MECHANICAL ENGINEERS' HANDBOOK

MECHANICAL ENGINEERS' HANDBOOK

Edited by

MYER KUTZ
John Wiley & Sons, Inc.

A Wiley-Interscience Publication
JOHN WILEY & SONS
New York · Chichester · Brisbane · Toronto · Singapore

TJ
151
m 395
1986

Library of Congress Cataloging in Publication Data:

Main entry under title:

Mechanical engineers' handbook.

"A Wiley-Interscience Publication."

   Includes index.
     1. Mechanical engineering—Handbooks, manuals, etc.
I. Kutz, Myer
TJ151.M395   1986   621   85-29468
ISBN 0-471-08817-X

Printed in the United States of America

10 9 8 7 6

For Mandy

CONTRIBUTORS

David M. Abels
Cuadra Associates, Inc.
Santa Monica, California

John E. Ahern
Aerojet Electrosystems Company
Azusa, California

Gordon A. Alison
The Aluminum Association, Inc.
Washington, D.C.

Dell K. Allen, Professor
Brigham Young University
Provo, Utah

R. Arvikar
Bell Laboratories
North Andover, Massachusetts

James W. Barr
The Aluminum Association, Inc.
Washington, D.C.

T. H. Bassford, R & D Fellow
Huntington Alloys, Inc.
Huntington, West Virginia

Alan Basting
Owens Corning Fiberglas
Toledo, Ohio

Andrew L. Bastone
Owens Corning Fiberglas
Toledo, Ohio

Adrian Bejan, Professor
University of Colorado
Boulder, Colorado

William E. Biles, Professor
Louisiana State University
Baton Rouge, Louisiana

Peter D. Blair
United States Congress
Office of Technology Assessment
Washington, D.C.

Carl Blumstein
University of California
Berkeley, California

John L. Boyen
Consulting Engineer
P.O. Box 8527
Emeryville, California

William Brett
New York, New York

Philip S. Brumbaugh, President
Quality Assurance, Inc.
St. Louis, Missouri

David A. Burge
David A. Burge Company, L.P.A.
Cleveland, Ohio

Robert S. Busk
International Magnesium Consultants
    Inc.
Hilton Head, South Carolina

Shun Chen
Westinghouse Electric Corporation
Philadelphia, Pennsylvania

Dale B. Cherchas, Associate Professor
Department of Mechanical Engineering
University of British Columbia
Vancouver, British Columbia, Canada

Jack A. Collins, Professor
Department of Mechanical Engineering
The Ohio State University
Columbus, Ohio

Carroll Cone
Professional Engineer
Toledo, Ohio

Kenneth Cooper
Borg Warner Corporation
York, Pennsylvania

Douglas A. Cornell
The Aluminum Association, Inc.
Washington, D.C.

Robert L. Crane
Wright Patterson Air Force Base
Dayton, Ohio

Balbir S. Dhillon, Associate Professor
University of Ottawa
Ottawa, Ontario, Canada

George M. Diehl
Consulting Engineer
Phillipsburg, New Jersey

Charles H. Drummond, III
Associate Professor
Ohio State University
Columbus, Ohio

Fritz Dusold
Science and Business Department
Mid-Manhattan Library
New York, New York

Seymour G. Epstein
The Aluminum Association, Inc.
Washington, D.C.

Franklin E. Fisher, Professor
Department of Mechanical Engineering
Loyola Marymount University
Los Angeles, California

Henry O. Fuchs, Professor
Mechanical Engineering Department
Stanford University
Stanford, California

Bernard J. Hamrock
National Aeronautics and Space
    Administration
Lewis Research Center
Cleveland, Ohio

Richard Harman
Department of Mechanical Engineering
University of Canterbury
Christchurch, New Zealand

K. E. Hickman
Borg Warner Corporation
York, Pennsylvania

E. L. Hixson, Professor
Department of Electrical Engineering
University of Texas
Austin, Texas

Byron W. Jones, Professor
Department of Mechanical Engineering
Kansas State University
Manhattan, Kansas

James G. Keppeler
Electric Fuels Corporation
St. Petersburg, Florida

William Kerr, Professor
Department of Nuclear Engineering
University of Michigan
Ann Arbor, Michigan

Robert J. King
United States Steel Corporation
Pittsburgh, Pennsylvania

Donald Knittel
Cabot Corporation
Kokomo, Indiana

Daniel T. Koenig
Canadian General Electric Co., Ltd.
Toronto, Ontario, Canada

Allan D. Kraus
Naval Post Graduate School
Monterey, California

Jan F. Kreider
Consulting Engineer
Boulder, Colorado

Peter Kuhn
Kuhn and Kuhn
Industrial Energy Consultants
Sausalito, California

Dudley Lesser
RCA, Government Systems Division
Morrestown, New Jersey

Peter E. Liley, Professor
School of Mechanical Engineering
Purdue University
West Lafayette, Indiana

Joseph A. Maciariello, Professor
Business Administration
Claremont Graduate School
Claremont, California

Ronald Douglas Matthews, Professor
University of Texas
Austin, Texas

F. C. McQuiston
School of Mechanical and Aerospace
  Engineering
Oklahoma State University
Stillwater, Oklahoma

Howard Mendenhall
Olin Corporation
New Haven, Connecticut

Benjamin W. Niebel, Professor
Department of Industrial and
  Management Systems Engineering
The Pennsylvania State University
University Park, Pennsylvania

Reuben M. Olson, Professor
Ohio University
Athens, Ohio

Carl C. Osgood
Mechanical and Metallurgical Engineer
Cranbury, New Jersey

Joseph Palen, HTRI
Alhambra, California

William J. Palm III, Professor
Mechanical Engineering Department
University of Rhode Island
Kingston, Rhode Island

Jerald D. Parker, Professor
School of Mechanical and Aerospace
  Engineering
Oklahoma State University
Stillwater, Oklahoma

Edward N. Peters
Plastics Division
General Electric Corporation
Pittsfield, Massachusetts

Peter Pollak
The Aluminum Association, Inc.
Washington, D.C.

W. William Pritsky
The Aluminum Association, Inc.
Washington, D.C.

Michael J. Rabins
Mechanical Engineering Department
Wayne State University
Detroit, Michigan

A. Ravindran, Professor
School of Industrial Engineering
University of Oklahoma
Norman, Oklahoma

Richard, J. Reed, Technical Director
North American Manufacturing
  Company
Cleveland, Ohio

G. V. Reklaitis
School of Chemical Engineering
Purdue University
West Lafayette, Indiana

J. B. ReVelle
Hughes Aircraft Company
Fullerton, California

E. A. Ripperger, Professor
College of Engineering
University of Texas
Austin, Texas

Andrew P. Sage, Professor
George Mason University
Fairfax, Virginia

Robert F. Schmidt
Colonial Metals
Columbia, Pennsylvania

Ali Seireg, Professor
Mechanical Engineering Department
University of Wisconsin
Madison, Wisconsin

Ronald Selner
Wright Patterson Air Force Base
Dayton, Ohio

George Silvestri, Jr.
Steam Turbine Generation Division
Westinghouse Electric Corporation
Orlando, Florida

Paul Smith, Professor
Department of Manufacturing
    Technology
Brigham Young University
Provo, Utah

William A. Smith
College of Engineering
University of South Florida
Tampa, Florida

Richard Stephenson, Professor
Department of Chemical
    Engineering
The University of Connecticut
Storrs, Connecticut

Rodney Stewart
Mobile Data Services
Huntsville, Alabama

Hans Thamhain
Associate Professor of Management
Worcester Polytechnic Institute
Worcester, Massachusetts

Wayne Tustin, President
Tustin Institute of Technology
Santa Barbara, California

Judith Wanger
Cuadra Associates, Inc.
Santa Monica, California

Leonard A. Wenzel, Professor
Department of Chemical Engineering
Lehigh University
Bethlehem, Pennsylvania

K. Preston White, Professor
Department of Engineering Science and
    Systems
University of Virginia
Charlottesville, Virginia

James B. C. Wu
Cabot Corporation
Kokomo, Indiana

Victor John Yannacone, Jr.
Patchogue, New York

Magd E. Zohdi, Professor
Louisiana State University
Baton Rouge, Louisiana

PREFACE

A handbook's contents portray the interests of the editor. As I pass back and forth over these chapters I find that they reflect my own career.

As an undergraduate in mechanical engineering at MIT I concentrated on mechanical design and actually thought, as I once told my father when he asked me how I intended to make a living, that I would be happy if someone would pay me to solve design problems. I was very young. While still an undergraduate I had a part-time job at the MIT Instrumentation Laboratory, where I did my Bachelor's thesis, *Design of a Bearing Torque Tester*. This was in the late 1950s, in the early days of such new and alternative materials as titanium and composites, covered in detail in this handbook, and when testing and failure analysis were much less sophisticated than they are today. (See the chapters on failure analysis and nondestructive testing.) As things turned out, however, when I went to work at the laboratory full-time, I found myself in a thermal design group working on temperature control for the guidance system of the Polaris missile. Most of my career as a mechanical engineer was focused on thermal design and temperature control. In fact, the title of my 1967 book was just that, *Temperature Control*. So the sections of this handbook on systems analysis, automatic control, thermodynamics, and heat transfer cover disciplines in which I was very much involved professionally throughout the 1960s. (I managed to employ systems engineering terminology in an article in *How Things Work in the Home*, published by Time-Life Books.)

Perhaps it was the Rockefellers' past identification with oil that suggested them as the topic for a book I wrote in the early 1970s (*Rockefeller Power*, published by Simon and Schuster in 1974), and possibly it is stretching the point to say that the chapters in this handbook on energy resources and power reflect what turned out to be a career transition. In 1976 I turned my professional activities to publishing as an acquisitions editor for professional and reference books in mechanical engineering and related disciplines at Wiley. Among the most successful titles in my editorial program were books on manufacturing, products liability, patents, and occupational safety and health—all areas where I had no prior professional experience. Sales of these books demonstrated the needs of the marketplace, of course. I would be neglecting the concerns of many mechanical engineers if I did not include these topics in the handbook. Hence, the major sections on manufacturing and management.

More recently my professional life has focused on two major areas—computers and management. For several years I ran the Electronic Publishing Division at Wiley, with responsibility for distributing publications online via telecommunications. The chapters on computers, sources of mechanical engineering information, and online searching all reflect my concerns of this period.

As chairman of the Publications Committee of the American Society of Mechanical Engineers and, currently, as Executive Publisher of the Scientific and Technical Division at Wiley, I have focused on financial matters. Several chapters of the handbook directly address this work. Worth noting here is that the ASME Board of Governors has identified the interaction of management with finance (and marketing) as being of key strategic importance.

In the past several decades, while my career was undergoing change, mechanical engineering was growing and evolving into a variety of different professions. This handbook, with its wide-ranging emphases, reflects this evolution—just as it must reflect the career growth and change of many mechanical engineers. It was Dick Zeldin, then head of Wiley's handbook program, who suggested that I undertake the editing of this handbook. At the outset, I intended it to be a one-volume updating of the old two-volume Kent's *Mechanical Engineers' Handbook*, which was first copyrighted in 1895. The most recent edition, the 12th, was published in 1950. It still continues to sell despite the fact that some of it is outdated and many topics that have become important over the past 35 years are not included. Not surprisingly, the 1950 edition of Kent contains little or nothing on computers and microprocessors, modern techniques of failure analysis, titanium, plastics and composites, modern techniques of nondestructive testing, group technology in manufacturing, computer integrated manufacturing, control systems design, modern financial techniques, safety engineering, online searching, energy, cogeneration, heat recovery, and nuclear power. The world of Kent, 12th edition, was a different world, a world in which engineering could afford large safety factors. Who cared how much anything weighed, or how much energy was consumed? So it is interesting to look back at the 12th edition and note the emphasis on steel. Steel must have been a relatively happy industry in the immediate postwar period—much happier than today. Obtaining an article on steel was an agonizing process. The industry is so demoralized, it seems, that when the Iron and Steel Institute moved from New York to Washington, it did not take along its library. I finally resolved the difficulty by borrowing part of the article that Wiley published in the Kirk–Othmer *Encyclopedia of Chemical Technology.*

Much of the 12th edition of Kent seemed to defy updating. How could you improve on the section on the efficiency of splices and knots? A revision of the chapter on woodworking (which was combined with a chapter on plastics molding) seemed better left in other hands—and in other publications.

In fact, the break between the 12th edition of Kent and a handbook edited in the 1980s was so severe that it seemed inappropriate to (1) attempt an updating, and (2) call a new mechanical engineers' handbook the 13th edition of Kent. I recommend, therefore, that you not clear space on your shelf for this volume by throwing out your old Kent, but that you keep both.

One thing you may not need the old Kent for, however, is its mathematical tables. With an inexpensive scientific calculator, most numbers you need are at your fingertips. You will not find mathematical tables in these pages. Why destroy forests to make paper for printing tables when ubiquitous microelectronics are so cheap and powerful?

They are changing the profession of mechanical engineering. In fact, CAD/ CAM or CIM (computer integrated manufacturing), and robotics, for example, have made mechanical engineering a "hot field." So this handbook begins with chapters on computers and microprocessors (and, later, has a chapter on robots and computer-aided manufacturing).

This order was the recommendation of the handbook's editorial board. I am indebted to them for this wisdom—and for their good sense in confirming the subject matter of the handbook. The final decision on what actually went into the handbook was mine alone—as were the choices of contributors. As it turned out, this was a massive undertaking. Although I have many connections within the mechanical engineering fraternity, not all of the 78 articles came easily. Some did not come forth at all, and I have had to borrow about 10 percent of the articles from other sources. These borrowings have been edited for this handbook, of course, often by the original authors.

The wonder is that I was able to obtain original articles from such a wide variety of authors. Some of them I knew because I had edited their books. Others were colleagues on ASME boards and committees. Others I found after exhaustive correspondence and telephone calls. All of them, no matter what their prior relationship to me, have exhibited great dedication. Writing an article for a handbook is its own reward. A desire to impart information, knowledge, and experience to others for little personal gain is a truly wonderful thing.

My thanks to colleagues Martin Grayson, Thurman Poston, Wiley's excellent production staff, particularly Ed Cantillon, Margaret Comaskey, Douglas Elam, and the unknown copyeditor and proofreaders. Thanks, too, to my secretary, Meryl Weiner, who put up with an awesome amount of last minute typing and organizational work. At this point, authors and editors generally mention spouses and other family members, thanking them for their patience and understanding. I have read hundreds of these citations to late nights in offices and Sunday afternoons spent with galleys spread over the dining room table. I know from personal experience that these statements are not pro forma; they are true, and they come from the heart. So does this: My wife Mandy deserves special thanks, particularly because this handbook was so long in the making.

MYER KUTZ

New York, New York
February 1986

CONTENTS

PART 4   SYSTEMS, CONTROLS, AND INSTRUMENTATION

PART 5   MANAGEMENT AND RESEARCH

PART 6   ENERGY AND POWER

PART 1
DIGITAL COMPUTERS

CHAPTER 1

INTRODUCTION TO MICROCOMPUTERS

ALI SEIREG

University of Wisconsin
Madison, Wisconsin

1.1 HISTORY

Although many ingenious calculating devices date back to ancient times, electronic computers are less than four decades old. Eniac, generally regarded as the first electronic computer, was funded by the Army Ballistic Research Laboratory at the University of Pennsylvania and was completed in 1946. It contained 18,000 vacuum tubes and 1500 relays and took up 1000 ft² of floor space. Operators programmed it by hand-setting thousands of switches and plugs.

The IBM pioneer computer SSEC, installed in January 1948, was the first machine to combine electronic computation with a stored program and, hence, to operate on its own instructions. The SSEC required a special room with over 3000 ft² of floor space for its 12,500 vacuum tubes, 21,400 relays, and 40,000 pluggable connections. Its memory size was 8 words in the arithmetic unit, 150 words in relays, and 2000 words on the tapes that were used for input and for long-term memory. The SSEC's tape drive consisted of a 400-lb reel of card stock that had to be lifted with a chain hoist. Sixteen-foot-high racks of wire contact relays formed the core of central memory.

In the short 40 year history of the technology, computers have shrunk from room-size behemoths to desktop devices that are faster and much friendlier. The size continues to diminish. Consider the Grid Compass, which measures 15 in. × 11.5 in. × 2 in. and weighs 9 lb. It is battery operated and has a nonvolatile memory. It has a powerful processor that is enhanced greatly by its ability to communicate with IBM mainframes, in order to store and retrieve information and update programs.

The Epson HX-20 Notebook Computer weighs less than 4 lb and has a full-size keyboard, built-in printer and liquid crystal display (LCD), and 48K of combined RAM and ROM. Its internal power supply will keep it running for over 50 hr. Radio Shack and Nippon Electric Co. (NEC) have recently introduced notebook-size computers of their own.

Currently, the consideration that is limiting personal computers from becoming even smaller is the size of input and output devices. (The microprocessor and memory chips combined are already smaller than a matchbox.) However, the bulky input and output components—the cathode ray tube and the disk drive—are giving way to the flat display (CRT, LCD, and, soon, plasma) and the bubble memory, which holds information magnetically when the power is off. The 16- and 32-bit processors allow not only enormously faster computation but also the creation of simpler and more versatile software for word processing and other tasks.

1.2 COMPUTERS IN MECHANICAL ENGINEERING

Advances in microelectronics are slowly changing the face of mechanical engineering and profoundly affecting the work of mechanical engineers. Computer technology has provided them with new tools, challenges, and opportunities that are unparalleled since the early period of mechanization. Its many manifestations are evident in equipment design, planning, production, and quality control. Computer-based software and graphic display hardware are routinely used in creating fully detailed designs that can be routed via digital links to numerically controlled machines, plotters, disk storage, microfilm devices, or hard copy devices as needed. Future computer systems are expected to automatically create task plans for part manufacture and to be used for material handling, inspection, and assembly systems capable of assembling a mix of parts. The engineer will be expected to design the program that produces the product rather than optimizing the product design.

1.2.1 Geometric Modeling

An important element of computer application in mechanical engineering is the ability to model accurately the geometry of mechanical parts. The early geometric modelers were based on "wire-frame" models used primarily for the production of drawings. The next generation of modelers were the surface modelers, which provide a more rigorous definition of geometry. The latest approach is the solid modelers, which contain sufficient information to construct totally and unambiguously the solid being modeled. An interesting development for producing high-quality realistic images is a ray-tracing algorithm, which enhances the visualization of virtual parts. It consists of tracing mathematical rays of light from an observer's viewpoint into three-dimensional space. Reflective and transparent surfaces as well as shadows can be simulated by recursively tracing rays propagated from points on the model's surface.

1.2.2 Finite Element Analysis

One of the most commonly used computer analysis techniques by mechanical engineers is the finite element method. Since its basic formulation was introduced in 1956, it has been widely accepted as one of the most powerful tools for calculating stress and deformation in structures. Its use has been extensive in dynamic analysis of solids with complex geometries and nonlinear material properties. It has been applied effectively to compressible as well as incompressible fluids and to heat transfer, thermal stress, wave propagation, fracture mechanics, metal working, and acoustical response problems. Numerous mapping methods are available for automated generation of two- and three-dimensional meshes. They are extensively used in practice and continue to improve in reliability and computational efficiency.

1.2.3   Integration of Computers and Machines

Mechanical system designers are frequently confronted with the integration of computers and other microelectronic devices in the system they develop.

There are many simple or repetitive tasks that existing machine intelligence technology is capable of dealing with more reliably and less expensively than if human beings were in the loop. This in turn frees human experts for more difficult judgmental tasks.

The integration of machine models and material handling equipment under computer control for producing a wide variety of parts in a flexible manufacturing system is beginning to replace job shops. The decision making function is computerized in such a system to allow for changes in design and part mix.

Computers are also playing an ever increasing role in the process industries in order to improve performance and safety. The issue of data and information management and its condensation and allocations are slowly being assigned to the process computer. It collects and interprets data from hundreds of different sensing devices and relays the proper information to the proper individual at the proper time.

1.2.4   Human–Machine Interface

In spite of many significant advances, the human–machine interfaces are areas that are undergoing continuous development. Many studies are directed toward better understanding of the nature of the human–machine system and developing human-compatible machines and computer languages. Other research efforts focus on improving operator guidance techniques for seldom-encountered emergency conditions and for augmenting the capabilities of the human operator and the handicapped operator, as well as developing general techniques for incorporating the human in the total system. Human–machine interface currently reflects the requirements of the job rather than the limitations of the computer.

Some formerly extravagant or impossible techniques are now highly cost effective. For example, just a few years ago speech output required complex custom-designed electromechanical players. Applications were once limited to telephone time/temperature services and aircraft warning systems. Today we have talking cars, elevators, toys, calculators, pinball machines, etc.

1.3   BASIC STRUCTURE OF DIGITAL COMPUTERS

The basic elements of a digital computer are illustrated in Fig. 1.1.

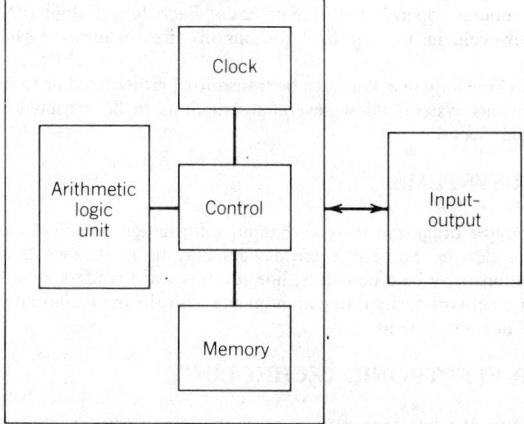

Fig. 1.1   Basic structure of a digital computer.

1.3.1 Input–Output Device

An input–output device (I/O) provides the interface between the user and the central processor unit. These devices provide the communication channels between the operators and the computer and are generally called terminals. They are available in a variety of forms, which are continually undergoing evolutionary change to simplify, enhance, and speed communications. Some of the devices in common use are keyboards, printers, teletype, and cathode ray tubes. The latter is the most commonly used at this time. Voice communication is receiving considerable attention because of its great potential. It is currently used in an elementary form in limited applications.

1.3.2 Central Processing Unit

A central processing unit (CPU) directs and controls the computer. It receives instructions and data and sends out commands on which instructions to execute next. It has two main components:

- The arithmetic logic unit (ALU), which performs the arithmetic and logic operations.
- The control unit, which oversees the program flow. It incorporates a synchronous device that performs the different instruction steps in timing with pulses from an internal clock whose speed determines the speed of the computer. The clock generates a continuous stream of precisely timed signals that coordinates the digital signal flow among the different components of the computer system. The control unit is in effect the nerve center of the entire machine. It is seldom a physically separable element of the computer. The control circuitry is often distributed throughout the computer and connected by numerous control lines that transmit the timing and synchronization signals to control the different events.

1.3.3 Computer Memory

Computer memory is composed of the storage devices into which the instructions and data are written for subsequent retrieval. There are two types of memory, which are categorized by the type of access they provide:

- Random access memories, which provide immediate access to all stored information at any time. These are generally made of ferrite cores or semiconductor integrated circuits. The programmer can distribute the stored information over the device in any chosen form.
- Serial access memories, where the access to information is dependent on its position relative to the last item stored or retrieved. Devices of this type are magnetic tapes, drums, or disks.

The first type of memory is considerably faster and more expensive and is usually used for storing the instructions and data needed immediately in the execution of the program. The second type is much slower and less expensive and is generally used for storage of programs and data that are not needed immediately.

1.3.4 Hardware Bus

The hardware bus is the conduit by which data are transmitted among the different elements of the computer. Digital computers operate with binary data. Each binary digit (or bit) is represented by either a high or a low voltage; the width of the bus in wires defines the width of the transmitted word in bits.

In a single-bus machine only one word can be transmitted either from or to its different components at any time. Multiple-bus systems allow several instructions to be executed at the same time and, consequently, are much faster.

1.4 COMPUTER SYSTEMS

Computer systems may be organized to receive input information and send output information from numerous and diverse devices. Some of these devices may be at stations that are remote from the main system. The stations may be typewriters, line printers, card readers, or other satellite computers that are connected in a network fashion to communicate with the main computer. Figure 1.2 illustrates the general structure of such systems.

1.5 COMPUTER ELECTRONIC TECHNOLOGY

Computers evolved over the last four decades with the innovations in electronic technology. There are now four generations of computer systems, and the fifth is expected to be developed within a

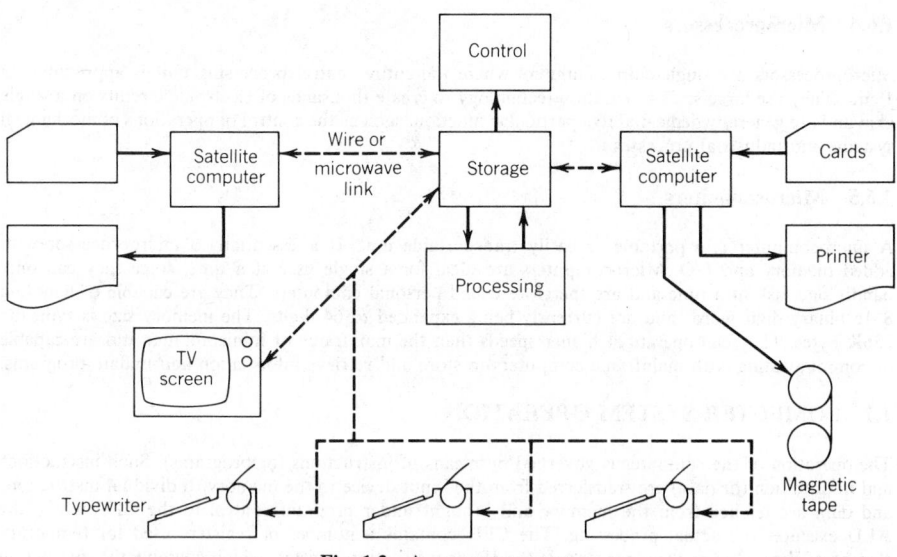

Fig. 1.2 A computer system.

decade. The first-generation machines used vacuum tubes, the second used the transistor, the third used integrated circuits, and the fourth used very-large-scale integrated circuits (VLSI). Fifth-generation computers are also expected to use VLSI technology but with the chips arranged in parallel. Parallel processing is expected to increase the speed and power on the order of 1000 times that of the previous generation.

These ultra-high-speed machines will be using programming in logic (Prolog) languages, will hold intricate catalogs of knowledge, and will aspire to mimic the thinking of skilled experts. They are envisioned to be able to recognize complex patterns, read human handwriting, make assumptions, use rules of thumb, and draw inferences to reach conclusions.

1.6 TYPES OF DIGITAL COMPUTERS

Computers are generally compared based on physical size, speed, cost, accuracy, memory size, time of run, and the input–output capabilities. Because of the continuous exponential improvements in microelectronic circuit technology and manufacturing techniques, classification criteria are becoming difficult and the boundaries between types of computers are not well defined. The following are the currently accepted categories.

1.6.1 Supercomputers

Supercomputers are characterized by very-high-speed processing and large storage capacity. Such computers are generally used on a shared time basis with remote access by many users, sometimes on a global basis. They are ideally suited for very large problems requiring extensive computations with great numerical accuracy.

1.6.2 Mainframe

Mainframes are room-size computers that are capable of handling 32–64 binary digit words and have a storage capacity on the order of 12,000K bytes. They are suited for an entire organization.

1.6.3 Minicomputers

Minicomputers typically are desk sized. They can handle 16–32 binary digits and have a storage capacity of 4000K bytes. They are ideally suited for scientific and engineering computations and are capable of serving multiple users.

1.6.4 Microprocessors

Microprocessors are single-chip computers where the entire central processing unit is approximately 1 cm². They use large-scale-integration technology to create thousands of electronic circuits on a single chip and are generally dedicated to a particular function, such as the control of operations of mechanical systems and industrial processes.

1.6.5 Microcomputers

A microcomputer is a portable or easily transportable unit. It is essentially a microprocessor with added memory and I/O. Microcomputers are ideal for a single user at a time, since they can only handle one task at a time and are therefore called personal computers. They are capable of handling 8–16 binary digit words and are currently being expanded to 64 digits. The memory size is typically 256K bytes. They can operate at higher speeds than the mainframe or minicomputer and are capable of communicating with mainframe computers to store and retrieve information and update programs.

1.7 COMPUTER SYSTEM OPERATION

The operation of the computer is governed by means of instructions (or programs). Such instructions and information (or data) are transferred from the input device to the memory. Individual instructions and data are fetched from the memory and brought under program control to the CPU where the ALU executes the actual processing. The CPU contains a number of registers used for temporary storage of data. One of these registers is the IR or instruction register, which contains the instruction that is being executed. Its output allows the control unit to generate the timing signals needed for the control of the execution of the instruction. Another register is the PC register or program counter, which keeps track of the execution of the instructions. During the execution of a particular instruction the contents of the PC are updated to correspond to the address of the next instruction that is to be transferred from the memory for execution.

The MAR or memory address register is used to hold the address of the location where the data are located and where they are to be transferred. The MDR or the memory data register contains the data to be written in or read out of the addressed location.

1.8 PROGRAM EXECUTION STEPS

The first step in the program execution is to set the PC to point out the first instruction. This is then transferred to the MAR and a "Read" control signal is sent to the memory. After a certain memory access time the addressed instruction is read out of the memory and loaded into the MDR. The instruction is subsequently transferred to the IR and it is then ready to be decoded and executed and the PC is updated to point at the next instruction. In order for the computer operation to be efficient, the transfer of data among the different units is generally done in parallel. This requires a large number of lines (called a bus) to transfer the data and control signals among the different locations.

Although many computers have several distinct buses, some have only a single bus. Since the bus is used for one transfer at a time, single-bus computers operate at slower speeds than multibus machines. The single-bus structure is commonly used in smaller computers, since it is considerably less expensive and provides more flexibility for connecting peripheral devices.

CHAPTER 2
COMPUTER PROGRAMS

ALI SEIREG

University of Wisconsin
Madison, Wisconsin

Computers are machines designed to perform tasks requested by the human operator. The set of instructions (or program) devised to describe in detail the different steps that the computer is required to execute in order to accomplish the task is the means of communication between the computer and the user. A well thought out, precise, and concise program is needed for the efficient use of a particular machine.

Execution speed is highly dependent on the suitability of the program to the machine used. Different instructions run at different speeds in different machines. A computer program generally contains a number of functionally different steps. These correspond to the following classes of operation:

- Input–output transfer.
- Data transfers among the main memory and the different registers in the CPU.
- Program sequencing and control.
- Arithmetic and logic operations on data.

The instruction format, the instruction sequencing, and the capability for testing conditions and choosing one of a set of alternatives is fundamental to the programming function.

2.1 MACHINE LANGUAGES

The processor unit in the computer can only execute instructions in machine code, which is based on detailed knowledge of the computer system architecture. Programming in machine code is the most efficient way to communicate instructions to the computer, but is very difficult to assemble. Besides being a long and tedious process, it requires detailed understanding of the different components of the machine and their interconnections.

2.2 ASSEMBLY LANGUAGES

Assemblers allow the user to program low-level languages (or assembler languages) where the conversion and arrangement of the different parts of each instruction are undertaken by the assembler. The use of the assembler reduces the effort of programming in machine languages, but it also requires intimate knowledge of the computer architecture. In general, the number of instructions are approximately the same when programming in machine code or assembler language.

2.3 HIGH-LEVEL PROGRAMMING LANGUAGES

There are several high-level languages that are oriented toward problem solutions rather than the machine structure. Such programs are much faster to write and easier to document. They are translated into machine code with the aid of special software called compilers. Examples of high-level languages are FORTRAN, COBOL, BASIC, ALGOL-60, ALGOL-68, PL/1, BCPL, LISP, PASCAL, and ADA. Some of these are briefly discussed in the following.

2.3.1 FORTRAN

One of the most common among these languages is FORTRAN, where the statements of a program are analogous to the grammatical and syntactical rules of English. The FORTRAN (or Formula Translator) language was developed for use in engineering and scientific applications requiring numerical computations. It is relatively easy to learn, and most computing machines can accept FORTRAN programs.

2.3.2 BASIC

BASIC is a relatively crude language, which is easy to learn and use. It is extensively used in microcomputers. It is usually translated into computer code line by line, and errors are reported back to the programmer for immediate correction. Although it is not a very efficient language, it is very useful for design problems where considerable human–machine interaction takes place.

2.3.3 ALGOL

ALGOL is a language with efficient facilities for program control and is oriented more toward problem solving than mere numerical computations.

2.3.4 PROLOG

PROLOG is a language for programming in logic developed for use in problem solving, game playing, and artificial intelligence.

2.3.5   PASCAL

PASCAL is a language that is particularly suited for teaching and conversational programs. It is widely used in developing future interactive programming and CAD research.

2.3.6   PL/1

PL/1 is a language developed by IBM for use in commercial and scientific programming.

2.3.7   LISP, BCPL, and ADA

LISP, BCPL, and ADA are languages that were developed for CAD systems and interactive programming.

2.4   LANGUAGE PROCESSORS

Computers are usually provided with language processors known as compilers and interpreters. A compiler is an algorithm that translates high-level-language programs into sequences of instructions that can be executed directly by the computer. An interpreter is an algorithm that allows computers to execute programs written in high-level languages line-by-line as if they were machine-code instructions. Although interpreters allow the user to deal directly with the machine, compilers allow programs to be executed more rapidly.

Language processors are designed to provide the following programming conveniences:

- They make a program easy to write, check, and verify.
- They provide brevity of style without creating confusion.
- They allow expression of the intent of the task in a manner that is easy to grasp.
- They optimize the programming and computing time for a particular computer and language.

2.5   COMPUTER ALGORITHMS AND FLOW CHARTS

The first step in any program is a detailed description of the task to be performed in sequential steps. Each step leads to another in a logical and orderly fashion. This precise description is called an algorithm. The second step is to represent the algorithm diagrammatically by a flow chart, which illustrates the sequential nature of the different steps of the task. Flow charts are composed of blocks representing certain operations or program statements. Each box has a conventional meaning. The standard symbols adopted by the American National Standards Institute (ANSI X3.5-1970) that relate to programming in FORTRAN are given in Table 2.1.

Table 2.1   Flow Chart Symbols

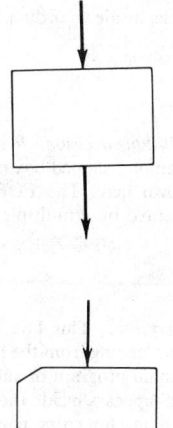

1. *Processing.* The basic rectangular box indicates a processing step, that is, a change in the value, form, or location of data. For FORTRAN flow charts, this box ordinarily encloses one or more assignment statements.

2. *Punched card input.* This box represents input from punched cards. We use this form for all input steps since, conceptually, the input is from punched cards (although, actually, the input data may be stored on some other medium, such as magnetic tape). Inside the box we write the list of items whose values are to be read.

Table 2.1 *(Continued)*

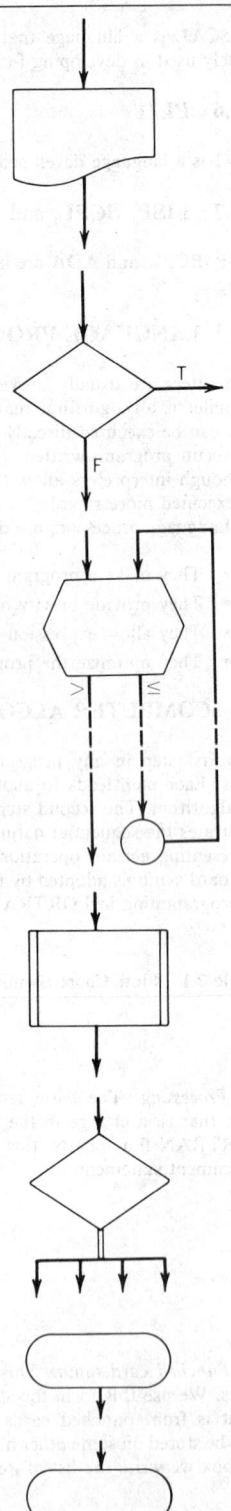

3. *Document output.* This box represents output in the form of a printed document. We use this form for output to be displayed on a line printer.

4. *Decision.* The diamond-shaped box represents a conditional control statement in FORTRAN. For the logical IF statement, the *assertion* to be tested is written inside the box, and two flowlines leaving the box (labeled "T" for *true* and "F" for *false*) indicate the alternative paths that the computation may take. For the arithmetic IF statement, the *expression* to be tested is written inside the box, and three flowlines (labeled "<0," "=0," and ">0") leave the box.

5. *Preparation.* We use this shape exclusively for DO statements in this text. At the upper right, inside the box, we indicate the *Label* of the terminal statement. Inside the box we also give the *Index,* the *Initial Value,* the *Limit,* and the *Increment* (if required). The flowline leaving the box at the lower right (labeled "<") leads to the group of statements to be repeated under control of the DO statement; this path will be taken initially, and thereafter so long as the *Index* value is less than or equal to the *Limit.* When the *Index* value exceeds the *Limit,* the flowline at the lower left (labeled ">") will be followed. The *Label* of the terminal statement also appears in a *Connector* symbol at the end of the group of instructions controlled by the DO statement; a flowline leads from this Connector symbol directly back to the box representing the DO statement.

6. *Predefined process.* This box, distinguished by vertical striping, is used to indicate the activation of a subroutine. The name of the subroutine and the list of arguments are shown inside the box. The activation of a function, since it occurs as a part of the evaluation of an expression, is shown in the context in which it occurs, for example, inside an ordinary rectangular box for the assignment statement.

7. *Multiple decision.* Where the number of branches exceeds three, the diamond-shaped box is supplemented with a system of flowlines, as shown here. The FORTRAN computed GO TO statement is represented by a multiple-decision box of this form.

8. *Terminal.* This box is used in two ways in this text. With a flowline leading from the box, it indicates the beginning of a program unit (main program or subprogram). For a main program the word "Start" appears inside the box. For a subprogram entry the subprogram name (or entry name) and the list of parameters are shown inside the box.

   With a flowline leading to the box, it is used for the STOP statement, or for the RETURN statement of a subprogram.

Table 2.1 *(Continued)*

9. *Flowline.* This is simply a line with an arrow, indicating the sequence in which the statements of a program will be executed.

10. *Connector.* This symbol, enclosing a label, often indicates a CONTINUE statement. It may indicate any other point in the program where we wish to draw attention to the label of a statement to be executed next.

11. *Annotation.* Descriptive information (sometimes indicating declarations) may be included in a flow chart by means of a symbol of this form.

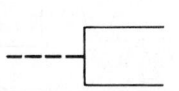

A program can be stated in a relatively small number of basic statements such as the following:

- An assignment statement is a fundamental procedural step by which the values of the variables are defined.
- A conditional statement or IF statement is the fundamental decision step of the FORTRAN language.
- Declarative or dimension statements supply information to the processor about the algorithm in order to reserve sufficient space in storage for expected lists.
- Iterative or DO statements are central statements that exercise control over the repeated execution of succeeding statements.
- CONTINUE statement acts as junction point, marking the last of the statements under control of the iterative (or DO) statement.
- STOP statement indicates the termination of an action sequence.
- END statement indicates that no other statements follow and accordingly marks the end of a program.

2.6 SAMPLE PROGRAM AND FLOW CHART

The programming process is illustrated by the following simple problem using FORTRAN as the programming language.

2.6.1 Task

Find the largest value of a set of numbers.

2.6.2 Mathematical Formulation

Given a set of numbers

$$a_j, \qquad j = 1 \rightarrow N, \qquad N < 100$$

find the largest number.

2.6.3 Algorithm and Flow Chart

1. Reserve space for storing up to 100 numbers
    $a_j, \qquad J = 1 \rightarrow N, \qquad N < 100$.
2. Start with $Z = a_1$.
3. $J = 2$.
4. Check if $Z < a_j$; if true, let $Z = A(J)$.
5. If false continue for $J = J + 1$.
6. Check if $J \leq n$; if true repeat step, if false stop.
7. End program.

The flow chart illustrating the different steps in this algorithm is shown in Fig. 2.1.

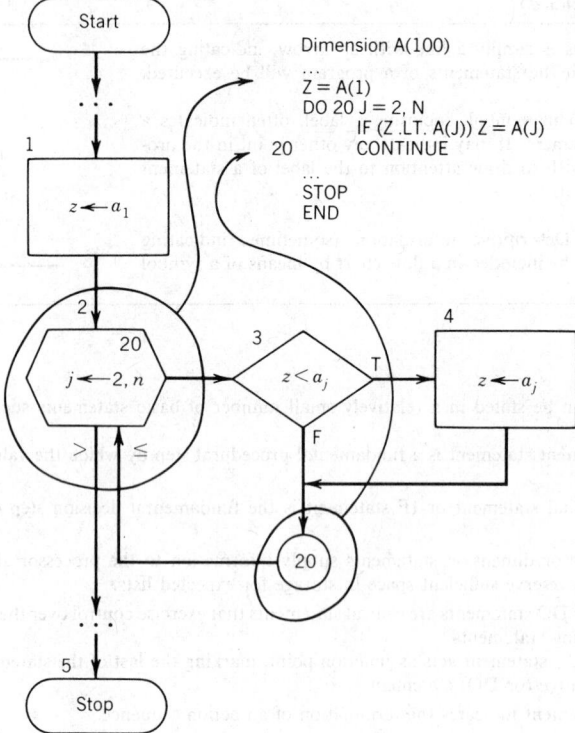

Dimension A(100)
...
Z = A(1)
DO 20 J = 2, N
   IF (Z .LT. A(J)) Z = A(J)
20   CONTINUE
...
STOP
END

Fig. 2.1 Flow chart.

2.6.4 FORTRAN Program

```
DIMENSION A(100)
Z = A(1)
DO 20 J = 2,N
IF (Z·LT·A(J)) Z=A(J)
20 CONTINUE
 . . .
STOP
END
```

The correspondence between the above program statement and the different steps of the flow chart is shown in Fig. 2.1.

2.7 SUBPROGRAMS AND MACROINSTRUCTIONS

An important reason for using subprograms is to divide complex programs into smaller, more manageable units. In such cases the program can be organized into a hierarchy of subprograms of different levels linked together and supervised by a main program as illustrated in Fig. 2.2. Each of the subprograms represents an independent unit that can be checked separately and incorporated if needed into other programs. The main program provides the macroinstructions that interconnect and coordinate the operation of the different subroutines needed to accomplish the task. (Subprograms can be classified as subroutines or functions.)

    A subroutine is an independent group of statements that can be called by a special CALL statement and is in effect incorporated into the main program by this statement. A computation that is repeated over and over can be made into a subroutine in order to shorten the main program. On the other hand, a function subprogram is supplied to the computer to compute a specific function and is activated by simply writing the name of the function as part of an expression. The function subprograms may be defined by the programmer or predefined as library functions (i.e., trigonometric or algebraic functions).

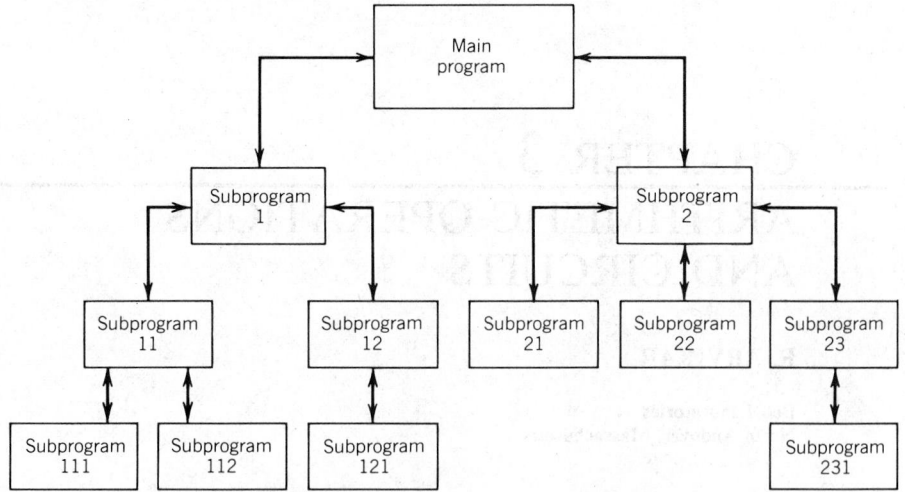

Fig. 2.2 Hierarchial structure of a complex program.

CHAPTER 3

ARITHMETIC OPERATIONS AND CIRCUITS

R. ARVIKAR

Bell Laboratories
North Andover, Massachusetts

In this chapter we will review how the basic mathematical operations are performed in the digital computer, and why the conventional decimal notation system would be impractical from a computer-system-design viewpoint. The representation of numbers in various numbering systems such as binary, octal, and decimal, and conversions between them are also explained. This is followed by some representative computational circuits, such as adders, subtractors, dividers, and multiplers, used to perform the basic binary arithmetic in a computer.

3.1 NUMBER SYSTEMS

3.1.1 Decimal System

The common decimal system consists of ten digits, 0–9. For a digital machine to be able to process these ten digits and perform arithmetic operations, one would have to devise a system where each of the digits would be represented by distinct "states" or "levels" (such as current or voltage levels). On the other hand, if we used a numbering system that had fewer number of digits, for example, a binary notation with two digits, 0 and 1, then the machine need understand only two "states," representing a 0 and a 1. The digits could be assigned the states "off" or "on" and should result in a machine that would be much less complex or unwieldy since there would be fewer states or levels to process. The user input to the computer, however, need not be in binary; prior to and after the actual computations, special processors can convert the data to the required binary format and reconvert to the format convenient to the user, usually decimal. As we shall see later, even coded numbering systems (such as binary-coded decimal) would be more suitable and are used widely in computers. In any numbering system, a number can be represented in terms of the "base" or "radix" of the system and a series of digits valid for that system. In general, any number can be formed as

$$N = C_n B^n + C_{n-1} B^{n-1} + \cdots + C_1 B^1 + C_0 B^0$$

where N is the number, B is the base, and C is any valid digit that can be used. For example, in the decimal notation the valid digits are 0–9, and the base is 10, so that the number 24 implies $+2$ tens plus 4. In other words,

$$24 = 2(10)^1 + 4(10)^0$$

Similarly,

$$746|_{10} = 7(10)^2 + 4(10)^1 + 6(10)^0$$

It is clear that the magnitude of the number is positional, that is, the weight of a digit is B times the weight of the digit to its right. For example, in a number in the base 3, each digit has three times the weight of the preceding (right-hand) digit. Hence, the number $122|_3$ (the suffix following the number identifies the base of the numbering system) is equivalent to decimal 17:

$$122|_3 = 1(3)^2 + 2(3)^1 + 2(3)^0 = 17|_{10}$$

3.1.2 Binary System

The binary system consists of two digits, 0 and 1, which are commonly called "bits." The base of the system is 2. A binary number will take the general form

$$N = C_n 2^n + C_{n-1} 2^{n-1} + \cdots + C_1 2^1 + C_0 2^0$$

where C can be either a 0 or a 1. For example, $11001|_2$ is equivalent to $25|_{10}$:

$$11001 = 1(2)^4 + 1(2)^3 + 0(2)^2 + 0(2)^1 + 1(2)^0 = 25|_{10}$$

Therefore, binary numbers are powers of 2. The exponent can be either positive or negative, as in

$$100.101|_2 = 1(2)^2 + 0(2)^1 + 0(2)^0 + 1(2)^{-1} + 0(2)^{-2} + 1(2)^{-3}$$
$$= 4.625|_{10}$$

3.1.3 Binary Arithmetic

The adaptability of the binary system for digital computer application stems from the fact that the switching hardware generally functions in two states, on–off or open–closed. Therefore, by assigning the two digits to the two functional states, the binary system can be conveniently implemented on a digital computer. Furthermore, the fundamental arithmetic operations are relatively simple when performed in binary. The basic addition and multiplication rules are summarized below:

+	0	1
0	0	1
1	1	10

×	0	1
0	0	0
1	0	1

Addition   Multiplication

Since the multiplier is either a 0 or a 1, the result is either a 0 or the multiplicand itself, which can be used to form the partial product. Therefore, the process of multiplication of two binary numbers consists of a series of shifts and additions. Consider, for example, the multiplication of 1101 (multiplicand) by 101 (multiplier):

$$
\begin{array}{r}
1101 = 13 \\
\times \quad 101 = 5 \\
\hline
1101 \\
0000 \qquad \text{Partial products} \\
1101 \\
\hline
1000001 = 65
\end{array}
$$

Note that whenever a 0 is encountered in the multiplier, the partial product is 0; whenever a 1 is encountered, the partial product is the multiplicand itself, shifted to the left. If we were to ignore the 0 in the multiplier, the result can be rewritten as

$$
\begin{array}{r}
1101 \\
101 \\
\hline
1101 \\
1101 \qquad \text{Shifted to left} \\
\hline
1000001
\end{array}
$$

This means that the final result can be obtained by simply shifting and adding the multiplicands several times as needed.

3.1.4 Octal Notation

The base of the octal system is 8, and the valid digits are 0–7. The usefulness of the octal system is its compact form compared to the binary; since one octal digit is always equal to three binary digits and vice versa, it is possible to convert between binary and octal by mere inspection. For example, 110011101 in binary can be converted to octal by simply marking off groups of three digits starting from the right and replacing each group by its octal equivalent:

$$
\begin{aligned}
110011101 &= 110 \quad 011 \quad 101 & \text{(binary)} \\
&= 6 \qquad 3 \qquad 5 & \text{(octal)} \\
&= 6(8)^2 + 3(8)^1 + 5(8)^0 = 413|_{10} & \text{(decimal)}
\end{aligned}
$$

Since the conversion from octal to binary and vice versa is simple, octal addition can be performed readily by converting to binary and then adding. For example,

$$
\begin{array}{r}
124|_8 = 001 \quad 010 \quad 100 \ |_2 \\
+ \ 52|_8 = \qquad\quad 101 \quad 010 \ |_2 \\
\hline
176|_8 = \ 1 \quad 111 \quad 110 \ |_2
\end{array}
$$

3.1.5 Hexadecimal System

The hexadecimal system has a base of 16 and a total of 16 digits, 0–9 and A–F, with the letters A–F standing for the decimal numbers 10–15, respectively. Thus, the decimal equivalent of 3A4B in hexadecimal would be

$$
\begin{aligned}
3A4B &= 3(16)^3 + 10(16)^2 + 4(16)^1 + 11(16)^0 \\
&= 14923|_{10}
\end{aligned}
$$

Since each hexadecimal digit is equal to four binary digits, one can convert binary to hexadecimal by marking off groups of four digits and replacing them by their hexadecimal equivalents. The decimal numbers 0–20 in the binary, octal, and hexadecimal systems are shown in Table 3.1

Table 3.1 Numbers Using Base 2, 8, and 16

Decimal (base = 10)	Binary (base = 2)	Octal (base = 8)	Hexadecimal (base = 16)
00	00000	000	0
01	00001	001	1
02	00010	002	2
03	00011	003	3
04	00100	004	4
05	00101	005	5
06	00110	006	6
07	00111	007	7
08	01000	010	8
09	01001	011	9
10	01010	012	A
11	01011	013	B
12	01100	014	C
13	01101	015	D
14	01110	016	E
15	01111	017	F
16	10000	20	10
17	10001	21	11
18	10010	22	12
19	10011	23	13
20	10100	24	14

3.2 CONVERSION BETWEEN NUMBER SYSTEMS

It is often necessary to convert among the octal, the binary, and the decimal systems. We have seen earlier that conversion between binary and octal is trivial and can be performed by mere inspection. Conversions to decimal (for both integers as well as fractions) can also be done simply by assigning proper value to each digit depending on its position and summing. For example,

$$1011|_2 = 1(2)^3 + 0(2)^2 + 1(2)^1 + 1(2)^0 = 11|_{10}$$
$$3406|_8 = 3(8)^3 + 4(8)^2 + 0(8)^1 + 6(8)^0 = 1792|_{10}$$
$$1.1011|_2 = 1(2)^0 + 1(2)^{-1} + 0(2)^{-2} + 1(2)^{-3} + 1(2)^{-4} = 1.6875|_{10}$$

Conversions from decimal to binary or octal is less trivial and is covered below.

3.2.1 Decimal to Octal

To convert a decimal number to octal, successively divide by 8 until the decimal integer is reduced to zero and then combine the remainders to form the octal equivalent. For example,

```
                        Remainder
        8 | 234            2
        8 | 29             5
        8 | 3              3
            0
```

Therefore,

$$234|_{10} = 352|_8$$

Decimal fractions can be converted to octal by successive multiplications by 8, with the integers produced becoming the octal digits. The process is continued until the fraction becomes zero or until the desired accuracy is attained. The conversion of $.234|_{10}$ to octal is shown below:

```
    Integer   .234
                  8
         1    .872
                  8
         6    .976
```

$$
\begin{array}{cc}
 & \underline{8} \\
7 & .808 \\
 & \underline{8} \\
6 & .464 \\
 & \underline{8} \\
3 & .712 \\
\end{array}
$$

Therefore, $.234|_{10} = .16763 + |_8$, which is also binary $.001110111110011$.

3.2.2 Decimal to Binary

To convert a decimal number to binary, it is divided successively by 2 until the decimal integer is reduced to zero and then combining the remainders (which are either a 0 or a 1) to form the binary equivalent. For example,

$$
\begin{array}{rl|c}
 & & \text{Remainder} \\
2 & \underline{46} & 0 \\
2 & \underline{23} & 1 \\
2 & \underline{11} & 1 \\
2 & \underline{5} & 1 \\
2 & \underline{2} & 0 \\
2 & \underline{1} & 1 \\
 & 0 \\
\end{array}
$$

Therefore,

$$46|_{10} = 101110|_2$$

If the decimal number is a fraction, it can be converted to binary by successive multiplication by 2. If the product is less than 1, a 0 is recorded, and if the product is greater than 1, a 1 is recorded. The multiplication process is continued until the answer becomes exactly 1 or when the desired accuracy is reached. To convert $.234|_{10}$ to binary,

$$
\begin{array}{cc}
\text{Integer} & \\
 & .234 \\
 & \underline{2} \\
0 & .468 \\
 & \underline{2} \\
0 & .936 \\
 & \underline{2} \\
1 & .982 \\
 & \underline{2} \\
1 & .744 \\
 & \underline{2} \\
1 & .488 \\
 & \underline{2} \\
0 & .976 \\
 & \underline{2} \\
1 & .952 \\
 & \underline{2} \\
1 & .904 \\
 & \underline{2} \\
1 & .808 \\
 & \underline{2} \\
1 & .616 \\
 & \underline{2} \\
1 & .232 \\
 & \underline{2} \\
0 & .464 \\
 & \vdots \\
\end{array}
$$

Therefore, $.234|_{10} = .001110111110+$, which is also equivalent to $.1676|_8$. As a quick check,

$$1676|_8 = 1(8)^{-1} + 6(8)^{-2} + 7(8)^{-3} + 6(8)^{-4} = 0.23389$$

3.3 BINARY CODES

3.3.1 BCD Notation

As was previously mentioned, digital computers perform arithmetic using the binary notation. It turns out that the pure binary form is impractical to implement since this would involve manipulating large amounts of data. The decimal 234, for example, consisting of three digits would become 11101010 containing eight bits in the binary form. To alleviate this problem, several alternate methods have been devised that involve using a binary-coded form rather than pure binary. The binary-coded decimal (BCD) notation results from using the binary equivalent of each decimal digit in a number. Thus, $234|_{10}$ would be written 0010 0011 0100, where each group of four binary digits is the binary equivalent of the numbers 2, 3, and 4, respectively. Note that this makes the conversion process simple, and the information then can be placed in or removed from the computer directly without having to convert the entire number to binary and vice versa.

Within the BCD notation itself several types of codes are popular such as the nonweighted excess-three code (or XS-3), the weighted 7421 and 8421 codes, and the Gray code (see Table 3.2). In a weighted code, weights can be assigned to the four binary digit positions so that the sum of the weights corresponding to the binary 1's results in the decimal digit being represented. In the 7421 code, for example, the weights of the respective positions from the right are 1, 2, 4, and 7 so that the number 1011 using this code would be equivalent to decimal $1(7) + 0(4) + 1(2) + 1(1)$ or 10, and the number 1000 would be simply 7. Similarly, the 8421 code has weights of 1, 2, 4, and 8 for the four positions starting from the right (i.e., same as a pure binary notation), so that number 0111 in this code corresponds to decimal 7. In a nonweighted code a constant decimal value cannot be assigned to each binary digit position. In the XS-3 code, for example, when the weights corresponding to the digits are added, the sum is $d + 3$, where d is the decimal digit the code represents. In this code, binary 0011 (decimal 3) is added to each digit to generate the code. Thus, the decimal 8 in this code would be 1011, formed by adding 0011 to 1000. An advantage of this code is that it is self-complementing; that is, inversion of all 1's and 0's for any number produces its "9's complement." For example, decimal digits 0 and 9 are represented by 0011 and 1100, respectively, and decimal digits 3 and 6 by 0110 and 1001, respectively. This feature is useful for performing subtraction in the computer.

Table 3.2 Binary Codes

Decimal	BCD Binary 8421	Excess-3	84-2-1	Biquinary 5-0-4-3-2-1-0	2-out-of-5 Code 8421P	Reflected Code
0	0000	0011	0000	0100001	10100	0000
1	0001	0100	0111	0100010	00011	0001
2	0010	0101	0110	0100100	00101	0011
3	0011	0110	0101	0101000	00110	0010
4	0100	0111	0100	0110000	01001	0110
5	0101	1000	1011	1000001	01010	0111
6	0110	1001	1010	1000010	01100	0101
7	0111	1010	1001	1000100	11000	0100
8	1000	1011	1000	1001000	10001	1100
9	1001	1100	1111	1010000	10010	1101

3.3.2 Biquinary Code

The biquinary code is a seven-bit decimal code with bits divided into a group of two (bi) and a group of five (quinary). In the "bi" part, the positions have weights of 5 and 0, respectively; in the "quinary" part, the weights are 4, 3, 2, 1, and 0, respectively. It should be noted that each valid number in this code has only two binary 1's, one each in the two groups. This feature facilitates error detection, since the occurrence of more than one binary 1 in either of the two groups indicates an error (see Table 3.2).

3.3.3 · Gray Code

The Gray code (reflected binary; cyclic) is useful in those applications where an analog signal is to be converted to a digital signal and the successive decimal representations (the next greater or lesser number) differ in only one digit position. A frequent use of this code is for shaft encoders and decoders where the shaft rotation (analog) is converted to digital data. The name "reflected" is derived from the fact that the code digits are obtained by changing the 0 to 1 at each power of 2 (i.e., at 2, 4, 8, 16, etc.) and repeating the previously used combinations for the other digits in reverse sequence.

3.3.4   Error-Detecting Codes

Since errors can occur in computers in the control section or in transmission of data or in networking, some means for error checking and detecting are necessary. A common method is by means of the parity code ("odd–even"), where an extra bit called the parity or check bit is added to each character code. The parity bit is always assigned so that the total number of 1's is always odd. For example, for 0101 ($5|_{10}$), a parity bit 1 is added as the first bit, so that it appears as 10101; for 0001, no bit is needed. Similarly, for even parity codes, a bit is added to each character code so that the total number of 1's is always even. With this approach, 0101 remains the same, while 0100 becomes 10100. Other five-bit decimal error-detection codes, called "two-out-of-five" codes, can also be devised where two of the five bits are always 1's. With five bits, it turns out that there are exactly 10 different possible combinations with two of the bits as 1's, and these can be assigned to the 10 decimal digits in any pattern to create a code. If the group is seen to contain less than or more than two 1's due to changing of 1 to 0 or 0 to 1, then an error has occurred. The various binary codes for decimal digits 0–9 are shown in Table 3.2.

3.3.5   Alphanumeric Codes

In many cases, computers are called upon to handle data containing not only numbers but also letters or characters. It therefore becomes necessary to develop a binary code for the alphabet. The code must also be able to represent decimal numbers as well as special characters. To be able to code the 26 letters of the alphabet, the 10 decimal digits, and the special characters ($+,-,*$, etc.), one would require at least six bits. This is because with six bits, 2^6 (=64) distinct combinations are possible. An alphanumeric code using six bits that is commonly used for internal representation inside computers is shown in Table 3.3. To be able to cover more than 64 characters (such as both uppercase and lowercase letters, special characters), seven- and eight-bit codes have also been created. One such code is the ASCII (American Standard Code for Information Interchange), which is a seven-bit code but with a parity bit added it becomes an eight-bit code. The EBCDIC (Extended BCD Interchange Code) is an eight-bit code. The standard 12-row, 80-column punch code is also shown in Table 3.3. A character is represented on the card by punching the appropriate rows in a column (a hole is treated as 1 and no hole is treated as 0). The rows are marked 12, 11, 0, 1,. . .,9 from top to bottom, so that to denote the letter A in a column, there will be two holes, one in row 12 and one in row 1, using the code shown in the table.

3.4   COMPLEMENTS

The arithmetic of addition and subtraction of numbers in a computer can be done conveniently by using complements. For a base R system there can be two types of complements: (1) the R's complement and (2) the $(R - 1)$'s complement. Thus, for base 10 we have the 10's complement and the 9's complement; for the binary base 2, we have the 2's complement and the 1's complement. The complement of a number is that quantity which, when added to the number, produces a power of the base of the number (for R's complement) or a power of the base minus one [for $(R - 1)$'s complement]. For example, in the decimal system, 3 is a 10's complement of 7 since $7 + 3 = 10$, or 74 is a 10's complement of 26 since $26 + 74 = 100 = 10^2$.

In general, for a positive N in base R with an integer part of n digits, the R's complement is $R^n - N$, so that

$$10\text{'s complement of } 3437|_{10} \text{ is } 10^4 - 3437 = 6563$$
$$10\text{'s complement of } 0.3437|_{10} \text{ is } 10^0 - .3437 = .6563$$
$$10\text{'s complement of } 26.3437|_{10} \text{ is } 10^2 - 26.3437 = 73.6563$$
$$2\text{'s complement of } 10011|_2 \text{ is } 2^5 - 10011 = 1101$$
$$2\text{'s complement of } .1011 \text{ is } 2^0 - .1011 = .0101$$

Similarly, for a positive N in base R with integer part of n digits and fraction part of m digits, the $(R - 1)$'s complement is $R^n - R^{-m} - N$. Thus,

$$9\text{'s complement of } 3437|_{10} \text{ is } 10^4 - 10^{-0} - 3437 = 6562$$
$$9\text{'s complement of } 0.3437|_{10} \text{ is } 10^0 - 10^{-4} - .3437 = .6562$$
$$1\text{'s complement of } 10011|_2 \text{ is } 2^5 - 2^{-0} - 10011 = 1100$$
$$1\text{'s complement of } .1011|_2 \text{ is } 2^0 - 2^{-4} - .1011 = .0100$$

Table 3.3 Alphanumeric Codes

Character	6-Bit Internal Code		7-Bit ASCII Code		8-Bit EBCDIC Code		12-Bit Card Code
A	010	001	100	0001	1100	0001	12,1
B	010	010	100	0010	1100	0010	12,2
C	010	011	100	0011	1100	0011	12,3
D	010	100	100	0100	1100	0100	12,4
E	010	101	100	0101	1100	0101	12,5
F	010	110	100	0110	1100	0110	12,6
G	010	111	100	0111	1100	0111	12,7
H	011	000	100	1000	1100	1000	12,8
I	011	001	100	1001	1100	1001	12,9
J	100	001	100	1010	1101	0001	11,1
K	100	010	100	1011	1101	0010	11,2
L	100	011	100	1100	1101	0011	11,3
M	100	100	100	1101	1101	0100	11,4
N	100	101	100	1110	1101	0101	11,5
O	100	110	100	1111	1101	0110	11,6
P	100	111	101	0000	1101	0111	11,7
Q	101	000	101	0001	1101	1000	11,8
R	101	001	101	0010	1101	1001	11,9
S	110	010	101	0011	1110	0010	0,2
T	110	011	101	0100	1110	0011	0,3
U	110	100	101	0101	1110	0100	0,4
V	110	101	101	0110	1110	0101	0,5
W	110	110	101	0111	1110	0110	0,6
X	110	111	101	1000	1110	0111	0,7
Y	111	000	101	1001	1110	1000	0,8
Z	111	001	101	1010	1110	1001	0,9
0	000	000	011	0000	1111	0000	0
1	000	001	011	0001	1111	0001	1
2	000	010	011	0010	1111	0010	2
3	000	011	011	0011	1111	0011	3
4	000	100	011	0100	1111	0100	4
5	000	101	011	0101	1111	0101	5
6	000	110	011	0110	1111	0110	6
7	000	111	011	0111	1111	0111	7
8	001	000	011	1000	1111	1000	8
9	001	001	011	1001	1111	1001	9
blank	110	000	010	0000	0100	0000	no punch
.	011	011	010	1110	0100	1011	12,8,3
(	111	100	010	1000	0100	1101	12,8,5
+	010	000	010	1011	0100	1110	12,8,6
$	101	011	010	0100	0101	1011	11,8,3
*	101	100	010	1010	0101	1100	11,8,4
)	011	100	010	1001	0101	1101	11,8,5
-	100	000	010	1101	0110	0000	11
/	110	001	010	1111	0110	0001	0,1
,	111	011	010	1110	0110	1011	0,8,3
=	001	011	011	1101	0111	1000	8,6

It is clear that to form the 9's complement of a decimal number, simply subtract each digit from 9. The 1's complement of a binary number is obtained by changing the 1's to 0's and the 0's to 1's. Note, also, that since it is easier to obtain the $(R - 1)$'s complement compared to R's, the latter can be derived from the former by adding R^{-m} to the least-significant digit. The complements serve a useful purpose when subtracting two numbers in the computer. For example, to subtract two positive numbers $A - B$, both of base R, the procedure is to add A to the R's complement of B. When an end carry occurs, it is rejected; when an end carry does not occur, the R's complement of the number with a negative sign is the required result. The procedure with $(R - 1)$'s complement is similar. In this case, the $(R - 1)$'s complement is added; if an end carry occurs, 1 is added to the least-significant digit; if it does not occur, the $(R - 1)$'s complement of the number with a negative sign is the result.

3.5 RADIX-POINT OPERATIONS—FIXED-POINT AND FLOATING-POINT NOTATIONS

While performing numerical computations, the decimal or the binary point in a number must be kept track of. The decimal point locates the unit's order in a number (i.e., 10's, 100's, etc.) and separates the positive and negative powers of the base. For example, in the number 345.721, the positive powers of the base are to the left of the decimal point and the negative powers are to the right of the decimal point, so that

$$345.721 = 3 \times 10^2 + 4 \times 10^1 + 5 \times 10^0 + 7 \times 10^{-1} + 2 \times 10^{-2} + 1 \times 10^{-3}$$

For computer operations, however, the coefficients of the powers are usually placed in predetermined locations, because the exact location of the decimal point is not important. This arrangement is known as fixed-point representation, and the decimal point for all numbers is commonly placed to the left of the most-significant digit. This would clearly limit the range of numbers to be used and would require scaling or normalizing of the numbers should a mathematical operation result in a number larger than the ability of the computer to handle. In scaling, the digits of the number are moved with respect to the assumed decimal point location, which implies either multiplication or division of the number by a power of the base depending on whether the decimal point is shifted to the right or to the left.

The problems involved in scaling can be avoided using the floating-point notation. In this case, the number is stored in a fixed format including the number itself and the power of the base. The general form of the number will be NR^m or NR^{-m}, where R is the radix and the number has its point m positions to the right or to the left, respectively, of the zero position. Either N (called mantissa) or m (called exponent) or both may be negative, and the number of digits for N or m need not be the same. The zero position of the point is customarily at the left end of N, with the left-hand digit for both N and m being the sign digit. For example, the number 705.8 can be written as .7058 $\times$ 10^3, with $N = .7058$, $m = 3$. Generally, the point is moved to some finite number of positions to the left depending on the storage space allocated for numbers within the computer. When performing addition or subtraction, the two numbers are shifted so that they line up with respect to the point position to make the exponents the same. Arithmetic operations are performed separately for the mantissa and the exponent; during multiplication, the exponents are added, while in division they are subtracted.

3.6 DIGITAL ARITHMETIC OPERATION CIRCUITS

Arithmetic operations in digital computers involving binary numbers or BCD numbers are performed using logic circuits (see Chapter 4). Some simple circuits that perform basic arithmetic such as addition, subtraction, multiplication, and division are presented here. Obviously, the same operation could be implemented using a variety of logic circuits; however, practical considerations such as hardware simplicity, economy, and speed should be kept in mind.

3.6.1 Adders

When two binary digits 0, 1 are added, the sum is either a 0, a 1, or a 10. The sum 10 occurs when both digits are 1; in this case 0 is the sum digit and 1 is the carry digit. The basic circuit to implement an adder of two bits is called a half-adder (HA). In terms of Boolean algebra, the sum (S) and carry (C) signals for a half-adder of two binary variables A, B can be represented by $C = AB$ and $S = AB' + A'B$. For adding two binary numbers of more than one digit, the carry from the lower-order bits must be added to the next-higher-order bit with the resulting carry added to the next higher order and so on. In other words, by combining more than one half-adder, numbers with more than one bit can be added as illustrated schematically in Fig. 3.1a. The numbers added are $...A_8A_4A_2A_1$ and $...B_8B_4B_2B_1$ producing a sum $...S_8S_4S_2S_1$, where the subscripts denote the weight of the binary position. In each HA, the half-carry and half-sum are shown by C and S, respectively. OR devices are used to combine signals from two HAs, because a carry of digit 1 occurs only if both digits to the adder are 1, in which case the half-sum from that adder will be 0. Note that in the addition process three digits are added, two from the corresponding positions of the numbers being added and the third is the carry from the preceding lower order. The equipment for adding three digits is known as a full-adder (FA). The box outlined in Fig. 3.1a consisting of two HAs and one OR device thus represents an FA. The HA circuits of Fig. 3.1a can be replaced by an FA circuit as shown in

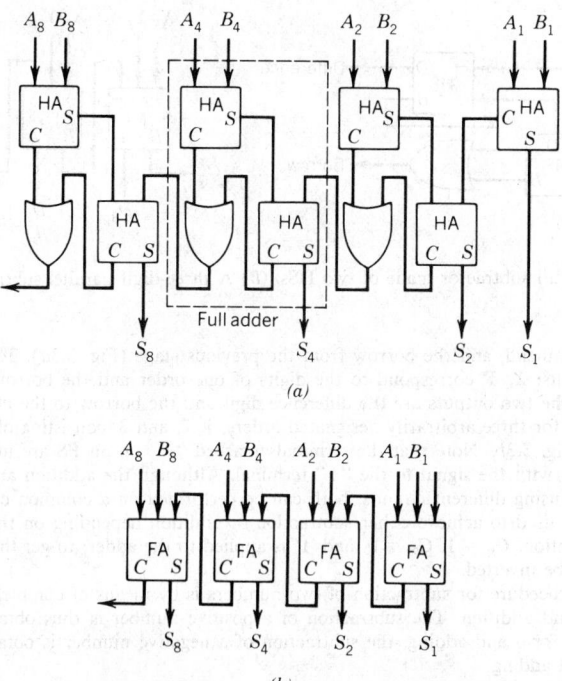

(a)

(b)

Fig. 3.1 (a) Half-adders for binary addition. (b) Full-adders for binary addition.

Fig. 3.1b. There are several alternatives for designing the HA circuitry (Fig. 3.2) with identical outputs. Note that the outputs of the HA circuits for adding A, B are two variables, C and S.

A deficiency of the described adder circuits is the length of computation time due to the "ripple" effect. The carry digits are generated in succession in going from the low to the high order, so that the total time for producing the overall sum will depend on the time required to travel from right to left as successive digits in the two numbers are added. To alleviate this problem, the basic hardware scheme can be modified in several ways, such as using A-O-I modules, carry bypass, or look-ahead carry schemes (see Ref. 1).

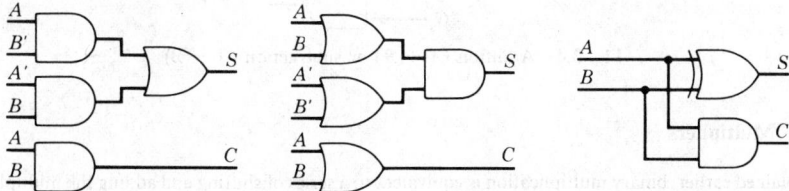

Fig. 3.2 Implementation of half-adders: $S = AB' + A'B$, $C = AB$.

3.6.2 Subtractors

The subtraction operation is similar to the addition operation provided the complement of the subtrahend is added to the minuend. However, direct subtraction using half- and full-subtractors is also possible, where the subtrahend bit is subtracted from the minuend and a 1 is borrowed from the next significant position in case the minuend bit is smaller. A half-subtractor (HS), thus, will have two inputs X, Y and two outputs, one that generates the difference (D) and the other that generates the binary signal (B) to inform the next stage that a 1 has been borrowed. The output B is 0 if $X \geqslant Y$, otherwise B is 1 when $X = 0$ and $Y = 1$. The Boolean expressions that represent this process are $B = X'Y$ and $D = X'Y + XY'$. A full-subtractor (FS) is similar to an HS except that it has three inputs: the

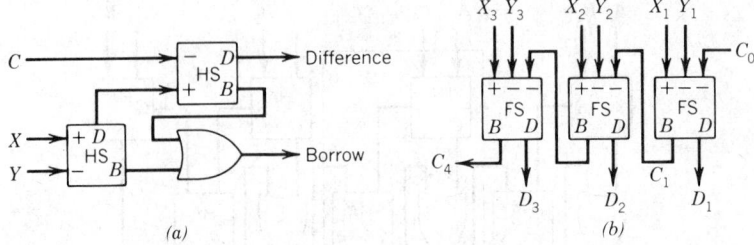

Fig. 3.3 (a) A full subtractor made of two HSs. (b) A three-digit parallel subtractor using FSs.

minuend, the subtrahend, and the borrow from the previous stage (Fig. 3.3a). The FS has two HSs and has three inputs: X, Y correspond to the digits of one order and the borrow digit C from the next-lower order; the two outputs are the difference digit and the borrow to the next-higher order. A parallel subtractor for three arbitrarily designated orders, 1, 2, and 3, consisting of a cascade of three FSs is shown in Fig. 3.3b. Note that the terminals marked "−" on an FS are interchangeable with each other but not with the signal to the "+" terminal. Although the addition and subtraction have been implemented using different circuits, both can be performed on a common circuit (Fig. 3.4). In this case an FA is used to achieve either subtraction or addition depending on the control signal G_s or G_a. For subtraction, $G_s = 1$, $G_a = 0$, and A' is applied to the adder; to get the correct difference the S signal must be inverted.

An alternate procedure for subtraction of two numbers is by means of complements, either 1's or 2's complement, and addition. The subtraction of a positive number is thus obtained by converting to its complement form and adding; the subtraction of a negative number is obtained by converting it to true form and adding.

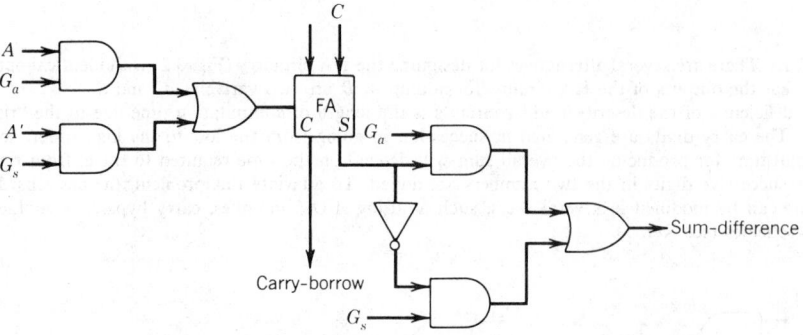

Fig. 3.4 Addition $(A + B)$ or subtraction $(A - B)$

3.6.3 Multipliers

As explained earlier, binary multiplication is equivalent to a series of shifting and adding the multiplicand depending on whether the multiplier digit is 0 or 1. In general, multiplication of two n-digit numbers produces a product with $2n$ digits. For example, to multiply two three-bit numbers $X_2X_1X_0$ and $Y_2Y_1Y_0$, the procedure is

		X_2	X_1	X_0
		Y_2	Y_1	Y_0
		X_2Y_0	X_1Y_0	X_0Y_0
	X_2Y_1	X_1Y_1	X_0Y_1	
X_2Y_2	X_1Y_2	X_0Y_2		
P_5	P_4	P_3	P_2	P_1	P_0

A multiplier network of FAs that implements the above logic for the product is shown in Fig. 3.5. In general, for n-bit numbers, n^2 two-input AND gates are required to form the bit products, which are then input to FAs to form the partial products. The time required for multiplication depends

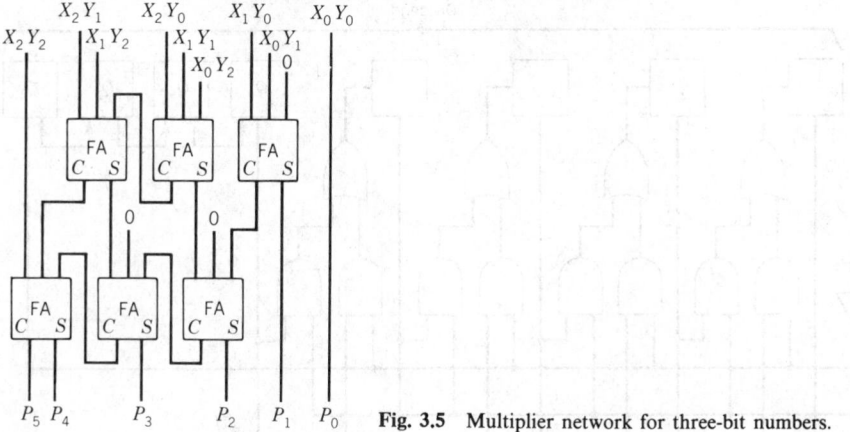

Fig. 3.5 Multiplier network for three-bit numbers.

on the time required for a carry from the right-hand end of the top adder to propagate to the P_5 output line. Owing to the excessive number of connections required, modified versions of this basic arrangement to improve speed and economy have been designed. For example, a scheme where a partial product is formed and added to the sum of previous partial products is shown in Fig. 3.6a. Note that this represents a 3×1 multiplier cell that forms the sum of

$$
\begin{array}{cccc}
 & S_2 & S_1 & S_0 \\
 & X_2 Y_j & X_1 Y_j & X_0 Y_j \\
\hline
P_3 & P_2 & P_1 & P_0 \\
\end{array}
$$

Thus, for a three-bit-number multiplication, one would need a combination of three cells, as shown in Fig. 3.6b. Another scheme consists of dividing the multiplier and/or the multiplicand digits in groups of three or more and taking action in accordance with the digit combination in each group rather than acting on the multiplier or the multiplicand digit individually.

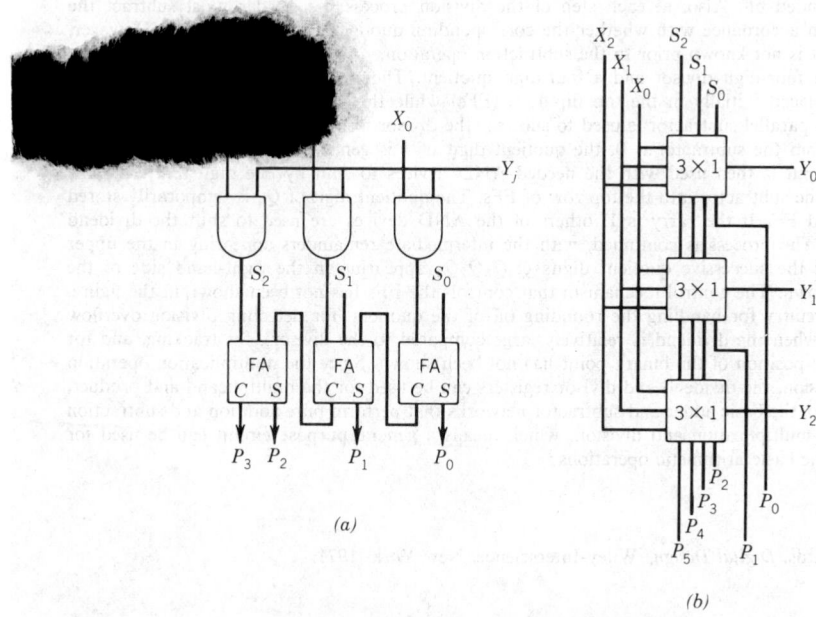

(a)

(b)

Fig. 3.6 (a) A 3×1 multiplier cell for forming partial products. (b) Three 3×1 multiplier cells for multiplying two three-bit numbers.

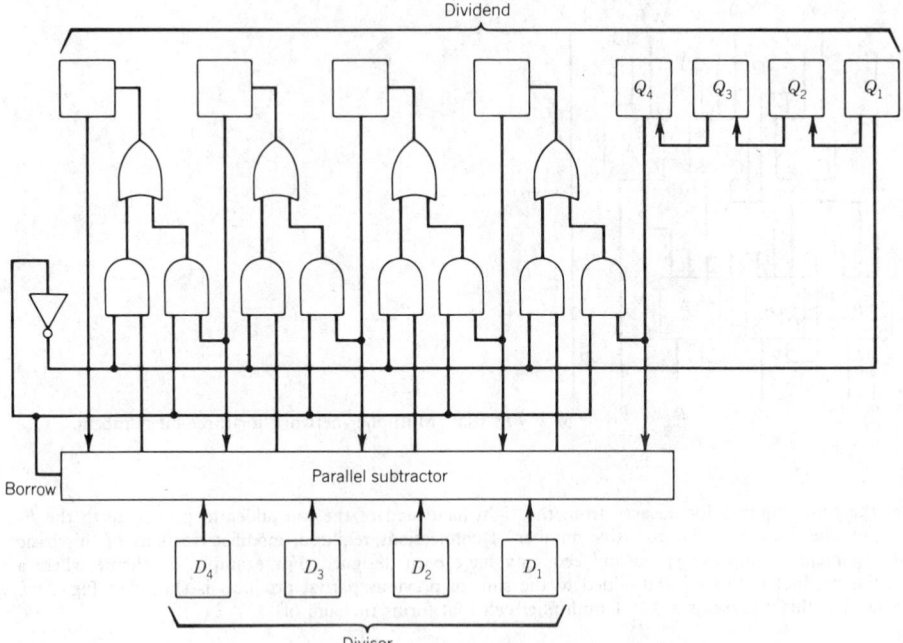

Fig. 3.7 Four-bit binary division network.

3.6.4 Dividers

Binary division is basically the inverse of multiplication. The division circuit networks tend to be more elaborate and time consuming than multipliers. This is because in division, the quotient may not be even, and a nonzero remainder may result. Where the remainder is not of interest, the quotient needs to be rounded off. Also, at each step of the division process the divider must subtract the divisor or zero in accordance with whether the corresponding quotien̶ ̶ ̶ ̶ ̶ ̶ ̶ ̶ ̶ ̶ However, the quotient digit is not known prior to the subtraction operation. A ̶ ̶ ̶ ̶ ̶ ̶ ̶ ̶ ̶ ̶ ̶ ̶ in Fig. 3.7 for a four-digit divisor and a four-digit quotient. The ̶ ̶ ̶ ̶ ̶ ̶ ̶ ̶ ̶ ̶ ̶ eight digits, is placed initially in the top flip-flops (FFs) while th̶ ̶ ̶ ̶ ̶ ̶ ̶ ̶ ̶ ̶ FFs. A five-digit parallel subtractor is used to subtract the divisor f̶.̶ ̶ ̶ ̶ ̶ ̶ If the borrow from the subtractor is 0, the quotient digit of 1 is gen̶e̶ ̶ ̶ ̶ ̶ ̶ The inverter output is then used with the needed AND devices to shift by one digit ̶a̶ ̶ ̶ ̶ ̶ difference from the subtractor into the top row of FFs. The quotient digit of Q_4 is temporarily stored in the right-hand FF. If the carry is 1, others of the AND devices are used to shift the dividend digits left by 1. The process is continued, with the intermediate remainders appearing in the upper row of FFs and the successive quotient digits $Q_4Q_3Q_2Q_1$ appearing in the right-hand side of the upper FFs as shown. The control mechanism that controls the FFs has not been shown in the figure. Additionally, circuitry for handling the rounding off of the quotient, for detecting division overflow (as can happen when the dividend is relatively large compared to the divisor), for tracking, and for manipulating the position of the binary point has not been shown. Since the multiplication operation is similar to division, the dividend and divisor registers can be used for the multiplicand and product, respectively. Also, the basic adder and subtractor networks that perform pure addition and subtraction can be used for multiplication and division, which means a general-purpose circuit can be used for performing all the basic arithmetic operations.

REFERENCE

1. R. K. Richards, *Digital Design*, Wiley-Interscience, New York, 1971.

CHAPTER 4
BOOLEAN ALGEBRA AND LOGIC CIRCUITS

R. ARVIKAR

Bell Laboratories
North Andover, Massachusetts

The advantage of the binary notation for mathematical operations in a digital machine is discussed in Chapter 3. The processing and storing of information is accomplished by means of registers comprised of binary cells that can assume either a 0 or a 1 value. The basic organization of a computer generally consists of memory registers for storing the information, process registers that manipulate the information (i.e., perform the actual computation), and control registers that control the sequence of the various operations. This chapter is devoted to an understanding of the basic electronic circuits used to accomplish these functions, and, in particular, how digital logic circuits manipulate the individual bits of information.[1,2] The behavior of these circuits that perform binary arithmetic, store data, and control the computational process can be most conveniently understood in terms of Boolean algebra.

4.1 BINARY LOGIC

Binary logic deals with binary variables that can have only two possible distinct values, 0 or 1 and three basic logic operations:

AND   The logical operation AND between two variables A and B is denoted by $A \cdot B$ and means that the result is 1 only if both A and B are 1, otherwise it is 0.

OR   The operation OR between two variables A and B is denoted by $A + B$ and means that the result is 1 if either A or B is 1; if both are 0, the result is 0.

NOT   The NOT operation results in the inverse of the variable so that $\overline{A}$ (or A') implies that the result is 1 if A is 0 and vice versa.

Electronic digital circuits that employ binary signals and binary logic are called switching circuits or logic circuits. The AND function is equivalent to the output of the two switches in series in Fig. 4.1a, while the OR function can be simulated by two switches in parallel as shown in Fig. 4.1b. The switches represent the variables A and B, with the "on" condition being logic 1 and the "off" condition being "0." In the AND circuit, the output is 1 only if both switches are closed or "on," while in the OR circuit of Fig. 4.1b the output is 1 if either switch is on.

Digital logic circuits that perform the basic operations are called gates (Fig. 4.2) and can be easily combined to implement complex expressions or functions of binary variables. For example, the function $\overline{A \cdot (B + C)}$ represents B, C as inputs to an OR gate with the resulting output along with the variable A being fed to an AND gate followed by an inverter gate to produce the function. Networks of logical gates are employed to generate desired logical manipulation paths for computations and control in computers. The design and analysis of logical circuitry can be performed using Boolean algebra.

4.2 BOOLEAN ALGEBRA—THEOREMS AND POSTULATES

Boolean algebra, in particular as applied to binary elements using binary operators, follows some of the same laws (such as closure, association, commutation, etc.) that govern simple algebra and arithmetic covering real numbers. Boolean algebra is very useful in the analysis of electrical switching circuits or logic gates that operate in either the "off" or "on" condition and that are employed to perform binary arithmetic in digital computers. The two-valued Boolean algebra is defined on a set of two elements (0, 1) and uses the basic operators $+$ and $\cdot$ (similar to the AND and OR operators described

Input A B Output   Input A B Output

(a)   (b)

Fig. 4.1   Switching circuits showing (a) AND and (b) OR operations.

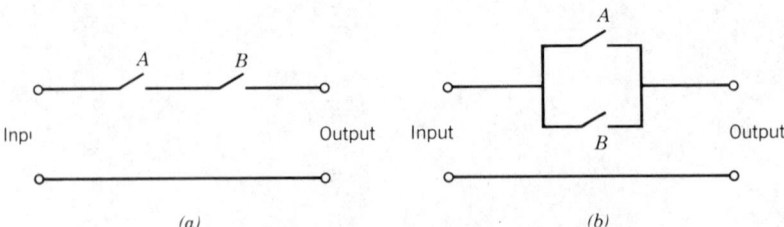

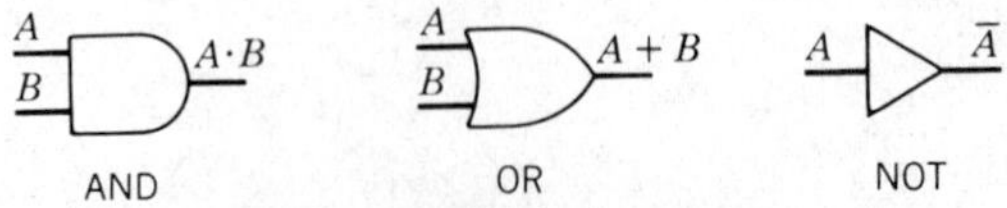

AND   OR   NOT

Fig. 4.2   Logical operators.

earlier) and a complement operator (similar to the inverter NOT). Some of the basic theorems and postulates associated with Boolean algebra are shown in Table 4.1. (Some · symbols have been omitted.) An important property of the postulates of Boolean algebra is duality; that is, every algebraic expression remains valid if the operators and the identity elements are interchanged. For example, the associative theorem states that $A + (B + C) = (A + B) + C$. Its dual-counterpart expression states that $A \cdot (B \cdot C) = (A \cdot B) \cdot C$. The dual of an algebraic expression is obtained simply by interchanging the OR (+) and AND (·) operators and replacing 1's by 0's and 0's by 1's.

DeMorgan's theorem shown for two variables in Table 4.1 is more general, and it can be shown that

$$(A + B + C)' = A'B'C'$$

or

$$(A_1 + A_2 + \cdots + A_n)' = A'_1 A'_2 \cdots A'_n$$

Similarly,

$$(A_1 \cdot A_2 \cdot \cdots A_n)' = A'_1 + A'_2 + \cdots + A'_n$$

This shows that the complement of a function can be obtained by interchanging + by · and complementing each literal, which is the same as forming the dual of the function and complementing each literal.

The theorems listed previously can be derived from the basic postulates. For example, to prove $A + AB = A$:

$$A + AB = A \cdot 1 + A \cdot B = A \cdot (1 + B) = A \cdot 1 = A$$

Another technique to prove the theorems, as well as for simplifying complex algebraic functions or relationships, is by means of truth tables. Simplification of complex functions is often required since in devising logic circuits that simulate or represent Boolean expressions, a simpler expression would require lesser circuitry for implementation. In evaluating Boolean expressions, the operator precedence to be followed is: parentheses, NOT, AND, and OR, which means expressions inside parentheses must be evaluated first.

4.3 TRUTH TABLES

The truth table is an exhaustive technique whereby variables in the expressions are assigned the 1 or 0 value and all possible combinations are individually evaluated. For example, the theorem $(A + B)' = A' \cdot B'$ can be proved using the truth table below. The two sides of the equation are shown to have the same values for all possible values of A and B:

A	B	$(A + B)$	$(A + B)'$	A'	B'	$A'B'$
0	0	0	1	1	1	1
1	0	1	0	0	1	0
0	1	1	0	1	0	0
1	1	1	0	0	0	0

Table 4.1 Theorems and Postulates of Boolean Algebra

		"+" Operator	"·" Operator
Postulates	Identity elements	$A + 0 = A$	$A \cdot 1 = A$
	Complements	$A + A' = 1$	$A \cdot A' = 0$
	Commutation	$A + B = B + A$	$A \cdot B = B \cdot A$
	Distribution	$A(B + C) = AB + AC$	$A + BC = (A + B)(A + C)$
Theorems		$A + A = A$	$A \cdot A = A$
		$A + 1 = 1$	$A \cdot 0 = 0$
	Involution	$(A')' = A$	
	Association	$A + (B + C) = (A + B) + C$	$A(BC) = (AB)C$
	DeMorgan's	$(A + B)' = A' \cdot B'$	$(A \cdot B)' = A' + B'$
	Absorption	$A + AB = A$	$A(A + B) = A$

In general, for an expression containing n different variables, the truth table will contain 2^n rows. In the example above, there are two variables, A, B (each of which can assume one of two values, 0 or 1); that is, $n = 2$ and so there are $2^2 = 4$ rows.

4.4 LOGIC SIMPLIFICATION

Boolean algebraic expressions consisting of binary variables and operators are implemented in computers using logic circuits. A Boolean function can only have either a 0 or a 1 value and can be represented by several forms such as algebraic expression, Venn diagram, truth tables, Karnaugh maps, and logic gates. For example, the function $F = XY' + YX'$ is a Boolean function in algebraic form. We could use a truth table using four rows to list the values the function will take for all combinations of X, Y values.

X	Y	X'	Y'	XY'	$X'Y$	$F = XY' + X'Y$
0	0	1	1	0	0	0
0	1	1	0	0	1	1
1	0	0	1	1	0	1
1	1	0	0	0	0	0

It is evident that there are only two cases—$X = 0$, $Y = 1$ or $X = 1$, $Y = 0$—when this function is true; that is, it is equal to 1.

We could also express the function in terms of the logic gates shown in Fig. 4.2, and this is shown in Fig. 4.3. From an economic standpoint, the implementation of a logic circuit is facilitated by simple expressions because it means fewer logic gates. Therefore, very often the original function or expression needs to be put in an alternate form with fewer literals and operations. Although the final form of the function is changed, it will yield the same value as the original function for the same values of the variables. Logic simplification can be accomplished in many ways, the most general being algebraic methods and Karnaugh maps.

4.4.1 Algebraic Methods

Algebraic methods are based on the theorems and postulates described earlier. There are no hard and fast rules, and the same result can be arrived at in many different ways. For example,

$$(X + Y)'(YZ)' = (X + Y + YZ)' = (X + Y(1 + Z))' = (X + Y)'$$
$$(X + Y)(X' + Z)(Y + Z) = (Y + XZ)(X' + Z)$$
$$= YX' + X'XZ + YZ + XZZ$$
$$= YX' + YZ + XZ + XX'$$
$$= (X' + Z)(X + Y)$$

The principle of duality is also useful in deriving alternate forms. The expression $X(X' + Y) = XY$ means that its dual is also true, that is, $X + X' \cdot Y = X + Y$. Similarly, $XY + X'Z + YZ = XY + X'Z$ and its dual $(X + Y)(X' + Z)(Y + Z) = (X + Y)(X' + Z)$ are both true.

Boolean expressions can also be represented by minterms (or standard product) and maxterms (or standard sum). The minterms are all the possible combinations of variables and their complements. For the two variables X, Y there are four minterms: XY, XY', $X'Y$, $X'Y'$. The eight minterms for the three variables X, Y, Z are shown in Table 4.2. The minterms are denoted by m_j, $j = 0$ to 7, where the subscript j represents the decimal value of the binary combination with the primed and unprimed variables being 0 and 1, respectively. The maxterms are formed in a similar way but using OR instead of AND and are denoted by m_j, $j = 0$ to 7 with j being the decimal value of the combination with the primed and unprimed variables being 1 and 0, respectively. In general, for n variables there will be 2^n minterms and maxterms.

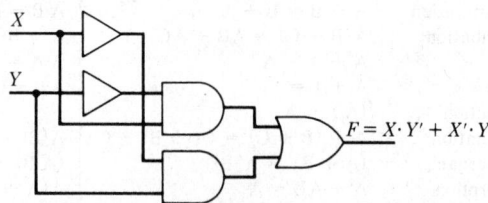

Fig. 4.3   Implementation of $F = X \cdot Y' + X' \cdot Y$.

Table 4.2 Minterms and Maxterms for Three Variables

Variables			Minterms		Maxterms	
X	Y	Z	Term	Notation	Term	Notation
0	0	0	X'Y'Z'	m_0	X + Y + Z	M_0
0	0	1	X'Y'Z	m_1	X + Y + Z'	M_1
0	1	0	X'YZ'	m_2	X + Y' + Z	M_2
0	1	1	X'YZ	m_3	X + Y' + Z'	M_3
1	0	0	XY'Z'	m_4	X' + Y + Z	M_4
1	0	1	XY'Z	m_5	X' + Y + Z'	M_5
1	1	0	XYZ'	m_6	X' + Y' + Z	M_6
1	1	1	XYZ	m_7	X' + Y' + Z'	M_7

Any Boolean function can be expressed as a sum of minterms (i.e., minterms combined with OR) or a product of maxterms (i.e., maxterms combined with AND). If a function is available as a truth table, then it is relatively simple to convert the table into an algebraic expression using maxterms and minterms. Consider the function of three variables shown in Table 4.3. The function is 1 for four combinations of the variables and 0 for the remaining four. The combinations for which f is 1 corresponds to the minterms m_3, m_4, m_6, and m_7, and, so, the function can be expressed as

$$f = X'YZ + XY'Z' + XYZ' + XYZ$$
$$= m_3 + m_4 + m_6 + m_7 = \sum (3,4,6,7)$$

By noting the maxterms corresponding to the combinations for which f is 0, the alternate form, in terms of maxterms, can be written as

$$f = M_0 \cdot M_1 \cdot M_2 \cdot M_5 = \prod (0,1,2,5)$$

The maxterm form could have been derived from the minterm form by expressing f as a sum of minterms that result in f being 0 and complementing it:

$$f = (f')' = (m_0 + m_1 + m_2 + m_5)'$$
$$= (X'Y'Z' + X'Y'Z + X'YZ' + XY'Z)'$$
$$= (X'Y'Z')' \cdot (X'Y'Z)' \cdot (X'YZ')' \cdot (XY'Z)'$$
$$= (X + Y + Z) \cdot (X + Y + Z') \cdot (X + Y' + Z) \cdot (X' + Y + Z')$$
$$= M_0 \cdot M_1 \cdot M_2 \cdot M_5$$

The conversion of a given Boolean expression to the sum of minterms or product of maxterms form is done readily. To obtain the sum-of-minterms form, the expression is first simplified as a sum of products of variables, and the missing terms are added successively by ANDing the missing variable,

Table 4.3 Truth table for function f.

X	Y	Z	f
0	0	0	0
0	0	1	0
0	1	0	0
0	1	1	1
1	0	0	1
1	0	1	0
1	1	0	1
1	1	1	1

say X, with the expression $(X + X')$. The function of $f = Y + X'Z$ is equivalent to $\Sigma(1,2,3,6,7)$ as follows:

$$
\begin{aligned}
f = Y + X'Z &= Y(X + X') + X'Z(Y + Y') \quad \text{(since } X + X' = Y + Y' = 1) \\
&= YX + YX' + X'YZ + X'Y'Z \\
&= YX(Z + Z') + YX'(Z + Z') + X'YZ + X'Y'Z \\
&= XYZ + XYZ' + \underline{X'YZ} + X'YZ' + \underline{X'YZ} + X'Y'Z \quad \text{(since } A + A = A) \\
&= XYZ + XYZ' + X'YZ + X'YZ' + X'Y'Z \\
&= m_7 + m_6 + m_3 + m_2 + m_1
\end{aligned}
$$

To obtain the product-of-maxterms form, the procedure is the same except that the function must be expressed as a product (OR) of terms and a missing variable, say X, is provided as XX'. Thus,

$$
\begin{aligned}
f = Y + X'Z &= (Y + X')(Y + Z) \quad \text{(using distribution)} \\
&= (Y + X' + ZZ')(XX' + Y + Z) \\
&= \underline{(Y + X' + Z)\,(Y + X' + Z')\,(X + Y + Z)(X' + Y + Z)} \quad \text{(since } A \cdot A = A) \\
&= (Y + X' + Z)(Y + X' + Z')(X + Y + Z) \\
&= M_4\, M_5\, M_0
\end{aligned}
$$

Note that, in general, the product of maxterms for a function can also be obtained from the sum of minterms by noting the numbers (subscripts) of the missing minterms and writing them as maxterms. Thus, $f = \Sigma(1,3,4,6) = \Pi(0,2,5,7)$. It is also evident that the complement of a function in minterms form is obtained by summing the missing minterms. That is, if $f = \Sigma(1,3,4,6)$, then $f' = \Sigma(0,2,5,7)$.

Algebraic functions expressed as minterms or maxterms are known as "canonical" forms. Another way to express an algebraic function is in "standard" form, either as a sum of products (sop) or as a product of sums (pos). The sop form is an expression containing OR terms, where each term is expressed as a product of the variables. Similarly, the pos form is an expression containing AND terms with each term expressed as a sum of the variables. For example, $f = X + XY + XY'Z$ is in sop form, whereas $f = X(X + Z)(X' + Y + Z)$ is in pos form. These forms can be thought of as reduced canonical forms, since not all the literals required to make it the equivalent sum-of-minterms of product-of-maxterms form of the function are present.

4.4.2 Karnaugh Maps (K-Maps)

A Karnaugh map (also known as Veitch diagram) is a graphical technique to represent and simplify a Boolean function. The map basically consists of squares, each representing a minterm. Therefore, any Boolean function expressed as a sum of minterms can be formed readily by noting the appropriate squares on the map. The map also offers a way to simplify or reduce a complex Boolean function to a unique form. A two-variable K-map is shown in Fig. 4.4. One variable's values appear as the labels for the vertical edge of the squares ($B = 0$ and $B = 1$), and the other variable's values correspond to the horizontal edges ($A = 0$ and $A = 1$). Each square contains the minterms reflected by the values of the variables. For example, the minterm $m_2 = AB'$ appears in the upper right square. Note, also, that A appears primed in the left column and unprimed in the right column, while B is primed in the top row and unprimed in the bottom row. Any function of two variables expressed as a sum of minterms can be marked on the map by putting a 1 in the appropriate square corresponding to the minterm. For example, $f = A + B$ appears as three 1's corresponding to m_2, m_1, and m_3 (see Fig. 4.4b, since $f = A + B = \Sigma(1,2,3)$.

A three-variable map can be constructed along similar lines and will require eight squares to represent the eight minterms of three variables (Fig. 4.5). There are four squares for each variable where it appears unprimed (i.e., where it is equal to 1) and four squares where the same variable is primed. The variable with its symbol is written under the four squares where it is unprimed, last two columns for A, middle two columns for B, and the bottom row for C. It is interesting to note that in any two adjacent squares along any row or column only one variable changes its values. A four-variable map, which requires $2^4 = 16$ squares, is shown in Fig. 4.6.

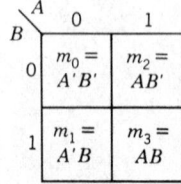

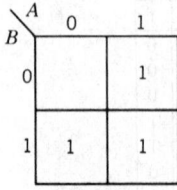

(a)   (b)

Fig. 4.4 (a) A two-variable Karnaugh map., (b) Map of $f = A + B$.

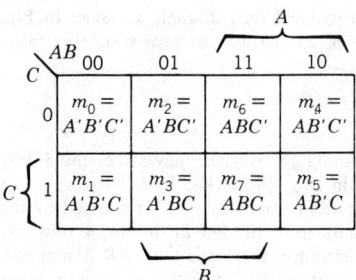

Fig. 4.5   A three-variable K-map.

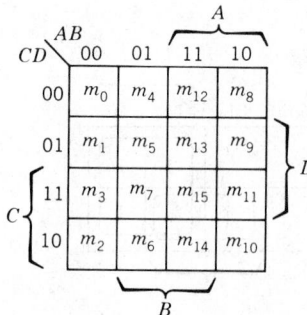

Fig. 4.6   A four-variable map.

The fact that only one variable changes in two adjacent squares is helpful in simplifying a Boolean function. When two terms from adjacent squares are combined using OR, the result is a single term containing only the unchanged variables. For example, in Fig. 4.5, if m_2 and m_6 were to be ORed, the result is $m_2 + m_6 = A'BC' + ABC' = BC'(A' + A) = BC'$. This procedure can be generalized by mapping the function, that is, by marking the appropriate squares on the map by 1 and then enclosing the adjacent squares that contain 1's. The simplification of $f = A'B'C + A'BC + ABC' + AB'C'$ is shown in Fig. 4.7a. The first term corresponds to the bottom left square, since $A'B'C$ means a square that does not belong to either A or B (which rules out the last two columns and the middle two columns) but belongs to C (the bottom row). Similarly, the term $AB'C'$ corresponds to the top right square since it belongs to A but not to B (middle two columns) or C (bottom row). The complete function is thus shown by 1's in four of the eight squares. The next step is to enclose adjacent squares as shown in the figure. The top two squares simplify to AC' while the two bottom squares simplify to $A'C$. Thus f can be simplified to $f = AC' + A'C$. The squares do not have to be physically adjacent to be considered adjacent. The first and the last column in the three-variable map are actually "adjacent" since combining two squares from these columns in the top row would result in the function $B'C'$ or $B'C$ in the bottom row. This is illustrated in Fig. 4.7b where $f = A'B'C' + A'BC + A'BC' + AB'C'$ is mapped and, by enclosing adjacent squares, is shown to be the same as $f = B'C' + A'B$. In a three-variable map, while two adjacent squares can be combined to give a term containing

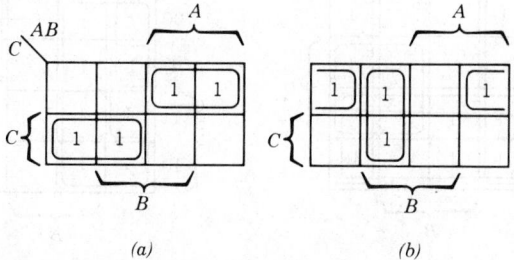

(a)   (b)

Fig. 4.7   (a) $f = A'B'C + A'BC + ABC' + AB'C' = A'C + AC'$. (b) $f = A'B'C' + A'BC + A'BC' + AB'C' = B'C' + A'B$.

two variables, four properly placed squares can be combined to yield only a single variable. In Fig. 4.8, $f = m_2 + m_6 + m_3 + m_7$ is shown to be equal to $f = B$, as can also be proved algebraically:

$$f = A'BC' + ABC' + A'BC + ABC$$
$$= A'B(C' + C) + AB(C' + C) = A'B + AB$$
$$= B(A' + A) = B$$

In enclosing the proper squares, if the adjacent squares containing 1's overlap, they are counted only once. For example, $f = BC + AC + AB + AB'C'$ is shown in Fig. 4.9 marked by five squares. The term BC represents the middle two adjacent squares in the bottom row, while AC is mapped by the last two squares in the same row. In this case, the third square from the left in the lower row (m_7 square), which is overlapped, contains only one 1 in it. Furthermore, when the term AB is mapped, it is shown by the third column from the left filled with 1's, again overlapping the m_7 square. Since this square already contains a 1 in it, another 1 need not be placed in the square again. Note that the function mapped in Fig. 4.9 can be simplified to be $f = A + BC$.

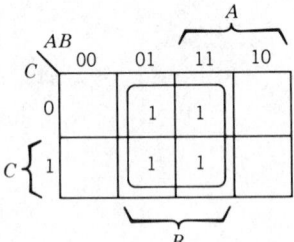

Fig. 4.8 $f = A'BC' + ABC' + A'BC + ABC = B.$

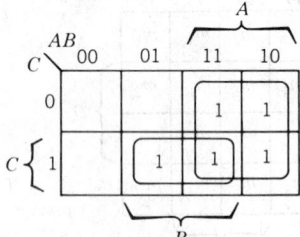

Fig. 4.9 $f = BC + AC + AB + AB'C' = A + BC.$

The simplification of a four-variable map is similar. In this case, two adjacent squares result in a term with three variables (because each individual square represents a four-variable term), four adjacent squares result in a two-variable term and eight adjacent squares result in a one-variable term. The functions $f = B' + C + AD$ and $f = A' + B'D + B'C'$ are shown mapped in Fig. 4.10a and 4.10b. Note the rule of overlapping squares containing no more than one 1 as explained earlier. Adjacent squares in a four-variable map are not only those that are physically adjacent in a row or column but also those in the upper and lower edges or the right and left edges.

Five- and six-variable maps are possible, but are not as convenient to use. A five-variable map (Fig. 4.11) will require $2^5 = 32$ squares. The minterm assigned to each square is obtained by reading the binary numbers for the row and the columns and obtaining the decimal equivalent. The adjacent square for a minterm is not only its neighbor but also its mirror image about the double centerline. The function $f = A'D'E' + BE + AB'E'$ is shown in Fig. 4.11. A six-variable map requiring $2^6 = 64$ squares representing m_0 through m_{63} minterms is shown in Fig. 4.12.

In general, when a function is expressed in its equivalent algebraic form by summing the minterms

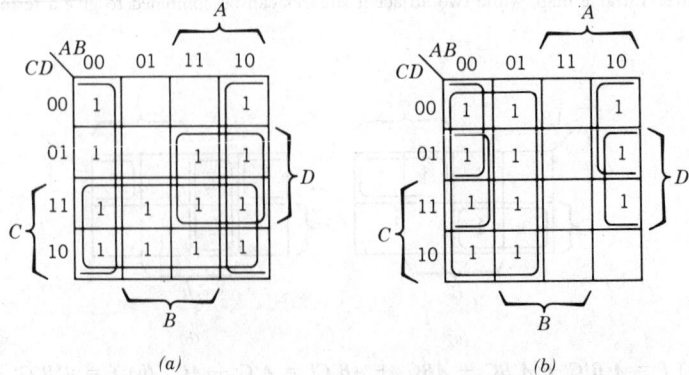

(a)   (b)

Fig. 4.10 (a) Map of $f = B' + C + AD$. (b) Map of $f = A' + B'D + B'C'$.

CDE
AB
000 001 011 010 110 111 101 100
00 1 1
01 1 1 1 1 1 1
11 1 1 1 1
10 1 1 1 1
C
B
A
D
E
E

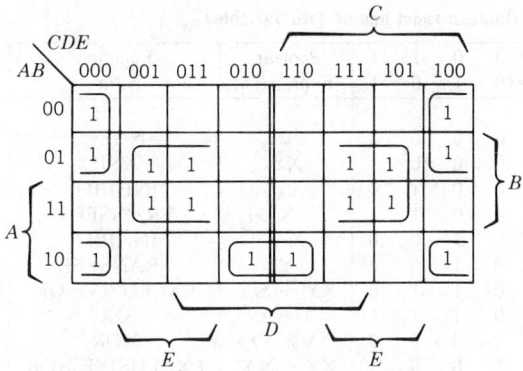

Fig. 4.11 A five-variable map showing $f = A'D'E' + BE + AB'E'$.

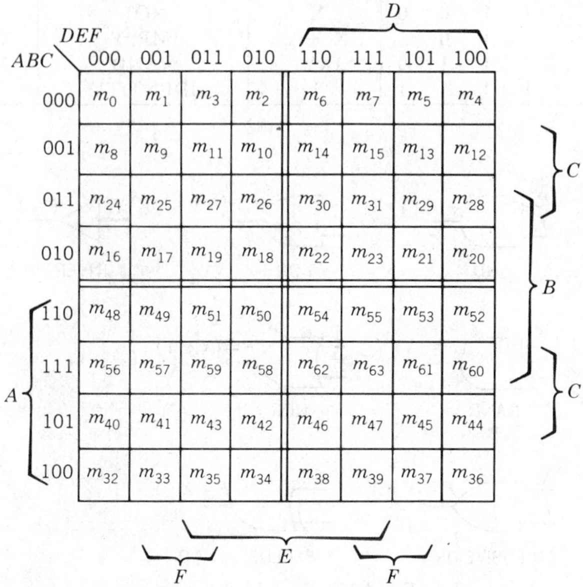

DEF
ABC
000 001 011 010 110 111 101 100
000 m_0 m_1 m_3 m_2 m_6 m_7 m_5 m_4
001 m_8 m_9 m_{11} m_{10} m_{14} m_{15} m_{13} m_{12}
011 m_{24} m_{25} m_{27} m_{26} m_{30} m_{31} m_{29} m_{28}
010 m_{16} m_{17} m_{19} m_{18} m_{22} m_{23} m_{21} m_{20}
110 m_{48} m_{49} m_{51} m_{50} m_{54} m_{55} m_{53} m_{52}
111 m_{56} m_{57} m_{59} m_{58} m_{62} m_{63} m_{61} m_{60}
101 m_{40} m_{41} m_{43} m_{42} m_{46} m_{47} m_{45} m_{44}
100 m_{32} m_{33} m_{35} m_{34} m_{38} m_{39} m_{37} m_{36}
D
C
B
C
A
E
F
F

Fig. 4.12 A six-variable map.

(i.e., the squares containing 1's) it will be in the sop form. The same function can also be expressed as a pos by combining the squares that do not contain a 1. The result, however, will be the complement of the function (since it combines squares that cause the function to be 0 rather than 1) from which the desired functions can be obtained merely by complementing. For example, the function $f = A' + B'D + B'C'$ as mapped by 1's in Fig. 4.10*b* is the sop form; its complement is $f' = AB + ACD'$, which represents the 0's map. Complement of f' results in $f = (A' + B')\cdot(A' + C' + D)$, the pos form.

4.5 LOGIC CIRCUIT IMPLEMENTATION

4.5.1 Basic Logic Gates

Boolean functions are implemented using logic gates. Boolean functions or expressions contain either a + or · operator, which is equivalent to the OR or AND operation using logic gates. While the AND, OR, and NOT gates represent the basic Boolean functions of a pair of binary variables X, Y, there are other functions that can be formed (out of a possible $2^{2^2} = 16$) as listed in Table 4.4. The truth tables for the functions and their operator notations are also shown therein. The gate symbols for the commonly implemented functions are shown in Fig. 4.13. The AND gate performs the operation

Table 4.4   Boolean functions of Two variables

F	X = 0	0	1	1	Boolean Expression	Function Name	Operator Notation
	Y = 0	1	0	1			
F_0	0	0	0	0	0	NULL	
F_1	0	0	0	1	XY	AND	$X \cdot Y$
F_2	0	0	1	0	XY'	INHIBIT	X/Y
F_3	0	0	1	1	X	TRANSFER	
F_4	0	1	0	0	X'Y	INHIBIT	Y/X
F_5	0	1	0	1	Y	TRANSFER	
F_6	0	1	1	0	XY' + X'Y	EXCLUSIVE OR	$X \oplus Y$
F_7	0	1	1	1	X + Y	OR	X + Y
F_8	1	0	0	0	(X + Y)'	NOR	$X \downarrow Y$
F_9	1	0	0	1	XY + X'Y'	EXCLUSIVE NOR (EQUIVALENCE)	$X \odot Y$
F_{10}	1	0	1	0	Y'	NOT	Y'
F_{11}	1	0	1	1	X + Y'	IMPLY	$Y \rightarrow X$
F_{12}	1	1	0	0	X'	NOT	X'
F_{13}	1	1	0	1	X' + Y	IMPLY	$X \rightarrow Y$
F_{14}	1	1	1	0	(XY)'	NAND	$X \uparrow Y$
F_{15}	1	1	1	1	1	IDENTITY	

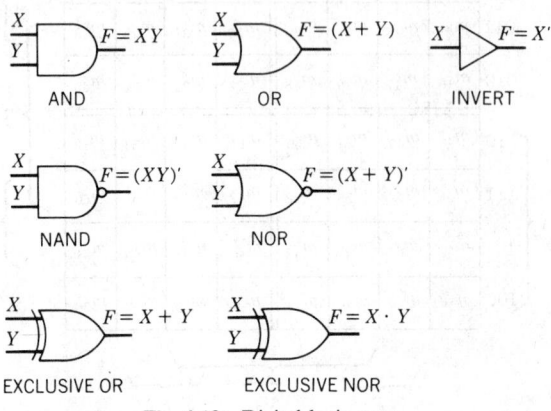

Fig. 4.13   Digital logic gates.

$F = X \cdot Y$, while the OR gate represents $F = X + Y$. The NOR gate (NOT OR) is the complement of OR function, that is, it represents $F = (X + Y)'$, while the NAND gate (NOT AND) is the complement of AND function resulting in $F = (X \cdot Y)'$. These gates are easily constructed using transistor circuits and are widely used in digital circuit design. The EXCLUSIVE OR is similar to OR but the function has a value 1 if either variable is 1 but not when both are equal, either 0 or 1. The equivalence function is 1 only when both variables are equal, either 0 or 1. Thus, the EXCLUSIVE OR and the equivalence functions are complements of each other. Although the gates shown are two-input gates, multiple input gates are also used (Fig. 4.14).

Complex functions are implemented by arranging the basic gates to generate the result. For example, the logic circuits in Figs. 4.15a and 4.15b represent the functions $F = XY'Z + X'YZ + X'Y'$ and $F = ABC + DE$, respectively.

It was described earlier that Boolean expressions can be simplified to the sop or pos form. These forms can be readily implemented by AND and OR gates. The sop form can be simulated by AND gates, representing the product, that are connected to OR gates to perform the sum. In pos form, OR gates are connected to a single AND gate. These arrangements are known as two-level implementation. However, transistor logic circuits are more frequently constructed with NAND and NOR gates rather than AND and OR gates. This means that Boolean expressions should be first converted to forms that render the NAND and NOR gate implementation feasible.

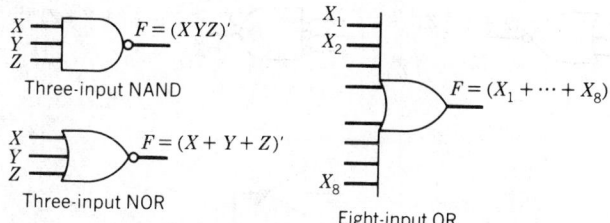

Fig. 4.14 Multiple input gates.

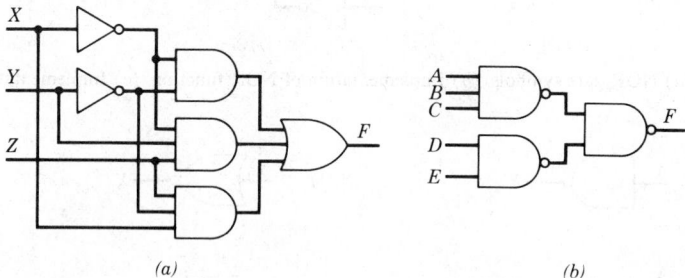

Fig. 4.15 (a) Implementation of $F = X'Y' + X'YZ + XY'Z$. (b) Implementation of $F = ((ABC)' \cdot (DE)')' = ABC + DE$.

4.5.2 NAND and NOR Gates

The NAND gate, along with its alternate graphic symbol, is shown in Fig. 4.16a. The same function can be obtained by ANDing the inputs and inverting it (Fig. 4.16b), or by inverting the inputs and forming the OR (Fig. 4.16c). Similarly, the NOR function shown symbolically in Fig. 4.17a can be obtained by ORing the inputs and inverting (Fig. 4.17b) or by inverting the inputs and forming the AND (Fig. 4.17c). Both the NAND and NOR gates are universal gates because any logic function can be expressed in terms of these. For example, the basic AND, OR, and NOT operations can be implemented by using only NAND and NOR gates (Fig. 4.18). This means that any circuit expressed with AND/OR gates can be converted to NAND or NOR implementation by replacing the basic AND/OR gates by means of equivalent gates from Fig. 4.18.

The general procedure to implement a Boolean function by NAND gates is to first express the function in sop form; each product is then represented by NAND gates with their outputs combined in second-level NAND gates. Consider the function $F = XY'Z' + YZ$ shown in map form and in NAND circuit form in Fig. 4.19b. In order to express the function using NOR gates, the procedure is similar. The function must first be expressed as pos form with each sum term represented by input

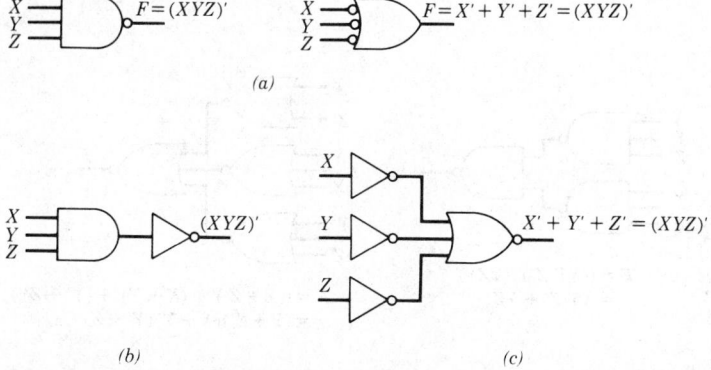

Fig. 4.16 (a) NAND gate symbols. (b) Implementation of NAND function. (c) Implementation of NAND function.

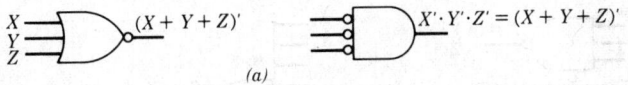

(a)

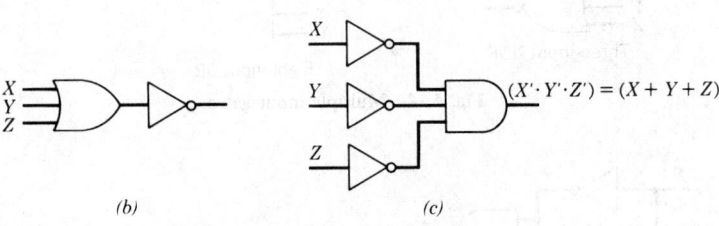

(b)   (c)

Fig. 4.17 (a) NOR gate symbols. (b) Implementation of NOR function. (c) Implementation of NOR function.

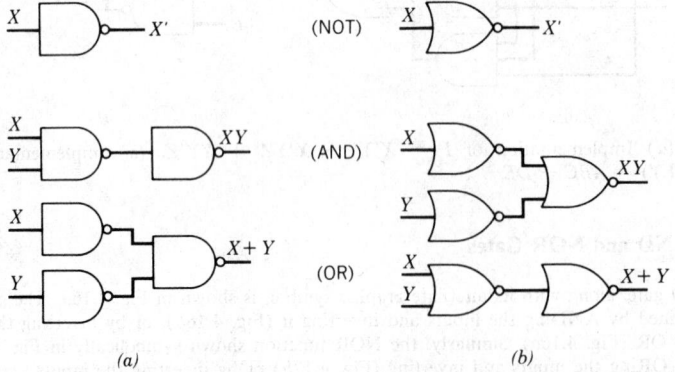

(a)   (b)

Fig. 4.18 (a) Implementation of NOT, AND, and OR using NAND gates. (b) Implementation of NOT, AND, and OR using NOR gates.

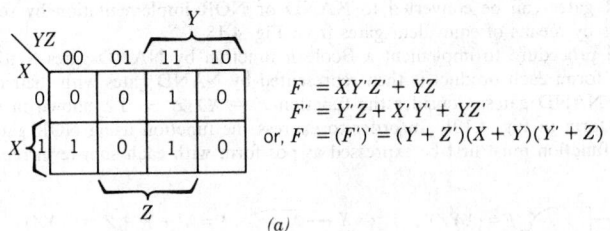

$F = XY'Z' + YZ$
$F' = Y'Z + X'Y' + YZ'$
or, $F = (F')' = (Y + Z')(X + Y)(Y' + Z)$

(a)

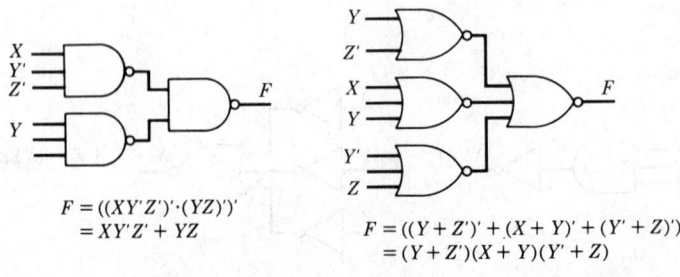

$F = ((XY'Z')' \cdot (YZ)')'$
$= XY'Z' + YZ$

$F = ((Y + Z')' + (X + Y)' + (Y' + Z)')'$
$= (Y + Z')(X + Y)(Y' + Z)$

(b)   (c)

Fig. 4.19 (a) Map for $F = XY'Z' + YZ$. (b) NAND circuit for $F = XY'Z' + YZ$. (c) NOR circuit for $F = (Y + Z')(X + Y)(Y' + Z)$.

to a first-level NOR gate followed by a second-level NOR gate to produce the product. For example, the function $F = XY'Z' + YZ$ can be expressed as pos form by forming the complement by using the "0" squares from the map and complementing, that is, $F' = (F')' = (X'Y' + Y'Z + Z'Y)' = (X + Y)(Y + Z')(Z + Y')$. Its implementation by NOR gates is shown in Fig. 4.19c.

4.5.3 AND-OR-INVERT and OR-AND-INVERT Gates

The four types of gates (AND, OR, NAND, and NOR) can be arranged to create 16 combinations for two-level implementation. However, there are only eight nondegenerate combinations of which AND-OR, OR-AND, NAND-NAND, and NOR-NOR have been discussed earlier. The remaining four are used to create two types of circuits, AND-OR-INVERT (A-O-I) and OR-AND-INVERT (O-A-I) functions. The A-O-I results from NAND-AND and AND-NOR two-level implementation (Fig. 4.20). Similarly, the O-A-I circuits are obtained from OR-NAND and NOR-OR types of arrangement (Fig. 4.21). For example, the A-O-I function $(AB + CD)'$ is done by a AND-NOR circuit or NAND-AND circuit; the O-A-I $((A + B) \cdot (C + D))'$ is done by OR-NAND- or NOR-OR-type arrangement. In the former, the expression must be in sop form; for the latter the function must be in pos form. As an example, the function $F = XY'Z' + YZ$ is implemented in Fig. 4.22a as an A-O-I by

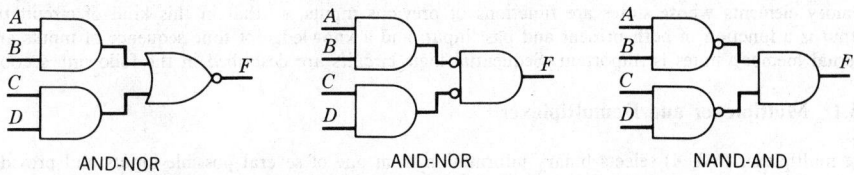

Fig. 4.20 AND-OR-INVERT circuits for $F = (A \cdot B + C \cdot D)'$.

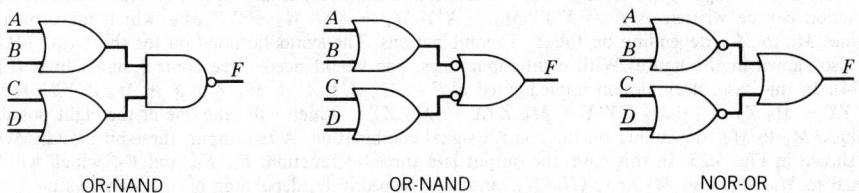

Fig. 4.21 OR-AND-INVERT circuits for $F + ((A + B) \cdot (C + D))'$.

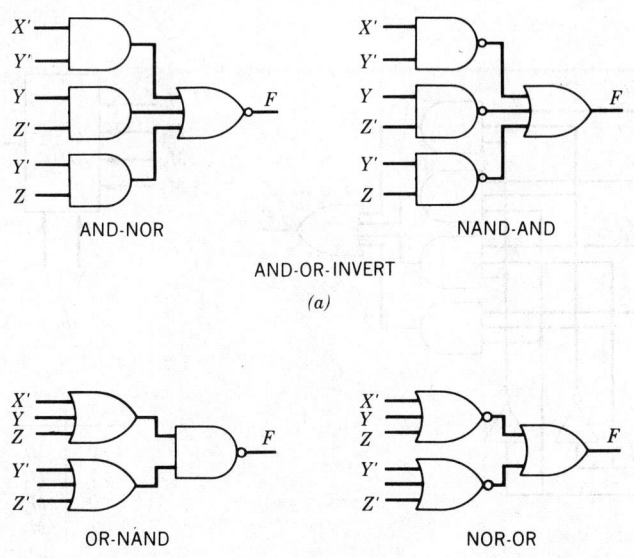

Fig. 4.22 A-O-I and O-A-I implementation of $F = XY'Z' + YZ$.

using the complement $F' = Y'Z + YZ' + X'Y'$ in sop form. Similarly, the same function is shown as an O-A-I circuit by using its complement as a pos (Fig. 4.22b), $F = (X' + Y + Z)(Y' + Z')$.

4.6 COMBINATIONAL LOGIC CIRCUITS

While any operation in a digital computer can be implemented using algebraic functions consisting of the basic AND, OR, and NOT gates, design synthesis at such low level would be too unwieldy. Instead, integrated building blocks or circuit packages that employ these fundamental gates have been devised which perform many of the common functions in the computer. Some of these functions involve transferring of data, selection of data from and transmitting data to one or more sources, conversion of data to different representation, arithmetic operations such as addition, subtraction, etc. Logic gates usually come prepackaged as integrated circuits (ICs) or chips and, depending on the number of gates on the chip, can be classified as small-, medium-, or large-scale integrated (SSI, MSI, LSI) circuits. Although an absolute differentiation is not possible, an SSI device usually has up to 10 gates, an MSI has from 10 to 100 gates, and an LSI has more than 100 gates. In this section some of the basic building blocks employed to devise combinational logic circuits are described. A "combinational" logic circuit is a circuit whose output is dependent solely on the present values of the input without any regard to past inputs. On the other hand, "sequential" logic circuits employ memory elements whose states are functions of previous inputs, so that in this kind of circuit the output is a function of both present and past input, and a knowledge of time sequence of inputs and internal memory states is important. Sequential logic circuits are described in the following section.

4.6.1 Multiplexer and Demultiplexer

The multiplexer (MUX) selects binary information from one of several possible inputs and provides it to a single output line. The particular output to be selected depends on a set of binary control signals, so that with n control signals any one of up to 2^n input lines can be selected. A four-way MUX with four inputs, M_0 to M_3, and two selection lines, X, Y, is shown in Fig. 4.23. The resulting function can be written as $F = X'Y' M_0 + X'Y M_1 + XY' M_2 + XY M_3$, which takes on the values M_0 to M_3, depending on the X, Y combinations. The symbolic notation for the 4-to-1 MUX is also shown in the figure. With eight input lines, one would need three control signal lines (Fig. 4.24). In this case, the function implemented is $F = M_0 Z'Y'X' + M_1 Z'Y'X + M_2 Z'YX' + M_3 Z'YX + M_4 ZY'X' + M_5 ZY'X + M_6 ZYX' + M_7 ZYX$, which will take one of the eight possible values, M_0 to M_7, depending on the control signal combination. A two-input, three-bit digital MUX is shown in Fig. 4.25. In this case, the output is a three-bit function, F_0, F_1, and F_2, which will be equal to M_0, M_1, and M_2 or to $(N_0$, N_1, and $N_2)$, respectively, depending on the signal value X. In general, for a m-input, k-bit MUX, one would need n control lines, where $m = 2^n$. Multiplexer ICs may have "enable" input to control the operation of the unit. Depending on the binary state of the

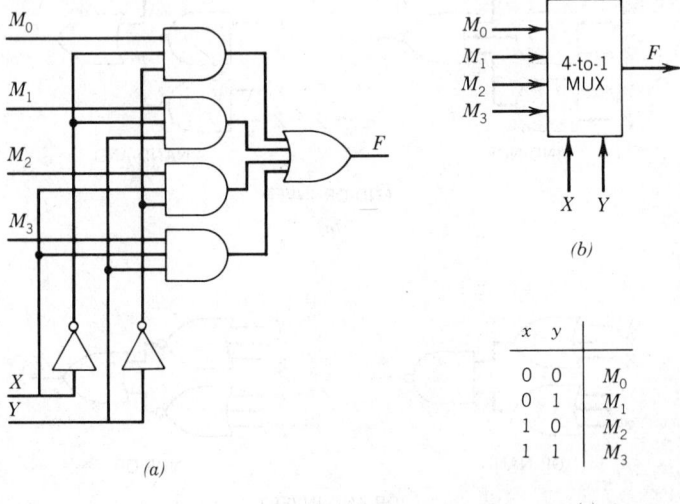

Fig. 4.23 A 4-to-1 multiplexer: (a) gate circuit; (b) symbol; (c) truth table.

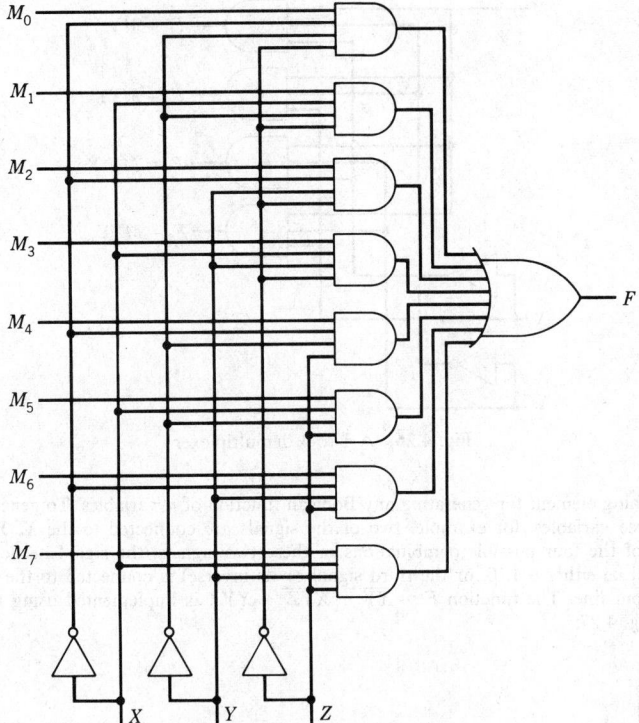

Fig. 4.24 A 8-to-1 digital multiplexor.

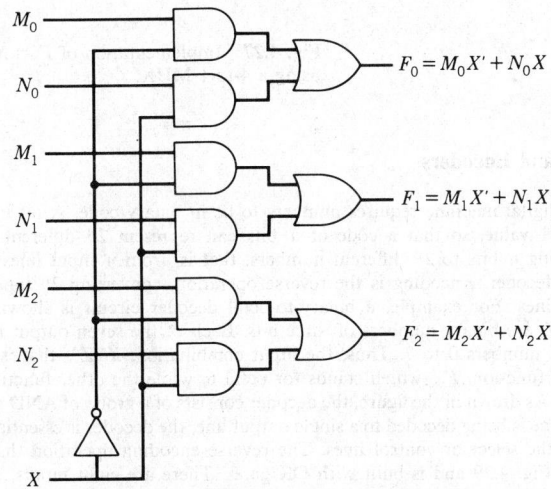

Fig. 4.25 Two-input, three-bit multiplexer.

enable input, the unit may be either operative or disabled. Several MUXs may be enclosed within one IC chip with common control and enable lines.

A demultiplexer performs the opposite function of a MUX and takes a single input and applies it to any one of the several possible outputs. A 1-to-4 demultiplexer is shown in Fig. 4.26, where the input is M and the outputs are F_0 to F_3, which will be equal to M or 0 depending on the control X, Y combination.

MUXs and demultiplexers are useful where many data lines are to be transmitted to one line from one end and to be received at many lines at the other end. Another use for a MUX is as a

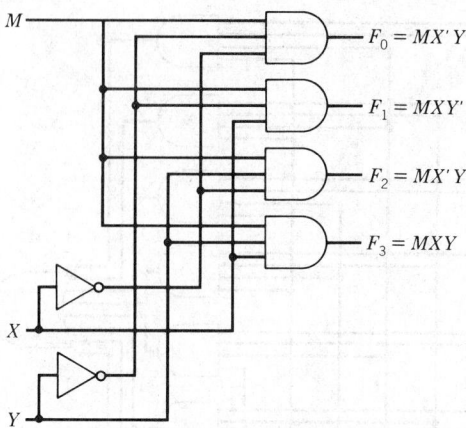

Fig. 4.26 A 1-to-4 demultiplexer.

universal switching element for generating any Boolean function of n variables. To generate a Boolean function of three variables, for example, two of the signals are connected to the X, Y control lines and, for each of the four possible combinations of these two signals, the signal needed at the input line (which will be either a 1, 0, or the third signal or its inverse) is connected to the corresponding multiplexed input line. The function $F = \overline{XY} + X\overline{YZ} + XYZ$ is implemented using a 4-to-1 MUX as shown in Fig. 4.27.

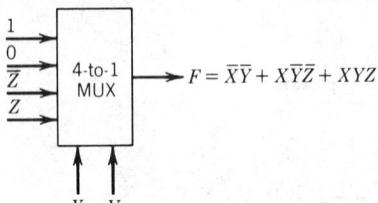

Fig. 4.27 Implementation of $F = \overline{XY} + X\overline{YZ} + XYZ$ using a 4-to-1 MUX.

4.6.2 Decoders and Encoders

Computation in a digital machine requires numbers to be in binary code. A bit in a binary code can take either a 0 or 1 value, so that a code of n bits can represent 2^n different combinations. The function of converting n bits to 2^n different numbers, that is, from n input lines to 2^n output lines, is performed by a decoder. Encoding is the reverse operation, converting 2^n input lines to a binary code of n output lines. For example, a binary-to-octal decoder circuit is shown in Fig. 4.28. The three inputs X, Y, Z are binary numbers of three bits. Each of the seven output functions, F_0 to F_7, represents the octal numbers 0 to 7. Thus, the input combination, XYZ', that is, binary 110 would result in the output function, F_6, which stands for octal 6, while the other functions, F_0 through F_5 and F_7, would be 0. As drawn in the figure, the decoder consists of a group of AND gates. Conceptually, since a three-input line is being decoded to a single output line, the decoder is essentially a demultiplexer, with the inputs as the select or control lines. The reverse encoding operation that converts octal to binary is shown in Fig. 4.29 and is built with OR gates. There are eight inputs, one for each of the eight digits 0 through 7 and three outputs that, in proper combination, represent the binary equivalent code. Although with eight inputs, 2^8 possible combinations can be represented, only eight of these are useful. A typical MSI decoder (74LS42) chip is shown schematically in Fig. 4.30. It is a 4-line to 10-line decoder, which will decode a decimal number 0 through 9 expressed as a four-bit binary code into 1 of 10 individual outputs. Note that with four binary input lines there will be 16 outputs; however, the six binary combinations 1010 through 1111 do not have decimal symbolic representation. Therefore, six of the output lines can be eliminated, resulting in a 10-pin chip. Sometimes, one of the input lines, say the D line, can also be used as an enable line, in which case the chip will act as an octal 3-to-8 decoder with eight outputs. A BCD-to-decimal decoder is shown in Fig. 4.31. The inputs are the 10-decimal digits expressed in BCD code with the output being the 10-decimal digits.

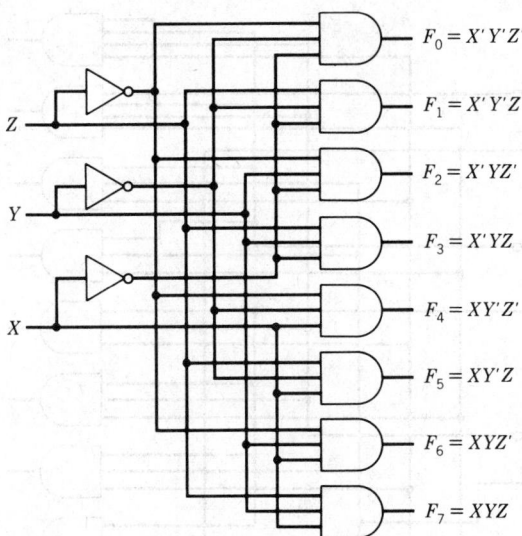

Fig. 4.28 A decoder for converting binary to octal.

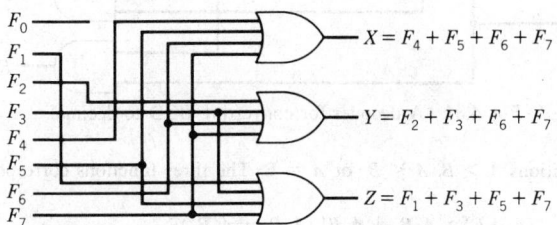

Fig. 4.29 An encoder for converting octal to binary.

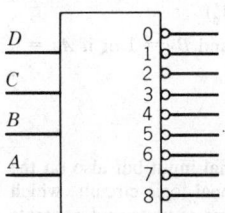

Fig. 4.30 The 74LS42 decoder.

Besides being used in decoding and encoding applications, these combinational logic modules are also used for displaying information stored in memory registers and for generating timing and sequencing signals for control purposes.

4.6.3 Comparators

To compare and determine the relationship among a set of binary numbers, a comparator is used. The output of a comparator for two numbers A, B can be $A > B$, $A = B$, or $A < B$. If A and B are two-bit numbers, for example, then the truth table that lists the possible combinations for the four bits will contain 16 rows. However, suitable algorithms can be developed for higher-bit numbers to simplify the truth table analysis and create the proper circuits. A three-bit comparator for A, B $(A_2 \, A_1 \, A_0 \, B_2 \, B_1 \, B_0)$ is shown in Fig. 4.32. The output is one of three binary functions, which

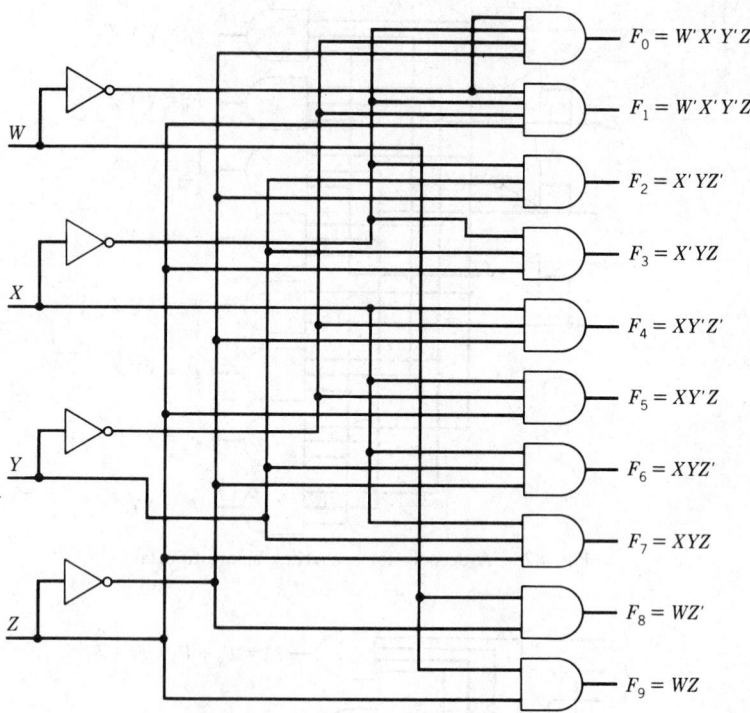

Fig. 4.31 A decoder for converting BCD to decimal.

represents the conditions $A > B$, $A < B$, or $A = B$. The three functions corresponding to the three conditions are

$$(A > B) = A_2 B_2' + A_1 B_1' (A_2 B_2 + A_2' B_2')$$
$$+ A_0 B_0' (A_2 B_2 + A_2' B_2') (A_1 B_1 + A_1' B_1')$$
$$(A < B) = A_2' B_2 + A_1' B_1 (A_2 B_2 + A_2' B_2')$$
$$+ A_0' B_0 (A_2 B_2 + A_2' B_2') (A_1 B_1 + A_1' B_1')$$
$$(A = B) = (A_2 B_2 + A_2' B_2') (A_1 B_1 + A_1' B_1') (A_0 B_0 + A_0' B_0')$$

Thus, when $A < B$, the output is logic 1, and this occurs when $A_2 = 0$ and $B_2 = 1$ or if $A_1 = 0$ and $B_1 = 1$ (with $A_2 = B_2$) or if $A_0 = 0$ and $B_0 = 1$ (with $A_2 = B_2$, $A_1 B_1$).

4.7 SEQUENTIAL LOGIC CIRCUITS

In sequential logic circuits the output depends not only on the present external input but also on the past inputs. Sequential logic circuits thus perform the function of combinational logic circuits, which accept the external binary inputs to be stored as well as the memory element outputs and generate the external output signals and memory element input signals. The state of memory elements depends not only on the external input but also on the present state. Sequential circuits are of two types depending on the timing of the signals, asynchronous and synchronous. Asynchronous circuits are "feedback" type, where memory elements utilize the propagation delay in logic gates to affect the output signals which, in turn, are fed back as input. In synchronous circuits, the behavior of the circuits can be determined based on the knowledge of the signals at discrete instants of time. This is accomplished by using a timed pulse of limited duration representing two logic states. The pulses are synchronized by a master clock generator and affect the state of the memory elements upon arrival. These are known as clocked sequential circuits and utilize AND gates for the signals and the clocked pulse. Memory elements used in clocked sequential circuits are called flip-flops and store one bit of information. A flip-flop (FF) circuit has two outputs, one for the normal value and one for the complement value of the bit stored in it. The common types of FFs are described in the following sections.

4.7.1 Basic Flip-Flop

The basic FF (also called asynchronous RS FF) consists of two NAND or two NOR gates (Fig. 4.33a). It is a bistable device, so that in the absence of any inputs it can assume either of two stable

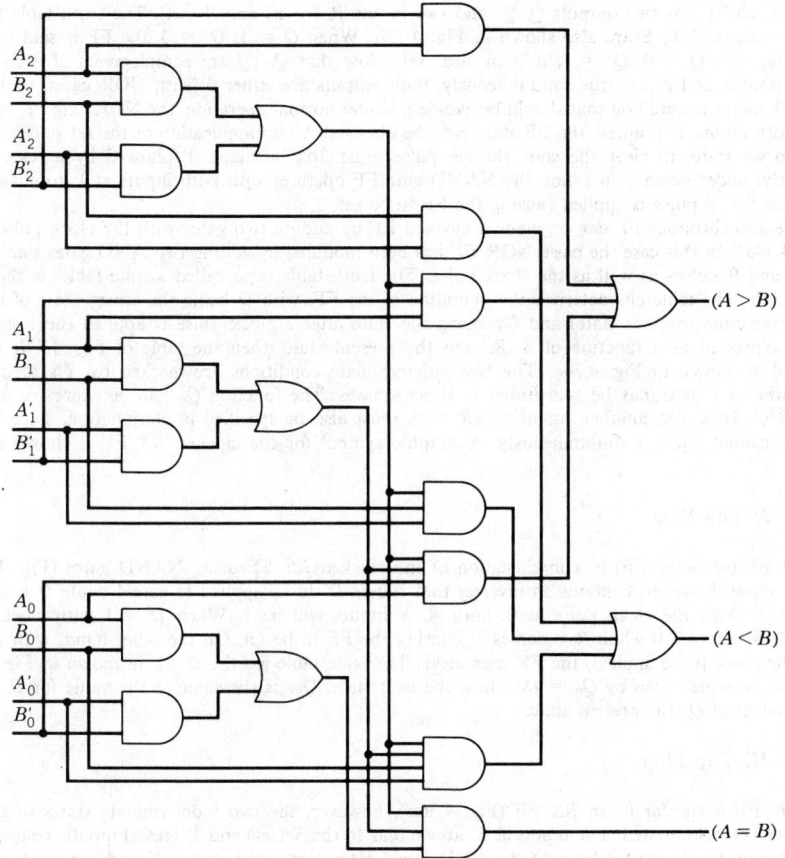

Fig. 4.32 A three-bit comparator.

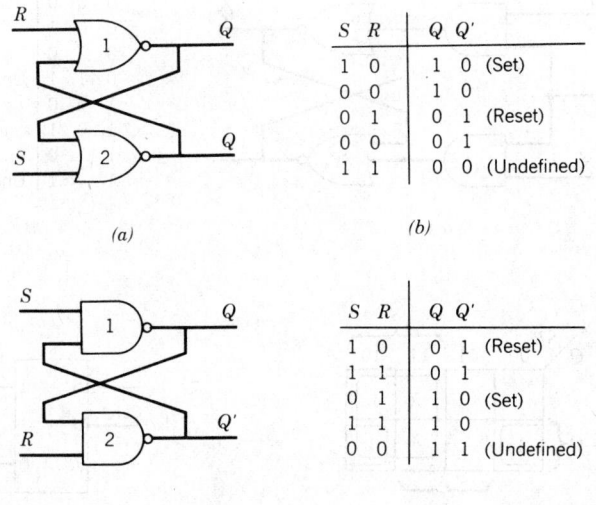

S	R	Q	Q'	
1	0	1	0	(Set)
0	0	1	0	
0	1	0	1	(Reset)
0	0	0	1	
1	1	0	0	(Undefined)

(a)   *(b)*

S	R	Q	Q'	
1	0	0	1	(Reset)
1	1	0	1	
0	1	1	0	(Set)
1	1	1	0	
0	0	1	1	(Undefined)

(c)   *(d)*

Fig. 4.33 (*a*) The basic FF using NOR gates. (*b*) Truth table for (*a*). (*c*) The basic FF using NAND gates. (*d*) Truth table for (*b*).

states. Each FF has two outputs Q, Q' and two inputs R (reset) and S (set). The truth table for the various values of R, S are also shown in Fig. 4.33b. When $Q = 1$, $Q' = 0$, the FF is said to be in set state; with $Q = 0$, $Q' = 1$, it is in clear set. Note that Q, Q' are complements of each other. When both S and R are true simultaneously, both outputs are either 0 (with NOR gates) or 1 (with NAND gates), a condition that should be avoided. Under normal operation, the NOR gate FF operates with both inputs at 0 unless the FF state is to be changed. Upon application of the set pulse, the FF goes to set state; to clear the state the set pulse must first be made 0 followed by a reset pulse. Similarly, under normal operation the NAND gate FF operates with both inputs at 1; to change the state, an $S = 0$ pulse is applied causing the FF to be set.

The asynchronous FF can be made a clocked FF by adding two gates with the clock pulse input (Fig. 4.34a). In this case the basic NOR FF has been modified by adding two AND gates that accept the R and S pulses as well as the clock pulse. The truth table (also called a state table) is shown in Fig. 4.34b. The table characterizes the operation of the FF, with Q being the binary state of the FF at a given time (present state) and Q_n being the state after a clock pulse is applied (next state). If Q_n is expressed as a function of S, R, and the present state, then the table of Fig. 4.34b can be mapped as shown in Fig. 4.34c. The two indeterminate conditions are marked by X's to indicate that either a 1 or 0 may be substituted in these squares. The function Q_n can be expressed as $Q = S + R'Q$. However, another equation, $SR = 0$, must also be specified in conjunction, since both S and R cannot equal 1 simultaneously. A graphic symbol for the clocked RS FF is shown in Fig. 4.34d.

4.7.2 D Flip-Flop

The D FF (or delay FF) is a modification of the clocked RS FF using NAND gates (Fig. 4.35a). The R input, however, contains an inverter that causes D' to be applied to gate 4 while D is applied to gate 3. With the clock pulse at 0, both R, S inputs will be 1. When $D = 1$ with clock pulse applied, S becomes 0 while R becomes 1, causing the FF to be set. On the other hand, with $D = 0$ with the clock pulse applied, the FF goes clear. The state table for the D FF is shown in Fig. 4.35b and can be represented by $Q_n = D$. Thus, the next state, Q_n, is the same as the value for D and is independent of Q, the present state.

4.7.3 JK Flip-Flop

The JK FF is similar to an RS FF (Fig. 4.36a), however, the two indeterminate states of the RS FF have been eliminated. The inputs J, K are similar to the S (set) and R (reset) inputs, respectively. The outputs Q, Q' are fed back to the AND input gates along with the J, K and the clock signals.

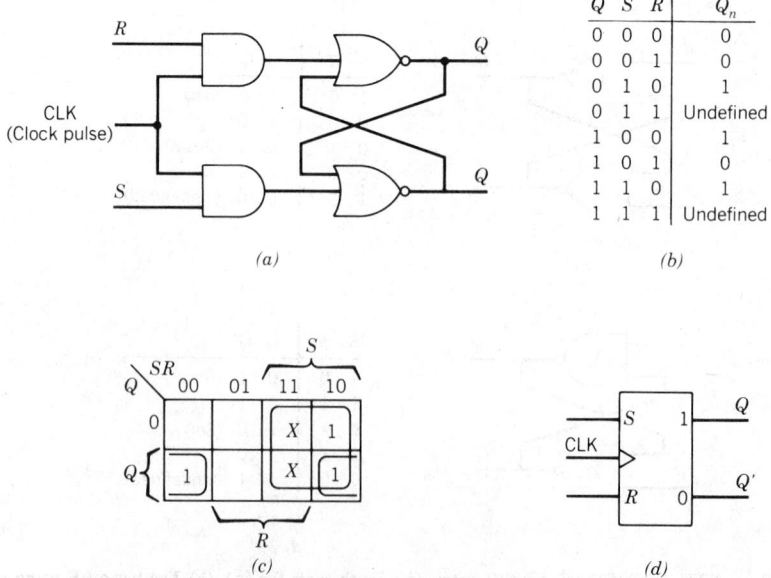

Q	S	R	Q_n
0	0	0	0
0	0	1	0
0	1	0	1
0	1	1	Undefined
1	0	0	1
1	0	1	0
1	1	0	1
1	1	1	Undefined

(a)   *(b)*

(c)   *(d)*

Fig. 4.34 (*a*) The clocked RS FF. (*b*) State table. (*c*) Map. (*d*) Symbol.

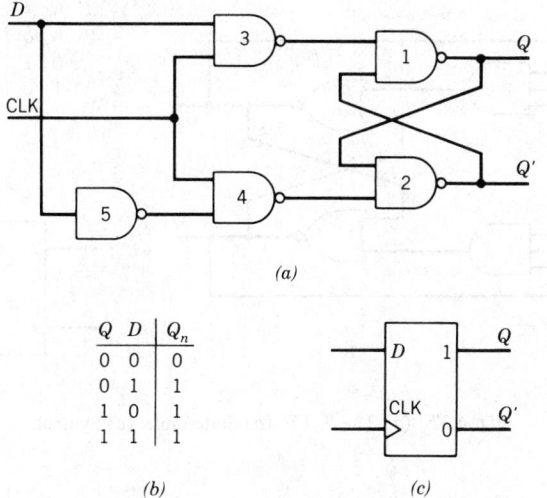

Q	D	Q_n
0	0	0
0	1	1
1	0	1
1	1	1

(b)

(c)

Fig. 4.35 (a) The D FF. (b) State table. (c) Symbol.

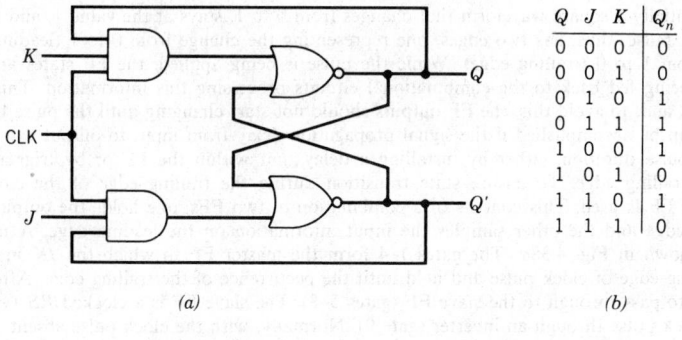

Q	J	K	Q_n
0	0	0	0
0	0	1	0
0	1	0	1
0	1	1	1
1	0	0	1
1	0	1	0
1	1	0	1
1	1	1	0

(a)

(b)

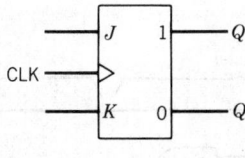

(c)

Fig. 4.36 (a) The clocked JK FF. (b) State table. (c) Symbol.

This means that during a clock pulse, the FF is cleared if Q was 1 prior to the application of the pulse, or it is set if Q' was 1. When both J and K are applied simultaneously, the flip-flop state is complemented, that is, if it was set initially, then it is cleared and vice versa. The state table is shown in Fig. 4.36b, and, when mapped, the function Q_n can be shown to be $Q_n = K'Q + JQ'$.

4.7.4 T Flip-Flop

The T FF (trigger FF) is shown in Fig. 4.37a and is seen to be the same as the JK FF except that there is only one input. The T FF can toggle between states, that is, it reverses its state at each clock pulse that the trigger is applied. The stable table in Fig. 4.37b can be mapped to yield $Q_n = T'Q + TQ'$.

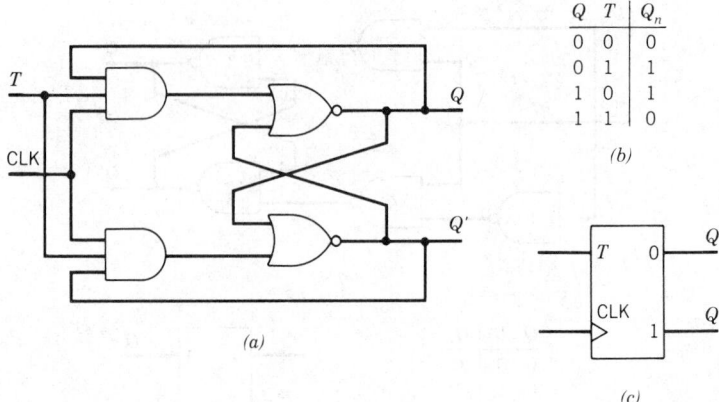

Q	T	Q_n
0	0	0
0	1	1
1	0	1
1	1	0

(b)

(a)

(c)

Fig. 4.37 (a) The T FF. (b) State table. (c) Symbol.

4.7.5 Triggering of Flip-Flops

The state of an FF can be altered (or "triggered") by the input signals. For an asynchronous FF it means changing the input signal levels, whereas for clocked FF it implies application of pulses. The pulse is essentially a square waveform that changes from 0 to 1, stays at the value 1, and then reverts back to 0. A pulse, thus, has two edges, one representing the change from 0 to 1 (leading edge) and the other from 1 to 0 (trailing edge). While the pulse is being applied, the FF states are changing, and this is being fed back to the combinational circuits processing this information. This can result in instability, and, to avoid this, the FF outputs should not start changing until the pulse has returned to 0. This can be accomplished if the signal propagation delay from input to output can be made to exceed the pulse duration, either by installing a delay unit within the FF or by triggering the FF during the trailing edge. To ensure state transition during the trailing edge of the clock pulse, a master–slave FF is used. This consists of a combination of two FFs, one holds the output state until the trailing edge and the other samples the input information on the leading edge. A master–slave JK FF is shown in Fig. 4.38a. The gates 1–4 form the master FF to which the JK inputs are fed on the leading edge of clock pulse and held until the occurrence of the trailing edge. After that they are allowed to pass through to the slave FF (gates 5–8). The slave FF is a clocked RS type fed from the same clock pulse through an inverter (gate 9). Normally, with the clock pulse absent, the outputs Q, Q' correspond to the outputs of the master–slave. When the clock pulse is 1, the master FF is affected but the slave FF is not because the output of gate 9 is 0. With the clock at 0, master FF is

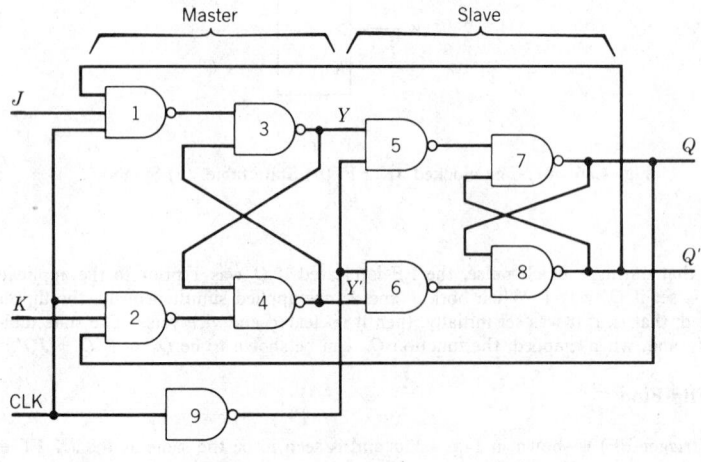

(a)

Fig. 4.38a The JK master–slave FF.

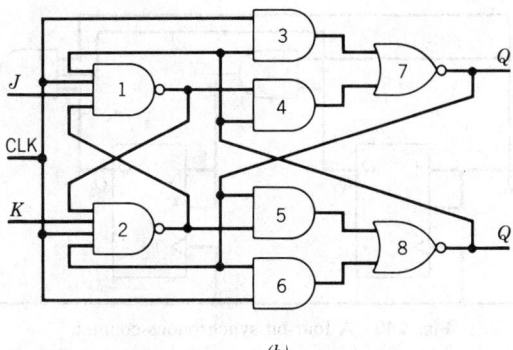

(b)

Fig. 4.38b *JK* edge-triggered FF.

isolated and the slave FF attains the same state as the master FF. The FF described is also known as "pulse-triggered" type, since the data are entered on the leading edge of the clock pulse and then transferred to the output on the trailing edge. One disadvantage of this type is that both *J, K* must be held constant during the clock pulse, otherwise, if either is allowed to change, the output is flipped. This is overcome by using an "edge-triggered" FF, which is activated on one of the edges of the pulse (see Fig. 4.38b). The *JK* edge-triggered FF shown therein triggers on the trailing edge. Suppose $Q = 0$, $J = 1$, $K = 0$ when the CLK is 0. When CLK goes to 1, output of gate 1 will be 0 and gate 2 will be 1. With Q' as 1, gate 3 will result in $Q = 0$. With Q' as 1, gate 3 will result in $Q = 0$. Thus, the result of clock activation is setting of the 1, 2 FFs. When CLK goes to 0, gates 3, 4 cause Q to become 1 and gate 5 causes Q' to be 0. Therefore, the output of the FFs 7, 8 depends on the state of FFs 1, 2 prior to the trailing edge of the CLK pulse. The data input to the FFs 1, 2 through *J, K* must occur before this time, called the setup time.

4.7.6 Counters

In digital design, the operation of counting is performed by "counters," which are sequential circuits utilizing FFs. Upon the application of an input pulse, counters react by going through a predetermined sequence of states; for example, a four-bit binary counter will generate the binary sequence from 0000 to 1111 and back to 0000. A two-bit synchronous *JK* FF counter is shown in Fig. 4.39a. With sequential clock pulses, the FF outputs Q_1, Q_0 (Q_0 is the least-significant digit) go through the sequence 010101. . . . This is a synchronous counter, since both FFs are triggered simultaneously by the clock pulse. In an asynchronous or ripple counter, each FF triggers the next FF, and so the counter states are changed in a ripple or staggered fashion (Fig. 4.39b). Ripple counters can be easily cascaded to produce higher-order counters. However, the change of state during a signal transition from 1 to 0 or 0 to 1 along with the timing delay due to signal propagation can result in "hazards" or "glitches" in ripple counters. Moreover, synchronous counters operate faster than ripple counters, since all the FFs change state simultaneously. A four-bit synchronous counter (for counting from 0000 to 1111) with the same clock pulse to all the four FFs is shown in Fig. 4.40. Using the extra AND gates the input to each FF is manipulated to cause the FFs to trigger at the proper time.

4.7.7 Shift Registers

Registers are sequential circuits using FFs that are used to store binary data in digital machines. An *n*-bit register is a group of *n* FFs that can store *n* bits of data. Shift registers perform the shifting

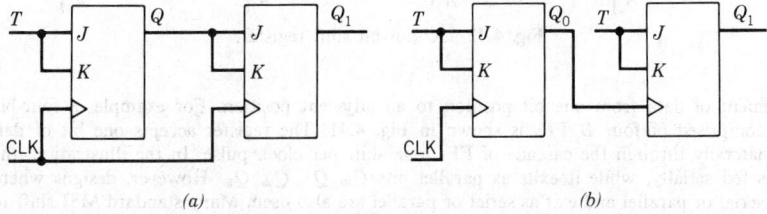

(a)   *(b)*

Fig. 4.39 A two-bit *JK* FF counter: (*a*) synchronous: (*b*) asynchronous or ripple.

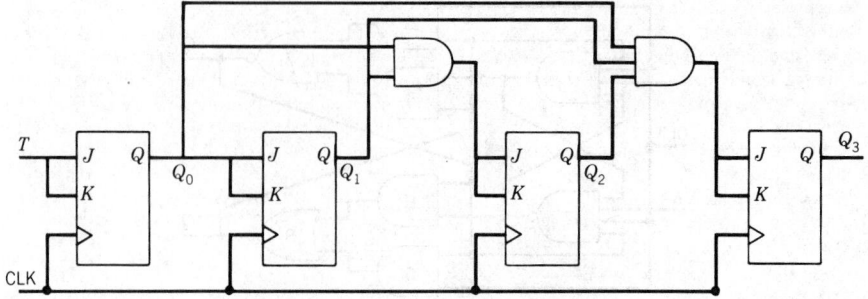

Fig. 4.40 A four-bit synchronous counter.

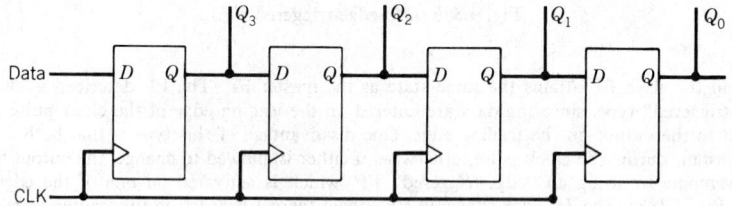

Fig. 4.41 A four-bit shift register using D FFs.

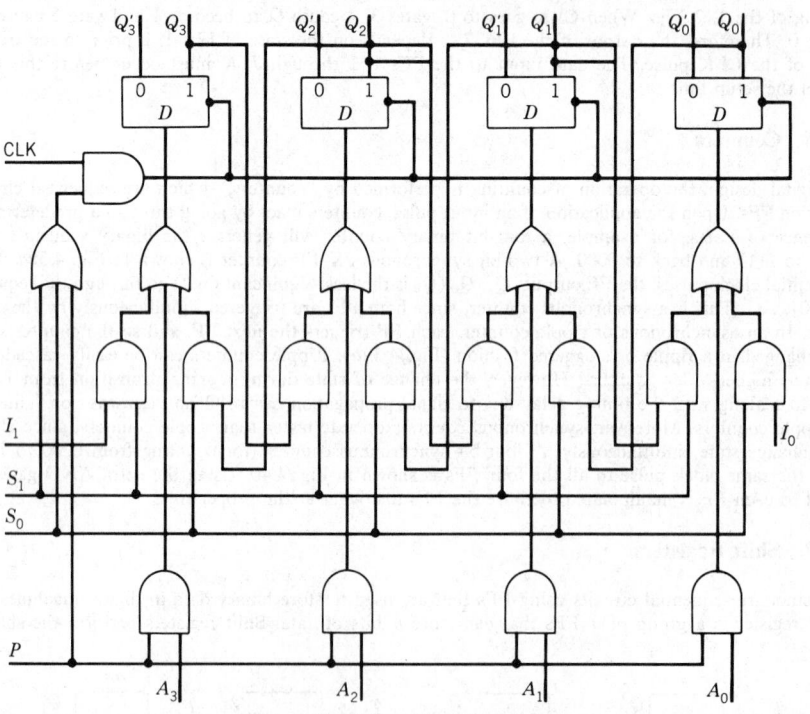

Fig. 4.42 A four-bit shift register.

or movement of data from one bit position to an adjacent position. For example, a four-bit shift register comprised of four D FFs is shown in Fig. 4.41. The register accepts one bit of data and shifts it laterally through the cascade of FFs, one shift per clock pulse. In the illustration, the data stream is fed serially, while it exits as parallel bits Q_0, Q_1, Q_2, Q_3. However, designs where data enter in serial or parallel and exit as serial or parallel are also used. Many standard MSI shift register chips are available, for example, the 74LS194, which is a four-bit, bidirectional, universal shift register.

In general, a shift register should be capable of loading data in serially or in parallel, shift the bits either left or right, and store or retain the present value of data. Thus, there are four operations: load new data, shift right, shift left, and store. In a four-bit register, this can be performed by using two control signals to derive the proper input to each of the four FFs. In other words, we can use a four-input MUX arrangement with two control signals to provide input to each FF to correspond to the operation desired (Fig. 4.42). The parallel-transfer, left-shift, and right-shift operations are controlled by the signals marked P, S_0, and S_1, respectively. When $S_1 = S_0 = 0$ and $P = 1$, the register stores the incoming data A_0 through A_3 as Q_0 through Q_3. For shift-left operation, $S_0 = 1$, P and $S_1 = 0$, so that, with the serial input at I_0, each FF receives its input from the adjacent FF on its right outputting at Q_3. Similarly, for shift-right operation, $S_1 = 0$, P and $S_0 = 0$, so that, with the data input serially at I_1, they are outputted at Q_0.

REFERENCES

1. D. L. Dietmeyer, *Logic Design of Digital Systems,* Allyn and Bacon, Boston, MA, 1979.
2. M. M. Mano, *Digital Logic and Computer Design,* Prentice-Hall, Englewood Cliffs, NJ, 1979.

CHAPTER 5
COMPUTER MEMORIES

ALI SEIREG

University of Wisconsin
Madison, Wisconsin

5.1 MEMORY DEVICES

Memory devices are an essential element of a computer. The size and method of operation of a computer memory is an important factor in rating it. These devices are necessary to store data and instructions so they can be retrieved as needed.

5.2 STORAGE CATEGORIES

Computer memories are classified into two basic categories: internal or primary and external or secondary storage. The former is essential to the operation of the computer. It represents its main memory where programs and data are stored during processing. Internal memories can operate at electronic speeds and therefore provide immediate access to the stored information. Data in such devices can be stored and retrieved at random at any position at the discretion of the programmer. This type of memory is usually expensive and consists of a large number of storage locations. Each location is called a storage cell and can hold a single binary digit.

In many computers, a storage space much larger than the available internal memory is necessary. The extra storage can be provided by external devices. This secondary memory is generally used to store very large amounts of data, which need not be accessed frequently. It provides cheaper storage with inherently slower access. Storage devices for secondary memories cover a wide variety of systems ranging from punch cards and paper tapes to magnetic tapes, disks, and drums.

5.3 INTERNAL MEMORIES (RAM)

The internal memory or the main computer memory consists of a large number of storage cells. Each of these cells stores one bit. The memory cells are generally organized so that a group of n bits defining a word are written or read in one single operation. Each word location has a distinct address for identification. The maximum size of the main memory for a 16-bit machine is 2^{16} locations and for a 24-bit computer is 2^{24} locations.

The main memory holds the data and computing instructions during program execution. The speed of the computer is determined by the access time (the time elapsing between the read instruction and delivery of the stored information) and the cycle time (the time elapsing between two successive read operations).

Main computer memories are designed so that information can be written or read in a fixed amount of time independent of the storage location. They are therefore called random access memories (RAMs). Main memories are constructed either of magnetic (ferrite) cores or semiconductor integrated circuits. Ferrite core memories have high reliability and low power consumption. They were the first practical means of providing large storage for digital computers economically. They are now being replaced in many computers by semiconductor integrated circuits.

5.3.1 Magnetic-Core Memories

Core Structure and Operation

The basic elements of this type are ferrite rings made of ceramic iron. Each ring, as illustrated in Fig. 5.1, is approximately 0.4 mm in diameter and has wires passing through the core to magnetize it in a clockwise or counterclockwise direction depending on the direction of the current in the wires. The resulting states of magnetization of the ring correspond to 1 and 0, respectively.

Write Operation

The writing operation is accomplished by passing a current I_m with sufficient magnitude in the drive wire to magnetize the coil to a state of 1 or 0 as required. The writing operation destroys the previous state of the core.

Read Operation

Reading a stored value in the core is accomplished by passing a current $-I_m$ in the drive wires in the opposite direction. If the core has a state of 1, a pulse is therefore induced in the sense wire as a result of the reversed magnetization of the coil. If the core is in a state of 0, no change in the direction of magnetization will occur and no pulse is therefore induced in the sense wire. It should be noted that reading is also a destructive process, since the core will be set to 0 immediately after the operation independent of its initial state. Since the stored value in the memory may be needed for later use, a rewriting operation would be necessary to reinstate it.

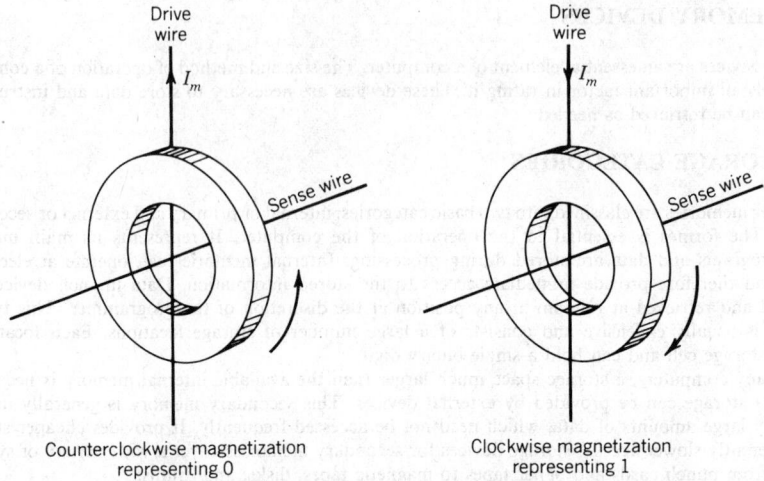

Counterclockwise magnetization
representing 0

Clockwise magnetization
representing 1

Fig. 5.1 Magnetic or ferrite core.

Memory Arrays and Multiple Module Memories

A computer memory is constructed by organizing the magnetic cores into arrays as shown in Fig.
5.2. The figure illustrates a 16-word memory with eight bits per word. Figure 5.2a shows the core
organization in one of the eight arrays, which corresponds to one of the eight bits. A word consists
of cores in corresponding positions in each array. Each core has an x-drive wire and a y-drive wire.
A stored value at any core location can therefore be read by passing a current $\frac{1}{2} I_m$ in the corresponding

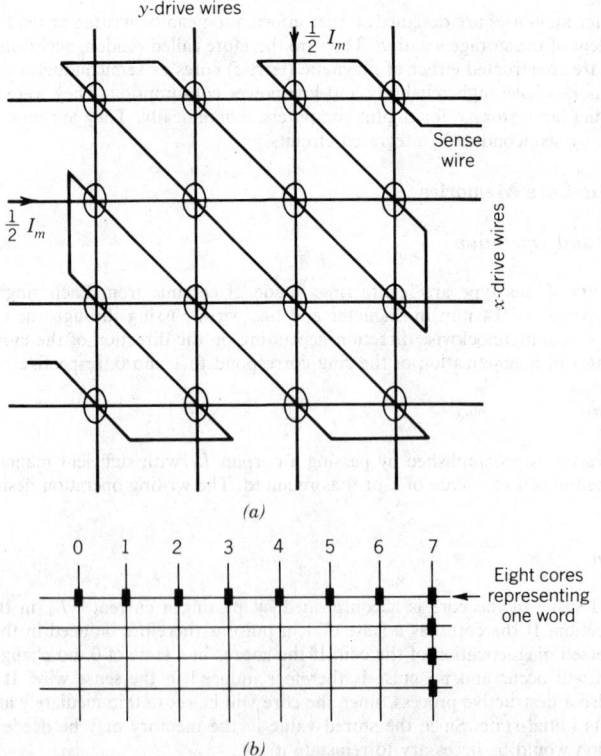

Fig. 5.2 Memory arrays: (a) core organization in one plane: (b) memory planes.

x and y drive wires (where I_m is the magnitude of the current necessary to magnetize the coil). Only the core to be accessed is magnetized by the appropriate selection for the x- and y-drive lines. The drive lines are threaded into the eight planes as illustrated in Fig. 5.2b, and the selection of all eight bits of a word is accomplished by specifying the corresponding x- and y-drive lines.

The rate of transmission of words to and from the main memory can be significantly increased by constructing the main memory from separate modules capable of performing read and write operations at the same time. The distribution of the individual addresses over the modules determines the number of modules that are operating during the computing process. Additional controls are required with multiple-module memories.

5.3.2 Semiconductor RAM Memories

Types of Semiconductor Memories

Semiconductor memories utilize integrated circuits as storage cells. Early memories of this type were faster but more expensive than magnetic core memories. Advances in large-scale integration (LSI) made the use of large semiconductor memories economically feasible in many computers. Based on the technology used, these memories are divided into two types: bipolar and metal oxide semiconductor (MOS). These are variations of the basic integrated circuit technology.

Bipolar Storage Cell

A bipolar memory cell is essentially a basic flip-flop incorporating two transistor inverters. The operation of the cell can be explained by the use of the block diagram given in Fig. 5.3.

- In order to write 0 in the cell the following steps are implemented:
  Set: x-address line voltage high
    y-address line voltage high
    data line 1 voltage low
    data line 2 voltage low
  "0" will be written in the cell as long as the supply voltage is maintained.
- In order to write 1 in the cell the following steps are implemented:
  Set: x-address line voltage low
    y-address line voltage low
    data line 1 voltage low
    data line 2 voltage high
  "1" will be written in the cell as long as the supply voltage is maintained. The high and low voltages are normally 2.5 V and 1.6 V, respectively.
- In order to read the stored value in a cell the following steps are implemented:
  Set: x-address line voltage high
    y-address line voltage high
    data line 1 voltage low
    data line 2 voltage low
  The state of the cell is determined by the data line through which the current flows.

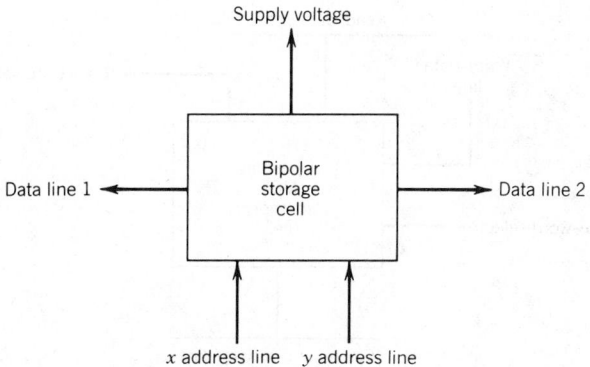

Fig. 5.3 Bipolar cell operation.

Metal Oxide Semiconductor (MOS) Memories

MOSs are higher impedance devices and thus dissipate less power than bipolar memories. They are easier to manufacture and allow higher bit densities on integrated circuit chips. They are however slower to operate than bipolar devices. The simplest form of a MOS cell is a flip-flop circuit similar to bipolar cells. Transistors are used to replace the resistors and diode switches and both cells operate in the same basic way.

Static, Dynamic, and Volatile Memories

The bipolar and MOS memories previously described require that the electrical supply to the cells is maintained in order to preserve the stored information. As long as the supply voltage is maintained, they are capable of storing the information indefinitely. Such cells are therefore called "static" memories.

There is another type of memory which is designated "dynamic," where the information has to be rewritten or "refreshed" every few milliseconds and every time a read operation is performed in order to guarantee that the stored information is always available in the memory. A separate control circuit is generally provided as a part of the memory chip for the refreshing process.

The use of dynamic memories reduces the electric power consumption and allows higher bit density. Dynamic memory cells incorporate a capacitor in their circuits. The charge in the capacitor determines whether a 1 or 0 is stored. Since charge leaks from the capacitor are to be expected, it is necessary to make provisions for reading and rewriting the stored information every few milliseconds. The high impedance of MOS cells allows their use for dynamic memories. A block diagram for a MOS dynamic memory is shown in Fig. 5.4. To write in the memory cell the voltage of the write-select line is set high to make it possible to use the write-data line to access the capacitor. On the other hand, to read from the cell the read-select wire voltage is set high, which produces a connection to the capacitor discharge line and, consequently, the read-data line to be discharged.

Semiconductor memories in general are called "volatile" since they are only capable of maintaining the stored information indefinitely as long as the supply power is on. When the power is switched off, all the stored information is lost. If it is required that the information be preserved when the power is off, magnetic-core memories should be used.

5.3.3   Read Only Memory (ROM)

The random access memories described in the previous section can be read at will. They can be used to read and write and can also be overwritten. There is another type of semiconductor memory, called read only memories (ROMs), that is used to store the control instructions for the central processing unit. ROMs contain the fixed programs or data necessary for the operation of the computer. Once written, this type of memory can be accessed at will but cannot be overwritten. Data in a ROM are usually programmed by the computer manufacturer. Some ROM designs allow the user to write data into the memory. These are called programmable ROMs (PROMs), and once programmed, they cannot be altered.

Another type of ROM allows the stored data to be erased and reprogrammed. The erasure is usually accomplished by exposure to ultraviolet light for a period of several minutes to an hour. This type is called an erasable PROM (EPROM). Another type of ROM allows the user to alter and reprogram individual words by applying a high voltage to its connections. This is called an electrically alterable ROM (EAROM).

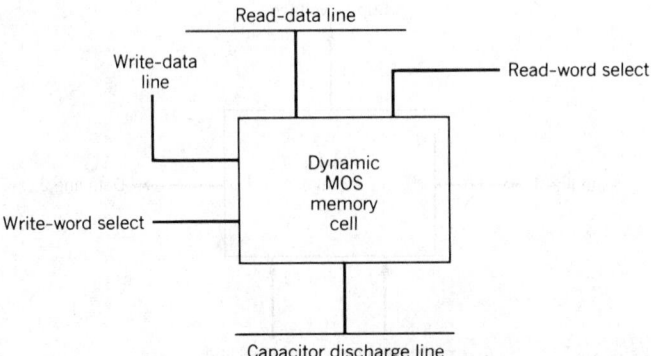

Fig. 5.4   Dynamic MOS memory cell.

5.4 EXTERNAL MEMORY

Examples of external memory devices used for external storage are discussed in the following.

5.4.1 Punch Cards

Punch cards are one of the early methods of external storage. Programmers punch program statements and data on an off-line card punch. The cards are then assembled in decks and become available to the computer through an on-line card reader connected to a direct memory access. Each card measures approximately 3.25 in. × 7.25 in. and has a grid with 12 rows and 80 columns. At each grid location a hole may be punched to represent 1. No hole represents 0. The code used includes the digits 0–9, the letters A–Z, and punctuation marks. A card reader, directly connected to the computer, utilizes a photoelectric device to translate the hole patterns into electric signals, which are passed on to the controller that translates the hole pattern into computer code. The transfer of information by the hole reading process to the main memory of the computer is accomplished at a speed of approximately 1200 cards per minute, which utilizes approximately 0.05% of the main memory cycle time. Most computers have the capability of off-line card storage in machine language by directly operating a card punch. Card punching is a much slower process than card reading, since the former is a mechanical process while the latter is accomplished photoelectrically.

5.4.2 Paper Tape

Paper tape is not often used as an input medium, since it is inconvenient to handle and operate. It is very slow to punch and read. A single error or damaged location in the tape renders the entire tape useless.

5.4.3 Magnetic-Disk Storage

Digital information can be stored on thin magnetic films deposited on both sides of disks placed on rotary drives as illustrated in Fig. 5.5. The drives move at a constant speed and the information is written or read by read/write heads, each of which incorporates a magnetizing coil. The same head is used for storing and reading. Both operations occur when the disk rotates at constant speed. Disks are not started or stopped during the storage or access operations in order to ensure their constant speed rotation.

The surfaces of each disk are divided into concentric tracks divided into sectors. Identification of the storage area is accomplished by specifying the sector track and surface numbers. The heads can be fixed or movable. The former system is usually designed with one head for each track and consequently provides much faster access to the information, since the latter system requires complex electromechanical devices to move the heads to the appropriate tracks. These disks can be designed to rotate at

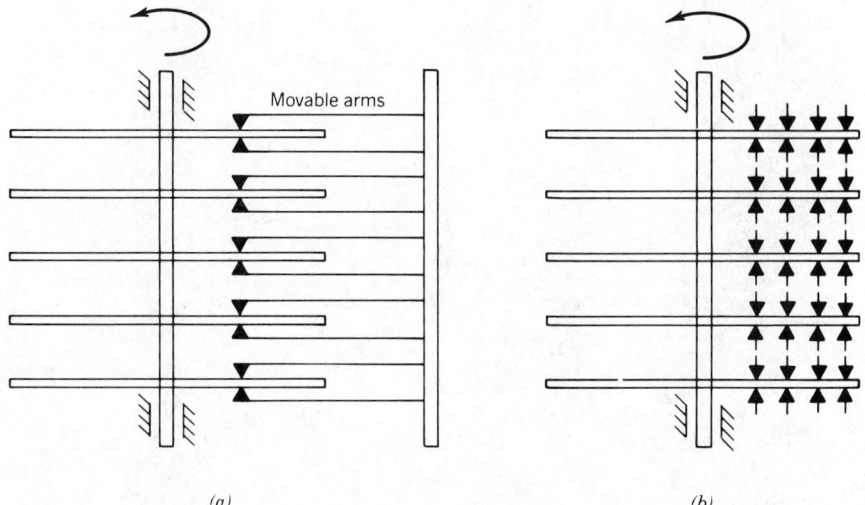

(a)   *(b)*

Fig. 5.5 Magnetic disk storage: (*a*) movable heads; (*b*) fixed heads.

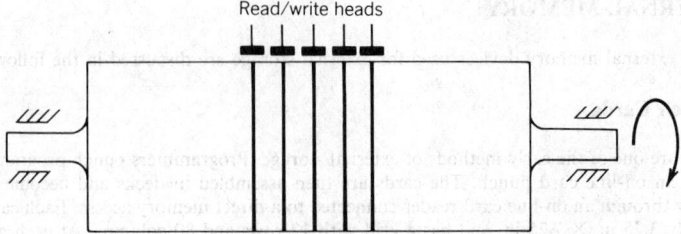

Read/write heads

Fig. 5.6 Magnetic drum.

speeds from 1500 to 6000 rpm. There are usually 16 sectors per track, 256 tracks per surface, and, consequently, an eight disk system has a total storage capacity of 30 million bytes.

Magnetic disks can be designed for over 300 million (3×10^8) character capacity. Transfer rates can be as high as 3 million (3×10^6) characters per second with an access rate of approximately 2.5 msec.

5.4.4 Magnetic-Drum Memories

Magnetic-drum memories operate on the same principles as the magnetic-disk system. The circular tracks in this system are organized on the thin film deposited on the surface of the rotating drum in closely spaced planes perpendicular to the axis of rotation. A set of read/write heads, which move parallel to the drum axis, can be used in this type of system, as shown in Fig. 5.6.

The storage capacity of drum systems is generally larger than disk systems and can be designed for over 3000 million (3×10^9) characters. The access time can be less than 10 msec and the transfer rate can be over 1 million (10^6) characters per second.

5.4.5 Magnetic-Tape Memories

Magnetic-tape memories are particularly suited for large off-line storage and for transfer of information between machines that are not directly linked. Tapes operate on the same principle as the previously described systems; in this system the thin magnetic film is deposited on a thin plastic tape. Data are usually organized on a tape in the form of files separated by long gaps. A file mark, composed of a single or multicharacter record, at the end of the gap serves to identify the file. Magnetic-tape systems can be designed for up to nine channel recordings with speeds of up to 200 inches per second and transfer rates of over 1 million (10^6) characters per second.

CHAPTER 6
SYSTEM OPERATION

University of Wisconsin
Madison, Wisconsin

6.1 GENERAL OUTLINE OF SYSTEM OPERATION

Computer system operation requires a set of programs designed to perform the following functions:

- Translate the user's program into machine language.
- Load the machine language program into the main memory of the computer.
- Optimize the sharing of the different computing resources in the system [input/output (I/O) devices, external storage, central processing unit (CPU), main memory, etc.] among a variety of users in order to provide efficient service with minimum cost.

A computing task requires the formulation of a suitable program incorporating a concise set of instructions written in a user-oriented high-level language. The program is fed into the computer through an appropriate input device. An assembler, interpreter, or compiler then translates the program into machine code. A loader moves the machine language program into the computer memory and prepares it for execution.

In addition to the instructions the data to be operated on are supplied to the computer and are stored in the memory in a manner that can be easily fetched, processed in the CPU, and then outputed in the desired format using an appropriate output device. Registers in the CPU provide temporary storage for the instructions, data, memory addresses, and program progress during the execution phase. Buffer registers are also commonly used to hold the information during transfer from one component of the computing system to another.

6.2 OPERATING SYSTEM PROGRAMS FOR INPUT OF INSTRUCTIONS AND DATA

As was previously stated, special system programs are necessary to translate the user's instructions and to properly prepare and store them for execution in the computer memory. Such programs are briefly described in the following sections.

6.2.1 Assembler

An assembler is a program that translates instructions written in assembly language into machine language. Since assembly language is close to machine language, the assembler is considerably less sophisticated than other translation programs. Some assemblers produce a machine language program that can be run immediately. Most, however, produce an output file (object module) that requires further processing by a loader or linkage editor to convert them into executable form.

6.2.2 Compiler

Compiler programs are system programs that translate programs from high-level languages into a machine-oriented form. Compilers are sophisticated programs that must not only be able to translate complex statements into machine instructions but must also resolve symbols, develop memory management based on the activities of the program being compiled, and diagnose and report errors in the program as written. A compiler is dedicated exclusively to a specific language and produces code for a specific CPU type. The input data for a compiler are a source program in a high-level language, and its output is an object module of machine instructions in a form executable by the CPU.

6.2.3 Interpreters

Interpreter programs are system programs that read the source program, translate its instructions, and execute them immediately. An interpreter is different from a compiler in that it translates a program every time it is run. The interpreter program is usually started by the command RUN.

6.2.4 Loader

A system control program moves a program into memory and prepares it for execution. The machine language program capable of being brought into memory and executed without further preparation is called a load module.

6.2.5 Linkage Editor

Linkage editor is a system program that combines two or more related computer programs into one. The inputs to a linkage editor are the object modules produced by a compiler or assembler. The outputs are machine language programs in a form suitable to be executed on a computer. Large programs are usually developed in modules which refer to each other. The linkage editor resolves these references by replacing them with the actual memory addresses it assigns in the process of arranging the modules into a monolithic program. Upon completion of the linking phase, a "linkage loader" loads the program into the memory for execution.

6.3 SEQUENCING AND TIMING

6.3.1 Sequencing

The statements of a program are normally executed in sequence. Execution of a given instruction consists of a two-phase procedure:

- In the first phase, called instruction fetch, the instruction is fetched from the main memory location whose address is in the program counter (PC). This instruction is placed in the instruction register (IR) in the CPU.
- At the start of the second phase, called instruction execute, the operation field of the instruction in IR is examined to determine which operation is to be performed. The specific operation is then performed in the CPU. This may involve fetching operands from the main memory, transferring them to the CPU, performing an arithmetic or logic operation, and storing the results into the main memory.

Sometimes during this two-phase procedure the contents of the PC are incremented by 1 to update them to point at the next instruction. Therefore, when the execute phase of an instruction is completed, the PC contains the address of the next instruction, and a new instruction fetch phase can begin.

The sequence of operations required to execute one instruction can be therefore summarized as

- Fetch instruction.
- Fetch first operand (contents of memory location pointed at by the address field of instruction).
- Perform the arithmetic operation.
- Load results in the register.
- A new fetch cycle is started by going back to the first step.

6.3.2 Branching

The conditional branch instruction is an essential part of loop control. Conditional branches are used in any program situation in which one of the two possible paths for continuing the computation must be chosen. This choice is usually based on an arithmetic or logic property or condition, or on the result of a recently performed operation.

Branching is accomplished by replacing the current contents of the PC by the address of the instruction to which branching is required. This is done by adding an offset that gives the address of the branch instruction to the current value of the PC. Execution starts as usual until the instruction is loaded in the IR.

6.3.3 Timing

The required timing for controlling the flow of the signals involved in transferring the data between registers as well as gating data to and from the CPU is an important factor in determining the speed of the computer. Timing aspects of the transfer of signals inside the CPU should take into consideration the finite time delay encountered for the register gate to open and for the data to travel along the bus to the input of the arithmetic logic unit (ALU) circuits. Also, for the result to be properly stored in the register the signal should be maintained on the bus for an additional period equal to the step time of the register.

6.4 PATHS AND DATA STORAGE

6.4.1 Transfer of Data Between the I/O Devices and CPU

An interface is needed to coordinate the transfer of data between the I/O devices and the CPU. The interface serves the following purposes:

1. Stores the device status for use by the computer when needed. A "status register" is used for this purpose.

2. Provides storage space for at least one data character to be used for transferring the data to and from the computer. A "data buffer register" is used for this purpose.
3. Recognizes the device address when it appears on the I/O address lines.
4. Provides appropriate timing and gating signals to accomplish the transfer of data or status as required.

The I/O transfer bus is composed of three lines:

* data lines
* address lines
* control lines

The control lines handle requests for interrupt and direct memory access operations as well as handle information related to initialization of the system, to power failure, and so on.

The control signals are used to specify the type of operation and the timing of the read or write data transferred over the bus.

6.4.2 Timing of Events During Input and Output Operations

At the initial time t_0, the CPU places the device address on the address lines and sets the mode control lines to indicate input operation.

This information travels over the bus. The clock pulse width $t_1 - t_0$ should be greater than the maximum propagation delay between the CPU and any of the devices connected to the bus. Another delay time $t_2 - t_1$ is needed for loading the data into its input buffer; this should be greater than the maximum bus propagation time plus the set-up time for the input buffer of the CPU. After t_2 the bus lines are cleared in preparation for a new bus cycle. A similar sequence exists for the output operation.

In the input operation the controller starts by placing a command on the data line that contains the device address and the mode of the transfer. The device interface then begins transmitting data. When all data are transferred, this is indicated by an END message. The controller then transmits a new command, placing another device in the required mode.

6.4.3 Sequence of the Arithmetic or Logic Operation

The ALU does not have any internal storage. All numbers needed for an operation should be available to the appropriate circuits in the unit at the same time. The sequence for the operation is as follows:

* The contents of the registers needed for the operation are transferred to the unit.
* The results are transferred to the destination register.

6.4.4 Control Sequence and Microprograms

A microprogram is a system program that incorporates a sequence of control words representing the control sequence of a set of machine instructions. The individual bits of a control word represent the various control signals called "microinstruction." Each step of the control sequence defines a unique set of instructions.

A microprogram counter (μPC) is used to read the control line sequentially from the microprogram memory. A "starting address generator" is responsible for loading the starting address of the microprogram into the μPC every time a new instruction is loaded in the IR. The control signals are delivered to various parts of the CPU in the correct sequence.

When a branch microinstruction is encountered and the branch condition is satisfied, the μPC is loaded with the branch address.

Microprograms are stored in ROM since they are changed very infrequently. The microprogram memory and the CPU constitute two separate entities. A separate bus is usually provided for the transfer between them. Execution of any machine instruction involves a number of fetches from the microprogram memory. Faster microprogram memory means a faster computer.

6.5 LIFO STACKS AND FIFO QUEUES

Data are usually organized in the computer memory in a structured form to speed writing and retrieval. Last-in first-out (LIFO) stack is a commonly used data storage structure. It represents a list of data elements (words or bytes) where access to the data is restricted to the element at the top of the stack. Consequently, the last element written in the stack is the first element to be retrieved.

A stack is generally stored in the main computer memory with its successive elements placed in memory locations with successively increasing memory addresses. A register is used as a stack pointer and contains the address of the current top element of the stack. The first instruction moves the new element from the memory location to the top of the stack and decreases the address number before the move. The second instruction moves the top element from the stack to the memory location and increments the address number in order that it points to the new top element.

This stack data structure is generally used to provide efficient implementation of computing tasks. It can allow arithmetic operations to be performed by one instruction by removing the two top elements from the stack, performing the arithmetic operation, and then replacing the result on the top of the stack.

First-in first-out (FIFO) queues are a form of data structure used in the main memory as data buffers during input and output operations. In this case an in-pointer and an out-pointer are used to monitor the locations of the words to be added or removed from the queue, respectively. Whenever an operation is prepared, the state of the queue should be inspected and updated.

6.6 OPERATING SYSTEMS

Large computing systems are invariably designed to serve many users. An operating system is therefore necessary to manage the utilization of the entire computing facility as efficiently as possible. The objective is to increase the system throughput and therefore reduce the computation cost for the individual user.

6.6.1 Batch Processing

In batch processing the different computing jobs are generally processed sequentially with a chance for some overlap in I/O with computation in individual jobs. This overlap is constrained by the fact that some steps on a particular component of the computing facility cannot proceed until the related steps are completed in other components. Significant increase in operating efficiency can be achieved by input and output spooling. A spool is a file maintained by systems software where an input or output issued by the program is written after being routed by the operating system until the particular computer component is freed to handle them. In large systems where several programs are running concurrently, several inputs or outputs may be spooled at the same time, and a system of priorities determines the order in which they are routed back. System programs called "spoolers/despoolers" control this type of operation.

Another process called "multiprogramming" is also used to improve the throughput efficiency in batch processing. This is accomplished by an operating system routine that controls the transfer of I/O information among the spooling disks, the main memory, and the CPU. Multiprogramming serves the goal of freeing the CPU from having to wait for disk transfer and, consequently, the CPU can be more effectively utilized. Multiprogramming therefore allows several selected jobs to be processed at the same time. The grouping of jobs is done by another operating system routine called "long-term scheduling." When job grouping is accomplished, a "memory management" system allocates specific parts in the main memory for loading the different jobs. A "short-term scheduling" routine controls the access of these jobs to the CPU.

The users in batch processing are generally isolated from the I/O spooling and multiprogramming. These functions are undertaken by the operating system according to job-process priorities and a criterion for the efficient utilization of the shared resources.

6.6.2 Interactive Computing

The primary goal in interactive computing is to achieve fast response to a large number of individual users. This type of computing is gaining wide popularity among scientific and business users. Scheduling techniques are used to assign the CPU to serve each user for a short time called a "quantum" after which the access is interrupted and the control returns back to the operating system. In cases when interactive use does not utilize all the CPU availability, the spare time is usually made available for batch processing with a lower priority.

CHAPTER 7
DIGITAL COMPUTER INTERFACING

DALE B. CHERCHAS

The University of British Columbia
Vancouver, British Columbia, Canada

7.1 INTERFACING

The methodology used to interface digital computers to other physical devices is of vital importance to the mechanical engineer when implementing digital computers in applications. The main function of an interface is to allow signals and information to flow between the computer bus and the external device, as shown in Fig. 7.1. The computer data bus is the source of data to be transmitted to an external device and also the bus onto which incoming data from a peripheral device is placed. Interfacing to the data bus requires bus drivers and receivers. The computer address bus is used to identify the correct peripheral device for data transfer, thus, address bus interfacing requires address decoding logic. The control bus lines are used to coordinate the data transfer operations using input–output (I/O) read, I/O write requests. A typical interface with its major logic devices is shown in Fig. 7.2.

The actual data transmission/reception to and from the peripheral device can be done through serial data transmission/reception or parallel data transmission/reception.

Serial interfacing is done by transmitting high- or low-voltage levels between a single transmission line and ground or sensing similar levels between a single receiving line and ground. The voltage levels represent logical 1 or logical 0. Baud rate is the number of unit-time intervals per second associated with serial transmission, and since the interval signal in this case is logical 1 or 0, the baud rate is the data transfer rate in bits per second. Logical data can also be transmitted/received by using a current loop. Two standards are accepted for serial data communication. For voltage-level serial communication, the Electrical Industry Association or EIA RS-232-C is the latest adopted standard. The primary information in the standard is given in Table 7.1.

The current loop standard is the 20 mA current loop standard. A current flow represents logical 1, and zero current flow represents a logical 0.

The simplest and least costly logical form of serial data transmission is asynchronous, that is, the receiver and transmitter clocks have approximately the same frequency, both much higher than the baud rate, but they have separate clocks. By using a start bit for each character transmitted, the receiver clock can be resynchronized at the beginning of each character. A full transmission of a character in asynchronous mode requires a series sequence of one start bit, five to eight data bits, an optional even or odd parity bit and one, one and one-half, or two stops bits as shown in Fig. 7.3.

Since data on the computer data bus are in parallel form, a serial interface must be able to convert between serial and parallel data formats as well as insert or recognize start and stop bits and detect errors. Today these operations are done by a single chip called a Universal Asynchronous Receiver

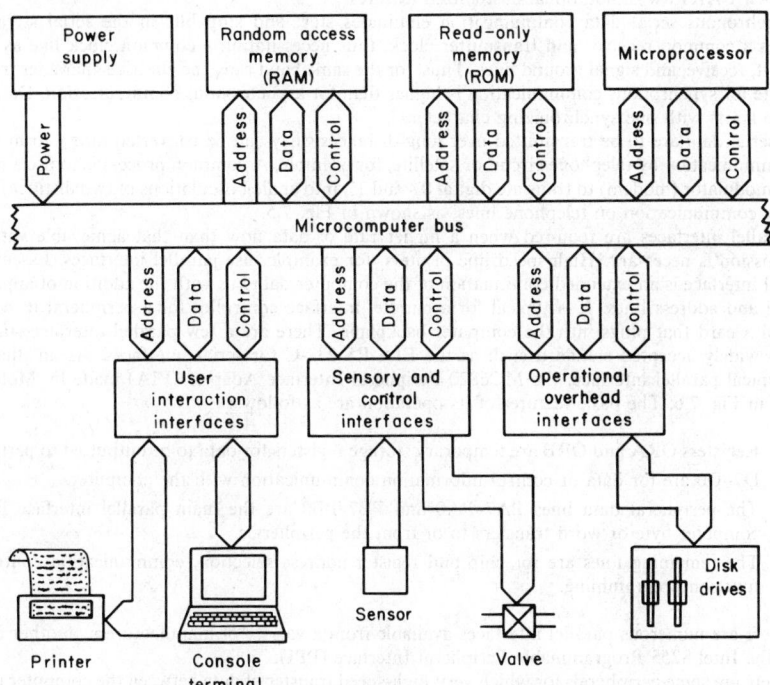

Fig. 7.1 Computer system interfaces.

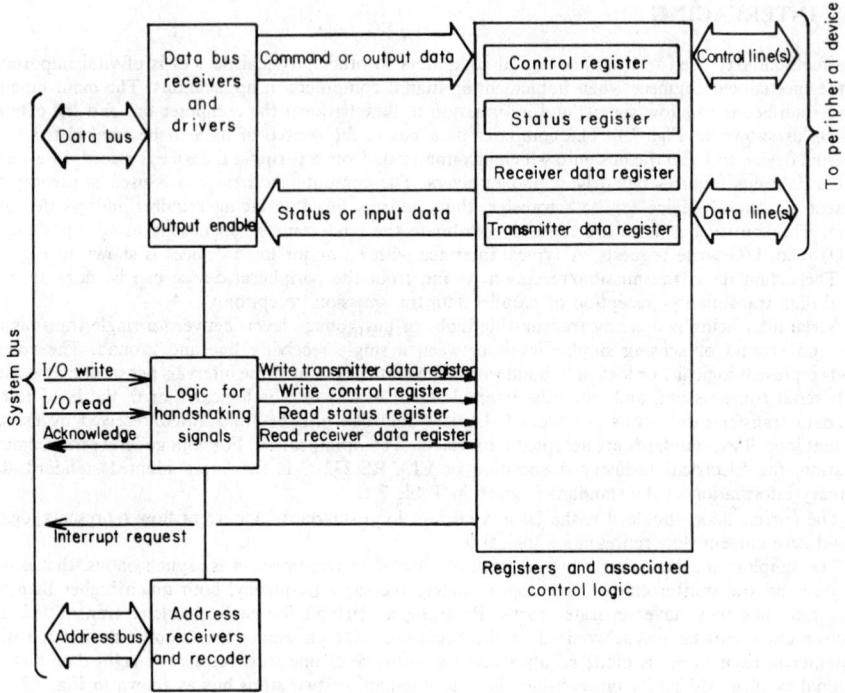

Fig. 7.2 Typical interface logic.

and Transmitter (UART), diagrammed in Fig. 7.4. A large number of manufacturers have designed their own UARTs with additional customized features.

Synchronous serial data communication eliminates start and stop bits in the serial stream, but requires a common receiver and transmitter clock, thus necessitating a common clock line as well as transmit, receive, and signal ground lines. Thus, for the same baud rate, the effective character transmission rate for synchronous communication is higher than for asynchronous communication. Data transmission starts with one synchronizing character.

If serial data are to be transmitted over long distances, they can be converted into a form suitable for communication by telephone circuit or satellite, for example. A common process is to use a modulator–demodulator (modem) to translate digital 0's and 1's into analog oscillations of two distinct frequencies for communication on telephone lines, as shown in Fig. 7.5.

Parallel interfaces are required when a higher rate of data flow than that achievable with serial transmission is necessary. High-speed line printers, for example, use parallel interfaces. Essentially, a parallel interface is an extended continuation of the computer data bus with the addition of appropriate control and address lines. It is typical for a parallel interface controller for a peripheral to be in the form of a card that plugs into the computer backplane. There are a few parallel interface standards, but no widely accepted standard such as the EIA RS-232-C for serial interfaces. As an illustration of a typical parallel interface, the MC6820 Peripheral Interface Adapter (PIA) made by Motorola is shown in Fig. 7.6. The basic features of its operation are as follows:

1. Registers ORA and ORB are temporary storage registers for data to be outputted to peripherals.
2. D7–D0 are for data or control information communication with the computer.
3. The peripheral data lines PA7–PA0 and PB7–PB0 are the main parallel interface lines for computer byte or word transfers to or from the peripherals.
4. The remaining lines are for chip and register address selection, communication control, and function programming.

There are numerous parallel interfaces available from a variety of manufacturers, another example being the Intel 8255 Programmable Peripheral Interface (PPI).

There are some peripherals for which very-high-speed transfer of data between the computer memory and the peripheral is necessary. Using the computer central processing unit (CPU) as an intermediate device in the transfer, that is, putting the data in the CPU accumulator or other general purpose

Table 7.1 EIA RS-232-C Standard[a]

Pin	Description
1	Protective ground
2	Transmitted data
3	Received data
4	Request to send
5	Clear to send
6	Data set ready
7	Signal ground (common return)
8	Received line signal detector
9	(Reserved for data set testing)
10	(Reserved for data set testing)
11	Unassigned (see Section 3.2.3 of the Standard)
12	Secondary received line signal detector
13	Secondary clear to send
14	Secondary transmitted data
15	Transmission signal element timing (DCE source)
16	Secondary received data
17	Receiver signal element timing (DCE source)
18	Unassigned
19	Secondary request to send
20	Data terminal ready
21	Signal quality detector
22	Ring indicator
23	Data signal rate selector (DTE/DCE source)
24	Transmit signal element timing (DTE source)
25	Unassigned

[a] Standard data communication baud rates: 50, 75, 110, 150, 300, 600, 1200, 2400, 4800, 9600, 19200.

Signal voltages (at interface point):

Minimum, $+5 \text{ V} = 1$, $-5 \text{ V} = 0$
Maximum, $+15 \text{ V} = 1$, $-15 \text{ V} = 0$

Current and voltage limitations:
1. Drivers must be able to withstand open or short circuits between any pins in the interface.
2. Voltages below -3 V shall be called *mark* potentials; above 3 V, *space* voltages.
3. Maximum transfer rate is 20,000 bits per second.
4. The load impedance of the terminator side of the interface must be between 3000 and 7000 Ω, and not more than 2500 pF.

register would greatly reduce the availability of the CPU for other tasks. Hence, a special device called a direct memory access (DMA) controller has been developed. The DMA controller is capable of controlling the computer bus and the computer memories for write/read operations to or from memory. If the DMA controller wishes to perform such an operation, it requests use of the computer bus from the CPU, and when granted, performs a memory read or write starting at a previously specified memory address. The DMA controller is placed between the peripheral interface and the computer bus. DMA controllers are currently available as single large-scale integrated (LSI) chips, an example being the Intel 8257.

7.2 OPERATOR DEVICE INTERFACING

The operator devices that require interfacing to a computer to allow user interaction are:

1. Computer terminal displays, that is, cathode ray tubes (CRTs) and storage tubes.
2. Keyboards.

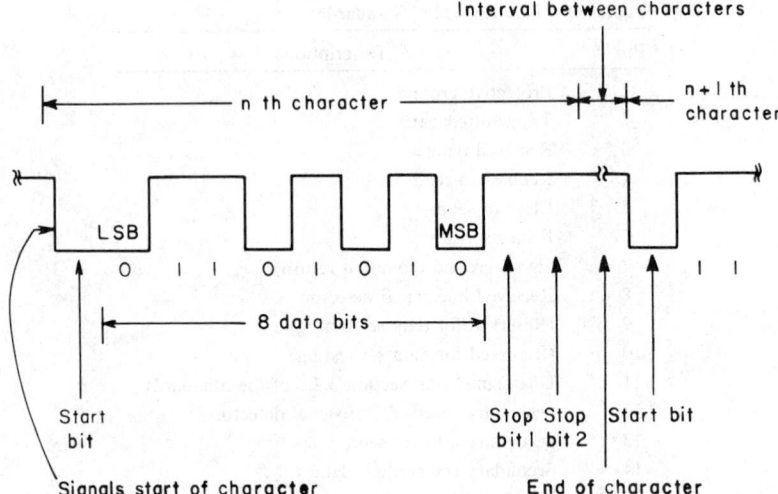

Fig. 7.3   Data format for asynchronous serial transmission.

3. Special graphic input devices, that is, digitizing tablets and light pens.
4. Printers.
5. Plotters.
6. Voice input and output devices.

    Since a variety of data, that is, characters, numbers, commands, must be communicated between the computer and the peripheral and since a string of 0's and 1's, that is, a binary number, is exactly that, a number, a numeric code that can be used to represent not only numbers but also alphabetic characters and special control characters is needed. Of the variety of alphanumeric codes developed, there are only two that are in extensive use. The first is the American Standard Code for Information Interchange (ASCII) and the second is the External Binary-Coded-Decimal Internal Code (EBCDIC). EBCDIC is almost exclusively used by IBM, while ASCII is most popular with other manufacturers and is in fact the most widely used code. Figure 7.7 gives the ASCII code characters and the corresponding binary, hexadecimal, and decimal numbers.

    Thus, there are two tasks that an operator device interface must do, irrespective of whether it is a serial or parallel interface. The first is to perform a speed conversion between the rate at which the operator and/or device can transmit or display characters and the rate at which the computer can receive or transmit the characters. For example, a computer can transmit characters by either a serial or parallel interface, at a faster rate than a low-speed character printer can print them. The second task is to translate the ASCII code, for example, if the letter "A" is pressed on a keyboard, then a conversion device (either hardware or software) in the keyboard generates (see Fig. 7.7) 1000001, which is then transmitted, either serially or in parallel, to the computer. Normally, the peripheral device handles the code translation and the computer handles the necessary character communication speed conversion. The most common operator interface device is a CRT display combined with a keyboard, using a serial communications interface. Such an arrangement is shown in Fig. 7.8. The keyboard decodes keystrokes into ASCII-coded binary numbers, which are then transferred to the keyboard UART for serial transmission to the computer serial interface ACIA. At the computer, two things happen to these characters: (1) any displayable alphanumeric character is immediately transmitted to the CRT display to allow the operator to visually verify what he or she has typed, and (2) the character or string of characters is directed to the computer operating system for analysis when an end of line or carriage return control character is typed by the operator. It should be noted that in what is typically local mode, output from the keyboard decoder is sent directly to the CRT. When the computer wishes to send a character to the CRT, the character is transmitted in serial form from the computer serial interface ACIA to the CRT UART where it is then put into the video display memory for the CRT display processor to then put on the screen.

    Printers and plotters receive ASCII-coded data in either serial or parallel form from the computer. After any necessary conversion to parallel form and decoding, the characters are then put into the printer/plotter buffer and are processed by the controller for the print head or plotter head. Since

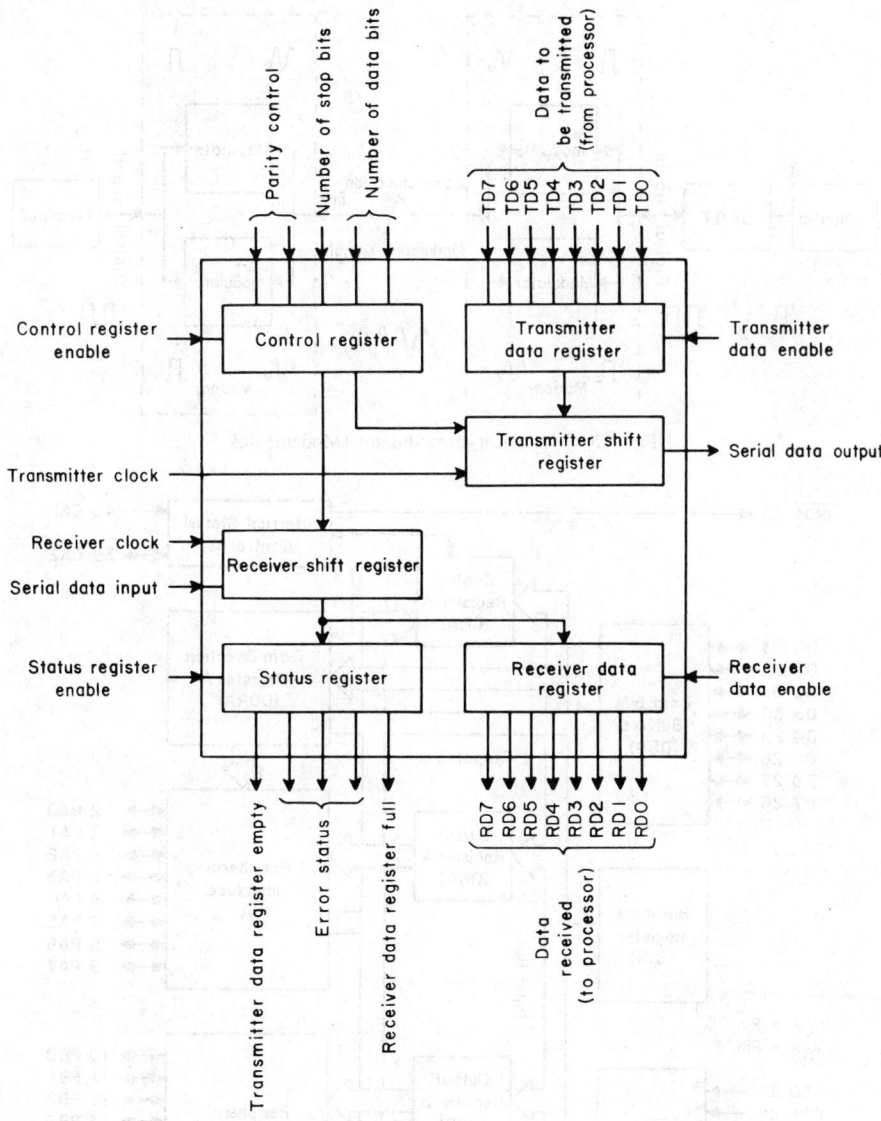

Fig. 7.4 UART logic diagram.

printers and plotters are electromechanical devices, they are generally not able to process data in their buffer as fast as the buffer can be filled by data coming from the transmitter. To prevent buffer overflow, controls are used to signal the main computer to start and stop transmission. With serial interfaces, this is accomplished using the "XON XOFF" control technique, and with parallel interfaces, this is accomplished by separate control lines. For line printers, up to 600 lines per minute, serial RS-232-C interfaces can be used. For higher printing rates, a parallel interface is required. Plotters generally use an RS-232-C serial interface.

Special graphics input devices such as digitizing tablets and light pens essentially convert analog position coordinates into digital numbers and then typically code the digital coordinate numbers into ASCII code for serial transmission to the computer. Voice I/O devices perform analog/digital conversion of data representing movement of microphone or speaker components; however, once the data are in digital form, they can be coded and communicated in serial or parallel form as for any other device.

One other parameter to be considered, particularly for serial communication interfaces, is the directionality of the communication. The directionality is identified as below:

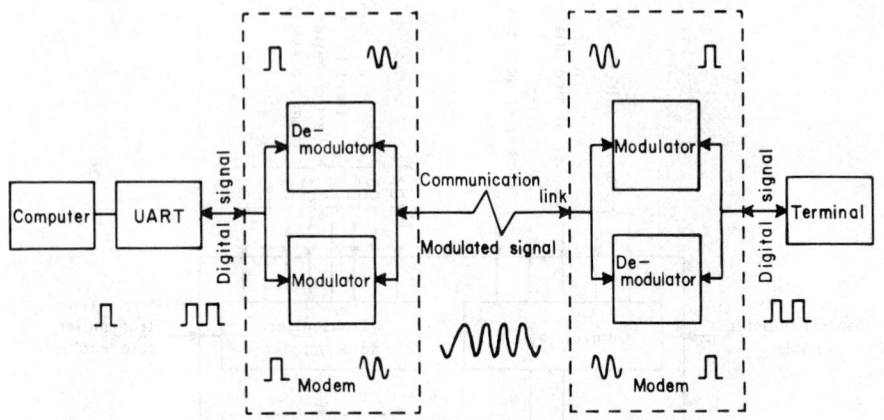

Fig. 7.5 Modulator–demodulator (Modem) link.

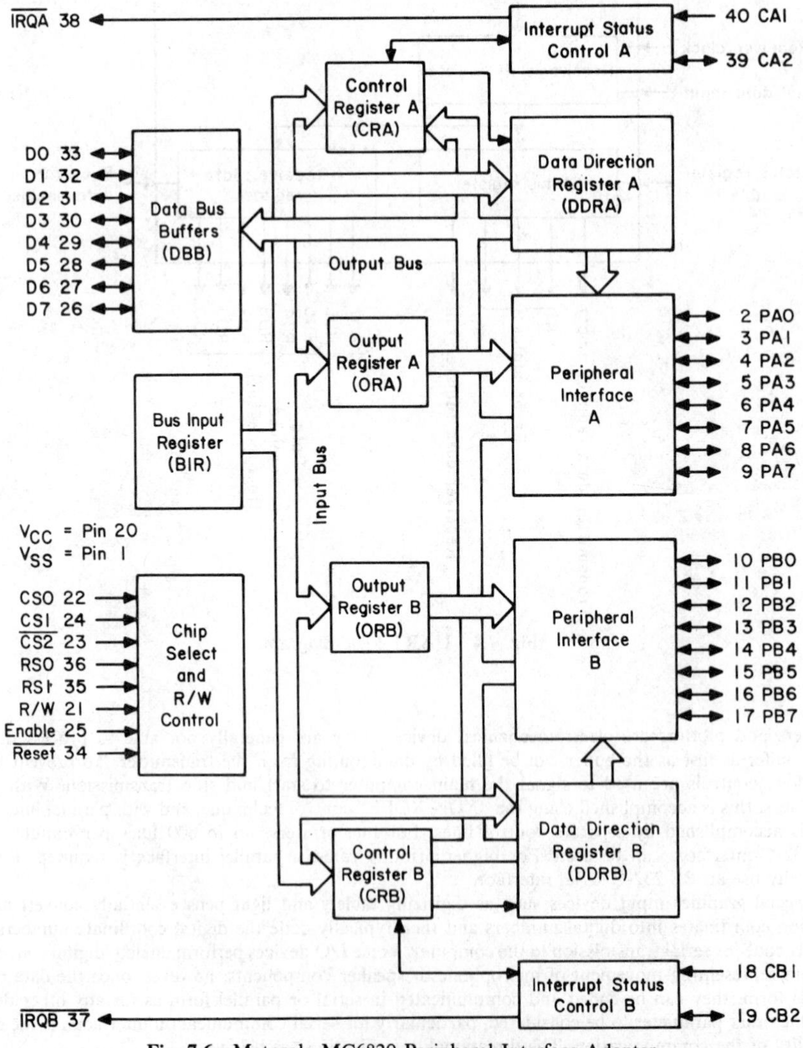

$\overline{\text{IRQA}}$ 38

Interrupt Status
Control A

40 CAI

39 CA2

Control
Register A
(CRA)

D0 33
D1 32
D2 31
D3 30
D4 29
D5 28
D6 27
D7 26

Data Bus
Buffers
(DBB)

Data Direction
Register A
(DDRA)

Output Bus

Output
Register A
(ORA)

Peripheral
Interface
A

2 PA0
3 PA1
4 PA2
5 PA3
6 PA4
7 PA5
8 PA6
9 PA7

Bus Input
Register
(BIR)

Input Bus

V_{CC} = Pin 20
V_{SS} = Pin 1

CS0 22
CS1 24
$\overline{\text{CS2}}$ 23
RS0 36
RS1 35
R/W 21
Enable 25
$\overline{\text{Reset}}$ 34

Chip
Select
and
R/W
Control

Output
Register B
(ORB)

Peripheral
Interface
B

10 PB0
11 PB1
12 PB2
13 PB3
14 PB4
15 PB5
16 PB6
17 PB7

Data Direction
Register B
(DDRB)

Control
Register B
(CRB)

Interrupt Status
Control B

18 CB1

19 CB2

$\overline{\text{IRQB}}$ 37

Fig. 7.6 Motorola MC6820 Peripheral Interface Adapter.

72

Character	Hex	Decimal	Binary	Character	Hex	Decimal	Binary	Character	Hex	Decimal	Binary	Character	Hex	Decimal	Binary
NUL	0	0	0000000	SP	20	32	0100000	@	40	64	1000000	`	60	96	1100000
SOH	1	1	0000001	!	21	33	0100001	A	41	65	1000001	a	61	97	1100001
STX	2	2	0000010	"	22	34	0100010	B	42	66	1000010	b	62	98	1100010
ETX	3	3	0000011	#	23	35	0100011	C	43	67	1000011	c	63	99	1100011
EOT	4	4	0000100	$	24	36	0100100	D	44	68	1000100	d	64	100	1100100
ENQ	5	5	0000101	%	25	37	0100101	E	45	69	1000101	e	65	101	1100101
ACK	6	6	0000110	&	26	38	0100110	F	46	70	1000110	f	66	102	1100110
BEL	7	7	0000111	'	27	39	0100111	G	47	71	1000111	g	67	103	1100111
BS	8	8	0001000	(	28	40	0101000	H	48	72	1001000	h	68	104	1101000
HT	9	9	0001001	)	29	41	0101001	I	49	73	1001001	i	69	105	1101001
LF	0A	10	0001010	*	2A	42	0101010	J	4A	74	1001010	j	6A	106	1101010
VT	0B	11	0001011	+	2B	43	0101011	K	4B	75	1001011	k	6B	107	1101011
FF	0C	12	0001100	,	2C	44	0101100	L	4C	76	1001100	l	6C	108	1101100
CR	0D	13	0001101	-	2D	45	0101101	M	4D	77	1001101	m	6D	109	1101101
SO	0E	14	0001110	.	2E	46	0101110	N	4E	78	1001110	n	6E	110	1101110
SI	0F	15	0001111	/	2F	47	0101111	O	4F	79	1001111	o	6F	111	1101111
DLE	10	16	0010000	0	30	48	0110000	P	50	80	1010000	p	70	112	1110000
DC1	11	17	0010001	1	31	49	0110001	Q	51	81	1010001	q	71	113	1110001
DC2	12	18	0010010	2	32	50	0110010	R	52	82	1010010	r	72	114	1110010
DC3	13	19	0010011	3	33	51	0110011	S	53	83	1010011	s	73	115	1110011
DC4	14	20	0010100	4	34	52	0110100	T	54	84	1010100	t	74	116	1110100
NAK	15	21	0010101	5	35	53	0110101	U	55	85	1010101	u	75	117	1110101
SYN	16	22	0010110	6	36	54	0110110	V	56	86	1010110	v	76	118	1110110
ETB	17	23	0010111	7	37	55	0110111	W	57	87	1010111	w	77	119	1110111
CAN	18	24	0011000	8	38	56	0111000	X	58	88	1011000	x	78	120	1111000
EM	19	25	0011001	9	39	57	0111001	Y	59	89	1011001	y	79	121	1111001
SUB	1A	26	0011010	:	3A	58	0111010	Z	5A	90	1011010	z	7A	122	1111010
ESC	1B	27	0011011	;	3B	59	0111011	[	5B	91	1011011	{	7B	123	1111011
FS	1C	28	0011100	<	3C	60	0111100	\	5C	92	1011100	|	7C	124	1111100
GS	1D	29	0011101	=	3D	61	0111101	]	5D	93	1011101	}	7D	125	1111101
RS	1E	30	0011110	>	3E	62	0111110	^	5E	94	1011110	~	7E	126	1111110
US	1F	31	0011111	?	3F	63	0111111	_	5F	95	1011111	del	7F	127	1111111

Fig. 7.7 ASCII code.

1. *Simplex.*   Operator device can only send or receive data.
2. *Half-duplex.*   Operator device can both send and receive data but cannot send and receive simultaneously.
3. *Full-duplex.*   Operator device can simultaneously send and receive data.

7.3 EXTERNAL MEMORY INTERFACING

Digital computer external memories are used to provide mass storage of data and programs at a lower cost, but also at a slower access speed, than that available using the digital computer registers, cache memory, or main memory. External memory devices include the following:

1. Hard magnetic disks.
2. Flexible magnetic diskettes.
3. Magnetic tape.
4. Magnetic bubble memory.

In order to interface successfully with such devices, a digital computer must have the following items:

1. A serial or parallel interface to communicate data, address, and control information.
2. An integrated circuit (IC) or set of ICs that act as a controller to operate, through the serial or parallel interface, the electromechanical devices in the external memory unit.
3. A special piece of software called a "driver," normally a subroutine that can be called from a main program, that handles the passage of data, address, and control information from the digital computer main memory to the interface and, as well, operates the controller in item 2, which subsequently operates the electromechanical devices in the external memory unit.

For high-speed external memory devices such as hard disks, data can be transferred in block form at a high rate by adding a direct memory access (DMA) controller to the controller and communications

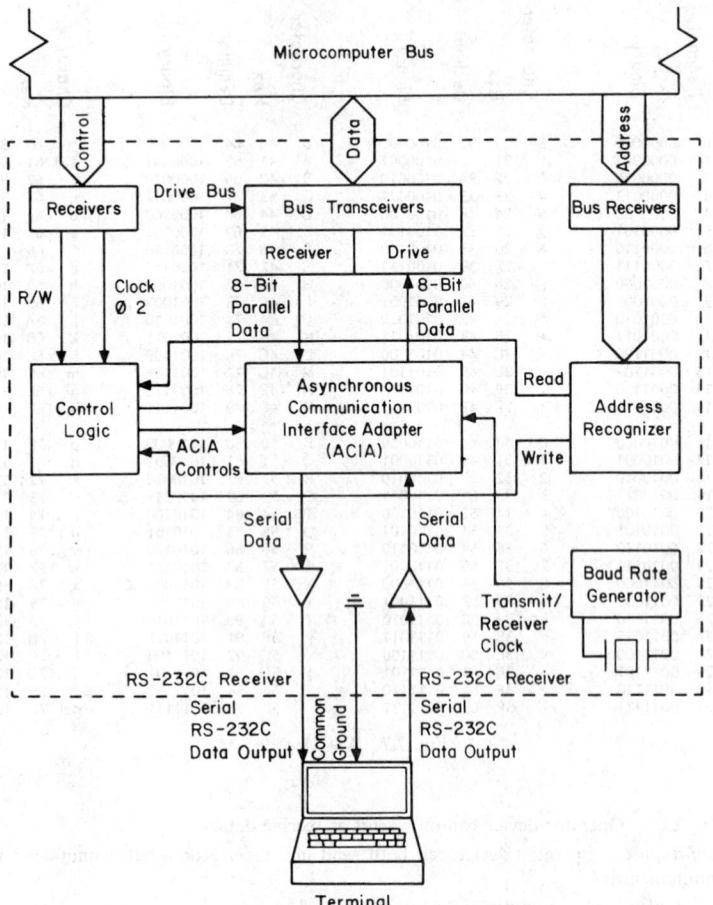

Fig. 7.8 CRT and keyboard serial interface.

interface. For hard disk drives, floppy diskette drives, and magnetic reel tape drives, the high rate of data transfer generally necessitates the use of parallel interfaces.

Cassette tape drives, with lower data transfer rates, can be interfaced with a serial interface.

7.4 INSTRUMENTATION INTERFACING

Instrumentation refers to the use of sensing devices to measure the operating state of a process or machine and/or the use of actuating devices to *control* the operating state of a process or machine. A digital computer can be used to receive and process data from the sensors and/or to determine and send control signals to the actuators.

In order to understand how a digital computer is interfaced, it is necessary to distinguish and describe four different types of signals, that is, continuous, discrete, analog, and digital. Given a signal $r(t)$, where r is the value of the signal and t is the time variable, the signals, referring to Fig. 7.9, are defined as follows:

1. *Continuous Time Signal.* t can take on a continuum of values.
2. *Discrete Time Signal.* t takes on a finite or infinitely countable set of discrete values, that is, t_k, $k = 0, \pm 1, \pm 2, \pm 3. \ldots$

Referring to Fig. 7.10,

3. *Analog Signal.* r can take on any value within a given range.
4. *Digital Signal.* r can take on any one of a specific set of values within a given range.

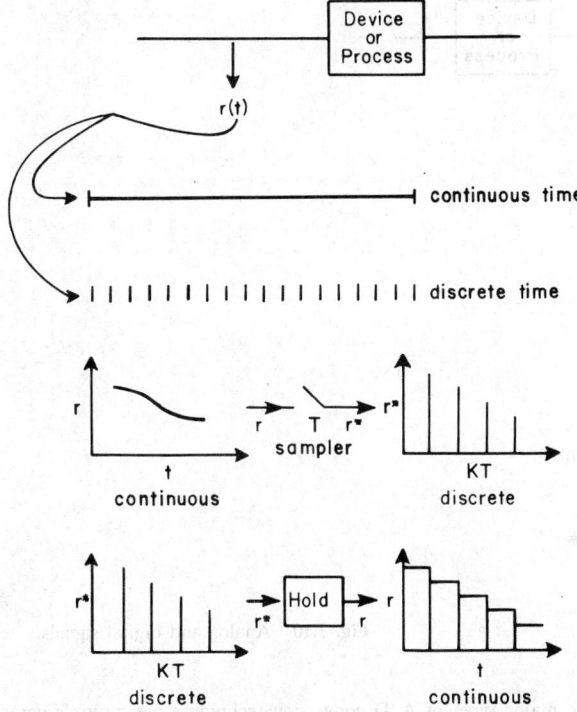

Fig. 7.9 Continuous and discrete time signals.

The necessary devices to convert between these signals are shown in Figs. 7.9 and 7.10. A *sampler* converts a continuous signal to a discrete time signal. A *hold* device converts a discrete time signal to a continuous signal. An analog-to-digital (A/D) converter and a digital-to-analog (D/A) converter are devices to convert between analog and digital signals.

Depending on the application, a signal may have to be converted to the correct form. For example, a digital computer utilizes discrete digital signals, however, in a control loop where the computer may act as the controller, the input and output may be of the continuous analog form.

The possible required conversions for a computer control loop are shown in Fig. 7.11. Thus for digital computer interfacing to instrumentation, it will be necessary to use one or more of four major subsystems, that is, an analog-input subsystem, an analog-output subsystem, a digital-input subsystem, and a digital-output subsystem.

7.4.1 Analog-Input Subsystem

Figure 7.12 shows an analog-input subsystem. The sensor(s) convert(s) a physical process variable, for example, temperature or pressure, into a sensor variable value that bears a known relationship to the process variables. If necessary, the sensor variable is then converted by means of a transducer into an electrical variable, typically a voltage. The continuous time electrical variable is then sampled by a sample and hold device through a multiplexer to allow sharing of the sample and hold and A/D converter. The value of the variable is then maintained in the hold, amplified through a programmable amplifier and made available to be converted to digital form. An active and/or passive filter may be inserted at this point. The A/D converter converts the held voltage into a digital word, which depending on the accuracy of the converter, can have from 6 to 16 bits. This digital word is latched into a buffer and then transmitted to the computer memory through a serial interface if the computer is remotely located or a parallel interface if the sample and hold and A/D converter can be located close to the computer. In fact it is very common to have the entire analog-input subsystem starting from the multiplexer on a single card that is plugged directly into the computer bus.

Returning to Fig. 7.12, the multiplexer switch(s), sample and hold, active filter, amplifier, A/D converter, and buffer are all available as separate IC chips. As noted earlier, completely assembled analog-input subsystems with all ICs assembled on a single printed circuit (PC) board can be purchased, or designers may construct their own. Of the major ICs, the most critical and expensive is the A/D

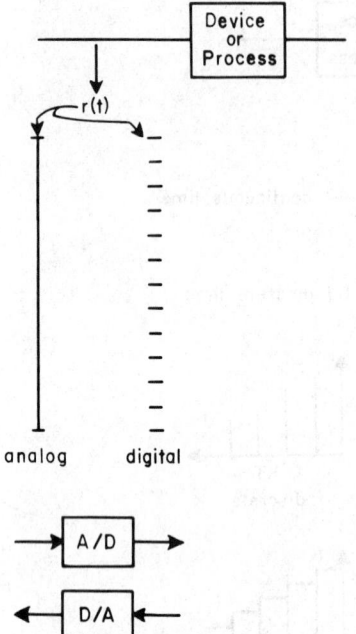

Fig. 7.10 Analog and digital signals.

converter chip. The major types of A/D conversion techniques are multiple comparator, successive approximation register (SAR), voltage-to-frequency, and ramp and dual ramp integration. Buchsbaum[1] gives a thorough discussion of these ICs. The SAR technique is a common and relatively inexpensive approach.

The five most important parameters of a sample and hold circuit are:

1. *Acquisition Time.*   The time between start of sampling and stable output—analogous to digital propagation delay.
2. *Aperture Time.*   The time it takes for the sampling switch to fully open.
3. *Aperture Uncertainty Time.*   The variation characteristics of the aperture time.

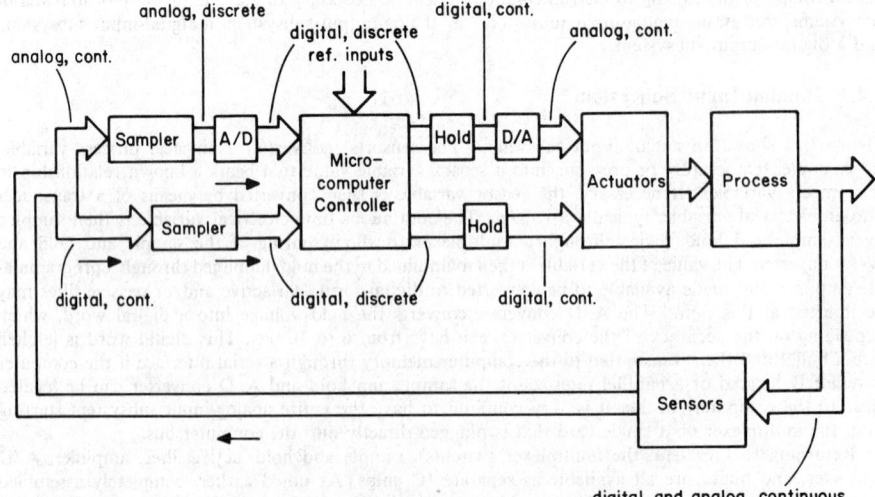

Fig. 7.11 Direct digital control loop.

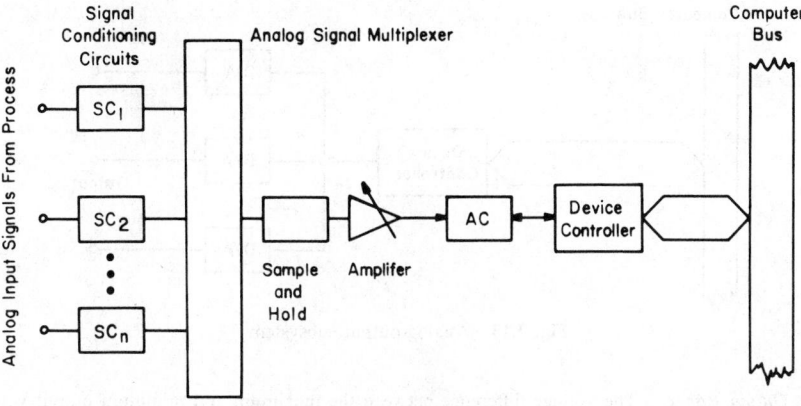

Fig. 7.12 Analog-input subsystem.

4. *Decay Rate.* The change in output voltage resulting from capacitor discharge per period of time.
5. *Feedthrough.* The amount of input signal that manages to leak through to the output when in the hold mode.

Selection Criteria

When choosing an A/D converter, there are a few important parameters to consider:

1. *Range.* The voltage difference between the converter's minimum and maximum input voltages.
2. *Resolution.* The size of the voltage steps. Resolution is expressed as the number of bits in the output code and the percentage of the range covered. A 10-bit A/D is a 0.1% A/D.
3. *Linearity.* The maximum difference between the voltage steps. Ideally, all voltage steps should be precisely the same size.
4. *Monotonicity.* The property of having an increasing output for an increasing input over the converter's entire range.
5. *Missing Codes.* The property of a converter skipping a code due to nonlinearity.
6. *Quantizing Error.* The maximum voltage error due to the converter's noninfinite resolution.
7. *Relative Accuracy.* The full-scale error in output voltage for any two input voltages across the entire range.
8. *Absolute Accuracy.* The full-scale error in output voltage for full-scale input voltage.
9. *Offset Error.* The value output by the converter when a 0 V input is applied.

7.4.2 Analog-Output Subsystem

Figure 7.13 shows an analog-output subsystem. The value to be outputted is stored as a digital word in the computer memory. The sequence of bits in the word is then transmitted either by parallel or serial interface (and this may be a remote transmission) to the buffer of the D/A converter. The D/A converter then converts the word to an analog voltage. The voltage is then adjusted for sign, conditioned and amplified, and made available to the control actuator. D/A converters are relatively inexpensive and typically one is dedicated to each analog output point.

D/A converters generally are a resistor network, the R-$2R$ network being the most common, to convert the digital word to an analog voltage. These converters are available very inexpensively as single IC chips. An excellent discussion of these ICs is given in Buchsbaum.[1]

The following features should be evaluated before purchasing a D/A converter:

1. *Resolution.* The number of bits converted.
2. *Accuracy.* Typically measured as percentage error in voltage output.
3. *Settling Time.* The time it takes for the digital input to be converted to an analog output with the rated specification characteristics.
4. *Linearity.* The maximum error in the conversion between adjacent input codes.

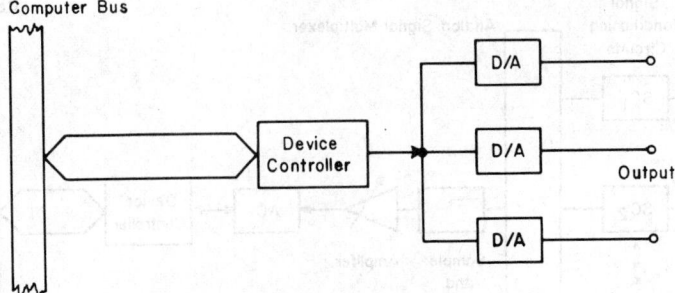

Fig. 7.13   Analog-output subsystem.

5. *Output Range.*   The voltage difference between the maximum and minimum output voltage.
6. *Input Coding.*   The binary of binary-coded-decimal format of input code.

7.4.3   Digital-Input Subsystem

A considerable amount of information on the operation of a process or machine can be provided to a digital computer in the form of digital input. A digital input is binary, that is, it is one that can take on only one of two values, a high voltage or a zero or low voltage. These two states correspond to the state of a device on the sensed process or machine, for example, a switch being open or closed, a threshold level being exceeded or not. Figure 7.14 shows a digital-input subsystem.

The sensors and signal conditioners sense the operating state of a device, adjust the signal to be acceptable to the register, and deliver the signal to the register where a single logical bit is set in the register. Thus, the register will contain a digital word with bits set corresponding to a number of sensed digital inputs. The digital word in the register is then transmitted by either a serial (for remote transmission) or a parallel interface to the computer.

The register is generally a single IC chip and can be latching or nonlatching. The register operation will depend on the chip technology. For example, a TTL chip will detect an input voltage of less than ~0.2 V as being a logical 0 and voltage greater than ~2 V (with 6 V being the allowed maximum) as a logical 1. Reference 1 gives an excellent discussion of digital-input registers.

7.4.4   Digital-Output Subsystem

A digital subsystem is shown in Fig. 7.15. The digital output subsystem provides one or more binary outputs determined by the computer. These binary outputs can be used to control, through current or voltage, an electrical device. For example, a digital subsystem can be used to open or close a valve, or to start or stop an electric motor.

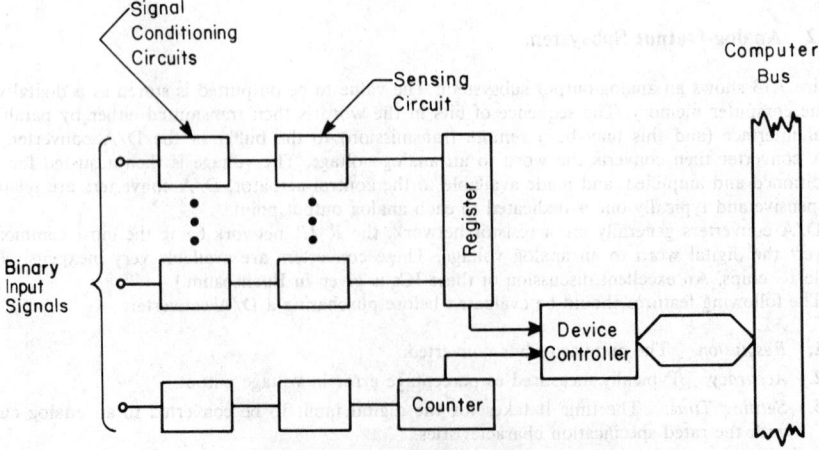

Fig. 7.14   Digital-input subsystem.

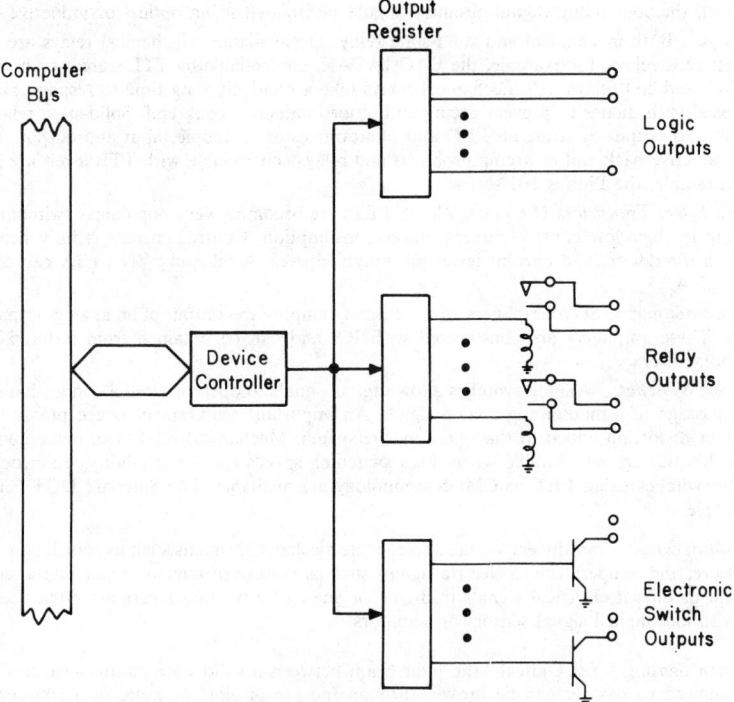

Fig. 7.15 Digital-output subsystem.

The computer sets the status of an individual digital output by setting a bit in a digital word in the computer memory. This word then contains the desired status for a number of digital outputs. The digital word is transmitted by a parallel or a serial (can be used for remote transmission) interface to the output register, typically a latched IC register. The outputs from the register then correspond to the individual desired digital outputs. An individual output can then be read directly by another digital device, used to operate a semiconductor switch or used to operate a relay.

7.4.5 Sensors and Actuators

Provided the correct transducers and signal conditioners are implemented, virtually any sensor or actuator can be interfaced to a digital computer using the subsystems described in the previous sections. However, certain sensors and actuators have been developed that are particularly well suited to use with digital computers. These will be discussed in three categories, that is, power switching devices, transducers, and optoelectronic devices.

1. *Power Switching Devices.* The limited current-handling capabilities of TTL gates (~16 mA) have necessitated the development of devices, both solid state and mechanical, for switching higher currents:

- *Buffer-Interface Gates.* These gates are similar in package and pinout to standard TTL gates but capable of handling up to three times the sink current. A good example is the 7438 NAND gate.
- *Power Transistors.* Power transistors can be used to switch high current levels by varying the base current directly, or through an interface gate, from a computer discrete output point. An example is the 2N222 *npn* transistor with a drive current up to 800 mA and a 40 V breakdown voltage. Darlington power transistors combine two transistors in one and allow control with very small base currents.
- *Thyristors.* These silicon-controlled rectifiers allow very large ratios (>1000) of load current to gate control current. They are switching devices, not linear amplifiers. The device will not turn off, even if the gate current is removed, until the load current is removed and reapplied; hence, they are best suited to controlling alternating current loads. When large loads are con-

trolled, the controlling digital circuitry should be isolated using optical or inductive coupling.

- *Relays.* Both mechanical and solid-state relays are available. Mechanical relays are generally either reed relays, for example, the EAC 1ACAH, controllable by TTL signals or miniaturized switch and coil relays. All mechanical relays take a relatively long time to respond and should be used with diodes to prevent arcing and absorb inductive back emf. Solid-state relays isolate input and output by using an LED and phototransistor to couple input and output. There are no inductive back emf or arcing problems and relays controllable with TTL levels are available, for example, the Philips 501 series.
- *Field Effect Transistors (FETs).* VMOS FETs are becoming very popular as switching devices owing to their low control current power consumption. Control current is only necessary to switch the device, and current levels are microamperes. A Siliconix VN 84GA can control up to 12.5 A.
- *Servoamplifiers.* Servoamplifiers will accurately amplify the output of an analog-output subsystem. These amplifiers are constructed with ICs and can be obtained from industrial control manufacturers.
- *Analog Switches.* Analog switches allow digital signals to open or close a connection to permit the passage of a modulating analog signal. An important requirement is acceptably low noise and/or distortion added to the signal by the switch. Mechanical relays can make good analog switches but are not suitable where high switching speeds and/or reliability are critical. Solid-state switches using FET or CMOS technology are available. The Siliconix DG175 is a good example.

2. *Transducers.* Transducers are used to convert mechanical signals such as translation, rotation, force, pressure, and temperature to electric signals such as voltage or current. Since digital computers either accept or output electrical signals in digital or analog form, transducers are often necessary to interface with mechanical signal sensors or actuators.

- *Motion Sensing.* (a) Optical—the light beam between a solid-state emitter–detector pair can be allowed to pass or can be broken through the use of slots or holes in a rotating disk or translating plate. The pulses from the emitter–detector pair are conditioned, typically using monostable multivibrators, and are then counted by the computer to determine the amount of motion. (b) Magnetic—induction coils are used to generate current pulses in the coils as a rotating or translating magnet moves past the coils. The pulses can be conditioned and counted or integrated with a capacitor and the capacitor voltage inputted through an analog-input subsystem. For slow motion, Hall-effect semiconductors should be used to detect the magnetic field strength. (c) Electromechanical Switches—a switch or switches are opened and closed at a prescribed point in the motion. The switch closing can be sensed, and the rate of this occurrence can be used by the computer to determine the rate of motion.
- *Pressure Sensing.* (a) Capacitive Devices—a compressible dielectric material is placed between two metal sheets. The application of pressure changes the capacitance; this change is then converted to a voltage change detectable by an analog-input subsystem. (b) Piezoelectric Devices—piezoelectric material is material that generates electricity when compressed and is particularly useful when pressure changes are being sensed.

Integrated silicon pressure transducers are available, an example being the National Semiconductor LX Series.

- *Temperature Sensing.* A number of transducers are available, for example, thermocouples, resistance detectors, thermistors, thyristor temperature switches, and, most recently, silicon diode temperature sensors. Regarding the silicon diode sensors, they are inexpensive, have a range of ~-40 to $+150°C$, require no trimming or calibration, and have a large output voltage swing. However, they are not suitable for temperature measurement above $\sim150°C$.
- *Solenoid Actuators.* A solenoid consists of an electromagnet that moves a metal plunger. Thus, electrical current in the magnet results in linear motion of the plunger. The plunger can be returned by a permanent magnet or a spring. Solenoids are used to control valves or switches.
- *Stepping Motor Actuators.* A popular device for achieving accurately controlled rotary motion from computer digital outputs. A stepping motor consists of a rotor that aligns itself with the one of a number of stators that is currently actuated. Figure 7.16 shows a stepping motor. The motor can be stepped forward or backward, at quite a high rate if necessary, by switching current between the stator coils. Stepping motor control ICs are available to convert digital words or pulse trains from the computer into correctly conditioned signals to the motor.
- *Servomechanism Actuators.* A servomechanism allows modulated control of translational or rotary motion. A servo converts an electrical voltage into controlled linear or rotary motion

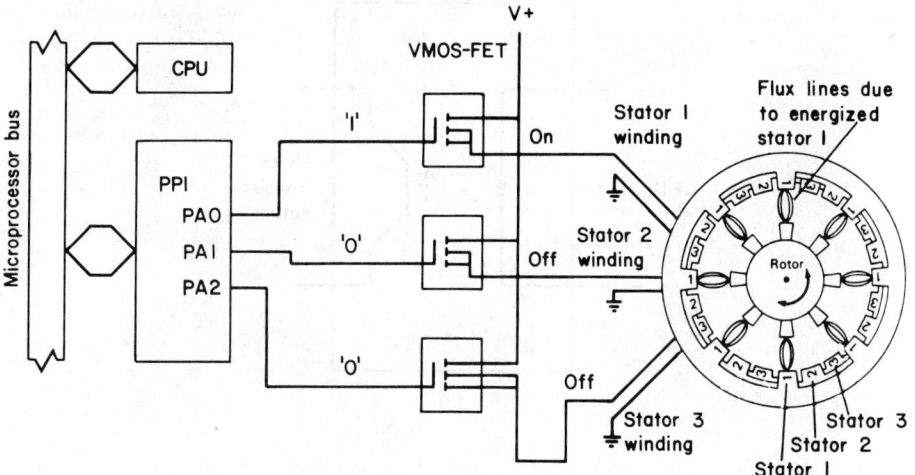

Fig. 7.16 Stepping motor system.

through the use of feedback control. The main components are an actuator (e.g., electric motor, hydraulic or pneumatic pistons), a motion sensor, and a feedback controller.

3. *Optoelectronic Devices.* Optoelectronic devices have become popular for interfacing with a digital computer. In switching applications, they are fast, interface easily with IC logic, have zero hysteresis, and exhibit virtually no arc or wear problems.

- *Light Emitting Diodes (LEDs).* LEDs will generate highly visible light with very low control current and voltage levels. The major colors available are red, orange, yellow, and green as well as infrared. LEDs generate light in very narrow color ranges. The voltage drop across an LED ranges from 1.6 to 3 V. LED response time is very fast, ranging from 90 to 200 nsec. LEDs are generally used with current limiting resistors as shown in Fig. 7.17.
- *Photocells.* Cadmium sulfide (CdS) photocells have been available for many years, are relatively inexpensive, but slow. Varying the intensity of light falling on the photocell varies the electrical resistance of the cell. In computer interfacing, the cell can be used as a switch or a variable resistor. Lead sulfide cells are sensitive to infrared light and selenium cells are sensitive to blue light. All photocells take up to 1 or 2 sec to turn off.

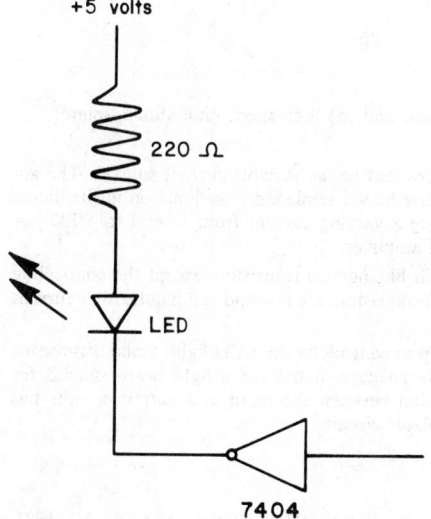

Fig. 7.17 Light-emitting diode circuit.

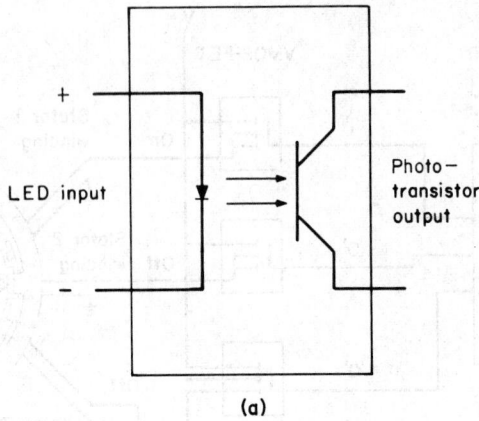

(a)

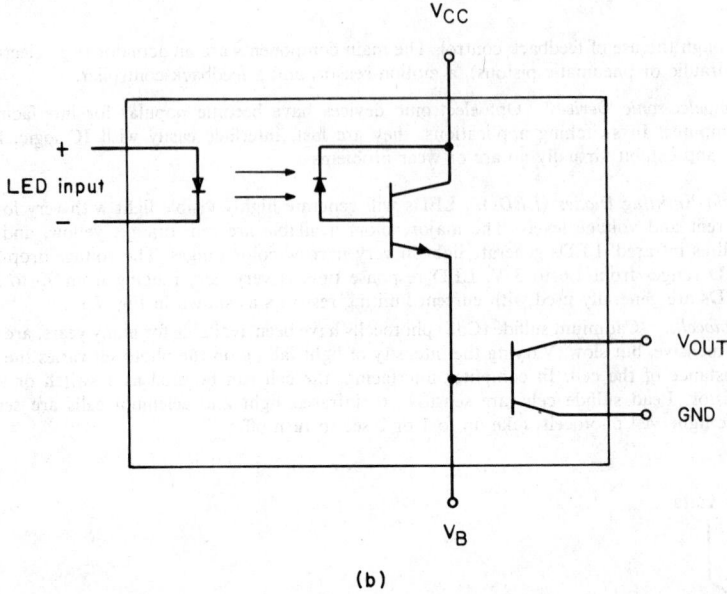

(b)

Fig. 7.18 Optoisolators: (a) phototransistor isolator and (b) high-speed photodiode isolator.

- *Photodiodes.* Photodiodes are high-speed devices that act as variable current sources. The applied light level changes the properties of a reverse-biased semiconductor junction in the diode. The reverse-biased diode (10–35 V) will generate a varying current from several to ~200 μA that can then be used to control an operational amplifier.
- *Phototransistors.* Phototransistors operate much like normal transistors except the controlling base voltage is replaced by a light window. Phototransistors are fast and can handle high current loads.
- *Optoisolators.* An optoisolator is an LED set up to control, by the LED light, a phototransistor or photodiode. The two devices are in a single package, and since a light beam is used for signal communication, effective electrical isolation between the input and output circuits has been achieved. Figure 7.18 shows some optoisolator circuits.

REFERENCE

1. W. H. Buchsbaum, *Encyclopedia of Integrated Circuits,* Prentice-Hall, Englewood Cliffs, NJ, 1981.

CHAPTER 8
DIGITAL COMPUTER SELECTION

DALE B. CHERCHAS

The University of British Columbia
Vancouver, British Columbia, Canada

The selection of a digital computer, in particular a microcomputer for use in a given mechanical engineering application, merits very careful consideration of a number of technical factors as well as price. Many of the technical features of a small digital computer are determined by the central processing unit (CPU), that is, for most small computers the microprocessor chip itself. The key considerations are discussed below.

8.1 MANUFACTURER

The capabilities and reputation of the manufacturer(s) of the main integrated circuit (IC) chips in the computer and the computer itself should be assessed. The manufacturer of the CPU or microprocessor chip should be a large, well-established, and reputable firm. Since most memory chips of a given word size are of a similar design, it is reasonable and prudent to require that at least two manufacturers be able to supply the memory chips used. The designer and manufacturer of the computer itself may not be an IC chip manufacturer, although many computer manufacturers will have a controlling interest in firms that make IC chips in order to ensure a chip supply. Manufacturers of small computers range from large, long-established firms down to small companies that have been in business a short time. The small manufacturer should not be overlooked, since they can often provide a highly innovative product at a relatively low price. A major consideration in selecting a computer manufacturer is service. All computers can be expected to fail at some point in their service life. Unless the user is sophisticated in digital electronics, service calls will be necessary. There is, therefore, a real advantage to purchasing a computer from a firm that has been and is expected to be in business for a long time and also has a capable and available service organization.

8.2 TECHNOLOGY

"Technology" refers to the basic fabrication process that is used to produce an IC chip. The type of technology used for the main ICs in a computer will have a major effect on the suitability of the computer for a given application. The type of technology will affect the following properties of a chip:

1. Power consumption.
2. Operating speed.
3. Density, that is, number of logic gates on the chip.
4. Reliability.
5. Noise sensitivity.
6. Interface requirements.
7. Cost.

The two main processes used for chip fabrication are bipolar [including conventional, Schottky, and integrated injection logic (I²L)] and metal oxide semiconductor (MOS) (including PMOS, NMOS, CMOS, SOS, VMOS, and HMOS). Table 8.1 summarizes the main characteristics of each technology. Regarding the speed–power product, the lower the value of the product, the higher the speed and the lower the power consumed by the chip.

8.3 INSTRUCTION SET

This parameter refers to the number of basic instructions in the assembly language instruction set that can be decoded by the CPU. The instructions can be categorized as one of transfer, arithmetic, logical, shift and rotate, index and count, bit manipulation, looping, branching, and input/output (I/O) communication and transfer. A measure of the sophistication of a microprocessor chip is the number of these operations that can be executed with only one instruction. In general, the suitability of a microprocessor to a given task is determined by the instruction set. For example, I/O instructions are very important if the processor is going to be used in an automatic control application. For data processing, arithmetic instructions are important. Owing to the lack of a uniform definition for an instruction set, the number of instructions is not a good index of the versatility of a processor; more important is what the instructions can do.

8.4 BIT WIDTH

Bit width or data bus size is the number of data lines in the processor data bus. Bit width, or data word length as it is often called, partially determines the accuracy, speed, and memory cost of a small digital computer. The earliest microprocessors were four-bit devices, for example, the Intel 4004 in 1971, which found extensive use in calculators and control applications. The most common micropro-

Table 8.1 Feature of IC Chip Fabrication Technologies

Characteristics	Technology							
	Schottky Bipolar	PMOS	NMOS Silicon Gate	HMOS	VMOS	CMOS	SOS	I²L
Logic voltage range (high–low)	5 V	10 V	5 V	2–5 V	5 V	Supply voltage	Supply voltage	0.6–0.7 V
Speed–power product (pj)	10–100	50–100	5–50	0.5–1	0.5–1	2–40	0.5–30	0.2–2
Gate propagation delay (nsec)	7	75	4–25	2	2	10–35	4–20	5–50
Gate power dissipation	1–5 mW	1.7 mW	1 mW	0.4–1 mW	1 mW	1 mW to 1 W	0.05 mW	1 nW to 100 μW
Gate current	0.2–2 mA	50 μA to 1 mA	50 μA to 1 mA	0.2–1 mA	0.5–1 mA	Nanoamperes	Nanoamperes	1 nA to 1 mA
Power supply voltage	4 V typically	−15—−5 V	5–15 V	2–5 V	5 V	3–18 V	3–18 V	1–15 V
Gate area (mils²)	20	10	5	2–3	2–3	10–30	15	5
Gates/mm²	20–40	70–120	100–180	250	250	40–90	100–200	75–150
Comments	High speed, less dense bit slice processor technology	High yield, proven technology, low speed, good packing density	High yield, higher density, and higher speed than PMOS, dominant microprocessor technology	Scaled-down NMOS resulting in higher performance, greater density, and lower cost; one of dominant technologies of the 1980s	More complex processing than NMOS; high speed and high density NMOS process; can sink higher currents than HMOS	Low power static operation, complementary n and p channel structure, wide operating range over temperature and power supply variations, less dense	Features of CMOS, higher speed, and density	Programmable speed and power characteristics, eliminates discrete load resistors; has high speed capability of bipolar, high density of MOS

cessor bit width today is eight bits. Eight-bit microprocessors have been used for data processing, text manipulation, and control. Sixteen-bit microprocessors are now available in small computers, with 32-bit processors still being used only for superminicomputers and mainframe computers. Smaller bit width computers require a greater number of instructions to be executed to achieve the same accuracy in arithmetic operations as larger bit width computers and cannot directly address as much memory. Thus, in general, as the bit width increases, for the same data processing accuracy, the execution speed of the computer increases as well as the size of program it can run without sophisticated memory management.

8.5 CLOCK FREQUENCY

Clock frequency is the number of clock pulses per second that the processor is operating at. The clock pulses are generated by an external circuit and are used as the basic timing reference signal for all processor operations. In general, the higher the maximum clock frequency the processor will run at, the faster the operation of the processor, although the true data processing speed depends on a number of additional factors such as the instruction set and the number of microcycles per instruction. The lower limit on clock frequency can be very small, that is, several hertz for some microprocessors to allow program step through. The typical maximum clock frequency for microprocessors is 2–5 mHz.

8.6 INSTRUCTION EXECUTION TIME

The time to execute a given instruction is the number of clock cycles in the instruction divided by the clock frequency. The number of clock cycles required is determined by the addressing mode and the complexity of the instruction. For a fixed clock frequency, decreased instruction execution time will reflect in increased program processing speed for programs using the high-speed instructions.

8.7 REGISTERS

The main processor registers, that is, program counter, accumulator, and instruction register, have previously been discussed. Some processors have additional programmable registers that provide the programmer with a great deal of flexibility, and since they act as high-speed memory, the appropriate use of programmable registers can reduce program execution time.

8.8 MEMORY

The amount of main memory both directly addressable and provided with the computer is very important. The common unit of memory is "K," referring to the number 2^{10} or 1024. Thus a 48K byte memory contains 49,152 bytes of addressable space. Normally memory is measured in terms of bytes, for example, 48K or 64K bytes, but it is often measured in terms of words, for example, 48K or 64K words, where a word may be 8, 16, or 32 bits, usually depending on the bit width of the processor.

Permanent main memory is required for permanent resident executive software such as bootstrap loaders, operating systems, language translators, etc. Permanent memory is normally in the form of ROM or PROM semiconductor memory. Also, random access storage memory (RAM), normally volatile memory, is required for user-written application-oriented software and executive software loaded from mass storage such as tape, floppy diskette, or hard disk. A typical total minimum amount of memory for a small computer with a text editor, language translator, small operating system, and space for user programs is 32K bytes. If more memory is needed, it can usually be purchased and retrofitted, the upper limit on the amount that can be added being determined by the bit width of the processor unless special memory management techniques are used.

8.9 INTERRUPTS

The ability of the computer processor to scan for and react to interrupt signals can be important particularly in automatic control applications where critical process or plant conditions must be detected and appropriate alarms and action programs executed quickly. A program to regularly scan sensors and react to changes of state can of course be written for a processor that does not have interrupt capability. However, excessive machine time may be required and the reaction may be slow. True interrupt handling can be of the software form, that is, an immediate jump to an interrupt handling routine or of the hardware form. In either case, if multiple interrupt conditions are expected to be present, the processor must have priority levels for the interrupts and a means of resolving simultaneous interrupts of the same priority level.

8.10 SOFTWARE

When selecting a digital computer for a given application, the system and application software that are available for the computer and the software tools to develop customized application software should be carefully considered. The need for the following system software should be determined:

- Monitor or operating system and I/O device drivers.
- Compiler.
- Assembler.
- Interpreter.
- Linker–loader.
- Editor.

A restricted amount of systems software for a microcomputer, for example, monitor, interpreter, and editor, is typically contained in ROM chips and becomes available as soon as the computer is powered up. Larger systems software packages are typically provided on mass storage media such as floppy diskettes or hard disks and are read into RAM after the computer is powered up. Systems software may be needed each time the computer is operated or may only be needed and run, usually in another computer, to develop the object code for the application computer. The object code is then permanently stored in ROM or PROM in the application computer and the system software is no longer needed.

The language translators—compiler, assembler, interpreter—will be selected based on the languages that the user wishes to use. The sophistication of the editor should be assessed based on the nature of language source and data file to be created.

The most critical component of the systems software is the operating system, which should be chosen based on needs such as:

- Multiuser capability.
- Multitask capability.
- Task priority assignments.
- Real time environment features.
- I/O devices.

In selecting a digital computer, the required systems software, its source, and reliability should be known. Often systems software is available from a source other than the computer manufacturer.

CHAPTER 9

DIGITAL COMPUTER APPLICATIONS

DALE B. CHERCHAS

The University of British Columbia
Vancouver, British Columbia, Canada

Digital computers, particularly small digital computers, are being implemented in many mechanical engineering applications and new applications are constantly being found.

9.1 ADVANTAGES

The major advantages of using a small digital computer in specific mechanical engineering applications are, in general:

1. Economy.
2. Computational capability.
3. Accuracy.
4. Flexibility.
5. Reliability.
6. Speed.

9.2 PROCESS CONTROL

The full potential of the digital computer as a tool for automatic control has not been realized; however, its use in a wide variety of control applications is growing rapidly. Process control is the automatic control of industrial processes such as paper machines, blast furnaces, chemical plants, electric power generating stations, and sewage treatment plants. The first replacement of analog controllers by a digital control system for process control was in the late 1950s.

Digital computers are used in various ways for process control:

1. As direct digital control (DDC) devices acting as the controller in single or multivariable feedback control loops. Figure 9.1 shows such a loop.
2. As a supervisory or set point controller determining the set points for DDC loops in an optimum manner to minimize plant operating costs. Figure 9.2 shows a supervisory computer control system.
3. As sequence controllers to control the open/close of valves and start/stop of motors in correct sequence to start up or shut down a system or subsystem. This operation today is best done by using a low-level logic device called a programmable logic controller (PLC). These devices are inexpensive and more easily maintained by technicians than standard computers.
4. As management information processing systems.

Figure 9.3 shows a complete process plant computer control hierarchy.

There is no lack of commercially available digital computer control systems for process control. Process control is a very big business in the developed countries of the world, and most major process control firms offer digital computer control systems from Level 1 DDC controllers to complete supervisory systems. Detailed technical specifications are available in the suppliers' equipment catalogs; however, a good guide to suppliers is Ref. 1.

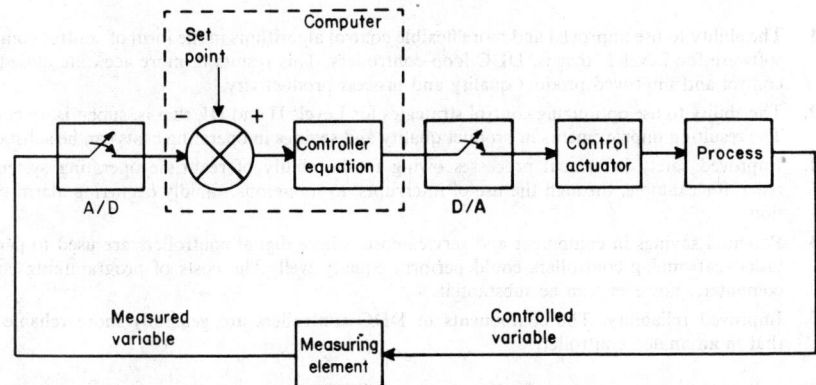

Fig. 9.1 Direct digital control computer.

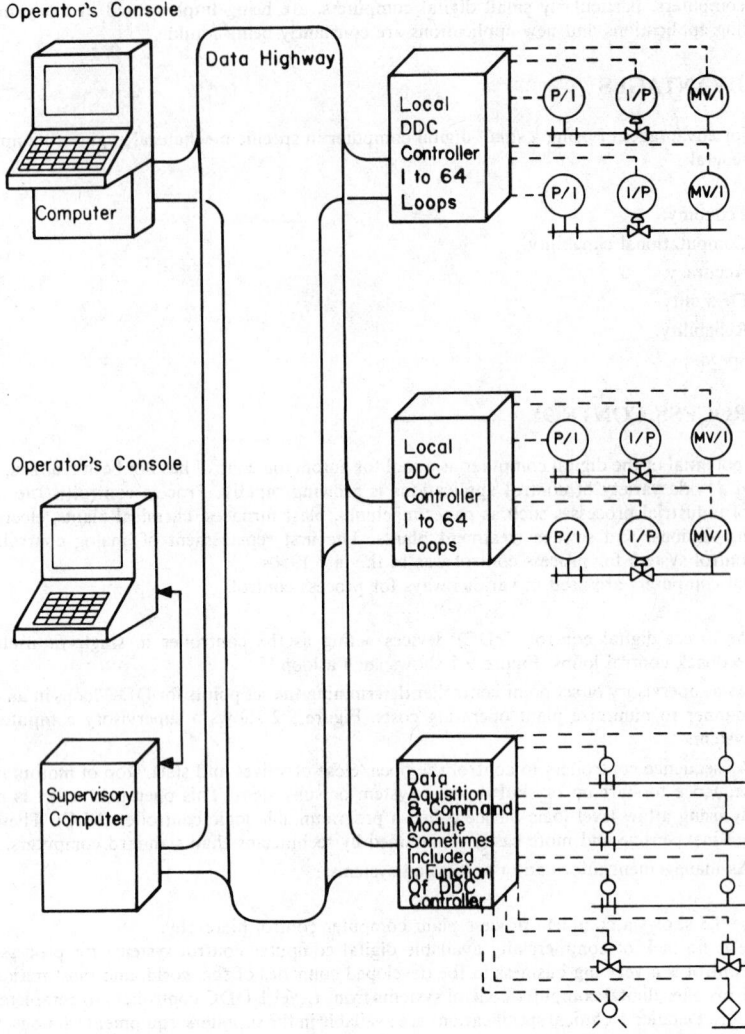

Fig. 9.2 Supervisory computer system.

The principal advantages in using computers in process control are:

1. The ability to use improved and more flexible control algorithms in the form of control computer software for Level 1, that is, DDC loop controllers. This results in more accurate closed-loop control and improved product quality and process productivity.

2. The ability to use optimizing control strategies for Levels II and III, that is, supervisory control. The resulting improvements in product quality and savings in operating costs can be substantial.

3. Improved safety in critical processes owing to the ability of real-time operating systems to react, for example, through the use of interrupts, to numerous, rapidly occurring alarm conditions.

4. Potential savings in equipment and service costs where digital controllers are used to perform tasks that analog controllers could perform equally well. The costs of programming control computers, however, can be substantial.

5. Improved reliability. The components in DDC controllers are generally more reliable than that in an analog controller.

Since digital computers process in discrete time, all control strategies must be programmed into the control computer software using discrete-time algorithms. For Level 1, that is, DDC loop control, the most common algorithms are given below. Referring to Fig. 9.1:

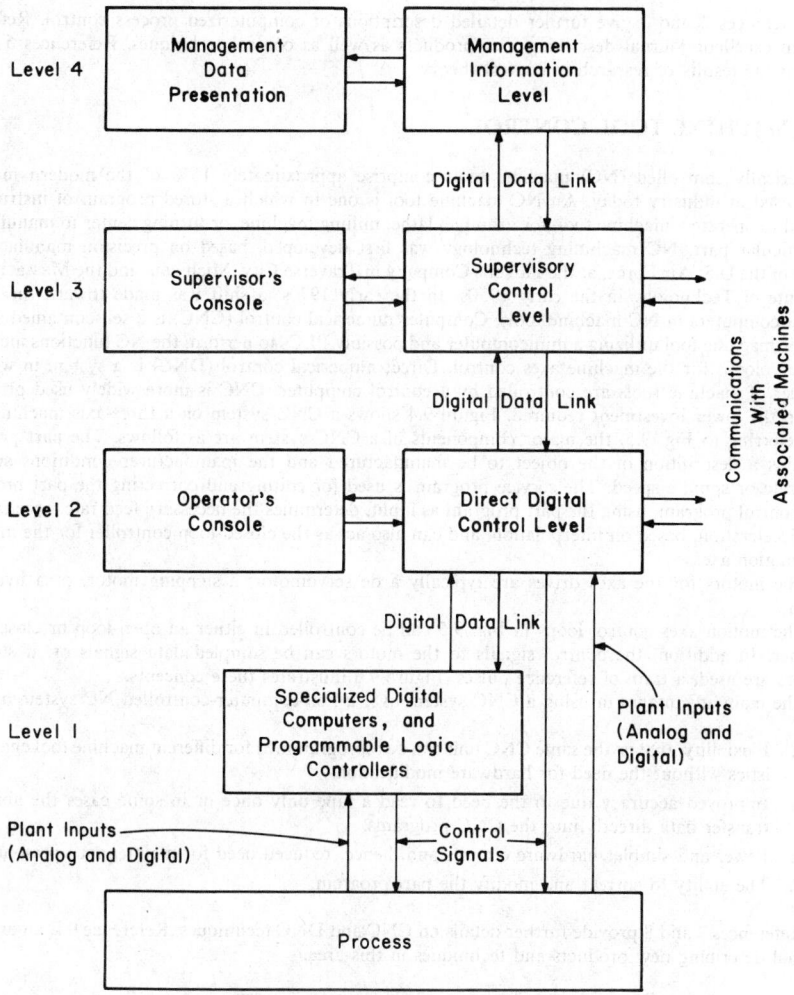

Fig. 9.3 Plant computer control hierarchy.

1. Proportional plus integral plus derivative (PID) control:

$$m(kT) = K_p e(kT) + K_I e_I(kT) + K_D \frac{e(kT) - e[(k-1)T]}{T} \qquad (9.1)$$

where K_p = proportional gain
K_I = integral gain
K_D = derivative gain
T = sample time (constant)
and

$$e_I(kT) = e_I[(k-1)T] + T e(kT) \qquad (9.2)$$

Equations (9.1) and (9.2) can easily be programmed directly using a high level language such as FORTRAN, BASIC, PASCAL, or FORTH. K_p, K_I, and K_D are chosen to "tune" the loop as for analog controllers. Allowing T to become too large can result in instability. The theory of discrete time control system stability is described in Ref. 2. PID algorithms can also of course be implemented using analog controllers.

2. Finite settling time, minimum risk time, zero steady-state error, and dead time (Dahlin) algorithms. These algorithms can only be implemented using a digital computer and are described in Ref. 2.

The principles of interfacing and computer selection as described in Chapters 7 and 8, respectively, apply to the use of digital computers in process control.

References 2 and 3 give further detailed descriptions of computerized process control. Reference 4 is an excellent journal describing new products as well as control techniques. References 5 and 6 present the results of research in control theory.

9.3 MACHINE TOOL CONTROL

Numerically controlled (NC) machine tools comprise approximately 15% of the modern machine tools used in industry today. An NC machine tool is one in which a stored program of instructions is used to operate a machine tool, for example, lathe, milling machine, or turning center to manufacture a particular part. NC machining technology was first developed, based on precision manufacturing needs of the U.S. Air Force, at the Parsons Company in Traverse City, Michigan, and the Massachusetts Institute of Technology in the early 1950s. In the early 1970s, a shift was made toward the use of digital computers in NC machine tools. Computer numerical control (CNC) is a self-contained system for one machine tool utilizing a minicomputer and possibly PLCs to perform the NC functions including the servoloops for the machine axes control. Direct numerical control (DNC) is a system in which a number of machine tools are controlled by a control computer. CNC is more widely used primarily due to the lower investment required. Figure 9.4 shows a CNC system on a three-axis machine tool.

Referring to Fig. 9.5, the major components of a CNC system are as follows. The part program contains a description of the object to be manufactured and the manufacturer conditions such as feed rate or spindle speed. The service program is used for editing and correcting the part program. The control program, using the part program as input, determines the necessary feed rate, acceleration, and deceleration, based on interpolation, and can also act as the closed-loop controller for the machine tool motion axes.

The motors for the axes drives are typically a dc servomotor, a stepping motor, or a hydraulic motor.

The motion axes control loops in Fig. 9.5 can be controlled in either an open-loop or closed-loop manner. In addition, the control signals to the motors can be sampled data signals or, if stepping motors are used, a train of reference pulses. Figure 9.6 illustrates these concepts.

The main advantages in using a CNC system over a non-computer-controlled NC system are:

1. Flexibility, that is, the same CNC unit can be reprogrammed for different machine tool characteristics without the need for hardware modifications.

2. Improved accuracy due to the need to read a tape only once or in some cases the ability to transfer data directly into the CNC programs.

3. Fewer and simpler hardware circuits and, hence, reduced need for maintenance personnel.

4. The ability to correct and modify the part program.

References 7 and 8 provide further details on CNC and DNC techniques. Reference 9 is an excellent journal describing new products and techniques in this area.

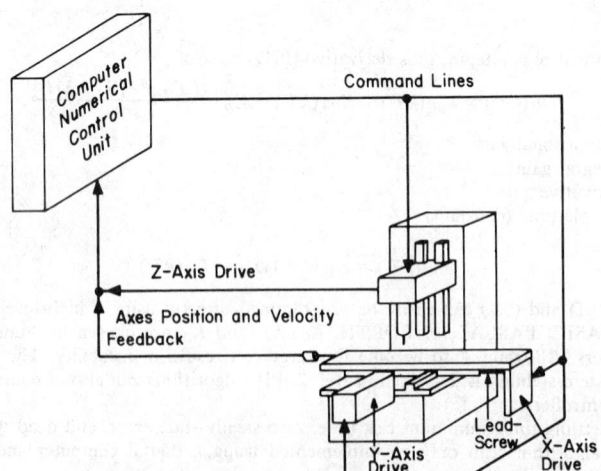

Fig. 9.4 Computer numerical control for a three-axis machine tool.

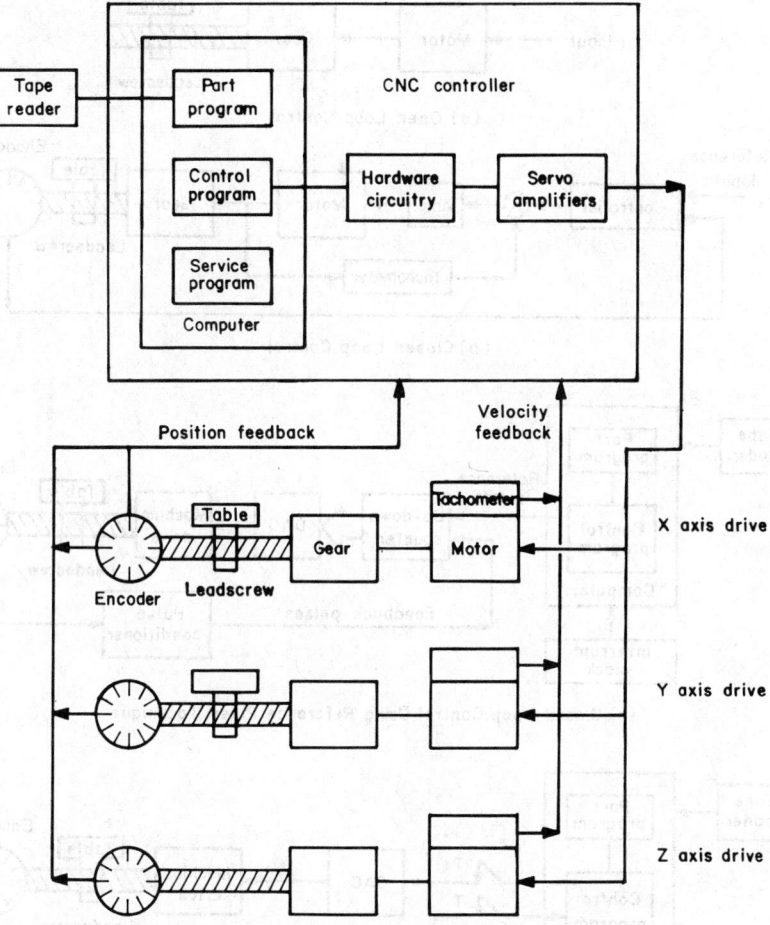

Fig. 9.5 Computer numerical control system components.

9.4 INDUSTRIAL ROBOTS

The Robot Institute of America's definition of a robot is "a reprogrammable, multifunctional manipulator designed to move material, parts, tools or specialized devices, through variable programmed motions for the performance of a variety of tasks." A manipulator is a mechanism, usually consisting of a series of segments, jointed or sliding relative to one another, for the purpose of grasping and moving objects, usually in several degrees of freedom. It may be remotely controlled by a computer or by a human. A manipulator can be operated directly by a person, usually through mechanical or electrical servo devices. Only when a manipulator is automatically operated by a controller that can store a preprogrammed task in its memory, either mechanical, electrical, or electronic, does the manipulator become a robot. Figure 9.7 shows the generic form of an industrial robot and controller.

A digital computer is commonly used in a robot to:

1. Store the task in computer memory when the robot is operating and in mass storage such as magnetic diskette for a permanent task record.
2. Facilitate the creation of a task program through direct programming in a high-level language, and store the robot endpiece location history as the robot is manually led through the desired program, and allow editing of the program.
3. Perform the necessary calculations to relate desired endpiece motion to required motion of the manipulator joints, that is, coordinate transformations.
4. Process incoming sensor data, for example, tactile data vision.
5. Control, in an open-loop or feedback control manner, the joints of the robot arm.

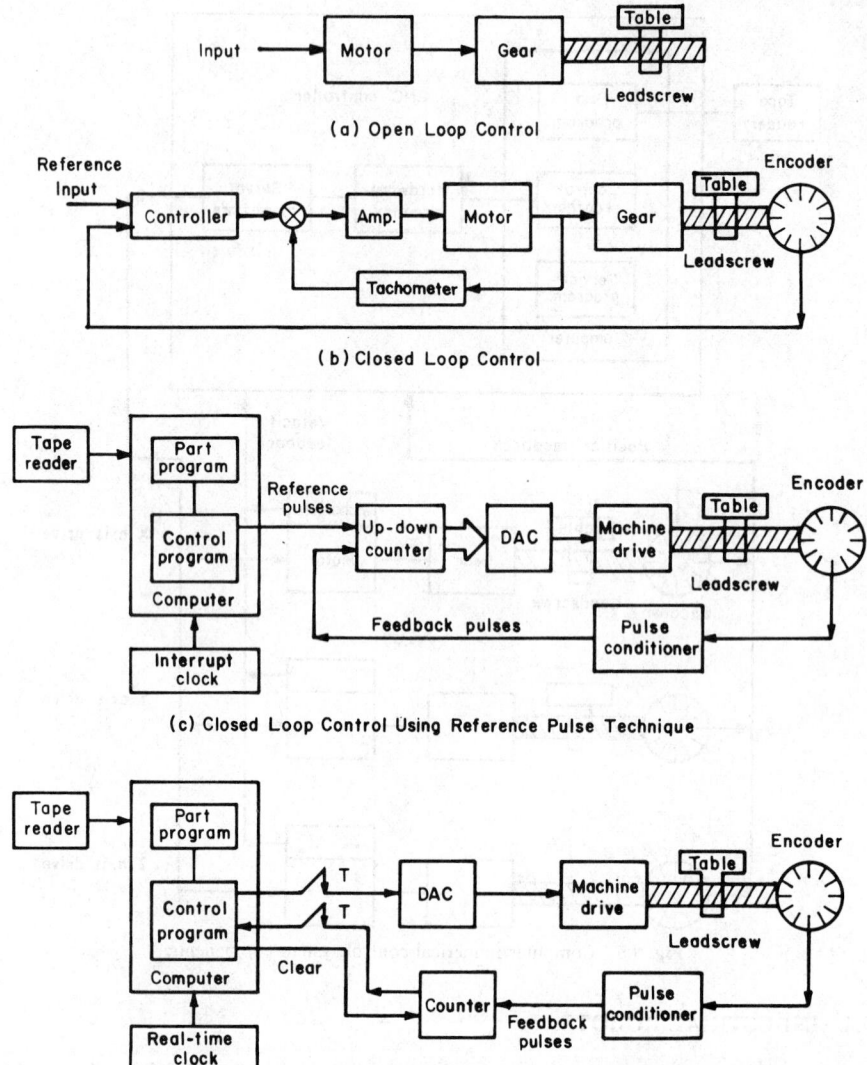

(a) Open Loop Control

(b) Closed Loop Control

(c) Closed Loop Control Using Reference Pulse Technique

(d) Closed Loop Control Using a Sampled Data System

Fig. 9.6 Open loop and closed loop motion axes control.

Figure 9.8 shows the general structure of a robot computer control system. The main computer monitors the keyboard/CRT terminal and the teaching pendant, which allows the robot joints to be manually controlled. It accepts input from sensors such as TV cameras and pressure sensors, and from other computers that may download data or commands. The main computer controls data flow with a mass storage device, typically a floppy diskette. An operating system allows the creation, editing, and execution of programs and monitoring of external sensory inputs and control of external actuators as well as the robot arm from these programs. The servomechanism control of the joints can be accomplished with separate analog controllers for each joint or a microprocessor based on digital controllers for each joint.

The joint drive units are generally stepping motors for very small robots, dc servomotors for medium-sized robots, and hydraulic power units for large robots. However, the general trend is to use electric motors for larger robots in new designs.

Robot control systems are expected to evolve into a hierarchical form as shown in Fig. 9.9 allowing decomposition of complex tasks into servomechanism commands and the use of high-level sensor

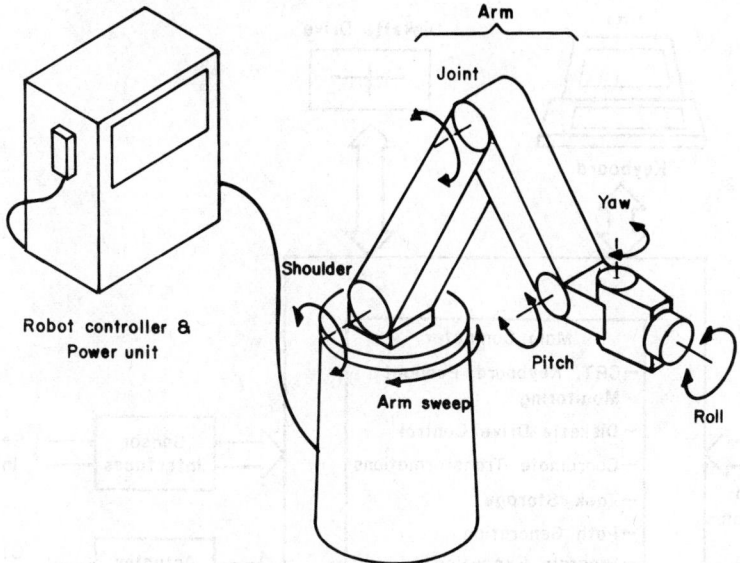

Fig. 9.7 Industrial robot and controller.

input, for example, visual input to plan tasks. A robot control system of this kind will make extensive use of that area of computer science referred to as "artificial intelligence." The significant integration of artificial intelligence into robot control systems will have a major impact on robot use. Multiprocessors are expected to be used.

A number of high-level robot control languages are in use in industry and in research labs, for example,

1. AL (Stanford University).
2. PAL (Purdue University).
3. VAL (Unimation Incorporated).
4. AUTOPASS, AML (IBM).
5. RPL (SRI International).

At the time of writing, a standard robot control language is not developed or in use comparable to FORTRAN or BASIC for data processing.

Robots can be programmed in a number of ways:

1. Typing the program on a keyboard.
2. Downloading a program from another computer where the program has been generated automatically. Such "automatic programming" is currently in the research stage.
3. "Walkthrough" using the teaching pendant to manually move the robot endpiece through the desired path and orientation history and record the path and/or path points in the computer memory.
4. "Leadthrough," that is, an operator physically grasps the endpiece and moves it in the desired manner while the path and/or path points are being recorded.

Reference 10 describes the use of robots in industry. Reference 11 analyzes the kinematics, dynamics, and control foundations of robotics. References 7 and 12 describe the use of computers in industrial robots. The potential use of artificial intelligence in robotics is described in Ref. 13. Reference 14 discusses robot control theory including discrete-time control. Reference 15 is a relatively new journal presenting research in robotics and Ref. 16 focuses on technology and applications.

9.5 BUILDING AUTOMATION SYSTEMS

A building automation system is a computer-based system providing information and heating, ventilating, and air conditioning (HVAC) equipment control functions. Building automation systems utilizing digital

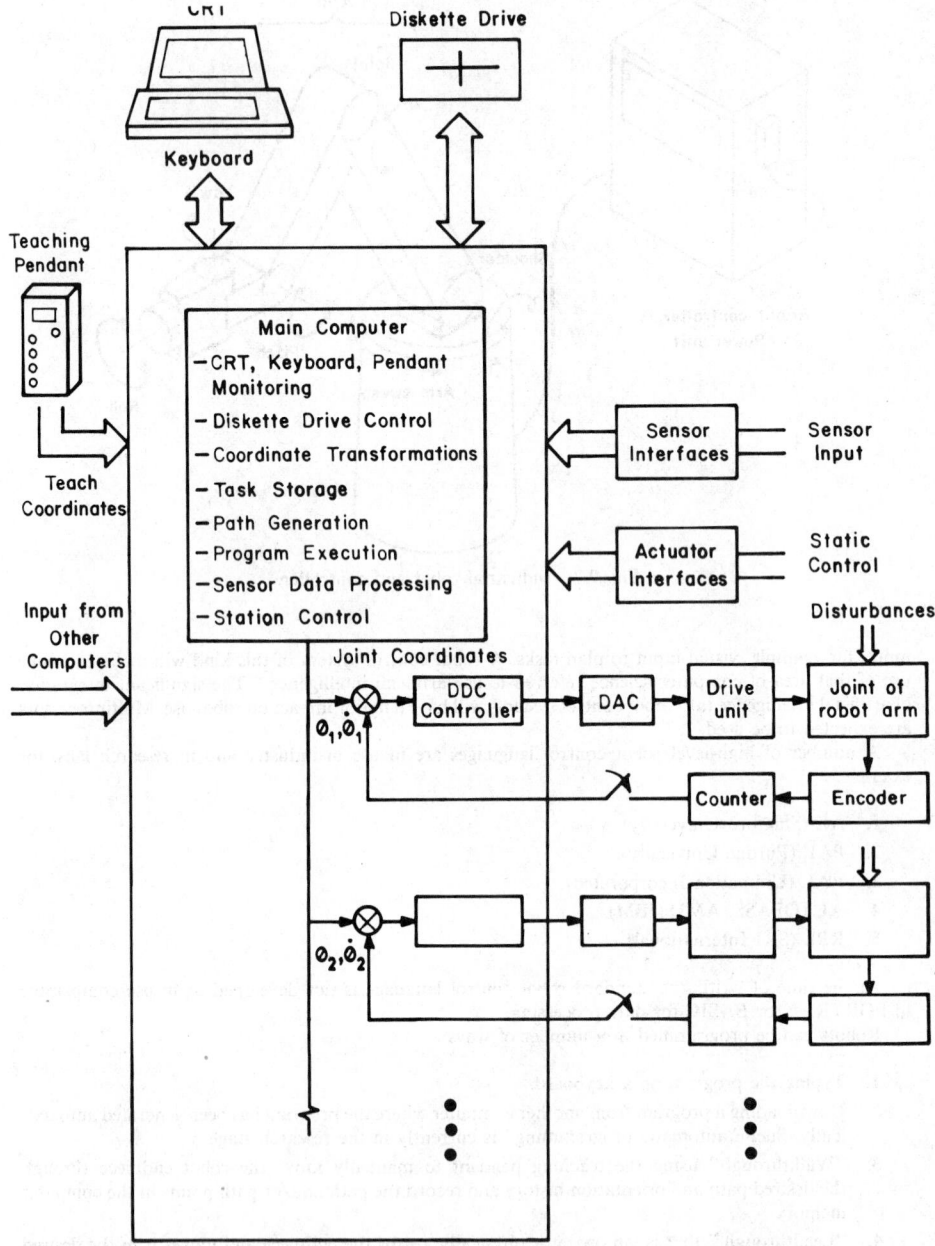

Fig. 9.8 Industrial robot computer control.

computers have primarily been used for providing information functions such as equipment monitoring and security. Recently, however, building automation systems have been designed and installed to provide not only information functions, but control functions also. That is, the building automation system plays a direct role in the operation of HVAC equipment. A building automation system providing both information and control functions consists of a number of specific components:

1. *Computer Hardware.* Usually the hardware consists of one central host minicomputer with a full or stripped down backup and one or more distributed remote processing units, typically microprocessors, sensor and actuator signal conditioners, and converters and required communi-

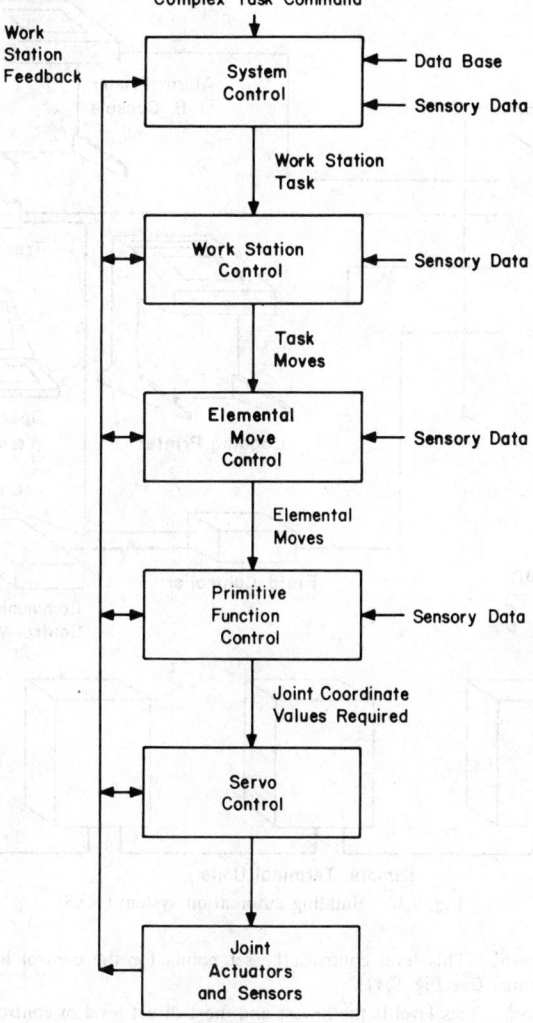

Fig. 9.9 Robot hierarchical control.

cation devices such as processor interface translators, modems, terminals, printers, and graphic displays (see Fig. 9.10).

2. *Computer Software.*   The software consists of operating systems or monitor software, processor communication software, I/O device software, and applications software.

3. *Field Hardware.*   Field hardware consists of sensors, actuators, and communication lines.

A given building automation system installation may use the digital computer in a very limited manner to perform control functions, or the computers may be used to perform virtually every required control function. Irrespective of the degree of application of computer control in the building automation system, the general task of *controlling* the building HVAC equipment can be viewed in the following manner:

• *Level III Control.*   Control of the operating states of HVAC equipment subsystems, for example, cooling tower fan and water pump on or off. The states are selected by turning drivers such as electric motors for pumps or fans on or off and opening or closing valves or dampers fully. The change of a subsystem from one state to another requires correct sequencing of the state selection devices (see Fig. 9.11).

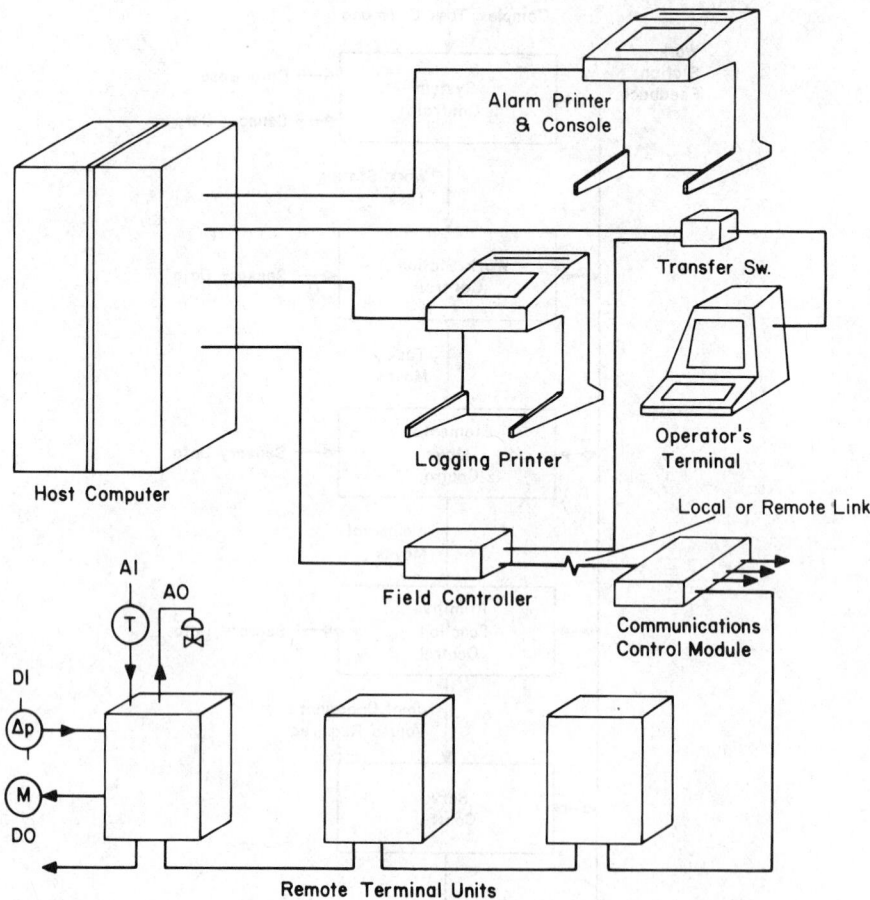

Fig. 9.10 Building automation system (BAS).

- *Level II Control.* This level controls the set points for the control loops of level I in the optimum manner (see Fig. 9.12).
- *Level I Control.* This level is the lowest and most direct level of control, consisting of closed control loops maintaining controlled process variable(s) as close as possible to the loop set points established by level II (see Fig. 9.13).

Depending on the building automation system design objectives, the three levels of control can be implemented entirely with human operators and analog controllers, entirely with digital computers, or with a combination of human and analog control and computer control. A number of new building automation system installations are achieving all three levels of control with digital computer processing.

The advantages of digital computer based building automation system control systems are:

1. Optimum selection of heating, ventilating, and air conditioning states.
2. Optimum selection of HVAC equipment set points, that is, operating points.
3. Increased accuracy and flexibility in feedback control loops.
4. Reduced HVAC equipment capital and running costs.
5. Trend to lower control system installation costs.

Table 9.1 lists the software typically implemented in a building automation system.

9.6 VEHICLE MOTION CONTROL

Computers are being used for motion control on air, sea, and ground vehicles. Figure 9.14 shows the general configuration for a computer-based vehicle control system.

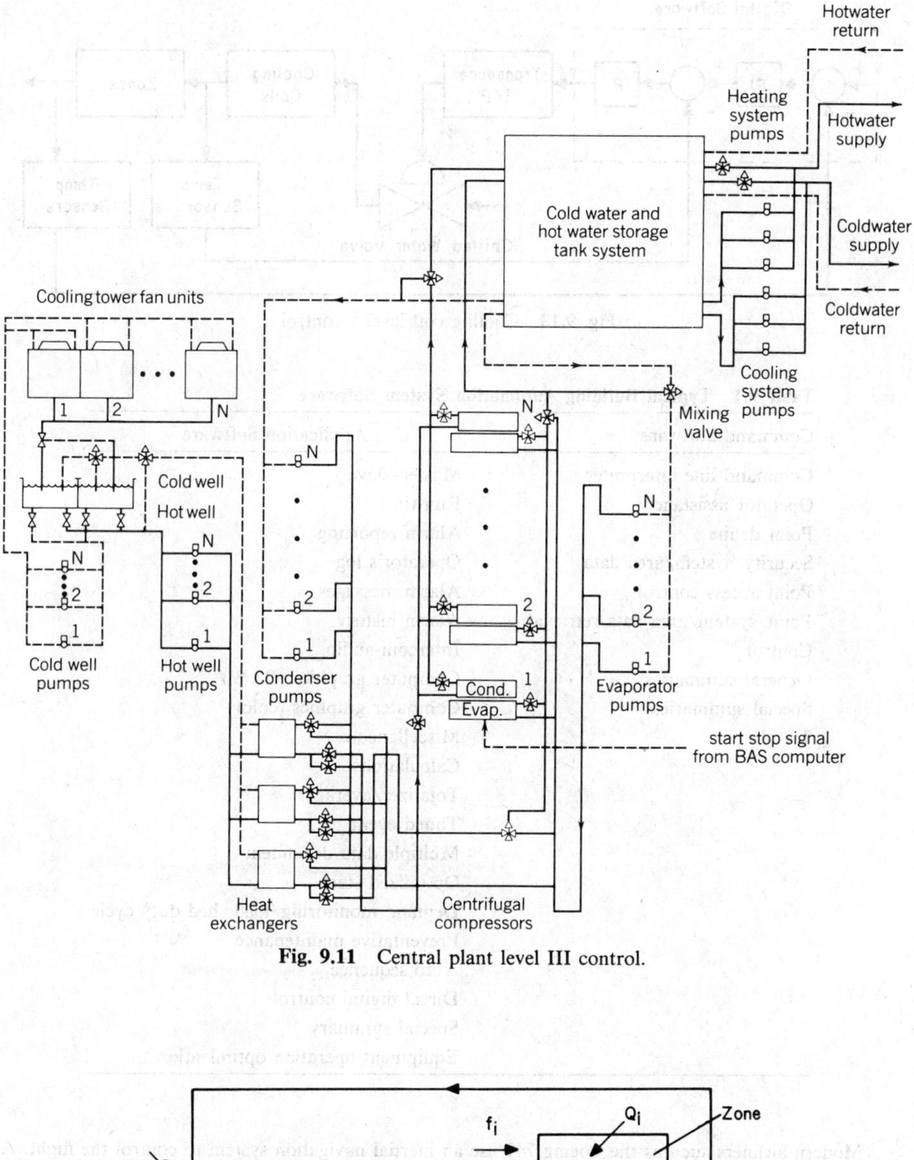

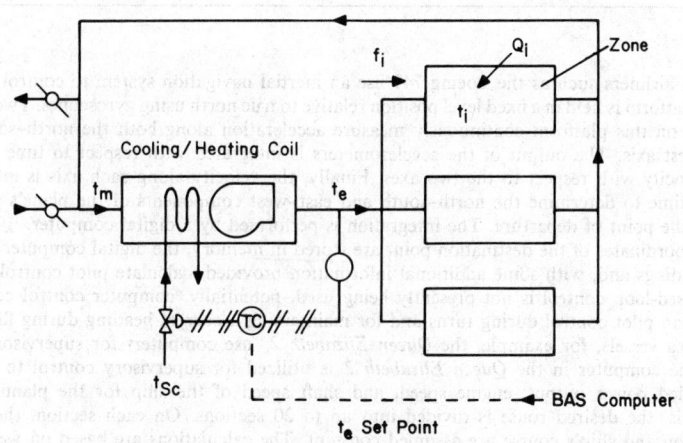

Fig. 9.11 Central plant level III control.

Fig. 9.12 Heating/cooling coil level II control.

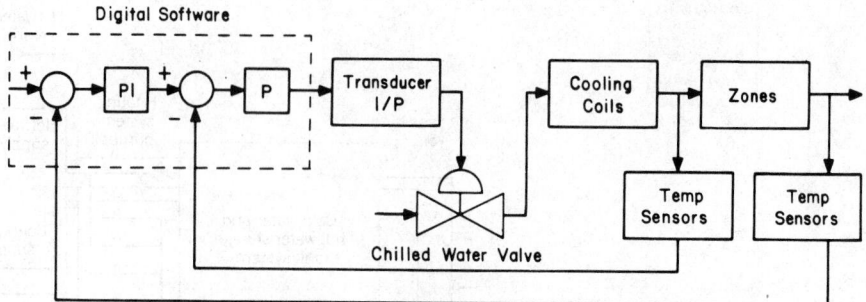

Fig. 9.13 Cooling coil level I control.

Table 9.1 Typical Building Automation System Software

Command Software	Application Software
Command line interpreter	Master–slave
Operator assistance	Run time
Point define	Alarm reporting
Security/system/area data	Operator's log
Point access control	Alarm messages
Point/system/area data retrieval	Alarm history
Control	Intercom-audio
General summaries	Computer graphics (B&W)
Special summaries	Computer graphics (color)
Trend	Miscellaneous
	Calculations
	Totalizer/averager
	Timed event
	Multiple data dependent
	Optimized start
	Demand monitoring/load shed duty cycle
	Preventative maintenance
	Auto sequence
	Direct-digital control
	Special summary
	Equipment operation optimization

Modern airliners such as the Boeing 747 use an inertial navigation system to control the flight. A reference platform is held in a fixed level position relative to true north using gyroscopes. Two accelerometers placed on this platform continuously measure acceleration along both the north–south axis and the east–west axis. The output of the accelerometers is integrated with respect to time to determine aircraft velocity with respect to the two axes. Finally, the velocity along each axis is integrated with respect to time to determine the north–south and east–west components of the plane's position with respect to the point of departure. The integration is performed by a digital computer.

If the coordinates of the destination point are stored in memory, the digital computer can compute correct headings and, with some additional information provided, calculate pilot control settings. Although closed-loop control is not presently being used, potentially, computer control could be more accurate than pilot control during turns and for maintaining a correct heading during flight.

Some sea vessels, for example, the *Queen Elizabeth 2*, use computers for supervisory and direct control. The computer in the *Queen Elizabeth 2* is utilized for supervisory control to calculate the recommended power output, engine speed, and shaft speed of the ship for the planned route. To perform this, the desired route is divided into up to 20 sections. On each section, the wind speed and direction and ship's course are assumed constant. The calculations are based on weather reports and forecasts, scheduled arrival time and the ship's load. Feedback control is used aboard the *Queen*

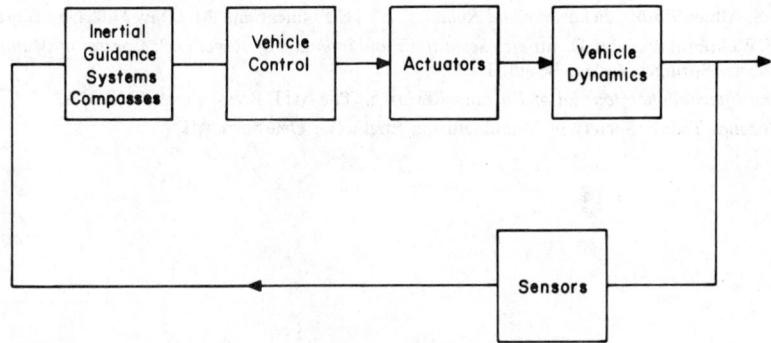

Fig. 9.14   Computer control of vehicle motion.

Elizabeth 2 to regulate circulating water scoop pumps for the condensers in order to increase the efficiency of the engine and decrease fuel consumption. The recommended relationship between engine shaft-horsepower and condenser vacuum is stored in the computer. Periodic computations are made to determine the shaft horsepower of the engine and the corresponding recommended condenser vacuum. The actual vacuum and recommended condenser vacuum are compared, and the change in vane angle of the scoop pump necessary to give the recommended vacuum is computed. Finally, the computer sends a signal to an actuator to make the correct change in vane angle.

Computers are used for tracked ground vehicles to control suspension systems, drive systems, attitude control systems, and, most recently, for headway control of a fleet of urban mass transit vehicles on tracked guideways. The intermediate capacity transit system developed by the Urban Transportation Development Corporation in Ontario, Canada, features driverless trains controlled by a central headway control system.

9.7   OTHER APPLICATIONS

Digital computers are used in numerous other mechanical engineering applications:

- Computer-aided drafting and design (CADD) systems.
- Engine control.
- Consumer products.

REFERENCES

1. *ISA Directory of Instrumentation,* Instrument Society of America, Research Triangle Park, NC (published annually).

2. P. B. Deshponde and R. H. Ash, *Elements of Computer Process Control,* Instrument Society of America, Research Triangle Park, NC, 1981.

3. T. J. Harrison (ed.), *Minicomputers in Industrial Control,* Instrument Society of America, Research Triangle Park, NC, 1978.

4. *Instrumentation Technology (Intech),* Instrument Society of America, Research Triangle Park, NC.

5. *Automatica,* Journal of the International Federation of Automatic Control, Pergamon Press, Elmsford, NY.

6. *Journal of Dynamic Systems, Measurement and Control,* Transactions of the American Society of Mechanical Engineers, New York, NY.

7. Y. Koren, *Computer Control of Manufacturing Systems,* McGraw-Hill, New York, 1983.

8. M. P. Groover, *Automation, Production Systems and Computer-Aided Manufacturing,* Prentice-Hall, Englewood Cliffs, NJ, 1980.

9. *Manufacturing Engineering,* Society of Manufacturing Engineers, Dearborn, MI.

10. J. F. Engelberger, *Robotics in Practice,* Amacom, New York, 1980.

11. R. P. Paul, *Robot Manipulators: Mathematics, Programming and Control,* The MIT Press, Cambridge, MA, 1981.

12. R. C. Dorf, *Robotics and Automated Manufacturing,* Reston Publishing Company (Prentice-Hall), Reston, VA, 1983.

13. J. S. Albus, *Brains, Behaviour and Robotics,* BYTE Publications, McGraw-Hill, New York, 1981.

14. M. Vukobratovic and D. Stokic, *Scientific Fundamentals of Robotics 2; Control of Manipulation Robots,* Springer-Verlag, Berlin, 1982.

15. *The International Journal of Robotics Research,* The MIT Press, Cambridge, MA.

16. *Robotics Today,* Society of Manufacturing Engineers, Dearborn, MI.

PART 2
MATERIALS AND MECHANICAL DESIGN

CHAPTER 10
STRUCTURE OF SOLIDS

CHARLES H. DRUMMOND III

Department of Ceramic Engineering
The Ohio State University
Columbus, Ohio

10.1 INTRODUCTION

10.1.1 Effects of Structure on Properties

Physical properties of metals, ceramics, and polymers such as ductility, thermal expansion, heat capacity, elastic modulus, electrical conductivity, and dielectric and magnetic properties are a direct result of the structure and bonding of the atoms and ions in the material. An understanding of the origin of the differences in these properties is of great engineering importance.

In single crystals, a physical property such as thermal expansion varies with direction, reflecting the crystal structure; whereas in polycrystalline and amorphous materials, a property does not vary with direction, reflecting the average property of the individual crystals or the randomness of the amorphous structure. Most engineering materials are polycrystalline, composed of many grains, and, thus, an understanding of the properties requires not only a knowledge of the structure of the single grains but also a knowledge of grain size and orientation, grain boundaries, and other phases present, that is, a knowledge of the microstructure of this material.

10.1.2 Atomic Structure

Atoms consist of electrons, protons, and neutrons. The central nucleus consists of positively charged protons and electrically neutral neutrons. Negatively charged electrons are in orbits about the nucleus in different energy levels, occupying a much larger volume than the nucleus.

In an atom the number of electrons equals the number of protons and, hence, an atom is neutral. The atomic number of an element is given by the number of protons, and the atomic weight is given by the total number of protons and neutrons. (The weight of the electrons is negligible.) Thus, hydrogen, H, with one proton and one electron has an atomic number of 1 and an atomic weight of 1 and is the first element in the periodic chart. Oxygen, O, with atomic number 8 has eight protons and eight neutrons and, hence, an atomic weight of 16.

Completed electronic shells have a lower energy than partially filled orbitals when bonded to other atoms. As a result of this energy reduction, atoms share electrons to complete the shells, or gain or lose electrons to form completed shells. In the latter case ions are formed in which the number of electrons is not equal to the number of protons. Thus O by gaining two electrons has a charge of -2 and forms the oxygen ion O^{2-}.

The periodic chart arranges elements in columns of the same electronic configuration. The first column consists of the alkalis Li, Na, K, Cs, Rb; each has one electron in the outer shell which can be lost. Similarly, the second column of alkaline-earths can form Mg^{2+}, Ca^{2+}, Sr^{2+}, Ba^{2+} by losing two electrons. The seventh column consists of the halogens Fl, Cl, Br, I, which by gaining one electron become the halides, all with a charge of -1. The eighth column consists of the inert gases He, Ne, Ar, K, Xe with completed shells. The bonding of the elements and ions with similar electronic configurations is similar. Moving down in a column increases the number of electrons and, hence, the atom's size increases even though the electronic configuration remains the same.

The outer electrons that are lost, gained, or shared are called valence electrons, and the inner electrons are called core electrons. For the most part, the valence electrons are important in determining the nature of the bonding and, hence, the structure and properties of the materials.

10.1.3 Bonding

When two atoms or ions are within atomic distances of each other, distances of 0.5–3.0 Å, bonding may occur between the atoms or ions. The resulting reduction in energy due to an attractive force leads to the formation of polyatomic gas molecules, liquids, and solids. If the energy of the bonds is large (75–275 kcal/mol), primary bonds are formed—metallic, ionic, or covalent. If the energy of the bond is smaller (1–10 kcal/mol), secondary bonds are formed—van der Waals and hydrogen. In addition, combinations of bond types, such as a mixture of ionic and covalent bonds, may occur.

Metallic Bonding

In a metallic crystal an ordered arrangement of nuclei and their electrons are embedded in a cloud of valence electrons, which are shared throughout the lattice. The resulting bonding is a nondirectional primary bond. Since the binding energy of the valence electrons is relatively small, the mobility of these electrons is high and creates high electrical and thermal conductivity. The atoms are approximately spherical in shape as a result of the shape of completed inner shell. Examples of metals are Cu, Au, Ag, and Na.

Ionic Bonding

The strongest type of bonding between two oppositely charged particles is called ionic bonding. The positively charged ions (cations) attract as many negatively charged ions (anions) as they can and form ionic bonds. The primary bond formed is nondirectional if the bonding is purely ionic. Li^+ and F^- in LiF form predominantly ionic bonds. In general, since the electrons are strongly bonded, electrical and thermal conductivities are much smaller than in metals, and, thus, ionic bonded materials are classified as insulators or dielectrics.

Covalent Bonding

Covalent bonding results from an overlap or sharing and not from gain or loss of valence electrons. A net reduction of energy as a result of each atom's completing the other's orbital also results in a primary bond, but it is directional. The directionality is a result of the shape of the orbitals involved in the bonding. When C is covalently bonded to four other C's in diamond, the bonding is purely covalent, and the configuration of these four bonds is tetrahedral. When B, however, is bonded to three other B's, a triangular configuration is formed. Organic polymers and diatomic gases such as Cl_2 are typical examples of covalent bonding. As a result of the strong bonding of the valence electrons, these materials for the most part have low electrical and thermal conductivity.

Van der Waals and Hydrogen Bonding

Van der Waals bonds are secondary bonds, the result of fluctuating dipoles due to the fact that at an instant of time the centers of positive and negative charge do not coincide. An example is an inert gas such as Ar, which below $-190°C$ forms a solid as a result of these weak attractive forces. Similar weak forces exist in molecules and solids. Hydrogen bonds are also secondary bonds, but they are the result of permanent dipoles. For example, the water molecule, H_2O, is nonlinear, and the bonding between H and an adjacent O in water results in H_2O being a liquid above $0°C$ at 1 atm pressure rather than a gas as is the case for other molecules of comparable molecular weight.

10.1.4   Simple Structures

If atoms or ions are considered to be spheres, then the most efficient packing of the spheres in space will form their most stable structure. However, the type of bonding, in particular, directional bonding, may affect the structure formed. In two dimensions, there is only one configuration that most efficiently fills space, the close packed layer (see Fig. 10.1). If similar layers are stacked to form a three-dimensional structure, an infinite number of configurations is possible. Two are important. In both, the first two layers are the same. In the first layer (A), the point at the center of three spheres provides a hollow for a fourth sphere to rest. A second close packed layer (B) then can be placed on the first layer with each sphere occupying the hollow. With the addition of a third layer to these two layers, two choices are possible. A sphere in the third layer can be placed above a sphere in the first layer in the spaces marked (•) in Fig. 10.2 or above a hollow not occupied by a sphere spaces marked (×) in the second layer. If the first stacking arrangement is continued, that is, the first and third layers on top of each other (denoted $ABABA$ . . .), the hexagonal-close-packed (hcp) structure is generated, so called because of the hexagonal symmetry of the structure. If the second stacking arrangement is continued, that is, the first and third layers are not on top of each other (denoted $ABCABC$ . . .), the cubic-close-packed or face-centered-cubic (fcc) structure is generated, so called because the lattice formed is a face-centered cube. Both structures are shown in Fig. 10.3. In both structures 74% of the volume is occupied, and each sphere is surrounded by 12 spheres (or 12 nearest neighbors), although

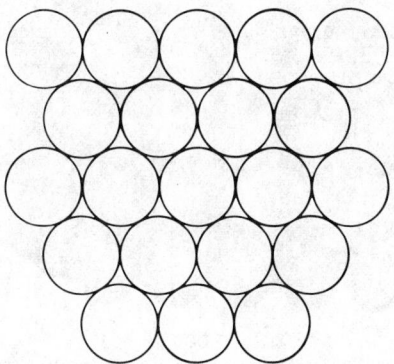

Fig. 10.1   Close packed layer.

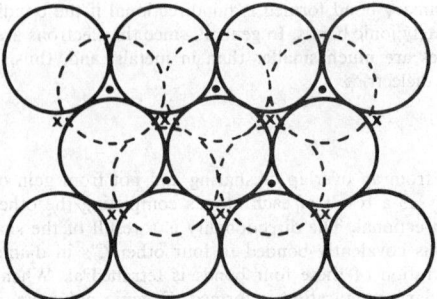

Fig. 10.2 Two possible sites for sphere in hcp and fcc structures: $\times$ and $\cdot$ (from D. M. Adams, *Inorganic Solids,* Wiley, New York, 1974).

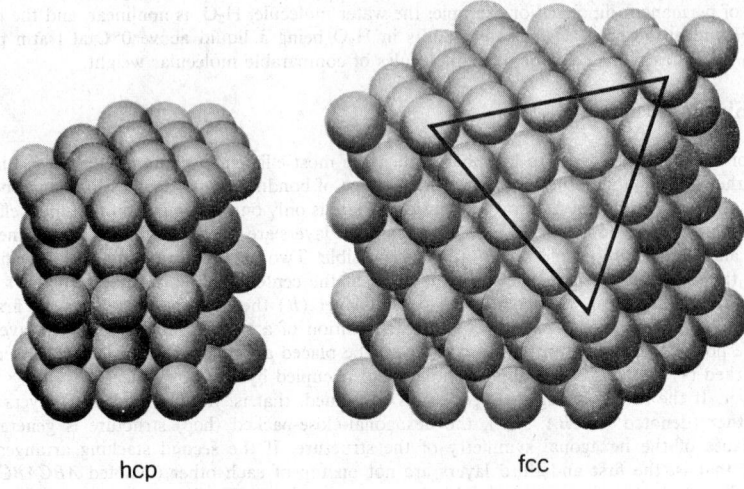

hcp

fcc

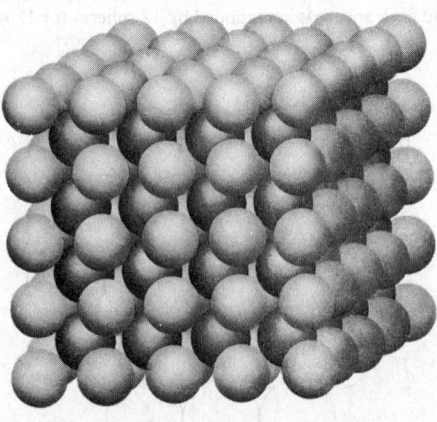

bcc

Fig. 10.3 hcp, fcc, and bcc structures (from W. G. Moffatt, G. W. Pearsall, and J. Wulff, *The Structure and Properties of Materials,* Wiley, New York, 1964, Vol. I, p. 51).

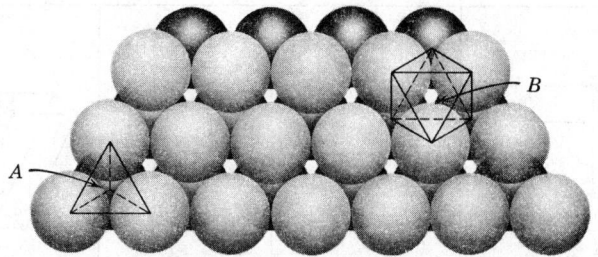

Fig. 10.4 Tetrahedral and octahedral sites (from G. W. Moffatt, G. W. Pearsall, and J. Wulff, *The Structure and Properties of Materials,* Wiley, New York, 1964, Vol. I, p. 58).

the arrangement is different. Another common structure is the body-centered-cubic (bcc) structure shown in Fig. 10.3. Here, each sphere has eight nearest neighbors with another six at a slightly greater distance. The volume fraction occupied is 68%. In the hcp and fcc structures, the stacking of a fourth sphere on top of three in any close packed layer generates a tetrahedral site or void, as shown in Fig. 10.4. Into such a site a smaller sphere with a coordination number of four could fit. Three spheres from each of two layers generate an octahedral site or void, as shown in Fig. 10.4. Into such a site a smaller sphere with a coordination number of six could fit. In the hcp and fcc structures, there are two tetrahedral and one octahedral sites per packing sphere; however, the configuration of these sites is different.

10.1.5 Crystallography

All possible crystallographic structures are described in terms of 14 Bravais space lattices—only 14 different ways of arranging points in space. These are shown in Fig. 10.5. Each of the positions in a given space lattice is equivalent, and an atom or ion or group of atoms or ions can be centered on each position. Each of the lattices is described by a unit cell, as shown in Fig. 10.5. The seven crystallographic classes are also shown in Fig. 10.5.

10.1.6 States of Matter

Matter can be divided into gases, liquids, and solids. In gases and liquids the positions of the atoms are not fixed with time, whereas in solids they are. Distances between atoms in gases are an order of magnitude or greater than the size of the atoms, whereas in solids and liquids distances between atoms are only several times the size of the atoms. Almost all engineering materials are solids, either crystalline or noncrystalline.

Crystalline Solids

In crystalline solids the atoms or ions occupy lattice points and vibrate about these equilibrium positions. The arrangement of the positions is some periodic array as discussed in Section 10.1.5. At $0°K$, except for a small zero-point vibration, the oscillation of the atoms is zero. With increasing temperature the amplitude and frequency of vibration increases up to the melting point. At the melting point the crystalline lattice is destroyed, and the material melts to form a liquid. For a particular single crystal the external shape is determined by the symmetry of the crystal class to which it belongs. Most engineering materials are not single crystals but polycrystalline, consisting of many small crystals. These crystals are randomly oriented and may be of the same composition or of different composition or of different structures. There may be small voids between these grains. Typical sizes of grains in such polycrystalline materials range from 0.01 to 10 mm in diameter.

Noncrystalline Solids

Noncrystalline solids (glasses) are solids in which the arrangement of atoms is aperiodic (random) and lacks any long-range order. The external shape is without form and has no defined external faces like a crystal. This is not to say that there is no structure. A local or short-range order exists in the structure. Since the bonding between atoms or ions in a glass is similar to that of the corresponding crystalline solid, it is not surprising that the local coordination, number of neighbors, configuration, and distances are similar for a glass and crystal of the same composition. In fused SiO_2, for example, four O's surround each Si in a tetrahedral coordination, the same as in crystalline SiO_2.

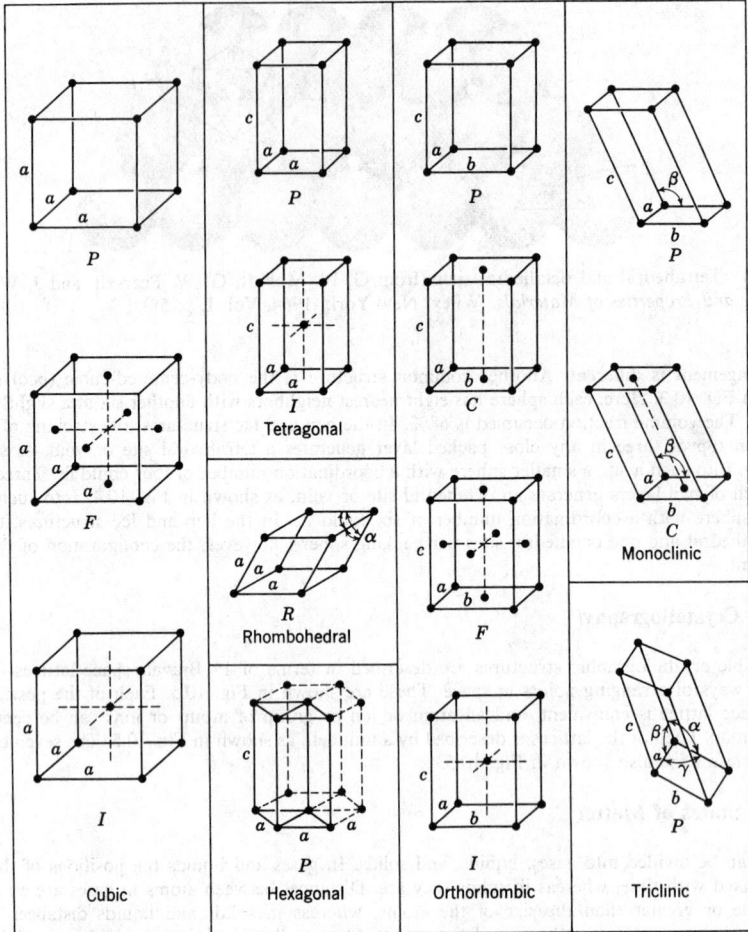

Fig. 10.5 Bravais lattices (from W. G. Moffatt, G. W. Pearsall, and J. Wulff, *The Structure and Properties of Materials*, Wiley, New York, 1964, Vol. I, p. 47).

Glasses do not have a definite melting point like crystals. Instead, they gradually soften to form a supercooled liquid at temperatures below the melting point of the corresponding crystal. Glass formation results when a liquid is cooled sufficiently rapidly to avoid crystallization. This behavior is summarized in Fig. 10.6 where the volume V is plotted as a function of temperature T.

10.1.7 Polymorphism

Crystalline materials of the same composition in many cases exhibit more than one crystalline structure called polymorphs. Fe, for example, exists in three different structures α, γ, and δ Fe. The α phase, ferrite, a bcc structure, transforms at 910°C to the γ phase, austenite, an fcc structure, and then at 1400°C changes back to bcc structures, δ-iron or δ-ferrite. The addition of C to Fe and the reaction and transformations that occur are extremely important in determining the properties of steel.

SiO$_2$ exhibits many polymorphs including α- and β-quartz, α- and β-tridymite, and α- and β-cristobalite. The SiO$_4$ tetrahedron is common to all the structures, but the arrangement or linking of these tetrahedra varies, leading to different structures. The $\alpha \rightarrow \beta$ transitions involve only a slight change in the Si–O–Si bond angle, are rapid, and are an example of a phase transformation that is called displacive. The quartz $\rightarrow$ tridymite $\rightarrow$ cristobalite transformations require the reformation of the new structure, are much slower than displacive transformations, and are called reconstructive phase transformations. The $\alpha \rightarrow \gamma \rightarrow \delta$ Fe transformations are other examples of reconstructive transformations.

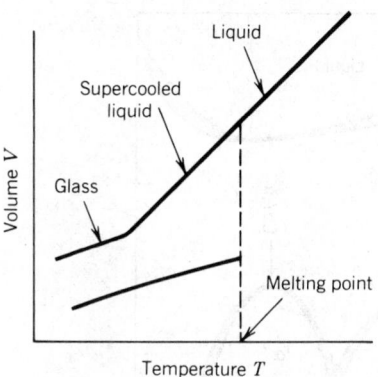

Fig. 10.6 Glass formation.

A phase diagram gives the equilibrium phases as a function of temperature, pressure, and composition. More commonly, the pressure is fixed at 1 atm and only the temperature and composition are varied. The Fe–C diagram is shown in Fig. 10.7.

10.1.8 Defects

The discussion of crystalline structures assumes that the lattices are perfect, with each lattice site occupied by the correct atom. In real materials at temperatures greater than 0°K, defects in the crystalline structure will exist. These defects may be formed by the substitution of atoms different from those normally occupying the site, vacancies on the site, atoms in sites not normally occupied (interstitials), geometrical alterations of the structure in the form of dislocations, twin boundaries, or grain boundaries.

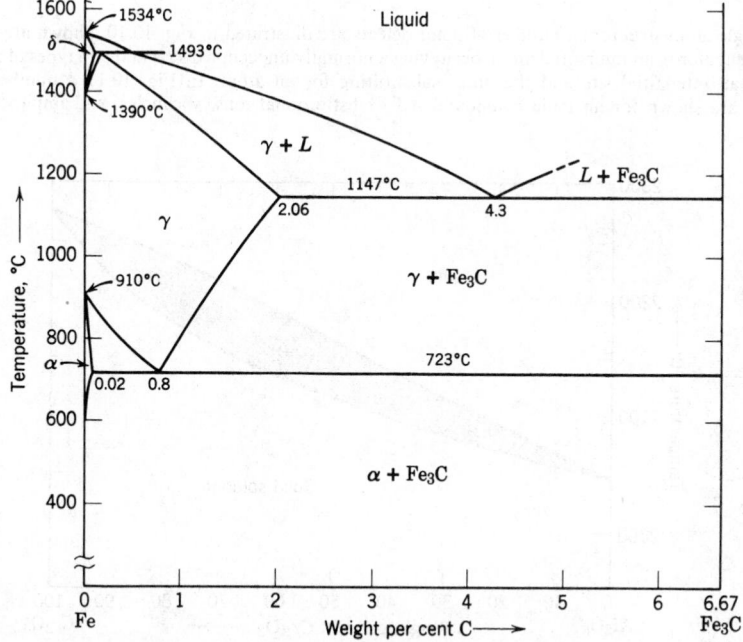

Fig. 10.7 Fe–C phase diagram (from W. G. Moffatt, G. W. Pearsall, and J. Wulff, *The Structure and Properties of Materials,* Wiley, New York, 1964, Vol. I, p. 185).

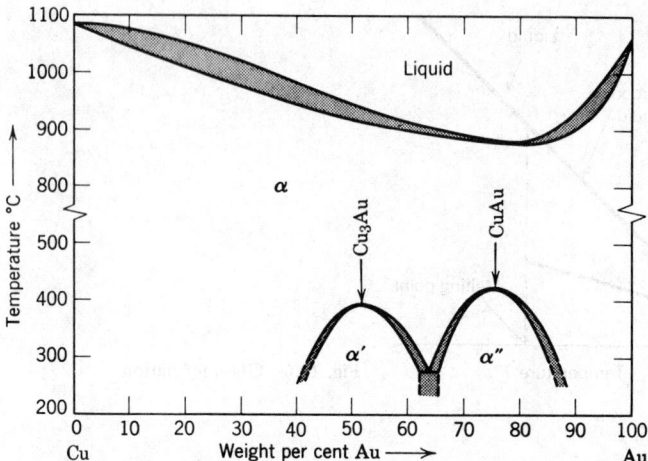

Fig. 10.8 Cu–Au system (from W. G. Moffatt, G. W. Pearsall, and J. Wulff, *The Structure and Properties of Materials,* Wiley, New York, 1964, Vol. I, p. 230).

Solid Solution

When atoms or ions are approximately the same size, they may substitute for another in the structure. For example, Cu and Au have similar radii and at high temperature form a complete solid solution, as shown in Fig. 10.8. A ceramic example is the Cr_2O_3–Al_2O_3 system shown in Fig. 10.9, where Cr and Al substitute for each other. Cr^{3+} has a radius of 0.76 Å and Al has a radius of 0.67 Å. Complete solid solution is not possible if the size difference between atoms or ions is too large, if the structures are different, or if there are charge differences between ions being substituted. In the latter case, substitution is possible only if the charge is compensated for by the creation of vacancies or by oxidation or reduction of ions.

Point Defects

For single-atom structure a number of point defects are illustrated in Fig. 10.10. Shown are a vacancy (an absent atom); an interstitial atom, occupying a normally unoccupied site; and two types of impurities, one in an interstitial site and the other substituting for an atom. In Fig. 10.11 a number of point defects are shown for an ionic compound *AB*. Substitutional ions, vacancies, and impurity ions are

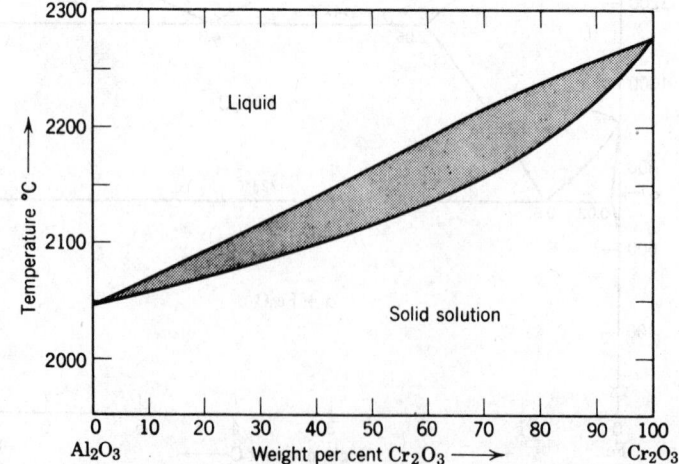

Fig. 10.9 Cr_2O_3–Al_2O_3 system (from W. G. Moffatt, G. W. Pearsall, and J. Wulff, *The Structure and Properties of Materials,* Wiley, New York, 1964, Vol. I, p. 229).

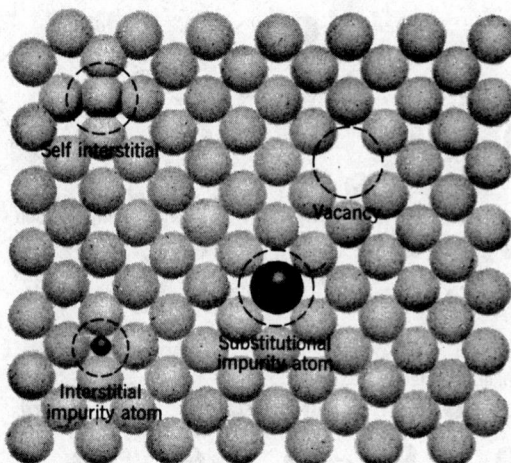

Fig. 10.10 Point defects (from W. G. Moffatt, G. W. Pearsall, and J. Wulff, *The Structure and Properties of Materials*, Wiley, New York, 1964, Vol. I, p. 77).

shown. In ionic compounds, because charges must be balanced, when a cation is removed, an anion is also removed. The resulting vacancy and interstitial point defects are called a Schottky pair. A Frenkel defect occurs when an ion is removed from its normal site and is placed in an interstitial site.

Dislocations

Two basic types of dislocations exist in solids—edge and screw dislocations. An edge dislocation consists of an extra plane of atoms as shown in Fig. 10.12. It is represented by the symbol ⊥ and has associated compression and tension. A screw dislocation is formed by the atom planes spiraling and is shown in Fig. 10.13. Combinations of screw and edge dislocations also exist, which are called mixed dislocations.

Dislocations are important because of their effect on the properties, in particular, the mechanical properties, of engineering materials. The slip of a metal is the result of the movement of dislocations; plastic deformation is the result of the generation of dislocations; the increased strength and brittleness as a result of cold working is due to a generation and pileup of dislocations; and creep in a material is the result of dislocation climb.

Grain Boundaries

Grain boundaries are the regions that occur when there is no alignment between grains in a polycrystalline material. Grain boundaries are important in determining the bulk properties of a material. Impurities segregate at grain boundaries if they reduce the surface energy. Diffusion is usually faster along grain boundaries than through the bulk of the material. Deformation of a material can occur by relative movement of grains.

10.2 METALS

Of the 104 elements in the periodic chart, most are metals, many of which are important technologically. The structure of metals can be considered the packing of spheres that most efficiently fills space. Three basic structures will be considered: face-centered cubic (fcc), hexagonal close packed (hcp), and body-centered cubic (bcc). An introductory discussion of these structures is given in Section 10.1.4.

10.2.1 Structures

Face-Centered Cubic (fcc)

The fcc structure is shown in Fig. 10.3. The *ABCABC* . . . layers, of which there are four sets, are perpendicular to the body diagonals of the cube. The 12 nearest neighbors at a distance D (the diameter of a sphere) form a cubo-octahedron about each sphere, as shown in Fig. 10.14. There are six next nearest neighbors at a distance $\sqrt{2}\,D$ and 24 third nearest neighbors at a distance $\sqrt{3}\,D$. The symmetry of the structure is cubic F in Fig. 10.5. The following metals adopt the fcc structures as one of their polymorphs: Al, Ca, Fe, Co, Ni, Cu, Sr, Y, Rh, Pd, Ag, Ir, Pt, Au, and Pb.

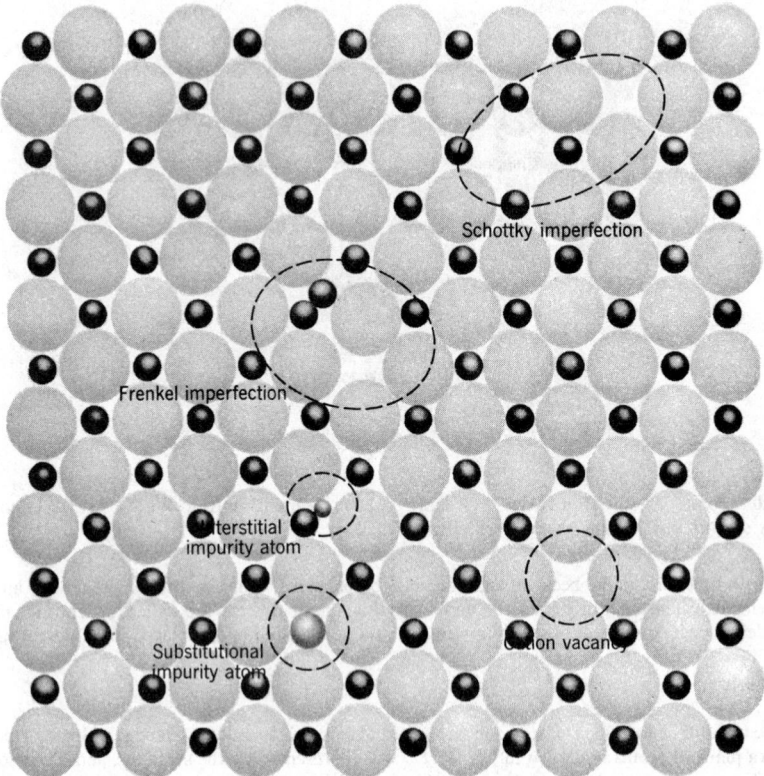

Fig. 10.11 Point defects in a compound AB (from W. G. Moffatt, G. W. Pearsall, and J. Wulff, *The Structure and Properties of Materials,* Wiley, New York, 1964, Vol. I, p. 78).

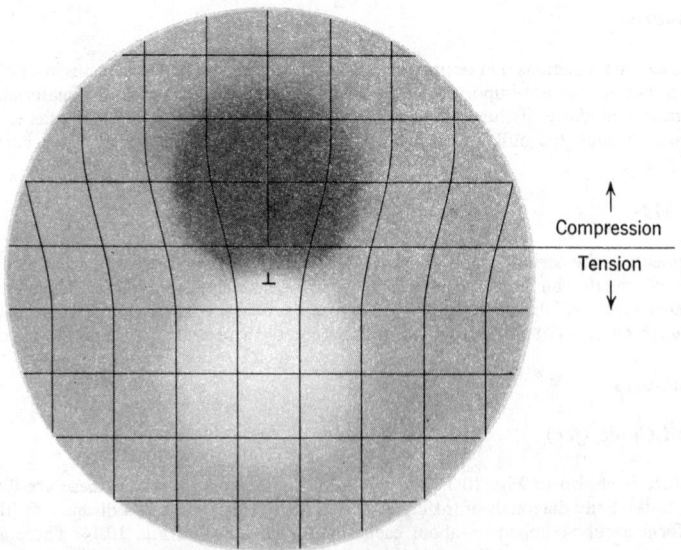

Fig. 10.12 Edge dislocations (from W. G. Moffatt, G. W. Pearsall, and J. Wulff, *The Structure and Properties of Materials,* Wiley, New York, 1964, Vol. I, p. 85).

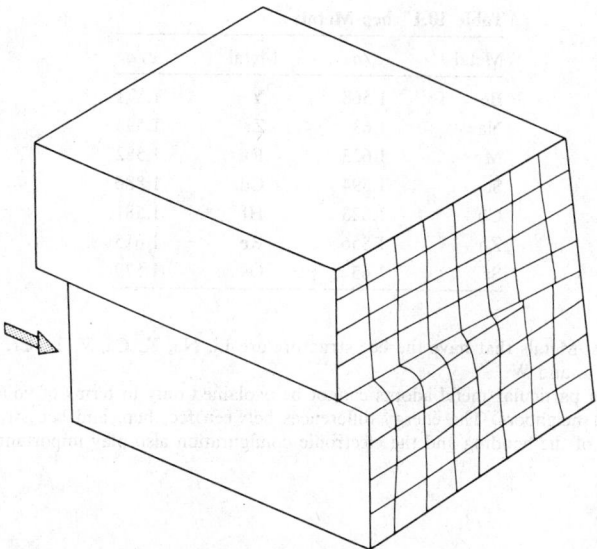

Fig. 10.13 Screw dislocation (from W. D. Kingery, H. K. Bowen, and D. R. Uhlmann, *Introduction to Ceramics,* Wiley, New York, 1976).

hcp   fcc

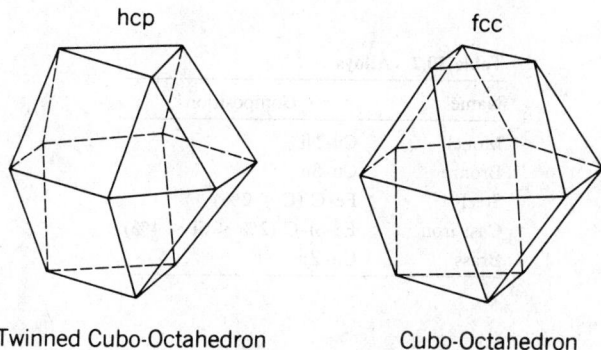

Twinned Cubo-Octahedron   Cubo-Octahedron

Fig. 10.14 Configuration of nearest neighbors in fcc and hcp structures (from W. G. Moffatt, G. W. Pearsall, and J. Wulff, *The Structure and Properties of Materials,* Wiley, New York, 1964, Vol. I, p. 38).

Hexagonal Close Packed (hcp)

The hcp structure is shown in Fig. 10.3. There is only one close-packed direction with a packing sequence *ABAB. . . .* The hexagonal symmetry is shown in Fig. 10.5. As in the fcc structure there are 12 nearest neighbors, but their configuration is different in the form of a twinned cubo-octahedron, as shown in Fig. 10.14. There are six next nearest neighbors as in the fcc structure, but only two third nearest neighbors at a distance $\sqrt{8/3} D$ or 1.633D, the distance from one sphere to the spheres in the second layer above or below the given sphere. The c/a ratio = 1.633 is defined in Fig. 10.5 for the hexagonal lattice. If the shape of the atoms is ellipsoidal rather than spherical, then the c/a ratio deviates from the 1.633 value. Metals with the hcp structure and their c/a ratio are given in Table 10.1.

Body-Centered Cubic (bcc)

The bcc structure is shown in Fig. 10.3. There the distance of the next nearest neighbors is close to the nearest-neighbor distance. Thus, the effective coordination number is 14, comparable to the fcc

Table 10.1   hcp Metals

Metal	c/a	Metal	c/a
Be	1.568	Y	1.571
Na	1.63	Zr	1.593
Mg	1.623	Ru	1.582
Sc	1.594	Cd	1.886
Co	1.623	Hf	1.581
Zn	1.856	Re	1.615
Sr	1.63	Os	1.579

and hcp structures. Metals that have the bcc structure are Li, Na, K, Ca, V, Ti, Cr, Fe, Rb, Sr, Nb, Mo, Cs, Ba, Hf, Ta, and W.

The structure a particular metal adopts cannot be explained only in terms of volume occupied or number of nearest neighbors. The energy differences between fcc, hcp, and bcc structures are very small. The nature of the bonding and the electronic configuration also play important roles.

10.2.2   Alloys

The addition of a second element (or more) to a metal results in an alloy, which may have improved engineering properties. Some examples of alloys are given in Table 10.2. The extent of solid solution and phases formed is given by the appropriate phase diagram. The extent of solid solution is determined by the relative sizes of the atoms.

Table 10.2   Alloys

Name	Composition
Monel	Cu–Ni
Bronze	Cu–Sn
Steel	Fe–C (C < 2%)
Cast iron	Fe–Si–C (2% < Si < 4%)
Brass	Cu–Zn

10.2.3   Noncrystalline Metals

Noncrystalline or amorphous metals can be prepared in the form of ribbon or film by rapid quenching techniques (cooling rate $> 10^5$ deg/sec) such as splat cooling or vapor, electrolytic, or chemical deposition. Compositions include a metal and metalloid such as Si, Ge, P, Sb, or C with generally 80 wt% metal. Typical compositions are Ni_3P, Au_3Si, and Pd–Fe–Si. The structure of these materials consists of a dense, random packing of the metal in which approximately 64% of the volume is occupied and in which the metalloid occupies irregularly shaped tetrahedra, octahedra, and other sites and stabilizes the structure. Improved mechanical properties including higher strengths, greater ductility, improved corrosion resistance, and interesting magnetic properties make these promising engineering materials.

10.3   NONMETALS

10.3.1   Ceramics

Ceramics are nonmetallic inorganic materials. Thus, the oxides of all metals such as Fe_2O_3, TiO_2, Al_2O_3, and SiO_2 and materials such as diamond, SiN, SiC, and Si are considered to be ceramics.

Crystalline Ceramics

Most oxides can be considered packings of oxygen ions with the cations occupying the tetrahedral and/or octahedral sites in the structure. As an example, α-alumina (α-Al_2O_3) consists of an hcp packing of O^{2-} with two-thirds of the octahedral sites occupied by Al^{3+} in an orderly fashion. Since for each O^{2-} there exists one octahedral and two tetrahedral sites, in Al_2O_3 there would be three octahedral sites in which two Al^{3+} are placed, thus two-thirds of the octahedral and none of the tetrahedral sites are filled. The compound is electrically neutral since $2\times(3+)$ (Al) $= 3\times(2-)$ (O). If the Al is shared by six O's, then $\frac{3}{6} = \frac{1}{2}$ of its charge is contributed to each O. For the charge on each O to be satisfied, four Al's need to be coordinated to each O since $4(\frac{1}{2}) = 2$. A notation to indicate the coordination scheme for α-Al_2O_3 is $6:4$—each Al is coordinated to six O's and each oxygen is coordinated to four Al's.

Table 10.3   Common Ceramic Structures

Structure	Examples	Coordination	Packing of Anion
Rock salt	MgO, CaO, SrO, FeO	6:6	ccp
Zincblende	SiC	4:4	ccp
Rutile	TiO_2, GeO_2	6:3	Distorted hcp
Perovskite	$SrTiO_3$, $BaTiO_3$	12:6:6	ccp
Spinel	$MgAl_2O_4$, $FeAl_2O_4$	4:6:4	ccp
Corundum	α-Al_2O_3, Fe_2O_3	6:4	hcp
Fluorite	UO_2, ZrO_2	8:4	Simple cube

A summary of common ceramic structures is given in Table 10.3. The structure of silicates is complicated, but the basic unit is the SiO_4 tetrahedron. The three polymorphs of SiO_2—quartz, tridymite and crystobalite—have different arrangements for the linking of all four vertices of the tetrahedron. Each Si is bonded to four O's and each O is bonded to two Si's. In the layer silicates such as micas, clays, and talc only three of the vertices are linked. The result is a laminar structure in which the bonding between layers is a weaker ionic bonding, hydrogen bonding, or van der Waals bonding, respectively, for mica, clay, and talc.

Of particular importance in semiconductors is the diamond structure. In this structure each atom is tetrahedrally coordinated to four other atoms. The predominant covalent bonding of the structure is manifested by the high degree of directionality in the bonding. In addition to diamond, Si and Ge have this structure as do other semiconductors that have been doped with other elements.

Noncrystalline Ceramics

Common glass compositions are fused SiO_2, soda-lime silica, soda-borosilicate, and alkali-lead silicate. Glass formers such as SiO_2 and B_2O_3 are characterized by a high viscosity at the melting point and readily form glasses when cooled. Network modifiers such as Na_2O and CaO do not form glasses unless quenched at extremely high rates. Intermediates such as Al_2O_3 and PbO, while not readily forming glasses by themselves, can be present in high concentrations when combined with glass formers.

Amorphous or fused SiO_2 has Si tetrahedrally coordinated to four O's with each O bonded to two Si's. Thus, SiO_4 tetrahedra are linked, sharing all four vertices in a continuous three-dimensional network. The structure has short-range order but no long-range order. The introduction of network modifiers results in the formation of nonbridging oxygen—oxygen bonded to only one Si and thus negatively charged. The cation, such as Na^+, is in the interstitial sites balancing the charge. The result is an increase in density, a large decrease in viscosity, and a decrease in thermal expansion with increasing alkali content. The alkaline earths behave in a similar manner. Commercial soda–lime–silica glass ($\sim$ 72 wt% SiO_2, 12–15 wt% Na_2O, 10–15 wt% CaO) has a broken up silica network with Na and Ca ions in the holes.

Glass–Ceramics

Glass–ceramics are materials that have been fabricated as glasses and then crystallized as a result of controlled nucleation and growth. In most cases nucleating agents such as TiO_2, P_2O_5, Pt, or ZrO_2 are added to aid in crystallization. The microstructure of many glass–ceramics consists of 95–98% crystalline phase with a grain size < 1 μm embedded in a pore-free glassy phase. Typical composition systems are Li_2O–Al_2O_3–SiO_2, MgO–Al_2O_3–SiO_2, and Na_2O–BaO–Al_2O_3–SiO_2. Some of the desirable properties of various glass–ceramic systems are zero or very low thermal expansion, high mechanical strength, high electrical resistivity, and machinability.

10.3.2   Polymers

Polymers are organic materials that consist of chains of C and H. The intrachain bonding is covalent, while the interchain bonding is van der Waals. The repeating structural units, monomers, are linked together to form the polymer.

Isomers are organic compounds of the same composition but with a different arrangement of the atoms. Copolymerization is the process of linking different polymers together. Many polymers are noncrystalline because the long chains become entangled or because of side groups attached to the chain, particularly if they are large or irregularly placed. Both of these factors make it difficult to crystallize the chains. The addition of plasticizers—low-molecular-weight compounds that separate the chains—also helps to prevent crystallization.

The manner in which the polymers are formed affects the final structure. Bifunctional monomers result in two bonds that form linear chains, whereas trifunctional or tetrafunctional monomers result in network or framework polymers. This results in cross-linking and in increased structural rigidity and less elasticity. The shape of the linear polymers can be altered by the addition of side groups; not only does the packing become less ordered, but the interbonding becomes stronger. Branching, the splitting of the polymer chain, is another way to introduce three dimensions into the polymer structure.

Since polymers are organic materials when compared to metals and ceramics, they tend to have low melting or softening temperatures and are flammable. The elastic moduli are lower by several orders of magnitude, and they serve as electrical insulators.

10.4 COMPOSITES AND COATINGS

Many modern engineering materials have been developed by combining two or more materials into a single material, a composite, or by coating one material with another. The structure of such composites and coatings will be discussed in general, but the specifics will not be covered.

10.4.1 Fiberglass

Fiberglass is formed from glass fibers impregnated in an ordered or random manner in a plastic material. The fibers are usually of a composition known as E glass (SiO_2, 54 wt%; Al_2O_3, 14 wt%; B_2O_3, 10 wt%; MgO, 4.5 wt%; and CaO, 17.5 wt%), are typically 0.00023–0.00053 in. in diameter, and are woven together to form continuous fibers or woven together to form cloth.

10.4.2 Coatings

Various coatings used in engineering applications are summarized in Table 10.4. Coatings can serve as a protective layer for the substrate and/or alter the appearance of the surface. The structures of the coating and the substrate have previously been discussed. Of great importance is the bonding and structure at the interface between the coating and the substrate.

Table 10.4 Coatings

Coating	Composition	Substrate
Enamel	Inorganic glass	Metal
Glaze	Inorganic glass	Ceramic
Paint	Organic	Metal, polymer
Galvanized and plating	Metal	Metal

In general, the bonding will be affected by atomistic and microscopic considerations of the surfaces. The bonding in the material, whether metallic, ionic, or covalent, may be continued or altered in the interface. Such factors as surface roughness, porosity, and oxidation/reduction, and the presence of impurities will affect the bonding at the interface.

Enamels are used on metals to protect the surface from oxidation and to change the color and appearance of the surface. The vitreous enamel is fused to the surface of the metal. The bonding changes from metallic to ionic–covalent on the enamel. The thermal expansion of the enamel is usually less than that of the metal substrate, so that the enamel surface is in compression, thus improving the mechanical properties of the enamel. Glazes are used to decrease the porosity of the ceramic substrate and to alter the appearance of the surface.

BIBLIOGRAPHY

C. S. Barrett and T. B. Massalski, *Structure of Metals,* 3rd ed., Pergamon Press, Oxford, 1980.

C. R. Barrett, W. D. Nix, and A. S. Tetelman, *The Principles of Engineering Materials,* Prentice-Hall, Englewood Cliffs, NJ, 1973.

R. A. Flinn and P. K. Trojan, *Engineering Materials and Their Applications,* 2nd ed., Houghton Mifflin, Boston, MA, 1981.

A. G. Guy, *Essentials of Materials Science,* McGraw-Hill, New York, 1976.

W. D. Kingery, H. K. Bowen, and D. R. Uhlmann, *Introduction to Ceramics,* 2nd ed., Wiley, New York, 1976.

The Structure and Properties of Materials, 4 Vols., Wiley, New York. Volume 1, *Structures,* W. G. Moffatt, G. W. Pearsall, and J. Wulff (eds.), 1964; Volume 2, *Thermodynamics of Structure,* J. H. Brophy, R. M. Rose, and J. Wulff (eds.), 1964; Volume 3, *Mechanical Behavior,* H. W. Hayden, W. G. Moffatt, and J. Wulff (eds.), 1965; Volume 4, *Electronic Properties,* R. M. Rose, L. A. Shepard, and J. Wulff (eds.), 1966.

L. H. Van Vlack, *Elements of Materials Science,* 4th ed., Addison-Wesley, Reading, MA, 1980.

L. H. Van Vlack, *Physical Ceramics for Engineers,* Addison-Wesley, Reading, MA, 1964.

CHAPTER 11

STEEL

ROBERT J. KING

**United States Steel Corporation
Pittsburgh, Pennsylvania**

Reprinted from *Kirk–Othmer Encyclopedia of Chemical Technology*, 3rd ed., Wiley, New York, 1983, Vol. 21, by permission of the publisher.

11.1 METALLOGRAPHY AND HEAT TREATMENT

The great advantage of steel as an engineering material is its versatility, which arises from the fact that its properties can be controlled and changed by heat treatment.[1-3] Thus, if steel is to be formed into some intricate shape, it can be made very soft and ductile by heat treatment; on the other hand, heat treatment can also impart high strength.

The physical and mechanical properties of steel depend on its constitution, that is, the nature, distribution, and amounts of its metallographic constituents as distinct from its chemical composition. The amount and distribution of iron and iron carbide determine the properties, although most plain carbon steels also contain manganese, silicon, phosphorus, sulfur, oxygen, and traces of nitrogen, hydrogen, and other chemical elements such as aluminum and copper. These elements may modify, to a certain extent, the main effects of iron and iron carbide, but the influence of iron carbide always predominates. This is true even of medium-alloy steels, which may contain considerable amounts of nickel, chromium, and molybdenum.

The iron in steel is called ferrite. In pure iron–carbon alloys, the ferrite consists of iron with a trace of carbon in solution, but in steels it may also contain alloying elements such as manganese, silicon, or nickel. The atomic arrangement in crystals of the allotropic forms of iron is shown in Fig. 11.1.

Cementite, the term for iron carbide in steel, is the form in which carbon appears in steels. It has the formula Fe_3C, and consists of 6.67% carbon and 93.33% iron. Little is known about its properties, except that it is very hard and brittle. As the hardest constituent of plain carbon steel, it scratches glass and feldspar but not quartz. It exhibits about two-thirds the induction of pure iron in a strong magnetic field.

Austenite is the high-temperature phase of steel. Upon cooling, it gives ferrite and cementite. Austenite is a homogeneous phase, consisting of a solid solution of carbon in the γ form of iron. It forms when steel is heated above 790°C. The limiting temperatures for its formation vary with composition and are discussed below. The atomic structure of austenite is that of γ iron, fcc; the atomic spacing varies with the carbon content.

When a plain carbon steel of ~ 0.80% carbon content is cooled slowly from the temperature range at which austenite is stable, ferrite and cementite precipitate together in a characteristically lamellar structure known as pearlite. It is similar in its characteristics to a eutectic structure but, since it is formed from a solid solution rather than from a liquid phase, it is known as a eutectoid structure. At carbon contents above and below 0.80%, pearlite of ~ 0.80% carbon is likewise formed on slow cooling, but excess ferrite or cementite precipitates first, usually as a grain-boundary network, but occasionally also along the cleavage planes of austenite. The excess ferrite or cementite rejected by the cooling austenite is known as a proeutectoid constituent. The carbon content of a slowly cooled steel can be estimated from the relative amounts of pearlite and proeutectoid constituents in the microstructure.

Bainite is a decomposition product of austenite consisting of an aggregate of ferrite and cementite. It forms at temperatures lower than those where very fine pearlite forms and higher than those at which martensite begins to form on cooling. Metallographically, its appearance is feathery if formed in the upper part of the temperature range, or acicular (needlelike) and resembling tempered martensite if formed in the lower part.

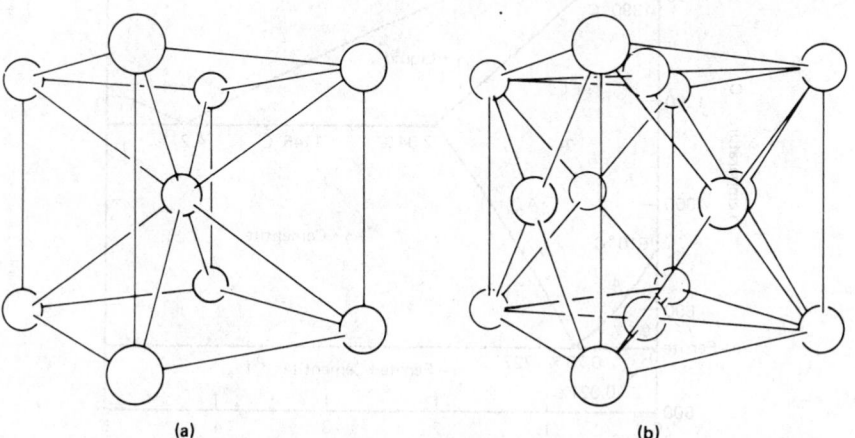

(a)   (b)

Fig. 11.1 Crystalline structure of allotropic forms of iron. Each white sphere represents an atom of (a) α and δ iron in bcc form, and (b) γ iron, in fcc (from Ref. 1).

Martensite in steel is a metastable phase formed by the transformation of austenite below the temperature called the M_s temperature, where martensite begins to form as austenite is cooled continuously. Martensite is an interstitial supersaturated solid solution of carbon in iron with a body-centered tetragonal lattice. Its microstructure is acicular.

11.2 IRON–IRON CARBIDE PHASE DIAGRAM

The iron–iron carbide phase diagram (Fig. 11.2) furnishes a map showing the ranges of compositions and temperatures in which the various phases such as austenite, ferrite, and cementite are present in slowly cooled steels. This diagram covers the temperature range from 600°C to the melting point of iron, and carbon contents from 0 to 5%. In steels and cast irons, carbon can be present either as iron carbide (cementite) or as graphite. Under equilibrium conditions, only graphite is present because iron carbide is unstable with respect to iron and graphite. However, in commercial steels, iron carbide is present instead of graphite. When a steel containing carbon solidifies, the carbon in the steel usually solidifies as iron carbide. Although the iron carbide in a steel can change to graphite and iron when the steel is held at ~900°C for several days or weeks, iron carbide in steel under normal conditions is quite stable.

The portion of the iron–iron carbide diagram of interest here is that part extending from 0 to 2.01% carbon. Its application to heat treatment can be illustrated by considering the changes occurring on heating and cooling steels of selected carbon contents.

Iron occurs in two allotropic forms, α or δ (the latter at a very high temperature) and γ (see Fig. 11.1). The temperatures at which these phase changes occur are known as the critical temperatures, and the boundaries in Fig. 11.2 show how these temperatures are affected by composition. For pure iron, these temperatures are 910°C for the α–γ phase change and 1390°C for the γ–δ phase change.

11.2.1 Changes on Heating and Cooling Pure Iron

The only changes occurring on heating or cooling pure iron are the reversible changes at ~910°C from bcc α iron to fcc γ iron and from the fcc δ iron to bcc γ iron at ~1390°C.

11.2.2 Changes on Heating and Cooling Eutectoid Steel

Eutectoid steels are those that contain 0.8% carbon. The diagram shows that at and below 727°C the constituents are α ferrite and cementite. At 600°C, the α-ferrite may dissolve as much as 0.007% carbon. Up to 727°C, the solubility of carbon in the ferrite increases until, at this temperature, the ferrite contains about 0.02% carbon. The phase change on heating an 0.8% carbon steel occurs at

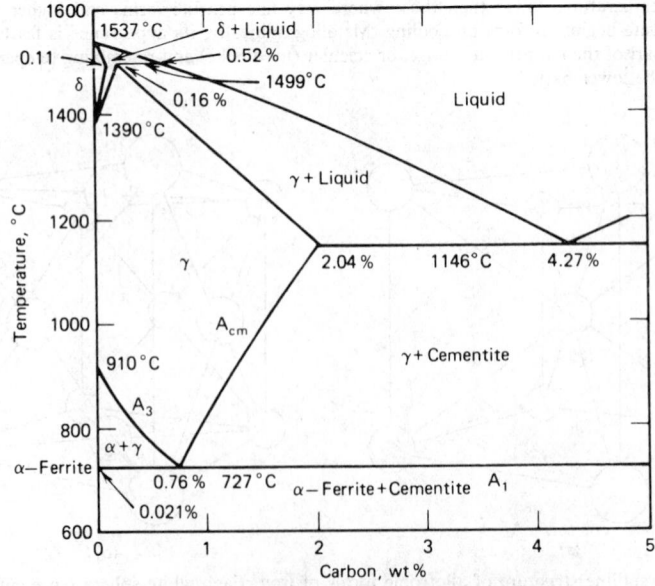

Fig. 11.2 Iron–iron carbide phase diagram (from Ref. 1).

727°C which is designated as A_1, as the eutectoid or lower critical temperature. On heating just above this temperature, all ferrite and cementite transform to austenite, and on slow cooling the reverse change occurs.

When a eutectoid steel is slowly cooled from ~738°C, the ferrite and cementite form in alternate layers of microscopic thickness. Under the microscope at low magnification, this mixture of ferrite and cementite has an appearance similar to that of a pearl and is therefore called pearlite.

11.2.3   Changes on Heating and Cooling Hypoeutectoid Steels

Hypoeutectoid steels are those that contain less carbon than the eutectoid steels. If the steel contains more than 0.02% carbon, the constituents present at and below 727°C are usually ferrite and pearlite; the relative amounts depend on the carbon content. As the carbon content increases, the amount of ferrite decreases and the amount of pearlite increases.

The first phase change on heating, if the steel contains more than 0.02% carbon, occurs at 727°C. On heating just above this temperature, the pearlite changes to austenite. The excess ferrite, called proeutectoid ferrite, remains unchanged. As the temperature rises further above A_1, the austenite dissolves more and more of the surrounding proeutectoid ferrite, becoming lower and lower in carbon content until all the proeutectoid ferrite is dissolved in the austenite, which now has the same average carbon content as the steel.

On slow cooling the reverse changes occur. Ferrite precipitates, generally at the grain boundaries of the austenite, which becomes progressively richer in carbon. Just above A_1, the austenite is substantially of eutectoid composition, 0.8% carbon.

11.2.4   Changes on Heating and Cooling Hypereutectoid Steels

The behavior on heating and cooling hypereutectoid steels (steels containing >0.80% carbon) is similar to that of hypoeutectoid steels, except that the excess constituent is cementite rather than ferrite. Thus, on heating above A_1, the austenite gradually dissolves the excess cementite until at the A_{cm} temperature the proeutectoid cementite has been completely dissolved and austenite of the same carbon content as the steel is formed. Similarly, on cooling below A_{cm}, cementite precipitates and the carbon content of the austenite approaches the eutectoid composition. On cooling below A_1, this eutectoid austenite changes to pearlite and the room-temperature composition is, therefore, pearlite and proeutectoid cementite.

Early iron–carbon equilibrium diagrams indicated a critical temperature at ~768°C. It has since been found that there is no true phase change at this point. However, between ~768 and 790°C there is a gradual magnetic change, since ferrite is magnetic below this range and paramagnetic above it. This change, occurring at what formerly was called the A_2 change, is of little or no significance with regard to the heat treatment of steel.

11.2.5   Effect of Alloys on the Equilibrium Diagram

The iron–carbon diagram may, of course, be profoundly altered by alloying elements, and its application should be limited to plain carbon and low-alloy steels. The most important effects of the alloying elements are that the number of phases that may be in equilibrium is no longer limited to two as in the iron–carbon diagram; the temperature and composition range, with respect to carbon, over which austenite is stable may be increased or reduced; and the eutectoid temperature and composition may change.

Alloying elements either enlarge the austenite field or reduce it. The former include manganese, nickel, cobalt, copper, carbon, and nitrogen and are referred to as austenite formers.

The elements that decrease the extent of the austenite field include chromium, silicon, molybdenum, tungsten, vanadium, tin, niobium, phosphorus, aluminum, and titanium; they are known as ferrite formers.

Manganese and nickel lower the eutectoid temperature, whereas chromium, tungsten, silicon, molybdenum, and titanium generally raise it. All these elements seem to lower the eutectoid carbon content.

11.2.6   Grain Size—Austenite

A significant aspect of the behavior of steels on heating is the grain growth that occurs when the austenite, formed on heating above A_3 or A_{cm}, is heated even higher; A_3 is the upper critical temperature and A_{cm} is the temperature at which cementite begins to form. The austenite, like any metal composed of a solid solution, consists of polygonal grains. As formed at a temperature just above A_3 or A_{cm}, the size of the individual grains is very small but, as the temperature is increased above the critical temperature, the grain sizes increase. The final austenite grain size depends, therefore, on the temperature above the critical temperature to which the steel is heated. The grain size of the austenite has a marked influence on transformation behavior during cooling and on the grain size of the constituents of the final microstructure. Grain growth may be inhibited by carbides that dissolve slowly or by dispersion of nonmetallic inclusions. Hot working refines the coarse grain formed by reheating steel to the relatively high temperatures used in forging or rolling, and the grain size of hot-worked steel is determined largely by the temperature at which the final stage of the hot-working process is carried out. The general effects of austenite grain size on the properties of heat-treated steel are summarized in Table 11.1.

11.2.7 Microscopic-Grain-Size Determination

The microscopic grain size of steel is customarily determined from a polished plane section prepared in such a way as to delineate the grain boundaries. The grain size can be estimated by several methods. The results can be expressed as diameter of average grain in millimeters (reciprocal of the square root of the number of grains per mm²), number of grains per unit area, number of grains per unit volume, or a micrograin-size number obtained by comparing the microstructure of the sample with a series of standard charts.

11.2.8 Fine- and Coarse-Grain Steels

As mentioned previously, austenite-grain growth may be inhibited by undissolved carbides or nonmetallic inclusions. Steels of this type are commonly referred to as fine-grained steels, whereas steels that are free from grain-growth inhibitors are known as coarse-grained steels.

The general pattern of grain coarsening when steel is heated above the critical temperature is as follows: Coarse-grained steel coarsens gradually and consistently as the temperature is increased, whereas fine-grained steel coarsens only slightly, if at all, until a certain temperature known as the coarsening temperature is reached, after which abrupt coarsening occurs. Heat treatment can make any type of steel either fine or coarse grained; as a matter of fact, at temperatures above its coarsening temperature, the fine-grained steel usually exhibits a coarser grain size than the coarse-grained steel at the same temperature.

Making steels that remain fine grained above 925°C involves the judicious use of deoxidation with aluminum. The inhibiting agent in such steels is generally conjectured to be a submicroscopic dispersion of aluminum nitride or, perhaps at times, aluminum oxide.

11.2.9 Phase Transformations—Austenite

At equilibrium, that is, with very slow cooling, austenite transforms to pearlite when cooled below the A_1 temperature. When austenite is cooled more rapidly, this transformation is depressed and occurs at a lower temperature. The faster the cooling rate, the lower the temperature at which transformation occurs. Furthermore, the nature of the ferrite–carbide aggregate formed when the austenite transforms varies markedly with the transformation temperature, and the properties are found to vary correspondingly. Thus, heat treatment involves a controlled supercooling of austenite, and in order to take full advantage of the wide range of structures and properties that this treatment permits, a knowledge of the transformation behavior of austenite and the properties of the resulting aggregates is essential.

11.2.10 Isothermal Transformation Diagram

The transformation behavior of austenite is best studied by observing the isothermal transformation at a series of temperatures below A_1. The transformation progress is ordinarily followed metallographi-

Table 11.1 Trends in Heat-Treated Products[57]

Property	Coarse-grain austenite	Fine-grain austenite
Quenched and tempered products		
hardenability	increasing	decreasing
toughness	decreasing	increasing
distortion	more	less
quench cracking	more	less
internal stress	higher	lower
Annealed or normalized products		
machinability		
rough finish	better	inferior
fine finish	inferior	better

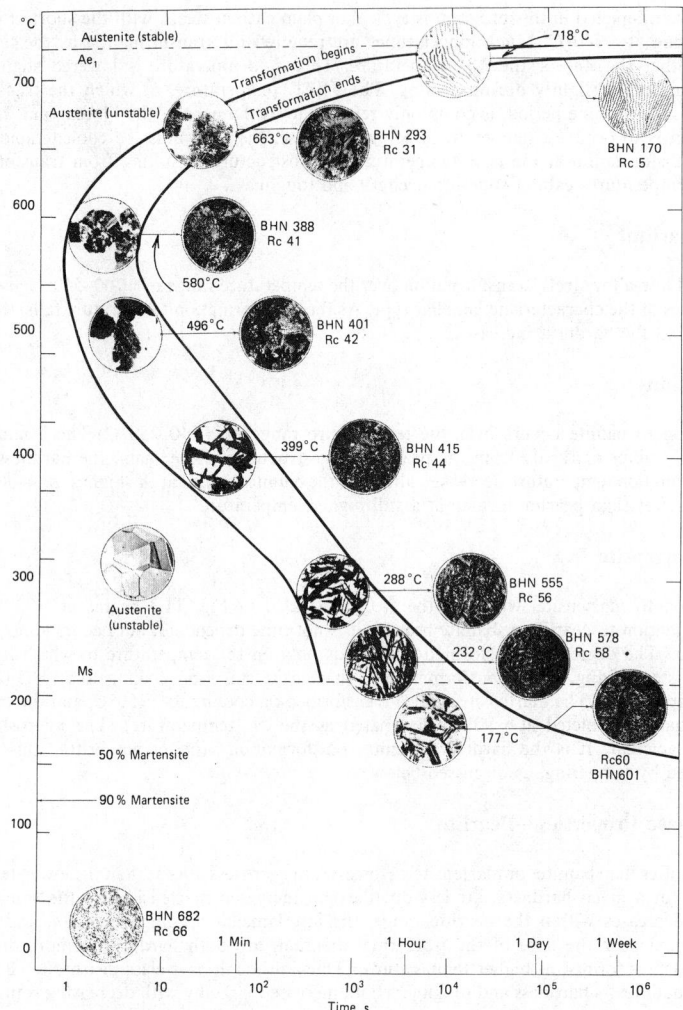

Fig. 11.3 Isothermal transformation diagram for a plain carbon eutectoid steel 1; $Ae_1 = A_1$ temperature at equilibrium; BHN = Brinell hardness number; Rc = Rockwell hardness scale C. C,0.89%; Mn, 0.29% austenitized at 885°C; grain size, 4–5; photomicrographs originally ×2500.

cally in such a way that both the time–temperature relationships and the manner in which the microstructure changes are established. The times at which transformation begins and ends at a given temperature are plotted, and curves depicting the transformation behavior as a function of temperature are obtained by joining these points (Fig. 11.3). Such a diagram is referred to as an isothermal transformation (IT) diagram, a time–temperature-transformation (TTT) diagram, or, an S curve.[4]

The IT diagram for a eutectoid carbon steel is shown in Fig. 11.3. In addition to the lines depicting the transformation, the diagram shows microstructures at various stages of transformation and hardness values. Thus, the diagram illustrates the characteristic subcritical austenite transformation behavior, the manner in which microstructure changes with transformation temperature, and the general relationship between these microstructural changes and hardness.

As the diagram indicates, the characteristic isothermal transformation behavior at any temperature above the temperature at which transformation to martensite begins (the M_s temperature) takes place over a period of time, known as the incubation period, in which no transformation occurs, followed by a period of time during which the transformation proceeds until the austenite has been transformed completely. The transformation is relatively slow at the beginning and toward the end, but much more rapid during the intermediate period in which ~25–75% of the austenite is transformed. Both the incubation period and the time required for completion of the transformation depend on the temperature.

The behavior depicted in this program is typical of plain carbon steels, with the shortest incubation period occurring at ~540°C. Much longer times are required for transformation as the temperature approaches either the Ae_1 or the M_s temperature. This A_1 temperature is lowered slightly during cooling and increased slightly during heating. The 540°C temperature, at which the transformation begins in the shortest time period, is commonly referred to as the nose of the IT diagram. If complete transformation is to occur at temperatures below this nose, the steel must be cooled rapidly enough to prevent transformation at the nose temperature. Microstructures resulting from transformation at these lower temperatures exhibit superior strength and toughness.

11.2.11 Pearlite

In carbon and low-alloy steels, transformation over the temperature range of ~700–540°C gives pearlite microstructures of the characteristic lamellar type. As the transformation temperature falls, the lamellae move closer and the hardness increases.

11.2.12 Bainite

Transformation to bainite occurs over the temperature range of ~540–230°C. The acicular bainite microstructures differ markedly from the pearlite microstructures. Here again, the hardness increases as the transformation temperature decreases, although the bainite formed at the highest possible tempera-ture is often softer than pearlite formed at a still higher temperature.

11.2.13 Martensite

Transformation to martensite, which in the steel illustrated in Fig. 11.3 begins at ~230°C, differs from transformation to pearlite or bainite because it is not time dependent, but occurs almost instantly during cooling. The degree of transformation depends only on the temperature to which it is cooled. Thus, in this steel of Fig. 11.3, transformation to martensite starts on cooling to 230°C (designated as the M_s temperature). The martensite is 50% transformed on cooling to ~150°C, and the transforma-tion is essentially completed at ~90°C (designated as the M_f temperature). The microstructure of martensite is acicular. It is the hardest austenite transformation product but brittle; this brittleness can be reduced by tempering, as discussed below.

11.2.14 Phase Properties—Pearlite

Pearlites are softer than bainites or martensites. However, they are less ductile than the lower-temperature bainites and, for a given hardness, far less ductile than tempered martensite. As the transformation temperature decreases within the pearlite range, the interlamellar spacing decreases, and these fine pearlites, formed near the nose of the isothermal diagram, are both harder and more ductile than the coarse pearlites formed at higher temperatures. Thus, although as a class pearlite tends to be soft and not very ductile, its hardness and toughness both increase markedly with decreasing transformation temperatures.

11.2.15 Phase Properties—Bainite

In a given steel, bainite microstructures are generally found to be both harder and tougher than pearlite, although less hard than martensite. Bainite properties generally improve as the transformation tempera-ture decreases and lower bainite compares favorably with tempered martensite at the same hardness or exceeds it in toughness. Upper bainite, on the other hand, may be somewhat deficient in toughness as compared with fine pearlite of the same hardness.[4]

11.2.16 Phase Properties—Martensite

Martensite is the hardest and most brittle microstructure obtainable in a given steel. The hardness of martensite increases with increasing carbon content up to the eutectoid composition, and, at a given carbon content, varies with the cooling rate.

Although for some applications, particularly those involving wear resistance, the hardness of martens-ite is desirable in spite of the accompanying brittleness, this microstructure is mainly important as starting material for tempered martensite structures, which have definitely superior properties.

11.2.17 Tempered Martensite

Martensite is tempered by heating to a temperature ranging from 170 to 700°C for 30 min to several hours. This treatment causes the martensite to transform to ferrite interspersed with small particles

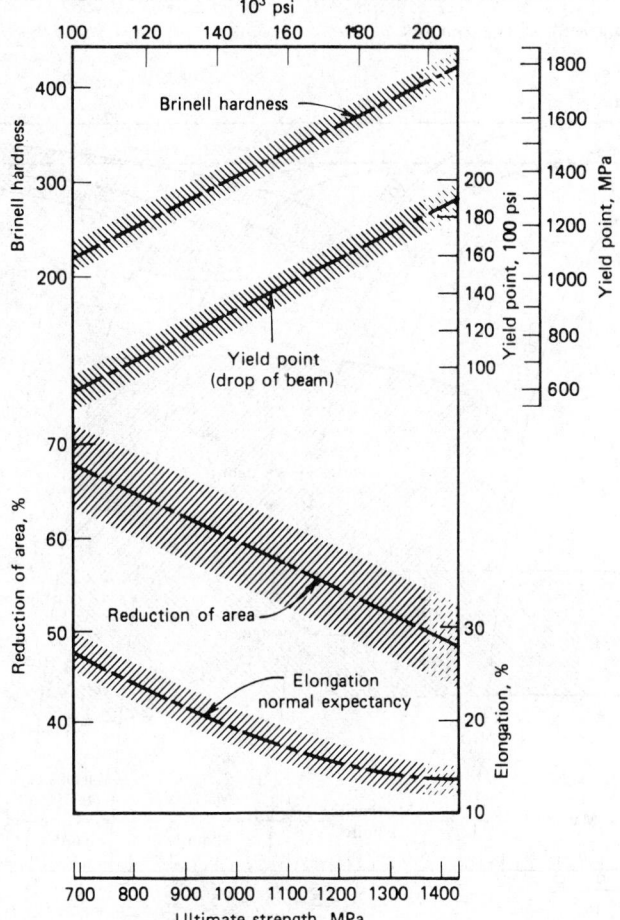

Fig. 11.4   Properties of tempered martensite (from Ref. 1). Fully heat-treated miscellaneous analyses, low-alloy steels; 0.30–0.50% C.

of cementite. Higher temperatures and longer tempering periods cause the cementite particles to increase in size and the steel to become more ductile and lose strength. Tempered martensitic structures are, as a class, characterized by toughness at any strength. The diagram of Fig. 11.4 describes, within ±10%, the mechanical properties of tempered martensite, regardless of composition. For example, a steel consisting of tempered martensite, with an ultimate strength of 1035 MPa (150,000 psi), might be expected to exhibit elongation of 16–20%, reduction of area of between 54 and 64%, yield point of 860–980 MPa (125,000–142,000 psi), and Brinell hardness of about 295–320. Because of its high ductility at a given hardness, this is the structure that is preferred.

11.2.18   Transformation Rates

The main factors affecting transformation rates of austenite are composition, grain size, and homogeneity. In general, increasing carbon and alloy content as well as increasing grain size tend to lower transformation rates. These effects are reflected in the isothermal transformation curve for a given steel.

11.2.19   Continuous Cooling

The basic information depicted by an IT diagram illustrates the structure formed if the cooling is interrupted and the reaction is completed at a given temperature. The information is also useful for interpreting behavior when the cooling proceeds directly without interruption, as in the case of annealing, normalizing, and quenching. In these processes, the residence time at a single temperature is generally

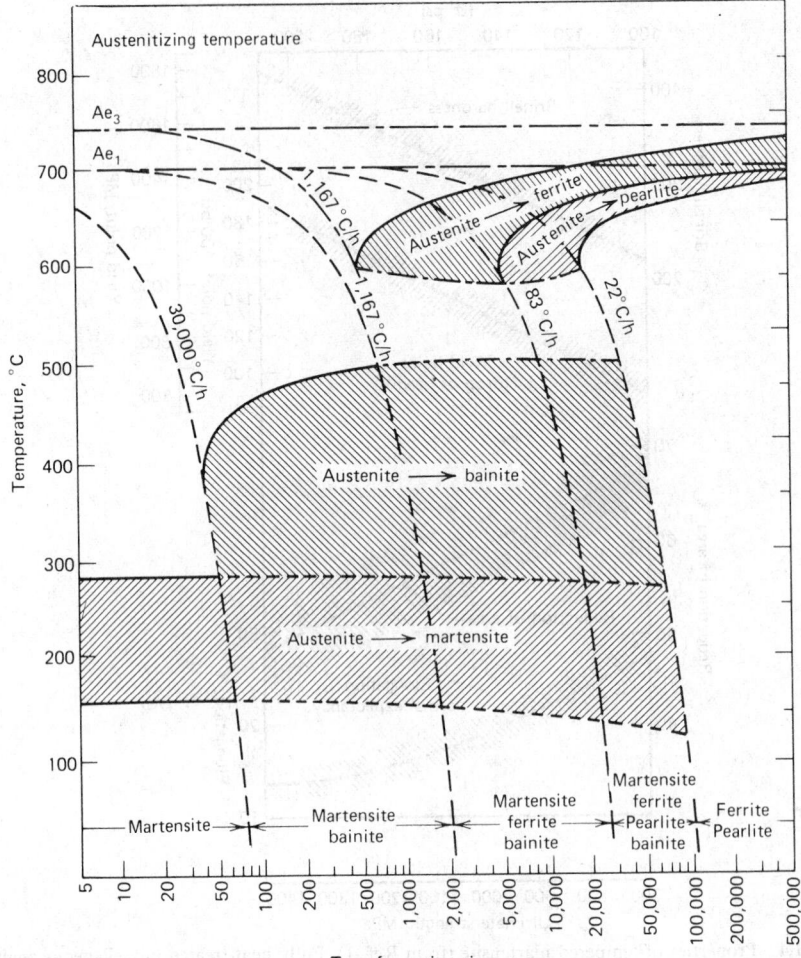

Fig. 11.5 Continuous-cooling transformation diagram for a type 4340 alloy steel, with superimposed cooling curves illustrating the manner in which transformation behavior during continuous cooling governs final microstructure (from Ref. 1). Ae_3 = critical temperature at equilibrium.

insufficient for the reaction to go to completion; instead, the final structure consists of an association of microstructures which were formed individually at successively lower temperatures as the piece cooled. However, the tendency to form several structures is still explained by the isothermal diagram.[5,6]

The final microstructure after continuous cooling depends on the times spent at the various transformation-temperature ranges through which a piece is cooled. The transformation behavior on continuous cooling thus represents an integration of these times by constructing a continuous-cooling diagram at constant rates similar to the isothermal transformation diagram (see Fig. 11.5). This diagram lies below and to the right of the corresponding IT diagram if plotted on the same coordinates; that is, transformation on continuous cooling starts at a lower temperature and after a longer time than the intersection of the cooling curve and the isothermal diagram would predict. This displacement is a function of the cooling rate, and increases with increasing cooling rate.

Several cooling-rate curves have been superimposed on Fig. 11.5. The changes occurring during these cooling cycles illustrate the manner in which diagrams of this nature can be correlated with heat-treating processes and used to predict the resulting microstructure.

Considering, first, the relatively low cooling rate ($<22°C/hr$), the steel is cooled through the regions in which transformations to ferrite and pearlite occur which constitute the final microstructure. This cooling rate corresponds to a slow cooling in the furnace such as might be used in annealing.

At a higher cooling rate ($22–83°C/hr$), such as might be obtained on normalizing a large forging, the ferrite, pearlite, bainite, and martensite fields are traversed and the final microstructure contains all these constituents.

At cooling rates of 1167–30,000°C/hr, the microstructure is free of proeutectoid ferrite and consists largely of bainite and a small amount of martensite. A cooling rate of at least 30,000°C/hr is necessary to obtain the fully martensitic structure desired as a starting point for tempered martensite.

Thus, the final microstructure, and therefore the properties of the steel, depend upon the transformation behavior of the austenite and the cooling conditions, and can be predicted if these factors are known.

11.3 HARDENABILITY

Hardenability refers to the depth of hardening or to the size of a piece that can be hardened under given cooling conditions, and not to the maximum hardness that can be obtained in a given steel.[7,8] The maximum hardness depends almost entirely upon the carbon content, whereas the hardenability (depth of hardening) is far more dependent on the alloy content and grain size of the austenite. Steels whose IT diagrams indicate a long time interval before the start of transformation to pearlite are useful when large sections are to be hardened, since if steel is to transform to bainite or martensite, it must escape any transformation to pearlite. Therefore, the steel must be cooled through the high-temperature transformation ranges at a rate rapid enough for transformation not to occur even at the nose of the IT diagram. This rate, which just permits transformation to martensite without earlier transformation at a higher temperature, is known as the critical cooling rate for martensite. It furnishes one method for expressing hardenability; for example, in the steel of Fig. 11.5, the critical cooling rate for martensite is 30,000°C/hr or 8.3°C/sec.

Although the critical cooling rate can be used to express hardenability, cooling rates ordinarily are not constant but vary during the cooling cycle. Especially when quenching in liquids, the cooling rate of steel always decreases as the steel temperature approaches that of the cooling medium. It is therefore customary to express hardenability in terms of depth of hardening in a standardized quench. The quenching condition used in this method of expression is a hypothetical one in which the surface of the piece is assumed to come instantly to the temperature of the quenching medium. This is known as an ideal quench; the diameter of a round steel bar, which is quenched to the desired microstructure, or corresponding hardness value, at the center in an ideal quench, is known as the ideal diameter for which the symbol D_I is used. The relationships between the cooling rates of the ideal quench and those of other cooling conditions are known. Thus, the hardenability values in terms of ideal diameter are used to predict the size of round or other shape that has the same cooling rate when cooled in actual quenches whose cooling severities are known. The cooling severities (usually referred to as severity of quench) which form the basis for these relationships are called H values. The H value for the ideal quench is infinity; those for some commonly used cooling conditions are given in Table 11.2.

Hardenability is most conveniently measured by a test in which a steel sample is subjected to a continuous range of cooling rates. In the end-quench or Jominy test, a round bar, 25 mm in diameter and 102 mm long, is heated to the desired austenitizing temperature and quenched in a fixture by a stream of water impinging on only one end. Hardness measurements are made on flats that are ground along the length of the bar after quenching. The results are expressed as a plot of hardness versus distance from the quenched end of the bar. The relationships between the distance from the quenched end and cooling rates in terms of ideal diameter (D_I) are known, and the hardenability can be evaluated in terms of D_I by noting the distance from the quenched end at which the hardness corresponding to the desired microstructure occurs and using this relationship to establish the corresponding cooling rate or D_I value. Published heat-flow tables or charts relate the ideal-diameter value to cooling rates in quenches or cooling conditions whose H values are known. Thus, the ideal-diameter value can be used to establish the size of a piece in which the desired microstructure can be obtained under the quenching conditions of the heat treatment to be used. The hardenability of steel is such an important property that it has become common practice to purchase steels to specified hardenability limits. Such steels are called H steels.

Table 11.2 H Values Designating Severity of Quench for Commonly Used Cooling Conditions[57]

Degree of agitation of medium	Quenching medium		
	Oil	Water	Brine
none	0.25–0.30	0.9–1.0	2
mild	0.30–0.35	1.0–1.1	2.0–2.2
moderate	0.35–0.40	1.2–1.3	
good	0.40–0.50	1.4–1.5	
strong	0.50–0.80	1.6–2.0	
violent	0.80–1.1	4.0	5.0

[a] H values are proportional to the heat-extracting capacity of the medium.

11.4 HEAT-TREATING PROCESSES

In heat-treating processes, steel is usually heated above the A_3 point and then cooled at a rate that results in the microstructure that gives the desired properties.[9,10]

11.4.1 Austenitization

The steel is first heated above the temperature at which austenite is formed. The actual austenitizing temperature should be high enough to dissolve the carbides completely and take advantage of the hardening effects of the alloying elements. In some cases, such as tool steels or high-carbon steels, undissolved carbides may be retained for wear resistance. The temperature should not be high enough to produce pronounced grain growth. The piece should be heated long enough for complete solution; for low-alloy steels in a normally loaded furnace, 1.8 min/mm of diameter or thickness usually suffices.

Excessive heating rates may create high stresses, resulting in distortion or cracking. Certain types of continuous furnaces, salt baths, and radiant-heating furnaces provide very rapid heating, but preheating of the steel may be necessary to avoid distortion or cracking, and sufficient time must be allowed for uniform heating throughout. Unless special precautions are taken, heating causes scaling or oxidation, and may result in decarburization; controlled-atmosphere furnaces or salt baths minimize these effects.

11.4.2 Quenching

The primary purpose of quenching is to cool rapidly enough to suppress all transformation at temperatures above the M_s temperature. The cooling rate required depends on the size of the piece and the hardenability of the steel. The preferred quenching media are water, oils, and brine. The temperature gradients set up by quenching create high thermal and transformational stresses which may lead to cracking and distortion; a quenching rate no faster than necessary should be employed to minimize these stresses. Agitation of the cooling medium accelerates cooling and improves uniformity. Cooling should be long enough to permit complete transformation to martensite. Then, in order to minimize cracking from quenching stresses, the article should be transferred immediately to the tempering furnace (Fig. 11.6).

11.4.3 Tempering

Quenching forms very hard, brittle martensite with high residual stresses. Tempering relieves these stresses and improves ductility, although at some expense of strength and hardness. The operation consists of heating at temperatures below the lower critical temperature (A_1).

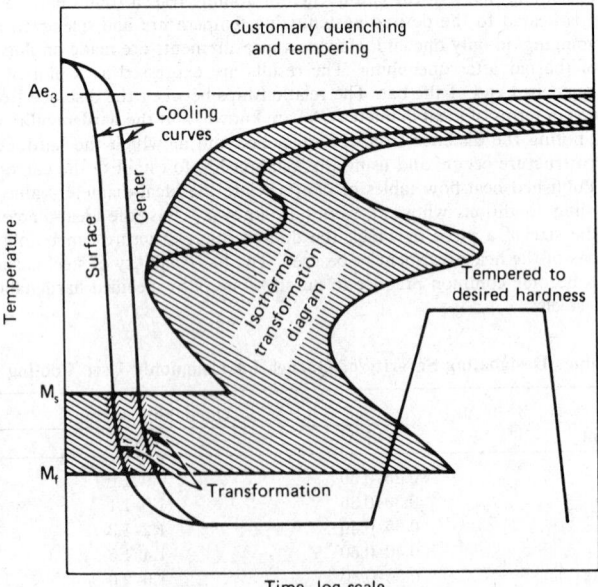

Fig. 11.6 Transformation diagram for quenching and tempering martensite; the product is tempered martensite (from Ref. 1).

Measurements of stress relaxation on tempering indicate that, in a plain carbon steel, residual stresses are significantly lowered by heating to temperatures as low as 150°C, but that temperatures of 480°C and above are required to reduce these stresses to very low values. The times and temperatures required for stress relief depend on the high-temperature yield strength of the steel, since stress relief results from the localized plastic flow that occurs when the steel is heated to a temperature at which yield strength decreases. This phenomenon may be affected markedly by composition, and particularly by alloy additions. The toughness of quenched steel, as measured by the notch impact test, first increases on tempering up to 200°C, then decreases on tempering between 200 and 310°C, and finally increases rapidly on tempering at 425°C and above. This behavior is characteristic and, in general, temperatures of 230–310°C should be avoided.

In order to minimize cracking, tempering should follow quenching immediately. Any appreciable delay may promote cracking.

The tempering of martensite results in a contraction, and if the heating is not uniform, stresses result. Similarly, heating too rapidly may be dangerous because of the sharp temperature gradient set up between the surface and the interior. Recirculating-air furnaces can be used to obtain uniform heating. Oil or salt baths are commonly used for low-temperature tempering; lead or salt baths are used at higher temperatures.

Some steels lose toughness on slow cooling from ~540°C and above, a phenomenon known as temper brittleness; rapid cooling after tempering is desirable in these cases.

11.4.4 Martempering

A modified quenching procedure known as martempering minimizes the high stresses created by the transformation to martensite during the rapid cooling characteristic of ordinary quenching (see Fig. 11.7). In practice, it is ordinarily carried out by quenching in a molten-salt bath just above the M_s temperature. Transformation to martensite does not begin until the piece reaches the temperature of the salt bath and is removed to cool relatively slowly in air. Since the temperature gradient characteristic of conventional quenching is absent, the stresses produced by the transformation are much lower and a greater freedom from distortion and cracking is obtained. After martempering, the piece may be tempered to the desired strength.

11.4.5 Austempering

As discussed earlier, lower bainite is generally as strong as and somewhat more ductile than tempered martensite. Austempering, which is an isothermal heat treatment that results in lower bainite, offers an alternative heat treatment for obtaining optimum strength and ductility.

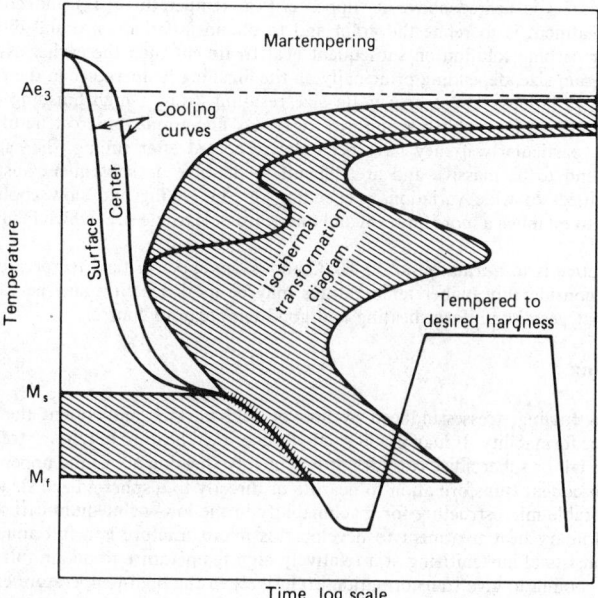

Fig. 11.7 Transformation diagram for martempering; the product is tempered martensite (from Ref. 57).

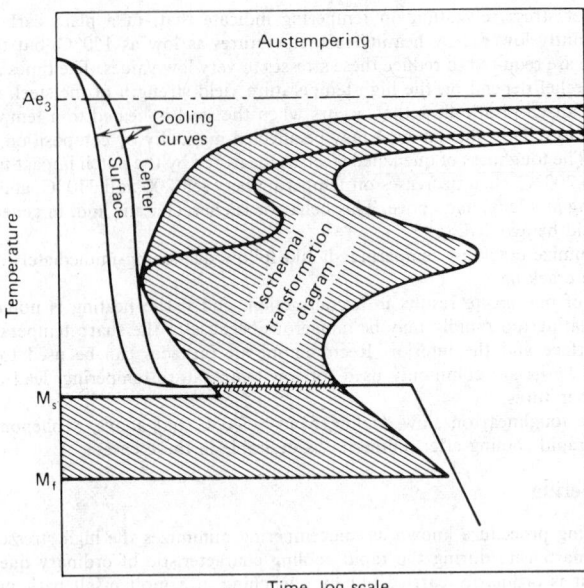

Fig. 11.8 Transformation diagram for austempering; the product is bainite (from Ref. 57).

In austempering the article is quenched to the desired temperature in the lower bainite region, usually in molten salt, and kept at this temperature until transformation is complete (see Fig. 11.8). Usually, it is held twice as long as the period indicated by the IT diagram. The article may be quenched or air cooled to room temperature after transformation is complete, and may be tempered to lower hardness if desired.

11.4.6 Normalizing

In this operation, steel is heated above its upper critical temperature (A_3) and cooled in air. The purpose of this treatment is to refine the grain and to obtain a carbide size and distribution that is more favorable for carbide solution on subsequent heat treatment than the earlier as-rolled structure.

The as-rolled grain size, depending principally on the finishing temperature in the rolling operation, is subject to wide variations. The coarse grain size resulting from a high finishing temperature can be refined by normalizing to establish a uniform, relatively fine-grained microstructure.

In alloy steels, particularly if they have been slowly cooled after rolling, the carbides in the as-rolled condition tend to be massive and are difficult to dissolve on subsequent austenitization. The carbide size is subject to wide variations, depending on the rolling and slow cooling. Here again, normalizing tends to establish a more uniform and finer carbide particle size, which facilitates subsequent heat treatment.

The usual practice is to normalize at 50–80°C above the upper critical temperature; however, for some alloy steels considerably higher temperatures may be used. Heating may be carried out in any type of furnace that permits uniform heating and good temperature control.

11.4.7 Annealing

Annealing relieves cooling stresses induced by hot- or cold-working and softens the steel to improve its machinability or formability. It may involve only a subcritical heating to relieve stresses, recrystallize cold-worked material, or spheroidize carbides; it may involve heating above the upper critical temperature (A_3) with subsequent transformation to pearlite or directly to a spheroidized structure on cooling.

The most favorable microstructure for machinability in the low- or medium-carbon steels is coarse pearlite. The customary heat treatment to develop this microstructure is a full annealing, illustrated in Fig. 11.9. It consists of austenitizing at a relatively high temperature to obtain full carbide solution, followed by slow cooling to give transformation exclusively in the high-temperature end of the pearlite range. This simple heat treatment is reliable for most steels. It is, however, rather time consuming since it involves slow cooling over the entire temperature range from the austenitizing temperature to a temperature well below that at which transformation is complete.

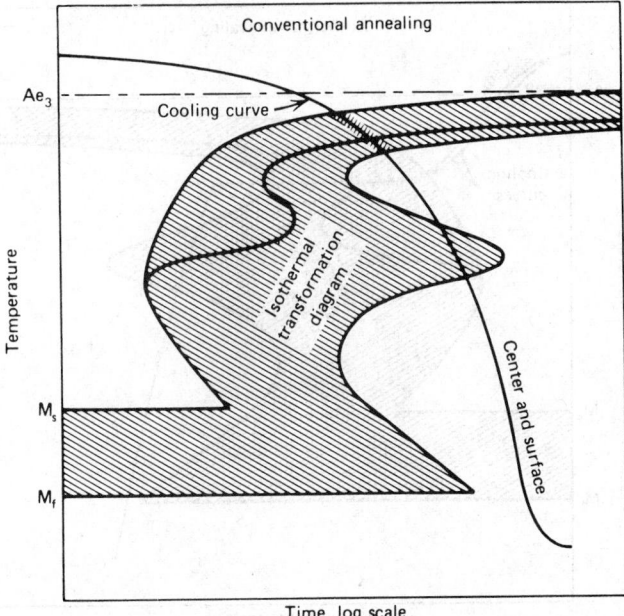

Fig. 11.9 Transformation diagram for full annealing; the product is ferrite and pearlite (from Ref. 57).

11.4.8 Isothermal Annealing

Annealing to coarse pearlite can be carried out isothermally by cooling to the proper temperature for transformation to coarse pearlite and holding until transformation is complete. This method, called isothermal annealing, is illustrated in Fig. 11.10. It may save considerable time over the full-annealing process described previously, since neither the time from the austenitizing temperature to the transformation temperature, nor from the transformation temperature to room temperature, is critical; these may be shortened as desired. If extreme softness of the coarsest pearlite is not necessary, the transformation may be carried out at the nose of the IT curve, where the transformation is completed rapidly and the operation further expedited; the pearlite in this case is much finer and harder.

Isothermal annealing can be conveniently adapted to continuous annealing, usually in specially designed furnaces, when it is commonly referred to as cycle annealing.

11.4.9 Spheroidization Annealing

Coarse pearlite microstructures are too hard for optimum machinability in the higher carbon steels. Such steels are customarily annealed to develop spheroidized microstructures by tempering the as-rolled, slowly cooled, or normalized materials just below the lower critical temperature range. Such an operation is known as subcritical annealing. Full spheroidization may require long holding times at the subcritical temperature and the method may be slow, but it is simple and may be more convenient than annealing above the critical temperature.

The annealing procedures described above to produce pearlite can, with some modifications, give spheroidized microstructures. If free carbide remains after austenitizing, transformation in the temperature range where coarse pearlite ordinarily would form proceeds to spheroidized rather than pearlite microstructures. Thus, heat treatment to form spheroidized microstructures can be carried out like heat treatment for pearlite, except for the lower austenitizing temperatures. Spheroidization annealing may thus involve a slow cooling similar to the full-annealing treatment used for pearlite, or it may be a treatment similar to isothermal annealing. An austenitizing temperature not more than 55°C above the lower critical temperature is customarily used for this supercritical annealing.

11.4.10 Process Annealing

Process annealing is the term used for subcritical annealing of cold-worked materials. It customarily involves heating at a temperature high enough to cause recrystallization of the cold-worked material

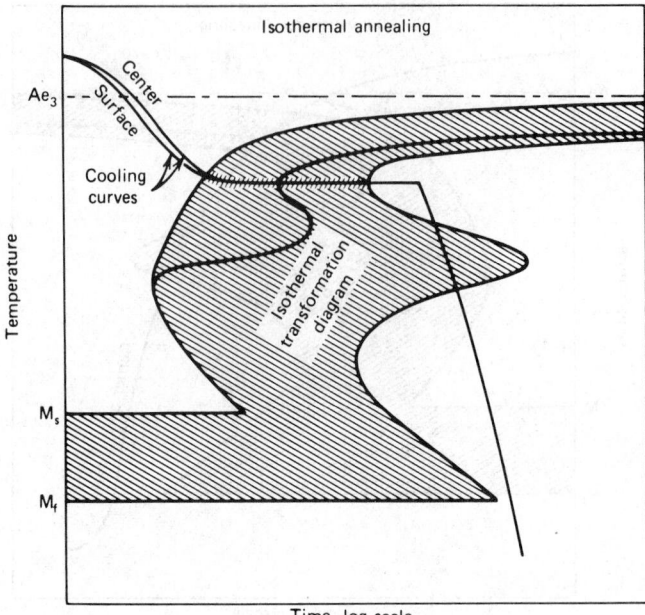

Fig. 11.10 Transformation diagram for isothermal annealing; the product is ferrite and pearlite (from Ref. 57).

and to soften the steel. The most important example of process annealing is the box annealing of cold-rolled low-carbon sheet steel. The sheets are enclosed in a large box that can be sealed to permit the use of a controlled atmosphere to prevent oxidation. Annealing is usually carried out between 590 and 700°C. The operation usually takes ~24 hr, after which the charge is cooled slowly within the box; the entire process takes ~40 hr.

11.4.11 Carburizing

In carburizing, low-carbon steel acquires a high-carbon surface layer by heating in contact with carbonaceous materials. On quenching after carburizing, the high-carbon skin hardens, whereas the low-carbon core remains comparatively soft. The result is a highly wear-resistant exterior over a very tough interior. This material is particularly suitable for gears, camshafts, etc. Carburizing is most commonly carried out by packing the steel in boxes with carbonaceous solids, sealing to exclude the atmosphere, and heating to about 925°C for a period of time depending on the depth desired; this method is called pack carburizing. Alternatively, the steel may be heated in contact with carburizing gases in which case the process is called gas carburizing; or, least commonly, in liquid baths of carburizing salts, in which case it is known as liquid carburizing.

11.4.12 Nitriding

The nitrogen case-hardening process, termed nitriding, consists of subjecting machined and (preferably) heat-treated parts to the action of a nitrogenous medium, commonly ammonia gas, under conditions whereby surface hardness is imparted without requiring any further treatment. Wear resistance, retention of hardness at high temperatures, and resistance to certain types of corrosion are also imparted by nitriding.

11.5 CARBON STEELS

The plain carbon steels represent by far the largest volume produced, with the most diverse applications of any engineering material, including castings, forgings, tubular products, plates, sheet and strip, wire and wire products, structural shapes, bars, and railway materials (rails, wheels, and axles). Carbon steels are made by all modern steelmaking processes and, depending on their carbon content and intended purpose, may be rimmed, semikilled, or fully killed.[11-15]

Table 11.3 Standard Numerical Designations of Plain Carbon and Constructional Alloy Steels (AISI–SAE Designations)[57]

Series designation[a]	Types	Series designation[a]	Types
10xx	nonresulfurized carbon-steel grades	47xx	1.05% Ni–0.45% Cr–0.20% Mo
11xx	resulfurized carbon-steel grades	48xx	3.50% Ni–0.25% Mo
12xx	rephosphorized and resulfurized carbon-steel grades	50xx	0.28 or 0.40% Cr
13xx	1.75% Mn	51xx	0.80, 0.90, 0.95, 1.00, or 1.05% Cr
23xx	3.50% Ni	5xxxx	1.00% C–0.50, 1.00, or 1.45% Cr
25xx	5.00% Ni	61xx	0.80 or 0.95% Cr–0.10 or 0.15% V
31xx	1.25% Ni–0.65% Cr	86xx	0.55% Ni–0.50 or 0.65% Cr–0.20% Mo
33xx	3.50% Ni–1.55% Cr	87xx	0.55% Ni–0.50% Cr–0.25% Mo
40xx	0.25% Mo	92xx	0.85% Mn–2.00% Si
41xx	0.50 or 0.95% Cr–0.12 or 0.20% Mo	93xx	3.25% Ni–1.20% Cr–0.12% Mo
43xx	1.80% Ni–0.50 or 0.80% Cr–0.25% Mo	98xx	1.00% Ni–0.80% Cr–0.25% Mo
46xx	1.55 or 1.80% Ni–0.20 or 0.25% Mo		

[a] The first figure indicates the class to which the steel belongs; 1xxx indicates a carbon steel, 2xxx a nickel steel, and 3xxx a nickel–chromium steel. In the case of alloy steels, the second figure generally indicates the approximate percentage of the principal alloying element. Usually, the last two or three figures (represented in the table by x) indicate the average carbon content in points or hundredths of 1 wt %. Thus, a nickel steel containing ca 3.5% nickel and 0.30% carbon would be designated as 2330.

The American Iron and Steel Institute has published standard composition ranges for plain carbon steels, which in each composition range are assigned an identifying number according to a method of classification (see Table 11.3). In this system, carbon steels are assigned to one of three series: 10xx (nonresulfurized), 11xx (resulfurized), and 12xx (rephosphorized and resulfurized). The 10xx steels are made with low phosphorus and sulfur contents, 0.04% max and 0.050% max, respectively. Sulfur in amounts as high as 0.33% max may be added to the 11xx and as high as 0.35% max to the 12xx steels to improve machinability. In addition, phosphorus up to 0.12% max may be added to the 12xx steels to increase stiffness.

In identifying a particular steel, the letters x are replaced by two digits representing average carbon content; for example, an AISI No. 1040 steel would have an average carbon content of 0.40%, with a tolerance of ±0.03%, giving a range of 0.37 to 0.43% carbon.

11.5.1 Properties

The properties of plain carbon steels are governed principally by carbon content and microstructure. The fact that properties can be controlled by heat treatment has been discussed in Section 11.1. Most plain carbon steels, however, are used without heat treatment.

The properties of plain carbon steels may be modified by residual elements other than the carbon, manganese, silicon, phosphorus, and sulfur that are always present, as well as gases, especially oxygen, nitrogen, and hydrogen, and their reaction products. These incidental elements are usually acquired from scrap, deoxidizers, or the furnace atmosphere. The gas content depends mostly on melting, deoxidizing, and pouring procedures; consequently, the properties of plain carbon steels depend heavily on the manufacturing techniques.

The average mechanical properties of as-rolled 2.5-cm bars of carbon steels as a function of carbon contents are shown in Fig. 11.11. This diagram is an illustration of the effect of carbon content when microstructure and grain size are held approximately constant.

11.5.2 Microstructure and Grain Size

The carbon steels with relatively low hardenability are predominantly pearlitic in the cast, rolled, or forged state. The constituents of the hypoeutectoid steels are, therefore, ferrite and pearlite, and of the hypereutectoid steels, cementite and pearlite. As discussed earlier, the properties of such pearlitic steels depend primarily on the interlamellar spacing of the pearlite and the grain size. Both hardness and ductility increase as the interlamellar spacing or the pearlite-transformation temperature decreases, whereas the ductility increases with decreasing grain size. The austenite-transformation behavior in

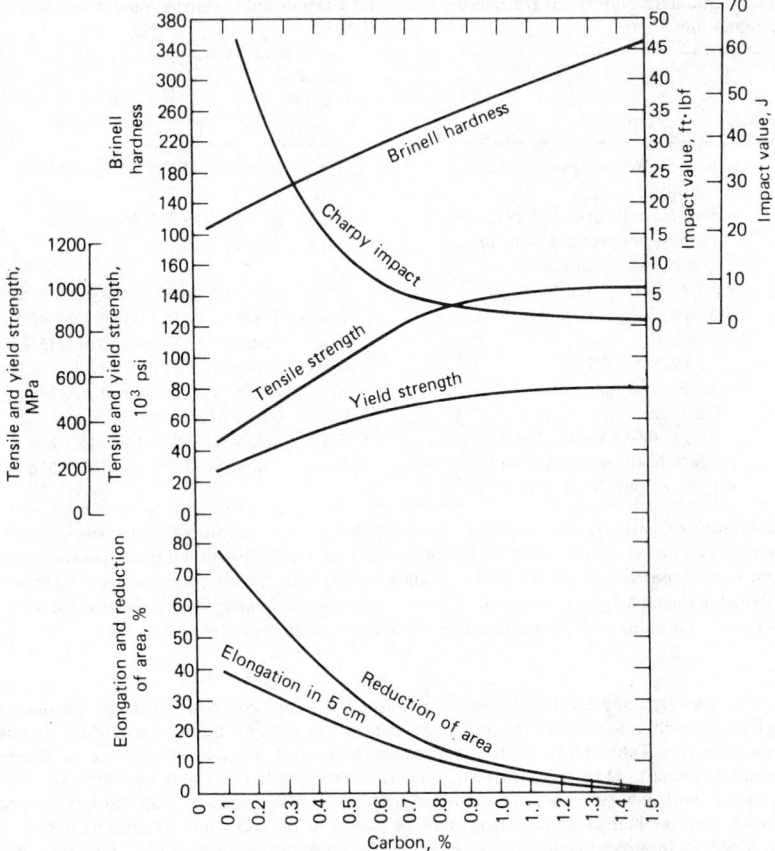

Fig. 11.11 Variations in average mechanical properties of as-rolled 2.5-cm bars of plain carbon steels, as a function of carbon content (from Ref. 1).

carbon steel is determined almost entirely by carbon and manganese content; the effects of phosphorus and sulfur are almost negligible; and the silicon content is normally so low as to have no influence. The carbon content is ordinarily chosen in accordance with the strength desired, and the manganese content selected to produce suitable microstructure and properties at that carbon level under the given cooling conditions.

11.5.3 Microstructure of Cast Steels

Cast steel is generally coarse grained, since austenite forms at high temperature and the pearlite is usually coarse, in as much as cooling through the critical range is slow, particularly if the casting is cooled in the mold. In hypoeutectoid steels, ferrite ordinarily precipitates at the original austenite grain boundaries during cooling. In hypereutectoid steels, cementite is similarly precipitated. Such mixtures of ferrite or cementite and coarse pearlite have poor strength and ductility properties, and heat treatment is usually necessary to obtain suitable microstructures and properties in cast steels.

11.5.4 Hot Working

Many carbon steels are used in the form of as-rolled finished sections. The microstructure and properties of these sections are determined largely by composition, rolling procedures, and cooling conditions. The rolling or hot working of these sections is ordinarily carried out in the temperature range in which the steel is austenitic, with four principal effects: Considerable homogenization occurs during the heating for rolling, tending to eliminate dendrite segregation present in the ingot; the dendritic structure is broken up during rolling; recrystallization takes place during rolling, with final austenitic grain size determined by the temperature at which the last passes are made (the finishing temperature); and dendrites and inclusions are reoriented, with markedly improved ductility, in the rolling direction.

Thus, homogeneity and grain size of the austenite is largely determined by the rolling technique. However, the recrystallization characteristics of the austenite and, therefore, the austenite grain size characteristic at a given finishing temperature, may be affected markedly by the steelmaking technique, particularly with regard to deoxidation.

The distribution of the ferrite or cementite and the nature of the pearlite are determined by the cooling rate after rolling. Since the usual practice is air cooling, the final microstructure and the properties of as-rolled sections depend primarily on composition and section size.

11.5.5 Cold Working

The manufacture of wire, sheet, strip, and tubular products often includes cold working, with effects that may be eliminated by annealing; however, some products, particularly wire, are used in the cold-worked condition. The most pronounced effects of cold working are increased strength and hardness and decreased ductility. The effect of cold working on the tensile strength of plain carbon steel is shown in Fig. 11.12.

Upon reheating cold-worked steel to the recrystallization temperature (400°C) or above, depending on composition, extent of cold work, and other variables, the original microstructure and properties may be restored.

11.5.6 Heat Treatment

Although most wrought (rolled or forged) carbon steels are used without a final heat treatment, it may be employed to improve the microstructure and properties for specific applications.

Annealing is applied when better machinability or formability is required than would be obtained

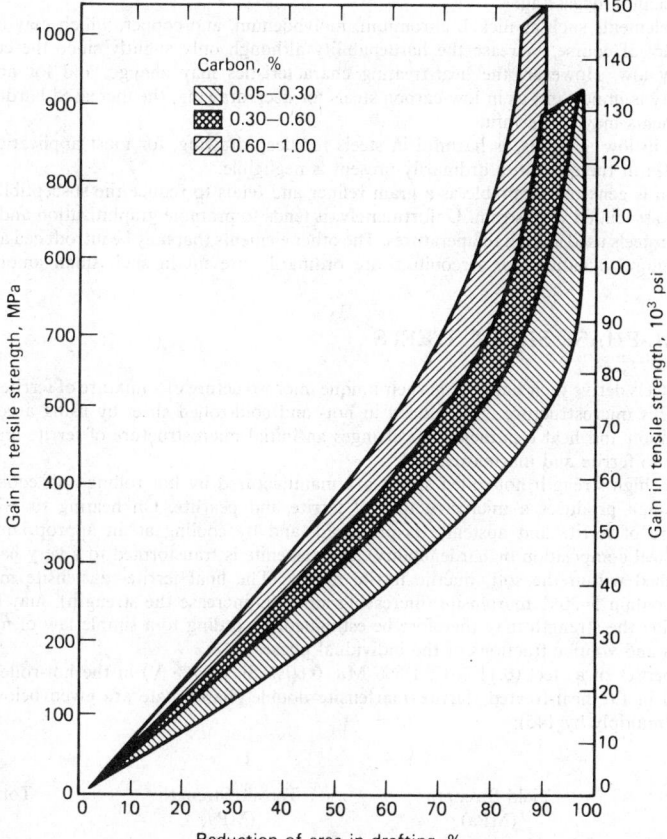

Fig. 11.12 Increase of tensile strength of plain carbon steel with increased cold working (from Ref. 57).

with the as-rolled microstructure. A complete annealing is generally employed to form coarse pearlite, although a subcritical annealing or spheroidizing treatment is occasionally used. Process annealing for optimum formability is universal with cold-rolled strip and sheet and cold-worked tubing.

The grain size of as-rolled products depends largely on the finishing temperature but is difficult to control. A final normalizing treatment from a relatively low temperature may establish a fine, uniform grain size for applications in which ductility and toughness are critical.

Quenching and tempering of plain carbon steels are being more frequently applied. In one type of treatment, the steel is heat treated to produce tempered martensite, but because of relatively low hardenability, the operation is limited to section sizes of not more than 10–13 mm. In the other type, large sections of plain carbon steels are quenched and tempered to produce fine pearlite microstructures with much better strength and ductility than those of the coarse pearlite microstructures in as-rolled or normalized products.

Thin sections of carbon steels (≤ 5 mm) are particularly suitable for the production of parts requiring toughness at high hardness by austempering.

11.5.7 Residual Elements

In addition to the carbon, manganese, phosphorus, sulfur, and silicon that are always present, carbon steels may contain small amounts of gases, such as hydrogen, oxygen, or nitrogen, introduced during the steelmaking process; nickel, copper, molybdenum, chromium, and tin, which may be present in the scrap; and aluminum, titanium, vanadium, or zirconium, which may be introduced during deoxidation.

Oxygen and nitrogen cause the phenomenon called aging, manifested as a spontaneous increase in hardness at room temperature and believed to be a precipitation effect.

An embrittling effect, the mechanism of which is not completely understood, is caused by a hydrogen content of more than ~3 ppm. As discussed earlier, the content of hydrogen and other gases can be reduced by vacuum degassing.

Alloying elements such as nickel, chromium, molybdenum, and copper, which may be introduced with scrap, do, of course, increase the hardenability although only slightly since the concentrations are ordinarily low. However, the heat-treating characteristics may change, and for applications in which ductility is important, as in low-carbon steels for deep drawing, the increased hardness imparted by these elements may be harmful.

Tin, even in low amounts, is harmful in steels for deep drawing; for most applications, however, the effect of tin in the quantities ordinarily present is negligible.

Aluminum is generally desirable as a grain refiner and tends to reduce the susceptibility of carbon steel to aging associated with strain. Unfortunately, it tends to promote graphitization and is, therefore, undesirable in steels used at high temperatures. The other elements that may be introduced as deoxidizers, such as titanium, vanadium, or zirconium, are ordinarily present in such small amounts as to be ineffective.

11.6 DUAL-PHASE SHEET STEELS

Dual-phase steels derive their name from their unique microstructure of a mixture of ferrite and martensite phases. This microstructure is developed in hot- and cold-rolled sheet by using a combination of steel composition and heat treatment that changes an initial microstructure of ferrite and pearlite (or iron carbide) to ferrite and martensite.[16-22]

Normally, high-strength hot-rolled sheets are manufactured by hot rolling and cooling on a hot-strip mill, which produces a microstructure of ferrite and pearlite. On heating to ~750–850°C, a microstructure of ferrite and austenite is produced, and by cooling at an appropriate rate (which depends on steel composition or hardenability), the austenite is transformed to a very hard martensite phase contained within the soft, ductile ferrite matrix. The final ferrite–martensite microstructure, which may contain 5–30% martensite (increasing amounts increase the strength), may be considered as a composite; the strength may therefore be estimated, according to a simple law of mixtures, from the strengths and volume fractions of the individual phases.

The properties of a steel (0.11% C, 1.6% Mn, 0.60% Si, 0.04% V) in the hot-rolled (ferrite and pearlite) and in the heat-treated (ferrite–martensite double-phase) state are given below (to convert MPa to psi, multiply by 145):

	Yield Strength (MPa)	Tensile Strength (MPa)	Total Elongation (%)
Hot rolled	480	4275	24
Dual phase	345	4516	32

Although the tensile (ultimate) strength of the steel is little affected by heat treating, the yield strength is substantially reduced and the ductility markedly improved. The low yield strength allows for the easy initiation of plastic deformation during press forming of dual-phase sheet material. However, dual-phase steels have the unique capacity to strain harden rapidly so that after a few percent deformation (3–5%) the yield strength exceeds 550 MPa (80,000 psi).

Dual-phase steels have found application in automotive bumpers and wheels where high ductility is required to form the complex shapes. The development of very high strength of ~550 MPa (80,000 psi) allows thinner, lighter weight sheet to be used, instead of steels having strengths of only 200–350 MPa (30,000–50,000 psi). However the heat-treating step increases production costs.

11.7 ALLOY STEELS

As a class, alloy steels may be defined as steels having enhanced properties owing to the presence of one or more special elements or larger proportions of elements (such as silicon and manganese) than are ordinarily present in carbon steel. Steels containing alloying elements are classified into high-strength low-alloy (HSLA) steels; AISI alloy steels; alloy tool steels; stainless steels; heat-resistant steels; and electrical steels (silicon steels). In addition, there are numerous steels, some with proprietary compositions, with exceptional properties developed to meet unusually severe requirements. The relatively small production of such steels does not reflect their engineering importance.[23,24]

11.7.1 Functions of Alloying Elements

In the broadest sense, alloy steels may contain up to ~50% of alloying elements which directly enhance properties. Examples are the increased corrosion resistance of high chromium steels, the enhanced electric properties of silicon steels, the improved strength of the HSLA steels, and the improved hardenability and tempering characteristics of the AISI alloy steels.

11.7.2 Thermomechanical Treatment

The conventional method of producing high-strength steels has been to add alloy elements such as Cr, Ni, and Mo to the liquid steel. The resulting alloy steels are often heat treated after rolling to develop the desired strength without excessive loss of toughness (resistance to cracking upon impact). In the 1970s, a less expensive method was developed to produce HSLA steels with improved toughness and yield strength ranging from 400 to 600 MPa (60,000 to 85,000 psi). In this thermomechanical treatment, the working of the steel is controlled while its temperature is changing and it is being hot rolled between 1300 and 750°C to its final thickness.[25-29] The HSLA steels that are commonly strengthened by thermomechanical treatment, also called controlled rolling, generally contain 0.05–0.20% carbon, 0.40–1.60% manganese, 0.05–0.50% silicon, plus 0.01–0.30% of one or more of the following elements: aluminum, molybdenum, niobium, titanium, and vanadium. Thermomechanical treatment usually involves a substantial degree of rolling, such as a 50–75% decrease in thickness in the last rolling passes; temperature maintained between 750 and 950°C; and a controlled rate of cooling after hot rolling. This procedure gives a very fine steel grain size and imparts strength and toughness. Steels so treated are increasingly used in automobiles and oil and gas pipelines.

11.7.3 High-Strength Low-Alloy (HSLA) Steels

HSLA steels are categorized according to mechanical properties, particularly the yield point; for example, within certain thickness limits they have yield points ranging from 310 to 450 MPa (45,000 to 65,000 psi) as compared with 225 to 250 MPa (33,000 to 36,000 psi) for structural carbon steel. This classification is in contrast to the usual classification into plain carbon or structural-carbon steels, alloy steels, and stainless steels on the basis of alloying elements.

The superior mechanical properties of HSLA steels are obtained by the addition of alloying elements (other than carbon), singly and in combination. Each steel must meet similar minimum mechanical requirements. They are available for structural use as sheets, strips, bars, and plates, and in various other shapes. They are not to be considered as special-purpose steels or requiring heat treatment.

To be of commercial interest, HSLA steels must offer economic advantages. They should be much stronger and often tougher than structural carbon steel. In addition, they must have sufficient ductility, formability, and weldability to be fabricated by customary techniques. Improved resistance to corrosion is often required. The abrasion resistance of these steels is somewhat higher than that of structural carbon steel containing 0.15–0.20% carbon. Superior mechanical properties permits the use of HSLA steels in structures with a higher unit working stress; this generally permits reduced section thickness with corresponding decrease in weight. Thus, HSLA steels may be substituted for structural carbon steel without change in section, resulting in a stronger and more durable structure without weight increase.

11.7.4 AISI Alloy Steels

The American Iron and Steel Institute defines alloy steels as follows: "By common custom steel is considered to be alloy steel when the maximum of the range given for the content of alloying elements exceeds one or more of the following limits: manganese, 1.65%; silicon, 0.60%; copper, 0.60%; or in which a definite range or a definite minimum quantity of any of the following elements is specified or required within the limits of the recognized field of constructional alloy steels: aluminum, boron, chromium up to 3.99%, cobalt, columbium (niobium), molybdenum, nickel, titanium, tungsten, vanadium, zirconium, or any other alloying element added to obtain a desired alloying effect,"[30] Steels that contain 4.00% or more of chromium are included by convention among the special types of alloy steels known as stainless steels subsequently discussed.[31-37]

Steels that fall within the AISI definition have been standardized and classified jointly by AISI and SAE as shown in Table 11.3. They represent by far the largest alloy steel production and are generally known as AISI alloy steels. They are also commonly referred to as constructional alloy steels.

The effect of the alloying elements on AISI steels is indirect since alloying elements control microstructure through their effect on hardenability. They permit the attainment of desirable microstructures and properties over a much wider range of sizes and sections than is possible with carbon steels.

11.7.5 Alloy Tool Steels

Alloy tool steels are classified roughly into three groups: Low-alloy tool steels, to which alloying elements have been added to impart hardenability higher than that of plain carbon tool steels; accordingly, they may be hardened in heavier sections or with less drastic quenches to minimize distortion; intermediate-alloy tool steels usually contain elements such as tungsten, molybdenum, or vanadium, which form hard, wear-resistant carbides; high-speed tool steels contain large amounts of carbide-forming elements that serve not only to furnish wear-resisting carbides, but also promote the phenomenon known as secondary hardening and thereby increase resistance to softening at elevated temperatures.

11.7.6 Stainless Steels

Stainless steels are more resistant to rusting and staining than plain carbon and low-alloy steels.[38-46] This superior corrosion resistance is due to the addition of chromium. Although other elements, such as copper, aluminum, silicon, nickel, and molybdenum, also increase corrosion resistance, they are limited in their usefulness.

No single nation can claim credit for the development of the stainless steels; Germany, the United Kingdom, and the United States share alike in their development. In the United Kingdom in 1912, during the search for steel that would resist fouling in gun barrels, a corrosion-resistant composition was reported of 12.8% chromium and 0.24% carbon. It was suggested that this composition be used for cutlery. In fact, the composition of AISI type 420 steel (12–14% chromium, 0.15% carbon) is similar to that of the first corrosion-resistant steel.

The higher chromium–iron alloys were developed in the United States from the early 20th century on, when the effect of chromium on oxidation resistance at 1090°C was first noticed. Oxidation resistance increased markedly as the chromium content was raised above 20%. Even now and with steels containing appreciable quantities of nickel, 20% chromium seems to be the minimum amount necessary for oxidation resistance at 1090°C.

The austenitic iron–chromium–nickel alloys were developed in Germany around 1910 in a search for materials for use in pyrometer tubes. Further work led to the versatile 18% chromium–8% nickel steels, so-called 18–8, which are widely used today.

The chromium content seems to be the controlling factor and its effect may be enhanced by additions of molybdenum, nickel, and other elements. The mechanical properties of the stainless steels, like those of the plain carbon and lower-alloy steels, are functions of structure and composition. Thus, austenitic steels possess the best impact properties at low temperatures and the highest strength at high temperatures, whereas martensitic steels are the hardest at room temperature. Thus, stainless steels, which are available in a variety of structures, exhibit a range of mechanical properties which, combined with their excellent corrosion resistance, makes these steels highly versatile from the standpoint of design.

The standard AISI and SAE types are identified in Table 11.4.

11.7.7 Martensitic Stainless Steels

Martensitic stainless steels are iron–chromium alloys that are hardenable by heat treatment. They include types 403, 410, 414, 416, 420, 431, 440A, 440B, 440C, 501, and 502 (see Table 11.4). The most widely used is type 410, containing 11.50–13.50% chromium and <0.15% carbon. In the annealed

Table 11.4 Standard Stainless and Heat-Resisting Steel Products[57]

AISI type number	SAE type[a] number	Chemical composition, %			
		Carbon	Chromium	Nickel	Other
201	30201	0.15 max	16.00–18.00	3.50–5.50	Mn 5.50–7.50[b] P 0.06 max[c] N 0.25 max
202	30202	0.15 max	17.00–19.00	4.00–6.00	Mn 7.50–10.00 P 0.06 max N 0.25 max
301	30301	0.15 max	16.00–18.00	6.00–8.00	
302	30302	0.15 max	17.00–19.00	8.00–10.00	
302B	30302B	0.15 max	17.00–19.00	8.00–10.00	Si 2.00–3.00[d] P 0.20 max S 0.15 min[e]
303	30303	0.15 max	17.00–19.00	8.00–10.00	Mo 0.60 max
303Se	30303Se	0.15 max	17.00–19.00	8.00–10.00	P 0.20 max S 0.06 max Se 0.15 min
303SeA		0.08 max	17.25–18.75	11.50–13.00	Se 0.15–0.35
304	30304	0.08 max	18.00–20.00	8.00–10.00	
304L		0.030 max	18.00–20.00	8.00–10.00	
305	30305	0.12 max	17.00–19.00	10.00–13.00	
307		0.07–0.15	19.50–21.50	9.00–10.50	Mo residual only
308	30308	0.08 max	19.00–21.00	10.00–12.00	
308 Mod		0.07–0.15	19.50–21.50	9.00–10.50	Mo residual only
309	30309	0.20 max	22.00–24.00	12.00–15.00	
309S	30309S	0.08 max	22.00–24.00	12.00–15.00	
309SCb		0.08 max	22.00–24.00	12.00–15.00	NbTa min, 10 times carbon Ta 0.10 max
309SCbTa		0.08 max	22.00–24.00	12.00–15.00	NbTa min, 10 times carbon
310	30310	0.25 max	24.00–26.00	19.00–22.00	
314	30314	0.25 max	23.00–26.00	19.00–22.00	
316	30316	0.08 max	16.00–18.00	10.00–14.00	Mo 2.00–3.00
316L	30316L	0.030 max	16.00–18.00	10.00–14.00	Mo 2.00–3.00
317	30317	0.08 max	18.00–20.00	11.00–15.00	Mo 3.00–4.00
318		0.10 max	16.00–18.00	10.00–14.00	Mo 2.00–3.00 NbTa min, 10 times carbon
D319		0.07 max	17.50–19.50	11.00–15.00	Mn 2.00 max Si 1.00 max Mo 2.25–3.00
321	30321	0.08 max	17.00–19.00	9.00–12.00	Ti min, 5 times carbon
330		0.25 max	14.00–16.00	33.00–36.00	
347	30347	0.08 max	17.00–19.00	9.00–13.00	NbTa min, 10 times carbon
348	30348	0.08 max	17.00–19.00	9.00–13.00	NbTa min, 10 times carbon Ta 0.10 max Co 0.20 max
403	51403	0.15 max	11.50–13.00		
405	51405	0.08 max	11.50–14.50		Al 0.10–0.30
410	51410	0.15 max	11.50–13.50		
410Mo		0.15 max	11.50–13.50		Mo 0.40–0.60
414	51414	0.15 max	11.50–13.50	1.25–2.50	
416	51416	0.15 max	12.00–14.00		P 0.06 max S 0.15 min Mo 0.60 max

Table 11.4 (Continued)

AISI type number	SAE type[a] number	Chemical composition, %			
		Carbon	Chromium	Nickel	Other
410Se	51410Se	0.15 max	12.00–14.00		P 0.06 max S 0.06 max Se 0.15 min
420	51420	>0.15	12.00–14.00		
420F	51420F	>0.15	12.00–14.00		S[f]
430	51430	0.12 max	14.00–18.00		
430F	51430F	0.12 max	14.00–18.00		P 0.06 max S 0.15 min Mo 0.60 max
430Ti		0.10 max	16.00–18.00		Ti 0.30–0.70
431	51431	0.20 max	15.00–17.00	1.25–2.50	
434A		0.05–0.10	15.00–17.00		Cu 0.75–1.10
442	51442	0.25 max	18.00–23.00		
446	51446	0.20 max	23.00–27.00		N 0.25 max
501	51501	>0.10	4.00–6.00		Mo 0.40–0.65
502	51502	0.10 max	4.00–6.00		Mo 0.40–0.65

[a] SAE chemical composition (ladle) ranges may differ slightly in certain elements from AISI limits.

[b] Manganese: All steels of AISI Type 300 series—2.00% max. All steels of AISI Type 400 and 500 series—1.00% max except 416, 416Se, 430F, and 430Se (1.25% max) and Type 446 (1.50% max).

[c] Phosphorus: All steels of AISI Type 200 series—0.060% max. All steels of AISI Type 300 series—0.045% max except Types 303 and 303Se (0.20% max). All steels of AISI Type 400 and 500 series—0.040% max except Types 416, 416Se, 430F, and 430FSe (0.060% max).

[d] Silicon: All steels of AISI Type 200, 300, 400, and 500 series—1.00% max except where otherwise indicated.

[e] Sulfur: All steels of AISI Type 200, 300, 400 and 500 series—0.30% max except Types 303, 416, and 430F (0.15% min) and Types 303Se, 416Se, and 430FSe (0.060% max).

[f] No restriction.

condition, this grade may be drawn or formed. It is an air-hardening steel, affording a wide range of properties by heat treatment. In sheet or strip form, type 410 is used extensively in the petroleum industry for ballast trays and liners. It is also used for parts of furnaces operating below 650°C, and for blades and buckets in steam turbines.

Type 420, with ~0.35% carbon and a resultant increased hardness, is used for cutlery. In bar form, it is used for valves, valve stems, valve seats, and shafting where corrosion and wear resistance are needed. Type 440 may be employed for surgical instruments, especially those requiring a durable cutting edge. The necessary hardness for different applications can be obtained by selecting grade A, B, or C, with increasing carbon content in that order.

Other martensitic grades are types 501 and 502, the former with >0.10% and the latter <0.10% carbon; both contain 4.6% chromium. These grades are also air hardening, but do not have the corrosion resistance of the 12% chromium grades. Types 501 and 502 have wide application in the petroleum industry for hot lines, bubble towers, valves, and plates.

11.7.8 Ferrite Stainless Steels

These steels are iron–chromium alloys that are largely ferritic and not hardenable by heat treatment (ignoring the 475°C embrittlement). They include types 405, 430, 430F, and 446 (see Table 11.4).

The most common ferritic grade is type 430, containing 0.12% carbon or less and 14–18% chromium. Because of its higher chromium content, the corrosion resistance of type 430 is superior to that of the martensitic grades. Furthermore, type 430 may be drawn, formed, and, with proper techniques, welded. It is widely used for automotive and architectural trim. It is employed in equipment for the manufacture and handling of nitric acid to which it is resistant. Type 430 does not have high creep strength but is suitable for some types of service up to 815°C and thus has application in combustion chambers for domestic heating furnaces.

The high chromium content of type 446 (23–27% chromium) imparts excellent heat resistance, although its high-temperature strength is only slightly better than that of carbon steel. Type 446 is used in sheet or strip form up to 1150°C. This grade does not have the good drawing characteristics of type 430, but it may be formed. Accordingly, it is widely used for furnace parts such as muffles, burner sleeves, and annealing baskets. Its resistance to nitric and other oxidizing acids makes it suitable for chemical-processing equipment.

11.7.9 Austenitic Stainless Steels

These steels are iron–chromium–nickel alloys not hardenable by heat treatment and predominantly austenitic. They include types 301, 302, 302B, 303, 304, 304L, 305, 308, 309, 310, 314, 316, 316L, 317, 321, and 347. In some recently developed austenitic stainless steels, all or part of the nickel is replaced by manganese and nitrogen in proper amounts, as in one proprietary steel and types 201 and 202 (see Table 11.4).

The most widely used austenitic stainless steel is type 302, known as 18–8; it has excellent corrosion resistance and, because of its austenitic structure, excellent ductility. It may be deep drawn or strongly formed. It can be readily welded, but carbide precipitation must be avoided in and near the weld by cooling rapidly enough after welding. Where carbide precipitation presents problems, types 321, 347, or 304L may be used. The applications of type 302 are wide and varied, including kitchen equipment and utensils; dairy installations; transportation equipment; and oil-, chemical-, paper-, and food-processing machinery.

The low nickel content of type 301 causes it to harden faster than type 302 because of reduced austenite stability. Accordingly, although type 301 can be drawn successfully, its drawing properties are not as good as those of type 302. For the same reason, type 301 can be cold rolled to very high strength.

Type 301, because of its lower carbon content, is not as prone as type 302 to give carbide precipitation problems in welding. In addition, its somewhat higher chromium content makes it slightly more resistant to corrosion. It is used to withstand severe corrosive conditions in the paper, chemical, and other industries.

The austenitic stainless steels are widely used for high-temperature service.

Types 321 and 347, with additions of titanium and niobium, respectively, are used in welding applications and high-temperature service under corrosive conditions. Type 304L may be used as an alternative for types 321 and 347 in welding and stress-relieving applications below 426°C.

The addition of 2–4% molybdenum to the basic 18–8 composition produces types 316 and 317 with improved corrosion resistance. These grades are employed in the textile, paper, and chemical industries where strong sulfates, chlorides, and phosphates and reducing acids such as sulfuric, sulfurous, acetic, and hydrochloric acids are used in such concentrations that the use of corrosion-resistant alloys is mandatory. Types 316 and 317 have the highest creep and rupture strengths of any commercial stainless steels.

The austenitic stainless steels most resistant to oxidation are types 309 and 310. Because of their high chromium and nickel contents, these steels resist scaling at temperatures up to 1090 and 1150°C and, consequently, are used for furnace parts and heat exchangers. They are somewhat harder and not as ductile as the 18–8 types, but they may be drawn and formed. They can be welded readily and have increasing use in the manufacture of jet-propulsion motors and industrial-furnace equipment.

For applications requiring good machinability, type 303 containing sulfur or selenium may be used.

11.7.10 High-Temperature Service, Heat-Resisting Steels

The term high-temperature service comprises many types of operations in many industries. Conventional high-temperature equipment includes steam boilers and turbines, gas turbines, cracking stills, tar stills, hydrogenation vessels, heat-treating furnaces, and fittings for diesel and other internal-combustion engines. Numerous steels are available from which to select the one best suited for each of the foregoing applications. Where unusual conditions occur, modification of the chemical composition may adapt an existing steel grade to service conditions. In some cases, however, entirely new alloy combinations must be developed to meet service requirements. For example, the aircraft and missile industries have encountered design problems of increased complexity, requiring metals of great high-temperature strength for both power plants and structures, and new steels are constantly under development to meet these requirements.[47,48]

A number of steels suitable for high-temperature service are given in Table 11.5.

The design of load-bearing structures for service at room temperature is generally based on the yield strength or for some applications on the tensile strength. The metal behaves essentially in an elastic manner, that is, the structure undergoes an elastic deformation immediately upon load application and no further deformation occurs with time; when the load is removed, the structure returns to its original dimensions.

At high temperature, the behavior is different. A structure designed according to the principles employed for room-temperature service continues to deform with time after load application, even

Table 11.5 Alloy Composition of High-Temperature Steels[57]

Ferritic steels	Austenitic steels	AISI type
0.5% Mo	18% Cr–8% Ni	304
0.5% Cr–0.5% Mo	18% Cr–8% Ni with Mo	316
1% Cr–0.5% Mo	18% Cr–8% Ni with Ti	321
2% Cr–0.5% Mo	18% Cr–8% Ni with Nb	347
2.25% Cr–1% Mo	25% Cr–12% Ni	309
3% Cr–0.5% Mo–1.5% Si	25% Cr–20% Ni	310
5% Cr–0.5% Mo–1.5% Si		
5% Cr–0.5% Mo, with Nb added		
5% Cr–0.5% Mo, with Ti added		
9% Cr–1% Mo		
12% Cr		410
17% Cr		430
27% Cr		446

though the design data may have been based on tension tests at the temperature of interest. This deformation with time is called creep, since at the design stresses at which it first was recognized it occurred at a relatively low rate.

In spite of the fact that plain carbon steel has lower resistance to creep than high-temperature alloy steels, it is widely used in such applications up to 540°C, where rapid oxidation commences and a chromium-bearing steel must be employed. Low-alloy steels containing small amounts of chromium and molybdenum have higher creep strengths than carbon steel and are employed where materials of higher strength are needed. Above ~540°C, the amount of chromium required to impart oxidation resistance increases rapidly. The 2% chromium steels containing molybdenum are useful up to ~620°C, whereas 10–14% chromium steels may be employed up to ~700–760°C. Above this temperature, the austenitic 18–8 steels are commonly used; their oxidation resistance is considered adequate up to ~815°C. For service between 815 and 1090°C, steels containing 25% chromium and 20% nickel, or 27% chromium are used.

The behavior of steels at high temperature is quite complex, and only a few design considerations have been mentioned here.

11.7.11 Quenched and Tempered Low-Carbon Constructional Alloy Steels

A class of quenched and tempered low-carbon constructional alloy steels has been very extensively used in a wide variety of applications such as pressure vessels, mining and earth-moving equipment, and large steel structures.[49-51]

As a general class, these steels are referred to as low-carbon martensites to differentiate them from constructional alloy steels of higher carbon content, such as AISI alloy steels, that develop high-carbon martensite upon quenching. They are characterized by a relatively high strength, with minimum yield strengths of 690 MPa (100,000 psi), toughness down to −45°C, and weldability with joints showing full joint efficiency when welded with low-hydrogen electrodes. They are most commonly used in the form of plates, but also sheet products, bars, structural shapes, forgings, or semifinished products.

Several steel-producing companies manufacture such steels under various tradenames; their compositions are proprietary.

11.7.12 Maraging Steels

A group of high-nickel martensitic steels called maraging steels contain so little carbon that they are referred to as carbon-free iron–nickel martensites.[52,53]

Iron–carbon martensite is hard and brittle in the as-quenched condition and becomes softer and more ductile when tempered. Carbon-free iron–nickel martensite, on the other hand, is relatively soft and ductile and becomes hard, strong, and tough when subjected to an aging treatment at 480°C.

The first iron–nickel martensitic alloys contained ~0.01% carbon, 20 or 25% nickel, and 1.5–2.5% aluminum and titanium. Later an 18% nickel steel containing cobalt, molybdenum, and titanium was developed, and still more recently a series of 12% nickel steels containing chromium and molybdenum came on the market.

By adjusting the content of cobalt, molybdenum, and titanium, the 18% nickel steel can attain yield strengths of 1380–2070 MPa (200,000–300,000 psi) after the aging treatment. Similarly, yield strengths of 12% nickel steel in the range of 1035–1380 MPa (150,000–200,000 psi) can be developed by adjusting its composition.

11.7.13 Silicon-Steel Electrical Sheets

The silicon steels are characterized by relatively high permeability, high electrical resistance, and low hysteresis loss when used in magnetic circuits. First patented in the United Kingdom around 1900, the silicon steels permitted the development of more powerful electrical equipment and have furthered the rapid growth of the electrical power industry. Steels containing 0.5–5% silicon are produced in sheet form for the laminated magnetic cores of electrical equipment and are referred to as electrical sheets.[54-56]

The grain-oriented steels, containing ~3.25% silicon, are used in the highest efficiency distribution and power transformers and in large turbine generators. They are processed in a special way to give them directional properties related to orientation of the crystals making up the structure of the steel in a preferred direction.

The nonoriented steels are subdivided into low-silicon steels, containing ~0.5–1.5% silicon, used mainly in rotors and stators of motors and generators. Steels containing ~1% silicon are used for reactors, relays, and small intermittent-duty transformers.

Intermediate-silicon steels (2.5–3.5% Si) are used in motors and generators of average to high efficiency and in small- to medium-size intermittent-duty transformers, reactors, and motors.

High-silicon steels (~3.75–5.00% Si) are used in power transformers and high-efficiency motors, generators, and transformers, and in communications equipment.

REFERENCES

1. G. L. Kehl, *The Principles of Metallographic Laboratory Practice,* McGraw-Hill, New York, 1949.

2. *Applications of Modern Metallographic Techniques,* STP 480, American Society for Testing Materials, Philadelphia, PA, 1970.

3. W. C. Leslie, *The Physical Metallurgy of Steels,* Hemisphere Publishing, McGraw-Hill, New York, 1981.

4. E. C. Bain and H. W. Paxton, *Alloying Elements in Steel,* American Society for Metals, Metals Park, OH, 1961.

5. *Heat Treatment '79,* Metals Society Publication 261, Metals Society, London, 1980.

6. K. E. Thelning, *Steel and Its Heat Treatment: Bofors Handbook,* Butterworths, Boston, 1975.

7. D. V. Doane and J. S. Kirkaldy, *Hardenability Concepts with Applications to Steel,* The Metallurgical Society–AIME, Warrendale, PA, 1978.

8. C. A. Siebert, D. V. Doane, and D. H. Breen, *The Hardenability of Steels,* American Society for Metals, Metals Park, OH, 1977.

9. G. Krauss, *Principles of Heat Treatment of Steel,* American Society for Metals, Metals Park, OH, 1980.

10. M. Atkins, *Atlas of Continuous Cooling Transformation Diagrams for Engineering Steels,* British Steel Corp., Sheffield, UK, 1978.

11. J. S. Blair, *The Profitable Way: Carbon Sheet Steel Specifying and Purchasing Handbook,* General Electric Technology Marketing, Schenectady, NY, 1978.

12. J. S. Blair, *The Profitable Way: Carbon Strip Steel Specifying and Purchasing Handbook,* General Electric Technology Marketing, Schenectady, NY, 1978.

13. J. D. Jevons, *The Metallurgy of Deep Drawing and Pressing,* Wiley, New York, 1942.

14. *Low Carbon Structural Steels for the Eighties,* Institution of Metallurgists, London, 1977.

15. J. S. Blair, *The Profitable Way: Carbon Plate Steel Specifying and Purchasing Handbook,* General Electric Technology Marketing, Schenectady, NY, 1978.

16. R. A. Kot and J. W. Morris (eds.), *Structure and Properties of Highly Formable Dual-Phase Steels,* The Metallurgical Society–AIME, Warrendale, PA, 1979.

17. A. T. Davenport (ed.), *Formable HSLA and Dual-Phase Steels,* The Metallurgical Society–AIME, Warrendale, PA, 1979.

18. A. B. Rothwell and J. M. Gray (eds.), *Welding of HSLA (Microalloyed) Structural Steels,* American Society for Metals, Metals Park, OH, 1978.

19. R. A. Kot and B. L. Bramfitt, *Fundamentals of Dual-Phase Steels,* The Metallurgical Society–AIME, Warrendale, PA, 1981.

20. P. E. Repas, *Iron Steelmaker* 7, 12 (1980).

21. M. D. Baughman, K. L. Fetters, G. Perrault, Jr., and K. Toda, *Iron Steel Eng.* 56, 52 (1979).

22. A. P. Coldren and G. T. Eldis, *J. Met.* 32, 41 (1980).

23. B. P. Bardes (ed.), *Metals Handbook,* 9th ed., American Society for Metals, Metals Park, OH, 1978, Vol. 1, p. 127.

24. *Nickel Alloy Steels Data Book,* International Nickel Co., Inc., New York, 1967.

25. *Micro Alloying 75: Proceedings of an International Symposium on High-Strength Low-Alloy Steels,* Union Carbide Corp., New York, 1977.

26. F. B. Pickering, *Physical Metallurgy and the Design of Steels,* Applied Science Publishers, London, 1978.

27. G. R. Speich and D. S. Dabkowski, in *The Hot Deformation of Austenite,* J. B. Ballance (ed.), The Metallurgical Society–AIME, Warrendale, PA, 1977.

28. *Iron Age* **214,** MP9 (Dec. 9, 1974).

29. A. T. Davenport, D. R. DiMicco, and D. W. Dickinson, *J. Met.* **32,** 28 (1980).

30. *Steel Products Manual; Strip Steel,* American Iron and Steel Institute, Washington, DC, 1978.

31. *Alloy Cross Index,* Mechanical Properties Data Center, Battelle's Columbus Laboratories, Columbus, OH, 1981.

32. P. M. Unterweiser, *Worldwide Guide to Equivalent Irons and Steels,* American Society for Metals, Metals Park, OH, 1979.

33. *Unified Numbering System for Metals and Alloys,* Society of Automotive Engineers, Warrendale, PA, 1977.

34. *Handbook of Comparative World Steel Standards,* International Technical Information Institute, Tokyo, Japan, 1980.

35. R. B. Ross, *Metallic Materials Specification Handbook,* E. and F. N. Spon Ltd., New York, 1980.

36. C. W. Wegst, *Key to Steel (Stahlschluessel),* Verlag Stahlschluessel Wegst KG, Marbach/Neckar, Federal Republic of Germany, 1974.

37. M. J. Wahll and R. F. Frontani, *Handbook of Soviet Alloy Compositions,* Metals and Ceramics Information Center, Battelle's Columbus Laboratories, Columbus, OH, 1976.

38. *Source Book on Stainless Steels,* American Society for Metals, Metals Park, OH, 1976.

39. K. G. Brickner and co-workers, *Selection of Stainless Steels,* American Society for Metals, Metals Park, OH, 1968.

40. D. Peckner and I. M. Bernstein, *Handbook of Stainless Steels,* McGraw-Hill, New York, 1977.

41. W. F. Simmons and R. B. Gunia, *Compilation and Index of Trade Nanes, Specifications, and Producers of Stainless Alloys and Superalloys,* Data Series, DS45A, American Society for Testing Materials, Philadelphia, PA, 1972.

42. G. E. Rowan and co-workers, *Forming of Stainless Steels,* American Society for Metals, Metals Park, OH, 1968.

43. R. A. Lula (ed.), *Toughness of Ferritic Stainless Steels,* STP 706, American Society for Testing Materials, Philadelphia, PA, 1980.

44. C. R. Brinkman and H. W. Garvin (eds.), *Properties of Austenitic Stainless Steels and Their Weld Metals,* STP 679, American Society for Testing Materials, Philadelphia, PA, 1978.

45. J. J. Demo (ed.), *Structure, Constitution, and General Characteristics of Wrought Ferritic Stainless Steels,* STP 619, American Society for Testing Materials, Philadelphia, PA, 1977.

46. F. B. Pickering (ed.), *The Metallurgical Evolution of Stainless Steels,* American Society for Metals, Metals Park, OH, 1979.

47. E. F. Bradley (ed.), *Source Book on Materials for Elevated-Temperature Application,* American Society for Metals, Metals Park, OH, 1979.

48. G. V. Smith (ed.), *Ductility and Toughness Considerations in Elevated Temperature Service,* American Society of Mechanical Engineers, New York, 1978.

49. R. L. Brockenbrough and B. G. Johnston, *USS Steel Design Manual,* ADUSS 27-3400-03, U.S. Steel Corp., Pittsburgh, PA, 1974.

50. J. H. Gross, *Trans. ASME; J. Pressure Vessel Technology* **96,** 9 (1974).

51. *Annual Book of ASTM Standards, Part 4—Steel,* American Society for Testing Materials, Philadelphia, PA, 1982, p. 465.

52. S. Floreen and G. R. Speich, *Trans. ASM* **57,** 714 (1964).

53. *Third Maraging Steel Project Review,* AD 425 299, Air Force Systems Command Technical Documentary Report No. RTD-TDR-63-4048, Nov. 1963, available from National Technical Information Service of the United States Department of Commerce, Springfield, VA.

54. A. E. DeBarr, *Soft Magnetic Materials Used in Industry,* Reinhold, New York, 1953.

55. F. Brailsford, *Magnetic Materials,* Wiley, New York, 1960.

56. R. M. Bozorth, *Ferromagnetism,* Van Nostrand, New York, 1951.

57. R. J. King, "Steel," in *Kirk–Othmer Encyclopedia of Chemical Technology,* 3rd ed., Wiley, New York, 1983, Vol. 21.

CHAPTER 12

ALUMINUM AND ITS ALLOYS

GORDAN A. ALISON
JAMES W. BARR
DOUGLAS A. CORNELL
SEYMOUR G. EPSTEIN
PETER POLLAK
W. WILLIAM PRITSKY

The Aluminum Association, Inc.
Washington, D.C.

12.1 INTRODUCTION

Aluminum is the most abundant metal and the third most abundant chemical element in the earth's crust, comprising over 8% of its weight. Only oxygen and silicon are more prevalent. Yet, until about 150 years ago aluminum in its metallic form was unknown to man. The reason for this is that aluminum, unlike iron or copper, does not exist as a metal in nature. Because of its chemical activity and its affinity for oxygen, aluminum is always found combined with other elements, mainly as aluminum oxide. As such it is found in nearly all clays and many minerals. Rubies and sapphires are aluminum oxide colored by trace impurities, and corundum, also aluminum oxide, is the second hardest naturally occurring substance on earth—only a diamond is harder.

It was not until 1886 that scientists learned how to economically extract aluminum from aluminum oxide. Yet in the 100 years since that time, aluminum has become the second most widely used of the approximately 60 naturally occurring metals, behind only iron.

12.2 PROPERTIES OF ALUMINUM

Let us consider the properties of aluminum that lead to its wide use.

One property of aluminum that everyone is familiar with is its light weight or, technically, its low specific gravity. The specific gravity of aluminum is only 2.7 times that of water, and roughly one-third that of steel or copper. An easy number to remember is that 1 in.^{3} of aluminum weighs 0.1 lb; 1 ft^3 weighs 170 lb compared to 62 lb for water and 490 lb for steel. The following are some other properties of aluminum and its alloys that will be examined in more detail in later sections:

Formability. Aluminum can be formed by every process in use today and in more ways than any other metal. Its relatively low melting point, 1220°F, while restricting high-temperature applications to about 500–600°F, does make it easy to cast, and there are over 1000 foundries casting aluminum in this country.

Mechanical Properties. Through alloying, naturally soft aluminum can attain strengths twice that of mild steel.

Strength to Weight Ratio. Some aluminum alloys are among the highest strength to weight materials in use today, in a class with titanium and superalloy steels.

Cryogenic Properties. Unlike most steels, which tend to become brittle at cryogenic temperatures, aluminum alloys actually get tougher at low temperatures and hence enjoy many cryogenic applications.

Corrosion Resistance. Aluminum possesses excellent resistance to corrosion by natural atmospheres and by many foods and chemicals.

High Electrical and Thermal Conductivity. On a volume basis the electrical conductivity of pure aluminum is roughly 60% of the International Annealed Copper Standard, but pound for pound aluminum is a better conductor of heat and electricity than copper and is surpassed only by sodium, which is a difficult metal to use in everyday situations.

Reflectivity. Aluminum can accept surface treatment to become an excellent reflector and it does not dull from normal oxidation.

Finishability. As we will also see later, aluminum can be finished in more ways than any other metal used today.

12.3 ALUMINUM ALLOYS

While commercially pure aluminum (defined as at least 99% aluminum) does find application in electrical conductors, chemical equipment, and sheet metal work, it is a relatively weak material, and its use is restricted to applications where strength is not an important factor. Some strengthening of the pure metal can be achieved through cold working, called strain hardening. However, much greater strengthening is obtained through alloying with other metals, and the alloys themselves can be further strengthened through strain hardening or heat treating. Other properties, such as castability and machinability, are also improved by alloying. Thus, aluminum alloys are much more widely used than is the pure metal, and many times when aluminum is mentioned, it is actually an aluminum alloy that is being referred to.

The principal alloying additions to aluminum are copper, manganese, silicon, magnesium, and zinc; other elements are also added in smaller amounts for metallurgical purposes. Since there have been literally hundreds of aluminum alloys developed for commercial use, the Aluminum Association formulated special designation systems to distinguish and classify the alloys in a meaningful manner.

Table 12.1 Designation System for Wrought Aluminum Alloys

Alloy Series	Description or Major Alloying Element
1xxx	99.00% minimum aluminum
2xxx	Copper
3xxx	Manganese
4xxx	Silicon
5xxx	Magnesium
6xxx	Magnesium and silicon
7xxx	Zinc
8xxx	Other element
9xxx	Unused series

Table 12.2 Designation System for Cast Aluminum Alloys

Alloy Series	Description or Major Alloying Element
1xx.x	99.00% minimum aluminum
2xx.x	Copper
3xx.x	Silicon plus copper and/or magnesium
4xx.x	Silicon
5xx.x	Magnesium
6xx.x	Unused series
7xx.x	Zinc
8xx.x	Tin
9xx.x	Other element

12.4 ALLOY DESIGNATION SYSTEMS

First the aluminum alloys are divided into two classes according to how they are formed for use: wrought and cast. The wrought category is a broad one, since aluminum is formed by virtually every known process including rolling, extruding, drawing, forging, and other special processes. Cast alloys are those that are poured molten into a sand or steel mold and are allowed to solidify to produce the desired shape. The wrought and cast alloys are quite different in composition; wrought alloys must be ductile for fabrication, while cast alloys must be fluid for castability.

About 20 years ago the Association published a designation system for wrought aluminum alloys that classifies the alloys by major alloying additions. More recently, a similar system for casting alloys was introduced.

Each wrought or cast aluminum alloy is designated by a number to distinguish it as a wrought or cast alloy and to categorize the alloy. A wrought alloy is given a four-digit number. The first digit classifies the alloy by alloy series, or principal alloying element. The second digit, if different than 0, denotes a modification in the basic alloy. The third and fourth digits form an arbitrary number which identifies the specific alloy in the series.* A cast alloy is assigned a three-digit number followed by a decimal. Here again the first digit signifies the alloy series or principal addition; the second and third digits identify the specific alloy; the decimal indicates whether the alloy composition is for the final casting (0.0) or for ingot (0.1 or 0.2). A capital letter prefix (A, B, C, etc.) indicates a modification of the basic alloy.

The designation systems for wrought and cast aluminum alloys are shown in Tables 12.1 and 12.2, respectively.

---

* An exception is for the 1xxx series alloys, where the last two digits indicate the minimum aluminum percentage. For example, alloy 1060 contains a minimum of 99.60% aluminum.

Specification of an aluminum alloy is not complete without designating the metallurgical condition, or temper, of the alloy. A temper designation system, unique for aluminum alloys, was developed by the Aluminum Association and is used for all wrought and cast alloys. The temper designation follows the alloy designation, the two being separated by a hyphen. Basic temper designations consist of letters; subdivisions, where required, are indicated by one or more digits following the letter. The basic tempers are:

F—As-Fabricated. Applies to the products of shaping processes in which no special control over thermal conditions or strain hardening is employed. For wrought products, there are no mechanical property limits.

O—Annealed. Applies to wrought products that are annealed to obtain the lowest strength temper, and to cast products that are annealed to improve ductility and dimensional stability. The O may be followed by a digit other than zero.

H—Strain-Hardened (Wrought Products Only). Applies to products that have their strength increased by strain hardening, with or without supplementary thermal treatments to produce some reduction in strength. The H is always followed by two or more digits.

W—Solution Heat Treated. An unstable temper applicable only to alloys that spontaneously age at room temperature after solution heat treatment. This designation is specific only when the period of natural aging is indicated; for example: W ½ hr.

T—Thermally Treated to Produce Stable Tempers Other than F, O, or H. Applies to products that are thermally treated, with or without supplementary strain hardening, to produce stable tempers. The T is always followed by one or more digits. (See Tables 12.3 and 12.4.)

12.5 MECHANICAL PROPERTIES OF ALUMINUM ALLOYS

Wrought alloys are divided into two categories: nonheat treatable and heat treatable. Nonheat-treatable alloys are those that derive strengths from solid solution or dispersion hardening and are further strengthened by strain hardening; they include the 1xxx, 3xxx, 4xxx, and 5xxx series alloys. Heat-treatable alloys are strengthened by controlling aging, and they include the 2xxx, some 4xxx, 6xxx, and 7xxx series alloys. Castings are not normally strain hardened but many casting alloys are heat treatable.

In Table 12.5 typical mechanical properties are shown for several representative nonheat-treatable alloys in the annealed, half-hard and full-hard tempers; values for super purity aluminum (99.99%) are included for comparison. Typical properties are usually higher than minimum, or guaranteed, properties and are not meant for design purposes but are useful for comparisons. It should be noted that pure aluminum can be substantially strain hardened, but a mere 1% alloying addition produces a comparable tensile strength to that of fully hardened pure aluminum with much greater ductility in the alloy. And the alloys can then be strain hardened to produce even greater strengths. Thus, the alloying effect is compounded. Note also that, while strain hardening increases both tensile and yield strengths, the effect is more pronounced for the yield strength so that it approaches the tensile strength in the fully hardened temper. Ductility and workability are reduced as the material is strain hardened, and most alloys have limited formability in the fully hardened tempers.

Table 12.6 lists typical mechanical properties and nominal compositions of some representative heat-treatable aluminum alloys. One can readily see that the strengthening effect of the alloying ingredients in these alloys is not reflected in the annealed condition to the same extent as in the nonheat-treatable alloys, but the true value of the additions can be seen in the aged condition. Presently, heat-treatable alloys are available with tensile strengths approaching 100,000 psi.

Table 12.3 Subdivisions of H Temper: Strain Hardened

First digit indicates basic operations:

    H1—Strain hardened only

    H2—Strain hardened and partially annealed

    H3—Strain hardened and stabilized

Second digit indicates degree of strain hardening:

    HX2—Quarter hard

    HX4—Half hard

    HX8—Full hard

    HX9—Extra hard

Third digit indicates variation of two-digit temper.

Table 12.4 Subdivisions of T Temper: Thermally Treated

First digit indicates specific sequence of treatments:

T1—Cooled from an elevated-temperature shaping process and naturally aged to a substantially stable condition

T2—Cooled from an elevated-temperature shaping process, cold worked, and naturally aged to a substantially stable condition

T3—Solution heat-treated, cold worked, and naturally aged to a substantially stable condition

T4—Solution heat-treated and naturally aged to a substantially stable condition

T5—Cooled from an elevated-temperature shaping process and then artificially aged

T6—Solution heat-treated and then artificially aged

T7—Solution heat-treated and overaged/stabilized

T8—Solution heat-treated, cold worked, and then artificially aged

T9—Solution heat-treated, artificially aged, and then cold worked

T10—Cooled from an elevated-temperature shaping process, cold worked, and then artificially aged

Second digit indicates variation in basic treatment:

Examples:

T42 or T62—Heat treated to temper by user

Additional digits indicate stress relief:

Examples:

TX51 or TXX51—Stress relieved by stretching

TX52 or TXX52—Stress relieved by compressing

TX54 or TXX54—Stress relieved by combination of stretching and compressing

Table 12.5 Typical Mechanical Properties of Representative Nonheat-Treatable Aluminum Alloys (Not for Design Purposes)

Alloy	Nominal Composition	Temper	Tensile Strength (ksi)	Yield Strength (ksi)	Elongation (% in 2 in.)	Hardness (BHN)
1199	99.99+% Al	O	6.5	1.5	50	—
		H18	17	16	5	—
1100	99+% Al	O	13	5	35	23
		H14	18	17	9	32
		H18	24	22	5	44
3003	1.2% Mn	O	16	6	30	28
		H14	22	21	8	40
		H18	29	27	4	55
5005	0.8% Mg	O	18	6	25	28
		H14	23	22	6	41
		H18	29	28	4	51
3004	1.2% Mn	O	26	10	20	45
	1.0% Mg	H34	35	29	9	63
		H38	41	36	5	77
5052	2.5% Mg	O	28	13	25	47
		H34	38	31	10	68
		H38	42	37	7	77
5456	5.1% Mg	O	45	23	24	70
	0.8% Mn	H321,H116	51	37	16	90
B443.0	5.0% Si	F[a]	19	8	8	40
		F[b]	23	9	10	45
514.0	4.0% Mg	F[a]	25	12	9	50

[a] Sand cast.

[b] Permanent mold cast.

Table 12.6 Typical Mechanical Properties of Representative Heat-Treatable Aluminum Alloys (Not for Design Purposes)

Alloy	Nominal Composition	Temper	Tensile Strength (ksi)	Yield Strength (ksi)	Elongation (% in 2 in.)	Hardness (BHN)
2024	4.4% Cu	O	27	11	20	47
	1.5% Mg	T4	68	47	20	120
	0.6% Mn	T6	69	57	10	125
		T86	75	71	6	135
6061	1.0% Mg	O	18	8	25	30
	0.6% Si	T4	35	21	22	65
		T6	45	40	12	95
7005	4.5% Zn	O	28	12	20	—
	1.4% Mg	T6	51	42	13	—
7075	5.6% Zn	O	33	15	17	60
	2.5% Mg	T6	83	73	11	150
	1.6% Cu	T73	73	63	13	—
356.0	7.0% Si	F[a]	24	18	6	—
	0.3% Mg	T6[a]	33	24	3.5	70
		F[b]	26	18	5	—
		T6[b]	37	27	5	80

[a] Sand cast.

[b] Permanent mold cast.

Again, casting alloys cannot be work hardened and are either used in as-cast or heat-treated conditions. Typical mechanical properties for commonly used casting alloys ranged from 20 to 50 ksi for ultimate tensile strength, from 15 to 50 ksi tensile yield strength and up to 20% elongation. The range of strengths available with wrought aluminum alloys is shown graphically in Fig. 12.1.

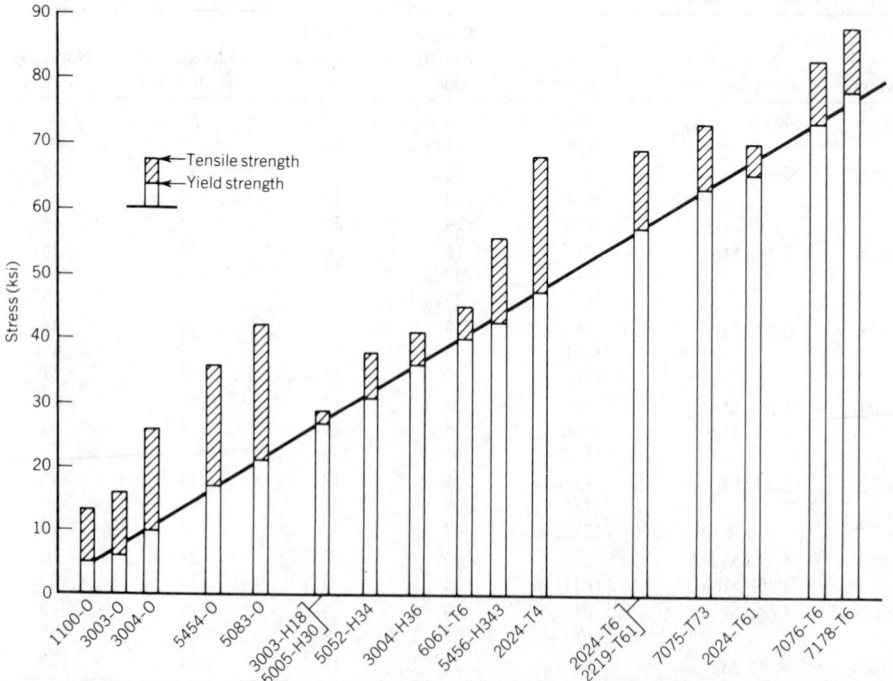

Fig. 12.1 Comparison of strengths of wrought aluminum alloys.

12.6 WORKING STRESSES

Aluminum is used in a wide variety of structural applications. These range from curtain walls on buildings to tanks and piping for handling cryogenic liquids. In establishing appropriate working stresses the factors of safety applied to the ultimate strength and yield strength of the aluminum alloy varies with the specific application. For building and similar type structures a factor of safety of 1.95 is applied to the tensile ultimate strength and 1.65 on the yield strength. For bridges and similar type structures the factors of safety are 2.20 on tensile ultimate strength and 1.85 on yield strength. For other types of applications the factors of safety may differ.

Selection of the working stresses for a particular application should be based on codes, specifications, and standards covering that application published by agencies of government or nationally recognized trade and professional organizations.

For building and bridge design, reference should be made to *Specifications for Aluminum Structures* published by the Aluminum Association. For boiler and pressure vessel design, reference should be made to the *Boiler and Pressure Vessel Code* published by the American Society of Mechanical Engineers.

For information on available codes, standards and specifications for other applications, the Aluminum Association may be consulted at 818 Connecticut Avenue, NW, Washington, DC 20006.

12.7 CHARACTERISTICS

In addition to strength, the combination of alloy and temper determine other characteristics such as corrosion resistance, workability, machinability, etc. Some of the more important characteristics of representative aluminum alloys are compared in Table 12.7. The ratings A through E are relative ratings to compare wrought and cast aluminum alloys *within each category* and are explained below. Where a range of ratings is given, the first rating applies to the alloy in the annealed condition and the second rating is for the alloy when fully hardened. Alloys shown are representative and other alloys of the same type generally have comparable ratings.

12.7.1 Resistance to General Corrosion

Ratings are based on exposures to sodium chloride solution by intermittent spraying or immersion. In general, alloys with A and B ratings can be used in industrial and seacoast atmospheres and in many applications without protection. Alloys with C, D, and E ratings generally should be protected, at least on faying surfaces.

12.7.2 Workability

Ratings A through D for workability (cold) are relative ratings in decreasing order of merit.

Table 12.7 Comparative Characteristics of Representative Aluminum Alloys

Alloy	Resistance to General Corrosion	Work-ability[c]	Machin-ability	Braze-ability	Weldability (Arc)
1100	A	A–C	E–D	A	A
2024	D[a]	C–D	B	D	B–C
3003	A	A–C	E–D	A	A
3004	A	A–C	D–C	B	A
5005	A	A–C	E–D	B	A
5052	A	A–C	D–C	C	A
5456	A[b]	B–C	D–C	D	A
6061	B	A–C	D–C	A	A
7075	C[a]	D	B	D	C
356.0	B	A	C	—	A
B443.0	B	A	E	—	A
514.0	A	D	A	—	C

[a] E in thick sections.

[b] May differ if material heated for long periods.

[c] Castability for casting alloys.

12.7.3 Weldability and Brazeability

Aluminum alloys can be joined by most fusion and solid-state welding processes as well as by brazing and soldering. Fusion welding is commonly done by gas metal-arc welding (GMAW) and gas tungsten-arc welding (GTAW).

The relative weldability and brazeability of representative aluminum alloys is covered in Table 12.7, where ratings A through D are defined as follows:

A = Generally weldable by all commercial procedures and methods.
B = Weldable with special techniques or for specific applications that justify preliminary trials or testing to develop welding procedures and weld performance.
C = Limited weldability because of crack sensitivity or loss in resistance to corrosion and mechanical properties.
D = No commonly used welding methods have been developed.

Table 12.8 gives practical thickness or cross-sectional areas that can be joined by various processes.

Table 12.8ᵃ Practical Aluminum Thickness Ranges for Various Joining Processes

Joining Process	Thickness (in.) [or Area (in.²)] Minimum	Maximum
Gas metal-arc welding	0.12	No limit
Gas tungsten-arc welding	0.02	1
Resistance spot welding	Foil	0.18
Resistance seam welding	0.01	0.18
Flash welding	0.05	(12)
Stud welding	0.02	No limit
Cold welding—butt joint	(0.0005)	(0.2)
Cold welding—lap joint	Foil	0.015
Ultrasonic welding	Foil	0.12
Electron beam welding	0.02	6
Brazing	0.006	No limit

ᵃ Reprinted from the American Welding Society, *Welding Handbook,* 7th ed., Miami, FL, 1982.

12.7.4 Typical Applications

Typical applications of commonly used wrought aluminum alloys are listed in Tables 12.9 and 12.10. By comparing these with Tables 12.5, 12.6, and 12.7, one can readily see that application is based on properties such as strength, corrosion resistance, weldability, etc. Where one desired property, such as high strength, is the prime requisite, then steps must be taken to overcome a possible undesirable characteristic, such as relatively poor corrosion resistance. In this case, the high-strength alloy would be protected by a protective coating such as alcladding, which will be described in a later section. Conversely, where resistance to attack is the prime requisite, then one of the more corrosion-resistant alloys would be employed and assurance of adequate strengths would be met through proper design. The best combination of strength and corrosion resistance for consumer applications in wrought products is found among the 5xxx and 6xxx series alloys. Several casting alloys have good corrosion resistance, and aluminum castings are widely used as cooking utensils and components of food processing equipment as well as for valves, fittings, and other components in various chemical applications.

12.8 MACHINING ALUMINUM

Aluminum alloys are readily machined and offer such advantages as almost unlimited cutting speed, good dimensional control, low cutting force, and excellent life. Relative machinability of commonly used alloys are classified as A, B, C, D, or E (see Table 12.7).

Table 12.9 Typical Applications of Wrought Nonheat-Treatable Aluminum Alloys

Alloy Series	Typical Alloys	Typical Applications
1xxx	1350	Electrical conductor
	1060	Chemical equipment, tank cars
	1100	Sheet metal work, cooking utensils, decorative
3xxx	3003, 3004	Sheet metal work, chemical equipment, storage tanks, beverage cans
4xxx	4043	Welding electrodes
	4343	Brazing alloy
5xxx	5005, 5050, 5052, 5657	Decorative and automotive trim, architectural and anodized, sheet metal work, appliances
5xxx (>3% Mg)	5083, 5086, 5182, 5454, 5456	Marine, welded structures, storage tanks, pressure vessels, armor plate, cryogenics, beverage can easy open ends

Table 12.10 Typical Applications of Wrought Heat-Treatable Aluminum Alloys

Alloy Series	Typical Alloys	Typical Applications
2xxx (Al–Cu)	2011	Screw machine products
	2219	Structural, high temperature
2xxx (Al–Cu–Mg)	2014, 2024, 2618	Aircraft structures and engines, truck frames and wheels
6xxx	6061, 6063	Marine, truck frames and bodies, structures, architectural, furniture
7xxx (Al–Zn–Mg)	7004, 7005	Structural, cryogenic, missile
7xxx (Al–Zn–Mg–Cu)	7001, 7075, 7178	High-strength structural and aircraft

12.8.1 Cutting Tools

Cutting tool geometry is described by seven elements: top or back rake angle, side rake angle, end relief angle, side relief angle, end cutting edge angle, and nose radius.

The depth of cut may be in the range of $\frac{1}{16}$–$\frac{1}{4}$ in. for small work up to $\frac{1}{2}$–$1\frac{1}{2}$ in. for large work. The feed depends on finish. Rough cuts vary from 0.006 to 0.080 in. and finishing cuts from 0.002 to 0.006 in. Speed should be as high as possible, up to 15,000 fpm.

Cutting forces for an alloy such as 6061-T651 are 0.30–0.50 hp/in.^{3}/min for a 0° rake angle and 0.25–0.35 hp/in.^{3}/min for a 20° rake angle.

Lubrication such as light mineral or soluble oil is desirable for high production. Alloys with a machinability rating of A or B may not need lubrication.

The main types of cutting tool materials include water-hardening steels, high-speed steels, hard-cast alloys, sintered carbides and diamonds:

1. Water-hardening steels (plain carbon or with additions of chromium, vanadium, or tungsten) are lowest in first cost. They soften if cutting edge temperatures exceed 300–400°F; have low resistance to edge wear; and are suitable for low cutting speeds and limited production runs.

2. High-speed steels are available in a number of forms, are heat treatable, permit machining at rapid rates, allow cutting edge temperatures of over 1000°F, and resist shock better than hard-cast or sintered carbides.

3. Hard-cast alloys are cast closely to finish size, are not heat treated, and lie between high-speed steels and carbides in terms of heat resistance, wear, and initial cost. They will not take severe shock loads.

4. Sintered carbide tools are available in solid form or as inserts. They permit speeds 10–30 times faster than for high-speed steels. They can be used for most machining operations. They should be used only when they can be supported rigidly and when there is sufficient power and speed. Many types are available.

   5. Mounted diamonds are used for finishing cuts where an extremely high-quality surface is required.

12.8.2 Single-Point Tool Operations

   1. *Turning.* Aluminum alloys should be turned at high speeds with the work held rigidly and supported adequately to minimize distortion.
   2. *Boring.* All types of tooling are suitable. Much higher speeds can be employed than for boring ferrous materials. Carbide tips are normally used in high-speed boring in vertical or horizontal boring machines.
   3. *Planing and Shaping.* Aluminum permits maximum table speeds and high metal removal rates. Tools should not strike the work on the return stroke.

12.8.3 Multipoint Tool Operations

Milling

Removal rate is high with correct cutter design, speed and feed, machine rigidity, and power. When cutting speeds are high, the heat developed is retained mostly in the chips, with the balance absorbed by the coolant. Speeds are high with cutters of high-speed and cast alloys, and very high with sintered carbide cutters.

All common types of solid-tooth, high-carbon, or high-speed steel cutters can be employed. High-carbon cutters operating at a maximum edge temperature of 400°F are preferred for short run production. For long runs, high-speed steel or inserted-tooth cutters are used.

Speeds of 15,000 fpm are not uncommon for carbide cutters. Maximum speeds for high-speed and high-carbon-steel cutters are around 5000 fpm and 600 fpm, respectively.

Drilling

General purpose drills with bright finishes are satisfactory for use on aluminum. Better results may be obtained with drills having a high helix angle. Flute areas should be large; the point angle should be 118° (130°–140° for deeper holes). Cutting lips should be equal in size. Lip relief angles are between 12° and 20°, increasing toward the center to hold the chisel angle between 130° and 145°.

No set rule can be given for achieving the correct web thickness. Generally, for aluminum, it may be thinner at the point without tool breakage.

A 1/8-in. drill at 6000 rpm has a peripheral speed of 2000 fpm. For drilling aluminum, machines are available with speeds up to 80,000 rpm.

If excessive heat is generated, hole diameter may be reduced even below drill size. With proper drills, feeds, speeds, and lubrication, no heat problem should occur.

For a feed of 0.008 ipr, and a depth to diameter ratio of 4:1, the thrust value is 170 lb and the torque value is 10 lb-in. for a 1/4-in. drill with alloy 6061-T651. Aluminum alloys can be counterbored, tapped, threaded by cutting or rolling, and broached. Machining fluid should be used copiously.

Grinding

Resin-bounded silicon carbide wheels of medium hardness are used for rough grinding of aluminum. Finish grinding requires softer, vitrified-bonded wheels. Wheel speeds can vary from 5500 to 6000 fpm. Abrasive belt grinding employs belt speeds from 4600 to 5000 sfpm. Grain size of silicon carbide abrasive varies from 36 to 80 for rough cuts and from 120 to 180 for finishing cuts. For contact wheel abrasive belt grinding, speeds are 4500–6500 sfpm. Silicon carbide or aluminum oxide belts (24–80 grit) are used for rough cuts.

Sawing, Shearing, Routing, and Arc Cutting Aluminum

Correct tooth contour is most important in *circular sawing*. The preferred saw blade has an alternate hollow ground side—rake teeth at about 15°. Operating speeds are 4000–15,000 fpm. Lower speeds are recommended for semi-high-speed steel, intermediate speeds for high-speed inserted-tooth steel blades, and high speeds for carbide-tipped blades.

Band sawing speeds should be between 2000 and 5000 fpm. Spring-tempered blades are recommended for sheet and soft blades with hardened teeth for plate. Tooth pitch should not exceed material thickness: four to five teeth to the inch for spring tempered, six to eight teeth to the inch for flexible backed. Contour sawing is readily carried out. Lubricant should be applied to the back of the blade.

Shearing of sheet may be done on guillotine shears. The clearance between blades is generally 10–12% of sheet thickness down to 5–6% for light gauge soft alloy sheet. Hold-down pads, shear beds, and tables should be covered to prevent marring. Routing can also be used with 0.188–0.50 in. material routed at feeds of 10–30 ipm. Plates of 3-in.-thick heat-treated material can be routed at feeds up to 10 ipm.

Chipless machining of aluminum can be carried out using shear spinning rotary swaging, internal swaging, thread rolling, and flame cutting.

12.9 CORROSION BEHAVIOR

Although aluminum is a chemically active metal, its resistance to corrosion is attributable to an invisible oxide film that forms naturally and is always present unless it is deliberately prevented from forming. Scratch the oxide from the surface and, in air, the oxide immediately reforms. Once formed, the oxide effectively protects the metal from chemical attack and also from further oxidation. Some properties of this natural oxide are:

1. It is very thin—200–400 billionths of an inch thick.
2. It is tenacious. Unlike iron oxide or rust which spalls from the surface leaving a fresh surface to oxidize, aluminum oxide adheres tightly to aluminum.
3. It is hard. Aluminum oxide is one of the hardest substances known.
4. It is relatively stable and chemically inert.
5. It is transparent and does not detract from the metal's appearance.

12.9.1 General Corrosion

The general corrosion behavior of aluminum alloys depends basically on three factors: (1) the stability of the oxide film, (2) the environment, and (3) the alloying elements; these factors are not independent of one another. The oxide film is considered stable between pH 4.5 and 9.0; however, aluminum can be attacked by certain anions and cations in neutral solutions, and it is resistant to some acids and alkalies.

In general, aluminum alloys have good corrosion resistance in the following environments: atmosphere, most fresh waters, seawater, most soils, most foods, and many chemicals. Since "good corrosion resistance" is intended to mean that the material will give long service life without surface protection, in support of this rating is the following list of established applications of aluminum in various environments:

In Atmosphere.   Roofing and siding, truck and aircraft skin, architectural.

With Most Fresh Waters.   Storage tanks, pipelines, heat exchangers, pleasure boats.

In Seawater.   Ship hulls and superstructures, buoys, pipelines.

In Soils.   Pipelines and drainage pipes.

With Foods.   Cooking utensils, tanks and equipment, cans and packaging.

With Chemicals.   Storage tanks, processing and transporting equipment.

It is generally true that the higher the aluminum purity, the greater is its corrosion resistance. However, certain elements can be alloyed with aluminum without reducing its corrosion resistance and in some cases an improvement actually results. Those elements having little or no effect include Mn, Mg, Zn, Si, Sb, Bi, Pb, Ti; those having a detrimental effect include Cu, Fe, Ni:

Al–Mn Alloys.   Al–Mn alloys (3xxx series) have good corrosion resistance and may possibly be better than 1100 alloy in marine environments and for cooking utensils because of a reduced effect by Fe in these alloys.

Al–Mg Alloys.   Al–Mg alloys (5xxx series) are as corrosion resistant as 1xxx alloys and even more resistant to salt water and some alkaline solutions. In general, they offer the best combination of strength and corrosion resistance of all aluminum alloys.

Al–Mg–Si Alloys.   Al–Mg–Si alloys (6xxx series) have good resistance to atmospheric corrosion, but generally slightly lower resistance than Al–Mg alloys. They can be used unprotected in most atmospheres and waters.

Alclad Alloys.   Alclad alloys are composite wrought products comprised of an aluminum alloy core with a thin layer of corrosion—protective pure aluminum or aluminum alloy metallurgically bonded to one or both surfaces of the core. As a class, alclad alloys have a very high resistance to corrosion. The cladding is anodic to the core and thus protects the core.

12.9.2 Pitting Corrosion

Pitting is the most common corrosive attack on aluminum alloy products. Pits form at localized discontinuities in the oxide film on aluminum exposed to atmosphere, fresh water, or saltwater, or other
neutral electrolytes. Since in highly acidic or alkaline solutions the oxide film is usually unstable,
pitting generally occurs in a pH range of about 4.5–9.0. The pits can be minute and concentrated or
can vary in size and be widely scattered, depending on alloy composition, oxide film quality, and the
nature of the corrodent.

The resistance of aluminum to pitting depends significantly on its purity; the purest metal is the
most resistant. The presence of other elements in aluminum, except Mn, Mg, and Zn, increases its
susceptibility to pitting. Copper and iron have the greatest effect on susceptibility. Alclad alloys have
greatest resistance to penetration since any pitting is confined to the more anodic cladding until the
cladding is consumed.

12.9.3 Galvanic Corrosion

Aluminum in contact with a dissimilar metal in the presence of an electrolyte tends to corrode more
rapidly than if exposed by itself to the same environment; this is referred to as galvanic corrosion.
The tendency of one metal to cause galvanic corrosion of another can be predicted from a "galvanic
series," which depends on environments. Such a series is listed below; the anodic metal is usually
corroded by contact with a more cathodic one:

	Magnesium and zinc	Protect aluminum
Anodic	Aluminum, cadmium, and chromium	Neutral and safe in most environments
	Steel and iron	Cause slow action on aluminum except in marine environments
	Lead	Safe except for severe marine or industrial atmospheres
	Copper and nickel	Tend to corrode aluminum
Cathodic	Stainless steel	Safe in most atmospheres and fresh water; tends to corrode aluminum in severe marine atmospheres

Since galvanic corrosion is akin to a battery and depends on current flow, several factors determine
the severity of attack. These are:

Electrolyte Conductivity. The higher the electrical conductivity, the greater the corrosive effect.

Polarization. Some couples polarize strongly to reduce the current flow appreciably. For example,
stainless steel is highly cathodic to aluminum, but because of polarization the two can safely be
used together in many environments.

Anode/Cathode Area Ratios. A high ratio minimizes galvanic attack; a low ratio tends to cause
severe galvanic corrosion.

12.10 FINISHING ALUMINUM

The aluminum surface can be finished in more different ways than any other metal. The normal
finishing operations fall into four categories: mechanical, chemical, electrochemical, and applied. They
are usually performed in the order listed although one or more of the processes can be eliminated
depending on the final effect desired.

12.10.1 Mechanical Finishes

This is an important starting point since even with subsequent finishing operations, a rough, smooth,
or textured surface may be retained and observed. For many applications, the as-fabricated finish
may be good enough. Or this surface can be changed by grinding, polishing, buffing, abrasive or shot
blasting, tumbling or burnishing, or even hammering for special effects. Rolled surfaces of sheet or
foil can be made highly specular by use of polishing rolls; one side or two sides bright and textures
can be obtained by using textured rolls.

12.10.2 Chemical Finishes

A chemical finish is often applied after the mechanical finish. The most widely used chemical finishes
include caustic etching for a matte finish, design etching, chemical brightening, conversion coatings,

and immersion coatings. Conversion coatings chemically convert the natural oxide coating on aluminum to a chromate, a phosphate, or a combination chromate–phosphate coating, and they are principally used and are the recommended ways to prepare the aluminum surface for painting.

12.10.3 Electrochemical Finishes

These include electrobrightening for maximum specularity, electroplating of another metal such as nickel or chromium for hardness and wear resistance, and, most importantly, anodizing. Anodizing is an electrochemical process whereby the natural oxide layer is increased in thickness over a thousand times and made more dense for increased resistance to corrosion and abrasion resistance.

The anodic oxide forms by the growth of cells, each cell containing a central pore. The pores are sealed by immersing the metal into very hot or boiling water. Sealing is an important step and will affect the appearance and properties of the anodized coating. There are several varieties of anodized coatings.

12.10.4 Clear Anodizing

On many aluminum alloys a thick, transparent oxide layer can be obtained by anodizing in a sulfuric acid solution—this is called clear anodizing. The thickness of the layer depends on the current density and the time in solution, and is usually between 0.1 and 1 mil in thickness.

12.10.5 Color Anodizing

Color can be added to the film simply by immersing the metal immediately after anodizing and before sealing into a vat containing a dye or metallic coloring agents and then sealing the film. A wide range of colors have been imparted to aluminum in this fashion for many years. However, the colors imparted in this manner tend to fade from prolonged exposure to sunlight.

12.10.6 Integral Color Anodizing

More lightfast colors for outdoor use are achieved through integral color anodizing. These are proprietary processes utilizing electrolytes containing organic acids and, in some cases, small amounts of impurities are added to the metal itself to bring about the desired colors. This is usually a one-stage process and the color forms as an integral part of the anodize. Colors are for the most part limited to golds, bronzes, grays, and blacks.

12.10.7 Electrolytically Deposited Coloring

Electrolytically deposited coloring is another means of imparting lightfast colors. Following sulfuric acid anodize, the parts are transferred to a second solution containing metallic pigments that are driven into the coating by an electric current.

12.10.8 Hard Anodizing

Hard anodizing, or hardcoating as it is sometimes called, usually involves anodizing in a combination of acids and produces a very dense coating, often 1–5 mils thick. It is very resistant to wear and is normally intended for engineering applications rather than appearance.

12.10.9 Electroplating

In electroplating, a metal such as chromium or nickel is deposited on the aluminum surface from a solution containing that metal. This usually is done for appearance or to improve the hardness or abrasion resistance of the surface. Electroplating has a "smoothing out" effect, whereas anodized coatings follow the contours of the base metal surface thus preserving a matte or a polished surface as well as any other patterns applied prior to the anodize.

12.10.10 Applied Coatings

Applied coatings include porcelain enamel, paints and organic coatings, and laminates such as plastic, paper, or wood veneers. Probably as much aluminum produced today is painted as is anodized. Adhesion can be excellent when the surface has been prepared properly. For best results paint should be applied over a clean conversion-coated or anodized surface.

CHAPTER 13
COPPER AND ITS ALLOYS

HOWARD MENDENHALL

Olin Brass
East Alton, Illinois

ROBERT F. SCHMIDT

Colonial Metals
Columbia, Pennsylvania

13.1 COPPER

Howard Mendenhall

13.1.1 Composition of Commercial Copper

Specifications for copper, generally accepted by industry, are the ASTM standard specifications. These also cover silver-bearing copper.

Low-resistance copper, used for electrical purposes, may be electrolytically or fire refined. It is required to have a content of copper plus silver not less than 99.90%. Maximum permissible resistivities in international ohms (meter, gram) are: copper wire bars, 0.15436; ingots and ingot bars, 0.15694.

Mechanical Properties of Copper

	Annealed	Cold Rolled or Drawn	Cast
Tensile strength			
psi	30,000–40,000	50,000–70,000	20,000–30,000
MPa	210–280	350–490	140–210
Elongation in 2 in.	25–40%	2–35%	25–45%
Reduction of area	40–60%	2–4%	—
Rockwell F hardness	65 max	54–100	—
Rockwell 30T hardness	31 max	18–70	—

Physical Properties of Copper

Density	0.323 lb/in.^{3}		8.94 g/cm^3	
Melting point	1981°F		1083°C	
Coefficient of linear	0.0000094/°F	(68–212°F)	0.0000170/°C	(20–100°C)
thermal expansion	0.0000097/°F	(68–392°F)	0.0000174/°C	(20–200°C)
	0.0000099/°F	(68–572°F)	0.0000178/°C	(20–300°C)
Pattern shrinkage	¼ in./ft		2%	
Thermal conductivity	226 Btu/ft^2/ft/hr/°F		398 W/m/°C	
	at 68°F		at 27°C	
Electric resistivity	10.3 ohms (circular mil/ft)		1.71 microhm/cm	
	at 68°F		at 20°C	
Temperature coefficient	0.023 ohms/°F		0.0068/°C	
of electric resistivity	at 68°F		at 20°C	
Specific heat	—		0.386 J/g/°C	
			at 20°C	
Magnetic property			Diamagnetic	
Optical property			Selectively reflecting	
Young's modulus	17,300,000 psi		119,300 MPa	

ASTM Specification B216-78, *Fire-Refined Copper for Wrought Products and Alloys,* calls for the following analysis: Cu + Ag, min 99.88%; As, max 0.012%; Sb, max 0.003%; Se + Te, max 0.025%; Ni, max 0.05%; Bi, max 0.003%; Pb, max 0.004%.

Oxygen-free high-conductivity copper is a highly ductile material, made under conditions that prevent the entrance of oxygen and the formation of copper oxide. It is utilized in deep-drawing, spinning, and edge-bending operations, and in welding, brazing, and other hot-working operations where embrittlement must be avoided. It has the same conductivity and tensile properties as tough pitch electrolytic copper.

Deoxidized copper containing silver has been utilized to increase softening resistance of copper. It does not affect oxygen level. A number of elements that reduce oxygen in copper, such as Zr, Cr, B, P, can also provide some softening resistance.

13.1.2 Hardening Copper

There are three methods for hardening copper: grain-size control, cold working, or alloying. When copper is hardened with tin, silicon, or aluminum, it generally is called bronze; when hardened with zinc, it is called brass.

13.1.3 Corrosion

Copper is resistant to the action of seawater and to atmospheric corrosion. It is not resistant to the common acids, and is unsatisfactory in service with ammonia and with most compounds of sulfur. Manufacturers should be consulted in regard to its use under corrosive conditions.

13.1.4 Fabrication

Copper may be hot forged, hot or cold rolled, hot extruded, hot pierced, and drawn, stamped, or spun cold. It can be silver-soldered, brazed, and welded. For brazing in reducing atmosphere or for welding by the oxyacetylene torch or electric arc, deoxidized copper will give more satisfactory joints than electrolytic or silver bearing copper. High-temperature exposure of copper containing oxygen, in reducing atmosphere, leads to decomposition of copper oxide and formation of steam with resulting embrittlement. Copper is annealed from 480 to 1400°F, depending on the properties desired. Ordinary commercial annealing is done in the neighborhood of 1100°F. Inert or reducing atmospheres give best surface quality; however, high temperature annealing of oxygen-containing coppers in reducing atmosphere can cause embrittlement. Copper may be electrodeposited from the alkaline cyanide solution, or from the acid sulfate solution.

13.2 SAND-CAST COPPER-BASE ALLOYS

Robert F. Schmidt

13.2.1 Introduction

The information required for selection of cast copper-base alloys for various types of applications can be found in Table 13.1. The principal data required by engineers and designers for castings made of copper-base alloys are given in Table 13.2. A cross-reference chart is shown in Table 13.3 for

Table 13.1 Application for Copper-Base Alloys

Uses	Types of Alloys	Alloy Number
Andirons	Leaded yellow brass	C85200
Architectural trim	Leaded red brass	C83600
	Leaded yellow brass	C85400
	Leaded nickel silver	C97400
Ball bearing races	Manganese bronze	C86200
	Aluminum bronze	C95400
	Leaded yellow brass	C85200
Bearings, high speed, low load	High-leaded tin bronze	C93200
		C93800
		C93700
Bearings, low speed, heavy load	Tin bronze	C91300
		C91000
	Manganese bronze	C86300
	Aluminum broonze	C95400
Bearings, medium speed	High-leaded tin bronze	C93700
		C93800
Bells	Tin bronze	C91300
	Silicon bronze	C87200
Carburetors	Leaded red brass	C83600
	Leaded tin bronze	C92200
Cocks and faucets	Leaded semired brass	C84400
		C84800
	Leaded yellow brass	C85200
Corrosion resistance to acids	Aluminum bronze	C95400
	Leaded nickel bronze	C97600
alkalies	Silicon bronze	C87200
seawater	Nickel aluminum bronze	C95800

Table 13.1 (*Cont.*)

water	Leaded red brass	C83600
	Leaded semired brass	C84400
Electrical hardware	Leaded red brass	C83300
	Silicon bronze	C87200
	Aluminum bronze	C95400
Fittings	Leaded semired brass	C84400
Food-handling equip-ment	Leaded nickel bronze	C97600 C97800
Gears	Tin bronze	C90700 C91600
	Aluminum bronze	C95400
General hardware	Leaded red brass	C83600
Gun mounts	Manganese bronze	C86200
	Aluminum bronze	C95300
High-strength alloy	Manganese bronze	C86300
Impellers	Tin bronze	C90300
	Leaded red brass	C83600
	Aluminum bronze	C95400
	Silicon brass	C87200
Landing gear parts	Aluminum bronze	C95400
Lever arms	Manganese bronze	C86500
Marine castings and fittings	Manganese bronze	C86500 C86200
	Aluminum bronze	C95800
Marine propellers	Aluminum bronze	C95800
	Manganese bronze	C86500
Musical instruments	Leaded nickel bronze	C97800
Ornamental bronze	Leaded yellow brass	C85200
Pickling baskets	Aluminum bronze	C95300
Piston rings	Tin bronze	C90500 C91300
Plumbing fixtures	Leaded semired brass	C84400 C84800
Pump bodies	Tin bronze	C90300
	Leaded tin bronze	C93800
	Aluminum bronze	C95800
Steam fittings and valves	Leaded tin bronze	C92200 C92300
Valves, high pressure	Leaded tin bronze	C92200 C92600
Valves, low pressures	Leaded red brass	C83600
	Leaded semired brass	C84400
Valve seats for elevated temperature	Leaded nickel bronze	C97800
Valve stems	Silicon brass	C87500
	Silicon bronze	C87200
Wear parts	High-leaded tin bronze	C93700 C93800
Weldability	Tin bronze	C90700
	Manganese bronze	C86500
	Aluminum bronze	All grades
	Silicon bronze	C87200
Welding jaws	Aluminum bronze	C95300
Wormwheels	Aluminum bronze	C95500

Table 13.2 Sand-Cast Copper-Base Alloys

Copper–tin–lead–zinc alloys (Nominal Composition columns: Cu, Sn, Pb, Zn, Others)

UNS Number	Ingot Number	Cu	Sn	Pb	Zn	Others	Yield Strength[a] ksi (MPa)	Tensile Strength[a] ksi (MPa)	Elongation[a] (%)	Brinell Hardness (500 kg)	Impact Strength (Izod) (ft-lb)	Electrical Conductivity (% IACS)	Pattern Shrinkage (in./ft)
C83600	115	85	5	5	5		14 (97)	30 (207)	20	65	9	15	11/64
C83800	120	83	4	6	7		13 (90)	30 (207)	20	60	8	15.2	11/64
C84400	123	81	3	7	9		13 (90)	29 (200)	18	55	8	16.7	11/64
C84800	130	76	3	6	15		12 (83)	28 (193)	16	55	12[b]	16.6	11/64
C85200	400	72	1	3	24		12 (83)	35 (241)	25	46		18.6	3/16
C85400	403	67	1	3	29		11 (76)	30 (207)	20	53		19.6	3/16
C85700	405.2	61	1	1	37.3	0.3 Al	14 (97)	40 (276)	15	76		21.8	7/32

Manganese and silicon bronzes (Nominal Composition columns: Cu, Sn, Fe, Zn, Al, Mn, Others)

UNS Number	Ingot Number	Cu	Sn	Fe	Zn	Al	Mn	Others	Yield Strength[a] ksi (MPa)	Tensile Strength[a] ksi (MPa)	Elongation[a] (%)	Brinell Hardness (500 kg)	Impact Strength (Izod) (ft-lb)	Electrical Conductivity (% IACS)	Pattern Shrinkage (in./ft)
C86200	423	64		3	26	4	3		45 (310)	90 (621)	18	180[c]	12	7.4	1/4
C86300	424	62		3	26	6	3		60 (414)	110 (758)	12	225[c]	15	8.0	9/32
C86400	420	58		1	38	0.75	0.25	0.75 Pb	20 (138)	60 (414)	15	105[c]	30	19.3	1/4
C86500	421	58		1	39	1	1		25 (172)	65 (448)	20	130[c]	32[b]	20.6	1/4
C87200	500	92						4 Si	18 (124)	45 (310)	20	87	33	6.1	1/4
C87200	500	95						1 Mn, 4 Si	18 (124)	45 (310)	20	88	33	5.9	1/4
C87500	500	82			14			4 Si	24 (165)	60 (414)	16	115[c]	32[b]	6.1	15/64

Leaded tin bronzes (Nominal Composition columns: Cu, Sn, Pb, Zn, Others)

UNS Number	Ingot Number	Cu	Sn	Pb	Zn	Others	Yield Strength[a] ksi (MPa)	Tensile Strength[a] ksi (MPa)	Elongation[a] (%)	Brinell Hardness (500 kg)	Impact Strength (Izod) (ft-lb)	Electrical Conductivity (% IACS)	Pattern Shrinkage (in./ft)
C90300	225	88	8	0	4		18 (124)	40 (276)	30	70	14[b]	12.4	3/16
C90500	210	88	10	0	2		18 (124)	40 (276)	30	75	10	10.9	3/16
C92200	245	86	6	1 1/2	4 1/2		16 (110)	34 (234)	24	64	19[b]	14.3	3/16
C92300	230	87	8	1	4		16 (110)	36 (248)	18	70	14	12.3	3/16
C92600	215	87	10	1	2		18 (124)	40 (276)	20	72	7	10.0	3/16
C93200	315	83	7	7	3		14 (97)	30 (207)	15	67	5	12.4	7/32
C93700	305	80	10	10			12 (83)	30 (207)	15	67	5	10.1	1/8
C93800	319	78	7	15			14 (97)	26 (179)	12	58	5	11.6	5/32

Aluminum bronzes (Nominal Composition columns: Cu, Fe, Ni, Al, Others)

UNS Number	Ingot Number	Cu	Fe	Ni	Al	Others	Yield Strength[a] ksi (MPa)	Tensile Strength[a] ksi (MPa)	Elongation[a] (%)	Brinell Hardness (500 kg)	Impact Strength (Izod) (ft-lb)	Electrical Conductivity (% IACS)	Pattern Shrinkage (in./ft)
C95200	415	88	3		9		25 (172)	65 (448)	20	120[c]	35	12.2	7/32
C95300	415	89	1		10		25 (172)	65 (448)	20	140[c]	30	15.3	7/32

164

Alloy	No.[a]	Cu	Fe (Sn)	Ni (Pb)	Al (Zn)	Others	Yield strength, ksi (MPa)[a]	Tensile strength, ksi (MPa)[a]	Elongation, %	Hardness	Impact, Charpy (ft-lb)[b]	Elec. cond., % IACS	Section, in.
C95400	415	86	3½	—	10½	—	30 (207) / 36 (248)	75 (517) / 92 (634)	12 / 18	156[c]	15	13	7/32
C95410	415	84	4	2	10	—	30 (207) / 36 (248)	75 (517) / 96 (662)	12 / 15	176[c]	15	13	9/32
C95500	415	81	4	4	11	—	40 (276) / 44 (303)	90 (621) / 102 (703)	6 / 12	200[c]	13	8.8	3/16
C95800	415	81½	4	4½	9	1 Mn	35 (241) / 37 (255)	85 (586) / 96 (662)	15 / 25	160[c]	20	7.0	3/16
C96400	—	68	1	30	—	1 Nb	32 (221) / 37 (255)	60 (414) / 68 (469)	20 / 28	140[b]	78[b]	5.0	3/16

Composition header for the following group: **Cu | Sn | Pb | Zn | Others**

Alloy	No.[a]	Cu	Sn	Pb	Zn	Others	Yield strength, ksi (MPa)[a]	Tensile strength, ksi (MPa)[a]	Elongation, %	Hardness	Impact, Charpy (ft-lb)[b]	Elec. cond., % IACS	Section, in.
C97300	410	57	2	9	20	12 Ni	15 (103) / 17 (117)	30 (207) / 36 (248)	8 / 25	60		5.9	1/8
C97400	411	60	3	5	16	16 Ni	16 (110) / 17 (117)	30 (207) / 38 (262)	8 / 20	70		5.5	1/8
C97600	412	64	4	4	8	20 Ni	17 (117) / 25 (172)	40 (276) / 47 (324)	22 / 22	85		4.8	1/8
C97800	413	66	5	2	2	25 Ni	22 (151) / 30 (207)	50 (345) / 55 (379)	10 / 15	130[b]	11[b]	4.5	3/16

Composition header for the following group: **Cu | Sn | Pb | Zn | Others**

Alloy	No.[a]	Cu	Sn	Pb	Zn	Others	Yield strength, ksi (MPa)[a]	Tensile strength, ksi (MPa)[a]	Elongation, %	Hardness	Impact, Charpy (ft-lb)[b]	Elec. cond., % IACS	Section, in.
C81100	—	99.7	—	—	—	1 Cr	6 (41)	20 (138)	50	40	84	92	1/4
C81400	—	99	—	—	—	2 Be, 0.5 Cr	36 (248)[d]	53 (365)[d]	11[d]	B69[e]		60	1/4
C82500	—	97½	—	—	—	0.25 Si	45 (310)	80 (551) / 160[d]	14 / 20	B82.5[e] / C40[d][e]		20[d]	
C83300	131	93	1	2	4	—	10 (69) / 14 (97)	30 (207) / 32 (220)	25 / 35	35	12	32	3/16
C83450	—	88	2½	2	6½	1 Ni	15 (103) / 18 (124)	35 (241) / 37 (255)	10 / 34	55		20	3/16
C90700	205	89	11	—	—	—	22 (152)	44 (303)	20	80		9.6	3/16
C91100	194	84	16	—	—	—	25 (172)	35 (241)	2	135[c]		8.5	3/16
C91300	—	81	19	—	—	—	30 (207)	35 (241)	0.5	170[c]		7.0	3/16
C91600	205A	88	10½	—	—	1½ Ni	17 (117) / 22 (152)	35 (241) / 44 (303)	16	85		10.0	3/16
C92900	206A	84	10	2½	0	3½ Ni	25 (172) / 26 (179)	45 (310) / 47 (324)	20	80		9.2	3/16

Composition header for the following group: **Cu | Fe | Ni | Al | Others**

Alloy	No.[a]	Cu	Fe	Ni	Al	Others	Yield strength, ksi (MPa)[a]	Tensile strength, ksi (MPa)[a]	Elongation, %	Hardness	Impact, Charpy (ft-lb)[b]	Elec. cond., % IACS	Section, in.
C99400	—	90.5	2	2.2	1.2	3 Zn, 1.2 Si	30 (207) / 34 (234)	60 (414) / 66 (445)	20 / 25	125[c]		16.8	3/16
C99500	—	88	4	4.5	1.2	1.2 Zn, 1.2 Si	40 (276)	70 (483)	12	145[c]		13.7	3/16
C99700	—	58	—	5	—	13 Mn, 23 Zn	25 (172)	55 (379)		110[c]		3.0	1/4
C99750	—	58	—	—	1	20 Mn, 20 Zn, 1 Pb	32 (220)	65 (448)		110[c]		2.0	1/4

[a] Left column is minimum; right column is typical; yield strength is 0.5% extension under load.

[b] Impact strength, Charpy (ft-lb).

[c] Brinell hardness (3000 kg).

[d] Heat treated.

[e] Rockwell.

quick reference in locating the specifications applying to these alloys. Additional information in regard to special alloys, such as high conductivity copper, chromium–copper, and beryllium copper, is covered in Section 13.2.5.

13.2.2 Selection of Alloy

Table 13.1 is an outline of the various types of alloys generally used for the purposes shown. When specifying a specific alloy for a new application, the foundry or ingot maker should be consulted. This is particularly important where corrosion resistance is involved or specific mechanical properties are required. While all copper-base alloys have good general corrosion resistance, specific environments, especially chemical, can cause corrosive attack or stress corrosion cracking. An example of this is the stress corrosion cracking that occurs when a manganese bronze alloy (high-strength yellow brass) is placed under load in certain environments.

The typical and minimum properties shown in Table 13.2 for the various alloys are for room temperature. The effect of elevated temperature on mechanical properties should be considered for any given application. The ingot maker or foundry should be consulted for this information.

Since copper-base alloy castings are often used for pressure-tight value and pump parts, caution should be exercised in alloy selection. In general, when small-sized, thin-wall castings are used, such as valve bodies with up to 3-in. openings, with all sections up to 1 in., the leaded red brass and leaded tin bronze alloys should be specified. When heavy-wall valves and pump bodies over 1-in. thickness are used, the castings should be made of nickel aluminum bronze or 70/30 cupronickel. These alloy preferences are based on differences in solidification behavior.

13.2.3 Fabrication

All sand-cast copper-base alloys can be machined, although some are far more machinable than others. The alloys containing lead, such as the leaded red brasses, leaded tin bronzes, and high-leaded tin bronzes, are very easily machined. On the other hand, aluminum and manganese bronzes do not machine easily. However, use of carbide tooling, proper tool angles, and coolants permit successful machining. In regard to weldability, no leaded alloys should be welded. In general, the aluminum bronzes, silicon bronzes, and α–β manganese bronzes can be welded successfully. This also applies to tin bronzes and 70/30 cupronickel. These alloys not only can be joined to other materials by welding, but can also be repaired by welding if exhibiting casting defects such as shrinkage porosity. All copper-base alloys can be joined by brazing.

13.2.4 Mechanical and Physical Properties

The mechanical and physical properties of the most widely used copper-base casting alloys are given in Table 13.2. Alloy numbers used are the UNS numbers developed by the Copper Development Association (CDA) and now adopted by the American Society for Testing Materials (ASTM), Society for Automotive Engineers (SAE), and the U.S. Government. Also shown for reference purposes are the ingot numbers still used by the ingot makers. Much of the data shown in Table 13.2 were taken from *Standards Handbook,* Part 7, Alloy Data, published by CDA. Table 13.2 not only shows the typical properties that can be attained, but also the minimum values called for in the various specifications listed in Table 13.3. These properties, of course, can only be attained when care is taken toward proper melting, gating, feeding, and venting of casting molds.

The CDA *Standards Handbook,* Part 7, contains a very complete list of physical properties on not only the alloys shown in Table 13.2, but also other alloys less widely used.

13.2.5 Special Alloys

There are a number of alloys shown in Table 13.2 that are used for special purposes and amount to much less tonnage than the red brasses, leaded red brasses, tin bronzes, manganese bronzes, and aluminum bronzes. The following sections mention the more widely used of the special alloy families.

Gear Bronzes

High-tin alloys such as C90700 (89% copper, 11% tin), C91600 (88% copper, 10% tin, 2% nickel), and C92900 (84% copper, 10% tin, 2½% lead, 3½% nickel) are widely used for cast bronze gears. In addition to these tin bronze alloys, aluminum bronze, such as C95400 (86% copper, 4% iron, and 10% aluminum) is also used for gear applications.

Bridge Bearing Plates

These castings are made almost entirely to ASTM B22 specification and are generally made from copper–tin alloys like C91300 (81% copper, 19% tin) and C91100 (86% copper, 14% tin). Three other alloys, specified under ASTM B22 are C86300 high-tensile manganese bronze, C90500 tin bronze, and C93700 high-leaded tin bronze.

Piston Rings

Tin bronzes, such as C91300 and C91100, are commonly used for piston rings. These castings are usually made by the centrifugal casting process.

High Conductivity

When the electrical conductivity of pure copper is required, it can be melted and deoxidized and poured into casting molds. Care must be taken to avoid contamination by elements usually present in cast copper-base alloys, such as phosphorous, iron, zinc, tin, and nickel. Electrical conductivity values of 85% to 90% IACS can be attained with low level impurities present. This alloy is C81100.

Moderate Conductivity, High Strength

All of the alloys shown in Table 13.2 have electrical conductivity less than 25% IACS. However, there are additional copper-base alloys available with higher electrical conductivity. Beryllium copper and low-tin bronzes are examples of alloys in the 25–35% IACS range. C83300, which has 32% IACS, has a composition of 93% copper, 1% tin, 2% lead, and 4% zinc. A typical beryllium copper casting alloy with around 25% IACS is C82500, which has as-cast typical properties of 80,000 psi tensile strength and 20% elongation in 2 in., and after heat treatment has a tensile strength of 155,000 psi and elongation of 1% in 2 in. Hardness of this alloy is typically Rockwell C40 in the heat-treated condition and Rockwell B82 when as-cast. This alloy has a composition of 2% beryllium, 0.5% cobalt, 0.25% silicon, and 97.20% copper.

When some strength is required in addition to high electrical conductivity, the best casting alloy is chromium copper, alloy C81400. This alloy is made up of 0.9% chromium, 0.1% silicon, and 99% copper. It is heat treatable and maintains an electrical conductivity of 85% IACS, a tensile strength of 51,000 psi, a yield strength of 40,000 psi, and an elongation of 17%. The hardness value for this alloy is 105 under a 500-kg load.

BIBLIOGRAPHY

Books

ASTM Book of Standards, Part 2.01, American Society for Testing Materials, Philadelphia, PA, 1983, Table 11-3.
Copper-Base Alloys Foundry Practice, 3rd ed., American Foundrymen's Society, Des Plaines, IL, 1965, Section 11.3.
Metals Handbook, 9th ed., American Society for Metals, Metals Park, OH, 1979, Vol. 2, Sections 11.1 and 11.2.
SAE Handbook, Society of Automotive Engineers, Warrendale, PA, 1982, Table 11-3.
Standards Handbook, Part 7, Cast Products, Copper Development Association, Greenwich, CT, 1978, Table 11-2 and Section 11.4.
Standards Handbook, Part 6, Specifications Index, Copper Development Association, Greenwich, CT, 1983, Table 11-3.

Periodicals

Foundry, Penton/IPC, Cleveland, OH.
Modern Castings, American Foundrymen's Society, Des Plaines, IL.

Transaction

Transaction, American Foundrymen's Society, Des Plaines, IL.

Table 13.3 Copper-Base Alloy Casting Specifications

Alloy Number	Commercial Designation	American Society for Testing Materials		Federal		Military	Society of Automotive Engineers	
		Specification Number	Alloy Number	QQ-C-390A Alloy Designation	Former Specification		Current	Former
C83600	85-5-5-5	B62,B584 B271,B505	C83600	836	QQ-L-225(2)	MIL-C-11866(25) MIL-C-15345(1) MIL-C-22087(2) MIL-C-22229(836)	836	40
C83800	83-4-6-7	B271,B584 B505	C83800	838	QQ-L-225(17)			
C84400	81-3-7-9	B271,B584 B505	C84400	844	QQ-L-225(11)	MIL-B-11553(11) MIL-B-18343		
C84800	76-2½-6½-15	B271,B584 B505	C84800					
C85200	72-1-3-24	B271 B584	C85200	852	QQ-B-621(C)			
C85400	67-1-3-29	B271 B584	C85400	854	QQ-B-621(B)		854	41
C85700	61-1-1-37	B271 B584	C85700	857	QQ-B-621(A)			
C86200	90,000 tensile manganese bronze	B271,B584 B505	C86200	862	QQ-B-726(B)	MIL-C-15345(3) MIL-C-11866(27) MIL-C-11866(20) MIL-C-22087(7) MIL-C-22229(862)	862	430A
C86300	110,000 tensile manganese bronze	B22,B505 B271,B584	C86300	863	QQ-B-726(C)	MIL-C-11866(21) MIL-C-15345(6) MIL-C-22087(9) MIL-C-22229(863)	863	430B
C86400	60,000 tensile manganese bronze	B271 B584	C86400	864	QQ-B-726(D)			
C86500	65,000 tensile manganese bronze	B271,B584 B505	C86500	865	QQ-B-726(A)	MIL-C-15345(4) MIL-C-22087(5) MIL-C-22229(865)	865	43
C87200	5% zinc max silicon bronze	B271 B584	C87200	872	QQ-C-593(B)	MIL-C-11866(19) MIL-C-22229(872)		
C87500	82-14-4 silicon brass	B271 B584	C87500		QQ-C-593(A)			
C90300	88-8-0-4	B271,B584 B505	C90300	903	QQ-L-225(5)	MIL-C-11866(26) MIL-C-15345(8)	903	620

UNS No.	Nominal composition / name	ASTM	SAE	Federal	Military		
C90500	88-10-0-2	B22,B505 B271,B584	905	QQ-L-225(16)	MIL-C-22087(3) MIL-C-22229(903)	905	62
C92200	88-6-1½-4½	B61,B505	922	QQ-L-225(1)	MIL-C-15345(9) MIL-B-16541	922	622
C92300	87-8-1-4	B271,B584	923	QQ-L-225(6-6X)	MIL-C-15345(10)	923	621
C93200	83-7-7-3	B271, B505,B584	932	QQ-L-225(12)	MIL-B-11553(12) MIL-B-16261(6)	932	660
C93500	85-5-9-1	B271,B584 B505	935	QQ-L-225(14)		935	66
C93700	80-10-10	B271,B584 B505	937		MIL-B-13506(792,797)	937	64
C93800	78-7-15	B22,B505 B271,B584	938	QQ-L-225(7)		938	67
C95200	88-3-9 aluminum bronze	B66,B271, B144,B505,B584	952	QQ-B-671(1)	MIL-C-22087(6) MIL-C-22229(952) MIL-C-11866(22)	952	68A
C95300	89-1-10 aluminum bronze	B148,B505 B271	953	QQ-B-671(2)	MIL-C-11866(23) MIL-C-15345(13) MIL-C-11866(24) MIL-C-15345(14) MIL-C-22087(8) MIL-C-22229(955) MIL-C-15345(28) MIL-B-21230(1) MIL-B-24480 MIL-B-22229(958)	953	68B
C95400	85-4-11 aluminum bronze	B148,B505 B271	954	QQ-B-671(3)			
C95500	81-4-11-4 aluminum bronze	B148,B505 B271	955	QQ-B-671(4)			
C95800	81-4-9-5-1Mn aluminum bronze	B148 B271	958				
C96400	70-30 cupronickel	B369 B505	964		MIL-C-15345(24) MIL-C-20159(1) MIL-C-15345(7)		
C97300	12% nickel nickel silver	B271 B584					
C94700	16% nickel nickel silver						
C97600	20% nickel nickel bronze	B271 B584					
C97800	25% nickel nickel bronze	B271 B584					

CHAPTER 14
NICKEL AND ITS ALLOYS

T. H. BASSFORD

Huntington Alloys, Inc.
Huntington, West Virginia

14.1 INTRODUCTION

Nickel, the 24th element in abundance, has an average content of 0.016% in the outer 10 miles of the earth's crust. This is greater than the total for copper, zinc, and lead. However, few of these deposits scattered throughout the world are of commercial importance. Oxide ores commonly called laterites are largely distributed in the tropics. The igneous rocks contain high magnesium contents and have been concentrated by weathering. Of the total known ore deposits, more than 80% is contained in laterite ores. The sulfide ores found in the northern hemispheres do not easily concentrate by weathering. The sulfide ores in the Sudbury district of Ontario, which contain important by-products such as copper, cobalt, iron, and precious metals are the world's greatest single source of nickel[1].

Nickel has an atomic number of 28 and is one of the transition elements in the fourth series in the periodic table. The atomic weight is 58.71 and density is 8.902 g/cm^3. Useful properties of the element are the modulus of elasticity and its magnetic and magnetostrictive properties, and high thermal and electrical conductivity. Hydrogen is readily adsorbed on the surface of nickel. Nickel will also adsorb other gases such as carbon monoxide, carbon dioxide, and ethylene. It is this capability of surface adsorption of certain gases without forming stable compounds that makes nickel an important catalyst.[2]

As an alloying element, nickel is used in hardenable steels, stainless steels, special corrosion-resistant and high-temperature alloys, copper–nickel, "nickel–silvers," and aluminum–nickel. Nickel imparts ductility and toughness to cast iron.

Approximately 10% of the total annual production of nickel is consumed by electroplating processes. Nickel can be electrodeposited to develop mechanical properties of the same order as wrought nickel; however, special plating baths are available that will yield nickel deposits possessing a hardness as high as 450 Vickers (425 BHN). The most extensive use of nickel plate is for corrosion protection of iron and steel parts and zinc-base die castings used in the automotive field. For these applications, a layer of nickel, 0.0015–0.003 in. thick, is used. This nickel plate is then finished or covered with a chromium plate consisting in thickness of about 1% of the underlying nickel plate thickness in order to maintain a brilliant, tarnish-free, hard exterior surface.

14.2 NICKEL ALLOYS

Most of the alloys listed and discussed are in commercial production. However, producers from time to time introduce improved modifications that make previous alloys obsolete. For this reason, or economic reasons, they may remove certain alloys from their commercial product line. Some of these alloys have been included to show how a particular composition compares with the strength or corrosion resistance of currently produced commercial alloys.

14.2.1 Classification of Alloys

Nickel and its alloys can be classified into the following groups on the basis of chemical composition.[3]

Nickel

(1) Pure nickel, electrolytic (99.56% Ni), carbonyl nickel powder and pellet (99.95% Ni); (2) commercially pure wrought nickel (99.6–99.97% nickel); and (3) anodes (99.3% Ni).

Nickel and Copper

(1) Low-nickel alloys (2–13% Ni); (2) cupronickels (10–30% Ni); (3) coinage alloy (25% Ni); (4) electrical resistance alloy (45% Ni); (5) nonmagnetic alloys (up to 60% Ni); and (6) high-nickel alloys, Monel (over 50% Ni).

Nickel and Iron

Wrought alloy steels (0.5–9% Ni); (2) cast alloy steels (0.5–9% Ni); (3) alloy cast irons (1–6 and 14–36% Ni); (4) magnetic alloys (20–90% Ni): (a) controlled coefficient of expansion (COE) alloys (29.5–32.5% Ni) and (b) high-permeability alloys (49–80% Ni); (5) nonmagnetic alloys (10–20% Ni); (6) clad steels (5–40% Ni); (7) thermal expansion alloys: (a) low expansion (36–50% Ni) and (b) selected expansion (22–50% Ni).

Iron, Nickel, and Chromium

(1) Heat-resisting alloys (40–85% Ni); (2) electrical resistance alloys (35–60% Ni); (3) iron-base superalloys (9–26% Ni); (4) stainless steels (2–25% Ni); (5) valve steels (2–13% Ni); (6) iron-base superalloys (0.2–9% Ni); (7) maraging steels (18% Ni).

Nickel, Chromium, Molybdenum, and Iron

(1) Nickel-base solution-strengthened alloys (40–70% Ni); (2) nickel-base precipitation-strengthened alloys (40–80% Ni).

Powder-Metallurgy Alloys

(1) Nickel-base dispersion strengthened (78–98% Ni); (2) nickel-base mechanically alloyed oxide-dispersion-strengthened (ODS) alloys (69–80% Ni).

The nominal chemical composition of nickel-base alloys is given in Table 14.1. This table does not include alloys with less than 30% Ni, cast alloys, or welding products. For these and those alloys not listed, the chemical composition and applicable specifications can be found in the *Unified Numbering System for Metals and Alloys,* published by the Society of Automotive Engineers, Inc.

14.2.2 Discussion and Applications

The same grouping of alloys used in Tables 14.1, 14.2, and 14.3, which give chemical composition and mechanical properties, will be used for discussion of the various attributes and uses of the alloys as a group. Many of the alloy designations are registered trademarks of producer companies.

Nickel Alloys

The corrosion resistance of nickel makes it particularly useful for maintaining product purity in the handling of foods, synthetic fibers, and caustic alkalies, and also in structural applications where resistance to corrosion is a prime consideration. It is a general-purpose material used when the special properties of the other nickel alloys are not required. Other useful features of the alloy are its magnetic and magnetostrictive properties; high thermal and electrical conductivity; low gas content; and low vapor pressure.[4]

Typical *nickel 200* applications are food-processing equipment, chemical shipping drums, electrical and electronic parts, aerospace and missile components, caustic handling equipment and piping, and transducers.

Nickel 201 is preferred to nickel 200 for applications involving exposure to temperatures above 316°C (600°F). Nickel 201 is used as coinage, plater bars, and combustion boats in addition to some of the applications for Nickel 200.

Nickel 270, a high-purity product, is essentially free from nonmetallic inclusions. The alloy exhibits high thermal conductivity and permits heavy cold deformation without reannealing. It is used in electronic devices, heat exchangers, and heat shields.

Permanickel alloy 300 by virtue of the magnesium content is age-hardenable. But, because of its low alloy content, alloy 300 retains many of the characteristics of nickel. Typical applications are grid lateral winding wires, magnetostriction devices, thermostat contact arms, solid-state capacitors, grid side rods, diaphragms, springs, clips, and fuel cells.

Duranickel alloy 301 is another age-hardenable high-nickel alloy, but is made heat treatable by aluminum and titanium additions. The important features of alloy 301 are high strength and hardness, good corrosion resistance, and good spring properties up to 316°C (600°F); and it is on these mechanical considerations that selection of the alloy is usually based. Typical applications are extrusion press parts, molds used in the glass industry, clips, diaphragms, and springs.

Nickel–Copper Alloys

Nickel–copper alloys are characterized by high strength, weldability, excellent corrosion resistance, and toughness over a wide temperature range. They have excellent service in seawater or brackish water under high-velocity conditions, as in propellers, propeller shafts, pump shafts, and impellers and condenser tubes, where resistance to the effects of cavitation and erosion are important. Corrosion rates in strongly agitated and aerated seawater usually do not exceed 1 mil/year.

Monel alloy 400 has low corrosion rates in chlorinated solvents, glass-etching agents, sulfuric and many other acids, and practically all alkalies, and it is resistant to stress-corrosion cracking. Alloy 400 is useful up to 538°C (1000°F) in oxidizing atmospheres, and even higher temperatures may be used if the environment is reducing. Springs of this material are used in corrosive environments up to 232°C (450°F). Typical applications are valves and pumps; pump and propeller shafts; marine fixtures and fasteners; electrical and electronic components; chemical processing equipment; gasoline and freshwater tanks; crude petroleum stills, process vessels, and piping; boiler feedwater heaters and other heat exchangers; and deaerating heaters.

Monel alloy 404 is characterized by low magnetic premeability and excellent brazing characteristics. Residual elements are controlled at low levels to provide a clean wettable surface even after prolonged firing in wet hydrogen. Alloy 404 has a low Curie temperature and its magnetic properties are not appreciably affected by processing or fabrication. This magnetic stability makes alloy 404 particularly suitable for electronic applications. Much of the strength of alloy 404 is retained at outgassing temperatures. Thermal expansion of alloy 404 is sufficiently close to that of many other alloys as to permit the firing of composite metal tubes with negligible distortion. Typical applications are waveguides, metal-to-ceramic seals, transistor capsules, and power tubes.

Monel alloy R-405 is a free-machining material intended almost exclusively for use as stock for automatic screw machines. It is similar to alloy 400 except that a controlled amount of sulfur is added for improved machining characteristics. The corrosion resistance of alloy R-405 is essentially the same as that of alloy 400, but the range of mechanical properties differs slightly. Typical applications are water meter parts, screw machine products, fasteners for nuclear applications, and valve seat inserts.

Monel alloy K-500 is an age-hardenable alloy that combines the excellent corrosion resistance characteristics of the Monel nickel–copper alloys with the added advantage of increased strength and hardness. Age hardening increases its strength and hardness. Still better properties are achieved when the alloy is cold-worked prior to the aging treatment. Alloy K-500 has good mechanical properties over a wide temperature range. Strength is maintained up to about 649°C (1200°F), and the alloy is strong, tough, and ductile at temperatures as low as −253°C (−423°F). It also has low permeability and is nonmagnetic to −134°C (−210°F). Alloy K-500 has low corrosion rates in a wide variety of environments. Typical applications are pump shafts and impellers, doctor blades and scrapers, oil-well drill collars and instruments, electronic components, and springs.

Nickel–Chromium–Iron Alloys

This family of alloys was developed for high-temperature oxidizing environments. These alloys typically contain 50–80% nickel, which permits the addition of other alloying elements to improve strength and corrosion resistance while maintaining toughness.

Inconel alloy 600 is a standard engineering material for use in severely corrosive environments at elevated temperatures. It is resistant to oxidation at temperatures up to 1177°C (2150°F). In addition to corrosion and oxidation resistance, alloy 600 presents a desirable combination of high strength and workability, and is hardened and strengthened by cold-working. This alloy maintains strength, ductility, and toughness at cryogenic as well as elevated temperatures. Because of its resistance to chloride-ion stress-corrosion cracking and corrosion by high-purity water, it is used in nuclear reactors. For this service, the alloy is produced to exacting specifications and is designated Inconel alloy 600T. Typical applications are furnace muffles, electronic components, heat-exchanger tubing, chemical- and food-processing equipment, carburizing baskets, fixtures and rotors, reactor control rods, nuclear reactor components, primary heat-exchanger tubing, springs, and primary water piping. Alloy 600, being one of the early high-temperature, corrosion-resistant alloys, can be thought of as being the basis of many of our present day special-purpose high-nickel alloys, as illustrated in Fig. 14.1.

Inconel alloy 601 has shown very low rates of oxidation and scaling at temperatures as high as 1093°C (2000°F). The high chromium content (nominally 23%) gives alloy 601 resistance to oxidizing, carburizing, and sulfur-containing environments. Oxidation resistance is further enhanced by the aluminum and chromium contents. Typical applications are heat-treating baskets and fixtures, radiant furnace tubes, strand-annealing tubes, thermocouple protection tubes, and furnace muffles and retorts.

Inconel alloy 690 is a high-chromium nickel alloy having very low corrosion rates in many corrosive aqueous media and high-temperature atmospheres. In various types of high-temperature water, alloy 690 also displays low corrosion rates and excellent resistance to stress-corrosion cracking—desirable attributes for nuclear steam-generator tubing. In addition, the alloy's resistance to sulfur-containing gases makes it a useful material for such applications as coal-gasification units, burners and ducts for processing sulfuric acid, furnaces for petrochemical processing, and recuperators and incinerators.

Inconel alloy 702 has a high aluminum content, which helps maintain low oxidation rates at temperatures up to 1316°C (2400°F). Alloy 7₀2 has good mechanical strength at high temperatures; age hardening improves the strength of the alloy up to approximately 816°C (1500°F). Typical applications are afterburner liners, furnace components, and fixtures.

Inconel alloy 706 is a precipitation-hardenable alloy with characteristics similar to alloy 718, except that alloy 706 has considerably improved machinability. It also has good resistance to oxidation and corrosion over a broad range of temperatures and environments. Like alloy 718, alloy 706 has excellent resistance to postweld strain-age cracking. Typical applications are gas-turbine components and other parts that must have high strength combined with good machinability and weldability.

Inconel alloy 718 is an age-hardenable high-strength alloy suitable for service at temperatures from −253°C (−423°F) to 704°C (1300°F). The fatigue strength of alloy 718 is high, and the alloy exhibits high stress-rupture strength up to 704°C (1300°F) as well as oxidation resistance up to 982°C (1800°F). It also offers good corrosion resistance to a wide variety of environments. The outstanding characteristic of alloy 718 is its slow response to age hardening. The slow response enables the material to be welded and annealed with no spontaneous hardening unless it is cooled slowly. Alloy 718 can

Table 14.1 Nominal Chemical Composition (wt%)

Material	Ni	Cu	Fe	Cr	Mo	Al	Ti	Cb	Mn	Si	C	Other Elements
Nickel												
Nickel 200	99.6	—	—	—	—	—	—	—	0.23	0.03	0.07	—
Nickel 201	99.7	—	—	—	—	—	—	—	0.23	0.03	0.01	—
Nickel 270	99.98	0.001[b]	0.003	0.001[b]	—	—	—	—	0.001[b]	0.001[b]	0.01	—
Permanickel alloy 300	98.7	—	0.02	—	—	—	0.49	—	0.11	0.04	0.29	0.38 Mg
Duranickel alloy 301	94.3	—	0.08	—	—	4.44	0.44	—	0.25	0.39	0.16	—
Nickel–Copper												
Monel alloy 400	65.4	31.5	1.7	—	—	—	—	—	1.0	0.19	0.12	—
Monel alloy 404	54.6	45.3	0.03	—	—	—	—	—	0.01	0.04	0.07	—
Monel alloy R-405	65.3	31.6	1.63	—	—	0.1	—	—	1.0	0.17	0.15	0.04 S
Monel alloy K-500	65.0	30	0.64	—	—	2.94	0.48	—	0.70	0.12	0.17	—
Nickel–Chromium–Iron												
Inconel alloy 600	76[a]	0.25	8.0	15.5	—	—	—	—	0.5	0.25	0.08	—
Inconel alloy 601	60.5	0.50	14.1	23.0	—	1.35	—	—	0.5	0.25	0.05	—
Inconel alloy 690	60	—	9.5	30	—	—	—	—	—	—	0.03	—
Inconel alloy 702	79.5[a]	0.25	1.0	15.5	—	3.25	0.63	—	0.50	0.35	0.05	—
Inconel alloy 706	41.5	0.15	40	16	—	0.20	1.8	2.9	0.18	0.18	0.03	—
Inconel alloy 718	52.5	0.15	18.5	19	3.0	0.5	0.9	5.1	0.18	0.18	0.04	—
Inconel alloy X-750	73[a]	0.25	7	15.5	—	0.70	2.5	1	0.50	0.25	0.04	—
Nickel–Iron–Chromium												
Incoloy alloy 800	32.5	0.38	46	21	—	0.38	0.38	—	0.75	0.50	0.05	—
Incoloy alloy 800H	32.5	0.38	46	21	—	0.38	0.38	—	0.75	0.50	0.07	—
Incoloy alloy 802	32.5	0.38	46	21	—	0.58	0.75	—	—	—	0.35	—
Incoloy alloy 825	42	2.2	30	21.5	3	0.10	0.90	—	0.50	0.25	0.03	—
Incoloy alloy 901	42.7	—	34	13.5	6.2	0.20	2.5	—	0.40	0.40	0.05	—
Incoloy alloy 925	43.2	1.8	28	20.8	2.70	0.35	2.10	—	0.60	0.22	0.03	—
Pyromet 860	44	—	Bal	13	6	1.0	3.0	—	0.25	0.10	0.05	4.0 Co
Refractaloy 26	38	—	Bal	18	3.2	0.2	2.6	—	0.8	1.0	0.03	20 Co

Nickel–Iron

Alloy	Ni	Cu	Fe	Cr	Mo	Nb	Ti	Al	Mn	Si	C	Other
Nilo alloy 42	41.6	—	57.4	—	—	—	—	—	1.00	0.06	0.03	—
Ni-Span-C alloy 902	42.3	0.05	48.5	5.33	—	—	2.6	0.55	0.40	0.50	0.03	—
Incoloy alloy 903	38	—	41.5	—	—	2.9	1.40	0.90	0.09	0.17	0.02	14.9 Co
Incoloy alloy 907	37.6	0.10	41.9	—	—	4.70	1.5	—	0.05	0.08	0.02	14.1 Co

Nickel–Chromium–Molybdenum

Alloy	Ni	Cu	Fe	Cr	Mo	Nb	Ti	Al	Mn	Si	C	Other
Hastelloy alloy X	Bal[c]	—	18	22	9	—	—	—	—	—	0.10	—
Hastelloy alloy G	Bal	2	19.5	22	6.5	2.1	—	—	<1	1.5	<0.05	<1 W, <2.5 Co
Hastelloy alloy C-276	Bal	—	5.5	15.5	16	—	—	—	<0.08	<1	<0.01	2.5 Co, 4 W, 0.35 V
Hastelloy alloy C	Bal	—	<3	16	16	—	<0.7	—	<0.08	<1	<0.01	<2 Co
Inconel alloy 617	54	—	—	22	9	—	—	1	—	—	0.07	12.5 Co
Inconel alloy 625	Bal	—	2.5	21.5	9	3.6	<0.4	<0.4	—	—	0.03	—
MAR-M-252	Bal	—	—	19	10	—	2.6	1	<0.5	<0.5	0.15	10 Co, 0.005 B
Rene' 41	Bal	—	—	19	10	—	3.1	1.5	—	—	0.09	11 Co, <0.010 B
Rene' 95	Bal	—	—	14	3.5	3.5	2.5	3.5	—	—	0.15	8 Co, 3.5 W, 0.01 B, 0.05 Zr
Astroloy	Bal	—	—	15	5.3	—	3.5	4.4	—	—	0.06	15 Co
Udimet 500	Bal	—	<0.5	19	4	—	3.0	3.0	—	—	0.08	18 Co, 0.007 B
Udimet 520	Bal	—	—	19	6	—	3.0	2.0	—	—	0.05	12 Co, 1 W, 0.005 B
Udimet 600	Bal	—	<4	17	4	—	2.9	4.2	—	—	0.04	16 Co, 0.02 B
Udimet 700	Bal	—	—	15	5.0	—	3.5	4.4	—	—	0.07	18.5 Co, 0.025 B
Udimet 1753	Bal	—	9.5	16.3	1.6	—	3.2	1.9	0.1	0.05	0.24	7.2 Co, 8.4 W, 0.008 B, 0.06 Zr
Waspaloy	Bal	<0.1	<2	19	4.3	—	3	1.5	—	—	0.08	14 Co, 0.006 B, 0.05 Zr

Nickel-Powder Alloys (Dispersion Strengthened)

Alloy	Ni	Cu	Fe	Cr	Mo	Nb	Ti	Al	Mn	Si	C	Other
TD-nickel	98	—	—	—	—	—	—	—	—	—	—	2 ThO2
TD-NiCr	Bal	—	—	20	—	—	—	—	—	—	—	1.7 ThO2

Nickel-Powder Alloys (Mechanically Alloyed)

Alloy	Ni	Cu	Fe	Cr	Mo	Nb	Ti	Al	Mn	Si	C	Other
Inconel alloy MA 754	78	—	1.0	20	—	—	0.5	0.3	—	—	0.05	0.6 Y2O3
Inconel alloy MA 6000	69	—	—	15	2	—	2.5	4.5	—	—	0.05	4 W, 2 Ta, 1.1 Y2O3

[a] Minimum. [b] Maximum. [c] Bal = balance.

Table 14.2 Mechanical Properties of Nickel Alloys

Material	0.2% Yield Strength (ksi)[a]	Tensile Strength (ksi)[a]	Elongation (%)	Rockwell Hardness
Nickel				
Nickel 200	21.5	67	47	55 Rb
Nickel 201	15	58.5	50	45 Rb
Nickel 270	16	50	50	35 Rb
Permanickel alloy 300	38	95	30	79 Rb
Duranickel alloy 301	132	185	28	36 Rc
Nickel–Copper				
Monel alloy 400	31	79	52	73 Rb
Monel alloy 404	31	69	40	68 Rb
Monel alloy R-405	56	91	35	86 Rb
Monel alloy K-500	111	160	24	25 Rc
Nickel–Chromium–Iron				
Inconel alloy 600	50	112	41	90 Rb
Inconel alloy 601	35	102	49	81 Rb
Inconel alloy 690	53	106	41	97 Rb
Inconel alloy 702	82	142	38	25 Rc
Inconel alloy 706	158	193	21	40 Rc
Inconel alloy 718	168	205	20	46 Rc
Inconel alloy X-750	102	174	25	33 Rc
Nickel–Iron–Chromium				
Incoloy alloy 800	48	88	43	84 Rb
Incoloy alloy 800H	29	81	52	72 Rb
Incoloy alloy 802	38	93	46	81 Rb
Incoloy alloy 825	44	97	53	84 Rb
Incoloy alloy 901	128	171	13	—
Incoloy alloy 925	119	176	24	34 Rc
Pyromet 860	115	180	21	37 Rc
Refractaloy 26	100	170	18	—
Nickel–Iron				
Nilo alloy 42	37	72	43	80 Rb
Ni-Span-C alloy 902	137	150	12	33 Rc
Incoloy alloy 903	174	198	14	39 Rc
Incoloy alloy 907	163	195	15	42 Rc
Nickel–Chromium–Molybdenum				
Hastelloy alloy X	52	114	43	—
Hastelloy alloy G	56	103	48.3	86 Rb
Hastelloy alloy C	64	130	38	—
Inconel alloy 617	43	107	70	81 Rb
Inconel alloy 625	63	140	51	96 Rb
MAR-M-252	122	180	16	—
Rene' 41	120	160	18	—
Rene' 95	190	235	15	—

Table 14.2 (Cont.)

Material	0.2% Yield Strength (ksi)[a]	Tensile Strength (ksi)[a]	Elongation (%)	Rockwell Hardness
Astroloy	152	205	16	—
Udimet 500	122	190	32	—
Udimet 520	125	190	21	—
Udimet 600	132	190	13	—
Udimet 700	140	204	17	—
Udimet 1753	130	194	20	39 Rc
Waspaloy	115	185	25	—
Nickel-Powder Alloys (Dispersion Strengthened)				
TD–Nickel	45	65	15	—
TD–NiCr	89	137	20	—
Nickel-Powder Alloys (Mechanically Alloyed)				
Inconel alloy MA 754	85	140	21	—
Inconel alloy MA 6000	187	189	3.5	—

[a] MPa = ksi × 6.895.

also be repair-welded in the fully aged condition. Typical applications are jet engine components, pump bodies and parts, rocket motors and thrust reversers, and spacecraft.

Inconel alloy X-750 is an age-hardenable nickel–chromium–iron alloy used for its corrosion and oxidation resistance and high creep-rupture strength up to 816°C (1500°F). The alloy is made age-hardenable by the addition of aluminum, columbium, and titanium, which combine with nickel, during proper heat treatment, to form the intermetallic compound $Ni_3(Al, Ti)$. Alloy X-750, originally developed for gas turbines and jet engines, has been adopted for a wide variety of other uses because of its favorable combination of properties. Excellent relaxation resistance makes alloy X-750 suitable for springs operating at temperatures up to about 649°C (1200°F). The material also exhibits good strength and ductility at temperatures as low as −253°C (−423°F). Alloy X-750 also exhibits high resistance to chloride-ion stress-corrosion cracking even in the fully age-hardened condition. Typical applications are gas-turbine parts (aviation and industrial), springs (steam service), nuclear reactors, bolts, vacuum envelopes, heat-treating fixtures, extrusion dies, aircraft sheet, bellows, and forming tools.

Nickel–Iron–Chromium Alloys

This series of alloys typically contains 30–45% Ni and is used in elevated- or high-temperature environments where resistance to oxidation or corrosion is required.

Incoloy alloy 800 is a widely used material of construction for equipment that must resist corrosion, have high strength, or resist oxidation and carburization. The chromium in the alloy imparts resistance to high-temperature oxidation and general corrosion. Nickel maintains an austenitic structure so that the alloy remains ductile after elevated-temperature exposure. The nickel content also contributes resistance to scaling, general corrosion, and stress-corrosion cracking. Typical applications are heat-treating equipment and heat exchangers in the chemical, petrochemical, and nuclear industries, especially where resistance to stress-corrosion cracking is required. Considerable quantities are used for sheathing on electric heating elements.

Incoloy alloy 800H is a version of Incoloy alloy 800 having significantly higher creep and rupture strength. The two alloys have the same chemical composition with the exception that the carbon content of alloy 800H is restricted to the upper portion of the standard range for alloy 800. In addition to a controlled carbon content, alloy 800H receives an annealing treatment that produces a coarse grain size—an ASTM number of 5 or coarser. The annealing treatment and carbon content are responsible for the alloy's greater creep and rupture strength.

Alloy 800H is useful for many applications involving long-term exposure to elevated temperatures or corrosive atmospheres. In chemical and petrochemical processing, the alloy is used in steam/hydrocarbon reforming for catalyst tubing, convection tubing, pigtails, outlet manifolds, quenching-system piping, and transfer piping; in ethylene production for both convection and cracking tubes; in oxo-alcohol

Table 14.3 1000-hr Rupture Stress (ksi)[a]

	1200°F	1500°F	1800°F	2000°F
Nickel–Chromium–Iron				
Inconel alloy 600	14.5	3.7	1.5	—
Inconel alloy 601	28	6.2	2.2	1.0
Inconel alloy 690	16	—	—	—
Inconel alloy 702	43	9	2.5	—
Inconel alloy 706	85	—	—	—
Inconel alloy 718	85	—	—	—
Inconel alloy X-750	68	17	—	—
Nickel–Iron–Chromium				
Incoloy alloy 800	20	—	—	—
Incoloy alloy 800H	23	6.8	1.9	0.9
Incoloy alloy 802	25	12.5	3.5	1.2
Incoloy alloy 825	26	6.0	1.3	—
Incoloy alloy 901	85	18	2.5	—
Pyromet 860	81	17	—	—
Refractaloy 26	65	15.5	—	—
Nickel–Chromium–Molybdenum				
Hastelloy alloy X	31	9.5	—	—
Hastelloy alloy C	44	12	—	—
Inconel alloy 617	52	14	3.8	1.5
Inconel alloy 625	60	7.5	—	—
MAR-M-252	79	22.5	—	—
Rene' 41	102	29	—	—
Rene' 95	125	—	—	—
Astrology	112	42	8	—
Udimet 500	110	30	—	—
Udimet 520	85	33	—	—
Udimet 600	—	37	—	—
Udimet 700	102	43	7.5	—
Udimet 1753	98	34	6.5	—
Waspaloy	89	26	—	—
Nickel-Powder Alloys (Dispersion Strengthened)				
TD–Nickel	21	15	10	7
TD–NiCr	—	—	8	5
Nickel-Powder Alloys (Mechanically Alloyed)				
Inconel alloy MA 754	38	—	19	14
Inconel alloy MA 6000	—	—	22	15

[a] MPa = ksi × 6.895.

production for tubing in hydrogenation heaters; in hydrodealkylation units for heater tubing; and in production of vinyl chloride monomer for cracking tubes, return bends, and inlet and outlet flanges.

Industrial heating is another area of wide usage for alloy 800H. In various types of heat-treating furnaces, the alloy is used for radiant tubes, muffles, retorts, and assorted furnace fixtures. Alloy 800H is also used in power generation for steam superheater tubing and high-temperature heat exchangers in gas-cooled nuclear reactors.

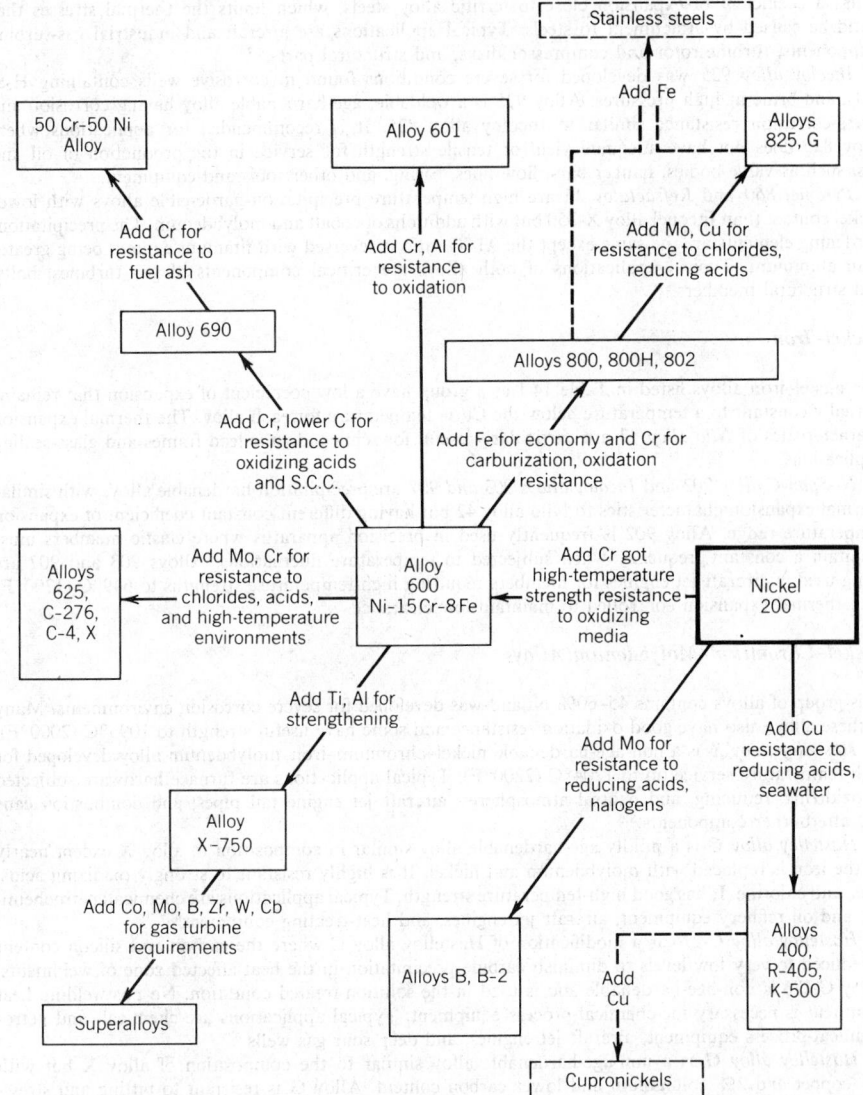

Fig. 14.1 Some compositional modifications of nickel and its alloys to produce special properties.

Incoloy alloy 802 is a higher-carbon modification of alloy 800; it has a typical carbon content of 0.35%. Alloy 802 has still higher creep-rupture strength than alloy 800H. The high-temperature strength results from a fine dispersion of carbide both within the grains and at the grain boundaries. It has the high-temperature corrosion and oxidation resistance of alloy 800. Typical applications are hot sizing dies for the aerospace industry, connecting pins for cast-link heat-treating furnace belts, and deck plates for iron ore sintering plants.

Incoloy alloy 825 was developed for use in aggressively corrosive environments. The nickel content of the alloy is sufficient to make it resistant to chloride-ion stress-corrosion cracking, and, with molybdenum and copper, alloy 825 has resistance to reducing acids. Chromium confers resistance to oxidizing chemicals. The alloy also resists pitting and intergranular attack when heated in the critical sensitization temperature range. Alloy 825 offers exceptional resistance to corrosion by sulfuric acid solutions, phosphoric acid solutions, and seawater. Typical applications are phosphoric acid evaporators, pickling-tank heaters, pickling hooks and equipment, chemical-process equipment, spent nuclear fuel element recovery, propeller shafts, tank trucks, and oil-country cold-worked tubulars.

Incoloy alloy 901 was designed for high creep-rupture strength in the temperature range of 538–760°C (1000–1400°F) for use in aircraft and land-based gas turbines. Owing to its nickel–iron content, it has a coefficient of expansion close to ferritic alloy steels, which limits the thermal stresses that could be caused by attachment to steels. Typical applications are aircraft and industrial gas-turbine components, turbine rotor and compressor disks, and structural parts.[6,7]

Incoloy alloy 925 was developed for severe conditions found in corrosive wells containing H_2S, CO_2, and brine at high pressures. Alloy 925 is a weldable, age-hardenable alloy having corrosion and stress-corrosion resistance similar to Incoloy alloy 825. It is recommended for applications where alloy 825 does not have adequate yield or tensile strength for service in the production of oil and gas, such as valve bodies, hanger bars, flow lines, casing, and other tools and equipment.

Pyromet 860 and *Refractaloy 26* are high-temperature precipitation-hardenable alloys with lower nickel content than Inconel alloy X-750 but with additions of cobalt and molybdenum. The precipitation-hardening elements are the same except the Al/Ti ratio is reversed with titanium content being greater than aluminum. Typical applications of both alloys are critical components of gas turbines, bolts, and structural members.[8]

Nickel–Iron

The nickel–iron alloys listed in Table 14.1 as a group have a low coefficient of expansion that remains virtually constant to a temperature below the Curie temperature for each alloy. The thermal expansion characteristics of *Nilo alloy 42* are particularly useful for semiconductor lead frames and glass-sealing applications.

Ni-Span-C alloy 902 and *Incoloy alloys 903 and 907* are precipitation-hardenable alloys with similar thermal expansion characteristics to Nilo alloy 42 but having different constant coefficient of expansion temperature range. Alloy 902 is frequently used in precision apparatus where elastic members must maintain a constant frequency when subjected to temperature fluctuations. Alloys 903 and 907 are being used in aircraft jet engines for members requiring high-temperature strengths to 649°C (1200°F) with thermal expansion controlled to maintain low clearance.

Nickel–Chromium–Molybdenum Alloys

This group of alloys contains 45–60% Ni and was developed for severe corrosion environments. Many of these alloys also have good oxidation resistance and some have useful strength to 1093°C (2000°F).

Hastelloy alloy X is a non-age-hardenable nickel–chromium–iron–molybdenum alloy developed for high-temperature service up to 1204°C (2200°F). Typical applications are furnace hardware subjected to oxidizing, reducing, and neutral atmospheres; aircraft jet engine tail pipes; and combustion cans and afterburner components.[5,6]

Hastelloy alloy C is a mildly age-hardenable alloy similar in composition to alloy X except nearly all the iron is replaced with molybdenum and nickel. It is highly resistant to strongly oxidizing acids, salts, and chlorine. It has good high-temperature strength. Typical applications are chemical, petrochemical, and oil refinery equipment; aircraft jet engines; and heat-treating equipment.[6,7]

Hastelloy alloy C-276 is a modification of Hastelloy alloy C where the carbon and silicon content is reduced to very low levels to diminish carbide precipitation in the heat-affected zone of weldments. Alloy C-276 is non-age-hardenable and is used in the solution-treated condition. No postwelding heat treatment is necessary for chemical-process equipment. Typical applications are chemical- and petrochemical-process equipment, aircraft jet engines, and deep sour gas wells.[6,7]

Hastelloy alloy G is a non-age-hardenable alloy similar to the composition of alloy X but with 2% copper and 2% columbium and lower carbon content. Alloy G is resistant to pitting and stress-corrosion cracking. Typical applications are paper and pulp equipment, phosphate fertilizer, and synthetic fiber processing.[6,7]

Inconel alloy 617 is a solid-solution-strengthened alloy containing cobalt that has an exceptional combination of high-temperature strength and oxidation resistance which makes alloy 617 a useful material for gas-turbine aircraft engines and other applications involving exposure to extreme temperatures, such as, steam generator tubing and pressure vessels for advanced high-temperature gas-cooled nuclear reactors.

Inconel alloy 625, like alloy 617, is a solid-solution-strengthened alloy but containing columbium instead of cobalt. This combination of elements is responsible for superior resistance to a wide range of corrosive environments of unusual severity as well as to high-temperature effects such as oxidation and carburization. The properties of alloy 625 that make it attractive for seawater applications are freedom from pitting and crevice corrosion, high corrosion fatigue strength, high tensile strength, and resistance to chloride-ion stress-corrosion cracking. Typical applications are wire rope for mooring cables; propeller blades; submarine propeller sleeves and seals; submarine snorkel tubes; aircraft ducting, exhausts, thrust-reverser, and spray bars; and power plant scrubbers, stack liners, and bellows.

MAR-M-252, Rene' 41, Rene' 95, and *Astroloy* are a group of age-hardenable nickel-base alloys containing 10–15% cobalt designed for highly stressed parts operating at temperatures from 871 to 982°C (1600 to 1800°F) in jet engines. MAR-M-252 and Rene' 41 have nearly the same composition but Rene' 41 contains more of the age-hardening elements allowing higher strengths to be obtained.

Rene' 95, of similar base composition but in addition containing 3.5% columbium and 3.5% tungsten, is used at temperatures between 371 and 649°C (700 and 1200°F). Its primary use is as disks, shaft retaining rings, and other rotating parts in aircraft engines of various types.[6-8]

Udimet 500, 520, 600, and 700 and *Unitemp 1753* are age-hardenable, nickel-base alloys having high strength at temperatures up to 982°C (1800°F). All contain a significant amount of cobalt. Applications include jet engine gas-turbine blades, combustion chambers, rotor disks, and other high-temperature components.[6-8]

Waspaloy is an age-hardenable nickel-base alloy developed to have high strength up to 760°C (1400°F) combined with oxidation resistance to 871°C (1600°F). Applications are jet engine turbine buckets and disks, air frame assemblies, missile systems, and high-temperature bolts and fasteners.[6,8]

Nickel Powder Alloys (Dispersion Strengthened)

These oxide dispersion strengthened (ODS) alloys are produced by a proprietary powder metallurgical process using thoria as the dispersoid. The mechanical properties to a large extent are determined by the processing history. The preferred thermomechanical processing results in an oriented texture with grain aspect ratios of about 3:1 to 6:1.

TD–nickel and *TD–NiCr* are dispersion-hardened nickel alloys developing useful strengths up to 1204°C (2200°F). These alloys are difficult to fusion weld without reducing the high-temperature strength. Brazing is used in the manufacture of jet engine hardware. Applications are jet engine parts, rocket nozzles, and afterburner liners.[6,8]

Nickel Powder Alloys (Mechanically Alloyed)

Inconel alloy MA 754 and *Iconel alloy MA 6000* are ODS nickel-base alloys produced by mechanical alloying.[9,10] An yttrium oxide dispersoid imparts high creep-rupture strength up to 1149°C (2100°F). MA 6000 is also age-hardenable, which increases strength at low temperatures up to 760°C (1400°F) These mechanical alloys like the thoria-strengthened alloys described are difficult to fusion weld without reducing high-temperature strength. Useful strength is obtained by brazing. MA 754 is being used as aircraft gas-turbine vanes and bands. Applications for MA 6000 are aircraft gas turbine buckets and test grips.

14.3 CORROSION

It is well recognized that the potential saving is very great by utilizing available and economic practices to improve corrosion prevention and control. Not only should the designer consider initial cost of materials, but he or she should also include the cost of maintenance, length of service, downtime cost, and replacement costs. This type of cost analysis can frequently show that more highly alloyed, corrosion-resistant materials are more cost effective. The National Commission on Materials Policy concluded that one of the "most obvious opportunities for material economy is control of corrosion."

Studies have shown that the total cost of corrosion is astonishing. The overall cost of corrosion in the United States has been estimated by the National Bureau of Standards in 1978 to be 4.2% of the gross national product. A study by a task group of The Federation of Materials Society in 1972 estimated the United States losses because of corrosion to be on the order of $15 billion per year. It was further estimated that approximately $5 billion of the total loss was recoverable through application of corrosion-control techniques already developed and available for use.

Since becoming commercially available shortly after the turn of the century, nickel has become very important in combating corrosion. It is a major constituent in the plated coatings and claddings applied to steel, corrosion-resistant stainless steels, copper–nickel and nickel–copper alloys, high-nickel alloys, and commercially pure nickel alloys. Not only is nickel a corrosion-resistant element in its own right, but, owing to its high tolerance for alloying, it has been possible to develop many metallurgically stable, special-purpose alloys.[11]

Figure 14.1 shows the relationship of these alloys and the major effect of alloying elements. Alloy 600 with 15% chromium, one of the earliest of the nickel–chromium alloys, can be thought of as the base for other alloys. Chromium imparts resistance to oxidizing environments and high-temperature strength. Increasing chromium to 30%, as in alloy 690, increases resistance to stress-corrosion cracking, nitric acid, steam, and oxidizing gases. Increasing chromium to 50% increases resistance to melting sulfates and vanadates found in fuel ash. High-temperature oxidation resistance is also improved by alloying with aluminum in conjunction with high chromium (e.g., alloy 601).

Without chromium, nickel by itself is used as a corrosion-resistant material in food processing and in high-temperature caustic and gaseous chlorine or chloride environments.

Of importance for aqueous reducing acids, oxidizing chloride environments, and seawater are alloy 625 and alloy C-276, which contain 9% and 16% molybdenum, respectively, and are among the most resistant alloys currently available. Low-level titanium and aluminum additions provide γ' strengthening while retaining good corrosion resistance, as in alloy X-750. Cobalt and other alloying element additions provide jet engine materials (superalloys) that combine high-temperature strength with resistance to gaseous oxidation and sulfidation.

Another technologically important group of materials are the higher-iron alloys, which were originally developed to conserve nickel and are often regarded as intermediate in performance and cost between nickel alloys and stainless steels. The prototype, alloy 800 (Fe/33% Ni/21% Cr), is a general purpose alloy with good high-temperature strength and resistance to steam and oxidizing or carburizing gases. Alloying with molybdenum and chromium, as in alloy 825 and alloy G, improves resistance to reducing acids and localized corrosion in chlorides.

Another important category is the nickel–copper alloys. At the higher-nickel end are the Monel alloys (30–45% Cu, balance Ni) used for corrosive chemicals such as hydrofluoric acid, and severe marine environments. At the higher-copper end are the cupronickels (10–30% Ni, balance Cu), which are widely used for marine applications because of their fouling resistance.

Nickel alloys exhibit high resistance to attack under nitriding conditions (e.g., in dissociated ammonia) and in chlorine or chloride gases. Corrosion in the latter at elevated temperatures proceeds by the formation and volatilization of chloride scales, and high-nickel contents are beneficial since nickel forms one of the least volatile chlorides. Conversely, in sulfidizing environments, high-nickel alloys without chromium can exhibit attack due to the formation of a low-melting-point $Ni–Ni_3S_2$ eutectic. However, high chromium contents appear to limit this form of attack.[5]

Friend explains corrosion reactions as wet or dry:[11]

> The term wet corrosion usually refers to all forms of corrosive attack by aqueous solutions of electrolytes, which can range from pure water (a weak electrolyte) to aqueous solutions of acids or bases or of their salts, including neutral salts. It also includes natural environments such as the atmosphere, natural waters, soils, and others, irrespective or whether the metal is in contact with a condensed film or droplets of moisture or is completely immersed. Corrosion by aqueous environments is electrochemical in nature, assuming the presence of anodic and cathodic areas on the surface of the metal even though these areas may be so small as to be indistinguishable by experimental methods and the distance between them may be only of atomic dimensions.
>
> The term dry corrosion implies the absence of water or an aqueous solution. It generally is applied to metal/gas or metal/vapor reactions involving gases such as oxygen, halogens, hydrogen sulfide, and sulfur vapor and even to "dry" steam at elevated temperatures. . . . High-temperature oxidation of metals has been considered to be an electrochemical phenomenon since it involves the diffusion of metal ions outward, or of reactant ions inward, through the corrosion product film, accompanied by a flow of electrons.

The decision to use a particular alloy in a commercial application is usually based on past corrosion experience and laboratory or field testing using test spools of candidate alloys. Most often weight loss is measured to rank various alloys; however, many service failures are due to localized attack such as pitting, crevice corrosion, intergranular corrosion, and stress-corrosion cracking, which must be measured by other means.

A number of investigations have shown the effect of nickel on the different forms of corrosion. Figure 14.2 shows the galvanic series of many alloys in flowing seawater. This series gives an indication of the rate of corrosion between different metals or alloys when they are electrically coupled in an electrolyte. The metal close to the active end of the chart will behave as an anode and corrode, and the metal closer to the noble end will act as a cathode and be protected. Increasing the nickel content will move an alloy more to the noble end of the series. There are galvanic series for other corrosive environments, and the film-forming characteristics of each material may change this series somewhat. Seawater is normally used as a rough guide to the relative positions of alloys in solution of good electrical conductivity such as mineral acids or salts.

Residual stresses from cold rolling or forming do not have any significant effect on the general corrosion rate. However, many low-nickel-containing steels are subject to stress-corrosion cracking in chloride-containing environments. Figure 14.3 from work by LaQue and Copson[12] shows that nickel–chromium and nickel–chromium–iron alloys containing about 45% Ni or more are immune from stress-corrosion cracking in boiling 42% magnesium chloride.[11]

When localized corrosion occurs in well-defined areas, such corrosion is commonly called *pitting attack*. This type of corrosion typically occurs when the protective film is broken or is penetrated by a chloride ion and the film is unable to repair itself quickly. The addition of chromium and particularly molybdenum makes nickel-base alloys less susceptible to pitting attack, as shown in Fig. 14.4, which shows a very good relationship between critical[11] pitting temperature in a salt solution. Along with significant increases in chromium and/or molybdenum, the iron content must be replaced with more nickel in wrought alloys to resist the formation of embrittling phases.[12,13]

Air *oxidation* at moderately high temperatures will form an intermediate subsurface layer between the alloy and gas quickly. Alloying of the base alloy can affect this subscale oxide and, therefore, control the rate of oxidation. At constant temperature, the resistance to oxidation is largely a function of chromium content. Early work by Eiselstein and Skinner has shown that nickel content is very beneficial under cyclic temperature conditions as shown in Fig. 14.5.[14]

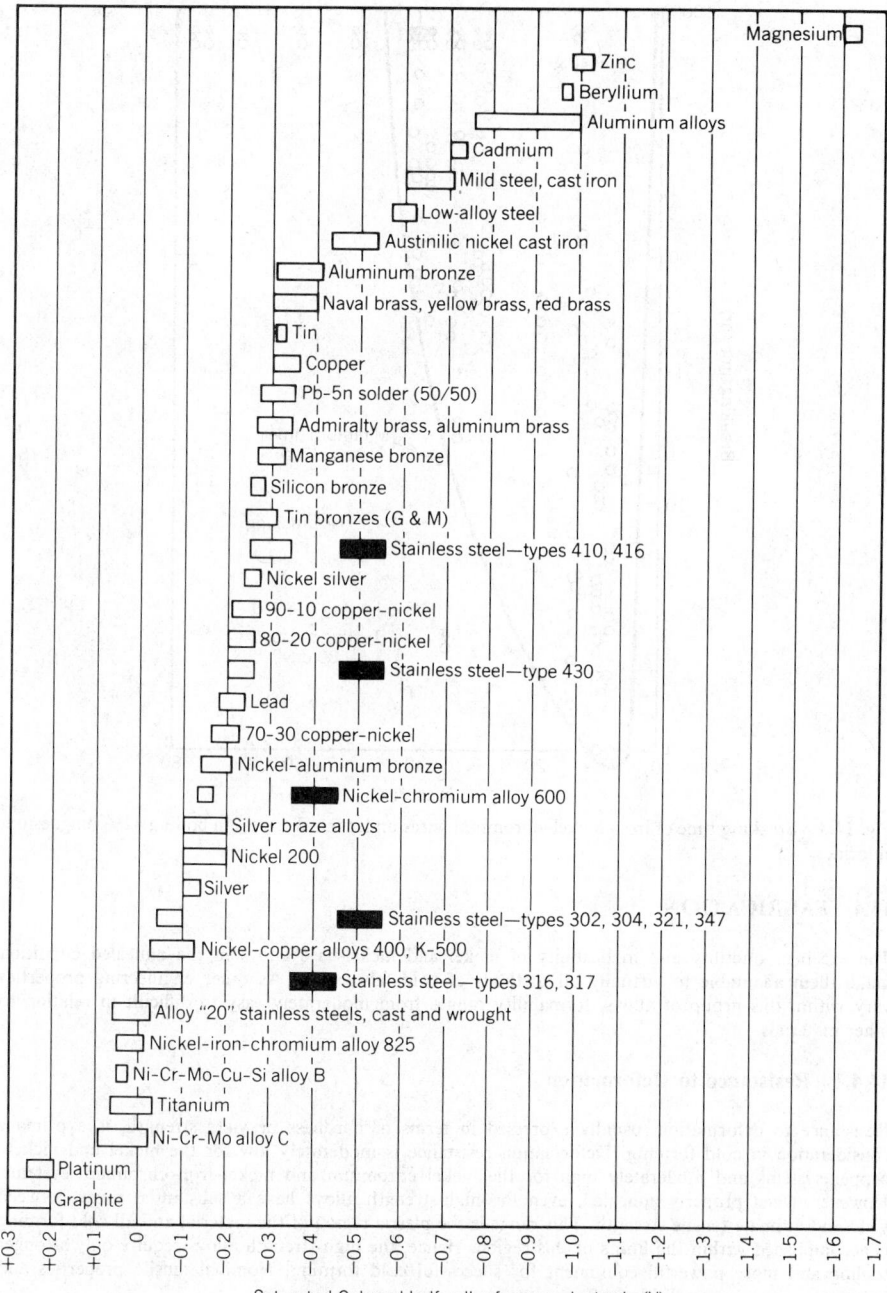

Fig. 14.2 Corrosion potentials in flowing seawater (8–13 ft/sec), temperature range 50–80°F. Alloys are listed in the order of the potential they exhibit in flowing seawater. Certain alloys, indicated by solid boxes, in low velocity or poorly aerated water, and at shielded areas, may become active and exhibit a potential near −0.5 V.

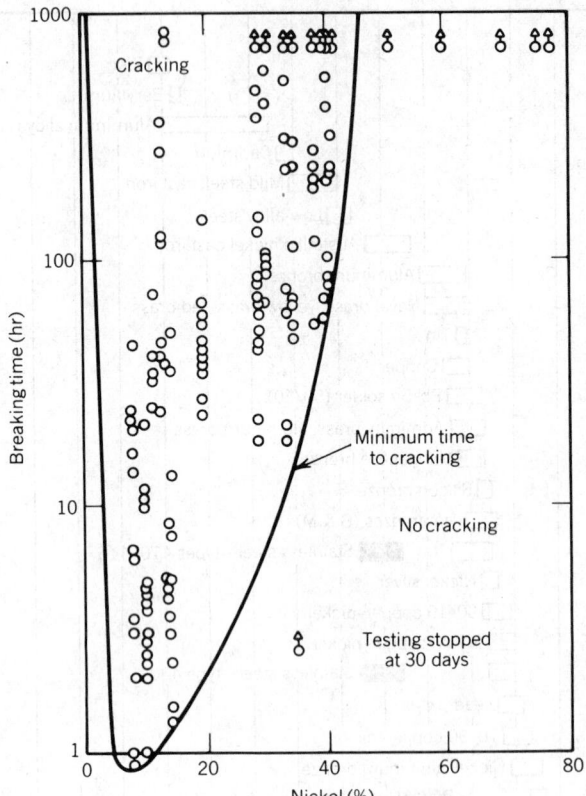

Fig. 14.3 Breaking time of iron–nickel–chromium wires under tensile stress in boiling 42% magnesium chloride.

14.4 FABRICATION

The excellent ductility and malleability of nickel and nickel-base alloys in the annealed condition make them adaptable to virtually all methods of cold fabrication. As other engineering properties vary within this group of alloys, formability ranges from moderately easy to difficult in relation to other materials.

14.4.1 Resistance to Deformation

Resistance to deformation, usually expressed in terms of hardness or yield strength, is a primary consideration in cold forming. Deformation resistance is moderately low for the nickel and nickel–copper systems and moderately high for the nickel–chromium and nickel–iron–chromium systems. However, when properly annealed, even the high-strength alloys have a substantial range between yield and ultimate tensile strength. This range is the plastic region of the material and all cold forming is accomplished within the limits of this region. Hence, the high-strength alloys require only stronger tooling and more powerful equipment for successful cold forming. Nominal tensile properties and hardnesses are given in Table 14.2.

14.4.2 Strain Hardening

A universal characteristic of the high-nickel alloys is that they have face-centered-cubic crystallographic structures, and, consequently, are subject to rapid strain hardening. This characteristic is used to advantage in increasing the room-temperature tensile properties and hardness of alloys that otherwise would have low mechanical strength, or in adding strength to those alloys that are hardened by a precipitation heat treatment. Because of this increased strength, large reductions can be made without rupture of the material. However, the number of reductions in a forming sequence will be limited before annealing is required, and the percentage reduction in each successive operation must be reduced.

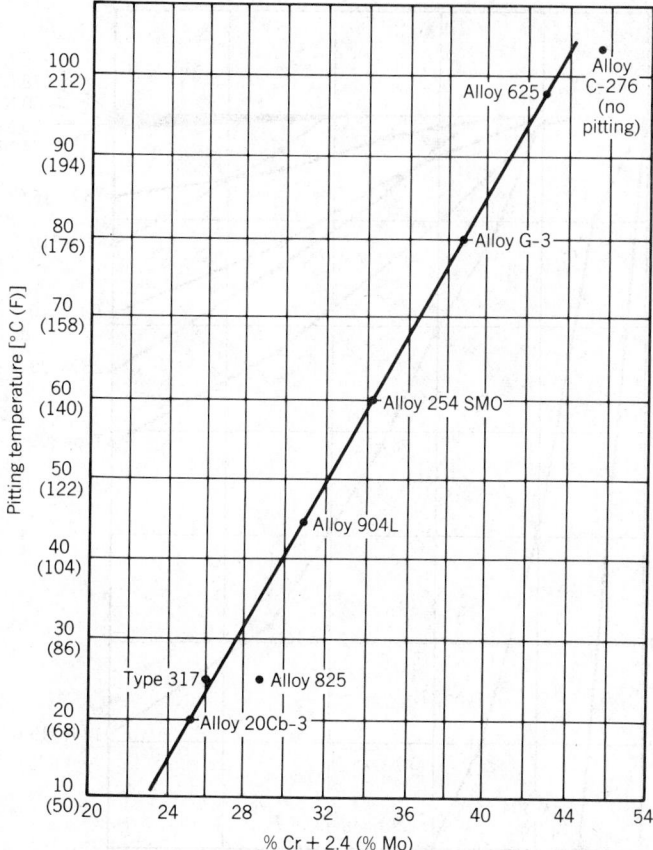

Fig. 14.4   Critical temperature for pitting in 4% NaCl + 1% $Fe_2(SO_4)_3$ + 0.01 M HCl versus composition for Fe–Ni–Cr–Mo alloys.

Since strain hardening is related to the solid-solution strengthening of alloying elements, the strain-hardening rate generally increases with the complexity of the alloy. Accordingly, strain-hardening rates range from moderately low for nickel and nickel–copper alloys to moderately high for nickel–chromium and nickel–iron–chromium alloys. Similarly, the age-hardenable alloys have higher strain-hardening rates than their solid-solution equivalents. Figure 14.6 compares the strain-hardening rates of some nickel alloys with those of other materials as shown by the increase in hardness with increasing cold reduction.

Laboratory tests have indicated that the shear strength of the high-nickel alloys in double shear averages about 65% of the ultimate tensile strength (see Table 14.4). These values, however, were obtained under essentially static conditions using laboratory testing equipment having sharp edges and controlled clearances. Shear loads for well-maintained production equipment can be found in Table 14.5. These data were developed on a power shear having a 31 mm/m (³⁄₈ in./ft) rake.

14.5   HEAT TREATMENT

High-nickel alloys are subject to surface oxidation unless heating is performed in a protective atmosphere or under vacuum. A protective atmosphere can be provided either by controlling the ratio of fuel and air to minimize oxidation or by surrounding the metal being heated with a prepared atmosphere.

Monel alloy 400, Nickel 200, and similar alloys will remain bright and free from discoloration when heated and cooled in a reducing atmosphere formed by the products of combustion. The alloys that contain chromium, aluminum, or titanium form thin oxide films in the same atmosphere and, therefore, require prepared atmospheres to maintain bright surfaces.

Regardless of the type of atmosphere used, it must be free of sulfur. Exposure of nickel alloys to sulfur-containing atmospheres at high temperatures can cause severe sulfidation damage.

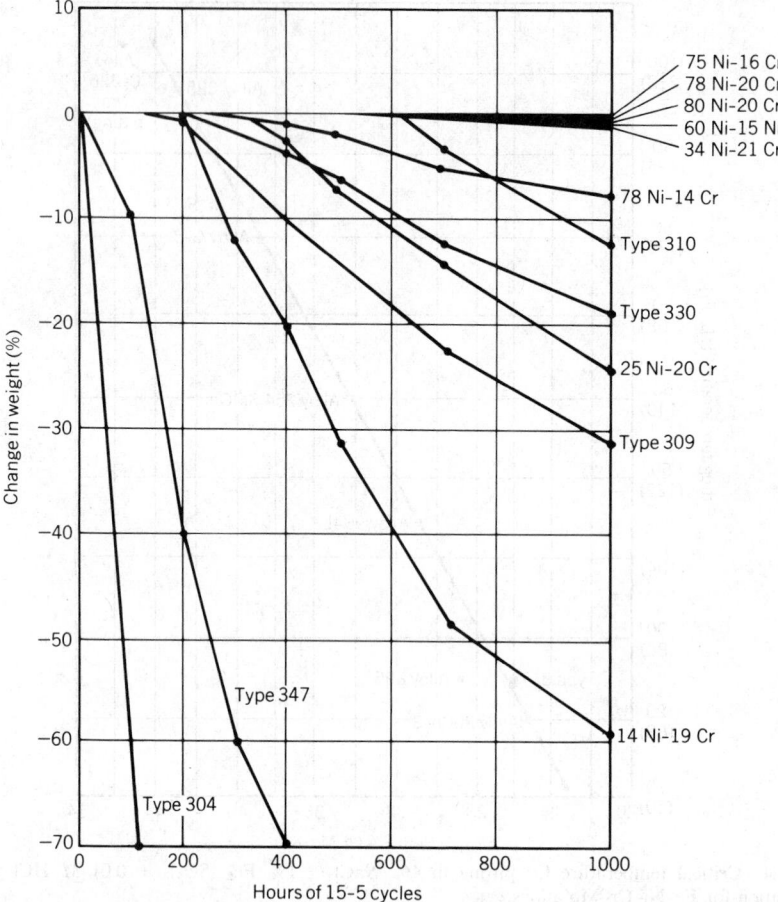

Fig. 14.5 Effect of nickel content on air oxidation of alloys. Each cycle consisted of 15 min at 1800°F followed by a 5-min air cooling.

   The atmosphere of concern is that in the immediate vicinity of the work, that is, the combustion gases that actually contact the surface of the metal. The true condition of the atmosphere is determined by analyzing gas samples taken at various points about the metal surface.

   Furnace atmospheres can be checked for excessive sulfur by heating a small test piece of the material, for example, 13 mm (½ in.) diameter rod or 13 mm × 25 mm (½ in. × 1 in.) flat bar, to the required temperature and holding it at temperature for 10–15 min. The piece is then air cooled or water quenched and bent through 180° flat on itself. If heating conditions are correct, there will be no evidence of cracking.

14.5.1   Reducing Atmosphere

The most common protective atmosphere used in heating the nickel alloys is that provided by controlling the ratio between the fuel and air supplied to the burners. A suitable reducing condition can be obtained by using a slight excess of fuel so that the products of combustion contain at least 4%, preferably 6%, of carbon monoxide plus hydrogen. The atmosphere should not be permitted to alternate from reducing to oxidizing; only a slight excess of fuel over air is needed.

   It is important that combustion takes place before the mixture of fuel and air comes into contact with the work, otherwise the metal may be embrittled. To ensure proper combustion, ample space should be provided to burn the fuel completely before the hot gases contact the work. Direct impingement of the flame can cause cracking.

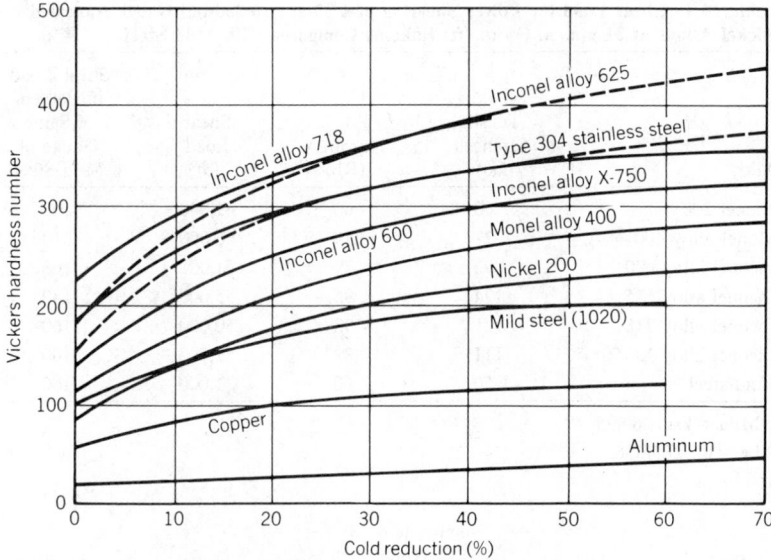

Fig. 14.6   Effect of cold work on hardness.

Table 14.4   Strength in Double Shear of Nickel and Nickel Alloys

Alloy	Condition	Shear Strength (ksi)[a]	Tensile Strength (ksi)	Hardness
Nickel 200	Annealed	52	68	46 Rb
	Half-hard	58	79	84 Rb
	Full-hard	75	121	100 Rb
Monel alloy 400	Hot-rolled, annealed	48	73	65 Rb
	Cold-rolled, annealed	49	76	60 Rb
Inconel alloy 600	Annealed	60	85	71 Rb
	Half-hard	66	98	98 Rb
	Full-hard	82	152	31 Rc
Inconel alloy X-750	Age hardened[b]	112	171	36 Rc

[a] MPa = ksi × 6.895.

[b] Mill-annealed and aged 1300°F (705°C)/20 hr.

14.5.2   Prepared Atmosphere

Various prepared atmospheres can be introduced into the heating and cooling chambers of furnaces to prevent oxidation of nickel alloys. Although these atmospheres can be added to the products of combustion in a directly fired furnace, they are more commonly used with indirectly heated equipment. Prepared protective atmospheres suitable for use with the nickel alloys include dried hydrogen, dried nitrogen, dried argon or any other inert gas, dissociated ammonia, and cracked or partially reacted natural gas. For the protection of pure nickel and nickel–copper alloys, cracked natural gas should be limited to a dew point of −1 to 4°C (30 to 40°F).

Figure 14.7 indicates that at a temperature of 1093°C (2000°F), a hydrogen dew point of less than −30°C (−20°F) is required to reduce chromium oxide to chromium; at 815°C (1500°F) the dew point must be below −50°C (−60°F). The values were derived from the thermodynamic relationships of pure metals with their oxides at equilibrium, and should be used only as a guide to the behavior

Table 14.5 Shear Load for Power Shearing of 6.35-mm (0.250-in.) Gauge Annealed Nickel Alloys at 31 mm/m (⅜ in./ft) Rake as Compared with Mild Steel

Alloy	Tensile Strength (ksi)[a]	Hardness (Rb)	Shear Load (lb)[b]	Shear Load in Percent of Same Gauge of Mild Steel
Nickel 200	60	60	61,000	200
Monel alloy 400	77	75	66,000	210
Inconel alloy 600	92	79	51,000	160
Inconel alloy 625	124	95	55,000	180
Inconel alloy 718	121	98	50,000	160
Inconel alloy X-750	111	88	57,000	180
Mild steel	50	60	31,000	100

[a] MPa = ksi × 6.895.

[b] kg = lb × 0.4536.

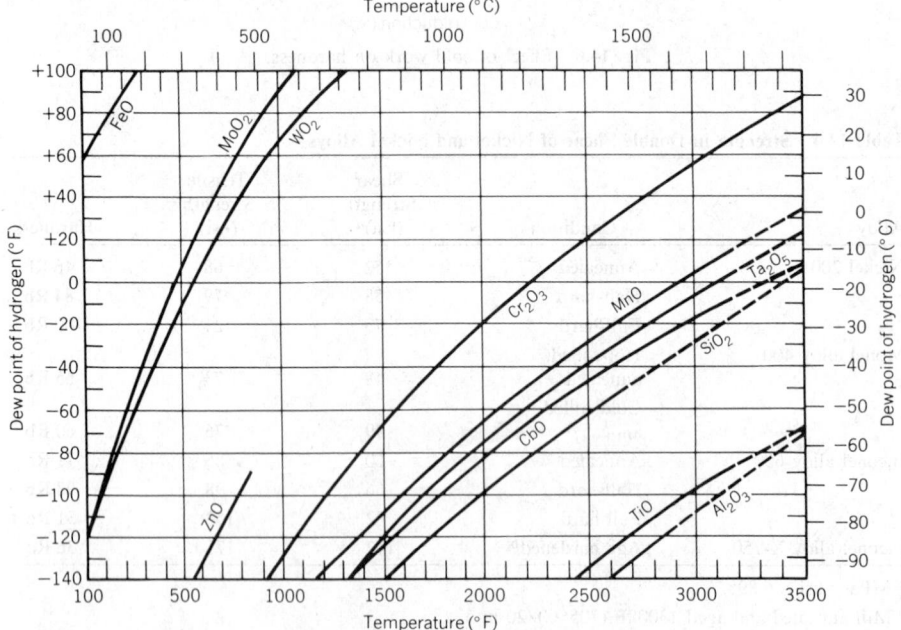

Fig. 14.7 Metal/metal oxide equilibria in hydrogen atmospheres.

of complex alloys under nonequilibrium conditions. However, these curves have shown a close correlation with practical experience. For example, Inconel alloy 600 and Incoloy alloy 800 are successfully bright-annealed in hydrogen having a dew point of −35 to −40°C (−30 to −40°F).

As indicated in Fig. 14.7, lower dew points are required as the temperature is lowered. To minimize oxidation during cooling, the chromium-containing alloys must be cooled rapidly in a protective atmosphere.

14.6 WELDING

Cleanliness is the single most important requirement for successful welded joints in nickel alloys. At high temperatures, nickel and its alloys are susceptible to embrittlement by sulfur, phosphorus, lead, and other low-melting-point substances. Such substances are often present in materials used in normal manufacturing/fabrication processes; some examples are grease, oil, paint, cutting fluids, marking crayons and inks, processing chemicals, machine lubricants, and temperature-indicating sticks, pellets, or lacquers. Since it is frequently impractical to avoid the use of these materials during processing and fabrication of the alloys, it is mandatory that the metal be thoroughly cleaned prior to any welding operation or other high-temperature exposure.

Before maintenance welding is done on high-nickel alloys that have been in service, products of corrosion and other foreign materials must be removed from the vicinity of the weld. Clean, bright base metal should extend 50–75 mm (2–3 in.) from the joint on both sides of the material. This prevents embrittlement by alloying of corrosion products during the welding process. Cleaning can be done mechanically by grinding with a fine grit wheel or disk, or chemically by pickling.

14.7 MACHINING

Nickel and nickel-base alloys can be machined by the same techniques used for iron-base alloys. However, higher loads will be imparted to the tooling requiring heavy-duty equipment to withstand the load and coolants to dissipate the heat generated. The cutting tool edge must be maintained sharp and have the proper geometry.

14.8 CLOSURE

There has been a vast amount of nickel-alloy developments since the 1950 edition of *Kent's Mechanical Engineer's Handbook*. It has not been possible to give the composition and discuss each commercial alloy and, therefore, one should refer to publications like Refs. 6–8 for alloy listings, which are revised periodically to include the latest alloys available. (See Table 14.6 for the producer companies of some of the alloys mentioned in this chapter.)

Table 14.6 Registered Trademarks of Producer Company

Trademark	Owner
Duranickel	Inco family of companies
Hastelloy	Cabot
Incoloy	Inco family of companies
Inconel	Inco family of companies
MAR-M	Martin Marietta Corp.
Monel	Inco family of companies
Nilo	Inco family of companies
Ni-Span-C	Inco family of companies
Permanickel	Inco family of companies
Pyromet	Carpenter Technology Corp.
Rene	General Electric Co.
Rene' 41	Allvac Metals Corp.
Udimet	Special Metals Corp.
Waspaloy	United Aircraft Corp.

REFERENCES

1. Joseph R. Boldt, Jr., *The Winning of Nickel,* Van Nostrand, New York, 1967.

2. *Nickel and Its Alloys,* NBS Monograph 106, May, 1968.

3. *Kent's Mechanical Engineer's Handbook,* 1950 edition, pp. 4–50 to 4–60.

4. Huntington Alloys, Inc., *Alloy Handbook,* and *Bulletins.*

5. Inco internal communication by A. J. Sedriks.

6. *Alloy Digest,* Engineering Alloy Digest, Inc., 1983.

7. *Aerospace Structural Metals Handbook,* 1983.

8. *Materials and Processing Databook,* 1983 Metals Progress.

9. J. S. Benjamin, *Met. Trans. AIME,* **1,** 2943 (1970).

10. J. P. Morse and J. S. Benjamin, *J. Met.,* **29** (12), 9 (1977).
11. Wayne Z. Friend, *Corrosion of Nickel and Nickel-Base Alloy,* Wiley, New York, 1980.
12. F. L. LaQue and H. R. Copson, *Corrosion Resistance of Metals and Alloys,* 2nd ed., Reinhold, New York, 1963.
13. J. Kolts et al., "Highly Alloyed Austenitic Materials for Corrosion Service," *Metal Prog.,* 25–36 (September, 1983).
14. *High Temperature Corrosion in Refinery and Petrochemical Service,* Inco Publication, 1960.

CHAPTER 15
TITANIUM AND ITS ALLOYS

DONALD KNITTEL

JAMES B. C. WU

Cabot Corporation
Kokomo, Indiana

Reprinted with additions from *Kirk–Othmer Encyclopedia of Chemical Technology*, 3rd ed., Wiley, New York, 1983, Vol. 23, by permission of the publisher.

15.1 INTRODUCTION

Titanium was first identified as a constituent of the earth's crust in the late 1700s. In 1790, William Gregor, an English clergyman and mineralogist, discovered a black magnetic sand (ilmenite), which he called menaccanite after his local parish. In 1795, a German chemist found that a Hungarian mineral, rutile, was the oxide of a new element he called titan, after the mythical Titans of ancient Greece. In the early 1900s, a sulfate purification process was developed to commercially obtain high-purity TiO_2 for the pigment industry, and titanium pigment became available in both the United States and Europe. During this period, titanium was also used as an alloying element in irons and steels. In 1910, 99.5% pure titanium metal was produced at General Electric from titanium tetrachloride and sodium in an evacuated steel container. Since the metal did not have the desired properties, further work was discouraged. However, this reaction formed the basis for the commercial sodium reduction process. In the 1920s, ductile titanium was prepared with an iodide dissociation method combined with Hunter's sodium reduction process.

In the early 1930s, a magnesium vacuum reduction process was developed for reduction of titanium tetrachloride to metal. Based on this process, the U.S. Bureau of Mines (BOM) initiated a program in 1940 to develop commercial production. Some years later, the BOM publicized its work on titanium and made samples available to the industrial community. By 1948, the BOM produced batch sizes of 104 kg. In the same year, E. I. du Pont de Nemours & Co., Inc., announced commercial availability of titanium, and the modern titanium metals industry began.[1]

By the mid-1950s, this new metals industry had become well established, with six producers, two other companies with tentative production plans, and more than 25 institutions engaged in research projects. Titanium, termed the wonder metal, was billed as the successor to aluminum and stainless steels. When, in the 1950s, the DOD (titanium's most staunch supporter) shifted emphasis from aircraft to missiles, the demand for titanium sharply declined. Only two of the original titanium metal plants are still in use, the Titanium Metals Corporation of America's (TMCA) plant in Henderson, Nevada, and National Distillers & Chemical Corporation's two-stage sodium reduction plant built in the late 1950s at Ashtabula, Ohio, which now houses the sponge production facility for RMI Corporation (formerly Reactor Metals, Inc.).

Overoptimism followed by disappointment has characterized the titanium-metals industry. In the late 1960s, the future again appeared bright. Supersonic transports and desalination plants were intended to use large amounts of titanium. Oregon Metallurgical Corporation, a titanium melter, decided at that time to become a fully integrated producer (i.e., from raw material to mill products). However, the supersonic transports and the desalination industry did not grow as expected. Nevertheless, in the late 1970s and early 1980s, the titanium-metal demand again exceeded capacity and both the United States and Japan expanded capacities. This growth was stimulated by greater acceptance of titanium in the chemical-process industry, power-industry requirements for seawater cooling, and commercial and military aircraft demands. However, with the economic recession of 1981–1983, the demand dropped well below capacity and the industry was again faced with hard times.

15.2 ALLOYS

Titanium alloy systems have been studied extensively. A single company evaluated over 3000 compositions in 8 years. Alloy development has been aimed at elevated-temperature aerospace applications, strength for structural applications, and aqueous corrosion resistance. The principal effort has been in aerospace applications to replace nickel- and cobalt-base alloys in the 500–900°C ranges. To date, titanium alloys have replaced steel in the 200–500°C range. The useful strength and corrosion-resistance temperature limit is ~550°C.

The addition of alloying elements alters the α–β transformation temperature. Elements that raise the transformation temperature are called α stabilizers; elements that depress the transformation temperature are called β stabilizers; the latter are divided into β-isomorphous and β-eutectoid types. The β-isomorphous elements have limited α solubility, and increasing additions of these elements progressively depresses the transformation temperature. The β-eutectoid elements have restricted beta solubility and form intermetallic compounds by eutectoid decomposition of the β phase. The binary phase diagram illustrating these three types of alloy systems is shown in Fig. 15.1.

The important α-stabilizing alloying elements include aluminum, tin, zirconium, and the interstitial alloying elements (i.e., elements that do not occupy lattice positions) oxygen, nitrogen, and carbon. Small quantities of interstitial alloying elements, generally considered to be impurities, have a very great effect on strength and ultimately embrittle the titanium at room temperature.[3] The effects of oxygen, nitrogen, and carbon on the ultimate tensile properties and elongation are shown in Table 15.1. These elements are always present and are difficult to control. Nitrogen has the greatest effect, and commercial alloys specify its limit to be less than 0.05 wt %. It may also be present as nitride (TiN) inclusions, which are detrimental to critical aerospace structural applications. Oxygen additions increase strength and serve to identify several commercial grades. This strengthening effect diminishes at elevated temperatures and under creep conditions at room temperature. For cryogenic service, low

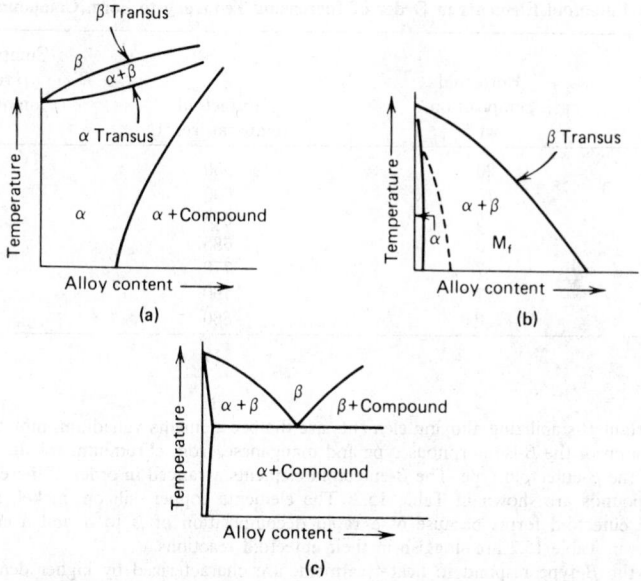

Fig. 15.1 The effect of alloying elements on the phase diagram of titanium: (a) α-stabilized system, (b) β-isomorphous system, and (c) β-eutectoid system.[2]

oxygen content is specified (<1300 ppm) because high concentrations of interstitial impurities increase sensitivity to cracking, cold brittleness, and fracture temperatures. Alloys with low interstitial content are identified as ELI (extra-low interstitials) after the alloy name. Carbon does not affect strength at concentration above 0.25 wt % because carbides (TiC) are formed. Carbon content is usually specified at 0.08 wt % max.[4]

The most important alloying element is aluminum, an α stabilizer. It is not expensive, and its atomic weight is less than that of titanium; hence, aluminum additions lower the density. The mechanical strength of titanium can be increased considerably by aluminum additions. Even though the solubility range of aluminum extends to 27 wt %, above 7.5 wt % the alloy becomes too difficult to fabricate and embrittles. The embrittlement is caused by a coherently ordered phase based on Ti₃Al. Other α-stabilizing elements also cause phase ordering. An empirical relationship below which ordering does not occur is[5]

$$\text{wt \% Al} + \frac{\text{wt \% Sn}}{3} + \frac{\text{wt \% Zr}}{6} + 10 \times \text{wt \% O} \leq 9$$

Table 15.1 Effects of O, N, and C on the Ultimate Tensile Strength.[2]

Concentration of impurity, wt %	Oxygen[c,d] UT, MPa[e]	Elong., %	Nitrogen[c,d] UT, MPa[e]	Elong., %	Carbon[c,d] UT, MPa[e]	Elong., %
0.025	330	37	380	35	310	40
0.05	365	35	460	28	330	39
0.1	440	30	550	20	370	36
0.15	490	27	630	15	415	32
0.2	545	25	700	13	450	26
0.3	640	23	embrittles		500	21
0.5	790	18			520	18
0.7	930	8			525	17

[a] Ref. 3.
[b] Tests were conducted using titanium produced by the iodide process.
[c] UT = ultimate tensile stress.
[d] Elongation on 2.54 cm.
[e] To convert MPa to psi, multiply by 145.

Table 15.2   β-Eutectoid Elements in Order of Increasing Tendency to Form Compounds.[2]

Element	Eutectoid composition, wt %	Eutectoid temperature, °C	Composition for β retention on quenching, wt %
manganese	20	550	6.5
iron	15	600	4.0
chromium	15	675	8.0
cobalt	9	685	7.0
nickel	7	770	8.0
copper	7	790	13.0
silicon	0.9	860	

[a] Ref. 6.

The important β-stabilizing alloying elements are the bcc elements vanadium, molybdenum, tantalum, and niobium of the β-isomorphous type and manganese, iron, chromium, cobalt, nickel, copper, and silicon of the β-eutectoid type. The β-eutectoid elements arranged in order of increasing tendency to form compounds are shown in Table 15.2. The elements copper, silicon, nickel, and cobalt are termed active eutectoid forms because of a rapid decomposition of β to α and a compound. The other elements in Table 15.2 are sluggish in their eutectoid reactions.

Alloys of the β type respond to heat treatment, are characterized by higher density than pure titanium, and are easily fabricated. The purpose of β alloying is to form an all-β-phase alloy with commercially useful qualities, form alloys with duplex α and β structure to enhance heat-treatment response (i.e., changing the α and β volume ratio), or use β-eutectoid elements for intermetallic hardening. The most important commercial β-alloying element is vanadium.

15.2.1   Aerospace Alloys

The alloys of titanium for aerospace use can be divided into three categories: an all-α structure, a mixed α–β structure, and an all-β structure. The α–β structure alloys are further divided into near-α alloys (<2% β stabilizers). Most of the approximately 100 commercially available alloys (approximately 30 in the United States, 40 in the USSR, and 10 in Europe and Japan) are of the α–β structure

Table 15.3   Properties, specifications,

Nominal composition, wt %	CAS Registry No.	ASTM B-265	CLTE[b], µm/(m·K) 21–100°C	21–538°C	Average physical properties Modulus of elasticity[c], GPa[d]	Modulus of rigidity[c], GPa[d]	Poisson's[c] ratio	Density, g/cm³	Condition
commercially pure									
99.5 Ti		grade 1	8.7	9.8	102	39	0.34	4.5	annealed
99.2 Ti		grade 2	8.7	9.8	102	39	0.34	4.5	annealed
99.1 Ti		grade 3	8.7	9.8	103	39	0.34	4.5	annealed
99.0 Ti		grade 4	8.7	9.8	104	39	0.34	4.5	annealed
99.2 Ti[h]		grade 7	8.7	9.8	102	39	0.34	4.5	annealed
98.9 Ti[i]								4.5	annealed
Ti–5 Al–2.5 Sn[j]	[11109-19-6]	grade 6	9.4	9.6	110			4.5	annealed
Ti–8 Al–1 Mo, 1 V[j]	[39303-55-4]		8.5	10.1	124	47	0.32	4.4	duplex annealed
Ti–6 Al–2 Sn, 4 Zr–2 Mo[j]	[11109-15-2]		7.8	8.1	114			4.5	annealed
Ti–3 Al–2.5 V[j]	[11109-23-2]		9.6	9.9	107			4.5	annealed
Ti–6 Al–4 V[j]	[12743-70-3]	grade 5	8.7	9.6	114	42	0.342	4.4	annealed
Ti–6 Al–6 V, 2 Sn[j]	[12606-77-8]		9.0	9.6	110			4.5	annealed
Ti–10 V–2 Fe, 3 Al[j]	[51809-47-3]				112			4.6	solution and age

[a] Refs. 9, 10.
[b] CLTE = coefficient of linear thermal expansion.
[c] Room temperature.
[d] To convert GPa to psi, multiply by 145,000.
[e] To convert MPa to psi, multiply by 145.

type.[7] Some of these, produced in the United States, are given in Table 15.3 along with some wrought properties.[8–10] The most important commercial alloy is Ti–6 Al–4 V, an α–β alloy with a good combination of strength and ductility. It can be age-hardened and has moderate ductility, and an excellent record of successful applications. It is mostly used for compressor blades and disks in aircraft gas-turbine engines, and also in lower-temperature engine applications such as rotating disks and fans. It is also used for rocket-motor cases, structural forgings, steam-turbine blades, and cryogenic parts for which ELI grades are usually specified.

Other commercially important α–β alloys are Ti–3 Al–2.5 V, Ti–6 Al–6 V–2 Sn, and Ti–10 V–2 Fe–3 Al (see Table 15.3). As a group, these alloys have good strength, moderate ductility, and can be age-hardened.[10,11] Weldability becomes more difficult with increasing β constituents, and fabrication of strip, foil, sheet, and tubing may be difficult. Temperature tolerances are lower than those of the α or near-α alloys. The alloy Ti–3 Al–2.5 V (called one-half Ti–6 Al–4 V) is easier to fabricate than Ti–6 Al–4 V and is used primarily as seamless aircraft-hydraulic tubing. The alloy Ti–6 Al–6 V–2 Sn is used for some aircraft forgings because it has a higher strength than Ti–6 Al–4 V. The alloy Ti–10 V–2 Fe–3 Al is easier to forge at lower temperatures than Ti–6 Al–4 V because it contains more β-alloying constituents and has good fracture toughness. This alloy can be hardened to high strengths [1.24–1.38 GPa or (1.8–2) $\times$ 10^5 psi] and is expected to be used as forgings for airframe structures to replace steel below temperatures of 300°C.[12]

The only α alloy of commercial importance is Ti–5 Al–2.5 Sn. It is weldable, has good elevated-temperature stability, and good oxidation resistance to about 600°C. It is used for forgings and sheet-metal parts such as aircraft-engine compressor cases because of weldability.

The commercially important near-α alloys are Ti–8 Al–1 Mo–1 V and Ti–6 Al–2 Sn–4 Zr–2 Mo. They exhibit good creep resistance and the excellent weldability and high strength of α alloys; the temperature limit is ~500°C. Alloy Ti–8 Al–1 Mo–1 V is used for compressor blades because of its high elastic modules and creep resistance; however, it may suffer from ordering embrittlement. Alloy Ti–6 Al–2 Sn–4 Zr–2 Mo is also used for blades and disks in aircraft engines. The service temperature limit of 470°C is ~70°C higher than that of Ti–8 Al–1 Mo–1 V.[5]

Commercialization of β alloys has not been very successful. Even though alloys with high strength [up to 1.5 GPa (217,500 psi)] were made, they suffered from intermetallic and ω-phase embrittlement. These alloys are metallurgically unstable and have little practical use above 250°C. They are fabricable but welds are not ductile. This alloy type is used in the cold-drawn or cold-rolled condition and finds application in spring manufacture (alloy Ti–13 V–11 Cr–3 Al).[13] There is one commercially available alloy of the β-eutectoid type (Ti–2.5 Cu) that uses a true precipitation-hardening mechanism to increase strength. The precipitate is Ti$_2$Cu. This alloy is only slightly heat treatable; it is used in engine castings and flanges.[5]

and Applications of Wrought Titanium Alloys.[2]

Average mechanical properties										
Room temperature					Extreme temperatures				Charpy	
Tensile strength, MPa[e]	Yield strength, MPa[e]	Elonga-tion, %	Reduction in area, %	Test temperature, °C	Tensile strength, MPa[e]	Yield strength, MPa[e]	Elonga-tion, %	Reduction in area, %	impact strength, J/m[f]	Hardness[g]
331	241	30	55	315	152	97	32	80		HB 120
434	346	28	50	315	193	117	35	75	43	HB 200
517	448	25	45	315	234	138	34	75	38	HB 225
662	586	20	40	315	310	172	25	70	20	HB 265
434	346	28	50	315	186	110	37	75	43	HB 200
517	448	25	42	315	324	207	32			
862	807	16	40	315	565	448	18	45	26	HRC 36
1000	952	15	28	540	621	517	25	55	33	HRC 35
979	896	15	35	540	648	490	26	60		HRC 32
690	586	20		315	483	345	25			
993	924	14	30	540	531	427	35	50	19	HRC 36
1069	1000	14	30	315	931	807	18	42	18	HRC 38
1276	1200	10	19	315	1103	979	13	42		

[f] To convert J/m to ft·lb/in., divide by 53.38.

[g] HB = Brinnell, HRC = Rockwell (C-scale).

[h] Also contains 0.2 Pd.

[i] Also contains 0.8 Ni and 0.3 Mo.

[j] Numerical designations = wt % of element.

15.2.2 Nonaerospace Alloys

The nonaerospace alloys are used primarily in industrial applications. The four grades (ASTM grade 1 through grade 4) differ primarily in oxygen and iron content (see Table 15.4). ASTM grade 1 has the highest purity and the lowest strength (strength is controlled by impurities). The two other alloys of this group are ASTM grade 7, Ti–0.2 Pd, and ASTM grade 12, Ti–0.8 Ni–0.3 Mo. The alloys in this group are distinguished by excellent weldability, formability, and corrosion resistance. The strength, however, is not maintained at elevated temperatures (see Table 15.3). The primary use of alloys in this group is in industrial-processing equipment (i.e., tanks, heat exchangers, pumps, electrodes, etc.), even though there is some use in airframes and aircraft engines. The ASTM grade 1 is used where higher purity is desired, for example, as weld wire for grade 2 fabrication and as sheet for explosive bonding to steel. Grade 1 is manufactured from high-purity sponge. The ASTM grade 2 is the most commonly used grade of commercially pure titanium. The chemistry for this grade is easy to meet with most sponge. The ASTM grades 3 and 4 are higher strength versions of grade 2; grades 7 and 12 have better corrosion resistance than grade 2 in reducing acids and acid chlorides. However, grade 7 is expensive and grade 12 is not readily available.

15.2.3 Other Alloys

Other alloying ranges include the aluminides (TiAl and Ti₃Al), the superconducting alloys (Ti–Nb type), the shape-memory alloys (Ni–Ti type), and the hydrogen-storage alloys (Fe–Ti). The aluminides TiAl and Ti₃Al have excellent high-temperature strengths, comparable to those of nickel- and cobalt-base alloys, with less than half the density. These alloys exhibit ultimate strengths of 1 GPa (145,000 psi), and 800 MPa (116,000 psi) yield, respectively, 4–5% elongation, and 7% reduction in area. Strengths are maintained to 800–900°C. The modulus of elasticity is high [125–165 GPa, (18–24) × 10⁶ psi], and oxidation resistance is good.[8] The aluminides are intended for both static and rotating parts in the turbine section of gas-turbine aircraft engines.

Titanium alloyed with niobium exhibits superconductivity, and a lack of electrical resistance below 10 K. Composition ranges from 25 to 50 wt % Ti. These alloys are β-phase alloys with superconducting transitional temperatures at ~10 K. Their use is of interest for power generation, propulsion devices, fusion research, and electronic devices.[14]

Titanium alloyed with nickel exhibits a memory effect, that is, the metal form switches from one specific shape to another in response to temperature changes. The group of Ti–Ni alloys (nitinol) was developed by the Navy in the early 1960s for F-14 fighter jets. The compositions are typically Ti with 55 wt % Ni. The transition temperature ranges from −100°C to >100°C and is controlled by additional alloying elements. These alloys are of interest for thermostats, recapture of waste heat, pipe joining, etc. The nitinols have not been extensively used because of high price and fabrication difficulties.[15]

Titanium alloyed with iron is a leading candidate for solid-hydride energy-storage material for automotive fuel. The hydride, $FeTiH_2$, absorbs and releases hydrogen at low temperatures. This hydride stores 0.9 kW·hr/kg. To provide the energy equivalent to a tank of gasoline would require about 800 kg $FeTiH_2$.[8]

Table 15.4   ASTM Requirements for Different Titanium Grades.[2]

Element	Grade 1	Grade 2	Grade 3	Grade 4	Grade 7	Grade 12
nitrogen, max	0.03	0.03	0.05	0.05	0.03	0.03
carbon, max	0.10	0.10	0.10	0.10	0.10	0.08
hydrogen, max	0.015	0.015	0.015	0.015	0.015	0.015
iron, max	0.20	0.30	0.30	0.50	0.30	0.30
oxygen, max	0.18	0.25	0.35	0.40	0.25	0.25
palladium					0.12–0.25	
molybdenum						0.2–0.4
nickel						0.6–0.9
residuals, max						
each	0.1	0.1	0.1	0.1	0.1	0.1
total	0.4	0.4	0.4	0.4	0.4	0.4
titanium	remainder	remainder	remainder	remainder	remainder	remainder

[a] Ref. 4.

Table 15.5 Physical Properties of Titanium.[2]

Property	Value
melting point, °C	1668 ± 5
boiling point, °C	3260
density, g/cm³	
α phase at 20°C	4.507
β phase at 885°C	4.35
allotropic transformation, °C	882.5
latent heat of fusion, kJ/kg[a]	440
latent heat of transition, kJ/kg[a]	91.8
latent heat of vaporization, MJ/kg[a]	9.83
entropy at 25°C, J/mol[a]	30.3
thermal expansion coefficient at 20°C per °C ·	8.41×10^{-6}
thermal conductivity at 25°C, W/(m·K)	21.9
emissivity	9.43
electrical resistivity at 20°C, nΩ·m	420
magnetic susceptibility, mks	180×10^{-6}
modulus of elasticity, GPa[b]	
tension	ca 101
compression	103
shear	44
Poisson's ratio	~0.41
lattice constants, nm	
α, 25°C	$a_0 = 0.29503$
	$c_0 = 0.46531$
β, 900°C	$a_0 = 0.332$
vapor pressure, kPa[c]	$\log P_{kPa} = 5.7904 - 24644/T - 0.000227\ T$
specific heat, J/(kg·K)[d]	$C_p = 669.0 - 0.037188\ T - 1.080 \times 10^7/T^2$

[a] To convert J to cal, divide by 4.184.
[b] To convert GPa to psi, multiply by 145,000.
[c] To convert $\log P_{kPa}$ to $\log P_{atm}$, add 2.0056 to the constant.
[d] $T > 298$ K.

15.3 PHYSICAL PROPERTIES

The physical properties of titanium are given in Table 15.5. The most important physical property of titanium from a commercial viewpoint is the ratio of its strength [ultimate strength > 690 MPa (100,000 psi)] at a density of 4.507 g/cm³. Titanium alloys have a higher yield strength-to-density rating between −200 and 540°C than either aluminum alloys or steel.[6,16] Titanium alloys can be made with strength equivalent to high-strength steel, yet with density ~60% that of iron alloys. At ambient temperatures, titanium's strength-to-weight ratio is equal to that of magnesium, one and one-half times greater than that of aluminum, two times greater than that of stainless steel, and three times greater than that of nickel. Alloys of titanium have much higher strength-to-weight ratios than alloys of nickel, aluminum, or magnesium, and stainless steel. Because of its high melting point, titanium can be alloyed to maintain strength well above the useful limits of magnesium and aluminum alloys. This property gives titanium a unique position in applications between 150 and 550°C where the strength-to-weight ratio is the sole criterion.

Solid titanium exists in two allotropic crystalline forms. The α phase, stable below 882.5°C, is a hexagonal closed-packed structure, whereas the β phase, a bcc crystalline structure, is stable between 882.5°C and the melting point of 1668°C. The high-temperature β phase can be found at room temperature when β-stabilizing elements are present as impurities or additions (see Section 15.2). The α and β phases can be distinguished by examining an unetched polished mount with polarized light. The α is optically active and changes from light to dark as the microscope stage is rotated. The microstructure of titanium is difficult to interpret without knowledge of the alloy content, working temperature, and thermal treatment.[6,17,18]

The heat-transfer qualities of titanium are characterized by the coefficient of thermal conductivity. Even though this is low, heat transfer in service approaches that of admiralty brass (thermal conductivity seven times greater) because titanium's greater strength permits thinner-walled equipment, relative absence of corrosion scale, erosion–corrosion resistance permitting higher operating velocities, and inherently passive film.

Table 15.6 Corrosion Data for ASTM Grade 2 Titanium.[2]

Media	Conc, wt %	Temperature, °C	Corrosion rate, mm/yr
acetaldehyde	100	149	0.0
acetic acid	5–99.7	124	0.0
adipic acid	67	232	0.0
aluminum chloride, aerated	10	100	0.002
	10	150	0.03
	20	149	16
	25	20	0.001
	25	100	6.6
	40	121	109
ammonia + 28% urea + 20.5% H_2O + 19% CO_2 + 0.3% inerts + air	32.2	182	0.08
ammonium carbamate	50	100	0.0
ammonium perchlorate aerated	20	88	0.0
aniline hydrochloride	20	100	0.0
aqua regia	3:1	RT	0.0
	3:1	79	0.9
barium chloride, aerated	5–20	100	<0.003
bromine–water solution		RT	0.0
calcium chloride		RT	0.0
	5	100	0.005
	10	100	0.007
	20	100	0.02
	55	104	0.0005
	60	149	<0.003
	62	154	0.05–0.4
	73	177	2.1
calcium hypochlorite	6	100	0.001
chlorine gas, wet	>0.7 H_2O	RT	0.0
	>1.5 H_2O	200	0.0
chlorine gas, dry	<0.5 H_2O	RT	may react
chlorine dioxide in steam	5	99	0.0
chloracetic acid	100	189	<0.1
chromic acid	50	24	0.01
citric acid	25	100	0.0009
copper sulfate + 2% H_2SO_4	saturated	RT	0.02
cupric chloride, aerated	1–20	100	<0.01
cyclohexane (plus traces of formic acid)		150	0.003
ethylene dichloride	100	boiling	0.005–0.1
ferric chloride	10–30	100	<0.1
formic acid, nonaerated	10	100	2.4
hydrochloric acid, aerated	5	35	0.04
	20	35	4.4
HCl, chlorine saturated	5	190	<0.03
HCl + 10% HNO_3	5	38	0.0
HCl + 1% CrO_3	5	93	0.03
hydrofluoric acid	1–48	RT	rapid
hydrogen peroxide	3	RT	<0.1
hydrogen sulfide, steam and 0.077% mercaptans	7.65	93–110	0.0
hypochlorous acid + Cl_2O and Cl_2	17	38	0.00003
lactic acid	10	boiling	<0.1
manganous chloride, aerated	5–20	100	0.0
magnesium chloride	5–40	boiling	0.0
mercuric chloride, aerated	1	100	0.0003
	5	100	0.01
	10	100	0.001
	55	102	0.0

Table 15.6 (Continued)

Media	Conc, wt %	Temperature, °C	Corrosion rate, mm/yr
mercury	100	RT	0.0
nickel chloride, aerated	5–20	100	0.0004
nitric acid	17	boiling	0.08–0.1
	70	boiling	0.05–0.9
nitric acid, red fuming	<about 2% H_2O RT		ignition sensitive
	>about 2% H_2O RT		nonignition sensitive
oxalic acid	1	37	0.3
oxygen, pure			ignition sensitive
phenol	saturated	21	0.1
phosphoric acid	10–30	RT	0.02–0.05
	10	boiling	10
potassium chloride	saturated	60	<0.0002
potassium dichromate			0.0
potassium hydroxide	50	27	0.01
	50	boiling	2.7
seawater, ten year test			0.0
sodium chlorate	saturated	boiling	0.0
sodium chloride	saturated	boiling	0.0
sodium chloride, titanium in contact with Teflon	23	boiling	crevice attack
sodium dichromate	saturated	RT	0.0
sodium hypochlorite + 12–15% sodium chloride + 1% sodium hydroxide + 1–2% sodium carbonate	1.5–4	66–93	0.03
stannic chloride	5	100	0.003
	24	boiling	0.04
sulfuric acid	1	boiling	2.5
sulfuric acid + 0.25% $CuSO_4$	5	93	0.0
terephthalic acid	77	218	0.0
urea–ammonia reaction mass		elevated temperature and pressure	no attack
zinc chloride	20	104	0.0
	50	150	0.0
	75	200	0.5
	80	200	203

[a] Refs. 16, 21.

15.4 CORROSION RESISTANCE

Titanium is immune to corrosion in all naturally occurring environments. It does not corrode in air, even if polluted or moist with ocean spray. It does not corrode in soil or even the deep salt-mine-type environments where nuclear waste might be buried. It does not corrode in any naturally occurring water and most industrial wastewater streams. For these reasons, titanium has been termed the metal for the earth, and 20–30% of consumption is used in corrosion-resistance applications.

Even though titanium is an active metal, it resists decomposition because of a tenacious protective oxide film. This film is insoluble, repairable, and nonporous in many chemical media and provides excellent corrosion resistance. However, where this oxide film is broken, the corrosion rate is very rapid. However, usually the presence of a small amount of water is sufficient to repair the damaged oxide film. In a seawater solution, this film is maintained in the passive region from ~ −0.2 to 10 V versus the saturated calomel electrode.[19,20]

Titanium is resistant to corrosion attack in oxidizing, neutral, and inhibited reducing conditions. Examples of oxidizing environments are nitric acid, oxidizing chloride ($FeCl_3$ and $CuCl_2$) solutions, and wet chlorine gas. Neutral conditions include all neutral waters (fresh, salt, and brackish), neutral salt solutions, and natural soil environments. Examples of inhibited reducing conditions are in hydrochloric or sulfuric acids with oxidizing inhibitors and in organic acids inhibited with small amounts of water. Corrosion resistances to a variety of media are given in Table 15.6.[22] Titanium resistance

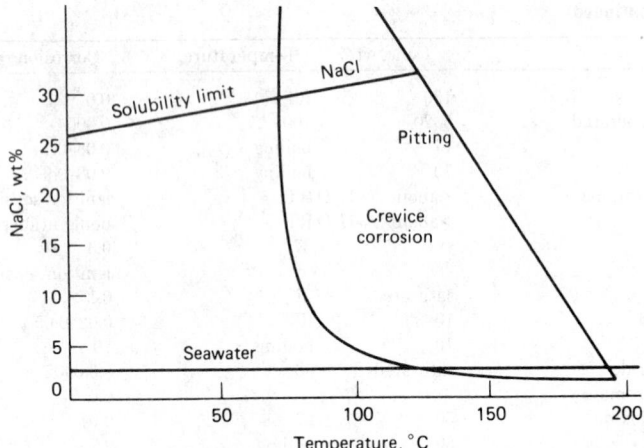

Fig. 15.2   Corrosion characteristics of titanium in aqueous NaCl solution.[23]

to aqueous chloride solutions and chlorine account for most of its use in corrosion-resistant applications.

Titanium corrodes very rapidly in acid fluoride environments. The degree of attack generally increases with the acidity and the fluoride content. It is attacked in boiling HCl or H_2SO_4 at acid concentrations >1% or in ~10 wt % acid concentration at room temperature. Titanium is also attacked by hot caustic solutions, phosphoric acid solutions (concentrations above 25 wt %), boiling $AlCl_3$ (concentrations >15 wt %), dry chlorine gas, anhydrous ammonia above 150°C, and dry hydrogen–dihydrogen sulfide above 150°C.

Titanium is susceptible to pitting and crevice corrosion in aqueous chloride environments. The area of susceptibility is shown in Fig. 15.2 as a function of temperature and sodium chloride content.[22] The susceptibility also depends on pH. The susceptibility temperature increases parabolically from 65°C as pH is increased from zero. With ASTM grades 7 or 12, crevice-corrosion attack is not observed above pH 2 until ~270°C. Noble alloying elements shift the equilibrium potential into the passive region where a protective film is formed and maintained.

Titanium does not stress crack in environments that cause stress cracking of other metal alloys (i.e., boiling 42% $MgCl_2$, NaOH, sulfides, etc.). Some of the alloys are susceptible to hot-salt stress cracking; however, this is a laboratory observation and has not been confirmed in service. Titanium stress cracks in methanol containing acid chlorides or sulfates, red fuming nitric acid, nitrogen tetroxide, and trichloroethylene.

Titanium is susceptible to failure by hydrogen embrittlement. Hydrogen attack initiates at sites of surface iron contamination or when titanium is galvanically coupled with iron.[23] In hydrogen-containing environments, titanium absorbs hydrogen above 80°C or in areas of high stress. If the surface oxide is removed by vacuum annealing or abrasion, pure dry hydrogen reacts at lower temperatures. Small amounts of oxygen or water vapor repair the oxide film and prevent this occurrence. Molybdenum-containing alloys are less susceptible to hydrogen attack. Titanium resists oxidation in air up to 650°C. Noticeable scale forms and embrittlement occurs at higher temperatures. Surface contaminants accelerate oxidation. In the presence of oxygen, the metal does not react significantly with nitrogen. Spontaneous ignition occurs in gas mixtures containing more than 40% oxygen under impact loading or abrasion. Ignition also occurs in dry halogen gases.

Titanium resists erosion–corrosion by fast-moving sand-laden water. In a high-velocity sand-laden seawater test (8.2 m/sec) for a 60-day period, titanium performed more than 100 times better than 18 Cr–8 Ni stainless steel, Monel, or 70 Cu–30 Ni. Resistance to cavitation (i.e., corrosion on surfaces exposed to high-velocity liquids) is better than by most other structural metals.[21,22]

In galvanic coupling, titanium is usually the cathode metal and, consequently, is not attacked. The galvanic potential in flowing seawater in relation to other metals is shown in Table 15.7.[21] Since titanium is the cathode metal, hydrogen attack may be of concern, as it occurs with titanium coupled to iron.

15.5   FABRICATION

Titanium can be fabricated similarly to nickel-base alloys and stainless steels. However, the characteristics of titanium have to be taken into account. Compared to these materials, titanium has:

Table 15.7 Galvanic Series in Flowing Seawater 4 m/sec at 24° C[2]

Metal	Potential, V[b]
T304 stainless steel, passive	0.08
Monel alloy	0.08
Hastelloy alloy C	0.08
unalloyed titanium	0.10
silver	0.13
T410 stainless steel, passive	0.15
nickel	0.20
T430 stainless steel, passive	0.22
70–30 copper–nickel	0.25
90–10 copper–nickel	0.28
admiralty brass	0.29
G bronze	0.31
aluminum brass	0.32
copper	0.36
naval brass	0.40
T410 stainless steel, active	0.52
T304 stainless steel, active	0.53
T430 stainless steel, active	0.57
carbon steel	0.61
cast iron	0.61
aluminum	0.79
zinc	1.03

[a] Ref. 23.
[b] Steady-state potential, negative to saturated calomel half-cell.

1. Lower modulus of elasticity.
2. Lower ductility.
3. Higher melting point.
4. Lower thermal conductivity.
5. Smaller strain-hardening coefficient, thereby, lower uniform elongation.
6. Greater tendency to cold weld, thereby, greater tendency to gall or sieze.
7. Greater tendency to be contaminated by oxygen, nitrogen, hydrogen, and carbon.

15.5.1 Boiler Code

The allowable stress values as determined by the Boiler and Pressure Vessel Committee of the American Society of Mechanical Engineers are listed in Tables 15.8 and 15.9 for various titanium grades and product forms.

15.5.2 Drawing

Commercially pure titanium can be cold drawn by tools required for austenitic stainless steels. Alpha–beta alloys, such as Ti–6 Al–4 V, are difficult to draw at room temperature. The following considerations should be given to drawing of titanium:

1. Slow drawing speeds are recommended.
2. Blanks should be profiled and cleaned.
3. Cleanliness should be maintained on the dies and blanks.
4. Room-temperature deformations should be held to 8% maximum on a single draw before annealing.
5. Proper lubrication, preferably using dry-film types with antigalling constituents, should be applied to blanks.
6. The large amount of springback should be considered.

Hot drawing of titanium is capable of resulting in deeper draws, lower loads, and less distortion. The recommended drawing temperature ranges are 200–300°C for commercially pure titanium and 500–650°C for titanium alloys, such as Ti–6 Al–4 V.

Table 15.8 Maximum Allowable Stress Values in Tension for Titanium and Its Alloys[a]

Form and Specification Number	Grade	Condition	Maximum Allowable Stress (ksi) for Metal Temperature (°F) Not Exceeding										
			100	150	200	250	300	350	400	450	500	550	600
Sheet, strip, plate, SB-265	1	Annealed	8.8	8.1	7.3	6.5	5.8	5.2	4.8	4.5	4.1	3.6	3.1
	2	Annealed	12.5	12.0	10.9	9.9	9.0	8.4	7.7	7.2	6.6	6.2	5.7
	3	Annealed	16.3	15.6	14.3	13.0	11.7	10.4	9.3	8.3	7.5	6.7	6.0
	7	Annealed	12.5	12.0	10.9	9.9	9.0	8.4	7.7	7.2	6.6	6.2	5.7
Bar, billet, SB-348	12		17.5	17.5	16.4	15.2	14.2	13.3	12.5	11.9	11.4	11.1	10.8
Pipe, SB-337	1	Seamless annealed	8.8	8.1	7.3	6.5	5.8	5.2	4.8	4.5	4.1	3.6	3.1
	2	Seamless annealed	12.5	12.0	10.9	9.9	9.0	8.4	7.7	7.2	6.6	6.2	5.7
Tubing, SB-338	3	Seamless annealed	16.3	15.6	14.3	13.0	11.7	10.4	9.3	8.3	7.5	6.7	6.0
	7	Seamless annealed	12.5	12.0	10.9	9.9	9.0	8.4	7.7	7.2	6.6	6.2	5.7
	12	Seamless annealed	17.5	17.5	16.4	15.2	14.2	13.3	12.5	11.9	11.4	11.1	10.8
Pipe, SB-337	1	Weld, annealed[b]	7.5	6.9	6.2	5.5	4.9	4.4	4.1	3.8	3.5	3.1	2.6
	2	Weld, annealed[b]	10.6	10.2	9.3	8.4	7.7	7.1	6.5	6.1	5.6	5.3	4.8
Tubing, SB-338	3	Weld, annealed[b]	13.9	13.3	12.2	11.1	10.0	8.8	7.9	7.1	6.4	5.7	5.1
	7	Weld, annealed[b]	10.6	10.2	9.3	8.4	7.7	7.1	6.5	6.1	5.6	5.3	4.8
	12	Weld, annealed[b]	14.9	14.9	13.9	12.9	12.1	11.3	10.6	10.1	9.7	9.4	9.2
Forgings, SB-381	1		Same as Grade 1 of sheet, strip, and plate										
	F 2	Annealed	Same as Grade 2 of sheet, strip, and plate										
	F 3	Annealed	Same as Grade 3 of sheet, strip, and plate										
	F 7		Same as Grade 7 of sheet, strip, and plate										
	F 12		Same as Grade 12 of sheet, strip, and plate										

[a] Boiler and Pressure Vessel Code, Section VIII–Division 1.

[b] 85% joint efficiency has been used in determining the allowable stress values for welded pipe and tube. Filler metal shall not be used in the manufacture of welded tubing or pipe.

Table 15.9 Design Stress Intensity Values in Tension for Titanium and Its Alloys

Form and Specification Number	Grade	Condition	Design Stress Intensity (ksi) for Metal Temperature (°F) Not Exceeding										
			100	150	200	250	300	350	400	450	500	550	600
Sheet, strip, plate, SB-265	1	Annealed	11.7	10.8	9.7	8.6	7.7	6.9	6.4	6.0	5.3	4.7	4.2
	2	Annealed	16.7	16.7	16.7	13.7	12.3	10.9	9.8	8.8	8.0	7.5	7.3
	3	Annealed	21.7	20.8	19.0	17.3	15.6	13.9	12.3	11.1	9.9	8.9	8.0
	7	Annealed	16.7	16.7	16.7	13.7	12.3	10.9	9.8	8.8	8.0	7.5	7.3
Bar, billet, SB-348													
Pipe, SB-337	1	Seamless annealed	11.7	10.8	9.7	8.6	7.7	6.9	6.4	6.0	5.3	4.7	4.2
	2	Seamless annealed	16.7	16.7	16.7	13.7	12.3	10.9	9.8	8.8	8.0	7.5	7.3
Tubing, SB-338	3	Seamless annealed	21.7	20.8	19.0	17.3	15.6	13.9	12.3	11.1	9.9	8.9	8.0
	7	Seamless annealed	16.7	16.7	16.7	13.7	12.3	10.9	9.8	8.8	8.0	7.5	7.3
	1	Weld, annealed[b]	9.9	9.2	8.3	7.3	6.5	5.9	5.4	5.1	4.5	4.0	3.6
	2	Weld, annealed[b]	14.2	14.2	14.2	11.6	10.5	9.3	8.3	7.5	6.8	7.4	6.2
	3	Weld, annealed[b]	18.4	17.7	16.2	14.7	13.3	11.8	10.5	9.4	8.4	7.6	6.8
	7	Weld, annealed[b]	14.2	14.2	14.2	11.6	10.5	9.3	8.3	7.5	6.8	7.5	6.2
Forgings, SB-381	F 1	Annealed	11.7	10.8	9.7	8.6	7.7	6.9	6.4	6.0	5.3	4.7	4.2
	F 2	Annealed	16.7	16.7	16.7	13.7	12.3	10.9	9.8	8.8	8.0	7.5	7.3
	F 3	Annealed	21.7	20.8	19.0	17.3	15.6	13.9	12.9	11.1	9.9	8.9	8.0
	F 7	Annealed	16.7	16.7	16.7	13.7	12.3	10.9	9.8	8.8	8.0	7.5	7.3

[a] Boiler and Pressure Vessel Code, Section VIII–Division 2.
[b] A quality factor of 0.85 has been applied in arriving at the design intensity values for this material. Filler metal shall not be used in the manufacture of welded tubing or pipe.

203

15.5.3 Bending

Titanium can be bent with press brake equipment used for cold-forming stainless steels. The minimum bend radii for bending various titanium alloys through an angle of 105° are given in Table 15.10. More severe bends can be accomplished at 200°C or higher temperatures depending on the alloy and the bend required. The amount of springback decreases with increasing temperature. For bending operations above 550°C, a descaling operation may be necessary to remove the surface oxide layer.

Table 15.10 Minimum Bend Radius for Annealed Titanium Sheet (ASTM B-265)

Grade	Minimum Bend Radius[a]	
	Under 1.8 mm Thick	1.8–4.75 mm Thick
1	1.5t	2.0t
2	2.0t	2.5t
3	2.0t	2.5t
4	2.5t	3.0t
5	4.5t	5.0t
6	4.0t	4.5t
7	2.0t	2.5t
10	3.0t	3.0t
11	1.5t	2.0t
12	2.0t	2.5t

[a] In multiples of sheet thickness, t, for room-temperature bending through an angle of 105°.

Titanium tubes (<25 mm outside diameter) can be bent at a radius equal to two to three times the outside diameter of the tube. For tubes with an outside diameter larger than 25 mm, larger bend radii are recommended for room-temperature bending. Tighter bends can be obtained by heating the tube above 200°C. Owing to the low modulus, titanium has a tendency to buckle under compressive stress. Therefore, it is recommended that both the inside and outside surfaces of the bend be subjected to tension to avoid buckling.

15.5.4 Cutting and Grinding

Titanium can be sheared, flame cut, saw cut, and abrasive cut. Sheared edges should be examined for cracks for plates over 9.5 mm thick. Flame-cut edges are contaminated with oxygen and carbon. It is recommended that at least 1.6 mm below the surface (measured from the lowest point of the cut roughness) be removed by grinding or machining. Thick plates may require removal of additional thickness. Generous amounts of coolant should be used in saw cutting and abrasive cutting to keep the workpiece cool and to minimize sparking. Water or water-soluble oil is recommended.

Abrasive grinding can be used on titanium, but care should be taken to avoid excessive heat buildup and contamination. The temperature of the grinding sparks is very high and precautionary measures should be taken accordingly.

15.5.5 Welding

Commercially pure titanium and most titanium alloys can be readily welded using the gas metal-arc (GMA) or gas tungsten-arc (GTA) welding process. Owing to titanium's highly reactive nature, the welding processes involving a noninert gas or a flux, such as, oxyacetylene-shielded metal arc, flux-cored arc, and submerged arc welding, are not suitable. For the same reason, welding titanium requires a clear environment and good inert-gas shielding in either GMA or GTA welding.

Common joint designs can be used for welding titanium as long as the design allows proper inert-gas shielding. The joint surfaces must be clean and free of grease, oil, moisture, visible oxides, and other contaminants. The oxides can be removed by grinding, brushing with a stainless-steel wire brush, or pickling in a room-temperature solution containing 30% nitric acid and 3% hydrofluoric acid, by weight.

A primary shield is needed for the molten weld puddle and a trailing secondary shield for the solidified weld deposit and the heat-affected zone. In addition, a backup shield is also needed for the backside of the weld and the heat-affected zone. Argon is generally preferred to helium for primary shielding because of better arc stability. Argon–helium mixtures can be used in some conditions where

Table 15.11 Comparison in Tensile Properties Between Weld and Parent Metal of Titanium Alloys[a]

Alloy	Condition	Yield Strength (MPa)	Tensile Strength (MPa)	Elongation (%)
Grade 1	Parent metal	215	315	50.4
	Single-bead metal	255	345	37.5
	Multiple-bead weld	270	365	37.7
Grade 2	Parent metal	325	460	26.2
	Single-bead weld	380	505	18.3
	Multiple-bead weld	385	510	13.3
Grade 3	Parent metal	395	545	25.9
	Single-bead weld	475	605	15.5
	Multiple-bead weld	480	615	14.7
Grade 4	Parent metal	530	660	22.3
	Single-bead weld	580	695	16.4
	Multiple-bead weld	585	710	16.0
Grade 5	Parent metal	945	1000	11.0
	Single-bead weld	920	1060	3.5
	Multiple-bead weld	945	1090	3.2
Grade 6	Parent metal	805	850	15.7
	Single-bead weld	770	920	9.8
	Multiple-bead weld	820	935	7.5
Grade 9	Parent metal	670	705	15.2
	Single-bead weld	600	705	12.7
	Multiple-bead weld	625	745	11.2
Ti–8 Al–1 Mo–1 V	Parent metal	1020	1060	15.0
	Single-bead weld	930	1085	5.5
	Multiple-bead weld	960	1115	3.2
Ti–6 Al–6 V–2 Sn	Parent metal	1005	1060	9.8
	Single-bead weld	1255	1295	0.3
	Multiple-bead weld	—	1280	0.1

[a] *Metals Handbook,* 9th ed., American Society for Metals, Metals Park, OH, 1980, Vol. 3.

high voltage and deep penetration are desired. Either argon or helium can be used in the secondary and backup shielding.

The mechanical properties of the welds depend on the alloying elements. The welds generally have higher strength but less ductility than the parent metals as shown in Table 15.11.

Other than the GTA and GMA welding processes, titanium can also be welded by electron beam, resistance, plasma arc, and friction welding. In general, titanium cannot be welded with a dissimilar metal owing to the fact that it forms brittle intermetallic compounds with most other metals. Mechanical joining is recommended when joining titanium with a dissimilar metal.

15.6 SPECIFICATIONS, STANDARDS, AND QUALITY CONTROL

The alloys of titanium have compositional specifications tabulated by ASTM. The ASTM specification number is given in Table 15.3 for the commercially important alloys. Military specifications are found under MIL-T-9046 and MIL-T-9047, and aerospace material specifications for bar, sheet, tubing, and wire under specification numbers 4900–4980. Each aircraft company has its own set of alloy specifications.

The alloy name in the United States usually includes a company name or trademark in conjunction with the composition for alloyed titanium or the strength (ultimate tensile strength for TMCA and yield strength for other U.S. producers) for unalloyed titanium. The common alloys and company designations are shown in Table 15.12.

Since titanium alloys are used in a variety of applications, several different material and quality standards are specified. Among them are ASTM, ASME, ASM, U.S. military, and a number of proprietary sources. The correct chemistry is basic to obtaining mechanical and other properties required for a given application. Minor elements controlled by specification include carbon, iron, hydrogen, nitrogen, and oxygen. For more stringent applications, yttrium may also be specified. In addition, control of the thermomechanical processing and subsequent heat treatment is vital to obtaining desired

Table 15.12 Company Names of Common Titanium Alloys[2]

Alloys	ASTM	Cabot	IMI[b]	RMI	Timet	USSR
99.5 Ti	grade 2	CABOT Ti 40	IMI-125	RMI 40	Ti–50 A	VT1-0
99.2 Ti	grade 3	CABOT Ti 55	IMI-130	RMI 55	Ti–65 A	VT1
99.0 Ti	grade 4	CABOT Ti 70	IMI-155	RMI 70	Ti–75 A	VT1-1
Ti–5 Al–2.5 Sn	grade 6		IMI-315	RMI 5 Al–2.5 Sn	Ti–5 Al–2.5 Sn	VT5-1
Ti–6 Al–4 V	grade 5	CABOT Ti–6 Al–4 V	IMI-317	RMI 6 Al–4 V	Ti–6 Al–4 V	VT6

[a] Refs. 7, 9.
[b] IMI = IMI Limited, Witton, Birmingham, UK.

properties. For extremely critical applications, such as rotating parts in aircraft gas turbines, raw materials, melting parameters, chemistry, thermomechanical processing, heat treatment, test, and finishing operations must be carefully and closely controlled at each step to ensure that required characteristics are present in the products supplied.

15.7 HEALTH AND SAFETY FACTORS

Titanium and its corrosion products are nontoxic. A safety problem does exist with titanium grindings, turnings, and some corrosion products which are pyrophoric. Grindings and turnings should be stored in a closed container and not left on the floor. Smoking must be prohibited in areas where titanium is ground or turned and, if a fire occurs, it must be extinguished with a class D extinguisher (for use against metal fires). The larger the surface area, the more pyrophoric the titanium fines. When titanium equipment is being worked on, all flammable products and corrosive products must be removed, and the area must be well ventilated. A pyrophoric corrosion product has been observed in environments of dry Cl_2 gas and in dry red fuming nitric acids.

15.8 USES

Titanium is primarily used in the form of high-purity titanium oxide. Although the principal application of high-purity (pigment-grade) TiO_2 is in paint pigments, other important uses are in plastics (for color in floor-covering products and to help protect plastic products and foodstuffs contained in plastic bags from ultraviolet radiation deterioration), in paper (as a filler and whitener), and in rubber. Future application areas include TiO_2 single-crystal electrodes for water decomposition for the production of hydrogen fuel, flue-gas denitrification catalysts, and high-purity TiO_2 to make barium titanate thermistors.

Titanium metal was first established as a material for aerospace, "metal-for-air" applications. In the late 1970s, it was developed as "metal-for-sea" uses. The metal-for-air and metal-for-sea designations characterize Japanese market development goals. In terms of volume, the U.S. titanium industry is still in the metal-for-air development stage; the statements about metal-for-earth and -sea reflect an optimistic outlook.

In the United States, the high strength-to-weight ratio of titanium accounts for approximately 70% of its uses. Before 1970, the high strength-to-weight ratio was the basis of over 90% of applications, such as engines, where the advantage of light weight is translated to higher flying, faster planes. Aerospace applications have shaped and controlled the titanium-metal industry.

The use of titanium in aircraft is divided about equally between engines and airframes. For engine components, titanium is limited because of temperature constraints at the compressor area where it is used as blades, casings, and disks. In the frame, it is used in bulkheads, firewall, flap tracks, landing-gear parts, wiring pivot structures, fasteners, rotor hubs, and hot-area skins. In the F-15, titanium accounts for about 32% of the structural weight. Design changes and weight savings owing to the use of titanium in Pratt and Whitney's JT3D engine, employed to power Boeing 707 and Douglas DC 8 aircraft, resulted in 42% more takeoff thrust, 13% lower specific fuel consumption, and 18% less weight than the prior JT3C engine.[3]

The other outstanding property of titanium metal is its corrosion resistance, although its use in corrosion-resistance applications in 1980 in the United States was a mere 5000 tons or ~0.001% of the metal used in corrosion-resistance markets. The largest application was heat-exchanger pipes and tubing (~800 μm or 22 gauge welded) for the power industry, and marine and desalination applications, where titanium provides protection against corrosion by seawater, brackish water, and other estuary waters containing high concentrations of chlorides and industrial wastes.

Titanium metal is especially utilized in environments of wet chlorine gas and bleaching solutions, that is, in the chlor-alkali industry and the pulp and paper industries. Here, titanium is used as anodes for chlorine production, chlorine–caustic scrubbers, pulp washers, and Cl_2, ClO_2, and $HClO_4$ storage and piping equipment.

In the chemical industry, titanium is used in heat-exchanger tubing for salt production, in the production of ethylene glycol, ethylene oxide, propylene oxide, and terephthalic acid, and in industrial wastewater treatment. Titanium is used in environments of aqueous chloride salts ($ZnCl_2$, NH_4Cl, $CaCl_2$, $MgCl_2$, etc.), chlorine gas, chlorinated hydrocarbons, and nitric acid. In metal recovery, titanium is used for ore-leaching solutions and as racks for metal plating. The leaching solutions contain HCl or H_2SO_4 with enough ferric or cupric ions to inhibit the corrosion of titanium. In metal-plating applications, titanium is cathodically protected against H_2SO_4 and chrome-plating solution corrosion. An important factor in using titanium for metal-plating applications is that the minute amount of dissolved titanium ions does not plate out as an impurity in the coatings.

In oil and gas refinery applications, titanium is used as protection in environments of H_2S, SO_2, CO_2, NH_3, caustic solutions, steam, and cooling water. It is used in heat-exchanger condensers for the fractional condensation of crude hydrocarbons, NH_3, propane, and desulfurization products using seawater or brackish water for cooling.

Other application areas include nuclear-waste storage canisters, pacemaker castings, implantations, geothermal equipment, automotive connection rods, and ordnance.

REFERENCES

1. S. C. Williams, *Report on Titanium*, J. W. Edwards, Inc., Ann Arbor, MI, 1965.
2. D. Knittel, "Titanium and Titanium Alloys," in *Kirk–Othmer Encyclopedia of Chemical Technology*, 3rd ed., Wiley, New York, 1983, Vol. 23.
3. A. D. McQuillan and M. K. McQuillan, in *Metallurgy of the Rarer Metals*, H. M. Finniston (ed.), Academic, New York, 1956, p. 335.
4. *ASTM Standard Specification for Titanium and Titanium Alloy Strip, Sheet, and Plate*, ANSI–ASTM B265-79, American Society for Testing and Materials, Philadelphia, PA, Oct. 1980.
5. R. M. Duncan, P. A. Blenkinsop, and R. E. Goosey, in *The Development of Gas Turbine Materials*, G. W. Meethan (ed.), Wiley, New York, 1981, p. 63.
6. *Facts About the Metallography of Titanium*, RMI Company, Niles, OH, 1975.
7. R. A. Wood, *The Titanium Industry in the Mid-1970's*, Battelle Report MCIC-75-26, Battelle Memorial Institute, Columbus, OH, June 1975.
8. R. I. Jaffee, in *Titanium '80 Science and Technology*, H. Kimura and O. Izumi (eds.), The Metallurgical Society/American Institute of Mining, Metallurgical and Petroleum Engineers, Warrendale, PA, 1980, p. 53.
9. H. Hucek and M. Wahll, *Handbook of International Alloy Compositions and Designations*, Battelle Report MCIC-HB-09, Battelle Memorial Institute, Columbus, OH, Nov. 1976, Vol. 1.
10. *Metals Prog. Databook* **110**, 94 (June 1976).
11. S. G. Glazunov, in *Titanium Alloys for Modern Technology*, N. P. Sazhin and co-workers (eds.), NASA TT F-596, National Aeronautics and Space Administration, Washington, DC, March 1970, p. 11.
12. C. C. Chen and R. R. Boyer, *J. Met.* **31**, 33 (1979).
13. E. L. Hayman, D. W. Greenwood, and B. G. Martin, *Exp. Mech.* **17**, 161 (May 1977).
14. E. M. Savitskiy, M. I. Bychkova, and V. V. Baron, Ref. 8, p. 735.
15. C. M. Wayman, *J. Met.* **32**, 129 (1980).
16. *How to Use Titanium Properties and Fabrication of Titanium Mill Products*, Titanium Metals Corporation of America, Pittsburgh, PA, 1975.
17. H. R. Ogden and F. C. Holden, *Metallography of Titanium Alloys*, TML Report No. 103, Battelle Memorial Institute, Columbus, OH, May 29, 1958.
18. *Metals Handbook*, American Society for Metals, Metals Park, OH, 1972, Vol. 7.
19. T. R. Beck, in *Localized Corrosion*, R. W. Staehle, B. F. Brown, J. Kruger, and A. Agarwal (eds.), National Association of Corrosion Engineers, Houston, TX, 1974, Vol. Nace-3, p. 644.
20. E. E. Millaway, *Mater. Prot.* **4**, 16 (1965).
21. L. C. Covington, R. W. Schultz, and I. A. Fronson, *Chem. Eng. Prog.* **74**, 67 (1978).
22. *Titanium for Industrial Brine and Sea Water Service*, Titanium Metal Corporation of America, Pittsburgh, PA, 1968.
23. L. C. Covington and R. W. Schultz, in *Industrial Applications of Titanium and Zirconium*, STP 728, E. W. Kleefisch (ed.), American Society for Testing and Materials, Philadelphia, PA, 1981, p. 163.

CHAPTER 16

MAGNESIUM AND ITS ALLOYS

ROBERT S. BUSK

International Magnesium Consultants, Inc.
Hilton Head, South Carolina

16.1 INTRODUCTION

Magnesium, with a specific gravity of only 1.74, is the lowest-density metal available for engineering use. It is produced either by electrolytic reduction of $MgCl_2$ or by chemical reduction of MgO by Si in the form of ferrosilicon. $MgCl_2$ is obtained from seawater, brine deposits, or salt lakes; MgO is obtained principally from seawater or dolomite. Because of the widespread, easy availability of magnesium ores (e.g., the ocean), the ore supply is, in human terms, inexhaustible.

16.2 USES

Magnesium is used both as a structural, load-bearing material and in applications that exploit its chemical and metallurgical properties

16.2.1 Nonstructural Applications

Because of its high place in the electromotive series, magnesium is used as a sacrificial anode to protect steel from corrosion; some examples are the protection of buried pipe lines and the prolongation of the life of household hot-water tanks. Magnesium is added to gray cast iron to produce ductile iron, an alloy that has many of the producibility advantages of cast iron but is ductile and strong. A rapidly growing use for magnesium is its addition to the iron tapped from blast furnaces to remove sulfur prior to converting to steel, thereby increasing the efficiency of the blast furnace and improving the toughness of the steel. Magnesium is used to produce the Grignard reagent, an organic intermediate used in turn to produce fine chemicals and pharmaceuticals. Magnesium is used in sheet and extruded form as a material to produce photoengravings. Magnesium is one of the principal alloying additions to aluminum, imparting improved strength and corrosion resistance to that metal.

16.2.2 Structural Applications

Magnesium structures are made from sand, permanent-mold, investment, and die castings, and from sheet, plate, extrusions, and forgings. The base forms produced in these ways are fabricated into finished products by machining, forming, and joining. Finishing for protective or decorative purposes is by chemical-conversion coatings, painting, or electroplating.

Table 16.1 is a short listing of some typical uses for magnesium with the property or properties of particular significance for each use indicated.

16.3 ALLOYS AND PROPERTIES

Many alloys (Table 16.2) have been developed to provide a range of properties and characteristics to meet the needs of a wide variety of applications. There are two major classes—one containing aluminum as the principal alloying ingredient, the other containing zirconium. Those containing aluminum are strong and ductile and have excellent resistance to atmospheric corrosion. Since zirconium is a potent grain refiner for magnesium alloys but is incompatible with the presence of aluminum in magnesium, it is added to all alloys not containing aluminum. Within this class, those alloys containing rare-earth or thorium additions are especially suited to applications at temperatures ranging to as high as 300°C. Those not containing rare-earth elements have zinc as a principal alloying element and are strong, ductile, and tough.

16.3.1 Mechanical Properties of Castings

Magnesium castings are produced in sand, permanent, investment, and pressure die-casting molds.

Castings produced in sand molds range in size from a few pounds to a few thousand pounds and can be very simple to extremely complex in shape. If production runs are large enough to justify higher tooling costs, then permanent instead of sand molds are used. The use of low pressure to fill a permanent mold is a low-cost method that is being used increasingly for magnesium. Investment casting is a specialized technique that permits the casting of very thin and intricate sections with excellent surface and high mechanical properties. Die casting is a process for the production of castings with good dimensional tolerances, good surface, and acceptable properties at quite low cost.

Mechanical properties of cast alloys are given in Table 16.3.

16.3.2 Mechanical Properties of Wrought Products

Wrought products are produced as forgings, extrusions, sheet, and plate. Mechanical properties are given in Table 16.4.

Table 16.1   Representative Uses for Magnesium Alloys

Use	Light Weight	Machin-ability	Damping	Corrosion Resistance	Dent Resistance	Non-crocking	Produci-bility
Automotive							
Brackets	X						X
Covers	X						X
Housings	X						X
Wheels	X				X		X
Portable tools							
Lawn mowers					X		X
Chain saws	X						X
Levels	X						
Cement working				X			
Aerospace							
Gear casings	X						
Housings	X						
Covers	X						
Wheels	X						
Instrument housings	X		X				
Engine parts	X						
Sports							
Catcher's masks	X				X		
Archery bows	X						
Manufacturing aids							
Jigs		X					
Fixtures		X					
Vibration testing		X	X				
Miscellaneous							
Luggage	X					X	X
Bakery racks						X	X
Garment racks						X	X

a Light weight: The lowest density of any structural metal. Machinability: The easiest of any metal to machine. Damping: The greatest damping capacity of any metal used for structural purposes. Corrosion resistance: Very resistant to alkaline environments. Dent resistance: Greater resistance to denting than most metals. Noncrocking: Bare magnesium does not mark clothing or skin. Producibility: Particularly suited to the die-casting process.

16.3.3   Physical Properties

A selection of physical properties of pure magnesium is given in Table 16.5. Most of these are insensitive to alloy addition, but melting point, density, and electrical resistivity vary enough that these properties are listed for alloys in Table 16.6.

16.4   FABRICATION

16.4.1   Machining

Magnesium is the easiest of all metals to machine; it requires only low power and produces clean, broken chips, resulting in good surfaces, even with heavy cuts.

16.4.2   Joining

All standard methods of joining can be used, including welding, riveting, brazing, and adhesive bonding.

Welding is by inert-gas-shielded processes using either helium or argon, and either MIG or TIG. Alloys containing more than 1.5% aluminum should be stress relieved after welding in order to prevent stress-corrosion cracking due to residual stresses associated with the weld joint. Rivets for magnesium are of aluminum rather than magnesium. Galvanic attack is minimized or eliminated by using aluminum rivets made of an alloy high in magnesium such as 5056. Brazing is used, but not extensively, since

Table 16.2 Magnesium Alloys in Common Use

ASTM Designation	Ag	Al	Cu (max)	Fe (max)	Mn	Ni (max)	Rare Earth	Si	Th	Zn	Zr	Forms[a]
AM60A		6.0			0.15							DC
AM100A		10.0			0.2							SC, PM, IC
AS41A		4.2			0.35			1.0				DC
AZ31B		3	0.04	0.005	0.3	0.005				1		S, P, F, E
AZ31C		3	0.10		0.3	0.03				1		S, P, F, E
AZ61A		6.5	0.05	0.005	0.2	0.005				0.9		F, E
AZ80A		8.5	0.05	0.005	0.15	0.005				0.5		F, E
AZ81A		7.6			0.15					0.7		SC, PM, IC
AZ91A		9	0.1		0.15					0.7		DC
AZ91C		8.7			0.15					0.7		SC, PM, IC
AZ91B		9	0.35		0.15					0.7		DC
AZ92A		9.2			0.15					2		SC, PM, IC
EZ33A							3.2			2.5	0.7	SC, PM, IC
HK31A									3.2		0.7	SC, PM, IC, S, P
HM21A					0.7				2			S, P, F
HZ32A									3.2	2.1	0.7	SC
K1A											0.7	SC, PM, IC
M1A					1.5							E
QE22A	2.5						2.2[b]				0.7	S, PM, IC
QH21A	2						2.0[b]		1		0.7	S, PM, IC
ZE41A							1.2			4.2	0.7	SC, PM, IC
ZE63A							2.6			5.8	0.7	SC, PM, IC
ZH62A									1.8	5.7	0.7	SC, PM, IC
ZK40A										4.0	0.7	E
ZK51A										4.6	0.7	SC, PM, IC
ZK60A										5.5	0.7	F, E
ZK61A										6.0	0.7	SC, IC

[a] DC = die casting; E = extrusion; F = forging; IC = investment casting; P = plate; PM = permanent mold casting; S = sheet; SC = sand casting.
[b] Rare earth is present in didymium.

Table 16.3 Typical Mechanical Properties for Castings

Alloy	Temper	Tensile Strength (MPa)	Yield Strength (MPa)	Elongation in 2 in. (%)
Sand and Permanent Mold Castings[a]				
AM100A*	T6	275	150	1
AZ81A*	T4	275	80	15
AZ91C*	F	200	100	3
	T4	275	100	12
	T5	200	110	3
	T6	275	195	6
AZ92A*	F	200	100	2
	T4	275	100	11
	T5	200	110	2
	T6	275	150	3
EZ33A*	T5	160	110	2
HK31A*	T6	220	105	8
HZ32A	T5	185	90	4
K1A	F	180	55	20
QE22A*	T6	260	195	3
QH21A	T6	275	205	4
ZE41A	T5	205	140	4
ZE63A	T6	300	190	10
ZH62A	T5	240	170	4
ZK51A	T5	205	165	4
ZK61A	T6	310	195	10
Investment Mold Castings				
AM100A	F	180	100	2
	T4	275	100	11
	T6	275	130	3
	T7	275	145	1
AZ81A	T4	275	100	12
AZ91C	F	165	100	2
	T4	275	100	12
	T5	180	100	3
	T7	275	140	5
AZ92A	F	180	100	2
	T4	275	100	11
	T5	180	100	2
	T6	275	150	1
EZ33A	T5	255	110	4
HK31A	T6	210	100	10
K1A	F	175	60	22
QE22A	T6	260	185	4
ZK61A	T6	300	185	9
Die Castings				
AM60A	F	220	130	8
AZ91A&B	F	230	160	3
AS41A	F	210	140	6

[a] Only starred alloys are produced in permanent mold.

212

Table 16.4 Typical Mechanical Properties of Wrought Products

Alloy	Temper	Tensile Strength (MPa)	Yield Strength (MPa) Tensile	Compressive	Elongation in 2 in. (%)
Sheet and Plate					
AZ31B&C	O	255	150	110	21
	H24	290	220	180	15
HK31A	O	230	130	90	21
	H24	255	200	160	9
HM21A	T8	235	170	130	11
Extrusions					
AZ31B	F	260	200	100	15
AZ61A	F	310	230	130	16
AZ80A	F	340	250	140	11
	T5	380	275	240	7
M1A	F	255	180	80	12
ZK40A	T5	275	255	140	4
ZK60A	F	315	235	170	12
	T5	365	305	250	11
Forgings					
AZ31B	F	260	195	80	9
AZ61A	F	300	180	115	12
AZ80A	F	320	215	170	8
	T5	345	235	190	6
	T6	345	250	185	5
HM21A	T5	235	150	110	9
ZK60A	T5	300	210	190	16
	T6	325	270	170	11

Table 16.5 Physical Properties of Pure Magnesium

Density	1.738 g/cm^3 at 20°C (Ref. 1)
Melting point	650°C (Ref. 2)
Boiling point	1107°C (Ref. 2)
Thermal expansion	25.2×10^{-6}/K (Ref. 3)
Specific heat	1.025 kJ/kg · K at 20°C (Ref. 4)
Latent heat of fusion	360–377 kJ/kg (Ref. 4)
Latent heat of sublimation	6113–6238 kJ/kg at 25°C (Ref. 2)
Latent heat of vaporization	5150–5400 kJ/kg (Ref. 2)
Heat of combustion	24,900–25,200 kJ/kg
Electrical resistivity	44.5 nΩ · m^3
Crystal structure	Close-packed hexagonal: $a = 0.32087$ nm; $c = 0.5209$ nm; $c/a = 1.6236$ (Ref. 5)
Young's modulus	45 GPa
Modulus of rigidity	16.5 GPa
Poisson's ratio	0.35

Table 16.6 Physical Properties of Alloys

Alloy	Density (g/cm³)	Melting Point (°C)		Electrical Resistivity (nΩ · m)
		Liquidus	Solidus	
AZ31B&C	1.77	632	605	92
AZ61A	1.8	620	525	125
AZ80A	1.8	610	490	145
HK31A	1.79	650	590	61
HM21A	1.78	650	605	52
M1A	1.76	649	648	50
ZK40A	1.82			
ZK60A	1.83	635	520	
AM60A	1.79	615	540	
AM100A	1.81	595	465	140
AS41A	1.77	620	565	130
AZ81A	1.80	610	490	
AZ91A,B,C	1.81	595	470	170
AZ92A	1.82	595	445	140
EZ33A	1.83	645	545	70
HK31A	1.79	650	590	77
HZ32A	1.83	650	550	65
K1A	1.74	649	648	57
QE22A	1.81	645	545	68
QH21A	1.82	640	535	68
ZE41A	1.82	645	525	60
ZE63A	1.87	635	510	56
ZH62A	1.82	630	520	65
ZK51A	1.81	640	560	62
ZK61A	1.83	635	530	

it can be done only on alloys with a high melting point such as AZ31B or K1A. Adhesive bonding is straightforward, and no special problems related to magnesium are encountered.

16.4.3 Forming

Magnesium alloys are formed by all the usual techniques, such as deep drawing, bending, spinning, rubber forming, stretch forming, and dimpling.

In general, it is preferable to form magnesium in the temperature range of 150–300°C. While this requires more elaborate tooling, there is some compensation in the ability to produce deeper draws (thus fewer tools), and in the elimination or minimizing of springback. Hydraulic rather than mechanical presses are preferred.

16.5 CORROSION AND FINISHING

Magnesium is highly resistant to alkalies, and to chromic and hydrofluoric acids. In these environments no protection is usually necessary. On the other hand, magnesium is less resistant to other acidic or salt-laden environments. While most magnesium alloys can be exposed without protection to dry atmosphere, it is in general desirable to provide a protective finish.

Magnesium is anodic to any other structural metal and will be preferentially attacked in the presence of an electrolyte. Therefore, galvanic contact must be avoided by separating magnesium from other metals by the use of films and tapes. These precautions do not apply in the case of 5056 aluminum alloy, since the galvanic attack in this case is minimal.

Because magnesium is not resistant to acid attack, standing water (which will become acidic by absorption of CO_2 from the atmosphere) must be avoided by providing drain holes.

16.5.1 Chemical-Conversion Coatings

There are a large number of chemical-conversion processes based on chromates or fluorides. These are simple to apply and provide good protection themselves in addition to being a good paint base.

16.5.2 Anodic Coatings

There are a number of good anodic coatings that offer excellent corrosion protection and also provide a good paint base.

16.5.3 Painting

If a good chemical-conversion or anodic coating is present, any paint will provide protection. Best protection results from the use of baked, alkaline-resistant paints.

16.5.4 Electroplating

Once a zinc coating is deposited chemically followed by a copper strike, standard electroplating procedures can be applied to magnesium to give decorative and protective finishes.

REFERENCES

1. R. S. Busk, *Trans. AIME* **194,** 207 (1952).
2. D. R. Stull and G. C. Sinke, *Thermodynamic Properties of the Elements* (Advances in Chemistry, Vol. 18), American Chemical Society, Washington, DC, 1956.
3. P. Hidnert and W. T. Sweeney, *J. Res. Nat. Bur. St.* **1,** 771 (1955).
4. R. A. McDonald and D. R. Stull, *J. Am. Chem. Soc.* **77,** 5293 (1955).
5. R. S. Busk, *Trans. AIME* **188,** 1460 (1950).

BIBLIOGRAPHY

M. R. Bothwell, *The Corrosion of Light Metals,* Wiley, New York, 1967.

E. F. Emley, *Principles of Magnesium Technology,* Pergamon Press, New York, 1966.

Fabricating with Magnesium, The Dow Chemical Company, Form 141-477-82, 1982.

Machining Magnesium, The Dow Chemical Company, Form 141-480-82, 1982.

Operations in Magnesium Finishing, The Dow Chemical Company, Form 141-479-583, 1983.

"Properties and Selection: Nonferrous Alloys and Pure Metals," *ASM Metals Handbook,* 9th ed., American Society for Metals, Metals Park, OH, 1979, Vol. 2.

C. S. Roberts, *Magnesium and Its Alloys,* Wiley, New York, 1960.

CHAPTER 17
PLASTICS

EDWARD N. PETERS

General Electric Company
Pittsfield, Massachusetts

The use of plastics has increased almost 10-fold in the last 20 years. Plastics have come on the scene as the result of a continual search for man-made substances that can perform better or can be produced at a lower cost than natural materials such as wood, glass, and metals, which require mining, refining, processing, milling, and machining. Plastics can also increase productivity by producing finished parts and by consolidating parts. These increases in productivity have led to fantastic growth. For example, the total volumes of raw steel and plastics production in the United States in 1978 were 5.67×10^8 and 5.77×10^8 ft^3, respectively. Thus, we are producing a larger volume of plastics than of steel.

Plastics can be classified several ways. From a performance standpoint they can be either commodity or engineering plastics. During fabrication a polymer will behave as a thermoplastic or a thermoset. Thermoplastics can be repeatedly softened by heat and shaped. By contrast, a thermoset can be shaped and cured by heat only once. Other classification schemes include crystalline/amorphous or synthetic/natural. These classifications will be useful in differentiating plastics in this chapter.

17.1 COMMODITY THERMOPLASTICS

The commodity thermoplastics include polyolefins and side-chain-substituted vinyl polymers.

17.1.1 Polyethylene

Polyethylenes (PEs) have the largest volume use of any plastic. They are prepared by the catalytic polymerization of ethylene. Depending on the mode of polymerization, one can obtain a high-density (HDPE) or a low-density polyethylene (LDPE) polymer. LDPE is prepared under more vigorous conditions, which result in short-chain branching. Linear low-density polyethylene (LLDPE) is prepared by introducing short-branching via copolymerization of PE with a small amount of long-chain olefin.

Polyethylenes are crystalline thermoplastics that exhibit toughness, near-zero moisture absorption, excellent chemical resistance, excellent electrical insulating properties, low coefficient of friction, and ease of processing. Their heat deflection temperatures are reasonable but not high. HDPE exhibits greater stiffness, rigidity, improved heat resistance, and increased resistance to permeability than LDPE. Some typical properties of PEs are listed in Table 17.1.

Table 17.1 Typical Property Values of Polyethylenes

	HDPE	LDPE
Density (Mg/m³)	0.96–0.97	0.91–0.93
Tensile modulus (GPa)	0.76–1.0	—
Tensile strength (MPa)	25–32	4–20
Elongation at break (%)	500–700	275–600
Flexural modulus (GPa)	0.8–1.0	0.21
Vicat soft point (°C)	120–129	80–96
Brittle temperature (°C)	−100 to −70	−85 to −35
Hardness (Shore)	D60–D69	D45–D52
Dielectric constant (10^6 Hz)	—	2.3
Dielectric strength (MV/m)	—	9–21
Dissipation factor (10^6 Hz)	—	0.0002
Linear mold shrinkage (in./in.)	0.007–0.009	0.015–0.035

Uses. HDPE's major use is in blow-molded bottles, automotive gas tanks, drums, and carboys; injection-molded material-handling pallets, trash and garbage cans, and household and automotive parts; and extruded pipe.

LDPEs find major applications in film form for food packaging, as a vapor barrier film; for extruded wire and cable insulation; and for bottles, closures, and toys.

17.1.2 Polypropylene

Polypropylene (PP) is prepared by the catalyzed polymerization of propylene. PP is a highly crystalline thermoplastic that exhibits low density, rigidity, excellent chemical resistance, negligible water absorption, and excellent electrical properties. Its properties appear in Table 17.2.

Table 17.2 Typical Property Values of Polypropylene

Density (Mg/m³)	0.90–0.91
Tensile modulus (GPa)	1.8
Tensile strength (MPa)	37
Elongation at break (%)	10–50
Heat deflection temperature at 0.45 MPa (°C)	100–105
Heat deflection temperature at 1.81 MPa (°C)	60–65
Vicat soft point (°C)	130–148
Linear thermal expansion (mm/mm·K)	3.8×10^{-5}
Hardness (Shore)	D76
Volume resistivity (Ω·cm)	1.0×10^{17}
Linear mold shrinkage (in./in.)	0.01–0.02

Uses. End uses for PP are in blow-molded bottles, closures, automotive parts, appliances, housewares, and toys. PP can be extruded into fibers and filaments for use in carpets, rugs, and cordage.

17.1.3 Polystyrene

Catalytic polymerization of styrene yields polystyrene (PS), a clear, amorphous polymer with a moderately high heat deflection temperature. PS has excellent electrical insulating properties; however, it is brittle under impact and exhibits very poor resistance to surfactants and solvents. Its properties appear in Table 17.3.

Table 17.3 Typical Properties of Styrene Thermoplastics

Property	PS	SAN	IPS/HIPS	ABS
Density (Mg/m³)	1.050	1.080	1.020–1.040	1.050–1.070
Tensile modulus (GPa)	2.76–3.10	3.4–3.9	2.0–2.4	2.5–2.7
Tensile strength (MPa)	41.4–51.7	65–76	26–40	36–39
Yield elongation (%)	1.5–2.5	—	—	2.2–2.6
Heat deflection temperature at 264 psi (°C)	82–93	101–103	81	80–95
Vicat soft point (°C)	98–107	110	88–101	90–99
Notched Izod (kJ/m)	0.02	0.02	0.10–0.32	0.09–0.48
Linear thermal expansion (10^{-5} mm/mm·K)	5–7	6.4–6.7	7.0–7.5	7.5–9.5
Hardness (Rockwell)	M60–M75	M80–M83	M45, L55	R69–R115
Linear mold shrinkage (in./in.)	0.007	0.003–0.004	0.007	0.0055

Uses. Ease of processing, rigidity, clarity, and low cost combine to support applications in toys, displays, and housewares. PS foams can readily be prepared and are characterized by excellent low thermal conductivity, high strength-to-weight ratio, low water absorption, and excellent energy absorption. These attributes have made PS foam of special interest as insulation boards for construction, protective packaging materials, insulated drinking cups, and flotation devices.

17.1.4 SAN (Styrene/Acrylonitrile Copolymer)

Copolymerization of styrene with a moderate amount of acrylonitrile provides a clear, amorphous polymer (SAN) with increased heat deflection temperature and chemical resistance compared to polystyrene. However, the impact resistance is still poor.

Uses. SAN is utilized in typical PS-type applications where a slight increase in the heat deflection temperature and/or chemical resistance are needed, such as housewares and appliances.

17.1.5   Impact Polystyrene

Copolymerization of styrene with butadiene can reduce the brittleness of PS, but only at the expense of rigidity and heat deflection temperature. These opaque materials (IPS and HIPS) generally exhibit poor weathering characteristics.

Uses. IPS and HIPS are used for small appliance, radio, and TV housings.

17.1.6   ABS

ABS is a terpolymer prepared from the combination of acrylonitrile, butadiene, and styrene monomers. Compared to PS, ABS exhibits good impact strength, improved chemical resistance, and similar heat deflection temperature and rigidity. ABS is also opaque.

Uses. The previously mentioned properties of ABS are suitable for tough consumer products; automotive parts; business machine housings; telephones; luggage; and pipe, fittings, and conduit.

17.1.7   Polyvinyl Chloride

The catalytic polymerization of vinyl chloride yields polyvinyl chloride. It is commonly referred to as PVC or vinyl and is second only to polyethylene in volume use. Normally, PVC has a low degree of crystallinity and good transparency. The high chlorine content of the polymer produces advantages in flame resistance, fair heat deflection temperature, good electrical resistance, and chemical resistance. However, PVC is difficult to process. The chlorine atoms also have a tendency to split out under the influence of heat during processing and heat and light during end use in finished products, producing discoloration and embrittlement. Therefore, special stabilizer systems are often used with PVC to retard degradation.

There are two major subclassifications of PVC, rigid and flexible. Properties appear in Table 17.4.

Table 17.4   Typical Property Values for Polyvinyl Chloride Materials

Property	General Purpose	Rigid	Rigid Foam	Plasticized	Copolymer
Linear mold shrinkage (in./in.)	0.003	—	—	—	—
Density (Mg/m^3)	1.400	1.340–1.390	0.750	1.290–1.340	1.370
Tensile modulus (GPa)	3.45	2.41–2.45	—	—	3.15
Tensile strength (MPa)	8.7	37.2–42.4	>13.8	14–26	52–55
Elongation at break (%)	113	—	>40	250–400	—
Notched Izod (kJ/m)	0.53	0.74–1.12	>0.06	—	0.02
Heat deflection temperature at 1.81 MPa (°C)	77	73–77	65	—	65
Brittle temperature (°C)	—	—	—	−60 to −30	—
Hardness	D85 (Shore)	R107–R112 (Rockwell)	D55 (Shore)	A71–A96 (Shore)	—
Linear thermal expansion (10^{-5} mm/mm·K)	7.00	5.94	5.58	—	—

Rigid PVC

PVC alone is a fairly good rigid polymer, but is difficult to process and has low impact strength. Both of these properties are improved by the addition of elastomers or impact-modified graft copolymers—such as ABS and impact acrylic polymers. These improve the melt flow during processing and improve the impact strength without seriously lowering the rigidity or the heat deflection temperature. With this improved balance of properties, rigid PVCs are used in such applications as door and window frames; pipe, fittings, and conduit; building panels and siding; rainwater gutters and downspouts; credit cards; and flooring.

Plasticized PVC

Flexible PVC is a plasticized material. Thus PVC is softened by the addition of compatible, nonvolatile, liquid plasticizers. The plasticizers, which are usually used in >20 parts per hundred (pph) resin level, lower the crystallinity in PVC and act as internal lubricants to give a clear, flexible plastic. Plasticized PVC is used for wire and cable insulation, outdoor apparel, rainwear, flooring, interior wall coverings, upholstery, automotive seat coverings, garden hose, toys, clear tubing, shoes, tablecloths, and shower curtains.

PVC is also available in liquid formulations known as plastisols or organosols. These materials are used in coating fabrics, paper, and metal; and are rotationally cast into dolls, balls, etc.

Foamed PVC

Rigid PVC can be foamed to a low-density cellular material that is used for decorative moldings and trim.

Foamed plastisols add greatly to the softness and energy absorption already inherent in plasticized PVC, giving richness and warmth to leatherlike upholstery, clothing, shoe fabrics, handbags, luggage, and auto door panels; and energy absorption for quiet and comfort in flooring, carpet backing, auto headliners, etc.

PVC Copolymer

Copolymerization of vinyl chloride with 10–15% vinyl acetate gives a vinyl copolymer with improved flexibility and less crystallinity than PVC, making such copolymers easier to process without detracting seriously from the rigidity and heat deflection temperature. These copolymers find primary applications in phonograph records, flooring, and solution coatings.

17.1.8 Poly(vinylidene Chloride)

Poly(vinylidene chloride) is prepared by the catalytic polymerization of 1,1-dichloroethylene. This crystalline polymer exhibits high strength, abrasion resistance, high melting point, better than ordinary heat resistance (100°C maximum service temperature), and outstanding impermeability to oil, grease, water vapor, oxygen, and carbon dioxide. It is used for packaging films, coatings, and monofilaments.

When the polymer is extruded into film, quenched, and oriented, the crystallinity is fine enough to produce high clarity and flexibility. These properties contribute to widespread use in packaging film, especially for food products that require impermeable barrier protection.

Poly(vinylidene chloride) and/or copolymers with vinyl chloride, alkyl acrylate, or acrylonitrile are used in coating paper, paperboard, or other films to provide more economical, impermeable materials.

A small amount of poly(vinylidene chloride) is extruded into monofilament and tape that is used in outdoor furniture upholstery.

17.1.9 Poly(methyl Methacrylate)

The catalytic polymerization of methylmethacrylate yields poly(methyl methacrylate) (PMMA), a strong, rigid, clear, amorphous polymer. PMMA has excellent resistance to weathering, low water absorption, and good electrical resistivity. PMMA properties appear in Table 17.5.

Uses. PMMA is used for glazing, lighting diffusers, skylights, outdoor signs, and automobile tail lights.

17.1.10 Poly(ethylene Terephthalate)

Poly(ethylene terephthalate) (PET) is prepared from the condensation polymerization of dimethyl terephthalate and ethylene glycol. It is a crystalline polymer that exhibits high modulus, high strength, high melting point, good electrical properties, and moisture and solvent resistance.

Uses. Primary applications of PET include fibers for wash and wear, wrinkle-resistant fabrics, and films that are used in food packaging, electrical applications (capacitors, etc.), magnetic recording tape, and graphic arts.

Table 17.5 Typical Properties of Poly(methyl Methacrylate)

Density (Mg/m³)	1.180–1.190
Tensile modulus (GPa)	3.10
Tensile strength (MPa)	72
Elongation at break (%)	5
Notched Izod (kJ/m)	0.4
Heat deflection temperature at 1.81 MPa (°C)	96
Hardness (Rockwell)	M90–M100
Linear thermal expansion (10^{-5} mm/mm·K)	6.3
Continuous service temperature (°C)	88
Linear mold shrinkage (in./in.)	0.002–0.008

17.2 ENGINEERING THERMOPLASTICS

Engineering polymers comprise a special high-performance segment of synthetic plastic materials that offer premium properties. When properly formulated, they may be shaped into mechanically functional, semiprecision parts or structural components. Mechanically functional implies that the parts may be subject to mechanical stress, impact, flexure, vibration, sliding friction, temperature extremes, hostile environments, etc., and continue to function.

As substitutes for metal in the construction of mechanical apparatus, engineering plastics offer advantages such as transparency, light weight, self-lubrication, and economy in fabricating and decorating. Replacement of metals by plastic is favored as the physical properties and operating temperature ranges of plastics improve and the cost of metals and their fabrication increases.

17.2.1 Polyesters (Thermoplastic)

Poly(butylene terephthalate) (PBT) is prepared from the condensation polymerization of butanediol with dimethyl terephthalate. PBT is a crystalline polymer. It seems to have a unique and favorable balance of properties between nylons and acetal resins. PBT has lower moisture absorption, extremely good self-lubrication, fatigue resistance, solvent resistance, and good maintenance of mechanical properties at elevated temperatures. Properties appear in Table 17.6.

Table 17.6 Typical Properties of Poly(butylene Terephthalate)

Property	PBT	PBT + 40% Glass Fiber
Density (Mg/m³)	1.300	1.600
Flexural modulus (GPa)	2.4	9.0
Flexural strength (MPa)	88	207
Elongation at break (%)	300	3
Notched Izod (kJ/m)	0.06	0.12
Heat deflection temperature at 1.81 MPa (°C)	54	232
Heat deflection temperature at 0.45 MPa (°C)	154	232
Hardness (Rockwell)	R117	M86
Linear thermal expansion (10^{-5} mm/mm·K)	9.54	1.89
Linear mold shrinkage (in./in.)	0.020	<0.007

Uses. Applications of PBT include gears, rollers, bearings, housings for pumps and appliances, impellers, pulleys, switch parts, automotive components, and electrical/electronic components.

17.2.2 Polyamides

The two major types of polyamides are nylon 6 and nylon 66. Nylon 6, or polycaprolactam, is prepared by the polymerization of caprolactam. Poly(hexamethylene adipate), or nylon 66, is derived from the condensation of hexamethylene diamine with adipic acid. These polyamides are crystalline polymers. Their key features include a high degree of solvent resistance, toughness, and fatigue resistance. Nylons do exhibit a tendency to creep under applied load. Their properties appear in Table 17.7.

Table 17.7 Typical Properties of Nylons

Property	Nylon 6	Nylon 6 + 40% Glass Fiber	Nylon 66	Nylon 66 + 40% Glass Fiber
Density (Mg/m³)	1.130	1.460	1.140	1.440
Flexural modulus (GPa)	2.8	10.3	2.8	9.3
Flexural strength (MPa)	113	248	—	219
Elongation at break (%)	150	3	60	4
Notched Izod (kJ/m)	0.06	0.16	0.05	0.14
Heat deflection temperature at 1.81 MPa (°C)	64	216	90	250
Heat deflection temperature at 0.45 MPa (°C)	170	218	235	260
Hardness (Rockwell)	R119	M92	R121	R119
Linear thermal expansion (10⁻⁵ mm/mm·K)	8.28	2.16	8.10	3.42
Linear mold shrinkage (in./in.)	0.013	0.003	0.0150	0.0025

Uses. The largest application of nylons is in fibers. Molded applications include automotive components, related machine parts (gears, cams, pulleys, rollers, boat propellers, etc.), appliance parts, and electrical insulation.

17.2.3 Polyacetals

Polyacetals are prepared via the polymerization of formaldehyde or its copolymerization with ethylene oxide. Polyacetals are crystalline polymers that exhibit rigidity, high strength, solvent resistance, fatigue resistance, toughness, self-lubricity, and cold-flow resistance. They also exhibit a tendency to thermally unzip and, hence, are difficult to flame retard. Properties appear in Table 17.8.

Table 17.8 Typical Properties of Polyacetals

Property	Polyacetal	Polyacetal + 40% Glass Fiber
Density (Mg/m³)	1.420	1.740
Flexural modulus (GPa)	2.7	11.0
Flexural strength (MPa)	107	117
Elongation at break (%)	75	1.5
Notched Izod (kJ/m)	0.12	0.05
Heat deflection temperature at 1.81 MPa (°C)	124	164
Heat deflection temperature at 0.45 MPa (°C)	170	167
Hardness (Rockwell)	M94	R118
Linear thermal expansion (10⁻⁵ mm/mm·K)	10.4	3.2
Linear mold shrinkage (in./in.)	0.02	0.003

Uses. Applications of polyacetals include moving parts in appliances and machines (gears, bearings, bushings, etc.), in automobiles (door handles, etc.), and in plumbing (valves, pumps, faucets, etc.).

17.2.4 Polyphenylene Sulfide

The condensation polymerization of dichlorobenzene and sodium sulfide yields a crystalline polymer, polyphenylene sulfide. It is characterized by high heat resistance, rigidity, excellent chemical resistance, low friction coefficient, good abrasion resistance, and electrical properties. Polyphenylene sulfides are somewhat difficult to process due to their very high melting temperature, relatively poor flow characteristics, and some tendency for slight cross-linking during processing. Properties appear in Table 17.9.

Table 17.9 Typical Properties of Polyphenylene Sulfide

Property	Polyphenylene Sulfide + 40% Glass Fiber
Density (Mg/m³)	1.640
Tensile modulus (GPa)	7.7
Tensile strength (MPa)	135
Flexural modulus (GPa)	11.7
Flexural strength (MPa)	200
Elongation at break (%)	1.3
Notched Izod (kJ/m)	0.08
Heat deflection temperature at 1.81 MPa (°C)	>260
Hardness (Rockwell)	R123
Linear thermal expansion (10^{-5} mm/mm·K)	4.0
Linear mold shrinkage (in./in.)	0.004
Constant service temperature (°C)	232

Uses. The unreinforced resin is used only in coatings. The reinforced materials are used in aerospace applications, pump components, electrical/electronic components, appliance parts, and automotive applications.

17.2.5 Polycarbonates

Most commercial polycarbonates are derived from bisphenol A and phosgene. Polycarbonates are transparent, amorphous polymers. They are among the stronger, tougher, and more rigid thermoplastics. Polycarbonates also show resistance to creep and excellent electrical-insulating characteristics. Polycarbonate properties appear in Table 17.10.

Uses. Applications of polycarbonates include safety glazing, safety shields, nonbreakable windows, automobile tail lights, electrical relay covers, various appliance parts and housings, power tool housings, automotive fender extensions, and blow-molded bottles.

17.2.6 Polysulfone

Polysulfone is prepared from the condensation polymerization of bisphenol A and dichlorodiphenyl sulfone. This transparent, amorphous resin is characterized by excellent thermooxidative resistance, hydrolytic stability, and creep resistance. Polysulfone properties appear in Table 17.11.

Uses. Typical applications of polysulfones include microwave cookware, medical equipment where sterilization by steam is required, coffee makers, and electrical/electronic components.

17.2.7 Modified Polyphenylene Ether

Blends of poly(2,6-dimethyl phenylene ether) with styrenics (i.e., HIPS, ABS, etc.) form a family of modified polyphenylene-ether-based resins. Depending on the blend, these materials have a broad tem-

Table 17.10   Typical Properties of Polycarbonates

Property	Polycarbonate	Polycarbonate + 40% Glass Fiber
Density (Mg/m³)	1.200	1.520
Tensile modulus (GPa)	2.4	11.6
Tensile strength (MPa)	65	158
Flexural modulus (GPa)	2.3	9.7
Flexural strength (MPa)	93	186
Elongation at break (%)	110	4
Notched Izod (kJ/m)	0.86	0.13
Heat deflection temperature at 1.81 MPa (°C)	132	146
Heat deflection temperature at 0.45 MPa (°C)	138	154
Hardness (Rockwell)	M70	M93
Linear thermal expansion (10^{-5} mm/mm·K)	6.75	1.67
Linear mold shrinkage (in./in.)	0.006	0.0015
Constant service temperature (°C)	121	135

Table 17.11   Typical Properties of Polysulfone

Density (Mg/m³)	1.240
Tensile modulus (GPa)	2.48
Tensile strength (MPa)	70
Flexural modulus (GPa)	2.69
Flexural strength (MPa)	106
Elongation at break (%)	75
Notched Izod (kJ/m)	0.07
Heat deflection temperature at 1.81 MPa (°C)	174
Hardness (Rockwell)	M69
Linear thermal expansion (10^{-5} mm/mm·K)	5.6
Linear mold shrinkage (in./in.)	0.007
Constant service temperature (°C)	150

perature use range. They are characterized by outstanding dimensional stability at elevated temperatures, outstanding hydrolytic stability, long-term stability under load, and excellent dielectric properties over a wide range of frequencies and temperatures. Their properties appear in Table 17.12.

Uses. Modified polyphenylene ether applications include automotive applications (dashboards, trim, etc.), TV cabinets, electrical connectors, pumps, plumbing fixtures, and small appliance and business machine housings.

17.2.8   Polyimides

Polyimides are a class of polymers prepared from the condensation reaction of a carboxylic acid dianhydride with a diamine. Thermoplastic and thermoset grades of polyimides are available. The thermoset polyimides are among the most heat resistant polymers; for example, they can withstand temperatures up to 250°C. Thermoplastic polyimides, which can be processed by standard techniques, fall into two main categories—polyetherimides and polyamideimides.

Table 17.12 Typical Properties of Modified Polyphenylene Ethers

Property	190 Grade	225 Grade	300 Grade
Density (Mg/m³)	1.080	1.090	1.060
Tensile modulus (GPa)	2.5	2.4	—
Tensile strength (MPa)	48	55	76
Flexural modulus (GPa)	2.2	2.4	2.4
Flexural strength (MPa)	56.5	76	104
Elongation at break (%)	35	35	—
Notched Izod (kJ/m)	0.37	0.32	0.53
Heat deflection temperature at 1.81 MPa (°C)	88	107	149
Heat deflection temperature at 0.45 MPa (°C)	96	118	157
Hardness (Rockwell)	R115	R116	R119
Linear thermal expansion (10^{-5} mm/mm·K)	—	—	5.9
Linear mold shrinkage (in./in.)	0.006	0.006	0.006
Constant service temperature (°C)	—	95	—

In general, polyimides have very good electrical properties, very good wear resistance, superior dimensional stability, outstanding flame resistance and high strength and rigidity. Polyimide properties appear in Table 17.13.

Table 17.13 Typical Properties of Polyimides

Property	Polyimide	Polyetherimide Unfilled	Polyetherimide 30% Glass Reinforced	Polyamideimide Unfilled	Polyamideimide 30% Glass Reinforced
Density (Mg/m³)	—	1.27	1.51	1.38	1.57
Tensile modulus (GPa)	2.65	0.30	0.90	—	1.15
Tensile strength (MPa)	196.2	104.8	168.9	117.2	195.2
Elongation at break (%)	90	60	3	10	5
Notched Izod (kJ/m)	—	0.6	0.11	0.13	0.11
Heat deflection temperature at 1.81 MPa (°C)	—	392	410	260	274
Heat deflection temperature at 0.45 MPa (°C)	—	410	414	—	—
Hardness (Rockwell)	—	R109	M125	E78	E94
Linear thermal expansion (10^{-5} mm/mm·K)	—	5.6	2.0	3.60	1.80
Linear mold shrinkage (in./in.)	—	0.5	0.2	—	0.25

Uses. Polyimide applications include gears, bushings, bearings, seals, insulators, electrical/ electronic components (printed wiring boards, connectors, etc.), microwave oven components, and structural components.

17.3 FLUORINATED THERMOPLASTICS

In general, fluoropolymers, fluoroplastics, or fluorocarbons are a family of fluorine-containing thermoplastics that exhibit some unusual properties. These properties include inertness to most chemicals, resistance to high temperature, extremely low coefficient of friction, and excellent dielectric properties. Properties appear in Table 17.14.

Mechanical properties are normally low, but can be improved when reinforced with glass or carbon fibers or molybdenum disulfide fillers.

Table 17.14   Typical Properties of Fluoropolymers

Property	PTFE	CTFE	FEP	ETFE	ECTFE
Density (Mg/m³)	2.160	2.100	2.150	1.700	1.680
Tensile modulus (GPa)	—	14.3	—	—	—
Tensile strength (MPa)	27.6	39.4	20.7	44.8	48.3
Elongation at break (%)	~275	~150	~300	100–300	200
Notched Izod (kJ/m)	—	0.27	0.15	—	—
Heat deflection temperature at 1.81 MPa (°C)	—	75	—	71	77
Heat deflection temperature at 0.45 MPa (°C)	—	126	—	104	116
Hardness	D55–65 (Shore)	D75–80 (Shore)	D55 (Shore)	D75 (Shore)	R93 (Rockwell)
Linear thermal expansion (10^{-5} mm/mm·K)	9.9	4.8	9.3	13.68	—
Dielectric strength (MV/m)	23.6	19.7	82.7	7.9	19.3
Dielectric constant at 10^2 Hz	2.1	3.0	2.1	2.6	2.5
Dielectric constant at 10^3 Hz	2.1	2.7	—	2.6	2.5
Constant service temperature (°)	260	199	204	—	150–170
Linear mold shrinkage (in./in.)	0.033–0.053	0.008	—	—	<0.025

The difficulty of fluorochemical synthesis makes the price of fluoropolymers relatively high, and their uses are largely restricted to critical specialty applications.

17.3.1   Poly(tetrafluoroethylene)

Prepared from tetrafluoroethylene, poly(tetrafluoroethylene) (PTFE) is a crystalline, very heat resistant (up to 500°F), and outstanding chemical resistant polymer, and it has the lowest coefficient of friction of any polymer. PTFE does not soften like other thermoplastics, and has to be processed by nonconventional techniques (PTFE powder is compacted to the desired shape and sintered).

Uses.   PFTE applications include nonstick coatings on cookware; nonlubricated bearings; chemical-resistant pipe fittings, valves, and pump parts; high-temperature electrical parts; and gaskets, seals, and packings.

17.3.2   Poly(chlorotrifluoroethylene)

Poly(chlorotrifluoroethylene) (CTFE) is less crystalline and exhibits higher rigidity and strength than PTFE; it is chemical resistant and has heat resistance up to 390°F. Unlike PTFE, it can be molded and extruded by conventional processing techniques.

Uses.   CTFE applications include electrical insulation, cable jacketing, electrical and electronic coil forms, pipe and pump parts, valve diaphragms, and coatings for corrosive process industries and other industrial parts.

17.3.3   Fluorinated Ethylene–Propylene

Copolymerization of tetrafluoroethylene with some hexafluoropropylene produces a polymer with less crystallinity, lower melting point, and improved impact strength than PTFE. This copolymer can be molded by thermoplastic molding techniques.

Uses.   Fluorinated ethylene–propylene applications include wire insulation and jacketing, high-frequency connectors, coils, gaskets, and tube sockets.

17.3.4 Polyvinylidene Fluoride

Polyvinylidene fluoride has high tensile strength, better ability to be processed but less thermal and chemical resistance than the previous fluoropolymers.

Uses. Polyvinylidene fluoride applications include insulation, seals and gaskets, diaphragms, and piping.

17.3.5 Poly(ethylene Trifluoroethylene)

Copolymerization of ethylene with trifluoroethylene produces poly(ethylene trifluoroethylene (ETFE) with good high-temperature and chemical resistance; ETFE can be processed by conventional techniques.

Uses. ETFE applications include molded labware, valve liners, electrical connectors, and coil bobbins.

17.3.6 Poly(ethylene Chlorotrifluoroethylene)

The copolymer of ethylene and chlorotrifluoroethylene, poly(ethylene chlorotrifluoroethylene) (ECTFE), is strong and chemical and impact resistant. It can be processed by conventional techniques.

Uses. ECTFE applications include wire and cable coatings, chemical-resistant coatings and linings, molded labware, and medical packaging.

17.3.7 Poly(vinyl Fluoride)

Poly(vinyl fluoride) films exhibit excellent outdoor durability.

Uses. Poly(vinyl fluoride) uses include glazing, lighting, and coatings on presurfaced exterior building panels.

17.4 THERMOSETS

Thermosetting polymers are used in molded and laminated plastics. They are first polymerized to a low-molecular-weight polymer, which is still soluble, fusible, and highly reactive during final processing. Thermosets are generally catalyzed and/or heated to finish the polymerization reaction, cross-linking them to almost infinite molecular weight. This step is often referred to as cure. Such cured polymers cannot be reprocessed or reshaped.

17.4.1 Phenolic

Phenolic resins combine the high reactivity of phenol and formaldehyde to form prepolymers and oligomers called resoles and novalaks. These materials are combined with fibrous fillers to give a phenolic resin, which when heated provides rapid complete cross-linking to highly cured structures. The high cross-linked aromatic structure has high hardness, rigidity, strength, heat resistance, chemical resistance, and good electrical properties.

Uses. Phenolic applications include automotive uses—distributor caps, rotors, brake linings; appliance parts—pot handles, knobs, bases; electrical/electronic components—connectors, circuit breakers, switches; and adhesive in laminates (e.g., plywood).

17.4.2 Epoxy Resins

Epoxy resins are low-molecular-weight materials that are liquid either at room temperature or on warming. Each polymer chain usually contains two or more epoxide groups. The high reactivity of the epoxide groups with amines, anhydrides, and other curing agents provides facile conversion into highly cross-linked materials. Cured epoxy resins exhibit hardness, strength, heat resistance, electrical resistance, and broad chemical resistance.

Uses. Epoxy applications include glass-reinforced, high-strength composites used in aerospace, pipes, tanks, pressure vessels; encapsulation or casting of various electrical and electronic components; adhesives; protective coatings in appliances, flooring, and industrial equipment; and sealants.

17.4.3 Unsaturated Polyesters

Properly formulated glass-reinforced unsaturated polyesters are commonly referred to as sheet molding compound (SMC) or reinforced plastics. Unsaturated polyesters are thermosets and are quite distinct from thermoplastic polyesters. In general, unsaturated polyesters are dissolved in styrene monomer to produce any viscosity material desired for impregnation and lamination of glass fibers. The low-molecular-weight polyesters have fumarate ester units that provide easy reactivity with styrene monomer.

In combination with reinforcing materials like glass fibers, cured resins offer outstanding strength, high rigidity, high weight-to-strength ratio, impact resistance, and chemical resistance.

Uses. The prime use of unsaturated polyesters is in combination with glass fibers in high-strength composites; these include transportation markets (body parts and components for automobiles, trucks, trailers, buses, and aircraft), marine uses (small- to medium-size boat hulls and associated marine equipment), building panels, housings, bathroom components (bathtubs and shower stalls), appliances, and electronic/electrical components.

17.4.4 Alkyd Resins

Alkyd resins are based on branched prepolymers from glycerol, phthalic anhydride, and glyceryl esters of fatty acids. Alkyds have excellent heat resistance, are dimensionally stable at high temperatures, and have excellent dielectric strength (>14 MV/m), high resistance to electrical leakage, and excellent arc resistance.

Uses. Alkyd resin applications include drying oils in enamel paints; lacquers for automobiles and appliances; and molding compounds when formulated with reinforcements for electrical applications (circuit breaker insulation, encapsulation of capacitors and resistors, and coil forms).

17.4.5 Diallyl Phthalate

Diallyl phthalate (DAP) is the most widely used compound in the allylic family. These low-molecular-weight prepolymers can be reinforced and compression molded into highly cross-linked, completely cured products.

.The most outstanding properties of DAP are excellent dimension stability and high insulation resistance. In addition, DAP has high dielectric strength, excellent arc resistance, and chemical resistance.

Uses. DAP applications include electronic parts, electrical connectors, bases, and housings. DAP is also used as a coating and impregnating material.

17.4.6 Amino Resins

The two main members of the amino family of thermosets are the melamine and urea resins. They are prepared from the reaction of melamine and urea with formaldehyde. In general, these materials exhibit extreme hardness, scratch resistance, electrical resistance, and chemical resistance.

Uses. Melamine resins find use in colorful, rugged dinnerware; decorative laminates (countertops, tabletops, and furniture surfacing); electrical applications (switchboard panels, circuit breaker parts, arc barriers, and armature and slot wedges); and adhesives and coatings.

Urea resins are used in particle board binders, decorative housings, closures, electrical parts, coatings, and paper and textile treatment.

17.5 GENERAL-PURPOSE ELASTOMERS

Elastomers are materials that can be stretched substantially beyond their original length and can retract rapidly and forcibly to essentially their original dimensions (on release of the force).

The optimum properties and/or economics of many rubbers are obtained through formulating with reinforcing agents, fillers, extending oils, vulcanizing agents, antioxidants, pigments, etc. End-use markets for formulated rubbers include automobile tire products (including tubes, retread applications, valve stems, and innerliners), adhesives, cements, caulks, sealants, latex foam products, hose (automotive, industrial, and consumer applications), belting (V-conveyor and trimming), footwear (heels, soles, slab stock, boots, and canvas), and molded, extruded, and calendered products (athletic goods, flooring, gaskets, household products, O-rings, blown sponge, thread, and rubber sundries). Properties appear in Table 17.15.

Table 17.15 Properties of General-Purpose Elastomers

Rubber	ASTM Nomenclature	Outstanding Property	Property Deficiency	Temperature Use Range (°C)
Butadiene rubber	BR	Very flexible; resistance to abrasive wear	Sensitive to oxidation; poor resistance to fuels and oil	−100 to 90
Natural rubber	NR	Similar to BR but less resilient	Similar to BR	−50 to 80
Isoprene rubber	IR	Similar to BR but less resilient	Similar to BR	−50 to 80
Isobutylene–isoprene rubber (butyl rubber)	IIR	High flexibility; low permeability to air		−45 to 150
Chloroprene	CR	Flame resistant; fair fuel and oil resistance; increased resistance toward oxygen, ozone, heat, light	Poor low-temperature flexibility	−40 to 115
Nitrile–butadiene rubber	NBR	Good resistance to fuels, oils, solvents; improved abrasion resistance	Lower resilience; higher hysteresis; poor electrical properties; poorer low temperature flexibility	−50 to 80
Styrene–butadiene rubber	SBR	Relatively low cost	Less resilience; higher hysteresis; limited low temperature flexibility	−50 to 80
Ethylene–propylene copolymer	EPM	Resistance to ozone and weathering	Poor hydrocarbon and oil resistance	−50 to <175
Ethylene–propylene–diene terpolymer	EPDM	Resistance to ozone and weathering	Poor hydrocarbon and oil resistance	−50 to <175
Polysulfide	T	Chemical resistance; resistance to ozone and weathering	Creep; low resilience	−45 to 120

Table 17.16 Properties of Specialty Rubbers

Rubber	ASTM Nomenclature	Temperature Use Range (°C)	Outstanding Property	Typical Applications
Silicones (polydimethylsiloxane)	MQ	−100 to 300	Wide temperature range; resistance to aging, ozone, sunlight; very high gas permeability	Seals, specialty molded and extruded goods, adhesives, biomedical
Fluoroelastomers	CFM	−40 to 200	Resistance to heat, oils, chemicals	Seals such as O-rings, corrosion-resistant coatings
Acrylic	AR	−40 to 200	Oil, oxygen, ozone, and sunlight resistance	Seals, hose
Epichlorohydrin	ECO	−18 to 150	Resistance to oils, fuels; some flame resistance; low gas permeability	Hose, tubing, coated fabrics, vibration isolators
Chlorosulfonated	CSM	−40 to 150	Resistance to oils, ozone weathering, oxidizing chemicals	Automotive hose, wire and cable, linings for reservoirs
Chlorinated polyethylene	CM	−40 to 150	Resistance to oils, ozone, chemicals	Impact modifier, automotive applications
Ethylene–acrylic		−40 to 175	Resistant to ozone, weathering	Seals, insulation vibration damping
Propylene oxide		−60 to 150	Low-temperature properties	Motor mounts

17.6 SPECIALTY ELASTOMERS

Specialty rubbers are more costly and, hence, are produced in smaller volume than the general-purpose rubbers. They fill the need for high-performance materials. Properties and uses are summarized in Table 17.16.

CHAPTER 18
FIBROUS COMPOSITES

ANDREW L. BASTONE
ALAN L. BASTING

Owens Corning Fiberglas
Toledo, Ohio

18.1 INTRODUCTION

Fibrous composites, sometimes referred to as fiberglass-reinforced plastics or glass-fiber-reinforced plastics, represent a family of engineering materials offering combinations of mechanical and physical properties not found in any other single-material system. Typically, fibrous composites are composed of reinforcing fibers, a polymer resin matrix, and various additives. Composite mechanical properties, such as strength and dimensional stability, are derived primarily from the reinforcement material—its type, form, amount, and orientation. Physical properties, such as chemical, electrical, and finish, are usually related to the choice of resin matrix. The resin matrix can be of either the thermosetting or thermoplastic classification. A number of other additives can be employed to provide specific product or process properties. Choice of composition usually depends on the property requirements of the finished product. The selected composition, the general shape considerations, and production volume, in turn, help determine the manufacturing process to be employed. Regardless of product size, complexity, production volume, or property requirements, there is an appropriate composite process available to deliver a product with favorable cost and performance.

The basic benefits offered by fibrous composites include such qualities as:

- A wide range of mechanical properties.
- High strength per unit of weight.
- Broad design flexibility.
- Dimensional stability.
- High dielectric strength.
- Broad chemical and corrosion resistance.
- Moderate tooling costs.
- Reduced finishing requirements.
- The potential for consolidating parts of an assembly.

Over 40,000 specific applications for composites have been identified, including a variety of applications for the automotive, agricultural, appliance, aviation/aerospace, business machine, chemical-processing, construction, electrical/electronic, home, marine, materials-handling, recreational, and transportation industries. Products range in complexity and size from a 200-ft boat hull to the intricate gears and pinions for a clock.

Critical to the successful use of fibrous composites is an integrated design approach, which depends on the design engineer's understanding of the relationships between materials, fabrication processes, and the service environment. When these relationships are considered early in a component or product's design development, fibrous composites offer the engineer nearly unlimited opportunities to develop an effective, cost/performance design.

18.2 FIBROUS REINFORCEMENTS

Fibrous composites can be described as materials constructed with a discrete fiber component embedded in a continuous resin matrix. The fibers can be very short and randomly dispersed, or long or continuous in length with specific orientations. They can also be combinations of both short, long, and/or continuous in almost any relationship and orientation. These fibers can be organic or inorganic, and either natural or synthetic in origin. Examples include:

Natural Organic	Natural Inorganic
Cellulose	Asbestos
Cotton	Wollastonite
Jute	
Sisal	

Synthetic Organic	Synthetic Inorganic
Aramid	Boron
Polyacrylonitrile	Ceramic
Polyamide	Glass
Polyester	Graphite–carbon
Polyvinyl alcohol	Metallic (steel, aluminum)
Rayon	Silicon carbide

Although all of these fibers can be used for composites, glass represents the largest percentage of the reinforcing-fiber market in the United States at this time. For this reason, the remainder of this chapter will be devoted primarily to glass-fiber reinforcements, although it must be recognized that the processes, matrix resins, and design principles would be essentially applicable to other fibrous reinforcement.

Glass fibers are available in a variety of forms to accommodate both the reinforcement needs of specific applications and facilitate handling in the fabrication processes. Regardless of their final form, all glass-fiber reinforcements originate with the attenuation or drawing of monofilaments from a furnace containing molten glass. A large number of these filaments are formed simultaneously and are gathered into a strand. A chemical surface treatment is applied to facilitate processing, maintain fiber integrity, and establish compatibility with specific resin systems. These treated strands are then further processed into basic reinforcing forms.

18.2.1   Continuous-Strand Roving

The continuous-strand roving form is a ropelike arrangement of untwisted strands wound into a cylindrical package from 35 to 400 lb in weight. One of the lowest cost reinforcement forms, continuous strand roving provides good overall processing through fast wet-out (penetration of resin into the strand), even tension, and abrasion resistance during processing. Strands can be cut cleanly and dispersed evenly throughout the resin matrix during fabrication. They provide excellent mechanical properties.

18.2.2   Woven Roving

Woven roving is a heavy drapable fabric made from continuous strand roving. Available in various widths, thicknesses, weights, and strength orientations, woven roving imparts high strengths to large molded parts.

18.2.3   Woven Fabrics

Woven fabrics are made from glass yarns and are available in a broad range of properties, widths, and lengths. Weights and thickness vary, and a number of strength orientations are possible.

18.2.4   Surfacing Mats (or Veil)

Used in conjunction with reinforcing mats and fabrics to provide a good surface finish, surfacing mat is effective in blocking out the fiber pattern of the underlying mat or fabric. Surfacing mats can also be used on the inside of fiber-reinforced plastic products to provide a smooth, resin-rich surface.

18.2.5   Reinforcing Mats

Made of either chopped strands or continuous strands of glass laid in a swirl pattern, mats are generally held together by resinous binders to allow handling as a sheet material. They are used for medium strength parts with uniform cross section. Both chopped and continuous-strand reinforcing mats are available in varying weights and widths.

18.2.6   Chopped Strands

Chopped strands are short strands, $1/8$–2 in. in length, which are blended with resins and other additives to prepare molding compounds for compression or injection molding and other processes. The shorter chopped-strand reinforcements are best suited for blending with thermoplastic resin systems for injection molding. Longer chopped strands are blended with thermosetting resins for compression or transfer molding.

18.2.7   Milled Fibers

Milled fibers, very short lengths of fibrous glass, $1/32$–$1/8$ in., are used for reinforcing thermoplastic parts where strength requirements are low and for reinforcing fillers and adhesives.

18.2.8   Glass Content

The mechanical-strength properties of fibrous composites depend primarily on the combined effect of the amount, type, and arrangement (orientation) of the glass-fiber reinforcements. In any composite made of glass fiber, strength will essentially increase directly and proportionally in relation to the percentage of glass fiber in the composite. The more glass, the stronger the part.

18.2.9 Fiber Orientation

In any component made of glass-fiber composite, the arrangement of fibers—the way in which the individual strands are positioned—determines the direction and level of strength that is achieved. There are three general types of fiber orientation:

1. *Unidirectional.* This orientation provides the greatest strength in the direction of the fibers. Up to 80% reinforcement loading, by weight, is possible.
2. *Bidirectional.* Bidirectional orientation is where some fibers are positioned at an angle to the remaining fibers, as with woven fabric. This provides different strength levels in each direction of fiber orientation. Up to 75% glass-fiber loadings, by weight, can be obtained with woven reinforcements.
3. *Multidirectional or Isotropic.* Multidirectional arrangement provides essentially equal strength in all directions in the finished part. From 10 to 50% reinforcement loading, by weight, can be obtained.

The relationship between the amount of glass-fiber reinforcement and the arrangement of the reinforcement can easily be seen: the more oriented the fibers, the greater the reinforcement loadings possible and, therefore, the greater strength obtainable in the finished part (in the direction of the fibers). The more random the arrangement, the lower the reinforcement loading, and, therefore, the lower the resulting strength.

Figure 18.1 illustrates how part strength is affected by glass-fiber-reinforcement percentage by weight, as well as by fiber orientation. Also indicated are reinforcement product types that yield the directional arrangements indicated, together with the widely used processing methods with which they are compatible.

The selection of the specific reinforcement for a particular composite application should not be wholly the result of mechanical-property requirements. The size and shape of the product being produced, and the expected method of production, are both integral elements of the design development process.

Generally, there are several reinforcement combinations to satisfy the structural needs for a given application. Careful consideration of mechanical properties, reinforcement cost, and process cost and requirements will yield the most effective, cost/performance reinforcement system.

18.3 MATRIX RESINS AND ADDITIVES

Of equal importance to the glass reinforcement is the selection of the matrix resin and additives that comprise the total fibrous composite system. The matrix resin is the primary determinant of the chemical, electrical, and thermal properties of the composite. A wide range of thermosetting and thermoplastic resins are available to produce composites that will meet the requirements for almost any application.

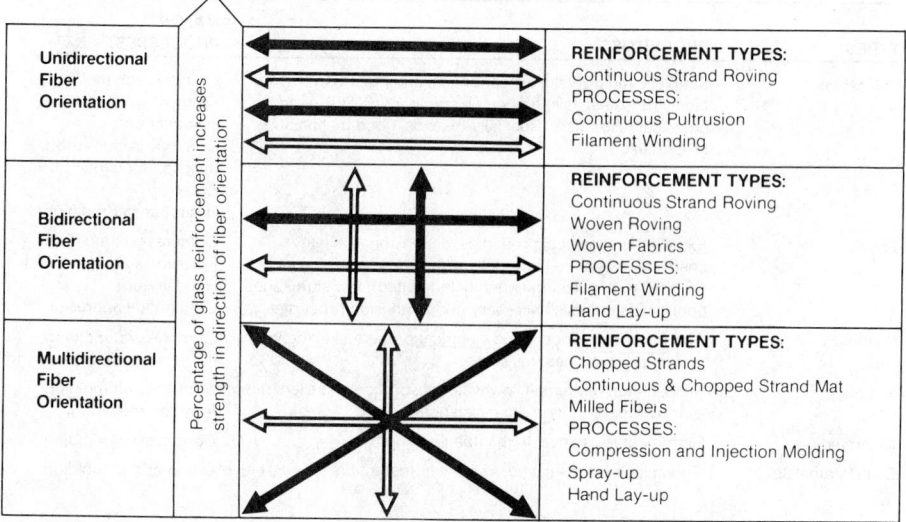

Fig. 18.1   Fiber orientation.

Thermosets undergo a chemical reaction (polymerization) during molding, forming a hardened material that cannot be remelted. Thermoplastics will repeatedly soften when heated and harden when cooled.

Additives are available for use with many resin systems to provide a variety of characteristics such as color, flame retardance, weather resistance, surface smoothness, toughness, lower density, low shrinkage, and reduced material costs. There are also additives used as effective processing aids to control cure rates, modify viscosity, and provide for ease in mold release. The optimum composite requires the proper selection and combination of all the various constituent materials, keeping in mind the desired end-product properties and the fabrication process to be used.

18.3.1 Thermosetting Resins

Thermosetting resins represent the most widely used matrix system for fibrous composites. A thermosetting resin undergoes an irreversible chemical reaction (called polymerization) when it is transformed from its typically liquid state to a solid form during the fabrication process. Generally speaking, the thermosetting composites have higher mechanical properties and greater heat resistance than thermoplastic composites.

Polyesters are the predominant class of thermosetting resins, offering good mechanical, chemical, and electrical properties, as well as dimensional stability, ease of handling, and reasonable cost. There are polyesters specifically designed for low- and high-temperature processes, for room- or high-temperature cure, and for flexible or rigid products. Additives are easily incorporated into polyester resin systems to produce a variety of specific properties and aid in fabrication. Figure 18.2 shows the major thermosetting resins and their characteristic properties. It also indicates the fabrication processes generally employed for the various resins.

18.3.2 Thermoplastic Resins

Thermoplastic resins are finding increased use in fibrous composites because they are easily fabricated and demonstrate a potential for high-production-rate processing. Also, newer high-performance engineering polymers have become increasingly available. Thermoplastic resins differ from thermosetting types in that they begin as solid materials that are heated and become liquid during fabrication, and then solidify again when cooled. This process, unlike thermosetting processes, is reversible. Figure 18.3 shows the major thermoplastic resins and their characteristic properties. Accompanying fabrication processes are also indicated.

The actual selection of the resin for a particular application, as with the selection of a reinforcement, requires consideration of not only the resin properties, but also the reinforcement system and the fabrication process to be used. Working with reinforced composites means working with an integrated systems approach—from choice of materials, through production and final application.

TYPES	PROPERTIES	MAJOR PROCESSES
Polyesters	Simplest, most versatile, economical and most widely used family of resins, having good electrical properties, good chemical resistance, especially to acids, good mechanical properties	Compression molding Filament winding Hand lay-up Continuous pultrusion Injection molding Spray-up Resin transfer molding
Epoxies	Excellent mechanical properties, dimensional stability, chemical resistance (especially alkalis), low water absorption, self-extinguishing (when halogenated), low shrinkage, good abrasion resistance, very good adhesion properties	Compression molding Filament winding Hand lay-up Continuous pultrusion
Phenolics	Good acid resistance, good electrical properties (except arc resistance), high heat resistance	Compression molding Continuous laminating
Silicones	Highest heat resistance, low water absorption, excellent dielectric properties, high arc resistance	Compression molding Injection molding
Melamines	Good heat resistance, high impact strength	Compression molding
Diallyl phthalate	Good electrical insulation, low water absorption	Compression molding

Fig. 18.2 Thermosetting resins.

TYPES	PROPERTIES	MAJOR PROCESSES
Polystyrene	Low cost, moderate heat distortion, good dimensional stability, good stiffness, good impact strength	Injection molding Continuous laminating
Nylon	High heat distortion, low water absorption, low elongation, good impact strength, good tensile and flexural strength	Injection molding Stamping
Polycarbonate	Self-extinguishing, high dielectric strength, high mechanical properties	Injection molding
Styrene-acrylo-nitrile (SAN)	Good solvent resistance, good long term strength, good appearance	Injection molding
Acrylics	Good gloss, weather resistance, optical clarity, and color; excellent electrical properties	Injection molding Continuous laminating
Polyvinyl chloride (PVC)	Excellent weatherability, superior electrical properties, excellent moisture and chemical resistance, self-extinguishing	Injection molding Continuous laminating
Acetals	Very high tensile strength and stiffness, exceptional dimensional stability, high chemical and abrasion resistance, no known room temperature solvent	Injection molding
Polyethylene	Good toughness, light weight, low cost, good flexibility, good chemical resistance; can be "welded"	Injection molding
Fluorocarbons	Very high heat and chemical resistance, non-burning, lowest coefficient of friction, high dimensional stability	Injection molding
Polyphenylene oxide (PPO) modified	Very tough engineering plastic, superior dimensional stability, low moisture absorption, excellent chemical resistance	Injection molding Continuous pultrusion
Polypropylene	Excellent resistance to stress or flex cracking, very light weight, hard, scratch-resistant surface, can be electroplated; good chemical and heat resistance; exceptional impact strength; good optical qualities	Injection molding Stamping
Polysulfone	Good transparency, high mechanical properties, heat resistance, electrical properties at high temperatures; can be electroplated	Injection molding
Acrylonitrile-butadiene-styrene (ABS)	Heat stability, chemical resistance, low temperature property retention, toughness, high impact strength, high gloss surface, rigidity, and good processing	Injection molding Extrusion
Polyphenylene-sulfide (PPS)	Outstanding high-temperature stability, inherent flame resistance, excellent chemical resistance, good property and shape retention at high temperature, high impact strength, high electrical arc resistance and low track rate	Injection molding
Thermoplastic Polyester (PBT Polybutylene Terephthalate)	Fast molding cycle, good chemical resistance, good high temperature properties, high dimensional stability, low coefficient of static and dynamic friction, good creep resistance at elevated temperature, good electrical properties	Injection molding
Urethane Elastomer	Tough, abrasion resistance, good mechanical strength properties, good impact strength	Injection molding

Fig. 18.3 Thermoplastic resins.

18.4 FABRICATION PROCESSES

There are many fabrication processes available to produce fibrous composite products. Each process has its own characteristics, as well as limitations accruing to product size, shape, production rate, compatible reinforcement type, and suitable resin system. The illustrated descriptions characterize each of the major processes and provide general guidance on which process best fits a particular application.

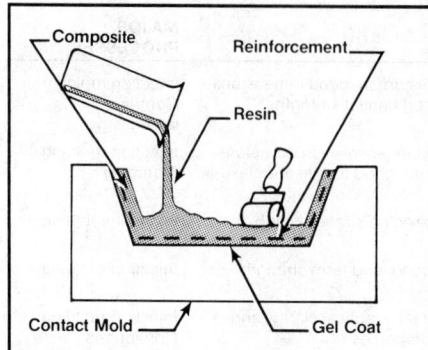

PROCESS CHARACTERISTICS
- offers low to medium production volume
- offers capability for large part production
- involves low tooling cost
- is labor intensive
- offers one finished surface
- involves minimum equipment investment

RESINS
- polyesters
- epoxies

REINFORCEMENTS
- mats
- woven roving
- fabrics

TYPICAL APPLICATIONS
- boats
- tanks
- housings
- building panels
- truck parts
- corrosion duct work

Fig. 18.4   Hand lay-up.

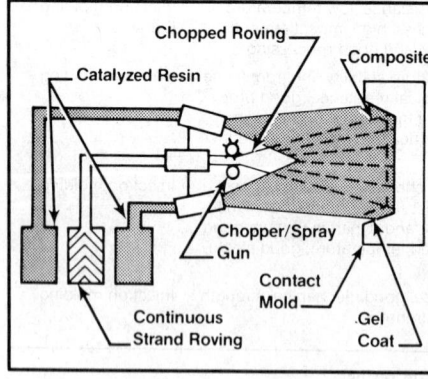

PROCESS CHARACTERISTICS
- offers low to medium production volume
- offers capability for large part production
- involves low tooling cost
- is labor intensive
- offers one finished surface
- is automatable
- provides extreme shape flexibility
- involves low investment costs

RESINS
- polyesters

REINFORCEMENTS
- roving
- woven roving

TYPICAL APPLICATIONS
- boats
- tanks
- tub/shower units
- truck parts
- motor home components
- custom automotive parts

Fig. 18.5   Spray-up.

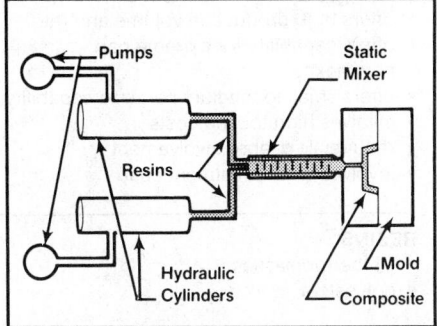

PROCESS CHARACTERISTICS
- offers intermediate production volume
- offers capability for large part production
- involves moderate tooling costs
- requires moderate labor involvement
- offers two finished surfaces
- involves moderate equipment investment

RESINS
- polyesters
- epoxies

REINFORCEMENTS
- mats, chopped and continuous
- woven roving

TYPICAL APPLICATIONS
- equipment housings
- automotive components
- camper and motor home parts

Fig. 18.6   Resin transfer molding.

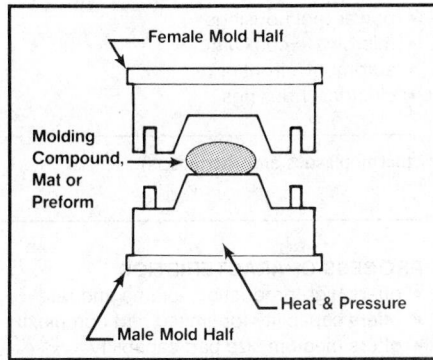

PROCESS CHARACTERISTICS
- offers high production volume
- offers moderate size part capability
- offers excellent surface finish
- involves high tooling costs
- requires low labor involvement
- involves high investment cost

PROCESS OPTIONS
- sheet molding compounds
- mat or preform
- bulk molding compounds

RESINS
- polyesters
- phenolics
- melamines
- silicones
- epoxies
- diallylphthalate

REINFORCEMENTS
- roving
- chopped strand
- continuous strand mats
- chopped strand mats
- fabrics

TYPICAL APPLICATIONS
WITH SHEET MOLDING COMPOUND
- automotive front ends
- appliance housings
- furniture
- business machine parts
- bath tubs

WITH MAT OR PREFORMS
- electrical sheet
- truck parts
- small fishing boats
- chair shells
- helmets

WITH BULK MOLDING COMPOUND
- complex housings
- electrical circuit breakers
- air conditioner parts

Fig. 18.7   Compression molding.

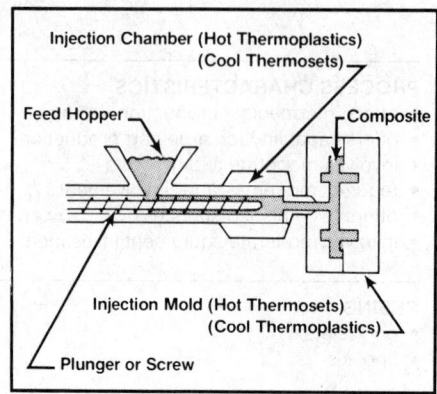

PROCESS CHARACTERISTICS
- offers high production volume and rate
- offers capability for extreme part complexity
- offers small to medium part size capability
- involves high tooling costs
- requires low labor involvement
- involves high investment cost

RESINS
- all thermoplastics
- polyesters

REINFORCEMENTS
- roving
- chopped strand

TYPICAL APPLICATIONS
THERMOPLASTIC
- household appliance parts
- gears
- pump housings
- automotive instrument panels
- laundry tubs

THERMOSETS
- power tool housings
- microwave cookware
- automotive front ends
- electrical housings

Fig. 18.8 Injection molding (includes thermoplastics and thermosets).

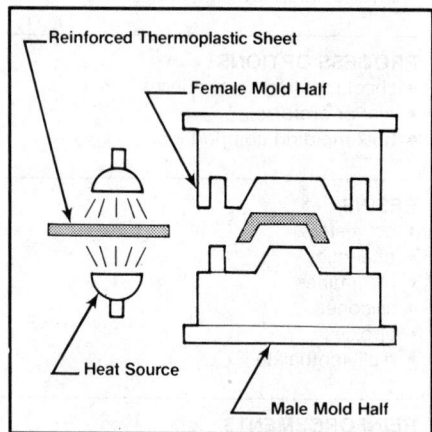

PROCESS CHARACTERISTICS
- offers high production volume and rate
- offers capability for limited part complexity
- offers medium size part capability
- requires low labor involvement
- involves high tooling costs
- involves high investment cost

RESINS
- thermoplastics
- polypropylene
- nylon

REINFORCEMENTS
- mats
- chopped strand
- continuous strand

TYPICAL APPLICATIONS
- automotive components
- musical instrument cases

Fig. 18.9 Cold stamping.

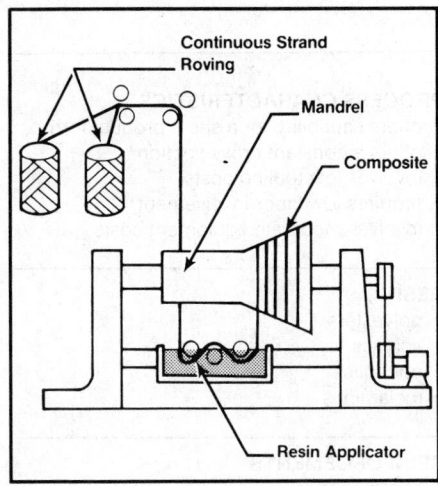

PROCESS CHARACTERISTICS
- shapes determined by surfaces of revolution
- offers accurate fiber orientation
- offers high reinforcement content
- requires low labor involvement
- involves moderately expensive equipment cost
- involves moderate tooling costs
- is automatable

RESINS
- polyesters
- epoxies
- thermoplastics

REINFORCEMENTS
- continuous roving
- tapes

TYPICAL APPLICATIONS
- pipe
- chemical tanks
- pressure vessels
- rocket motor cases
- chimney stack liners

Fig. 18.10 Filament winding.

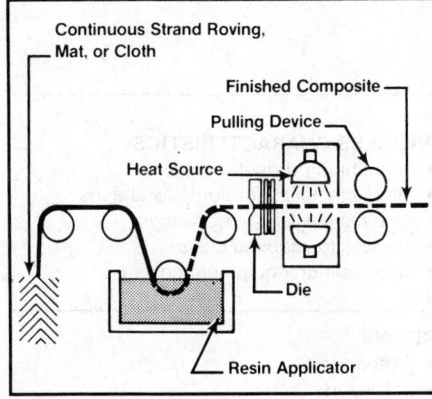

PROCESS CHARACTERISTICS
- offers a constant cross-section
- offers high fiber orientation
- requires low labor involvement
- involves moderate equipment cost
- involves low tooling costs
- is automatable

RESINS
- polyesters
- epoxies
- thermoplastics

REINFORCEMENTS
- continuous roving
- mats
- woven roving

TYPICAL APPLICATIONS
- fishing rods
- structural shapes
- pipe
- oil well sucker rods
- hammer and axe handles
- ladder siderails

Fig. 18.11 Continuous pultrusion.

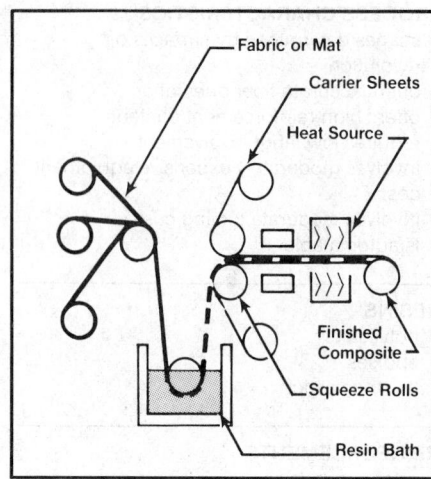

PROCESS CHARACTERISTICS
- offers capability for a sheet product form
- offers a constant cross-section
- involves low tooling costs
- requires low labor involvement
- involves moderate equipment costs

RESIN
- polyesters
- acrylics
- phenolics
- melamines

REINFORCEMENTS
- continuous roving
- mats
- fabrics

TYPICAL APPLICATIONS
- flat and corrugated glazing panels
- electrical insulating sheet
- truck liner panels
- highway sign stock

Fig. 18.12 Continuous laminating.

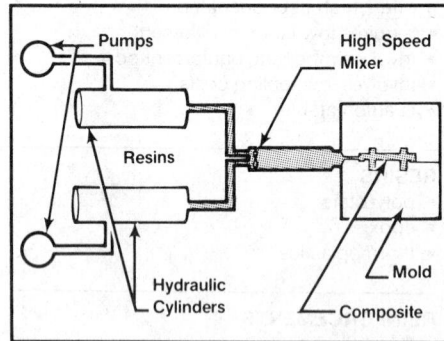

PROCESS CHARACTERISTICS
- offers high production volume
- offers capability for complex shapes
- requires low labor involvement
- involves low tooling costs
- involves high equipment costs

RESINS
- urethanes
- polyesters
- epoxies

REINFORCEMENTS
- milled fibers
- glass flake

TYPICAL APPLICATIONS
- automotive bumper covers and sight shields
- flexible fenders

Fig. 18.13 Reinforced reaction injection molding.

These fabrication processes represent the major processes used to produce composite parts. There are variations and combinations of these processes for specific applications. More complete information can be provided by the various molders and material suppliers of the composite industry.

18.5 DESIGN IN COMPOSITES

Design in composites is similar to design in other structural materials, providing consideration is given to such factors as the interrelationship of materials and process, the ability to vary mechanical properties, and the effects of service environment, strength orientation, and other factors. Composite materials, when considered early in the development phase, can provide unique solutions to the most critical design problems. Composites can be fashioned to satisfy a variety of specific end-use requirements, including wide temperature range, electrical-insulation properties, corrosion resistance, and a spectrum of mechanical properties. Surfaces can be smooth or textured, and in many cases colors can be molded in. Parts can be large or small, simple or complex. Consideration of these end-use requirements leads the designer to the proper choice of reinforcement, resin system, and process.

The general design approach includes 10 basic steps. Each step is critical to the successful development of a composite application.

18.5.1 Determining Functional Requirements

As noted, it is important to determine the functional end-use requirements at the outset of the design process. A thorough analysis must be made of the mechanical, thermal, electrical, chemical, and other requirements, such as surface finish, color, appearance, machinability, and properties that may be sought to meet standard or code requirements. This will provide the information necessary to select the proper composite material and process. An application development checklist, which provides typical predesign considerations for functional requirements, is shown in Figure 18.14, in Section 18.5.11. Composite fabricators and material suppliers are also excellent resources and can be helpful during early inquiry.

18.5.2 Determining Economic Requirements

This step establishes the necessary cost targets that ultimately will be used to determine if the composite part is economically feasible. When considering the economics of a given end use, a "systems" approach should be taken so that the composite application is viewed in the context of its total contribution to the end use. Such "systems" considerations should include tooling cost, finishing cost, and other secondary operations, such as assembling, machining, warehousing, shipping, packaging, quality and inventory control, and other related application costs. An evaluation of this magnitude will take the designer beyond a simple comparison of materials and labor costs. In most cases, life-cycle costing is advisable because of the long-term durability of composite materials.

18.5.3 Determining Material and Process

As is generally the case, choosing material and process involves a single selection because of the relationship they share. With functional and economic requirements at hand, the designer can select the type of reinforcement that will give the mechanical properties needed within appropriate economic parameters. Second, a resin formulation can be selected that will provide the chemical, electrical, and thermal performance required. The selected reinforcement and resin formulation will help determine the process and tooling that will produce the most cost-effective results.

General shape, size, and production volume will usually limit process selection to one or two choices. Process descriptions shown in Section 18.4 (Figs. 18.4–18.13) provide some general characteristics to aid selection of candidate processes; however, these figures are not meant to be all-inclusive for either materials or applications. (There are over 40,000 applications developed for glass-fiber composites alone.) Final process selection will be based on the overall cost/performance that each system will provide. Once the material and process selection has been completed, typical material properties can be determined using tables, such as those provided in the Appendix to this chapter. The tables and the information required to generate them are available from composite material suppliers and fabricators.

Also supplied in the Appendix is a table for test methods for plastics, which includes a list of properties and both the Federal specifications and ASTM standardized testing methods for determining these properties.

18.5.4 Creating the Initial Design Concept

Using information generated by such sources as the Application Development Checklist (Fig. 18.14), in conjunction with the general characteristics of the selected process, the initial design configuration is defined and general proportions are established. This can be done empirically, based on similar applications, or by detailed stress analysis using engineering formulas and the mechanical properties of the selected material system (refer to a mechanical properties table, such as Table 18.2 or other reference source). Computer-aided-design codes now exist that provide sophisticated analysis using finite-element methods. These computer-aided-design techniques are very useful in comparing various design approaches to optimize cost, deflection, weight, or other specific performance factors. Computers can also utilize the directional properties that result from fiber orientation with relative ease, greatly simplifying the design process.

Other design details such as mounting locations, inserts, stiffening ribs, bosses, and molded-in holes must also be considered. Design guides for molding processes (such as Fig. 18.15) are available from material suppliers and fabricators who normally are involved with this phase of design to add efficiency and cost effectiveness to the design process and ensure that the resultant design will be producible. Detailed design drawings are then made.

18.5.5 Prototype Construction

With new applications it is always desirable to build a prototype of the part before the final commitment is made for production tooling. The prototype serves several functions. Generally, it is used to show what the part will really look like in full scale. Many times, a design "on paper" will appear quite differently when it is translated into a three-dimensional part. The prototype can also be used to determine how the part fits with other components when the application is part of a total assembly. Finally, the prototype can be subjected to the same mechanical loading that the production part must withstand in actual use. This allows refinement of the design before the actual production tooling is completed.

There are various ways of building prototypes, depending on the production process that is to be used for the actual part. Sometimes the prototype can be produced on partially completed production tools, allowing some modification of tool design after the prototype has been evaluated. With the lower volume and tooling cost processes, the first production mold can be used for prototypes.

18.5.6 Performance Evaluation

Before final commitment to production tooling, the prototype part is tested to confirm that each of the established functional requirements has been satisfied. Static load and deflection requirements can be checked easily. Dynamic performance, involving fatigue or vibration resistance, requires more elaborate testing. In some cases, strain gages are used to determine the actual stress developed at a critical point. A comparison of this stress and the ultimate strength of the material will define the safety factor that exists. Long-term properties, such as chemical and weather resistance, are determined by comparisons to similar applications of the same material system or by specific exposure testing.

18.5.7 Design Optimization

Utilizing the information gained from performance evaluation, problems of high stress can be resolved by increasing wall thickness or modifying the type of reinforcement. High deflections may require adding ribs or other stiffeners. Each composite application must be considered individually. Results of this effort are used to modify the production tools. Through molding and testing of prototypes, the design is refined and the final configuration is established, resulting in a composite part with maximum function and minimum material.

18.5.8 Pilot Production

As soon as the first production tool is completed, several parts should be molded to confirm that the tool works properly and that the part meets the design requirements. Critical measurements are confirmed to ensure proper fit.

18.5.9 Final Evaluation

Parts produced in the production mold should be evaluated in the same way that the prototype parts were evaluated (see Sec. 18.5.6) to confirm that the final part is adequate. The part should also be checked to confirm that the required thicknesses have been achieved. This can be done by cutting the part into strips or by drilling holes and making actual micrometer measurements. Any changes required at this point are more expensive to make because the tool must be modified. However, if the design is found faulty, the changes must be made.

DIMENSION AND SHAPE REQUIREMENTS
1. Dimensions of existing or proposed part.
2. Limitations or critical dimensions.
3. Present material, part thickness, and weight.
4. Weight limitation.
5. Secondary operations of present part.
   - machining
   - drilling
   - tapping
   - assembly
   - welding
   - other
6. Finishing operations of present part.
   - plating
   - polishing
   - painting
   - other
7. Present part cost or cost objective.

ENVIRONMENTAL REQUIREMENTS
1. Temperature service range.
   - inside
   - outside
   - cycles-high to low
   - cycles-freeze/thaw
2. Liquid, moisture, or vapor exposure.
3. Chemical exposure.
   - concentration
   - temperature
4. Fire safety.
   - incombustibility
   - flame spread
   - toxic gases
   - fuel content
5. Light transmission.
6. Surface.
   - texture
   - color
   - abrasion resistance
7. Thermal insulation.
8. Electrical insulation
   - dielectric properties
   - arc and track resistance
9. Weather exposure.
   - moisture
   - sun
   - sand
   - pollutants

MECHANICAL REQUIREMENTS
1. General function of part.
2. Loads to be applied.
   - Gravity — Fixed weight / Live weight
   - Pressure — Fluid / Earth / Wind
   - Dynamic — Impact / Seismic / Cyclic / Handling and shipping
3. Deflection limitations.

MAINTENANCE REQUIREMENTS
1. Special cleaning agents and frequency.
2. Lubricant or oil exposure.
3. Abrasive exposure.

PRODUCTION REQUIREMENTS
1. Yearly volume required.
2. Production rate (monthly-highest level).
3. Total volume (design life).
4. Shipping range.

SPECIAL REQUIREMENTS
1. Codes and standards.
   - Underwriters Laboratory
   - Federal Drug Administration
   - National Safety Foundation
   - building codes
   - others
2. Shipping and handling.

Fig. 18.14 Application development checklist.

18.5.10 Full Production

When final evaluation is complete and any required corrections are made, the part is ready for full production. Normal production quality control procedures are then necessary to ensure that a consistent product is produced.

It should be recognized that these steps outline a simplified procedure that can generally be related to a wide variety of end-use composite applications. Obviously, some applications will require a much more extensive design approach. However, the same general procedures remain applicable. More detailed information concerning design with composites can be found in the reference materials cited in the Appendix.

18.5.11 Application Development Checklist

An application development checklist is given in Fig. 18.14.

18.6 APPENDIX

Various properties and design guides are listed in Tables 18.1–18.3 and Fig. 18.15.

Table 18.1 Test Methods for Plastics

Property	Federal Spec. 6P-4066	ASTM
Flexural Strength	1031	D790
Flexural Modulus	1031	D790
Tensile Strength	1011	D638/D651
Tensile Modulus	1011	D638
Compressive Strength	1021	D695
Elongation	1011	D638
Izod Impact	1071	D256
Thermal Conductivity	N.T.	C177
Specific Heat	N.T.	N.T.
Flammability	2021	D635
Rockwell Hardness	1081	D785
Dielectric Strength	4031	D149
Specific Gravity	5011/5012	D792
Density	5011/5012	D792
Heat Distortion Temperature	2011	D648
Continuous Heat Resistance	N.T.	N.T.
Thermal Coefficient of Expansion	2031	D696

N.T.—means no standardized method from these sources.

Table 18.2 Typical Mechanical Properties[a]

	% Glass Fiber By Wt.	Flexural Strength psi $\times 10^3$	Flexural Strength M Pa	Flexural Modulus psi $\times 10^6$	Flexural Modulus G Pa	Tensile Strength at Yield psi $\times 10^3$	Tensile Strength at Yield M Pa	Tensile Modulus psi $\times 10^5$	Tensile Modulus G Pa	Compressive Strength psi $\times 10^3$	Compressive Strength M Pa	Ultimate Tensile Elongation %	Impact Strength Izod Ft-lb/in of notch	Impact Strength Izod NM/cm of notch
THERMOSETTING RESIN PROCESSES														
Spray Up—Mat Hand Lay Up—Polyester	30-50	16-28	110-190	10-12	7-8	9-18	60-120	8-18	6-12	15-25	100-170	1.0-12	4-12	2-6
Resin Transfer Molding—Mat Polyester	20-30	22-37	150-250	13-19	9-13	12-20	80-140	—	—	—	—	1-2	9-12	5-6
Compression Molding—Sheet Molding Compound (SMC)—Polyester	15-30	18-30	120-210	14-20	10-14	8-20	60-140	16-25	11-17	15-30	100-210	3-1.5	8-22	4-12
Compression Molding—Mat or Preform—Polyester	25-50	10-40	70-280	13-18	9-12	25-30	170-210	9-20	6-14	15-30	100-210	1-2	10-20	5-11
Compression Molding—Bulk Molding Compound (BMC)—Polyester	15-35	10-20	70-140	14-20	10-14	4-10	30-70	16-25	11-17	20-30	140-210	3-.5	2-10	1-5
Compression Molding—Bulk Molding Compound (BMC)—Phenolic	5-25	18-24	120-170	30	21	7-17	50-120	26-29	18-20	14-35	97-240	.25-.6	1-6	.5-3
Filament Winding—Epoxy	30-80	100-270	690-1860	50-70	34-48	80-250	550-1720	40-90	28-62	45-70	310-480	1.6-2.8	40-60	21-32
Pultrusion—Polyester	40-80	100-180	690-1240	40-60	28-41	60-180	410-1240	40-60	28-41	30-70	210-480	1.6-2.5	45-60	24-32
THERMOPLASTIC RESIN INJECTION MOLDING														
Acetal	20-40	15-28	100-190	8-13	6-9	9-18	62-124	8-15	6-10	11-17	76-117	2	.8-2.8	0.4-1.5
Nylon	6-60	7-50	50-345	2-26	1-18	13-33	90-228	2-20	1-14	13-24	90-165	2-10	.8-4.5	0.4-2.4
Polycarbonate	20-40	17-30	120-210	7.5-15	5.2-10	12-25	83-172	7.5-17	5.2-12	14-24	97-165	2	1.5-3.5	0.8-1.9
Polyethylene (High Density)	10-40	7-12	50-83	2.1-6	1.4-4	6.5-11	45-76	4-9	3-6	4-8	28-55	1.5-3.5	1.2-4.0	
Polypropylene	20-40	7-11	50-76	3.5-8.2	2.4-5.6	5.5-10.5	38-72	4.5-9	3.1-6	6-8	41-55	1-3	1-4	0.5-2
Polystyrene (High Impact)	20-35	10-17	70-120	8-12	6-8	10-15	69-103	8.4-12.1	5.8-8.34	14-19	96-131	1.0-1.4	1.0-4.5	0.2-2.4
Polysulfone	20-40	21-27	140-190	8-15	6-10	13-20	90-138	15	10	21-26	145-179	2-3	1.3-2.5	0.7-1.3
ABS (Acrylonitrile Butadiene Styrene)	20-40	23-26	160-180	9.2-15	6.3-10	11-16	76-110	6-10	4-7	12-22	80-152	3-3.4	1-2.4	0.5-1.3
PVC (Polyvinyl Chloride)	15-35	20-25	140-170	9-16	6-11	14-18	97-124	10-18	7-12	13-17	90-117	2.4	0.8-1.6	0.4-0.9
Polyphenylene Oxide (Modified)	20-40	17-31	120-210	8-15	6-10	15-22	103-152	9.5-15	6.6-10	18-20	120-140	1.7-5	1.6-22	0.9-12
SAN (Styrene Acrylonitrile)	20-40	15-21	100-140	8.0-18	5.5-12	13-18	90-124	9-18.5	6-12.8	12-23	80-160	1.1-1.6	0.4-4.5	0.2-1.3
Thermoplastic Polyester	20-35	19-29	130-200	8.7-15	6.0-10	14-19	97-131	13-15.5	9-10.7	16-18	110-120	1-5	1.0-2.7	0.5-1.4
CONVERSION FACTOR TO SI[①]		6.894 757E-03		6.894 757E-06		6.894 757E-03		6.894 757E-06		6.894 757E-03			5.337 800E-01	

PER ANSI Z210.1-1976

ASTM E380-76

IEEE 268-1976

[a] These values are the values most often obtained, but based on material combinations, fabrication processes, and test techniques, it is possible to derive numbers higher or lower than shown.

① Example
Flexural strength (16 x 10³ psi) (6.894 757E-03) (16,000) (0.006 894 757) : 110.316 112 or rounded : 110 M Pa
Impact strength (1.2 ft-lbs in of notch) (5.337 800 E-01) (1.2) (0.533 780) : 0.640 536 or rounded to 2 significant places : 0.6 NM/cm of notch

Table 18.3 Typical Physical Properties

	% Glass Fiber By Wt.	Thermal Conductivity Btu·in/hr·ft²·°F (k Value)	W/m K	Specific Heat Btu/lb/°F (c Value)	J/kg K	Dielectric Strength Volts/mil	kV/mm	Specific Gravity	Density lb/in³	kg/m³	Heat Distortion °F @ 264 psi	°C @ 1820k Pa	Continuous Heat Resistance °F	°C	Thermal Coeff. of Expansion in/in/°F x 10⁶	cm/cm/°C x 10⁶
THERMOSETTING RESIN PROCESSES																
Spray Up—Mat Hand Lay Up—Polyester	30-50	1.2-1.6	0.17-0.23	31-34	1300-1400	200-400	7.87-15.7	1.4-1.6	.050-.058	1380-1600	350-400	180-200	150-350	66-180	12-20	7-11
Resin Transfer Molding—Mat—Polyester	20-30	1.3-1.8	0.19-0.26	30-33	1300-1400	300-600	11.8-23.6	1.5-1.7	.054-.061	1500-1700	350-400	180-200	150-400	66-200	10-18	6-10
Compression Molding—Sheet Molding Compound (SMC)—Polyester	15-30	1.3-1.7	0.19-0.25	30-35	1300-1500	300-450	11.8-17.7	1.7-2.1	.061-.075	1700-2100	400-500	200-260	300-400	150-200	8-12	4-7
Compression Molding—Mat or Preform—Polyester	25-50	1.3-1.8	0.19-0.26	30-33	1300-1400	300-600	11.8-23.6	1.5-1.7	.054-.061	1500-1700	350-400	180-200	150-400	66-200	10-18	6-10
Compression Molding—Bulk Molding Compound (BMC)—Polyester	15-35	1.3-1.7	0.19-0.25	30-35	1300-1500	300-450	11.8-17.7	1.8-2.1	.065-.075	1800-2100	400-500	200-260	300-400	150-200	8-12	4-7
Compression Molding—Bulk Molding Compound (BMC)—Phenolic	5-25	1.1-2.0	0.16-0.29	20-30	800-1300	150-370	5.91-14.6	1.7-1.9	.061-.069	1700-1900	400-500	200-260	325-350	160-180	4.5-9	2.5-5
Filament Winding—Epoxy	30-80	1.9-2.3	0.27-0.33	23-25	960-1050	300-400	11.8-15.7	1.7-2.2	.061-.079	1700-2200	350-400	180-200	500	260	2-6	1-3
Pultrusion-Polyester	40-80	1.9-2.3	0.27-0.33	22-25	920-1050	200-400	7.87-15.7	1.6-2.0	.058-.072	1600-2000	325-375	160-190	150-500	66-260	3-8	2-4
THERMOPLASTIC RESIN INJECTION MOLDING																
Acetal	20-40	—	—	—	—	500-600	20-24	1.6-17	—	—	315-335	157-168	185-220	85-100	19-35	11-19
Nylon	6-60	—	—	30-35	1256-1465	400-500	16-20	1.5-1.7	.049	1360	300-500	149-260	300-400	150-200	11-21	6.1-12
Polycarbonate	20-40	—	—	—	—	450	18	1.2-1.5	—	—	285-300	141-149	275	135	12-18	6.7-10
Polyethylene (High Density)	10-40	—	—	—	—	450-500	18-20	1.2-1.3	—	—	150-200	66-93	280-300	140-150	17-27	9.4-15
Polypropylene	20-40	—	—	—	—	500-600	20-24	1.0-1.2	—	—	230-300	110-149	300-320	150-160	16-24	8.9-13
Polystyrene (High Impact)	20-35	—	—	23-35	963-1465	350-425	14-17	1.2-1.3	.045-.048	1250-1330	200-220	93-104	180-200	80-90	17-22	9.4-12
Polysulfone	20-40	—	—	—	—	—	—	1.4-1.6	—	—	333-350	167-177			12-17	6.7-9.4
ABS (Acrylonitrile Butadiene Styrene)	20-40	—	—	—	—	—	—	1.2-1.4	—	—	215-240	102-116	200-230	90-110	16-20	8.9-11
PVC (Polyvinyl Chloride)	15-35	—	—	—	—	500-550	20-22	1.5-1.6	—	—	155-165	68-74	—	—	12	6.7
Polyphenylene Oxide (Modified)	20-40	—	—	—	—	—	—	1.2-1.4	—	—	220-315	104-157	240-265	120-130	10-20	5.6-11
SAN (Styrene Acrylonitrile)	20-40	—	—	—	—	—	—	1.2-1.4	—	—	210-230	99-110	200-220	90-100	16-21	8.9-12
Thermoplastic Polyester	20-35	1.3	0.19	23-35	920-1050	560-750	22-30	1.5-1.6	—	—	380-470	193-243	275-375	140-190	24-33	13-18
CONVERSION FACTOR TO SI PER ANSI Z210.1-1976			1.442 279E-01		4.186 800E+03		3.937 008E-02			2.767 990E+04	°C = (°F-32)/1.8 (264 psi)(6.894 757E00) = 1820k Pa		°C = (°F-32)/1.8		cm/cm = in/in	cm/cm /1.8

ASTM E380-76

IEEE 268-1976

		COMPRESSION MOLDING			INJECTION MOLDING (THERMO-PLASTICS)	RESIN TRANSFER MOLDING (RTM)	SPRAY-UP AND HAND LAY-UP
		SHEET MOLDING COMPOUND	BULK MOLDING COMPOUND	PRE-FORM MOLDING			
Minimum inside radius, inches (mm)	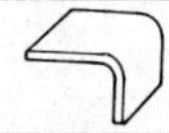	1/16" (1.6)	1/16" (1.6)	1/8" (3.2)	1/16" (1.6)	1/4" (6.4)	1/4" (6.4)
Molded-in holes		yes*	yes*	yes*	yes*	no	large
Trimmed in mold		yes	yes	yes	no	yes	no
Core pull & slides		yes	yes	no	yes	no	no
Undercuts		yes	yes**	no	yes	no	yes**
Minimum recommended draft, in./° (mm/°)	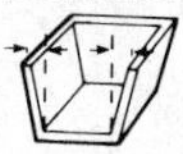	1/4" to 6" (6mm to 150mm) depth: 1° to 3° 6" (150mm) depth and over: 3° or as required				2° 3°	0°
Minimum practical thickness, in. (mm)	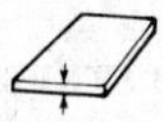	.050" (1.3)	.060" (1.5)	.030" (0.8)	.035" (0.89)	.080" (2.1)	.060" (1.5)
Maximum practical thickness, in. (mm)	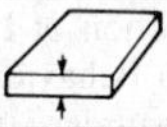	1" (25)	1" (25)	.250" (6.4)	.500" (13)	.500" (13)	no limit
Normal thickness Variation, inches (mm)	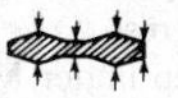	±.005" (±0.13)	±.005" (±0.13)	±.008" (±0.20)	±.005 (±0.13)	±.010" (±0.25)	±.020" (±0.51)
Maximum thickness build-up, heavy build-up & increased cycle	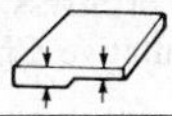	as req'd.	as req'd.	2-to-1 max.	as req'd.	2-to-1 max.	as req'd.
Corrugated sections		yes	yes	yes	yes	yes	yes
Metal inserts		yes	yes	not recommended	yes	no	yes
Bosses		yes	yes	yes	yes	not recommended	yes
Ribs	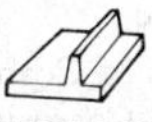	as req'd.	yes	not recommended	yes	not recommended	yes
Molded-in labels		yes	yes	yes	no	yes	yes
Raised numbers		yes	yes	yes	yes	yes	yes
Finished surfaces (reproduces mold surface)	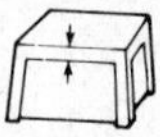	two	two	two	two	two	one

*Parallel or perpendicular to ram action only.
**With slides in tooling, or split mold.

Fig. 18.15 Design guides. For other molding processes, consult a fabricator as material supplier.

CHAPTER 19

FAILURE CONSIDERATIONS

JACK A. COLLINS

The Ohio State University
Columbus, Ohio

19.1 CRITERIA OF FAILURE

Any change in the size, shape, or material properties of a structure, machine, or machine part that renders it incapable of performing its intended function must be regarded as a mechanical failure of the device. It should be carefully noted that the key concept here is that *improper functioning* of a machine part constitutes failure. Thus, a shear pin that does *not* separate into two or more pieces upon the application of a preselected overload must be regarded as having failed as surely as a drive shaft has failed if it *does* separate into two pieces under normal expected operating loads.

Failure of a device or structure to function properly might be brought about by any one or a combination of many different responses to loads and environments while in service. For example, too much or too little elastic deformation might produce failure. A fractured load-carrying structural member or a shear pin that does not shear under overload conditions each would constitute failure. Progression of a crack due to fluctuating loads or aggressive environment might lead to failure after a period of time if resulting excessive deflection or fracture interferes with proper machine function.

A primary responsibility of any mechanical designer is to ensure that his or her design functions as intended for the prescribed design lifetime and, at the same time, that it be competitive in the marketplace. Success in designing competitive products while averting premature mechanical failures can be achieved consistently only by recognizing and evaluating all potential modes of failure that might govern the design. To recognize potential failure modes a designer must be acquainted with the array of failure modes observed in practice, and with the conditions leading to these failures. The following section summarizes the mechanical failure modes most commonly observed in practice, followed by a brief description of each one.

19.2 FAILURE MODES

A failure mode may be defined as the physical process or processes that take place or that combine their effects to produce a failure, as just discussed. In the following list of commonly observed failure modes it may be noted that some failure modes are unilateral phenomena, whereas others are combined phenomena. For example, fatigue is listed as a failure mode, corrosion is listed as a failure mode, and corrosion fatigue is listed as still another failure mode. Such combinations are included because they are commonly observed, important, and often *synergistic*. In the case of corrosion fatigue, for example, the presence of active corrosion aggravates the fatigue process and at the same time the presence of a fluctuating load accelerates the corrosion process.

The following list is not presented in any special order but it includes all commonly observed modes of mechanical failure:[1]

1. Force and/or temperature-induced elastic deformation.
2. Yielding.
3. Brinnelling.
4. Ductile rupture.
5. Brittle fracture.
6. Fatigue:
   a. High-cycle fatigue
   b. Low-cycle fatigue
   c. Thermal fatigue
   d. Surface fatigue
   e. Impact fatigue
   f. Corrosion fatigue
   g. Fretting fatigue
7. Corrosion:
   a. Direct chemical attack
   b. Galvanic corrosion
   c. Crevice corrosion
   d. Pitting corrosion
   e. Intergranular corrosion
   f. Selective leaching
   g. Erosion corrosion
   h. Cavitation corrosion
   i. Hydrogen damage
   j. Biological corrosion
   k. Stress corrosion

8. Wear:
    a. Adhesive wear
    b. Abrasive wear
    c. Corrosive wear
    d. Surface fatigue wear
    e. Deformation wear
    f. Impact wear
    g. Fretting wear
9. Impact:
    a. Impact fracture
    b. Impact deformation
    c. Impact wear
    d. Impact fretting
    e. Impact fatigue
10. Fretting:
    a. Fretting fatigue
    b. Fretting wear
    c. Fretting corrosion
11. Creep.
12. Thermal relaxation.
13. Stress rupture.
14. Thermal shock.
15. Galling and seizure.
16. Spalling.
17. Radiation damage.
18. Buckling.
19. Creep buckling.
20. Stress corrosion.
21. Corrosion wear.
22. Corrosion fatigue.
23. Combined creep and fatigue.

As commonly used in engineering practice, the failure modes just listed may be defined and described briefly as follows. It should be emphasized that these failure modes only produce failure when they generate a set of circumstances that interferes with the proper functioning of a machine or device.

Force and/or temperature-induced elastic deformation failure occurs whenever the elastic (recoverable) deformation in a machine member, brought about by the imposed operational loads or temperatures, becomes large enough to interfere with the ability of the machine to perform its intended function satisfactorily.

Yielding failure occurs when the plastic (unrecoverable) deformation in a ductile machine member, brought about by the imposed operational loads or motions, becomes large enough to interfere with the ability of the machine to perform its intended function satisfactorily.

Brinnelling failure occurs when the static forces between two curved surfaces in contact result in local yielding of one or both mating members to produce a permanent surface discontinuity of significant size. For example, if a ball bearing is statically loaded so that a ball is forced to indent permanently the race through local plastic flow, the race is brinnelled. Subsequent operation of the bearing might result in intolerably increased vibration, noise, and heating; and, therefore, failure would have occurred.

Ductile rupture failure occurs when the plastic deformation, in a machine part that exhibits ductile behavior, is carried to the extreme so that the member separates into two pieces. Initiation and coalescence of internal voids slowly propagate to failure, leaving a dull, fibrous rupture surface.

Brittle fracture failure occurs when the elastic deformation, in a machine part that exhibits brittle behavior, is carried to the extreme so that the primary interatomic bonds are broken and the member separates into two or more pieces. Preexisting flaws or growing cracks form initiation sites for very rapid crack propagation to catastrophic failure, leaving a granular, multifaceted fracture surface.

Fatigue failure is a general term given to the sudden and catastrophic separation of a machine part into two or more pieces as a result of the application of fluctuating loads or deformations over a period of time. Failure takes place by the initiation and propagation of a crack until it becomes unstable and propagates suddenly to failure. The loads and deformations that typically cause failure by fatigue are far below the static failure levels. When loads or deformations are of such magnitude that more than about 10,000 cycles are required to produce failure, the phenomenon is usually termed *high-cycle fatigue*. When loads or deformations are of such magnitude that less than about 10,000 cycles are required to produce failure, the phenomenon is usually termed *low-cycle fatigue*. When load or strain cycling is produced by a fluctuating temperature field in the machine part, the process is usually termed *thermal fatigue*. *Surface fatigue* failure, usually associated with rolling surfaces in contact, manifests itself as pitting, cracking, and spalling of the contacting surfaces as a result of the cyclic Hertz contact stresses that result in maximum values of cyclic shear stresses slightly below the surface. The cyclic subsurface shear stresses generate cracks that propagate to the contacting surface, dislodging particles in the process to produce surface pitting. This phenomenon is often viewed as a type of wear. Impact fatigue, corrosion fatigue, and fretting fatigue are described later.

Corrosion failure, a very broad term, implies that a machine part is rendered incapable of performing its intended function because of the undesired deterioration of the material as a result of chemical or electrochemical interaction with the environment. Corrosion often interacts with other failure modes such as wear or fatigue. The many forms of corrosion include the following. *Direct chemical attack*, perhaps the most common type of corrosion, involves corrosive attack of the surface of the machine part exposed to the corrosive media, more or less uniformly over the entire exposed surface. *Galvanic corrosion* is an accelerated electrochemical corrosion that occurs when two dissimilar metals in electrical contact are made part of a circuit completed by a connecting pool or film of electrolyte or corrosive medium, leading to current flow and ensuing corrosion. *Crevice corrosion* is the accelerated corrosion process highly localized within crevices, cracks, or joints where small volume regions of stagnant solution are trapped in contact with the corroding metal. *Pitting corrosion* is a very localized attack that leads to the development of an array of holes or pits that penetrate the metal. *Intergranular corrosion* is the localized attack occurring at grain boundaries of certain copper, chromium, nickel, aluminum, magnesium, and zinc alloys when they are improperly heat treated or welded. Formation of local galvanic cells that precipitate corrosion products at the grain boundaries seriously degrades the material strength because of the intergranular corrosive process. *Selecting leaching* is a corrosion process in which one element of a solid alloy is removed, such as in dezincification of brass alloys or graphitization of gray cast irons. *Erosion corrosion* is the accelerated chemical attack that results when abrasive or viscid material flows past a containing surface, continuously baring fresh, unprotected material to the corrosive medium. *Cavitation corrosion* is the accelerated chemical corrosion that results when, because of differences in vapor pressure, certain bubbles and cavities within a fluid collapse adjacent to the pressure-vessel walls, causing particles of the surface to be expelled, baring fresh, unprotected surface to the corrosive medium. *Hydrogen damage*, while not considered to be a form of direct corrosion, is induced by corrosion. Hydrogen damage includes hydrogen blistering, hydrogen embrittlement, hydrogen attack, and decarburization. *Biological corrosion* is a corrosion process that results from the activity of living organisms, usually by virtue of their processes of food ingestion and waste elimination, in which the waste products are corrosive acids or hydroxides. *Stress corrosion*, an extremely important type of corrosion, is described separately later.

Wear is the undesired cumulative change in dimensions brought about by the gradual removal of discrete particles from contacting surfaces in motion, usually sliding, predominantly as a result of mechanical action. Wear is not a single process, but a number of different processes that can take place by themselves or in combination, resulting in material removal from contacting surfaces through a complex combination of local shearing, plowing, gouging, welding, tearing, and others. *Adhesive wear* takes place because of high local pressure and welding at asperity contact sites, followed by motion-induced plastic deformation and rupture of asperity junctions, with resulting metal removal or transfer. *Abrasive wear* takes place when the wear particles are removed from the surface by the plowing, gouging, and cutting action of the asperities of a harder mating surface or by hard particles entrapped between the mating surfaces. When the conditions for either adhesive wear or abrasive wear coexist with conditions that lead to corrosion, the processes interact synergistically to produce *corrosive wear*. As described earlier, *surface fatigue wear* is a wear phenomenon associated with curved surfaces in rolling or sliding contact, in which subsurface cyclic shear stresses initiate microcracks that propagate to the surface to spall out macroscopic particles and form wear pits. *Deformation wear* arises as a result of repeated *plastic* deformation at the wearing surfaces, producing a matrix of cracks that grow and coalesce to form wear particles. Deformation wear is often caused by severe impact loading. *Impact wear* is impact-induced repeated *elastic* deformation at the wearing surface that produces a matrix of cracks that grows in accordance with the surface fatigue description just given. Fretting wear is described later.

Impact failure results when a machine member is subjected to nonstatic loads that produce in the part stresses or deformations of such magnitude that the member no longer is capable of performing its function. The failure is brought about by the interaction of stress or strain waves generated by dynamic or suddenly applied loads, which may induce local stresses and strains many times greater than would be induced by the static application of the same loads. If the magnitudes of the stresses and strains are sufficiently high to cause separation into two or more parts, the failure is called *impact fracture*. If the impact produces intolerable elastic or plastic deformation, the resulting failure is called *impact deformation*. If repeated impacts induce cyclic elastic strains that lead to initiation of a matrix of fatigue cracks, which grows to failure by the surface fatigue phenomenon described earlier, the process is called *impact wear*. If fretting action, as described in the next paragraph, is induced by the small lateral relative displacements between two surfaces as they impact together, where the small displacements are caused by Poisson strains or small tangential "glancing" velocity components, the phenomenon is called *impact fretting*. *Impact fatigue* failure occurs when impact loading is applied repetitively to a machine member until failure occurs by the nucleation and propagation of a fatigue crack.

Fretting action may occur at the interface between any two solid bodies whenever they are pressed together by a normal force and subjected to small-amplitude cyclic relative motion with respect to each other. Fretting usually takes place in joints that are not intended to move but, because of vibrational loads or deformations, experience minute cyclic relative motions. Typically, debris produced by fretting action is trapped between the surfaces because of the small motions involved. *Fretting fatigue* failure is the premature fatigue fracture of a machine member subjected to fluctuating loads or strains together with conditions that simultaneously produce fretting action. The surface discontinuities and microcracks generated by the fretting action act as fatigue crack nuclei that propagate to failure under conditions of fatigue loading that would otherwise be acceptable. Fretting fatigue failure is an insidious failure mode because the fretting action is usually hidden within a joint where it cannot be seen and leads to premature, or even unexpected, fatigue failure of a sudden and catastrophic nature. *Fretting wear* failure results when the changes in dimensions of the mating parts, because of the presence of fretting action, become large enough to interfere with proper design function or large enough to produce geometrical stress concentration of such magnitude that failure ensues as a result of excessive local stress levels. *Fretting corrosion* failure occurs when a machine part is rendered incapable of performing its intended function because of the surface degradation of the material from which the part is made, as a result of fretting action.

Creep failure results whenever the plastic deformation in a machine member accrues over a period of time under the influence of stress and temperature until the accumulated dimensional changes interfere with the ability of the machine part to perform satisfactorily its intended function. Three stages of creep are often observed: (1) transient or primary creep during which time the rate of strain decreases, (2) steady-state or secondary creep during which time the rate of strain is virtually constant, and (3) tertiary creep during which time the creep strain rate increases, often rapidly, until rupture occurs. This terminal rupture is often called creep rupture and may or may not occur, depending on the stress–time–temperature conditions.

Thermal relaxation failure occurs when the dimensional changes due to the creep process result in the relaxation of a prestrained or prestressed member until it no longer is able to perform its intended function. For example, if the prestressed flange bolts of a high-temperature pressure vessel relax over a period of time because of creep in the bolts, so that, finally, the peak pressure surges exceed the bolt preload to violate the flange seal, the bolts will have failed because of thermal relaxation.

Stress rupture failure is intimately related to the creep process except that the combination of stress, time, and temperature is such that rupture into two parts is ensured. In stress rupture failures the combination of stress and temperature is often such that the period of steady-state creep is short or nonexistent.

Thermal shock failure occurs when the thermal gradients generated in a machine part are so pronounced that differential thermal strains exceed the ability of the material to sustain them without yielding or fracture.

Galling failure occurs when two sliding surfaces are subjected to such a combination of loads, sliding velocities, temperatures, environments, and lubricants that massive surface destruction is caused by welding and tearing, plowing, gouging, significant plastic deformation of surface asperities, and metal transfer between the two surfaces. Galling may be thought of as a severe extension of the adhesive wear process. When such action results in significant impairment to intended surface sliding or in seizure, the joint is said to have failed by galling. *Seizure* is an extension of the galling process to such severity that the two parts are virtually welded together so that relative motion is no longer possible.

Spalling failure occurs whenever a particle is spontaneously dislodged from the surface of a machine part so as to prevent the proper function of the member. Armor plate fails by spalling, for example, when a striking missile on the exposed side of an armor shield generates a stress wave that propagates across the plate in such a way as to dislodge or spall a secondary missile of lethal potential on the protected side. Another example of spalling failure is manifested in rolling contact bearings and gear teeth because of the action of surface fatigue as described earlier.

Radiation damage failure occurs when the changes in material properties induced by exposure to a nuclear radiation field are of such a type and magnitude that the machine part is no longer able to perform its intended function, usually as a result of the triggering of some other failure mode, and often related to loss in ductility associated with radiation exposure. Elastomers and polymers are typically more susceptible to radiation damage than are metals, whose strength properties are sometimes enhanced rather than damaged by exposure to a radiation field, although ductility is usually decreased.

Buckling failure occurs when, because of a critical combination of magnitude and/or point of load application, together with the geometrical configuration of a machine member, the deflection of the member suddenly increases greatly with only a slight change in load. This nonlinear response results in a buckling failure if the buckled member is no longer capable of performing its design function.

Creep buckling failure occurs when, after a period of time, the creep process results in an unstable combination of the loading and geometry of a machine part so that the critical buckling limit is exceeded and failure ensues.

Stress corrosion failure occurs when the applied stresses on a machine part in a corrosive environment generate a field of localized surface cracks, usually along grain boundaries, that render the part incapable of performing its function, often because of triggering some other failure mode. Stress corrosion is a very important type of corrosion failure mode because so many different metals are susceptible to it. For example, a variety of iron, steel, stainless-steel, copper, and aluminum alloys are subject to stress corrosion cracking if placed in certain adverse corrosive media.

Corrosion wear failure is a combination failure mode in which corrosion and wear combine their deleterious effects to incapacitate a machine part. The corrosion process often produces a hard, abrasive corrosion product that accelerates the wear, while the wear process constantly removes the protective corrosion layer from the surface, baring fresh metal to the corrosive medium and thus accelerating the corrosion. The two modes combine to make the result more serious than either of the modes would have been otherwise.

Corrosion fatigue is a combination failure mode in which corrosion and fatigue combine their deleterious effects to cause failure of a machine part. The corrosion process often forms pits and surface discontinuities that act as stress raisers which in turn accelerate fatigue failure. Furthermore, cracks in the usually brittle corrosion layer also act as fatigue crack nuclei that propagate into the base material. On the other hand, the cyclic loads or strains cause cracking and flaking of the corrosion layer, which bares fresh metal to the corrosive medium. Thus, each process accelerates the other, often making the result disproportionately serious.

Combined creep and fatigue failure is a combination failure mode in which all of the conditions for both creep failure and fatigue failure exist simultaneously, each process influencing the other to produce failure. The interaction of creep and fatigue is probably synergistic but is not well understood.

19.3 ELASTIC DEFORMATION AND YIELDING

Small changes in the interatomic spacing of a material, brought about by applied forces or changing temperatures, are manifested macroscopically as elastic strain. Although the maximum elastic strain in crystalline solids, including engineering metals, is typically very small, the force required to produce the small strain is usually large; hence, the accompanying stress is large. On the other hand, certain other noncrystalline materials such as elastomers may exhibit recoverable (but not necessarily linear) strains of several hundred percent. For uniaxial loading of a machine or structural element, the total elastic deformation of the member may be found by integrating the elastic strain over the length of the element. Thus, for a uniform bar subjected to uniaxial loading the total deformation of the bar in the axial direction is

$$\Delta l = l \epsilon \tag{19.1}$$

where Δl is total axial deformation of the bar, l is the original bar length, and ϵ is the axial elastic strain. If Δl exceeds the design-allowable axial deformation, failure will occur. For example, if the blade-axial deformation of an aircraft gas-turbine blade, due to the centrifugal force field, exceeds the tip clearance gap, failure will occur because of force-induced elastic deformation. Likewise, if thermal expansion of the blade produces a blade-axial deformation that exceeds the tip clearance gap, failure will occur because of temperature-induced elastic deformation.

When the state of stress is more complicated, it becomes necessary to calculate the elastic strains induced by the multiaxial states of stress in three mutually perpendicular directions through the use of the generalized Hooke's law equations given by

$$\epsilon_x = \frac{1}{E} \left[\sigma_x - \nu(\sigma_y + \sigma_z) \right]$$

$$\epsilon_y = \frac{1}{E} \left[\sigma_y - \nu(\sigma_x + \sigma_z) \right] \tag{19.2}$$

$$\epsilon_z = \frac{1}{E} \left[\sigma_z - \nu(\sigma_x + \sigma_y) \right]$$

where σ_x, σ_y, and σ_z are the normal stresses in the three coordinate directions, E and ν are Young's modulus and Poisson's ratio, respectively, and ϵ_x, ϵ_y, and ϵ_z are the elastic strains in the three coordinate directions. Again, total elastic deformation of a member in any of the coordinate directions may be found by integrating the strain over the member's length in that direction. If the change in length of the member in any direction exceeds the design-allowable deformation in that direction, failure will occur.

If applied loads reach certain critical levels, the atoms within the microstructure may be moved into new equilibrium positions and the induced strains are not fully recovered upon release of the loads. Such permanent strains, usually the result of slip, are called plastic strains, and the macroscopic permanent deformation due to plastic strain is called yielding. If applied loads are increased even more, the plastic deformation process may be carried to the point of instability where *necking* begins: internal voids form and slowly coalesce to finally produce a ductile rupture of the loaded member.

After plastic deformation has been initiated, the Hooke's law equations (19.2) are no longer valid and the predictions of plastic strains and deformations under multiaxial states of stress are more difficult. If a designer can tolerate a prescribed plastic deformation without experiencing failure, and if the loading produces a multiaxial state of stress, the techniques of *proportional deformation theory* may be used to generate a set of plastic true-stress, true-strain relationships somewhat similar in structure to the elastic Hooke's law relationships. These equations for plastic true strains in the three principal axes directions are

$$\delta_1 = \left[\frac{\sigma_1'}{k}\right]^{1/n} [\alpha^2 + \beta^2 + 1 - \alpha\beta - \alpha - \beta]^{(1-n)/2n} \left[1 - \frac{\alpha}{2} - \frac{\beta}{2}\right]$$

$$\delta_2 = \left[\frac{\sigma_2'}{k}\right]^{1/n} [\alpha^2 + \beta^2 + 1 - \alpha\beta - \alpha - \beta]^{(1-n)/2n} \left[\alpha - \frac{\beta}{2} - \frac{1}{2}\right] \qquad (19.3)$$

$$\delta_3 = \left[\frac{\sigma_3'}{k}\right]^{1/n} [\alpha^2 + \beta^2 + 1 - \alpha\beta - \alpha - \beta]^{(1-n)/2n} \left[\beta - \frac{\alpha}{2} - \frac{1}{2}\right]$$

where σ_1', σ_2', and σ_3' are principal true stresses; k is the strength coefficient for the material and n is its strain-hardening exponent; δ_1, δ_2, and δ_3 are the principal true strains in the three principal directions; and α and β are the ratios

$$\alpha = \frac{\sigma_2'}{\sigma_1'}, \qquad \beta = \frac{\sigma_3'}{\sigma_1'} \qquad (19.4)$$

The total plastic deformation of a member in any of the three principal directions may be found by integrating the plastic strain over the member's length in that direction. If the plastic change in length of the member in any direction exceeds the design-allowable plastic deformation in that direction, failure will occur. For some cases, for example, when the ratio of principal shearing strains to principal shearing stresses do not remain constant, Eqs. (19.3) may not be sufficiently accurate, and an incremental strain technique becomes necessary. Otherwise, the use of Eqs. (19.3) gives good results well into the plastic range, and in many cases the results are valid all the way up to the instability point.

For the case of simple uniaxial loading, the onset of yielding may be accurately predicted to occur when the uniaxial maximum normal stress reaches a value equal to the yield point strength of the material read from an engineering stress–strain curve. If the loading is more complicated, and a multiaxial state of stress is produced by the loads, the onset of yielding may no longer be predicted by comparing any one of the normal stress components with uniaxial material yield strength, not even the maximum principal normal stress. Onset of yielding for multiaxially stressed critical points in a machine or structure is more accurately predicted through the use of a *combined stress theory of failure*, which has experimentally been validated for the prediction of yielding. The two most widely accepted theories for predicting the onset of yielding are the distortion energy theory (also called the octahedral shear stress theory or the Huber–von Mises–Hencky theory) and the maximum shearing stress theory. The distortion energy theory is somewhat more accurate while the maximum shearing stress theory may be slightly easier to use.

In words, the distortion energy theory may be expressed as follows:

Failure is predicted to occur in the multiaxial state of stress when the distortion energy per unit volume becomes equal to or exceeds the distortion energy per unit volume at the time of failure in a simple uniaxial stress test using a specimen of the same material.

Mathematically, the distortion energy theory may be formulated as

Failure is predicted by the distortion energy theory to occur if

$$\tfrac{1}{2}[(\sigma_1 - \alpha_2)^2 + (\sigma_2 - \sigma_3)^2 + (\sigma_3 - \sigma_1)^2] \geq \sigma_f^2 \qquad (19.5)$$

The maximum shearing stress theory may be stated in words as:

Failure is predicted to occur in the multiaxial state of stress when the maximum shearing stress magnitude becomes equal to or exceeds the maximum shearing stress magnitude at the time of failure in a simple uniaxial stress test using a specimen of the same material.

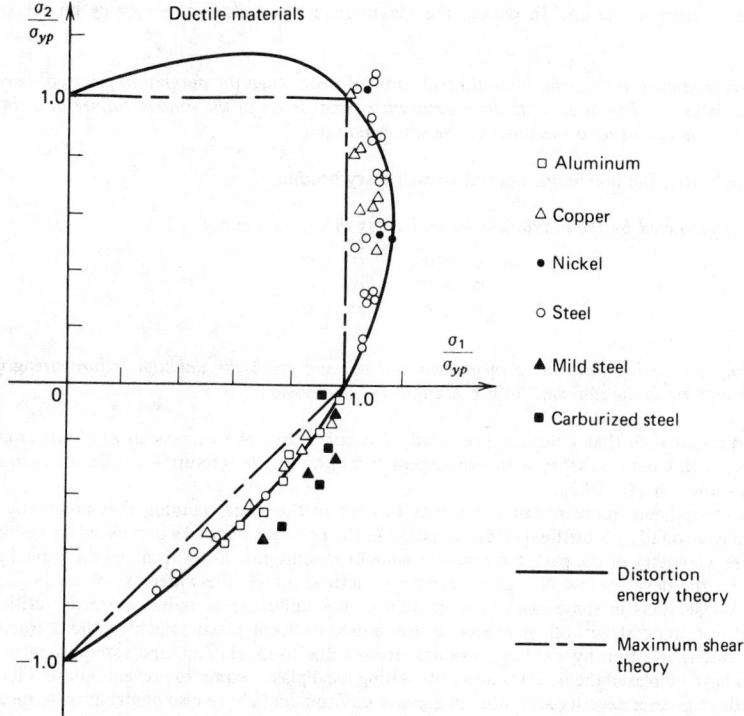

Fig. 19.1 Comparison of biaxial yield strength data with theories of failure for a variety of ductile materials.

Mathematically, the maximum shearing stress theory becomes:

Failure is predicted by the maximum shearing stress theory to occur if

$$|\tau_1| \geq |\tau_f|$$
$$|\tau_2| \geq |\tau_f| \qquad (19.6)$$
$$|\tau_3| \geq |\tau_f|$$

or, written in terms of principal normal stresses:

Failure is predicted by the maximum shearing stress theory to occur if

$$|\sigma_1 - \sigma_2| \geq |\sigma_f|$$
$$|\sigma_2 - \sigma_3| \geq |\sigma_f| \qquad (19.7)$$
$$|\sigma_3 - \sigma_1| \geq |\sigma_f|$$

where σ_1, σ_2, and σ_3 are the principal normal stresses and σ_f is the uniaxial failure strength in tension.

It is important to note that failure is predicted to occur if any one expression of (19.6) or (19.7) is satisfied.

Comparisons of these two failure theories with experimental data on yielding are shown in Fig. 19.1 for a variety of materials and different biaxial states of stress.

19.4   FRACTURE MECHANICS AND CRACK PROPAGATION

When the material behavior is brittle rather than ductile, the mechanics of the failure process are much different. Instead of the slow coalescence of voids associated with ductile rupture, brittle fracture proceeds by the high-velocity propagation of a crack across the loaded member. If the material behavior is clearly brittle, fracture may be predicted with reasonable accuracy through use of the maximum normal stress theory of failure. In words, the maximum normal stress theory may be expressed as follows:

Failure is predicted to occur in the multiaxial state of stress when the maximum principal normal stress becomes equal to or exceeds the maximum normal stress at the time of failure in a simple uniaxial stress test using a specimen of the same material.

Mathematically, the maximum normal stress theory becomes:

Failure is predicted by the maximum normal stress theory to occur if

$$
\begin{aligned}
\sigma_1 &\geq \sigma_t, & \sigma_1 &\leq \sigma_c \\
\sigma_2 &\geq \sigma_t, & \sigma_2 &\leq \sigma_c \\
\sigma_3 &\geq \sigma_t, & \sigma_3 &\leq \sigma_c
\end{aligned}
\tag{19.8}
$$

where σ_1, σ_2, *and* σ_3 *are the principal normal stresses;* σ_t *is the uniaxial failure strength in tension; and* σ_c *is the uniaxial failure strength in compression.*

It is important to note that failure is predicted to occur if any one expression of (19.8) is satisfied. Comparison of this failure theory with experimental data on brittle fracture for different biaxial states of stress is shown in Fig. 19.2.

On the other hand, more recent experience has led to the understanding that nominally ductile materials may also fail by a brittle fracture response in the presence of cracks or flaws if the combination of crack size, geometry of the part, temperature, and/or loading rate lies within certain critical regions. Furthermore, the development of higher-strength structural alloys, the wider use of welding, and the use of thicker sections in some cases have combined their influence to reduce toward a critical level the capacity of some structural members to accommodate local plastic strain without fracture. At the same time, fabrication by welding, residual stresses due to machining, and assembly mismatch in production have increased the need for accommodating local plastic strain to prevent failure. Fluctuating service loads of greater severity and more aggressive environments have also contributed to unexpected fractures. From the study of all these factors the basic concepts of *fracture control* were conceived and developed. Fracture control consists, simply, of controlling the nominal stress and crack size so that the combination always lies below a critical level for the material being used in a given design application.

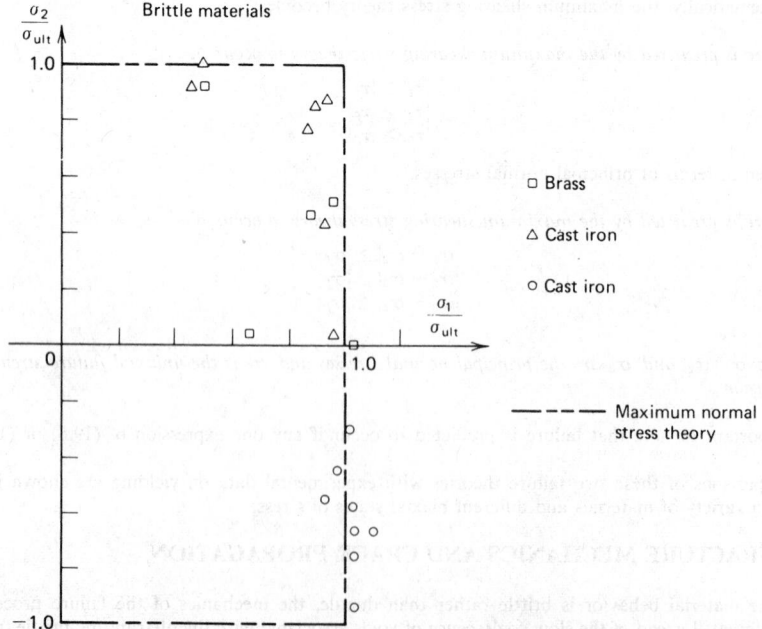

Fig. 19.2 Comparison of biaxial brittle fracture strength data with maximum normal stress theory for several brittle materials.

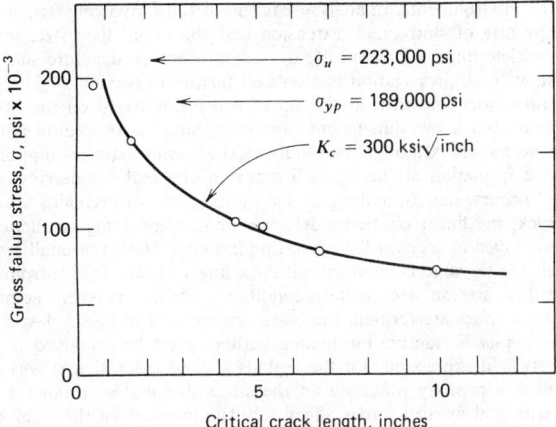

Fig. 19.3 Influence of crack length on gross failure stress for center cracked steel plate, 36 in. wide, 0.14 in. thick, room temperature, 4330 M steel, longitudinal direction. (After Ref. 2, copyright ASTM; adapted with permission.)

An important observation in studying fracture behavior is that the magnitude of the nominal applied stress that causes fracture is related to the size of the crack or cracklike flaw within the structure.[2] For example, observations of the behavior of central through-the-thickness cracks, oriented normal to the applied tensile stress, in steel and aluminum plates, yielded the results shown in Figs. 19.3 and 19.4. In these tests, as the tensile loading on the precracked plates was slowly increased, the crack extension slowly increased for a time and then abruptly extended to failure by rapid crack propagation. Slow stable crack growth was characterized by speeds of the order of fractions of an inch per minute. Rapid crack propagation was characterized by speeds of the order of hundreds of feet per second. The data of Figs. 19.3 and 19.4 indicate that for longer initial crack length the fracture stress, that is, the stress corresponding to the onset of rapid crack extension, was lower. For the aluminum alloy the fracture stress was less than the yield strength for cracks longer than about 0.75 in. For the steel alloy the fracture stress was less than the yield strength for cracks longer than about 0.5 in. In both cases, for shorter cracks the fracture stress approaches the ultimate strength of the material determined from a conventional uniaxial tension test.

Experience has shown that the abrupt change from slow crack propagation to rapid crack propagation establishes an important material property termed *fracture toughness*. The fracture toughness may be used as a design criterion in fracture prevention, just as the yield strength is used as a design criterion in prevention of yielding of a ductile material under static loading.

In many cases slow crack propagation is also of interest, especially under conditions of fluctuating

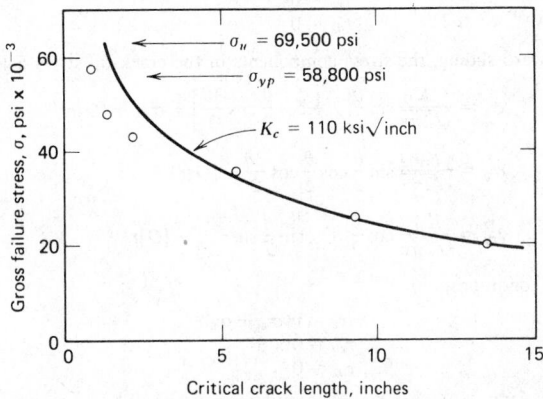

Fig. 19.4 Influence of crack length on gross failure stress for center cracked aluminum plate, 24 in. wide, 0.1 in. thick, room temperature, 2219–T87 aluminum alloy, longitudinal direction. (After Ref. 2, copyright ASTM, adapted with permission.)

loads and/or aggressive environments. In analyses and predictions involving fatigue failure phenomena, characterization of the rate of slow crack extension and the initial flaw size, together with critical crack size, are used to determine the useful life of a component or structure subjected to fluctuating loads. The topic of slow crack propagation is discussed further in Section 19.5.

The simplest useful model for stress at the tip of a crack is based on the assumptions of linear elastic material behavior and a two-dimensional analysis; thus, the procedure is often referred to as linear elastic fracture mechanics. Although the validity of the linear elastic assumption may be questioned in view of plastic zone formation at the tip of a crack in any real engineering material, as long as "small-scale yielding" occurs, that is, as long as the plastic zone size remains small compared to the dimensions of the crack, the linear elastic model gives good engineering results, especially if a small correction factor is employed to account for crack-tip plasticity. Thus, the small-scale yielding concept implies that the small plastic zone is confined within a linear elastic field surrounding the crack tip. If the material properties, section size, loading conditions, and environment combine in such a way that "large-scale" plastic zones are formed, the basic assumptions of linear elastic fracture mechanics are violated, and elastic–plastic fracture mechanics methods must be employed.

In developing stress field expressions for the regions around crack tips, it was recognized that the free crack surfaces have a primary influence on the stress distribution around a crack tip, whereas other remote boundaries and applied forces affect only the intensity of the local stress field near the crack tip.

Three basic types of stress fields can be defined for crack-tip stress analysis, each one associated with a distinct mode of crack deformation, as illustrated in Fig. 19.5. The crack-opening mode, mode I, is associated with local displacement in which the crack surfaces move directly apart, as shown in Fig. 19.5a. The edge sliding or forward sliding mode, mode II, is developed when crack surfaces slide over each other in a direction perpendicular to the leading edge of the crack, as shown in Fig. 19.5b. The side sliding, or tearing, or parallel shear mode, mode III, is characterized by crack surfaces sliding with respect to each other in a direction parallel to the leading edge of the crack, as shown in Fig. 19.5c. Superposition of these three modes will fully describe the most general three-dimensional case of local crack-tip deformation and stress field.

Based on the methods developed by Westergaard,[3] Irwin[4] developed the two-dimensional stress field and displacement field equations for each of the three modes depicted in Fig. 19.5, expressing them in terms of the coordinates shown in Fig. 19.6.

For mode I, crack opening, the stress components in the crack-tip stress field are

$$\sigma_x = \frac{K_I}{\sqrt{2\pi r}} \cos\frac{\theta}{2}\left[1 - \sin\frac{\theta}{2}\sin\frac{3\theta}{2}\right] + \sigma_{x0} + [O]r^{1/2} \tag{19.9}$$

$$\sigma_y = \frac{K_I}{\sqrt{2\pi r}} \cos\frac{\theta}{2}\left[1 + \sin\frac{\theta}{2}\sin\frac{3\theta}{2}\right] + [O]r^{1/2} \tag{19.10}$$

$$\tau_{xy} = \frac{K_I}{\sqrt{2\pi r}} \sin\frac{\theta}{2}\cos\frac{\theta}{2}\cos\frac{3\theta}{2} + [O]r^{1/2} \tag{19.11}$$

For conditions of plane strain, where displacements in the z direction are constrained to be zero (thick members), the remaining three stress components are

$$\sigma_z = \nu(\sigma_x + \sigma_y) \tag{19.12}$$
$$\tau_{xz} = 0 \tag{19.13}$$
$$\tau_{yz} = 0 \tag{19.14}$$

For mode II, forward sliding, the stress components in the crack-tip stress field are

$$\sigma_x = \frac{-K_{II}}{\sqrt{2\pi r}} \sin\frac{\theta}{2}\left[2 + \cos\frac{\theta}{2}\cos\frac{3\theta}{2}\right] + \sigma_{x0} + [O]r^{1/2} \tag{19.15}$$

$$\sigma_y = \frac{K_{II}}{\sqrt{2\pi r}} \sin\frac{\theta}{2}\cos\frac{\theta}{2}\cos\frac{3\theta}{2} + [O]r^{1/2} \tag{19.16}$$

$$\tau_{xy} = \frac{K_{II}}{\sqrt{2\pi r}} \cos\frac{\theta}{2}\left[1 - \sin\frac{\theta}{2}\sin\frac{3\theta}{2}\right] + [O]r^{1/2} \tag{19.17}$$

and, for plane strain conditions,

$$\sigma_z = \nu(\sigma_x + \sigma_y) \tag{19.18}$$
$$\tau_{xy} = 0 \tag{19.19}$$
$$\tau_{yz} = 0 \tag{19.20}$$

For mode III, side sliding, the stress components in the crack-tip stress field are

$$\tau_{xz} = \frac{K_{III}}{\sqrt{2\pi r}} \sin\frac{\theta}{2} + \tau_{xz0} + [O]r^{1/2} \tag{19.21}$$

$$\tau_{yz} = \frac{K_{III}}{\sqrt{2\pi r}} \cos\frac{\theta}{2} + [O]r^{1/2} \tag{19.22}$$

$$\sigma_x = \sigma_y = \sigma_z = \tau_{xy} = 0 \tag{19.23}$$

In expressions (19.9) through (19.23), higher-order terms such as the uniform stresses parallel to cracks σ_{x0} and τ_{xz0}, and terms of the order of $r^{1/2}$, that is, $[O]r^{1/2}$, are indicated. These terms are usually neglected as higher order compared to the leading $1/\sqrt{r}$ term. The parameters K_I, K_{II}, and K_{III} are called crack-tip stress field intensity factors, or simply *stress intensity factors*. They represent the *strength* of the stress field surrounding the tip of the crack. Physically, K_I, K_{II}, and K_{III} may be interpreted as the intensity of load transmittal through the crack-tip region, induced by the introduction of a crack into a flaw-free body. Since fracture is induced by the crack-tip stress field, the stress intensity factors are primary correlation parameters in current practice.

For remote loadings, in general, the expressions for the stress intensity factor are of the form

$$K = C\sigma\sqrt{\pi a} \tag{19.24}$$

where C is dependent on the type of loading and the geometry away from the crack. Much work has been completed in determining values of C for a wide variety of conditions. (See, for example, Ref. 5.)

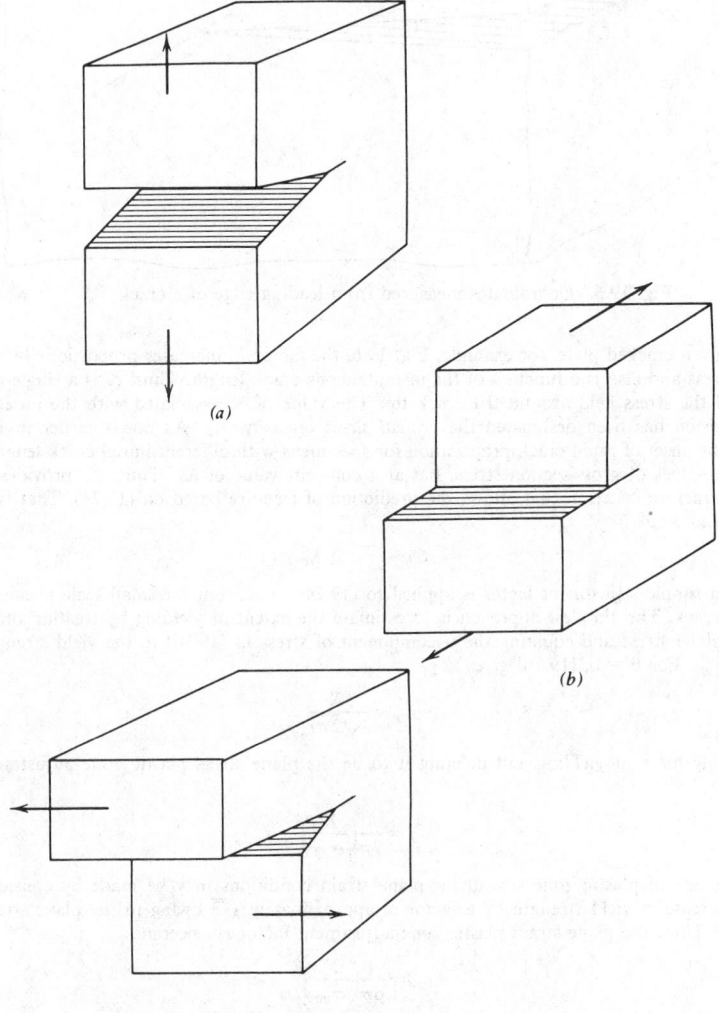

Fig. 19.5 Basic modes of crack displacement: (*a*) mode I; (*b*) model II; (*c*) mode III.

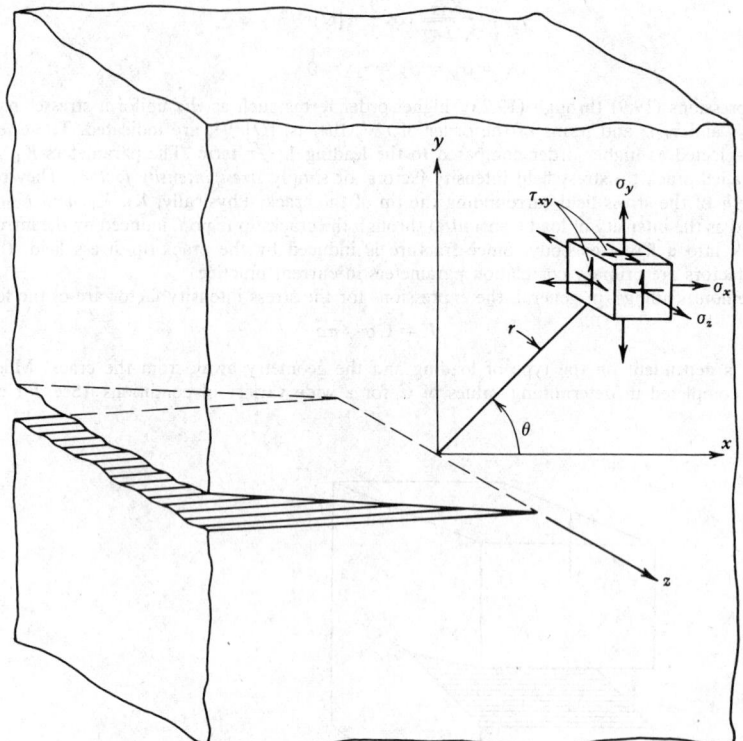

Fig. 19.6   Coordinates measured from leading edge of a crack.

For a given cracked plate, for example, Fig. 19.6, the factor K increases proportionally with gross nominal stress and also is a function of the instantaneous crack length. Thus, K is a single-parameter measure of the stress field around the crack tip. The value of K associated with the onset of rapid crack extension has been designated the *critical stress intensity*, K_c. As noted earlier in Figs. 19.3 and 19.4, the onset of rapid crack propagation for specimens with different initial crack lengths occurs at different values of gross-section stress, but at a constant value of K_c. Thus, K_c provides a single-parameter fracture criterion that allows the prediction of fracture based on (19.24). That is, *fracture is predicted to occur if*

$$C\sigma\sqrt{\pi a} \geq K_c \qquad (19.25)$$

Often, a simple adjustment factor is applied to (19.24) to account for small-scale plasticity at the tip of the crack. The simplest approach is to estimate the extent of yielding by treating the problem as one of plane stress and equating the y component of stress in (19.10) to the yield strength of the material, σ_{yp}. For $\theta = 0$, (19.10) gives

$$\sigma_{yp} = \frac{K}{\sqrt{2\pi r}} \qquad (19.26)$$

Then, solving for r at yielding and defining it to be the plane stress plastic zone adjustment factor $r_{Y\,\sigma}$,

$$r_{Y\,\sigma} = \frac{1}{2\pi}\left(\frac{K}{\sigma_{yp}}\right)^2 \qquad (19.27)$$

An estimate of plastic zone size under plane strain conditions may be made by considering the effective increase in yield strength by a factor of approximately $\sqrt{3}$ owing to the plane strain elastic constraint.[6] Thus, the plane strain plastic zone adjustment factor $r_{Y\,\epsilon}$ becomes

$$r_{y\,\epsilon} = \frac{1}{6\pi}\left(\frac{K}{\sigma_{yp}}\right)^2 \qquad (19.28)$$

Stress redistribution accompanies plastic yielding and causes the plastic zone to extend approximately $2r_Y$ ahead of the real crack tip to satisfy equilibrium conditions. Because of the crack-tip plastic deformation, the crack is "blunted," as indicated in Fig. 19.7, and the stresses in the surrounding elastic medium "focus" on a virtual crack tip within the plastic zone as the "effective" elastic crack tip. To account for this effect, then, an effective crack length a' is inserted into (19.24) where

$$a'1 = a + r_Y \tag{19.29}$$

to yield

$$K = C\sigma\sqrt{\pi(a + r_Y)} \tag{19.30}$$

It should be noted that (19.30) requires an iterative solution since K depends on the magnitude of r_Y, and r_Y depends on the magnitude of K. For stress levels and crack sizes well below critical values, the magnitude of r_Y is typically very small and often neglected in practice. For cases in which the stress intensity factor K approaches its critical magnitude K_c, the plastic zone correction factor becomes significant and should be included. However, it should be pointed out that if the plastic zone becomes "large" relative to the crack size, that is, if r_Y is more than a few percent of a, the accuracy of linear elastic fracture mechanics techniques becomes questionable, and elastic–plastic fracture mechanics procedures should be considered.

In studying material behavior, one finds that for a given material, depending on the state of stress at the crack tip, the critical stress intensity K_c decreases to a lower limiting value as the state of strain approaches the condition of plane strain. This lower limiting value defines a basic material property K_{Ic}, the *plane strain fracture toughness* for the material. Standard test methods have been established for the determination of K_{Ic} values.[7] A few data are shown in Table 19.1.

For the plane strain fracture toughness K_{Ic} to be a valid failure prediction criterion for a specimen or a machine part, plane strain conditions must exist at the crack tip; that is, the material must be *thick* enough to ensure plane strain conditions. It has been estimated empirically that for plane strain conditions the minimum material thickness B must be

$$B \geq 2.5\left(\frac{K_{Ic}}{\sigma_{yp}}\right)^2 \tag{19.31}$$

If the material is not thick enough to meet the criterion of (19.31), plane stress is a more likely state of stress at the crack tip; and K_c, the critical stress intensity factor for failure prediction under plane stress conditions, may be as much as 10 times higher than the lower limiting value of K_{Ic}. Under such conditions, although it would be on the safe side from the failure prediction standpoint to use the value K_{Ic} as a failure criterion, it would be more efficient, and a better design approach, to try to employ elastic–plastic fracture mechanics methods.

It should be pointed out that the concepts of linear elastic fracture mechanics may be developed independently either in terms of stress intensity factor K, as has been done in the preceding presentation, or in terms of "crack extension force" or "strain energy release rate" G, the strain energy released

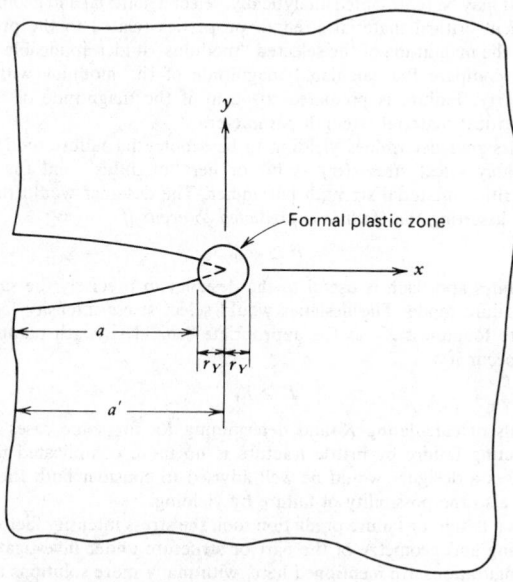

Fig. 19.7 Definition of formal plastic zone at crack tip.

Table 19.1 Yield Strength and Plane Strain Fracture Toughness Data for Selected Engineering Alloys[8,10]

Alloy	Form	Test Temperature °F	°C	σ_{yp} ksi	MPa	K_{Ic} ksi$\sqrt{\text{in}}$	MPa$\sqrt{\text{m}}$
4340 (500°F temper) steel	Plate	70	21	217–238	1495–1640	45–57	50–63
4340 (800°F temper) steel	Forged	70	21	197–211	1360–1455	72–83	79–91
D6AC (1000°F temper) steel	Plate	70	21	217	1495	93	102
D6AC (1000°F temper) steel	Plate	−65	−54	228	1570	56	62
A 538 steel				250	1722	100	111
2014-T6 aluminum	Forged	75	24	64	440	28	31
2024-T351 aluminum	Plate	80	27	54–56	370–385	28–40	31–44
7075-T6 aluminum				85	585	30	33
7075-T651 aluminum	Plate	70	21	75–81	515–560	25–28	27–31
7075-T7351 aluminum	Plate	70	21	58–66	400–455	28–32	31–35
Ti-6Al-4V titanium	Plate	74	23	119	820	96	106

by an incremental increase in crack length. There are cases in which the strain energy release rate approach may be more useful, as, for example, in cases where different crack displacement modes are simultaneously induced, in compliance testing techniques, or in certain elastic–plastic fracture mechanics methods. The concept of a critical value of strain energy release rate, G_c, at which a crack becomes unstable or self-propagating is discussed in the literature (see, for example, Refs. 8 or 9) and may be related directly to the concept of critical stress intensity factor K_c. In any case, the stress intensity factor K and strain energy release rate G are directly related as follows:

$$G = \frac{K^2}{E} \quad \text{(for plane stress)} \tag{19.32}$$

and

$$G = \frac{K^2(1 - \nu^2)}{E} \quad \text{(for plane strain)} \tag{19.33}$$

In predicting failure for designing a part so that failure will not occur, a designer must, at an early stage, identify the probable mode of failure, employ a suitable "modulus" by which severity of loading and environment may be represented analytically, select a material and geometry for the proposed part, and obtain pertinent critical material strength properties related to the probable failure mode. He must next calculate the magnitude of the selected "modulus" under applicable loading and environmental conditions and compare the calculated magnitude of the modulus with the proper critical material strength property. Failure is predicted to occur if the magnitude of the selected modulus equals or exceeds the critical material strength parameter.

For example, if a designer determines yielding to be a potential failure mode for his or her part, he or she would probably select stress (σ) as his or her "modulus" and the uniaxial yield point strength (σ_{yp}) as the critical material strength parameter. The designer would then assess the quality of his or her design by asserting that *failure is predicted to occur if*

$$\sigma \geq \sigma_{yp} \tag{19.34}$$

The fracture mechanics approach is useful to the designer in precisely the same way when brittle fracture is a possible failure mode. The designer would select stress intensity factor K as his or her "modulus" and fracture toughness K_c as the appropriate critical strength parameter and assert that failure is predicted to occur if

$$K \geq K_c \tag{19.35}$$

Although the details of calculating K and determining K_c for some cases may be difficult, the basic concept of predicting failure by brittle fracture is no more complicated than this. It is worth noting that in most cases a designer would be well advised to consider both the possibility of failure by brittle fracture and also the possibility of failure by yielding.

To utilize (19.35) as a design or failure prediction tool, the stress intensity factor must be determined for the particular loading and geometry of the part or structure under investigation. To illustrate the procedure, several configurations are mentioned here, with many more solutions available in the literature. (See, for example, Refs. 5, 10, and 11.)

For *central through-the-thickness cracks, double edge through-the-thickness cracks,* and *single edge through-the-thickness cracks* under *direct tension loading* or *shear loading,* the form of the stress intensity factor is

$$K = C\sigma_t\sqrt{\pi a} \qquad\qquad (19.36)$$

or

$$K = C\tau\sqrt{\pi a} \qquad\qquad (19.37)$$

where C is a function of geometry and crack displacement mode, as given in Figs. 19.8, 19.9, and 19.10.

For a *beam* with a *single through-the-thickness edge crack* under a *pure bending moment,* the form of the stress intensity factor is

$$K_I = C_1\sigma_b\sqrt{\pi a} \qquad\qquad (19.38)$$

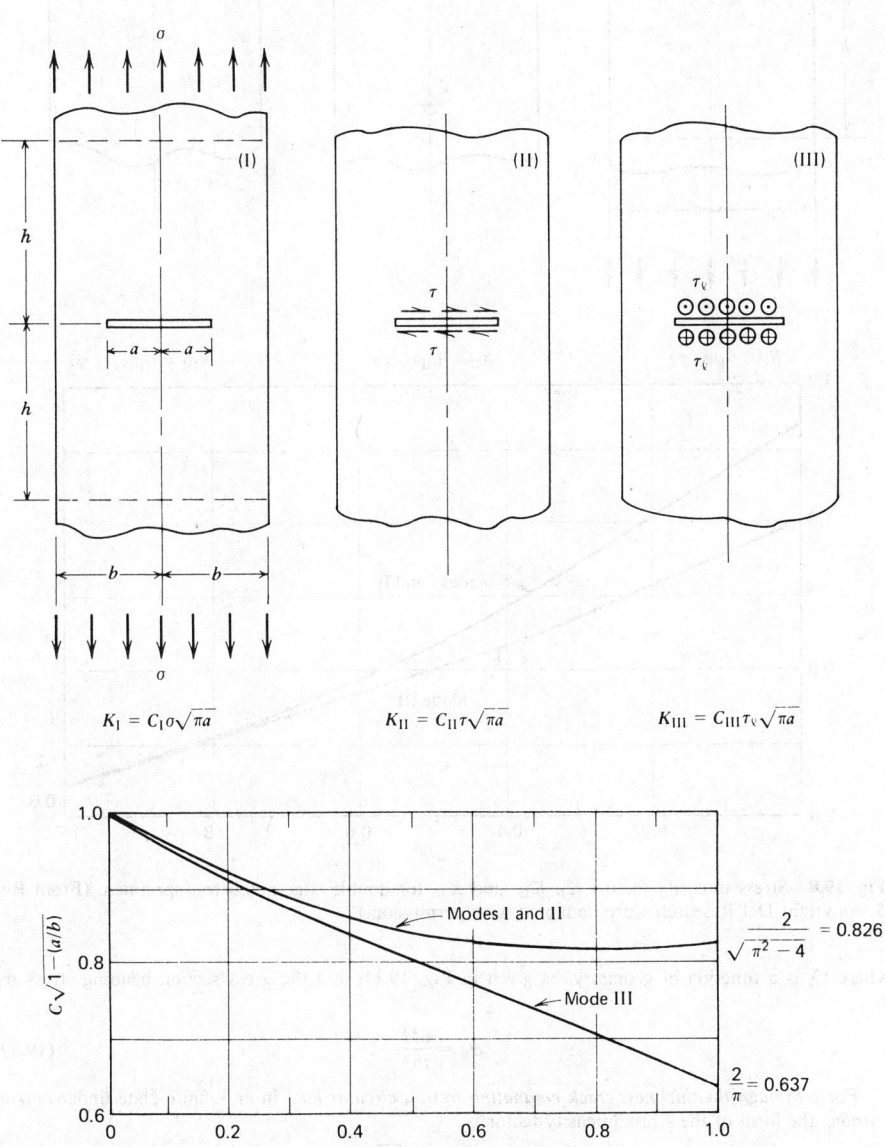

Fig. 19.8   Stress intensity factors K_I, K_{II}, and K_{III} for center cracked test specimen. (From Ref. 5, copyright Del Research Corp.; adapted with permission.)

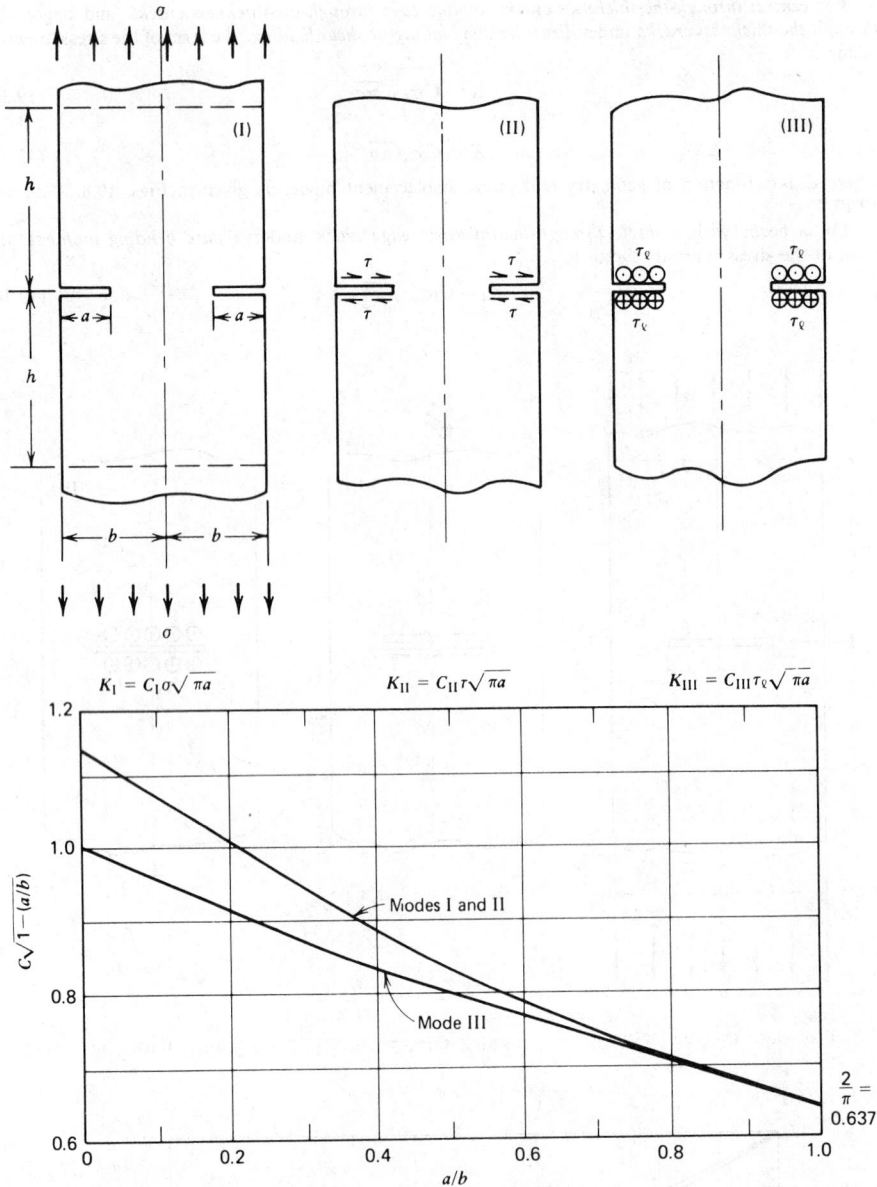

Fig. 19.9 Stress intensity factors K_I, K_{II}, and K_{III} for double edge notch test specimen. (From Ref. 5, copyright Del Research Corp.; adapted with permission.)

where C_I is a function of geometry, as given in Fig. 19.11, and the gross section bending stress σ_b is

$$\sigma_b = \frac{6M}{tb^2} \tag{19.39}$$

For a *through-the-thickness crack emanating from a circular hole* in an infinite plate under *biaxial tension*, the form of the stress intensity factor is

$$K_I = C_I \sigma \sqrt{\pi a} \tag{19.40}$$

where C_I is a function of geometry and the ratio of biaxial stress components, as shown in Fig. 19.12.

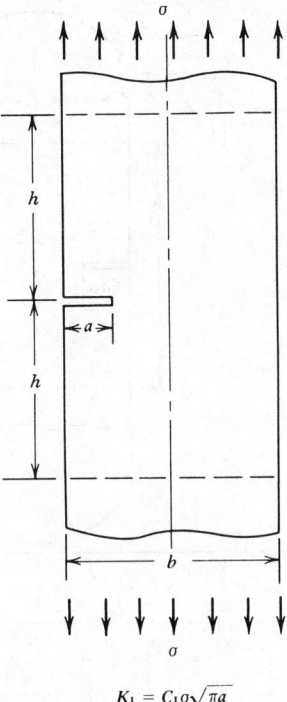

$$K_I = C_I \sigma \sqrt{\pi a}$$

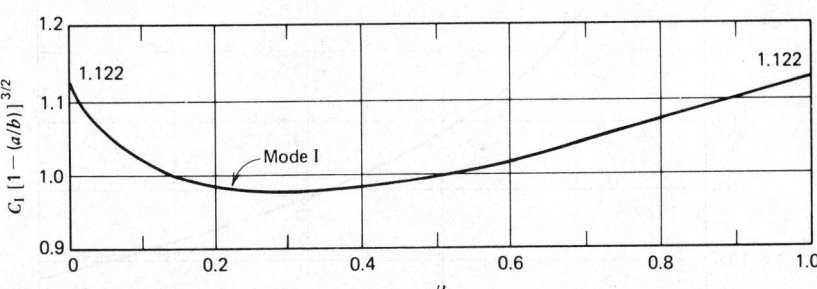

Fig. 19.10 Stress intensity factor K_I for single edge notch test specimen. (From Ref. 5, copyright Del Research Corp.; reprinted with permission.)

For a *part-through thumbnail surface crack* in a plate subjected to *uniform tension loading*, the form of the stress intensity factor is

$$K_I = \frac{1.12}{\sqrt{Q}} \sigma_t \sqrt{\pi a} \qquad (19.41)$$

where Q is a surface flaw shape parameter that depends on the ratio of crack depth to length and the ratio of nominal applied stress to yield strength of the material, as shown in Fig. 19.13.

Utilizing (19.36) through (19.41) and other similar expressions from the literature, together with fracture toughness properties for the material of interest, a designer may utilize (19.35) to predict failure or, more important, to design a part so that failure will not occur under service loading. It should be reiterated that fracture toughness is not only a function of metallurgical factors such as alloy composition and heat treatment, but a function of service temperature, loading rate, and state of stress in the vicinity of the crack tip as well. In many practical applications the plastic zone size ahead of the crack tip becomes so large that the assumption of small-scale yielding is no longer valid,

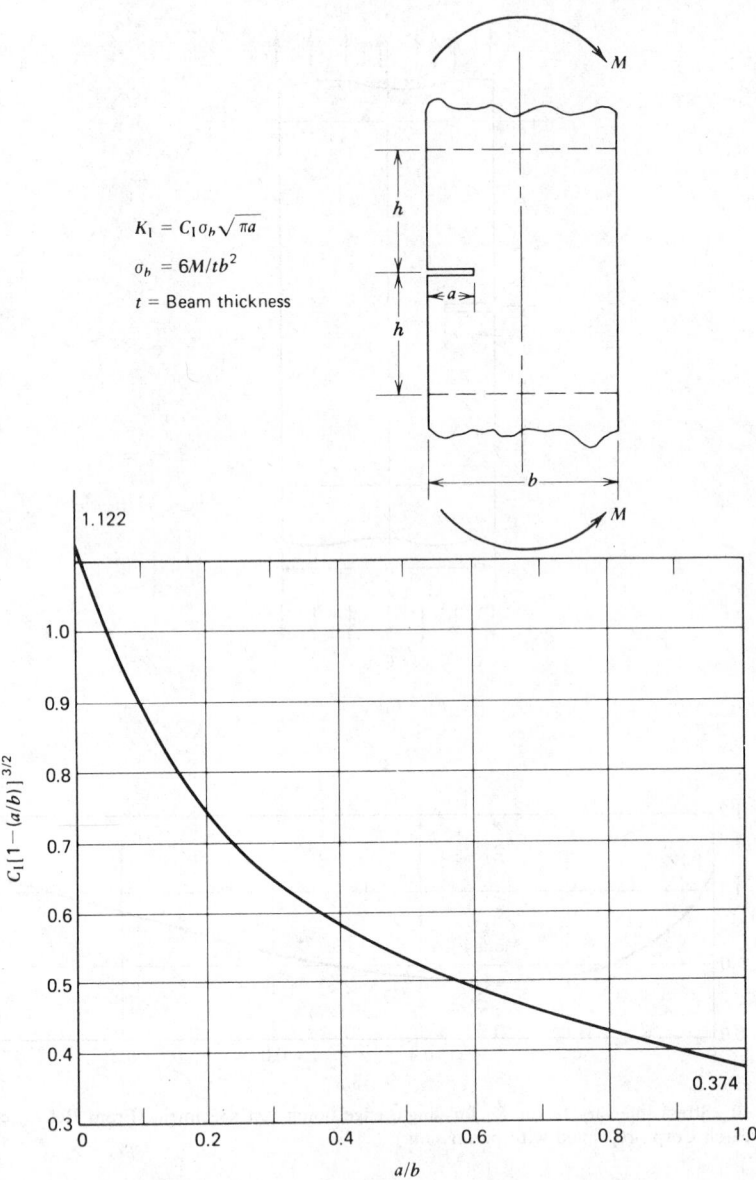

$$K_1 = C_1 \sigma_b \sqrt{\pi a}$$

$$\sigma_b = 6M/tb^2$$

t = Beam thickness

Fig. 19.11 Stress intensity factor K_1 for single through-the-thickness edge crack under pure bending moment. (From Ref. 5, copyright Del Research Corp.; adapted with permission.)

and the linear elastic analyses just described are no longer appropriate. A large percentage of low-strength and medium-strength steels used in the fabrication of modern ships, bridges, pressure vessels, structures, and machine parts are typically used in section sizes that are too thin to maintain plane strain conditions at the crack tip. The use of linear elastic fracture mechanics under such circumstances and the use of K_{Ic} for the failure criterion are, in these cases, not valid. Efforts to extend the techniques of fracture mechanics into this elastic–plastic regime have led to several promising procedures (see, for example, Ref. 9, Chap. 4), including (1) crack opening displacement (COD) methods, (2) R-curve methods, and (3) J-integral methods. The basic concept of the COD method is that fracture behavior in the region of a sharp crack may be described in terms of the opening displacement of the crack faces. The J-integral technique characterizes the stress and strain fields at the tip of a crack by a

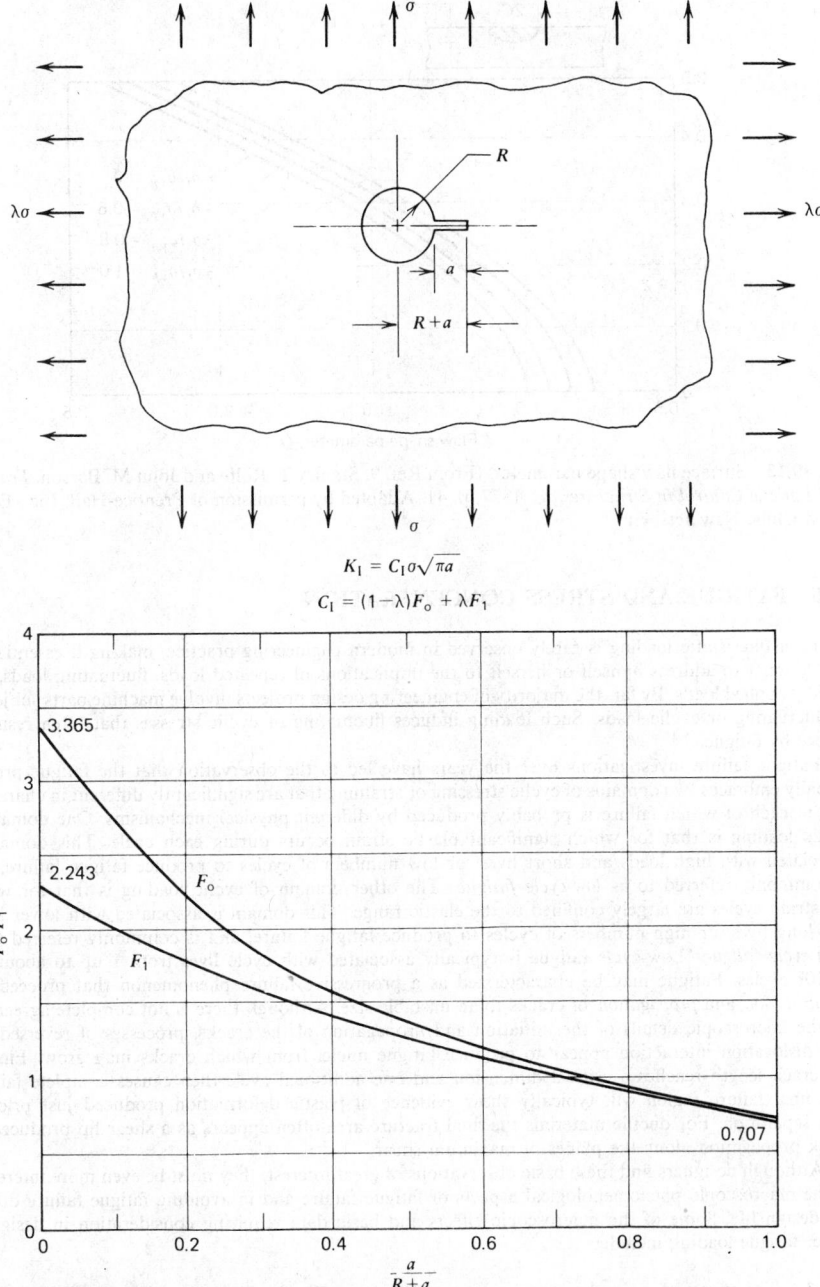

$$K_1 = C_1 \sigma \sqrt{\pi a}$$

$$C_1 = (1 - \lambda) F_o + \lambda F_1$$

Fig. 19.12 Stress intensity factor K_I for a through-the-thickness crack emanating from a circular hole in an infinite plate under biaxial tension. (From Ref. 5, copyright Del Research Corp.; adapted with permission.)

path-independent line integral close to the crack tip, which is evaluated by substituting an integration path that lies a distance away from the crack-tip yield zone. Behavior in the crack-tip region is then inferred by analyzing the region away from the crack tip. Resistance curves, or R curves, are a means of characterizing the resistance of materials to failure during the period of slow stable crack extension under the influence of increasing applied loads. Each of these methods is being used by fracture mechanics specialists with some success.

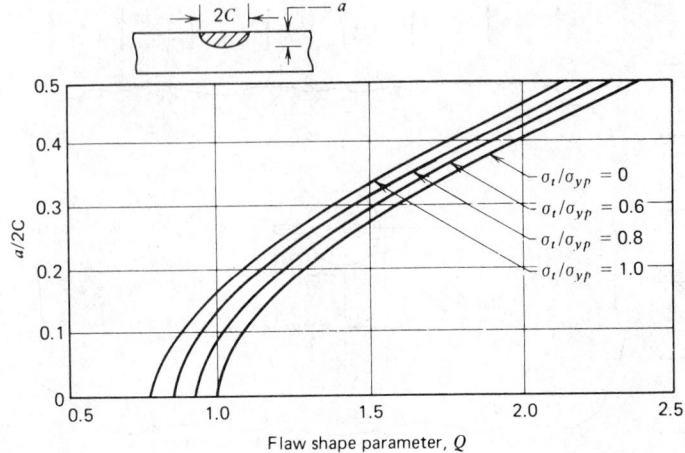

Fig. 19.13   Surface flaw shape parameter. (From Ref. 9, Stanley T. Rolfe and John M. Barson, *Fracture and Fatigue Control in Structures,* @ 1977, p. 41. Adapted by permission of Prentice-Hall, Inc., Englewood Cliffs, New Jersey.)

19.5   FATIGUE AND STRESS CONCENTRATION

Static or quasistatic loading is rarely observed in modern engineering practice, making it essential for the designer to address himself or herself to the implications of repeated loads, fluctuating loads, and rapidly applied loads. By far, the majority of engineering design projects involve machine parts subjected to fluctuating or cyclic loads. Such loading induces fluctuating or cyclic stresses that often result in failure by fatigue.

Fatigue failure investigations over the years have led to the observation that the fatigue process actually embraces two domains of cyclic stressing or straining that are significantly different in character, and in each of which failure is probably produced by different physical mechanisms. One domain of cyclic loading is that for which significant plastic strain occurs during each cycle. This domain is associated with high loads and short lives, or low numbers of cycles to produce fatigue failure, and is commonly referred to as *low-cycle fatigue.* The other domain of cyclic loading is that for which the strain cycles are largely confined to the elastic range. This domain is associated with lower loads and long lives, or high numbers of cycles to produce fatigue failure, and is commonly referred to as *high-cycle fatigue.* Low-cycle fatigue is typically associated with cycle lives from 1 up to about 10^4 or 10^5 cycles. Fatigue may be characterized as a progressive failure phenomenon that proceeds by the *initiation* and *propagation* of cracks to an unstable size. Although there is not complete agreement on the microscopic details of the initiation and propagation of the cracks, processes of reversed slip and dislocation interaction appear to produce fatigue nuclei from which cracks may grow. Finally, the crack length reaches a critical dimension and one additional cycle then causes complete failure. The final failure region will typically show evidence of plastic deformation produced just prior to final separation. For ductile materials the final fracture area often appears as a shear lip produced by crack propagation along the planes of maximum shear.

Although designers find these basic observations of great interest, they must be even more interested in the macroscopic phenomenological aspects of fatigue failure and in avoiding fatigue failure during the design life. Some of the macroscopic effects and basic data requiring consideration in designing under fatigue loading include:

1. The effects of a simple, completely reversed alternating stress on the strength and properties of engineering materials.
2. The effects of a steady stress with superposed alternating component, that is, the effects of cyclic stresses with a nonzero mean.
3. The effects of alternating stresses in a multiaxial state of stress.
4. The effects of stress gradients and residual stresses, such as imposed by shot peening or cold rolling, for example.
5. The effects of stress raisers, such as notches, fillets, holes, threads, riveted joints, and welds.
6. The effects of surface finish, including the effects of machining, cladding, electroplating, and coating.

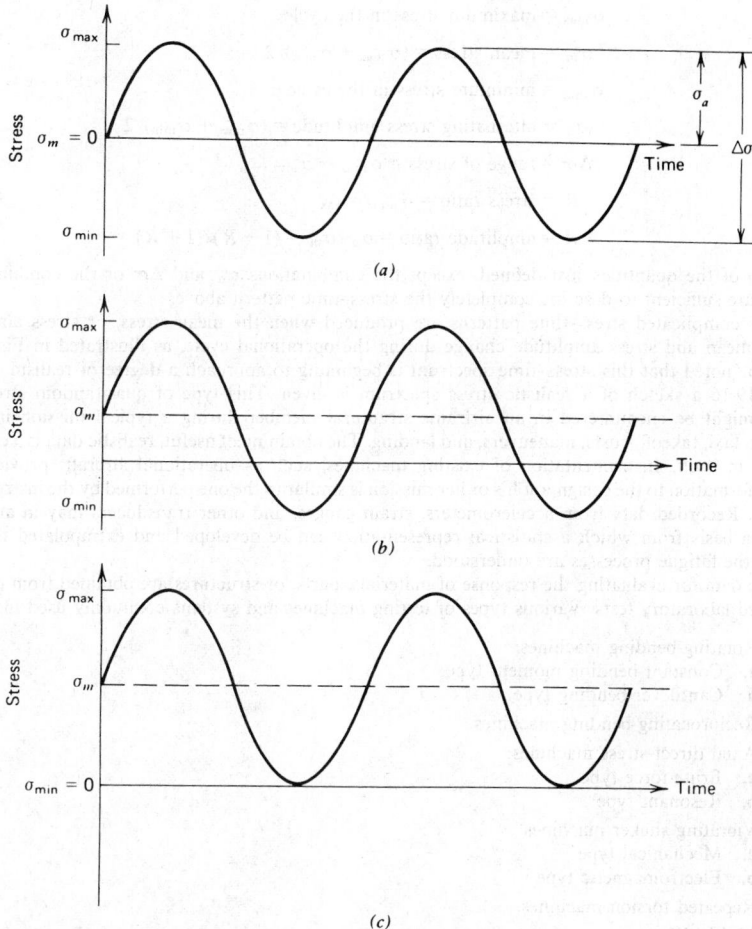

Fig. 19.14 Several constant-amplitude stress–time patterns of interest: (a) completely reversed, $R = -1$; (b) nonzero mean stress; (c) released tension, $R = 0$.

7. The effects of temperature on fatigue behavior of engineering materials.
8. The effects of size of the structural element.
9. The effects of accumulating cycles at various stress levels and the permanence of the effect.
10. The extent of the variation in fatigue properties to be expected for a given material.
11. The effects of humidity, corrosive media, and other environmental factors.
12. The effects of interaction between fatigue and other modes of failure, such as creep, corrosion, and fretting.

19.5.1 Fatigue Loading and Laboratory Testing

Faced with the design of a fatigue-sensitive element in a machine or structure, a designer is very interested in the fatigue response of engineering materials to various loadings that might occur throughout the design life of the machine under consideration. That is, the designer is interested in the effects of various *loading spectra* and associated *stress spectra*, which will in general be a function of the design configuration and the operational use of the machine.

Perhaps the simplest fatigue stress spectrum to which an element may be subjected is a zero-mean sinusoidal stress–time pattern of constant amplitude and fixed frequency, applied for a specified number of cycles. Such a stress–time pattern, often referred to as a completely reversed cyclic stress, is illustrated in Fig. 19.14a. Utilizing the sketch of Fig. 19.14, we can conveniently define several useful terms and symbols; these include:

$$\sigma_{max} = \text{maximum stress in the cycle}$$

$$\sigma_m = \text{mean stress} = (\sigma_{max} + \sigma_{min})/2$$

$$\sigma_{min} = \text{minimum stress in the cycle}$$

$$\sigma_a = \text{alternating stress amplitude} = (\sigma_{max} + \sigma_{min})/2$$

$$\Delta\sigma = \text{range of stress} = \sigma_{max} - \sigma_{min}$$

$$R = \text{stress ratio} = \sigma_{min}/\sigma_{max}$$

$$A = \text{amplitude ratio} = \sigma_a/\sigma_m = (1 - R)/(1 + R)$$

Any two of the quantities just defined, except the combinations σ_a and $\Delta\sigma$ or the combination A and R, are sufficient to describe completely the stress–time pattern above.

More complicated stress–time patterns are produced when the mean stress, or stress amplitude, or both mean and stress amplitude change during the operational cycle, as illustrated in Fig. 19.15. It may be noted that this stress–time spectrum is beginning to approach a degree of realism. Finally, in Fig. 19.16 a sketch of a realistic stress spectrum is given. This type of quasirandom stress–time pattern might be encountered in an airframe structural member during a typical mission including refueling, taxi, takeoff, gusts, maneuvers, and landing. The obtaining of useful, realistic data is a challenging task in itself. Instrumentation of existing machines, such as operational aircraft, provide some useful information to the designer if his or her mission is similar to the one performed by the instrumented machine. Recorded data from accelerometers, strain gauges, and other transducers may in any event provide a basis from which a statistical representation can be developed and extrapolated to future needs if the fatigue processes are understood.

Basic data for evaluating the response of materials, parts, or structures are obtained from carefully controlled laboratory tests. Various types of testing machines and systems commonly used include:

1. Rotating-bending machines:
   a. Constant bending moment type
   b. Cantilever bending type
2. Reciprocating-bending machines.
3. Axial direct-stress machines:
   a. Brute-force type
   b. Resonant type
4. Vibrating shaker machines:
   a. Mechanical type
   b. Electromagnetic type
5. Repeated torsion machines.
6. Multiaxial stress machines.
7. Computer-controlled closed-loop machines.
8. Component testing machines for special applications.
9. Full-scale or prototype fatigue testing systems.

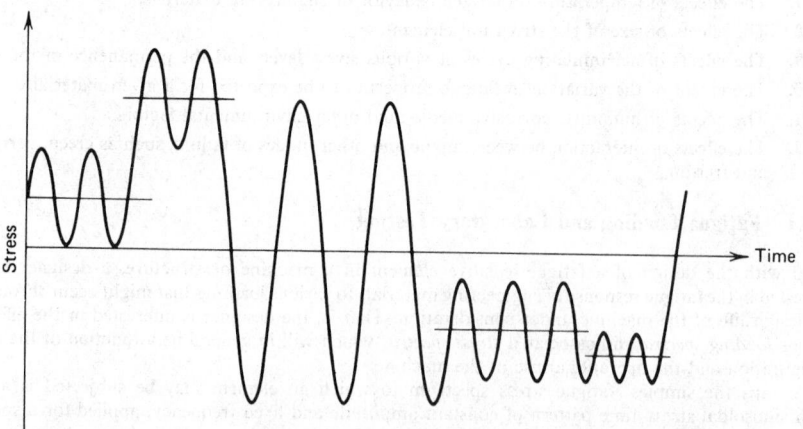

Fig. 19.15 Stress–time pattern in which both mean and amplitude change to produce a more complicated stress spectrum.

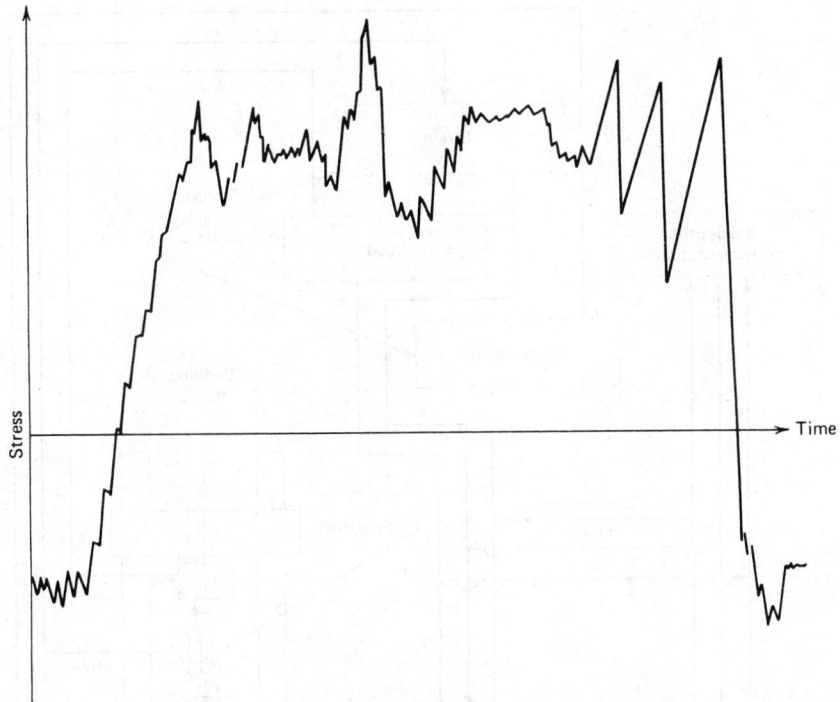

Fig. 19.16 A quasirandom stress–time pattern that might be typical of an operational aircraft during any given mission.

Computer-controlled fatigue testing machines are widely used in all modern fatigue testing laboratories. Usually such machines take the form of precisely controlled hydraulic systems with feedback to electronic controlling devices capable of producing and controlling virtually any strain–time, load–time, or displacement–time pattern desired. A schematic diagram of such a system is shown in Fig. 19.17.

Special testing machines for component testing and full-scale prototype testing systems are not found in the general fatigue testing laboratory. These systems are built up especially to suit a particular need, for example, to perform a full-scale fatigue test of a commercial jet aircraft.

It may be observed that fatigue testing machines range from very simple to very complex. The very complex testing systems, used, for example, to test a full-scale prototype, produce very specialized data applicable only to the particular prototype and test conditions used; thus, for the particular prototype and test conditions the results are very accurate, but extrapolation to other test conditions and other pieces of hardware is difficult, if not impossible. On the other hand, simple smooth-specimen laboratory fatigue data are very general and can be utilized in designing virtually any piece of hardware made of the specimen material. However, to use such data in practice requires a quantitative knowledge of many pertinent differences between the laboratory and the application, including the effects of nonzero mean stress, varying stress amplitude, environment, size, temperature, surface finish, residual stress pattern, and others. Fatigue testing is performed at the extremely simple level of smooth specimen testing, the extremely complex level of full-scale prototype testing, and everywhere in the spectrum between. Valid arguments can be made for testing at all levels.

19.5.2 The S–N–P Curves—A Basic Design Tool

Basic fatigue data in the high-cycle life range can be conveniently displayed on a plot of cyclic stress level versus the logarithm of life, or alternatively, on a log–log plot of stress versus life. These plots, called S–N curves, constitute design information of fundamental importance for machine parts subjected to repeated loading. Because of the scatter of fatigue life data at any given stress level, it must be recognized that there is not only one S–N curve for a given material, but a family of S–N curves with probability of failure as the parameter. These curves are called the S–N–P curves, or curves of constant probability of failure on a stress-versus-life plot.

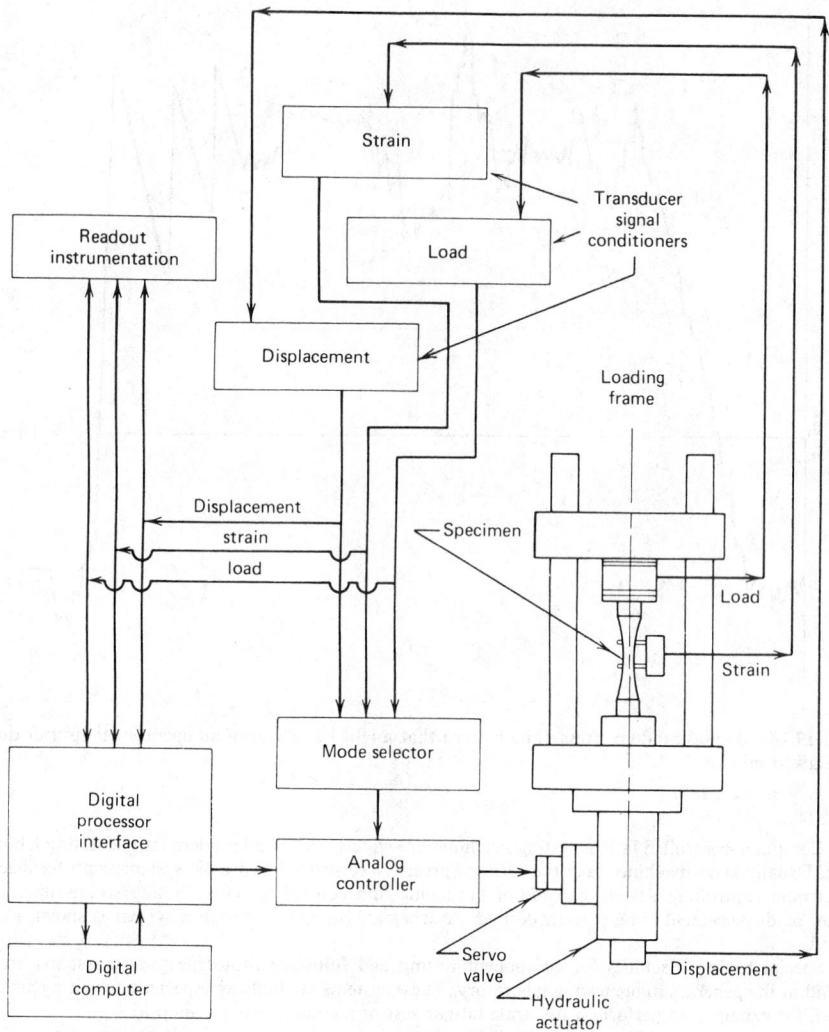

Fig. 19.17 Schematic diagram of a computer-controlled closed-loop fatigue testing machine.

To develop an *S–N–P* plot in the fatigue laboratory by "standard" methods, one would proceed in the following way:

1. Select a large group of carefully prepared, polished fatigue specimens of the material of interest and subdivide them into four or five smaller groups of at least 15 specimens each.

2. Select four or five stress levels, perhaps judged by a few exploratory tests, that span the stress range of the *S–N* curve.

3. Run an entire subgroup at each of the selected stress levels following the procedures to be outlined here.

4. To make each test run, mount a specimen in the testing machine, using due care to avoid spurious stresses. Set the machine for the desired stress amplitude, with cycle counter set on zero.

5. Start the machine and run at constant stress amplitude until the specimen fails or the machine reaches a predetermined *runout* criterion, for example, 5×10^7 cycles.

6. Record the stress amplitude used and the cycle count at the time of failure or runout.

7. Using a new specimen, repeat the procedure, again recording the stress level and life at failure or runout. Continue to repeat this procedure until all specimens designated for the selected stress level have been tested.

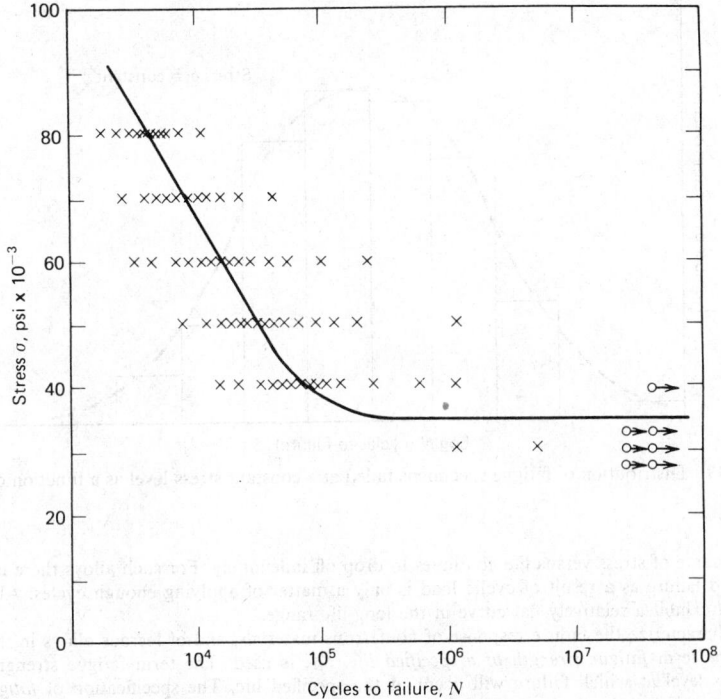

Fig. 19.18 Plot of stress–cycle (S–N) data as it might be collected by laboratory fatigue testing of a new alloy.

8. Change to a new stress level and repeat the preceding procedure until all specimens designated for this second stress level have been tested. Repeat this procedure until all selected stress levels have been tested. Note that the entire output from a complete fatigue test is a single point on the S–N plot.

9. Plot all data collected on a stress-versus-log-of-life coordinate system as shown in Fig. 19.18. Runouts, or points for which fatigue failure was not observed during the test, are indicated by a small arrow to the right.

Considering the data plotted in Fig. 19.18, one could simply construct a visual mean curve through the data. Doing this, it becomes clear that a substantial scatter of data about the mean clouds the design usefulness of such a mean curve. A better approach would be to construct for each stress level a *histogram*, such as the one shown in Fig. 19.19, which shows the *distribution* of failures as a function of the log of life for the sample tested. Computation of the sample mean and variance permits the estimation of population mean and variance if the form of the distribution is known for fatigue tests at a constant stress level. Extensive testing of large samples has indicated that a log-normal distribution of life at a constant stress level is a good estimate. Using a Weibull distribution gives a better estimate, but it is slightly more complicated to analyze. Assuming the life distribution to be log-normal, the sample mean and variance can be used to specify any desired probability of failure. Repeating the analysis at all test stress levels, we can connect points of equal probability of failure to obtain curves of constant probability of failure on the S–N plot. Such a family of S–N–P curves is shown in Fig. 19.20. It is also of interest to note that the "reliability" R is defined to be one minus the probability of failure; hence, $R = 1 - P$. Thus, the 5% probability of failure curve may alternatively be designated as the 95% reliability curve ($R = 0.95$), for example. Thus, some reference will be found in the literature to these so-called R–S–N curves. It should also be noted that references to the "S–N curve" in the literature generally refer to the *mean* curve unless otherwise specified.

The mean S–N curves sketched in Fig. 19.21 distinguish two types of material response to cyclic loading commonly observed. The ferrous alloys and titanium exhibit a steep branch in the relatively short life range, leveling off to approach a stress asymptote at longer lives. This stress asymptote is called the *fatigue limit* (formerly called endurance limit) and is the stress level below which an infinite number of cycles can be sustained without failure. The nonferrous alloys do not exhibit an asymptote,

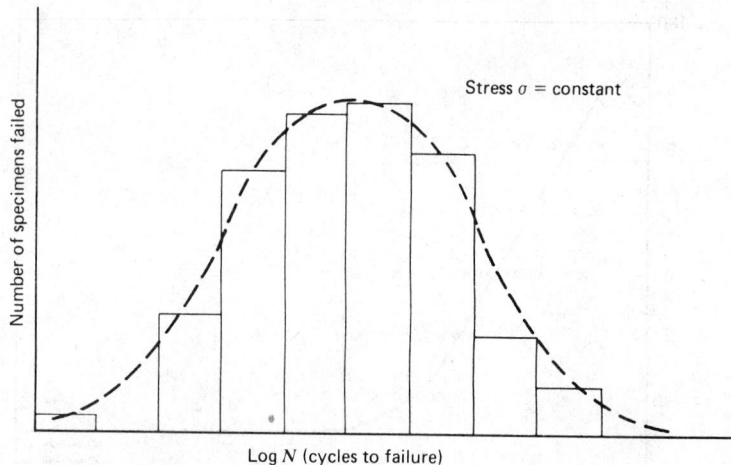

Fig. 19.19   Distribution of fatigue specimens failed at a constant stress level as a function of logarithm of life.

and the curve of stress versus life continues to drop off indefinitely. For such alloys there is no fatigue limit, and failure as a result of cyclic load is only a matter of applying enough cycles. All materials, however, exhibit a relatively flat curve in the long-life range.

To characterize the failure response of nonferrous materials, and of ferrous alloys in the finite-life range, the term *fatigue strength at a specified life*, S_N, is used. The term fatigue strength identifies the stress level at which failure will occur at the specified life. The specification of *fatigue strength* without specifying the corresponding life is meaningless. The specification of a *fatigue limit* always implies infinite life.

Many different fatigue testing procedures have been devised to provide information suitable for meeting a variety of different objectives. For example, it may be desired to obtain life distribution data at a constant stress level, strength distribution data at a constant life level, significant fatigue design data with the smallest possible sample size or in the shortest possible time, or data to meet

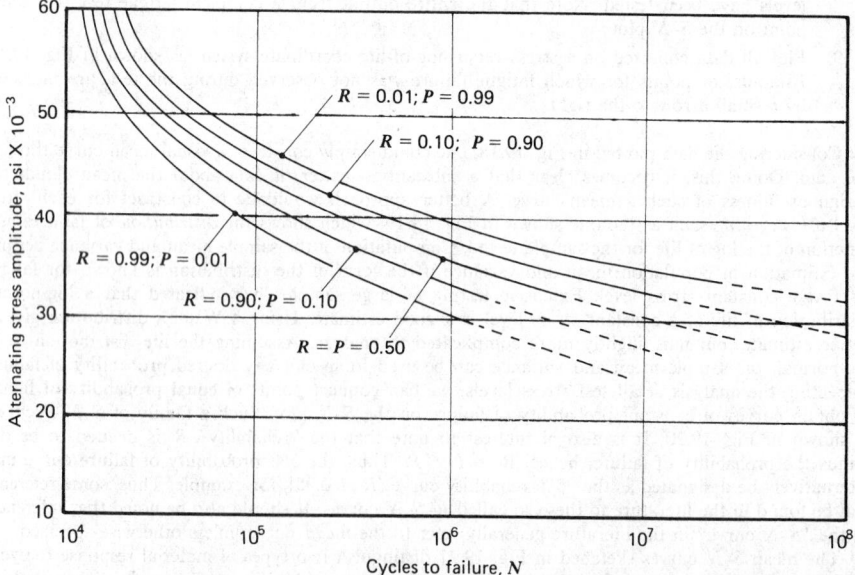

Fig. 19.20   Family of *S–N–P* curves, or *R–S–N* curves, for 7075-T6 aluminum alloy. Note: $P =$ probability of failure; $R =$ reliability $= 1 - P$. (Adapted from Ref. 12, p. 117; with permission from John Wiley & Sons, Inc.)

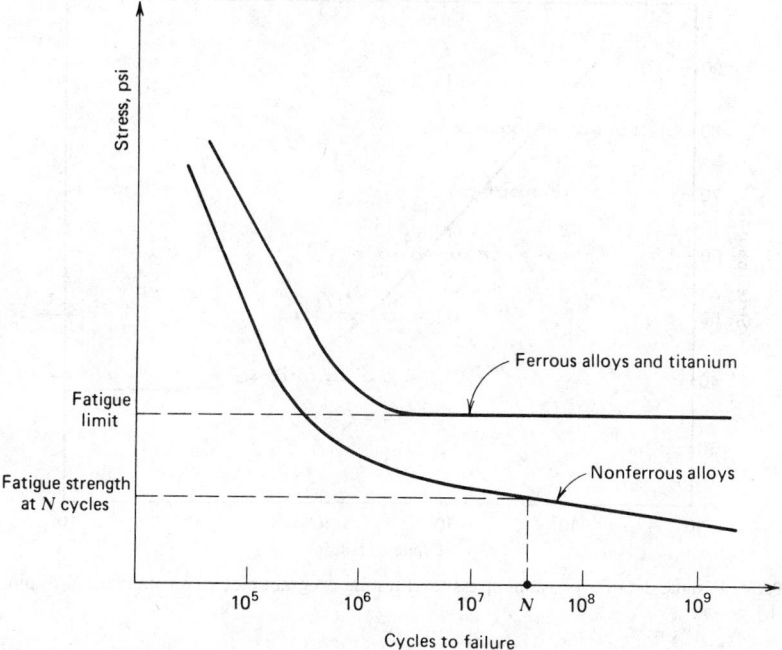

Fig. 19.21 Two types of material response to cyclic loading.

other specific needs. Several fatigue testing procedures are outlined in the following paragraphs to meet a variety of objectives. It should be noted that the procedures are applicable to all types of laboratory and field service fatigue data, no matter what types of testing machines or methods are used. Testing machines and methods have been widely discussed in the literature.[13]

When only a few parts or specimens are available for testing, and an estimate of the whole *S–N* curve is needed, the testing method often used is the so-called "standard" method. The standard method simply involves running one or two specimens at each of several different stress amplitude levels and recording the stress amplitude and number of cycles to failure. If specimens "run out," they are sometimes run again at a higher stress level, taking due precaution to note that damage or coaxing effects may influence the final failure lives of these specimens. The data are typically plotted on a standard *S–N* plot as shown in Fig. 19.22. Scatter of the data is to be expected, and it is doubtful that much useful statistical information can be developed from the "standard" method, even under the best conditions, because the sample size is so small.

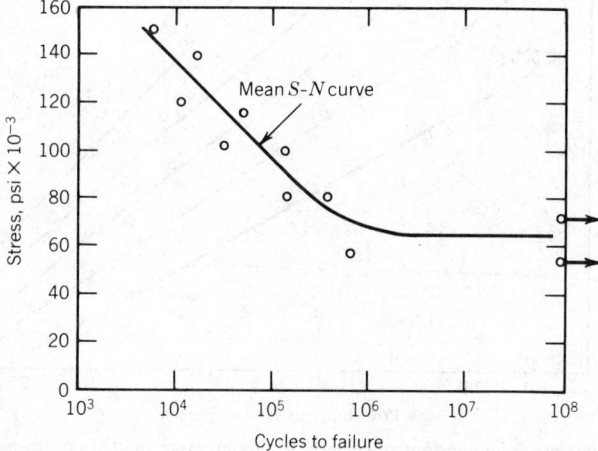

Fig. 19.22 *S–N* curves resulting from "standard" fatigue testing method.

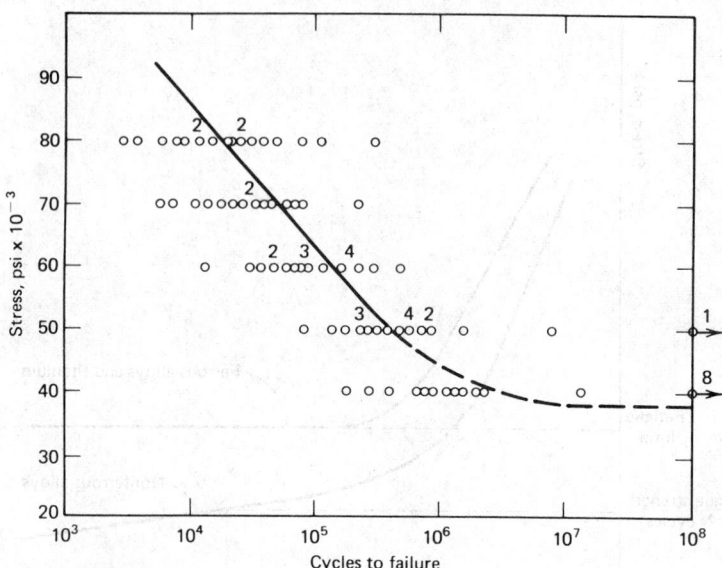

Fig. 19.23 Fatigue data for constant stress level testing program plotted on standard *S–N* plot. (After Ref. 14.)

The constant stress level procedure involves testing groups of approximately 15 or more specimens at each of four or more different constant stress levels in the stress range between the fatigue limit and the yield strength of the material, recording failure lives for all tests. Experience has indicated that the distributions of fatigue failure lives at constant stress levels higher than the fatigue limit are log-normal to a good approximation. All the data taken at each constant stress level are therefore plotted on log-normal probability paper to verify the distribution and to determine the mean and variance for log of life at that stress level. Figure 19.23 shows an example of the results of a constant stress level testing program plotted on a standard *S–N* plot, and Fig. 19.24 shows the same data plotted on log-normal probability paper. It may be noted that the log-normal assumption is accurate

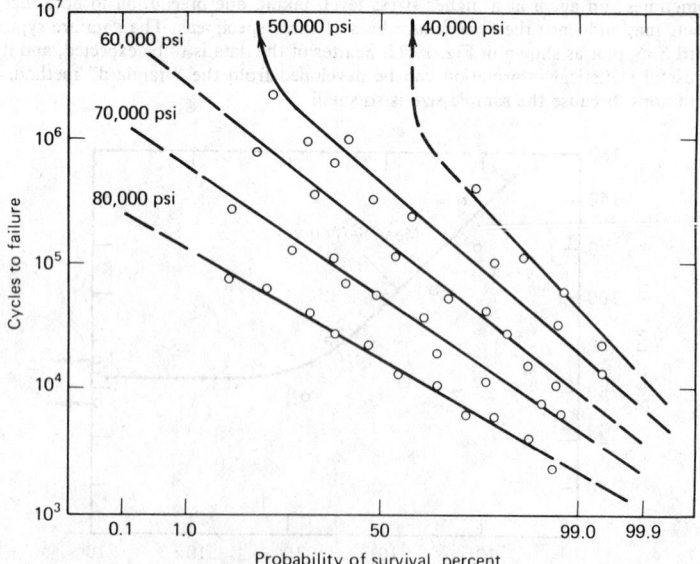

Fig. 19.24 Fatigue data from constant stress level testing program plotted on log–normal probability paper. (After Ref. 14.)

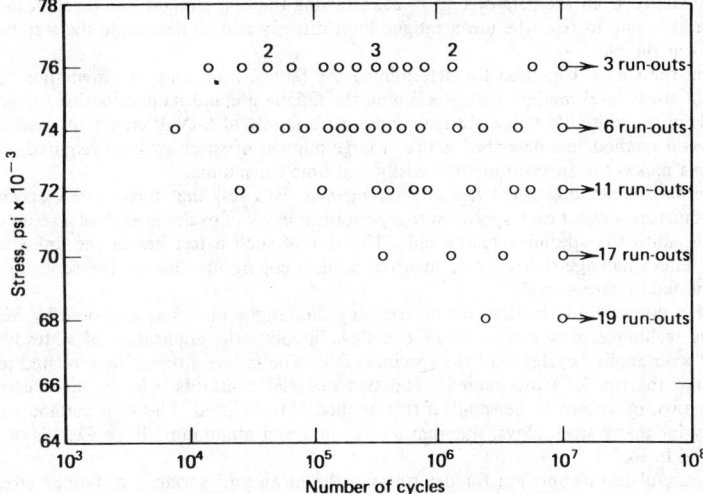

Fig. 19.25   S–N plot showing results of a survival test for determination of fatigue limit.

at the higher stress levels, but at stress levels near the fatigue limit is not valid. A much more useful method for determining the fatigue limit is the *survival method,* which is also called the *mortality method, quantal response method,* or *all-or-nothing method.* This technique may be used to determine the mean and variance of the fatigue limit or the mean and variance of fatigue strength at any specified life.

The survival method involves testing several groups of specimens at closely spaced stress levels spanning the range from about two standard deviations below the fatigue limit to two standard deviations above it. For example, preliminary testing may indicate a fatigue limit around 72,000 psi. On this basis five stress levels might be selected, ranging from 68,000 to 76,000 psi at 2000 psi intervals. If 20 specimens were tested at each stress level, the data might plot on S–N coordinates as shown in Fig. 19.25. It may be observed that at the lowest selected stress level most specimens are runouts, whereas at the highest selected stress level only a few specimens are runouts, a result that might reasonably be expected. It is important to select the stress levels judiciously because if the stress level increments are too large, many groups may have all failures and other groups may have no failures, with little or no intermediate response. Usually, however, if the groups span the stress range of about two standard deviations above and below the mean, the response will be satisfactory.

The results of the survival test shown in Fig. 19.25 may next be plotted on normal probability paper, as shown in Fig. 19.26 where stress level is the random variable plotted against probability of

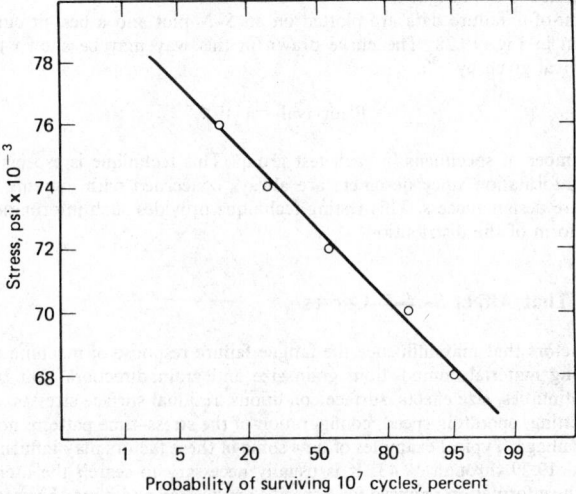

Fig. 19.26   Survival test data plotted on normal probability paper.

survival calculated from the data of Fig. 19.25. Drawing the best straight line through the probability plot then allows one to read the mean fatigue limit directly and to determine the standard deviation in stress, from the plot.

Utilizing the survival method for determining the fatigue limit and its distribution, and utilizing the constant stress level method for determining the fatigue life and its distribution at each of several stress levels in the finite life range, we may construct a family of S–N–P curves. A clear disadvantage of the survival method just described is that a large number of specimens are required, usually about 60–100. This makes the survival method costly and time consuming.

An alternative method, called the *step-test method*, is a test that forces every specimen to fail. The technique is to subject each specimen to a prescribed block of cycles at each of a series of increasing stress levels, until the specimen finally fails. The risk of such a test lies in the unknown effects of coaxing or latent damage that may be incurred while stepping up through the series of stress levels to reach the failure stress levels.

In 1948 a unique rapid method for determining the fatigue limit was proposed by Marcel Prot.[15] This testing technique, now called the *Prot method,* involves the application of a steadily increasing stress level with applied cycles until the specimen fails. The failure stress is then related to the fatigue limit through the rate of stress increase and two material constants. Clearly the effects of coaxing must be known, or known to be small, if this method is to be used. The Prot method has been used successfully for many steel alloys, titanium alloys, and even aluminum alloys. Details of Prot testing are discussed in Ref. 1.

A very useful testing method for determining the mean and variance of fatigue strength at any specified life is the *up-and-down testing method.* The method is equally valid for determining fatigue limit, since the fatigue limit is simply the fatigue strength at infinite life.

To perform an up-and-down test a group of at least 15 specimens is selected to evaluate the fatigue strength at the life of interest. The first specimen is tested at a stress level a little higher than the estimated fatigue strength until it either fails or runs out at the life of interest. If the specimen fails before reaching the life of interest, the stress level is *decreased* by a preselected increment and the second specimen is tested at this new lower stress level. If the first specimen runs out, the stress level is *increased* by the preselected increment and the second specimen is tested at this new higher stress level. The test is continued in this manner in sequence, with each succeeding specimen being tested at a stress level that is one stress increment above or below its predecessor, depending on whether the predecessor ran out or failed. Thus the test is sequential in nature, has an up-and-down character, and tends to center on the mean value of failure for the life of interest. A typical up-and-down test is illustrated in Fig. 19.27. Details of this procedure for analyzing up-and-down data may be found in Ref. 1.

Finally, a method sometimes used to establish extreme value probability S–N curves is the so-called *extreme value method* or *least-of-n method.* For this type of test the technique is to select a group of n specimens to be run simultaneously in n different identical fatigue testing machines, all at the same stress level. When the first specimen of the group fails, the stress level and number of cycles to failure are recorded; then all other machines are stopped and the specimens are discarded.

Next, a second group of n specimens is run at a new stress level, again recording data for the first failure and discarding all the other specimens. The procedure is repeated for several different stress levels at and above the fatigue limit.

Finally, the first-of-n failure data are plotted on an S–N plot and a best-fit curve drawn through the data, as shown in Fig. 19.28. The curve drawn in this way may be shown to correspond to a probability of survival given by

$$P\{\text{survival}\} = (\tfrac{1}{2})^{1/n} \tag{19.42}$$

where n is the number of specimens in each test group. This technique is especially appropriate for providing design information since designers are always concerned with extreme value probabilities of survival to ensure design success. This testing technique provides such information without specific knowledge of the form of the distribution.

19.5.3 Factors That Affect S–N–P Curves

There are many factors that may influence the fatigue failure response of machine parts or laboratory specimens, including material composition, grain size and grain direction, heat treatment, welding, geometrical discontinuities, size effects, surface conditions, residual surface stresses, operating temperature, corrosion, fretting, operating speed, configuration of the stress–time pattern, nonzero mean stress, and prior fatigue damage. Typical examples of how some of these factors may influence fatigue response are shown in Figs. 19.29 through 19.43. It is usually necessary to search the literature and existing data bases to find the information required for a specific application and it may be necessary to undertake experimental testing programs to produce data where they are unavailable.

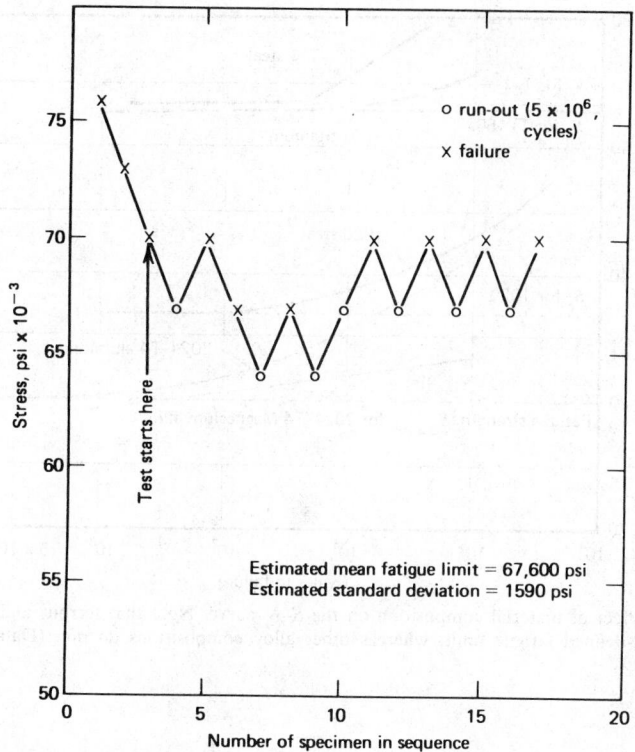

Fig. 19.27 Up-and-down fatigue test used to determine median (mean) fatigue strength at 5×10^6 cycles for a 4340 steel alloy.

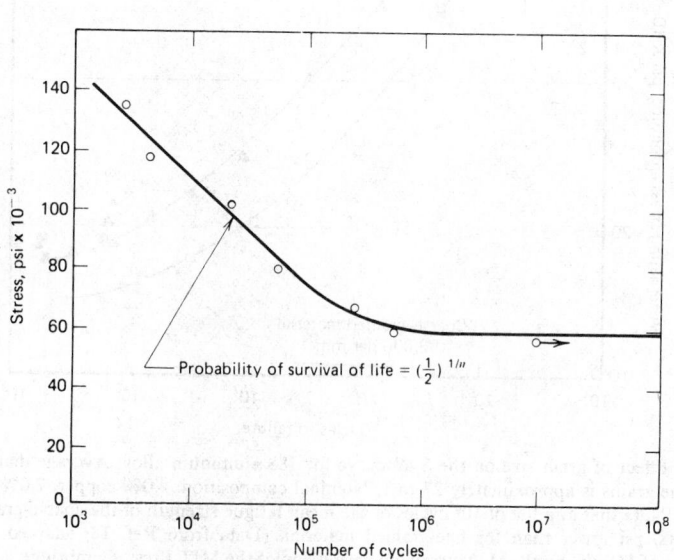

Fig. 19.28 Extreme value probability S–N plot based on least-of-n testing. (After Ref. 14.)

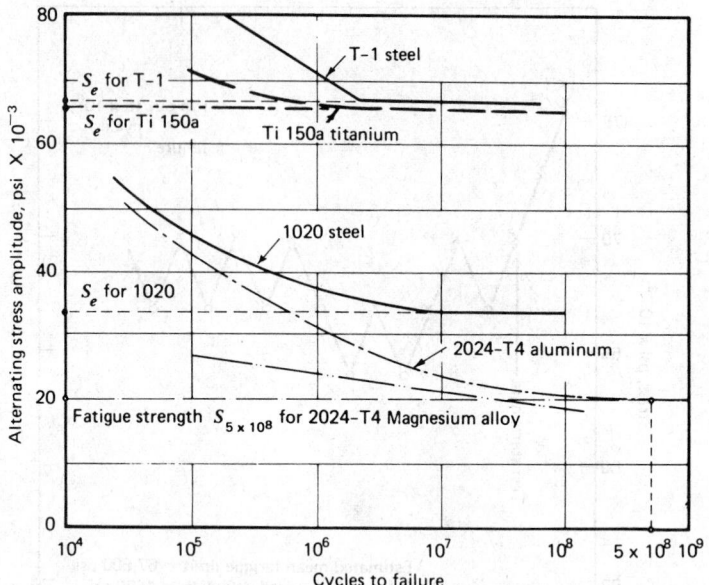

Fig. 19.29 Effect of material composition on the *S–N* curve. Note that ferrous and titanium alloys exhibit a well-defined fatigue limit, whereas other alloy compositions do not. (Data from Refs. 16 and 17.)

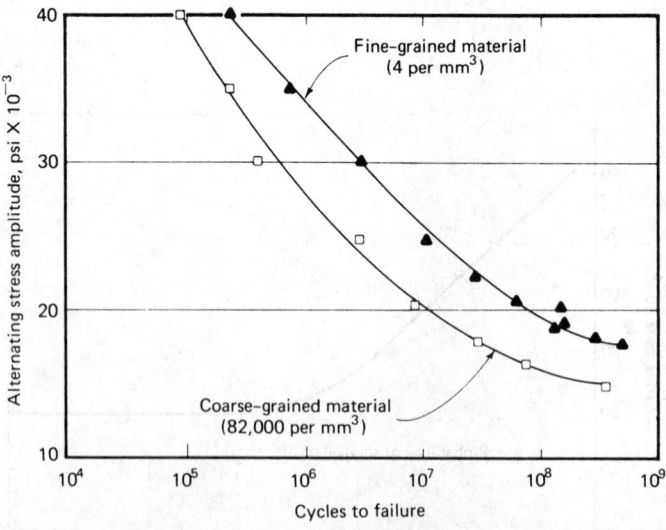

Fig. 19.30 Effect of grain size on the *S–N* curve for 18S aluminum alloy. Average diameter ratio of coarse to fine grains is approximately 27 to 1. Nominal composition: 4.0% copper, 2.0% nickel, 0.6% magnesium. Note that at a life of 10^8 cycles of the mean fatigue strength of the coarse-grained material is about 3000 psi lower than for fine-grained material. (Data from Ref. 18; adapted from *Fatigue and Fracture of Metals,* by W. M. Murray, by permission of the MIT Press, Cambridge, Massachusetts, copyright, 1952.)

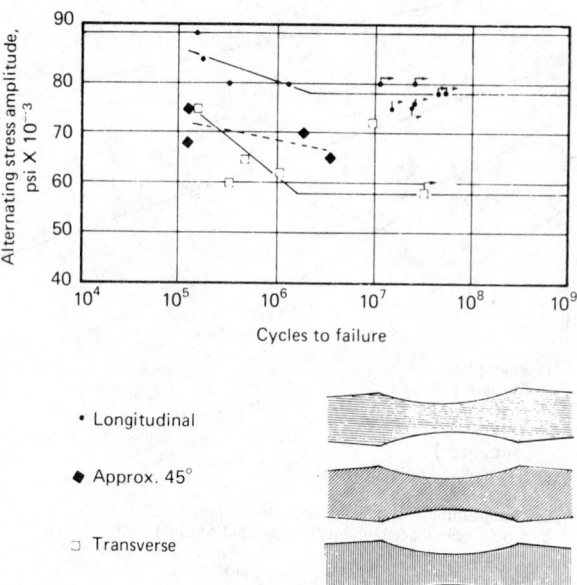

Fig. 19.31 Effect on the *S–N* curve of grain flow direction relative to longitudinal loading direction for specimens machined from crankshaft forgings. Nominal composition: 0.41% carbon, 0.47% manganese, 0.01% silicon, 0.04% phosphorous, 1.8% nickel. $S_u = 139,000$ psi, $S_{yp} = 115,000$ psi, e (2.0 in.) = 20%. (Data from Ref. 19.)

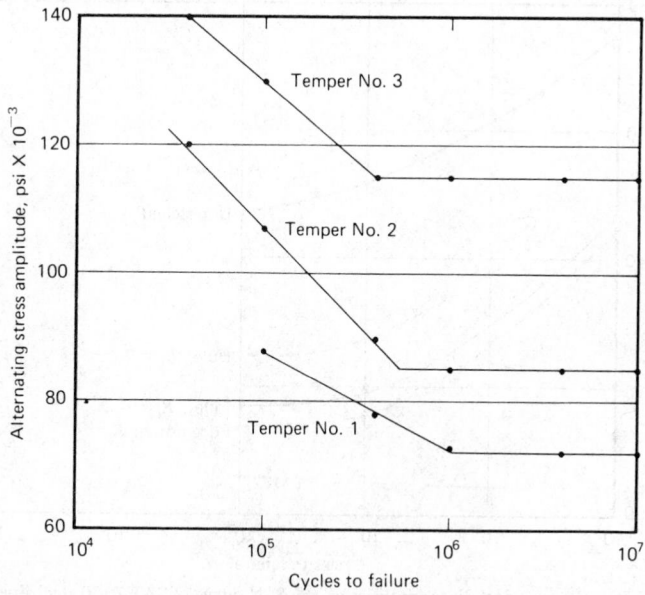

Fig. 19.32 Effects of heat treatment on the *S–N* curve of SAE 4130 steel, using 0.19-in.-diameter rotating bending specimens cut from ⅜-in. plate, 1625°F, oil quenched, followed by three different tempers. Temper No. 1: $S_u = 129,000$ psi; $S_{yp} = 118,000$ psi. Temper No. 2: $S_u = 150,000$ psi; $S_{yp} = 143,000$ psi. Temper No. 3: $S_u = 206,000$ psi; $S_{yp} = 194,000$ psi.

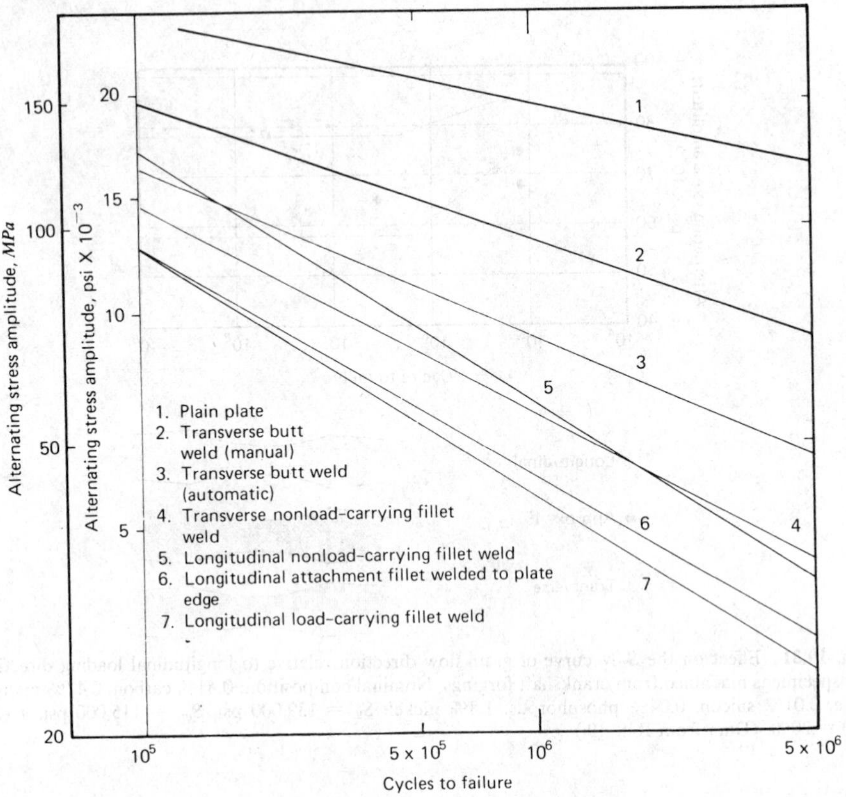

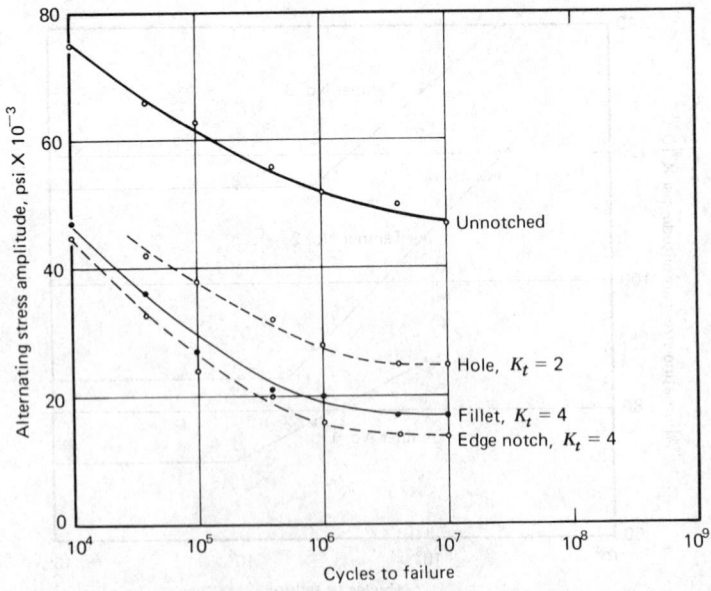

Fig. 19.33 Effects of welding detail on the *S–N* curve of structural steel, with yield strength in the range 30,000–52,000 psi. Tests were released tension ($\sigma_{\min} = 0$). (Data from Ref. 20.)

Fig. 19.34 Effects of geometrical discontinuities on the *S–N* curve of SAE 4130 steel sheet, normalized, tested in completely reversed axial fatigue test. Specimen dimensions (t = thickness, w = width, r = notch radius): Unnotched: $t = 0.075$ in., $w = 1.5$ in. Hole: $t = 0.075$ in., $w = 4.5$ in., $r = 1.5$ in. Fillet: $t = 0.075$ in., $n_{\text{net}} = 1.5$ in., $w_{\text{gross}} = 2.25$ in., $r = 0.0195$ in. Edge notch: $t = 0.075$ in., $w_{\text{net}} = 1.5$ in., $w_{\text{gross}} = 2.25$ in., p = 0.057 in. (Data from Ref. 16.)

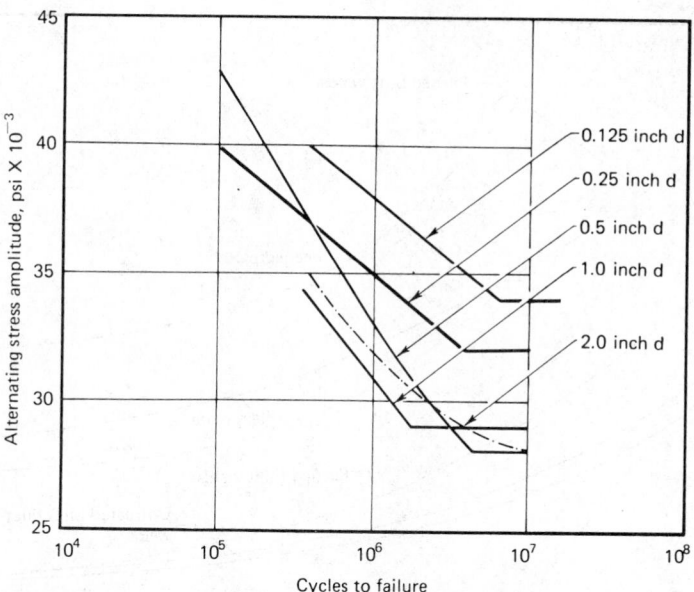

Fig. 19.35 Size effects on the *S–N* curve of SAE 1020 steel specimens cut from a 3½-in.-diameter hot-rolled bar, testing in rotating bending. (Data from Ref. 21.)

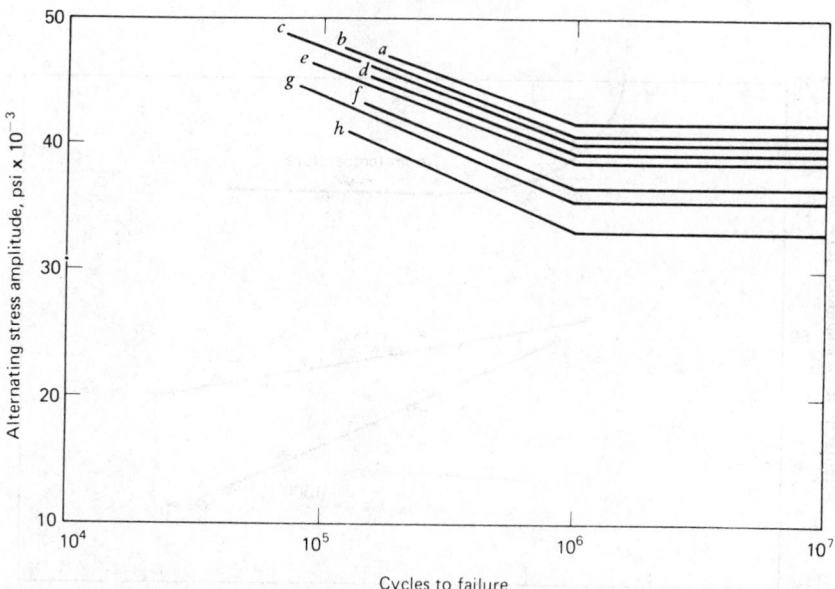

Fig. 19.36 Effect of surface finish on the *S–N* curve of 0.33% carbon steel specimens, testing in a rotating cantilever beam machine: (*a*) high polish, longitudinal direction; (*b*) FF emery finish; (*c*) No. 1 emery finish; (*d*) coarse emery finish; (*e*) smooth file; (*f*) as-turned; (*g*) bastard file; (*h*) coarse file. (Data from Ref. 22.)

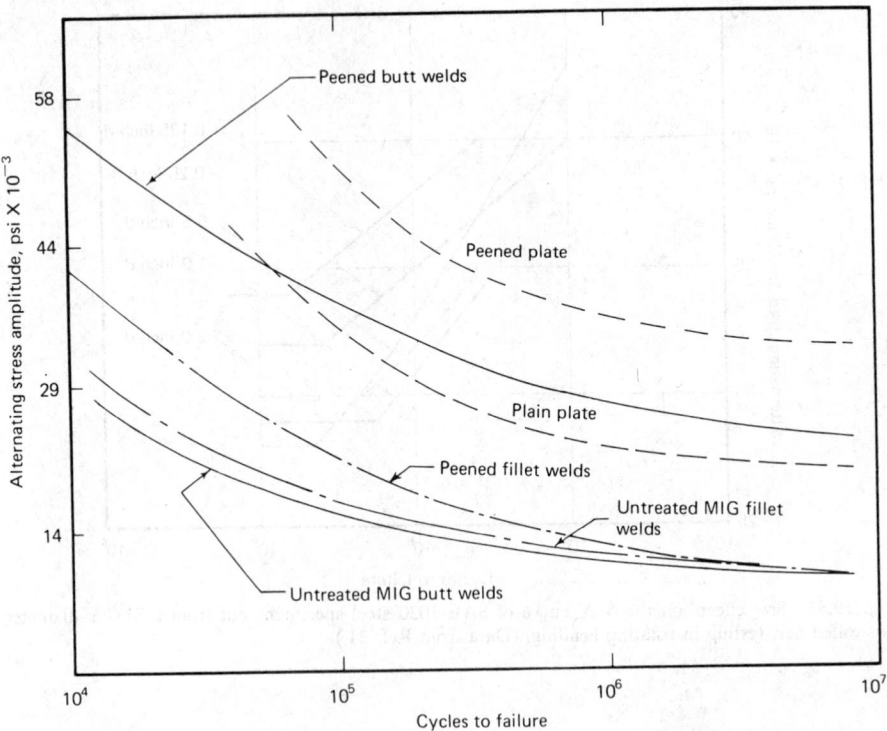

Fig. 19.37 Effect of shot peening on the *S–N* curves for welded and unwelded steel plate. (Data from Ref. 20.)

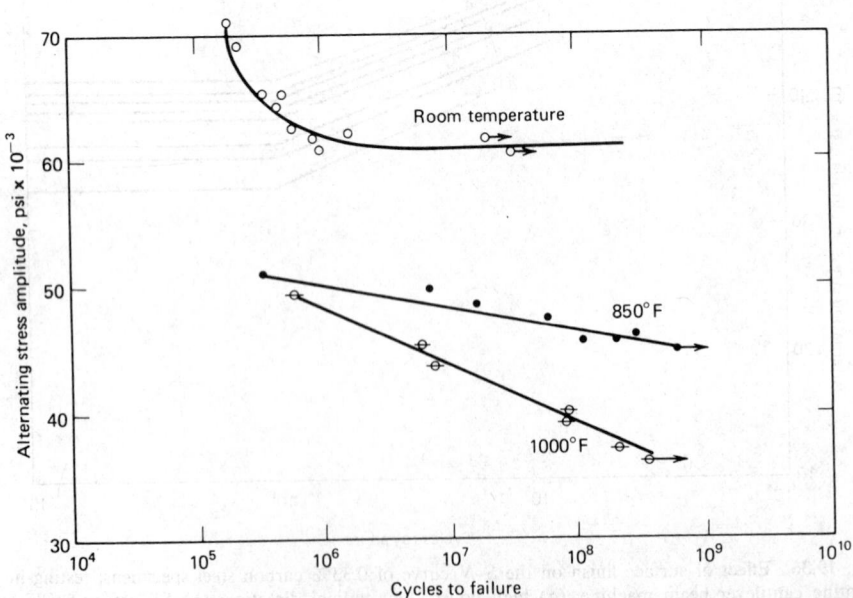

Fig. 19.38 Effect of operating temperature on the *S–N* curve of a 12% chromium steel alloy. Alloy composition = 0.10% C, 0.45% Mn, 0.21% Ni, 12.3% Cr, and 0.38% Mo. (Data from Ref. 23.)

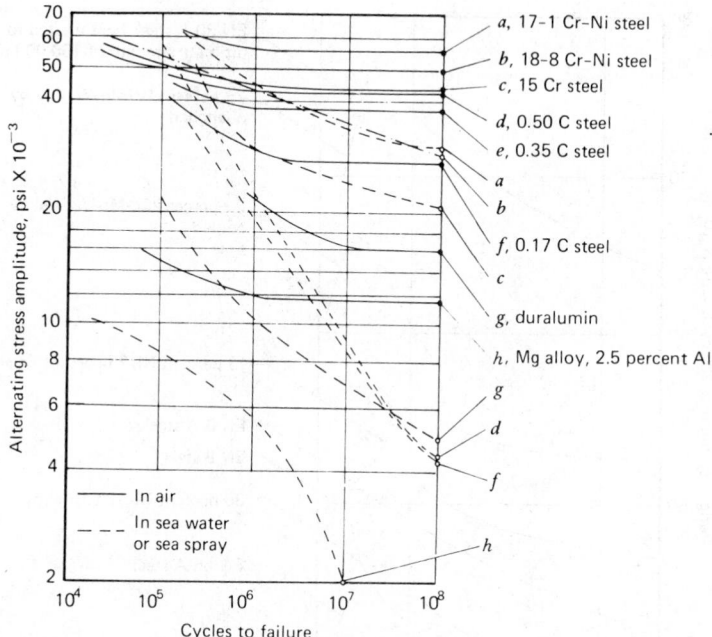

Fig. 19.39 Effects of corrosion on the *S–N* curves of various aircraft materials tested in push–pull loading in seawater or sea spray. (Data from Ref. 24.)

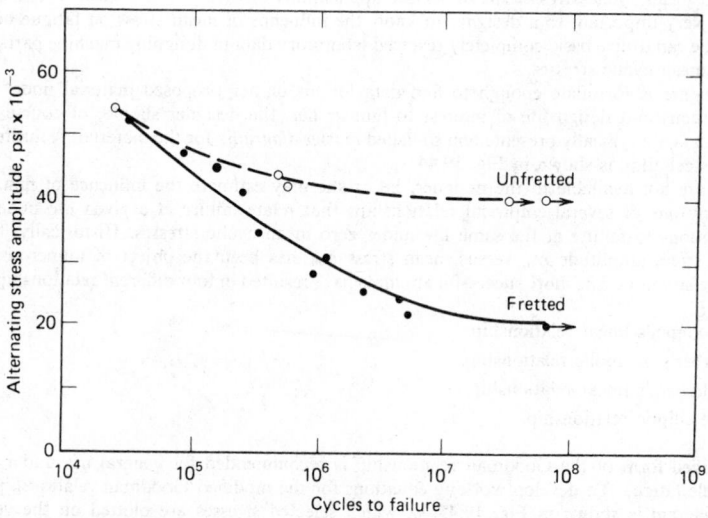

Fig. 19.40 Effect of fretting on the *S–N* curve of a forged 0.24% steel. (Data from Ref. 25; reprinted with permission from McGraw-Hill Book Company.)

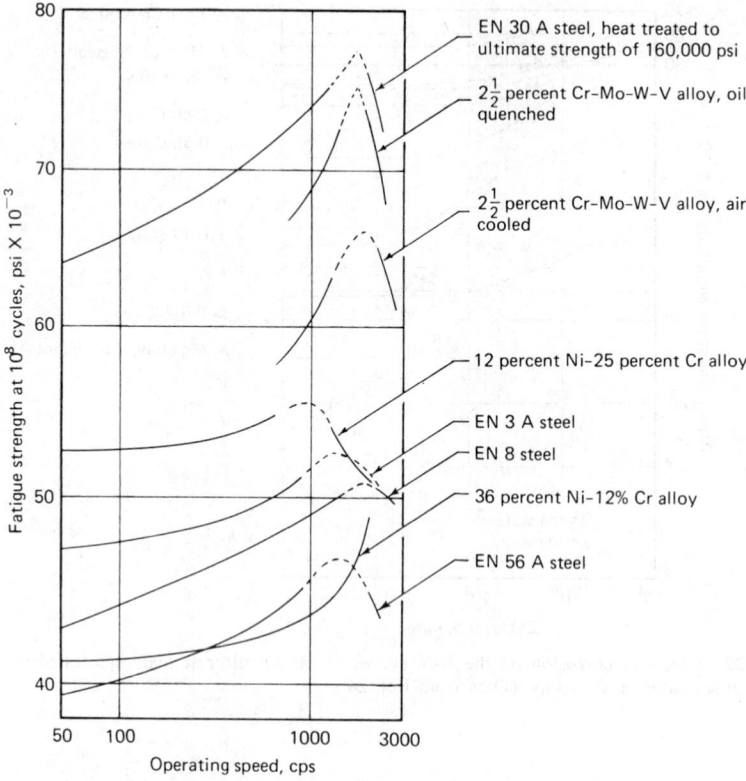

Fig. 19.41 Effect of operating speed on the fatigue strength at 10^8 cycles for several different ferrous alloys. (Data from Ref. 26, p. 381).

19.5.4 Nonzero Mean and Multiaxial Fatigue Stresses

Most basic fatigue data collected in the laboratory are for completely reversed alternating stresses, that is, zero mean cyclic stresses. Most service applications involve nonzero mean cyclic stresses. It is therefore very important to a designer to know the influence of mean stress on fatigue behavior so that he or she can utilize basic completely reversed laboratory data in designing machine parts subjected to nonzero mean cyclic stresses.

If a designer is fortunate enough to find data for his or her proposed material under the mean stress conditions and design life of interest to him or her, the designer should, of course, use these data. Such data are typically presented on so-called *master diagrams* for the material. A master diagram for a 4340 steel alloy is shown in Fig. 19.44.

If data are not available to the designer, he or she may estimate the influence of nonzero mean stress by any one of several empirical relationships that relate failure at a given life under nonzero mean conditions to failure at the same life under zero mean cyclic stresses. Historically, the plot of alternating stress amplitude σ_a versus mean stress σ_m has been the object of numerous empirical curve-fitting attempts. The more successful attempts have resulted in four different relationships, namely:

1. Goodman's linear relationship.
2. Gerber's parabolic relationship.
3. Soderberg's linear relationship.
4. The elliptic relationship.

A modified form of the Goodman relationship is recommended for general use under conditions of high-cycle fatigue. To develop working equations for the modified Goodman relationships a range-of-stress diagram is shown in Fig. 19.45, in which selected stresses are plotted on the vertical axis versus mean stress on the horizontal axis. The entire range-of-stress diagram in Fig. 19.45 is constructed for a single fatigue lifetime of N cycles.

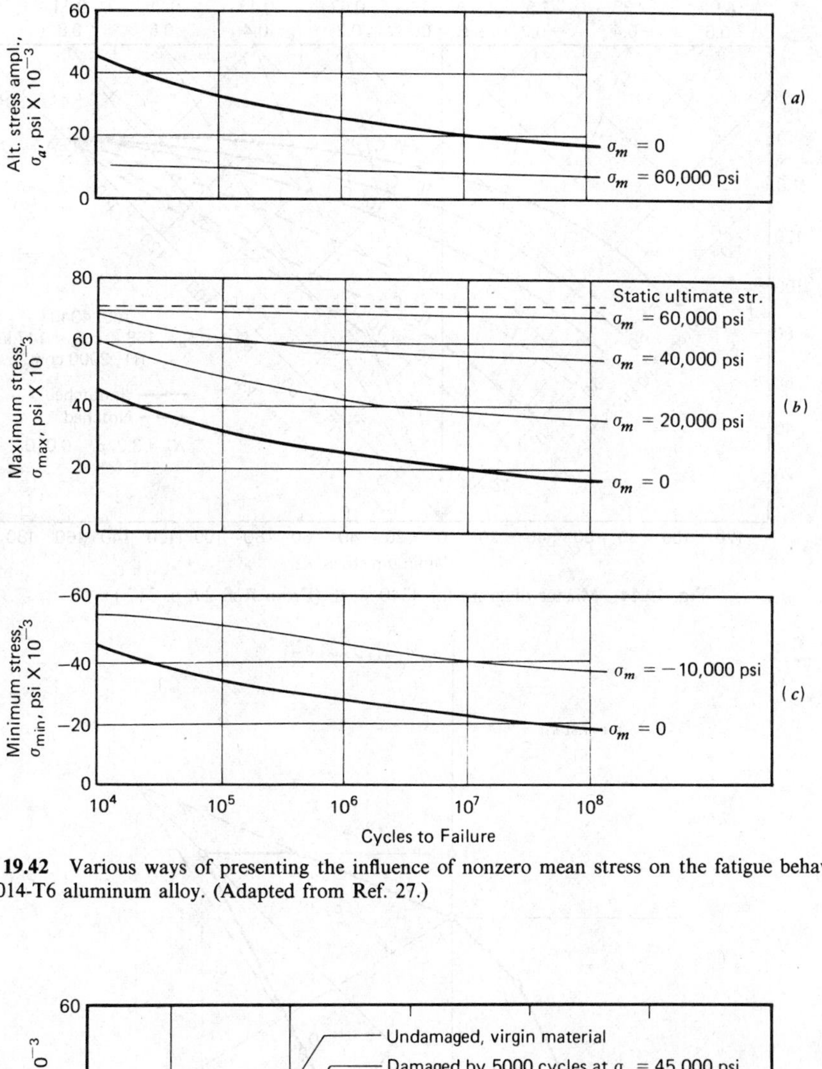

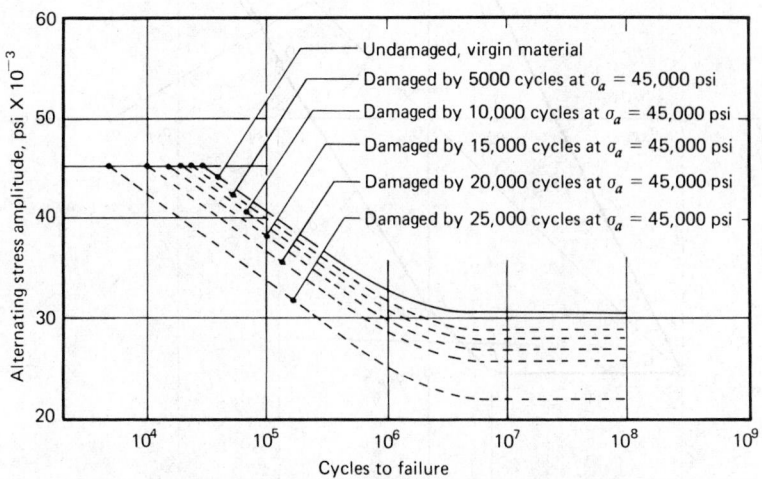

Fig. 19.42 Various ways of presenting the influence of nonzero mean stress on the fatigue behavior of 2014-T6 aluminum alloy. (Adapted from Ref. 27.)

Fig. 17.43 Illustration of the influence of accumulated fatigue damage on subsequent fatigue behavior of carbon steel. Note: Life of virgin material at $\sigma_a = 45,000$ psi is approximately 30,000 cycles. (Data from Ref. 28; reprinted with permission from John Wiley & Sons, Inc.)

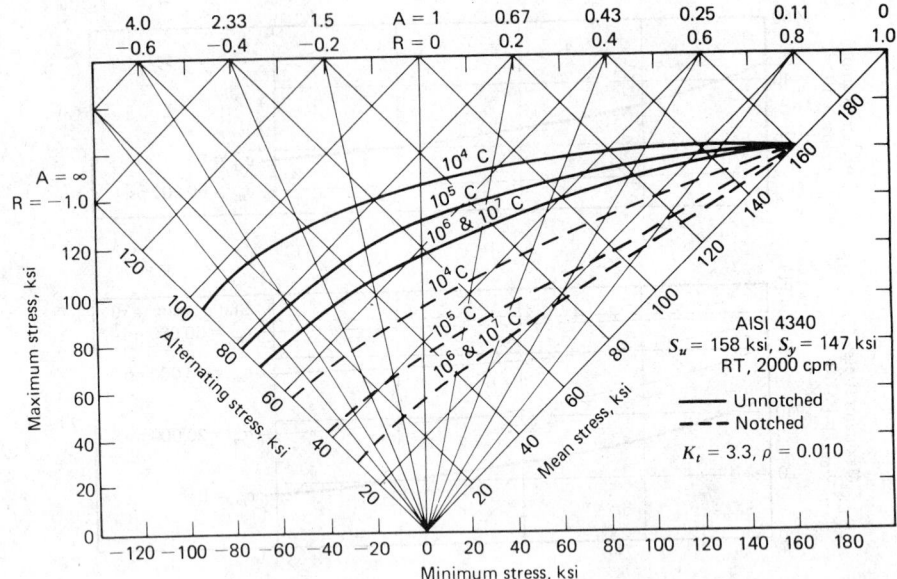

Fig. 19.44 Master diagram for 4340 steel. (From Ref. 29, p. 317.)

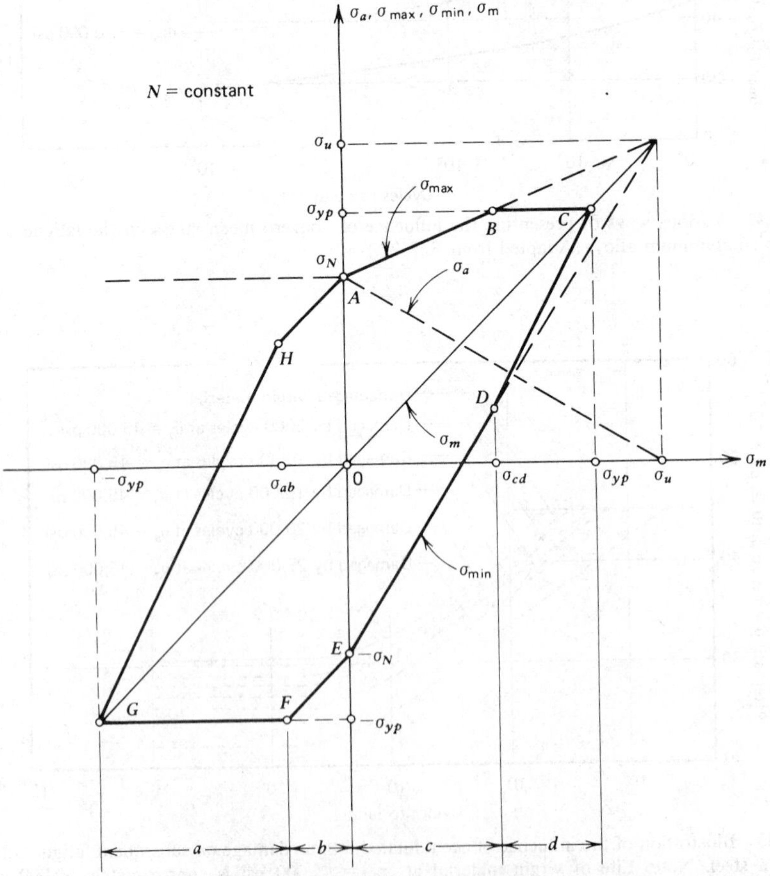

Fig. 19.45 Modified Goodman range-of-stress diagram for fatigue failure in N cycles.

Table 19.2 Failure Prediction Equations for Nonzero Mean Cyclic Stressing

Region	Failure Equation (failure is predicted to occur if:)	Limits on Validity of Equation
a	$\sigma_{max} - 2\sigma_m \geq \sigma_{yp}$	$-\sigma_{yp} \leq \sigma_m \leq (\sigma_N - \sigma_{yp})$
b	$\sigma_{max} - \sigma_m \geq \sigma_N$	$(\sigma_N - \sigma_{yp}) \leq \sigma_m \leq 0$
c	$\sigma_{max} - (1-r)\sigma_m \geq \sigma_N$	$0 \leq \sigma_m \leq \left(\dfrac{\sigma_{yp} - \sigma_N}{1-r}\right)$
d	$\sigma_{max} \geq \sigma_{yp}$	$\left(\dfrac{\sigma_{yp} - \sigma_N}{1-r}\right) \leq \sigma_m \leq \sigma_{yp}$

$$\text{where } r \equiv \frac{\sigma_N}{\sigma_u}$$

The modified Goodman range-of-stress diagram is a failure locus for the case of *uniaxial* fatigue stressing. Any cyclic loading that produces a stress amplitude that exceeds the bounds of the locus will cause failure in fewer than N cycles. Any stress amplitude that lies within the locus will result in more than N cycles without failure. Stress amplitudes that just touch the locus produce failure in exactly N cycles. Writing equations for the four regions of the diagram, and transforming them into failure prediction equations, yields the results given in Table 19.2.

Utilizing the results of Table 19.2, a designer may estimate whether failure will occur under any nonzero mean stress condition at the design life if he or she knows the ultimate strength of the material, the yield strength, and the simple completely reversed fatigue strength for the material at the design life. All these properties are usually available.

If the machine part under consideration is subjected not only to a nonzero mean stress, but also a multiaxial state of stress, the results of Table 19.2 may be integrated with appropriate combined stress theories of failure. Thus the maximum normal stress multiaxial fatigue failure theory represents a combination of the maximum normal stress theory for static stresses, shown in Section 19.4, and the modified Goodman relationships, shown in Table 19.2. Writing the results of this combination in complete detail results in a total of 12 failure equations, which, when written together with their regions of validity, are as follows:

Failure is predicted to occur if:

$$\sigma_{1max} - 2\sigma_{1m} - \sigma_{yp} \geq 0, \qquad -\sigma_{yp} \leq \sigma_{1m} \leq \sigma_N - \sigma_{yp} \qquad (19.43)$$
$$\sigma_{1max} - \sigma_{1m} - \sigma_N \geq 0, \qquad (\sigma_N - \sigma_{yp}) \leq \sigma_{1m} \leq 0 \qquad (19.44)$$

$$\sigma_{1max} - \sigma_{1m}(1-r) - \sigma_N \geq 0, \qquad 0 \leq \sigma_{1m} \leq \frac{\sigma_{yp} - \sigma_N}{1-r} \qquad (19.45)$$

$$\sigma_{1max} - \sigma_{yp} \geq 0, \qquad \frac{\sigma_{yp} - \sigma_N}{1-r} \leq \sigma_{1m} \leq \sigma_{yp} \qquad (19.46)$$

or if

$$\sigma_{2max} - 2\sigma_{2m} - \sigma_{yp} \geq 0, \qquad -\sigma_{yp} \leq \sigma_{2m} \leq \sigma_N - \sigma_{yp} \qquad (19.47)$$
$$\sigma_{2max} - \sigma_{2m} - \sigma_N \geq 0, \qquad \sigma_N - \sigma_{yp} \leq \sigma_{2m} \leq 0 \qquad (19.48)$$

$$\sigma_{2max} - \sigma_{2m}(1-r) - \sigma_N \geq 0, \qquad 0 \leq \sigma_{2m} \leq \frac{\sigma_{yp} - \sigma_N}{1-r} \qquad (14.49)$$

$$\sigma_{2max} - \sigma_{yp} \geq 0, \qquad \frac{\sigma_{yp} - \sigma_N}{1-r} \leq \sigma_{2m} \leq \sigma_{yp} \qquad (19.50)$$

or if

$$\sigma_{3max} - 2\sigma_{3m} - \sigma_{yp} \geq 0, \qquad -\sigma_{yp} \leq \sigma_{3m} \leq \sigma_N - \sigma_{yp} \qquad (19.51)$$
$$\sigma_{3max} - \sigma_{3m} - \sigma_N \geq 0, \qquad \sigma_N - \sigma_{yp} \leq \sigma_{3m} \leq 0 \qquad (19.52)$$

$$\sigma_{3max} - \sigma_{3m}(1-r) - \sigma_N \geq 0, \qquad 0 \leq \sigma_{3m} \leq \frac{\sigma_{yp} - \sigma_N}{1-r} \qquad (19.53)$$

$$\sigma_{3max} - \sigma_{yp} \geq 0, \qquad \frac{\sigma_{yp} - \sigma_N}{1-r} \leq \sigma_{3m} \leq \sigma_{yp} \qquad (19.54)$$

Whichever one of these 12 equations proves to be most critical will govern the design, and the geometry is determined in accordance with this most critical equation. It should also be noted that the subscripts 1, 2, and 3 in these equations denote the three principal normal stresses.

The maximum shearing stress multiaxial fatigue theory represents a combination of the maximum shearing stress theory for static stresses, from Section 19.4, and the modified Goodman relationships, shown in Table 19.2. It may be recalled that for a uniaxial state of stress the largest principal shearing stress is equal to one-half the largest principal normal stress. This relationship is utilized by substituting 2τ for σ everywhere that it appears in the modified Goodman relationships. Writing the results of this substitution for the first principal shearing stress τ_1 yields the following set of equations:

Failure is predicted to occur if:

$$2\tau_{1max} - 4\tau_{1m} - \sigma_{yp} \geq 0, \qquad -\sigma_{yp} \leq 2\tau_{1m} \leq \sigma_N - \sigma_{yp} \tag{19.55}$$

$$2\tau_{1max} - 2\tau_{1m} - \sigma_N \geq 0, \qquad \sigma_N - \sigma_{yp} \leq 2\tau_{1m} \leq 0 \tag{19.56}$$

$$2\tau_{1max} - 2\tau_{1m}(1-r) - \sigma_N \geq 0, \qquad 0 \leq 2\tau_{1m} \leq \frac{\sigma_{yp} - \sigma_N}{1-r} \tag{19.57}$$

$$2\tau_{1max} - \sigma_{yp} \geq 0, \qquad \frac{\sigma_{yp} - \sigma_N}{1-r} \leq 2\tau_{1m} \leq \sigma_{yp} \tag{19.58}$$

In addition to this set of equations for τ_1, there are two other sets of equations, identical in structure, that must be written for τ_2 and τ_3. This results in a total of 12 failure equations that must be examined to determine the most critical equation that governs the design. The values of τ_1, τ_2, and τ_3 to be used in these failure equations are the multiaxial principal shearing stresses as defined by the expressions

$$\tau_1 = \pm\left(\frac{\sigma_2 - \sigma_3}{2}\right) \tag{19.59}$$

$$\tau_2 = \pm\left(\frac{\sigma_3 - \sigma_1}{2}\right) \tag{19.60}$$

$$\tau_3 = \pm\left(\frac{\sigma_1 - \sigma_2}{2}\right) \tag{19.61}$$

where σ_1, σ_2, and σ_3 are the principal normal stresses. The geometry of the machine part would then be determined in accordance with the most critical of the 12 equations. The distortion energy multiaxial fatigue failure theory represents a combination of the distortion energy theory for a static state of stress, discussed in Section 19.3, and the modified Goodman relationships, shown in Table 19.2. The expression for distortion energy per unit volume is

$$u_d = \left[\frac{1+v}{3E}\right]\left[\frac{(\sigma_1 - \sigma_2)^2}{2} + \frac{(\sigma_2 - \sigma_3)^2}{2} + \frac{(\sigma_3 - \sigma_1)^2}{2}\right] \tag{19.62}$$

Evaluating the distortion energy per unit volume for a uniaxial state of stress, expressions for σ_{max} and σ_m become

$$\sigma_{max} = \sqrt{\frac{3E}{1+v}}\sqrt{u_{d\,max}} \tag{19.63}$$

$$\sigma_m = \sqrt{\frac{3E}{1+v}}\sqrt{u_{dm}} \tag{19.64}$$

and substituting these expressions in the modified Goodman relationships yields the following result:

Failure is predicted to occur if:

$$\left[\frac{3E}{1+v}\right]^{1/2}[(u_{d\,max})^{1/2} - 2(u_{dm})^{1/2}] - \sigma_{yp} \geq 0 \tag{19.65}$$

$$\left[\frac{3E}{1+v}\right]^{1/2}[(u_{d\,max})^{1/2} - (u_{dm})^{1/2}] - \sigma_N \geq 0 \tag{19.66}$$

$$\left[\frac{3E}{1+v}\right]^{1/2}[(u_{d\,max})^{1/2} - (u_{dm})^{1/2}(1-r)] - \sigma_N \geq 0 \tag{19.67}$$

$$\left[\frac{3E}{1+v}\right]^{1/2}[u_{d\,max}]^{1/2} - \sigma_{yp} \geq 0 \tag{19.68}$$

where $r \equiv \sigma_N/\sigma_u$.

Values of $u_{d\,max}$ and u_{dm} are determined for actual multiaxial loading ranges using (19.62) when possible. If this is impractical, the maximum and mean values for distortion energy per unit volume are often approximated using

$$u_{d\,max} = \left[\frac{1+v}{3E}\right]\left[\frac{(\sigma_{1max} - \sigma_{2max})^2}{2} + \frac{(\sigma_{2max} - \sigma_{3max})^2}{2} + \frac{(\sigma_{3max} - \sigma_{1max})^2}{2}\right] \qquad (19.69)$$

$$u_{dm} = \left[\frac{1+v}{3E}\right]\left[\frac{(\sigma_{1m} - \sigma_{2m})^2}{2} + \frac{(\sigma_{2m} - \sigma_{3m})^2}{2} + \frac{(\sigma_{3m} - \sigma_{1m})^2}{2}\right] \qquad (19.70)$$

Each of the fatigue failure theories provides a means of predicting failure, or designing to prevent failure, under conditions of nonzero mean cyclic multiaxial states of stress as long as the stress amplitude remains constant throughout the life of the part. Although there is only a relatively small amount of multiaxial fatigue data available, the existing experimental evidence leads to the following observations:

1. For brittle materials, the maximum normal multiaxial fatigue failure theory is the best theory to use.

2. For ductile materials, the distortion energy multiaxial fatigue failure theory is the best theory to use.

3. For ductile materials, the maximum shearing stress multiaxial fatigue failure theory is almost as good as the distortion energy multiaxial fatigue failure theory.

4. As a rule of thumb, materials that exhibit a ductility of less than 5% elongation in 2 in. may be regarded as brittle, whereas materials that exhibit a ductility of 5% or more elongation in 2 in. may be regarded as ductile.

Some data illustrating the validity of these observations are plotted in Fig. 19.46.

19.5.5   Spectrum Loading and Cumulative Damage

In virtually every engineering application where fatigue is an important failure mode, the alternating stress amplitude may be expected to vary or change in some way during the service life. Such variations and changes in load amplitude, often referred to as *spectrum loading*, make the direct use of standard *S–N* curves inapplicable because these curves are developed and presented for constant stress amplitude operation. Therefore, it becomes important to a designer to have available a theory or hypothesis, verified by experimental observations, that will permit good design estimates to be made for operation under conditions of spectrum loading using the standard constant amplitude *S–N* curves.

The basic postulate adopted by all fatigue investigators working with spectrum loading is that operation at any given cyclic stress amplitude will produce *fatigue damage*, the seriousness of which will be related to the number of cycles of operation at that stress amplitude and also related to the total number of cycles that would be required to produce failure of an undamaged specimen at that stress amplitude. It is further postulated that the damage incurred is permanent and operation at several different stress amplitudes in sequence will result in an accumulation of total damage equal to the sum of the damage increments accrued at each individual stress level. When the total accumulated damage reaches a critical value, fatigue failure occurs. Although the concept is simple in principle, much difficulty is encountered in practice because the proper assessment of the amount of damage incurred by operation at any given stress level S_i for a specified number of cycles n_i is not straightforward. Many different *cumulative damage* theories have been proposed for the purposes of assessing fatigue damage caused by operation at any given stress level and the addition of damage increments to properly predict failure under conditions of spectrum loading. The first cumulative damage theory was proposed by Palmgren in 1924 and later developed by Miner in 1945. This linear theory, which is still widely used, is referred to as the *Palmgren–Miner hypothesis* or the *linear damage rule*. The theory may be described using the *S–N* plot shown in Fig. 19.47.

By definition of the *S–N* curve, operation at a constant stress amplitude S_1 will produce complete damage, or failure, in N_1 cycles. Operation at stress amplitude S_1 for a number of cycles n_1 smaller than N_1 will produce a smaller fraction of damage, say D_1. D_1 is usually termed the damage fraction. Operation over a spectrum of different stress levels results in a damage fraction D_1 for each of the different stress levels S_i in the spectrum. When these damage fractions sum to unity, failure is predicted; that is,

Failure is predicted to occur if:

$$D_1 + D_2 + \cdots + D_{i-1} + D_i \geq 1 \qquad (19.71)$$

The Palmgren–Miner hypothesis asserts that the damage fraction at any stress level S_i is linearly proportional to the ratio of number of cycles of operation to the total number of cycles that would produce failure at that stress level; that is

$$D_i = \frac{n_i}{N_i} \qquad (19.72)$$

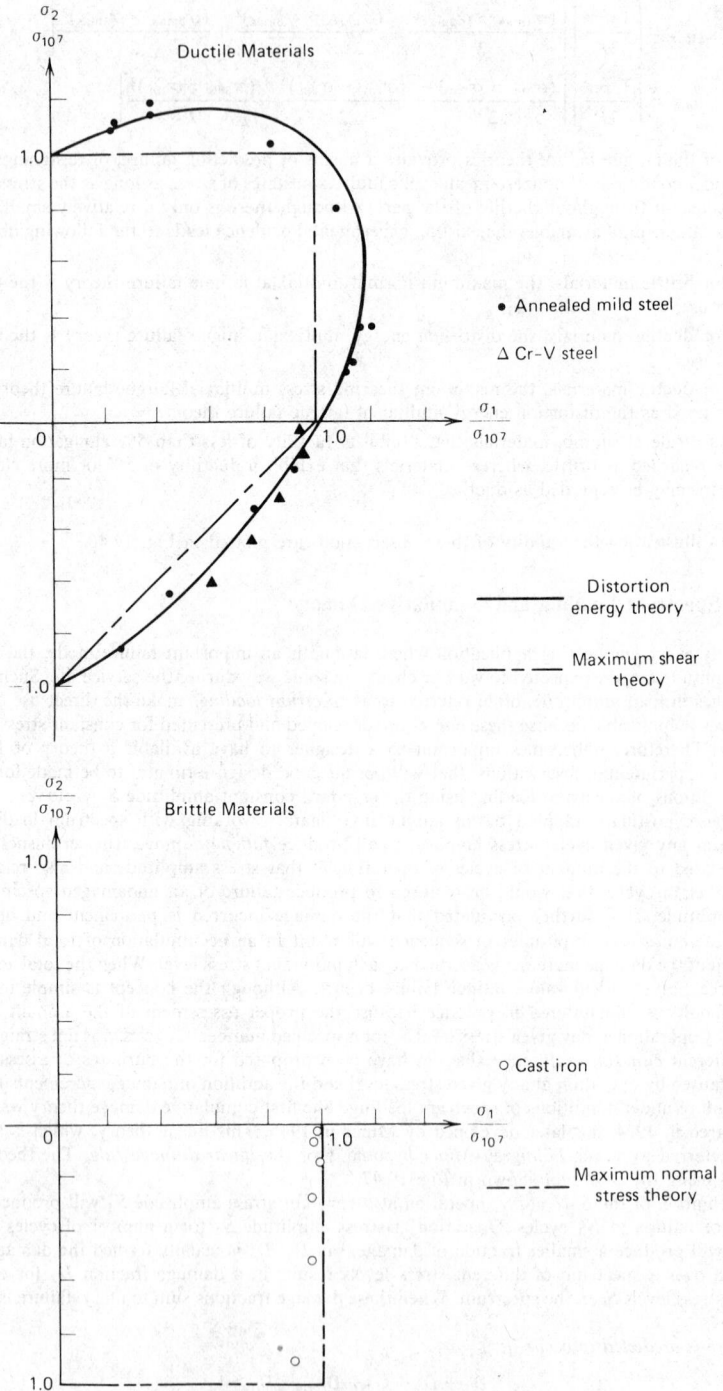

Fig. 19.46 Comparison of biaxial fatigue strength data with multiaxial fatigue failure theories for ductile and brittle materials. (See Chap. 7 of Ref. 30.)

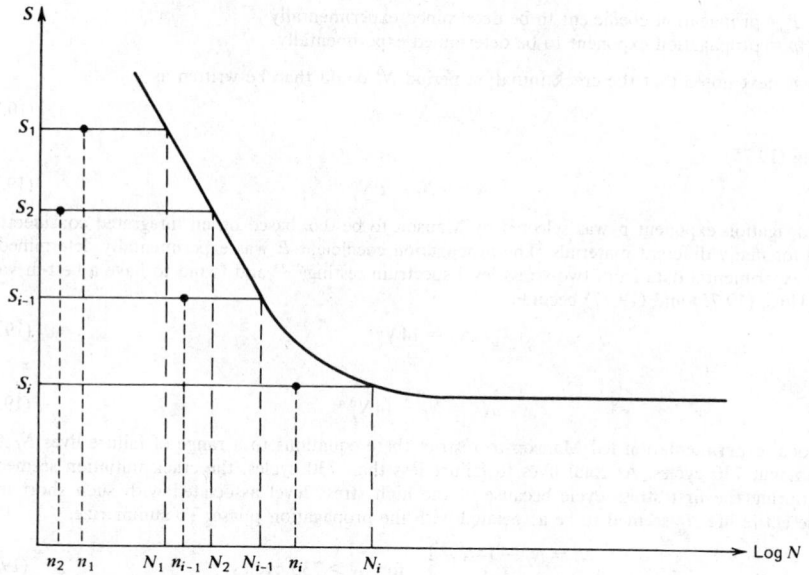

Fig. 19.47   Illustration of spectrum loading where n_i cycles of operation are accrued at each of the different corresponding stress levels S_i, and the N_i are cycles to failure at each S_i.

By the Palmgren–Miner hypothesis, then, utilizing (19.72), we may write (19.71) as
   Failure is predicted to occur if:

$$\frac{n_1}{N_1} + \frac{n_2}{N_2} + \cdots + \frac{n_{i-1}}{N_{i-1}} + \frac{n_i}{N_i} \geq 1 \qquad (19.73)$$

or
   Failure is predicted to occur if:

$$\sum_{j=1}^{i} \frac{n_j}{N_j} \geq 1 \qquad (19.74)$$

This is a complete statement of the Palmgren–Miner hypothesis or the linear damage rule. It has one important virtue, namely, *simplicity;* and for this reason it is widely used. It must be recognized, however, that in its simplicity certain significant influences are unaccounted for, and failure prediction errors may therefore be expected. Perhaps the most significant shortcomings of the linear theory are that no influence of the order of application of various stress levels is recognized, and damage is assumed to accumulate at the same rate at a given stress level without regard to past history. Experimental data indicate that the order in which various stress levels are applied does have a significant influence and also that damage rate at a given stress level is a function of prior cyclic stress history. Experimental values for the Miner's sum at the time of failure often range from about ¼ to about 4, depending on the type of decreasing or increasing cyclic stress amplitudes used. If the various cyclic stress amplitudes are mixed in the sequence in a quasi-random way, the experimental Miner's sum more nearly approaches unity at the time of failure, with values of Miner's sums corresponding to failure in the range of about 0.6 to 1.6. Since many service applications involve quasi-random fluctuating stresses, the use of the Palmgren–Miner linear damage rule is often satisfactory for failure protection.

Of the many attempts (see, for example, Ref. 1, Chap. 7) to develop more accurate cumulative damage predictions through the use of nonlinear models, only the Manson bilinear damage rule will be given here. The Manson rule was based on a proposal by Grover[31] that cumulative damage estimates might be improved by breaking the fatigue process down into a crack initiation phase and a crack propagation phase and applying a linear damage rule to each phase separately. No basis was suggested for quantitatively defining the ranges for these two phases, however, until Manson et al.[32,33] presented an empirical technique for establishing the ranges and the damage equations for these two phases. Manson suggested that the crack propagation period could be expressed as

$$N_p = PN_f^p \qquad (19.75)$$

where N_p = number of cycles to propagate a crack to failure after it has been initiated
      N_f = total number of cycles to failure

P = propagation coefficient to be determined experimentally
p = propagation exponent to be determined experimentally

It was next noted that the crack initiation period N' could then be written as

$$N' = N_f - N_p \tag{19.76}$$

or, using (19.75)

$$N' = N_f - PN_f^p \tag{19.77}$$

The propagation exponent p was selected by Manson to be 0.6, based on an integrated consideration of data for many different materials. The propagation coefficient P was experimentally determined to best fit experimental data from two-stress-level spectrum testing[32,33] and found to have a best-fit value of 14. Thus, (19.75) and (19.77) become

$$N_p = 14N_f^{0.6} \tag{19.78}$$

and

$$N' = N_f - 14N_f^{0.6} \tag{19.79}$$

Additional experimentation led Manson to restrict these equations to a range of failure lives N_f that exceed about 730 cycles. At total lives to failure less than 730 cycles, the crack initiation seemed to occur during the first stress cycle because of the high stress level associated with such short lives, and the entire life N_f seemed to be associated with the propagation phase. To summarize,

$$\left. \begin{array}{l} N' = N_f - 14N_f^{0.6} \\ N_p = 14N_f^{0.6} \end{array} \right\} \quad \text{for } N_f > 730 \text{ cycles} \tag{19.80}$$

and

$$\left. \begin{array}{l} N' = 0 \\ N_p = N_f \end{array} \right\} \quad \text{for } N_f \leq 730 \text{ cycles} \tag{19.81}$$

where N' = number of cycles to initiate a crack
N_p = number of cycles to propagate a crack from the initiation stage to failure
N_f = total number of cycles to failure

Utilizing these empirical expressions, we apply a linear damage rule to each phase individually to yield the prediction of initiation and failures as follows:
Fatigue nuclei of critical size are initiated when

$$\sum_{i=1}^{m} \frac{n_i}{N_i'} = 1 \tag{19.82}$$

and *fatigue cracks are propagated to failure if cracks of critical size have been initiated and then*

$$\sum_{j=1}^{q} \frac{n_j}{(N_p)_j} = 1 \tag{19.83}$$

where in each case n is the number of cycles applied at the ith or jth stress level.

Thus, (19.82) must be applied until a critical crack is initiated, and then (19.83) is used until failure is predicted. Use of this double linear damage rule appears to give relatively good agreement with two-stress-level tests for several materials.[33] Figure 19.48 shows the experimental results for a variety of two-stress-level tests on SAE 4130 steel compared to predictions made by Manson's double linear damage rule and also the Palmgren–Miner hypothesis predictions.

19.5.6 Stress Concentration

Failures in machines and structures almost always initiate at sites of local stress concentration caused by geometrical or microstructural discontinuities. These *stress concentrations,* or *stress raisers,* often lead to local stresses many times higher than the nominal net section stress that would be calculated without considering stress concentration effects. An intuitive appreciation of the stress concentration associated with a geometrical discontinuity may be developed by thinking in terms of "force flow" through a member as it is subjected to external loads. The sketches of Fig. 19.49 illustrate the concept. The rectangular flat plate of width w and thickness t is fixed at the lower edge and subjected to a total force F uniformly distributed along the upper edge. The dashed lines each represent a fixed quantum of force, and the local spacing between lines is therefore an indication of the local force intensity, or stress. In Fig. 19.49a the lines are uniformly spaced throughout the plate, and the stress σ is uniform and calculable as

$$\sigma = \frac{F}{wt} \qquad (19.84)$$

In the sketch of Fig. 19.49b a flat rectangular plate of the same thickness has been subjected to the same total force F, but the plate has been made wider and notched to provide the same net section width w at the site of the notch. The lines of force flow may be visualized in very much the

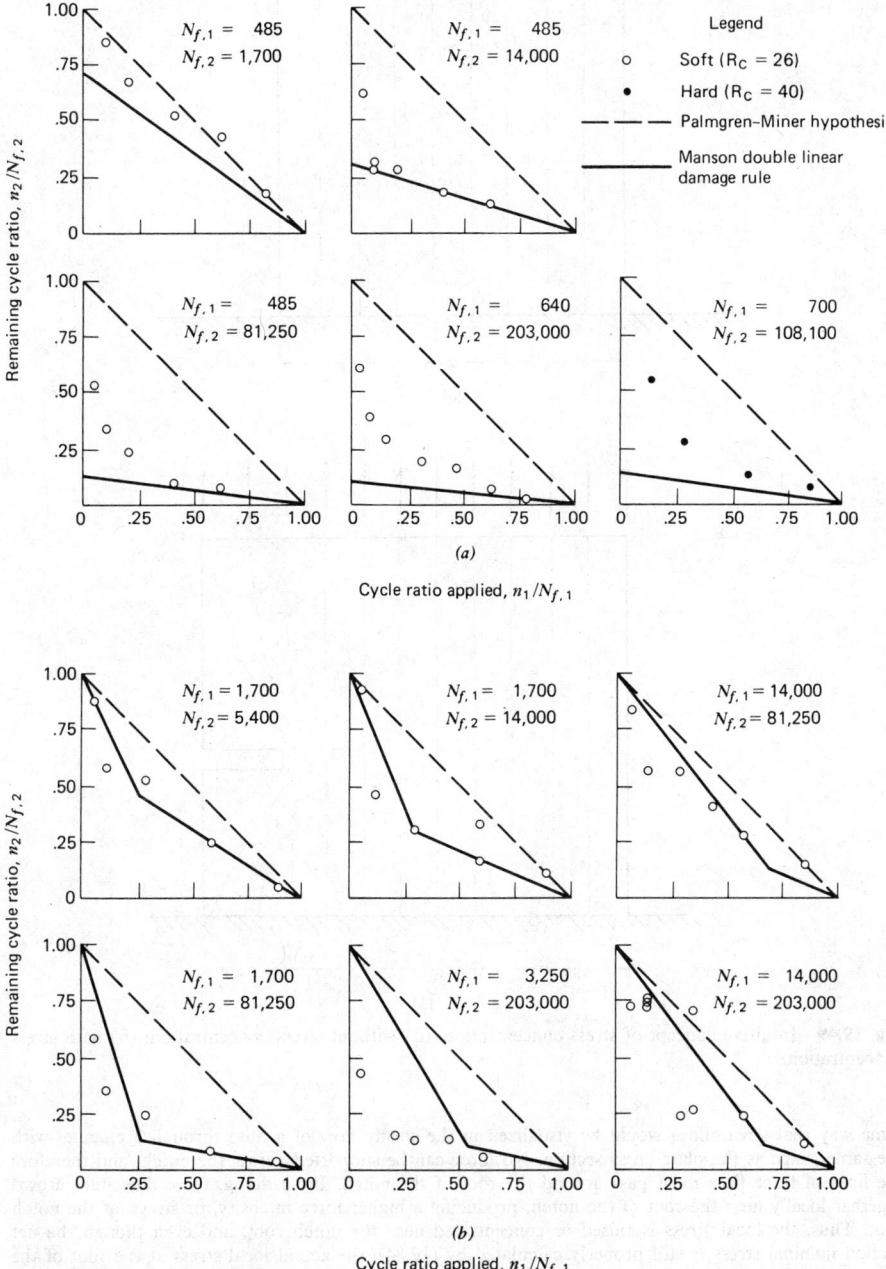

Fig. 19.48 Comparison of predicted fatigue behavior by Manson's double linear damage rule and Palmgren–Miner linear damage rule with experimental data for two-stress-level tests on SAE 4130 steel. R. R. Moore rotating–bending: (a) high to low stress with low initial life; (b) high to low stress with relatively high initial life. (After Ref. 33, copyright ASTM; adapted with permission)

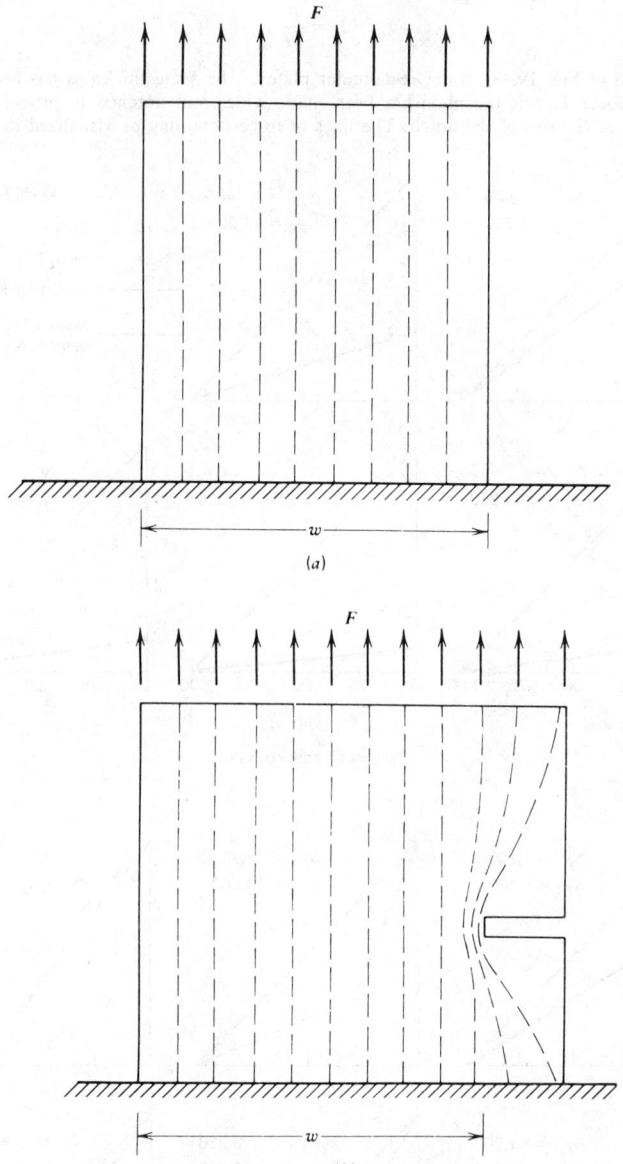

Fig. 19.49 Intuitive concept of stress concentration: (*a*) without stress concentration; (*b*) with stress concentration.

same way that streamlines would be visualized in the steady flow of a fluid through a channel with the same shape as the plate cross section. No force can be supported across the notch, and therefore the lines of force flow must pass around the root of the notch. In so doing, force flow lines crowd together locally near the root of the notch, producing a higher force intensity, or stress, at the notch root. Thus, the local stress is raised or concentrated near the notch root, and even though the net section nominal stress is still properly calculated by (19.84), the actual local stress at the root of the notch may be many times higher than the calculated nominal stress. Many common examples of stress concentration may be cited, some of which are illustrated in Fig. 19.50. Discontinuities at the roots of gear teeth, at the corners of keyways in shafting, at the roots of screw threads, at the fillets of shaft shoulders, around rivet holes and bolt holes, and in the neighborhood of welded joints all constitute stress raisers that usually must be considered by a designer. The seriousness of the stress

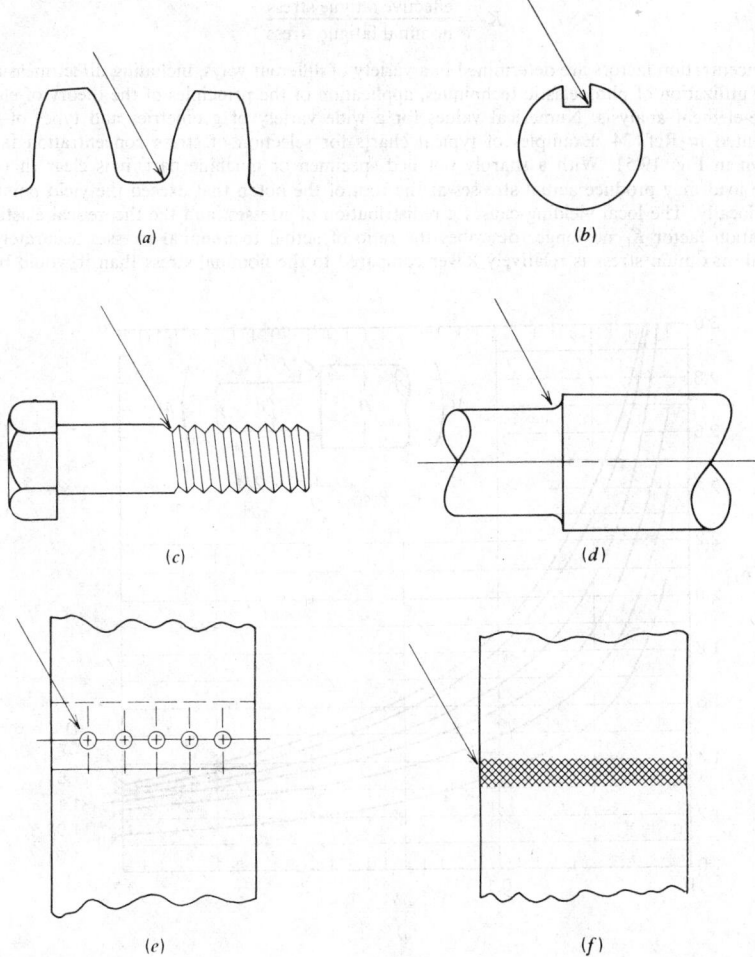

Fig. 19.50 Some common examples of stress concentration: (a) gear teeth; (b) shaft keyway; (c) bolt threads; (d) shaft shoulder; (e) riveted or bolted joint; (f) welded joint.

concentration depends on the type of loading, the type of material, and the size and shape of the discontinuity.

Stress raisers may be classified as being either *highly local* or *widely distributed*. Highly local stress raisers are those for which the volume of material containing the concentration of stress is negligibly small compared to the overall volume of the stressed member. Widely distributed stress raisers are those for which the volume of material containing the concentration of stress is a significant portion of the overall volume of the stressed member.

The theoretical elastic stress concentration factor, K_t, is defined to be the ratio of the actual maximum local stress in the region of the discontinuity to the nominal net section stress calculated by simple theory as if the discontinuity exerted no stress concentration effect; that is,

$$K_t = \frac{\text{actual maximum stress}}{\text{nominal stress}} \qquad (19.85)$$

and the magnitude of K_t is found to be a function of geometry and type of loading, but not a function of material. It should be noted that this definition of K_t is valid only for stress levels within the elastic range, and must be suitably modified if stresses are in the plastic range.

The fatigue stress concentration factor, K_f, is defined to be the ratio of the effective fatigue stress at the root of the discontinuity to the nominal fatigue stress calculated as if the notch has no stress concentrating effect. This definition is made for the high-cycle fatigue range and must be suitably modified for the low-cycle fatigue range. Thus, the fatigue stress concentration factor may be defined as

$$K_f = \frac{\text{effective fatigue stress}}{\text{nominal fatigue stress}} \qquad (19.86)$$

Stress concentration factors are determined in a variety of different ways, including direct measurement of strain, utilization of photoelastic techniques, application of the principles of the theory of elasticity, and finite-element analysis. Numerical values for a wide variety of geometries and types of loading are presented in Ref. 34. Examples of typical charts for selection of stress concentration factor K_t are shown in Fig. 19.51. With a sharply notched specimen or machine part, it is clear that even a moderate load may produce actual stresses at the root of the notch that exceed the yield point of the material locally. The local yielding causes a redistribution of stresses, and the theoretical elastic stress concentration factor K_t no longer describes the ratio of actual to nominal stresses accurately, since the actual maximum stress is relatively lower compared to the nominal stress than it would be if the

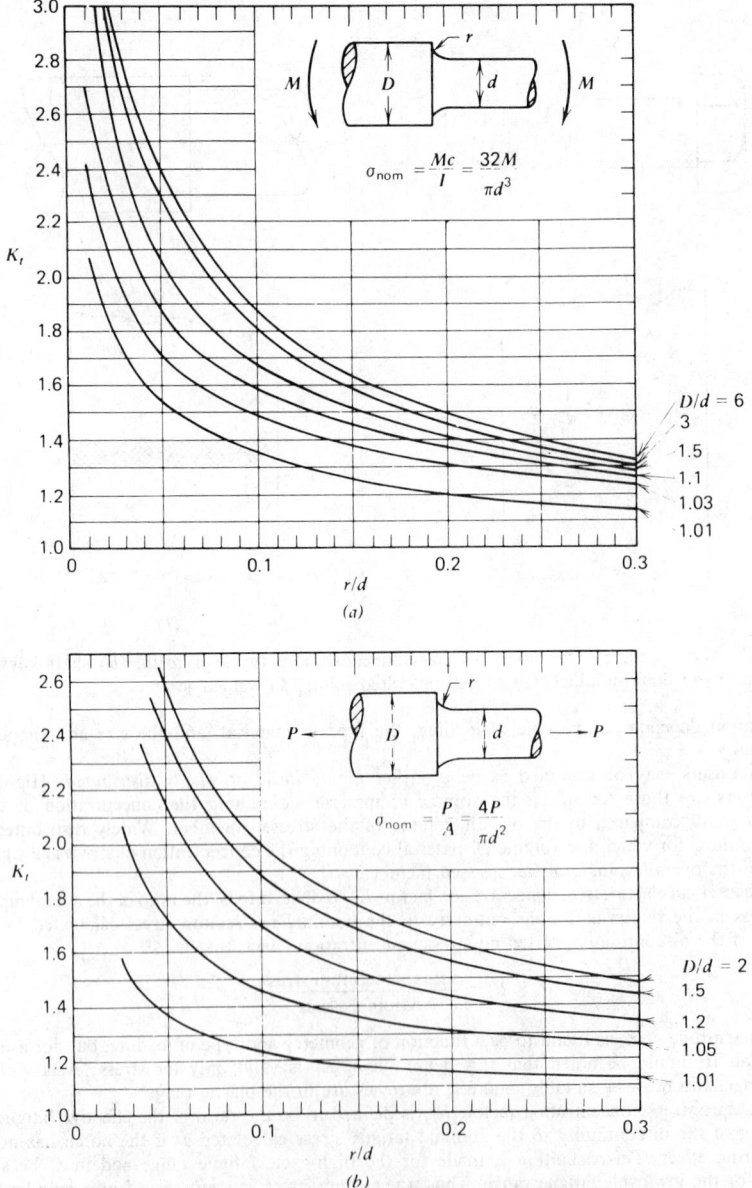

Fig. 19.51 Stress concentration factors for a shaft with a fillet subjected to (a) bending, (b) axial load, or (c) torsion. (From Ref. 34, adapted with permission from John Wiley & Sons, Inc.)

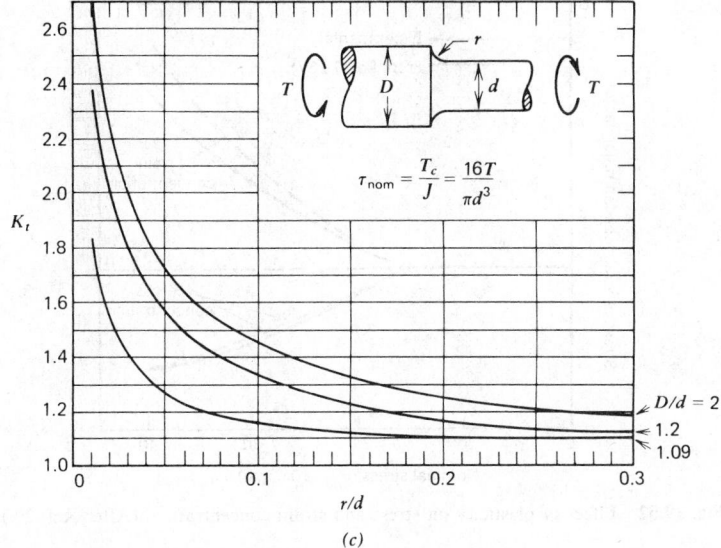

$$\tau_{\text{nom}} = \frac{T_c}{J} = \frac{16T}{\pi d^3}$$

(c)

Fig. 19.51 (*Continued*)

material remained elastic. That is, the stress concentration factor is diminished in magnitude by local plastic flow, whereas the local strain is made larger than would be predicted by elastic theory.

Mathematical solutions of elastic–plastic stress and strain distributions around notches are relatively difficult to obtain, even using numerical solutions and digital computer techniques. One of the more successful approximations for stress concentration due to a circular hole in a very wide plate under tension has been given as[35]

$$K = 1 + 2\frac{E_s}{E} \tag{19.87}$$

where E = Young's modulus
E_s = secant modulus
K = stress concentration factor

Figure 19.52 illustrates values of stress concentration factor and strain concentration factor computed for a specific case by (19.87) and compared with values measured using very small strain gages. The agreement between calculated and measured values is good. Unlike the theoretical stress concentration factor K_t, the fatigue stress concentration factor K_f is a *function of the material* as well as geometry and type of loading. To account for the influence of material characteristics, a *notch sensitivity index* q has been defined to relate the actual effect of a notch on fatigue strength of a material to the effect that might be predicted solely on the basis of elastic theory. The definition of notch sensitivity index q is given by

$$q = \frac{K_f - 1}{K_t - 1} \tag{19.88}$$

where K_f = fatigue stress concentration factor
K_t = theoretical stress concentration factor
q = notch sensitivity index valid for high-cycle fatigue range

The reason for subtracting unity from the numerator and denominator in this definition is to provide a scale for q that ranges from zero for no notch effect to unity for full notch effect. That is, for full notch effect K_f is equal to K_t. The notch sensitivity index has been found to be a function of both material and notch radius. Scatter in the experimental data is a serious problem in evaluating notch sensitivity index, as may be seen in the experimental results shown in Fig. 19.53. The notch sensitivity index for a range of steels and an aluminum alloy are shown in Fig. 19.54 for axial, bending, and torsional loading. These curves provide sufficient accuracy for most design applications and clearly

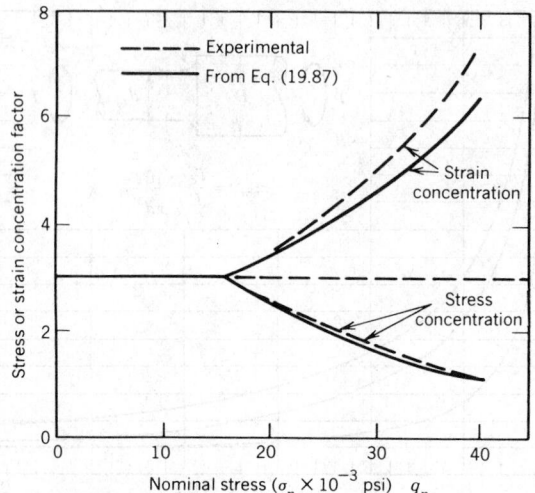

Fig. 19.52 Effect of plasticity on stress and strain concentration. (After Ref. 29.)

demonstrate that the notch sensitivity index is a function of both the material and the notch root radius. An expression for fatigue stress concentration factor may be written from (19.88) as

$$K_f = q(K_t - 1) + 1 \qquad (19.89)$$

where the theoretical elastic stress concentration factor K_t may be determined on the basis of geometry and loading from handbook charts such as those depicted in Fig. 19.51. The notch sensitivity index q may be read from charts, such as the one shown in Fig. 19.54.

For uniaxial states of cyclic stress it is sometimes convenient to use K_f as a "strength reduction factor" rather than as a "stress concentration factor." That is, for uniaxial stressing only, a designer may choose to divide the fatigue limit by K_f rather than multiplying the applied nominal cyclic stress by K_f. Although, conceptually, it is clearly more correct to think of K_f as a stress concentration factor, computationally, it is equivalent, and often simpler, to use K_f as a strength reduction factor. For multiaxial states of stress, however, K_f must be used as a stress concentration factor, since an appropriate value of K_f used as a strength reduction factor would be undefined.

The fatigue stress concentration factor (or strength reduction factor) determined from (19.89) is strictly applicable only in the high-cycle fatigue range, that is, for cycle lives of 10^5–10^6 cycles and larger. For ductile materials and static loads, effects of stress concentration may usually be neglected. Thus, in the intermediate- and low-cycle life range from a quarter cycle (static load) up to about 10^5–10^6 cycles, the stress concentration factor changes from unity to K_f. As shown in Fig. 19.55, the notched and unnotched S–N curves converge as they approach the low-cycle end of the range and coincide at the quarter cycle point A. Many materials exhibit fatigue stress concentration factors very near unity for lives less than 1000 cycles. Estimates of fatigue stress concentration factor are often made by constructing a straight line on a semilogarithmic S–N plot from the ultimate strength at a life of 1 cycle to the unnotched fatigue strength divided by K_f at a life of 10^6 cycles. Such a straight line construction is shown in Fig. 19.55. The ratio of unnotched to notched fatigue strength values read at a specified life then becomes the estimate of fatigue stress concentration factor to be used for that life.

Finally, it should be noted that experimental investigations have indicated that for fatigue of *ductile* materials the fatigue stress concentration factor should be applied *only* to the *alternating component* of stress and not to the steady component of stress that exists in any nonzero mean cyclic stress. For fatigue loading of *brittle* materials, the stress concentration factor should be applied to the steady component as well.

19.5.7 Low-Cycle Fatigue

Two domains of cyclic loading were identified in the introduction to this section. One domain is that for which the cyclic loads are relatively low, strain cycles are confined largely to the elastic range, and long lives or high numbers of cycles to failure are exhibited. This behavior has traditionally been called high-cycle fatigue. The other domain is that for which the cyclic loads are relatively high, significant amounts of plastic strain are induced during each cycle, and short lives or low numbers

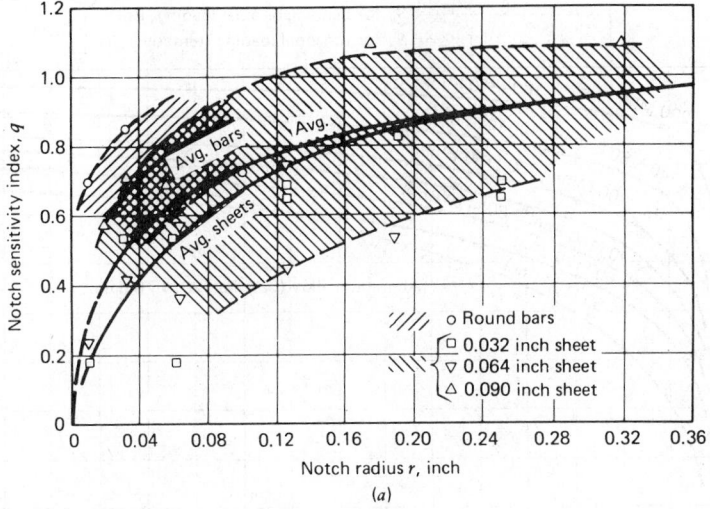

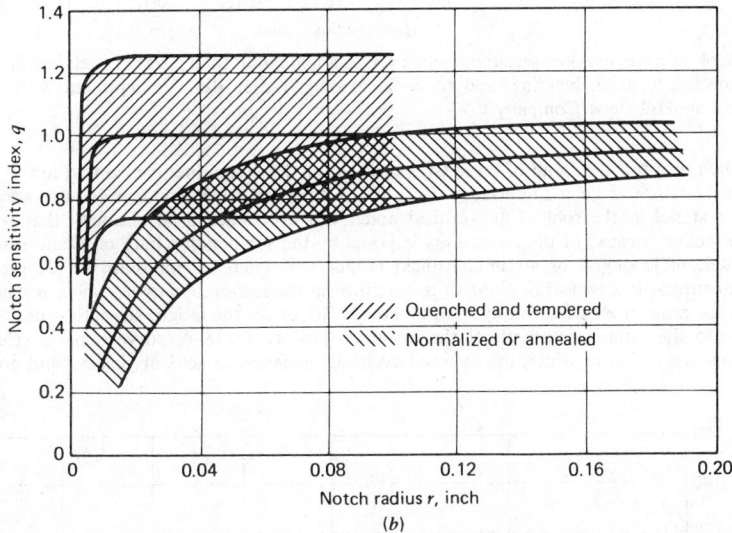

Fig. 19.53 An indication of scatter in the experimental determination of notch sensitivity index for alloys of aluminum and steel. (*a*) 24ST aluminum under completely reversed axial loading. (*b*) Steel alloys under alternating bending. (After Ref. 30; reprinted with permission from McGraw-Hill Book Company.)

of cycles to failure are exhibited if these relatively high loads are repeatedly applied. This type of behavior has been commonly called *low-cycle fatigue* or, more recently, cyclic strain-controlled fatigue. The transition from low-cycle-fatigue behavior to high-cycle-fatigue behavior generally occurs in the range from about 10^4 to 10^5 cycles, and some investigators define the low-cycle-fatigue range to be failure in 50,000 cycles or less.[37] Although the usual objective of an engineering designer is to provide long life, there are several circumstances in which the low-cycle fatigue or strain-controlled life response is of great importance. For example, in the design of high-performance devices such as missiles and rockets, the total design lifetime may be only a few hundred or a few thousand cycles from launch to delivery, and low-cycle-fatigue analysis and design methods are of direct interest. In the design of other high-performance devices, such as aircraft gas-turbine blades and wheels, nuclear pressure vessels and fuel elements, or steam turbine rotors and shells, the occurrence of occasional large mechanical or thermal transients during operation may give rise to significant damage accumulation due to a

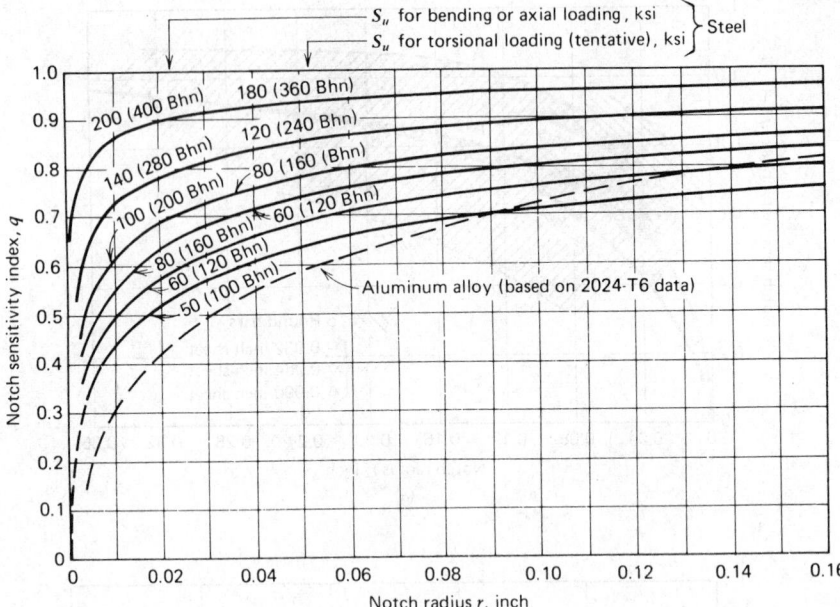

Figure 19.54 Curves of notch sensitivity index versus notch radius for a range of steels and an aluminum alloy subjected to axial, bending, and torsional loading. (After Ref. 36; reprinted with permission from McGraw-Hill Book Company.)

few hundred or a few thousand of these large cycles over the design lifetime, so that low-cycle-fatigue design methods are of great importance. Even if the loads on a machine or structure are nominally low, the material at the root of any critical notch will experience local plasticity that is cyclically strain controlled because of the constraints imposed by the surrounding bulk of elastic material, and the methods of low-cycle or strain-controlled fatigue will again be important in life prediction of such components. If a typical *S–N* curve is examined in the low-cycle-fatigue region, it will be found that over the range from a quarter cycle up to around 10^3 cycles the fatigue strength is nearly constant and close to the ultimate strength of the material. That is, the *S–N* curve remains relatively flat throughout this region in which the material cyclically experiences general yielding and gross plastic

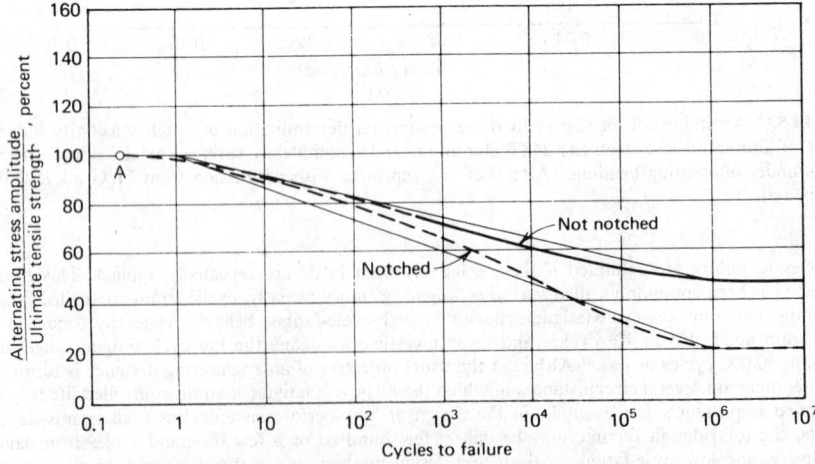

Fig. 19.55 *S–N* curves for notched and unnotched specimens subjected to completely reversed axial loading. (After Ref. 18, *Fatigue and Fracture of Metals,* by W. M. Murray, by permission of The MIT Press, Cambridge, Massachusetts, Copyright 1952.)

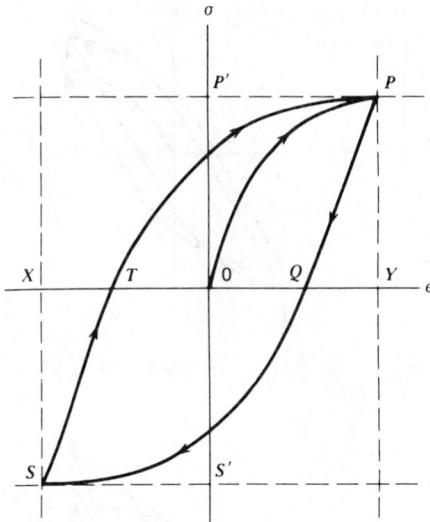

Fig. 19.56 Hysteresis loop associated with cyclic loading that produces low-cycle-fatigue damage.

deformation. In this region of macroscopic plastic behavior the fatigue life is much more accurately described as a function of the cyclic *strain amplitude* rather than the cyclic stress amplitude. Stress–strain behavior under these circumstances is characterized by a stress–strain hysteresis loop, as shown in Fig. 19.56, with evidence of a measurable plastic strain in the specimen or machine part. This plastic behavior is typically nonlinear and history dependent, and it has been observed that the stress–strain response of most materials changes significantly with cyclic straining into the plastic range. Some materials exhibit cyclic strain hardening and others exhibit cyclic strain softening, as illustrated in Fig. 19.57. The stress–strain response of most materials changes significantly with applied cyclic strains early in life, but typically the hysteresis loops tend to stabilize so that the stress amplitude remains reasonably constant under strain control over the remaining large portion of the fatigue life. Based on the stable hysteresis loops for a family of different constant strain amplitudes, a curve passed through the tips of these hysteresis loops, as shown in Fig. 19.58, defines a "cyclic" stress–strain curve for the material. In Fig. 19.59 cyclic stress–strain curves are compared with static or monotonic stress–strain curves for several different materials.

The usual method of displaying the results of low-cycle fatigue tests is to plot the logarithm of strain amplitude or strain range versus the logarithm of number of cycles (or reversals) to failure. Sometimes the plastic strain amplitude or strain range is plotted, and sometimes the total strain amplitude or strain range is plotted as the ordinate. Early experimental investigations had indicated that if the plastic strain amplitude were plotted versus cycles to failure on a log–log plot, the data would approximate a straight line with a slope of about −0.5. Subsequent investigations have indicated that the slope ranges from about −0.7 to −0.5. Such plots seem to be remarkably similar for a wide range of materials,[40] as indicated in Fig. 19.60. Experimental evidence accumulated by various investigators in recent years seems to indicate that the cyclic life is better related to total strain than to plastic strain, especially at the longer life end of the low-cyclic range. An example of a plot of strain amplitude versus life is shown in Fig. 19.61, separately showing the plastic strain amplitude and the total strain amplitude for a nickel–steel alloy.

Data such as those shown in Fig. 19.60 led to the proposal of an empirical equation relating plastic strain range $\Delta \epsilon_p$ to failure life N_f for completely reversed strain cycling under uniaxial stress conditions in the low-cycle-fatigue regime. The relationship, independently proposed by Manson[41] and Coffin,[42,43] may be expressed as

$$\frac{\Delta \epsilon_p}{2} = \epsilon_f'(2N_f)^c \tag{19.90}$$

where $\Delta \epsilon_p / 2$ = plastic strain amplitude
ϵ_f' = fatigue ductility coefficient, defined as strain intercept at one load reversal, that is, at $2N_f = 1$ (see Fig. 19.62)
$2N_f$ = total reversals to failure
c = fatigue ductility exponent defined as slope of plastic strain amplitude versus reversals to failure curve on log–log plot (see Fig. 19.62)

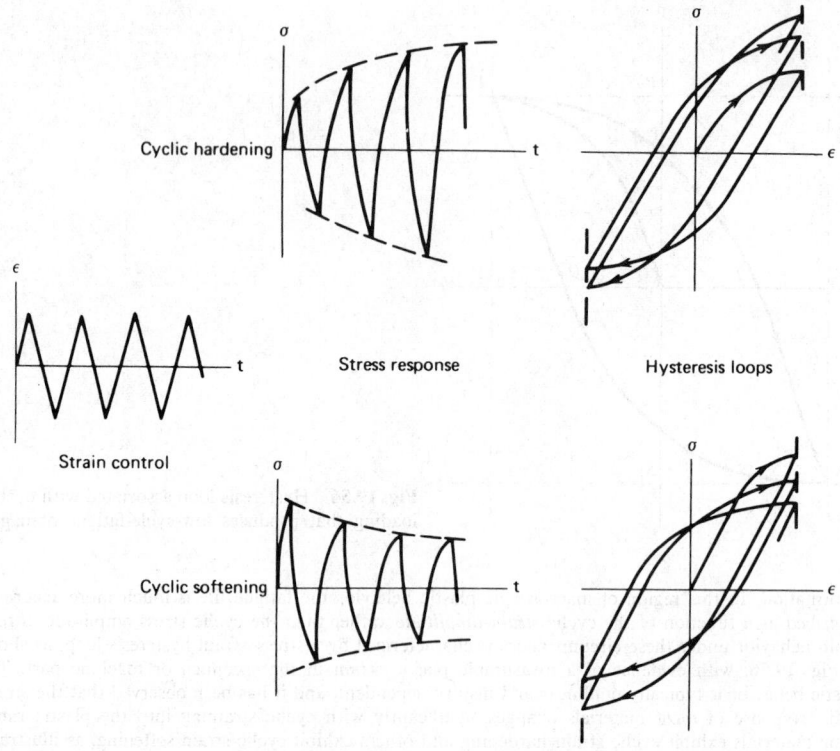

Fig. 19.57 Illustration of cyclic strain hardening and cyclic strain softening phenomena under strain control. (From Ref. 38, copyright ASTM; reprinted with permission.)

Later work by many investigators, capitalizing on the Manson–Coffin equation of (19.62), has indicated that total strain amplitude, the sum of elastic strain amplitude plus plastic strain amplitude, may be better correlated to life. As shown in Fig. 19.62, the total strain amplitude is the sum of elastic plus plastic components. This has been modeled mathematically by Morrow et al.[44] as

$$\frac{\Delta\epsilon}{2} = \frac{\sigma_f'}{E}(2N_f)^b + \epsilon_f'(2N_f)^c \qquad (19.91)$$

where constants b and σ_f'/E are the slope and one-reversal intercept of the elastic curve in Fig. 19.62, and constants c and ϵ_f' are the slope and one-reversal intercept of the plastic curve in Fig. 19.62. Typically,[38] b ranges from about -0.05 to -0.15 and c ranges from about -0.5 to -0.8. Although these constants are best evaluated from cyclic testing, they may be approximated from static properties, if fatigue data are unavailable. This may be done by taking σ_f' equal to true fracture strength σ_f, ϵ_f' equal to true fracture ductility ϵ_f, c equal to -0.6, and b equal to $-0.16 \log 2\sigma_f/\sigma_u$. However, actual fatigue data should be used where available.

From the schematic representation of (19.63) in Fig. 19.62, it may be noted that at short lives the plastic strain amplitude component dominates, whereas at longer lives the elastic strain amplitude component dominates. The point at which the elastic and plastic curves intersect has been called the "transition life." A thoughtful consideration of (19.91) and Fig. 19.62 leads to the observation that for design lives less than the transition life, materials with high fatigue ductility (high fracture ductility) are superior, whereas design lives greater than the transition life demand materials with high values of true fracture strength. This contradictory set of requirements, namely, high strength with high ductility, requires careful consideration on the part of the designer to match the appropriate material to the application, with careful attention to the magnitude of operating strain amplitude. This point is emphasized in Fig. 19.63 where three idealized materials are shown, one very high strength material (strong), one very ductile material (ductile), and one whose properties lie between the two extremes (tough). These curves cross at about 10^3 cycles (2×10^3 reversals) at a total cyclic strain amplitude of about 0.01. Thus, one would select the "strong" material for design life requirements greater than about 10^3 cycles, pick the "ductile" material for design life requirements shorter than about 10^3 cycles, and "optimize" with the "tough" material for spectrum loading of a more complicated nature. It is

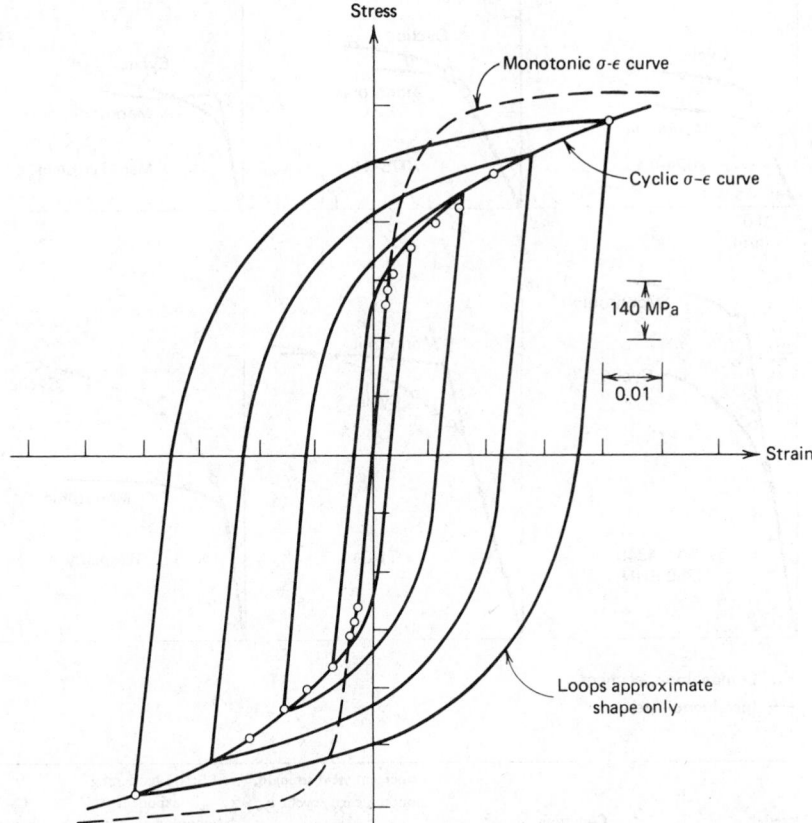

Fig. 19.58 Cyclic stress–strain curve compared to monotonic stress–strain curve for SAE 4340 steel. (From Ref. 38, copyright ASTM; reprinted with permission.)

interesting to note that all types of materials seem to have about the same fatigue resistance for total strain amplitude around 0.01, corresponding to a failure life of about 10^3 cycles. Presently attainable combinations of true fracture strength and true fracture ductility are illustrated in Fig. 19.64.

The effects of nonzero mean *strain* under low-cycle fatigue conditions have been studied by relatively few investigators, principally for the case of tensile mean strain. The experimental results of these few investigations[45-48] indicate that the effect of a compressive mean strain on low-cycle fatigue life is essentially the same as the effect of a tensile mean strain if their magnitudes are the same. These results also indicate that mean strain effects are of primary importance only in the operating range where the *plastic* strain component dominates, that is, at design lives less than the transition life for the material.

The effects of nonzero mean *stress* are of primary importance only in the operating range where the *elastic* strain component dominates, that is, at design lives greater than the transition life of the material.

Just as in the case of high-cycle fatigue, the assessment of low-cycle-fatigue damage under conditions where the cyclic strain amplitude ranges over a spectrum of values requires the use of a cumulative damage theory. Many investigators have studied cumulative damage effects in low-cycle fatigue with the general conclusion that a linear damage rule of the Palmgren–Miner type yields acceptable results if local stress–strain behavior can be accurately determined as a function of applied loads and if cycle counting is properly conducted.

If, for example, the fatigue-modified Neuber rule described in Section 19.5.8 is utilized to determine local stress and strain amplitude history and if the rain-flow cycle counting method, also described in Section 19.5.8, is properly applied to the local strain–time spectrum, results of using the Palmgren–Miner linear damage rule of (19.74) have been found to give acceptable life predictions. The range of values for $\Sigma(n/N)$ corresponding to failure under a variety of spectrum loading conditions has been reported to be from about 0.6 to about 1.6 for a variety of materials.[45] This is a narrower range for

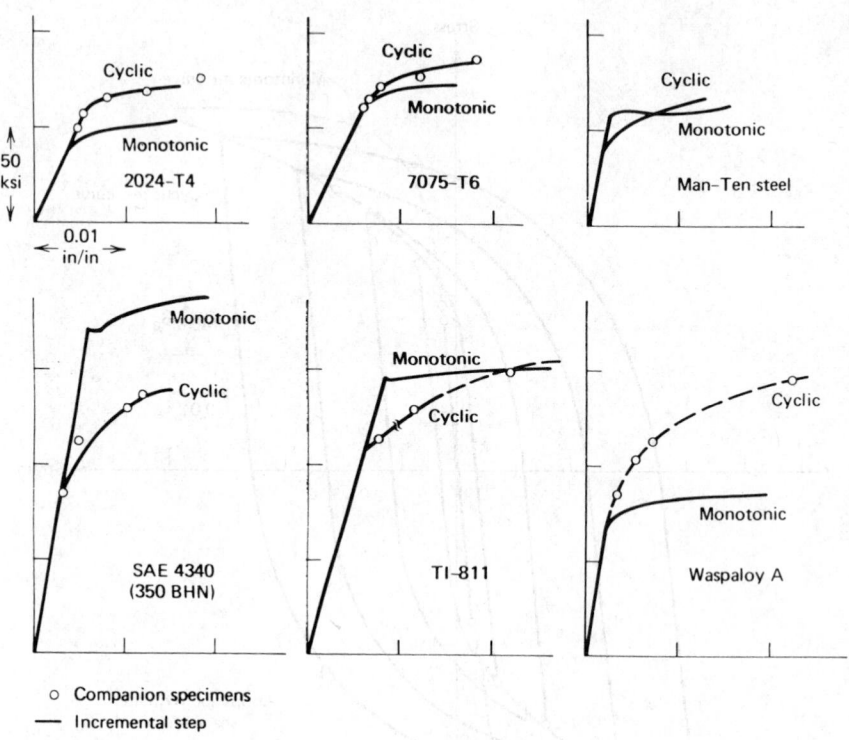

○ Companion specimens

— Incremental step

Material	Condition	0.2 percent yield strength, monotonic σ_{yp}/cyclic σ_{yp} ksi	Strain-hardening exponent n(monotonic)/n'(cyclic)	Cyclic behavior
OFHC copper	Annealed	3/20	0.40/0.15	Hardens
	Partial annealed	37/29	0.13/0.16	Stable
	Cold worked	50/34	0.10/0.12	Softens
2024 aluminum alloy	T4	44/65	0.20/0.11	Hardens
7075 aluminum alloy	T6	68/75	0.11/0.11	Hardens
Man-Ten steel	As-received	55/50	0.15/0.16	Softens and hardens
SAE 4340 steel	Quenched and tempered, 350 BHN	170/110	0.066/0.14	Softens
Ti-8Al-1Mo-1V	Duplex annealed	145/115	0.078/0.14	Softens and hardens
Waspaloy		79/102	0.11/0.17	Hardens
SAE 1045 steel	Quenched and tempered, 595 BHN	270/250	0.071/0.14	Stable
	Quenched and tempered, 500 BHN	245/185	0.047/0.12	Softens
	Quenched and tempered, 450 BHN	220/140	0.041/0.15	Softens
	Quenched and tempered, 390 BHN	185/110	0.044/0.17	Softens
SAE 4142 steel	As-quenched, 670 BHN	235/...	0.14/...	Hardens
	Quenched and tempered, 560 BHN	245/250	0.092/0.13	Stable
	Quenched and tempered, 475 BHN	250/195	0.048/0.12	Softens
	Quenched and tempered, 450 BHN	230/155	0.040/0.17	Softens
	Quenched and tempered, 380 BHN	200/120	0.051/0.18	Softens

Fig. 19.59 Cyclic stress–strain behavior of several materials. (From Ref. 39, copyright ASTM; reprinted with permission.)

low-cycle fatigue than has been observed for high-cycle fatigue. It should be noted, however, that for multiaxial states of stress, the linear damage rule may be much less reliable.[49]

A detailed treatment of the effects of multiaxial states of stress on low-cycle-fatigue behavior is beyond the scope of this discussion, but proposed techniques have been presented for estimating low-cycle-fatigue life under these conditions. (See Refs. 45, 50, and p. 165ff of 51.) The proposed technique involves the definition of an *equivalent stress* and an *equivalent total strain range*, both calculable from the multiaxial states of stress and strain. Life estimates based on the equivalent total strain

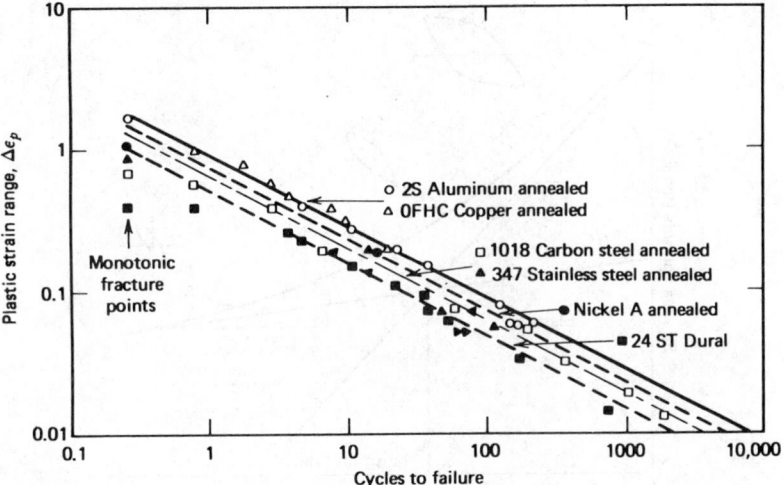

Fig. 19.60 Low-cycle-fatigue plot for several different materials. (From Ref. 40, copyright American Society for Metals, 1959; reprinted with permission.)

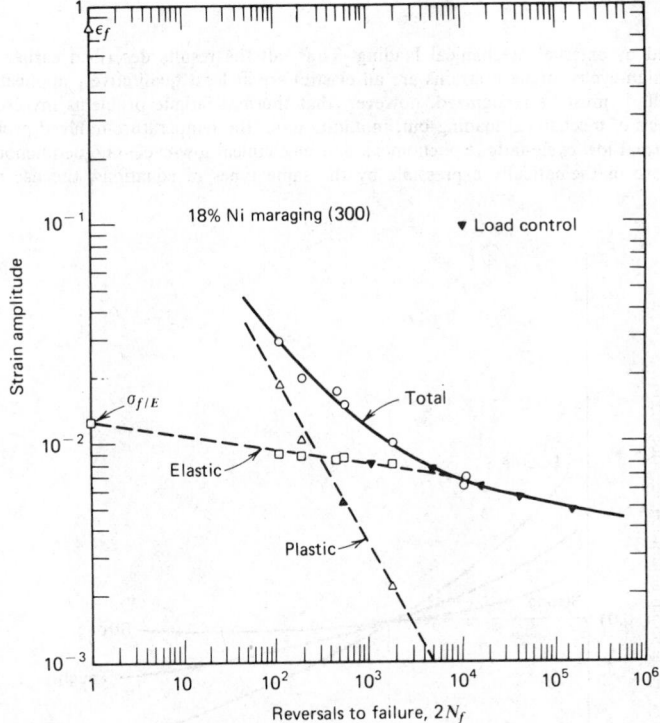

Fig. 19.61 Strain amplitude versus life for 18% Ni maraging steel, separately showing elastic, plastic, and total components of the strain. (From Ref. 38, copyright ASTM; reprinted with permission.)

range under conditions of a multiaxial state of stress can then be made from uniaxial state-of-stress low-cycle-fatigue data expressed as total strain range versus cycles to failure.

Although strains are often produced mechanically, it is perhaps even more usual to find the cyclic strains produced by a cyclic thermal field. If a machine part undergoes cyclic temperature changes and if the natural thermal expansions and contractions are either wholly or partially constrained, cyclic strains and stresses result. These cyclic strains produce fatigue failure just as if the strains

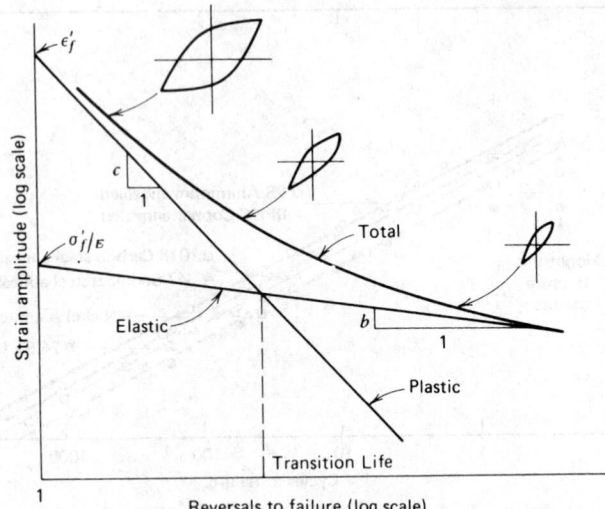

Fig. 19.62 Schematic representation of elastic, plastic, and total strain amplitude versus fatigue life. (From Ref. 38, copyright ASTM; reprinted with permission.)

were produced by external mechanical loading. Thus, all the results described earlier for low-cycle fatigue (and high-cycle fatigue if strains are all elastic) are at least qualitatively applicable to thermal fatigue as well. It must be recognized, however, that thermal fatigue problems involve not only all the complexities of mechanical loading but, in addition, all the temperature-induced problems as well.

While thermal low-cycle-fatigue phenomena and mechanical low-cycle-fatigue phenomena are very similar, and are mathematically expressible by the same types of equations, the use of mechanical

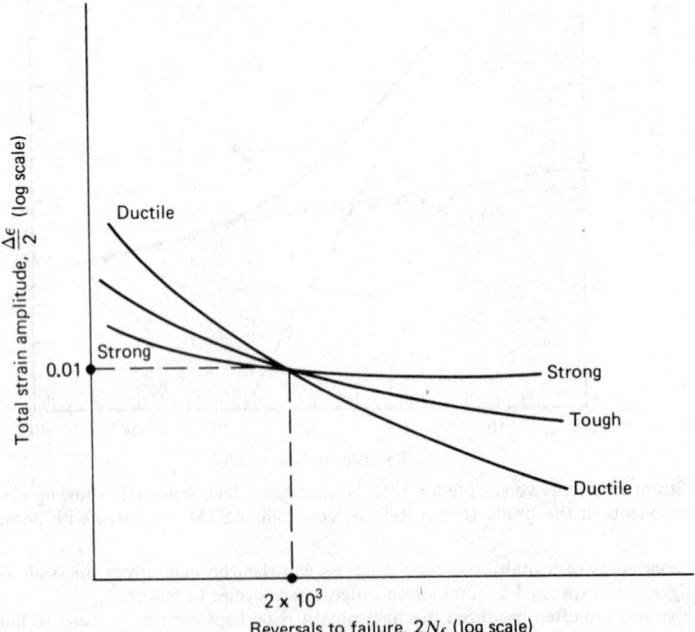

Fig. 19.63 Idealized representation of cyclic strain failure resistance of various types of materials. (After Ref. 38, copyright ASTM; reprinted with permission.)

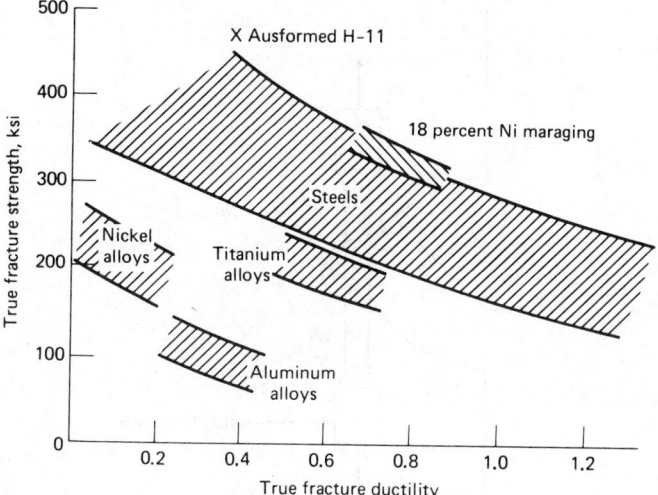

Fig. 19.64 Monotonic fracture strength–ductility combinations presently attainable for various alloy classes. (From Ref. 38, copyright ASTM; reprinted with permission.)

low-cycle-fatigue results to predict thermal low-cycle-fatigue performance must be undertaken with care.

Some of the differences between thermally and mechanically induced low-cycle fatigue that give rise to apparent discrepancies in low-cycle-fatigue data for these two cases are (see pp. 270–272 of Ref. 51):

1. In thermal fatigue the plastic strain tends to become concentrated in the hottest regions of the body, since the yield point is locally reduced at these hottest regions.

2. In thermal fatigue there is often a localized region of strain developed by virtue of plastic flow during the compressive branch of the strain cycle to produce a bulging at the hottest region. This is followed by a necking tendency adjacent to the bulge caused by plastic flow during the tensile branch of the strain cycle upon cooling.

3. Cyclic variations in temperature may, in and of themselves, have important effects on the material properties and ability to resist low-cycle-fatigue failure.

4. There may be interaction effects caused by superposition of simultaneous variations in temperatures and strains.

5. Rates at which the strain cycling is induced may have an important effect, since the testing speeds in thermal fatigue tests are often greatly different from the rates used in mechanical low-cycle-fatigue tests.

For these reasons caution is necessary in the prediction of thermal low-cycle-fatigue behavior from mechanical low-cycle-fatigue results or vice versa.

19.5.8 Fracture Mechanics Methods for Fatigue Life Prediction

In recent years it has been recognized that the fatigue failure process involves three phases. A crack initiation phase occurs first, followed by a crack propagation phase; finally, when the crack reaches a critical size, the final phase of unstable rapid crack growth to fracture completes the failure process. The modeling of each of these phases has been under intense scrutiny. Although excellent progress has been made in modeling the propagation phase and final fracture phase, progress has been much slower in modeling the crack initiation phase. The most promising approach seems to be the "local stress–strain method."

The basic premise of the local stress–strain approach is that the local fatigue response of the material at the critical point, that is, the site of crack initiation, is analogous to the fatigue response of a small, smooth specimen subjected to the same cyclic strains and stresses.[52] This concept is illustrated schematically in Fig. 19.65 for a simple notched plate under cyclic loading. The cyclic stress–strain response of the critical material may be determined from the characterizing smooth specimen through appropriate laboratory testing. To properly perform such laboratory tests, the local cyclic stress–strain history at the critical point in the structure must be determined, by either analytical or experimental

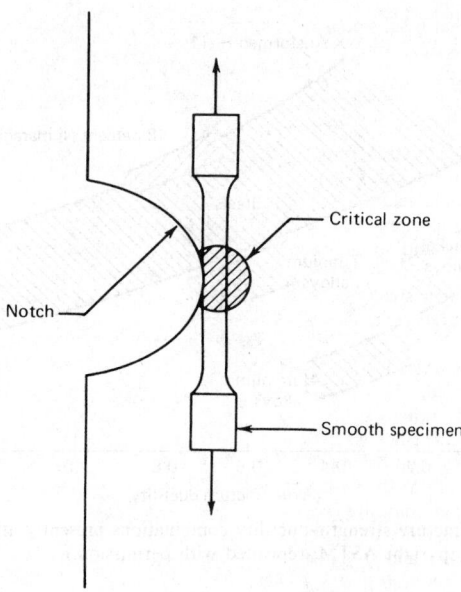

Fig. 19.65 Smooth specimen analog of material at critical point in the structure. (See Ref. 52.)

means. Thus, valid stress analysis procedures, finite-element modeling, or experimental strain measurements are necessary, and the ability to properly account for plastic behavior must be included. In performing smooth specimen tests of this type it must be recognized that the phenomena of cyclic hardening, cyclic softening, and cycle-dependent stress relaxation, as well as sequential loading effects and residual stress effects, may be experienced by the specimen as it accumulates fatigue damage presumed to be the same as at the critical point in the structural member being simulated. Some data have been accumulated to support the validity of this postulate.[53-55]

Digital computer simulation of the smooth specimen simulation has also been shown to be feasible.[55-57] To successfully utilize this computer simulation technique, it is necessary to have access to both monotonic and cyclic material properties, since the stress–strain response of most materials changes significantly with cyclic straining into the plastic range as already noted in Fig. 19.59. If cyclic materials response data are not available from the literature or an accessible data bank, it is necessary to perform enough smooth specimen testing to characterize the cyclic stress–strain response and fracture resistance of the material. With cyclic materials properties available, the computer simulation model for prediction of crack initiation must contain the following abilities:

1. To compute local stresses and strains, including means and ranges, from the applied loads and geometry of the structure.
2. To count cycles and associate mean and range values of stress and strain with each cycle.
3. To convert nonzero mean cycles to equivalent completely reversed cycles.
4. To compute fatigue damage in each cycle from stress and/or strain amplitudes and cyclic materials properties.
5. To compute damage cycle by cycle and sum the damage to give desired prediction of crack initiation.

To compute local stresses and strains from the external loading and geometry, Neuber's rule is often used in conjunction with the cyclic stress–strain properties and fatigue stress concentration factor. Neuber's rule may be modified for application to fatigue loading[58] by utilizing the fatigue stress concentration factor K_f together with nominal stress *range* ΔS, local stress *range* $\Delta \sigma$, and local strain *range* $\Delta \epsilon$ to develop the expression

$$\frac{(K_f \, \Delta S)^2}{E} = \Delta \sigma \, \Delta \epsilon \tag{19.92}$$

All the terms on the left-hand side of (19.92) are known from the geometry and from loading and material properties of the structure. To resolve the right-hand term, an empirical expression for the cyclic stress–strain curve that is satisfactory for most engineering metals may be obtained by separating the cyclic strain amplitude $\Delta\epsilon/2$ into elastic and plastic components to yield (from p. 7 of Ref. 52)

$$\frac{\Delta\epsilon}{2} = \frac{\Delta\epsilon_e}{2} + \frac{\Delta\epsilon_p}{2} = \frac{\Delta\sigma}{2E} + \left[\frac{\Delta\sigma}{2k'}\right]^{1/n'} \qquad (19.93)$$

where k' and n' are the *cyclic* strength coefficient and *cyclic* strain-hardening exponent, respectively, determined from the intercept and slope of a log–log plot of cyclic stress amplitude versus cyclic plastic strain amplitude. Some values for n' are shown in Fig. 19.59. The use of (19.93), then, provides a means for computing local stresses and strains, when used in conjunction with (19.92). The need to include cyclic hardening and softening in the prediction model is dependent on the accuracy of other parts of the model, and in some cases these transient phenomena may be regarded as second-order effects. Cycle-dependent stress relaxation may be important in cases in which occasional large overload cycles produce large residual stresses in the structure that relax with additional cycles of local plastic strain.

To properly interpret complex load, stress, or strain versus time histories requires that an appropriate cycle counting method be used. The *rain-flow cycle counting method,* illustrated in Fig. 19.66, is probably more widely used than any other method. The strain–time history is plotted so that the time axis is vertically downward, and the lines connecting the strain peaks are imagined to be a series of roofs. Several rules are imposed on rain dripping down these roofs so that cycles and half-cycles are defined. Rain flow begins successively at the inside of each strain peak. The rain flow initiating at each peak is allowed to drip down and continue except that, if it initiates at a minimum, it must stop when it comes opposite a minimum more negative than the minimum from which it initiated. For example, in Fig. 19.66 begin at peak 1 and stop opposite peak 9, peak 9 being more negative than peak 1. A half-cycle is thus counted between peaks 1 and 8. Similarly, if the rain flow initiates at a maximum, it must stop when it comes opposite a maximum more positive than the maximum from which it initiated. For example, in Fig. 19.66 begin at peak 2 and stop opposite peak 4, thus counting a half-cycle between peaks 2 and 3. A rain flow must also stop if it meets the rain from a roof above. For

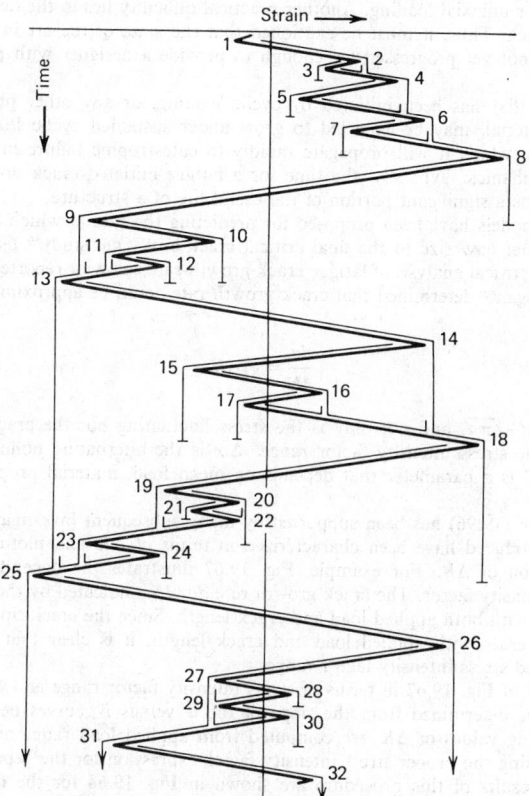

Fig. 19.66 Example of rain-flow cycle counting method. (After Ref. 59.)

example, in Fig. 19.66, the half-cycle beginning at peak 3 ends beneath peak 2. Note that every part of the strain–time history is counted once and only once.

If cycles are to be counted for a "duty cycle" or a "mission profile" that is repeated until failure occurs, one complete strain cycle should be counted between the most positive and most negative peaks in the sequence, and other smaller complete cycles that are interruptions of this largest cycle should also be counted. This will be accomplished by the rain-flow method if the cycle counting is started at either the most positive or most negative peak in the sequence. Using this procedure to identify the maximum and minimum strain (or stress), the range and mean may be tabulated for each cycle in the stress–strain history.

The nonzero mean stress cycles may be converted to equivalent completely reversed cycles by utilizing either the modified Goodman equations or some empirical expression based on specific material data. Using the modified Goodman results of Table 19.2, for a tensile mean stress the equivalent completely reversed stress σ_{eqC-R} is

$$\sigma_{eqC-R} = \frac{\sigma_a}{1 - \sigma_m/\sigma_u}, \qquad \sigma_m \geq 0 \tag{19.94}$$

and for a compressive mean stress

$$\sigma_{eqC-R} = \sigma_a, \qquad \sigma_m \leq 0 \tag{19.95}$$

To compute the fatigue damage in each cycle associated with the equivalent completely reversed stress and strain range, it is necessary to have available data for strain amplitude versus cycles to failure, N_f (or reversals to failure, $2N_f$), as illustrated in Fig. 19.62. An expression for total strain amplitude has already been given in Eq. (19.91) as a function of total number of cycles to failure.

Damage is summed by utilizing an appropriate cumulative damage theory. Using the procedure described in this section for prediction of crack initiation, the Palmgren–Miner linear damage hypothesis of (19.74) gives results that are as good as any other proposed technique. Thus, when the sum of cycle ratios becomes equal to unity, it is predicted that crack initiation has occurred. The prediction techniques described in this section are practical only with the help of a digital computer program designed to perform the tedious cycle-by-cycle analyses involved, and the validity of the method remains to be proven, even for uniaxial loading. Another practical difficulty lies in the definition and detection of an "initiated" crack. Thus, it must be cautioned that the state of the art in prediction of fatigue crack initiation has not yet progressed far enough to provide a designer with proven tools for such predictions.

A fatigue crack that has been initiated by cyclic loading, or any other preexisting flaw in the structure or the material, may be expected to grow under sustained cyclic loading until it reaches the critical size from which it will propagate rapidly to catastrophic failure in accordance with the laws of fracture mechanics. Typically, the time for a fatigue-initiated crack or a preexisting flaw to grow to critical size is a significant portion of the useful life of a structure.

Many different models have been proposed for predicting the rate at which a fatigue crack grows from a specified initial flaw size to the final critical crack size. One study[60] lists 33 proposed crack growth "laws." In a critical analysis of fatigue crack-growth-rate behavior reported by several investigators, Paris and Erdogan[61] determined that crack growth rate could be approximated by an expression of the form

$$\frac{da}{dN} = g(\Delta K) \tag{19.96}$$

where $\Delta K = C \, \Delta\sigma \, \sqrt{\pi a}$, and not only is the stress fluctuating but the crack is growing. In this expression ΔK is the stress intensity factor *range*, $\Delta\sigma$ is the alternating nominal stress range, a is crack length, and C is a parameter that depends on mean load, material properties, and secondary variables.

The expression of (19.96) has been supported by many subsequent investigations, and most crack propagation data produced have been characterized in terms of ΔK and plotted either as a log–log or log–linear function of ΔK. For example, Fig. 19.67 illustrates the dependence of fatigue crack growth on stress intensity factor. The crack growth rate da/dN, indicated by the slope of the a versus N curves, increases with both applied load and crack length. Since the crack-tip stress intensity factor range ΔK also increases with applied load and crack length, it is clear that crack growth rate is related to the applied stress intensity factor range.

To plot the data of Fig. 19.67 in terms of stress intensity factor range and crack growth rate, the crack growth rate is determined from the slope of the a versus N curves between successive data points. Corresponding values of ΔK are computed from applied load range and mean crack length for each interval, using the proper stress intensity factor expression for the geometry of the specimen being tested. The results of this procedure are shown in Fig. 19.68 for the data presented in Fig. 19.67. It should be noted that all the curves of Fig. 19.67 incorporate themselves into a single curve in Fig. 19.68 through use of the stress intensity-factor concept, and the curve of Fig. 19.68 is therefore

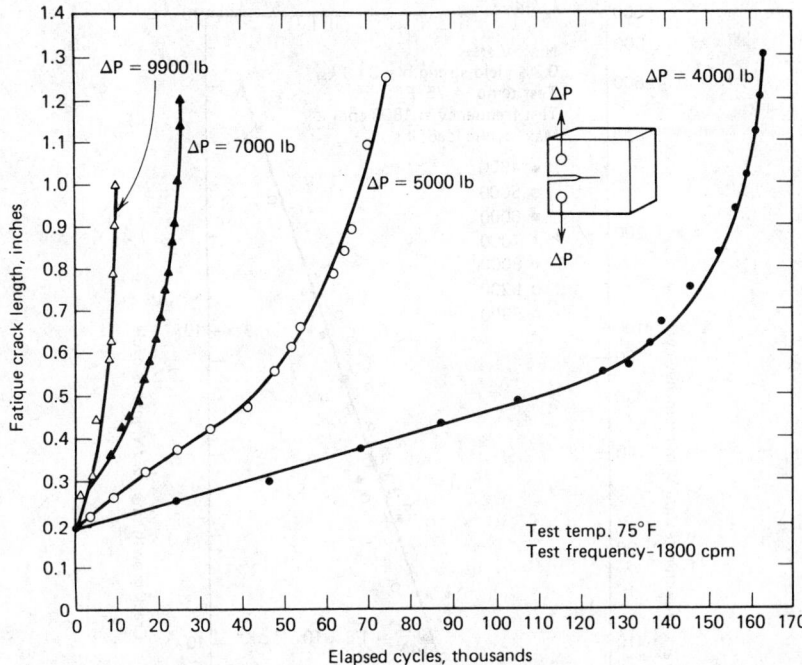

Fig. 19.67 Effect of cyclic-load range on crack growth in Ni–Mo–V alloy steel for released tension loading. (From Ref. 62, copyright Society for Experimental Stress Analysis, 1971; reprinted with permission.)

applicable to any combination of cyclic stress range and crack length for released loading on specimens of this geometry.

Fatigue crack-growth-rate data similar to that shown in Fig. 19.68 have been reported for a wide variety of engineering metals. Thus, it was concluded[61] that the fatigue crack-growth-rate for engineering metals could be generalized to the form

$$\frac{da}{dN} = C_{PE}(\Delta K)^n \qquad (19.97)$$

where n is the slope of the log da/dN versus log ΔK plot, such as shown in Fig. 19.68, and C_{PE} is an empirical parameter that depends on material properties, frequency, mean load, and perhaps other secondary variables. Thus, if the parameters C_{PE} and n are known for a particular application, the crack length a_N, after application of N cycles of loading, may be computed from the expression

$$a_N = a_i + \sum_{i=1}^{N} C_{PE}\,\Delta K^n \qquad (19.98)$$

or

$$a_N = a_i + \int_1^N C_{PE}\,\Delta K^n\,dN \qquad (19.99)$$

where a_i is the initial crack length and N is the total number of loading cycles. It must be emphasized that (19.97), (19.98), and (19.99) are applicable only to region II crack growth. Region I of Fig. 19.69 corresponds to the nucleation period, and region III corresponds to the transition into the unstable regime of rapid crack extension.

Modifications of the Paris–Erdogan expression have been developed to account for influence of magnitude of peak stress or the existence of a threshold value of ΔK as indicated in Fig. 19.69 at the boundary between regions I and II. One such expression is

$$\frac{da}{dN} = \frac{C_{HS}(\Delta K - \Delta K_{th})^m}{(1-R)K_c - \Delta K} \qquad (19.100)$$

where R is the stress ratio $\sigma_{min}/\sigma_{max}$, K_c is the fracture toughness for unstable crack growth during monotonic loading, and ΔK_{th} is the threshold stress intensity factor range for fatigue crack propagation.

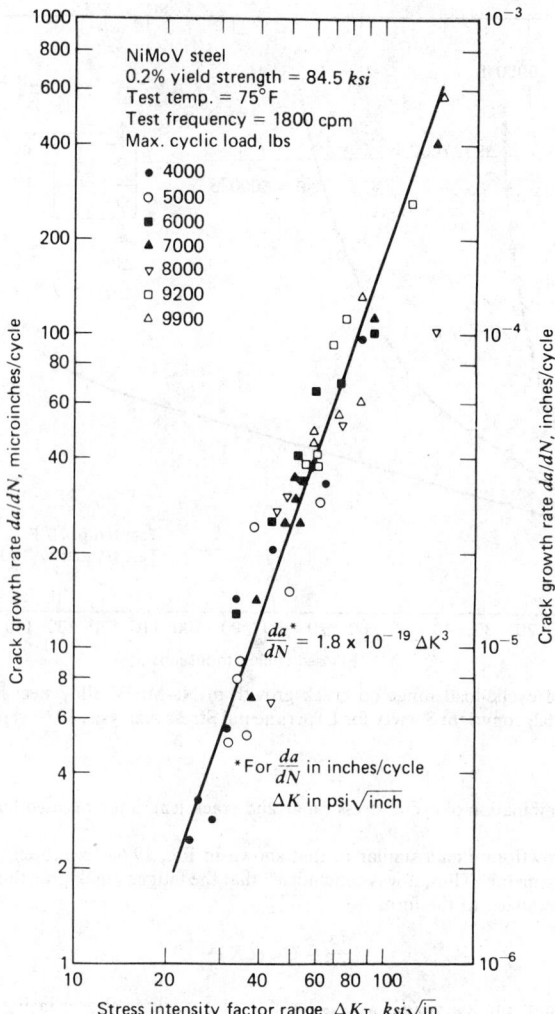

Fig. 19.68 Crack growth rate as a function of stress-intensity range for Ni–Mo–V steel. (From Ref. 62, copyright Society for Experimental Stress Analysis, 1971; reprinted with permission.)

Crack growth rates determined from constant-amplitude cyclic loading tests are approximately the same as for random loading tests in which the maximum stress is held constant but mean and range of stress vary randomly. However, in random loading tests where the maximum stress is also allowed to vary, the sequence of loading cycles may have a marked effect on crack growth rate, with the overall crack growth being significantly higher for random loading spectra.[64]

Several investigations have shown that in some circumstances a significant delay in crack propagation occurs following intermittent application of high stresses. That is, fatigue damage and crack extension are dependent on preceding cyclic load history. This dependence of crack extension on preceding history and the effects upon future damage increments are referred to as *interaction* effects. Most of the interaction studies conducted thus far have dealt with *retardation* of crack growth as a result of the application of occasional tensile overload cycles. Retardation can be characterized as a period of reduced crack growth rate following the application of a peak load or loads higher and in the same direction as those peaks that follow.

Although much progress has been made in crack-growth-rate prediction in the laboratory, it must be recognized that load sequence effects, environmental effects, frequency effects, multiaxial state-of-stress effect, and problems in determination of applicable ΔK values still make it essential to conduct full-scale fatigue tests to ensure proper resistance to failure by fatigue.

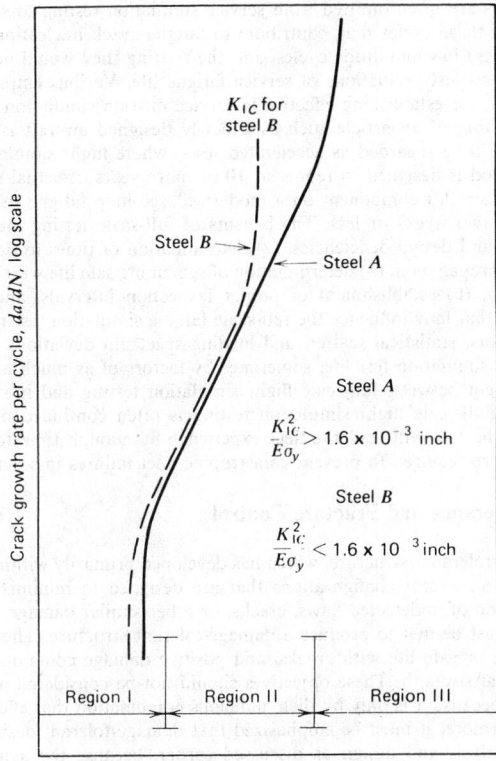

Fig. 19.69 Schematic representation of fatigue crack growth in steel. (See Ref. 63.)

19.5.9 Service Spectrum Simulation and Full-Scale Testing

To achieve a reliable fatigue-resistant design configuration, a realistic service loading simulation test or a full-scale fatigue test will generally be required. For service load simulation testing it is essential that both the test specimen and the loading spectrum, including sequence, be representative of actual service conditions. This means that the specimen should be an actual component, a complete structural subassembly, or a complete full-scale machine or structure. It also means that an exact simulation of the service load–time history would be best, if known. Usually, however, an estimated load–time history must be generated on the basis of a statistical representation of similar mission loading spectra obtained from similar instrumented articles already in actual use.

In designing service loading simulation spectra, special attention must be given to the highest load levels to be incorporated because they may exert a major influence on crack propagation and fatigue life. As noted in Section 19.5.8, the life-enhancing retardation phenomenon associated with large load peaks can be extremely important. It should be recognized that the largest peak loads in service will vary from article to article on a statistical basis, so that some articles will experience the maximum load peak more than once, whereas others will never see this load level. For these reasons it is usual to truncate the design loading spectrum, discarding all load levels that occur fewer than 10 times during the projected design lifetime. Although it may seem intuitively wrong to discard the high load peaks to achieve a more critical simulation test spectrum, it must be realized that the crack retardation effect of the occasional high loads gives a longer test life than if the high loads were omitted. Thus, it is conservative to omit the 10 highest loads from the simulation test spectrum. Likewise, truncation is utilized in analyses for establishing suitable inspection intervals for aircraft so that propagating cracks will have a higher probability of being detected before they approach critical size.

The application of failsafe or limit loads at regular intervals during a service spectrum simulation or full-scale fatigue test *must be avoided*, since they may contribute to crack growth retardation that is not typical of the actual flight spectrum. Limit loading should be applied only at the end of the fatigue test.

Low-amplitude cycles are often omitted from service simulation testing to save time. However, it must be recognized that these cycles may contribute to fatigue crack nucleation through the fretting process, and omitting these low-amplitude cycles, and the fretting they would normally produce, may result in unsafe simulation test predictions of service fatigue life. Various empirical procedures have been developed, however, for establishing effective in-service mission simulation spectra.[65]

Full-scale fatigue testing of an article such as a newly designed aircraft is extremely expensive. Such tests generally would be regarded as accelerated tests, where flight simulation testing over a 6- to 12-month testing period is designed to represent 10 or more years of actual service history. Flight simulation testing of an aircraft component on a modern closed-loop fatigue testing machine may be accomplished in one or two weeks or less. The benefits of full-scale testing include (1) discovery of fatigue-critical elements and design deficiencies, (2) determination of times to detectable cracking, (3) obtaining data on crack propagation, (4) determination of remaining safe life with cracks, (5) determination of residual strength, (6) establishment of proper inspection intervals, and (7) development of repair methods. Factors that may influence the full-scale fatigue simulation test results include loading rate, environmental factors, statistical scatter, and loading spectrum deviations. Actual service life is usually shorter than the simulation test life, sometimes by factors of as much as 2 to 4. However, in recent years the agreement between full-scale flight simulation testing and in-service experience has improved significantly. Full-scale flight simulation testing is often continued over the long term so that fatigue failures in the test will lead the fleet experience by enough time to redesign and install whatever modifications are required to prevent catastrophic fleet failures in service before they occur.

19.5.10 Damage Tolerance and Fracture Control

The concept of "damage-tolerant" structure, which has developed primarily within the aerospace industry, is characterized by structural configurations that are designed to minimize the loss of aircraft because of the propagation of undetected flaws, cracks, or other similar damage. There are two major design objectives that must be met to produce a damage-tolerant structure. These objectives are controlled safe flaw growth, or safe life with cracks, and positive damage containment, which implies a safe remaining or residual strength. These objectives should not be considered as separate or distinct requirements, however, because it is only by their judicious combination that effective fracture control can be achieved. Furthermore, it must be emphasized that damage-tolerant design is not a substitute for a careful fatigue analysis and design as discussed earlier, because the achievement of "fatigue quality" through careful stress analysis, geometry selection, detail design, material selection, surface finish, and workmanship is a necessary prerequisite to effective damage-tolerant design and fracture control.

The general goals of damage-tolerant design and fracture control include the selection of fracture-resistant materials and manufacturing processes, the design for inspectability, and the use of damage-tolerant structural configurations such as multiple load paths or crack stoppers, in addition to the usual rules of good design practice.

In the application of the fracture control philosophy, the basic assumption is made that flaws do exist even in new structures and that they may go undetected. The first major requirement for damage tolerance, therefore, is that any member in the structure, including each element of a redundant load path group, must have a safe life with assumed cracks present. For any specific application the primary factors influencing the design include the type or class of structure, the quality of the nondestructive inspection (NDI) techniques used in production assembly, the accessibility of the structure to inspection, the assurance that the member will be inspected on schedule when in service, and the probability that a flaw of subcritical size will go undetected even though periodic in-service inspections are made on schedule.

Most structural arrangements may be classified according to load path as class 1, single load path; class 2, single primary load path with auxiliary crack arrest features; or class 3, multiple or redundant load path. Each of these structural classes is illustrated in Fig. 19.70. Clearly, for the class 1 structure it is essential to satisfy the safe-life-with-cracks requirement, because failure is catastrophic. For class 2 structures, including pressurized cabins and pressure vessels, relatively large amounts of damage may be contained by providing tear straps or stiffeners. There is usually a high probability of damage detection for a class 2 structure because of fuel or pressure leakage, that is, "leak-before-break" design is characteristic of class 2 structures. Class 3 structures are usually designed to provide a specified percentage of the original strength, that is, a specified residual strength, during and subsequent to the failure of one element. This is often called "failsafe" type of structure. However, the preexisting flaw concept requires that all members, including every member of a multiple load path structure, be assumed to contain flaws. It is usual to assume a smaller initial flaw size for class 3 structures because it is appropriate to take a larger risk of operating with cracks if multiple load paths are available.

The development of inspection procedures is an important part of any fracture control program. Appropriate inspection procedures must be established for each structural element, and regions within elements may be classified with respect to required NDI sensitivity. Inspection intervals are established on the basis of crack growth information assuming a specified initial flaw size and a "detectable"

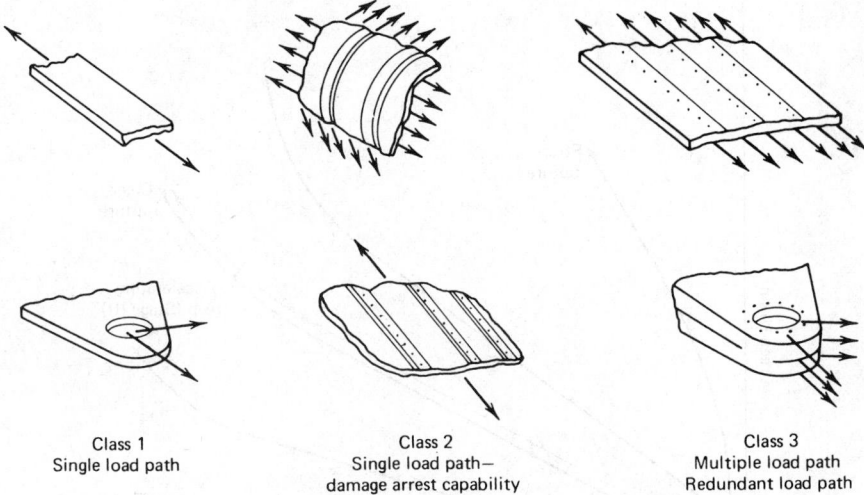

Class 1
Single load path

Class 2
Single load path—
damage arrest capability

Class 3
Multiple load path
Redundant load path

Fig. 19.70 Structural arrangements. (After Ref. 66.)

flaw size that depends on the NDI procedure. Inspection intervals are established to ensure that an undetected flaw will not grow to critical size before the next inspection, with a comfortable margin of safety. The intervals are usually picked so that two inspections will occur before any crack will reach critical size.

A good fracture control program should encompass and interact with design, materials selection, fabrication, inspection, and operational phases in the development of any high-performance engineering system.

19.6 CREEP AND STRESS RUPTURE

Creep in its simplest form is the progressive accumulation of plastic strain in a specimen or machine part under stress at elevated temperature over a period of time. Creep failure occurs when the accumulated creep strain results in a deformation of the machine part that exceeds the design limits. *Creep rupture* is an extension of the creep process to the limiting condition where the stressed member actually separates into two parts. *Stress rupture* is a term used interchangeably by many with creep rupture; however, others reserve the term stress rupture for the rupture termination of a creep process in which steady-state creep is never reached, and use the term creep rupture for the rupture termination of a creep process in which a period of steady-state creep has persisted. Figure 19.71 illustrates these differences. The interaction of creep and stress rupture with cyclic stressing and the fatigue process has not yet been clearly understood but is of great importance in many modern high-performance engineering systems.

Creep strains of engineering significance are not usually encountered until the operating temperatures reach a range of approximately 35–70% of the melting point on a scale of absolute temperature. The approximate melting temperature for several substances is shown in Table 19.3.

Not only is excessive deformation due to creep an important consideration, but other consequences of the creep process may also be important. These might include creep rupture, thermal relaxation, dynamic creep under cyclic loads or cyclic temperatures, creep and rupture under multiaxial states of stress, cumulative creep effects, and effects of combined creep and fatigue.

Creep deformation and rupture are initiated in the grain boundaries and proceed by sliding and separation. Thus, creep rupture failures are intercrystalline, in contrast, for example, to the transcrystalline failure surface exhibited by room-temperature fatigue failures. Although creep is a plastic flow phenomenon, the intercrystalline failure path gives a rupture surface that has the appearance of brittle fracture. Creep rupture typically occurs without necking and without warning. Current state-of-the-art knowledge does not permit a reliable prediction of creep or stress rupture properties on a theoretical basis. Furthermore, there seems to be little or no correlation between the creep properties of a material and its room-temperature mechanical properties. Therefore, test data and empirical methods of extending these data are relied on heavily for prediction of creep behavior under anticipated service conditions.

Metallurgical stability under long-time exposure to elevated temperatures is mandatory for good creep-resistant alloys. Prolonged time at elevated temperatures acts as a tempering process, and any improvement in properties originally gained by quenching may be lost. Resistance to oxidation and

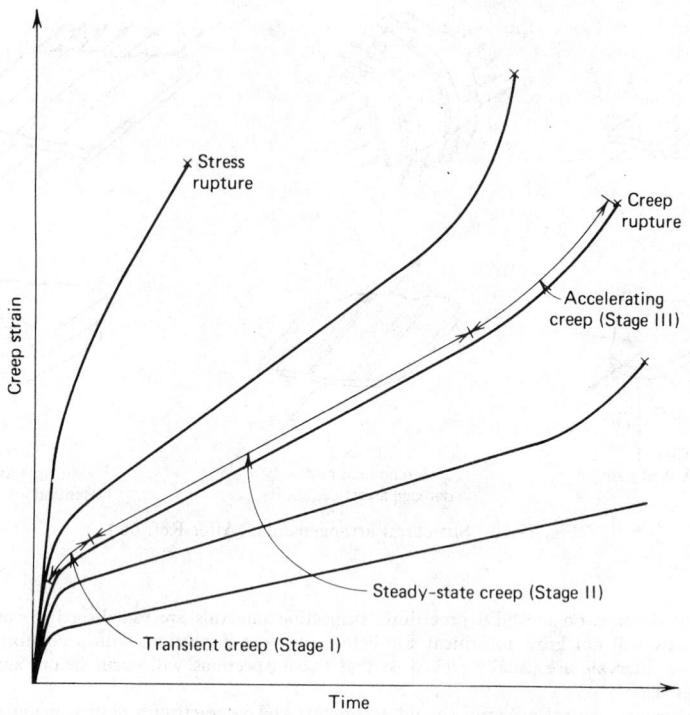

Fig. 19.71   Illustration of creep and stress rupture.

other corrosive media are also usually important attributes for a good creep-resistant alloy. Larger grain size may also be advantageous since this reduces the length of grain boundary, where much of the creep process resides.

19.6.1   Prediction of Long-Term Creep Behavior

Much time and effort has been expended in attempting to devise good short-time creep tests for accurate and reliable prediction of long-term creep and stress rupture behavior. It appears, however, that really reliable creep data can be obtained only by conducting long-term creep tests that duplicate actual

Table 19.3   Melting Temperatures[36]

Material	°F	°C
Hafnium carbide	7030	3887
Graphite (sublimes)	6330	3500
Tungsten	6100	3370
Tungsten carbide	5190	2867
Magnesia	5070	2800
Molybdenum	4740	2620
Boron	4170	2300
Titanium	3260	1795
Platinum	3180	1750
Silica	3140	1728
Chromium	3000	1650
Iron	2800	1540
Stainless steels	2640	1450
Steel	2550	1400
Aluminum alloys	1220	660
Magnesium alloys	1200	650
Lead alloys	605	320

service loading and temperature conditions as nearly as possible. Unfortunately, designers are unable to wait for years to obtain design data needed in creep failure analysis. Therefore, certain useful techniques have been developed for approximating long-term creep behavior based on a series of short-term tests. Data from creep testing may be cross plotted in a variety of different ways. The basic variables involved are stress, strain, time, temperature, and, perhaps, strain rate. Any two of these basic variables may be selected as plotting coordinates, with the remaining variables treated as parametric constants for a given curve. Three commonly used methods for extrapolating short-time creep data to long-term applications are the abridged method, the mechanical acceleration method, and the thermal acceleration method. In the abridged method of creep testing the tests are conducted at several different stress levels and at the contemplated operating temperature. The data are plotted as creep strain versus time for a family of stress levels, all run at constant temperature. The curves are plotted out to the laboratory test duration and then extrapolated to the required design life. In the mechanical acceleration method of creep testing, the stress levels used in the laboratory tests are significantly higher than the contemplated design stress levels, so the limiting design strains are reached in a much shorter time than in actual service. The data taken in the mechanical acceleration method are plotted as stress level versus time for a family of constant strain curves all run at a constant temperature. The thermal acceleration method involves laboratory testing at temperatures much higher than the actual service temperature expected. The data are plotted as stress versus time for a family of constant temperatures where the creep strain produced is constant for the whole plot.

It is important to recognize that such extrapolations are not able to predict the potential of failure by creep rupture prior to reaching the creep design life. In any testing method it should be noted that creep testing guidelines usually dictate that test periods of less than 1% of the expected life are not deemed to give significant results. Tests extending to at least 10% of the expected life are preferred where feasible.

Several different theories have been proposed in recent years to correlate the results of short-time elevated-temperature tests with long-term service performance at more moderate temperatures. The more accurate and useful of these proposals to date are the Larson–Miller theory and the Manson–Haferd theory.

The Larson–Miller theory[67] postulates that for each combination of material and stress level there exists a unique value of a parameter P that is related to temperature and time by the equation

$$P = (\theta + 460)(C + \log_{10} t) \tag{19.101}$$

where P = Larson–Miller parameter, constant for a given material and stress level
θ = temperature, °F
C = constant, usually assumed to be 20
t = time in hours to rupture or to reach a specified value of creep strain

This equation was investigated for both creep and rupture for some 28 different materials by Larson and Miller with good success. By using (19.101) it is a simple matter to find a short-term combination of temperature and time that is equivalent to any desired long-term service requirement. For example, for any given material at a specified stress level the test conditions listed in Table 19.4 should be equivalent to the operating conditions.

The Manson–Haferd theory[68] postulates that for a given material and stress level there exists a unique value of a parameter P' that is related to temperature and time by the equation

$$P' = \frac{\theta - \theta_a}{\log_{10} t - \log_{10} t_a} \tag{19.102}$$

where P' = Manson–Haferd parameter, constant for a given material and stress level
θ = temperature, °F
t = time in hours to rupture or to reach a specified value of creep strain
θ_a, t_a = material constants

In the Manson–Haferd equation values of the constants for several materials are shown in Table 19.5.

Table 19.4 Equivalent Conditions Based on Larson–Miller Parameter

Operating Condition	Equivalent Test Condition
10,000 hours at 1000°F	13 hours at 1200°F
1,000 hours at 1200°F	12 hours at 1350°F
1,000 hours at 1350°F	12 hours at 1500°F
1,000 hours at 300°F	2.2 hours at 400°F

Table 19.5 Constants for Manson–Haferd Equation[68]

Material	Creep or Rupture	θ_a	$\log_{10} t_a$
25–20 stainless steel	Rupture	100	14
18–8 stainless steel	Rupture	100	15
S-590 alloy	Rupture	0	21
DM steel	Rupture	100	22
Inconel X	Rupture	100	24
Nimonic 80	Rupture	100	17
Nimonic 80	0.2 percent plastic strain	100	17
Nimonic 80	0.1 percent plastic strain	100	17

19.6.2 Creep under Uniaxial State of Stress

Many relationships have been proposed to relate stress, strain, time, and temperature in the creep process. If one investigates experimental creep strain versus time data, it will be observed that the data are close to linear for a wide variety of materials when plotted on log strain versus log time coordinates. Such a plot is shown, for example, in Fig. 19.72 for three different materials. An equation describing this type of behavior is

$$\delta = At^a \tag{19.103}$$

where δ = true creep strain
t = time
A, a = empirical constants

Differentiating (19.102) with respect to time gives

$$\dot{\delta} = aAt^{(a-1)} \tag{19.104}$$

or, setting $aA = b$ and $(1 - a) = n$,

$$\dot{\delta} = bt^{-n} \tag{19.105}$$

This equation represents a variety of different types of creep strain versus time curves, depending on the magnitude of the exponent n. If n is zero, the behavior, characteristic of high temperatures, is termed *constant creep rate,* and the creep strain is given as

$$\delta = b_1 t + C_1 \tag{19.106}$$

If n lies between 0 and 1, the behavior is termed *parabolic creep,* and the creep strain is given by

$$\delta = b_3 t^m + C_3 \tag{19.107}$$

This type of creep behavior occurs at intermediate and high temperatures. The coefficient b_3 increases exponentially with stress and temperature, and the exponent m decreases with stress and increases with temperature. The influence of stress level σ on creep rate can often be represented by the empirical expression

$$\dot{\delta} = B\sigma^N \tag{19.108}$$

Assuming the stress σ to be independent of time, we may integrate (19.108) to yield the creep strain

$$\delta = Bt\sigma^N + C' \tag{19.109}$$

If the constant C' is small compared with $Bt\sigma^N$, as it often is, the result is called the *log–log stress–time creep law,* given as

$$\delta = Bt\sigma^N \tag{19.110}$$

As long as the instantaneous deformation on load application and the stage I transient creep are small compared to stage II steady-state creep, (19.110) is useful as a design tool.

If it is necessary to consider all stages of the creep process, the creep strain expression becomes much more complex. The most general expression for the creep process is (see p. 438 of Ref. 70)

$$\delta = \frac{\sigma}{E} + k_1\sigma^m + k_2(1 - e^{-qt})\sigma^n + k_3 t\sigma^p \tag{19.111}$$

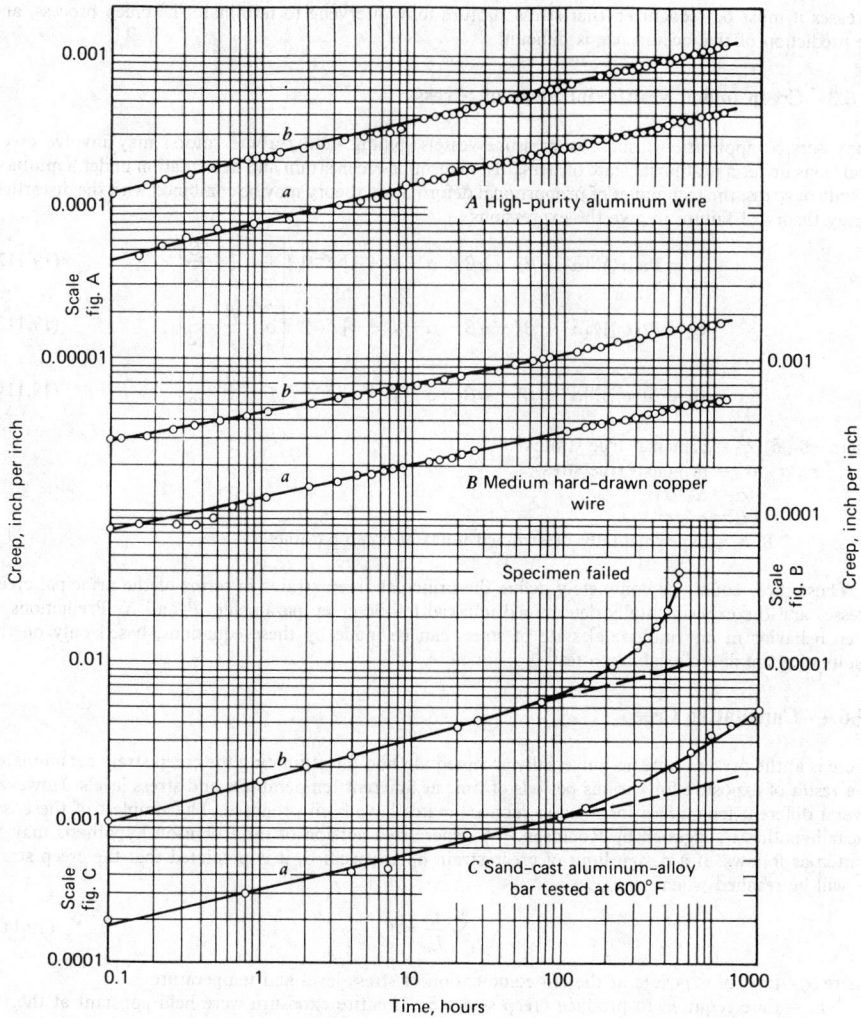

Fig. 19.72 Creep curves for three materials plotted on log–log coordinates. (From Ref. 69.)

where

$$\delta = \text{total creep strain}$$
$$\sigma/E = \text{initial elastic strain}$$
$$k_1\sigma^m = \text{initial plastic strain}$$
$$k_2(1 - e^{-qt})\sigma^n = \text{anelastic strain}$$
$$k_3t\sigma^p = \text{viscous strain}$$
$$\sigma = \text{stress}$$
$$E = \text{modulus of elasticity}$$
$$m = \text{reciprocal of strain-hardening exponent}$$
$$k_1 = \text{reciprocal of strength coefficient}$$
$$q = \text{reciprocal of Kelvin retardation time}$$
$$k_2 = \text{anelastic coefficient}$$
$$n = \text{empirical exponent}$$
$$k_3 = \text{viscous coefficient}$$
$$p = \text{empirical exponent}$$
$$t = \text{time}$$

To utilize this empirical nonlinear expression in a design environment requires specific knowledge of the constants and exponents that characterize the material and temperature of the application. In all cases it must be recognized that stress rupture may intervene to terminate the creep process, and the prediction of this occurrence is difficult.

19.6.3   Creep under Multiaxial State of Stress

Many service applications, such as pressure vessels, piping, and turbine rotors, may involve creep conditions under a multiaxial state of stress. To determine creep strain and deformation under a multiaxial state of stress, the techniques of proportional deformation theory may be combined with the distortion energy theory of failure to give the expressions

$$\delta_1 = Bt(\sigma_1')^N[\alpha^2 + \beta^2 - \alpha\beta - \alpha - \beta + 1]^{(N-1)/2}\left[1 - \frac{\alpha}{2} - \frac{\beta}{2}\right] \tag{19.112}$$

$$\delta_2 = Bt(\sigma_1')^N[\alpha^2 + \beta^2 - \alpha\beta - \alpha - \beta + 1]^{(N-1)/2}\left[\alpha - \frac{\beta}{2} - \frac{1}{2}\right] \tag{19.113}$$

$$\delta_3 = Bt(\sigma_1')^N[\alpha^2 + \beta^2 - \alpha\beta - \alpha - \beta + 1]^{(N-1)/2}\left[\beta - \frac{\alpha}{2} - \frac{1}{2}\right] \tag{19.114}$$

where $\delta_1, \delta_2, \delta_3$ = principal true strains
$\quad\quad\sigma_1', \sigma_2', \sigma_3'$ = principal true stresses
$\quad\quad\alpha = \alpha_2'/\sigma_1'$
$\quad\quad\beta = \sigma_3'/\sigma_1'$
$\quad\quad B, N$ = experimentally determined uniaxial creep parameters

These three equations completely define the principal creep strains in terms of the principal creep stresses and the experimentally determined uniaxial tensile creep parameters B and N. Predictions of creep behavior in any multiaxial state of stress can be made by these equations, based only on the results of a simple uniaxial creep test.

19.6.4   Cumulative Creep

There is at the present time no universally accepted method for estimating the creep strain accumulated as a result of exposure for various periods of time at different temperatures and stress levels. However, several different techniques for making such estimates have been proposed. The simplest of these is a linear hypothesis suggested by Robinson.[71] A generalized version of the Robinson hypothesis may be written as follows: If a design limit of creep strain δ_D is specified, it is predicted that the creep strain δ_D will be reached when

$$\sum_{i=1}^{k} \frac{t_i}{L_i} = 1 \tag{19.115}$$

where t_i = time of exposure at the ith combination of stress level and temperature
$\quad\quad L_i$ = time required to produce creep strain δ_D if entire exposure were held constant at the ith combination of stress level and temperature

Stress rupture may also be predicted by (19.115) if the L_i values correspond to stress rupture. This prediction technique gives relatively accurate results if the creep deformation is dominated by stage II steady-state creep behavior. Under other circumstances the method may yield predictions that are seriously in error.

Other cumulative creep prediction techniques that have been proposed include the time-hardening rule, the strain-hardening rule, and the life-fraction rule. The time-hardening rule is based on the assumption that the major factor governing the creep rate is the length of exposure at a given temperature and stress level, no matter what the past history of exposure has been. The strain-hardening rule is based on the assumption that the major factor governing the creep rate is the amount of prior strain, no matter what the past history of exposure has been. The life-fraction rule is a compromise between the time-hardening rule and the strain-hardening rule which accounts for influence of both time history and strain history. The life-fraction rule is probably the most accurate of these prediction techniques.

19.7   COMBINED CREEP AND FATIGUE

There are several important high-performance applications of current interest in which conditions persist that lead to combined creep and fatigue. For example, aircraft gas turbines and nuclear power reactors are subjected to this combination of failure modes. To make matters worse, the duty cycle in these applications might include a sequence of events including fluctuating stress levels at constant temperature, fluctuating temperature levels at constant stress, and periods during which both stress and temperature are simultaneously fluctuating. Furthermore, there is evidence to indicate that the fatigue and creep processes interact to produce a synergistic response.

It has been observed that interrupted stressing may accelerate, retard, or leave unaffected the time under stress required to produce stress rupture. The same observation has also been made with respect to creep rate. Temperature cycling at constant stress level may also produce a variety of responses, depending on material properties and the details of the temperature cycle.

No general law has been found by which cumulative creep and stress rupture response under temperature cycling at constant stress or stress cycling at constant temperature in the creep range can be accurately predicted. However, some recent progress has been made in developing life prediction techniques for combined creep and fatigue. For example, a procedure sometimes used to predict failure under combined creep and fatigue conditions for isothermal cyclic stressing is to assume that the creep behavior is controlled by the mean stress σ_m and that the fatigue behavior is controlled by the stress amplitude σ_a, with the two processes combining linearly to produce failure. This approach is similar to the development of the Goodman diagram described in Section 19.5.4 except that instead of an intercept of σ_u on the σ_m axis, as shown in Fig. 19.45, the intercept used is the *creep-limited static stress* σ_{cr}, as shown in Fig. 19.73. The creep-limited static stress corresponds either to the design limit on creep strain at the design life or to creep rupture at the design life, depending on which failure mode governs. The linear failure prediction rule then may be stated as

Failure is predicted to occur under combined isothermal creep and fatigue if

$$\frac{\sigma_a}{\sigma_N} + \frac{\sigma_m}{\sigma_{cr}} \geq 1 \qquad (19.116)$$

An elliptic relationship is also shown in Fig. 19.73, which may be written as

Failure is predicted to occur under combined isothermal creep and fatigue if

$$\left(\frac{\sigma_a}{\sigma_N}\right)^2 + \left(\frac{\sigma_m}{\sigma_{cr}}\right)^2 \geq 1 \qquad (19.117)$$

The linear rule is usually (but not always) conservative. In the higher-temperature portion of the creep range the elliptic relationship usually gives better agreement with data. For example, in Fig. 19.74a actual data for combined isothermal creep and fatigue tests are shown for several different temperatures using a cobalt-base S-816 alloy. The elliptic approximation is clearly better at higher temperatures for this alloy. Similar data are shown in Fig. 19.74b for 2024 aluminum alloy. Detailed studies of the relationships among creep strain, strain at rupture, mean stress, and alternating stress amplitude over a range of stresses and constant temperatures involve extensive, complex testing pro-

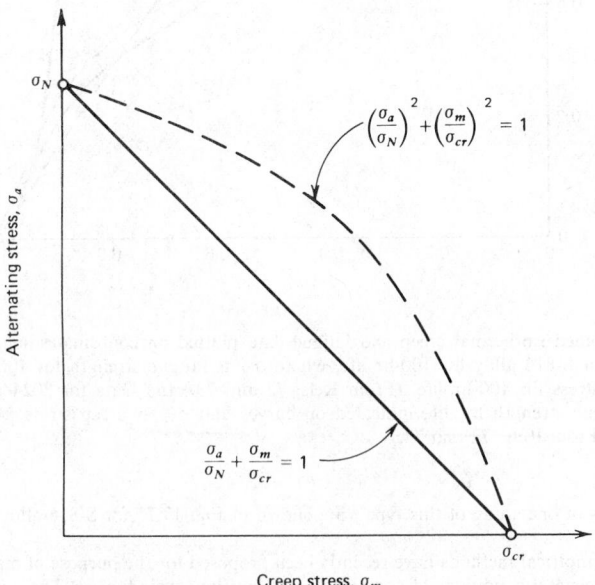

Fig. 19.73 Failure prediction diagram for combined creep and fatigue under constant-temperature conditions.

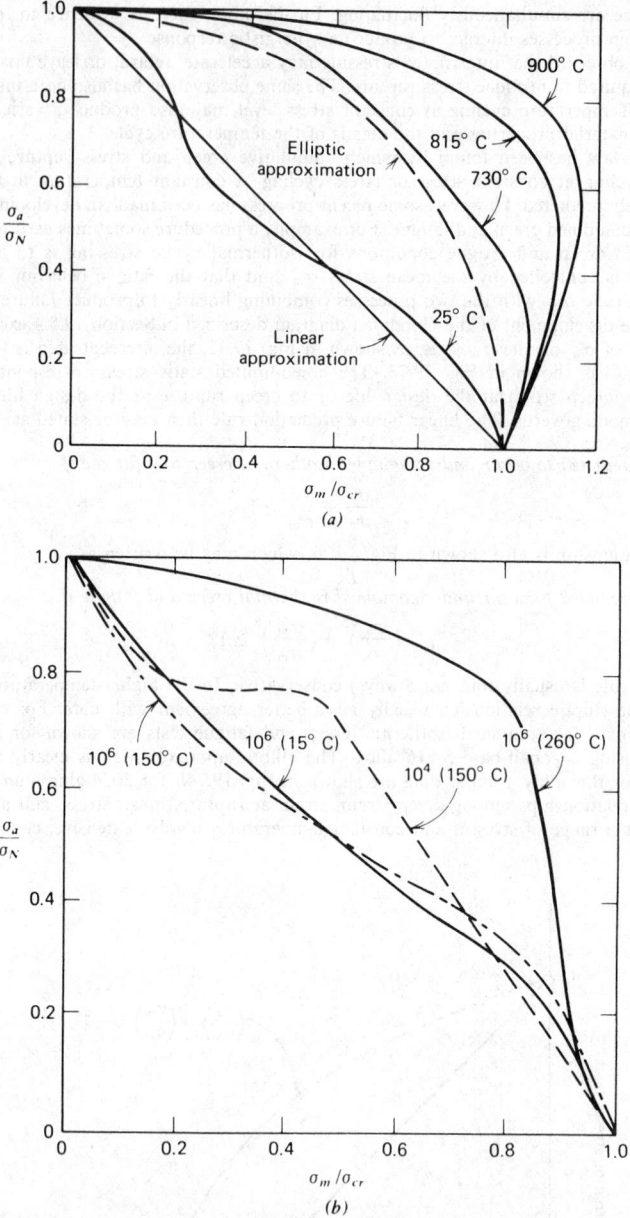

Fig. 19.74 Combined isothermal creep and fatigue data plotted on coordinates suggested in Figure 19.73. (*a*) Data for S-816 alloy for 100-hr life, where σ_N is fatigue strength for 100-hr life and σ_{cr} is creep rupture stress for 100-hr life. (From Refs. 72 and 73.) (*b*) Data for 2024 aluminum alloy, where σ_N is fatigue strength for life indicated on curves and σ_{cr} is creep stress for corresponding time to rupture. (From Refs. 72 and 74.)

grams. The results of one study of this type[74] are shown in Fig. 19.75 for S-816 alloy at two different temperatures.

Several other empirical methods have recently been proposed for the purpose of making life predictions under more general conditions of combined creep and low-cycle fatigue. These methods include:

1. Frequency-modified stress and strain-range method.[75]
2. Total time to fracture versus time-of-one-cycle method.[76]

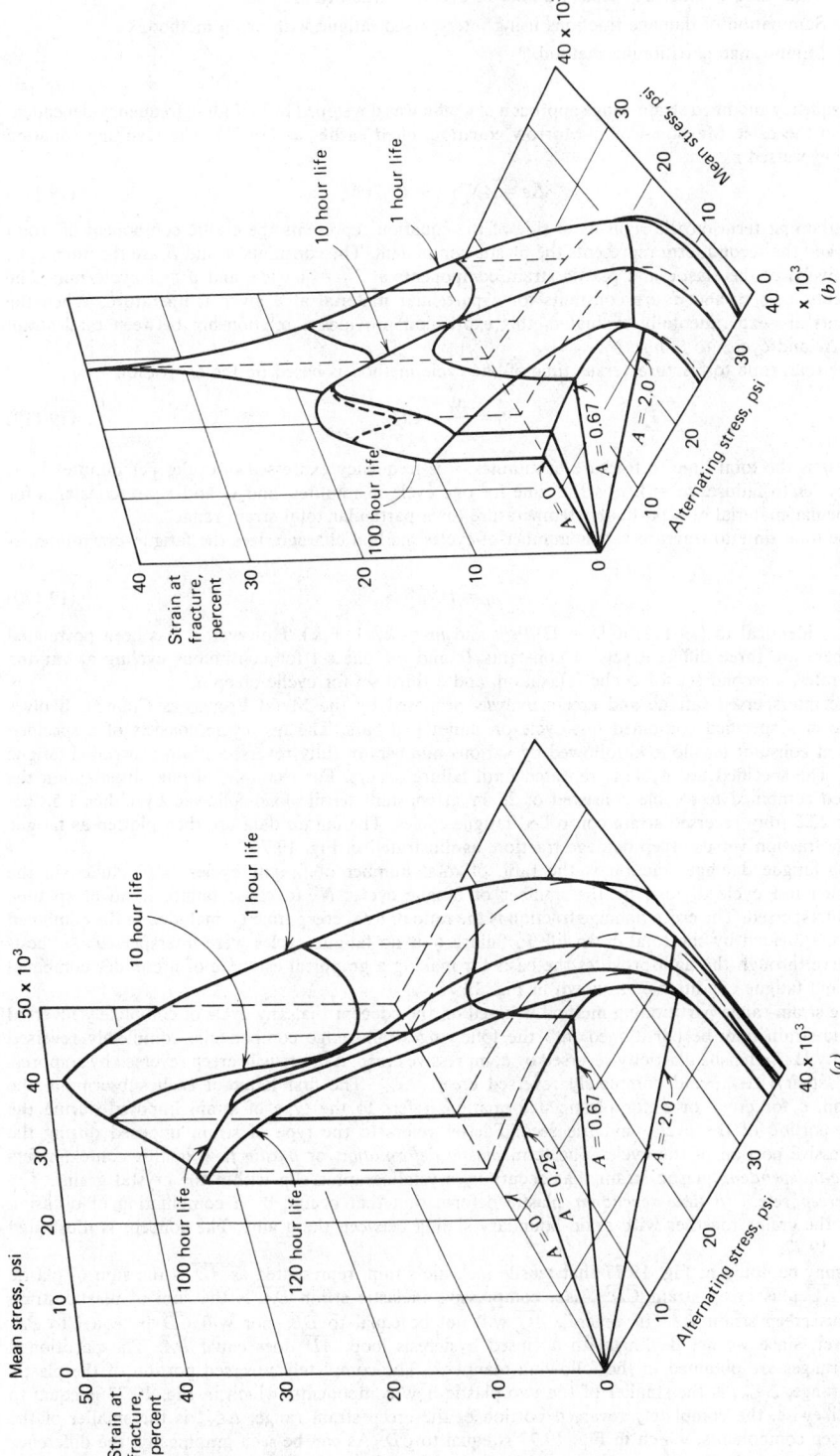

Fig. 19.75 Strain at fracture for various combinations of mean and alternating stresses in unnotched specimens of S-816 alloy. (a) Data taken at 900°C. (b) Data taken at 816°C. (From Refs. 72 and 73.)

3. Total time to fracture versus number of cycles to fracture method.[77]
4. Summation of damage fractions using interspersed fatigue with creep method.[78]
5. Strain-range partitioning method.[79]

The frequency-modified strain-range approach of Coffin was developed by including frequency-dependent terms in the basic Manson–Coffin–Morrow equation, cited earlier as (19.91). The resulting equation can be expressed as

$$\Delta\epsilon = AN_j^a \nu^b + BN_f^c \nu^d \qquad (19.118)$$

where the first term on the right-hand side of the equation represents the elastic component of strain range, and the second term represents the plastic component. The constants A and B are the intercepts, respectively, of the elastic and plastic strain components at $N_f = 1$ cycle and $\nu = 1$ cycle/min. The exponents a, b, c, and d are constants for a particular material at a given temperature. When the constants are experimentally evaluated, this expression provides a relationship between total strain range $\Delta\epsilon$ and cycles to failure N_f.

The total time to fracture versus time-of-one-cycle method is based on the expression

$$t_f = \frac{N_f}{\nu} = Ct_c^k \qquad (19.119)$$

where t_f is the total time to fracture in minutes, ν is frequency expressed in cycles per minute, N_f is total cycles to failure, $t_c = 1/\nu$ is the time for one cycle in minutes, and C and k are constants for a particular material at a particular temperature for a particular total strain range.

The total time to fracture versus number-of-cycles method characterizes the fatigue–creep interaction as

$$t_f = DN_f^{-m} \qquad (19.120)$$

which is identical to (19.119) if $D = C^{1/(1-k)}$ and $m = k/(1 - k)$. However, it has been postulated that there are three different sets of constants D and m: one set for continuous cycling at varying strain rates, a second set for cyclic relaxation, and a third set for cyclic creep.

The interspersed fatigue and creep analysis proposed by the Metal Properties Council involves the use of a specified combined test cycle on unnotched bars. The test cycle consists of a specified period at constant tensile load followed by various numbers of fully reversed strain-controlled fatigue cycles. The specified test cycle is repeated until failure occurs. For example, in one investigation the specified combined test cycle consisted of 23 hr at constant tensile load followed by either 1.5, 2.5, 5.5, or 22.5 fully reversed strain-controlled fatigue cycles. The failure data are then plotted as fatigue damage fraction versus creep damage fraction, as illustrated in Fig. 19.76.

The fatigue damage fraction is the ratio of total number of fatigue cycles N_f' included in the combined test cycle divided by the number of fatigue cycles N_f to cause failure if no creep time were interspersed. The creep damage fraction is the ratio of total creep time t_{cr} included in the combined test cycle divided by the total creep life to failure t_f if no fatigue cycles were interspersed. A "best-fit" curve through the data provides the basis for making a graphical estimate of life under combined creep and fatigue conditions, as shown in Fig. 19.76.

The strain-range partitioning method is based on the concept that any cycle of completely reversed inelastic strain may be partitioned into the following strain-range components: completely reversed plasticity, $\Delta\epsilon_{pp}$; tensile plasticity reversed by compressive creep, $\Delta\epsilon_{pc}$; tensile creep reversed by compressive plasticity, $\Delta\epsilon_{cp}$; and completely reversed creep, $\Delta\epsilon_{cc}$. The first letter of each subscript in the notation, c for creep or p for plastic deformation, refers to the type of strain imposed during the tensile portion of the cycle, and the second letter refers to the type of strain imposed during the compressive portion of the cycle. The term *plastic deformation* or *plastic flow* in this context refers to *time-independent* plastic strain that occurs by crystallographic slip within the crystal grains. The term *creep* refers to *time-dependent* plastic deformation that occurs by a combination of diffusion within the grains together with grain boundary sliding between the grains. The concept is illustrated in Fig. 19.77.

It may be noted in Fig. 19.77 that tensile inelastic strain, represented as $\overline{AD}$ is the sum of plastic strain $\overline{AC}$ plus creep strain $\overline{CD}$. Also, compressive inelastic strain $\overline{DA}$ is the sum of plastic strain $\overline{DB}$ plus creep strain $\overline{BA}$. In general, $\overline{AC}$ will not be equal to $\overline{DB}$, nor will $\overline{CD}$ be equal to $\overline{BA}$. However, since we are dealing with a closed hysteresis loop, $\overline{AD}$ does equal $\overline{DA}$. The partitioned strain ranges are obtained in the following manner:[81] The completely reversed portion of the plastic strain range, $\Delta\epsilon_{pp}$, is the smaller of the two plastic flow components, which in Fig. 19.77 is equal to $\overline{DB}$. Likewise, the completely reversed portion of the creep strain range, $\Delta\epsilon_{cc}$, is the smaller of the two creep components, which in Fig. 19.77 is equal to $\overline{CD}$. As can be seen graphically, the difference between the two plastic components must be equal to the difference between the two creep components, or $\overline{AC} - \overline{DB}$ must equal $\overline{BA} - \overline{CD}$. This difference then is either $\Delta\epsilon_{pc}$ or $\Delta\epsilon_{cp}$, in accordance with the notation just defined. For the case illustrated in Fig. 19.77, the difference is $\Delta\epsilon_{pc}$, since the tensile

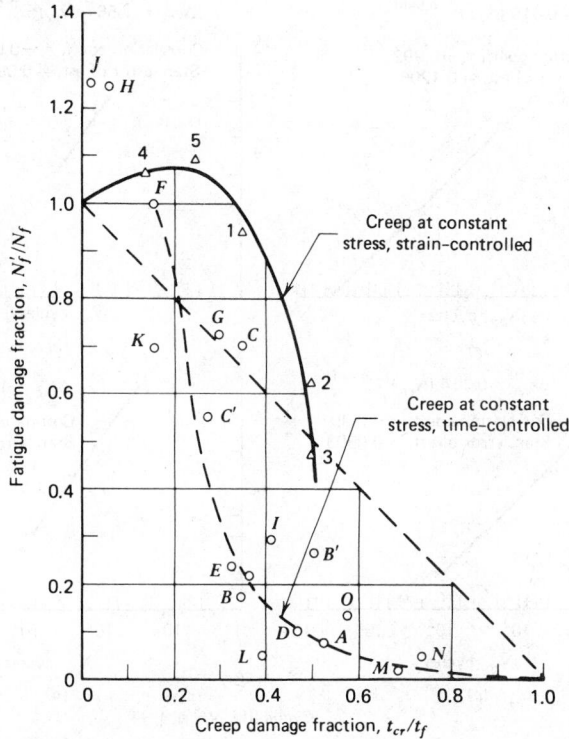

Fig. 19.76 Plot of fatigue damage fraction versus creep damage fraction for 1 Cr–1 Mo–¼ V rotor steel at 1000°F in air, using the method of the Metal Properties Council. (After Ref. 80, copyright Society for Experimental Stress Analysis, 1973; reprinted with permission.)

plastic strain component is greater than the compressive plastic strain component. It follows from this discussion that the sum of the partitioned strain ranges will necessarily be equal to the total inelastic strain range, or the width of the hysteresis loop.

It is next assumed that a unique relationship exists between cyclic life to failure and each of the four strain-range components listed. Available data indicate that these relationships are of the form of the basic Manson–Coffin–Morrow expression (19.91), as indicated, for example, in Fig. 19.78 for a

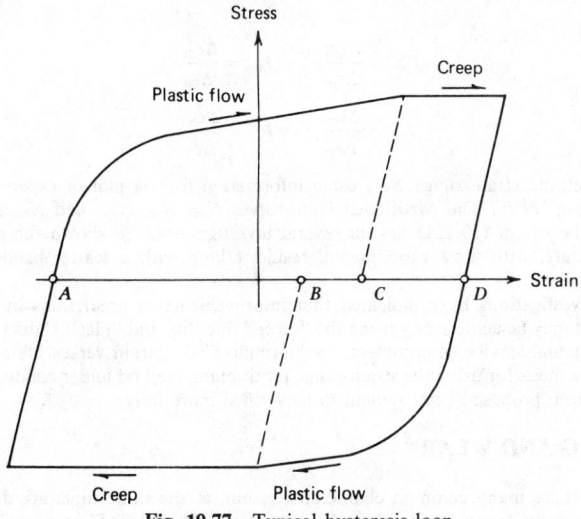

Fig. 19.77 Typical hysteresis loop.

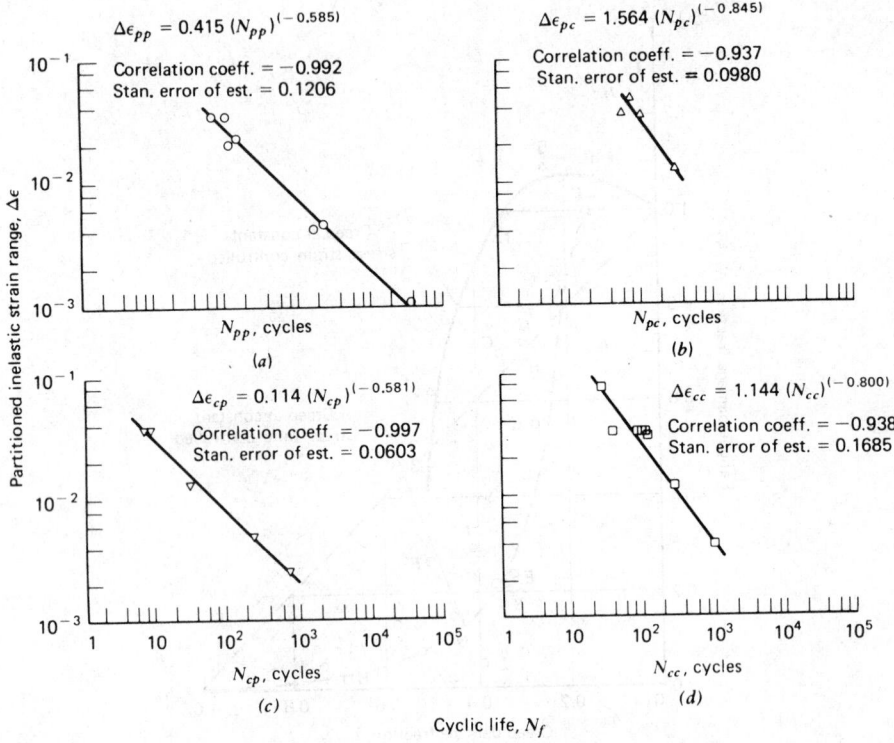

Fig. 19.78 Summary of partitioned strain-life relations for type 316 stainless steel at 1300°F (After Ref. 82): (a) pp-type strain range; (b) pc-type strain range; (c) cp-type strain range; (d) cc-type strain range.

type 316 stainless-steel alloy at 1300°F. The governing life prediction equation, or "interaction damage rule," is then postulated to be

$$\frac{1}{N_{\text{pred}}} = \frac{F_{pp}}{N_{pp}} + \frac{F_{pc}}{N_{pc}} + \frac{F_{cp}}{N_{cp}} + \frac{F_{cc}}{N_{cc}} \qquad (19.121)$$

where N_{pred} is the predicted total number of cycles to failure under the combined *straining* cycle containing all of the pertinent strain range components. The terms F_{pp}, F_{pc}, F_{cp}, and F_{cc} are defined as

$$F_{pp} = \frac{\Delta \epsilon_{pp}}{\Delta \epsilon_p}, \qquad F_{pc} = \frac{\Delta \epsilon_{pc}}{\Delta \epsilon_p}$$

$$(19.122)$$

$$F_{cp} = \frac{\Delta \epsilon_{cp}}{\Delta \epsilon_p}, \qquad F_{cc} = \frac{\Delta \epsilon_{cc}}{\Delta \epsilon_p}$$

for any selected inelastic strain range $\Delta \epsilon_p$, using information from a plot of experimental data such as that shown in Fig. 19.78. The partitioned failure lives N_{pp}, N_{pc}, N_{cp}, and N_{cc} are also obtained from Fig. 19.78. The use of (19.121) has, in several investigations,[82-87] shown the predicted lives to be acceptably accurate, with most experimental results falling with a scatter band of $\pm 2N_f$ of the predicted value.

More recent investigations have indicated that improvements in predictions by the strain-range partitioning method may be achieved by using the "creep" ductility and "plastic" ductility of a material determined in the actual service environment, to "normalize" the strain versus life equations prior to using (19.122). Procedures for using the strain-range partitioning method under conditions of multiaxial loading have also been proposed[86] but remain to be verified more fully.

19.8 FRETTING AND WEAR

Fretting and wear share many common characteristics but, at the same time, are distinctly different in several ways. Basically, fretting action has, for many years, been defined as a combined mechanical and chemical action in which contacting surfaces of two solid bodies are pressed together by a normal force and are caused to execute oscillatory sliding relative motion, wherein the magnitude of normal force is great enough and the amplitude of the oscillatory sliding motion is small enough to significantly restrict the flow of fretting debris away from the originating site.[88] More recent definitions of fretting action have been broadened to include cases in which contacting surfaces periodically separate and then reengage, as well as cases in which the fluctuating friction-induced surface tractions produce stress fields that may ultimately result in failure. The complexities of fretting action have been discussed by numerous investigators, who have postulated the combination of many mechanical, chemical, thermal, and other phenomena that interact to produce fretting. Among the postulated phenomena are plastic deformation caused by surface asperities plowing through each other, welding and tearing of contacting asperities, shear and rupture of asperities, friction-generated subsurface shearing stresses, dislodging of particles and corrosion products at the surfaces, chemical reactions, debris accumulation and entrapment, abrasive action, microcrack initiation, and surface delamination.[89-104]

Damage to machine parts due to fretting action may be manifested as corrosive surface damage due to fretting corrosion, loss of proper fit or change in dimensions due to fretting wear, or accelerated fatigue failure due to fretting fatigue. Typical sites of fretting damage include interference fits; bolted, keyed, splined, and riveted joints; points of contact between wires in wire ropes and flexible shafts; friction clamps; small-amplitude-of-oscillation bearings of all kinds; contacting surfaces between the leaves of leaf springs; and all other places where the conditions of fretting persist. Thus, the efficiency and reliability of the design and operation of a wide range of mechanical systems are related to the fretting phenomenon.

Wear may be defined as the undesired cumulative change in dimensions brought about by the gradual removal of discrete particles from contacting surfaces in motion, due predominantly to mechanical action. It should be further recognized that corrosion often interacts with the wear process to change the character of the surfaces of wear particles through reaction with the environment. Wear is, in fact, not a single process but a number of different processes that may take place by themselves or in combination. It is generally accepted that there are at least five major subcategories of wear (see p. 120 of Ref. 105; see also Ref. 106), including adhesive wear, abrasive wear, corrosive wear, surface fatigue wear, and deformation wear. In addition, the categories of fretting wear and impact wear[107-109] have been recognized by wear specialists. Erosion and cavitation are sometimes considered to be categories of wear as well. Each of these types of wear proceeds by a distinctly different physical process and must be separately considered, although the various subcategories may combine their influence either by shifting from one mode to another during different eras in the operational lifetime of a machine or by simultaneous activity of two or more different wear modes.

19.8.1 Fretting Phenomena

Although fretting fatigue, fretting wear, and fretting corrosion phenomena are potential failure modes in a wide variety of mechanical systems, and much research effort has been devoted to the understanding of the fretting process, there are very few quantitative design data available, and no generally applicable design procedure has been established for predicting failure under fretting conditions. However, even though the fretting phenomenon is not fully understood, and a good general model for prediction of fretting fatigue or fretting wear has not yet been developed, significant progress has been made in establishing an understanding of fretting and the variables of importance in the fretting process. It has been suggested that there may be more than 50 variables that play some role in the fretting process.[110] Of these, however, there are probably only eight that are of major importance; they are:

1. The magnitude of relative motion between the fretting surfaces.
2. The magnitude and distribution of pressure between the surfaces at the fretting interface.
3. The state of stress, including magnitude, direction, and variation with respect to time in the region of the fretting surfaces.
4. The number of fretting cycles accumulated.
5. The material, and surface condition, from which each of the fretting members is fabricated.
6. Cyclic frequency of relative motion between the two members being fretted.
7. Temperature in the region of the two surfaces being fretted.
8. Atmospheric environment surrounding the surfaces being fretted.

These variables interact so that a quantitative prediction of the influence of any given variable is very dependent on all the other variables in any specific application or test. Also, the combination of variables that produce a very serious consequence in terms of fretting fatigue damage may be quite different from the combinations of variables that produce serious fretting wear damage. No general techniques yet exist for quantitatively predicting the influence of the important variables of fretting fatigue and fretting wear damage, although many special cases have been investigated. However, it has been observed that certain trends usually exist when the variables just listed are changed. For example, fretting damage tends to increase with increasing contact pressure until a nominal pressure of a few thousand pounds per square inch is reached, and further increases in pressure seem to have relatively little direct effect. The state of stress is important, especially in fretting fatigue. Fretting damage accumulates with increasing numbers of cycles at widely different rates, depending on specific operating conditions. Fretting damage is strongly influenced by the material properties of the fretting pair—surface hardness, roughness, and finish. No clear trends have been established regarding frequency effects on fretting damage, and although both temperature and atmospheric environment are important influencing factors, their influences have not been clearly established. A clear presentation of the current state of knowledge relative to these various parameters is given, however, in Ref. 101.

Fretting fatigue is fatigue damage directly attributable to fretting action. It has been suggested that premature fatigue nuclei may be generated by fretting through either abrasive pit-digging action, asperity-contact microcrack initiation,[111] friction-generated cyclic stresses that lead to the formation of microcracks,[112] or subsurface cyclic shear stresses that lead to surface delamination in the fretting zone.[104] Under the abrasive pit-digging hypothesis, it is conjectured that tiny grooves or elongated pits are produced at the fretting interface by the asperities and abrasive debris particles moving under the influence of oscillatory relative motion. A pattern of tiny grooves would be produced in the fretted region with their longitudinal axes all approximately parallel and in the direction of fretting motion, as shown schematically in Fig. 19.79.

The asperity-contact microcrack initiation mechanism is postulated to proceed due to the contact force between the tip of an asperity on one surface and another asperity on the mating surface as the surfaces move back and forth. If the initial contact does not shear one or the other asperity from its base, the repeated contacts at the tips of the asperities give rise to cyclic or fatigue stresses in the region at the base of each asperity. It has been estimated[97] that under such conditions the region at the base of each asperity is subjected to large local stresses that probably lead to the nucleation of fatigue microcracks at these sites. As shown schematically in Fig. 19.80, it would be expected that the asperity-contact mechanism would produce an array of microcracks whose longitudinal axes would be generally perpendicular to the direction of fretting motion.

The friction-generated cyclic stress fretting hypothesis[99] is based on the observation that when

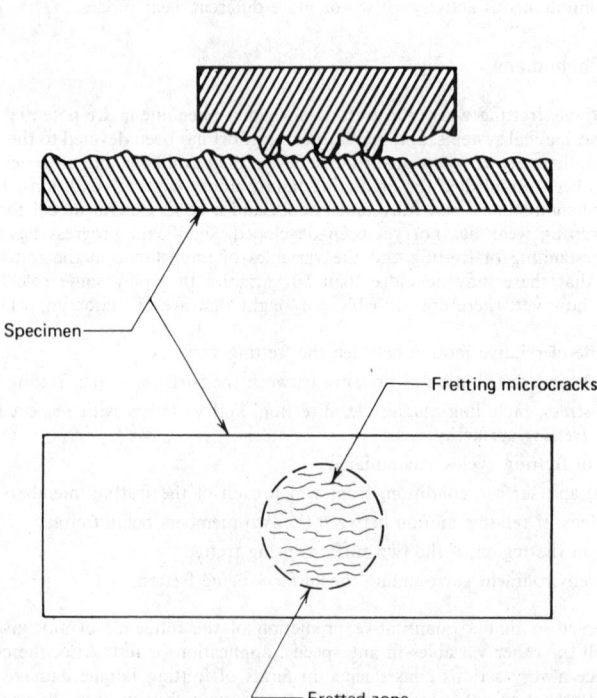

Fig. 19.79 Idealized schematic illustration of the stress concentrations produced by the abrasive pit-digging mechanism.

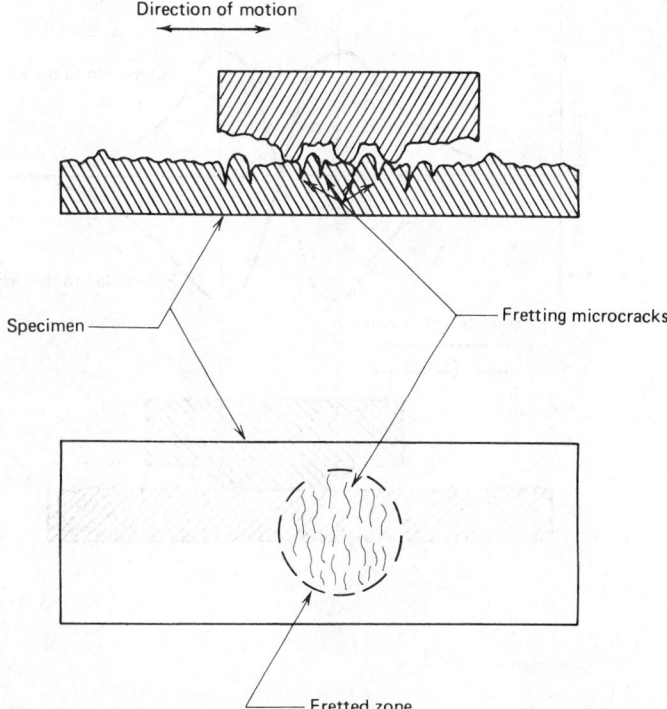

Direction of motion

Specimen

Fretting microcracks

Fretted zone

Fig. 19.80 Idealized schematic illustration of the stress concentrations produced by the asperity-contact microcrack initiation mechanism.

one member is pressed against the other and caused to undergo fretting motion, the tractive friction force induces a compressive tangential stress component in a volume of material that lies ahead of the fretting motion, and a tensile tangential stress component in a volume of material that lies behind the fretting motion, as shown in Fig. 19.81*a*. When the fretting direction is reversed, the tensile and compressive regions change places. Thus, the volume of material adjacent to the contact zone is subjected to a cyclic stress that is postulated to generate a field of microcracks at these sites. Furthermore, the geometrical stress concentration associated with the clamped joint may contribute to microcrack generation at these sites.[100] As shown in Fig. 19.81*c,* it would be expected that the friction-generated microcrack mechanism would produce an array of microcracks whose longitudinal axes would be generally perpendicular to the direction of fretting motion. These cracks would lie in a region adjacent to the fretting contact zone.

In the delamination theory of fretting[104] it is hypothesized that the combination of normal and tangential tractive forces transmitted through the asperity-contact sites at the fretting interface produce a complex multiaxial state of stress, accompanied by a cycling deformation field, which produces subsurface peak shearing stress and subsurface crack nucleation sites. With further cycling, the cracks propagate approximately parallel to the surface, as in the case of the surface fatigue phenomenon, finally propagating to the surface to produce a thin wear sheet, which "delaminates" to become a particle of debris.

Supporting evidence has been generated to indicate that under various circumstances each of the four mechanisms is active and significant in producing fretting damage.

The influence of the state of stress in the member during the fretting is shown for several different cases in Fig. 19.82, including static tensile and compressive mean stresses during fretting. An interesting observation in Fig. 19.82 is that fretting under conditions of compressive mean stress, either static or cyclic, produces a drastic reduction in fatigue properties. This, at first, does not seem to be in keeping with the concept that compressive stresses are beneficial in fatigue loading. However, it was deduced[113] that the compressive stresses during fretting shown in Fig. 19.82 actually resulted in local residual compressive stresses in the fretted region. Likewise, the tensile stresses during fretting shown in Fig. 19.82 actually resulted in local residual compressive stresses in the fretted region. The conclusion, therefore, is that local compressive stresses are beneficial in minimizing fretting fatigue damage.

Further evidence of the beneficial effects of compressive residual stresses in minimizing fretting fatigue damage is illustrated in Fig. 19.83, where the results of a series of Prot tests (see Section

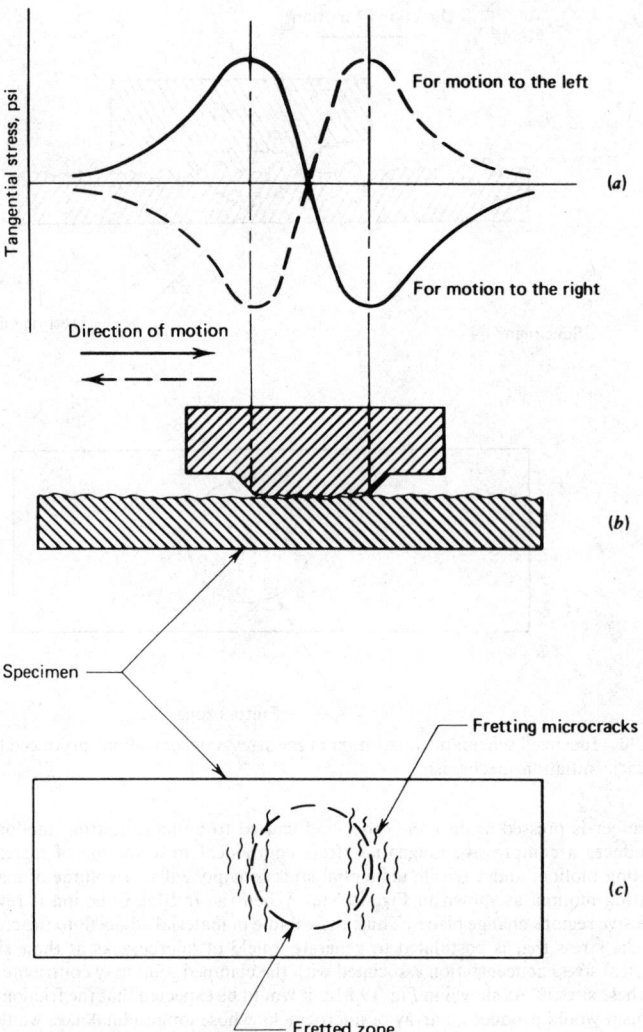

Fig. 19.81 Idealized schematic illustration of the tangential stress components and microcracks produced by the friction-generated microcrack initiation mechanism.

19.5.2) are reported for steel and titanium specimens subjected to various combinations of shot peening and fretting or cold rolling and fretting. It is clear from these results that the residual compressive stresses produced by shot peening and cold rolling are effective in minimizing the fretting damage. The reduction in scatter of the fretted fatigue properties for titanium is especially important to a designer because design stress is closely related to the lower limit of the scatter band.

Recent efforts to apply the tools of fracture mechanics to the problem of life prediction under fretting fatigue conditions have produced encouraging preliminary results that may ultimately provide designers with a viable quantitative approach.[114] These studies emphasize that the principal effect of fretting in the fatigue failure process is to accelerate crack initiation and the early stages of crack growth, and they suggest that when cracks have reached a sufficient length, the fretting no longer has a significant influence on crack propagation. At this point the fracture mechanics description of crack propagation described in Section 19.5.8 becomes valid.

In the final analysis, it is necessary to evaluate the seriousness of fretting fatigue damage in any specific design by running simulated service tests on specimens or components. Within the current state-of-the-art knowledge in the area of fretting fatigue, there is no other safe course of action open to the designer.

Fretting wear is a change in dimensions through wear directly attributable to the fretting process

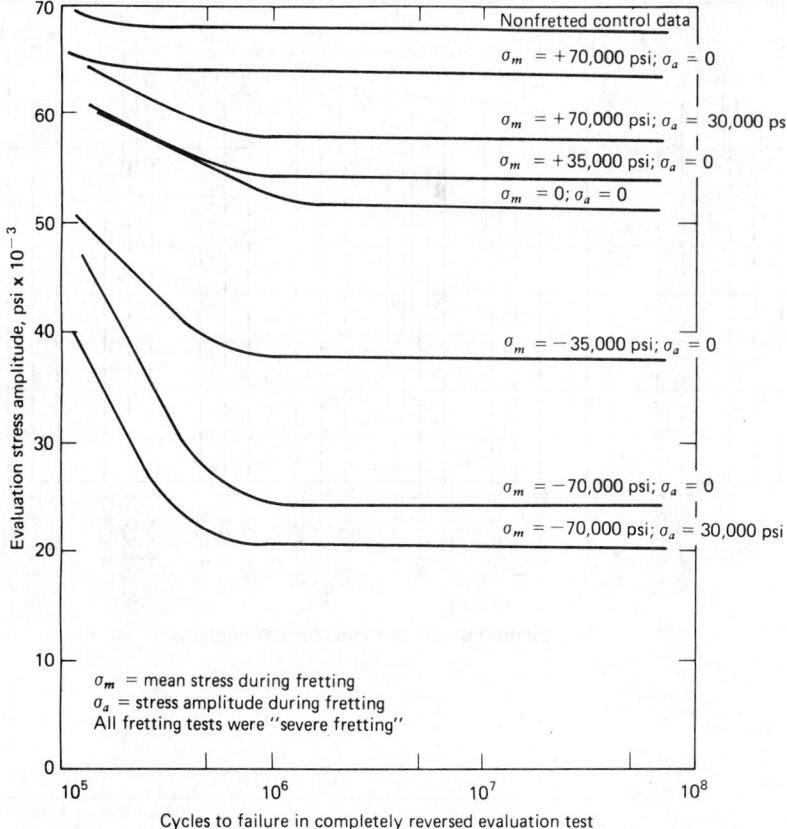

Fig. 19.82 Residual fatigue properties subsequent to fretting under various states of stress.

between two mating surfaces. It is thought that the abrasive pit-digging mechanism, the asperity-contact microcrack initiation mechanism, and the wear-sheet delamination mechanism may all be important in most fretting wear failures. As in the case of fretting fatigue, there has been no good model developed to describe the fretting wear phenomenon in a way useful for design. An expression for weight loss due to fretting has been proposed[94] as

$$W_{\text{total}} = (k_0 L^{1/2} - k_1 L)\frac{C}{F} + k_2 SLC \qquad (19.123)$$

where W_{total} = total specimen weight loss
L = normal contact load
C = number of fretting cycles
F = frequency of fretting
S = peak-to-peak slip between fretting surfaces
k_0, k_1, k_2 = constants to be empirically determined

This equation has been shown to give relatively good agreement with experimental data over a range of fretting conditions using mild steel specimens.[94] However, weight loss is not of direct use to a designer. Wear depth is of more interest. Prediction of wear depth in an actual design application must in general be based on simulated service testing.

Some investigators have suggested that estimates of fretting wear depth may be based on the classical adhesive or abrasive wear equations, in which wear depth is proportional to load and total distance slid, where the total distance slid is calculated by multiplying relative motion per cycle times number of cycles. Although there are some supporting data for such a procedure,[115] more investigation is required before it could be recommended as an acceptable approach for general application.

If fretting wear at a support interface, such as between tubes and support plates of a steam generator or heat exchanger or between fuel pins and support grids of a reactor core, produces loss of fit at a support site, impact fretting may occur. Impact fretting is fretting action induced by the small lateral

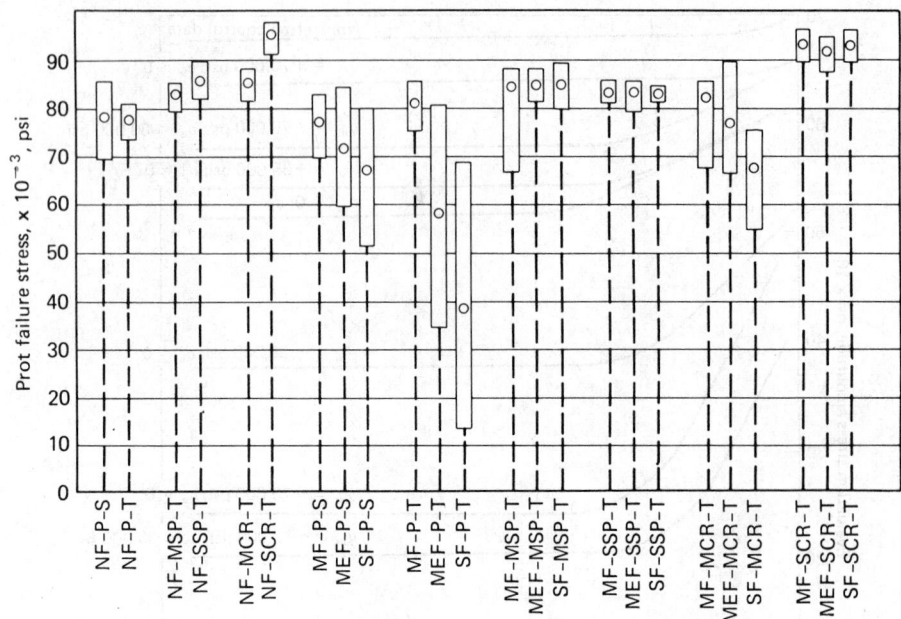

Test conditions used (see table for key symbols)

Test Condition Used	Code Designation	Sample Size	Mean Prot Failure Stress, psi	Unbiased Standard Deviation, psi
Nonfretted, polished, SAE 4340 steel	NF-P-S	15	78,200	5,456
Nonfretted, polished, Ti-140-A titanium	NF-P-T	15	77,800	2,454
Nonfretted, mildly shot-peened, Ti-140-A titanium	NF-MSP-T	15	83,100	1,637
Nonfretted, severely shot-peened, Ti-140-A titanium	NF-SSP-T	15	85,700	2,398
Nonfretted, mildly cold-rolled, Ti-140-A titanium	NF-MCR-T	15	85,430	1,924
Nonfretted, severely cold-rolled, Ti-140-A titanium	NF-SCR-T	15	95,400	2,120
Mildly fretted, polished, SAE 4340 steel	MF-P-S	15	77,280	4,155
Medium fretted, polished, SAE 4340 steel	MeF-P-S	15	71,850	5,492
Severely fretted, polished, SAE 4340 steel	SF-P-S	15	67,700	6,532
Mildly fretted, polished, Ti-140-A titanium	MF-P-T	15	81,050	3,733
Medium fretted, polished, Ti-140-A titanium	MeF-P-T	15	58,140	15,715
Severely fretted, polished, Ti-140-A titanium	SF-P-T	15	38,660	19,342
Mildly fretted, mildly shot-peened, Ti-140-A titanium	MF-MSP-T	15	84,520	5,239
Medium fretted, mildly shot-peened, Ti-140-A titanium	MeF-MSP-T	15	84,930	2,446
Severely fretted, mildly shot-peened, Ti-140-A titanium	SF-MSP-T	15	84,870	2,647
Mildly fretted, severely shot-peened, Ti-140-A titanium	MF-SSP-T	15	83,600	1,474
Medium fretted, severely shot-peened, Ti-140-A titanium	MeF-SSP-T	15	83,240	1,332
Severely fretted, severely shot-peened, Ti-140-A titanium	SF-SSP-T	15	83,110	1,280
Mildly fretted, mildly cold-rolled, Ti-140-A titanium	MF-MCR-T	15	82,050	4,313
Medium fretted, mildly cold-rolled, Ti-140-A titanium	MeF-MCR-T	15	76,930	8,305
Severely fretted, mildly cold-rolled, Ti-140-A titanium	SF-MCR-T	15	67,960	5,682
Mildly fretted, severely cold-rolled, Ti-140-A titanium	MF-SCR-T	15	93,690	1,858
Medium fretted, severely cold-rolled, Ti-140-A titanium	MeF-SCR-T	15	91,950	2,098
Severely fretted, severely cold-rolled, Ti-140-A titanium	SF-SCR-T	15	93,150	1,365

Fig. 19.83 Fatigue properties of fretted steel and titanium specimens with various degrees of shot peening and cold rolling. (see Ref. 98).

relative displacements between two surfaces when they impact together, where the small displacements are caused by Poisson strains or small tangential "glancing" velocity components. Impact fretting has only recently been addressed in the literature,[116] but it should be noted that under certain circumstances impact fretting may be a potential failure mode of great importance.

Fretting corrosion may be defined as any corrosive surface involvement resulting as a direct result of fretting action. The consequences of fretting corrosion are generally much less severe than for either fretting wear or fretting fatigue. Note that the term *fretting corrosion* is not being used here as a synonym for fretting, as in much of the early literature on this topic. Perhaps the most important single parameter in minimizing fretting corrosion is proper selection of the material pair for the application. Table 19.6 lists a variety of material pairs grouped according to their resistance to fretting corrosion.[117] Cross comparisons from one investigator's results to another's must be made with care because testing conditions varied widely. The minimization or prevention of fretting damage must be carefully considered as a separate problem in each individual design application because a palliative in one application may significantly accelerate fretting damage in a different application. For example, in a joint that is designed to have no relative motion, it is sometimes possible to reduce or prevent fretting by increasing the normal pressure until all relative motion is arrested. However, if the increase in normal pressure does not completely arrest the relative motion, the result may be significantly increasing fretting damage instead of preventing it.

Nevertheless, there are several basic principles that are generally effective in minimizing or preventing fretting. These include:

1. Complete separation of the contacting surfaces.
2. Elimination of all relative motion between the contacting surfaces.
3. If relative motion cannot be eliminated, it is sometimes effective to superpose a large unidirectional relative motion that allows effective lubrication. For example, the practice of driving the inner or outer race of an oscillatory pivot bearing may be effective in eliminating fretting.
4. Providing compressive residual stresses at the fretting surface; this may be accomplished by shot peening, cold rolling, or interference fit techniques.
5. Judicious selection of material pairs.
6. Use of interposed low-shear-modulus shim material or plating, such as lead, rubber, or silver.
7. Use of surface treatments or coatings as solid lubricants.
8. Use of surface grooving or roughening to provide debris escape routes and differential strain matching through elastic action.

Of all these techniques, only the first two are completely effective in preventing fretting. The remaining concepts, however, may often be used to minimize fretting damage and yield an acceptable design.

19.8.2   Wear Phenomena

The complexity of the wear process may be better appreciated by recognizing that many variables are involved, including the hardness, toughness, ductility, modulus of elasticity, yield strength, fatigue properties, and structure and composition of the mating surfaces, as well as geometry, contact pressure, temperature, state of stress, stress distribution, coefficient of friction, sliding distance, relative velocity, surface finish, lubricants, contaminants, and ambient atmosphere at the wearing interface. Clearance versus contact-time history of the wearing surfaces may also be an important factor in some cases. Although the wear processes are complex, progress has been made in recent years toward development of quantitative empirical relationships for the various subcategories of wear under specified operating conditions. Adhesive wear is often characterized as the most basic or fundamental subcategory of wear since it occurs to some degree whenever two solid surfaces are in rubbing contact and remains active even when all other modes of wear have been eliminated. The phenomenon of adhesive wear may be best understood by recalling that all real surfaces, no matter how carefully prepared and polished, exhibit a general waviness upon which is superposed a distribution of local protuberances or asperities. As two surfaces are brought into contact, therefore, only a relatively few asperities actually touch, and the *real* area of contact is only a small fraction of the *apparent* contact area. (See Chap. I of Ref. 118 and Chap. II of Ref. 119.) Thus, even under very small applied loads the local pressures at the contact sites become high enough to exceed the yield strength of one or both surfaces, and local plastic flow ensues. If the contacting surfaces are clean and uncorroded, the very intimate contact generated by this local plastic flow brings the atoms of the two contacting surfaces close enough together to call into play strong adhesive forces. This process is sometimes called *cold welding*. Then if the surfaces are subjected to relative sliding motion, the cold-welded junctions must be broken. Whether they break at the original interface or elsewhere within the asperity depends on surface conditions, temperature distribution, strain-hardening characteristics, local geometry, and stress distribution. If the junction is broken away from the original interface, a particle of one surface is transferred

Table 19.6 Fretting Corrosion Resistance of Various Material Pairs[117]

Material Pairs Having Good Fretting Corrosion Resistance

Sakmann and Rightmire	Lead	on	Steel
	Silver plate	on	Steel
	Silver plate	on	Silver plate
	'Parco-lubrized' steel	on	Steel
Gray and Jenny	Grit blasted steel plus lead plate	on	Steel (very good)
	1/16 in. nylon insert	on	Steel (very good)
	Zinc and iron phosphated (Bonderizing) steel	on	Steel (good with thick coat)
McDowell	Laminated plastic	on	Gold plate
	Hard tool steel	on	Tool steel
	Cold-rolled steel	on	Cold-rolled steel
	Cast iron	on	Cast iron with phosphate coating
	Cast iron	on	Cast iron with rubber cement
	Cast iron	on	Cast iron with tungsten sulphide coating
	Cast iron	on	Cast iron with rubber insert
	Cast iron	on	Cast iron with Molykote lubricant
	Cast iron	on	Stainless steel with Molykote lubricant

Material Pairs Having Intermediate Fretting Corrosion Resistance

Sakmann and Rightmire	Cadmium	on	Steel
	Zinc	on	Steel
	Copper alloy	on	Steel
	Zinc	on	Aluminium
	Copper plate	on	Aluminium
	Nickel plate	on	Aluminium
	Silver plate	on	Aluminium
	Iron plate	on	Aluminium
Gray and Jenny	Sulphide coated bronze	on	Steel
	Cast bronze	on	"Parco-lubrized" steel
	Magnesium	on	"Parco-lubrized" steel
	Grit-blasted steel	on	Steel
McDowell	Cast iron	on	Cast iron (rough or smooth surface)
	Copper	on	Cast iron
	Brass	on	Cast iron
	Zinc	on	Cast iron
	Cast iron	on	Silver plate
	Cast iron	on	Copper plate
	Magnesium	on	Copper plate
	Zirconium	on	Zirconium
Sakmann and Rightmire	Steel	on	Steel
	Nickel	on	Steel
	Aluminium	on	Steel
	Al-Si alloy	on	Steel
	Antimony plate	on	Steel
	Tin	on	Steel
	Aluminium	on	Aluminium
	Zinc plate	on	Aluminium
Gray and Jenny	Grit blast plus silver plate	on	Steel*
	Steel	on	Steel
	Grit blast plus copper plate	on	Steel
	Grit blast plus tin plate	on	Steel
	Grit blast and aluminium foil	on	Steel
	Be-Cu insert	on	Steel
	Magnesium	on	Steel
	Nitrided steel	on	Chromium plated steel†

Table 19.6 (*Continued*)

		Material Pairs Having Poor Fretting Corrosion Resistance	
McDowell	Aluminium	on	Cast iron
	Aluminium	on	Stainless steel
	Magnesium	on	Cast iron
	Cast iron	on	Chromium plate
	Laminated plastic	on	Cast iron
	Bakelite	on	Cast iron
	Hard tool steel	on	Stainless steel
	Chromium plate	on	Chromium plate
	Cast iron	on	Tin plate
	Gold plate	on	Gold plate

*Possibly effective with light loads and thick (0.005 inch) silver plate.
†Some improvement by heating chromium plated steel to 538°C for 1 hour.

to the other surface, marking one event in the adhesive wear process. Later sliding interactions may dislodge the transferred particles as loose wear particles, or they may remain attached. If this adhesive wear process becomes severe and large-scale metal transfer takes place, the phenomenon is called *galling*. If the galling becomes so severe that two surfaces adhere over a large region so that the actuating forces can no longer produce relative motion between them, the phenomenon is called *seizure*. If properly controlled, however, the adhesive wear rate may be low and self-limiting, often being exploited in the "wearing-in" process to improve mating surfaces such as bearings or cylinders so that full film lubrication may be effectively used.

One quantitative estimate of the amount of adhesive wear is given as follows (see Ref. 105 and Chaps. 2 and 6 of Ref. 120):

$$d_{adh} = \frac{V_{adh}}{A_a} = \left(\frac{k}{9\sigma_{yp}}\right)\left(\frac{W}{A_a}\right)L_s \tag{19.124}$$

or

$$d_{adh} = k_{adh}p_m L_s \tag{19.125}$$

where d_{adh} is the average wear depth, A_a is the apparent contact area, L_s is the total sliding distance, V_{adh} is the wear volume, W is the applied load, $p_m = W/A_a$ is the mean nominal contact pressure between bearing surfaces, and $k_{adh} = k/9\sigma_{yp}$ is a wear coefficient that depends on the probability of formation of a transferred fragment and the yield strength (or hardness) of the softer material. Typical values of the wear constant k for several material pairs are shown in Table 19.7, and the influence of lubrication on the wear constant k is indicated in Table 19.8.

Noting from (19.125) that

$$k_{adh} = \frac{d_{adh}}{p_m L_s} \tag{19.126}$$

it may be observed that if the ratio $d_{adh}/p_m L_s$ is experimentally found to be constant, (19.125) should be valid. Experimental evidence has been accumulated (see pp. 124–125 of Ref. 105) to confirm that for a given material pair this ratio is constant up to mean nominal contact pressures approximately equal to the uniaxial yield strength. Above this level the adhesive wear coefficient increases rapidly, with attendant severe galling and seizure.

Table 19.7 **Archard Adhesive Wear Constant k for Various Unlubricated Material Pairs in Sliding Contact**[a]

Material Pair	Wear Constant k
Zinc on zinc	160×10^{-3}
Low-carbon steel on low-carbon steel	45×10^{-3}
Copper on copper	32×10^{-3}
Stainless steel on stainless steel	21×10^{-3}
Copper (on low-carbon steel)	1.5×10^{-3}
Low-carbon steel (on copper)	0.5×10^{-3}
Bakelite on bakelite	0.02×10^{-3}

[a] From Chap. 6 of Ref. 120, with permission of John Wiley & Sons.

Table 19.8 Order of Magnitude Values for Adhesive Wear Constant k Under Various Conditions of Lubrication[a]

Lubrication Condition	Metal (on metal)		Nonmetal (on metal)
	Like	Unlike	
Unlubricated	5×10^{-3}	2×10^{-4}	5×10^{-6}
Poorly lubricated	2×10^{-4}	2×10^{-4}	5×10^{-6}
Average lubrication	2×10^{-5}	2×10^{-5}	5×10^{-6}
Excellent lubrication	2×10^{-6} to 10^{-7}	2×10^{-6} to 10^{-7}	2×10^{-6}

[a] From Chap. 6 of Ref. 120, with permission of John Wiley & Sons.

In the selection of metal combinations to provide resistance to adhesive wear, it has been found that the sliding pair should be composed of mutually insoluble metals and that at least one of the metals should be from the B subgroup of the periodic table. (See p. 31 of Ref. 121.) The reasons for these observations are that the number of cold-weld junctions formed is a function of the mutual solubility, and the strength of the junction bonds is a function of the bonding characteristics of the metals involved. The metals in the B subgroup of the periodic table are characterized by weak, brittle covalent bonds. These criteria have been verified experimentally, as shown in Table 19.9, where 114 of 123 pairs tested substantiated the criteria.

In the case of abrasive wear, the wear particles are removed from the surface by the plowing and gouging action of the asperities of a harder mating surface or by hard particles trapped between the rubbing surfaces. This type of wear is manifested by a system of surface grooves and scratches, often called *scoring*. The abrasive wear condition in which the hard asperities of one surface wear away the mating surface is commonly called *two-body wear*, and the condition in which hard abrasive particles between the two surfaces cause the wear is called *three-body wear*.

An average abrasive wear depth d_{abr} may then be estimated as

$$d_{abr} = \frac{V_{abr}}{A_a} = \frac{(\tan \theta)_m}{3\pi \sigma_{yp}} \left(\frac{W}{A_a} \right) L_s \qquad (19.127)$$

or

$$d_{abr} = k_{abr} p_m L_s \qquad (19.128)$$

where W is total applied load, $(\tan \theta)_m$ is a weighted mean value for all asperities, L_s is total distance of sliding, σ_{yp} is the uniaxial yield point strength for the softer material, V_{abr} is abrasive wear volume, $p_m = W/A_a$ is mean nominal contact pressure between bearing surfaces, and $k_{abr} = (\tan \theta)_m/3\sigma_{yp}$ is an abrasive wear coefficient that depends on the roughness characteristics of the surface and the yield strength (or hardness) of the softer material.

Comparing (19.127) for abrasive wear volume with (19.124) for adhesive wear volume, we note that they are formally the same except the constant $k/3$ in the adhesive wear equation is replaced by $(\tan \theta)_m/\pi$ in the abrasive wear equation. Typical values of the wear constant $3(\tan \theta)_m/\pi$ for several materials are shown in Table 19.10. As indicated in Table 19.10, experimental evidence shows that k_{abr} for three-body wear is typically about an order of magnitude smaller than for the two-body case, probably because the trapped particles tend to roll much of the time and cut only a small part of the time.

In selecting materials for abrasive wear resistance, it has been established that both hardness and modulus of elasticity are key properties. Increasing wear resistance is associated with higher hardness and lower modulus of elasticity since both the amount of elastic deformation and the amount of elastic energy that can be stored at the surface are increased by higher hardness and lower modulus of elasticity.

Table 19.11 tabulates several materials in order of descending values of (hardness)/(modulus of elasticity). Well-controlled experimental data are not yet available, but general experience would provide an ordering of materials for decreasing wear resistance compatible with the array of Table 19.11. When the conditions for adhesive or abrasive wear exist together with conditions that lead to corrosion, the two processes persist together and often interact synergistically. If the corrosion product is hard and abrasive, dislodged corrosion particles trapped between contacting surfaces will accelerate the abrasive wear process. In turn, the wear process may remove the "protective" surface layer of corrosion product to bare new metal to the corrosive atmosphere, thereby accelerating the corrosion process. Thus, the corrosion wear process may be self-accelerating and may lead to high rates of wear.

On the other hand, some corrosion products, for example, metallic phosphates, sulfides, and chlorides, form as soft lubricative films that actually improve the wear rate markedly, especially if adhesive wear is the dominant phenomenon.

Three major wear control methods have been defined, as follows (see p. 36 of Ref. 121): *principle*

Table 19.9 Adhesive Wear Behavior of Various Pairs[a]

Description of Metal Pair	Material Combination				Remarks
	Al Disk	Steel Disk	Cu Disk	Ag Disk	
Soluble pairs with poor adhesive wear resistance	Be	Be	Be	Be	These pairs substantiate the criteria of solubility and B subgroup metals
	Mg	—	Mg	Mg	
	Al	Al	Al	—	
	Si	Si	Si	Si	
	Ca	—	Ca	—	
	Ti	Ti	Ti	—	
	Cr	Cr	—	—	
	—	Mn	—	—	
	Fe	Fe	—	—	
	Co	Co	Co	—	
	Ni	Ni	Ni	—	
	Cu	—	Cu	—	
	—	Zn	Zn	—	
	Zr	Zr	Zr	Zr	
	Nb	Nb	Nb	—	
	Mo	Mo	Mo	—	
	Rh	Rh	Rh	—	
	—	Pd	—	—	
	Ag	—	Ag	—	
	—	—	Cd	Cd	
	—	—	In	In	
	Sn	—	Sn	—	
	Ce	Ce	Ce	—	
	Ta	Ta	Ta	—	
	W	W	W	—	
	—	Ir	—	—	
	Pt	Pt	Pt	—	
	Au	Au	Au	Au	
	Th	Th	Th	Th	
	U	U	U	U	
Soluble pairs with fair or good adhesive wear resistance. (F) = Fair	—	Cu(F)	—		These pairs do not substantiate the stated criteria
	Zn(F)	—	—		
	—	—	Sb(F)		
Insoluble pairs, neither from the B subgroup, with poor adhesive wear resistance		Li			These pairs substantiate the stated criteria
		Mg			
		Ca			
		Ba			
Insoluble pairs, one from the B subgroup, with fair or good adhesive wear resistance. (F) = Fair	—	C(F)	—	—	These pairs substantiate the stated criteria
	—	—	—	Ti(F)	
	—	—	Cr(F)	Cr(F)	
	—	—	—	Fe(F)	
	—	—	—	Co(F)	
	—	—	Ge(F)	—	
	—	Se(F)	Se(F)	—	
	—	—	—	Nb(F)	
	—	Ag	—	—	
	Cd	Cd	—	—	
	In	In	—	—	
	—	Sn(F)	—	—	
	—	Sb(F)	Sb	—	
	Te(F)	Te(F)	Te(F)	—	
	Tl	Tl	Tl	—	
	Pb(F)	Pb	Pb	—	
	Bi(F)	Bi	Bi(F)	—	
Insoluble pairs, one from the B subgroup, with poor adhesive wear resistance	C	—	C	C	These pairs do not substantiate the stated criteria
	—	—	—	Ni	
	Se	—	—	—	
	—	—	—	Mo	

[a] See pp. 34–35 of Ref. 121.

341

Table 19.10   Abrasive Wear Constant $3(\tan \theta)_m / \pi$ for Various Materials in Sliding Contact as Reported by Different Investigators[a]

Materials	Wear type	Particle size, μ	$3(\tan \theta)_m / \pi$
Many	Two-body	–	180×10^{-3}
Many	Two-body	110	150×10^{-3}
Many	Two-body	40–150	120×10^{-3}
Steel	Two-body	260	80×10^{-3}
Many	Two-body	80	24×10^{-3}
Brass	Two-body	70	16×10^{-3}
Steel	Three-body	150	6×10^{-3}
Steel	Three-body	80	4.5×15^{-3}
Many	Three-body	40	2×10^{-3}

[a] See p. 169 of Ref. 120. Reprinted with permission from John Wiley & Sons.

of protective layers, including protection by lubricant, surface film, paint, plating, phosphate, chemical, flame-sprayed, or other types of interfacial layers; *principle of conversion,* in which wear is converted from destructive to permissible levels through better choice of metal pairs, hardness, surface finish, or contact pressure; and *principle of diversion,* in which the wear is diverted to an economical replaceable wear element that is periodically discarded and replaced as "wear out" occurs. When two surfaces operate in rolling contact, the wear phenomenon is quite different from the wear of sliding surfaces just described, although the "delamination" theory[122] is very similar to the mechanism of wear between rolling surfaces in contact as described here. Rolling surfaces in contact result in Hertz contact stresses that produce maximum values of shear stress slightly below the surface. (See, for example, p. 389 of Ref. 123.) As the rolling contact zone moves past a given location on the surface, the subsurface peak shear stress cycles from zero to a maximum value and back to zero, thus producing a cyclic stress field. Such conditions may lead to fatigue failure by the initiation of a subsurface crack that propagates under repeated cyclic loading and that may ultimately propagate to the surface to spall out a macroscopic surface particle to form a wear pit. This action, called *surface fatigue wear,* is a common failure mode in antifriction bearings, gears, and cams, and all machine parts that involve rolling surfaces in contact. Deformation wear arises as a result of repeated plastic deformations at the wearing surfaces; this wear may induce a matrix of cracks that grow and coalesce to form wear particles or may produce cumulative permanent plastic deformations that finally grow into an unacceptable surface indentation or wear scar. Deformation wear is generally caused by conditions that lead to

Table 19.11   Values of (Hardness/Modulus of Elasticity) for Various Materials [105] †

Material	Condition	BHN*/$(E \times 10^{-6})$ (in mixed units)
Alundum (Al_2O_3)	Bonded	143
Chrome plate	Bright	83
Gray iron	Hard	33
Tungsten carbide	9% Co	22
Steel	Hard	21
Titanium	Hard	17
Aluminum alloy	Hard	11
Gray iron	As cast	10
Structural steel	Soft	5
Malleable iron	Soft	5
Wrought iron	Soft	3.5
Chromium metal	As cast	3.5
Copper	Soft	2.5
Silver	Pure	2.3
Aluminum	Pure	2.0
Lead	Pure	2.0
Tin	Pure	0.7

†Reprinted from copyrighted work with permission; courtesy of Elsevier Publishing Company.

*Brinell hardness number.

impact loading between the two wearing surfaces. Although some progress has been made in deformation wear analysis, the techniques are highly specialized. Fretting wear, which has received renewed attention in the recent literature (see p. 55 of Ref. 124 and p. 75 of Ref. 120), has already been discussed. Impact wear is a term reserved for impact-induced repeated elastic deformations at the wearing surfaces that produce a matrix of cracks that grow in accordance with surface fatigue phenomena. Under some circumstances impact wear may be generated by purely normal impacts, and under other circumstances the impact may contain elements of rolling and/or sliding as well. The severity of the impact is generally measured or expressed in terms of the kinetic energy of the striking mass. The geometry of the impacting surfaces and the materials properties of the two contacting surfaces play a major role in the determination of severity of impact wear damage. The objective of a designer faced with impact wear as a potential failure mode is to predict the size of the wear scar, or its depth, as a function of the number of repetitive load cycles.

An *empirical* approach to the prediction of sliding wear has been developed,[125] and the pertinent empirical constants have been evaluated for a wide variety of materials and lubricant combinations for various operating conditions. This empirical development permits the designer to specify a design configuration to ensure "zero wear" during the specified design lifetime. *Zero wear* is defined to be wear of such small magnitude that the surface finish is not significantly altered by the wear process. That is, the wear depth for zero wear is of the order of one-half the peak-to-peak surface finish dimension.

If a *pass* is defined to be a distance of sliding W equal to the dimension of the contact area in the direction of sliding, N is the number of passes, τ_{max} is the maximum shearing stress in the vicinity of the surface, τ_{yp} is the shear yield point of the specified material, and γ_r is a constant for the particular combination of materials and lubricant, then the empirical model asserts that there will be "zero wear" for N passes if

$$\tau_{max} \leq \left[\frac{2 \times 10^3}{N}\right]^{1/9} \gamma_r \tau_{yp} \qquad (19.129)$$

or, to interpret it differently, the number of passes that can be accommodated without exceeding the zero wear level is given by

$$N = 2 \times 10^3 \left[\frac{\gamma_r \tau_{yp}}{\tau_{max}}\right]^9 \qquad (19.130)$$

It may be noted that the constant γ_r is referred to 2000 passes and must be experimentally determined. For quasihydrodynamic lubrication, γ_r ranges between 0.54 and 1. For dry or boundary lubrication, γ_r is 0.54 for materials with low susceptibility to adhesive wear and 0.20 for materials with high susceptibility to adhesive wear.

Calculation of the maximum shear stress τ_{max} in the vicinity of the contacting surface must include both the normal force and the friction force. Thus, for conforming geometries, such as a flat surface on a flat surface or a shaft in a journal bearing, a critical point at the contacting interface may be analyzed by the maximum shear stress theory to determine τ_{max}.

The number of passes will usually require expression as a function of the number of cycles, strokes, oscillations, or hours of operation in the design lifetime.

Utilizing these definitions and a proper stress analysis at the wear interface allows one to design for "zero wear" through use of Eqs. (19.129) or (19.130).

19.9 CORROSION AND STRESS CORROSION

Corrosion may be defined as the undesired deterioration of a material through chemical or electrochemical interaction with the environment, or destruction of materials by means other than purely mechanical action. Failure by corrosion occurs when the corrosive action renders the corroded device incapable of performing its design function. Corrosion often interacts synergistically with another failure mode, such as wear or fatigue, to produce the even more serious combined failure modes, such as corrosion wear or corrosion fatigue. Failure by corrosion and protection against failure by corrosion has been estimated to cost in excess of 8 billion dollars annually in the United States alone. (See p. 1 of Ref. 126.)

The complexity of the corrosion process may be better appreciated by recognizing that many variables are involved, including environmental, electrochemical, and metallurgical aspects. For example, anodic reactions and rate of oxidation; cathodic reactions and rate of reduction; corrosion inhibition, polarization, or retardation; passivity phenomena; effect of oxidizers; effect of velocity; temperature; corrosive concentration; galvanic coupling; and metallurgical structure all influence the type and rate of the corrosion process.

Corrosion processes have been categorized in many different ways. One convenient classification divides corrosion phenomena into the following types (see p. 28 of Ref. 126 and p. 85 of Ref. 127): direct chemical attack, galvanic corrosion, crevice corrosion, pitting corrosion, intergranular corrosion, selective leaching, erosion corrosion, cavitation corrosion, hydrogen damage, biological corrosion, and stress corrosion cracking. Depending on the types of environment, loading, and mechanical function of the machine parts involved, any of the types of corrosion may combine their influence with other failure modes to produce premature failures. Of particular concern are interactions that lead to failure by corrosion wear, corrosion fatigue, fretting wear, fretting fatigue, and corrosion-induced brittle fracture.

19.9.1 Types of Corrosion

Direct chemical attack is probably the most common type of corrosion. Under this type of corrosive attack the surface of the machine part exposed to the corrosive media is attacked more or less uniformly over its entire surface, resulting in a progressive deterioration and dimensional reduction of sound load-carrying net cross section. The rate of corrosion due to direct attack can usually be estimated from relatively simple laboratory tests in which small specimens of the selected material are exposed to a well-simulated actual environment, with frequent weight change and dimensional measurements carefully taken. The corrosion rate is usually expressed in mils per year (mpy) and may be calculated as (see p. 133 of Ref. 126)

$$R = \frac{534W}{\gamma At} \tag{19.131}$$

where R is rate of corrosion penetration in mils (1 mil = 0.001 in.) per year (mpy), W is weight loss in milligrams, A is exposed area of the specimen in square inches, γ is density of the specimen in grams per cubic centimeter, and t is exposure time in hours. Use of this corrosion rate expression in predicting corrosion penetration in actual service is usually successful if the environment has been properly simulated in the laboratory. Corrosion rate data for many different combinations of materials and environments are available in the literature.[128-130] Figure 19.84 illustrates one presentation of such data.

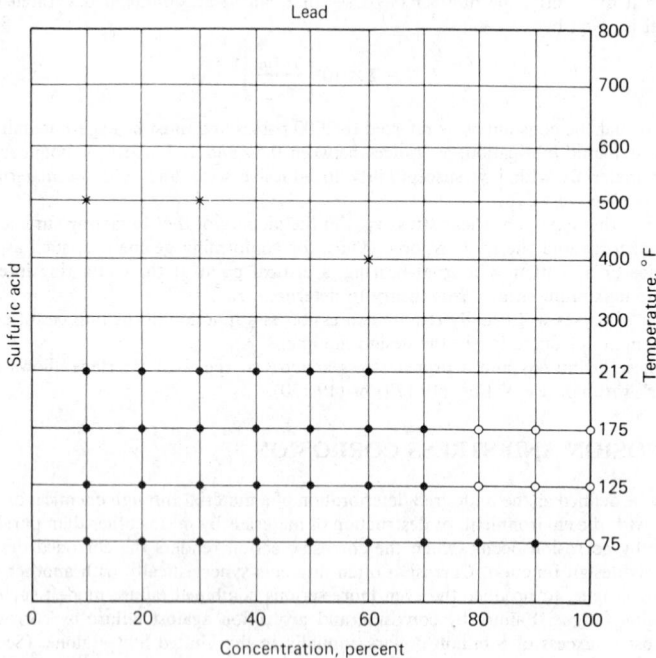

- Corrosion rate less than 2 mpy (mils/year)
- Corrosion rate less than 20 mpy
- Corrosion rate from 20 to 50 mpy
- Corrosion rate greater than 50 mpy

Fig. 19.84 Nelson's method for summarizing corrosion rate data for lead in sulfuric acid environment as a function of concentration and temperature. (See Ref. 128; reprinted with permission of McGraw-Hill Book Company.)

Direct chemical attack may be reduced in severity or prevented by any one or a combination of several means, including selecting proper materials to suit the environment; using plating, flame spraying, cladding, hot dipping, vapor deposition, conversion coatings, and organic coatings or paint to protect the base material; changing the environment by using lower temperature or lower velocity, removing oxygen, changing corrosive concentration, or adding corrosion inhibitors; using cathodic protection in which electrons are supplied to the metal surface to be protected either by galvanic coupling to a sacrificial anode or by an external power supply; or adopting other suitable design modifications.

Galvanic corrosion is an accelerated electrochemical corrosion that occurs when two dissimilar metals in electrical contact are made part of a circuit completed by a connecting pool or film of electrolyte or corrosive medium. Under these circumstances, the potential difference between the dissimilar metals produces a current flow through the connecting electrolyte, which leads to corrosion, concentrated primarily in the more anodic or less noble metal of the pair. This type of action is completely analogous to a simple battery cell. Current must flow to produce galvanic corrosion, and, in general, more current flow means more serious corrosion. The relative tendencies of various metals to form galvanic cells, and the probable direction of the galvanic action, are illustrated for several commercial metals and alloys in seawater in Table 19.12. (See p. 32 of Ref. 126 or p. 86 of Ref. 127.)

Ideally, tests in the actual service environment should be conducted; but, if such data are unavailable, the data of Table 19.12 should give a good indication of possible galvanic action. The farther apart the two dissimilar metals are in the galvanic series, the more serious the galvanic corrosion problem

Table 19.12 Galvanic Series of Several Commercial Metals and Alloys in Seawater[a]

↑	Platinum
	Gold
Noble or	Graphite
cathodic	Titanium
(protected	Silver
end)	⌈ Chlorimet 3 (62 Ni, 18 Cr, 18 Mo) ⌉
	⌊ Hastelloy C (62 Ni, 17 Cr, 15 Mo) ⌋
	⌈ 18-8 Mo stainless steel (passive) ⌉
	18-8 stainless steel (passive)
	⌊ Chromium stainless steel 11-30% Cr (passive) ⌋
	⌈ Inconel (passive)(80 Ni, 13 Cr, 7 Fe) ⌉
	⌊ Nickel (passive) ⌋
	Silver solder
	⌈ Monel (70 Ni, 30 Cu) ⌉
	Cupronickels (60-90 Cu, 40-10 Ni)
	Bronzes (Cu-Sn)
	Copper
	⌊ Brasses (Cu-Zn) ⌋
	⌈ Chlorimet 2 (66 Ni, 32 Mo, 1 Fe) ⌉
	⌊ Hastelloy B (60 Ni, 30 Mo, 6 Fe, 1 Mn) ⌋
	⌈ Inconel (active) ⌉
	⌊ Nickel (active) ⌋
	Tin
	Lead
	Lead-tin solders
	⌈ 18-8 Mo stainless steel (active) ⌉
	⌊ 18-8 stainless steel (active) ⌋
	Ni-Resist (high Ni cast iron)
	Chromium stainless steel, 13% Cr (active)
	⌈ Cast iron ⌉
	⌊ Steel or iron ⌋
Active or	2024 aluminum (4.5 Cu, 1.5 Mg, 0.6 Mn)
anodic	Cadmium
(corroded	Commercially pure aluminum (1100)
end)	Zinc
↓	Magnesium and magnesium alloys

[a] See p. 32 of Ref. 126. Reprinted with permission of McGraw-Hill Book Company.

may be. Material pairs within any bracketed group exhibit little or no galvanic action. It should be noted, however, that there are sometimes exceptions to the galvanic series of Table 19.12, so wherever possible corrosion tests should be performed with actual materials in the actual service environment.

The accelerated galvanic corrosion is usually most severe near the junction between the two metals, decreasing in severity at locations farther from the junction. The ratio of cathodic area to anodic area exposed to the electrolyte has a significant effect on corrosion rate. It is *desirable* to have a *small ratio* of cathode area to anode area. For this reason, if only *one* of two dissimilar metals in electrical contact is to be coated for corrosion protection, the *more* noble or more corrosion-resistant metal should be coated. Although this at first may seem the wrong metal to coat, the area effect, which produces anodic corrosion rate of 10^2–10^3 times cathodic corrosion rates for equal areas, provides the logic for this assertion.

Galvanic corrosion may be reduced in severity or prevented by one or a combination of several steps, including the selection of material pairs as close together as possible in the galvanic series, preferably in the same bracketed group; electrical insulation of one dissimilar metal from the other as completely as possible; maintaining as small a ratio of cathode area to anode area as possible; proper use and maintenance of coatings; the use of inhibitors to decrease the aggressiveness of the corroding medium; and the use of cathodic protection in which a third metal element anodic to both members of the operating pair is used as a sacrificial anode that may require periodic replacement.

Crevice corrosion is an accelerated corrosion process highly localized within crevices, cracks, and other small-volume regions of stagnant solution in contact with the corroding metal. For example, crevice corrosion may be expected in gasketed joints; clamped interfaces; lap joints; rolled joints; under bolt and rivet heads; and under foreign deposits of dirt, sand, scale, or corrosion product. Until recently, crevice corrosion was thought to result from differences in either oxygen concentration or metal ion concentration in the crevice compared to its surroundings. More recent studies seem to indicate, however, that the local oxidation and reduction reactions result in oxygen depletion in the stagnant crevice region, which leads to an excess positive charge in the crevice due to increased metal ion concentration. This, in turn, leads to a flow of chloride and hydrogen ions into the crevice, both of which accelerate the corrosion rate within the crevice. Such accelerated crevice corrosion is highly localized and often requires a lengthy incubation period of perhaps many months before it gets under way. Once started, the rate of corrosion accelerates to become a serious problem. To be susceptible to crevice corrosion attack, the stagnant region must be wide enough to allow the liquid to enter but narrow enough to maintain stagnation. This usually implies cracks and crevices of a few thousandths to a few hundredths of an inch in width.

To reduce the severity of crevice corrosion, or prevent it, it is necessary to eliminate the cracks and crevices. This may involve caulking or seal welding existing lap joints; redesign to replace riveted or bolted joints by sound, welded joints; filtering foreign material from the working fluid; inspection and removal of corrosion deposits; or using nonabsorbent gasket materials. Pitting corrosion is a very localized attack that leads to the development of an array of holes or pits that penetrate the metal. The pits, which typically are about as deep as they are across, may be widely scattered or so heavily concentrated that they simply appear as a rough surface. The mechanism of pit growth is virtually identical to that of crevice corrosion described, except that an existing crevice is not required to initiate pitting corrosion. The pit is probably initiated by a momentary attack due to a random variation in fluid concentration or a tiny surface scratch or defect. Some pits may become inactive because of a stray convective current, whereas others may grow large enough to provide a stagnant region of stable size, which then continues to grow over a long period of time at an accelerating rate. Pits usually grow in the direction of the gravity force field since the dense concentrated solution in a pit is required for it to grow actively. Most pits, therefore, grow downward from horizontal surfaces to ultimately perforate the wall. Fewer pits are formed on vertical walls, and very few pits grow upward from the bottom surface.

Measurement and assessment of pitting corrosion damage is difficult because of its highly local nature. Pit depth varies widely and, as in the case of fatigue damage, a statistical approach must be taken in which the probability of a pit of specified depth may be established in laboratory testing. Unfortunately, a significant size effect influences depth of pitting, and this must be taken into account when predicting service life of a machine part based on laboratory pitting corrosion data.

The control or prevention of pitting corrosion consists primarily of the wise selection of material to resist pitting or, since pitting is usually the result of stagnant conditions, imparting velocity to the fluid. Increasing its velocity may also decrease pitting corrosion attack.

Because of the atomic mismatch at the grain boundaries of polycrystalline metals, the stored strain energy is higher in the grain boundary regions than in the grains themselves. These high-energy grain boundaries are more chemically reactive than the grains. Under certain conditions depletion or enrichment of an alloying element or impurity concentration at the grain boundaries may locally change the composition of a corrosion-resistant metal, making it susceptible to corrosive attack. Localized attack of this vulnerable region near the grain boundaries is called intergranular corrosion. In particular, the austenitic stainless steels are vulnerable to intergranular corrosion if *sensitized* by heating into the temperature range from 950° to 1450°F, which causes depletion of the chromium near the grain boundaries as chromium carbide is precipitated at the boundaries. The chromium-poor regions then corrode because of local galvanic cell action, and the grains literally fall out of the matrix. A special case of intergranular corrosion, called "weld decay," is generated in the portion of the weld-affected zone, which is heated into the sensitizing temperature range.

To minimize the susceptibility of austenitic stainless steels to intergranular corrosion, the carbon content may be lowered to below 0.03%, stabilizers may be added to prevent depletion of the chromium near the grain boundaries, or a high-temperature solution heat treatment, called quench-annealing, may be employed to produce a more homogeneous alloy.

Other alloys susceptible to intergranular corrosion include certain aluminum alloys, magnesium alloys, copper-based alloys, and die-cast zinc alloys in unfavorable environments.

The corrosion phenomenon in which one element of a solid alloy is removed is termed selective leaching. Although the selective leaching process may occur in any of several alloy systems, the more common examples are *dezincification* of brass alloys and *graphitization* of gray cast iron. Dezincification may occur as either a highly local "plug-type" or a broadly distributed layer-type attack. In either case, the dezincified region is porous, brittle, and weak. Dezincification may be minimized by adding inhibitors such as arsenic, antimony, or phosphorus to the alloy; by lowering oxygen in the environment; or by using cathodic protection.

In the case of graphitization of gray cast iron, the environment selectively leaches the iron matrix to leave the graphite network intact to form an active galvanic cell. Corrosion then proceeds to destroy the machine part. Use of other alloys, such as nodular or malleable cast iron, mitigates the problem because there is no graphite network in these alloys to support the corrosion residue. Other alloy systems in adverse environments that may experience selective leaching include aluminum bronzes, silicon bronzes, and cobalt alloys.

Erosion corrosion is an accelerated, direct chemical attack of a metal surface due to the action of a moving corrosive medium. Because of the abrasive wear action of the moving fluid, the formation of a protective layer of corrosion product is inhibited or prevented, and the corroding medium has direct access to bare, unprotected metal. Erosion corrosion is usually characterized by a pattern of grooves or peaks and valleys generated by the flow pattern of the corrosive medium. Most alloys are susceptible to erosion corrosion, and many different types of corrosive media may induce erosion corrosion, including flowing gases, liquids, and solid aggregates. Erosion corrosion may become a problem in such machine parts as valves, pumps, blowers, turbine blades and nozzles, conveyors, and piping and ducting systems, especially in the regions of bends and elbows.

Erosion corrosion is influenced by the velocity of the flowing corrosive medium, turbulence of the flow, impingement characteristics, concentration of abrasive solids, and characteristics of the metal alloy surface exposed to the flow. Methods of minimizing or preventing erosion corrosion include reducing the velocity, eliminating or reducing turbulence, avoiding sudden changes in the direction of flow, eliminating direct impingement where possible, filtering out abrasive particles, using harder and more corrosion-resistant alloys, reducing the temperature, using appropriate surface coatings, and using cathodic protection techniques.

Cavitation often occurs in hydraulic systems, such as turbines, pumps, and piping, when pressure changes in a flowing liquid give rise to the formation and collapse of vapor bubbles at or near the containing metal surface. The impact associated with vapor bubble collapse may produce high-pressure shock waves that may plastically deform the metal locally or destroy any protective surface film of corrosion product and locally accelerate the corrosion process. Furthermore, the tiny depressions so formed act as a nucleus for subsequent vapor bubbles, which continue to form and collapse at the same site to produce deep pits and pockmarks by the combined action of mechanical deformation and accelerated chemical corrosion. This phenomenon is called cavitation corrosion. Cavitation corrosion may be reduced or prevented by eliminating the cavitation through appropriate design changes. Smoothing the surfaces, coating the walls, using corrosion-resistant materials, minimizing pressure differences in the cycle, and using cathodic protection are design changes that may be effective.

Hydrogen damage, although not considered to be a form of direct corrosion, is often induced by corrosion. Any damage caused in a metal by the presence of hydrogen or the interaction with hydrogen is called hydrogen damage. Hydrogen damage includes hydrogen blistering, hydrogen embrittlement, hydrogen attack, and decarburization.

Hydrogen blistering is caused by the diffusion of hydrogen atoms into a void within a metallic structure where they combine to form molecular hydrogen. The hydrogen pressure builds to a high level that, in some cases, causes blistering, yielding, and rupture. Hydrogen blistering may be minimized by using materials without voids, by using corrosion inhibitors, or by using hydrogen-impervious coatings.

Hydrogen embrittlement is also caused by the penetration of hydrogen into the metallic structure to form brittle hydrides and pin dislocation movement to reduce slip, but the exact mechanism is not yet fully understood. Hydrogen embrittlement is more serious at the higher-strength levels of susceptible alloys, which include most of the high-strength steels. Reduction and prevention of hydrogen embrittlement may be accomplished by "baking out" the hydrogen at relatively low temperatures for several hours, use of corrosion inhibitors, or use of less susceptible alloys.

Decarburization and hydrogen attack are both high-temperature phenomena. At high temperatures hydrogen removes carbon from an alloy, often reducing its tensile strength and increasing its creep rate. This carbon-removing process is called *decarburization*. It is also possible that the hydrogen may lead to the formation of methane in the metal voids, which may expand to form cracks, another form of hydrogen attack. Proper selection of alloys and coatings is helpful in prevention of these corrosion-related problems.

Biological corrosion is a corrosion process or processes that results from the activity of living organisms. These organisms may be microorganisms, such as aerobic or anaerobic bacteria, or they may be macroorganisms, such as fungi, mold, algae, or barnacles. The organisms may influence or produce corrosion by virtue of their processes of food ingestion and waste elimination. There are, for example, sulfate-reducing anaerobic bacteria, which produce iron sulfide when in contact with buried steel structures, and aerobic sulfur-oxidizing bacteria, which produce localized concentrations of sulfuric acid and serious corrosive attack on buried steel and concrete pipe lines. There are also iron bacteria, which ingest ferrous iron and precipitate ferrous hydroxide to produce local crevice corrosion attack. Other bacteria oxidize ammonia to nitric acid, which attacks most metals, and most bacteria produce carbon dioxide, which may form the corrosive agent carbonic acid. Fungi and mold assimilate organic matter and produce organic acids. Simply by their presence, fungi may provide the site for crevice corrosion attacks, as does the presence of attached barnacles and algae. Prevention or minimization of biological corrosion may be accomplished by altering the environment or by using proper coatings, corrosion inhibitors, bactericides or fungicides, or cathodic protection.

19.9.2 Stress Corrosion Cracking

Stress corrosion cracking is an extremely important failure mode because it occurs in a wide variety of different alloys. This type of failure results from a field of cracks produced in a metal alloy under the combined influence of tensile stress and a corrosive environment. The metal alloy is not attacked over most of its surface, but a system of intergranular or transgranular cracks propagates through the matrix over a period of time.

Stress levels that produce stress corrosion cracking are well below the yield strength of the material, and residual stresses as well as applied stresses may produce failure. The lower the stress level, the longer is the time required to produce cracking, and there appears to be a threshold stress level below which stress corrosion cracking does not occur. (See p. 96 of Ref. 126.)

The chemical compositions of the environments that lead to stress corrosion cracking are highly specific and peculiar to the alloy system, and no general patterns have been observed. For example, austenitic stainless steels are susceptible to stress corrosion cracking in chloride environments but not in ammonia environments, whereas brasses are susceptible to stress corrosion cracking in ammonia environments but not in chloride environments. Thus, the "season cracking" of brass cartridge cases in the crimped zones was found to be stress corrosion cracking due to the ammonia resulting from decomposition of organic matter. Likewise, "caustic embrittlement" of steel boilers, which resulted in many explosive failures, was found to be stress corrosion cracking due to sodium hydroxide in the boiler water.

Stress corrosion cracking is influenced by stress level, alloy composition, type of environment, and temperature. Crack propagation seems to be intermittent, and the crack grows to a critical size, after which a sudden and catastrophic failure ensues in accordance with the laws of fracture mechanics. Stress corrosion crack growth in a statically loaded machine part takes place through the interaction of mechanical strains and chemical corrosion processes at the crack tip. The largest value of plane strain stress intensity factor for which crack growth does not take place in a corrosive environment is designated K_{Iscc}. In many cases, corrosion fatigue behavior is also related to the magnitude of K_{Iscc}.[9]

Prevention of stress corrosion cracking may be attempted by lowering the stress below the critical threshold level, choice of a better alloy for the environment, changing the environment to eliminate the critical corrosive element, use of corrosion inhibitors, or use of cathodic protection. Before cathodic protection is implemented care must be taken to ensure that the phenomenon is indeed stress corrosion cracking because hydrogen embrittlement is accelerated by cathodic protection techniques.

19.10 FAILURE ANALYSIS AND RETROSPECTIVE DESIGN

In spite of all efforts to design and manufacture machines and structures to function properly without failure, failures do occur. Whether the failure consequences simply represent an annoying inconvenience, such as a "binding" support on the sliding patio screen, or a catastrophic loss of life and property, as in the crash of a jumbo jet, it is the responsibility of the designer to glean all of the information possible from the failure event so that similar events can be avoided in the future. Effective assessment of service failures usually requires the intense interactive scrutiny of a team of specialists, including at least a mechanical designer and a materials engineer trained in failure analysis techniques. The team might often include a manufacturing engineer and a field service engineer as well. The mission of the failure analysis team is to discover the initiating cause of failure, identify the best solution, and redesign the product to prevent future failures. Although the results of failure analysis investigations may often be closely related to product liability litigation, the legal issues will not be addressed in this discussion.

Techniques utilized in the failure analysis effort include the inspection and documentation of the event through direct examination, photographs and eyewitness reports; preservation of all parts, especially failed parts; and pertinent calculations, analyses, and examinations that may help establish and validate the cause of failure. The materials engineer may utilize macroscopic examination, low-power magnification, microscopic examination, transmission or scanning electron microscopic techniques, energy-dispersive X-ray techniques, hardness tests, spectrographic analysis, metallographic examination, or other techniques of determining the failure type, failure location, material abnormalities, and potential causes of failure. The designer may perform stress and deflection analyses, examine geometry, assess service loading, and environmental influences, reexamine the kinematics and dynamics of the application, and attempt to reconstruct the failure scenario. Other team members may examine the quality of manufacture, the quality of maintenance, the possibility of unusual or unconventional usage by the operator, or other factors that may have played a role in the service failure. Piecing all of this information together, it is the objective of the failure analysis team to identify as accurately as possible the probable cause of failure.

As undesirable as service failures may be, the results of a well-executed failure analysis may be transformed directly into improved product reliability by a designer who capitalizes on service failure data and failure analysis results. These techniques of retrospective design have become important working tools of the profession and are likely to continue to grow in importance.

REFERENCES

1. J. A. Collins, *Failure of Materials in Mechanical Design; Analysis, Prediction, Prevention,* Wiley, New York, 1981.

2. "Progress in Measuring Fracture Toughness and Using Fracture Mechanics," *Materials Research and Standards,* 103–119 (March, 1964).

3. H. M. Westergaard, "Bearing Pressures and Cracks," *Journal of Applied Mechanics* **66,** 49 (1939).

4. G. R. Irwin, "Analysis of Stresses and Strains Near the End of a Crack Traversing a Plate," *Journal of Applied Mechanics* **24,** 361 (1957).

5. H. Tada, P. C. Paris, and G. E. Irwin, *The Stress Analysis of Cracks Handbook,* Del Research Corporation, Hellertown, PA, 1973.

6. G. R. Irwin, "Linear Fracture Mechanics, Fracture Transition, and Fracture Control," *Engineering Fracture Mechanics,* No. 2, 1 (August 1968).

7. "Standard Method of Test for Plane-Strain Fracture Toughness of Metallic Materials," Designation: E399–72, Annual Book of ASTM Standards, Part 31, The American Society for Testing and Materials, Philadelphia, PA, 1972.

8. R. W. Hertzberg, *Deformation and Fracture Mechanics of Engineering Materials,* Wiley, New York, 1976.

9. S. T. Rolfe and J. M. Barsom, *Fracture and Fatigue Control in Structures,* Prentice-Hall, Englewood Cliffs, NJ, 1977.

10. J. E. Cambell, W. E. Berry, and C. E. Fedderson, *Damage Tolerant Design Handbook,* MCIC-HB-01, Battelle, Columbus, OH, September, 1973.

11. G. C. Sih, *Handbook of Stress Intensity Factors for Researchers and Engineers,* Institute of Fracture and Solid Mechanics, Lehigh University, Bethlehem, PA, 1973

12. A. F. Madayag, *Metal Fatigue, Theory and Design,* Wiley, New York, 1969.

13. S. R. Swanson (ed.), *Handbook of Fatigue Testing,* STP-655, American Society for Testing and Materials, Philadelphia, PA, 1974.

14. D. E. Hardenbergh (ed.), *Statistical Methods in Materials Research,* Proceedings of Short Course, Department of Engineering Mechanics, Pennsylvania State University, State College, PA, 1956.

15. E. M. Prot, *Fatigue Testing Under Progressive Loading; A New Technique for Testing Materials* (translation by E. J. Ward), WADC TR52-148, Wright-Patterson Air Force Base, OH, September 1952.

16. H. J. Grover, S. A. Gordon, L. R. Jackson, *Fatigue of Metals and Structures,* Government Printing Office, Washington, DC, 1954.

17. A. Higdon, E. H. Ohlsen, W. B. Stiles, and J. A. Weese, *Mechanics of Materials,* Wiley, New York, 1967.

18. W. M. Murray (ed.), *Fatigue and Fracture of Metals,* Wiley, New York, 1952.

19. J. B. Johnson, "Aircraft Engine Material," *SAE Journal* **40**, 153–162 (1937).

20. *Proceedings of the Conference on Welded Structures,* The Welding Institute, Cambridge, England, 1971, Vols. I and II.

21. H. F. Moore, "A Study of Size Effect and Notch Sensitivity in Fatigue Tests of Steel," *ASTM Proceedings* **45**, 507 (1945).

22. W. N. Thomas, "Effect of Scratches and Various Workshop Finishes upon the Fatigue Strength of Steel," *Engineering* **116**, 449ff (1923).

23. G. V. Smith, *Properties of Metals at Elevated Temperatures,* McGraw-Hill, New York, 1950.

24. H. J. Gough and D. G. Sopwith, *Journal of Iron and Steel Institute* **127**, 301 (1933).

25. O. J. Horger, *Metals Engineering Design* (ASME Handbook), American Society of Mechanical Engineers, McGraw-Hill, New York, 1953.

26. *Proceedings of International Conference on Fatigue,* American Society of Mechanical Engineers (jointly with Institution of Mechanical Engineers), New York, 1956.

27. F. M. Howell, and J. L. Miller, "Axial Stress Fatigue Strengths of Several Structural Aluminum Alloys," *ASTM Proceedings* **55**, 955 (1955).

28. Battelle Memorial Institute, *Prevention of Fatigue in Metals,* Wiley, New York, 1941

29. H. J. Grover, *Fatigue of Aircraft Structures,* Government Printing Office, Washington, DC, 1966.

30. G. Sines and J. L. Waisman, *Metal Fatigue,* McGraw-Hill, New York, 1959.

31. H. J. Grover, "An Observation Concerning the Cycle Ratio in Cumulative Damage," *Fatigue in Aircraft Structures,* STP-274, American Society for Testing Materials, Philadelphia, PA, 1960, pp. 120–124.

32. S. S. Manson, "Interfaces Between Fatigue, Creep, and Fracture," *Proceedings of International Conference on Fracture,* Vol. 1, Japanese Society for Strength and Fracture of Metals, Sendai, Japan, September, 1965, and *International Journal of Fracture Mechanics,* March 1966.

33. S. S. Manson, J. C. Frecke, and C. R. Ensign, "Applications of a Double Linear Damage Rule to Cumulative Fatigue," *Fatigue Crack Propagation,* STP-415, American Society for Testing and Materials, Philadelphia, PA, 1967, p. 384.

34. R. E. Peterson, *Stress Concentration Factors,* Wiley, New York, 1974.

35. E. Z. Stowell, "Stress and Strain Concentration at a Circular Hole in an Infinite Plate," NACA TN2073, National Advisory Committee for Aeronautics, Cleveland, OH, April 1950.

36. R. C. Juvinall, *Engineering Considerations of Stress, Strain, and Strength,* McGraw-Hill, New York, 1967.

37. *Manual on Low Cycle Fatigue Testing,* STP-465, American Society for Testing and Materials, Philadelphia, PA, 1969.

38. R. W. Landgraf, "The Resistance of Metal to Cyclic Deformation," *Achievement of High Fatigue Resistance in Metal Alloys,* STP-467, American Society for Testing and Materials, Philadelphia, PA, 1970.

39. B. M. Wundt, *Effects of Notches on Low Cycle Fatigue,* STP-490, American Society for Testing and Materials, Philadelphia, PA, 1972.

40. J. F. Tavernelli and L. F. Coffin, Jr., "A Compilation and Interpretation of Cyclic Strain Fatigue Tests on Metals," *ASM Transactions of American Society for Metals* **51**, 438–453 (1959).

41. S. S. Manson, "Behavior of Materials Under Conditions of Thermal Stress," NACA TN-2933, National Advisory Committee for Aeronautics, Cleveland, OH, 1954.

42. L. F. Coffin, Jr., "A Study of the Effects of Cyclic Thermal Stresses in a Ductile Metal," *ASME Transactions* **16**, 931–950 (1954).

43. L. F. Coffin, Jr., "Design Aspects of High Temperature Fatigue with Particular Reference to Thermal Stresses," *ASME Transactions* **78**, 527–532 (1955).

44. J. Morrow, J. F. Martin, and N. E. Dowling, "Local Stress–Strain Approach to Cumulative Fatigue Damage Analysis," Final Report, T. & A. M. Report No. 379, Department of Theoretical and Applied Mechanics, University of Illinois, Urbana, IL, January 1974.

45. K. Ohji, W. R. Miller, and J. Marin, "Cumulative Damage and Effect of Mean Strain in Low Cycle Fatigue of a 2024-T351 Aluminum Alloy," *Journal of Basic Engineering* **88**, 801 (1966).

46. G. Sachs, W. W. Gerberich, V. Weiss, and J. V. Latorre, "Low Cycle Fatigue of Pressure Vessel Materials," *ASTM Proceedings* **60**, 512–529 (1961).

47. J. G. Sessler and V. Weiss, "Low Cycle Fatigue Damage in Pressure Vessel Materials," *Journal of Basic Engineering* **85**, 539–547 (1963).

48. S. S. Manson, "Thermal Stresses in Design, Part 21—Effect of Mean Stress and Strain on Cyclic Life," *Machine Design* **32** (16), 129–135 (August 4, 1969).

49. A. A. Blatherwick and N. D. Viste, "Cumulative Damage Under Biaxial Fatigue Stress," *Materials Research and Standards,* American Society for Testing and Materials, Philadelphia, PA, 1967, Vol. 7, pp. 331–336.

50. E. Krempl, *The Influence of State of Stress on Low-Cycle Fatigue of Structural Materials: A Literature Survey and Interpretation Report,* STP-549, American Society for Testing and Materials, Philadelphia, PA, 1974.

51. S. S. Manson, *Thermal Stress and Low Cycle Fatigue,* McGraw-Hill, New York, 1966.

52. J. D. Morrow, J. F. Martin, and N. E. Dowling, "Local Stress–Strain Approach to Cumulative Fatigue Damage Analysis," Final Report, T. & A. M. Report No. 379, Department of Theoretical and Applied Mechanics, University of Illinois, Urbana, IL, January 1974.

53. S. J. Stadnick and J. Morrow, "Techniques for Smooth Specimen Simulation of the Fatigue Behavior of Notched Members," *Testing for Prediction of Material Performance in Structures and Components,* STP-515, American Society for Testing and Materials, Philadelphia, PA, 1972, pp. 229–252.

54. S. J. Stadnick, "Simulation of Overload Effects in Fatigue Based on Neuber's Analysis," T. & A. M. Report No. 325, Department of Theoretical and Applied Mechanics, University of Illinois, Urbana, IL, 1969.

55. T. H. Topper and J. Morrow, "Simulation of the Fatigue Behavior at the Notch Root in Spectrum Loaded Notched Members (U)," T. & A. M. Report No. 333, Department of Theoretical and Applied Mechanics, University of Illinois, Urbana, IL, January 1970 (final report for Aero Structures Department, Naval Air Development Center).

56. J. F. Martin, T. H. Topper, and G. M. Sinclair, "Computer Based Simulation of Cyclic Stress–Strain Behavior with Applications to Fatigue," *Materials Research and Standards* 11 (2), 23 (February 1971).

57. J. F. Martin, T. H. Topper, and G. M. Sinclair, "Computer Based Simulation of Cyclic Stress–Strain Behavior," T. & A. M. Report No. 326, Department of Theoretical and Applied Mechanics, University of Illinois, Urbana, IL, July 1969.

58. T. H. Topper, R. M. Wetzel, and J. Morrow, "Neuber's Rule Applied to Fatigue of Notched Specimens," *Journal of Materials* 4 (1), 200–209 (March 1969).

59. N. E. Dowling, "Fatigue Failure Predictions for Complicated Stress–Strain Histories," *Journal of Materials* 7 (1), 71–87 (March 1972). See also, N. E. Dowling, "Fatigue Failure Predictions for Complicated Stress–Strain Histories," T. & A. M. Report No. 337, Department of Theoretical and Applied Mechanics, University of Illinois, Urbana, IL, January 1971.

60. D. W. Hoeppner and W. E. Krupp, "Prediction of Component Life by Application of Fatigue Crack Growth Knowledge," *Engineering Fracture Mechanics* 6, 47–70 (1974).

61. P. C. Paris and F. Erdogan, "A Critical Analysis of Crack Propagation Laws," *Journal of Basic Engineering* 85 (4), 528–534 (1963).

62. W. G. Clark, Jr., "Fracture Mechanics in Fatigue," *Experimental Mechanics* (September 1971).

63. J. M. Barsom, "Effect of Cyclic Stress Form on Corrosive Fatigue Crack Propagation Below K_{ISCC} in a High Yield Strength Steel," *Corrosion Fatigue,* 424–436 (1972).

64. J. L. McMillan and R. M. N. Pelloux, "Fatigue Crack Propagation Under Program and Random Loads," STP-415, American Society for Testing and Materials, Philadelphia, PA, 1967, p. 505.

65. J. C. Ekvall and L. Young, "Converting Fatigue Loading Spectra for Flight-by-Flight Testing of Aircraft and Helicopter Components," Report of Lockheed California Company, Burbank, CA, 1974.

66. H. A. Wood, "Fracture Control Procedures for Aircraft Structural Integrity," AFFDL Report TR-21-89, Wright-Patterson Air Force Base, OH, July 1971.

67. F. R. Larson and J. Miller, "Time–Temperature Relationships for Rupture and Creep Stresses," *ASME Transactions* 74, 765 (1952).

68. S. S. Manson and A. M. Haferd, "A Linear Time–Temperature Relation for Extrapolation of Creep and Stress Rupture Data," NACA Technical Note 2890, National Advisory Committee for Aeronautics, Cleveland, OH, March 1953.

69. R. G. Sturm, C. Dumont, and F. M. Howell, "A Method of Analyzing Creep Data," *ASME Transactions* 58, A62 (1936).

70. N. H. Polakowski and E. J. Ripling, *Strength and Structure of Engineering Materials,* Prentice-Hall, Englewood Cliffs, NJ, 1966.

71. E. L. Robinson, "Effect of Temperature Variation on the Long-Time Rupture Strength of Steels," *ASME Transactions* 74, 777–781 (1952).

72. A. J. Kennedy, *Processes of Creep and Fatigue in Metals,* Wiley, New York, 1963.

73. F. W. Demoney and B. J. Lazan, WADC Tech Report 53-510, Wright-Patterson Air Force Base, OH, 1954.

74. F. Vitovec and B. J. Lazan, *Symposium on Metallic Materials for Service at Temperatures Above 1600°F*, STP-174, American Society for Testing and Materials, Philadelphia, PA, 1956.

75. L. F. Coffin, Jr., "The Effect of Frequency on the Cyclic Strain and Low Cycle Fatigue Behavior of Cast Udimet 500 at Elevated Temperature," *Metallurgical Transactions* **12**, 3105–3113 (November 1971).

76. J. B. Conway and J. T. Berling, "A New Correlation of Low-Cycle Fatigue Data Involving Hold Periods," *Metallurgical Transactions* **1** (1), 324–325 (January 1970).

77. J. R. Ellis and E. P. Esztergar, "Considerations of Creep-Fatigue Interaction in Design Analysis," *Symposium on Design for Elevated Temperature Environment*, ASME, New York, 1971, pp. 29–33.

78. R. M. Curran and B. M. Wundt, "A Program to Study Low-Cycle Fatigue and Creep Interaction in Steels at Elevated Temperatures," *Current Evaluation of 2¼ Chrome 1 Molybdenum Steel in Pressure Vessels and Piping*, ASME, New York, 1972, pp. 49–82.

79. S. S. Manson, G. R. Halford, and M. H. Hirschberg, "Creep-Fatigue Analysis by Strain-Range Partitioning," *Symposium on Design for Elevated Temperature Environment*, ASME, New York, 1971, pp. 12–24.

80. M. M. Leven, "The Interaction of Creep and Fatigue for a Rotor Steel," *Experimental Mechanics* **13** (9), 353–372 (September 1973).

81. S. S. Manson, G. R. Halford, and M. H. Hirschberg, NASA Technical Memo TMX-67838, Lewis Research Center, Cleveland, OH, May 1971.

82. J. F. Saltsman and G. R. Halford, "Application of Strain-Range Partitioning to the Prediction of Creep-Fatigue Lives of AISI Types 304 and 316 Stainless Steel," NASA Technical Memo TMX-71898, Lewis Research Center, Cleveland, OH, September 1976.

83. C. G. Annis, M. C. Van Wanderham, and R. M. Wallace, "Strain-Range Partitioning Behavior of an Automotive Turbine Alloy," Final Report NASA TR 134974, February 1976.

84. M. H. Hirschberg and G. R. Halford, "Use of Strain-Range Partitioning to Predict High-Temperature Low Cycle Fatigue Life," NASA TN D-8072, January 1976.

85. J. F. Saltsman and G. R. Halford, "Application of Strain-Range Partitioning to the Prediction of MPC Creep-Fatigue Data for 2¼ Cr-1 Mo Steel," NASA TMX-73474, December 1976.

86. S. S. Manson and G. R. Halford, "Treatment of Multiaxial Creep-Fatigue by Strain-Range Partitioning," NASA TMX-73488, December 1976.

87. *Characterization of Low Cycle High Temperature Fatigue by the Strain-Range Partitioning Method*, AGARD Conference Proceedings No. 243, distributed by NASA, Langley Field, VA, April 1978.

88. J. A. Collins, "Fretting-Fatigue Damage-Factor Determination," *Journal of Engineering for Industry* **87** (8), 298–302 (August 1965).

89. D. Godfrey, "Investigation of Fretting by Microscopic Observation," NACA Report 1009, Cleveland, OH, 1951 (formerly TN-2039, February 1950).

90. F. P. Bowden and D. Tabor, *The Friction and Lubrication of Solids*, Oxford University Press, Amen House, London, 1950.

91. D. Godfrey and J. M. Bailey, "Coefficient of Friction and Damage to Contact Area During the Early Stages of Fretting; I—Glass, Copper, or Steel Against Copper," NACA TN-3011, Cleveland, OH, September 1953.

92. M. E. Merchant, "The Mechanism of Static Friction," *Journal of Applied Physics* **11** (3), 232 (1940).

93. E. E. Bisson, R. L. Johnson, M. A. Swikert, and D. Godfrey, "Friction, Wear, and Surface Damage of Metals as Affected by Solid Surface Films," NACA TN-3444, Cleveland, OH, May 1955.

94. H. H. Uhlig, "Mechanism of Fretting Corrosion," *Journal of Applied Mechanics* **76**, 401–407 (1954).

95. I. M. Feng and B. G. Rightmire, "The Mechanism of Fretting," *Lubrication Engineering* **9**, 134ff (June 1953).

96. I. M. Feng, "Fundamental Study of the Mechanism of Fretting," Final Report, Lubrication Laboratory, Massachusetts Institute of Technology, Cambridge, 1955.

97. H. T. Corten, "Factors Influencing Fretting Fatigue Strength," T. & A. M. Report No. 88, Department of Theoretical and Applied Mechanics, University of Illinois, Urbana, IL, June 1955.

98. W. L. Starkey, S. M. Marco, and J. A. Collins, "Effects of Fretting on Fatigue Characteristics of Titanium–Steel and Steel–Steel Joints," ASME Paper 57-A-113, New York, 1957.

99. W. D. Milestone, "Fretting and Fretting-Fatigue in Metal-to-Metal Contacts," ASME Paper 71-DE-38, New York, 1971.

100. G. P. Wright and J. J. O'Connor, "The Influence of Fretting and Geometric Stress Concentrations on the Fatigue Strength of Clamped Joints," *Proceedings, Institution of Mechanical Engineers,* 186 (1972).

101. R. B. Waterhouse, *Fretting Corrosion,* Pergamon Press, New York, 1972.

102. "Fretting in Aircraft Systems," AGARD Conference Proceedings CP161, distributed through NASA, Langley Field, VA, 1974.

103. "Control of Fretting Fatigue," Report No. NMAB-333, National Academy of Sciences, National Materials Advisory Board, Washington, DC, 1977.

104. N. P. Suh, S. Jahanmir, J. Fleming, and E. P. Abrahamson, "The Delamination Theory of Wear—II," Progress Report, Materials Processing Lab, Mechanical Engineering Dept., MIT Press, Cambridge, MA, September 1975.

105. J. T. Burwell, Jr., "Survey of Possible Wear Mechanisms," *Wear* 1, 119–141 (1957).

106. M. B. Peterson, M. K. Gabel, and M. J. Derine, "Understanding Wear"; K. C. Ludema, "A Perspective on Wear Models"; E. Rabinowicz, "The Physics and Chemistry of Surfaces"; J. McGrew, "Design for Wear of Sliding Bearings"; R. G. Bayer, "Design for Wear of Lightly Loaded Surfaces," *ASTM Standardization News* 2 (9), 9–32 (September 1974).

107. R. B. Waterhouse, *Fretting Corrosion,* Pergamon Press, New York, 1972.

108. P. A. Engel, "Predicting Impact Wear," *Machine Design,* 100–105 (May 1977).

109. P. A. Engel, *Impact Wear of Materials,* Elsevier, New York, 1976.

110. J. A. Collins, "A Study of the Phenomenon of Fretting-Fatigue with Emphasis on Stress-Field Effects," Dissertation, Ohio State University, Columbus, 1963.

111. J. A. Collins and F. M. Tovey, "Fretting Fatigue Mechanisms and the Effect of Direction of Fretting Motion on Fatigue Strength," *Journal of Materials* 7 (4) (December 1972).

112. W. D. Milestone, "An Investigation of the Basic Mechanism of Mechanical Fretting and Fretting-Fatigue at Metal-to-Metal Joints, with Emphasis on the Effects of Friction and Friction-Induced Stresses," Dissertation, Ohio State University, Columbus, 1966.

113. J. A. Collins and S. M. Marco, "The Effect of Stress Direction During Fretting on Subsequent Fatigue Life," *ASTM Proceedings* 64, 547 (1964).

114. J. A. Alic, "Fretting Fatigue in the Presence of Periodic High Tensile or Compressive Loads," Final Scientific Report, Grant No. AFOSR77-3422, Wright-Patterson AFB, Ohio, April 1979.

115. H. Lyons, "An Investigation of the Phenomenon of Fretting-Wear and Attendant Parametric Effects Towards Development of Failure Prediction Criteria," Ph.D. Dissertation, Ohio State University, Columbus, 1978.

116. P. L. Ko, "Experimental Studies of Tube Fretting in Steam Generators and Heat Exchangers," ASME/CSME Pressure Vessels and Piping Conference, Nuclear and Materials Division, Montreal, Canada, June 1978.

117. R. B. Heywood, *Designing Against Fatigue of Metals,* Reinhold, New York, 1962.

118. F. P. Bowden and D. Tabor, *Friction and Lubrication of Solids,* Oxford University Press, London, 1950.

119. F. P. Bowden and D. Tabor, *Friction and Lubrication,* Methuen, London, 1967.

120. E. Rabinowicz, *Friction and Wear of Materials,* Wiley, New York, 1966.

121. C. Lipson, *Wear Considerations in Design,* Prentice-Hall, Englewood Cliffs, NJ, 1967.

122. N. P. Suh, "The Delamination Theory of Wear," *Wear* 25, 111–124 (1973).

123. J. E. Shigley, *Mechanical Engineering Design,* 2nd ed., McGraw-Hill, New York, 1972.

124. W. A. Glaeser, K. C. Ludema, and J. K. Rhee (eds.), *Wear of Materials,* American Society of Mechanical Engineers, New York, April 25–28, 1977.

125. C. W. MacGregor (ed.), *Handbook of Analytical Design for Wear,* Plenum Press, New York, 1964.

126. M. G. Fontana and N. D. Greene, *Corrosion Engineering,* McGraw-Hill, New York, 1967.

127. L. S. Seabright and R. J. Fabian, "The Many Faces of Corrosion," *Materials in Design Engineering* 57 (1) (January 1963).

128. G. Nelson, *Corrosion Data Survey,* National Association of Corrosion Engineers, Houston, TX, 1972.

129. H. H. Uhlig (ed.), *Corrosion Handbook,* Wiley, New York, 1948.

130. E. Rabald, *Corrosion Guide,* Elsevier, New York, 1951.

CHAPTER 20

FAULT TREE ANALYSIS

BALBIR S. DHILLON

University of Ottawa
Ottawa, Ontario, Canada

20.1 INTRODUCTION

Fault tree analysis is performed to evaluate the reliability of a system. The fault tree technique is widely known and is applied frequently in industry to analyze complex engineering systems. It may be said that the fault tree is a model and is used to bridge the gap between the world of mathematics and the real world. Furthermore, it is a graphical representation of the Boolean failure logic associated with the determination of the overall reliability of a system.

The starting point of the fault tree analysis is the identification of the undesirable event of the system known as the top event. An example of a top event is the "failure of the system." Here one may be concerned with calculating the failure probability of the system or the probability of the occurrence of the top event of the system.

Once the system undesirable event has been established, then a fault tree is generated by asking a question such as "How can this event occur?" The events that could make the top event occur are connected and generated by logic operators such as OR and AND. The fault tree analysis proceeds by generating fault events in a successive manner until one reaches those points where the basic events are encountered. In other words, those events for which there is no need for further development. In summary, a fault tree is simply the logic structure that relates the system undesirable event to the system's basic events.

The history of the fault tree technique goes back to 1961 when H. A. Watson of Bell Telephone Laboratories developed this concept and applied it to a safety analysis of the Minuteman Launch Control System.[1] The fault tree technique was further developed by Boeing Company analysts, including D. F. Haasel. In 1965, a system safety symposium was sponsored by the University of Washington and the Boeing Company. This symposium further stimulated the interest in the fault tree technique. Between 1965 and 1975 several more publications appeared on the subject. Another symposium partially devoted to the fault tree technique was given in 1975. This symposium was held at the University of California at Berkeley, and the proceedings were published[2] by the Society for Industrial and Applied Mathematics. The fault tree technique is described in detail in Ref. 3, where a comprehensive list of published references on the topic is also given. This chapter briefly presents the various aspects of the fault tree analysis and the basic mathematics associated with the technique.

20.2 REVIEW OF ESSENTIAL MATHEMATICS ASSOCIATED WITH THE FAULT TREE TECHNIQUE

This section briefly reviews the essential mathematics associated with the fault tree technique. Briefly, this section presents the Boolean algebra properties, basic probability concepts, and reliability measures related to the fault tree technique.

20.2.1 Boolean Algebra Properties

The following Boolean algebra properties find application in the fault tree analysis.

Associate Law

$$X + (Y + Z) = (X + Y) + Z \tag{20.1}$$
$$X(YZ) = (XY)Z \tag{20.2}$$

where X, Y, and Z are known as events or sets.

Identities

$$Y + Y = Y \tag{20.3}$$
$$Y \cdot Y = Y \tag{20.4}$$

Absorption Law

$$Y(Y + X) = Y \tag{20.5}$$
$$Y + (Y \cdot X) = Y \tag{20.6}$$
$$X(XY) = X \cdot Y \tag{20.7}$$

Commutative Law

$$Y + X = X + Y \tag{20.8}$$
$$YX = XY \tag{20.9}$$

Distributive Law

$$Y + X \cdot Z = (Y + X)(Y + Z) \tag{20.10}$$
$$Y(X + Z) = YX + YZ \tag{20.11}$$

20.2.2 Probability Concepts

This section presents two important probability properties used in the fault tree analysis.

Probability of a Union

The probability of a union of k events is given by

$$P(E_1 + E_2 + E_3 + E_4 + \cdots + E_k) = \{P(E_1) + P(E_2) + P(E_3) + \cdots + P(E_k)\}$$
$$- \{P(E_1 \cdot E_2) + P(E_1 \cdot E_3) + \cdots + P(E_j \cdot E_i)j \neq i\} \cdots \quad (20.12)$$
$$+ (-1)^{k-1}\{P(E_1 \cdot E_2 \cdot E_3 \cdot \cdots \cdot E_k)\}$$

where E_i is the ith event, for $i = 1, 2, 3, \ldots k$.

Probability of an Intersection

The probability of an intersection of k events is given by

$$P(E_1 \cdot E_2 \cdot E_3 \cdot \cdots \cdot E_k) = P(E_1)P(E_2/E_1) \cdots P(E_k/E_1 \cdot E_2 \cdot E_3 \cdots E_{k-1}) \quad (20.13)$$

For independent events, Eq. (20.13) reduces to

$$P(E_1 \cdot E_2 \cdot E_3 \cdot \cdots \cdot E_k) = P(E_1) \cdot P(E_2) \cdot P(E_3) \cdot \cdots \cdot P(E_k) \quad (20.14)$$

20.2.3 Reliability Measures

Generally to perform a fault tree analysis of engineering systems, it is assumed that the basic fault events or components failure and repair rates are constant. Therefore, this section discusses the failure and repair rates, reliability and availability of a basic component.

Constant Failure Rate

In reliability analysis it is well known that when component failure times are exponentially distributed, then its failure rate is constant. This is demonstrated by the following example.

Example 20.1

Component failure times are described by an exponential probability density function. Prove that its failure rate is constant.

Thus,

$$f(t) = \lambda e^{-\lambda t} \quad (20.15)$$

where $f(t) =$ the probability density function of the component
$\quad t =$ time
$\quad \lambda =$ the component's constant failure rate

By definition the component's cumulative distribution function, $F(t)$, is given by

$$F(t) = \int_0^t f(t) \cdot dt = \int_0^t \lambda e^{-\lambda t} \cdot dt$$
$$= 1 - e^{-\lambda t} \quad (20.16)$$

Thus, the component reliability $R(t)$ is

$$R(t) = 1 - F(t) = e^{-\lambda t} \quad (20.17)$$

Again by definition the component hazard rate $\lambda(t)$ is given by

$$\lambda(t) = \frac{f(t)}{R(t)} = \frac{\lambda e^{-\lambda t}}{e^{-\lambda t}} = \lambda \quad (20.18)$$

Therefore, the component failure rate is constant.

Constant Repair Rate

Similar to the constant failure rate of a component, it can easily be proven that the constant repair rate of such a component is associated with an exponential probability density function. In other words, the repair times are exponentially distributed.

Availability and Unavailability

From Ref. 4, the availability $A(t)$ and unavailability $U(t)$ of a repairable component with constant failure and constant repair rates are given by

$$A(t) = \frac{\mu}{\lambda + \mu} + \frac{\lambda}{\lambda + \mu} \cdot e^{-(\lambda + \mu)t} \tag{20.19}$$

and

$$U(t) = \frac{\lambda}{\lambda + \mu} - \frac{\lambda}{\lambda + \mu} \cdot e^{-(\lambda + \mu)t} \tag{20.20}$$

where λ = the constant failure rate of the component
$\quad\mu$ = the constant repair rate of the component
$\quad t$ = time

For a large value of t, the component's steady-state availability and unavailability from Eqs. (20.19) and (20.20) are

$$A = \frac{\mu}{\lambda + \mu} \tag{20.21}$$

$$U = \frac{\lambda}{\lambda + \mu} \tag{20.22}$$

where A = the component steady-state availability
$\quad U$ = the component steady-state unavailability

20.3 FAULT TREE SYMBOLS

This section presents the symbols commonly used to develop a fault tree. Other symbols used to develop fault trees are given in Refs. 5 and 6. Some of the symbols used to represent gates are shown in Fig. 20.1. In addition, Figs. 20.2 and 20.3 also show symbols frequently used to develop fault trees of systems. In Fig. 20.2, the symbols shown are used to represent various types of events, whereas Fig. 20.3 shows those symbols that represent neither gates nor events.

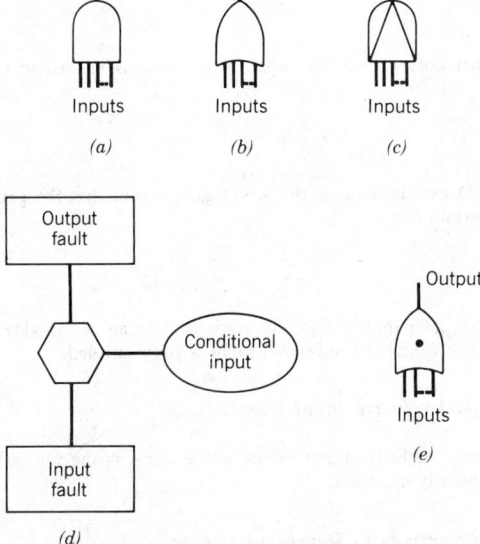

Fig. 20.1 Symbols used to represent gates in fault tree analysis: (*a*) AND gate; (*b*) OR gate; (*c*) priority AND gate; (*d*) inhibit gate; (*e*) OR gate with mutually exclusive input events.

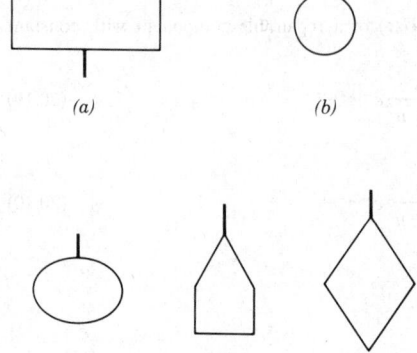

(a)   (b)

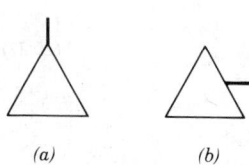

(c)   (d)   (e)

Fig. 20.2 Symbols used to represent events in fault tree anslysis: (a) rectangle; (b) circle; (c) ellipse; (d) house; (e) diamond.

(a)   (b)   **Fig. 20.3**   Triangle symbols: (a) type I; (b) type II.

20.3.1   Definitions of Symbols Used to Represent Gates

This section presents the definitions of the symbols shown in Fig. 20.1.

AND Gate

An AND gate signifies that all the input fault events must occur to produce an output fault event.

OR Gate

An OR gate signifies that only one of the input fault events is needed to produce an output fault event.

Priority AND Gate

Basically, a priority AND gate is same as the AND gate, except that the gate input fault events are required to occur in a certain order.

Inhibit Gate

An inhibit gate is another gate that is logically equivalent to an AND gate. The gate output fault event only occurs when the specified conditional input is fully satisfied.

OR Gate with Mutually Exclusive Input Events

An OR gate with mutually exclusive input events is the same as the OR gate, except that the gate input fault events are mutually exclusive.

20.3.2   Definitions of Symbols to Represent Events

This section defines the symbols for the five types of events shown in Fig. 20.2.

Rectangle

A rectangle is used to denote an event resulting from the combination of fault events acting through the input of a logic gate.

Circle

A circle denotes a basic fault event. For example, the failure of an elementary component of a system. The parameters of a basic fault event are failure rate, repair rate, occurrence probability, and unavailability. The values of such parameters are estimated from field data or through other means.

Ellipse

An ellipse is used to represent a condition that applies to a logic gate.

House

A house is used to represent a fault event whose occurrence is expected.

Diamond

A diamond denotes a fault event that could be further developed.

20.3.3 Triangle Symbols

Essentially there are two types of triangle symbols, which are frequently used to avoid repeating sections of a fault tree. Both triangle symbols are shown in Fig. 20.3.

Triangle Type I

A type I symbol is used to denote a "transfer in" situation. In other words, this indicates the input point for a certain portion of a fault tree.

Triangle Type II

A type II symbol denotes a "transfer out" situation. More clearly, this indicates that a certain portion of the fault tree provides an input to the fault tree at a specific point. Whenever a fault tree uses a "transfer out" symbol, then it also has to make use of a corresponding "transfer in" symbol.

20.4 FAULT TREE ANALYSIS PROCEDURE

This section describes briefly a procedure for performing fault tree analysis. This procedure is essentially composed of four steps,[1] outlined in Fig. 20.4.

These four steps are self-explanatory, therefore no description is given here. However, they are described in detail in Refs. 1, 3, and 7.

20.5 ANALYSIS OF LOGIC GATES

This section is concerned with the analysis associated with AND and OR logic gates.

20.5.1 AND Gate

A three-input AND gate is shown in Fig. 20.5. This figure represents a fault tree of an aircraft with three independent engines. At least one engine must function normally in order for the aircraft to fly. Thus, Fig. 20.5 can be used to compute the probability of the aircraft crashing because of engine failures. Thus, from Fig. 20.5

$$T = A_1 \cdot A_2 \cdot A_3 \tag{20.23}$$

where A_1, A_2, and A_3 denote failure events of engines A_1, A_2, and A_3, respectively.

With the aid of Eqs. (20.14) and (20.23) the probability $P(T)$ of the aircraft crashing is given by

$$P(T) = P(A_1) \cdot P(A_2) \cdot P(A_3) \tag{20.24}$$

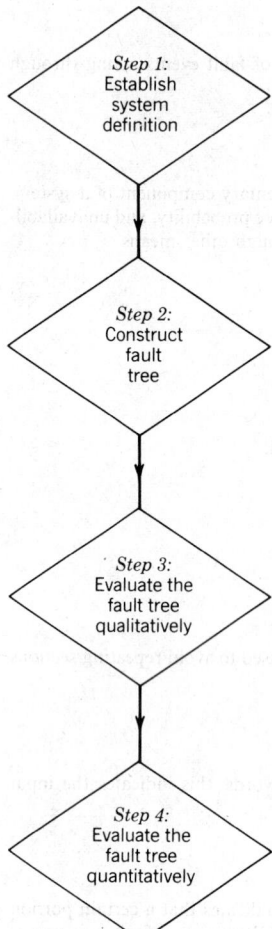

Fig. 20.4 Steps in fault tree analysis.

Step 1:
Establish
system
definition

Step 2:
Construct
fault
tree

Step 3:
Evaluate the
fault tree
qualitatively

Step 4:
Evaluate the
fault tree
quantitatively

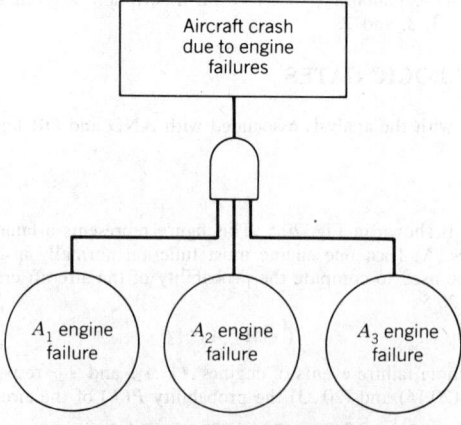

Fig. 20.5 A three-input AND gate.

Aircraft crash
due to engine
failures

A_1 engine
failure

A_2 engine
failure

A_3 engine
failure

where $P(A_1)$ = the probability of failure of engine A_1
$P(A_2)$ = the probability of failure of engine A_2
$P(A_3)$ = the probability of failure of engine A_3

If the aircraft had k engines, then in Fig. 20.5 there would have been k AND gate inputs. In this situation Eqs. (20.23) and (20.24) become

$$T = A_1 \cdot A_2 \cdot A_3 \cdot A_4 \cdot \cdots \cdot A_k \tag{20.25}$$

and

$$P(T) = P(A_1) \cdot P(A_2) \cdot P(A_3) \cdot P(A_4) \cdot \cdots \cdot P(A_k) \tag{20.26}$$

where $P(A_k)$ is the failure probability of engine A_k.

Example 20.2

Assume that from the past field data, the estimates for probabilities $P(A_1)$, $P(A_2)$, and $P(A_3)$ are 0.1, 0.15, and 0.05, respectively. Calculate the probability of the aircraft crashing.
Substituting the above data into Eq. (20.24) yields

$$
\begin{aligned}
P(T) &= P(A_1) \cdot P(A_2) \cdot P(A_3) \\
&= (0.1)(0.15)(0.05) \\
&= 0.00075
\end{aligned}
$$

Thus, the probability of the aircraft crashing is 0.00075.

20.5.2   OR Gate

A two-input OR gate is shown in Fig. 20.6. This figure represents a fault tree of an aircraft with two independent engines. The aircraft can fly only if both engines function normally. The single logic gate fault tree of Fig. 20.6 can be used to compute the probability of the aircraft crashing due to engine failure.
Thus, from Fig. 20.6 we get

$$T = A_1 + A_2 \tag{20.27}$$

The probability of occurrence of the top event T, calculated from Eqs. (20.27) and (20.12), is given by

$$P(T) = P(A_1) + P(A_2) - P(A_1) \cdot P(A_2) \tag{20.28}$$

If we assume that the aircraft has k engines, then in this situation the Fig. 20.6 OR gate will have k inputs. Under this condition, Eq. (27) can be generalized to the following form:

$$T = A_1 + A_2 + A_3 + A_4 + \cdots + A_k \tag{20.29}$$

From Eqs. (20.12) and (20.29), the probability of occurrence of top event T is given by

$$P(T) = 1 - \prod_{i=1}^{k} \{1 - P(A_i)\} \tag{20.30}$$

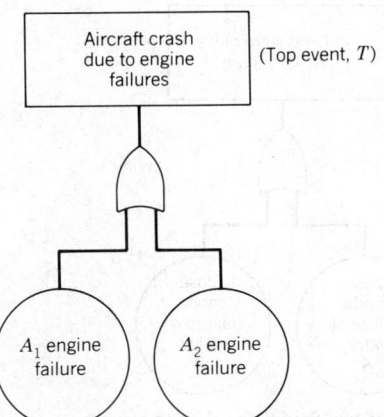

Fig. 20.6   A two-input OR gate.

OR Gate with Mutually Exclusive Inputs

An OR gate with mutually exclusive inputs is the same as an ordinary OR gate, except that its inputs are mutually exclusive. A two-input OR gate is shown in Fig. 20.7. This figure shows a fault tree of a fluid flow valve failure. Thus

$$T = E_1 + E_2 \qquad (20.31)$$

where E_1 = an event denoting the failure of the valve in open mode
  E_2 = an event denoting the failure of the valve in close mode

Since the events E_1 and E_2 are mutually exclusive, the probability of the top event, T, occurrence is given by

$$P(T) = P(E_1) + P(E_2) \qquad (20.32)$$

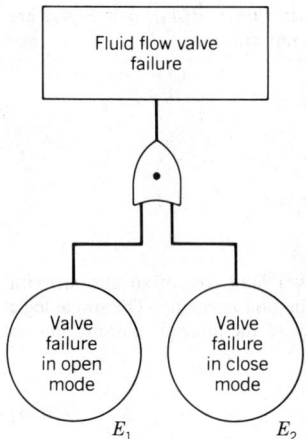

Fig. 20.7   A two-mutually-exclusive-input OR gate.

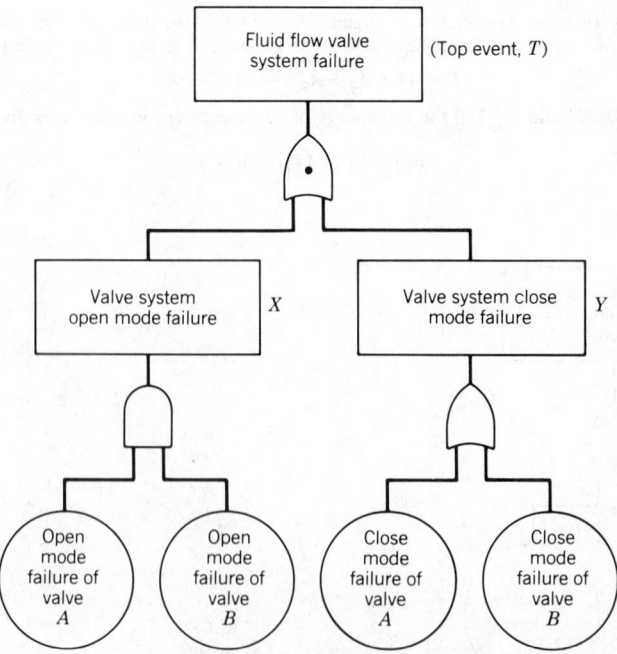

Fig. 20.8   A fault tree of a system composed of two fluid flow valves in series.

where $P(E_1)$ = the probability of valve failure in open mode
    $P(E_2)$ = the probability of valve failure in close mode

Example 20.3

A fluid flow valve close mode and open mode failure probabilities are 0.01 and 0.03, respectively. Calculate the overall failure probability of the valve.

Substituting this data into Eq. (20.32) yields

$$P(T) = P(E_1) + P(E_2)$$
$$= 0.03 + 0.01$$
$$= 0.04$$

The probability of fluid flow valve failure is 0.04.

20.6 EXAMPLES OF FAULT TREES

This section presents three different examples of fault trees. The first is concerned with the fault tree of a system composed of two independent fluid flow valves in series. Each valve can fail in either open or close mode. In this situation the top event of the fault tree is concerned with the fluid flow valve system failure. The fault tree of the system is shown in Fig. 20.8.

In the second example the top event is "no hot water supply in the kitchen of the house." The house uses gas to heat the water. The fault tree of such situation is shown in Fig. 20.9.

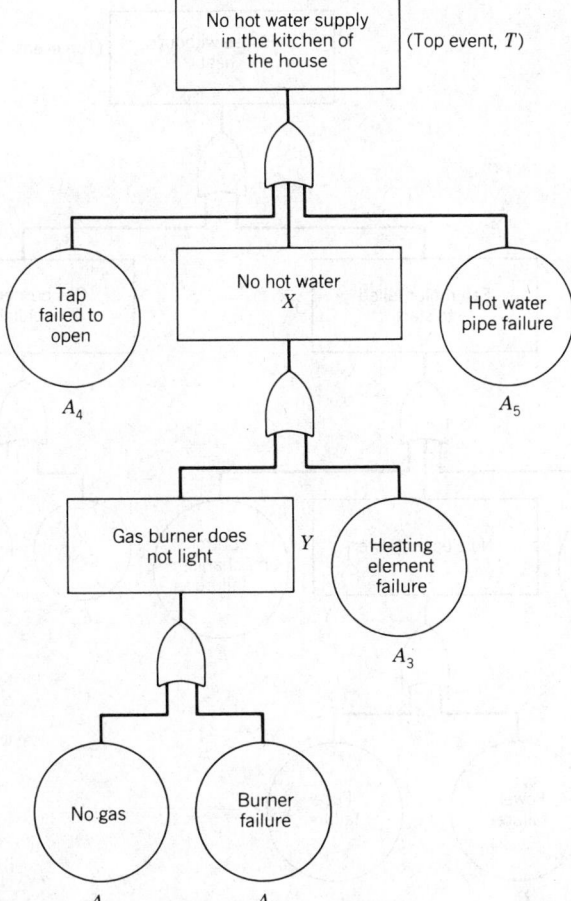

Fig. 20.9 A fault tree for no hot water supply in the kitchen of a house.

Finally, in the third example the top event is "house without heat." In this situation it is assumed that the house heating system is centralized and uses gas. In the heating system an electric fan is used to circulate the warm air to all the rooms of the house. The simplified fault tree is shown in Fig. 20.10.

20.7 PROBABILITY EVALUATION OF A FAULT TREE WITH A REPEATED EVENT

A fault tree with a repeated event is shown in Fig. 20.11. In this diagram the fault event E_1 is the repeated event.

The fault tree in Fig. 20.11 can be represented by the following Boolean expression:

$$T = Y \cdot X = (E_1 + E_3) \cdot (E_1 + E_2) \tag{20.33}$$

The above expression contains the repeated event E_1. Therefore, this expression has to be minimized with the application of Boolean algebra properties before taking the probability of the top event T.

Thus, with the aid of the Boolean algebra property stated by Eq. (20.10), expression (20.33) reduces to

$$T = E_1 + E_2 \cdot E_3 \tag{20.34}$$

The repeated-event-free fault tree of expression (20.34) is shown in Fig. 20.12.

Once a fault tree is repeated-event free, then the occurrence probability of the output fault event of each gate may be calculated. However, if the fault tree is not repeated-event free, then this approach will yield an incorrect result. To obtain minimal cut sets of a fault tree, an algorithm is described in Ref. 3.

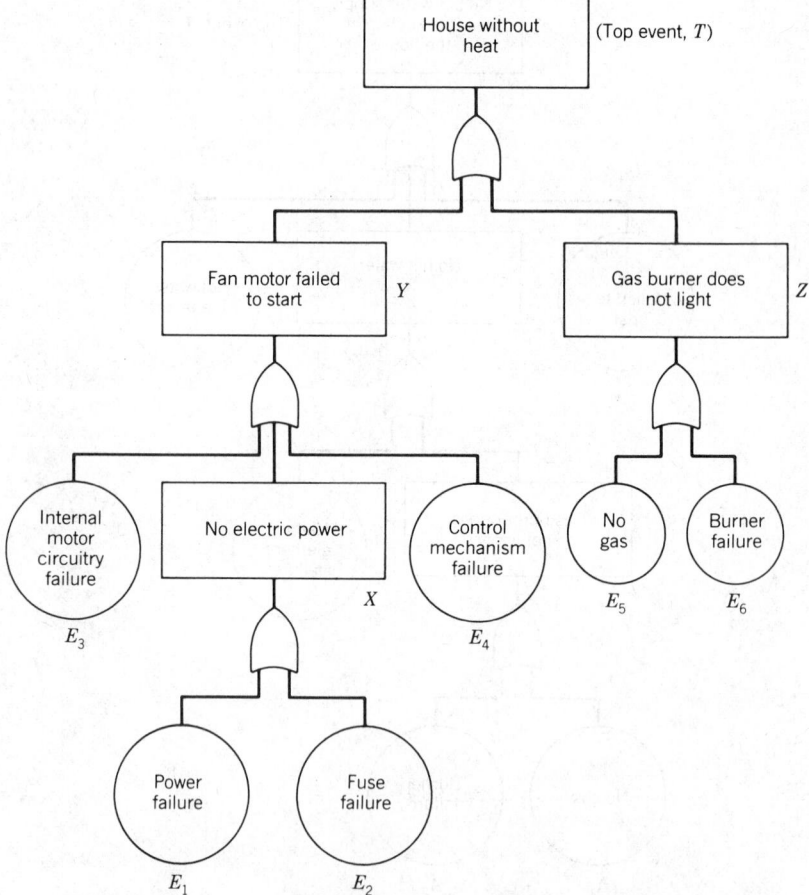

Fig. 20.10. A simplified fault tree of a house without heat.

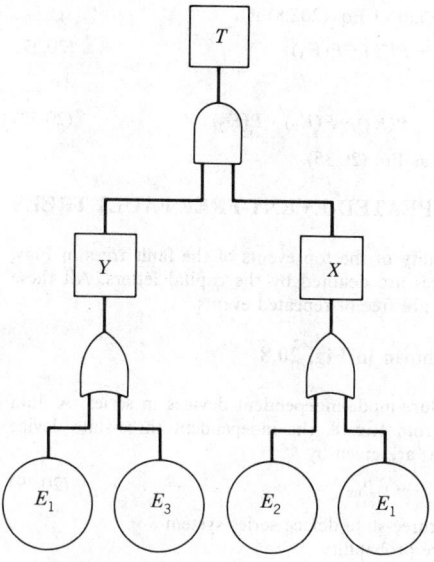

Fig. 20.11   A fault tree with a repeated event.

The occurrence probability of the top event T of Fig. 20.12 can be obtained by two different procedures.

Procedure I

Take the probability of the top fault event T. The Boolean expression for the top event is given by Eq. (20.34). Thus, with the aid of Eqs. (20.12) and (20.14),

$$P(T) = P(E_1 + E_2 \cdot E_3) = P(E_1) + P(E_2) \cdot P(E_3) - P(E_1) \cdot P(E_2) \cdot P(E_3) \qquad (20.35)$$

Procedure II

This procedure calls for taking the probability of occurrence of each output fault event of the fault tree's logic gates. Thus, the probability of occurrence of the fault event Z in Fig. 20.12, with the application of Eq. (20.26), is given by

$$P(Z) = P(E_2) \cdot P(E_3) \qquad (20.36)$$

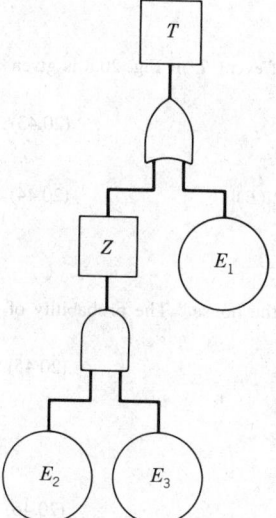

Fig. 20.12   A repeated-event-free fault tree.

Hence, the probability of the top event T with the aid of Eq. (20.28) is

$$P(T) = P(Z) + P(E_1) - P(Z) \cdot P(E_1) \tag{20.37}$$

Substituting Eq. (20.36) into Eq. (20.37) yields

$$P(T) = P(E_2) \cdot P(E_3) + P(E_1) - P(E_1) \cdot P(E_2) \cdot P(E_3) \tag{20.38}$$

We note here that Eq. (20.38) is exactly the same as Eq. (20.35).

20.8   PROBABILITY EVALUATION OF REPEATED-EVENT-FREE FAULT TREES

This section is concerned with evaluating the probability of the top events of the fault trees in Figs. 20.8 and 20.9. The fault events in all these fault trees are denoted by the capital letters. All these fault trees contain independent basic fault events and are free of repeated events.

20.8.1   Probability Evaluation of Fault Tree Shown in Fig. 20.8

This is a fault tree of a system composed of two failure-mode-independent devices in series. A fluid flow valve is a typical example of such a device. From Ref. 8, the independent three-state device series system open and close mode failure probabilities are given by

$$P_0 = P_{01} \cdot P_{02} \cdots P_{0k} \tag{20.39}$$

where P_0 = the open mode failure probability of the three-state device series system
$\quad P_{0k}$ = the kth fluid flow valve open mode failure probability

and

$$P_c = 1 - \prod_{i=1}^{k}(1 - P_{ci}) \tag{20.40}$$

where P_c = the close mode failure probability of the three-state device series system
$\quad P_{ci}$ = the ith fluid flow valve close mode failure probability

With the aid of Eq. (20.39), the probability of occurrence of event X in Fig. 20.8 is given by

$$P(X) = P_0(A) \cdot P_0(B) \tag{20.41}$$

where $P_0(A)$ = the open mode failure probability of fluid flow valve A
$\quad P_0(B)$ = the open mode failure probability of fluid flow valve B

Similarly, with the aid of Eq. (20.40), the probability of occurrence of event Y in Fig. 20.8 is given by

$$P(Y) = 1 - \{1 - P_c(A)\}\{1 - P_c(B)\} \tag{20.42}$$

where $P_c(A)$ = the close mode failure probability of fluid flow valve A
$\quad P_c(B)$ = the close mode failure probability of fluid flow valve B

Finally, with the aid of Eq. (20.32), the probability of occurrence of event T in Fig. 20.8 is given by

$$P(T) = P(X) + P(Y) \tag{20.43}$$

Substituting Eqs. (20.41) and (20.42) into Eq. (20.43) results in

$$P(T) = P_0(A) \cdot P_0(B) + 1 - \{1 - P_c(A)\}\{1 - P_c(B)\} \tag{20.44}$$

20.8.2   Probability Evaluation of Fault Tree Shown in Fig. 20.9

The top event in Fig. 20.9 is "no hot water supply in the kitchen of the house." The probability of occurrence of event Y with the aid of Eq. (20.28) is

$$P(Y) = P(A_1) + P(A_2) - P(A_1) \cdot P(A_2) \tag{20.45}$$

where $P(A_1)$ = the probability of having no gas
$\quad P(A_2)$ = the probability of the gas burner failure

Similarly, the probability of occurrence of event X is given by

$$P(X) = P(Y) + P(A_3) - P(Y) \cdot P(A_3) \tag{20.46}$$

where $P(A_3)$ = the probability of heating element failure.

Finally, the Fig. 20.9 top event probability of occurrence is, with the aid of Eq. (20.30),

$$P(T) = 1 - \{1 - P(A_4)\}\{1 - P(A_5)\} \cdot \{1 - P(X)\} \tag{20.47}$$

where $P(A_4)$ = the probability of tap failing to open
$P(A_5)$ = the probability of hot water pipe failure

20.9 UNAVAILABILITY EVALUATION OF FAULT TREES

In the earlier failure probability analysis, only the occurrence probabilities of the basic fault events were considered. The fault tree analysis covered in this section takes into consideration the constant failure and repair rates of the elementary components (basic fault events). Both these parameters can be used to compute the steady-state unavailability of each elementary component (basic fault event) of the fault tree. In this analysis it is assumed that the basic events (elementary components' failures) occur independently and the fault tree is repeated-event free.

From Eq. (20.22) the steady-state unavailability of an elementary component (basic event) is

$$U = \frac{\lambda}{\lambda + \mu} \tag{20.48}$$

This approach is demonstrated by evaluating the steady-state unavailability of the top event of fault tree shown in Fig. 20.10. This fault tree is again shown in Fig. 20.13. In the fault tree diagram

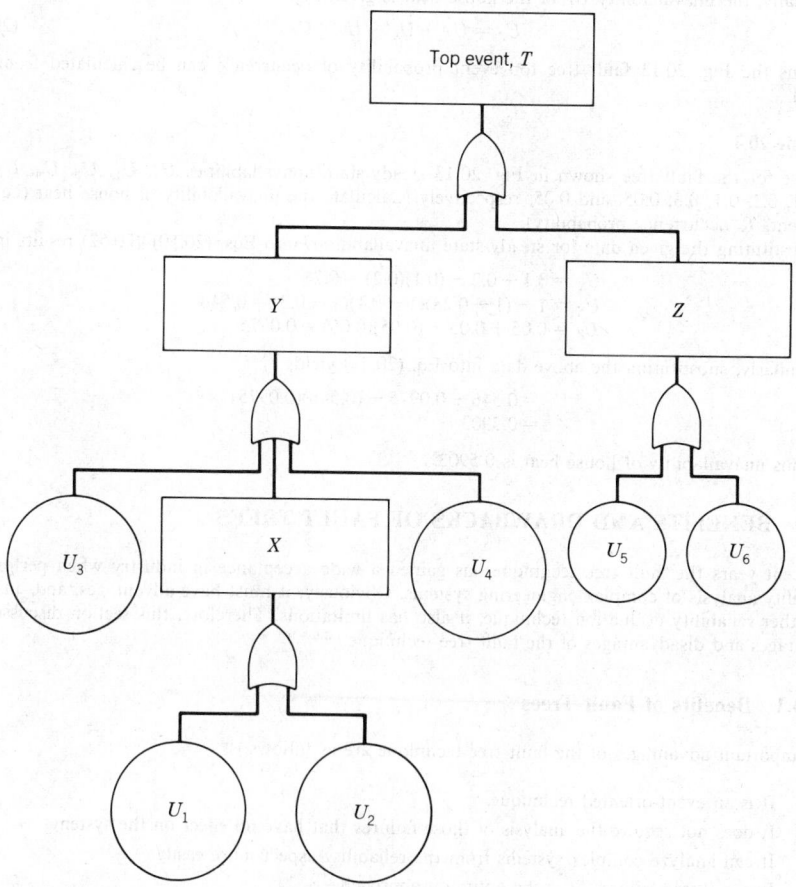

Fig. 20.13 A fault tree with specified basic events unavailabilities.

U_i denotes the unavailability associated with the ith basic event, for $i = 1, 2, 3, 4, 5, 6$. From Eq. (20.48) the U_i are defined as follows:

$$U_i = \frac{\lambda_i}{\lambda_i + \mu_i} \tag{20.49}$$

where λ_i = the ith fault event of the Fig. 20.10 (in this diagram E_i denotes event i, for $i = 1, 2, 3, 4, 5, 6$) constant failure rate; $i = 1$ (power failure), $i = 2$ (fuse failure), $i = 3$ (internal motor circuitry failure), $i = 4$ (control mechanism failure), $i = 5$ (no gas), $i = 6$ (burner failure)

μ_i = the ith fault event of the Fig. 20.10 (in this diagram E_i denotes event i, for $i = 1, 2, 3, 4, 5, 6$) constant repair rate; $i = 1$ (power failure), $i = 2$ (fuse failure), $i = 3$ (internal motor circuitry failure), $i = 4$ (control mechanism failure), $i = 5$ (no gas), $i = 6$ (burner failure)

With the aid of Eqs. (20.49) and (20.28) the unavailability U_X of electric power (fault event X in Fig. 20.13) is given by

$$U_X = U_1 + U_2 - U_1 \cdot U_2 \tag{20.50}$$

Similarly, the unavailabilities U_Y and U_Z of fan motor starting and gas burner lighting, respectively, are

$$U_Y = 1 - (1 - U_X)(1 - U_3)(1 - U_4) \tag{20.51}$$

and

$$U_Z = U_5 + U_6 - U_5 \cdot U_6 \tag{20.52}$$

Finally, the unavailability U_T of the house heat is given by

$$U_T = U_Y + U_Z - U_Y \cdot U_Z \tag{20.53}$$

Thus the Fig. 20.13 fault tree top event probability of occurrence can be calculated from Eq. (20.53).

Example 20.4

Suppose for the fault tree shown in Fig. 20.13 steady-state unavailabilities U_1, U_2, U_3, U_4, U_5, U_6 are 0.1, 0.2, 0.1, 0.3, 0.05, and 0.05, respectively. Calculate the unavailability of house heat (i.e., the top event, T, occurrence probability).

Substituting the given data for steady-state unavailabilities into Eqs. (20.50)–(20.52) results in

$$U_X = 0.1 + 0.2 - (0.1)(0.2) = 0.28$$
$$U_Y = 1 - (1 - 0.28)(1 - 0.1)(1 - 0.3) = 0.546$$
$$U_Z = 0.05 + 0.05 - (0.05)(0.05) = 0.0975$$

Similarly, substituting the above data into Eq. (20.53) yields

$$U_T = 0.546 + 0.0975 - (0.546)(0.0975)$$
$$= 0.5903$$

Thus unavailability of house heat is 0.5903.

20.10 BENEFITS AND DRAWBACKS OF FAULT TREES

In recent years the fault tree technique has gained a wide acceptance in industry when performing reliability analysis of complex engineering systems. Obviously, it must have advantages, and, just like any other reliability evaluation technique, it also has limitations. Therefore, this section discusses the advantages and disadvantages of the fault tree technique.[1,3,5,9,10]

20.10.1 Benefits of Fault Trees

The important advantages of the fault tree technique are as follows:

1. It is an event-oriented technique.
2. It does not require the analysis of those failures that have no effect on the system.
3. It can analyze complex systems from the reliability aspect more easily.
4. For multiple failures, it is the optimum approach.
5. It provides a visibility tool to both management and engineers.

6. It provides the reliability analyst with insight into the behavior of the system.
7. It helps the analyst thoroughly understand the system under consideration.
8. It allows the reliability analyst to focus on one fault event at a time.
9. It gives the reliability analyst the option of performing qualitative or quantitative analysis.

20.10.2 Drawbacks of Fault Trees

Like other methods, the fault tree technique has drawbacks. Some of them are as follows:

1. The major drawbacks of this technique are oversight and omission.[1]
2. It fails to take into consideration properly the mutual exclusiveness of fault events.
3. There is difficulty in checking the results obtained.
4. It is a costly method.
5. It is a time-consuming approach.
6. It is difficult to apply in many cases.[11]

REFERENCES

1. J. B. Fussell, G. J. Powers, and R. G. Bennetts, "Fault Trees—A State of the Art Discussion," *IEEE Transactions on Reliability* **R-23**, 51–55 (1974).
2. R. E. Barlow, J. B. Fussell, and N. D. Singpurwalla, *Reliability and Fault Analysis: Theoretical and Applied Aspects of System Reliability and Safety Assessment,* Society for Industrial and Applied Mathematics (SIAM), Philadelphia, PA, 1975.
3. B. S. Dhillon and C. Singh, *Engineering Reliability: New Techniques and Applications,* Wiley, New York, 1981.
4. B. S. Dhillon, *Systems Reliability, Maintainability and Management,* Petrocelli Books Inc., New York, 1983.
5. R. J. Schroder, "Fault Tree for Reliability Analysis," *Proceedings of the Annual Symposium on Reliability,* IEEE, New York, 1970, pp. 198–205.
6. Flow Research Report, *Risk Analysis Using the Fault Tree Technique,* Flow Research, Inc., New York, 1973.
7. J. B. Fussell, "Fault Tree Analysis—Concepts and Techniques," in *Proceedings of the NATO Advanced Study Institute on Generic Techniques of System Reliability Assessment,* Nordhoff, Leiden, Netherlands, 1975, pp. 133–162.
8. B. S. Dhillon, "The Analysis of the Reliability of Multi-State Device Networks," Ph.D. Dissertation, 1975. (Available from the National Library of Canada, Ottawa, Ontario.)
9. K. H. Eagle, "Fault Tree and Reliability Analysis Comparison," in *Proceedings of the Annual Symposium on Reliability,* IEEE, New York, 1969, pp. 12–17.
10. P. A. Crosetti and R. A. Bruce, "Commercial Application of Fault Tree Analysis," in *Proceedings of the Annual Reliability and Maintainability Symposium,* IEEE, New York, 1970, pp. 230–244.
11. R. E. Barlow and H. E. Lambert, "Introduction to Fault Tree Analysis," in *Reliability and Fault Tree Analysis: Theoretical and Applied Aspects of System Reliability and Safety Assessment,* Society for Industrial and Applied Mathematics, Philadelphia, PA, 1975, pp. 7–35.

CHAPTER 21

DESIGN OPTIMIZATION— AN OVERVIEW

A. RAVINDRAN

School of Industrial Engineering
University of Oklahoma

G. V. REKLAITIS

School of Chemical Engineering
Purdue University

INTRODUCTION

This chapter presents an overview of optimization theory and its application to problems arising in engineering. In the most general terms, optimization theory is a body of mathematical results and numerical methods for finding and identifying the best candidate from a collection of alternatives without having to enumerate and evaluate explicitly all possible alternatives. The process of optimization lies at the root of engineering, since the classical function of the engineer is to design new, better, more efficient, and less expensive systems, as well as to devise plans and procedures for the improved operation of existing systems. The power of optimization methods to determine the best case without actually testing all possible cases comes through the use of a modest level of mathematics and at the cost of performing iterative numerical calculations using clearly defined logical procedures or algorithms implemented on computing machines. Because of the scope of most engineering applications and the tedium of the numerical calculations involved in optimization algorithms, the techniques of optimization are intended primarily for computer implementation.

21.1 REQUIREMENTS FOR THE APPLICATION OF OPTIMIZATION METHODS

In order to apply the mathematical results and numerical techniques of optimization theory to concrete engineering problems it is necessary to delineate clearly the boundaries of the engineering system to be optimized, to define the quantitative criterion on the basis of which candidates will be ranked to determine the "best," to select the system variables that will be used to characterize or identify candidates, and to define a model that will express the manner in which the variables are related. This composite activity constitutes the process of *formulating* the engineering optimization problem. Good problem formulation is the key to the success of an optimization study and is to a large degree an art. It is learned through practice and the study of successful applications and is based on the knowledge of the strengths, weaknesses, and peculiarities of the techniques provided by optimization theory.

21.1.1 Defining the System Boundaries

Before undertaking any optimization study it is important to define clearly the boundaries of the system under investigation. In this context a system is the restricted portion of the universe under consideration. The system boundaries are simply the limits that separate the system from the remainder of the universe. They serve to isolate the system from its surroundings, because, for purposes of analysis, all interactions between the system and its surroundings are assumed to be frozen at selected, representative levels. Since interactions, nonetheless, always exist, the act of defining the system boundaries is the first step in the process of approximating the real system.

In many situations it may turn out that the initial choice of system boundary is too restrictive. In order to analyze a given engineering system fully it may be necessary to expand the system boundaries to include other subsystems that strongly affect the operation of the system under study. For instance, suppose a manufacturing operation has a paint shop in which finished parts are mounted on an assembly line and painted in different colors. In an initial study of the paint shop we may consider it in isolation from the rest of the plant. However, we may find that the optimal batch size and color sequence we deduce for this system are strongly influenced by the operation of the fabrication department that produces the finished parts. A decision thus has to be made whether to expand the system boundaries to include the fabrication department. An expansion of the system boundaries certainly increases the size and complexity of the composite system and thus may make the study much more difficult. Clearly, in order to make our work as engineers more manageable, we would prefer as much as possible to break down large complex systems into smaller subsystems that can be dealt with individually. However, we must recognize that this decomposition is in itself a potentially serious approximation of reality.

21.1.2 The Performance Criterion

Given that we have selected the system of interest and have defined its boundaries, we next need to select a criterion on the basis of which the performance or design of the system can be evaluated so that the "best" design or set of operating conditions can be identified. In many engineering applications, an economic criterion is selected. However, there is a considerable choice in the precise definition of such a criterion: total capital cost, annual cost, annual net profit, return on investment, cost to benefit ratio, or net present worth. In other applications a criterion may involve some technological factors, for instance, minimum production time, maximum production rate, minimum energy utilization, maximum torque, and minimum weight. Regardless of the criterion selected, in the context of optimization the "best" will always mean the candidate system with either the *minimum* or the *maximum* value of the performance index.

It is important to note that within the context of the optimization methods, only *one* criterion or performance measure is used to define the optimum. It is not possible to find a solution that, say, simultaneously minimizes cost and maximizes reliability and minimizes energy utilization. This again is an important simplification of reality, because in many practical situations it would be desirable to achieve a solution that is "best" with respect to a number of different criteria. One way of treating multiple competing objectives is to select one criterion as primary and the remaining criteria as secondary. The primary criterion is then used as an optimization performance measure, while the secondary criteria are assigned acceptable minimum or maximum values and are treated as problem constraints. However, if careful considerations were not given while selecting the acceptable levels, a feasible design that satisfies all the constraints may not exist. This problem is overcome by a technique called *goal programming*, which is fast becoming a practical method for handling multiple criteria. In this method, all the objectives are assigned target levels for achievement and a relative priority on achieving these levels. Goal programming treats these targets as goals to aspire for and not as absolute constraints. It then attempts to find an optimal solution that comes as "close as possible" to the targets in the order of specified priorities. Readers interested in multiple criteria optimizations are directed to recent specialized texts.[1,2]

21.1.3 The Independent Variables

The third key element in formulating a problem for optimization is the selection of the independent variables that are adequate to characterize the possible candidate designs or operating conditions of the system. There are several factors that must be considered in selecting the independent variables.

First, it is necessary to distinguish between variables whose values are amenable to change and variables whose values are fixed by external factors, lying outside the boundaries selected for the system in question. For instance, in the case of the paint shop, the types of parts and the colors to be used are clearly fixed by product specifications or customer orders. These are specified system parameters. On the other hand, the order in which the colors are sequenced is, within constraints imposed by the types of parts available and inventory requirements, an independent variable that can be varied in establishing a production plan.

Furthermore, it is important to differentiate between system parameters that can be treated as fixed and those that are subject to fluctuations which are influenced by external and uncontrollable factors. For instance, in the case of the paint shop, equipment breakdown and worker absenteeism may be sufficiently high to influence the shop operations seriously. Clearly, variations in these key system parameters must be taken into account in the production planning problem formulation if the resulting optimal plan is to be realistic and operable.

Second, it is important to include in the formulation all of the important variables that influence the operation of the system or affect the design definition. For instance, if in the design of a gas storage system we include the height, diameter, and wall thickness of a cylindrical tank as independent variables, but exclude the possibility of using a compressor to raise the storage pressure, we may well obtain a very poor design. For the selected fixed pressure we would certainly find the least cost tank dimensions. However, by including the storage pressure as an independent variable and adding the compressor cost to our performance criterion, we could obtain a design that has a lower overall cost because of a reduction in the required tank volume. Thus, the independent variables must be selected so that all important alternatives are included in the formulation. Exclusion of possible alternatives, in general, will lead to suboptimal solutions.

Finally, a third consideration in the selection of variables is the level of detail to which the system is considered. While it is important to treat all of the key independent variables, it is equally important not to obscure the problem by the inclusion of a large number of fine details of subordinate importance. For instance, in the preliminary design of a process involving a number of different pieces of equipment— pressure vessels, towers, pumps, compressors, and heat exchangers—one would normally not explicitly consider all of the fine details of the design of each individual unit. A heat exchanger may well be characterized by a heat-transfer surface area as well as shell-side and tube-side pressure drops. Detailed design variables such as number and size of tubes, number of tube and shell passes, baffle spacing, header type, and shell dimensions would normally be considered in a separate design study involving that unit by itself. In selecting the independent variables a good rule to follow is to include only those variables that have a significant impact on the composite system performance criterion.

21.1.4 The System Model

Once the performance criterion and the independent variables have been selected, then the next step in problem formulation is the assembly of the model that describes the manner in which the problem variables are related and the performance criterion is influenced by the independent variables. In principle, optimization studies may be performed by experimenting directly with the system. Thus, the independent variables of the system or process may be set to selected values, the system operated under those conditions, and the system performance index evaluated using the observed performance. The optimization methodology would then be used to predict improved choices of the independent variable values and the experiments continued in this fashion. In practice most optimization studies are carried out with the help of a *model,* a simplified mathematical representation of the real system. Models are used because it is too expensive or time consuming or risky to use the real system to carry out the study. Models are typically used in engineering design because they offer the cheapest and fastest way of studying the effects of changes in key design variables on system performance.

In general, the model will be composed of the basic material and energy balance equations, engineering design relations, and physical property equations that describe the physical phenomena taking place in the system. These equations will normally be supplemented by inequalities that define allowable operating ranges, specify minimum or maximum performance requirements, or set bounds on resource availabilities. In sum, the model consists of all of the elements that normally must be considered in calculating a design or in predicting the performance of an engineering system. Quite clearly the assembly of a model is a very time-consuming activity, and it is one that requires a thorough understanding of the system being considered. In simple terms, a model is a collection of equations and inequalities that define how the system variables are related and that constrain the variables to take on acceptable values.

From the preceding discussion, we observe that a problem suitable for the application of optimization methodology consists of a performance measure, a set of independent variables, and a model relating the variables. Given these rather general and abstract requirements, it is evident that the methods of optimization can be applied to a very wide variety of applications. We shall illustrate next a few engineering design applications and their model formulations.

21.2 APPLICATIONS OF OPTIMIZATION IN ENGINEERING

Optimization theory finds ready application in all branches of engineering in four primary areas:

1. Design of components of entire systems.
2. Planning and analysis of existing operations.
3. Engineering analysis and data reduction.
4. Control of dynamic systems.

In this section we briefly consider representative applications from the first three areas.

In considering the application of optimization methods in design and operations, the reader should keep in mind that the optimization step is but one step in the overall process of arriving at an optimal design or an efficient operation. Generally, that overall process will, as shown in Fig. 21.1, consist of an iterative cycle involving synthesis or definition of the structure of the system, model formulation, model parameter optimization, and analysis of the resulting solution. The final optimal design or new operating plan will be obtained only after solving a series of optimization problems, the solution to each of which will have served to generate new ideas for further system structures. In the interest of brevity, the examples in this section show only one pass of this iterative cycle and focus mainly on preparations for the optimization step. This focus should not be interpreted as an indication of the dominant role of optimization methods in the engineering design and systems analysis process. Optimization theory is but a very powerful tool that, to be effective, must be used skillfully and intelligently by an engineer who thoroughly understands the system under study. The primary objective of the following example is simply to illustrate the wide variety but common form of the optimization problems that arise in the design and analysis process.

21.2.1 Design Applications

Applications in engineering design range from the design of individual structural members to the design of separate pieces of equipment to the preliminary design of entire production facilities. For purposes of optimization the shape or structure of the system is assumed known and optimization problem reduces to the selection of values of the unit dimensions and operating variables that will yield the best value of the selected performance criterion.

Example 21.1 Design of an Oxygen Supply System

Description. The basic oxygen furnace (BOF) used in the production of steel is a large fed-batch chemical reactor that employs pure oxygen. The furnace is operated in a cyclic fashion: ore and flux are charged to the unit, are treated for a specified time period, and then are discharged. This cyclic operation gives rise to a cyclically varying demand rate for oxygen. As shown in Fig. 21.2, over each cycle there is a time interval of length t_1 of low demand rate, D_0, and a time interval $(t_2 - t_1)$ of

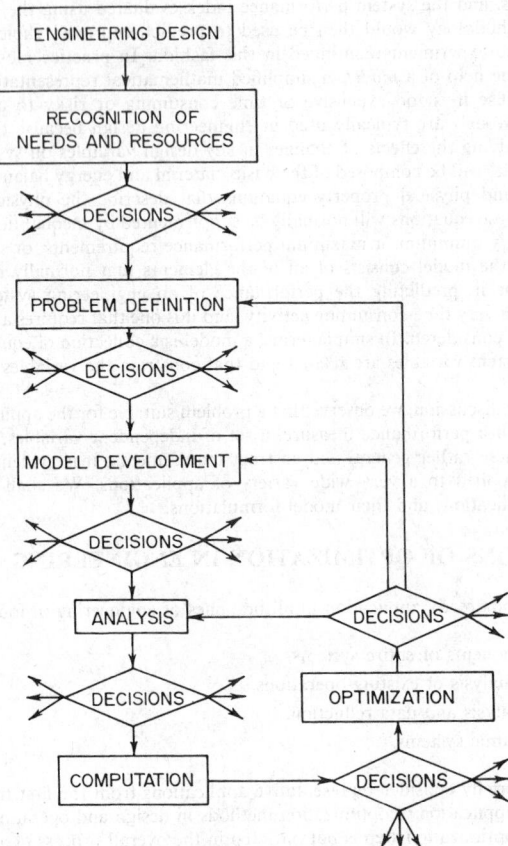

```
        ┌─────────────────────┐
        │  ENGINEERING DESIGN │
        └─────────────────────┘
                  │
        ┌─────────────────────┐
        │   RECOGNITION OF     │
        │ NEEDS AND RESOURCES  │
        └─────────────────────┘
                  │
             ◇ DECISIONS ◇
                  │
        ┌─────────────────────┐
        │ PROBLEM DEFINITION   │◄──
        └─────────────────────┘
                  │
             ◇ DECISIONS ◇
                  │
        ┌─────────────────────┐
        │  MODEL DEVELOPMENT   │◄──
        └─────────────────────┘
                  │
             ◇ DECISIONS ◇
                  │
   ┌──────────┐          ◇ DECISIONS ◇
   │ ANALYSIS │
   └──────────┘          ┌─────────────────┐
                         │  OPTIMIZATION   │
      ◇ DECISIONS ◇      └─────────────────┘
                                 │
   ┌─────────────┐      ◇ DECISIONS ◇
   │ COMPUTATION │─────►
   └─────────────┘
```

Fig. 21.1 Optimal design process.

high demand rate, D_1. The oxygen used in the BOF is produced in an oxygen plant. Oxygen plants are standard process plants in which oxygen is separated from air using a combination of refrigeration and distillation. These are highly automated plants, which are designed to deliver a fixed oxygen rate. In order to mesh the continuous oxygen plant with the cyclically operating BOF, a simple inventory system shown in Fig. 21.3 and consisting of a compressor and a storage tank must be designed. A number of design possibilities can be considered. In the simplest case, one could select the oxygen plant capacity to be equal to D_1, the high demand rate. During the low-demand interval the excess oxygen could just be vented to the air. At the other extreme, one could also select the oxygen plant capacity to be just enough to produce the amount of oxygen required by the BOF over a cycle. During the low-demand interval, the excess oxygen production would then be compressed and stored for use during the high-demand interval of the cycle. Intermediate designs could involve some combination of venting and storage of oxygen. The problem is to select the optimal design.

   Formulation. The system of concern will consist of the O_2 plant, the compressor, and the storage tank. The BOF and its demand cycle are assumed fixed by external factors. A reasonable performance index for the design is the total annual cost, which consists of the oxygen production cost (fixed and variable), the compressor operating cost, and the fixed costs of the compressor and of the storage vessel. The key independent variables are the oxygen plant production rate F (lb O_2/hr), the compressor and storage tank design capacities, H (hp) and V (ft³), respectively, and the maximum tank pressure, p (psia). Presumably the oxygen plant design is standard, so that the production rate fully characterizes the plant. Similarly, we assume that the storage tank will be of a standard design approved for O_2 service.

   The model will consist of the basic design equations that relate the key independent variables.

   If I_{max} is the maximum amount of oxygen that must be stored, then using the corrected gas law we have

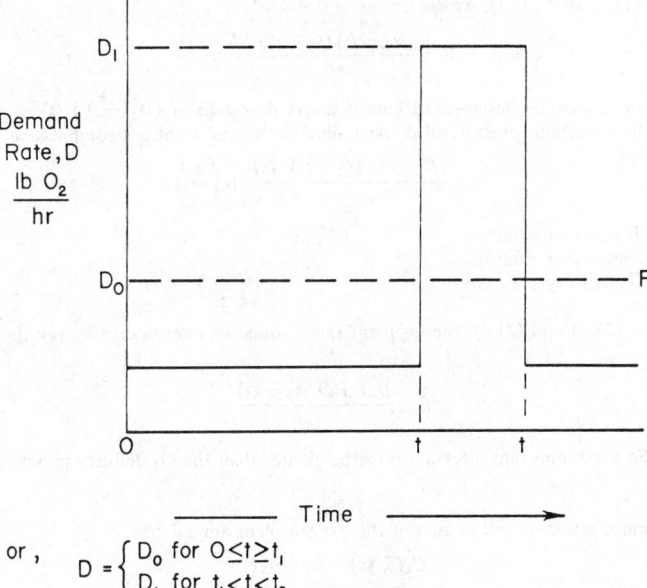

Fig. 21.2   Oxygen demand cycle.

or,

$$D = \begin{cases} D_0 & \text{for } 0 \le t \ge t_1 \\ D_1 & \text{for } t_1 \le t \le t_2 \end{cases}$$

$$V = \frac{I_{max}}{M} \frac{RT}{p} z \tag{21.1}$$

where R = the gas constant
T = the gas temperature (assume fixed)
z = the compressibility factor
M = the molecular weight of O_2

From Fig. 21.1, the maximum amount of oxygen that must be stored is equal to the area under the demand curve between t_1 and t_2 and D_1 and F. Thus,

$$I_{max} = (D_1 - F)(t_2 - t_1) \tag{21.2}$$

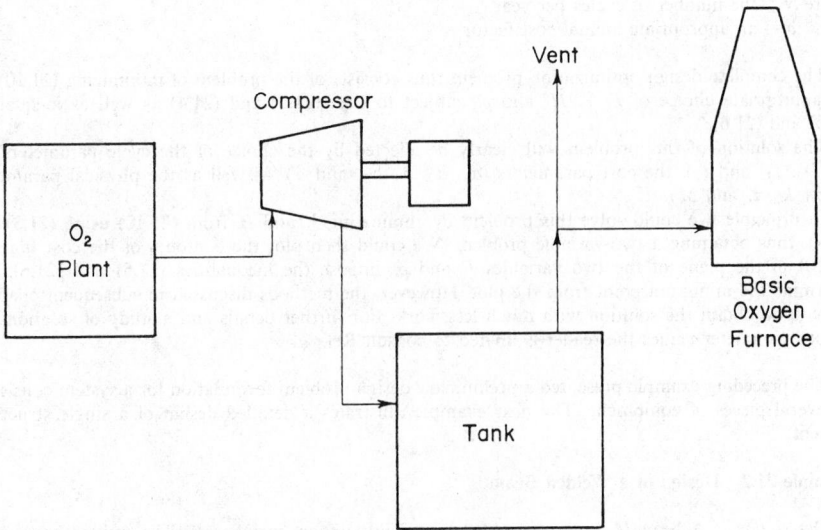

Fig. 21.3   Design of oxygen production system.

Substituting (21.2) into (21.1), we obtain

$$V = \frac{(D_1 - F)(t_2 - t_1)}{M} \frac{RT}{p} z \qquad (21.3)$$

The compressor must be designed to handle a gas flow rate of $(D_1 - F)(t_2 - t_1)/t_1$ and to compress it to the maximum pressure of p. Assuming isothermal ideal gas compression,[3]

$$H = \frac{(D_1 - F)(t_2 - t_1)}{t_1} \frac{RT}{k_1 k_2} \ln\left(\frac{p}{p_0}\right) \qquad (21.4)$$

where $k_1 =$ a unit conversion factor
$\quad k_2 =$ the compressor efficiency
$\quad p_0 =$ the O_2 delivery pressure

In addition to (21.3) and (21.4), the O_2 plant rate F must be adequate to supply the total oxygen demand, or

$$F \geq \frac{D_0 t + D_1(t_2 - t_1)}{t_2} \qquad (21.5)$$

Moreover, the maximum tank pressure must be greater than the O_2 delivery pressure,

$$p \geq p_0 \qquad (21.6)$$

The performance criterion will consist of the oxygen plant annual cost,

$$C_1(\$/\text{yr}) = a_1 + a_2 F \qquad (21.7)$$

where a_1 and a_2 are empirical constants for plants of this general type and include fuel, water, and labor costs.

The capital cost of storage vessels is given by a power-law correlation,

$$C_2(\$) = b_1 V^{b_2} \qquad (21.8)$$

where b_1 and b_2 are empirical constants appropriate for vessels of a specific construction.

The capital cost of compressors is similarly obtained from a correlation,

$$C_3(\$) = b_3 H^{b_4} \qquad (21.9)$$

The compressor power cost will, as an approximation, be given by

$$b_5 t_1 H$$

where b_5 is the cost of power.

The total cost function will thus be of the form,

$$\text{Annual cost} = a_1 + a_2 F + d\{b_1 V^{b_2} + b_3 H^{b_4}\} + N b_5 t_1 H \qquad (21.10)$$

where $N =$ the number of cycles per year
$\quad d =$ an appropriate annual cost factor

The complete design optimization problem thus consists of the problem of minimizing (21.10), by the appropriate choice of F, V, H, and p, subject to Eqs. (21.3) and (21.4) as well as inequalities (21.5) and (21.6).

The solution of this problem will clearly be affected by the choice of the cycle parameters (N, D_0, D_1, t_1, and t_2), the cost parameters (a_1, a_2, b_1–b_5, and d), as well as the physical parameters (T, p_0, k_2, z, and M).

In principle, we could solve this problem by eliminating V and H from (21.10) using (21.3) and (21.4), thus obtaining a two-variable problem. We could then plot the contours of the cost function (21.10) in the plane of the two variables F and p, impose the inequalities (21.5) and (21.6), and determine the minimum point from the plot. However, the methods discussed in subsequent chapters allow us to obtain the solution with much less work. For further details and a study of solutions for various parameter values the reader is invited to consult Ref. 4.

The preceding example presented a preliminary design problem formulation for a system consisting of several pieces of equipment. The next example illustrates a detailed design of a single structural element.

Example 21.2 Design of a Welded Beam

Description. A beam A is to be welded to a rigid support member B. The welded beam is to consist of 1010 steel and is to support a force F of 6000 lb. The dimensions of the beam are to be selected so that the system cost is minimized. A schematic of the system is shown in Fig. 21.4.

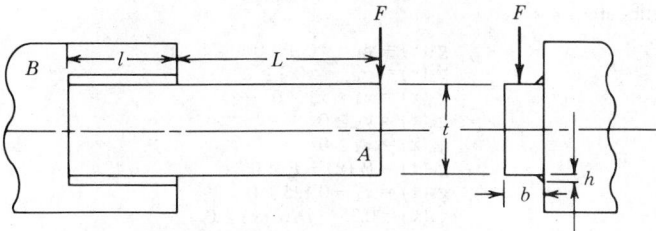

Fig. 21.4 Welded beam.

Formulation. The appropriate system boundaries are quite self-evident. The system consists of the beam A and the weld required to secure it to B. The independent or design variables in this case are the dimensions h, l, t, and b as shown in Fig. 21.4. The length L is assumed to be specified at 14 in. For notational convenience we redefine these four variables in terms of the vector of unknowns $\mathbf{x}$,

$$\mathbf{x} = [x_1,\, x_2,\, x_3,\, x_4]^T = [h,\, l,\, t,\, b]^T$$

The performance index appropriate to this design is the cost of a weld assembly. The major cost components of such an assembly are (a) set-up labor cost, (b) welding labor cost, and (c) material cost:

$$F(x) = c_0 + c_1 + c_2 \tag{21.11}$$

where $F(x)$ = cost function
$\quad\quad c_0$ = set-up cost
$\quad\quad c_1$ = welding labor cost
$\quad\quad c_2$ = material cost

Set-Up Cost: c_0. The company has chosen to make this component a weldment, because of the existence of a welding assembly line. Furthermore, assume that fixtures for set-up and holding of the bar during welding are readily available. The cost c_0 can, therefore, be ignored in this particular total cost model.

Welding Labor Cost: c_1. Assume that the welding will be done by machine at a total cost of $10 per hour (including operating and maintenance expense). Furthermore, suppose that the machine can lay down 1 in.³ of weld in 6 min. Therefore, the labor cost is

$$c_1 = \left(10\,\frac{\$}{\text{hr}}\right)\left(\frac{1\ \text{hr}}{60\ \text{min}}\right)\left(6\,\frac{\text{min}}{\text{in.}^3}\right)V_w = 1\left(\frac{\$}{\text{in.}^3}\right)V_w$$

where V_w = weld volume, in.³

Material Cost: c_2.

$$c_2 = c_3 V_w + c_4 V_B$$

where c_3 = \$/volume of weld material = $(0.37)(0.283)(\$/\text{in.}^3)$
$\quad\quad c_4$ = \$/volume of bar stock = $(0.17)(0.283)(\$/\text{in.}^3)$
$\quad\quad V_B$ = volume of bar A (in.³)

From the geometry,

$$V_w = 2(\tfrac{1}{2}h^2 l) - h^2 l \quad\quad\text{and}\quad\quad V_B = tb(L + l)$$

so

$$c_2 = c_3 h^2 l + c_4 tb(L + l)$$

Therefore, the cost function becomes

$$F(x) = h^2 l + c_3 h^2 l + c_4 tb(L + l) \tag{21.12}$$

or, in terms of the x variables

$$F(x) = (l + c_3)x_1^2 x_2 + c_4 x_3 x_4(L + x_2) \tag{21.13}$$

Note all combinations of x_1, x_2, x_3, and x_4 can be allowed if the structure is to support the load required. Several functional relationships between the design variables that delimit the region of feasibility must certainly be defined. These relationships, expressed in the form of inequalities, represent the design model. Let us first define the inequalities and then discuss their interpretation.

The inequities are:

$$g_1(x) = \tau_d - \tau(x) \geq 0 \tag{21.14}$$
$$g_2(x) = \sigma_d - \sigma(x) \geq 0 \tag{21.15}$$
$$g_3(x) = x_4 - x_1 \geq 0 \tag{21.16}$$
$$g_4(x) = x_2 \geq 0 \tag{21.17}$$
$$g_5(x) = x_3 \geq 0 \tag{21.18}$$
$$g_6(x) = P_c(x) - F \geq 0 \tag{21.19}$$
$$g_7(x) = x_1 - 0.125 \geq 0 \tag{21.20}$$
$$g_8(x) = 0.25 - DEL(x) \geq 0 \tag{21.21}$$

where
τ_d = design shear stress of weld
$\tau(x)$ = maximum shear stress in weld; a function of x
σ_d = design normal stress for beam material
$\sigma(x)$ = maximum normal stress in beam; a function of x
$P_c(x)$ = bar buckling load; a function of x
$DEL(x)$ = bar end deflection; a function of x

In order to complete the model it is necessary to define the important stress states.

Weld stress: $\tau(x)$. After Shigley,[5] the weld shear stress has two components, τ' and τ'', where τ' is the primary stress acting over the weld throat area and τ'' is a secondary torsional stress:

$$\tau' = F/\sqrt{2}x_1x_2 \quad \text{and} \quad \tau'' = MR/J$$

with $M = F[L + (x_2/2)]$
$R = \{(x_2^2/4) + [(x_3 + x_1)/2]^2\}^{1/2}$
$J = 2\{0.707x_1x_2[x_2^2/12 + (x_3 + x_1)/2)^2]\}$

where M = moment of F about the center of gravity of the weld group
J = polar moment of inertia of the weld group

Therefore, the weld stress τ becomes

$$\tau(x) = [(\tau')^2 + 2\tau'\tau'' \cos \theta + (\tau'')^2]^{1/2}$$

where $\cos \theta = x_2/2R$.

Bar Bending Stress: $\sigma(x)$. The maximum bending stress can be shown to be equal to

$$\sigma(x) = 6FL/x_4x_3^2$$

Bar Buckling Load: $P_c(x)$. If the ratio $t/b = x_3/x_4$ grows large, there is a tendency for the bar to buckle. Those combinations of x_3 and x_4 that will cause this buckling to occur must be disallowed. It has been shown[6] that for narrow rectangular bars, a good approximation to the buckling load is

$$P_c(x) = \frac{4.013\sqrt{EI\alpha}}{L^2}\left[1 - \frac{x_3}{2L}\sqrt{\frac{EI}{\alpha}}\right]$$

where E = Young's modulus = 30×10^6 psi
$I = \frac{1}{12} x_3x_4^3$
$\alpha = \frac{1}{3} Gx_3x_4^3$
G = shearing modulus = 12×10^6 psi

Bar deflection: $DEL(x)$. To calculate the deflection assume the bar to be a cantilever of length L. Thus,

$$DEL(x) = 4FL^3/Ex_3^3x_4$$

The remaining inequalities are interpreted as follows.

g_3 states that it is not practical to have the weld thickness greater than the bar thickness. g_4 and g_5 are nonnegativity restrictions on x_2 and x_3. Note that the nonnegativity of x_1 and x_4 are implied by g_3 and g_7. Constraint g_6 ensures that the buckling load is not exceeded. Inequality g_7 specifies that it is not physically possible to produce an extremely small weld.

Finally, the two parameters τ_d and σ_d in g_1 and g_2 depend on the material of construction. For 1010 steel τ_d = 13,600 psi and σ_d = 30,000 psi are appropriate.

The complete design optimization problem thus consists of the cost function (21.13) and the complex system of inequalities that results when the stress formulas are substituted into (21.14) through (21.21). All of these functions are expressed in terms of four independent variables.

This problem is sufficiently complex that graphical solution is patently infeasible. However, the optimum design can readily be obtained numerically using the methods of subsequent sections. For a further discussion of this problem and its solution the reader is directed to Ref. 7.

21.2.2 Operations and Planning Applications

The second major area of engineering application of optimization is found in the tuning of existing operations. We shall discuss an application of goal programming model for machinability data optimization in metal cutting.[8]

Example 21.3 An Economic Machining Problem with Two Competing Objectives

Consider a single-point, single-pass turning operation in metal cutting wherein an optimum set of cutting speed and feed rate is to be chosen which balances the conflict between metal removal rate and tool life as well as being within the restrictions of horsepower, surface finish, and other cutting conditions. In developing the mathematical model of this problem, the following constraints will be considered for the machining parameters:

Constraint 1: Maximum Permissible Feed.

$$f \leqq f_{max} \tag{21.22}$$

where f is the feed in inches per revolution. f_{max} is usually determined by a cutting force restriction or by surface finish requirements.[9]

Constraint 2: Maximum Cutting Speed Possible. If v is the cutting speed in surface feet per minute, then

$$v \leqq v_{max} \tag{21.23}$$

where

$$v_{max} = \frac{\pi D N_{max}}{12}$$

and

$$N_{max} = \text{maximum spindle speed available on the machine}$$

Constraint 3: Maximum Horsepower Available. If P_{max} is the maximum horsepower available at the spindle, then

$$vf^{\alpha} \leq \frac{P_{max}(33,000)}{c_t d_c^{\beta}}$$

where α, β, and c_t are constants.[9] d_c is the depth of cut in inches, which is fixed at a given value. For a given P_{max}, c_t, β, and d_c, the right-hand side of the above constraint will be a constant. Hence, the horsepower constraint can be written simply as

$$vf^{\alpha} \leqq \text{constant} \tag{21.24}$$

Constraint 4: Nonnegativity Restrictions on Feed Rate and Speed.

$$v, f \geqq 0 \tag{21.25}$$

In optimizing metal cutting there are a number of optimality criteria that can be used. Suppose we consider the following objectives in our optimization: (i) maximize metal removal rate (MRR), (ii) maximize tool life (TL). The expression for MRR is

$$\text{MRR} = 12vfd_c \text{ in.}^3/\text{min} \tag{21.26}$$

TL for a given depth of cut is given by

$$\text{TL} = \frac{A}{v^{1/n} f^{1/n_1}} \tag{21.27}$$

where A, n, and n_1 are constants. We note that the MRR objective is directly proportional to feed and speed, while the TL objective is inversely proportional to feed and speed. In general, there is no single solution to a problem formulated in this way, since MRR and TL are competing objectives and their respective maxima must include some compromise between the maximum of MRR and the maximum of TL.

A Goal Programming Model

Goal programming is a technique specifically designed to solve problems involving complex, usually conflicting multiple objectives. Goal programming requires the user to select a set of goals (which may or may not be realistic) that ought to be achieved (if possible) for the various objectives. It then uses preemptive weights or priority factors to rank the different goals and tries to obtain an optimal solution satisfying as many goals as possible. For this, it creates a single objective function that minimizes the deviations from the stated goals according to their relative importance.

Before we discuss the goal programming formulation of the machining problem, we should discuss the difference between the terms "real constraint" and "goal constraint" (or simply "goal") as used in goal programming models. The real constraints are absolute restrictions placed on the behavior of the design variables, while the goal constraints are conditions one would like to achieve but are not mandatory. For instance, a real constraint given by

$$x_1 + x_2 = 3$$

requires all possible values of $x_1 + x_2$ to always equal 3. As opposed to this, if we simply had a goal requiring $x_1 + x_2 = 3$, then this is not mandatory and we can choose values of x_1, x_2 such that $x_1 + x_2 \geq 3$ as well as $x_1 + x_2 \leq 3$. In a goal constraint positive and negative deviational variables are introduced as follows:

$$x_1 + x_2 + d_1^- - d_1^+ = 3, \qquad d_1^-, d_1^+ \geq 0$$

Note that if $d_1^- > 0$, then $x_1 + x_2 < 3$, and if $d_1^+ > 0$, then $x_1 + x_2 > 3$. By assigning suitable preemptive weights on d_1^- and d_1^+, the model will try to achieve the sum $x_1 + x_2$ as close as possible to 3.

Returning to the machining problem with competing objectives, suppose that management considers that a given single-point, single-pass turning operation will be operating at an acceptable efficiency level if the following goals are met as closely as possible:

1. The MRR must be greater than or equal to a given rate M_1 (in.^{3}/min).
2. The tool life must equal T_1 (min).

In addition, management requires that a higher priority be given to achieving the first goal than the second.

The goal programming approach may be illustrated by expressing each of the goals as goal constraints as shown below. Taking the MRR goal first,

$$12vfd_c + d_1^- - d_1^+ = M_1$$

where d_1^- represents the amount by which the MRR goal is underachieved, and d_1^+ represents any overachievement of the MRR goal. Similarly, the TL goal can be expressed as

$$\frac{A}{v^{1/n} f^{1/n_1}} + d_2^- - d_2^+ = T_1$$

Since the objective is to have an MRR of at least M_1, the objective function must be set up so that a high penalty will be assigned to the underachievement variable d_1^-. No penalty will be assigned to d_1^+. In order to achieve a tool life of T_1, penalties must be associated with both d_2^- and d_2^+ so that both of these variables are minimized to their fullest extent. The relative magnitudes of these penalties must reflect the fact that the first goal is considered to be more important than the second. Accordingly, the goal programming objective function for this problem is

$$\text{Minimize } z = P_1 d_1^- + P_2(d_2^- + d_2^+)$$

where P_1 and P_2 are nonnumerical preemptive priority factors such that $P_1 >>> P_2$ (i.e., P_1 is infinitely larger than P_2). With this objective function every effort will be made to satisfy completely the first goal before any attempt is made to satisfy the second.

In order to express the problem as a linear goal programming problem, M_1 is replaced by M_2, where

$$M_2 = \frac{M_1}{12d_c}$$

The goal T_1 is replaced by T_2, where

$$T_2 = \frac{A}{T_1}$$

and logarithms are taken of the goals and constraints. The problem can then be stated as follows:

$$\text{Minimize } z = P_1 d_1^- + P_2(d_2^- + d_2^+)$$

Subject to

(MRR goal)	$\log v + \log f + d_1^- - d_1^+ = \log M_2$
(TL goal)	$(1/n) \log v + (1/n_1) \log f + d_2^- - d_2^+ = \log T_2$
(f_{max} constraint)	$\log f \leq \log f_{max}$
(V_{max} constraint)	$\log v \leq \log v_{max}$
(Horsepower constraint)	$\log v + \alpha \log f \leq \log$ constant
	$\log v, \log f, d_1^-, d_1^+, d_2^-, d_2^+ \geq 0$

We would like to reemphasize here that the last three inequalities are real constraints on feed, speed, and horsepower that must be satisfied at all times, while the equations for MRR and TL are simply goal constraints. For a further discussion of this problem and its solution, see Ref. 8. An efficient algorithm and a computer code for solving linear goal programming problems is given in Ref. 10. Readers interested in other optimization models in metal cutting should see Ref. 11. The textbook by Lee[12] contains a good discussion of goal programming theory and its applications.

21.2.3 Analysis and Data Reduction Applications

A further fertile area for the application of optimization techniques in engineering can be found in nonlinear regression problems as well as in many analysis problems arising in engineering science. A very common problem arising in engineering model development is the need to determine the parameters of some semitheoretical model given a set of experimental data. This data reduction or regression problem inherently transforms to an optimization problem, because the model parameters must be selected so that the model fits the data as closely as possible.

Suppose some variable y is assumed to be dependent on an independent variable x and related to x through a postulated equation $y = f(x, \theta_1, \theta_2)$, which depends on two parameters θ_1 and θ_2. To establish the appropriate values of θ_1 and θ_2, we run a series of experiments in which we adjust the independent variable x and measure the resulting y. As a result of a series of N experiments covering the range of x of interest, a set of y and x values (y_i, x_i), $i = 1, \ldots, N$, is available. Using these data we now try to "fit" our function to the data by adjusting θ_1 and θ_2 until we get a "good fit." The most commonly used measure of a "good fit" is the *least squares criterion,*

$$L(\theta_1, \theta_2) = \sum_{i=1}^{N} [y_i - f(x_i, \theta_1, \theta_2)]^2 \tag{21.28}$$

The difference $y_i - f(x_i, \theta_1, \theta_2)$ between the experimental value y_i and the predicted value $f(x_i, \theta_1, \theta_2)$ measures how close our model prediction is to the data and is called the *residual.* The sum of the squares of the residuals at all the experimental points gives an indication of goodness of fit. Clearly, if $L(\theta_1, \theta_2)$ is equal to zero, then the choice of θ_1, θ_2 has led to a perfect fit; the data points fall exactly on the predicted curve. The data-fitting problem can thus be viewed as an optimization problem in which $L(\theta_1, \theta_2)$ is minimized by appropriate choice of θ_1 and θ_2.

Example 21.4 Nonlinear Curve Fitting

Description.   The pressure–molar–volume–temperature relationship of real gases is known to deviate from that predicted by the ideal gas relationship

$$Pv = RT$$

where P = pressure (atm)
v = molar volume (cm³/g·mol)
T = temperature (K)
R = gas constant (82.06 atm·cm³/g·mol·K)

The semiempirical Redlich–Kwong equation

$$P = \frac{RT}{v - b} - \frac{a}{T^{1/2}v(v + b)} \tag{21.29}$$

is intended to correct for the departure from ideality but involves two empirical constants a and b whose values are best determined from experimental data. A series of PvT measurements listed in Table 21.1 are made for CO_2, from which a and b are to be estimated using nonlinear regression.

Formulation.   Parameters a and b will be determined by minimizing the least squares function (21.28). In the present case, the function will take the form

$$\sum_{i=1}^{\delta} \left[P_i - \frac{RT_i}{v_i - b} + \frac{a}{T_i^{1/2}v_i(v_i + b)} \right]^2 \tag{21.30}$$

Table 21.1   *PVT* Data for CO_2

Experiment Number	P (atm)	v (cm³/g·mol)	T°(K)
1	33	500	273
2	43	500	323
3	45	600	373
4	26	700	273
5	37	600	323
6	39	700	373
7	38	400	273
8	63.6	400	373

where P_i is the experimental value at experiment i, and the remaining two terms correspond to the value of P predicted from Eq. (21.29) for the conditions of experiment i for some selected value of the parameters a and b. For instance, the term corresponding to the first experimental point will be

$$\left(33 - \frac{82.06(273)}{500 - b} + \frac{a}{(273)^{1/2}(500)(500 + b)}\right)^2$$

Function (21.30) is thus a two-variable function whose value is to be minimized by appropriate choice of the independent variables a and b. If the Redlich–Kwong equation were to precisely match the data, then at the optimum the function (21.30) would be exactly equal to zero. In general, because of experimental error and because the equation is too simple to accurately model the CO_2 nonidealities, Eq. (21.30) will not be equal to zero at the optimum. For instance, the optimal values of $a = 6.377 \times 10^7$ and $b = 29.7$ still yield a squared residual of 9.7×10^{-2}.

21.3   STRUCTURE OF OPTIMIZATION PROBLEMS

Although the application problems discussed in the previous section originate from radically different sources and involve different systems, at root they have a remarkably similar form. All four can be expressed as problems requiring the minimization of a real-valued function $f(x)$ of an N-component vector argument $x = (x_1, x_2, \ldots, x_N)$ whose values are restricted to satisfy a number of real-valued equations $h_k(x) = 0$, a set of inequalities $g_j(x) \geq 0$, and the variable bounds $x_i^{(U)} \geq x_i \geq x_i^{(L)}$. In subsequent discussions we will refer to the function $f(x)$ as the *objective function*, to the equations $h_k(x) = 0$ as the *equality constraints*, and to the inequalities $g_j(x) \geq 0$ as the *inequality constraints*. For our purposes, these problem functions will always be assumed to be real valued, and their number will always be finite.

The general problem,

$$\text{Minimize } f(x)$$
$$\text{Subject to } h_k(x) = 0 \qquad k = 1, \ldots, K$$
$$g_j(x) \geq 0 \qquad j = 1, \ldots, J$$
$$x_i^{(U)} \geq x_i \geq x_i^{(L)} \qquad i = 1, \ldots, N$$

is called the *constrained* optimization problem. For instance, Examples 21.1, 21.2, and 21.3 are all constrained problems. The problem in which there are no constraints, that is,

$$J = K = 0$$

and

$$x_i^{(U)} = -x_i^{(L)} = \infty, \qquad i = 1, \ldots, N$$

is called the *unconstrained* optimization problem. Example 21.4 is an unconstrained problem. Optimization problems can be classified further based on the structure of the functions f, h_k, and g_j and on the dimensionality of x. Figure 21.5 illustrates one such classification. The basic subdivision is between unconstrained and constrained problems. There are two important classes of methods for solving the unconstrained problems. The direct search methods require only that the objective function be evaluated at different points, at least through experimentation. Gradient-based methods require the analytical form of the objective function and its derivatives.

An important class of constrained optimization problems is *linear programming*, which requires both the objective function and the constraints to be linear functions. Out of all optimization models, linear programming models are the most widely used and accepted in practice. Professionally written software programs are available from all major computer manufacturers for solving very large linear

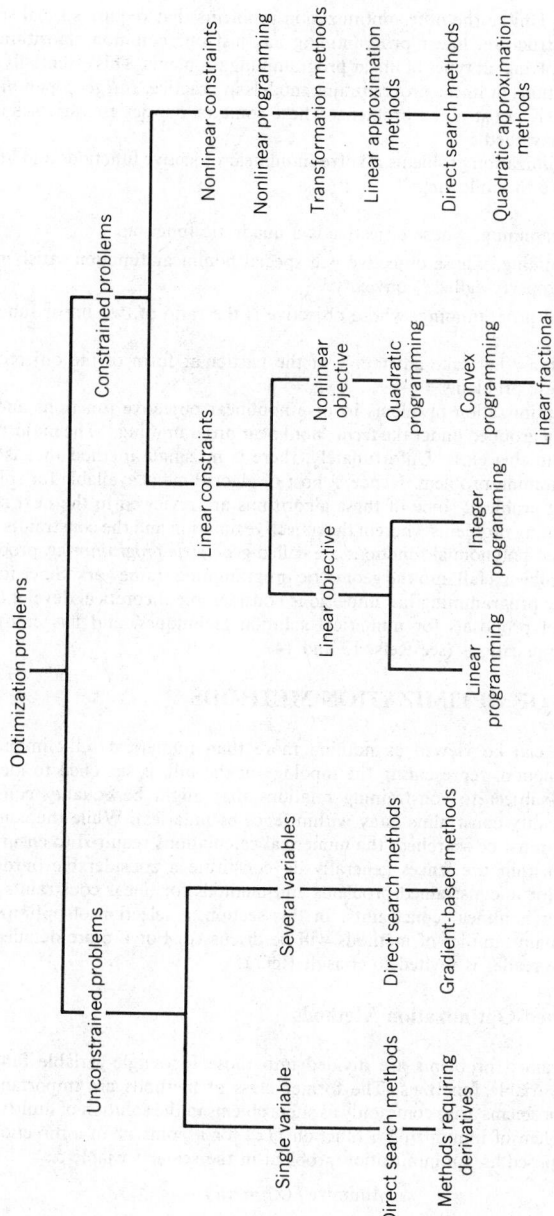

Fig. 21.5 Classification of optimization problems.

programming problems. Unlike the other optimization problems that require special solution methods based on the problem structure, linear programming has just one common algorithm, known as the "simplex method," for solving all types of linear programming problems. This essentially has contributed to the successful applications of linear programming models in practice. *Integer programming* is another important class of linearly constrained problems where some of the design variables are restricted to being discrete or integer valued.

The next class of optimization problems involves nonlinear objective functions and linear constraints. Under this class we have the following:

1. Quadratic programming, whose objective is a quadratic function.
2. Convex programming, whose objective is a special nonlinear function satisfying an important mathematical property called "convexity."
3. Linear fractional programming, whose objective is the ratio of two linear functions.

Special-purpose algorithms that take advantage of the particular form of the objective functions are available for solving these problems.

The most general optimization problems involve nonlinear objective functions and nonlinear constraints and are generally grouped under the term "nonlinear programming." The majority of engineering design problems fall into this class. Unfortunately, there is no single method that is best for solving every nonlinear programming problem. Hence, a host of algorithms is available for solving the general nonlinear programming problem, some of these algorithms are reviewed in the next section.

Nonlinear programming problems wherein the objective function and the constraints can be expressed as the sum of generalized polynomial functions are called *geometric programming* problems. A number of engineering design problems fall into the geometric programming framework. Since its earlier development in 1961, geometric programming has undergone considerable theoretical development, has experienced a proliferation of proposals for numerical solution techniques, and has enjoyed considerable practical engineering applications (see Refs. 13 and 14).

21.4 OVERVIEW OF OPTIMIZATION METHODS

Optimization methods can be viewed as nothing more than numerical hill-climbing procedures in which the objective function, representing the topology of the hill, is searched to identify the highest point—or maximum—subject to constraining relations that might be equality constraints (stay on winding path) or inequality constraints (stay within fence boundaries). While the constraints do serve to reduce the area that must be searched, the numerical calculations required to ensure that the search stays on the path or within the fences generally do constitute a considerable burden. Accordingly, optimization methods for unconstrained problems and methods for linear constraints are less complex than those designed for nonlinear constraints. In this section, a selection of optimization techniques representative of the main families of methods will be discussed. For a more detailed presentation of individual methods the reader is invited to consult Ref. 15.

21.4.1 Unconstrained Optimization Methods

Methods for unconstrained problems are divided into those for single-variable functions and those appropriate for multivariable functions. The former class of methods are important because single-variable optimization problems arise commonly as subproblems in the solution of multivariable problems. For instance, the problem of minimizing a function $f(x)$ for a point x^0 in a direction d (often called a *line search*) can be posed as a minimization problem in the scalar variable α:

$$\text{Minimize } f(x^0 + \alpha d)$$

Single Variable Methods

These methods are roughly divided into *region elimination methods* and *point estimation methods*. The former use comparison of function values at selected trial points to reject intervals within which the optimum of the function does not lie. The latter typically use polynomial approximating functions to estimate directly the location of the optimum. The simplest polynomial approximating function is the quadratic

$$\tilde{f}(x) = ax^2 + bx + c$$

whose coefficients a, b, c can be evaluated readily from those trial values of the actual function. The point at which the derivative of $\tilde{f}$ is zero is used readily to predict the location of the optimum of the true function

$$\tilde{x} = -b/2a$$

The process is repeated using successively improved trial values until the differences between successive estimates $\bar{x}$ become sufficiently small.

Multivariable Unconstrained Methods

These algorithms can be divided into *direct search methods* and *gradient-based methods*. The former methods only use direct function values to guide the search, while the latter also require the computation of function gradient and, in some cases, second derivative values. Direct search methods in widespread use in engineering applications include the simplex search, the pattern search method of Hooke and Jeeves, random-sampling-based methods, and the conjugate directions method of Powell (see Chap. 3 of Ref. 15). All but the last of these methods make no assumptions about the smoothness of the function contours and hence can be applied to both discontinuous and discrete-valued objective functions.

Gradient-based methods can be grouped into the classical methods of steepest descent (Cauchy) and Newton's method and the modern quasi-Newton methods such as the conjugate gradient, Davidson–Fletcher–Powell, and Broyden–Fletcher–Shanno algorithms. All gradient-based methods employ the first derivative or gradient of the function at the current best solution estimate $\bar{x}$ to compute a direction in which the objective function value is guaranteed to decrease (a *descent* direction). For instance, Cauchy's classical method used the direction.

$$d = -\nabla f(\bar{x})$$

followed by a line search from $\bar{x}$ in this direction. In Newton's method the gradient vector is premultiplied by the matrix of second derivatives to obtain an improved direction vector

$$d = -(\nabla^2 f(\bar{x}))^{-1} \nabla f(\bar{x})$$

which in theory at least yields very good convergence behavior. However, the computation of $\nabla^2 f$ is often too burdensome for engineering applications. Instead, in recent years quasi-Newton methods have found increased application. In these methods, the direction vector is computed as

$$d = -H \nabla f(\bar{x})$$

where H is a matrix whose elements are updated as the iterations proceed using only values of gradient and function value difference from successive solution estimates. Quasi-Newton methods differ in the details of H updating, but all use the general form

$$H^{n+1} = H^n + C^n$$

where H^n is the previous value of H and C^n is a suitable correction matrix. The attractive feature of this family of methods is that convergence rates approaching those of Newton's method are attained without the need for computing $\nabla^2 f$ or solving the linear equation set

$$\nabla^2 f(\bar{x}) \cdot d = -f(\bar{x})$$

to obtain d. Recent developments in these methods have focused on strategies for eliminating the need for detailed line searching along the direction vectors and on enhancements for solving very large problems. For a detailed discussion of quasi-Newton methods the reader is directed to Refs. 15 and 16.

21.4.2 Constrained Optimization Methods

Constrained optimization methods can be classified into those applicable to totally linear or at least linearly constrained problems and those applicable to general nonlinear problems. The linear or linearly constrained problems can be well solved using methods of linear programming and extensions, as discussed earlier. The algorithms suitable for general nonlinear problems comprise four broad categories of methods:

1. Direct search methods that use only objective and constraint function values.
2. Transformation methods that use constructions that aggregate constraints with the original objective function to form a single composite unconstrained function.
3. Linearization methods that use linear approximations of the nonlinear problem functions to produce efficient search directions.
4. Successive quadratic programming methods that use quasi-Newton constructions to solve the general problem via a series of subproblems with quadratic objective function and linear constraints.

Direct Search

The direct search methods essentially consist of extensions of unconstrained direct search procedures to accommodate constraints. These extensions are generally only possible with inequality constraints or linear equality constraints. Nonlinear equalities must be treated by implicit or explicit variable elimination. That is, each equality constraint is either explicitly solved for a selected variable and used to eliminate that variable from the search or the equality constraints are numerically solved for values of the dependent variables for each trial point in the space of the independent variables.

For example, the problem

$$\text{Minimize} \quad f(\mathbf{x}) = x_1 x_2 x_3$$
$$\text{Subject to} \quad h_1(\mathbf{x}) = x_1 + x_2 + x_3 - 1 = 0$$
$$h_2(\mathbf{x}) = x_1^2 x_3 + x_2 x_3^2 + x_2^{-1} - 2 = 0$$
$$0 \le (x_1, x_3) \le \tfrac{1}{2}$$

involves two equality constraints and hence can be viewed as consisting of two dependent variables and one independent variable. Clearly, h_1 can be solved for x_1 to yield

$$x_1 = 1 - x_2 - x_3$$

Thus, on substitution the problem reduces to

$$\text{Minimize} \quad (1 - x_2 - x_3) x_2 x_3$$
$$\text{Subject to} \quad (1 - x_2 - x_3)^2 x_3 + x_2 x_3^2 + x_2^{-1}(1 - x_2 - x_3) 2 = 0$$
$$0 \le 1 - x_2 - x_3 \le \tfrac{1}{2}$$
$$0 \le x_3 \le \tfrac{1}{2}$$

Solution of the remaining equality constraint for one variable, say x_3, in terms of the other is very difficult. Instead, for each value of the independent variable x_2, the corresponding value of x_3 would have to be calculated numerically using some root-finding method.

Some of the more widely used direct search methods include the adaptation of the simplex search due to Box (called the *complex method*), various direct random-sampling-type methods, and combined random sampling/heuristic procedures such as the combinatorial heuristic method[17] advanced for the solution of complex optimal mechanism design problems.

A typical direct sampling procedure is given by the formula,

$$x_i p = \bar{x}_i \times z_i (2r - 1)^k, \qquad \text{for each variable } x_i, \quad i = 1, \ldots, n$$

where $\bar{x}_i$ = the current best value of variable i
z_i = the allowable range of the variable i
r = a random variable uniformly distributed on the interval 0–1
k = an adaptive parameter whose value is adjusted based on past successes or failures in the search

For given $\bar{x}$, z, and k, r is sampled N times and the new point x^p evaluated. If x^p satisfies all constraints, it is retained; if it is infeasible, it is rejected and a new set of N r values is generated. If x^p is feasible, $f(x^p)$ is compared to $f(\bar{x})$, and if improvement is found, x^p replaces $\bar{x}$. Otherwise x^p is rejected. The parameter k is an adaptive parameter whose value will regulate the contraction or expansion of the sampling region. A typical adjustment procedure for k might be to increase k by 2 whenever a specified number of improved points is found or to decrease it by 2 when no improvement is found after a certain number of trials.

The general experience with direct search and especially random-sampling-based methods for constrained problems is that they can be quite effective for severely nonlinear problems that involve multiple local minima but are of low dimensionality.

Transformation Methods

This family consists of strategies for converting the general constrained problem to a parametrized unconstrained problem that is solved repeatedly for successive values of the parameters. The approaches can be grouped into the penalty/barrier function constructions, exact penalty methods, and augmented Lagrangian methods. The classical penalty function approach is to transform the general constrained problem to the form

$$P(x, R) = f(x) + \Omega(R, g(x), h(x))$$

where R = the penalty parameter
Ω = the penalty term

The ideal penalty function will have the property that

$$P(x, R) = \begin{cases} f(x), & \text{if } x \text{ is feasible} \\ \infty, & \text{if } x \text{ is infeasible} \end{cases}$$

Given this idealized construction, $P(x, R)$ could be minimized using any unconstrained optimization method, and, hence, the underlying constrained problem would have been solved. In practice such radical discontinuities cannot be tolerated from a numerical point of view, and, hence, practical penalty functions use penalty terms of the form

$$\Omega(R, g, h) = R\left(\sum_k h_k(x)\right)^2 + R(\Sigma(\min(0, g_j(x)))^2)$$

A series of unconstrained minimizations of $P(x, R)$ with different values of R are carried out beginning with a low value of R (say $R = 1$) and progressing to very large values of R. For low values of R, the unconstrained minima of $P(x, R)$ obtained will involve considerable constraint violations. As R increases, the violations decrease until in the limit as $R \to \infty$, the violations will approach zero. A large number of different forms of the Ω function have been proposed; however, all forms share the common feature that a sequence of problems must be solved and that, as the penalty parameter R becomes large, the penalty function becomes increasingly distorted and thus its minimization becomes increasingly more difficult. As a result the penalty function approach is best used for modestly sized problems (2–10 variables), few nonlinear equalities (2–5), and a modest number of inequalities. In engineering applications, the unconstrained subproblems are most commonly minimized using direct search methods, although successful use of quasi-Newton methods is also reported.

The exact penalty function and augmented Lagrangian approaches have been developed in an attempt to circumvent the need to force convergence by using increasing values of the penalty parameter. One typical representative of this type of method is the so-called method of multipliers.[18] In this method, once a sufficiently large value of R is reached, further increases are not required. However, the method does involve additional finite parameters that must be updated between subproblem solutions. Computational evidence reported to date suggests that, while augmented Lagrangian approaches are more reliable than penalty-function methods, they, as a class, are not suitable for larger dimensionality problems.

Linearization Methods

The common characteristic of this family of methods is the use of local linear approximations to the nonlinear problem functions to define suitable, preferably feasible, directions for search. Well-known members of this family include the method of feasible directions, the gradient projection method, and the generalized reduced gradient (GRG) method. Of these, the GRG method has seen the widest engineering application.

The key constructions of the GRG method are the following:

1. The calculation of the reduced objective function gradient $\nabla \bar{f}$.
2. The use of the reduced gradient to determine a direction vector in the space of the independent variables.
3. The adjustment of the dependent variable values using Newton's method so as to achieve constraint satisfaction.

Given a feasible point x^0, the gradients of the equality constraints are evaluated and used to form the constraint Jacobian matrix A. This matrix is partitioned into a square submatrix J and the residual rectangular matrix C where the variable associated with the columns of J are the dependent variables and those associated with C are the independent variables.

If J is selected to have nonzero determinant, then the reduced gradient is defined as

$$\nabla \bar{f}(x^0) = \nabla \bar{f} - \nabla \hat{f} J^{-1} C$$

where $\nabla \bar{f}$ is the subvector of objective function partial derivatives corresponding to the (dependent) variables and $\nabla \hat{f}$ is the corresponding subvector whose components correspond to the independent variables. The reduced gradient $\nabla \hat{f}$ provides an estimate of the rate of change of $f(x)$ with respect to the independent variables when the dependent variables are adjusted to satisfy the linear approximations to the constraints.

Given $\nabla \bar{f}$, in the simplest version of GRG algorithm, the direction subvector for the independent variables $\bar{d}$ is selected to be the reduced gradient descent direction

$$\bar{d} = -\nabla \bar{f}$$

For a given step α in that direction, the constraints are solved iteratively to determine the value of the dependent variables $\hat{x}$ that will lead to a feasible point. Thus, the system

$$h_k(\bar{x}^0 + \alpha\bar{d}, \hat{x}) = 0, \qquad k = 1, \ldots, K$$

is solved for the K unknown variables $\hat{x}$. The new feasible point is checked to determine whether an improved objective value has been obtained and, if not, α is reduced and the solution for $\hat{x}$ repeated. The overall algorithm terminates when a point is reached at which the reduced gradient is sufficiently close to zero.

The GRG algorithm has been extended to accommodate inequality constraints as well as variable bounds. Moreover, the use of efficient equation solving procedures, line search procedures for α, and quasi-Newton formulas to generate improved direction vectors $\bar{d}$ have been investigated. A commercial quality GRG code will incorporate such developments and thus will constitute a reasonably complex software package. Computational testing using such codes indicates that GRG implementations are among the most robust and efficient general purpose nonlinear optimization methods currently available.[19] One of the particular advantages of this algorithm, which can be critical in engineering applications, is that it generates *feasible* intermediate points; hence, it can be interrupted prior to final convergence to yield a feasible solution. Of course, this attractive feature and the general efficiency of the method are attained at the price of providing (analytically or numerically) the values of the partial derivatives of all of the model functions.

Successive Quadratic Programming (SQP) Methods

This family of methods seeks to attain superior convergence rates by employing subproblems constructed using higher-order approximating functions than those employed by the linearization methods. The SQP methods are still the subject of active research; hence, developments and enhancements are proceeding apace. However, the basic form of the algorithm is well established and can be sketched out as follows.

At a given point x^0, a direction finding subproblem is constructed, which takes the form of a quadratic programming problem:

$$\text{Minimize} \quad \nabla^T f \cdot d + \tfrac{1}{2}\, d^T H\, d$$
$$\text{Subject to} \quad h_k(x^0) + \nabla^T h_k(x^0)d = 0$$
$$g_j(x^0) + \nabla^T g_j(x^0)\, d \geq 0$$

The symmetric matrix H is a quasi-Newton approximation of the matrix of second derivatives of a composite function (the Lagrangian) containing terms corresponding to all of the functions f, h_k, and g_j. H is updated using only gradient differences as in the unconstrained case. The direction vector d is used to conduct a line search, which seeks to minimize a penalty function of the type discussed earlier. The penalty function is required because, in general, the intermediate points produced in this method will be infeasible. Use of the penalty function ensures that improvements are achieved in either the objective function values or the constraint violations or both. One major advantage of the method is that very efficient methods are available for solving large quadratic programming problems and, hence, that the method is suitable for large scale applications. Recent computational testing indicates that the SQP approach is very efficient, outperforming even the best GRG codes.[20] However, it is restricted to models in which infeasibilities can be tolerated and will produce feasible solutions only when the algorithm has converged.

21.4.3 Code Availability

With the exception of the direct search methods and the transformation-type methods, the development of computer programs implementing state-of-the-art optimization algorithms is a major effort requiring expertise in numerical methods in general and numerical linear algebra in particular. For that reason, it is generally recommended that engineers involved in design optimization studies take advantage of the number of good quality implementations now available through various public sources. A partial list of state-of-the-art software sources follows:

GRGA: GRG/Fletcher–Reeves
Contact: J. Abadie
          University of Paris VI
          Institut de Programmation
          4, Place Jussieu
          Paris, France

GRG2: GRG/BFS
Contact: L. S. Lasdon
          General Business
          University of Texas
          Austin, TX 78712

OPT: GRG/Fletcher–Reeves
Contact: K. M. Ragsdell
          University of Missouri
          Aerospace and Mechanical Engineering
          Columbia, MO 65211

MINOS/Augmented: Augmented Lagrangian/GRG for linearly constrained subproblems

Contact: M. Saunders
        Systems Optimization Laboratory
        Operations Research
        Stanford University
        Stanford, CA 94305

COMPUTE II: Penalty package with various penalty forms and unconstrained methods

Contact: A. B. King
        Gulf Research and Development
        P.O. Box 2038
        Pittsburgh, PA 15230

BIAS: MOM/DFP/automatic scaling

Contact: K. M. Ragsdell
        University of Missouri
        Aerospace and Mechanical Engineering
        Columbia, MO 65211

OPRQP: SQP/BFS

Contact: M. C. Bartholomew-Biggs
        Numerical Optimization Centre
        Hatfield Polytechnic
        19 St. Albans Road
        Hatfield, Herts, UK

VFO2AD: SQP/BFS +

Contact: M. J. D. Powell
        Computer Science and Systems Div.
        A.E.R.E.
        Harwell, Oxfordshire, UK

VMCON1:   Similar to VFO2AD

Contact: L. C. Ranzini
        National Energy Software Center
        Argonne National Laboratory
        9700 S. Cass Avenue
        Argonne, IL 60439

21.5 SUMMARY

In this chapter an overview was given of the elements and methods comprising design optimization methodology. The key element in the overall process of design optimization was seen to be the engineering model of the system constructed for this purpose. The assumptions and formulation details of the model govern the quality and relevance of the optimal design obtained. Hence, it is clear that design optimization studies cannot be relegated to optimization software specialists but are the proper domain of the well-informed design engineer.

The chapter also gave a structural classification of optimization problems and a broad brush review of the main families of optimization methods. Clearly this review can only hope to serve as entry point to this broad field. For a more complete discussion of optimization techniques with emphasis on engineering applications, guidelines for model formulation, practical solution strategies, and available computer software, the readers are referred to the recent text by Reklaitis, Ravindran, and Ragsdell.[15]

The Design Automation Committee of the Design Engineering Division of ASME has been sponsoring conferences devoted to engineering design optimization. Several of these presentations have subsequently appeared in the *Journal of Mechanical Design, ASME Transactions.* Ragsdell[21] presents a review of the papers published up to 1977 in the areas of machine design applications and numerical methods in design optimization. ASME published, in 1981, a special volume entitled *Progress in Engineering Optimization,* edited by Mayne and Ragsdell.[22] It contains several articles pertaining to advances in optimization methods and their engineering applications in the areas of mechanism design, structural design, optimization of hydraulic networks, design of helical springs, optimization of hydrostatic journal bearing, and others. Finally, the persistent and mathematically oriented reader may wish to pursue the fine exposition given by Avriel,[23] which explores the theoretical properties and issues of nonlinear programming methods.

REFERENCES

1. M. Zeleny, *Multiple Criteria Decision Making,* McGraw-Hill, New York, 1982.
2. T. L. Vincent and W. J. Grantham, *Optimality in Parametric Systems,* Wiley, New York, 1981.
3. K. E. Bett, J. S. Rowlinson, and G. Saville, *Thermodynamics for Chemical Engineers,* MIT Press, Cambridge, MA, 1975.
4. F. C. Jen, C. C. Pegels, and T. M. Dupuis, Optimal Capacities of Production Facilities," *Management Science* **14B,** 570–580 (1968).
5. J. E. Shigley, *Mechanical Engineering Design,* McGraw-Hill, New York, 1973, p. 271.
6. S. Timoshenko and J. Gere, *Theory of Elastic Stability,* McGraw-Hill, New York, 1961, p. 257.
7. K. M. Ragsdell and D. T. Phillips, "Optimal Design of a Class of Welded Structures Using Geometric Programming," *ASME J. Eng. Ind. Ser. B* **98** (3), 1021–1025 (1975).
8. R. H. Philipson and A. Ravindran, "Application of Goal Programming to Machinability Data Optimization," *Journal of Mechanical Design, Trans. of ASME* **100,** 286–291 (1978).
9. E. J. A. Armarego and R. H. Brown, *The Machining of Metals,* Prentice-Hall, Englewood Cliffs, NJ, 1969.
10. J. L. Arthur and A. Ravindran, "PAGP-Partitioning Algorithm for (Linear) Goal Programming Problems," *ACM Transactions on Mathematical Software* **6,** 378–386 (1980).
11. R. H. Philipson and A. Ravindran, "Application of Mathematical Programming to Metal Cutting," *Mathematical Programming Study* **11,** 116–134 (1979).
12. S. M. Lee, *Goal Programming for Decision Analysis,* Auerbach Publishers, Philadelphia, PA, 1972.
13. C. S. Beightler and D. T. Phillips, *Applied Geometric Programming,* Wiley, New York, 1976.
14. M. J. Rijckaert, "Engineering Applications of Geometric Programming," in *Optimization and Design,* M. Avriel, M. J. Rijckaert, and D. J. Wilde (eds.), Prentice-Hall, Englewood Cliffs, NJ, 1974.
15. G. V. Reklaitis, A. Ravindran, and K. M. Ragsdell, *Engineering Optimization: Methods and Applications,* Wiley, New York, 1983.
16. R. Fletcher, *Practical Methods of Optimization: Unconstrained Optimization,* Wiley, New York, 1980.
17. T. W. Lee and F. Freudenstein, "Heuristic Combinatorial Optimization in the Kinematic Design of Mechanisms: Part 1: Theory," *J. Eng. Ind. Trans. ASME,* 1277–1280 (1976).
18. S. B. Schuldt, G. A. Gabriele, R. R. Root, E. Sandgren, and K. M. Ragsdell, "Application of a New Penalty Function Method to Design Optimization," *J. Eng. Ind. Trans. ASME,* 31–36 (1977).
19. E. Sandgren and K. M. Ragsdell, "The Utility of Nonlinear Programming Algorithms: A Comparative Study—Parts 1 and 2," *Journal of Mechanical Design, Trans. of ASME* **102,** 540–541 (1980).
20. K. Schittkowski, *Nonlinear Programming Codes: Information, Tests, Performance,* Lecture Notes in Economics and Mathematical Systems, Vol. 183, Springer Verlag, New York, 1980.
21. K. M. Ragsdell, "Design and Automation," *Journal of Mechanical Design, Trans. of ASME* **102,** 424–429 (1980).
22. R. W. Mayne and K. M. Ragsdell (eds.), *Progress in Engineering Optimization,* ASME, New York, 1981.
23. M. Avriel, *Nonlinear Programming: Analysis and Methods,* Prentice-Hall, Englewood Cliffs, NJ, 1976.

CHAPTER 22

SOLID MECHANICS

FRANKLIN E. FISHER

Mechanical Engineering Department
Loyola Marymount University
Los Angeles, California
and
Senior Staff Engineer
Hughes Aircraft Company

Revised from Chapter 8, *Kent's Mechanical Engineer's Handbook*, 12th ed., by John M. Lessells and G. S. Cherniak.

22.1 STRESSES, STRAINS, STRESS INTENSITY

22.1.1 Fundamental Definitions

Static Stresses

TOTAL STRESS on a section *mn* through a loaded body is the resultant force *S* exerted by one part
of the body on the other part in order to maintain in equilibrium the external loads acting on
the part. Thus, in Figs. 22.1, 22.2, and 22.3 the total stress on section *mn* due to the external
load *P* is *S*. The units in which it is expressed are those of load, that is, pounds, tons, etc.

UNIT STRESS more commonly called stress σ, is the total stress per unit of area at section *mn*. In
general it varies from point to point over the section. Its value at any point of a section is the
total stress on an elementary part of the area, including the point divided by the elementary
area. If in Figs. 22.1, 22.2, and 22.3 the loaded bodies are one unit thick and four units wide,
then when the total stress *S* is uniformly distributed over the area, $\sigma = P/A = P/4$. Unit stresses
are expressed in pounds per square inch, tons per square foot, etc.

TENSILE STRESS OR TENSION is the internal total stress *S* exerted by the material fibers to resist
the action of an external force *P* (Fig. 22.1), tending to separate the material into two parts
along the line *mn*. For equilibrium conditions to exist, the tensile stress at any cross section
will be equal and opposite in direction to the external force *P*. If the internal total stress *S* is
distributed uniformly over the area, the stress can be considered as unit tensile stress $\sigma = S/A$.

COMPRESSIVE STRESS OR COMPRESSION is the internal total stress *S* exerted by the fibers to resist
the action of an external force *P* (Fig. 22.2) tending to decrease the length of the material. For
equilibrium conditions to exist, the compressive stress at any cross section will be equal and
opposite in direction to the external force *P*. If the internal total stress *S* is distributed uniformly
over the area, the unit compressive stress $\sigma = S/A$.

SHEAR STRESS is the internal total stress *S* exerted by the material fibers along the plane *mn* (Fig.
22.3) to resist the action of the external forces, tending to slide the adjacent parts in opposite
directions. For equilibrium conditions to exist, the shear stress at any cross section will be equal
and opposite in direction to the external force *P*. If the internal total stress *S* is uniformly distributed
over the area, the unit shear stress $\tau = S/A$.

NORMAL STRESS is the component of the resultant stress that acts normal to the area considered
(Fig. 22.4).

AXIAL STRESS is a special case of normal stress and may be either tensile or compressive. It is the
stress existing in a straight homogeneous bar when the resultant of the applied loads coincides
with the axis of the bar.

SIMPLE STRESS exists when either tension, compression, or shear is considered to operate singly on
a body.

TOTAL STRAIN on a loaded body is the total elongation produced by the influence of an external
load. Thus, in Fig. 22.5, the total strain is equal to δ. It is expressed in units of length, that is,
inches, feet, etc.

UNIT STRAIN or deformation per unit length is the total amount of deformation divided by the
original length of the body before the load causing the strain was applied. Thus, if the total

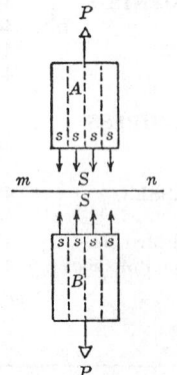

Fig. 22.1 Tensile stress.

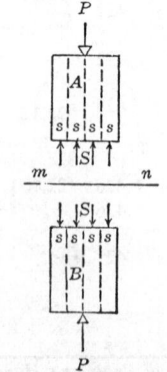

Fig. 22.2 Compressive stress.

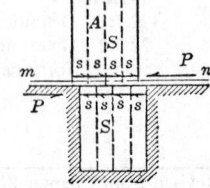

Fig. 22.3 Shear stress.

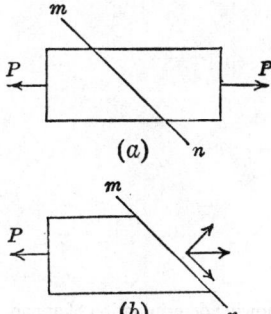

Fig. 22.4 Normal and shear stress components of resultant stress on section *mn*.

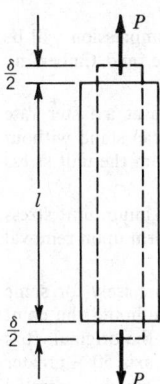

Fig. 22.5 Strain due to tension.

elongation is δ in an original gage length l, the unit strain $e = \delta/l$. Unit strains are expressed in inches per inch and feet per foot.

TENSILE STRAIN   is the strain produced in a specimen by tensile stresses, which in turn are caused by external forces.

COMPRESSIVE STRAIN   is the strain produced in a bar by compressive stresses, which in turn are caused by external forces.

SHEAR STRAIN   is a strain produced in a bar by the external shearing forces.

POISSON'S RATIO   is the ratio of lateral unit strain to longitudinal unit strain under the conditions of uniform and uniaxial longitudinal stress within the proportional limit. It serves as a measure of lateral stiffness. Average values of Poisson's ratio for the usual materials of construction are:

Material	Steel	Wrought Iron	Cast Iron	Brass	Concrete
Poisson's ratio	0.300	0.280	0.270	0.340	0.100

ELASTICITY   is that property of a material that enables it to deform or undergo strain and return to its original shape upon the removal of the load.

HOOKE'S LAW   states that within certain limits (not to exceed the proportional limit) the elongation of a bar produced by an external force is proportional to the tensile stress developed. Hooke's law gives the simplest relation between stress and strain.

PLASTICITY   is that state of matter where permanent deformations or strains may occur without fracture. A material is plastic if the smallest load increment produces a permanent deformation. A perfectly plastic material is nonelastic and has no ultimate strength in the ordinary meaning of that term. Lead is a plastic material. A prism tested in compression will deform permanently under a small load and will continue to deform as the load is increased, until it flattens to a thin sheet. Wrought iron and steel are plastic when stressed beyond the elastic limit in compression. When stressed beyond the elastic limit in tension, they are partly elastic and partly plastic, the degree of plasticity increasing as the ultimate strength is approached.

STRESS–STRAIN RELATIONSHIP   gives the relation between unit stress and unit strain when plotted on a stress–strain diagram in which the ordinate represents unit stress and the abscissa represents unit strain. Figure 22.6 shows a typical tension stress–strain curve for medium steel. The form

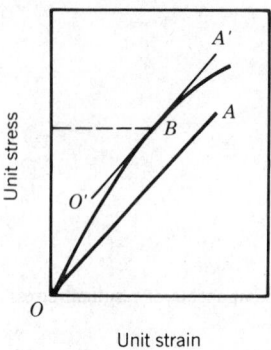

O

Unit strain

Fig. 22.6   Stress–strain relationship showing determination of apparent elastic limit.

of the curve obtained will vary according to the material, and the curve for compression will be different from the one for tension. For some materials like cast iron, concrete, and timber, no part of the curve is a straight line.

PROPORTIONAL LIMIT   is that unit stress at which unit strain begins to increase at a faster rate than unit stress. It can also be thought of as the greatest stress that a material can stand without deviating from Hooke's law. It is determined by noting on a stress–strain diagram the unit stress at which the curve departs from a straight line.

ELASTIC LIMIT   is the least stress that will cause permanent strain, that is, the maximum unit stress to which a material may be subjected and still be able to return to its original form upon removal of the stress.

JOHNSON'S APPARENT ELASTIC LIMIT.   In view of the difficulty of determining precisely for some materials the proportional limit, J. B. Johnson proposed as the "apparent elastic limit" the point on the stress–strain diagram at which the rate of strain is 50% greater than at the original. It is determined by drawing OA (Fig. 22.6) with a slope with respect to the vertical axis 50% greater than the straight-line part of the curve; the unit stress at which the line $O'A'$ which is parallel to OA is tangent to the curve (point B, Fig. 22.6) is the apparent elastic limit.

YIELD POINT   is the lowest stress at which strain increases without increase in stress. Only a few materials exhibit a true yield point. For other materials the term is sometimes used as synonymous with yield strength.

YIELD STRENGTH   is the unit stress at which a material exhibits a specified permanent deformation or state. It is a measure of the useful limit of materials, particularly of those whose stress–strain curve in the region of yield is smooth and gradually curved.

ULTIMATE STRENGTH   is the highest unit stress a material can sustain in tension, compression, or shear before rupturing.

RUPTURE STRENGTH OR BREAKING STRENGTH   is the unit stress at which a material breaks or ruptures. It is observed in tests on steel to be slightly less than the ultimate strength because of a large reduction in area before rupture.

MODULUS OF ELASTICITY   (Young's modulus) in tension and compression is the rate of change of unit stress with respect to unit strain for the condition of uniaxial stress within the proportional limit. For most materials the modulus of elasticity is the same for tension and compression.

MODULUS OF RIGIDITY   (modulus of elasticity in shear) is the rate of change of unit shear stress with respect to unit shear strain for the condition of pure shear within the proportional limit. For metals it is equal to approximately 0.4 of the modulus of elasticity.

TRUE STRESS   is defined as a ratio of applied axial load to the corresponding cross-sectional area. The units of true stress may be expressed in pounds per square inch, pounds per square foot, etc.,

$$\sigma = \frac{P}{A}$$

where σ is the true stress, pounds per square inch, P is the axial load, pounds, and A is the smallest value of cross-sectional area existing under the applied load P, square inches.

TRUE STRAIN   is defined as a function of the original diameter to the instantaneous diameter of the test specimen:

$$q = 2 \log_e \frac{d_0}{d} \quad \text{in./in.}$$

where q = true strain, inches per inch, d_0 = original diameter of test specimen, inches, and d = instantaneous diameter of test specimen, inches.

TRUE STRESS–STRAIN RELATIONSHIP   is obtained when the values of true stress and the corresponding true strain are plotted against each other in the resulting curve (Fig. 22.7). The slope of the nearly straight line leading up to fracture is known as the coefficient of strain hardening. It as well as the true tensile strength appear to be related to the other mechanical properties.

DUCTILITY   is the ability of a material to sustain large permanent deformations in tension, such as drawing into a wire.

MALLEABILITY   is the ability of a material to sustain large permanent deformations in compression, such as beating or rolling into thin sheets.

BRITTLENESS   is that property of a material that permits it to be only slightly deformed without rupture. Brittleness is relative, no material being perfectly brittle, that is, capable of no deformation before rupture. Many materials are brittle to a greater or less degree, glass being one of the most brittle of materials. Brittle materials have relatively short stress–strain curves. Of the common structural materials, cast iron, brick, and stone are brittle in comparison with steel.

TOUGHNESS   is the ability of the material to withstand high unit stress together with great unit strain, without complete fracture. The area *OAGH,* or *OJK,* under the curve of the stress–strain diagram (Fig. 22.8), is a measure of the toughness of the material. The distinction between ductility and

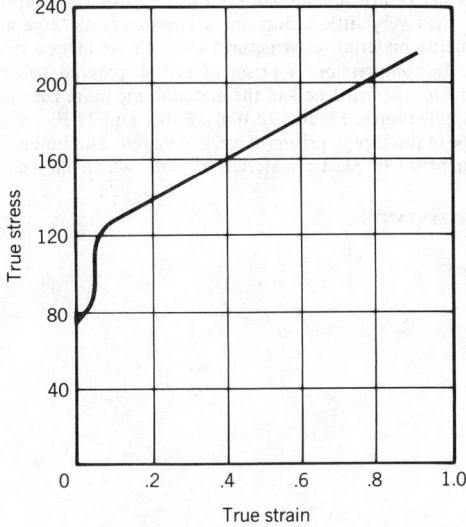

Fig. 22.7   True stress–strain relationship.

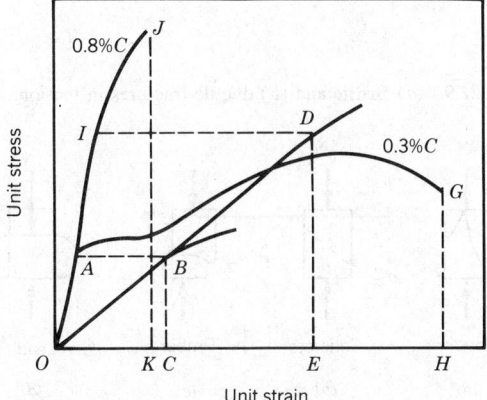

Fig. 22.8   Toughness comparison.

toughness is that ductility deals only with the ability to deform, whereas toughness considers both the ability to deform and the stress developed during deformation.

STIFFNESS is the ability to resist deformation under stress. The modulus of elasticity is the criterion of the stiffness of a material.

HARDNESS is the ability to resist very small indentations, abrasion, and plastic deformation. There is no single measure of hardness, as it is not a single property but a combination of several properties.

CREEP or flow of metals is a phase of plastic or inelastic action. Some solids, as asphalt or paraffin, flow appreciably at room temperatures under extremely small stresses; zinc, plastics, fiber-reinforced plastics, lead, and tin show signs of creep at room temperature under moderate stresses. At sufficiently high temperatures, practically all metals creep under stresses that vary with temperature, the higher the temperature the lower being the stress at which creep takes place. The deformation due to creep continues to increase indefinitely and becomes of extreme importance in members subjected to high temperatures, as parts in turbines, boilers, super-heaters, etc.

   Creep limit is the maximum unit stress under which unit distortion will not exceed a specified value during a given period of time at a specified temperature. A value much used in tests, and suggested as a standard for comparing materials, is the maximum unit stress at which creep does not exceed 1% in 100,000 hours.

TYPES OF FRACTURE. A bar of brittle material, such as cast iron, will rupture in a tension test in a clean sharp fracture with very little reduction of cross-sectional area and very little elongation (Fig. 22.9a). In a ductile material, as structural steel, the reduction of area and elongation are greater (Fig. 22.9b). In compression, a prism of brittle material will break by shearing along oblique planes; the greater the brittleness of the material, the more nearly will these planes parallel the direction of the applied force. Figures 22.10a, 22.10b, and 22.10c, arranged in order of brittleness, illustrate the type of fracture in prisms of brick, concrete, and timber. Figure 22.10d represents the deformation of a prism of plastic material, as lead, which flattens out under load without failure.

RELATIONS OF ELASTIC CONSTANTS

Modulus of elasticity, E:

$$E = \frac{Pl}{Ae}$$

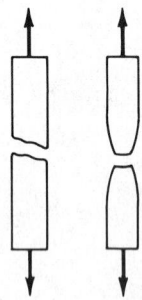

(a)   (b)   Fig. 22.9 (a) Brittle and (b) ductile fractures in tension.

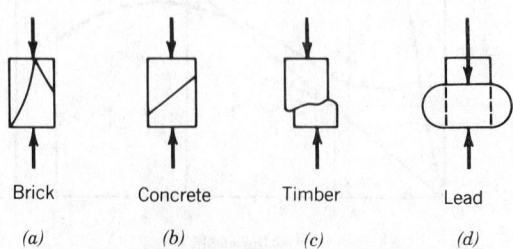

   Brick   Concrete   Timber   Lead

    (a)   (b)   (c)   (d)

Fig. 22.10 Fractures in compression.

where P = load, pounds, l = length of bar, inches, A = cross-sectional area acted on by the axial load P, and e = total strain produced by axial load P.

Modulus of rigidity, G:

$$G = \frac{E}{2(1 + \nu)}$$

where E = modulus of elasticity and ν = Poisson's ratio.

Bulk modulus, K, is the ratio of normal stress to the change in volume.

Relationships. The following relationships exist between the modulus of elasticity E, the modulus of rigidity G, the bulk modulus of elasticity K, and Poisson's ratio ν:

$$E = 2G(1 + \nu); \qquad G = \frac{E}{2(1 + \nu)}; \qquad \nu = \frac{E - 2G}{2G}$$

$$K = \frac{E}{3(1 - 2\nu)}; \qquad \nu = \frac{3K - E}{6K}$$

ALLOWABLE UNIT STRESS, also called allowable working unit stress, allowable stress, or working stress, is the maximum unit stress to which it is considered safe to subject a member in service. The term allowable stress is preferable to working stress, since the latter often is used to indicate the actual stress in a material when in service. Allowable unit stresses for different materials for various conditions of service are specified by different authorities on the basis of test or experience. In general, for ductile materials, allowable stress is considerably less than the yield point.

FACTOR OF SAFETY is the ratio of ultimate strength of the material to allowable stress. The term was originated for determining allowable stress. The ultimate strength of a given material divided by an arbitrary factor of safety, dependent on material and the use to which it is to be put, gives the allowable stress. In present design practice, it is customary to use allowable stress as specified by recognized authorities or building codes rather than an arbitrary factor of safety. One reason for this is that the factor of safety is misleading, in that it implies a greater degree of safety than actually exists. For example, a factor of safety of 4 does not mean that a member can carry a load four times as great as that for which it was designed. It also should be clearly understood that, even though each part of a machine is designed with the same factor of safety, the machine as a whole does not have that factor of safety. When one part is stressed beyond the proportional limit, or particularly the yield point, the load or stress distribution may be completely changed throughout the entire machine or structure, and its ability to function thus may be changed, even though no part has ruptured.

Although no definite rules can be given, if a factor of safety is to be used, the following circumstances should be taken into account in its selection:

1. When the ultimate strength of the material is known within narrow limits, as for structural steel for which tests of samples have been made, when the load is entirely a steady one of a known amount and there is no reason to fear the deterioration of the metal by corrosion, the lowest factor that should be adopted is 3.

2. When the circumstances of (1) are modified by a portion of the load being variable, as in floors of warehouses, the factor should not be less than 4.

3. When the whole load, or nearly the whole, is likely to be alternately put on and taken off, as in suspension rods of floors of bridges, the factor should be 5 or 6.

4. When the stresses are reversed in direction from tension to compression, as in some bridge diagonals and parts of machines, the factor should be not less than 6.

5. When the piece is subjected to repeated shocks, the factor should be not less than 10.

6. When the piece is subjected to deterioration from corrosion, the section should be sufficiently increased to allow for a definite amount of corrosion before the piece is so far weakened by it as to require removal.

7. When the strength of the material or the amount of the load or both are uncertain, the factor should be increased by an allowance sufficient to cover the amount of the uncertainty.

8. When the strains are complex and of uncertain amount, such as those in the crankshaft of a reversing engine, a very high factor is necessary, possibly even as high as 40.

9. If the property loss caused by failure of the part may be large or if loss of life may result, as in a derrick hoisting materials over a crowded street, the factor should be large.

Dynamic Stresses

DYNAMIC STRESSES occur where the dimension of time is necessary in defining the loads. They include creep, fatigue, and impact stresses.

CREEP STRESSES occur when either the load or deformation progressively vary with time. They are usually associated with noncyclic phenomena.

FATIGUE STRESSES occur when the cyclic variation of either load or strain is coincident with respect to time.

IMPACT STRESSES occur from loads which are transient with time. The duration of the load application is of the same order of magnitude as the natural period of vibration of the specimen.

22.1.2 Work and Resilience

EXTERNAL WORK. Let P = axial load, pounds, on a bar, producing an internal stress not exceeding the elastic limit; σ = unit stress produced by P, pounds per square inch; A = cross-sectional area, square inches; l = length of bar, inches; e = deformation, inches; E = modulus of elasticity; W = external work performed on bar, inch-pounds = $\frac{1}{2}Pe$. Then

$$W = \frac{1}{2}A\sigma\left(\frac{\sigma l}{E}\right) = \frac{1}{2}\left(\frac{\sigma^2}{E}\right)Al \qquad (22.1)$$

The factor $\frac{1}{2}(\sigma^2/E)$ is the work required per unit volume, the volume being Al. It is represented on the stress–strain diagram by the area ODE or area OBC (Fig. 22.11), in which DE and BC are ordinates representing the unit stresses considered.

RESILIENCE is the strain energy that may be recovered from a deformed body when the load causing the stress is removed. Within the proportional limit, the resilience is equal to the external work performed in deforming the bar, and may be determined by Eq. (22.1). When σ is equal to the proportional limit, the factor $\frac{1}{2}(\sigma^2/E)$ is the *modulus of resilience,* that is, the measure of capacity of a unit volume of material to store strain energy up to the proportional limit. Average values of the modulus of resilience under tensile stress are given in Table 22.1.

   The total resilience of a bar is the product of its volume and the modulus of resilience. These formulas for work performed on a bar, and its resilience, do not apply if the unit stress is greater than the proportional limit.

WORK REQUIRED FOR RUPTURE. Since beyond the proportional limit the strains are not proportional to the stresses, $\frac{1}{2}P$ does not express the mean value of the force acting. Equation (22.1), therefore, does not express the work required for strain after the proportional limit of the material has been passed, and cannot express the work required for rupture. The work required per unit volume to produce strains beyond the proportional limit or to cause rupture may be determined from the stress–strain diagram as it is measured by the area under the stress–strain curve up to the strain in question, as $OAGH$ or OJK (Fig. 22.11). This area, however, does not represent the

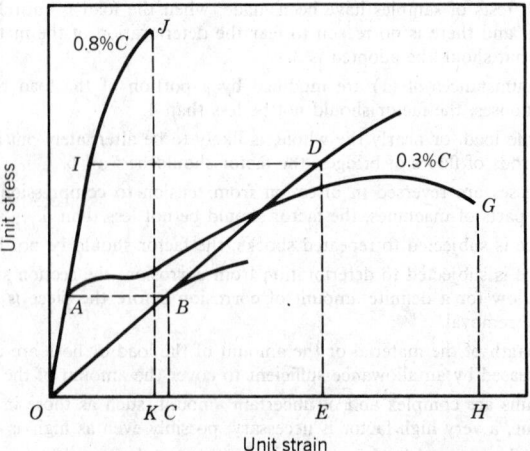

Fig. 22.11 Work areas on stress–strain diagram.

Table 22.1 Modulus of Resilience and Relative Toughness under Tensile Stress (Ave. Values)

Material	Modulus of Resilience (in.-lb/in.^{3})	Relative Toughness (Area under Curve of Stress-Deformation Diagram)
Gray cast iron	1.2	70
Malleable cast iron	17.4	3,800
Wrought iron	11.6	11,000
Low-carbon steel	15.0	15,700
Medium-carbon steel	34.0	16,300
High-carbon steel	94.0	5,000
Ni-Cr steel, hot-rolled	94.0	44,000
Vanadium steel, 0.98% C, 0.2% V, heat-treated	260.0	22,000
Duralumin, 17 ST	45.0	10,000
Rolled bronze	57.0	15,500
Rolled brass	40.0	10,000
Oak	2.3[a]	13[a]

[a] Bending.

resilience, since part of the work done on the bar is present in the form of hysteresis losses and cannot be recovered.

DAMPING CAPACITY (HYSTERESIS). Observations show that when a tensile load is applied to a bar, it does not produce the complete elongation immediately, but there is a definite time lapse which depends on the nature of the material and the magnitude of the stresses involved. In parallel with this it is also noted that, upon unloading, complete recovery of energy does not occur. This phenomenon is variously termed *elastic hysteresis* or, for vibratory stresses, damping. Figure 22.12 shows a typical hysteresis loop obtained for one cycle of loading. The area of this hysteresis loop, representing the energy dissipated per cycle, is a measure of the damping properties of the material. While the exact mechanism of damping has not been fully investigated, it has been found that under vibratory conditions the energy dissipated in this manner varies approximately as the cube of the stress.

22.2 DISCONTINUITIES, STRESS CONCENTRATION

The direct design procedure assumes no abrupt changes in cross section, discontinuities in the surface, or holes, through the member. In most structural parts this is not the case. The stresses produced at these discontinuities are different in magnitude from those calculated by various design methods. The effect of the localized increase in stress, such as that caused by a notch, fillet, hole, or similar *stress raiser*, depends mainly on the type of loading, the geometry of the part, and the material. As a result, it is necessary to consider a stress-concentration factor K_t, which is defined by the relationship

$$K_t = \frac{\sigma_{max}}{\sigma_{nominal}} \qquad (22.2)$$

In general σ_{max} will have to be determined by the methods of experimental stress analysis or the theory of elasticity, and $\sigma_{nominal}$ by a simple theory such as $\sigma = P/A$, $\sigma = Mc/I$, $\tau = Tc/J$ without

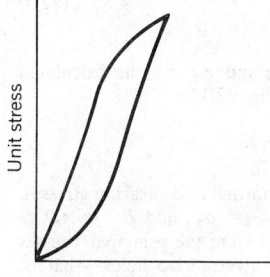

Unit strain   **Fig. 22.12** Hysteresis loop for loading and unloading.

Table 22.2 Stress-Concentration Factors[a]

Type	K_t Factors					
Circular hole in plate or rectangular bar	$\dfrac{h}{a} = 0.67$	0.77	0.91	1.07	1.29	1.56
	$k = 4.37$	3.92	3.61	3.40	3.25	3.16

Square shoulder with fillet for rectangular and circular cross sections in bending	$\dfrac{r}{d}$	0.05	0.10	0.20	0.27	0.50	1.0
$\dfrac{h}{r}$	0.5	1.61	1.49	1.39	1.34	1.22	1.07
	1.0	1.91	1.70	1.48	1.38	1.22	1.08
	1.5	2.00	1.73	1.50	1.39	1.23	1.08
	2.0		1.74	1.52	1.39	1.23	1.09
	3.5		1.76	1.54	1.40	1.23	1.10

[a] Adapted by permission from R. J. Roark and W. C. Young, *Formulas for Stress and Strain*, 5th ed., McGraw-Hill, New York, 1975.

taking into account the variations in stress conditions caused by geometrical discontinuities such as holes, grooves, and fillets. For ductile materials it is not customary to apply stress-concentration factors to members under static loading. For brittle materials, however, stress concentration is serious and should be considered.

Stress-Concentration Factors for Fillets, Keyways, Holes, and Shafts

In Table 22.2 selected stress-concentration factors have been given from a complete table in Refs. 1 and 2.

22.3 COMBINED STRESSES

Under certain circumstances of loading a body is subjected to a combination of tensile, compressive, and/or shear stresses. For example, a shaft that is simultaneously bent and twisted is subjected to combined stresses, namely, longitudinal tension and compression and torsional shear. For the purposes of analysis it is convenient to reduce such systems of combined stresses to a basic system of stress coordinates known as principal stresses. These stresses act on axes that differ in general from the axes along which the applied stresses are acting and represent the maximum and minimum values of the normal stresses for the particular point considered.

Determination of Principal Stresses

The expressions for the principal stresses in terms of the stresses along the x and y axes are

$$\sigma_1 = \frac{\sigma_x + \sigma_y}{2} + \sqrt{\left(\frac{\sigma_x - \sigma_y}{2}\right)^2 + \tau_{xy}^2} \tag{22.3}$$

$$\sigma_2 = \frac{\sigma_x + \sigma_y}{2} - \sqrt{\left(\frac{\sigma_x - \sigma_y}{2}\right)^2 + \tau_{xy}^2} \tag{22.4}$$

$$\tau_1 = \pm \sqrt{\left(\frac{\sigma_x - \sigma_y}{2}\right)^2 + \tau_{xy}^2} \tag{22.5}$$

where σ_1, σ_2, and τ_1 are the principal stress components and σ_x, σ_y, and τ_{xy} are the calculated stress components, all of which are determined at any particular point (Fig. 22.13).

Graphical Method of Principal Stress Determination—Mohr's Circle

Let the axes x and y be chosen to represent the directions of the applied normal and shearing stresses, respectively (Fig 22.14). Lay off to suitable scale distances $OA = \sigma_x$, $OB = \sigma_y$, and $BC = AD = \tau_{xy}$. With point E as a center construct the circle DFC. Then OF and OG are the principal stresses σ_1 and σ_2, respectively, and EC is the maximum shear stress τ_1. The inverse also holds—that is, given the principal stresses, σ_x and σ_y can be determined on any plane passing through the point.

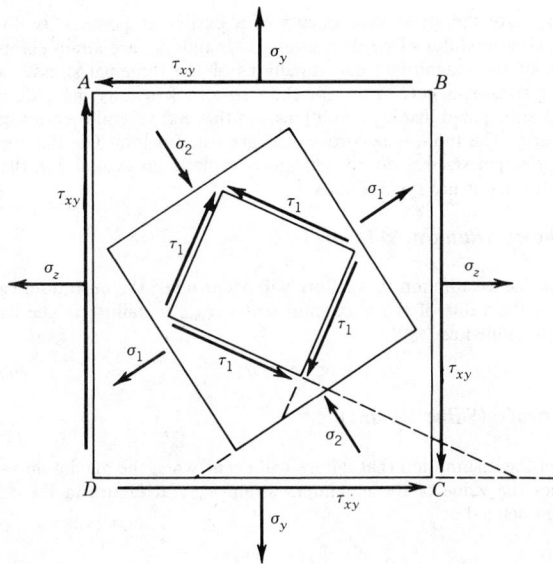

Fig. 22.13 Diagram showing relative orientation of stresses. (Reproduced by permission from J. Marin, *Mechanical Properties of Materials and Design,* McGraw-Hill, New York, 1942.)

Stress–Strain Relations

The linear relation between components of stress and strain is known as *Hooke's law.* This relation for the two-dimensional case can be expressed as

$$e_x = \frac{1}{E}(\sigma_x - \nu\sigma_y) \tag{22.6}$$

$$e_y = \frac{1}{E}(\sigma_y - \nu\sigma_x) \tag{22.7}$$

$$\gamma_{xy} = \frac{1}{G}\tau_{xy} \tag{22.8}$$

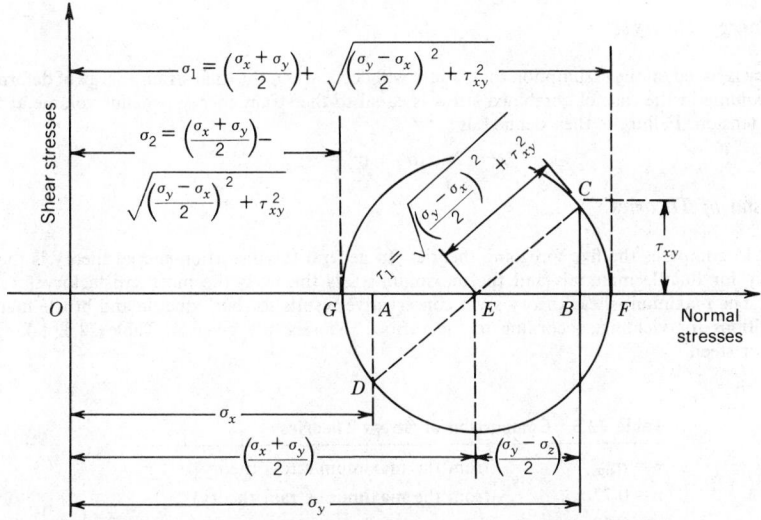

Fig. 22.14 Mohr's circle used for the determination of the principal stresses. (Reproduced by permission from J. Marin, *Mechanical Properties of Materials and Design,* McGraw-Hill, New York, 1942.)

where σ_x, σ_y, and τ_{xy} are the stress components of a particular point, ν is Poisson's ratio, E is modulus of elasticity, G is modulus of rigidity, and e_x, e_y, and γ_{xy} are strain components.

The determination of the magnitudes and directions of the principal stresses and strains and of the maximum shearing stresses is carried out for the purpose of establishing criteria of failure within the material under the anticipated loading conditions. To this end several theories have been advanced to elucidate these criteria. The more noteworthy ones are listed below. The theories are based on the assumption that the principal stresses do not change with time, an assumption that is justified since the applied loads in most cases are synchronous.

Maximum-Stress Theory (Rankine's Theory)

This theory is based on the assumption that failure will occur when the maximum value of the greatest principal stress reaches the value of the maximum stress σ_{max} at failure in the case of simple axial loading. Failure is then defined as

$$\sigma_1 \text{ or } \sigma_2 = \sigma_{max} \tag{22.9}$$

Maximum-Strain Theory (Saint Venant)

This theory is based on the assumption that failure will occur when the maximum value of the greatest principal strain reaches the value of the maximum strain e_{max} at failure in the case of simple axial loading. Failure is then defined as

$$e_1 \text{ or } e_2 = e_{max} \tag{22.10}$$

If e_{max} does not exceed the linear range of the material, Eq. (22.10) may be written as

$$\sigma_1 - \nu\sigma_2 = \sigma_{max} \tag{22.11}$$

Maximum-Shear Theory (Guest)

This theory is based on the assumption that failure will occur when the maximum shear stress reaches the value of the maximum shear stress at failure in simple tension. Failure is then defined as

$$\tau_1 = \tau_{max} \tag{22.12}$$

Distortion-Energy Theory (Hencky–Von Mises) (Shear Energy)

This theory is based on the assumption that failure will occur when the distortion energy corresponding to the maximum values of the stress components equals the distortion energy at failure for the maximum axial stress. Failure is then defined as

$$\sigma_1^2 - \sigma_1\sigma_2 + \sigma_2^2 = \sigma_{max}^2 \tag{22.13}$$

Strain-Energy Theory

This theory is based on the assumption that failure will occur when the total strain energy of deformation per unit volume in the case of combined stress is equal to the strain energy per unit volume at failure in simple tension. Failure is then defined as

$$\sigma_1^2 - 2\nu\sigma_1\sigma_2 + \sigma_2^2 = \sigma_{max}^2 \tag{22.14}$$

Comparison of Theories

Figure 22.15 compares the five foregoing theories. In general the distortion-energy theory is the most satisfactory for ductile materials and the maximum-stress theory is the most satisfactory for brittle materials. The maximum-shear theory gives conservative results for both ductile and brittle materials. The conditions for yielding, according to the various theories, are given in Table 22.3, taking $\nu = 0.300$ as for steel.

Table 22.3 Comparison of Stress Theories

$\tau = \sigma_{yp}$	(from the maximum-stress theory)
$\tau = 0.77\sigma_{yp}$	(from the maximum-strain theory)
$\tau = 0.50\sigma_{yp}$	(from the maximum-shear theory)
$\tau = 0.62\sigma_{yp}$	(from the maximum-strain-energy theory)

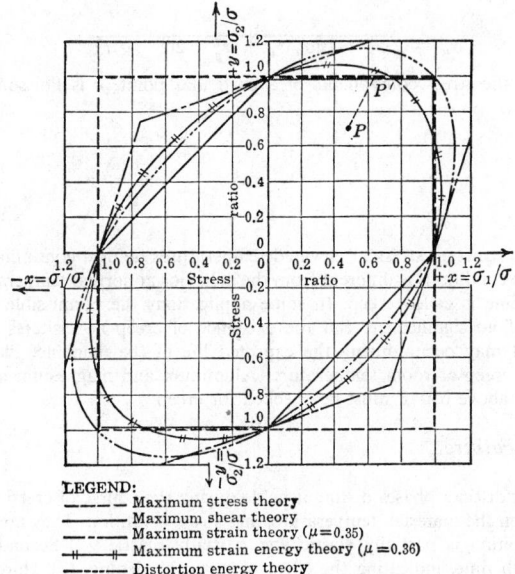

LEGEND:
———— Maximum stress theory
——— Maximum shear theory
——·—— Maximum strain theory ($\mu = 0.35$)
—++—++— Maximum strain energy theory ($\mu = 0.36$)
········· Distortion energy theory

Fig. 22.15 Comparison of five theories of failure. (Reproduced by permission from J. Marin, *Mechanical Properties of Materials and Design*, McGraw-Hill, New York, 1942.)

Static Working Stresses

Ductile Materials. For ductile materials the criteria for working stresses are

$$\sigma_w = \frac{\sigma_{yp}}{n} \quad \text{(tension and compression)} \tag{22.15}$$

$$\tau_w = \frac{1}{2}\frac{\sigma_{yp}}{n} \tag{22.16}$$

Brittle Materials. For brittle materials the criteria for working stresses are

$$\sigma_w = \frac{\sigma_{\text{ultimate}}}{K_t \times n} \quad \text{(tension)} \tag{22.17}$$

$$\sigma_w = \frac{\sigma_{\text{compressive}}}{K_t \times n} \quad \text{(compression)} \tag{22.18}$$

where K_t is the stress-concentration factor, n is the factor of safety, σ_w and τ_w are working stresses, and σ_{yp} is stress at the yield point.

Working-Stress Equations for the Various Theories.
Stress Theory

$$\sigma_w = \frac{\sigma_x + \sigma_y}{2} \pm \sqrt{\left(\frac{\sigma_x - \sigma_y}{2}\right)^2 + \tau_{xy}^2} \tag{22.19}$$

Shear Theory

$$\sigma_w = 2\sqrt{\left(\frac{\sigma_x - \sigma_y}{2}\right)^2 + \tau_{xy}^2} \tag{22.20}$$

Strain Theory

$$\sigma_w = (1 - \nu)\left(\frac{\sigma_x + \sigma_y}{2}\right) + (1 + \nu)\sqrt{\left(\frac{\sigma_x - \sigma_y}{2}\right)^2 + \tau_{xy}^2} \tag{22.21}$$

Distortion-Energy Theory

$$\sigma_w = \sqrt{\sigma_x^2 - \sigma_x\sigma_y + \sigma_y^2 + 3\tau_{xy}^2} \tag{22.22}$$

Strain-Energy Theory

$$\sigma_w = \sqrt{\sigma_x^2 - 2\nu\sigma_x\sigma_y + \sigma_y^2 + 2(1+\nu)\tau_{xy}^2} \qquad (22.23)$$

where σ_x, σ_y, τ_{xy} are the stress components of a particular point, ν is Poisson's ratio, and σ_w is working stress.

22.4   CREEP

Introduction

Materials subjected to a constant stress at elevated temperatures deform continuously with time, and the behavior under these conditions is different from the behavior at normal temperatures. This continuous deformation with time is called creep. In some applications the permissible creep deformations are critical, in others of no significance. But the existence of creep necessitates information on the creep deformations that may occur during the expected life of the machine. Plastic, zinc, tin, and fiber-reinforced plastics creep at room temperature. Aluminum and magnesium alloys start to creep at around 300°F. Steels above 650°F must be checked for creep.

Mechanism of Creep Failure

There are generally four distinct phases distinguishable during the course of creep failure. The elapsed time per stage depends on the material, temperature, and stress condition. They are: (1) Initial phase—where the total deformation is partially elastic and partially plastic. (2) Second phase—where the creep rate decreases with time, indicating the effect of strain hardening. (3) Third phase—where the effect of strain hardening is counteracted by the annealing influence of the high temperature which produces a constant or minimum creep rate. (4) Final phase—where the creep rate increases until fracture occurs owing to the decrease in cross-sectional area of the specimen.

Creep Equations

In conducting a conventional creep test, curves of strain as a function of time are obtained for groups of specimens; each specimen in one group is subjected to a different constant stress, while all of the specimens in the group are tested at one temperature.

In this manner families of curves like those shown in Fig. 22.16 are obtained. Several methods have been proposed for the interpretation of such data. (See Refs. 1 and 3.) Two frequently used expressions of the creep properties of a material can be derived from the data in the following form:

$$C = B\sigma^m$$
$$\epsilon = \epsilon_0 + Ct \qquad (22.24)$$

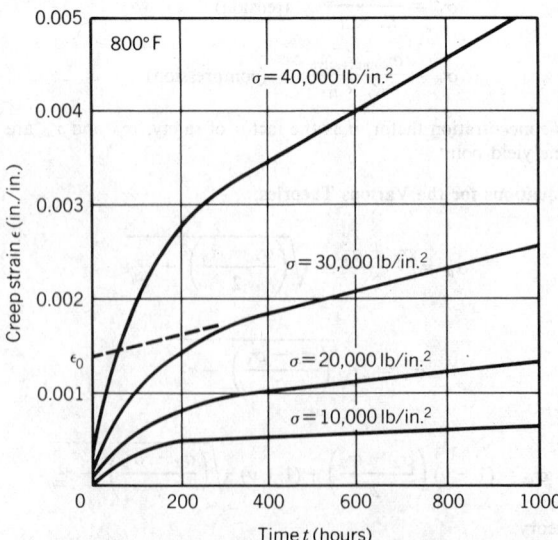

Fig. 22.16   Curves of creep strain for various stress levels.

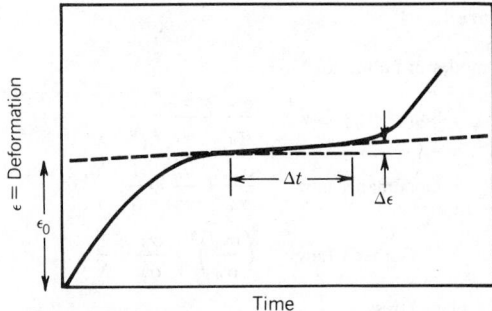

Fig. 22.17 Method of determining creep rate.

where C = creep rate, B, m = experimental constants, σ = stress, ϵ = creep strain at any time t, ϵ_0 = zero-time strain intercept, and t = time. See Fig. 22.17.

Stress Relaxation

Various types of bolted joints and shrink or press fit assemblies are applications of creep taking place with diminishing stress. This deformation tends to loosen the joint and produce a stress reduction or stress relaxation. The performance of a material to be used under diminishing creep-stress condition is determined by a tensile stress-relaxation test.

22.5 FATIGUE

Definitions

STRESS CYCLE. A stress cycle is the smallest section of the stress-time function that is repeated identically and periodically, as shown in Fig. 22.18.

MAXIMUM STRESS. σ_{max} is the largest algebraic value of the stress in the stress cycle, being positive for a tensile stress and negative for a compressive stress.

MINIMUM STRESS. σ_{min} is the smallest algebraic value of the stress in the stress cycle, being positive for a tensile stress and negative for a compressive stress.

RANGE OF STRESS. σ_r is the algebraic difference between the maximum and minimum stress in one cycle:

$$\sigma_r = \sigma_{max} - \sigma_{min} \tag{22.25}$$

For most cases of fatigue testing the stress varies about zero stress, but other types of variation may be experienced.

ALTERNATING-STRESS AMPLITUDE (VARIABLE STRESS COMPONENT). σ_a is one-half the range of stress, $\sigma_a = \sigma_r/2$.

MEAN STRESS (STEADY STRESS COMPONENT). σ_m is the algebraic mean of the maximum and minimum stress in one cycle:

$$\sigma_m = \frac{\sigma_{max} + \sigma_{min}}{2} \tag{22.26}$$

STRESS RATIO. R is the algebraic ratio of the minimum stress and the maximum stress in one cycle.

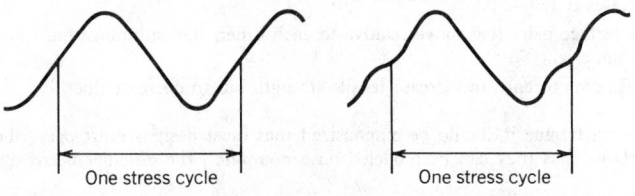

Fig. 22.18 Definition of one stress cycle.

22.5.1 Modes of Failure

The three most common modes of failure are*

$$\text{Soderberg's Law} \qquad \frac{\sigma_m}{\sigma_y} + \frac{\sigma_a}{\sigma_e} = \frac{1}{N} \tag{22.27}$$

$$\text{Goodman's Law} \qquad \frac{\sigma_m}{\sigma_u} + \frac{\sigma_a}{\sigma_e} = \frac{1}{N} \tag{22.28}$$

$$\text{Gerber's Law} \qquad \left(\frac{\sigma_m}{\sigma_u}\right)^2 + \frac{\sigma_a}{\sigma_e} = \frac{1}{N} \tag{22.29}$$

From distortion energy for plane stress

$$\sigma_m = \sqrt{\sigma_{xm}^2 - \sigma_{xm}\sigma_{ym} + \sigma_{ym}^2 + 3\tau_{xym}^2} \tag{22.30}$$

$$\sigma_a = \sqrt{\sigma_{xa}^2 - \sigma_{xa}\sigma_{ya} + \sigma_{ya}^2 + 3\tau_{xya}^2} \tag{22.31}$$

The stress concentration factor,[4] K_t or K_f, is applied to the individual stress for both σ_a and σ_m for brittle materials and only to σ_a for ductile materials. N is a reasonable factor of safety. σ_u is the ultimate tensile strength, and σ_y is the yield strength. σ_e is developed from the endurance limit σ_e' and reduced or increased depending on conditions and manufacturing procedures and to keep σ_e less than the yield strength:

$$\sigma_e = k_a k_b \cdots k_n \sigma_e'$$

where σ_e' (Ref. 1) for various materials is:

Steel	$0.5\sigma_u$ and never greater than 100 kpsi at 10^6 cycles
Magnesium	$0.35\sigma_u$ at 10^8 cycles
Nonferrous alloys	$0.35\sigma_u$ at 10^8 cycles
Aluminum alloys	$(0.16\text{–}0.3)\sigma_u$ at 5×10^8 cycles (see Military Handbook 5D)

and where the other k factors are affected as follows:

Surface Condition.   For surfaces that are from machined to ground, the k_a varies from 0.7 to 1.0. When surface finish is known, k_a can be found[1] more accurately.

Size and Shape.   If the size of the part is 0.30 in. or larger, the reduction is 0.85 or less, depending on the size.

Reliability.   The endurance limit and material properties are averages and both should be corrected. A reliability of 90% reduces values 0.897, while one of 99% reduces 0.814.

Temperature.   The endurance limit at −190°C increases 1.54–2.57 for steels, 1.14 for aluminums, and 1.4 for titaniums. The endurance limit is reduced approximately 0.68 for some steels at 1382°F, 0.24 for aluminum around 662°F, and 0.4 for magnesium alloys at 572°F.

Residual Stresses.   For steel, shot peening increases the endurance limit 1.04–1.22 for polished surfaces, 1.25 for machined surfaces, 1.25–1.5 for rolled surfaces, and 2–3 for forged surfaces. The shot-peening effect disappears above 500°F for steels and above 250°F for aluminum. Surface rolling affects the steel endurance limit approximately the same as shot peening, while the endurance limit is increased 1.2–1.3 in aluminum, 1.5 in magnesium, and 1.2–2.93 in cast iron.

Corrosion.   A corrosive environment decreases the endurance limit of anodized aluminum and magnesium 0.76–1.00, while nitrided steel and most materials are reduced 0.6–0.8.

Surface Treatments.   Nickel plating reduces the endurance limit of 1008 steel 0.01 and of 1063 steel 0.77, but, if the surface is shot peened after it is plated, the endurance limit can be increased over that of the base metal. The endurance limit of anodized aluminum is in general not affected. Flame and induction hardening as well as caburizing increases the endurance limit 1.62–1.85, while nitriding increases it 1.30–2.00.

Fretting.   In surface pairs that move relative to each other, the endurance limit is reduced 0.70–0.90 for each material.

Radiation.   Radiation tends to increase tensile strength but to decrease ductility.

In discussions on fatigue it should be emphasized that most designs must pass vibration testing. When sizing parts so that they can be modeled on a computer, the designer needs a starting point

---

* This section is condensed from Ref. 1, Chap. 12.

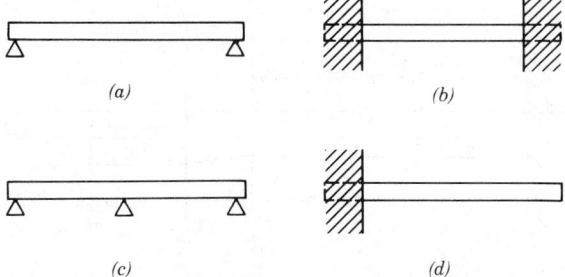

Fig. 22.19 (*a*) Simple, (*b*) constrained, (*c*) continuous, and (*d*) cantilever beams.

until he or she can receive feedback from the modeling. A helpful starting point is to estimate the static load to be carried, to find the level of vibration testing in *G* levels, to assume that the part vibrates with a magnification of 10, and to multiply these together to get an equivalent static load. The stress level should be $\sigma_u/4$, which should be less than the yield strength. When the design is modeled, changes can be made to bring the design within the required limits.

22.6 BEAMS

22.6.1 Theory of Flexure

Types of Beams

A beam is a bar or structural member subjected to transverse loads that tend to bend it. Any structural member acts as a beam if bending is induced by external transverse forces.

A **simple beam** (Fig. 22.19*a*) is a horizontal member that rests on two supports at the ends of the beam. All parts between the supports have free movement in a vertical plane under the influence of vertical loads.

A **fixed beam, constrained beam,** or **restrained beam** (Fig. 22.19*b*) is rigidly fixed at both ends or rigidly fixed at one end and simply supported at the other.

A **continuous beam** (Fig. 22.19*c*) is a member resting on more than two supports.

A **cantilever beam** (Fig. 22.19*d*) is a member with one end projecting beyond the point of support, free to move in a vertical plane under the influence of vertical loads placed between the free end and the support.

Phenomena of Flexure

When a simple beam bends under its own weight, the fibers on the upper or concave side are shortened, and the stress acting on them is compression; the fibers on the under or convex side are lengthened, and the stress acting on them is tension. In addition, shear exists along each cross section, the intensity of which is greatest along the sections at the two supports and zero at the middle section.

When a cantilever beam bends under its own weight, the fibers on the upper or convex side are lengthened under tensile stresses; the fibers on the under or concave side are shortened under compressive stresses, the shear is greatest along the section at the support, and zero at the free end.

The **neutral surface** is that horizontal section between the concave and convex surfaces of a loaded beam, where there is no change in the length of the fibers and no tensile or compressive stresses acting upon them.

The **neutral axis** is the trace of the neutral surface on any cross section of a beam. (See Fig. 22.20).

The **elastic curve** of a beam is the curve formed by the intersection of the neutral surface with the side of the beam, it being assumed that the longitudinal stresses on the fibers are within the elastic limit.

Reactions at Supports

The reactions, or upward pressures at the points of support, are computed by applying the following conditions necessary for equilibrium of a system of vertical forces in the same plane: (1) The algebraic sum of all vertical forces must equal zero; that is, the sum of the reactions equals the sum of the downward loads. (2) The algebraic sum of the moments of all the vertical forces must equal zero. Condition (1) applies to cantilever beams and to simple beams uniformly loaded, or with equal concen-

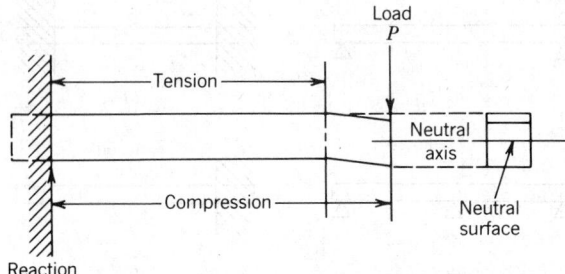

Fig. 22.20   Loads and stress conditions in a cantilever beam.

trated loads placed at equal distances from the center of the beam. In the cantilever beam, the reaction is the sum of all the vertical forces acting downward, comprising the weight of the beam and the superposed loads. In the simple beam each reaction is equal to one-half the total load, consisting of the weight of the beam and the superposed loads. Condition (2) applies to a simple beam not uniformly loaded. The reactions are computed separately, by determining the moment of the several loads about each support. The sum of the moments of the load around one support is equal to the moment of the reaction of the other support around the first support.

Example 22.1

Determine reactions R_1, R_2 of a simple beam 20 ft long, loaded as in Fig. 22.21. The weight of the beam may be taken as a concentrated load applied at the center of gravity of the beam. Then, moments around left-hand and right-hand supports are, respectively,

Left hand: $(2000 \times 5) + (3000 \times 9) + (1200 \times 10) - (R_2 \times 20) = 0$; $R_2 = 2450$ lb

Right hand: $(R_1 \times 20) - (2000 \times 15) - (3000 \times 11) - (1200 \times 10) = 0$; $R_1 = 3750$ lb

$R_1 + R_2 = 6200$ lb = sum of all loads acting downward

Conditions of Equilibrium

The fundamental laws for the stresses at any cross section of a beam in equilibrium are: (1) Sum of horizontal tensile stresses = sum of horizontal compressive stresses. (2) Resisting shear = vertical shear. (3) Resisting moment = bending moment.

Vertical Shear.   At any cross section of a beam the resultant of the external vertical forces acting on one side of the section is equal and opposite to the resultant of the external vertical forces acting on the other side of the section. These forces tend to cause the beam to shear vertically along the section. The value of either resultant is known as the vertical shear at the section considered. It is computed by finding the algebraic sum of the vertical forces to the left of the section; that is, it is equal to the left reaction minus the sum of the vertical downward forces acting between the left support and the section.

A **shear diagram** is a graphic representation of the vertical shear at all cross sections of the beam. Thus in the simple beam (Fig. 22.22) the ordinates to the line *AOB* represent to scale the intensity of the vertical shear at the corresponding sections of the beam. The vertical shear is greatest at the supports, where it is equal to the reactions, and it is zero at the center of the span. In the cantilever beam (Fig. 22.23) the vertical shear is greatest at the point of support, where it is equal to the reaction, and it is zero at the free end. Figure 22.24 shows graphically the vertical shear on all sections of a simple beam carrying two concentrated loads at equal distances from the supports, the weight of the beam being neglected.

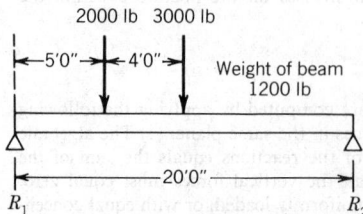

Fig. 22.21   Example of a simple loaded beam.

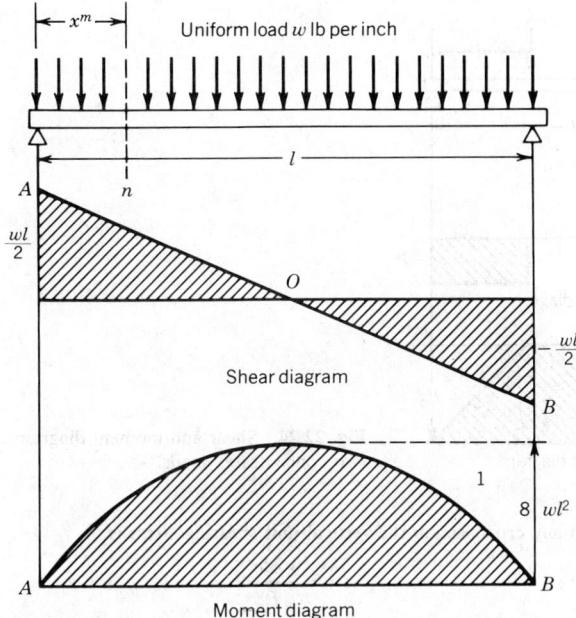

Fig. 22.22 Shear and moment diagrams, simple beam.

Resisting Shear. The tendency of a beam to shear vertically along any cross section, due to the vertical shear, is opposed by an internal shearing stress at that cross section known as the resisting shear; it is equal to the algebraic sum of the vertical components of all the internal stresses acting on the cross section.

If V = vertical shear, pounds; V_r = resisting shear, pounds; τ = average unit shearing stress, pounds per square inch; and A = area of the section, square inches, then at any cross section

$$V_r = V = \tau A; \qquad \tau = \frac{V}{A} \qquad (22.32)$$

The resisting shear is not uniformly distributed over the cross section, but the intensity varies from zero at the extreme fiber to its maximum value at the neutral axis.

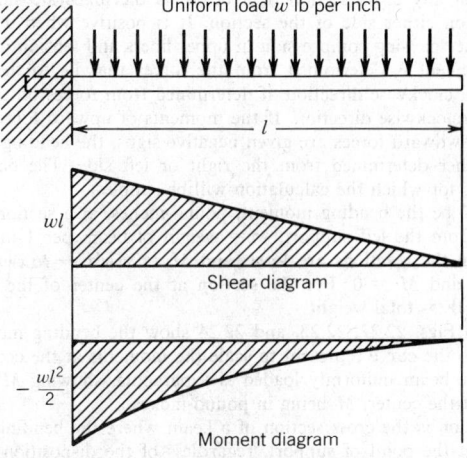

Fig. 22.23 Shear and moment diagrams, cantilever beam.

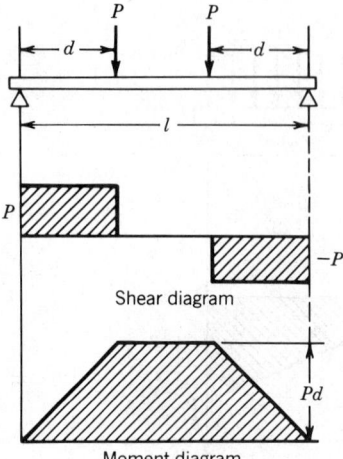

Shear diagram

Moment diagram

Fig. 22.24 Shear and moment diagrams, simple beam with two concentrated loads.

At any point in any cross section the vertical unit shearing stress is

$$\tau = \frac{VA'c'}{It} \tag{22.33}$$

where V = total vertical shear in pounds for section considered; A' = area in square inches of cross section between a horizontal plane through the point where shear is being found and the extreme fiber on the same side of the neutral axis; c' = distance in inches from neutral axis to center of gravity of area A'; I = moment of inertia of the section, inches[4]; t = width of section at plane of shear, inches. Maximum value of the unit shearing stress, where A = total area, square inches, of cross section of the beam, is

For a solid rectangular beam: $\quad \tau = \dfrac{3V}{2A} \tag{22.34}$

For a solid circular beam: $\quad \tau = \dfrac{4V}{3A} \tag{22.35}$

Horizontal Shear. In a beam, at any cross section where there is a vertical shearing force, there must be resultant unit shearing stresses acting on the vertical faces of particles that lie at that section. On a horizontal surface of such a particle, there is a unit shearing stress equal to the unit shearing stress on a vertical surface of the particle. Equation (22.33) therefore, also gives the horizontal unit shearing stress at any point on the cross section of a beam.

Bending moment, at any cross section of a beam, is the algebraic sum of the moments of the external forces acting on either side of the section. It is positive when it causes the beam to bend convex downward, hence causing compression in upper fibers and tension in lower fibers of the beam. When the bending moment is determined from the forces that lie to the left of the section, it is positive if they act in a clockwise direction; if determined from forces on the right side, it is positive if they act in a counterclockwise direction. If the moments of upward forces are given positive signs, and the moments of downward forces are given negative signs, the bending moment will always have the correct sign, whether determined from the right or left side. The bending moment should be determined for the side for which the calculation will be simplest.

In Fig. 22.22 let M be the bending moment, pound-inches, at a section mn of a simple beam at a distance x, inches, from the left support; w = weight of beam per 1 in. of length; l = length of the beam, inches. Then the reactions are $\frac{1}{2}wl$, and $M = \frac{1}{2}wlx - \frac{1}{2}xwx$. For the sections at the supports, $x = 0$ or l and $M = 0$. For the section at the center of the span $x = \frac{1}{2}l$ and $M = \frac{1}{8}wl^2 = \frac{1}{8}Wl$, where W = total weight.

A moment diagram Figs. 22.22, 22.23, and 22.24 show the bending moment at all cross sections of a beam. Ordinates to the curve represent to scale the moments at the corresponding cross sections. The curve for a simple beam uniformly loaded is a parabola, showing $M = 0$ at the supports and $M = \frac{1}{8}wl^2 = \frac{1}{8}Wl$ at the center, M being in pound-inches.

The dangerous section is the cross section of a beam where the bending moment is greatest. In a cantilever beam it is at the point of support, regardless of the disposition of the loads. In a simple beam it is that section where the vertical shear changes from positive to negative, and it may be located graphically by constructing a shear diagram or numerically by taking the left reaction and subtracting the loads in order from the left until a point is reached where the sum of the loads subtracted equals the reaction. For a simple beam, uniformly loaded, the dangerous section is at the center of the span.

The tendency to rotate about a point in any cross section of a beam is due to the bending moment at that section. This tendency is resisted by the **resisting moment,** which is the algebraic sum of the moments of all the horizontal stresses with reference to the same point.

Formula for Flexure

Let M = bending moment; M_r = resisting moment of the horizontal fiber stresses; σ = unit stress (tensile or compressive) on any fiber, usually that one most remote from the neutral surface; c = distance of that fiber from the neutral surface. Then

$$M = M_r = \frac{\sigma I}{c} \tag{22.36}$$

$$\sigma = \frac{Mc}{I} \tag{22.37}$$

where I = moment of inertia of the cross section with respect to its neutral axis. If σ is in pounds per square inch, M must be in pound-inches, I in inches^4 and c in inches.

Equation (22.37) is the basis of the design and investigation of beams. It is true only when the maximum horizontal fiber stress σ does not exceed the proportional limit of the material.

Moment of inertia is the sum of the products of each elementary area of the cross section multiplied by the square of the distance of that area from the assumed axis of rotation, or

$$I = \Sigma r^2 \, \Delta A = \int r^2 \, dA \tag{22.38}$$

where Σ is the sign of summation, ΔA is an elementary area of the section, and r is the distance of ΔA from the axis. The moment of inertia is greatest in those sections (such as I-beams) having much of the area concentrated at a distance from the axis. Unless otherwise stated, the neutral axis is the axis of rotation considered. I usually is expressed in inches^4. See Table 22.4 for values of moments of inertia of various sections.

Modulus of rupture is the term applied to the value of σ as found by Eq. (22.37), when a beam is loaded to the point of rupture. Since Eq. (22.37) is true only for stresses within the proportional limit, the value σ of the rupture strength so found is incorrect. However, the equation is used, as a measure of the ultimate load-carrying capacity of a beam. The modulus of rupture does not show the actual stress in the extreme fiber of a beam; it is useful only as a basis of comparison. If the strength of a beam in tension differs from its strength in compression, the modulus of rupture is intermediate between the two.

Section modulus, the factor I/c in flexure [Eq. (22.36)], is expressed in inches^3. It is the measure of a capacity of a section to resist a bending moment. For values of I/c for simple shapes, see Table 22.4. See Refs. 6 and 17 for properties of standard steel and aluminum structural shapes.

Elastic Deflection of Beams

When a beam bends under load, all points of the elastic curve except those over the supports are deflected from their original positions. The radius of curvature ρ of the elastic curve at any section is expressed as

$$\rho = \frac{EI}{M} \tag{22.39}$$

where E = modulus of elasticity of the material, pounds per square inch; I = moment of inertia, inches^4, of the cross section with reference to its neutral axis; M = bending moment, pound-inches, at the section considered. Where there is no bending moment, ρ is infinity and the curve is a straight line; where M is greatest, ρ is smallest and the curvature, therefore, is greatest.

If the elastic curve is referred to a system of coordinate axes in which x represents horizontal distances, y vertical distances, and l distances along the curve, the value of ρ is found, by the aid of the calculus, to be $d^3l/dx \cdot d^2y$. Differential equation (22.40) of the elastic curve which applies to all beams when the elastic limit of the material is not exceeded is obtained by substituting this value in the expression $\rho = EI/M$ and assuming that dx and dl are practically equal:

$$EI \frac{d^2y}{dx^2} = M \tag{22.40}$$

Equation (22.40) is used to determine the deflection of any point of the elastic curve, by regarding the point of support as the origin of the coordinate axis, taking y as the vertical deflection at any point on the curve and x as the horizontal distance from the support to the point considered. The values of E, I, and M are substituted and the expression is integrated twice, giving proper values to the constants of integration, and the deflection y is determined for any point. See Table 22.5.

For example, the cantilever beam shown in Fig. 22.20 has a length = l, inches, and carries a

Table 22.4   Elements of Sections

A = area of section	I/c = section modulus
I = moment of inertia about axis $I\text{-}I$	r = radius of gyration
c = distance from axis $I\text{-}I$ to remotest point of section	

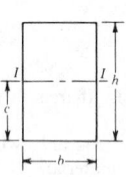

RECTANGLE

Axis through center

$A = bh$

$c = h/2$

$I = bh^3/12$

$I/c = bh^2/6$

$r = h/\sqrt{12} = 0.289h$

RECTANGLE

Axis any line through center of gravity

$A = bh$

$c = (b \sin \alpha + h \cos \alpha)/2$

$I = bh(b^2 \sin^2 \alpha + h^2 \cos^2 \alpha)/12$

$I/c = \dfrac{bh(b^2 \sin^2 \alpha + h^2 \cos^2 \alpha)}{6(b \sin \alpha + h \cos \alpha)}$

$r = \sqrt{(b^2 \sin^2 \alpha + h^2 \cos^2 \alpha)/12}$

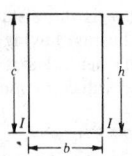

RECTANGLE

Axis on base

$A = bh$

$c = h$

$I = bh^3/3$

$I/c = bh^2/3$

$r = h/\sqrt{3} = 0.577h$

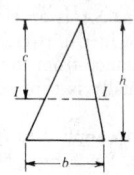

TRIANGLE

Axis through center of gravity

$A = bh/2$

$c = 2/3\, h$

$I = bh^3/36$

$I/c = bh^2/24$

$r = h/\sqrt{18} = 0.236h$

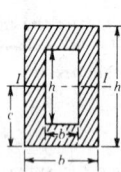

HOLLOW RECTANGLE

Axis through center

$A = bh - b_1 h_1$

$c = h/2$

$I = (bh^3 - b_1 h_1^3)/12$

$I/c = (bh^3 - b_1 h_1^3)/6h$

$r = \sqrt{\dfrac{bh^3 - b_1 h_1^3}{12(bh - b_1 h_1)}}$

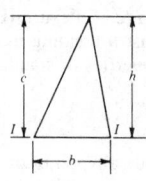

TRIANGLE

Axis through base

$A = bh/2$

$c = h$

$I = bh^3/12$

$I/c = bh^2/12$

$r = h/\sqrt{6} = 0.408h$

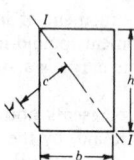

RECTANGLE

Axis on diagonal

$A = bh$

$c = bh/\sqrt{b^2 + h^2}$

$I = b^3 h^3/6(b^2 + h^2)$

$I/c = b^2 h^2/6\sqrt{(b^2 + h^2)}$

$r = bh/\sqrt{6(b^2 + h^2)}$

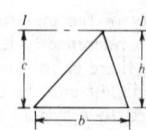

TRIANGLE

Axis through apex

$A = bh/2$

$c = h$

$I = bh^3/4$

$I/c = bh^2/4$

$r = h/\sqrt{2} = 0.707h$

Table 22.4 (*Continued*)

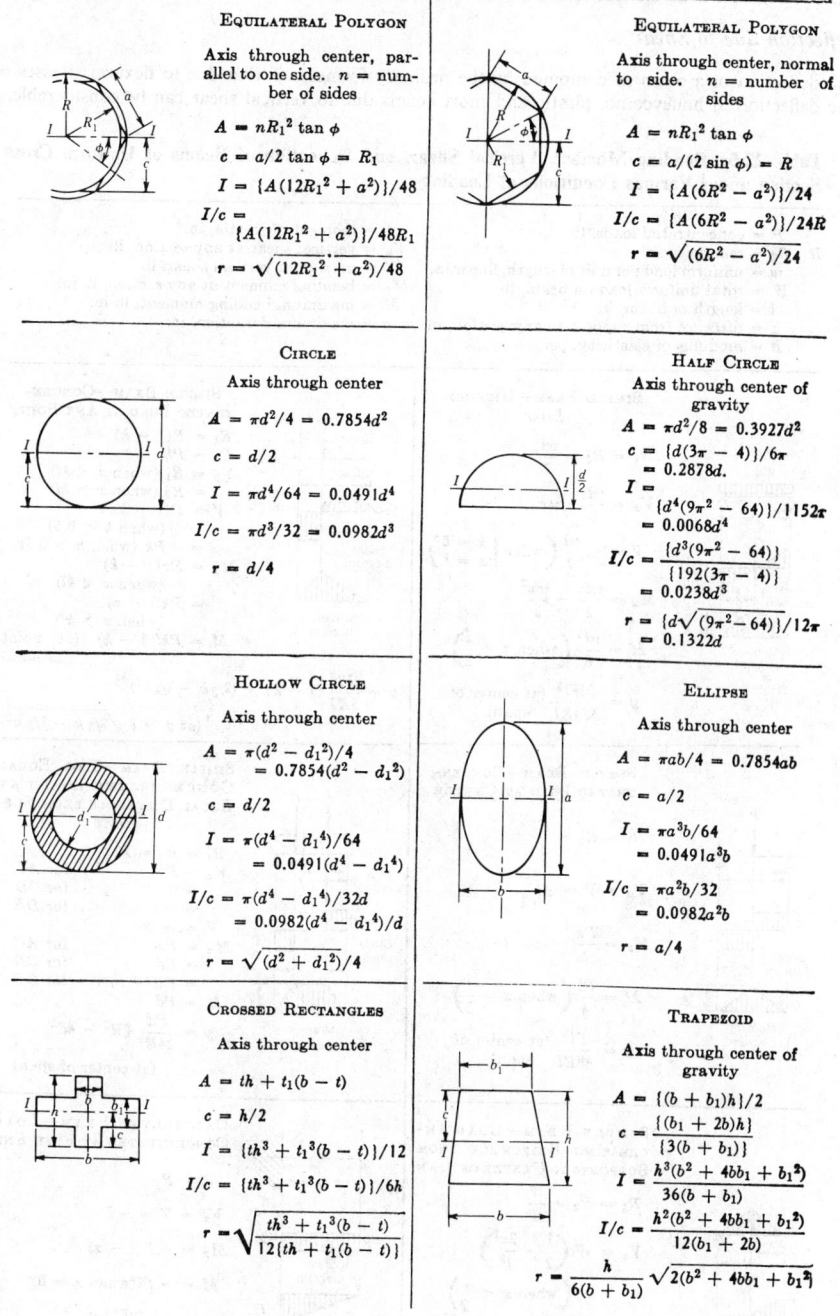

EQUILATERAL POLYGON	EQUILATERAL POLYGON
Axis through center, parallel to one side. n = number of sides	Axis through center, normal to side. n = number of sides
$A = nR_1^2 \tan \phi$	$A = nR_1^2 \tan \phi$
$c = a/2 \tan \phi = R_1$	$c = a/(2 \sin \phi) = R$
$I = \{A(12R_1^2 + a^2)\}/48$	$I = \{A(6R^2 - a^2)\}/24$
$I/c = \{A(12R_1^2 + a^2)\}/48R_1$	$I/c = \{A(6R^2 - a^2)\}/24R$
$r = \sqrt{(12R_1^2 + a^2)/48}$	$r = \sqrt{(6R^2 - a^2)/24}$

CIRCLE	HALF CIRCLE
Axis through center	Axis through center of gravity
$A = \pi d^2/4 = 0.7854d^2$	$A = \pi d^2/8 = 0.3927d^2$
$c = d/2$	$c = \{d(3\pi - 4)\}/6\pi$ $= 0.2878d.$
$I = \pi d^4/64 = 0.0491d^4$	$I = \{d^4(9\pi^2 - 64)\}/1152\pi$ $= 0.0068d^4$
$I/c = \pi d^3/32 = 0.0982d^3$	$I/c = \dfrac{\{d^3(9\pi^2 - 64)\}}{\{192(3\pi - 4)\}}$ $= 0.0238d^3$
$r = d/4$	$r = \{d\sqrt{(9\pi^2 - 64)}\}/12\pi$ $= 0.1322d$

HOLLOW CIRCLE	ELLIPSE
Axis through center	Axis through center
$A = \pi(d^2 - d_1^2)/4$ $= 0.7854(d^2 - d_1^2)$	$A = \pi ab/4 = 0.7854ab$
$c = d/2$	$c = a/2$
$I = \pi(d^4 - d_1^4)/64$ $= 0.0491(d^4 - d_1^4)$	$I = \pi a^3 b/64$ $= 0.0491a^3b$
$I/c = \pi(d^4 - d_1^4)/32d$ $= 0.0982(d^4 - d_1^4)/d$	$I/c = \pi a^2 b/32$ $= 0.0982a^2b$
$r = \sqrt{(d^2 + d_1^2)/4}$	$r = a/4$

CROSSED RECTANGLES	TRAPEZOID
Axis through center	Axis through center of gravity
$A = th + t_1(b - t)$	$A = \{(b + b_1)h\}/2$
$c = h/2$	$c = \dfrac{\{(b_1 + 2b)h\}}{\{3(b + b_1)\}}$
$I = \{th^3 + t_1^3(b - t)\}/12$	$I = \dfrac{h^3(b^2 + 4bb_1 + b_1^2)}{36(b + b_1)}$
$I/c = \{th^3 + t_1^3(b - t)\}/6h$	$I/c = \dfrac{h^2(b^2 + 4bb_1 + b_1^2)}{12(b_1 + 2b)}$
$r = \sqrt{\dfrac{th^3 + t_1^3(b - t)}{12\{th + t_1(b - t)\}}}$	$r = \dfrac{h}{6(b + b_1)}\sqrt{2(b^2 + 4bb_1 + b_1^2)}$

load P, pounds, at the free end. It is required to find the deflection of the elastic curve at a point distant x, inches, from the support, the weight of the beam being neglected.

The moment $M = -P(l - x)$. By substitution in Eq. (22.40), the equation for the elastic curve becomes $EI(d^2y/dx^2) = -Pl + Px$. By integrating and determining the constant of integration by the condition that $dy/dx = 0$ when $x = 0$, $EI(dy/dx) = -Plx + \frac{1}{2}Px^2$ results. By integrating a second time and determining the constant by the condition that $x = 0$ when $y = 0$, $EIy = -\frac{1}{2}Plx^2$

$+ \frac{1}{6}Px^3$, which is the equation of the elastic curve, results. When $x = l$, the value of y, or the deflection in inches at the free end, is found to be $-Pl^3/3EI$.

Deflection due to Shear

The deflection of a beam as computed by the ordinary formulas is that due to flexural stresses only. The deflection in honeycomb, plastic and short beams due to vertical shear can be considerable, and

Table 22.5   Bending Moment, Vertical Shear, and Deflection of Beams of Uniform Cross Section under Various Conditions of Loading

P = concentrated loads, lb
R_1, R_2 = reactions, lb
w = uniform load per unit of length, lb per in.
W = total uniform load on beam, lb
l = length of beam, in
x = distance from support to any section, in
E = modulus of elasticity, psi

I = moment of inertia, in.[4]
V_x = vertical shear at any section, lb
V = maximum vertical shear, lb
M_x = bending moment at any section, lb-in.
M = maximum bending moment, lb-in.
y = maximum deflection, in.

SIMPLE BEAM—UNIFORM LOAD

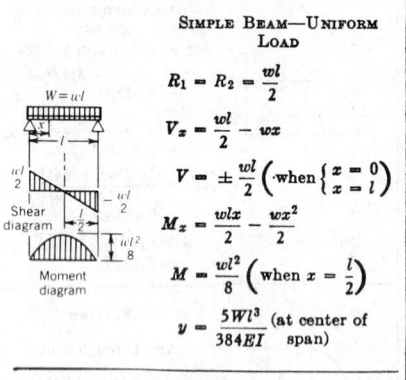

$$R_1 = R_2 = \frac{wl}{2}$$

$$V_x = \frac{wl}{2} - wx$$

$$V = \pm \frac{wl}{2} \left(\text{when } \begin{cases} x = 0 \\ x = l \end{cases} \right)$$

$$M_x = \frac{wlx}{2} - \frac{wx^2}{2}$$

$$M = \frac{wl^2}{8} \left(\text{when } x = \frac{l}{2} \right)$$

$$y = \frac{5Wl^3}{384EI} \quad \text{(at center of span)}$$

SIMPLE BEAM—CONCENTRATED LOAD AT ANY POINT

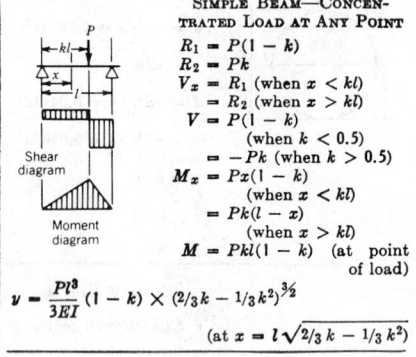

$$R_1 = P(1 - k)$$
$$R_2 = Pk$$
$$V_x = R_1 \text{ (when } x < kl)$$
$$= R_2 \text{ (when } x > kl)$$
$$V = P(1 - k)$$
$$\qquad \text{(when } k < 0.5)$$
$$= -Pk \text{ (when } k > 0.5)$$
$$M_x = Px(1 - k)$$
$$\qquad \text{(when } x < kl)$$
$$= Pk(l - x)$$
$$\qquad \text{(when } x > kl)$$
$$M = Pkl(1 - k) \quad \text{(at point of load)}$$

$$y = \frac{Pl^3}{3EI} (1 - k) \times (2/3 k - 1/3 k^2)^{3/2}$$

$$(\text{at } x = l \sqrt{2/3 k - 1/3 k^2})$$

SIMPLE BEAM—CONCENTRATED LOAD AT CENTER

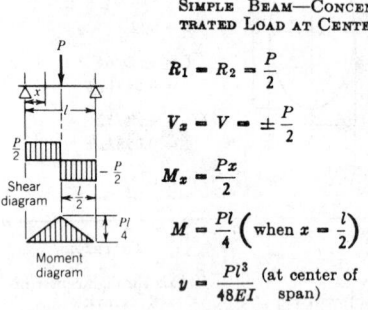

$$R_1 = R_2 = \frac{P}{2}$$

$$V_x = V = \pm \frac{P}{2}$$

$$M_x = \frac{Px}{2}$$

$$M = \frac{Pl}{4} \left(\text{when } x = \frac{l}{2} \right)$$

$$y = \frac{Pl^3}{48EI} \quad \text{(at center of span)}$$

SIMPLE BEAM—TWO EQUAL CONCENTRATED LOADS AT EQUAL DISTANCES FROM SUPPORTS

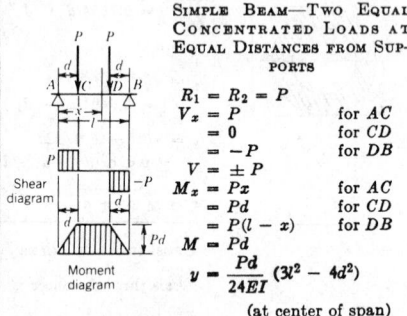

$$R_1 = R_2 = P$$
$$V_x = P \qquad \text{for } AC$$
$$= 0 \qquad \text{for } CD$$
$$= -P \qquad \text{for } DB$$
$$V = \pm P$$
$$M_x = Px \qquad \text{for } AC$$
$$= Pd \qquad \text{for } CD$$
$$= P(l - x) \quad \text{for } DB$$
$$M = Pd$$

$$y = \frac{Pd}{24EI} (3l^2 - 4d^2)$$

$$(\text{at center of span})$$

SIMPLE BEAM—LOAD INCREASING UNIFORMLY FROM SUPPORTS TO CENTER OF SPAN

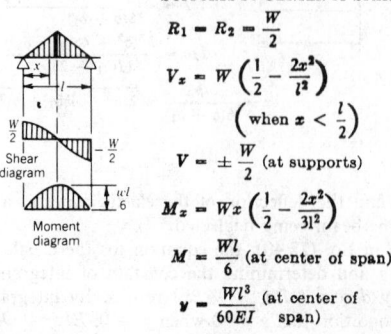

$$R_1 = R_2 = \frac{W}{2}$$

$$V_x = W \left(\frac{1}{2} - \frac{2x^2}{l^2} \right)$$

$$\left(\text{when } x < \frac{l}{2} \right)$$

$$V = \pm \frac{W}{2} \quad \text{(at supports)}$$

$$M_x = Wx \left(\frac{1}{2} - \frac{2x^2}{3l^2} \right)$$

$$M = \frac{Wl}{6} \quad \text{(at center of span)}$$

$$y = \frac{Wl^3}{60EI} \quad \text{(at center of span)}$$

CANTILEVER BEAM—LOAD CONCENTRATED AT FREE END

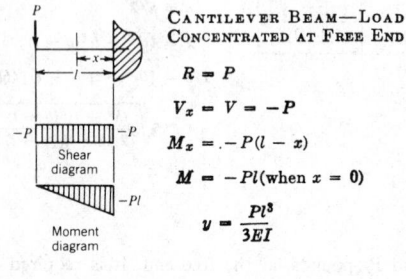

$$R = P$$

$$V_x = V = -P$$

$$M_x = -P(l - x)$$

$$M = -Pl \text{(when } x = 0)$$

$$y = \frac{Pl^3}{3EI}$$

should always be checked. Because of the nonuniform distribution of the shear over the cross section of the beam, computing the deflection due to shear by exact methods is difficult. It may be approximated by $y_s = M/AE_s$, where y_s = deflection, inches, due to shear; M = bending moment, pound-inches, at the section where the deflection is calculated; E_s = modulus of elasticity in shear, pounds per square inch; A = area of cross section of beam, square inches.[7] For a rectangular section, the ratio

Table 22.5 (*Continued*)

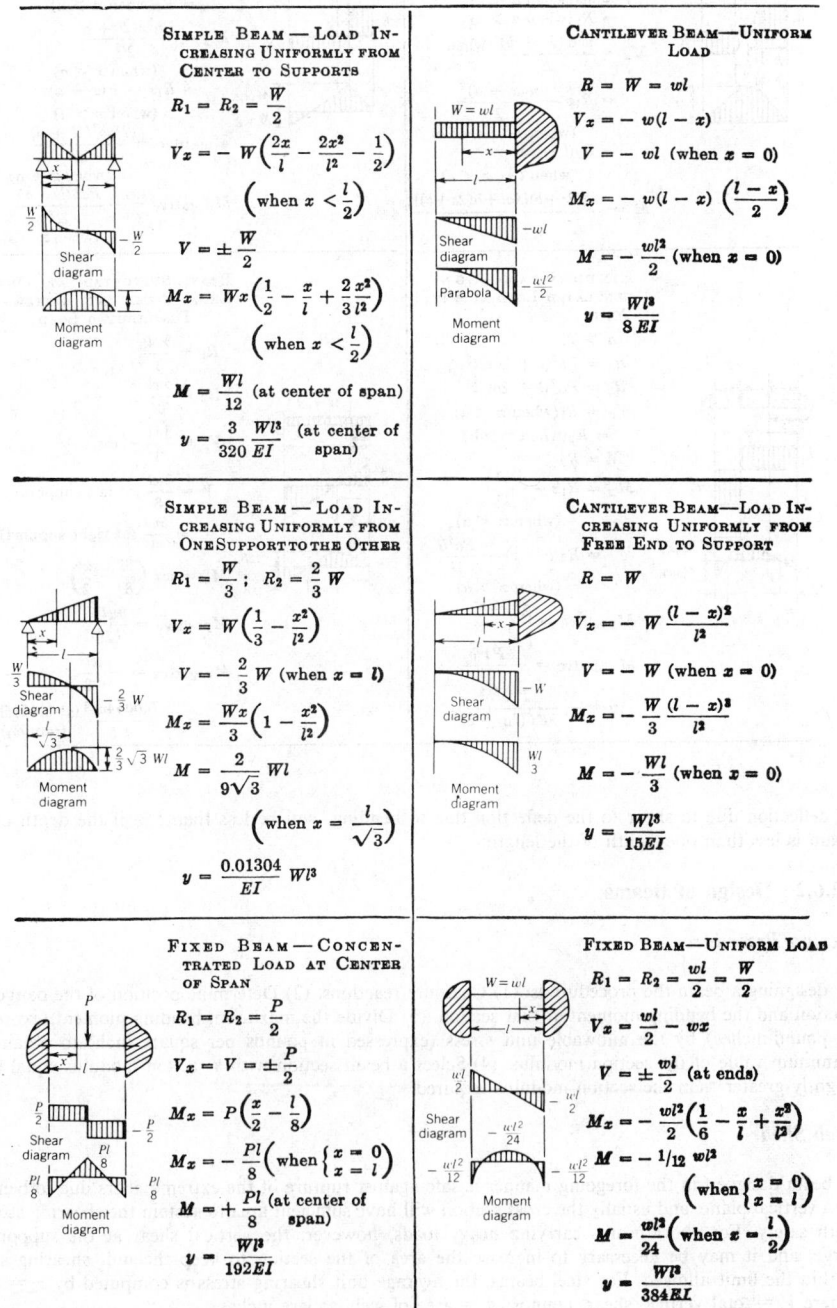

Simple Beam — Load Increasing Uniformly from Center to Supports

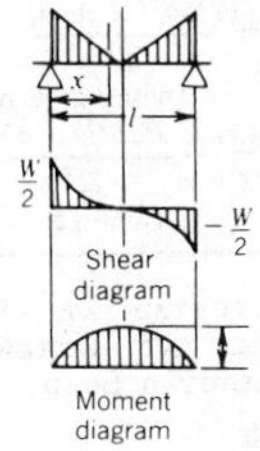

Shear diagram
Moment diagram

$$R_1 = R_2 = \frac{W}{2}$$

$$V_x = -W\left(\frac{2x}{l} - \frac{2x^2}{l^2} - \frac{1}{2}\right) \quad \left(\text{when } x < \frac{l}{2}\right)$$

$$V = \pm\frac{W}{2}$$

$$M_x = Wx\left(\frac{1}{2} - \frac{x}{l} + \frac{2}{3}\frac{x^2}{l^2}\right) \quad \left(\text{when } x < \frac{l}{2}\right)$$

$$M = \frac{Wl}{12} \text{ (at center of span)}$$

$$y = \frac{3}{320}\frac{Wl^3}{EI} \text{ (at center of span)}$$

Cantilever Beam—Uniform Load

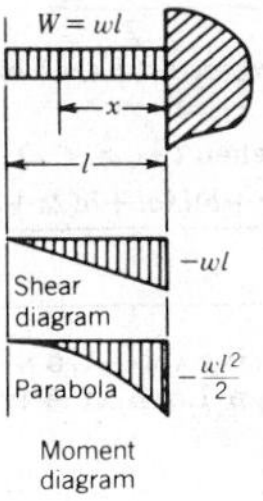

W = wl
Shear diagram
Parabola
Moment diagram

$$R = W = wl$$

$$V_x = -w(l - x)$$

$$V = -wl \text{ (when } x = 0)$$

$$M_x = -w(l - x)\left(\frac{l - x}{2}\right)$$

$$M = -\frac{wl^2}{2} \text{ (when } x = 0)$$

$$y = \frac{Wl^3}{8EI}$$

Simple Beam — Load Increasing Uniformly from One Support to the Other

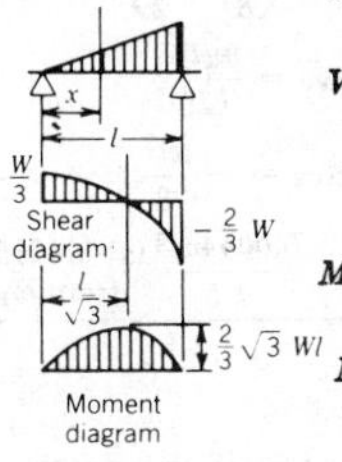

Shear diagram
Moment diagram

$$R_1 = \frac{W}{3};\ R_2 = \frac{2}{3}W$$

$$V_x = W\left(\frac{1}{3} - \frac{x^2}{l^2}\right)$$

$$V = -\frac{2}{3}W \text{ (when } x = l)$$

$$M_x = \frac{Wx}{3}\left(1 - \frac{x^2}{l^2}\right)$$

$$M = \frac{2}{9\sqrt{3}}Wl \quad \left(\text{when } x = \frac{l}{\sqrt{3}}\right)$$

$$y = \frac{0.01304}{EI}Wl^3$$

Cantilever Beam—Load Increasing Uniformly from Free End to Support

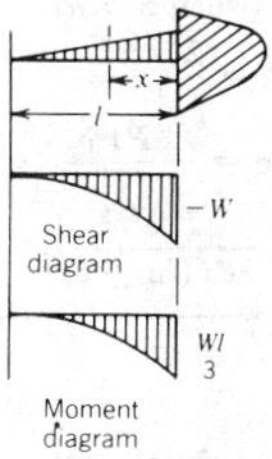

Shear diagram
Moment diagram

$$R = W$$

$$V_x = -W\frac{(l - x)^2}{l^2}$$

$$V = -W \text{ (when } x = 0)$$

$$M_x = -\frac{W}{3}\frac{(l - x)^3}{l^2}$$

$$M = -\frac{Wl}{3} \text{ (when } x = 0)$$

$$y = \frac{Wl^3}{15EI}$$

Fixed Beam — Concentrated Load at Center of Span

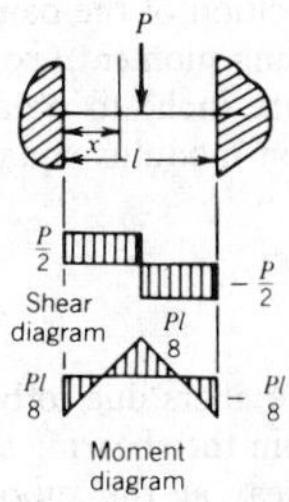

Shear diagram
Moment diagram

$$R_1 = R_2 = \frac{P}{2}$$

$$V_x = V = \pm\frac{P}{2}$$

$$M_x = P\left(\frac{x}{2} - \frac{l}{8}\right)$$

$$M_x = -\frac{Pl}{8}\left(\text{when } \begin{cases} x = 0 \\ x = l \end{cases}\right)$$

$$M = +\frac{Pl}{8} \text{ (at center of span)}$$

$$y = \frac{Wl^3}{192EI}$$

Fixed Beam—Uniform Load

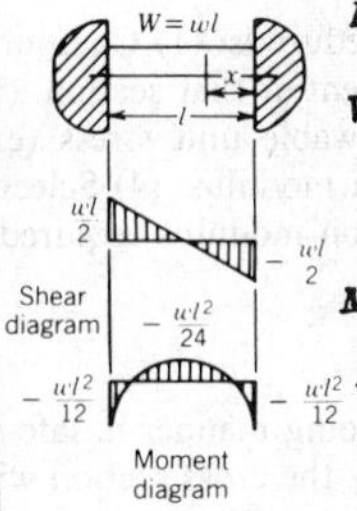

W = wl
Shear diagram
Moment diagram

$$R_1 = R_2 = \frac{wl}{2} = \frac{W}{2}$$

$$V_x = \frac{wl}{2} - wx$$

$$V = \pm\frac{wl}{2} \text{ (at ends)}$$

$$M_x = -\frac{wl^2}{2}\left(\frac{1}{6} - \frac{x}{l} + \frac{x^2}{l^2}\right)$$

$$M = -\frac{1}{12}wl^2 \left(\text{when } \begin{cases} x = 0 \\ x = l \end{cases}\right)$$

$$M = \frac{wl^2}{24}\left(\text{when } x = \frac{l}{2}\right)$$

$$y = \frac{Wl^3}{384EI}$$

Table 22.5 *(Continued)*

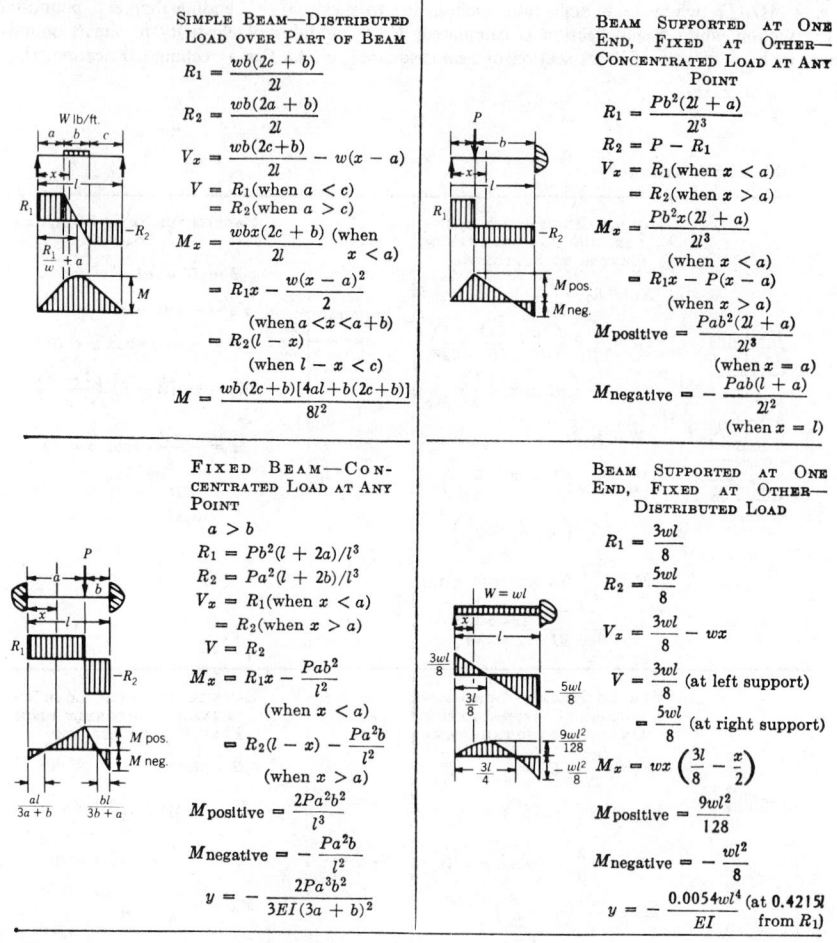

SIMPLE BEAM—DISTRIBUTED LOAD OVER PART OF BEAM

$$R_1 = \frac{wb(2c + b)}{2l}$$

$$R_2 = \frac{wb(2a + b)}{2l}$$

$$V_x = \frac{wb(2c+b)}{2l} - w(x - a)$$

$$V = R_1(\text{when } a < c)$$
$$\quad = R_2(\text{when } a > c)$$

$$M_x = \frac{wbx(2c + b)}{2l} \text{ (when } x < a)$$

$$\quad = R_1x - \frac{w(x - a)^2}{2}$$
(when $a < x < a+b$)
$$\quad = R_2(l - x)$$
(when $l - x < c$)

$$M = \frac{wb(2c+b)[4al+b(2c+b)]}{8l^2}$$

BEAM SUPPORTED AT ONE END, FIXED AT OTHER—CONCENTRATED LOAD AT ANY POINT

$$R_1 = \frac{Pb^2(2l + a)}{2l^3}$$

$$R_2 = P - R_1$$

$$V_x = R_1(\text{when } x < a)$$
$$\quad = R_2(\text{when } x > a)$$

$$M_x = \frac{Pb^2x(2l + a)}{2l^3}$$
(when $x < a$)
$$\quad = R_1x - P(x - a)$$
(when $x > a$)

$$M \text{ positive} = \frac{Pab^2(2l + a)}{2l^3}$$
(when $x = a$)

$$M \text{ negative} = -\frac{Pab(l + a)}{2l^2}$$
(when $x = l$)

FIXED BEAM—CONCENTRATED LOAD AT ANY POINT

$a > b$

$$R_1 = Pb^2(l + 2a)/l^3$$
$$R_2 = Pa^2(l + 2b)/l^3$$
$$V_x = R_1(\text{when } x < a)$$
$$\quad = R_2(\text{when } x > a)$$
$$V = R_2$$
$$M_x = R_1x - \frac{Pab^2}{l^2}$$
(when $x < a$)
$$\quad = R_2(l - x) - \frac{Pa^2b}{l^2}$$
(when $x > a$)

$$M \text{ positive} = \frac{2Pa^2b^2}{l^3}$$

$$M \text{ negative} = -\frac{Pa^2b}{l^2}$$

$$y = -\frac{2Pa^3b^2}{3EI(3a + b)^2}$$

BEAM SUPPORTED AT ONE END, FIXED AT OTHER—DISTRIBUTED LOAD

$$R_1 = \frac{3wl}{8}$$

$$R_2 = \frac{5wl}{8}$$

$$V_x = \frac{3wl}{8} - wx$$

$$V = \frac{3wl}{8} \text{ (at left support)}$$
$$\quad = \frac{5wl}{8} \text{ (at right support)}$$

$$M_x = wx\left(\frac{3l}{8} - \frac{x}{2}\right)$$

$$M \text{ positive} = \frac{9wl^2}{128}$$

$$M \text{ negative} = -\frac{wl^2}{8}$$

$$y = -\frac{0.0054wl^4}{EI} \text{ (at 0.4215}l \text{ from } R_1)$$

of deflection due to shear to the deflection due to bending, will be less than 5% if the depth of the beam is less than one-eighth of the length.

22.6.2 Design of Beams

Design Procedure

In designing a beam the procedure is: (1) Compute reactions. (2) Determine position of the dangerous section and the bending moment at that section. (3) Divide the maximum bending moment (expressed in pound-inches) by the allowable unit stress (expressed in pounds per square inch) to obtain the minimum value of the section modulus. (4) Select a beam section with a section modulus equal to or slightly greater than the section modulus required.

Web Shear

A beam designed in the foregoing manner is safe against rupture of the extreme fibers due to bending in a vertical plane, and usually the cross section will have sufficient area to sustain the shearing stresses with safety. For short beams carrying heavy loads, however, the vertical shear at the supports is large, and it may be necessary to increase the area of the section to keep the unit shearing stress within the limit allowed. For steel beams, the average unit shearing stress is computed by $\tau = V/A$, where V = total vertical shear, pounds; A = area of web, square inches.

Shear Center

Closed or solid cross sections with two axes of symmetry will have a shear center at the origin. If the loads are applied here, then the bending moment can be used to calculate the deflections and bending stress, which means there are no torsional stresses. The open section or unsymmetrical section generally has a shear center that is offset on one axis of symmetry and must be calculated.[2,8,9] The load applied at this location will develop bending stresses and deflections. If any sizable torsion is developed, then torsional stresses and rotations must be accounted for.

Miscellaneous Considerations

Other considerations which will influence the choice of section under certain conditions of loading are: (1) Maximum vertical deflection that may be permitted in beams coming in contact with plaster. (2) Danger of failure by sidewise bending in long beams, unbraced against lateral deflection. (3) Danger of failure by the buckling of the web of steel beams of short span carrying heavy loads. (4) Danger of failure by horizontal shear, particularly in wooden beams.

Vertical Deflection

If a beam is to support or come in contact with materials like plaster, which may be broken by excessive deflection, it is usual to select such a beam that the maximum deflection will not exceed ($\frac{1}{360}$ × span). It may be shown that for a simple beam, supported at the ends, with a total uniformly distributed load W, pounds, the deflection, inches, is

$$y = \frac{30\sigma L^2}{Ed} \tag{22.41}$$

where σ = allowable unit fiber stress, pounds per square inch; L = span of beam, feet; E = modulus of elasticity, pounds per square inch; d = depth of beam, inches.

If the deflection of a steel beam is to be less than $\frac{1}{360}$th of the span, it may be shown from Eq. (22.41) that, for a maximum allowable fiber stress of 18,000 psi, the limit of span in feet is approximately $1.8d$, where d = depth of the beam inches.

For the deflection due to the impact of a moving load falling on a beam, see Section 22.6.6.

Horizontal Shear in Timber Beams

In beams of a homogeneous material which can withstand equally well shearing stresses in any direction, vertical and horizontal shearing stresses are equally important. In timber, however, shearing strength along the grain is much less than that perpendicular to the grain. Hence, the beams may fail owing to horizontal shear. Short wooden beams should be checked for horizontal shear in order that allowable unit shearing stress along the grain shall not be exceeded. (See the example below.)

Restrained Beams

A beam is considered to be restrained if one or both ends are not free to rotate. This condition exists if a beam is built into a masonry wall at one or both ends, if it is riveted or otherwise fastened to a column, or if the ends projecting beyond the supports carry loads that tend to prevent tilting of the ends which would naturally occur as the beam deflects. The shears and moments given in Table 22.5 for fixed end conditions are seldom, if ever, attained, since the restraining elements themselves deform and reduce the magnitude of the restraint. This reduction of restraint decreases the negative moment at the support and increases the positive moment in the central portion of the span. The amount of restraint that exists is a matter which must be judged for each case in the light of the construction used, the rigidity of the connections, and the relative sizes of the connecting members.

Safe Loads on Simple Beams

Equation (22.42) gives the safe loads on simple beams. This formula is obtained by substituting in the flexure equation (22.36), the value of M for a simple beam, uniformly loaded, as given in Table 22.5. Let W = total load, pounds; σ = extreme fiber unit stress, pounds per square inch; S = section modulus, inches[3]; L = length of span, feet. Then

$$W = \frac{2}{3}\sigma\frac{S}{L} \tag{22.42}$$

If σ is taken as a maximum allowable unit fiber stress, this equation gives the maximum allowable load on the beam. Most building codes permit a value of $\sigma = 18,000$ psi for quiescent loads on steel. For this value of σ, Eq. (22.42) becomes

$$W = \frac{12,000\ S}{L} \tag{22.43}$$

If the load is concentrated at the center of the span, the safe load is one-half the value given by Eq. (22.43). If the load is neither uniformly distributed nor concentrated at the center of the span, the maximum bending moment must be used. The foregoing equations are for beams laterally supported and are for flexure only. The other factors which influence the strength of the beam, as shearing, buckling, etc., must also be considered.

Use of Tables in Design

The following is an example in the use of tables for the design of a wooden beam.

Example 22.2

Design a southern pine girder, of common structural grade, to carry a load of 9600 lb distributed uniformly over a 16-ft span in the interior of a building, the beam being a simple beam, freely supported at each end.

Solution. From Table 22.5, the bending moment of a simple beam uniformly loaded is $M = wl^2/8$. Since $W = wl$ and $l = 12L$,

$$M = 9600 \times 16 \times {}^{12}\!/\!_8 = 230,400 \text{ lb-in.}$$

If the allowable unit stress on yellow pine is 1200 psi,

$$\frac{I}{c} = \frac{230,400}{1200} = 192 \text{ in.}^3$$

From Table 22.4, the section modulus of a rectangular section is $bd^2/6$. Assume $b = 8$ in. Then $8d^2/6 = 192$, and $d = \sqrt{144} = 12.0$ in. A beam 8 by 12 in. is selected tentatively, and checked for shear.

Maximum shearing stress (horizontal and vertical) is at the neutral surface over the supports. Equation (22.34) for horizontal shear in a solid rectangular beam is $\tau = 3V/2A$; $V = 9600/2 = 4800$, and $A = 8 \times 12 = 96$, whence $\tau = (3 \times 4800)/(2 \times 96) = 75$ psi.

If the safe horizontal unit shearing stress for common-grade southern yellow pine is 88 psi, and since the actual horizontal unit shearing stress is less than 88 lb, the 8 by 12 in. beam will be satisfactory.

A beam of uniform strength is one in which the dimensions are such that the maximum fiber stress σ is the same throughout the length of the beam. The form of the beam is determined by finding the areas of various cross sections from the flexure formula $M = \sigma I/c$, keeping σ constant and making I/c vary with M. For a rectangular section of width b and depth d, the section modulus $I/c = \frac{1}{6}bd^2$, and, therefore, $M = \frac{1}{6}\sigma bd^2$. By making bd^2 vary with M, the dimensions of the various sections are obtained. Table 22.6 gives the dimensions b and d, at any section, the maximum unit fiber stress σ and the maximum deflection y, of some rectangular beams of uniform strength. In this table, the bending moment has been assumed to be the controlling factor. On account of the vertical shear near the ends of the beams, the area of the sections must be increased over that given by an amount necessary to keep the unit shearing stress within the allowable unit shearing stress. The discussion of beams of uniform strength, although of considerable theoretical interest, is of little practical value since the cost of fabrication will offset any economy in the use of the material. A plate girder in a bridge or a building is an approximation in practice to a steel beam of uniform strength.

22.6.3 Continuous Beams

As in simple beams, the expressions $M = \sigma I/c$ and $\tau = V/A$ govern the design and investigation of beams resting on more than two supports. In the case of continuous beams, however, the reactions cannot be obtained in the manner described for simple beams. Instead, the bending moments at the various sections must be determined, and from these values the vertical shears at the sections and the reactions at the supports may be derived.

Consider the second span of length l_2, inches, of the continuous beam (Fig. 22.25). Vertical shear V_x at any section distant x, inches, from the left support of the span is equal to the algebraic sum of all the vertical forces on one side of the section. Thus, if V_2 = vertical shear at a section to the right of, but infinitely close to, the left support, w_2x = uniform load, and ΣP_2 = sum of the concentrated loads along the distance x, applied at a distance kl_2 from the left support, k being a fraction less than unity, then

Table 22.6   Rectangular Beams of Uniform Strength*

σ = maximum fiber unit stress, pounds per square inch; E = modulus of elasticity; w = uniform load, pounds per inch; d = depth of beam, inches; b = width of beam, inches; y = maximum deflection, inches.   All other dimensions in inches.

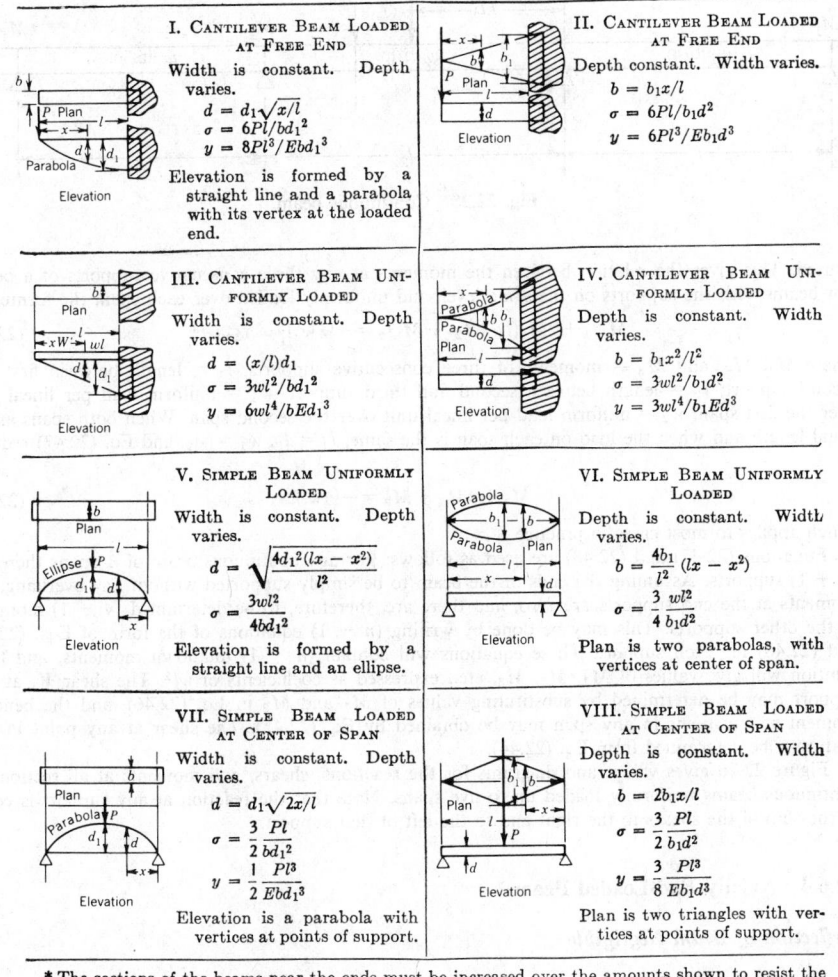

I. Cantilever Beam Loaded at Free End

Width is constant.   Depth varies.

$$d = d_1\sqrt{x/l}$$
$$\sigma = 6Pl/bd_1^2$$
$$y = 8Pl^3/Ebd_1^3$$

Elevation is formed by a straight line and a parabola with its vertex at the loaded end.

II. Cantilever Beam Loaded at Free End

Depth constant.   Width varies.

$$b = b_1x/l$$
$$\sigma = 6Pl/b_1d^2$$
$$y = 6Pl^3/Eb_1d^3$$

III. Cantilever Beam Uniformly Loaded

Width is constant.   Depth varies.

$$d = (x/l)d_1$$
$$\sigma = 3wl^2/bd_1^2$$
$$y = 6wl^4/bEd_1^3$$

IV. Cantilever Beam Uniformly Loaded

Depth is constant.   Width varies.

$$b = b_1x^2/l^2$$
$$\sigma = 3wl^2/b_1d^2$$
$$y = 3wl^4/b_1Ed^3$$

V. Simple Beam Uniformly Loaded

Width is constant.   Depth varies.

$$d = \sqrt{\dfrac{4d_1^2(lx - x^2)}{l^2}}$$
$$\sigma = \dfrac{3wl^2}{4bd_1^2}$$

Elevation is formed by a straight line and an ellipse.

VI. Simple Beam Uniformly Loaded

Depth is constant.   Width varies.

$$b = \dfrac{4b_1}{l^2}(lx - x^2)$$
$$\sigma = \dfrac{3}{4}\dfrac{wl^2}{b_1d^2}$$

Plan is two parabolas, with vertices at center of span.

VII. Simple Beam Loaded at Center of Span

Width is constant.   Depth varies.

$$d = d_1\sqrt{2x/l}$$
$$\sigma = \dfrac{3}{2}\dfrac{Pl}{bd_1^2}$$
$$y = \dfrac{1}{2}\dfrac{Pl^3}{Ebd_1^3}$$

Elevation is a parabola with vertices at points of support.

VIII. Simple Beam Loaded at Center of Span

Depth is constant.   Width varies.

$$b = 2b_1x/l$$
$$\sigma = \dfrac{3}{2}\dfrac{Pl}{b_1d^2}$$
$$y = \dfrac{3}{8}\dfrac{Pl^3}{Eb_1d^3}$$

Plan is two triangles with vertices at points of support.

* The sections of the beams near the ends must be increased over the amounts shown to resist the vertical shear expressed by the formula $\sigma = 3/2\ V/A$.

$$V_x = V_2 - w_2x - \Sigma P_2 \tag{22.44}$$

At any section, distant x from the left support, the bending moment is equal to the algebraic sum of the moments of all forces on one side of the section. If M_2 is the moment, pound-inches, at the support to the left,

$$M_x = M_2 + V_2x - \frac{w_2x^2}{2} - \Sigma P_2(x - kl_2) \tag{22.45}$$

Assume that $x = l_2$. Then M_x becomes the moment M_3 at the next support to the right, and the expression may be written

$$V_2l_2 = M_3 - M_2 + \frac{w_2l_2^2}{2} + \Sigma P_2(l_2 - kl_2) \tag{22.46}$$

From Eqs. (22.44), (22.45), and (22.46) it is evident that the bending moment M_x and the shear V_x at any section between two consecutive supports may be determined if the bending moments M_2 and M_3 at those supports are known.

To determine bending moments at the supports an expression known as the *theorem of three moments*

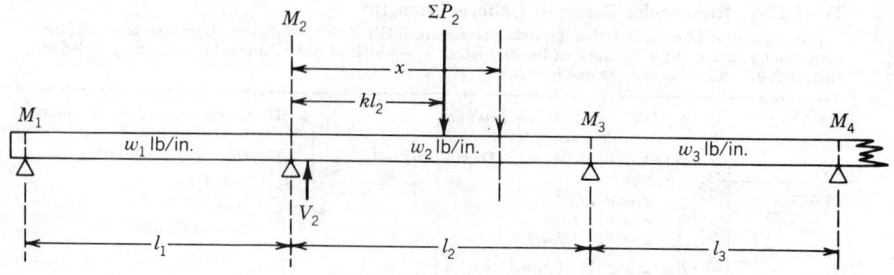

Fig. 22.25 Continuous beam.

is used. This gives the relation between the moments at any three consecutive supports of a beam. For beams with the supports on the same level, and uniformly loaded over each span, the formula is

$$M_1 l_1 + 2M_2(l_1 + l_2) + M_3 l_2 = -\tfrac{1}{4} w_1 l_1^3 - \tfrac{1}{4} w_2 l_2^3 \qquad (22.47)$$

where M_1, M_2, and M_3 = moments of three consecutive supports; l_1 = length between first and second support; l_2 = length between second and third support; w_1 = uniform load per lineal unit over the first span; w_2 = uniform load per lineal unit over the second span. When both spans are of equal length and when the load on each span is the same, $l_1 = l_2$, $w_1 = w_2$, and Eq. (22.47) reduces to

$$M_1 + 4M_2 + M_3 = -\tfrac{1}{2} w l^2 \qquad (22.48)$$

which applies to most cases in practice.

Equations (22.47) and (22.48) are used as follows: For any continuous beam of n spans there are $(n + 1)$ supports. Assuming the ends of the beam to be simply supported without any overhang, the moments at the end supports are zero, and there are, therefore, to be determined $(n - 1)$ moments at the other supports. This may be done by writing $(n - 1)$ equations of the form of Eqs. (22.47) and (22.48) for each support. These equations will contain $(n - 1)$ unknown moments, and their solution will give values of M_1, M_2, M_3, etc., expressed as coefficients of $w l^2$. The shear V_1 at any support may be determined by substituting values of M_1 and M_2 in Eq. (22.46), and the bending moment at any point in any span may be obtained by Eq. (22.45). The shear at any point in any span may be determined from Eq. (22.44).

Figure 22.26 gives values and diagrams for the reactions, shears, and moments at all sections of continuous beams uniformly loaded up to five spans. Note that the reaction at any support is equal to the sum of the shears to the right and to the left of that support.

22.6.4   Axially End-Loaded Beams*

Deflection of Beam Negligible

When a beam is subjected to an axial end load as well as to a transverse bending load, the resultant unit stress developed at any point in the beam is the algebraic sum of the unit stresses produced by each of the loads acting independently of each other. If the beam of cross-sectional area A, square inches, is considered so short that its deflection under load may be neglected, the unit stress produced at any cross section of the beam by an axial load P, pounds, is P/A. The maximum unit stress (tensile or compressive), pounds per square inch, due to the transverse or bending load is given by $\sigma = M/(I/c)$, where M, pound-inches, is the maximum bending moment due to this load; I, inches^4, is the moment of inertia of the cross section with respect to the neutral axis; c, inches, is the distance from the neutral axis to the extreme fiber. If the axial load is compressive, the maximum unit stress is compressive, occurring at the top of the beam, and is

$$\sigma = \frac{P}{A} + \frac{Mc}{I} \qquad (22.49)$$

The unit stress on the bottom fiber may be either tensile or compressive, according as Mc/I is larger or smaller than P/A. If the longitudinal load P were a tensile load, the maximum unit stress would occur on the bottom fiber and would be a tensile stress.

---

* Section 22.6.4 is condensed from Seely, *Resistance of Materials*, 3rd ed., Wiley, New York, 1947, Chap. VII.

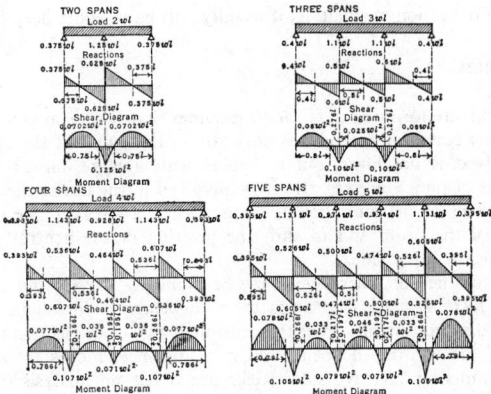

Fig. 22.26 Shear and moment diagrams of continuous beams.

Deflection of Beam Not Negligible

If the deflection of the beam is not negligible, the load P cannot be considered an axial load with respect to any cross sections except the end sections. The longitudinal load then has a moment arm equal to the deflection, and the stress due to this moment should be added algebraically to that caused by the other moments or loads. The stress due to deflection, y, inches (all other symbols as above), is Pyc/I. Thus, for a compressive longitudinal load, the maximum unit stress, pounds per square inch, is at the top of the beam and is

$$\sigma = \frac{P}{A} + (M + Py)\frac{c}{I} \qquad (22.50)$$

Unit tensile stress, pounds per square inch, at the bottom of the beam for the same load is

$$\sigma = -\frac{P}{A} + (M + Py)\frac{c}{I} \qquad (22.51)$$

Since the value of y depends on the total bending moment $(M + Py)$, and the bending moment depends, in turn, on the value of y, Eqs. (22.50) and (22.51) usually are solved by a method of approximation. The value of y that would be caused by the cross-bending moment M, considered acting alone, is found first, and then is used in the expression, $M + Py$. This new value of the total bending moment is used to find a closer approximation to y. This operation may be repeated as many times as desired. When the deflection is small, as in comparatively short beams, the value of y as found from the cross-bending moment above often is used without any further approximation.

Example 22.3

Find the stress at the bottom and at the top of a timber beam 6 in. wide, 8 in. deep, and 12 ft long, supported at each end, and carrying a distributed load of 400 lb/ft, and also an axial compressive load of 10,000 lb. The modulus of elasticity of the timber may be taken as 1,600,000.

Solution.   Maximum unit stress will be at the center of the beam.

$$\frac{P}{A} = \frac{10,000}{48} = 208 \text{ psi}$$

$$\frac{Mc}{I} = \frac{400 \times 12 \times 144}{8} \times \frac{6}{6 \times 8^2} = 1350 \text{ psi}$$

See Table 22.5,

$$y = \frac{5 \times 4800 \times 144^3 \times 12}{384 \times 6 \times 8^3 \times 1,600,000} = 0.456 \text{ in.}$$

$$\frac{Pyc}{I} = \frac{10,000 \times 0.456 \times 6}{6 \times 8^2} = 71 \text{ psi}$$

Total unit compressive stress (top) = 208 + 1350 + 71 = 1629 psi.
Total unit tensile stress (bottom) = −208 + 1350 + 71 = 1213 psi.

Here the stress due to deflection is small, as it usually will be in short deep beams.

22.6.5 Curved Beams

The derivation of the flexure formula, $\sigma = Mc/I$, assumes that the beam is initially straight; therefore, any deviation from this condition introduces an error in the value of the stress. If the curvature is slight, the error involved is not large, but in beams with a large amount of curvature, as hooks, chain links, and frames of punch presses, the error involved in the use of the ordinary flexure formula is considerable. The effect of the curvature is to increase the stress in the inside and to decrease it on the outside fibers of the beam and to shift the position of the neutral axis from the centroidal axis toward the concave or inner side.

The correct value for the unit fiber stress may be found by introducing a correction factor in the flexure formula, $\sigma = K(P/A \pm Mc/I)$; the factor K depends on the shape of the beam and on the ratio R/c, where R = distance, inches, from the centroidal axis of the section to the center of curvature of the central axis of the unstressed beam; and c = distance, inches, of centroidal axis from the extreme fiber on the inner or concave side. Reference 8 has an analysis of curved beams as does Table 22.7, which gives values of K for a number of shapes and ratios of R/c. For slightly different shapes or proportions K may be found by interpolation.

Deflection of Curved and Slender Curved Beams

The deflection of curved beams,[8,9] Fig. 22.27, in the curved portion can be found by

$$U = \int \frac{1}{2} \frac{P^2}{EA} \, ds + \int \frac{\phi V^2}{GA} \, ds + \int \frac{1}{2} \frac{M^2}{EAy_oR} \, ds + \int \frac{MP}{EAR} \, ds \tag{22.52}$$

$$\frac{\partial U}{\partial Q} = \delta_Q \tag{22.53}$$

where Q is a fictitious load of a couple where the deflection or rotation is desired or can be thought of as a 1 lb load or 1 in.-lb couple. y_0 is from Table 22.7, ϕ is a shape factor[2] often taken as 1, and ds is $R \, d\theta$. When $R/c > 4$, the last two terms condense to the integral of $(M^2/2EI) \, ds$. When the length of the curved portion to the depth of the beam is greater than 10, the second term of Eq. (22.52) can be dropped. When in doubt, include all terms.

When beams are not curved (Fig. 22.28), such as some clamps, the following equations (used by permission of McGraw-Hill from the 4th ed. of Ref. 2) are useful:

$$M = M_0 + HR \, [\sin(\theta - x) - s] - VR \, [\cos(\theta - x) - c] + pR^2 \, (1 - u) \tag{22.54}$$

Vertical deflection =

$$\frac{1}{EI} [M_0R^2 (s - \theta c) + VR^3 (\tfrac{1}{2} \, \theta + c^2\theta - \tfrac{3}{2} \, sc)$$

$$+ HR^3 (\tfrac{1}{2} - c + sc\theta + \tfrac{1}{2} \, c^2 - s^2) + pR^4 (s + sc - \tfrac{3}{2} \, \theta c - \tfrac{1}{2} \, s^3 - \tfrac{1}{2} \, c^2s)] \tag{22.55}$$

Horizontal deflection =

$$\frac{1}{EI} [M_0R^2 (1 - \theta s - c) + VR^3 (\tfrac{1}{2} - c + \theta sc + \tfrac{1}{2} \, c^2 - s^2) +$$

$$HR^3 (-2s + \theta s^2 + \tfrac{1}{2} \, \theta + \tfrac{3}{2} \, sc) + pR^4 (1 - \tfrac{3}{2} \, \theta s + s^2 - c)] \tag{22.56}$$

Rotation =

$$\frac{1}{EI} [M_0R \theta + VR^2 (s - \theta c) + HR^2 (1 - \theta s - c) + pR^3 (\theta - s)] \tag{22.57}$$

where $u = \cos x$, $s = \sin \theta$, and $c = \cos \theta$.

22.6.6 Impact Stresses in Bars and Beams

Effect of Sudden Loads

If a sudden load P is applied to a bar, it will cause a deformation el, and the work done by the load will be Pel. Since the external work equals the internal work, $Pel = \sigma^2Al/2E$, and since $e = \sigma/E$. $P = \sigma A/2$, or $\sigma = 2P/A$. The unit stress and also the unit strain are double those obtained by an equal load applied gradually. However, the bar does not maintain equilibrium at the point of maximum stress and strain. After a series of oscillations, however, in which the surplus energy is dissipated in

Table 22.7 Values of Constant K for Curved Beams

Section	$\frac{R}{c}$	Values of K Inside Fiber	Values of K Outside Fiber	$\frac{Y_0*}{R}$	Section	$\frac{R}{c}$	Values of K Inside Fiber	Values of K Outside Fiber	$\frac{Y_0*}{R}$
(circular)	1.2	3.41	.54	.224	(rectangular)	1.2	2.89	.57	.305
	1.4	2.40	.60	.151		1.4	2.13	.63	.204
	1.6	1.96	.65	.108		1.6	1.79	.67	.149
	1.8	1.75	.68	.084		1.8	1.63	.70	.112
	2.0	1.62	.71	.069		2.0	1.52	.73	.090
	3.0	1.33	.79	.030		3.0	1.30	.81	.041
	4.0	1.23	.84	.016		4.0	1.20	.85	.021
	6.0	1.14	.89	.0070		6.0	1.12	.90	.0093
	8.0	1.10	.91	.0039		8.0	1.09	.92	.0052
	10.0	1.08	.93	.0025		10.0	1.07	.94	.0033
(trapezoid)	1.2	3.01	.54	.336	(trapezoid)	1.2	3.09	.56	.336
	1.4	2.18	.60	.229		1.4	2.25	.62	.229
	1.6	1.87	.65	.168		1.6	1.91	.66	.168
	1.8	1.69	.68	.128		1.8	1.73	.70	.128
	2.0	1.58	.71	.102		2.0	1.61	.73	.102
	3.0	1.33	.80	.046		3.0	1.37	.81	.046
	4.0	1.23	.84	.024		4.0	1.26	.86	.024
	6.0	1.13	.88	.011		6.0	1.17	.91	.011
	8.0	1.10	.91	.0060		8.0	1.13	.94	.0060
	10.0	1.08	.93	.0039		10.0	1.11	.95	.0039
(triangle)	1.2	3.14	.52	.352	(triangle)	1.2	3.26	.44	.361
	1.4	2.29	.54	.243		1.4	2.39	.50	.251
	1.6	1.93	.62	.179		1.6	1.99	.54	.186
	1.8	1.74	.65	.138		1.8	1.78	.57	.144
	2.0	1.61	.68	.110		2.0	1.66	.60	.116
	3.0	1.34	.76	.050		3.0	1.37	.70	.052
	4.0	1.24	.82	.028		4.0	1.27	.75	.029
	6.0	1.15	.87	.012		6.0	1.16	.82	.013
	8.0	1.12	.91	.0060		8.0	1.12	.86	.0060
	10.0	1.11	.93	.0039		10.0	1.09	.88	.0039
(T-section)	1.2	3.63	.58	.418	(T-section)	1.2	3.55	.67	.409
	1.4	2.54	.63	.299		1.4	2.48	.72	.292
	1.6	2.14	.67	.229		1.6	2.07	.76	.224
	1.8	1.89	.70	.183		1.8	1.83	.78	.178
	2.0	1.73	.72	.149		2.0	1.69	.80	.144
	3.0	1.41	.79	.069		3.0	1.38	.86	.067
	4.0	1.29	.83	.040		4.0	1.26	.89	.038
	6.0	1.18	.88	.018		6.0	1.15	.92	.018
	8.0	1.13	.91	.010		8.0	1.10	.94	.010
	10.0	1.10	.92	.0065		10.0	1.08	.95	.0065
(channel)	1.2	2.52	.67	.408	(I-section)	1.2	2.37	.73	.453
	1.4	1.90	.71	.285		1.4	1.79	.77	.319
	1.6	1.63	.75	.208		1.6	1.56	.79	.236
	1.8	1.50	.77	.160		1.8	1.44	.81	.183
	2.0	1.41	.79	.127		2.0	1.36	.83	.147
	3.0	1.23	.86	.058		3.0	1.19	.88	.067
	4.0	1.16	.89	.030		4.0	1.13	.91	.036
	6.0	1.10	.92	.013		6.0	1.08	.94	.016
	8.0	1.07	.94	.0076		8.0	1.06	.95	.0089
	10.0	1.05	.95	.0048		10.0	1.05	.96	.0057
(hollow circle)	1.2	3.28	.58	.269	(hollow box)	1.2	2.63	.68	.399
	1.4	2.31	.64	.182		1.4	1.97	.73	.280
	1.6	1.89	.68	.134		1.6	1.66	.76	.205
	1.8	1.70	.71	.104		1.8	1.51	.78	.159
	2.0	1.57	.73	.083		2.0	1.43	.80	.127
	3.0	1.31	.81	.038		3.0	1.23	.86	.058
	4.0	1.21	.85	.020		4.0	1.15	.89	.031
	6.0	1.13	.90	.0087		6.0	1.09	.92	.014
	8.0	1.10	.92	.0049		8.0	1.07	.94	.0076
	10.0	1.07	.93	.0031		10.0	1.06	.95	.0048

* Y_0 is distance from centroidal axis to neutral axis, where beam is subjected to pure bending.

damping, the bar finally comes to rest with the same strain and stress as that due to the equal static load.

Stress due to Live Loads

In structural design two loads are considered, the dead load or weight of the structure and the live load or superimposed loads to be carried. The stresses due to the dead load and to the live load are

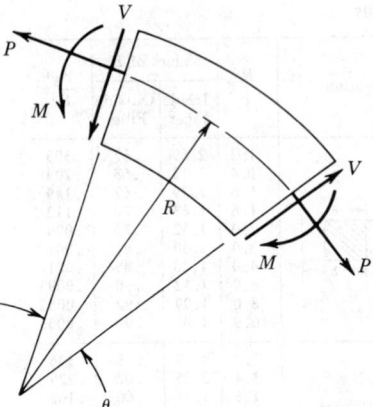

Fig. 22.27 Positive sign convention for curved beams.

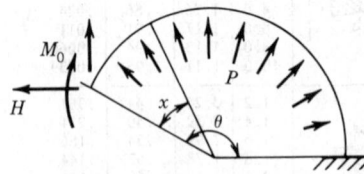

Fig. 22.28 Circular cantilever with end loading and uniform radial pressure p lb/linear in.

computed separately, each being regarded as a static load. It is obvious that the stress due to the live load may be greatly increased, depending on the suddenness with which the load is applied. It has been shown above that the stress due to a suddenly applied load is double the stress caused by a static load. The term *coefficient of impact* is used extensively in structural engineering to denote the number by which the computed static stress is multiplied to obtain the value of the increased stress assumed to be caused by the suddenness of the application of the live load. If σ = static unit stress computed from the live load, and i = coefficient of impact, then the increase of unit stress due to sudden loading is $i\sigma$, and the total unit stress due to live load is $\sigma + i\sigma$. The value of i has been determined by empirical methods and varies according to different conditions.

In the building codes of most cities, specified floor loadings for buildings include the impact allowance, and no increase is needed for live loads except for special cases of vibration or other unusual conditions. For railroad bridges, the value of i depends upon the proportion of the length of the bridge which is loaded. No increase in the static stress is needed when the mass of the structure, as in monolithic concrete, is great. For machinery and for unusual conditions, such as elevator machinery and its supports, each structure should be considered by itself and the coefficient assumed accordingly. It should be noted that the meaning of the word impact used above differs somewhat from its strict theoretical meaning and as it is used in the next paragraph. The use of the terms impact and coefficient of impact in connection with live load stresses is, however, very general.

Axial Impact on Bars

A load P dropped from a height h onto the end of a vertical bar of cross-sectional area A, rigidly secured at the bottom end, produces in the bar a unit stress which increases from 0 up to σ', with a corresponding total strain increasing from 0 up to e_1. The work done on the bar is $P(h + e_1)$, which, provided no energy is expended in hysteresis losses or in giving velocity to the bar, is equal to the energy $\frac{1}{2}\,\sigma' A e_1$ stored in the bar; that is,

$$P(h + e_1) = \frac{1}{2}\,\sigma' A e_1 \tag{22.58}$$

If e = strain produced by a static load P, within the proportional limit

$$\frac{e}{e_1} = \frac{P/A}{\sigma'} \tag{22.59}$$

Combining this with Eq. (22.58) gives

$$\sigma' = \sigma + \sigma \sqrt{1 + 2\frac{h}{e}} \tag{22.60}$$

$$e_1 = e + e \sqrt{1 + 2\frac{h}{e}} \tag{22.61}$$

A wrought-iron bar 1 in. square and 5 ft long under a static load of 5000 lb will be shortened about 0.012 in., assuming no lateral flexure to occur; but, if a weight of 5000 lb drops on its end from a height of 0.048 in., a stress of 20,000 lb will be produced.

Equations (22.60) and (22.61) give values of stress and strain that are somewhat high because part of the energy of the applied force is not effective in producing stress, but is expended in overcoming the inertia of the bar and in producing local stresses. For light bars they give approximately correct results.

If the bar is horizontal and is struck at one end by a weight P, moving with a velocity V, the strain produced is e_1. Then, as before, $\frac{1}{2}\,\sigma'Ae_1 = Ph$. In this case $h = V^2/2g =$ height from which P would have to fall to acquire velocity V ($g =$ acceleration due to gravity $= 32.16$ ft/sec²). Combining with Eq. (22.59),

$$\sigma' = \sigma \sqrt{2\frac{h}{e}} \tag{22.62}$$

$$e_1 = e \sqrt{2\frac{h}{e}} \tag{22.63}$$

Impact on Beams

If a weight P falls on a horizontal beam from a height h, producing a maximum deflection y and a maximum unit stress σ' in the extreme fiber, the values of σ' and y are given by

$$\sigma' = \sigma + \sigma \sqrt{1 + 2\frac{h}{y}} \tag{22.64}$$

$$y_1 = y + y \sqrt{1 + 2\frac{h}{y}} \tag{22.65}$$

where $\sigma =$ extreme fiber unit stress and $y =$ deflection due to P, considered as a static load. The value of σ may be obtained from the flexure formula [Eq. (22.37)]; that of y from the proper formula for deflection under static load.

If a weight P moving horizontally with a velocity V strikes a beam (the ends of which are secured against horizontal movement), the maximum fiber unit stress and the maximum lateral deflection are given by

$$\sigma' = \sigma \sqrt{2\frac{h}{y}} \tag{22.66}$$

$$y_1 = y \sqrt{2\frac{h}{y}} \tag{22.67}$$

where σ and y are as before and h is height through which P would have to fall to acquire the velocity V. These formulas, like those for axial impact on bars, give results higher than those observed in tests, particularly if the weight of the beam is great. For further discussion, see Ref. 7.

Rupture from Impact

Rupture may be caused by impact provided the load has the requisite velocity. The above formulas, however, do not apply since they are valid only for stresses within the proportional limit. It has been found that the dynamic properties of a material are dependent on volume, velocity of the applied load, and material condition. If the velocity of the applied load is kept within certain limiting values, the total energy values for static and dynamic conditions are identical. If the velocity is increased, the impact values are considerably reduced. For further information see Ref. 10.

22.6.7 Steady and Impulsive Vibratory Stresses

For steady vibratory stresses of a weight, W, supported by a beam or rod, the deflection of the bar, or beam, will be increased by the dynamic magnification factor. The relation is given by

$$\delta_{\text{dynamic}} = \delta_{\text{static}} \times \text{dynamic magnification factor}$$

An example of the calculating procedure for the case of no damping losses is

$$\delta_{\text{dynamic}} = \delta_{\text{static}} \times \frac{1}{1 - (\omega/\omega_n)^2} \tag{22.68}$$

where ω is the frequency of oscillation of the load and ω_n is the natural frequency of oscillation of a weight on the bar.

For the same beam excited by a single sine pulse of magnitude A in./sec² and a sec duration, then for $t < a$ a good approximation is

$$\sigma_{\text{dynamic}} = \frac{\delta_{\text{static}}(A/g)}{1 - \left(\dfrac{\omega}{4\pi\omega_n}\right)^2} \left[\sin \omega t - \frac{1}{4\pi^2}\left(\frac{\omega}{\omega_n}\right) \sin \omega_n t \right] \tag{22.69}$$

where A/g is the number of g's and ω is π/a.

22.7 SHAFTS, BENDING, AND TORSION

22.7.1 Definitions

TORSIONAL STRESS. A bar is under torsional stress when it is held fast at one end, and a force acts at the other end to twist the bar. In a round bar (Fig. 22.29) with a constant force acting, the straight line ab becomes the helix ad, and a radial line in the cross section, ob, moves to the position od. The angle bad remains constant while the angle bod increases with the length of the bar. Each cross section of the bar tends to shear off the one adjacent to it, and in any cross section the shearing stress at any point is normal to a radial line drawn through the point. Within the shearing proportional limit, a radial line of the cross section remains straight after the twisting force has been applied, and the unit shearing stress at any point is proportional to its distance from the axis.

TWISTING MOMENT, T, is equal to the product of the resultant, P, of the twisting forces, multiplied by its distance from the axis, p.

RESISTING MOMENT, T_r, in torsion, is equal to the sum of the moments of the unit shearing stresses acting along a cross section with respect to the axis of the bar. If dA is an elementary area of the section at a distance of z units from the axis of a circular shaft (Fig. 22.29b), and c is the distance from the axis to the outside of the cross section where the unit shearing stress is τ, then the unit shearing stress acting on dA is $(\tau z/c)\ dA$, its moment with respect to the axis is $(\tau z^2/c)\ dA$, and the sum of all the moments of the unit shearing stresses on the cross section is $\int (\tau z^2/c)\ dA$. In this expression the factor $\int z^2\ dA$ is the polar moment of inertia of the section with respect to the axis. Denoting this by J, the resisting moment may be written $\tau J/c$.

THE POLAR MOMENT OF INERTIA   of a surface about an axis through its center of gravity and perpendicular to the surface is the sum of the products obtained by multiplying each elementary area by the square of its distance from the center of gravity of its surface; it is equal to the sum of the moments of inertia taken with respect to two axes in the plane of the surface at right angles to each other passing through the center of gravity. It is represented by J, inches⁴. For the cross section of a round shaft,

$$J = \tfrac{1}{32}\pi d^4 \quad \text{or} \quad \tfrac{1}{2}\pi r^4 \tag{22.70}$$

For a hollow shaft,

$$J = \tfrac{1}{32}\pi(d^4 - d_1^4) \tag{22.71}$$

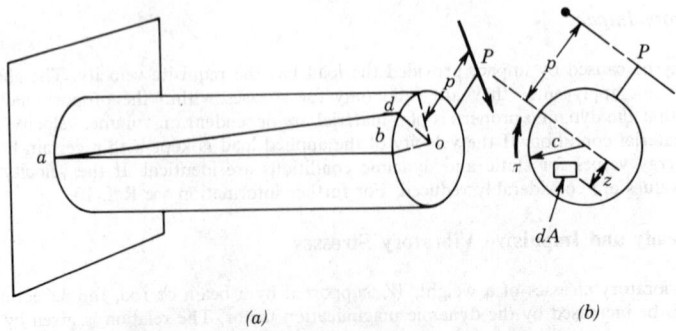

(a)   *(b)*

Fig. 22.29 Round bar subject to torsional stress.

where d is the outside and d_1 is the inside diameter, inches, or

$$J = \tfrac{1}{2}\,\pi(r^4 - r_1^4) \tag{22.72}$$

where r is the outside and r_1 the inside radius, inches.

THE POLAR RADIUS OF GYRATION, k_p, sometimes is used in formulas; it is defined as the radius of a circumference along which the entire area of a surface might be concentrated and have the same polar moment of inertia as the distributed area. For a solid circular section,

$$k_p^2 = \tfrac{1}{8}\,d^2 \tag{22.73}$$

For a hollow circular section,

$$k_p^2 = \tfrac{1}{8}\,(d^2 - d_1^2) \tag{22.74}$$

22.7.2 Determination of Torsional Stresses in Shafts

Torsion Formula for Round Shafts

The conditions of equilibrium require that the twisting moment, T, be opposed by an equal resisting moment, T_r, so that for the values of the maximum unit shearing stress, τ, within the proportional limit, the torsion formula for round shafts becomes

$$T_r = T = \tau\frac{J}{c} \tag{22.75}$$

If τ is in pounds per square inch, then T_r and T must be in pound-inches, J is in inches⁴, and c is in inches. For solid round shafts having a diameter, d, inches,

$$J = \tfrac{1}{32}\,\pi d^4 \quad\text{and}\quad c = \tfrac{1}{2}\,d \tag{22.76}$$

and

$$T = \tfrac{1}{16}\,\pi d^3\tau \quad\text{or}\quad \tau = \frac{16T}{\pi d^3} \tag{22.77}$$

For hollow round shafts,

$$J = \frac{\pi(d^4 - d_1^4)}{32} \quad\text{and}\quad c = \tfrac{1}{2}\,d \tag{22.78}$$

and the formula becomes

$$T = \frac{\tau\pi(d^4 - d_1^4)}{16d} \quad\text{or}\quad \tau = \frac{16Td}{\pi(d^4 - d_1^4)} \tag{22.79}$$

The torsion formula applies only to solid circular shafts or hollow circular shafts, and then only when the load is applied in a plane perpendicular to the axis of the shaft and when the shearing proportional limit of the material is not exceeded.

Shearing Stress in Terms of Horsepower

If the shaft is to be used for the transmission of power, the value of T, pound-inches, in the above formulas becomes $63,030H/N$, where H = horsepower to be transmitted and N = revolutions per minute. The maximum unit shearing stress, pounds per square inch, then is

$$\text{For solid round shafts:}\quad \tau = \frac{321,000H}{Nd^3} \tag{22.80}$$

$$\text{For hollow round shafts:}\quad \tau = \frac{321,000Hd}{N(d^4 - d_1^4)} \tag{22.81}$$

If τ is taken as the allowable unit shearing stress, the diameter, d, inches, necessary to transmit a given horsepower at a given shaft speed can then be determined. These formulas give the stress due to torsion only, and allowance must be made for any other loads, as the weight of shaft and pulleys, and tension in belts.

Angle of Twist

When the unit shearing stress τ does not exceed the proportional limit, the angle bod (Fig. 22.29) for a solid round shaft may be computed from the formula

$$\theta = \frac{Tl}{GJ} \tag{22.82}$$

where θ = angle in radians; l = length of shaft in inches; G = shearing modulus of elasticity of the material; T = twisting moment, pound-inches. Values of G for different materials are steel, 12,000,000; wrought iron, 10,000,000; and cast iron, 6,000,000.

When the angle of twist on a section begins to increase in a greater ratio than the twisting moment, it may be assumed that the shearing stress on the outside of the section has reached the proportional limit. The shearing stress at this point may be determined by substituting the twisting moment at this instant in the torsion formula.

Torsion of Noncircular Cross Sections

The analysis of shearing stress distribution along noncircular cross sections of bars under torsion is complex. By drawing two lines at right angles through the center of gravity of a section before twisting, and observing the angular distortion after twisting, it has been found from many experiments that in noncircular sections the shearing unit stresses are not proportional to their distances from the axis. Thus in a rectangular bar there is no shearing stress at the corners of the sections, and the stress at the middle of the wide side is greater than at the middle of the narrow side. In an elliptical bar the shearing stress is greater along the flat side than at the round side.

It has been found by tests (Refs. 5 and 11) as well as by mathematical analysis that the torsional resistance of a section, made up of a number of rectangular parts, is approximately equal to the sum of the resistances of the separate parts. It is on this basis that nearly all the formulas for noncircular sections have been developed. For example, the torsional resistance of an I-beam is approximately equal to the sum of the torsional resistances of the web and the outstanding flanges. In an I-beam in torsion the maximum shearing stress will occur at the middle of the side of the web, except where the flanges are thicker than the web, and then the maximum stress will be at the midpoint of the width of the flange. Reentrant angles, as those in I-beams and channels, are always a source of weakness in members subjected to torsion. Table 22.8[2] gives values of the maximum unit shearing stress τ and the angle of twist θ induced by twisting bars of various cross sections, it being assumed that τ is not greater than the proportional limit.

Torsion of thin-wall closed sections, Fig. 22.30,

$$T = 2qA \tag{22.83}$$

$$q = \tau t \tag{22.84}$$

$$\theta_i = \frac{\theta}{L} = \frac{T}{2A} \frac{1}{2AG} \frac{S}{t} = \frac{T}{GJ} \tag{22.85}$$

where S is the arc length around area A over which τ acts for a thin-wall section; shear buckling should be checked. When more than one cell is used[1,12] or if section is not constructed of a single material,[12] the calculations become more involved:

$$J = \frac{4A^2}{\oint ds/t} \tag{22.86}$$

Ultimate Strength in Torsion

In a torsion failure, the outer fibers of a section are the first to shear, and the rupture extends toward the axis as the twisting is continued. The torsion formula for round shafts has no theoretical basis after the shearing stresses on the outer fibers exceed the proportional limit, as the stresses along the section then are no longer proportional to their distances from the axis. It is convenient, however, to compare the torsional strength of various materials by using the formula to compute values of τ at which rupture takes place. These computed values of the maximum stress sustained before rupture are somewhat higher for iron and steel than the ultimate strength of the materials in direct shear. Computed values of the ultimate strength in torsion are found by experiment to be: cast iron, 30,000 psi; wrought iron, 55,000 psi; medium steel, 65,000 psi; timber, 2000 psi. These computed values of twisting strength may be used in the torsion formula to determine the probable twisting moment that will cause rupture of a given round bar or to determine the size of a bar that will be ruptured by a given twisting moment. In design, large factors of safety should be taken, especially when the stress is reversed as in reversing engines and when the torsional stress is combined with other stresses as in shafting.

22.7.3 Bending and Torsional Stresses

The stress for combined bending and torsion can be found from Eqs. (22.20), shear theory, and (22.22), distortion energy, with $\sigma_y = 0$:

$$\frac{\sigma_w}{2} = \sqrt{\left(\frac{Mc}{2I}\right)^2 + \left(\frac{Tr}{J}\right)^2} \tag{22.87}$$

For solid round rods, this equation reduces to

$$\frac{\sigma_w}{2} = \frac{16}{\pi d^3}\sqrt{M^2 + T^2} \tag{22.88}$$

Table 22.8   Formulas for Torsional Deformation and Stress

General formulas: $\theta = \dfrac{TL}{KG}$, $\tau = \dfrac{T}{Q}$, where $\theta =$ angle of twist, radians; $T =$ twisting moment, in.-lb; $L =$ length, in.; $\tau =$ unit shear stress, psi; $G =$ modulus of rigidity, psi; K, in.^{4}; and Q, in.^{3} are functions of the cross section.

Shape	Formula for K in $\theta = \dfrac{TL}{KG}$	Formula for Shear Stress
	$K = \dfrac{\pi d^4}{32}$	$\tau = \dfrac{16T}{\pi d^3}$
	$K = 1/32\,\pi(d^4 - d_1{}^4)$	$\tau = \dfrac{16Td}{\pi(d^4 - d_1{}^4)}$
	$K = 2/3\,\pi r t^3$	$\tau = \dfrac{3T}{2\pi r t^2}$
	$K = \dfrac{\pi a^3 b^3}{a^2 + b^2}$	$\tau = \dfrac{2T}{\pi a b^2}$
	$K = \dfrac{\pi a_1{}^3 b_1{}^3}{a_1{}^2 + b_1{}^2}[(1 + q)^4 - 1]$ $q = \dfrac{a - a_1}{a_1}$ $q = \dfrac{b - b_1}{b_1}$	$\tau = \dfrac{2T}{\pi a_1 b_1{}^2[(1 + q)^4 - 1]}$
	$K = \dfrac{b^4\sqrt{3}}{80}$	$\tau = \dfrac{20T}{b^3}$
	$K = 2.69b^4$	$\tau = \dfrac{1.09T}{b^3}$
	$K = \dfrac{ab^3}{16}\left[\dfrac{16}{3} - 3.36\dfrac{b}{a}\left(1 - \dfrac{b^4}{12a^4}\right)\right]$	$\tau = \dfrac{(3a + 1.8b)T}{a^2 b^2}$
	$K = \dfrac{2t_1 t_2(a - t_2)^2(b - t_1)^2}{at_2 + bt_1 - t_2{}^2 - t_1{}^2}$	$\tau = \dfrac{T}{2t_2(a - t_2)(b - t_1)}$
	$K = 0.1406b^4$	$\tau = \dfrac{4.8T}{b^3}$

Table 22.8 (*Continued*)

General formulas: $\theta = \dfrac{TL}{KG}$, $\tau = \dfrac{T}{Q}$, where θ = angle of twist, radians; T = twisting moment, in.-lb; L = length, in.; τ = unit shear stress, psi; G = modulus of rigidity, psi; K, in.^{4}; and Q, in.^{3} are functions of the cross section.

Shape	Formula for K in $\theta = \dfrac{TL}{KG}$	Formula for Shear Stress
	r = fillet radius D = diameter largest inscribed circle $K = 2K_1 + K_2 + 2\alpha D^4$ $K_1 = ab^3\left[\dfrac{1}{3} - 0.21\dfrac{b}{a}\left(1 - \dfrac{b^4}{12a^4}\right)\right]$ $K_2 = cd^3\left[\dfrac{1}{3} - 0.105\dfrac{d}{c}\left(1 - \dfrac{d^4}{192c^4}\right)\right]$ $\alpha = \dfrac{b}{d}\left(0.07 + 0.076\dfrac{r}{b}\right)$	For all solid sections of irregular form the maximum shear stress occurs at or very near one of the points where the largest inscribed circle touches the boundary, and of these, at the one where the curvature of the boundary is algebraically least. (Convexity represents positive, concavity negative, curvature of the boundary.) At a point where the curvature is positive (boundary of section straight or convex) this maximum stress is given approximately by:
	$K = 2K_1 + K_2 + 2\alpha D^4$ $K_1 = ab^3\left[\dfrac{1}{3} - 0.21\dfrac{b}{a}\left(1 - \dfrac{b^4}{12a^4}\right)\right]$ $K_2 = 1/3\,cd^3$ $\alpha = \dfrac{t}{t_1}\left(0.15 + 0.1\dfrac{r}{b}\right)$ $t = b$ if $b < d$ $t = d$ if $d < b$ $t_1 = b$ if $b > d$ $t_1 = d$ if $d > b$	$\tau = G\dfrac{\theta}{L}c$ or $\tau = \dfrac{T}{K}c$ where $c = \dfrac{D}{1 + \dfrac{\pi^2 D^4}{16A^2}} \times$ $\left[1 + 0.15\left(\dfrac{\pi^2 D^4}{16A^2} - \dfrac{D}{2r}\right)\right]$
	$K = K_1 + K_2 + \alpha D^4$ $K_1 = ab^3\left[\dfrac{1}{3} - 0.21\dfrac{b}{a}\left(1 - \dfrac{b^4}{12a^4}\right)\right]$ $K_2 = cd^3\left[\dfrac{1}{3} - 0.105\dfrac{d}{c}\left(1 - \dfrac{d^4}{192c^4}\right)\right]$ $\alpha = \dfrac{b}{d}\left(0.07 + 0.076\dfrac{r}{b}\right)$	where D = diameter of largest inscribed circle, r = radius of curvature of boundary at the point (positive for this case), A = area of the section.

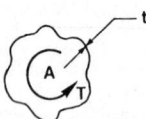

Fig. 22.30   Thin-walled tube.

From distortion energy

$$\sigma = \sqrt{\left(\frac{Mc}{I}\right)^2 + 3\left(\frac{Tr}{J}\right)^2} \tag{22.89}$$

For solid round rods, the equation yields

$$\sigma = \frac{32}{\pi d^3}\sqrt{M^2 + \tfrac{3}{4}\,T^2} \tag{22.90}$$

22.8   COLUMNS

22.8.1   Definitions

A COLUMN OR STRUT is a bar or structural member under axial compression, which has an unbraced length greater than about eight or ten times the least dimension of its cross section. On account of its length, it is impossible to hold a column in a straight line under a load; a slight sidewise bending always occurs, causing flexural stresses in addition to the compressive stresses induced directly by the load. The lateral deflection will be in a direction perpendicular to that axis of the cross section about which the moment of inertia is the least. Thus in Fig. 22.31a the column will bend in a direction perpendicular to *aa*, in Fig. 22.31b it will bend perpendicular to *aa* or *bb*, and in Fig. 22.31c it is likely to bend in any direction.

RADIUS OF GYRATION of a section with respect to a given axis is equal to the square root of the quotient of the moment of inertia with respect to that axis, divided by the area of the section, that is

$$k = \sqrt{\frac{I}{A}}; \qquad \frac{I}{A} = k^2 \qquad (22.91)$$

where I is the moment of inertia and A is the sectional area. Unless otherwise mentioned, an axis through the center of gravity of the section is the axis considered. As in beams, the moment of inertia is an important factor in the ability of the column to resist bending, but for purposes of computation it is more convenient to use the radius of gyration.

LENGTH OF A COLUMN is the distance between points unsupported against lateral deflection.

SLENDERNESS RATIO is the length l divided by the least radius of gyration k, both in inches. For steel, a *short column* is one in which $l/k < 20$ or 30, and its failure under load is due mainly to direct compression; in a *medium-length column*, $l/k =$ about 30–175, failure is by a combination of direct compression and bending; in a *long column*, $l/k >$ about 175–200, failure is mainly by bending. For timber columns these ratios are about 0–30, 30–90, and above 90 respectively. The load which will cause a column to fail decreases as l/k increases. The above ratios apply to round-end columns. If the ends are fixed (see below), the effective slenderness ratio is one-half that for round-end columns, as the distance between the points of inflection is one-half of the total length of the column. For flat ends it is intermediate between the two.

CONDITIONS OF ENDS. The various conditions which may exist at the ends of columns usually are divided into four classes: (1) Columns with round ends; the bearing at either end has perfect freedom of motion, as there would be with a ball-and-socket joint at each end. (2) Columns with hinged ends; they have perfect freedom of motion at the ends in one plane, as in compression members in bridge trusses where loads are transmitted through end pins. (3) Columns with flat ends; the bearing surface is normal to the axis of the column and of sufficient area to give at least partial fixity to the ends of the columns against lateral deflection. (4) Columns with fixed ends; the ends are rigidly secured, so that under any load the tangent to the elastic curve at the ends will be parallel to the axis in its original position.

Experiments prove that columns with fixed ends are stronger than columns with flat, hinged, or round ends, and that columns with round ends are weaker than any of the other types. Columns with hinged ends are equivalent to those with round ends in the plane in which they have free movement; columns with flat ends have a value intermediate between those with fixed ends and those with round ends. It often happens that columns have one end fixed and one end hinged, or some other combination. Their relative values may be taken as intermediate between those represented by the condition at either end. The extent to which strength is increased by fixing the ends depends on the length of column, fixed ends having a greater effect on long columns than on short ones.

22.8.2 Theory

There is no exact theoretical formula that gives the strength of a column of any length under an axial load. Formulas involving the use of empirical coefficients have been deduced, however, and they give results that are consistent with the results of tests.

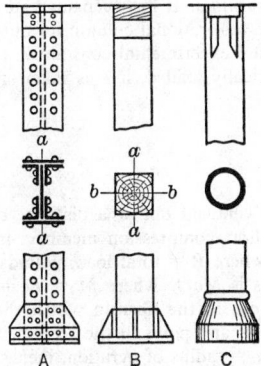

A   B   C   **Fig. 22.31** Column end designs.

Euler's Formula

Euler's formula assumes that the failure of a column is due solely to the stresses induced by sidewise bending. This assumption is not true for short columns, which fail mainly by direct compression, nor is it true for columns of medium length. The failure in such cases is by a combination of direct compression and bending. For columns in which $l/k > 200$, Euler's formula is approximately correct and agrees closely with the results of tests.

Let P = axial load, pounds; l = length of column, inches; I = least moment of inertia, inches^4; k = least radius of gyration, inches; E = modulus of elasticity; y = lateral deflection, inches, at any point along the column, that is caused by load P. If a column has round ends, so that the bending is not restrained, the equation of its elastic curve is

$$EI \frac{d^2y}{dx^2} = -Py \tag{22.92}$$

when the origin of the coordinate axes is at the top of the column, the positive direction of x being taken downward and the positive direction of y in the direction of the deflection. Integrating the above expression twice and determining the constants of integration give

$$P = \Omega \pi^2 \frac{EI}{l^2} \tag{22.93}$$

which is Euler's formula for long columns. The factor Ω is a constant depending on the condition of the ends. For round ends $\Omega = 1$; for fixed ends $\Omega = 4$; for one end round and the other fixed $\Omega = 2.05$. P is the load at which, if a slight deflection is produced, the column will not return to its original position. If P is decreased, the column will approach its original position, but if P is increased, the deflection will increase until the column fails by bending.

For columns with value of l/k less than about 150, Euler's formula gives results distinctly higher than those observed in tests. Euler's formula is now little used except for long members and as a basis for the analysis of the stresses in some types of structural and machine parts. It always gives an *ultimate* and never an allowable load.

Secant Formula

The deflection of the column is used in the derivation of the Euler formula, but if the load were truly axial it would be impossible to compute the deflection. If the column is assumed to have an initial eccentricity of load of e in. (see Ref. 7, for suggested values of e), the equation for the deflection y becomes

$$y_{max} = e \left(\sec \frac{l}{2} \sqrt{\frac{P}{EI}} - 1 \right) \tag{22.94}$$

The maximum unit compressive stress becomes

$$\sigma = \frac{P}{A} \left(1 + \frac{ec}{k^2} \sec \frac{l}{2} \sqrt{\frac{P}{EI}} \right) \tag{22.95}$$

where l = length of column, inches; P = total load, pounds; A = area, square inches; I = moment of inertia, inches^4; k = radius of gyration, inches; c = distance from neutral axis to the most compressed fiber, inches; E = modulus of elasticity; both I and k are taken with respect to the axis about which bending takes place. The ASCE indicates $ec/k^2 = 0.25$ for central loading. Because the formula contains the secant of the angle $(l/2) \sqrt{P/EI}$, it is sometimes called the *secant formula*. It has been suggested by the Committee on Steel-Column Research (Refs. 13 and 14) that the best rational column formula can be constructed on the secant type, although of course it must contain experimental constants.

The secant formula can be used also for columns that are eccentrically loaded, if e is taken as the actual eccentricity plus the assumed initial eccentricity.

Eccentric Loads on Short Compression Members

Where a direct push acting on a member does not pass through the centroid but at a distance e, inches, from it, both direct and bending stresses are produced. For short compression members in which column action may be neglected, the direct unit stress is P/A, where P = total load, pounds, and A = area of cross section, square inches. The bending unit stress is Mc/I, where $M = Pe$ = bending moment, pound-inches; c = distance, inches, from the centroid to the fiber in which the stress is desired; I = moment of inertia, inches^4. The total unit stress at any point in the section is $\sigma = P/A + Pec/I$, or $\sigma = (P/A)(1 + ec/k^2)$, since $I = Ak^2$, where k = radius of gyration, inches.

Eccentric Loads on Columns

Various column formulas must be modified when the loads are not balanced, that is, when the resultant of the loads is not in line with the axis of the column. If P = load, pounds, applied at a distance e in. from the axis, bending moment $M = Pe$. Maximum unit stress σ, pounds per square inch, due to this bending moment alone, is $\sigma = Mc/I = Pec/Ak^2$, where c = distance, inches, from the axis to the most remote fiber on the concave side; A = sectional area in square inches; k = radius of gyration in the direction of the bending, inches. This unit stress must be added to the unit stress that would be induced if the resultant load were applied in line with the axis of the column.

The secant formula, Eq. (22.95) also can be used for columns that are eccentrically loaded if e is taken as the actual eccentricity plus the assumed initial eccentricity.

Column Subjected to Transverse or Cross-Bending Loads

A compression member that is subjected to cross-bending loads may be considered to be (1) a beam subjected to end thrust as discussed (Eq. 22.51) or (2) a column subjected to cross-bending loads, depending on the relative magnitude of the end thrust and cross-bending loads, and on the dimensions of the member. The various column formulas may be modified so as to include the effect of cross-bending loads. In this form the modified secant formula for transverse loads is

$$\sigma = \frac{P}{A}\left[1 + (e+y)\frac{c}{k^2}\sec\frac{l}{2k}\sqrt{\frac{P}{AE}}\right] + \frac{Mc}{Ak^2} \tag{22.96}$$

In the formula, σ = maximum unit stress on concave side, pounds per square inch; P = axial end load, pounds; A = cross-sectional area, square inches; M = moment due to cross-bending load, pound-inches; y = deflection due to cross-bending load, inches; k = radius of gyration, inches; l = length of column, inches; e = assumed initial eccentricity, inches; c = distance, inches, from axis to the most remote fiber on the concave side.

22.8.3 Wooden Columns

Wooden Column Formulas

One of the principal formulas is that formerly used by the AREA, $P/A = \sigma_1(1 - l/60d)$, where P/A = allowable unit load, pounds per square inch; σ_1 = allowable unit stress in direct compression on short blocks, pounds per square inch; l = length, inches; d = least dimension, inches. This formula is being replaced rapidly by formulas recommended by the ASTM and AREA. Committees of these societies, working with the U.S. Forest Products Laboratory, classified timber columns in three groups (ASTM Standards, 1937, D245-37):

1. *Short Columns.* The ratio of unsupported length to least dimension does not exceed 11. For these columns, the allowable unit stress should not be greater than the values given in Table 22.9 under compression parallel to the grain.

2. *Intermediate-Length Columns.* Where the ratio of unsupported length to least dimension is greater than 10, Eq. (22.97), of the fourth power parabolic type, shall be used to determine allowable unit stress, until this allowable unit stress is equal to two-thirds of the allowable unit stress for short columns.

$$\frac{P}{A} = \sigma_1\left[1 - \frac{1}{3}\left(\frac{l}{Kd}\right)^4\right] \tag{22.97}$$

where P = total load, pounds; A = area, square inches; σ_1 = allowable unit compressive stress parallel to grain, pounds per square inch (see Table 22.9); l = unsupported length, inches; d = least dimension, inches; $K = l/d$ at the point of tangency of the parabolic and Euler curves, at which $P/A = \frac{2}{3}\sigma_1$. The value of K for any species and grade is $\pi/2\sqrt{E/6\sigma_1}$, where E = modulus of elasticity.

3. *Long Columns.* Where P/A as computed by Eq. (22.97) is less than $\frac{2}{3}\sigma_1$, Eq. (22.98) of the Euler type, which includes a factor of safety of 3, shall be used:

$$\frac{P}{A} = \frac{1}{36}\left[\frac{\pi^2 E}{(l/d)^2}\right] \tag{22.98}$$

Timber columns should be limited to a ratio of l/d equal to 50. No higher loads are allowed for square-ended columns. The strength of round columns may be considered the same as that of square columns of the same cross-sectional area.

Table 22.9   Basic Stresses for Clear Material*

Species	Extreme Fiber in Bending or Tension Parallel to Grain	Maximum Horizontal Shear	Compression Perpendicular to Grain	Compression Parallel to Grain $L/d = 11$ or Less	Modulus of Elasticity in Bending
SOFTWOODS					
Baldcypress (Southern cypress)	1900	150	300	1450	1,200,000
Cedars					
Redcedar, Western	1300	120	200	950	1,000,000
White-cedar, Atlantic (Southern white-cedar) and northern	1100	100	180	750	800,000
White-cedar, Port Orford	1600	130	250	1200	1,500,000
Yellow-cedar, Alaska (Alaska cedar)	1600	130	250	1050	1,200,000
Douglas-fir, coast region	2200	130	320	1450	1,600,000
Douglas-fir, coast region, close-grained	2350	130	340	1550	1,600,000
Douglas-fir, Rocky Mountain region	1600	120	280	1050	1,200,000
Douglas-fir, dense, all regions	2550	150	380	1700	1,600,000
Fir, California red, grand, noble, and white	1600	100	300	950	1,100,000
Fir, balsam	1300	100	150	950	1,000,000
Hemlock, Eastern	1600	100	300	950	1,100,000
Hemlock, Western (West Coast hemlock)	1900	110	300	1200	1,400,000
Larch, Western	2200	130	320	1450	1,500,000
Pine, Eastern white (Northern white), ponderosa, sugar, and Western white (Idaho white)	1300	120	250	1000	1,000,000
Pine, jack	1600	120	220	1050	1,100,000
Pine, lodgepole	1300	90	220	950	1,000,000
Pine, red (Norway pine)	1600	120	220	1050	1,200,000
Pine, southern yellow	2200	160	320	1450	1,600,000
Pine, southern yellow, dense	2550	190	380	1700	1,600,000
Redwood	1750	100	250	1350	1,200,000
Redwood, close-grained	1900	100	270	1450	1,200,000
Spruce, Engelmann	1100	100	180	800	800,000
Spruce, red, white, and Sitka	1600	120	250	1050	1,200,000
Tamarack	1750	140	300	1350	1,300,000
HARDWOODS					
Ash, black	1450	130	300	850	1,100,000
Ash, commercial white	2050	185	500	1450	1,500,000
Beech, American	2200	185	500	1600	1,600,000
Birch, sweet and yellow	2200	185	500	1600	1,600,000
Cottonwood, Eastern	1100	90	150	800	1,000,000
Elm, American and slippery (white or soft elm)	1600	150	250	1050	1,200,000
Elm, rock	2200	185	500	1600	1,300,000
Gums, blackgum, sweetgum (red or sap gum)	1600	150	300	1050	1,200,000
Hickory, true and pecan	2800	205	600	2000	1,800,000
Maple, black and sugar (hard maple)	2200	185	500	1600	1,600,000
Oak, commercial red and white	2050	185	500	1350	1,500,000
Tupelo	1600	150	300	1050	1,200,000
Yellow poplar	1300	120	220	950	1,100,000

* These stresses are applicable with certain adjustments to material of any degree of seasoning.

(For use in determining working stresses according to the grade of timber and other applicable factors
    All values are in pounds per square inch.   U. S. Forest Products Laboratory.)

Use of Timber Column Formulas

The values of E (modulus of elasticity) and σ_1 (compression parallel to grain) in the above formulas are given in Table 22.9. Table 22.10 gives the computed values of K for some common types of timbers. These may be substituted directly in Eq. (22.97) for intermediate-length columns, or may be used in conjunction with Table 22.11, which gives the strength of columns of intermediate length, expressed as a percentage of strength (σ_1) of short columns. In the tables, the term "continuously

Table 22.10   Values of K for Columns of Intermediate Length

ASTM Standards, 1937, D245–37

Species	Continuously Dry		Occasionally Wet		Usually Wet	
	Select	Common	Select	Common	Select	Common
Cedar, western red	24.2	27.1	24.2	27.1	25.1	28.1
Cedar, Port Orford	23.4	26.2	24.6	27.4	25.6	28.7
Douglas fir, coast region	23.7	27.3	24.9	28.6	27.0	31.1
Douglas fir, dense	22.6	25.3	23.8	26.5	25.8	28.8
Douglas fir, Rocky Mountain region	24.8	27.8	24.8	27.8	26.5	29.7
Hemlock, west coast	25.3	28.3	25.3	28.3	26.8	30.0
Larch, western	22.0	24.6	23.1	25.8	25.8	28.8
Oak, red and white	24.8	27.8	26.1	29.3	27.7	31.1
Pine, southern	....	27.3	....	28.6	....	31.1
Pine, dense	22.6	25.3	23.8	26.5	25.8	28.8
Redwood	22.2	24.8	23.4	26.1	25.6	28.6
Spruce, red, white, Sitka	24.8	27.8	25.6	28.7	27.5	30.8

Table 22.11   Strength of Columns of Intermediate Length, Expressed as a Percentage of Strength of Short Columns

ASTM Standards, 1937, D245–37

Values for expression $\{1 - 1/3(l/Kd)^4\}$ in eq. 33

K	Ratio of Length to Least Dimension in Rectangular Timbers, l/d																			
	12	13	14	15	16	17	18	19	20	21	22	23	24	25	26	27	28	29	30	31
22	97	96	95	93	91	88	85	81	77	72	67	...	...	...	...	...	...	...	...	...
23	98	97	95	94	92	90	87	84	81	77	72	67	...	...	...	...	...	...	...	...
24	98	97	96	95	93	92	89	87	84	80	76	72	67	...	...	...	...	...	...	...
25	98	98	97	96	94	93	91	89	86	83	80	76	72	67	...	...	...	...	...	...
26	99	98	97	96	95	93	92	91	89	86	83	80	76	72	67	...	...	...	...	...
27	99	98	98	97	96	95	93	92	90	88	85	82	79	74	71	67	...	...	...	...
28	99	98	98	97	96	95	94	93	91	89	87	85	82	79	75	71	67	...	...	...
29	99	99	98	98	97	96	95	94	92	91	89	87	84	82	79	75	71	67	...	...
30	99	99	98	98	97	97	96	95	94	92	90	88	86	84	81	78	75	71	67	...
31	99	99	99	98	98	97	97	96	95	94	93	92	90	88	86	84	81	78	75	67

Note. This table can also be used for columns not rectangular, the l/d being equivalent to $0.289l/k$, where k is the least radius of gyration of the section.

dry" refers to interior construction where there is no excessive dampness or humidity; "occasionally wet but quickly dry" refers to bridges, trestles, bleachers, and grandstands; "usually wet" refers to timber in contact with the earth or exposed to waves or tidewater.

22.8.4   Steel Columns

Types

Two general types of steel columns are in use: (1) rolled shapes and (2) built-up sections. The rolled shapes are easily fabricated, accessible for painting, neat in appearance where they are not covered, and convenient in making connections. A disadvantage is the probability that thick sections are of lower-strength material than thin sections because of the difficulty of adequately rolling the thick material. For the effect of thickness of material on yield point, see Ref. 14, p. 1377.

General Principles in Design

The design of steel columns is always a cut-and-try method, as no law governs the relation between area and radius of gyration of the section. A column of given area is selected, and the amount of load that it will carry is computed by the proper formula. If the allowable load so computed is less than that to be carried, a larger column is selected and the load for it is computed, the process being repeated until a proper section is found.

A few general principles should guide in proportioning columns. The radius of gyration should be approximately the same in the two directions at right angles to each other; the slenderness ratio of the separate parts of the column should not be greater than that of the column as a whole; the different parts should be adequately connected in order that the column may function as a single unit; the material should be distributed as far as possible from the centerline in order to increase the radius of gyration.

Steel Column Formulas

A variety of steel column formulas are in use, differing mostly in the value of unit stress allowed with various values of l/k. See Ref. 15, for a summary of the formulas.

Tests on Steel Columns

After the collapse of the Quebec Bridge in 1907 as a result of a column failure, the ASCE, the AREA, and the U.S. Bureau of Standards cooperated in tests of full-size steel columns. The results of these tests are reported in Ref. 16, pp. 1583–1688. The tests showed that, for columns of the proportions commonly used, the effect of variation in the steel, kinks, initial stresses, and similar defects in the column was more important than the effect of length. They also showed that the thin metal gave definitely higher strength, per unit area, than the thicker metal of the same type of section.

Example 22.4

Design a steel column 18 ft long to carry an axial load of 105,000 lb and an additional load of 43,000 lb which comes from a crane runway, and which is 1¼ in. beyond the outside edge of the column.

The formula used will be the AISC formula (see Ref. 17),

$$\frac{P}{A} = \frac{18,000}{1 + \left(\dfrac{l^2}{18,000k^2}\right)}$$

The shape selected will be a plate and four angles, Fig. 22.32.

Using 10,000 psi as a preliminary estimate for the first selection, one plate 12 × ⁵⁄₁₆ in. and four angles 5 × 3½ × ⅜ in. will have an area of 15.95 in.². The safe load for this column will now be

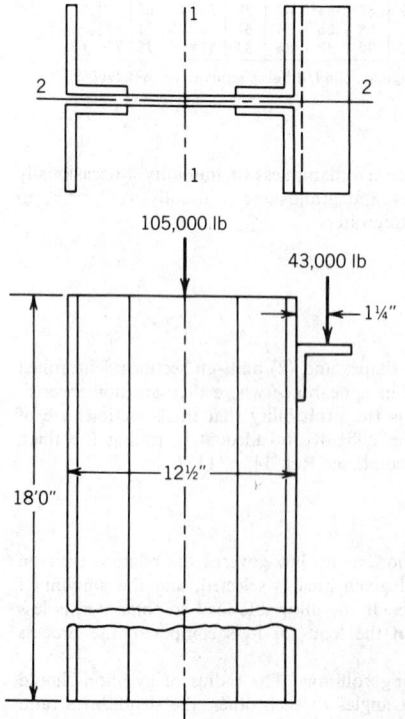

Fig. 22.32 Built-up column.

found. The moment of inertia I_{1-1} is 412 in.^{4}, the radius of gyration k_{1-1} is 5.08 in., and the radius of gyration k_{2-2} is 2.08 in.

Considering the safe load with respect to axis 2–2 about which neither load is eccentric,

$$P = \frac{15.95 \times 18,000}{1 + \dfrac{(18 \times 12)^2}{18,000 \times (2.08)^2}} = 179,400 \text{ lb}$$

Consider now the safe load about axis 1–1, with respect to which the load of 105,000 lb is axial, and the load of 43,000 lb has an eccentricity of $7\frac{1}{2}$ in. The load of 43,000 lb will produce a stress due to its eccentricity, and this stress will reduce by that amount the unit load that the column can carry. The safe load will therefore be

$$P = 15.95 \left(\frac{18,000}{1 + \dfrac{(18 \times 12)^2}{18,000 \times (5.08)^2}} - \frac{43,000 \times 7.5 \times 6.25}{412} \right) = 183,000 \text{ lb}$$

The area exceeds that needed by about 3 in.^{2}; the next smaller size is one plate $12 \times \frac{5}{16}$ in. and four angles $5 \times 3\frac{1}{2} \times \frac{5}{16}$ in. The area is 13.99 in.^{2}, moment of inertia, $I_{1-1} = 356$ in.^{4}, $k_{1-1} = 5.04$ in., and $k_{2-2} = 2.03$ in. With respect to the axis 2–2,

$$P = \frac{13.99 \times 18,000}{1 + \dfrac{(18 \times 12)^2}{18,000 \times (2.03)^2}} = 154,600 \text{ lb}$$

With respect to axis 1–1,

$$P = 13.99 \left(\frac{18,000}{1 + \dfrac{(18 \times 12)^2}{18,000 \times (5.04)^2}} - \frac{43,000 \times 7.5 \times 6.25}{356} \right) = 149,300 \text{ lb}$$

As the load to be carried is 148,000 lb this column is satisfactory.

Steel Column, Secondary Buckling

When thin-wall columns[12] are designed, plates making up the cross section must be checked as in ex. 22.4 to see if they buckle individually. In the column in Fig. 22.32 the flanges and web have the possibility of buckling individually. The flanges are simply supported on three sides and free on the fourth, long side. The buckling stress is

$$\sigma = 0.416 \frac{E}{1 - v^2} \left(\frac{t}{b} \right)^2 = 78,975 \text{ psi}$$

where $b = 5$ in., $t = 0.375$ in., $v = \frac{1}{3}$, and $E = 30 \times 10^6$ psi. For the web, which is simply supported on four sides,

$$\sigma = 3.29 \frac{E}{1 - v^2} \left(\frac{t}{b} \right)^2 = 75,302 \text{ psi}$$

where $b = 12$ in. and $t = 0.3125$ in. The right flange stress is

$$\sigma = \frac{105,000 \text{ lb}}{13.99 \text{ in.}^2} + \frac{43,000 \text{ lb}}{2 \times 5 \text{ in.} \times 0.375 \text{ in.}} = 18,298 \text{ psi}$$

The members of the column will not buckle individually. If the thickness is one-half of the present value, then difficulties will be encountered.

22.9 CYLINDERS, SPHERES, AND PLATES

22.9.1 Thin Cylinders and Spheres under Internal Pressure

A cylinder is regarded as thin when the thickness of the wall is small compared with the mean diameter, or $d/t > 20$. There are only tensile membrane stresses in the wall developed by the internal pressure p

$$\frac{\sigma_1}{R_1} + \frac{\sigma_2}{R_2} = \frac{p}{t} \qquad (22.99)$$

In the case of a cylinder where R_1, the curvature, is R and R_2 is infinite, and the hoop stress is

$$\sigma_1 = \sigma_h = \frac{pR}{t} \tag{22.100}$$

If the two equations are compared, it is seen that the resistance to rupture by circumferential stress [Eq. (22.100)] is one-half the resistance to rupture by longitudinal stress [Eq. (22.101)]. For this reason cyclindrical boilers are single riveted in the circumferential seams and double or triple riveted in the longitudinal seams.

From the equations of equilibrium, the longitudinal stress is

$$\sigma_2 = \sigma_L = \frac{pR}{2t} \tag{22.101}$$

For a sphere, using Eq. (22.99), $R_1 = R_2 = R$ and $\sigma_1 = \sigma_2$, making

$$\sigma_1 = \sigma_2 = \frac{pR}{2t} \tag{22.102}$$

In using the foregoing formulas to design cylindrical shells or piping, thickness t must be increased to compensate for rivet holes in the joints. Water pipes, particularly those of cast iron, require a high factor of safety, which results in increased thickness to provide security against shocks caused by water hammer or rough handling before they are laid. Equation (22.101) applies also to the stresses in the walls of a thin hollow sphere, hemisphere, or dome. When holes are cut, the tensile stresses must be found by the method used in riveted joints.

Thin Cylinders under External Pressure

Equations (22.100) and (22.101) apply equally well to cases of external pressure if P is given a negative sign, but the stresses so found are significant only if the pressure and dimensions are such that no buckling can occur.

22.9.2 Thick Cylinders and Spheres

Cylinders

When the thickness of the shell or wall is relatively large, as in guns, hydraulic machinery piping, and similar installations, the variation in stress from the inner surface to the outer surface is relatively large, and the ordinary formulas for thin wall cylinders are no longer applicable. In Fig. 22.33 the stresses, strains, and deflections are related[1,18,19] by

$$\sigma_t = \frac{E}{1 - \nu^2}(\epsilon_t + \nu\epsilon_r) = \frac{E}{1 - \nu^2}\left[\frac{u}{r} + \nu\frac{\partial u}{\partial r}\right] \tag{22.103}$$

$$\sigma_r = \frac{E}{1 - \nu^2}(\epsilon_r + \nu\epsilon_t) = \frac{E}{1 - \nu^2}\left[\frac{\partial u}{\partial r} + \nu\frac{u}{r}\right] \tag{22.104}$$

where E is the modulus and ν is Poisson's ratio. In a cylinder (Fig. 22.34) that has internal and external pressures, p_i and p_o; internal and external radii, a and b; $K = b/a$; the stresses are

$$\sigma_t = \frac{p_i}{K^2 - 1}\left(1 + \frac{b^2}{r^2}\right) - \frac{p_o K^2}{K^2 - 1}\left(1 + \frac{a^2}{r^2}\right) \tag{22.105}$$

$$\sigma_r = \frac{p_i}{K^2 - 1}\left(1 - \frac{b^2}{r^2}\right) - \frac{p_o K^2}{K^2 - 1}\left(1 - \frac{a^2}{r^2}\right) \tag{22.106}$$

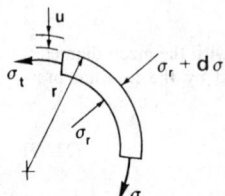

Fig. 22.33 Cylindrical element.

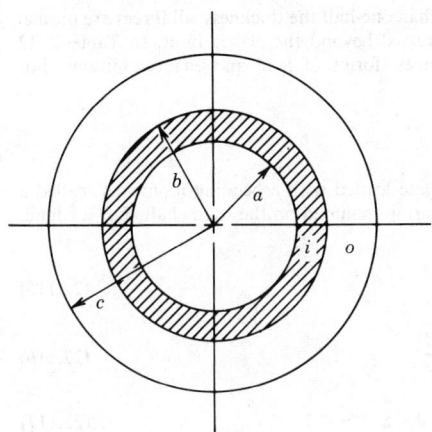

Fig. 22.34 Cylinder press fit.

if $p_o = 0$, and σ_t, σ_r are maximum at $r = a$; if $p_i = 0$, σ_t is maximum at $r = a$; and σ_r is maximum at $r = b$.

In shrinkage fits, Fig. 23.34, a hollow cylinder is pressed over a cylinder with a radial interference δ at $r = b$. p_f, the pressure between the cylinders, can be found from

$$\delta = \frac{bp_f}{E_o}\left(\frac{c^2 + b^2}{c^2 - b^2} + \nu_o\right) + \frac{bp_f}{E_i}\left(\frac{a^2 + b^2}{a^2 - b^2} - \nu_i\right) \qquad (22.107)$$

The radial deflection can be found at a which shrinks and c which expands by knowing σ_r is zero and using Eqs. (22.103) and (22.104):

$$u_a = \frac{\sigma_t}{E_i}a, \qquad u_c = \frac{\sigma_t}{E_o}c \qquad (22.108)$$

Spheres

The stress, strain, and deflections[19,20] are related by

$$\sigma_t = \frac{E}{1 - \nu - 2\nu^2}[\epsilon_t + \nu\epsilon_r] = \frac{E}{1 - \nu - 2\nu^2}\left[\frac{u}{r} + \nu\frac{\partial u}{\partial r}\right] \qquad (22.109)$$

$$\sigma_r = \frac{E}{1 - \nu - 2\nu^2}[2\nu\epsilon_t + (1 - \nu)\epsilon_r] = \frac{E}{1 - \nu - 2\nu^2}\left[2\nu\frac{u}{r} + (1 - \nu)\frac{\partial u}{\partial r}\right] \qquad (22.110)$$

The stresses for a thick wall sphere with internal and external pressure, p_i and p_o, and $K = b/a$ are

$$\sigma_t = \frac{p_i(1 + b^3/2r^3)}{K^3 - 1} - \frac{p_oK^3(1 + a^3/2r^3)}{K^3 - 1} \qquad (22.111)$$

$$\sigma_r = \frac{p_i(1 - b^3/r^3)}{K^3 - 1} - \frac{p_oK^3(1 - a^3/r^3)}{K^3 - 1} \qquad (22.112)$$

If $p_i = 0$, $\sigma_r = 0$ at $r = a$, then

$$u_a = (1 - \nu)\frac{\sigma_t}{E}a \qquad (22.113)$$

Conversely, if $p_o = 0$, $\sigma_r = 0$ at $r = b$, then

$$u_b = (1 - \nu)\frac{\sigma_t}{E}b \qquad (22.114)$$

22.9.3 Plates

The formulas that apply for plates are based on the assumptions that the plate is flat, of uniform thickness, and of homogeneous isotropic material, thickness is not greater than one-fourth the least transverse dimension, maximum deflection is not more than one-half the thickness, all forces are normal to the plane of the plate, and the plate is nowhere stressed beyond the elastic limit. In Table 22.12 are formulas for deflection and stress for various shapes, forms of load and edge conditions. For further information see Refs. 12 and 21.

22.9.4 Trunnion

A solid shaft (Fig. 22.35) on a round or rectangular plate loaded with a bending moment is called a trunnion. The loading generally is developed from a bearing mounted on the solid shaft. For a round, simply supported plate

$$\sigma_r = \frac{\beta M}{at^2} \tag{22.115}$$

$$\theta = \frac{\gamma M}{Et^3} \tag{22.116}$$

$$\left. \begin{array}{l} \beta = 10^{(0.7634 - 1.252x)} \\ \log \gamma = 0.248 - \pi x^{1.5} \end{array} \right\} \ 0 < x = \frac{b}{a} < 1 \tag{22.117}$$

For the fixed-end plate

$$\left. \begin{array}{l} \beta = 10^{(1 - 1.959x)} \\ \log \gamma = 0.179 - 3.75x^{1.5} \end{array} \right\} \ 0 < x = \frac{b}{a} < 1 \tag{22.118}$$

The equations for β, γ are derived from curve fitting of data (see, for example, Refs. 2, 4th ed., and 21).

22.9.5 Socket Action

In Fig. 22.36a, summation of moments in the middle of the wall yields

$$2\left[\left(\frac{\omega''}{2} \frac{l}{2} \right) \left(\frac{2}{3} \frac{l}{2} \right) \right] = F\left(a + \frac{l}{2} \right)$$

$$\omega'' = \frac{6}{l^2} \left[F\left(a + \frac{l}{2} \right) \right] \tag{22.119}$$

Summation of forces in the horizontal gives

$$\omega' = \frac{F}{l} \tag{22.120}$$

At B, the bearing pressure in Fig. 22.36c is

$$p_i = \frac{\omega' + \omega''}{d} \ \text{psi} \tag{22.121}$$

In Eq. (22.105) $p_o = 0$ and

$$\sigma_t = \frac{p_i}{R^2 - 1} \left[1 + \left(\frac{b}{d/2} \right)^2 \right]$$

At A in Fig. 22.36c

$$\sigma = \frac{\phi 8F}{\pi^2 bl}$$

where $2b/d = 2, 4$ and $\phi = 4.3, 4.4$;

$$F = (\omega' + \omega'')l$$

If a pin is pressed into the frame hole, σ_t created by p_f [Eq. (22.107)] must be added. Furthermore, if the pin and frame are different metals, additional σ_t will be created by temperature changes that vary p_f.

The stress in the pin can be found from the maximum moment developed by ω' and ω'', and then calculating the bending stress.

22.10 CONTACT STRESSES

The stresses caused by the pressure between elastic bodies are of importance in connection with the design or investigation of ball and roller bearings, trunnions, expansion rollers, track stresses, gear teeth, etc.

Table 22.12 Formulas for Flat Plates[a]

(By permission from *Formulas for Stress and Strain*, 2nd Ed., by R. J. Roark, copyrighted, 1943. McGraw-Hill Book Co., Inc.)

Notation: W = total applied load, lb; w = unit applied load, psi; t = thickness of plate, in.; σ = stress at surface of plate, psi; y = vertical deflection of plate from original position, in.; E = modulus of elasticity; m = reciprocal of ν, Poisson's ratio. q denotes any given point on the surface of plate; r denotes the distance of q from the center of a circular plate. Other dimensions and corresponding symbols are indicated on figures. Positive sign for σ indicates tension at upper surface and equal compression at lower surface; negative sign indicates reverse condition. Positive sign for y indicates upward deflection, negative sign downward deflection. Subscripts r, t, a, and b used with σ denote, respectively, radial direction, tangential direction, direction of dimension a, and direction of dimension b. All dimensions are in inches. All logarithms are to the base e ($\log_e x = 2.3026 \log_{10} x$).

Type of Load and Support	Formulas for Stress and Deflection

CIRCULAR FLAT PLATES

Outer edges supported. Uniform load over entire surface.

$w\pi a^2 = W$

At center:

$$\max \sigma_r = \sigma_t = \frac{-3W}{8\pi m t^2}(3m+1) \qquad \max y = -\frac{3W(m-1)(5m+1)a^2}{16\pi E m^2 t^3}$$

At q:

$$\sigma_r = -\frac{3W}{8\pi m t^2}\left[(3m+1)\left(1-\frac{r^2}{a^2}\right)\right] \qquad \sigma_t = -\frac{3W}{8\pi m t^2}\left[(3m+1)-(m+3)\frac{r^2}{a^2}\right]$$

$$y = -\frac{3W(m^2-1)}{8\pi E m^2 t^3}\left[\frac{(5m+1)a^2}{2(m+1)}+\frac{r^4}{2a^2}-\frac{(3m+1)r^2}{m+1}\right]$$

Outer edges fixed. Uniform load over entire surface.

$W = w\pi a^2$

At center:

$$\sigma_r = \sigma_t = -\frac{3W(m+1)}{8\pi m t^2} \qquad \max y = -\frac{3W(m^2-1)a^2}{16\pi E m^2 t^3}$$

At q:

$$\sigma_r = \frac{3W}{8\pi m t^2}\left[(3m+1)\frac{r^2}{a^2}-(m+1)\right] \qquad \sigma_t = \frac{3W}{8\pi m t^2}\left[(m+3)\frac{r^2}{a^2}-(m+1)\right]$$

$$y = \frac{-3W(m^2-1)}{16\pi E m^2 t^3}\left[\frac{(a^2-r^2)^2}{a^2}\right]$$

Outer edges supported. Uniform load over concentric circular area of radius r_0.

$W = w\pi r_0^3$

At q, $r < r_0$:

$$\sigma_r = -\frac{3W}{2\pi m t^2}\left[m+(m+1)\log\frac{a}{r_0}-(m-1)\frac{r_0^2}{4a^2}-(3m+1)\frac{r^2}{4r_0^2}\right]$$

$$\sigma_t = -\frac{3W}{2\pi m t^2}\left[m+(m+1)\log\frac{a}{r_0}-(m-1)\frac{r_0^2}{4a^2}-(m+3)\frac{r^2}{4r_0^2}\right]$$

$$y = -\frac{3W(m^2-1)}{16\pi E m^2 t^3}\left[4a^2-5r_0^2+\frac{r^4}{r_0^2}-(8r^2+4r_0^2)\log\frac{a}{r_0}-\frac{2(m-1)r_0^2(a^2-r^2)}{(m+1)a^2}\right.$$
$$\left.+\frac{8m(a^2-r^2)}{m+1}\right]$$

At q, $r > r_0$:

$$\sigma_r = -\frac{3W}{2\pi m t^2}\left[(m+1)\log\frac{a}{r}-(m-1)\frac{r_0^2}{4a^2}+(m-1)\frac{r_0^2}{4r^2}\right]$$

$$\sigma_t = -\frac{3W}{2\pi m t^2}\left[(m-1)+(m+1)\log\frac{a}{r}-(m-1)\frac{r_0^2}{4a^2}-(m-1)\frac{r_0^2}{4r^2}\right]$$

$$y = -\frac{3W(m^2-1)}{16\pi E m^2 t^3}\left[\frac{(12m+4)(a^2-r^2)}{m+1}-\frac{2(m-1)r_0^2(a^2-r^2)}{(m+1)a^2}\right.$$
$$\left.-(8r^2+4r_0^2)\log\frac{a}{r}\right]$$

At center:

$$\max \sigma_r = \sigma_t = -\frac{3W}{2\pi m t^2}\left[m+(m+1)\log\frac{a}{r_0}-(m-1)\frac{r_0^2}{4a^2}\right]$$

$$\max y = -\frac{3W(m^2-1)}{16\pi E m^2 t^3}\left[\frac{(12m+4)a^2}{m+1}-4r_0^2\log\frac{a}{r_0}-\frac{(7m+3)r_0^2}{m+1}\right]$$

[a] By permission from *Formulas for Stress and Strain,* 2nd ed., by R. J. Roark, copyrighted, 1943. McGraw-Hill Book Co., Inc.

Table 22.12 (*Continued*)

TYPE OF LOAD AND SUPPORT	FORMULAS FOR STRESS AND DEFLECTION

CIRCULAR FLAT PLATES

Outer edges supported. Uniform load on concentric circular ring of radius r_0.

At q, $r < r_0$:

$$\max \sigma_r = \sigma_t = -\frac{3W}{2\pi m t^2}\left[\frac{1}{2}(m-1)+(m+1)\log\frac{a}{r_0}-(m-1)\frac{r_0^2}{2a^2}\right]$$

$$y = -\frac{3W(m^2-1)}{2\pi E m^2 t^3}\left[\frac{(3m+1)(a^2-r^2)}{2(m+1)}-(r^2+r_0^2)\log\frac{a}{r_0}+(r^2-r_0^2)\right.$$
$$\left.-\frac{(m-1)r_0^2(a^2-r^2)}{2(m+1)a^2}\right]$$

At q, $r > r_0$:

$$\sigma_r = -\frac{3W}{2\pi m t^2}\left[(m+1)\log\frac{a}{r}+(m-1)\frac{r_0^2}{2r^2}-(m-1)\frac{r_0^2}{2a^2}\right]$$

$$\sigma_t = -\frac{3W}{2\pi m t^2}\left[(m-1)+(m+1)\log\frac{a}{r}-(m-1)\frac{r_0^2}{2r^2}-(m-1)\frac{r_0^2}{2a^2}\right]$$

$$y = -\frac{3W(m^2-1)}{2\pi E m^2 t^3}\left[\frac{(3m+1)(a^2-r^2)}{2(m+1)}-(r^2+r_0^2)\log\frac{a}{r}-\frac{(m-1)r_0^2(a^2-r^2)}{2(m+1)a^2}\right]$$

Outer edges fixed. Uniform load over concentric circular area of radius r_0.

At q, $r < r_0$:

$$\sigma_r = -\frac{3W}{2\pi m t^2}\left[(m+1)\log\frac{a}{r_0}+(m+1)\frac{r_0^2}{4a^2}-(3m+1)\frac{r^2}{4r_0^2}\right]$$

$$\sigma_t = -\frac{3W}{2\pi m t^2}\left[(m+1)\log\frac{a}{r_0}+(m+1)\frac{r_0^2}{4a^2}-(m+3)\frac{r^2}{4r_0^2}\right]$$

$$y = -\frac{3W(m^2-1)}{16\pi E m^2 t^3}\left[4a^2-(8r^2+4r_0^2)\log\frac{a}{r_0}-\frac{2r^2r_0^2}{a^2}+\frac{r^4}{r_0^2}-3r_0^2\right]$$

At q, $r > r_0$:

$$\sigma_r = -\frac{3W}{2\pi m t^2}\left[(m+1)\log\frac{a}{r}+(m+1)\frac{r_0^2}{4a^2}+(m-1)\frac{r_0^2}{4r^2}-m\right]$$

$$\sigma_t = -\frac{3W}{2\pi m t^2}\left[(m+1)\log\frac{a}{r}+(m+1)\frac{r_0^2}{4a^2}-(m-1)\frac{r_0^2}{4r^2}-1\right]$$

$$y = -\frac{3W(m^2-1)}{16\pi E m^2 t^3}\left[4a^2-(8r^2+4r_0^2)\log\frac{a}{r}-\frac{2r^2r_0^2}{a^2}-4r^2+2r_0^2\right]$$

$W = w\pi r_0^2$

At center:

$$\sigma_r = \sigma_t = -\frac{3W}{2\pi m t^2}\left[(m+1)\log\frac{a}{r_0}+(m+1)\frac{r_0^2}{4a^2}\right]=\max \sigma_r \text{ when } r_0 < 0.588a$$

$$\max y = -\frac{3W(m^2-1)}{16\pi E m^2 t^3}\left[4a^2-4r_0^2\log\frac{a}{r_0}-3r_0^2\right]$$

Outer edges fixed. Uniform load on concentric circular ring of radius r_0.

At q, $r < r_0$:

$$\sigma_r = \sigma_t = -\frac{3W}{4\pi m t^2}\left[(m+1)\left(2\log\frac{a}{r_0}+\frac{r_0^2}{a^2}-1\right)\right]=\max \sigma \text{ when } r < 0.31a$$

$$y = -\frac{3W(m^2-1)}{2\pi E m^2 t^3}\left[\frac{1}{2}\left(1+\frac{r_0^2}{a^2}\right)(a^2-r^2)-(r^2+r_0^2)\log\frac{a}{r_0}+(r^2-r_0^2)\right]$$

At q, $r > r_0$:

$$\sigma_r = -\frac{3W}{4\pi m t^2}\left[(m+1)\left(2\log\frac{a}{r}+\frac{r_0^2}{a^2}\right)+(m-1)\frac{r_0^2}{r^2}-2m\right]$$

$$\sigma_t = -\frac{3W}{4\pi m t^2}\left[(m+1)\left(2\log\frac{a}{r}+\frac{r_0^2}{a^2}\right)-(m-1)\frac{r_0^2}{r^2}-2\right]$$

$$y = -\frac{3W(m^2-1)}{2\pi E m^2 t^3}\left[\frac{1}{2}\left(1+\frac{r_0^2}{a^2}\right)(a^2-r^2)-(r^2+r_0^2)\log\frac{a}{r}\right]$$

At center:

$$\max y = -\frac{3W(m^2-1)}{2\pi E m^2 t^3}\left[\frac{1}{2}(a^2-r_0^2)-r_0^2\log\frac{a}{r_0}\right]$$

Table 22.12 (*Continued*)

Type of Load and Support	Formulas for Stress and Deflection
	Circular Flat Plates with Concentric Circular Hole

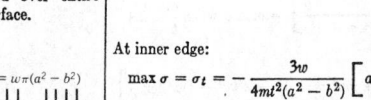

Outer edge supported. Uniform load over entire surface.

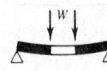

$W = w\pi(a^2 - b^2)$

At inner edge:

$$\max \sigma = \sigma_t = -\frac{3w}{4mt^2(a^2 - b^2)}\left[a^4(3m+1) + b^4(m-1) - 4ma^2b^2 - 4(m+1)a^2b^2\log\frac{a}{b}\right]$$

$$\max y = -\frac{3w(m^2-1)}{2m^2Et^3}\left[\frac{a^4(5m+1)}{8(m+1)} + \frac{b^4(7m+3)}{8(m+1)} - \frac{a^2b^2(3m+1)}{2(m+1)}\right.$$
$$\left. + \frac{a^2b^2(3m+1)}{2(m-1)}\log\frac{a}{b} - \frac{2a^2b^4(m+1)}{(a^2-b^2)(m-1)}\left(\log\frac{a}{b}\right)^2\right]$$

Outer edge supported. Uniform load along inner edge.

W

At inner edge:

$$\max \sigma = \sigma_t = -\frac{3W}{2\pi mt^2}\left[\frac{2a^2(m+1)}{a^2-b^2}\log\frac{a}{b} + (m-1)\right]$$

$$\max y = -\frac{3W(m^2-1)}{4\pi Em^2t^3}\left[\frac{(a^2-b^2)(3m+1)}{(m+1)} + \frac{4a^2b^2(m+1)}{(m-1)(a^2-b^2)}\left(\log\frac{a}{b}\right)^2\right]$$

Supported along concentric circle near outer edge. Uniform load along concentric circle near inner edge.

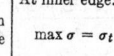

$d \mid d$

$c \leftarrow \mid \rightarrow c$

At inner edge:

$$\max \sigma = \sigma_t = -\frac{3W}{2\pi mt^2}\left[\frac{2a^2(m+1)}{a^2-b^2}\log\frac{c}{d} + (m-1)\frac{c^2-d^2}{a^2-b^2}\right]$$

Inner edge supported. Uniform load over entire surface.

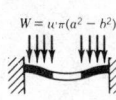

(reference figure)

$W = w\pi(a^2 - b^2)$

At inner edge:

$$\max \sigma = \sigma_t = \frac{3w}{4mt^2(a^2-b^2)}\left[4a^4(m+1)\log\frac{a}{b} + 4a^2b^2 + b^4(m-1) - a^4(m+3)\right]$$

At outer edge:

$$\max y = \frac{3w(m-1)}{16Em^2t^3}\left[a^4(7m+3) + b^4(5m+1) - a^2b^2(12m+4)\right.$$
$$\left. - \frac{4a^2b^2(3m+1)(m+1)}{(m-1)}\log\frac{a}{b} + \frac{16a^4b^2(m+1)^2}{(a^2-b^2)(m-1)}\left(\log\frac{a}{b}\right)^2\right]$$

Outer edge fixed and supported. Uniform load over entire surface.

$W = w\pi(a^2 - b^2)$

At outer edge:

$$\max \sigma_r = \frac{3w}{4t^2}\left[a^2 - 2b^2 + \frac{b^4(m-1) - 4b^4(m+1)\log\frac{a}{b} + a^2b^2(m+1)}{a^2(m-1) + b^2(m+1)}\right] = \nu \max \sigma$$

At inner edge:

$$\max \sigma_t = -\frac{3w(m^2-1)}{4mt^2}\left[\frac{a^4 - b^4 - 4a^2b^2\log\frac{a}{b}}{a^2(m-1) + b^2(m+1)}\right]$$

$$\max y = -\frac{3w(m^2-1)}{16m^2Et^3}\left[a^4 + 5b^4 - 6a^2b^2 + 8b^4\log\frac{a}{b}\right.$$
$$\left. + \frac{\left\{[-8b^6(m+1) + 4a^2b^4(3m+1) + 4a^4b^2(m+1)]\log\frac{a}{b} - 16a^2b^4(m+1)\left(\log\frac{a}{b}\right)^2\right.}{\left. + 4a^2b^4 - 2a^4b^2(m+1) + 2b^6(m-1)\right\}}{a^2(m-1) + b^2(m+1)}\right]$$

Table 22.12 (*Continued*)

Type of Load and Support	Formulas for Stress and Deflection

CIRCULAR FLAT PLATES WITH CONCENTRIC CIRCULAR HOLE

Outer edge fixed and supported. Uniform load along inner edge.

At outer edge:

$$\max \sigma_r = \frac{3W}{2\pi t^2}\left[1 - \frac{2mb^2 - 2b^2(m+1)\log\frac{a}{b}}{a^2(m-1)+b^2(m+1)}\right] = \max \sigma \text{ when } \frac{a}{b} < 2.4$$

At inner edge:

$$\max \sigma_t = \frac{3W}{2\pi m t^2}\left[1 + \frac{ma^2(m-1) - mb^2(m+1) - 2(m^2-1)a^2\log\frac{a}{b}}{a^2(m-1)+b^2(m+1)}\right]$$

$$= \max \sigma \text{ when } \frac{a}{b} > 2.4$$

$$\max y = -\frac{3W(m^2-1)}{4\pi m^2 E t^3} \times$$

$$\left[a^2 - b^2 + \frac{2mb^2(a^2-b^2) - 8ma^2b^2\log\frac{a}{b} + 4a^2b^2(m+1)\left(\log\frac{a}{b}\right)^2}{a^2(m-1)+b^2(m+1)}\right]$$

Outer edge fixed. Uniform moment along inner edge.

At inner edge:

$$\max \sigma_r = \frac{6M}{t^2}$$

$$\max y = \frac{6M(m^2-1)}{mEt^3}\left[\frac{a^2b^2 - b^4 - 2a^2b^2\log\frac{a}{b}}{a^2(m-1)+b^2(m+1)}\right]$$

At outer edge:

$$\sigma_r = -\frac{6M}{t^2}\left[\frac{2mb^2}{(m+1)b^2+(m-1)a^2}\right]$$

Outer edge supported. Unequal uniform moments along edges.

At q:

$$\sigma_r = \frac{6}{t^2(a^2-b^2)}\left[a^2M_a - b^2M_b - \frac{a^2b^2}{r^2}(M_a - M_b)\right]$$

$$\sigma_t = \frac{6}{t^2(a^2-b^2)}\left[a^2M_a - b^2M_b + \frac{a^2b^2}{r^2}(M_a - M_b)\right]$$

From outer edge level:

$$y = \frac{12(m^2-1)}{mEt^3(a^2-b^2)}\left[\frac{a^2-r^2}{2}\left(\frac{a^2M_a - b^2M_b}{m+1}\right) + \log\frac{a}{r}\left(\frac{a^2b^2(M_a - M_b)}{m-1}\right)\right]$$

Contact Stress Theory

H. Hertz[22] developed the mathematical theory for the surface stresses and the deformations produced by pressure between curved bodies, and the results of his analysis are supported by research. Formulas based on this theory give the maximum compressive stresses which occur at the center of the surfaces of contact, but do not consider the maximum subsurface shear stresses nor the maximum tensile stresses which occur at the boundary of the contact area. In Table 22.13 formulas are given for the elastic stress and deformation produced by bodies in contact. Numerous tests have been made to determine the bearing strength of balls and rollers, but there is difficulty in interpreting the results

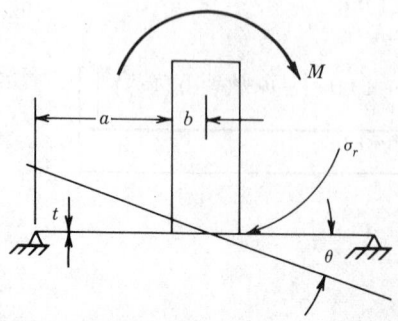

Fig. 22.35 Simply supported trunnion.

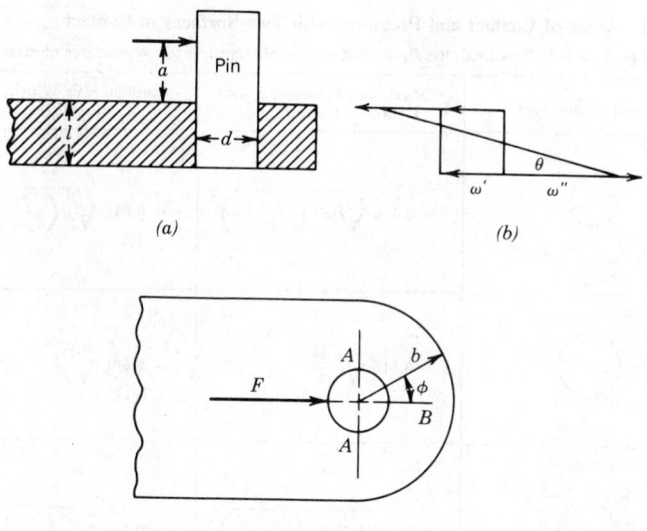

Fig. 22.36 Socket action near an edge.

for lack of a satisfactory criterion of failure. One arbitrary criterion of failure is the amount of allowable plastic yielding. For further information on contact stresses see Refs. 2, 23, and 24.

22.11 ROTATING ELEMENTS

22.11.1 Shafts

The stress[1] in the center of a rotating shaft or solid cylinder is

$$\sigma_r = \sigma_h = \frac{3 - 2\nu}{8(1 - \nu)} \left(\frac{\gamma \omega^2}{g} \right) r_o^2 \tag{22.122}$$

$$\sigma_z = \frac{\nu \gamma \omega^2}{4g(1 - \nu)} r_o^2 \tag{22.123}$$

where ν is Poisson's ratio, ω is in rad/sec, γ is the density in lb/in.[3], and g is 386 in./sec[2]. The limiting ω can be found by using distortion energy; however, most shafts support loads and are limited by critical speeds from torsional or bending modes of vibration. Holzer's method and Dunkerley's equation are used.

22.11.2 Disks

A rotating disk[1,9,19] of inside radius a and outside radius b has $\sigma_r = 0$ at a and b, while σ_t is

$$\sigma_{ta} = \frac{3 + \nu}{4g} \gamma \omega^2 \left(b^2 + \frac{1 - \nu}{3 + \nu} a^2 \right) \tag{22.124}$$

$$\sigma_{tb} = \frac{3 + \nu}{4g} \gamma \omega^2 \left(a^2 + \frac{1 - \nu}{3 + \nu} b^2 \right) \tag{22.125}$$

Substitution in Eq. (22.108) gives the outside and inside radial expansions.

The solid disk of radius b has stresses at the center

$$\sigma_t = \sigma_r = \frac{3 + \nu}{8g} \gamma \omega^2 b^2$$

Substitution into the distortion energy [Eq. (22.22)] can give one the limiting speed.

22.11.3 Blades

Blades attached to a rotating shaft will experience a tensile force at the attachment to the shaft. These can be found from dynamics of machinery texts; however, the forces developed from a fluid

Table 22.13   Areas of Contact and Pressures with Two Surfaces in Contact

Poisson's ratio = 0.3; P = load, lb; P_1 = load per in. of length, lb; E = modulus of elasticity.

Character of Surfaces	Maximum Pressure, s, at Center of Contact, psi	Radius, r, or Width, b, of Contact Area, in.
Two spheres	$s = 0.616 \sqrt[3]{PE^2 \left(\dfrac{d_1 + d_2}{d_1 d_2}\right)^2}$	$r = 0.881 \sqrt[3]{\dfrac{P}{E}\left(\dfrac{d_1 d_2}{d_1 + d_2}\right)}$
Sphere and plane	$s = 0.616 \sqrt[3]{\dfrac{PE^2}{d^2}}$	$r = 0.881 \sqrt[3]{\dfrac{Pd}{E}}$
Sphere and hollow sphere	$s = 0.616 \sqrt[3]{PE^2 \left(\dfrac{d_2 - d_1}{d_1 d_2}\right)^2}$	$r = 0.881 \sqrt[3]{\dfrac{P}{E}\left(\dfrac{d_1 d_2}{d_2 - d_1}\right)}$
Cylinder and plane	$s = 0.591 \sqrt{\dfrac{P_1 E}{d}}$	$b = 2.15 \sqrt{\dfrac{P_1 d}{E}}$
Two cylinders	$s = 0.591 \sqrt{P_1 E \left(\dfrac{d_1 + d_2}{d_1 d_2}\right)}$	$b = 2.15 \sqrt{\dfrac{P_1}{E}\left(\dfrac{d_1 d_2}{d_1 + d_2}\right)}$
General case of two bodies in contact	$s = \dfrac{1.5P}{\pi c d}$	$c = \alpha \sqrt[3]{\dfrac{P\delta}{K}}$ $d = \beta \sqrt[3]{\dfrac{P\delta}{K}}$ $\delta = \dfrac{4}{\dfrac{1}{R_1} + \dfrac{1}{R_2} + \dfrac{1}{R_1'} + \dfrac{1}{R_2'}}$ $K = \dfrac{8}{3}\dfrac{E_1 E_2}{E_2(1-\nu_1^2) + E_1(1-\nu_2^2)}$

$$\theta = \text{arc cos}\ \frac{1}{4}\,\delta \sqrt{\left\{ \left(\frac{1}{R_1} - \frac{1}{R_1'}\right)^2 + \left(\frac{1}{R_2} - \frac{1}{R_2'}\right)^2 + 2\left(\frac{1}{R_1} - \frac{1}{R_1'}\right)\left(\frac{1}{R_2} - \frac{1}{R_2'}\right)\cos 2\phi \right\}}$$

θ	0°	10°	20°	30°	40°	50°	60°	70°	80°	90°
α	∞	6.612	3.778	2.731	2.136	1.754	1.486	1.284	1.128	1.00
β	0	0.319	0.408	0.493	0.567	0.641	0.717	0.802	0.893	1.00

driven by the blades develop more problems. The blades, if not in the plane, will develop additional forces and moments from the driving force plus vibration of the blades on the shaft.

22.12   EXPERIMENTAL STRESS ANALYSIS

The determination of static and dynamic strains in specimens and machine parts by means of physical measurements is defined as *experimental stress analysis*. Direct strain measurements are made possible

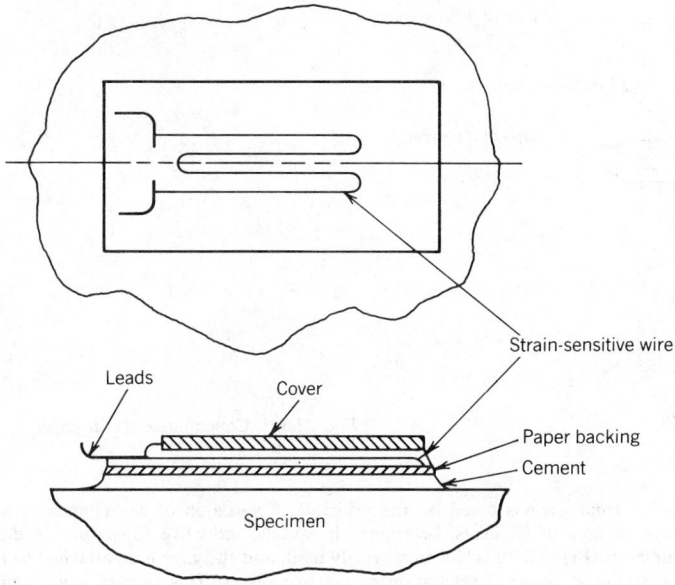

Fig. 22.37 Bonded wire strain gage.

by mechanical and electrical pickup devices. Mechanical and optical levers, brittle lacquers, photogrids, and cathetometers are mechanical systems. Electromagnetic, resistance, and capacitance methods are forms of electrical systems. Sometimes the nature of the installation and the operating conditions preclude direct strain measurements. Indirect methods such as high-speed photography of body displacements and velocity or acceleration measurements by means of moving coil pickups have proved to be acceptable.

Direct Strain Measurements

A *bonded wire strain gage* (Fig. 22.37) consists of a grid of strain-sensitive wire protected by a paper sheath and cemented to the specimen. A change in the length of the grid due to strains in the specimen changes the resistance of the wire. This change in resistance is readily measured by means of suitable electrical circuits. The gage is suitable for measurement of dynamic as well as static strains.

A *mechanical strain gage* is used to measure static strains in a specimen under load. The instrument employs either an optical or a mechanical lever system to multiply the strain, which then may be read from a suitable scale.

A *magnetic strain gage* is based on the principle of a change in either self- or mutual inductance of the gaging element. The simplest type is the variable inductance gage where the change in an air gap in a magnetic circuit varies the permeance of the circuit, thus giving an indication of the strains produced. See Fig. 22.38. This gage is used to measure both static and dynamic strains.

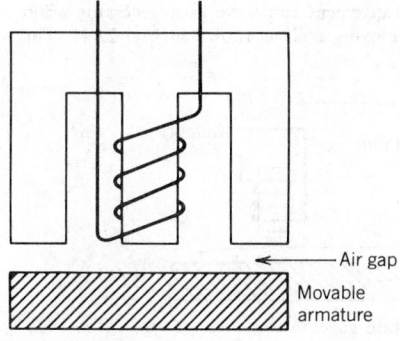

Fig. 22.38 Magnetic strain gage.

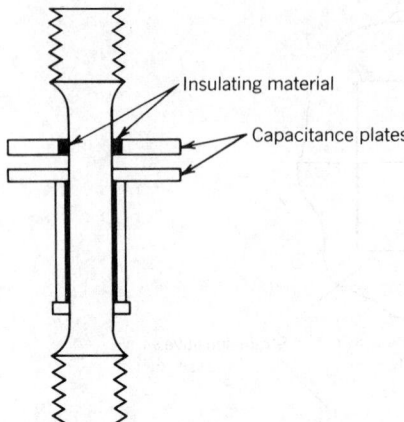

Fig. 22.39   Capacitance strain gage.

A *capacitance strain gage* is based on the principle of variation of capacitance by separation of elements, change in area of elements, or change in specific inductive capacity of a dielectric. The first of these methods (Fig. 22.39) is the most widely used, and the gage is so attached to the specimen that the strain produced causes a separation of the two plates. This type of gage is applicable for measuring both static and dynamic strains.

An *acoustic strain gage* employs a taut wire between two points mounted on the specimen, as shown in Fig. 22.40. The strain produced in the specimen alters the tension in the wire, thus changing its natural frequency. The natural frequency of the wire is determined by a variable frequency oscillator. Resonance is indicated by the output meter of the unit. By suitable methods the strain can be computed. This gage is suitable for measuring static and dynamic strains.

Brittle lacquer coatings are very convenient as they graphically present an overall strain picture, including the principal strain directions. They are revealed by cracks produced in the coating due to the strain in the specimen. The method consists of coating the specimen with lacquer, which becomes brittle on hardening. When the strains in the piece reach a prescribed magnitude, the lacquer cracks. This method, which gives quantitative results accurate to within ±10%, can be used to detect static and dynamic strains in tension or compression.

The *photogrid* method consists of coating the specimen with a photosensitive film on which a grid system is printed. The dimensional change in the grid system under load is a measure of the strain produced. This method is suitable for measuring permanent static strains only.

The *cathetometer* method consists of measuring the relative movement of two points on a test specimen under load by means of a microscope with a vernier scale attachment. Both static and dynamic strains can be measured by this method.

Indirect Strain and Displacement Measurements

Displacement pickups are used to detect the motion or displacement of the specimen. In general they consist of a seismic mass–spring system having suitable dynamic characteristics. The motion of the seismic element is used to produce an electrical signal by any of the electrical methods just described.

Velocity pickups usually employ the principle of a moving permanent magnet within a coil whose field strength is constant. The voltage induced in the coil is proportional to the relative velocity of the magnet and coil. By mechanical modification, the arrangement can serve as a generator where the voltage output is proportional to the velocity of the moving coil, as shown in Fig. 22.41. This

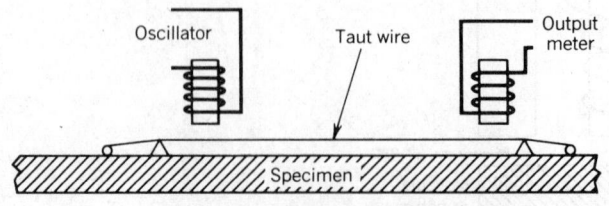

Fig. 22.40   Acoustic strain gage.

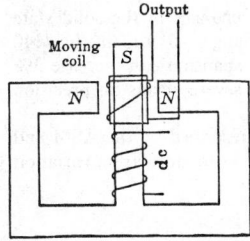

Fig. 22.41   Velocity pickup.

pickup can be used to measure dynamic strains. By electrically integrating the output, a signal proportional to displacement can be obtained.

Acceleration can be detected by pickups similar to those used for displacement and velocity. The output can be integrated for convenience. In some designs the system consists of a piezoelectric crystal which, by virtue of its high natural frequency, permits the detection of sharp transients and high-frequency accelerations.

Photoelastic methods involve the construction of a model of the prototype from an isotropic transparent material such as celluloid and glass. When these substances are subjected to stress, they become doubly refracting, and plane polarized light on passing through the material is changed into two component mutually perpendicular rays which are parallel and perpendicular to the directions of stress. The emergent rays will be out of phase, depending on the stress and the thickness of the material. A photograph of the emergent rays gives light and dark fringes indicative of the distribution of the principal stresses. A schematic diagram of one method is shown in Fig. 22.42. This procedure is used to measure static stress.

22.13   RESIDUAL STRESSES

22.13.1   Sources

Residual stresses can be used to increase fatigue life and performance of a design. Residual stresses[1,18,20] are developed by machining, heat treatments, welding, plating, and surface treatments. Residual stresses can be designed in from autofrettage in cylinders or from rolling, shot peening, and forming operations in manufacturing. Phase changes in materials in the solid state can induce plastic deformations and residual stresses. Loading of a part above its yield will cause residual stresses after the load is removed.

Heat treating tends to give surface compressive residual stresses (30–100 kpsi), while the core has tensile residual stresses (30–100 kpsi), with the overall part failing in tension at a higher average stress. Surface treatments (carburizing, nitriding, flame hardening, and induction hardening) develop surface compressive residual stresses (50–100 kpsi) much like heat treating but with sample surface hardnesses of around R_c 70. Rolling and shot peening also develop surface compression up to the yield strength of the part. Manufacturing can develop tensile and compressive residual stress. Rough machining develops tensile values, while a light grinding or honing can develop compressive residual stresses. Welding will produce tensile residual stresses with warping. Plating, generally applied at elevated temperatures, on a witness strip will cause bowing when the two metals are cooled to room temperature owing to differences in thermal expansion and chemical composition of the plating. The plating residual stresses are around 30 kpsi.

The residual stress condition and values described are the most commonly encountered; however, they are exceptions given in the literature. Furthermore, these stresses can fade or creep with temperature, time, and loading; in addition, they can change from compressive to tensile values, for example, a

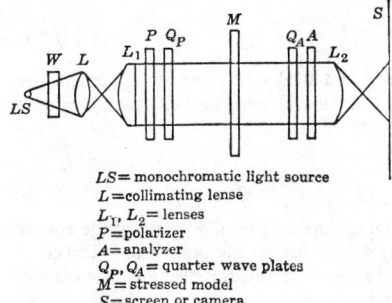

$LS=$ monochromatic light source
$L =$ collimating lense
$L_1, L_2 =$ lenses
$P =$ polarizer
$A =$ analyzer
$Q_p, Q_A =$ quarter wave plates
$M =$ stressed model
$S =$ screen or camera

Fig. 22.42   Polariscope.

mirror changing from convex to concave while on a storage shelf. Phase changes in the solid state after 15 years can cause this type of variation. Part dimensions[25] can change $+5$ μin./in. in 4140 steel to -15 μin./in. in 303 stainless steel, and the thermal coefficient of expansion can increase 3% with the elastic modulus decreasing 6%. These types of changes can have severe effects on precision mirrors, instruments, and parts requiring exact and unvarying dimensions.

Manufactured and welded parts should be stress relieved by methods prescribed by the ASM and AWS to reduce tensile residual stresses. Other parts may require stabilization to stop stress variation by using the following methods:

1. Stress relieving at 500–1000°F after a fabrication process.
2. Thermal cycling over the operating range.
3. Load cycling to above maximum levels.

These procedures are to develop dimensional stability. Where possible, the residual stress should be measured by X-ray diffraction or by experiment to see if the residual stress exists or vanishes before and after each stabilizing operation. Some machine tool manufacturers store large castings in open areas for years for seasoning and stabilization.

The values of residual stress can be determined by analytical and experimental means.[20]

22.13.2 Analytical Methods

Autofrettage

Cylinders, which are required to hold large internal pressures, can be designed[1,18] to have one or more sleeves pressed on the outside to allow a higher internal pressure. Wire and ribbon wound cylinders accomplish the same purpose. The basic equations are in Section 22.9.2.

Plastic Loading of Beams

Beams can be loaded into the plastic range, and using the stress–strain curve, the amount of residual stress can be estimated. Concrete beams are preloaded on the tensile side to increase the load capability.

Overspinning of Disks and Cylinders

Overspinning (meaning yielding at the inner radius) develops residual stresses[1] in the disks to increase the rotating speed.

Shrink Fits

Shrink fitting a disk on a shaft allows the disk to resist forces and torques because of the interference pressure and the friction between mating surfaces.

22.13.3 Experimental Methods

X-Ray-Diffraction Method

X-ray diffraction is a nondestructive means[26] of finding residual stress, which can work in areas as small as 1/64 in.[2]; it allows the determination of sharp stress gradients. This method works only on elastic strains, and is one of the most used methods.

Magnetic Method

The magnetic method is a nondestructive method[26] for nickel and steel where the reversible effective permeability is measured over a range of frequencies of an applied alternating field over the outer surface of a part.

Blind Hole Drilling Method

The blind hole drilling method is an ASTM STD E837,[26] which makes use of a strain gauge rosette into which a milled hole is drilled. The 0.122 in. (3.10 mm) drill is flat on the point and is drilled in 0.02 in. (0.51 mm) increments. From these data the residual stress is determined. Reflecting photoelasticity instead of strain gauges has also been used in this method.

Sach's Boring Method

In Sach's boring method residual stresses in cylinders[26] in the longitudinal, radial, and tangential directions can be found by successive removal of metal from the diameter and length. Split rings and longitudinal strips can be used.

Indentation Method

The indentation method[26] uses optical interference to measure the permanent surface deformation around a shallow spherical indentation in a polished area on a metal specimen. Calibrated specimens allow the determination of tensile stresses in parts made of similar material.

REFERENCES

1. J. H. Faupel and F. E. Fisher, *Engineering Design,* 2nd ed., Wiley, New York, 1981.
2. R. J. Roark and W. C. Young, *Formulas for Stress and Strain,* 5th ed., McGraw-Hill, New York, 1975.
3. J. Marin, *Mechanical Properties of Materials and Design,* McGraw-Hill, New York, 1942.
4. R. E. Peterson, *Stress Concentration Factors,* 2nd ed., Wiley, New York, 1974.
5. Young, *Bulletin 4,* School of Engineering Research, University of Toronto.
6. *Aluminum Standards and Data,* 3rd ed., The Aluminum Association, New York, 1972.
7. F. B. Seely and J. O. Smith, *Resistance of Materials,* 4th ed., Wiley, New York, 1957.
8. A. P. Boresi, O. Sidebottom, F. B. Seely, and J. O. Smith, *Advanced Mechanics of Materials,* 3rd ed., Wiley, New York, 1978.
9. S. P. Timoshenko, *Strength of Materials,* 3rd ed., Krieger, Melbourne, FL, 1958, Vols. I and II.
10. H. C. Mann, *Proc. Am. Soc. Testing Materials,* 1935, 1936, and 1937.
11. Bach, Elastizitäl u. Festigkeit.
12. R. M. Rivello, *Theory Analysis of Flight Structures,* McGraw-Hill, New York, 1969.
13. B. G. Johnston (Ed.), *Structural Research Council, Stability Design Criteria for Metal Structures,* 3rd ed., Wiley, New York, 1976.
14. *Trans. Am. Soc. Civil Engr.,* **xcviii** (1933).
15. S. P. Timoshenko and J. M. Gere, *Theory of Elastic Stability,* 2nd ed., McGraw-Hill, New York, 1961.
16. *Trans. Am. Soc. Civil Engr.,* **lxxxiii** (1919–20).
17. *AISC Handbook,* American Institute of Steel Construction, New York.
18. R. C. Juvinall, *Stress, Strain and Strength,* McGraw-Hill, New York, 1967.
19. S. P. Timoshenko and J. N. Goodier, *Theory of Elasticity,* 3rd ed., McGraw-Hill, New York, 1970.
20. M. Hetényi, *Handbook of Experimental Stress Analysis,* Wiley, New York, 1950.
21. W. Griffel, *Handbook of Formulas for Stress and Strain,* Frederick Ungar, New York, 1966.
22. H. Hertz, *Gesammelte Werke,* Vol. 1, Leipzig, 1895.
23. R. K. Allen, *Rolling Bearings,* Pitman and Sons, London, 1945.
24. A. Palmgren, *Ball and Roller Bearing Engineering,* SKF Industries, Philadelphia, PA, 1945.
25. M. Di Giovanni, *Flat and Corrugated Diaphragm Design Handbook,* Marcel Dekker, New York, 1982.
26. *Proceedings of the Society for Experimental Stress Analysis.*

GENERAL REFERENCES

J. O. Almen and P. H. Black, *Residual Stresses and Fatigue in Metals,* McGraw-Hill, New York, 1963.

W. Flügge (Ed.), *Handbook of Engineering Mechanics,* 1st ed., McGraw-Hill, New York, 1962.

W. R. Osgood (Ed.), *Residual Stresses in Metals and Metal Construction,* Reinhold, New York, 1954.

Symposium on Internal Stresses in Metals and Alloys, Institute of Metals, London, 1948.

CHAPTER 23

LUBRICATION OF MACHINE ELEMENTS

BERNARD J. HAMROCK

Department of Mechanical Engineering
The Ohio State University
Columbus, Ohio

By the middle of this century two distinct regimes of lubrication were generally recognized. The first of these was hydrodynamic lubrication. The development of the understanding of this lubrication regime began with the classical experiments of Tower,[1] in which the existence of a film was detected from measurements of pressure within the lubricant, and of Petrov,[2] who reached the same conclusion from friction measurements. This work was closely followed by Reynolds' celebrated analytical paper[3] in which he used a reduced form of the Navier–Stokes equations in association with the continuity equation to generate a second-order differential equation for the pressure in the narrow, converging gap of a bearing contact. Such a pressure enables a load to be transmitted between the surfaces with very low friction since the surfaces are completely separated by a film of fluid. In such a situation it is the physical properties of the lubricant, notably the dynamic viscosity, that dictate the behavior of the contact.

The second lubrication regime clearly recognized by 1950 was boundary lubrication. The understanding of this lubrication regime is normally attributed to Hardy and Doubleday,[4,5] who found that very thin films adhering to surfaces were often sufficient to assist relative sliding. They concluded that under such circumstances the chemical composition of the fluid is important, and they introduced the term "boundary lubrication." Boundary lubrication is at the opposite end of the lubrication spectrum from hydrodynamic lubrication. In boundary lubrication it is the physical and chemical properties of thin films of molecular proportions and the surfaces to which they are attached that determine contact behavior. The lubricant viscosity is not an influential parameter.

In the last 30 years research has been devoted to a better understanding and more precise definition of other lubrication regimes between these extremes. One such lubrication regime occurs in nonconformal contacts, where the pressures are high and the bearing surfaces deform elastically. In this situation the viscosity of the lubricant may rise considerably, and this further assists the formation of an effective fluid film. A lubricated contact in which such effects are to be found is said to be operating elastohydrodynamically. Significant progress has been made in our understanding of the mechanism of elastohydrodynamic lubrication, generally viewed as reaching maturity.

This chapter describes briefly the science of these three lubrication regimes (hydrodynamic, elastohydrodynamic, and boundary) and then demonstrates how this science is used in the design of machine elements.

SYMBOLS

A_p	total projected pad area, m²
a_b	groove width ratio
a_f	bearing-pad load coefficient
B	total conformity of ball bearing
b	semiminor axis of contact, m; width of pad, m
$\bar{b}$	length ratio, b_s/b_r
b_g	length of feed groove region, m
b_r	length of ridge region, m
b_s	length of step region, m
C	dynamic load capacity, N
C_l	load coefficient, $F/p_a Rl$
c	radial clearance of journal bearing, m
c'	pivot circle clearance, m
c_b	bearing clearance at pad minimum film thickness (Fig. 23.16), m
c_d	orifice discharge coefficient
D	distance between race curvature centers, m
$\bar{D}$	material factor
D_x	diameter of contact ellipse along x axis, m
D_y	diameter of contact ellipse along y axis, m
d	diameter of rolling element or diameter of journal, m
d_a	overall diameter of ball bearing (Fig. 23.76), m
d_b	bore diameter of ball bearing, m
d_c	diameter of capillary tube, m
d_i	inner-race diameter of ball bearing, m
d_o	outer-race diameter of ball bearing, m
$\bar{d}_o$	diameter of orifice, m
E	modulus of elasticity, N/m²

E' effective elastic modulus, $2\left(\dfrac{1-\nu_a^2}{E_a}+\dfrac{1-\nu_b^2}{E_b}\right)^{-1}$, N/m^2

$\tilde{E}$ metallurgical processing factor

$\mathscr{E}$ elliptic integral of second kind

e eccentricity of journal bearing, m

F applied normal load, N

F' load per unit length, N/m

$\tilde{F}$ lubrication factor

$\mathscr{F}$ elliptic integral of first kind

F_c pad load component along line of centers (Fig. 23.41), N

F_e rolling-element-bearing equivalent load, N

F_r applied radial load, N

F_s pad load component normal to line of centers (Fig. 23.41), N

F_t applied thrust load, N

f race conformity ratio

f_c coefficient dependent on materials and rolling-element bearing type (Table 23.19)

G dimensionless materials parameter

$\tilde{G}$ speed effect factor

G_f groove factor

g_e dimensionless elasticity parameter, $W^{8/3}/U^2$

g_v dimensionless viscosity parameter, GW^3/U^2

H dimensionless film thickness, h/R_x

$\tilde{H}$ misalignment factor

H_a dimensionless film thickness ratio, h_s/h_r

H_b pad pumping power, N m/sec

H_c power consumed in friction per pad, W

H_f pad power coefficient

H_{min} dimensionless minimum film thickness, h_{min}/R_x

$\hat{H}_{min}$ dimensionless minimum film thickness, $H_{min}(W/U)^2$

H_p dimensionless pivot film thickness, h_p/c

H_t dimensionless trailing-edge film thickness, h_t/c

h film thickness, m

$\bar{h}_i$ film thickness ratio, h_i/h_o

h_i inlet film thickness, m

h_l leading-edge film thickness, m

h_{min} minimum film thickness, m

h_o outlet film thickness, m

h_p film thickness at pivot, m

h_r film thickness in ridge region, m

h_s film thickness in step region, m

h_t film thickness at trailing edge, m

h_0 film constant, m

J number of stress cycles

K load deflection constant

$\bar{K}$ dimensionless stiffness coefficient, cK_p/p_aRl

K_a dimensionless stiffness, $-c\ \partial\bar{W}/\partial c$

K_p film stiffness, N/m

K_1 load-deflection constant for a roller bearing

$K_{1.5}$ load-deflection constant for a ball bearing

$\bar{K}_\infty$ dimensionless stiffness, cK_p/p_aRl

k ellipticity parameter, D_y/D_x

k_c capillary tube constant, m^3

k_o orifice constant, m^4/N$^{1/2}$ sec

L	fatigue life
L_a	adjusted fatigue life
L_{10}	fatigue life where 90% of bearing population will endure
L_{50}	fatigue life where 50% of bearing population will endure
l	bearing length, m
l_c	length of capillary tube, m
l_r	roller effective length, m
l_t	roller length, m
l_v	length dimension in stress volume, m
l_1	total axial length of groove, m
M	probability of failure
$\bar{M}$	stability parameter, $\bar{m}p_a h_t^5/2R^5 l\eta^2$
m	number of rows of rolling elements
$\bar{m}$	mass supported by bearing, N sec^2/m
m_p	preload factor
N	rotational speed, rps
N_R	Reynolds number
n	number of rolling elements or number of pads or grooves
P	dimensionless pressure, p/E'
P_d	diametral clearance, m
P_e	free endplay, m
p	pressure, N/m^2
p_a	ambient pressure, N/m^2
p_l	lift pressure, N/m^2
p_{max}	maximum pressure, N/m^2
p_r	recess pressure, N/m^2
p_s	bearing supply pressure, N/m^2
Q	volume flow of lubricant, m^3/sec
$\bar{Q}$	dimensionless flow, $3\eta Q/\pi p_a h_t^3$
Q_c	volume flow of lubricant in capillary, m^3/sec
Q_o	volume flow of lubricant in orifice, m^3/sec
Q_s	volume side flow of lubricant, m^3/sec
q	constant, $\pi/2 - 1$
q_f	bearing-pad flow coefficient
R	curvature sum on shaft or bearing radius, m
$\bar{R}$	groove length fraction, $(R_o - R_g)/(R_o - R_i)$
R_g	groove radius (Fig. 23.60), m
R_o	orifice radius, m
R_x	effective radius in x direction, m
R_y	effective radius in y direction, m
R_1	outer radius of sector thrust bearing, m
R_2	inner radius of sector thrust bearing, m
r	race curvature radius, m
r_c	roller corner radius, m
S	probability of survival
Sm	Sommerfeld number for journal bearings, $\eta N d^3 l/2Fc^2$
Sm_t	Sommerfeld number for thrust bearings, $\eta u b l^2/Fh_0^2$
s	shoulder height, m
T	tangential force, N
$\bar{T}$	dimensionless torque, $6 T_r/\pi p_a/(R_1^2 + R_2^2) h_r \Lambda_c$
T_c	critical temperature
T_r	torque, N m
U	dimensionless speed parameter, $u\eta_0/E'R_x$

u	mean surface velocity in direction of motion, m/sec
v	elementary volume, m³
N	dimensionless load parameter, $F/E'R_x^2$
$\overline{W}$	dimensionless load capacity, $F/p_a l(b_r + b_s + b_g)$
$\overline{W}_\infty$	dimensionless load, $1.5G_f F/\pi p_a(R_1^2 - R_2^2)$
X, Y	factors for calculation of equivalent load
x, y, z	coordinate system
$\bar{x}$	distance from inlet edge of pad to pivot, m
α	radius ratio, R_y/R_x
α_a	offset factor
α_b	groove width ratio, $b_s/(b_r + b_s)$
α_p	angular extent of pad, deg
α_r	radius ratio, R_2/R_1
β	contact angle, deg
β'	iterated value of contact angle, deg
β_a	groove angle, deg
β_f	free or initial contact angle, deg
β_p	angle between load direction and pivot, deg
Γ	curvature difference
γ	groove length ratio, l_1/l
Δ	rms surface finish, m
δ	total elastic deformation, m
ϵ	eccentricity ratio, e/c
η	absolute viscosity of lubricant, N sec/m²
η_k	kinematic viscosity, ν/ρ, m²/sec
η_0	viscosity at atmospheric pressure, N sec/m²
θ	angle used to define shoulder height, deg
$\bar{\theta}$	dimensionless step location, $\theta_i/(\theta_i + \theta_o)$
θ_g	angular extent of lubrication feed groove, deg
θ_i	angular extent of ridge region, deg
θ_o	angular extent of step region, deg
Λ	film parameter (ratio of minimum film thickness to composite surface roughness)
Λ_c	dimensionless bearing number, $3\eta\omega(R_1^2 - R_2^2)/p_a h_r^2$
Λ_j	dimensionless bearing number, $6\eta\omega R^2/p_a c^2$
Λ_t	dimensionless bearing number, $6\eta u l/p_a h_r^2$
λ	length-to-width ratio
λ_a	length ratio, $(b_r + b_s + b_g)/l$
λ_b	$(1 + 2/3\alpha)^{-1}$
μ	coefficient of friction, T/F
ν	Poisson's ratio
ξ	pressure–viscosity coefficient of lubricant, m²/N
ξ_p	angle between line of centers and pad leading edge, deg
ρ	lubricant density, N sec²/m⁴
ρ_0	density at atmospheric pressure, N sec²/m⁴
σ_{max}	maximum Hertzian stress, N/m²
τ	shear stress, N/m²
τ_0	maximum shear stress, N/m²
ϕ	attitude angle in journal bearings, deg
ϕ_p	angle between pad leading edge and pivot, deg
ψ	angular location, deg
ψ_l	angular limit of ψ, deg
ψ_s	step location parameter, $b_s/(b_r + b_s + b_g)$
ω	angular velocity, rad/sec

ω_B angular velocity of rolling-element race contact, rad/sec
ω_b angular velocity of rolling element about its own center, rad/sec
ω_c angular velocity of rolling element about shaft center, rad/sec
ω_d rotor whirl frequency, rad/sec
$\overline{\omega}_d$ whirl frequency ratio, ω_d/ω_j
ω_j journal rotational speed, rad/sec

Subscripts

a solid a
b solid b
EHL   elastohydrodynamic lubrication
e elastic
HL   hydrodynamic lubrication
i inner
iv   isoviscous
o outer
pv   piezoviscous
r rigid
x,y,z coordinate system

23.1 LUBRICATION FUNDAMENTALS

A lubricant is any substance that is used to reduce friction and wear and to provide smooth running and a satisfactory life for machine elements. Most lubricants are liquids (like mineral oils, the synthetic esters and silicone fluids, and water), but they may be solids (such as polytetrafluoroethylene) for use in dry bearings, or gases (such as air) for use in gas bearings. An understanding of the physical and chemical interactions between the lubricant and the tribological surfaces is necessary if the machine elements are to be provided with satisfactory life. To help in the understanding of this tribological behavior, the first section describes some lubrication fundamentals.

23.1.1 Conformal and Nonconformal Surfaces

Hydrodynamic lubrication is generally characterized by surfaces that are conformal; that is, the surfaces fit snugly into each other with a high degree of geometrical conformity, as shown in Fig. 23.1, so that the load is carried over a relatively large area. Furthermore, the load-carrying surface remains essentially constant while the load is increased. Fluid-film journal bearings (as shown in Fig. 23.1) and slider bearings exhibit conformal surfaces. In journal bearings the radial clearance between the shaft and bearing is typically one-thousandth of the shaft diameter; in slider bearings these inclination of the bearing surface to the runner is typically one part in a thousand. These converging surfaces, coupled with the fact that there is relative motion and a viscous fluid separating the surfaces, enable a positive pressure to be developed and exhibit a capacity to support a normal applied load. The magnitude of the pressure developed *is not* generally large enough to cause significant elastic deformation of the surfaces. The minimum film thickness in a hydrodynamically lubricated bearing is a function of applied load, speed, lubricant viscosity, and geometry. The relationship between the minimum film thickness $h_{\min}$ and the speed u and applied normal load F is given as

$$(h_{\min})_{HL} \propto \left(\frac{u}{F}\right)^{1/2} \tag{23.1}$$

More coverage of hydrodynamic lubrication can be found in Section 23.2.

Many machine elements have contacting surfaces that *do not* conform to each other very well, as

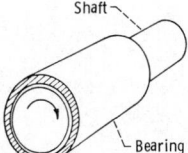

Shaft

Bearing

Fig. 23.1 Conformal surfaces. (From Ref. 6.)

shown in Fig. 23.2 for a rolling-element bearing. The full burden of the load must then be carried by a very small contact area. In general, the contact areas between nonconformal surfaces enlarge considerably with increasing load, but they are still smaller than the contact areas between conformal surfaces. Some examples of nonconformal surfaces are mating gear teeth, cams and followers, and rolling-element bearings (as shown in Fig. 23.2). The mode of lubrication normally found in these nonconformal contacts is elastohydrodynamic lubrication. The requirements necessary for hydrodynamic lubrication (converging surfaces, relative motion, and viscous fluid) are also required for elastohydrodynamic lubrication.

The relationship between the minimum film thickness and normal applied load and speed for an elastohydrodynamically lubricated contact is

$$(h_{min})_{EHL} \propto F^{-0.073} \tag{23.2}$$

$$(h_{min})_{EHL} \propto u^{0.68} \tag{23.3}$$

Comparing the results of Eqs. (23.2) and (23.3) with that obtained for hydrodynamic lubrication expressed in Eq. (23.1) indicates that:

1. The exponent on the normal applied load is nearly seven times larger for hydrodynamic lubrication than for elastohydrodynamic lubrication. This implies that in elastohydrodynamic lubrication the film thickness is only slightly affected by load while in hydrodynamic lubrication it is significantly affected by load.

2. The exponent on mean velocity is slightly higher for elastohydrodynamic lubrication than that found for hydrodynamic lubrication.

More discussion of elastohydrodynamic lubrication can be found in Section 23.3.

The load per unit area in conformal bearings is relatively low, typically averaging only 1 MN/m² and seldom over 7 MN/m². By contrast, the load per unit area in nonconformal contacts will generally exceed 700 MN/m² even at modest applied loads. These high pressures result in elastic deformation of the bearing materials such that elliptical contact areas are formed for oil-film generation and load support. The significance of the high contact pressures is that they result in a considerable increase in fluid viscosity. Inasmuch as viscosity is a measure of a fluid's resistance to flow, this increase greatly enhances the lubricant's ability to support load without being squeezed out of the contact zone. The high contact pressures in nonconforming surfaces therefore result in both an elastic deformation of the surfaces and large increases in the fluid's viscosity. The minimum film thickness is a function of the parameters found for hydrodynamic lubrication with the addition of an effective modulus of elasticity parameter for the bearing materials and a pressure–viscosity coefficient for the lubricant.

23.1.2 Bearing Selection

Ball bearings are used in many kinds of machines and devices with rotating parts. The designer is often confronted with decisions on whether a nonconformal bearing such as a rolling-element bearing or a conformal bearing such as a hydrodynamic bearing should be used in a particular application. The following characteristics make rolling-element bearings *more desirable* than hydrodynamic bearings in many situations:

1. Low starting and good operating friction.
2. The ability to support combined radial and thrust loads.
3. Less sensitivity to interruptions in lubrication.
4. No self-excited instabilities.
5. Good low-temperature starting.

Within reasonable limits changes in load, speed, and operating temperature have but little effect on the satisfactory performance of rolling-element bearings.

The following characteristics make nonconformal bearings such as rolling-element bearings *less desirable* than conformal (hydrodynamic) bearings:

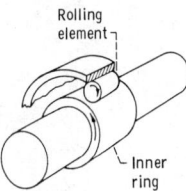

Rolling
element

Inner
ring

Fig. 23.2 Nonconformal surfaces. (From Ref. 6.)

1. Finite fatigue life subject to wide fluctuations.
2. Large space required in the radial direction.
3. Low damping capacity.
4. High noise level.
5. More severe alignment requirements.
6. Higher cost.

Each type of bearing has its particular strong points, and care should be taken in choosing the most appropriate type of bearing for a given application.

The Engineering Sciences Data Unit documents[7,8] provide an excellent guide to the selection of the type of journal or thrust bearing most likely to give the required performance when considering the load, speed, and geometry of the bearing. The following types of bearings were considered:

1. Rubbing bearings, where the two bearing surfaces rub together (e.g., unlubricated bushings made from materials based on nylon, polytetrafluoroethylene, also known as PTFE, and carbon).
2. Oil-impregnated porous metal bearings, where a porous metal bushing is impregnated with lubricant and thus gives a self-lubricating effect (as in sintered-iron and sintered-bronze bearings).
3. Rolling-element bearings, where relative motion is facilitated by interposing rolling elements between stationary and moving components (as in ball, roller, and needle bearings).
4. Hydrodynamic film bearings, where the surfaces in relative motion are kept apart by pressures generated hydrodynamically in the lubricant film.

Figure 23.3, reproduced from the Engineering Sciences Data Unit publication,[7] gives a guide to the typical load that can be carried at various speeds, for a nominal life of 10,000 hr at room temperature, by journal bearings of various types on shafts of the diameters quoted. The heavy curves indicate the preferred type of journal bearing for a particular load, speed, and diameter and thus divide the graph into distinct regions. From Fig. 23.3 it is observed that rolling-element bearings are preferred at lower speeds and hydrodynamic oil film bearings are preferred at higher speeds. Rubbing bearings and oil-impregnated porous metal bearings are not preferred for any of the speeds, loads, or shaft diameters considered. Also, as the shaft diameter is increased, the transitional point at which hydrodynamic bearings are preferred over rolling-element bearings moves to the left.

The applied load and speed are usually known, and this enables a preliminary assessment to be made of the type of journal bearing most likely to be suitable for a particular application. In many cases the shaft diameter will have been determined by other considerations, and Fig. 23.3 can be used to find the type of journal bearing that will give adequate load capacity at the required speed. These curves are based on good engineering practice and commercially available parts. Higher loads and speeds or smaller shaft diameters are possible with exceptionally high engineering standards or specially produced materials. Except for rolling-element bearings the curves are drawn for bearings with a width equal to the diameter. A medium-viscosity mineral oil lubricant is assumed for the hydrodynamic bearings.

Similarly, Fig. 23.4, reproduced from the Engineering Sciences Data Unit publication,[8] gives a guide to the typical maximum load that can be carried at various speeds for a nominal life of 10,000 hr at room temperature by thrust bearings of the various diameters quoted. The heavy curves again indicate the preferred type of bearing for a particular load, speed, and diameter and thus divide the graph into major regions. As with the journal bearing results (Fig. 23.3) the hydrodynamic bearing is preferred at lower speeds. A difference between Figs. 23.3 and 23.4 is that at very low speeds there is a portion of the latter figure in which the rubbing bearing is preferred. Also, as the shaft diameter is increased, the transitional point at which hydrodynamic bearings are preferred over rolling-element bearings moves to the left. Note also from this figure that oil-impregnated porous metal bearings are not preferred for any of the speeds, loads, or shaft diameters considered.

23.1.3 Lubricants

Both oils and greases are extensively used as lubricants for all types of machine elements over wide range of speeds, pressures, and operating temperatures. Frequently, the choice is determined by considerations other than lubrication requirements. The requirements of the lubricant for successful operation of nonconformal contacts such as in rolling-element bearings and gears are considerably more stringent than those for conformal bearings and therefore will be the primary concern in this section.

Because of its fluidity oil has several advantages over grease: It can enter the loaded conjunction most readily to flush away contaminants, such as water and dirt, and, particularly, to transfer heat from heavily loaded machine elements. Grease, however, is extensively used because it permits simplified designs of housings and enclosures, which require less maintenance, and because it is more effective in sealing against dirt and contaminants.

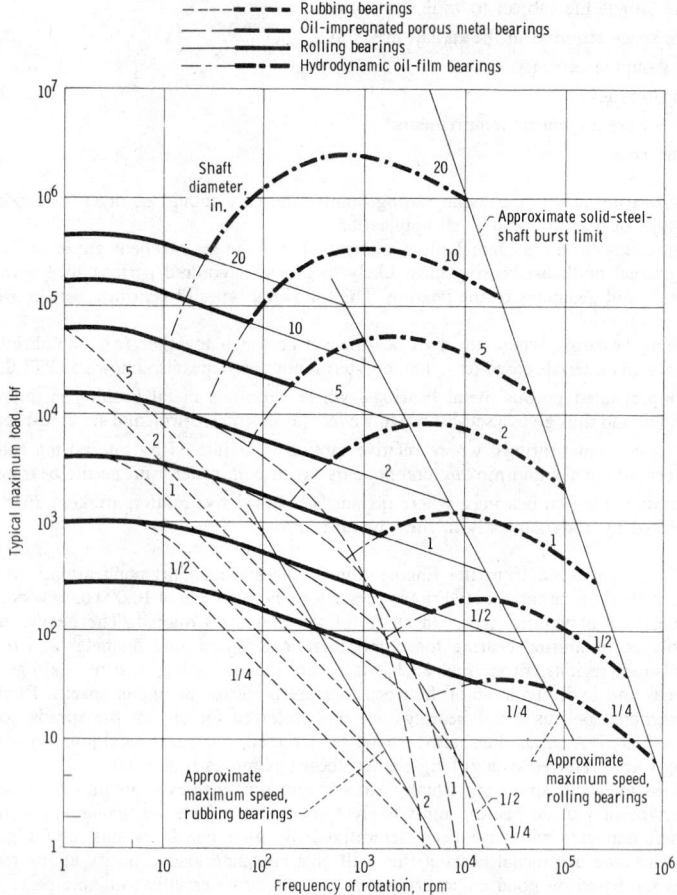

----■--- Rubbing bearings
─────── Oil-impregnated porous metal bearings
─────── Rolling bearings
──-──■-■ Hydrodynamic oil-film bearings

Fig. 23.3 General guide to journal bearing type. (Except for roller bearings, curves are drawn for bearings with width equal to diameter. A medium-viscosity mineral oil lubricant is assumed for hydrodynamic bearings.) (From Ref. 7.)

Viscosity

In hydrodynamic and elastohydrodynamic lubrication the most important physical property of a lubricant is its viscosity. The viscosity of a fluid may be associated with its resistance to flow, that is, with the resistance arising from intermolecular forces and internal friction as the molecules move past each other. Thick fluids, like molasses, have relatively high viscosity; they do not flow easily. Thinner fluids, like water, have lower viscosity; they flow very easily.

The relationship for internal friction in a viscous fluid (as proposed by Newton,[9] can be written as

$$\tau = \eta \frac{du}{dz} \tag{23.4}$$

where τ = internal shear stress in the fluid in the direction of motion

η = coefficient of absolute or dynamic viscosity or coefficient of internal friction

du/dz = velocity gradient perpendicular to the direction of motion (i.e., shear rate)

It follows from Eq. (23.4) that the unit of dynamic viscosity must be the unit of shear stress divided by the unit of shear rate. In the newton-meter-second system the unit of shear stress is the newton per square meter while that of shear rate is the inverse second. Hence the unit of dynamic viscosity will be newton per square meter multiplied by second, or N sec/m². In the SI system the unit of pressure or stress (N/m²) is known as pascal, abbreviated Pa, and it is becoming increasingly common to refer to the SI unit of viscosity as the pascal-second (Pa sec). In the cgs system, where the dyne

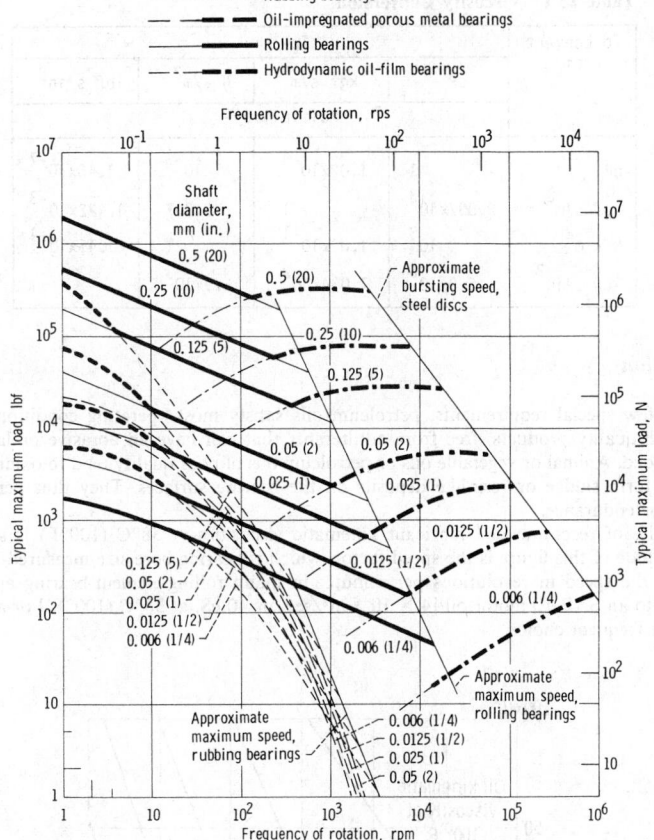

Fig. 23.4 General guide to thrust bearing type. (Except for roller bearings, curves are drawn for typical ratios of inside to outside diameter. A medium-viscosity mineral oil lubricant is assumed for hydrodynamic bearings.) (From Ref. 8.)

is the unit of force, dynamic viscosity is expressed as dyne-second per square centimeter. This unit is called the poise, with its submultiple the centipoise ($1 \text{ cP} = 10^{-2} \text{ P}$) of a more convenient magnitude for many lubricants used in practice.

Conversion of dynamic viscosity from one system to another can be facilitated by Table 23.1. To convert from a unit in the column on the left-hand side of the table to a unit at the top of the table, multiply by the corresponding value given in the table. For example, $\eta = 0.04 \text{ N sec/m}^2 = 0.04 \times 1.45 \times 10^{-4} \text{ lbf sec/in.}^2 = 5.8 \times 10^{-6} \text{ lbf sec/in.}^2$. One English and three metric systems are presented—all based on force, length, and time. Metric units are the centipoise, the kilogram force-second per square meter, and the newton-second per square meter (or Pa sec). The English unit is pound force-second per square inch, or reyn, in honor of Osborne Reynolds.

In many situations it is convenient to use the *kinematic viscosity* rather than the dynamic viscosity. The kinematic viscosity η_k is equal to the dynamic viscosity η divided by the density ρ of the fluid ($\eta_k = \eta/\rho$). The ratio is literally kinematic, all trace of force or mass cancelling out. The unit of kinematic viscosity may be written in SI units as square meters per second or in English units as square inches per second or, in cgs units, as square centimeters per second. The name stoke, in honor of Sir George Gabriel Stokes, was proposed for the cgs unit by Max Jakob in 1928. The centistoke, or one-hundredth part, is an everyday unit of more convenient size, corresponding to the centipoise.

The viscosity of a given lubricant varies within a given machine element as a result of the nonuniformity of pressure or temperature prevailing in the lubricant film. Indeed, many lubricated machine elements operate over ranges of pressure or temperature so extensive that the consequent variations in the viscosity of the lubricant may become substantial and, in turn, may dominate the operating characteristics of machine elements. Consequently, an adequate knowledge of the viscosity–pressure and viscosity–pressure–temperature relationships of lubricants is indispensable.

Table 23.1 Viscosity Conversion

To convert from-	To-			
	cP	kgf s/m^2	N s/m^2	lbf s/in^2
	Multiply by-			
cP	1	1.02×10^{-4}	10^{-3}	1.45×10^{-7}
kgf s/m^2	9.807×10^3	1	9.807	1.422×10^{-3}
N s/m^2	10^3	1.02×10^{-1}	1	1.45×10^{-4}
lbf s/in^2	6.9×10^6	7.034×10^2	6.9×10^3	1

Oil Lubrication

Except for a few special requirements, petroleum oils satisfy most operating conditions in machine elements. High-quality products, free from adulterants that can have an abrasive or lapping action, are recommended. Animal or vegetable oils or petroleum oils of poor quality tend to oxidize, to develop acids, and to form sludge or resinlike deposits on the bearing surfaces. They thus penalize bearing performance or endurance.

A composite of recommended lubricant kinematic viscosities at 38°C (100°F) is shown in Fig. 23.5. The ordinate of this figure is the speed factor, which is bearing bore size measured in millimeters multiplied by the speed in revolutions per minute. In many rolling-element-bearing applications an oil equivalent to an SAE-10 motor oil [4×10^{-6} m²/sec, or 40 cS, at 38°C (100°F)] or a light turbine oil is the most frequent choice.

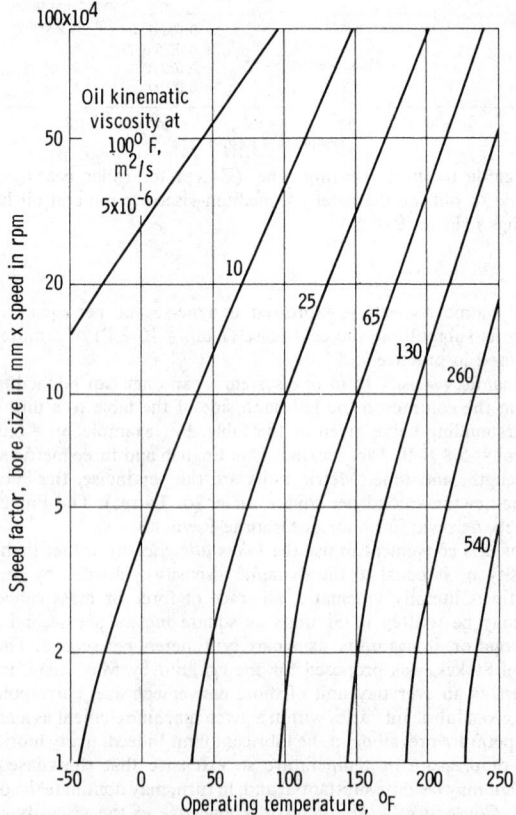

Fig. 23.5 Recommended lubricant viscosities for ball bearings. (From Ref. 10.)

For a number of military applications where the operational requirements span the temperature range −54 to 204°C (−65 to 400°F), synthetic oils are used. Ester lubricants are most frequently employed in this temperature range. In applications where temperatures exceed 260°C (500°F), most synthetics will quickly break down, and either a solid lubricant (e.g., MoS$_2$) or a polyphenyl ether is recommended. A more detailed discussion of synthetic lubricants can be found in Bisson and Anderson.[11]

Grease Lubrication

The simplest method of lubricating a bearing is to apply grease, because of its relatively nonfluid characteristics. The danger of leakage is reduced, and the housing and enclosure can be simpler and less costly than those used with oil. Grease can be packed into bearings and retained with inexpensive enclosures, but packing should not be excessive and the manufacturer's recommendations should be closely adhered to.

The major limitation of grease lubrication is that it is not particularly useful in high-speed applications. In general, it is not employed for speed factors over 200,000, although selected greases have been used successfully for higher speed factors with special designs.

Greases vary widely in properties depending on the type and grade or consistency. For this reason few specific recommendations can be made. Greases used for most bearing operating conditions consist of petroleum, diester, polyester, or silicone oils thickened with sodium or lithium soaps or with more recently developed nonsoap thickeners. General characteristics of greases are as follows:

1. Petroleum oil greases are best for general-purpose operation from −34 to 149°C (−30 to 300°F).
2. Diester oil greases are designed for low-temperature service down to −54°C (−65°F).
3. Ester-based greases are similar to diester oil greases but have better high-temperature characteristics, covering the range from −73 to 177°C (−100 to 350°F).
4. Silicone oil greases are used for both high- and low-temperature operation, over the widest temperature range of all greases [−73 to 232°C (−100 to 450°F)], but have the disadvantage of low load-carrying capacity.
5. Fluorosilicone oil greases have all of the desirable features of silicone oil greases plus good load capacity and resistance to fuels, solvents, and corrosive substances. They have a very low volatility in vacuum down to 10^{-7} torr, which makes them useful in aerospace applications.
6. Perfluorinated oil greases have a high degree of chemical inertness and are completely nonflammable. They have good load-carrying capacity and can operate at temperatures as high as 280°C (550°F) for long periods, which makes them useful in the chemical processing and aerospace industries, where high reliability justifies the additional cost.

Grease consistency is important since grease will slump badly and churn excessively when too soft and fail to lubricate when too hard. Either condition causes improper lubrication, excessive temperature rise, and poor performance and can shorten machine element life. A valuable guide to the estimation of the useful life of grease in rolling-element bearings has been published by the Engineering Sciences Data Unit.[12]

It has recently been demonstrated by Aihara and Dowson[13] and by Wilson[14] that the film thickness in grease-lubricated components can be calculated with adequate accuracy by using the viscosity of the base oil in the elastohydrodynamic equation (see Section 23.3). This enables the elastohydrodynamic lubrication film thickness formulas to be applied with confidence to grease-lubricated machine elements.

23.1.4 Lubrication Regimes

If a machine element is adequately designed and lubricated, the lubricated surfaces are separated by a lubricant film. Endurance testing of ball bearings, as reported by Tallian et al.,[15] has demonstrated that when the lubricant film is thick enough to separate the contacting bodies, fatigue life of the bearing is greatly extended. Conversely, when the film is not thick enough to provide full separation between the asperities in the contact zone, the life of the bearing is adversely affected by the high shear resulting from direct metal-to-metal contact.

To establish the effect of film thickness on the life of the machine element, we first introduce a relevant parameter Λ. The relationship between Λ and the minimum film thickness $h_{\min}$ is defined to be

$$\Lambda = \frac{h_{\min}}{(\Delta_a^2 + \Delta_b^2)^{1/2}} \qquad (23.5)$$

where Δ_a = rms surface finish of surface a
Δ_b = rms surface finish of surface b

Hence Λ is just the minimum film thickness in units of the composite roughness of the two bearing surfaces.

Hydrodynamic Lubrication Regime

Hydrodynamic lubrication occurs when the lubricant film is sufficiently thick to prevent the opposing solids from coming into contact. This condition is often referred to as the ideal form of lubrication since it provides low friction and a high resistance to wear. The lubrication of the contact is governed by the bulk physical properties of the lubricant, notably viscosity, and the frictional characteristics arise purely from the shearing of the viscous lubricant. The pressure developed in the oil film of hydrodynamically lubricated bearings is due to two factors:

1. The geometry of the moving surfaces produces a convergent film shape.
2. The viscosity of the liquid results in a resistance to flow.

The lubricant films are normally many times thicker than the surface roughness so that the physical properties of the lubricant dictate contact behavior. The film thickness normally exceeds 10^{-6} m. For hydrodynamic lubrication the film parameter Λ, defined in Eq. (23.5), is in excess of 10 and may even rise to 100. Films of this thickness are clearly also insensitive to chemical action in surface layers of molecular proportions.

For normal load support to occur in bearings, positive pressure profiles must develop over the length of the bearing. Three different forms of hydrodynamic lubrication are presented in Fig. 23.6. Figure 23.6a shows a slider bearing. For a positive load to be developed in the slider bearing shown in Fig. 23.6a the lubricant film thickness must be decreasing in the direction of sliding.

A squeeze film bearing is another mechanism of load support of hydrodynamic lubrication, and it is illustrated in Fig. 23.6b. The squeeze action is the normal approach of the bearing surfaces. The squeeze mechanism of pressure generation provides a valuable cushioning effect when the bearing surfaces tend to be pressed together. Positive pressures will be generated when the film thickness is diminishing.

An externally pressurized bearing is yet a third mechanism of load support of hydrodynamic lubrication, and it is illustrated in Fig. 23.6c. The pressure drop across the bearing is used to support the load. The load capacity is independent of the motion of the bearing and the viscosity of the lubricant.

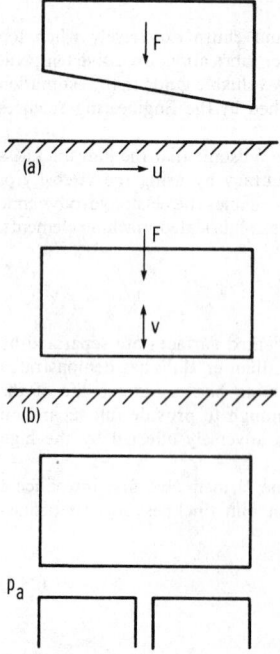

Fig. 23.6 Mechanisms of load support for hydrodynamic lubrication. (*a*) Slider bearing. (*b*) Squeeze film bearing. (*c*) Externally pressurized bearing.

There is no problem of contact at starting and stopping as with the other hydrodynamically lubricated bearings because pressure is applied before starting and is maintained until after stopping.

Hydrodynamically lubricated bearings are discussed further in Section 23.2.

Elastohydrodynamic Lubrication Regime

Elastohydrodynamic lubrication is a form of hydrodynamic lubrication where elastic deformation of the bearing surfaces becomes significant. It is usually associated with highly stressed machine components of low conformity. There are two distinct forms of elastohydrodynamic lubrication (EHL).

Hard EHL.   Hard EHL relates to materials of *high* elastic modulus, such as metals. In this form of lubrication both the elastic deformation and the pressure–viscosity effects are equally important. Engineering applications in which elastohydrodynamic lubrication are important for high-elastic-modulus materials include gears and rolling-element bearings.

Soft EHL.   Soft EHL relates to materials of *low* elastic modulus, such as rubber. For these materials the elastic distortions are large, even with light loads. Another feature of the elastohydrodynamics of low-elastic-modulus materials is the negligible effect of the relatively low pressures on the viscosity of the lubricating fluid. Engineering applications in which elastohydrodynamic lubrication are important for low-elastic-modulus materials include seals, human joints, tires, and a number of lubricated elastomeric material machine elements.

The common factors in hard and soft EHL are that the local elastic deformation of the solids provides coherent fluid films and that asperity interaction is largely prevented. Elastohydrodynamic lubrication normally occurs in contacts where the minimum film thickness is in the range 0.1 μm $<$ $h_{min} \leq 10$ μm and the film parameter Λ is in the range $3 \leq \Lambda < 10$. Elastohydrodynamic lubrication is discussed further in Section 23.3.

Boundary Lubrication Regime

If in a lubricated contact the pressures become too high, the running speeds too low, or the surface roughness too great, penetration of the lubricant film will occur. Contact will take place between the asperities. The friction will rise and approach that encountered in dry friction between solids. More importantly, wear will take place. Adding a small quantity of certain active organic compounds to the lubricating oil can, however, extend the life of the machine elements. These additives are present in small quantities ($<$1%) and function by forming low-shear-strength surface films strongly attached to the metal surfaces. Although they are sometimes only one or two molecules thick, such films are able to prevent metal-to-metal contact.

Some boundary lubricants are long-chain molecules with an active end group, typically an alcohol, an amine, or a fatty acid. When such a material, dissolved in a mineral oil, meets a metal or other solid surface, the active end group attaches itself to the solid and gradually builds up a surface layer. The surface films vary in thickness from 5×10^{-9} to 10^{-8} m depending on molecular size, and the film parameter Λ is less than unity ($\Lambda < 1$). Boundary lubrication is discussed further in Section 23.4.

Figure 23.7 illustrates the film conditions existing in hydrodynamic, elastohydrodynamic, and boundary lubrication. The surface slopes in this figure are greatly distorted for the purpose of illustration. To scale, real surfaces would appear as gently rolling hills rather than sharp peaks.

23.1.5   Relevant Equations

This section presents the equations frequently used in hydrodynamic and elastohydrodynamic lubrication theory. They are not relevant to boundary lubrication since in this lubrication regime bulk fluid effects are negligible. The differential equation governing the pressure distribution in hydrodynamically and

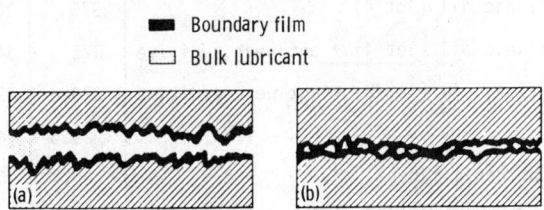

■ Boundary film
☐ Bulk lubricant

Fig. 23.7   Film conditions of lubrication regimes. (*a*) Hydrodynamic and elastohydrodynamic lubrication—surfaces separated by bulk lubricant film. (*b*) Boundary lubrication—performance essentially dependent on boundary film.

elastohydrodynamically lubricated machine elements is known as the Reynolds equation. For steady-state hydrodynamic lubrication the Reynolds equation normally appears as

$$\frac{\partial}{\partial x}\left(h^3 \frac{\partial p}{\partial x}\right) + \frac{\partial}{\partial y}\left(h^3 \frac{\partial p}{\partial y}\right) = 12\eta u \frac{\partial h}{\partial x} \tag{23.6}$$

where h = film shape measured in the z direction, m
   p = pressure, N/m²
   η = lubricant viscosity, N sec/m²
   u = mean velocity, $(u_a + u_b)/2$, m/sec

Solutions of Eq. (23.6) are rarely achieved analytically, and approximate numerical solutions are sought.
   For elastohydrodynamic lubrication the steady-state form of the Reynolds equation normally appears as

$$\frac{\partial}{\partial x}\left(\frac{\rho h^3}{\eta} \frac{\partial p}{\partial x}\right) + \frac{\partial}{\partial y}\left(\frac{\rho h^3}{\eta} \frac{\partial p}{\partial y}\right) = 12u \frac{\partial(\rho h)}{\partial x} \tag{23.7}$$

where ρ is lubricant density in N sec²/m². The essential difference between Eqs. (23.6) and (23.7) is that Eq. (23.7) allows for variation of viscosity and density in the x and y directions. Equations (23.6) and (23.7) allow for the bearing surfaces to be of finite length in the y direction. Side leakage, or flow in the y direction, is associated with the second term in Eqs. (23.6) and (23.7). The solution of Eq. (23.7) is considerably more difficult than that of Eq. (23.6); therefore, only numerical solutions are available.
   The viscosity of a fluid may be associated with the resistance to flow, with the resistance arising from the intermolecular forces and internal friction as the molecules move past each other. Because of the much larger pressure variation in the lubricant conjunction, the viscosity of the lubricant for elastohydrodynamic lubrication does not remain constant as is approximately true for hydrodynamic lubrication.
   As long ago as 1893, Barus[16] proposed the following formula for the isothermal viscosity–pressure dependence of liquids:

$$\eta = \eta_0 \, e^{\xi p} \tag{23.8}$$

where η_0 = viscosity at atmospheric pressure
   ξ = pressure–viscosity coefficient of lubricant

Table 23.2 Absolute Viscosities of Test Fluids at Atmospheric Pressure and Three Temperatures (From Ref. 17)

Test fluid	Temperature, °C		
	38	99	149
	Absolute viscosity, η, cP		
Advanced ester	25.3	4.75	2.06
Formulated advanced ester	27.6	4.96	2.15
Polyalkyl aromatic	25.5	4.08	1.80
Polyalkyl aromatic + 10 wt % heavy resin	32.2	4.97	2.03
Synthetic paraffinic oil (lot 3)	414	34.3	10.9
Synthetic paraffinic oil (lot 4)	375	34.7	10.1
Synthetic paraffinic oil (lot 4) + antiwear additive	375	34.7	10.1
Synthetic paraffinic oil (lot 2) + antiwear additive	370	32.0	9.93
C-ether	29.5	4.67	2.20
Superrefined naphthenic mineral oil	68.1	6.86	2.74
Synthetic hydrocarbon (traction fluid)	34.3	3.53	1.62
Fluorinated polyether	181	20.2	6.68

The pressure–viscosity coefficient ξ characterizes the liquid considered and depends in most cases only on temperature, not on pressure.

Table 23.2 lists the absolute viscosities of 12 lubricants at atmospheric pressure and three temperatures as obtained from Jones et al.[17] These values would correspond to η_0 to be used in Eq. (23.8) for the particular fluid and temperature to be used. The 12 fluids with manufacturer and manufacturer's designation are shown in Table 23.3. The pressure–viscosity coefficients ξ, expressed in square meters per newton, for these 12 fluids at three different temperatures are shown in Table 23.4.

For a comparable change in pressure the relative density change is smaller than the viscosity change. However, very high pressures exist in elastohydrodynamic films, and the liquid can no longer be considered as an incompressible medium. From Dowson and Higginson[18] the density can be written as

$$\rho = \rho_0 \left(1 + \frac{0.6p}{1 + 1.7p} \right) \tag{23.9}$$

where p is given in gigapascals.

The film shape appearing in Eq. (23.7) can be written with sufficient accuracy as

$$h = h_0 + \frac{x^2}{2R_x} + \frac{y^2}{2R_y} + \delta(x,y) \tag{23.10}$$

where h_0 = constant, m
$\delta(x,y)$ = total elastic deformation, m
R_x = effective radius in x direction, m
R_y = effective radius in y direction, m

The elastic deformation can be written, from standard elasticity theory, in the form

$$\delta(x,y) = \frac{2}{\pi E'} \iint_A \frac{p(x,y)\,dx_1\,dy_1}{[(x - x_1)^2 + (y - y_1)^2]^{1/2}} \tag{23.11}$$

Table 23.3 Fluids with Manufacturer and Manufacturer's Designation (From Ref. 17)

Test fluid	Manufacturer	Designation
Advanced ester	Shell Oil Co.	Aeroshell® turbine oil 555 (base oil)
Formulated advanced ester	Shell Oil Co.	Aeroshell® turbine oil 555 (WRGL-358)
Polyalkyl aromatic	Continental Oil Co.	DN-600
Synthetic paraffinic oil (lot 3)	Mobil Oil Corp.	XRM 109F3
Synthetic paraffinic oil (lot 4)		XRM 109F4
Synthetic paraffinic oil + antiwear additive (lot 2)		XRM 177F2
Synthetic paraffinic oil + antiwear additive (lot 4)		XRM 177F4
C-ether	Monsanto Co.	MCS-418
Superrefined naphthenic mineral oil	Humble Oil and Refining Co.	FN 2961
Synthetic hydrocarbon (traction fluid)	Monsanto Co.	MCS-460
Fluorinated polyether	DuPont Co.	PR 143 AB (Lot 10)

Table 23.4 Pressure–Viscosity Coefficients for Test Fluids at Three Temperatures (From Ref. 17)

Test fluid	Temperature, °C		
	38	99	149
	Pressure-viscosity coefficient, ξ, m^2/N		
Advanced ester	1.28×10^{-8}	0.987×10^{-8}	0.851×10^{-8}
Formulated advanced ester	1.37	1.00	.874
Polyalkyl aromatic	1.58	1.25	1.01
Polyalkyl aromatic + 10 wt % heavy resin	1.70	1.28	1.06
Synthetic paraffinic oil (lot 3)	1.77	1.51	1.09
Synthetic paraffinic oil (lot 4)	1.99	1.51	1.29
Synthetic paraffinic oil (lot 4) + antiwear additive	1.96	1.55	1.25
Synthetic paraffinic oil (lot 2) + antiwear additive	1.81	1.37	1.13
C-ether	1.80	.980	.795
Superrefined naphthenic mineral oil	2.51	1.54	1.27
Synthetic hydrocarbon (traction fluid)	3.12	1.71	.939
Fluorinated polyether	4.17	3.24	3.02

where

$$E' = 2\left(\frac{1-v_a^2}{E_a} + \frac{1-v_b^2}{E_b}\right)^{-1} \tag{23.12}$$

and $v =$ Poisson's ratio
$E =$ modulus of elasticity, N/m^2

Therefore, Eq. (23.6) is normally involved in hydrodynamic lubrication situations, while Eqs. (23.7)–(23.11) are normally involved in elastohydrodynamic lubrication situations.

23.2 HYDRODYNAMIC AND HYDROSTATIC LUBRICATION

Surfaces lubricated hydrodynamically are normally conformal as pointed out in Section 23.1.1. The conformal nature of the surfaces can take its form either as a thrust bearing or as a journal bearing, both of which will be considered in this section. Three features must exist for hydrodynamic lubrication to occur:

1. A viscous fluid must separate the lubricated surfaces.
2. There must be relative motion between the surfaces.
3. The geometry of the film shape must be larger in the inlet than at the outlet so that a convergent wedge of lubricant is formed.

If feature 2 is absent, lubrication can still be achieved by establishing relative motion between the fluid and the surfaces through external pressurization. This is discussed further in Section 23.2.3.

In hydrodynamic lubrication the entire friction arises from the shearing of the lubricant film so that it is determined by the viscosity of the oil: the thinner (or less viscous) the oil, the lower the friction. The great advantages of hydrodynamic lubrication are that the friction can be very low ($\mu \simeq 0.001$) and, in the ideal case, there is no wear of the moving parts. The main problems in hydrodynamic lubrication are associated with starting or stopping since the oil film thickness theoretically is zero when the speed is zero.

The emphasis in this section is on hydrodynamic and hydrostatic lubrication. This section is not intended to be all inclusive but rather to typify the situations existing in hydrodynamic and hydrostatic lubrication. For additional information the reader is recommended to investigate Gross et al.,[19] Reiger,[20] Pinkus and Sternlicht,[21] and Rippel.[22]

23.2.1 Liquid-Lubricated Hydrodynamic Journal Bearings

Journal bearings, as shown in Fig. 23.8, are used to support shafts and to carry radial loads with minimum power loss and minimum wear. The bearing can be represented by a plain cylindrical bush

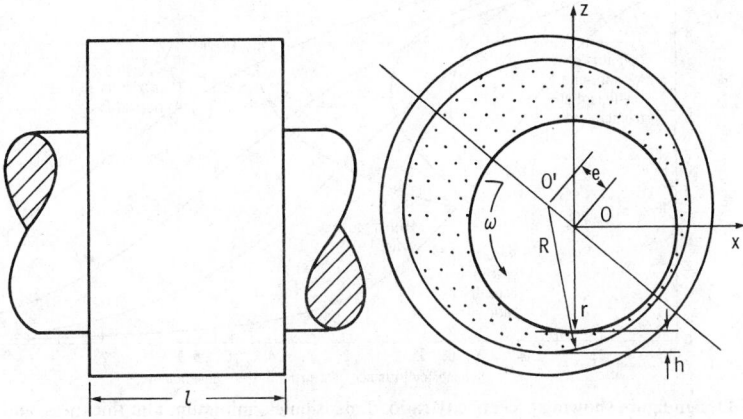

Fig. 23.8 Journal bearing.

wrapped around the shaft, but practical bearings can adopt a variety of forms. The lubricant is supplied at some convenient point through a hole or a groove. If the bearing extends around the full 360° of the shaft, the bearing is described as a full journal bearing. If the angle of wrap is less than 360°, the term "partial journal bearing" is employed.

Plain

Journal bearings rely on the motion of the shaft to generate the load-supporting pressures in the lubricant film. The shaft does not normally run concentric with the bearing center. The distance between the shaft center and the bearing center is known as the eccentricity. This eccentric position within the bearing clearance is influenced by the load that it carries. The amount of eccentricity adjusts itself until the load is balanced by the pressure generated in the converging portion of the bearing. The pressure generated, and therefore the load capacity of the bearing, depends on the shaft eccentricity e, the frequency of rotation N, and the effective viscosity of the lubricant η in the converging film, as well as the bearing dimensions l and d and the clearance c. The three dimensionless groupings normally used for journal bearings are:

1. The eccentricity ratio, $\epsilon = e/c$
2. The length-to-diameter ratio, $\lambda = l/d$
3. The Sommerfeld number, $Sm = \eta N d^3 l / 2Fc^2$

When designing a journal bearing, the first requirement to be met is that it should operate with an adequate minimum film thickness, which is directly related to the eccentricity ($h_{min} = c - e$). Figures 23.9, 23.10, and 23.11 show the eccentricity ratio, the dimensionless minimum film thickness, and the dimensionless Sommerfeld number for, respectively, a full journal bearing and partial journal bearings of 180° and 120°. In these figures a recommended operating eccentricity ratio is indicated as well as a preferred operational area. The left boundary of the shaded zone defines the optimum eccentricity ratio for minimum coefficient of friction, and the right boundary is the optimum eccentricity ratio for maximum load. In these figures it can be observed that the shaded area is significantly reduced for the partial bearings as compared with the full journal bearing. These plots were adapted from results given in Raimondi and Boyd.[23]

Figures 23.12, 23.13, and 23.14 show a plot of attitude angle ϕ (angle between the direction of the load and a line drawn through the centers of the bearing and the journal) and the bearing characteristic number for various length-to-diameter ratios for, respectively, a full journal bearing and partial journal bearings of 180° and 120°. This angle establishes where the minimum and maximum film thicknesses are located within the bearing. These plots were also adapted from results given in Raimondi and Boyd,[23] where additional information about the coefficient of friction, the flow variable, the temperature rise, and the maximum film pressure ratio for a complete range of length-to-diameter ratios as well as for full or partial journal bearings can be found.

Nonplain

As applications have demanded higher speeds, vibration problems due to critical speeds, imbalance, and instability have created a need for journal bearing geometries other than plain journal bearings.

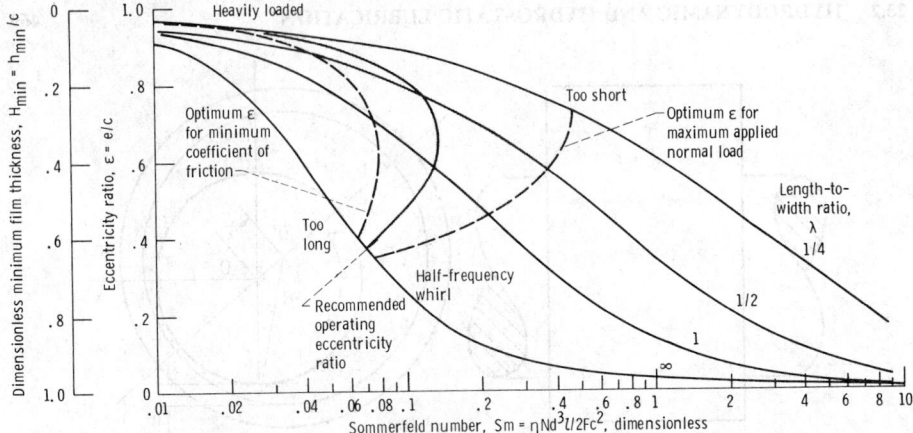

Fig. 23.9 Design figure showing eccentricity ratio, dimensionless minimum film thickness, and Sommerfeld number for full journal bearings. (Adapted from Ref. 23.)

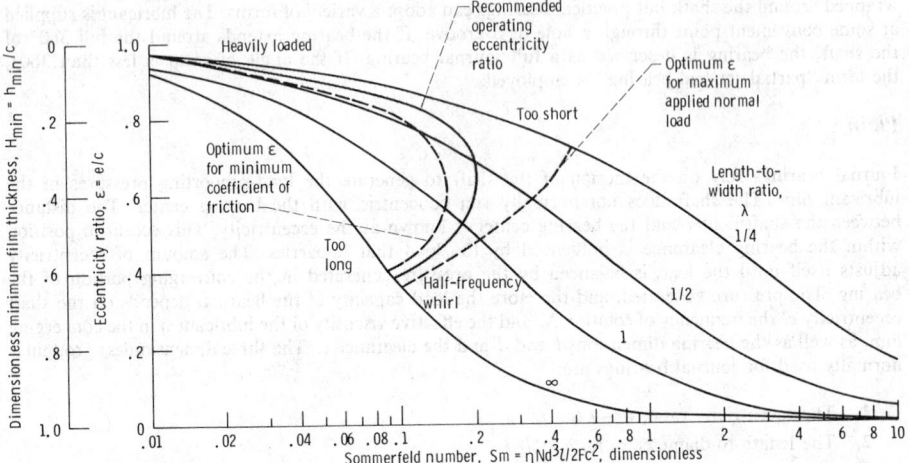

Fig. 23.10 Design figure showing eccentricity ratio, dimensionless minimum film thickness, and Sommerfeld number for 180° partial journal bearings, centrally loaded. (Adapted from Ref. 23.)

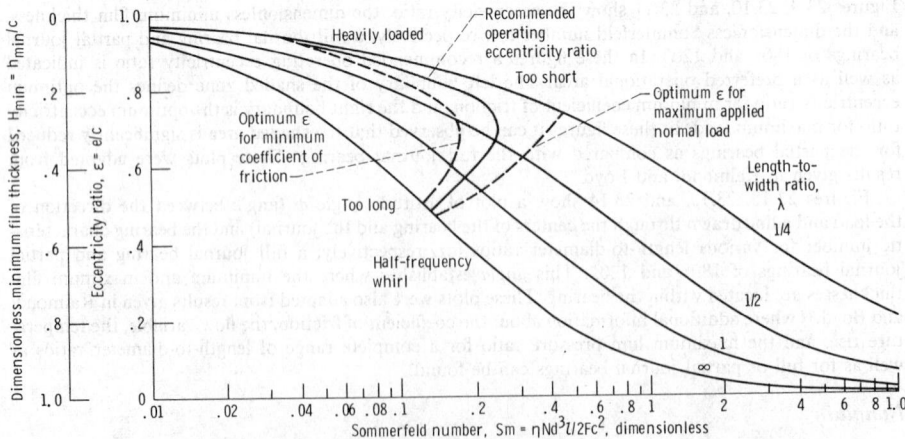

Fig. 23.11 Design figure showing eccentricity ratio, dimensionless minimum film thickness, and Sommerfeld number for 120° partial journal bearings, centrally loaded. (Adapted from Ref. 23.)

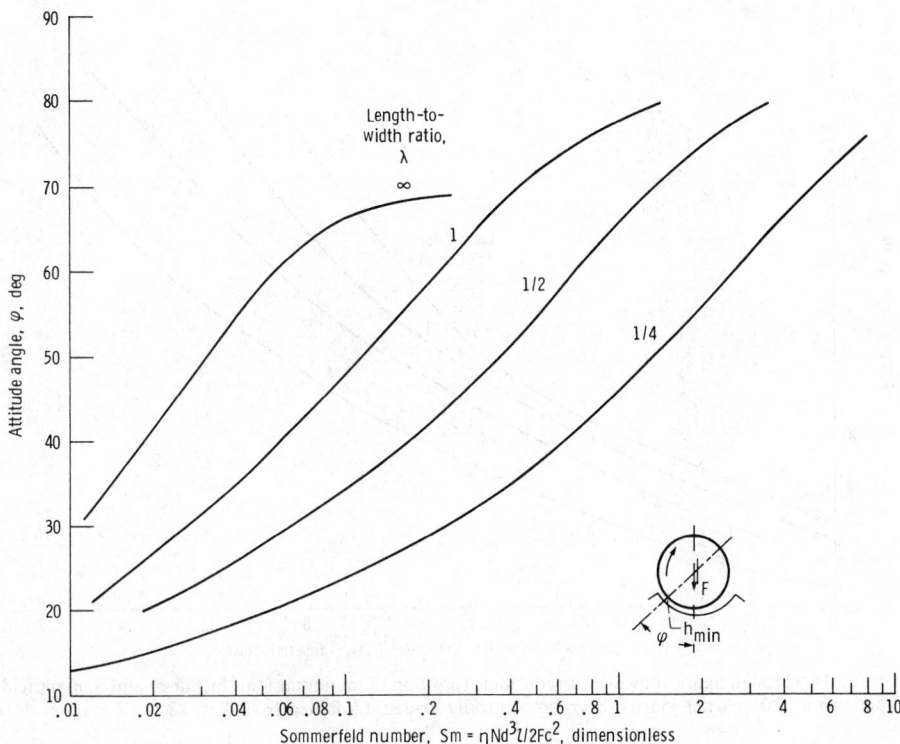

Fig. 23.12 Design figure showing attitude angle (position of minimum film thickness) and Sommerfeld number for full journal bearings, centrally loaded. (Adapted from Ref. 23.)

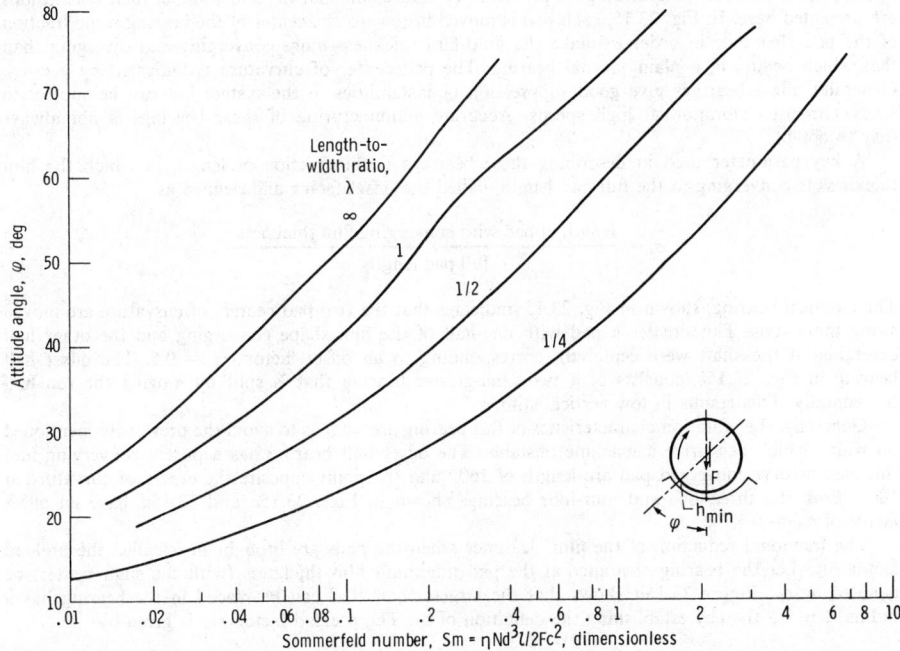

Fig. 23.13 Design figure showing attitude angle (position of minimum film thickness) and Sommerfeld number for 180° partial journal bearings, centrally loaded. (Adapted from Ref. 23.)

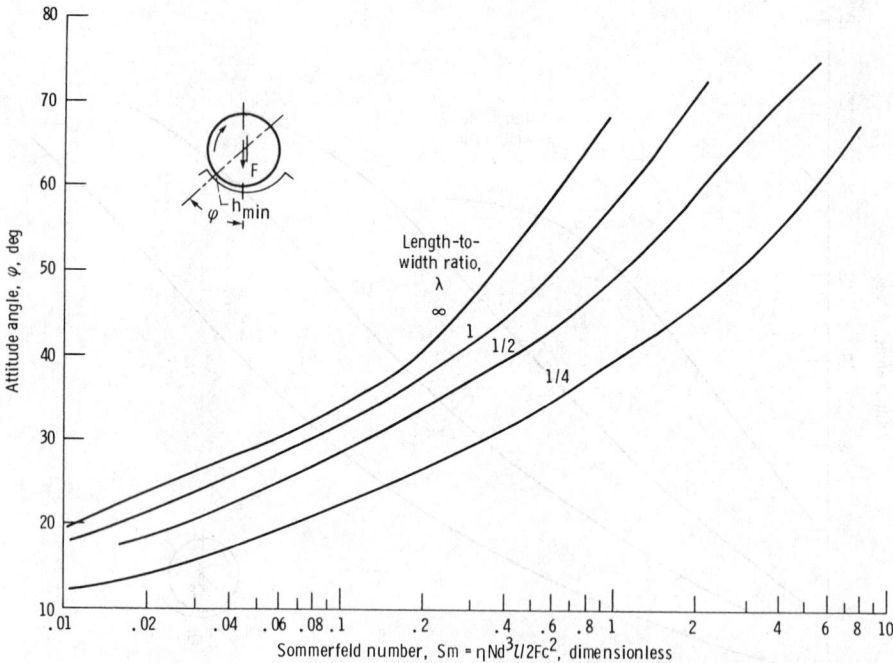

Fig. 23.14   Design figure showing attitude angle (position of minimum film thickness) and Sommerfeld number for 120° partial journal bearings, centrally loaded. (Adapted from Ref. 23.)

These geometries have various patterns of variable clearance so as to create pad film thicknesses that have more strongly converging and diverging regions. Figure 23.15 shows elliptical, offset half, three-lobe, and four-lobe bearings—bearings different from the plain journal bearing. An excellent discussion of the performance of these bearings is provided in Allaire and Flack,[24] and some of their conclusions are presented here. In Fig. 23.15, each pad is moved in toward the center of the bearing some fraction of the pad clearance in order to make the fluid-film thickness more converging and diverging than that which occurs in a plain journal bearing. The pad center of curvature is indicated by a cross. Generally, these bearings give good suppression of instabilities in the system but can be subject to subsynchronous vibration at high speeds. Accurate manufacturing of these bearings is not always easy to obtain.

A key parameter used in describing these bearings is the fraction of length in which the film thickness is converging to the full pad length, called the offset factor and defined as

$$\alpha_a = \frac{\text{length of pad with converging film thickness}}{\text{full pad length}}$$

The elliptical bearing, shown in Fig. 23.15, indicates that the two pad centers of curvature are moved along the y axis. This creates a pad with one-half of the film shape converging and the other half diverging (if the shaft were centered), corresponding to an offset factor $\alpha_a = 0.5$. The offset half bearing in Fig. 23.15b consists of a two-axial-groove bearing that is split by moving the top half horizontally. This results in low vertical stiffness.

Generally, the vibration characteristics of this bearing are such as to avoid the previously mentioned oil whirl, which can drive a machine unstable. The offset half bearing has a purely converging film thickness with a converged pad arc length of 160° and the point opposite the center of curvature at 180°. Both the three-lobe and four-lobe bearings shown in Figs. 23.15c and 23.15d have an offset factor of $\alpha_a = 0.5$.

The fractional reduction of the film clearance when the pads are brought in is called the preload factor m_p. Let the bearing clearance at the pad minimum film thickness (with the shaft centered) be denoted by c_b. Figure 23.16a shows that the largest shaft that can be placed in the bearing has a radius $R + c_b$, thereby establishing the definition of c_b. The preload factor m_p is given by

$$m_p = \frac{c - c_b}{c}$$

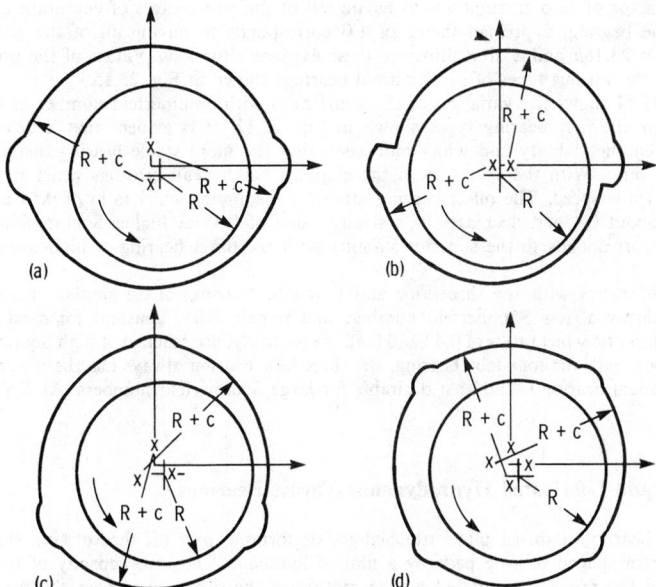

Fig. 23.15 Types of fixed-incline pad preloaded journal bearings. (From Ref. 24.) (a) Elliptical bore bearing ($\alpha_a = 0.5$, $m_p = 0.4$). (b) Offset half bearing ($\alpha_a = 1.125$, $m_p = 0.4$). (c) Three-lobe bearing ($\alpha_a = 0.5$, $m_p = 0.4$). (d) Four-lobe bearing ($\alpha_a = 0.5$, $m_p = 0.4$).

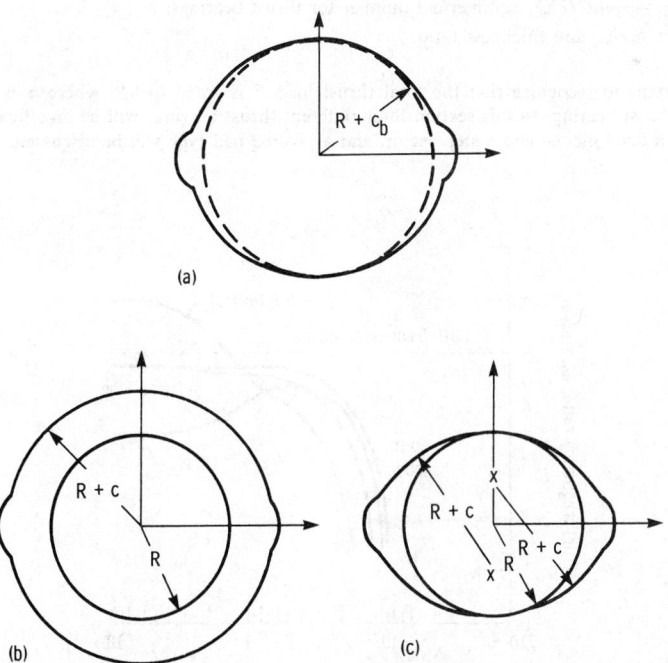

Fig. 23.16 Effect of preload on two-lobe bearings. (From Ref. 24.) (a) Largest shaft that fits in bearing. (b) $m = 0$, largest shaft $= R + c$, bearing clearance $c_b = c$. (c) $m = 1.0$, largest shaft $= R$, bearing clearance $c_b = 0$.

A preload factor of zero corresponds to having all of the pad centers of curvature coinciding at the center of the bearing; a preload factor of 1.0 corresponds to having all of the pads touching the shaft. Figures 23.16b and 23.16c illustrate these extreme situations. Values of the preload factor are indicated in the various types of fixed journal bearings shown in Fig. 23.15.

Figure 23.17 shows the variation of the whirl ratio with Sommerfeld number at the threshold of instability for the four bearing types shown in Fig. 23.15. It is evident that a definite relationship exists between the stability and whirl ratio such that the more stable bearing distinctly whirls at a lower speed ratio. With the exception of the elliptical bearing, all bearings whirl at speeds less than 0.48 of the rotor speed. The offset bearing attains a maximum whirl ratio of 0.44 at a Sommerfeld number of about 0.4 and decreases to a steady value of 0.35 at higher Sommerfeld numbers. This observation corresponds to the superior stability with the offset bearing at high-speed and light-load operations.

The whirl ratios with the three-lobe and four-lobe bearings share similar characteristics. They both rise sharply at low Sommerfeld numbers and remain fairly constant for most portions of the curves. Asymptotic whirl ratios of 0.47 and 0.48, respectively, are reached at high Sommerfeld numbers. In comparison with the four-lobe bearing, the three-lobe bearing always has the lower whirl ratio.

The elliptical bearing is the least desirable for large Sommerfeld numbers. At $Sm > 1.3$ the ratio exceeds 0.5.

23.2.2 Liquid-Lubricated Hydrodynamic Thrust Bearings

In a thrust bearing, a thrust plate attached to, or forming part of, the rotating shaft is separated from the sector-shaped bearing pads by a film of lubricant. The load capacity of the bearing arises entirely from the pressure generated by the motion of the thrust plate over the bearing pads. This action is achieved only if the clearance space between the stationary and moving components is convergent in the direction of motion. The pressure generated in, and therefore the load capacity of, the bearing, depends on the velocity of the moving slider $u = (R_1 + R_2)\omega/2 = \pi(R_1 + R_2)N$, the effective viscosity, the length of the pad l, the width of the pad b, the normal applied load F, the inlet film thickness h_i, and the outlet film thickness h_o. For thrust bearings three dimensionless parameters are used:

1. $\lambda = l/b$, pad length-to-width ratio;
2. $Sm_t = \eta u b l^2/F h_o^2$, Sommerfeld number for thrust bearings;
3. $\bar{h}_i = h_i/h_o$, film thickness ratio.

It is important to recognize that the total thrust load F is equal to nF, where n is the number of pads in a thrust bearing. In this section three different thrust bearings will be investigated. Two fixed-pad types, a fixed incline and a step sector, and a pivoted-pad type will be discussed.

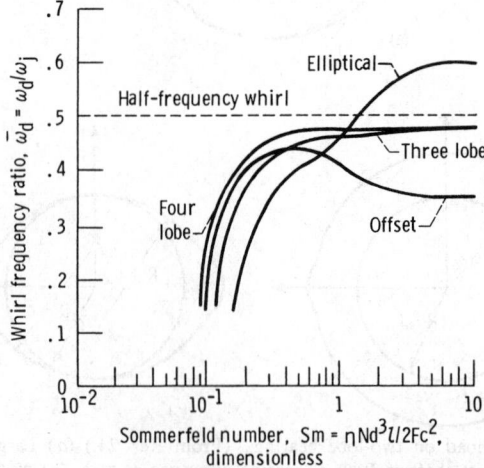

Fig. 23.17 Chart for determining whirl frequency ratio. (From Ref. 24.)

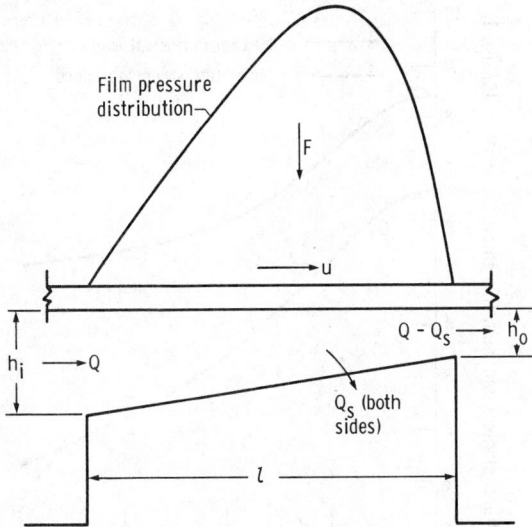

Fig. 23.18 Configuration of fixed-incline pad bearing. (From Ref. 25.)

Fixed-Incline Pad

The simplest form of fixed-pad thrust bearing provides only straight-line motion and consists of a flat surface sliding over a fixed pad or land having a profile similar to that shown in Fig. 23.18. The fixed-pad bearing depends for its operation on the lubricant being drawn into a wedge-shaped space and thus producing pressure that counteracts the load and prevents contact between the sliding parts. Since the wedge action only takes place when the sliding surface moves in the direction in which the lubricant film converges, the fixed-incline bearing, shown in Fig. 23.18, can only carry load for this direction of operation. If reversibility is desired, a combination of two or more pads with their surfaces sloped in opposite direction is required. Fixed-incline pads are used in multiples as in the thrust bearing shown in Fig. 23.19.

The following procedure assists in the design of a fixed-incline pad thrust bearing:

1. Choose a pad width-to-length ratio. A square pad ($\lambda = 1$) is generally felt to give good performance. From Fig. 23.20, if it is known whether maximum load or minimum power is most important in the particular application, a value of the film thickness ratio can be determined.

2. Within the terms in the Sommerfeld number the term least likely to be preassigned is the outlet film thickness. Therefore, determine h_o from Fig. 23.21. Since $\bar{h}_i$ is known from Fig. 23.20, h_i can be determined ($h_i = \bar{h}_i h_o$).

3. Check Table 23.5 to see if minimum (outlet) film thickness is sufficient for the preassigned surface finish. If not:

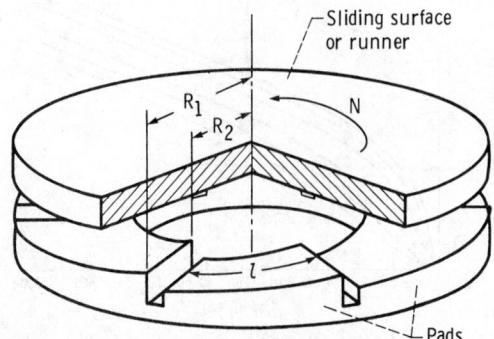

Fig. 23.19 Configuration of fixed-incline pad thrust bearing. (From Ref. 25.)

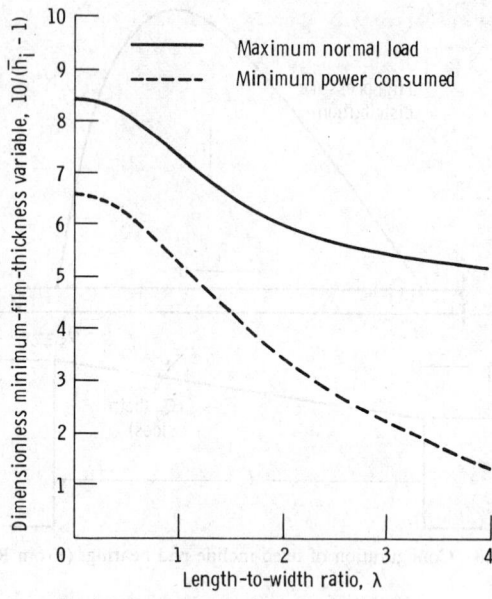

Fig. 23.20 Chart for determining minimum film thickness corresponding to maximum load or minimum power loss for various pad proportions—fixed-incline pad bearings. (From Ref. 25.)

    a. Increase the fluid viscosity or speed of the bearing.

    b. Decrease the load or the surface finish. Upon making this change return to step 1.

   4. Once an adequate minimum film thickness has been determined, use Figs. 23.22–23.24 to obtain, respectively, the coefficient of friction, the power consumed, and the flow.

Pivoted Pad

The simplest form of pivoted-pad bearing provides only for straight-line motion and consists of a flat surface sliding over a pivoted pad as shown in Fig. 23.25. If the pad is assumed to be in equilibrium under a given set of operating conditions, any change in these conditions, such as a change in load, speed, or viscosity, will alter the pressure distribution and thus momentarily shift the center of pressure and create a moment that causes the pad to change its inclination until a new position of equilibrium

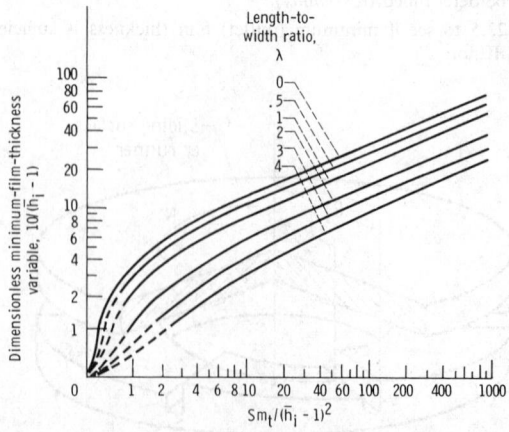

Fig. 23.21 Chart for determining minimum film thickness for fixed-incline pad thrust bearings. (From Ref. 25.)

Table 23.5 Allowable Minimum Outlet Film Thickness for a Given Surface Finish (From Ref. 8)

Surface finish		Description of surface	Examples of manufacturing methods	Approximate relative costs	Allowable minimum outlet film thickness[a], h_o	
Familiar British units, μin. CLA[b]	SI units, μm[c] CLA				Familiar British units, in.	SI units, m
4-8	0.1 - 0.2	Mirror-like surface without toolmarks, close tolerances	Grind, lap, and superfinish	17-20	0.00010	0.0000025
8-16	0.2 - 0.4	Smooth surface without scratches, close tolerances	Grind and lap	17-20	.00025	.0000062
16-32	0.4 - 0.8	Smooth surface, close tolerances	Grind, file, and lap	10	.00050	.0000125
32-63	0.8 - 1.6	Accurate bearing surface without toolmarks	Grind, precision mill, and file	7	.00100	.000025
63-125	1.6 - 3.2	Smooth surface without objectionable toolmarks, moderate tolerances	Shape, mill, grind, and turn	5	.00200	.000050

[a]The values of film thickness are given only for guidance. They indicate the film thickness required to avoid metal-to-metal contact under clean oil conditions with no misalignment. It may be necessary to take a larger film thickness than that indicated (e.g., to obtain an acceptable temperature rise). It has been assumed that the average surface finish of the pads is the same as that of the runner.
[b]CLA = centerline average.
[c]μm = micrometer; 40 μin. (microinch) = 1 μm.

is established. It can be shown that if the position of that pivot, as defined by the distance $\bar{x}$, is fixed by choosing $\bar{x}/l$, the ratio of the inlet film thickness to the outlet film thickness, h_i/h_o, also becomes fixed and is independent of load, speed, and viscosity. Thus the pad will automatically alter its inclination so as to maintain a constant value of h_i/h_o.

Pivoted pads are sometimes used in multiples as pivoted-pad thrust bearings, shown in Fig. 23.26. Calculations are carried through for a single pad, and the properties for the complete bearing are found by combining these calculations in the proper manner.

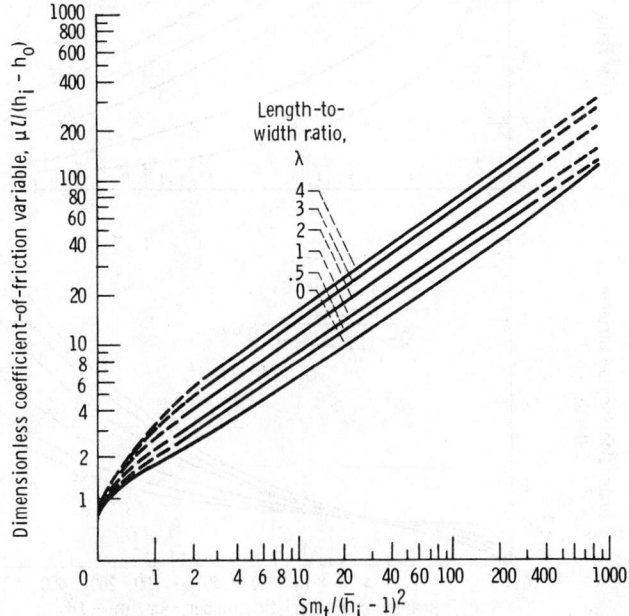

Fig. 23.22 Chart for determining coefficient of friction for fixed-incline pad thrust bearings. (From Ref. 25.)

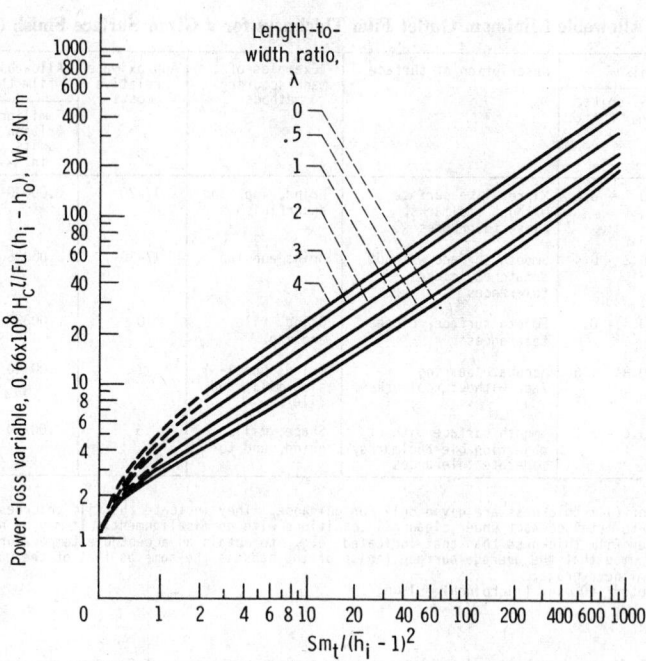

Fig. 23.23 Chart for determining power loss for fixed-incline pad thrust bearings. (From Ref. 25.)

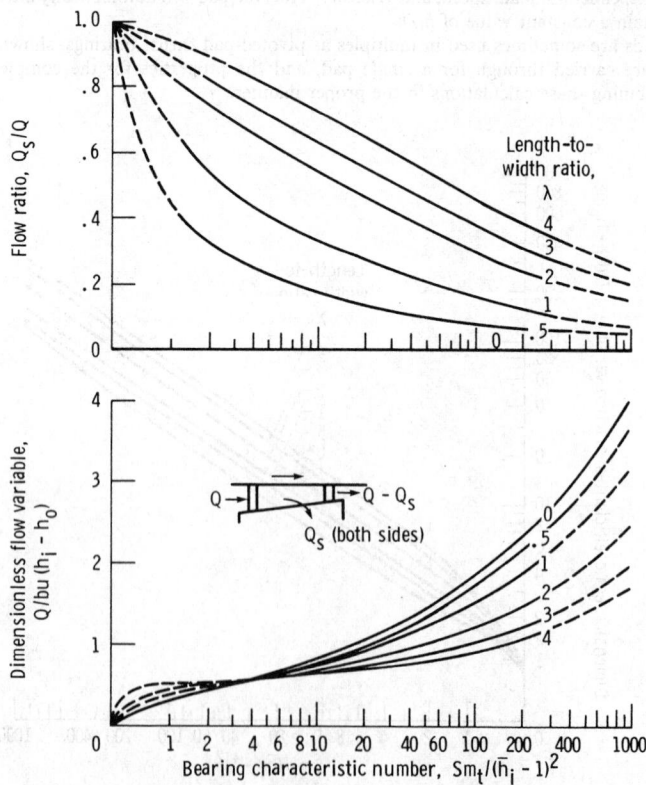

Fig. 23.24 Charts for determining lubricant flow for fixed-incline pad thrust bearings. (From Ref. 25.)

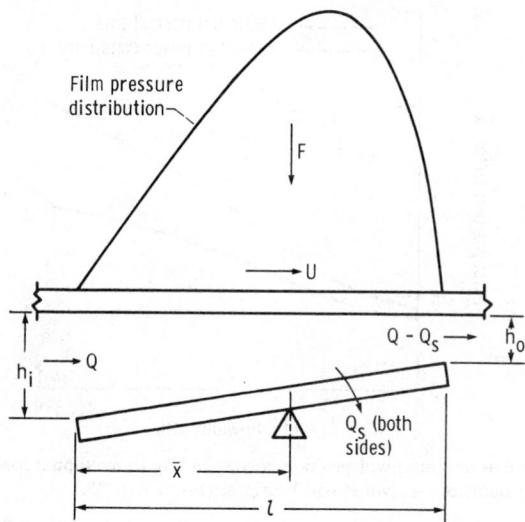

Fig. 23.25 Configuration of pivoted-pad bearings. (From Ref. 25.)

Normally, a pivoted pad will only carry load if the pivot is placed somewhere between the center of the pad and the outlet edge ($0.5 < \bar{x}/l \leq 1.0$). With the pivot so placed, the pad therefore can only carry load for one direction of rotation.

The following procedure helps in the design of pivoted-pad thrust bearings:

1. Having established if minimum power or maximum load is more critical in the particular application and chosen a pad length-to-width ratio, establish the pivot position from Fig. 23.27.

2. In the Sommerfeld number for thrust bearings the unknown parameter is usually the outlet or minimum film thickness. Therefore, establish the value of h_o from Fig. 23.28.

3. Check Table 23.5 to see if the outlet film thickness is sufficient for the preassigned surface finish. If sufficient, go on to step 4. If not, consider:

   a. Increasing the fluid viscosity.
   b. Increasing the speed of the bearing.
   c. Decreasing the load of the bearing.
   d. Decreasing the surface finish of the bearing lubrication surfaces.

Upon making this change return to step 1.

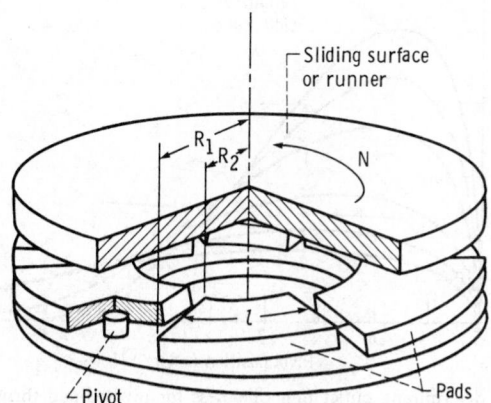

Fig. 23.26 Configuration of pivoted-pad thrust bearings. (From Ref. 25.)

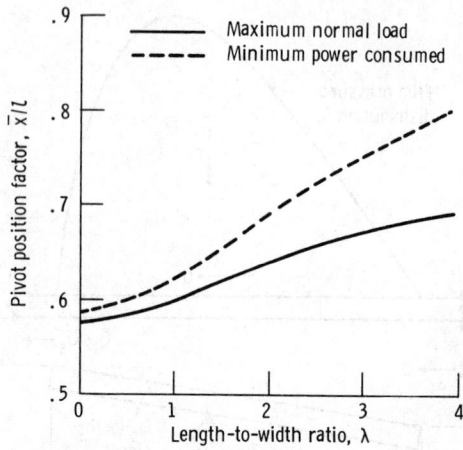

Fig. 23.27 Chart for determining pivot position corresponding to maximum load or minimum power loss for various pad proportions—pivoted-pad bearings. (From Ref. 25.)

    4. Once an adequate outlet film thickness is established, determine the film thickness ratio, power loss, coefficient of friction, and flow from Figs. 23.29–23.32.

Step Sector

The configuration of a step-sector thrust bearing is shown in Fig. 23.33. The parameters used to define the dimensionless load and stiffness are:

  1. $\bar{h}_i = h_i/h_o$, film thickness ratio.
  2. $\bar{\theta} = \theta_i/(\theta_i + \theta_o)$, dimensionless step location.
  3. n, number of sectors.
  4. $\alpha_r = R_2 R_1$, radius ratio.
  5. θ_g, angular extent of lubrication feed groove.

Note that the first four parameters are dimensionless and the fifth is dimensional and expressed in radians.

    The optimum parallel step-sector bearing for *maximum load capacity* for a given α_r and θ_g is

$$\bar{\theta}_{\text{opt}} = 0.558, \qquad (\bar{h}_i)_{\text{opt}} = 1.668, \qquad \text{and} \qquad n_{\text{opt}} = \frac{2\pi}{\theta_g + \dfrac{2.24(1 - \alpha_r)}{1 + \alpha_r}}$$

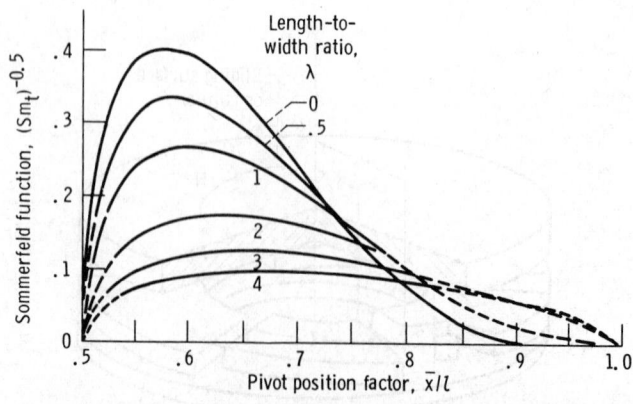

Fig. 23.28 Chart for determining outlet film thickness for pivoted-pad thrust bearings. (From Ref. 25.)

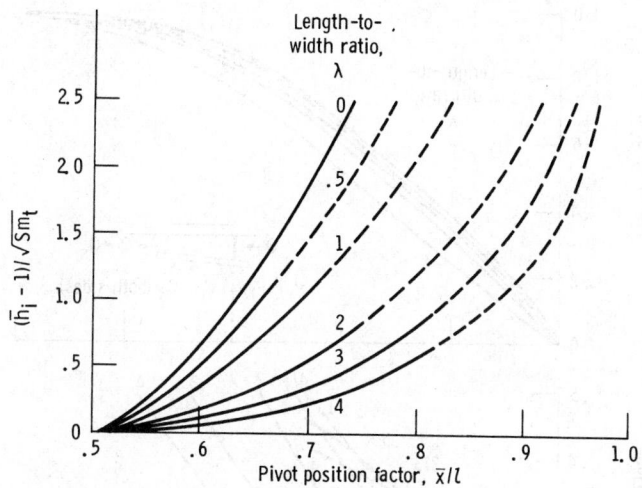

Fig. 23.29 Chart for determining film thickness ratio $\bar{h}_i$ for pivoted-pad thrust bearings. (From Ref. 25.)

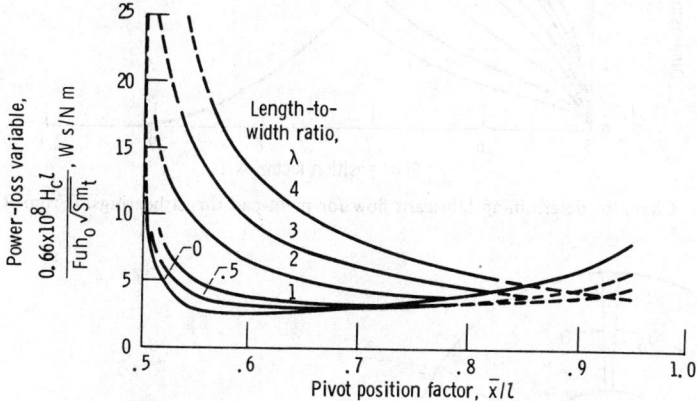

Fig. 23.30 Chart for determining power loss for pivoted-pad thrust bearings. (From Ref. 25.)

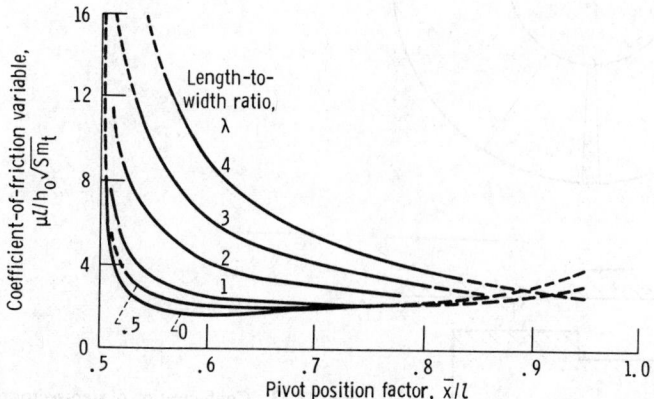

Fig. 23.31 Chart for determining coefficient of friction for pivot-pad thrust bearings. (From Ref. 25.)

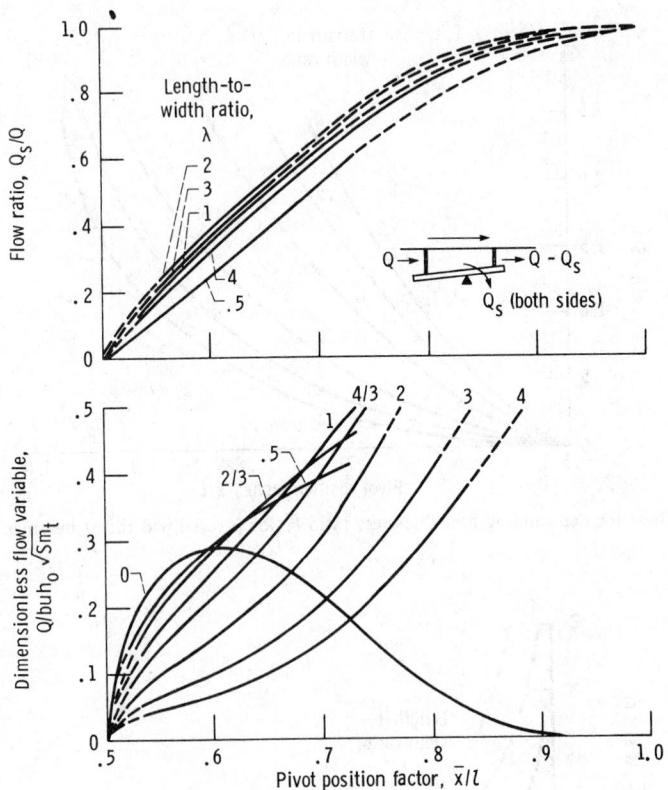

Fig. 23.32 Chart for determining lubricant flow for pivot-pad thrust bearings. (From Ref. 25.)

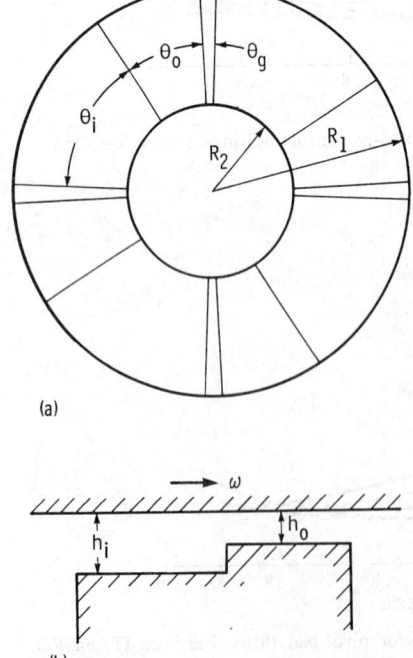

Fig. 23.33 Configuration of step-sector thrust bearing. (From Ref. 26.) (a) Top view. (b) Section through a sector.

where n_{opt} is rounded off to the nearest integer and its minimum value is 3. For *maximum stiffness*, results are identical to the above with the exception that $(\bar{h}_i)_{opt} = 1.467$. These results are obtained from Hamrock.[26]

23.2.3 Hydrostatic Bearings

In Sections 23.2.1 and 23.2.2 the load-supporting fluid pressure is generated by relative motion between the bearing surfaces. Thus its load capacity depends on the relative speeds of the surfaces. When the relative speeds of the bearing are low or the loads are high, the liquid-lubricated journal and thrust bearings may not be adequate. If full-film lubrication with no metal-to-metal contact is desired under such conditions, another technique, called hydrostatic or externally pressurized lubrication, may be used.

The one salient feature that distinguishes hydrostatic from hydrodynamic bearings is that the fluid is pressurized externally to the bearings and the pressure drop across the bearing is used to support the load. The load capacity is independent of the motion of bearing surfaces or the fluid viscosity. There is no problem of contact of the surfaces at starting and stopping as with conventional hydrodynamically lubricated bearings because pressure is applied before starting and maintained until after stopping. Hydrostatic bearings can be very useful under conditions of little or no relative motion and under extreme conditions of temperature or corrosivity, where it may be necessary to use bearing materials with poor boundary lubricating properties. Surface contact can be avoided completely, so material properties are much less important than in hydrodynamic bearings. The load capacity of a hydrostatic bearing is proportional to the available pressure.

Hydrostatic bearings do, however, require an external source of pressurization such as a pump. This represents an additional system complication and cost.

The chief advantage of hydrostatic bearings is their ability to support extremely heavy loads at slow speeds with a minimum of driving force. For this reason they have been successfully applied in rolling mills, machine tools, radio and optical telescopes, large radar antennas, and other heavily loaded, slowly moving equipment.

The formation of a fluid film in a hydrostatic bearing system is shown in Fig. 23.34. A simple bearing system with the pressure source at zero pressure is shown in Fig. 23.34a. The runner under the influence of a load F is seated on the bearing pad. As the source pressure builds up, Fig. 23.34b, the pressure in the pad recess also increases. The pressure in the recess is built up to a point, Fig. 23.34c, where the pressure on the runner over an area equal to the pad recess area is just sufficient to lift the load. This is commonly called the lift pressure. Just after the runner separates from the bearing pad, Fig. 23.34d, the pressure in the recess is less than that required to lift the bearing runner $(p_r < p_l)$. After lift, flow commences through the system. Therefore, a pressure drop exists between the pressure source and the bearing (across the restrictor) and from the recess to the exit of the bearing.

If more load is added to the bearing, Fig. 23.34e, the film thickness will decrease and the recess pressure will rise until pressure within the bearing clearance and the recess is sufficient to carry the increased load. If the load is now decreased to less than the original, Fig. 23.34f, the film thickness will increase to some higher value and the recess pressure will decrease accordingly. The maximum load that can be supported by the pad will be reached, theoretically, when the pressure in the recess is equal to the pressure at the source. If a load greater than this is applied, the bearing will seat and remain seated until the load is reduced and can again be supported by the supply pressure.

Pad Coefficients

To find the load-carrying capacity and flow requirements of any given hydrostatic bearing pad, it is necessary to determine certain pad coefficients. Since the selection of pad and recess geometries is up to the designer, the major design problem is the determination of particular bearing coefficients for particular geometries.

The load-carrying capacity of a bearing pad, regardless of its shape or size, can be expressed as

$$F = a_f A_p p_r \tag{23.13}$$

where a_f = bearing pad load coefficients
A_p = total projected pad area, m^2
p_r = recess pressure, N/m^2

The amount of lubricant flow across a pad and through the bearing clearance is

$$Q = q_f \frac{F}{A_p} \frac{h^3}{\eta} \tag{23.14}$$

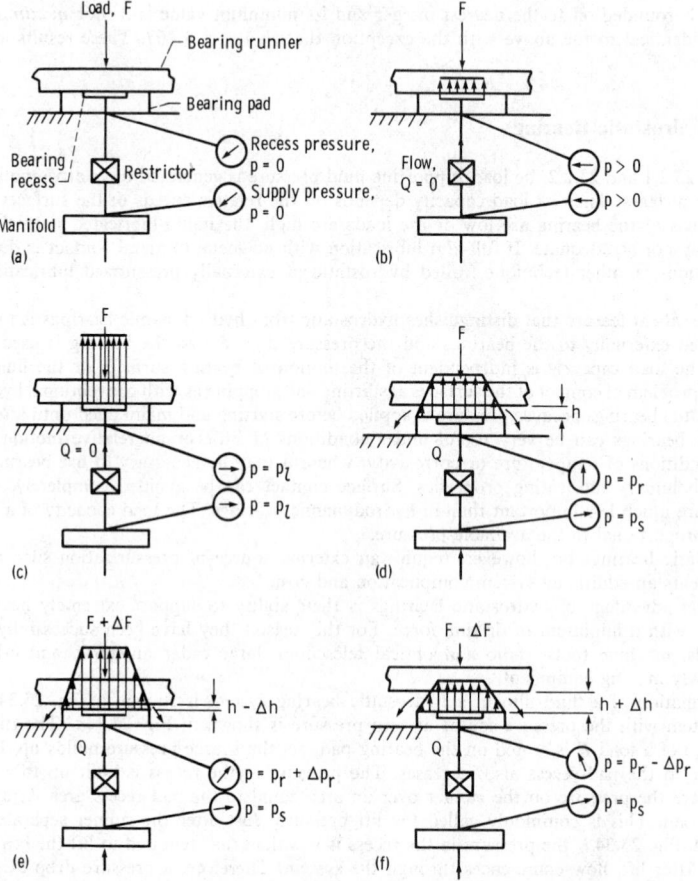

Fig. 23.34 Formation of fluid film in hydrostatic bearing system. (From Ref. 22.) (a) Pump off. (b) Pressure building up. (c) Pressure × recess area = F. (d) Bearing operating. (e) Increased load. (f) Decreased load.

where q_f = pad flow coefficient
   h = film thickness, m
   η = lubricant absolute viscosity, N sec/m²

The pumping power required by the hydrostatic pad can be evaluated by determining the product of recess pressure and flow:

$$H_b = p_r Q = H_f \left(\frac{F}{A_p}\right)^2 \frac{h^3}{\eta} \tag{23.15}$$

where $H_f = q_f/a_f$ is the bearing pad power coefficient. Therefore, in designing hydrostatic bearings the designer is primarily concerned with the bearing coefficients (a_f, q_f, and H_f) expressed in Eqs. (23.13)–(23.15).

Bearing coefficients are dimensionless quantities that relate performance characteristics of load, flow, and power to physical parameters. The bearing coefficients for two types of bearing pads will be considered, both of which exhibit pure radial flow and are flat, thrust-loaded types of bearings. For other types of hydrostatic bearings the reader is referred to Rippel.[22]

Circular Step Bearing Pad. The bearing coefficients for this type of pad are expressed as

$$a_f = \frac{1}{2}\left[\frac{1 - (R_o/R)^2}{\log_e (R/R_o)}\right] \tag{23.16}$$

$$q_f = \frac{\pi}{3}\left[\frac{1}{1 - (R_o/R)^2}\right] \tag{23.17}$$

$$H_f = \frac{2\pi \log_e (R/R_o)}{3 [1 - (R_o/R)^2]^2}$$ (23.18)

For this type of pad the total projected bearing pad area A_p is equal to πR^2.

Figure 23.35 shows the three bearing pad coefficients for various ratios of recess radius to bearing radius for a circular step thrust bearing. The bearing-pad load coefficient a_f varies from zero for extremely small recesses to unity for bearings having large recesses with respect to pad dimensions. In a sense, a_f is a measure of how efficiently the bearing uses the recess pressure to support the applied load.

In Fig. 23.35 we see that the pad flow coefficient q_f varies from unity for pads with relatively small recesses to a value approaching infinity for bearings with extremely large recesses. Physically, as the recess becomes larger with respect to the bearing, the hydraulic resistance to fluid flow decreases, and thus flow increases.

From Fig. 23.35, the power coefficient H_f approaches infinity for very small recesses, decreases to a minimum value as the recess size increases, and approaches infinity again for very large recesses. For this particular bearing the minimum value of H_f occurs at a ratio of recess radius to bearing radius R_o/R of 0.53. All bearing-pad configurations exhibit minimum values of H_f when their ratios of recess length to bearing length are approximately 0.4 to 0.6.

Annular Thrust Bearing.   Figure 23.36 shows an annular thrust pad bearing. In this bearing the lubricant flows from the annular recess over the inner and outer sills. For this type of bearing the pad coefficients are

$$a_f = \frac{1}{2(R_4^2 - R_1^2)} \left[\frac{R_4^2 - R_3^2}{\log_e (R_4/R_3)} - \frac{R_2^2 - R_1^2}{\log_e (R_2/R_1)} \right]$$ (23.19)

$$q_f = \frac{\pi}{6a_f} \left[\frac{1}{\log_e (R_4/R_3)} - \frac{1}{\log_e (R_2/R_1)} \right]$$ (23.20)

$$H_f = \frac{q_f}{a_f}$$ (23.21)

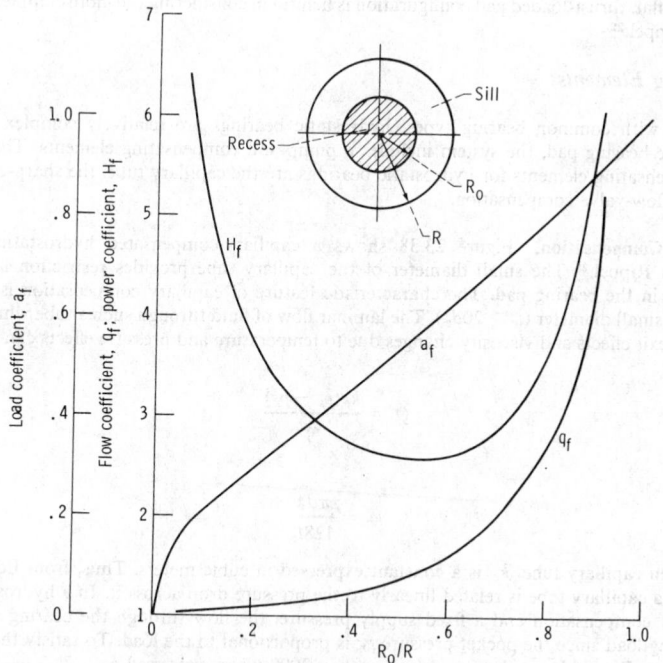

Fig. 23.35   Chart for determining bearing pad coefficients for circular step thrust bearing. (From Ref. 22.)

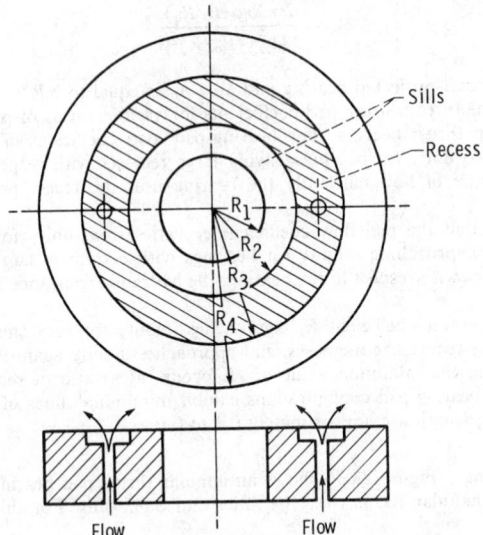

Fig. 23.36 Configuration of annular thrust pad bearing. (From Ref. 22.)

For this bearing the total projected bearing-pad area is

$$A_p = \pi(R_4^2 - R_1^2) \tag{23.22}$$

Figure 23.37 shows the bearing-pad load coefficient for an annular thrust pad bearing as obtained from Eqs. (23.19)–(23.21). For this figure it is assumed that the annular recess is centrally located within the bearing width; this therefore implies that $R_1 + R_4 = R_2 + R_3$. The curve for a_f applies for all R_1/R_4 ratios.

The hydrostatic bearings considered in this section have been limited to flat thrust-loaded bearings. Design information about other pad configurations can be obtained from Rippel.[22] The approach used for the simple, flat, thrust-loaded pad configuration is helpful in considering the more complex geometries covered by Rippel.[22]

Compensating Elements

As compared with common bearing types, hydrostatic bearings are relatively complex systems. In addition to the bearing pad, the system includes a pump and compensating elements. Three common types of compensating elements for hydrostatic bearings are the capillary tube, the sharp-edged orifice, and constant-flow-valve compensation.

Capillary Compensation. Figure 23.38 shows a capillary-compensated hydrostatic bearing as obtained from Rippel.[22] The small diameter of the capillary tube provides restriction and resultant pressure drop in the bearing pad. The characteristic feature of capillary compensation is a long tube or a relatively small diameter ($l_c > 20d_c$). The laminar flow of fluid through such a tube while neglecting entrance and exit effects and viscosity changes due to temperature and pressure effects can be expressed as

$$Q_c = \frac{k_c(p_s - p_r)}{\eta} \tag{23.23}$$

where

$$k_c = \frac{\pi d_c^4}{128 l_c} \tag{23.24}$$

For a given capillary tube, k_c is a constant expressed in cubic meters. Thus, from Eq. (23.23) the flow through a capillary tube is related linearly to the pressure drop across it. In a hydrostatic bearing with capillary compensation and a fixed supply pressure, the flow through the bearing will decrease with increasing load since the pocket pressure p_r is proportional to the load. To satisfy the assumption of laminar flow, Reynolds number must be less than 2000 when expressed as

$$N_R = \frac{4\rho Q_c}{\pi d_c \eta} < 2000 \tag{23.25}$$

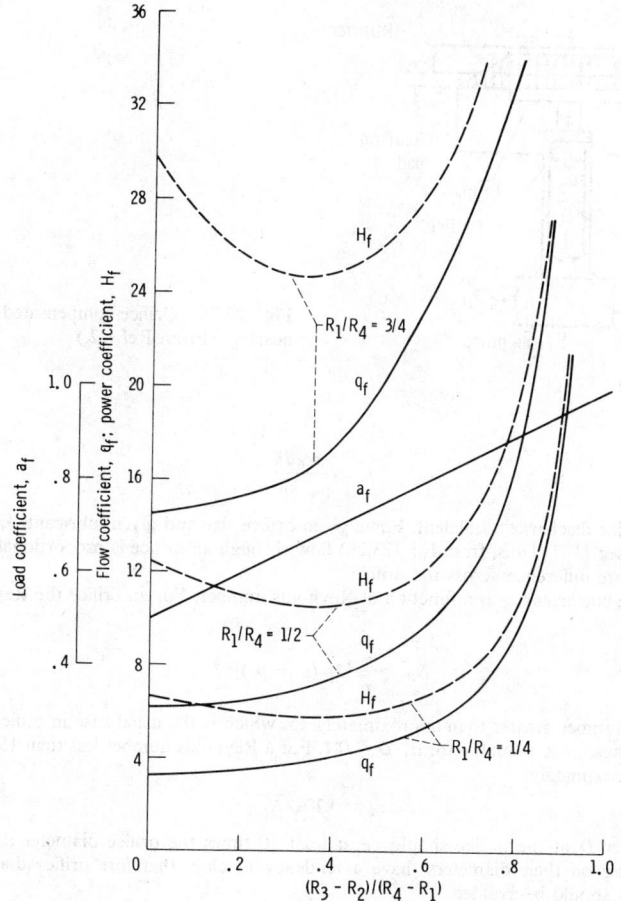

Fig. 23.37 Chart for determining bearing pad coefficients for annular thrust pad bearings. (From Ref. 22.)

where ρ is the mass density of the lubricant in N sec²/m⁴. Hypodermic needle tubing serves quite well as capillary tubing for hydrostatic bearings. Although very small diameter tubing is available, diameters less than 6×10^{-4} m should not be used because of their tendency to clog.

Orifice Compensation. Orifice compensation is illustrated in Fig. 23.39. The flow of an incompressible fluid through a sharp-edged orifice can be expressed as

$$Q_o = k_o (p_s - p_r)^{1/2} \tag{23.26}$$

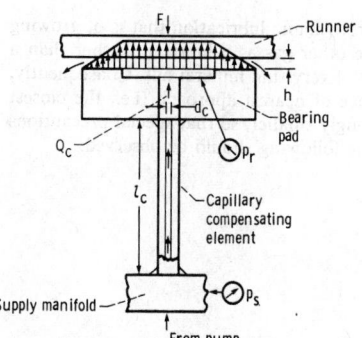

Fig. 23.38 Capillary-compensated hydrostatic bearing. (From Ref. 22.)

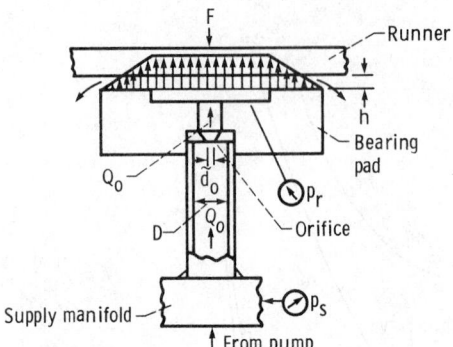

Fig. 23.39 Orifice-compensated hydrostatic bearing. (From Ref. 22.)

where

$$k_o = \frac{\pi c_d d_o^2}{\sqrt{8\rho}}$$

and c_d is the orifice discharge coefficient. For a given orifice size and given lubricant, k_o is a constant expressed in $m^4/\text{sec } N^{1/2}$. Thus, from Eq. (23.26) flow through an orifice is proportional to the square root of the pressure difference across the orifice.

The discharge coefficient c_d is a function of Reynolds number. For an orifice the Reynolds number is

$$N_R = \frac{d_o}{\eta} [2\rho (p_s - p_r)]^{1/2} \qquad (23.27)$$

For a Reynolds number greater than approximately 15, which is the usual case in orifice-compensated hydrostatic bearings, c_d is about 0.6 for $d_o/D < 0.1$. For a Reynolds number less than 15, the discharge coefficient is approximately

$$c_d = 0.20\sqrt{N_R} \qquad (23.28)$$

The pipe diameter D at the orifice should be at least 10 times the orifice diameter d_o. Sharp-edged orifices, depending on their diameters, have a tendency to clog, therefore orifice diameters d_o less than 5×10^{-4} m should be avoided.

Constant-Flow-Valve Compensation. Constant-flow-valve compensation is illustrated in Fig. 23.40. This type of restrictor has a constant flow regardless of the pressure difference across the valve. Hence, the flow is independent of recess pressure.

The relative ranking of the three types of compensating elements with regard to a number of considerations is given in Table 23.6. A rating of 1 in this table indicates best or most desirable. This table should help in deciding which type of compensation is most desirable in a particular application.

Basically, any type of compensating element can be designed into a hydrostatic bearing system if loads on the bearing never change. But if stiffness, load, or flow vary, the choice of the proper compensating element becomes more difficult and the reader is again referred to Rippel.[22]

23.2.4 Gas-Lubricated Hydrodynamic Bearings

A relatively recent (within the last 30 years) extension of hydrodynamic lubrication that is of growing importance is gas lubrication. It consists of using air or some other gas as a lubricant rather than a mineral oil. The viscosity of air is 1000 times smaller than that of very thin mineral oils. Consequently, the viscous resistance is very much less. However, the distance of nearest approach (i.e., the closest distance between the shaft and the bearing) is also correspondingly smaller, so that special precautions must be taken. To obtain full benefits from gas lubrication, the following should be observed:

1. Surfaces must have a very fine finish.
2. Alignment must be very good.
3. Dimensions and clearances must be very accurate.
4. Speeds must be high.
5. Loading must be relatively low.

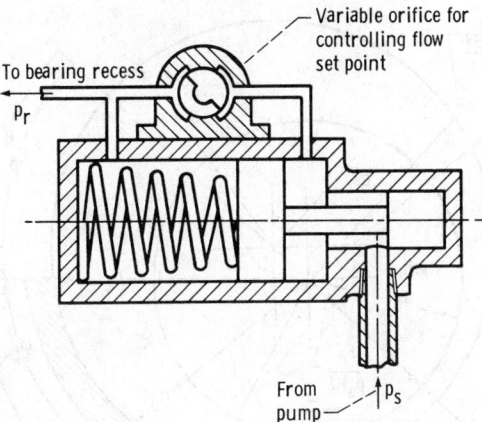

Fig. 23.40 Constant-flow-valve compensation in hydrostatic bearing. (From Ref. 22.)

Another main difference between the behavior of similar gas and liquid films besides that of viscosity is the compressibility of the gas. At low relative speeds it is reasonable to expect the gas-film density to remain nearly constant and the film therefore to behave as if it were imcompressible. At high speeds, however, the density change is likely to become of primary importance so that such gas-film properties must differ appreciably from those of similar liquid films.

Gas-lubricated bearings can also operate at very high temperatures since the lubricant will not degrade chemically. Furthermore, if air is used as the lubricant, it costs nothing. Gas bearings are finding increasing use in gas-cycle machinery where the cycle gas is used in the bearings, thus eliminating the need for a conventional lubrication system; in gyros, where precision and constancy of torque are critical; in food and textile processing machinery, where cleanliness and absence of contaminants are critical; and also in the magnetic recording tape industry.

Journal Bearings

Plain gas-lubricated journal bearings are of little interest because of their poor stability characteristics. Lightly loaded bearings that operate at low eccentricity ratios are subjected to fractional frequency whirl, which can result in bearing destruction. Two types of gas-lubricated journal bearings find widespread use, namely, the pivoted pad and the herringbone groove.

Table 23.6 Compensating-Element Considerations[a] (From Ref. 22)

Consideration	Compensating element		
	Capillary	Orifice	Constant-flow valve
Initial cost	2	1	3
Cost to fabricate and install	2	3	1
Space required	2	1	3
Reliability	1	2	3
Useful life	1	2	3
Commercial availability	2	3	1
Tendency to clog	1	2	3
Serviceability	2	1	3
Adjustability	3	2	1

[a]Rating of 1 is best or most desirable.

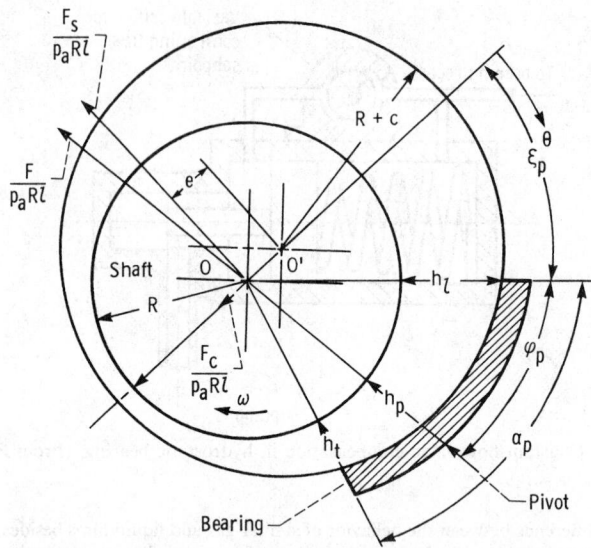

Fig. 23.41   Geometry of individual shoe-shaft bearing. (From Ref. 27.)

Pivoted Pad.   Pivoted-pad journal bearings are most frequently used as shaft supports in gas-bearing machinery because of their excellent stability characteristics. An individual pivot pad and shaft are shown in Fig. 23.41, and a three-pad pivoted-pad bearing assembly is shown in Fig. 23.42. Generally, each pad provides pad rotation degrees of freedom about three orthogonal axes (pitch, roll, and yaw). Pivoted-pad bearings are complex because of the many geometric variables involved in their design. Some of these variables are:

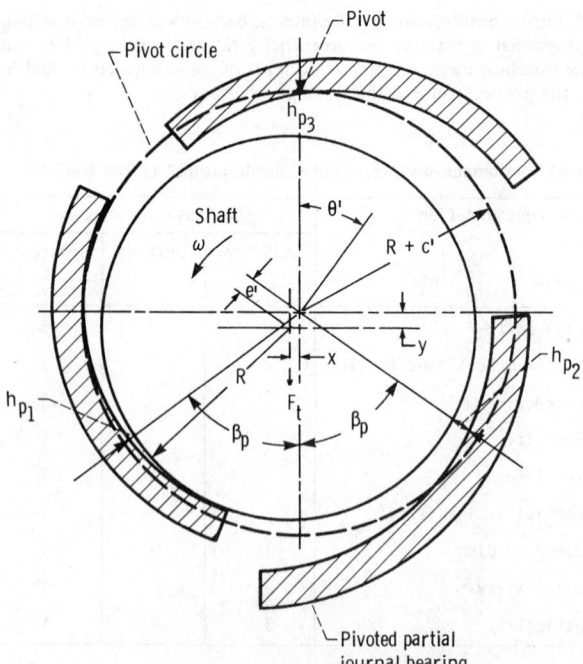

Fig. 23.42   Geometry of pivoted-pad journal bearing with three shoes. (From Ref. 27.)

1. Number of pads.
2. Circumferential extent of pads, α_p.
3. Aspect ratio of pad, R/l.
4. Pivot location, ϕ_p/α_p.
5. Machined-in clearance ratio, c/R.
6. Pivot circle clearance ratio, c'/R.
7. Angle between line of centers and pad leading edge, ξ_p.

Analysis is accomplished by first determining the characteristics of an individual pad. Both geometric and operating parameters influence the design of a pivoted pad. The operating parameter of importance is the dimensionless bearing number Λ_j, where

$$\Lambda_j = \frac{6\eta\omega R^2}{p_a c^2}$$

The results of computer solutions obtained from Gunter et al.[27] for the performance of a single pad are shown in Figs. 23.43–23.45. These figures illustrate load coefficient, pivot film thickness, and trailing-edge film thickness as functions of pivot location and eccentricity ratio. These field maps apply for a pad with a radius-to-width ratio of 0.606, a circumferential extent of 94.5° (an aspect ratio of 1), and $\Lambda_j = 3.5$. For other geometries and Λ values similar maps must be generated. Additional maps are given in Gunter et al.[27]

Figures 23.46–23.48 show load coefficient and stiffness coefficient for a range of Λ_j values up to 4. These plots are for a pivot position of $\frac{2}{3}$.

When the individual pad characteristics are known, the characteristics of the multipad bearing can be determined by using a trial-and-error approach. With the arrangement shown in Fig. 23.42, the load is directed between the two lower pivots. For this case the load carried by each of the lower pads is initially assumed to be $F \cos \beta$. The pivot film thicknesses h_{p_1} and h_{p_2} are then calculated. The upper-pad pivot film thickness h_{p_3}, eccentricity ratio ϵ, and load coefficient C_{l_3} can be determined. The additional load on the shaft due to the reaction of pad 3 is added to the system load. Calculations are repeated until the desired accuracy is achieved.

Pivoted-pad journal bearings are usually assembled with a pivot circle clearance c' somewhat less than the machined-in clearance c. When $c'/c < 1$, the bearing is said to be preloaded. Preload is usually given in terms of a preload coefficient, which is equal to $(c - c')/c$. Preloading is used to increase bearing stiffness and to prevent complete unloading of one or more pads. The latter condition can lead to pad flutter and possible contact of the pad leading edge and the shaft, which, in turn, can result in bearing failure.

Herringbone Groove.   A fixed-geometry bearing that has demonstrated good stability characteristics and thus promise for use in high-speed gas bearings is the herringbone bearing. It consists of a circular journal and bearing sleeve with shallow, herringbone-shaped grooves cut into either member. Figure 23.49 illustrates a partially grooved herringbone journal bearing. In this figure the groove and bearing parameters are also indicated. Figures 23.50–23.54 were obtained from Hamrock and Fleming[28] and are design charts that present curves for optimizing the design parameters for herringbone journal bearings for maximum radial load. The (a) portion of these figures is for the grooved member rotating and the (b) portion is for the smooth member rotating. The only groove parameter not represented in these figures is the number of grooves to be used. From Hamrock and Fleming[28] it was found that the *minimum* number of grooves to be placed around the journal can be represented by $n \geq \Lambda_j/5$.

More than any other factors, self-excited whirl instability and low-load capacity limit the usefulness of gas-lubricated journal bearings. The whirl problem is the tendency of the journal center to orbit the bearing center at an angular speed less than or equal to half that of the journal about its own center. In many cases the whirl amplitude is large enough to cause destructive contact of the bearing surfaces.

Figure 23.55, obtained from Fleming and Hamrock,[29] shows the stability attained by the optimized herringbone bearings. In this figure the stability parameter $\overline{M}$ is introduced, where

$$\overline{M} = \frac{\overline{m} p_a h_r^5}{2R^5 l \eta^2}$$

and $\overline{m}$ is the mass supported by the bearing.

In Fig. 23.55, the bearings with the grooved member rotating are substantially more stable than those with the smooth member rotating, especially at high compressibility numbers.

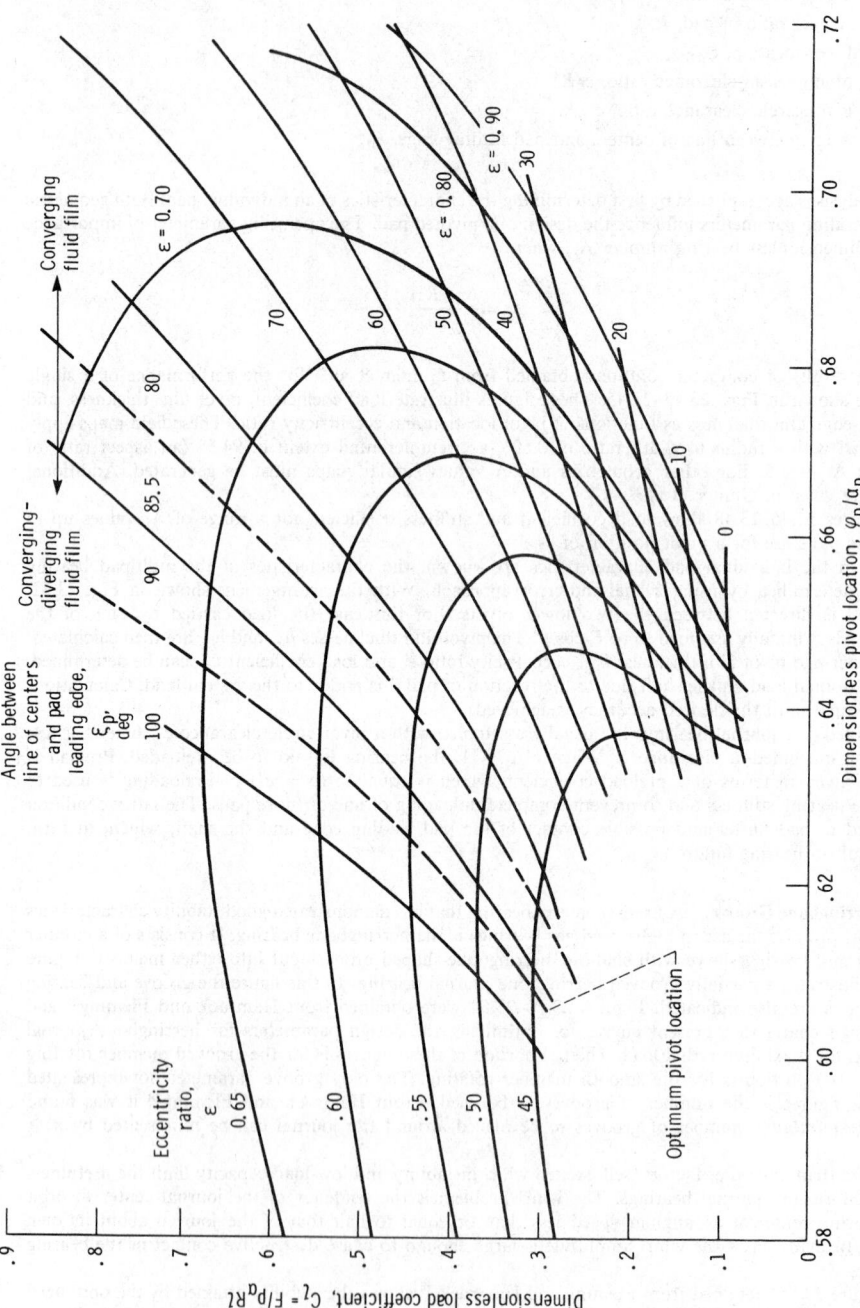

Fig. 23.43 Chart for determining load coefficient. Bearing radius-to-length ratio, R/l, 0.6061; angular extent of pad, α_p, 94.5°; dimensionless bearing number, Λ_j, 3.5. (From **Ref. 27**.)

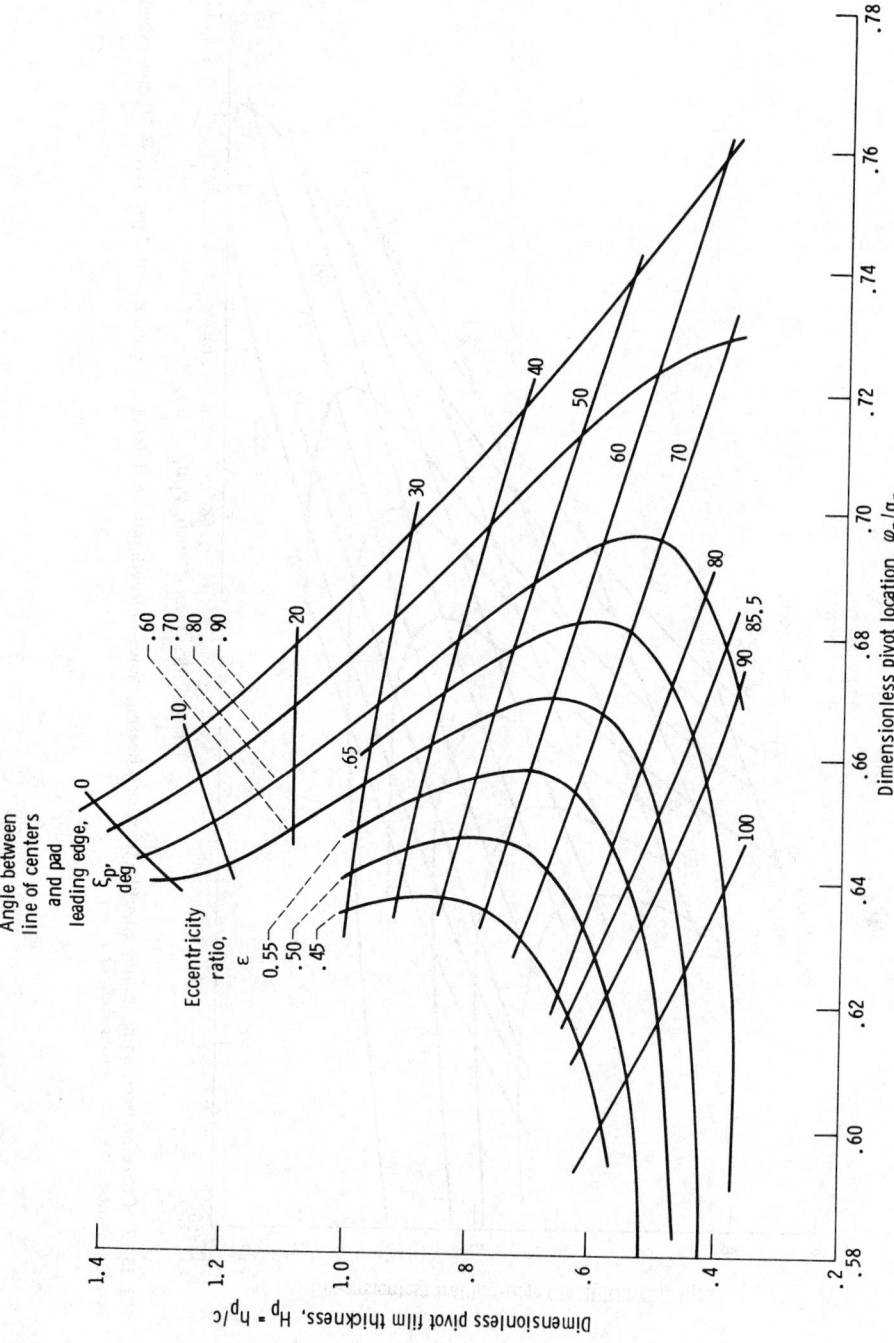

Fig. 23.44 Chart for determining pivot film thickness. Bearing radius-to-length ratio, R/l, 0.6061; angular extent of pad, α_p, 94.5°; dimensionless bearing number, Λ_J, 3.5. (From **Ref. 27**.)

Angle between line of centers and pad leading edge, ξ_p, deg

Eccentricity ratio, ε

Dimensionless pivot film thickness, $H_p = h_p/c$

Dimensionless pivot location, φ_p/α_p

493

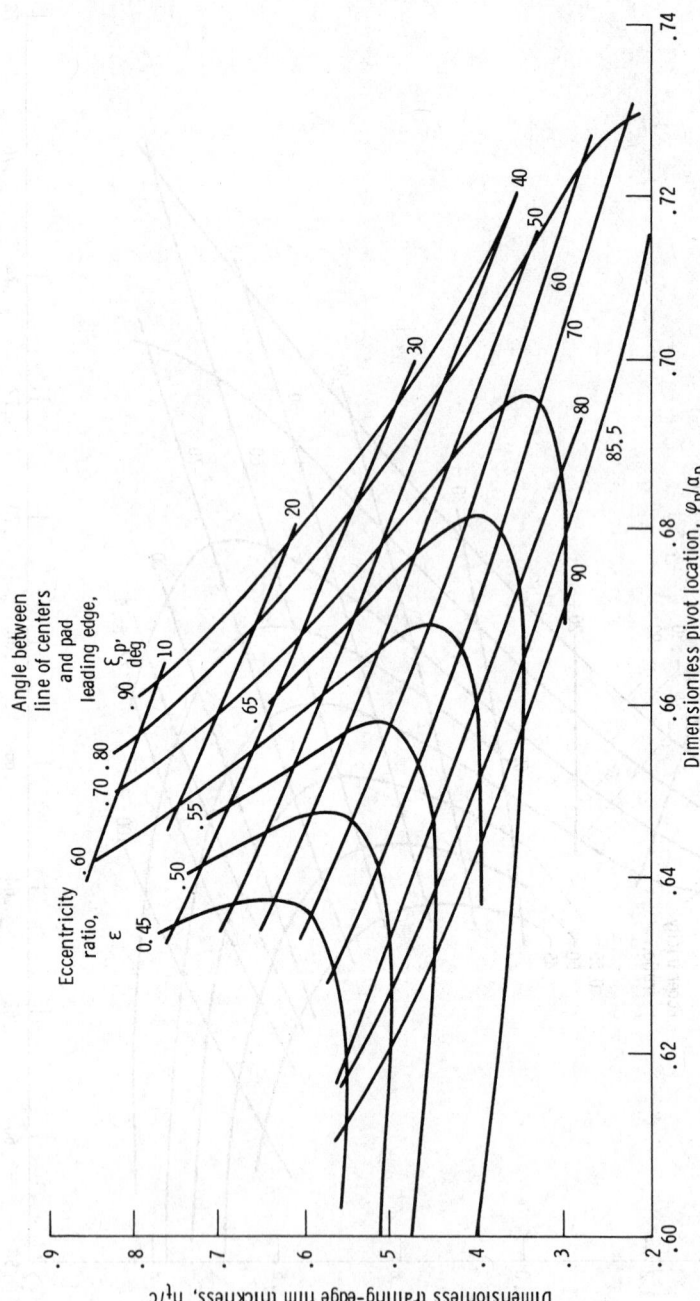

Fig. 23.45 Chart for determining trailing-edge film thickness. Bearing radius-to-length ratio, R/l, 0.6061; angular extent of pad, α_p, 94.5°; dimensionless bearing number, Λ_J, 3.5. (From Ref. 27.)

Dimensionless pivot location, φ_p/α_p

Dimensionless trailing-edge film thickness, h_t/c

Angle between line of centers and pad leading edge, ξ_p, deg

Eccentricity ratio, ε

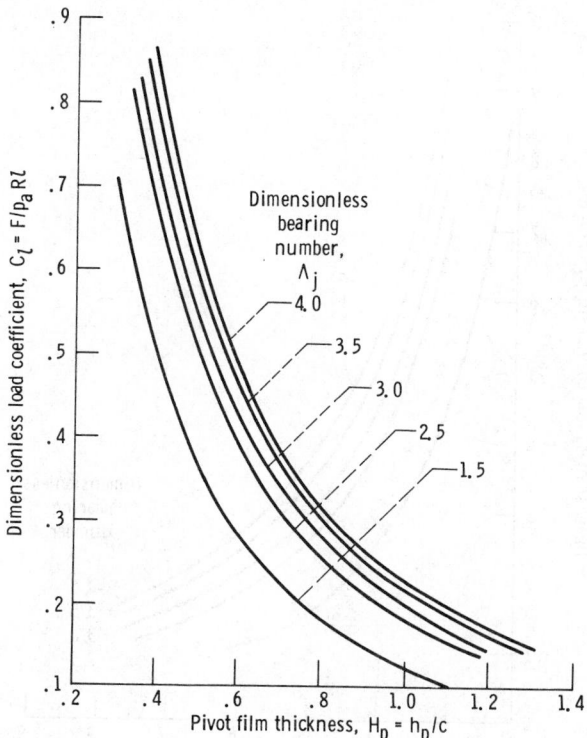

Fig. 23.46   Chart for determining load coefficient. Angular extent of pad, α_p, 94.5°; ratio of angle between pad leading edge and pivot to α_p, φ_p/α_p, $\frac{2}{3}$; length-to-width ratio, λ, 1.0. (From Ref. 27.)

Thrust Bearings

Two types of gas-lubricated thrust bearings have found the widest use in practical applications. These are the Rayleigh step and the spiral- or herringbone-groove bearings.

Rayleigh Step Bearing.   Figure 23.56 shows a Rayleigh step thrust bearing. In this figure the ridge region is where the film thickness is h_r and the step region is where the film thickness is h_s. The feed groove is the deep groove separating the end of the ridge region and the beginning of the next step region. Although not shown in the figure, the feed groove is orders of magnitude deeper than the film thickness h_r. A pad is defined as the section that includes ridge, step, and feed groove regions. The length of the feed groove is small relative to the length of the pad. It should be noted that each pad acts independently since the pressure profile is broken at the lubrication feed groove.

The load capacity and stiffness of a Rayleigh step thrust bearing are functions of the following parameters:

1. $\Lambda_t = 6\eta u l/p_a h_r^2$, the dimensionless bearing number.
2. $\lambda_a = (b_r + b_s + b_g)/l$, length ratio.
3. $H_a = h_s/h_r$, film thickness ratio.
4. $\psi_s = b_s/(b_s + b_r + b_g)$, step location parameter.

Figure 23.57a shows the effect of Λ_t on λ_a, H_a, and ψ_s for the *maximum-load-capacity condition*. The optimal step parameters λ_a, H_a, and ψ_s approach an asymptote as the dimensionless bearing number Λ_t becomes small. This asymptotic condition corresponds to the incompressible solution or $\lambda_a = 0.918$, $\psi_s = 0.555$, $H_a = 1.693$. For $\Lambda_t > 1$ it is observed that there is a different optimum value of λ_a, H_a, and ψ_s for each value of Λ_t.

Figure 23.57b shows the effect of Λ_t on λ_a, H_a, and ψ_s for the *maximum-stiffness condition*. As in Fig. 23.57a the optimal step parameters approach asymptotes as the incompressible solution is reached. The asymptotes are $\lambda_a = 0.915$, $\psi_s = 0.557$, and $H_a = 1.470$. Note that there is a difference

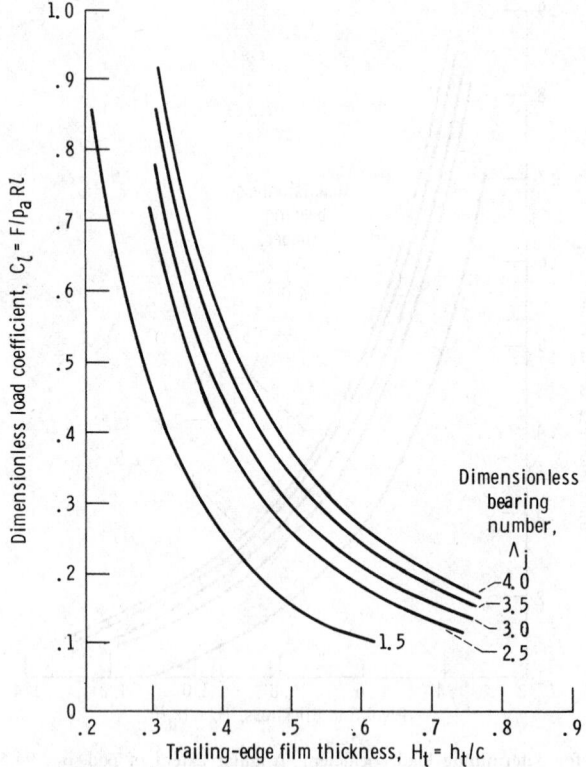

Fig. 23.47 Chart for determining load coefficient. Angular extent of pad, α_p, 94.5°; ratio of angle between pad loading edge and pivot to α_p, φ_p/α_p, $\frac{2}{3}$; length-to-width ratio, λ, 1.0. (From Ref. 27.)

in the asymptote for the film thickness ratio but virtually no change in λ_a and ψ_s when compared with the results obtained for maximum-load-capacity condition.

Figure 23.58 shows the effect of dimensionless bearing number Λ_t on dimensionless load capacity and stiffness. The difference in these figures is that the optimal step parameters are obtained in Fig. 23.58a for maximum load capacity and in Fig. 23.58b for maximum stiffness.

For optimization of a step-sector thrust bearing, parameters for the sector must be found that are analogous to those for the rectangular step bearing. The following substitutions accomplish this transformation:

$$l \rightarrow R_1 - R_2$$

$$n(b_s + b_r + b_g) \rightarrow \pi(R_1 + R_2)$$

$$u \rightarrow \frac{\omega}{2}(R_1 + R_2)$$

where n is the number of pads placed in the step sector. By making use of these equations, the dimensionless bearing number can be rewritten as

$$\Lambda_c = \frac{3\eta\omega\,(R_1^2 - R_2^2)}{p_a h_f^2}$$

The optimal number of pads to be placed in the sector is obtained from the formula

$$n = \frac{\pi(R_1 + R_2)}{(\lambda_a)_{\text{opt}}\,(R_1 - R_2)}$$

where $(\lambda_a)_{\text{opt}}$ is obtained from Fig. 23.57a or 23.57b for a given dimensionless bearing number Λ_t. Since n will not normally be an integer, rounding it to the nearest integer is required. Therefore, through the parameter transformation discussed above, the results presented in Figs. 23.57 and 23.58 are directly usable in designing optimal step-sector gas-lubricated thrust bearings.

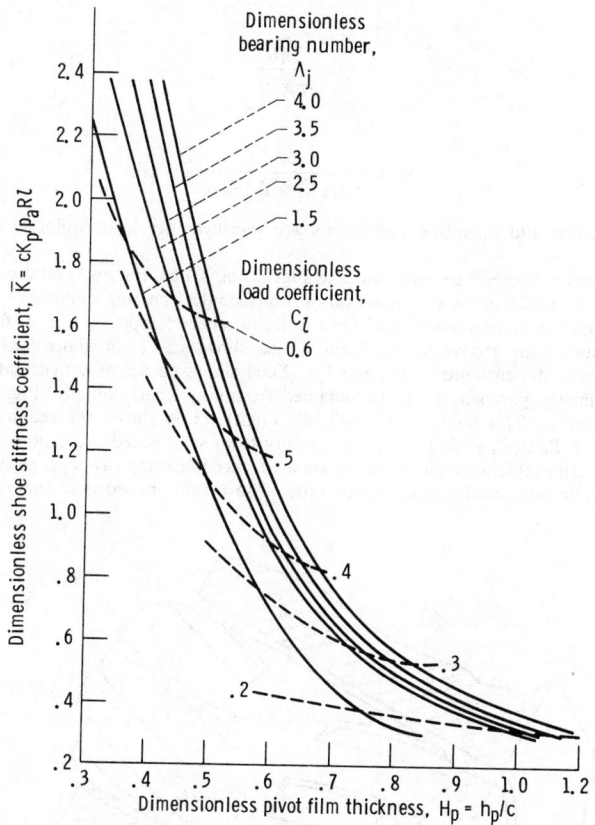

Fig. 23.48 Chart for determining shoe stiffness coefficient. (From Ref. 27.)

Spiral-Groove Thrust Bearings. An inward-pumping spiral-groove thrust bearing is shown in Fig. 23.59. An inward-pumping thrust bearing is somewhat more efficient than an outward-pumping thrust bearing and therefore is the only type considered here.

The dimensionless parameters normally associated with a spiral-groove thrust bearing are:

1. Angle of inclination, β_a.
2. Width ratio, $\bar{b} = b_s/b_r$.
3. Film ratio, $H_a = h_s/h_r$.
4. Radius ratio, $\alpha_r = R_2/R_1$.
5. Groove length fraction, $\bar{R} = (R_1 - R_g)/(R_1 - R_2)$.
6. Number of grooves, n.
7. Dimensionless bearing number, $\Lambda_c = 3\eta\omega(R_1^2 - R_2^2)/p_a h_r^2$.

The first six parameters are geometrical parameters and the last parameter is an operating parameter.

The performance of spiral-groove thrust bearings is represented by the following dimensionless parameters:

Load

$$\bar{W}_\infty = \frac{1.5 G_f F}{\pi p_a (R_1^2 - R_2^2)} \tag{23.29}$$

Stiffness

$$\bar{K}_\infty = \frac{1.5 h_r G_f K_p}{\pi p_a (R_1^2 - R_2^2)} \tag{23.30}$$

Flow

$$\bar{Q} = \frac{3\eta Q}{\pi p_a h_r^3} \qquad (23.31)$$

Torque

$$\bar{T} = \frac{6T_r}{\pi p_a (R_1^2 + R_2^2) h_r \Lambda_c} \qquad (23.32)$$

When the geometrical and operating parameters are specified, the load, stiffness, flow, and torque can be obtained.

The design charts of Reiger[20] are reproduced as Figs. 23.60–23.66. Figure 23.60 shows the dimensionless load for various radius ratios as a function of dimensionless bearing number Λ_c. This figure can be used to calculate the dimensionless load for a finite number of grooves; Fig. 23.61 can be used to determine the value of the groove factor. Figure 23.62 shows curves of dimensionless stiffness; Fig. 23.63 shows curves of dimensionless flow; and Fig. 23.64 shows curves of dimensionless torque. Optimized groove geometry parameters can be obtained from Fig. 23.65. Finally, Fig. 23.66 is used to calculate groove radius R_g (shown in Fig. 23.59). Figure 23.66 shows the required groove length fraction $\bar{R} = (R_o - R_g)/(R_o - R_i)$ to ensure stability from self-excited oscillations.

In a typical design problem the given factors are load, speed, bearing envelope, gas viscosity, ambient pressure, and an allowable radius-to-clearance ratio. The maximum value of the radius-to-clearance

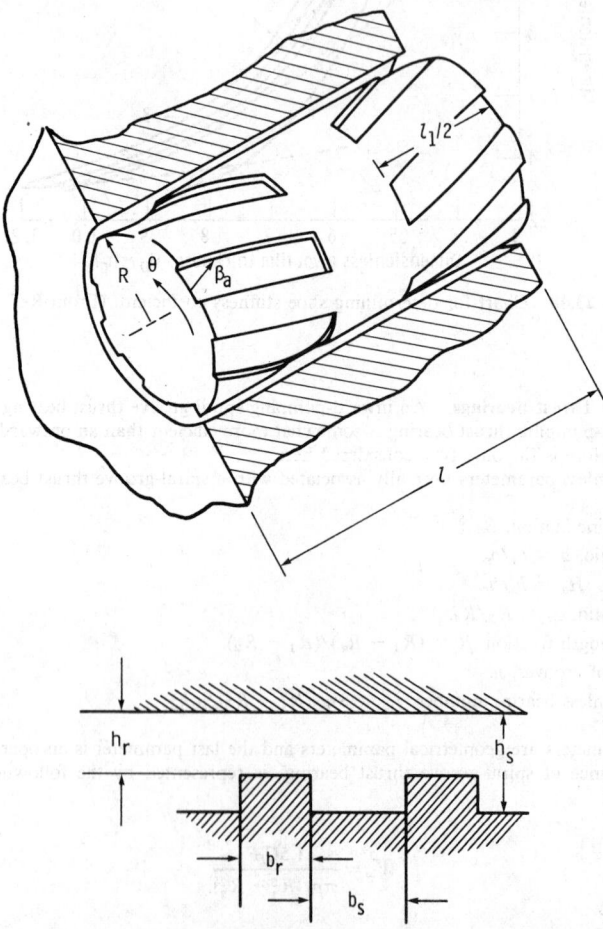

Fig. 23.49 Configuration of concentric herringbone-groove journal bearing. Bearing parameters: $\lambda = l/2R$; $\Lambda_j = 6\mu UR/p_a h_r^2$. Groove parameters: $H_a = h_s/h_r$; $\alpha_b = b_s/(b_r + b_s)$; β_a; $\gamma = l_1/l$; n. (From Ref. 28.)

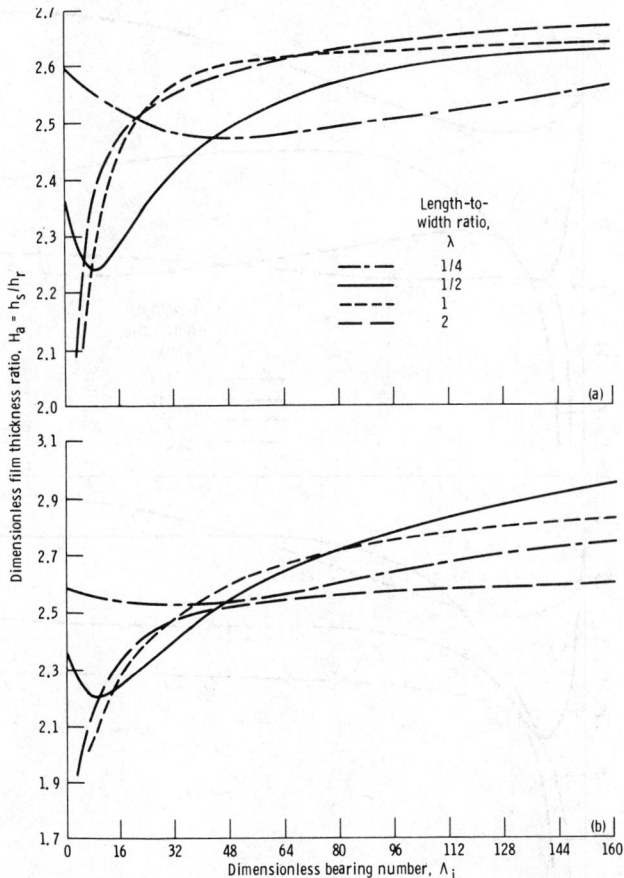

Fig. 23.50 Chart for determining optimal film thickness. (From Ref. 28.) (*a*) Grooved member rotating. (*b*) Smooth member rotating.

ratio is usually dictated by the distortion likely to occur to the bearing surfaces. Typical values are 5000–10,000. The procedure normally followed in designing a spiral-groove thrust bearing while using the design curves given in Figs. 23.60–23.66 is as follows:

1. Select the number of grooves n.
2. From Fig. 23.61 determine the groove factor G_f for given $\alpha_r = R_i/R_o$ and n.
3. Calculate $\overline{W}_\infty = 1.5 G_f F / \pi p_a (R_1^2 - R_2^2)$.
4. If $\overline{W}_\infty < 0.8$, R_1 must be increased. Return to step 2.
5. From Fig. 23.60, given $\overline{W}_\infty$ and α_r establish Λ_c.
6. Calculate

$$\frac{R_1}{h_r} = \left\{ \frac{\Lambda_c p_a}{3\eta(\omega_h - \omega_o)\left[1 - (R_2/R_1)^2\right]} \right\}^{1/2}$$

   If $R_1/h_r > 10,000$ (or whatever preassigned radius-to-clearance ratio), a larger bearing or higher speed is required. Return to step 2. If these changes cannot be made, an externally pressurized bearing must be used.
7. Having established what α_r and Λ_c should be, obtain values of $\overline{K}_\infty$, $\overline{Q}$, and $\overline{T}$ from Figs. 23.62, 23.63, and 23.64, respectively. From Eqs. (23.29), (23.30), and (23.31) calculate K_p, Q, and T_r.
8. From Fig. 23.65 obtain groove geometry (b, β_a, and H_a) and from Fig. 23.66 obtain R_g.

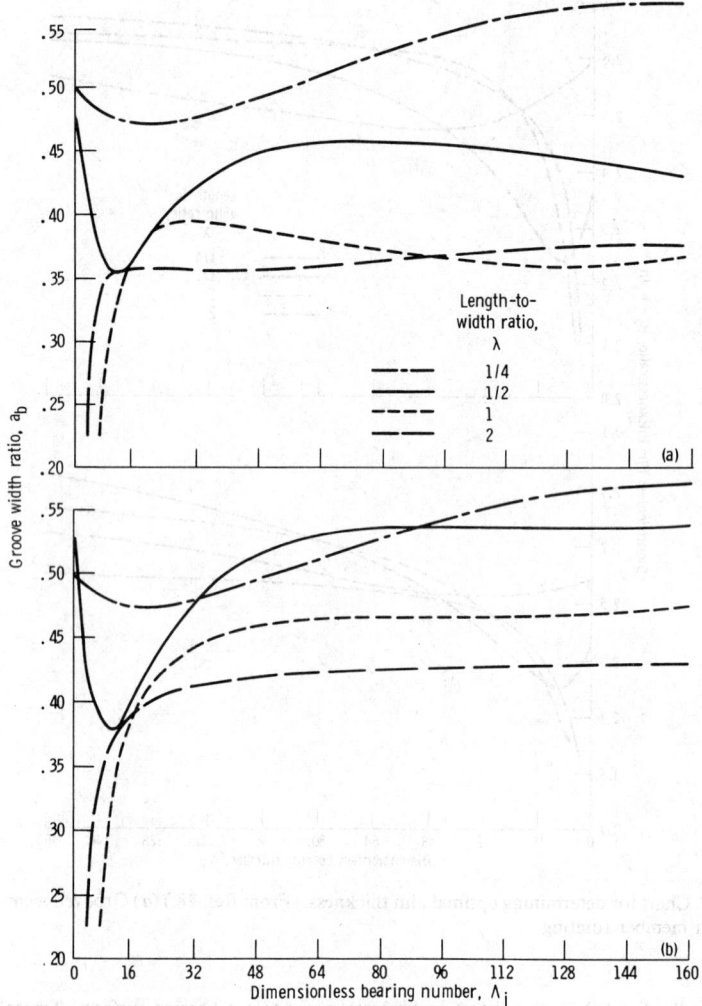

Fig. 23.51 Chart for determining optimal groove width ratio. (From Ref. 28.) (*a*) Grooved member rotating. (*b*) Smooth member rotating.

23.3 ELASTOHYDRODYNAMIC LUBRICATION

Dowson[31] defines elastohydrodynamic lubrication (EHL) as "the study of situations in which elastic deformation of the surrounding solids plays a significant role in the hydrodynamic lubrication process." Elastohydrodynamic lubrication implies complete fluid-film lubrication and no asperity interaction of the surfaces. There are two distinct forms of elastohydrodynamic lubrication.

1. *Hard EHL.* Hard EHL relates to materials of high elastic modulus, such as metals. In this form of lubrication not only are the elastic deformation effects important, but the pressure–viscosity effects are equally as important. Engineering applications in which this form of lubrication is dominant include gears and rolling-element bearings.

2. *Soft EHL.* Soft EHL relates to materials of low elastic modulus, such as rubber. For these materials the elastic distortions are large, even with light loads. Another feature is the negligible pressure–viscosity effect on the lubricating film. Engineering applications in which soft EHL is important include seals, human joints, tires, and a number of lubricated elastomeric material machine elements.

The recognition and understanding of elastohydrodynamic lubrication presents one of the major developments in the field of tribology in this century. The revelation of a previously unsuspected

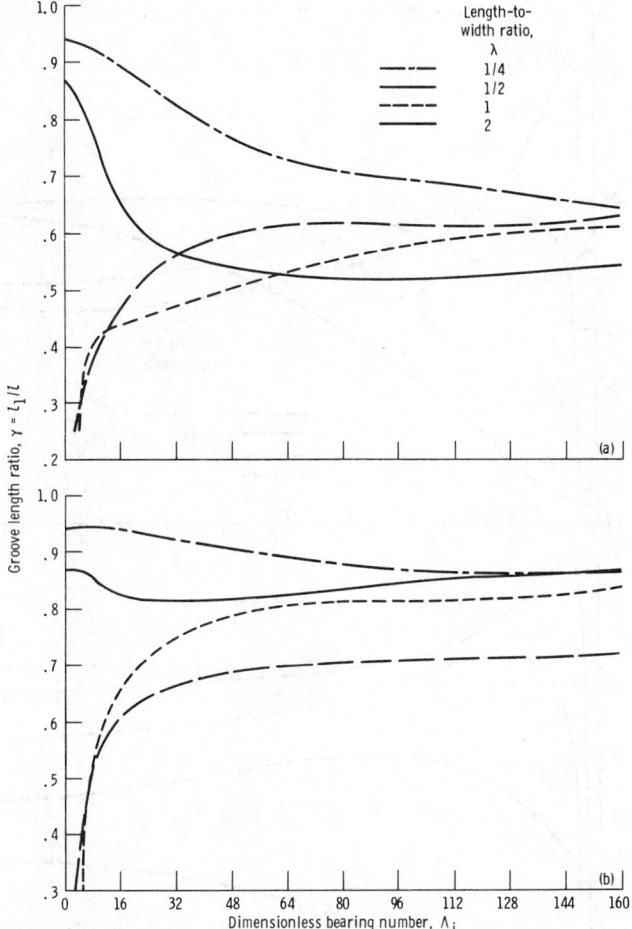

Fig. 23.52 Chart for determining optimal groove length ratio. (From Ref. 28.) (*a*) Grooved member rotating. (*b*) Smooth member rotating.

regime of lubrication is clearly an event of importance in tribology. Elastohydrodynamic lubrication not only explained the remarkable physical action responsible for the effective lubrication of many machine elements, but it also brought order to the understanding of the complete spectrum of lubrication regimes, ranging from boundary to hydrodynamic.

A way of coming to an understanding of elastohydrodynamic lubrication is to compare it to hydrodynamic lubrication. The major developments that have led to our present understanding of hydrodynamic lubrication[1,3] predate the major developments of elastohydrodynamic lubrication[32,33] by 65 years. Both hydrodynamic and elastohydrodynamic lubrication are considered as fluid-film lubrication in that the lubricant film is sufficiently thick to prevent the opposing solids from coming into contact. Fluid-film lubrication is often referred to as the ideal form of lubrication since it provides low friction and high resistance to wear.

This section highlights some of the important aspects of elastohydrodynamic lubrication while illustrating its use in a number of applications. It is not intended to be exhaustive but to point out the significant features of this important regime of lubrication. For more details the reader is referred to Hamrock and Dowson.[10]

23.3.1 Contact Stresses and Deformations

As was pointed out in Section 23.1.1, elastohydrodynamic lubrication is the mode of lubrication normally found in nonconformal contacts such as rolling-element bearings. A load–deflection relationship for nonconformal contacts is developed in this section. The deformation within the contact is calculated

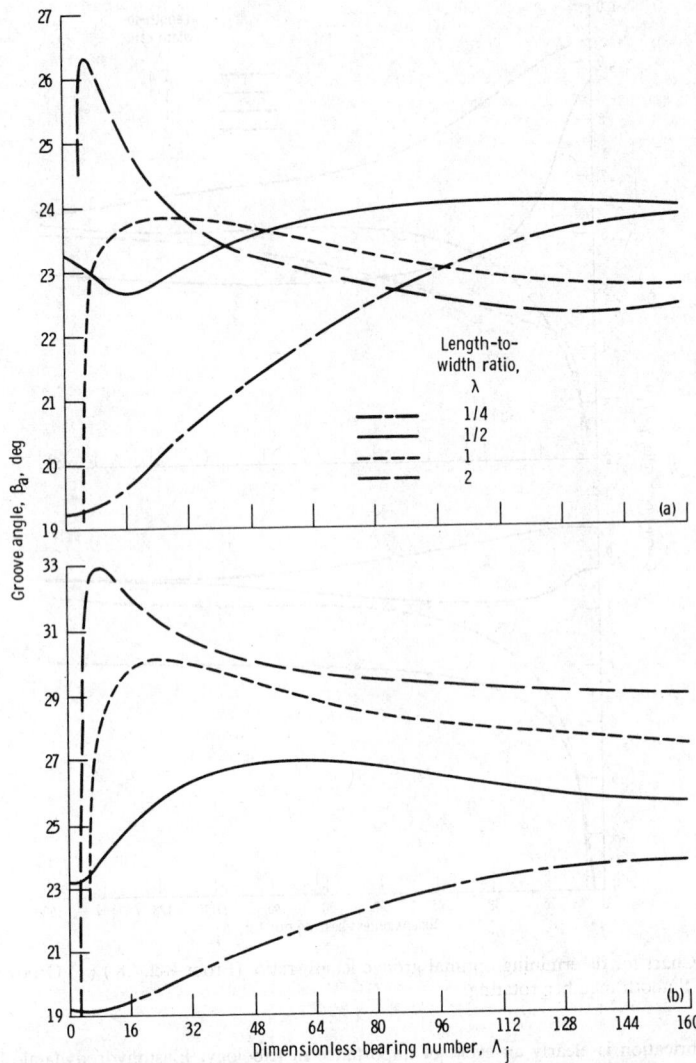

Fig. 23.53 Chart for determining optimal groove angle. (From Ref. 28.) (*a*) Grooved member rotating. (*b*) Smooth member rotating.

from, among other things, the ellipticity parameter and the elliptic integrals of the first and second kinds. Simplified expressions that allow quick calculations of the stresses and deformations to be made easily from a knowledge of the applied load, the material properties, and the geometry of the contacting elements are presented in this section.

Elliptical Contacts

The undeformed geometry of contacting solids in a nonconformal contact can be represented by two ellipsoids. The two solids with different radii of curvature in a pair of principal planes (*x* and *y*) passing through the contact between the solids make contact at a single point under the condition of zero applied load. Such a condition is called point contact and is shown in Fig. 23.67, where the radii of curvature are denoted by *r*'s. It is assumed that convex surfaces, as shown in Fig. 23.67, exhibit positive curvature and concave surfaces exhibit negative curvature. Therefore if the center of curvature lies within the solids, the radius of curvature is positive; if the center of curvature lies outside the solids, the radius of curvature is negative. It is important to note that if coordinates *x* and *y* are chosen such that

$$\frac{1}{r_{ax}} + \frac{1}{r_{bx}} > \frac{1}{r_{ay}} + \frac{1}{r_{by}} \tag{23.33}$$

coordinate x then determines the direction of the semiminor axis of the contact area when a load is applied and y determines the direction of the semimajor axis. The direction of motion is always considered to be along the x axis.

The curvature sum and difference, which are quantities of some importance in the analysis of contact stresses and deformations, are

$$\frac{1}{R} = \frac{1}{R_x} + \frac{1}{R_y} \tag{23.34}$$

$$\Gamma = R\left(\frac{1}{R_x} - \frac{1}{R_y}\right) \tag{23.35}$$

where

$$\frac{1}{R_x} = \frac{1}{r_{ax}} + \frac{1}{r_{bx}} \tag{23.36}$$

$$\frac{1}{R_y} = \frac{1}{r_{ay}} + \frac{1}{r_{by}} \tag{23.37}$$

$$\alpha = \frac{R_y}{R_x} \tag{23.38}$$

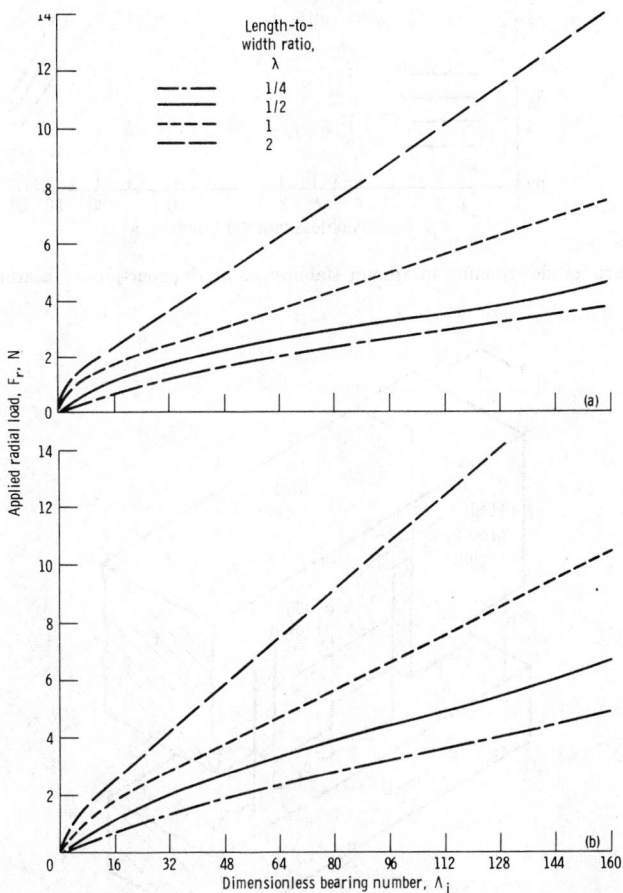

Fig. 23.54   Chart for determining maximum radial load capacity. (From Ref. 28.) (*a*) Grooved member rotating. (*b*) Smooth member rotating.

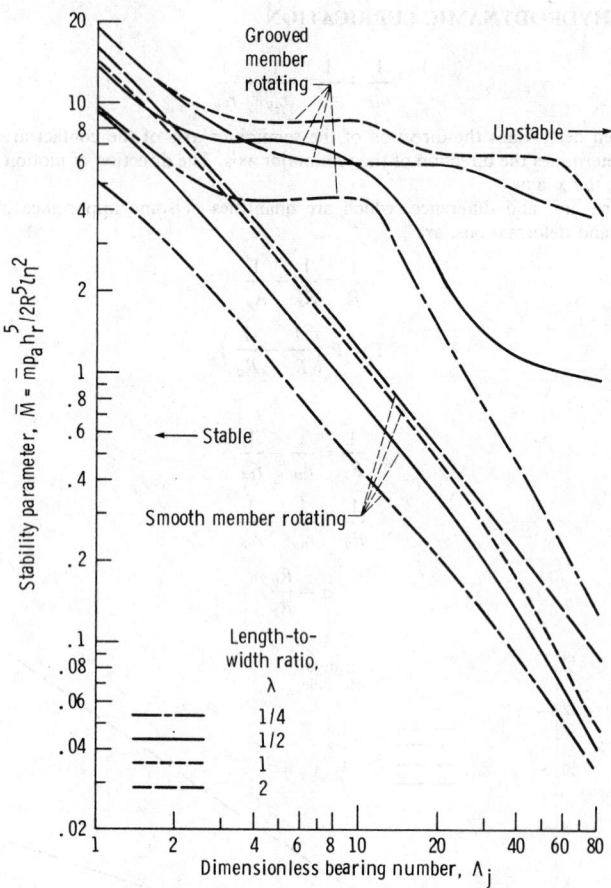

Fig. 23.55 Chart for determining maximum stability of herringbone-groove bearings. (From Ref. 29.)

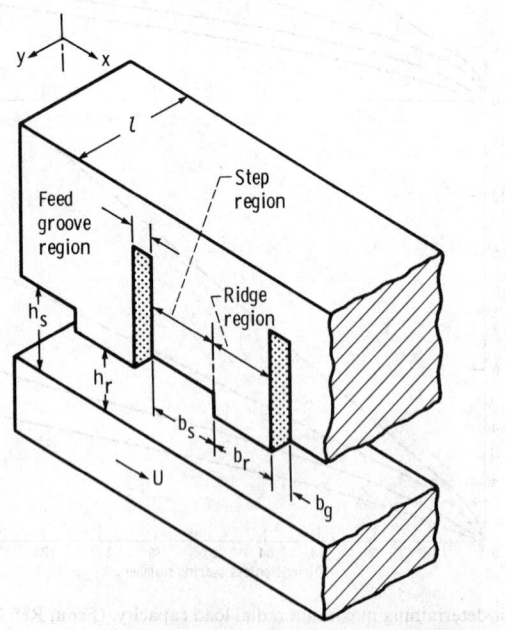

Fig. 23.56 Configuration of rectangular step thrust bearing. (From Ref. 30.)

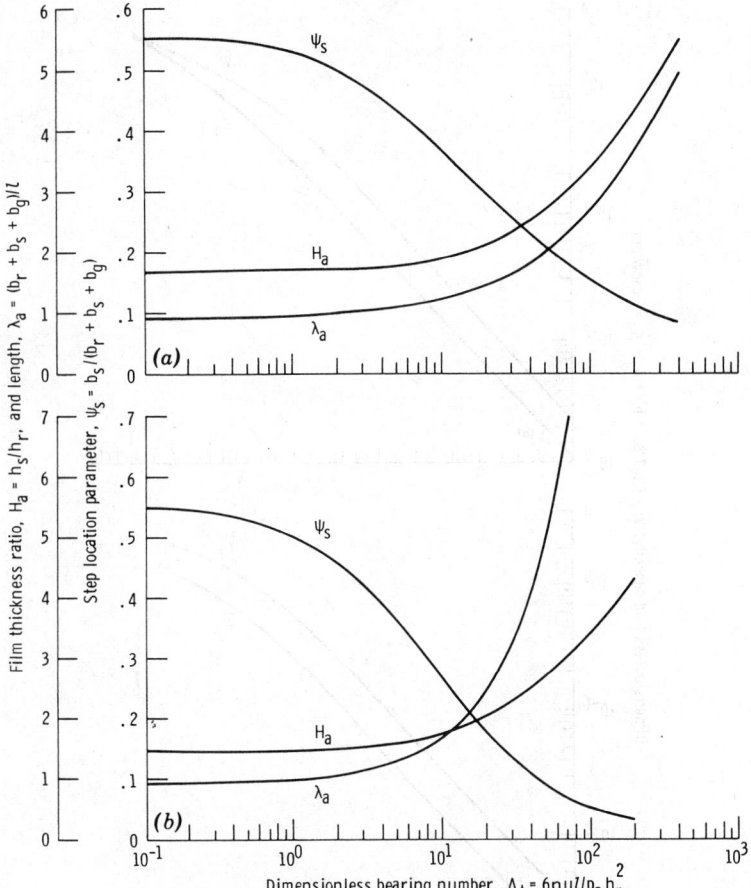

Fig. 23.57 Chart for determining optimal step parameters. (From Ref. 30.) (a) Maximum dimensionless load. (b) Maximum dimensionless stiffness.

Equations (23.36) and (23.37) effectively redefine the problem of two ellipsoidal solids approaching one another in terms of an equivalent ellipsoidal solid of radii R_x and R_y approaching a plane.

The ellipticity parameter k is defined as the elliptical-contact diameter in the y direction (transverse direction) divided by the elliptical-contact diameter in the x direction (direction of motion) or $k = D_y/D_x$. If Eq. (23.33) is satisfied and $\alpha \geq 1$, the contact ellipse will be oriented so that its major diameter will be transverse to the direction of motion, and, consequently, $k \geq 1$. Otherwise, the major diameter would lie along the direction of motion with both $\alpha \leq 1$ and $k \leq 1$. Figure 23.68 shows the ellipticity parameter and the elliptic integrals of the first and second kinds for a range of curvature ratios ($\alpha = R_y/R_x$) usually encountered in concentrated contacts.

Simplified Solutions for $\alpha > 1$. The classical Hertzian solution requires the calculation of the ellipticity parameter k and the complete elliptic integrals of the first and second kinds $\mathscr{F}$ and $\mathscr{E}$. This entails finding a solution to a transcendental equation relating k, $\mathscr{F}$, and $\mathscr{E}$ to the geometry of the contacting solids. Possible approaches include an iterative numerical procedure, as described, for example, by Hamrock and Anderson,[35] or the use of charts, as shown by Jones.[36] Hamrock and Brewe[34] provide a shortcut to the classical Hertzian solution for the local stress and deformation of two elastic bodies in contact. The shortcut is accomplished by using simplified forms of the ellipticity parameter and the complete elliptic integrals, expressing them as functions of the geometry. The results of Hamrock and Brewe's work[34] are summarized here.

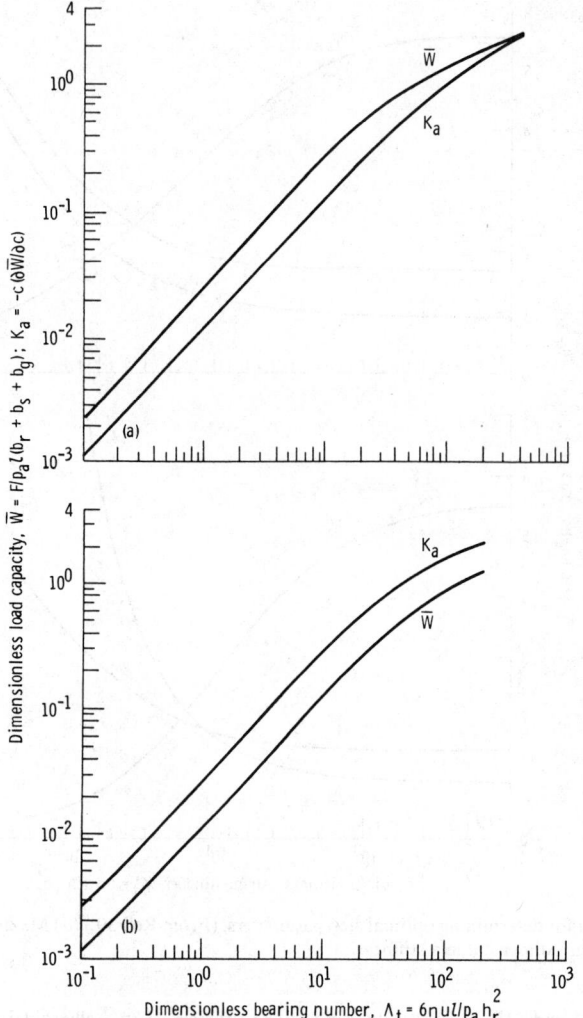

Fig. 23.58 Chart for determining dimensionless load capacity and stiffness. (From Ref. 30.) (a) Maximum dimensionless load capacity. (b) Maximum stiffness.

A power fit using linear regression by the method of least squares resulted in the following expression for the ellipticity parameter:

$$k = \alpha^{2/\pi}, \qquad \text{for} \quad \alpha \geq 1 \tag{23.39}$$

The asymptotic behavior of $\mathscr{E}$ and $\mathscr{F}$ ($\alpha \to 1$ implies $\mathscr{E} \to \mathscr{F} \to \pi/2$, and $\alpha \to \infty$ implies $\mathscr{F} \to \infty$ and $\mathscr{E} \to 1$) was suggestive of the type of functional dependence that $\mathscr{E}$ and $\mathscr{F}$ might follow. As a result, an inverse and a logarithmic fit were tried for $\mathscr{E}$ and $\mathscr{F}$, respectively. The following expressions provided excellent curve fits:

$$\mathscr{E} = 1 + \frac{q}{\alpha} \qquad \text{for} \quad \alpha \geq 1 \tag{23.40}$$

$$\mathscr{F} = \frac{\pi}{2} + q \ln \alpha \qquad \text{for} \quad \alpha \geq 1 \tag{23.41}$$

where

$$q = \frac{\pi}{2} - 1 \tag{23.42}$$

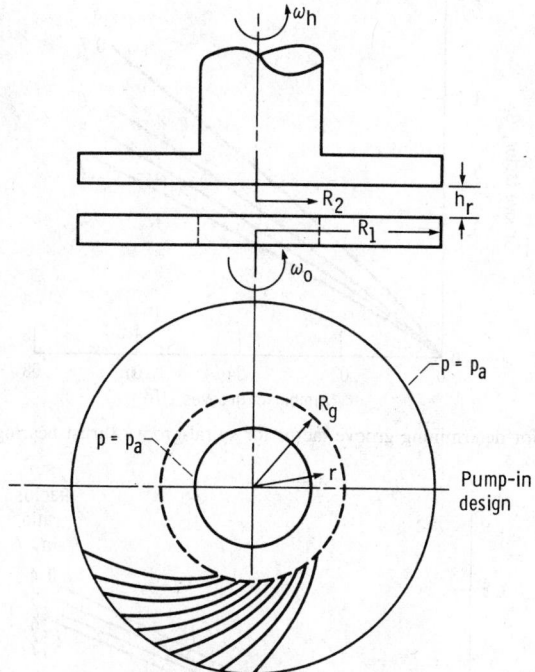

Fig. 23.59 Configuration of spiral-groove thrust bearing. (From Ref. 20.)

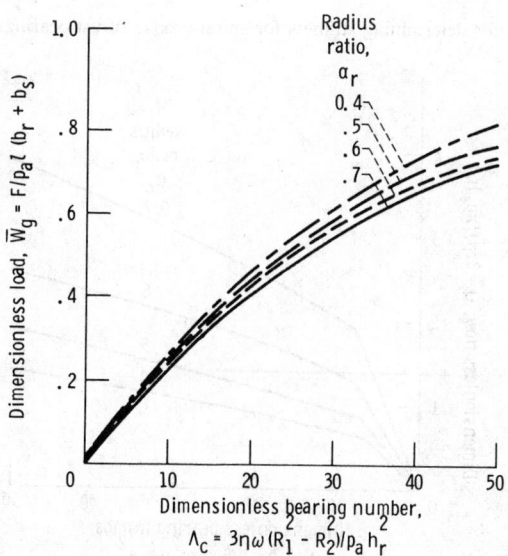

Fig. 23.60 Chart for determining load for spiral-groove thrust bearings. (From Ref. 20.)

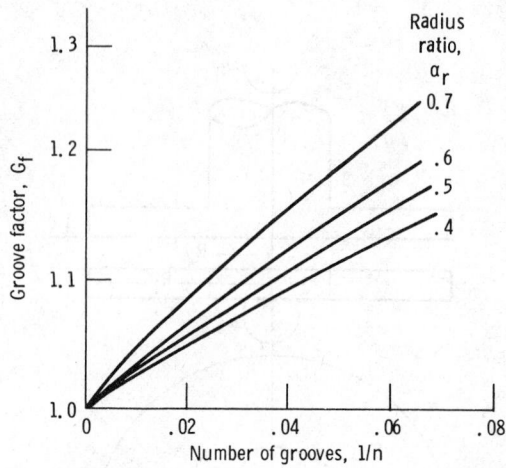

Fig. 23.61 Chart for determining groove factor for spiral-groove thrust bearings. (From Ref. 20.)

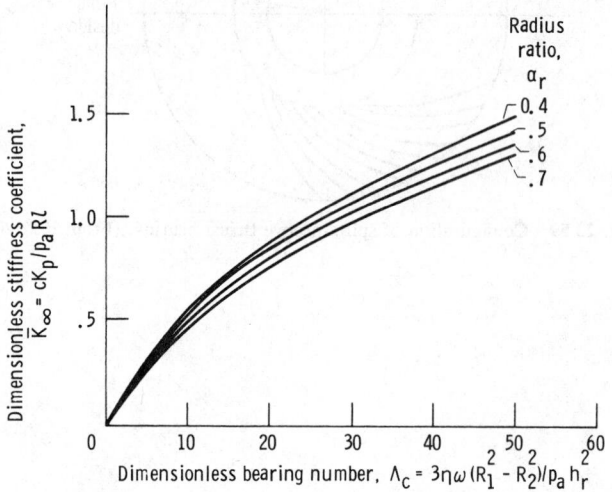

Fig. 23.62 Chart for determining stiffness for spiral-groove thrust bearings. (From Ref. 20.)

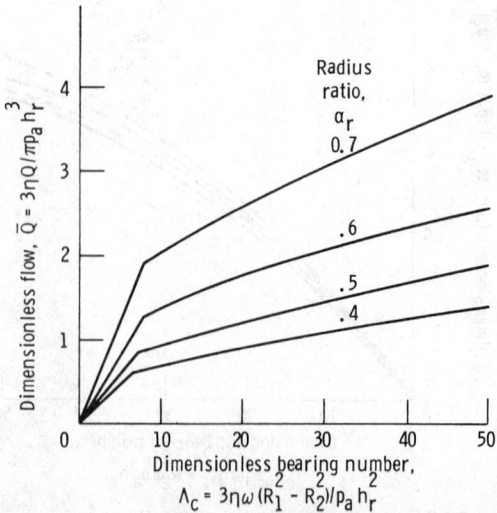

Fig. 23.63 Chart for determining flow for spiral-groove thrust bearings. (From Ref. 20.)

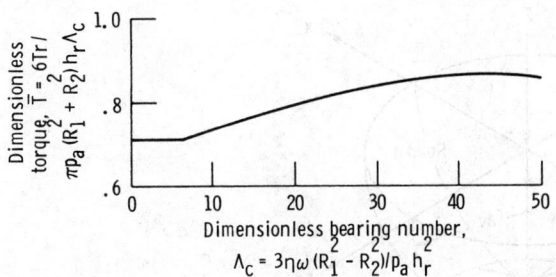

Fig. 23.64 Chart for determining torque for spiral-groove thrust bearings. (Curve is for all radius ratios. From Ref. 20.)

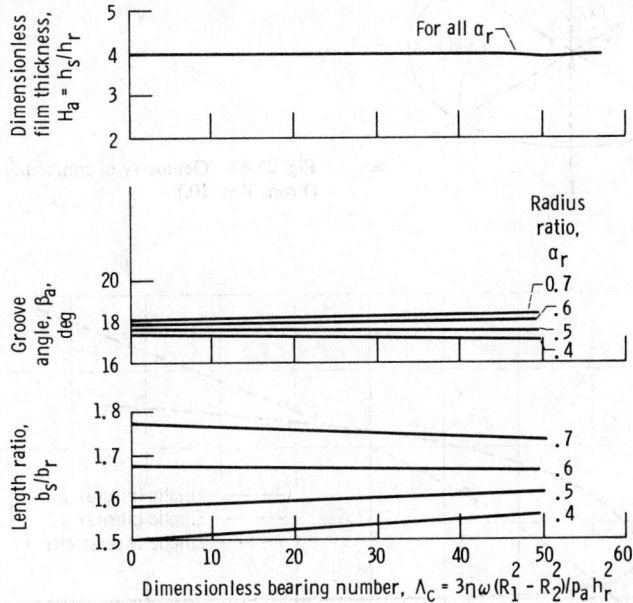

Fig. 23.65 Chart for determining optimal groove geometry for spiral-groove thrust bearings. (From Ref. 20.)

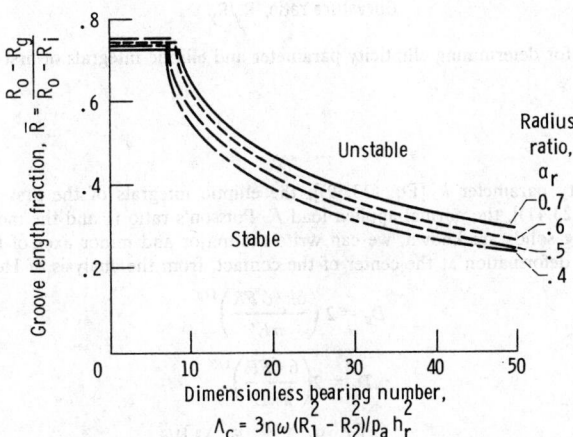

Fig. 23.66 Chart for determining groove length fraction for spiral-groove thrust bearings. (From Ref. 20.)

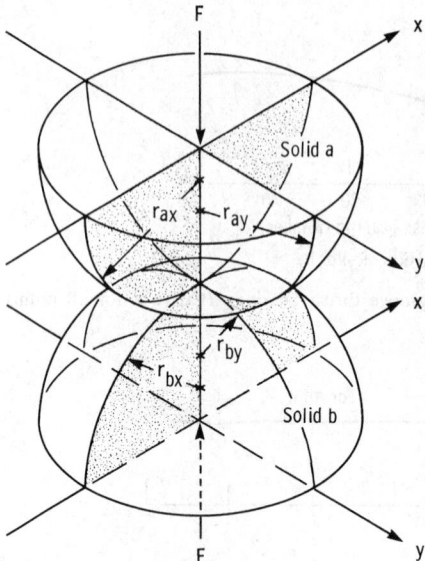

Fig. 23.67 Geometry of contacting elastic solids. (From Ref. 10.)

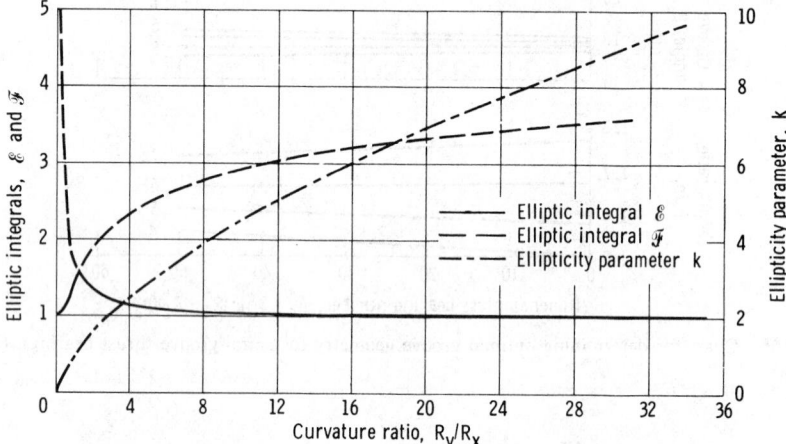

Fig. 23.68 Chart for determining ellipticity parameter and elliptic integrals of first and second kinds. (From Ref. 34.)

When the ellipticity parameter k [Eq. (23.39)], the elliptic integrals of the first and second kinds [Eqs. (23.40) and (23.41)], the normal applied load F, Poisson's ratio ν, and the modulus of elasticity E of the contacting solids are known, we can write the major and minor axes of the contact ellipse and the maximum deformation at the center of the contact, from the analysis of Hertz,[37] as

$$D_y = 2\left(\frac{6k^2 \mathscr{E} FR}{\pi E'}\right)^{1/3}$$

(23.43)

$$D_x = 2\left(\frac{6 \mathscr{E} FR}{\pi k E'}\right)^{1/3}.$$

(23.44)

$$\delta = F\left[\left(\frac{9}{2\mathscr{E}R}\right)\left(\frac{F}{\pi k E'}\right)^2\right]^{1/3}$$

(23.45)

where [as in Eq. (23.12)]

$$E' = 2\left(\frac{1 - \nu_a^2}{E_a} + \frac{1 - \nu_b^2}{E_b}\right)^{-1} \tag{23.46}$$

In these equations D_y and D_x are proportional to $F^{1/3}$ and δ is proportional to $F^{2/3}$.

The maximum Hertzian stress at the center of the contact can also be determined by using Eqs. (23.42) and (23.44)

$$\sigma_{max} = \frac{6F}{\pi D_x D_y} \tag{23.47}$$

Simplified Solutions for $\alpha < 1$. Table 23.7 gives the simplified equations for $\alpha < 1$ as well as for $\alpha \geq 1$. Recall that $\alpha \geq 1$ implies $k \geq 1$ and Eq. (23.33) is satisfied, and $\alpha < 1$ implies $k < 1$ and Eq. (23.33) is not satisfied. It is important to make the proper evaluation of α, since it has a great significance in the outcome of the simplified equations.

Table 23.7 Simplified Equations (From Ref. 6)

$\alpha \geq 1$	$\alpha < 1$
$\bar{k} = \alpha^{2/\pi}$	$\bar{k} = \alpha^{2/\pi}$
$\bar{\mathcal{F}} = \dfrac{\pi}{2} + q \ln \alpha$	$\bar{\mathcal{F}} = \dfrac{\pi}{2} + q \ln \alpha$
where $q = \dfrac{\pi}{2} - 1$	where $q = \dfrac{\pi}{2} - 1$
$\bar{\mathcal{E}} = 1 + \dfrac{q}{\alpha}$	$\bar{\mathcal{E}} = 1 + q\alpha$
$D_y = 2\left(\dfrac{6k^2 \mathcal{E} FR}{\pi E'}\right)^{1/3}$	$D_y = 2\left(\dfrac{6k \mathcal{E} FR}{\pi E'}\right)^{1/3}$
where $R^{-1} = R_x^{-1} + R_y^{-1}$	where $R^{-1} = R_x^{-1} + R_y^{-1}$
$D_x = 2\left(\dfrac{6 \mathcal{E} FR}{\pi k E'}\right)^{1/3}$	$D_x = 2\left(\dfrac{6 \mathcal{E} FR}{\pi E' k^2}\right)^{1/3}$
$\delta = \mathcal{F}\left[\left(\dfrac{4.5}{\mathcal{E} R}\right)\left(\dfrac{F}{\pi k E'}\right)^2\right]^{1/3}$	$\delta = \mathcal{F}\left[\left(\dfrac{4.5}{\mathcal{E} R}\right)\left(\dfrac{Fk}{\pi E'}\right)^2\right]^{1/3}$

Figure 23.69 shows three diverse situations in which the simplified equations can be usefully applied. The locomotive wheel on a rail (Fig. 23.69a) illustrates an example in which the ellipticity parameter k and the radius ratio α are less than 1. The ball rolling against a flat plate (Fig. 23.69b) provides pure circular contact (i.e., $\alpha = k = 1.0$). Figure 23.69c shows how the contact ellipse is formed in the ball–outer-race contact of a ball bearing. Here the semimajor axis is normal to the direction of rolling and, consequently, α and k are greater than 1. Table 23.8 shows how the degree of conformity affects the contact parameters for the various cases illustrated in Fig. 23.69.

Rectangular Contacts

For this situation the contact ellipse discussed in the preceding section is of infinite length in the transverse direction ($D_y \rightarrow \infty$). This type of contact is exemplified by a cylinder loaded against a plate, a groove, or another parallel cylinder or by a roller loaded against an inner or outer ring. In these situations the contact semiwidth is given by

$$b = R_x \left(\frac{8W}{\pi}\right)^{1/2} \tag{23.48}$$

where

$$W = \frac{F'}{E'R_x} \tag{23.49}$$

and F' is the load per unit length along the contact.

The maximum deformation due to the approach of centers of two cylinders can be written as[12]

$$\delta = \frac{2WR_x}{\pi}\left[\frac{2}{3} + \ln\left(\frac{2r_{ax}}{b}\right) + \ln\left(\frac{2r_{bx}}{b}\right)\right] \tag{23.50}$$

The maximum Hertzian stress in a rectangular contact can be written as

$$\sigma_{\max} = E'\left(\frac{W}{2\pi}\right)^{1/2} \tag{23.51}$$

23.3.2   Dimensionless Grouping

The variables appearing in elastohydrodynamic lubrication theory are

$$E' = \text{effective elastic modulus, N/m}^2$$
$$F = \text{normal applied load, N}$$
$$h = \text{film thickness, m}$$
$$R_x = \text{effective radius in } x \text{ (motion) direction, m}$$
$$R_y = \text{effective radius in } y \text{ (transverse) direction, m}$$
$$u = \text{mean surface velocity in } x \text{ direction, m/sec}$$
$$\xi = \text{pressure–viscosity coefficient of fluid, m}^2/\text{N}$$
$$\eta_0 = \text{atmospheric viscosity, N sec/m}^2$$

From these variables the following five dimensionless groupings can be established.
  Dimensionless film thickness

$$H = \frac{h}{R_x} \tag{23.52}$$

Ellipticity parameter

$$k = \frac{D_y}{D_x} = \left(\frac{R_y}{R_x}\right)^{2/\pi} \tag{23.53}$$

Dimensionless load parameter

$$W = \frac{F}{E'R_x^2} \tag{23.54}$$

Dimensionless speed parameter

$$U = \frac{\eta_0 u}{E'R_x} \tag{23.55}$$

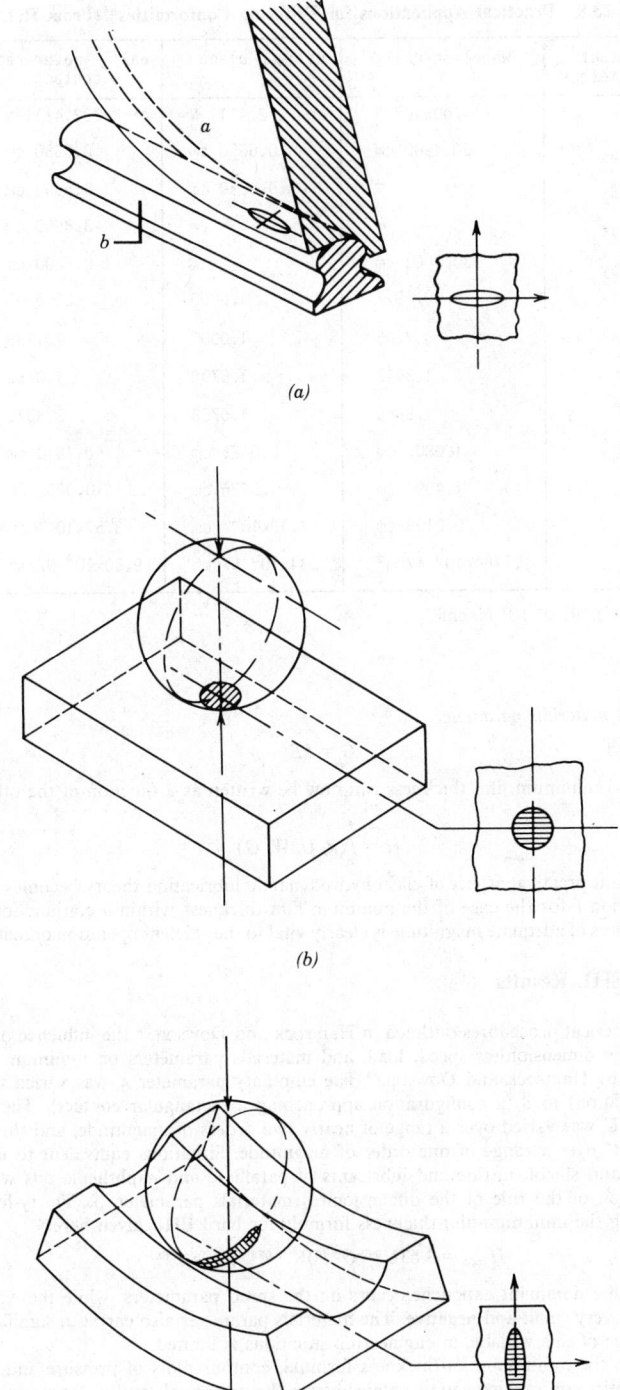

Fig. 23.69 Three degrees of conformity. (From Ref. 34.) (a) Wheel on rail. (b) Ball on plane. (c) Ball–outer-race contact.

Table 23.8   Practical Applications for Differing Conformities^a (From Ref. 34)

Contact parameters	Wheel on rail	Ball on plane	Ball - outer-race contact
F	1.00×10^5 N	222.4111 N	222.4111 N
r_{ax}	50.1900 cm	0.6350 cm	0.6350 cm
r_{ay}	∞	0.6350 cm	0.6350 cm
r_{bx}	∞	∞	-3.8900 cm
r_{by}	30.0000 cm	∞	-0.6600 cm
u	0.5977	1.0000	22.0905
k	0.7206	1.0000	7.1738
$\mathscr{E}$	1.3412	1.5708	1.0258
$\mathscr{F}$	1.8645	1.5708	3.3375
D_y	1.0807 cm	0.0426 cm	0.1810 cm
D_x	1.4997 cm	0.0426 cm	0.0252 cm
δ	0.0108 cm	7.13×10^{-4} cm	3.57×10^{-4} cm
σ_{max}	1.1784×10^5 N/cm^2	2.34×10^5 N/cm^2	9.30×10^4 N/cm^2

$^a E' = 2.197 \times 10^7$ N/cm².

Dimensionless materials parameter

$$G = \xi E' \tag{23.56}$$

The dimensionless minimum film thickness can now be written as a function of the other parameters involved:

$$H = f(k, U, W, G)$$

The most important practical aspect of elastohydrodynamic lubrication theory becomes the determination of this function f for the case of the minimum film thickness within a conjunction. Maintaining a fluid-film thickness of adequate magnitude is clearly vital to the efficient operation of machine elements.

23.3.3   Hard-EHL Results

By using the numerical procedures outlined in Hamrock and Dowson,[38] the influence of the ellipticity parameter and the dimensionless speed, load, and materials parameters on minimum film thickness was investigated by Hamrock and Dowson.[39] The ellipticity parameter k was varied from 1 (a ball-on-plate configuration) to 8 (a configuration approaching a rectangular contact). The dimensionless speed parameter U was varied over a range of nearly two orders of magnitude, and the dimensionless load parameter W over a range of one order of magnitude. Situations equivalent to using materials of bronze, steel, and silicon nitride and lubricants of paraffinic and naphthenic oils were considered in the investigation of the role of the dimensionless materials parameter G. Thirty-four cases were used in generating the minimum-film-thickness formula for hard EHL given here:

$$H_{min} = 3.63 U^{0.68} G^{0.49} W^{-0.073} (1 - e^{-0.68k}) \tag{23.57}$$

In this equation the dominant exponent occurs on the speed parameters, while the exponent on the load parameter is very small and negative. The materials parameter also carries a significant exponent, although the range of this variable in engineering situations is limited.

In addition to the minimum-film-thickness formula, contour plots of pressure and film thickness throughout the entire conjunction can be obtained from the numerical results. A representative contour plot of dimensionless pressure is shown in Fig. 23.70 for $k = 1.25$, $U = 0.168 \times 10^{-11}$, and $G = 4522$. In this figure and in Fig. 23.71, the + symbol indicates the center of the Hertzian contact zone. The dimensionless representation of the X and Y coordinates causes the actual Hertzian contact ellipse to be a circle regardless of the value of the ellipticity parameter. The Hertzian contact circle is shown by asterisks. On this figure is a key showing the contour labels and each corresponding value of dimensionless pressure. The inlet region is to the left and the exit region is to the right. The

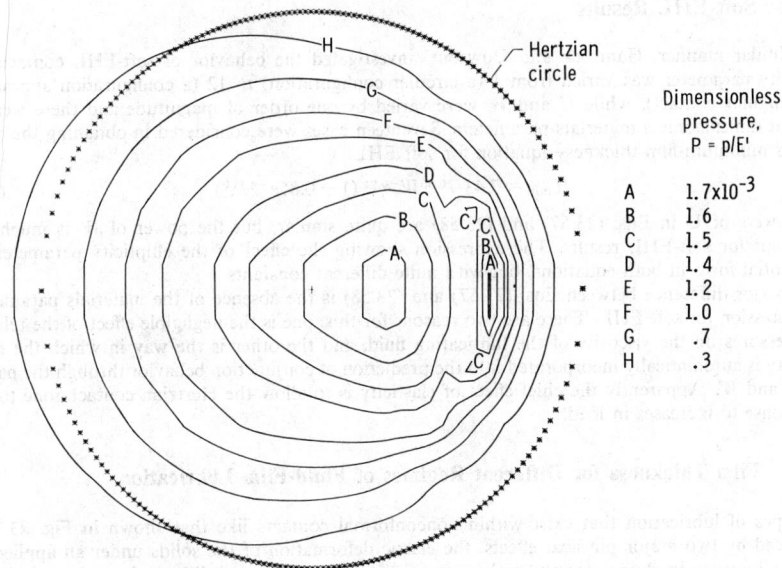

Fig. 23.70 Contour plot of dimensionless pressure. $k = 1.25$; $U = 0.168 \times 10^{-11}$; $W = 0.111 \times 10^{-6}$; $G = 4522$. (From Ref. 39.)

	Dimensionless pressure, $P = p/E'$
A	1.7×10^{-3}
B	1.6
C	1.5
D	1.4
E	1.2
F	1.0
G	.7
H	.3

pressure gradient at the exit end of the conjunction is much larger than that in the inlet region. In Fig. 23.70 a pressure spike is visible at the exit of the contact.

Contour plots of the film thickness are shown in Fig. 23.71 for the same case as Fig. 23.70. In this figure two minimum regions occur in well-defined lobes that follow, and are close to, the edge of the Hertzian contact circle. These results contain all of the essential features of available experimental observations based on optical interferometry.[40]

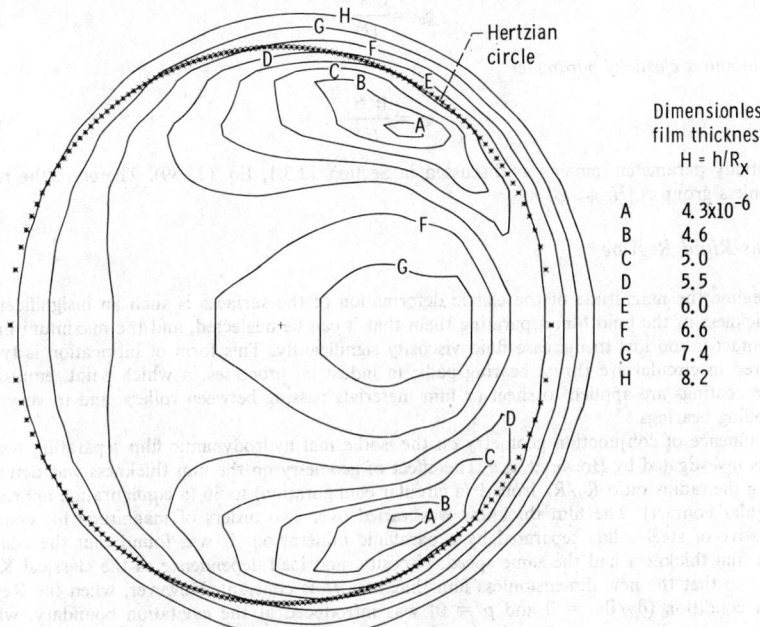

	Dimensionless film thickness, $H = h/R_x$
A	4.3×10^{-6}
B	4.6
C	5.0
D	5.5
E	6.0
F	6.6
G	7.4
H	8.2

Fig. 23.71 Contour plot of dimensionless film thickness. $k = 1.25$; $U = 0.168 \times 10^{-11}$; $W = 0.111 \times 10^{-6}$; $G = 4522$. (From Ref. 39.)

23.3.4 Soft-EHL Results

In a similar manner, Hamrock and Dowson[41] investigated the behavior of soft-EHL contacts. The ellipticity parameter was varied from 1 (a circular configuration) to 12 (a configuration approaching a rectangular contact), while U and W were varied by one order of magnitude and there were two different dimensionless materials parameters. Seventeen cases were considered in obtaining the dimensionless minimum-film-thickness equation for soft EHL:

$$H_{min} = 7.43 U^{0.65} W^{-0.21} (1 - 0.85 e^{-0.31k})$$
(23.58)

The powers of U in Eqs. (23.57) and (23.58) are quite similar, but the power of W is much more significant for soft-EHL results. The expression showing the effect of the ellipticity parameter is of exponential form in both equations, but with quite different constants.

A major difference between Eqs. (23.57) and (23.58) is the absence of the materials parameter in the expression for soft EHL. There are two reasons for this: one is the negligible effect of the relatively low pressures on the viscosity of the lubricating fluid, and the other is the way in which the role of elasticity is automatically incorporated into the prediction of conjunction behavior through the parameters U and W. Apparently the chief effect of elasticity is to allow the Hertzian contact zone to grow in response to increases in load.

23.3.5 Film Thickness for Different Regimes of Fluid-Film Lubrication

The types of lubrication that exist within nonconformal contacts like that shown in Fig. 23.70 are influenced by two major physical effects: the elastic deformation of the solids under an applied load and the increase in fluid viscosity with pressure. Therefore, it is possible to have four regimes of fluid-film lubrication, depending on the magnitude of these effects and on their relative importance. In this section because of the need to represent the four fluid-film lubrication regimes graphically, the dimensionless grouping presented in Section 23.3.2 will need to be recast. That is, the set of dimensionless parameters given in Section 23.3.2 $\{H, U, W, G, \text{ and } k\}$—will be reduced by one parameter without any loss of generality. Thus the dimensionless groupings to be used here are:

Dimensionless film parameter

$$\hat{H} = H \left(\frac{W}{U} \right)^2$$
(23.59)

Dimensionless viscosity parameter

$$g_v = \frac{GW^3}{U^2}$$
(23.60)

Dimensionless elasticity parameter

$$g_e = \frac{W^{8/3}}{U^2}$$
(23.61)

The ellipticity parameter remains as discussed in Section 23.3.1, Eq. (23.39). Therefore the reduced dimensionless group is $\{\hat{H}, g_v, g_e, k\}$.

Isoviscous-Rigid Regime

In this regime the magnitude of the elastic deformation of the surfaces is such an insignificant part of the thickness of the fluid film separating them that it can be neglected, and the maximum pressure in the contact is too low to increase fluid viscosity significantly. This form of lubrication is typically encountered in circular-arc thrust bearing pads; in industrial processes in which paint, emulsion, or protective coatings are applied to sheet or film materials passing between rollers; and in very lightly loaded rolling bearings.

The influence of conjunction geometry on the isothermal hydrodynamic film separating two rigid solids was investigated by Brewe et al.[42] The effect of geometry on the film thickness was determined by varying the radius ratio R_y/R_x from 1 (a circular configuration) to 36 (a configuration approaching a rectangular contact). The film thickness was varied over two orders of magnitude for conditions representative of steel solids separated by a paraffinic mineral oil. It was found that the computed minimum film thickness had the same speed, viscosity, and load dependence as the classical Kapitza solution,[43] so that the new dimensionless film thickness H is constant. However, when the Reynolds cavitation condition $(\partial p / \partial n = 0$ and $p = 0)$ was introduced at the cavitation boundary, where n represents the coordinate normal to the interface between the full film and the cavitation region, an additional geometrical effect emerged. According to Brewe et al.,[42] the dimensionless minimum-film-thickness parameter for the isoviscous-rigid regime should now be written as

$$(\hat{H}_{\min})_{\text{ir}} = 128\alpha\lambda_b^2 \left[0.131 \tan^{-1}\left(\frac{\alpha}{2}\right) + 1.683 \right]^2 \tag{23.62}$$

where

$$\alpha = \frac{R_y}{R_x} \approx (k)^{\pi/2} \tag{23.63}$$

and

$$\lambda_b = \left(1 + \frac{2}{3\alpha} \right)^{-1} \tag{23.64}$$

In Eq. (23.62) the dimensionless film thickness parameter $\hat{H}$ is shown to be strictly a function only of the geometry of the contact described by the ratio $\alpha = R_y/R_x$.

Piezoviscous-Rigid Regime

If the pressure within the contact is sufficiently high to increase the fluid viscosity within the conjunction significantly, it may be necessary to consider the pressure–viscosity characteristics of the lubricant while assuming that the solids remain rigid. For the latter part of this assumption to be valid, it is necessary that the deformation of the surfaces remain an insignificant part of the fluid-film thickness. This form of lubrication may be encountered on roller end-guide flanges, in contacts in moderately loaded cylindrical tapered rollers, and between some piston rings and cylinder liners.

From Hamrock and Dowson[44] the minimum-film-thickness parameter for the piezoviscous-rigid regime can be written as

$$(\hat{H}_{\min})_{\text{pvr}} = 1.66\, g_v^{2/3}\, (1 - e^{-0.68k}) \tag{23.65}$$

Note the absence of the dimensionless elasticity parameter g_e from Eq. (23.65).

Isoviscous-Elastic (Soft-EHL) Regime

In this regime the elastic deformation of the solids is a significant part of the thickness of the fluid film separating them, but the pressure within the contact is quite low and insufficient to cause any substantial increase in viscosity. This situation arises with materials of low elastic modulus (such as rubber), and it is a form of lubrication that may be encountered in seals, human joints, tires, and elastomeric material machine elements.

If the film thickness equation for soft EHL [Eq. (23.58)] is rewritten in terms of the reduced dimensionless grouping, the minimum-film-thickness parameter for the isoviscous-elastic regime can be written as

$$(\hat{H}_{\min})_{\text{ie}} = 8.70\, g_e^{0.67}\, (1 - 0.85 e^{-0.31k}) \tag{23.66}$$

Note the absence of the dimensionless viscosity parameter g_v from Eq. (23.66).

Piezoviscous-Elastic (Hard-EHL) Regime

In fully developed elastohydrodynamic lubrication the elastic deformation of the solids is often a significant part of the thickness of the fluid film separating them, and the pressure within the contact is high enough to cause a significant increase in the viscosity of the lubricant. This form of lubrication is typically encountered in ball and roller bearings, gears, and cams.

Once the film thickness equation [Eq. (23.57)] has been rewritten in terms of the reduced dimensionless grouping, the minimum film parameter for the piezoviscous-elastic regime can be written as

$$(\hat{H}_{\min})_{\text{pve}} = 3.42\, g_v^{0.49} g_e^{0.17}\, (1 - e^{-0.68k}) \tag{23.67}$$

An interesting observation to make in comparing Eqs. (23.65) through (23.67) is that in each case the sum of the exponents on g_v and g_e is close to the value of $\frac{2}{3}$ required for complete dimensional representation of these three lubrication regimes: piezoviscous-rigid, isoviscous-elastic, and piezoviscous-elastic.

Contour Plots

Having expressed the dimensionless minimum-film-thickness parameter for the four fluid-film regimes in Eqs. (23.62) to (23.67), Hamrock and Dowson[44] used these relationships to develop a map of the lubrication regimes in the form of dimensionless minimum-film-thickness parameter contours. Some of these maps are shown in Figs. 23.72–23.74 on a log-log grid of the dimensionless viscosity and elasticity parameters for ellipticity parameters of 1, 3, and 6, respectively. The procedure used to

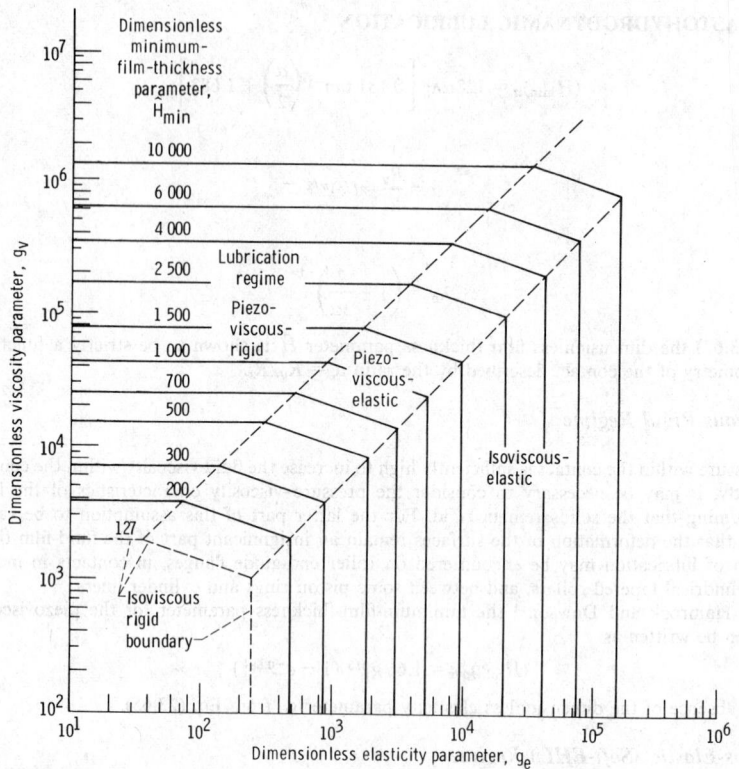

Fig. 23.72 Map of lubrication regimes for ellipticity parameter k of 1. (From Ref. 44.)

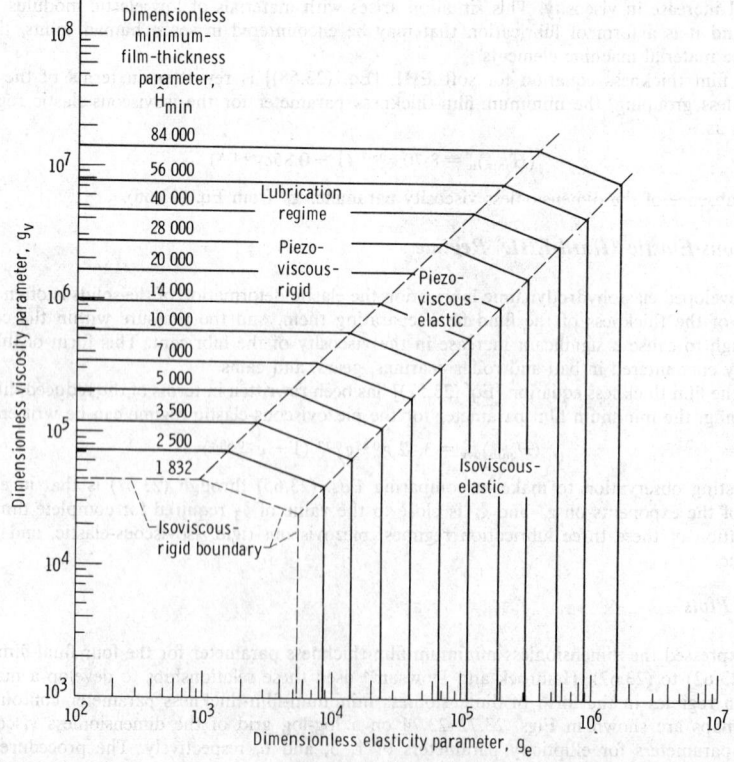

Fig. 23.73 Map of lubrication regimes for ellipticity parameter k of 3. (From Ref. 44.)

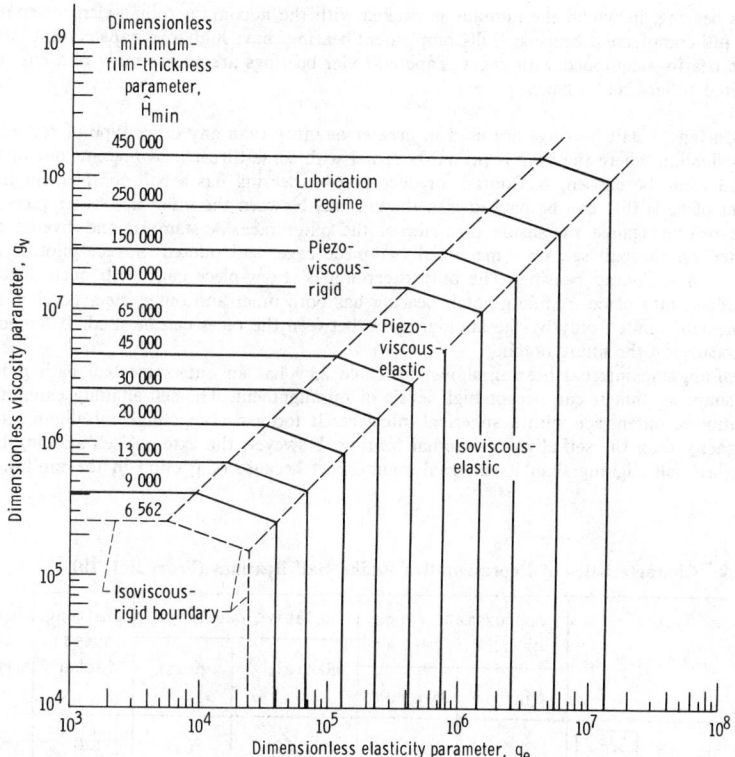

Fig. 23.74 Map of lubrication regimes for ellipticity parameter k of 6. (From Ref. 44.)

obtain these figures can be found in Ref. 44. The four lubrication regimes are clearly shown in Figs. 23.72–23.74. By using these figures for given values of the parameters k, g_v, and g_e, the fluid-film lubrication regime in which any elliptical conjunction is operating can be ascertained and the approximate value of $\hat{H}_{min}$ can be determined. When the lubrication regime is known, a more accurate value of $\hat{H}_{min}$ can be obtained by using the appropriate dimensionless minimum-film-thickness equation. These results are particularly useful in initial investigations of many practical lubrication problems involving elliptical conjunctions.

23.3.6 Rolling-Element Bearings

Rolling-element bearings are precision, yet simple, machine elements of great utility, whose mode of lubrication is elastohydrodynamic. This section describes the types of rolling-element bearings and their geometry, kinematics, load distribution, and fatigue life, and demonstrates how elastohydrodynamic lubrication theory can be applied to the operation of rolling-element bearings. This section makes extensive use of the work by Hamrock and Dowson[10] and by Hamrock and Anderson.[6]

Bearing Types

A great variety of both design and size range of ball and roller bearings is available to the designer. The intent of this section is not to duplicate the complete descriptions given in manufacturers' catalogs, but rather to present a guide to representative bearing types along with the approximate range of sizes available. Tables 23.9–23.17 illustrate some of the more widely used bearing types. In addition, there are numerous types of specialty bearings available for which space does not permit a complete cataloging. Size ranges are given in metric units. Traditionally, most rolling-element bearings have been manufactured to metric dimensions, predating the efforts toward a metric standard. In addition to bearing types and approximate size ranges available, Tables 23.9–23.17 also list approximate relative load-carrying capabilities, both radial and thrust, and, where relevant, approximate tolerances to misalignment.

Rolling bearings are an assembly of several parts—an inner race, an outer race, a set of balls or rollers, and a cage or separator. The cage or separator maintains even spacing of the rolling elements.

A cageless bearing, in which the annulus is packed with the maximum rolling-element complement, is called a full-complement bearing. Full-complement bearings have high load capacity but lower speed limits than bearings equipped with cages. Tapered-roller bearings are an assembly of a cup, a cone, a set of tapered rollers, and a cage.

Ball Bearings. Ball bearings are used in greater quantity than any other type of rolling bearing. For an application where the load is primarily radial with some thrust load present, one of the types in Table 23.9 can be chosen. A Conrad, or deep-groove, bearing has a ball complement limited by the number of balls that can be packed into the annulus between the inner and outer races with the inner race resting against the inside diameter of the outer race. A stamped and riveted two-piece cage, piloted on the ball set, or a machined two-piece cage, ball piloted or race piloted, is almost always used in a Conrad bearing. The only exception is a one-piece cage with open-sided pockets that is snapped into place. A filling-notch bearing has both inner and outer races notched so that a ball complement limited only by the annular space between the races can be used. It has low thrust capacity because of the filling notch.

The self-aligning internal bearing shown in Table 23.9 has an outer-race ball path ground in a spherical shape so that it can accept high levels of misalignment. The self-aligning external bearing has a multipiece outer race with a spherical interface. It too can accept high misalignment and has higher capacity than the self-aligning internal bearing. However, the external self-aligning bearing is somewhat less self-aligning than its internal counterpart because of friction in the multipiece outer race.

Table 23.9 Characteristics of Representative Radial Ball Bearings (From Ref. 10)

Type	Approximate range of bore sizes, mm		Relative capacity		Limiting speed factor	Tolerance to mis-alignment
	Minimum	Maximum	Radial	Thrust		
Conrad or deep groove	3	1060	1.00	[a]0.7	1.0	±0°15'
Maximum capacity or filling notch	10	130	1.2-1.4	[a]0.2	1.0	±0°3'
Magneto or counterbored outer	3	200	0.9-1.3	[b]0.5-0.9	1.0	±0°5'
Airframe or aircraft control	4.826	31.75	High static capacity	[a]0.5	0.2	0°
Self-aligning, internal	5	120	0.7	[b]0.2	1.0	±2°30'
Self-aligning, external	------	------	1.0	[a]0.7	1.0	High
Double row, maximum	6	110	1.5	[a]0.2	1.0	±0°3'
Double row, deep groove	6	110	1.5	[a]1.4	1.0	0°

[a]Two directions.
[b]One direction.

Representative angular-contact ball bearings are illustrated in Table 23.10. An angular-contact ball bearing has a two-shouldered ball groove in one race and a single-shouldered ball groove in the other race. Thus it is capable of supporting only a unidirectional thrust load. The cutaway shoulder allows assembly of the bearing by snapping over the ball set after it is positioned in the cage and outer race. This also permits use of a one-piece, machined, race-piloted cage that can be balanced for high-speed operation. Typical contact angles vary from 15° to 25°.

Angular-contact ball bearings are used in duplex pairs mounted either back to back or face to face as shown in Table 23.10. Duplex bearing pairs are manufactured so that they "preload" each other when clamped together in the housing and on the shaft. The use of preloading provides stiffer shaft support and helps prevent bearing skidding at light loads. Proper levels of preload can be obtained from the manufacturer. A duplex pair can support bidirectional thrust load. The back-to-back arrangement offers more resistance to moment or overturning loads than does the face-to-face arrangement.

Where thrust loads exceed the capability of a simple bearing, two bearings can be used in tandem, with both bearings supporting part of the thrust load. Three or more bearings are occasionally used in tandem, but this is discouraged because of the difficulty in achieving good load sharing. Even slight differences in operating temperature will cause a maldistribution of load sharing.

The split-ring bearing shown in Table 23.10 offers several advantages. The split ring (usually the inner) has its ball groove ground as a circular arc with a shim between the ring halves. The shim is then removed when the bearing is assembled so that the split-ring ball groove has the shape of a gothic arch. This reduces the axial play for a given radial play and results in more accurate axial positioning of the shaft. The bearing can support bidirectional thrust loads but must not be operated for prolonged periods of time at predominantly radial loads. This results in three-point ball–race contact and relatively high frictional losses. As with the conventional angular-contact bearing, a one-piece precision-machined cage is used.

Table 23.10   Characteristics of Representative Angular-Contact Ball Bearings (From Ref. 10)

Type	Approximate range of bore sizes, mm		Relative capacity		Limiting speed factor	Tolerance to mis-alignment
	Minimum	Maximum	Radial	Thrust		
One-directional thrust	10	320	[b]1.00–1.15	[a,b]1.5–2.3	[b]1.1–3.0	±0°2'
Duplex, back to back	10	320	1.85	[c]1.5	3.0	0°
Duplex, face to face	10	320	1.85	[c]1.5	3.0	0°
Duplex, tandem	10	320	1.85	[a]2.4	3.0	0°
Two-directional or split ring	10	110	1.15	[c]1.5	3.0	±0°2'
Double row	10	140	1.5	[c]1.85	0.8	0°
Double row, maximum	10	110	1.65	[a]0.5 [d]1.5	0.7	0°

[a]One direction.
[b]Depends on contact angle.
[c]Two directions.
[d]In other direction.

Ball thrust bearings (90° contact angle), Table 23.11, are used almost exclusively for machinery with vertically oriented shafts. The flat-race bearing allows eccentricity of the fixed and rotating members. An additional bearing must be used for radial positioning. It has low load capacity because of the very small ball–race contacts and consequent high Hertzian stress. Grooved-race bearings have higher load capacities and are capable of supporting low-magnitude radial loads. All of the pure thrust ball bearings have modest speed capability because of the 90° contact angle and the consequent high level of ball spinning and frictional losses.

Roller Bearings. Cylindrical roller bearings, Table 23.12, provide purely radial load support in most applications. An N or U type of bearing will allow free axial movement of the shaft relative to the housing to accommodate differences in thermal growth. An F or J type of bearing will support a light thrust load in one direction; and a T type of bearing will support a light bidirectional thrust load.

Cylindrical roller bearings have moderately high radial load capacity as well as high-speed capability. Their speed capability exceeds that of either spherical or tapered-roller bearings. A commonly used bearing combination for support of a high-speed rotor is an angular-contact ball bearing or duplex pair and a cylindrical roller bearing.

As explained in the following section on bearing geometry, the rollers in cylindrical roller bearings are seldom pure cylinders. They are crowned or made slightly barrel shaped to relieve stress concentrations of the roller ends when any misalignment of the shaft and housing is present.

Cylindrical roller bearings may be equipped with one- or two-piece cages, usually race piloted. For greater load capacity, full-complement bearings can be used, but at a significant sacrifice in speed capability.

Spherical roller bearings, Tables 23.13–23.15, are made as either single- or double-row bearings. The more popular bearing design uses barrel-shaped rollers. An alternative design employs hourglass-shaped rollers. Spherical roller bearings combine very high radial load capacity with modest thrust load capacity (with the exception of the thrust type) and excellent tolerance to misalignment. They find widespread use in heavy-duty rolling mill and industrial gear drives, where all of these bearing characteristics are requisite.

Tapered-roller bearings, Table 23.16, are also made as single- or double-row bearings with combinations of one- or two-piece cups and cones. A four-row bearing assembly with two- or three-piece cups and cones is also available. Bearings are made with either a standard angle for applications in which moderate thrust loads are present or with a steep angle for high thrust capacity. Standard and special cages are available to suit the application requirements.

Single-row tapered-roller bearings must be used in pairs because a radially loaded bearing generates a thrust reaction that must be taken by a second bearing. Tapered-roller bearings are normally set up with spacers designed so that they operate with some internal play. Manufacturers' engineering journals should be consulted for proper setup procedures.

Needle roller bearings, Table 23.17, are characterized by compactness in the radial direction and are frequently used without an inner race. In the latter case the shaft is hardened and ground to

Table 23.11   Characteristics of Representative Thrust Ball Bearings (From Ref. 10)

Type		Approximate range of bore sizes, mm		Relative capacity		Limiting speed factor	Tolerance to mis-alignment
		Minimum	Maximum	Radial	Thrust		
One directional, flat race		6.45	88.9	0	^{a}0.7	0.10	^{b}0°
One directional, grooved race		6.45	1180	0	^{a}1.5	0.30	0°
Two directional, grooved race		15	220	0	^{c}1.5	0.30	0°

^{a}One direction.
^{b}Accepts eccentricity.
^{c}Two directions.

Table 23.12 Characteristics of Representative Cylindrical Roller Bearings (From Ref. 10)

Type	Approximate range of bore sizes, mm		Relative capacity		Limiting speed factor	Tolerance to mis-alignment
	Minimum	Maximum	Radial	Thrust		
Separable outer ring, nonlocating (RN, RIN)	10	320	1.55	0	1.20	±0°5'
Separable inner ring, nonlocating (RU, RIU)	12	500	1.55	0	1.20	±0°5'
Separable outer ring, one-direction locating (RF, RIF)	40	177.8	1.55	[a]Locating	1.15	±0°5'
Separable inner ring, one-direction locating (RJ, RIJ)	12	320	1.55	[a]Locating	1.15	±0°5'
Self-contained, two-direction locating	12	100	1.35	[b]Locating	1.15	±0°5'
Separable inner ring, two-direction locating (RT, RIT)	20	320	1.55	[b]Locating	1.15	±0°5'
Nonlocating, full complement (RK, RIK)	17	75	2.10	0	0.20	±0°5'
Double row, separable outer ring, nonlocating (RD)	30	1060	1.85	0	1.00	0°
Double row, separable inner ring, nonlocating	70	1060	1.85	0	1.00	0°

[a]One direction.
[b]Two directions.

Table 23.13   Characteristics of Representative Spherical Roller Bearings (From Ref. 10)

Type		Approximate range of bore sizes, mm		Relative capacity		Limiting speed factor	Tolerance to mis-alignment
		Minimum	Maximum	Radial	Thrust		
Single row, barrel or convex		20	320	2.10	0.20	0.50	±2°
Double row, barrel or convex		25	1250	2.40	0.70	0.50	±1°30'
Thrust		85	360	[a]0.10 [b]0.10	[a]1.80 [b]2.40	0.35-0.50	±3°
Double row, concave		50	130	2.40	0.70	0.50	±1°30'

[a]Symmetric rollers.
[b]Asymmetric rollers.

serve as the inner race. Drawn cups, both open and closed end, are frequently used for grease retention. Drawn cups are thin walled and require substantial support from the housing. Heavy-duty roller bearings have relatively rigid races and are more akin to cylindrical roller bearings with long-length-to-diameter-ratio rollers.

Needle roller bearings are more speed limited than cylindrical roller bearings because of roller skewing at high speeds. A high percentage of needle roller bearings are full-complement bearings. Relative to a caged needle bearing, these have higher load capacity but lower speed capability.

There are many types of specialty bearings available other than those discussed here. Aircraft bearings for control systems, thin-section bearings, and fractured-ring bearings are some of the more widely used bearings among the many types manufactured. A complete coverage of all bearing types is beyond the scope of this chapter.

Angular-contact ball bearings and cylindrical roller bearings are generally considered to have the highest speed capabilities. Speed limits of roller bearings are discussed in conjunction with lubrication methods. The lubrication system employed has as great an influence on limiting bearing speed as does the bearing design.

Table 23.14   Characteristics of Standardized Double-Row, Spherical Roller Bearings (From Ref. 10)

Type		Roller design	Retainer design	Roller guidance	Roller-race contact
SLB		Symmetric	Machined, roller piloted	Retainer pockets	Modified line, both races
SC		Symmetric	Stamped, race piloted	Floating guide ring	Modified line, both races
SD		Asymmetric	Machined, race piloted	Inner-ring center rib	Line contact, outer; point contact, inner

Table 23.15   Characteristics of Spherical Roller Bearings (From Ref. 10)

Series	Types	Approximate range of bore sizes, mm		Approximate relative capacity[a]		Limiting speed factor
		Minimum	Maximum	Radial	Thrust	
202	Single-row barrel	20	320	1.0	0.11	0.5
203	Single-row barrel	20	240	1.7	.18	.5
204	Single-row barrel	25	110	2.1	.22	.4
212	SLB	35	75	1.0	.26	.6
213	SLB	30	70	1.7	.53	
22, 22K	SLB, SC, SD	30	320	1.7	.46	
23, 23K	SLB, SC, SD	40	280	2.7	1.0	
30, 30K	SLB, SC, SD	120	1250	1.2	.29	.7
31, 31K	SLB, SC, SD	110	1250	1.7	.54	.6
32, 32K	SLB, SC, SD	100	850	2.1	.78	.6
39, 39K	SD	120	1250	.7	.18	.7
40, 40K	SD	180	250	1.5	----	.7

[a]Load capacities are comparative within the various series of spherical roller bearings only. For a given envelope size, a spherical roller bearing has a radial capacity approximately equal to that of a cylindrical roller bearing.

Table 23.16   Characteristics of Representative Tapered Roller Bearings (From Ref. 10)

Type		Subtype	Approximate range of bore sizes, mm	
			Minimum	Maximum
Single row (TS)			8	1690
		TST - Tapered bore	24	430
		TSS - Steep angle	16	1270
		TS - Pin cage	----	
		TSE, TSK - keyway cones	12	380
		TSF, TSSF - flanged cup	8	1070
		TSG - steering gear (without cone)	----	
Two row, double cone, single cups (TDI)			30	1200
		TDIK, TDIT, TDITP - tapered bore	30	860
		TDIE, TDIKE - slotted double cone	24	690
		TDIS - steep angle	55	520
Two row, double cup, single cones, adjustable (TDO)			8	1830
		TDO		
		TDOS - steep angle	20	1430
Two row, double cup, single cones, nonadjustable (TNA)			20	60
		TNA		
		TNASW - slotted cones	30	260
		TNASWE - extended cone rib	20	305
		TNASWH - slotted cones, sealed	8	70
		TNADA, TNHDADX - self-aligning cup AD	---	----
Four row, cup adjusted (TQO)			70	1500
		TQO, TQOT - tapered bore	250	1500
Four row, cup adjusted (TQI)		TQIT - tapered bore	---	----

Table 23.17　Characteristics of Representative Needle Roller Bearings (From Ref. 10)

Type	Bore sizes, mm		Relative load capacity		Limiting speed factor	Misalignment tolerance
	Minimum	Maximum	Dynamic	Static		
Drawn cup, needle （Open end　Closed end）	3	185	High	Moderate	0.3	Low
Drawn cup, needle, grease retained	4	25	High	Moderate	0.3	Low
Drawn cup, roller （Open end　Closed end）	5	70	Moderate	Moderate	0.9	Moderate
Heavy-duty roller	16	235	Very high	Moderate	1.0	Moderate
Caged roller	12	100	Very high	High	1.0	Moderate
Cam follower	12	150	Moderate to high	Moderate to high	0.3-0.9	Low
Needle thrust	6	105	Very high	Very high	0.7	Low

Geometry

The operating characteristics of a rolling-element bearing depend greatly on the diametral clearance of the bearing. This clearance varies for the different types of bearings discussed in the preceding section. In this section, the principal geometrical relationships governing the operation of unloaded rolling-element bearings are developed. This information will be of vital interest when such quantities as stress, deflection, load capacity, and life are considered in subsequent sections. Although bearings rarely operate in the unloaded state, an understanding of this section is vital to the appreciation of the remaining sections.

Geometry of Ball Bearings

Pitch Diameter and Clearance.　The cross section through a radial, single-row ball bearing shown in Fig. 23.75 depicts the radial clearance and various diameters. The pitch diameter d_e is the mean of the inner- and outer-race contact diameters and is given by

$$d_e = d_i + \tfrac{1}{2}(d_o - d_i) \qquad \text{or} \qquad d_e = \tfrac{1}{2}(d_o + d_i) \qquad (23.68)$$

Also from Fig. 23.75, the diametral clearance denoted by P_d can be written as

$$P_d = d_o - d_i - 2d \qquad (23.69)$$

Diametral clearance may therefore be thought of as the maximum distance that one race can move diametrally with respect to the other when no measurable force is applied and both races lie in the same plane. Although diametral clearance is generally used in connection with single-row radial bearings, Eq. (23.69) is also applicable to angular-contact bearings.

Race Conformity.　Race conformity is a measure of the geometrical conformity of the race and the ball in a plane passing through the bearing axis, which is a line passing through the center of the bearing perpendicular to its plane and transverse to the race. Figure 23.76 is a cross section of a ball bearing showing race conformity, expressed as

$$f = \frac{r}{d} \qquad (23.70)$$

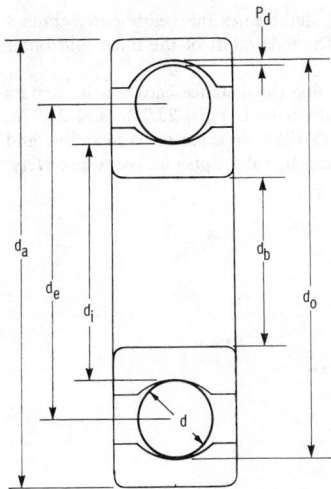

Fig. 23.75 Cross section through radial, single-row ball bearing. (From Ref. 10.)

For perfect conformity, where the radius of the race is equal to the ball radius, f is equal to ½. The closer the race conforms to the ball, the greater the frictional heat within the contact. On the other hand, open-race curvature and reduced geometrical conformity, which reduce friction, also increase the maximum contact stresses and, consequently, reduce the bearing fatigue life. For this reason, most ball bearings made today have race conformity ratios in the range $0.51 \leq f \leq 0.54$, with $f = 0.52$ being the most common value. The race conformity ratio for the outer race is usually made slightly larger than that for the inner race to compensate for the closer conformity in the plane of the bearing between the outer race and ball than between the inner race and ball. This tends to equalize the contact stresses at the inner- and outer-race contacts. The difference in race conformities does not normally exceed 0.02.

Contact Angle. Radial bearings have some axial play since they are generally designed to have a diametral clearance, as shown in Fig. 23.77. This implies a free-contact angle different from zero. Angular-contact bearings are specifically designed to operate under thrust loads. The clearance built

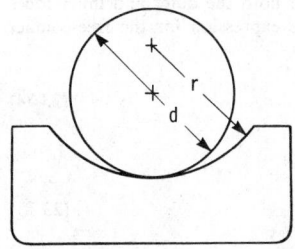

Fig. 23.76 Cross section of ball and outer race, showing race conformity. (From Ref. 10.)

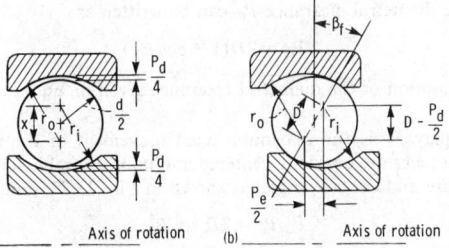

Fig. 23.77 Cross section of radial ball bearing, showing ball–race contact due to axial shift of inner and outer rings. (From Ref. 10.) (*a*) Initial position. (*b*) Shifted position.

into the unloaded bearing, along with the race conformity ratio, determines the bearing free-contact angle. Figure 23.77 shows a radial bearing with contact due to the axial shift of the inner and outer races when no measurable force is applied.

Before the free-contact angle is discussed, it is important to define the distance between the centers of curvature of the two races in line with the center of the ball in both Figs. 23.77a and 23.77b. This distance—denoted by x in Fig. 23.77a and by D in Fig. 23.77b—depends on race radius and ball diameter. Denoting quantities referred to the inner and outer races by subscripts i and o, respectively, we see from Figs. 23.77a and 23.77b that

$$\frac{P_d}{4} + d + \frac{P_d}{4} = r_o - x + r_i$$

or

$$x = r_o + r_i - d - \frac{P_d}{2}$$

and

$$d = r_o - D + r_i$$

or

$$D = r_o + r_i - d \tag{23.71}$$

From these equations, we can write

$$x = D - \frac{P_d}{2}$$

This distance, shown in Fig. 23.77, will be useful in defining the contact angle.

By using Eq. (23.70), we can write Eq. (23.71) as

$$D = Bd \tag{23.72}$$

where

$$B = f_o + f_i - 1 \tag{23.73}$$

The quantity B in Eq. (23.72) is known as the total conformity ratio and is a measure of the combined conformity of both the outer and inner races to the ball. Calculations of bearing deflection in later sections depend on the quantity B.

The free-contact angle β_f (Fig. 23.77) is defined as the angle made by a line through the points of contact of the ball and both races with a plane perpendicular to the bearing axis of rotation when no measurable force is applied. Note that the centers of curvature of both the outer and inner races lie on the line defining the free-contact angle. From Fig. 23.77, the expression for the free-contact angle can be written as

$$\cos \beta_f = \frac{D - P_d/2}{D} \tag{23.74}$$

By using Eqs. (23.69) and (23.71), we can write Eq. (23.74) as

$$\beta_f = \cos^{-1} \left[\frac{r_o + r_i - \frac{1}{2}(d_o - d_i)}{r_o + r_i - d} \right] \tag{23.75}$$

Equation (23.75) shows that if the size of the balls is increased and everything else remains constant, the free-contact angle is decreased. Similarly, if the ball size is decreased, the free-contact angle is increased.

From Eq. (23.74) the diametral clearance P_d can be written as

$$P_d = 2D(1 - \cos \beta_f) \tag{23.76}$$

This is an alternative definition of the diametral clearance given in Eq. (23.69).

Endplay.   Free endplay P_e is the maximum axial movement of the inner race with respect to the outer race when both races are coaxially centered and no measurable force is applied. Free endplay depends on total curvature and contact angle, as shown in Fig. 23.77, and can be written as

$$P_e = 2D \sin \beta_f \tag{23.77}$$

The variation of free-contact angle and free endplay with the ratio $P_d/2d$ is shown in Fig. 23.78 for four values of total conformity normally found in single-row ball bearings. Eliminating β_f in Eqs.

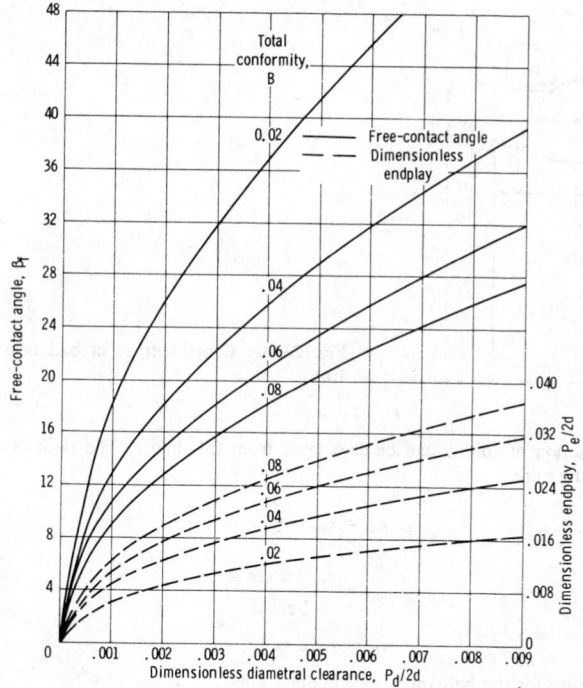

Fig. 23.78 Chart for determining free-contact angle and endplay. (From Ref. 10.)

(23.76) and (23.77) enables the establishment of the following relationships between free endplay and diametral clearance:

$$P_d = 2D - [(2D)^2 - P_e^2]^{1/2}$$

$$P_e = (4DP_d - P_d^2)^{1/2}$$

Shoulder Height. The shoulder height of ball bearings is illustrated in Fig. 23.79. Shoulder height, or race depth, is the depth of the race groove measured from the shoulder to the bottom of the groove and is denoted by s in Fig. 23.79. From this figure the equation defining the shoulder height can be written as

$$s = r(1 - \cos \theta) \qquad (23.78)$$

The maximum possible diametral clearance for complete retention of the ball–race contact within the race under zero thrust load is given by

$$(P_d)_{max} = \frac{2Ds}{r}$$

Curvature Sum and Difference. A cross section of a ball bearing operating at a contact angle β is shown in Fig. 23.80. Equivalent radii of curvature for both inner- and outer-race contacts in, and

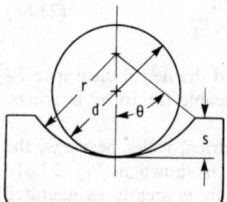

Fig. 23.79 Shoulder height in ball bearing. (From Ref. 10.)

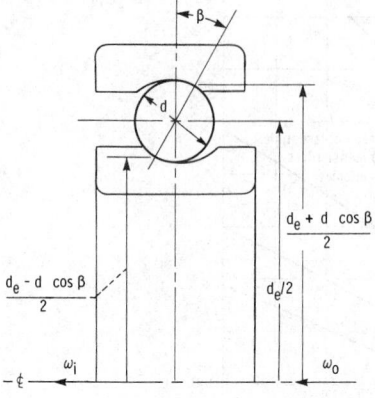

Fig. 23.80 Cross section of ball bearing. (From Ref. 10.)

normal to, the direction of rolling can be calculated from this figure. The radii of curvature for the ball–inner-race contact are

$$r_{ax} = r_{ay} = \frac{d}{2} \tag{23.79}$$

$$r_{bx} = \frac{d_e - d \cos \beta}{2 \cos \beta} \tag{23.80}$$

$$r_{by} = -f_i d = -r_i \tag{23.81}$$

The radii of curvature for the ball–outer-race contact are

$$r_{ax} = r_{ay} = \frac{d}{2} \tag{23.82}$$

$$r_{bx} = -\frac{d_e + d \cos \beta}{2 \cos \beta} \tag{23.83}$$

$$r_{by} = -f_o d = -r_o \tag{23.84}$$

In Eqs. (23.80) and (23.83), β is used instead of β_f since these equations are also valid when a load is applied to the contact. By setting $\beta = 0°$, Eqs. (23.79)–(23.84) are equally valid for radial ball bearings. For thrust ball bearings, $r_{bx} = \infty$ and the other radii are defined as given in the preceding equations.

Equations (23.36) and (23.37) effectively redefine the problem of two ellipsoidal solids approaching one another in terms of an equivalent ellipsoidal solid of radii R_x and R_y approaching a plane. From the radius-of-curvature expressions, the radii R_x and R_y for the contact example discussed earlier can be written for the ball–inner-race contact as

$$R_x = \frac{d(d_e - d \cos \beta)}{2d_e} \tag{23.85}$$

$$R_y = \frac{f_i d}{2f_i - 1} \tag{23.86}$$

and for the ball–outer-race contact as

$$R_x = \frac{d(d_e + d \cos \beta)}{2d_e} \tag{23.87}$$

$$R_y = \frac{f_o d}{2f_o - 1} \tag{23.88}$$

Roller Bearings. The equations developed for the pitch diameter d_e and diametral clearance P_d for ball bearings in Eqs. (23.68) and (23.69), respectively, are directly applicable for roller bearings.

Crowning. To prevent high stresses at the edges of the rollers in cylindrical roller bearings, the rollers are usually crowned as shown in Fig. 23.81. A fully crowned roller is shown in Fig. 23.81a and a partially crowned roller in Fig. 23.81b. In this figure the crown curvature is greatly exaggerated

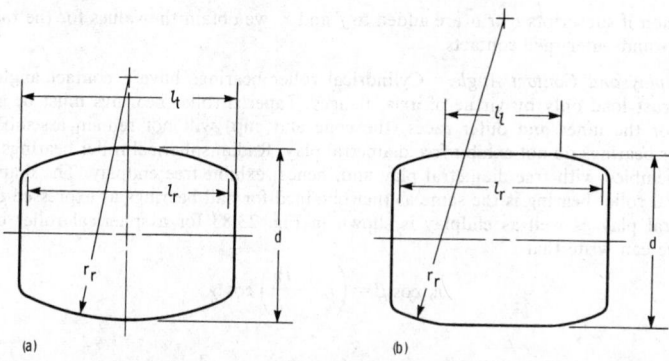

Fig. 23.81 Spherical and cylindrical rollers. (From Ref. 6.) (a) Spherical roller (fully crowned). (b) Cylindrical roller (partially crowned).

for clarity. The crowning of rollers also gives the bearing protection against the effects of slight misalignment. For cylindrical rollers, $r_{ay}/d \approx 10^2$. In contrast, for spherical rollers in spherical roller bearings, as shown in Fig. 23.81, $r_{ay}/d \approx 4$. In Fig. 23.81 it is observed that the roller effective length l_r is the length presumed to be in contact with the races under loading. Generally, the roller effective length can be written as

$$l_r = l_t - 2r_c$$

where r_c is the roller corner radius or the grinding undercut, whichever is larger.

Race Conformity.   Race conformity applies to roller bearings much as it applies to ball bearings. It is a measure of the geometrical conformity of the race and the roller. Figure 23.82 shows a cross section of a spherical roller bearing. From this figure the race conformity can be written as

$$f = \frac{r}{2r_{ay}}$$

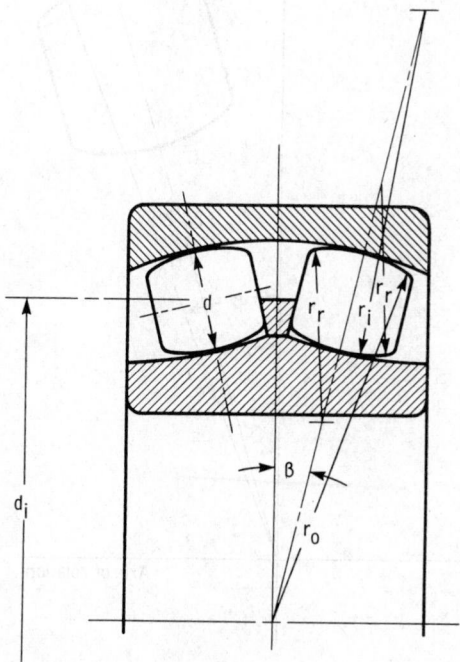

Fig. 23.82 Spherical roller bearing geometry. (From Ref. 6.)

In this equation if subscripts i or o are added to f and r, we obtain the values for the race conformity for the inner- and outer-race contacts.

Free Endplay and Contact Angle. Cylindrical roller bearings have a contact angle of zero and may take thrust load only by virtue of axial flanges. Tapered-roller bearings must be subjected to a thrust load or the inner and outer races (the cone and cup) will not remain assembled; therefore, tapered-roller bearings do not exhibit free diametral play. Radial spherical roller bearings are, however, normally assembled with free diametral play and, hence, exhibit free endplay. The diametral play P_d for a spherical roller bearing is the same as that obtained for ball bearings as expressed in Eq. (23.69). This diametral play as well as endplay is shown in Fig. 23.83 for a spherical roller bearing. From this figure we can write that

$$r_o \cos \beta = \left(r_o - \frac{P_d}{2} \right) \cos \gamma$$

or

$$\beta = \cos^{-1}\left[\left(1 - \frac{P_d}{2r_o} \right) \cos \gamma \right]$$

Also from Fig. 23.83 the free endplay can be written as

$$P_e = 2r_o (\sin \beta - \sin \gamma) + P_d \sin \gamma$$

Curvature Sum and Difference. The same procedure will be used for defining the curvature sum and difference for roller bearings as was used for ball bearings. For spherical roller bearings, as shown in Fig. 23.82, the radii of curvature for the roller–inner-race contact can be written as

$$r_{ax} = \frac{d}{2}, \qquad r_{ay} = f_i \left(\frac{r_i}{2} \right)$$

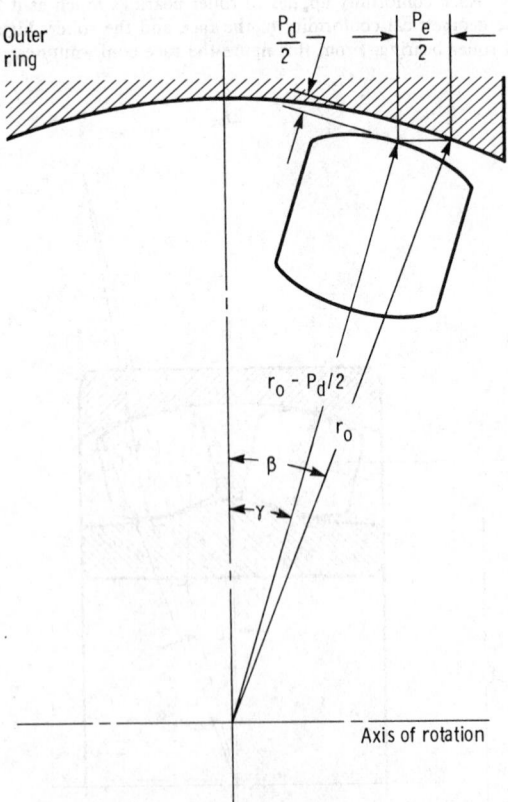

Fig. 23.83 Schematic diagram of spherical roller bearing, showing diametral play and endplay. (From Ref. 6.)

$$r_{bx} = \frac{d_e - d \cos \beta}{2 \cos \beta}, \qquad r_{by} = -2f_i r_{ay}$$

For the spherical roller bearing shown in Fig. 23.82 the radii of curvature for the roller–outer-race contact can be written as

$$r_{ax} = \frac{d}{2}, \qquad r_{ay} = f_o \left(\frac{r_o}{2} \right)$$

$$r_{bx} = -\frac{d_e + d \cos \beta}{2 \cos \beta}, \qquad r_{by} = -2f_o r_{ay}$$

Knowing the radii of curvature for the contact condition, we can write the curvature sum and difference directly from Eqs. (23.34) and (23.35). Furthermore, the radius-of-curvature expressions R_x and R_y for spherical roller bearings can be written for the roller–inner-race contact as

$$R_x = \frac{d(d_e - d \cos \beta)}{2d_e} \tag{23.89}$$

$$R_y = \frac{2r_{ay} f_i}{2f_i - 1} \tag{23.90}$$

and for the roller–outer-race contact as

$$R_x = \frac{d(d_e + d \cos \beta)}{2d_e} \tag{23.91}$$

$$R_y = \frac{2r_{ay} f_o}{2f_o - 1} \tag{23.92}$$

Kinematics

The relative motions of the separator, the balls or rollers, and the races of rolling-element bearings are important to understanding their performance. The relative velocities in a ball bearing are somewhat more complex than those in roller bearings, the latter being analogous to the specialized case of a zero- or fixed-value contact-angle ball bearing. For that reason the ball bearing is used as an example here to develop approximate expressions for relative velocities. These are useful for rapid but reasonably accurate calculation of elastohydrodynamic film thickness, which can be used with surface roughnesses to calculate the lubrication life factor.

When a ball bearing operates at high speeds, the centrifugal force acting on the ball creates a difference between the inner- and outer-race contact angles, as shown in Fig. 23.84, in order to maintain force equilibrium on the ball. For the most general case of rolling and spinning at both inner- and outer-race contacts, the rolling and spinning velocities of the ball are as shown in Fig. 23.85.

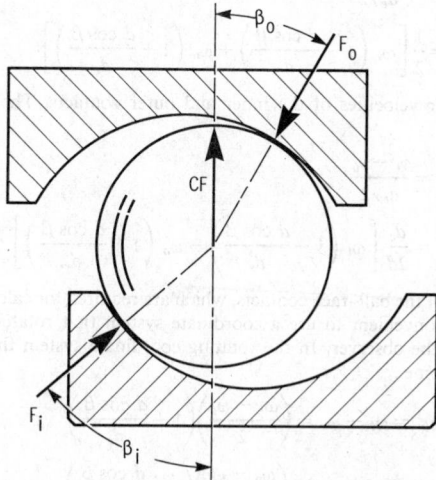

Fig. 23.84   Contact angles in a ball bearing at appreciable speeds. (From Ref. 6.)

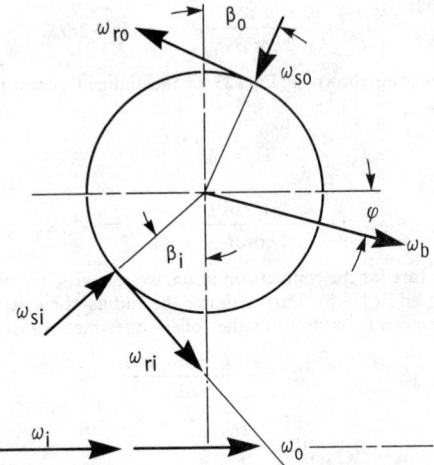

Fig. 23.85 Angular velocities of a ball. (From Ref. 6.)

The equations for ball and separator angular velocity for all combinations of inner- and outer-race rotation were developed by Jones.[45] Without introducing additional relationships to describe the elastohydrodynamic conditions at both ball–race contacts, however, the ball–spin-axis orientation angle ϕ cannot be obtained. As mentioned, this requires a long numerical solution except for the two extreme cases of outer- or inner-race control. These are illustrated in Fig. 23.86.

Race control assumes that pure rolling occurs at the controlling race, with all of the ball spin occurring at the other race contact. The orientation of the ball rotational axis is then easily determinable from bearing geometry. Race control probably occurs only in dry bearings or dry-film-lubricated bearings where Coulomb friction conditions exist in the ball–race contact ellipses. Pure rolling will occur at the race contact with the higher magnitude spin-opposing moment. This is usually the inner race at low speeds and the outer race at high speeds.

In oil-lubricated bearings in which elastohydrodynamic films exist in both ball–race contacts, rolling with spin occurs at both contacts. Therefore, precise ball motions can only be determined through use of a computer analysis. We can approximate the situation with a reasonable degree of accuracy, however, by assuming that the ball rolling axis is normal to the line drawn through the centers of the two ball–race contacts. This is shown in Fig. 23.80.

The angular velocity of the separator or ball set ω_c about the shaft axis can be shown to be

$$\omega_c = \frac{(v_i + v_o)/2}{d_e/2}$$

$$= \frac{1}{2}\left[\omega_i\left(1 - \frac{d\cos\beta}{d_e}\right) + \omega_o\left(1 + \frac{d\cos\beta}{d_e}\right)\right] \tag{23.93}$$

where v_i and v_o are the linear velocities of the inner and outer contacts. The angular velocity of a ball ω_b about its own axis is

$$\omega_b = \frac{v_i - v_o}{d_e/2}$$

$$= \frac{d_e}{2d}\left[\omega_i\left(1 - \frac{d\cos\beta}{d_e}\right) - \omega_o\left(1 + \frac{d\cos\beta}{d_e}\right)\right] \tag{23.94}$$

To calculate the velocities of the ball–race contacts, which are required for calculating elastohydrodynamic film thicknesses, it is convenient to use a coordinate system that rotates at ω_c. This fixes the ball–race contacts relative to the observer. In the rotating coordinate system the angular velocities of the inner and outer races become

$$\omega_{ir} = \omega_i - \omega_c = \left(\frac{\omega_i - \omega_o}{2}\right)\left(1 + \frac{d\cos\beta}{d_e}\right)$$

$$\omega_{or} = \omega_o - \omega_c = \left(\frac{\omega_o - \omega_i}{2}\right)\left(1 - \frac{d\cos\beta}{d_e}\right)$$

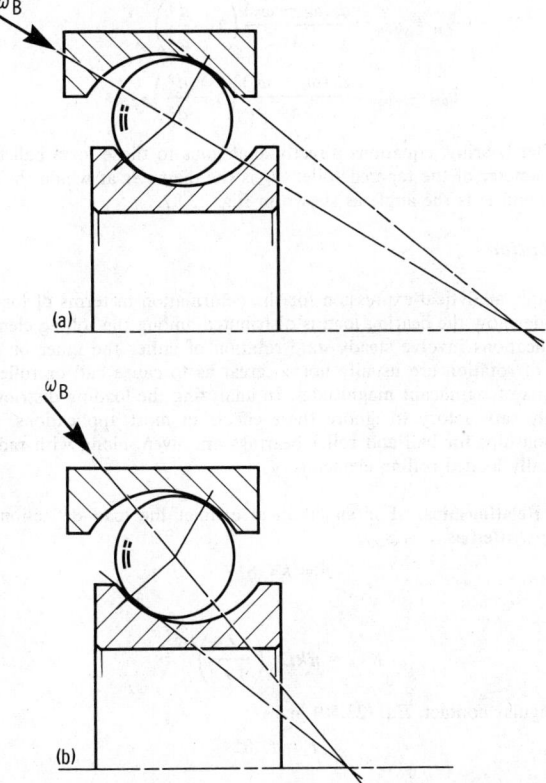

Fig. 23.86 Ball spin axis orientations for outer- and inner-race control. (From Ref. 6.) (*a*) Outer-race control. (*b*) Inner-race control.

The surface velocities entering the ball–inner-race contact for pure rolling are

$$u_{ai} = u_{bi} = \left(\frac{d_e - d \cos \beta}{2}\right) \omega_{ir} \tag{23.95}$$

or

$$u_{ai} = u_{bi} = \frac{d_e(\omega_i - \omega_o)}{4}\left(1 - \frac{d^2 \cos^2 \beta}{d_e^2}\right) \tag{23.96}$$

and those at the ball–outer-race contact are

$$u_{ao} = u_{bo} = \left(\frac{d_e + d \cos \beta}{2}\right) \omega_{or}$$

or

$$u_{ao} = u_{bo} = \frac{d_e(\omega_o - \omega_i)}{4}\left(1 - \frac{d^2 \cos^2 \beta}{d_e^2}\right) \tag{23.97}$$

For a cylindrical roller bearing $\beta = 0°$ and Eqs. (23.92), (23.94), (23.96), and (23.97) become, if d is roller diameter,

$$\omega_c = \frac{1}{2}\left[\omega_i\left(1 - \frac{d}{d_e}\right) + \omega_o\left(1 + \frac{d}{d_e}\right)\right]$$

$$\omega_R = \frac{d_e}{2d}\left[\omega_i\left(1 - \frac{d}{d_e}\right) + \omega_o\left(1 + \frac{d}{d_e}\right)\right]$$

$$u_{ai} = u_{bi} = \frac{d_e(\omega_i - \omega_o)}{4}\left(1 - \frac{d^2}{d_e^2}\right) \qquad (23.98)$$

$$u_{ao} = u_{bo} = \frac{d_e(\omega_o - \omega_i)}{4}\left(1 - \frac{d^2}{d_e^2}\right)$$

For a tapered-roller bearing, equations directly analogous to those for a ball bearing can be used if d is the average diameter of the tapered roller, d_e is the diameter at which the geometric center of the rollers is located, and ω is the angle as shown in Fig. 23.87.

Static Load Distribution

Having defined a simple analytical expression for the deformation in terms of load in Section 23.3.1, it is possible to consider how the bearing load is distributed among the rolling elements. Most rolling-element bearing applications involve steady-state rotation of either the inner or outer race or both; however, the speeds of rotation are usually not so great as to cause ball or roller centrifugal forces or gyroscopic moments of significant magnitudes. In analyzing the loading distribution on the rolling elements, it is usually satisfactory to ignore these effects in most applications. In this section the load–deflection relationships for ball and roller bearings are given, along with radial and thrust load distributions of statically loaded rolling elements.

Load–Deflection Relationships. For an elliptical contact the load–deflection relationship given in Eq. (23.45) can be written as

$$F = K_{1.5}\delta^{3/2} \qquad (23.99)$$

where

$$K_{1.5} = \pi k E'\left(\frac{2\mathscr{E}R}{9\mathscr{F}^3}\right)^{1/2} \qquad (23.100)$$

Similarly for a rectangular contact, Eq. (23.50) gives

$$F = K_1\delta$$

where

$$K_1 = \left(\frac{\pi l E'}{2}\right)\left[\frac{1}{\frac{2}{3} + \ln(2r_{ax}/b) + \ln(2r_{bx}/b)}\right] \qquad (23.101)$$

In general, then,

$$F = K_j\delta^j \qquad (23.102)$$

in which $j = 1.5$ for ball bearings and 1.0 for roller bearings. The total normal approach between two races separated by a rolling element is the sum of the deformations under load between the rolling element and both races. Therefore

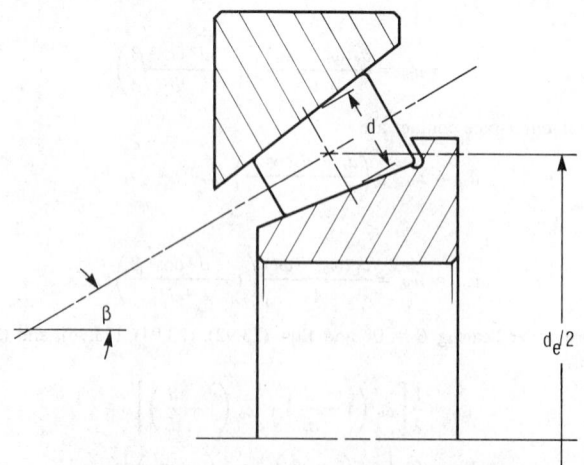

Fig. 23.87 Simplified geometry for tapered-roller bearing. (From Ref. 6.)

$$\delta = \delta_o + \delta_i \qquad (23.103)$$

where

$$\delta_o = \left[\frac{F}{(K_j)_o}\right]^{1/j} \qquad (23.104)$$

$$\delta_i = \left[\frac{F}{(K_j)_i}\right]^{1/j} \qquad (23.105)$$

Substituting Eqs. (23.103)–(23.105) into Eq. (23.102) gives

$$K_j = \frac{1}{\{[1/(K_j)_o]^{1/j} + [1/(K_j)_i]^{1/j}\}^j}$$

Recall that $(K_j)_o$ and $(K_j)_i$ are defined by Eq. (23.100) or (23.101) for an elliptical or rectangular contact, respectively. From these equations we observe that $(K_j)_o$ and $(K_j)_i$ are functions of only the geometry of the contact and the material properties. The radial and thrust load analyses are presented in the following two sections and are directly applicable for radially loaded ball and roller bearings and thrust-loaded ball bearings.

Radially Loaded Ball and Roller Bearings. A radially loaded rolling element with radial clearance P_d is shown in Fig. 23.88. In the concentric position shown in Fig. 23.88a, a uniform radial clearance between the rolling element and the races of $P_d/2$ is evident. The application of a small radial load to the shaft causes the inner race to move a distance $P_d/2$ before contact is made between a rolling element located on the load line and the inner and outer races. At any angle there will still be a radial clearance c that, if P_d is small compared with the radius of the tracks, can be expressed with adequate accuracy by

$$c = (1 - \cos \psi) P_d/2$$

On the load line where $\psi = 0$ the clearance is zero, but when $\psi = 90°$, the clearance retains its initial value of $P_d/2$.

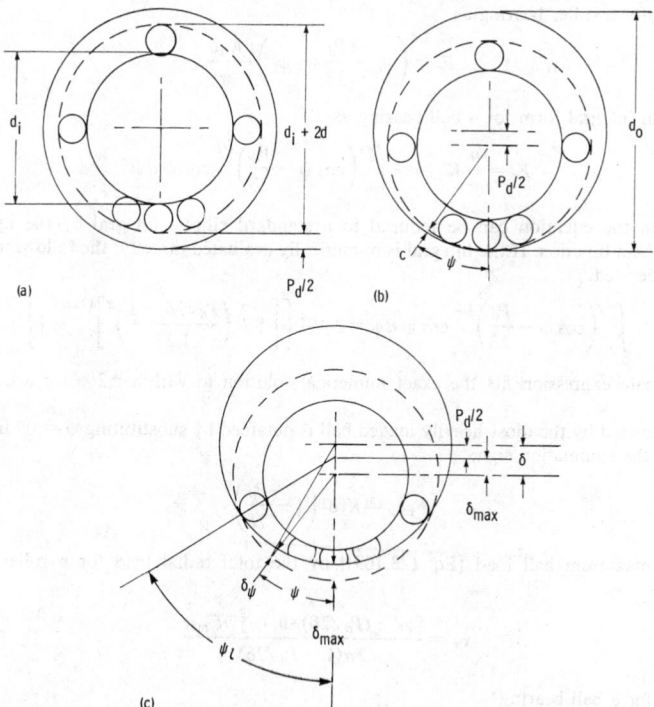

Fig. 23.88 Radially loaded rolling-element bearing. (From Ref. 10.) (a) Concentric arrangement. (b) Initial contact. (c) Interference.

The application of further load will cause elastic deformation of the balls and the elimination of clearance around an arc $2\psi_c$. If the interference or total elastic compression on the load is δ_{max}, the corresponding elastic compression of the ball δ_ψ along a radius at angle ψ to the load line will be given by

$$\delta_\psi = (\delta_{max} \cos \psi - c) = (\delta_{max} + P_d/2) \cos \psi - P_d/2$$

This assumes that the races are rigid. Now, it is clear from Fig. 23.88 that $(\delta_{max} + P_d/2)$ represents the total relative radial displacement of the inner and outer races. Hence,

$$\delta_\psi = \delta \cos \psi - P_d/2 \qquad (23.106)$$

The relationship between load and elastic compression along the radius at angle ψ to the load vector is given by Eq. (23.102) as

$$F_\psi = K_j \delta_\psi^j$$

Substituting Eq. (23.106) into this equation gives

$$F_\psi = K_j (\delta \cos \psi - P_d/2)^j$$

For static equilibrium the applied load must equal the sum of the components of the rolling-element loads parallel to the direction of the applied load:

$$F_r = \sum F_\psi \cos \psi$$

Therefore

$$F_r = K_j \sum \left(\delta \cos \psi - \frac{P_d}{2} \right)^j \cos \psi \qquad (23.107)$$

The angular extent of the bearing arc $2\psi_l$ in which the rolling elements are loaded is obtained by setting the root expression in Eq. (23.107) equal to zero and solving for ψ:

$$\psi_l = \cos^{-1}\left(\frac{P_d}{2\delta}\right)$$

The summation in Eq. (23.107) applies only to the angular extent of the loaded region. This equation can be written for a roller bearing as

$$F_r = \left(\psi_l - \frac{P_d}{2\delta} \sin \psi_l \right) \frac{nK_1\delta}{2\pi} \qquad (23.108)$$

and similarly in integral form for a ball bearing as

$$F_r = \frac{n}{\pi} K_{1.5} \delta^{3/2} \int_0^{\psi_l} \left(\cos \psi - \frac{P_d}{2\delta} \right)^{3/2} \cos \psi \, d\psi$$

The integral in the equation can be reduced to a standard elliptic integral by the hypergeometric series and the beta function. If the integral is numerically evaluated directly, the following approximate expression is derived:

$$\int_0^{\psi_l} \left(\cos \psi - \frac{P_d}{2\delta} \right)^{3/2} \cos \psi \, d\psi = 2.491 \left\{ \left[1 + \left(\frac{P_d/2\delta - 1}{1.23} \right)^2 \right]^{1/2} - 1 \right\}$$

This approximate expression fits the exact numerical solution to within $\pm 2\%$ for a complete range of $P_d/2\delta$.

The load carried by the most heavily loaded ball is obtained by substituting $\psi = 0°$ in Eq. (23.107) and dropping the summation sign:

$$F_{max} = K_j \delta^j \left(1 - \frac{P_d}{2\delta} \right)^j$$

Dividing the maximum ball load [Eq. (23.109)] by the total radial load for a roller bearing [Eq. (23.108)] gives

$$F_r = \frac{[\psi_l - (P_d/2\delta) \sin \psi_l] \, nF_{max}}{2\pi(1 - P_d/2\delta)} \qquad (23.109)$$

and similarly for a ball bearing

$$F_r = \frac{nF_{max}}{Z} \qquad (23.110)$$

where

$$Z = \frac{\pi(1 - P_d/2\delta)^{3/2}}{2.491\left\{\left[1 + \left(\frac{1 - P_d/2\delta}{1.23}\right)^2\right]^{1/2} - 1\right\}} \tag{23.111}$$

For *roller bearings* when the diametral clearance P_d is zero, Eq. (23.105) gives

$$F_r = \frac{nF_{max}}{4} \tag{23.112}$$

For *ball bearings* when the diametral clearance P_d is zero, the value of Z in Eq. (23.110) becomes 4.37. This is the value derived by Stribeck[46] for ball bearings of zero diametral clearance. The approach used by Stribeck was to evaluate the finite summation for various numbers of balls. He then derived the celebrated Stribeck equation for static load-carrying capacity by writing the more conservative value of 5 for the theoretical value of 4.37:

$$F_r = \frac{nF_{max}}{5} \tag{23.113}$$

In using Eq. (23.113), it should be remembered that Z was considered to be a constant and that the effects of clearance and applied load on load distribution were not taken into account. However, these effects were considered in obtaining Eq. (23.110).

Thrust-Loaded Ball Bearings. The static-thrust-load capacity of a ball bearing may be defined as the maximum thrust load that the bearing can endure before the contact ellipse approaches a race shoulder, as shown in Fig. 23.89, or the load at which the allowable mean compressive stress is reached, whichever is smaller. Both the limiting shoulder height and the mean compressive stress must be calculated to find the static-thrust-load capacity.

The contact ellipse in a bearing race under a load is shown in Fig. 23.89. Each ball is subjected to an identical thrust component F_t/n, where F_t is the total thrust load. The initial contact angle before the application of a thrust load is denoted by β_f. Under load, the normal ball thrust load F acts at the contact angle β and is written as

$$F = \frac{F_t}{n}\sin\beta \tag{23.114}$$

A cross section through an angular-contact bearing under a thrust load F_t is shown in Fig. 23.90. From this figure the contact angle after the thrust load has been applied can be written as

$$\beta = \cos^{-1}\left(\frac{D - P_d/2}{D + \delta}\right) \tag{23.115}$$

The initial contact angle was given in Eq. (23.74). Using that equation and rearranging terms in Eq. (23.115) give, solely from geometry (Fig. 23.90),

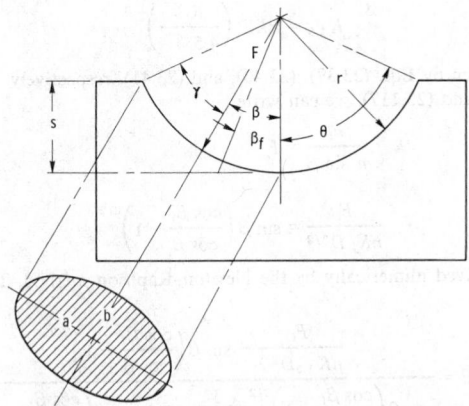

Fig. 23.89 Contact ellipse in bearing race. (From Ref. 10.)

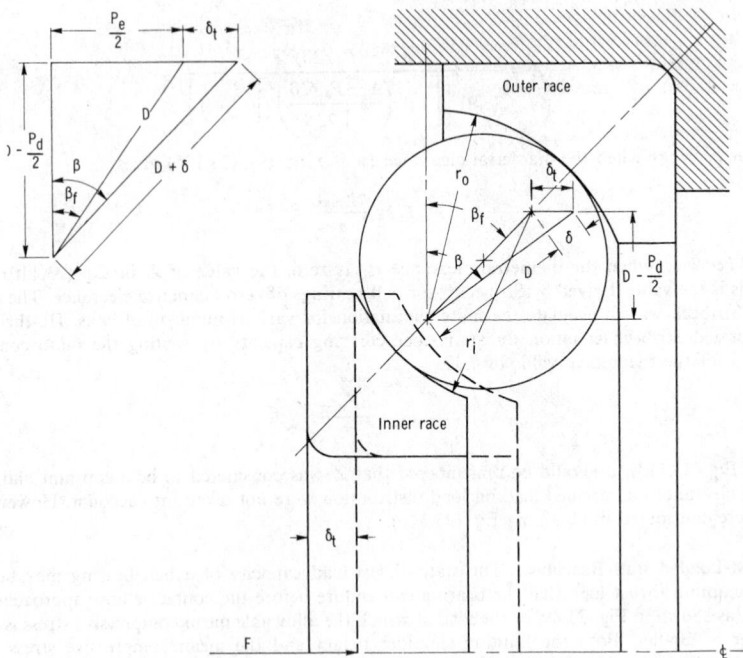

Fig. 23.90 Angular-contact ball bearing under thrust load. (From Ref. 10.)

$$\delta = D\left(\frac{\cos \beta_f}{\cos \beta} - 1\right) = \delta_o + \delta_i$$

$$= \left[\frac{F}{(K_j)_o}\right]^{1/j} + \left[\frac{F}{(K_j)_i}\right]^{1/j}$$

$$K_j = 1 \Bigg/ \left\{\left[\frac{1}{(K_j)_o}\right]^{1/j} + \left[\frac{1}{(K_j)_i}\right]^{1/j}\right\}^j$$

$$K_j = 1 \Bigg/ \left\{\left[\frac{4.5\,\mathscr{F}_o^3}{\pi k_o E_o'(R_o\,\mathscr{E}_o)^{1/2}}\right]^{2/3} + \left[\frac{4.5\,\mathscr{F}_i^3}{\pi k_i E_i'(R_i\,\mathscr{E}_i)^{1/2}}\right]^{2/3}\right\} \tag{23.116}$$

$$F = K_j D^{3/2}\left(\frac{\cos \beta_f}{\cos \beta} - 1\right)^{3/2} \tag{23.117}$$

where

$$K_{1.5} = \pi k E'\left(\frac{R\,\mathscr{E}}{4.5\,\mathscr{F}^3}\right)^{1/2} \tag{23.118}$$

and k, $\mathscr{E}$, and $\mathscr{F}$ are given by Eqs. (23.39), (23.40), and (23.41), respectively.
   From Eqs. (23.114) and (23.117), we can write

$$\frac{F_t}{n \sin \beta} = F$$

$$\frac{F_t}{nK_j D^{3/2}} = \sin \beta\left(\frac{\cos \beta_f}{\cos \beta} - 1\right)^{3/2} \tag{23.119}$$

This equation can be solved numerically by the Newton–Raphson method. The iterative equation to be satisfied is

$$\beta' - \beta = \frac{\dfrac{F_t}{nK_{1.5}D^{3/2}} - \sin \beta\left(\dfrac{\cos \beta_f}{\cos \beta} - 1\right)^{3/2}}{\cos \beta\left(\dfrac{\cos \beta_f}{\cos \beta} - 1\right)^{3/2} + \dfrac{3}{2}\cos \beta_f \tan^2 \beta\left(\dfrac{\cos \beta_f}{\cos \beta} - 1\right)^{1/2}} \tag{23.120}$$

In this equation convergence is satisfied when $\beta' - \beta$ becomes essentially zero.

When a thrust load is applied, the shoulder height is limited to the distance by which the pressure–contact ellipse can approach the shoulder. As long as the following inequality is satisfied, the pressure–contact ellipse will not exceed the shoulder height limit:

$$\theta > \beta + \sin^{-1}\left(\frac{D_y}{fd}\right)$$

From Fig. 23.79 and Eq. (23.68), the angle used to define the shoulder height θ can be written as

$$\theta = \cos^{-1}\left(\frac{1-s}{fd}\right)$$

From Fig. 23.77 the axial deflection δ_t corresponding to a thrust load can be written as

$$\delta_t = (D + \delta)\sin\beta - D\sin\beta_f \qquad (23.121)$$

Substituting Eq. (23.116) into Eq. (23.121) gives

$$\delta_t = \frac{D\sin(\beta - \beta_f)}{\cos\beta}$$

Having determined β from Eq. (23.120) and β_f from Eq. (23.103), we can easily evaluate the relationship for δ_t.

Preloading. The use of angular-contact bearings as duplex pairs preloaded against each other is discussed in the first subsection in Section 23.3.6. As shown in Table 23.10 duplex bearing pairs are used in either back-to-back or face-to-face arrangements. Such bearings are usually preloaded against each other by providing what is called "stickout" in the manufacture of the bearing. This is illustrated in Fig. 23.91 for a bearing pair used in a back-to-back arrangement. The magnitude of the stickout and the bearing design determine the level of preload on each bearing when the bearings are clamped together as in Fig. 23.91. The magnitude of preload and the load–deflection characteristics for a given bearing pair can be calculated by using Eqs. (23.74), (23.99), (23.114), and (23.116)–(23.119).

The relationship of initial preload, system load, and final load for bearings a and b is shown in Fig. 23.92. The load–deflection curve follows the relationship $\delta = KF^{2/3}$. When a system thrust load F_t is imposed on the bearing pairs, the magnitude of load on bearing b increases while that on bearing a decreases until the difference equals the system load. The physical situation demands that the change in each bearing deflection be the same ($\Delta a = \Delta b$ in Fig. 23.92). The increments in bearing load,

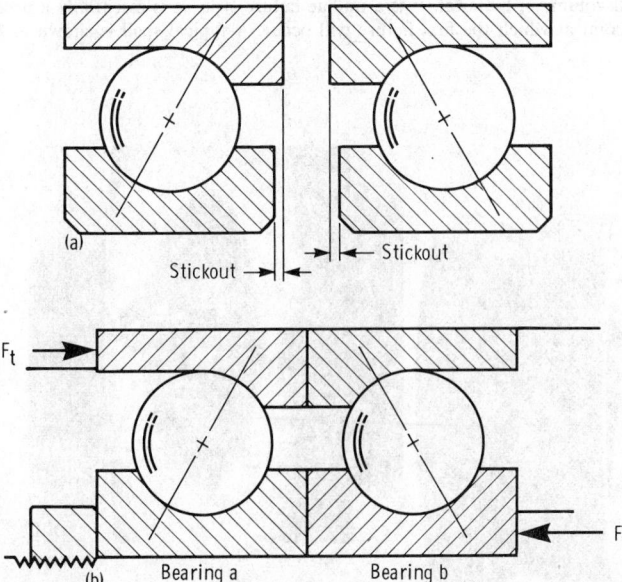

Fig. 23.91 Angular-contact bearings in back-to-back arrangement, shown individually as manufactured and as mounted with preload. (From Ref. 6.) (a) Separated. (b) Mounted and preloaded.

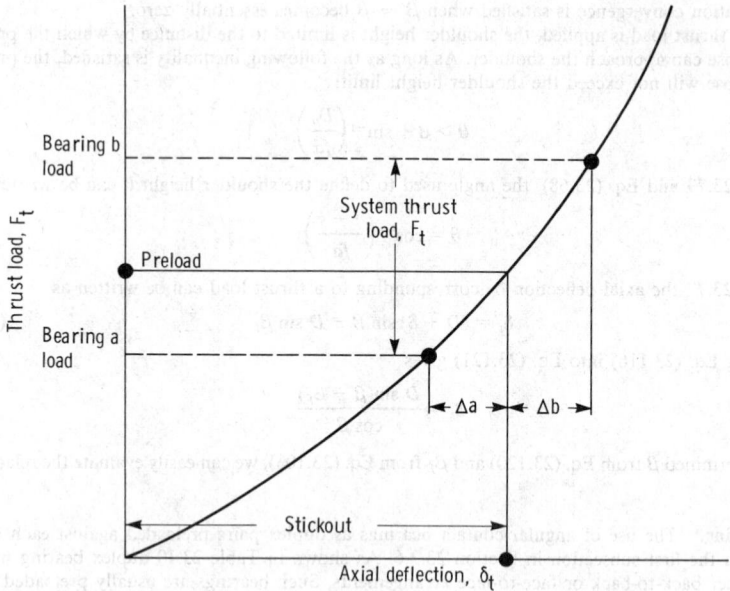

Fig. 23.92 Thrust-load–axial-deflection curve for a typical ball bearing. (From Ref. 6.)

however, are not the same. This is important because it always requires a system thrust load far greater than twice the preload before one bearing becomes unloaded. Prevention of bearing unloading, which can result in skidding and early failure, is an objective of preloading.

Rolling Bearing Fatigue Life

Contact Fatigue Theory. Rolling fatigue is a material failure caused by the application of repeated stresses to a small volume of material. It is a unique failure type. It is essentially a process of seeking out the weakest point at which the first failure will occur. A typical spall is shown in Fig. 23.93. We

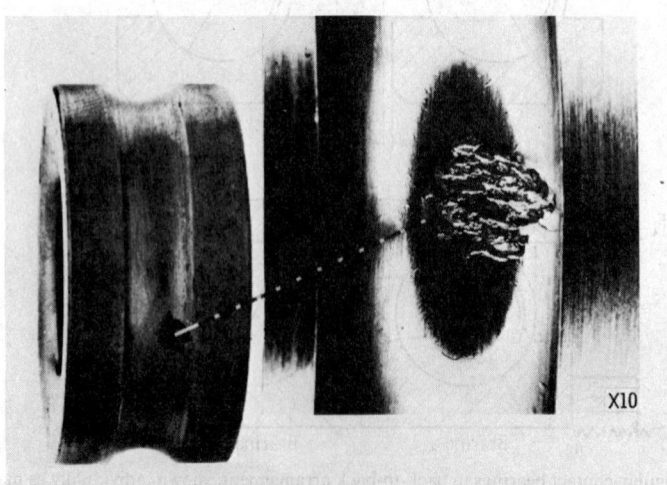

Fig. 23.93 Typical fatigue spall.

can surmise that on a microscale there will be a wide dispersion in material strength or resistance to fatigue because of inhomogeneities in the material. Because bearing materials are complex alloys, we would not expect them to be homogeneous nor equally resistant to failure at all points. Therefore, the fatigue process can be expected to be one in which a group of supposedly identical specimens exhibit wide variations in failure time when stressed in the same way. For this reason it is necessary to treat the fatigue process statistically.

To be able to predict how long a typical bearing will run under a specific load, we must have the following two essential pieces of information:

1. An accurate, quantitative estimate of the life dispersion or scatter.
2. The life at a given survival rate or reliability level. This translates into an expression for the "load capacity," or the ability of the bearing to endure a given load for a stipulated number of stress cycles or revolutions. If a group of supposedly identical bearings is tested at a specific load and speed, there will be a wide scatter in bearing lives, as shown in Fig. 23.94.

The Weibull Distribution. Weibull[47] postulates that the fatigue lives of a homogeneous group of rolling-element bearings are dispersed according to the following relation:

$$\ln \ln \frac{1}{S} = e_1 \ln L/A$$

where S is the probability of survival, L is the fatigue life, and e_1 and A are constants. The Weibull distribution results from a statistical theory of strength based on probability theory, where the dependence of strength on volume is explained by the dispersion in material strength. This is the "weakest link" theory.

Consider a volume being stressed that is broken up into m similar volumes:

$$s_1 = 1 - M_1 \qquad S_2 = 1 - M_2 \qquad S_3 = 1 - M_3 \qquad \cdots \qquad S_m = 1 - M_m$$

The M's represent the probability of failure and the S's represent the probability of survival. For the entire volume we can write

$$S = S_1 \cdot S_2 \cdot S_3 \cdot \cdots \cdot S_m$$

Then

$$1 - M = (1 - M_1)(1 - M_2)(1 - M_3) \cdots (1 - M_m)$$

$$1 - M = \prod_{i=1}^{m} (1 - M_i)$$

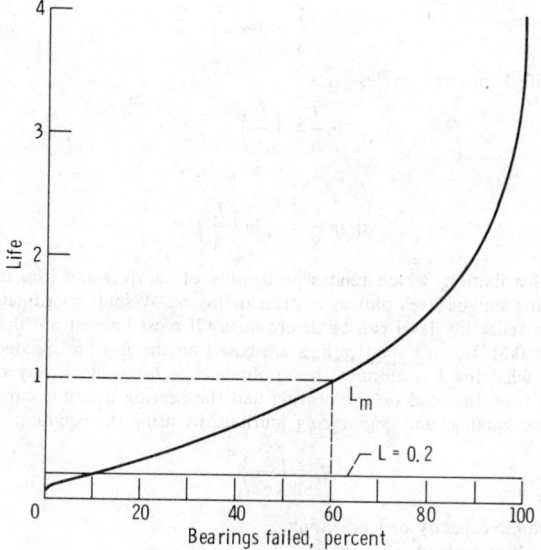

Fig. 23.94   Distribution of bearing fatigue failures. (From Ref. 6.)

$$S = \prod_{i=1}^{m} (1 - M_i)$$

The probability of a crack starting in the ith volume is

$$M_i = f(x)v_i$$

where $f(x)$ is a function of the stress level, the number of stress cycles, and the depth into the material where the maximum stress occurs and v_i is the elementary volume. Therefore,

$$S = \prod_{i=1}^{m} [1 - f(x)v_i]$$

$$\ln S = \sum_{i=1}^{m} \ln[1 - f(x)v_i]$$

Now if $f(x)v_i \ll 1$, then $\ln[1 - f(x)v_i] = -f(x)v_i$ and

$$\ln S = -\sum_{i=1}^{m} f(x)v_i$$

Let $v_i \to 0$; then

$$\sum_{i=1}^{m} f(x)v_i = \int (x)\, dv = f(x)V$$

Lundberg and Palmgren[48] assume that $f(x)$ could be expressed as a power function of shear stress τ_0, number of stress cycles J, and depth to the maximum shear stress Z_0:

$$f(x) = \frac{\tau_0^{c_1} J^{c_2}}{Z_0^{c_3}} \tag{23.122}$$

They also choose as the stressed volume

$$V = D_y Z_0 l_v$$

Then

$$\ln S = -\frac{\tau_0^{c_1} J^{c_2} D_y l_v}{Z_0^{c_3 - 1}}$$

or

$$\ln \frac{1}{S} = \frac{\tau_0^{c_1} J^{c_2} D_y l_v}{Z_0^{c_3 - 1}}$$

For a specific bearing and load (e.g., stress) τ_0, D_y, l_v, and Z_0 are all constant, so that

$$\ln \frac{1}{S} \approx J^{c_2}$$

Designating J as life L in stress cycles gives

$$\ln \frac{1}{S} = \left(\frac{L}{A}\right)^{c_2}$$

or

$$\ln \ln \frac{1}{S} = c_2 \ln \left(\frac{L}{A}\right) \tag{23.123}$$

This is the Weibull distribution, which relates probability of survival and life. It has two principal functions. First, bearing fatigue lives plot as a straight line on Weibull coordinates (log log vs log), so that the life at any reliability level can be determined. Of most interest are the L_{10} life ($S = 0.9$) and the L_{50} life ($S = 0.5$). Bearing load ratings are based on the L_{10} life. Second, Eq. (23.123) can be used to determine what the L_{10} life must be to obtain a required life at any reliability level. The L_{10} life is calculated, from the load on the bearing and the bearing dynamic capacity or load rating given in manufacturers' catalogs and engineering journals, by using the equation

$$L = \left(\frac{C}{F_e}\right)^m$$

where C = basic dynamic capacity or load rating
      F_e = equivalent bearing load
      m = 3 for elliptical contacts and 10/3 for rectangular contacts

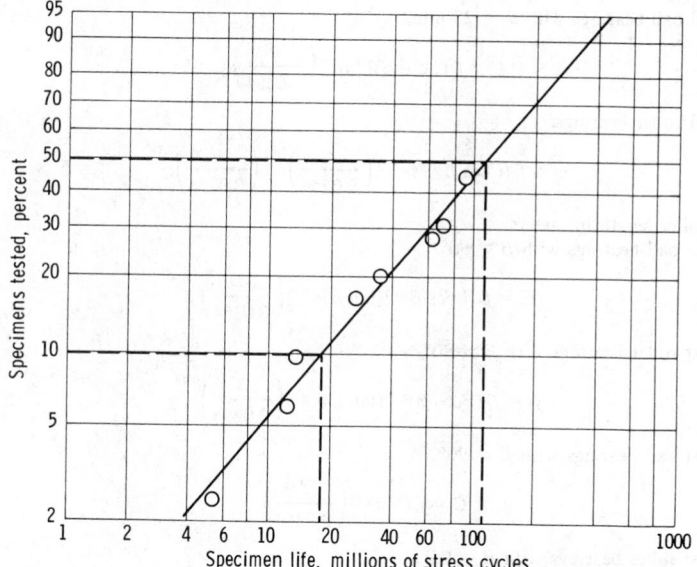

Fig. 23.95 Typical Weibull plot of bearing fatigue failures. (From Ref. 10.)

A typical Weibull plot is shown in Fig. 23.95.

Lundberg–Palmgren Theory. The Lundberg–Palmgren theory, on which bearing ratings are based, is expressed by Eq. (23.122). The exponents in this equation are determined experimentally from the dispersion of bearing lives and the dependence of life on load, geometry, and bearing size. As a standard of reference, all bearing load ratings are expressed in terms of the specific dynamic capacity C, which, by definition, is the load that a bearing can carry for 10^6 inner-race revolutions with a 90% chance of survival.

Factors on which specific dynamic capacity and bearing life depend are:

1. Size of rolling element.
2. Number of rolling elements per row.
3. Number of rows of rolling elements.
4. Conformity between rolling elements and races.
5. Contact angle under load.
6. Material properties.
7. Lubricant properties.
8. Operating temperature.
9. Operating speed.

Only factors 1–5 are incorporated in bearing dynamic capacities developed from the Lundberg–Palmgren theory. The remaining factors must be taken into account in the life adjustment factors discussed later.

The formulas for specific dynamic capacity as developed by Lundberg and Palmgren[48,49] are as follows:

For radial ball bearings with $d \leq 25$ mm,

$$C = f_c (i \cos \beta)^{0.7} n^{2/3} \left(\frac{d}{0.0254} \right)^{1.8}$$

where d = diameter of rolling element, m
$\quad i$ = number of rows of rolling elements
$\quad n$ = number of rolling elements per row
$\quad \beta$ = contact angle
$\quad f_c$ = coefficient dependent on material and bearing type

For radial ball bearings with $d \geq 25$ mm,

$$C = f_c (i \cos \beta)^{0.7} n^{2/3} \left(\frac{d}{0.0254} \right)^{1.4}$$

For radial roller bearings,

$$C = f_c (i \cos \beta)^{0.78} n^{3/4} \left(\frac{d}{0.0254} \right)^{1.07} \left(\frac{l_t}{0.0254} \right)^{0.78}$$

where l_t is roller length in meters.

For thrust ball bearings with $\beta \neq 90°$,

$$C = f_c (i \cos \beta)^{0.7} (\tan \beta) n^{2/3} \left(\frac{d}{0.0254} \right)^{1.8}$$

For thrust roller bearings with $\beta \neq 90°$,

$$C = f_c (i \cos \beta)^{0.78} (\tan \beta) n^{3/4} \left(\frac{l_t}{0.0254} \right)^{0.78}$$

For thrust ball bearings with $\beta = 90°$,

$$C = f_c i^{0.7} n^{2/3} \left(\frac{d}{0.0254} \right)^{1.8}$$

For thrust roller bearings with $\beta = 90°$,

$$C = f_c i^{0.78} n^{3/4} \left(\frac{d}{0.0254} \right)^{1.07} \left(\frac{l_t}{0.0254} \right)^{0.78}$$

For ordinary bearing steels such as SAE 52100 with mineral oil lubrication, f_c can be evaluated by using Tables 23.18 and 23.19, but a more convenient method is to use tabulated values from the most recent Antifriction Bearing Manufacturers Association (AFBMA) documents on dynamic load ratings and life.[50] The value of C is calculated or determined from bearing manufacturers' catalogs. The equivalent load F_e can be calculated from the equation

$$F_e = X F_r + Y F_t$$

Factors X and Y are given in bearing manufacturers' catalogs for specific bearings.

In addition to specific dynamic capacity C, every bearing has a specific static capacity, usually designated as C_0. Specific static capacity is defined as the load that, under static conditions, will result in a permanent deformation of 0.0001 times the rolling-element diameter. For some bearings C_0 is less than C, so it is important to avoid exposing a rolling to a static load that exceeds C_0. Values of C_0 are also given in bearing manufacturers' catalogs.

The AFBMA Method. Shortly after publication of the Lundberg–Palmgren theory, the AFBMA began efforts to standardize methods for establishing bearing load ratings and making life predictions. Standardized methods of establishing load ratings for ball bearings[51] and roller bearings[52] were devised, based essentially on the Lundberg–Palmgren theory. These early standards are published in their entirety in Jones.[45] In recent years significant advances have been made in rolling-element bearing material quality and in our understanding of the role of lubrication in bearing life through the development of elastohydrodynamic theory. Therefore the original AFBMA standards in AFBMA[51,52] have been updated with life adjustment factors. These factors have been incorporated into ISO,[50] which is discussed in the following section.

Life Adjustment Factors. A comprehensive study of the factors affecting the fatigue life of bearings, which were not taken account of in the Lundberg–Palmgren theory, is reported in Bamberger et al.[53] In that reference it was assumed that the various environmental or bearing design factors are multiplicative in their effect on bearing life. The following equation results:

$$L_A = (\tilde{D})(\tilde{E})(\tilde{F})(\tilde{G})(\tilde{H}) L_{10}$$

or

$$L_A = (\tilde{D})(\tilde{E})(\tilde{F})(\tilde{G})(\tilde{H})(C/F_e)^m$$

where $\tilde{D}$ = materials factor
$\tilde{E}$ = metallurgical processing factor
$\tilde{F}$ = lubrication factor
$\tilde{G}$ = speed effect factor
$\tilde{H}$ = misalignment factor
F_e = bearing equivalent load
m = load-life exponent; either 3 for ball bearings or 10/3 for roller bearings

Table 23.18 Capacity Formulas for Rectangular and Elliptic Contacts[a] (From Ref. 6)

Function	Elliptical contact of ball bearings			Rectangular contact of roller bearings		
C	$f_c f_a i^{0.7} N^{2/3} d^{1.8}$			$f_c f_a i^{7/9} N^{3/4} d^{29/27} \ell_{t,i}^{7/9}$		
f_c	$g_c f_1 f_2 \left(\dfrac{d_i}{d_i - d}\right)^{0.41}$			$g_c f_1 f_2$		
g_c	$\left[1 + \left(\dfrac{C_i}{C_o}\right)^{10/8}\right]^{-0.8}$			$\left[1 + \left(\dfrac{C_i}{C_o}\right)^{9/2}\right]^{-2/9}$		
C_i/C_o	$f_3 \left[\dfrac{d_i(d_o - d)}{d_o(d_i - d)}\right]^{0.41}$			$f_3 \left(\dfrac{\ell_{t,i}}{\ell_{t,o}}\right)^{7/9}$		
	Radial	Thrust		Radial	Thrust	
		$\beta \neq 90°$	$\beta = 90°$		$\beta \neq 90°$	$\beta = 90°$
γ	$\dfrac{d \cos \beta}{d_e}$	$\dfrac{d}{d_e}$		$\dfrac{d \cos \beta}{d_e}$	$\dfrac{d}{d_e}$	
f_a	$(\cos \beta)^{0.7}$	$(\cos \beta)^{0.7} \tan \beta$	1	$(\cos \beta)^{7/9}$	$(\cos \beta)^{7/9} \tan \beta$	1
f_1	3.7-4.1	6-10		18-25	36-60	
f_2	$\dfrac{\gamma^{0.3}(1-\gamma)^{1.39}}{(1+\gamma)^{1/3}}$		$\gamma^{0.3}$	$\dfrac{\gamma^{2/9}(1-\gamma)^{29/27}}{(1+\gamma)^{1/3}}$		$\gamma^{2/9}$
f_3	$104 f_4$	f_4	1	$1.14 f_4$	f_4	1
f_4	$\left(\dfrac{1-\gamma}{1+\gamma}\right)^{1.72}$			$\left(\dfrac{1-\gamma}{1+\gamma}\right)^{38/37}$		

[a] Units in kg and mm.

Factors $\bar{D}$, $\bar{E}$, and $\bar{F}$ are briefly reviewed here. The reader is referred to Bamberger et al.[53] for a complete discussion of all five life adjustment factors.

Materials Factors $\bar{D}$ and $\bar{E}$. For over a century, AISI 52100 steel has been the predominant material for rolling-element bearings. In fact, the basic dynamic capacity as defined by AFBMA in 1949 is based on an air-melted 52100 steel, hardened to at least Rockwell C 58. Since that time, better control of air-melting processes and the introduction of vacuum remelting processes have resulted in more homogeneous steels with fewer impurities. Such steels have extended rolling-element bearing fatigue lives to several times the AFBMA or catalog life. Life improvements of 3–8 times are not uncommon. Other steel compositions, such as AISI M-1 and AISI M-50, chosen for their higher-

Table 23.19 Capacity Formulas for Mixed Rectangular and Elliptical Contacts[a] (From Ref. 6)

Function	Radial bearing	Thrust bearing $\beta \neq 90°$	Thrust bearing $\beta = 90°$	Radial bearing	Thrust bearing $\beta \neq 90°$	Thrust bearing $\beta = 90°$
	Inner race			Outer race		
γ	$\dfrac{d\cos\beta}{d_e}$		$\dfrac{d}{d_e}$	$\dfrac{d\cos\beta}{d_e}$		$\dfrac{d}{d_e}$
	Rectangular contact C_i			Elliptical contact C_o		
C_i or C_o	$f_1 f_2 f_a i^{7/9} N^{3/4} D^{29/27} \ell_{t,i}^{7/9}$			$f_1 f_2 f_a \left(\dfrac{2R}{D}\dfrac{r_o}{r_o - R}\right)^{0.41} i^{0.7} N^{2/3} D^{1.8}$		
f_a	$(\cos\beta)^{7/9}$	$(\cos\beta)^{7/9}\tan\beta$	1	$(\cos\beta)^{0.7}$	$(\cos\beta)^{0.7}\tan\beta$	1
f_1	18-25	36-60		3.5-3.9	6-10	
f_2	$\dfrac{\gamma^{2/9}(1-\gamma)^{29/27}}{(1+\gamma)^{1/3}}$		$\gamma^{3/9}$	$\dfrac{\gamma^{0.3}(1+\gamma)^{1.39}}{(1-\gamma)^{1/3}}$		$\gamma^{0.3}$
	Point contact C_i			Line contact C_o		
C_i or C_o	$f_1 f_2 f_a \left(\dfrac{2R}{D}\dfrac{r_i}{r_i - R}\right)^{0.41} i^{0.7} n^{2/3} d^{1.8}$			$f_1 f_2 f_a i^{7/9} n^{3/4} d^{29/27} \ell_{t,o}^{7.9}$		
f_a	$(\cos\alpha)^{0.7}$	$(\cos\alpha)^{0.7}\tan\alpha$	1	$(\cos\alpha)^{7/9}$	$(\cos\alpha)^{7/9}\tan\alpha$	1
f_1	3.7-4.1	6-10		15-22	36-60	
f_2	$\dfrac{\gamma^{0.3}(1-\gamma)^{1.39}}{(1+\gamma)^{1/3}}$		$\gamma^{0.3}$	$\dfrac{\gamma^{2/9}(1+\gamma)^{29/27}}{(1-\gamma)^{1/3}}$		$\gamma^{2/9}$

[a] $C' = C_i[1 + (C_i/C_o)^4]^{1/4}$ units in kg and mm.

temperature capabilities and resistance to corrosion, also have shown greater resistance to fatigue pitting when vacuum melting techniques are employed. Case-hardened materials, such as AISI 4620, AISI 4118, and AISI 8620, used primarily for roller bearings, have the advantage of a tough, ductile steel core with a hard, fatigue-resistant surface.

The recommended $\tilde{D}$ factors for various alloys processed by air melting are shown in Table 23.20. Insufficient definitive life data were found for case-hardened materials to recommend $\tilde{D}$ factors for them. It is recommended that the user refer to the bearing manufacturer for the choice of a specific case-hardened material.

The metallurgical processing variables considered in the development of the $\tilde{E}$ factor included melting practice (air and vacuum melting) and metal working (thermomechanical working). Thermomechanical working of M-50 has also been shown to result in improved life, but it is costly and still not fully developed as a processing technique. Bamberger et al.[53] recommend an $\tilde{E}$ factor of 3 for consumable-electrode-vacuum-melted materials.

Table 23.20 Material Factor for Through-Hardened Bearing Materials[a] (From Ref. 53)

Material	$\widetilde{D}$-Factor
52100	2.0
M-1	.6
M-2	.6
M-10	2.0
M-50	2.0
T-1	.6
Halmo	2.0
M-42	.2
WB 49	.6
440C	0.6-0.8

[a] Air-melted materials assumed.

The translation of factors into a standard[50] is discussed later.

Lubrication factor $\bar{F}$.   Until approximately 1960 the role of the lubricant between surfaces in rolling contact was not fully appreciated. Metal-to-metal contact was presumed to occur in all applications with attendant required boundary lubrication. The development of elastohydrodynamic lubrication theory showed that lubricant films of thicknesses of the order of microinches and tens of microinches occur in rolling contact. Since surface finishes are of the same order of magnitude as the lubricant film thicknesses, the significance of rolling-element bearing surface roughnesses to bearing performance became apparent. Tallian[54] first reported on the importance on bearing life of the ratio of elastohydrodynamic lubrication film thickness to surface roughness. Figure 23.96 shows life as a percentage of calculated L_{10} life as a function of Λ, where

$$\Lambda = \frac{h_{\min}}{(\Delta_a^2 + \Delta_b^2)^{1/2}}$$

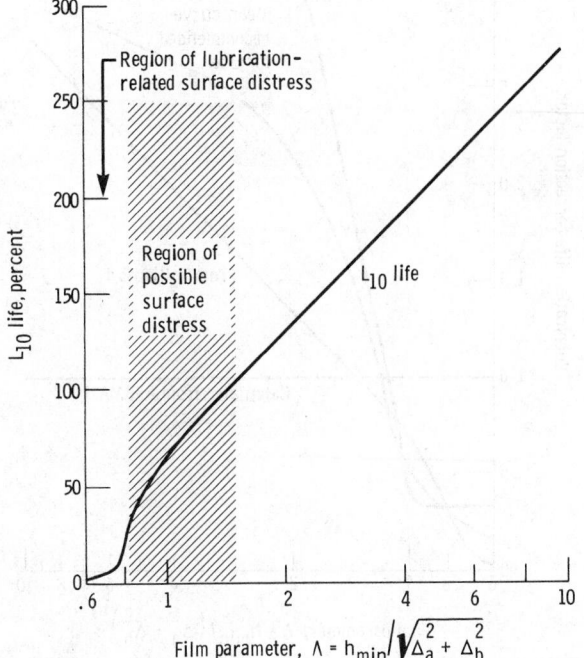

Fig. 23.96 Chart for determining group fatigue life L_{10}. (From Ref. 54.)

Figure 23.97, from Bamberger et al.,[53] presents a curve of the recommended $\bar{F}$ factor as a function of the Λ parameter. A mean of the curves presented in Tallian[54] for ball bearings and in Skurka[55] for roller bearings is recommended for use. A formula for calculating the minimum film thickness h_{min} in the hard-EHL regime is given in Eq. (23.57).

The results of Bamberger et al.[53] have not been fully accepted into the current AFBMA standard represented by ISO.[50] The standard presents the following:

1. Life and dynamic load rating formulas for radial and thrust ball bearings and radial and thrust roller bearings.
2. Tables of f_c for all cases.
3. Tables of X and Y factors for calculating equivalent loads.
4. Load rating formulas for multirow bearings.
5. Life correction factors for high-reliability levels a_1, materials a_2, and lubrication or operating conditions a_3.

Procedures for calculating a_2 and a_3 are less than definitive, reflecting the need for additional research, life data, and operating experience.

Applications

In this section two applications of the film thickness equations developed throughout this chapter are presented to illustrate how the fluid-film lubrication conditions in machine elements can be analyzed. Specifically, a typical roller bearing and a typical ball bearing problem are considered.

Cylindrical-Roller-Bearing Problem. The equations for elastohydrodynamic film thickness that have been developed earlier relate primarily to elliptical contacts, but they are sufficiently general to allow them to be used with adequate accuracy in line-contact problems, as would be found in a cylindrical

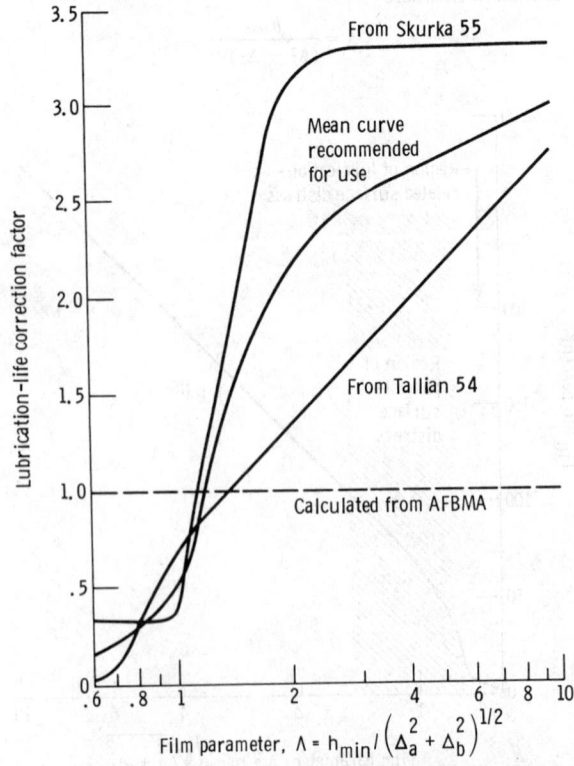

Fig. 23.97 Chart for determining lubrication-life correction factor. (From Ref. 53.)

roller bearing. Therefore, the minimum elastohydrodynamic film thicknesses on the inner and outer races of a cylindrical roller bearing with the following dimensions are calculated:

Inner-race diameter, d_i, mm (m)	64 (0.064)
Outer-race diameter, d_o, mm (m)	96 (0.096)
Diameter of cylindrical rollers, d, mm (m)	16 (0.016)
Axial length of cylindrical rollers, l, mm (m)	16 (0.016)
Number of rollers in complete bearing, n	9

A bearing of this kind might well experience the following operating conditions:

Radial load, F_r, N	10,800
Inner-race angular velocity, ω_i, rad/sec	524
Outer-race angular velocity, ω_o, rad/sec	0
Lubricant viscosity at atmospheric pressure at operating temperature of bearings, η_0, N sec/m^2	0.01
Viscosity–pressure coefficient, ξ, m^2/N	2.2×10^{-8}
Modulus of elasticity for both rollers and races, E, N/m^2	2.075×10^{11}
Poisson's ratio, ν	0.3

Calculation. From Eq. (23.124), the most heavily loaded roller can be expressed as

$$F_{max} = \frac{4F_r}{n} = \frac{4(10,800 \text{ N})}{9} = 4800 \text{ N} \tag{23.124}$$

Therefore, the radial load per unit length on the most heavily loaded roller is

$$F'_{max} = \frac{4800 \text{ N}}{0.016 \text{ m}} = 0.3 \text{ MN/m} \tag{23.125}$$

From Fig. 23.98 we can write the radii of curvature as

$$r_{ax} = 0.008 \text{ m}, \quad r_{ay} = \infty$$
$$r_{bx,i} = 0.032 \text{ m}, \quad r_{by,i} = \infty$$
$$r_{bx,o} = 0.048 \text{ m}, \quad r_{by,o} = \infty$$

Then

$$\frac{1}{R_{x,i}} = \frac{1}{0.008} + \frac{1}{0.032} = \frac{5}{0.032}$$

giving $R_{x,i} = 0.0064$ m,

$$\frac{1}{R_{x,o}} = \frac{1}{0.008} - \frac{1}{0.048} = \frac{5}{0.048} \tag{23.126}$$

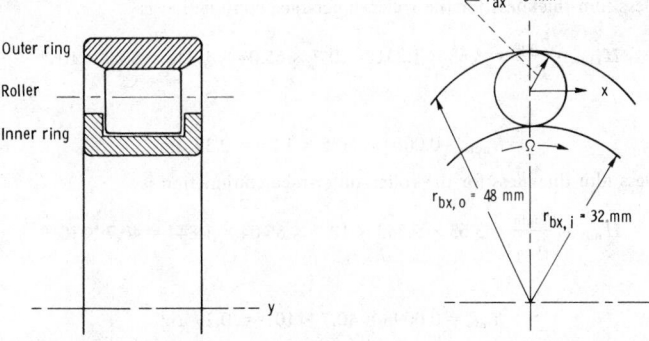

Fig. 23.98 Roller bearing example: $r_{ay} = r_{by,i} = r_{by,o} = \infty$.

giving $R_{x,o} = 0.0096$ m, and

$$\frac{1}{R_{y,i}} = \frac{1}{R_{y,o}} = \frac{1}{\infty} + \frac{1}{\infty} = 0 \qquad (23.127)$$

giving $R_{y,i} = R_{y,o} = \infty$.

From the input information, the effective modulus of elasticity can be written as

$$E' = 2\left(\frac{1 - \nu_a^2}{E_a} + \frac{1 - \nu_b^2}{E_b}\right) = 2.28 \times 10^{11} \text{ N/m}^2 \qquad (23.128)$$

For pure rolling, the surface velocity u relative to the lubricated conjunctions for a cylindrical roller is

$$u = |\omega_i - \omega_o| \frac{d_e^2 - d^2}{4d_e} \qquad (23.129)$$

where d_e is the pitch diameter and d is the roller diameter.

$$d_e = \frac{d_o + d_i}{2} = \frac{0.096 + 0.064}{2} = 0.08 \text{ m} \qquad (23.130)$$

Hence,

$$u = \frac{0.08^2 - 0.16^2}{4 \times 0.08} |524 - 0| = 10.061 \text{ m/sec} \qquad (23.131)$$

The dimensionless speed, materials, and load parameters for the inner- and outer-race conjunctions thus become

$$U_i = \frac{\eta_0 u}{E' R_{x,i}} = \frac{0.01 \times 10.061}{2.28 \times 10^{11} \times 0.0064} = 6.895 \times 10^{-11} \qquad (23.132)$$

$$G_i = \xi E' = 5016 \qquad (23.133)$$

$$W_i = \frac{F}{E'(R_{x,i})^2} = \frac{4800}{2.28 \times 10^{11} \times (0.0064)^2} = 5.140 \times 10^{-4} \qquad (23.134)$$

$$U_o = \frac{\eta_0 u}{E' R_{x,o}} = \frac{0.01 \times 10.061}{2.28 \times 10^{11} \times 0.0096} = 4.597 \times 10^{-11} \qquad (23.135)$$

$$G_o = \xi E' = 5016 \qquad (23.136)$$

$$W_o = \frac{F}{E'(R_{x,o})^2} = \frac{4800}{2.28 \times 10^{11} \times (0.0096)^2} = 2.284 \times 10^{-4} \qquad (23.137)$$

The appropriate elliptical-contact elastohydrodynamic film thickness equation for a fully flooded conjunction is developed in Section 23.3.3 and recorded as Eq. (23.138):

$$H_{min} = \frac{h_{min}}{R_x} = 3.63 U^{0.68} G^{0.49} W^{-0.073}(1 - e^{-0.68k}) \qquad (23.138)$$

For a roller bearing, $k = \infty$ and this equation reduces to

$$H_{min} = 3.63 U^{0.68} G^{0.49} W^{-0.073}$$

The dimensionless film thickness for the roller–inner-race conjunction is

$$H_{min} = \frac{h_{min}}{R_{x,i}} = 3.63 \times 1.231 \times 10^{-7} \times 65.04 \times 1.738 = 50.5 \times 10^{-6}$$

and hence

$$h_{min} = 0.0064 \times 50.5 \times 10^{-6} = 0.32 \ \mu\text{m}$$

The dimensionless film thickness for the roller–outer-race conjunction is

$$H_{min} = \frac{h_{min}}{R_{x,o}} = 3.63 \times 9.343 \times 10^{-8} \times 65.04 \times 1.844 = 40.7 \times 10^{-6}$$

and hence

$$h_{min} = 0.0096 \times 40.7 \times 10^{-6} = 0.39 \ \mu\text{m}$$

It is clear from these calculations that the smaller minimum film thickness in the bearing occurs at the roller–inner-race conjunction, where the geometrical conformity is less favorable. It was found that if the ratio of minimum film thickness to composite surface roughness is greater than 3, an adequate elastohydrodynamic film is maintained. This implies that a composite surface roughness of <0.1 μm is needed to ensure that an elastohydrodynamic film is maintained.

Radial Ball Bearing Problem. Consider a single-row, radial, deep-groove ball bearing with the following dimensions:

Inner-race diameter, d_i, m	0.052291
Outer-race diameter, d_o, m	0.077706
Ball diameter, d, m	0.012700
Number of balls in complete bearing, n	9
Inner-groove radius, r_i, m	0.006604
Outer-groove radius, r_o, m	0.006604
Contact angle, β, deg	0
rms surface finish of balls, Δ_b, μm	0.0625
rms surface finish of races, Δ_a, μm	0.175

A bearing of this kind might well experience the following operating conditions:

Radial load, F_r, N	8900
Inner-race angular velocity, ω_i, rad/sec	400
Outer-race angular velocity, ω_o, rad/sec	0
Lubricant viscosity at atmospheric pressure and effective operating temperature of bearing, η_0, N sec/m²	0.04
Viscosity–pressure coefficient, ξ, m²/N	2.3×10^{-8}
Modulus of elasticity for both balls and races, E, N/m²	2×10^{11}
Poisson's ratio for both balls and races, ν	0.3

The essential features of the geometry of the inner and outer conjunctions (Figs. 23.75 and 23.76) can be ascertained as follows:

Pitch diameter [Eq. (23.68)]:

$$d_e = 0.5(d_o + d_i) = 0.065 \text{ m}$$

Diametral clearance [Eq. (23.69)]:

$$P_d = d_o - d_i - 2d = 1.5 \times 10^{-5} \text{ m}$$

Race conformity [Eq. (23.70)]:

$$f_i = f_o = \frac{r}{d} = 0.52$$

Equivalent radius [Eq. (23.85)]:

$$R_{x,i} = \frac{d(d_e - d)}{2d_e} = 0.00511 \text{ m}$$

Equivalent radius [Eq. (23.87)]:

$$R_{x,o} = \frac{d(d_e + d)}{2d_e} = 0.00759 \text{ m}$$

Equivalent radius [Eq. (23.86)]:

$$R_{y,i} = \frac{f_i d}{2f_i - 1} = 0.165 \text{ m}$$

Equivalent radius [Eq. (23.88)]:

$$R_{y,o} = \frac{f_o d}{2f_o - 1} = 0.165 \text{ m}$$

The curvature sum

$$\frac{1}{R_i} = \frac{1}{R_{x,i}} + \frac{1}{R_{y,i}} = 201.76 \tag{23.139}$$

gives $R_i = 4.956 \times 10^{-3}$ m, and the curvature sum

$$\frac{1}{R_o} = \frac{1}{R_{x,o}} + \frac{1}{R_{y,o}} = 137.81 \qquad (23.140)$$

gives $R_o = 7.256 \times 10^{-3}$ m. Also, $\alpha_i = R_{y,i}/R_{x,i} = 32.35$ and $\alpha_o = R_{y,o}/R_{x,o} = 21.74$.
The nature of the Hertzian contact conditions can now be assessed.
Ellipticity parameters:

$$k_i = \alpha_i^{2/\pi} = 9.42, \qquad k_o = \alpha_o^{2/\pi} = 7.09 \qquad (23.141)$$

Elliptic integrals:

$$q = \frac{\pi}{2} - 1$$

$$\mathscr{E}_i = 1 + \frac{q}{\alpha_i} = 1.0188, \qquad \mathscr{E}_o = 1 + \frac{q}{\alpha_o} = 1.0278 \qquad (23.142)$$

$$\mathscr{F}_i = \frac{\pi}{2} + q \ln \alpha_i = 3.6205, \qquad \mathscr{F}_o = \frac{\pi}{2} + q \ln \alpha_o = 3.3823 \qquad (23.143)$$

The effective elastic modulus E' is given by

$$E' = 2\left(\frac{1 - v_a^2}{E_a} + \frac{1 - v_b^2}{E_b}\right) - 1 = 2.198 \times 10^{11} \text{ N/m}^2$$

To determine the load carried by the most heavily loaded ball in the bearing, it is necessary to adopt an iterative procedure based on the calculation of local static compression and the analysis presented in the fourth subsection in Section 23.3.6. Stribeck[46] found that the value of Z was about 4.37 in the expression

$$F_{\max} = \frac{ZF_r}{n}$$

where $F_{\max}$ = load on most heavily loaded ball
F_r = radial load on bearing
n = number of balls

However, it is customary to adopt a value of $Z = 5$ in simple calculations in order to produce a conservative design, and this value will be used to begin the iterative procedure.

Stage 1. Assume $Z = 5$. Then

$$F_{\max} = \frac{5F_r}{9} = \frac{5}{9} \times 8900 = 4944 \text{ N} \qquad (23.144)$$

The maximum local elastic compression is

$$\delta_i = \mathscr{F}_i\left[\left(\frac{9}{2\mathscr{E}_i R_i}\right)\left(\frac{F_{\max}}{\pi k_i E'}\right)^2\right]^{1/3} = 2.902 \times 10^{-5} \text{ m}$$

$$\delta_o = \mathscr{F}_o\left[\left(\frac{9}{2\mathscr{E}_o R_o}\right)\left(\frac{F_{\max}}{\pi k_o E'}\right)^2\right]^{1/3} = 2.877 \times 10^{-5} \text{ m} \qquad (23.145)$$

The sum of the local compressions on the inner and outer races is

$$\delta = \delta_i + \delta_o = 5.779 \times 10^{-5} \text{ m}$$

A better value for Z can now be obtained from

$$Z = \frac{\pi(1 - P_d/2\delta)^{3/2}}{2.491\left\{\left[1 + \left(\frac{1 - P_d/2\delta}{1.23}\right)^2\right]^{1/2} - 1\right\}}$$

since $P_d/2\delta = (1.5 \times 10^{-5})/(5.779 \times 10^{-5}) = 0.1298$. Thus

$$Z = 4.551$$

Stage 2.

$$Z = 4.551$$

$$F_{max} = (4.551 \times 8900)/9 = 4500 \text{ N}$$

$$\delta_i = 2.725 \times 10^{-5} \text{ m}, \qquad \delta_o = 2.702 \times 10^{-5} \text{ m}$$

$$\delta = 5.427 \times 10^{-5} \text{ m}$$

$$\frac{P_d}{2\delta} = 0.1382$$

Thus

$$Z = 4.565$$

Stage 3.

$$Z = 4.565$$

$$F_{max} = \frac{4.565 \times 8900}{9} = 4514 \text{ N}$$

$$\delta_i = 2.731 \times 10^{-5} \text{ m}, \qquad \delta_o = 2.708 \times 10^{-5} \text{ m}$$

$$\delta = 5.439 \times 10^{-5} \text{ m}$$

$$\frac{P_d}{2\delta} = 0.1379$$

and hence

$$Z = 4.564$$

This value is very close to the previous value from stage 2 of 4.565, and a further iteration confirms its accuracy.

Stage 4.

$$Z = 4.564$$

$$F_{max} = \frac{4.564 \times 8900}{9} = 4513 \text{ N}$$

$$\delta_i = 2.731 \times 10^{-5} \text{ m}, \qquad \delta_o = 2.707 \times 10^{-5} \text{ m}$$

$$\delta = 5.438 \times 10^{-5} \text{ m}$$

$$\frac{P_d}{2\delta} = 0.1379$$

and hence

$$Z = 4.564$$

The load on the most heavily loaded ball is thus 4513 N.

Elastohydrodynamic Minimum Film Thickness. For pure rolling

$$u = |\omega_o - \omega_i| \frac{d_e^2 - d^2}{4d_e} = 6.252 \text{ m/sec} \tag{23.146}$$

The dimensionless speed, materials, and load parameters for the inner- and outer-race conjunctions thus become

$$U_i = \frac{\eta_o u}{E' R_{x,i}} = \frac{0.04 \times 6.252}{2.198 \times 10^{11} \times 5.11 \times 10^{-3}} = 2.227 \times 10^{-10} \tag{23.147}$$

$$G_i = \xi E' = 2.3 \times 10^{-8} \times 2.198 \times 10^{11} = 5055 \tag{23.148}$$

$$W_i = \frac{F}{E'(R_{x,i})^2} = \frac{4513}{2.198 \times 10^{11} \times (5.11)^2 \times 10^{-6}} = 7.863 \times 10^{-4} \tag{23.149}$$

$$U_o = \frac{\eta_o u}{E' R_{x,o}} = \frac{0.04 \times 6.252}{2.198 \times 10^{11} \times 7.59 \times 10^{-3}} = 1.499 \times 10^{-10} \tag{23.150}$$

$$G_o = \xi E' = 2.3 \times 10^{-8} \times 2.198 \times 10^{11} = 5055 \qquad (23.151)$$

$$W_o = \frac{F}{E'(R_{x,o})^2} = \frac{4513}{2.198 \times 10^{11} \times (7.59)^2 \times 10^{-6}} = 3.564 \times 10^{-4} \qquad (23.152)$$

The dimensionless minimum elastohydrodynamic film thickness in a fully flooded elliptical contact is given by

$$H_{\min} = \frac{h_{\min}}{R_x} = 3.63 U^{0.68} G^{0.49} W^{-0.073} (1 - e^{-0.68k}) \qquad (23.153)$$

For the ball-inner-race conjunction it is

$$(H_{\min})_i = 3.63 \times 2.732 \times 10^{-7} \times 65.29 \times 1.685 \times 0.9983$$
$$= 1.09 \times 10^{-4} \qquad (23.154)$$

Thus

$$(h_{\min})_i = 1.09 \times 10^{-4} R_{x,i} = 0.557 \ \mu m$$

The lubrication factor Λ discussed in the fifth subsection of Section 23.3.6 was found to play a significant role in determining the fatigue life of rolling-element bearings. In this case

$$\Lambda_i = \frac{(h_{\min})_i}{(\Delta_a^2 + \Delta_b^2)^{1/2}} = \frac{0.557 \times 10^{-6}}{[(0.175)^2 + (0.06225)^2]^{1/2} \times 10^{-6}} = 3.00 \qquad (23.155)$$

Ball–outer-race conjunction is given by

$$(H_{\min})_o = \frac{(h_{\min})_o}{R_{x,o}} = 3.63 U_o^{0.68} G^{0.49} W^{-0.073} (1 - e^{-0.68k_o}) \qquad (23.156)$$

$$= 3.63 \times 2.087 \times 10^{-7} \times 65.29 \times 1.785 \times 0.9919$$

$$= 0.876 \times 10^{-4}$$

Thus

$$(h_{\min})_o = 0.876 \times 10^{-4} R_{x,o} = 0.665 \ \mu m$$

In this case, the lubrication factor Λ is given by

$$\Lambda_o = \frac{0.665 \times 10^{-6}}{[(0.175)^2 + (0.0625)^2]^{1/2} \times 10^{-6}} = 3.58 \qquad (23.157)$$

Once again, it is evident that the smaller minimum film thickness occurs between the most heavily loaded ball and the inner race. However, in this case the minimum elastohydrodynamic film thickness is about three times the composite surface roughness, and the bearing lubrication can be deemed to be entirely satisfactory. Indeed, it is clear from Fig. 23.97 that very little improvement in the lubrication factor $\tilde{F}$ and thus in the fatigue life of the bearing could be achieved by further improving the minimum film thickness and hence Λ.

23.4 BOUNDARY LUBRICATION

If the pressures in fluid-film-lubricated machine elements are too high, the running speeds are too low, or the surface roughness is too great, penetration of the lubricant film will occur. Contact will take place between asperities, leading to a rise in friction and wear rate. Figure 23.99 (obtained from Bowden and Tabor[56] shows the behavior of the coefficient of friction in the different lubrication regimes. It is to be noted in this figure that in boundary lubrication, although the friction is much higher than in the hydrodynamic regime, it is still much lower than for unlubricated surfaces. As the running conditions are made more severe, the amount of lubricant breakdown increases, until the system scores or seizes so badly that the machine element can no longer operate successfully.

Figure 23.100 shows the wear rate in the different lubrication regimes as determined by the operating load. In the hydrodynamic and elastohydrodynamic lubrication regimes, since there is no asperity contact, there is little or no wear. In the boundary lubrication regime the degree of asperity interaction and wear rate increases as the load increases. The transition from boundary lubrication to an unlubricated condition is marked by a drastic change in wear rate. Machine elements cannot operate successfully in the unlubricated region. Together Figs. 23.99 and 23.100 show that both friction and wear can be greatly decreased by providing a boundary lubricant to unlubricated surfaces.

Understanding boundary lubrication depends first on recognizing that bearing surfaces have asperities that are large compared with molecular dimensions. On the smoothest machined surfaces these asperities may be 25 nm (0.025 μm) high; on rougher surfaces they may be ten to several hundred times higher.

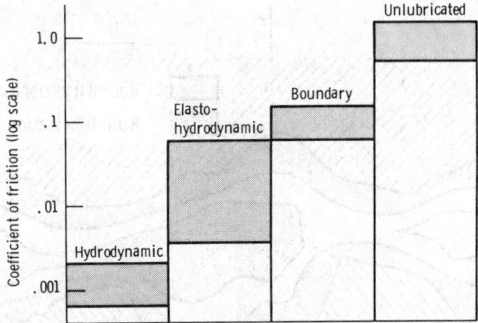

Fig. 23.99 Schematic drawing showing how type of lubrication shifts from hydrodynamic to elastohydrodynamic to boundary lubrication as the severity of running conditions is increased. (From Ref. 56.)

Figure 23.101 illustrates typical surface roughness as a random distribution of hills and valleys with varying heights, spacing, and slopes. In the absence of hydrodynamic or elastohydrodynamic pressures these hills or asperities must support all of the load between the bearing surfaces. Understanding boundary lubrication also depends on recognizing that bearing surfaces are often covered by boundary lubricant films such as are idealized in Fig. 23.101. These films separate the bearing materials and, by shearing preferentially, provide some control of friction, wear, and surface damage.

Many mechanisms, such as door hinges, operate totally under conditions (high load, low speed) of boundary lubrication. Others are designed to operate under full hydrodynamic or elastohydrodynamic lubrication. However, as the oil film thickness is a function of speed, the film will be unable to provide complete separation of the surfaces during startup and rundown, and the condition of boundary lubrication will exist. The problem from the boundary lubrication standpoint is to provide a boundary film with the proper physical characteristics to control friction and wear. The work of Bowden and Tabor,[56] Godfrey,[59] and Jones[60] was relied upon in writing the sections that follow.

23.4.1 Formation of Films

The most important aspect of boundary lubrication is the formation of surface films that will protect the contacting surfaces. There are three ways of forming a boundary lubricant film: physical adsorption, chemisorption, and chemical reaction. The surface action that determines the behavior of boundary lubricant films is the energy binding the film molecules to the surface, a measure of the film strength. The formation of films is presented in the order of such a film strength, the weakest being presented first.

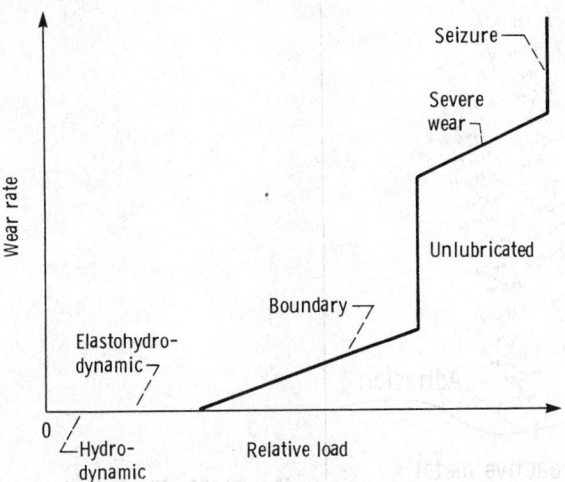

Fig. 23.100 Chart for determining wear rate for various lubrication regimes. (From Ref. 57.)

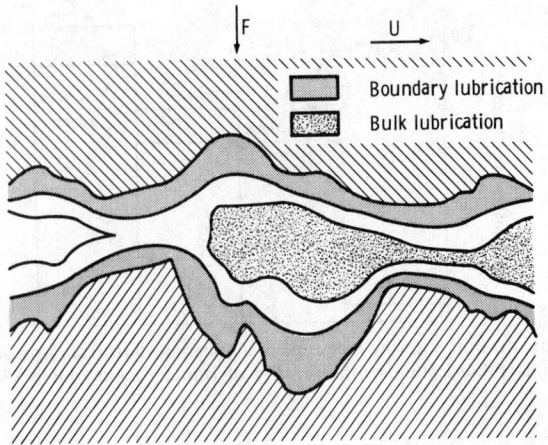

Fig. 23.101   Lubricated bearing surfaces. (From Ref. 58.)

Physical Adsorption

Physical adsorption involves intermolecular forces analogous to those involved in condensation of vapors to liquids. A layer of lubricant one or more molecules thick becomes attached to the surfaces of the solids, and this provides a modest protection against wear. Physical adsorption is usually rapid, reversible, and nonspecific. Energies involved in physical adsorption are in the range of heats of condensations. Physical adsorption may be monomolecular or multilayer. There is no electron transfer in this process. An idealized example of physical adsorption of hexadecanol on an unreactive metal is shown in Fig. 23.102. Because of the weak bonding energies involved, physically adsorbed species are usually not very effective boundary lubricants.

Chemical Adsorption

Chemically adsorbed films are generally produced by adding animal and vegetable fats and oils to the base oils. These additives contain long-chain fatty acid molecules, which exhibit great affinity for

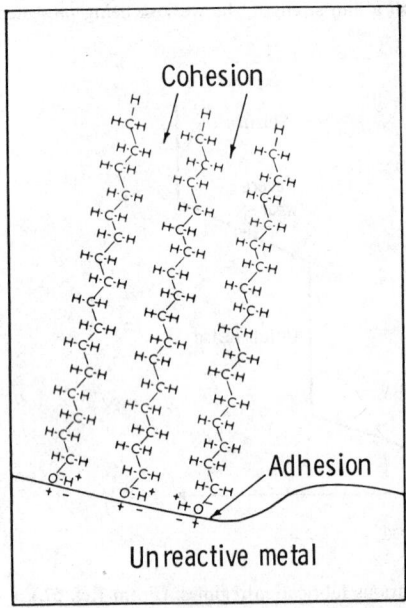

Fig. 23.102   Physical adsorption of hexadecanol. (From Ref. 59.)

metals at their active ends. The usual configuration of these polar molecules resembles that of a carpet pile with the molecules standing perpendicular to the surface. Such fatty acid molecules form metal soaps that are low-shear-strength materials with coefficients of friction in the range 0.10–0.15. The soap film is dense because of the preferred orientation of the molecules. For example, on a steel surface stearic acid will form a monomolecular layer of iron stearate, a soap containing 10^{14} molecules/ cm^2 of surface. The effectiveness of these layers is limited by the melting point of the soap ($180°C$ for iron stearate). It is clearly essential to choose an additive that will react with the bearing metals, so that less reactive, inert metals like gold and platinum are not effectively lubricated by fatty acids.

Examples of fatty acid additives are stearic, oleic, and lauric acid. The soap films formed by these acids might reduce the coefficient of friction to 50% of that obtained by a straight mineral oil. They provide satisfactory boundary lubrication at moderate loads, temperatures, and speeds and are often successful in situations showing evidence of mild surface distress.

Chemisorption of a film on a surface is usually specific, may be rapid or slow, and is not always reversible. Energies involved are large enough to imply that a chemical bond has formed (i.e., electron transfer has taken place). In contrast to physical adsorption, chemisorption may require an activation energy. A film may be physically adsorbed at low temperatures and chemisorbed at higher temperatures. In addition, physical adsorption may occur on top of a chemisorbed film. An example of a film of stearic acid chemisorbed on an iron oxide surface to form iron stearate is shown in Fig. 23.103.

Chemical Reaction

Films formed by chemical reaction provide the greatest film strength and are used in the most severe operating conditions. If the load and sliding speeds are high, significant contact temperatures will be developed. It has already been noted that films formed by physical and chemical adsorption cease to be effective above certain transition temperatures, but some additives start to react and form new high-melting-point inorganic solids at high temperatures. For example, sulfur will start to react at about $100°C$ to form sulfides with melting points of over $1000°C$. Lubricants containing additives like sulfur, chlorine, phosphorous, and zinc are often referred to as extreme-pressure (EP) lubricants, since they are effective in the most arduous conditions.

The formation of a chemical reaction film is specific; may be rapid or slow (depending on temperature, reactivity, and other conditions); and is irreversible. An idealized example of a reacted film of iron sulfide on an iron surface is shown in Fig. 23.104.

23.4.2 Physical Properties of Boundary Films

The two physical properties of boundary films that are most important in determining their effectiveness in protecting surfaces are melting point and shear strength. It is assumed that the film thicknesses involved are sufficient to allow these properties to be well defined.

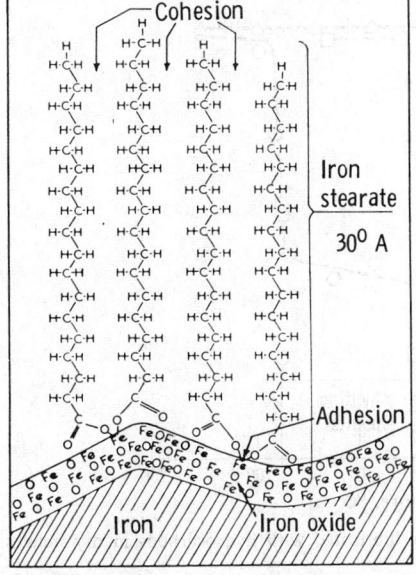

Fig. 23.103 Chemisorption of stearic acid on iron surface to form iron stearate. (From Ref. 59.)

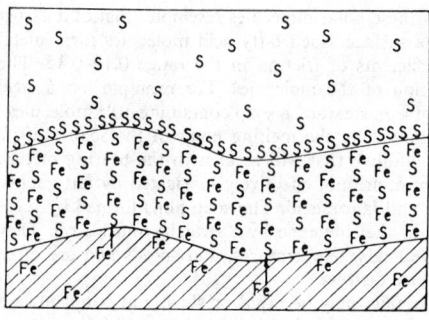

Fig. 23.104 Formation of inorganic film by reaction of sulfur with iron to form iron sulfide. (From Ref. 59.)

Melting Point

The melting point of a surface film appears to be one discriminating physical property governing failure temperature for a wide range of materials including inorganic salts. It is based on the observation that only a surface film that is solid can properly interfere with potentially damaging asperity contacts. Conversely, a liquid film allows high friction and wear. Under practical conditions, physically adsorbed additives are known to be effective only at low temperatures, and chemisorbed additives at moderate temperatures. High-melting-point inorganic materials are used for high-temperature lubricants.

The correlation of melting point with failure temperature has been established for a variety of organic films. An illustration is given in Fig. 23.105 (obtained from Russell et al.[61]) showing the friction transition for copper lubricated with pure hydrocarbons. Friction data for two hydrocarbons (mesitylene and dotriacontane) are given in Fig. 23.105 as a function of temperature. In this figure the boundary film failure occurs at the melting point of each hydrocarbon.

In contrast, chemisorption of fatty acids on reactive metals yields failure temperature based on the softening point of the soap rather than the melting point of the parent fatty acid.

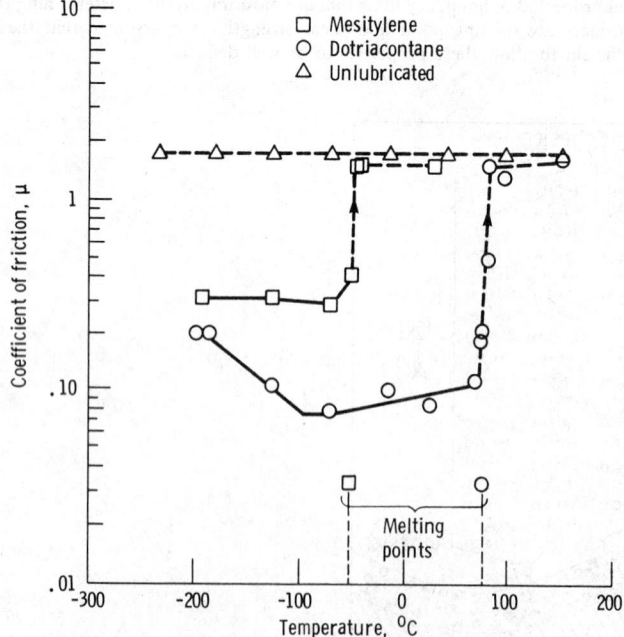

Fig. 23.105 Chart for determining friction of copper lubricated with hydrocarbons in dry helium. (From Ref. 61.)

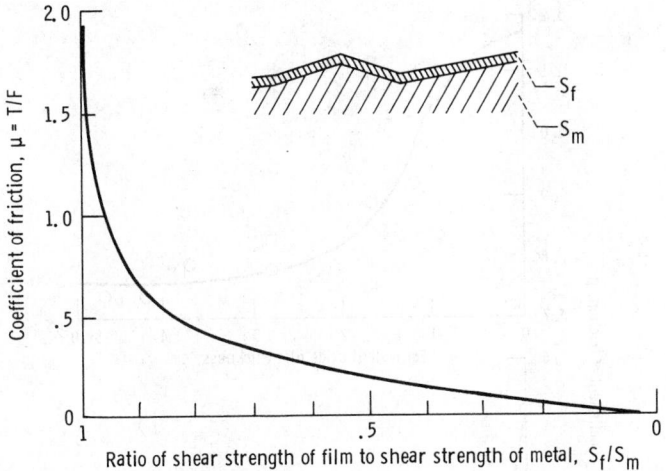

Fig. 23.106 Chart for determining friction as function of shear strength ratio. (From Ref. 59.)

Shear Strength

The shear strength of a boundary lubricating film should be directly reflected in the friction coefficient. In general, this is true with low-shear-strength soaps yielding low friction and high-shear-strength salts yielding high friction. However, the important parameter in boundary friction is the ratio of shear strength of the film to that of the substrate. This relationship is shown in Fig. 23.106, where the ratio is plotted on the horizontal axis with a value of 1 at the left and zero at the right. These results are in agreement with experience. For example, on steel an MoS_2 film gives low friction and Fe_2O_3 gives high friction. The results from Fig. 23.106 also indicate how the same friction value can be obtained with various combinations provided that the ratio is the same. It is important to recognize that shear strength is also affected by pressure and temperature.

23.4.3 Film Thickness

Boundary film thickness can vary from a few angstroms (adsorbed gas) to thousands of angstroms (chemical reaction films). In general, as the thickness of a boundary film increases, the coefficient of friction decreases. This effect is shown in Fig. 23.107a, which shows the coefficient of friction plotted against oxide film thickness formed on a copper surface. However, continued increases in thickness may result in an increase in friction. This effect is shown in Fig. 23.107b, which shows the coefficient of friction plotted against indium film thickness on copper surface. It should also be pointed out that the shear strengths of all boundary films decrease as their thicknesses increase, which may be related to the effect seen in Fig. 23.107b.

For physically adsorbed or chemisorbed films, surface protection is usually enhanced by increasing film thickness. The frictional transition temperature of multilayers also increases with increasing number of layers.

For thick chemically reacted films there is an optimum thickness for minimum wear that depends on temperature, concentration, or load conditions. The relationship between wear and lubricant (or additive) reactivity is shown in Fig. 23.108. Here, if reactivity is not great enough to produce a thick enough film, adhesion wear occurs. On the other hand, if the material is too reactive, very thick films are formed and corrosive wear ensues.

23.4.4 Effect of Operating Variables

The effect of load, speed, temperature, and atmosphere can be important for the friction and wear of boundary lubrication films. Such effects are considered in this section.

On Friction

Load. The coefficient of friction is essentially constant with increasing load.

Speed. In general, in the absence of viscosity effects, friction changes little with speed over a sliding speed range of 0.005 to 1.0 cm/sec. When viscosity effects do come into play, two types of

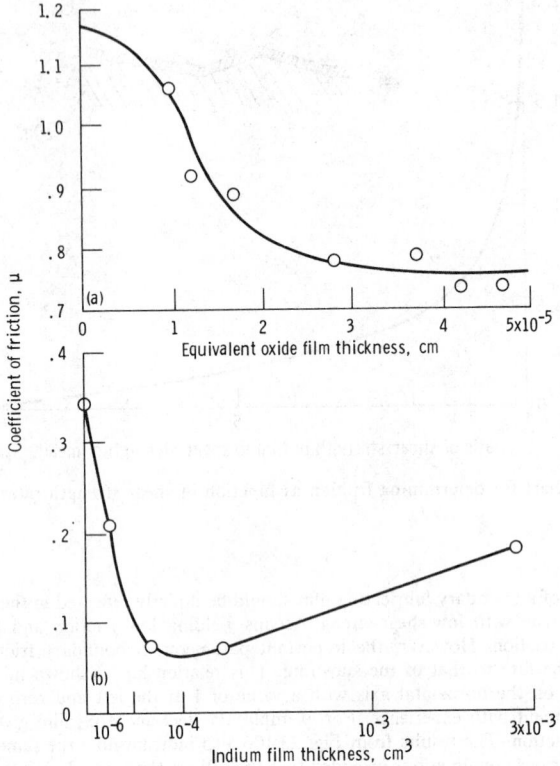

Fig. 23.107 Chart for determining relationship of friction and thickness of films on copper surfaces. (From Ref. 62.)

behavior are observed, as shown in Fig. 23.109. In this figure relatively nonpolar materials such as mineral oils show a decrease in friction with increasing speed, while polar fatty acids show the opposite trend. At higher speeds viscous effects will be present, and increases in friction are normally observed.

Temperature. It is difficult to make general comments on the effect of temperature on boundary friction since so much depends on the other conditions and the type of materials present. Temperature can cause disruption, desorption, or decomposition of boundary films. It can also provide activation energy for chemisorption or chemical reactions.

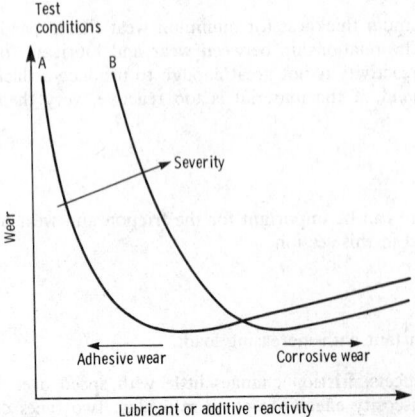

Fig. 23.108 Relationship between wear and lubricant reactivity. (From Ref. 63.)

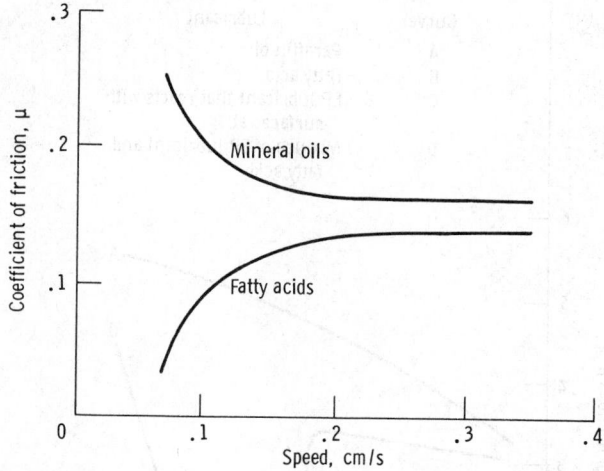

Fig. 23.109 Effect of speed on coefficient of friction. (From Ref. 64.)

Atmosphere. The presence of oxygen and water vapor in the atmosphere can greatly affect the chemical processes that occur in the boundary layer. These processes can, in turn, affect the friction coefficient.

On Wear

Load. It is generally agreed that wear increases with increasing load, but no simple relationship seems to exist, at least before the transition to severe wear occurs. At this point a discontinuity of wear versus load is often like that illustrated in Fig. 23.100.

Speed. For practical purposes, wear rate in a boundary lubrication regime is essentially independent of speed. This assumes no boundary film failure due to contact temperature rise.

Temperature. As was the case for friction, there is no way to generalize the effect of temperature on wear. The statement that pertains to friction also pertains to wear.

Atmosphere. Oxygen has been shown to be an important ingredient in boundary lubrication experiments involving load-carrying additives. The presence of oxygen or moisture in the test atmosphere has a great effect on the wear properties of lubricants containing aromatic species.

23.4.5 Extreme-Pressure (EP) Lubricants

The best boundary lubricant films cease to be effective above 200–250°C. At these high temperatures the lubricant film may iodize. For operation under more severe conditions, EP lubricants might be considered.

Extreme-pressure lubricants usually consist of a small quantity of an EP additive dissolved in a lubricating oil, usually referred to as the base oil. The most common additives used for this purpose contain phosphorus, chlorine, or sulfur. In general, these materials function by reacting with the surface to form a surface film that prevents metal-to-metal contact. If, in addition, the surface film formed has a low shear strength, it will not only protect the surface, but it will also give a low coefficient of friction. Chloride films give a lower coefficient of friction ($\mu = 0.2$) than sulfide films ($\mu = 0.5$). Sulfide films, however, are more stable, are unaffected by moisture, and retain their lubricating properties to very high temperatures.

Although EP additives function by reacting with the surface, they must not be too reactive, otherwise chemical corrosion may be more troublesome than frictional wear. They should only react when there is a danger of seizure, usually noted by a sharp rise in local or global temperature. For this reason it is often an advantage to incorporate in a lubricant a small quantity of a fatty acid that can provide effective lubrication at temperatures below those at which the additive becomes reactive. Bowden and Tabor[56] describe this behavior in Fig. 23.110, where the coefficient of friction is plotted against temperature. Curve A is for paraffin oil (the base oil) and shows that the friction is initially high and increases as the temperature is raised. Curve B is for a fatty acid dissolved in the base oil: it reacts with the

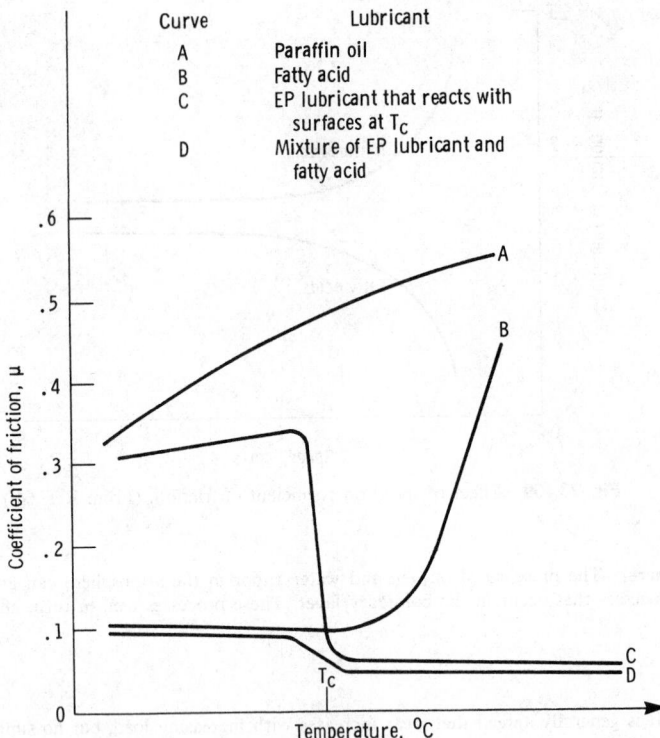

Fig. 23.110 Graph showing frictional behavior of metal surfaces with various lubricants. (From Ref. 56.)

surface to form a metallic soap, which provides good lubrication from room temperature up to the temperature at which the soap begins to soften. Curve C is for a typical EP additive in the base oil: this reacts very slowly below the temperature T_c, so that in this range the lubrication is poor, while above T_c the protective film is formed and effective lubrication is provided to a very high temperature. Curve D is the result obtained when the fatty acid is added to the EP solution. Good lubrication is provided by the fatty acid below T_c, while above this temperature the greater part of the lubrication is due to the additive. At still higher temperatures, a deterioration of lubricating properties will also occur for both curves C and D.

REFERENCES

1. B. Tower, "First Report on Friction Experiments (Friction of Lubricated Bearings)," *Proc. Inst. Mech. Eng., London*, 632–659 (1883).

2. N. P. Petrov, "Friction in Machines and the Effect of the Lubricant," *Inzh. Zh. St-Petreb.* **1**, 71–140 (1883); **2**, 227–279 (1883); **3**, 377–436 (1883); **4**, 535–564 (1883).

3. O. Reynolds, "On the Theory of Lubrication and Its Application to Mr. Beauchamp Tower's Experiments, Including an Experimental Determination of the Viscosity of Olive Oil," *Philos. Trans. R. Soc. London* **177**, 157–234 (1886).

4. W. B. Hardy and I. Doubleday, "Boundary Lubrication—The Temperature Coefficient," *Proc. R. Soc.* **A101**, 487–492 (1922).

5. W. B. Hardy and I. Doubleday, "Boundary Lubrication—The Paraffin Series," *Proc. R. Soc.* **A104**, 25–39 (1922).

6. B. J. Hamrock and W. J. Anderson, *Rolling-Element Bearings*, NASA RP-1105, 1983.

7. ESDU, "General Guide to the Choice of Journal Bearing Type," Engineering Sciences Data Unit, Item 65007, Institution of Mechanical Engineers, London, 1965.

8. ESDU, "General Guide to the Choice of Thrust Bearing Type," Engineering Sciences Data Unit, Item 67033, Institution of Mechanical Engineers, London, 1967.

9. I. Newton, *Philosophiae Naturalis Principia Mathematica,* 1687. Imprimature S. Pepys. Reg. Soc. Praess, 5 Julii 1866. Revised and supplied with a historical and explanatory appendix by F. Cajori, edited by R. T. Crawford, 1934. Published by the University of California Press, Berkeley and Los Angeles, 1966.

10. B. J. Hamrock and D. Dowson, *Ball Bearing Lubrication—The Elastohydrodynamics of Elliptical Contacts,* Wiley, New York, 1981.

11. E. E. Bisson and W. J. Anderson, *Advanced Bearing Technology,* NASA SP-38, 1964.

12. ESDU, "Contact Stresses," Engineering Sciences Data Unit, Item 78035, Institution of Mechanical Engineers, London, 1978.

13. S. Aihara and D. Dowson, "A Study of Film Thickness in Grease Lubricated Elastohydrodynamic Contacts," in *Proceedings of Fifth Leeds–Lyon Symposium on Tribology on "Elastohydrodynamics and Related Topics,"* D. Dowson, C. M. Taylor, M. Godet, and D. Berthe (Eds.), Mechanical Engineering Publications, Bury St. Edmunds, Suffolk, 1979, pp. 104–115.

14. A. R. Wilson, "The Relative Thickness of Grease and Oil Films in Rolling Bearings," *Proc. Inst. Mech. Eng., London,* **193**(17), 185–192 (1979).

15. T. Tallian, L. Sibley, and R. Valori, "Elastohydrodynamic Film Effect on the Load-Life Behavior of Rolling Contacts," ASME Paper 65-LUB-11, 1965.

16. C. Barus, "Isotherms, Isopeistics and Isometrics Relative to Viscosity," *Am. J. Sci.* **45**, 87–96 (1893).

17. W. R. Jones, R. L. Johnson, W. O. Winer, and D. M. Sanborn, "Pressure-Viscosity Measurements for Several Lubricants to 5.5×10^8 Newtons Per Square Meter (8×10^4 psi) and 149° C (300° F)," *ASLE Trans.* **18**(4), 249–262 (1975).

18. D. Dowson and G. R. Higginson, *Elastohydrodynamic Lubrication, the Fundamentals of Roller and Gear Lubrication,* Pergamon, Oxford, 1966.

19. W. A. Gross, L. A. Matsch, V. Castelli, A. Eshel, and M. Wildmann, *Fluid Film Lubrication,* Wiley, New York, 1980.

20. N. F. Reiger, *Design of Gas Bearings,* Mechanical Technology, Inc., Latham, New York, 1967.

21. O. Pinkus and B. Sternlicht, *Theory of Hydrodynamic Lubrication,* McGraw-Hill, New York, 1961.

22. H. C. Rippel, *Cast Bronze Hydrostatic Bearing Design Manual,* Cast Bronze Bearing Institute, Inc., Cleveland, OH, 1963.

23. A. A. Raimondi and J. Boyd, "A Solution for the Finite Journal Bearing and Its Application to Analysis and Design; III," *Trans. ASLE* **1**(1), 194–209 (1959).

24. P. E. Allaire and R. D. Flack, "Journal Bearing Design for High Speed Turbomachinery," *Bearing Design—Historical Aspects, Present Technology and Future Problems,* W. J. Anderson (Ed.), American Society of Mechanical Engineers, New York, 1980, pp. 111–160.

25. A. A. Raimondi and J. Boyd, "Applying Bearing Theory to the Analysis and Design of Pad-type Bearings," *Trans. ASME,* 287–309 (Apr. 1955).

26. B. J. Hamrock, "Optimum Parallel Step-Sector Bearing Lubricated with an Incompressible Fluid," NASA TM-83356, 1983.

27. E. J. Gunter, J. G. Hinkle, and D. D. Fuller, "Design Guide for Gas-Lubricated Tilting-Pad Journal and Thrust Bearings with Special Reference to High Speed Rotors," Franklin Institute Research Laboratories Report I-A2392-3-1, 1964.

28. B. J. Hamrock and D. P. Fleming, "Optimization of Self-Acting Herringbone Grooved Journal Bearings for Minimum Radial Load," in *Proceedings of Fifth International Gas Bearing Symposium,* University of Southampton, Southampton, England, 1971, Paper 13.

29. D. P. Fleming and B. J. Hamrock, "Optimization of Self-Acting Herringbone Journal Bearings for Maximum Stability," *6th International Gas Bearing Symposium,* University of Southampton, Southampton, England, 1974, Paper c1, pp. 1–11.

30. B. J. Hamrock, "Optimization of Self-Acting Step Thrust Bearings for Load Capacity and Stiffness," *ASLE Trans.* **15**(3), 159–170 (1972).

31. D. Dowson, "Elastohydrodynamic Lubrication—An Introduction and a Review of Theoretical Studies," *Institution of Mechanical Engineers, London, Proceedings,* Vol. 180, Pt. 3B, 1965, pp. 7–16.

32. A. N. Grubin, "Fundamentals of the Hydrodynamic Theory of Lubrication of Heavily Loaded Cylindrical Surfaces," in *Investigation of the Contact Machine Components,* Kh. F. Ketova (Ed.), translation of Russian Book No. 30, Central Scientific Institute of Technology and Mechanical Engineering, Moscow, 1949, Chap. 2. (Available from Dept. of Scientific and Industrial Research, Great Britain, Transl. CTS-235, and from Special Libraries Association, Chicago, Transl. R-3554.)

33. A. J. Petrusevich, "Fundamental Conclusion from the Contact-Hydrodynamic Theory of Lubrication," *Zv. Akad, Nauk, SSSR (OTN)* **2**, 209 (1951).

34. B. J. Hamrock and D. Brewe, "Simplified Solution for Stresses and Deformation," *J. Lubr. Technol.* **105**(2), 171–177 (1983).

35. B. J. Hamrock and W. J. Anderson, "Analysis of an Arched Outer-Race Ball Bearing Considering Centrifugal Forces," *J. Lubr. Techn.* **95**(3), 265–276 (1973).

36. A. B. Jones, "Analysis of Stresses and Deflections," New Departure Engineering Data, General Motors Corporation, Bristol, CT, 1946.

37. H. Hertz, "The Contact of Elastic Solids," *J. Reine Angew. Math.* **92,** 156–171 (1881).

38. B. J. Hamrock and D. Dowson, "Isothermal Elastohydrodynamic Lubrication of Point Contacts, Part I—Theoretical Formulation," *J. Lubr. Technol.* **98**(2), 223–229 (1976).

39. B. J. Hamrock and D. Dowson, "Isothermal Elastohydrodynamic Lubrication of Point Contacts, Part III—Fully Flooded Results," *J. Lubr. Technol.* **99**(2), 264–276 (1977).

40. A. Cameron and R. Gohar, "Theoretical and Experimental Studies of the Oil Film in Lubricated Point Contacts," *Proc. R. Soc. London, Ser. A,* **291**, 520–536 (1966).

41. B. J. Hamrock and D. Dowson, "Elastohydrodynamic Lubrication of Elliptical Contacts for Materials of Low Elastic Modulus, Part I—Fully Flooded Conjunction," *J. Lubr. Technol.* **100**(2), 236–245 (1978).

42. D. E. Brewe, B. J. Hamrock, and C. M. Taylor, "Effect of Geometry on Hydrodynamic Film Thickness," *J. Lubr. Technol.,* **101**(2), 231–239 (1979).

43. P. L. Kapitza, "Hydrodynamic Theory of Lubrication During Rolling," *Zh. Tekh. Fig.* **25**(4), 747–762 (1955).

44. B. J. Hamrock and D. Dowson, "Minimum Film Thickness in Elliptical Contacts for Different Regimes of Fluid-Film Lubrication," in *Proceedings of Fifth Leeds–Lyon Symposium on Tribology on Elastohydrodynamics and Related Topics,* D. Dowson, C. M. Taylor, M. Godet, and D. Berthe (Eds.), Mechanical Engineering Publications, Bury St. Edmunds, Suffolk, 1979, pp. 22–27.

45. A. B. Jones, "The Mathematical Theory of Rolling Element Bearings," in *Mechanical Design and Systems Handbook,* H. A. Rothbart (Ed.), McGraw-Hill, New York, 1964, pp. 13-1–13-76.

46. R. Stribeck, "Kugellager fur beliebige Belastungen," *Z. VDI-Zeitschrift* **45**(3), 73–125 (1901).

47. W. Weibull, "A Statistical Representation of Fatigue Failures in Solids," *Trans. Roy. Inst. Tech., Stockholm,* **27** (1949).

48. G. Lundberg and A. Palmgren, "Dynamic Capacity of Rolling Bearings," *Acta Polytechnica,* Mechanical Engineering Series, Vol. I, No. 3, 1947.

49. G. Lundberg and A. Palmgren, "Dynamic Capacity of Rolling Bearings," *Acta Polytechnica,* Mechanical Engineering Series, Vol. II, No. 4, 1952.

50. ISO, "Rolling Bearings, Dynamic Load Ratings and Rating Life," ISO/TC4/JC8, Revision of ISOR281. Issued by International Organization for Standardization, Technical Committee ISO/TC4, 1976.

51. AFBMA, *Method of Evaluating Load Ratings for Ball Bearings,* AFBMA Standard System No. 9, Revision No. 4, Anti-Friction Bearing Manufacturers Association, Inc., Arlington, VA, 1960.

52. AFBMA, *Method of Evaluating Load Ratings for Roller Bearings,* AFBMA Standard System No. 11, Anti-friction Bearing Manufacturers Association, Inc., Arlington, VA, 1960.

53. E. N. Bamberger, T. A. Harris, W. M. Kacmarsky, C. A. Moyer, R. J. Parker, J. J. Sherlock, and E. V. Zaretsky, *Life Adjustment Factors for Ball and Roller Bearings—An Engineering Design Guide,* American Society for Mechanical Engineers, New York, 1971.

54. T. E. Tallian, "On Competing Failure Modes in Rolling Contact," *Trans. ASLE* **10**, 418–439 (1967).

55. J. C. Skurka, "Elastohydrodynamic Lubrication of Roller Bearings," *J. Lubr. Technol.* **92**(2), 281–291 (1970).

56. F. P. Bowden and D. Tabor, *Friction—An Introduction to Tribology,* Heinemann, London, 1973.

57. A. Beerbower, "Boundary Lubrication," *Scientific and Technical Application Forecasts,* Department of the Army, DAC-19-69-C-0033, 1972.

58. R. S. Fein and F. J. Villforth, "Lubrication Fundamentals," *Lubrication,* Texaco Inc., New York, 1973, pp. 77–88.

59. D. Godfrey, "Boundary Lubrication," in *Interdisciplinary Approach to Friction and Wear,* P. M. Ku (Ed.), NASA SP-181, 1968, pp. 335–384.

60. W. R. Jones, *Boundary Lubrication—Revisited,* NASA TM-82858, 1982.

61. J. A. Russell, W. C. Campbell, R. A. Burton, and P. M. Ku, "Boundary Lubrication Behavior of Organic Films at Low Temperatures," *ASLE Trans.* **8**(1), 48 (1965).

62. I. V. Kragelski, *Friction and Wear,* Butterworth, London, 1965, pp. 158–163.

63. C. N. Rowe, "Wear Corrosion and Erosion," *Interdisciplinary Approach to Liquid Lubricant Technology,* P. M. Ku (Ed.), NASA SP-237, 1973, pp. 469–527.

64. D. Clayton, "An Introduction to Boundary and Extreme Pressure Lubrication," *Physics of Lubrication, Br. J. Appl. Phys.* **2,** Suppl. 1, 25 (1951).

CHAPTER 24
DESIGN OF JOINTS

CARL C. OSGOOD

Cranbury, New Jersey

24.0 INTRODUCTION

The importance of proper joint design cannot be overemphasized. The dependency of the fatigue behavior of an assembly on that of the joints within it has generally been known, and Ref. 1 provides a splendid example showing that the fatigue life of an airplane is just the fatigue life of the joints in its primary structure. The conventional approach to the fatigue design of a multimember structure treats the members as first in importance, with the joints as inevitable complications, possibly only as necessary evils. In some cases it may be more profitable to consider the assembly as a collection of joints with the members as simple connections among them.

A joint is here defined as an assembly of two or more massive pieces of material, having been mechanically transported to, and being restrained in a position of contact or of close proximity. Such a definition is intended to exclude the interfaces formed by such methods as plating, painting, and vapor deposition. It also reduces the scope of the topic to joints made by the four time-proven processes: bolting, riveting, welding, and adhesive bonding. Some of the newer methods such as joining by explosive forces, by high-velocity impact, and by the transfer of momentum to heat are being developed rapidly, but the available data have not yet been validated in a sufficiently universal manner to warrant their inclusion here.

It is evident that, in addition to the material, the shape of the parts and the modes of loading are very important. The designer may not always have control of the external envelope, but he or she must consider the part details and the location of the joint along the load path. If, for example, the bending component on a bolt could be eliminated and the loading simplified to shear by changing the joint's position or orientation, its overall size and weight may be reduced. The shape of the parts within the joint is, of course, highly significant, since it may transform the external loading to a different mode, and it is inevitably concerned with stress concentrations. "Shape" here refers to any geometric characteristic such as the thread form on bolts, the groove profile on weld plate edges, or the radius/thickness ratio at a clevis throat. Design for resistance to fatigue failure, at any lifetime, requires close attention to physical detail, especially for joints.

The overall envelope of the joint may often be dictated by functional, loading, and displacement requirements, but within the envelope considerable latitude exists for the arrangement of bolts, rivets, or weld beads. By applying one of the first principles of design in the form of drawing the load lines, in all three dimensions, across the joint, it is immediately seen that as usual the straighter the lines the greater the strength. Both static and fatigue loading capacities are improved by the elimination or reduction of transverse components. While the use of lugs or protruding bosses is often unavoidable, their offset from the load line should be minimized. Drawing the analogy between load lines and fluid flow results in the above conclusion, and also that there should be no sudden changes in section of the connected parts, with the obvious effect on stress concentration factors. If one considers a shear joint with two or more rows of bolts, it follows that the "flow" would be improved—made smoother—with the rows placed in tandem along the lines, rather than staggered.

The least-weight joint or splice is that having the shortest overlap or cover plate, meaning that the rivet or bolt spacing should be the minimum, practically and structurally. The minimum structural spacing is of course that calculated on the appropriate criteria of shear, bearing, or tension, while the minimum practical spacing depends on clearances for wrenches or dimpling tools. The latter is a particularly cogent item for flush riveting; in any event, a practical spacing must be held to ensure well-formed heads and an undistorted sheet. Splices in structural shapes are becoming less necessary because of the increased sizes of extrusions and forgings. But should such a joint be required, the shape may be divided into strip elements, as the web and flanges, and the splice designed such that each element has the same stress distribution as in the structural shape. If it is not possible or practical to splice each element, as in the case of an angle with a small return flange, the return may be considered to be spliced by the rivets in a closely adjacent element; the eccentricity so developed is usually negligible. But the splicing member should be flanged and in the same direction as the flange of the structural shape for continuity. There appears to be no restriction on the application of this approach to any shape under any combination of loads.

24.1 MECHANICALLY FASTENED JOINTS

24.1.1 Bolted Joints

Many geometric patterns of mechanical connections have come into use through necessity and experience: splices, gussets, hangars. Figure 24.1 shows a few of those most frequently found and, of course, many patterns are dictated by the function of the device, as the flange plan of a pump or turbine.

Basically, the envelope of the joint and the fastener pattern need to be considered as an entity in order to determine the effective width of the pattern and to identify the critical sections. Typical cases are shown in Figs. 24.2 and 24.3, and Refs. 2, 3, and 4 provide methods for finding the nominal loads in each fastener, including treatment of the eccentrically loaded pattern.

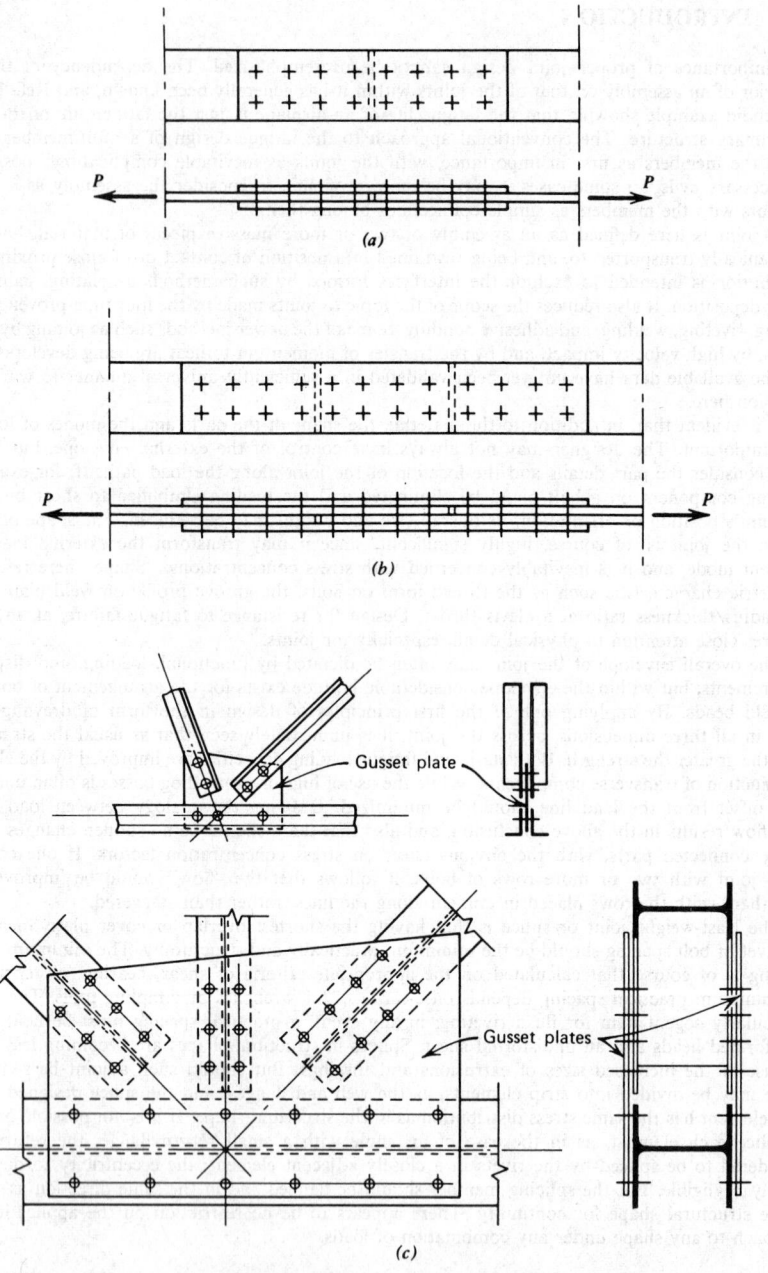

Fig. 24.1 Typical riveted and bolted connections. (a) Symmetric butt splice; (b) shingle splice; (c) single-plane construction (top), double-plane construction (bottom); (d) lap splice; (e) bracket connection; (f) girder web splice; (g) hanger connection.

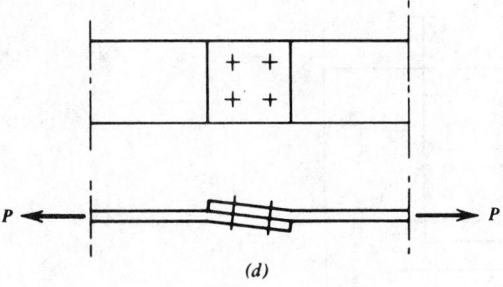

(d)

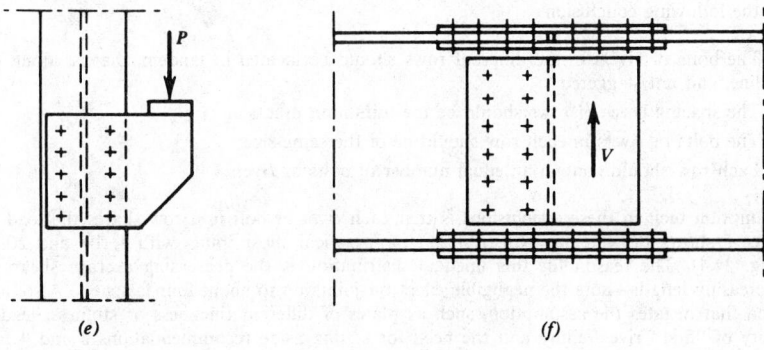

(e)   (f)

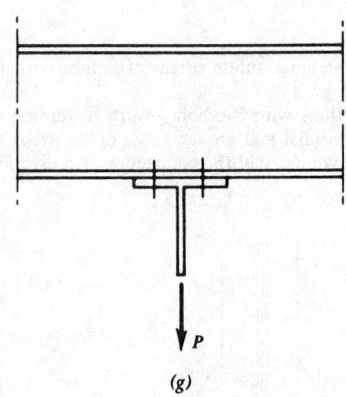

(g)

Fig. 24.1 (*Continued*)

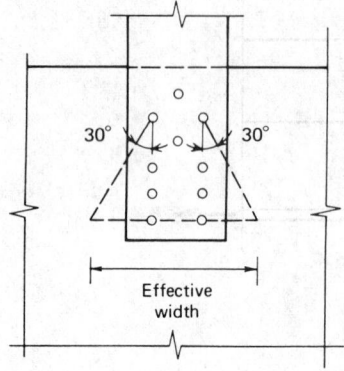

Fig. 24.2 Effective width for a fastener pattern.

Many tests have been conducted on bolt and rivet patterns for maximum efficiency resulting, generally, in the following conclusions:

1. The bolts or rivets in the different rows should be located in tandem, that is, along the load line, and not staggered.
2. The spacing between rows should be the minimum practical.
3. The bolts or rivets in each row should be of the same size.
4. Each row should contain an equal number of bolts or rivets.

An assumption tacit in these conclusions is that each rivet or bolt in a row shares the load equally. Reference 5 shows the calculated distribution for practical shear joints with 4, 10, and 20 bolts in line (Fig. 24.4). The reason for this unequal distribution is the decreasing average shear strength with increasing length—note the negligible effect on joints up to about four fasteners. Also, any local condition that negates the assumption, such as plates of different thickness or stiffness, leads to the possibility of "first" rivet failure and the need for setting aside recommendations 3 and 4 for equal size and number. For patterns using fasteners of unequal size in shear, it is usually considered that the load on the row will be distributed in proportion to the shear strength of each fastener; when the plate stresses are purely elastic, the load distribution is probably more nearly proportional to the fastener bearing area.

Prying Action

A particularly frequent and sometimes subtle origin of failure is that from the increased force from prying on the bolts in a tension group.

One of the simplest connections with the bolt groups in tension is the symmetric T-stub hanger with a single line of fasteners parallel and on each side of the web. The fasteners are assumed to be stressed equally because of the symmetry of the connection. An external tensile load on the connection

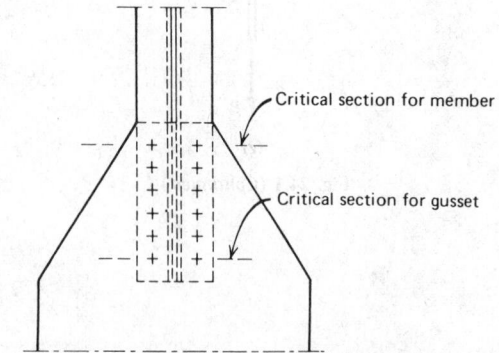

Fig. 24.3 Critical sections for a joint subject to fatigue loading.

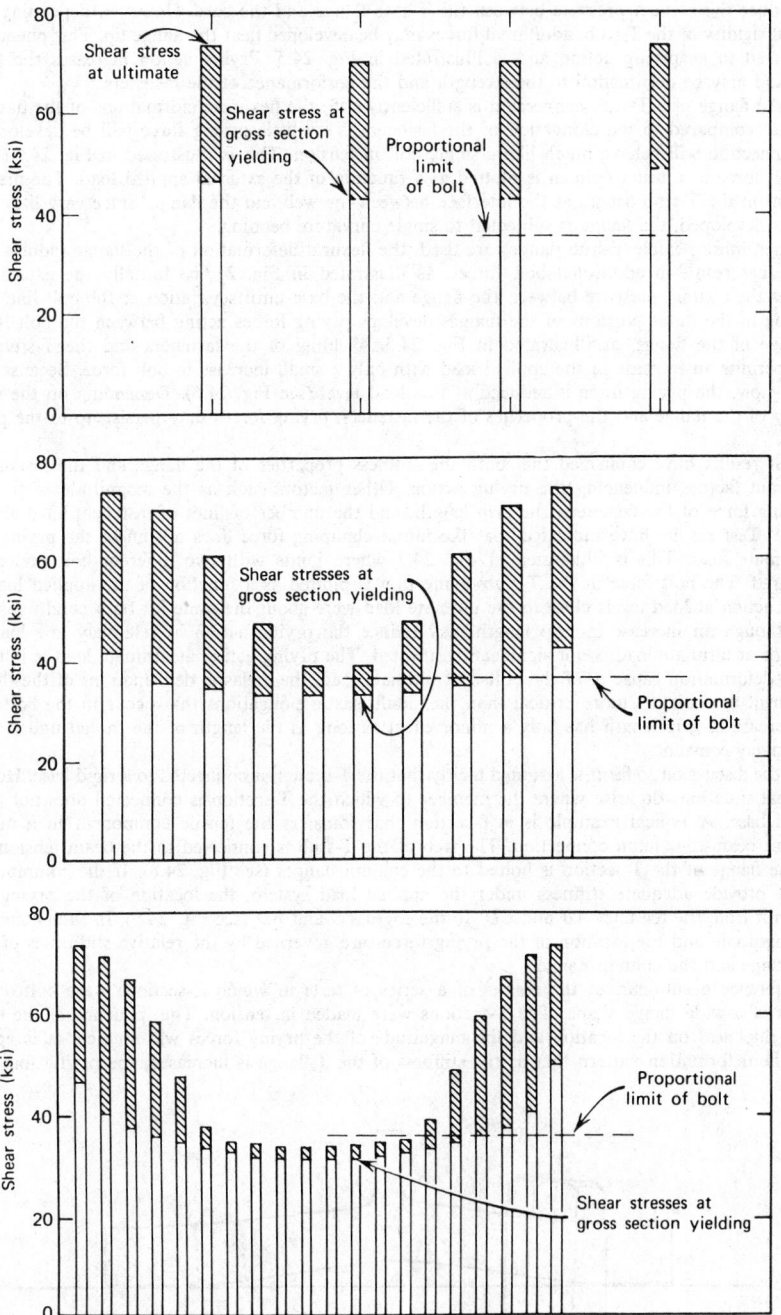

Fig. 24.4 Load distribution in joints with multiple fasteners. Plate material $S_{yp} = 36$ ksi. Bolts are ⅞ in. dia. A325: (a) 4 bolts in line; (b) 10 bolts in line; (c) 20 bolts in line.

will reduce the contact pressure between the T-stub flange and the base. However, depending on the flexural rigidity of the T-stub, additional forces may be developed near the flange tip. This phenomenon is referred to as prying action and is illustrated in Fig. 24.5. Prying action increases the fastener force and may be detrimental to the strength and the performance of the fasteners.

If the flange of a T-stub connection is sufficiently stiff, the flexural deformations of the flange will be small compared to the elongation of the fasteners. Very little prying force will be developed and the connection will behave much like a single bolt in tension. This is illustrated in Fig. 24.6a, where the bolt force in a test specimen is plotted as a function of the external applied load. The maximum moment in the T-stub occurs at the interface between the web and the flange. Since very little prying force is developed, the flange is subjected to single curvature bending.

When more flexible T-stub flanges are used, the flexural deformation of the flange induces prying forces that result in additional bolt forces, as illustrated in Fig. 24.6b. Initially the external load reduces the contact pressure between the flange and the base until separation at the bolt line occurs. Bending in the outer portions of the flanges develops prying forces acting between the bolt line and the edge of the flange, as illustrated in Fig. 24.5. Yielding of the fasteners and the T-stub flange often permits an increase in the applied load with only a small increase in bolt force. Because of this plastic flow, the prying force is reduced at this load level (see Fig. 24.6). Depending on the flexural rigidity of the flange and the properties of the fasteners, prying forces may persist up to the point of failure.

Test results have confirmed that both the stiffness properties of the flange and the fasteners are significant factors influencing the prying action. Other factors such as the magnitude of the initial clamping force of the fasteners, the grip length, and the number of lines of fasteners have also been studied. Test results have indicated that the initial clamping force does not affect the prying action at ultimate load. This is illustrated in Fig. 24.7 where joints with two different bolt preloads are compared. The bolt force in the T-stub connection is plotted as a function of the applied load. The prying action at load levels close to the ultimate load were about the same for both conditions.

Although an increase in grip length may reduce the prying action at relatively low loads, the behavior at ultimate load is not significantly affected. The prying action at ultimate load is influenced by the deformation capacity of the bolts. At ultimate load, the inelastic deformations of the threaded portion of the bolt are more critical than the small elastic elongations that occur in the bolt shank. An increase in grip length has only a minor effect as long as the length of the thread under the nut is relatively constant.

In the discussion so far it is assumed tacitly that the T-section is connected to a rigid base. However, practical situations do arise where the member to which the T-section is connected does not provide a rigid base. A typical example is a T-section that transfers the tensile component in a moment-resistant beam-to-column connection. The web of the T-stub is connected to the beam tension flange and the flange of the T-section is bolted to the column flanges (see Fig. 24.8). If the column flanges do not provide adequate stiffness under the applied load system, the location of the prying forces may shift from the toe lines AB and CD, to the edges AC and BD (see Fig. 24.9). In such connections the magnitude and the location of the prying forces are governed by the relative stiffnesses of the T-stub flange and the column flange.

Reference 6 summarizes the results of a series of tests in which T-sections were bolted to the flanges of a wide flange shape. The T-sections were loaded in tension. The influence of the column flange thickness on the location and the magnitude of the prying forces was studied. It is apparent from the deformation pattern that as the stiffness of the T-flange is increased, the prying forces tend

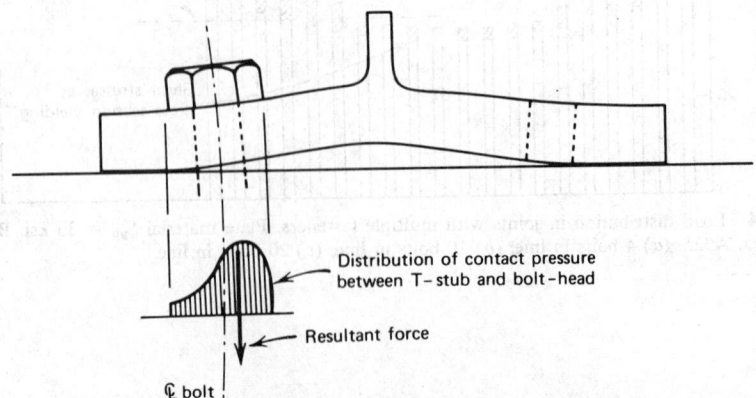

Distribution of contact pressure
between T-stub and bolt-head

Resultant force

℄ bolt

Fig. 24.5 Influence of flange deformations on location of resultant bolt force.

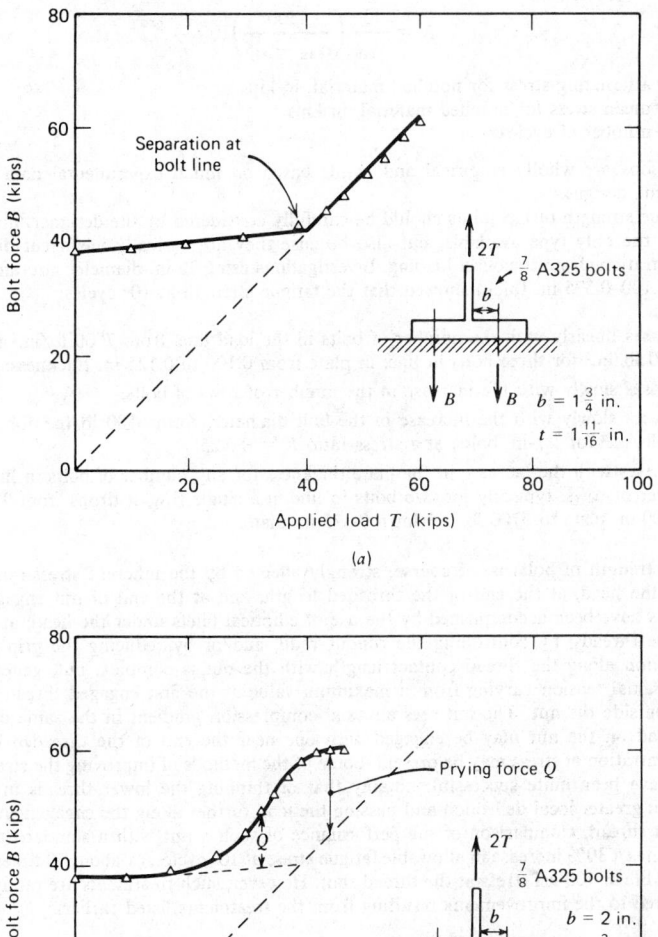

Fig. 24.6 Influence of plate thickness on applied load.

to concentrate in the areas near the corners of the T-section. When the stiffness of the T-stub flange is much greater than the stiffness of the column flange, the T-section provides the rigid base, and prying forces are developed because of deformations of the column flange.

Fatigue Behavior

The general expressions for fatigue behavior are

$$\sigma_{an} = 1.5 + 1500 \left(\frac{13}{N \sigma_{mn}} \right)^{1/2}$$

or

$$N = \frac{13}{\sigma_{mn}} \left(\frac{1500}{\sigma_{an} - 1.5} \right)^2$$

where σ_{an} = alternating stress for notched material, in kips

σ_{mn} = mean stress for notched material, in kips

N = number of cycles

Both expressions are wholly empirical and, while based on much experimental data (Ref. 7), may fail for extreme designs.

The fatigue strength of lap joints should be carefully considered by the designer, not only because they may be the only type available, but also because they are basically inefficient due to the high stress concentrations from the offset loading. Investigations using $\frac{3}{8}$-in.-diameter aircraft bolts in 2024-T3 Alclad (0.100–0.375 in. thick) showed that the fatigue strength at 10^6 cycles:

1. Increases linearly with the number of bolts in the load line: from 7000 lb/in.² for one bolt to 13,000 lb/in.² for three bolts in line; in plate from 0.100 to 0.125 in. thickness.
2. Increases slowly with the increase in the number of rows of bolts.
3. Increases slowly with the increase in the bolt diameter, from 6500 lb/in.² for ¼-in. bolts to 8500 lb/in.² for $\frac{7}{16}$-in. bolts, at a stress ratio $R = +0.25$.
4. Decreases with the increase in the plate thickness for any number of bolts in line and for any number of rows; typically for two bolts in line in a single row, it drops from 9700 lb/in.² for a 0.100-in. plate to 5700 lb/in.² for a 0.375-in. plate.

The fatigue strength of bolts is, of course, strongly affected by the inherent stress-raisers, especially those under the head, at the end of the threaded length, and at the end of nut engagement. Major improvements have been accomplished by the use of elliptical fillets under the head; at thread-runout by rolling the threads; by controlling the runout radii; and/or by reducing the grip diameter. The load distribution along the thread contact length with the nut is complex, but, generally speaking, the bolt sees axial tension varying from a maximum value at the first engaged thread to zero at the first thread outside the nut. The nut sees an axial compression gradient in the same direction. Since the first thread on the nut may be engaged with one near the end of the threaded bolt section, a serious concentration of stress may be present. Some of the methods of improving the stress distribution in the nut have been quite successful, notably that of thinning the lower threads in the nut, thus allowing them greater local deflection and passing the load further along the engagement length, away from the first thread. Comparison of the performance of such a nut with a standard type on a 220-ksi bolt indicates a 30% increase in allowable fatigue stress at 10^5 cycles, or about 50 times improvement in the fatigue life at 100-ksi stress at the thread root. However, such treatments are rarely cost-effective when compared to the improvements resulting from the treatments listed earlier.

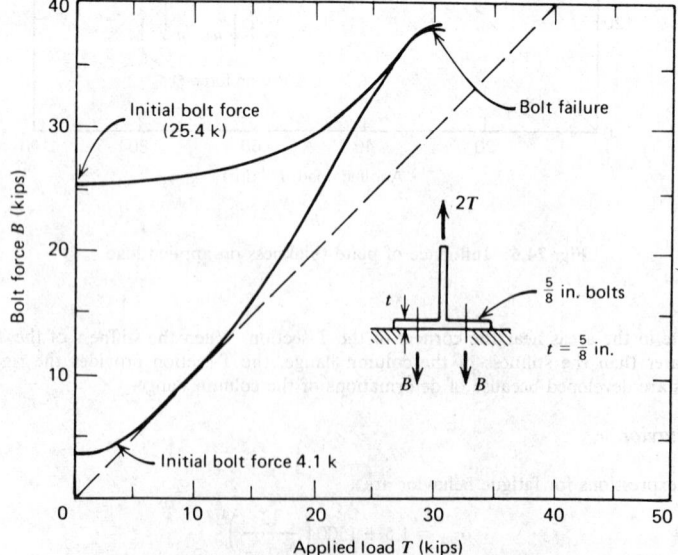

Fig. 24.7 Influence of initial bolt preload on prying action.

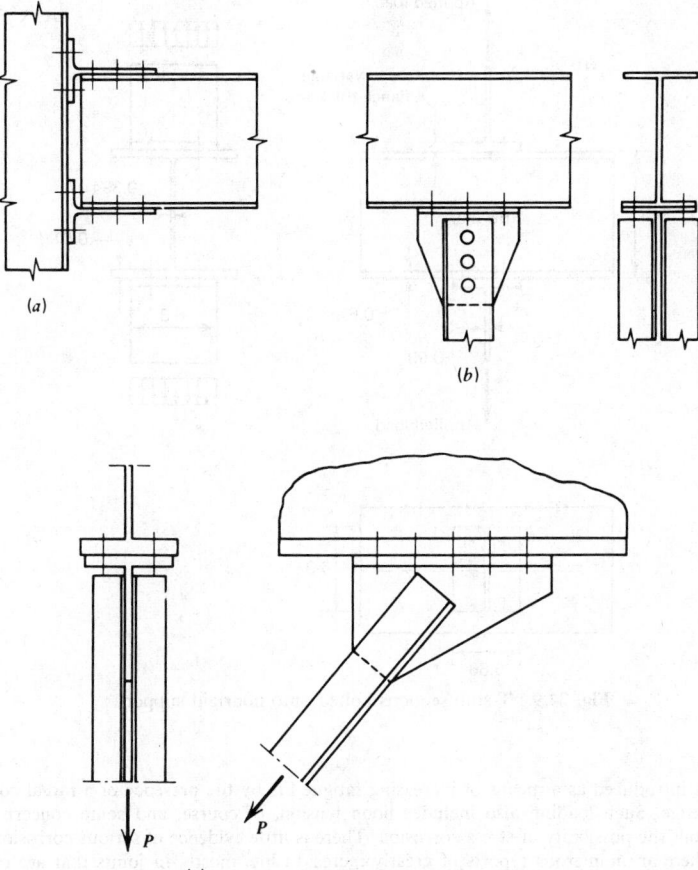

Fig. 24.8 Typical uses of T-type structural connections: (a) beam-to-column connection; (b) hanger connection; (c) diagonal brace connection.

Both bolted and riveted joints inevitably contain holes in the basic sheet or plate; and it has often been observed that a loaded hole may be the most important of all sources of stress concentrations. By the contrasting nature of their geometry, weldments and adhesively bonded joints avoid this particular source of stress concentration. For a single hole in an infinite sheet under tension, the K_t is 3.0, and Fig. 24.10 shows the redistribution, in the sheet adjacent to the hole, of the load that would have been supported by the hole area. It is seen that the peak stress is not much affected by either trimming the sheet to a width of three or four hole diameters or by adding holes in a row at a pitch of three or four hole diameters. Thus, for multiple holes at a normal pitch of three hole diameters, for example, the accumulative effect of the K_t's for each hole will be much less than their simple product and may be determined by superposition. Considering the original hole plus a new hole on the right and one on the left, the peak stress relative to the gross area stress can be approximated by $3(1.02)(1.02)S_t$ $= 3.13S_t$, an increase of only 4% over that for a single hole, although one-third of the sheet has been removed along this section. Had the stress calculation been based on the net section, the value of K_t would have been 27% greater than that for a single hole.

For designs using a hole pitch less than three diameters, the static strength becomes the governing criterion. The fatigue strength is less critical because the low gross-area stress dictated by the static strength greatly reduces the peak stresses. Any bearing stress that may be present will, of course, add to the peak tension stress through a corresponding K_t, but this effect, too, is very highly localized. Thus, for practical design calculations using conventional hole spacings, the peak stress is a function primarily of gross-area stress and the type of hole; the bearing-stress/hole-spacing ratio has little effect on the peak stress.

The foregoing discussion pertains to joints with straight-shank fasteners. Tapered shank bolts have

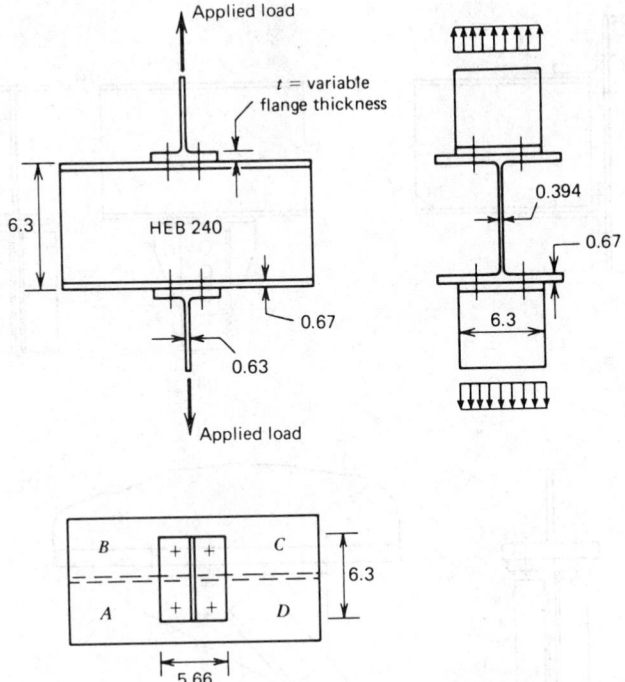

Fig. 24.9   T-stub sections bolted onto nonrigid support.

recently been introduced as a means of increasing fatigue life by the presence of a radial compression in the hole edge. Such loading also includes hoop tension, of course, and some concern had been expressed about the possibility of stress corrosion. There is little evidence of serious corrosion reported as yet, but there are numerous reports of greatly increased life, mostly in joints that are rather high loaded (Ref. 8), in either constant amplitude or spectrum tests.

Tension Bolts. For bolts in tension, with or without bending, the most suitable criterion for failure is separation of the joint, a condition allowing any number of undesirable results such as loss of pressure, excessive distortion, and impact. Separation of the connected members may occur at loads and bolt stresses lower than those for fracture of the bolt, the latter being considered as ultimate, catastrophic failure. Thus a major objective in the design of these joints is to show that the bolt can prevent separation while sustaining the maximum design load for the required number of cycles.

The first requirement for fatigue as well as for static design is the determination of the maximum

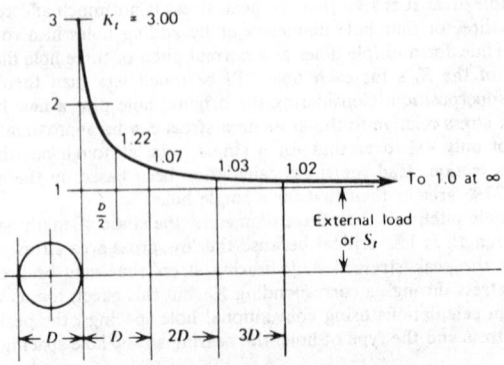

Fig. 24.10   Stress distribution near a hole.

bolt load, a value that has been the subject of much consideration, resulting generally in the correct but only qualitative conclusion that it is minimized by making the bolt more "elastic" than the bolted parts. Thus, spring constants arise as necessary parameters in the analytical considerations, but unfortunately, the k of the bolted parts is usually calculable for only the simplest shapes (i.e., a pair of plates). For complex shapes, such as connecting rod bosses, one must resort to direct tests. However, a general expression for spring constants may be formed as

$$k = F/\delta = AE/L$$

The ratio of the (area) × (modulus) products for the bolt and for the parts

$$(A_b E_b / L_b)/(A_p E_p / L_p)$$

may then be taken as a quantitative description of the joint stiffness. The question of the effective compression area of the plates is considered below.

In establishing an expression for the total load on the bolt, the assembly is considered as a pair of springs in parallel, with the bolt in tension, and compression in the annular volumes of the plates between the head and nut. To maintain equilibrium, the forces in the bolts and plate must be equal and opposite. The deflections at preload would be equal only if the stiffnesses were equal. As the external or working load is applied via the plates, the total load on the bolt increases, the limit to this action being defined as the condition of impending separation. If the applied load is increased beyond that at separation, then the total bolt load will, of course, continue to increase. This action is shown schematically in Fig. 24.11a. During loading between preload and separation, the *changes* in bolt and plate deflection are equal.

The load expressions are obtained for a conventional assembly of multiple plates or flanges by first writing the force equality under preload alone:

$$\text{Tensile force in bolt} = \text{Compressive force in plate(s)} \tag{24.1}$$

$$P_i = C$$

or

$$\frac{P_i L_b}{A_b E_b} = \frac{CL_{p1}}{A_{p1} E_{p1}} + \frac{CL_{p2}}{A_{p2} E_{p2}} + \cdots + \frac{CL_{pn}}{A_{pn} E_{pn}} \tag{24.2}$$

The relations between the preload and the working load at separation are derived from this expression and the condition that the compressive preload and deflection become zero, leading eventually to:

$$F_{\max} = P_i + W \left[\left(\frac{A_b E_b / L_b}{A_p E_p / L_p} \right) \middle/ \left(\frac{A_b E_b / L_b}{A_p E_p / L_p} \right) + 1 \right] \tag{24.3}$$

which is simplified through the use of $r = (A_b E_b / L_b)/(A_p E_p / L_p)$ to

$$F_{\max} = P_i + Wr/(r + 1) \tag{24.3a}$$

Similarly, for the total load in the plates, we get

$$F_p = P_i - W/(r + 1) \tag{24.4}$$

The full derivation is given in Ref 9, and Ref. 10 provides a discussion of the behavior of bolts during torque-up.

In view of these relationships Fig. 24.11b reveals that:

1. Separation occurs only at $W = P_i$ or $W/P_i = 1$ for all values of r. Prior to separation the total bolt load cannot exceed $2P_i$ for any value of r; $F_{\max}$ approaches $2P_i$ as r approaches ∞, at $W/P_i = 1$. Although $W_s = P_i$ at separation, $F_{\max} \neq 2P_i$ because a portion of W_s is being reacted by compression in the plates. Only when $r = \infty$ (or $E_p = 0$) does $F_{\max} = 2P_i$.

2. After separation, $F_{\max}$ increases with both W and r, the latter indicating, quantitatively, lower loads in the lower modulus bolts or lower AE joints.

3. Although the general desirability of "low" bolt loads is not debated, a difficulty should be noted with low-load, low-r joints in that sufficient clamping force, P_i, may be obtainable only with very large bolt elongations.

The above considerations have, through Eq. (24.3a) related the three prime parameters: total bolt load, $F_{\max}$; preload, P_i; and working load, W. A further relation would be highly useful, that between the critical preload, P_c, and the alternating component of W, which is W_a. If the initial preload is made equal to P_c, the force compressing the plates changes between the limits of P_c and 0 during reversal of the working load between 0 and W. But part of W is used up reacting the preload, so that the remainder is the alternating component

$$W_a + P_c = W$$

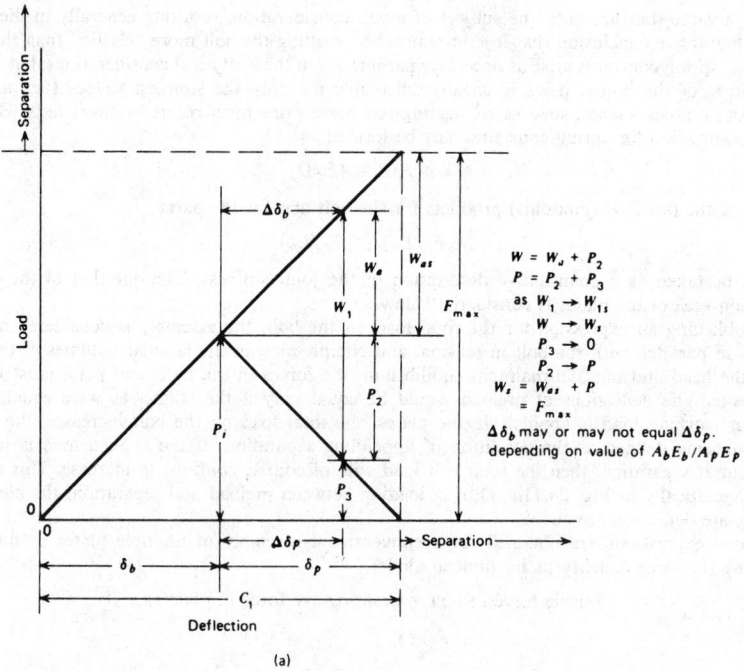

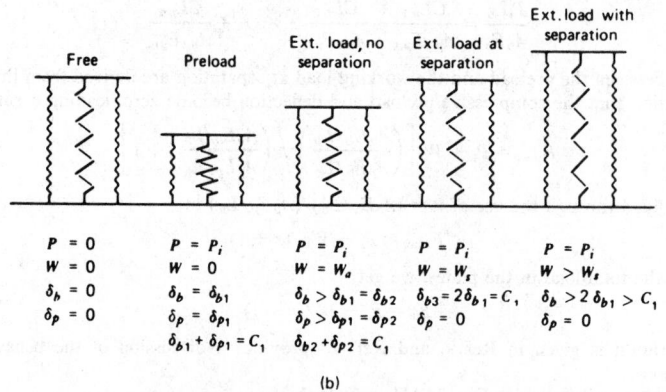

Fig. 24.11 Force–deflection conditions for a bolted joint.

After some manipulation,

$$P_c = W \left(\frac{1}{r+1} \right) \qquad (24.4a)$$

For practical reasons, and to provide a margin of safety, the actual preload is usually made somewhat greater than P_c:

$$P_{act} = cP_c$$

where c is often given a popular engineering factor of safety value. In fatigue design it is imperative to recognize the effect of the range of alternating loads, and joint design is no exception to the rule. A load ratio is defined here as

$$R = \frac{F_{min}}{F_{max}}$$

Then

$$F_{min} = P_c = W\left(\frac{1}{r+1}\right)$$

$$F_{max} = P_c + W_a$$

$$= W\left(\frac{1}{r+1}\right) + \left(\frac{r}{r+1}\right)$$

and Eq. (24.3a) reappears, modified as

$$F_{max} = cW\left(\frac{1}{r+1}\right) + W\left(\frac{r}{r+1}\right) \tag{24.5}$$

In fatigue terms the preload, P or P_c, represents the mean stress, and the variable part of the applied load, W_a, represents the alternating stress. Thus the latter is the more critical and should be decreased, at the expense of increasing P_c whenever possible, the relationship being dependent on the relative stiffnesses: the r value.

In stress terms, preloaded bolts in a joint under cyclic stressing are subject to that combination of steady and alternating stresses characteristic of fatigue designs. Much has been written on the desirability of protecting the bolt from the effect of the alternating component by increasing the mean stress, or preload. A glance at the master life diagrams shows immediately that the slopes of the lifelines are such that, in general, much life can be gained by boosting the mean stress. However, recalling that the major objective of bolted design is to prevent separation, it follows that the mean stress (in the fatigue sense) from the preload must be sufficient to accommodate the maximum working load. Then the total load on the bolt is, as given by Eq. (24.5), composed of the preload mean stress and the working load alternating stress. The suggestion has been made that the effect of the alternating component, S_s, could be canceled by making $P_i = F_{max}$, but this is irrelevant, since S_a arises only from W_a and always adds to P_i leading to a new, higher value of F_{max}. Not only is there no benefit in increasing P_i beyond that value needed to prevent separation, but such action results in a lower fatigue life because of an unnecessarily high mean stress. The preceding argument indicates that there is a proper value of preload for each design condition; insufficient preload allows joint separation with its large increase in the alternating component; too high a preload leads to lowered fatigue life because of the high mean stress.

A major result of this derivation, Eq. (24.3a) is plotted as Fig. 24.12 for practical values of r,

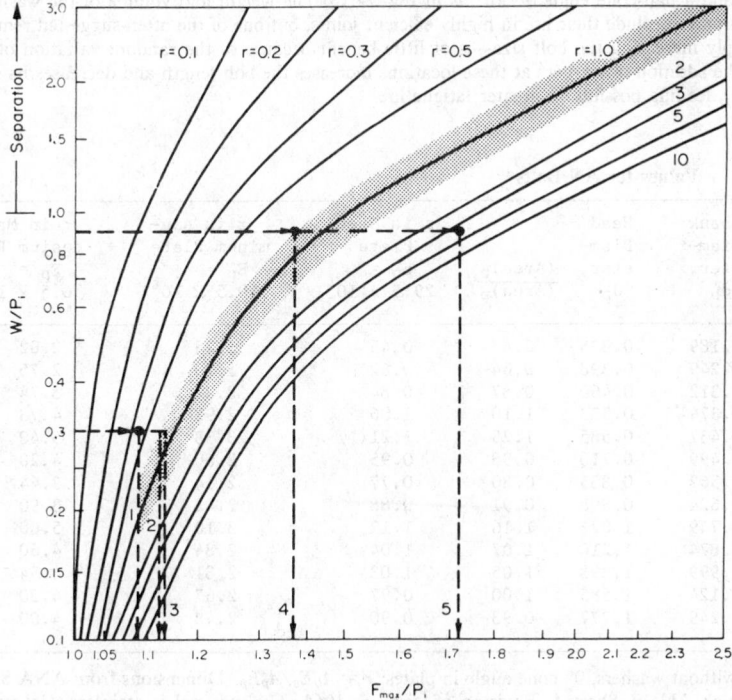

Fig. 24.12 Relationship of total bolt load to the external loads and preloads.

Table 24.1   r Values for NAS Bolts[a]

	Shank Diameter, d_b	Head Diameter, d_p	$\dfrac{\text{(Area)}_b}{\text{(Area)}_p}$	r in Steel Plate $E_p = 29.5 \times 10^6$	r in Aluminum Plate, $E_p = 10.5 \times 10^6$	r in Magnesium Plate, $E_p = 6.5 \times 10^6$
NAS-3	0.189	0.375	0.25	0.24	0.94	1.51
NAS-4	0.249	0.438	0.47	0.45	1.25	2.03
NAS-5	0.312	0.500	0.64	0.62	1.70	2.76
NAS-6	0.374	0.563	0.81	0.78	2.18	3.50
NAS-7	0.437	0.625	0.97	0.94	2.57	4.18
NAS-8	0.499	0.750	0.80	0.77	2.13	3.45
NAS-9	0.562	0.875	0.69	0.67	1.84	2.98
NAS-10	0.624	0.938	0.80	0.77	2.13	3.45
NAS-12	0.749	1.063	0.98	0.95	2.62	4.23
NAS-14	0.874	1.250	0.95	0.92	2.54	4.10
NAS-16	0.999	1.438	0.93	0.90	2.49	4.00
NAS-18	1.124	1.625	0.92	0.89	2.46	3.95
NAS-20	1.249	1.812	0.92	0.89	2.46	3.95

[a] r values without washers, 0° cone angle in plates; $r = A_b E_b / A_p E_p$. Dimensions from NAS-501 Standard Sheet 1, Revision of January 31, 1956. The material is stainless-steel type 321, AMS 5645; $E_b = 28 \times 10^6$ psi.

and is normalized on the preload. A study of this plot, together with Tables 24.1 and 24.2 yield several useful conclusions. The total bolt load, F_{max}, increases sharply with increasing slope, or r value, a condition that must be controlled in attempts to design for any given F_{max}. Note that, of the two basic parameters in the ratio, E is controllable only by choice of material, while A is adjustable by the details of the joint design. For example, $r = 2.6$ for an AN-8 bolt in aluminum plates with no washers, giving an F_{max}/P_i of about 1.7 at separation. This value could be reduced to about 1.4 by bringing r into the range 0.8–1.0 through the use of conical steel washers with a large end diameter of 1.15 in. and a half-cone angle of 30°, as in Fig. 24.13. The weight and volume of the washers and longer bolts may preclude their use in highly efficient joints, but one of the often-suggested remedies—that of simply increasing the bolt size—is of little benefit because of the random variation of r with bolt size. The addition of washers at these locations increases the bolt length and decreases its stiffness ($k = AE/l$), leading possibly to greater fatigue life.

Table 24.2   r Values for AN Bolts[a]

	Shank Diameter, d_b	Head Diameter, d_p	$\dfrac{\text{(Area)}_b}{\text{(Area)}_p}$	r in Steel Plate $E_p = 29.5 \times 10^6$	r in Aluminum Plate $E_p = 10.5 \times 10^6$	r in Magnesium Plate, $E_p = 6.5 \times 10^6$
AN-3	0.189	0.335	0.47	0.45	1.25	2.02
AN-4	0.249	0.398	0.64	0.62	1.70	2.75
AN-5	0.312	0.460	0.87	0.84	2.32	3.74
AN-6	0.374	0.523	1.10	1.06	2.92	4.73
AN-7	0.437	0.585	1.25	1.21	3.33	5.40
AN-8	0.499	0.710	0.98	0.95	2.60	4.20
AN-9	0.562	0.835	0.80	0.77	2.14	3.44
AN-10	0.624	0.898	0.91	0.88	2.42	3.90
AN-12	0.749	1.023	1.16	1.12	3.10	5.00
AN-14	0.874	1.210	1.07	1.04	2.84	4.60
AN-16	0.999	1.398	1.05	1.03	2.81	4.54
AN-18	1.124	1.585	1.00	0.97	2.67	4.30
AN-20	1.249	1.772	0.93	0.90	2.48	4.00

[a] r values without washers, 0° cone angle in plates; $r = A_b E_b / A_p E_p$. Dimensions from ANA Standard AN-3 through AN-20, Sheet 1, Revision of April 6, 1964. The material is stainless-steel type 321, AMS 5645; $E_b = 28 \times 10^6$ psi.

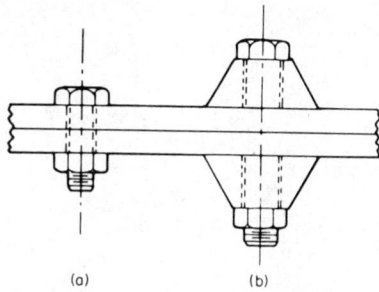

(a)   (b)

Fig. 24.13   Variation of r and F_{max} for steel bolts and washers in aluminum alloy plates: (a) $r = 2.5$, $F_{max}/P_i = 1.7$; (b) $r = 0.85$, $F_{max}/P_i = 1.4$.

More extreme examples may be noted in Tables 24.1 and 24.2, the maximum r value of 5.4 indicating a total bolt load of nearly twice the preload at separation. As expected, the higher r's are associated with the higher values of E_b/E_p; thus, the higher bolt load conditions appear for steel bolts in magnesium plates. The stiffness r varies little with the type of steel in the bolts, but it is, of course, a strong function of the modulus for nonferrous bolts.

Tables 24.1 and 24.2 were developed to show the variation in r with the *size* of standard bolts. The ratio of shank area to annulus area is neither constant nor uniformly varying with bolt size, a condition that leads to wide variation in r and the bolt load. When heavy washers are used to control the bolt load, their outside diameter should be sized to minimize r to the greatest extent practical. The compression area in the plate, A_p, was calculated as that of a cylindrical annulus (i.e., of $0°$ cone angle), proportional to $d_a^2 - d_b^2$. Thus, the tabulated values of r are conservative to a degree dependent on L, the plate thickness. For $L < \frac{1}{8}$ in., the values are essentially as given. A relevant A_p in each design case should be calculated by Eq. (24.6) from the plate thickness.

Example 24.1:   Find F_{max} for Tension Bolts

Consider the bolting for a circular steel header plate on a compressed air-storage tank, the total force on the plate being 8000 lb, and the design objective being to prevent loss of pressure.

1. Assume a number of bolts and a pattern—here take 10 bolts in an 8-in.-dia. circle.
2. Assume a load distribution to determine the maximum working load—here take a uniform distribution, so $W = 800$ lb per bolt.
3. Assume a trial bolt size, and grade or class. Here take NAS-4, SAE Grade 2, for which the maximum allowable preload, or clamping load = 1320 lb (from Table 4 of Ref. 11; or Class T-1 ultimate load = 1837 lb [from Table 8.1.2(b) of Ref. 12].
4. From Table 24.1 obtain the r value for the trial size as 0.45.
5. Choose a value for W/P_i—here try 0.3.
6. Calculate the preload: $P_i = (1/0.3)W = 3(800) = 2400$ lb.
7. Enter the curves of Fig. 24.12 on line 1 at $r = 0.45$ and $W/P_i = 0.3$ to find $F_{max}/P_i = 1.09$.
8. Calculate the total bolt load: $F_{max} = 1.09(2400) = 2620$ lb.

It is immediately seen that F_{max} exceeds the capacity of the trial bolt size and grade; recourse to Refs. 11 and 12 indicates that a Grade 8, or a Grade 5 fine thread, or a Class T-4 would be required. While any of these three choices form a correct solution, questions of cost and availability may make a further trial with larger, low-grade bolts more interesting.

Consider a NAS-5, Grade 2 bolt:

1. As before.
2. As before.
3. Clamping load = 2160 lb. Ultimate load = 2952 lb.
4. From Table 24.1, $r = 0.62$.
5. $W/P_i = 0.3$.
6. Preload = 3(800) = 2400 lb, as before.
7. $F_{max}/P_i = 1.12$ (line 2, Fig. 24.12).
8. $F_{max} = 1.12(2400) = 2700$ lb, which also indicates insufficient capacity.

For a NAS-6, Grade 2 bolt:

1. As before.
2. As before.
3. Clamping load = 3200 lb. Ultimate load = 4530 lb.
4. From Table 24.1, $r = 0.78$.
5. $W/P_i = 0.3$.
6. As before: preload = 3(800) = 2400 lb.
7. $F_{max}/P_i = 1.13$ (from line 3, Fig. 24.12).
8. $F_{max} = 1.13(2400) = 2725$ lb.

Comparison of this 2725 lb load with either the 3200 lb clamping or the 4530 lb ultimate capacities indicates a positive margin.

One of the most important factors affecting the total bolt load is the ratio of working load to preload, W/P_i. In the example so far, the arbitrary value of 0.3 was used. The general trend is that as W/P_i increases, the total load comes down, but by definition, so also does the margin between the working (or external) load and the preload (or clamping force). Thus, use of the more efficiently designed bolts with the higher values of W/P_i requires an increasingly accurate knowledge of both W and P_i (line 4 in Fig. 24.12). This ratio is determined primarily by engineering judgment: in many

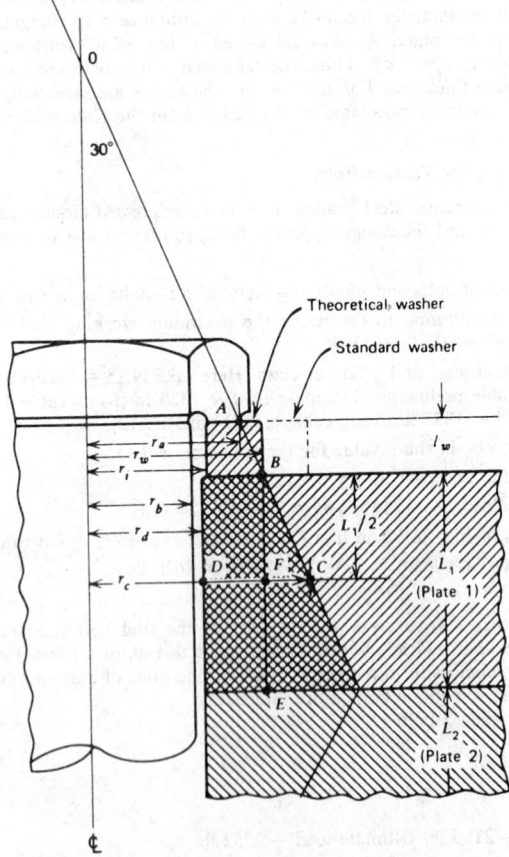

Fig. 24.14 Sizing of washers. *Graphical Construction:* (1) Point A is located at maximum radius of bearing annulus under bolt head. (2) Draw the 30° line through A, establishing points O, B, and C. C is placed at $L_1/2$. (3) Drop the vertical BE. Then: $r_c = r_w = 0.25L_1$ and the compression area, A_p, is the annulus at CD, equal to:

$$A_p = (\pi/4)[(d_w + 0.50L_1)^2 - d_d^2].$$

instances it would appear desirable to build in the safety of a wide margin between the external load and the preload (low W/P_i), but this condition is obtainable only at relatively higher total loads.

As noted earlier, joint stiffness has a pronounced effect on total load. Consider the same example again, but with magnesium alloy plates instead of steel. Now, from Table 24.1 the r value is 3.45, leading to higher F_{max} for all W/P_i; for the more efficient joints, the change in plate material raises F_{max} by about 50%; see line 5 in Fig. 24.12.

In the steel bolt–steel plate combination, the r value increases with increasing bolt size (in the range considered); thus, one would expect the total load to go up, but the bolt capacity fortunately increases faster with size than does the r value. This condition may not hold for all sizes; examination of Tables 24.1 and 24.2 reveals much randomness, especially for the AN group. For joint designs using bolts of other standards, the r values should be calculated for each size and material combination.

The sizing of washers to ensure that a sufficient compression area of the plate is brought into action may be done with reference to Fig. 24.14. If we assume that plate thickness $= L_1$, bolt shank radius $= r_b$, head bearing radius $= r_a$, hole radius $= r_d$, and r_c represents the radius of the effective compression area, then it follows that the washer outer radius, r_w, and thickness, l_w, must be such as to place points B and C on the line OA extended. The compression area is properly taken as the annulus at C–D, since the results of studies in photoelasticity show that the volume in compression takes on the profile of a truncated cone with an apex angle between 25° and 33° depending on plate material, thickness, and stress level, and that the maximum stress appears near the central plane of the plate. For simplicity in the construction of Fig. 24.14, BF is made equal to $L_1/2$ and the angle is taken as 30°, from which a general expression for the compression area is derived as

$$A_p = \pi/4[(d_w + 0.50L_1)^2 - d_d^2] \tag{24.6}$$

for d_b/L_1 in the range 0.68–1.4. Recent work[13] indicates that the angle is probably greater than 30°, thus making the use of Eq. (24.6) somewhat conservative.

A highly useful relationship between thickness and effective radius in flanges or plates is given in Fig. 24.15 for several bolt sizes. It can also be seen that the washer might be decreased considerably from the proportions shown before any bearing area under the head is lost, but this procedure often leads to failure by local crushing. It is important to develop the full areas in compression at both washer interfaces to avoid this type of failure.

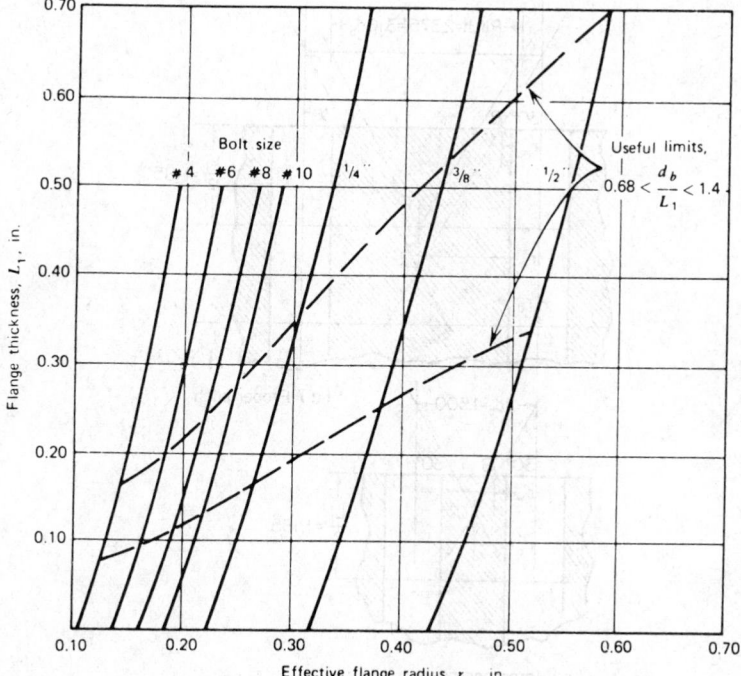

Fig. 24.15 Thickness versus effective radius in bolted flanges.

Example 24.2

A Class T-8, AN-6 steel bolt is applied to ½-in.-thick plates of 2024-T851. From Ref. 12, Table 8.1.2(b), the ultimate tensile load on this bolt is 15,100 lb. Determine the (steel) washer dimensions necessary to develop the ultimate compressive stress in the plates for a factor of safety of 1.5 on the load. The allowable load then is $15,100/1.5 = 10,000$ lb. The ultimate compressive stress, F_c, for 2024-T851 is 60,000 psi from Table 3.2.3.0(b) of Ref. 12. The minimum compression area to prevent yielding in the plate is then $A_p = 10,000/60,000 = 0.16$ in.^{2}. First check whether any washer is needed. The area of the bearing annulus under the bolt head is $\pi/4[(0.523)^2 - (0.374)^2] = 0.102$ in.^{2}; the dimensions are from Table 24.2. Since this area is less than A_p, a washer is needed to prevent local crushing. As a trial washer size, take $l_w = 0.063$, then $r_w - r_a = l_w \tan 30° = 0.036$ in. Now A_p becomes $\pi/4[(0.595)^2 - (0.374)^2] = 0.17$ in.^{2}, sufficiently greater than the 0.16 in.^{2} required.

The nearest standard size plain washer is a type B, narrow, 0.406 in. inside diameter; $^{47}\!\!/_{64}$ in. outside diameter; 0.963 in. nominal thickness (from ANSI B27.2-1958). Applying these dimensions to Fig. 24.14, d_c turns out to be 0.978 in., and d_i is 0.406 in., from which the plate compression area A_c is 0.61 in.^{2}.

This relatively large value for A_c just means that the maximum compressive stress in the plate will be 0.16/0.61, or about one-quarter of the plate yield strength, thus allowing for a stress concentration factor up to 4, a more than ample margin. The fact that the standard washer has a considerably greater outside diameter than that required is neglected—it would be unduly conservative to calculate A_p on a 30° line struck from the standard washer outside diameter.

Checking the effect of this washer on the ratio of F_{max}/P_i: from Table 24.2, r without any washer was 2.92; for these dimensions, r changes to

$$r = \frac{\pi/4(0.374)^2(29 \times 10^6)}{0.61(10.5 \times 10^6)} = 0.55$$

From Fig. 24.14, for $r = 2.92$ and a W/P_i somewhat less than 1, say 0.8, $F_{max}/P_i = 1.61$. At the same W/P_i and $r = 0.55$, F_{max}/P_i drops to 1.23, indicating that, as usual, the use of a properly

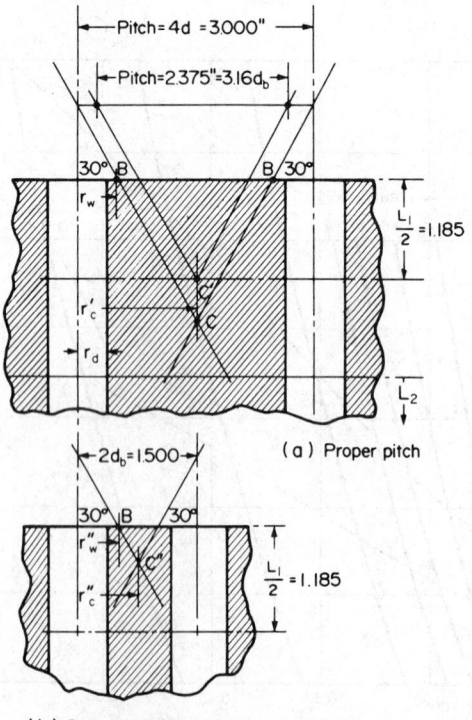

(a) Proper pitch

(b) Improper pitch

Fig. 24.16 Bolt-pitch considerations.

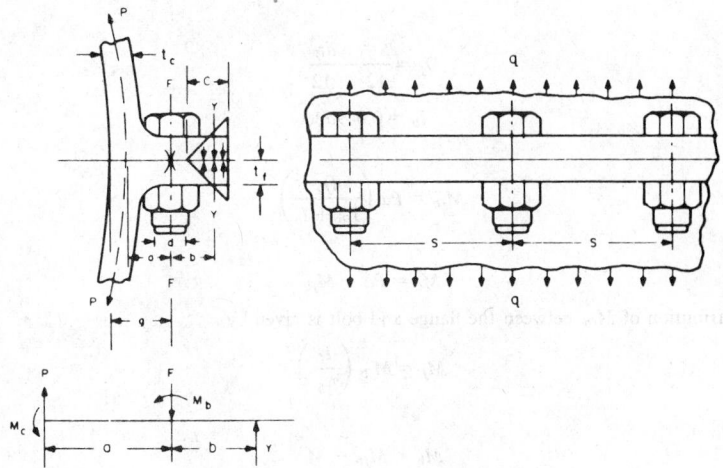

Fig. 24.17 Typical bolted flange.

sized washer enhances the fatigue life of the bolt by decreasing the r value of the joint, making the bolt "more elastic."

In the derivation above for A_p, [Eq. (24.6)], the pitch or bolt spacing was assumed to be "proper," that is, in the approximate range of 2.5–5 times the bolt diameter. A brief procedure for finding the most efficient pitch is given in Fig. 24.16a. Here the pitch was originally chosen at $4d$ for ¾-in.-diam. bolts and, in laying out, the intersection point "C" fell considerably off the midplane of the plate. Projection of C to this plane as point C', and thence along the 30° lines, results in a spacing of 2.375 in. (measured) or a pitch of $3.16d$. This value could be rounded off to $3d$ for convenience. The numerical dimensions were arbitrarily chosen.

Figure 24.16b illustrates the danger of too small a pitch; point C'' approaches B with consequent overstressing in the region of the hole edge. Both the short-column and shear-out forces are active in the same region. Comparing the compression stress for a proper pitch of $3.16d$ with that at a pitch of $2d$ indicates an increase by the factor of $(r_c'/r_c'')^2$, or for the dimensions shown $(1.5/0.75)^2 = 4$. Thus, the ever-present problem of shear-out or bearing failure is easily compounded by choosing too low a value for the pitch. Although the compressive and bearing stresses always add, it is probably adequate to use a pitch such that C' does not depart from the midplane of the plate by more than $\pm 25\%$ of $L_1/2$. Considerations of pitch should also include the ratio of bolt diameter to plate thickness, a practical range being from 1.0 down to about 0.25 min.

For types of threaded fasteners other than conventional, headed bolts, the variation in r due to different design dimensions may be rather wide, sometimes beneficial. For example, a typical blind bolt assembly of nominal diameter ⅜ in. (core bolt diameter ¼ in.) used in aluminum plates has an r value of only 45% of that for a ¼-in. AN bolt, leading to a reduction in bolt load of about 30% for the same preload. Recent designs of very-high-strength bolts have tended to show smaller heads; but some designs are of the washer type, which means that they have a larger area of the annulus under the head, resulting in a reduction of both r and the bolt load. For proper control of the bolt load, an r value should be found for every *head size* in combination with the bolt and plate materials.

The problem of finding the maximum bolt load for an actual case is illustrated by a calculation for bolts in the split-line flange of an axial-flow compressor casing. The loading here involves both direct tension and bending. Figure 24.17 shows typical sections and gives the nomenclature; the term "flange" has the same meaning as the earlier term "plate," $E_f = E_p$, and so forth.

In the calculation of the moment distribution between the flange and case, it is considered that the flange and bolts act together, that is, undergo the same bending deflection, and the conservative assumption made is that the maximum moment occurs at the bolt. The moment Pa is distributed between the flange and case in proportion to their moments of inertia, thus:

$$M = Pa = \text{total moment} \tag{24.7}$$

$$I_c = \frac{St_c^3}{12} \tag{24.8}$$

and

$$I_{fb} = I_f + I_b \tag{24.9}$$

where

$$I_f = \frac{St_f^3}{12} - \frac{dt_f^3}{12} \tag{24.10}$$

$$I_b = 0.049d_2^4 \tag{24.11}$$

then

$$M_{fb} = Pa \left(\frac{I_{fb}}{I_{fb} + I_c} \right) \tag{24.12}$$

and

$$M_c = Pa - M_{fb} \tag{24.13}$$

The distribution of M_{fb} between the flange and bolt is given by

$$M_f = M_{fb} \left(\frac{I_f}{I_{fb}} \right) \tag{24.14}$$

and

$$M_b = M_{fb} - M_f \tag{24.15}$$

Taking moments about Y, we obtain

$$M_y = P(a + b) - T_b b - M_c - M_b = 0 \tag{24.16}$$

from which

$$F = \frac{P(a + b) - M_c - M_b}{b} \tag{24.17}$$

The total static bolt stress is the sum of the tension and bending stresses:

$$f_{bs} = \frac{4F}{\pi d_2^2} + \frac{M_b d_2}{2I_b} \tag{24.18}$$

Using AN-5 bolts and the dimensions of the compressor section of a typical aircraft turbine, we obtain a static stress because of the internal pressure of $f_{vs} = 11,500$ psi, compared to a yield strength of 100,000 psi. If detailed calculations involve a factor of safety, it would be applied in the load analysis on the P term in Eq. (24.7), a value of 1.15 being commonly used where yielding is the final failure mode.

Before examining the alternating component, the preload and total bolt load are found. Equation (24.19) is an expression for W/P_i, at separation, for this loading condition of tension plus bending:

$$\frac{W}{P_i} = \frac{1/E_f A_f + [(g - a)(0 - a)]/[E_f(I_b + I_f)]}{1/(E_f A_f + E_b A_b) - [(a - t_c/2)(0 - 2a)]/[E_f(I_b + I_f)]} \tag{24.19}$$

The substitution of typical dimensions for steel bolts in steel flanges yields $W/P_i \sim 0.3$. The working load W corresponding to the static stress of Eq. (24.18) is 600 lb; the approximation above indicates that the preload must be $1/0.3$ times this value. The total bolt load F_{max} is found by the following procedure:

1. Obtain r for an AN-5 bolt in steel plates as 0.84, from Table 24.2.
2. Enter the curves, Fig. 24.12, at $W/P_i = 0.3$ and $r = 0.84$ to find $F_{max}/P_i = 1.2$.
3. $P_i = (1/0.3)W = 3(600) = 1800$ lb.
4. Then $F_{max} = 1.2(1800) = 2160$ lb.

The latter discussion might have been left in stress terms and in Fig. 24.12 the stress could be read for load.

The working load on the bolts in this example has been treated solely as a static load arising from the internal pressure, its cyclic nature from the turbine start–stop cycle being neglected. The chief source of the alternating component of load comes from the residual imbalance in the rotor at the once-around frequency of 12,000 rpm or 200 Hz. The load paths are sufficiently complex so that calculation of this component is impractical; resort has been made to direct experiment. Strain gage data indicate that the peak amplitude of this cyclic force is about 120 lb, or $S_a = 2300$ psi. This stress is added to the value of f_{bs} above for a total working stress of 13,800 psi or $W = 720$ lb.

Now the appropriate steps in the above sequence are repeated to obtain the total bolt load:

3. $P_i = 3W = 3(720) = 2160$ lb. corresponding to a mean stress of 41,500 psi.
4. $F_{max} = 1.2P_i = 1.2(2160) = 2600$ lb, corresponding to a peak stress of 50,000 psi.

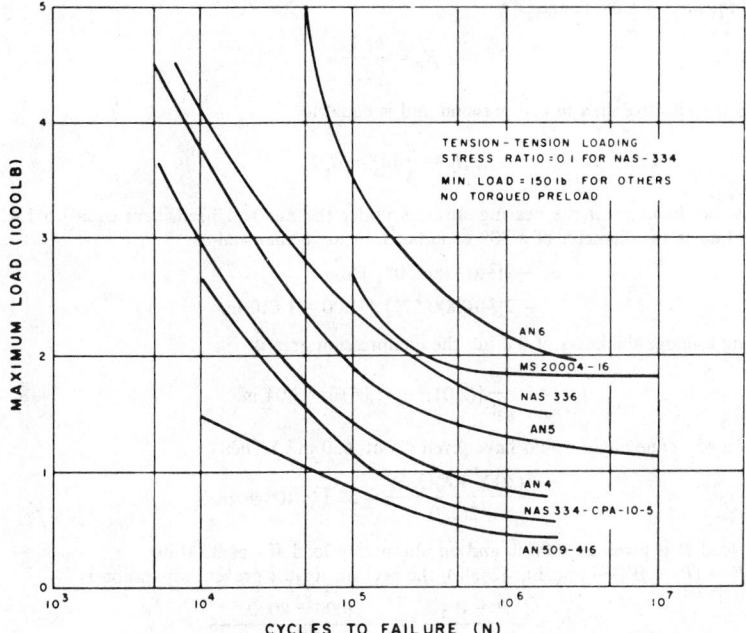

Fig. 24.18 S–N curves for various bolt–nut assemblies.

An approximate bolt life may be determined by entering a Master Diagram for the bolt material with S_a and the new value of P_i as the mean stress. Care should be taken to use a diagram for material with the proper stress concentration factor; if curves for smooth material only are available, the K_t should be calculated and applied to S_a.

A somewhat more accurate, but still not completely rigorous, method is the use of curves for bolted assemblies, which comprise at least the bolt and its nut, Fig. 24.18. Here a life of about 1×10^5 cycles is indicated, but unfortunately the curves were obtained with various low stress ratios, below about $S_{min}/S_{max} = 0.2$, while the data above give a value of 0.8 approximately (41,500/50,000). From the life diagrams of other alloy steels of the same ultimate strength, it is known that an increase in the stress ratio from 0.2 to 0.8 (at the same mean stress) would increase the life by about one order of magnitude, leading to an approximate life of the bolts in this example of 1.0×10^6 cycles. Since a practical life requirement is 20×10^6, it is concluded that these particular bolts are overstressed, but it is seen that a reduction of only 600 or 700 lb in F_{max} would yield this life. Despite the extensive and useful increase in available information on fatigue, very often the designer is still unable to find just what he needs. He should have curves that apply to the particular problem with the data based on:

1. Same material.
2. Same strength, ultimate or yield.
3. Same shapes, that is, the same stress concentration factors.
4. Same stress ratio.
5. Same test temperature.

Figure 24.18 represented the best data available for this problem, but it requires further information to account for the mismatch in item 4.

Another flange example is given to illustrate the calculation method; this example uses spring constants directly. Consider a pair of standard steel pipe flanges of equal thickness, 0.8 in., with no gasket. The proposed bolting consists of four NAS-16 steel bolts, shank diameter = 0.8 in. = D_s. $A_b = 0.502$ in.^{2}; basic major thread diameter = 1 in. = D, minor diameter = 0.846 in. = D_k; and thread root area = 0.562 in.^{2} = A_k. The spring constant for the bolt is then

$$k_b = \frac{E_b A_b}{L} = \frac{(30 \times 10^6)(0.502)}{2(0.8)} = 9.43 \times 10^6 \text{ lb/in.}$$

The spring constant for the flange is

$$K_p = \frac{E_p A_p}{L}$$

where A_p is the effective area in compression and is equal to

$$A_p = \frac{\pi}{4}(d_p^2 - d_h^2)$$

where d_h is the diameter of the bearing surfaces under the nut and head, here equal to 1.4 in. (no washer), and d_p is the diameter of a 30° cone taken at its median radius:

$$d_p = 2[\tfrac{2}{3}(t)\tan 30°] + d_h$$
$$= 2[\tfrac{2}{3}(0.8)0.577] + 1.00 = 1.610 \text{ in.}$$

Assuming a flange thickness of 0.8 in., the compression area is

$$A_p = \frac{\pi}{4}[(2.01)^2 - (1.00)^2] = 3.04 \text{ in.}^2$$

(The use of a 45° cone angle would have given about 3.30 in.^{2}.) Then

$$k_p = \frac{(30 \times 10^6)(3.04)}{2(0.8)} = 57.1 \times 10^6 \text{ lb/in.}$$

The steady load P is given as 1000 lb and an alternating load W_1 as 8000 lb.
From $W = (P + W_1)/4$ and Eq. (24.4a), the preload to just prevent separation is

$$P_c = \frac{P + W_1}{4[1/(r + 1)]} = \frac{1000 + 8000}{4[1/(0.502/3.04 + 1)]}$$
$$= 2200 \text{ lb per bolt}$$

The additional alternate force, W_a, is, from Eq. (24.4a),

$$W_a = \frac{W_1}{4[r/(r + 1)]} = \frac{8000}{4(0.17/1.17)}$$
$$= 290 \text{ lb per bolt}$$

and, similarly, the additional constant force is

$$P_a = \frac{P}{4[r/(r + 1)]} = \frac{1000}{4(0.17/1.17)}$$
$$= 36 \text{ lb per bolt}$$

The highest stresses will occur at the minimum section or at the thread root, where the area in tension is A_k, and

$$S_{\min} = \frac{P_c + P_a - W_a}{A_k} = \frac{2220 + 36 - 290}{0.562} = 3500 \text{ psi}$$

$$S_{\max} = \frac{P_c + P_a + W_a}{A_k} = \frac{2562}{0.562} = 4570 \text{ psi}$$

The stress ratio R is $S_{\min}/S_{\max} = 3500/4570 = 0.755$. The mean stress is then $(S_{\max} + S_{\min})/2 = 4010$ psi, and the alternating stress is $(S_{\max} - S_{\min})/2 = 560$ psi, both in tension.
The shear stress developed by the preload torque is

$$S_s = 16M/\pi D_k^3$$

where $M = \beta P_c D$ and β is the coefficient of friction, taken as 0.15. Then $M = 0.15(2220)(1.00) = 332$ in.-lb, and $S_s = 16(332)/\pi(0.846)^3 = 2800$ psi. The preload shear stress then combines with the $S_{\max}$ above to give the maximum resultant tension:

$$S_{t\,\max} = \tfrac{1}{2}(S_{\max} + \sqrt{S_{\max}^2 + 4S_s^2})$$
$$= \tfrac{1}{2}[4570 + \sqrt{(4570)^2 + 4(2800)^2}] = 5530 \text{ psi}$$

Thus the benefit of a small back-off, say 5°, after torquing to preload is readily observed. (If S_s tends to zero, $S_{t\,\max}$ drops about 20% to 4570 psi.)

Shear-Tension and Shear-Bearing Joints. Lap joints and butts with cover plates have long been designed on the basis of shear in the fastener and bearing in the plates as the major modes of loading. Particularly with rivets, however, the axial tension or clamping force is variable from unit to unit and is difficult to determine. During recent years the results of observation and of many investigations have been leading to a consensus that the highly clamped joint was superior in both static and fatigue loading. This conclusion is, of course, generally in agreement with the analytical knowledge that the stress concentration factors for holes in plates are fairly high, $K_t \sim 2$ or 3 for round holes, and tending very high as the hole becomes elliptical, a K_t of 25 being possible.

The clamping force has a dual role: (1) to develop frictional resistance to slip on the contact planes of the joint and (2) to reduce the stress concentration at the edge of the hole. Any designed increase in the tension/shear ratio is generally beneficial to fatigue life (see the discussion below for Fig. 24.19). However, the effect of clamping force on stress concentrations is complex, a major result being the transfer of the concentration from the hole edge to the washer edge. In joints designed with high ratios of tension to shear, failures have been observed to start in the plate at the edge of the washer. Thus the details of washer design take on an added importance; particularly, the use of sharp-edged, carburized washers is to be avoided. Properly profiled washers should hold the stress concentration factors in the plates near the theoretical minimums of 2 or 3; and with sufficient clamping force to carry the load by friction, the fatigue life of such joints will be relatively long. Coefficients of friction up to 0.35 had been observed on joint surfaces not specially prepared, which, together with the high-strength bolting now available (F_{tu} up to 300 ksi), make a tension-friction joint an achievable design goal. This general benefit from the clamping force has led to more extensive consideration of tension bolts for joints originally designed as shear-bearing types.

For those joints whose overall restrictions do not permit the elimination of shear bearing, the bolts may be designed for combined loading in tension and shear by an interaction equation of the form

$$R^3 + R^2 = 1$$

or

$$\frac{x^3}{a^3} + \frac{y^2}{b^2} = 1$$

where x = shear load
  y = tensile load
  a = shear allowable load
  b = tensile allowable load

and the curves of Fig. 24.19. Starting with the known shear x and tension y loads, point a is located. The extrapolation of a line from the origin through a locates b on the next outer (larger) bolt curve, determining the P_s and P_t allowable loads for that bolt. Then the margin of safety is

$$MS = \left(\frac{P_s}{x}\right) - 1$$

or

$$MS = \left(\frac{P_t}{y}\right) - 1$$

whichever is the smaller. Plots for other types of bolts are easily constructed; the extrapolation procedure is discussed more fully in Ref. 12. The older elliptical form $x^2/a^2 + y^2/b^2 = 1$ has been superseded by that above; the latter is recommended by Ref. 12. In both expressions, however, the allowable stresses are for static loading, while the requirement is for fatigue allowables. These values are derived, somewhat approximately, by replotting S–N information in the form of Fig. 24.19. The approximation arises because of the absence of corresponding tension-fatigue and shear-fatigue data on specific bolts. It is reasonably conservative, however, to lower the shear allowable as a function of cycles in greater proportion than the tension allowable, because of the much greater increase in stress concentration factors for shear, as mentioned above.

The effect of varying the ratio of areas in tension and in shear is illustrated in Fig. 24.20, using A325 and A354 bolts. These data, typical of the structural steels, follow the interaction equation with the cubic terms very well, but it should be noted that, for those slightly less than optimum joints with threads in the shear plane, the curves trend more closely to the unmodified elliptic form. On this plot, the lines for the various ratios of tension shear area provide information corresponding to those lines for the ratio of S_{min}/S_{max} on the constant life diagrams.

The results of many investigations of joints in the heavier structural steels indicate that, if the stresses are such as to allow survival for approximately 2×10^6 cycles, the life of the joint may be considered as indefinitely long.

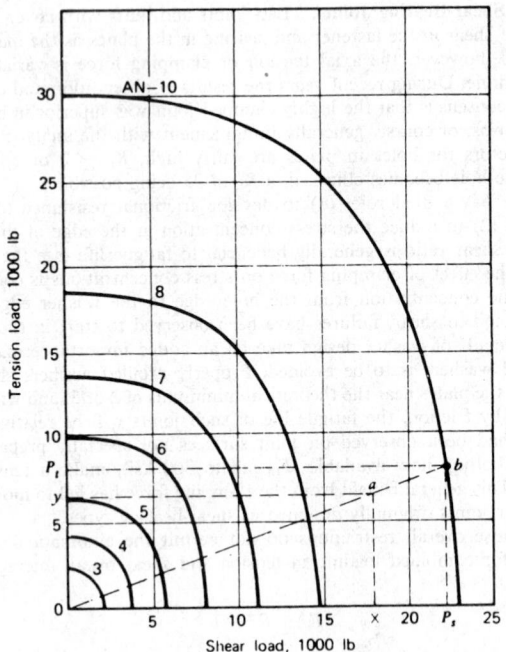

Fig. 24.19 Combined static shear and tension loading for AN steel bolts. ($F_{tu} = 125$ ksi, $F_{su} = 75$ ksi.)

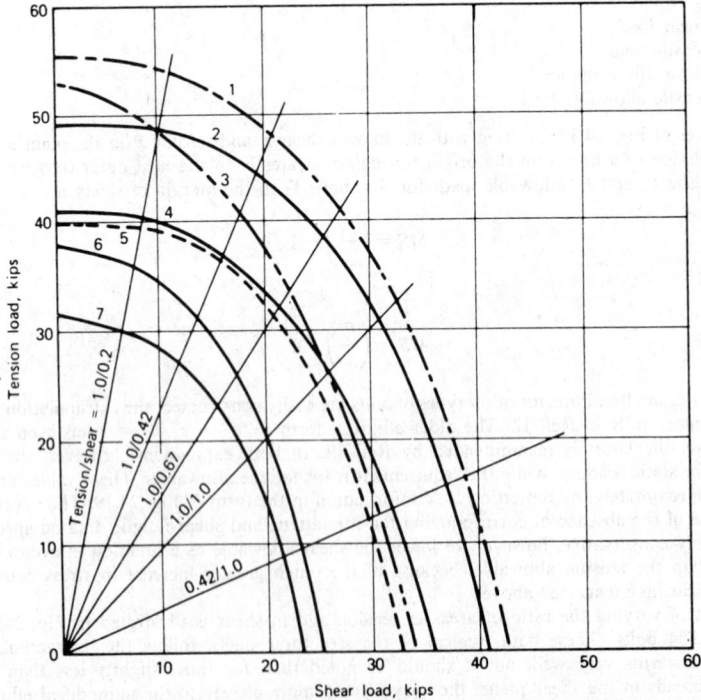

Fig. 24.20 Interaction curces for A325 and A354-BD bolts. (1) A354-BD bolt in T1 plates, shank in shear plane; (2) A325-HH bolts in A36 plates, shank in shear plane; (3) A354-BD, A242 plates, threads in shear plane; (4) A325, A36 plates, shank in shear plane; (5) A325 in T1 plates, shank in shear plane; (6) A325 in A36 plates, threads in shear plane; (7) rivet average. Nominal diam. = ¾ in.

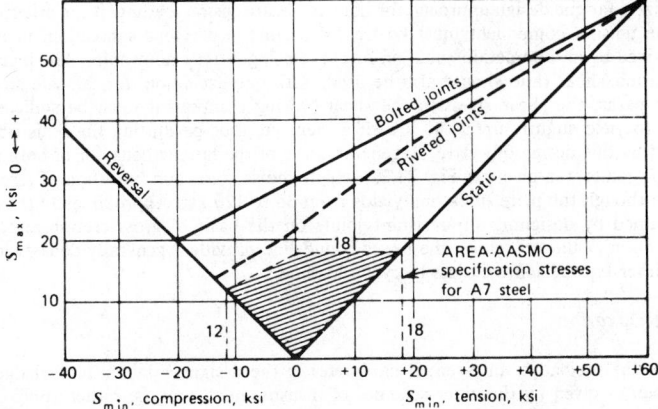

Fig. 24.21 Modified Goodman diagram for A325 and A354 BD bolts at 2×10^6 cycles.

Upon this basis, a modified Goodman diagram, Fig. 24.21, has been constructed for the joints of Fig. 24.20. A comparison of these results with design specifications for A36 steel shows that the bolted joints provided a fatigue life of 2×10^6 or more at stresses approximately 50% greater than those permitted by the specifications, thus the plot is very conservative.

In regard to the effect of clamping force or bolt tension on the fatigue life of these joints, measurements of bolt elongation before and at various intervals during the test showed that the joints providing a life of 2×10^6 cycles or more had a loss of elongation of less than 20%. For the group of failed joints, the loss of elongation increased very sharply as the life decreased. These conditions combined with the fact that the fatigue interaction curves for A325 and A354 bolts must appear similar to those of Fig. 24.20 sharply illustrate the benefits of using the highest practical ratio of tension to shear, in terms of force, stress, or area, and Fig. 24.22 provides a general performance summary.

Many designs of shear joints have been developed: plain and double-scarfed, serrated, keyed, stepped, etc. Test results combined with economic considerations indicate that, generally, the single shear joint with its very high K_t of 13 is unacceptable, at least for high-performance applications; the double-shear type at $K_t = 4.1$ is probably the best all-around choice, and the double-scarf type at $K_t = 3.2$ is rarely worth the effort. Many of the others are unsuitable, for both their relatively high K_t's and the expense of their manufacture.

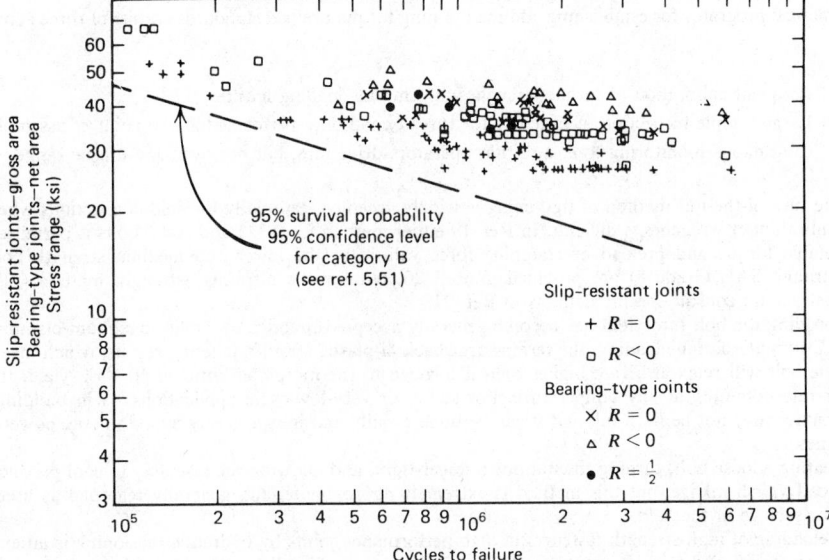

Fig. 24.22 Fatigue behavior of bolted joints.

To summarize a fatigue design approach for bolts in shear-tension loading: if the criterion of separation applies, the tension component must be treated as in the previous subsection; in all cases, the bolt should be sized by an interaction equation or curve using fatigue allowables. The highest practical value of the tension/shear ratio should also be used, with one exception: for fail-safe structures with redundant load paths, the shear stress in the shear-bearing component must be sufficiently high to allow the joint to yield slightly upon the loss of a member, thus permitting the redistribution of the load as required by this design objective. A general curve of the fatigue behavior of both slip-resistant and bearing-type joints was given as Fig. 24.22, which shows some concentration of stresses between 25 and 40 ksi, although the plate stock had yields from 36 to 120 ksi. Approximately the same fatigue strength is obtained by designing slip-resistant joints on the basis of gross section and bearing-type joints on net section. Values taken on the lower-bound line provide a generally conservative basis for the design of either type of joint subject to cyclic loading.

Preload and Relaxation

As noted above, the preload is an essential parameter in the design of joints. Knowledge of the "as-assembled" value, as given by the older schemes of measurement, may be rather approximate, while the newer ones can provide improved accuracy at an increased cost. The general *reliability* dictates the level of accuracy needed: pressure vessel headers and high-speed machinery often require hydraulic tensioners and microwave transducers, while domestic appliances can be handled with an air wrench at a given pressure. Production rates and levels usually set the degree of use of robots and automation, but considerable human judgment and monitoring are required, except for the simplest operations in very high quantities.

Probably the most accurate method of determining actual preload is by measurement of the elongation in the bolt with a micrometer, but this procedure may not always be possible because of inaccessibility, and is clumsy and time-consuming at best. Bolts with built-in elongation indicators are available[14]; various deformable or crushable devices, usually in washer form, have been designed,[15,16] but in most cases an additional gage and measurement are required. Break-off portions of threaded nuts have been sized such that the driving torque develops the correct preload at the ultimate shear stress on this section.[17] Pull-and-swage-type serrated fasteners use a tension break-off load to control the preload.[18] This type has some advantage in fatigue loading in that the soft "nut" or collar is swaged around the harder "bolt" or pin serrations, developing a somewhat more uniform distribution of working load along the nut length.

Tightening to preload by means of a torque wrench is perhaps the simplest and cheapest but is apt to be rather inaccurate. The inevitable variation in friction from one set of threads to the next, plus the random presence of lubricant and/or dirt, can result in widely varying preloads. Proper control in manufacturing, handling, and assembly, in addition to frequent calibration of the wrenches, can make this method acceptable but rigid control is necessary to get top performance out of properly designed, top quality fasteners. While the preload may be much smaller than the maximum bolt load, the corresponding stresses may be high, depending on the margins of safety used.

Practical programs for establishing and maintaining torque-preload standards consist of three activities:

1. Frequent calibration of torque wrenches in a master loading fixture.
2. Large-sample inspection of fasteners for thread condition, performed at the point of assembly.
3. Continuous monitoring to ensure that operators drive into, but not over, the torque range.

The turn-of-the-nut method of tightening is widely accepted, especially for field connections where the calibration of wrenches is difficult. In Ref. 19 either method for A325 and A490 bolts is permitted. Acceptable torque and preload or clamping force values for the lower and medium strength bolts (up through SAE Grade 8) are provided in Ref. 20, but for the ultra-high-strength materials, the designer should consult specific data, as in Ref. 21.

Torquing the bolt into yield has become generally accepted practice to obtain maximum clamping force. But, under fatigue loading, the serious drawback of plastic strain-ratcheting results, which means that the bolt will relax at a rate higher than if torqued to the more conventional 60–80% yield, this higher rate obtaining at any temperature. For static, or very-low-cycle applications, as in buildings, the practice may not be harmful, and it can be done rapidly and inexpensively by automatic, powered machines.

Heating a loose bolt, turning the nut down finger-tight, and allowing the assembly to cool produces a preload via shrinkage, but this method is extremely approximate, and generally regarded as useful only for salvage in the field.

Preloading of high-strength fasteners in high-performance joints by hydraulic tensioning is an eminently successful and fairly accurate method, especially for larger diameters where the torque would become prohibitively high. Tensioning combined with one of the sophisticated methods of measuring bolt elongation, such as ultrasonic reflection, form a highly accurate, although expensive, way to find the actual preload.

Loss of preload, or relaxation, occurs in essentially all bolted joints, at any temperature, owing to seating or embedment; the smoothing of the peaks in the surface finish; and more gross, local yielding at the high spots, especially on the threads. The degree of loss depends on accuracy of alignment and surface finish, and tests have generally indicated that, for well put-up joints with finish below about 6 rms, the loss increases asymptotically to 3–5% of the initial preload during the first 10 cycles of applied stress. Thus, there is considerable advantage to a few cycles of "shakedown" loading, followed by retightening.

Effects of Hole Sizing and Interference

In spite of the best efforts to design tension-friction joints with high-strength fastener materials and high clamping forces, bearing stresses very often exist, and, as discussed above, they can be high because of extreme values of K_t. Among the methods of reducing these stresses is that of introducing residual compression on the hole surface. The effects of "sizing" or "drifting"—the operation of forcing a smooth, oversize pin or ball through the hole to cause a small plastic deformation—are generally beneficial. The residual tangential compression absorbs part of the tangential tension from the active load, thus increasing fatigue life to failure by tear-out. This type of "preload" is of opposite sign to that in a tension bolt, wherein both the preload and active load add in tension. The results of hole sizing for some specimens testing in tension–tension with open holes are listed in Table 24.3.

Such results show numerically the expected trend; that the presence of residual compression (below compression yield) improves the fatigue limit in tension. Tests on similar specimens having the holes filled with a self-sizing Huckbolt (interference of 0.002–0.003 in.) gave a fatigue limit of 21.5 ksi for all three hole conditions. This single value indicates that, although the benefits of the sizing are present, the differences in the three hole conditions were masked by the apparently excessive interference. While the benefits of the hole sizing are not debated, the operation is very difficult to control in the shop because of the sharp dependence of the value of residual compression on dimensional tolerances, especially for small diameters.

The relatively large increase, with interference, of radial compression at the expense of a small increase in hoop tension is the basis for the improved fatigue behavior of such joints, particularly those using tapered bolts. The latter provide a far more uniform bearing on the elements of the joint than do the straight-shank interference bolts. Figure 24.23 provides some comparative life data, but it should be noted that the benefits of the Taper-lok are available only at the relatively greater expense of hole preparation and higher cost of the fastener itself. The "stress pin" is a new item; it is basically a refined rivet completely filling the hole to the extent of inducing radial compression, plus some plastic compression in the plate edges under the holding collars. Its use forms a joint that is not normally demountable.

It is especially important to distinguish among the states of stress for sized and nonsized holes, with clearance bolt or interference bolt and for open hole conditions. See Table 24.4.

The problem of producing holes with a smooth, damage-free bore is severe; reamer or drill removal marks, punching burrs, etc., act as initial defects, especially in aluminum alloys. Thus, unless the local tension can be made low enough to make the crack-growth rate insignificant, the "life" of a joint is better calculated by the procedures of fracture mechanics. Joints may fail by excessive cracking or fracture of either the fastener or the flange.

Bolting Materials and Specifications

The tensile and shear strengths of a material are not necessarily good indications of its behavior as a bolt. The only valid strength data come from tests of an assembly of a fully finished bolt and nut;

Table 24.3 Hole Sizing Results

	Hole Condition in 7075–T6 Alclad Plate	Fatigue Limit at 10^7 cycles, ksi
1.	Hole predrilled; size drilled; edges honed.	8.5
2.	Repeat (1), plus sizing 0.001 in. to 0.002 in. with drift pin.	10.5
3.	Repeat (1), plus sizing 0.005 in. to 0.006 in. with drift pin.	13.5

Source: Huck Manufacturing Co.

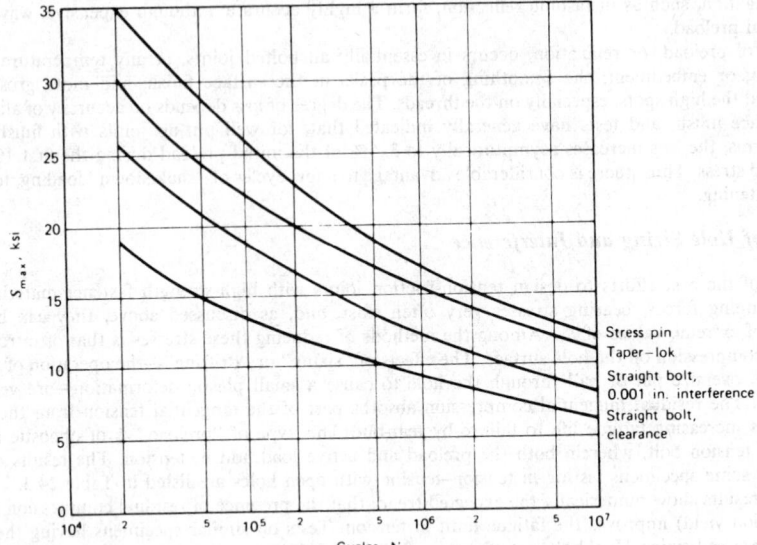

Fig. 24.23 *S–N* curves for various titanium fasteners and interferences: No. 10's in two 0.080-in. Alclad sheets; *R* = +0.10.

then other items in the working environment must be accounted for, such as corrosion and combined loads, and particularly in the high-strength steels, stress corrosion and hydrogen embrittlement. One useful base for indexing the materials is their resistance to the loss of strength or toughness as a function of temperature. See Table 24.5.

For a reasonable range around room temperature, about −100 to +250°F, the designer has a wide choice of materials, so that many parameters must be balanced: strength, corrosion resistance, the degree of magnetism, cost, etc. Outside this approximate range, the number of material choices decreases rather sharply to the superalloys and ultra-high-strength steels.

On a unit strength basis at normal temperatures, the designer may use the SAE graded (0–8) steels, beryllium, titanium, cold-formed aluminum, copper-base alloys, or any of the three types of stainless steels. On the basis of corrosion resistance, there are the austenitic and ferritic stainless steels; copper and nickel alloys; and many nonmetallics such as nylon, Teflon, Delrin, and polycarbonates. An optimum choice for both strength and corrosion resistance is often the 316-type stainless steel in the 5 c condition (min F_{ty} = 100,000 psi), but, of course, the particulars of the application must be satisfied.

Bolts, studs, and other threaded fasteners are available in a great variety of sizes, shapes, materials, and finishes, but not in all possible combinations. Generally, it is more prudent and economical to design the joint with a standard fastener than to require a special fastener with all the associated testing, which, however extensive, will result in a relatively narrow data base.

The available range of quality is also rather wide; for domestic use, the higher levels probably cannot be justified economically, but for critical applications such as automotive wheel studs and connecting rod bolts, the consequences of failure can be extensive. High quality and reliability can be achieved by specifying to a unique Special Lot Requirement or by choice of the appropriate sections of one of the very comprehensive commercial specifications such as Ref. 22. The cost of items to either of the latter specifications will of course be higher than if they came off-the-shelf.

Some 15 organizations in the United States write bolting specifications, predominantly the SAE and ASTM, plus the military and the federal government.

Flange Design

The design of the flanges for a bolted connection is not trivial. The major task is the proportioning of the flange section and the determination of the number and size of the bolts. Analysis generally considers the flanges to be continuous, annular plates whose flexural characteristics can be approximated by beam theory. For identical pairs, the flanges are assumed to be discrete radial beams and the moments are balanced.

For nonidentical pairs, the theory of rigid body rotations is also considered, and the moments are unbalanced. Maintenance of a pressure-tight joint, gasketed or not, depends on the flange/bolt

Table 24.4 Stresses on Hole Surface (Prior to External Loading)[a]

	STRESS	Hole Not Sized	Hole Sized
Open hole		Zero	Small; depends on dimensions
		Zero	Zero
		Zero	Zero
		Zero	Small; depends on Δdiameter
		Zero	Small; depends on Δdiameter
		Zero	Zero
With straight clearance bolt or rivet	HT	Zero, or increased, depending on fit-up	Small; depends on Δdiameter
	HC	Zero	Zero
	RT	Zero	Zero
	RC	Zero, or increased, depending on fit-up	Small; depends on Δdiameter
	LT	Zero	Zero
	LC	Depends on clamping force	Depends on clamping force
With straight interference bolt	HT	Increased, amount depends on fit-up	Small increase, depends on initial compression
	HC	Zero	Decreased from initial value
	RT	Zero	Zero
	RC	Increased, amount depends on fit-up	Considerable increase
	LT	Zero	Zero
	LC	Depends on clamping force	Depends on clamping force
With tapered bolt	HT	Depends on fit-up and pull-in	Not applicable
	HC	Zero	Not applicable
	RT	Zero	Not applicable
	RC	Greatly increased, amount depends on fit-up and pull-in	Not applicable
	LT	Zero	Not applicable
	LC	Depends on clamping force	Not applicable

[a] HT = hoop tension. RT = radial tension. LT = longitudinal tension. HC = hoop compression. RC = radial compression. LC = longitudinal compression.

stiffness and on the bolt preload, the earlier comments being applicable here. The predominant stress is bending, but if a membrane stress occurs in conjunction with the bending, the combination represents the maximum absolute value and may be either tension or compression.

A comprehensive treatment of flange design is given in Section VIII, Division 1 of Ref. 23, Appendicies II, S, and Y. Whether the device in question is being designed to Code or not, the practice described in the reference is recommended as conservative.

Table 24.5 Working Temperatures for Bolting Materials (°F)[a]

−400	−200	0	+200	+400	+600	+800	+1000	+1200	+1400	+1600
		←— Plastics ——→								
		←—SAE Graded Steels—→								
		←—Aluminum 2024-T4—→								
		←—Brass———→								
		←—Bronze———→								
		←—Beryllium QMV ——————→								
←----←		—Stainless Steel Type 302———				→				
←		—Stainless Steel Type 316———			→- --→				
←		—Stainless Steel Type 431—				→				
←		— Stainless Steel Type 17-4PH—			→					
←		—Stainless Steel Type 17-7PH—			→- ------→				
		←—Titanium 4Al-4Mn ——————→								
←		— Titanium 6Al-4V ———————→- ----→							
←		← Titanium 7Al-12Zr —→- ----- ►							
←		← Titanium 1Al-8V-5Fe———→								
←		—Maraging Steel 300 ———————→							
←		—AISI 4340 ————————→							
◄		—A-286 ——→								
		←—AM-350 (AMS 5745)——————→							
◄		—AMS-5615 (Greek Ascoloy)—			→					
		←—AFC-77 ————→- ---→							
◄		—H-11 ————————→							
◄		—Unitemp 212 ———→								
◄		—Inconel 718 —————————————————————————————————→						
◄		— René 41 ————————————————————→						
◄		— U-212 ——→						
◄		—Waspalloy————————————————————————————————————→						

[a] Data from *Metal Progress*, July 1966, by permission ASM.

Lugs or Clevises

The single-pin joint or lug represents a simple form of bolted joint, and its fatigue strength may well govern the fatigue behavior of the entire structure. Proper design may be achieved by the following procedure, which is based on a "standard" lug with a 1-in.-diam. hole and a K_t of 3.0.

It has been observed from data that the fatigue strength of the lug assembly is extremely low in comparison to that of the materials. For the failure of steel lugs in 10^7 cycles, S_a is only about ± 4% of the tensile strength, and for high-strength aluminums it is only ± 2.5%. This condition points to the theoretical stress concentration factor as the primary influence on the fatigue life. The stress concentration factor as the ratio of the maximum stress to the average stress on the minimum cross-sectional area at the hole:

$$K_t = \frac{\sigma_{max}}{P/(D - d)t}$$

Considering K_t as a function of geometry, we note that as d/D approaches a value of 1, the lug tends to act as a flexible strap around the pin, thus reducing the stress concentration factor toward a value of 1 also. One immediate problem that often arises is that of determining the optimum strength of lugs for a given overall width.

The stress concentration factor for this condition may be redefined as the ratio of maximum stress to the average stress acting on the full section:

$$C_t = \frac{K_t}{1 - d/D} = \frac{\sigma_{max}}{P/Dt}$$

The curves in Fig. 24.24 show a distinct saddle at d/D of about 0.4, indicating a minimum stress concentration factor or a maximum strength at this ratio.

A number of other parameters may affect the stress concentration factor: (1) the effect of "waisting" the lug as shown in Fig. 24.25, (2) pin bending, (3) clearance or interference fits, (4) stress distribution around the hole, and (5) pin material and lubrication.

The effect of bending and shear of the pin is to increase the maximum stress near the faces of the lug. The increase depends primarily on the size of the pin compared to the thickness of the lug, d/t; photoelastic tests have indicated that the increase is fairly small. For a semicircular-ended lug

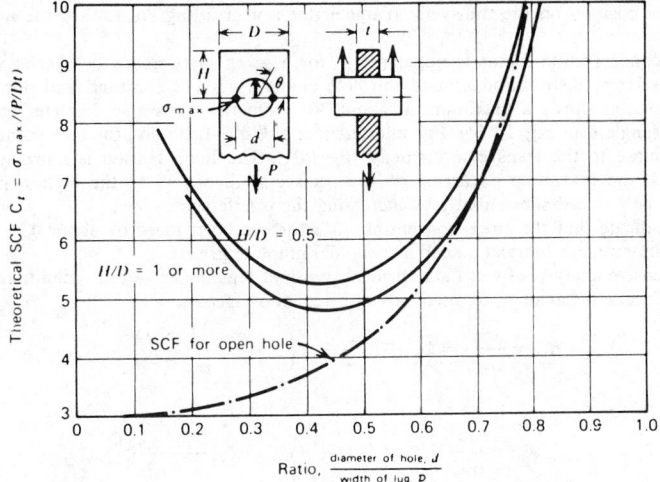

Fig. 24.24 Theoretical stress concentration factors for loaded lugs based on the gross area of the lug. Small clearance between hole and pin.

having ratios of overhung length to width, H/D, of 0.5 and a pin diameter to width, d/D, of ½, the results below were obtained.

Ratio d/t	0.51	0.72	1.27
Increase in maximum (shear) stress due to pin-bending (%)	17	10	7.5

The evidence appears to confirm the comparatively small influence of the pin-bending effect on fatigue. For a very thick lug it is an advantage to use a large diameter pin, and it has been suggested that the optimum strength is obtained when the ratio of the pin diameter to the width of the lug, d/D, is about 0.6.

Greater stress concentration factors are obtained with larger clearances between the pin and the hole. The data show that the factor is dependent not only on clearance, but also on the magnitude of the load applied, with higher loads reducing the factor. The points of maximum stress are shifted

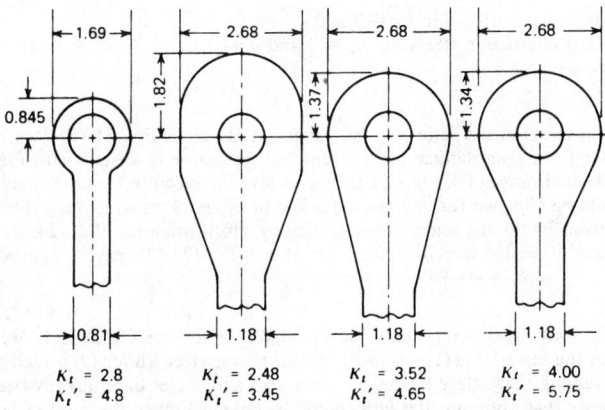

Fig. 24.25 Stress concentration factors for "waisted" lugs. For all cases, the diameter of the hole is 1 in.; for K_t, a neat fit pin is used; for K_t', diametral clearances are 0.019 in. for the first case, and 0.015 in. for other cases.

from their usual position on the transverse diameter for a neat fitting pin toward the loaded side of the pin.

The stress concentration factor is not constant for a given interference but varies with the pin load. The shear stress distribution around the hole boundary is not constant, and, for the push-fit case, the distribution shows a maximum at about 30° from the transverse diameter on the loaded side of the pin (angle θ in Fig. 24.24). For interferences of 0.003–0.006 in., the maximum shear stress occurred very close to the transverse diameter ($\theta \sim 0°$). The hoop tension is more symmetrically distributed, with the maximum occurring on the transverse diameter. At the highest pin load this maximum may be reduced substantially by increasing the interference.

Data also indicate that the stress concentration factor can be reduced by about 0.25 through the use of a high-film-strength lubricant, such as colloidal graphite grease.

A comprehensive analysis of test data on many types of lugs under various conditions resulted in expressions of fatigue behavior in terms of strength reduction factors:

$$K_m = \frac{\sigma_m}{\sigma_{mn}} = K_s + (K_t - K_s)\left(1 - \frac{\sigma_{mn} + \sigma_{an}}{\sigma_{tn}}\right)^2 \tag{24.20}$$

and

$$K_a = \frac{\sigma_a}{\sigma_{an}} = K_s + [\beta(1+d)K_t - K_s]\frac{n^4}{b + n^4} \tag{24.21}$$

where K_m, K_a = the maximum values of the strength reduction factors, for 10^7 cycles or more
σ_m = mean stress, smooth specimen
σ_{mn} = mean stress, notched specimen
K_s = static stress concentration factor, here equal to 1
d = diameter of hole, in.
b = a constant, $\cong 1000$
β = a constant, ~2.7 for steel and 1.85 for aluminum

Both expressions apply for notched specimens, but Eq. (24.21) is modified to include the maximum stress concentration factor by the term $\beta(1 + d)K_t$, the fretting factor effect.

A correlation has been made by interpreting the various results for different designs to a standard condition of $K_t = 3$, representing a near-optimum strength design, and a hole diameter of 1 in. The standard σ_{mn} and σ_{an} are derived from the observed σ_{mn} and σ_{an} of any lug from Eqs. (24.20) and (24.21), for the condition of equality between the peak σ_{mn} and σ_{an}. It is also assumed that the static stress concentration factor, K_s, equals 1, leading to

$$\frac{\sigma'_{mn}}{\sigma'_t} = \frac{\sigma_{mn}}{\sigma_t}\left\{\frac{1 + (K_t - 1)[1 - (\sigma_{mn} + \sigma_{an})/\sigma_t]^2}{1 + 2[1 - (\sigma'_{mn} + \sigma'_{an})/\sigma'_t]^2}\right\} \tag{24.22}$$

and

$$\frac{\sigma'_{mn}}{\sigma'_t} = \frac{\sigma_{mn}}{\sigma_t}\left[\frac{1000 + K_t(1+d)\beta n^4}{1000 + 6\beta n^4}\right] \tag{24.23}$$

where σ'_{mn} = standard mean stress for $K_t = 3$ and $d = 1$ in.
σ'_{an} = standard alternating stress for $K_t = 3$ and $d = 1$ in.
σ'_t = tensile strength
$n = \log N$

Results are plotted in terms of the ratio σ'_{an}/σ'_t for the steels, as in Fig. 24.26.

In designing a lug for a given fatigue life, the simplest procedure is to start with Fig. 24.26, taking into account the indicated range of scatter. These curves give the magnitude of the standard alternating stress σ'_{an} that would be obtained for any given number of cycles. This stress may then be interpreted to the alternating stress for the lug under consideration by substitution into Eq. (24.23). An allowance for mean stress is hardly needed if $\sigma_{mn} > \sigma_{an}$, but, if not, Eq. (24.22) may be applied.

Example 24.3

A lug is required to withstand a repeated load of 10,000 lb for 100,000 cycles. What dimensions should be used when the material is (1) a steel of 160 ksi tensile strength, or (2) a high-tensile-strength aluminum alloy, assuming that safety factors for both load and cycles have already been applied?

Figure 24.24 shows that optimum strength should be obtained when the ratio of hole diameter to the width of lug is about $d/D = 0.43$. This ratio is reduced slightly if one takes account of the size effect implied in Eq. (24.21), giving an optimum ratio of, say, $d/D = 0.4$. The associated theoretical K_t is 2.9 for a value of H/D of 1 or more. For repeated loading the nominal mean and alternating stresses are of equal value, or

$$\sigma_{mn} = \sigma_{an} = \frac{P}{1(D-d)} = \frac{5}{1.5dt} \text{ ksi}$$

for $d/D = 0.4$ and $P = 5000$ lb.

For steel, Fig. 24.26 indicates that a representative value of the standard ratio stress at 10^5 cycles is $\sigma'_{an}/\sigma_t = \pm 0.07$. Substitution into Eq. (24.23) gives

$$\sigma_{an} = 0.07 \times 160 \left[\frac{1000 + 6 \times 2.7 \times 5^4}{1000 + 2.7(1+d)2.9 \times 5^4} \right] = \frac{21.1}{1+0.83d}$$

Equating the two expressions for σ_{an} we find that

$$\frac{5}{1.5dt} = \frac{21.1}{1+0.83d}$$

giving

$$t = \frac{0.158}{d} + 0.131$$

This expression shows that there is an infinite series of solutions of values for t and d that satisfy the conditions. Typical solutions are listed in Table 24.6.

It is not advisable to use lug thicknesses appreciably greater than the diameter of the hole, since strength becomes adversely affected by pin-bending effects. Thus the pin diameter should exceed 0.5 in. The weight of the lug probably depends on the cross-sectional area Dt, and for this case the above figures show that the smallest diameter pin gives the least weight. The best design is thus represented approximately by the dimensions with ½ in. pin diameter. All lugs in Table 24.6 should have equal strengths, and it is interesting to note that the nominal stress σ_{an} varies with the size of the pin, since a size effect is involved in Eq. (24.21).

Use of Bushings. From a fatigue-strength point of view, the pin–bushing combination is roughly equivalent to a pin of the same outside diameter as the bushing when there is clearance at the bushing outside diameter. For lugs with interference-fit bushings (0.003–0.010 in. per in. of diameter) tests have shown consistently longer lives than with interference-fit pins—up to four times as long.

Interference-Fit Pins. Interference-fit pins also show greater lives than clearance-fit pins. Nonfailure at 10^7 cycles can usually be obtained for all aluminum pin–lug combinations at interferences of at least 0.004 in. per in. of diameter, and similarly for steels with about 0.003 in. per in. maximum.

Hole Surface Conditions. These conditions appear to have approximately the same effect on lug life and strength as they do for bolted joints, but the scatter in both lives and strengths is very much greater. It is generally worthwhile to use a lubricant. The effect of the method of machining the hole is inconclusive: drilled holes were superior to reamed holes for steel lugs, the reverse was true for aluminum. Anodizing the bores in aluminum lugs is quite harmful to life, but has little effect on

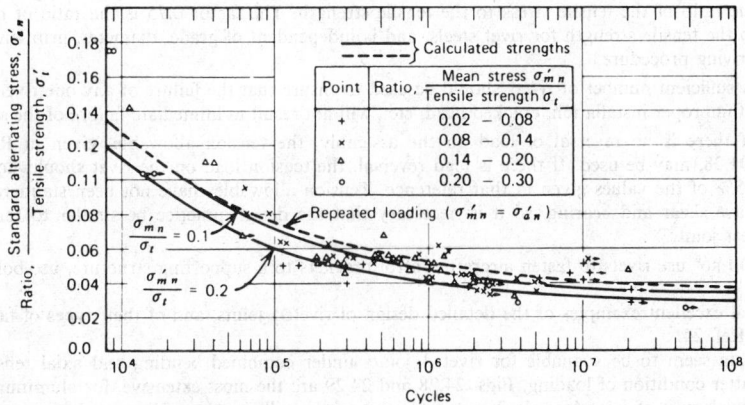

Fig. 24.26 "Standard" fatigue strength of steel lugs as a function of mean stress. $K_t = 3$; hole diameter $= 1$ in. Clearance pin.

Table 24.6   Typical Solutions for Example 24.3

d	0.25	0.5	0.75	1
t	0.76	0.44	0.34	0.29
D	0.625	1.25	1.87	2.5
Area Dt	0.47	0.56	0.64	0.72
$\sigma_{an} = [5/(D-d)t]$ (ksi)	17.6	15.0	13.1	11.6
Bearing stress (maximum) = $10/dt$ (ksi)	53	45	39	35

static strength. Cadmium or zinc plating the pin gives some improvement in life, but much more improvement if both pin and lug are plated, particularly with zinc. Small-sample tests indicate chromium plating may be very harmful, depending on the degree of relief of embrittlement. While these statements may appear uselessly qualitative, attempts at correlation of present data do not yield more rigorous conclusions. Designs of optimum geometry and minimum stress concentration factors, together with interference bushing fits, seem to be the best approach to long life.

24.1.2   Riveted Joints

The design of both butt splices and lap joints is based necessarily on the tension–shear–bearing ratio, the value of which may vary widely, Fig. 24.27. Riveted joints are infrequently designed to carry any significant part of the sheet load by friction induced by tension in the driven rivet because it is essentially impossible to specify this tensile force. Measurements of the tensile force in hot-driven steel rivets indicate variable consistency from lot to lot, and, as expected, an increasing force with increasing grip. "Proper" clamping force or tension in the driven rivet is usually obtained by trials on scrap stock with correctly dimensioned rivets to determine a proper driving pressure. Rivets are, of course, loaded in tension whenever the condition is unavoidable in the joint design; with the application of conservative judgment in the form of the restrictions below, success is usually attained:

1. The tension on rivets should be restricted to conditions in which the tension load is incidental to the major shear-carrying purpose of the rivet. When it is difficult to determine if the tension component is incidental or major, a bolt should be used. The tensile capacity of a rivet is

$$P_{ult} = AS_{ult}$$

    where A is the undriven area. S_{ult} is usually taken as 60 ksi for A502 grade 1 steel and 80 ksi for grade 2.

2. Rivets loaded in both shear and tension should be checked for combined stresses using the interaction equation:

$$X^2/(0.75)^2 + Y^2 = 1.0$$

    where X is the ratio of the shear stress on the shear plane to the tensile strength and Y is the ratio of the tensile stress to the tensile strength. The factor 0.75 is the ratio of the shear to the tensile strength for rivet steels, and is independent of grade, diameter, grip length, and driving procedure.

3. A sufficient number of rivets should be used to ensure that the failure of any one rivet because of improper installation, cracked head, etc., will not result in immediate failure of the structure.

4. If there is no reversal of load on the assembly, the tension allowable, given in Ref. 3, p. D1.28, may be used. If there is load reversal, the tension load on the rivet should not exceed 25% of the values given in that reference. Tension allowables have not been standardized, as have shear and bearing, so it is necessary that the design practice be similar to that of the test joint.

5. Do not use rivets to fasten aircraft control brackets to a supporting structure, use bolts.

Several excellent examples of the detailed design of riveted joints, and of their types of failure are given in Ref. 24.

No data seem to be available for riveted joints under combined bending and axial tension, but for the latter condition of loading, Figs. 24.28 and 24.29 are the most extensive, for aluminum alloys. The material, type of joint (double-shear), and joint size are illustrative of best modern practice. The data are further useful not only from the wide variety of alloys included, but also for the comparison between the lives of plain specimens and those of actual joints made up of the same material.

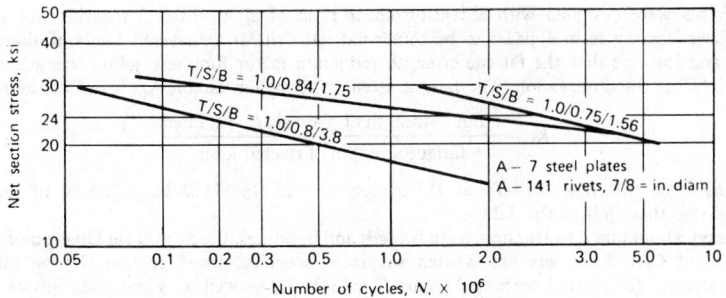

Fig. 24.27 Fatigue life as a function of tension–shear–bearing ratio.

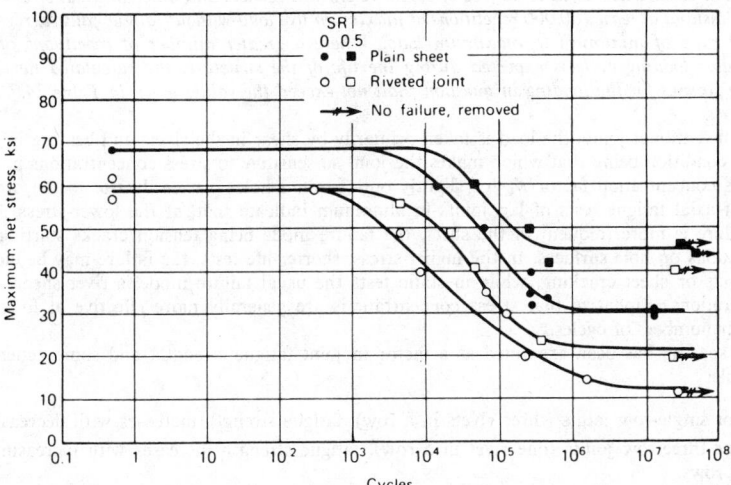

Fig. 24.28 Axial-stress fatigue strengths of 2024-T3 plain sheet specimens riveted joints.

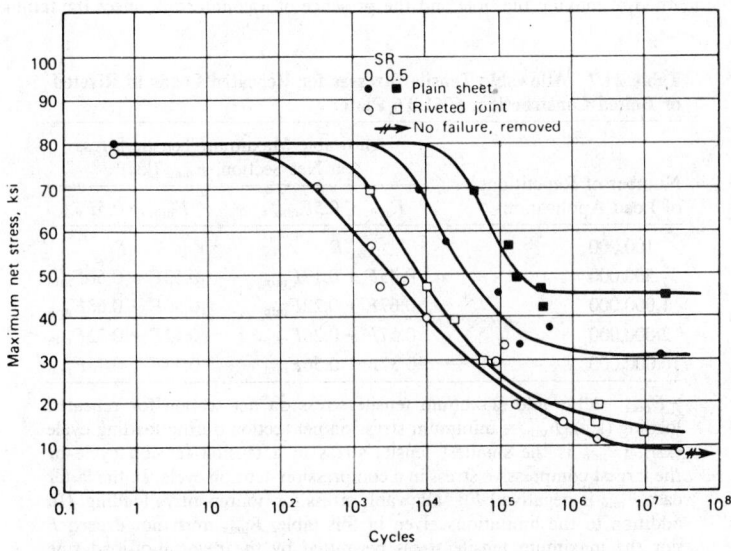

Fig. 24.29 Axial-stress fatigue strengths of 7075-T6 plain sheet specimens and riveted joints.

The joints were designed with a tension/shear ratio of about 1.0/1.0 (bearing not given). The tension/shear/bearing ratio appears to be somewhat less critical for riveted joints of aluminum than for steel, and the fact that the fatigue strength reduction factor for these joints was generally below the K_t reinforces the observation. The fatigue strength reduction factor, K_t, is defined here as

$$K_t = \frac{\text{fatigue strength of plain sheet specimen}}{\text{fatigue strength of riveted joint}}$$

and ranged from 1.04 (for 2014-T6 at 10^3 cycles) to 4.16 (for 7178-T6 at 2.5×10^7 cycles). The value of K_t for these joints was 3.2.

For heavy aluminum construction, as in bridges and buildings, the Structural Division of the American Society of Civil Engineers has written suggested specifications.[25] Section G-4 on fatigue gives allowable stresses for riveted or bolted joints (Table 24.7), as well as some good advice on fatigue design:

Tests indicate that riveted members designed in accordance with the requirements of these specifications and constructed so as to be free of severe re-entrant corners and other unusual stress raisers will withstand at least 100,000 repetitions of maximum live load without fatigue failure, regardless of the ratio of minimum to maximum load. Where a greater number of repetitions of some particular loading cycle is expected during the life of the structure, the calculated net section tensile stresses for the loading in question shall not exceed the values given in Table 24.7.

For conventional joints the load is taken primarily by shear in the rivets and bearing in the sheet, the latter condition being that which makes the joint so sensitive to stress concentrations in the sheet. The stress concentration factor K_t for slightly out-of-round holes can easily rise to the order of 10. Results of axial fatigue tests of lap joints in aluminum indicate that, at the lower-stress, longer-life levels, failure is more frequent in the sheet, the failure mode being tension cracks starting at stress concentrations on hole surfaces. In the higher-stress, shorter-life tests, the failure may be either shear of the rivets or sheet cracking, while in static tests the usual failure mode is rivet shear. Together, these conditions emphasize that stress concentrations are generally more effective at lower stresses and higher numbers of cycles.

Rivet spacing has been examined as a factor in joint fatigue strength, and some generalizations are possible:

1. For single-row joints (three rivets in a row), fatigue strength increases with decreasing pitch.
2. For three-row joints (one rivet in a row), fatigue strength increases with increasing spacing of rows.
3. For multirow joints (three in a row), fatigue strength increases with increasing numbers of rows.

Table 24.8 gives more specific results.

Both the method of making the hole and the presence of a countersink affect the fatigue strength

Table 24.7　Allowable Tensile Stresses for Repeated Loads in Riveted or Bolted Construction, 6061-T6 Plate

Number of Repetitions of Load Application	Allowable Maximum Tensile Stress on Net Section, F_{max} (ksi)[a]	
	$F_{min} \leq 0.5 F_{max}$	$F_{min} > 0.5 F_{max}$
100,000	F	F
500,000	$0.75F + 0.17F_{min}$	$0.53F + 0.66F_{min}$
1,000,000	$0.67F + 0.22F_{min}$	$0.50F + 0.68F_{min}$
2,000,000	$0.57F + 0.26F_{min}$	$0.42F + 0.72F_{min}$
10,000,000	$0.37F + 0.56F_{min}$	$0.33F + 0.73F_{min}$

[a] F_{max} = allowable maximum tensile stress on net section for repeated loading (ksi). F_{min} = minimum stress on net section during loading cycle (ksi). (F_{min} is the smallest tensile stress in a tension–tension cycle or the largest compressive stress in a compression–tension cycle. In the latter case F_{min} is negative.) F = allowable stress for nonrepetitive loading. (In addition to the limitations given in this table, F_{max} must not exceed F nor the maximum tensile stress permitted by the reversal-of-load rule in Specification G-1 (Ref. 25).

Table 24.8 Fatigue Strength of Riveted Joints in 2024-T3 Alclad Sheet[a]

Number of Rows	Spacing of Rows	Rivet Position	Average Static Strength (lb/rivet)	Fatigue Strength (max load) (lb/rivet)	
				10^5 cycles	10^7 cycles
1	—	—	692	282	170
2	$3/16$	Staggered	668	222	156
2	$3/8$	Staggered	690	266	133
2	$11/16$	Staggered	705	290	137
2	$3/4$	In line	725	288	135

[a] Rivets, AN426-AD5. Pitch $= 3/4$ in. Edge distance $= 3/8$ in. $R = 0.25$. Specimen width $= 5$ in.

of joints in thin sheet. Joints with flush-head rivets have generally shown poorer fatigue behavior than those with protruding heads, because the stress concentration factor is higher for the countersunk hole than for the plain hole, by about 20% typically. The detailed dimensions of the hole are important to the life of the joint, especially those in the top sheet. The taper should not extend through the sheet to produce a feather edge with its extremely high K_t.

Variations in the surface finish of the hole seem to have little effect on fatigue life, but the method of hole production must be carefully chosen and controlled to prevent corner cracks. The order of decreasing fatigue strength is: coil-dimpling; spin dimpling; drilling plus reaming; drilling only; and machine-countersinking.

For joints in heavy plate of both plain carbon and alloy steels, a summary of many tests emphasizes the crucial point that, despite the wide range of static strengths available, from about 60 to 100 ksi, the fatigue strength of the joints at 2,000,000 cycles was the same, about 26 ksi. This condition may well raise the question of the economics of specifying the higher strength, and higher cost, nickel or silicon steels, unless some other property such as corrosion resistance is required. Because of the importance of the tension/shear/bearing ratio, Fig. 24.27 was given to show the wide variation in life for different ratios. Note particularly the lower life for the higher bearing values, another indication of the tremendous effect of the stress concentrations at the hole surface on the allowable bearing strength. This condition is most effective in promoting the use of high-strength bolts instead of rivets for such work. In a bolted joint of proper design it is possible to carry the shear load by friction, thus avoiding these particular difficulties.

24.2 WELDED JOINTS

24.2.1 Arc and Gas Welds

In contrast to the bolted or riveted joint, a weld constitutes a continuous load path without any mechanical interface. This absence of an interface makes the region monolithic and thus not capable of stopping a running crack. While there is no reason why a welded joint cannot be satisfactorily designed for its application, this potential crackpath makes it imperative that the design and manufacture adhere to the requirements of fracture mechanics; that is, the size, location, and orientation of the inevitable defects must be controlled. For supercritical applications, an artificial bolted or bonded joint may be introduced to act as a "crackstopper"; see Ref. 9, p. 30.

The fatigue life of weldments depends not only on the size of the initial defect but also on the joint geometry and on the stress range. The bead shape and the defects are controlled by the technique or procedure used, and, for acceptable results, an applicable qualified procedure must be followed. The *AWS Welding Handbook* and other references[26-32] provide information as to the important details for standard work; while for other cases, especially thick sections, Ref. 30 and the publications of the Welding Research Council (WRC) and the Heavy Section Steel Technology Program (HSST) are useful.

As compared with manual welding, semiautomatic (gas metal arc) and automatic (submerged arc and electroslag) welding processes generally provide greater fatigue resistance, because they produce welds with fewer internal discontinuities and with a smoother surface. Electroslag welding has been observed to produce transverse butt welds with up to 90% of the base metal fatigue strength, but this value is sensitive to procedural parameters.

Gas tungsten arc welding (GTAW) of the toes and roots of welds laid by other processes increases the fatigue resistance by up to 100%. This increase is probably due to reduction of both the residual stresses and of the stress raiser at the toe or root radius. The various processes and procedures are discussed at length in Refs. 26 and 30.

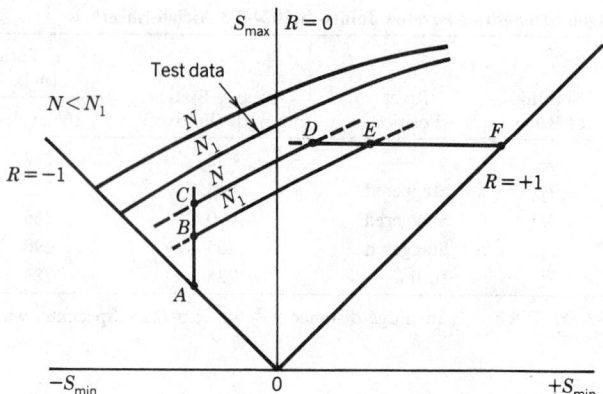

Fig. 24.30 Design criteria. (Reprinted with permission of ASTM, 1916 Race St., Philadelphia, PA 19103.)

One particular source of toe cracking is physical restraint of the thermal contraction during cooling, the restraint being provided by alignment fixtures, jacks, or greatly disparate thermal masses being joined. Such restraint should be minimized, and so also should the total energy input and its rate, the "heat rate." To this end the procedure should use the smallest weld section, the lowest current or gas flow, and the highest travel speed practical. Additionally, to prevent cracking in austenitic steels (300 series stainless steels) these parameters must be correlated to produce at least 3–4% delta-ferrite in the weld deposit; see the Schaeffler Diagram, Ref. 30, p. 322. (A modification known as the DeLong diagram is also useful.) A detailed description of the general problem of residual stresses, distortion, and cracking may be found in Ref. 33.

To achieve any reasonable degree of acceptability, qualification is required of:

The design of the weldment, of its structure and of its joint details.

The materials: parent metal, filler rod, and flux if any.

The process and procedure, especially the heat rate.

The operators, by their making acceptable sample welds.

Allowable Stresses and Fatigue Design Criteria*

Allowable stresses for welded joints are established by the AWS,[27] the AISC,[28] and the ASME.[29] Fatigue design criteria for weldments consider the effects of maximum stress, stress range, material tensile strength, and stress concentration on the service life of a structural element.

A few criteria, principally those for vehicle and machine design show constant-life diagrams for each type of weld joint and material. In Fig. 24.30, for example, the allowable stress envelope is defined by the allowable static tensile stress (line DF), the allowable compressive stress (line AC), the allowable cyclic stress (line CD or BE) and the ray $R = +1$. The allowable cyclic stress is a function of life and generally represents the cases: $N = 100,000$; $100,000 \leq N = 5000$: $500,000 \leq N = 2,000,000$, and $N > 2,000,000$ cycles. The constant-life lines are derived from S–N test curves; the allowable cyclic stress lines are offset from them by suitable factors of safety, providing such results as Fig. 24.31.

The criteria governing land and marine structures in the United States and Great Britain assume weldment fatigue resistance to be independent of mean stress and cite only the stress range. In such criteria the lines CD and BE would be parallel to the ray $R = +1$ in Fig. 24.30. However, these criteria do not show constant-life diagrams for each detail. Instead, they show sketches of the various weld-joint configurations from which the designer selects the stress category closest to his or her detail. The designer then enters a tabular array with this stress category and the desired service life, and selects the allowable stress range for his or her particular case.

All allowable stresses cited in the various criteria are nominal net section normal or shear stresses. Many of the structural design criteria furnish guidelines for handling spectrum loading via the Miner–Palmgren linear cumulative damage hypothesis. None of the criteria consider the effects of environment.

The general design of welded joints, as outlined below, assumes a normal qualified procedure

---

* Adapted from Ref. 34 by permission of ASTM.

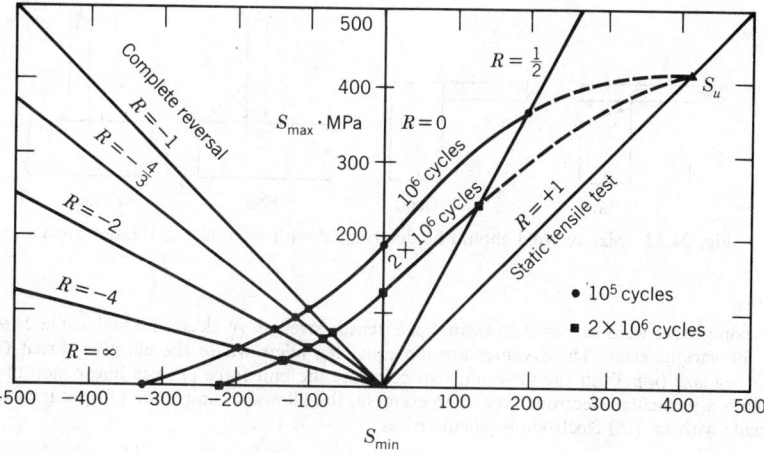

Fig. 24.31 Constant-life diagram, carbon-steel butt welds. (Reprinted with permission of ASTM, 1916 Race St., Philadelphia, PA 19103.)

resulting in a "normal" content of defects. Thus, to ensure the highest integrity, a fracture mechanics analysis should also be performed. The normal joint and weld types are shown in Fig. 24.32. The fillet or corner weld is one of those most commonly used, because it requires no groove preparation; however, it frequently lacks the integrity of the penetration weld. The size of the weld should always be designed with reference to the thinner member. The joint cannot be made any stronger by using the thicker member for the weld size, and more weld metal will be required (see Fig. 24.33). In the United States a fillet weld is measured by the leg size of the largest right triangle that may be inscribed in the cross-sectional area (Fig. 24.34). The throat, a better index of strength, is the shortest distance between the root of the joint and the face of the diagrammatical weld. As Fig. 24.34 shows, the leg size used may be shorter than the actual leg of the weld. With convex fillets, as they all should be, the actual throat may be longer than the throat of the inscribed triangle. Reference 35 provides an excellent scheme for optimizing weld design by determining the size for which failure would occur in the weld or in the plate (p. 131).

The AISC and AWS have established the allowable shear value for weld metal in a fillet or partial-

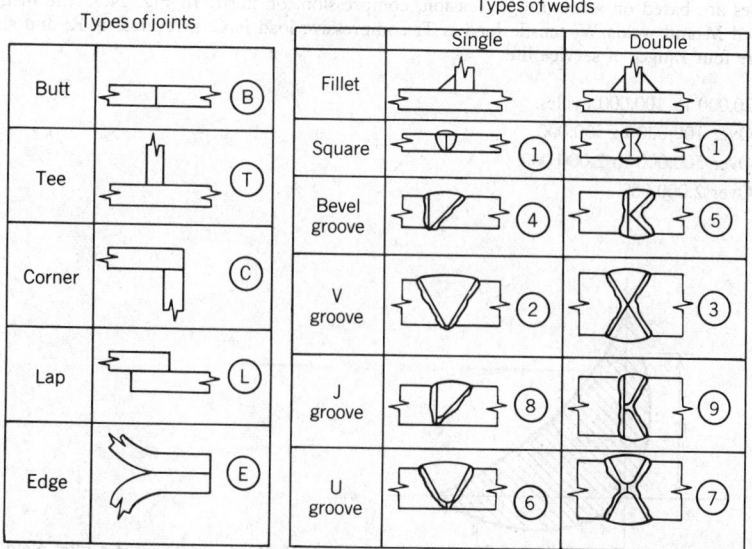

Fig. 24.32 Joint designs (left); weld grooves (right).

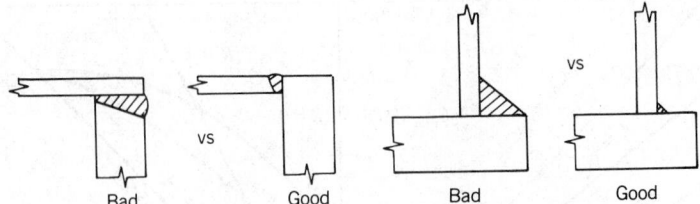

Fig. 24.33 Size of weld should be determined with reference to thinner member.

penetration groove weld as $\tau = 0.30$ (Min. Spec. tensile strength of electrode) and Table 24.9 provides values for various sizes. These values are for equal leg fillets where the effective throat (t_e) equals $0.707 \times$ leg size (w). With the table one can calculate the unit force (f) per linear inch of weld size made with a particular electrode type. For example, the allowable unit force (f) per inch of a ½-in. fillet made with an E70 electrode is calculated as

$$f = 0.707 \ (w) \ (\tau)$$
$$= 0.707 \ (w) \ (0.30) \ (EXX)$$
$$= 0.707 \ (0.5) \ (0.30) \ (70{,}000)$$
$$= 7420 \ \text{psi/linear inch}$$

Note that the table lists the allowable shear stress (τ) for the commonly used E70 weld metal as 21,000 psi (0.30 × 70,000). *Caution:* These values are for *shear only*—fillet welds should not be subjected to normal, that is, transverse, forces or moments because even the most minor defects (toe cracks) act as high stress raisers, leading to early failure. Prudence dictates that some other arrangement of load paths and joint details be found. Groove welds with full penetration are usually preferable.

Reference Table for Allowable Stresses

Table 24.10 summarizes the AWS structural welding code and the AISC allowables for weld metal. It is intended to provide a ready reference for picking the proper strength levels for various types of steels. Once this selection has been made, the allowables for various types of welds may be found.

Fatigue Allowables—AISC

The AISC specifications include fatigue allowables which are also accepted by AWS section 8, Buildings; section 10, Bridges, sets its fatigue allowables on a 10% lower allowable design stress. The AISC allowables are based on stress range: tension, compression, or shear. In Fig. 24.35 the members are designated M and welds W; tensile load is T; compressive load is C; a reversal is R; and shear is S. There are four ranges of service life:

1. 20,000 to 100,000 cycles.
2. Over 100,000 to 500,000.
3. Over 500,000 to 2,000,000.
4. Over 2,000,000.

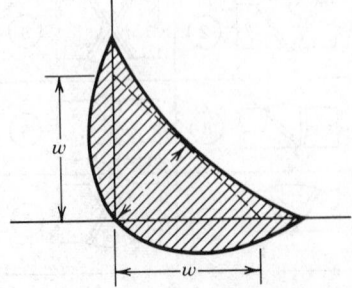

Fig. 24.34 Leg size w of a fillet weld.

Table 24.9 Allowable Load for Various Sizes of Fillet Welds

	Strength Level of Weld Metal (EXX)					
	60	70	80	90	100	110
	Allowable Shear Stress on Throat of Fillet Weld or Partial-Penetration Groove Weld (1000 psi)					
$\tau =$	18.0	21.0	24.0	27.0	30.0	33.0
	Allowable Unit Force on Fillet Weld (1000 psi/linear in.)					
$f =$	12.73w	14.85w	16.97w	19.09w	21.21w	23.33w
Leg Size w (in.)	Allowable Unit Force for Various Sizes of Fillet Welds (1000 lb/linear in.)					
1	12.73	14.85	16.97	19.09	21.21	23.33
7/8	11.14	12.99	14.85	16.70	18.57	20.41
3/4	9.55	11.14	12.73	14.32	15.92	17.50
5/8	7.96	9.28	10.61	11.93	13.27	14.58
1/2	6.37	7.42	8.48	9.54	10.61	11.67
7/16	5.57	6.50	7.42	8.35	9.28	10.21
3/8	4.77	5.57	6.36	7.16	7.95	8.75
5/16	3.98	4.64	5.30	5.97	6.63	7.29
1/4	3.18	3.71	4.24	4.77	5.30	5.83
3/16	2.39	2.78	3.18	3.58	3.98	4.38
1/8	1.59	1.86	2.12	2.39	2.65	2.92
1/16	0.795	0.930	1.06	1.19	1.33	1.46

There are also eight categories representing the type of joint and detail of member. The figure provides the allowable range in stress (σ_{sr} or τ_{sr}) that may be used in the conventional fatigue equations:

$$\frac{\sigma_{sr}}{1-K} \quad \text{or} \quad \frac{\tau_{sr}}{1-K}$$

where K = min. stress/max. stress or min. force/max. force or min. moment/max. moment. No circumstance can allow the maximum fatigue value of σ or τ to exceed that for steady loading.

An alternate use of the allowable range of stress—taken from Table 24.10—is to divide it into the range of applied load. This will provide the required property of the section—area or section modulus. The section, as determined, must additionally be large enough to support the total load (dead and live load) at steady allowable stresses.

Effect of Geometry, Defects, and Penetration of Life of Welds

For welds of high criticality or where a service life warrantee may be involved, the importance of these parameters cannot be overemphasized.

Geometry

Within the joint geometry, the flank angle (Fig. 24.36) is probably the most effective factor and, for useful lives, it should be minimized by careful control of the bead deposition for a smooth, low profile, or preferably by machining the reinforcement off flush. Typical values of life versus angle are given in Fig. 24.36.

Penetration

Incomplete penetration may occasionally be unavoidable; however, it degrades the design since it constitutes a sharp crack normal to the tensile stress (Fig. 24.37). As a matter of good design and fabrication practice, partial-penetration welds should not be used; some Codes permit them but only under restricted conditions; see AWS Paragraph 9.12, etc., and ASME Paragraph NB 3337.3, 3352.4, etc.

Table 24.10 Permissible Stress in Weld

Type of Weld and Stress	Permissible Stress	Required Strength Level[a,b]
Complete Penetration Groove Welds		
Tension normal to the effective throat.	Same as base metal.	Matching weld metal must be used.[f]
Compression normal to the effective throat.[c]	Same as base metal.	
Tension or compression parallel to the axis of the weld.	Same as base metal.	Weld metal with a strength level equal to or less than matching weld metal may be used.
Shear on the effective throat.	$0.30 \times$ nominal tensile strength of weld metal (ksi) except stress on base metal shall not exceed $0.40 \times$ yield stress of base metal.	
Partial Penetration Groove Welds		
Compression normal to effective throat.[c]	Same as base metal.	
Tension or compression parallel to axis of the weld.	Same as base metal.	
Shear parallel to axis of weld.	$0.30 \times$ nominal tensile strength of weld metal (ksi) except stress on base metal shall not exceed $0.40 \times$ yield stress of base metal.	Weld metal with a strength level equal to or less than matching weld metal may be used.
Tension normal to effective throat.[d]	$0.30 \times$ nominal tensile strength of weld metal (ksi) except stress on base metal shall not exceed $0.60 \times$ yield stress of base metal.	
Fillet Welds[e]		
Stress on effective throat, regardless of direction of application of load.	$0.30 \times$ nominal tensile strength of weld metal (ksi) except stress on base metal shall not exceed $0.40 \times$ yield stress of base metal.	Weld metal with a strength level equal to or less than matching weld metal may be used.
Tension or compression parallel to the axis of weld.	Same as base metal.	

Plug and Slot Welds

Shear parallel to faying surfaces.	$0.30 \times$ nominal tensile strength of weld metal (ksi) except stress on base metal shall not exceed $0.40 \times$ yield stress of base metal.	Weld metal with a strength level equal to or less than matching weld metal may be used.

[a] For matching weld metal, see table below.

[b] Weld metal, one strength level stronger than matching weld metal, will be permitted.

[c] AISC allows lower strength weld metal to be used.

[d] Cannot be used in tension normal to their axis under fatigue loading (AWS 2.5). AWS Bridge prohibits their use on any butt joint (9.12.1.1), or any splice in a tension or compression member (9.17), or splice in beams or girders (9.21), however, are allowed on corner joints parallel to axial force of components of built up members (9.12.1.2(2)): Cannot be used in girder splices (AISC 1.10.8).

[e] Fillet welds and partial penetration groove welds joining the component elements of built up members (except flange to web welds) may be designed without regard to the axial tensile or compressive stress applied to them.

[f] Matching weld metal and base metal: see above, left.

Weld Metal	60 (or 70)	70	80	90	100	110
Type of steel	A36 A53 Gr B	A242 A441	A537 Class 2		A514 A517 2½" & over	A514 A517
	A106 Gr B A131	A537 Class 1	A572 Gr 65			
	A139 Gr B A375	A516 Gr 65, 70				
	A381 Gr Y35 A500	A572 Gr 42–60				
	A501 A516 Gr 55, 60	A588 A618				
	A524 A529	API 5LX Gr 42				
	A570 Gr D,E A573 Gr 65	ABS Gr AH, DH, EH				
	API 5L Gr B					
	ABS Gr A,B,C,CS,D,E,R					

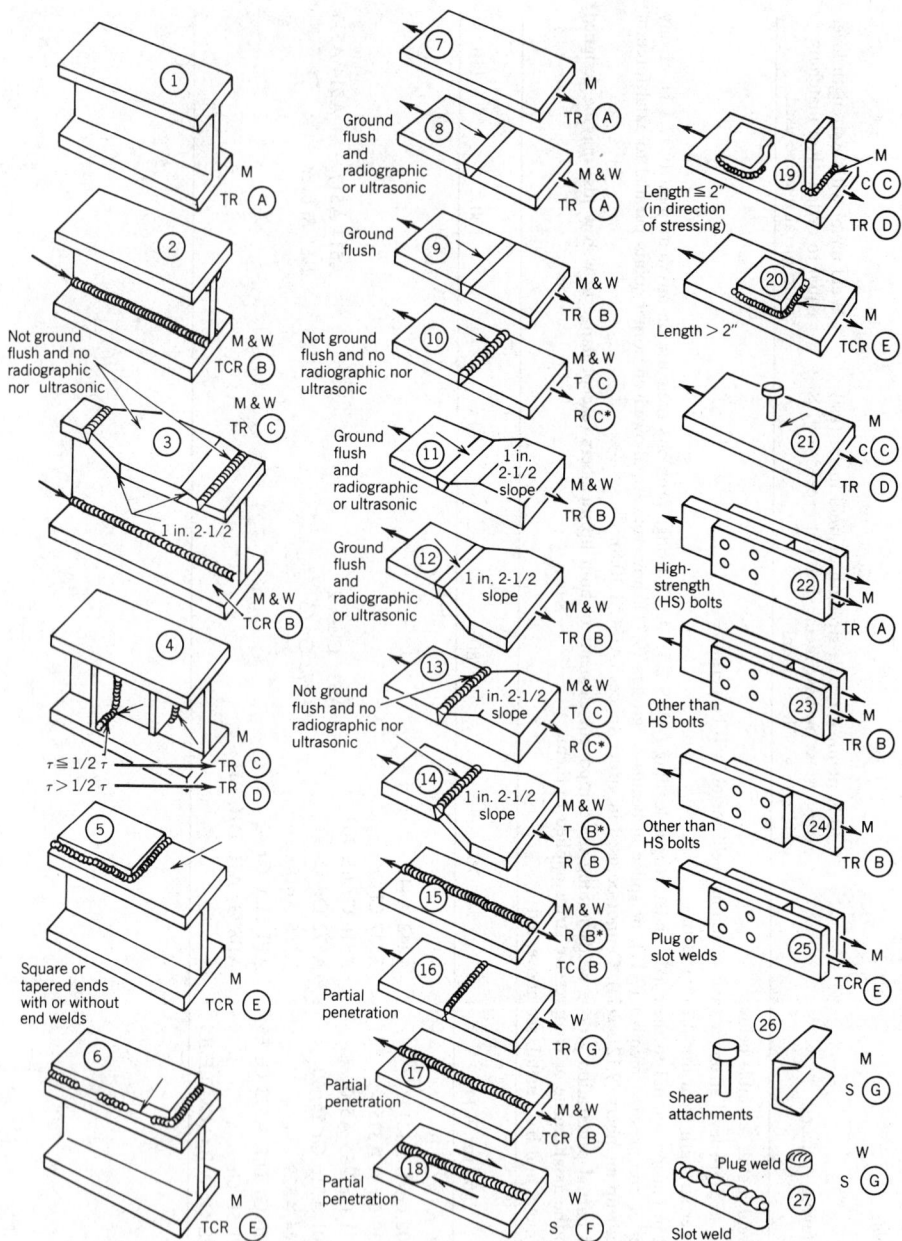

Fig. 24.35a AISC allowable range of stress.

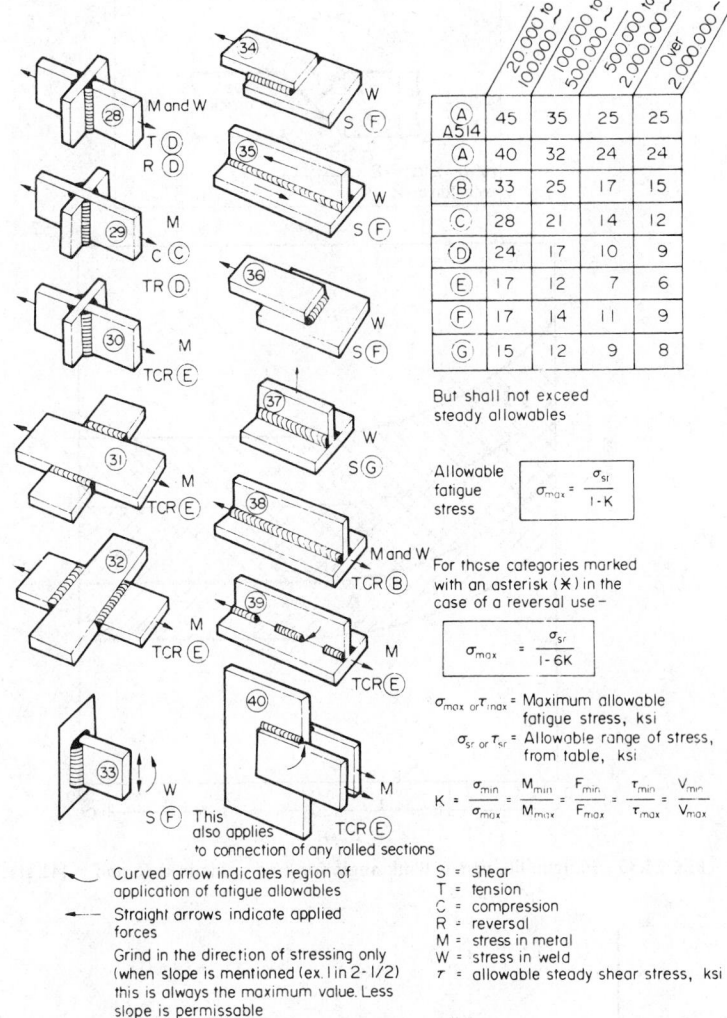

		20,000 to 100,000	100,000 to 500,000	500,000 to 2,000,000	Over 2,000,000
Ⓐ	A514	45	35	25	25
Ⓐ		40	32	24	24
Ⓑ		33	25	17	15
Ⓒ		28	21	14	12
Ⓓ		24	17	10	9
Ⓔ		17	12	7	6
Ⓕ		17	14	11	9
Ⓖ		15	12	9	8

But shall not exceed
steady allowables

Allowable
fatigue $\sigma_{max} = \dfrac{\sigma_{sr}}{1-K}$
stress

For those categories marked
with an asterisk (✱) in the
case of a reversal use –

$$\sigma_{max} = \frac{\sigma_{sr}}{1-6K}$$

σ_{max} or τ_{max} = Maximum allowable
fatigue stress, ksi
σ_{sr} or τ_{sr} = Allowable range of stress,
from table, ksi

$$K = \frac{\sigma_{min}}{\sigma_{max}} = \frac{M_{min}}{M_{max}} = \frac{F_{min}}{F_{max}} = \frac{\tau_{min}}{\tau_{max}} = \frac{V_{min}}{V_{max}}$$

This
also applies
to connection of any rolled sections

Curved arrow indicates region of
application of fatigue allowables
Straight arrows indicate applied
forces
Grind in the direction of stressing only
(when slope is mentioned (ex. 1 in 2-1/2)
this is always the maximum value. Less
slope is permissable

S = shear
T = tension
C = compression
R = reversal
M = stress in metal
W = stress in weld
τ = allowable steady shear stress, ksi

Fig. 24.35b AISC allowable range of stress.

Defects (Shrinkage Cracks)

Figure 24.38 illustrates an inescapable conclusion: if steel weldments are to have a useful "long" life—above about 10^5 cycles—the stress must be below about 8 ksi, since the best of techniques will still produce cracks up to about 0.040 in. The value of 8 ksi holds without regard for the static strength, that is, simply using a steel of higher static strength does not improve the fatigue performance, in fact, the performance may be degraded because of the lower toughness (strength and toughness are reciprocal properties). This conclusion is further borne out by Fig. 24.39, which results from extensive testing of conventional joints and techniques.

Thus, while the earlier methods of determining allowable stresses are acceptable for static loads, the preceding parameters must be satisfied for successful designs under dynamic (repeated) loads.

Figures 24.40–24.42 illustrate attainable fatigue performance for several weldments.

24.2.2 Spot Welds

The fatigue behavior of spot-welded joints is similar to that of riveted joints, the strength being influenced by the type of joint (lap or butt), the spot pattern, and the weld quality. The mode of failure in

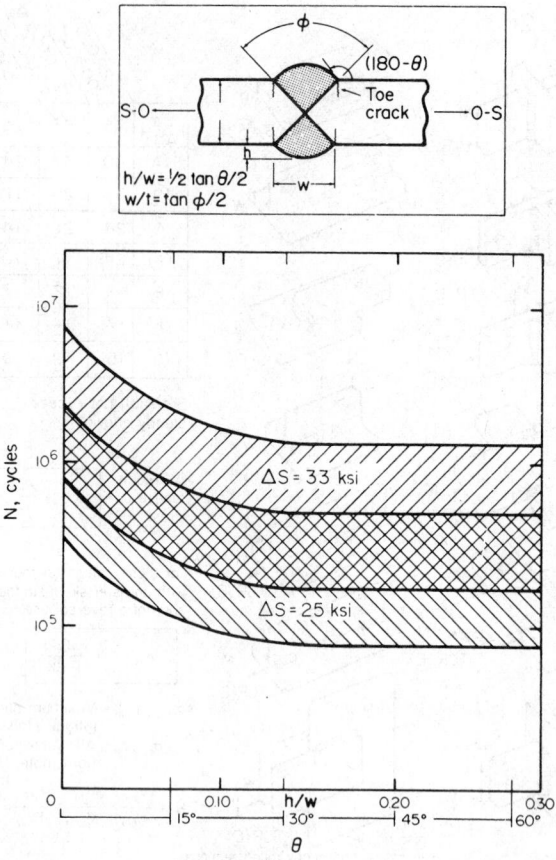

Fig. 24.36 Fatigue life versus flank angle for butt welds in A36 and A441 steels.

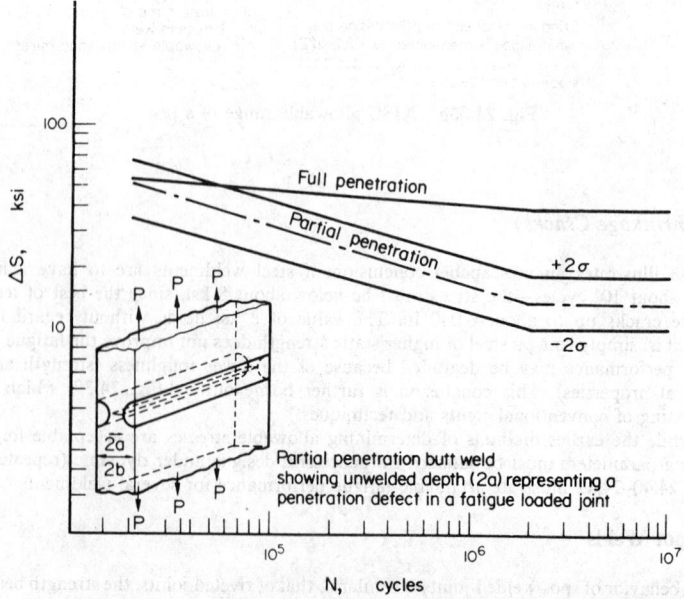

Fig. 24.37 Fatigue life versus penetration but butt welds in A36 steel.

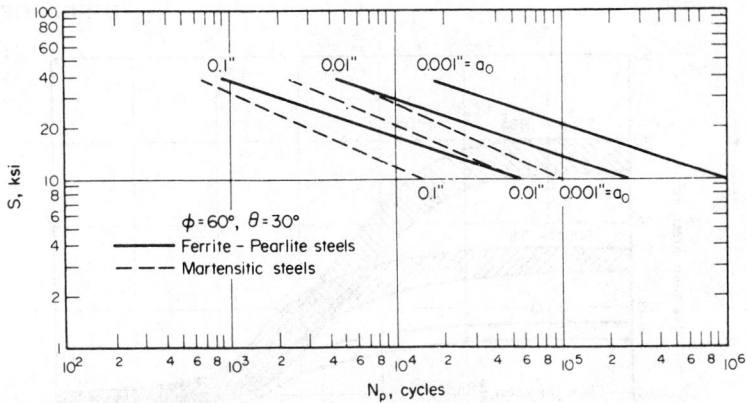

Fig. 24.38 Effect of initial flaw size on fatigue crack propagation life.

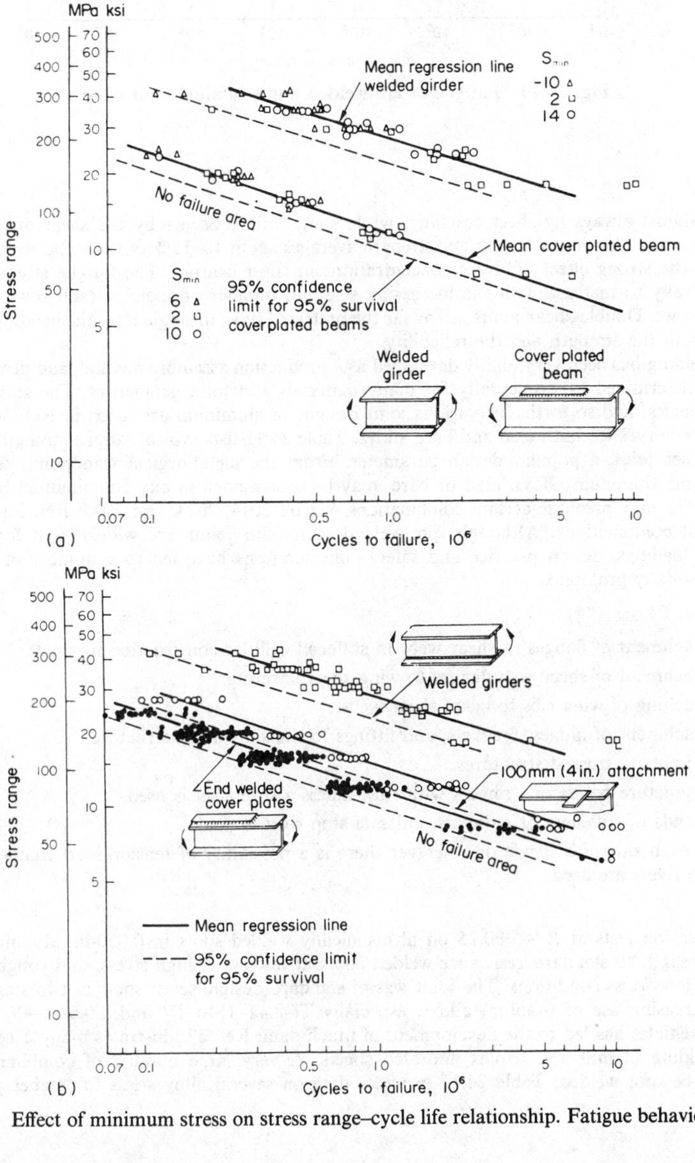

(a)

(b)

Fig. 24.39 Effect of minimum stress on stress range–cycle life relationship. Fatigue behavior of welded beams.

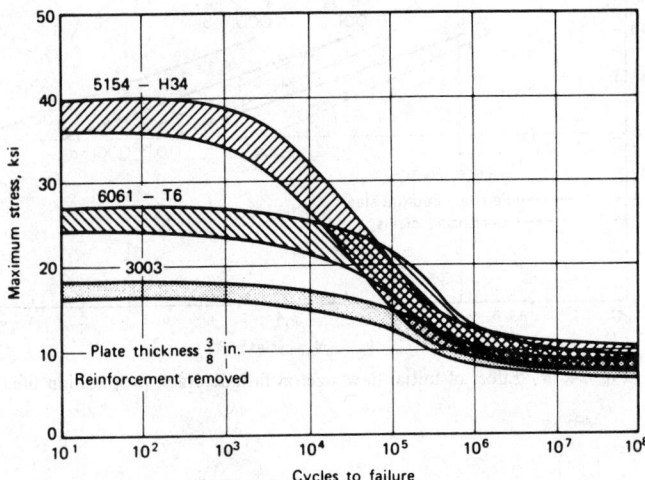

Fig. 24.40 Fatigue of arc-welded joints in aluminum plate.

fatigue is almost always by sheet cracking, while static failure occurs by the shear of the spot. For joints in steel and aluminum, the fatigue strength averages about 10–15% of the static strength, indicating, again, the strong effect of stress concentrations in sheet bearing. The fatigue strength has been found generally to increase with the increasing size and number of spots in each row and with the number of rows. Double-shear joints are by far the preferred type; in single laps, the bending component decreases both the strength and the reliability.

Spot welding has been very highly developed as a production assembly method, and design strengths have been determined experimentally for many materials and joint geometries. The static strengths, sheet efficiencies, and so forth, for various joint designs in aluminum are listed in Ref. 36, as well as specific *S–N* curves for both clad and bare alloys. Table 24.11 lists typical fatigue strengths for Alclad in pounds per joint, a popular design parameter. From the metallurgical standpoint, essentially all the structural aluminum alloys, clad or bare, may be spot-welded in any combination; but corrosion considerations may preclude certain combinations of bare 2014, 2024, and 7075; Ref. 3 gives a list of all practical combinations. Although spot-welded aluminum joints are widely used for both static and cyclic loadings, design practice and safety considerations have led to a number of applications that are normally prohibited:

1. Attachment of flanges to shear webs in stiffened cellular construction in wings.
2. Attachment of shear web flanges to wing sheet covering.
3. Attaching of wing ribs to beam shear webs.
4. Attachment of hinges, brackets, and fittings to the supporting structure.
5. At joints in trussed structures.
6. At juncture points of stringers with ribs, unless a stop rivet is used.
7. At ends of stiffeners or stringers, unless a stop rivet is used.
8. On each side of a joggle or wherever there is a possibility of tension load component, unless stop rivets are used.

Axial tension tests at $R = +0.25$ on ultrasonically welded spots in 0.050-in. aluminum indicate superior strength to standard resistance-welded spots at low-cycle, high-stress, and rough equality at high-cycle, low-stress conditions. The joint was of standard design: seven spots in two staggered rows.

The increasing use of titanium alloys, especially Ti–8A1–1Mo–1V and Ti–6A1–4V, for aircraft and space vehicles has led to the development of much data; Ref. 37 illustrates many excellent results of spot welding in mill and triplex annealed sheets. A very large number of combinations of the steels may be spot welded; Table 24.12 provides data on several alloy steels for higher performance joints.

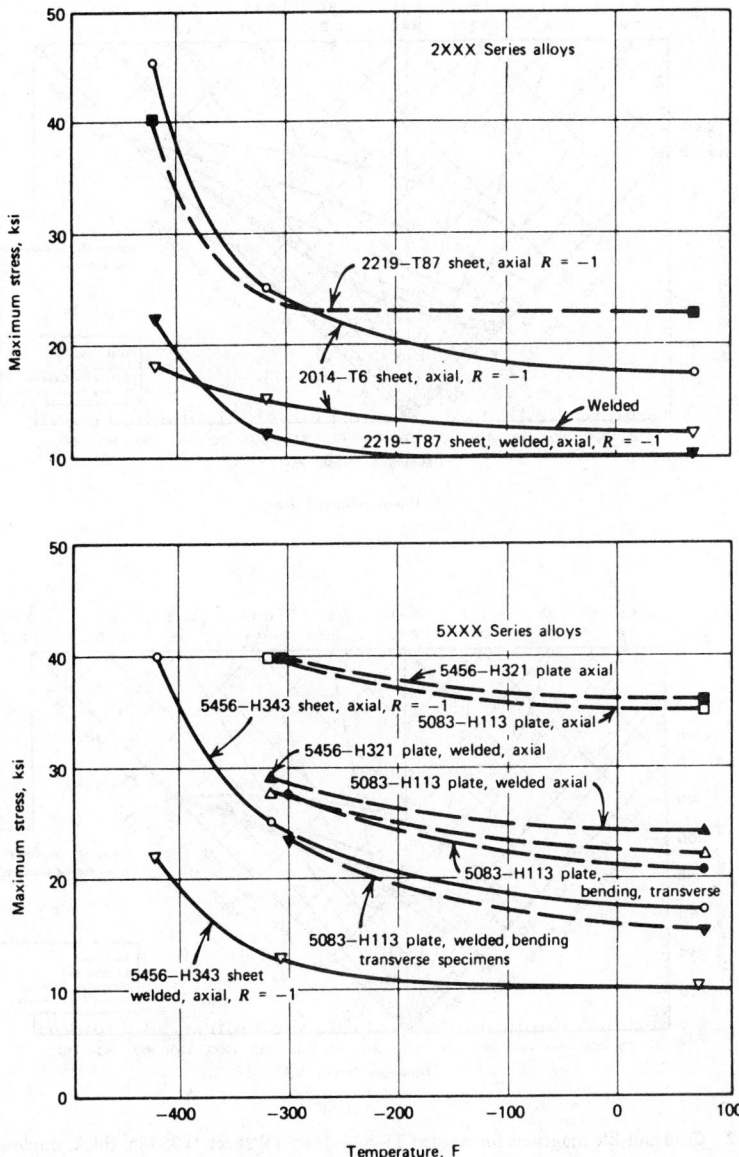

Fig. 24.41 Low-temperature fatigue strengths of welded aluminum alloys, at 10^6 cycles.

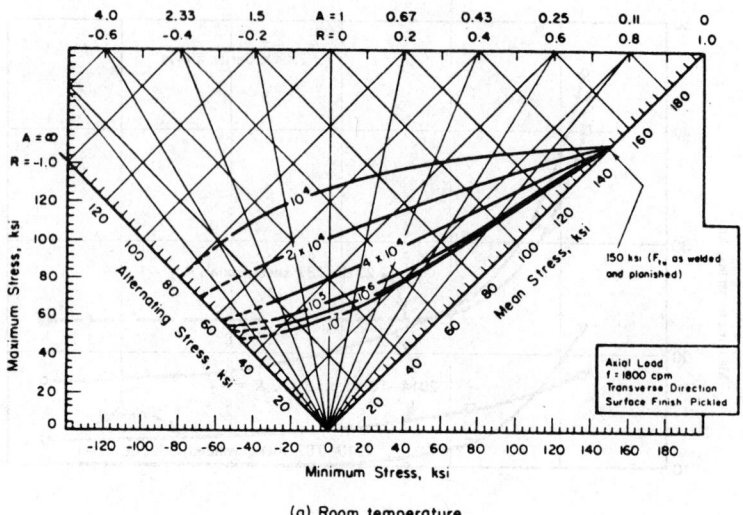

(a) Room temperature

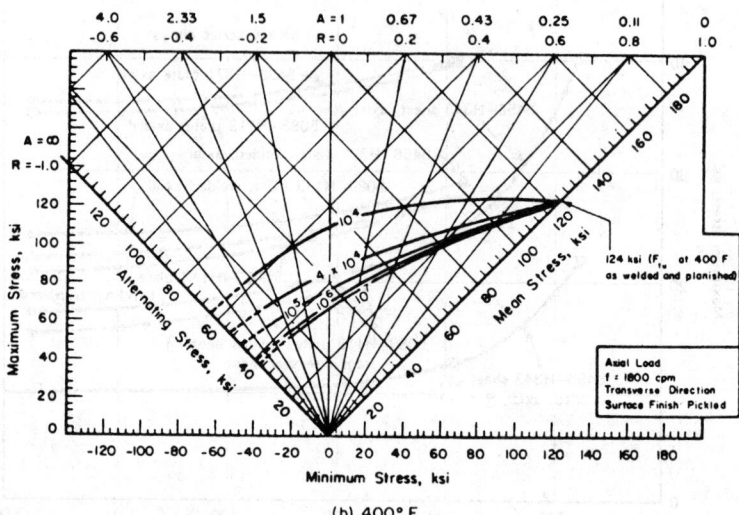

(b) 400° F

Fig. 24.42 Constant-life diagrams for welded Ti–8Al–1Mo–IV sheet, 0.050-in. thick, duplex-annealed.

Table 24.11 Typical Fatigue Data from Spot Welds in Alclad Sheet[a]

Sheet Material	Spots per Joint	Sheet Gage, in.	Static Strength, lb/joint	Type of Test	Loading conditions (R)	Fatigue Strength of Joint (at Given Number of Cycles), lb/joint			
						10^4	10^5	10^6	10^7
24S-T Alclad	4[a]	0.025	335	Axial	+0.25	210	140	110	100
24S-T Alclad	4[a]	0.025	335	Axial	+0.75	---	225	160	140
24S-T Alclad	6[b] (1)	0.025	313	Axial	+0.25	185	150	100	74
24S-T Alclad	6[b]	0.025	313	Axial	+0.75	---	200	155	145
24S-T Alclad	4[a]	0.032	312	Axial	+0.25	245	170	135	125
24S-T Alclad	6[a]	0.032	325	Axial	+0.25	210	180	120	90
24S-T81 Alclad	1	0.064	1060	Reversed Shear	-1.0	440	350	160	100
24S-T3 Alclad	1	0.064	1220	Reversed Shear	-1.0	450	390	200	110
24S-T86 Alclad	1	0.068	1100	Reversed Shear	-1.0	510	360	160	85
75S-T6 Alclad	1	0.064	1320	Reversed Shear	-1.0	530	380	180	120
14S-T4 Alclad	1	0.064	1210	Reversed Shear	-1.0	480	320	170	100

(Rows 1–6, columns 10^4–10^7 are bracketed with the note (2).)

[a] (1) Specimens marked [a] had a 1¼-in. pitch; those marked [b] had a ¾-in. pitch. (2) Strength values are in pounds per spot for these specimens.

24.3 BONDED JOINTS

The successful design of adhesively bonded joints is dependent on the relationship between the shape of the joint section and the loading mode. In addition to the *shape* of the *section* its proportions as described by the joint factor are critical in determining the fatigue properties.

For joining sheet or strip the inexpensive offset lap, Fig. 24.43, is often used, but it has relatively little resistance to bending. By comparison the in-line type, Fig. 24.44, generally provides both higher static strength and superior fatigue performance owing to its lower stress concentrations. Tapering or beveling of the adherend in both types is beneficial, since this makes the stress distribution more uniform. The double lap is frequently best because its symmetrical load path (in tension) eliminates bending.

It should be noted for all types that to reduce the stress concentration below the major category requires preparatory shaping, additional pieces, or other complications. For tees and corners, Fig. 24.45, the same conditions apply generally, and the type offers the opportunity for improvement through the use of extruded strips or integrally molded flanges.

Both the static and fatigue strengths are functions of the joint factor, defined as $\sqrt{t/l}$, where t is the sheet thickness and l is the length of overlap. The fatigue relationship is given in Fig. 24.46 for standard lap joints, with the alternating stress in the sheet expressed as

$$S_a = \frac{1.15/\sqrt{t}}{1 + 1.5\sqrt{t}\,(l)} \text{ ksi}$$

based on the fatigue strength of the bond material, not of the sheet. Thus, for thin sheets and large overlap, the fatigue load capacity of the bond would be very high, as might be expected; but for extreme values of the joint factor, the equation does not hold, and failure of the sheet could occur at a lower stress than indicated. At the other extreme of joint geometry—thick sheets and short overlap—failure would occur in the bond at lower stresses, as given by the preceding equation. A typical joint of $l = 1.0$ in. and $t = 0.050$ in., bonded with an epoxy resin (Araldite), would show a fatigue strength in axial tension of 3.6–4.0 ksi, or about 8% of the static strength. For large values of joint factor (small overlaps), the effect of the offset in a single lap becomes important; the adherends bend and reduce the achievable shear stress. For large overlap (small joint factor), the yield strength of the adherend may be exceeded, leading to a metal (adherend) failure.

Low-cycle fatigue (below about 10^6 cycles) tends to occur in the bond, but for lower loadings and longer lives, the failure is usually in the adherend. Figure 24.47 shows that generally better performance is obtained by the use of small joint factors (longer overlaps). The stress concentrations at the ends of the (metal) adherends are slightly reduced by increasing the overlap. A typical specification,

Table 24.12 Typical Fatigue Data from Spot Welds in Alloy Steel Sheet

Sheet Material	Condition	Spots per Joint	Sheet gage, in.	Static Strength, lb./joint	Type of Test	Loading condi- tions(R)	Fatigue strength of Joint (at Given Numbers of Cycles), lb./joint			
							10^4	10^5	10^6	10^7
SAEX4130 steel	As welded	4	0.025	1387	Axial	0	850	670	500	---
SAEX4130 steel	Welded, stress relieved 950°F.	4	0.025	2665	Axial	0	960	760	420	---
SAEX4130 steel	Welded, stress relieved 1100°F.	4	0.025	2742	Axial	0	(1100)	880	560	---
SAE 1010 steel	---	---	0.010	2192	Axial	0	900	550	225	---
18-8 stainless steel	---	---	0.020	2783	Axial	0	825	425	---	---
18-8 stainless steel	---	2	0.030	2916	Axial	0	920	500-530	230-290	---
Austenitic Mn-Cr steel	---	2	0.035	2844	Axial	0	920	500-530	230-290	---
Austenitic Mn steel	---	2	0.035	2443	Axial	0	920	500-530 5×10^5	230-290	---
Republic Cor-ten steel	---	1	0.050	2590	Bending	-1.0	---	32.5	27.0	17
Republic Cor-ten steel	---	1	0.050	3020	Bending	-1.0	---	28.0	22.5	16
18-8 stainless steel	---	1	0.050	1185	Bending	-1.0	---	28.5	24.5	20
18-8 stainless steel	---	1	0.050	3225	Bending	-1.0	---	24.0	20.5	17

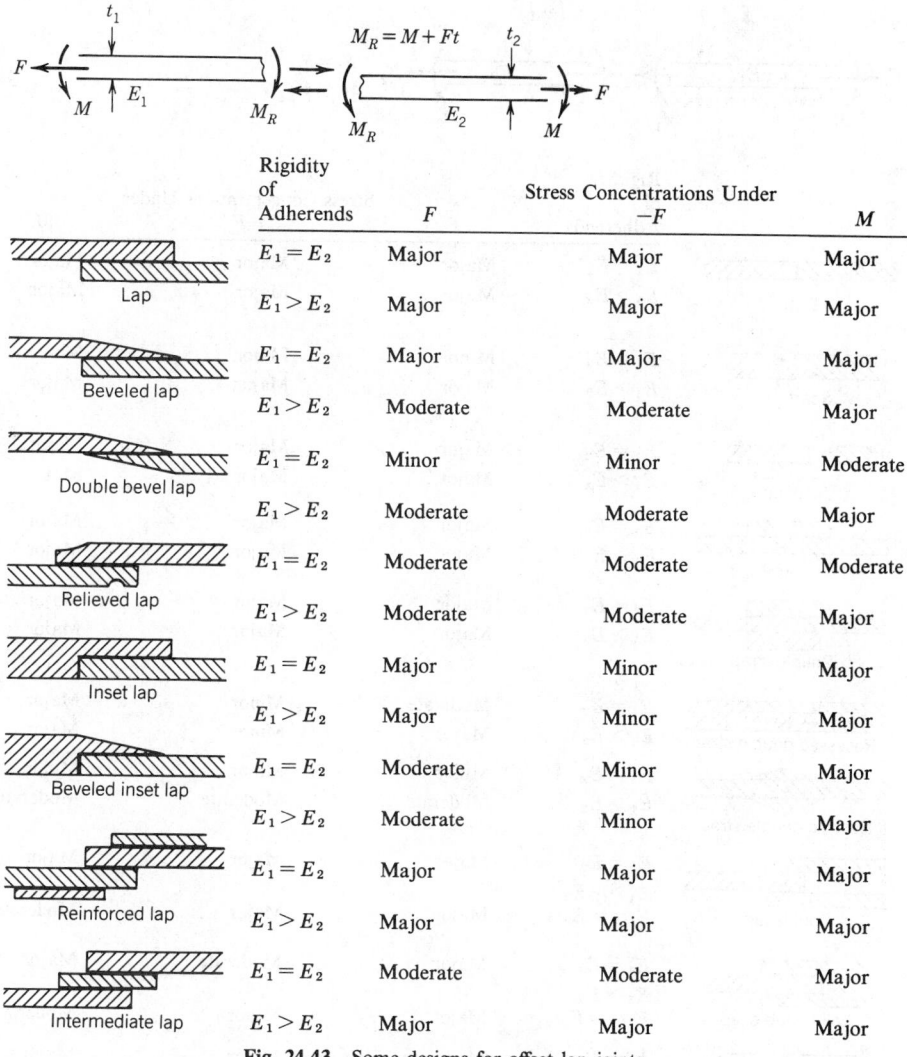

Fig. 24.43 Some designs for offset lap joints.

MIL-A-8331, requires a fatigue strength of 650 psi minimum at 10^7 cycles, irrespective of the joint factor value.

For joints in which the adherends are of different thickness, the thinner of the two should be used to determine the joint factor. This practice is conservative because the thicker sheet takes more of the bending moment, reducing the offset of the thinner sheet.

In the design of panels and similar structural elements using conventional sheet metal construction, the local buckling stress of a flat sheet is considered as proportional to the square of the thickness-to-width ratio, $(t/l)^2$. In bonded construction, it has been found that, over the width of the bond, two sheets act as one of twice the thickness, so that the allowable buckling stress is quadrupled. Compared to riveted or spot-welded assemblies, this condition provides much greater stiffness and reduces considerably the effective width for buckling of adjacent bonded sheets in the same plane. This condition holds for shear webs and compression panels; Fig. 24.48 shows the superior performance of the bonded panel over that of a riveted one. Typical strength requirements are, for epoxy-bonded aluminum, 2500 psi static and 600 psi at 10^6 cycles, with the test conditions as in Mil Spec MMM-A-134, Para. 4.5.5.4.

To ensure adequate integrity and proper process control the ASTM recommended practices are

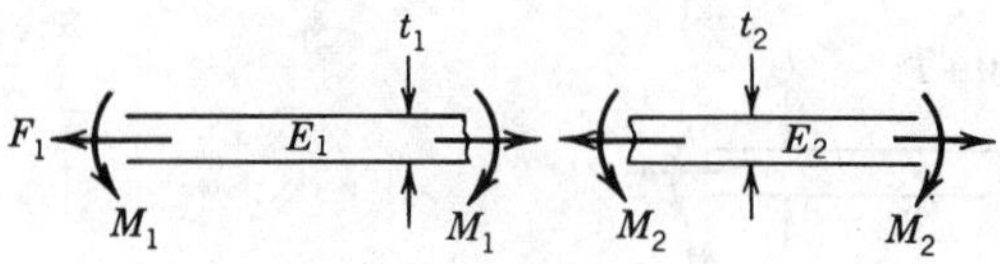

$$F_1 = F_2$$
$$M_1 = M_2$$

	Rigidity of Adherends	Stress Concentrations Under		
		F	$-F$	M
 Butt	$E_1 = E_2$	Major	Major	Major
	$E_1 > E_2$	Major	Major	Major
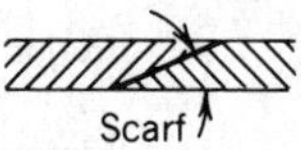 Scarf	$E_1 = E_2$	Minor	Minor	Major
	$E_1 > E_2$	Major	Major	Major
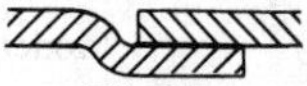 Offset lap	$E_1 = E_2$	Major	Major	Major
	$E_1 > E_2$	Major	Major	Major
 Strap	$E_1 = E_2$	Major	Major	Major
	$E_1 > E_2$	Major	Major	Major
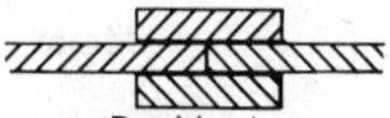 Double strap	$E_1 = E_2$	Major	Major	Major
	$E_1 > E_2$	Major	Major	Major
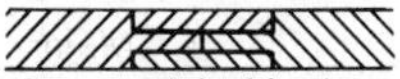 Recessed double strap	$E_1 = E_2$	Moderate	Minor	Major
	$E_1 > E_2$	Major	Minor	Major
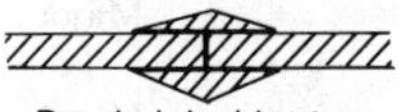 Beveled double strap	$E_1 = E_2$	Minor	Minor	Minor
	$E_1 > E_2$	Moderate	Moderate	Moderate
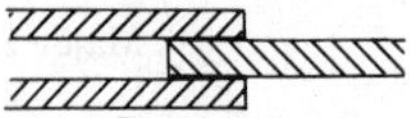 Double lap	$E_1 = E_2$	Major	Major	Major
	$E_1 >> E_2$ $E_1 t_2 = E_1 t_2$	Major	Major	Moderate
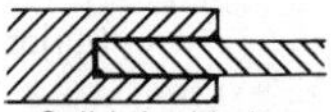 Solid double lap	$E_1 = E_2$	Major	Moderate	Major
	$E_1 > E_2$ $E_1 t_1 = E_2 t_2$	Major	Minor	Moderate
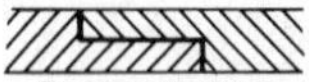 Double butt lap	$E_1 = E_2$	Moderate	Minor	Major
	$E_1 > E_2$	Major	Minor	Major
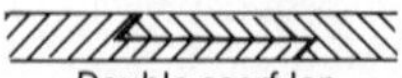 Double scarf lap	$E_1 = E_2$	Moderate	Minor	Major
	$E_1 > E_2$	Major	Minor	Major

Fig. 24.44 Some designs for collinear lap joints.

often used, especially D2918 Peel Strength; D2919 Shear Strength; D3528 Tensile Strength; and D3762 Surface Durability.

A properly bonded adhesive joint will often outperform a mechanical welded joint because it contains an inherent crackstopper, and because its lower local stiffness usually leads to lower local stresses, especially in bending.

Table 24.13 lists the types of structural adhesives; and the general fatigue behavior of a variety of joints, adhesives, and adherends is illustrated in Figs. 24.49 and 24.50.

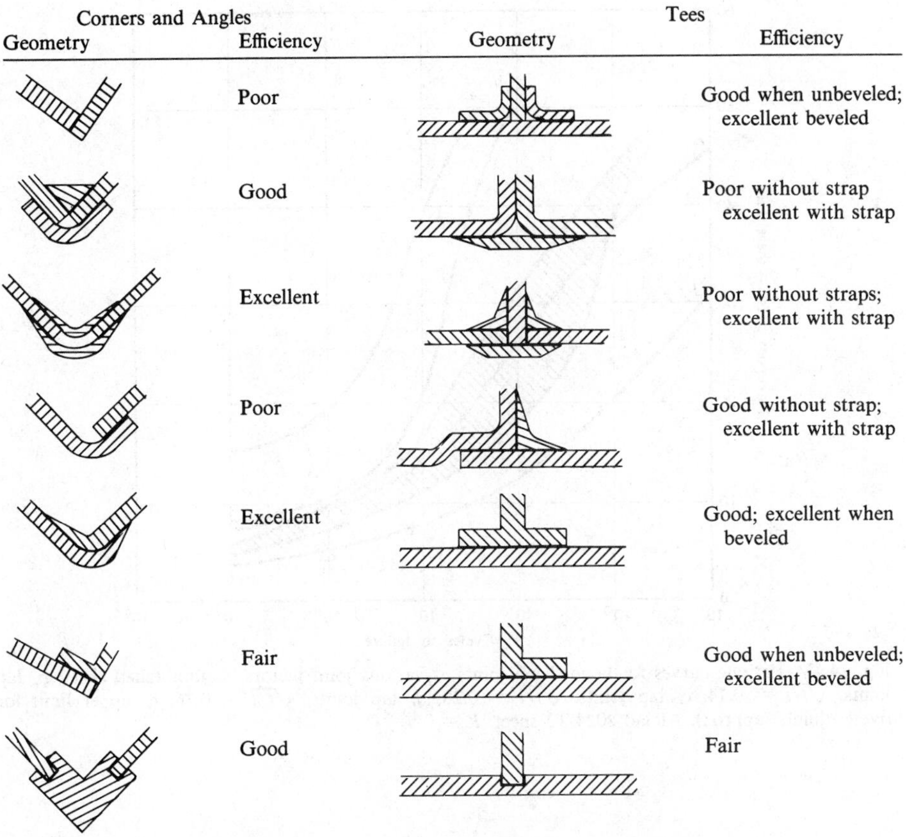

| Corners and Angles | | Tees | |
Geometry	Efficiency	Geometry	Efficiency
	Poor		Good when unbeveled; excellent beveled
	Good		Poor without strap excellent with strap
	Excellent		Poor without straps; excellent with strap
	Poor		Good without strap; excellent with strap
	Excellent		Good; excellent when beveled
	Fair		Good when unbeveled; excellent beveled
	Good		Fair

Fig. 24.45 Some designs for edge, angle, and tee joints.

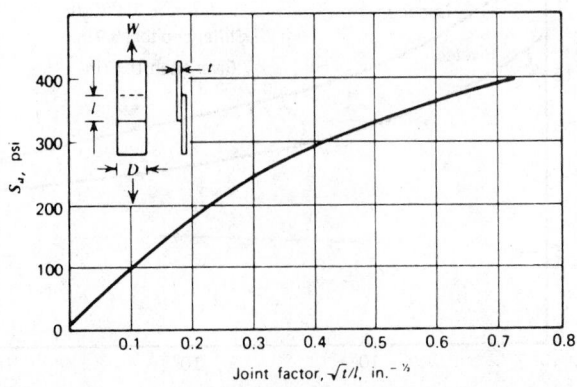

Joint factor, $\sqrt{t/l}$, in.$^{-\frac{1}{2}}$

Fig. 24.46 Fatigue on bonded lap joints in axial tension. Aluminum sheets, Araldite, or Redux Bond. 10^6 cycles; $R = 0.5$–0.9.

623

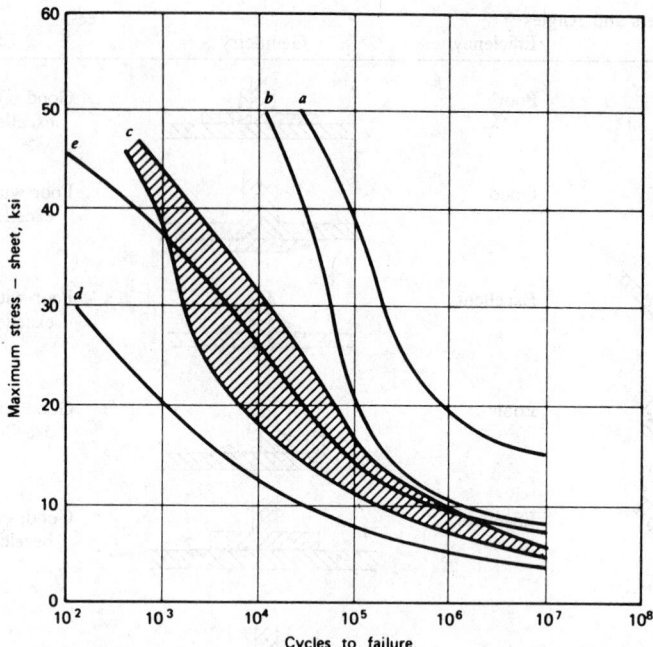

Fig. 24.47 Fatigue curves for bonded lap joints at various joint factors: *a,* unnotched sheet; *b,* lap joints, $\sqrt{t/l}$ = 0.14; *c,* lap joints, $\sqrt{t/l}$ = 0.28; *d,* lap joints, $\sqrt{t/l}$ = 0.36; *e,* upper limit for riveted joints (approx). Alclad 2024-T3 sheet. $R = 0$.

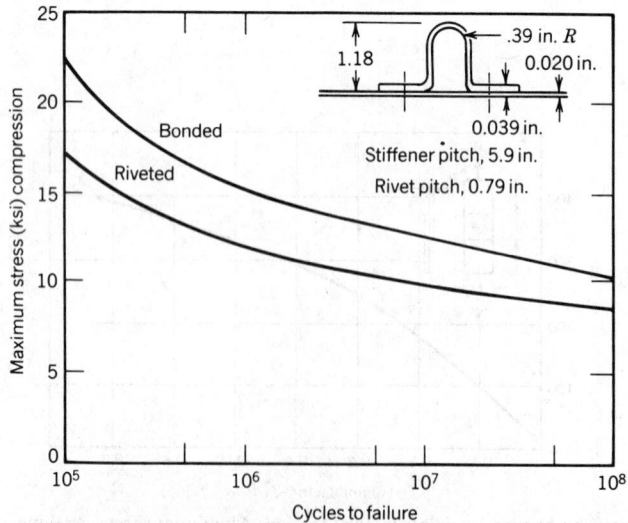

Fig. 24.48 Fatigue of bonded and riveted panels. Static failure load—riveted: 22.1 ksi; static failure load—bonded: 26.3. Minimum mean load = 1.1 ksi compression.

Table 24.13 Types of Structural Adhesives

Hot-Melt Adhesives

Thermoplastic resins—100% solids
Fast application and fast set
Rigid to flexible bonds
Bonds to permeable or impermeable surfaces
Nonpolar, adhesion poor, and wetting poor
Low cost in materials and labor
Requires special dispensing equipment
Insensitive to moisture and solvent attack
Low heat resistance—degrades as temperature rises
Poor creep resistance—5 psi @ 125° F

Dispersion or Solution Adhesives

Thermoplastic resins—20-50% solids
Very easy to apply—long shelf life
Slow or fast setting by evaporation
Low material cost and solvent loss
Excellent wetting and penetration
No special equipment required
Moderate clamp pressure needed
Low strength under load poor creep resistance
Requires two permeable surfaces
Sensitive to heat, solvent, and moisture

Cyanoacrylate Adhesives

Thermosetting—100% solids—liquid
Easiest to apply and cure
Fast cure—30 sec to 5 min
Low viscosity—mated surfaces needed
Good tensile strength—poor impact strength
No special equipment needed
Will not bond permeable materials
High material cost—poor shelf life
Poor solvent resistance
Handling hazards—bonds to skin

"Second-Generation" Modified Acrylic Adhesives

Thermoset—100% reactive—liquids and pastes
Fast room-temperature cure—3-60 sec
One or two pack
High strength bonds with metal and plastics—good peel and shear strength
Tolerant of dirty surfaces
Good gap filling properties
Limited open time
Versatile application techniques
Variable curing time and pot life
Excellent for engineered and reinforced plastics
Characteristic acrylic odor
Rigid and flexible bonds are possible

Epoxy Adhesives

Thermoset—100% solids—liquid
Variable pot life and cure time
Long shelf life—one to two components
Rigid or flexible bonds—highly polar
Poor creep under load—little shrinkage
Resists most chemicals and solvents
Wide variety of formulations available
Requires careful controls and clean surfaces
Need equipment to weigh, mix, dispense, and clamp
Requires clean surfaces

Polyurethane Adhesives

Thermoset—100% solids—liquid
Flexible sealant/adhesive—low creep
Bonds permeable or impermeable surfaces
One or two components—room or oven cure
Good flexibility at low temperature
Sensitive to moisture both when unmixed or cured
May undergo reversion with heat and moisture
Moderate material cost
Some formulations are toxic
Special equipment to mix, and dispense

Table 24.13 (*Continued*)

Silicone Adhesives	*Phenolic or Urea Adhesives*	*Anaerobic Adhesives*
Thermoset—100% solids—liquid	Thermoset resins—100% solids or solution	Thermoset—100% solids—liquid
Rubberlike—high impact and peel strength	Gases during cure—solvent evaporates	One component—moderate cost
Retains properties in 100–400°F temperature range	High tensile strength—low impact strength	Easiest application—simple cure
Excellent sealant for low stress applications	Highly polar—excellent adhesion	Machinery and structural grades available
One component—short shelf life	Long shelf life—application easy	Good cohesive, but low-adhesive strength
Easy application—1–5-day cure	Requires high pressure—oven cure	Excellent for bolts, nuts, and static joints
Resists most chemicals and solvents	Very low material cost	Requires primer for many materials
Simple equipment needed	Excellent wetting and penetration	Not suitable for permeable surfaces
Reversion possible	High shrinkage stresses—brittle	Will not cure where air contacts
High material cost		Crazes some plastics

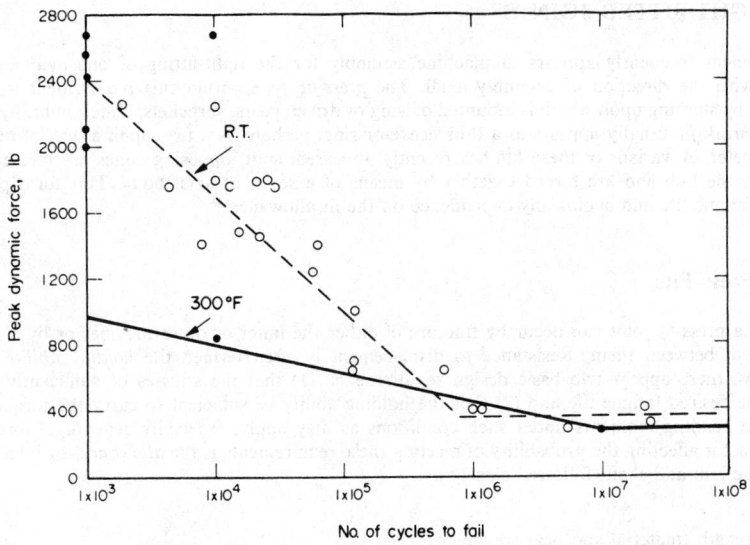

Fig. 24.49 Fatigue strength of Imidite 850 joints at various temperatures.

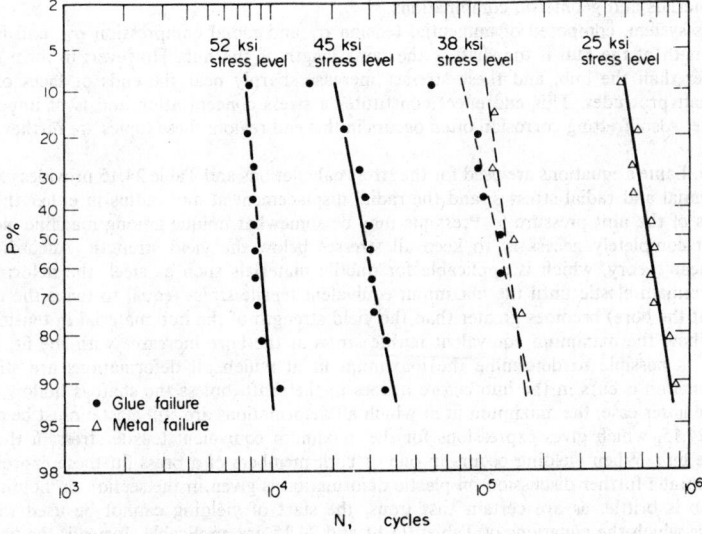

Fig. 24.50 *S–N–P* curves for 2024-T3 Alclad. Bonded with FM-123-2 adhesive; double-strap joints; tapered adherends; log–normal probability.

24.4   TIGHT-FITTED JOINTS

A requirement frequently appears in machine assembly for the tight-fitting of one cylinder inside another, with the direction of assembly axial. The *press-fit* type within this requirement is readily illustrated by shafting upon which is mounted driving or driven gears, sprockets, clutch hubs, flywheels, etc. The *shrink fit* usually appears as a thin sleeve or ring, perhaps as a tire, upon a disk of relatively large diameter. A variant of these fits has recently appeared: split, opposing cones are located in an annulus in the hub and are forced together by means of a series of axial bolts. This form provides excellent fatigue life and avoids any dependence on the fit allowance.[38]

24.4.1   Press Fits

Failure of a press-fit joint can occur by fracture of either the inner or outer member or by a relative displacement between them. Resistance to displacement is often termed the *holding ability* of the joint. Thus, there appear two basic design requirements: (1) that the stresses be sufficiently low to provide the desired fatigue life and (2) that the holding ability be sufficient to carry the torque, axial forces, and bending moments under such conditions as may apply, especially centrifugal force. The principal factor affecting the probability of meeting these requirements is the *allowance*, or interference between the hub and shaft, followed closely by:

1. Strength (material and heat treatment).
2. Type of loading cycle.
3. Surface finish, including out-of-roundness and taper.
4. Lubricant and the coefficient of friction.
5. Shape of the end of the shaft and of the bore.
6. Number of times a pair of members have been assembled.

General experience has led to the use of an allowance, e, of 0.001 in. per in. of shaft diameter as a safe specification value for the machine-tool, automotive, and other mass-production industries. However, for the make-up of wheels and axles, the railroads prefer to specify the maximum force (tonnage) to assemble. Because, in any case, the assembly force may be relatively large, press fitting is usually considered to be a shop rather than a field operation. By contrast, shrink fitting can be done wherever heating and/or cooling facilities are provided.

For control of the stresses and displacements at the design stage, the unit pressure p is needed. Figures 24.51 and 24.52 provide values of p in terms of e and d/D for two popular combinations: steel shaft/steel hub and steel shaft/cast-iron hub.[39] Table 24.14 gives the general expressions for p for many materials and geometric combinations.

The stress system, composed of tangential tension σ_t and radial compression σ_c, and is uniform within the length of the hub if the shaft is the same length as the hub. However, in most cases, the shaft is longer than the hub, and these stresses increase sharply near the ends or faces of the hub where the shaft protrudes. This end effect constitutes a stress concentration and is of importance to the fatigue life. Also, fretting corrosion often occurs in this end region; these topics are further discussed below.

Generally, Lamé's equations are used for the stress calculations and Table 24.15 provides expressions for the tangential and radial stresses, and the radial displacement at any radius in either the shaft or hub, in terms of the unit pressure p. Press fits may be somewhat unique among machine elements in that it is not completely necessary to keep all stresses below the yield strength. According to the maximum shear theory, which is applicable for ductile materials such as steel, the deformations of the hub will remain elastic until the maximum equivalent tensile stress (equal to twice the maximum shear stress at the bore) becomes greater than the yield strength of the hub material in tension. Figure 24.53 shows how the maximum equivalent tensile stress at the bore increases with the fit, and from this figure it is possible to determine the maximum fit at which all deformations are still elastic. Plastic deformation occurs in the hub before it does in the shaft, unless the shaft is hollow and thin-walled. In the latter case, the maximum fit at which all deformations are still elastic must be calculated from Table 24.15, which gives expressions for the maximum equivalent tensile stress in the shaft as well as in the hub. When yielding occurs in one or both members of a press fit, these expressions are no longer accurate; further discussion on plastic deformation is given in the section on holding ability.

If the hub is brittle, as are certain cast irons, the start of yielding cannot be used to find the limiting fit for which the equations of Tables 24.14 and 24.15 are applicable. Instead, the fit at which the hub will fracture becomes important; the maximum stress theory is used to predict the failure of brittle materials. Figure 24.54 shows how the maximum tensile stress in a hub, or the tangential

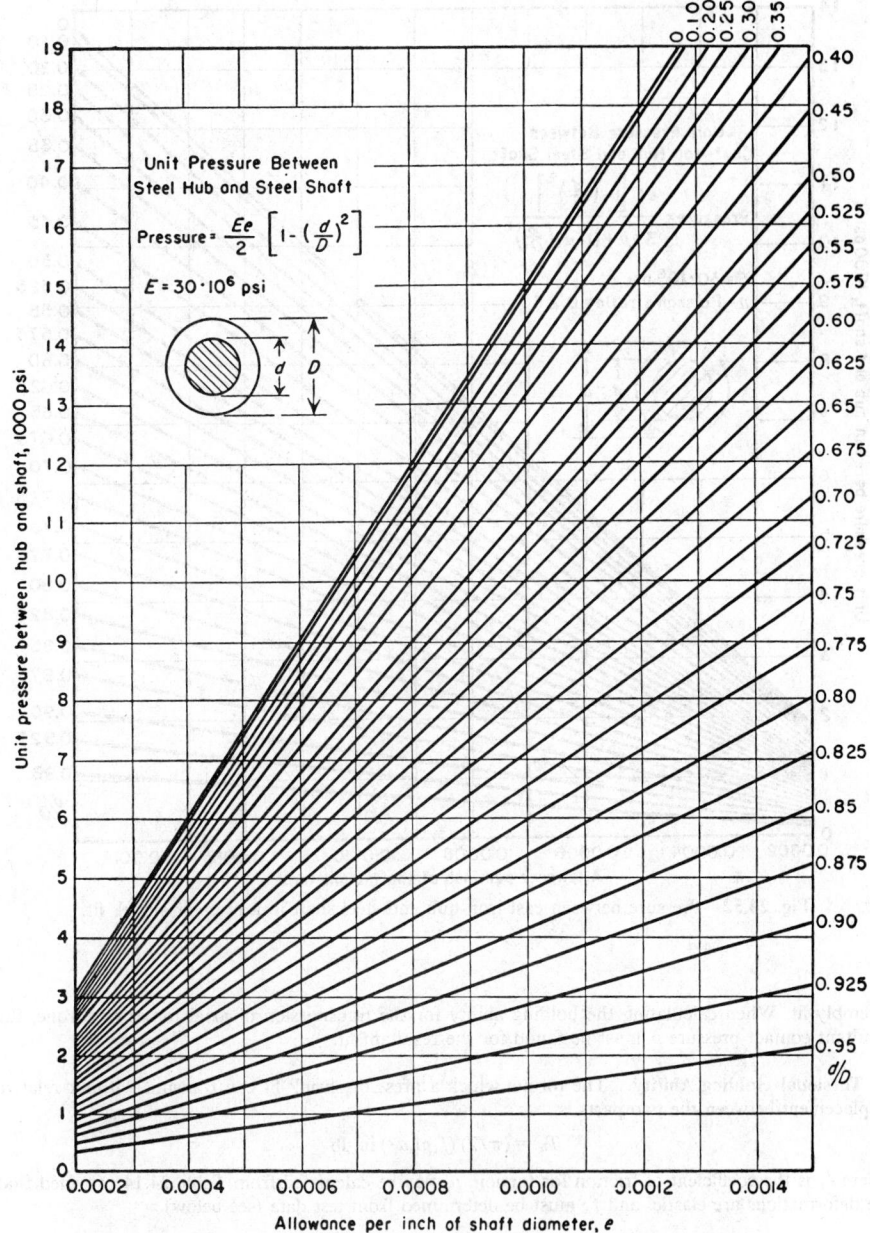

Fig. 24.51 Pressure between steel hub and steel shaft in a press or shrink fit.

stress at the bore, increases with the fit allowance. When this stress becomes equal to the ultimate tensile strength of the hub material, fracture occurs. Fatigue failure is discussed below.

Holding Ability as Affected by Service Conditions

During service a press-fitted shaft and hub will be subjected to forces in addition to those from the fit allowance.

Centrifugal Forces. In high-speed machinery the deformations due to centrifugal force cause a decrease in the effective fit. This decrease is given in Table 24.16 and must always be less than the

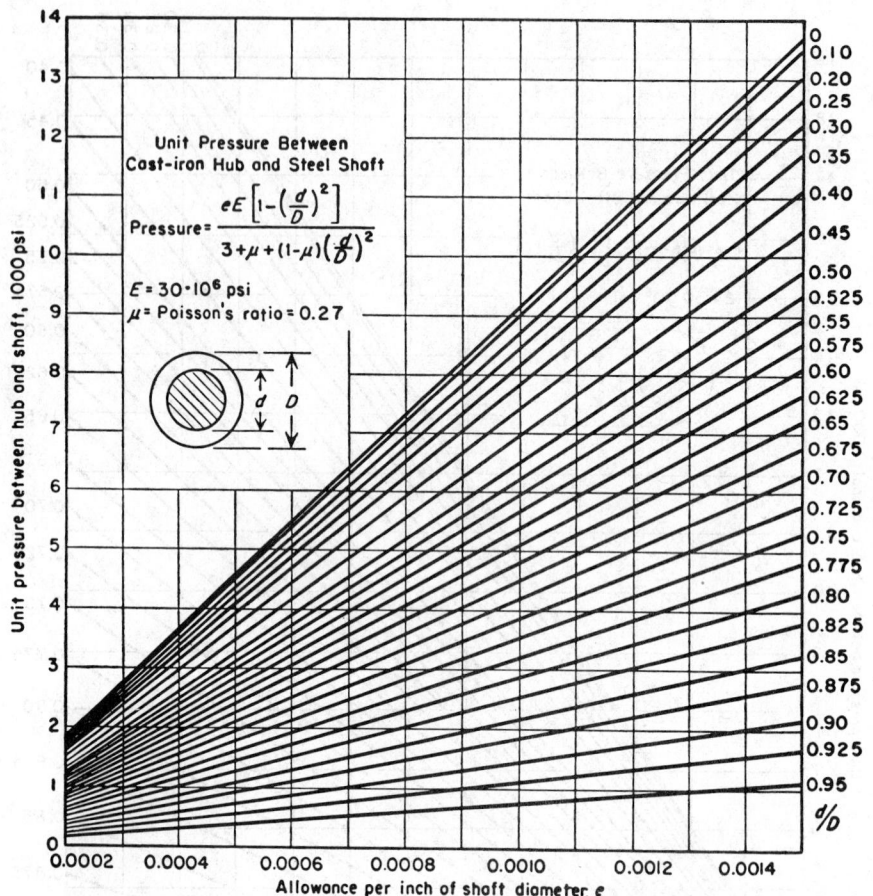

Fig. 24.52   Pressure between cast-iron hub and steel shaft in a press or shrink fit.

assembly fit. When calculating the holding ability for the transmission of axial forces or torque, the resultant contact pressure p must be found for the resultant fit.

Torsional Holding Ability.   The torque which a press or shrink fit can transmit without relative displacement between the two parts is

$$T_h = (\pi/2)\,(f_1 p L d^2)\ \text{in.-lb}$$

where f_1 is the coefficient of friction for torsion. p may be calculated from Table 24.14 provided that the deformations are elastic, and f_1 must be determined from test data (see below).

Axial Holding Ability.   The axial force that a press or shrink fit can transmit without relative displacement of the two parts is

$$F_h = \pi f p L d\ \text{lb}$$

where L is the length of the fit; d is the diameter of the shaft; p is the unit pressure from Table 24.14; and f is the coefficient of friction. Figure 24.55 shows that the value of p and, therefore, the holding ability can be increased by increasing the wall thickness of the hub. However, after the hub thickness becomes equal to the shaft radius, only a small increase in p can be obtained by further thickening of the hub. According to Fig. 24.53 yielding should begin at a fit of 0.0017 in. per in. of shaft diameter, but Ref. 40 has shown that, for a plain carbon steel (yield strength of 52 ksi), both the mounting pressure and holding ability continue to increase up to a limiting fit of 0.003 in. per in. of shaft diameter. Beyond this value the assembly force decreases until the fit is 0.007 in. per in., then becomes greater again for still tighter fits; the holding ability decreases asymptotically and never

Table 24.14 Unit Pressure between Hub and Shaft, Expressed in Terms of e, the Press Fit per Inch Shaft Diameter

Press-fitted Members	p
Hub and hollow shaft of different materials	$\dfrac{e}{\dfrac{1}{E_1}\left[\left(\dfrac{D^2+d^2}{D^2-d^2}\right)+\mu_1\right]+\dfrac{1}{E}\left[\left(\dfrac{d^2+d_i^2}{d^2-d_i^2}\right)-\mu\right]}$
Hub and hollow shaft of same material	$\dfrac{(D^2-d^2)(d^2-d_i^2)Ee}{2d^2(D^2-d_i^2)}$
Cast-iron hub and hollow steel shaft $E_1=E/2$; $\mu_1=\mu$	$\dfrac{Ee}{\mu+\dfrac{3d^2(D^2-d_i^2)=D^2d_i^2+d^4}{(D^2-d^2)(d^2-d_i^2)}}$
Hub and solid shaft of different materials	$\dfrac{e}{\dfrac{1}{E_1}\left[\left(\dfrac{D^2+d^2}{D^2-d^2}\right)+\mu_1\right]+\dfrac{1-\mu}{E}}$
Hub and solid shaft of same material	$\dfrac{Ee}{2}\left[\dfrac{D^2-d^2}{D^2}\right]$
Cast-iron hub and solid steel shaft $E_1=E/2$; $\mu_1=\mu$	$\dfrac{Ee(D^2-d^2)}{(3+\mu)D^2+(1-\mu)d^2}$

p = unit pressure between hub and shaft
e = press fit per inch of shaft diameter
D = OD of hub
d = nominal outside diameter of shaft
   = diameter of bore of hub
d_i = diameter of bore of hollow shaft
E_1 = modulus of elasticity for hub material
E = modulus of elasticity for shaft material
μ_1 = Poisson's ratio for hub material
μ = Poisson's ratio for shaft material

Source: Ref. 40, by permission of the ASME.

rises again. These findings indicate a good basis for permitting the bore stresses to exceed the yield strength criterion expressed in Fig. 24.53.

Holding Ability for Transverse Loading. In most cases the shaft projects from the hub and is loaded by transverse forces, which cause crushing stresses in both the hub and shaft along the fit. The stress level depends on the bending moment M and the total transverse shearing force V at the hub face from which the shaft projects. The equations given in Table 24.15 are in such form that they are applicable not only for crankpin fits but also for any case of a projecting shaft that is loaded by transverse forces all on the same side of the hub. If the joint is not to loosen, the crushing stresses (Table 24.15) should not exceed the unit pressure (Table 24.14) due to the fit allowance. Thus, the ratio of crushing stress to pressure is a measure of the ability of the joint to transmit bending moment M and total transverse shearing force V, which act over the cross section of the shaft at the hub face. Usually the maximum M is known fairly accurately because the fatigue strength of the shaft is known. Then the crushing stress can be calculated from

$$(\text{crushing stress})/(\text{bending stress}) = \tfrac{3}{8}(d/l)^2$$

which is sufficiently accurate provided that the shear stress at the hub face is less than one-eighth the maximum bending stress, that is, $M/Vd > 1$.

Factors Affecting Holding Ability and Assembly

Coefficient of Friction, Lubricant, and Surface Finish. As might be expected there is considerable variation in friction depending on the lubricant and finish, but enough data exist to make reasonable approximations. Table 24.17 summarizes the results of the large number of tests from Ref. 41. In these tests the fit surfaces were ground and the lubricant was a mixture of white lead and machine oil. From the torsion tests on 2.5-in.-diam. shafts it was found that complete slip was obtained for coefficients of friction from 0.25 to 0.33. Initial slip from the hub face from which the shaft projects begins at a torque that is much lower than the torsional holding ability, that is, the ultimate torque required to produce complete slip. The torque to produce initial slip is essentially independent of the

Table 24.15 Stresses and Displacements in a Hub and Shaft that Are Press-fitted Together[a]

	For hub with: $D = $ OD $d = $ ID $L = $ hub length	For shaft with: $d = $ OD $d_i = $ ID	For solid shaft with: $d = $ OD
Tangential stress at any radius r	$\dfrac{d^2}{(D^2-d^2)}\left(\dfrac{D^2}{4r^2}+1\right)p$	$-\dfrac{d^2}{(d^2-d_i^2)}\left(1+\dfrac{d_i^2}{4r^2}\right)p$	$-p$
Tangential stress at OD	$\dfrac{2d^2}{(D^2-d^2)}p$	$-\dfrac{(d^2+d_i^2)}{(d^2-d_i^2)}p$	$-p$
Tangential stress at bore	$\dfrac{(D^2+d^2)}{(D^2-d^2)}p$	$-\dfrac{2d^2}{(d^2-d_i^2)}p$	
Radial stress at any radius r	$-\dfrac{d^2}{(D^2-d^2)}\left(\dfrac{D^2}{4r^2}-1\right)p$	$-\dfrac{d^2}{(d^2-d_i^2)}\left(1-\dfrac{d_i^2}{4r^2}\right)p$	$-p$
Radial stress at OD	0	$-p$	$-p$
Radial stress at bore	$-p$	0	
Max equivalent tensile stress at any radius = $2 \times$ max shearing stress at same radius	$\dfrac{d^2D^2}{2(D^2-d^2)r^2}p$	$\dfrac{d^2}{(d^2-d_i^2)}\left(1+\dfrac{d_i^2}{4r^2}\right)p$	p
Max equivalent tensile stress at OD	$\dfrac{2d^2}{(D^2-d^2)}p$	$\dfrac{(d^2+d_i^2)}{(d^2-d_i^2)}p$	p
Max equivalent tensile stress at bore	$\dfrac{2D^2}{(D^2-d^2)}p$	$\dfrac{2d^2}{(d^2-d_i^2)}p$	
Radial displacement at any radius	$\dfrac{[(1-\mu_1)r+(1+\mu_1)(D^2/4r)]}{[(D^2/d^2)-1]}\dfrac{p}{E_1}$	$-\dfrac{[(1-\mu)r+(1+\mu)(d_i^2/4r)]}{[1-(d_i^2/d^2)]}\dfrac{p}{E}$	$-(1-\mu)r\dfrac{p}{E}$
Radial displacement at OD	$\dfrac{d^2D}{(D^2-d^2)}\dfrac{p}{E_1}$	$-\left[\dfrac{d(d^2+d_i^2)}{(d^2-d_i^2)}+\mu d\right]\dfrac{p}{2E}$	$-(1-\mu)d\dfrac{p}{2E}$
Radial displacement at bore	$\left[\dfrac{d(D^2+d^2)}{(D^2-d^2)}+\mu_1 d\right]\dfrac{p}{2E_1}$	$-\dfrac{d^2d_i}{(d^2-d_i^2)}\dfrac{p}{E}$	
Crushing stress at hub face from which shaft projects	$\dfrac{4}{\pi dL^2}(3M+2VL)$		
Crushing stress at other hub face	$\dfrac{4}{\pi dL^2}(3M+VL)$		

Source: Ref. 40, by permission of the ASME.

[a] $r = $ any radius measured from center line or shaft or hub

$L = $ length of fit

$M = $ bending moment in the shaft of the hub face from which the shaft projects

$V = $ total transverse shearing force at the same cross section of the shaft $\begin{Bmatrix}\text{positive}\\\text{negative}\end{Bmatrix}$ when M $\begin{Bmatrix}\text{increases}\\\text{decreases}\end{Bmatrix}$ as hub face is approached

For other notations, see Table 24.14.

length of the fit (except for very short fits), and is affected by the radius of the fillet that joins the hub seat on the shaft to the projecting portion of the shaft.

Reference 42 assembled 200 press fits: the coefficients of friction ranged from 0.077 to 0.33 for steel shafts in steel hubs, and from 0.053 to 0.30 for cast-iron hubs; the average for all tests was 0.17.

Reference 43 pressed $\frac{5}{16}$–8-in. shafts in hubs, both parts being hardened ground steel, using rape oil as a lubricant. The calculated coefficients of friction for these tests were 0.054–0.22, with an average of 0.11.

Reference 44 pressed 1.5-in.-diam. steel pins into steel collars with a total allowance of 0.0008 in. The ram pressure at assembly, per inch of shaft diameter, was 3.39 tons for rape oil; 3.95 tons for cutting oil; 4.25 tons for graphite in engine oil; 6.38 tons for Texaco oil; and 9.00 tons for Bayonne oil. When these parts were shrunk together with a 0.001-in. fit and dry surface, 19 tons were required to break the fit; whereas if rape oil were used in the shrink fit, only 5.67 tons were necessary. Thus the dry shrink-fit joint had a holding ability of 3.35 times the lubricated one. By cooling the pin with liquid oxygen before assembly, demounting pressures of 3.37 tons for an etched surface, 2.28 tons for rape oil, and 13.9 tons dry were obtained. Tests in torsion indicated that rape oil gave about the same coefficients of friction for both press and shrink fits, but dry-shrunk joints gave values about 2.75 times as great.

The comparison of ground versus rough-turned surfaces is inconclusive in the rigorous sense; results from Refs. 41 and 43 indicate, typically, an insignificant difference: $f = 0.086$ for ground and 0.091 for turned. The hardness, composition, and method of manufacture (cast, rolled, or forged) appear to have little effect on the coefficient of friction, at least for the steels and cast irons.

Shape of the Shaft End and of the Hub Bore. It is often necessary to improve the entering conditions for the shaft into the bore and/or to increase the holding ability. These objectives can

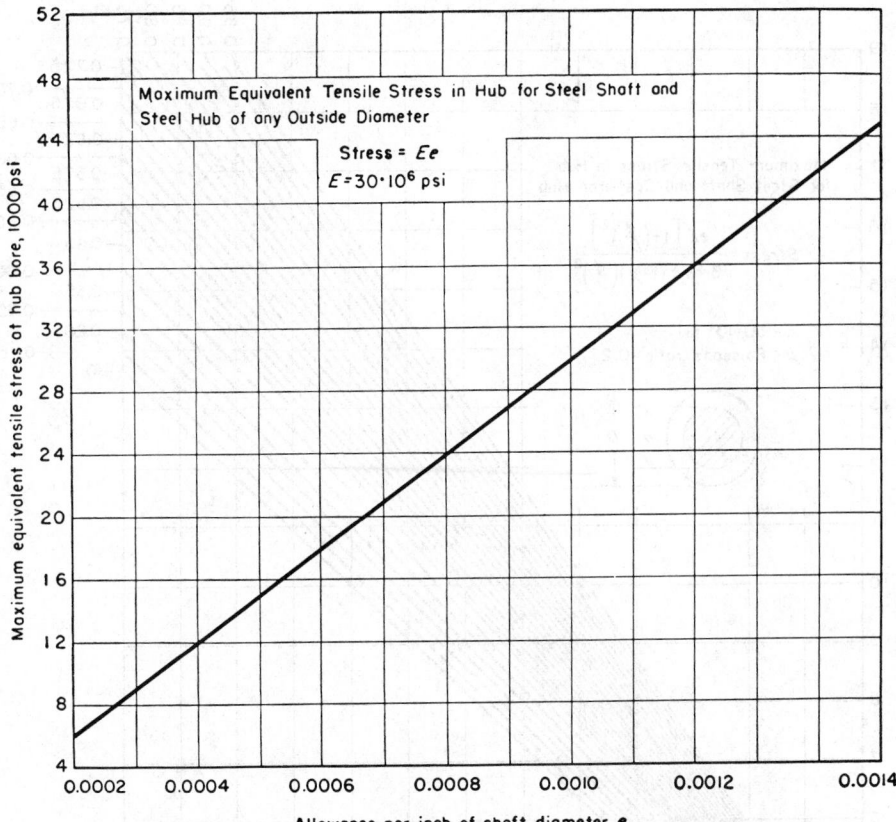

Fig. 24.53 Maximum equivalent tensile stress in hub (steel hub and steel shaft.

usually be accomplished by chamfering or radiusing the entering end of the shaft. Sharp edges cause scraping or wearing away of the bore and also a permanent set resulting in a bellmouth shape on the entering end of the bore. During assembly, the bore opposite the entering end actually closes in so that the major portion of the hub not yet in contact with the shaft restrains the expansion of the entering end, and causes the hub stresses there to be higher than those calculated on the fit allowance. For this reason, it is often found that the hub, when finally mounted, is not in contact with the shaft at the entering end for a considerable distance, as much as an inch for heavy press fits. Even if contact is present, the pressure may be very low.

Type of Loading Cycle. The generally available data do not indicate that variation of the type of cycle was a test parameter. Since the great majority of uses of press and shrink fits is in rotating–bending, the stress ratio R turns out to be -1.0 ($R = S_{min}/S_{max}$), and the only variable is the stress level. No data have been found to indicate directly the effect of *defects* on the life of press fits, but such fracture mechanics calculations should be made, especially in the consideration of the quality of cast hubs; see Ref. 45.

Fatigue Failure of Shaft

The fatigue strength of a press or shrink fit approaches that of a monolithic machined piece of the same geometry. The weakness due to shape is intensified by fretting corrosion in the end regions, evidence of which is found by the presence of iron oxide particles near the ends. When fretting is present, as it usually is, fatigue cracks can initiate at stresses far below the endurance limit. Then the nominally high service stresses will propagate these cracks to fracture. If the net tension in service can be made sufficiently low, such that the stress-intensity range maximum is *very* approximately 4–8 ksi $\sqrt{in.}$, the crack-growth rate should be low enough to give an adequate life. As noted above, crack growth estimates should be made for both the shaft and hub.

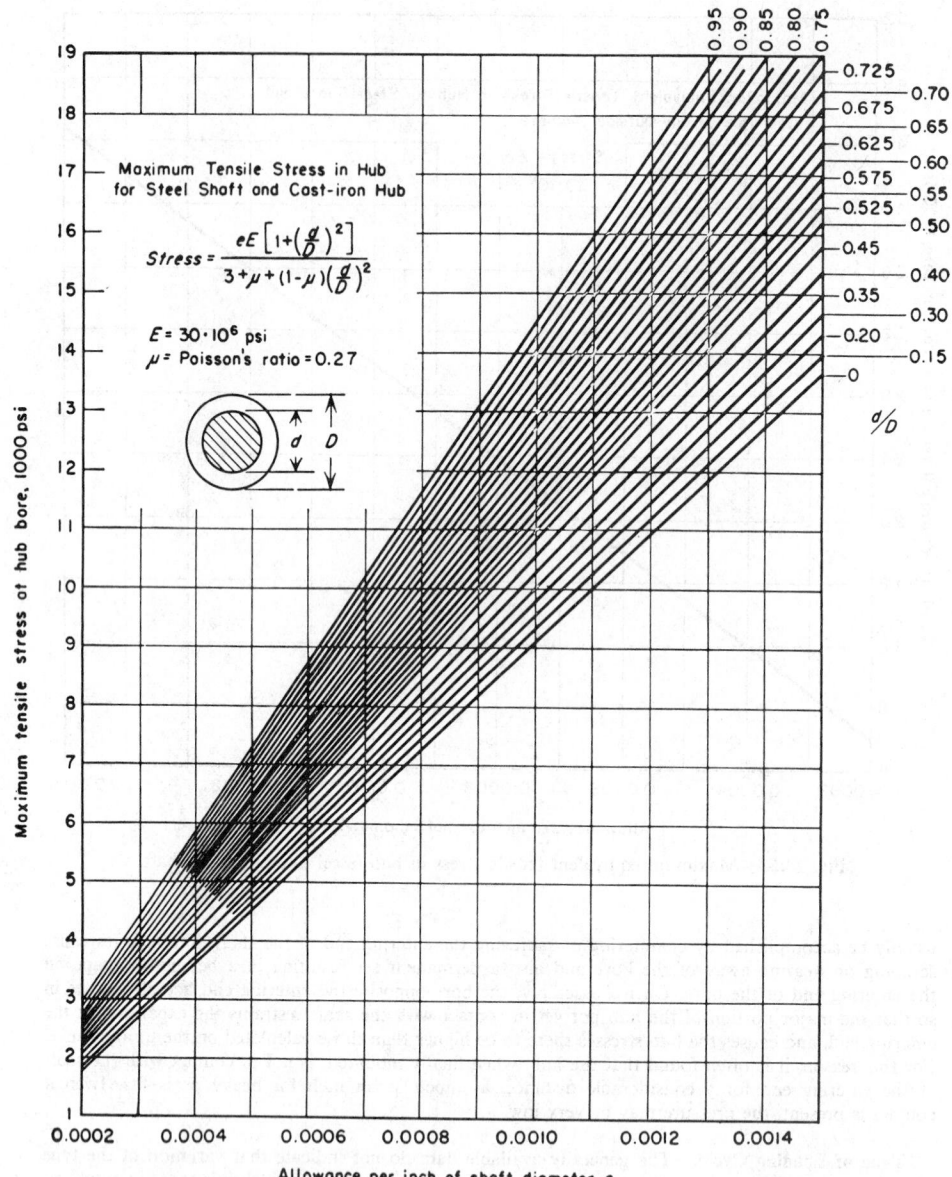

Fig. 24.54 Maximum tensile stress in hub (cast-iron hub and steel shaft).

Improvement in fatigue strength can be obtained by attention to the geometric details of the shaft; changes of diameter with small radii are apt to be hazardous, but the raised seat, as in Fig. 24.56, is excellent. These curves are based on tests of full-sized members for approximately 10^8 stress reversals. The upper curve predicts complete fracture, while the lower curve indicates the maximum nominal stress that will not produce fatigue cracks large enough to be visible by the dry-magnetic-particle method of inspection. Figure 24.57 shows four typical designs that have been tested in the form of full-sized (9-in.-dia.) shafts for 10^8 cycles at $R = -1.0$. For Fig. 24.57a a nominal stress of 11,000 psi caused fracture; the other three designs required about 15,000 psi to fracture. There is some opinion that these last four designs show lower resistance to crack initiation due to the low value of L/D, than does the raised seat design. Permissible design stresses for small shafts run higher than those

Table 24.16 Decrease in Fit per Inch of Shaft Diameter between Hub and Shaft Due to Centrifugal Forces[a]

Press-fitted Members	Decrease in e
Hub and hollow shaft of different materials	$\dfrac{U^2}{4g}\left\{\dfrac{\gamma_1}{E_1}\left[(3+\mu_1)+(1-\mu_1)\dfrac{d^2}{D^2}\right]-\dfrac{\gamma}{E}\left[(3+\mu)\dfrac{d_i^2}{D^2}+(1-\mu)\dfrac{d^2}{D^2}\right]\right\}$
Hub and hollow shaft of same material	$\dfrac{U^2\gamma}{4g}\dfrac{(3+\mu)}{E}\left(1-\dfrac{d^2}{D^2}\right)$
Cast-iron hub and hollow steel shaft, $E_1=E/2$, $\mu_1=\mu$, $\gamma_1=0.92\gamma$	$\dfrac{U^2\gamma}{4gE}\left[1.84(3+\mu)-(3+\mu)\dfrac{d_i^2}{D^2}+0.84(1-\mu)\dfrac{d^2}{D^2}\right]$
Hub and solid shaft of different materials	$\dfrac{U^2}{4g}\left\{\dfrac{\gamma_1}{E_1}\left[(3+\mu_1)+(1-\mu_1)\dfrac{d^2}{D^2}\right]-\gamma(1-\mu)\dfrac{d^2}{D^2}\right\}$
Hub and solid shaft of same material	$\dfrac{U^2\gamma}{4g}\dfrac{(3+\mu)}{E}$
Cast-iron hub and solid steel shaft, $E_1=E/2$, $\mu_1=\mu$, $\gamma_1=0.92\gamma$	$\dfrac{U^2\gamma}{4gE}\left[1.84(3+\mu)+0.84(1-\mu)\dfrac{d^2}{D^2}\right]$

Source: Ref. 40, by permission of the ASME.

[a] γ_1 = weight per unit volume of hub material
γ = weight per unit volume of shaft material
g = acceleration due to gravity = 384 in. per sec per sec
U = linear velocity at the OD of the hub. For other notations, see Table 24.14.

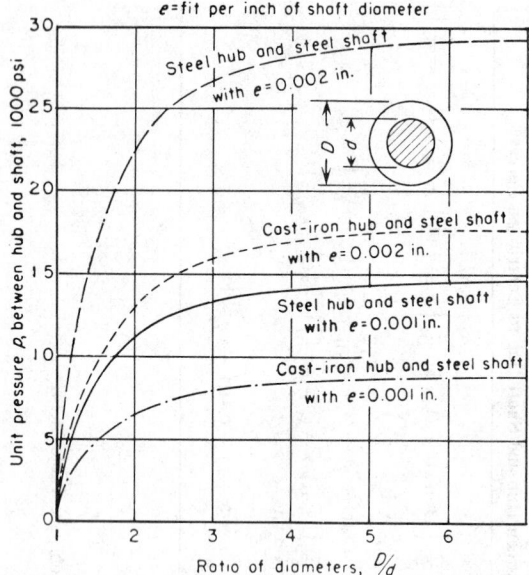

Fig. 24.55 Effect of increasing the hub thickness on the unit pressure between hub and shaft.

for the large shafts previously described, but in any case, it is necessary to use the strength level characteristic of the diameter involved.

Another very effective method of improving the fatigue strength of shafting is the introduction of a surface stress in compression. This may be accomplished by cold-working (use of cold-drawn shafting) or local heat treatment of a hardenable alloy by induction or flame heating followed by the appropriate cooling rate. Carburizing and nitriding are possibilities, but the loss of toughness or ductility may well result in a net loss of strength or life.

Table 24.17 Coefficients of Friction for Pressed and Shrunk Assemblies

	Coefficient of friction		
	Min	*Max*	*Av*
a. Assembly of 62 press fits; lower values due to hub yielding as a result of a too large fit allowance...	0.030	0.250	0.086
b. Pressing four steel axles of 5¼ in. diam and 175 Brinell hardness into a cast-steel spider...............	0.100	0.140	0.115
Pressing off the pressed-on spider..	0.120	0.150	0.137
Pressing off the shrunk-on spider..	0.160	0.170	0.165
c. Pressing 15 axles of 6¹⁄₁₆ in. diameter into gears, both of which were steel having a hardness of 170 Brinell...	0.075	0.250	0.150
d. Pressing steel gears off steel shafts having tapered fits, where the mean diameter was 6½ in.: When the parts were pressed-together.....................	0.150	0.210	0.177
When the parts were shrunk together.....................	0.210	0.230	0.220

Source: Ref. 40, by permission of the ASME.

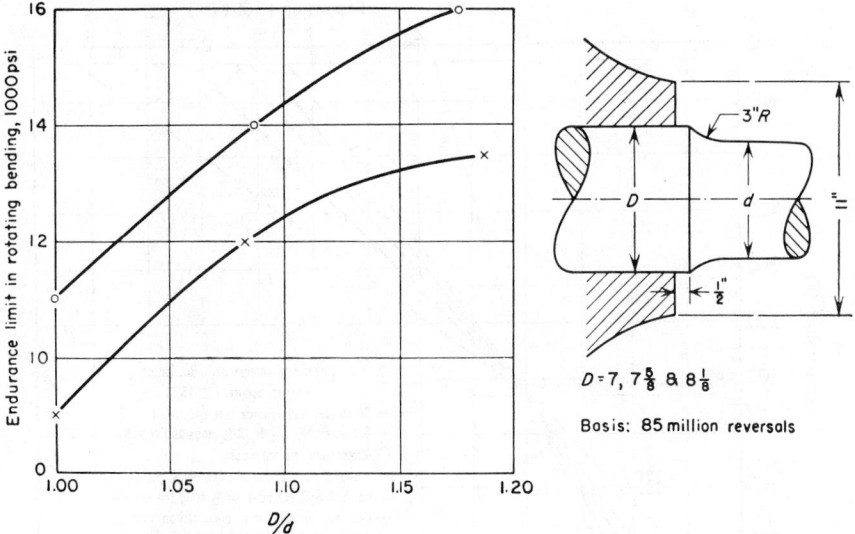

Fig. 24.56   Effect of height of raised wheel seat on endurance limit.

24.4.2   Shrink Fits

The purposes of shrink-fit construction are many: to increase the fatigue life of a wheel and axle, or as a tire, to increase the resistance to contact stresses and wear; to increase the elastic pressure range of a pressure vessel (either as a liner or an external sleeve); or to strengthen a forging or extrusion die.

Shrink fits may be set up as two-shell or as multishell constructions, the former being the more frequent. Generally, thin-ring theory can be used for uniform stresses in the outer member, provided that r/t exceeds about 30. Below this approximate value the members are regarded as thick-walled and their stresses and displacements must be found by appropriate analyses, typically Ref. 46. General stress distributions are shown in Fig. 24.58.

In producing a shrink-fit type of construction, it is common to heat the outer component in order to expand it beyond the interference on the diameter, and then slip it over the inner component while expanded. The temperature differential through which the outer part must be heated in order to obtain the required expansion is

$$t_f - t_i = (D_f - D_i)/\alpha D_i$$

where t_i and t_f are initial and final temperatures, D_i and D_f are initial and final diameters of the cylinder, and α is the coefficient of thermal expansion. To increase the temperature differential or decrease the high temperature of the outer component, the inner component of the assembly may also be cooled so that it contracts. If cooling only is used, then

$$-(t_i - t_f) = (D_f - D_i)/\alpha D_i$$

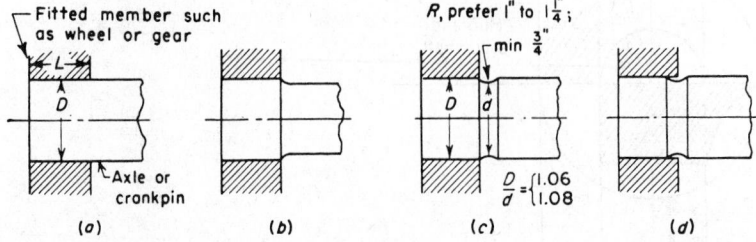

Fig. 24.57   Use of raised wheel seat to improve fatigue resistance. (*a*) Conventional practice giving low fatigue resistance; (*b*), (*c*), (*d*) improved practice giving increased fatigue resistance.

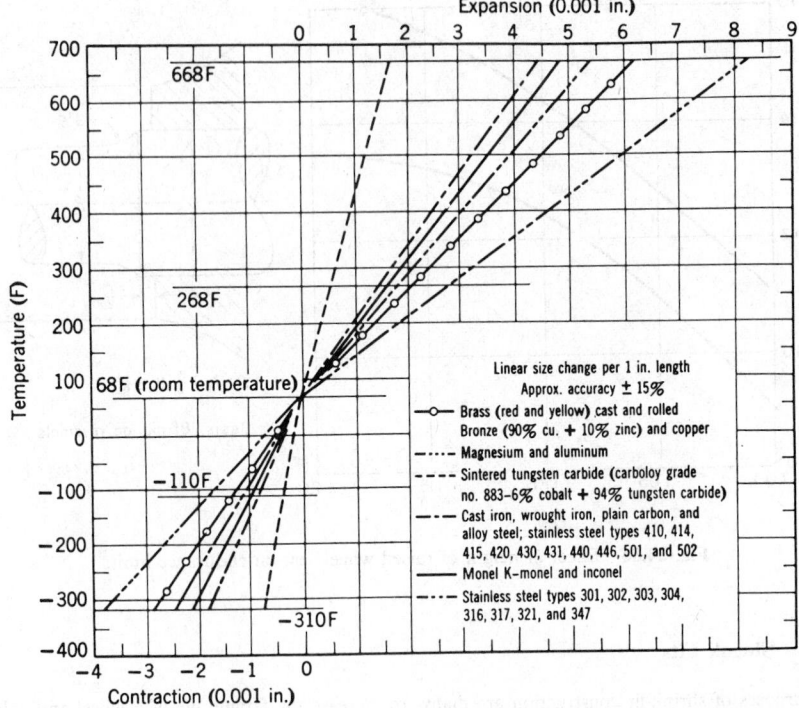

Fig. 24.58 Size change due to heating or cooling.

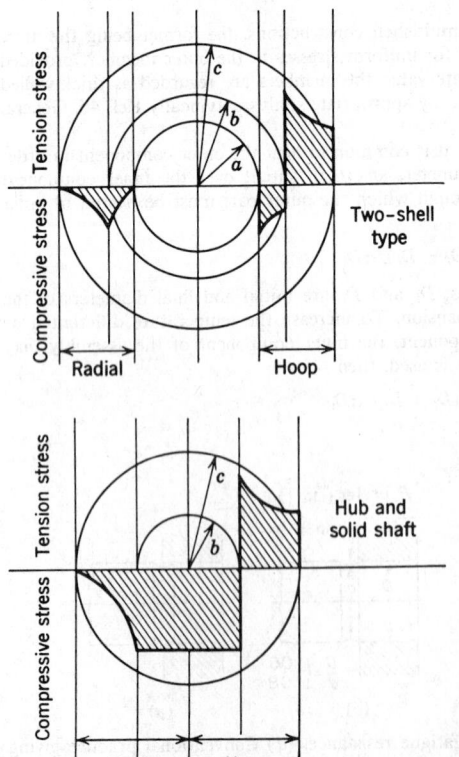

Two-shell type

Hub and solid shaft

Fig. 24.59 Stress distributions in assemblies subjected to shrink-fit pressure.

The approximate size change, per inch of length or diameter, which occurs when parts are heated or cooled is shown in Fig. 24.59. To determine size change for a given material on heating or cooling, start at the temperature specified, follow across the graph horizontally to the material in question, then follow the vertical lines to read the size change per inch of dimension. The size change per inch multiplied by the length or diameter of the part gives the total size change. For example, suppose a 4-in.-diam. steel shaft is to be cooled from room temperature to $-310°F$. Start at $-310°F$, follow horizontally across to the line for steel, then down to read approximately -0.0018 in./in. Multiply this size change per inch by the diameter of the shaft. The result, $-0.0018 \times 4 = -0.0072$ in., is the total size change cooled from room temperature. To determine the diameter of this shaft at $-310°F$, the assembly temperature, subtract the total size change from the diameter at room temperature, result, $6.0 - 0.0072 = 5.9928$.

Figure 24.59 may also be used to determine the temperature to which a part must be heated or cooled to effect a specified size change. For example, to shrink a 5-in.-diam. steel shaft 0.005 in. for assembly, divide the total shrink required by the diameter; the shrinkage must be $0.005/5 = 0.001$ in./in. of diameter. Using the curve in Fig. 24.59, start at -0.001 in., and follow up to the line for steel, then across to the temperature scale. The chart shows that the part must be cooled from room temperature to approximately $-150°F$.

The chart and equations do not account for residual stresses. If the stress relieving or tempering were done at a temperature of $1000°F$ or higher, the residuals would be insignificant, but calculation of the fit allowance must accommodate these stresses, especially in heat-treated stock. Shrinkage temperatures over $500°F$ may have an annealing effect on certain steels tempered to a high hardness level. High thermal gradients at the instant of contact of the components may induce high thermal stress and accompanying plastic flow if the stresses are above the yield strength of the material. Seizing of shells on liners or shafts before complete assembly is also a frequently encountered inconvenience. Dimensional tolerances must be accurately controlled because stresses set up in shrink fitting are directly dependent on the accuracy of machining (Fig. 24.59).

REFERENCES

1. P. G. Forrest, *Fatigue of Metals,* Addison-Wesley, Reading, MA, 1962.

2. *Manual of Steel Construction,* 2nd ed., AISC, Chicago, IL, 1973.

3. E. F. Bruhn, *Analysis and Design of Flight Vehicle Structures,* Tri-State Offset Co., Cincinnati, OH, 1965.

4. J. E. Shigley, *Mechanical Engineering Design,* 3rd ed., McGraw-Hill, New York, 1977.

5. J. W. Fisher and J. H. A. Struik, *Guide to Design Criteria for Bolted and Riveted Joints,* Wiley–Interscience, New York, 1974.

6. J. deBack et al., *High Strength Bolted Beam-to-Column Connections,* Report 6-72-13, Stevin Lab., Delft University of Technology, Delft, The Netherlands, 1972.

7. R. B. Heywood, *Designing against Fatigue of Metals,* Reinhold, New York, 1962.

8. C. R. Smith, "The Effect of Tapered Holes on Structural Integrity," *Assembly Engineering* (July 1967).

9. C. C. Osgood, *Fatigue Design,* 2nd ed., Pergamon, Oxford, U.K. 1982.

10. D. S. Kimball and J. H. Barr, *Elements of Machine Design,* Wiley, New York, 1935.

11. *Machine Design,* **41,** 20 (Sept. 11, 1969).

12. MIL-HDBK-5a, *Metallic Materials and Elements for Flight Vehicle Structures,* DOD, 1968.

13. N. Motosh, "Determination of Joint Stiffness in Bolted Connections," *Trans. ASME* (Aug. 1976).

14. Bland Burner Co., Precision Threaded Products Div., *Tru-load Fasteners,* Hartford, CT, 1963.

15. Standard Pressed Steel Co., Jenkintown, PA, 1962–1963, advertisements.

16. Cooper and Turner Ltd., Sheffield, England, advertisement in *Machine Design,* **35**(6) (Mar. 14, 1963).

17. Hi-Shear Corp., Torrance, CA, Report 4-1502, Sept. 20, 1957.

18. Huck Mfg. Co., Detroit, MI, Brochure No. 132, 1962.

19. *Specification for Structural Joints Using ASTM A325 or A490 Bolts,* Research Council on Riveted and Bolted Structural Joints, Eng. Foundation, March, 1962.

20. Fasteners Reference Issue, *Machine Design,* **39** (June 15, 1967).

21. *Evaluation of a 250,000 psi High-Fatigue Fastener System,* Standard Pressed Steel Lab. Report 1325, Jenkintown, PA, 1966.

22. *Bolt Procurement Specification: SPS-B-640, Fatigue Rated,* Standard Pressed Steel, Jenkintown, PA, 1971.

23. *ASME Boiler and Pressure Vessel Code,* ANSI/ASME BPV-I-XI, ASME, New York, 1980.

24. *Machinery's Handbook,* 22nd ed., The Industrial Press, New York, 1984, pp. 1188 et seq.
25. *Suggested Specifications for Structures of Aluminum Alloys 6061-T6 and 6062-T6,* Amer. Soc. Civil Engrs. Proc. Papers Nos. 3341 and 3342, Vol. 88, No. ST6, Dec. 1962.
26. *Welding Handbook,* American Welding Society, New York, 1981, Vols 3, 4, and 5.
27. *Structural Welding Code,* American Welding Society, D1.1–82 et seq., Miami, FL.
28. *Manual of Steel Construction,* American Institute of Steel Construction, New York, 1982.
29. *ASME Boiler and Pressure Vessel Code,* in Ref. 23, Sec. III, Div. 1; Sec. VIII, Div. 1, Part UW.
30. "Welding and Brazing," *Metals Handbook,* American Society for Metals, Metals Park, OH, 1983, Vol. 6.
31. R. Bakish and S. S. White, *Handbook of Electron Beam Welding,* Wiley, New York, 1964.
32. *The Procedure Handbook of Arc Welding,* Lincoln Electric Co., Cleveland, OH.
33. K. Masubuchi, *Analysis of Welded Structures,* Pergamon, Elmsford Park, NY, 1980.
34. H. S. Reemsnyder, *Development and Application of Fatigue Data for Structural Steel Weldments,* American Society for Testing and Materials, Spec. Tech. Publ. 648, 1977.
35. T. R. Gurney, *Fatigue of Welded Structures,* Cambridge University Press, Cambridge, 1979.
36. MIL-HDBK-5A, *Metallic Materials and Elements for Flight, Vehicle Structures,* Department of Defense, 1982.
37. R. A. Wood, *The Ti-8A1-1Mo-1V Alloy,* DMIC Report No. S-10, Battelle Memorial Institute, Columbus, OH, 1965.
38. Catalogue No. S226E, Ringfelder Corp., Westwood, NJ, 1972.
39. O. Horger (Ed.), *Metals Engineering Design,* ASME Handbook, McGraw-Hill, New York, 1953.
40. S. Worth, *Austauschbare Langspressitze,* Forschungschaft 383, V.D.E., 1937.
41. J. W. Baugher et al., *Transmission of Torque by Press-fits,* ASME Paper No. MSP-53-10, ASME, New York.
42. C. F. MacGill, "A Record of Press-fits," *Trans. ASME,* **819** (1913).
43. N. J. Savin, "Research on Force Fits," *American Machinist,* **68** (1928).
44. R. Russell, "Factors Affecting the Grip in Force, Shrink and Expansion Fits," *Proc. Inst. Mech. Engrs.,* **125** (1933).
45. C. C. Osgood, *Fatigue Design,* 2nd ed., Pergamon, Oxford, U.K. 1982.
46. S. Timoshenko, *Strength of Materials,* Van Nostrand, New York, 1966, Vol. II.

CHAPTER 25
VIBRATION AND SHOCK

WAYNE TUSTIN

Tustin Institute of Technology
Santa Barbara, California

25.1 VIBRATION

In any structure or assembly, certain whole-body motions and certain deformations are more common than others: the most likely (easiest to excite) motions will occur at certain natural frequencies. Certain exciting or forcing frequencies may coincide with the natural frequencies (resonance) and give relatively severe vibration responses.

We will now discuss the much-simplified system shown in Fig. 25.1. It includes a weight W (it is technically preferred to use mass M here, but weight W is what people tend to think about), a spring of stiffness K, and a viscous damper of damping constant C. K is usually called the spring rate; a static force of K newtons will statically deflect the spring by δ mm, so that spring length l becomes $\delta + l$. (In "English" units, a force of K lb will statically deflect the spring by 1 in.) This simplified system is constrained to just one motion—vertical translation of the mass. Such single-degree-of-freedom (SDF) systems are not found in the real world, but the dynamic behavior of many real systems approximate the behavior of SDF systems over small ranges of frequency.

Suppose that we pull weight W down a short distance further and then let it go. The system will oscillate with W moving up-and-down at natural frequency f_N, expressed in cycles per second (cps) or in hertz (Hz); this condition is called "free vibration." Let us here ignore the effect of the damper, which acts like the "shock absorbers" or dampers on your automobile's suspension—using up vibratory energy so that oscillations die out. f_N may be calculated by

$$f_N = \frac{1}{2\pi} \sqrt{\frac{Kg}{W}} \tag{25.1}$$

It is often convenient to relate f_N to the static deflection δ due to the force caused by earth's gravity, $F = W = Mg$, where $g = 386$ in./sec² $= 9807$ mm/sec², opposed by spring stiffness K expressed in either lb/in. or N/mm. On the moon, both g and W would be considerably less (about one-sixth as large as on earth). Yet f_N will be the same. Classical texts show Eq. (25.1) as

$$f_N = \frac{1}{2\pi} \sqrt{\frac{K}{M}}$$

In the "English" System:

$$\delta = \frac{F}{K} = \frac{W}{K}$$

Then

$$f_N = \frac{1}{2\pi} \sqrt{\frac{g}{\delta}} = \frac{1}{2\pi} \sqrt{\frac{386}{\delta}}$$
$$= \frac{19.7}{2\pi\sqrt{\delta}} = \frac{3.13}{\sqrt{\delta}} \tag{25.2a}$$

In the International System:

$$\delta = \frac{F}{K} = \frac{Mg}{K}$$

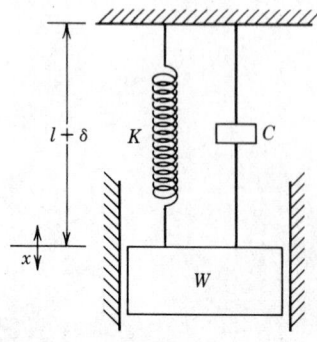

Fig. 25.1 Single-degree-of-freedom system.

Then

$$f_N = \frac{1}{2\pi} \sqrt{\frac{g}{\delta}} = \frac{1}{2\pi} \sqrt{\frac{9807}{\delta}}$$

$$= \frac{99.1}{2\pi\sqrt{\delta}} = \frac{15.76}{\sqrt{\delta}}$$

(25.2b)

Relationships (25.2a) and (25.2b) often appear on specialized "vibration calculators." As increasingly large mass is supported by a spring, δ becomes larger and f_N drops.

Let $K = 1000$ lb/in. and vary W:

W (lb)	$\delta = \dfrac{W}{K}$ (in.)	f_N (Hz)
0.001	0.000 001	3 130
0.01	0.000 01	990
0.1	0.000 1	313
1.	0.001	99
10.	0.01	31.3
100.	0.1	9.9
1 000.	1.	3.13
10 000.	10.	0.99

Let $K = 1000$ N/mm and vary M:

M (kg)	$\delta = \dfrac{9.81M}{K}$	f_N (Hz)
0.00102	10 nm	4 980
0.0102	100 nm	1 576
0.102	1 μm	498
1.02	10 μm	157.6
10.2	100 μm	49.8
102.	1 mm	15.76
1 020.	10 mm	4.98
102 000.	1 m	0.498

Note, from Eqs. (25.2a) and (25.2b), that f_N depends on δ, and thus on both M and K (or W and K). As long as both load and stiffness change proportionately, f_N does not change.

The peak-to-peak or double displacement amplitude D will remain constant if there is no damping to use up energy. The potential energy we put into the spring becomes zero each time the mass passes through the original position and becomes maximum at each extreme. Kinetic energy becomes maximum as the mass passes through zero (greatest velocity) and becomes zero at each extreme (zero velocity). Without damping, energy is continually transferred back and forth from potential to kinetic energy. But with damping, motion gradually decreases; energy is converted to heat. A vibration pickup on the weight would give oscilloscope time history patterns like Fig. 25.2; more damping was present for the lower pattern and motion decreased more rapidly.

Assume that the "support" at the top of Fig. 25.1 is vibrating with a constant D of, say, 1 in. Its frequency may be varied. How much vibration will occur at weight W? The answer will depend on

1. The frequency of "input" vibration.
2. The natural frequency and damping of the system.

Let us assume that this system has an f_N of 1 Hz while the forcing frequency is 0.1 Hz, one-tenth the natural frequency, Fig. 25.3. We will find that weight W has about the same motion as does the input, around 1 in. D. Find this at the left edge of Fig. 25.3; transmissibility, the ratio of response vibration divided by input vibration, is $1/1 = 1$. As we increase the forcing frequency, we find that

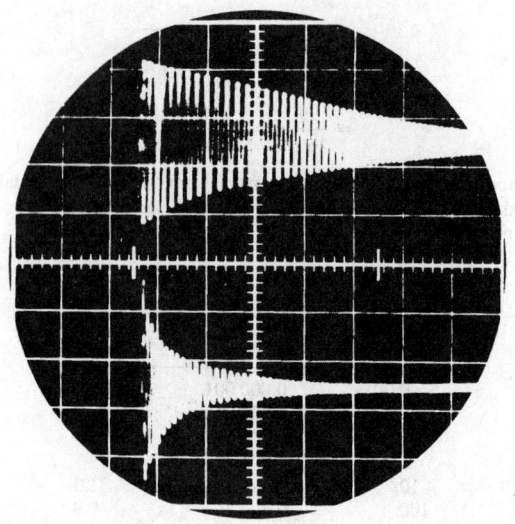

Fig. 25.2 Oscilloscope time history patterns of damped vibration.

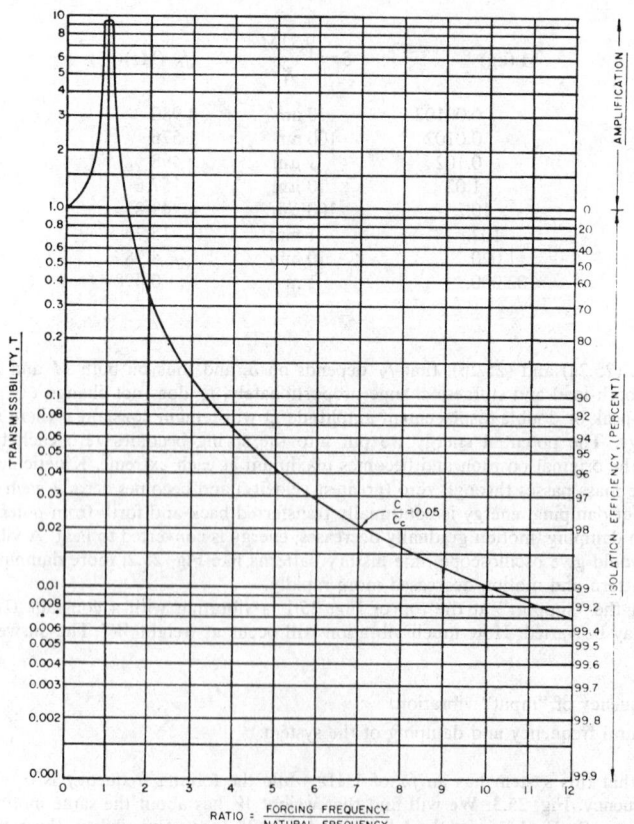

Fig. 25.3 Transmissibility of a lightly damped system.

the response increases. How much? It depends on the amount of damping in the system. Let us assume that our system is lightly damped, that it has a ratio C/C_c of 0.05 (ratio of actual damping to "critical" damping is 0.05). When our forcing frequency reaches 1 Hz (exactly f_N), weight W has a response D of about 10 in., 10 times as great as the input D. At this "maximum response" frequency, we have the condition of "resonance"; the exciting frequency is the same as the f_N of the load. As we further increase the forcing frequency (see Fig. 25.3), we find that response decreases. At 1.414 times f_N, the response has dropped so that D is again 1 in. As we further increase the forcing frequency, the response decreases further. At a forcing frequency of 2 Hz, the response D will be about 0.3 in. and at 3 Hz it will be about 0.1 in.

Note that the abscissa of Fig. 25.3 is "normalized"; that is, the transmissibility values of the preceding paragraph would be found for another system whose natural frequency is 10 Hz, when the forcing frequency is, respectively, 1, 10, 14.14, 20, and 30 Hz. Note also that the vertical scale of Fig. 25.3 can represent (in addition to ratios of motion) ratios of force, where force can be measured in pounds or newtons.

The region above 1.414 times f_N (where transmissibility is less than 1) is called the region of "isolation." That is, weight W has less vibration than the input; it is *isolated*. This illustrates the use of vibration isolators—rubber elements or springs that reduce the vibration input to delicate units in aircraft, missiles, ships, and other vehicles, and on certain machines. We normally try to set f_N (by selecting isolators) considerably below the expected forcing frequency of vibration. Thus, if our support vibrates at 50 Hz, we might select isolators whose K makes f_N 25 Hz or less. According to Fig. 25.3, we will have best isolation at 50 Hz if f_N is as low as possible. (However, we should not use too-soft isolators because of instabilities that could arise from too-large static deflections, and because of need for excessive clearance to any nearby structures.)

Imagine a system with a weight supported by a spring whose stiffness K is sufficient that $f_N = $ 10 Hz. At an exciting frequency of 50 Hz, the frequency ratio will be 50/10 or 5, and we can read transmissibility = 0.042 from Fig. 25.3. The weight would "feel" only 4.2% as much vibration as if it were rigidly mounted to the support. We might also read the "isolation efficiency" as being 96%. However, as the source of 50-Hz vibration comes up to speed (passing slowly through 10 Hz), the isolated item will "feel" about 10 times as much vibration as if it were rigidly attached, without any isolators. Here is where damping is helpful: to limit the "Q" or "mechanical buildup" at resonance. Observe Fig. 25.4, plotted for several different values of damping. With little damping present, there is much resonant magnification of the input vibration. With more damping, maximum transmissibility is not so high. For instance, when C/C_c is 0.01, "Q" is about 40. Even higher Q values are found with certain structures having little damping; Q's over 1000 are sometimes found.

Most structures (ships, aircraft, missiles, etc.) have Q's ranging from 10 to 40. Bonded rubber vibration isolating systems often have Q's around 10; if additional damping is needed (to keep Q lower), dashpots or rubbing elements may be used. Note that there is less buildup at resonance, but that isolation is not as effective when damping is present.

Figure 25.1 shows guides or constraints that restrict the motion to up-and-down translation. A single measurement on an SDF system will describe the arrangement of its parts at any instant. Another SDF system is a wheel attached to a shaft. If the wheel is given an initial twist, that system will also oscillate at a certain f_N, which is determined by shaft stiffness and wheel inertia. This imagined rotational system is an exact counterpart of the SDF system shown in Fig. 25.1.

If weight W in Fig. 25.1 did not have the guides shown, it would be possible for weight W to move in five other motions—five additional degrees of freedom. Visualize the six possibilities:

Vertical translation.

North–south translation.

East–west translation.

Rotation about the vertical axis.

Rotation about the north–south axis.

Rotation about the east–west axis.

Now this solid body has six degrees of freedom—six measurements would be required in order to describe the various whole-body motions that may be occurring.

Suppose now that the system of Fig. 25.1 were attached to another mass, which in turn is supported by another spring and damper, as shown in Fig. 25.5. The reader will recognize that this is more typical of many actual systems than is Fig. 25.1. Machine tools, for example, are seldom attached directly to bedrock, but rather to other structures that have their own vibration characteristics. Weight W_2 will introduce six additional degrees of freedom, making a total of 12 for the system of Fig. 25.5. That is, in order to describe all of the possible solid-body motions of W_1 and W_2, it would be necessary to consider all 12 motions and to describe the instantaneous positions of the two masses.

The reader can extend that reasoning to include additional masses, springs, and dampers, and additional degrees of freedom—possible motions. Finally, consider a continuous beam or plate, where

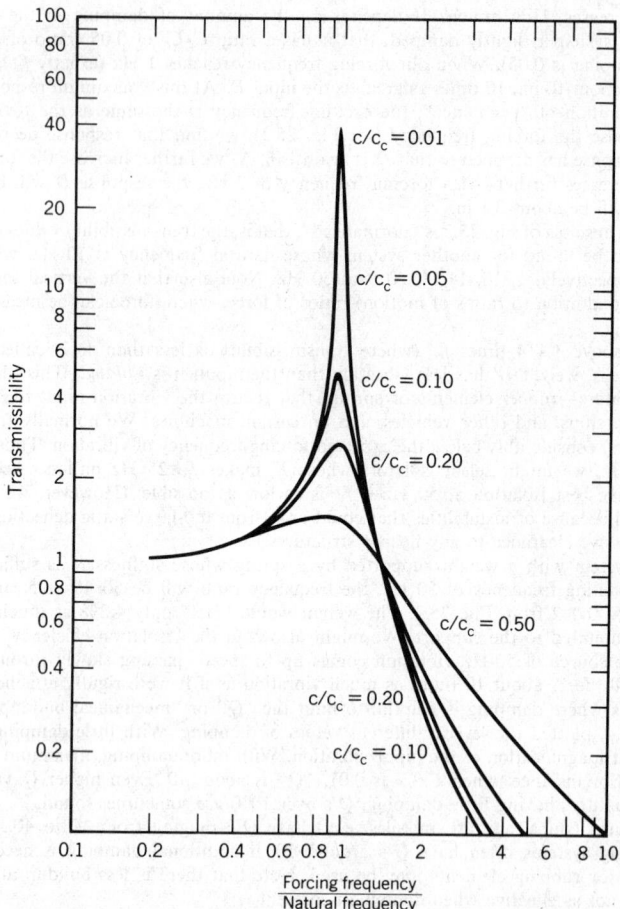

Fig. 25.4 Transmissibility for several different values of damping.

mass, spring, and damping are distributed rather than being concentrated as in Fig. 25.1 and 25.5. Now we have an infinite number of possible motions, depending on the exciting frequency, the distribution of mass and stiffness, the point or points at which vibration is applied, etc. Fortunately, only a few of these motions are likely to occur in any reasonable range of frequencies.

Figure 25.6 consists of three photographs, each taken at a different forcing frequency. Up-and-down shaker motion is coupled through a holding fixture to a pair of beams. (Let us only consider

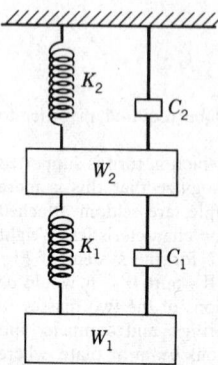

Fig. 25.5 System with 12 degrees of freedom.

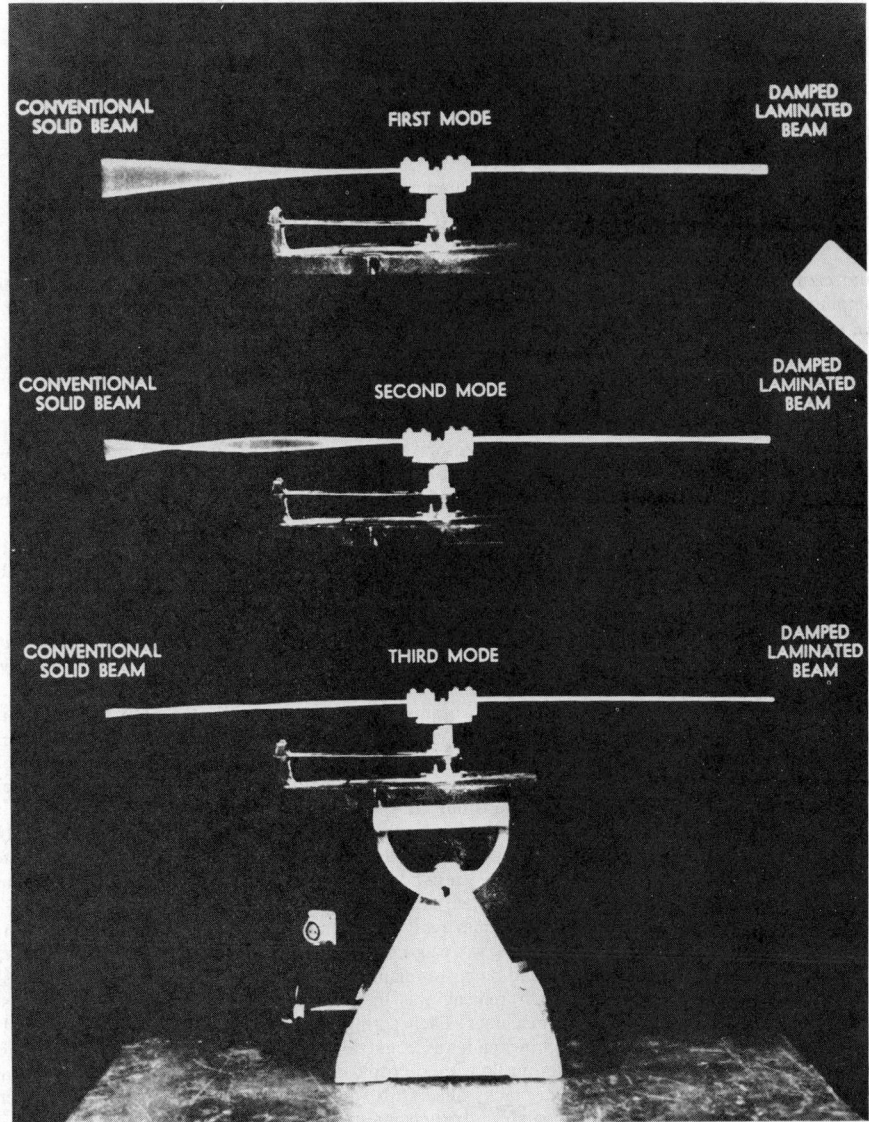

Fig. 25.6 Pair of beams excited at three different forcing frequencies.

the left-hand beams for now.) Frequency is adjusted first to excite the fundamental shape (also called the "first mode" or "fundamental mode") of beam response. Then frequency is readjusted higher to excite the second and third modes. There will be an infinite number of such resonant frequencies, an infinite number of modes, if we go to an infinitely high test frequency. We can mentally extend this reasoning to various structures found in vehicles, machine tools, appliances, etc.; an infinite number of resonances could exist in any structure. Fortunately, there are usually limits to the exciting frequencies that must be considered. Whenever the frequency coincides with one of the f_N's of a structure, we will have a resonance; high stresses, large forces, and large motions result.

All remarks about free vibration, f_N, and damping apply to continuous systems. We can "pluck" the beam of Fig. 25.6 and cause it to respond in one or more of the patterns shown. We know that vibration will gradually die out, as indicated by the upper trace of Fig. 25.2, because there is internal (or hysteresis) damping of the beam. Stress reversals create heat, which uses up vibratory energy. With more damping, vibration would die out faster, as in the lower trace of Fig. 25.2. The right-

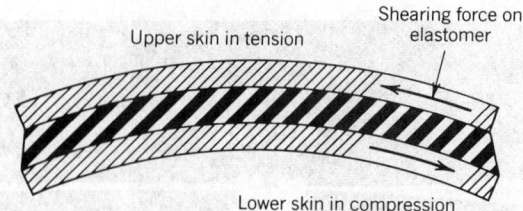

Fig. 25.7   Detail of laminated beam (Courtesy of Lord Manufacturing Co.)

hand cantilever beam of Fig. 25.6 is made of thin metal layers joined by a layer of a viscoelastic damping material, as in Fig. 25.7. Shear forces between the layers use up vibratory energy faster and free vibration dies out more quickly.

Statements about forced vibration and resonance also apply to continuous systems. At very low frequencies, the motion is the same (transmissibility = 1) at all points along the beams. At certain f_N's, large motion results.

The points of minimum D are called "nodes" and the points having maximum D are called "antinodes." With a strobe light and/or vibration sensors we could show that (in a pure mode) all points along the beam are moving either in-phase or out-of-phase and that phase reverses from one side of a node to the other. Also bending occurs at the attachment point and at the antinodes; these are the locations where fatigue failures usually occur.

It is possible for several modes to occur at once, if several natural frequencies are present as in complex or in broad-band random vibration. It is also possible for several modes to simultaneously be excited by a shock pulse (and then to die out). Certain modes are most likely to cause failure in a particular installation; the frequencies causing such critical modes are called "critical frequencies."

Intense sounds can cause modes to be excited, especially in thin panels; stress levels in aircraft and missile skins may cause fatigue failures. Damping treatments on the skin are often very effective in reducing such vibration.

Figure 25.6 shows forced vibration of two beams whose length is adjusted until their f_N's are identical. The conventional solid beam responds more violently than the damped laminated beam. We say that the maximum transmissibility (often called mechanical "Q") of the solid beam is much greater than that of the damped beam. Assuming that the vibration continues indefinitely, which beam will probably fail in fatigue first? Here is one reason for using damping.

Resonance is sometimes helpful and desirable; at other times it is harmful. Some readers will be familiar with deliberate applications of vibration to move bulk materials, to compact materials, to remove entrapped gases, or to perform fatigue tests. Maximum vibration is achieved (assuming the vibratory force is limited) by operating the system at resonance.

A resonant vibration absorber can sometimes reduce motion, if the vibration input to a structure is at a fixed frequency. Imagine, for example, weaving machines in a relatively soft, multistory factory building. They happen to excite an up-and-down resonant motion of their floor. (Some "old timers" claim they saw D's of several inches.) A remedy was to attach springs to the undersides of those floors, directly beneath the offending machines. Each spring supported a pail which was gradually filled with sand until the f_N of the spring/pail was equal to the exciting frequency of the weaving machine above it. A dramatic reduction in floor motion told maintenance people that the spring was correctly loaded. Similar methods (using tanks filled with water) have been used on ships. However, any change in exciting frequency necessitates a bothersome readjustment.

25.2   ROTATIONAL IMBALANCE

Where rotating engines are used in ships, automobiles, aircraft, or other vehicles; where turbines are used in vehicles or electrical power generating stations; where propellers are used in ships and aircraft—in all of these and many other varied applications, imbalance of the rotating members causes vibration.

Consider first the simple disk shown in Fig. 25.8. This disk has some extra material on one side so that the center of gravity is not at the rotational center. If we attach this disk to a shaft and allow the shaft to rotate on knife-edges, we observe that the system comes to rest with the heavy side of the disk downward. This type of imbalance is called *static* imbalance, since it can be detected statically. It can be measured statically, also, by determining some weight W at some radius r that must be attached to the side opposite the heavy side, in order to restore the center of gravity to the rotational center and thus to bring the system into static balance; that is, so that the disk will have no preferred position and will rest in any angular position. The product Wr is the value of the original imbalance. It is often expressed in units of ounce-inches, gram-millimeters, etc.

Static balancing is the simplest technique of balancing, and is often used for the wheels on automobiles, for instance. It locates the center of gravity at the center of the wheel. But we will show that

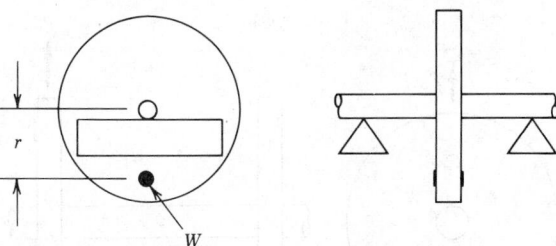

Fig. 25.8 Static balancing of a disk.

this compensation is not completely satisfactory. Let us now support the disc and shaft of Fig. 25.8 by a bearing at each end and cause the disk and shaft to spin. A rotating vector force of $Mr\omega^2$ lb results, in phase with the center of gravity of the rotating system, as shown in Fig. 25.9. W is the total weight, ω is the angular velocity in radians per second, and r is the radial distance (inches or mm) from the shaft center to the center of gravity.

The shaft and bearings must absorb and transmit not only the weight of the rotor but also a new force, one which rotates; one which, at high rotational speeds, may be greater than the weight of the rotor.

Though your automotive mechanic may only statically balance the wheels of your car by adding wheel weights on the light side of each wheel, this static balancing results in noticeable improvement in car ride, in passenger comfort, and in tire wear, because the force $Mr\omega^2$ is greatly reduced.

A numerical example may interest the reader. Imbalance can be measured in ounce-inches (or, in metric units, gram-millimeters). One ounce-inch means that an excess or deficiency of weight of one ounce exists at a radius of 1 in. How big is an ounce-inch? It sounds quite small, but at high rotational speeds (since force is proportional to the square of rotational speed) this "small" imbalance can cause very high forces. You will recall that centrifugal force may be calculated by

$$F = \frac{Mv^2}{r}$$

where M is the mass in kilograms (or weight in pounds divided by 386 in./sec^2, the acceleration due to the earth's gravity); r is the radius in inches; and v is the tangential velocity in inches per second. We can calculate $v = 2\pi fr$, where f is the frequency of rotation in hertz and r is the radius in inches or mm. Then

$$F = \frac{M}{r}(2\pi fr)^2 = 4\pi^2 Mf^2 r = \frac{4\pi^2}{386} Wrf^2 = 0.1023 Wrf^2$$

Let us calculate the force that results from 1 oz-in. of imbalance on a member rotating at 8000 rpm or 133 rps:

$$1 \text{ oz} = \frac{1}{16} \text{ lb}$$

$$r = 1 \text{ in.}$$

Then

$$F = 0.1023(\tfrac{1}{16})(1)(133)^2 = 114 \text{ lb}$$

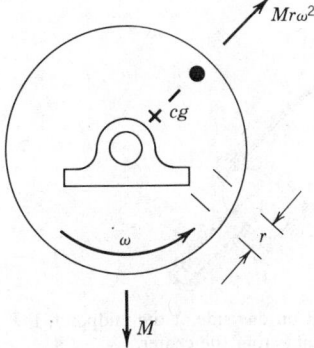

Fig. 25.9 Unbalanced disk in rotation.

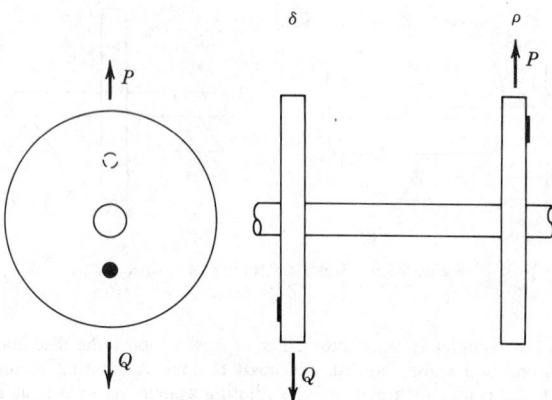

Fig. 25.10 Schematic of unbalanced shaft.

If this centrifugal force of 114 lb occurred in an electrical motor, for instance, whose weight was less than 114 lb, the imbalance force acting through the bearings would lift the motor off its supports once each revolution, or 8000 times a minute. If the motor were fastened to some framework, a vibratory force would be apparent.

In most rotating elements, such as motor armatures or engine crankshafts, the mass of the rotor is distributed along the shaft rather than being concentrated in a disk as shown in Figs. 25.8 and 25.9. If we test such a rotor as we tested in Fig. 25.8, we may find that we have static balance, that the rotating element has no preferred angular position, and that the center of gravity coincides with the shaft center. But when we spin such a unit, we may find severe forces being transmitted by shaft and bearings. Obviously we are not truly balanced; since this new imbalance is apparent only when the system is rotated, we call it *dynamic imbalance.*

As a simplified example of such a system, consider Fig. 25.10. If the two imbalances P and Q are exactly equal, if they are exactly 180° apart, and if the two disks are otherwise uniform and identical, this system will be statically balanced. But if we rotate the shaft, each disk will have a rotating centrifugal force similar to Fig. 25.9. These two forces are out-of-phase with each other. The result is dynamic imbalance forces in our simple two-disk system; they must be countered by two rotating forces rather than by one rotating force as before, in static balancing.

If we again consider one wheel of our automobile, having spent our money for static balancing only, we may still have unbalanced forces at high speeds. We may find it necessary to both statically and dynamically balance the wheel to reduce the forces to zero. Few automotive mechanics, doing this work every day, are aware of this. You will find it quite difficult to find repair shops that both statically and dynamically balance the wheels of your automobile, but the results, in increased comfort and tire wear, often repay one for the effort and expense.

Imagine that we have a perfectly homogeneous and balanced rotor. Now we add a weight on one side at the midpoint. If we now spin this rotor, but do not rigidly restrain its movement, the motion will resemble the left sketch in Fig. 25.11; the centerline of the shaft will trace out a cylinder. On

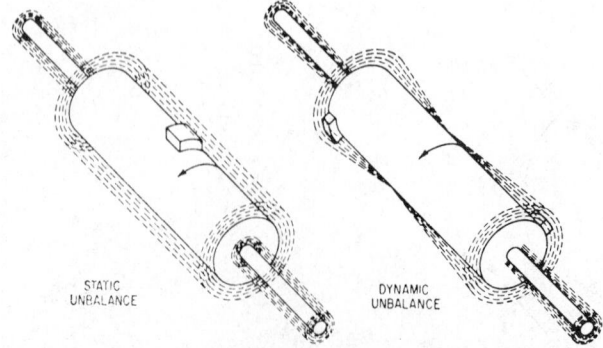

Fig. 25.11 Imbalance in a rotor; the rotor on the left has a weight on one side at the midpoint; the rotor on the right has two equal weights on opposite sides, equidistant from the center.

the other hand, suppose that we had added two equal weights on opposite sides, equidistant from the center, so that we have static balance. If we spin the rotor, its centerline will trace out two cones, as shown in the right sketch; the apex of each cone will be at the center of gravity of the rotor. In practical unbalanced rotors, the motion will be some complex combination of these two movements.

Imbalance in machinery rotors can come from a number of sources. One is lack of symmetry; the configuration of the rotor may not be symmetrical in design, or a core may have shifted in casting, or a rough cast or forged area may not be machined. Another source is lack of homogeneity in the material due, perhaps, to blowholes in a casting or to some other variation in density. The rotor (a fan blade, for example) may distort at operating rpm. The bearings may not be aligned properly.

Generally, manufacturing processes are the major source of imbalance; this includes manufacturing tolerances and processes that permit any unmachined portions, any eccentricity or lack of squareness with the shaft, or any tolerances that permit parts of the rotor to shift during assembly. When possible, rotors should be designed for inherent balance. If operating speeds are low, balancing may not be necessary; today's trends are all toward higher rpm and toward lighter-weight assemblies; balancing is more required than it was formerly.

Figure 25.12 shows two unbalanced disks on a shaft; they represent the general case of any rotor, but the explanation is simpler if the weight is concentrated into two disks. The shaft is supported by two bearings, a distance l apart. At a given rotational speed, one disk generates a centrifugal force P, while the other disk generates a centrifugal force Q. In the plane of bearing 1, forces P and Q may be resolved into two forces by setting the sum of the moments about plane 2 equal to zero; the force diagram is shown in Fig. 25.12. Similarly, in the plane of bearing 2, forces P and Q may be resolved into two forces by setting the sum of the moments about plane 1 equal to zero; this force diagram is also shown in Fig. 25.12. Both force diagrams represent rotating vectors with angular velocity ω radians per second. Force P has been replaced by two forces, one at each bearing plane: $P(l_1/l)$ and $P(l_2/l)$. Similarly, force Q may be replaced by two forces, one at each bearing plane: $Q(l_3/l)$ and $Q(l_4/l)$.

We may combine the two forces at bearing plane 1 into one resultant force R_1. We may combine the two forces at bearing plane 2 into one resultant force R_2. If we can somehow apply rotating counterforces at the bearings, namely, $-R_1$ and $-R_2$, we will achieve complete static and dynamic balance. No force will be transmitted from our rotor to its bearings.

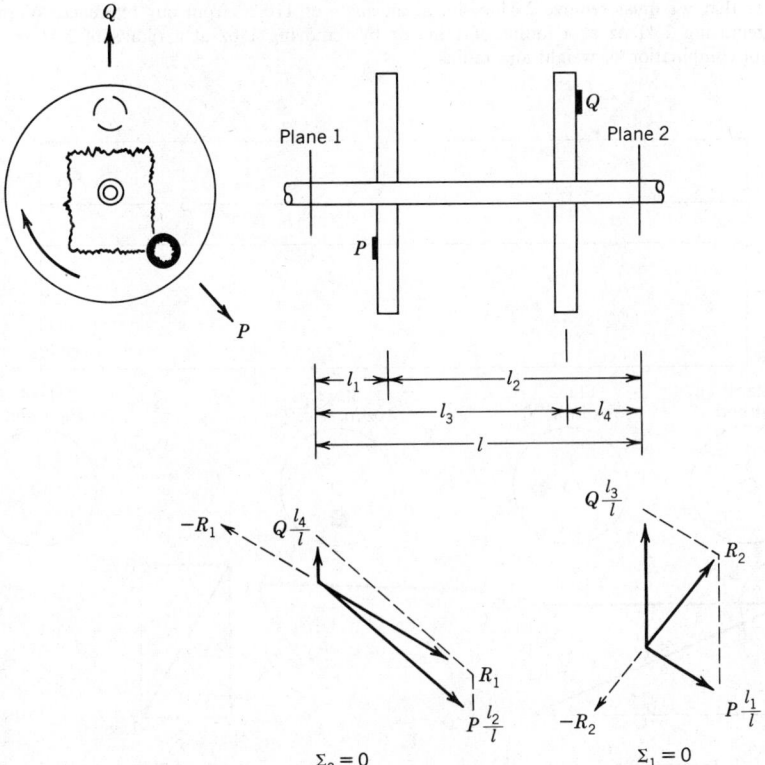

Fig. 25.12 Representation of the general case of an unbalanced rotor.

In practice, of course, we do not actually use forces at the bearings. We add balancing weights at two locations along the shaft and at the proper angles around the shaft. These weights each generate a centrifugal force, which is also proportional to the square of shaft speed. Or we may remove weight. Either way, we must achieve forces $-R_1$ and $-R_2$. The amount of each weight added (or subtracted) depends on where along the shaft we can conveniently perform the physical operation.

When we have our automobile wheels balanced, to use this simple example for the last time, only the weight and angular position of the counterweights may be selected. The weights are always attached to either the inner or outer rim of the wheel. In early automobile wheels, with their large diameter/thickness ratio, static balancing was usually sufficient. Modern automobiles have smaller, thicker wheels (turning at higher speeds) and they approach Fig. 25.11. Dynamic balancing is definitely better.

In the case of a rotating armature we may subtract weight at a convenient point along the length of the armature (usually at the end) and at the proper angular position by simply drilling out a bit of the armature material. This operation is repeated at the other end. Then the armature is both statically and dynamically balanced.

As an example, consider the rotor shown in Fig. 25.13. This rotor has 3 oz-in. of imbalance at station 2, located 3 in. from the left end, and at an angular position of 90° from an arbitrary reference. Another imbalance of 2 oz-in. exists at station 3, located 5 in. from the right end, and at an angle of 180° from the same reference. We want to statically and dynamically balance the rotor, by means of corrections at the two ends, at stations 1 and 4.

Let us now draw a vector diagram of the forces at station 1, as shown in Fig. 25.13. First we take a summation of forces about station 4, just as we did in Fig. 25.12. The imbalance at station 2, when sensed at station 1, is $3 \times \frac{9}{12} = \frac{9}{4} = 2.25$ or a vector 2.25 at 90°. The imbalance at station 3, when sensed at station 1, is $2 \times \frac{5}{12} = \frac{5}{6} = 0.83$ or a vector 0.83 at 180°. By taking the square root of 2.25^2 plus 0.83^2, we find that we must remove 2.41 oz-in. Now at what angle shall we remove weight?

$$\phi_1 = 90° + \tan^{-1}\frac{0.83}{2.25}$$

$$= 90° + 20.3° = 110.3°$$

So we see that we must remove 2.41 oz-in. at an angle of 110.3° from our reference. We may do this by removing 2.41 oz at a radius of 1 in., or by removing 1 oz at a radius of 2.41 in. or any convenient combination of weight and radius.

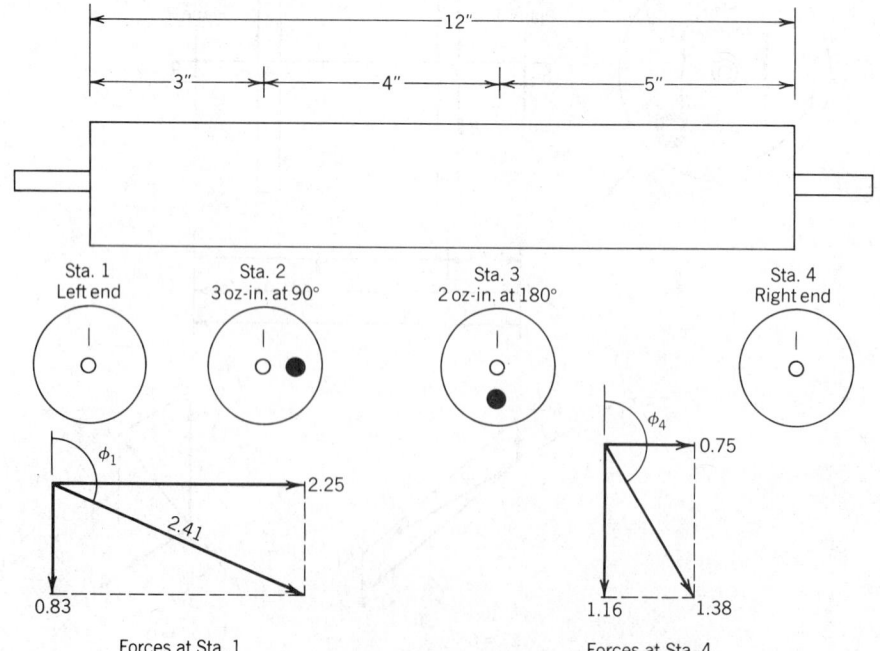

Fig. 25.13 Forces on an unbalanced rotor.

Let us now draw a vector diagram of forces at station 4, summing forces about station 1. The imbalance at station 2, when sensed at station 4, is $3 \times \frac{3}{12} = \frac{3}{4} = 0.75$ or a vector 0.75 at 90°. The imbalance at station 3, when sensed at station 4, is $2 \times \frac{7}{12} = \frac{7}{6} = 1.16$ or a vector of 1.16 at 180°. By taking the square root of 0.75^2 plus 1.16^2, we get a resultant of 1.38 oz-in. Now to get the proper angle:

$$\phi_4 = 180° - \tan^{-1}\frac{0.75}{1.16}$$

$$= 180° - 32.8° = 147.2°$$

So we see that we must remove 1.38 oz-in. at an angle of 147.2° from our reference. We may again select any convenient combination of weight and radius along this 147.2° radius line.

The reader must be cautioned that elastic bodies cannot be balanced in this manner; imbalance must be removed in the plane where it exists, rather than in some plane that happens to be convenient.

We have reviewed the meanings of several terms and shown how balancing may be accomplished. Now we will discuss the test equipment used in learning the amount and location of imbalance.

The simplest balancing machines make use of gravity. A rotor, on its shaft, may rest on hard knife-edges, which are straight and level. The heavy side of the disk will seek out its lowest level and thus automatically indicate the angle at which a balancing weight must be added. Various weights are added until the disk has no preferred position. Or the disk may be horizontal with its center pivoted on a point, or attached to a string. The heavy side will seek its lowest level, thus indicating the angle at which a correcting weight should be added. Weight is added until the disk is level. With either of these gravity machines, which accomplish static balancing only (the disk is stationary, not rotating), we recognize that balancing could have been accomplished by removing material from the heavy side, rather than by adding material to the light side.

Static balancing is occasionally helpful as a first step in dynamic balancing. A rotor may be so badly unbalanced that it cannot be brought up to speed in a dynamic balancing machine. It must be given a preliminary static balance.

There are many variations on the basic idea of centrifugal (dynamic) balancing machines. We will not attempt to describe any one unit; rather we will mention rather briefly some of the variations that a reader may encounter.

Bearings may be supported in a soft, flexible manner, to approximate the unrestricted rotor of Fig. 25.11. Bearings are supported by thin rods or wires which are soft in a single plane; the resulting motion is primarily in that plane. Motion of the flexible bearing supports is measured as an indication of imbalance force. The earliest machines measured motion mechanically. Subsequent machines measured motion electrically, with pickups that sensed either displacement or velocity. Newer machines employ force-sensing transducers to generate a signal for the electronic system. A readout informs the operator how much imbalance exists and where, and usually tells this in terms of how much material must be removed (or added) and where. Some machines require that the work-piece be removed to, say, a drill press. Others combine the balancing operation with drilling, welding, soldering, etc., so the part can be balanced and rechecked without removing it from the machine. On some machines the entire operation is automatic.

The reader may be interested to learn that transducers and meters of some balancing machines are capable of measuring peak-to-peak displacements of about 0.08 μin. This extreme sensitivity makes it possible to detect very small imbalances.

Table 25.1 indicates the balancing accuracy of some common commercial items. Gyroscopes are held to even closer tolerances, on the order of 0.000 000 25 oz-in., since even slight vibration causes long-time drifting.

Table 25.1 Selected Balancing Accuracies

Part	Accuracy of Balance (oz-in.)
Vacuum sweeper armature	0.00005–0.0001
Automobile crankshaft	0.0001–0.0002
Automobile flywheel	0.0002–0.0005
Automobile clutch	0.001–0.002
Aircraft crankshaft	0.0001–0.0004
Aircraft supercharger	0.00055–0.0001
Electric motor armatures, 1 hp	
1800 rpm	0.0001–0.0002
3600 rpm	0.00005–0.0001
Small instrument motor armatures	0.000005–0.00001

In-place dynamic balancing of rotating machinery is often needed when it is impossible or inconvenient to remove a rotor for balancing. A stroboscopic light is synchronized to the rpm of the rotor
to be balanced; this is usually done with a vibration pickup on the frame of the machine. It may be
necessary to "tune out" interfering vibrations from other machines by means of a filter between pickup
and strobe light; when this is accomplished, the offending rotor will appear to be "frozen" and not
moving. A reference mark on the rotor and the angle at which the mark appears indicate the angle
of imbalance. Readout from the vibration pickup is proportional to the amount of imbalance.

The rotor is stopped, and a compensating weight is attached at some location. The rotor is again
spun at operating rpm and any improvement (or degradation) is noted in the vibration level, along
with the new reference angle. Different weights and radial and angular locations are tried until satisfactory balance is achieved. The "cut-and-try" process is greatly speeded by drawing vector diagrams
representing imbalance forces or from use of a specially programmed calculator. Skill in this work
comes from wide experience; a wide range of problems arises, as compared to production-line balancing
work.

25.3 VIBRATION MEASUREMENT

Let us first discuss the measurement of vibratory *displacement*, normally the peak-to-peak displacement
or double-amplitude D measured in inches or in millimeters. If a structure is steadily vibrating with
a large enough D, we can estimate motion by merely holding a ruler alongside. The reader is requested
to mentally estimate the smallest D which would be read with $\pm 10\%$ accuracy. ½ or ¼ in.? 10 or
5 mm? If we could hold the ruler steady, if the vibration remained constant, and if we took several
readings and averaged them, we might get an accuracy of $\pm 10\%$ at ¼ in. or 10 mm D.

Optical techniques for measuring vibratory displacement D are limited in accuracy, especially at
the small displacements we find in vibration measurement and testing. Let us calculate D under the
test conditions of 10g (peak) acceleration at a frequency of 1000 Hz by using the following relationships:
English unit example:

$$A = 0.0511 f^2 D$$

or

$$D = \frac{A}{0.0511 f^2}$$

$$= \frac{10}{.0511(1000)(1000)}$$

$$= 0.0001957 \text{ in.}$$

International System example

$$A = 0.002\,02 f^2 D$$

or

$$D = \frac{A}{0.002\,02 f^2}$$

$$= \frac{10}{0.002\,02(1000)(1000)}$$

$$= 0.00495 \text{ mm}$$

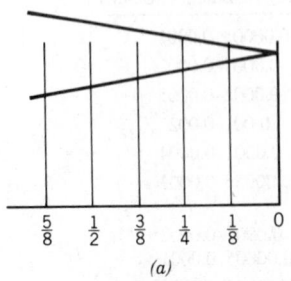

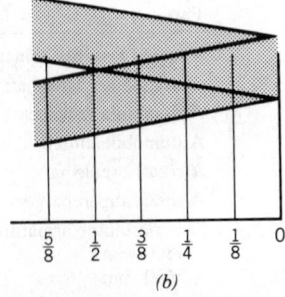

Fig. 25.14 Vibration measurement by "optical wedge" device.

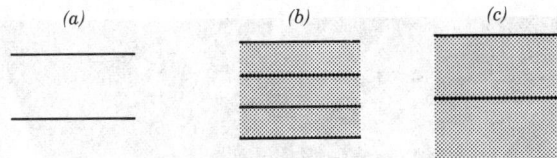

Fig. 25.15 Vibration calibration.

This is about 200 μin., an example of the extremely small displacements we find in much vibration work. At higher frequencies, D is still smaller; for example, at 10g (peak), 2000 Hz, D would be about 50 μin. or 1.3 μm.

A very popular optical technique uses the "optical wedge" device, shown in Fig. 25.14a, which is cemented onto a structure that will be vibrating. This device depends on our eyes' persistence of vision; it works best above 15 Hz. Velocity is zero at the extremes of position. There we get a crisp image. In between, where the pattern is moving, we let a lighter, gray image. The result *appears* to be two images, overlapping each other as in Fig. 25.14b. The ends of the images will be separated by exactly D. D is read by noting where the two crisp images intersect, then by looking directly below this intersection on the ruled scale. In Fig. 25.17b, D is 0.50 in. With great care, accuracies of ±10% or even ±5% are possible, particularly at larger amplitudes. This technique is most useful when an "unknown" D is to be measured.

When one wants an optical indication that D has reached a particular value, the scheme of Fig. 25.15 is useful. Two parallel lines, perpendicular to the direction of motion, are drawn ½ in. apart. When the target is stationary, the lines appear as in Fig. 25.15a. To adjust the D of a shaker to, say, ¼ in., increase shaker power until two gray bands separated by a white band, all ¼ in. high, appear as in Fig. 25.15b. For a D of ½ in., increase the power to the shaker until the two gray bands just merge, as in Fig. 25.15c.

Suppose you wish to calibrate an accelerometer/cable/amplifier/meter system at 10g (peak) and 20 Hz. Carefully rule lines 0.49 in. apart, then adjust D until the pattern resembles Fig. 25.15c. Some laboratories use lines scribed through dull paint on the surface of a metal block bolted to the table of a shaker. Lines are scribed with spacings of 0.100, 0.200, 0.036, 0.49, etc., in. These D's, at known frequencies, give useful levels of velocity and acceleration to check velocity pickups and accelerometers. Use a magnifying glass or a low-power microscope to determine when areas merge.

Accuracy statements on these techniques are not conclusive. Factors include accuracy of original construction of the device used, accuracy and judgment of the user, the D being read (greatest accuracy at large D's), etc. Only with all factors aiding the operator do optical pattern techniques give ±5% accuracy.

For measuring D's smaller than 0.1 in., use optical magnification. Use a 40× or 50× microscope with internal "cross hairs" and with cross-feeds that permit moving the entire microscope body. A stroboscopic light is helpful. Focus the microscope on some well-defined point or edge on the vibrating structure (no motion); with the structure vibrating, move the microscope body until the cross hairs line up with the extreme excursion of the point or edge being observed. Read a dial wheel on the cross-feed. Move the microscope until the cross hairs line up with the other extreme position, and take another reading. D equals the distance the microscope body was moved—the difference in readings. In practice, this technique is difficult to use: the "neutral" position of the vibrating structure may shift and lead to large errors. "Filar" elements inside the microscope are helpful.

A better microscope technique, generally used with about a 40× or 50× instrument, requires a calibrated reticule or eyepiece, having regularly spaced lines, usually 0.001 in. apart. Before the structure

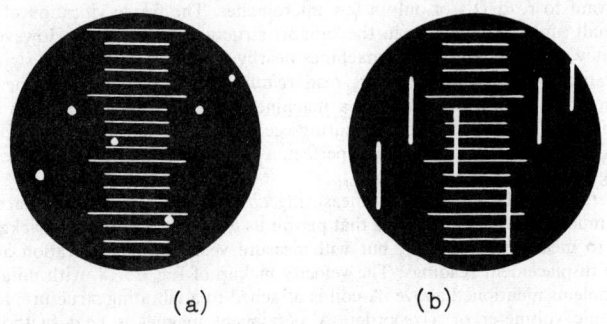

Fig. 25.16 Vibration measurement with a microscope.

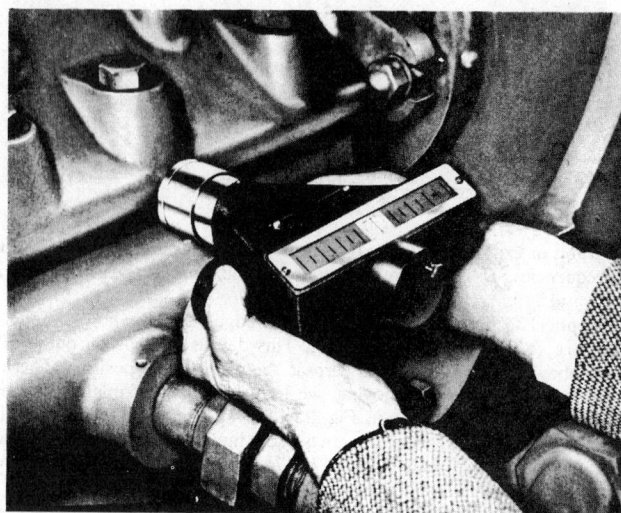

Fig. 25.17 Hand-held vibration measurement instrument.

commences to vibrate, focus the microscope on a bit of Scotchlite tape, fine garnet, or emery paper. You will see the grains of the target; with a strong light from the side, each grain will reflect a bright point of light, as in Fig. 25.16a. When the structure vibrates, each point of light "stretches" into a line as in Fig. 25.16b. The length of any line is estimated by comparing it with the rulings. In Fig. 25.16b, the rulings are 0.001 in. apart, and D is 0.005 in.

Optical (and mechanical) magnification is also used in hand-held instruments typified by Fig. 25.17. A probe maintains contact with a vibrating surface. Motion results in a bright pattern appearing on a scale; the pattern width represents D. A similar unit permanently records waveforms of instantaneous displacement versus time on a paper strip. As paper speed is known, frequency can be calculated. These instruments are satisfactory when a vibrating machine is so heavy that its motion is not affected by the mass of the probe. The instrument case must not move. Thus the operator must stand on a nonvibrating support and must hold the instrument steady against forces transmitted through it. These limitations are never fully met; consequently, these instruments are seldom used for measuring the tiny displacements found in most vibration situations, originating in faulty bearings, gears, etc. They are mainly used in measuring the result of rotational imbalance upon heavy machines such as engines, pumps, etc.

There are a number of displacement-sensing pickups that send an electrical signal to remote amplifiers, meters, oscilloscopes, oscillographs, tape recorders, analyzers, etc. Several different physical methods can convert changes in position into electrical signals: these include linkages from vibrating structures to the sliders of stationary variable resistances; also noncontacting variations in capacitance, inductance, eddy currents, etc., which are caused by changes in distance. Note that some portion of each instrument must be held stationary, if one is to gain accurate information about the *absolute* motion of a vibrating structure. This is very difficult to do. The "background noise" of vibration in many industrial buildings is 0.005 in. D or more; this will certainly prevent accurate measurement of, say, 0.001 in. D. And many times we want to read D's of only a few microinches. The frame vibration of a gas turbine is normally very small (unless resonances in the support structure are excited). However, there may be large, low-frequency motion due to other machines nearby; this may "mask" the signal of interest.

On the other hand, displacement sensors read *relative* displacements well. One type is inserted into a drilled, threaded hole in the frame of a machine. Its sensing end is close to a rotating shaft, either perpendicular to the shaft axis for measuring eccentric shaft motion or parallel to the shaft for measuring axial shaft motion. If motion is not perfect, a signal proportional to relative motion between shaft and housing will result.

Since the greatest trouble with directly measuring *absolute* D's is in the presence of background vibration, we are much interested in systems that permit us to largely ignore such background vibration. We will not try to measure D directly, but will measure velocity or acceleration and then convert those signals into displacement readings. The velocity pickup of Fig. 25.18, with minor modifications, eliminates the problems mentioned above. A coil is attached to a vibrating structure; leads are brought out to an electronic voltmeter or a recorder. A permanent magnet is held (without any motion!) close to the coil; the electrical signal generated by the coil is proportional to the relative velocity

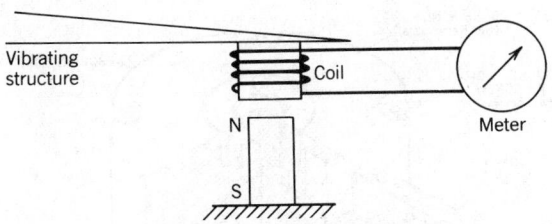

Fig. 25.18 A velocity pickup.

between coil and magnet, so any motion of the magnet creates an error in the measurement. This system can be calibrated in volts per unit of velocity.

If the magnet can be held still, the reading of the meter will be proportional to actual velocity of the moving structure. Such a pickup is too bulky to be practical for much vibration measurement work. A number of firms build self-contained seismic velocity pickups, as exemplified by Fig. 25.19. In this unit, the sensing coil is attached to an arm; the arm in turn is attached to the frame by pivots and delicate springs. The coil moves relative to the frame and to a magnetic field which is supplied by Alnico permanent magnets. The entire pickup is attached to the structure being investigated. At the pickup's f_N of about 5 Hz, the arm and coil move more than does the pickup frame. This unit is only used at frequencies above 10 Hz where the arm and coil are "isolated" from vibration; thus they remain stationary (seismic support). The magnetic field sweeps up-and-down across the coil; this generates a voltage proportional to the velocity of the pickup.

Functions of the coil and magnet may be reversed. The permanent magnet of Fig. 25.20 is seismically suspended. When the case vibration is well above resonance (operation of this unit starts at 45 Hz), the magnet remains stationary. The coil, attached to the frame, sweeps through the magnetic field so that a voltage is induced. The frequency range is 45–1500 Hz. Stroke is limited; the maximum D is 0.1 in. Some velocity pickups are built for lower frequencies and longer strokes. Accelerations greater than $50g$ can cause damage.

The velocity pickup shown in Fig. 25.21 is similar to the unit of Fig. 25.19. However, a length of 0.030 in. drill rod, connected to the sensing coil, passes out through the housing. This unit is normally held by hand, with the "probe" firmly touching a vibrating surface. Motion is transmitted to the coil, in which a voltage proportional to velocity is generated. This unit is very useful for hand-scanning over a vibrating surface, to locate points of maximum and minimum motion.

Velocity pickups have many advantages: they are self-contained devices requiring no external source of power—no dc or ac excitation. Because of their low internal impedance, they can be used at great distances from the readout instrument. They are most used in the frequency range 10–500 Hz. A typical sensitivity is 100 millivolts (peak) per inch per second (peak) velocity.

Velocity pickups have certain disadvantages, of course. They are generally quite large and heavy,

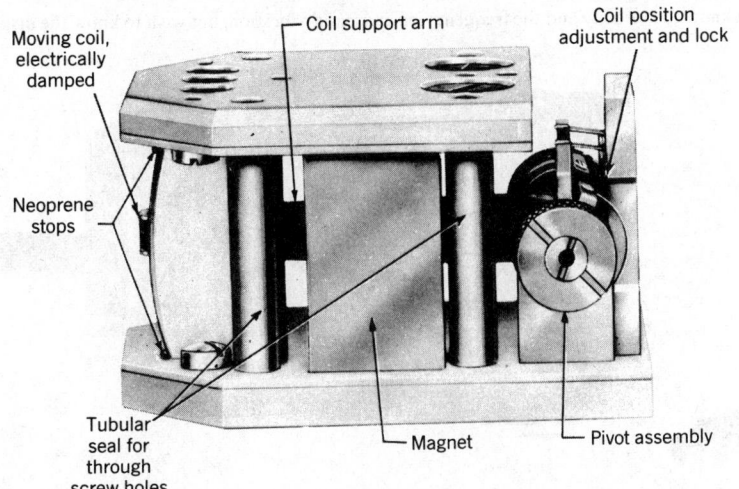

Fig. 25.19 A self-contained seismic velocity pickup. (Courtesy of Vibra-Metrics Inc.)

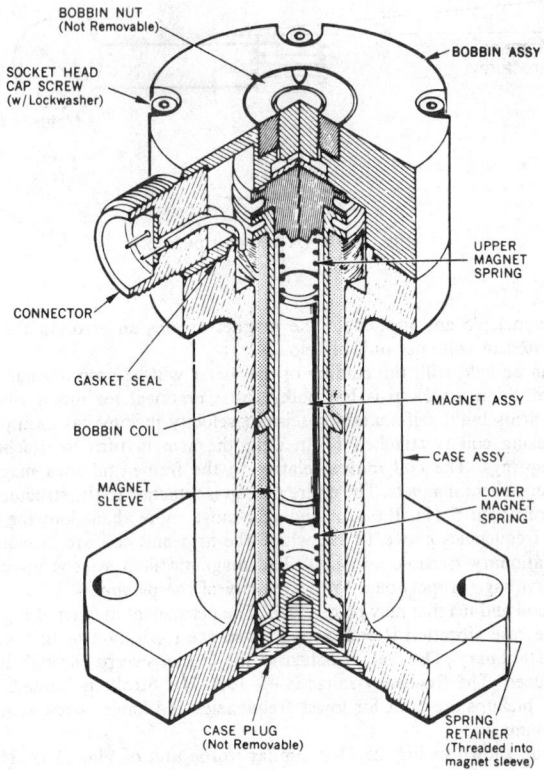

BOBBIN NUT
(Not Removable)

BOBBIN ASSY

SOCKET HEAD
CAP SCREW
(w/Lockwasher)

UPPER
MAGNET
SPRING

CONNECTOR

GASKET SEAL

MAGNET ASSY

BOBBIN COIL

CASE ASSY

MAGNET
SLEEVE

LOWER
MAGNET
SPRING

CASE PLUG
(Not Removable)

SPRING
RETAINER
(Threaded into
magnet sleeve)

Fig. 25.20 Velocity pickup with permanent magnet suspended seismically. (Courtesy of CEC.)

especially as compared with some accelerometers. They cannot be used at very large displacements, because of limited stroke of their moving parts. Since they will sometimes be used close to their natural frequencies, some form of damping (usually oil or eddy current) must be used; this introduces certain problems in measuring any nonsinusoidal motion.

Readout from velocity pickups is very simple, as in Fig. 25.18. The pickup sensitivity is known in terms of millivolts per unit of velocity. An electronic voltmeter reading in terms of millivolts can be interpreted in terms of velocity.

If you know the velocity and the frequency of *sinusoidal* vibration, but wish to know the displacement

Fig. 25.21 Hand-held velocity pickup used for locating points of minimum and maximum motion.

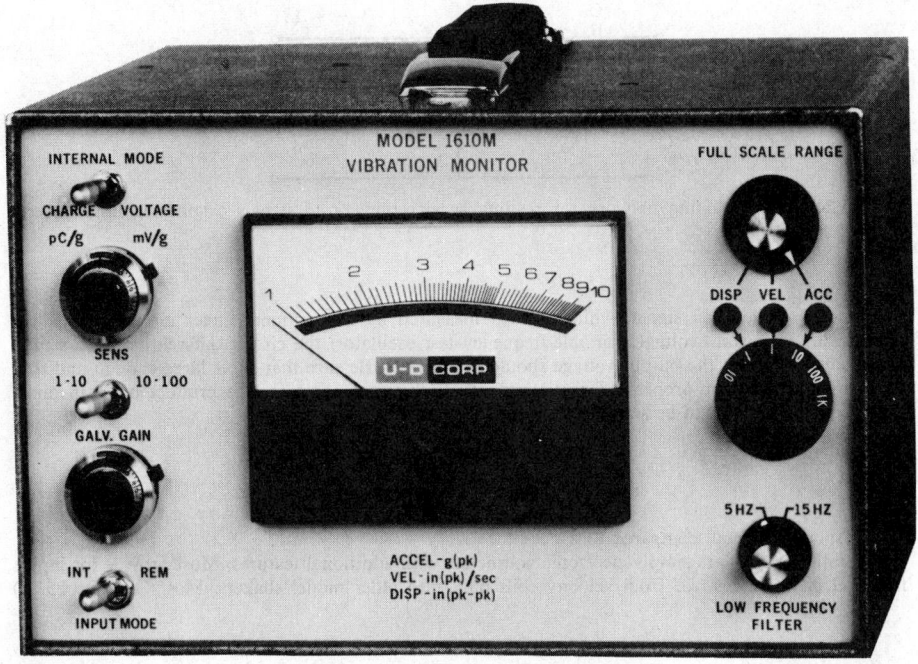

Fig. 25.22 A specialized vibration meter.

or acceleration, you could calculate these. If you did not need extreme accuracy, you could use one of the cardboard vibration calculators that are available from some shaker and accelerometer manufacturers. One of the advantages of velocity pickups when used with a common electronic voltmeter: values of velocity, displacement, and/or acceleration may easily be determined. This advantage was very important around 1945–1955; funds for purchase of more sophisticated vibration instruments were not readily available.

Specialized vibration meters such as the unit in Fig. 25.22 were soon developed. When one wishes to read D on a meter whose input signal comes from a velocity pickup, the meter must contain an integrating network similar to Fig. 25.23. The velocity signal is electronically integrated to form a displacement signal, which is then measured by conventional meter circuitry. Test the circuit with a constant-voltage, variable-frequency test oscillator; the circuit integrates the signal. If frequency doubles, the output voltage should drop to ½, etc. Be sure that R is large enough and that X_c is small enough for proper integrating action; the "time constant," the product of R in ohms times C in farads, should be large compared to the period T, where T is

$$T = \frac{1}{\text{lowest frequency at which integrator is used}}$$

and R should be large compared to X_c.

Most meters also provide a differentiating network, something like Fig. 25.24, so that A may be read from a signal that originates in a velocity pickup. The velocity signal is electronically differentiated

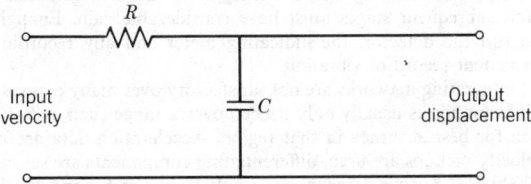

Fig. 25.23 Integrating network for reading displacements when input signals come from a velocity pickup.

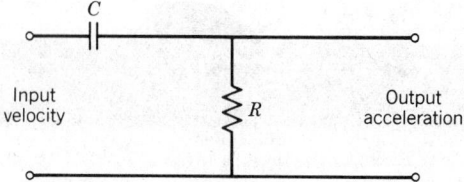

Fig. 25.24 Differentiating network for reading accelerations from signals originating in a velocity pickup.

to form an acceleration signal, which is then measured by conventional meter circuitry. Test the circuit with a constant-voltage, variable-frequency test oscillator; the circuit differentiates the signal. If frequency doubles, the output voltage should double, etc. Be sure that X_c is large enough and that R is small enough for proper differentiating action; the time constant, the product of R in ohms times C in farads, should be small compared to

$$T = \frac{1}{\text{highest frequency at which differentiator is used}}$$

and R should be small compared to X_c.

Vibration meters are simply electronic voltmeters with additional features. Most provide for inputs from velocity pickups and from velocity coils built into older model shakers. Most (see Fig. 25.25)

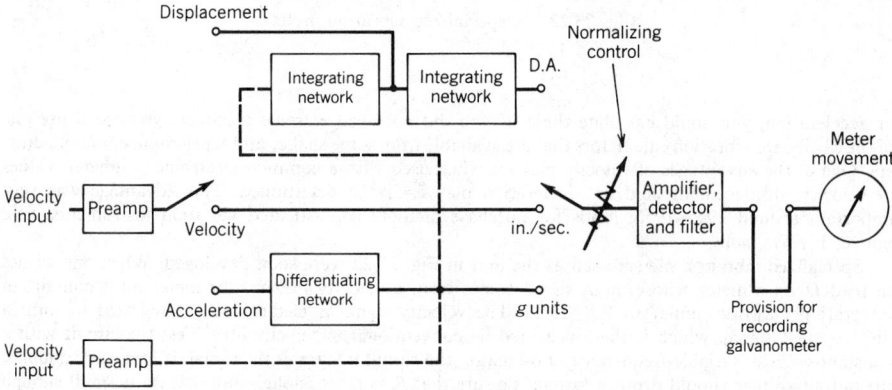

Fig. 25.25 A vibration meter circuit.

have a three-position switch marked Displacement–Velocity–Acceleration. In the Velocity position, the unit is a simple ac voltmeter. The variable-gain "normalizing" control is set for the particular pickup sensitivity being used. When such a meter is used to read displacement, an integrating network is inserted into the signal path, as in Fig. 25.25, by switching to Displacement. For reading acceleration, the signal is switched through a differentiating network, as shown, by switching to Acceleration.

All meters today provide for acclerometer inputs, as in Fig. 25.25. If the incoming signal is twice integrated, it becomes proportional to displacement. Integration and differentiation "waste" much signal; therefore preceding and subsequent stages must have considerable gain. Enough voltage and power must be developed to run the detector, the indicating meter and any recording galvanometer that may be used for a permanent record of vibration.

Differentiating and integrating networks are not satisfactory over many octaves of frequency. Fortunately, displacement information is usually only needed over a range such as 5–100 Hz, so integrating components are chosen for best accuracy in that region. Acceleration data are usually needed above about 50 Hz; when velocity pickups are used, differentiating components are selected for best operation from 50 to perhaps 2000 Hz. Serious inaccuracies result from use beyond the frequencies indicated in Table 25.2. Table 25.2 gives typical tolerances for the vibration meter only, not including the velocity pickup or accelerometer.

Table 25.2   Vibration Meter Data

Function	Range	Frequency Limits	Tolerance
Velocity	0.01–100 in./sec	5–500 Hz	±2%
Displacement	0.001–10 in. D	5–5 000 Hz	±2%
Acceleration	0.1–1 000 g	10–5 000 Hz	±1%
		5–10 000 Hz	±2%
		2–25 000 Hz	±5%

25.4 ACCELERATION MEASUREMENT

Let us now take up the subject of accelerometers—units whose instantaneous output voltage is proportional to the instantaneous value of acceleration. Most vibration and shock measurements today are made with accelerometers.

At high frequencies, accelerometers generate a larger signal than do velocity pickups. This may be shown by calculating the peak velocity that would exist if the peak acceleration were 1 g at a frequency of 2000 Hz:

English units:

$$V = \frac{A}{0.0162f}$$

$$V = 0.031 \text{ in./sec (peak)}$$

SI units:

$$V = \frac{A}{0.000\,642f}$$

$$V = 0.78 \text{ mm/sec (peak)}$$

Let us assume that a velocity pickup has a sensitivity of 105 mV/in. sec. It will generate about 3.22 mV (peak) or about 2.28 mV (rms). (At 200 Hz, 1g, the velocity would be 10 times greater, of course, and so would be the voltage.) A signal of only 2 or 3 mV rms is difficult to measure under some conditions. We will show that accelerometers built for these higher frequencies have a number of advantages; one advantage is higher output signal at higher test frequencies found in modern vibration tests and field measurements.

A reasonable sensitivity figure for a crystal accelerometer is 10 mV (peak) per g (peak); this sensitivity is relatively constant up to, say, 10,000 Hz. In the situation discussed above (1g at 2000 Hz) the accelerometer would generate 10 mV (peak) or 7.07 mV rms. This is about three times as much signal as the velocity pickup generates. This advantage would double with each doubling of frequency (octave).

Accelerometers always operate *below* their natural frequencies. Thus the natural frequency must be high; the useful "flat" range is to about one-fifth of the natural frequency, where the sensitivity is about 4% higher than it is at low frequencies. A 50-kHz unit thus permits operation to 10 kHz.

How might we build an instrument that would sense acceleration? The automobile enthusiast senses acceleration by "feeling" his or her body sink into the seat cushions. As the car and seat gain a high velocity (accelerate), a deflection proportional to the intensity of acceleration is noted. His or her body has inertia and tends to keep its former velocity. If we could measure the amount of deflection, we would have a crude accelerometer. In Fig. 25.27 we see a crude accelerometer and how acceleration could be measured on a meter. A cantilever beam supports a weight W. On the top and bottom are cemented strain gages, elements that change resistance when stretched or compressed. When the structure being observed accelerates upward, the weight W, having inertia, tends to stay behind; the top side of the beam stretches while the bottom side shortens.

This results in R_1 increasing while R_2 decreases. The electrical signal to the meter (which had been zero before, when $R_1 = R_2$) becomes negative. The larger the acceleration, the larger the voltage and deflection of the meter pointer. On a downward acceleration, all these actions reverse.

For very low frequencies, we can use a dc meter, as shown in Fig. 25.27. The pointer could be adjusted to the center of the scale for zero acceleration; it will swing left for upward acceleration and right for downward acceleration. If we gradually increase the vibration frequency, the meter will not "follow" but will just quiver about the zero position. We will have to switch to an ac meter or use an oscilloscope. The ac meter will probably display the rms value of the alternating voltage input. Its reading will remain constant as long as the acceleration peak or rms value of each cycle of vibration remains the same. The oscilloscope will show a sinusoidal pattern, with peaks representing peak acceleration.

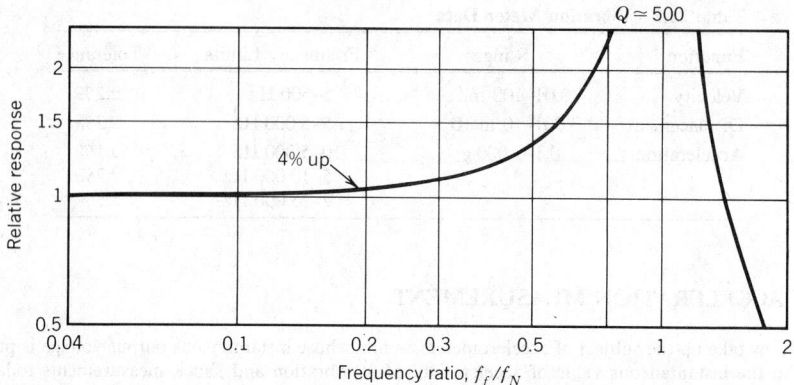

Fig. 25.26 Useful "flat" range for an accelerometer.

Let us assume that our crude accelerometer has a natural frequency of 20 Hz; this is 1.0 on the frequency scale of Fig. 25.26. As we slowly increase the frequency of vibration, holding the peak acceleration constant, we will observe higher and higher readings as we approach 20 Hz, as indicated by Fig. 25.26. Thus we can use our crude accelerometer only to 4 or 5 Hz; at higher frequencies it gives false readings. We could put some kind of dash pot or other damper on the beam. This would prevent so much buildup and would permit testing to perhaps 10 or 15 Hz. However, this would introduce phase-shift problems if we tried to measure nonsinusoidal waveforms or shock. We could use a stiffer beam (or a lighter weight) in order to get a higher natural frequency. While this would permit testing to higher frequencies, it would reduce the bending and thus the electrical response per g, the sensitivity.

Accelerometers using this principle are used in laboratories and in navigational equipment. But the configuration of Fig. 25.27 would be hard to package. Figure 25.28 shows a practical instrument. A mass is free to slide in the direction shown; it is restrained by guides against other motions. The mass is "sprung" from the case by four wire springs, which are also electrical strain gages; these change resistance when deformed. If the case accelerates to the right (due to acceleration of the structure to which the unit is attached), R_3 and R_4 increase while R_1 and R_2 decrease. Electrical polarity reverses when acceleration reverses. Output voltage is proportional to the intensity of acceleration. In place of a battery or other direct-current supply, as shown in Fig. 25.28, excitation could be an alternating current or "carrier" supply. Variations in resistance would amplitude modulate the carrier; this has some advantages in that electronic amplification is simpler. Strain gage accelerometers generally have low natural frequencies. Sensitivity has to be sacrificed if higher natural frequencies are desired. Typical natural frequencies are 90, 155, and 250 Hz with full-scale ranges of $\pm1g$, $\pm2.5g$, and $\pm5g$, respectively.

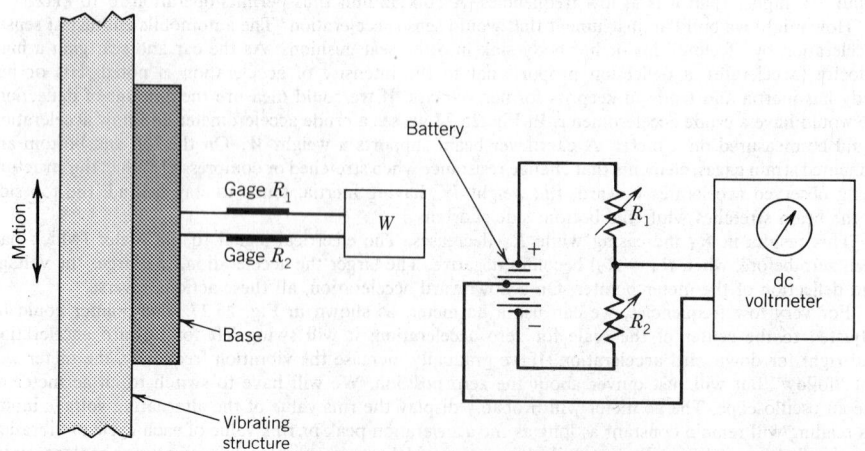

Fig. 25.27 A crude accelerometer.

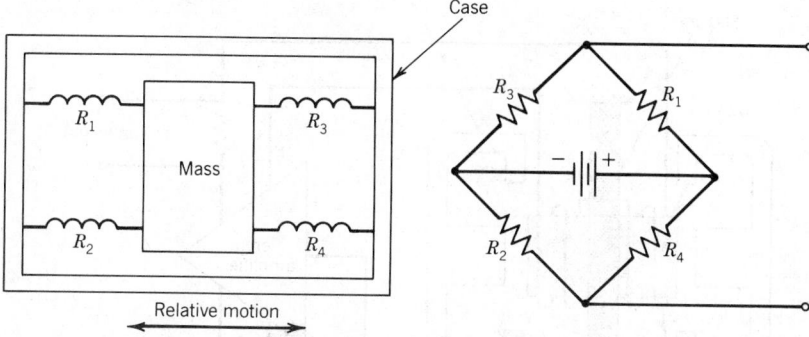

Fig. 25.28 A practical accelerometer.

One way to get a higher frequency range without great loss of sensitivity is to use piezoresistive units, suggested by Fig. 25.29. Note that the spring rate (and thus the natural frequency) is determined by how much material is cut away from the mass. Semiconductor strain gages (higher gage factor than wire) sense any deflection of the mass that may result from an input acceleration. A typical piezoresistive accelerometer has a natural frequency of 10,000 Hz with a useful frequency range of 0–2000 Hz; its full-scale range is ±250g, while its sensitivity is about 1 mV/g.

The most sensitive (capable of sensing distant earthquakes in micro-g's) and most precise (used in navigation of spacecraft) accelerometer is the servo or force-rebalance type, suggested by Fig. 25.30. Acceleration of the case results in very slight relative motion between mass and case; this is sensed by an internal position pickup. The resulting signal is amplified, then it passes through a coil on the mass, restoring the mass to its original position. The restoring current also acts as the output signal from the accelerometer.

The piezoresistive, servo, and strain-gage types have one important advantage over most piezoelectric crystal types: they can be easily calibrated by merely turning them over; this is a static calibration at ±1g, which is very useful. (Some piezoelectrics used with some charge amplifiers may also be calibrated in this manner.) Imagine that the accelerometer of Fig. 25.28 were turned onto its left end; then the mass would stretch R_3 and R_4 while it shortens R_1 and R_2, giving a dc output for 1g. Turning the accelerometer on its other end would reverse the electrical signal. This calibration can only be performed on accelerometers having response down to zero frequency. Piezoresistive and strain-gage types can be used and calibrated on centrifuges where the acceleration is one-directional and constant at 1g,

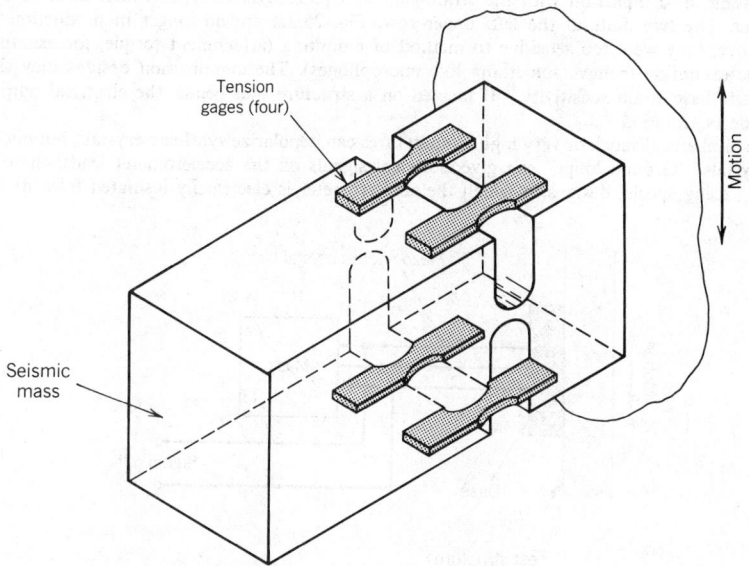

Fig. 25.29 A piezoresistive accelerometer.

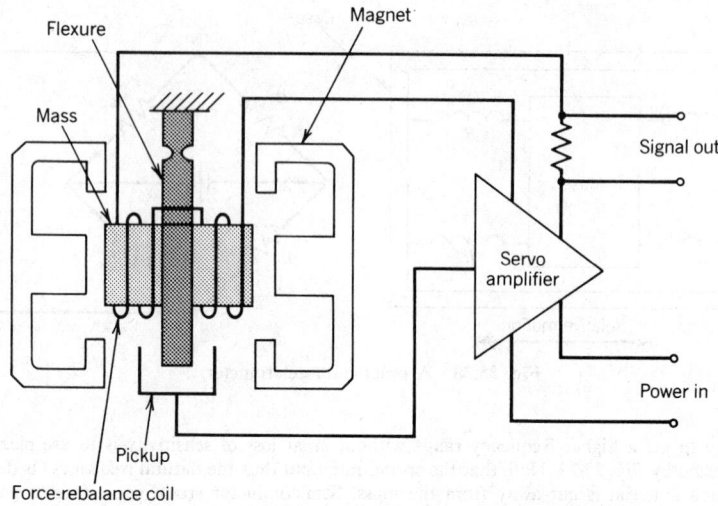

Fig. 25.30 A servo or force-rebalance accelerometer.

10g, etc. For some vibration measurements, however, it is best to suppress electrical signals due to static accelerations; in missile testing, for example, we may be interested only in vibratory loadings on structures. We can insert a blocking capacitor to suppress zero-frequency signals if they are troublesome.

A different technique of generating a signal proportional to acceleration is shown by Fig. 25.31. This impractical, but possible, device includes a bar made from a natural crystalline material such as a quartz or from a synthetic crystal such as barium titanate. Such a bar will develop an electrical charge Q when it is bent. (This is quite similar to some phonograph pickups.) Upward acceleration of a test structure and the inertia-caused lagging of the mass will cause such bending and charge Q. Downward acceleration will cause reversed polarity. We can electrically measure either charge Q or voltage E. If the measuring system permits, static acceleration will give zero-frequency charge and voltage, but usually there is some low-frequency limitation on the system response.

The design of Fig. 25.31 would be difficult to package, whereas the basic designs of Fig. 25.32 can give high natural frequencies, have wide dynamic range (fraction of 1g to several thousand g's), are small and rugged, and are quite simple to mount mechanically. All crystal accelerometers are self-generating, in comparison with the strain-gage and piezoresistive types which must be supplied with power. The two units at the left, upper row, Fig. 25.32, are no longer in production by any manufacturer; they were too sensitive to method of mounting (attachment torque, for example) and to acoustic excitation (behave something like microphones). The compression designs may show an effect called "base strain sensitivity." If located on a structure that bends, the electrical output may masquerade as vibration.

Severe mechanical shock or very high temperatures can depolarize synthetic crystals, but not natural quartz crystals. "Ground loops" can give electrical signals on the accelerometer leads; these can be avoided by using special designs in which the accelerometer is electrically insulated from its case, or

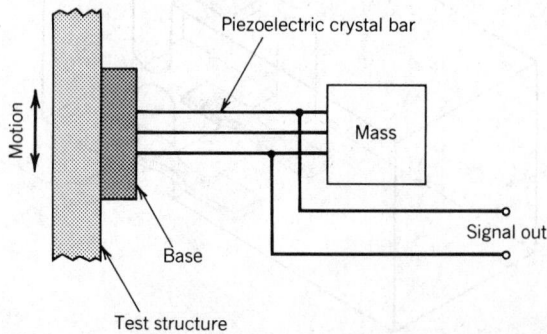

Fig. 25.31 A device for generating a signal proportional to acceleration.

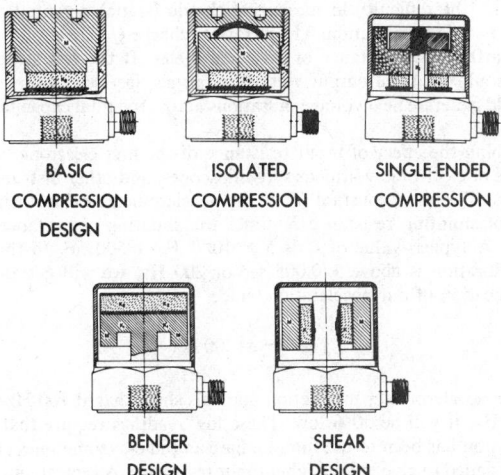

BASIC   ISOLATED   SINGLE-ENDED
COMPRESSION   COMPRESSION   COMPRESSION
DESIGN

BENDER   SHEAR
DESIGN   DESIGN

Fig. 25.32 Basic designs for accelerometers.

by electrically insulating the case from the vibrating structure. The latter method can change the frequency response of the unit, however, and any insulation should be present at the time of calibration.

The compression types, Fig. 25.32, generate electrical charge when the crystal element is compressed due to upward acceleration or decompressed due to downward acceleration. Similarly, a crystal plate bends in the bender type and a crystal ring or flat sandwich shears in the shear type.

An analogy for readers not inclined toward understanding of electronics: accelerometers behave something like a sponge. When the crystal is squeezed (or bent or sheared), it puts out "juice" (electrons). If we remove the squeeze, the juice returns. If we measure the amount of juice squeezed out (electrical charge), we have a measure of acceleration. Electrically, this requires a *charge amplifier*, a storage-measuring device for electrons.

Until recently, voltage was measured. Voltage E is proportional to charge Q divided by shunt capacitance C; see Fig. 25.33. An accelerometer's sensitivity is given in terms of charge (picocoulombs or 10^{-12} C) per g or in terms of voltage per g when used with a specified shunt capacitance or with a specified cable:

$$E = \frac{Q}{C_{\text{total}}} = \frac{Q}{C_a + C_c}$$

Part of the shunting capacitance is found in the accelerometer itself; a typical value is 500 pF. Part of the shunting capacitance is found in the signal cable; a typical value is 30 pF (30×10^{-12} F) per foot or 300 pF for a 10-ft cable. If an accelerometer has a voltage sensitivity, open circuit, of 10 mV per g with 500 pF capacitance, it will drop to

$$\frac{10\,\text{mV}}{g} \times \frac{500}{500 + 300} = 10(0.625) = 6.24\,\text{mV/g}$$

This is a severe loss of signal for only 10 ft of cable. More important, there is always danger that the user will forget to apply correction factors for the particular cable he or she uses. Laboratory calibrations are usually done with a standard value of external capacitance, such as 300 pF. The capacitances of test cables used for environmental or service tests affect the voltage per g; test personnel must remember to correct for the cables used or they will make errors in their measurements. With charge amplifiers (storage devices), the amount of shunting capacitance used is not so important. Sensitivity is essentially the same for any length cable.

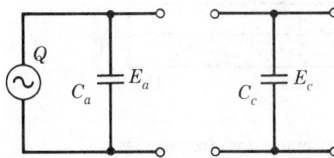

Fig. 25.33 Voltage measurement in an accelerometer.

We mentioned earlier the difficulty in measuring the dc (static) potential generated when crystal accelerometers receive a static acceleration. The electrical charge Q "leaks off" rather quickly through the approximately 10 mΩ input resistance of a typical meter. If the test structure were vibrating at perhaps 1000 Hz, we could read the output voltage on an ac electronic voltmeter, because successive cycles of vibration would generate new voltage pulsations across the total capacitance before the previous pulsation leaked off.

Here we are describing the effect of input resistance of the first electronic unit in the signal path. The shunting resistance of electronic voltmeters, oscilloscopes, and other instruments affects their readings at low frequencies. The time constant T of the accelerometer–cable–instrument combination is equal to the product of shunting resistance R times the shunting capacitance C; that is, $T = RC$. Typically, R is 10^7 Ω. A typical value of C is 5×10^{-10} F, or 500 pF. In this instance $T = RC = 0.005$ sec. When the vibration is above $1/0.005$ sec or 200 Hz, we will get an accurate (error about -1.5% at 200 Hz) indication of our acceleration, since

$$f \approx \frac{1}{RC} \approx 200 \text{ Hz}$$

Graphs published in accelerometer instruction manuals show that at 100 Hz the instrument reading will be 5% low; at 18 Hz, it will be 50% low. These low readings require that data be corrected.

The traditional solution has been to use intermediate amplifiers (sometimes called cathode followers or emitter followers) designed to give much higher input resistance. A typical unit has an input resistance of 10^8 Ω. Now, with the same 500 pF system, $T = RC = 0.05$ sec. Vibration at all frequencies above 20 Hz will be accurately indicated. At 10 Hz, the response will be 5% low. At 1.8 Hz, the response will be 50% low. For many applications, this is satisfactory. For very-low-frequency performance, intermediate amplifiers with input resistances of 10^9 Ω have been used, but there are many difficulties with connector contamination (skin oils, salt, moisture, etc.) that may change system sensitivity; if such a change occurred without the operator's knowledge, the results could be most serious!

Another approach to voltage amplification, sometimes used to eliminate need for intermediate amplifiers, has employed accelerometers with very large internal shunting capacitance. However, these generally have a very low voltage sensitivity. Since $E = Q/C$, small accelerations are hard to measure.

One of the major advantages of charge amplifiers or charge converters over voltage amplifiers is that these low-frequency errors are avoided. The RC factor has no effect on sensitivity. Moderate contamination of cables and connectors has no effect. Response *can* go to extremely low frequencies as suggested by the solid line of Fig. 25.34. (Most stop at 0.2 or 0.5 Hz to avoid responding to thermal gradients, etc.) The two dashed lines show how sensitivity falls off with loads of different resistance, when using voltage amplification.

Ideally, there should be a constant ratio (called sensitivity) between the electrical output from an accelerometer and the mechanical input. Then the electrical signal is an accurate measure of the mechanical input. Whenever that ratio changes, measurement errors can creep in; data must be corrected. Not only is this true for measurement of sinusoidal vibration, but also for measurement of shock (especially long-duration shock pulses) and of random vibration.

If the measurement system (accelerometer, *all* amplifiers in the system *and* the readout device) responds clear down to zero frequency, the electrical signal will be a duplicate of the mechanical input. If the mechanical input is one of the shock pulses shown by solid lines of Fig. 25.35, the electrical signal *should* duplicate it. This requires that RC be much greater than T, where T is the time duration of the mechanical shock pulse. For instance, with a rectangular pulse, if $RC = 50T$, there will only be a 2% error at the end of the pulse, whereas if $RC = 20T$, there will be a 5% error. The magnitude of these errors is shown by Fig. 25.35 for three common shock pulses for $RC = 6T$, $3T$, $1.5T$, and $0.6T$.

The cable from an accelerometer to the first electronic unit is important in that it must transmit the signal without any distortion or introducing any "noise." Coaxial cable is used to minimize pickup from strong electrical or magnetic fields. However, when this cable is overly flexed or mechanically distorted, it can generate unwanted electrical signals, or "noise." See Fig. 25.36. If the shield parts from the dielectric, a voltage is generated by "triboelectricity." The excess of electrons on the dielectric

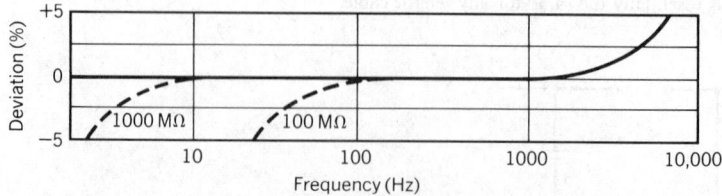

Fig. 25.34 Deviation in frequency measurement at extremely low frequencies.

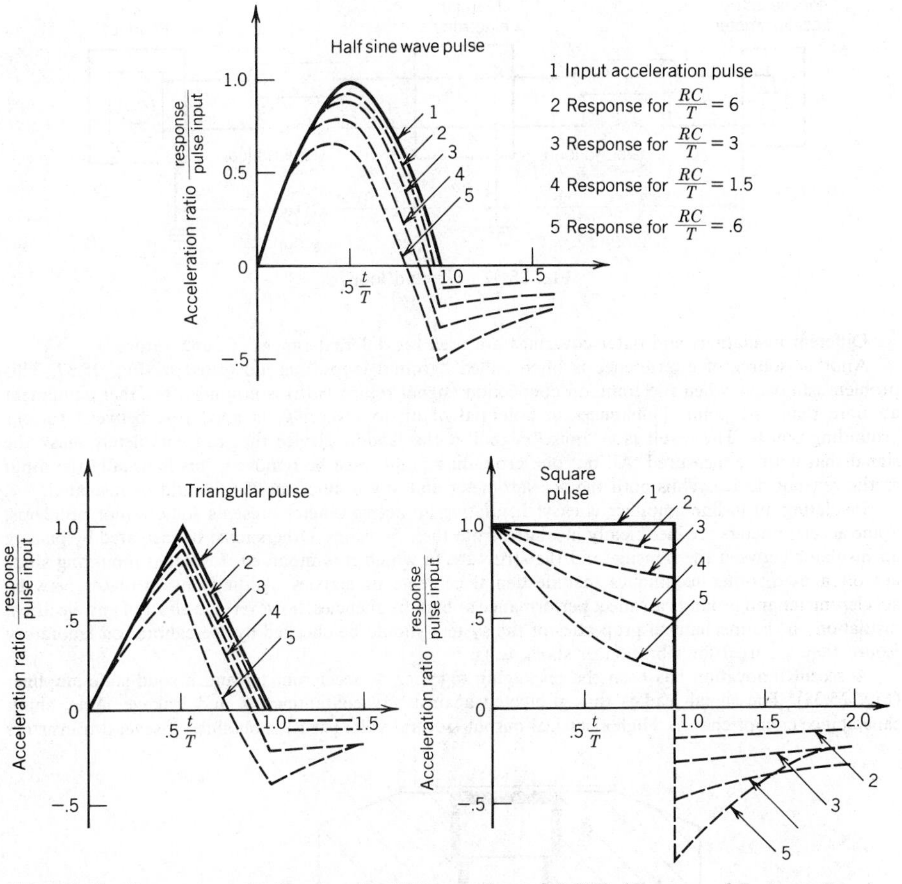

Half sine wave pulse

1 Input acceleration pulse

2 Response for $\dfrac{RC}{T} = 6$

3 Response for $\dfrac{RC}{T} = 3$

4 Response for $\dfrac{RC}{T} = 1.5$

5 Response for $\dfrac{RC}{T} = .6$

Triangular pulse

Square pulse

Fig. 25.35 Response to shock pulses for different values of RC/T. (Courtesy of Endevco Corp.)

is transferred to the center conductor at that point; excess electrons flow along the center conductor to the input resistance of the first electronic unit then back along the shield. The amplifier cannot distinguish between this brief flow of electrons and an accelerometer signal. When the cable distortion is relieved, the opposite occurs. Untreated cables used with early crystal accelerometers frequently generated noise signals greater than accelerometer signals! This problem has largely been solved through treating the surfaces of the dielectric with a conductive coating, which permits the triboelectric charges to redistribute themselves locally rather than through the amplifier input.

One should watch out for large-amplitude resonant motions of accelerometer cables, especially at low frequencies. These can easily be seen with the naked eye, and are especially troublesome when cables are secured with a "strain-relief loop," a few inches from the accelerometer. Shock pulses often cause severe whipping of the cable and can thus give faulty signals. Note that cable must be small, light, and flexible so that cable resonance and whipping do not distort the accelerometer case, thus generating stresses on the crystal and further noise.

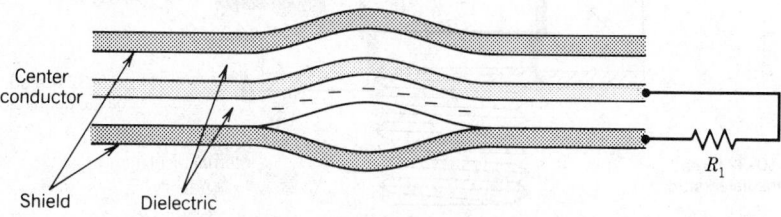

Center conductor

Shield

Dielectric

R_1

Fig. 25.36 Distortion in accelerometer cable.

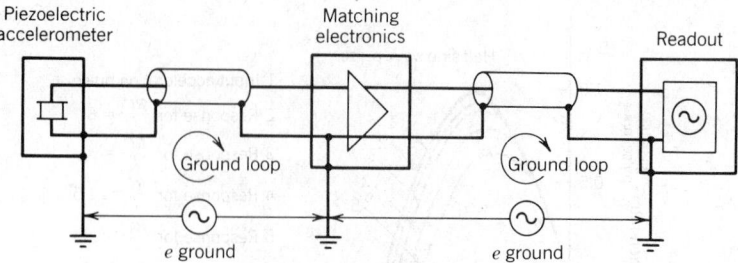

Fig. 25.37 Ground loops.

Different insulations and outer coverings are used for different ranges of temperature.

Another source of interference is often called "ground loops," as illustrated by Fig. 25.37. This problem can occur when the common connection (signal return path) is grounded to other equipment at more than one point. Differences in potential of up to several volts may exist between various grounding points. The result is a "noise" signal at the readout device that can completely mask the signal that is to be measured. All but one grounding point must be removed; this is usually the input to the readout device. Thus both the accelerometer and any in-line amplifier should be insulated.

Insulating an in-line amplifier is easy. Insulating an accelerometer presents some minor problems. Some accelerometers are electrically insulated from their housings. Others must be insulated by placing an insulator between the housing and the structure to which it is mounted. Insulated mounting studs are often used; other techniques include dental cements or epoxies. Adding any insulator between accelerometer and ground can affect performance at high frequencies, however; the effect of any electrical insulation on the mechanical properties of the system should be checked by the calibration laboratory *before* they are used for vibration or shock tests.

A recent innovation has been the packaging together of accelerometer and a solid-state amplifier (Fig. 25.38). The signal lead is thus protected against any contamination and leakage. Also, shunt capacitance cannot change. High electrical output (several volts per g) is obtained if several transistors

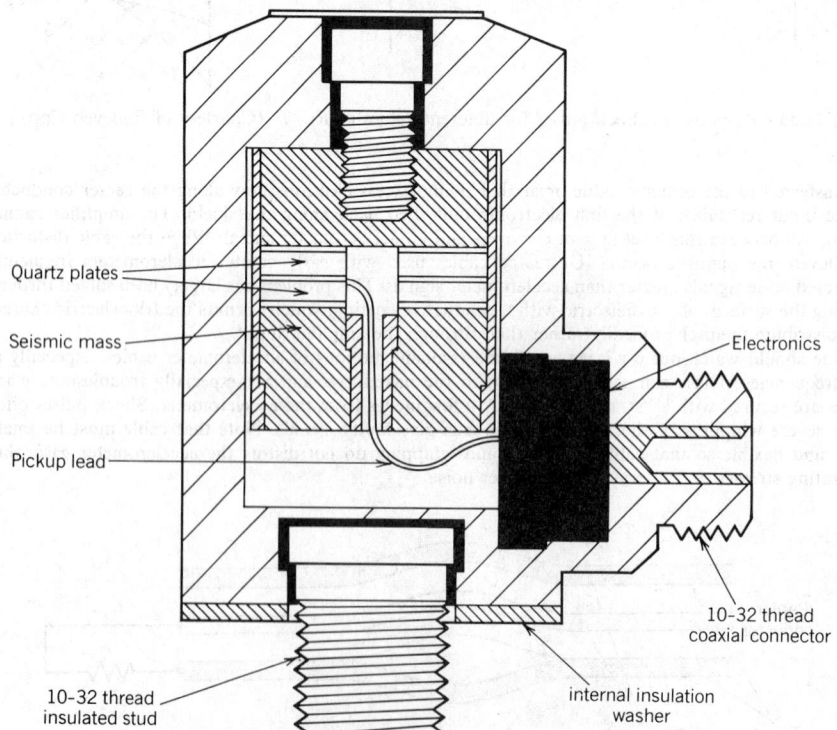

Fig. 25.38 Packaging together of accelerometer and solid-state amplifier.

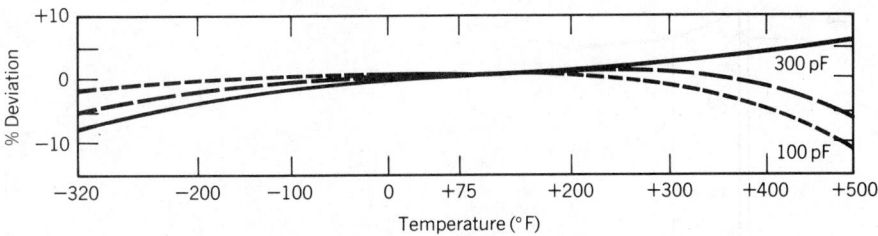

Fig. 25.39 Effect of temperature on sensitivity of an accelerometer.

are used. Some use only one transistor, mainly to give very low output impedance (typically only 100 Ω). Long cables may be used without much electrical interference. Ground loops do not present much difficulty, nor does cable noise. This innovation offers large advantages, though at an increase in cost (partly offset by simpler external electronics). All manufacturers of accelerometers offer them.

We would like vibration pickups such as accelerometers to respond only to vibration, and not at all to other environments such as acoustical fields, pressure fields, temperature, etc. Let us briefly discuss the effect of varying ambient temperature on the sensitivity of an acclerometer. Figure 25.39 shows such an effect upon a particular crystal accelerometer. The solid line shows how sensitivity varies when charge-sensing electronics is used. The dashed lines show how sensitivity varies when voltage-sensing electronics is used. (Two different values of shunt capacitance were used.) Although, as we can see, sensitivity does vary with temperature, there is a broad flat region in which these sensitivity changes are quite small. (Today's accelerometers are far superior to early day units in this regard.) Curves such as Fig. 25.39 often differ widely between voltage and charge response; it is important to use charge-sensing electronics with accelerometers having flat charge–temperature characteristics and to use voltage-sensing electronics with accelerometers having flat voltage–temperature characteristics.

One manufacturer offers four usable temperature ranges: −65 to +185°F, −65 to +350°F, −320 to +500°F, and −452 to +750°F. Another manufacturer offers units said to work well to +1100°F. If a unit is briefly exposed to too high a temperature, it may still be usable (after a calibration check). But if the crystal reaches its Curie temperature, the accelerometer becomes completely unusable. Some accelerometer materials change resistance and/or capacitance with extremes of temperature; this can greatly affect their frequency responses, mainly at low test frequencies.

Not only is temperature a factor, but also the *rate of change of temperature*. With some accelerometers, temperature transients can produce several volts signal due to stresses in the crystal element and/or case. Thermal shielding is often helpful, as is blocking of amplifier response to extremely low frequencies.

Accelerometers may be attached in various ways to the structure being investigated. The best way is by means of a machine screw or stud; this exerts a compression preload between the accelerometer base and the supporting structure. This preload should be greater than the maximum tension force that can occur, so that there will be no "chattering" between accelerometer base and the structure. At very high test frequencies (5000 Hz and higher), a little oil or grease assists in getting a solid mechanical connection. A mounting torque value is usually stated; this ensures a firm connection. If a screw connection cannot be made, or if one is only interested in comparative vibration values, he or she may choose to cement his or her accelerometer to the vibrating structure. Dental cement or Eastman 910 cement may be used. Some (usually quite small) accelerometers have a flat base for easy cementing. However, calibration data (taken in a standards laboratory where only screw connections are used for mounting) will not apply, since the cement introduces compliance that affects the accelerometer's sensitivity at certain frequencies. Insulated studs, which are often used to mount accelerometers when "ground loops" are expected to give trouble, also introduce compliance.

Some types of accelerometers, particularly the compression types, exhibit an effect called base-strain sensitivity. As suggested by Fig. 25.40, an accelerometer may be located at a point of small motion. Yet the readout instrument may show large motion. Why? Sometimes the difficulty is distortion of the base caused by bending of the structure being measured, as would be shown by strain measurements; this distortion can cause an alterating force on the crystal and a resulting electrical signal. One solution is to select accelerometers that do not have much such sensitivity. Another is to reduce the effect by mounting the accelerometer on an insulated mounting stud that will transmit vibration but not bending force.

Both in industrial and in aerospace applications, displacement and velocity pickups have been losing favor since about 1955, when rugged piezoelectric accelerometers first became available. Let us discuss a few relative advantages of these accelerometers. First, they are far more rugged; some accelerometers for shock measurements are routinely calibrated at levels greater than 10,000g. In comparison, few velocity pickups will withstand 100g accelerations and are easily damaged by rough

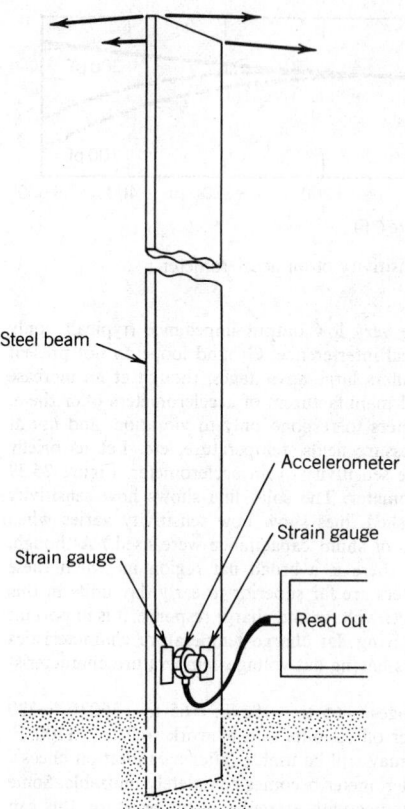

Steel beam

Accelerometer

Strain gauge

Strain gauge

Read out

Fig. 25.40 Location of accelerometer at point of small motion.

handling. Noncontacting displacement sensors are no doubt rugged enough to withstand any ordinary industrial application.

We might note that aircraft engine vibrations (in aircraft and test stands) were traditionally measured with velocity sensors mounted close to main bearings; output was integrated and read on a meter as inches D. This practice carried into jet engines; a typical overall displacement limit is 0.003 in. Recently, however, many people have become convinced that acceleration measurements are better indications

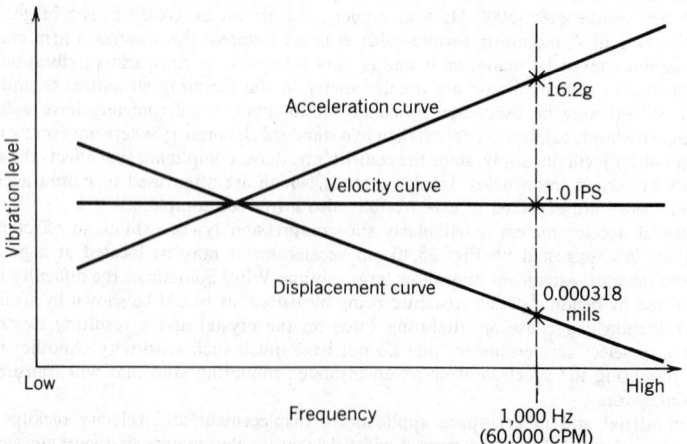

Fig. 25.41 With constant peak velocity maintained over a wide range of sinusoidal vibration frequencies, displacement voltage is inversely proportional to frequency and acceleration voltage is proportional to frequency.

of internal forces due to failed gears, bad bearings, imbalance due to failed turbine blades, etc. The giant Lockheed C5A was the first to carry accelerometers on each engine, with a centralized vibration monitoring and alarm system.

The principal advantage of accelerometers is their far wider flat frequency range, often to beyond 10,000 Hz. Many vibration components are that high in frequency. Many velocity pickups have flat response only to about 100 Hz, although this rolloff can be compensated to perhaps 2000 Hz by special amplifier characteristics.

If constant peak velocity were maintained over a wide range of sinusoidal vibration frequencies, the user would note a constant voltage from a velocity pickup. From a companion displacement sensor, he or she would note voltage dropping, inversely proportional to frequency, as in Fig. 25.41. From a companion accelerometer, he or she would note voltage rising proportional to frequency.

If the vibration of some machine happened to be concentrated at high frequencies, it would be much less observable on the signal from a displacement pickup than on the signal from an accelerometer. Such high-frequency vibration is characteristic of the rubbing of machine parts that are supposed to slide smoothly. High-frequency vibration caused by rough bearings, rough gears, etc., then, is most easily observed when sensed by an accelerometer, the preferred pickup. An accelerometer can be located nearly anywhere on such a machine (although it is preferable to place it close to a critical bearing, gear, etc.) and high-frequency vibration will find its way through the structure to be sensed. If, on the other hand, the vibration of concern will show up as whole-body vibration at relatively low shaft frequencies, then displacement or velocity sensing is better.

25.5   SHOCK MEASUREMENT AND ANALYSIS

Shock may be defined as a transmission of kinetic energy (energy of motion) to a system that takes place in a relatively short period of time compared to the natural period of the system. Shock is followed by a natural decay of the oscillatory motion that has been given to the system. The reader will recognize that there can be no hard and fast division between what we casually call vibration and what we casually call shock.

Consider Fig. 25.42, which shows three oscilloscope views of the same shock pulse. They appear

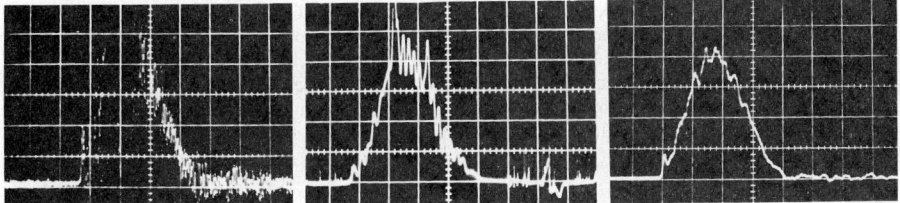

Fig. 25.42   Three oscilloscope views of the same shock pulse.

different because the accelerometers were different (left to right: $f_N = 35,000$ Hz, undamped piezoelectric; $f_N = 2000$ and 850 Hz, both damped strain gage types). The trace at the left of Fig. 25.42 gives us much more information about the shock pulse (mainly its high-frequency content) than do the others. The others are not "wrong"; they just give a less complete picture. However, for loads that are not rigid, the trace at the right may give sufficient information.

We can get very much the same result by using the same electrical pulse (Fig. 25.42 at the left) and by passing it through low-pass electrical filters to decrease the amount of high-frequency "hash." There is a real temptation to filter shock pulses (this temptation should be resisted) and there is considerable past history and existing practice to be overcome. Observe Fig. 25.43, oscilloscope views

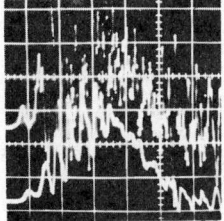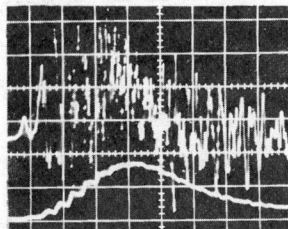

Fig. 25.43   High-frequency hash due to rattling of carriages in shock-testing machines.

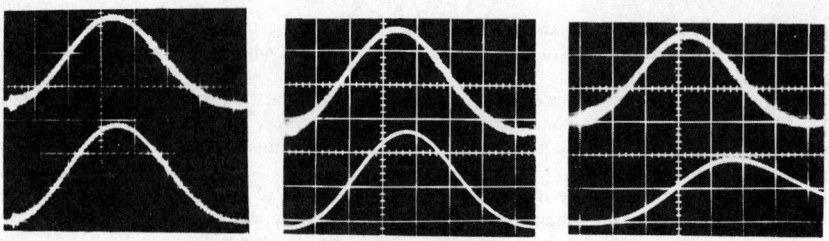

Fig. 25.44   Lower traces filtered; upper traces unfiltered.

of shock pulses originating in a piezoelectric accelerometer during a shock test on a very old type of shock test machine. Test personnel were told to expect a shock pulse approximating a half-sine wave. The upper traces contain so much hash that any possible half-sine pulse is obscured. But by a little filtering, the half-sine pulse becomes visible; with more filtering, it becomes easy to see. However, note how the height of the right-hand lower-trace pulse is lowered. Also note how it appears to be stretched out in time. If the operator had previously calibrated the oscilloscope (using sinusoidal excitation) at Y_1 g/in. on the vertical scale, and at X_1 msec/in. on the horizonal scale, he or she would be expecting a certain Y_2 deflection and a certain X_2 deflection on his or her oscilloscope. If Y_2 were too low, and if X_2 were too large, he or she would probably readjust his or her shock test machine for a more severe, shorter shock pulse. Then he or she would *think* he or she was meeting the test specification, but in reality would not be. For this reason, accelerometer manufacturers, in their instruction manuals, often say, "For shock measurements, use *no* filtering at all."

The reason for the high-frequency hash in the upper traces of Fig. 25.43 is that early designed shock test machines had carriages that rattled violently during the shock. More modern types often have cast carriages that do not rattle; clean shock pulses can be obtained without filtering. But to illustrate answers to the question "How much filtering *can* I use?," consider Fig. 25.44. In all three cases, taken on the same machine, the upper traces are unfiltered. The frequency ranges for the lower traces were, respectively, 0.06–60,000 Hz, 0.06–600 Hz, and 0.06–60 Hz. Why only 60 Hz? It may be shown that this shock pulse, approximately 0.011 sec in duration (11 msec) contains frequencies that are multiples of 45 Hz. An old "rule of thumb" in such cases was to set the cutoff frequency of the filter at 1.5 × the lowest frequency present; thus 60 Hz. It was felt that this would show how much motion was present at the lowest frequency, but would block motion signals at other frequencies. True enough, but look at the result. A better "rule of thumb" would be 15 × lowest frequency present; in this case 600 Hz.

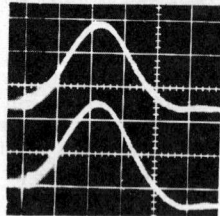

Fig. 25.45   Lower trace filtered with low-frequency cutoff of 6 Hz.

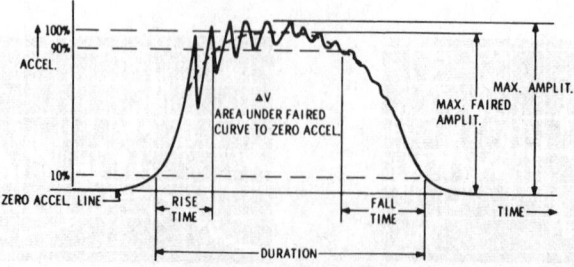

Fig. 25.46   Half-sine pulse.

The low-frequency response must extend nearly to zero. The lower trace of Fig. 25.45 here shows what happens with a low-frequency cutoff of 6 Hz; note the negative undershoot of the trace after the shock pulse was over. There was also a reduction in apparent pulse height, but this is harder to see.

Since so-called half-sine pulses more closely approximate what is called a haversine pulse, there is a little difficulty in establishing, from a record, just when the shock pulse starts and when it stops. One method is illustrated by Fig. 25.46. Some specifications also give the "rise time" and the "fall time" of a shock pulse; these are also shown by Fig. 25.46.

Accelerometer systems used for shock measurement and analysis should be shock calibrated. An accelerometer might be shown to have a sensitivity of, say, 10 mV/g under sinusoidal vibration conditions at vibration intensities of 50g or 100g, the maximum available from most shakers. But what will the sensitivity be at 1000g or 10,000g? Is the accelerometer linear? It is sometimes possible to sinusoidal vibration-calibrate an accelerometer on a tuning fork or other resonant structure (driven by a shaker) at 500g or 1000g, but only at one frequency per structure. Shock calibration as suggested by Figs. 25.47 and 25.48 gives sensitivity at pulses to perhaps 15,000g and at various pulse time durations.

In the drop-ball calibrator of Fig. 25.47, the accelerometer is at rest, attached to an anvil which is struck by a falling steel ball. Padding on the impact surface varies pulse intensity and duration. In the machine shown in Fig. 25.48, the accelerometer being calibrated is attached to a falling carriage and is stopped by a target, which contains a force gage (previously calibrated by dead weights).

Another advantage of shock calibration (in addition to linearity checks of accelerometer sensitivity) is that the electronic portions of the measurement system are checked for performance with short, severe input pulses. Some amplifiers, detectors, and readout devices work satisfactorily on sinusoidal input signals but fail miserably with pulse inputs.

Readout of shock pulses is most often done on an oscilloscope. Pulse height (acceleration intensity), also pulse shape and duration, can be read if the oscilloscope is a storage-type unit. Alternately, a Poiaroid or other camera may be used for a permanent record. Oscillographs running at high speed,

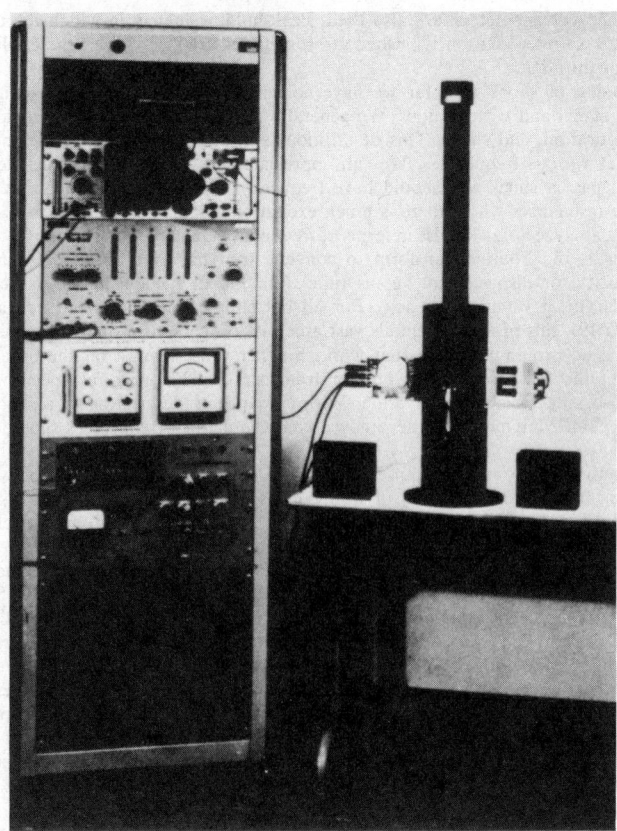

Fig. 25.47 Drop-ball calibrator.

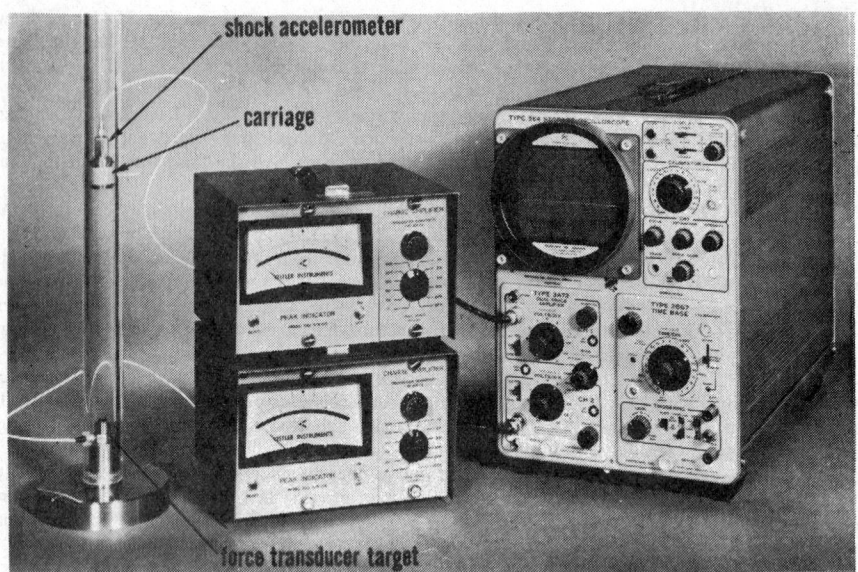

Fig. 25.48 Calibration of accelerometer: accelerometer is attached to a falling carriage and is stopped by a target containing a force gage previously calibrated by dead weights.

with high-frequency galvanometers, are also used. Peak-holding meters in which the maximum pulse height is held until a "reset" button is pushed are sometimes used. However, these tell nothing about the pulse shape or duration.

In our discussion of shock thus far we have concerned ourselves only with converting motion into an electrical signal and with readout. We generally use accelerometers that provide a description of pulse height, duration, and shape. This description tells us very little about the energy content of the shock pulse at various frequencies. We can appreciate that a designer responsible for building a product that will pass a shock test should have frequency information in order that his design will avoid having any resonances where there is much excitation.

How might we analyze a shock pulse in terms of frequency? One method is to employ a mathematical procedure known as the Fourier transform to convert from the time domain (instantaneous force, acceleration, velocity, or displacement versus time) into the frequency domain (force, acceleration, velocity, or displacement versus frequency). For all but simple time histories (such as the left-hand column of Fig. 25.49), this procedure entails vast amounts of work. If a digital computer is available, one can use a special program to save that work. Further time savings, on the order of 300:1, are possible by use of the fairly recent fast Fourier transform (FFT) program. Special purpose hybrid computers can also make this transformation. With a frequency description of a shock that is likely to occur in service, an equipment designer knows what the natural frequencies should be, in order to avoid damage.

Another way to analyze a shock pulse in terms of frequency is by means of what is called a *response spectrum*. Imagine that an array of spring–mass–damper systems (such as a group of reeds), as suggested by Fig. 25.50, each with a different natural frequency (but the same mechanical buildup Q) is attached to a structure. A single unit from that array is shown in Fig. 25.51. A shock pulse occurs. Its displacement time history is recorded as x, the shock input. Some method, such as the pen making trace y, is used to measure the maximum response motion of each reed. Some reeds will respond more than others, depending on the amount of energy in the shock motion at different frequencies. A graph of maximum response is made versus reed natural frequency; this graph is called a response spectrum. It gives information about relative amounts of energy at various frequencies. Note that it does not describe the shock pulse itself; it only describes what the pulse does to a series of idealized mechanical systems. The right-hand column of Fig. 25.49 indicates the response spectra of the several common shock time histories shown in the left-hand column. The residual spectrum deals only with events during the decay period. The primary spectrum (not shown) deals only with events during the shock. The maximum spectrum is the larger of the two at any instant of time. The maximum value of the maximum spectrum (maximax value) is often considered.

Note that the symmetrical time histories (half-sine, square, and triangular shock pulses) have wide variations from one frequency to another, as opposed to the sawtooth pulse, which is relatively constant

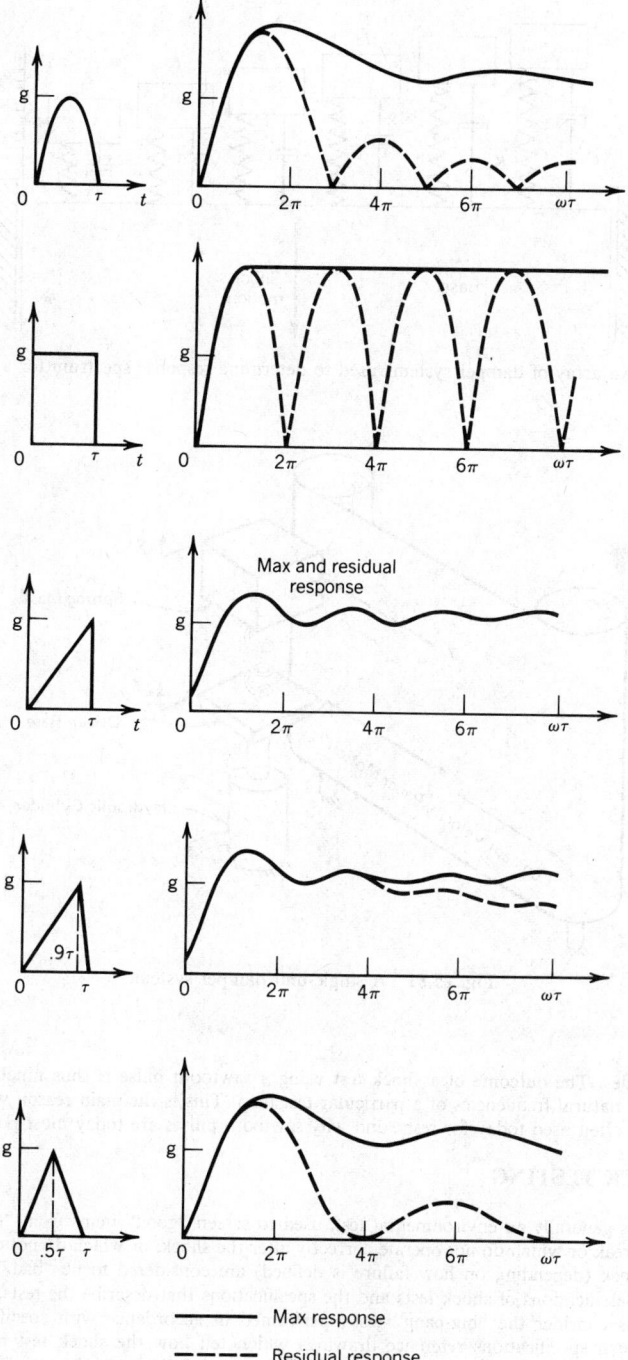

Max response

- - - - Residual response

Fig. 25.49 Conversion of shock pulses from time domain to frequency domain.

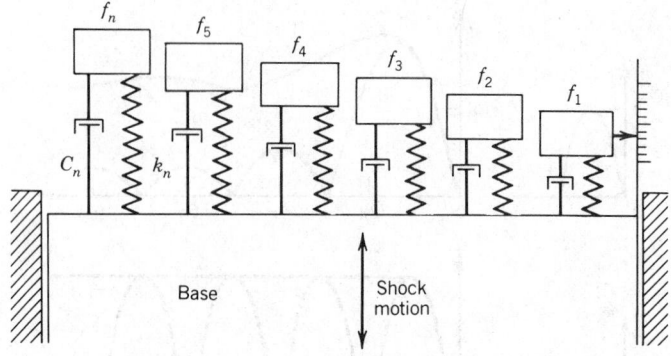

Fig. 25.50 An array of damper systems used to determine response spectrum for a shock pulse.

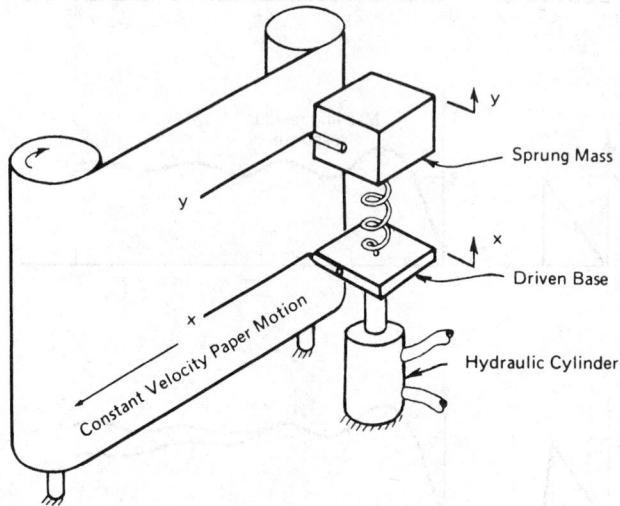

Fig. 25.51 A single-unit damper system.

at all frequencies. The outcome of a shock test using a sawtooth pulse is thus much less dependent on the specific natural frequencies of a particular test item. This is the main reason why symmetrical pulses are less often used today for tests, and why sawtooth pulses are today most popular.

25.6 SHOCK TESTING

A shock test is generally an environmental test, used to screen "good" items from "bad" items. Test items which break or which do not operate correctly after the shock, or which do not operate correctly during the shock (depending on how failure is defined) are considered to be "bad." Let us look at three major classifications of shock tests and the specifications that describe the tests.

First let us consider the "big-bang" tests, performed in accordance with specifications such as MIL-S-901. Such specifications reference drawings which tell how the shock test machine is to be built. The smaller Navy machine, Fig. 25.52, tests loads to 400 lb fastened to an 800 lb anvil plate (vertical) and struck by a swinging 400 lb hammer, which is dropped from a specified height. The larger Navy machine, Fig. 25.53, tests 250–6000 lb loads fastened to a 3500 lb anvil plate (horizontal) and struck by a 3000 lb hammer, which is dropped from a specified height. Still larger Navy loads (to 400,000 lb) are fastened to special steel barges and are tested by setting off underwater explosives nearby.

These test machines and their operation are defined by the test specification. Specifications tell nothing about the shock pulse that results or about the effect of the shock pulse on test items (real

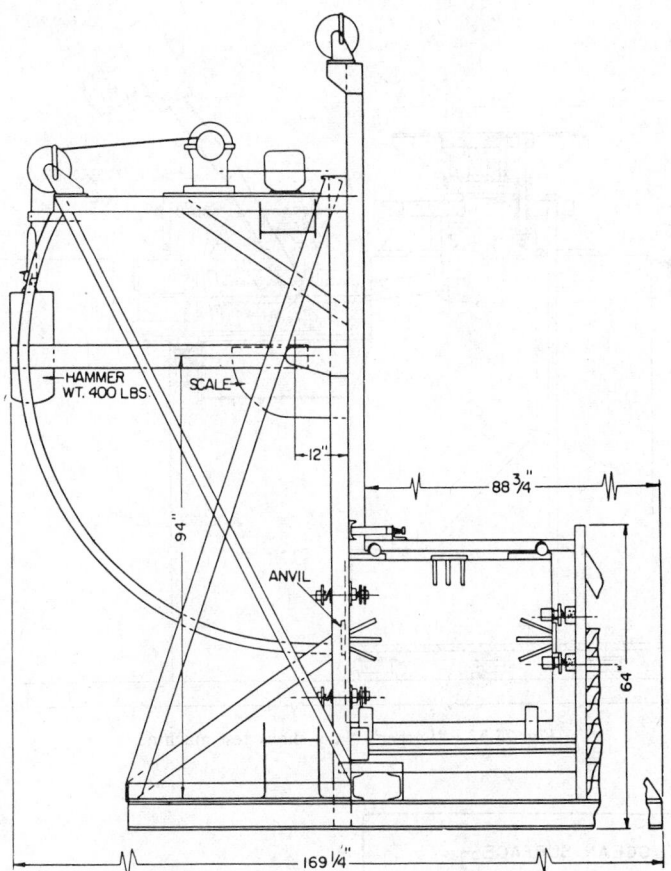

Fig. 25.52   "Smaller" Navy shock test machine.

or idealized). In fact, with these specifications and machines, there is no need for any instrumentation other than a ruler! The Navy has quite a bit of experience indicating that equipment items which have passed such tests are rugged and capable of withstanding the effects of a ship's own gunfire and of enemy explosions.

Consider Fig. 25.54, which shows several paths for energy transmission from an underwater explosion to the hull of a submarine. Depending on the relative amounts of energy following the various paths, the shock motion of the hull will be different. No one simple specification could duplicate the effects of the variety of shock pulses that the hull could receive.

Consider next Fig. 25.55, which gives the shock spectra at three different levels in a surface ship. Note, curve A, that high-frequency energy is found only at or near the ship's bottom. Energy at B, higher in the hull, shows less high-frequency energy. Energy at C, in the structure higher than the hull, shows still less high-frequency energy but accentuated energy at low frequencies. No one simple specification could duplicate the effects of even a simple standard pulse received by the hull, when that pulse is so much modified by hull resonances. And, of course, there are wide variations between ships of a class and of different classes.

Thus there is some justification for the Navy "big-bang" approach to shock testing. The drawback is that one cannot tell, in advance of a test, what the effect will be on a new test item. Furthermore, there is no way to compare that these machines give the "same" pulse; in fact response spectrum investigations indicate that no two machines are alike.

Let us next consider tests for which the motion is specified, such as MIL-STD-810C. Here a definite waveform (intensity, time duration, and pulse shape) is specified. Test personnel select any machine that will give the desired pulse on the test item.

Most test laboratories buy machines such as those shown in Figs. 25.56, 25.57, and 25.58, although a few labs have built their own. The unit shown in Fig. 25.56 is rather typical of the common free-fall gravity drop testers. If the fall is arrested by impacting a rubber cushion, a haversine deceleration

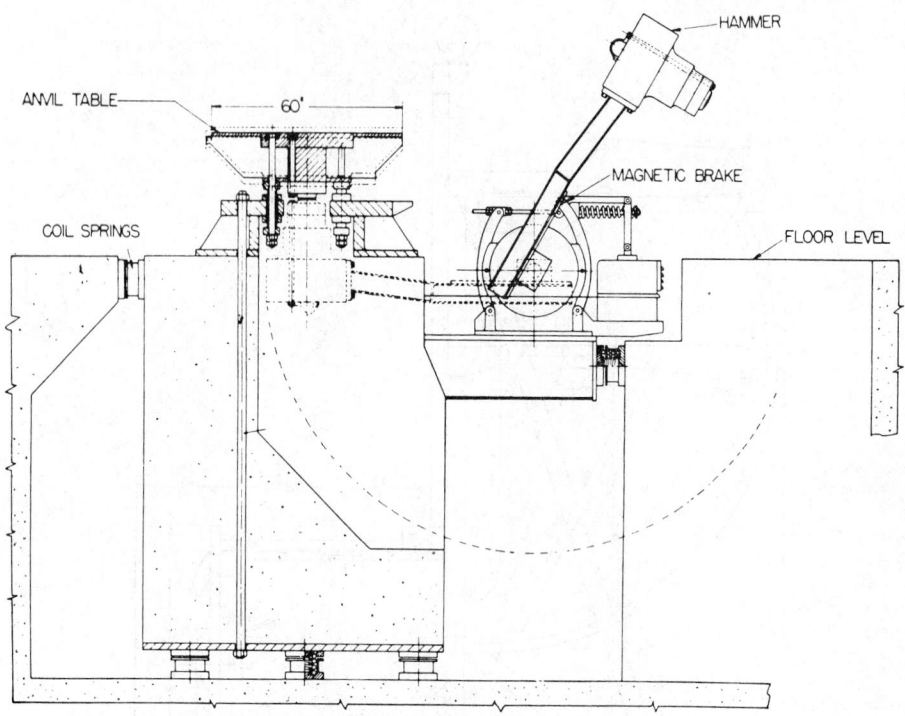

Fig. 25.53 "Larger" Navy shock test machine.

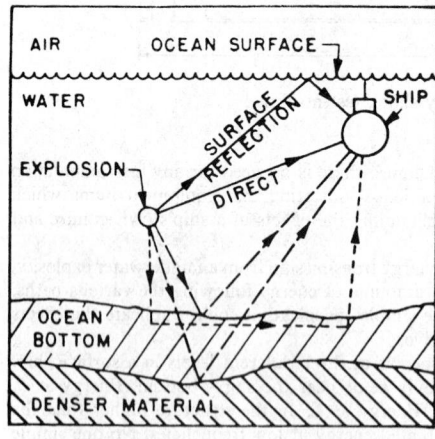

Fig. 25.54 Paths for energy transmission from underwater explosion to hull of submarine.

pulse results. If another pulse shape is desired, the fall can be arrested· by impacting a shaped soft lead casting; the shape of the casting dictates the deceleration versus time curve. Lead is often used for sawtooth and triangular pulses. Crushable honeycomb materials are often used for square pulses. Some machines use a fluid/orifice system to arrest the motion; various pulse shapes may be programmed. Whatever the stopping method, just before the carriage stops it closes a switch which initiates or triggers the sweep of the beam of an oscilloscope, thus a picture of deceleration versus time is obtained.

Some machines similiar to Fig. 25.57 use stretched cords (such as Bungee or aircraft shock cord) to give additional velocity to the downward-falling carriage, thus a greater impact. The machine shown by Fig. 25.58 uses shop compressed air for operation. The carriage is driven downward and impacts an arresting device. The Hyge machine illustrated by Fig. 25.58 uses dry nitrogen under high pressure to accelerate a carriage for test; a braking system later arrests the carriage gently.

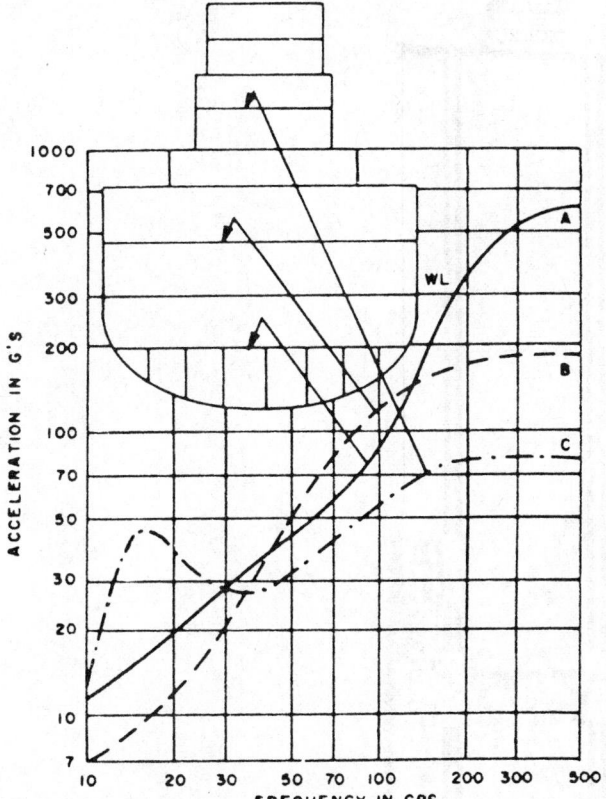

ACCELERATION IN G'S

FREQUENCY IN CPS

Fig. 25.55 Shock spectra at three different levels in a surface ship.

Still another way to generate a specified pulse of acceleration versus time is to use a shaker. Electrohydraulic shakers have been used; one advantage is their long stroke, required for long-duration shock pulses. A few electromagnetic shakers have been built with a long stroke for the same purpose. In general, though, the 1-in. peak-to-peak motion of typical shakers restricts their usefulness for testing with specified (half-sine, haversine, square, triangular, sawtooth, etc.) pulse shapes, especially if pulses are more than a very few milliseconds long.

There are several advantages in using a shaker for shock testing: the cost of a shock test machine may be saved; the cost of an extra test fixture to adapt a test item to a shock test machine may be saved if shock testing may be done on an existing shaker; reversal of the direction of a shock is as simple as reversing a switch, whereas a complex attachment to a shock test machine would be required; setup time may be greatly reduced, compared to transferring a load from vibration test to a shock test machine.

Figure 25.59a suggests older analog equipment. Usually, a switch on the pulse generator is closed once for each pulse. Proper shaping of the pulse is done with the gain of the power amplifier reduced; by means of trial and error the pulse is shaped until it meets the requirements of the test specification. Then the power amplifier gain is increased for the actual shock test. Figure 25.59b shows that the electrical pulse (when viewed on an oscilloscope) looks much different from the motion pulse. Digital computers control most of these tests now.

Simple pulse shapes (half-sine, sawtooth, square, etc.) are not found in the real world and many newer testing specifications avoid them. Instead, they call for much more typical time histories such as the one set into Fig. 25.60, brief intervals of oscillatory behavior. These cannot be duplicated by machines typified by Figs. 25.56 and 25.57, but can be duplicated by tape recorders and by shakers operating under control of digital computers.

Finally, let us consider tests in which the *response spectrum* is dictated as in MIL-STD-810D. Let us imagine that an in-flight shock has been measured and analyzed in terms of its frequency content, as in the lower line of Fig. 25.60. Shock measured during a static firing is also shown. In

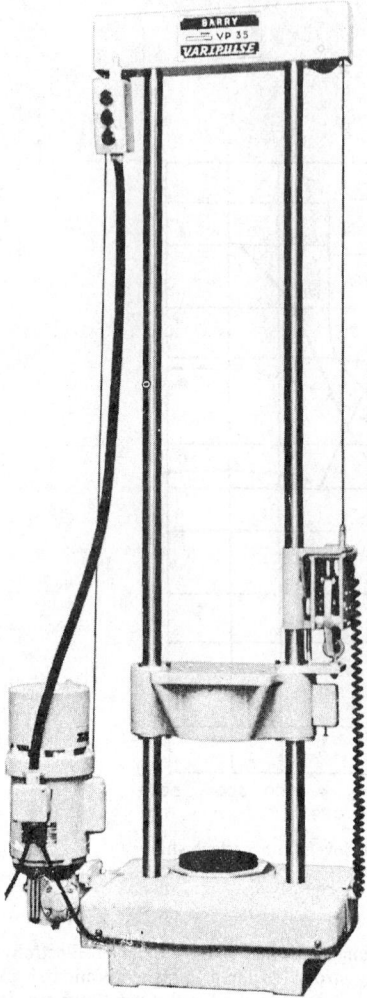

Fig. 25.56   A free-fall gravity drop tester.

order to specify a test somewhat more severe than these shocks, the test requirement envelope is described by the following points:

Intensity *(g)*	Frequency (Hz)
80	50
200	200
200	700
320	1000
320	2000

These coordinate points are connected by straight lines on a semilog plot, as in the upper line of Fig. 25.60. A shock pulse of mechanical motion must be generated. Its shape and duration are less important than that the response spectrum be as specified. A signal from the control accelerometer will be analyzed by a shock response spectrum analyzer.

Early shock response spectrum testing (1960s and early 1970s) was done as follows: A single

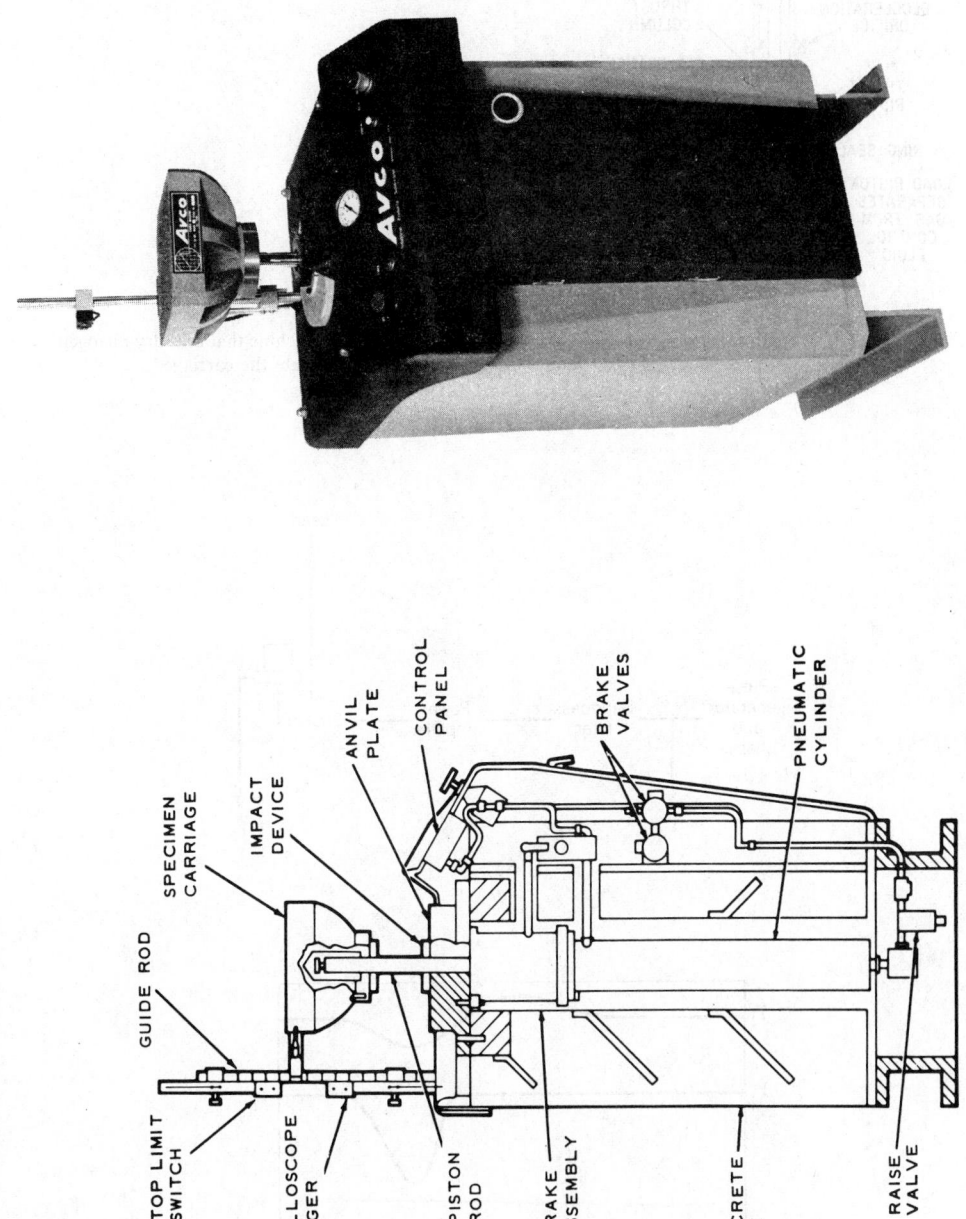

TOP LIMIT
SWITCH

GUIDE ROD

SPECIMEN
CARRIAGE

IMPACT
DEVICE

ANVIL
PLATE

CONTROL
PANEL

BRAKE
VALVES

OSCILLOSCOPE
TRIGGER

PISTON
ROD

BRAKE
ASSEMBLY

PNEUMATIC
CYLINDER

CONCRETE
BASE

RAISE
VALVE

Fig. 25.57 A shock testing machine that uses shop compressed air.

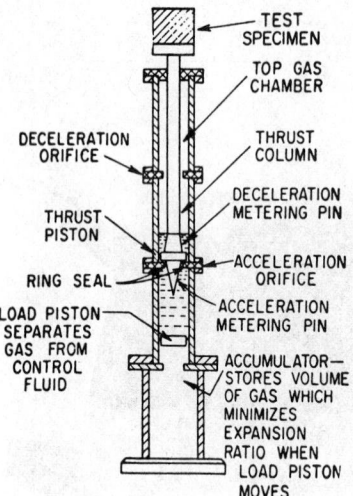

TEST SPECIMEN

TOP GAS CHAMBER

DECELERATION ORIFICE

THRUST COLUMN

DECELERATION METERING PIN

THRUST PISTON

ACCELERATION ORIFICE

RING SEAL

ACCELERATION METERING PIN

LOAD PISTON SEPARATES GAS FROM CONTROL FLUID

ACCUMULATOR— STORES VOLUME OF GAS WHICH MINIMIZES EXPANSION RATIO WHEN LOAD PISTON MOVES

Fig. 25.58 A shock testing machine that uses dry nitrogen under high pressure to accelerate the carriage.

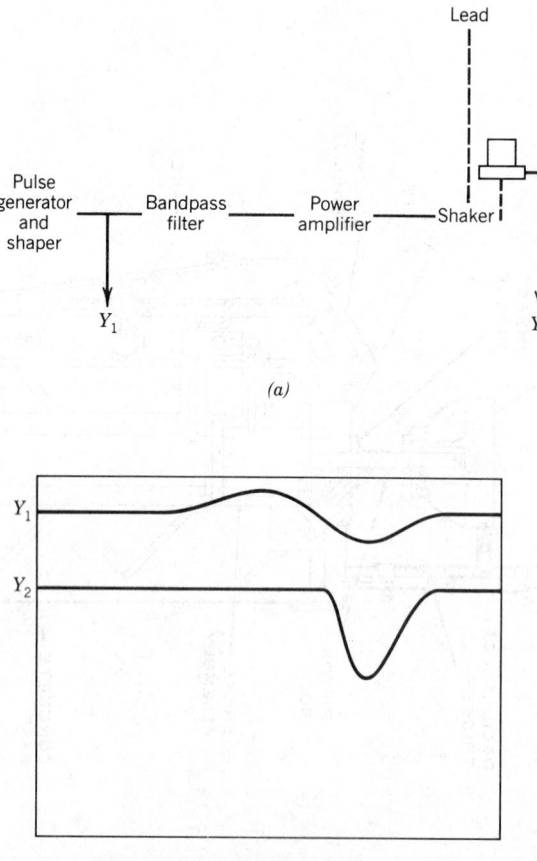

Fig. 25.59 (a) Equipment needed for a shock test. (b) Y_1: electrical pulse; Y_2: motion pulse.

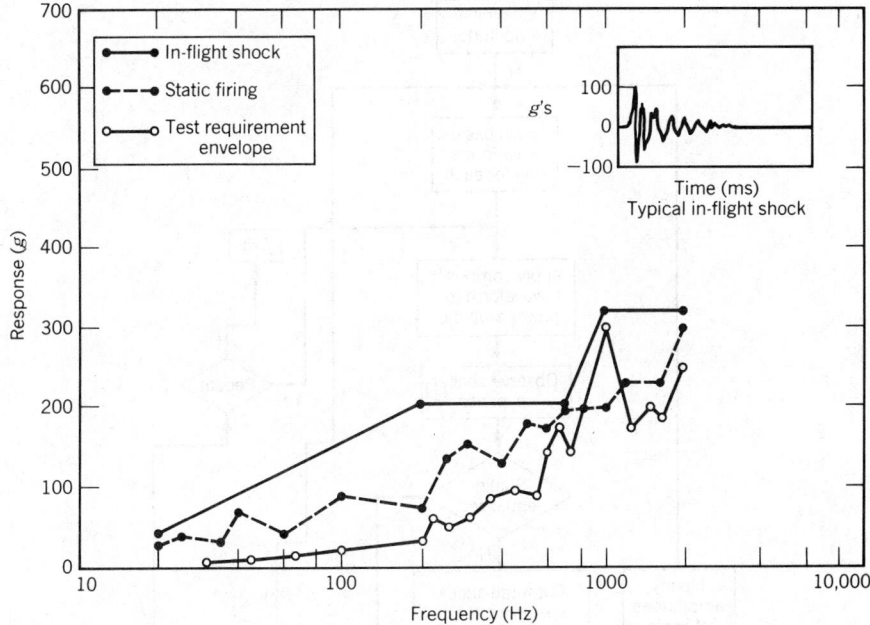

Fig. 25.60 Analysis of in-flight shock in terms of its frequency content.

pulse (short, with much high-frequency content) was generated, then frequency divided by a bank of filters. Each filter had an adjustable attenuator to adjust the energy content in its frequency "window" of the final pulse. The synthesized pulse then went to the power amplifier and shaker. The response spectrum was analyzed with an identical filter bank. Energy in the various frequency windows could be adjusted so that the response spectrum met requirements.

Since the energy content of the shock pulse at various frequencies is specified, the test operator will have to vary the shaker's motion in order to meet the specification. This process closely resembles the process of equalizing a shaker and test load for random vibration testing. Note in Fig. 25.61 the similarity between the block diagrams of equipment needed for random vibration testing and for shock spectrum testing.

With the introduction of digital control techniques, it became feasible to reproduce very complex time histories, not only for long-time vibration tests but also for short-time shock tests.

One method of synthesizing the shock spectrum is known as the "parallel filters" method. It is a direct digital implementation of the analog method. The required shock spectrum is divided into frequency windows. The operator may select the spacing of these windows (one-half, one-third, and one-sixth octave are typical). At one-sixth octave spacing, 53 windows is typical. The minicomputer chooses a basic waveform for each window and adds these waveforms to form a composite waveform.

The minicomputer will ask certain questions that must be answered by the test operator. These include the date, test title, shock duration, shock peak amplitude, damping coefficient, number of

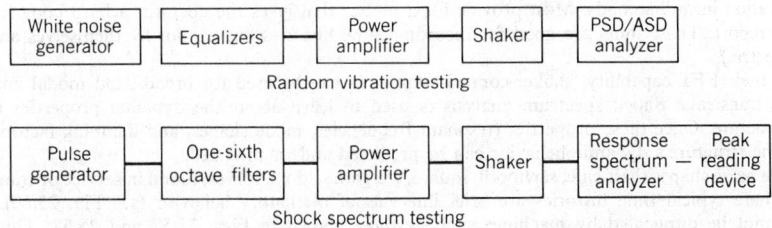

Fig. 25.61 Flow charts for random vibration and shock spectrum testing.

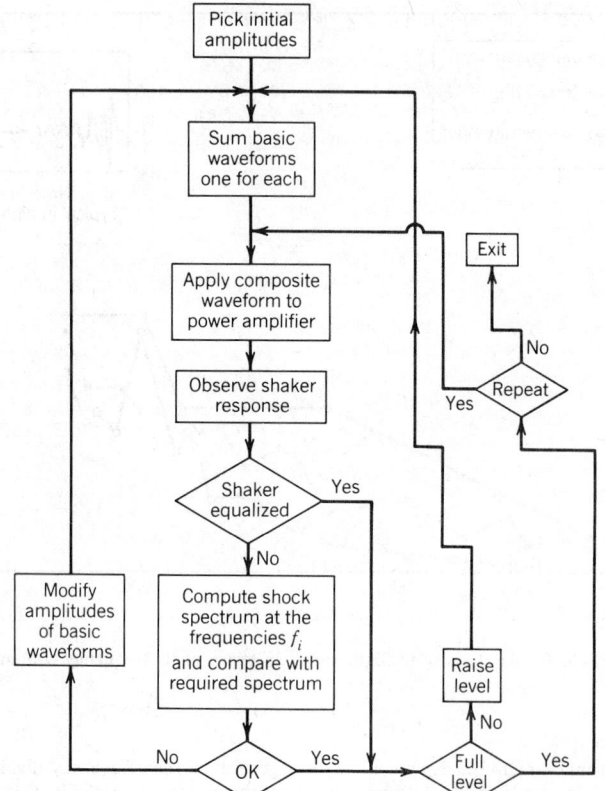

Fig. 25.62 Flow chart for shock spectrum testing to required spectrum.

shocks to be applied (see "repeat" in Fig. 25.62) and wave type. Typical wave types can include half-sine, square, sawtooth, triangular, or damped sinusoidal.

Consider the flow chart of Fig. 25.62. Initial shock amplitudes at specific frequencies are generated by the computer, summed, and applied to the shaker at low level (one-fourth or one-eighth of full level is typical). The computer compares the shock response spectrum (SRS) of the low-level shock to the desired SRS. The display CRT will present any error between the actual and desired SRS's as a percentage. If the error is within tolerances, the test can proceed to full level. If not, the computer produces a corrected signal to again drive the shaker. After this equalization process is complete, the operator proceeds to test at full level.

Certain limitations apply. Maximum acceleration is limited by the momentary force capability of the shaker and the ability of the amplifier to deliver high peak current. Pulse duration is limited by several factors, including shaker displacement and power-amplifier low-frequency response. When test parameters exceed shaker system limits, testing must be performed by one of the older methods (Figs. 25.56, 25.57, and 25.58). (Even when a conventional shock machine is used to generate a shock pulse, one can measure and record pulses digitally. Several manufacturers offer units that digitize a shock pulse signal and store it in memory. Most provide readouts that display peak amplitude in g's and pulse duration in milliseconds. Most provide for a plotter that gives the operator a hard copy for the final test report. These units are good for recording time histories but cannot by themselves analyze shock spectra.)

With their FFT capability, shaker-control computers can be used for broad-band modal analysis of shock transients. Shock spectrum analysis is used to learn about the dynamic properties of an elastic structure. Once these properties (resonant frequencies, mode shapes, and damping factors) are known, the structure's dynamic behavior can be predicted and/or modified.

Simple pulse shapes (half-sine, sawtooth, square, etc.) are seldom, if ever, found in service or transportation, where typical time histories are brief intervals of oscillatory behavior (see Fig. 25.60). The latter cannot be duplicated by machines such as those shown in Figs. 25.57 and 25.58. Thus the newer test specifications (Fig. 25.61) must be performed under the control of a digital shaker system.

25.7 SHAKE TESTS FOR ELECTRONIC ASSEMBLIES*

Not all electronic gear enjoys the protected environment of a computer, business machine, or bench instrument. A substantial portion of electronic assemblies ends up in vehicles, aircraft, military hardware, or hand-held devices where service life entails a fair amount of shock and vibration. And in some applications, such as consumer electronics, mere shipping and handling often impose severe mechanical stresses.

For these reasons, both prototype and production electronic hardware normally is subjected to some sort of vibration testing. Typically, the vibration frequency is swept through some specified spectrum. Although this procedure ostensibly seems logical, it actually may induce failures that would not occur in field service. Moreover, this type of test sometimes fails to uncover true weaknesses that are not found until the product is in a customer's hands.

Experience has shown, however, that equipment proof-tested with random vibrations rarely fails in service. And random testing has proved to be effective for disclosing flaws even in equipment that is not subject to in-service vibration.

Electronic equipment generally is mounted on shake-tables for vibration testing. For many years, shake-tables were mechanically vibrated at frequencies up to about 60 Hz. But in the late 1940s, electromagnetic shakers operating at frequencies up to 500 Hz were developed. Later, test vibration frequencies were increased to over 2000 Hz.

Early electromagnetic shakers generally were powered by adjustable-frequency alternators. During a test, the vibration frequency was swept back and forth over a frequency band, and vibration amplitude was set somewhat higher than that expected in service. This is known as sine-wave testing, and many vibration tests were performed in this manner.

Investigations made following a number of missile launch failures in the 1950s indicated that sine-wave testing had some deficiencies. These investigations showed, much to everyone's surprise, that vibrations during launch were much less severe than those which had been used for sine-wave testing the failed components.

Additional studies showed, moreover, that vibrations produced by a missile launch contain a wide band of randomly changing frequencies. This is much different than the vibrations used for sine-wave tests. Investigators concluded that missile components should be tested by vibrations closely matching those actually occurring during launch and flight.

The first random vibration test specifications were issued in 1955. Electromagnetic shakers were powered by large audio amplifiers because adjustable-frequency alternators could not produce random-vibration power. These new specifications and test equipment were important factors in improving missile reliability. And this type of test now is applied to a wide variety of electronic equipment for other applications.

Sine-wave and other periodic vibrations are easy to visualize because they can be described graphically or by simple mathematical equations (see Fig. 25.63). And they can be completely defined by specifying frequency, amplitude, and phase characteristics. Thus, maximum acceleration, an important parameter for test purposes, is calculable.

Random vibrations, however, are difficult to picture because they have nonperiodic waveforms that change magnitude unpredictably. A random vibration, though often limited to a specific frequency band, contains many frequencies simultaneously. And all frequencies in the band are possible.

Neither instantaneous velocities nor instantaneous displacements can be predicted for random vibrations. Thus, maximum acceleration is indeterminate, and statistical theory must be used to calculate rms acceleration and to predict the probability for any specific instantaneous acceleration.

A random vibration can be defined for test purposes by specifying an acceleration spectral density (ASD) value for all frequencies included in the test. (Acceleration spectral density sometimes is known as power spectral density or PSD.) The concept for acceleration spectral density is derived from statistical theory. It is typically specified in g^2/Hz units and is the square root of the vibrational power contained in a 1-Hz-wide slice of the defined random vibration. The rms acceleration for a 1-Hz-wide slice can be calculated by extracting the square root of its ASD value (Fig. 25.64).

An ASD chart can be produced by plotting ASD values versus frequency on log-log graph paper. This type of chart, by showing upper and lower frequency limits as well as ASD values, enables one to visualize frequency and vibrational power distribution.

Neither instantaneous nor maximum acceleration values can be determined from an ASD chart. That information can be obtained only by using probability techniques. But rms acceleration can be calculated by extracting the square root of the area under the ASD curve.

The simplest type of ASD chart has a constant ASD value for an entire test-frequency band, the ASD value dropping sharply to zero at the maximum and minimum frequencies. This is known as "white" random vibration (Fig. 25.65).

Based on probability theory, instantaneous acceleration rates are less than the rms value 68% of

* From *Machine Design,* January 22, 1981, used by permission.

Sine-wave vibrations can be represented by a simple equation. And the instantaneous acceleration produced by a sine wave can be determined by taking the second derivative of the sine-wave equation:

$$a = 4\pi^2 f^2 X \sin 2\pi f t$$

where a = instantaneous acceleration, f = frequency, X = peak amplitude, and t = time. From this, maximum acceleration can be calculated:

$$A = 0.511 f^2 D$$

where A = maximum acceleration in in./sec, f = frequency in Hz, and $D = 2X$ = total displacement in in.

Random vibrations, however, cannot be represented by a simple mathematical expression. Moreover, maximum acceleration for random vibrations is indeterminant. But other useful acceleration values can be obtained from statistical principles. Thus, if the instantaneous acceleration produced by a random vibration is observed over a long period, the mean, the variance, and the standard deviation can be measured and calculated:

$$\bar{a} = \frac{I}{T} \int_0^T a(t)\, dt$$

$$\overline{a^2} = \frac{I}{T} \int_0^T a(t)^2\, dt$$

$$\sigma = \sqrt{\lim_{T \to \infty} \frac{I}{T} \int_0^T a(t)^2\, dt}$$

where $\bar{a}$ = mean acceleration, $\overline{a^2}$ = acceleration standard variance, σ = acceleration deviation, and T = measurement period.

The mean acceleration (usually zero) has little value for test purposes, but variance does. Variance, or mean square acceleration, is proportional to the total vibrational power. And standard deviation is a useful figure because it is equal to rms, or effective, acceleration.

Acceleration variance, or mean square acceleration, for a 1-Hz-wide slice of a random vibration is known as acceleration spectral density (ASD) and is represented by

$$\text{ASD} = \lim_{\Delta f \to 1} \frac{a(f)^2}{\Delta f}$$

Acceleration spectral density is a useful value because it can be used for presenting random vibrations graphically. And ASD can be measured easily.

Figure 25.63   Statistical Acceleration

the time, less than twice the rms value 95% of the time, and over three times the rms value only 0.3% of the time. Theoretically, the acceleration rate can occasionally be infinitely high. However, test equipment is not capable of producing infinitely high acceleration rates, even briefly. Thus, maximum acceleration rates for random vibrations typically are limited to about three times the rms value.

Test specifications for products subject to in-service vibrations typically are based on measurements and analysis of vibration that the product will experience. These specifications then may be suitable only for those particular conditions. However, some test specifications have been widely used for many different products and applications.

Random Vibration is widely used for stress screening electronic products that are not subject to heavy vibration in service. A vibration stress screen can uncover design weaknesses as well as faulty components, nicked wires, cold-soldered joints, and other workmanship flaws. Products that will be stress screened in production generally should be subjected to the same test during development.

Stress screening is a technique for identifying flaws in subassemblies or in complete products before they are placed in service. Stress screening typically includes burn-in procedures and temperature cycling as well as vibration.

Stress screening is not a test in the generally accepted sense of the word. The word test implies that a product must meet certain performance criteria. But in stress screening, the only real failure is one where a flawed item is approved for use.

Sine-wave vibrations have been and still are widely used for stress screening electronic products. But experience has shown that random vibrations are more effective for the purpose. As in pure performance testing, sine-wave vibrations can cause failures that would not occur in service, and they can leave some flaws hidden. Random vibrations, however, rarely trigger unwarranted failures, and they often expose design and workmanship faults that are undiscovered by sine waves.

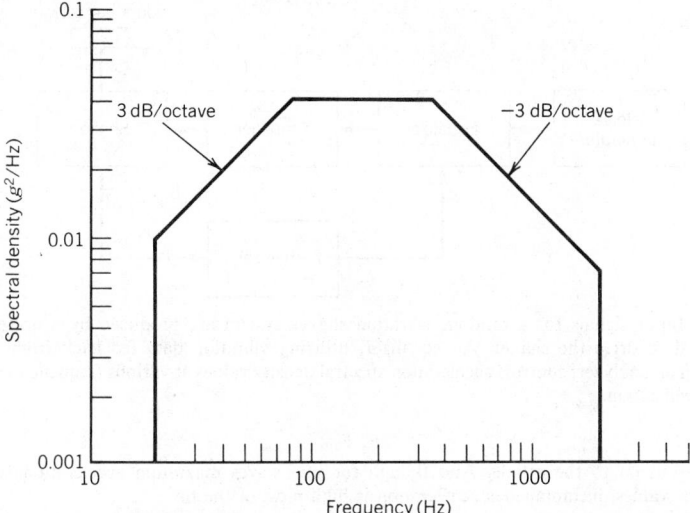

Fig. 25.64 This typical spectral density chart defines a random vibration test that is widely used for stress-screening electronic equipment. Spectral density, measured in g^2/Hz, is proportional to the vibrational power contained in a 1-Hz-wide slice of a random vibration spectrum. The square root of the area under the curve is an rms $6g$, the effective acceleration value for the entire spectrum. The use of dB/octave for defining the sloping portions of the diagram is based on filter rolloff terminology.

When a printed circuit board (PCB), for example, is subjected to sine-wave vibrations, the amplitude of the board vibration typically is far greater than that of the shaker table. This occurs when a test frequency coincides with the board's natural frequency. The ratio of the board amplitude to the shaker table amplitude is known as the Q of the board. The value of Q is a function of an article's damping characteristic.

If a PCB has a Q of 25, its amplitude, when subjected to a sine-wave vibration having a frequency equal to the board's natural frequency, will be 25 times the shaker amplitude. However, if the PCB is subjected to random vibrations, the PCB amplitude will be much lower to the square root of Q, or in this case, 5.

In sine-wave testing, moreover, all vibrational energy is concentrated in one frequency. Thus, when that frequency briefly coincides with the natural frequency of an article under test, all the vibrational

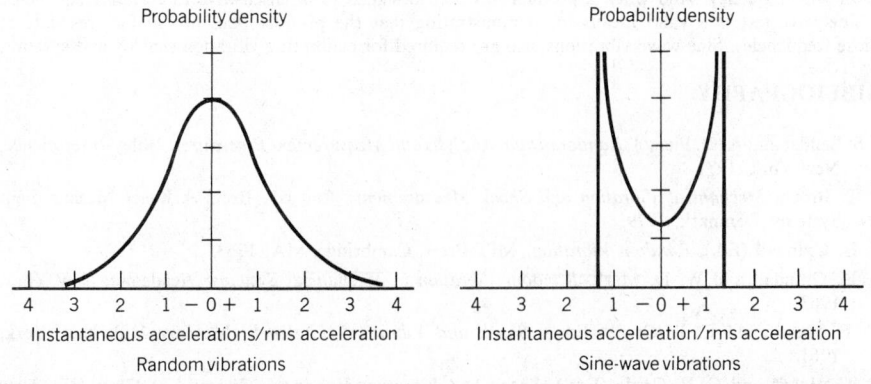

Fig. 25.65 Instantaneous acceleration rates for wide-band random vibrations are defined by a Gaussian probability density curve as shown on the left. This curve indicates that acceleration is close to zero most of the time. High rates do occur, but only rarely. Instantaneous acceleration rates for a sine wave are described by the curve on the right. This curve shows that acceleration is near the rms value most of the time and only rarely close to zero.

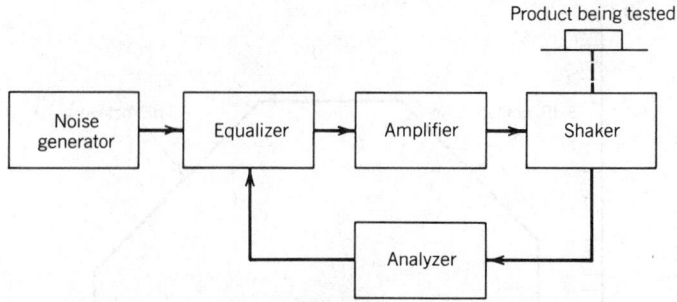

Fig. 25.66 Input signals for a random vibration shaker system are produced by a noise generator and amplified to drive the shaker. An equalizer, utilizing vibration data fed back from the shaker table through an analyzer, controls acceleration spectral density values at various frequencies as required by a test specification.

energy is passed on to the article. And though for sine waves maximum acceleration is only $\sqrt{2}$ times the rms values, instantaneous acceleration is high most of the time.

Random vibrations briefly can produce very high acceleration rates. But instantaneous acceleration rates are less than the rms value most of the time. And vibrational energy is distributed in both time and frequency domains. Thus, any one narrow frequency band contains only a small portion of the total energy available. And the narrower the band, the less energy it contains. Items having a high Q are affected only by frequencies quite close to their natural frequency. Thus, random vibrations typically are gentler than sinusoidal vibrations. And random vibration tests are more representative of transportation and in-service vibrations (Fig. 25.66).

Random vibration screens for products that are not subject to vibration in service must, of necessity, be developed empirically. One widely used screen is described in NAVMAT P-9492, a Navy publication on stress screening. This calls for rms 6g random vibration having a 20–2000-Hz frequency band. The test period is 10 min for products tested on one axis. Test period for additional axes is five minutes each.

An rms 6g screen may seem excessively severe for stress screening products that are not subject to severe vibration in service. But this screen is not intended for qualification or acceptance. The purpose is to weed out flawed products. And a brief period of an rms 6g random vibration will not harm properly designed and manufactured electronic products.

The NAVMAT P-9492 test often is run in conjunction with a temperature cycling test. The temperature range is −65 to 131°F with 40°F/min ramps. Temperature extremes are held until internal parts have stabilized to within 5°F of the specific temperature. Products are cycled 1–10 times through this temperature range depending on product complexity. Those containing 4000 components are cycled 10 times, but those containing 100 or less are cycled only once.

Sine-wave vibrations should be used for fatigue testing products that are subject to in-service vibration of known frequency. And when a product has been designed to be insensitive to certain frequencies, a sine-wave test is a valid means for demonstrating that the product does not in fact resonate at those frequencies. Sine-wave vibrations also are required for calibrating vibration sensors and systems.

BIBLIOGRAPHY

J. S. Bendat and A. G. Piersol, *Random Data: Analysis and Measurement Procedures,* Wiley-Interscience, New York, 1971.

J. T. Broch, *Mechanical Vibration and Shock Measurements,* 2nd ed., Bruel & Kjaer Measurement Systems, Denmark, 1979.

S. H. Crandall (Ed.), *Random Vibration,* MIT Press, Cambridge, MA, 1963.

S. H. Crandall and W. D. Mark, *Random Vibration in Mechanical Systems,* Academic, New York, 1963.

C. E. Harris and C. E. Crede (Eds.), *Shock and Vibration Handbook,* McGraw-Hill, New York, 1961.

C. E. Harris and C. E. Crede (Eds.), *Shock and Vibration Handbook,* 2nd ed., McGraw-Hill, New York, 1976.

D. N. Keast, *Measurements in Mechanical Dynamics,* McGraw-Hill, New York, 1967.

R. Morrison, *Grounding and Shielding Techniques in Instrumentation,* Wiley, New York, 1967.

C. T. Morrow, *Shock and Vibration Engineering,* Wiley, New York, 1963, Vol. 1.

J. P. Salter, *Steady-State Vibration,* Kenneth Mason, London, 1969.

D. S. Steinberg, *Vibration Analysis for Electronic Equipment,* Wiley, New York, 1973.

W. T. Thomson, *Theory of Vibration with Applications,* Prentice-Hall, Englewood Cliffs, NJ, 1981.

The following texts are printed in volume form by the Shock and Vibration Information Center, U.S. Dept. of Defense (Washington D.C.):

R. H. Lyon, *Random Noise and Vibration in Space Vehicles* (SVM-1), 1967.

G. S. Mustin, *Theory and Practice of Cushion Design* (SVM-2), 1968.

L. D. Enochson, *Programming and Analysis for Digital Time Series Data* (SVM-3), 1968.

R. G. Loewy and V. J. Piarulli, *Dynamics of Rotating Shafts* (SVM-4), 1969.

R. D. Kelly and G. Richman, *Principles and Techniques of Shock Data Analysis* (SVM-5), 1969.

E. Sevin and W. D. Pilkey, *Optimum Shock and Vibration Isolation* (SVM-6), 1971.

J. E. Ruzicka and T. R. Derby, *Influence of Damping in Vibration Isolation* (SVM-7), 1971.

A. J. Curtis, N. G. Tinling, and H. T. Abstein, Jr., *Selection and Performance of Vibration Tests* (SVM-8), 1971.

W. C. Fackler, *Equivalence Techniques for Vibration Testing* (SVM-9), 1972.

B. Pilkey and W. Pilkey (Eds.), *Shock and Vibration Computer Programs* (SVM-10), 1975.

R. R. Bouche, *Calibration of Shock and Vibration Measuring Transducers* (SVM-11), 1979.

CHAPTER 26

NOISE MEASUREMENT AND CONTROL

GEORGE M. DIEHL, P.E.

Consulting Engineer
Machinery Acoustics
Phillipsburg, NJ

26.1 SOUND CHARACTERISTICS

Sound is a compressional wave. The particles of the medium carrying the wave vibrate longitudinally, or back and forth, in the direction of travel of the wave, producing alternating regions of compression and rarefaction. In the compressed zones the particles move forward in the direction of travel, whereas in the rarefied zones they move opposite to the direction of travel. Sound waves differ from light waves in that light consists of transverse waves, or waves that vibrate in a plane normal to the direction of propagation.

26.2 FREQUENCY AND WAVELENGTH

Wavelength, the distance from one compressed zone to the next, is the distance the wave travels during one cycle. Frequency is the number of complete waves transmitted per second. Wavelength and frequency are related by the equation

$$v = f\lambda$$

where v = velocity of sound, in meters per second
f = frequency, in cycles per second or hertz
λ = wavelength, in meters

26.3 VELOCITY OF SOUND

The velocity of sound in air depends on the temperature, and is equal to

$$v = 20.05 \sqrt{273.2 + C°} \text{ m/sec}$$

where $C°$ is the temperature in degrees Celsius.
  The velocity in the air may also be expressed as

$$v = 49.03 \sqrt{459.7 + F°} \text{ ft/sec}$$

where $F°$ is the temperature in degrees Fahrenheit.
  The velocity of sound in various materials is shown in Tables 26.1, 26.2, and 26.3.

26.4 SOUND POWER AND SOUND PRESSURE

Sound power is measured in watts. It is independent of distance from the source, and independent of the environment. Sound intensity, or watts per unit area, is dependent on distance. Total radiated sound power may be considered to pass through a spherical surface surrounding the source. Since the radius of the sphere increases with distance, the intensity, or watts per unit area, must also decrease with distance from the source.
  Microphones, sound-measuring instruments, and the ear of a listener respond to changing pressures in a sound wave. Sound power, which cannot be measured directly, is proportional to the mean-square sound pressure, p^2, and can be determined from it.

26.5 DECIBELS AND LEVELS

In acoustics, sound is expressed in decibels instead of watts. By definition, a decibel is 10 times the logarithm, to the base 10, of a ratio of two powers, or powerlike quantities. The reference power is 1 pW, or 10^{-12} W. Therefore,

$$L_W = 10 \log \frac{W}{10^{-12}} \tag{26.1}$$

where L_W = sound power level in dB
W = sound power in watts
$\log$ = logarithm to base 10

  Sound pressure level is 10 times the logarithm of the pressure ratio squared, or 20 times the logarithm of the pressure ratio. The reference sound pressure is 20 μPa, or 20×10^{-6} Pa. Therefore,

$$L_p = 20 \log \frac{p}{20 \times 10^{-6}} \tag{26.2}$$

where L_p = sound pressure level in dB
p = root-mean-square sound pressure in Pa
$\log$ = logarithm to base 10

Table 26.1   Velocity of Sound in Solids

Material	Longitudinal Bar Velocity		Plate (Bulk) Velocity	
	cm/sec	fps	cm/sec	fps
Aluminum	5.24×10^5	1.72×10^4	6.4×10^5	2.1×10^4
Antimony	3.40×10^5	1.12×10^4	—	—
Bismuth	1.79×10^5	5.87×10^3	2.18×10^5	7.15×10^3
Brass	3.42×10^5	1.12×10^4	4.25×10^5	1.39×10^4
Cadmium	2.40×10^5	7.87×10^3	2.78×10^5	9.12×10^3
Constantan	4.30×10^5	1.41×10^4	5.24×10^5	1.72×10^4
Copper	3.58×10^5	1.17×10^4	4.60×10^5	1.51×10^4
German silver	3.58×10^5	1.17×10^4	4.76×10^5	1.56×10^4
Gold	2.03×10^5	6.66×10^3	3.24×10^5	1.06×10^4
Iridium	4.79×10^5	1.57×10^4	—	—
Iron	5.17×10^5	1.70×10^4	5.85×10^5	1.92×10^4
Lead	1.25×10^5	4.10×10^3	2.40×10^5	7.87×10^3
Magnesium	4.90×10^5	1.61×10^4	—	—
Manganese	3.83×10^5	1.26×10^4	4.66×10^5	1.53×10^4
Nickel	4.76×10^5	1.56×10^4	5.60×10^5	1.84×10^4
Platinum	2.80×10^5	9.19×10^3	3.96×10^5	1.30×10^4
Silver	2.64×10^5	8.66×10^3	3.60×10^5	1.18×10^4
Steel	5.05×10^5	1.66×10^4	6.10×10^5	2.00×10^4
Tantalum	3.35×10^5	1.10×10^4	—	—
Tin	2.73×10^5	8.96×10^3	3.32×10^5	1.09×10^4
Tungsten	4.31×10^5	1.41×10^4	5.46×10^5	1.79×10^4
Zinc	3.81×10^5	1.25×10^4	4.17×10^5	1.37×10^4
Cork	5.00×10^4	1.64×10^3	—	—
Crystals				
Quartz X cut	5.44×10^5	1.78×10^4	5.72×10^5	1.88×10^4
Rock salt X cut	4.51×10^5	1.48×10^4	4.78×10^5	1.57×10^4
Glass				
Heavy flint	3.49×10^5	1.15×10^4	3.76×10^5	1.23×10^4
Extra heavy flint	4.55×10^5	1.49×10^4	4.80×10^5	1.57×10^4
Heaviest crown	4.71×10^5	1.55×10^4	5.26×10^5	1.73×10^4
Crown	5.30×10^5	1.74×10^4	5.66×10^5	1.86×10^4
Quartz	5.37×10^5	1.76×10^4	5.57×10^5	1.81×10^4
Granite	3.95×10^5	1.30×10^4	—	—
Ivory	3.01×10^5	9.88×10^3	—	—
Marble	3.81×10^5	1.25×10^4	—	—
Slate	4.51×10^5	1.48×10^4	—	—
Wood				
Elm	1.01×10^5	3.31×10^3	—	—
Oak	4.10×10^5	1.35×10^4	—	—

26.6   COMBINING DECIBELS

It is often necessary to combine sound levels from several sources. For example, it may be desired to estimate the combined effect of adding another machine in an area where other equipment is operating. The procedure for doing this is to combine the sounds on an energy basis, as follows:

$$L_p = 10 \log[10^{0.1L_1} + 10^{0.1L_2} + \cdots + 10^{0.1L_n}] \tag{26.3}$$

where L_p = total sound pressure level in dB
    L_1 = sound pressure level of source No. 1
    L_n = sound pressure level of source No. n
    log = logarithm to base 10

Table 26.2   Velocity of Sound in Liquids

Material	Temperature		Velocity	
	°C	°F	cm/sec	fps
Alcohol, ethyl	12.5	54.5	1.21×10^5	3.97×10^3
	20	68	1.17×10^5	3.84×10^3
Benzene	20	68	1.32×10^5	4.33×10^3
Carbon bisulfide	20	68	1.16×10^5	3.81×10^3
Chloroform	20	68	1.00×10^5	3.28×10^3
Ether, ethyl	20	68	1.01×10^5	3.31×10^3
Glycerine	20	68	1.92×10^5	6.30×10^3
Mercury	20	68	1.45×10^5	4.76×10^3
Pentane	20	68	1.02×10^5	3.35×10^3
Petroleum	15	59	1.33×10^5	4.36×10^3
Turpentine	3.5	38.3	1.37×10^5	4.49×10^3
	27	80.6	1.28×10^5	4.20×10^3
Water, fresh	17	62.6	1.43×10^5	4.69×10^3
Water, sea	17	62.6	1.51×10^5	4.95×10^3

Table 26.3   Velocity of Sound in Gases

Material	Temperature		Velocity	
	°C	°F	cm/sec	fps
Air	0	32	3.31×10^4	1.09×10^3
	20	68	3.43×10^4	1.13×10^3
Ammonia gas	0	32	4.15×10^4	1.48×10^3
Carbon dioxide	0	32	2.59×10^4	8.50×10^2
Carbon monoxide	0	32	3.33×10^4	1.09×10^3
Chlorine	0	32	2.06×10^4	6.76×10^2
Ethane	10	50	3.08×10^4	1.01×10^3
Ethylene	0	32	3.17×10^4	1.04×10^3
Hydrogen	0	32	1.28×10^5	4.20×10^3
Hydrogen chloride	0	32	2.96×10^4	9.71×10^2
Hydrogen sulfide	0	32	2.89×10^4	9.48×10^2
Methane	0	32	4.30×10^4	1.41×10^3
Nitric oxide	10	50	3.24×10^4	1.06×10^3
Nitrogen	0	32	3.34×10^4	1.10×10^3
	20	68	3.51×10^4	1.15×10^3
Nitrous oxide	0	32	2.60×10^4	8.53×10^2
Oxygen	0	32	3.16×10^4	1.04×10^3
	20	68	3.28×10^4	1.08×10^3
Sulfur dioxide	0	32	2.13×10^4	6.99×10^2
Water vapor	0	32	1.01×10^4	3.31×10^2
	100	212	1.05×10^4	3.45×10^2

26.7   SOUND PRODUCED BY SEVERAL MACHINES OF THE SAME TYPE

The total sound produced by a number of machines of the same type can be determined by adding $10 \log n$ to the sound produced by one machine alone. That is,

$$L_p(n) = L_p + 10 \log n$$

where $L_p(n)$ = sound pressure level of n machines
L_p = sound pressure level of one machine
n = number of machines of the same type

In practice, the increase in sound pressure level measured at any location seldom exceeds 6 dB, no matter how many machines are operating. This is because of the necessary spacing between machines, and the fact that sound pressure level decreases with distance.

26.8 AVERAGING DECIBELS

There are many occasions when the average of a number of decibel readings must be calculated. One example is when sound power level is to be determined from a number of sound pressure level readings. In such cases the average may be calculated as follows:

$$\bar{L}_p = 10 \log \left\{ \frac{1}{n} [10^{0.1 L_1} + 10^{0.1 L_2} + \cdots + 10^{0.1 L_n}] \right\} \tag{26.4}$$

where $\bar{L}_p$ = average sound pressure level in dB
$\quad\quad L_1$ = sound pressure level at location No. 1
$\quad\quad L_n$ = sound pressure level at location No. n
$\quad\quad n$ = number of locations
$\quad\quad \log$ = logarithm to base 10

The calculation may be simplified if the difference between maximum and minimum sound pressure levels is small. In such cases arithmetic averaging may be used instead of logarithmic averaging, as follows:

If the difference between the maximum and minimum of the measured sound pressure levels is 5 dB or less, average the levels arithmetically.

If the difference between maximum and minimum sound pressure levels is between 5 and 10 dB, average the levels arithmetically and add 1 dB.

The results will usually be correct within 1 dB when compared to the average calculated by Eq. (26.4).

26.9 SOUND-LEVEL METER

The basic instrument in all sound measurements is the sound-level meter. It consists of a microphone, a calibrated attenuator, an indicating meter, and weighting networks. The meter reading is in terms of root-mean-square sound pressure level.

The A-weighting network is the one most often used. Its response characteristics approximate the response of the human ear, which is not as sensitive to low-frequency sounds as it is to high-frequency sounds. A-weighted measurements can be used for estimating annoyance caused by noise and for estimating the risk of noise-induced hearing damage. Sound levels read with the A-network are referred to as dBA.

26.10 SOUND ANALYZERS

The octave-band analyzer is the most common analyzer for industrial noise measurements. It separates complex sounds into frequency bands one octave in width, and measures the level in each of the bands.

An octave is the interval between two sounds having a frequency ratio of two. That is, the upper cutoff frequency is twice the lower cutoff frequency. The particular octaves read by the analyzer are identified by the center frequency of the octave. The center frequency of each octave is its geometric mean, or the square root of the product of the lower and upper cutoff frequencies. That is,

$$f_0 = \sqrt{f_1 f_2}$$

where f_0 = the center frequency, in Hz
$\quad\quad f_1$ = the lower cutoff frequency, in Hz
$\quad\quad f_2$ = the upper cutoff frequency, in Hz

f_1 and f_2 can be determined from the center frequency. Since $f_2 = 2f_1$ it can be shown that $f_1 = f_0/\sqrt{2}$ and $f_2 = \sqrt{2} f_0$.

Third-octave band analyzers divide the sound into frequency bands one-third octave in width. The upper cutoff frequency is equal to $2^{1/3}$, or 1.26, times the lower cutoff frequency.

When unknown frequency components must be identified for noise control purposes, narrow-band analyzers must be used. They are available with various bandwidths.

26.11 CORRECTION FOR BACKGROUND NOISE

The effect of ambient or background noise should be considered when measuring machine noise. Ambient noise should preferably be at least 10 dB below the machine noise. When the difference is less than 10 dB, adjustments should be made to the measured levels as shown in Table 26.4.

Table 26.4 Correction for Background Sound

Level Increase due to the Machine (dB)	Value to be Subtracted from Measured Level (dB)
3	3.0
4	2.2
5	1.7
6	1.3
7	1.0
8	0.8
9	0.6
10	0.5

If the difference between machine octave-band sound pressure levels and background octave-band sound pressure levels is less than 6 dB, the accuracy of the adjusted sound pressure levels will be decreased. Valid measurements cannot be made if the difference is less than 3 dB.

26.12 MEASUREMENT OF MACHINE NOISE

The noise produced by a machine may be evaluated in various ways, depending on the purpose of the measurement and the environmental conditions at the machine. Measurements are usually made in overall A-weighted sound pressure levels, plus either octave-band or third-octave-band sound pressure levels. Sound power levels are calculated from sound pressure level measurements.

26.13 SMALL MACHINES IN A FREE FIELD

A free field is one in which the effects of the boundaries are negligible, such as outdoors, or in a very large room. When small machines are sound tested in such locations, measurements at a single location are often sufficient. Many sound test codes specify measurements at a distance of 1 m from the machine.

Sound power levels, octave band, third-octave band, or A-weighted, may be determined by the following equation:

$$L_W = L_p + 20 \log r + 7.8 \qquad (26.5)$$

where L_W = sound power level, in dB
$\qquad L_p$ = sound pressure level, in dB
$\qquad r$ = distance from source, in m
$\qquad \log$ = logarithm to base 10

26.14 MACHINES IN SEMIREVERBERANT LOCATIONS

Machines are almost always installed in semireverberant environments. Sound pressure levels measured in such locations will be greater than they would be in a free field. Before sound power levels are calculated adjustments must be made to the sound pressure level measurements.

There are several methods for determining the effect of the environment. One uses a calibrated reference sound source, with known sound power levels, in octave or third-octave bands. Sound pressure levels are measured on the machine under test, at predetermined microphone locations. The machine under test is then replaced by the reference sound source, and measurements are repeated. Sound power levels can then be calculated as follows:

$$L_{Wx} = \bar{L}_{px} + (L_{Ws} - \bar{L}_{ps}) \qquad (26.6)$$

where L_{Wx} = band sound power level of the machine under test
$\quad\quad \overline{L}_{px}$ = average sound pressure level measured on the machine under test
$\quad\quad L_{Ws}$ = band sound power level of the reference source
$\quad\quad \overline{L}_{ps}$ = average sound pressure level on the reference source

Another procedure for qualifying the environment uses a reverberation test. High-speed recording equipment and a special noise source are used to measure the time for the sound pressure level, originally in a steady state, to decrease 60 dB after the special noise source is stopped. This reverberation time must be measured for each frequency, or each frequency band of interest.

Unfortunately, neither of these two laboratory procedures is suitable for sound tests on large machinery, which must be tested where it is installed. This type of machinery usually cannot be shut down while tests are being made on a reference sound source, and reverberation tests cannot be made in many industrial areas because ambient noise and machine noise interfere with reverberation time measurements.

26.15 TWO-SURFACE METHOD

A procedure that can be used in most industrial areas to determine sound pressure levels and sound power levels of large operating machinery is called the two-surface method. It has definite advantages over other laboratory-type tests. The machine under test can continue to operate. Expensive, special instrumentation is not required to measure reverberation time. No calibrated reference source is needed; the machine is its own sound source. The only instrumentation required is a sound level meter and an octave-band analyzer. The procedure consists of measuring sound pressure levels on two imaginary surfaces enclosing the machine under test. The first measurement surface, S_1, is a rectangular parallelepiped 1 m away from a reference surface. The reference surface is the smallest imaginary rectangular parallelepiped that will just enclose the machine, and terminate on the reflecting plane, or floor. The area, in square meters, of the first measurement surface is given by the formula

$$S_1 = ab + 2ac + 2bc \tag{26.7}$$

where $a = L + 2$
$\quad\quad b = W + 2$
$\quad\quad c = H + 1$

and L, W, and H are the length, width, and height of the reference parallelepiped, in meters.

The second measurement surface, S_2, is a similar but larger, rectangular parallelepiped, located at some greater distance from the reference surface. The area, in square meters, of the second measurement surface is given by the formula

$$S_2 = de + 2df + 2ef \tag{26.8}$$

where $d = L + 2x$
$\quad\quad e = W + 2x$
$\quad\quad f = H + x$

and x is the distance in meters from the reference surface to S_2.

Microphone locations are usually those shown on Fig. 26.1.

First, the measured sound pressure levels should be corrected for background noise as shown in Table 26.4. Next, the average sound pressure levels, in each octave band of interest, should be calculated as shown in Eq. (26.4)

Octave-band sound pressure levels, corrected for both background noise and for the semireverberant environment, may then be calculated by the equations

$$\overline{L}_p = \overline{L}_{p1} - C \tag{26.9}$$

$$C = 10 \log\left\{ \left[\frac{K}{K-1} \right] \left[1 - \frac{S_1}{S_2} \right] \right\} \tag{26.10}$$

$$K = 10^{0.1(\overline{L}_{p1} - \overline{L}_{p2})} \tag{26.11}$$

where $\overline{L}_p$ = average octave-band sound pressure level over area S_1, corrected for both background sound and environment
$\quad\quad \overline{L}_{p1}$ = average octave-band sound pressure level over area S_1, corrected for background sound only
$\quad\quad C$ = environmental correction
$\quad\quad \overline{L}_{p2}$ = average octave-band sound pressure level over area S_2, corrected for background sound

As an alternative, the environmental correction C may be obtained from Fig. 26.2.

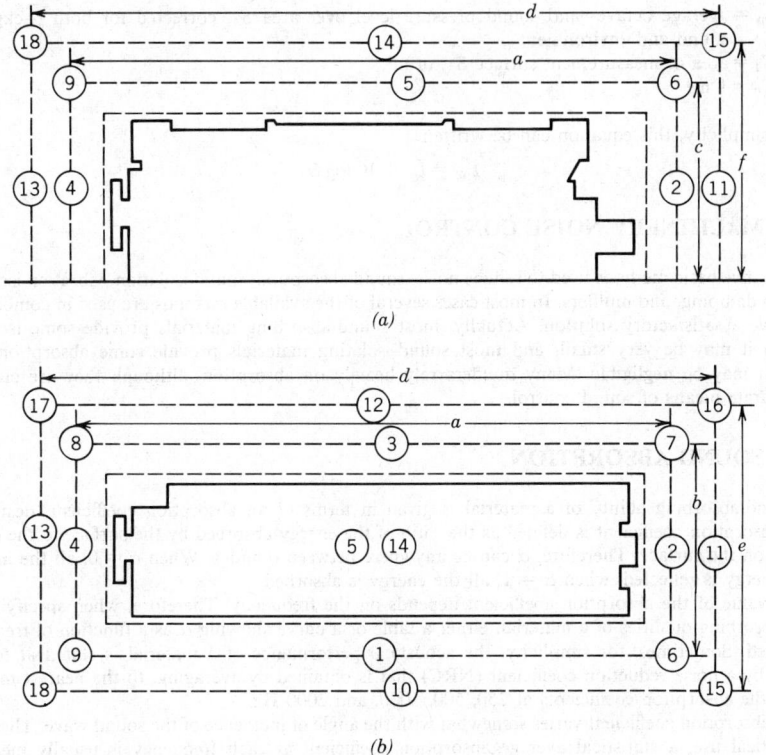

Fig. 26.1 Microphone locations. *(a)* Side view; *(b)* plan view.

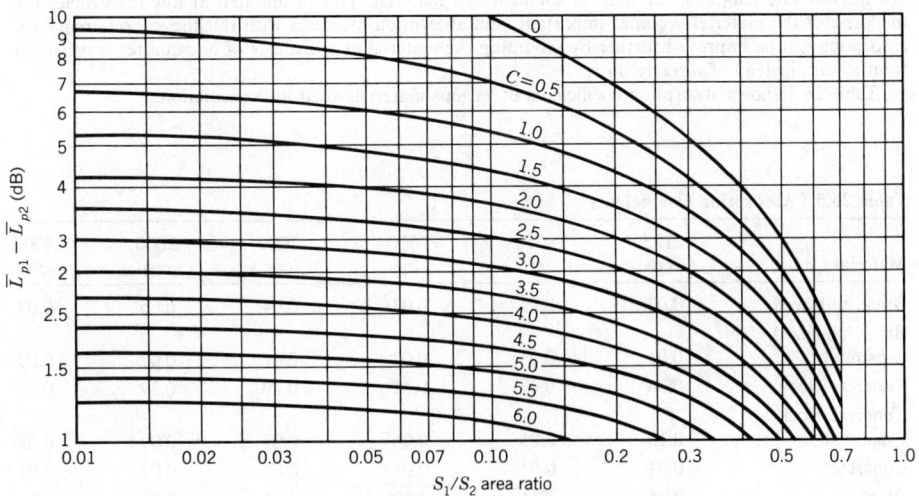

Fig. 26.2 $S_1 S_2$ area ratio.

Sound power levels, in each octave band of interest, may be calculated by the equation

$$L_W = \bar{L}_p + 10 \log \left[\frac{S_1}{S_0} \right] \qquad (26.12)$$

where L_W = octave-band sound power level, in dB

$\overline{L_p}$ = average octave-band sound pressure level over area S_1, corrected for both background sound and environment

S_1 = area of measurement surface S_1, in m²

$S_0 = 1$ m²

For simplicity, this equation can be written

$$L_W = \overline{L_p} + 10 \log S_1$$

26.16   MACHINERY NOISE CONTROL

There are five basic methods used to reduce noise: sound absorption, sound isolation, vibration isolation, vibration damping, and mufflers. In most cases several of the available methods are used in combination to achieve a satisfactory solution. Actually, most sound-absorbing materials provide some isolation, although it may be very small; and most sound-isolating materials provide some absorption, even though it may be negligible. Many mufflers rely heavily on absorption, although they are classified as a separate means of sound control.

26.17   SOUND ABSORPTION

The sound-absorbing ability of a material is given in terms of an absorption coefficient, designated by α. Absorption coefficient is defined as the ratio of the energy absorbed by the surface to the energy incident on the surface. Therefore, α can be anywhere between 0 and 1. When α = 0, all the incident sound energy is reflected; when α = 1, all the energy is absorbed.

The value of the absorption coefficient depends on the frequency. Therefore, when specifying the sound-absorbing qualities of a material, either a table or a curve showing α as a function of frequency is required. Sometimes, for simplicity, the acoustical performance of a material is stated at 500 Hz only, or by a noise reduction coefficient (NRC) that is obtained by averaging, to the nearest multiple of 0.05, the absorption coefficients at 250, 500, 1000, and 2000 Hz.

The absorption coefficient varies somewhat with the angle of incidence of the sound wave. Therefore, for practical use, a statistical average absorption coefficient at each frequency is usually measured and stated by the manufacturer. It is often better to select a sound-absorbing material on the basis of its characteristics for a particular noise rather than by its average sound-absorbing qualities.

Sound absorption is a function of the length of path relative to the wavelength of the sound, and not the absolute length of the path of sound in the material. This means that at low frequencies the thickness of the material becomes important, and absorption increases with thickness. Low-frequency absorption can be improved further by mounting the material at a distance of one-quarter wavelength from a wall, instead of directly on it.

Table 26.5 shows absorption coefficients of various materials used in construction.

Table 26.5   Absorption Coefficients

Material	125 cps	250 cps	500 cps	1000 cps	2000 cps	4000 cps
Brick, unglazed	0.03	0.03	0.03	0.04	0.05	0.07
Brick, unglazed, painted	0.01	0.01	0.02	0.02	0.02	0.03
Concrete block	0.36	0.44	0.31	0.29	0.39	0.25
Concrete block, painted	0.10	0.05	0.06	0.07	0.09	0.08
Concrete	0.01	0.01	0.015	0.02	0.02	0.02
Wood	0.15	0.11	0.10	0.07	0.06	0.07
Glass, ordinary window	0.35	0.25	0.18	0.12	0.07	0.04
Plaster	0.013	0.015	0.02	0.03	0.04	0.05
Plywood	0.28	0.22	0.17	0.09	0.10	0.11
Tile	0.02	0.03	0.03	0.03	0.03	0.02
6 lb/ft³ fiberglass	0.48	0.82	0.97	0.99	0.90	0.86

The sound absorption of a surface, expressed in either square feet of absorption, or sabins, is equal to the area of the surface, in square feet, times the absorption coefficient of the material on the surface.

Average absorption coefficient, $\overline{\alpha}$, is calculated as follows:

$$\overline{\alpha} = \frac{\alpha_1 S_1 + \alpha_2 S_2 + \cdots + \alpha_n S_n}{S_1 + S_2 + \cdots + S_n} \tag{26.13}$$

where $\overline{\alpha}$ = the average absorption coefficient
$\alpha_1, \alpha_2, \alpha_n$ = the absorption coefficients of materials on various surfaces
S_1, S_2, S_n = the areas of various surfaces

26.18 NOISE REDUCTION DUE TO INCREASED ABSORPTION IN ROOM

A machine in a large room radiates noise that decreases at a rate inversely proportional to the square of the distance from the source. Soon after the machine is started the sound wave impinges on a wall. Some of the sound energy is absorbed by the wall, and some is reflected. The sound intensity will not be constant throughout the room. Close to the machine the sound field will be dominated by the source, almost as though it were in a free field, while farther away the sound will be dominated by the diffuse field, caused by sound reflections. The distance where the free field and the diffuse field conditions control the sound depends on the average absorption coefficient of the surfaces of the room and the wall area. This critical distance can be calculated by the following equation:

$$r_c = 0.2\sqrt{R} \tag{26.14}$$

where r_c = distance from source, in m
R = room constant of the room, in m²

Room constant is equal to the product of the average absorption coefficient of the room and the total internal area of the room divided by the quantity one minus the average absorption coefficient. That is,

$$R = \frac{S_t \overline{\alpha}}{1 - \overline{\alpha}} \tag{26.15}$$

where R = the room constant, in m²
$\overline{\alpha}$ = the average absorption coefficient
S_t = the total area of the room, in m²

Essentially free-field conditions exist farther from a machine in a room with a large room constant than they do in a room with a small room constant.

The distance r_c determines where absorption will reduce noise in the room. An operator standing close to a noisy machine will not benefit by adding sound absorbing material to the walls and ceiling. Most of the noise heard by the operator is radiated directly by the machine, and very little is reflected noise. On the other hand, listeners farther away, at distances greater than r_c, will benefit from the increased absorption.

The noise reduction in those areas can be estimated by the following equation:

$$NR = 10 \log \frac{\overline{\alpha}_2 S}{\overline{\alpha}_1 S} \tag{26.16}$$

where NR = far field noise reduction, in dB
$\overline{\alpha}_1 S$ = room absorption before treatment
$\overline{\alpha}_2 S$ = room absorption after treatment

Equation (26.16) shows that doubling the absorption will reduce noise by 3 dB. It requires another doubling of the absorption to get another 3 dB reduction. This is much more difficult than getting the first doubling, and considerably more expensive.

26.19 SOUND ISOLATION

Noise may be reduced by placing a barrier or wall between a noise source and a listener. The effectiveness of such a barrier is described by its transmission coefficient.

Sound transmission coefficient of a partition is defined as the fraction of incident sound transmitted through it.

Sound transmission loss is a measure of sound-isolating ability, and is equal to the number of decibels by which sound energy is reduced in transmission through a partition. By definition, it is 10 times the logarithm to the base 10 of the reciprocal of the sound transmission coefficient. That is,

$$TL = 10 \log \frac{1}{\tau} \qquad (26.17)$$

where TL = the transmission loss, in dB
τ = the transmission coefficient

Transmission of sound through a rigid partition or solid wall is accomplished mainly by the forced vibration of the wall. That is, the partition is forced to vibrate by the pressure variations in the sound wave.

Under certain conditions porous materials can be used to isolate high-frequency sound, and, in general, the loss provided by a uniform porous material is directly proportional to the thickness of the material. For most applications, however, sound absorbing materials are very ineffective sound isolators because they have the wrong characteristics. They are porous, instead of airtight, and they are lightweight, instead of heavy. The transmission loss of nonporous materials is determined by weight per square foot of surface area, and how well all cracks and openings are sealed. Transmission loss is affected also by dynamic bending stiffness and internal damping. Table 26.6 shows the transmission loss of various materials used in construction.

Table 26.6   Transmission Loss of Building Materials

Item	TL
Hollow-core door ($\frac{3}{16}$-in. panels)	15
$1\frac{3}{4}$-in. solid-core oak door	20
$2\frac{1}{2}$-in. heavy wood door	25–30
4-in. cinder block	20–25
4-in. cinder block, plastered	40
4-in. cinder slab	40–45
4-in. slab-suspended concrete, plastered	50
Two 4-in. cinder blocks—4-in. air space	55
4-in. brick	45
4-in. brick, plastered	47
8-in. brick, plastered	50
Two 8-in. cinder block—4-in. air space	57

26.20   SINGLE PANEL

The simplest type of sound-isolating barrier is a single, homogeneous, nonporous partition. In general, the transmission loss of a single wall of this type is proportional to the logarithm of the mass. Its isolating ability also increases with frequency, and the approximate relationship is given by the following equation:

$$TL = 20 \log W + 20 \log f - 33 \qquad (26.18)$$

where TL = the transmission loss, in dB
W = the surface weight, in pounds per square foot
f = the frequency, in Hz

This means that the transmission loss increases 6 dB each time the weight is doubled, and 6 dB each time the frequency is doubled. In practice, these numbers are each about 5 dB, instead of 6 dB.

In general, a single partition or barrier should not be counted on to provide noise reduction of more than about 10 dB.

26.21   COMPOSITE PANEL

Many walls or sound barriers are made of several different materials. For example, machinery enclosures are commonly constructed of sheet steel, but they may have a glass window to observe instruments inside the enclosure. The transmission coefficients of the two materials are different. Another example is when there are necessary cracks or openings in the enclosure where it fits around a rotating shaft. In this case, the transmission loss of the opening is zero, and the transmission coefficient is 1.0.

The effectiveness of such a wall is related to both the transmission coefficients of the materials in

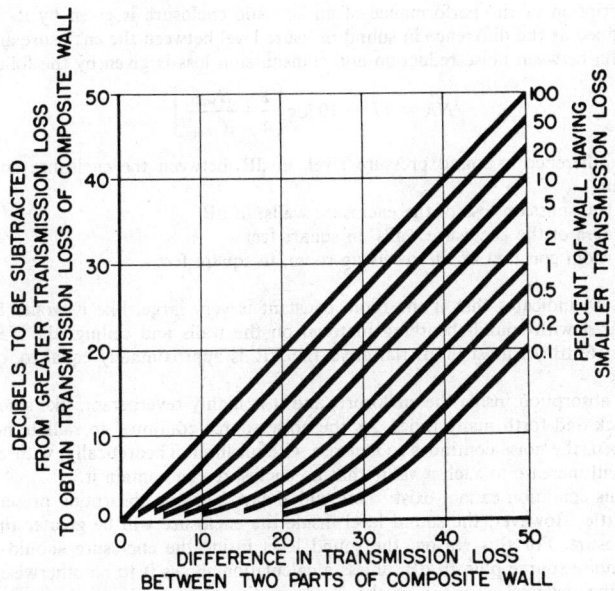

Fig. 26.3 Decibels to be subtracted from greater transmission loss to obtain transmission loss of composite wall. The difference in transmission loss between two parts of composite wall. Percentage of wall having smaller transmission loss.

the wall and the areas of the sections. A large area transmits more noise than a small one made of the same material. Also, more noise can be transmitted by a large area with a relatively small transmission coefficient than by a small one with a comparatively high transmission coefficient. On the other hand, a small area with a high transmission coefficient can ruin the effectiveness of an otherwise excellently designed enclosure. Both transmission coefficients and areas must be controlled carefully.

The average transmission coefficient of a composite panel is

$$\bar{\tau} = \frac{\tau_1 S_1 + \tau_2 S_2 + \tau_3 S_3 + \cdots + \tau_n S_n}{S_1 + S_2 + S_3 + \cdots + S_n} \quad (26.19)$$

where $\bar{\tau}$ = the average transmission coefficient

$\tau_1, \ldots, \tau_n$ = the transmission coefficients of the various areas

$S_1, \ldots, S_n$ = the various areas

Figure 26.3 shows how the transmission loss of a composite wall or panel may be determined from the transmission loss values of its parts. It also shows the damaging effect of small leaks in the enclosure. In this case the transmission loss of the leak opening is zero.

A leak of only 0.1% in an expensive, high-quality door or barrier, constructed of material with a transmission loss of 50 dB, would reduce the transmission loss by 20 dB, resulting in a *TL* of only 30 dB instead of 50 dB. A much less expensive 30 dB barrier would be reduced by only 3 dB, resulting in a *TL* of 27 dB. This shows that small leaks are more damaging to a high-quality enclosure than to a lower-quality one.

26.22 ACOUSTIC ENCLOSURES

When machinery noise must be reduced by 20 dB or more, it is usually necessary to use complete enclosures. It must be kept in mind that the actual decrease in noise produced by an enclosure depends on other things as well as the transmission loss of the enclosure material. Vibration resonances must be avoided, or their effects must be reduced by damping; structural and mechanical connections must not be permitted to short circuit the enclosure; and the enclosure must be sealed as well as possible to prevent acoustic leaks. In addition, the actual noise reduction depends on the acoustic properties of the room in which the enclosure is located. For this reason, published data on transmission loss of various materials should not be assumed to be the same as the noise reduction that will be obtained when using those materials in enclosures. How materials are used in machinery enclosures is just as important as which materials are used.

A better description of the performance of an acoustic enclosure is given by its noise reduction NR, which is defined as the difference in sound pressure level between the enclosure and the receiving room. The relation between noise reduction and transmission loss is given by the following equation:

$$NR = TL - 10 \log\left[\frac{1}{4} + \frac{S_{wall}}{R_{room}}\right] \qquad (26.20)$$

where NR = the difference in sound pressure level, in dB, between the enclosure and the receiving room

TL = the transmission loss of the enclosure walls, in dB

S_{wall} = the area of the enclosure walls, in square feet

R_{room} = the room constant of the receiving room, in square feet

Equation (26.20) indicates that if the room constant is very large, like it would be outdoors, or in a very large area with sound-absorbing material on the walls and ceiling, the NR could exceed the TL by almost 6 dB. In most industrial areas the NR is approximately equal to, or is several dB less than, the TL.

If there is no absorption inside the enclosure, and it is highly reverberant, like smooth sheet steel, sound reflects back and forth many times. As the noise source continues to radiate noise, with none of it being absorbed, the noise continues to increase without limit. Theoretically, with zero absorption, the sound level will increase to such a value that no enclosure can contain it.

Practically, this condition cannot exist; there will always be some absorption present, even though it may be very little. However, the sound level inside the enclosure will be greater than it would be without the enclosure. For this reason, the sound level inside the enclosure should be assumed to equal the actual noise source plus 10 dB, unless a calculation shows it to be otherwise.

When absorbing material is added to the inside of the enclosure, sound energy decreases each time it is reflected.

An approximate method for estimating the noise reduction of an enclosure is

$$NR = 10 \log\left[1 + \frac{\bar{\alpha}}{\bar{\tau}}\right] \qquad (26.21)$$

where NR = the noise reduction, in dB

$\bar{\alpha}$ = the average absorption coefficient of the inside of the enclosure

$\bar{\tau}$ = the average transmission coefficient of the enclosure

Equation (26.21) shows that in the theoretical case where there is no sound absorption, there is no noise reduction.

26.23 DOUBLE WALLS

A 4-in.-thick brick wall has a transmission loss of about 45 dB. An 8-in.-thick brick wall, with twice as much weight, has a transmission loss of about 50 dB. After a certain point has been reached it is found to be impractical to try to obtain higher isolation values simply by doubling the weight, since both the weight and the cost become excessive, and only a 5 dB improvement is gained for each doubling of weight.

An increase can be obtained, however, by using double-wall construction. That is, two 4-in.-thick walls separated by an air space are better than one 8-in. wall. However, noise radiated by the first panel can excite vibration of the second one and cause it to radiate noise. If there are any mechanical connections between the two panels, vibration of one directly couples to the other, and much of the benefit of double-wall construction is lost.

There is another factor that can reduce the effectiveness of double-wall construction. Each of the walls represents a mass, and the air space between them acts as a spring. This mass–spring–mass combination has a series of resonances that greatly reduce the transmission loss at the corresponding frequencies. The effect of the resonances can be reduced by adding sound-absorbing material in the space between the panels.

26.24 VIBRATION ISOLATION

There are many instances where airborne sound can be reduced substantially by isolating a vibrating part from the rest of the structure. A vibration isolator, in its simplest form is some type of resilient support. The purpose of the isolator may be to reduce the magnitude of force transmitted from a vibrating machine or part of a machine to its supporting structure. Conversely, its purpose may be to reduce the amplitude of motion transmitted from a vibrating support to a part of the system that is radiating noise due to its vibration.

Vibration isolators can be in the form of steel springs, cork, felt, rubber, plastic, or dense fiberglass.

Steel springs can be calculated quite accurately and can do an excellent job of vibration isolation. However, they also can have resonances, and high-frequency vibrations can travel through them readily, even though they are effectively isolating the lower frequencies. For this reason, springs are usually used in combination with elastomers or similar materials. Elastomers, plastics, and materials of this type have high internal damping and do not perform well below about 15 Hz. However, this is below the audible range, and, therefore, it does not limit their use in any way for effective sound control.

The noise reduction that can be obtained by installing an isolator depends on the characteristics of the isolator and the associated mechanical structure. For example, the attenuation that can be obtained by spring isolators depends not only on the spring constant, or spring stiffness (the force necessary to stretch or compress the spring one unit of length), but also on the mass load on the spring, the mass and stiffness of the foundation, and the type of excitation.

If the foundation is very massive and rigid, and if the mounted machine vibrates at constant amplitude, the reduction in force on the foundation is independent of frequency. If the machine vibrates at a constant force, the reduction in force depends on the ratio of the exciting frequency to the natural frequency of the system.

When a vibrating machine is mounted on an isolator, the ratio of the force applied to the isolator by the machine, to the force transmitted by the isolator, to the foundation is called the "transmissibility." That is,

$$\text{transmissibility} = \frac{\text{transmitted force}}{\text{impressed force}}$$

Under ideal conditions this ratio would be zero. In practice, the objective is to make it as small as possible. This can be done by designing the system so that the natural frequency of the mounted machine is very low compared to the frequency of the exciting force.

If no damping is present, the transmissibility can be expressed by the following equation:

$$T = \frac{1}{1 - (\omega/\omega_n)^2} \tag{26.22}$$

where T = the transmissibility, expressed as a fraction
$\quad \omega$ = the circular frequency of the exciting force, in radians per second
$\quad \omega_n$ = the circular frequency of the mounted system, in radians per second

When $\omega/\omega_n = 0$, the transmissibility equals 1.0. That is, there is no benefit obtained from the isolator.

If ω/ω_n is greater than zero but less than 1.41, the isolator actually increases the magnitude of the transmitted force. This is called the "region of amplification." In fact, when ω/ω_n equals 1.0, the theoretical amplitude of the transmitted force goes to infinity, since this is the point where the frequency of the disturbing force equals the system natural frequency.

Equation (26.22) indicates that the transmissibility becomes negative when ω/ω_n is greater than 1.0. The negative number is simply due to the phase relation between force and motion, and it can be disregarded when considering only the amount of transmitted force.

Since vibration isolation is achieved only when ω/ω_n is greater than 1.41, the equation for transmissibility can be written so that T is positive:

$$T = \frac{1}{(\omega/\omega_n)^2 - 1}$$

Also, since $\omega = 2\pi f$,

$$T = \frac{1}{(f/f_n)^2 - 1} \tag{26.23}$$

The static deflection of a spring when stretched or compressed by a weight is related to its natural frequency by the equation

$$f_n = 3.14 \sqrt{\frac{1}{d}} \tag{26.24}$$

where f_n = the natural frequency, in hertz
$\quad d$ = the deflection, in inches

When this is substituted in the equation for transmissibility, Eq. (26.23), it can be shown that

$$d = \left(\frac{3.14}{f}\right)^2 \left(\frac{1}{T} + 1\right) \tag{26.25}$$

This shows that the transmissibility can be determined from the deflection of the isolator due to its supported load.

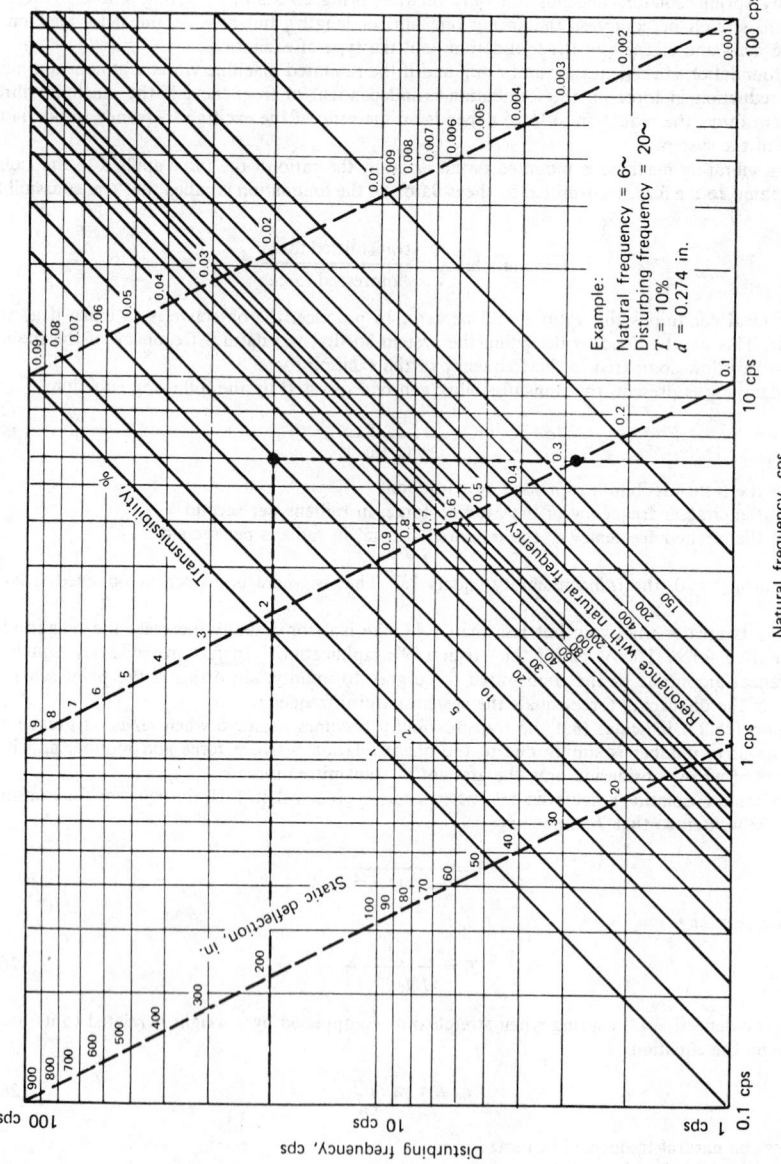

Fig. 26.4 Transmissibility of flexible mountings: $T = 1/[(\omega/\omega_n)^2 - 1]$.

Example:
Natural frequency = 6~
Disturbing frequency = 20~
T = 10%
d = 0.274 in.

Transmissibility, %

Resonance with natural frequency

Static deflection, in.

Natural frequency, cps

Disturbing frequency, cps

Equations (26.23) and (26.25) can be plotted, as shown in Fig. 26.4, for convenience in selecting isolator natural frequencies or deflections.

For critical applications, the natural frequency of the isolator should be about one-tenth to one-sixth of the disturbing frequency. That is, the transmissibility should be between 1 and 3%. For less critical conditions, the natural frequency of the isolator should be about one-sixth to one-third of the driving frequency, with transmissibility between 3 and 12%.

26.25 VIBRATION DAMPING

Complex mechanical systems have many resonant frequencies, and whenever an exciting frequency is coincident with one of the resonant frequencies, the amplitude of vibration is limited only by the amount of damping in the system. If the exciting force is wide band, several resonant vibrations can occur simultaneously, thereby compounding the problem. Damping is one of the most important factors in noise and vibration control.

There are three kinds of damping. Viscous damping is the type that is produced by viscous resistance in a fluid, for example, a dashpot. The damping force is proportional to velocity. Dry friction, or Coulomb damping, produces a constant damping force, independent of displacement and velocity. The damping force is produced by dry surfaces rubbing together, and it is opposite in direction to that of the velocity. Hysteresis damping, also called material damping, produces a force that is in phase with the velocity but is proportional to displacement. This is the type of damping found in solid materials, such as elastomers, widely used in sound control.

A large amount of noise radiated from machine parts comes from vibration of large areas or panels. These parts may be integral parts of the machine or attachments to the machine. They can be flat or curved, and vibration can be caused by either mechanical or acoustic excitation. The radiated noise is a maximum when the parts are vibrating in resonance.

When the excitation is mechanical, vibration isolation may be all that is needed. In other instances, the resonant response can be reduced by bonding a layer of energy-dissipating polymeric material to the structure. When the structure bends, the damping material is placed alternately in tension and compression, thus dissipating the energy as heat.

This extensional or free-layer damping is remarkably effective in reducing resonant vibration and noise in relatively thin, lightweight structures such as panels. It becomes less effective as the structure stiffness increases, because of the excessive increase in thickness of the required damping layer.

In a vibrating structure, the amount of energy dissipated is a function of the amount of energy necessary to deflect the structure, compared to that required to deflect the damping material. If 99% of the vibration energy is required to deflect the structure, and 1% is required to deflect the damping layer, then only 1% of the vibration energy is dissipated.

Resonant vibration amplitude in heavier structures can be controlled effectively by using constrained-layer damping. In this method, a relatively thin layer of viscoelastic damping material is constrained between the structure and a stiff cover plate. Vibration energy is removed from the system by the shear motion of the damping layer.

26.26 MUFFLERS

Silencers, or mufflers, are usually divided into two categories: absorptive and reactive. The absorptive type, as the name indicates, removes sound energy by the use of sound-absorbing materials. They have relatively wide-band noise-reduction characteristics, and are usually applied to problems associated with continuous spectra, such as fans, centrifugal compressors, jet engines, and gas turbines. They are also used in cases where a narrow-band noise predominates, but the frequency varies because of a wide range of operating speed.

A variety of sound-absorbing materials are used in many different configurations, determined by the level of the unsilenced noise and its frequency content, the type of gas being used, the allowable pressure drop through the silencer, the gas velocity, gas temperature and pressure, and the noise criterion to be met.

Fiberglass or mineral wool with density approximately 0.5–6.0 lb/ft^3 is frequently used in absorptive silencers. These materials are relatively inexpensive and have good sound-absorbing characteristics. They operate on the principle that sound energy causes the material fibers to move, converting the sound energy into mechanical vibration and heat. The fibers do not become very warm since the sound energy is actually quite low, even at fairly high decibel levels.

The simplest kind of absorptive muffler is a lined duct, where the absorbing material is either added to the inside of the duct walls or the duct walls themselves are made of sound-absorbing material. The attenuation depends on the duct length, thickness of the lining, area of the air passage, type of absorbing material, and frequency of the sound passing through.

The acoustical performance of absorptive mufflers is improved by adding parallel or annular baffles to increase the amount of absorption. This also increases pressure drop through the muffler, so that spacing and area must be carefully controlled.

Reactive mufflers have a characteristic performance that does not depend to any great extent on the presence of sound-absorbing material, but utilizes the reflection characteristics and attenuating properties of conical connectors, expansion chambers, side branch resonators, tail pipes, and so on, to accomplish sound reduction.

Expansion chambers operate most efficiently in applications involving discrete frequencies rather than broad-band noise. The length of the chamber is adjusted so that reflected waves cancel the incident waves, and since wavelength depends on frequency, expansion chambers should be tuned to some particular frequency. When a number of discrete frequencies must be attenuated, several expansion chambers can be placed in series, each tuned to a particular wavelength.

An effective type of reactive muffler, called a Helmholtz resonator, consists of a vessel containing a volume of air, that is connected to a noise source, such as a piping system. When a pure-tone sound wave is propagated along the pipe, the air in the vessel expands and contracts. By proper design of the area and length of the neck, and volume of the chamber, sound wave cancellation can be obtained, thereby reducing the tone. This type of resonator produces maximum noise reduction over a very narrow frequency range, but it is possible to combine several Helmholtz resonators on a piping system so that not only will each cancel out at its own frequency, but they can be made to overlap so that noise is attenuated over a wider range instead of at sharply tuned points.

Helmholtz resonators are normally located in side branches, and for this reason they do not affect flow in the main pipe.

The resonant frequency of these devices can be calculated by the equation

$$f = \frac{C}{2\pi}\sqrt{\frac{A}{LV}} \tag{26.26}$$

where f = the resonant frequency, in hertz
$\quad C$ = the speed of sound in the fluid, in feet per second
$\quad A$ = the cross-sectional area of the neck, in square feet
$\quad L$ = the length of the neck, in feet
$\quad V$ = the volume of the chamber, in cubic feet

The performance of all types of mufflers can be stated in various ways. Not everyone uses the same terminology, but in general, the following definitions apply:

INSERTION LOSS   is defined as the difference between two sound pressure levels measured at the same point in space before and after a muffler is inserted in the system.

DYNAMIC INSERTION LOSS   is the same as insertion loss, except that it is measured when the muffler is operating under rated flow conditions. Therefore, the dynamic insertion loss is of more interest than ratings based on no-flow conditions.

TRANSMISSION LOSS   is defined as the ratio of sound power incident on the muffler to the sound power transmitted by the muffler. It cannot be measured directly, and it is difficult to calculate analytically. For these reasons the transmission loss of a muffler has little practical application.

ATTENUATION   is used to describe the decrease in sound power as sound waves travel through the muffler. It does not convey information about how a muffler performs in a system.

NOISE REDUCTION   is defined as the difference between sound pressure levels measured at the inlet of a muffler and those at the outlet.

26.27 SOUND CONTROL RECOMMENDATIONS

Sound control procedures should be applied during the design stages of a machine, whenever possible. A list of recommendations for noise reduction follows:

1. Reduce horsepower. Noise is proportional to horsepower. Therefore, the machine should be matched to the job. Excess horsepower means excess noise.
2. Reduce speed. Slow-speed machinery is quieter than high-speed machinery.
3. Keep impeller tip speeds low. However, it is better to keep the rpm low and the impeller diameter large than to keep the rpm high and the impeller diameter small, even though the tip speeds are the same.
4. Improve dynamic balance. This decreases rotating forces, structure-borne sound, and the excitation of structural resonances.
5. Reduce the ratio of rotating masses to fixed masses.
6. Reduce mechanical run-out of shafts. This improves the initial static and dynamic balance.
7. Avoid structural resonances. These are often responsible for many unidentified components in the radiated sound. In addition to being excited by sinusoidal forcing frequencies, they can be excited by impacting parts and sliding and rubbing contacts.

8. Eliminate or reduce impacts. Either reduce the mass of impacting parts or their striking velocities.

9. Reduce peak acceleration. Reduce the rate of change of velocity of moving parts by using the maximum time possible to produce the required velocity change, and by keeping the acceleration as nearly constant as possible over the available time period.

10. Improve lubrication. Inadequate lubrication is often the cause of bearing noise, structure-borne noise due to friction, and the excitation of structural resonances.

11. Maintain closer tolerances and clearances in bearings and moving parts.

12. Install bearings correctly. Improper installation accounts for approximately half of bearing noise problems.

13. Improve alignment. Improper alignment is a major source of noise and vibration.

14. Use center of gravity mounting whenever feasible. When supports are symmetrical with respect to the center of gravity, translational modes of vibration do not couple to rotational modes.

15. Maintain adequate separation between operating speeds and lateral and torsional resonant speeds.

16. Consider the shape of impeller vanes from an acoustic standpoint. Some configurations are noisier than others.

17. Keep the distance between impeller vanes and cutwater or diffuser vanes as large as possible. Close spacing is a major source of noise.

18. Select combinations of rotating and stationary vanes that are not likely to excite strong vibration and noise.

19. Design turning vanes properly. They are a source of self-generated noise.

20. Keep the areas of inlet passages as large as possible and their length as short as possible.

21. Remove or keep at a minimum any obstructions, bends, or abrupt changes in fluid passages.

22. Pay special attention to inlet design. This is extremely important in noise generation.

23. Item 22 applies also to the discharge, but the inlet is more important than the discharge from an acoustic standpoint.

24. Maintain gradual, not abrupt, transition from one area to the next in all fluid passages.

25. Reduce flow velocities in passages, pipes, and so on. Noise can be reduced substantially by reducing flow velocities.

26. Reduce jet velocities. Jet noise is proportional to the eighth power of the velocity.

27. Reduce large radiating areas. Surfaces radiating certain frequencies can often be divided into smaller areas with less radiating efficiency.

28. Disconnect possible sound radiating parts from other vibrating parts by installing vibration breaks to eliminate metal to metal contact.

29. Provide openings or air leaks in large radiating areas so that air can move through them. This reduces pressure build-up and decreases radiated noise.

30. Reduce clearances, piston weights, and connecting rod weights in reciprocating machinery to reduce piston impacts.

31. Apply additional sound control devices, such as inlet and discharge silencers and acoustic enclosures.

32. When acoustic enclosures are used, make sure that all openings are sealed properly.

33. Install machinery on adequate mountings and foundations to reduce structure-borne sound and vibration.

34. Take advantage of all directivity effects whenever possible by directing inlet and discharge openings away from listeners or critical areas.

35. When a machine must meet a particular sound specification, purchase driving motors, turbines, gears, and auxiliary equipment that produce 3- to 5-dB lower sound levels than the machine alone. This ensures that the combination is in compliance with the specification.

CHAPTER 27

NONDESTRUCTIVE TESTING

ROBERT L. CRANE
RONALD SELNER

AFWAL/MLLP
Wright Patterson Air Force Base
Dayton, Ohio

27.1 INTRODUCTION

Nondestructive evaluation (NDE) encompasses those physical and chemical tests that are used to determine if a component or structure can perform its intended function without the test methods impairing the component's performance. Until recently NDE has been thought to be limited to the detection of physical flaws and the quantification of their dimensions. These data are commonly used to determine if the part should be scrapped or repaired based on a damage tolerance criterion. Such traditional definitions are being expanded as requirements for high-reliability cost-effective NDE tests are increasing. In addition, NDE tests are changing as they become an integral part of the automated manufacturing process. This chapter is but a brief review of the more commonly used NDE methods. Those who require more detailed information should consult Refs. 1 through 15 at the end of the chapter.

The NDE methods reviewed here consist of the five classical techniques—penetrants, ultrasonic methods, radiography, magnetic particle tests, and eddy current methods. Additionally thermal inspection methods are briefly covered. Table 27.1 contains a very brief definition of each of these methods and the types of flaws for which they are used for both detection and quantification of size.

27.2 LIQUID PENETRANT

The liquid penetrant method is a nondestructive method for finding surface discontinuities in solid and essentially nonporous materials. The method uses a penetrating liquid, which is applied to the surface of parts. The liquid eventually enters the discontinuity if the part and discontinuity are clean. After the surface penetrant has been removed, penetrant from the discontinuity indicates the presence and location of the discontinuity.

27.2.1 Penetrant Classification

Both technical societies and the government (military specifications) have developed classification systems for penetrants. Society documents (typically ASTM E165) categorize penetrants into two methods (visible and fluorescent) and three types (water washable, post-emulsifiable, and solvent removable). In MIL-I-6866, the categories are reversed, that is, two types of dyes and three methods of surface penetrant removal. Amendment 4 of MIL-I-25135C subdivided Group VI post-emulsifiable penetrants into two groups of high and ultrahigh sensitivity. Penetrants, then, are classified by type of dye, rinse process, and sensitivity.

27.2.2 Penetrant Procedure

The first step in penetrant inspection or testing (PT) is to clean the part (Fig. 27.1). Many times this critical step is the most neglected phase of the inspection. Since PT only detects flaws that are open to the surface, the flaw and part surface must, prior to inspection, be free of dirt, grease, oil, water, chemicals, and other foreign materials. Typical cleaning procedures use vapor degreasers, ultrasonic cleaners, alkaline cleaners, or solvents.

After the surface is clean, penetrant is applied to the part by dipping, spraying, or brushing. Step 2 in Fig. 27.1 shows the penetrant on the part surface and in the flaw. In the case of tight surface openings, such as fatigue cracks, the penetrant is allowed to dwell for a minimum of 30 min to enhance the probability of complete flaw filling. A high-sensitivity penetrant is used for crack detection.

At the conclusion of the minimum dwell time, the surface penetrant is removed by one of three processes, depending on the characteristics of the inspection penetrant. Ideally, only the surface penetrant is removed and the penetrant in the flaw is left undisturbed (step 3 in Fig. 27.1).

The final step in a basic penetrant inspection is the application of a developer, wet or dry, to the part surface. The developer aids in the withdrawal of penetrant from the flaw and provides a suitable background for flaw detection (step 4 in Fig. 27.1). The part is then viewed with either black or white light after a suitable development time, which can be up to one-half of the penetrant dwell time. White light is used for visible penetrants while black light is used for fluorescent penetrants.

27.2.3 Surface Penetrant Removal

As mentioned earlier, surface penetrant removal falls into three basic categories: water washable, post-emulsifiable, and solvent removable. Water-washable penetrants contain an emulsifier that mixes with penetrant oils and permits direct removal by a water spray immediately after the penetrant dwell. Post-emulsifiable penetrants do not contain an emulsifier and cannot be removed by plain water. Removal is made possible by applying an emulsifier as a separate process step. This step converts the surface penetrant into an emulsifiable mixture that can be removed with a water spray. Solvent-removable

Table 27.1 Capabilities of the Common NDE Methods

Method	Typical Flaws Detected	Typical Application	Advantages	Disadvantages
Radiography	Voids, porosity, inclusions, and cracks	Castings, forgings, weldments, and structural assemblies	Detects internal flaws; useful on a wide variety of geometric shapes; portable; provides a permanent record	High cost; insensitive to thin laminar flaws, such as tight fatigue cracks and delaminations; potential health hazard
Liquid penetrants	Cracks, gouges, porosity, laps, and seams open to a surface	Castings, forgings, weldments, and components subject to fatigue or stress-corrosion cracking	Inexpensive; easy to apply; portable; easily interpreted	Flaw must be open to an accessible surface; level of detectability operator dependent
Eddy current testing	Cracks, and variations in alloy composition or heat treatment, wall thickness, dimensions	Tubing, local regions of sheet metal, alloy sorting, and coating thickness measurement	Moderate cost; readily automated; portable	Detects flaws which change conductivity of metals; shallow penetration; geometry sensitive
Magnetic particles	Cracks, laps, voids, porosity, and inclusions	Castings, forgings, and extrusions	Simple; inexpensive; detects shallow subsurface flaws as well as surface flaws	Useful for ferromagnetic materials only; surface preparation required; irrelevant indications often occur; operator dependent
Thermal testing	Voids or disbonds in both metallic and nonmetallic materials, location of hot or cold spots in thermally active assemblies	Laminated structures, honeycomb, and electronic circuit boards	Produces a thermal image that is easily interpreted	Difficult to control surface emissivity; poor discrimination
Ultrasonic testing	Cracks, voids, porosity, inclusions and delaminations	Composites, forgings, castings, and weldments	Excellent depth penetration; good sensitivity and resolution; can provide permanent record	Requires acoustic coupling to component; slow; interpretation is often difficult

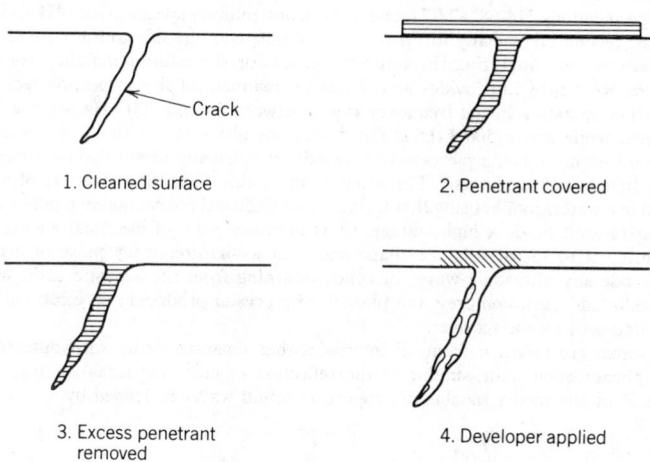

Fig. 27.1 Liquid penetrant examination procedure.

penetrants are usually standard post-emulsifiable penetrants that can be removed with solvents. This process is generally used in field applications where water-removal techniques are undesirable.

27.2.4 Reference Standards

Several types of reference standards are used to check the effectiveness of liquid-penetrant systems. One of the oldest and most often used methods involves chromium-cracked panels, which are available in sets containing fine, medium, and coarse cracks.[3] A typical set of these standards is shown in Fig. 27.2. These indications are from a high-sensitivity, post-emulsifiable, fluorescent-penetrant system. The panels are capable of classifying penetrant materials by sensitivity and identifying changes in the penetrant process.

27.2.5 Limitations

The major limitation of liquid-penetrant inspection is that it can only detect flaws that are open to the surface. Some other method must be used for detecting subsurface flaws.

Another factor that may inhibit the effectiveness of liquid-penetrant inspection is the surface roughness of the part being inspected. Extremely rough surfaces are likely to produce excessive background or false indications during inspection.

Although the liquid-penetrant method is used to inspect some porous parts, such as powder metallurgy parts, the process generally is not well suited for the inspection of porous materials, because the background penetrant from pores obscures flaw indications.

27.3 ULTRASONIC METHODS

Ultrasonic methods use sound waves to inspect the interior of components that possess a continuous path for the propagation of sound waves. This process is quite analogous to the use of sonar to detect

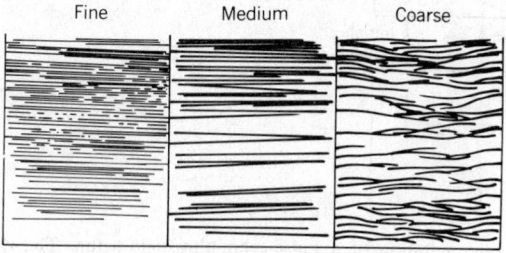

Fig. 27.2 Chromium-cracked panels.

schools of fish or map the ocean floor. Both government and industry have developed standards to regulate ultrasonic inspections. These include but are not limited to the American Society for Testing and Materials Specifications 214-68, 428-71, and 494-75, and military specification MIL-I-8950B. Acoustic and ultrasonic testing takes many forms from the simple coin tapping to the transmission of sonic waves into a material and analyzing the returning echos for the information they contain about its internal structure. Reference 13 provides an exhaustive treatment of this inspection technique.

Instrumentation operating in the frequency range between 20 and 500 kHz are usually defined as sonic instruments, while above 500 kHz is the domain of ultrasonic methods. In order to generate and receive the ultrasonic wave, a piezoelectric transducer is usually used to convert electrical signals to sound wave signals and vice versa. The usual form of this transducer consists of a piezoelectric crystal mounted in a waterproof housing that facilitates its electrical connection to a pulser (transmitter)/receiver. In the transmit mode a high-voltage, short-duration pulse of electrical energy is applied to the crystal, causing it to rapidly change shape and emit a high-frequency pulse of acoustic energy. In the receive mode any ultrasonic waves or echos returning from the acoustic path, which includes the coupling media and part, compress the piezoelectric crystal producing an electrical signal that is amplified and processed by the receiver.

Ultrasonic waves are reflected from all interfaces that separate media with different mechanical impedances, a phenomenon quite similar to the reflection of electrical signal in transmission lines. The impedance Z of any media capable of supporting sound waves is defined by

$$Z = \rho c$$

where ρ is the density of the media and c is its acoustic velocity.

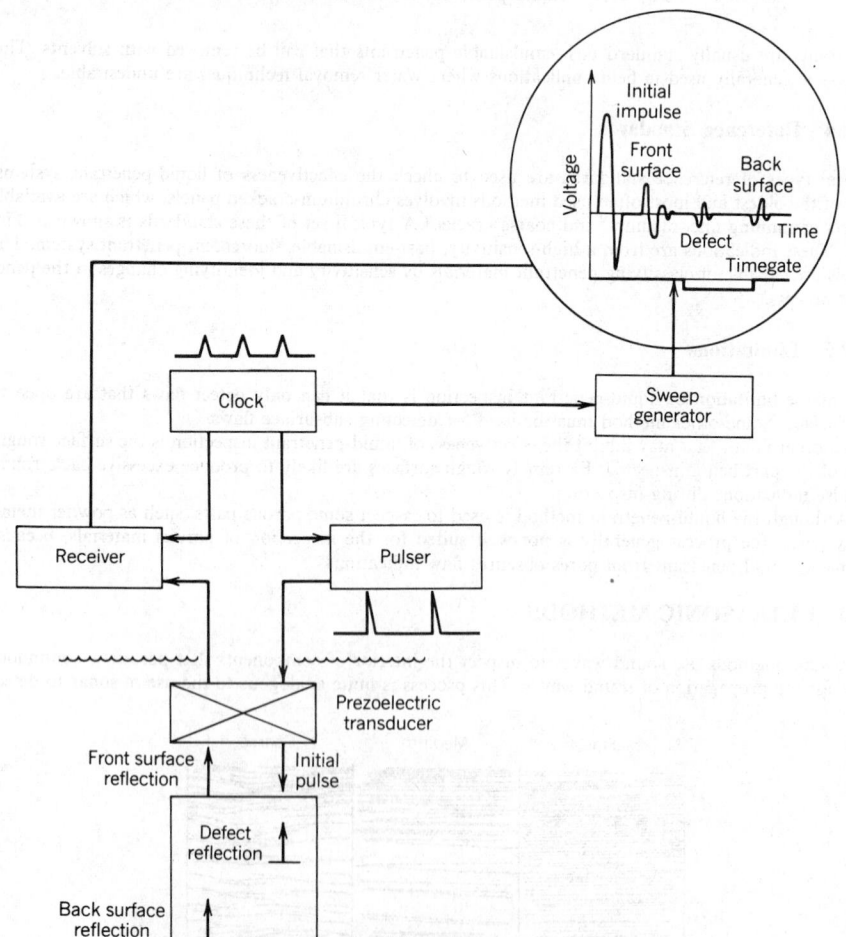

Fig. 27.3 Schematic representation of a pulse-echo ultrasonic setup. The display mode is called A-scan. The time gate is used to key alarm circuitry when a defect signal is detected in its time window.

Since very nearly all the acoustic energy incident on air/solid interfaces is reflected because of the large impedance mismatch of these two media, a coupling media with an impedance closer to that of the part is needed to transmit ultrasonic energy into the part under examination. A liquid couplant has obvious advantages for components with complex external geometries, and water is the couplant of choice for most inspection situations. The receiver, in addition to amplifying the returning echos, also time gates echos that return between the front surface and rear surfaces of the component. Thus, any unusually occurring echo can either be displayed separately or used to set off an alarm.

A schematic diagram of a typical ultrasonic pulse echo setup is shown in Fig. 27.3. This display of voltage amplitude versus time or depth (if acoustic velocity is known) is known as an A-scan. The defect echo is the result of a reflection from an interface between the front and back surface of the component. The portion of sound energy that is reflected from or transmitted through each interface is a function of the impedances of media on each side of the interface. The reflection coefficient R and transmission coefficient T for an acoustic wave incident from the normal on to an interface are

$$R = \frac{I_r}{I_i} = \left(\frac{Z_2 - Z_1}{Z_2 + Z_1}\right)^2$$

$$T = \frac{I_t}{I_i} = \frac{4Z_2 Z_1}{(Z_2 + Z_1)^2}$$

where I_i, I_r, and I_t are the incident, reflected, and transmitted acoustic field intensities, respectively. Z_1 is the acoustic impedance of the media from which the sound wave is incident and Z_2 is impedance into which the wave is transmitted. From these equations it is apparent that for a cracklike flaw containing air, $Z_1 = 45$ g/cm^2 · sec, located in, say, a piece of steel, $Z_2 = 46 \times 10^5$ g/cm^2 · sec, the reflection coefficient for the flaw is practically -1.0. The minus sign indicates a phase change of $180°$ for the reflected pulse (note that the defect signal in Fig. 27.3 is inverted or phase shifted $180°$ from the front surface signal).

Effectively no acoustic energy is transmitted across an air gap, necessitating the use of water as a coupling media in ultrasonic testing. The acoustic properties of several common materials are shown in Table 27.2. These data are useful for a number of simple, yet informative, calculations.

Thus far the discussion has involved only compression or longitudinal waves. This type of wave motion is the only type that can travel through fluids such as air and water. This wave motion is quite similar to the motion one would observe in a spring, or a Slinky toy, where the displacement and wave motion are collinear. Compressional waves are the type of sound used for human speech and so are quite familiar to everyone. However, several other types of acoustic wave motions are possible and two commonly used in ultrasonic nondestructive testing include, but are not limited to, shear waves and Rayleigh or surface waves. Shear waves have a wave motion that is analogous to

Table 27.2 Acoustic Properties

Material	Wave Speeds (10^5 cm/sec)		Density (g/cm^3)
	Longitudinal	Transverse	
Liquids			
Acetic acid	1.173		1.049
Acetone	1.192		0.792
Amyl acetate	1.168		0.879
Aniline	0.656		1.022
Benzene	1.326		0.879
Blood (horse)	1.571		—
Bromoform	0.928		2.890
n-Butyl alcohol	1.268		0.810
Caprylic acid	1.331		0.910
Carbon disulfide	1.158		1.263
Carbon tetrachloride	0.938		1.595
Chloroform	1.005		1.498
Formaldehyde (25°C)	1.587		0.815
Gasoline (34°C)	1.25		0.803
Glycerin	1.923		1.261
Kerosene (25°C)	1.315		0.82
Mercury	1.451		13.546
Methyl alcohol	1.123		0.796

Table 27.2 Acoustic Properties (Continued)

Material	Wave Speeds (10^5 cm/sec)		Density (g/cm³)
	Longitudinal	Transverse	
Liquids (Cont.)			
Oil			
Campor (25°C)	1.390	—	
Castor	1.500		0.969
Condenser	1.432		—
Olive (22°C)	1.440		0.918
SAE 20	1.74		0.87
Sperm (32°C)	1.411		—
Transformer	1.38		0.92
Oleic acid	1.333		0.873
n-Pentane	1.044		—
Silicon tetrachloride (30°C)	0.766		1.483
Toluene	1.328		0.867
Water (distilled)	1.482		1.00
m-Xylene	1.340		0.864
Solids—Metals			
Aluminum	6.37	3.11	2.70
Al 1100	6.31	3.08	2.71
Al 2014	6.37	3.07	2.80
Al 2024-T4	6.37	3.16	2.77
Al 2117-T4	6.50	3.12	2.80
Al 6061-T6	6.31	3.14	2.70
Bearing babbit	2.30	—	10.1
Beryllium	12.890	8.880	1.82
Bismuth	2.18	1.10	9.80
Brass (70% Cu–30% Zn)	4.37	2.10	8.50
Brass (naval)	4.43	2.12	8.42
Bronze (phosphor 5%)	3.53	2.23	8.86
Cadmium	2.780	1.50	8.64
Cerium	2.424	1.415	6.77
Chromium	6.608	4.005	7.2
Cobalt	5.88	3.10	8.9
Columbium	4.92	2.10	8.57
Constantan	5.177	2.625	8.88
Copper	4.759	2.325	8.93
Copper 110	4.70	2.26	8.9
Dysprosium	2.296	1.733	8.53
Erbium	2.064	1.807	9.06
Europium	1.931	1.237	5.17
Gadolinium	2.927	1.677	7.89
Germanium	5.18	3.10	5.47
Gold	3.24	1.20	19.32
Hafnium	2.84	—	13.3
Hastelloy C	5.84	2.90	8.94
Hastelloy X	5.79	2.74	8.23
Holmium	3.089	1.729	8.80
Indium	2.56	0.74	7.30
Invar	4.657	2.658	—
Lanthanum	2.362	1.486	6.16
Lead	2.160	0.700	11.34

Table 27.2 Acoustic Properties (*Continued*)

Material	Wave Speeds (10^5 cm/sec) Longitudinal	Transverse	Density (g/cm^3)
Solids—Metals (Cont.)			
Lead (94Pb–6Sb)	2.16	0.81	10.88
Lutetium	2.765	1.574	9.85
Magnesium	5.823	3.163	1.74
Magnesium AM-35	5.79	3.10	1.74
Magnesium FS-1	5.47	3.03	1.69
Magnesium J-1	5.67	3.01	1.70
Magnesium M1A	5.74	3.10	1.76
Magnesium O-1	5.80	3.04	1.82
Manganese	4.66	2.35	7.39
Manganin	4.66	2.35	8.40
Molybdenum	6.29	3.35	10.2
Neodynium	2.751	1.502	7.01
Nickel			
German Silver	4.76	2.16	8.40
Inconel	5.82	3.02	8.5
Inconel (X-750)	5.94	3.12	8.3
Inconel (wrought)	7.82	3.02	8.25
Monel	5.35	2.72	8.83
Monel (wrought)	6.02	2.72	8.83
Pure	5.63	2.96	8.88
Silver-nickel (18%)	4.62	2.32	8.75
Platinum	3.96	1.67	21.4
Potassium	2.47	1.22	0.862
Praseodymium	2.639	1.437	6.75
Samarium	2.875	1.618	7.48
Silver	3.60	1.59	10.5
Sodium	3.03	1.70	0.97
Steel			
1020	5.89	3.24	7.71
1095	5.90	3.19	7.80
4150, Rc14	5.86	2.79	7.84
4150, Rc18	5.88	3.18	7.82
4150, Rc43	5.87	3.20	7.81
4150, Rc64	5.83	2.77	7.80
4340	5.85	3.24	7.80
Iron (cast)	3.5–5.6	2.2–3.2	6.95–7.35
52100 steel			
Annealed	5.99	3.27	7.83
Hardened	5.89	3.20	7.8
D6 tool steel			
Annealed	6.14	3.31	7.7
Hardened	6.01	3.22	7.7
Stainless steel			
Type 302	5.66	3.12	7.9
Type 304L	5.64	3.07	7.9
Type 347	5.74	3.10	7.91
Type 410	5.39	2.99	7.67
Type 430	6.01	3.36	7.7
Tantalum	4.10	2.90	16.6
Thorium	2.94	1.56	11.3
Thulium	3.009	1.809	9.29
Tin	3.32	1.67	7.29
Titanium	6.07	3.11	4.50
Titanium (6A1, 4V)	6.18	3.29	4.50

Table 27.2 Acoustic Properties (*Continued*)

Material	Wave Speeds (10^5 cm/sec)		Density (g/cm³)
	Longitudinal	Transverse	
Solids—Metals (*Cont.*)			
Tungsten			
Annealed	5.221	2.887	19.25
Drawn	5.410	2.640	19.25
Uranium	3.37	1.98	18.7
Vanadium	6.023	2.774	6.03
Ytterbium	1.946	1.193	6.99
Yttrium	4.10	2.38	4.34
Zinc	4.187	2.421	7.10
Zirconium	4.65	2.25	6.48
Solids—Ceramics			
Aluminum oxide	10.84	6.36	3.98
Barium nitrate	4.12	2.28	3.24
Barium titanate	5.65	3.03	5.5
Bone (human tibia)	4.00	1.97	1.7–2.0
Cobalt oxide (CoO)	6.56	3.32	6.39
Concrete	4.25–5.25	—	2.60
Glass			
Crown	5.66	3.42	2.50
Flint	4.26	2.56	3.60
Lead	3.760	2.220	4.6
Plate	5.77	3.43	2.51
Pyrex	5.57	3.44	2.23
Soft	5.40	—	2.40
Granite	3.95	—	2.75
Graphite	4.21	2.03	2.25
Ice ($-16°C$)	3.83	1.92	0.94
Indium antimonide	3.59	1.91	—
Lead nitrate	3.28	1.47	4.53
Lithium fluoride	6.56	3.84	2.64
Magnesium oxide	9.32	5.76	3.58
Manganese oxide	6.68	3.59	5.37
Nickel oxide	6.60	3.68	6.79
Porcelain	5.34	3.12	2.41
Quartz			
Fused	5.57	3.52	2.60
Natural	5.73	—	2.65
Polycrystaline	5.75	3.72	2.65
Rock salt	4.60	2.71	2.17
Rutile (TiO_2)	8.72	4.44	4.26
Sandstone	2.920	1.840	2.2–2.4
Sapphire (*c*-axis)	11.91	7.66	3.97
Slate	4.50	—	2.6–3.3
Titanium carbide	8.27	5.16	5.15
Tourmaline (*Z*-cut)	7.54	—	3.10
Tungsten carbide	6.66	3.98	10.15
Ytterium iron garnet	7.29	4.41	5.17
Zinc oxide	6.00	2.84	5.61
Zinc sulfide	5.17	2.42	4.02

Table 27.2 Acoustic Properties (*Continued*)

Material	Wave Speeds (10^5 cm/sec) Longitudinal	Transverse	Density (g/cm³)
Solids—Polymers			
Acrylic resin	2.67	1.12	1.18
Bakelite	2.59		1.40
Butyl rubber	1.990		1.13
Cellulose acetate	2.45		1.30
Cork	0.5		0.2
Delrin (acetalhomopolymer) (0°C)	2.515		1.42
Ebonite	2.500		1.15
Lexan (polycarbonate) (0°C)	2.280		1.19
Neoprene	1.730		1.42
Nylon	2.680		—
Nylon 6,6	1.68		—
Paraffin	2.20		0.83
Perspex	2.70	1.33	1.29
Phenolic	1.42		1.34
Plexiglas			
UVA I	2.76		1.27
UVA II	2.73	1.43	1.18
Polyacrylonitrile–butadiene–styrene I	2.160	0.930	1.041
Polyacrylonitrile–butadiene–styrene II	2.020	0.810	1.022
Polybutadiene rubber	1.570		1.10
Polycaprolactam	2.700	1.120	1.146
Polycarboranesiloxane	1.450		1.041
Polydimethylsiloxane	1.020		1.045
Polyepoxide + glass spheres I	2.220	1.170	0.691
Polyepoxide + glass spheres II	2.400	1.280	0.718
Polyepoxide + glass spheres III	2.100	1.020	0.793
Polyepoxide + MPDA	2.820	1.230	1.205
Polyester + water	1.840	0.650	1.042
Polyethylene	2.67		1.10
Polyhexamethylene adipamide	2.710	1.120	1.147
Polymethacrylate	2.690	1.344	1.191
Polyoxymethylene	2.440	1.000	1.425
Polypropylene	2.650	1.300	0.913
Polystyrene	2.400	1.150	1.052
Polysulfane resin	2.297		1.24
Polytetrafluoroethylene (Teflon)	1.380		2.177
Polyvinylbutyral	2.350		1.107
Polyvinyl chloride	2.300		—
Polyvinylidene chloride	2.400		—
Polyvinylidene fluoride	1.930		1.779
Rubber			
India	1.48		0.90
Natural	1.550		1.12
Rubber/carbon (100/40)	1.680	—	—
Silicon rubber	0.948		1.48

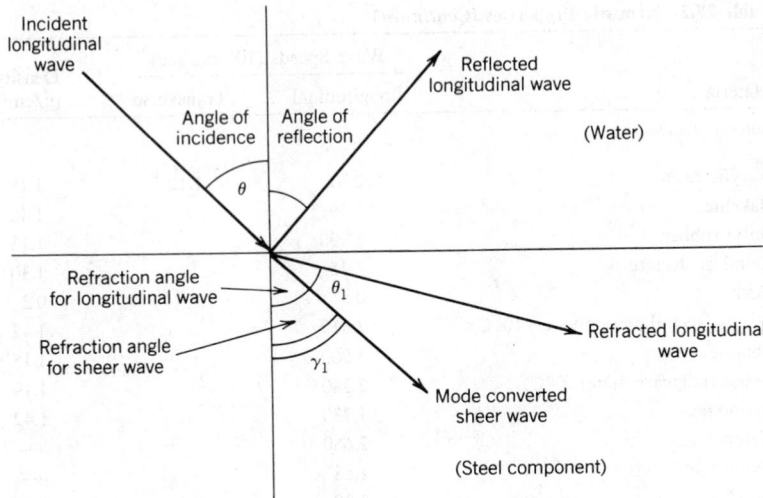

Fig. 27.4 Schematic diagram of incident, reflected, and refracted acoustic waves at a typical water/metal interface.

the motion one gets by snapping a rope, that is, the displacement of the rope is perpendicular to the direction of wave travel along the rope. The velocity of this wave motion is about one-half that of compressional waves and is only supported by solid media as shown in Table 27.2. Shear waves can be generated when a longitudinal wave is incident on a fluid/solid interface at angles of incidence other than 0°. Rayleigh or surface waves are used to detect surface and very near surface flaws. Their generation and utilization require special techniques and will not be covered here (see Ref. 13 for more details).

The direction of propagation of acoustic waves is described by the acoustic equivalent of Snell's law. Referring to Fig. 27.4, the directions of propagation are determined with the following equation:

$$\frac{\sin \theta}{c} = \frac{\sin \theta_1}{c_1} = \frac{\sin \gamma_1}{b_1}$$

where c is the longitudinal wave speed of the incident fluid media and c_1 and b_1 are the longitudinal and shear wave speeds of the steel, respectively. Since the water has a lower wave speed than either the compressional or shear wave speeds of the steel, the acoustic waves in the metal are refracted away from the normal. There will be critical angle at which point the refracted wave travels along the interface and does not enter the solid. Above this angle the acoustic wave will not propagate in the metal. A computer drawn curve is shown in Fig. 27.5 in which the normalized acoustic energy reflected and refracted at a water/steel interface are plotted as a function of angle of incidence. Note the longitudinal or first critical angle occurs at 14.5°. Likewise the shear or second critical angle occurs at 27°. If the angle of incidence is increased, above the first critical angle, then only a shear wave is generated in the metal and propagates at an angle of refraction determined by Snell's law. Angles of incidence above the second critical angle produce a complete reflection of the incident acoustic wave, that is, no acoustic energy enters the solid. However, between the two critical angles, only the shear wave is present in the material. In this region, shear wave testing is performed, which has two advantages. First, with only one type of wave present, the ambiguity that would exist concerning where a reflected wave originates is not present. Second, the lower wave speed of the shear wave means that about twice the time is available for resolving distances within the part under examination. These advantages mean shear wave inspection is often chosen for inspection of thin metallic structures such as those in aircraft.

Using only Snell's law and the reflection and transmission coefficient relationships a great deal of information can be deduced about any ultrasonic inspection situation in which the acoustic wave is incident at 90° to the surface. For other angles of incidence or for a component containing one or more thin layers, a computer program is used to analyze the acoustic interactions. For more complicated materials or structures such as fiber-reinforced composite any analytical predictions require computer computation for even relatively simple situations. In these cases it is often much simpler to determine empirically the inspection parameters.

Once the type of inspection has been determined, it remains only to choose the mode of presentation of the data. If the size of the flaw is small compared to the transducer, then the A-scan method can be chosen, as shown in Fig. 27.3. In this mode, the voltage output of the transducer is displayed

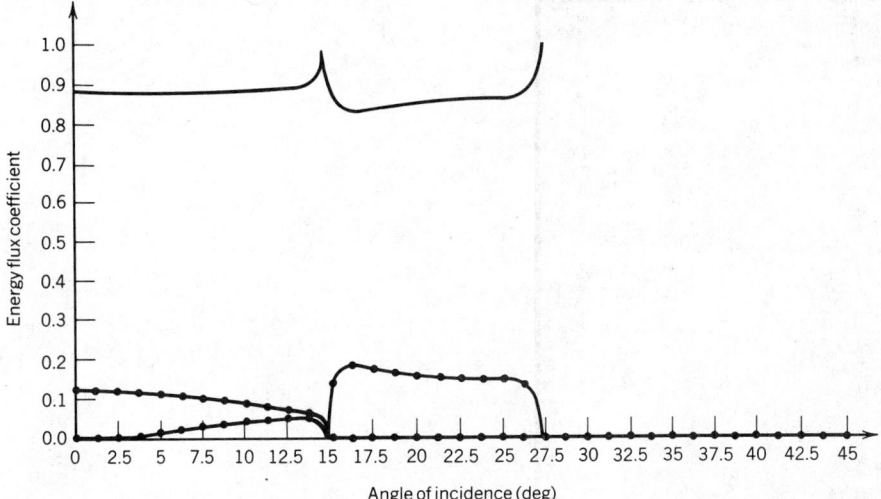

Fig. 27.5 Acoustic energy flux coefficients versus angle of incidence for a water/steel interface. Angles of incidence above the first critical angle, at 14.5°, do not permit the refraction of a longitudinal wave into the steel. Likewise, angles of incidence above the second critical angle, at 27°, do not permit the generation of shear waves in the steel. (Dotted line, transmitted waves; solid line, longitudinal wave; dashed line, shear wave.)

versus time or depth in the part. The size of the flaw is often inferred by comparing the size of the defect signal to a set of standard calibration blocks, which have varying sizes of flat bottom holes drilled in one end. For a specific transducer, the magnitude of the signal from the flat bottom hole is an increasing function of hole diameter. In this way an equivalent flat-bottom-hole size can be given for a flaw signal obtained from a defect in a component. The equivalent size is meaningful only for flaws that are nearly perpendicular to the ultrasonic signal path.

If the flaw size is larger than the transducer or if a number of flaws are expected, then the C-scan mode is usually selected. In this inspection mode, as shown in Fig. 27.6, the transducer is rastered back and forth across the part and when a flaw signal is detected between the front and back surface signals, then a line the size of intersection of the raster scan with the flaw is left blank on a piece of paper or CRT screen. In this manner a planar projection of each flaw is viewed and

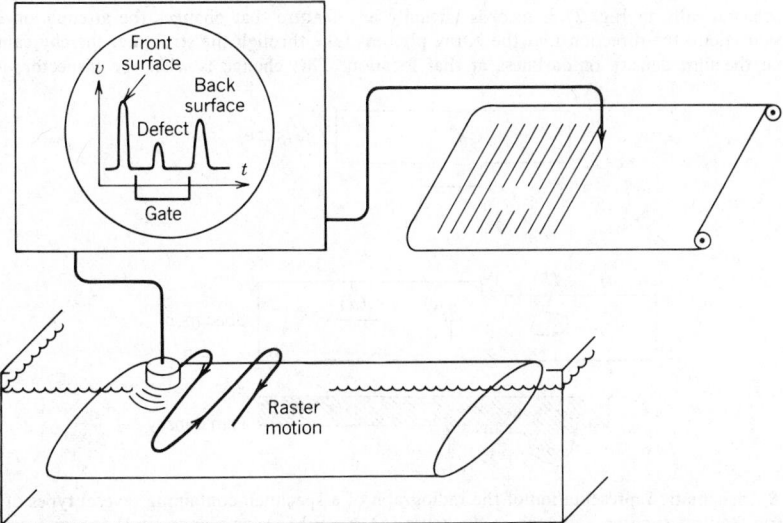

Fig. 27.6 Schematic representation of the C-scan mode of displaying defect information obtained from raster scanning a part. The electronic gate shown on the pulser/receiver module is used to change the printing of the image in such a manner that a projected image of the flow is displayed.

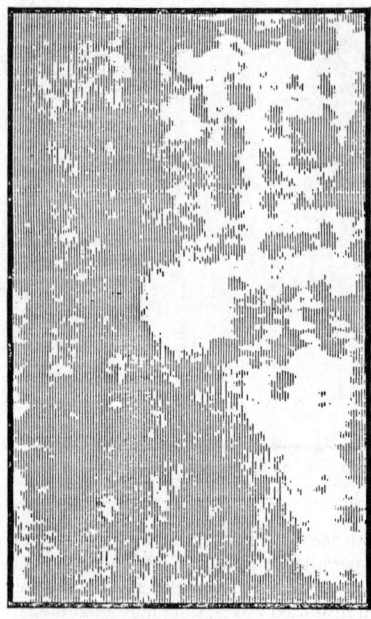

Fig. 27.7 A typical C-scan image of an adhesively bonded metal specimen. The circular indication in the center of the image is an intentionally implanted flaw. All other white areas represent unintentional porosity.

its positional relationships to others and to the part boundaries are easily assessed. Unfortunately in this mode the depth information for each flaw is frequently lost, therefore, this mode is mostly used for thin, layered aircraft structure.

Depending on the structural complexity and the attenuation of the signal by the material and electronic instrumentation, flaws as small as 0.015 in. can be reliably detected and quantified using this ultrasonic method. A typical C-scan printout of an adhesively bonded test panel is shown in Fig. 27.7. While the panel was fabricated with a Teflon void-simulating implant in the center, the numerous white areas indicate the presence of a great deal of porosity in the adhesive. For a much more extensive treatment of this inspection technique, the reader is referred to Ref. 13.

27.4 RADIOGRAPHY

Radiography is an NDE method in which the projected X-ray attenuation for various paths through a specimen are recorded as a two-dimensional image on some type of recording media. This process, shown schematically in Fig. 27.8, records virtually any feature that changes the attenuation of the X-ray beam along the direction that the X-ray photons take through the structure, thereby causing a change in the film density or darkness at that location. This change is what the inspector uses to

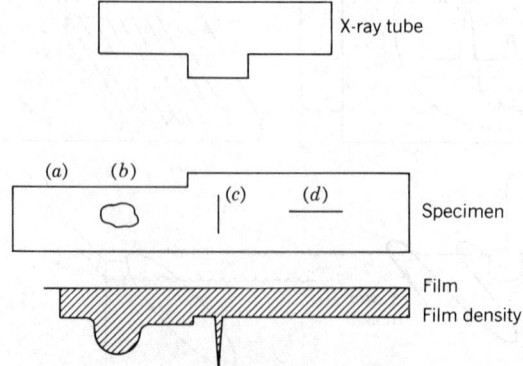

Fig. 27.8 Schematic representation of the radiograph of a specimen containing several types of flaws. Geometric features such as a change in thickness (a), a void (b), or planar circle (c), are visible on the film because they alter the attenuation of the X-ray beam as it traverses the specimen. Flaw d does not alter the attenuation and is, therefore, not visible.

detect internal anomalies. The inspector is greatly aided in detecting and quantifying specific flaw conditions by the geometric relationships in the image. For example, the inspector can often determine a great deal about a flaw, such characteristics as its size, location in the part, and its probable nature. It should be noted in Fig. 27.8 that only flaws which change the attenuation of the X-ray beam on passage through the part are recorded. For example, a delamination which is not the result of missing material and is located perpendicular to the path of the X ray through the part does not alter the attenuation of the X-ray beam and therefore is not detectable with this method. Therefore, small cracklike defects must be placed such that a portion of the X-ray beam passes along their length in order for them to be visible on the film.

X radiation can be produced via a number of processes. The most common form of producing X rays is with an electron tube in which a beam of energetic electrons impact a metal target and are rapidly decelerated producing a wide band of X radiation referred to as bremsstrahlung or breaking radiation. Higher-energy electrons produce shorter-wavelength or more energetic X rays. The relationship between the shortest-wavelength X radiation produced and the highest voltage applied to the tube is given by

$$\lambda_{min} = \frac{12,336}{voltage}$$

where λ is the shortest wavelength of the X radiation produced in the tube in angstroms. Figure 27.9 shows schematically the relationship between the intensity of the radiation and its wavelength, for various accelerating voltages, for a tube with a tungsten target. High-energy radiation is generally used to inspect dense materials, for example, welds in steels, and large structures, for example, large rocket motors, both of which significantly attenuate the X-ray beam. For these purposes equipment capable of producing several million electron volts for accelerating the electron beam are commonly used. On the other hand, the inspection of thin, low-attenuating materials such as composites require the use of very-low-energy X rays in the range between 10 and 50 kV.

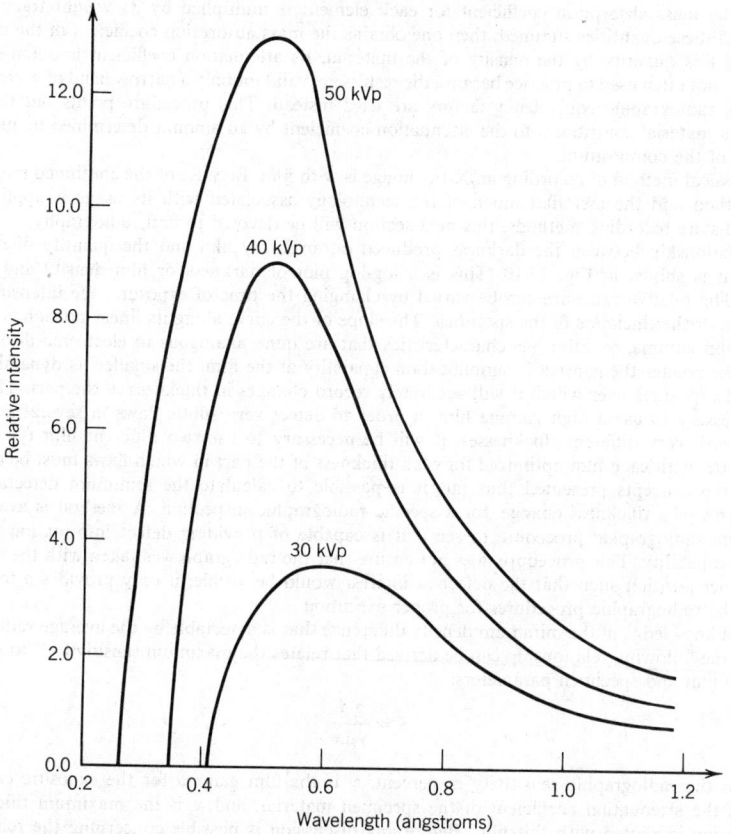

Fig. 27.9 The relative intensity of the white radiation from a tungsten target as functions of wavelength and maximum accelerating voltage (kilovolts-peak).

X radiation can also be obtained from the decay of radioactive sources. In this case the radiation is usually referred to as gamma radiation. These types of X-ray sources have several distinct characteristics that distinguish them from the aforementioned tubes. First, their radiation is either monochromatic or nearly so, that is, the radiation spectrum contains only one or two dominant characteristic energies. Second, the energies of most gamma ray sources is in the million volt range, making such a source ideal for inspecting highly attenuating structures. Third, their small size permits them to be used in situations where an X-ray tube would be unusable. Fourth, since the gamma-ray source is continually decaying, adjustments to the exposure time must be made in order to achieve consistent results over time. Finally, the operator must constantly be aware that the source is always on and therefore a potential safety hazard. In use, gamma radiography differs little from standard practice with X-ray tubes, therefore no further distinction between the two will be mentioned. For more information on the usage of a particular type of source the reader is referred to any one of the standard NDE references.

Since the interpretation of a radiograph requires some understanding of the fundamentals of the absorption of the X-ray beam, the basic relationship governing this phenomenon is given:

$$I = I_0 \exp(-\mu x)$$

where I and I_0 are the transmitted and incident X-ray beam intensities, respectively, μ is the attenuation coefficient of the material in reciprocal cm, and x is the thickness of the specimen, in cm, through which the X-ray beam passes. Since the attenuation coefficient is a function of both the composition of the specimen and the wavelength of the X ray, it is necessary to calculate or measure μ for each inspection situation. It is possible to calculate the attenuation coefficient of a material for a specific energy X-ray beam with the following procedure involving the use of the mass absorption coefficient, μ_m, as defined below. Mass absorption coefficients for most elements are readily available[1] for a variety of X-ray energies:

$$\mu_m = \mu/\rho$$

where μ is the attenuation coefficient of a specific element, in reciprocal cm and ρ is its density in g/cm^3. If the mass absorption coefficient for each element is multiplied by its weight fraction of a material and these quantities summed, then one obtains the mass absorption coefficient of the material. Multiplying this quantity by the density of the material, its attenuation coefficient is obtained. This procedure is not often used in practice because the results are valid for only a narrow band of wavelengths. In practice, radiographic equivalency factors are used instead. This procedure points out that each element in a material contributes to the attenuation coefficient by an amount determined by its atomic percentage of the composition.

The classical method of recording an X-ray image is with film. Because of the continued importance of this method and the fact that much of the technology associated with its usage is applicable to newer solid-state recording methods, this next section will be devoted to film radiography.

The relationship between the darkness produced on an X-ray film and the quantity of radiation falling on it is shown in Fig. 27.10. This is a log-log plot of darkness or film density and relative exposure. The relative exposure can be varied by changing the time of exposure, the intensity of the X-ray beam or the thickness of the specimen. The slope of the curve along its linear portion is referred to as the film gamma, γ. Film has characteristics that are quite analogous to electronic devices. For example, the greater the gamma or amplification capability of the film, the smaller its dynamic range, or range of exposures over which it will accurately record changes in thickness of the part. Therefore, if it is necessary to use a high gamma film in order to detect very subtle flaws in several areas of a specimen with very different thicknesses, it will be necessary to use two different film types in the same cassette, with each film optimized for each thickness of the part in which flaws must be detected.

Using the concepts presented thus far, it is possible to calculate the minimum detectable flaw size, in terms of a thickness change, for a specific radiographic inspection. A method is available to test the film radiographic procedure to see if it is capable of providing defect information with the calculated capability. This procedure does not ensure that the radiograph was taken with the specimen in the proper position such that the defect of interest would be visible; it only provides a method of checking the radiographic procedures for proper execution.

Using a knowledge of the minimum density difference that is detectable by the average radiographic inspector, the following relationship can be derived that relates the maximum sensitivity S to thickness changes to film and specimen parameters:

$$S = \frac{2.3}{\gamma \mu x}$$

where S is the radiographic sensitivity in percent, γ is the film gamma for the exposure conditions used, μ is the attenuation coefficient of the specimen material, and x is the maximum thickness of the part being inspected with this film. Before any discussion is possible concerning the relationship between the radiographic sensitivity and the minimum detectable thickness, that is, size of a correctly oriented flaw, it is necessary to discuss the measurement device for sensitivity, the penetrameter.

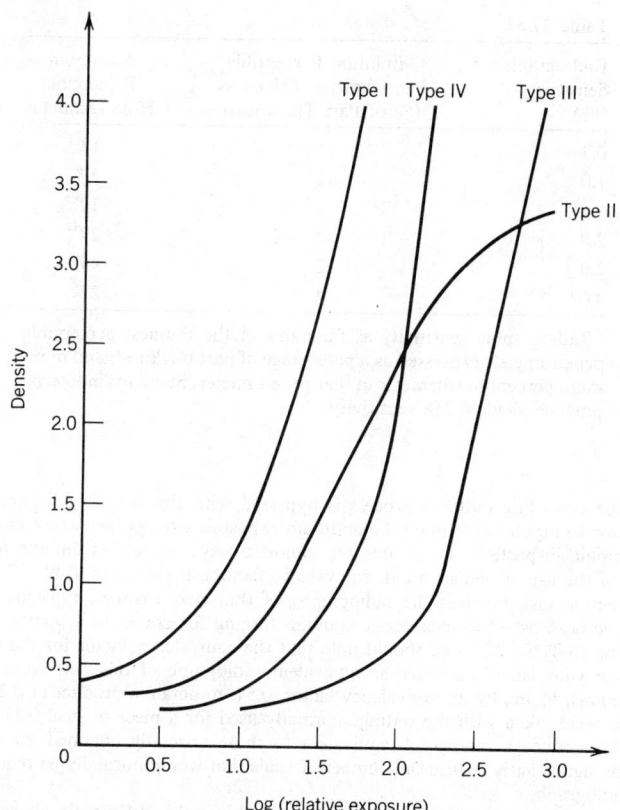

Fig. 27.10 The relationship between the darkness or density of an X-ray film and the amount of radiation incident upon it. The response of four film types is shown.

One example of a penetrameter is shown schematically in Fig. 27.11. This particular type was chosen because it is easily related to the following discussion of radiographic sensitivity. This device is quite simply a thin strip of metal in which three holes of varying sizes are punched and upon which an identifying mark is placed so that the penetrameter can be easily identified in a radiograph. The penetrameter is made of the same material as the specimen being inspected and has a thickness either 1%, 2%, or 4% of its maximum thickness. The holes in the penetrameter have diameters which are 1, 2, and 4 times its thickness. The sensitivity achieved for each radiographic can be easily determined by noting the smallest hole just visible in the thinnest penetrameter on a film. By referring to Table 27.3 the radiographic sensitivity achieved for that particular radiographic can be determined. Therefore, using the formula for sensitivity and then noting if that level was achieved in practice the radiographic process can be quantitatively controlled from a processing standpoint. While this procedure does not offer any guarantee of detecting flaws, it is quite useful in controlling the mechanics of the inspection.

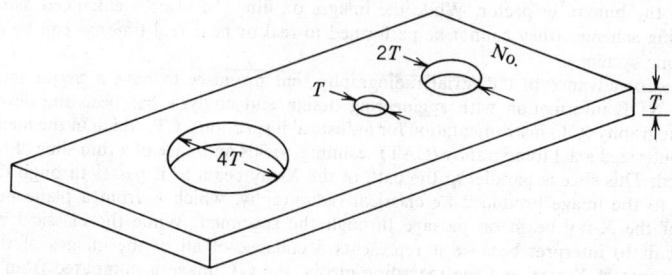

Fig. 27.11 Schematic representation of one type of penetrameter. T is the thickness of the penetrameter.

Table 27.3[a]

Radiographic Sensitivity (%)	Minimum Perceptible Penetrameter Thickness (% of Part Thickness)	Minimum Perceptible Hole Diameter
0.7	1	$1T$
1.0	1	$2T$
1.4	2	$1T$
2.0	2	$2T$
2.8	2	$4T$
4.0	4	$2T$

[a]Radiographic sensitivity as functions of the thinnest perceptible penetrameter, expressed as a percentage of part thickness, and minimum perceptible diameter in that penetrameter. Standard industrial practice yields a 2% sensitivity.

In practice the preceding outlined process is bypassed with the aid of tables and graphs which allow the inspector to rapidly determine the optimum exposure settings for almost any specimen that requires radiographic inspection. These aids are almost always available from the film supplier. A simple example of the use of one such aid, equivalence factors, is shown in Table 27.4. Suppose that one's usual inspection task involves the radiography of thin steel castings, typically ¼–1 in. thick, and that you have been asked to examine a titanium forging for cracks in a section which is ½ in. thick. By referring to Table 27.4 one should note that the equivalency factor for the titanium is 0.54 for the 220 kV in your lab. To achieve an equivalent radiograph of this new specimen you would multiply its thickness, ½ in., by its equivalency factor, 0.54, to obtain a product of 0.27 in. Therefore, if the radiograph were taken with the settings normally used for a piece of steel 0.27 or about ¼ in. thick, you would produce a radiograph equivalent to those normally obtained for steel specimens. These simple aids significantly reduce the number of trials that would normally be required to produce an acceptable radiograph.

Other variables of the radiographic process may also be easily and rapidly changed with the aid of tables, graphs, and nomograms, which are usually provided by film manufacturer free of charge. For more information in this regard the reader is referred to the commercial literature.

While film radiography represents the bulk of the radiographic NDE performed at this time, new methods of both recording the data and analyzing it are rapidly becoming commonplace. For example, filmless radiography (FR) uses solid-state detectors or television detection and image processing methods instead of film to record the image. These methods have several advantages along with some disadvantages. For example, FR permits the viewing of a radiographic image while the specimen is being moved. This often permits the detection of flaws that would normally be missed in conventional film radiography because of the limited number of views or exposures usually taken—remember that the X-ray beam must pass along a cracklike flaw in order to be detectable. Additionally, the motion of some flaws enhances their detectability because they present the inspector with a different image as a function of time than he or she is accustomed to viewing. The price to be paid for these advantages is the lower resolution of the FR system when compared to film. Typical resolution capabilities of an FR system are in the range from 4 to perhaps 12 line pairs/mm, while film has resolution capabilities in the range from 10 to 100 line pairs/mm. This means some types of very fine flaws cannot be detected with FR and the inspector must resort to film. Another fact that must be considered when comparing these two methods is the possibility of digital image processing that is easily implemented with FR systems. These methods can be of enormous benefit in enhancing flaws that might otherwise be invisible to the human inspector. While the images on film can also be enhanced using the same image-processing schemes, they cannot be performed in real or near real time, as can be done with a totally electronic system.

Another recent advance in industrial radiography that promises to have a major impact on the interaction of NDE information with engineering design and analysis has been the development of computed tomography (CT) instrumentation for industrial inspections. CT, which in the medical context is called computerized axial tomography (CAT) scanning, yields an image of a thin slice of the specimen being examined. This slice is parallel to the path of the X-ray beam as it passes through the specimen as contrasted to the image produced by classical radiography, which is from a plane perpendicular to the path of the X-ray beam on passage through the specimen. While the classical radiographic image is difficult to interpret because it represents a collapse of all of the images of the specimen between the source of X rays and the recording media, the CT image is computed from transmitted X-ray intensity data and does not contain image information from planes outside the thin slice produced.

Table 27.4 Approximate Radiograph Equivalence Factors

Metal	Energy Level									
	100 kV	150 kV	220 kV	250 kV	400 kV	1 MeV	2 MeV	4–25 MeV	^{192}Ir	^{60}Co
Magnesium	0.05	0.05	0.08							
Aluminum	0.08	0.12	0.18						0.35	0.35
Aluminum alloy	0.10	0.14	0.18						0.35	0.35
Titanium		0.54	0.54		0.71	0.9	0.9	0.9	0.9	0.9
Iron/all steels	1.0	1.0	1.0	1.0	1.0	1.0	1.0	1.0	1.0	1.0
Copper	1.5	1.6	1.4	1.4	1.4	1.1	1.1	1.2	1.1	1.1
Zinc		1.4	1.3		1.3			1.2	1.1	1.0
Brass		1.4	1.3		1.3	1.2	1.1	1.0	1.1	1.0
Inconel X		1.4	1.3		1.3	1.3	1.3	1.3	1.3	1.3
Monel	1.7		1.2							
Zirconium	2.4	2.3	2.0	1.7	1.5	1.0	1.0	1.0	1.2	1.0
Lead	14.0	14.0	12.0			5.0	2.5	2.7	4.0	2.3
Hafnium			14.0	12.0	9.0	3.0				
Uranium			20.0	16.0	12.0	4.0		3.9	12.6	3.4

This comparison is best explained with actual images from these two modalities. Figure 27.12 shows a conventional radiograph of a small rocket motor with its typical confusion of multiple images, and it also shows a CT slice through the same motor. The contrast in available information to the inspector is rather striking. First, the detectability of a defect is independent of its position in the image. This is certainly not the case with the classical radiograph where the defect detectability decreases significantly with depth in the rocket motor because the defect represents a smaller change in the attenuation of the X-ray beam as its depth increases. While this situation is particularly severe in cylindrical-shaped objects, it is also common in other geometric shapes. Second, the defect detectability is very nearly independent of its orientation. This again is clearly not the case with classical radiography. The extent to which CT will alter radiographic NDE is only now being realized. New applications where inspection for anomalous conditions was not possible a short time ago are being discovered every day. For example, with the computed digital image of CT it is possible to search for various flaw conditions using computer analysis and relieve the inspector of much of the tedium of examining acceptable structure for the odd flaw. In addition, it is possible to link the digital CT image with engineering analysis software such as a finite element program and examine precisely how the flaws present within the cross section imaged will affect such parameters as the stress distribution and heat flow. With little effort one could analyze the full three-dimensional performance of many engineering structures. Thus, this relatively new NDE capability promises to provide the engineering community with quantitative information about flaw-sensitive structures and to change completely how defect information is used.

27.5 EDDY CURRENT INSPECTION

Eddy current methods are used to inspect electrically conducting components for flaw conditions such as surface connected and near-surface cracks and voids; heat-treatment condition; external dimensions and wall thickness of thin-walled tubing; and thickness of nonconducting coatings on a metal

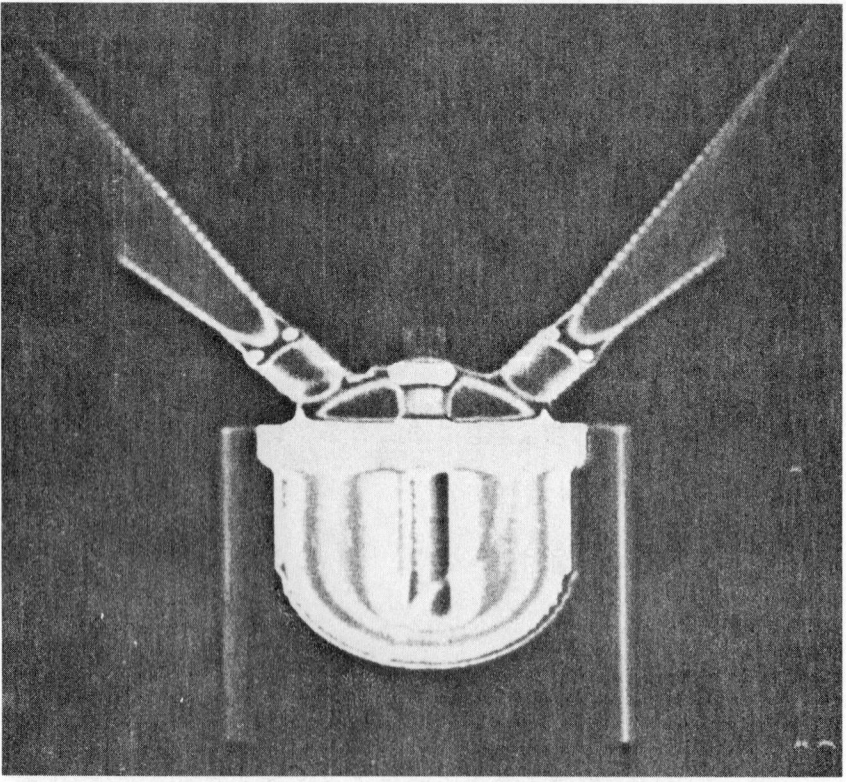

(a)

Fig. 27.12 (a) Standard radiographic image of a small solid-fuel rocket motor. (b) A tomographic slice taken through the body of the rocket motor. Note the increased detail in this image; most, if not all, the flaws clearly visible in the solid fuel would not be detectable in standard radiographic images.

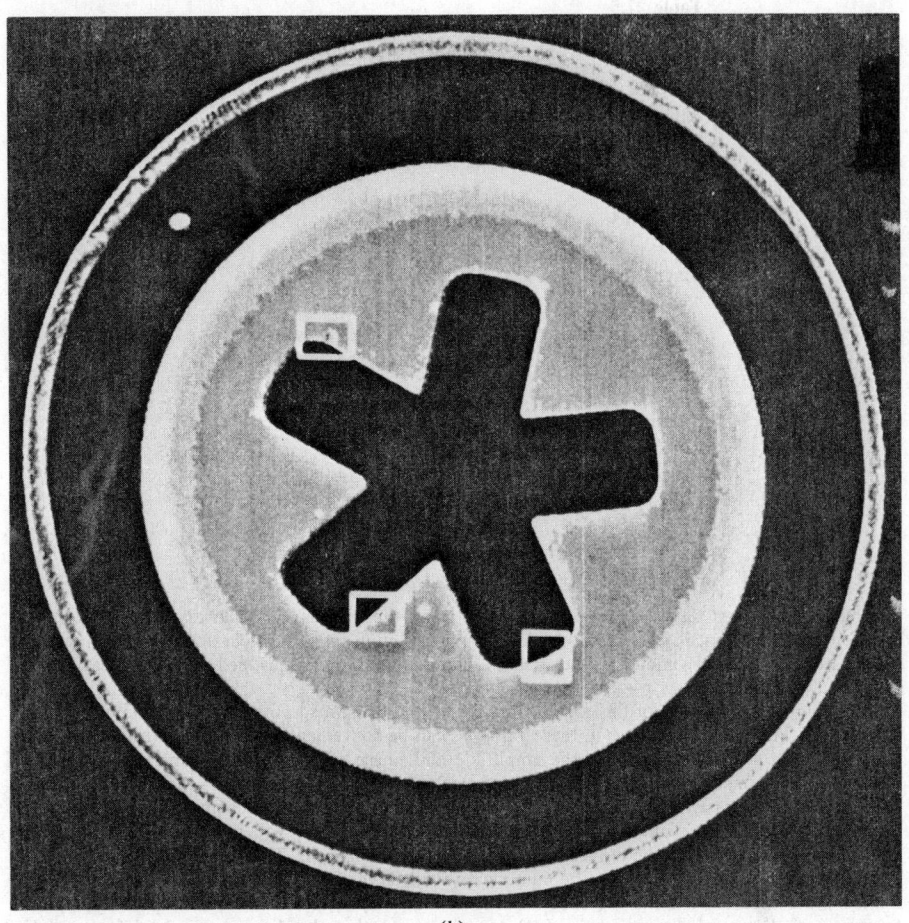

(b)

Fig. 27.12 (*Continued*)

substrate. Quite often several of these conditions can be monitored simultaneously if instrumentation capable of operating at multiple frequencies is used and the results are analyzed using modern computer algorithms.

This NDE method is based on the principle that eddy currents are induced in a conducting material when a coil with an alternating current is placed in close proximity to the surface of a conducting material. The induced currents produce a magnetic field that opposes the field of the inducing coil in accordance with Lenz's law. The eddy currents circulate in the part in closed, continuous paths, and their magnitude depends on the following variables: the magnitude and frequency of the current in the inducing coil; the coil's shape and position relative to the surface of the part; the electrical conductivity, magnetic permeability, and shape of the part; and the presence of discontinuities or inhomogeneities within the material.

Since alternating currents are necessary to perform this type of inspection, information from the inspection is limited to the near-surface region by the skin effect. The density of the eddy current field falls off exponentially with depth and diminishes to a value of about 37% of the at-surface value at a depth referred to as the standard depth of penetration (SDP). The SDP in meters can be calculated with the simple formula

$$SDP = 1/(\pi f \sigma \mu)^{1/2}$$

where f is the test frequency in Hz, σ is the test material's electrical conductivity in mho/m (see Table 27.5 for a list of conductivities for several common materials), and μ is its permeability in H/m. This latter quantity is the product of the relative permeability of the specimen, 1.0 for nonmagnetic materials, and the permeability of free space which is 4×10 H/m. While the SDP is used to give an indication of the depth from which useful information can be obtained, the choice of the independent

Table 27.5

Metal	Conductivity—(mho/m $\times$ 10^{-7})
Copper	5.7
Aluminum	3.4
Aluminum alloys	
6061-T6	2.4
7075-T6	1.89
2024-T4	1.92
Lead	0.46
Magnesium	2.17
70-30 Brass	1.6
Phosphor bronze	0.63
Zircaloy	0.14
Titanium	0.18
Titanium alloy	
GA1-4V	0.058
Stainless steel	0.14
Inconel 600	0.087
Hastelloy X	0.083

variables in most test situations is usually made using the impedance plane diagram suggested by Förster.[14] It is theoretically possible to calculate the optimum inspection parameters from numerical codes based on Maxwell's equations, but this is a laborious task that is justified only in critical applications.

The complex impedance of the coil used in the inspection of a specimen is a function of a number of variables. The effect of changes in these variables can be conveniently displayed with the impedance diagram which shows variations in the amplitude and phase of the coil impedance as functions of the dependent variables—specimen conductivity, thickness, and distance between the coil and specimen or lift-off. For the case of an encircling coil on a solid cylinder, shown schematically in Fig. 27.13, the complex impedance plane is displayed in Fig. 27.14. The reader will note that the ordinate and abscissa are normalized by the inductive reactance of the empty coil. This eliminates the effect of the geometry of the coil and specimen. The numerical values on the large curve, which are called reference numbers, are used to combine the effects of the conductivity and size of the test specimen, and the frequency of the measurement into a single parameter, so that the diagram is useful for most test conditions. The reference numbers shown on the outermost curve are obtained with the following relationship, for nonmagnetic materials:

$$\text{reference number} = r\sqrt{2\pi f \mu \sigma}$$

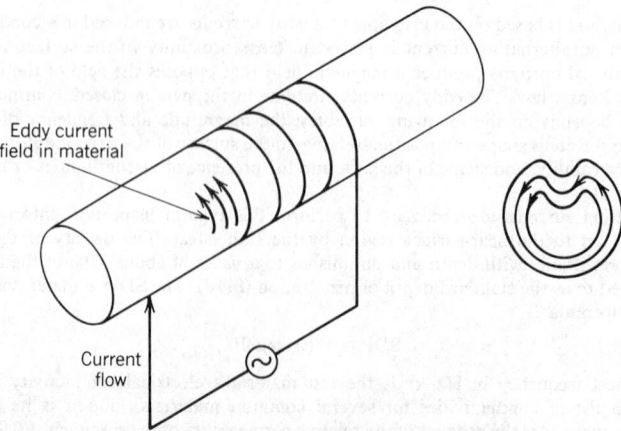

Fig. 27.13 Schematic representation of simplified eddy current inspection of a solid cylinder. Also shown are the eddy current flow paths in a solid cylinder with a surface flaw.

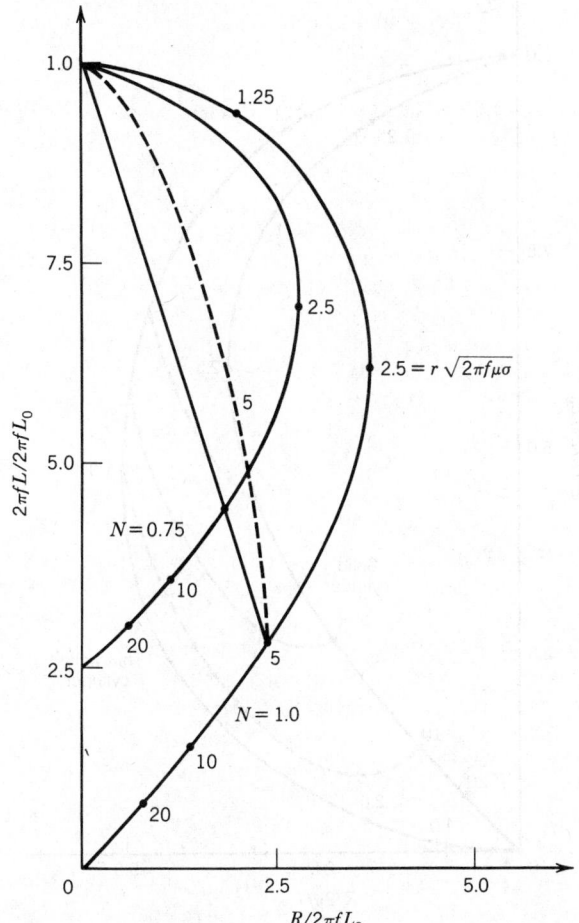

Fig. 27.14 Normalized impedance diagram for a long encircling coil on solid, nonferromagnetic cylinders. For $N = 1$, the coil and cylinder have the same diameter; while for $N = 0.75$, the coil is approximately 1.155 times larger than the cylinder.

where r is the radius of the bar in m, f is the frequency of the test in Hz, μ is the magnetic permeability of free space $(4 \times 10^{-7}$ H/m), and σ is the conductivity of the specimen in mho/m. The outer curve in Fig. 27.14 is useful only for the case that the coil is the same size as the solid cylinder under test. For those cases where the coil is larger than the test specimen, the usual case, a coil-filling factor must be calculated for this test situation and the appropriate point on a new curve located. This is quite easily accomplished using the fill factor N of the coil, which is defined as

$$N = \left[\frac{\text{diam (specimen)}}{\text{diam (coil)}} \right]^2$$

In Fig. 27.14 the curve for a specimen/coil inspection geometry with a fill factor of 0.75 is shown. Note that the reference numbers on the curves representing the different fill factors can be determined by projecting a straight line from the point 1.0 on the ordinate to the reference number of interest, as is shown for the reference number 5.0. Both the fill factor and the reference number change when the size of the specimen or coil changes. Assume that a reference number of 3.0 is appropriate to a specific test with $N = 1.0$; if the coil diameter is changed so that the fill factor is now 0.75, then the new reference number would be equal to $\sqrt{0.75} \times 3.0 = 4.33$. The inspection geometry discussed thus far has been for a solid cylinder. The other geometry of general interest is the thin-walled tube—remember that the skin effect limits the thickness that can be effectively inspected.

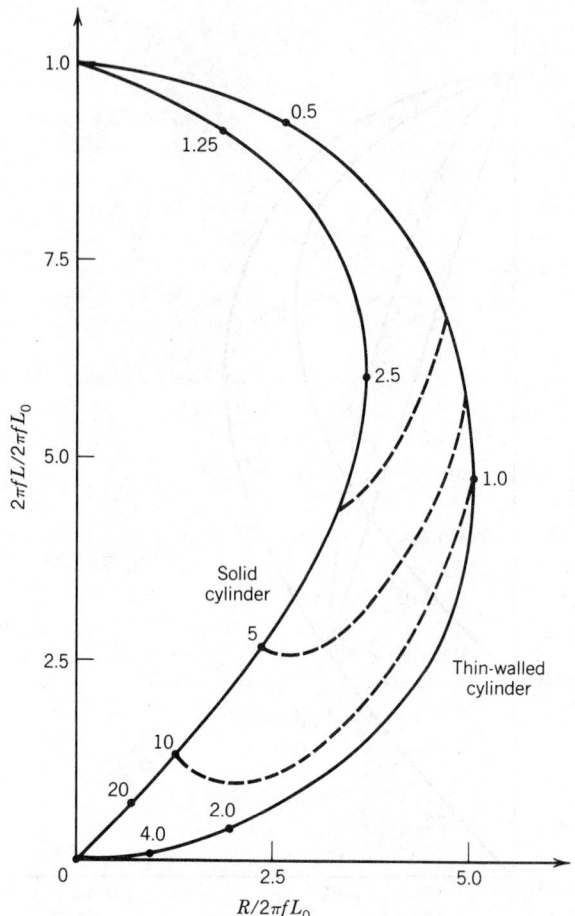

Fig. 27.15 Normalized impedence diagram for a long encircling coil on solid- and thin-walled cylinders of varying thicknesses. The dashed curves represent the effect of varying wall thickness. The numerical values given along each curve are for $r\sqrt{2\pi f\sigma\mu}$.

For a infinitely thin-walled tube the impedance plane is shown in Fig. 27.15. Also included in this figure is the curve for a solid cylinder. The curve segments that connect these two extreme cases are for thin-walled cylinders of varying thicknesses. The semicircular curve for the thin cylinder is used in the same manner as described above for the solid cylinder.

In most inspection situations the only independent variables are frequency and lift-off. While it is also possible to change the coil shape and measurement configuration, the reader is referred to the literature for a discussion of the effects of these more complex variables. Therefore, the relationships discussed thus far would seem of little practical usefulness. However, if a small section of Fig. 27.15 is expanded as shown in Fig. 27.16, then their utility will become apparent. In this figure changes in thickness, lift-off, or external dimension of a tube or cylinder, and conductivity are represented by small vectors. Since these vectors all point in different directions, the signals that they represent have different phases. Thus, instrumentation with phase detection circuitry can differentiate among changes in these different conditions. Changes in conductivity can arise from several different conditions, the most important of which are due to material inhomogeneity, for example, aluminum alloys can have quite different conductivities depending on their heat treatment, or the changes due to the presence of cracks or voids. These flaws decrease the apparent conductivity of the specimen because the eddy currents must travel a longer distance to complete their circuit within the material. Thus, many different flaw conditions can be rapidly detected. There are other inspection situations that cannot be covered in this brief description of this important NDE tool. These include the inspection of ferromagnetic alloys, plate and sheet stock, and the measurement of film thicknesses on metal substrates. For a

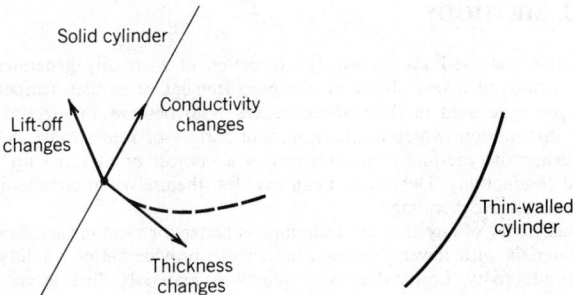

Fig. 27.16 The effects of various changes in inspection conditions on the local signal changes in the impedance plane of Fig. 27.15. Phase differentiation instrumentation can easily distinguish between these conditions.

treatment of these and other special applications of eddy current NDE, the reader is referred to Refs. 14 and 15.

There are numerous methods of making eddy current NDE measurements. Two of the more common generic methods are shown schematically in Fig. 27.17. In the absolute coil arrangement very accurate measurements can be made with the differences between the two samples easily amplified. In the differential coil method it is the differences between the two variables at two slightly different locations that are measured. For this arrangement slightly varying changes in dimensions and conductivity are not sensed, while singularities such as cracks or voids are highlighted even in the presence of other slowly changing variables. Since the specific electronic circuitry used to accomplish this task can vary dramatically depending on the specific inspection situation, this aspect of eddy current testing will not be covered here, and the reader is referred to the current NDE and instrumentation literature listed in the references.

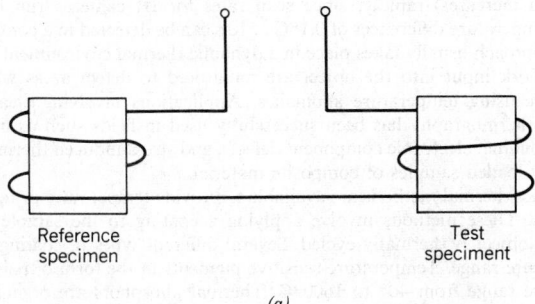

Reference   Test
specimen   specimen

(a)

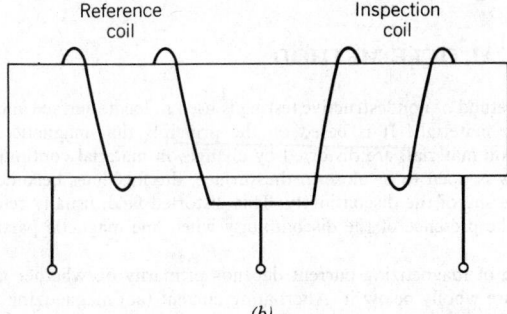

Reference   Inspection
coil   coil

(b)

Fig. 27.17 *(a)* Absolute and *(b)* differential configurations of the coils and specimens used in eddy current inspection.

27.6   THERMAL METHODS

Thermal nondestructive test methods involve the detection of thermally generated electromagnetic radiation from the surface of a test object or the measurement of surface temperature by another method. The basic principle used in thermal techniques is to observe and record anomalies in the surface temperature distribution, which are indicators of surface or near-surface defects in an object. Properties that influence the thermal characteristics of an object or material are the specific heat, density, and thermal conductivity. Defects that can manifest themselves in changes in these properties include porosity, cracks, and debonding.

In general, the sensitivity of any thermal technique is better for near-surface flaws than for deeply embedded flaws. Materials with lower thermal conductivity provide better resolution than materials with high thermal conductivity. Lower-thermal-conductivity materials allow larger thermal gradients to be obtained, and the thermal signature of a defect is a temperature gradient in the material.

Infrared thermography detects the thermal radiation which is given off by any object whose temperature is above absolute zero. For objects at moderate temperatures, the thermal radiation is predominately infrared and measurements are concentrated in the 8–14 m wavelength region. IR cameras are available that view large areas by scanning the field of view over a single, liquid-nitrogen-cooled detector using rotating prisms to obtain an image similar to a video image. Since the IR image can be stored in digital form, further image processing is easily accomplished and permanent records can be produced.

For many applications, a thermal image which reveals the relative temperature across an object is sufficient to detect flaws and defects. However, if absolute temperatures are required (to compare with analytical predictions), the IR camera must be calibrated to account for the surface emissivity of the test object.

IR thermography's ability to detect flaws is significantly affected by the type of flaw and its orientation with respect to the camera. The flaw must perturb the surface temperature of the object in order for the camera to find it. The important parameter is the projected area of the flaw in the field of view of the camera. Subsurface flaws such as cracks parallel to the surface of the object, porosity, and debonding of a surface layer are easily detected. Cracks that are perpendicular to the object surface, even if they have dangerous lengths, can present an infinitesimal area to the camera and are very difficult to find using thermography.

Where other NDE methods can be used, they generally provide better spatial resolution than thermal methods. The greatest advantage to thermography is that it is a noncontact, remote technique requiring only line-of-sight access to one side of an object. Large areas can be viewed (by sacrificing resolution as the area increases) rapidly, since scan rates for IR cameras run between 16 and 30 frames per second. Temperature differences of 0.1°C or less can be detected in a controlled environment.

The best testing approach usually takes place in a dynamic thermal environment where the transient effects of a heat or work input into the object are monitored to detect areas where different heat-transfer rates occur, causing temperature anomalies. Applications involving steady-state conditions are more limited. IR thermography has been successfully used in fields such as medical diagnosis of tumors, detecting debonding, electronic component defects, and stress-induced thermal gradients around defects in dynamically loaded samples of composite material.

A number of contact thermal methods are available to provide temperature or temperature-distribution data for surfaces. These methods involve applying a coating to the sample and observing the coating change as the object is thermally cycled. Several different types of coatings are available that cover a wide temperature range. Temperature-sensitive pigments in the form of paints have been made to cover a temperature range from 40° to 1600°C. Thermal phosphors are organic components that emit visible light when exposed to UV radiation. (The amount of visible light is an inverse function of the temperature.) Thermochromic compounds and liquid crystals change color at a specific surface temperature. The advantages are the simplicity of the test and the relatively low cost if small areas are involved.

27.7   MAGNETIC PARTICLE METHOD

The magnetic particle method of nondestructive testing is used to locate surface and subsurface discontinuities in ferromagnetic materials. It is based on the principle that magnetic lines of force, when present in a ferromagnetic material, are distorted by changes in material continuity, caused by cracks or inclusions. If the flaw is open to or close to the surface, the flux lines become distorted and some escape the surface at the site of the discontinuity. This distorted field, usually referred to as a leakage field, is used to reveal the presence of the discontinuity when fine magnetic particles are attracted to it.

Selection of the type of magnetizing current depends primarily on whether the defects are either open to the surface or are wholly below it. Alternating current (ac) magnetizing currents are best for the detection of surface discontinuities, because the current is concentrated in the near-surface region of the part. Direct current (dc) magnetizing currents are best suited for subsurface discontinuities, because of the current's much deeper penetration of the part. While dc can be obtained from batteries or dc generators, it is usually produced by half-wave or full-wave rectification of commercial power. Rectified current is classified as half-wave direct current (HWDC) or full-wave direct current (FWDC). Alternating current fields are usually obtained from conventional power mains, but it is supplied to the part at reduced voltage, for reasons of safety and the high-current requirements of the magnetizing process.

Two general types of magnetic particles are available. One type of particle is low-carbon steel with high permeability and low retentivity, which is used dry and consists of different sizes and shapes to respond to both weak and strong leakage fields. The other type of particle, in general use, consists of extremely fine particles of magnetic iron oxide that are suspended in a liquid (either a petroleum distillate or water). These particles are smaller and have a lower permeability than the dry particles. Their small mass permits them to be held by the weak leakage fields at very fine surface cracks. Magnetic particles are available in several colors to increase their contrast against different surfaces. Dry powders are typically gray, red, yellow, and black, while wet particles are usually red, black, or fluorescent.

Because the field is always stronger while the magnetizing current is on, the continuous magnetizing method is generally preferred, and if the part has low retentivity, the continuous method must be used. In the continuous method, the current can be applied in short pulses, typically 0.5 sec. The magnetic particles are applied to the surface during this interval and are free to move to the site of the leakage fields. In this case, the use of liquid-suspended fluorescent particles yields the most sensitive detection method. For field inspections, the magnetizing current is usually continuously on during the test to give time for the powder to migrate to the defect site.

In the residual method, the particles are applied after the magnetizing current is removed. This method is particularly suited for production inspection of multiple parts.

The choice of direction of the magnetizing field within the part involves the nature of the flaw and its direction with respect to the surface and the major axis of the part. In circular magnetization, the field runs circumferentially around the part. It is induced into the part by passing current through it between two contacting electrodes. Since only flaws perpendicular to the magnetizing lines are readily detectable, circular magnetization is used to detect flaws that are parallel or less than 45° to the surface of the part. Longitudinal magnetization is usually produced by placing the part in a coil, creates a field running lengthwise through the part, and is used to detect transverse discontinuities to the axis of the part.

The surface of the part to be examined should be essentially clean, dry, and free of contaminants such as oil, grease, loose rust, loose sand, loose scale, lint, thick paint, welding flux, and weld splatter. Cleaning of the test part may be accomplished by detergents, organic solvents, or mechanical means.

Portable and stationary equipment are available. Selection of the specific type depends on the nature and location of testing. Portable equipment is available in lightweight units (35–90 lb), which can be readily taken to the inspection site. Generally, these units operate off of 115, 230, or 460 V ac and supply current outputs of 750–1500 A in half-wave or ac.

The magnetic particle flaw indication is very similar in appearance to the penetrant indication, therefore, no separate figures of this inspection mode are presented. Magnetic particle inspection will work only on ferromagnetic materials. Other limitations include:

1. The magnetic field must be in a direction that will intercept the principal plane of the flaw.
2. Demagnetization after inspection is often necessary.
3. Exceedingly large currents, which are required for very large parts, can locally overheat or burn the finish of a part.

REFERENCES

1. R. C. McMaster (Ed.), *Nondestructive Testing Handbook*, Ronald Press, New York, 1959.
2. R. C. McMaster (Ed.), *Nondestructive Testing Handbook*, 2nd ed., The American Society for Nondestructive Testing, Columbus, OH, 1983, Vols. 1 and 2.
3. W. J. McGonnagle, *Nondestructive Testing*, Gordon and Breach, New York, 1961.
4. *Metals Handbook*, American Society for Metals, Metals Park, OH, 1976, Vol. 11.
5. *Annual Book of ASTM Standards*, Part II, *Metallography and Nondestructive Testing*, American Society of Testing and Materials, Philadelphia, PA.
6. *Nondestructive Testing: A Survey*, NASA SP-5113, NASA Technology Utilization Office, Washington, DC, 1973.
7. *Materials Evaluation*, American Society for Nondestructive Testing, Columbus, OH.
8. *British Journal of Nondestructive Testing*, British Institute of Nondestructive Testing, Northampton, UK.
9. *NDT International*, IPC Business Press Ltd, Sussex, UK.
10. *The Soviet Journal of Nondestructive Testing*, translated by Consultants Bureau, New York.

11. *Journal of Nondestructive Evaluation,* Plenum Press, New York.

12. *Nondestructive Testing Communications,* Gordon and Breach Science Publishers, London, UK.

13. J. Krautkramer and H. Krautkramer, *Ultrasonic Testing of Materials,* 3rd ed., Springer-Verlag, New York, 1983.

14. F. Förster, "Theoretische und experimentelle Grundlagen der zerstörungsfreien Werkstoffprufung mit Wirbelstromverfahren. I. Das Tastpulverfahren," *Z. Metallk.,* **43,** 163–171 (1952).

15. H. L. Libby, *Introduction to Electromagnetic Nondestructive Test Methods,* Wiley-Interscience, New York, 1971.

PART 3
MANUFACTURING ENGINEERING

CHAPTER 28
THE MANUFACTURING SYSTEM

DANIEL T. KOENIG

General Electric Co.
Bridgeport, Connecticut

The best ideas are useless if they cannot be implemented. Likewise, the most elegant design is worthless if it cannot be produced. The engineer in Manufacturing lives with these self-evident truths. This breed of engineer is responsible for conceiving and implementing ways of producing in physical form the ideas formalized in design. This is at least as creative an exercise, and sometimes even more so, than the process of defining the design itself. The engineer in Manufacturing, always working with the severe constraints of the design, costs, and time must create a product. He or she is the end of the entire chain of the engineering activity. His or her output—his or her ability to manufacture the design—is the endpoint of the engineering effort. In this section, the various manufacturing processes and control techniques will be defined, explained, and illustrated. Before starting the detailed discussions of the various management methods and techniques used to machine, mold, fabricate, etc., raw materials into finished products, it is necessary to describe the Manufacturing System, to explain the process of Manufacturing, in an overview sense, so that we can begin to understand its nature and how it is controlled.

28.1 THE SYSTEMS APPROACH TO MANUFACTURING

A modern factory is an organized way of building a product, be it one of a kind or one of a multi-thousand order. The purpose of the factory organization is to perform the work of Manufacturing in as efficient manner as possible. Therefore, the use of systems to achieve coordination of the various aspects of Manufacturing is as old as the industrial revolution itself.

A scheme or a system cannot evolve in a vacuum. The scheme for a factory to produce a product, therefore, has to be in harmony with the nature of the product. It should be obvious that the nature of a factory making wearing apparel would be quite different from a factory making large electric motors. The machinery would be far different. The method of work would be different. The way materials are moved about the factory would be different. The two environments would be dissimilar. In short, the observable schemes for production in these two factories are different. We can say the specific system used for manufacturing a product is dictated by its nature, which, in turn, specifies the types of equipment and processes that are used.

Let us now look beyond what is easily observable. For our two dissimilar factories we know that the processes are completely alien to each other. We also know that each system of manufacturing had to be designed to fit the respective product. But so far we have not discussed the underlying concepts of what is really happening in the manufacturing process.

Let us take a very simple look at that. First, product specification has to be accepted. Then a method of producing has to be conceived. Based on the method, equipment and processes to make the products have to be assembled and put in place. A schedule for producing must be determined, again based on the method. Raw materials will be ordered and scheduled for delivery as needed. Materials will be dispatched to the starting operation and work commences on the factory floor. Supervisors will monitor conformance to methods and schedules. Quality assurance personnel measure interim results for compliance with product specifications. Materials will be moved from one workstation to the next workstation. Finally, the completed product will be tested and shipped to the customer. The manager will then record cost and quality results as compared to the schedules. The cycle will repeat itself by going on to the next production plan.

Take a look at what has been described. It is a scheme, or a system, for manufacturing a product. Is it for the wearing apparel or the large electric motors? It can be for either. Take the starting operations. In a clothing factory they would perhaps be cutting cloth to basic sizes using patterns. Similarly, in a large electric motor factory, a starting operation could be the winding of copper wire in bundles to make motor coils. Therefore, the scheme described is a basic manufacturing system, repeated thousands of times. In fact, it is "the" manufacturing system. There are many modifications to it, many variances in use, but essentially the steps described are the steps used to make all types of products, ranging from black iron anvils, through automobiles, to computers.

For easy reference purposes the steps of the Manufacturing Systems are as follows:

1. Obtain product specification.
2. Design a production method, including design and purchase of equipment and processes, if required.
3. Schedule to produce.
4. Purchase raw materials in accordance with the schedule.
5. Produce in the factory.
6. Monitor results, technical compliance, and costs control.
7. Ship the completed product to the customer.

No matter how we vary them, these seven steps are the Manufacturing System. We engineers in Manufacturing have made tremendous strides in optimizing each of the seven steps and tremendous strides undoubtedly will be achieved in the future. But, no matter how we look at it, we always

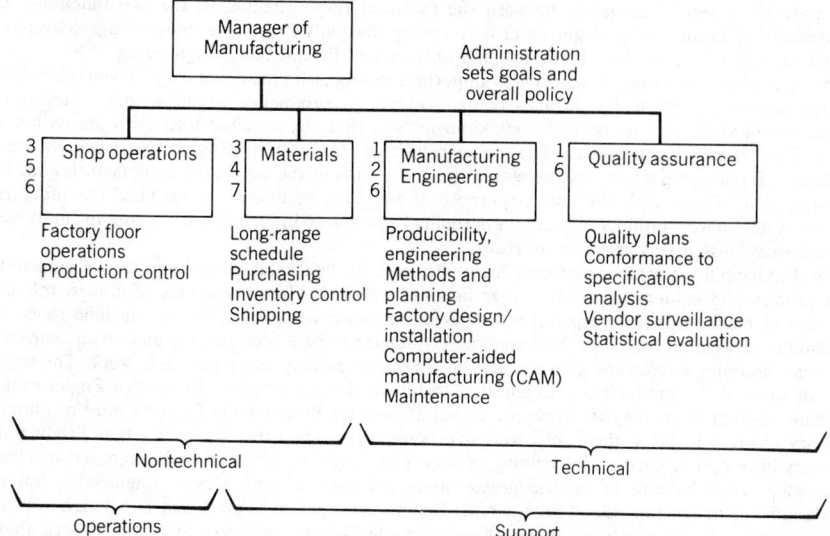

Fig. 28.1 Typical Manufacturing organization.

come back to these seven basic steps and all engineers in Manufacturing are continuously involved with optimizing these steps.

This leads to the next item of explanation, what engineers actually do in Manufacturing. For overview purposes we must first understand the nature of the Manufacturing organizations, then we can begin to see how engineers fit into it.

The seven steps of the Manufacturing System dictate a basic split in the Manufacturing organization: Operations functions and Support functions. Of course, there are times when the lines of demarcation become blurred, but mostly they remain sufficiently distinct. The definitions of the distinctions are readily apparent.

Operations is that function that actually makes the product. To put it another way, Operations adds value to the product during the steps from raw materials to finished manufactured product.

Support means providing everything necessary to Operations personnel so that they can make a product. Support can include supplying methods of manufacture, ensuring that raw materials are purchased and arrive on time, or determining via inspections that a product is being manufactured in accordance with specifications.

A typical Manufacturing organization with assigned responsibilities is shown in Fig. 28.1. The numbers to the left of each subfunction box of Fig. 28.1 represent the steps each of the Manufacturing subfunctions are involved in with respect to the seven steps of the Manufacturing System. As can be seen, some steps are shared among different subfunctions. For example, the Materials subfunction is responsible for long-range scheduling, that is, a master schedule of when each order is to be produced, while the Shop Operations subfunction is responsible for the detailed workstation by workstation schedule during the time a specific shop order is being produced.

Note in Fig. 28.1 the basic split within Manufacturing between Operations and Support. Noting the assigned responsibilities between all the subfunctions, it is evident that all subfunctions are involved in activities that support the Shop Operations task of making the product.

The final pair of brackets of Fig. 28.1 indicates the technical, nontechnical split in the Manufacturing organization and shows where the engineers are employed within the typical Manufacturing organization. This is not to say that there are no engineers in the Materials and Shop Operations subfunctions; in fact, there usually are. The brackets indicate which subfunctions are primarily technically based and which are commercially based (but dependent on technology to do their job). Many industries prefer engineers in all management positions within Manufacturing due to the nature of their products. It is often advantageous to have technically trained people in the so-called nontechnical areas of Manufacturing for the obvious reason that they relate fully to the products, while their nontechnical counterparts may not. Within Manufacturing Engineering and Quality Assurance, however, the work is all technically oriented and only engineers would suffice.

28.2 PRODUCIBILITY ENGINEERING

Producibility Engineering is that first intersection between the Design Engineering and Manufacturing functions. Producibility Engineering resides in the Manufacturing Engineering subfunction and, as such, provides a very close liaison between the technical responsibilities of the two functions. The responsibility of Producibility Engineering is to ensure that only "workable designs" are delivered to Manufacturing to produce. This is the one and only task of Producibility Engineering.

The concept of a workable design is very important to successful manufacturing. A workable design is simply one that is producible in a particular factory. By producible we mean that a factory can produce the product with its normal methodologies and that any required tolerances are within the means of the factory's capabilities. The actual results of Producibility Engineering work should be that Design Engineering either designs within the constraints of the associated manufacturing facility, or, failing that, design with the known purpose of requiring additional or modified manufacturing facilities. A design for a product that cannot normally be produced in their factory creates the unpleasant surprises that cause delays and cost overruns.

The Producibility Engineer performs his or her task by first cataloging the ranges of capabilities of all process and equipment in his or her factory, including lifting capacities of cranes, tolerance capacities of machine tools, temperature ranges of furnaces, work capacities of machine tools, and availabilities of welding processes. This catalogue, intended to be a wide ranging and all encompassing document, becomes a reference guide to Design Engineers as they carry out their work. The second phase of work of the Producibility Engineer is to review designs proposed by Design Engineering to determine whether or not they are workable. If they are not, the Producibility Engineer must recommend necessary changes to bring them into workable boundaries. The process requires that Producibility Engineers have both a good understanding of their company's products and a background in Design Engineering. They have to be knowledgeable about the goals of both Design Engineering function and Manufacturing function. Good Producibility Engineers are perhaps more attuned to design workability than Design Engineers, because they have to guide Design Engineers to produce work that is suitable for the Manufacturing function.

As can be seen from Fig. 28.2, Producibility Engineering is the sole Engineering activity within Manufacturing that precedes the seven steps of the Manufacturing System. This implies that Producibility Engineers are hybrids, for example, they perform their tasks both outside and within the Manufacturing System. All the other Engineering tasks within Manufacturing are involved with the seven operations of the Manufacturing System. Therefore, Producibility Engineering is a true ambassadorial function.

The Producibility Engineer, always keeping in mind the need for workable designs, must be skilled

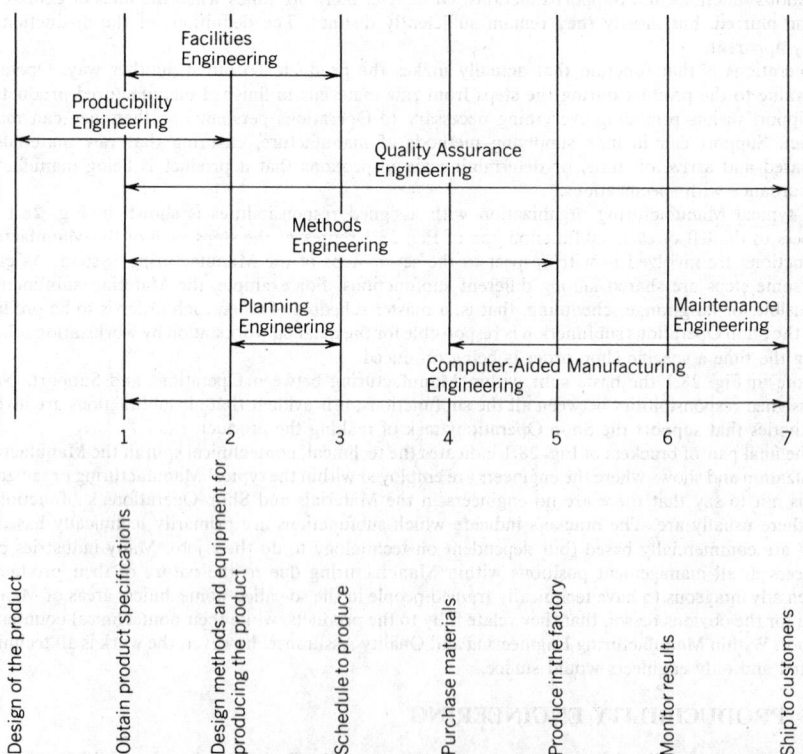

Fig. 28.2 The Manufacturing System: influence areas—engineers in Manufacturing.

in design processes and must continuously guide and persuade his Design Engineering colleagues in producing specifications acceptable to Manufacturing.

28.3 QUALITY ASSURANCE

Quality Assurance is that unique subfunction within Manufacturing that acts as a policing agent for a design, by ensuring that manufactured products comply with specifications. Quality Assurance always starts with the premise that the processes being monitored throughout the Manufacturing System's seven steps are capable of producing the design called for. Quality Assurance then develops the procedures to ensure that proper inspections and certifications are performed parallel to the various steps in the manufacturing procedures.

Inspection and certification plans are traditionally called Quality Plans. They usually closely follow the sequence of events Manufacturing Engineering defines as the procedure to make a product. A typical Quality Plan will require specific actions to be taken for all process steps. These actions could range from certification of pedigree of a raw material's physical and chemical properties, to Operators' affidavits that proper procedures were followed during manufacturing, and an inspector's verifying dimensions and tests. The exact nature of the actions called for in the Quality Plan will be based on the actual product specification that the Manufacturing function is working on.

Before preparing a Quality Plan the Quality Assurance Engineer has to be thoroughly familiar with the workable design, the customer's product specification (or in the case where products are not made for identified customers, industry standards or the company's own standards), and the procedure which Manufacturing Engineering will dictate for producing the part. The Quality Plan, therefore, is a checklist to be followed by all subfunctions of Manufacturing during production.

Theoretically, then, if the Quality Plan is followed explicitly, the finished product will meet all design specifications.

Unfortunately no plan is perfect. One hundred percent conformance with designated manufacturing procedure cannot be guaranteed. So there is need for monitoring of results to ensure that a certain threshold of conformance is maintained. This is usually achieved through a series of Inspection/Operator data reports required by the Quality Plan. The exact nature of such reports depends on the type of product. For example, if a factory is fabricating a pressure vessel, data reports probably will include dimensional data to ensure compliance with size requirements and assembly procedures, physical and chemical analyses to verify design stress calculations, and records of nondestructive testing of welds as well as actual temperatures and cycle times of heat treatments. These are only a few of the normal data reports one could require for pressure vessels, but these examples point out the potential for complexity of the Quality Plan.

The Quality Plan also has provisions for reporting off-standard conditions as they occur. These are called defect reports. They describe in quantitative terms off-standard conditions and frequently recommend corrections. The Quality Assurance Engineer, who is familiar with design specifications, then determines whether or not a recommended correction is sufficient or if another solution to the problem is required. Only rarely will a problem be submitted to Design Engineering for resolution. Quality Assurance Engineers should possess the necessary educational background to be able to converse effectively with Design Engineers.

Quality Assurance Engineers use both required normal processing data and defect report data to make statistical inferences. The degree to which inferences can be made depends on the type of product being made, the size of each production lot, and the degree of control maintained and desired during the manufacturing process. Ranges of control vary widely from industry to industry. Obviously, the key factors are cost and customer needs. For example, a pencil manufacturer would have less need for tight control than a semiconductor manufacturer. For the former, tight control would needlessly increase cost, while the latter needs tight control to ensure product acceptance.

Much has been written about statistical quality control. Putting it in perspective, it is a tool for controlling costs. It would be unwise to forego opportunities to use all the statistical techniques that can apply to specific products. Likewise, it is foolish to try to force-fit statistical techniques when they result in disruptions and difficulties in data interpretation. Therefore, the prudent Quality Assurance Engineer will continuously test for the appropriateness of statistical concepts. To give exact guidelines as to where statistical concepts should be used is virtually impossible. In general terms, Quality Assurance Engineers should attempt to use statistical concepts for long production runs where tooling is rigidly controlled. Statistical concepts are usually not successful where job shop activities are employed, that is, where general purpose machines are making parts in small lot size (under 50 per order).

The final item the Quality Assurance Engineer has to be concerned with is the matter of sufficiency: What is good enough? We often hear the phrase zero defects. Exactly what does that mean? Does it mean that a finished part must be in exact accordance with the corresponding drawing? Or must it be only within the design tolerance band? The answer is neither. Zero defects means that a part will work, that it will function as designed. Sufficiency, then, is the decision making process that a Quality Assurance Engineer goes through to determine whether or not a part will work regardless of whether or not a defect report has been issued against the part or family of parts. Here, again, the Quality

Assurance Engineer uses his or her skills as a statistician (evaluating tolerance buildups on the plus or minus side) and as a pragmatic engineer (knowing a product well enough to determine whether or not an off-standard condition will interfere with proper performance). Sufficiency skills are mandatory for successful Quality Assurance Engineers. More important, they must be in place within Manufacturing, otherwise a company could be at hazard of absorbing considerable additional manufacturing costs through rejection of actually usable products. At the other extreme is the danger of allowing actual deficient parts to be sent to customers. As can be seen, the "art of engineering" still exists for Quality Assurance Engineers as they monitor manufacturing processes (step 6 of the Manufacturing System).

28.4   THE BASIC TECHNIQUES OF PRODUCTION CONTROL AND PLANNING

While the work of production control is seldom carried out by engineers in Manufacturing, the technique is an engineering development dating back to the beginning of the industrial revolution. It is necessary for engineers to understand production control techniques so that they may gain better insight to the requirements for computer-assisted techniques and for adequate planning.

The basic reason for production control is to monitor shop operations in such a way that allows management to deploy resources in a tactical manner. The manager has certain equipment, processes, and personnel available. He or she has to use these resources in as optimal a manner as possible to economically make a product. The Production Control process allows the manager to do just that by forcing the establishment of an overall plan for a particular time period.

Let us look at various items of Production Control through an example of producing a shop order of 10,000 hex head 3-in.-long by ⅝-in.-dia. machine screws. The Production Control unit receives planning that dictates how the machine screws are to be made with planned times assigned to each operation. The Production Control Specialists review the sequence and planned times. In our example let us say the planning calls for the following sequence:

1. Obtain $11\!/_{16}$-in.-dia. round bar stock—15.00 min.
2. Set up cutoff saw—15.00 min per lot.
3. Cut to length—50 pieces per 1.00 min.
4. Set up in chucking lathe A—25.00 min per lot; use NC tape No. 83672.
5. Machine shank and threads—one piece per 1.00 min.
6. Set up on chucking lathe B—25.00 min per lot; use NC tape No. 83673.
7. Machine hex head—one piece per 0.15 min.
8. Conduct first piece inspection—15.00 min per lot.
9. Conduct sample inspection—20 pieces per 1000, 20.00 min per 1000.

The first action the Production Control Specialist undertakes is to convert planned times to effective times. Since planned times are ideal times determined by scientific principles of motion economy, where the human/machine team performs at a constant 100% efficiency, these times have to be converted to realistic times where unavoidable delays are anticipated. The simplest way to do this is by constantly measuring actual times versus planned times, thus achieving an efficiency ratio. This value is then multiplied by the planned times to give production scheduling times.

The second action the Production Control Specialist does is convert each planned step into a sequenced chronological elapsed time. This is shown as a Gantt chart representation (see Fig. 28.3).

The Production Control Specialist has determined that it will take 4.25 hr to saw cut, 185.65 hr for lathe A, 46.75 hr for lathe B, and 4.00 hr for inspection.

The Gantt chart of Fig. 28.3 is single product oriented, which is fine for a one product factory. But the vast majority of factories in North America are multiproduct, so the next step in Production Control is to mortgage time against a workstation for a particular product. This is now done by

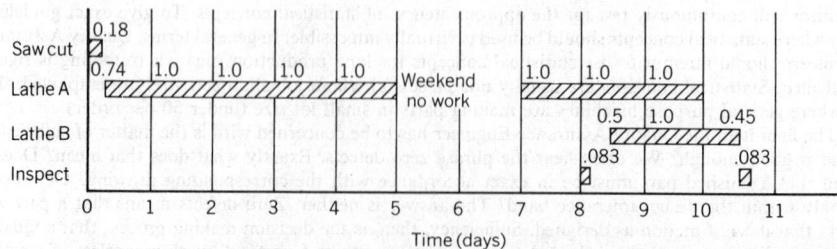

Fig. 28.3   Planned times for producing 10,000 3 in × five-eighths in. hex head machine screws.

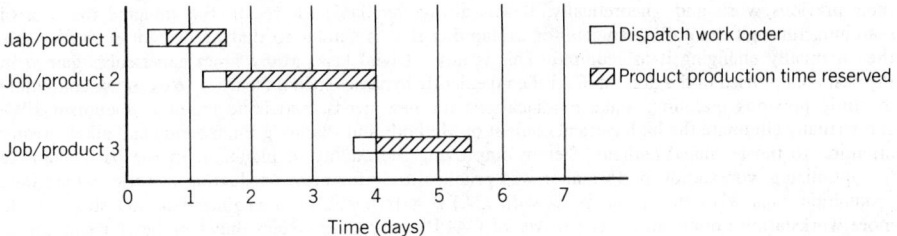

Fig. 28.4 Workstation gantt chart.

computer-based scheduling, but the concept used is the workstation Gantt chart (Fig. 28.4). This is the basic control mechanism by which the Production Control Specialist monitors progress, that is, when instructions have to be issued to commence work, when work is to begin, and when it is to be completed. Simply by monitoring progress against scheduled dates it is possible to determine if the proper results are being obtained.

Now we can list the basic technique of Production Control:

1. Receive planning from Manufacturing Engineering.
2. Assign estimated times for total lots.
3. Mortgage workstation time by schedule required chronological sequences.
4. Dispatch work to workstations as per item 3.
5. Monitor results.
6. Reevaluate progress, adjust schedules as required.

Production Control drives the factory, but it cannot drive it efficiently if Planning is not optimum. This is where the engineer in industry comes to the forefront. A workable design is usually delivered to the Planning and Methods activity of Manufacturing Engineering from the Producibility Engineering activity. If the workable design has been developed effectively, the Planning and Methods Engineers start their tasks with the knowledge that the product has been designed to match the capabilities of the factory.

With this knowledge it is now a matter of sequencing operations such that, for example, high locked-in stresses are not induced and processes used do not destroy the desired design surface finish or tolerance results. This implies and actually does require a sound knowledge of engineering mechanics, thermodynamics, and, in fact, all the classic mechanical engineering subject matter disciplines.

Planning and Methods Engineering are actually two distinct but interactive activities. Planning deals with sequencing and has to take into account the effects of such sequencing on the finished products. Methods has to do with the proper work actions at the workstation. This includes tool selections, fixturing, and the actual sequencing at the workstation as opposed to sequencing done on a macro sense in the planning phase. Methods also attends to the needs of developing performance time standards used in the planning phase to establish sequence times.

As we have seen, the Production Control function has a very definite need for computers to handle the large amounts of data required to manipulate schedules. One of the outstanding developments in recent years has been the work done by engineers in Manufacturing to develop and implement computer-assisted production control. The basic problem of production control is how to readjust schedules when unexpected events cause delays and bottlenecks. Parts pile up at a problem workstation, while workstations for future operations are starved for work. For a typical factory with 50 workstations and 50 shop orders being produced at any one time, imagine the enormity of the task of rescheduling the factory in a manual mode when the factory flow is imbalanced. In fact, unless the factory has a vast army of planners, this cannot be done effectively. Therefore, it is stuck with a static schedule and must hope that no significant unexpected delays occur. Usually buffer time is built into schedules to mitigate the effects of delays. With the advent of computer-assisted production scheduling it is possible to readjust schedules relatively easily and even to do simulations for various alternatives. This is indeed a significant development because it allows dynamic scheduling instead of static scheduling and the minimization of buffer time.

Another equally significant use of the computer is that of computer-aided process planning and group technology. Computer-aided process planning (CAPP) does two very important things. First, it reduces clerical drudgery to a bare minimum thus allowing the planner time for producing optimized plans. Second, it allows the planner instant recall of the best plans of the past, which is like having an acknowledged expert always on immediate call for consultation. The computer accomplishes this task by its capability of storing information and its quick recall. In the past, engineers filed copies of

744

THE MANUFACTURING SYSTEM

their previous work and, theoretically, it was always available for recall. But imagine the task of searching through many file cabinets for a planning that is similar to that needed for a current job, then manually changing it as required. This is not a trivial task. Many times experiences gained in the past are not used in the present. CAPP conveniently bypasses such difficulties. We can use computers to study previous planning, make modifications for new needs, and issue required documents. We can virtually eliminate the high clerical content of Methods and Planning Engineering and allow greater attention to the technical content, thereby improving the quality of planning. In the past, planning for optimizing workstation performance was practical only for mass production factories where long production runs were the norm. Now with CAPP it is possible for engineers in job shops to do more workstation optimization. The driver of CAPP is a coding system that searches for similarities between current and previous projects. Since the mid-1960s a type of classification and coding known as group technology has been used for classifying characteristics used in manufacturing. Elaborate codes have been developed by national productivity groups and private concerns to identify families of parts or products having similar characteristics. In essence, group technology allows job shops to approach the efficiency levels of continuous flow shops. Some examples follow.

For the case of CAPP, group technology allows for grouping of parts by either design or physical characteristics. Obviously, if a previously manufactured part is coded, it is simple for the computer to find it and then compare it with a current part for which planning is to be developed. This allows the planner to simply enter the group technology code for the current part and recall all planning of other parts with similar or identical codes. Coding in essence dictates planning, and becomes a very powerful tool in running a factory, since if it focuses on planning, it also will focus on development of necessary machine tools and process equipment.

In continuous-flow manufacturing (mass production) setups are minimized and fixtures are optimized for long production runs. Unfortunately, continuous-flow manufacturing accounts for only approximately 30% of all manufacturing activities. Job shops deal with many small lot sizes that require their own setups and fixturing, hence a much greater proportion of time is required for nonproductive work. With group technology, it is possible to minimize this work by finding parts from various shop orders that use similar manufacturing processes, then grouping them for longer production runs. Usually this results in only minimal design changes in fixtures to make them more universal, hence setup times are minimized. This leads to the development of manufacturing cells, that is, sets of dedicated machines for making specific geometric shapes called families of parts. Although group technology is relatively new, it is making an impact far beyond the obvious classification and coding benefits. An example is the development of the computer numerical control (CNC) machining centers, where one machine tool is capable of making an entire complex part.

Finally, the group technology concept forces the engineer in manufacturing to view manufacturing as more of an overall systems approach to solving problems. No longer do we look at particular parts and match them with particular machines. We look for the fit of a part within the manufacturing overview—the way the making of the one part fits into the making of all the parts bounded by time and cost constraints. Group technology is symbolic of the overall systems approach the Manufacturing System has always implied, but until the advent of the computer was only possible on a macro sense. Now with computers programmed in accordance with classification and coding concepts it is possible to achieve systems use in a micro sense.

28.5 SELECTING PRODUCTION PROCESSES AND EQUIPMENT TO MAKE THE PRODUCT

Let us now look at how production processes are selected. This is an essential step in making any product and is intimately related to the other engineering functions previously discussed.

In some aspects this is a "chicken and egg" question. The answer for production processes and equipment selection is neither, it is an interactive, ongoing action. The selection of equipment naturally depends on the product. Conversely, a company's choice of product depends on the processes and equipment the company has available. Therefore, selecting equipment is more of a slowly evolving process than a revolutionary one.

We start again with the workable design. Ideally it can be made with existing facilities. But often, owing to competitive pressure or to the promise of better performance with a new design that is not producible, new equipment is required. It is the responsibility of Manufacturing Engineering to develop the specifications for new equipment; obtain competitive quotes; justify its purchase; purchase, install, and debug it; and finally turn the equipment over to Shop Operations for use.

Recognizing that no factory's capital equipment can remain static, most companies dedicate funds annually to purchase new equipment. It is the responsibility of Facilities Engineering to prepare the budget and justify the expenditures for new equipment. This is done through continuous consultation with Design Engineering and Producibility Engineering. It can be thought of as the method of introducing the physical reality to new product development. Design Engineering develops the concepts, Producibility Engineering reviews them against current manufacturing capabilities, and Facilities Engineering develops whatever new machines are needed to supplement current capabilities. This often means very close cooperation between Design Engineering and Manufacturing Engineering. Of course, here, as a matter of practicality, much engineering intercourse occurs between the functions. This intercourse further strengthens the viability of the new or modified product.

Selection of new processes and equipment, even for a new product, is evolutionary. It is not practical from a cost and time viewpoint to embark on radically different technology. All engineers would readily agree that extrapolation from known technology is always the prudent course for production equipment. Therefore, virtually all new equipment for new designs are developed this way. This is not to say that Manufacturing Engineering does not embark on radically new ways of producing a family of products. But this approach is usually taken through research and development activities. It would be foolhardy to allow a producing factory to become an R&D facility. Therefore, evolutionary equipment selection activity is supplemented with laboratory or off-site R&D efforts which should lead to the gradual introduction of new techniques into the factory in a way that does not put the entire manufacturing capability of the company in jeopardy. It would be unfortunate for a company to have to cease production because of an unforeseen problem that occurred during the introduction of an improperly developed process. Engineers in Manufacturing must always keep in mind that their number one priority is to keep their factories producing. The purpose of introducing new technology into a factory is to keep it competitive. If new technology has significant development problems during implementation, it is then counterproductive and at odds with the number one priority.

Maintenance Engineering is a specialized subbranch of Manufacturing Engineering that deals primarily with equipment and processes previously selected, that is, making sure the equipment continues functioning in accordance with their specifications. Maintenance Engineering interfaces with all phases of the Manufacturing function: with Shop Operations, which must be assured that equipment is functioning; with Planning, which has to know that equipment is available to mortgage space/capacity against; and with engineers responsible for writing specifications for new equipment, who need to be able to take advantage of experiences gained on existing equipment.

The Maintenance function can be divided into two components, Preventive Maintenance and Rapid Response. The latter is the "fire brigade" of Manufacturing. It requires engineers who can diagnose problems quickly, devise short-term fixes to get equipment operating again, and then develop permanent fixes. The Preventive Maintenance activity includes periodic checks, replacement activities, and tests to make sure catastrophic failures do not occur. Also included in Preventive Maintenance is the requirement to ensure that the factory complies with safety and environmental regulations. The extent of the need for Rapid Response indicates the relative success of a Preventive Maintenance program. Design of a good Preventive Maintenance program must take into account a thorough understanding of the design theory employed to make a factory's equipment. It should be quite evident that Maintenance Engineering is a very sophisticated branch of applied engineering.

Another responsibility of Maintenance Engineering is to consult with those responsible for purchasing new equipment. Here the Maintenance Engineer is charged with making sure that specifications call for all the safeguards needed to ensure that the new equipment is not prone to failure. A good Maintenance Engineer will provide histories of failures of previous similar machines and equipment such that the engineer writing the specifications will know what to guard against. Maintenance Engineering shows once more that Engineering in Manufacturing is an integrated function.

28.6 CONCLUSION

We can see from this chapter that Engineering within the Manufacturing function is a complex activity. It ranges from the very theoretical—Producibility Engineering—to the very pragmatic—Maintenance Engineering. It encompasses traditional activities such as Methods and Planning, but also includes very advanced uses of computers, such as CAPP and group technology. Above all, Engineering in Manufacturing requires the use of technology to the application of the seven steps of the Manufacturing System. Engineers in Manufacturing need to apply virtually all aspects of all mechanical engineering disciplines. While they can practice the full range of the theoretical disciplines to which all engineers are trained for, they also have the advantage of pursuing their work from theory to practicality and thus are in a position to advance and refine the various theories making up the discipline we call mechanical engineering.

CHAPTER 29

PRODUCT DESIGN FOR ECONOMIC PRODUCTION

BENJAMIN W. NIEBEL

The Pennsylvania State University
University Park, Pennsylvania

29.1 OBJECTIVES OF DESIGNING FOR ECONOMIC PRODUCTION

The principal objective of designing for economic production is to produce a design that will satisfy both functional and physical requirements at a cost that is acceptable to the user. That is, the design itself must be producible at a cost that will permit the product to be introduced competitively at the marketplace. Designing for economic production implies the incorporation of producibility not only during the conception and development of the design but also during the production phases and throughout the entire life cycle of the product. It is during the functional design of the product that producibility should be incorporated in concert as the design progressively takes shape.

A corollary to the principal objective of designing for production is to minimize unit costs, tooling costs, elaborate test systems, use of high cost and/or critical processes, design changes in production, use of limited availability items, and the time for delivery.

Effective designing for economic production will maximize design simplicity, standardization of materials and components, ease of product inspectability and testing, safety in production, and competitive procurement.

29.2 ENGINEERING DRAWINGS

The principal communication tool between the functional designer and the producers of the design is the engineering drawing. Whether the drawing is prepared by the draftsman or is the plotter output from computer-aided design, it is the engineering drawings alone that control and completely delineate shape, form, fit, finish, function, and interchangeability requirements that lead to the most competitive procurement. An accurate engineering drawing together with reference specifications and standards will permit a qualified manufacturer to produce the design within the dimensional and surface tolerance specifications provided. The drawing will convey to the creative manufacturing planner how the design can best be produced. Frequently, in view of the space constraints, product specifications such as quality assurance checkpoints and inspection procedures will be separately summarized but should always be cross-referenced on the engineering drawing.

The following points are characteristic of sound functional design that should be clearly presented on the engineering drawings:

1. The functional design as depicted on the engineering drawings is conducive to the application of economic processing.
2. The functional design has applied the group technology concept on all the components that comprise the design and thus capitalizes on the "sameness" of the various components with similar or like components produced in the past. Interchangeability of parts is maximized.
3. Every component embracing the design can be produced with the current manufacturing state of the art.
4. The design specifications are compatible with the performance specifications.
5. Critical location surfaces and/or points are identified on the drawing.
6. The design should endeavor to avoid the use of proprietary items. Standard parts are specified when possible that are available from several suppliers and producers.
7. All specifications relating to the design should be definitive. For example, if chamfers are indicated, they must be dimensioned, and if desired, they must be shown.
8. All dimensions and associated tolerances, parallelism, perpendicularity, etc., should be shown so that they lend to easy measurement and/or inspectability.
9. Tolerances are no more restrictive than necessary.
10. The design permits easy and rapid assembly of making parts.
11. The design should be as simple as possible—overdesign should be avoided.

29.3 THE PRODUCT DESIGN PROCESS

To help ensure successful designs, the design process should be undertaken systematically. Figure 29.1 illustrates the six steps associated with the product design process: initial conception and evaluation, analysis, development of general design, development of detailed design, hardware development, and development of test and production models. Although this systematic procedure will be simplified in the case of certain elementary product designs, and will be extended in the case of more complex designs, the steps enumerated are characteristic of the procedure that should be followed in most cases. It should be noted that feedback paths must exist throughout the design continuance. Consideration is given to the classic manufacturing resources: people, facilities, technology, raw material, capital, and time. Throughout the product design process thought should be given to producibility so that the final design will reflect a design that can be economically produced. Perhaps the most common

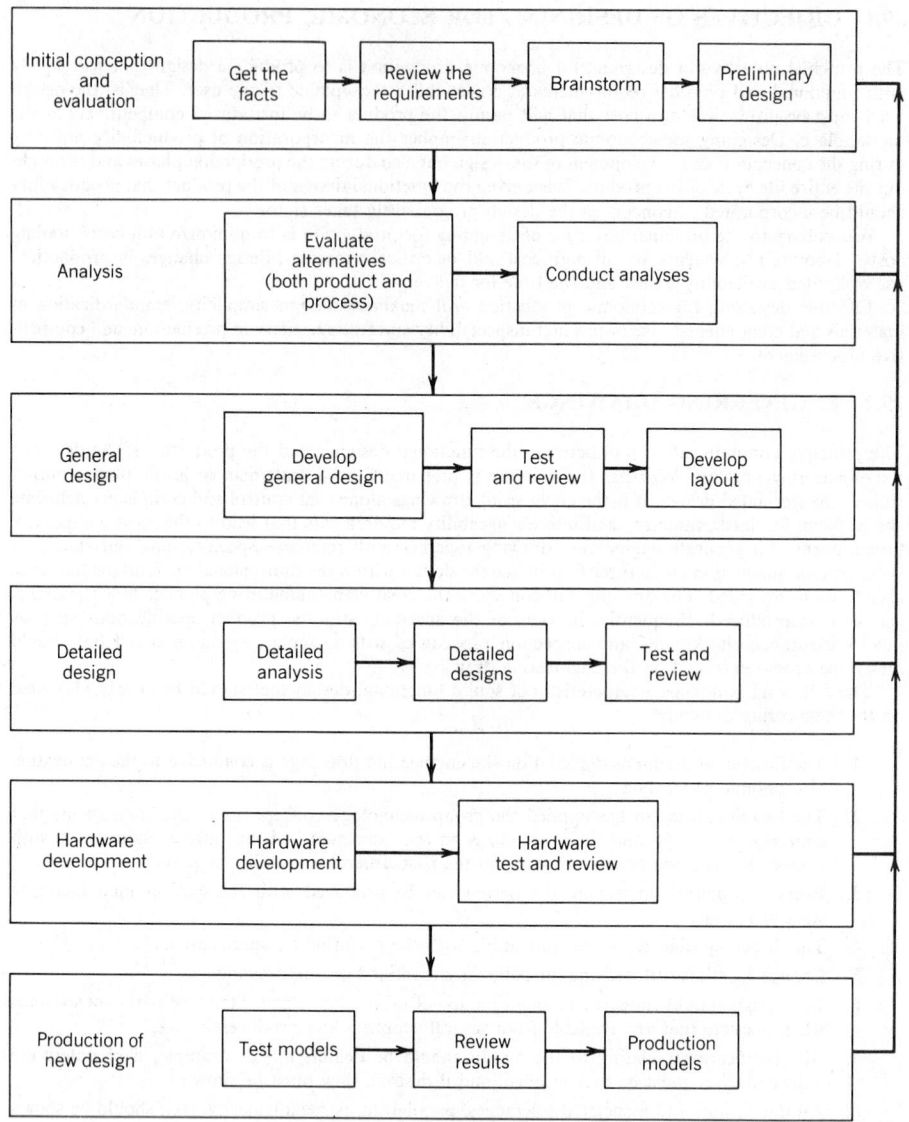

Fig. 29.1 The product design process.

deficiency of a new design is excessive complexity. Simpler designs invariably lead to more economic production, are more reliable, are more easily maintained, and, typically, will exceed the performance of the more complex design. Simplicity is often incorporated by the elimination of components of an assembly by building their function into other components. The talents of the manufacturing engineer should be utilized by the functional designer throughout the design process to avoid production restrictions. Typical errors in design that should be avoided so as to promote economical production include inconsistent or double dimensioning, specifying a threaded inside diameter to the bottom of the blind hole, specifying a ground tolerance on an outside diameter to a shoulder, etc.

The selection of the most appropriate materials and processes for economic production while maintaining product quality and reliability is a continuing activity that takes place throughout the design life cycle from initial conception to production. These considerations need to be made in concert, since it is only after materials are processed that the total cost (cost of materials and cost of processing) is known. An inexpensive raw material that is difficult to process can be an expensive finished material.

29.3.1 Selection of Materials

Usually there are several alternatives for the functional designer to consider in the selection of materials for a specific application. To simplify the selection process, the designer should organize information related to possible materials into three categories: properties, specifications, and required data for procurement.

The property category usually will provide the information that suggests the most desirable material from the standpoint of product quality and reliability over a predicted life. Alternative materials under consideration should be compared by their property profiles. Such important properties as yield point, modulus of elasticity, resistance to corrosion, conductivity, toughness, ductility, and machinability should be compared. Those materials that qualify because of their properties will stand out.

The second category where material information is organized will include unique specifications, standard grades, and cost. This information will allow the designer to make a final decision between two or more materials that have survived an analysis based on the data from the properties category. Finally, the third category will be the data needed to order the material. This includes sources of supply, availability, quantity break points, minimum order size, and economic order quantity.

29.3.2 Selection of Processes

When designing for economic production, we are concerned with the lowest processed material costs, which involves costs for all processing steps including set-up and production time costs along with the preprocessed material cost and the cost of scrap, chips, and waste generation.

Liquid Material Basic Processes

In the development of the preliminary design, consideration as to whether to start with a basic process that uses materials in the liquid state such as a casting or in the solid state such as a forging need be given. If the designer makes the decision to use a casting, he or she then must decide which alloy and what casting process can most nearly meet the required geometrical configuration, mechanical properties, dimensional tolerances, reliability, and production rate at the least cost.

The family of processes designated as castings have several distinct assets: the ability to conform to a complex shape, economy, and a wide choice of alloys. The principal inherent problems that may exist include internal porosity, dimensional variations resulting from shrinkage upon cooling, and solid or gaseous inclusions that result from the molding operation. These problems can be minimized by sound detailed design.

All casting processes are basically similar in that the metal being formed initially is in a liquid or highly viscous state and is poured or injected into a cavity of a desired configuration.

The following casting design guides will lead to economic production:

1. When changes in section are required, use smoothly tapered sections to reduce stress concentration. Where sections join, use generous radii that blend in with joined sections.
2. The casting drawing should show machining allowances so that patterns are produced to ensure adequate stock and yet avoid excessive machining.
3. Avoid concave notches in casting design, because the metal mold or die is difficult to produce. Convex product designs permit easy milling in the making of the mold or die.
4. Specify raised lettering rather than depressed lettering in the casting.
5. Avoid thin sections in the design, since they are difficult to fill completely.
6. Cored holes that will subsequently be drilled and/or reamed and/or tapped should have countersinking on both ends to facilitate the secondary operations.
7. To avoid warpage and distortion, break up large plain surfaces with ribs or serrations.
8. For maximum strength, keep material away from the neutral axis. Endeavor to keep plates in tension and ribs in compression.

Table 29.1 provides the important design parameters associated with the principal casting processes and provides those limitations and constraints that should be observed by the functional designer to assure economic production.

Solid Material Basic Processes

The functional designer may consider using a forging rather than a casting because of improved mechanical properties associated with forgings. The hot forging process will break up the large dendritic grain structure, characteristic of castings, thus giving the metal a refinement of the grain structure and

Table 29.1 Casting Design Parameters

Design Parameter	Sand Casting		Shell	Plaster (Preheated Mold)	Investment (Preheated Mold)	Permanent Mold (Preheated Mold)	Die (Preheated Mold)	Centrifugal
	Green	Dry/Cold Set						
Weight	3.5 oz to 440 tons (100 g to 400 metric tons)	3.5 oz to 440 tons (100 g to 400 metric tons)	3.5 oz to 225 lb (100 g to 100 kg)	3.5 oz to 225 lb (100 g to 100 kg)	Fraction of oz to 110 lb (fraction of g to 50 kg)	3.5 oz to 55 lb (100 g to 25 kg)	Fraction of oz to 65 lb (fraction of g to 30 kg)	0.25 oz to 440 lb (7 g to 200 kg)
Minimum section thickness	0.120 in. (3 mm)	0.120 in. (3 mm)	0.06 in. (1.5 mm)	0.04 in. (1 mm)	0.02 in. (0.5 mm)	0.12 in. (3 mm)	0.03 in. (0.75 mm)	0.25 in. (6 mm)
Allowance for machining	Ferrous—0.10-0.375 in. (2.5-9.5 mm) Nonferrous—0.06-0.25 in. (1.5-6.5 mm)	Ferrous—0.10-0.375 in. (2.5-9.5 mm) Nonferrous—0.06 to 0.25 in. (1.5-6.5 mm)	Often not required; when required 0.10-0.375 in. (2.5-6.5 mm)	0.03 in. (0.75 mm)	0.01-0.03 in. (0.25-0.75 mm)	0.030-0.12 in. (0.80-3 mm)	0.030-0.060 in. (0.80-1.60 mm)	0.030-0.060 in. (0.75-1.5 mm)
General tolerance	±0.015-±0.25 in. (±0.4-±6.4 mm)	±0.015-±0.25 in. (±0.4-±6.4 mm)	±0.003-±0.063 in. (±0.08-±1.60 mm)	±0.005-±0.010 in. (±0.13-±0.26 mm)	±0.002-±0.006 in. (±0.05-±1.5 mm)	±0.010-±0.06 in. (±0.25-±1.5 mm)	±0.001-±0.005 in. (±0.025-±0.125 mm)	±0.031-±0.138 in. (±0.80-±3.5 mm)
Surface finish (µm rms)	6.0-24.0	6.0-24.0	1.25-6.35	0.8-1.3	0.5-2.2	2.5-6.35	0.8-2.25	2.5-13.0
Process reliability	90	90	90	90	90	90	95	90
Cored holes (diameter)	Holes as small as 0.25 in. (6 mm)	Holes as small as 0.25 in. (6 mm)	Holes as small as 0.25 in. (6 mm)	Holes as small as 0.50 in. (12 mm)	Holes as small as 0.02 in. (0.5 mm)	Holes as small as 0.20 in. (5 mm)	Holes as small as 0.031 in. (0.80 mm)	Holes as small as 1 in. (25 mm); no undercuts
Minimum lot size	1	1	100	1	20	1000	3000	100
Draft allowances	1-3°	1-3°	¼-1°	½-2°	0-½°	2-3°	2-5°	0-3°

inclusions will be stretched out in the direction that plastic flow occurs. In order to acquire the best load-carrying ability, forgings should be designed so that the flow lines run in the direction of the greatest load during service.

Guidelines that the designer should observe in the design of forgings in order to help ensure economic production and reliability include the following:

1. The unsupported length of bar should not be longer than three times the diameter of the bar or distance across the flats. This is the maximum length of bar that can be upset in a single stroke without some buckling of the unsupported portion.

2. Recesses perpendicular to the forging plane with depth up to the diameter of the recess can be incorporated in either or both sides of a section. Secondary piercing operations to remove the residual web should be used on through-hole designs.

3. Draft should be added to all surfaces perpendicular to the forging plane to permit easy removal of the forged part. Outside draft can be less than inside draft, since the outside surfaces will shrink toward bosses in the die.

4. Draft for difficult forging materials such as titanium and nickel-base alloys should be larger than for easy forging materials. Similarly, deep die cavities need more draft than shallow ones.

5. Uniform draft results in more economic dies. One draft should be specified on all outside surfaces and one, larger draft on all inside surfaces.

6. Corner and fillet radii should be large in order to facilitate metal flow and provide long die life. Usually 0.25 in. (6 mm) is the minimum radius for parts forged from high-temperature alloys, stainless steels, and titanium alloys.

7. In order to simplify the die design and keep its cost of construction low, endeavor to keep the parting line in one plane.

8. To avoid deep impressions with accompanying high die wear and difficulty in removing the forged part from the die, locate the parting line on a central element of the part.

Table 29.2 provides helpful design information for the economic production of quality forgings.

Other Basic Processes

There are several other basic metal-forming processes in addition to casting and forging that may be considered in order to impart the approximate finished geometry to material that is in powdered, sheet, rod, or shape form. The more important of these include powder metallurgy, cold heading, extrusion, roll forming, press forming, spinning, electroforming, and automatic screw machine work.

Powdered metallurgy may be considered to be a basic process, although parts from this process often do not require any secondary operations. In this process, powdered metal is placed in a die and compressed under high pressure. The resulting cold formed part is then sintered in a furnace to a point below the melting point of its major constituent.

Cold heading involves striking a segment of cold material, up to 1 in. (25 mm) in diameter, in a die so that it is plastically deformed to the die configuration. Cold heading, like powdered metallurgy, frequently does not require secondary operations.

Extrusion is performed by forcing heated metal through a die having an aperture of the desired shape. The extruded lengths are then cut into the desired length. The following design features applicable to extrusion should be incorporated to ensure sound design for economic production and reliability:

1. Thin sections with large circumscribed area should be avoided.

2. Any heavy wedge section that tapers to a thin edge should be avoided.

3. Thin sections that have close space tolerance should be avoided.

4. Do not include sharp corners in the design.

5. Do not design semiclosed shapes that necessitate dies with long thin projections. Such dies are expensive to make and have a short life.

6. When a thin member is attached to a heavy section, the length of the thin member should not exceed 10 times its thickness.

Roll forming is a process where strip metal passes through a series of rolls that progressively change the shape of the metal by stretching it beyond its yield point. In designing for the amount of bend caused by each roll, allowance must be made for springback. The final rolls will bring the material to its desired geometrical shape. As in extrusion, roll formed lengths are cut into the desired length.

Forming in presses (hydraulic, pneumatic, and mechanical) is a process applied to sheet metal

Table 29.2 Forging Design Parameters

Design Parameter	Open Die	Conventional Utilizing Preblocked	Closed Die	Upset	Precision Die
Size or weight	1.1–11000 lb (500 g to 5000 kg)	0.5 oz to 45 lb (14 g to 20 kg)	0.5 oz to 45 lb (14 g to 20 kg)	0.75–10 in. bar (20–250 mm bar)	0.5 oz to 45 lb (14 g to 20 kg)
Allowance for finish machining	0.08–0.40 in. (2–10 mm)	0.08–0.40 in. (2–10 mm)	0.04–0.20 in. (1–5 mm)	0.20–0.40 in. (5–10 mm)	0–0.12 in. (0–3 mm)
Thickness tolerance	+0.024–0.008 in. to +0.12–0.04 in. (+0.6–0.2 mm to +3.00–1.00 mm)	+0.016–0.008 in. to +0.08–0.03 in. (+0.4–0.2 mm to +2.00–0.75 mm)	+0.012–0.006 in. to +0.06–0.02 in. (+0.3–0.15 mm to +1.5–0.5 mm)	—	+0.008–0.004 in. to +0.04–0.008 in. (+0.2–0.1 mm to +1–0.2 mm)
Fillet and corners	0.2–0.28 in. (5–7 mm)	0.12–0.20 in. (3–5 mm)	0.078–0.16 in. (2–4 mm)	—	0.04–0.079 in. (1–2 mm)
Surface finish (μm rms)	3.8–4.5	3.8–4.5	3.2–3.8	4.5–5.0	1.25–2.25
Process reliability	95	95	95	95	95
Minimum lot size	25	1000	1500	25	2000
Draft allowance	5–10°	3–5°	2–5°	—	0–3°
Die wear tolerance	±0.0014 in./lb (±0.075 mm/kg weight of forging)	±0.0014 in./lb (±0.075 mm/kg weight of forging)	±0.0014 in./lb (±0.075 mm/kg weight of forging)	—	±0.0014 in./lb (±0.075 mm/kg weight of forging)
Mismatching tolerance	±0.0015–+0.00006 in./lb (±0.25–+0.01 mm per 3 kg weight of forging)	±0.0015–+0.00006 in./lb (±0.25–+0.01 mm per 3 kg weight of forging)	±0.0015–+0.0006 in./lb (±0.25–+0.01 mm per 3 kg weight of forging)	—	±0.0015–+0.00006 in./lb (±0.25–+0.01 mm per 3 kg weight of forging)
Shrinkage tolerance	±0.003 in. (±0.08 mm)	±0.003 in. (±0.08 mm)	±0.003 in. (±0.08 mm)	—	±0.003 in. (±0.08 mm)

where the material is stressed beyond its yield point in a metal die. The original material may be reduced in thickness in view of drawing and/or ironing. The process is based on two principles:

1. Compressing and stretching a material beyond its elastic limit on the inside and outside of a bend.
2. Compressing a material beyond the elastic limit without stretching or stretching the material beyond the elastic limit without compression.

Press forming, although classified as a basic forming operation here, will produce parts that require no additional work.

Spinning is a process applied to sheet metal where the work is formed over a symmetrical pattern, often made of hard wood, plastic, or metal. The pattern and material are spun while a tool is forced against the material bringing it in contact with the pattern and finally taking shape conforming to the shape of the pattern. The process will result in thinning the material as it is formed over the pattern. This process is limited to symmetrical shapes and often requires no secondary operations.

Automatic screw machine work will often produce components without the necessity of performing any secondary operations. Here bar stock is fed and cut with multiple tools to obtain the desired configuration.

Table 29.3 outlines important design information for the utilization of these processes in an economic fashion.

29.4 DESIGNING FOR SECONDARY METAL-FORMING AND -SIZING OPERATIONS

Once the basic process is selected for producing a given functional design, consideration need be given to the selection of the most favorable secondary forming and sizing operations. Invariably there is more than one competing way to produce a part. The factors relating to a given design that need to be considered in secondary process selection include: the shape desired, the material being processed, the tolerances required, the surface finishes required, the quantity to be produced, and the cost.

Some design considerations that should be observed so that secondary operations may be economically performed include:

1. Provide flat surfaces for the entering and exit of the drill on holes to be drilled.
2. On long members, design so that male threads can be machined between centers as opposed to female threads, where it would be difficult to support the work.
3. Parts to be machined should be designed with gripping surfaces so that the work can be held securely while machining takes place.
4. Parts that are held during machining must be sufficiently rigid to withstand the forces of machining.
5. In the design of mating parts, avoid double fits. It is easier to maintain close tolerances when a single fit is specified.
6. To minimize tooling costs, avoid the design of special contours that would require a form tool.
7. When shearing sheet metal work, avoid feather edges. Internal edges should be rounded, while corners along the edge of the strip stock should be sharp.
8. The flat blanks from which press-formed parts are made should have straight edges.
9. Allow a distance of at least 1.5 times the thread pitch between the last thread and the bottom of the tapped blind hole.
10. All blind holes should end with a conical geometry in order to allow the use of standard drills.
11. Work should be designed so that diameters of external features increase from the exposed face, while diameters of internal features decrease.
12. Internal corners of the workpiece should be dimensioned with a radius machinable by a simple cutting tool.
13. Consider designing the work so that all secondary operations can be performed while holding the work in a single jig or fixture.

Table 29.4 provides pertinent information in connection with the principal secondary operations performed by machine tools.

Table 29.3 Metal-Forming Design Parameters

Design Parameter	Powder Metallurgy	Cold Heading	Extrusion
Size	Diam.—0.06–12 in. (1.5–300 mm) Length—0.12–9 in. (3–225 mm)	Diam.—0.03–0.75 in. (0.75–20 mm) Length—0.06–10 in. (1.50–250 mm)	Diam.—0.06–10 in. (1.5–250 mm)
Minimum thickness	0.040 in. (1 mm)	—	0.040 in. (1 mm)
Allowance for finish machining	To size	To size	To size
Tolerance	Diam.—±0.001–±0.005 in. (±0.025–0.125 mm) Length—±0.01–±0.02 in. (±0.25–0.50 mm)	Diam.—±0.002–0.005 in. (±0.05–0.125 mm) Length—±0.03–±0.09 in. (±0.75–2.25 mm)	Flatness—±0.0004 in. (±0.01 mm) per inch of width Wall thickness—±0.006–±0.01 in. (±0.15–±0.25 mm) Cross section—±0.006–±0.008 in. (±0.15–±0.20 mm)
Surface finish µin. (µm)	5–50 (0.125–1.25)	85–100 (2.2–2.6)	100–120 (2.6–3)
Process reliability	95	99	99
Minimum lot size	1000	5000	1500 ft
Draft allowance	0°	—	—
Bosses permitted	Yes	Yes	Yes
Undercuts permitted	No	Yes	Yes
Inserts permitted	Yes	No	No
Holes permitted	Yes	No	Yes

29.5 DESIGNING FOR THE SHAPING OF PLASTIC COMPONENTS

When considering plastics as a product material, the functional designer has more than 30 distinct families to consider. From these families evolve thousands of specific formulations that could be considered in the case of each component of the design. Although the number of plastic material alternatives is large, the number of basic processes utilized in the shaping of plastics is limited. The principal production processes used to shape plastic materials include compression molding, transfer molding, injection molding, extrusion, casting, cold molding, thermoforming, and blow molding. In order to design effectively for economic production, the functional designer must consider the capabilities of competing processes in concert with the unique physical and mechanical properties of each plastic material under consideration.

29.5.1 Compression Molding

This process is compatible for both thermosetting and thermoplastic materials. Here an appropriate amount of the plastic compound (usually in powder form) is placed into a heated metallic mold. The mold is closed under pressure and the molding material becomes soft by heat and is formed into a continuous mass having the shape of the mold cavity. If the material being processed is thermoplastic, the mold is cooled and the hardened product is removed. If the material is thermosetting, further heating will result in the hardening of the material.

Roll Forming	Press Forming	Spinning	Electro-forming	Automatic Screw Machine
To 80 in. (to 2000 mm)	Up to 20 ft (6 m) in diameter	0.25 in. to 13 ft in diameter (6–4000 mm)	Limited to size of plating tanks	0.031 in. (0.8 mm) diam. by 0.06 in. (1.5 mm) length to 8 in. (200 mm) diam. by 35 in. (900 mm) length
0.003 in. (0.075 mm) To size	0.003 in. (0.075 mm) To size	0.004 in. (0.1 mm) To size	0.0001 in. (0.0025 mm) To size	— To size
Cross section— ±0.002–0.014 in. (±0.050–0.35 mm) Length—±0.06 in. (±1.5 mm)	±0.01 in. (±0.25 mm)	Length—±0.005 in. (±0.12 mm) Thickness—±0.002 in. (±0.05 mm)	Wall thicknesses— ±0.001 in. (±0.025 mm) Dimension— ±0.0002 in. (±0.005 mm)	Diam.—±0.0004 to 0.002 in. (±0.01 to 0.06 mm) Length— ±0.0016–0.004 in. (±0.04–0.10 mm) Concentricity— ±0.002 in. (0.06 mm)
85–100 (2.2–2.6)	85–160 (2.2–4.0)	15–85 (0.4–2.2)	5–10 (0.125–0.250)	12–100 (0.30–2.5)
99	99	90–95	99	98
10,000 ft	1500	5	25	1000
—	0–¼°	—	—	—
Yes	Yes	No	Yes	Yes
Yes	Yes	Yes	Yes	Yes
No	No	No	No	No
Yes	Yes	No	Yes	Yes

Some considerations for the selection of compression molding in order to design for economic production include:

1. Parts with thin walls (less than 0.060 in.) can be molded with complete filling of the mold cavity, little part warpage, and good dimensional stability.
2. Since the material is placed directly into the mold cavity, no gate markings will be made.
3. Less material shrinkage and a more uniform shrinkage is characteristic of compression molding.
4. The process is economical for large parts (parts weighing more than 2 lb).
5. Mold costs are usually less than either transfer or injection molds designed to produce the same part.
6. When reinforcing fibers are used, the end product is usually stronger and tougher than when either transfer or injection molding is used since in these closed mold processes, the fibers are broken up.

29.5.2   Transfer Molding

In this process, plastic material temporarily stored in an auxiliary chamber is transferred under pressure through an orifice into the closed mold. The formed part along with the residue, referred to as the

Table 29.4 Secondary Operations

Process	Shape Produced	Machine	Cutting Tool	Tolerance	Surface Finish, μin. (μm)	Relative Motion Tool	Relative Motion Work
Turning (external)	Surface of revolution (cylindrical)	Lathe, boring machine	Single point	±0.001–±0.003 in. (±0.025–±0.075 mm)	30–250 (0.8–6.4)		
Boring (internal)	Cylindrical (enlarges holes)	Boring machine	Single point	±0.001–±0.003 in. (±0.025–±0.075 mm)	15–200 (0.4–5.0)		
Shaping and planning	Flat surfaces or slots	Shaper planner	Single point	±0.002–±0.004 in. (±0.050–±0.100 mm)	30–250 (0.8–6.4)		
Milling (end, form, slab)	Flat and contoured surfaces and slots	Milling machine: horizontal and vertical bed types	Multiple points	±0.001–±0.003 in. (±0.025–±0.075 mm)	30–250 (0.8–6.4)		
Drilling	Cylindrical (originating holes 0.1–100 mm in diameter	Drill press	Twin-edged drill	±0.002–±0.004 in. (±0.050–±0.100 mm)	100–250 (2.5–6.4)		or Fixed
Grinding (cylindrical surface, plunge)	Cylindrical, flat, and formed	Grinding machine Cylindrical Surface Thread	Multiple points	±0.0002–±0.001 in. (±0.005–±0.025 mm)	8–100 (0.2–2.5)		or Fixed
Reaming	Cylindrical (enlarging and improving finish of holes)	Drill press, Turret lathe	Multiple points	±0.0005–±0.002 in. (±0.0125–±0.0500 mm)	30–100 (0.8–2.5)		Fixed
Broaching	Cylindrical, flat, slots	Broaching machine, Press	Multiple points	±0.0002–±0.0006 in. (±0.005–±0.0150 mm)	30–100 (0.8–2.5)		Fixed
Electric discharge machining	Variety of shapes depending on shape of electrode	Electric discharge machine	Single-point electrode	±0.002 in. (±0.050 mm)	30–200 (0.8–5.0)	Spark	Fixed

Process	Typical applications	Machine	Tolerance	Surface finish, μin (μm)	Cutting action	Material-removal mechanism	Tool/workpiece motion
Electrochemical machining	Variety of shapes; usually odd shaped cavities of hard material	Electrochemical machine	±0.002 in. (±0.050 mm)	12–60 (0.3–1.5)	Dissolution process	Anodic dissolution; tool is cathode	Workpiece is anode
Chemical machining	Variety of shapes; usually blanking of intricate shapes, printed circuit etching, or shallow cavities	Chemical machining machine	±0.002 in. (±0.050 mm)	24–70 (0.6–1.8)	Chemical attack of exposed surfaces	Chemical attack of exposed surfaces	Fixed
Laser machining	Cylindrical holes as small as 5 μm	Laser beam machine	Holes are reproducible within ±3% of diam. (for thin materials)	24–100 (0.6–2.5)	Single-wavelength beam of collimated light	Vaporization or melting	Fixed
Ultrasonic machining	Same shape as tool	Machine equipped with magnetostrictive transducer, generator power supply	±0.001 in. (±0.025 mm)	12–35 (0.3–0.9)	Shaped tool and abrasive powder	Sonic erosion	Fixed (↔ reciprocating)
Electron beam machining	Cylindrical slots	Electron beam machine equipped with vacuum of 10^{-4} mm of mercury	±0.001 in. (±0.025 mm)	24–70 (0.6–1.8)	High-velocity electrons focus on workpiece	Vaporization or melting	Fixed
Gear generating	Eccentric cams, ratchets, gears	Gear shaper	±0.0005–±0.001 in. (±0.013–±0.025 mm)	70–150 (1.8–3.8)	Single-point reciprocating	(rotary/reciprocating motion ↻)	(↻)
Hobbing	Any form that regularly repeats itself on periphery of circular part	Hobbing machine	±0.0005–±0.001 in. (±0.013–±0.025 mm)	70–150 (1.8–3.8)	Multiple points	(rotary motion ↻)	(↻)
Trepanning	Large through-holes, circular grooves	Lathelike	±0.005 in. (±0.13 mm)	100–250 (2.5–6.4)	One or more single-point cutters revolving around a center	(rotary + linear motion ↻ →)	(↻)

cull, are removed when the mold is opened. Unlike compression molding, there is no flash to trim. Only the cull and the runners between the mold cavities in multiple cavity molds need to be removed.

29.5.3 Injection Molding

More thermoplastic compounds are molded by this process than any other. Here the plastic material in the form of pellets or grains is placed in a hopper above a heated cylinder called the barrel. From the hopper, an appropriate amount of material is metered into the barrel every cycle. The plastic material, under pressures up to 600 psi, is forced into the closed mold. The cooled mold, when molding thermoplastics, is opened and the molded parts are ejected. When molding thermosets, such as phenolic resins, the barrel temperatures are considerably lower than when molding thermoplastics. Thermoset barrel temperatures range from 150 to 250°F, while thermoplastic barrel temperatures typically range from 350 to 600°F.

29.5.4 Extrusion

This process provides a continuous cross-sectional shape by forcing softened plastic material through an orifice having approximately the geometrical profile of the cross section of the work. The extruded plastic form now passes through a cooling cycle where it is hardened. Such products as filaments, tubes, rod, and uniform cross-sectional shapes are economically produced using the continuous extrusion process.

29.5.5 Casting

In this process, plastic materials in the liquid form are introduced into a mold that has the geometrical configuration of the desired part. The mold material is often flexible, such as rubber, or at times made of nonflexible materials, such as plaster or metal. Typical plastic families utilizing casting include the epoxies, phenolics, and polyesters.

29.5.6 Cold Molding

Cold molding is a process that is used for the production of thermosetting parts. The molding material is placed into a mold at room temperature (unlike compression molding). The mold is then closed under pressure. Subsequently, the mold is opened and the formed part is removed to a heating oven where it is baked until the thermosetting material becomes hard.

29.5.7 Thermoforming

Under thermoforming, sheets of thermoplastic material are heated to a plastic condition and then drawn over a mold contour where, upon cooling, the sheet takes the shape of the mold. This process can also use a sequence of rolls to produce the desired form from plastic sheet. Typically, the plastic sheet is brought to the forming temperature (275–425°F) by infrared radiant heat; electrical resistant heating; or ovens using gas, fuel oil, or coal.

29.5.8 Calendering

This process involves the passing of thermoplastic compounds between a series of heated rolls in order to produce a sheet of material of uniform thickness. The thickness of the sheet is governed by adjusting the distance between mating rolls. After the sheet passes through the final sizing rolls, it is cooled in order to solidify the material preparatory to winding into large rolls.

29.5.9 Blow Molding

This process is used in the economic production of hollow, bottle-shaped thermoplastic products. Here a tube of plastic material called the parison is extruded over an apparatus called the blow pipe and then is encased in a split metallic mold. Air is now injected into this hot plastic section of extruded stock through the blow pipe. The parison is now blown outward against the contour of the mold. The part is then cooled (in heavy sections liquid nitrogen or carbon dioxide may be used to hasten the cooling), the mold is opened, and the solidified part is removed.

29.6   FACTORS AFFECTING THE SELECTION OF THE MOST FAVORABLE PROCESS IN THE MOLDING OF PLASTICS

The selection of the most favorable molding process for plastic components will have an impact on the quality, the reliability, and the cost of the design. Those parameters that need to be evaluated for the process selection include the specific plastic material used in the design, the geometry of the part, the quantity to be produced, and the cost of the process including the die or mold cost.

Usually, the functional designer, in view of the physical and/or mechanical property requirements, is able to identify whether a thermoplastic or thermosetting resin will be used in the design. This information will be helpful in reducing the number of molding processes to be considered. If a thermosetting resin is selected, both thermoforming and blow molding can generally be eliminated, since these processes are usually restricted to thermoplastics. If the quantity to be produced is large, for economic reasons, compression and transfer molding should be restricted to thermosetting resins. In both simple and complex shapes, where the quantity is large and the material is thermoplastic, injection molding will generally prove to be the most competitive.

After considering the material to be used, the functional designer will study the geometry of the part in order to help identify the most favorable process. If the part has a continuous cross section, extrusion would be the logical selection. If the design is bottle-shaped and thin-walled, it would probably be blow molded. Flat sheet designs would be limited to calendering.

Quantity also has an important impact on the process choice. Compression molding is economical for many thermoplastic and thermosetting designs where the quantity required is small or modest, yet would not be economical if the quantity required were large.

In order to design plastic components for producibility, the following guidelines should be observed:

1. Small holes (less than $\frac{1}{16}$ in. diameter) should be drilled after molding.
2. Molded blind holes should be limited in depth to twice their diameter.
3. All holes should be located perpendicular to the parting line to allow easy removal of the part from the mold.
4. Undercuts should be avoided since they require a more costly mold. The mold must either be split or have a removable core section.
5. The distance between adjacent holes should be larger than $\frac{1}{8}$ in.
6. The height of bosses should be less than twice their diameter.
7. In the design of bosses, taper of at least 5° on each side should be incorporated to ensure easy withdrawal from the mold.
8. Radii at both the top and the base should be included in the design of bosses and ribs. These radii should not be less than $\frac{1}{32}$ in.
9. Ribs should be designed with 2–5° taper on each side and the rib width at the base should be at least one-half of the wall thickness; the height should be limited to 1½ times the wall thickness.
10. At the parting line, outside edges should be designed without a radius.
11. The design should permit both ends of inserts to be supported in the mold.
12. Inserts should be at right angles to the parting line.
13. A taper of 1–2° should be specified on those vertical surfaces parallel to the direction of mold pressure.
14. Concave numbers and lettering should be engraved in the mold. Letters should be approximately $\frac{3}{32}$ in. high and 0.007 in. deep.
15. Threads that are less than $\frac{5}{16}$ in. in diameter should be cut after molding.

Table 29.5 provides helpful design information related to the principal processes used in the fabrication of thermoplastic and thermosetting resins.

29.7   GROUP TECHNOLOGY IN FUNCTIONAL DESIGN

Group technology implies the classification and coding of the various components utilized in a company's products so that parts similar in shape and processing sequence are numerically identified. When all of a company's products are so classified and coded, it is an easy matter to retrieve similar designs for review and study when beginning a new design. In this way not only will design time be reduced by capitalizing on existing designs, but design quality including designing for economic production will be enhanced. New designs can be developed to take advantage of existing facilities equipped with universal-type quick-acting jigs and fixtures.

Where group technology is practiced, frequently the functional designer will find it unnecessary

Table 29.5 Basic Processes for Fabricating Plastics and Their Principal Parameters

Process	Shape Produced	Machine	Mold or Tool	Material	Typical Tolerance	Minimum Wall Thickness	Ribs	Draft	Inserts	Minimum Quantity Requirements
Calendering	Continuous sheet or film	Multiple roll calender	None	Thermoplastic	0.002–0.008 in. (0.05–0.200 mm) depending on material	None	None	None	None	Low
Extrusion	Continuous form as rods, tubes, filaments, and simple shapes	Extrusion press	Hardened steel die	Thermoplastic	0.004–0.012 in. (0.10–0.30 mm) depending on material	None	None	None	Possible to extrude over or around wire insert	Low (tooling is inexpensive)
Compression molding	Simple outlines and plain cross sections	Compression press	Hardened tool steel mold	Thermoplastic or thermosetting	0.0016–0.01 in. (0.04–0.25 mm) depending on material	0.05 in. (1.25 mm)	None	None	Yes	Low
Transfer molding	Complex geometries possible	Transfer press	Hardened tool steel mold	Thermosetting	0.0016–0.01 in. (0.04–0.25 mm) depending on material	0.06 in. (1.5 mm)	3–5° taper Height larger than three wall thickness	½–5°	Yes	High
Injection molding	Complex geometries possible	Injection press	Hardened tool steel mold	Thermoplastic or thermosetting	0.0016–0.01 in. (0.04–0.25 mm) depending on material	0.05 in. (1.25 mm)	2–5° taper Height = 1½ wall thickness Width = ½ wall thickness	¼–4°	Yes	High

Process	Shape	Equipment	Mold	Material	Thickness	Max thickness	Reinforcement	Cost
Casting	Simple outlines and plain cross sections	None	Metal mold or epoxy mold	Thermosetting	0.004–0.02 in. (0.10–0.50 mm) depending on material	0.08 in. (2.0 mm)	Yes	Low to medium depending on mold
Cold molding	Simple outlines and plain cross sections	None	Mold of wood, plaster, or steel	Thermosetting	0.004–0.020 in. (0.10–0.50 mm) depending on material	0.08 in. (2.0 mm)	Yes	Low
Thermoforming	Thin walled and cup shaped	Thermoforming machine	Suitable form	Thermoplastic			No	Low
Blow molding	Thin walled and bottle shaped	Pneumatic blow molding machine	Tool steel mold	Thermoplastic	0.012–0.024 in. (0.30–0.60 mm)	0.12 in. (3.0 mm)	No	High
Rotational molding	Full or semi-enclosures (hollow objects)	Rotomolding system	Cast aluminum or fabricated metal	Thermoplastic, limited thermosetting	0.012–0.024 in. (0.30–0.60 mm)	0.12 in. (3.0 mm)	No	Medium
Filament winding	Tubes, piping, tanks	Filament winding machine	Must have axis about which the filament can be wound	Single-end continuous strand glass fiber and thermoplastic	0.008–0.020 in. (0.20–0.50 mm)	0.12 in. (3.0 mm)	No	Medium

to design a completely new part. He or she will have three choices for design rationalization: use a former design as it currently is, modify an existing design, and design a new part.

29.8   COMPUTER-AIDED MANUFACTURING PLANNING

Manufacturing planning is an important step in the transformation of a new functional design into a marketable product. Manufacturing planning involves the identification of the best ways to produce the product so that quality and reliability are ensured at a competitive profitable price. Good manufacturing planning will help ensure the success of the enterprise, while poor manufacturing planning may result in costs that are so prohibitive that the product may not sell.

Manufacturing planning must consider several alternatives in order to select the best procedure to produce a part. Furthermore, there is seldom much time to perform this function. The product usually is required by the customer shortly after design approval. The computer permits process planning rapidly while considering all possible alternatives.

To plan the complete processing of a product, four main classes of processes should be considered: basic processes, secondary processes, finishing processes, and quality control processes. Different parameters need to be considered in connection with each of these four families.

1. *Basic Processes.*   Those processes that are initially used to approximate the geometrical configuration of the component. Typical basic processes include casting, forging, extrusion, and roll forming.

2. *Secondary Processes.*   Those processes that follow the basic processes to bring the product to its final dimensional accuracy and form exclusive of protective and/or aesthetic coatings. Representative secondary processes include drilling, reaming, tapping, turning, facing, grooving, grinding, broaching, honing, and heat treating.

3. *Finishing Processes.*   Those processes applied, usually after the completion of all secondary processes, to bring the product to its final specifications by the application of a protective or decorative coating. Finishing processes include plating, anodizing, enameling, and painting.

4. *Quality Control Processes.*   Those processes applied to ensure the quality and reliability of the product. Typical quality control processes include inspection and packing for shipment.

With reference to the selection of the optimum basic process, the computer program would need to consider the following parameters: size of part, geometry or configuration of part, material being used, microstructure resulting from the process, relative cost of secondary process that will need to be performed, quantity to be produced, and cost.

All of the preceding parameters can serve as constraints. Size has a limiting effect on several basic processes. For metals, die castings larger than 75 lb are seldom produced, sand castings less than 2 oz usually are not made, and extrusions with cross-sectional areas of more than 140 in.^{2} are generally not produced.

Geometry, too, serves as an important constraint. For example, complex geometries can be cast but cannot be forged, nonsymmetrical bowl-shaped parts cannot be spun, and designs with undercut or reentrant angles are not acceptable to the powdered metal process.

Material is a very important parameter related to process selection. Ferrous metals generally are not pressure die cast because of economics. Plaster mold casting is limited to nonferrous metals. Compression molding is the usual technique for producing thermoplastic plates less than 0.25 in. thick.

The microstructure resulting from the process may be a consideration. For example, controlled grain flow may be required for end use products subject to dynamic or impact loading.

Quantity is a most important consideration. A permanent mold such as a die casting die would not be made for the production of a small number of parts. Similarly, a dozen parts from a given bar stock design would be produced on a lathe not on an automatic screw machine.

The cost of the secondary operations must also be considered in selecting the most favorable basic operations. Secondary operation cost will vary considerably depending upon what basic operation was used. For example, the secondary operations required on a ferrous metal gear made of powdered metal would be much less than a similar gear made as a ferrous sand casting.

Of course, cost of the basic operation must be considered. If the quality and product reliability resulting from competing processes are equal and assuming the delivery time is the same, we will want to select the process that will result in the lowest product cost.

Software can be developed that considers the preceding parameters so that all alternatives are considered and the most favorable process may be selected for new designs during initial design phases. In this way the best process can be selected well in advance of completion of the design, and the functional designer will be able to incorporate design requirements characteristic of the processes selected.

BIBLIOGRAPHY

L. Alting, *Manufacturing Engineering Processes,* Marcel Dekker, New York, 1982.

Defense Systems Management College, *Manufacturing Management Handbook for Program Managers,* Fort Belvoir, VA, 1982.

E. P. DeGarmo, *Materials and Processes in Manufacturing,* 5th ed., MacMillan, New York, 1979.

L. E. Doyle, *Manufacturing Processes and Materials for Engineers,* Prentice-Hall, Englewood Cliffs, NJ, 1969.

D. D. Greenwood, *Mechanical Details for Product Design,* McGraw-Hill, New York, 1964.

Headquarters, U.S. Army Material Command, *Engineering Design Handbook, Design Guidance for Producibility,* Washington, DC, 1971.

R. LeGrand, *Manufacturing Engineers' Manual,* McGraw-Hill, New York, 1971.

B. W. Niebel and A. B. Draper, *Product Design and Process Engineering,* McGraw-Hill, New York, 1974.

B. Niebel and E. N. Baldwin, *Designing for Production,* Richard D. Irwin, Inc., Homewood, IL, 1963.

H. E. Trucks, *Designing for Economical Production,* Society of Manufacturing Engineers, Dearborn, MI, 1974.

H. W. Yankee, *Manufacturing Processes,* Prentice-Hall, Englewood Cliffs, NJ, 1979.

CHAPTER 30
CLASSIFICATION SYSTEMS

DELL K. ALLEN
PAUL SMITH

CAM Software Laboratory
Brigham Young University
Provo, Utah

30.1 PART FAMILY CLASSIFICATION AND CODING

30.1.1 Introduction

History

Classification and coding practices are as old as the human race. They were used by Adam in the Bible to classify and name plants and animals, by Aristotle to identify basic elements of the earth, and in more modern times to classify concepts, books, and documents. But the classification and coding of manufactured pieceparts is relatively new. Early pioneers associated with workpiece classification are Mitrafanov of the USSR, Gombinski and Brisch both of the United Kingdom, and Opitz of Germany. In addition, there are many who have espoused the principles developed by these men and have adapted them, have enlarged upon them, and have created comprehensive workpiece classification systems. It has been reported that over 100 such systems have been created—specifically for machined parts, others for castings or forgings, others for sheet metal parts, etc. In the United States there are at least six workpiece classification systems that are commercially available. These include the Brisch–Bern system, MDSI's "CODE," TNO's "MICLASS," the "Analog" system, NAMCO's "UNIDEX," and the Opitz code.

Why are there so many different systems? In attempting to answer this question, it should be pointed out that different workpiece classification systems were initially developed for different purposes. Mitrafanov apparently developed his system to aid in formulating group production cells and in facilitating the design of standard tooling packages. Opitz developed his system for ascertaining the workpiece shape/size distribution to aid in designing suitable production equipment. The Brisch system is oriented toward design retrieval. More recent systems are production oriented.

Thus, the intended application perceived by those who have developed workpiece classification systems has been a major factor in their proliferation. Another significant factor has been personal preferences in identification of attributes and relationships. Few system developers totally agree as to what should or should not be the basis of classification. For example: Is it better to classify a workpiece by function as "standard" or "special" or by geometry as "rotational" or "nonrotational"? Either of these choices makes a significant impact on how a classification system will be developed.

Most classification systems are hierarchal, going from the general to the very specific. This has been referred to by the Brisch people as a monocode system. In an attempt to derive a workpiece code that addressed the question of how to include several related, but nonhierarchal workpiece features, the feature code or "polycode" concept was developed. Some classification systems now include both polycode and monocode concepts.

A few classification systems are quite simple and yield a short code of five or six digits. Other classification systems are very comprehensive and yield codes of up to 32 digits. Some codes are numeric and some are alphanumeric. The combination of such factors as application, identified attributes and relationships, hierarchal versus feature coding, comprehensiveness, and code format and length have resulted in a proliferation of classification systems.

30.1.2 Application

Identification of intended applications for a workpiece classification system are critical to the selection, development, or tailoring of a system.

It is not likely that any given system can readily satisfy both known present applications and unknown future applications. Nevertheless, a classification system can be developed in such a way as to minimize problems of adaptation. To do this, present and anticipated applications must be identified. It should be pointed out that development of a classification system for a narrow, specific application is relatively straightforward. Creation of a classification system for multiple applications, on the other hand, can become very complex and costly.

Figure 30.1 is a matrix illustrating this principle. As the applications increase, the number of required attributes also generally increases. Consequently, system complexity also increases but often at a geometric or exponential rate owing to the increased number of combinations possible. Therefore, it is important to establish reasonable application requirements first while avoiding unnecessary requirements and, at the same time, to make provision for adaptation to future needs.

In general, a classification system can be used to aid (1) design, (2) process planning, (3) materials control, and (4) management planning. A brief description of selected applications follows.

Design Retrieval

Before new workpieces are introduced into the production system, it is important to retrieve similar designs to see if a suitable one already exists or if an existing design may be slightly altered to accommo-

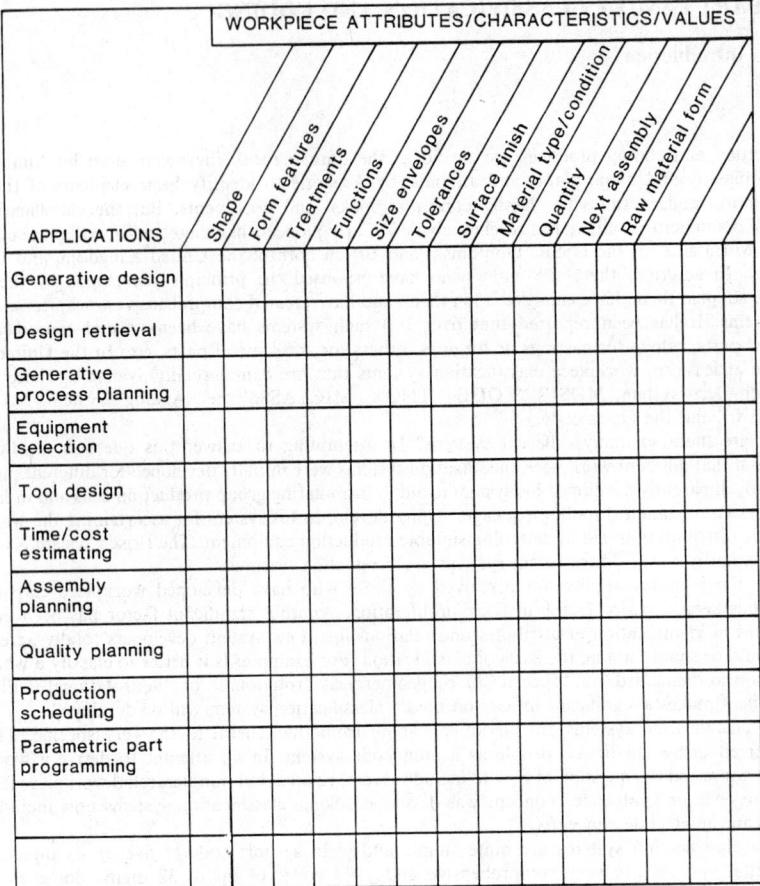

Fig. 30.1 Attribute selection matrix.

date new requirements. Potential savings from avoiding redundant designs range in the thousands of dollars.

Design retrieval also provides an excellent starting point for standardization and modularization. It has been stated that "only 10–20% of the geometry of most workpieces relates to the product function." The other 80–90% of the geometric features are often a matter of individual designer taste or preference. It is usually in this area that standardization could greatly reduce production costs, improve product reliability, increase ease of maintenance, and provide a host of other benefits.

One potential benefit of classification is in meeting the product liability challenge. If standard analytic tools are developed for each part family, and if product performance records are kept for those families, then the chances of negligent or inaccurate design is greatly reduced.

The most significant production savings in manufacturing enterprise begin with the design function. The function must be carefully integrated with the other functions of the company, including materials requisition, production, marketing, and quality assurance. Otherwise, suboptimization will likely occur with its attendant frequent redesign, rework, scrap, excess inventory, employee frustration, low productivity, and high costs.

Generative Process Planning

One of the most challenging and yet potentially beneficial applications of workpiece classification is that of process planning. The workpiece class code can provide the information required for logical, consistent process selection and operation planning.

The various segments of the part family code may be used as keywords on a comprehensive process classification taxonomy. Candidate processes are those that satisfy the conditions of the given basic shape *and* the special features *and* the size *and* the precision *and* the material type *and* the form *and* the quantity/time requirements.

After outputting the suitable processes, economic or other considerations may govern final process selection. When the suitable process has been selected, the codes for form features, heat treatments, coatings, surface finish, and part tolerance govern computerized selection of fabrication and inspection operations. The result is a generated process plan.

Production Estimating

Estimating of production time and cost is usually an involved and laborious task. Often the results are questionable because of unknown conditions, unwarranted assumptions, or shop deviations from the operation plan. The part family code can provide an index to actual production times and costs for each part family. A simple regression analysis can then be used to provide an accurate predictor of costs for new parts falling in a given part family. Feedback of these data to the design group could provide valuable information for evaluating alternative designs prior to their release to production.

Parametric and Generative Design

Once the product mix of a particular manufacturing enterprise has been established, high-cost, low-profit items can be singled out. During this sorting and characterization process, it is also possible to establish tabular or parametric designs for each basic family. Inputting of dimensional values and other data to a computer graphics system can result in the automatic production of a drawing for a given part. Taking this concept back one more step, it is conceivable that merely inputting a product name, specifications, functional requirements, and some dimensional data would result in the generation of a finished design drawing. Workpiece classification offers many exciting opportunities for productivity improvement in the design arena.

Parametric Part Programming

A logical extension of parametric design is that of parametric part programming. Although parametric part programming or family of parts programming has been employed for some time in advanced numerical control (NC) work, it has not been tied effectively to the design data base. It is believed that workpiece classification and coding can greatly assist with this integration. Parametric part programming provides substantial productivity increases by permitting the use of common program modules and reduction of tryout time.

Tool Design Standardization

The potential savings in tooling costs are astronomical when part families are created and when form features are standardized. The basis for this work is the ability to adequately characterize component pieceparts through workpiece classification and coding.

30.1.3 Classification Theory

This section outlines the basic premises and conventions underlying the development of a Part Family Classification and Coding System.

Basic Premises

The first premise underlying the development of such a system was that a workpiece may be best characterized by its most apparent and permanent attribute, which is its basic shape. The second premise was that each basic shape may have many special features (e.g., holes, slots, threads, coatings) superimposed upon it while retaining membership in its original part family. The third premise was that a workpiece may be completely characterized by (1) basic shape, (2) special features, (3) size, (4) precision, and (5) material type, form, and condition. The fourth premise was that code segments could be linked to provide a humanly recognizable code, and that these code segments could provide pointers to more detailed information. A fifth premise was that a short code would be adequate for human monitoring, and linking to other classification trees, but that a bitstring (0's, 1's) that is computer recognizable would best provide the comprehensive and detailed information required for retrieval and planning purposes. Each bit in the bitstring represents the presence or absence of a given feature and provides a very compact, computer-processible representation of a workpiece without an excessively long code. The sixth premise was that mutually exclusive workpiece characteristics should provide unique basic shape families for the classification, and that common elements (e.g., special features, size, precision, and materials) should be included only once but accessed by all families.

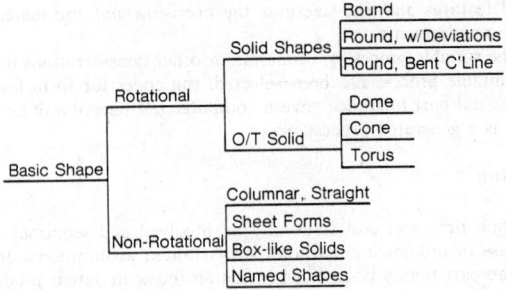

Fig. 30.2 E-tree concept applied to basic shape classification.

E-Tree Concept

Hierarchal classification trees with mutually exclusive data (E-trees) provide the foundation for establishing the basic part shape (Fig. 30.2). Although a binary-type hierarchal tree is preferred because it is easy to use, it is not uncommon to find three or more branches. It should be pointed out, however, that because the user must select only one branch, more than two branches require a greater degree of discrimination. With two branches the user may say, "Is it this or that?" With five branches the user must consider, "Is it this or this or this or this or this?" The reading time and error rate likely increase with the number of branches at each node. The E-tree is very useful for dividing a large collection of items into mainly exclusive families or sets.

N-Tree Concept

The N-tree concept is based on a hierarchal tree with nonmutually exclusive paths (i.e., all paths may be selected concurrently). This type of tree (Fig. 30.3) is particularly useful for representing the common attributes mentioned earlier (e.g., form features, heat treatments, surface finish, size, precision, and material type, form, and condition).

In the example shown in Fig. 30.3, the keyword is Part Number (P/N) 101. The attributes selected are shown by means of an asterisk (*). In this example the workpiece is characterized as having a "bevel," a "notch," and a "tab."

Bitstring Representation

During the traversal of either an E-tree or an N-tree a series of 1's and 0's are generated, depending on the presence or absence of particular characteristics or attributes. The keyword (part number) and its associated bitstring might look something like this:

$$\text{P/N-101} = 100101 \cdots 010$$

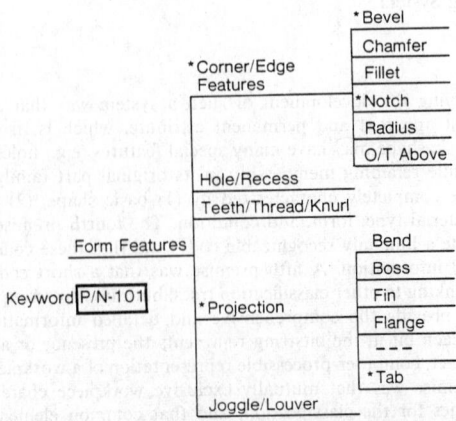

Fig. 30.3 N-tree concept applied to form features.

The significance of the bitstring is twofold. First, one 16-bit computer word can contain as many as 16 different workpiece attributes. This represents a significant reduction in computer storage space as compared with conventional representation. Second, the bitstring is in the proper format for rapid computer processing and information retrieval. The conventional approach is to use lists and pointers. This requires relatively large amounts of computation and usually a large computer to achieve reasonable response time.

Keywords

A keyword is an alphanumeric label with its associated bitstring. The label may be descriptive of a concept (e.g., stress, speed, feed, chip-thickness ratio), or it may be descriptive of an entity (e.g., cutting tool, vertical mill, 4340 steel, P/N-101). In conjunction with the Part Family Classification and Coding System, a number of standard keywords are provided. To conserve space and facilitate data entry, some of these keywords consist of one to three character alphanumeric codes. For example, the keyword code for a workpiece that is rotational and concentric, with two outside diameters and one bore diameter is "B11." The keyword code for a family of low-alloy low-carbon steels is "A1." These codes are easy to use and greatly facilitate concise communication. These codes may be used as output keys or input keys to provide the very powerful capability of linking to other types of hierarchal information trees, such as those used for process selection, equipment selection, or automated time standard setting.

Standard keyword codes for the basic shape families, form features, size, precision, and material type are given in Appendix 30.1.

30.1.4 Part Family Code

Purpose

Part classification and coding is a prerequisite to the introduction of group technology, computer-aided process planning, design retrieval, and many other manufacturing activities. Part classification and coding is aimed at improving productivity, reducing unnecessary variety, improving product quality, and reducing direct and indirect cost.

Code Format and Length

The part family code shown in Fig. 30.4 is composed of a five-section alphanumeric code. The first section of the code gives the basic shape. Other sections provide for form features, size, precision, and material. Each section of the code may be used as a pointer to more detailed information or as an output key for subsequent linking with related decision trees. The code length is eight digits. Each digit place has been carefully analyzed so that a compact code would result that is suitable for human communication and yet sufficiently comprehensive for generative process planning. The three-digit basic shape code provides for 240 standard families, 1160 custom families, and 1000 functional or named families. In addition, the combination of 50 form features, 9 size ranges, 5 precision classes, and 79 material types makes possible 2.5×10^{71} unique combinations! This capability should satisfy even the most sophisticated user.

Basic Shape

The basic shapes may be defined as those created from primitive solids and their derivatives (Fig. 30.5) by means of a basic founding process (cast, mold, machine). Primitives have been divided into rotational shapes and nonrotational shapes. Rotational primitives include the cylinder, sphere, cone, ellipsoid, hyperboloid, and toroid. The nonrotational primitives include the cube (parallelepiped), polyhedron, warped (contoured) surfaces, free forms, and named shapes. The basic shape families are subdivided

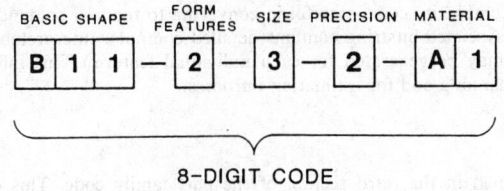

Fig. 30.4 Part family code.

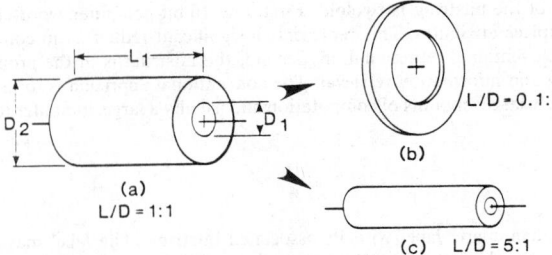

Fig. 30.5 Permutations of concentric cylinders.

on the basis of predominant geometric characteristics, including external and internal characteristics. (See Appendix 30.1.)

The derivative concentric cylinder shown in Fig. 30.5 may have several permutations. Each permutation is created by merely changing dimensional ratios as illustrated or by adding form features. The rotational cylindrical shape shown may be thought of as being created from the intersection of a negative cylinder with a positive cylinder.

Figure 30.5a, with length/diameter (L/D) ratio of $1:1$, could be a spacer; Fig. 30.5b, with an L/D ratio of $0.1:1$, would be a washer; and Fig. 30.5c, with an L/D ratio of $5:1$ could be a thin-walled tube. If these could be made using similar processes, equipment, and tooling, they could be said to constitute a family of parts.

Name or Function Code

Some geometric shapes are so specialized that they may serve only one function. For example, a crankshaft has the major function of transmitting reciprocating motion to rotary motion. It is difficult to use a crankshaft for other purposes. For design retrieval and process planning purposes, it would probably be well to classify all crankshafts under the code name "crankshaft." Of course, it may still have a geometric code such as "P75," but the descriptive code will aid in classification and retrieval. A controlled glossary of function codes with cross references, synonyms, and preferred labels would aid in using name and function codes and avoid unnecessary proliferation.

Special Features

To satisfy product design requirements, the designer creates the basic shape of a workpiece and selects the engineering material of which it is to be made. The designer may also require special processing treatments to enhance properties of a given material. In other words, the designer adds special features. Special features of a workpiece include form features, heat treatments, and special surface finishes.

Form features may include holes, notches, splines, threads, etc. The addition of a form feature does not change the basic part shape (family) but does enable it to satisfy desired functional requirements. Form features are normally imparted to the workpiece subsequent to the basic founding process.

Heat treatments are often given to improve strength, hardness, and wear resistance of a material. Heat treatments such as stress relieving or normalizing may also be given to aid in processing the workpiece.

Surface finishing treatments such as plating, painting, and anodizing, are given to enhance corrosion resistance, improve appearance, or meet some other design requirement.

The special features are contained in an N-tree format with an associated complexity evaluation and classification feature. This permits the user to select many special features while still maintaining a relatively simple code. Basically, nine values (1–9) have been established as the special feature complexity codes. As the user classifies the workpiece and identifies the special features required, the number of features is tallied and an appropriate complexity code is stored. Figure 30.6 shows the number count for special features and the associated feature code.

The special feature complexity code is useful in conveying to the user some idea of the complexity of the workpiece. The associated bitstring contains detailed computer-interpretable information on all features. (Output keys may be generated for each individual feature.) This information is valuable for generative process planning and for estimating purposes.

Size Code

The size code is contained in the third section of the part family code. This code consists of one numeric digit. Values range from 1 to 9, with 9 representing very large parts (Fig. 30.7). The main

FEATURE COMPLEXITY CODE	NO. SPECIAL FEATURES
1	1
2	· 2
3	3
4	5
5	8
6	13
7	21
8	34
9	GT 34

Fig. 30.6 Complexity code for special features.

purpose of the size code is to give the code user a feeling for the overall size envelope for the coded part. The size code is also useful in selecting production equipment of the appropriate size.

Precision Class Code

The precision class code is contained in the fourth segment of the part family code. It consists of a single numeric digit with values ranging from 1 to 5. Precision in this instance represents a composite of tolerance and surface finish. Class 1 precision represents very close tolerances and a precision-ground or lapped-surface finish. Class 5, on the other hand, represents a rough cast or flame-cut surface with a tolerance of greater than $\frac{1}{32}$ in. High precision is accompanied by multiple processing operations and careful inspection operations. Production costs increase rapidly as closer tolerances and finer surface finishes are specified. Care is needed by the designer to ensure that high precision is warranted. The precision class code is shown in Fig. 30.8.

Material Code

The final two digits of the part family code represent the material type. The material form and condition codes are captured in the associated bitstring.

Seventy-nine distinct material families have been coded (Fig. 30.9). Each material family or type is identified by a two-digit code consisting of a single alphabetic character and a single numeric digit.

PART FAMILY SIZE CLASSIFICATION

SIZE CODE	MAXIMUM DIMENSION		DESCRIPTION	EXAMPLES
	ENGLISH (in.)	METRIC (mm)		
1	.5	10	Sub−miniature	Capsules
2	2	50	Miniature	Paper clip box
3	4	100	Small	Large match box
4	10	250	Medium−small	Shoe box
5	20	500	Medium	Bread box
6	40	1000	Medium−large	Washing machine
7	100	2500	Large	Pickup truck
8	400	10000	Extra−large	Moving van
9	1000	25000	Giant	Railroad box−car

Fig. 30.7 Part family size classification.

PRECISION CLASS CODE

CLASS CODE	TOLERANCE	SURFACE FINISH
1	LE .0005″	LE 4 RMS
2	.0005″–.002″	4–32 RMS
3	.002″–.010″	32–125 RMS
4	.010″–.030″	125–500 RMS
5	GT .030″	GT 500 RMS

Fig. 30.8 Precision class code.

The stainless-steel family, for example, is coded "A6." The tool steel family is "A7." This code provides a pointer to specification sheets containing comprehensive data on material properties, availability, and processability.

The material code provides a set of standard interface codes to which may be appended a given industry class code when appropriate. For example, the stainless-steel code may have appended to it a specific material code to uniquely identify it as follows: "A6-430" represents a chromium-type, ferritic, nonhardenable stainless steel.

30.1.5 Tailoring the System

It has been found that nearly all classification systems must be customized to meet the needs of each individual company or user. This effort can be greatly minimized by starting with a general system and then tailoring it to satisfy unique user needs. The Part Family Classification and Coding System would permit this customizing. It is easy to add new geometric configurations to the existing family of basic shapes. It is likewise simple to add additional special features or to modify the size or precision class ranges. New material codes may be readily added if necessary.

The ability to modify easily an existing classification system without extensively reworking the system is one test of its design.

30.2 ENGINEERING MATERIALS TAXONOMY

30.2.1 Introduction

Serious and far-reaching problems exist with traditional methods of engineering materials selection. The basis for selecting a material is often tenuous and unsupported by defensible selection criteria and methods. A taxonomy of engineering materials accompanied by associated property files can greatly assist the designer in choosing materials to satisfy a design's functional requirements as well as procurement and processing requirements.

Material Varieties

The number of engineering materials from which a product designer may choose is staggering. It is estimated that over 40,000 metals and alloys are available, plus 250,000 plastics, uncounted composites, ceramics, rubbers, wood products, etc. From this list, the designer must select the one for use with the new product. Each of these materials can exhibit a wide range of properties, depending on its form and condition. The challenge faced by the designer in selecting optimum materials can be reduced by a classification system to aid in identifying suitable material families.

Material Shortages

Dependency on foreign nations for certain key alloying elements such as chromium, cobalt, tungsten, and tin points up the critical need for conserving valuable engineering materials, and for selecting less strategic materials wherever possible. The recycleability of engineering materials has become another selection criterion.

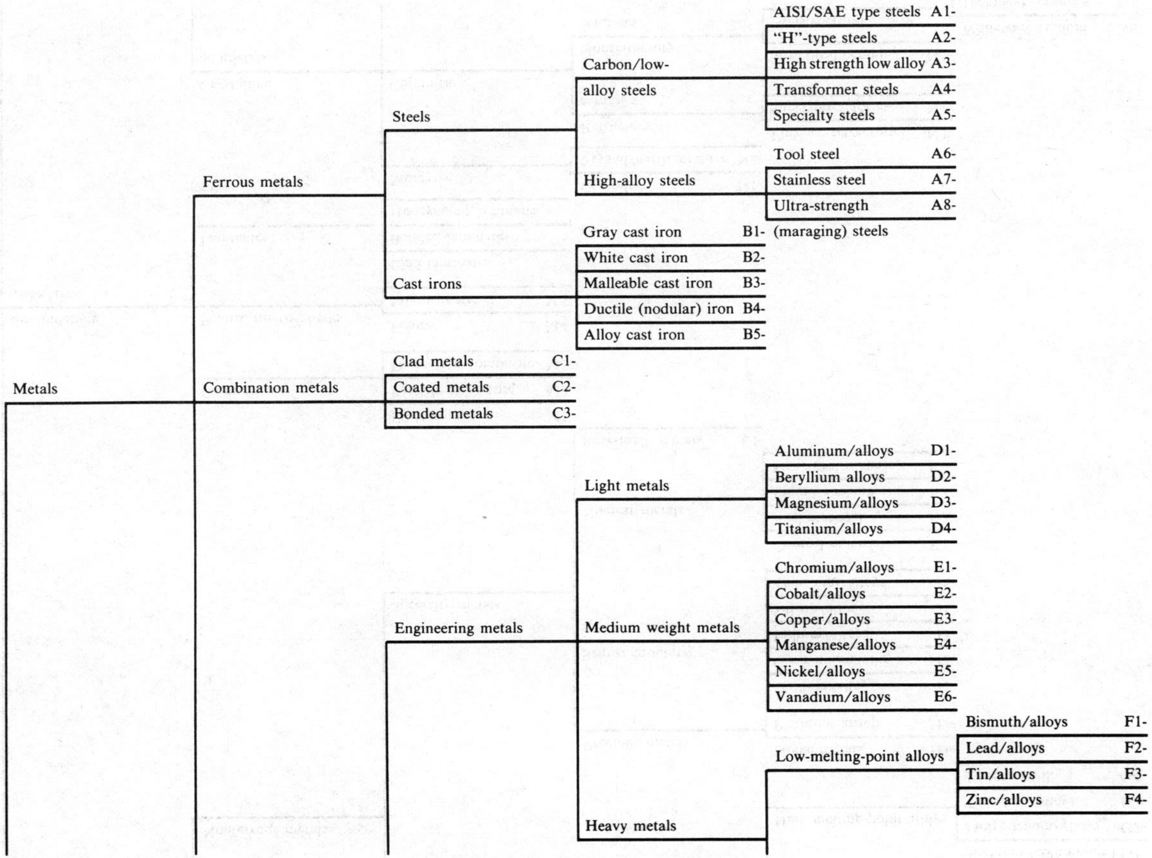

Fig. 30.9 Engineering materials.

Engineering materials classification:

- **Nonferrous metals**
  - **Specialty metals**
    - High-melting-point alloys
      - Niobium (columbium) G1-
      - Molybdenum/alloys G2-
      - Tantalum/alloys G3-
      - Tungsten/alloys G4-
    - Precious metals
      - Noble metals H1-
      - Platinum group H2-
    - Semiconductor/specialty metals
      - Gallium/alloys J1-
      - Germanium/alloys J2-
      - Indium/alloys J3-
      - Silicon/alloys J4-
      - Tellurium/alloys J5-
    - Nuclear metals
      - Control materials K1-
      - Fuel material K2-
      - Liquid coolants K3-
      - Structural materials K4-
    - Rare-earth metals L1-

- **Combination materials**
  - **Composites**
    - Fiber composite M1-
    - Particle composite M2-
    - Dispersion composite M3-
  - **Foams, microspheres**
    - Foams M4-
    - Microspheres M5-
  - **Laminates**
    - Clad laminates
    - Bonded laminates M6-
    - Honeycomb laminates

- **Crystalline nonmetals**
  - **Minerals**
    - Crystals N1-
    - Crystal/earth mixture N2-
  - **Ceramics**
    - Refractory ceramics
      - Furnace refractories N3-
      - Super-refractories N4-
    - Nonrefractory ceramics
      - Structural ceramics N5-
      - Nonstructural
        - Whiteware ceramics N6-
        - Technical ceramics N7-

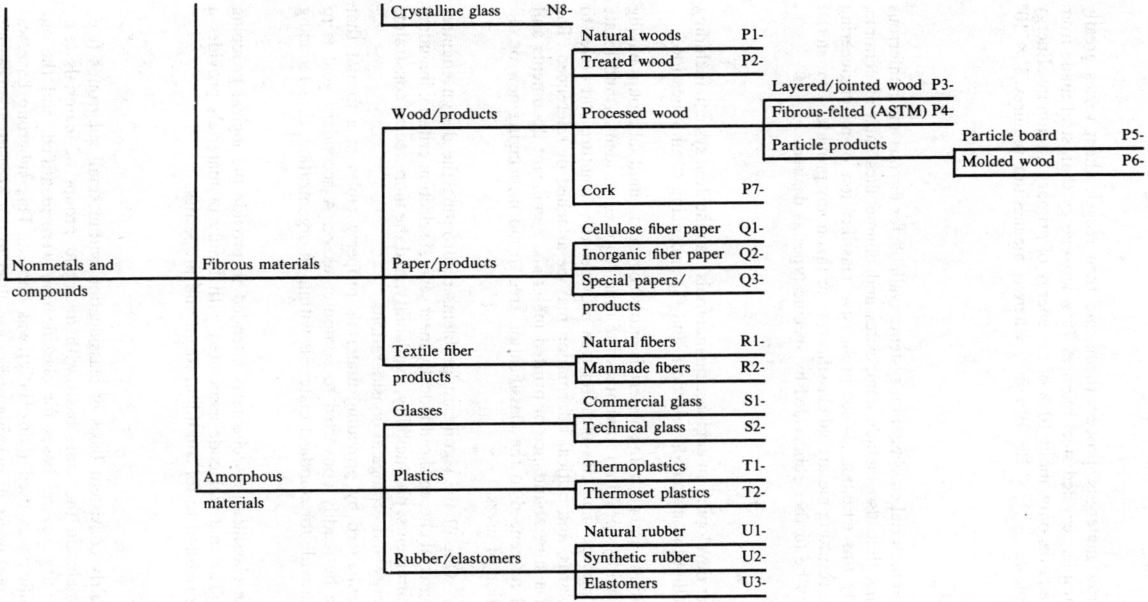

Crystalline glass — N8-

Nonmetals and compounds

Fibrous materials

Wood/products
- Natural woods — P1-
- Treated wood — P2-
- Processed wood
  - Layered/jointed wood — P3-
  - Fibrous-felted (ASTM) — P4-
  - Particle products
    - Particle board — P5-
    - Molded wood — P6-
- Cork — P7-

Paper/products
- Cellulose fiber paper — Q1-
- Inorganic fiber paper — Q2-
- Special papers/products — Q3-

Textile fiber products
- Natural fibers — R1-
- Manmade fibers — R2-

Amorphous materials

Glasses
- Commercial glass — S1-
- Technical glass — S2-

Plastics
- Thermoplastics — T1-
- Thermoset plastics — T2-

Rubber/elastomers
- Natural rubber — U1-
- Synthetic rubber — U2-
- Elastomers — U3-

Fig. 30.9 (*Continued*)

Energy Requirements

The energy required to produce raw materials, process them, and then recycle them varies greatly from material to material. For example, recycled steel requires 75% less energy than steel made from iron ore, and recycled aluminum requires only about 10% of the energy of primary aluminum. Energy on a per volume basis for ABS plastic is 2×10^6 Btu/in.^{3}, whereas magnesium requires 8×10^6 Btu/in.^{3}.

30.2.2   Material Classification

Although there are many specialized material classification systems available for ferrous and nonferrous metals, there are no known systems that also include composites and nonmetalics such as ceramic, wood, plastic, or glass. To remedy this situation, a comprehensive classification of all engineering materials has been undertaken. The resulting hierarchal classification or taxonomy provides 79 material families. Each of these families may be further subdivided by specific types as desired.

Objectives

Three objectives were established for developing an engineering materials classification system, including (1) minimizing search time, (2) facilitating materials selection, and (3) enhancing communication.

Minimize Search Time.   Classifying and grouping materials into recognized, small subgroups having similar characteristic properties (broadly speaking) minimizes the time required to identify and locate other materials having similar properties. The classification tree provides the structure and codes to which important procedures, standards, and critical information may be attached or referenced. The information explosion has brought a superabundance of printed materials. Significant documents and information may be identified and referenced to the classification tree to aid in bringing new or old reference information to the attention of users.

Facilitate Materials Selection.   One of the significant problems confronting the design engineer is that of selecting materials. The material chosen should ideally meet several selection criteria, including satisfying the design functional requirements, producibility, availability, and the more recent constraints for life-cycle costing, including energy and ecological considerations.

Materials selection is greatly enhanced by providing materials property tables in a format that can be used manually or that can be readily converted to computer usage. A secondary goal is to reduce material proliferation and provide for standard materials within an organization, thus reducing unnecessary materials inventory.

Enhance Communication.   The classification scheme is intended to provide the logical grouping of materials for coding purposes. The material code associated with family of materials provides a pointer to the specific material desired and to its condition, form, and properties.

Basis of Classification

Although it is possible to use a fairly consistent basis of classification within small subgroups (e.g., stainless steels), it is difficult to maintain the same basis with divergent groups of materials (e.g., nonmetals). Recognizing this difficulty, several bases for classification were identified, and the one which seemed most logical (or which was used industrially) was chosen. This subgroup base was then cross-examined relative to its usefulness in meeting objectives cited in the preceding subsection.

The various bases for classification considered for the materials taxonomy are shown in Fig. 30.10.

The particular basis selected for a given subgroup depends on the viewpoint chosen. The overriding viewpoint for each selection was (1) will it facilitate material selection for design purposes, and (2) does it provide a logical division that will minimize search time in locating materials with a predominant characteristic or property?

Taxonomy of Engineering Materials

An intensive effort to produce a taxonomy of engineering materials has resulted in the classification shown in Fig. 30.11. The first two levels of this taxonomy classifies all engineering materials into the broad categories of metals, nonmetals and compounds, and combination materials. Metals are further subdivided into ferrous, nonferrous, and combination metals. Nonmetals are classified as crystalline, fibrous, and amorphous.

Combination materials are categorized as composites, foams, microspheres, and laminates. Each of these groups is further subdivided until a relatively homogeneous materials family is identified. At this final level a family code is assigned.

Base		Example
A.	State	Solid–liquid–gas
B.	Structure	Fibrous–crystalline–amorphous
C.	Origin	Natural–synthetic
D.	Application	Adhesive–paint–fuel–lubricant
E.	Composition	Organic–inorganic
F.	Structure	Metal–nonmetal
G.	Structure	Ferrous–nonferrous
H.	Processing	Cast–wrought
I.	Processing response	Water-hardening–oil-hardening–air-hardening, etc.
J.	Composition	Low alloy–high alloy
K.	Application	Nuclear–semiconducting–precious
L.	Property	Light weight–heavy
M.	Property	Low melting point–high melting point
N.	Operating environment	Low-temperature–high-temperature
O.	Operating environment	Corrosive–noncorrosive

Fig. 30.10 Basis for classifying engineering materials.

Customizing

The Engineering Materials Taxonomy may be easily modified to fit a unique user's needs. For example, if it were desirable to further subdivide "fiber-reinforced composites," it could easily be done on the basis of type of filament used (e.g., boron, graphite, glass) and further by matrix employed (polymer, ceramic, metal). The code "M1" representing fiber-reinforced composites could have appended to it a dash number uniquely identifying the specific material desired. Many additional material families may also be added if desired.

30.2.3 Material Code

As was mentioned earlier, there are many material classification systems, each of which covers only a limited segment of the spectrum of engineering materials available. The purpose of the Engineering Materials Taxonomy is to overcome this limitation. Furthermore, each of the various materials systems has its own codes. This creates additional problems. To solve this coding compatibility problem a two-character alphanumeric code is provided as a standard interface code to which any industry or user code may be appended. This provides a very compact standard code so that any user will recognize the basic material family even though perhaps not recognizing a given industry code.

Material Code Format

The format used for the material code is shown in Fig. 30.12. The code consists of four basic fields of information. The first field contains a two-character interface code signifying the material family. The second field is to contain the specific material type based on composition or property. This code may be any five-character alphanumeric code. The third field contains a two-digit code containing the material condition (e.g., hot-worked, as-cast, ¾-hard). The fourth and final field of the code contains a one-digit alphabetic code signifying the material form (e.g., bar, sheet, structural shape).

Material Families

Of the 79 material families identified, 13 are ferrous metals, 30 are nonferrous metals, 6 are combination materials (composites, foams, laminates), and 26 are nonmetals and compounds. These materials are shown in Appendix 30.2.

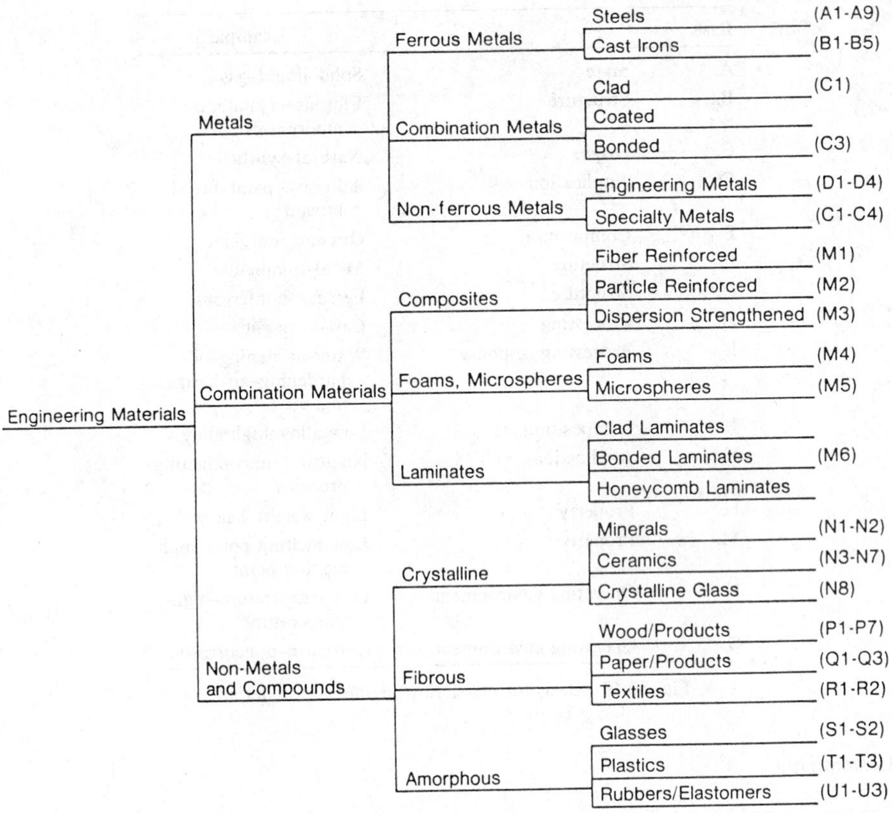

Engineering Materials
Metals
Combination Materials
Non-Metals and Compounds
Ferrous Metals
Combination Metals
Non-ferrous Metals
Composites
Foams, Microspheres
Laminates
Crystalline
Fibrous
Amorphous
Steels (A1-A9)
Cast Irons (B1-B5)
Clad (C1)
Coated
Bonded (C3)
Engineering Metals (D1-D4)
Specialty Metals (C1-C4)
Fiber Reinforced (M1)
Particle Reinforced (M2)
Dispersion Strengthened (M3)
Foams (M4)
Microspheres (M5)
Clad Laminates
Bonded Laminates (M6)
Honeycomb Laminates
Minerals (N1-N2)
Ceramics (N3-N7)
Crystalline Glass (N8)
Wood/Products (P1-P7)
Paper/Products (Q1-Q3)
Textiles (R1-R2)
Glasses (S1-S2)
Plastics (T1-T3)
Rubbers/Elastomers (U1-U3)

Fig. 30.11 Engineering materials taxonomy—three levels.

Material Type

The five-digit code space reserved for material type is sufficient to accommodate the UNS (Unified Numbering System) recently developed by ASTM, SAE, and others for metals and alloys. It will also accommodate industry or user-developed codes for nonmetals or combination materials. An example of the code (Fig. 30.10) for an open-hearth, low-carbon steel would be "A1-C1020," with the first two digits representing the steel family and the last five digits the specific steel alloy.

Material Condition

The material condition code consists of a two-digit code derived for each material family. The intent of this code is to reflect processes to which the material has been subjected and its resultant structure. Because of the wide variety of conditions that do exist for each family of materials, the creation of a D-tree for each of the 79 families seems to be the best approach. The D-tree can contain processing treatments along with resulting grain size, microstructure, or surface condition if desired. Typical material condition codes for steel family "A1" are given in Fig. 30.13.

MATERIAL FAMILY
MATERIAL TYPE
CONDITION
FORM
A 1 — C 1 0 2 0 — 3 A — A
10–DIGIT CODE

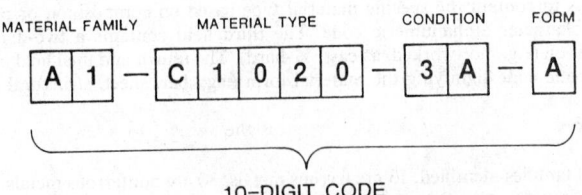

Fig. 30.12 Format for engineering materials code.

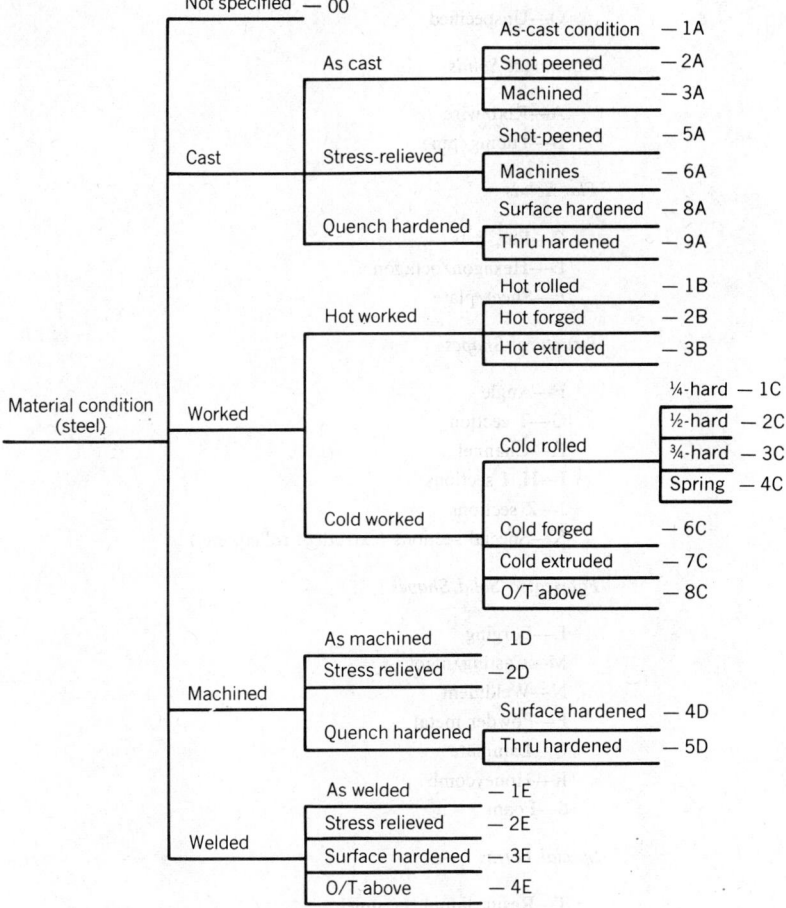

Fig. 30.13 Material condition for steel family "A1."

Material Form

Material form code consists of a single alphabetic character to represent this raw material form (e.g., rod, bar, tubing, sheet, structural shape). Typical forms are shown in Fig. 30.14.

30.2.4 Material Properties

Material properties have been divided into three broad classes: (1) mechanical properties, (2) physical properties, and (3) chemical properties. Each of these will be discussed briefly.

Mechanical Properties

The mechanical properties of an engineering material describe its behavior or quality when subjected to externally applied forces. Mechanical properties include strength, hardness, fatigue, elasticity, and plasticity. Figure 30.15 shows representative mechanical properties. Note that each property has been identified with a unique code number; the purpose of this code number is to reduce confusion in communicating precisely which property is intended. Confusion often arises because of the multiplicity of testing procedures that have been divised to assess the value of a desired property. For example, there are at least 15 different penetration hardness tests in common usage, each of which yields different numerical results from the others. The code uniquely identifies the property and the testing method used to ascertain it.

Each property of a material is intimately related to its composition, surface condition, internal

O—Unspecified

Rotational Solids

    A—Rod/wire
    B—Tubing/pipe

Flat Solids

    C—Bar, flats
    D—Hexagon/octagon
    E—Sheet/plate

Structural Shapes

    F—Angle
    G—T section
    H—Channel
    I—H, I sections
    J—Z sections
    K—Special sections (extruded, rolled, etc.)

Fabricated Solid Shapes

    L—Forging
    M—Casting/ingot
    N—Weldment
    P—Powder metal
    Q—Laminate
    R—Honeycomb
    S—Foam

Special Forms

    T—Resin, liquid, granules
    U—Fabric, roving, filament
    V—Putty, clay
    W—Other
    Y—Reserved
    Z—Reserved

Fig. 30.14 Raw material forms.

condition, and material form. These factors are all included in the material code. A modification of any of these factors either by itself or in combination can result in quite different mechanical properties. Thus, each material code combination is treated as a unique material. As an example of this, consider the tensile strength of a heat-treated 6061 aluminum alloy: in the wrought condition, the ultimate tensile strength is 19,000 psi; with the T4-temper, the ultimate tensile strength is 35,000 psi; and in the T913 condition, the ultimate tensile strength is 68,000 psi.

Physical Properties

The physical properties of an engineering material have to do with the intrinsic or structure-insensitive properties. These include melting point, expansion characteristics, dielectric strength, and density. Figure 30.16 shows representative physical properties.

Again, each property has been coded to aid in communication. Magnetic properties and electrical properties are included in this section for the sake of simplicity.

| Mechanical Properties | D | 1 | - | 0 | 6 | 0 | 6 | 1 | - | 1 | B | - | C |

Material Family/Type: Aluminum 6061-T6

Prepared by: Date: Approved by: Date:

Revision No./Date:

Code	Description	Value	Units
11.02	Brinell hardness number	95	HB
12.06	Yield strength, 0.2% offset	40,000	psi
12.11	Ultimate tensile strength	45,000	psi
12.20	Ultimate shear (bearing) strength	30,000	psi
12.30	Impact energy (Charpy V-notch)		ft-lb
12.60	Fatigue (endurance limit)	14,000	psi
12.70	Creep strength		psi
13.01	Modulus of elasticity (tensile)	10.0×10^6	psi
13.02	Modulus of elasticity (compressive)	10.2×10^6	psi
13.20	Poisson's ratio	—	—
14.02	Elongation	15	%
14.10	Reduction of area	—	%
14.30	Strain hardening coefficient	—	%
14.40	Springback	—	%

Fig. 30.15 Representative mechanical properties.

Chemical Properties

The chemical properties of an engineering material deal with its reactance to other materials or substances, including its operating environment. These properties include chemical reactivity, corrosion characteristics, and chemical compatibility.

Atomic structure factors, chemical valence, and related factors useful in predicting chemical properties may also be included in the broad category of chemical properties. Figure 30.17 shows representative chemical properties.

30.2.5 Material Availability

The availability of an engineering material is a prime concern in materials selection and use. Material availability includes such factors as stock shapes, sizes, and tolerances; material condition and finish; delivery; and price.

Other factors of increasing significance are energy requirements for winning the material from nature and recycleability. Figure 30.18 shows representative factors for accessing material availability.

30.2.6 Material Processability

Relative processability ratings for engineering materials in conjunction with material properties and availability can greatly assist the engineering designer in selecting materials that will meet essential design criteria. All too often the processability of a selected engineering material is unknown to the designer. As likely as not, the materials may warp during welding or heat treatment and be difficult to machine, which may result in undesirable surface stresses because of tearing or cracking during drawing operations. Many of these problems could be easily avoided if processability ratings of various materials were ascertained, recorded, and used by the designer during the material selection process. Figure 30.19 shows relative processability ratings. These ratings include machinability, weldability, castability, moldability, formability, and heat-treatability. Relative ratings are established through experience for each family. Ratings must not be compared between families. For example, the machinability rating of two steels may be compared, but they should not be evaluated against brass or aluminum.

Physical Properties	D	1	-	0	6	0	6	1	

Material Family/Type: Aluminum 6061-T6

Prepared by:	Date:	Approved by:	Date:

Revision No./Date:

Code	Description	Value	Units
21.01	Coefficient of linear expansion	13×10^{-6}	in./in./°F
21.05	Thermal conductivity	1070	Btu/in./ft²/°F/hr
21.40	Minimum service temperature	−320	°F
21.50	Maximum service temperature	700	°F
21.66	Melting range	1080–1200	°F
21.80	Recrystallization temperature	650	°F
21.90	Annealing temperature	775	°F, 2–3 hr
21.92	Stress-relieving temperature	450	°F, 1 hr
21.95	Solution heat treatment	970	°F
21.96	Precipitation heat treatment	350	°F, 6–10 hr
22.01	Electrical conductivity (weight)	40	%
22.02	Electrical conductivity (volume)	135	%
22.10	Electrical resistivity (volume)	26	ohms mil, ft
26.01	Specific weight	0.098	lb/in.³
26.03	Specific gravity	270	gm/cm³
26.35	Crystal (lattice) system	f.c.c.	—
26.70	Damping index	0.03	Very low
26.71	Strength-to-weight ratio		
26.72	Basic refining energy	100,000	Btu/lb
26.73	Recycling energy	10,000	Btu/lb

Fig. 30.16   Representative physical properties.

30.3   FABRICATION PROCESS TAXONOMY

30.3.1   Introduction

Purpose

The purpose for classifying manufacturing processes is to form families of related processes to aid in process selection, documentation of process capabilities, and information retrieval. A taxonomy or classification of manufacturing processes can aid in process selection by providing a display of potential manufacturing options available to the process planner.

Documentation of process capabilities can be improved by providing files containing the critical attributes and parameters for each classified process. Information *retrieval* and communication relative to various processes can be enhanced by providing a unique code number for each process. Process information can be indexed, stored, and retrieved by this code.

Classification and coding is an art, and, as such, it is difficult to describe the steps involved, and even more difficult to maintain consistency in the results. The anticipated benefits to users of a well-planned process classification outweighs the anticipated difficulties, and, thus, the following plan is being formulated to aid in uniform and consistent classification and coding of manufacturing processes.

Primary Objectives

There are three primary objectives for classifying and coding manufacturing processes. These objectives include (1) facilitating process planning, (2) improving process capability assessment, and (3) aiding in information retrieval.

Chemical Properties	D	1	-	0	6	0	6	1	

Material Family/Type: Aluminum 6061-T6

Prepared by:	Date:	Approved by:	Date:

Revision No./Date:

Code	Description	Value[a]	Units
32.01	Resistance to high-temperature corrosion	C	
32.02	Resistance to stress corrosion cracking	C	
32.03	Resistance to corrosion pitting	B	
32.04	Resistance to intergranular corrosion	B	
32.10	Resistance to fresh water	A	
32.11	Resistance to salt water	A	
32.15	Resistance to acids	A	
32.20	Resistance to alkalies	C	
32.25	Resistance to petrochemicals	A	
32.30	Resistance to organic solvents	A	
32.35	Resistance to detergents	B	
33.01	Resistance to weathering	A	

[a] *Key:* A = fully resistant; B = slightly attacked; C = unsatisfactory.

Fig. 30.17 Representative chemical properties.

Facilitate Process Selection.   One of the significant problems confronting the new process planner is that of process selection. The planner must choose, from many alternatives, the basic process, equipment, and tooling required to produce a given product of the desired quality and quantity in the specified time.

Although there are many alternative processes and subprocesses from which to choose, the process planner may be well acquainted with only a small number of them. The planner may thus continue to select these few rather than become acquainted with many of the newer and more competitive processes. The proposed classification will aid in bringing to the attention of the process planner all the processes suitable for modifying the *shape* of a material or for modifying its *properties.*

Improve Process Capability Assessment.   One of the serious problems facing manufacturing managers is that they can rarely describe their process capabilities. As a consequence, there is commonly a mismatch between process capability and process needs. This may result in precision parts being produced on unsuitable equipment with consequent high scrap rates, or parts with no critical tolerances being produced on highly accurate and expensive machines, resulting in high manufacturing costs.

Process capability files may be prepared for each family of processes to aid in balancing capacity with need.

Aid Information Retrieval.   The classification and grouping of manufacturing processes into subgroups having similar attributes will minimize the time required to identify and retrieve similar processes. The classification tree will provide a structure and branches to which important information may be attached or referenced regarding process attributes, methods, equipment, and tooling.

The classification tree will provide a logical arrangement for coding existing processes as well as a place for new processes.

30.3.2 Process Divisions

Manufacturing processes, as shown in Appendix 30.3, can be broadly grouped into two categories: (1) shaping processes and (2) nonshaping processes. *Shaping processes* are concerned primarily with modifying the shape of the plan material into the desired geometry of the finished part. *Nonshaping processes* are primarily concerned with modifying material properties.

Availability	D	1	-	0	6	0	6	1	

Material Family/Type: Aluminum 6061-T6

Prepared by:		Date:		Approved by:		Date:

Revision No./Date:

Surface Condition
  Cold worked
  Hot worked
  Cast
  Clad
  Peened
  Chromate
  Anodized
  Machined

Internal Condition
  Annealed
  Solution treated—naturally aged
  Solution treated—artificially aged
  Stress relieved
  Cold worked

Forms Available
  Sheet
  Plate
  Bar
  Tubing
  Wire
  Rod
  Extrusions
  Ingot

Fig. 30.18 Factors relating to material availability.

Shaping Processes

Processes available for shaping the raw material to produce a desired geometry may be classified into three subdivisions: (1) mass-reducing processes, (2) mass-conserving processes, and (3) mass-increasing or joining processes. These processes may then be further subdivided into mechanical, thermal, and chemical processes.

Mass-reducing processes include cutting, shearing, melting or vaporizing, and dissolving or ionizing processes. Mass-conserving processes include casting, molding, compacting, deposition, and laminating processes. Mass-increasing or, more commonly, joining, processes include pressure and thermal welding, brazing, soldering, and bonding. The joining processes are those that produce a megalithic structure not normally disassembled.

Nonshaping Processes

Nonshaping processes that are available for modifying material properties or appearance may be classified into two broad subdivisions: (1) heat-treating processes and (2) surface-finishing processes.

Processability		D	1	-	0	6	0	6	1	-	1	B	-	C

Material Family/Type: Aluminum 6061-T6

Prepared by:	Date:	Approved by:	Date:

Revision No./Date:

Processability Type	Rating			
	Poor 1	Fair 2	Good 3	Excellent 4
Machinability				X
Grindability (silicon carbide adhesive)			X	
Shear behavior				X
EDM rating	X			
Chemical etch factor		X		
Forgeability			X	
Extrudability				
Formability			X	
Weldability				X
Heat-treatability				X

Fig. 30.19 Relative processability ratings.

Heat-treating processes are designed primarily to modify mechanical properties or the processability of engineering materials. Heat-treating processes may be subdivided into (1) annealing (softening) processes, (2) hardening processes, and (3) other processes. The "other" category includes sintering, firing/ glazing, curing/bonding, and cold treatments. Annealing processes are designed to soften the work material, relieve internal stresses, or change the grain size. Hardening treatments, on the other hand, are often designed to increase strength and resistance to surface wear or penetration. Hardening treatments may be applied to the surface of a material or the treatments may be designed to change material properties throughout the section.

Surface-finishing processes are those used to prepare the workpiece surface for subsequent operations, to coat the surface, or to modify the surface. Surface-preparation processes include descaling, deburring, and degreasing. Surface coatings include organic and inorganic; metallic coatings applied by spraying, electrostatic methods, vacuum deposition, and electroplating; and coatings applied through chemical-conversion methods.

Surface-modification processes include burnishing, brushing, peening, and texturing. These processes are most often used for aesthetic purposes, although some peening processes are used to create warped surfaces or to modify surface stresses.

30.3.3 Process Taxonomy

There are many methods for classifying production processes and each method may serve unique purposes. The Fabrication Process Taxonomy is the first known comprehensive classification of all processes used for the fabrication of discrete parts for the durable goods manufacturing industries.

Basis of Classification

The basis for process classification may be the source of *energy* (i.e., mechanical, electrical, or chemical); the *temperature* at which the processing is carried out (i.e., hot-working, cold-working); the type of *material* to be processed (i.e., plastic, steel, wood, zinc, or powdered metal); or another basis of classification.

The main purpose of the hierarchy is to provide functional groupings without drastically upsetting recognized and accepted families of processes within a given industry. For several reasons, it is difficult to select only one basis for classification and apply it to all processes and achieve usable results. Thus, it will be noted that the fabrication process hierarchy has several bases for classification, each depending on the level of classification and on the particular family of processes under consideration.

Classification Rules and Procedures

In an effort to create a uniform classification of processes, the following rules were developed:

Rule 1. Processes were classified as either shaping or nonshaping, with appropriate mutually exclusive subdivisions.

Rule 2. Processes were classified as independent of materials and temperature as possible.

Rule 3. Critical attributes of various processes were identified early to aid in forming process families.

Rule 4. Processes were subdivided at each level to show the next options available.

Rule 5. Each process definition is in terms of relevant critical attributes.

Rule 6. Shaping process attributes include:

    6.1 Geometric shapes produced

    6.2 Form features or treatments imparted to the workpiece

    6.3 Size, weight, volume, or perimeter of parts that can be produced

    6.4 Precision of parts produced

    6.5 Production rates

    6.6 Set-up time tooling costs

    6.7 Relative labor costs

    6.8 Scrap and wastes material costs

    6.9 Unit costs versus quantities of 10; 100; 1000; 10,000; 100,000

Rule 7. All processes are characterized by:

    7.1 Prerequisite processes

    7.2 Materials that can be processed, including initial form construction

    7.3 Basic energy model: mechanical, thermal, and chemical

    7.4 Influence on mechanical properties such as strength, hardness, toughness

    7.5 Influence on physical properties such as conductivity, resistance, change in density, color

    7.6 Influence on chemical properties involving corrosion resistance

Rule 8. At the operational level, the process may be fully described by the operation description and sequence, equipment, tooling, processing parameters, operating instructions, and standard time.

The procedure followed in creating the taxonomy was first to identify all the processes that were used in fabrication processes. These processes were then grouped on the basis of relevant attributes. Next, most prominent attributes were selected as the parent node label. Through a process of selection, grouping, and classification, the taxonomy was developed. The taxonomy was evaluated and found to be quite good for accommodating new processes that subsequently were identified. This aided in verifying the generic design of the system. The process taxonomy was further cross-checked with the equipment and tooling taxonomies to see if related categories existed. In several instances, small modifications were required to ensure that the various categories were compatible. Following this method for cross-checking among processes, equipment, and tooling, the final taxonomy was prepared. As the taxonomy was subsequently typed and checked, and large charts were developed for printing, small remaining discrepancies were noted and corrected. Thus, the process taxonomy is presented as the best that is currently available. Occasionally, a process is identified that could be classified in one or more categories. In this case, the practice is to classify it with the preferred group and cross-reference it in the second group.

30.3.4   Process Code

The process taxonomy is used as a generic framework for creating a unique and unambiguous numeric code to aid in communication and information retrieval.

The process code consists of a three-digit numeric code. The first digit indicates the basic process division and the next two digits indicate the specific process group. The basic process divisions are as follows:

    000   Material identification and handling

    100   Material removal processes

    200   Consolidation processes

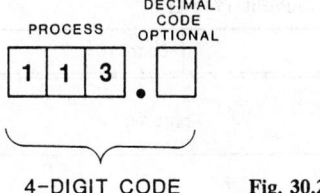

4-DIGIT CODE   **Fig. 30.20**   Basic process code.

300   Deformation processes
400   Joining processes
500   Heat-treating processes
600   Surface-finishing processes
700   Inspection
800   Assembly
900   Testing

The basic process code may be extended with the addition of an optional decimal digit similar to the Dewey Decimal System. The process code is organized as shown in Fig. 30.20.

The numeric process code provides a unique, easy-to-use shorthand communication symbol that may be used for manual or computer-assisted information retrieval. Furthermore, the numeric code can be used on routing sheets, in computer data bases, for labeling of printed reports for filing and retrieval purposes, and for accessing instructional materials, process algorithms, appropriate mathematical and graphical models, and the like.

Appendix 30.3 contains an alphabetic listing of all the fabrication processes along with their process codes. This listing may be used to identify quickly desired processes and also to provide a cross-reference index.

30.3.5   Process Capabilities

Fundamental to process planning is an understanding of the capabilities of various fabrication processes. This understanding is normally achieved through study, observation, and industrial experience. Because each planner has different experiences and observes processes through different eyes, there is considerable variability in derived process plans.

The taxonomy of fabrication processes in Appendix 30.3 is developed to aid the new process planner in learning about the various processes used in discrete part production.

Fabrication processes have been grouped into families having certain common attributes. A study of these common attributes will enable the prospective planner to learn quickly the significant characteristics of the process without becoming confused from the large amount of factual data that may be available about the given process.

Also, a knowledge about other processes in a given family will help the prospective planner learn about a specific process by inference. For example, if the planner understands that "turning" and "boring" are part of the family of single-point cutting operations and has learned about cutting-speed calculations for turning processes, the planner may correctly infer that cutting speeds for boring operations would be calculated in a similar manner. It is important at this point to let the prospective planner know the boundaries or exceptions for such generalizations.

A study of the common attributes and processing "clues" associated with each of these various processes will aid the planner. For example, an understanding of the attributes of a given process and recognition of process clues such as "feed marks," "ejector-pin marks," or "parting lines" can help the prospective planner to quickly identify how a given part was produced.

Figures 30.21 and 30.22 show a process capability sheet that has been designed for capturing information relative to each production process.

30.4   FABRICATION EQUIPMENT CLASSIFICATION

30.4.1   Introduction

Control of Capital Resources

One of the primary purposes for equipment classification systems is to better control capital resources. The amount of capital equipment and tooling per manufacturing employee has been reported to range

PROCESS CAPABILITY SHEET

Process: Turning/Facing	Code: 101

Prepared by: Date:	Approved by: Date:

Revision No. & Date:

Schematic:

Attributes:

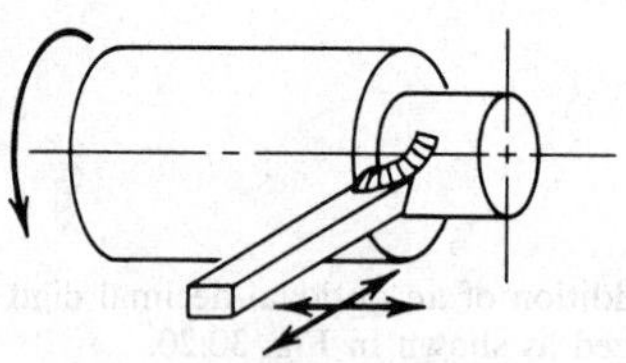

- Single point cutting tool
- Chips removed from external surface
- Helical or annular (tree-ring) feed marks are present.

Basic Shapes Produced:
  Surfaces of revolution (cylindrical, tapered, spherical) or flat shoulders or ends.
  May have discontinuities in surfaces (interrupted cut).

Form Features or Treatments:
  Bead, boss, chf'r, groove , lip, radius, thread

Size Range:
  1-6

Precision Class:
  1-4

Raw Material Type: Steel, cast iron, light metals, non-ferrous engineering metals,
  low-m.p. metals, refractory metals, nuclear metals, composites, refractories,
  wood, polymers, rubbers and elastometers

Fig. 30.21 Process capability sheet.

from $30,000 to $50,000. An equipment classification system can be a valuable aid in capacity planning, equipment selection, equipment maintenance scheduling, equipment replacement, elimination of unnecessary equipment, tax depreciation, and amortization.

Equipment Selection

A key factor in equipment selection is a knowledge of the various types of equipment and their capabilities. This knowledge may be readily transmitted through the use of an equipment classification tree showing the various types of equipment and through equipment specification sheets that capture significant information regarding production capabilities.

Equipment selection may be regarded as matching—the matching of production needs with equipment capabilities. Properly defined needs based on current and anticipated requirements when coupled with an equipment classification system provide a logical, consistent strategy for equipment selection.

Process: Turning/Facing				Code: 101		

Raw Material Condition:
  Hot-rolled, cold-rolled, forged, cast

Raw Material Form:
  rod, tubing, forgings, castings

Production Rate		1 A	10 B	100 C	1,000 D	10,000 E	100,000 F
Tooling Costs	High-3						
	Med-2						
	Low-1						
Set-up Time	High-3						
	Med-2						
	Low-1						
Labor Costs	High-3						
	Med-2						
	Low-1						
Scrap & Waste Material Costs	High-3						
	Med-2						
	Low-1						
Unit Costs	High-3						
	Med-2						
	Low-1						

Prerequisite Processes:
  Hot-rolling, cold rolling, forging, casting, p/m compacting

Influence on Mechanical Properties:
  Creates very thin layer of stressed work material. Grains may be slightly
  deformed, and built-up edge may be present on work surface.

Influence on Physical Properties:
  N/A

Influence on Chemical Properties:
  Highly stressed work surface may promote corrosion.

Fig. 30.22 Process capability sheet.

Manufacturing Engineering Services.

Some of the manufacturing engineering services that can be greatly benefited by the availability of an equipment classification system include process planning, tool design, manufacturing development, industrial engineering, and plant maintenance. The equipment classification code can provide an index pointer to performance records, tooling information, equipment specification sheets mentioned previously, and other types of needed records.

Quality Assurance Activities

Acceptance testing, machine tool capability assessment, and quality control are three important functions that can be enhanced by means of an equipment classification and coding system. As before, the derived code can provide a pointer to testing and acceptance procedures appropriate for the given family of machines.

30.4.2 Standard and Special Equipment

The classification system described below can readily accommodate both standard and special equipment. *Standard fabrication equipment* includes catalog items such as lathes, milling machines, drills, grinders, presses, furnaces, and welders. Furthermore, they can be used for making a variety of products. Although these machines often have many options and accessories, they are still classified as standard machines.

Special fabrication machines, on the other hand, are custom designed for a special installation or application. These machines are usually justified for high-volume production or special products that are difficult or costly to produce on standard equipment. Examples of special machines include transfer machines, special multistation machines, and the like.

30.4.3 Equipment Classification

The relationship among the fabrication process, equipment, and tooling is shown graphically in Fig. 30.23. The term *process* is basically a concept and requires equipment and tooling for its physical embodiment. For example, the grinding process cannot take place without equipment and tooling. In some instances the process can be implemented without equipment, as in "hand-deburring." The hierarchal relationships shown in Fig. 30.23 between the process, equipment, and tooling provide a natural linkage for generative process planning. Once the required processes have been identified for reproducing a given geometric shape and its associated form features and special treatments, the selection of equipment and tooling is quite straightforward.

Rationale

Two major functions of an equipment classification system are for process planning and tool design. These functions are performed each time a new product or piecepart is manufactured. Consequently, the relationships between processes, equipment, and tooling have been selected as a primary one in development of the equipment classification system.

The equipment taxonomy parallels the process taxonomy as far as possible. Primary levels of classification include those processes whose *intent* is to change the *form* of the material, for example, shaping processes, and those processes whose *intent* is to modify or enhance the material *properties*. These nonshaping processes include heat treatments and coating processes, along with attendant cleaning and deburring.

As each branch of a process tree is traversed, it soon becomes apparent that there is a point at which an equipment branch must be grafted in. It is at this juncture that the basis for equipment classification must be carefully considered. There are a number of possible bases for classification of equipment. Some of these bases include:

1. Form change (shaping, nonshaping).
2. Mass change (reduction, consolidation, joining).
3. Basic process (machine, cast, forge).
4. Basic subprocess (deep hole drill, precision drill).
5. Machine type (gang drill, radial drill).
6. Energy source (chemical, electrical, mechanical).
7. Energy transfer mechanism (mechanical, hydraulic, pneumatic).
8. Raw material form (sheet metal, forging, casting).
9. Shape produced (gear shaper, crankshaft lathe).
10. Speed of operation (high speed, low speed).
11. Machine orientation (vertical, horizontal).
12. Machine structure (open-side, two-column).
13. General purpose/special purpose (universal mill, spar mill).
14. Kinematics/motions (moving head, moving bed).
15. Control type (automatic, manual NC).

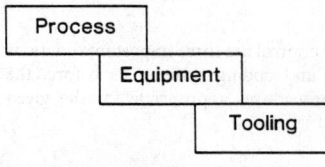

Fig. 30.23 Relationships between process, equipment, and tooling.

16. Feature machined (surface, internal).
17. Operating temperature (cold rolling, hot rolling).
18. Material composition (plastic molding, aluminum die casting).
19. Machine size (8 in. chucker, 12 in. chucker).
20. Machine power (600 ton, 100 ton).
21. Manufacturer (Landis, Le Blond, Gisholt).

In reviewing these bases of classification, it is apparent that some describe fundamental characteristics for dividing the equipment population into families, and others are simply attributes of a given family. For example, the features of "shaping," "consolidation," and "die casting" are useful for subdivision of the population into families (E-tree), whereas attributes such as "automatic," "cold-chamber," "aluminum," "100 ton," and "Reed–Prentice" are useful for characterizing equipment within a given family (N-tree). "Automatic" is an attribute of many machines; likewise, "100 ton" could apply to general purpose presses, forging presses, powder-metal compacting presses, etc. Similarly, the label "Reed–Prentice" could be applied equally well to lathes, die casting machines, or injection molders. In other words, these terms are not very useful for development of a taxonomy but are useful for characterizing a family.

Equipment Taxonomy

The taxonomy for fabrication equipment is given in Appendix 30.4. The first major division, paralleling the processes, is shaping or nonshaping. The second level for shaping is (1) mass reducing, (2) mass conserving, and (3) joining (mass increasing). The intent of this subdivision, as was mentioned earlier, is to classify equipment whose intent is to change the form or shape of the workpiece. The second level for nonshaping equipment includes (1) heat treating and (2) surface finishing. The intent of this subdivision is to classify equipment designed to modify or enhance material properties or appearance. The existing taxonomy identifies 257 unique families of fabrication equipment.

Customizing

As with other taxonomies described herein, equipment taxonomy is designed to accommodate readily new classes of machines. This may be accomplished by traversing the tree until a node point is reached where the new class or equipment must appropriately fit. The new equipment with its various subclasses may be grafted in at this point and an appropriate code number assigned. It should be noted that code numbers have been intentionally reserved for this purpose.

30.4.4 Equipment Code

The code number for fabrication equipment consists of a nine-character code. Three of the digits identify the basic process, leaving the remaining six characters to identify uniquely any given piece of equipment. As can be seen in Fig. 30.24, the code consists of four fields. Each of these fields will be briefly described in the following paragraphs.

Process Code

The process code is a three-digit code that refers to one of the 222 fabrication processes currently classified. For example, code 111 = drilling and 121 = grinding. Appended to this code is a code for the specific type of equipment required to implement the given process.

Equipment Family Code

The code for equipment type consists of a one-character alphabetic code. For example, this provides for up to 26 types of turning machines, 101-A through 101-Z, 26 types of drilling machines, etc. Immediately following the code for equipment type is a code that uniquely identifies the manufacturer.

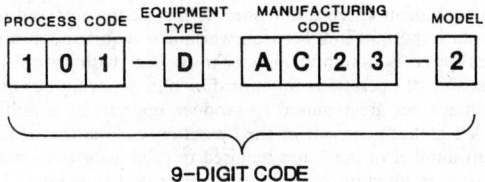

Fig. 30.24 Fabrication equipment code.

Manufacturer Code

The manufacturer code consists of a four-digit alphanumeric code. The first character in the code is an alphabetic character (A–Z) representing the first letter in the name of the manufacturing company. The next three digits are used to identify uniquely a given manufacturing company. In-house developed equipment would receive a code for your own company.

Model Number

The final character in the code is used to identify a particular manufacturer model number. Thus, the nine-digit code is designed to provide a shorthand designation as to the basic process, type of equipment, manufacturer, and model number. The code can serve as a pointer to more detailed information as might be contained on specification sheets, installation instructions, maintenance procedures, etc.

30.4.5 Equipment Specification Sheets

In the preceding subsections the rationale for equipment classification was discussed along with the type of information useful in characterizing a given piece of equipment. It has been found that this characterization information is best captured by means of a series of equipment specification sheets.

The philosophy has been that each family must have a tailored list of features or specifications to characterize it adequately. For example, the terms *swing, center-distance, rpm,* and *feed per revolution* are appropriate for a lathe family but not for forming presses, welding machines, electrical discharge machines, or vibratory deburring machines. Both common attributes and selected ones are described in the following paragraphs and shown in Figs. 30.25 and 30.26.

Equipment Family/Code

The equipment family label consists of the generic family name, subgroup name (e.g., "drill," "radial"), and the nine-digit equipment code.

Equipment Identification

Equipment name, make, model, serial number, and location provide a unique identification description for a given piece of equipment. The equipment name may be the familiar name given to the piece of equipment and may differ from the generic equipment family name.

Acquisition Data

The acquisition data include capital cost, date acquired, estimated life, and year of manufacture. This information can be used for amortization and depreciation purposes.

Facilities

Facilities required for installing and operating the equipment include power (voltage, current, phases, frequency), and other connections, floor space, height, and weight.

Specifications

The specifications for each family of machines must be carefully selected to meet needs of intended applications.

Operation Codes

A special feature of the equipment specification sheet is the section reserved for operation codes. The operation codes provide an important link between workpiece requirements and equipment capability. For example, if the workpiece is a rotational, machined part with threads, grooves, and a milled slot, and if a lathe is capable of operations for threading and grooving but not a milling slot, then it follows that either two machines are required to produce the part or a milling attachment must be installed on the lathe. A significant benefit of the operation code is that it can aid process planners in selecting the minimum number of machines required to produce a given workpiece. This fact must, of course, be balanced with production requirements and production rates. The main objective is to reduce transportation and waiting time and minimize cost. (See Fig. 30.26.)

Equipment Family: Mill, NC, vertical	Code: 113-K

Identification:

Name: Cintimatic, Single Spindle
Make/Model: Cincinnati Milling Machine Co.

Serial No.: 364
Location: 115 SNLB

Acquisition:

Capital Cost: $25,000
Date Acquired: 15 Aug. 1963

Estimated Life: 15 yr
Year of Manufacture: 1963

Maintenance:

Condition: U2
Date: 15 Aug. 1963

Reevaluation Date: 15 January 1981

Facilities:

Voltage: 230 volts, 3 ph, 60 Hz
Current: 3 hp
Other Connections: Air, 40 psi

Floor Space: 63 in. × 74 in.
Height: 101 in.
Weight:

Specifications:

Working surface . . . . 22 in. × 36 in.
Throat . . . . 16¼ in.
Table top to spindle . . . . 14–24 in. (8 in. travel)
Weight capacity, max . . . . 1000 lb
Spindle:
    Axes . . . . One axis
    Range . . . . 85–3800 rpm
    Rate . . . . 1–40 ipm
Table:
    Axes . . . . Two axes
    Range . . . . 15 in. × 25 in.
    Rate . . . . Feed—0–40 ipm, rapid travel 200 ipm
Motor hp:
    Drive motor . . . . 3.0 hp
    Feed . . . . Hydraulic servo motors
    Coolant . . . . Air mist spray
Spindle taper . . . . #40 NMTB
T-slots . . . . 3 in X axis, $1\frac{1}{16}$ in. wide
Accuracy . . . . ±0.001 in. in 24 in.
Control type . . . . Accramatic Series 200 control

Fig. 30.25 Equipment specifications.

Photograph or Sketch

A photograph or line drawing provides considerable data to aid in plant layout, tool design, process planning, and other production planning functions. Line drawings often provide information regarding T-slot size, spindle arbor size, limits of machine motion, and other information useful in interfacing the machine with tooling, fixtures, and the workpiece.

30.5 FABRICATION TOOL CLASSIFICATION AND CODING

30.5.1 Introduction

Because standard and special tooling represent a sizable investment, it is prudent to minimize redundant tooling, to evaluate performance of perishable tooling, and to provide good storage and retrieval practices to avoid excessive tool-float and loss of valuable tooling.

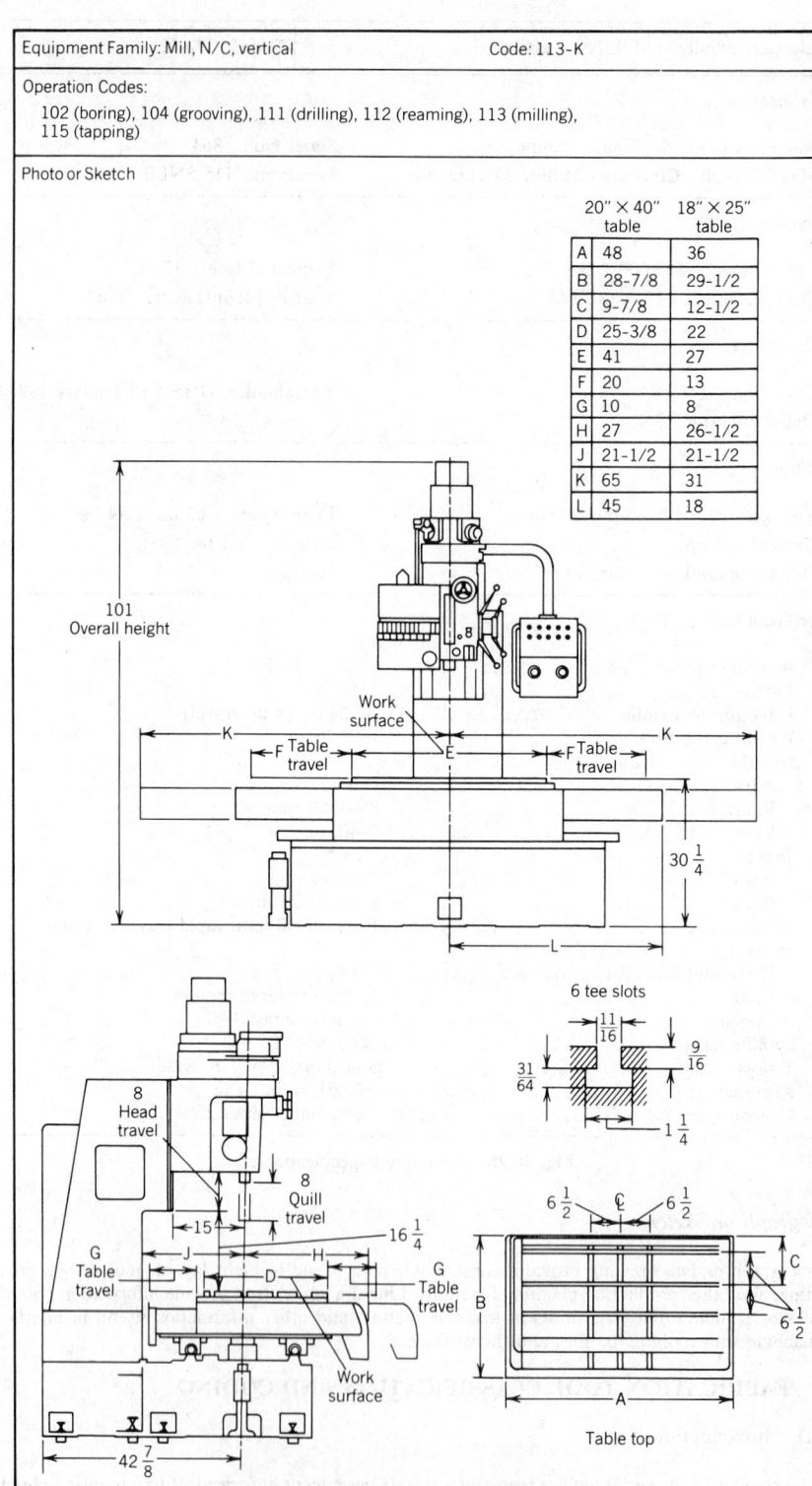

Equipment Family: Mill, N/C, vertical   Code: 113-K

Operation Codes:
   102 (boring), 104 (grooving), 111 (drilling), 112 (reaming), 113 (milling),
   115 (tapping)

Photo or Sketch

	20" × 40" table	18" × 25" table
A	48	36
B	28-7/8	29-1/2
C	9-7/8	12-1/2
D	25-3/8	22
E	41	27
F	20	13
G	10	8
H	27	26-1/2
J	21-1/2	21-1/2
K	65	31
L	45	18

Fig. 30.26 Operation codes and machine illustration.

The use of a standard tool classification system could provide many benefits for both the supplier and the user. The problem is to derive a comprehensive tool classification system that is suitable for the extremely wide variety of tools available to industry and that is agreeable to all suppliers and users. Although no general system exists, most companies have devised their own proprietary tool classification schemes. This has resulted in much duplicate effort. Because of the difficulty of developing general systems, which are expandable to accommodate new tooling categories, many of these existing schemes for tool classification are found to be inadequate.

This section describes a new classification and coding system for fabrication tooling. Assembly, inspection, and testing tools are not included. This new system for classifying fabrication tools is a derivative of the work on classifying fabrication processes and fabrication equipment. Furthermore, special tooling categories are directly related through a unique coding system to the basic shape of the workpieces they are used to fabricate.

Investment

The investment a manufacturing company must make for standard and special tooling is usually substantial. Various manufacturing companies may carry in stock from 5000 to 10,000 different tools and may purchase several thousand special tools as required. As a rule of thumb, the investment in standard tooling for a new machine tool is often 20–30% of the basic cost of the machine. Special tooling costs may approach or even exceed the cost of certain machines. For instance, complex die-casting molds costing from $50,000 to $250,000 are quite commonplace.

The use of a tool classification system can aid a manufacturing enterprise by helping to get actual cost data for various tooling categories and thus begin to monitor and control tooling expenditures.

The availability of good tooling is essential for economical and productive manufacturing. The intent of monitoring tooling costs should be to ensure that funds are available for such needed tooling and that these funds are wisely used. The intent should not be the miserly allocation of tooling money.

Tool Control

Tool control is a serious challenge in almost all manufacturing enterprises. Six important aspects of any good tool control system include*

Tool procurement.

Tool storage.

Tool identification and marking.

Tool dispensing.

Performance measurement.

Tool maintenance.

The availability of a standard, comprehensive tool classification and coding system can greatly aid each of these elements of tool control. For example, tool procurement data may be easily cross-indexed with a standard tool number, thus reducing problems in communication between the user and the purchasing department.

Tool storage and retrieval may be enhanced by means of standard meaningful codes to identify tools placed in a given bin or at a given location. This problem is especially acute with molds, patterns, fixtures, and other special tooling.

Meaningful tool identification markings aid in preventing loss or misplacement of tools. Misplaced tools can quickly be identified and returned to their proper storage locations. Standard tool codes may also be incorporated into bar codes or other machine-readable coding systems if desirable.

The development of an illustrated tooling manual can be a great asset to both the user and to the tool crib personnel in identifying and dispensing tools. The use of a cross-referenced standard tool code can provide an ideal index to such a manual.

Tool performance measures require the use of some sort of coding system for each type of tool to be evaluated. Comparison of tools within a given tool family may be facilitated by means of expanded codes describing the specific application and the various types of failures. Such extended codes may be easily tied to the standard tool family code.

Tool maintenance and repair costs can be best summarized when they are referenced to a standard tool family code. Maintenance and acquisition costs could be easily reduced to obtain realistic life cycle costs for tools of a given family or type.

In summary, tool control in general may be enhanced with a comprehensive, meaningful tool classification and coding system.

---

* "Small Tools Planning and Control," Section 3. *Tool Engineers Handbook*, McGraw-Hill, New York, 1959.

30.5.2 Standard and Special Tooling

Although tooling may be classified in many ways such as "durable or perishable tooling," "fabrication or assembly tooling," and "company-owned or customer-owned tooling," the fabrication tooling system described below basically classifies tooling as "standard tooling" or "special tooling."

Definition of Terms

STANDARD TOOLING is defined as that which is basically off-the-shelf and which may be used by different users or a variety of products. Standard tooling is usually produced in quantity, and the cost is relatively low.

SPECIAL TOOLING is that which is designed and built for a specific application, such as a specific product or family of products. Delivery on such tooling may be several weeks, and tooling costs are relatively high.

Examples

Examples of standard tooling are shown in Fig. 30.27. Standard tooling usually includes cutting tools, die components, nozzles, certain types of electrodes, rollers, brushes, tool holders, laps, chucks, mandrels, collets, centers, adaptors, arbors, vises, step-blocks, parallels, angle-plates, and the like.

Examples of special tooling are shown in Fig. 30.28. This usually includes dies, molds, patterns, jigs, fixtures, cams, templates, NC programs, and the like. Some standard tooling may be modified to perform a special function. When this modification is performed in accordance with a specified design, then the tooling is classified as special tooling.

30.5.3 Tooling Taxonomy

The tooling taxonomy is based on the same general classification system used for fabrication processes and for fabrication equipment. The first-level divisions are "shaping" and "nonshaping." The second-level divisions for shaping are "mass reduction," "mass conserving," and "joining" (or "mass increasing"). Second-level divisions for nonshaping are "heat treatment" and "surface finishing." Third-level subdivisions are more variable but include "mechanical," "chemical," and "thermal" among other criteria for subdivisions.

Rationale

The basic philosophy has been to create a tooling classification system that is related to fabrication processes, to fabrication equipment, and to fabrication products insofar as possible. The statement "Without this process, there is no product" has aided in clarifying the importance of a process classification to all phases of manufacturing. It was recognized early that the term *process* is really a concept and that only through the application of "equipment and tooling" could a process ever be implemented. Thus, this process taxonomy was used as the basis of both equipment classification and this tooling classification. Most tooling is used in conjunction with given families of equipment and in that way is related to the equipment taxonomy. Special tooling is also related to the workpiece geometry through a special coding system that will be explained later. Standard tooling may be applied to a number of product families and may be used on a variety of different machine tools.

Tool Taxonomy Charts

A series of convenient wall charts and notebook-size charts have been prepared showing the various types of fabrication tooling and their relationships. The complete tool taxonomy is given in Appendix 30.5. As described above, the primary divisions on the charts parallel those for fabrication processes and equipment to ensure linkage between processes, equipment, and tooling.

During the development of the various tooling classification charts, three points became apparent: First, there are various distinct types of tools, each of which constitutes quite different categories and each of which requires different criteria for subdivision. Second, the process node is the focal point for subdivision of equipment and tooling (e.g., at the node "turning-101"). The next subdivision for equipment is type of turning machine, whereas the subdivision for turning tooling includes tool insert and tool holder. Third, there are some processes that do not require equipment but which do require tooling (e.g., manual deburring).

Customizing

The tooling charts may be readily customized to meet unique company needs. Additional branches may be added or deleted from the tree, or the coding scheme may be modified as desired.

CONVENTIONAL

SUBLAND (Multi-Diameter)

CARBIDE TIPPED (For hard materials)

INSERT
INSERT HEAD BORING TOOL

CORE (For drilling out existing holes)

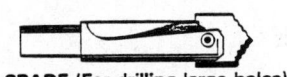
SPADE (For drilling large holes)

BORING BAR FOR HIGH SPEED
STEEL TOOLS

END MILL

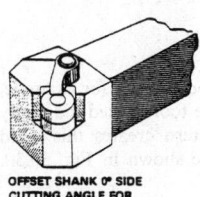
OFFSET SHANK 0° SIDE
CUTTING ANGLE FOR
STRAIGHT TURNING AND
FACING.

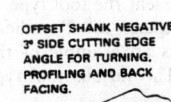
OFFSET SHANK NEGATIVE
3° SIDE CUTTING EDGE
ANGLE FOR TURNING,
PROFILING AND BACK
FACING.

PERIPHERAL MILL

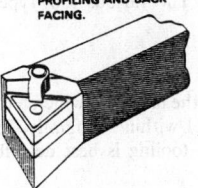

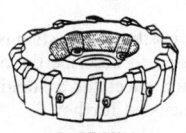
FACE MILL

Fig. 30.27 Standard tooling.

30.5.4 Tool Coding

The tool code is a shorthand notation used for identification and communication purposes. This tool code has been designed to provide the maximum amount of information in a short, flexible code. Complete tool information may be held in a computer data base or charts and tables. The code provides a pointer to this information.

Tool Code Format

The format used for the tool code is shown in Fig. 30.29. The code consists of three basic fields of information. The first field contains a three-digit process code that identifies the process for which the tooling is to be used. The second field consists of a one-digit code that indicates the tool type. Tool types are explained in the next subsection. The last field consists of either a three-digit numeral code for standard tooling or a three-digit alphanumeric code for special tooling. *Standard tool codes* have been designed to accommodate further subdivision of tool families if so desired. For example, the tool code for single-point turning inserts is "101-1-020." The last three digits could be amplified for given insert geometry (e.g., triangular, -021; square, -022; round, -023). A dash number may further

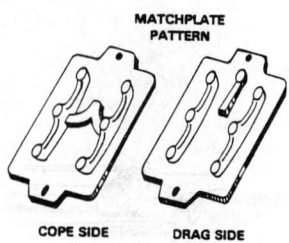

MATCHPLATE
PATTERN

COPE SIDE   DRAG SIDE

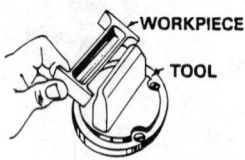

WORKPIECE

TOOL

Fig. 30.28 Special tooling.

be appended to these codes to uniquely identify a given tool, as shown on the tool specification sheets. *Special tool codes* are identified in the charts by a box containing three small squares. It is intended that the first three digits of this part family code will be inserted in this box, thus indicating the basic shape family for which the tool is designed. This way it will be possible to identify tool families and benefit from the application of group technology principles.

Tool Types

A single-digit alphanumeric code is used to represent the tool type. A code type of "-1," for example, indicates that the tool actually contacts the workpiece, a "-2" indicates that the tool is used indirectly in shaping the workpiece. A foundry pattern is an example of this: the pattern creates this mold cavity into which molten metal is introduced. The various tool type codes are shown in Fig. 30.30.

30.5.5 Tool Specification Sheets

The tool classification system is used to identify the family to which a tool belongs. The tool specification sheet is used to describe the attributes of a tool within the family. Figure 30.31 shows a sample tool specification sheet for a standard tool. Special tooling is best described with tool drawings and will not be discussed further.

Tool Identification

Tool identification consists of the tool name and the tool code number. The general tool family name is written first, followed by a specific qualifying label if applicable (e.g., "drill, subland, straight-shank"). The tool code consists of the seven-digit code described previously.

Acquisition Information

Information contained in this acquisition section may contain identifying codes for approved suppliers. This section may also contain information relating to the standard quantity per package if applicable, special finish requirements, or other pertinent information.

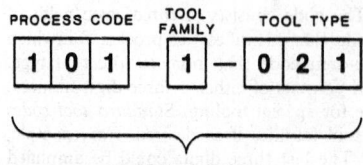

PROCESS CODE   TOOL   TOOL TYPE
                   FAMILY

1 0 1 — 1 — 0 2 1

7-DIGIT CODE   **Fig. 30.29** Fabrication tool code format.

TOOL FAMILY	
-1-	TOOL/DIE (CONTACTS WORKPIECE)
-2-	MOLDS, PATTERNS, NEGATIVES
-3-	TIPS, NOZZLES
-4-	ELECTRODES
-5-	OTHER TOOLS
-6-	RESERVED
-7-	TOOL HOLDERS
-8-	WORK HOLDERS
-9-	RESERVED
-A-	N/C PROGRAMS
-B-	CAMS
-C-	TRACING TEMPLATES (2-D)
-D-	TRACING PATTERNS (3-D)

Fig. 30.30 Tool types.

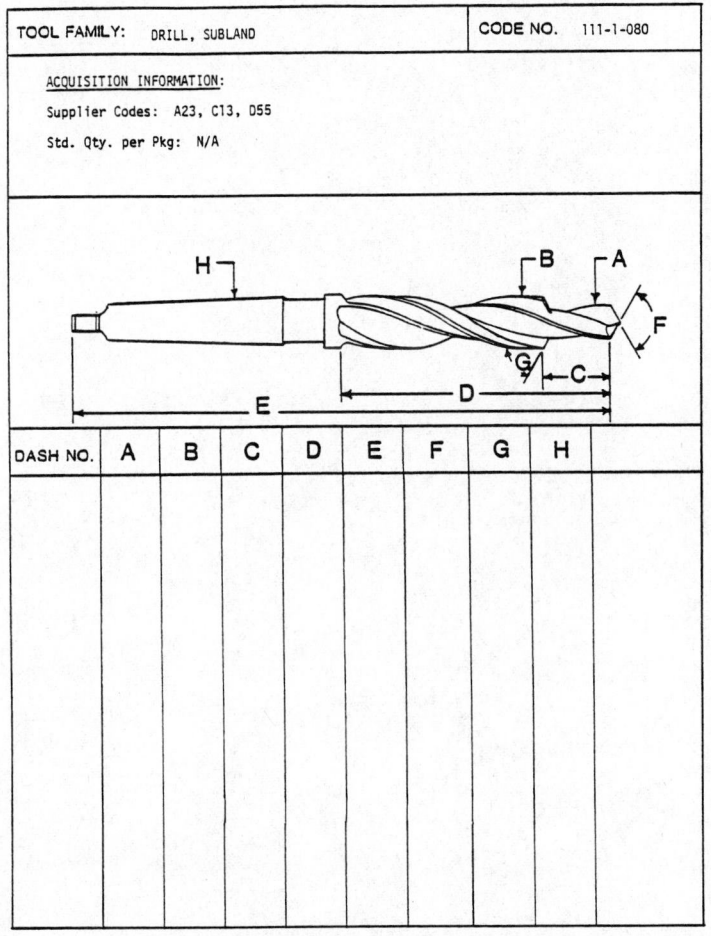

TOOL FAMILY: DRILL, SUBLAND	CODE NO. 111-1-080

ACQUISITION INFORMATION:
Supplier Codes: A23, C13, D55
Std. Qty. per Pkg: N/A

DASH NO.	A	B	C	D	E	F	G	H	

Fig. 30.31 Sample tool specification sheet.

Tool Sketches

Tool sketches are a valuable feature of the tool specification sheet. Prominent geometric relationships and parameters are shown in the sketch. Information relative to interfacing this tool to other devices or adaptors should also be shown. This may include type and size or shape, keyslot size, and tool capacity.

Tool Parameters

Tool parameters are included on the tool specification sheet to aid selection of the most appropriate tools. Because it is expensive to stock all possible tools, the usual practice is to identify preferred tools and to store these. Preferred tools may be so indicated by a special symbol such as an asterisk (*) in the dash-number column.

Tool parameters must be selected that are appropriate for each tool family, and that match product requirements. Tool parameters must also be identified that will aid in interfacing tools with fabrication equipment, as was explained earlier.

Typical tool parameters for a subland drill are shown in Fig. 30.31.

APPENDIX 30.1   BASIC SHAPE TAXONOMY

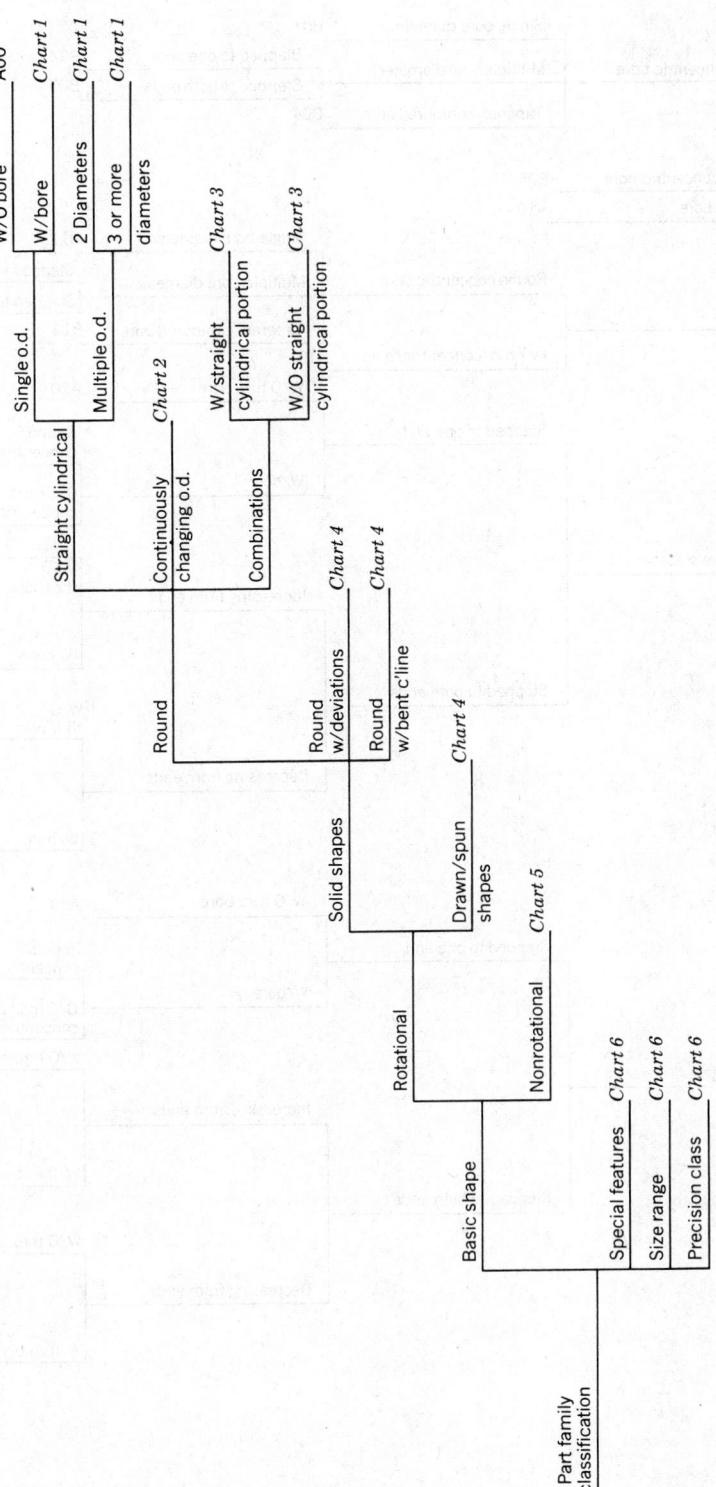

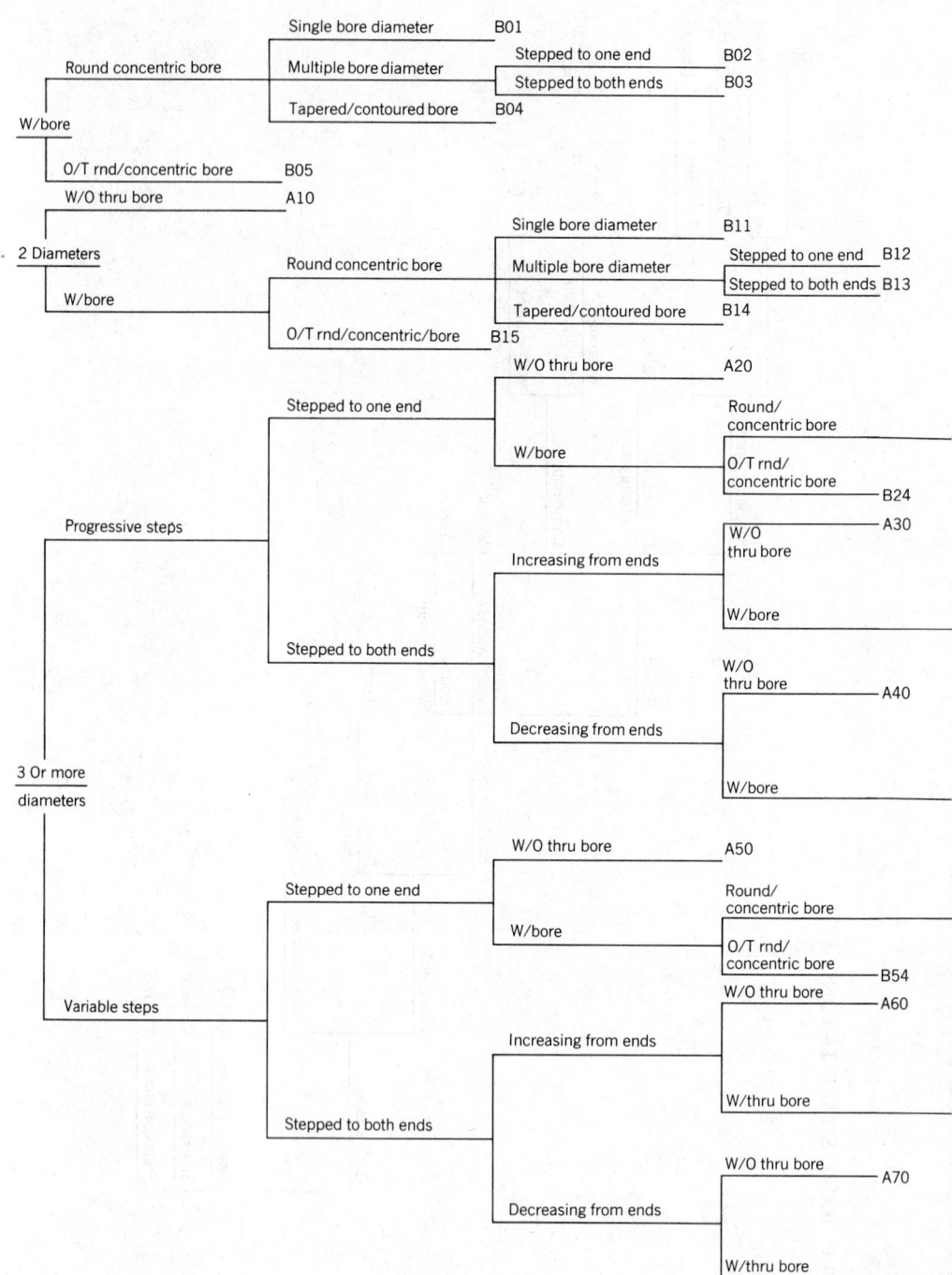

CHART 1

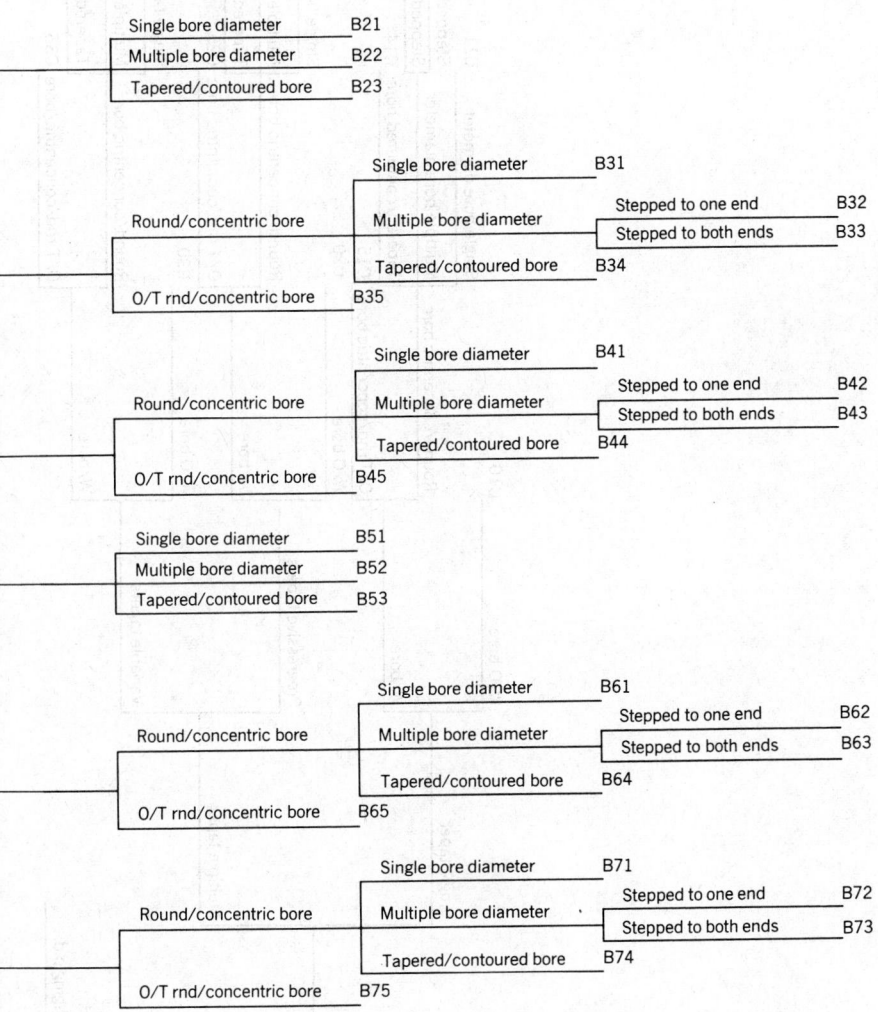

Single bore diameter	B21
Multiple bore diameter	B22
Tapered/contoured bore	B23

Round/concentric bore

Single bore diameter	B31
Multiple bore diameter	Stepped to one end	B32
	Stepped to both ends	B33
Tapered/contoured bore	B34

O/T rnd/concentric bore   B35

Round/concentric bore

Single bore diameter	B41
Multiple bore diameter	Stepped to one end	B42
	Stepped to both ends	B43
Tapered/contoured bore	B44

O/T rnd/concentric bore   B45

Single bore diameter	B51
Multiple bore diameter	B52
Tapered/contoured bore	B53

Round/concentric bore

Single bore diameter	B61
Multiple bore diameter	Stepped to one end	B62
	Stepped to both ends	B63
Tapered/contoured bore	B64

O/T rnd/concentric bore   B65

Round/concentric bore

Single bore diameter	B71
Multiple bore diameter	Stepped to one end	B72
	Stepped to both ends	B73
Tapered/contoured bore	B74

O/T rnd/concentric bore   B75

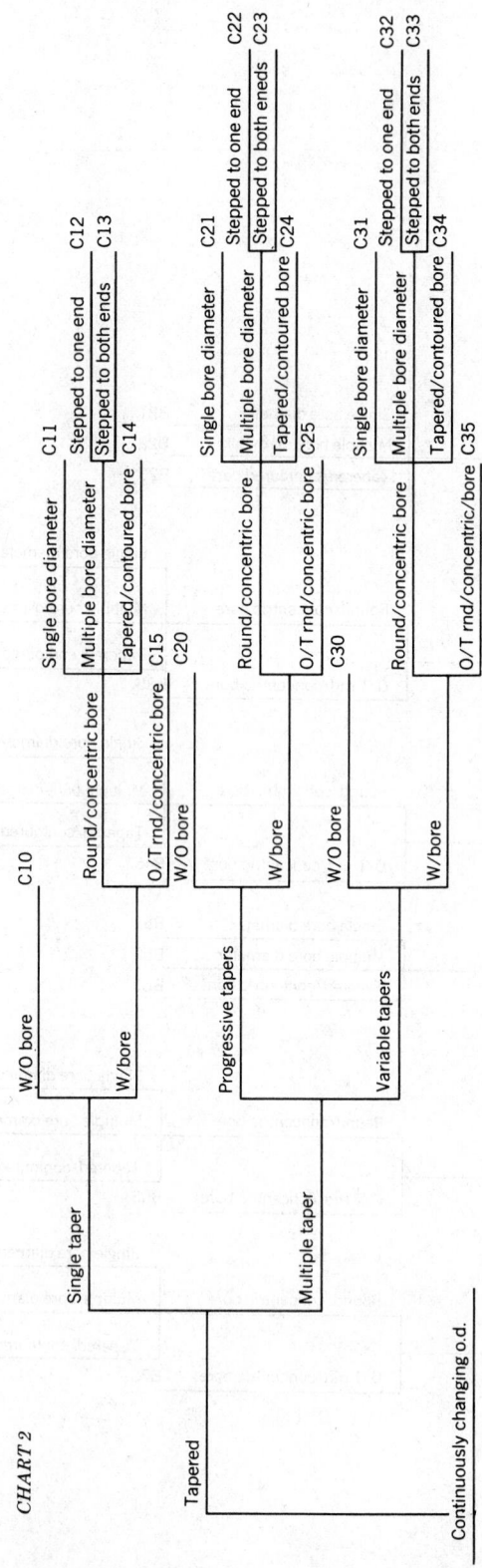

CHART 2

Tapered

Continuously changing o.d.

Curved/spherical

- **Concave**
  - W/O bore — D10
  - W/bore
    - Round/concentric bore
      - Single bore diameter — D11
      - Multiple bore diameter
        - Stepped to one end — D12
        - Stepped to both ends — D13
      - Tapered/contoured bore — D14
    - O/T rnd/concentric bore — D15
- **Convex**
  - W/O bore — D20
  - W/bore
    - Round/concentric bore
      - Single bore diameter — D21
      - Multiple bore diameter
        - Stepped to one end — D22
        - Stepped to both ends — D23
      - Tapered/contoured bore — D24
    - O/T rnd/concentric bore — D25
- **Spherical**
  - W/O bore — D30
  - W/bore — D31
- **Combination curved**
  - W/O bore — D40
  - W/bore
    - Round/concentric bore
      - Single bore diameter — D41
      - Multiple bore diameter
        - Stepped to one end — D42
        - Stepped to both ends — D43
      - Tapered/contoured bore — D44
    - O/T rnd/concentric bore — D45

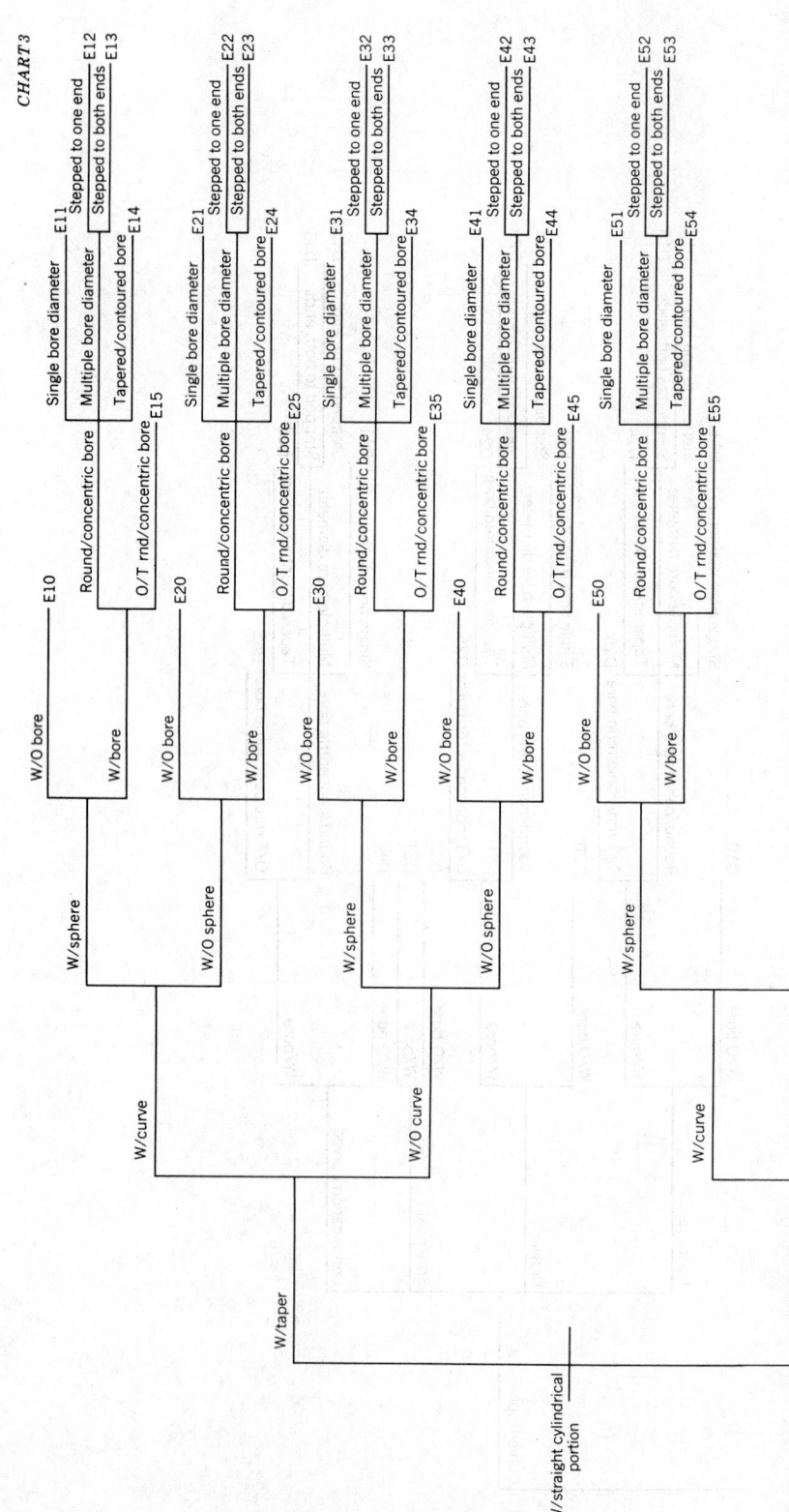

CHART 3

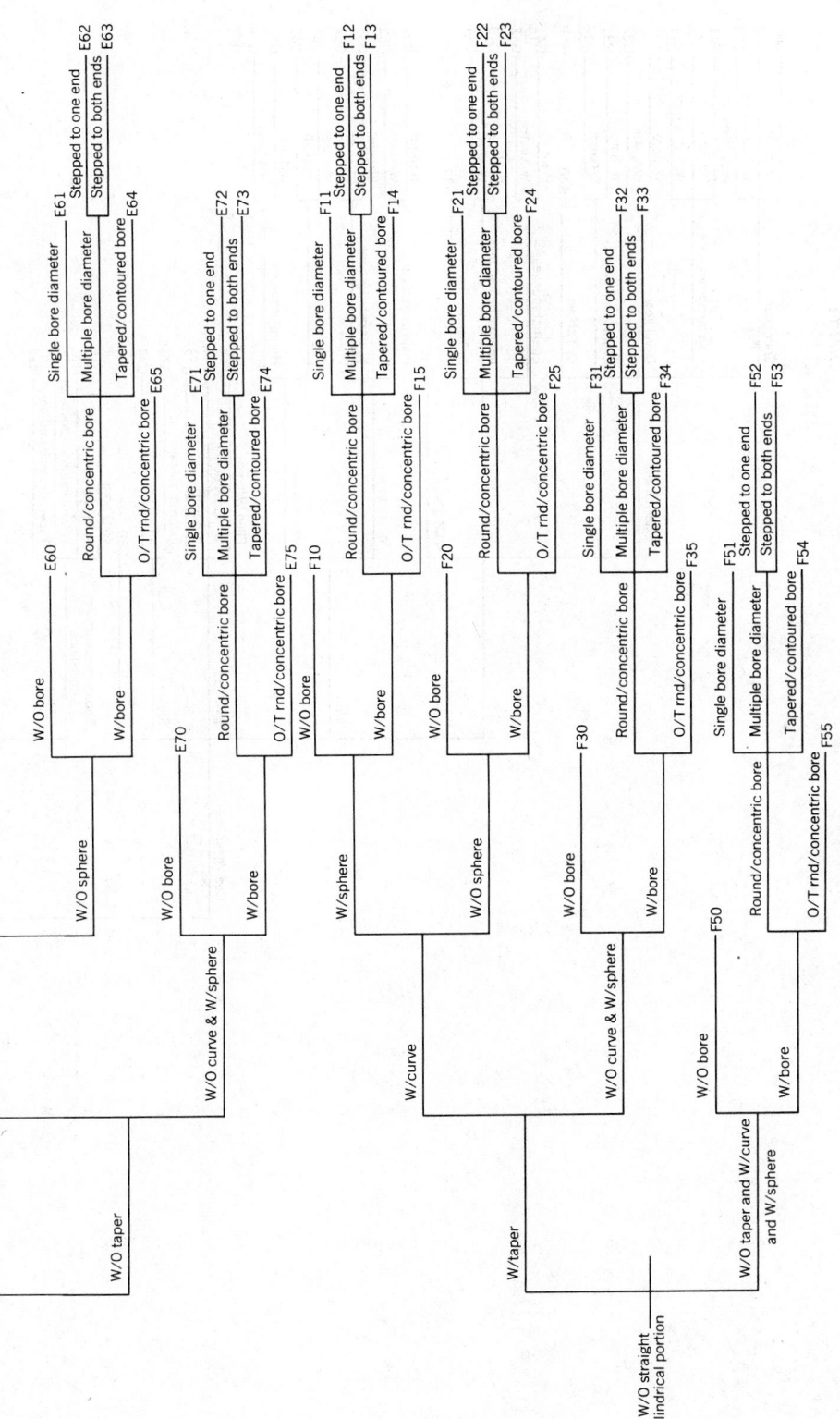

CHART 4

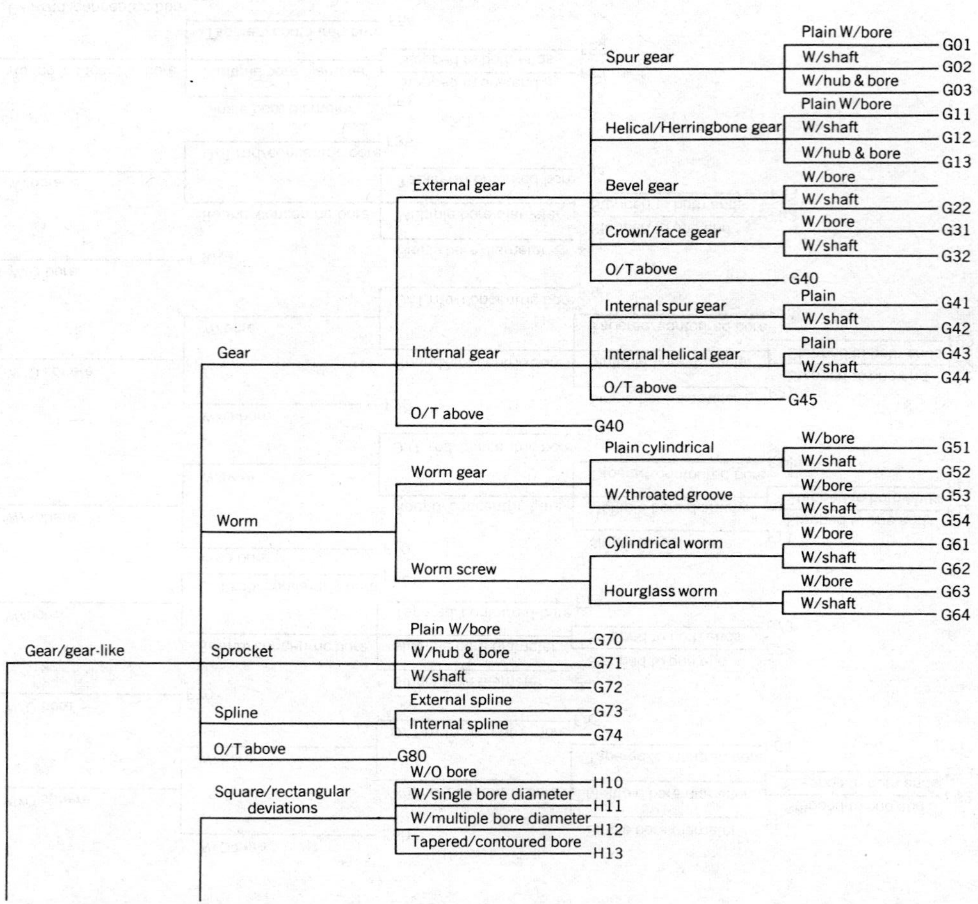

Gear/gear-like

- **Gear**
  - **External gear**
    - **Spur gear**
      - Plain W/bore — G01
      - W/shaft — G02
      - W/hub & bore — G03
    - **Helical/Herringbone gear**
      - Plain W/bore — G11
      - W/shaft — G12
      - W/hub & bore — G13
    - **Bevel gear**
      - W/bore — G22
      - W/shaft
    - **Crown/face gear**
      - W/bore — G31
      - W/shaft — G32
    - O/T above — G40
  - **Internal gear**
    - **Internal spur gear**
      - Plain — G41
      - W/shaft — G42
    - **Internal helical gear**
      - Plain — G43
      - W/shaft — G44
    - O/T above — G45
  - O/T above — G40
- **Worm**
  - **Worm gear**
    - **Plain cylindrical**
      - W/bore — G51
      - W/shaft — G52
    - **W/throated groove**
      - W/bore — G53
      - W/shaft — G54
  - **Worm screw**
    - **Cylindrical worm**
      - W/bore — G61
      - W/shaft — G62
    - **Hourglass worm**
      - W/bore — G63
      - W/shaft — G64
- **Sprocket**
  - Plain W/bore — G70
  - W/hub & bore — G71
  - W/shaft — G72
- **Spline**
  - External spline — G73
  - Internal spline — G74
- O/T above — G80
- **Square/rectangular deviations**
  - W/O bore — H10
  - W/single bore diameter — H11
  - W/multiple bore diameter — H12
  - Tapered/contoured bore — H13

Drawn/spun shapes

- **Round W/deviations**
  - **Regular deviations**
    - Hexagonal deviations
      - W/O bore — H20
      - W/single bore diameter — H21
      - W/multiple bore diameter — H22
      - Tapered/contoured bore — H23
    - Octagonal deviations
      - W/O bore — H30
      - W/single bore diameter — H31
      - W/multiple bore diameter — H32
      - Tapered/contoured bore — H33
    - O/T above — H40
  - **Irregular deviations**
    - Cams/eccentrics — J10
    - Lobes
      - Single lobe — J20
      - Multiple-camshaft-like — J21
    - Crankshaft-like — J30
    - O/T above — J40
- **Round/bent c'line**
  - **Uniform radius**
    - Bent-one plane
      - Solid — K10
      - Hollow — K11
    - Bent-2, 3 planes
      - Solid — K20
      - Hollow — K21
  - **Non-uniform radius**
    - Bent-one plane
      - Solid — K30
      - Hollow — K31
    - Bent-2, 3 planes
      - Solid — K40
      - Hollow — K41
- **Cylinder**
  - Straight sides
    - W/O flange — L10
    - W/flange — L11
  - Convex/bulged
    - W/O flange — L12
    - W/flange — L13
  - Concave
    - W/O flange — L14
    - W/flange — L15
  - O/T above
    - Plain sidewalls
      - W/O flange — L20
      - W/flange — L21
    - Convex sidewalls (ojive)
      - W/O flange — L22
      - W/flange — L23
    - Concave sidewalls
      - W/O flange — L24
      - W/flange — L25
- **Cone shape**
  - O/T above — L27
- **Dome shape**
  - W/O flange — L30
  - W/flange — L31
- **Torus** — L40
- **O/T above** — L50

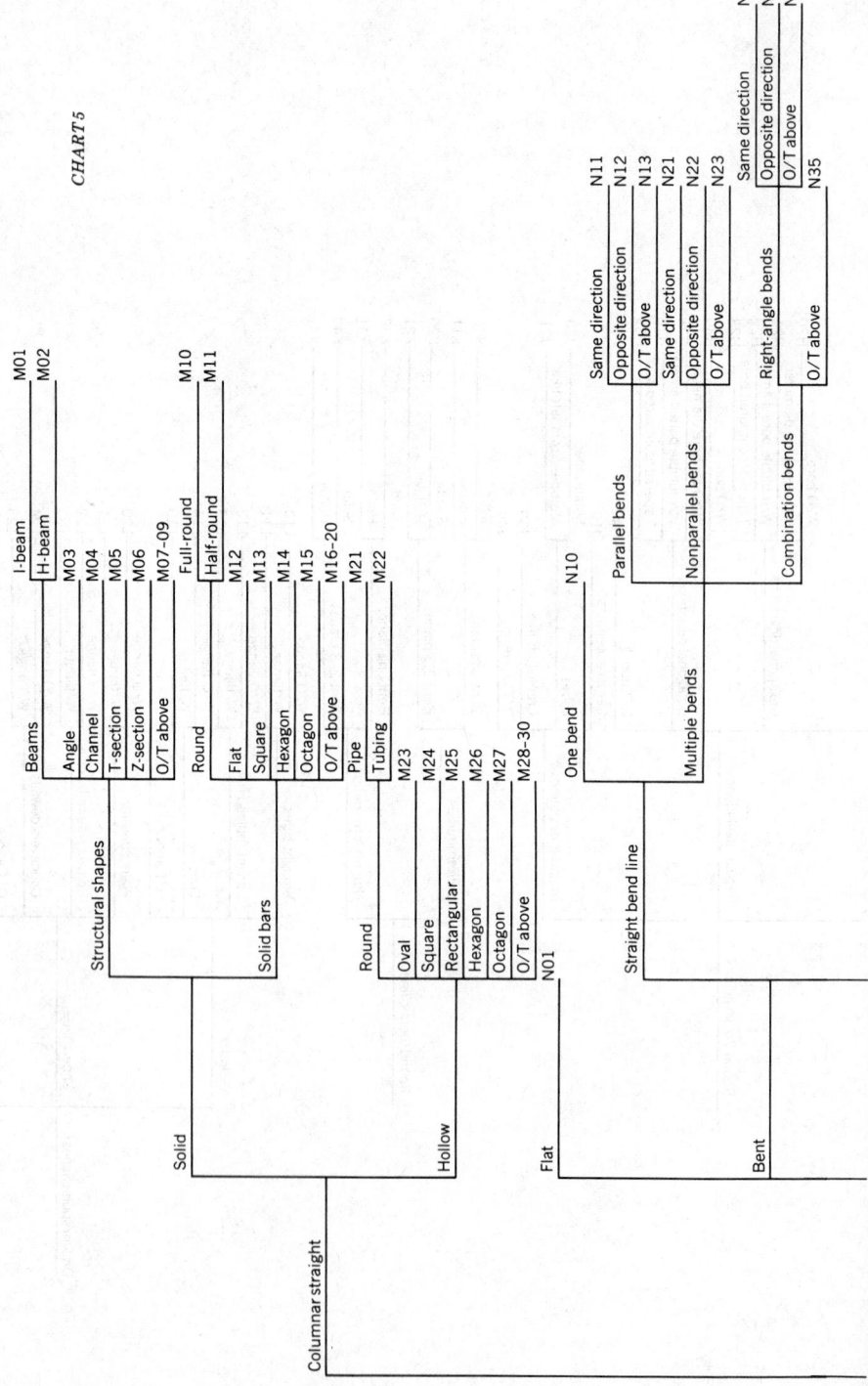

CHART 5

Columnar straight

- **Solid**
  - **Structural shapes**
    - Beams
      - I-beam — M01
      - H-beam — M02
    - Angle — M03
    - Channel — M04
    - T-section — M05
    - Z-section — M06
    - O/T above — M07–09
  - **Solid bars**
    - Round
      - Full-round — M10
      - Half-round — M11
    - Flat — M12
    - Square — M13
    - Hexagon — M14
    - Octagon — M15
    - O/T above — M16–20
- **Hollow**
  - Pipe — M21
  - Tubing — M22
  - Round
    - Oval — M23
    - Square — M24
    - Rectangular — M25
    - Hexagon — M26
    - Octagon — M27
    - O/T above — M28–30

Flat — N01

Bent
- **Straight bend line**
  - One bend — N10
  - Multiple bends
    - Parallel bends
      - Same direction — N11
      - Opposite direction — N12
      - O/T above — N13
    - Nonparallel bends
      - Same direction — N21
      - Opposite direction — N22
      - O/T above — N23
    - Combination bends
      - Right-angle bends
        - Same direction — N31
        - Opposite direction — N32
        - O/T above — N33
      - O/T above — N35

810

Sheet form
- Curved bend line
  - Flanged
    - Stretch flange — N41
    - Shrink flange — N42
    - Curved angles — N45
    - Curved channels — N46
    - Curved hat-section — N47
    - O/T above — N48
  - Curved section
    - Regular shallow cross-section
      - W/O flange — N61
      - W/flange — N62
    - Irregular shallow cross-section
      - W/O flange — N63
      - W/flange — N64
    - Regular deep cross-section
      - W/O flange — N65
      - W/flange — N66
    - Irregular deep cross-section
      - W/O flange — N67
      - W/flange — N68
- Warped
  - Simple curved — N51
  - Compound curved — N52
  - Reverse curved — N53
- Drawn
  - Shallow drawn
  - Deep drawn
- Bulged — N70
- O/T above — N80

Nonrotational

Box-like solids
- Plane surfaces
  - Parallel sides
    - Cube — P01
    - O/T above — P02
  - Nonparallel sides
    - Pyramid shape — P10
    - O/T above — P11
  - Combination parallel/non-parallel
    - Prism
      - Regular — P20
      - Irregular — P21
    - Frustum — P30
    - O/T above — P40
- Curved surfaces
  - Convex surfaces — P51
  - Concave surfaces — P52
  - Combination convex/concave surfaces — P53

Named shapes — Q-Z

CHART 6

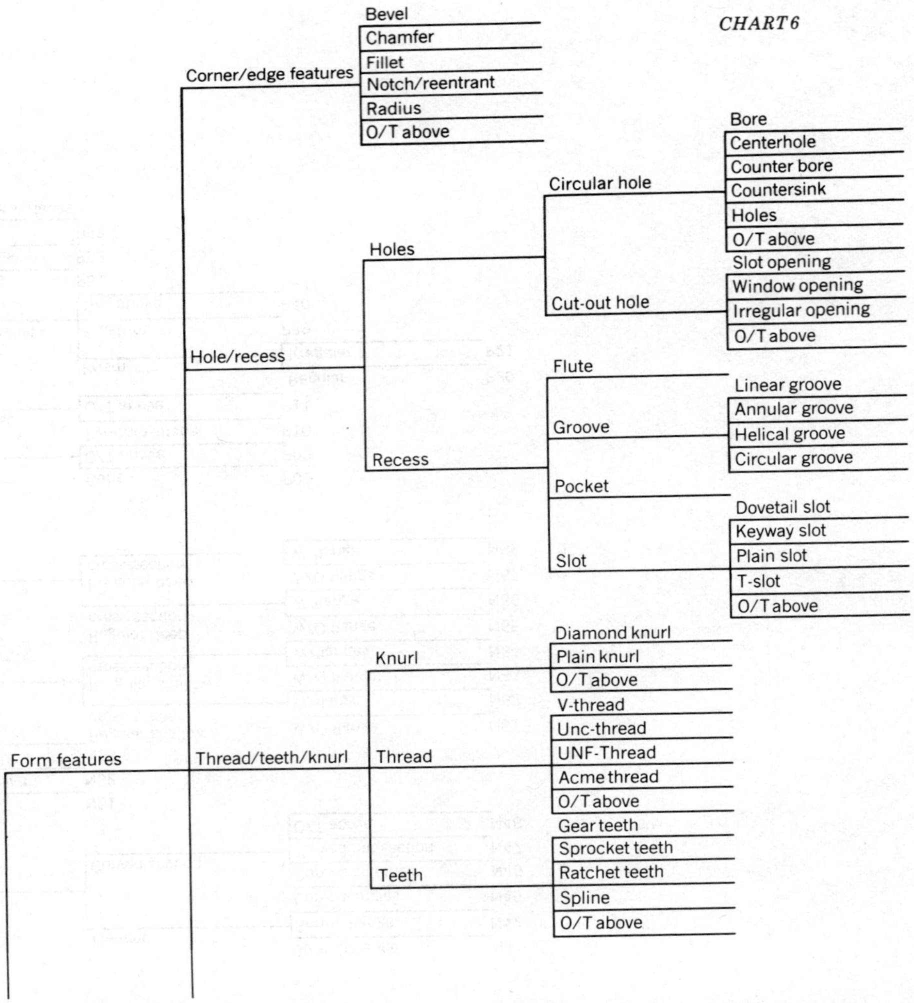

Form features

- **Corner/edge features**
  - Bevel
  - Chamfer
  - Fillet
  - Notch/reentrant
  - Radius
  - O/T above

- **Hole/recess**
  - **Holes**
    - **Circular hole**
      - **Bore**
        - Centerhole
        - Counter bore
        - Countersink
      - Holes
      - O/T above
      - Slot opening
      - Window opening
      - Irregular opening
      - O/T above
    - **Cut-out hole**
  - **Recess**
    - **Flute**
      - Linear groove
      - Annular groove
      - Helical groove
      - Circular groove
    - **Groove**
    - **Pocket**
      - Dovetail slot
      - Keyway slot
      - Plain slot
      - T-slot
      - O/T above
    - **Slot**

- **Thread/teeth/knurl**
  - **Knurl**
    - Diamond knurl
    - Plain knurl
    - O/T above
  - **Thread**
    - V-thread
    - Unc-thread
    - UNF-Thread
    - Acme thread
    - O/T above
  - **Teeth**
    - Gear teeth
    - Sprocket teeth
    - Ratchet teeth
    - Spline
    - O/T above

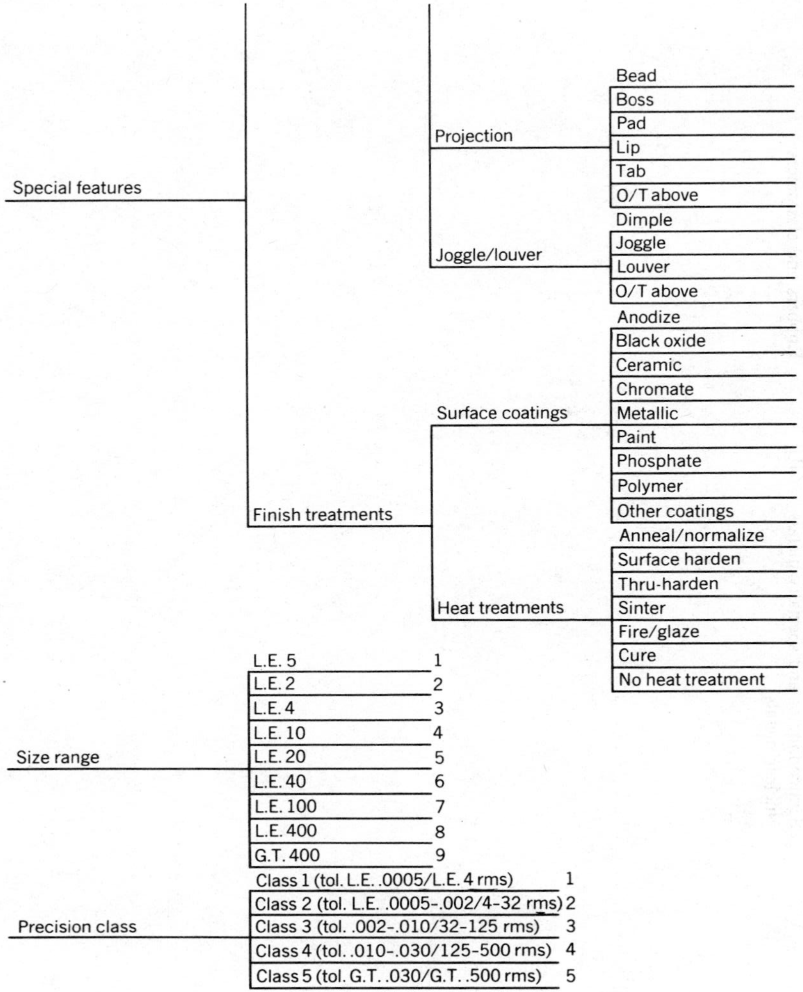

Special features

Finish treatments

Projection
- Bead
- Boss
- Pad
- Lip
- Tab
- O/T above

Joggle/louver
- Dimple
- Joggle
- Louver
- O/T above

Surface coatings
- Anodize
- Black oxide
- Ceramic
- Chromate
- Metallic
- Paint
- Phosphate
- Polymer
- Other coatings

Heat treatments
- Anneal/normalize
- Surface harden
- Thru-harden
- Sinter
- Fire/glaze
- Cure
- No heat treatment

Size range
- L.E. 5 — 1
- L.E. 2 — 2
- L.E. 4 — 3
- L.E. 10 — 4
- L.E. 20 — 5
- L.E. 40 — 6
- L.E. 100 — 7
- L.E. 400 — 8
- G.T. 400 — 9

Precision class
- Class 1 (tol. L.E. .0005/L.E. 4 rms) — 1
- Class 2 (tol. L.E. .0005–.002/4–32 rms) — 2
- Class 3 (tol. .002–.010/32–125 rms) — 3
- Class 4 (tol. .010–.030/125–500 rms) — 4
- Class 5 (tol. G.T. .030/G.T. .500 rms) — 5

APPENDIX 30.2   ENGINEERING MATERIAL FAMILIES

Family	Descriptive Title	Family	Descriptive Title
A	Steel	L	Rare-earth metals
B	Cast iron	M	Composites
C	Clad, coated, bonded metals	N	Minerals, refractories, and ceramics
D	Light metals	P	Wood and wood products
E	Nonferrous engineering metals	Q	Paper and paper products
F	Low-melting-point metals	R	Textile fibers and products
G	Refractory metals	S	Glass
H	Precious metals	T	Polymers
J	Semiconductor and specialty metals	U	Rubbers and elastomers
K	Nuclear metals		

APPENDIX 30.3   FABRICATION PROCESSES

PRODUCTION PROCESSES CODE	
DIVISION	DESCRIPTION
000	MATERIAL HANDLING
100	MATERIAL REMOVAL
200	CONSOLIDATION
300	DEFORMATION
400	JOINING
500	HEAT-TREATING
600	SURFACE FINISHES
700	INSPECTION
800	ASSEMBLY
900	TESTING

Shaping processes - 1

Mechanical reducing

Reduction (chips)

Single point cutting
- 101 Turning/facing
- 102 Boring
- 103 Shaping/planning
- 104 Parting/grooving
- 105 Threading (S.P.)

Multipoint cutting
- 111 Drilling
- 112 Reaming
- 113 Milling/routing
- 114 Broaching
- 115 Threading (M.P.)
- 116 Filing
- 117 Sawing
- 118 Gear cutting

Abrasive machining
- 121 Grinding
- 122 Honing
- 123 Lapping
- 124 Superfinishing
- 125 Ultrasonic machining
- 126 Jet machining

Shearing
- 131 Squaring
- 132 Slitting
- 133 Rotary shear
- 134 Nibbling

Blanking
- 141 Conventional blanking
- 142 Steel-rule-die blanking
- 143 Fine blanking

Mass reducing

Separation (shear)
- Shaving/trimming — 144
- Dinking — 145
- Piercing
  - Punching — 146
  - Perforating — 147
  - Lancing — 148
  - Notching

Thermal reducing

- Torch cutting
  - Air-arc cutting — 161
  - Gas cutting — 162
  - Plasma arc cutting — 163
- Electrical discharge mach
  - Cavity-type E.D.M. — 171
  - E.D.M. grinding — 172
  - E.D.M. sawing — 173
- High energy beam machining
  - Electron-beam cutting — 181
  - Laser-beam cutting — 182
  - Ion-beam cutting — 183

Chemical reducing

- Chemical milling
  - Immersion chem. milling — 191
  - Spray chemical milling — 192
- Electrochemical milling
  - Cavity-type E.C.M. — 194
  - Grinder-type E.C.M. — 195
- Photochemical milling
  - Photoetching — 197
  - Photomilling — 198

- **Casting**
  - Nonreusable mold
    - Cermaic mold casting
      - Investment casting — 201
      - Plaster mold casting — 202
    - Sand mold casting
      - Sand casting — 211
      - Shell mold casting — 212
      - No-bake mold casting — 213
      - Full-mold casting — 214
  - Reusable mold
    - Die casting — 221
    - Permanent mold casting — 222
    - Flexible mold casting — 223
    - Continuous casting — 224
- **Molding**
  - Cermaic molding
    - Wet forming — 231
    - Dry pressing — 232
  - Polymer molding
    - Injection molding — 241
    - Blow molding — 242
    - Transfer molding — 243
    - Compression molding — 244
    - Extrusion molding — 245
    - Thermoform molding — 246
    - Rotational molding — 247
- **Compacting**
  - Continuous
    - P/M extrusion — 251
    - P/M rolling — 252
  - Noncontinuous
    - Pressing (compacting) — 261
    - Centrifugal compacting — 262
    - Explosive compacting — 263
    - Isostatic pressing — 264
    - Slip casting (compacting) — 265
- **Deposition**
  - Electroforming — 271
- **Laminating**
  - Filament winding — 281
  - Sheet laminating — 282
  - Bulk laminating
    - Spray lay-up — 284
    - Hand lay-up — 285
  - Pultrusion — 287
- **Consolidation**
  - Hot forging
    - Hammer forging — 301
    - Drop forging — 302
    - Press forging — 303
    - Upset forging — 304
    - Roll forging — 305

Mass conserving

- Deformation
  - Total deformation
    - Forge
      - Cold forging
        - Swaging (solids) — 307
        - Cold heading — 308
    - Extrude
      - Direct extrusion — 311
      - Indirect extrusion — 312
      - Impact extrusion — 313
    - Draw
      - Wire drawing — 321
      - Tube drawing — 322
    - Roll
      - Sheet rolling — 331
      - Foil rolling — 332
      - Structural rolling — 333
      - Pierce rolling — 334
    - Shearspin — 341
    - Coining/sizing/hobbing — 342
    - Thread forming — 343
    - Knurling — 344
    - Bending
      - Straight angle bending — 351
      - Corrugation bending — 352
      - Joggle bending — 353
      - Curling — 354
      - Seaming — 355
      - Tube bending — 356
      - Roll forming — 357
  - Local deformation
    - Conventional forming
      - Die forming
        - Matched die drawing
          - Simple rigid die — 361
          - Compound die — 362
          - Progressive die — 363
        - Rubber die drawing
          - Guerin process
          - Marform process — 366
          - Hydroform process — 367
      - Sheet forming
        - Conventional spinning — 371
        - Stretch forming — 372
        - Embossing — 373
      - Tube forming
        - Swaging — 381
        - Flaring — 382
        - Intraforming — 383
    - High energy rate forming
      - Explosive forming — 391
      - Electromagnetic forming — 392
      - Electrohydraulic forming — 393

Mechanical
- Pressure (cold) welding — 401
- Friction (inertia) welding — 402
- Ultrasonic welding — 403
- Explosive welding — 404

Electric arc welding
- Shielded metal arc welding — 411
- Gas metal arc (Mig) welding — 412
- Gas tungsten arc (Tig) welding — 413
- Submerged arc welding — 414
- Carbon arc welding — 415
- Stud welding — 416

Electrical resistance welding
- Spot welding — 421
- Seam welding — 422
- Projection welding — 423
- Butt welding — 424
- Percussion welding — 425
- Electroslag welding — 426

Gas/chemical welding

Thermal welding
- Combustible gas welding — 431
- Atomic hydrogen welding — 432

Braze welding
- Gas brazing — 441
- Carbon arc brazing — 442

Joining

Thermal

- Diffusion bonding 451
- High energy beam welding
  - Electron beam welding 461
  - Laser beam welding 462
  - Plasma arc welding 463

Brazing
- Infrared brazing 471
- Resistance brazing 472
- Torch brazing 473
- Dip brazing 474
- Furnace brazing 475
- Induction brazing 476

Soldering
- Friction/ultrasonic soldering 481
- Induction soldering 482
- Infrared soldering 483
- Dip soldering 484
- Iron soldering 485
- Resistance soldering 486
- Torch soldering 487
- Wave soldering 488

Chemical
- Adhesive bonding 491

Nonshaping processes

Heat treatment

- Annealing
  - Recovery
    - Stress relieve — 501
    - Tempering — 502
  - Recrystallization
    - Full anneal — 511
    - Process anneal — 512
    - Short-cycle anneal — 513
- Hardening
  - Surface hardening
    - Carburizing — 521
    - Chromizing — 522
    - Carbonitriding — 523
    - Cyaniding — 524
    - Nitriding — 525
    - Diffusion hardening — 526
    - Flame hardening — 527
    - Induction hardening — 528
  - Through hardening
    - Water quench hardening — 531
    - Oil quench hardening — 532
    - Air quench hardening — 533
    - Martempering — 534
    - Austempering — 535
    - Age hardening — 536
  - Other thermal treatments
    - Sintering
      - Solid-phase sintering — 541
      - Liquid-phase sintering — 542
    - Cold treatment (subzero) — 544
    - Firing/glazing — 545
    - Curing/bonding — 546

Nonshaping processes

Surface preparation

- Descaling
  - Mechanical descaling
    - Abrasive blasting — 601
    - Belt sanding — 602
    - Shot peening preparation — 603
    - Wire brushing — 604
    - Grinding — 605
  - Thermal descaling
    - Flame cleaning — 607
  - Chemical descaling
    - Chemical pickling — 609
- Deburring
  - Mechanical deburring
    - Barrel tumbling — 611
    - Vibratory finishing — 612
    - Knife deburring — 613
  - Thermal deburring
    - Thermochemical deburring — 615
  - Chemical deburring
    - Electrochemical deburring — 617

- Surface finishing
  - Surface coating
    - Degreasing
      - Mechanical degreasing
        - Ultrasonic degreasing — 621
      - Chemical degreasing
        - Vapor degreasing — 623
        - Solvent degreasing — 624
        - Alkali degreasing — 625
    - Mechanical coating
      - Spray coating
        - Pressure transferred
          - Air gun spraying — 631
          - High pressure airless spray — 632
        - Charged transferred
          - Electrostatic coatings — 634
          - Vacuum coatings — 635
      - Dip/flow coating
        - Cold dip coating — 641
        - Hot dip coating — 642
        - Electrocoating — 643
        - Fluidized bed coating — 644
        - Curtain coating — 645
      - Dust coating — 647
      - Roll coating
        - Calendering — 648
        - Roller coating — 649
    - Thermal coating
      - Flame spraying
        - Combustion flame spraying — 651
        - Plasma arc spraying — 652
        - Detonation gun spraying — 653
      - Vaporized metal coating
        - Vacuum metallizing — 661
        - Sputtering — 662
        - Chemical vapor—phase deposition — 663
      - Heat tinting — 665
    - Chemical coating
      - Electroplate — 671
      - Chemical conversion
        - Anodize
          - Alkaline oxide — 674
          - Fused nitrate — 675
          - Proprietary treatment — 676
        - Phosphate — 677
        - Chromate — 678
  - Surface modification
    - Burnishing — 681
    - Peening
      - Shot peening mod. — 683
      - Hammer peening — 684
    - Texturing
      - Wire brush finishing — 686
      - Buffing/polishing — 687

Alphabetical Listing of Processes and Codes

Abrasive blasting—601
Abrasive machining—121, 122, 123, 124, 125, 126
Adhesive bonding—491
Age hardening—536
Air-gun spraying—631
Air quench hardening—533
Airless spray, high-pressure—632
Alkali degreasing—625
Alkaline oxide—674
Anneal, full—511
Anneal, process—512
Anneal, short-cycle—513
Anodize—673
Arc cutting—161
Arc spraying, plasma—652
Arc (Mig) welding, gas metal—412
Arc (Tig) welding, gas tungsten—413
Arc welding, electric—411, 412, 413, 414, 415, 416
Arc welding, plasma—463
Arc welding, shielded metal—411
Arc welding, submerged—414
Atomic hydrogen welding—432
Austemper hardening—535

Barrell tumbling—611
Beam cutting, electron—181
Beam cutting, ion—183
Beam cutting, laser—182
Beam machining, high-energy—181, 182, 183
Beam welding, electron—461
Belt welding, high-energy—461, 462, 463
Beam welding, laser—462
Belt sanding—602
Bending—351, 352, 353, 354, 355, 356, 357
Bending, corregation—352
Bending, curling—354
Bending, jobble—353
Bending, straight-angle—351
Bending, tube—356
Blanking—141, 142, 143, 144, 145
Blanking, conventional—141
Blanking, fine—143
Blanking, shaving—144
Blanking, steel-rule die—142
Blanking, trimming/cut-off—144
Blasting, abrasive—601
Blow molding—242
Bonding, adhesive—491
Bonding, diffusion—451
Bonding/curing—546
Boring—102
Braze welding—441, 442
Brazing—471–476
Brazing, carbon arc—442
Brazing, dip—474
Brazing, furnace—475
Brazing, gas—441
Brazing, induction—476
Brazing, infrared—471
Brazing, resistance—472

Brazing, torch—473
Broaching—114
Brush finishing, wire—686
Brushing, wire—604
Buffing/polishing—687
Bulk laminating—284, 285
Burnishing—681
Butt welding—424

Calendering—648
Carbon arc brazing—442
Carbon arc welding—415
Carburizing—521
Carbonitriding—523
Casting, ceramic mold—201, 202
Casting, continuous—224
Casting, die—221
Casting, flexible mold—223
Casting, full-mold—214
Casting, investment—201
Casting, no-bake mold—213
Casting, permanent mold—222
Casting, plaster mold—202
Casting, sand—211
Casting, sand mold—211, 212, 213, 214
Casting, shell mold—212
Centrifugal compacting—262
Ceramic mold casting—201, 202
Ceramic molding—231, 232
Charged transferred—634, 635
Chemical conversion—673, 674, 675, 676, 677, 678
Chemical deburring—617
Chemical degreasing—623, 624, 625
Chemical descaling—609
Chemical joining—491
Chemical milling—191, 192
Chemical milling, immersion—191
Chemical milling, spray—192
Chemical pickling—609
Chemical vapor-phase deposition—663
Chemical/gas welding—431, 432
Chromizing—522
Chromate—678
Cleaning, flame—607
Coating, cold dip—641
Coating, curtain—645
Coating, dust—647
Coating, electrostatic—634
Coating, fluidized bed—644
Coating, hot dip—642
Coating, roller—649
Coating, vaporized metal—661, 662, 663
Coatings, vacuum—635
Coining/sizing/hobbing—342
Cold dip coating—641
Cold forging—307, 308
Cold heading—308
Cold treatment (subzero)—544
Combustible gas welding—431
Combustion flame spraying—651
Compacting, centrifugal—262

APPENDIX 30.4 EQUIPMENT TAXONOMY

Equipment
1. Mass reducing

Reduction (chips)

- **Single point cutting**
  - **Turning**
    - Manual engine lathes — 101-A
    - Semiautomatic lathes
      - Turret lathes — 101-D
      - Hand screw lathes — 101-E
    - Automatic lathes
      - Bar lathes — 101-H
      - Chucking lathes — 101-J
      - Screw lathes — 101-K
      - N.C. & C.N.C. lathes — 101-L
      - Special lathes — 101-M
  - **Boring**
    - Boring machines — 103-A
    - Jig borers — 103-B
  - **Shaping/planing**
    - Planers — 104-A
    - Shapers — 104-B
- **Multipoint cutting**
  - **Drills**
    - General purpose
      - Multiple operation
        - Gang drills — 111-A
        - Multiple spindle drills — 111-B
        - Turret drills — 111-C
        - N.C. drills — 111-D
      - Single operation
        - Vertical drills — 111-G
        - Horizontal drills — 111-H
    - Special purpose
      - Gun drills — 111-L
      - Deep-hole drills — 111-M
      - Other drills — 111-N
  - **Mills**
    - Manual mills
      - Vertical — 113-A
      - Horizontal
        - Plain hor. mill — 113-D
        - Universal hor. mill — 113-E
    - Automatic mills
      - N.C. mills
        - Spar mills — 113-H
        - Skin mills — 113-J
        - Machining centers — 113-K
        - Bed-type mills — 113-L
        - Partsmaker — 113-M
      - Tracer mills — 113-Q
      - Other automatic mills — 113-R
  - **Milling/routing**
    - Routers — 113-U
  - **Broaching**
    - Surface broaches
      - Cont. sur. broaching mach. — 114-A
      - Intermittant broaching mach — 114-B
    - Int. broaching machine — 114-E
    - Sur. & int. broaching mach. — 114-F
    - Spec. broaching mach. — 114-G
  - **Threading**
    - Internal
      - Tapping mach. — 115-A
      - Other int. threading mach. — 115-B
    - External
      - Die cutting mach. — 115-E
      - Thread milling mach. — 115-F
  - **Filing**
    - Band filing machine — 116-A
    - Recip. filing mach. — 116-B
    - Disk filing mach. — 116-C

Mass reducing equipment

Mechanical reducing

- **Separation (shear)**
  - Sawing
    - Band saws
      - Straight cut band saws 117-A
      - Contour cutting band saws 117-B
    - Circular saws 117-E
    - Reciprocating saws 117-F
  - Gear cutting eq.
    - Gear hobbing mach. 118-A
    - Gear shapers 118-B
  - Abrasive machining
    - Grinding
      - Cylindrical grinder
        - Int. cylindrical grinder 212-A
        - Ext. cylindrical grinder
          - Centerless grinder 121-D
          - Centershape grinder 121-E
      - Surface grinders 121-H
      - Thread grinding mach. 121-J
    - Honing
      - Cylindrical hone
        - Internal cylindrical hone 122-A
        - External cylindrical hone 122-B
      - Flat surface hone 122-E
    - Lapping
      - Cylindrical lap 123-A
      - Flat surface lap 123-B
    - Superfinishing
      - Cyl. superfinishing mach. 124-A
      - Flat sur. superfinishing mach. 124-B
    - Ultrasonic mach. eq. 125-A
    - Jet mach. eq. 126-A
  - Shearing eq.
    - Blade shears 131-A
    - Butting machine 132-A
    - Rotary shears 133-A
    - 134-A
  - Stamping press
    - Tryout 140-A
    - Production 140-B

Thermal reducing

- Torch cutting
  - Arc cutting
    - Plasma arc eq. 163-A
    - Arc air eq. 161-A
  - Gas cutting
    - Inert gas eq. 162-A
    - Oxy-acetylene eq. 162-B
- Elec. discharge mach.
  - Cavity-type E.D.M. eq. 171-A
  - E.D.M. grinder 172-A
  - E.D.M. saw (traveling wire) 173-A
- High energy beam mach.
  - Electron beam cutting eq. 181-A
  - Laser cutting eq. 182-A
  - Ion beam cutting eq. 183-A

Chemical reducing

- Chemical milling
  - Immersion chem. milling eq. 191-A
  - Spray chem. milling eq. 192-A
- Electrochemical milling
  - Cavity-type E.C.M. eq. 194-A
  - Grinder-type E.C.M. eq. 195-A
- Photochemical milling
  - Photoetching eq. 197-A
  - Photomilling eq. 198-A

Consolidation

- **Casting**
  - **Remelting furnaces**
    - General purpose
      - Cupola furnace — 200-A
      - Crucible furnace — 200-B
      - Direct arc furnace — 200-C
      - Indirect arc furnace — 200-D
      - Electric furnace — 200-E
      - Induction furnace — 200-F
    - Special purpose
      - Vacuum furnace — 200-J
      - Special atmosphere furnace — 200-K
  - **Nonreusable mold**
    - Ceramic mold casting
      - Investment casting eq. — 201-A
      - Plaster mold casting eq. — 202-A
    - Sand mold casting
      - Sand casting eq. — 211-A
      - Shell mold casting eq. — 212-A
      - No-bake mold casting eq. — 213-A
      - Full-mold casting eq. — 214-A
  - **Reusable mold**
    - Die casting machines — 221-A
    - Permanent mold casting eq. — 222-A
    - Flexible mold casting eq. — 223-A
    - Centrifugal casting mach. — 222-B
    - Continuous casting mach. — 224-A

- **Molding**
  - **Ceramic molding**
    - Wet forming eq. — 231-A
    - Dry forming eq. — 232-A
  - **Polymer molding**
    - Injection molding mach. — 241-A
    - Blow molding mach. — 242-A
    - Transfer molding press — 243-A
    - Compression molding press — 244-A
    - Extrusion molding mach. — 245-A
    - Thermoform molding mach. — 246-A
    - Rotational molding mach. — 247-A

- **Compacting**
  - **Continuous**
    - Powder metal extruder — 251-A
    - Powder metal rolling mill — 252-A
  - **Noncontinuous**
    - Compacting press — 261-A
    - Compacting centrifuge — 262-A
    - Explosive compacting eq. — 263-A
    - Isostatic press — 264-A
    - Slip casting eq. — 265-A

- **Deposition**
  - Electroforming eq. — 271-A

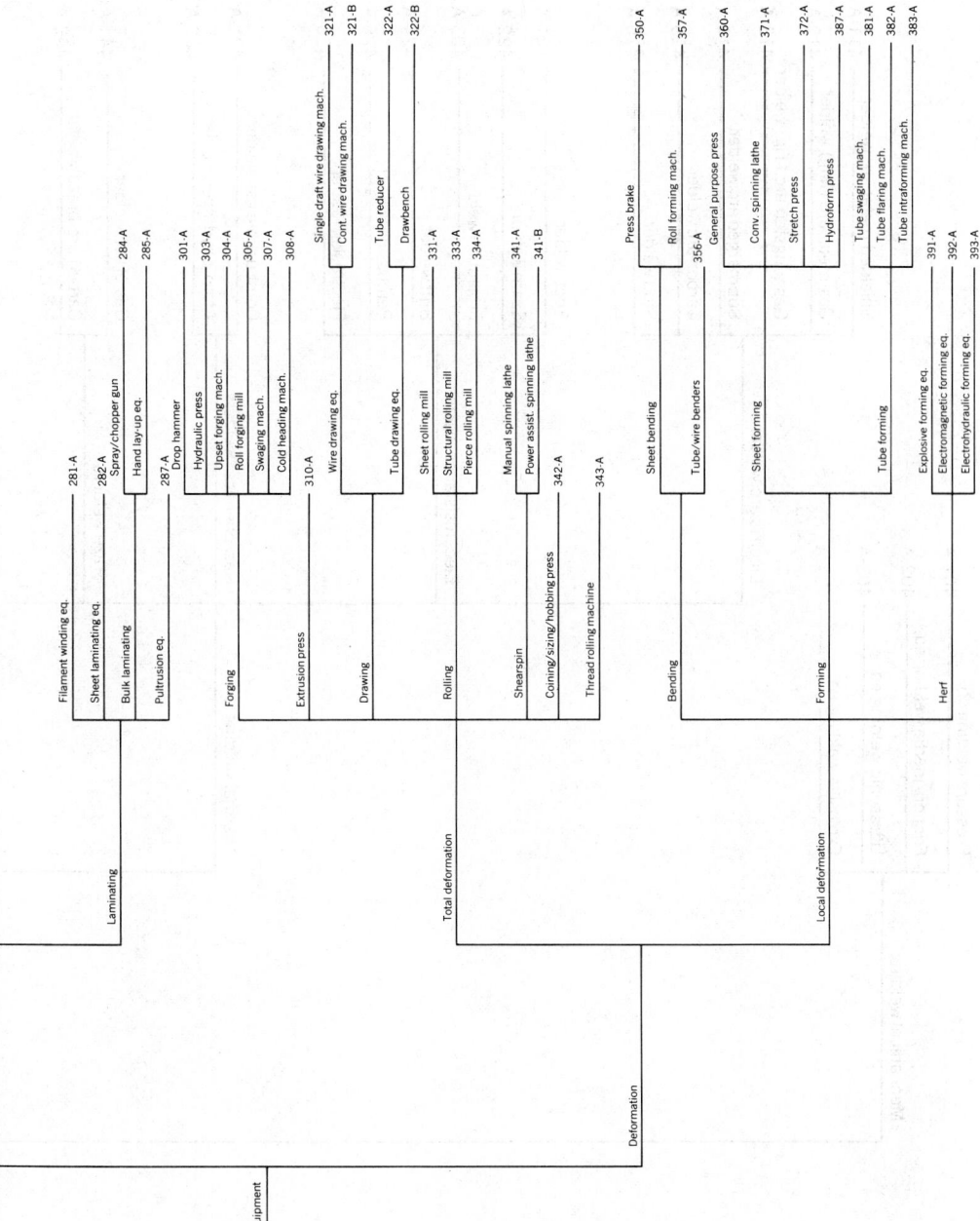

Mass conserving equipment

Deformation

Laminating
- Filament winding eq. — 281-A
- Sheet laminating eq. — 282-A
- Bulk laminating
  - Spray/chopper gun — 284-A
  - Hand lay-up eq. — 285-A
- Pultrusion eq. — 287-A

Total deformation

- Forging
  - Drop hammer — 301-A
  - Hydraulic press — 303-A
  - Upset forging mach. — 304-A
  - Roll forging mill — 305-A
  - Swaging mach. — 307-A
  - Cold heading mach. — 308-A
- Extrusion press — 310-A
- Drawing
  - Wire drawing eq.
    - Single draft wire drawing mach. — 321-A
    - Cont. wire drawing mach. — 321-B
  - Tube drawing eq.
    - Tube reducer — 322-A
    - Drawbench — 322-B
- Rolling
  - Sheet rolling mill — 331-A
  - Structural rolling mill — 333-A
  - Pierce rolling mill — 334-A
- Shearspin
  - Manual spinning lathe — 341-A
  - Power assist. spinning lathe — 341-B
- Coining/sizing/hobbing press — 342-A
- Thread rolling machine — 343-A

Local deformation

- Bending
  - Sheet bending
    - Press brake — 350-A
    - Roll forming mach. — 357-A
  - Tube/wire benders — 356-A
- Forming
  - General purpose press — 360-A
  - Sheet forming
    - Conv. spinning lathe — 371-A
    - Stretch press — 372-A
    - Hydroform press — 387-A
  - Tube forming
    - Tube swaging mach. — 381-A
    - Tube flaring mach. — 382-A
    - Tube intraforming mach. — 383-A
- Herf
  - Explosive forming eq. — 391-A
  - Electromagnetic forming eq. — 392-A
  - Electrohydraulic forming eq. — 393-A

Mechanical welding

- Pressure welding eq. — 401-A
- Friction/inertia welding eq. — 402-A
- Ultrasonic welding eq. — 403-A
- Explosive welding eq. — 404-A
- Electric arc welder
  - Shielded metal arc welder — 411-A
  - Gas metal arc (Mig) welder — 412-A
  - Gas tungsten arc (Tig) welder — 413-A
  - Submerged arc welder — 414-A
  - Carbon arc welder — 415-A
  - Stud welder — 416-A
- Elec. resistance welders
  - Spot welder — 421-A
  - Seam welder — 422-A
  - Projection welder — 423-A
  - Butt welder — 424-A
  - Percussion welder — 425-A
  - Electroslag welder — 426-A

Thermal welding

- Gas/chemical welders
  - Combustible gas welder — 431-A
  - Atomic hydrogen welder — 432-A
- Braze welders
  - Gas braze welder — 441-A
  - Carbon arc braze welder — 442-A

Joining

- Thermal
  - Diffusion bonding — 451-A
  - High energy beam welders
    - Electron beam welder — 461-A
    - Laser beam welder — 462-A
    - Plasma arc welder — 463-A
  - Brazing
    - Infrared brazing eq. — 471-A
    - Resistance brazing eq. — 472-A
    - Torch brazing eq. — 473-A
    - Dip brazing eq. — 474-A
    - Furnace brazing eq. — 475-A
    - Induction brazing eq. — 476-A
  - Soldering
    - Friction/ultrasonic sold. eq. — 481-A
    - Induction soldering eq. — 482-A
    - Infrared soldering eq. — 483-A
    - Dip soldering eq. — 484-A
    - Iron soldering eq. — 485-A
    - Resistance soldering eq. — 486-A
    - Torch soldering eq. — 487-A
    - Wave soldering eq. — 488-A
- Chemical
  - Adhesive bonding — 491-A

835

Equipment
4. Nonshaping

Nonshaping processes equipment

- Heat treatments
  - Heating eq.
    - Furnaces
      - Batch-type furnace
        - Horizontal furnace
          - box-type furnace — 500-A
          - Car bottom box-type furnace — 500-B
        - Vertical furnace
          - Pit furnace — 500-E
          - Bell furnace — 500-F
          - Elevator furnace — 500-G
          - Salt bath furnace — 500-H
      - Continuous furnace — 500-L
    - Ovens — 500-P
    - Flame hardening eq. — 527-A
    - Induction hardening eq. — 528-A
  - Cooling eq.
    - Quench tanks — 530-A
    - Refrigeration eq. — 544-A
- Surface preparation
  - Descaling
    - Mechanical descaling
      - Abrasive blastep — 601-A
      - Belt sander — 602-A
      - Shot peening eq. — 603-A
      - Wire brush eq. — 604-A
      - Grinder — 605-A
    - Thermal descaling
      - Flame cleaning eq. — 607-A
    - Chemical descaling
      - Chemical pickling eq. — 609-A
  - Deburring
    - Mechanical deburring
      - Barrel tumbler
        - Cov. barrel tumbling eq. — 611-A
        - Centrifugal barrel tumbling eq. — 611-B
      - Vibratory finisher — 612-A
    - Thermal deburring
      - Thermochemical deburring eq. — 615-A
    - Chemical deburring
      - Electrochemical deburring eq. — 617-A
  - Degreasing
    - Mechanical degreasing
      - Ultrasonic degreaser — 621-A
    - Chemical degreasing
      - Vapor degreasing eq. — 623-A
      - Solvent degreasing eq. — 624-A
      - Alkali degreasing eq. — 625-A

Surface finishing

- **Surface coating**
  - Mechanical coating
    - Spray coating
      - Pressure transferred
        - Air-gun sprayer — 631-A
        - High pressure airless sprayer — 632-A
      - Charged transferred
        - Electrostatic coating eq. — 634-A
        - Vacuum coating eq. — 635-A
    - Dip/flow coating
      - Cold dip coating eq. — 641-A
      - Hot dip coating eq. — 642-A
      - Electrocoating eq. — 643-A
      - Fluidized bed coating — 644-A
      - Curtain coating eq. — 645-A
    - Dust coating eq. — 647-A
    - Roll coating eq.
      - Calendering eq. — 648-A
      - Roller coating eq. — 649-A
  - Thermal coating
    - Flame spraying eq.
      - Combustion flame sprayer — 551-A
      - Plasma arc sprayer — 652-A
      - Detonation gun sprayer — 653-A
    - Vaporized metal coating eq.
      - Vacuum metallizing eq. — 661-A
      - Sputtering eq. • — 662-A
      - Vapor—phase deposition eq. — 663-A
    - Heat tinting eq. — 665-A
  - Chemical coating
    - Electroplating eq. — 671-A
    - Chemical conversion
      - Chemical bath eq. — 670-A
      - Anodizing eq. — 673-A
      - Fused bath eq. — 675-A
- **Surface modification**
  - Burnishing eq. — 681-A
  - Peening
    - Shot peening mach. — 682-A
    - Hammer peening mach. — 684-A
  - Texturing
    - Wire brush — 686-A
    - Buffing/polishing eq. — 687-A

APPENDIX 30.5 TOOL TAXONOMY CHARTS

Reduction (chips)

- **Single point cutting**
  - **Turning/facing**
    - Inserts — 101-1-020
    - Solid tool blanks — 101-1-040
  - **Boring**
    - Inserts — 102-1-020
    - Solid boring bars — 102-1-040
  - **Shaping/planing**
    - Standard tools — 103-1-020
    - Special form tools — 103-1-040
  - **Parting/grooving**
    - Parting/cut-off tools
      - Inserts — 104-1-020
      - Solid tool blanks — 104-1-040
    - Grooving tools
      - Inserts — 104-1-080
      - Solid tool blanks — 104-1-100
  - **Single point threading**
    - Inserts — 105-1-020
    - Solid bars — 105-1-040

- **Multiple point threading**
  - **Drills**
    - Twist drills
    - Helical
      - Center drills — 111-1-020
      - Core drills — 111-1-040
      - Subland drills — 111-1-060
      - Step drills — 111-1-080
      - Counter sinks — 111-1-100
      - Counter bores — 111-1-120
      - Spot facers — 111-1-140
    - W/O helix
      - Deep hole drills — 111-1-160
      - Gun drills — 111-1-200
      - Space drill inserts — 111-1-220
      - Prepanning drills — 111-1-240
      - Half round drills — 111-1-260
      - — 111-1-280
  - **Reamers**
    - Chucking reamers — 112-1-020
    - Rose chucking reamers — 112-1-040
    - Shell reamers — 112-1-060
    - Expansion reamers — 112-1-080
    - Adjustable reamer inserts — 112-1-100
    - Paper reamers — 112-1-120
  - **Milling/routing**
    - **Milling cutters**
      - General purpose mills
        - Face mills — 113-1-020
        - Plain mills — 113-1-040
        - Blade mills — 113-1-060
        - Staggered tooth mills — 113-1-080
        - Cutting saws — 113-1-100
        - End mills
          - Solid end mills — 113-1-140
          - Shell end mills — 113-1-160
          - Ball nose end mills — 113-1-180
          - Hollow end mills — 113-1-200
          - Corner rounding mills — 113-1-220
          - Taper end mills — 113-1-240
          - Stub end mills — 113-1-260
        - Angle mills — 113-1-300
      - Special purpose mills
        - T-slot mills — 113-1-340
        - Woodruff-byseat mills — 113-1-360
        - Form mills
          - Concave mills — 113-1-400
          - Convex mills — 113-1-420
          - Special form mills — 113-1-440
        - Thread milling cutters — 113-1-480
    - Routing cutters — 113-1-520

838

Mass reduction tooling

- **Mechanical reduction**
  - **Broaches**
    - Internal broaches
      - Spline broach — 114-1-020
      - Spiral broach — 114-1-040
      - Keyway broach — 114-1-060
      - Rifling broach — 114-1-080
      - Form broach — 114-1-100
      - Burnishing broach — 114-1-120
    - External broaches
      - Slab broach — 114-1-160
      - Ring broach — 114-1-180
      - Special external broach — 114-1-200
  - **Multiple point cutting**
    - Internal threading
      - Standard taps — 115-1-020
      - Machine taps — 115-1-040
    - External threading
      - Self-opening dies — 115-1-060
      - Standard dies — 115-1-100
  - **Files**
    - Straight files
      - Machine files — 116-1-020
      - Hand files — 116-1-040
    - Band segment files — 116-1-080
    - Disc files — 116-1-100
  - **Saws**
    - Hacksaw blades — 117-1-020
    - Bandsaw blades
      - Standard tooth blades — 117-1-060
      - Friction bands — 117-1-080
    - Circular saw blades
      - Gold saw blades — 117-1-120
      - Steel friction disk — 117-1-140
  - **Gear cutting tools**
    - Gear cutters — 118-1-020
    - Gear hobs — 118-1-040
  - **Abrasive machining**
    - Grinding
      - Plain wheels — 121-1-020
      - Contoured wheels
        - Thread grinding wheels — 121-1-060
        - Form grinding wheels — 121-1-080
      - Abrasive disks — 121-1-120
      - Abrasive belts — 121-1-140
    - Honing sticks
      - Bonded sticks — 124-1-040
    - Superfinishing
      - Cup wheels
    - Ultrasonic maching tips — 125-1-020
    - Jet machining nozzles — 125-3-020
  - **Separation (shear)**
    - Shears
      - Blades — 131-1-020
      - Cutting rolls — 132-1-020
      - Shearing rollers — 133-1-020
      - 134-1-020
    - Dies
      - Matched die — 140-1-020
      - Sinking die — 145-1-020
- **Thermal reduction**
  - Torch cutting — 152-1-020
  - E.D.M. electrodes
- **Chemical reduction**
  - E.C.M. electrodes

839

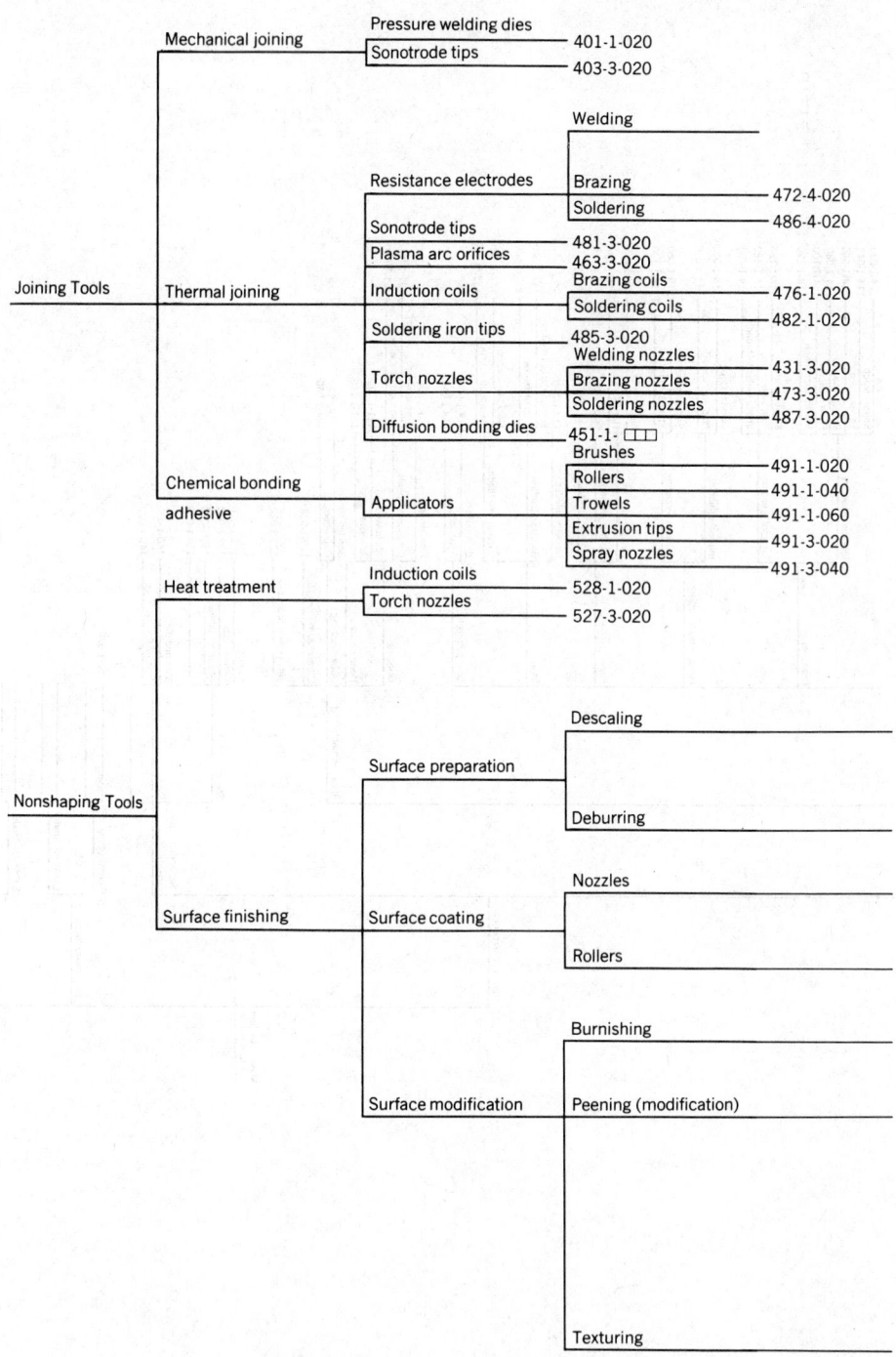

TOOL TAXONOMY
Joining Tools
Nonshaping Tools

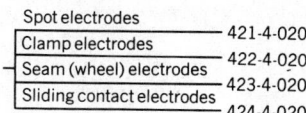

Spot electrodes — 421-4-020
Clamp electrodes — 422-4-020
Seam (wheel) electrodes — 423-4-020
Sliding contact electrodes — 424-4-020

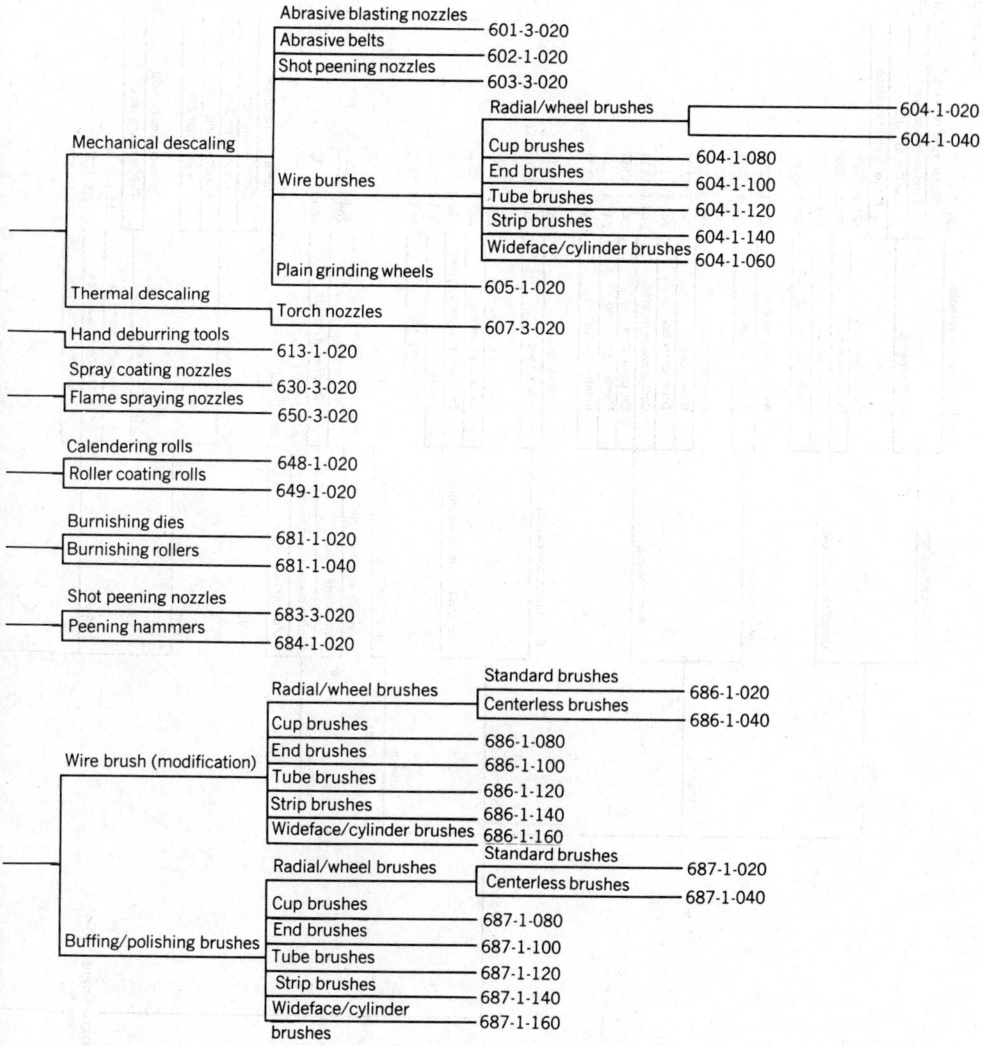

Mechanical descaling

Abrasive blasting nozzles — 601-3-020
Abrasive belts — 602-1-020
Shot peening nozzles — 603-3-020

Wire burshes
Radial/wheel brushes — 604-1-020 / 604-1-040
Cup brushes — 604-1-080
End brushes — 604-1-100
Tube brushes — 604-1-120
Strip brushes — 604-1-140
Wideface/cylinder brushes — 604-1-060

Plain grinding wheels — 605-1-020

Thermal descaling
Torch nozzles — 607-3-020

Hand deburring tools — 613-1-020

Spray coating nozzles — 630-3-020
Flame spraying nozzles — 650-3-020

Calendering rolls — 648-1-020
Roller coating rolls — 649-1-020

Burnishing dies — 681-1-020
Burnishing rollers — 681-1-040

Shot peening nozzles — 683-3-020
Peening hammers — 684-1-020

Wire brush (modification)
Radial/wheel brushes
Standard brushes — 686-1-020
Centerless brushes — 686-1-040
Cup brushes — 686-1-080
End brushes — 686-1-100
Tube brushes — 686-1-120
Strip brushes — 686-1-140
Wideface/cylinder brushes — 686-1-160

Buffing/polishing brushes
Radial/wheel brushes
Standard brushes — 687-1-020
Centerless brushes — 687-1-040
Cup brushes — 687-1-080
End brushes — 687-1-100
Tube brushes — 687-1-120
Strip brushes — 687-1-140
Wideface/cylinder brushes — 687-1-160

Tool Taxonomy
Mass Conserving

Mass conserving tooling

- **Consolidation**
  - **Casting**
    - Indirect contact molds
      - Investment wax molds — 201-2-
      - Mold patterns
        - Plaster mold patterns — 202-2-
        - Sand mold patterns — 211-2-
        - Shell mold patterns — 212-2-
        - No-bake mold patterns — 213-2-
    - Direct contact molds
      - Die casting dies — 221-1-
      - Permanent mold — 222-1-
      - Flexible molds — 223-1-
      - Continuous casting dies — 224-1-
  - **Molding**
    - Ceramic molding
      - Wet forming molds — 231-1-
      - Dry pressing molds — 232-1-
    - Polymer molding
      - Injection molds — 241-1-
      - Blow molds — 242-1-
      - Transfer molds — 243-1-
      - Compression molds — 244-1-
      - Extrusion dies — 245-1-
      - Thermoform mold — 246-1-
      - Rotational mold — 247-1-
  - **Compacting**
    - Continuous compacting
      - P/M extrusion — 251-1-
      - P/M rollers — 252-2-020
    - Noncontinuous compacting
      - Matched die — 261-1-
      - Centrifugal molds — 262-1-
      - Explosive compacting dies — 263-1-
      - Isostatic mold — 264-1-
      - Slip casting mold — 265-1-
  - **Deposition**
    - Electroforming molds/mandrels — 271-1-
  - **Laminating**
    - Filament winding mandrels — 281-1-
    - Machine lay-up molds — 284-1-
    - Hand lay-up molds — 285-1-
    - Laminating molds
    - Spray/chopper gun nozzles — 284-3-020
    - Pultrusion dies — 287-1-
  - **Forge**
    - Hot forge
      - Hammer forging dies — 301-1-020
      - Drop forging dies — 302-1-
      - Press forging dies — 303-1-
      - Upset forging dies — 304-1-
      - Shaped forging rolls — 305-1-
    - Cold forge
      - Swaging dies (solids) — 307-1-020
      - Cold heading dies — 308-1-
  - **Extrude**
    - Direct extrusion dies — 311-1-
    - Indirect extrusion dies — 312-1-
    - Impact extrusion dies — 313-1-

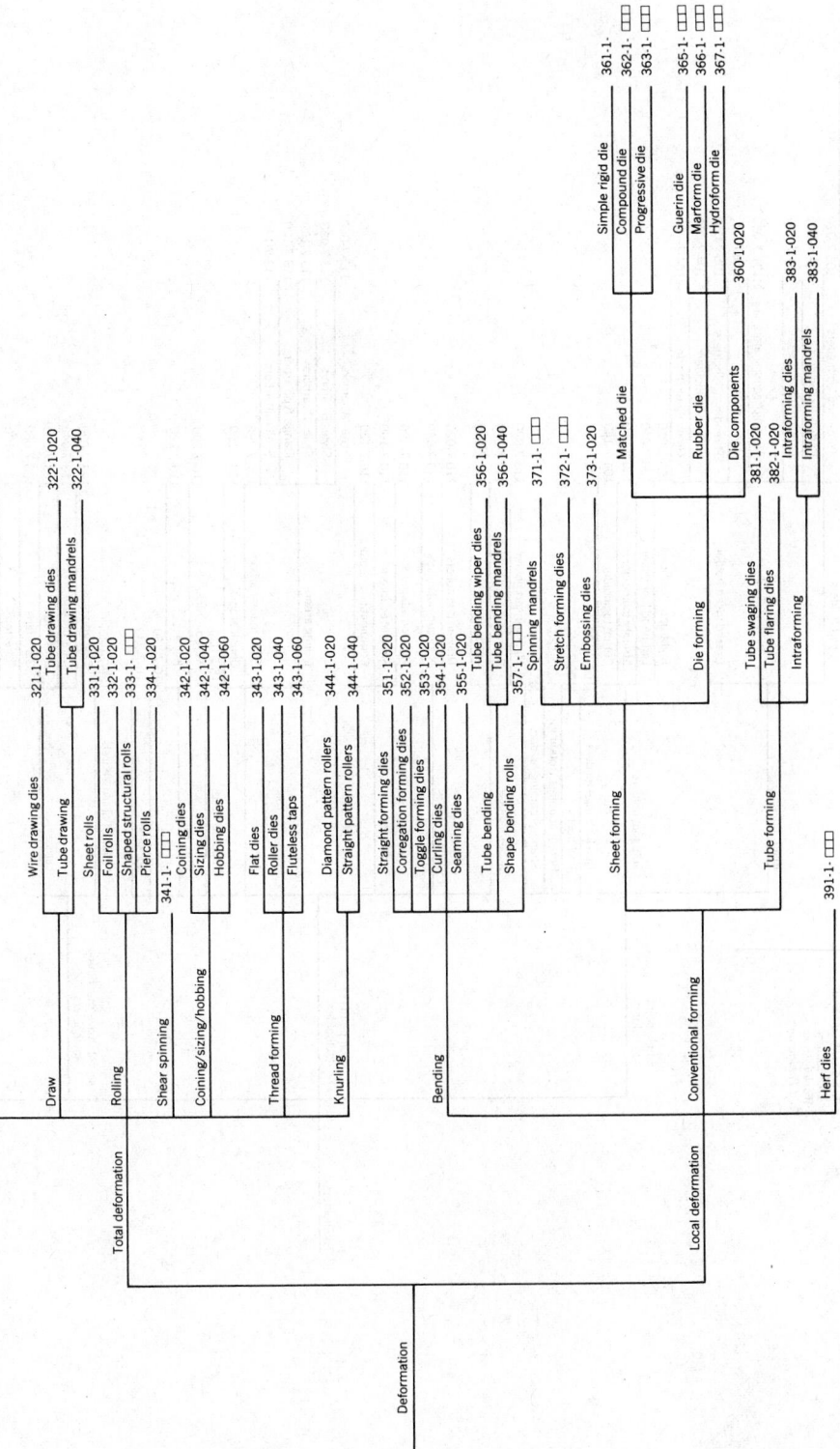

Deformation

Total deformation

Draw
Wire drawing dies — 321-1-020
Tube drawing
Tube drawing dies — 322-1-020
Tube drawing mandrels — 322-1-040

Rolling
Sheet rolls — 331-1-020
Foil rolls — 332-1-020
Shaped structural rolls — 333-1-☐☐☐
Pierce rolls — 334-1-020

Shear spinning — 341-1-☐☐☐

Coining/sizing/hobbing
Coining dies — 342-1-020
Sizing dies — 342-1-040
Hobbing dies — 342-1-060

Thread forming
Flat dies — 343-1-020
Roller dies — 343-1-040
Fluteless taps — 343-1-060

Knurling
Diamond pattern rollers — 344-1-020
Straight pattern rollers — 344-1-040

Bending
Straight forming dies — 351-1-020
Corregation forming dies — 352-1-020
Toggle forming dies — 353-1-020
Curling dies — 354-1-020
Seaming dies — 355-1-020
Tube bending
Tube bending wiper dies — 356-1-020
Tube bending mandrels — 356-1-040
Shape bending rolls — 357-1-☐☐☐

Local deformation

Conventional forming

Sheet forming
Spinning mandrels — 371-1-☐☐☐
Stretch forming dies — 372-1-☐☐☐
Embossing dies — 373-1-020
Die forming
Matched die
Simple rigid die — 361-1-☐☐☐
Compound die — 362-1-☐☐☐
Progressive die — 363-1-☐☐☐
Rubber die
Guerin die — 365-1-☐☐☐
Marform die — 366-1-☐☐☐
Hydroform die — 367-1-☐☐☐
Die components — 360-1-020

Tube forming
Tube swaging dies — 381-1-020
Tube flaring dies — 382-1-020
Intraforming
Intraforming dies — 383-1-020
Intraforming mandrels — 383-1-040

Herf dies — 391-1-☐☐☐

843

TOOL TAXONOMY
Toolholders
Workholders
Tool Positioners

- Toolholders
  - Single point toolholders
    - Insert holders
      - Mechanical clamp insert holders
        - Turning/facing holder
          - Tool shanks — 101-7-020
          - Tool cartridges — 101-7-040
        - Boring toolholders
          - Boring bars — 102-7-020
          - Boring heads — 102-7-040
        - Threading toolholders — 105-7-020
        - Parting toolholders — 104-7-020
        - Grooving toolholders — 104-7-040
      - Brazed insert holder
        - Turning/facing shanks — 101-7-080
        - Boring bars — 102-7-080
        - Threading toolholders — 105-7-060
        - Parting toolholders — 104-7-080
        - Grooving toolholders — 104-7-100
    - HSS tool holders
      - Right-hand toolholder — 100-7-020
      - Left-hand toolholder — 100-7-040
      - Straight toolholder — 100-7-060
      - Cut-off toolholder — 104-7-140
    - Quick change holders — 100-7-100
    - Tool posts — 100-7-140
    - Boring adapters (rotating) — 1-2-7-120
    - Shaping/planing holders — 1-3-7-020
  - Multipoint toolholders
    - Insert holders
      - Milling insert holder — 113-7-020
      - Reaming insert holder — 112-7-020
      - Spade drill shanks — 111-7-020
    - Collets
      - Draw-in bolt collets — 110-7-020
      - Non-pull-out collets — 110-7-040
    - Adaptors
      - Tapered adaptors — 110-7-080
      - Straight adaptors — 110-7-100
      - Combination adaptors — 110-7-120
    - Arbors
      - Milling arbors
        - Style "A" arbors — 113-7-060
        - Style "B" arbors — 113-7-080
        - Style "C" arbors — 113-7-100
        - Flange-type arbors — 113-7-120
        - Shell mill arbors — 113-7-140
      - Shell reamer arbors — 112-7-060
      - Chuck arbors — 110-7-160
    - Tapping heads — 115-7-020
    - Chucks
      - Keyed chucks — 110-7-200
      - Keyless chucks — 110-7-220
    - Sleeves — 110-7-260
    - Sockets — 110-7-280
  - Abrasive machining toolholders
    - Honing bars/mandrels — 122-7-020
    - Laps
      - Cast iron laps — 123-7-020
      - Ceramic laps — 123-7-040
      - Brass laps — 123-7-060
      - Glass laps — 123-7-080
      - Other laps — 123-7-100
  - Pressworking
    - Die shoes — 690-7-020
    - Die component holders — 690-7-040

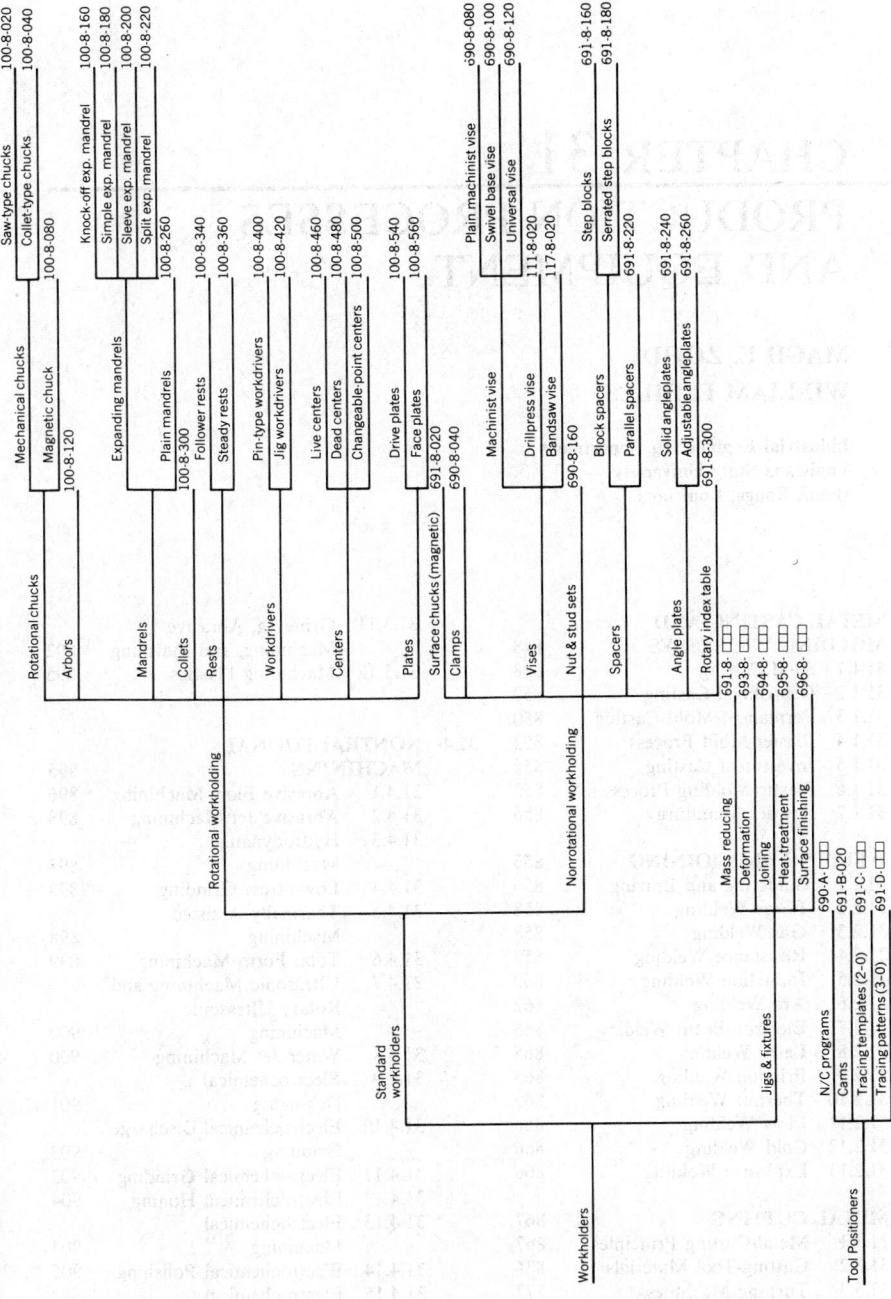

CHAPTER 31

PRODUCTION PROCESSES AND EQUIPMENT

MAGD E. ZOHDI
WILLIAM E. BILES

Industrial Engineering Department
Louisiana State University
Baton Rouge, Louisiana

31.1 METAL CASTING AND MOLDING PROCESSES

Casting provides a versatility and flexibility that have maintained casting's position as a primary production method of machine elements.

Methods of Casting

Casting processes are divided according to the specific type of molding method used in casting, as follows:

1. Sand
2. Centrifugal
3. Permanent
4. Die
5. Plaster-mold
6. Investment

31.1.1 Sand Casting

Sand casting consists basically of pouring molten metal into appropriate cavities formed in a sand mold, Fig. 31.1. The sand used may be natural, synthetic, or an artificially blended material.

Molds

The two common types of sand molds are the dry sand mold and the green sand molds. In the dry sand mold, the mold is dried thoroughly prior to closing and pouring, while the green sand mold is used without any preliminary drying. The dry sand mold is more firm and resistant to collapse than the green sand mold, and, consequently, core pieces for molds are usually made in this way. Cores are placed in mold cavities to form the interior surfaces of castings.

Patterns

To produce a mold for a conventional sand cast part it is necessary to make a pattern of the part. Patterns are made from wood or metal to suit a particular design with allowances to compensate for such factors as natural metal shrinkage and contraction characteristics. These and other effects, such as mold resistance, distortion, casting design, and mold design, which are not entirely within the range of accurate prediction, generally make it necessary to adjust the pattern in order to produce castings of the required dimensions.

Access to the mold cavity for entry of the molten metal is provided by sprues, runners, and gates.

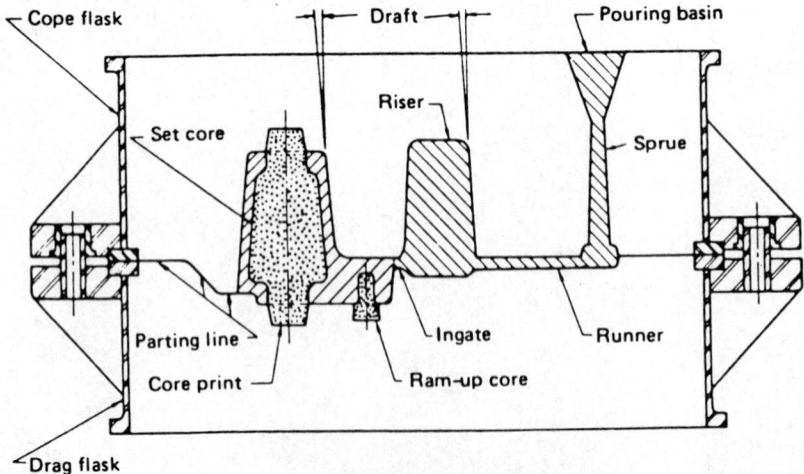

Fig. 31.1 Sectional view of a casting mold

Shrinkage

Allowances must be made on patterns to counteract the contraction in size as the metal cools. The amount of shrinkage is dependent on the design of the coating, type of metal used, solidification temperature, and mold resistance. Table 31.1 gives average shrinkage allowance values used in sand casting. Smaller values apply generally to large or cored castings of intricate design. Larger values apply to small to medium, simple castings designed with unrestrained shrinkage.

Table 31.1 Pattern Shrinkage Allowance (in./ft)

Metal	Shrinkage
Aluminum alloys	$\frac{1}{10}$ to $\frac{5}{32}$
Beryllium copper	$\frac{1}{8}$
Copper alloys	$\frac{3}{16}$ to $\frac{7}{32}$
Everdur	$\frac{3}{16}$
Gray irons	$\frac{1}{8}$
Hastelloy alloys	$\frac{1}{4}$
Magnesium alloys	$\frac{1}{8}$ to $\frac{11}{64}$
Malleable irons	$\frac{1}{16}$ to $\frac{3}{16}$
Meehanite	$\frac{1}{10}$ to $\frac{5}{32}$
Nickel and nickel alloys	$\frac{1}{4}$
Steel	$\frac{1}{8}$ to $\frac{1}{4}$
White irons	$\frac{3}{16}$ to $\frac{1}{4}$

Machining

Allowances are required in many cases because of unavoidable surface impurities, warpage, and surface variations. Average machining allowances are given in Table 31.2. Good practice dictates use of minimum section thickness compatible with the design. The normal minimum section recommended for various metals is shown in Table 31.3.

Table 31.2 Machining Allowances for Sand Castings (in.)

Metal	Casting Size	Finish Allowance
Cast irons	Up to 12 in.	$\frac{3}{32}$
	13–24 in.	$\frac{1}{8}$
	25–42 in.	$\frac{3}{16}$
	43–60 in.	$\frac{1}{4}$
	61–80 in.	$\frac{5}{16}$
	81–120 in.	$\frac{3}{8}$
Cast steels	Up to 12 in.	$\frac{1}{8}$
	13–24 in.	$\frac{3}{16}$
	25–42 in.	$\frac{5}{16}$
	43–60 in.	$\frac{3}{8}$
	61–80 in.	$\frac{7}{16}$
	81–120 in.	$\frac{1}{2}$
Malleable irons	Up to 8 in.	$\frac{1}{16}$
	9–12 in.	$\frac{3}{32}$
	13–24 in.	$\frac{1}{8}$
	25–36 in.	$\frac{3}{16}$
Nonferrous metals	Up to 12 in.	$\frac{1}{16}$
	13–24 in.	$\frac{1}{8}$
	25–36 in.	$\frac{5}{32}$

**Table 31.3   Minimum Sections
for Sand Castings (in.)**

Metal	Section
Aluminum alloys	$\frac{3}{16}$
Copper alloys	$\frac{3}{32}$
Gray irons	$\frac{1}{8}$
Magnesium alloys	$\frac{5}{32}$
Malleable irons	$\frac{1}{8}$
Steels	$\frac{1}{4}$
White irons	$\frac{1}{8}$

31.1.2   Centrifugal Casting

Centrifugal casting consists of having a sand, metal, or ceramic mold that is rotated at high speeds. When the molten metal is poured into the mold it is thrown against the mold wall, where it remains until it cools and solidifies. The process is increasingly being used for such products as cast-iron pipes, cylinder liners, gun barrels, pressure vessels, brake drums, gears, and flywheels. The metals used include almost all castable alloys. Most dental tooth caps are made by a combined lost-wax process and centrifugal casting.

Advantages and Limitations

Because of the relatively fast cooling time, centrifugal castings have a fine grain size. There is a tendency for the lighter nonmetallic inclusions, slag particles, and dross to segregate toward the inner radius of the castings (Fig. 31.2), where it can be easily removed by machining. Owing to the high purity of the outer skin, centrifugally cast pipes have a high resistance to atmospheric corrosion. Figure 31.2 shows a schematic sketch of how a pipe would be centrifugally cast in a horizontal mold.

   Parts that have diameters exceeding their length are produced by vertical-axis casting; see Fig. 31.3.

31.1.3   Permanent-Mold Casting

As demand for quality castings in production quantities increased, the attractive possibilities of metal molds brought about the development of the permanent-mold process. Although not as flexible regarding design as sand casting, the use of metal-mold casting made possible the continuous production of quantities of casting from a single mold as compared to batch production of individual sand molds.

Metal Molds and Cores

In permanent-mold casting both metal molds and cores are used, the metal being poured into the mold cavity with the usual gravity head as in sand casting. Molds are normally made of dense iron

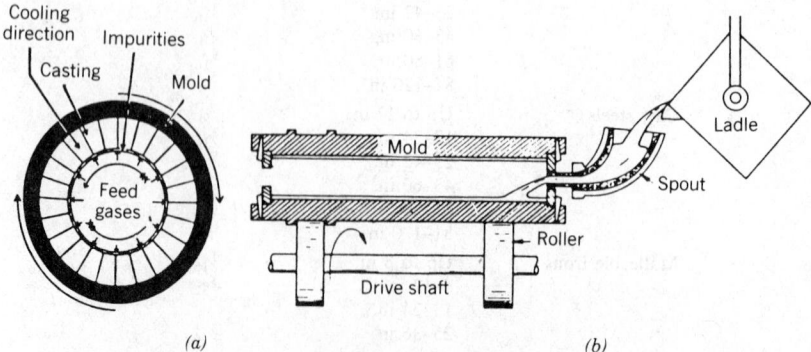

(a)   *(b)*

Fig. 31.2   The principle of centrifugal casting is to produce high-grade metal by throwing the heavier metal outward and forcing the impurities to congregate inward (*a*). Shown at (*b*) is a schematic of how a horizontal-bond centrifugal casting is made.

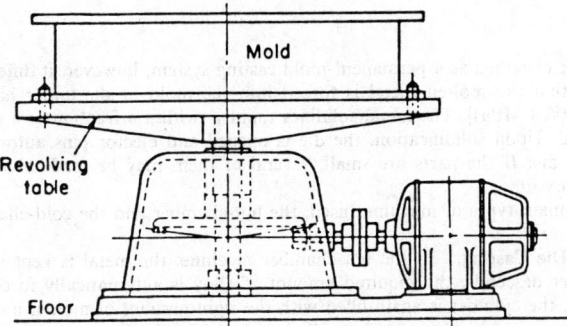

Fig. 31.3 Floor-type vertical centrifugal casting machine for large-diameter parts.

or meehanite, large cores are made of cast iron, and small or collapsible cores are made of alloy steel. All necessary sprues, runners, gates, and risers must be machined into the mold, and the mold cavity itself is made with the usual metal-shrinkage allowances. The mold is usually composed of one, two, or more parts, which may swing or slide for rapid operation. Whereas in sand casting the longest dimension is always placed in a horizontal position, with permanent-mold casting the longest dimension of a part is normally placed in a vertical position.

Production Quantities

Wherever quantities are in the range of 500 pieces or more, permanent-mold casting becomes competitive in cost with sand casting, and if the design is simple, runs as small as 200 pieces are often economical. Production runs of 1000 pieces or more will generally produce a favorable cost difference. High rates of production are possible, and multiple-cavity dies with as many as 16 cavities can be used. In casting gray iron in multiple molds as many as 50,000 castings per cavity are common with small parts. With larger parts of gray iron, weighing from 12 to 15 lb, single-cavity molds normally yield 2000–3000 pieces per mold on an average. Up to 100,000 parts per cavity or more is not uncommon with nonferrous metals, magnesium providing the longest die life. Low-pressure permanent mold casting is economical for quantities up to 40,000 pieces (Fig. 31.4).

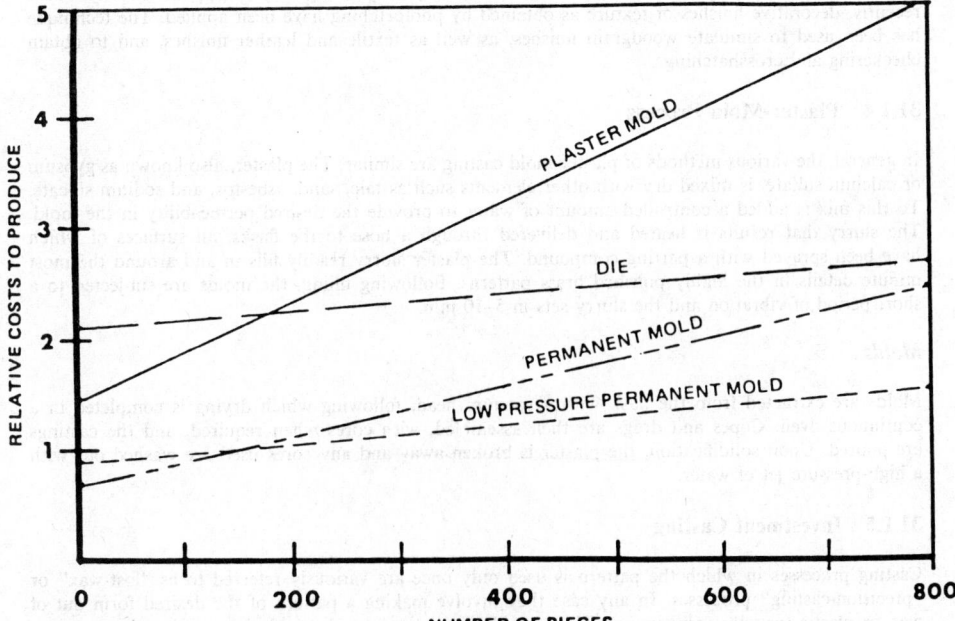

Fig. 31.4 Cost comparison of various casting systems. At approximately 40,000 pieces die casting crosses and becomes more economical than LPPM casting.

Die Casting

Die casting may be classified as a permanent-mold casting system; however, it differs from the process just described in that the molten metal is forced into the mold or die under high pressure [1000–30,000 psi (6.89–206.8 MPa)]. The metal solidifies rapidly (within a fraction of a second) because the die is water cooled. Upon solidification, the die is opened and ejector pins automatically knock the casting out of the die. If the parts are small, several of them may be made at one time in what is termed a multicavity die.

There are two main types of machines used: the hot-chamber and the cold-chamber types.

Hot-Chamber Die Casting. In the hot-chamber machine, the metal is kept in a heated holding pot. As the plunger descends, the required amount of alloy is automatically forced into the die. As the piston retracts, the cylinder is again filled with the right amount of molten metal. Metals such as aluminum, magnesium, and copper tend to alloy with the steel plunger and cannot be used in the hot chamber.

Cold-Chamber Die Casting. This process gets its name from the fact that the metal is ladled into the cold chamber for each shot. This procedure is necessary to keep the molten-metal contact time with the steel cylinder to a minimum. Iron pickup is prevented, as is freezing of the plunger in the cylinder.

Advantages and Limitations

Die-casting machines can produce large quantities of parts with close tolerances and smooth surfaces. The size is limited only by the capacity of the machine. Most die castings are limited to about 75 lb (34 kg) zinc; 65 lb (30 kg) aluminum; and 44 lb (20 kg) of magnesium. Die castings can provide thinner sections than any other casting process. Wall thickness as thin as 0.015 in. (0.38 mm) can be achieved with aluminum in small items. However, a more common range on larger sizes will be 0.105–0.180 in. (2.67–4.57 mm).

Some difficulty is experienced in getting sound castings in the larger capacities. Gases tend to be entrapped, which results in low strength and annoying leaks. Of course, one way to reduce metal sections without sacrificing strength is to design in ribs and bosses. Another approach to the porosity problem has been to operate the machine under vacuum. This process is now being developed.

The surface quality is dependent on that of the mold. Parts made from new or repolished dies may have a surface roughness of 24 μin. (0.61 μm). The high surface finish available means that, in most cases, coatings such as chrome-plating, anodizing, and painting may be applied directly. More recently, decorative finishes of texture as obtained by photoetching have been applied. The technique has been used to simulate woodgrain finishes, as well as textile and leather finishes, and to obtain checkering and crosshatching.

31.1.4 Plaster-Mold Process

In general, the various methods of plaster-mold casting are similar. The plaster, also known as gypsum or calcium sulfate, is mixed dry with other elements such as talc, sand, asbestos, and sodium silicate. To this mix is added a controlled amount of water to provide the desired permeability in the mold. The slurry that results is heated and delivered through a hose to the flasks, all surfaces of which have been sprayed with a parting compound. The plaster slurry readily fills in and around the most minute details in the highly polished brass patterns. Following filling, the molds are subjected to a short period of vibration and the slurry sets in 5–10 min.

Molds

Molds are extracted from the flask with a vacuum head, following which drying is completed in a continuous oven. Copes and drags are then assembled, with cores when required, and the castings are poured. Upon solidification, the plaster is broken away and any cores used are washed out with a high-pressure jet of water.

31.1.5 Investment Casting

Casting processes in which the pattern is used only once are variously referred to as "lost-wax" or "precision-casting" processes. In any case they involve making a pattern of the desired form out of wax or plastic (usually polystyrene). The expendable pattern may be made by pressing the wax into a split mold or by the use of an injection-molding machine. The patterns may be gated together so that several parts can be made at once. A metal flask is placed around the assembled patterns and a refractory mold slurry is poured in to support the patterns and form the cavities. A vibrating table equipped with a vacuum pump is used to eliminate all the air from the mold. Formerly, the standard procedure was to dip the patterns in the slurry several times until a coat was built up. This was called the investment process. After the mold material has set and dried, the pattern material is melted and allowed to run out of the mold.

The completed flasks are heated slowly to dry the mold and to melt out the wax, plastic, or whatever pattern material was used. When the molds have reached a temperature of 100°F (37.8°C), they are ready for pouring. Vacuum may be applied to the flasks to ensure complete filling of the mold cavities.

When the metal has cooled, the investment material is removed by vibrating hammers or by tumbling. As with other castings, the gates and risers are cut off and ground down.

Ceramic Process

The ceramic process is somewhat similar to the investment-casting process in that a creamy ceramic slurry is poured over a pattern. In this case, however, the pattern, made out of plastic, plaster, wood, metal, or rubber, is reusable. The slurry hardens on the pattern almost immediately and becomes a strong green ceramic of the consistency of vulcanized rubber. It is lifted off the pattern while it is still in the rubberlike phase. The mold is ignited with a torch to burn off the volatile portion of the mix. It is then put in a furnace and baked at 1800°F (982°C), resulting in a rigid refractory mold. The mold can be poured while still hot.

Full-Mold Casting

Full-mold casting may be considered a cross between conventional sand casting and the investment technique of using lost wax. In this case, instead of a conventional pattern of wood, metals, or plaster, a polystyrene foam or Styrofoam is used. The pattern is left in the mold and is vaporized by the molten metal as it rises in the mold during pouring. Before molding, the pattern is usually coated with a zirconite wash in an alcohol vehicle. The wash produces a relatively tough skin separating the metal from the sand during pouring and cooling. Conventional foundry sand is used in backing up the mold.

31.1.6 Plastic-Molding Processes

Plastic molding is similar in many ways to metal molding. For most molding operations, plastics are heated to a liquid or a semifluid state and are formed in a mold under pressure. Some of the most common molding processes are discussed.

Injection Molding

The largest quantity of plastic parts are made by injection molding. Plastic compound is fed in powdered or granular form from a hopper through metering and melting stages and then injected into a mold. After a brief cooling period, the mold is opened and the solidified part is ejected.

Coinjection Molding

Coinjection molding makes it possible to mold articles with a solid skin of one thermoplastic and a core of another thermoplastic. The skin material is usually solid while the core material contains blowing agents.

The basic process may be one-, two-, or three-channel technology. In one-channel technology, the two melts are injected into the mold, one after the other. The skin material cools and adheres to the colder surface; a dense skin is formed under proper parameter settings. The thickness of the skin can be controlled by adjustment of injection speed, stock temperature, mold temperature, and flow compatibility of the two melts.

In the two- and three-channel techniques, both plastic melts may be introduced simultaneously. This allows for better control of wall thickness of the skin, especially in gate areas on both sides of the part.

Injection-Molded Carbon-Fiber Composites

By mixing carbon or glass fibers in injection-molded plastic parts, they can be lightweight yet stiffer than steel.

Rotomolding

In rotational molding, the product is formed inside a closed mold that is rotated about two axes as heat is applied. Liquid, or powdered thermoplastic or thermosetting plastic is poured into the mold, either manually or automatically.

Expandable-Bead Molding

The expandable-bead process consists of placing small beads of polystyrene along with a small amount of blowing agent in a tumbling container. The polystyrene beads soften under heat, which allows a blowing agent to expand them. When the beads reach a given size, depending on the density required, they are quickly cooled. This solidifies the polystyrene in its larger foamed size. The expanded beads are then placed in a mold until it is completely filled. The entrance port is then closed and steam is injected, resoftening the beads and fusing them together. After cooling, the finished, expanded part is removed from the mold.

Extruding

Plastic extrusion is similar to metal extrusion in that a hot material (plastic melt) is forced through a die having an opening shaped to produce a desired cross section. Depending on the material used, the barrel is heated anywhere from 250 to 600°F (121 to 316°C) to transform the thermoplastic from a solid to a melt. At the end of the extruder barrel is a screen pack for filtering and building back pressure. A breaker plate serves to hold the screen pack in place and straighten the helical flow as it comes off the screen.

Blow Molding

Blow molding is a process that is used extensively to make bottles and other lightweight, hollow, plastic parts. Two methods are used: injection blow molding and extrusion blow molding.

Injection blow molding is used primarily for small containers. The parison or tube is formed by the injection of plasticized material around a hollow mandrel. While the material is still molten and still on the mandrel, it is transferred into the blowing mold where air is used to inflate it. Accurate threads may be formed at the neck.

In extrusion-type blow molding, molten-plastic pipe (parison) is inflated under relatively low pressure inside a split-metal mold. The die closes, pinching the end and closing the top around the mandrel. Air enters through the mandrel and inflates the tube until the plastic contacts the cold wall, where it solidifies. The mold opens, the bottle is ejected, and the tailpiece falls off.

Thermoforming

Thermoforming refers to heating a sheet of plastic material until it becomes soft and pliable and then forming it either under vacuum, by air pressure, or between matching mold halves.

Reinforced-Plastic Molding

Reinforced plastics generally refer to polymers that have been reinforced with glass fibers. Other materials used are asbestos, sisal, synthetic fibers such as nylon and polyvinyl chloride, and cotton fibers. High-strength composites using graphite fibers are now commercially available with moduli of 50,000,000 psi (344,700,000 MPa) and tensile strengths of about 300,000 psi (2,068,000 MPa). They are as strong as or stronger than the best alloy steels and are lighter than aluminum.

Forged-Plastic Parts

The forging of plastic materials is a relatively new process. It was developed to shape materials that are difficult or impossible to mold and is used as a low-cost solution for small production runs.

The forging operation starts with a blank or billet of the required shape and volume for the finished part. The blank is heated to a preselected temperature and transferred to the forging dies, which are closed to deform the work material and fill the die cavity. The dies are kept in the closed position for a definite period of time, usually 15–60 sec. When the dies are opened, the finished forging is removed. Since forging involves deformation of the work material in a heated and softened condition, the process is applicable only to thermoplastics.

31.1.7 Powder Metallurgy

Powder metallurgy is the name given to a process wherein fine metal powders are pressed into a desired shape, usually in a metal die and under high pressure, and the compacted powder is then heated (sintered), with a protective atmosphere. The density of sintered compacts may be increased by repressing. Repressing is also performed to improve the dimensional accuracy, either concurrently or subsequently, for a period of time at a temperature below the melting point of the major constituent. The process (commonly designated as P/M) has a number of distinct advantages that account for its rapid growth in recent years, including (1) no material is wasted, (2) usually no machining is required, (3) only semiskilled labor is required, and (4) some unique properties can be obtained, such as controlled degrees of porosity and built-in lubrication.

A crude form of powder metallurgy appears to have existed in Egypt as early as 3000 BC, using particles of sponge iron. In the 19th century P/M was used for producing platinum and tungsten wires. However, its first significant use related to general manufacturing was in Germany, following World War I, for making tungsten carbide cutting-tool tips. Since 1945 the process has been highly developed, and large quantities of a wide variety of P/M products are made annually, many of which could not be made by any other process. Most are under 2 in. (50.8 mm) in size, but many are larger, some weighing up to 50 lb (22.7 kg) and measuring up to 20 in. (508 mm).

The powder metallurgy normally consists of four basic steps:

1. Producing a fine metallic powder.
2. Mixing and preparing the powder for use.
3. Pressing the powder into the desired shape.
4. Heating (sintering) the shape at an elevated temperature.

Other operations can be added to obtain special results.

The pressing and sintering operations are of special importance. The pressing and repressing greatly affect the density of the product, which has a direct relationship to the strength properties. Sintering strips contaminants from the surface of the powder particles, permitting diffusion bonding to occur and resulting in a single piece of material. Sintering usually is done in a controlled, inert atmosphere, but sometimes it is done by the discharge of a spark through the powder while it is under compaction in the mold.

Properties of P/M Products

Because the strength properties of powder metallurgy products depend on so many variables—type and size of powder, pressing pressure, sintering temperature, finishing treatments, and so on—it is difficult to give generalized information. In general, the strength properties of products that are made from pure metals (unalloyed) are about the same as those made from the same wrought metals. As alloying elements are added, the resulting strength properties of P/M products fall below those of wrought products by varying, but usually substantial, amounts. The ductility usually is markedly less, as might be expected because of the lower density. However, tensile strengths of 40,000–50,000 psi (275.8–344.8 MPa) are common, and strengths above 100,000 psi (689.5 MPa) can be obtained. Table 31.4 gives a few P/M materials and their strength properties, with some wrought metals shown for comparison. As larger presses and forging combined with P/M preforms are used, to provide greater density, the strength properties of P/M materials will more nearly equal those of wrought materials. Coining can also be used to increase the strength properties of P/M products and to improve their dimensional accuracy.

31.2 WELDING AND JOINING

Welding is a metal-joining process in which coalescence is obtained by heat and/or pressure. It may also be defined as a metallurgical bond accomplished by the attracting force between atoms.

Welding processes differ widely in the manner in which heat is applied and in the equipment used. These processes are:

1. Braze welding
2. Forge welding
3. Gas welding
   a. Oxycetylene
   b. Oxyhydrogen
   c. Air–acetylene
   d. Pressure
4. Resistance welding
   a. Spot
   b. Projection
   c. Seam
   d. Butt
   e. Flash
   f. Percussion

Table 31.4 Typical Properties of Some P/M Materials, with Some Similar Wrought Materials

Material	Composition	Theoretical Density (%)	Condition	Tensile Strength (MPa)	Tensile Strength ×10³ (psi)	Elongation in 2 in. (%)
Iron[a]			Wrought hot rolled	331	48	30
Iron		84	As-sintered	214	31	2
		95	Repressed	283	41	25
Steel[a]	AISI 1025		Hot rolled	586	85	25
Steel	0.25% C; 99.75% Fe	84	As-sintered	234	34	2
Stainless steel[a]	Type 303		Annealed	621	90	50
Stainless steel	Type 303	75	As-sintered	241	35	2
Alloy steel[a]	AISI 4620		Annealed	552	80	30
Alloy steel	0.75% Mo; 0.35% Ni; 0.40% Mn; 0.25% Si; 0.5% C	85	As-sintered	352	51	Nil
Aluminum alloy[a]	7075		T6	572	83	11
Aluminum alloy	75	90	Heat-treated	310	45	3
Brass[a]	90% Cu; 10% Zn		Annealed	255	37	45
Brass	90% Cu; 10% Zn	89	As-sintered	234	34	29

[a] Shown for comparison.

5. Induction welding	7. Electron beam
6. Arc welding	8. Laser welding
a. Carbon electrode	9. Friction welding
i. Shielded	10. Thermit welding
ii. Unshielded	11. Flow welding
b. Metal electrode	12. Cold welding
i. Atomic hydrogen	a. Pressure
ii. Inert gas	b. Ultrasonic
iii. Arc apot	13. Explosion welding
iv. Submerged arc	
v. Stud	
vi. Plasma torch	
vii. Electroslag	

31.2.1 Soldering and Brazing

Soldering and brazing are processes that unite metals with a third joining metal, which is introduced into the joint in a liquid state and allowed to solidify. These processes have wide commercial use in uniting of small assemblies and electrical parts.

Soldering

Soldering is the uniting of two pieces of metal with a different metal, which is applied between the two in a molten state at a temperature not exceeding 800°F (430°C). In this process a little alloying with the base metal takes place, and additional strength is obtained by mechanical bonding. Lead and tin alloys having a melting range of 350–700°F (180–270°C) are used principally, and the strength of the joint is largely determined by the adhesive quality of the alloy. Cleaning is important in all soldering operations, and flux is necessary.

Brazing

In brazing, a nonferrous alloy is introduced in a liquid state between the pieces of metal to be joined and allowed to solidify. The filler metal, having a melting temperature of over 800°F (340°C), but lower than the melting temperature of the parent metal, is distributed between the surfaces by capillary attraction. The brazing metals and alloys commonly used are as follows:

1. Copper: melting point 1982°F (1083°C).
2. Copper alloys: brass and bronze alloys having melting points ranging from 1600 to 2250°F (870 to 2130°C).
3. Silver alloys: melting temperatures ranging from 1165 to 1550°F (630 to 845°C).
4. Aluminum alloys: melting temperatures ranging from 1060 to 1185°F (570 to 640°C).

The processes that have wide commercial use in the uniting of small assemblies and electrical parts are listed in Table 31.5.

Table 31.5 Brazing and Braze Welding

Dipping	Furnace	Torch	Electric	Welding
1. Metal	1. Gas	1. Oxyacetylene	1. Resistance	1. Torch
2. Chemical	2. Electric	2. Oxyhydrogen	2. Induction	2. Arc

To be efficient, a welded joint must be properly designed according to the service for which it is intended. A few of the common types may be seen in Fig. 31.5.

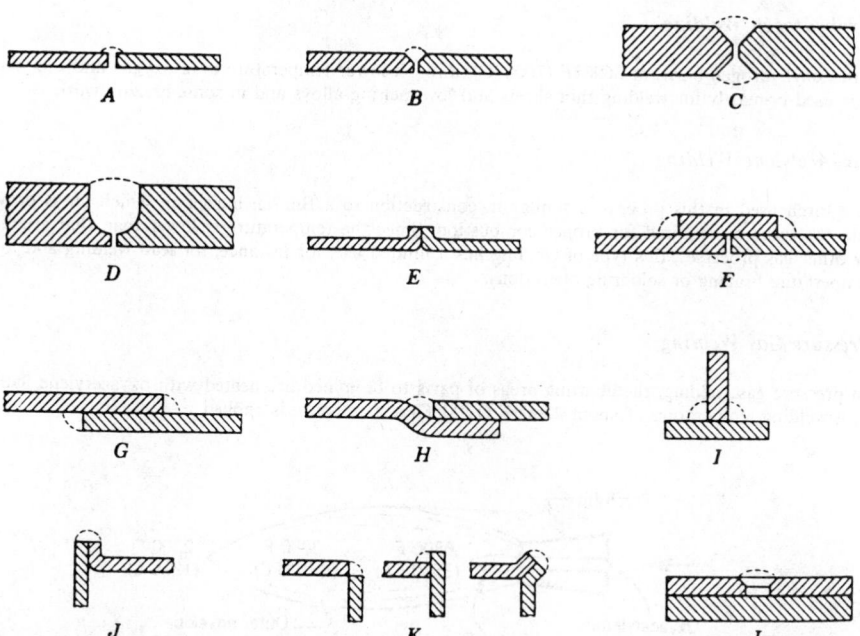

Fig. 31.5 Types of welded joints: (*a*) butt weld; (*b*) single vee; (*c*) double vee (heavy plates); (*d*) U-shaped (heavy casting); (*e*) flange weld (thin metal); (*f*) single-strap butt joint; (*g*) lap joint (single or double fillet weld); (*h*) joggled lap joint (single or double weld); (*i*) tee joint (fillet welds); (*j*) edge weld (used on thin plates); (*k*) corner welds (thin metal); (*l*) plug or rivet butt joint.

31.2.2 Forge Welding

Forge welding consists of heating the metal in a forge to a plastic condition and then uniting it by pressure. The manual process is naturally limited to light work, because all forming and welding is accomplished with a hand sledge.

Forge welding is rather slow, and there is considerable danger of an oxide scale forming on the surface. For this type of welding, low-carbon steel and wrought iron are recommended, since they have a broad welding-temperature range. The welding range decreases as the carbon content increases.

31.2.3 Gas Welding

Gas welding includes all the processes in which gases are used in combination to obtain a hot flame. Those commonly used are acetylene, natural gas, and hydrogen in combination with oxygen. Oxyhydrogen welding was the first gas process to be commercially developed. The maximum temperature developed by this process is 3600°F (1980°C). The most used combination is the oxyacetylene process, which has a flame temperature of 6300°F (3500°C).

Oxyacetylene Welding

An oxyacetylene weld is produced by heating with a flame obtained from the combustion of oxygen and acetylene, with or without the use of a filler metal. In most cases the joint is heated to a state of fusion; as a rule, no pressure is used.

Three types of flames that can be obtained are reducing, neutral, and oxidizing. Of the three, the neutral flame shown in Fig. 31.6 has the widest application in welding and cutting operations.

When there is excess of acetylene used, there is a decided change in the appearance of the flame. This flame, known as a reducing or carburizing flame, is used in the welding of Monel metal, nickel, certain alloy steels, and many of the nonferrous, hard-surfacing materials.

If the torch is adjusted to give excess oxygen, a flame similar in appearance to the neutral flame is obtained, except that the inner luminous cone is much shorter and the outer envelope appears to have more color. This oxidizing flame may be used in fusion welding of brass and bronze, but it is undesirable in other applications.

Oxyhydrogen Welding

Since oxyhydrogen burns at 3600°F (2000°C), a much lower temperature than oxygen and acetylene, it is used primarily for welding thin sheets and low-melting alloys and in some brazing work.

Air–Acetylene Welding

The torch used in this process is similar in construction to a Bunsen burner in which air is drawn into the torch as required for proper combustion. Since the temperature is lower than that attained by other gas processes, this type of welding has a limited use, for instance, for lead welding and low-temperature brazing or soldering operations.

Pressure Gas Welding

In pressure gas welding, the abutting areas of parts to be joined are heated with oxyacetylene flames to a welding temperature of about 2200°F (1200°C) and pressure is applied.

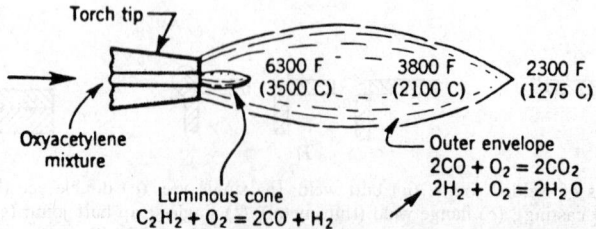

Fig. 31.6 Sketch of neutral flame showing temperatures attained.

Oxyacetylene Torch Cutting

Cutting steel with a torch is an important production process. A simple hand torch for flame cutting differs from the welding torch in that it has several small holes for preheating flames surrounding a central hole through which pure oxygen passes. The preheating flames are exactly like the welding flames and are intended only to preheat the steel before the cutting operation. The principle on which flame cutting operates is that oxygen has an affinity for iron and steel. At ordinary temperatures this action is slow, but eventually an oxide in the form of rust materializes. As the temperature of the steel is increased, this action becomes much more rapid. If the steel is heated to a red color and a jet of pure oxygen is blown on the surface, the action is almost instantaneous and the steel is actually burned into an iron oxide. About 1.3 ft³ (0.04 m³) of oxygen are required to burn up 1 in.³ (16.380 mm³) of iron. Metal up to 30 in. (762 mm) in thickness can be cut by this process.

31.2.4 Resistance Welding

In this process, a large electric current is passed through the metals causing local heating at the joint, the weld being completed by the application of pressure. A transformer in the welding machine reduces the alternating current voltage from either 120 or 240 V to around 4–12 V and raises the amperage sufficiently to produce a good heating current. When the current passes through the metal, most of the heating takes place at the point of greatest resistance in the electrical path, which is at the interface of the two sheets. It is here that the weld forms.

In all resistance welding the three factors that must be given consideration are expressed in the formula: heat $= A^2\, \Omega t$, where A is the welding current in amperes, Ω is the resistance between electrodes being used, and t is the time.

Resistance welding processes include (1) spot welding, (2) projection welding, (3) seam welding, (4) butt welding, (5) flash welding, and (6) percussion welding.

Spot Welding

In this form of resistance welding, two or more sheets of metal are held between metal electrodes, as shown in Fig. 31.7. The welding cycle is started with the electrodes contacting the metal under pressure before the current is applied, for a period known as the squeeze time. A low-voltage current of sufficient amperage is then passed between the electrodes, causing the metal in contact to be raised rapidly to a welding temperature. As soon as the temperature is reached, the pressure between the electrodes squeezes the metal together and completes the weld. This period, usually 3–30 cycles per second (Hz), is known as the weld time.

Projection Welding

Projection welding, a process similar to spot welding, is illustrated in Fig. 31.8. Projection welds are produced at localized points in workpieces held under pressure between suitable electrodes.

Seam Welding

Seam welding consists of a continuous weld on two overlapping pieces of sheet metal. Coalescence is produced by heat obtained from the resistance of current. The current passes through the overlapping

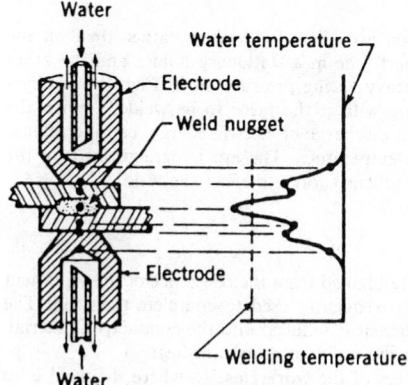

Fig. 31.7 Temperature distribution in spot welding.

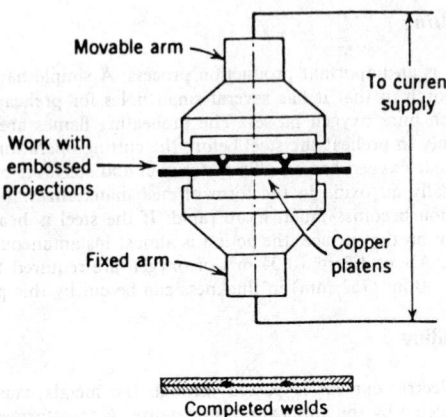

Fig. 31.8 Projection welding.

sheets, which are held together under pressure between two circular electrodes. This method is, in effect, a continuous spot-welding process, since the current is not usually on continuously but is regulated by the timer on the machine. In high-speed seam welding using continuous current, the frequency of the current acts as an interrupter.

The three types of seam welds used in industry are illustrated in Fig. 31.9.

Butt Welding

This form of welding, illustrated in Fig. 31.10, is the gripping together of two pieces of metal that have the same cross section, and pressing them together while heat is being generated in the contact surface by electrical resistance. Although pressure is maintained while the heating takes place, at no time is the temperature sufficient to melt the metal. The joint is upset somewhat by the process, but this defect can be eliminated by subsequently rolling or grinding.

Figure 31.11 illustrates a special type of butt seam welding used in pipe manufacturing. Two rolling electrodes bring a high-amperage current across the joint; the current generates the heat in the contact surfaces by electrical resistance. For thin-walled tubes high-frequency current from an induction coil can be used to generate the heat in place of electrodes. Otherwise, the procedure is the same.

Flash Welding

Butt and flash-butt welding, although similar in application, differ somewhat in the manner of heating the metal. Again, in Fig. 31.10 it is noted that for flash welding the parts must be brought together in a very light contact. A high voltage starts a flashing action between the two surfaces and continues as the parts advance slowly and the forging temperature is reached. The weld is completed by the application of a sufficient forging pressure of 5000–25,000 psi (35–170 MPa) to effect a weld.

Percussion Welding

Like the flash-weld process, percussion welding relies on arc effect for heating rather than on the resistance in the metal. Pieces to be welded are held apart, one in a stationary holder and the other in a clamp mounted in a slide and backed up against heavy spring pressure, as in Fig. 31.10. When the movable clamp is released, it moves rapidly carrying with it the piece to be welded. When the pieces are about $\frac{1}{16}$ in. (1.6 mm) apart, there is a sudden discharge of electric energy, causing intense arcing over the surfaces and bringing them to a high temperature. The arc is extinguished by the percussion blow of the two parts coming together with sufficient force to effect the weld.

31.2.5 Induction Welding

Coalescence in induction welding is produced by the heat obtained from the resistance of the weldment to the flow of an induced electrical current. Pressure is frequently used to complete the weld. The inductor coil is not in contact with the weldment; the current is induced into the conductive material. Resistance of the material to this current flow results in the rapid generation of heat.

In operation, a high current is induced into both edges of the work close to where the weld is to

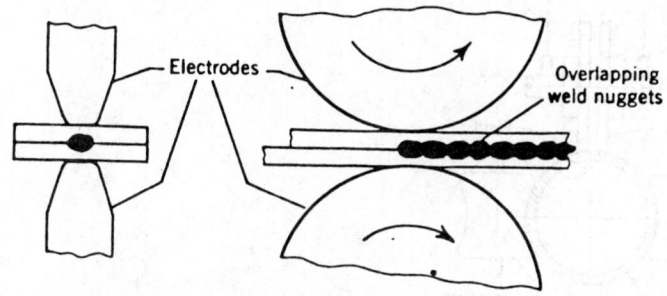

Lap Seam Weld

Electrodes

Overlapping
weld nuggets

Mash Seam Weld

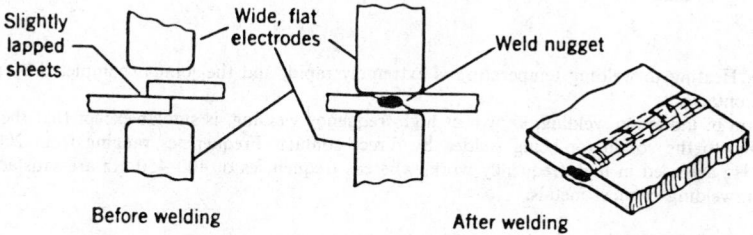

Slightly
lapped
sheets

Wide, flat
electrodes

Weld nugget

Before welding

After welding

Metal Finish Seam Weld

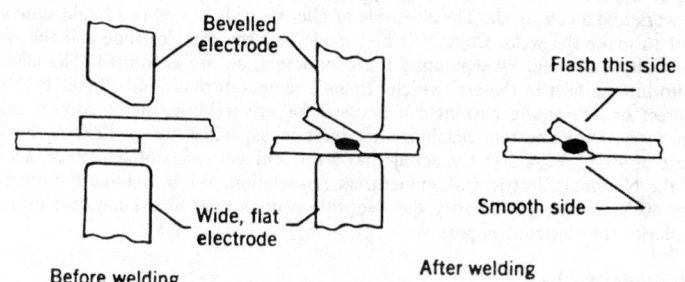

Bevelled
electrode

Flash this side

Wide, flat
electrode

Smooth side

Before welding

After welding

Fig. 31.9 Types of seam welds.

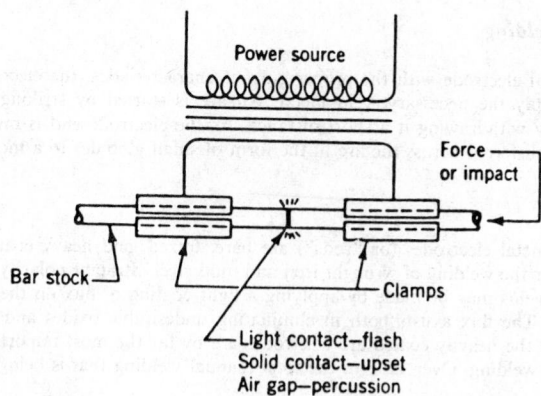

Power source

Force
or impact

Bar stock

Clamps

Light contact—flash
Solid contact—upset
Air gap—percussion

Fig. 31.10 Sketch illustrating forms of butt-welding bar stock.

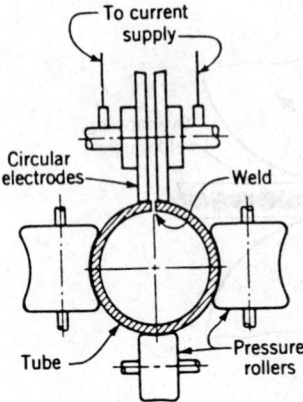

Fig. 31.11   Continuous resistance butt-welding of steel pipe.

be made. Heating to welding temperature is extremely rapid, and the joint is completed by pressure rolls or contacts.

A form of induction welding, known as high-frequency welding, is similar except that the current is supplied to the conductor being welded by direct contact. Frequencies ranging from 200,000 to 500,000 Hz are used in high-frequency work, whereas frequencies of 400–450 Hz are satisfactory for induction welding of most metals.

31.2.6   Arc Welding

Arc welding is a process in which coalescence is obtained by heat produced from an electric arc between the work and an electrode. The electrode or filler metal is heated to a liquid state and deposited into the joint to make the weld. Contact is first made between the electrode and the work to create an electric circuit, and, then, by separating the conductors, an arc is formed. The electric energy is converted into intense heat in the arc, which attains a temperature around 10,000°F (5500°C).

Either direct or alternating current can be used for arc welding, direct current being preferred for most purposes. Direct-current machines are built in capacities up to 1000 A, having an open-circuit voltage of 40–95 V. A 200-A machine has a rate current range of 40–250 A, according to the standard of the National Electrical Manufacturers Association. While welding is proceeding, the arc voltage is 18–40 V. In straight polarity the electrode is connected to the negative terminal, whereas in reverse polarity the electrode is positive.

Carbon Electrode Welding

The first methods of arc welding, which used only carbon electrodes, are still in use to some extent for both manual and machine operation. The carbon arc is used only as a source of heat, and the torch is handled in a fashion similar to that used in gas welding. Filler rods supply weld metal if additional metal is necessary.

Metal Electrode Welding

By the use of a metal electrode with the proper current characteristics, the electrode itself could be melted down to supply the necessary weld metal. An arc is started by striking the work with an electrode and quickly withdrawing it a short distance. As the electrode end is melted by the intense heat, most of it is transferred across the arc in the form of small globules to a molten pool.

Electrodes

The three types of metal electrodes (or "rods") are bare, fluxed, and heavy coated. Bare electrodes have a limited use for the welding of wrought iron and mild steel. Straight polarity is generally recommended. Improved welds may be made by applying a light coating of flux on the rod with a dusting or washing process. The flux assists both in eliminating undesirable oxides and in preventing their formation. However, the heavily coated arc electrodes are by far the most important ones used in all types of commercial welding. Over 95% of the total manual welding that is being done today is with coated electrodes.

Figure 31.12 is a diagrammatic sketch showing the action of an arc using a heavy-coated electrode.

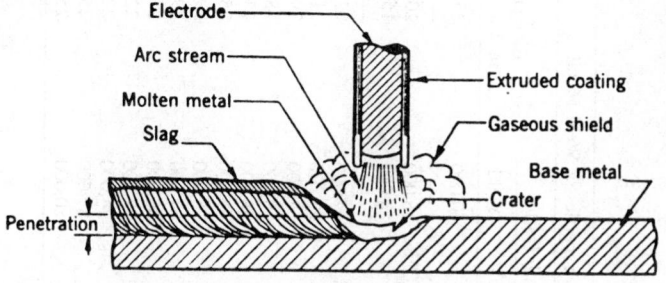

Fig. 31.12   Diagrammatic sketch of arc flame.

Coating compositions may be classified as organic and inorganic, although in some cases both types might be used. Inorganic coatings can be further subdivided into flux compounds and slag-forming compounds. The following list is of some of the principle constituents used:

1. Slag-forming constituents: SiO_2, MnO_2, and FeO. Al_2O_3 is sometimes used, but it makes the arc less stable.
2. Constituents to improve arc characteristics: Na_2O, CaO, MgO, and TiO_2.
3. Deoxidizing constituents: graphite, aluminum, and wood flour.
4. Binding material: sodium silicate, potassium silicate, and asbestos.
5. Alloying constituents to improve strength of weld: vanadium, cesium, cobalt, molybdenum, aluminum, zirconium, chromium, nickel, manganese, and tungsten.

Typical electrode performance and mechanical properties of weld metals are given in Table 31.6.

Atomic-Hydrogen Arc Welding

In this process a single-phase ac arc is maintained between two tungsten electrodes, and hydrogen is introduced into the arc. As the hydrogen enters the arc, the molecules are broken up into atoms, which recombine into molecules of hydrogen outside the arc. This reaction is accompanied by the liberation of an intense heat, attaining a temperature of about 11,000°F (6100°C). Weld metal may be added to the joint in the form of welding rod.

Inert-Gas-Shielded Arc Welding

In this process, coalescense is produced by heat from an arc between a metal electrode and the work, which is shielded by an atmosphere of argon, helium, carbon dioxide, or a mixture of gases. Two methods are employed: one using a tungsten electrode with filler metal added as in gas welding (TIG welding, tungsten inert gas), and the other using a consumable metal wire as the electrode (MIG welding, metal inert gas). Both methods are adaptable to either manual or automatic machine welding, and no flux or wire coating is required for protection of the weld.

Arc Spot Welding

An interesting application of inert-gas arc welding is the making of spot, plug, or tack welds by an argon-shielded electric arc using consumable electrodes. To effect a weld, a small welding gun with pistol grip is held tightly against the work to be welded. As the trigger is released, the argon valve is opened, and the current is allowed to pass through the electrode for a preset interval (2–5 sec), and then both are shut off.

Submerged-Arc Welding

This process is so named because the metal arc is shielded by a blanket of granular, fusible flux during the welding operation. Aside from this feature, its operation is quite similar to other automatic arc-welding methods. A bare electrode is fed through the welding head into the granular material. This metal is laid down along the seam to be welded, and the entire welding action takes place beneath it.

Table 31.6 Typical Electrode Performance and Mechanical Properties of Weld Metal

Electrode Class	Electrode Size (in.)	Current (ac) (A)	Deposition Rate (100% arc time) (oz/hr)	Deposition Efficiency (%)	Spatter Loss (% of core wire)	Properties of Weld Metal		
						Strength (psi)		Elongation in 2 in. (%)
						Tensile	Yield	
E6011	3/16	200 max	88	76	16	70,000	60,000	25
	1/4	300 max	112	72	—	—	—	—
E6012	3/16	225	66	68	14	79,000	66,000	20
	1/4	380	120	78	16	71,000	58,000	15
E6013	5/16	475	178	78	—	—	—	—
	3/16	225	67	67	13	76,000	67,000	22
E6020	3/16	225	92	68	11	69,000	59,000	31
	1/4	380	177	69	11	65,000	52,000	29
E7014	5/16	450	215	69	9	63,000	50,000	28
	3/16	260	87	68	—	74,000	68,000	23
E7016	1/4	340	119	69	6	73,000	67,000	25
	3/16	225	63	70	—	87,000	75,000	29
E7018	3/16	240	83	69	—	80,000	67,000	29
	1/4	270	134	68	—	90,000	80,000	22
E7024	1/4	360	186	69	—	89,000	77,000	22
	5/16	475	250	73	—	83,000	74,000	22
E7028	3/16	300	125	68	—	90,000	81,000	25
	1/4	390	192	70	—	88,000	80,000	25

Stud Arc Welding

Stud welding is a dc arc welding process developed to end-weld metal studs to flat surfaces. It is accomplished with a pistol-shaped welding gun that holds the stud or fastener to be welded. When the trigger of the gun is pressed, the stud is lifted to create an arc and then forced against the molten pool by a backing spring.

Transferred Arc Cutting

In a plasma torch, a gas is heated by a tungsten arc to such a high temperature that it becomes ionized and acts as a conductor of electricity. In this state the arc gas is known as plasma. The torch is generally designed so that the gas is closely confined to the arc column through a small orifice. This increases the temperature of the plasma and concentrates its energy on a small area of the workpiece which rapidly melts the metal. As the gas jet stream leaves the nozzle, it expands rapidly removing the molten metal continuously as the cut progresses. Since the heat obtained does not depend on a chemical reaction, this torch can be used to cut any metal. Temperatures approach 60,000°F (33,000°C), which is roughly 10 times that possible by the reaction of oxygen and acetylene.

Electroslag Welding

Electroslag welding, another metal-arc welding process, has its best use in the welding of heavy plates, the joint to be welded being in a vertical position. Heat is obtained from the resistance of current in an electrically conductive molten flux. One or more electrodes are continuously fed into a molten pool of slag which maintains a temperature in excess of 3200°F (1800°C).

31.2.7 Electron-Beam Welding

In electron-beam welding, coalescence is produced by bombarding the workpiece with a dense beam of high-velocity electrons. The metal is joined by melting the edges of the workpiece or by penetrating through the material, and usually no filler metal is added. This process may be used not only to join common metals, but also refractory metals, highly oxidizable metals, and various superalloys that have previously been impossible to weld.

31.2.8 Laser Welding

Lasers produce high heat intensity, which can be used for welding. Since the laser delivers its energy in the form of light, it can be operated in any transparent medium without contact with the workpiece.

In welding, the power is delivered in pulses rather than as a continuous beam. The beam is focused on the workpiece, where the weld is to be made, and the intense heat produces a fusion weld. Laser welding is slow and is used only for special jobs involving small weldments. Its greatest use is found in the electronics industry.

31.2.9 Friction Welding

In friction welding, coalescence is produced by the heat of friction generated by rotating one piece against another under controlled axial pressure. The two surfaces in contact are heated to a melting temperature; the adjacent material becomes plastic. The relative motion between the two is then stopped and a forging pressure is applied, which upsets the joint slightly.

31.2.10 Thermit Welding

Thermit welding is the only welding process employing an exothermal chemical reaction for the purpose of developing a high temperature and is based on the fact that aluminum has a great affinity for oxygen and can be used as a reducing agent for many oxides. The usual Thermit mixture or compound consists of finely divided aluminum and iron oxide, mixed in a ratio of about 1 to 3 by weight. The iron oxide is usually in the form of roll mill scale. This mixture is not explosive and can be ignited only at a temperature of about 2800°F (1500°C). A special ignition powder is used to start the reaction. The chemical reaction requires only about 30 sec and attains a temperature around 4500°F (2500°C). The mixture reacts according to the chemical equation

$$8Al + 3Fe_3O_4 = 9Fe + 4Al_2O_3$$

The resultant products are a highly purified iron (actually steel) and an aluminum oxide slag, which floats on top and is not used. Other reactions also take place, as most Thermit metal is alloyed with manganese, nickel, or other elements.

There is no limit to the size of welds that can be made by Thermit welding. It is used primarily for the repairs of large parts that would be difficult to weld by other processes.

31.2.11   Flow Welding

Flow welding is defined as a welding process in which coalescence is produced by heating with molten filler metal poured over the surfaces to be welded until the welding temperature is attained and the required filler metal has been added. Flow welding is used in the joining of thick sections of nonferrous metals using a filler of the same composition as the base metal.

In operation, the weld area is first properly prepared and preheated. Molten filler metal is then poured between the ends of the material until melting starts. At this point, the flow is stopped and the joint, filled with metal, is allowed to cool slowly. To ensure the fusion of the top edges and accomplishments of a good weld, the level of the molten filler metal must be kept higher than the surfaces being welded.

31.2.12   Cold Welding

Cold welding is a method of joining metals at room temperature by the application of pressure alone. The pressure applied causes the surface metals to flow, which produces the weld. It is a solid-state bonding process in which no heat is supplied from an external source. Butt welds of wire and rods can also be made by clamping the ends in special dies and bringing them together under a load sufficient to produce plastic flow at the joint. Before a weld is made, the surfaces or parts to be joined must be wire-brushed thoroughly at a surface speed of around 3000 fpm (15 m/sec) to remove oxide films on the surface. Other methods of cleaning seem to be unsatisfactory. In making a weld, the pressure is applied over a narrow strip so that the metal can flow away from the weld on both sides. It may be applied either by impact or with a slow squeezing action, both methods being equally effective. The pressure required for aluminum is 25,000–35,000 psi (170–240 MPa). Spot welds are rectangular in shape and in terms of gage thickness are approximately $t \times 5t$ in area. In addition, both ring welds and continuous seam welds can be made. This method of welding is used with aluminum and copper; however, lead, nickel, zinc, and Monel can also be joined by cold pressure.

Ultrasonic Welding

Ultrasonic welding is a solid-state bonding process for the joining of similar or dissimilar metals, generally with an overlap-type joint. High-frequency vibratory energy is introduced into the weld area in a plane parallel to the surface of the weldment. The forces involved set up oscillating shear stresses at the weld interface, which break up and expel surface oxides. This interfacial slip results in metal-to-metal contact permitting the intermingling of the metal and the forming of a sound weld nugget. No external heat is applied, although the weld metal does undergo a modest temperature rise.

31.2.13   Explosive Welding

Explosive welding or cladding, as it is often called, is the bringing together of two metal surfaces with sufficient impact and pressure to cause them to bond. Pressure is developed by a high-explosive shot placed in contact with, or in close proximity to, the metals, as illustrated in Fig. 31.13.

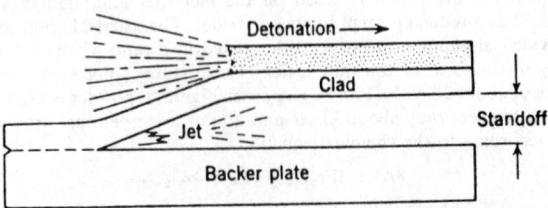

Fig. 31.13 Explosion-bonding process showing high-velocity jet emanating from the collision point because of upstream pressure.

31.3 METAL CUTTING

31.3.1 Metal-Cutting Principles

Mechanics of Metal Cutting

Material removal by chipping process began as early as 4000 BC when the Egyptians used a rotating bowstring device to drill holes in stones. Scientific work developed starting about the mid-19th century. The basic chip-type machining operations are shown in Fig. 31.14. The mechanics of the metal-cutting

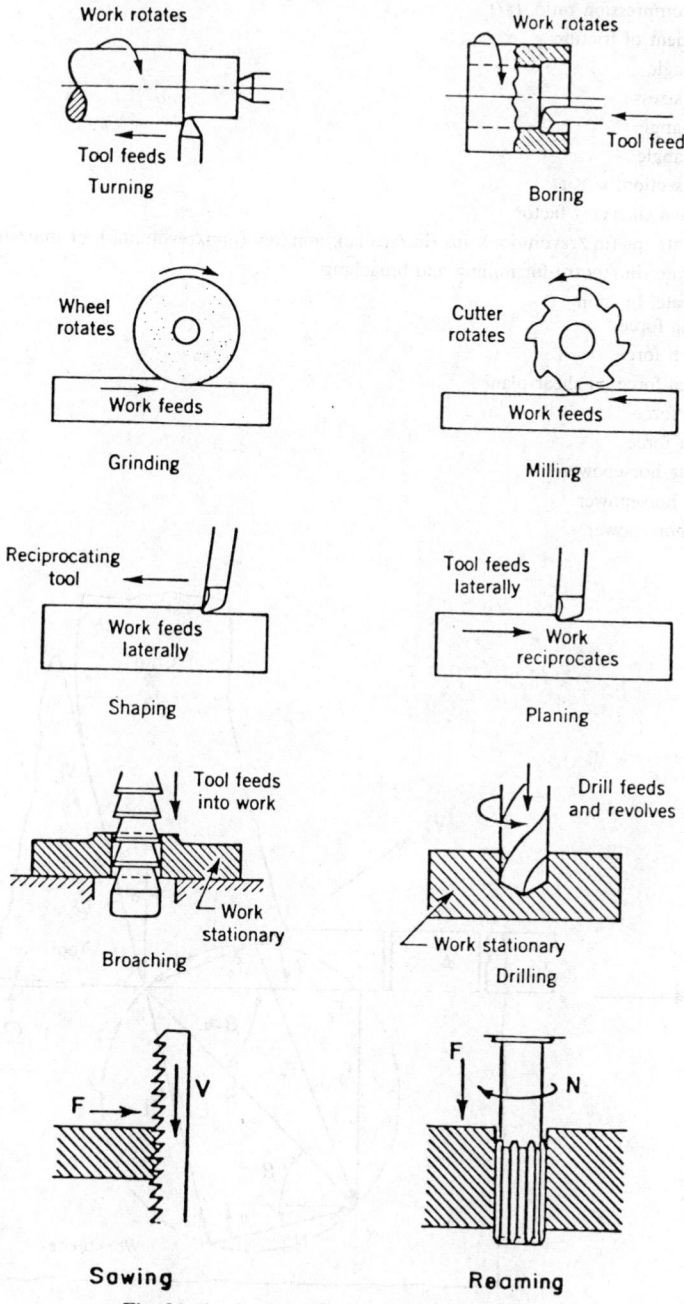

Fig. 31.14 Conventional machining processes.

process are shown in Fig. 31.15. Figure 31.15 shows a two-dimensional type of cutting in which the cutting edge is perpendicular to the cut, known as *orthogonal* cutting, as contrasted with the three-dimensional *oblique* cutting shown in Fig. 31.16. The main three cutting velocities are shown in Fig. 31.17. The metal cutting factors are defined as follows:

α	rake angle
β	friction angle
γ	strain
λ	chip compression ratio, t_2/t_1
μ	coefficient of friction
ψ	tool angle
τ	shear stress
ϕ	shear angle
Ω	relief angle
A_0	cross section, $w \times t_1$
e_m	machine efficiency factor
f	feed rate ipr (in./revolution), ips (in./stroke), mm/rev (mm/revolution), or mm/stroke
f_t	feed rate (in./tooth) for milling and broaching
F	feed rate, in./min
F_c	cutting force
F_f	friction force
F_n	normal force on shear plane
F_s	shear force
F_t	thrust force
HP_c	cutting horsepower
HP_g	gross horsepower
HP_u	unit horsepower

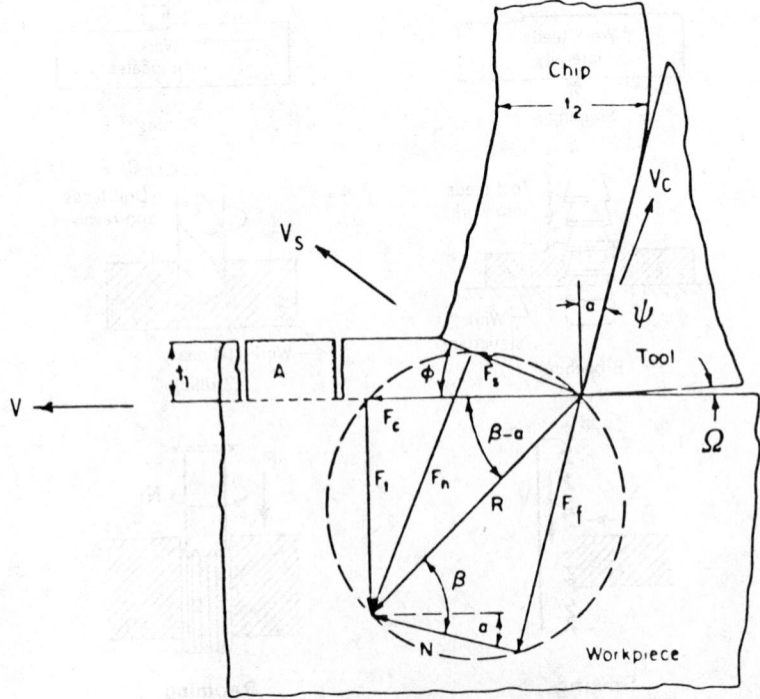

Fig. 31.15 Mechanics of metal cutting process.

N	revolutions per minute
Q	rate of metal removal, in.³/min
R	resultant force
T	tool life in minutes
t_1	depth of cut
t_2	chip thickness
V	cutting speed, ft/min
V_c	chip velocity
V_s	shear velocity

The *shear angle* ϕ controls the thickness of the chip and is given by

$$\tan \phi = \frac{\cos \alpha}{\lambda - \sin \alpha}$$

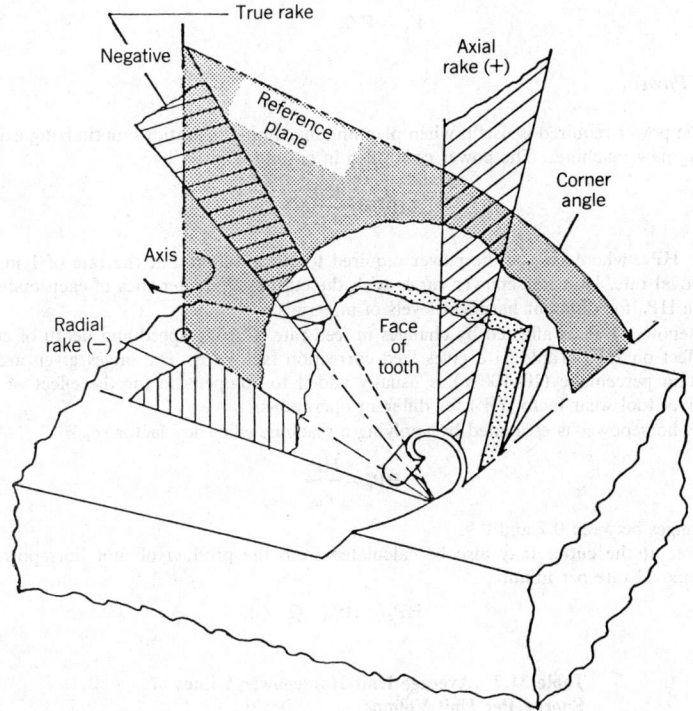

Fig. 31.16 Oblique cutting.

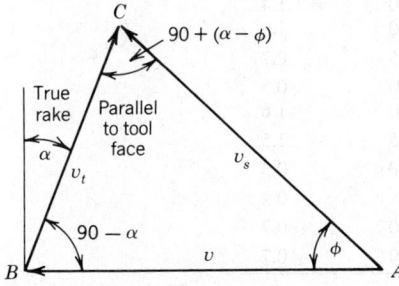

Fig. 31.17 Cutting velocities.

The *strain* γ that the material undergoes in shearing is given by

$$\gamma = \cot \phi + \tan(\phi - \alpha)$$

The *coefficient of friction* μ on the face of the tool is

$$\mu = \frac{F_t + F_c \tan \alpha}{F_c - F_t \tan \alpha}$$

The *friction force* F_f along the tool is given by

$$F_f = F_t \cos \alpha + F_c \sin \alpha$$

Cutting forces are usually measured with dynamometers and/or wattmeters. The shear stress τ in the shear plane is

$$\tau = \frac{F_c \sin \phi \cos \phi - F_t \sin^2 \phi}{A}$$

The speed relationships are

$$\frac{V_c}{V} = \frac{\sin \phi}{\cos(\phi - \alpha)}$$

$$V_c = V / \lambda$$

Machining Power

Estimating the power required is useful when planning machining operations, optimizing existing ones, and specifying new machines. The power consumed in cutting is given by

$$HP_c = \frac{F_c V}{33,000}$$

or $HP_c = Q \cdot HP_u$, where HP_u is the power required to cut a material at the rate of 1 in.^{3}/min and Q is the removal rate. Unit horsepower varies with the cutting characteristics of each material. Table 31.7 gives the HP_u for different hardness levels of materials.

Unit horsepower HP_u is affected by changes in feed rate. Cutting speed and depth of cut have no significant effect on HP_u. Table 31.8 gives feed correction factor (c). The tables given are for sharp tools. A certain percentage (10–100%) is usually added to compensate for the effect of tool wear. Table 31.9 gives tool wear factor (W) for different operations.

The gross horsepower is estimated by applying a machine efficiency factor (e_m):

$$HP_g = \frac{HP_c}{e_m}$$

e_m usually ranges between 0.7 and 0.9.

Horsepower at the cutter may also be calculated from the product of unit horsepower and the cubic inch removal rate per minute.

$$HP_c = HP_u \cdot Q$$

Table 31.7 Average Unit Horsepower Values of Energy Per Unit Volume

Material	BHN	Unit Power
Carbon steels	150–200	1.0
	200–250	1.4
	250–350	1.6
Leaded steels	150–175	0.7
Cast irons	125–190	0.5
	190–250	1.6
Stainless steels	135–275	1.5
Aluminum alloys	50–100	0.3
Magnesium alloys	40–90	0.2
Copper	125–140	0.7
Copper alloys	100–150	0.7

Table 31.8 Horsepower Feed Correction Factors for Turning, Planing, and Shaping

Feed (ipr or ips)	mm/rev or mm/stroke	Factor
0.002	0.05	1.4
0.005	0.12	1.2
0.008	0.20	1.05
0.012	0.30	1.0
0.020	0.50	0.9
0.030	0.75	0.80
0.040	1.00	0.80
0.050	1.25	0.75

Table 31.9 Tool Wear Factors W

Type of Operation[a]	W
Turning	
Finish turning (light cuts)	1.10
Normal rough and semifinish turning	1.30
Extra-heavy-duty rough turning	1.60–2.00
Milling	
Slab milling	1.10
End milling	1.10
Light and medium face milling	1.10–1.25
Extra-heavy-duty face milling	1.30–1.60
Drilling	
Normal drilling	1.30
Drilling hard-to-machine materials and drilling with a very dull drill	1.50
Broaching	
Normal broaching	1.05–1.10
Heavy-duty surface broaching	1.20–1.30

[a] For all operations with sharp cutting tools.

For turning, planing, and shaping this becomes

$$HP_c = (HP_u)12CWVfd$$

For milling,

$$HP_c = (HP_u)CWFwd$$

For drilling,

$$HP_c = (HP_u)CW(\text{rpm})f\left(\frac{\pi D^2}{4}\right)$$

For broaching

$$HP_c = (HP_u)12CWVn_c d_t$$

where V = cutting speed, fpm
$\quad C$ = feed correction factor
$\quad f$ = feed, ipr (turning and drilling), ips (planing and shaping)
$\quad F$ = feed, ipm
$\quad d$ = depth of cut
$\quad d_t$ = maximum depth of cut per tooth
$\quad n_c$ = number of teeth engaged in work
$\quad w$ = width of cut
$\quad W$ = tool wear factor

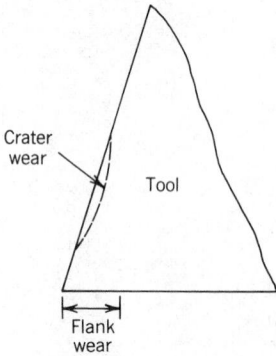

Fig. 31.18 Types of tool wear.

The four basic types of chips formed during machining processes are:

1. Continuous chips result from cutting ductile materials, such as mild steel and aluminum.
2. Discontinuous chips are produced when cutting most brittle materials, such as cast iron, cast brass, and bronze, or in cutting ductile materials at very low speeds or low rake angles (0–10°) for feeds greater than 0.008 in. (0.2 mm).
3. Built-up edge is a localized, highly deformed zone of material attached or welded to the tool face. This type of chip results from high friction between chip and tool, poor lubrication, reduced rake angles, and low speeds. Built-up edge produces rough machined surface and reduces the tool life. Built-up edge can be reduced by decreasing chip thickness, increasing rake angle, increasing tool sharpness and cutting speed, use of ceramic tools, and application or chlorinated or sulfurized cutting fluid.
4. Inhomogeneous chips result from machining metals whose yield strength decreases sharply with temperature, such as titanium alloys.

Tool Life

Tool life is a measure of the length of time a tool will cut satisfactorily, and may be measured in different ways. Tool wear as in Fig. 31.18 is a measure of tool failure if it reaches a certain limit. These limits are usually 0.062 in. (1.58 mm) for high-speed tools and 0.030 in. (0.76 mm) for carbide tools. In some cases the life is determined by surface finish deterioration and an increase in cutting forces. The cutting speed is the variable that has the greatest effect on tool life. The relationship between tool life and cutting speed is given by the Taylor equation

$$VT^n = C$$

where V = cutting speed, fpm (m/sec)
T = tool life, min (sec)
n = exponent depending on cutting condition
C = constant, the cutting speed for a tool life of 1 min

Table 31.10 gives the approximate ranges for the exponent n. Taylor's equation is equivalent to

$$\log V = C - n \log T$$

which when plotted on log–log paper gives a straight line, as shown in Fig. 31.19.

Table 31.10 Average Values of n

Tool Material	Work Material	n
High-speed steel	Steel	0.125
	Cast iron	0.14
Cemented carbide	Steel	0.20
	Cast iron	0.25
Sintered oxide	Steel	0.50

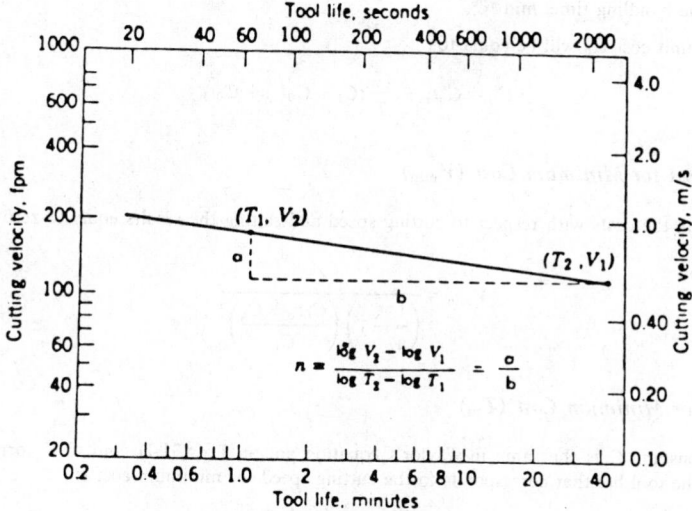

Fig. 31.19 Cutting-speed–tool-life relationship.

Metal-Cutting Economics

The efficiency of machine tools increases as cutting speeds increase, but tool life is reduced. To achieve the optimum conditions, that is, the minimum cost while considering the principal individual costs—machining cost, tool cost, tool-changing cost, and handling cost—is the main objective of metal-cutting economics. Figure 31.20 shows the relationships among these four factors.

$$\text{Machining cost} = C_0 t_m$$

where C_0 = the operating cost per minute which is equal to the machine operator's rate plus appropriate overhead

t_m = the machine time in minutes and is equal to $L/(fN)$, where L is the axial length of cut

$$\text{Tool cost per operation} = C_t \frac{t_m}{T}$$

where C_t = the tool cost per cutting edge

T = the tool life and is equal to $(C/V)^{1/n}$

$$\text{Tool changing cost} = C_0 t_c (t_m/T)$$

where t_c = the tool changing time, min

$$\text{Handling cost} = C_0 t_h$$

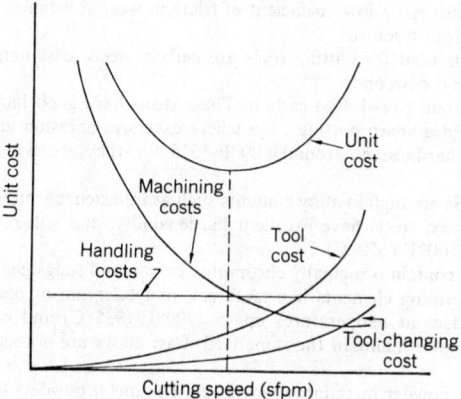

Fig. 31.20 Cost factors.

where t_h = the handling time, min

The average unit cost C_u will be equal to

$$C_u = C_0 t_m + \frac{t_m}{T}(C_t + C_0 t_c) + C_0 t_h$$

Cutting Speed for Minimum Cost (V_{min})

Differentiating the costs with respect to cutting speed and setting the results equal to zero will result in V_{min}:

$$V_{min} = \frac{C}{\left(\dfrac{1}{n} - 1\right)\left(\dfrac{C_0 t_c + C_t}{C_0}\right)^n} \qquad (31.1)$$

Tool Life for Minimum Cost (T_m)

Since the constant C is the same in Taylor's equation and in Eq. (31.1), and if V corresponds to V_{min}, then the tool life that corresponds to the cutting speed for minimum cost is

$$T_{min} = \left(\frac{1}{n} - 1\right)\left(\frac{C_0 t_c + C_t}{C_0}\right) \qquad (31.2)$$

Cutting Speed for Maximum Production (V_{max})

This speed can be determined from Eq. (31.1) for the cutting speed for minimum cost by assuming that the tool cost is negligible, that is, by setting $C_1 = 0$:

$$V_{max} = \frac{C}{\left[\left(\dfrac{1}{n} - 1\right)t_c\right]^n} \qquad (31.3)$$

Tool Life for Maximum Production (T_{max})

By analogy to Taylor's equation, the tool life that corresponds to the maximum production rate is given by

$$T_{max} = \left(\frac{1}{n} - 1\right)t_c \qquad (31.4)$$

31.3.2   Cutting-Tool Materials

The desirable properties for any tool material include the ability to resist softening at high temperature, which is known as red hardness; a low coefficient of friction; wear resistance; and sufficient toughness and shock resistance to avoid fracture.

The principal materials used for cutting tools are carbon steels, cast nonferrous alloys, carbides, ceramic tools or oxides, and diamonds.

High-carbon steels contain (0.8–1.2%) carbon. These steels have good hardening ability, and with proper heat treatment hold a sharp cutting edge where excessive abrasion and high heat are absent. Because these tools lose hardness at around 600°F (315°C), they are not suitable for high speeds and heavy-duty work.

High-speed steels (HSS) are high in alloy contents such as tungsten, chromium, vanadium, molybdenum, and cobalt. High-speed steels have excellent hardenability, and will retain a keen cutting edge to temperatures around 1200°F (650°C).

Cast nonferrous alloys contain principally chromium, cobalt, and tungsten, with smaller percentages of one or more carbide-forming elements like tantalum, molybdenum, or boron. Cast-alloy tools can maintain good cutting edges at temperatures up to 1700°F (935°C) and can be used at twice the cutting speed as HSS and still maintain the same feed. Cast alloys are not as tough as HSS and have less shock resistance.

Carbides are made by powder metallurgy technique. The metal powders used are tungsten carbide (WC), cobalt (Co), titanium carbide (TiC), and tantalum carbide (TaC) in different ratios. Carbide tools may be coated with a bonded layer of titanium carbide, aluminum oxide, or titanium nitride (TiN) to provide additional wear resistance while maintaining the same strength and toughness. Carbide tools will maintain a keen cutting edge at temperatures over 2200°F (1210°C) and can be used at speeds two to three times those of cast alloy tools. Cemented carbides are used in the form of inserts or tips which are clamped to a steel shank. *Micrograin carbide* is made of fine-grain tungsten carbide and has high strength and hardness qualities. It is used when cutting speeds are too low for regular carbides, and is used for form and cutoff tools. Machines using carbide tools must be rigidly built, have ample power, and have a range of feeds and speeds suitable to the machined materials. Cemented carbides are classified from C1 to C8. Those with lower numbers are of straight tungsten carbide–cobalt type and the higher the grade number, the greater its wear-resistance. C5 to C8 carbides are recommended for steel cutting, with titanium and tantalum added. Table 31.11 indicates carbide cutting tool application.

Table 31.11 Carbide Cutting-Tool Classification

Cast Iron, Nonferrous, and Nonmetallic Materials	Steel and Steel Alloys
C-1 Roughing	C-5 Roughing
C-2 General purpose	C-6 General purpose
C-3 Finishing	C-7 Finishing
C-4 Precision finishing	C-8 Precision finishing

Ceramic or oxide tool inserts are made from fine aluminum oxide (Al_2O_3) grains with minor additions of titanium, magnesium, or chromium oxide by powder-metallurgy techniques. These inserts have an extremely high abrasion resistance and compressive strength, lack affinity for metals being cut, resistance to cratering, and poor heat conductivity. They are harder than cemented carbides but lack impact toughness. The ceramic tool softening point is above 2000°F (1090°C) and these tools can be used at high speeds (1500–2000 ft/min) with large depth of cuts. Ceramic tools have tremendous potential because they are composed of materials that are abundant in the earth's crust. Optimum cutting conditions can be achieved by applying negative rank angles (5–7°), rigid tool mountings, and rigid machine tools.

Diamonds are used as a single-point tool for light cuts at high speeds of 1000–5000 fpm to achieve good surface finish and dimensional accuracy. They are used also for hard materials difficult to cut with other tool material.

Cutting-Tool Geometry

The shape and position of the tool relative to the workpiece have a very important effect in metal cutting. There are six single-point tool angles critical to the machining process. These can be divided into three groups.

Rake angles affect the direction of chip flow, the characteristics of chip formation, and tool life. Positive rake angles reduce the cutting forces and direct the chip flow away from the material. Negative rake angles increase cutting forces but provide greater strength as is recommended for hard materials.

Relief angles avoid excessive friction between the tool and workpiece and allow better access of coolant to tool–work interface.

The side cutting-edge angle allows the full load of the cut to be built up gradually. The *end cutting-edge angle* allows sufficient clearance so that the surface of the tool behind the cutting point will not rub over the work surface.

The purpose of the *nose radius* is to give a smooth surface finish and to increase the tool life by increasing the strength of the cutting edge. The elements of the single-point tool are written in the following order: back rake angle, side rake angle, end relief angle, side relief angle, end cutting-edge angle, side cutting-edge angle and nose radius. Figure 31.21 shows the basic tool geometry.

Cutting tools used in various machining operations often appear to be very different from the single-point tool in Fig. 31.21. Often they have several cutting edges as in the case of drills, broaches, saws, and milling cutters. Simple analysis will show that such tools are comprised of a number of single-point cutting edges arranged so as to cut simultaneously or sequentially.

Cutting Fluids

The major roles of the cutting fluids—liquids or gases—are: (1) removal of the heat of friction and deformation; (2) reduction of friction among chip, tool, and workpiece; (3) washing away chips; (4)

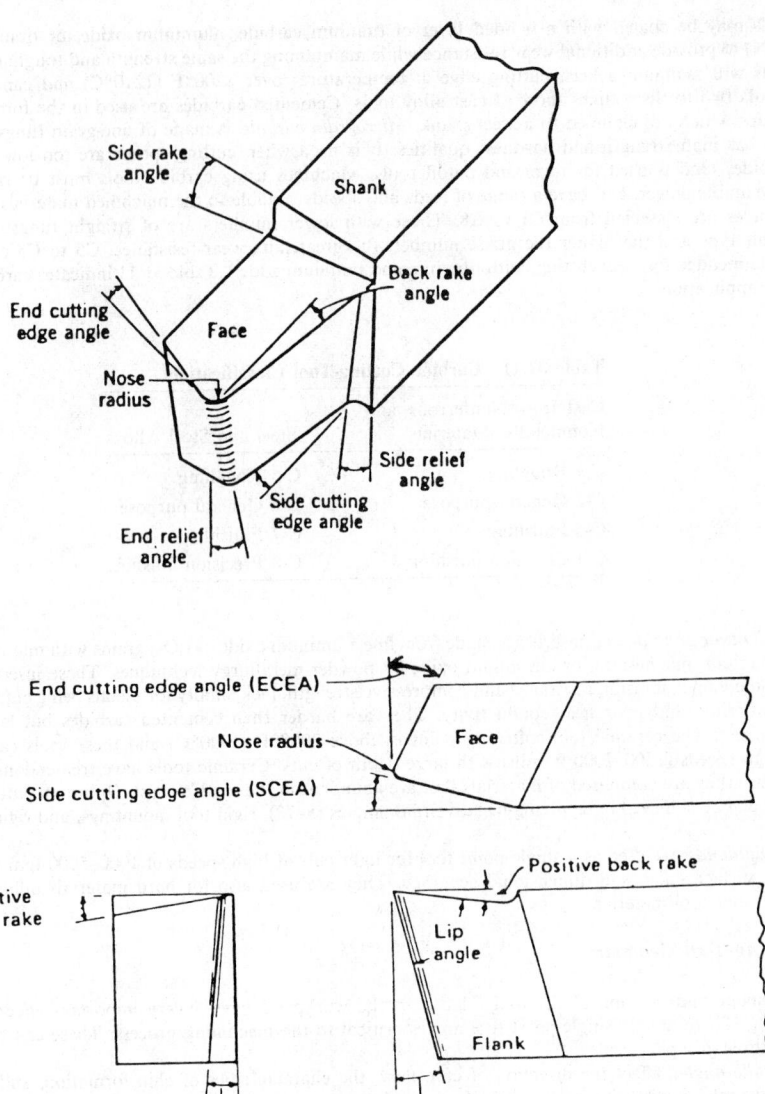

Side rake angle

Shank

Back rake angle

End cutting edge angle

Face

Nose radius

Side relief angle

Side cutting edge angle

End relief angle

End cutting edge angle (ECEA)

Nose radius

Side cutting edge angle (SCEA)

Face

Positive side rake

Positive back rake

Lip angle

Flank

Side relief   End relief

Fig. 31.21 Basic tool geometry.

reduction of possible corrosion on both workpiece and machine; and (5) prevention of built up edges. Cutting fluids work as coolants and lubricants. Cutting fluids applied depend primarily on the kind of material being used and the type of operation. The four major types of cutting fluids are: (1) soluble oil emulsions with water to oil ratios of 20:1 to 80:1; (2) oils; (3) chemicals and synthetics; and (4) air. At low cutting speeds (40 ft/min and below) oils are highly recommended, especially in tapping, reaming, and gear and thread machining. Cutting fluids with the maximum specific heat, such as soluble oil emulsions, are recommended at high speeds.

Machinability

Machinability refers to a system for rating materials on the basis of their relative ability to be machined easily, long tool life, low cutting forces, and acceptable surface finish. Additives such as lead, manganese sulfide, or sodium sulfide with percentages less than 3%, can improve the machinability of steel and copper-based alloys such as brass and bronze. In aluminum alloys, additions up to 1–3% of zinc and magnesium improve their machinability.

Cutting Speeds and Feeds

Cutting speed is expressed in feet per minute (m/sec) and is the relative surface speed between the cutting tool and the workpiece. It may be expressed by the simple formula $CS = \pi DN/12$ fpm, where D is the diameter of the workpiece in inches in case of turning or the diameter of the cutting tool in case of drilling, reaming, boring, and milling; N is the revolutions per minute. If D is given in millimeters, the cutting speed $CS = \pi DN/60,000$ m/sec.

Feed refers to the rate at which a cutting tool advances along or into the surface of the workpiece. For machines in which either the workpiece or the tool turns, feed is expressed in inches per revolution, ipr (mm/rev). For reciprocating tool or workpiece, feed is expressed in inches per stroke, ips (mm/stroke).

The recommended cutting speeds, feeds, and depth of cut that resulted from extensive research, for different combinations of tools and materials under different cutting conditions, can be found in many references, such as Society of Manufacturing Engineers (SME), publications such as *Tool and Manufacturing Engineers Handbook, Machining Data Handbook, Journal of Manufacturing Engineers, Manufacturing Engineering Transactions; American Society of Metals (ASM) Handbook; American Machinist's Handbook; Machinery's Handbook;* American Society of Mechanical Engineering (ASME) publications; Society of Automotive Engineers (SAE) Publications, and *International Journal of Machine Tool Design and Research.*

31.3.3 Turning Machines

Turning is a machining process for generating external surfaces of revolution by the action of a cutting tool on a rotating workpiece usually held in a lathe. Figure 31.22 shows some of the external operations that can be done on a lathe. When the same action is applied to internal surfaces of revolution, the process is termed boring. Operations that can be performed on a lathe are turning, facing, drilling, reaming, boring, chamfering, tapering, grinding, threading, tapping, and knurling.

The primary factors involved in turning are speed, feed, depth of cut, and tool geometry. Figure 31.23 shows the tool geometry along with the feed (f) and depth of cut (d). Table 31.12 gives the recommended cutting angles for HSS tools. The cutting speed (CS) is the surface speed in feet per minute (sfpm). The feed (f) is expressed in inches of tool advance per revolution of the spindle (ipr). The depth of cut (d) is expressed in inches. Table 31.13 gives the recommended speeds while using HSS tools and carbides for the case of finishing and rough machining. The cutting speed (fpm) is calculated by

$$CS = \frac{\pi DN}{12} \text{ fpm}$$

where D is the workpiece diameter in inches and N is the spindle revolutions per minute.

The tool advancing rate is $F = f \times N$ ipm. The machining time (T_1) required to turn a workpiece of length L in. is calculated from

$$T_1 = \frac{L}{F} \quad \text{min}$$

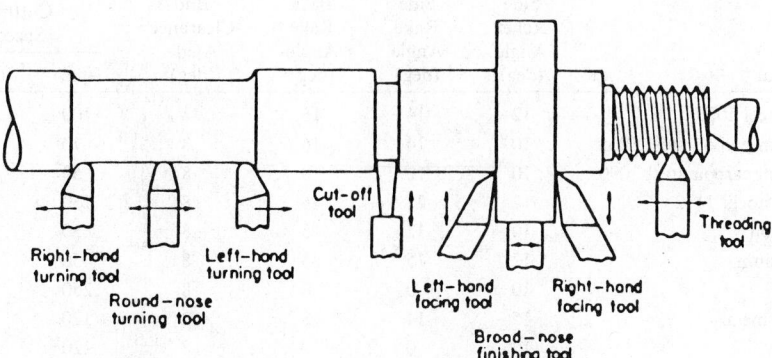

Fig. 31.22 Common lathe operations.

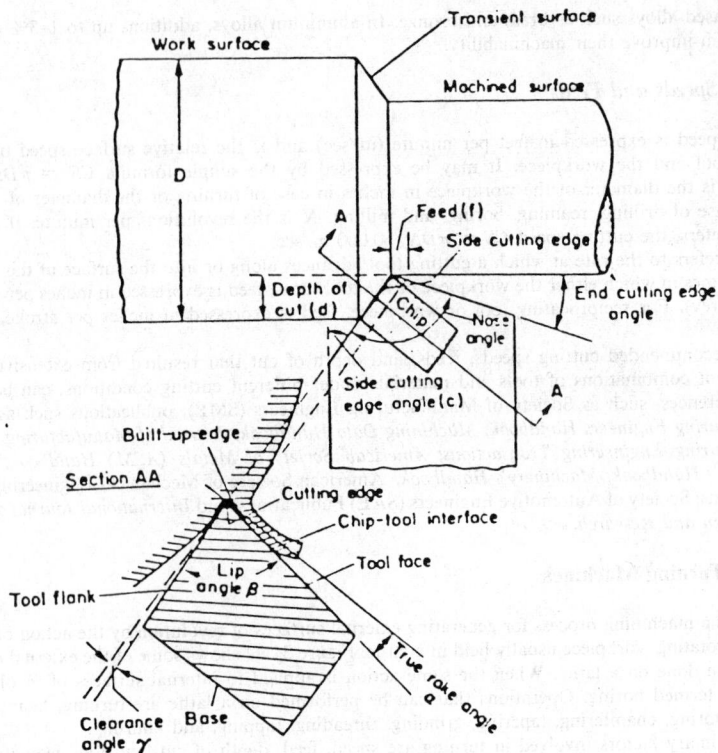

Fig. 31.23 Tool geometry.

The machining time (T_2) required to face a workpiece of diameter D is given by

$$T_2 = \frac{D/2}{F} \quad \text{min}$$

The rate of metal removal is given by

$$Q = 12f \times d \times \text{CS} \quad \text{in.}^3/\text{min}$$

Lathe Size

The size of a lathe is specified in terms of the diameter of the work it will swing and the workpiece length it can accommodate.

Table 31.12 Tool Angles and Cutting Speeds for HSS Tools

Material	Side Relief Angle (deg)	Side Rake Angle (deg)	Back Rake Angle (deg)	End Clearance Angle (deg)	Cutting Speed	
					(fpm)	(m/sec)
Mild steel 1020	12	14	16	8	100	0.5
Medium carbon steel 1035	10	14	16	8	70	0.4
Medium carbon steel 1090	10	12	8	8	50	0.3
Screw stock, 1112	12	22	16	8	150	0.8
Cast iron	10	12	5	8	50	0.3
Aluminum	12	15	35	8	450	2.3
Brass	10	0	0	8	250	1.3
Monel metal	15	14	8	12	120	0.6
Plastics	12	0	0	8	120	0.6
Fiber	15	0	0	12	80	0.4

Table 31.13 Typical Cutting Speeds

Material	High-Speed Steel		Carbide	
	Finish[a]	Rough[b]	Finish[a]	Rough[b]
Free cutting steels, 1112, 1315	250–350 (1.3–1.8)	80–150 (0.4–0.8)	600–750 (3.0–3.8)	350–450 (1.8–2.3)
Carbon steels, 1010, 1025	225–300 (1.1–1.5)	75–125 (0.4–0.6)	550–700 (2.8–3.5)	300–400 (1.5–2.0)
Medium steels, 1030, 1050	200–275 (1.0–1.4)	70–120 (0.4–0.6)	450–600 (2.3–3.0)	250–350 (1.3–1.8)
Nickel steels, 2330	200–275 (1.0–1.4)	70–110 (0.4–0.6)	425–550 (2.1–2.8)	225–325 (1.1–1.6)
Chromium nickel, 3120, 5140	150–200 (0.8–1.0)	50–75 (0.3–0.4)	325–425 (1.7–2.1)	175–260 (0.9–1.3)
Soft gray cast iron	120–150 (0.6–0.8)	75–90 (0.4–0.5)	350–450 (1.8–2.3)	200–250 (1.0–1.3)
Brass, normal	275–350 (1.4–1.8)	150–225 (0.8–1.1)	600–700 (3.0–3.5)	400–500 (2.0–2.5)
Aluminum	225–350 (1.1–1.8)	100–150 (0.5–0.8)	450–700 (2.3–3.5)	200–300 (1.0–1.5)
Plastics	300–500 (1.5–2.5)	100–200 (0.5–1.0)	400–650 (2.0–3.3)	150–250 (0.8–1.3)

[a] Cut depth, 0.015–0.095 in. (0.38–2.39 mm); feed, 0.005–0.015 ipr (0.13–0.38 mm/rev).
[b] Cut depth, 0.187–0.375 in. (4.75–9.53 mm); feed, 0.030–0.050 ipr (0.75–1.27 mm/rev).

The main types of lathes are engine, turret, single-spindle automatic, automatic screw machine, multispindle automatic, multistation machines, boring, vertical, and tracer. The level of automation can range from semiautomatic to tape-controlled machining centers. The *engine lathe* is considered to be the basic turning machine. Its main components are the head stock, bed, carriage, and quick-change gear box. The *turret lathe* has a turret mounted in place of the tailstock. Tools are mounted on the turret and can be indexed so that the appropriate tool is brought to the work when required. The *automatic screw machine* has five radially mounted tools that are cam-controlled. The stock can be made to feed as it is being cut, thus any desired cylindrical shape may be generated. The *tracer lathes* are basically the same as engine or turret lathes, but a two- or three-dimensional template is used to control the path of the cutting tool. The *numerically controlled (NC) machines* are capable of producing complicated design with great degree of accuracy and repetition. The control provides automatic functioning of speed, feed, depth of cut, tool path, turret indexing, oil application, and other necessary functions.

Break-Even (BE) Conditions

The selection of a specific machine for the production of a required quantity q must be done in a way to achieve minimum cost per unit produced. If the incremental setup cost is given by ΔC_t, C_1 is the machining cost per unit on the first machine, and C_2 is the machining cost for the second machine, the break-even point will be calculated as follows:

$$BE = \Delta C_t / (C_1 - C_2)$$

31.3.4 Drilling Machines

Drills are used as the basic method of producing holes in a wide variety of materials. Figure 31.24 indicates the nomenclature of a standard twist drill and its comparison with a single-point tool. Knowledge of the thrust force and torque developed in the drilling process is important for design consideration. Figure 31.25 shows the forces developed during the drilling process. From the force diagram the thrust force must be greater than $2Py + Py$ to include the friction on the sides and to be able to penetrate in the metal. The torque required is equal to $P_z X$. It is reported in the *Tool and Manufacturing Engineers Handbook* that the following relations reasonably estimate the torque and thrust requirements of sharp twist drills of various sizes and designs.

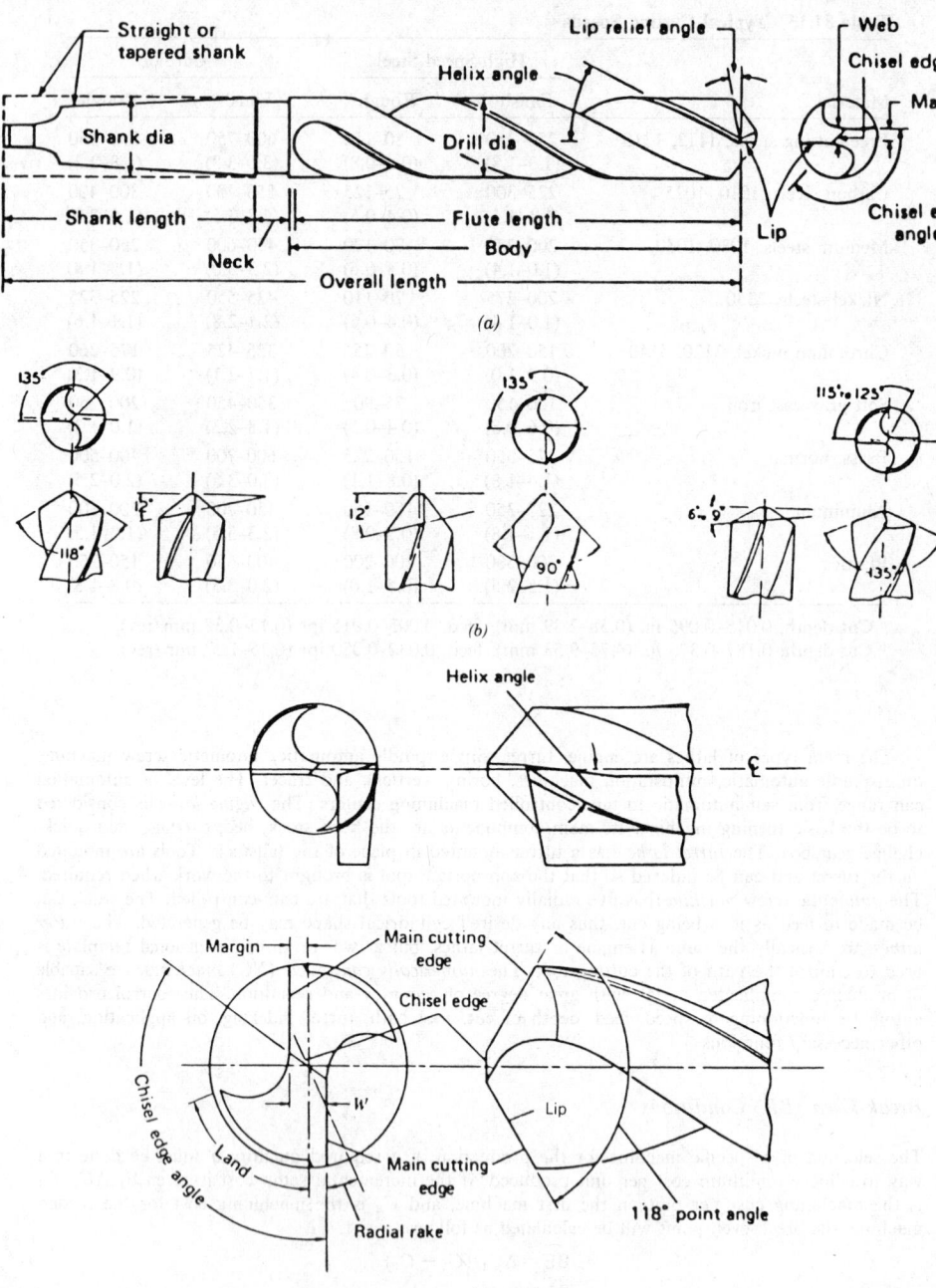

Fig. 31.24   Drill geometry.

For torque

$$M = kf^{0.8}d^{1.8}A \quad \text{in.-lb}$$

For thrust

$$T = 2kf^{0.8}d^{0.8}B + kd^2E \quad \text{lb}$$

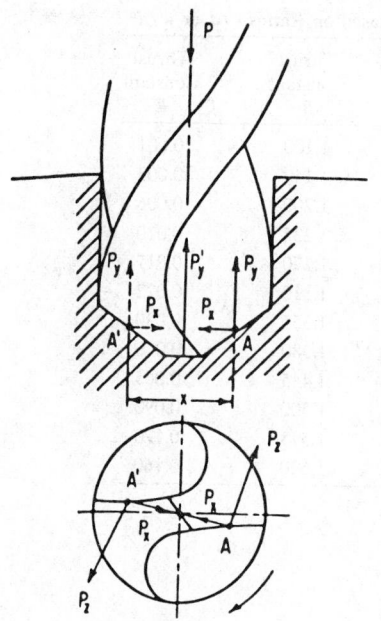

Fig. 31.25 Thrust forces and torque produced by a twist drill.

For horsepower

$$\text{hp} = MN/63.025$$

where k = work-material constant
  f = drill feed, ipr
  d = drill diameter, in.
A,B,E = design constants
  N = drill speed, rpm

Tables 31.14 and 31.15 give the constants used with the previous equations. Cutting speed at the surface is usually taken as 80% of turning speeds and is given by

$$\text{CS} = \frac{\pi d N}{12} \quad \text{fpm}$$

or

$$\text{CS} = \frac{\pi d_1 N}{60,000} \quad \text{m/sec}$$

where d_1 is given in millimeters.

Table 31.14 Work-Material Constants for Calculating Torque and Thrust

Work Material	K
Steel, 200 Bhn	24,000
Steel, 300 Bhn	31,000
Steel, 400 Bhn	34,000
Most aluminum alloys	7,000
Most magnesium alloys	4,000
Most brasses	14,000
Leaded brass	7,000

Table 31.15   Torque and Thrust Constants Based on Ratios c/d or w/d^a

c/d	Approximate w/d	Torque Constant A	Thrust Constant B	Thrust Constant E
0.03	0.025	1.000	1.100	0.001
0.05	0.045	1.005	1.140	0.003
0.08	0.070	1.015	1.200	0.006
0.10	0.085	1.020	1.235	0.010
0.13	0.110	1.040	1.270	0.017
0.15	0.130	1.080	1.310	0.022
0.18	0.155	1.085	1.355	0.030
0.20	0.175	1.105	1.380	0.040
0.25	0.220	1.155	1.445	0.065
0.30	0.260	1.235	1.500	0.090
0.35	0.300	1.310	1.575	0.120
0.40	0.350	1.395	1.620	0.160

^{a} c = chisel-edge length, in.
  d = drill diameter, in.
  w = web thickness, in.

Along the cutting edge of the drill the cutting speed is reduced toward the center as the diameter is reduced. The cutting speed is actually zero at the center. To avoid the region of very low speed and to reduce high thrust forces that might affect the alignment of the finished hole, a pilot hole is usually drilled before drilling holes of medium and large sizes.

The advancing rate is

$$F = f \times N \quad \text{ipm}$$

The recommended feeds are given in Table 31.16.

Table 31.16   Recommended Feeds for Drills

Diameter		Feed	
(in.)	(mm)	(ipr)	(mm/rev)
Under ⅛	3.2	0.001–0.002	0.03–0.05
⅛–¼	3.2–6.4	0.002–0.004	0.05–0.10
¼–½	6.4–12.7	0.004–0.007	0.10–0.18
½–1	12.7–25.4	0.007–0.015	0.18–0.38
Over 1	25.4	0.015–0.025	0.38–0.64

The time (T) required to drill a hole of depth h is given by

$$T = \frac{h + 0.3d}{F} \quad \text{min}$$

The extra distance $0.3d$ is approximately equal to the distance from the tip to the effective diameter of the tool. The rate of metal removal in case of blind holes is given by

$$Q = \left(\frac{\pi d^2}{4}\right) F \quad \text{in.}^3/\text{min}$$

For the case of drilling with a pilot hole

$$Q = \frac{\pi}{4}(d^2 d_p^2) F$$

$$= \frac{\pi}{4}(d + d_p)(d - d_p) F \quad \text{in.}^3/\text{min}$$

where d_p is pilot hole diameter in inches. When torque is unknown, the horsepower requirement can be calculated by

$$HP_c = QCW(HP_u) \quad hp$$

C, W, HP_u are given in previous sections.

Accuracy of Drills

The accuracy of holes drilled with a two-fluted twist drill is influenced by many factors, which include the accuracy of the drill point; the size of the drill; the chisel edge; the jigs used; the workpiece material; the cutting fluid used; the rigidity and accuracy of the machine used; and the cutting speed. Usually, when drilling most materials, the diameter of the drilled holes will be oversize. Table 31.17 provides the results of tests reported by The Metal Cutting Tool Institute for holes drilled in steel and cast iron.

Table 31.17   Oversize Diameters in Drilling

Drill Diameter (in.)	Amount Oversize (in.)		
	Average Max.	Mean	Average Min.
1/16	0.002	0.0015	0.001
1/8	0.0045	0.003	0.001
1/4	0.0065	0.004	0.0025
1/2	0.008	0.005	0.003
3/4	0.008	0.005	0.003
1	0.009	0.007	0.004

Gun drills differ from conventional drills in that they are usually made with a single flute. A hole provides a passageway for pressurized coolant, which serves as a means of both keeping the cutting edge cool and flushing out the chips, especially in deep cuts.

Spade drills (Fig. 31.26) are made by inserting a spade-shaped blade into a shank. Some advantages of spade drills are (1) efficiency in making holes up to 15 in. in diameter; (2) low cost, only the insert is replaced; (3) deep hole drilling; and (4) easiness of chip breaking and removal.

Trepanning is a machining process for producing a circular hole, groove, disk, cylinder, or tube from solid stock. The process is accomplished by a tool containing one or more cutters, usually single

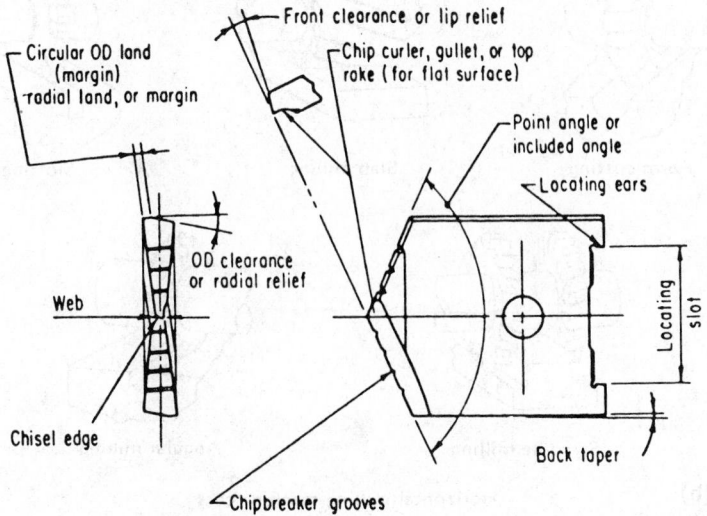

Fig. 31.26 Spade-drill blade elements.

point, revolving around a center. The advantages of trepanning are (1) the central core left is solid material, not chips, which can be used in later work, and (2) the power required to produce a given hole diameter is highly reduced because only the annulus is actually cut.

Reaming, boring, counterboring, centering and countersinking, spotfacing, tapping, and chamfering processes can be done on drills. Microdrilling and submicrodrilling achieve holes in the range of 0.000025 to 0.20 in. diameter.

Drilling machines are usually classified in the following manner: (1) bench: plain or sensitive; (2) upright: single-spindle or turret; (3) radial; (4) gang; (5) multispindle; (6) deep-hole: vertical or horizontal; and (7) transfer.

31.3.5 Milling Processes

The milling machines use a rotary multitooth cutter that can be designed to mill flat or irregularly shaped surfaces, cut gears, generate helical shapes, drill, bore, or do slotting work. Milling machines are classified broadly as vertical or horizontal. Figure 31.27 shows some of the operations that are done on both types. Figure 31.28 indicates the nomenclature of the milling cutter.

Feed in milling (F) is specified in inches per minute, but it is determined from the amount each tooth can remove or feed per tooth. (f_t). The feed in in./min is calculated from

$$F = f_t \cdot n \cdot N \quad \text{in./min}$$

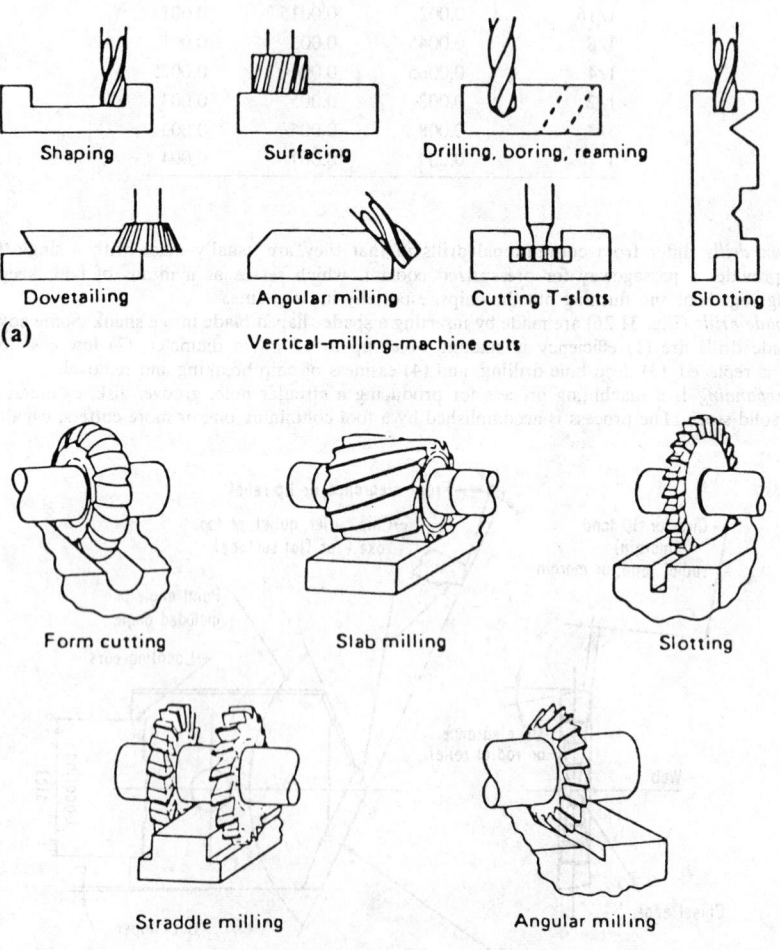

(a)

Vertical-milling-machine cuts

(b)

Horizontal-milling-machine cuts

Fig. 31.27 Application of (a) vertical and (b) horizontal milling machines.

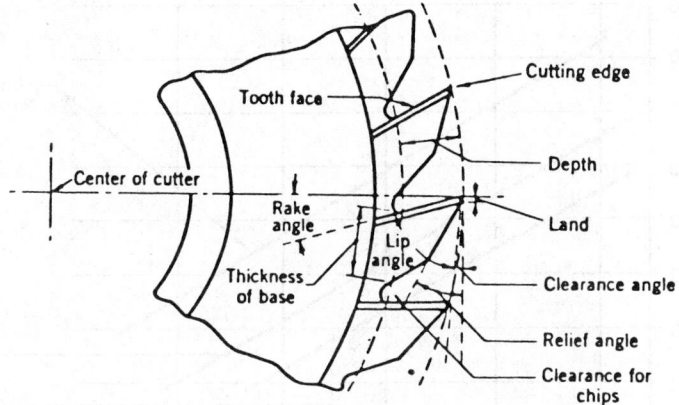

Fig. 31.28 Milling cutter with nomenclature indicated.

where n is number of teeth in cutter, and N is the rpm. Table 31.18 gives the recommended f_t for carbides and HSS tools. The cutting speed CS is calculated from

$$CS = \frac{\pi DN}{12} \text{ fpm}$$

where D is the tool diameter in inches.

Table 31.18 Recommended Feeds per Tooth for Milling Steel with Carbide and HSS Cutters

	Feed per Tooth	
Type of Milling	Carbides	HSS
Face	0.008–0.015	0.010
Side or straddle	0.008–0.012	0.006
Slab	0.008–0.012	0.008
Slotting	0.006–0.010	0.006
Slitting saw	0.003–0.006	0.003

Figure 31.29 shows the recommended cutting speed while using carbide and Table 31.19 gives the recommended cutting speeds while using HSS. The relationship between cutter rotation and feed direction is shown in Fig. 31.30. In climb milling or down milling, the chips are cut to maximum thickness at initial engagement and decrease to zero thickness at the end of engagement. In conventional or up milling, the reverse occurs. Because of the initial impact, climb milling requires rigid machines with backlash eliminators.

The rate of metal removal $Q = F \cdot w \cdot d$, where w is the width of cut and d is the depth of cut. The horsepower required for milling is given by

$$HP_c = HP_u \times Q$$

Machine horsepower is determined by

$$HP_m = \frac{HP_c}{\eta} + HP_i$$

where η = efficiency of machine
      HP_i = idle horsepower

The time required for milling is equal to distance required to be traveled by the cutter to complete the cut (L_1) divided by the feed rate F. L_1 is equal to the length of cut (L) plus cutter approach A and the overtravel OT. The machining time T is calculated from

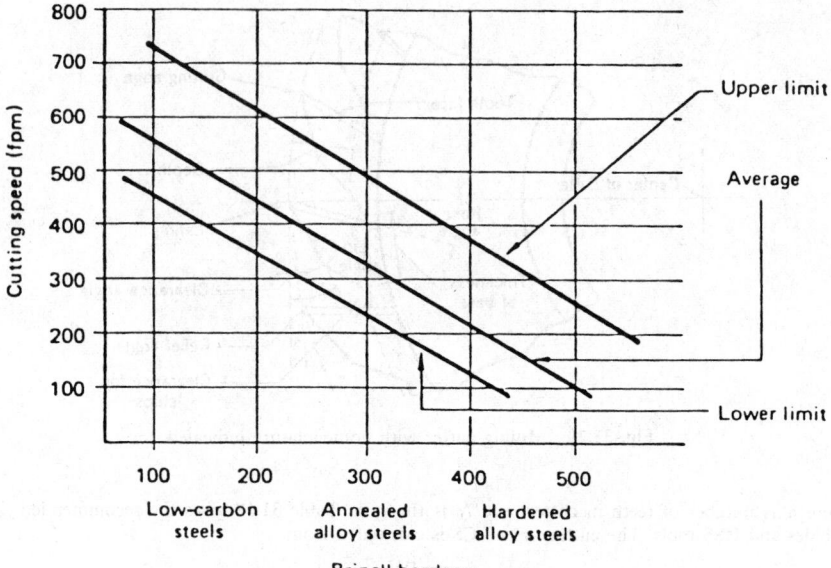

Fig. 31.29 Suggested cutting speeds for carbide milling cutters.

$$T = \frac{L + A + OT}{F} \quad \text{min}$$

As indicated in Fig. 31.31, in peripheral milling A is equal to $\sqrt{dD - d^2}$, and in face milling is $D/2 - \sqrt{(D^2 - W^2)}/4$. Overtravel (OT) depends on the specific milling operation.

Table 31.19 Table of Cutting Speeds (sfpm)

	HSS Tools		Carbide-Tipped Tools	
Work Material	Rough Mill	Finish Mill	Rough Mill	Finish Mill
Cast iron	50–60	80–110	180–200	350–400
Semisteel	40–50	65–90	140–160	250–300
Malleable iron	80–100	110–130	250–300	400–500
Cast steel	45–60	70–90	150–180	200–250
Copper	100–150	150–200	600	1000
Brass	200–300	200–300	600–1000	600–1000
Bronze	100–150	150–180	600	1000
Aluminum	400	700	800	1000
Magnesium	600–800	1000–1500	1000–1500	1000–5000
SAE steels				
1020 (coarse feed)	60–80	60–80	300	300
1020 (fine feed)	100–120	100–120	450	450
1035	75–90	90–120	250	250
X-1315	175–200	175–200	400–500	400–500
1050	60–80	100	200	200
2315	90–110	90–110	300	300
3150	50–60	70–90	200	200
4150	40–50	70–90	200	200
4340	40–50	60–70	200	200
Stainless steel	60–80	100–120	240–300	240–300
Titanium	30–70		200–350	

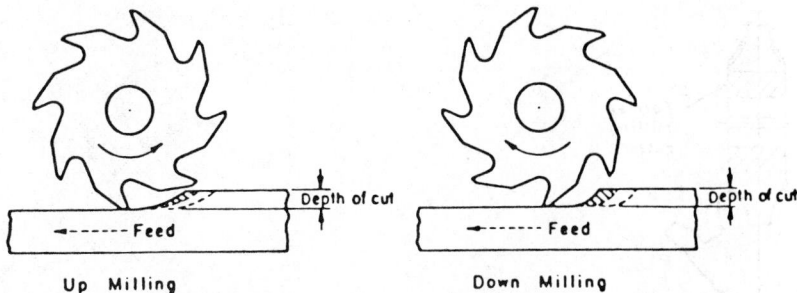

Fig. 31.30 Cutting action in up milling and down milling.

The milling machine is designated according to the longitudinal table travel. Milling machines are built in different types, some of which are (1) column-and-knee: vertical, horizontal, universal, and ram; (2) bed type, multispindle, (3) planer, (4) special, turret, profilers, and duplicators; and (5) numerically controlled.

31.3.6 Gear Manufacturing

Gears are made by various methods such as machining, rolling, extrusion, blanking, powder metallurgy, casting, or forging. Machining still is the unsurpassed method of producing gears of all types and sizes with high accuracy. Roll forming can be used only on ductile materials; however, it has been highly developed and widely adopted in recent years.

Machining Methods

There are three basic methods for machining gears: form cutting, template machining, and generating processes. Form cutting uses the principle illustrated in Fig. 31.32. The equipment and cutters required are relatively simple, and standard machines, usually milling, are often used. Theoretically, there should be different shaped cutters for each size of gear for a given pitch, as there is a slight change in the curvature of the involute. However, one cutter can be used for several gears having different numbers of teeth without much sacrifice in their operating action. The eight standard involute cutters are listed in Table 31.20. On the milling machine, the index or dividing head is used to rotate the gear blank through a certain number of degrees after each cut. The rule to use is: turns of index handle = 40/ *N*, where *N* is the number of teeth. Form cutting is usually slow. *Template machining* utilizes a simple, single-point cutting tool that is guided by a template. However, the equipment is specialized, and the method is seldom used except for making large bevel gears. The *generating process* is used to produce most high-quality gears. This process is based on the principle that any two involute gears, or any gear and a rack, of the same diametral pitch will mesh together. Applying this principle, one of the gears (or the rack) is made into a cutter by proper sharpening and is used to cut into a mating gear blank and thus generate teeth on the blank. Gear shapers, gear-hobbing machines, and bevel-gear generating machines are good examples of the gear generating machines.

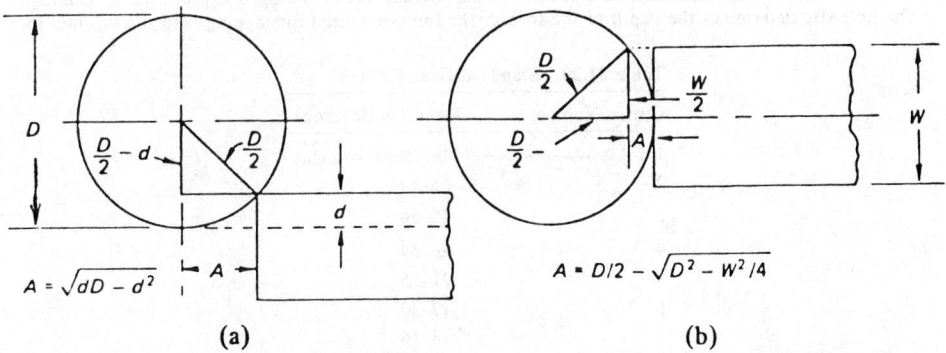

Fig. 31.31 Allowance for approach on (*a*) plain or slot milling and (*b*) face milling.

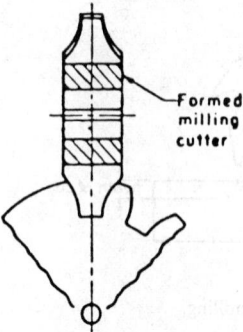

Formed
milling
cutter

Fig. 31.32   Basic method of maching a gear by form cutting.

Cold-Roll Forming of Gears

This process employs hardened forming rolls that are fed inward gradually during several revolutions. Owing to the rapidity of the process, its ease of being mechanized, and less material needed, and because skilled labor is not required, roll-formed gears are rapidly replacing machined gears whenever the process can be used. Small gears often are made by rolling a length of shaft and then slicing the individual gear units.

Gear Finishing

To operate efficiently and have satisfactory life, gears must have accurate tooth profile, and smooth and hard faces. Gears are usually produced from relatively soft blanks and are subsequently heat treated to obtain greater hardness, if it is required. Such heat treatment usually results in some slight distortion and surface roughness. *Gear shaving* is the most commonly used method for gear finishing. The gear is run, at high speed, in contact with a shaving tool. Such a tool is a very accurate, hardened, and ground gear that contains a number of peripheral gashes, or grooves, thus forming a series of sharp cutting edges on each tooth. This process usually requires less than a minute to produce a very accurate tooth profile. *Gear burnishing* is limited to unhardened gears. The unhardened gear is rolled under pressure contact with three hardened, accurately formed, burnishing gears. *Grinding and lapping* are used to obtain very accurate teeth on hardened gears.

31.3.7   Thread Cutting and Forming

Three basic methods are used for the manufacturing of threads: cutting, rolling, and casting. Die casting and molding of plastics are good examples of casting. The largest number of threads are made by rolling, even though it is restricted to standardized and simple parts, and ductile materials. Large number of threads are cut by the following methods: (1) on an engine lathe; (2) with a stock and die (manual, external); (3) with a tap and holder (internal); (4) with automatic dies (external) and collapsible taps (internal); (5) by milling; and (6) by grinding.

Internal Threads

In most cases the hole that must be made before an internal thread is tapped is produced by drilling. The hole size determines the depth of the thread, the forces required for tapping, and the tap life. In

Table 31.20   Standard Gear Cutters

Cutter Number	Gear Tooth Range
1	135 teeth to rack
2	55–134
3	35–54
4	26–34
5	21–25
6	17–20
7	14–16
8	12–13

most applications a drill size is selected that will result in a thread having about 75% of full thread depth. This practice makes tapping much easier, increases the tap's life, and the resulting strength is only slightly reduced. Table 31.21 gives the drill sizes used to produce 75% thread depth for several sizes of UNC threads. The feed of a tap depends on the lead of the screw and is equal to l/lead ipr.

Table 31.21 Recommended Tap-Drill Sizes for Standard Screw-Thread Pitches (American National Coarse-Thread Series)

Number or Diameter	Threads per Inch	Outside Diameter of Screw	Tap Drill Sizes	Decimal Equivalent of Drill
6	32	0.138	36	0.1065
8	32	0.164	29	0.1360
10	24	0.190	25	0.1495
12	24	0.216	16	0.1770
¼	20	0.250	7	0.2010
⅜	16	0.375	⁵⁄₁₆	0.3125
½	13	0.500	²⁷⁄₆₄	0.4219
¾	10	0.750	²¹⁄₃₂	0.6562
1	8	1.000	⅞	0.875

Cutting speeds depend on many factors such as (1) material hardness; (2) depth of cut; (3) thread profile; (4) tooth depth; (5) hole depth; (6) fineness of pitch; and (7) cutting fluid. Cutting speeds can range from 3 ft/min (1 m/min) for high strength steels to 150 ft/min (45 m/min) for aluminum alloys. Long screws with different configurations can be cut successfully on milling machines, as in Fig. 31.33. The feed per tooth is given by the following equation:

$$f_t = \frac{\pi d S}{n N}$$

where d = diameter of thread
n = number of teeth in cutter
N = rpm of cutter
S = rpm of work

Thread Rolling

In thread rolling, the metal on the cylindrical blank is cold-forged under considerable pressure by either rotating cylindrical dies or reciprocating flat dies. The advantages of thread rolling include (1) improved strength, (2) smooth surface finish, (3) less material used (~19%), and (4) high production rate. The limitations of thread rolling processes are (1) blank tolerance must be close, (2) economical only for large quantities, (3) limited to external threads, and (4) applicable only for ductile materials, less than Rockwell C37.

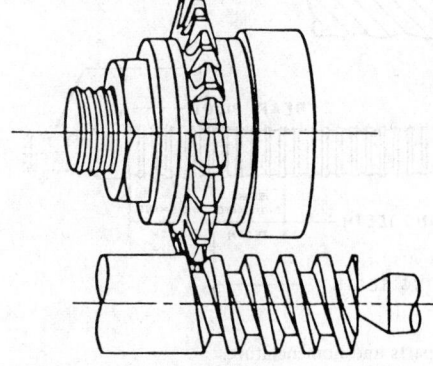

Fig. 31.33 Single-thread milling cutter.

31.3.8   Broaching

Broaching is unique in that it is the only one of the basic machining processes in which the feed of the cutting edges is built into the tool. The machined surface always is the inverse of the profile of the broach. The process usually is completed in a single, linear stroke. A broach is composed of a series of single-point cutting edges projecting from a rigid bar, with successive edges protruding farther from the axis of the bar. Figure 31.34 illustrates the parts and nomenclature of the broach. Most broaching machines are driven hydraulically, and they are of the pull or push type.

The maximum force an internal pull broach can withstand without damage is given by

$$P = \frac{A_r F_y}{s} \quad \text{lb}$$

where A_r = minimum tool section, in.^{2}

$\quad\quad F_y$ = tensile yield strength of tool steel, psi

$\quad\quad s$ = safety factor

The maximum push force is determined by the minimum tool diameter D_r, the length of the broach L, and the minimum compressive yield strength F_y. The ratio L/D_r should be less than 25 so that the tool will not bend under load. The maximum allowable pushing force is given by

$$P = \frac{A_r F_y'}{s} \quad \text{lb}$$

where F_y' is minimum compressive yield strength. If the L/D_r ratio is greater than 25 (long broach), the *Tool and Manufacturing Engineers Handbook* gives the following formula:

$$P = \frac{5.6 \times 10^7 \, D_r^4}{s L^2} \quad \text{lb}$$

D_r and L are given in inches.

Broaching speeds are relatively low, seldom exceeding 50 fpm, but, because a surface usually is completed in one stroke, the productivity is high.

31.3.9   Shaping, Planing, and Slotting

The shaping and planing operations generate surfaces with a single-point tool by a combination of a reciprocating motion along one axis and a feed motion normal to that axis (Fig. 31.35). Slots and limited inclined surfaces can also be produced. In shaping the tool is mounted on a reciprocating ram and the table is fed at each stroke of the ram. Planers handle large heavy workpieces. In planing the workpiece reciprocates and the feed increment is provided by moving the tool at each reciprocation. To reduce the lost time on the return stroke they are provided with a quick-return mechanism. For

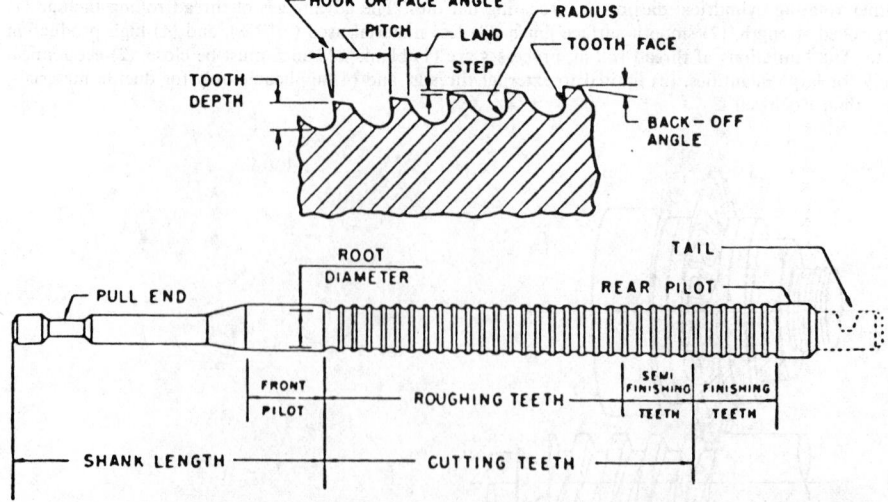

Fig. 31.34   Standard broach parts and nomenclature.

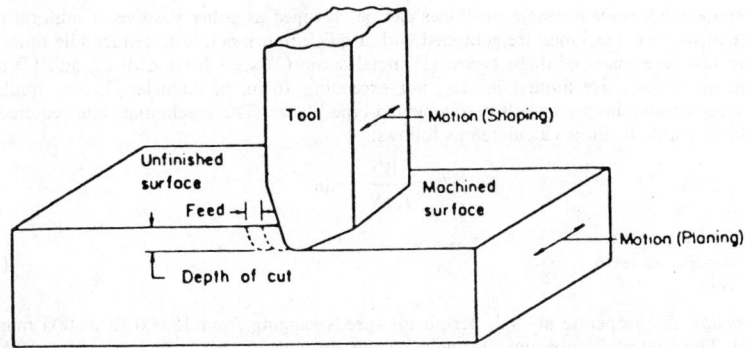

Fig. 31.35 Basic relationships of tool motion, feed, and depth of cut in shaping and planing.

mechanically driven shapers, the ratio of cutting time to return stroke averages 3:2, and for hydraulic shapers the ratio is 2:1. The average cutting speed may be determined by the following formula:

$$CS = \frac{LN}{12C} \quad \text{fpm}$$

where N = strokes per minute
$\quad\;\; L$ = stroke length in inches
$\quad\;\; C$ = cutting time ratio, cutting time divided by total time

For mechanically driven shapers, the cutting speed reduces to

$$CS = \frac{LN}{7.2} \quad \text{fpm}$$

or

$$CS = \frac{L_1 N}{600} \quad \text{m/min}$$

when L_1 is the stroke length in millimeters. For hydraulically driven shapers

$$CS = \frac{LN}{8} \quad \text{fpm}$$

or

$$CS = \frac{L_1 N}{666.7} \quad \text{m/min}$$

when L_1 is the stroke length in millimeters. The time T required to machine a workpiece of width W is calculated by

$$T = \frac{W}{N \times f}$$

when f is the feed in inches per stroke. The number of strokes (S) required to complete a job is then

$$S = \frac{W}{f}$$

The power required can be approximated by

$$HP_c = Kdf(CS)$$

where d = depth of cut, in.
$\quad\;\; CS$ = cutting speed, fpm
$\quad\;\; K$ = cutting constant, for medium cast iron, 3; free-cutting steel, 6; and bronze, 1.5

31.3.10 Sawing, Shearing, and Cutting-Off

Saws are among the most common of machine tools, even though the surfaces they produce often require further finishing operation. Saws have two general areas of application: (1) contouring and (2) cutting-off. There are three basic types of saws: (1) hacksaw; (2) circular; and (3) band saw.

The *reciprocating power hacksaw* machines can be classified as either positive or uniform-pressure feeds. Most of the new machines are equipped with a quick-return action to reduce idle time.

Circular saws are made of three types: (1) metal saws; (2) steel friction disks; and (3) abrasive disks. Solid metal saws are limited in size, not exceeding 16 in. in diameter. Large circular saws have either replaceable inserted teeth or segmented-type blades. The machining time required to cut a workpiece of width W in. is calculated as follows:

$$T = \frac{W}{f_t n N} \quad \text{min}$$

where f_t = feed per tooth
$\quad n$ = number of teeth
$\quad N$ = rpm

Steel friction disks operate at high peripheral speeds ranging from 18,000 to 25,000 fpm (90 to 125 m/sec). The heat of friction quickly melts a path through the part being cut. About 0.5 min is required to cut through a 24-in. I-beam.

Abrasive disks are mainly aluminum oxide grains or silicon carbide grains bonded together. They will cut ferrous or nonferrous metals. The finish and accuracy is better than steel friction blades, but they are limited in size compared to steel friction blades.

Band saw blades are of the continuous type. Band sawing can be used for cutting and contouring. Band-sawing machines operate with speeds that range from 50 to 1500 fpm. The time required to cut a workpiece of width W in. can be calculated as follows:

$$T = \frac{W}{12 f_t n V}$$

where f_t = feed, inches per tooth
$\quad n$ = number of teeth per inch
$\quad V$ = cutting speed, fpm

Cutting can also be achieved by band-friction cutting blades with a surface speed up to 15,000 fpm. Other band tools include band filing, diamond bands, abrasive bands, spiral bands, and special purpose bands.

31.3.11 Grinding, Abrasive Machining, and Finishing

Abrasive machining is the basic process in which chips are removed by very small edges of abrasive particles, usually synthetic. In many cases the abrasive particles are bonded into wheels of different shapes and sizes. When wheels are used mainly to produce accurate dimensions and smooth surfaces, the process is called *grinding*. When the primary objective is rapid metal removal to obtain a desired shape or approximate dimensions, it is termed *abrasive machining*. When fine abrasive particles are used to produce very smooth surfaces and to improve the metallurgical structure of the surface, the process is called *finishing*.

Synthetic Abrasives

Aluminum oxide (Al_2O_3) performs best on carbon and alloy steels, annealed malleable iron, hard bronze, and similar metals. Al_2O_3 wheels are not used in grinding very hard materials such as tungsten carbide because the grains will get dull prior to fracture. Common trade names for aluminum oxide abrasives are Alundum and Aloxite.

Silicon carbide (SiC) crystals are very hard, being about 9.5 on the Moh's scale, where diamond hardness is 10. SiC crystals are brittle, and this limits their use. Silicon carbide wheels are recommended for materials of low tensile strength, such as cast iron, brass, stone, rubber, leather, and cemented carbides.

Cubic boron nitride (CBN) is the second hardest natural or manmade substance. It is good for grinding hard and tough-hardened tool-and-die steels.

Diamonds may be classified as both natural and synthetic. Commercial diamonds are now manufactured in high, medium, and low impact strength.

Grain Size

To have uniform cutting action, abrasive grains are graded into various sizes, indicated by the numbers 4 to 600. The number indicates the number of openings per linear inch in a standard screen through which most of the particles of a particular size would pass. Grain sizes from 4 to 24 are termed coarse; 30 to 60, medium; and 70 to 600, fine. Fine grains produce smoother surfaces than coarse ones, but cannot remove as much metal.

Bonding materials have the following effects on the grinding process: (1) they determine the strength of the wheel and its maximum speed; (2) they determine whether the wheel is rigid or flexible; and (3) they determine the force available to pry the particles loose. If only a small force is needed to release the grains, the wheel is said to be soft. Hard wheels are recommended for soft materials, and soft wheels for hard materials. The bonding materials used are vitrified, silicate, rubber, resinoid, shellac, and oxychloride.

Structure or Grain Spacing

Structure relates to the spacing of the abrasive grain. Soft, ductile materials require a wide spacing to accommodate the relatively large chips. A fine finish requires a wheel with a close spacing. Figure 31.36 shows the standard system of grinding wheels as adopted by the American National Standards Institute.

Speeds

Wheel speed depends on the wheel type, bonding material, and operating conditions. Wheel speeds range between 4500 and 18,000 sfpm (1400 and 5500 m/min). Work speeds depend on type of material, grinding operation, and machine rigidity. Work speeds range between 15 and 200 fpm.

Feeds

Cross feed depends on the width of grinding wheel. For rough grinding the range is one-half to three-quarters of the width of the wheel. Finer feed is required for finishing, and it ranges between one-tenth and one-third of the width of the wheel.

Depth of Cut

Rough grinding conditions will dictate the maximum depth of cut. In the finishing operation the depth of cut is usually small, 0.0005–0.001 in. (0.013–0.025 mm). Good surface finish and close tolerance can be achieved by "sparking out" or letting the wheel run over the workpiece without increasing the depth of cut till sparks die out. *The grinding ratio, G-ratio,* refers to the ratio of the cubic inches of stock removed to the cubic inches of grinding wheel worn away. G-ratio is important in calculating grinding and abrasive machining cost, which may be calculated by the following formula:

$$C = \frac{C_a}{G} + \frac{L}{tq}$$

where C = specific cost of removing a cubic inch of material
C_a = cost of abrasive, \$/in.^{3}
G = grinding ratio
L = labor and overhead charge, \$/hr
q = machining rate, in.^{3}/hr
t = fraction of time the wheel is in contact with workpiece

Grinding and abrasive machines include (1) surface grinders, reciprocating or rotating table; (2) cylindrical grinders, work between centers, centerless, crankshaft, thread and gear form work, and internal and other special applications; (3) jig grinders; (4) tool and cutter grinders; (5) snagging, foundry rough work; (6) cutting off and profiling; (7) abrasive grinding, belt, disk, and loose grit; and (8) mass media, barrell tumbling and vibratory.

Surface Finishing

Finishing processes produce an extra fine surface finish; in addition, tool marks are removed and very close tolerances are achieved. Some of these processes are discussed briefly.

Honing is a low-velocity abrading process. It uses fine abrasive stones to remove very small amounts of metals usually left from previous grinding processes. The amount of metal removed usually is less than 0.005 in. (0.13 mm). Because of low cutting speeds, heat and pressure are minimized, resulting in excellent sizing and metallurgical control.

Lapping is an abrasive surface-finishing process wherein fine abrasive particles are charged in some sort of a vehicle, such as grease, oil, or water, and are embedded into a soft material, called a lap. Metal laps must be softer than the work and are usually made of close-grained gray cast iron. Other materials such as steel, copper, and wood are used where cast iron is not suitable. As the charged

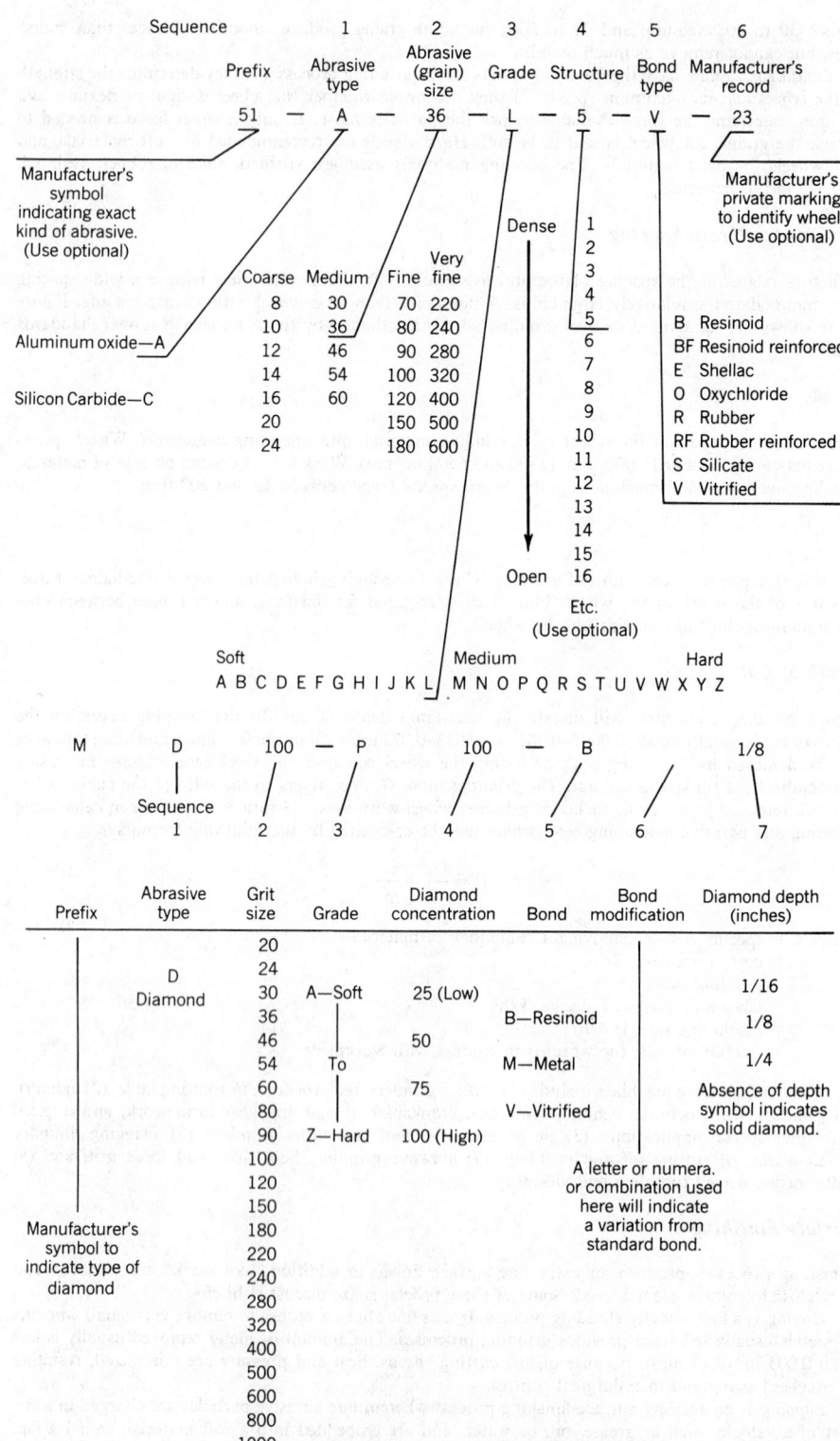

Fig. 31.36 Standard system for grinding wheels.

lap is rubbed against a surface, small amounts of material are removed from the harder surface. The amount of material removed is usually less than 0.001 in. (0.03 mm).

Superfinishing is a surface-improving process that removes undesirable fragmentation, leaving a base of solid crystalline metal. It uses fine abrasive stones like honing, but differs in the type of motion. Very rapid short strokes, very light pressure, and low-viscosity lubricant–coolant are used in superfinishing. It is essentially a finishing process and not a dimensional one, and can be superimposed on other finishing operations.

Ultrasonic Machining

In ultrasonic machining, material is removed from the workpiece through high-velocity bombardment by abrasive particles, in the form of a slurry, through the action of an ultrasonic transducer.

Buffing

Buffing wheels are made from a variety of soft materials. The most widely used is muslin, however flannel, canvas, sisal, and heavy paper are used for special applications. Buffing is usually divided into two operations: (1) cutting down and (2) coloring. The first is used to smooth the surface and the second to produce a high luster. The abrasives used are extremely fine powders of aluminum oxide, tripoli (an amorphous silicon), crushed flint or quartz, silicon carbide, and red rouge (iron oxide). Buffing speeds range between 6000 and 12,000 fpm.

Electropolishing is the reverse of electroplating; that is, the work is the anode instead of the cathode and metal is removed rather than added.

31.3.12 Machining Plastics

Although most plastics are readily machined, because their properties vary so greatly, it is impossible to give instructions that are exactly correct for all. It is very important to remember some general characteristics that affect their machinability. First, all plastics are poor heat conductors. Consequently, little of the heat that results from chip formation will be conducted away through the material or carried away in the chips. As a result, cutting tools run very hot and may fail more rapidly than when cutting metal. Carbide tools frequently are more economical to use than HSS tools if cuts are of moderately long duration or if high-speed cutting is to be done.

Second, because considerable heat and high temperatures do develop at the point of cutting, thermoplastics tend to soften, swell, and bind or clog the cutting tool. Thermosetting plastics give less trouble in this regard.

Third, cutting tools should be kept very sharp at all times. Drilling is best done by means of straight-flute drills or by "dubbing" the cutting edge of a regular twist drill to produce a zero rank angle. Rotary files and burs, saws, and milling cutters should be run at high speeds in order to improve cooling, but with feed carefully adjusted to avoid jamming the gullets. In some cases coolants can be used advantageously if they do not discolor the plastic or cause gumming. Water, soluble oil and water, and weak solutions of sodium silicate in water are used. In turning and milling plastics diamond tools provide the best accuracy, surface finish, and uniformity of finish. Surface speeds of 500 to 600 fpm with feeds of 0.002–0.005 in. are typical.

Fourth, filled and laminated plastics usually are quite abrasive and may produce a fine dust that may be a health hazard.

31.4 NONTRADITIONAL MACHINING

Nontraditional machining is a generic designation applied to those material-removal processes that have recently emerged, have not been used extensively heretofore, or are new to the user. These processes are sometimes labeled nonconventional, layless, or nonmechanical.

They have been grouped for discussion according to their primary energy mode; that is, mechanical, electrical, thermal, or chemical, as shown in Table 31.22.

Nontraditional processes provide manufacturing engineers with additional choices or alternatives to the traditional or conventional mechanical and abrasive material-removal processes. Figure 31.37 and Table 31.23 demonstrate the relationships among the conventional and the nontraditional machining processes with respect to surface roughness, dimensional tolerance, and metal removal rate. It is important to note that carefully selected and properly applied nontraditional machining processes offer some unique capabilities and fresh opportunities for cost improvement in the field of material removal.

The Machinery Handbook, Vol. 2, published by Machinability Data Center, Department of Defense Information Analysis Center, is an excellent reference for nontraditional machining process values, ranges, and limitations.

Table 31.22 Current Commercially Available Nontraditional Material Removal Process

Mechanical

AFM	Abrasive flow machining
AJM	Abrasive jet machining
HDM	Hydrodynamic machining
LSG	Low-stress grinding
RUM	Rotary ultrasonic machining
TAM	Thermally assisted machining
TFM	Total form machining
USM	Ultrasonic machining
WJM	Water jet machining

Electrical

ECD	Electrochemical deburring
ECDG	Electrochemical discharge grinding
ECG	Electrochemical grinding
ECH	Electrochemical honing
ECM	Electrochemical machining
ECP	Electrochemical polishing
ECS	Electrochemical sharpening
ECT	Electrochemical turning
ES	Electro-stream™
STEM™	Shaped tube electrolytic machining

Thermal

EBM	Electron-beam machining
EDG	Electrical discharge grinding
EDM	Electrical discharge machining
EDS	Electrical discharge sawing
EDWC	Electrical discharge wire cutting
LBM	Laser-beam machining
LBT	Laser-beam torch
PBM	Plasma-beam machining

Chemical

CHM	Chemical machining
ELP	Electropolish
PCM	Photochemical machining
TCM	Thermochemical machining (or TEM, thermal energy method)

31.4.1 Abrasive Flow Machining

Abrasive flow machining (AFM) is the removal of material by a viscous, abrasive media flowing, under pressure, through or across a workpiece. Figure 31.38 contains a schematic presentation of the AFM process. The grit-loaded media of polymeric-base material is selected for its viscosity and type and proportion of grit, in order to suit the part shape and the intended action—deburring, polishing, and radiusing. Generally, the puttylike media is extruded through or over the workpiece with motion usually in both directions using from 1 to 100 flow reversals per fixture load. Aluminum oxide, silicon carbide, boron carbide, or diamond abrasives are used. The velocity of the extruded media is dependent on the principal parameters of viscosity, pressure, passage size, and length.

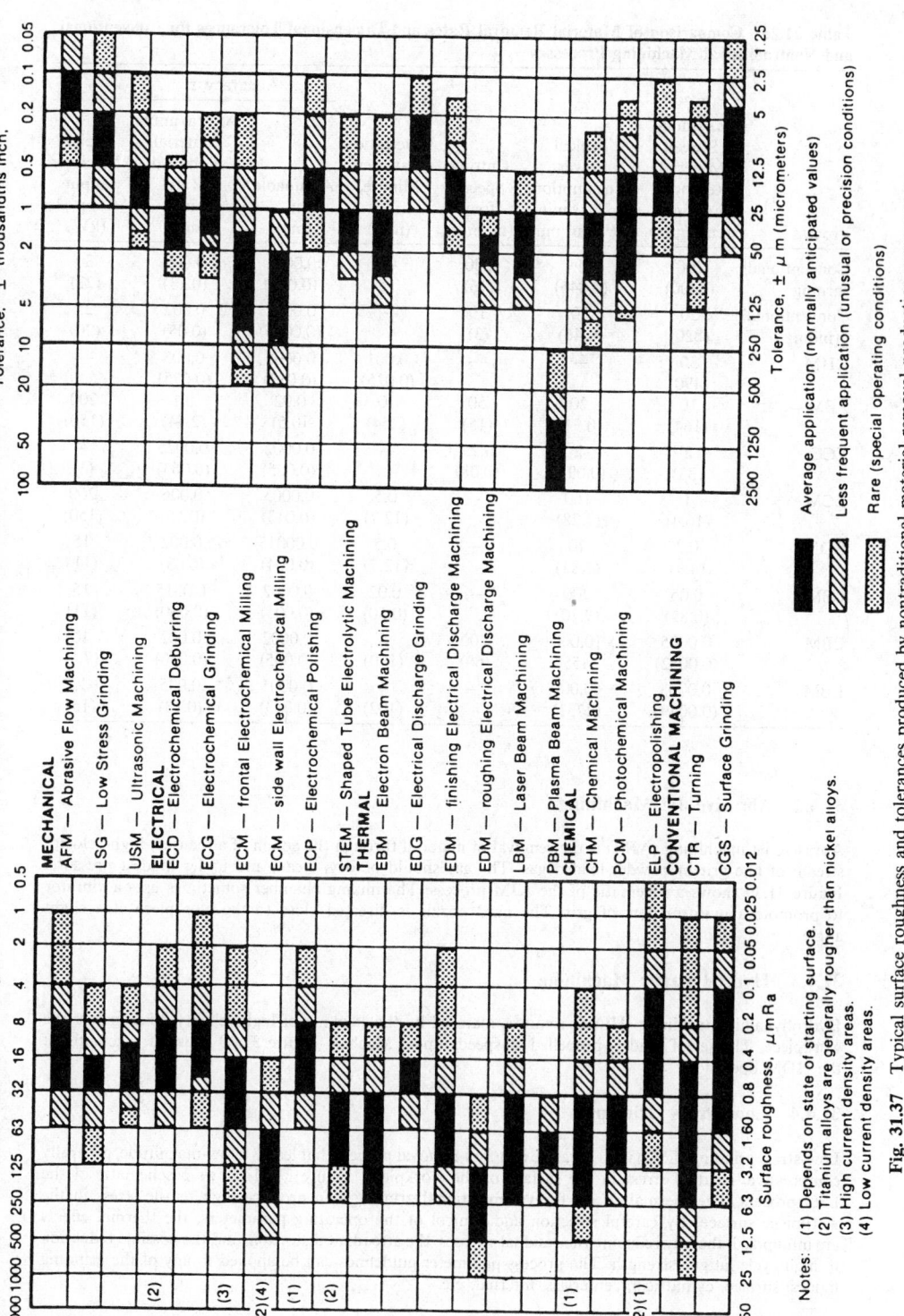

Surface roughness, μin Ra

Tolerance, ± thousandths inch.

MECHANICAL	
AFM	— Abrasive Flow Machining
LSG	— Low Stress Grinding
USM	— Ultrasonic Machining
ELECTRICAL	
ECD	— Electrochemical Deburring
ECG	— Electrochemical Grinding
ECM	— frontal Electrochemical Milling
ECM	— side wall Electrochemical Milling
ECP	— Electrochemical Polishing
STEM	— Shaped Tube Electrolytic Machining
THERMAL	
EBM	— Electron Beam Machining
EDG	— Electrical Discharge Grinding
EDM	— finishing Electrical Discharge Machining
EDM	— roughing Electrical Discharge Machining
LBM	— Laser Beam Machining
PBM	— Plasma Beam Machining
CHEMICAL	
CHM	— Chemical Machining
PCM	— Photochemical Machining
ELP	— Electropolishing
CONVENTIONAL MACHINING	
CTR	— Turning
CGS	— Surface Grinding

Notes: (1) Depends on state of starting surface.
(2) Titanium alloys are generally rougher than nickel alloys.
(3) High current density areas.
(4) Low current density areas.

Tolerance, ± μ m (micrometers)

■ Average application (normally anticipated values)

▨ Less frequent application (unusual or precision conditions)

▩ Rare (special operating conditions)

Fig. 31.37 Typical surface roughness and tolerances produced by nontraditional material removal production processes.

Table 31.23 Comparison of Material Removal Rates and Dimensional Tolerances for Conventional and Nontraditional Machining Processes

Process	Maximum Rate of Material Removal, in.³/min (cm³/min)	Typical Power Consumption, hp/in.³ min (kW/cm³ min)	Cutting Speed, fpm (m/min)	Penetration Rate per Minute, in. (mm)	Accuracy ± Attainable, in. (mm)	Accuracy ± At Maximum Material Removal Rate in. (mm)	Typical Machine Input hp (kW)
Conventional turning	200 (3300)	1 (0.046)	250 (76)	—	0.0002 (0.005)	0.005 (0.13)	30 (22)
Conventional grinding	50 (820)	10 (0.46)	10 (3)	—	0.0001 (0.0025)	0.002 (0.05)	25 (20)
CHM	30 (490)	—	—	0.001 (0.025)	0.0005 (0.013)	0.003 (0.075)	—
PBM	10 (164)	20 (0.91)	50 (15)	10 (254)	0.02 (0.5)	0.1 (2.54)	200 (150)
ECG	2 (33)	2 (0.091)	0.25 (0.08)	—	0.0002 (0.005)	0.0025 (0.063)	4 (3)
ECM	1 (16.4)	160 (7.28)	—	0.5 (12.7)	0.0005 (0.013)	0.006 (0.15)	200 (150)
EDM	0.3 (4.9)	40 (1.82)	—	0.5 (12.7)	0.00015 (0.004)	0.002 (0.05)	15 (11)
USM	0.05 (0.82)	200 (9.10)	—	0.02 (0.50)	0.0002 (0.005)	0.0015 (0.040)	15 (11)
EBM	0.0005 (0.0082)	10,000 (455)	200 (60)	6 (150)	0.0002 (0.005)	0.002 (0.050)	10 (7.5)
LBM	0.0003 (0.0049)	60,000 (2731)	—	4 (102)	0.0005 (0.013)	0.005 (0.13)	20 (15)

31.4.2 Abrasive Jet Machining

Abrasive jet machining (AJM) is the removal of material through the action of a focused, high-velocity stream of fine grit or powder-loaded gas. The gas should be dry, clean, and under modest pressure. Figure 31.39 shows a schematic of the AJM process. The mixing chamber sometimes uses a vibrator to promote a uniform flow of grit. The hard nozzle is directed close to the workpiece at a slight angle.

31.4.3 Hydrodynamic Machining

Hydrodynamic machining (HDM) removes material by the stroking of high-velocity fluid against the workpiece. The jet of fluid is propelled at speeds up to Mach 3. Figure 31.40 shows a schematic of the HDM operation.

31.4.4 Low-Stress Grinding

Low-stress grinding (LSG) is an abrasive material-removal process that leaves a low-magnitude, generally compressive, residual stress in the surface of the workpiece. Figure 31.41 shows a schematic of the LSG process. The thermal effects from conventional grinding can produce high tensile stress in the workpiece surface. By careful selection and control of the operating parameters, the thermal effects are minimized, thereby reducing the residual stresses, the attendant distortion, and the possible reduction in high-cycle fatigue strength. The process parameter guidelines can be applied to any of the grinding modes: surface, cylindrical, centerless, internal, etc.

31.4.5 Thermally Assisted Machining

Thermally assisted machining (TAM) is the addition of significant amounts of heat to the workpiece immediately prior to single-point cutting so that the material is softened, but the strength of the tool

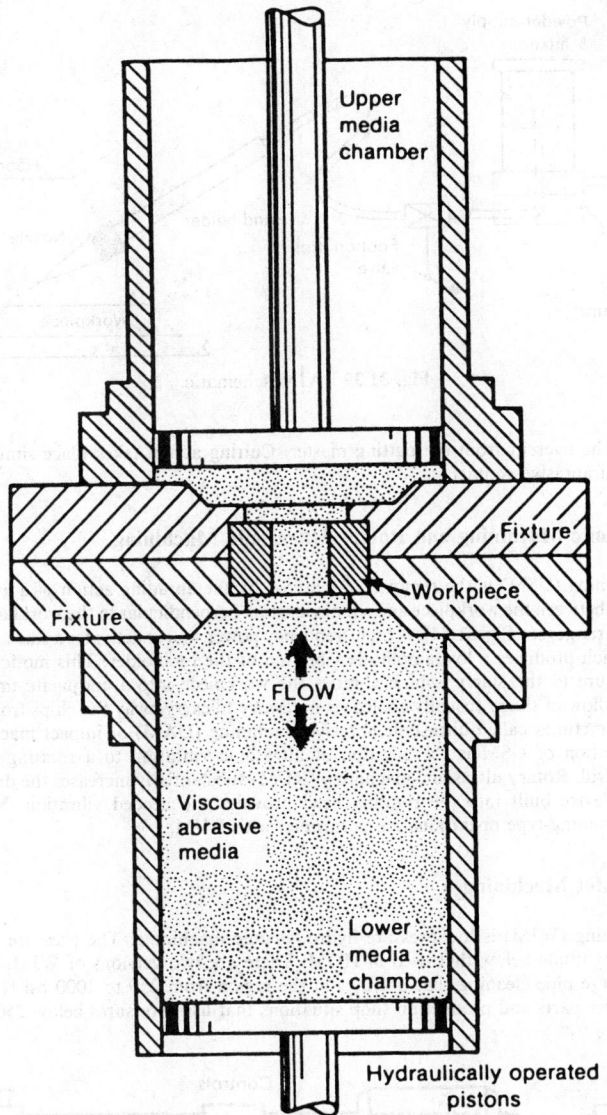

Fig. 31.38 AFM schematic.

bit is unimpaired (see Fig. 31.42). While resistive heating and induction heating offer possibilities, the plasma arc offers the most practical and effective method for external heating at the required rates. The plasma arc has a core temperature of 14,500°F (8000°C) and a surface temperature of 6500°F (3600°C). The torch can produce 2000°F (1100°C) in the workpiece in approximately one-quarter revolution of the workpiece between the point of application of the torch and the cutting tool.

31.4.6 Total Form Machining

Total form machining (TFM) is a process in which an abrasive master abrades its full three-dimensional shape into the workpiece by the application of force while a full-circle, orbiting motion is applied to the workpiece via the worktable (see Fig. 31.43). The cutting master is advanced into the work until the desired depth of cut is achieved. Uniformity of cutting is promoted by the fluid that continuously transports the abraded particles out of the working gap. Adjustment of the orbiting cam drive controls

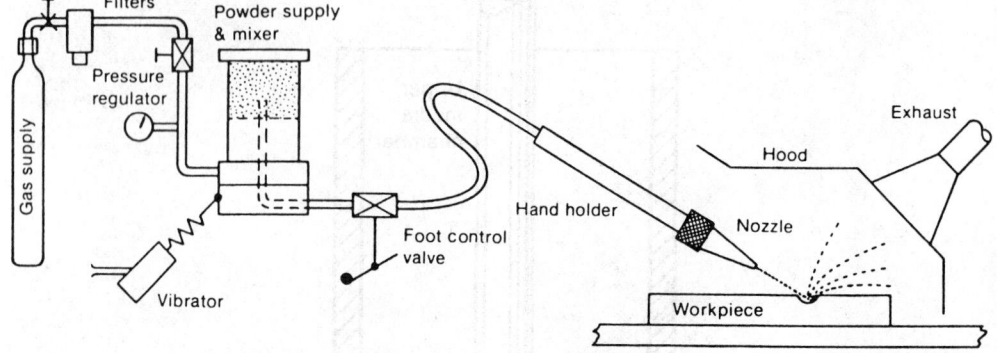

Fig. 31.39   AJM schematic.

the precision of the overcut from the cutting master. Cutting action takes place simultaneously over the full surface of abrasive contact.

31.4.7   Ultrasonic Machining and Rotary Ultrasonic Machining

Ultrasonic machining (USM) is the removal of material by the abrading action of a grit-loaded liquid slurry circulating between the workpiece and a tool vibrating perpendicular to the workface at a frequency above the audible range (see Fig. 31.44). A high-frequency power source activates a stack of magnetostrictive material, which produces a low-amplitude vibration of the toolholder. This motion is transmitted under light pressure to the slurry, which abrades the workpiece into a conjugate image of the tool form. A constant flow of slurry (usually cooled) is necessary to carry away the chips from the workface. The process is sometimes called ultrasonic abrasive machining (UAM) or impact machining.

A prime variation of USM is the addition of ultrasonic vibration to a rotating tool—usually a diamond-plated drill. Rotary ultrasonic machining (RUM) substantially increases the drilling efficiency. A piezoelectric device built into the rotating head provides the needed vibration. Milling, drilling, threading, and grinding-type operations are performed with RUM.

31.4.8   Water-Jet Machining

Water jet machining (WJM) is low-pressure hydrodynamic machining. The pressure range for WJM is an order of magnitude below that used in HDM. There are two versions of WJM: one for mining, tunneling, and large pipe cleaning that operates in the region from 250 to 1000 psi (1.7 to 6.9 MPa); and one for smaller parts and production shop situations that uses pressures below 250 psi (1.7 MPa).

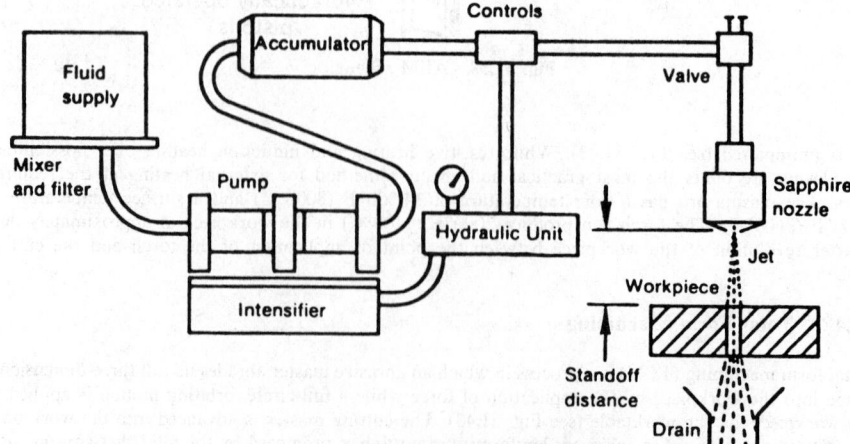

Fig. 31.40   HDM schematic.

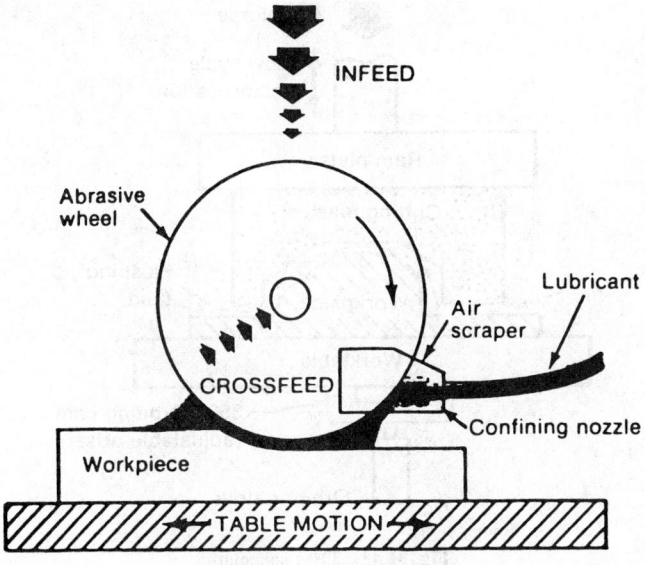

Fig. 31.41 LSG schematic.

The first version, or high-pressure range, is characterized by use of a pumped water supply with hoses and nozzles that generally are hand-directed. In the second version, more production-oriented and controlled equipment, such as that shown in Fig. 31.45, is involved. In some instances, abrasives are added to the fluid flow to promote rapid cutting. Single- or multiple-nozzle approaches to the workpiece depend on the size and number of parts per load. The principle is that WJM is high volume, not high pressure.

31.4.9 Electrochemical Deburring

Electrochemical deburring (ECD) is a special version of ECM (see Fig. 31.46). ECD was developed to remove burrs and fins or to round sharp corners. Anodic dissolution occurs on the workpiece burrs in the presence of a closely placed cathodic tool whose configuration matches the burred edge. Normally, only a small portion of the cathode is electrically exposed so a maximum concentration of the electrolytic action is attained. The electrolyte flow usually is arranged to carry away any burrs

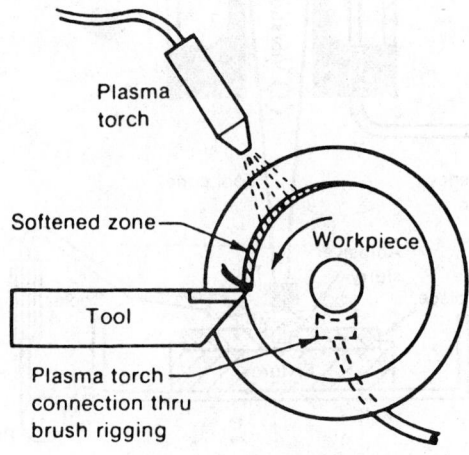

Fig. 31.42 TAM schematic.

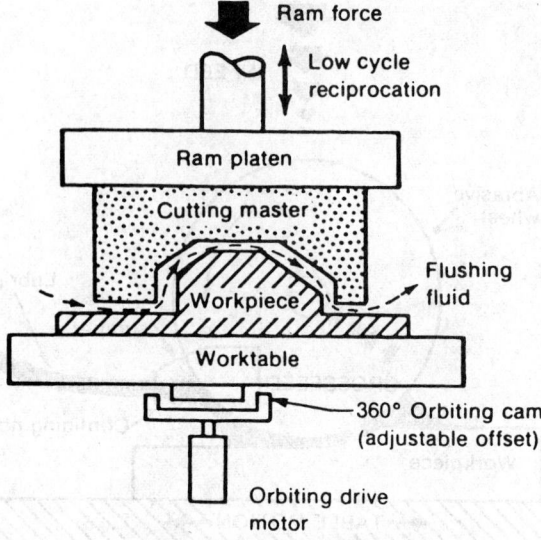

Fig. 31.43 TFM schematic.

that may break loose from the workpiece during the cycle. Voltages are low, current densities are high, electrolyte flow rate is modest, and electrolyte types are similar to those used for ECM. The electrode (tool) is stationary, so equipment is simpler than that used for ECM. Cycle time is short for deburring. Longer cycle time produces a natural radiusing action.

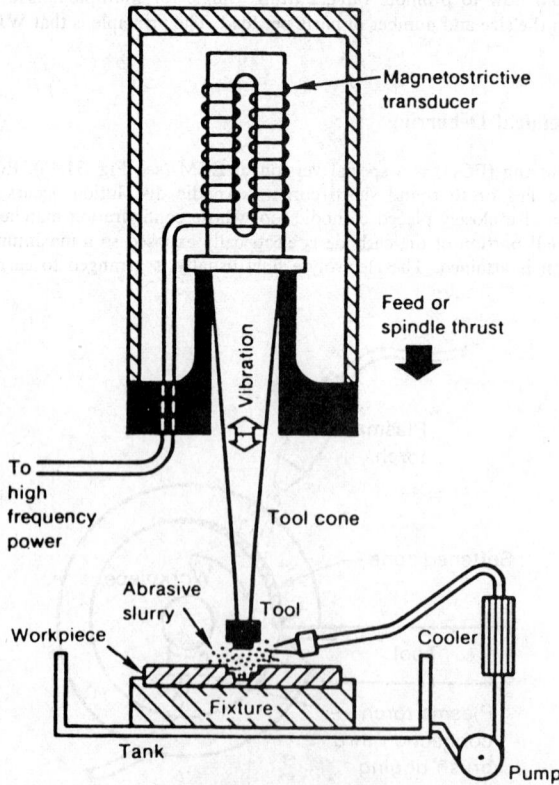

Fig. 31.44 USM schematic.

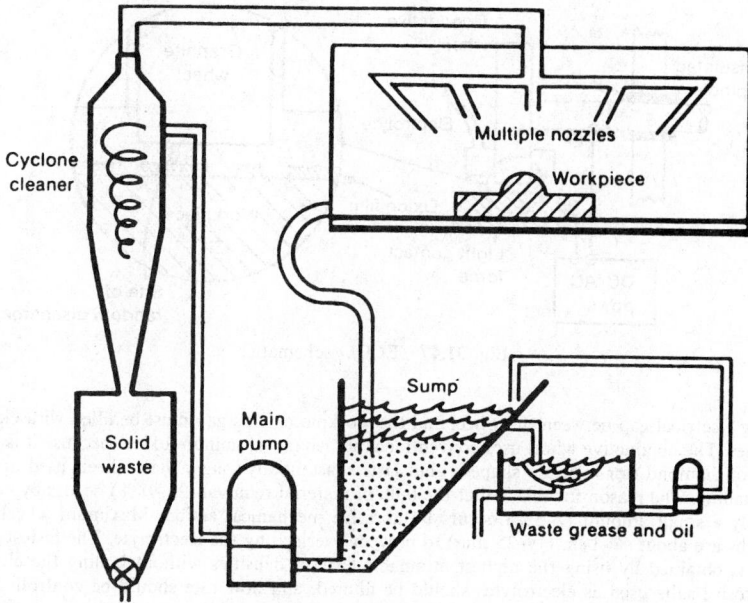

Fig. 31.45   WJM schematic.

31.4.10   Electrochemical Discharge Grinding

Electrochemical discharge grinding (ECDG) combines the features of both electrochemical and electrical discharge methods of material removal (see Fig. 31.47). ECDG has the arrangement and electrolytes of electrochemical grinding (ECG), but uses a graphite wheel without abrasive grains. The random spark discharge is generated through the insulating oxide film on the workpiece by the power generated in an ac source or by a pulsating dc source. The principal material removal comes from the electrolytic action of the low-level dc voltages. The spark discharges erode the anodic films to allow the electrolytic action to continue.

31.4.11   Electrochemical Grinding

Electrochemical grinding (ECG) is a special form of electrochemical machining in which the conductive workpiece material is dissolved by anodic action, and any resulting films are removed by a rotating, conductive, abrasive wheel (see Fig. 31.48). The abrasive grains protruding from the wheel form the

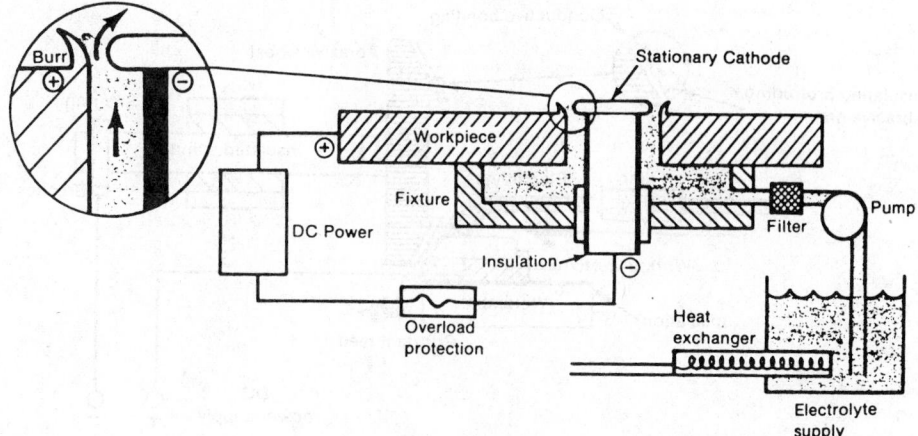

Fig. 31.46   ECD schematic.

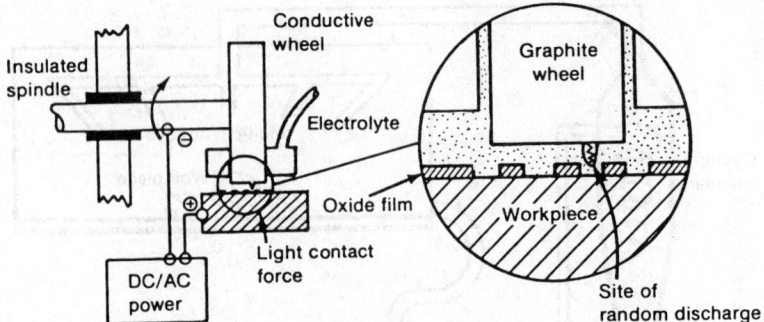

Fig. 31.47   ECDG schematic.

insulating electrical gap between the wheel and the workpiece. This gap must be filled with electrolyte at all times. The conductive wheel uses conventional abrasives—aluminum oxide (because it is nonconductive) or diamond (for intricate shapes)—but lasts substantially longer than wheels used in conventional grinding. The reason for this is that the bulk of material removal (95–98%) occurs by deplating, while only a small amount (2–5%) occurs by abrasive mechanical action. Maximum wheel contact arc lengths are about ¾–1 in. (19–25 mm) to prevent overheating the electrolyte. The fastest material removal is obtained by using the highest attainable current densities without boiling the electrolyte. The corrosive salts used as electrolytes should be filtered, and flow rate should be controlled for the best process control.

31.4.12   Electrochemical Honing

Electrochemical honing (ECH) is the removal of material by anodic dissolution combined with mechanical abrasion from a rotating and reciprocating abrasive stone (carried on a spindle, which is the cathode) separated from the workpiece by a rapidly flowing electrolyte (see Fig. 31.49). The principal material removal action comes from electrolytic dissolution. The abrasive stones are used to maintain size and to clean the surfaces to expose fresh metal to the electrolyte action. The small electrical gap is maintained by the nonconducting stones that are bonded to the expandable arbor with cement. The cement must be compatible with the electrolyte and the low dc voltage. The mechanical honing action uses materials, speeds, and pressures typical of conventional honing.

31.4.13   Electrochemical Machining

Electrochemical machining (ECM) is the removal of electrically conductive material by anodic dissolution in a rapidly flowing electrolyte, which separates the workpiece from a shaped electrode (see Fig. 31.50). The filtered electrolyte is pumped under pressure and at controlled temperature to bring a

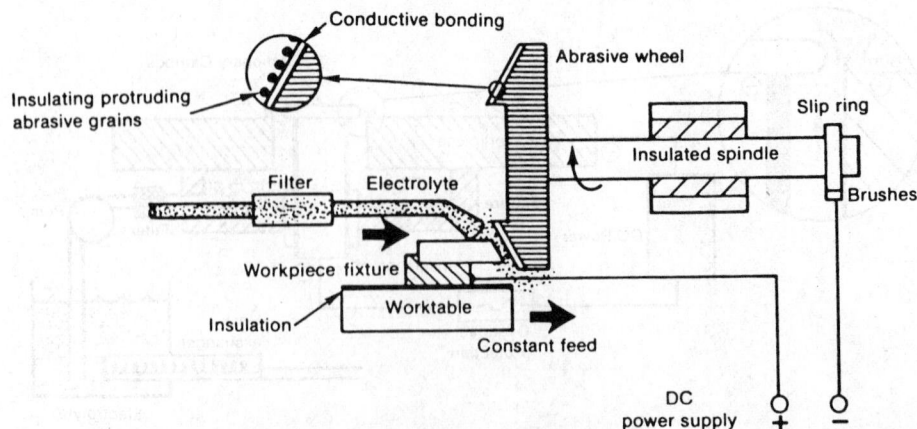

Fig. 31.48   ECG schematic.

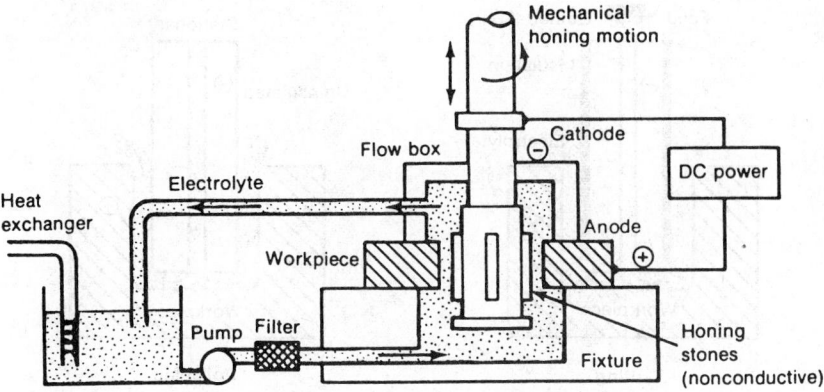

Fig. 31.49   ECH schematic.

controlled-conductivity fluid into the narrow gap of the cutting area. The shape imposed on the workpiece is nearly a mirror or conjugate image of the shape of the cathodic electrode. The electrode is advanced into the workpiece at a constant feed rate that exactly matches the rate of dissolution of the work material. The current density is the chief factor in setting feed rates and in attaining smoothness.

31.4.14   Electrochemical Polishing

Electrochemical polishing (ECP) is a special form of electrochemical machining arranged for cutting or polishing a workpiece (see Fig. 31.51). Polishing parameters are similar in range to those for cutting, but without the feed motion. ECP generally uses a larger gap and a lower current density than does ECM. This requires modestly higher voltages. [In contrast, electropolishing (ELP) uses still lower current densities, lower electrolyte flow, and more remote electrodes.]

31.4.15   Electrochemical Sharpening

Electrochemical sharpening (ECS) is a special form of electrochemical machining arranged to accomplish sharpening or polishing by hand (see Fig. 31.52). A portable power pack and electrolyte reservoir supply a finger-held electrode with a small current and flow. The fixed gap incorporated on the several styles of shaped electrodes controls the flow rate. A suction tube picks up the used electrolyte for recirculation after filtration.

31.4.16   Electrochemical Turning

Electrochemical turning (ECT) is a special form of electrochemical machining designed to accommodate rotating workpieces (see Fig. 31.53). The rotation provides additional accuracy but complicates the

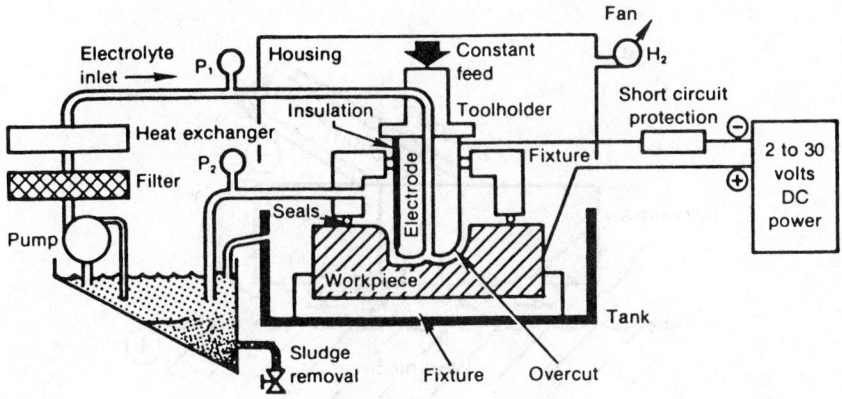

Fig. 31.50   ECM schematic.

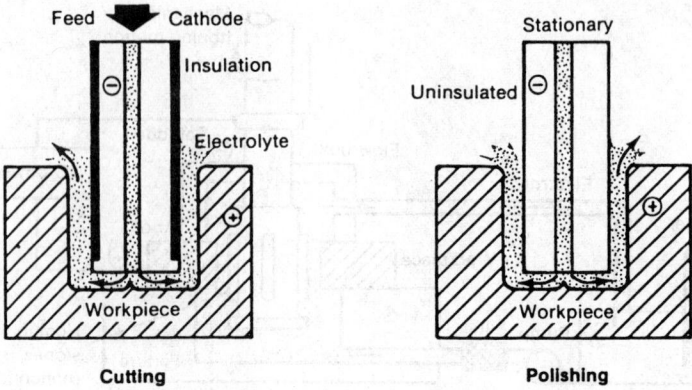

Fig. 31.51 ECP schematic.

equipment with the method of introducing the high currents to the rotating part. Electrolyte control may also be complicated because rotating seals are needed to direct the flow properly. Otherwise, the parameters and considerations of electrochemical machining apply equally to the turning mode.

31.4.17 Electro-Stream

Electro-stream (ES™) is a special version of electrochemical machining adapted for drilling very small holes using high voltages and acid electrolytes (see Fig. 31.54). The voltages are more than 10 times

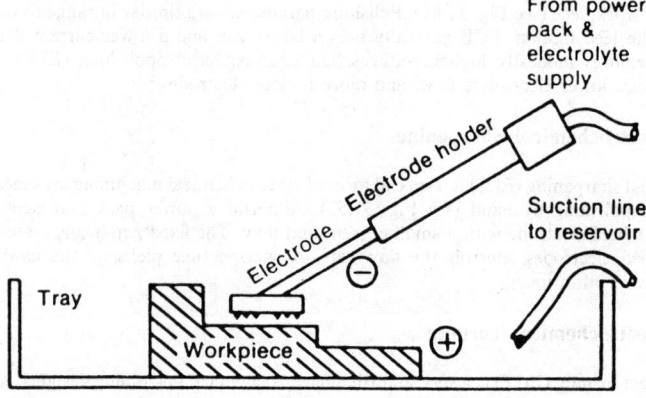

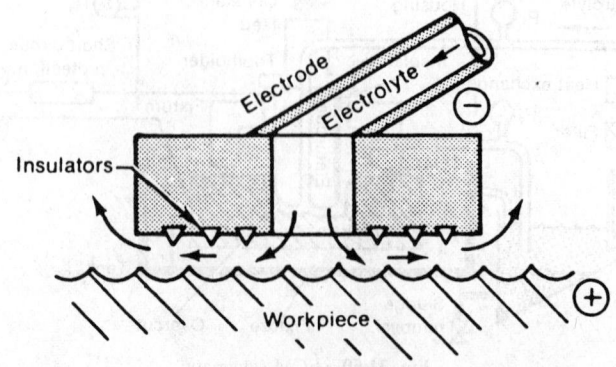

Fig. 31.52 ECS schematic.

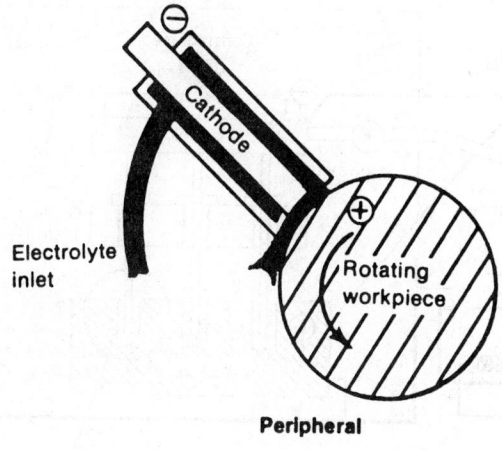

Peripheral

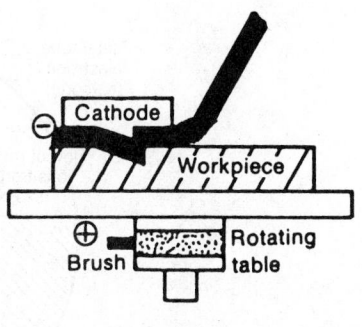

Face

Fig. 31.53 ECT schematic.

those employed in ECM or STEM, so special provisions for containment and protection are required. The feed must be controlled to match exactly the rate of dissolution of the workpiece material. The actual drilling takes place in a plastic chamber, suitably vented, with controls for automatic rapid advance of the tool, close programming of feed, dwell at hole breakthrough, if desired, and rapid retraction prior to indexing to the next location. The tool is a drawn-glass nozzle, 0.001–0.002 in. smaller than the desired hole size. An electrode inside the nozzle or the manifold ensures electrical contact with the acid. Multiple hole drilling predominates.

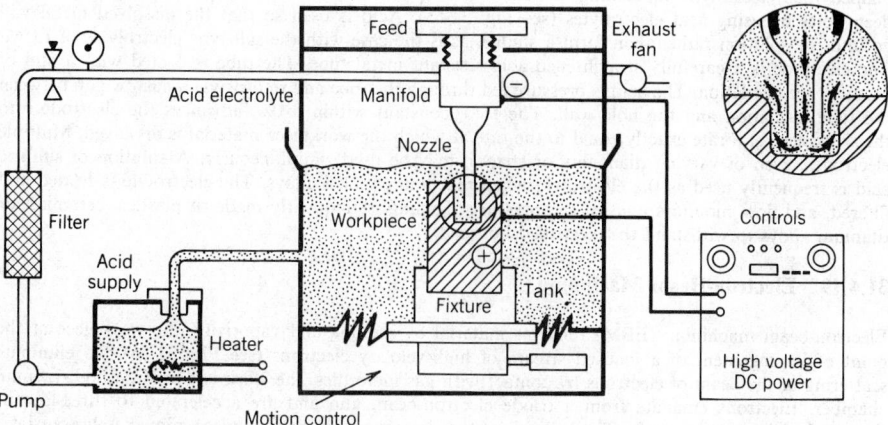

Fig. 31.54 ES schematic.

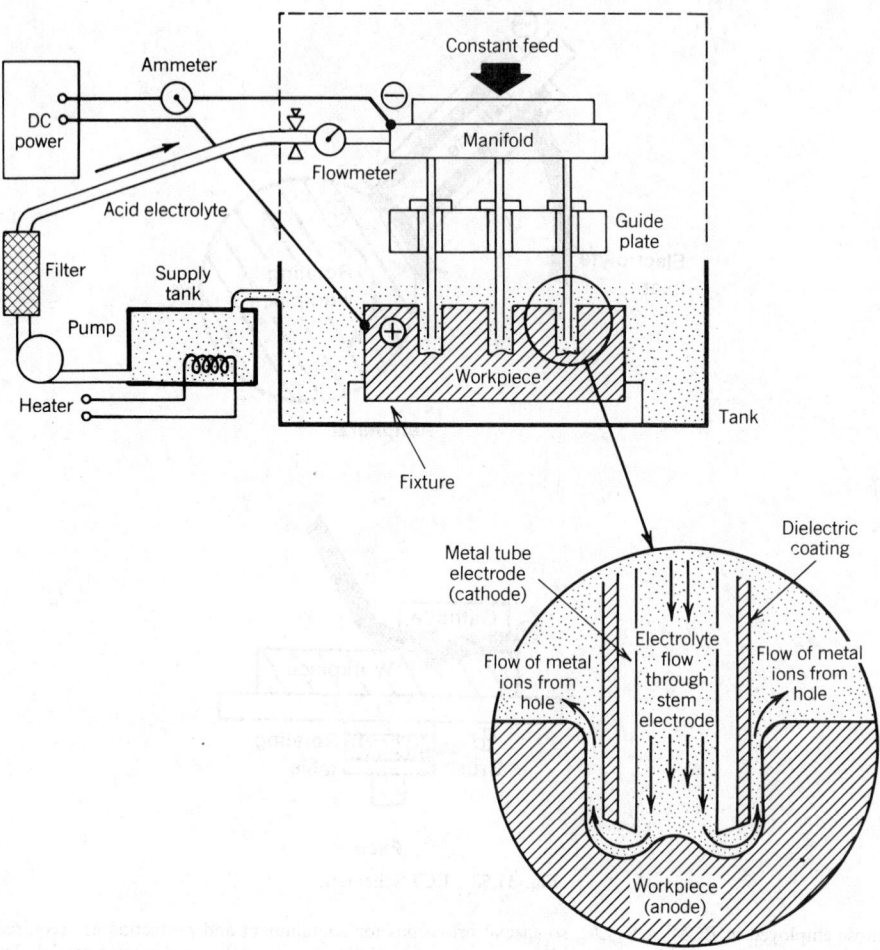

Fig. 31.55 STEM schematic.

31.4.18 Shaped-Tube Electrolytic Machining

Shaped tube electrolytic machining (STEM™) is a specialized ECM technique for "drilling" small, deep holes by using acid electrolytes (see Fig. 31.55). Acid is used so that the dissolved metal will go into the solution rather than form a sludge, as is the case with the salt-type electrolytes of ECM. The electrode is a carefully straightened acid-resistant metal tube. The tube is coated with a film of enamel-type insulation. The acid is pressure fed through the tube and returns via a narrow gap between the tube insulation and the hole wall. The feed, constant within ±1%, advances the electrode into the workpiece at a rate exactly equal to the rate at which the workpiece material is dissolved. Multiple electrodes, even of varying diameters or shapes, may be used simultaneously. A solution of sulfuric acid is frequently used as the electrolyte when machining nickel alloys. The electrolyte is heated and filtered, and flow monitors control the pressure. Tooling is frequently made of plastics, ceramics, or titanium alloys to withstand the electrified hot acid.

31.4.19 Electron-Beam Machining

Electron-beam machining (EBM) removes material by melting and vaporizing the workpiece at the point of impingement of a focused stream of high-velocity electrons (see Fig. 31.56). To eliminate scattering of the beam of electrons by contact with gas molecules, the work is done in a high-vacuum chamber. Electrons emanate from a triode electron-beam gun and are accelerated to three-fourths the speed of light at the anode. The collision of the electrons with the workpiece immediately translates

Electron Beam Gun

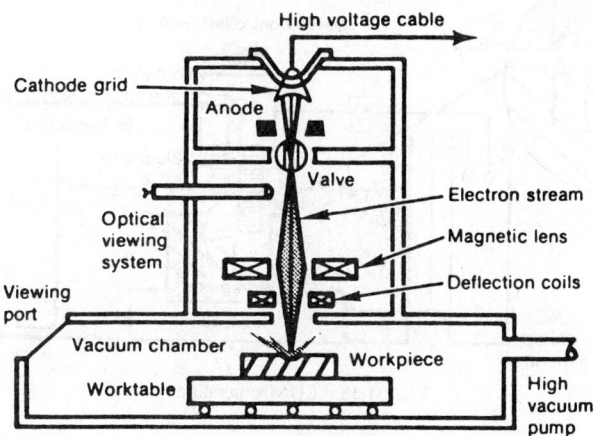

Fig. 31.56 EBM schematic.

their kinetic energy into thermal energy. The low-inertia beam can be simply controlled by electromagnetic fields. Magnetic lenses focus the electron beam on the workpiece, where a 0.001-in. (0.025-mm) diameter spot can attain an energy density of up to 10^9 W/in.^{2} (1.55×10^8 W/cm^2) to melt and vaporize any material. The extremely fast response time of the beam is an excellent companion for three-dimensional computer control of beam deflection, beam focus, beam intensity, and workpiece motion. View ports and optical tracking systems help guide the beam or the workpiece. Deflection coils permit magnetic control and programming of the beam to any desired pattern over a small area—about ¼ in. (6.4 mm) square.

31.4.20 Electrical-Discharge Grinding

Electrical discharge grinding (EDG) is the removal of a conductive material by rapid, repetitive spark discharges between a rotating tool and the workpiece, which are separated by a flowing dielectric fluid (see Fig. 31.57). (EDG is similar to EDM except the electrode is in the form of a grinding wheel and the current is usually lower.) The spark gap is servo controlled. The insulated wheel and the worktable are connected to the dc pulse generator. A positive charge on the workpiece is "standard." Higher currents produce faster cutting, rougher finishes, and deeper heat-affected zones in the workpiece.

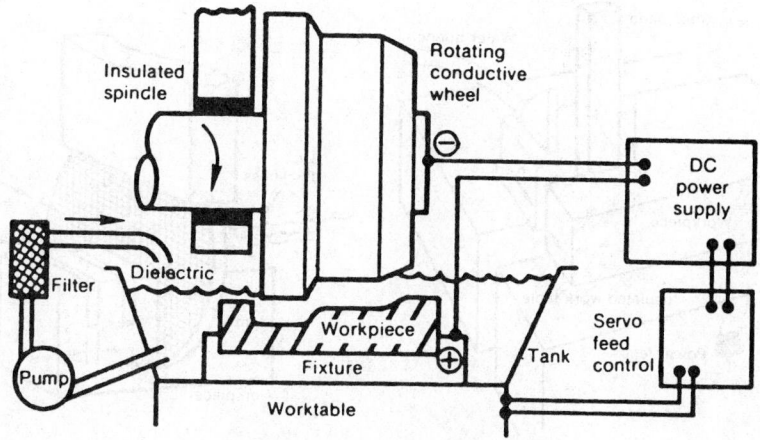

Fig. 31.57 EDG schematic.

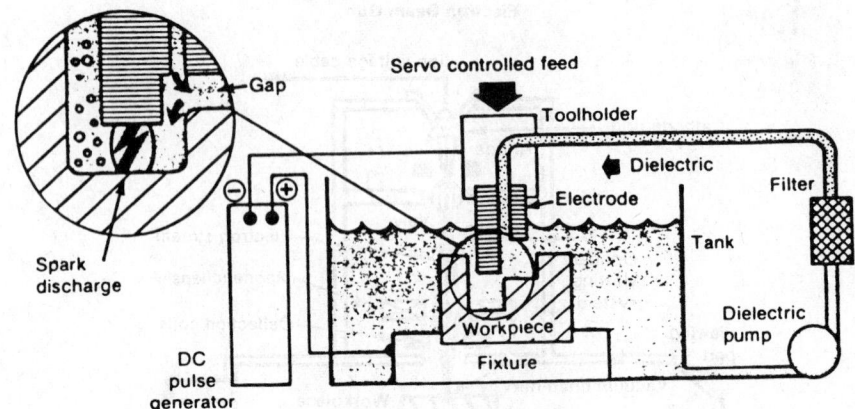

Fig. 31.58 EDM schematic.

31.4.21 Electrical-Discharge Machining

Electrical discharge machining (EDM) removes electrically conductive material by means of rapid, repetitive spark discharges from a pulsating dc power supply with dielectric flowing between the work-piece and the tool (see Fig. 31.58). The shaped tool (electrode) is fed into the workpiece under servo control. A spark discharge then breaks down the dielectric fluid. The frequency and energy per spark are set and controlled with a dc power source. The servo control maintains a constant gap between the tool and the workpiece while advancing the electrode. The dielectric coil cools and flushes out the vaporized and condensed material while reestablishing insulation in the gap.

31.4.22 Electrical-Discharge Sawing

Electrical discharge sawing (EDS) is a variation of electrical discharge machining (EDM) that combines the motion of either a band saw or a circular disk saw with electrical erosion of the workpiece (see Fig. 31.59). The rapid-moving, untoothed, thin, special steel band or disk is guided into the workpiece by carbide-faced inserts. A kerf only 0.002–0.005-in. (0.050–0.13-mm) wider than the blade or disk is formed as they are fed into the workpiece. The feed can be (1) uniform with continuous arcing and constant feed or (2) servo controlled to maintain a positive gap and arc control. The low-voltage dc power supplies high currents to the gap, and water is used as a cooling quenchant for the tool, swarf, and workpiece. Circular cutting is usually performed under water, thereby reducing noise and fumes. While the work is power-fed into the band (or the disk into the work), it is not subjected to appreciable forces because the arc does the cutting, so fixturing can be minimal. Precision adjustment of the feed rate is made to be in exact balance with the arc erosion rate.

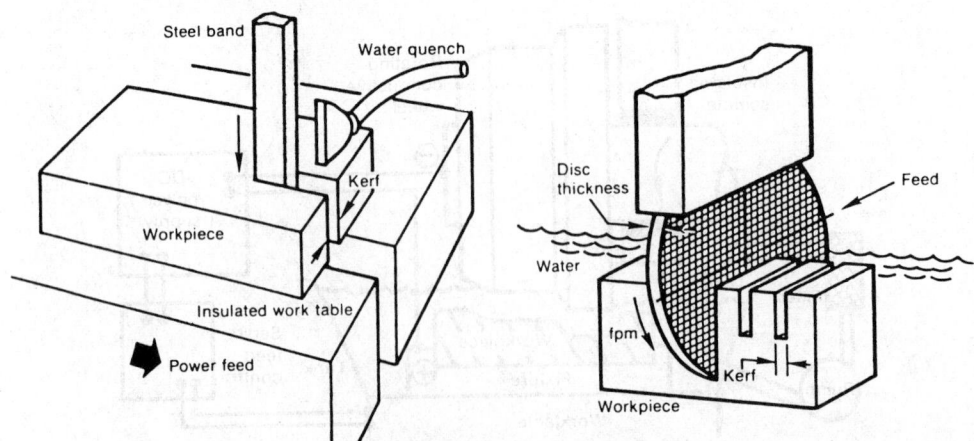

Fig. 31.59 EDS schematic. Left, band saw; right, disc saw.

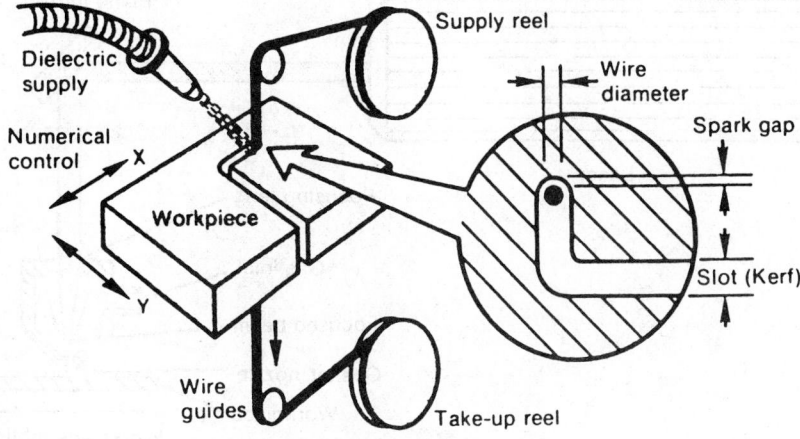

Fig. 31.60 EDWC schematic.

31.4.23 Electrical-Discharge Wire Cutting

Electrical discharge wire cutting (EDWC) is a special form of electrical discharge machining wherein the electrode is a continuously moving conductive wire (see Fig. 31.60). EDWC is often called traveling wire EDM. A small-diameter tension wire is guided to produce a straight, narrow-kerf size control. Usually, a programmed or numerically controlled motion guides the cutting, while the width of the kerf is maintained by the wire size and discharge controls. The dielectric is oil or deionized water carried into the gap by motion of the wire. The wire is inexpensive enough to be used only once. The cutting is basically a two-axis process.

31.4.24 Laser-Beam Machining

Laser-beam machining (LBM) removes material by melting, ablating, and vaporizing the workpiece at the point of impingement of a highly focused beam of coherent monochromatic light (see Fig. 31.61). Laser is an acronym for "light amplification by stimulated emission of radiation." The electromagnetic radiation operates at wavelengths from the visible to the infrared. The principal lasers used for material removal are the neodymium–glass, the Nd–YAD (neodymium–yttrium aluminum garnet), the ruby, and the carbon dioxide (CO_2). The last is a gas laser (most frequently used as a torch with an assisting gas—see LBT, laser-beam torch), while others are solid-state lasing materials.

For pulsed operation, the power supply produces short, intense bursts of electricity into the flash

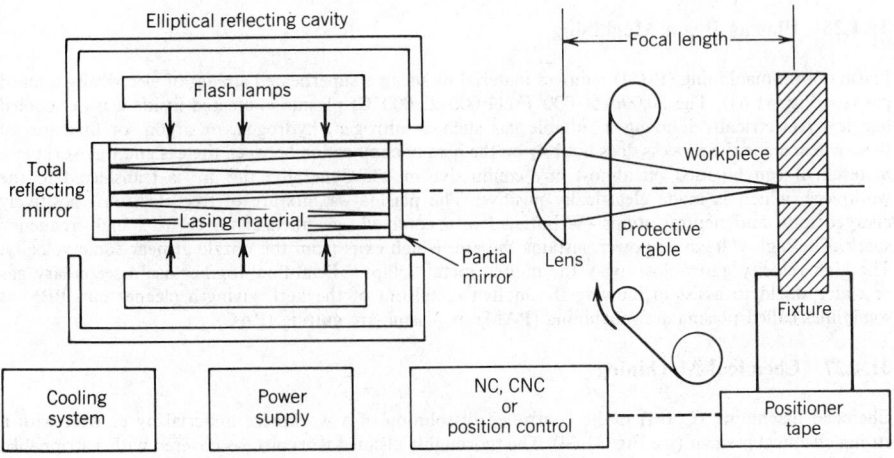

Fig. 31.61 LBM schematic.

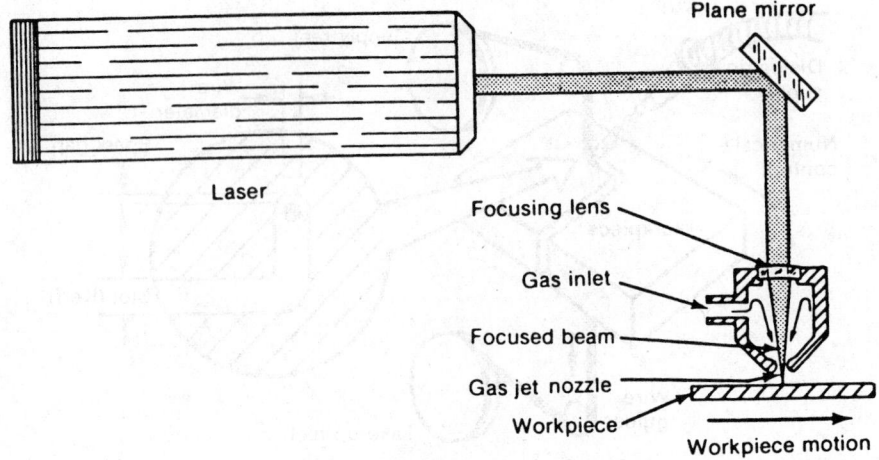

Fig. 31.62   LBT schematic.

lamps, which concentrate their light flux on the lasing material. The resulting energy from the excited atoms is released at a characteristic, constant frequency. The monochromatic light is amplified during successive reflections from the mirrors. The thoroughly collimated light exits through the partially reflecting mirror to the lens, which focuses it on or just below the surface of the workpiece. The small beam divergence, high peak power, and single frequency provide excellent, small-diameter spots of light with energy densities up to 3×10^{10} W/in.^{2} (4.6×10^9 W/cm^2), which can sublime almost any material. Cutting requires energy densities of 10^7–10^9 W/in.^{2} (1.55×10^6–1.55×10^8 W/cm^2), at which rate the thermal capacity of most materials cannot conduct energy into the body of the workpiece fast enough to prevent melting and vaporization.

31.4.25   Laser-Beam Torch

Laser-beam torch (LBT) is a process in which material is removed by the simultaneous focusing of a laser beam and a gas stream on the workpiece (see Fig. 31.62). A continuous-wave (CW) laser or a pulsed laser with more than 100 pulses per second is focused on or slightly below the surface of the workpiece, and the absorbed energy causes localized melting. An oxygen gas stream promotes an exothermic reaction and purges the molten material from the cut. Argon or nitrogen gas is sometimes used to purge the molten material while also protecting the workpiece.

Argon or nitrogen gas is often used when organic or ceramic materials are being cut. Close control of the spot size and the focus on the workpiece surface is required for uniform cutting. The type of gas used has only a modest effect on laser penetrating ability. Typically, short laser pulses with high peak power are used for cutting and welding. The CO$_2$ laser is the laser most often used for cutting.

31.4.26   Plasma-Beam Machining

Plasma-beam machining (PBM) removes material by using a superheated stream of electrically ionized gas (see Fig. 31.63). The 20,000–50,000°F (11,000–28,000°C) plasma is created inside a water-cooled nozzle by electrically ionizing a suitable gas such as nitrogen, hydrogen, or argon, or mixtures of these gases. Since the process does not rely on the heat of combustion between the gas and the workpiece material, it can be used on almost any conductive metal. Generally, the arc is transferred to the workpiece, which is made electrically positive. The plasma—a mixture of free electrons, positively charged ions, and neutral atoms—is initiated in a confined, gas-filled chamber by a high-frequency spark. The high-voltage dc power sustains the arc, which exits from the nozzle at near sonic velocity. The high-velocity gases blow away the molten metal "chips." Dual-flow torches use a secondary gas or water shield to assist in blowing the molten metal out of the kerf, giving a cleaner cut. PBM is sometimes called plasma-arc machining (PAM) or plasma-arc cutting (PAC).

31.4.27   Chemical Machining

Chemical machining (CHM) is the controlled dissolution of a workpiece material by contact with a strong chemical reagent (see Fig. 31.64). The thoroughly cleaned workpiece is covered with a strippable, chemically resistant mask. Areas where chemical action is desired are outlined on the workpiece with

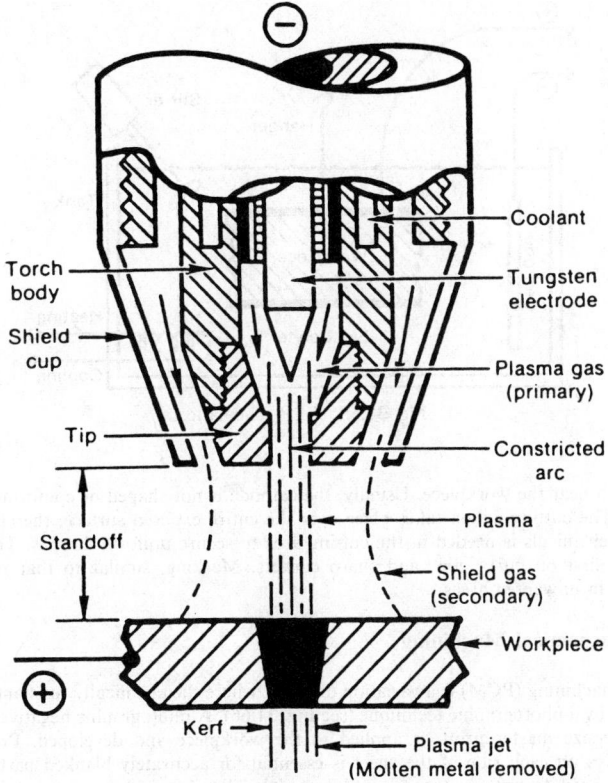

Fig. 31.63 PBM schematic.

the use of a template and then stripped off the mask. The workpiece is then submerged in the chemical reagent to remove material simultaneously from all exposed surfaces. The solution should be stirred or the workpiece should be agitated for more effective and more uniform action. Increasing the temperatures will also expedite the action. The machined workpiece is then washed and rinsed, and the remaining mask is removed. Multiple parts can be maintained simultaneously in the same tank.

31.4.28 Electropolishing

Electropolishing (ELP) is a specialized form of chemical machining that uses an electrical deplating action to enhance the chemical action (see Fig. 31.65). The chemical action from the concentrated heavy acids does most of the work, while the electrical action smooths or polishes the irregularities. A metal cathode is connected to a low-voltage, low-amperage, dc power source and is installed in

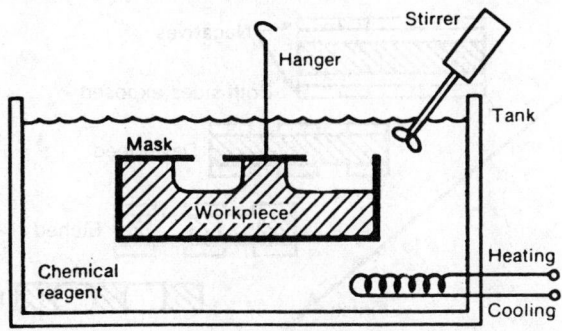

Fig. 31.64 CHM schematic.

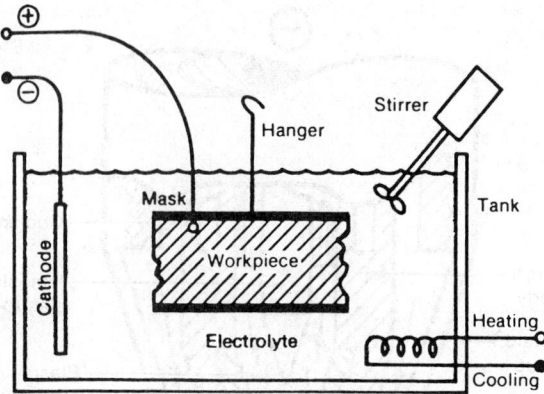

Fig. 31.65 ELP schematic.

the chemical bath near the workpiece. Usually, the cathode is not shaped or conformed to the surface being polished. The cutting action takes place over the entire exposed surface; therefore, a good flow of heated, fresh chemicals is needed in the cutting area to secure uniform finishes. The cutting action will concentrate first on burrs, fins, and sharp corners. Masking, similar to that used with CHM, prevents cutting in unwanted areas.

31.4.29 Photochemical Machining

Photochemical machining (PCM) is a variation of CHM where the chemically resistant mask is applied to the workpiece by a photographic technique (see Fig. 31.66). A photographic negative, often a reduced image of an oversize master print, is applied to the workpiece and developed. Precise registry of duplicate negatives on each side of the sheet is essential for accurately blanked parts. Immersion or spray etching is used to remove the exposed material. The chemicals used must be active on the workpiece, but inactive against the photoresistant mask. There will be some undercutting behind the mask, however, which limits the use of PCM to thin materials—up to $\frac{3}{32}$ in. (2.4 mm) (see Fig. 31.67).

31.4.30 Thermochemical Machining

Thermochemical machining (TCM) removes the workpiece material—usually only burrs and fins— by exposure of the workpiece to hot, corrosive gases (see Fig. 31.68). The process is sometimes called combustion machining, thermal deburring, or thermal energy method (TEM). The workpiece is exposed

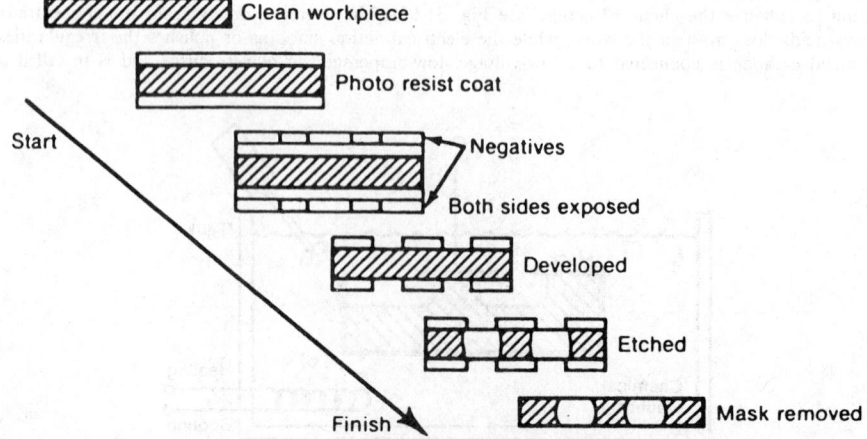

Fig. 31.66 PCM schematic.

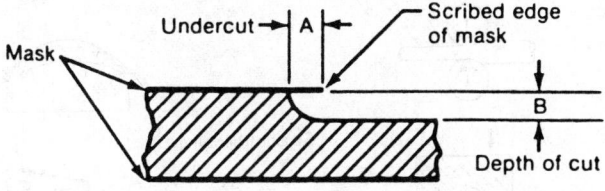

$$\text{Etch factor} = \frac{\text{Depth of cut}}{\text{Undercut}} = \frac{B}{A}$$

Fig. 31.67 Etch factor in photomechanical machining. (Note: Inverse of that used in chemical machining.)

for a very short time to extremely hot gases, which are formed by detonating an explosive mixture. The ignition of the explosive—usually hydrogen or natural gas and oxygen—creates a transient thermal wave that vaporizes the burrs and fins. The main body of the workpiece remains unaffected and relatively cool because of its low surface-to-mass ratio and the brevity of the exposure to high temperatures.

31.5 METAL FORMING

Metal-forming processes use a remarkable property of metals—their ability to flow plastically in the solid state without concurrent deterioration of properties. Moreover, by simply moving the metal to the desired shape, there is little or no waste. Figure 31.69 shows some of the metal-forming processes. Metal-forming processes are classified into two categories: hot-working processes and cold-working processes.

31.5.1 Hot-Working Processes

Hot working is defined as the plastic deformation of metals above their recrystallization temperature. Here it is important to note that the crystallization temperature varies greatly with different materials. Lead and tin are hot worked at room temperature, steels require temperatures of 2000°F (1100°C), and tungsten is still in a cold or warm working state at 2000°F (1100°C). Hot working does not necessarily imply high absolute temperatures.

Hot working can produce the following improvements:

1. Production of randomly oriented, spherical-shaped grain structure, which results in a net increase not only in the strength but also in ductility and toughness.

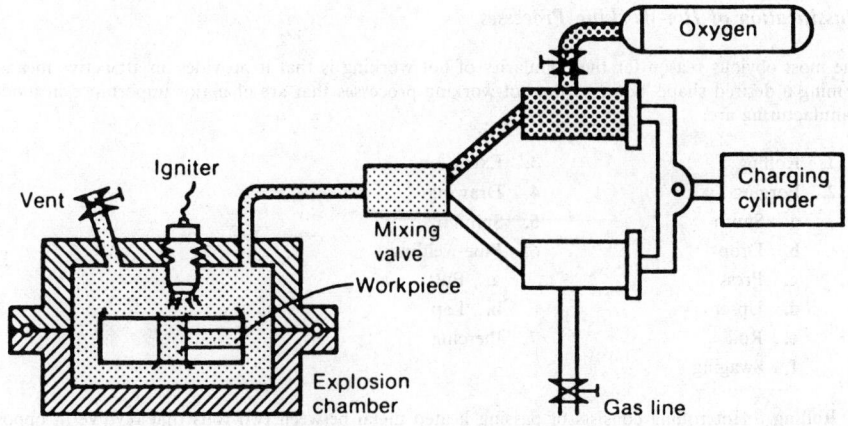

Fig. 31.68 TCM schematic.

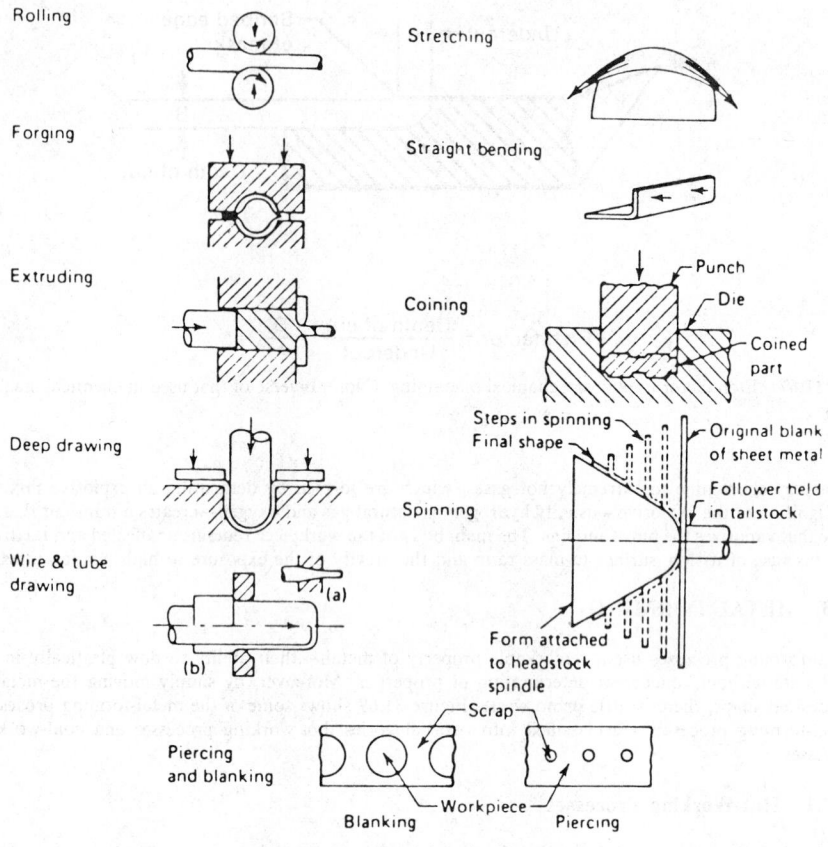

Fig. 31.69   Metal forming processes.

2. The reorientation of inclusions or impurity material in metal. The impurity material often distorts and flows along with the metal.

This material, however, does not recrystallize with the base metal and often produces a fiber structure. Such a structure clearly has directional properties, being stronger in one direction than in another. Moreover, an impurity originally oriented so as to aid crack movement through the metal is often reoriented into a "crack-arrestor" configuration perpendicular to crack propagation.

Classification of Hot-Working Processes

The most obvious reason for the popularity of hot working is that it provides an attractive means of forming a desired shape. Some of the hot-working processes that are of major importance in modern manufacturing are:

1. Rolling
2. Forging
   a. Smith
   b. Drop
   c. Press
   d. Upset
   e. Roll
   f. Swaging

3. Extrusion
4. Drawing
5. Spinning
6. Pipe welding
   a. Butt
   b. Lap
7. Piercing

Rolling.   Hot rolling consists of passing heated metal between two rolls that revolve in opposite directions, the space between the rolls being somewhat less than the thickness of the entering metal.

Many finished parts, such as hot-rolled structural shapes, are completed entirely by hot rolling. More often, however, hot-rolled products, such as sheets, plates, bars, and strips, serve as input material for other processes, such as cold forming or machining.

In hot rolling, as in all hot working, it is very important that the metal be heated uniformly throughout to the proper temperature before processing. This usually requires prolonged heating at the desired temperature, a procedure known as soaking. If the temperature is not uniform, the subsequent deformation will also be nonuniform, the hotter exterior flowing in preference to the cooler and, therefore, stronger, interior. Cracking, tearing, and associated problems may result.

Isothermal Rolling.   The ordinary rolling of some high-strength metals, such as titanium and stainless steels, particularly in thicknesses below about 0.150 in. (3.8 mm), is difficult because the heat in the sheet is transferred rapidly to the cold and much more massive rolls. This has been overcome by isothermal rolling. Localized heating is accomplished in the area of deformation by the passage of a large electrical current between the rolls, through the sheet. Reductions up to 90% per roll have been achieved. The process usually is restricted to widths below 2 in. (50 mm).

Forging.   Forging is the plastic working of metal by means of localized compressive forces exerted by manual or power hammers, presses, or special forging machines.

Various types of forging have been developed to provide great flexibility, making it economically possible to forge a single piece or to mass produce thousands of identical parts. The metal may be (1) drawn out, increasing its length and decreasing its cross section; (2) upset, increasing the cross section and decreasing the length; or (3) squeezed in closed impression dies to produce multidirectional flow. The state of stress in the work is primarily uniaxial or multiaxial compression.

The common forging processes are:

1. Open-die hammer or smith forging
2. Impression-die drop forging
3. Press forging
4. Upset forging
5. Roll forging
6. Swaging.

Open-Die Hammer Forging.   Open-die forging does not confine the flow of metal, the hammer and anvil often being completely flat. The desired shape is obtained by manipulating the workpiece between blows. Specially shaped tools or a slightly shaped die between the workpiece and the hammer or anvil are used to aid in shaping sections (round, concave, or convex), making holes, or performing cutoff operations.

Impression-Die Drop Forging.   In impression-die or closed-die drop forging, the heated metal is placed in the lower cavity of the die and struck one or more blows with the upper die. This hammering causes the metal to flow so as to fill the die cavity. Excess metal is squeezed out between the die faces along the periphery of the cavity to form a flash. When forging is completed, the flash is trimmed off by means of a trimming die.

Press Forging.   Press forging employs a slow-squeezing action that penetrates throughout the metal and produces a uniform metal flow. In hammer or impact forging, metal flow is a response to the energy in the hammer–workpiece collision. If all the energy can be dissipated through flow of the surface layers of metal and absorption by the press foundation, the interior regions of the workpiece can go undeformed. Therefore, when the forging of large sections is required, press forging must be employed.

Upset Forging.   Upset forging involves increasing the diameter of the end or central portion of a bar of metal by compressing its length. Upset-forging machines are used to forge heads on bolts and other fasteners, valves, couplings, and many other small components.

Roll Forging.   Roll forging, in which round or flat bar stock is reduced in thickness and increased in length, is used to produce such components as axles, tapered levers, and leaf springs.

Swaging.   Swaging involves hammering or forcing a tube or rod into a confining die to reduce its diameter, the die often playing the role of the hammer. Repeated blows cause the metal to flow inward and take the internal form of the die.

Extrusion.   In the extrusion process, metal is compressively forced to flow through a suitably shaped die to form a product with reduced cross section. Although it may be performed either hot or cold, hot extrusion is employed for many metals to reduce the forces required, to eliminate cold-working effects, and to reduce directional properties. The stress state within the material is triaxial compression.

Lead, copper, aluminum, and magnesium, and alloys of these metals are commonly extruded, taking advantage of the relatively low yield strengths and extrusion temperatures. Steel is more difficult to extrude. Yield strengths are high, and the metal has a tendency to weld to the walls of the die and confining chamber under the conditions of high temperature and pressures. With the development and use of phosphate-based and molten glass lubricants, substantial quantities of hot steel extrusions are now produced. These lubricants adhere to the billet and prevent metal-to-metal contact throughout the process.

Almost any cross-section shape can be extruded from the nonferrous metals. Hollow shapes can be extruded by several methods. For tubular products, the stationary or moving mandrel process is often employed. For more complex internal cavities, a spider mandrel or torpedo die is used. Obviously, the cost for hollow extrusions is considerably greater than for solid ones, but a wide variety of shapes can be produced that cannot be made by any other process.

Drawing. Drawing is a process for forming sheet metal between an edge-opposing punch and a die (draw ring) to produce a cup, cone, box, or shell-like part. The work metal is bent over and wrapped around the punch nose. At the same time, the outer portions of the blank move rapidly toward the center of the blank until they flow over the die radius as the blank is drawn into the die cavity by the punch. The radial movement of the metal increases the blank thickness as the metal moves toward the die radius; as the metal flows over the die radius, this thickness decreases, because of the tension in the shell wall between the punch nose and the die radius, and (in some instances) because of the clearance between the punch and the die.

Wall thinning is more pronounced in stainless-steel parts than in those made of low-carbon steel. One or more redraws, with or without intermediate annealing, are often required in order to complete a deep drawn part. Figure 31.70 illustrates the progression of metal flow in drawing a tub from a flat blank. Draw ratio or the reduction in drawing cylindrical shells is generally expressed in terms of the diameters of the blank and the cup. Strain depends only slightly on the blank thickness. The drawability of a metal is expressed either as the ratio of the blank diameter to the punch diameter, D/d, or as the percentage reduction from the blank diameter to the cup diameter, $100(1 - d/D)$.

The force (load) required for drawing a round cup is expressed by the following empirical equation:

$$L = \pi dt S \left(\frac{D}{d} - k \right)$$

where L is press load, in pounds; d is cup diameter, in inches; D is blank diameter, in inches; t is work-metal thickness, in inches; S is tensile strength, in pounds per square inch; and k, a constant that takes into account frictional and bending forces, is usually 0.6–0.7.

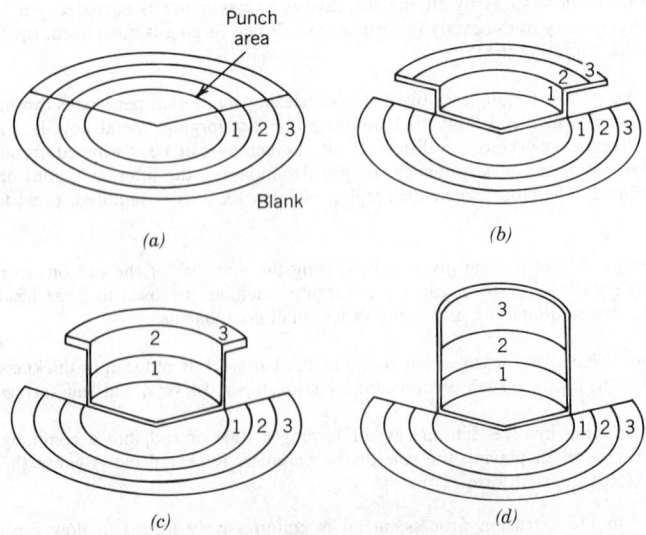

Fig. 31.70 Progression of metal flow in drawing a cup from a flat blank.

The force (load) required for drawing a rectangular cup can be calculated from the following equation:

$$L = tS(2\pi R k_A + l k_B)$$

where L is press load, in pounds; t is work-metal thickness, in inches; S is tensile strength, in pounds per square inch; R is corner radius of the cup, in inches; l is the sum of the lengths of straight sections of the sides, in inches; and k_A and k_B are constants. Values for k_A range from 0.5 (for a shallow cup) to 2.0 (for a cup of depth five to six times the corner radius). Values for k_B range from 0.2 (for easy draw radius, ample clearance, and no blankholding force) and 0.3 (for similar free flow and normal blankholding force of about $L/3$) to a maximum of 1.0 (for metal clamped too tightly to flow).

Figure 31.71 can be used as a general guide for computing maximum drawing load for a round shell. These relations are based on a free draw with sufficient clearance so that there is no ironing, using a maximum reduction of 50%. The nomograph gives the load required to fracture the cup.

Spinning. Spinning is a method of forming sheet metal or tubing into seamless hollow cylinders, cones, hemispheres, or other circular shapes by a combination of rotation and force. On the basis of techniques used, applications, and results obtainable, the method may be divided into two categories: manual spinning (with or without mechanical assistance to increase the force) and power spinning.

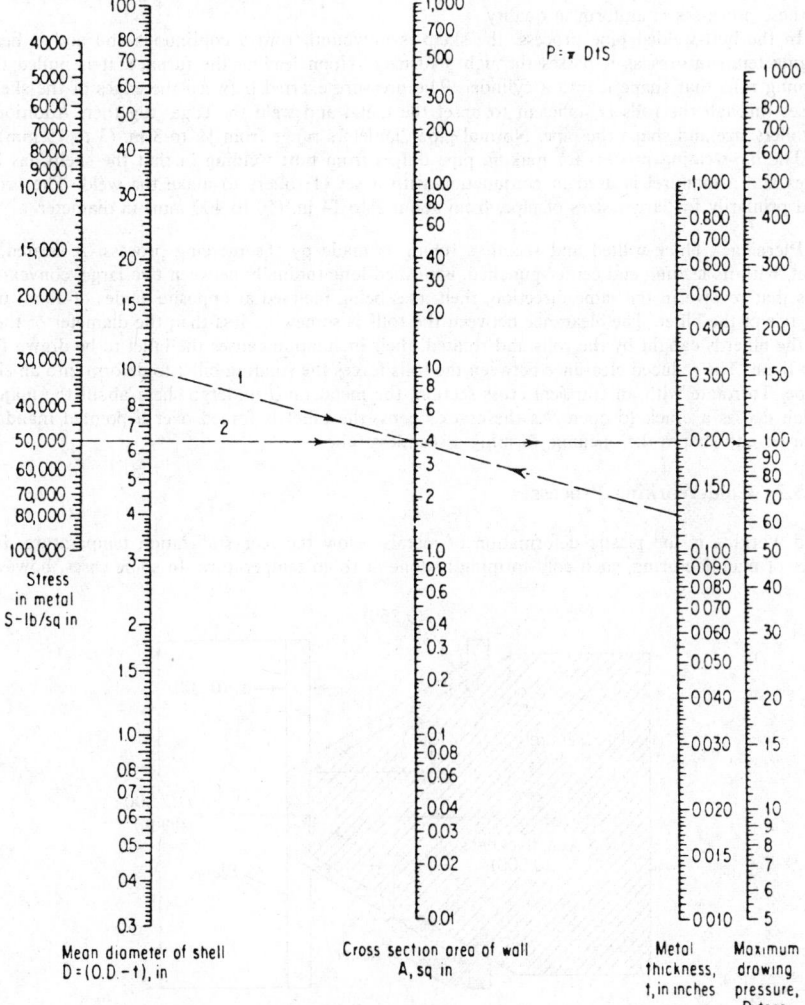

Fig. 31.71 Nomograph for estimating drawing pressures.

Manual spinning entails no appreciable thinning of metal. The operation ordinarily done in a lathe, consists of pressing a tool against a circular metal blank that is rotated by the headstock.

Power spinning is also known as shear spinning because in this method metal is intentionally thinned, by shear forces. In power spinning, forces as great as 400 tons are used.

The application of shear spinning to conical shapes is shown schematically in Fig. 31.72. The metal deformation is such that forming is in accordance with the sine law, which states that the wall thickness of the starting blank and that of the finished workpiece are related as

$$t_2 = t_1 (\sin \alpha)$$

where t_1 is the thickness of the starting blank, t_2 is the thickness of the spun workpiece, and α is one-half the apex angle of the cone.

Tube Spinning. Tube spinning is a rotary-point method of extruding metal, much like cone spinning, except that the sine law does not apply. Because the half-angle of a cylinder is zero, tube spinning follows a purely volumetric rule, depending on the practical limits of deformation that the metal can stand without intermediate annealing.

Pipe Welding. Large quantities of small-diameter steel pipe are produced by two processes that involve hot forming of metal strip and welding of its edges through utilization of the heat contained in the metal. Both of these processes, butt welding and lap welding of pipe, utilize steel in the form of skelp—long and narrow strips of the desired thickness. Because the skelp has been previously hot rolled and the welding process produces further compressive working and recrystallization, pipe welding by these processes is uniform in quality.

In the butt-welded pipe process, the skelp is unwound from a continuous coil and is heated to forging temperatures as it passes through a furnace. Upon leaving the furnace, it is pulled through forming rolls that shape it into a cylinder. The pressure exerted between the edges of the skelp as it passes through the rolls is sufficient to upset the metal and weld the edges together. Additional sets of rollers size and shape the pipe. Normal pipe diameters range from ⅛ to 3 in. (3 to 75 mm).

The lap-welding process for making pipe differs from butt welding in that the skelp has beveled edges and a mandrel is used in conjunction with a set of rollers to make the weld. The process is used primarily for larger sizes of pipe, from about 2 to 14 in. (50 to 400 mm) in diameter.

Piercing. Thick-walled and seamless tubing is made by the piercing process. A heated, round billet, with its leading end center-punched, is pushed longitudinally between two large, convex-tapered rolls that revolve in the same direction, their axes being inclined at opposite angles of about 6° from the axis of the billet. The clearance between the rolls is somewhat less than the diameter of the billet. As the billet is caught by the rolls and rotated, their inclination causes the billet to be drawn forward into them. The reduced clearance between the rolls forces the rotating billet to deform into an elliptical shape. To rotate with an elliptical cross section, the metal must undergo shear about the major axis, which causes a crack to open. As the crack opens, the billet is forced over a pointed mandrel that enlarges and shapes the opening, forming a seamless tube.

31.5.2 Cold-Working Processes

Cold working is the plastic deformation of metals below the recrystallization temperature. In most cases of manufacturing, such cold forming is done at room temperature. In some cases, however, the

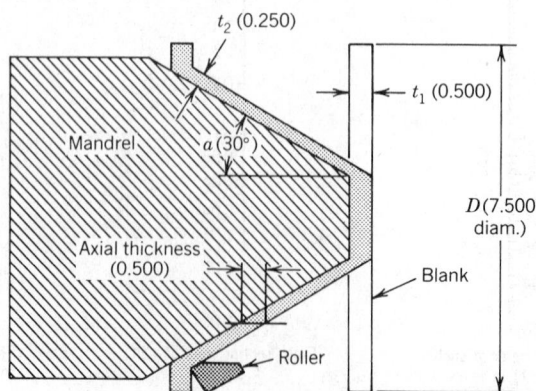

Fig. 31.72 Setup and dimensional relations for one-operation power spinning of a cone. See text for application of sine law in relation to this illustration.

working may be done at elevated temperatures that will provide increased ductility and reduced strength, but will be below the recrystallization temperature.

When compared to hot working, cold-working processes have certain distinct advantages:

1. No heating is required.
2. Better surface finish is obtained.
3. Superior dimension control.
4. Better reproducibility and interchangeability of parts.
5. Improved strength properties.
6. Directional properties can be imparted.
7. Contamination problems are minimized.

Some disadvantages associated with cold-working processes include:

1. Higher forces are required for deformation.
2. Heavier and more powerful equipment is required.
3. Less ductility is available.
4. Metal surfaces must be clean and scale-free.
5. Strain hardening occurs (may require intermediate anneals).
6. Imparted directional properties may be detrimental.
7. May produce undesirable residual stresses.

Classification of Cold-Working Operations

The major cold-working operations can be classified basically under the heading of squeezing, bending, shearing, and drawing as follows:

Squeezing		Bending	
1.	Rolling	1.	Angle
2.	Swaging	2.	Roll
3.	Cold forging	3.	Roll forming
4.	Sizing	4.	Seaming
5.	Extrusion	5.	Flanging
6.	Riveting	6.	Straightening
7.	Staking		
8.	Coining		
9.	Peening		
10.	Burnishing		
11.	Die hobbing (hubbing)		
12.	Thread rolling		

Shearing		Drawing	
1.	Shearing	1.	Bar and tube drawing
	Slitting	2.	Wire drawing
2.	Blanking	3.	Spinning
3.	Piercing	4.	Embossing
	Lancing	5.	Stretch forming
	Perforating	6.	Shell drawing
4.	Notching	7.	Ironing
	Nibbling		
5.	Shaving		
6.	Trimming		
7.	Cutoff		
8.	Dinking		

Squeezing Processes

Most of the cold-working squeezing processes have identical hot-working counterparts or are extensions of them. The primary reasons for deforming cold rather than hot are to obtain better dimensional accuracy and surface finish. In many cases the equipment is basically the same, except that it must be more powerful.

Cold Rolling Cold rolling accounts for by far the greatest tonnage of cold-worked products. Sheets, strip, bars, and rods are cold-rolled to obtain products that have smooth surfaces and accurate dimensions.

Swaging. Swaging basically is a process for reducing the diameter, tapering, or pointing round bars or tubes by external hammering. A useful extension of the process involves the formation of internal cavities. A shaped mandrel is inserted inside a tube, and the tube is then collapsed around it by swaging.

Cold Forging. Extremely large quantities of products are made by cold forging, in which the metal is squeezed into a die cavity that imparts the desired shape. Cold heading is used for making enlarged sections on the ends of a piece or rod or wire, such as the heads on bolts, nails, rivets, and other fasteners.

Sizing. Sizing involves squeezing areas of forgings or ductile castings to a desired thickness. It is used principally on bosses and flats, with only enough deformation to bring the region to a desired dimension.

Extrusion. This process is often called impact extrusion and was first used only with the low-strength ductile metals such as lead, tin, and aluminum for producing such items as collapsible tubes for toothpaste, medications, and so forth; small "cans" such as are used for shielding in electronics and electrical apparatus; and larger cans for food and beverages. In recent years, cold extrusion has been used for forming mild steel parts, often being combined with cold heading.

Another type of cold extrusion, known as hydrostatic extrusion, uses high fluid pressure to extrude a billet through a die, either into atmospheric pressure or into a lower-pressure chamber. The pressure-to-pressure process makes possible the extrusion of relatively brittle materials such as molybdenum, beryllium, and tungsten. Billet–chamber friction is eliminated, billet–die lubrication is enhanced by the pressure, and the surrounding pressurized atmosphere suppresses crack initiation and growth.

Riveting. In riveting, a head is formed on the shank end of a fastener to provide a permanent method of joining sheets or plates of metal together. Although riveting usually is done hot in structural work, in manufacturing it almost always is done cold.

Staking. Staking is a commonly used cold-working method for permanently fastening two parts together where one protrudes through a hole in the other. A shaped punch is driven into one of the pieces, deforming the metal sufficiently to squeeze it outward.

Coining. Coining involves cold working by means of positive displacement punch while the metal is completely confined within a set of dies.

Surface Improvement by Cold Working. Two cold-working methods are used extensively for improving or altering the surface of metal products. Peening involves striking the surface repeated blows by impelled shot or a round-nose tool. The highly localized blows deform and tend to stretch the metal surface. Because the surface deformation is resisted by the metal underneath, the result is a surface layer under residual compression. This condition is highly favorable to resist cracking under fatigue conditions, such as repeated bending, because the compressive stresses are subtractive from the applied tensile loads. For this reason, shafting, crankshafts, gear teeth, and other cyclic-loaded components are frequently peened.

Burnishing involves rubbing a smooth hard object under considerable pressure over the minute surface protrusions that are formed on a metal surface during machining or shearing, thereby reducing their depth and sharpness through plastic flow.

Hobbing. Hobbing is a cold-working process that is used to form cavities in various types of dies, such as those used for molding plastics. A male hob is made with the contour of the part that ultimately will be formed by the die. After the hob is hardened, it is slowly pressed into an annealed die block by means of hydraulic press until the desired impression is produced.

Thread Rolling. Threads can be rolled in any material sufficiently plastic to withstand the forces of cold working without disintegration. Threads can be rolled by flat or roller dies. Thread rolling was discussed in Section 31.3.7.

Bending

Bending is the method of producing shapes by stressing material beyond its yield strength but below its ultimate strength. Figure 31.73 illustrates bending terms. To estimate the press capacity needed for bending in V-dies, the bending load in tons can be computed from

$$F = \frac{LT^2 kS}{s}$$

where F is press load, in tons; L is length of bend (parallel to bend axis), in inches; T is work-metal thickness, in inches; k is a die-opening factor (varying from 1.2 for a die opening of $16T$ to 1.33 for a die opening of $8T$); S is tensile strength of the work metal, in tons per square inch; and s is width of die opening, in inches.

For U-dies, the constant k should be twice the value shown above. Ultimate strengths S for various materials are

Metal	Ultimate Strength (ton/in.^{2})
Aluminum and alloys	6.5–38.0
Brass	19.0–38.0
Bronze	31.5–47.0
Copper	16.0–25.0
Steel	22.0–40.0
Tin	1.1– 1.4
Zinc	9.7–13.5

Several factors must be considered when designing parts that are to be made by bending. Of primary importance is the minimum radius that can be bent successfully without metal cracking. This, of course, is related to the ductility of the metal.

Angle Bending. Angle bends up to 150° in sheet metal under about $\frac{1}{16}$ in. (1.5 mm) in thickness may be made in a bar folder. Heavier sheet metal and more complex bends in thinner sheets are made on a press brake.

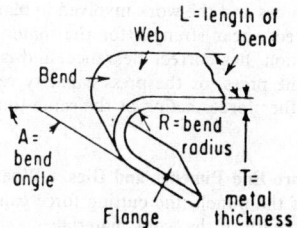

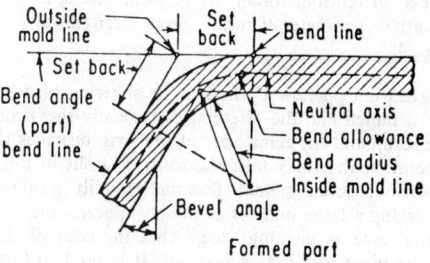

Fig. 31.73 Bend terms.

Roll Bending. Plates, heavy sheets, and rolled shapes can be bent to a desired curvature on forming rolls. These usually have three rolls in the form of a pyramid, with the two lower rolls being driven and the upper roll adjustable to control the degree of curvature. Supports can be swung clear to permit removal of a closed shape from the rolls. Bending rolls are available in a wide range of sizes, some being capable of bending plate up to 6 in. (150 mm) thick.

Cold-Roll Forming. This process involves the progressive bending of metal strip as it passes through a series of forming rolls. A wide variety of moldings, channeling, and other shapes can be formed on machines that produce up to 10,000 ft (3000 m) of product per day.

Seaming. Seaming is used to join ends of sheet metal to form containers such as cans, pails, and drums. The seams are formed by a series of small rollers on seaming machines that range from small hand-operated types to large automatic units capable of producing hundreds of seams per minute in the mass production of cans.

Flanging. Flanges can be rolled on sheet metal in essentially the same manner as seaming is done. In many cases, however, the forming of flanges and seams involves drawing, since localized bending occurs on a curved axis.

Straightening. Straightening or flattening has as its objective the opposite of bending and often is done before other cold-forming operations to ensure that flat or straight material is available. Two different techniques are quite common. Roll straightening or roller leveling involves a series of reverse bends. The rod, sheet, or wire is passed through a series of rolls having decreased offsets from a straight line. These bend the metal back and forth in all directions, stressing it slightly beyond its previous elastic limit and thereby removing all previous permanent set.

Sheet may also be straightened by a process called stretcher leveling. The sheets are grabbed mechanically at each end and stretched slightly beyond the elastic limit to remove previous stresses and thus produce the desired flatness.

Shearing

Shearing is the mechanical cutting of materials in sheet or plate form without the formation of chips or use of burning or melting. When the two cutting blades are straight, the process is called shearing. Other processes, in which the shearing blades are in the form of the curved edges of punches and dies, are called by other names, such as blanking, piercing, notching, shaving, and trimming. These all are basically shearing operations however.

Blanking. A blank is a shape cut from flat or preformed stock. Ordinarily, a blank serves as a starting workpiece for a formed part; less often, it is a desired end product. Figure 31.74 shows the blanking operation.

Calculation of the forces and the work involved in blanking gives average figures that are applicable only when (a) the correct shear strength for the material is used; and (b) the die is sharp and the punch is in good condition, has correct clearance, and is functioning properly.

The total load on the press, or the press capacity required to do a particular job, is the sum of the cutting force and other forces acting at the same time, such as the blankholding force exerted by a die cushion.

Cutting Force: Square-End Punches and Dies. When punch and die surfaces are flat and at right angles to the motion of the punch, the cutting force can be found by multiplying the area of the cut section by the shear strength of the work material:

$$L = S_8 tl$$

where L is load on the press, lb (cutting force); S_8 is shear strength of the stock, psi; t is stock thickness, in.; and l is the length or perimeter of cut, in. Shear strengths of various steels and nonferrous metals are given in Table 31.24.

Piercing. Piercing is a shearing operation wherein the shearing blades take the form of closed, curved lines on the edges of a punch and die. Piercing is basically the same as blanking except that the piece punched out is the scrap and the remainder of the strip becomes the desired workpiece.

Lancing is a piercing operation that may take the form of a slit in the metal or an actual hole. The purpose of lancing is to permit adjacent metal to flow more readily in subsequent forming operations.

Perforating consists of piercing a large number of closely spaced holes.

Notching is essentially the same as piercing except that the edge of the sheet of metal forms a portion of the periphery of the piece that is punched out. It is used to form notches of any desired shape along the edge of a sheet.

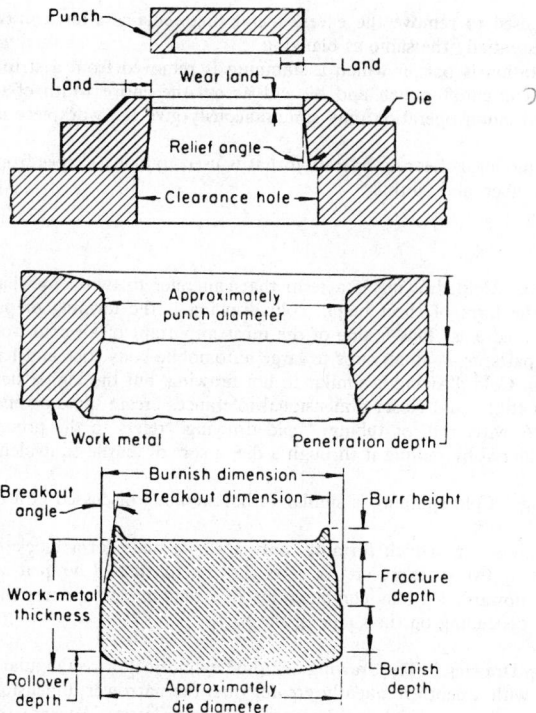

Fig. 31.74 Blanking die and characteristics of the sheared edges of a blank. Curvature and angles are exaggerated for emphasis.

Nibbling is a variation of notching in which a special machine makes a series of overlapping notches, each further into the sheet of metal.

Shaving is a finished operation in which a very small amount of metal is sheared away around the edge of a blanked part. Its primary use is to obtain greater dimensional accuracy, but it also may be used to obtain a square or smoother edge.

Table 31.24 Shear Strengths of Various Steels on Nonferrous Metals at Room Temperature

Metal	Shear Strength (psi)
Carbon steels	
0.10% C	35,000–43,000
0.20% C	44,000–55,000
0.30% C	52,000–67,000
High-strength, low-alloy steels	45,000–63,700
Silicon steels	60,000–70,000
Stainless steels	57,000–129,000
Nonferrous Metals	
Aluminum alloys	7000–46,000
Copper and bronze	22,000–70,000
Lead alloys	1825–5870
Magnesium alloys	17,000–29,000
Nickel alloys	35,000–116,000
Tin alloys	2900–11,100
Titanium alloys	60,000–70,000
Zinc alloys	14,000–38,000

Trimming is used to remove the excess metal that remains after a drawing, forging, or casting operation. It is essentially the same as blanking.

A cutoff operation is one in which a stamping is removed from a strip of stock by means of a punch and die. The cutoff punch and die cut across the entire width of the strip. Frequently, an irregularly shaped cutoff operation may simultaneously give the workpiece all or part of the desired shape.

Dinking is a modified shearing operation that is used to blank shapes from low-strength materials, primarily rubber, fiber, and cloth.

Drawing

Cold Drawing. Cold drawing is a term that can refer to two somewhat different operations. If the stock is in the form of sheet metal, cold drawing is the forming of parts wherein plastic flow occurs over a curved axis. This is one of the most important of all cold-working operations because a wide range of parts, from small cups to large automobile body tops and fenders, can be drawn in a few seconds each. Cold drawing is similar to hot drawing, but the higher deformation forces, thinner metal, limited ductility, and closer dimensional tolerances create some distinctive problems.

If the stock is wire, rod, or tubing, "cold drawing" refers to the process of reducing the cross section of the material by pulling it through a die, a sort of tensile equivalent to extrusion.

Cold Spinning. Cold spinning is similar to hot spinning discussed in a previous subsection.

Stretch Forming. In stretch forming, only a single male form block is required. The sheet of metal is gripped by two or more sets of jaws that stretch it and wrap it around the form block as the latter raises upward. Various combinations of stretching, wrapping, and upward motion of the blocks are used, depending on the shape of the part.

Shell or Deep Drawing. The drawing of closed cylindrical or rectangular containers, or a variation of these shapes, with a depth frequently greater than the narrower dimension of their opening, is one of the most important and widely used manufacturing processes. Because the process had its earliest uses in manufacturing artillery shells and cartridge cases, it is sometimes called shell drawing. When the depth of the drawn part is less than the diameter, or minimum surface dimension, of the blank, the process is considered to be shallow drawing. If the depth is greater than the diameter, it is considered to be deep drawing.

The design of complex parts that are to be drawn has been aided considerably by computer techniques, but is far from being completely and successfully solved. Consequently, such design still involves a mix of science, experience, empirical data, and actual experimentation. The body of known information is quite substantial, however, and is being used with outstanding results.

Forming with Rubber or Fluid Pressure. Several methods of forming use rubber or fluid pressure to obtain the desired information and thereby eliminate either the male or female member of the die set. Blanks of sheet metal are placed on top of form blocks, which usually are made of wood. The upper ram, which contains a pad of rubber 8–10 in. (200–250 mm) thick in a steel container, then descends. The rubber pad is confined and transmits force to the metal, causing it to bend to the desired shape. Since no female die is used and form blocks replace the male die, die cost is quite low.

The hydroform process or "rubber bag forming" replaces the rubber pad with a flexible diaphragm back by controlled hydraulic pressure. Deeper parts can be formed with truly uniform fluid pressure.

In bulging, oil or rubber is used for applying an internal bulging force to expand a metal blank or tube outward against a female mold or die, and thereby eliminates the necessity for a complicated, multiple-piece male die member.

Ironing. Ironing is the name given to the process of thinning the walls of a drawn cylinder by passing it between a punch and a die where the separation is less than the original wall thickness. The walls are elongated and thinned while the base remains unchanged. The most common example of an ironed product is the thin-walled all-aluminum beverage can.

Embossing. Embossing is a method for producing lettering or other designs in thin sheet metal. Basically, it is a very shallow drawing operation, usually in open dies, with the depth of the draw being from one to three times the thickness of the metal.

High-Energy-Rate Forming. A number of methods have been developed for forming metals through the release and application of large amounts of energy in a very short interval. These processes are called high-energy-rate forming processes, commonly designated HERF. Many metals tend to deform more readily under the ultrarapid rates of load application used in these processes, a phenomenon apparently related to the relative rates of load application and the movement of dislocations through the metal. As a consequence, HERF makes it possible to form large workpieces and difficult-to-form metals with less-expensive equipment and tooling than would otherwise be required.

The high energy-release rates are obtained by five methods: (1) underwater explosions, (2) underwater spark discharge (electrohydraulic techniques), (3) pneumatic–mechanical means, (4) internal combustion of gaseous mixtures, and (5) rapidly formed magnetic fields (electromagnetic techniques).

31.6 SURFACE TREATMENT

Products that have been completed to their proper shape and size frequently require some type of surface finishing to enable them to satisfactorily fulfill their function. In some cases, it is necessary to improve the physical properties of the surface material for resistance to penetration or abrasion.

Surface finishing may sometimes become an intermediate step in processing. For instance, cleaning and polishing are usually essential before any kind of plating process. Another important need for surface finishing is for corrosion protection in a variety of environments. The type of protection provided will depend largely on the anticipated exposure, with due consideration to the material being protected and the economic factors involved.

Satisfying the above objectives necessitates the use of many surface-finishing methods that involve chemical change of the surface; mechanical work affecting surface properties; cleaning by a variety of methods; and the application of protective coatings, organic and metallic.

31.6.1 Cleaning

Few, if any, shaping and sizing processes produce products that are usable without some type of cleaning unless special precautions are taken. Figure 31.75 indicates some of the cleaning methods available.

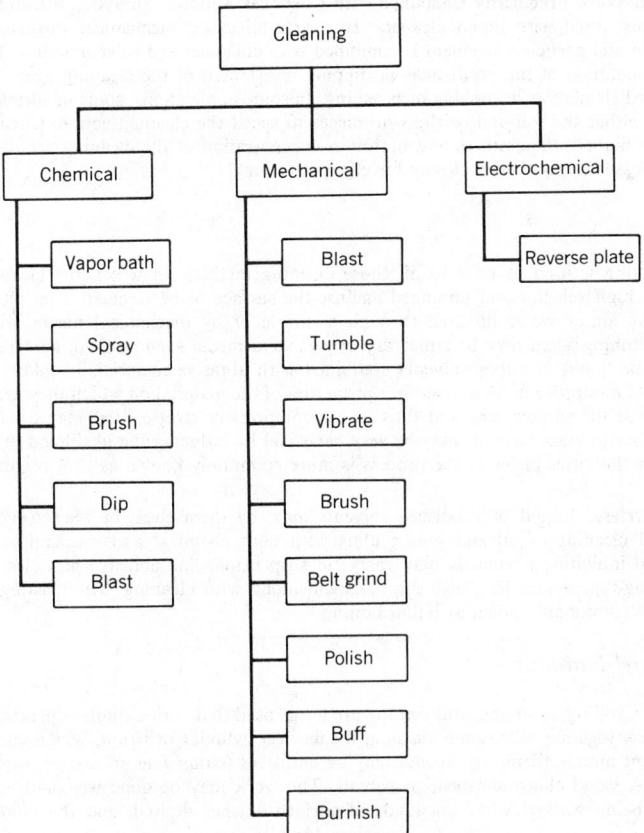

Fig. 31.75 Cleaning methods.

Some cleaning methods provide multiple benefits. Cleaning and finish improvements are often combined. Probably of even greater importance is the combination of corrosion protection with finish improvement, although corrosion protection is more often a second step that involves coating an already cleaned surface with some other material or chemical conversion.

Liquid and Vapor Baths

Liquid and Vapor Solvents. The most widely used cleaning methods make use of a cleaning medium in liquid or vapor form. These methods depend on a solvent or chemical action between the surface contaminants and the cleaning material.

Petroleum Solvents. Among the more common cleaning jobs required is the removal of grease and oil deposited during manufacturing or intentionally coated on the work to provide protection. One of the most efficient ways to remove this material is by use of solvents that dissolve the grease and oil but have no effect on the base metal. Petroleum derivatives such as Stoddard solvent and kerosene are common for this purpose, but, since they introduce some danger of fire, chlorinated solvents, such as trichlorethylene, that are free of this fault are sometimes substituted.

Conditioned Water. One of the most economical cleaning materials is water. However, it is seldom used alone even if the contaminant is fully water soluble, because the impurity of the water itself may contaminate the work surface. Depending on its use, water is treated with various acids and alkalies to suit the job being performed.

Pickling. Water containing sulfuric acid in a concentration from about 10 to 25% and at a temperature of approximately 149°F (65°C) is commonly used in a process called pickling for removal of surface oxides or scale on iron and steel.

Mechanical Work Frequently Combined with Chemical Action. Spraying, brushing, and dipping methods are also used with liquid cleaners. In nearly all cases, mechanical work to cause surface film breakdown and particle movement is combined with chemical and solvent action. The mechanical work may be agitation of the product as in dipping, movement of the cleaning agent as in spraying, or use of a third element as in rubbing or brushing. In some applications, sonic or ultrasonic vibrations are applied to either the solution or the workpieces to speed the cleaning action. Chemical activity is increased with higher temperatures and optimum concentration of the cleaning agent, both of which must in some cases be controlled closely for efficient action.

Blasting

The term blasting is used to refer to all those cleaning methods in which the cleaning medium is accelerated to high velocity and impinged against the surface to be cleaned. The high velocity may be provided by air or water directed through a nozzle or by mechanical means with a revolving slinger. The cleaning agent may be either dry or wet solid media such as sand, abrasive, steel grit, or shot, or may be liquid or vapor solvents combined with abrasive material. In addition to cleaning, solid particles can improve finish and surface properties of the material on which they are used. Blasting tends to increase the surface area and thus set up compressive stresses that may cause a warping of thin sections, but in other cases, it may be very beneficial by reducing the likelihood of fatigue failure. When used for the latter purpose, the process is more commonly known as shot peening.

Water Slurries. Liquid or vaporized solvents may, by themselves, be blasted against a surface for high-speed cleaning of oil and grease films with both chemical and mechanical action. Water containing rust-inhibiting chemicals may carry, in suspension, fine abrasive particles that provide a grinding cutting-type action for finish improvement along with cleaning. The blasting method using this medium is commonly known as liquid honing.

Abrasive Barrel Finishing

Barrel finishing, rolling, tumbling, and rattling are terms used to describe similar operations that consist of packing parts together with some cleaning media in a cylinder or drum, which can be rotated to cause movement among them. The media may be abrasives (either fine or coarse); metal stars, slugs, or balls; stones; wood chips; sawdust; or cereals. The work may be done wet or dry, depending on the materials being worked with, the kinds of surface finishes desired, and the kind of equipment available.

Wire Brushing

A number of cleaning operations can be quickly and easily performed by use of a high-speed rotating wire brush. In addition to cleaning, the contact and rubbing of the wire ends across the work surface produce surface improvement by a burnishing-type action. Sharp edges and burs can be removed.

Abrasive Belt Finishing

Continuous fabric belts coated with abrasive can be driven in several kinds of machines to provide a straight-line cutting motion for grinding, smoothing, and polishing work surfaces. Plane surfaces are the most common surfaces worked on with fabric belts.

Polishing

The term polishing may be interpreted to mean any nonprecision procedure providing a glossy surface but is most commonly used to refer to a surface-finishing process using a flexible abrasive wheel. The wheels may be constructed of felt or rubber with an abrasive band, of multiple coated abrasive discs, of leaves of coated abrasive, of felt or fabric to which loose abrasive is added as needed, or of abrasives in a rubber matrix.

Buffing

About the only difference between buffing and polishing is that, for buffing, a fine abrasive carried in wax or a similar substance is charged on the surface of a flexible wheel.

Electropolishing

If a workpiece is suspended in an electrolyte and connected to the anode in an electrical circuit, it will supply metal to the electrolyte in a reverse plating process. Material will be removed faster from the high spots of the surface than from the depressions and will thereby increase the average smoothness. The cost of the process is prohibitive for very rough surfaces because larger amounts of metal must be removed to improve surface finish than would be necessary for the same degree of improvement by mechanical polishing. Electropolishing is economical only for improving a surface that is already good or for polishing complex and irregular shapes, the surfaces of which are not accessible to mechanical polishing and buffing equipment.

31.6.2 Coatings

Many products, in particular those exposed to view and those subject to change by the environment with which they are in contact, need some type of coating for improved appearance or for protection from chemical attack. The need for corrosion protection for maintenance and appearance is important. In addition to change of appearance, loss of actual material, change of dimensions, and decrease of strength, corrosion may be the cause of eventual loss of service or failure of a product. Material that must carry loads in structural applications, especially when the loads are cyclic in nature, may fail with fatigue if corrosion is allowed to take place. Corrosion occurs more readily in highly stressed material where it attacks grain boundaries in such a way as to form points of stress concentration that may be nuclei for fatigue failure.

Hardness and wear resistance, however, can be provided on a surface by plating with hard metals. Chromium plating of gages and other parts subject to abrasion is frequently used to increase their wear life. Coatings of plastic material and asphaltic mixtures are sometimes placed on surfaces to provide sound deadening. The additional benefit of protection from corrosion is usually acquired at the same time.

Plastics of many kinds, mostly of the thermoplastic type because they are easier to apply and also easier to remove later if necessary, are used for mechanical protection. Highly polished material may be coated with plastic, which may be stripped off later, to prevent abrasion and scratches during processing. It is common practice to coat newly sharpened cutting edges of tools by dipping them in thermoplastic material to provide mechanical protection during handling and storage.

Organic Coatings

Organic coatings are used to provide pleasing colors, to smooth surfaces, to provide uniformity in both color and texture, and to act as a protective film for control of corrosion. Organic resin coatings do not ordinarily supply any chemical-inhibiting qualities. Instead, they merely provide a separating film between the surface to be protected and the corrosive environment. The important properties, therefore, are continuity, permeability, and adhesion characteristics.

Paints, Varnishes, and Enamels

Paints. Painting is a generic term that has come to mean the application of almost any kind of organic coating by any method. Because of this interpretation, it is also used generally to describe a broad class of products. As originally defined and as used most at present, paint is a mixture of pigment in a drying oil. The oil serves as a carrier for the pigment and in addition creates a tough continuous film as it dries. Drying oils, one of the common ones of which is linseed oil, become solid when large surface areas are exposed to air. Drying starts with a chemical reaction of oxidation. Nonreversible polymerization accompanies oxidation to complete the change from liquid to solid.

Varnish. Varnish is a combination of natural or synthetic resins and drying oil, sometimes containing volatile solvents as well. The material dries by a chemical reaction in the drying oil to a clear or slightly amber-colored film.

Enamel. Enamel is a mixture of pigment in varnish. The resins in the varnish cause the material to dry to a smoother, harder, and glossier surface than is produced by ordinary paints. Some enamels are made with thermosetting resins that must be baked for complete dryness. These baking enamels provide a toughness and durability not usually available with ordinary paints and enamels.

Lacquers

The term lacquer is used to refer to finishes consisting of thermoplastic materials dissolved in fast-drying solvents. One common combination is cellulose nitrate dissolved in butyl acetate. Present-day lacquers are strictly air drying and form films very quickly after being applied, usually by spraying. No chemical change occurs during the hardening of lacquers, consequently, the dry film can be redissolved in the thinner. Cellulose acetate is used in place of cellulose nitrate in some lacquers because it is nonflammable. Vinyls, chlorinated hydrocarbons, acrylics, and other synthetic thermoplastic resins are also used in the manufacture of lacquers.

Vitreous Enamels

Vitreous, or porcelain, enamel is actually a thin layer of glass fused onto the surface of a metal, usually steel or iron. Shattered glass, ball milled in a fine particle size, is called frit. Frit is mixed with clay, water, and metal oxides, which produce the desired color, to form a thin slurry called slip. This is applied to the prepared metal surface by dipping or spraying and, after drying, is fired at approximately 1470°F (800°C) to fuse the material to the metal surface.

Metallizing

Metal spraying, or metallizing, is a process in which metal wire or powder is fed into an oxyacetylene heating flame and, then, after melting, is carried by high-velocity air to be impinged against the work surface. The small droplets adhere to the surface and bond together to build up a coating.

Vacuum Metallizing

Some metals can be deposited in very thin films, usually for reflective or decorative purposes, as a vapor deposit. The metal is vaporized in a high vacuum chamber containing the parts to be coated. The metal vapor condenses on the exposed surfaces in a thin film that follows the surface pattern. The process is cheap for coating small parts, considering the time element only, but the cost of special equipment needed is relatively high.

Aluminum is the most used metal for deposit by this method and is used frequently for decorating or producing a mirror surface on plastics. The thin films usually require mechanical protection by covering with lacquer or some other coating material.

Hot-Dip Plating

Several metals, mainly zinc, tin, and lead, are applied to steel for corrosion protection by a hot-dip process. Steel in sheet, rod, pipe, or fabricated form, properly cleansed and fluxed, is immersed in molten plating metal. As the work is withdrawn, the molten metal that adheres solidifies to form a protective coat. In some of the large mills, the application is made continuously to coil stock that is fed through the necessary baths and even finally inspected before being recoiled or cut into sheets.

Electroplating

Coatings of many metals can be deposited on other metals, and on nonmetals when suitably prepared, by electroplating. The objectives of plating are to provide protection against corrosion, to improve appearance, to establish wear- and abrasion-resistant surfaces, to add material for dimensional increase, and to serve as an intermediate step of multiple coating. Some of the most common metals deposited in this way are copper, nickel, cadmium, zinc, tin, silver, and gold. The majority are used to provide some kind of corrosion protection but appearance also plays a strong part in their use.

Temporary Corrosion Protection

It is not uncommon in industry for periods of time, sometimes quite long periods, to elapse between manufacture, assembly, shipment, and use of parts. Unless a new processing schedule can be worked out, about the only cure for the problem is corrosion protection suitable for the storage time and exposure. The coatings used are usually nondrying organic materials called slushing compounds that can be removed easily. The two principal types of compounds used for this purpose are petroleum-based materials, varying from extremely light oils to semisolids, and thermoplastics. The most common method of application of slushing compounds for small parts is by dipping. Larger parts that cannot be handled easily may be sprayed, brushed, or flow coated with the compound.

31.6.3 Chemical Conversions

A relatively simple and often fully satisfactory method for protection from corrosion is by conversion of some of the surface material to a chemical composition that resists attack from the environment. These converted metal surfaces consist of relatively thin [seldom more than 0.001 in. (0.025 mm) thick] inorganic films that are formed by chemical reaction with the base material. One important feature of the conversion process is that the coatings have little effect on the product dimensions.

Anodizing

Aluminum, magnesium, and zinc can be treated electrically in a suitable electrolyte to produce a corrosion-resistant oxide coating. The metal being treated is connected to the anode in the circuit, which provides the name anodizing for the process. Aluminum is commonly treated by anodizing that produces an oxide film thicker than, but similar to, that formed naturally with exposure to air. Anodizing of zinc has very limited use. The coating produced on magnesium is not as protective as that formed on aluminum, but does provide some protective value and substantially increases protection when used in combination with paint coatings.

Chromate Coatings

Zinc is usually considered to have relatively good corrosion resistance. This is true when the exposure is to normal outdoor atmosphere where a relatively thin corrosion film forms. Contact with either highly aerated water films or immersion in stagnant water containing little oxygen causes uneven corrosion and pitting. The corrosion products of zinc are less dense than the base material so that heavy corrosion not only destroys the product appearance but also may cause malfunction by binding moving parts. Corrosion of zinc can be substantially slowed by the production of chromium salts on its surface. The corrosion resistance of magnesium alloys can be increased by immersion or anodic treatment in acid baths containing dichromates. Chromate treatment of both zinc and magnesium improves corrosion resistance, but is used also to improve adhesion of paint.

Phosphate Coatings

Phosphate coatings, used mostly on steel, result from a chemical reaction of phosphoric acid with the metal to form a nonmetallic coating that is essentially phosphoric salts. The coating is produced by immersing small items or spraying large items with the phosphating solution. Phosphate surfaces may be used alone for corrosion resistance, but their most common application is as a base for paint coatings. Two of the most common application methods are called parkerizing and bonderizing.

Chemical Oxide Coatings

A number of proprietary blacking processes, used mainly on steel, produce attractive black oxide coatings. Most of the processes involve the immersing of steel in a caustic soda solution, heated to about 300°F (150°C) and made strongly oxidizing by the addition of nitrites or nitrates. Corrosion resistance is rather poor unless improved by application of oil, lacquer, or wax. As in the case of most of the other chemical-conversion procedures, this procedure also finds use as a base for paint finishes.

31.7 MEASUREMENTS AND QUALITY CONTROL

Metric and English measuring systems are the two measuring systems commonly used throughout the world. The metric system is universally used in most scientific applications, but, for manufacturing in the United States, has been limited to a few specialties, mostly items that are related in some way to products manufactured abroad.

31.7.1 Dimension and Tolerance

In dimensioning a drawing, the numbers placed in the dimension lines represent dimensions that are only approximate and do not represent any degree of accuracy unless so stated by the designer. To specify the degree of accuracy, it is necessary to add tolerance figures to the dimension. Tolerance is the amount of variation permitted in the part or the total variation allowed in a given dimension.

Dimensions given close tolerances mean that the part must fit properly with some other part. Both must be given tolerances in keeping with the allowance desired, the manufacturing processes available, and the minimum cost of production and assembly that will maximize profit. Generally speaking, the cost of a part goes up as the tolerance is decreased.

Allowance, which is sometimes confused with tolerance, has an altogether different meaning. It is the minimum clearance space intended between mating parts and represents the condition of tightest permissible fit.

31.7.2 Quality Control

When parts must be inspected in large numbers, 100% inspection of each part is not only slow and costly, but, in addition, does not eliminate all of the defective pieces. Mass inspection tends to be careless; operators become fatigued; and inspection gages become worn or out of adjustment more frequently. The risk of passing defective parts is variable and of unknown magnitude, whereas in a planned sampling procedure the risk can be calculated. Many products such as bulbs cannot be 100% inspected, since any final test made on one results in the destruction of the product. Inspection is costly and nothing is added to a product that has been produced to specifications.

Quality control enables an inspector to sample the parts being produced in a mathematical manner and to determine whether or not the entire stream of production is acceptable, provided that the company is willing to allow up to a certain known number of defective parts. This number of acceptable defectives is usually taken as 3 out of 1000 parts produced. Other values might be used.

To use quality control techniques in inspection, the following steps must be taken:

1. Sample the stream of products by taking N samples, each of size n.
2. Measure the desired dimensions in the sample, mainly the central tendency.
3. Calculate the deviations of the dimensions.
4. Construct a control chart.
5. Plot succeeding data on the control chart.

The arithmetic mean of the set of n units is the main measure of central tendency. The symbol $\bar{X}$ is used to designate the arithmetic mean of the sample and may be expressed in algebraic terms as

$$\bar{X} = (X_1 + X_2 + X_3 + \cdots + X_n)/n$$

where X_1, X_2, X_3, etc., represent the specific dimensions in question. The most useful measure of dispersion of a set of numbers is the standard deviation σ. It is defined as the root-mean-square deviation of the observed numbers from their arithmetic mean. The standard deviation σ is expressed in algebraic terms as

$$\sigma = \sqrt{\frac{(X_1 - \bar{X})^2 + (X_2 - \bar{X})^2 + \cdots + (X_n - \bar{X})^2}{n}}$$

Another important measure of dispersion, used particularly in control charts, is the range R. The range is the difference between the largest observed value and the smallest observed in a specific sample. Even though the distribution of the X values in the universe can be of any shape, the distribution of the $\bar{X}$ values tends to be close to the normal distribution. The larger the sample size and the more nearly normal the universe, the closer will the frequency distribution of the average $\bar{X}$'s approach the normal curve, as in Fig. 31.76.

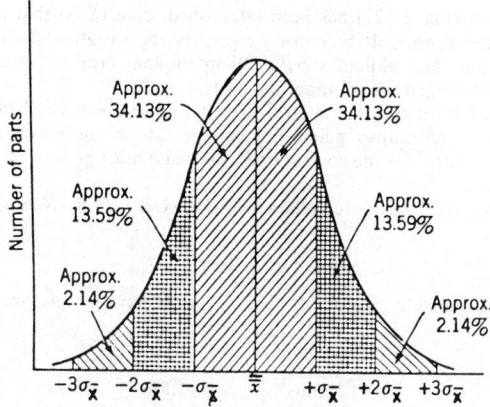

Fig. 31.76 Normal distribution and percentage of parts that will fall within σ limits.

According to the statistical theory—the Central Limit Theory—in the long run, the average of the $\overline{X}$ values will be the same as $\overline{X}'$, the average of the universe. And in the long run, the standard deviation of the frequency distribution of $\overline{X}$ values, $\sigma_{\overline{X}}$ will be given by

$$\sigma_{\overline{X}} = \frac{\sigma'}{\sqrt{n}}$$

where σ' is the standard deviation of the universe. To construct the control limits the following steps are taken:

1. Calculate the average of the average $\overline{X}$ as follows:

$$\overline{\overline{X}} = \sum_{1}^{N} \overline{X}_i/N, \qquad i = 1, 2, \ldots, N$$

2. Calculate the average deviation, $\overline{\sigma}$, where

$$\overline{\sigma} = \sum_{1}^{N} \sigma_i/N, \qquad i = 1, 2, \ldots, N$$

Statistical theory predicts the relationship between $\overline{\sigma}$ and $\sigma_{\overline{X}}$. The relationship for the $3\sigma_{\overline{X}}$ limits on the 99.73% limits is

$$A_1 \overline{\sigma} = 3\sigma_{\overline{X}}$$

This means that control limits are set so that only 0.27% of the produced units will fall outside the limits. The value of $3\sigma_{\overline{X}}$ is an arbitrary limit that has found acceptance in industry.

The value of A_1 calculated by probability theory is dependent on the number of pieces, as follows:

Sample Size	A_1
2 units	3.76
3 units	2.39
4 units	1.88
5 units	1.60
10 units	1.03

The formulas for 3-σ control limits using this factor are

Upper Control Limit

$$\text{UCL } \overline{X} = \overline{\overline{X}} + A_1 \overline{\sigma}$$

Lower Control Limit

$$\text{LCL } \overline{X} = \overline{\overline{X}} - A_1 \overline{\sigma}$$

Once the control chart (Fig. 31.77) has been established, data ($\bar{X}_i$'s) that result from samples of the same size n are recorded on it. It becomes a record of the variation of the inspected dimension over a period of time. The data plotted should fall in random fashion between the control limits 99.73% of the time if a stable pattern of variation exists.

So long as the points fall between the control lines, no adjustments or changes in the process are necessary. If five to seven consecutive points fall on one side of the mean, the process should be checked. When points fall outside of the control lines, the cause must be located and corrected immediately.

Statistical theory also gives the expected relationship between $\bar{R}$ ($\Sigma R_i/N$) and $\sigma_{\bar{x}}$. The relationship for the $3\sigma_{\bar{x}}$ limits is

$$A_2\bar{R} = 3\sigma_{\bar{x}}$$

The values for A_2 calculated by probability theory, for different sample sizes, are given as follows:

Sample Size	A_2
2 units	1.88
3 units	1.02
4 units	0.73
5 units	0.58
10 units	0.31

The formulas for 3-σ control limits using this factor are

Upper Control Limit

UCL $\bar{X} = \bar{\bar{X}} + A_2\bar{R}$

Lower Control Limit

LCL $\bar{X} = \bar{\bar{X}} - A_2\bar{R}$

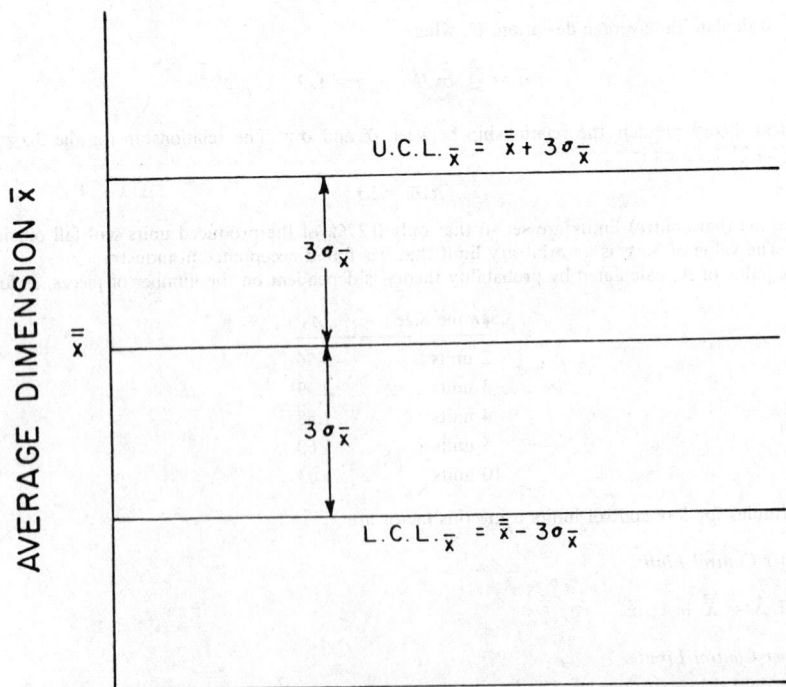

Fig. 31.77 Control chart $\bar{X}$.

In control chart work, the ease of calculating R is usually much more important than any slight theoretical advantage that might come from the use of σ. However, in some cases where the measurements are costly and it is necessary that the inferences from a limited number of tests be as reliable as possible, the extra cost of calculating σ is justified. It should be noted that, because Fig. 31.77 shows averages rather than individual values, it would have been misleading to indicate the tolerance limits on this chart. It is the individual article that has to meet the tolerances, not the average of a sample. Tolerance limits should be compared to the machine capability limits. Capability limits are the limits on a single unit and it can be calculated by

$$\text{Capability limits} = \overline{\overline{X}} \pm 3\sigma'$$

Since $\sigma' = \sqrt{n}\ \sigma_{\overline{x}}$, the capability limits can be given by

$$\text{Capability limits} = \overline{\overline{X}} \pm 3\sqrt{n}\ \sigma_{\overline{x}}$$

Figure 31.78 shows the relationships among the control limits, the capability limits, and assumed tolerance limits for a machine that is capable of producing the product with this specified tolerance. Capability limits indicate that the production facility can produce 99.73% of its products within these limits. If the specified tolerance limits are greater than the capability limits, the production facility is capable of meeting the production requirement. If the specified tolerance limits are tighter than the

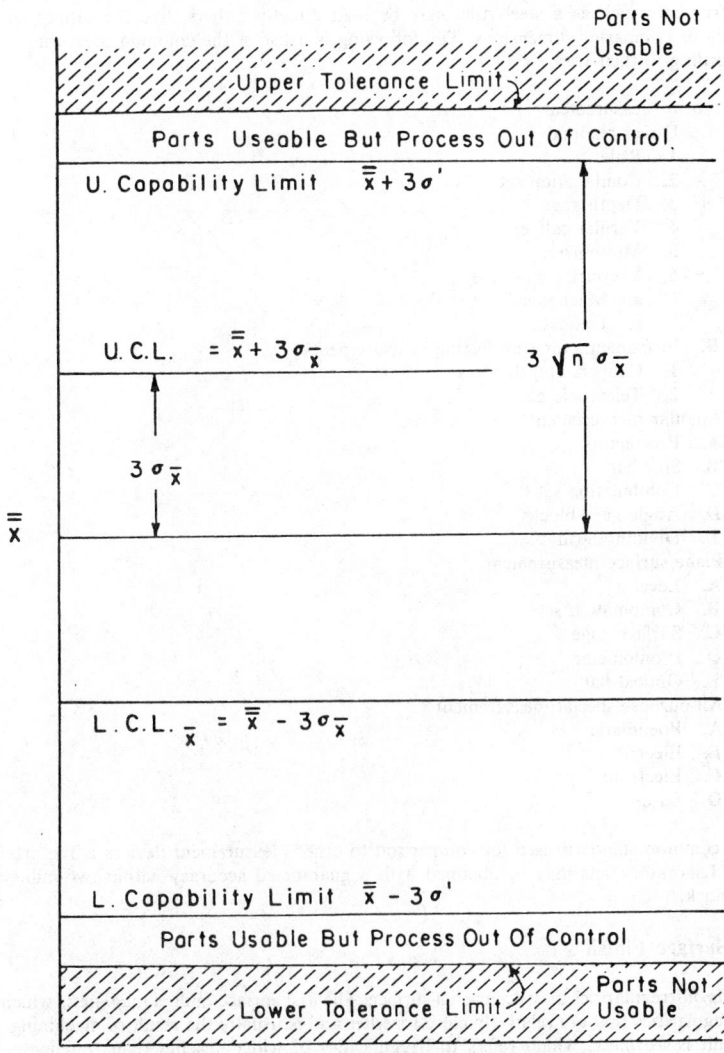

Fig. 31.78 Control, capability, and tolerance limits.

capability limits, a certain percentage of the production will not be usable and 100% inspection will be required to detect the products outside the tolerance limits.

31.7.3 Interrelationship of Tolerances of Assembled Products

Mathematical statistics states that the dimension on an assembled product may be the sum of the dimensions of the several parts that make the product. It states also that the standard deviation of the sum of any number of independent variables is the square root of the sum of the squares of the standard deviations of the independent variables:

$$\sigma'_{sum} = \sqrt{(\sigma'_1)^2 + (\sigma'_2)^2 + \cdots + (\sigma'_n)^2}$$

Whenever it is reasonable to assume that the tolerance ranges of the parts are proportional to their respective σ' values, such tolerance ranges may be combined by taking the square root of the sum of the squares:

$$T = \sqrt{T_1^2 + T_2^2 + T_3^2 + \cdots + T_n^2}$$

31.7.4 Measuring Instruments

A measuring instrument is any device that may be used to obtain a dimensional or angular measurement. Some instruments such as a steel rule, may be read directly; others, like the caliper, are used for transferring or comparing dimensions. The following is a list of the common measuring instruments listed according to their use:

    I. Linear measurement
        A. Direct reading
            1. Rule
            2. Combination set
            3. Depth gage
            4. Vernier caliper
            5. Micrometer
            6. Measuring machine
                a. Mechanical
                b. Optical
        B. Instruments for transferring measurements
            1. Calipers and dividers
            2. Telescopic gages
    II. Angular measurement
        A. Protractors
        B. Sine bar
        C. Combination set
        D. Angle gage blocks
        E. Dividing head
    III. Plane surface measurement
        A. Level
        B. Combination set
        C. Surface gage
        D. Profilometer
        E. Optical flat
    IV. All-purpose special measurement
        A. Pneumatic
        B. Electric
        C. Electronic
        D. Laser

The most common standard used for comparison to other measurement devices is the precision gage block set. Laboratory sets may be obtained with a guaranteed accuracy within two-millionths of an inch per block.

31.7.5 Surface Finish

Three kinds of irregularities may occur on a surface. The first surface fault is *roughness,* which describes surface irregularities that are relatively close together and is usually the result of machining. A second surface fault is *waviness,* which refers to irregularities of wider spacing than roughness. Waviness

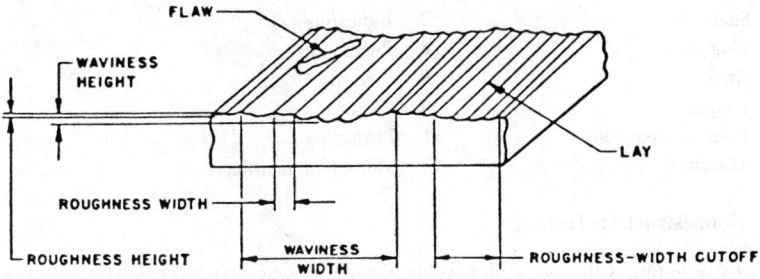

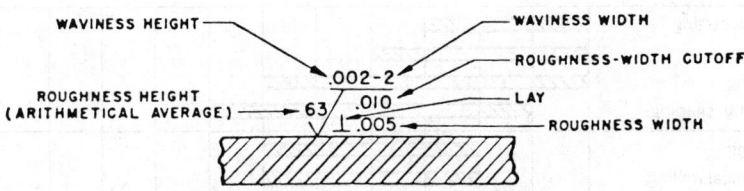

Fig. 31.79   Surface roughness terminology.

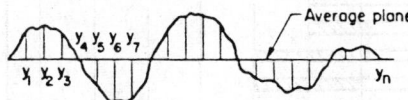

Fig. 31.80   Measurement of surface roughness.

may be the result of warping, deflection, or springing while the workpiece is being worked on, or the result of a tool movement pattern while the workpiece is being cut. The third fault is an irregularity called a *flaw* or imperfection, which is relatively infrequent and usually randomly located. Flaws consist of such things as scratches, holes, ridges, and cracks.

These surface irregularities, as well as the symbol for specifying surface finish, are shown in Fig. 31.79. Surface finish is measured in microinches either by the arithmetical average (AA) or root-mean-square (rms) average. These measurements are done from a reference line or a theoretical surface as shown in Fig. 31.80. The algebraic relation for the arithmetic average (AA) and the root-mean-square (rms) are

$$AA = \frac{\Sigma y_i}{n} \quad \mu\text{in.}$$

$$rms = \sqrt{\frac{\Sigma y_i^2}{n}} \quad \mu\text{in.}$$

Surface roughnesses available by common production methods are indicated in Fig. 31.81. The cost of production increases significantly as the requirement of surface roughness is reduced.

31.7.6   Gaging

A gage is used to determine whether a part has been produced within the tolerances required. It does not usually give a specific dimension. Gaging requires a minimum of time, and many times it is done while production equipment is in operation with no loss in production time. The most commonly used gages in production can be classified as follows:

<table>
<tr><td>

1. Snap
2. Plug
3. Ring
4. Length
5. Form comparison
6. Thickness

</td><td>

7. Indicating
8. Pneumatic
9. Electric
10. Electronic
11. Projecting
12. Multiple dimension

</td></tr>
</table>

31.7.7 Nondestructive Testing

Nondestructive testing is the use of all possible methods of detection and measurement of properties or performance capabilities, of materials, parts, assemblies, structures, and machines, which do not damage or destroy their serviceability.

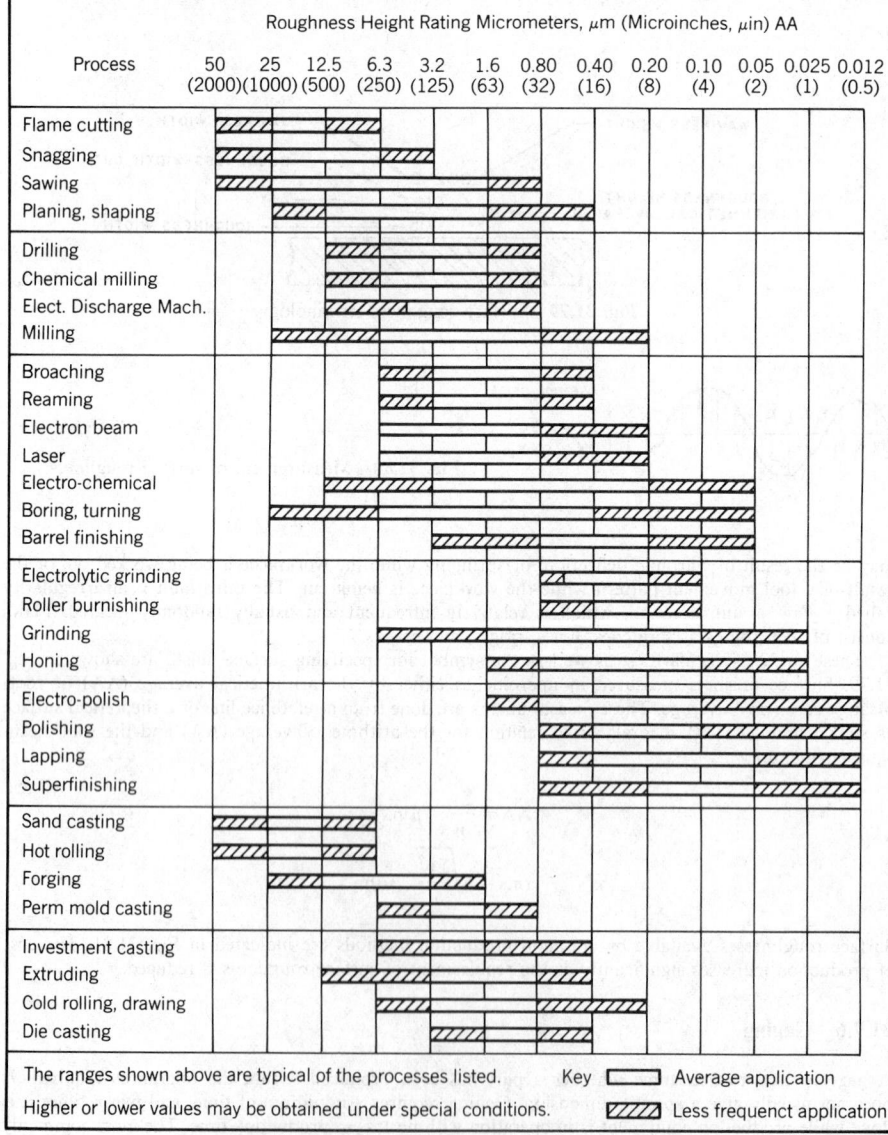

The ranges shown above are typical of the processes listed.   Key ⬚ Average application

Higher or lower values may be obtained under special conditions.   ▨ Less frequenct application

Extracted from General Motors Drafting Standards, June 1973 revision

Fig. 31.81 Surface roughness from common production methods.

Typical nondestructive test uses are as follows:

1. Identifying and sorting raw material.
2. Checking chemical composition.
3. Detecting variations in structure.
4. Detecting conditions of mechanical stress.
5. Detecting internal voids or discontinuities.
6. Detecting internal inclusions.
7. Detecting surface cracks.
8. Gaging difficult dimensions.
9. Detecting properties or presence of specific materials.

Nondestructive Tests

Radiography includes X-ray, gamma-ray, or radioisotope examination. Penetrating rays are absorbed in proportion to the density of the material through which they pass. Radiography is used to examine internal flaws in materials and welds and to measure thicknesses of materials.

Beta rays—electrons—projected through a moving sheet of steel, paper, or other relatively thin material can serve to measure its thickness. This type of gage can feed back information for automatic correction of errors in the system.

Electric or magnetic fields are applied to the material in such a way that variation of properties in the work will disturb the field, allowing comparison with results from known samples.

Magnetic particle inspection applies an intense magnetic field in the ferromagnetic material subjected for inspection. Fine magnetic powders are then applied to the work. Cracks, voids, and material discontinuities cause the lines of the magnetic flux to be distorted. Magnetic particle inspection will show up both external and internal flaws.

Charged particle inspection applies the triboelectric principle in inspecting very fine cracks and flaws. Powdered chalk dust is sprayed over the article inspected. Positively charged particles collect on the surface, revealing cracks and flaws invisible to the eye.

Electric sparks are used to inspect pinholes, cracks, and other imperfections in coatings of plastic or other nonconducting materials. A relatively high-voltage electric power is applied across the coating. Cracks or holes will be revealed by breakdown and electric sparks.

Fluorescent penetrants are applied by dipping, spraying, or brushing on the surface of material to be tested. The penetrant is washed and a powder is applied to absorb the penetrant remaining in cracks and voids revealing the flaws.

Low-frequency mechanical vibration is increasingly used in nondestructive testing. Sonic frequency that results from its application to a product reveals its resonance condition.

Ultrasonic mechanical vibrations are capable of detecting defects smaller than might be found by radiography. Pulses of vibration are transmitted into the work where they are reflected back from any discontinuity, hole, inclusion, or change of geometry. Ultrasonic inspection may be used for a wide range of large work sizes.

Eddy current testing is useful for flaw detection, sorting by metallurgical properties such as hardness, and thickness of measurement. Discontinuities in the part change the amplitude and direction of flow of the induced eddy current. The change of magnitude and phase difference can be used to sort parts according to alloy, temper, and other metallurgical properties.

31.7.8 Thermal and Infrared Testing

The human eye is a sensitive image detector between 0.1 and 0.75 μm wavelengths. This is the visible range of the spectrum, and in it we can see colors from violet to red. Most of the objects on earth are too cool to emit radiation within the visible light region. Objects like people, plants, and structures emit radiation in the region of 10 μm, which is in the far-infrared region and can be detected by the infrared-sensitive image detector.

Thermal nondestructive testing is used in two ways. In some cases it is only necessary to find the absolute temperature of the spot in the specimen. In others, the objective is to get the temperature difference of an area in contrast to the surrounding areas. Thermal methods can be used to get either thermographic or thermometric information, Infrared scanning is very fast, so that flicker-free pictures can be produced.

BIBLIOGRAPHY

L. Alting, *Manufacturing Engineering Processes,* Marcel Dekker, New York, 1982.

B. H. Amstead, P. F. Ostwald, and M. L. Begeman, *Manufacturing Processes,* 7th ed. Wiley, New York, 1977.

E. P. DeGarmo, J. T. Black, and R. A. Kohser, *Material and Processes in Manufacturing,* 6th ed. Macmillan, New York, 1984.

L. E. Doyle, G. F. Schrader, and M. B. Singer, *Manufacturing Processes and Materials for Engineers,* Prentice-Hall, Englewood Cliffs, NJ.

R. LeGrand (ed.), *American Machinist's Handbook,* McGraw-Hill, New York, 1973.

S. Kalpakjian, *Manufacturing Processes for Engineering Materials,* Addison-Wesley, Chicago, IL, 1984.

M. Kronenberg, *Machining Science and Application,* Pergamon, London, 1966.

R. A. Lindberg, *Processes and Materials of Manufacture,* 2nd ed., Allyn and Bacon, Boston, MA, 1977.

Machining Data Handbook, 3rd ed., Machinability Data Center, Cincinnati, OH, 1980.

Machinery's Handbook, 21st ed., Industrial Press, New York, 1979.

Metals Handbook, 8th ed., American Society of Metals, 1976.

H. D. Moore and D. R. Kibbey, *Manufacturing Materials and Processes,* 3rd ed., Wiley, New York, 1982.

B. W. Niebel and A. B. Draper, *Product Design and Process Engineering,* McGraw-Hill, New York, 1974.

J. A. Schey, *Introduction to Manufacturing Processes,* McGraw-Hill, New York, 1977.

M. C. Shaw, *Metal Cutting Principles,* Oxford University Press, Oxford, 1984.

Society of Manufacturing Engineers, *Tool and Manufacturing Engineers Handbook,* McGraw-Hill, New York.

CHAPTER 32
SHOT PEENING

HENRY O. FUCHS

Stanford University
Stanford, California

32.1 INTRODUCTION

Shot peening is a process for treating metal parts by controlled high-speed impact of many balls (shot). It is used mainly to increase the resistance of metal parts against fatigue, fretting fatigue, stress corrosion, and corrosion fatigue. It is also used to form slender metal parts to desired shapes, to produce a desired surface topography, to harden surfaces, to close pores, and to test bonds.

Practically all automobile valve springs and suspension springs are shot peened to prevent fatigue failures and most modern airplanes have wing skins formed by shot peening and shot peened to prevent stress corrosion.

The process is extremely flexible: Parts of any shape and of any size can be treated. Control of the process is essential, but the process is more tolerant of minor deviations from the specified parameters than heat treating or precision machining are.

Table 32.1 shows data on nine typical applications. Although each application must be considered individually, the following points can be made.

In general, the process works by plastically deforming a shallow surface layer. The shot impacts must be strong enough to achieve this, and they must cover the surface well enough to subject all of it to the effects of the plastic deformation.

Plastic deformation that extends too deeply into the part decreases the beneficial effect and may be detrimental. This also applies to the presence of sharp corners and to fins on thicker parts.

Peening produces a small extension of the peened surfaces and an extension or curvature of the entire peened part. This must be considered when high precision of dimensions is required.

When used to increase fatigue resistance, shot peening is more effective on harder materials than on softer materials (Example 6 of Table 32.1); is more effective on notched parts and on weakened surfaces than on smooth specimens (Examples 2 and 5 of Table 32.1); is more effective against applied tensile stresses and against shear stresses than against compressive stresses; and is more effective against applied stresses cycling from around zero to a maximum than against fully reversed stresses.

To prevent stress corrosion, shot peening must completely cover the surface.

When forming parts, shot peening is more effective for forming slender parts, for instance, aircraft skins. No dies are required, only checking fixtures.

The reasons for these rules and the methods for predicting effects on fatigue will be discussed in Section 32.5.

32.2 EQUIPMENT USED FOR SHOT PEENING

The equipment used for shot peening consists basically of eight components:

1. The shot, usually steel or glass. Its material, size, and shape are closely specified.[9,10]
2. Devices to remove shot worn beyond specification tolerances—for instance, screens.
3. Machinery to impel the shot—for instance, centrifugal wheels or compressed air.
4. Means to direct the shot stream—for instance, nozzles.
5. Means to expose parts to the shot stream—for instance, conveyors.
6. Means to collect the shot after it has been thrown against the part—for instance, enclosures with rubber curtains.
7. Means to recirculate shot to the impelling device—for instance, bucket elevators.
8. Controls for shot flow, shot velocity, and time of treatment, and means for inspection.

To expose all desired areas of a part to the shot stream it is necessary either to move the part under a stationary stream or to move the shot stream over a stationary part, or to move both parts and shot stream in a required pattern. This can be done by a conscientious operator, but to obtain consistently good results the motions are mechanized. Large wing skins, for instance, may be peened in machines that travel lengthwise over the skins while nozzles reciprocate in the machines transversely to the direction of travel. Suspension coil springs are conveyed through the shot stream in a helical path, rotating while they advance.

Large structures can be peened by machines that are brought to the structure, rather than bringing the structure to the machine (Example 4 of Table 32.1). This method (portable peening machines) has also been used when it was necessary to peen parts of aircraft that were in service.

Effects similar to those obtained by shot peening have also been obtained by peening with air hammers or with rotating impact tools. These substitutes are more difficult to control and are used only when shot peening is not feasible.

Equipment and methods of peening are described in more detail in the literature.[11-13]

Table 32.1 Typical Examples of Shot Peening Applications

	Parts, Sizes in in. (mm)	Peening—Intensity, Shot Size in in. (mm)	Testing	Results — Not Peened	Results — Peened ksi (MPa) <cycles>	Increase	Reference
1	Helical compression springs ⅛–¼ (3–6) wire diameter (1a) alloy steel (1b) stainless steel (1c) phosphor bronze	10A, 0.016 (0.4); also other intensities and shot sizes	Stress range to give 10^7 million cycles	(1a) 70 (48) (1b) 45 (310) (1c) 15 (103)	115 (795) 90 (620) 30 (205)	60% 100% 100%	Zimmerli[1]
2	Tractor axle shafts 3.0 (76) diameter, RC50 hard (2a) clean steel (2b) (2c) } with stringers	(2a) 23A, 230 (2b) 23A, 230 (2c) 13C, 550	Torsion stress zero to 113 ksi (780 MPa)	(2a) <1700> min (2b) <50> min (2c) <50> min	<80,000> <6000> <75,000>	47 times 120 times 1500 times	Eckert[2]
3	Thick tubes, steel, RC36 hard 1¼ (31.8) ID, 3¼ (82.6) OD	8–10A, 170	Shear stress amplitude to give 3×10^5 cycles (internal hydraulic pressure)	16.3 (113)	33 (228)	102%	Findley and Reed[3]
4	Induced draft fans welded T-1 steel plates	Heavy peening at welds on overhaul	Service on 750 MW coal-fired boilers	Cracks 12 in. (300 mm) long after 2 years	No cracks after 6 years		Takhar and Fryzel[4]
5	Notched specimens steel 120 ksi (825 MPa) ultimate tensile strength $K_T = 1.76$ dia. = 0.24 (6)	50 psi (345 kPa) air, 0.031 (0.8) shot	Rotating beam fatigue limit at 7×10^6 cycles	37.4 (258)	62 (428)	65%	Harris[5]
6	Bending specimens 0.35 (9) thick (6a) aluminum alloy (6b) titanium alloy (6c) maraging steel	12A steel shot followed by 2N glass beads	Flight spectrum S_{max} = 0.9 ultimate tensile strength S_{min} = −0.22 ultimate tensile strength			(6a) 2.7 times (6b) 10 times (6c) over 11 times	Schütz[6]
7	Reactor piping 321 stainless 7½ (190) OD	11A, 230	Boiling solution 42% $MgCl_2$	22 hr	No cracks after 264 hr	Over 10 times	Friske and Page[7]
8	Tensile specimens stainless 26% Cr, 4% N, 0.2% Mo; 0.177 (4.5) diameter	8A, 230	Axial pull in NaCl solution (pH = 2); fatigue strength at 10^7 cycles	40 (280)	65 (450)	60%	Müller et al.[8]
9	Aircraft wing skins	To form contour	Continued use by Lockheed, Boeing, etc.				

32.3 SHOT PEENING SPECIFICATIONS

The peening effect depends on the number of impacts per unit area and on the masses, velocities, sizes, shapes, rheological properties, and angles of impingement of the different particles that produce the impact. It may also depend on temperature, on surface films, and on the sequence of impacts (large after small or small after large).

The Almen intensity integrates all these variables as far as their effect on the fatigue resistance of medium hard steel parts such as springs is concerned.[14-16] It is the most widely used, most practical, and most clearly defined method of specifying shot peening. For parts made of materials other than medium hard steel (RC 35-55) a more complete specification should include the shot to be used (defined by size, material, and hardness).[16] The degree of coverage is also specified when double or triple coverage is required.

Details of the design and use of Almen gages and test strips are given in Military Specification MIL-S-13165[9] and in an SAE standard.[17]

A typical intensity call-out may read "12-14A." This means a reading of 12-14 units on the Almen gage; it measures the curvature produced by peening one side of test strip type A. The curvature is proportional to the bending moment of the compressive self-stresses generated by peening. Thorough saturation, so that additional peening produces little additional curvature, is implied. Test strips of standard size and material, free from stresses before being peened, are also implied.

Test strips type A are used for measuring and specifying medium intensities. They are 0.051 in. (1.29 mm) thick. Heavier peening is specified and checked with type C strips 0.094 in. (2.39 mm) thick, and lighter peening with type N strips, which are 0.031 in. (0.79 mm) thick. All types are made of cold rolled spring steel, tempered to RC 44-50, and hot pressed for at least 2 hr to remove self-stresses.

Older specifications may read "0.012 A" instead of 12A because the dial on the Almen gage reads in increments of 0.001 in. Still older specifications may say "0.012A2," meaning the second model Almen gage, which is now universally used. In Europe the gage is usually equipped with a dial indicator reading increments of 0.01 mm.

A European specification "0.2 mm A" or sometimes "20A" is equivalent to 8A of the SAE standard. According to the SAE standard[17] the Almen intensity is to be used as a number, for instance, like a Rockwell hardness number, not as a measurement in inches.

A more complete call-out might read "Peen areas marked P to 18-20A with 390 shot RC 60-65 hard. 200% coverage required. Mask areas marked M."

Shot sizes like "390" as well as sizes of cut wire shot and of glass beads used for peening, and some of their properties, are defined in the Military Specification[9] and in a number of SAE Recommended Practices.[10]

The nominal size "390" means 0.039 in. (1 mm); the actual size distribution of the pellets is defined by tolerances on percentages passed through or retained on standard screens.

Two hundred percent coverage means that the parts be must be exposed to the peening treatment twice as long as required to achieve full coverage at the required intensity. The terms coverage and saturation and their relation are explained in more detail in Section 32.6.

Masking is called for when certain areas must be protected against shot impacts. When masking is not required, the part is either peened all over, or thoroughly peened where required, with scattered peening dimples permitted in other areas. No damage by scattered dimples (other than their presence) has ever been observed.

The angle of impingement is seldom specified. It usually is as close to perpendicular as feasible.

32.4 SELECTION OF SHOT PEENING PARAMETERS

Figure 32.1 gives indications of intensities specified for shot peening steel parts. It shows that, in general, thicker parts are peened with greater intensities. Section 32.5 discusses the reasons for this practice.

Higher intensities may require the use of larger size shot. For otherwise equal conditions the number of balls thrown per minute is inversely proportional to the cube of their diameter. Treatment with larger shot sizes is therefore more expensive. The tractor axle shafts, Examples 2b and 2c of Table 32.1, were test specimens. In production such shafts were made of steel without magnetic indication of stringers and peened to 23A. Tests of such shafts showed a minimum fatigue life of 80,000 cycles, about equal to that of shafts with stringers peened to 13C. Lower than optimum intensity, together with cleaner steel, was the more economical choice for production shot peening. For these 3 in. (76 mm) diameter shafts the production treatment (23A) is at the lower edge of the band shown in Fig. 32.1, and the best experimental treatment (13C) is near the middle of the band.

Excessively high intensities on thin parts can be damaging. Zimmerli[14] reported fatigue limits of 110 ksi (760 MPa) on compression coil springs of 0.148 in. (3.76 mm) diameter steel wire peened to intensities from 5A to 15A with shot of various hardness, from RC 62 down to RC 26. With intensities

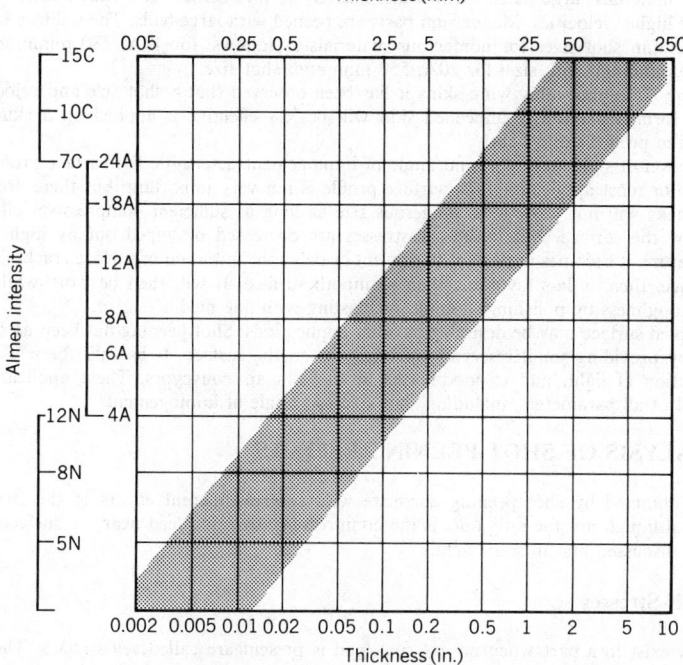

Fig. 32.1 Peening intensities commonly used on steel parts of different thickness.

about 9C the fatigue limit dropped to less than 100 ksi (690 MPa). Unpeened springs had a fatigue limit of 74 ksi (510 MPa).

Another factor may limit the shot size, and thereby the intensity: The shot radius should be much less than the radius of fillets or grooves that must be peened, so that numerous good impacts can be obtained in the grooves. The military specification[9] requires that, unless otherwise specified, the nominal size shot used on fillet surfaces shall not be greater than one-half the fillet radius. (See Fig. 32.2.) For torsion bar springs, approximately 2.4 in. (61 mm) diameter, with splined ends, the SAE recommended[18] intensities of 10C minimum on the body, which required 550 shot, and 11A minimum on the splines, where the shot cannot be larger than 230 because of the radius at the root of the splines. A user of such torsion bars specifies 200% coverage, obtained by passing each bar twice through the process.

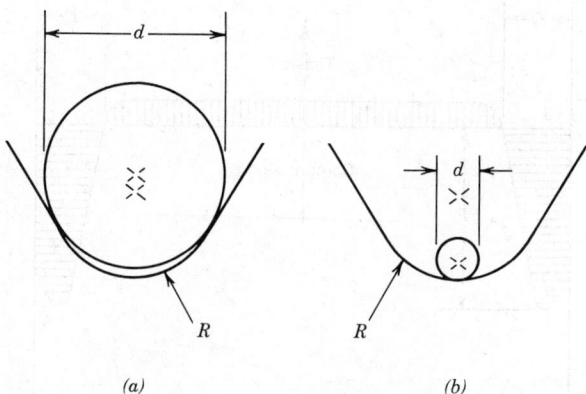

(a)   *(b)*

Fig. 32.2 Shot diameter d and groove radius R: (a) shot size too large, $d = 2.2\,R$; (b) maximum shot size permitted by Ref. 9, $d = \frac{1}{2}R$.

On some materials large balls, thrown at relatively low velocities, give much better results than small balls at higher velocities. Magnesium parts are peened with large balls. The military specification[9] requires minimum shot sizes for nonferrous materials as follows: for 12A, 280 minimum shot size; for 16A, 390 minimum shot size; for 20A, 550 minimum shot size.

In forming aluminum alloy wing skins it has been observed that a shot size and velocity that are sufficient to form a previously unpeened skin will be less effective if applied on a skin which has previously been peened more lightly.

The selection of shot, intensity, and angle of impingement determine the surface profiles obtained by peening. For most applications the surface profile is not very important, but there are exceptions.

Small cracks will not grow to a dangerous size as long as sufficient compressive self-stresses are present below the surface. But if the self-stresses are decreased or wiped out by high loads or by high temperature, it becomes important to prevent or delay the initiation of surface cracks. The dimpled peened surface, then, is less favorable than a smooth surface. It will then be worthwhile to remove the surface roughness by polishing, honing, or blasting with fine media.

The dimpled surface may be desirable in other applications. Shot peening has been used to improve lubrication by providing small reservoirs of lubricant on the surface. It has also been used to obtain diffuse reflection of light, and to modify the air flow in air conveyors. These applications require exact control of all parameters, including shot size and angle of impingement.

32.5 ANALYSIS OF SHOT PEENING EFFECTS

The results obtained by shot peening correlate with several different effects of the treatment. The chief cause, although not the only one, is the compressive stress induced near the surface by peening. This will be discussed first in some detail.

32.5.1 Self-Stresses

Stresses that exist in a part when no external load is present are called self-stresses. They have also been called self-equilibrating stresses, residual stresses, internal stresses, trapped stresses, and locked-up stresses.

Shot peening produces self-stresses in the skin of parts by plastically stretching a relatively shallow layer (the skin) near the surface. The skin by itself is made longer, wider, and thinner than it was before peening. The bulk of the material in the interior restrains the expansion of the skin. This produces substantial compressive stresses in the skin, balanced by smaller tensile stresses in the interior. Figure 32.3 shows a typical distribution of self-stress in a shot-peened thin plate.

Equilibrium demands that the tensile stress in the core becomes greater if the compressively stressed layer extends more deeply. This limits the intensities that should be used on thin parts.

If one would peen a thin sheet so heavily that all the material is stretched plastically, there would be only very small self-stresses. This might happen at sharp edges, which should be avoided on peened parts or which should be peened only lightly.

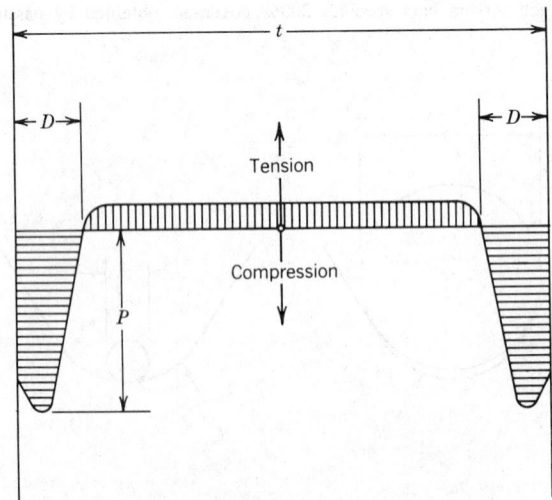

Fig. 32.3 Typical distribution of self-stress in a shot-peened plate: thickness, t; depth of compressive stress, D; and peak magnitude of compressive stress, P.

As the interior material is not exposed to atmosphere or accidental damage, it is more resistant than the surface to fatigue. Fatigue fractures starting below the surface are not likely unless the tensile stress below the surface is higher than the fatigue limit of smooth specimens of the same material.

When axial loads are applied, peening can bring the fatigue strength of parts with poor surfaces up to that of smooth polished specimens. For bending and torsion it can do much more because the applied stresses decrease toward the inside where peening produces small tensile self-stresses. For notched parts it can do still more because of the greater stress gradient, as, for instance, in Example 5 of Table 32.1 where the increase by peening overcomes the decrease by notching. In most cases peening can bring the fatigue strength of notched parts up to the strength of smooth specimens. It can overcome the notch effect by preventing small cracks that originate in the notch root from propagating toward the interior of parts.

The peak value of the compressive self-stress (P in Fig. 32.2) depends mainly on the material of the peened part and on the restraints imposed on the part during the peening process. In a part peened without restraints the value of P is of the order of half the yield strength of the material, usually somewhat more. Note that the yield strength may be raised by the cold work of peening, and further raised or lowered by repeated working of skin elements in continued peening. The compressive self-stress can be increased by applying tension to the surface while it is being peened.[15,19] The compressive self-stress at the surface itself is almost always somewhat less than the peak value P at some depth in the skin.

Changing the peening parameters (shot velocity and shot size) will change mainly the width of the peak and the depth (D in Fig. 32.3) to which the compressive self-stress extends. It will have only a minor effect on the peak value. The depth D is roughly equal to the diameter of the peening dimples.[16] It also is roughly proportional to the peening "intensity" defined in Section 32.3.

The compressive self-stress has two different effects on fatigue resistance: It increases the resistance to the formation of small cracks, and it slows down or prevents the growth of cracks. The latter effect is far more important than the former. It explains the very large improvement obtained in some cases and the substantial, but smaller, improvements in other cases, as shown below.

The compressive self-stress is also an important element in preventing stress corrosion, and the only factor active in changing the shape of parts to form them.

In analysis, self-stresses are added to mean load stresses, but there is one important difference between load stresses and self-stresses. When the total stress caused by mean load, alternating load, and self-stress reaches the yield strength, then the self-stress will decrease so that the sum equals the yield strength. Load stresses that exceed the yield strength wipe out the original self-stress produced by peening.

32.5.2 Constant Life Diagrams

Haigh diagrams are an excellent means of analyzing and visualizing the effects of self-stresses. They show the yield strength and the fatigue strength plotted in terms of alternating stress versus mean stress.

Haigh diagrams for a smooth part are shown in Figs. 32.4a and 32.4b and for a notched part in Fig. 32.4c.

Figure 32.4 is based on data for G104500 steel (SAE 1045), quenched and tempered to 450 Brinell hardness, 230 ksi (1585 MPa) ultimate tensile strength. Its monotonic yield strength is 220 ksi (1515 MPa), its cyclic yield strength is 140 ksi (965 MPa), and its smooth push–pull fatigue limit 84 ksi (580 MPa) for zero mean stress. For other values of mean stress the fatigue strength of smooth unpeened specimens for the same number of cycles is assumed to increase or decrease linearly, reaching zero at 305 ksi (2100 MPa) tensile stress. The self-stress produced by peening in this material is 125 ksi (860 MPa) compression.

Figure 32.4a shows that the fatigue strength for fully reversed loading can be increased to 100 ksi (690 MPa) by peening. The self-stress is reduced from 125 to 60 ksi (860 to 415 MPa) so that the sum of load stress plus self-stress equals 160 ksi (1100 MPa), which is a value of yield strength linearly interpolated between the monotonic and cyclic values. The strength increase obtained by peening is 19%.

A much greater increase (41%) is observed for loading from zero to maximum ($S_m = S_a$) as shown in Fig. 32.4b. Without peening the fatigue strength in terms of alternating stress is less than that for fully reversed loading, only 66 ksi (455 MPa). In terms of maximum stress it is 132 ksi (910 MPa). With peening the fatigue strength is increased to 93 ksi (640 MPa) in terms of alternating stress, 186 ksi (1280 MPa) in terms of maximum stress, or a 41% increase. The applied tensile stress does not wipe out any compressive self-stress.

Much greater improvement of long-life fatigue strength (122%) is seen in Fig. 32.4c for a notched part with 2.1 notch factor. The Haigh diagram for this part shows the same yield lines as Fig. 32.4a because we are not concerned with the onset of yielding at the notch root, but with yielding far below the notch root. The fully reversed fatigue limit is lowered to $84/2.1 = 40$ ksi (280 MPa). The line that indicates the formation of small cracks at the root of the notch has the same slope as the

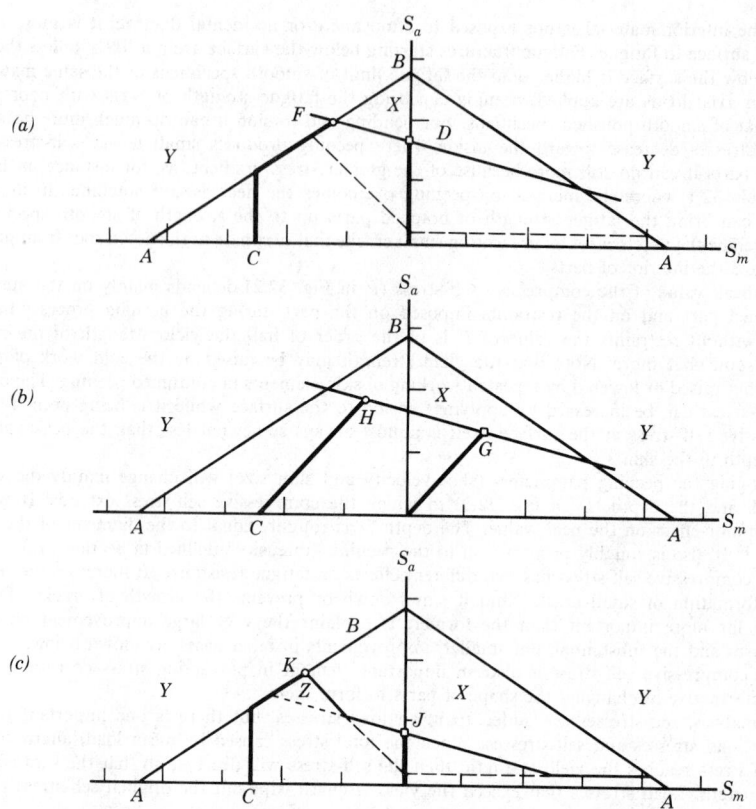

Fig. 32.4 Haigh diagrams for fatigue limit, peened and unpeened, for G104500 steel at 450 Brinell hardness: (*a*) smooth, fully reversed ($K = 1$, $R = -1$); (*b*) smooth, zero to maximum ($K = 1$, $R = 0$); (*c*) notched, fully reversed ($K = 2.1$, $R = -1$). In these figures,

A = monotonic yield stress	220 ksi	(1540 MPa)
B = cyclic yield stress	140 ksi	(980 MPa)
C = self-stress by peening	−125 ksi	(−875 MPa)
X = region of fatigue fractures		
Y = region of yielding		
D = fatigue limit, unpeened	84 ksi	(590 MPa)
F = fatigue limit, peened	100 ksi	(700 MPa)
G = fatigue limit (in terms of maximum load stress), unpeened	132 ksi	(925 MPa)
H = fatigue limit (in terms of maximum load stress), peened	186 ksi	(1300 MPa)
J = fatigue limit, unpeened	40 ksi	(280 MPa)
K = fatigue limit, peened	89 ksi	(625 MPa)
Z = region of nonpropagating cracks		

fatigue strength line in Fig. 32.4*a*. But such cracks at the notch root will not propagate to the interior of the part unless a small alternating tensile stress is present. It is calculated as a nominal stress, analogous to the threshold stress intensity factor of fracture mechanics. In Fig. 32.4*c* this limiting value is conservatively assumed to be 5 ksi (35 MPa). In fully reversed loading the peening stress of 125 ksi (860 MPa) will be decreased by yielding to 80 ksi (560 MPa), but it will still stop cracks at alternating load stresses of 89 ksi (614 MPa), corresponding to an increase of 122%. The long-life fatigue strength of the notched part is slightly higher than the fatigue strength of the unpeened smooth part. The notch damage is overcome. This corresponds to many tests on production parts, for instance, Example 2 in Table 32.1, where the stringers acted like notches, and to Example 5 in Table 32.1. Table 32.2 is a summary of the readings in Fig. 32.4. For short fatigue lives the strength improvements obtainable by shot peening will be smaller because the fully reversed fatigue limit, without peening,

**Table 32.2 Fatigue Limits
Predicted by Fig. 32.4 for G104500 Steel at 450 BHN**

Condition	Strength, ksi (MPa)		Strength Increase
	Unpeened	Peened	
Smooth, fully reversed	84 (580)	100 (690)	19%
Smooth, zero to maximum	132 (925)	186 (1280)	41%
Notched, $K = 2.1$, fully reversed	40 (280)	89 (614)	122%

will be higher but the yield limitation remains the same. The improvement in fatigue life can still be very worthwhile, as in Example 2 of Table 32.1.

Occasional stress peaks greater than yield strength, imposed in service, may also decrease the effectiveness of shot peening. Comparison of Example 6a with Examples 6b and 6c in Table 32.1 illustrates this effect. The service load spectra for the three materials were in proportion to their ultimate tensile strengths (UTS). The load stress of 90% UTS, applied a few times, reduced the remaining self-stress in the aluminum alloy to a value where it became less effective. For the specimens made of steel or titanium alloy the self-stress remained more effective.

Haigh diagrams are very useful for estimating the improvements obtainable by shot peening. To improve parts with notches or with weakened surfaces, the compressive self-stress should extend to the depth at which the load stresses are less than at the surface, as in the case of notches, or where the material is stronger, as in the case of decarburized steel. To determine the minimum required depth, or to determine the intensity for maximum strength, requires either experience or experiments or a detailed study of the distributions of load stresses and self-stresses.[15,20]

32.5.3 Metallurgical Effects of Peening

The plastic deformations produced by peening change the internal structures of the deformed grains. This can be important in preventing corrosion fatigue, intergranular corrosion, and fatigue at high temperatures, or in producing a strain-hardened surface.

Friske and Page[7] have shown that breaking down surface grains by peening can prevent intergranular corrosion. Müller et al.[8] explain the increased resistance to corrosion fatigue by more densely distributed smaller slip steps. These do not break the passive film, which is ruptured by the larger slip steps in unpeened material. Hornbogen et al.[21] found that the more homogeneous distribution of slip delays the appearance of cracks, but favors their propagation in a precipitation hardening steel. They consider this to be the cause of the superiority of shot peening over cold rolling. Even at 1544°F (840°C), where the self-stress is annealed, they found that the fatigue life of peened specimens was 80 times the life of electropolished specimens.

Wang et al.[22] tested jet engine superalloys at elevated temperatures. Even at temperatures which relieved self-stresses, they found improvements in fatigue strength. The improvements were substantial, around 20%, for notched specimens, and much less or in some cases zero for smooth specimens. They did not venture explanations for this observation.

32.6 QUALITY CONTROL

When shot peening is used to change the shape or the surface of parts, it is checked by the usual methods, such as templates, gages, and profilometers.

Inspection becomes more difficult when peening is used to improve resistance against fatigue or stress corrosion. The self-stresses and the metallurgical changes are not visible on the surface. Control of the process itself, in addition to inspection of the finished part, is then necessary. It is done by checking the qualities of the shot stream and the manner of exposing the part.

The shot stream is checked by inspecting the shot and its peening power. The size distribution is measured by screen analysis according to specifications.[9,10] Shot is also inspected for shape. Only a small percentage of broken or nonspherical particles are allowed.

The peening power is checked by Almen strips. It is important to distinguish the Almen intensity specified for the peened part from the Almen intensity that checks the peening power of the shot stream. The part intensity is obtained by exposing strips in the same manner as a critical surface of the part, and for the same time as that surface of the part. The shot stream intensity might be checked in the same manner, but it can also be checked by exposing strips in another, possibly simpler or quicker, manner which might show a higher reading.

It is important that the intensity reading corresponds to a strip saturated by peening. Saturation means that the curvature of the test strip cannot be substantially increased by longer exposure to the shot stream. Ideally one might want to say "cannot be increased at all," but one finds that this is not feasible. Therefore, one defines the time to saturation as that time which produces a reading that is increased no more than $X\%$ if a test strip is exposed for double that time. For many applications the agreed percentage X is 20%. The military specification[9] requires 10%.

Proof that saturation is obtained requires at least two test strips. Once this has been established, one can rely on a single test strip to show that conditions have not changed, provided the shot itself remains in the same condition.

Saturation is important because scattered heavy impacts can produce the same curvature of the strip without providing protection to the areas that have not been hit by shot.

If all balls are of the same size and hit the surface at equal angles, a visual inspection can show whether every point of the surface has been hit at least once, and this "coverage" will correspond to saturation, provided cyclic stress–strain relations that govern plastic deformation for often repeated impacts are the same as the monotonic relations that govern it on the first impact.

Because of the greater ease of inspection, coverage is used to specify and check the exposure to the shot stream; 200% coverage means double the exposure required to obtain full visual coverage. If the shot is a mixture of small and large balls, full (visual) coverage is obtained by the impact of small balls long before saturation by self-stress is achieved by the larger balls. Use of visual coverage for inspection is useful only if uniformity of shot size is also controlled.

Coverage is often checked by use of a 10-power magnifying glass. It can be done more quickly by coating the part with Dyescan liquid,[9] peening after the liquid has dried, and checking under a black light that all of the dried liquid coating has been removed by the impacts of peening. When the shot is not harder than the peened surface, it may produce good effects on the part and saturation of the test strip without leaving visible dimples. Coverage can then be verified by checking with Dyescan, but not by looking through a magnifying glass.

Coverage refers to impacts on the surface. Saturation refers to the bending moment of self-stresses on the test strip. To verify the actual condition below the surface of the peened part one needs other methods.

X-ray diffraction can measure the self-stresses in a very shallow layer beneath a surface. The quality of peening is determined by conditions in deeper layers of the skin. With present commercially available methods, those conditions can be determined only by cutting into the skin.

X-ray diffraction measurements made after removing some skin are one method of determining self-stress profiles. Use of strain gages together with drilling small holes in the center of the gage rosette, or with trepanning a circular channel around the gage, are other "semidestructive" methods of inspection.[23,24]

Checking metallurgical transformations in the deeper layers of the skin also requires cutting specimens.

Efforts to develop nondestructive methods of checking shot peening continue. Acoustic methods are in the laboratory development stage.

Because of the difficulties in checking more than the surface of finished parts, one must have assurance that the process of peening is carried out in a well-controlled repeatable manner, so that peened production parts will perform as well as the prototypes that led to the specification of peening.

REFERENCES

1. F. P. Zimmerli, "Heat Treating, Setting, and Shot Peening of Mechanical Springs," *Metal Progress,* **61**(6), 97–106 (1952).

2. E. J. Eckert, "Torsional Fatigue Testing of Axle Shafts," in *Large Fatigue Testing Machines and Their Results,* STP216, ASTM, Philadelphia, PA, 1958, pp. 21–36.

3. W. N. Findley and R. M. Reed, "Fatigue of Thick Tubes," *Trans. ASME, J. Eng. Mat. Techn.,* **104**(3), 174–179 (1982).

4. J. S. Takhar and R. A. Fryzel, "Use of Strain Gages for Equipment Reliability Improvement in the Electric-Utility Industry," *Experimental Mechanics,* **22**(4), 121–131 (1982), and personal communication by R. A. Fryzel.

5. W. J. Harris, *Metallic Fatigue,* Pergamon, London, 1961, p. 48.

6. W. Schütz, "Fatigue Life Improvement of High Strength Materials by Shot Peening," *Preceedings of the First International Conference on Shot Peening,* Pergamon, Oxford, 1982, pp. 423–433.

7. W. H. Friske and J. P. Page, "Shot Peening to Prevent Stress Corrosion of Austenitic Stainless Steels," *Proceedings of the First International Conference on Shot Peening,* Pergamon, Oxford, 1982, pp. 485–491.

8. M. P. Müller, C. Verpoort, and H. Gessinger, "The Influence of Shot Peening on the Fatigue and Corrosion Fatigue Behavior of an Austenitic-Ferritic Stainless Steel," *Proceedings of the First International Conference on Shot Peening,* Pergamon, Oxford, 1982, pp. 479–484.

9. *Shot Peening of Metal Parts,* MIL-S-13165B, amended 1979.

10. Recommended Practices: J 441, Cut wire shot; J 1444a, Cast shot and grit size specifications; J 827, Cast steel shot; J 1173, Size classification and characteristics of glass beads for peening; *SAE Handbook*, Society of Automotive Engineers, Warrendale, PA, 1982, pp. 9.05–9.12.

11. *Manual on Shot Peening*, Society of Automotive Engineers, Warrendale, PA, 1967.

12. H. J. Plaster, *Blast Cleaning and Allied Processes*, Industrial Newspapers Ltd., London, 1972.

13. "Machines and Materials," *Proceedings of the First International Conference on Shot Peening*, Pergamon, Oxford, 1982, pp. 83–135.

14. F. P. Zimmerli, "Shot Quality," *Steel*, 126–129 (18 October 1948).

15. R. L. Mattson and W. S. Coleman, "Effects of Shot Peening Variables and Residual Stresses on the Fatigue Life of Leaf Spring Specimens," *Trans. SAE*, **62**, 546–556 (1954).

16. H. O. Fuchs, "Shot Peening Effects and Specifications," STP 196, ASTM, Philadelphia, PA, 1957, pp. 22–32.

17. "Test Strip, Holder and Gage for Shot Peening," Standard J 442, August 1979, *SAE Handbook*, SAE, Warrendale, PA, 1982, pp. 9.05–9.06.

18. *Manual on the Design and Manufacture of Torsion Bar Springs*, SAE, Warrendale, PA, 1947.

19. J. O. Almen and P. H. Black, *Residual Stresses and Fatigue in Metals*, McGraw-Hill, New York, 1963, pp. 70–80.

20. H. O. Fuchs, "The Effect of Self Stresses on High Cycle Fatigue," *J. Testing and Evaluation*, **10**(4), 168–178 (1982).

21. E. Hornbogen, M. Thumann, and C. Verpoort, "Influence of Shot Peening on the Fatigue Behavior of a Precipitation Hardenable Austenitic Stainless Steel," *Proceedings of the First International Conference on Shot Peening*, Pergamon, Oxford, 1982, pp. 381–387.

22. R. Wang, X. Zhang, D. Song, and Y. Yin, "Shot Peening of Superalloys and Its Fatigue Properties at Elevated Temperatures," *Proceedings of the First International Conference on Shot Peening*, Pergamon, Oxford, 1982, pp. 395–403.

23. L. J. Vande Walle (Ed.), *Residual Stress for Designers and Metallurgists*, ASM, Metals Park, Ohio, 1981, pp. 189–210, 211–222, and 223–243.

24. E. B. Evans, *Residual Stresses: Origin, Measurement, Control*, Marcel Dekker, New York (to be published).

CHAPTER 33

MATERIAL HANDLING

WILLIAM E. BILES
MAGD E. ZOHDI

Industrial Engineering Department
Louisiana State University
Baton Rouge, Louisiana

33.1 INTRODUCTION

Material handling may be defined as the movement and storage of parts, material, and finished products so that they are on hand the moment they are needed in a manufacturing process or service operation. This definition establishes five aspects of material handling. These are:

1. Motion—parts, material, and finished products must be moved from location to location, in an efficient manner and at minimum cost.
2. Time—supplies must be on hand the moment they are needed.
3. Place—material must be positioned in the proper location for use.
4. Quantity—rate of demand varies between steps in process operations. Thus, material handling means that each location continually receives the correct weight, volume, or number of items of material.
5. Space—storage space, and its efficient utilization, are key factors in the overall cost of an operation or process.

The science and engineering of material handling generally falls into two categories, depending on the form of the material. *Bulk solids handling* involves the storage and movement of solids that are flowable, such as fine, free-flowing materials (e.g., wheat flour or sand), pelletized materials (e.g., soybeans or soap flakes), or lumpy materials (e.g., coal or wood bark). *Unit handling* refers to the storage and movement of items that have individual form, such as fabricated metal parts, or that are placed in boxes, in bins, or on pallets (so-called unit loads). The handling of liquids and gases is usually left in the domain of fluid mechanics, although the storage and movement of drums of liquid material properly falls into the domain of unit material handling.

33.2 BULK MATERIAL HANDLING

The handling of bulk solids involves four main areas: (1) conveying; (2) storage; (3) packaging; and (4) transportation.

33.2.1 Conveying of Bulk Solids

The selection of the proper equipment for conveying bulk solids depends on a number of interrelated factors. First, alternative types of conveyors must be evaluated, and the correct model and size must be chosen. Because standardized equipment designs and complete engineering data are available for many types of conveyors, their performance can be accurately predicted when they are used with materials having well-known conveying characteristics. Some of the primary factors involved in conveyor equipment selection are as follows:

1. *Capacity Requirement.* The rate at which material must be transported (e.g., tons per hour). For instance, belt conveyors can be manufactured in relatively large sizes, operate at high speeds, and deliver large weights and volumes of material economically. On the other hand, screw conveyors can become very cumbersome in large sizes, and cannot be operated at high speeds without severe abrasion problems.
2. *Length of Travel.* The distance material must be moved from origin to destination. For instance, belt conveyors can span miles, whereas pneumatic and vibrating conveyors are limited to hundreds of feet.
3. *Lift.* The vertical distance material must be transported. Vertical bucket elevators are commonly applied in those cases in which the angle of inclination exceeds 30°.
4. *Material Characteristics.* The chemical and physical properties of the bulk solids to be transported, particularly flowability.
5. *Processing Requirements.* The treatment material incurs during transport, such as heating, mixing, and drying.
6. *Life Expectancy.* The period of performance before equipment must be replaced; typically, the economic life of the equipment.
7. *Comparative Costs.* The installed first cost and annual operating costs of competing conveyor systems must be evaluated in order to select the most cost effective configuration.

Table 33.1 lists various types of conveyor equipment for certain common industrial functions. Table 33.2 provides information on the various types of conveyor equipment used with materials having certain characteristics.

Table 33.1 Types of Conveyor Equipment and Their Functions

Function	Conveyor Type
Conveying materials horizontally	Apron, belt, continuous flow, drag flight, screw, vibrating, bucket, pivoted bucket, air
Conveying materials up or down an incline	Apron, belt, continuous flow, flight, screw, skip hoist, air
Elevating materials	Bucket elevator, continuous flow, skip hoist, air
Handling materials over a combination horizontal and vertical path	Continuous flow, gravity-discharge bucket, pivoted bucket, air
Distributing materials to or collecting materials from bins, bunkers, etc.	Belt, flight, screw, continuous flow, gravity-discharge bucket, pivoted bucket, air
Removing materials from railcars, trucks, etc.	Car dumper, grain-car unloader, car shaker, power shovel, air

Table 33.2 Material Characteristics and Feeder Type

Material Characteristics	Feeder Type
Fine, free-flowing materials	Bar flight, belt, oscillating or vibrating, rotary vane, screw
Nonabrasive and granular materials, materials with some lumps	Apron, bar flight, belt, oscillating or vibrating, reciprocating, rotary plate, screw
Materials difficult to handle because of being hot, abrasive, lumpy, or stringy	Apron, bar flight, belt, oscillating or vibrating, reciprocating
Heavy, lumpy, or abrasive materials similar to pit-run stone and ore	Apron, oscillating or vibrating, reciprocating

The choice of the conveyor itself is not the only task involved in selecting a conveyor system. Conveyor drives, motors, and auxiliary equipment must also be chosen. Conveyor drives comprise from 10 to 30% of the total cost of the conveyor system. Fixed-speed drives and adjustable speed drives are available, depending on whether changes in conveyor speed are needed during the course of normal operation. Motors for conveyor drives are generally three-phase, 60-cycle, 220-V units; 220/440-V units; 550-V units; or four-wire, 208-V units. Also available are 240-V and 480-V ratings. Auxiliary equipment includes such items as braking or arresting devices on vertical elevators to prevent reversal of travel, torque-limiting devices or electrical controls to limit power to the drive motor, and cleaners on belt conveyors.

33.2.2 Screw Conveyors

A screw conveyor consists of a helical shaft mounted within a pipe or trough. Power may be transmitted through the helix, or in the case of a fully enclosed pipe conveyor through the pipe itself. Material is forced through the channel formed between the helix and the pipe or trough. Screw conveyors are generally limited to rates of flow of about 10,000 ft³/hr. Figure 33.1 shows a chute-fed screw conveyor, one of several types in common use. Table 33.3 gives capacities and loading conditions for screw conveyors on the basis of material classifications.

33.2.3 Belt Conveyors

Belt conveyors are widely used in industry. They can traverse distances up to several miles at speeds up to 1000 ft/min and can handle thousands of tons of material per hour. Belt conveyors are generally

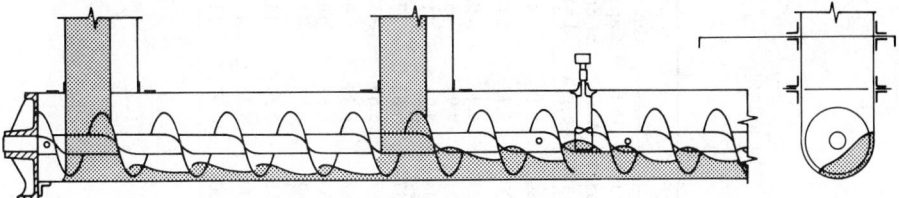

Fig. 33.1 Chute-fed screw conveyor.

placed horizontally or at slopes ranging from 10° to 20°, with a maximum incline of 30°. Direction changes can occur readily in the vertical plane of the belt path, but horizontal direction changes must be managed through such devices as connecting chutes and slides between different sections of belt conveyor.

Belt-conveyor design depends largely on the nature of the material to be handled. Particle-size distribution and chemical composition of the material dictate selection of the width of the belt and the type of belt. For instance, oily substances generally rule out the use of natural rubber belts. Conveyor-belt capacity requirements are based on peak load rather than average load. Operating conditions which affect belt-conveyor design include climate, surroundings, and period of continuous service. For instance, continuous service operation will require higher-quality components than will intermittent service, which allows more frequent maintenance. Belt width and speed depend on the bulk density of the material and lump size. The horsepower to drive the belt is a function of the following factors: (1) power to drive an empty belt; (2) power to move the load against the friction of the rotating parts; (3) power to elevate and lower the load; (4) power to overcome inertia in placing material in motion; and (5) power to operate a belt-driven tripper. Table 33.4 provides typical data for estimating belt-conveyor design requirements. Figure 33.2 illustrates a typical belt-conveyor loading arrangement.

33.2.4 Bucket Elevators

Bucket elevators are used for vertical transport of bulk solid materials. They are available in a wide range of capacities and may operate in the open or totally enclosed. They tend to be acquired in highly standardized units, although specifically engineered equipment can be obtained for use with special materials, unusual operating conditions, or high capacities. Figure 33.3 shows a common type of bucket elevator, the spaced-bucket centrifugal-discharge elevator. Other types include spaced-bucket positive-discharge elevators, V-bucket elevators, continuous-bucket elevators, and supercapacity continuous-bucket elevators. The latter handle high tonnages and are usually operated at an incline to improve loading and discharge conditions.

Bucket elevator horsepower requirements can be calculated for space-bucket elevators by multiplying the desired capacity (tons per hour) by the lift and dividing by 500. Table 33.5 gives bucket elevator specifications for spaced-bucket centrifugal-discharge elevators.

33.2.5 Vibrating or Oscillating Conveyors

Vibrating conveyors are usually directional-throw devices that consist of a spring-supported horizontal pan or trough, vibrated by an attached arm or rotating weight. The motion imparted to the material particles abruptly tosses them upward and forward so that the material travels in the desired direction. The conveyor returns to a reference position, which gives rise to the term "oscillating conveyor." The capacity of the vibrating conveyor is determined by the magnitude and frequency of trough displacement, angle of throw, and slope of the trough, and the ability of the material to receive and transmit through its mass the directional "throw" of the trough. Classifications of vibrating conveyors include (1) mechanical, (2) electrical, and (3) pneumatic and hydraulic vibrating conveyors. Capacities of vibrating conveyors are very broad, ranging from a few ounces or grams for laboratory-scale equipment to thousands of tons for heavy industrial applications. Figure 33.4 depicts a leaf-spring mechanical vibrating conveyor, and provides a selection chart for this conveyor.

33.2.6 Continuous-Flow Conveyors

The continuous-flow conveyor is a totally enclosed unit that operates on the principle of pulling a surface transversely through a mass of bulk solids material, such that it pulls along with it a cross section of material that is greater than the surface of the material itself. Figure 33.5 illustrates a typical configuration for a continuous-flow conveyor.

Three common types of continuous-flow conveyors are (1) closed-belt conveyors, (2) flight conveyors,

Table 33.3 Capacity and Loading Conditions for Screw Conveyors

Capacity tons/hr	Capacity ft³/hr	Diam. of Flights (in.)	Diam. of Pipe (in.)	Diam. of Shafts (in.)	Hanger Centers (ft)	Max. Size Lumps — All Lumps	Max. Size Lumps — Lumps 20–25%	Max. Size Lumps — Lumps 10% or Less	Speed (rpm)	Max. Torque Capacity (in.-lb)	Feed Section Diam. (in.)	Hp at Motor 15 ft Max. Length	30 ft Max. Length	45 ft Max. Length	60 ft Max. Length	75 ft Max. Length	Max. Hp. Capacity at Speed Listed
5	200	9	2½	2	10	¾	1½	2¼	40	7,600	6	0.43	0.85	1.27	1.69	2.11	4.8
10	400	10	2½	2	10	¾	1½	2½	55	7,600	9	0.85	1.69	2.25	3.00	3.75	6.6
15	600	10	2½	2	10	¾	1½	2½	80	7,600	9	1.27	2.25	3.38	3.94	4.93	9.6
		12	2½	2		1	2	3	45	7,600	10						5.4
		12	3½	3		1	2	3	45	16,400	10						11.7
20	800	12	2½	2	12	1	2	3	60	7,600	10	1.69	3.00	3.94	4.87	5.63	7.2
		12	3½	3		1	2	3	60	16,400	10						15.6
25	1000	12	2½	2	12	1	2	3	75	7,600	10	2.12	3.75	4.93	5.63	6.55	9.0
		14	3½	3		1¼	2½	3½	75	16,400	10						19.5
		14	3½	3		1¼	2½	3½	45	16,400	12						11.7
30	1200	14	3½	3	12	1¼	2½	3½	55	16,400	12	2.25	3.94	5.05	6.75	7.50	14.3
35	1400	14	3½	3	12	1¼	2½	3½	65	16,400	12	2.62	4.58	5.90	7.00	8.75	16.9
40	1600	16	3½	3	12	1½	3	4	50	16,400	14	3.00	4.50	6.75	8.00	10.00	13.0

Table 33.4 Data for Estimating Belt Conveyor Design Requirements

Belt Width (in.)	Cross-Sectional Area of Load (ft²)	Belt Speed — Normal Operating Speed (ft/min)	Belt Speed — Max. Advisable Speed (ft/min)	Belt Plies Min.	Belt Plies Max.	Max. Size Lump (in.) — Sized Material 80% Under	Max. Size Lump (in.) — Unsized Material Not Over 20%	Belt Speed (ft/min)	50 lb/ft³ Material — Capacity (tons/hr)	50 lb/ft³ — hp 10-ft Lift	50 lb/ft³ — hp 100-ft Centers	100 lb/ft³ Material — Capacity (tons/hr)	100 lb/ft³ — hp 10-ft Lift	100 lb/ft³ — hp 100-ft Centers	Add hp for Tripper
14	0.11	200	300	3	5	2	3	100	16	0.17	0.22	32	0.34	0.44	1.00
								200	32	0.34	0.44	64	0.68	0.88	
								300	48	0.52	0.66	96	1.04	1.32	
16	0.14	200	300	3	5	2½	4	100	22	0.23	0.28	44	0.46	0.56	1.25
								200	44	0.45	0.56	88	0.90	1.12	
								300	66	0.68	0.84	132	1.36	1.68	
18	0.18	250	350	4	6	3	5	100	27	0.29	0.35	54	0.58	0.7	1.50
								250	67	0.71	0.88	134	1.42	1.76	
								350	95	1.00	1.21	190	2.00	2.42	
20	0.22	250	350	4	6	3½	6	100	33	0.35	0.42	66	0.70	0.84	1.60
								250	82	0.86	1.03	164	1.72	2.06	
								350	115	1.22	1.45	230	2.44	2.9	
24	0.33	300	400	4	7	4½	8	100	49	0.51	0.51	98	1.02	1.02	1.75
								300	147	1.53	1.52	294	3.06	3.04	
								400	196	2.04	2.02	392	4.08	4.04	
30	0.53	300	450	4	8	7	12	100	79	0.80	0.75	158	1.60	1.5	2.50
								300	237	2.40	2.25	474	4.80	4.5	
								450	355	3.60	3.37	710	7.20	6.74	
36	0.78	400	600	4	9	8	15	100	115	1.22	0.80	230	2.44	1.59	3.53
								400	460	4.87	3.18	920	9.74	6.36	
								600	690	7.30	4.76	1380	14.6	9.52	
42	1.09	400	600	4	10	10	18	100	165	1.75	1.14	330	3.50	2.28	4.79
								400	660	7.00	4.56	1320	14.0	9.12	
								600	990	11.6	6.84	1980	23.2	13.68	
48	1.46	400	600	4	12	12	21	100	220	2.33	1.52	440	4.66	3.04	6.42
								400	880	9.35	6.07	1760	18.7	12.14	
								600	1320	14.0	9.10	2640	28.0	18.2	
54	1.90	450	600	6	12	14	24	100	285	3.02	1.97	570	6.04	3.94	10.56
								450	1282	13.6	8.85	2564	27.2	17.7	
								600	1710	18.1	11.82	3420	36.2	23.6	
60	2.40	450	600	6	13	16	28	100	360	3.82	2.49	720	7.64	4.98	
								450	1620	17.2	11.20	3240	34.4	22.4	
								600	2160	22.9	14.95	4320	45.8	29.9	

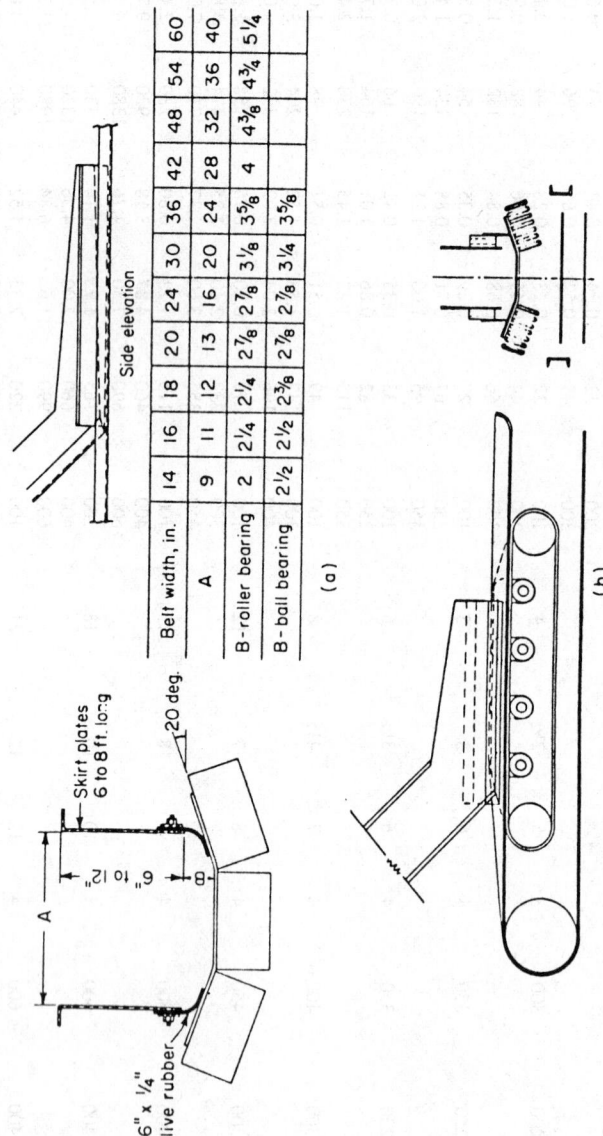

Belt width, in.	14	16	18	20	24	30	36	42	48	54	60
A	9	11	12	13	16	20	24	28	32	36	40
B-roller bearing	2	2¼	2¼	2⅞	2⅞	3⅛	3⅝	4	4⅜	4¾	5¼
B-ball bearing	2½	2½	2⅜	2⅞	2⅞	3¼	3⅝				

Side elevation

(a)

(b)

Skirt plates
6 to 8 ft. long

20 deg.

6" to 12"

A

B

6" x ¼"
live rubber

Fig. 33.2 A typical belt conveyor loading arrangement.

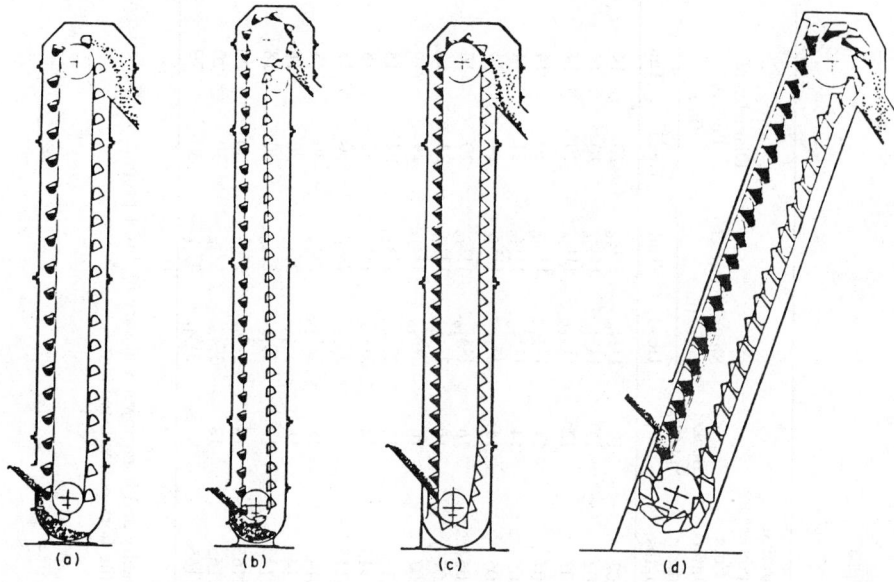

Fig. 33.3 Bucket elevators.

and (3) apron conveyors. These conveyors employ a chain-supported transport device, which drags through a totally enclosed boxlike tunnel.

33.2.7 Pneumatic Conveyors

Pneumatic conveyors operate on the principle of transporting bulk solids suspended in a stream of air over vertical and horizontal distances ranging from a few inches or centimeters to hundreds of feet or meters. Materials in the form of fine powders are especially suited to this means of conveyance, although particle sizes up to a centimeter in diameter can be effectively transported pneumatically. Materials with bulk densities from one to more than 100 lb/ft³ can be transported through pneumatic conveyors.

The capacity of a pneumatic conveying system depends on such factors as the bulk density of the product, energy within the conveying system, and the length and diameter of the conveyor.

There are four basic types of pneumatic conveyor systems: (1) pressure, (2) vacuum, (3) combination pressure and vacuum, and (4) fluidizing. In pressure systems, the bulk solids material is charged into an air stream operated at higher-than-atmospheric pressures, such that the velocity of the air stream maintains the solid particles in suspension until it reaches the separating vessel, usually an air filter or cyclone separator. Vacuum systems operate in much the same way, except that the pressure of the system is kept lower than atmospheric pressure. Pressure-vacuum systems combine the best features of these two techniques, with a separator and a positive-displacement blower placed between the vacuum "charge" side of the system and the pressure "discharge" side. One of the most common applications of pressure-vacuum systems is with combined bulk vehicle (e.g., hopper car) unloading and transporting to bulk storage. Fluidizing systems operate on the principle of passing air through a porous membrane, which forms the bottom of the conveyor, thus giving finely divided, non-free-flowing bulk solids the characteristics of free-flowing material. This technique, commonly employed in transporting bulk solids over short distances (e.g., from a storage bin to the charge point to a pneumatic conveyor), has the advantage of reducing the volume of conveying air needed, thereby reducing power requirements. Figure 33.6 illustrates these four types of pneumatic conveyor systems.

33.3 BULK MATERIALS STORAGE

33.3.1 Storage Piles

Open-yard storage is a commonplace approach to the storage of bulk solids. Belt conveyors are most often used to transport to and from such a storage area. Cranes, front-end loaders, and draglines are the types of equipment employed in moving material at the storage site.

Enclosed storage piles are employed where the bulk solids materials can erode or dissolve in rainwater,

Table 33.5 Bucket Elevator Specifications

Size of Bucket (in.)[a]	Elevator Centers (ft)	Capacity (tons/hr) Material Weighing 100 lb/ft³	Size Lumps Handled (in.)[c]	Bucket Speed (ft/min)	rpm Head Shaft	Horsepower[b] Required at Head Shaft	Additional Horsepower[b] per Foot for Intermediate Lengths	Bucket Spacing (in.)	Shaft Diameter (in.) Head	Shaft Diameter (in.) Tail	Diameter of Pulleys (in.) Head	Diameter of Pulleys (in.) Tail	Belt Width (in.)
6 × 4 × 4¼	25	14	¾	225	43	1.0	0.02	12	1¹⁵⁄₁₆	1¹¹⁄₁₆	20	14	7
	50	14	¾	225	43	1.6	0.02	12	1¹⁵⁄₁₆	1¹¹⁄₁₆	20	14	7
	75	14	¾	225	43	2.1	0.02	12	1¹⁵⁄₁₆	1¹¹⁄₁₆	20	14	7
8 × 5 × 5½	25	27	1	225	43	1.6	0.04	14	1¹⁵⁄₁₆	1¹¹⁄₁₆	20	14	9
	50	30	1	260	41	3.5	0.05	14	1¹⁵⁄₁₆	1¹¹⁄₁₆	24	14	9
	75	30	1	260	41	4.8	0.05	14	2⁷⁄₁₆	1¹¹⁄₁₆	24	14	9
10 × 6 × 6¼	25	45	1¼	225	43	3.0	0.063	16	1¹⁵⁄₁₆	1¹⁵⁄₁₆	20	16	11
	50	52	1¼	260	41	5.2	0.07	16	2⁷⁄₁₆	1¹⁵⁄₁₆	24	16	11
	75	52	1¼	260	41	7.2	0.07	16	2¹⁵⁄₁₆	1¹⁵⁄₁₆	24	16	11
12 × 7 × 7¼	25	75	1½	260	41	4.7	0.1	18	2⁷⁄₁₆	1¹⁵⁄₁₆	24	18	13
	50	84	1½	300	38	8.9	0.115	18	2¹⁵⁄₁₆	1¹⁵⁄₁₆	30	18	13
	75	84	1½	300	38	11.7	0.115	18	3⁷⁄₁₆	2⁷⁄₁₆	30	18	13
14 × 7 × 7¼	25	100	1¾	300	38	7.3	0.14	18	2¹⁵⁄₁₆	2⁷⁄₁₆	30	18	15
	50	100	1¾	300	38	11.0	0.14	18	3⁷⁄₁₆	2⁷⁄₁₆	30	18	15
	75	100	1¾	300	38	14.3	0.14	18	3⁷⁄₁₆	2⁷⁄₁₆	30	18	15
16 × 8 × 8½	25	150	2	300	38	8.5	0.165	18	2¹⁵⁄₁₆	2⁷⁄₁₆	30	20	18
	50	150	2	300	38	12.6	0.165	18	3⁷⁄₁₆	2⁷⁄₁₆	30	20	18
	75	150	2	300	38	16.7	0.165	18	3¹⁵⁄₁₆	2⁷⁄₁₆	30	20	18

[a] Size of buckets given: width × projection × depth.
[b] Capacities and horsepowers given for materials weighing 100 lb/ft³. For materials of other weights, capacity and horsepower will vary in direct proportion. For example, an elevator handling coal weighing 50 lb/ft³ will have half the capacity and will require approximately half the horsepower listed above.
[c] If volume of lumps averages less than 15% of total volume, lumps of twice size listed may be handled.

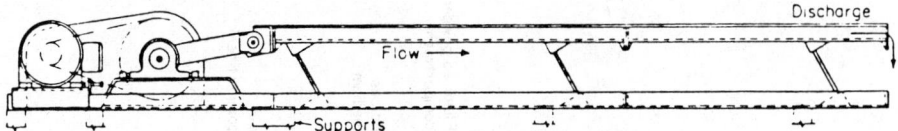

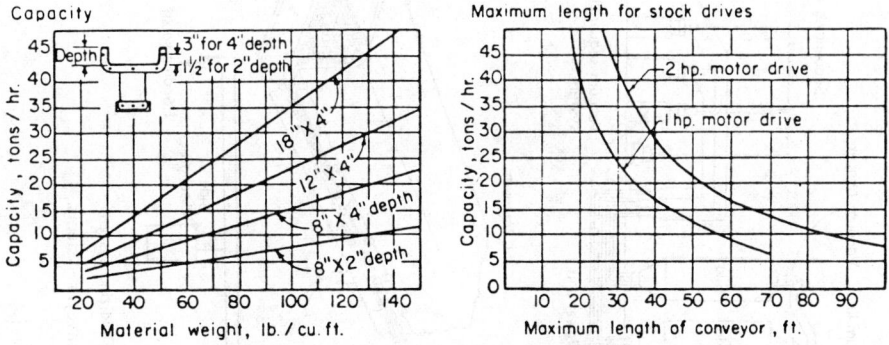

Fig. 33.4   Leaf-spring mechanical vibrating conveyor.

as in the case of salt for use on icy roads. The necessary equipment for one such application, the circular storage facility, are (1) feed conveyor, (2) central support column, (3) stacker, (4) reclaimer, (5) reclaim conveyor, and (6) the building or dome cover.

33.3.2   Storage Bins, Silos, and Hoppers

A typical storage vessel for bulk solids materials consists of two components—a bin and a hopper. The bin is the upper section of the vessel and has vertical sides. The hopper is the lower part of the vessel, connecting the bin and the outlet, and must have at least one sloping side. The hopper serves as the means by which the stored material flows to the outlet channel. Flow is induced by opening the outlet port and using a feeder device to move the material, which drops through the outlet port.

   If all material stored in the bin moves whenever material is removed from the outlet port, mass flow is said to prevail. However, if only a portion of the material moves, the condition is called funnel flow. Figure 33.7 illustrates these two conditions.

   Many flow problems in storage bins can be reduced by taking the physical characteristics of the bulk material into account. Particle size, moisture content, temperature, age, and oil content of the stored material affect flowability. Flow-assisting devices and feeders are usually needed to overcome flow problems in storage bins.

33.3.3   Flow-Assisting Devices and Feeders

To handle those situations in which bin design alone does not produce the desired flow characteristics, flow-assisting devices are available. Vibrating hoppers are one of the most important types of flow-

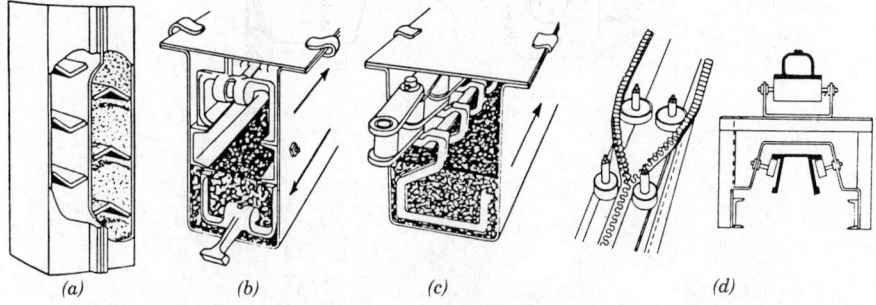

    (a)   *(b)*   *(c)*   *(d)*

Fig. 33.5   Continuous-flow conveyor.

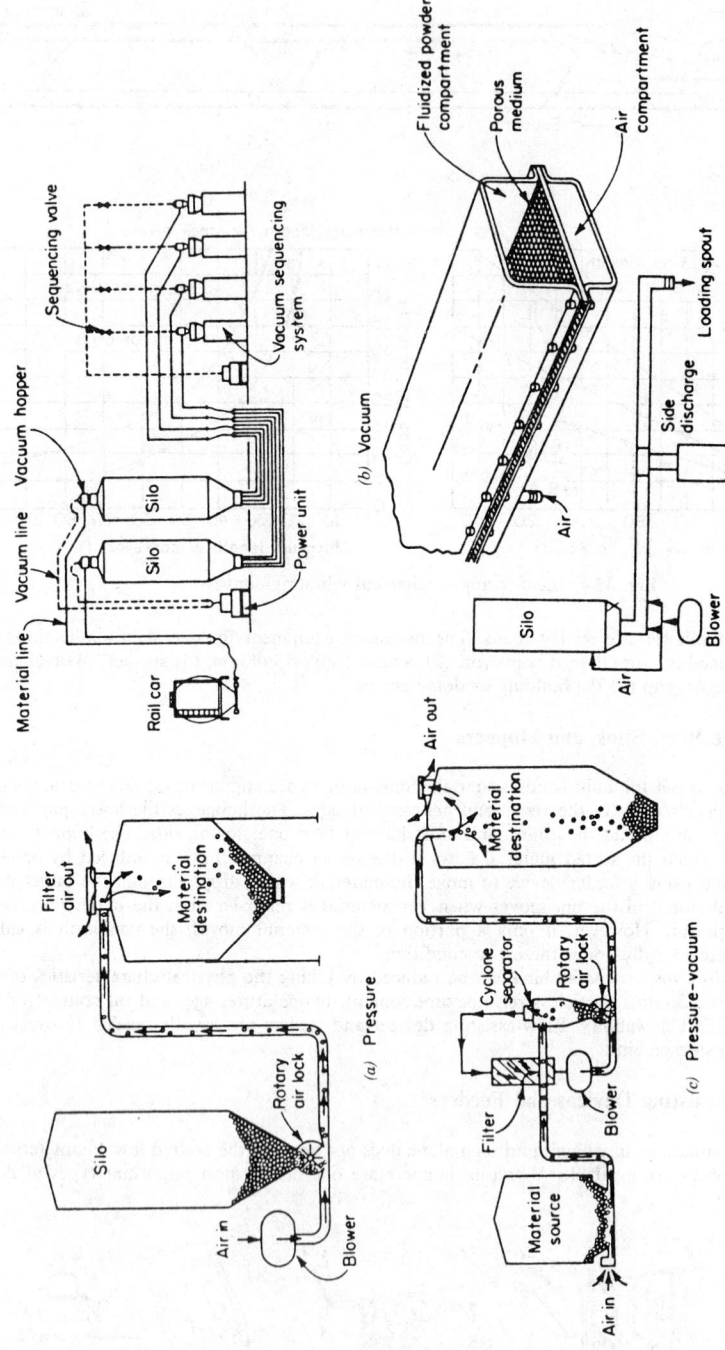

Fig. 33.6 Four types of pneumatic conveyor systems.

(a) Pressure

(b) Vacuum

(c) Pressure–vacuum

(d) Fluidizing system

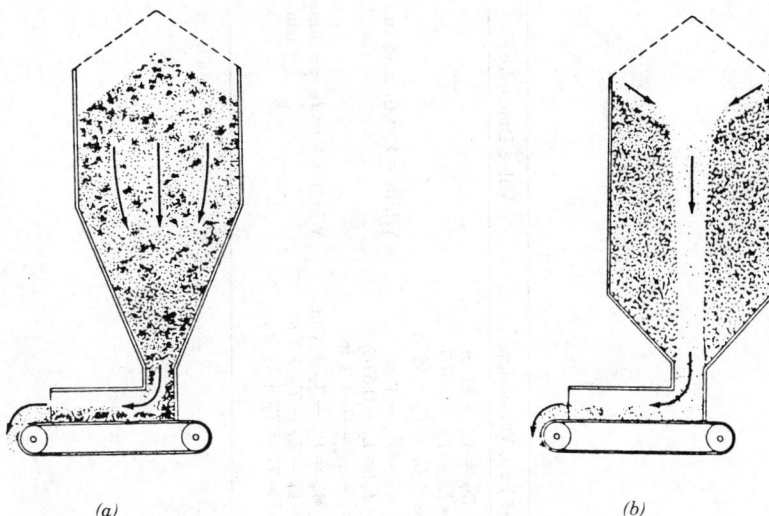

Fig. 33.7 Mass-flow (*a*) and funnel-flow (*b*) in storage bins.

assisting devices. These devices fall into two categories: (1) gyrating devices, in which vibration is applied perpendicular to the flow channel; and (2) whirlpool devices, which apply a twisting motion and a lifting motion to the material, thereby disrupting any bridges that might tend to form. Screw feeders are used to assist in bin unloading by removing material from the hopper opening.

33.3.4 Packaging of Bulk Materials

Bulk materials are often transported and marketed in containers, such as bags, boxes, and drums. Packaged solids lend themselves to material handling by means of unit material handling.

Bags

Paper, plastic, and cloth bags are common types of containers for bulk solids materials. Multiwall paper bags are made from several plies of kraft paper. Bag designs include valve and open-mouth designs. Valve-type bags are stitched or glued at both ends prior to filling, and are filled through a valve opening at one corner of the bag. Open-mouth bags are sealed at one end during manufacture, and at the open end after filling. Valve bags more readily lend themselves to automated filling than open-mouth bags, yielding higher packing rates.

Bag size is determined by the weight or volume of material to be packed and its bulk density. Three sets of dimensions must be established in bag sizing: (1) tube—outside length and width of the bag tube before closures are fabricated; (2) finished face—length, width, and thickness of the bag after fabrication; and (3) filled face—length, width, and thickness of the bag after filling and closure. Figure 33.8 shows the important dimensions of multiwall paper bags, and Table 33.6 gives their relationships to tube, finished face, and filled face dimensions.

Boxes

Bulk boxes are fabricated from corrugated kraft paper. They are used to store and ship bulk solid materials in quantities ranging from 50 lb to several hundred pounds. A single-wall corrugated kraft

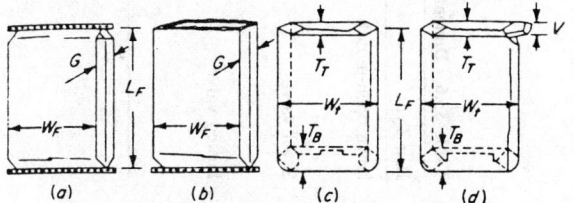

Fig. 33.8 Dimensions of multiwall paper bags.

Table 33.6 Dimensions of Multiwall Paper Bags

Bag Type	Tube Dimensions	Finished-Face Dimensions	Filled-Face Dimensions	Valve Dimensions
Sewn open-mouth	Width = $W_t = W_f + G_f$ Length = $L_t = L_f$	Width = $W_f = W_t - G_f$ Length = $L_f = L_t$ Gusset = G_f	Width = $W_F = W_f + \frac{1}{2}$ in. Length = $L_F = L_f - 0.67G_f$ Thickness = $G_F = G_f + \frac{1}{2}$ in.	
Sewn valve	Width = $W_t = W_f + G_f$ Length = $L_t = L_f$	Width = $W_f = W_t - G_f$ Length = $L_f = L_t$ Gusset = G_f	Width = $W_F = W_f + 1$ in. Length = $L_F = L_f - 0.67G_f$ Thickness = $G_F = G_f + 1$ in.	Width = $V = G_f \pm \frac{1}{2}$ in.
Pasted valve	Width = $W_t = W_f$ Length = $L_t = L_f$	Width = $W_f = W_t$ Length = $L_f = L_t - (T_T + T_B)/2 - 1$ Thickness at top = T_T Thickness at bottom = T_B	Width = $W_F = W_f - T_T + 1$ in. Length = $L_F = L_f - T_T + 1$ in. Thickness = $T_F = T_T + \frac{1}{2}$ in.	Width = $V = T_T \begin{cases} +0 \text{ in.} \\ -1 \text{ in.} \end{cases}$

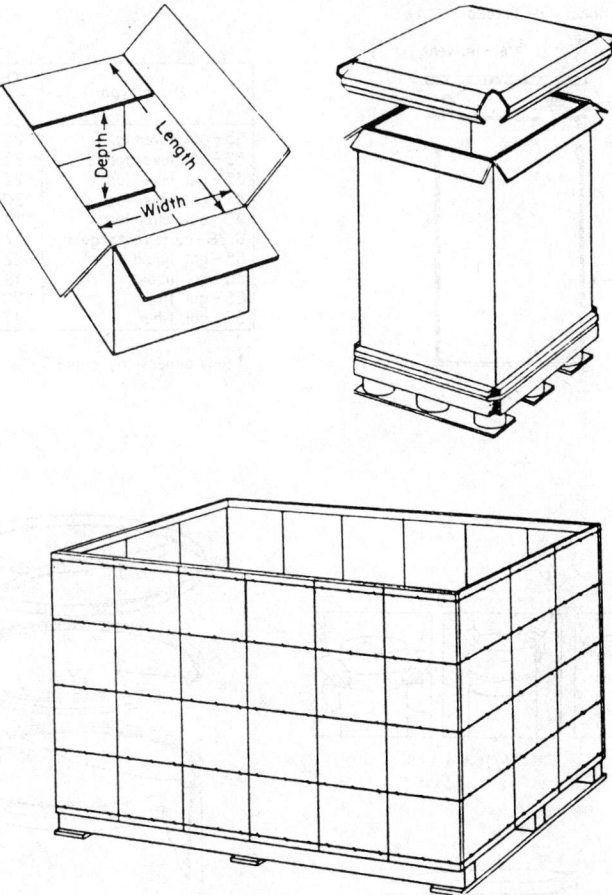

Fig. 33.9 Bulk boxes and cartons.

board consists of an outside liner, a corrugated medium, and an inside liner. A double-wall board has two corrugated mediums sandwiched between three liners. The specifications for bulk boxes depend on the service requirements; 600 lb/in.^{2} is common for loads up to 1000 lb, and 200 lb/in.^{2} for 100-lb loads. Bulk boxes have the advantages of reclosing and of efficient use of storage and shiping space, called "cube." Disadvantages include the space needed for storage of unfilled boxes and limited reusability. Figure 33.9 shows important characteristics of bulk boxes.

Folding cartons are used for shipping bulk solids contained in individual bottles, bags, or folding boxes. Cartons are of less sturdy construction than bulk boxes, because the contents can assist in supporting vertically imposed loads.

Drums

As illustrated in Fig. 33.10, drums can be made of either steel or fiber, although the less expensive fiber drums are most often used with bulk solids. Fiber drums are available in only a few sizes, ranging in volume from ½ to 50 gal, and for weights ranging from 50 to 500 lb. Steel drums are preferred for liquids.

33.3.5 Transportation of Bulk Materials

The term transportation of bulk materials refers to the movement of raw materials, fuels, and bulk products by land, sea, and air. A useful definition of a bulk shipment is any unit greater than 4000 lb or 40 ft^3.

The most common bulk carriers are railroad hopper cars, highway hopper trucks, portable bulk bins, van-type containers, barges, and ships. Factors affecting the choice among these alternative modes

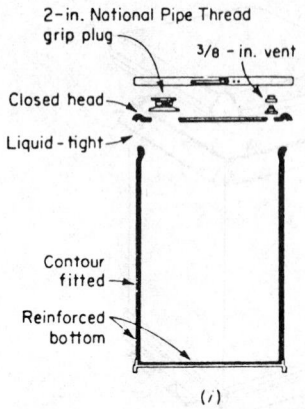

Drum type	Outside dimensions	
	Dia., in.	Height, in.
55 – gal. lever top	21	40 3/4
55 – gal. lever top	23 1/2	30 3/4
55 – gal. lever top	22	34 3/4
41 – gal. lever top	20 1/2	30 1/4
30 – gal. lever top	19	26 1/4
6.28 – cu. ft. rectangular	17 5/8*	37 1/2
55 – gal. liquid	22	37 1/2
30 – gal. liquid	19	28
55 – gal. fiber	20 3/8	40 3/4
30 – gal. fiber	17 3/8	30 3/4

* Side dimension, square

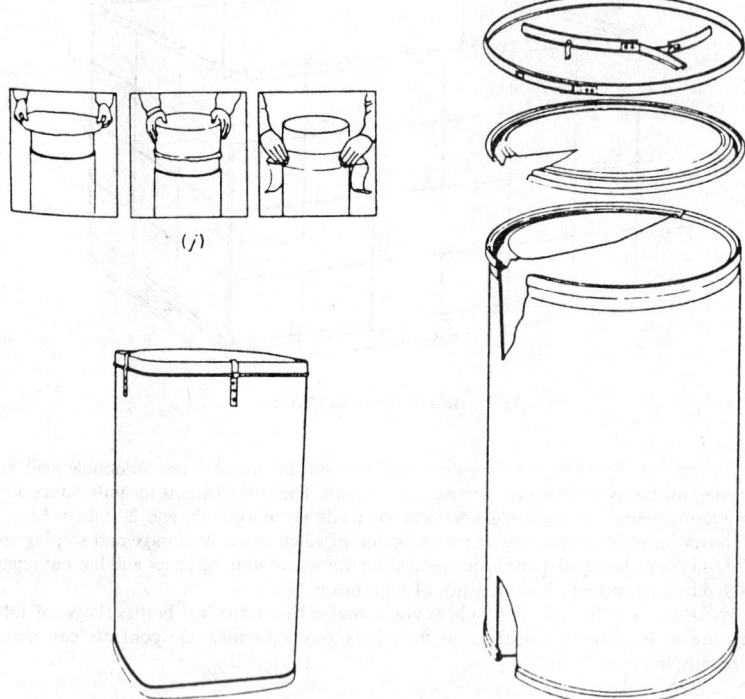

Fig. 33.10 Storage drums.

of transportation include the characteristics of the material, size of shipment, available transportation routes from source to destination (e.g., highway, rail, water), and the time available for shipment.

Railroad Hopper Cars

Railroad hopper cars are of three basic designs: (1) covered, with bottom unloading ports; (2) open, with bottom unloading ports; and (3) open, without unloading ports. Gravity, pressure differential, and fluidizing unloading systems are available with railroad hopper cars.

Loading of hopper cars can be done with most types of conveyors: belt, screw, pneumatic, etc. Unloading of bottom unloading hopper cars can be managed by constructing a special dumping pit beneath the tracks, equipped with screw or belt takeaway conveyors.

Hopper Trucks

Hopper trucks are used for highway transportation of bulk solids materials. The most common types include (1) closed type with a pneumatic conveyor unloading system, and (2) the open dump truck. With the first type, a truck can discharge its cargo directly into a storage silo. The shipment weights carried by trucks depend on state highway load limits, usually from 75,000 to 125,000 lb.

33.4 UNIT MATERIAL HANDLING

33.4.1 Introduction

Unit material handling involves the movement and storage of separate, individual parts, goods, or assemblies. Examples include automobile body components, engine blocks, bottles, cans, bags, pallets of boxes, bins of loose parts, etc. The word "unit" refers to the single entity that is handled; that entity can consist of a single item or numerous items that have been unitized for purposes of movement and storage.

This section discusses various categories of material handling equipment used in unit material handling, and describes the procedures employed in material handling systems design.

33.4.2 Conveyors

Conveyors transport materials along fixed paths. Their functions may be solely the movement of items from one location in a process or facility to another point, or they may move items through various stages of receiving, processing, assembly, inspection, packaging, and shipping.

Conveyors used in material handling are of two basic types:

1. *Gravity Conveyors.* These include chutes, slides, and gravity wheel or roller conveyors that essentially exploit the principle of gravity to move items from a point at higher elevation to another point at a lower elevation. As is subsequently discussed, one should maximize the use of the gravity principle in designing material handling systems.
2. *Powered Conveyors.* These use electric motors to drive belts, chains, or rollers in a variety of floor-mounted, in-floor, and overhead types of conveyors.

In general, conveyors are employed in unit material handling when:

1. Loads are uniform.
2. Materials move continuously.
3. Routes do not vary.
4. Load is constant.
5. Movement rate is relatively fixed.
6. Cross traffic can be by-passed.
7. Path is relatively fixed.
8. Movement is point-to-point.
9. Automatic counting, sorting, weighing, or dispatching is needed.
10. In-process storage is required.
11. In-process inspection is required.
12. Production pacing is necessary.
13. Process control is required.
14. Controlled flow is needed.
15. Materials are handled at extreme temperatures, or other adverse conditions.
16. Handling is required in a hazardous area.
17. Hazardous materials are handled.
18. Machines are integrated into a system.
19. Robots are integrated into a system.
20. Materials are moved between workplaces.
21. Manual handling and/or lifting is undesirable.
22. Changes in production volume or pace are needed.
23. Visual surveillance of a production process is required.
24. Floor space can be saved by utilizing overhead space.

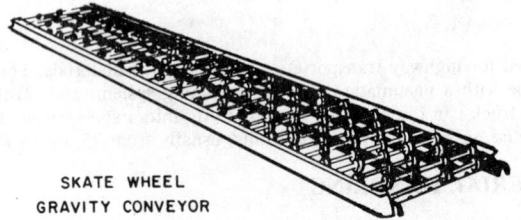

SKATE WHEEL
GRAVITY CONVEYOR

Fig. 33.11   Gravity conveyors.

25. Flexibility is required to meet changes in production processes.

26. Integration between computer-aided design and computer-aided manufacturing is required.

This section provides essential information about four main classes of conveyors used in unit material handling: (1) gravity conveyors; (2) powered conveyors; (3) chain-driven conveyors; and (4) power-and-free conveyors.

Gravity Conveyors

Gravity conveyors exploit gravity to move material without the use of other forms of energy. Chutes, skate wheel conveyors, and roller conveyors are the most common forms of gravity conveyors. Figure 33.11 illustrates wheel and roller conveyors. Advantages of gravity conveyors are low cost, relatively low maintenance, and negligible breakdown rate. The main requirement for using gravity conveyors is the ability to provide the necessary gradient in the system configuration at the point at which gravity units are placed.

Powered Conveyors

The two principal types of powered conveyors are (1) belt conveyors and (2) roller conveyors, as shown in Fig. 33.12. Electric motors provide the energy to drive the belt or rollers on these conveyors.

Belt conveyors are used either in the horizontal plane or with inclines up to 30°. They can range in length from a few feet to hundreds of feet, and are unidirectional. Changes in direction must be managed through the use of connecting chutes.

Roller conveyors are used for heavier loads than can be moved with belt conveyors, and are of sturdier construction. Roller conveyors can be used to provide spacing between items. Inclines are possible to about 10°; declines of about 15° are possible.

Chain-Driven Conveyors

Chain conveyors are those in which closed-loop systems of chain, usually driven by electric motors, are used to pull items or carts along a specified path. The three principal types of chain-driven conveyors used in unit material handling are (1) flight conveyors, (2) overhead towlines and monorails, and (3) in-floor towlines. Figure 33.13 illustrates each of these types of chain-driven conveyors.

Flight conveyors consist of one or more endless strands of chain with spaced transverse flights or scrapers attached, which push the material along through a trough. Used primarily in bulk material handling, its primary use in unit material handling includes movement of cans or bottles in food canning and bottling. A flight conveyor is limited to speeds up to 120 ft/min.

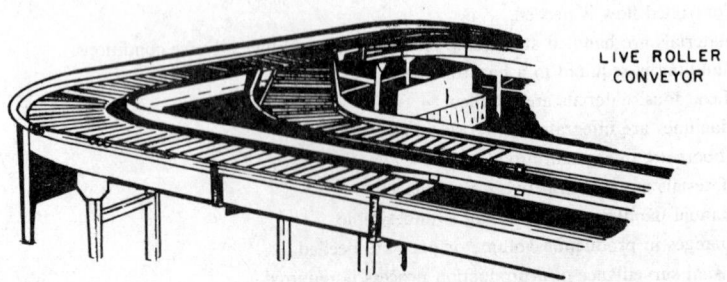

LIVE ROLLER
CONVEYOR

Fig. 33.12   Roller conveyors.

Fig. 33.13 Chain-driven conveyor.

Fig. 33.14 Power-and-free conveyor.

Overhead towlines consist of a track mounted 8–9 ft above the floor. Carts on the floor are attached to the chain, which moves through the overhead track. Overhead towlines free the floor for other uses, and are cheaper and more flexible than in-floor towlines.

In-floor towlines consist of chain tracks mounted in the floor. A cart is pulled along the track by attaching a handle to the chain. In-floor towlines are capable of greater speeds than overhead towlines and have a smoother pickup action. They are more difficult to maintain, and lack flexibility for rerouting.

Power-and-Free Conveyors

Power-and-free conveyors are a combination of powered trolley conveyors and unpowered monorail-type free conveyors. Two sets of tracks are used, one positioned above the other. The upper track carries the powered trolley or monorail-type conveyor, which is chain-driven. The lower track is the free, unpowered monorail. Load-carrying free trolleys are engaged by pushers attached to the powered trolley conveyors. Load trolleys can be switched to and from adjacent unpowered free tracks.

Interconnections on power-and-free conveyors may be manually or automatically controlled. Track switches may divert trolleys from "power" to "free" tracks. Speeds may vary from one "power" section to another. Computers can be used to control power-and-free conveyors.

Power-and-free conveyors, shown in Fig. 33.14, are relatively expensive, and costly to relocate.

33.4.3 Hoists, Cranes, and Monorails

Hoists, cranes, and monorails are used for a variety of overhead handling tasks. A hoist typically consists of a hook, a lifting medium—a chain, cable or rope—and a drum for reeling in unused chain, cable, or rope. Hoists may be manually operated or pneumatically driven or they may be driven by an electric motor.

A monorail is a single-beam overhead track whose lower flange serves as a runway for a trolley-mounted hoist. Through the use of switches, turntables, and other path-changing devices, an overhead monorail can be made to follow a predetermined path, carrying a series of trolleys through various stations in processing or assembly. A chain-driven overhead monorail is very similar to the overhead towline in its configuration, except that it carries uniformly spaced trolleys overhead instead of pulling carts along the floor. The monorail can be made to dip down at specified points to deliver items to machines or other processing stations.

A crane also involves a hoist traveling on a trolley. Frequently, the trolley may also be transported, as in the case of the bridge crane shown in Fig. 33.15. Cranes may be manually, electrically, or pneumatically powered. The jib crane has a horizontal beam on which a hoist trolley rides. The beam is cantilevered from a vertical mast. The jib crane has a broad range of coverage within its rotary path.

33.4.4 Industrial Trucks

Industrial trucks provide flexible handling of materials along variable flow paths. One of the most familiar types is the forklift truck, which uses a pair of forks—capable of variable spacing—riding on a vertical mast to engage, lift, lower, and move loads. Lift trucks may be manually propelled or powered by electric motors, gasoline, liquefied propane, or diesel-fueled engines. With some models, the operator walks behind the truck. On others, he or she rides on the truck, in either a standing or sitting position. Figure 33.16 depicts several types of forklift trucks.

Lift trucks are very effective in lifting, stacking, and unloading materials from storage racks, highway vehicles, railroad cars, and other equipment. Some lift trucks are designed for general-purpose use, while others are designed for specific tasks such as narrow-aisle or high-rack handling.

33.4.5 Automated Storage and Retrieval Systems

An automated storage and retrieval system consists of a machine—equipped with code-reading scanners and microprocessor controls—that retracts desired coded bins from specified storage locations and brings them to an operator station. While the operator is selecting items from one bin at the operator station, the machine is returning a previously used bin to storage. The system can be operator directed through a keyboard entry terminal, or it may be under computer control. Figure 33.17 shows an automated storage and retrieval system.

33.4.6 Rack Storage

A rack is a framework designed to facilitate the storage of loads. This framework consists of upright columns and horizontal members for supporting the loads, and diagonal bracing for stability. The structural members may be bolted into place or lock-fitted. Figure 33.18 illustrates rack storage.

Racks may be used to store pallet loads of material or bins or individual items. They may be

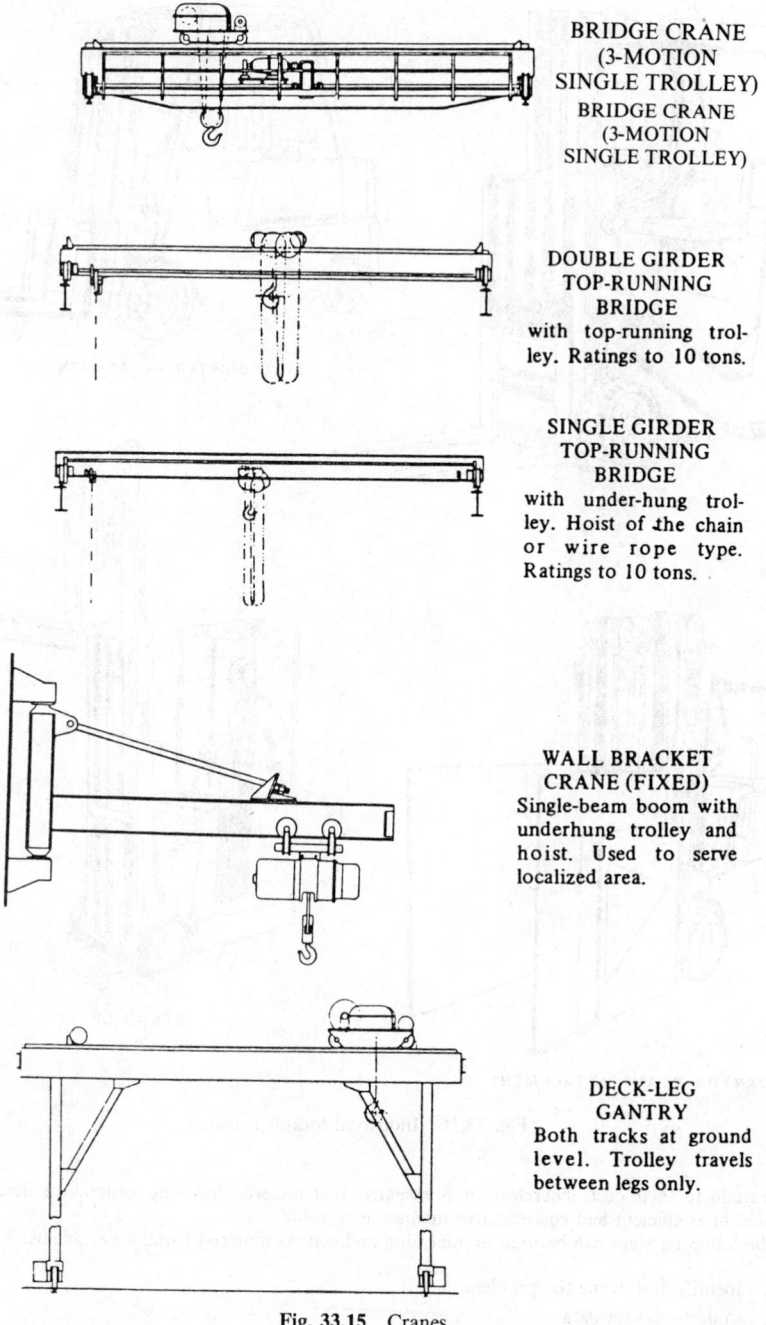

BRIDGE CRANE
(3-MOTION
SINGLE TROLLEY)
BRIDGE CRANE
(3-MOTION
SINGLE TROLLEY)

DOUBLE GIRDER
TOP-RUNNING
BRIDGE
with top-running trolley. Ratings to 10 tons.

SINGLE GIRDER
TOP-RUNNING
BRIDGE
with under-hung trolley. Hoist of the chain or wire rope type. Ratings to 10 tons.

WALL BRACKET
CRANE (FIXED)
Single-beam boom with underhung trolley and hoist. Used to serve localized area.

DECK-LEG
GANTRY
Both tracks at ground level. Trolley travels between legs only.

Fig. 33.15 Cranes.

served by forklift trucks, by stacker cranes, or by hand. They may be fixed into position or made to slide along track for denser storage. The primary purpose in using rack storage is to provide a highly organized unit material storage system that facilitates highly efficient operations in either manufacturing or distribution.

33.5 ANALYSIS OF MATERIAL HANDLING SYSTEMS

Material handling is an indispensable element in most production and distribution systems. Yet material handling adds nothing to the value of the materials and products that flow through the system, but

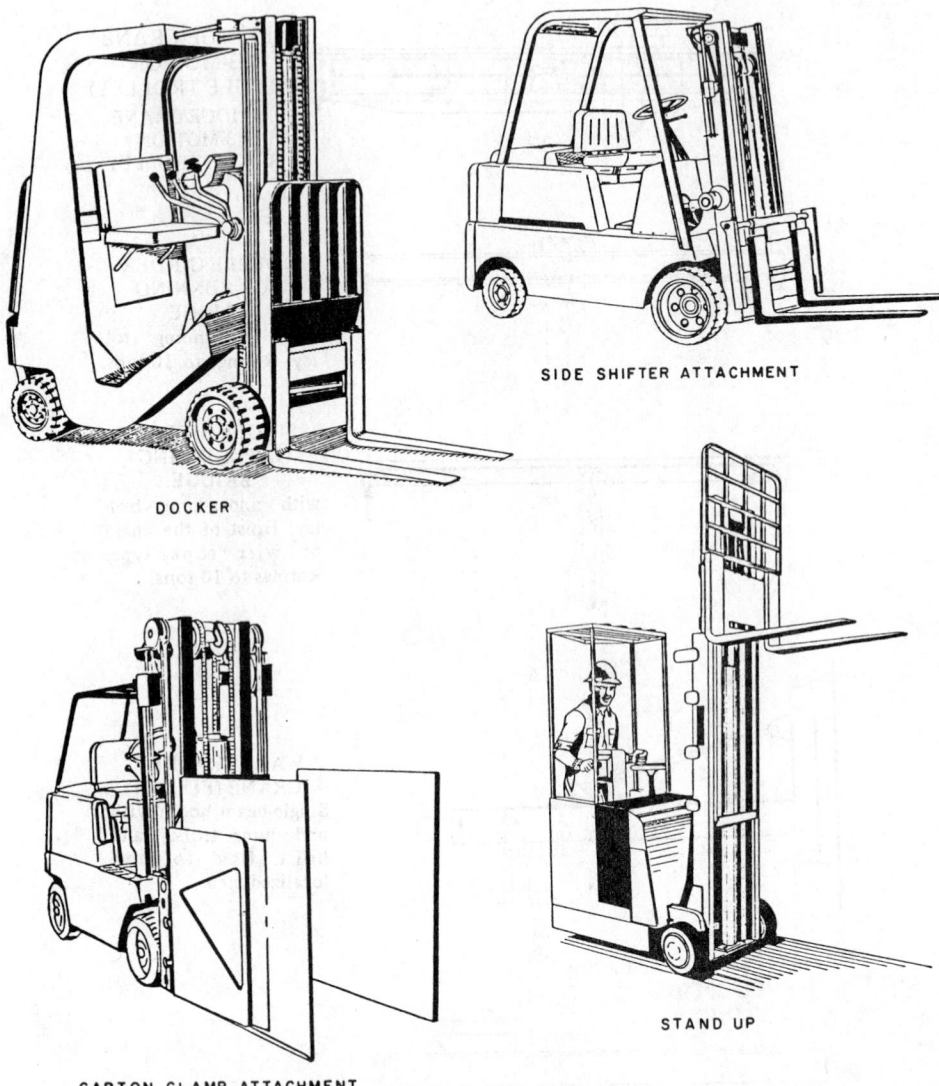

SIDE SHIFTER ATTACHMENT

DOCKER

STAND UP

CARTON CLAMP ATTACHMENT

Fig. 33.16   Industrial forklift trucks.

it does add to their cost. Therefore, it is essential that material handling systems be designed and operated in as efficient and cost effective manner as possible.

The following steps can be used in analyzing and solving material handling problems:

1. Identify and define the problem.
2. Collect relevant data.
3. Develop a plan.
4. Implement the solution.

The following sections explain how to organize an analysis of a material handling system according to this four-step procedure.

33.5.1   Identifying and Defining the Problem

In an existing plant, the best way to begin the process of identifying and defining the problem is to tour the facility, looking for the material handling aspects of the various processes observed. It is a

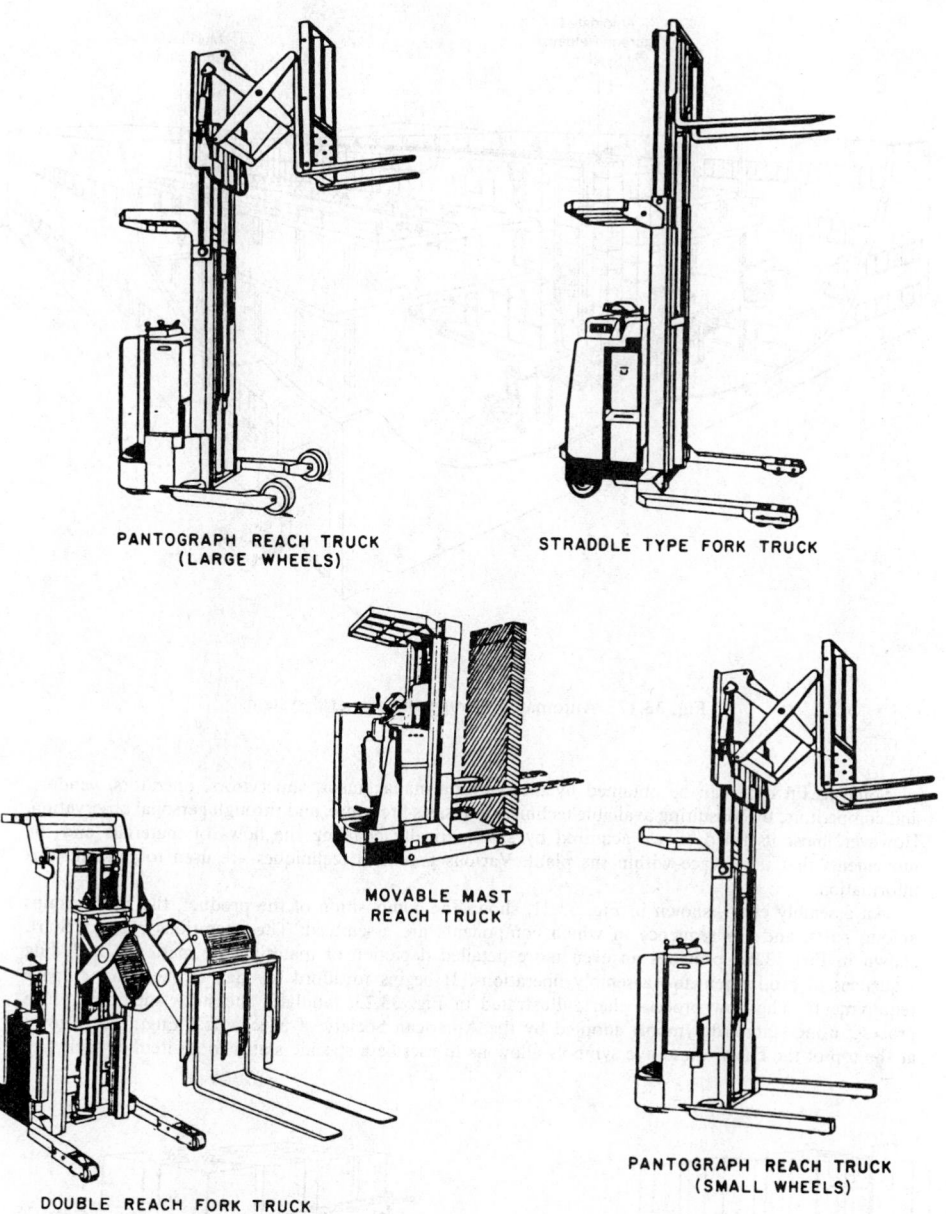

PANTOGRAPH REACH TRUCK
(LARGE WHEELS)

STRADDLE TYPE FORK TRUCK

MOVABLE MAST
REACH TRUCK

DOUBLE REACH FORK TRUCK

PANTOGRAPH REACH TRUCK
(SMALL WHEELS)

good idea to take along a checklist, such as that shown in Fig. 33.19. Another useful guide is the listing of "The Twenty Principles of Material Handling," as given in Fig. 33.20.

Once the problem has been identified, its scope must be defined. For example, if most of the difficulties are found in one area of the plant, such as shipping and receiving, the study can be focused there. Are the difficulties due to lack of space? Or, is part of the problem due to poor training of personnel in shipping and receiving? In defining the problem, it is necessary to answer the basic questions: who? what? when? where? why?

33.5.2 Collecting Data

Sometimes the problem cannot be fully defined until all relevant data are collected and analyzed. Data collection and analysis are also necessary before a solution to the problem can be found and implemented.

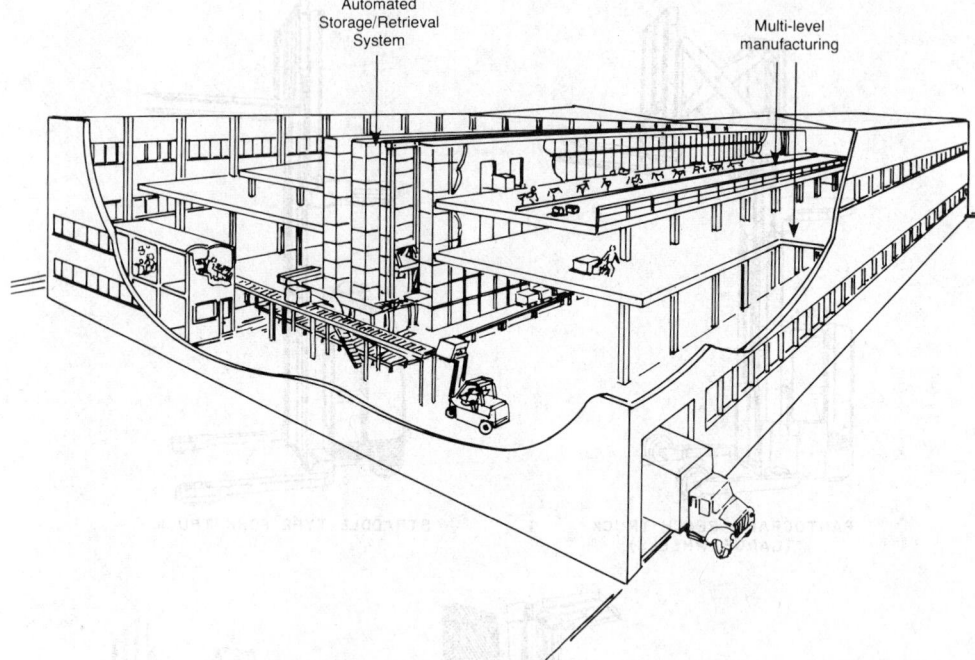

Fig. 33.17 Automated storage and retrieval system.

Some useful data can be obtained by interviewing management, supervisors, operators, vendors, and competitors, by consulting available technical and sales literature, and through personal observation. However, most useful data are acquired by systematically charting the flows of materials and the movements that take place within the plant. Various graphical techniques are used to generate this information.

An assembly chart, shown in Fig. 33.21, shows the composition of the product, the relationships among parts, and the sequence in which components are assembled. The operations process chart, shown in Fig. 33.22, provides an even more detailed depiction of material flow patterns, including sequences of production and assembly operations. It begins to afford an idea of the relative space requirements. The flow process chart, illustrated in Fig. 33.23, tabulates the steps and moves in a process, using standard symbols adopted by the American Society of Mechanical Engineers. Shown at the top of the chart, these five symbols allow us to ascribe a specific status to an item at each step

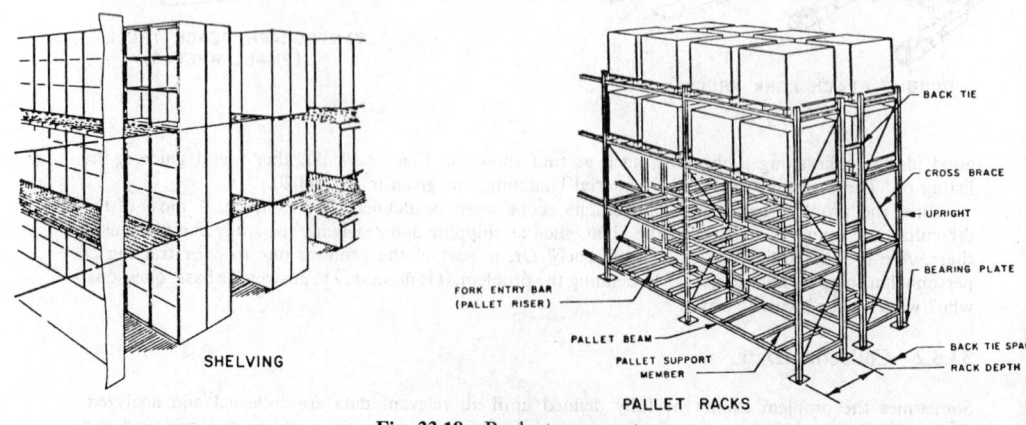

Fig. 33.18 Rack storage systems.

Material Handling Checklist

☐ Is the material handling equipment more than 10 years old?

☐ Do you use a wide variety of makes and models which require a high spare parts inventory?

☐ Are equipment breakdowns the result of poor preventive maintenance?

☐ Do the lift trucks go too far for servicing?

☐ Are there excessive employee accidents due to manual handling of materials?

☐ Are materials weighing more than 50 pounds handled manually?

☐ Are there many handling tasks that require 2 or more employees?

☐ Are skilled employees wasting time handling materials?

☐ Does material become congested at any point?

☐ Is production work delayed due to poorly scheduled delivery and removal of materials?

☐ Is high storage space being wasted?

☐ Are high demurrage charges experienced?

☐ Is material being damaged during handling?

☐ Do shop trucks operate empty more than 20% of the time?

☐ Does the plant have an excessive number of rehandling points?

☐ Is power equipment used on jobs that could be handled by gravity?

☐ Are too many pieces of equipment being used, because their scope of activity is confined?

☐ Are many handling operations unnecessary?

☐ Are single pieces being handled where unit loads could be used?

☐ Are floors and ramps dirty and in need of repair?

☐ Is handling equipment being overloaded?

☐ Is there unnecessary transfer of material from one container to another?

☐ Are inadequate storage areas hampering efficient scheduling of movement?

☐ Is it difficult to analyze the system because there is no detailed flow chart?

☐ Are indirect labor costs too high?

Fig. 33.19 Material handling checklist.

The 20 Principles of Material Handling

1. Planning Principle. Plan all material handling and storage activities to obtain maximum overall operating efficiency.

2. Systems Principle. Integrate as many handling activities as is practical into a coordinated system of operations, covering vendor, receiving, storage, production, inspection, packaging, warehousing, shipping, transportation, and customer.

3. Material Flow Principle. Provide an operation sequence and equipment layout optimizing material flow.

4. Simplification Principle. Simplify handling by reducing, eliminating, or combining unnecessary movements and/or equipment.

5. Gravity Principle. Utilize gravity to move material wherever practical.

6. Space Utilization Principle. Make optimum utilization of building cube.

7. Unit Size Principle. Increase the quantity, size, or weight of unit loads or flow rate.

8. Mechanization Principle. Mechanize handling operations.

9. Automation Principle. Provide automation to include production, handling, and storage functions.

10. Equipment Selection Principle. In selecting handling equipment consider all aspects of the material handled — the movement and the method to be used.

11. Standardization Principle. Standardize handling methods as well as types and sizes of handling equipment.

12. Adaptability Principle. Use methods and equipment that can best perform a variety of tasks and applications where special purpose equipment is not justified.

13. Dead Weight Principle. Reduce ratio of dead weight of mobile handling equipment to load carried.

14. Utilization Principle. Plan for optimum utilization of handling equipment and manpower.

15. Maintenance Principle. Plan for preventive maintenance and scheduled repairs of all handling equipment.

16. Obsolescence Principle. Replace obsolete handling methods and equipment when more efficient methods or equipment will improve operations.

17. Control Principle. Use material handling activities to improve control of production, inventory and order handling.

18. Capacity Principle. Use handling equipment to help achieve desired production capacity.

19. Performance Principle. Determine effectiveness of handling performance in terms of expense per unit handled.

20. Safety Principle. Provide suitable methods and equipment for safe handling.

Fig. 33.20 Twenty principles of material handling.

in processing. The leftmost column in the flow process chart lists the identifiable activities comprising the process, in sequential order. In the next column, one of the five standard symbols is selected to identify the activity as an operation, transportation, inspection, delay, or storage. Other columns allow us to record more detailed information. Note, for example, that, in the flow process chart in Fig. 33.23, for each step shown as a "transportation," a distance (in feet) is recorded. In the column headed "possibilities," improvement opportunities can be noted.

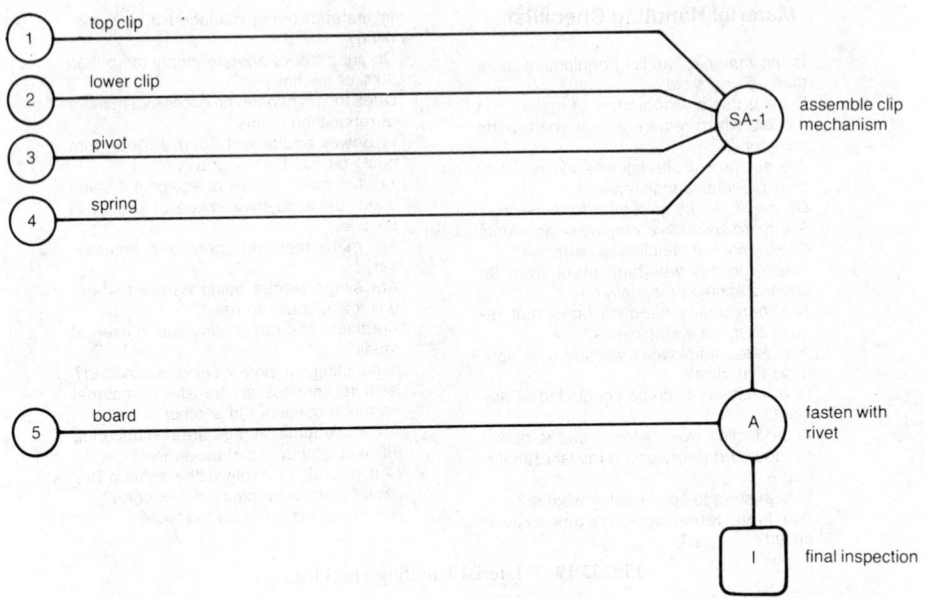

Fig. 33.21 Assembly chart.

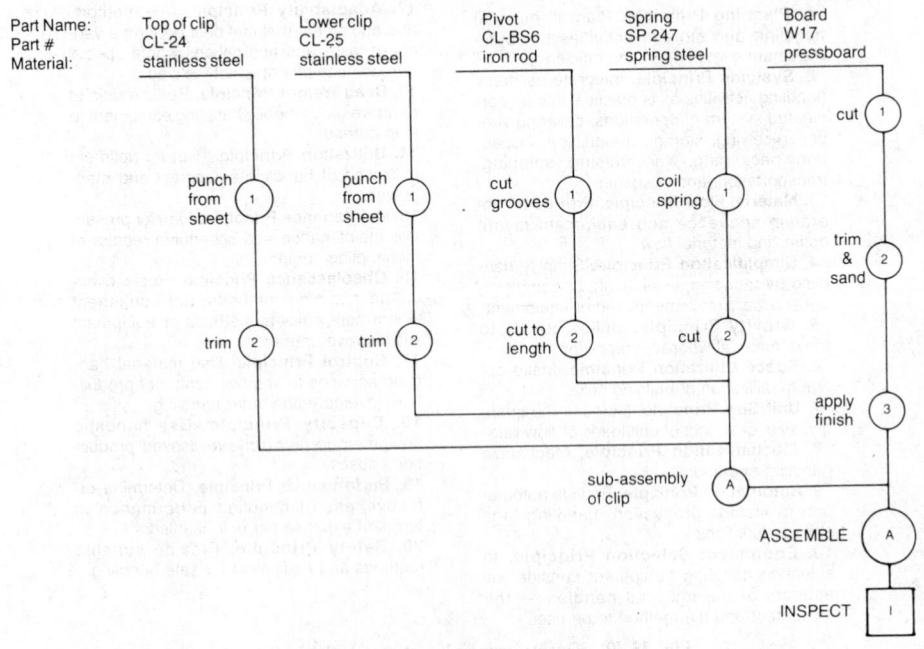

Fig. 33.22 Operations process chart.

The flow diagram, depicted in Fig. 33.24, provides a graphical record of the sequence of steps taken in the process, superimposed on an area layout. This graphical technique augments the flow process chart. The from–to chart, illustrated in Fig. 33.25, uses a matrix representation of material movement. The activity relationship chart, shown in Fig. 33.26, can be used to lay out the flow of materials between departments in a plant. Read like a highway mileage table that indicates distance between cities, the activity relationship chart allows the analyst to note information about the type

Symbol	Name	Results
○	Operation	Produces, prepares, and accomplishes
◇	Transportation	Moves
□	Inspection	Verifies
D	Delay	Interfere, waits
▽	Storage	Keeps, retains

DETAILS OF (PRESENT / PROPOSED) METHOD	OPERATION	TRANSPORT	INSPECTION	DELAY	STORAGE	DISTANCE IN FEET	QUANTITY	TIME	ELIMINATE	COMBINE	SEQUENCE	PLACE	PERSON	IMPROVE	SAFER?	$ SAVED?	NOTES
1. Receive raw materials	○	◇	□	D	▽	50											
2. Inspect	○	◇	□	D	▽												
3. Move by fork lift	○	◇	□	D	▽	40											
4. Store	○	◇	□	D	▽												
5. Move by fork lift	○	◇	□	D	▽	45											
6. Set up and print	○	◇	□	D	▽												
7. Moved by printer	○	◇	□	D	▽	120											
8. Stack at end of printer	○	◇	□	D	▽												
9. Move to stripping	○	◇	□	D	▽	165											
10. Delay	○	◇	□	D	▽												
11. Being stripped	○	◇	□	D	▽												
12. Move to temp. storage	○	◇	□	D	▽	150											
13. Storage	○	◇	□	D	▽												
14. Move to folders	○	◇	□	D	▽	200											
15. Delay	○	◇	□	D	▽												
16. Set up, fold, glue	○	◇	□	D	▽												
17. Mechanically moved	○	◇	□	D	▽	90											
18. Stack, count, crate	○	◇	□	D	▽												
19. Move by fork lift	○	◇	□	D	▽	525											
20. Storage	○	◇	□	D	▽												

SUMMARY

	PRESENT NO.	PRESENT TIME	PROPOSED NO.	PROPOSED TIME	DIFERENCE NO.	DIFERENCE TIME
○ OPERATIONS	5					
◇ TRANSPORTATIONS	9					
□ INSPECTIONS	1					
D DELAYS	2					
▽ STORAGES	3					
Distance Traveled	1485 FT.		FT.		FT.	

JOB: Manufacture of a tissue box

□ OPERATOR
☒ MATERIAL
CHART BEGINS: Receiving (raw materials)
CHART ENDS: Shipping (finished product)
CHARTED BY: T.P.C.

QUESTION EACH DETAIL	ANALYSIS	
	WHAT?	WHEN?
	WHY?	WHO?
	WHERE?	HOW?

DATE
NUMBER
PAGE 1 OF 1

Fig. 33.23 Flow process chart.

of relationship that should exist among departments, whether closeness is desired or important (using an alphabetic symbol) and the reason for that rating (using a numeric symbol). Together these charting techniques provide the analyst extensive, quantitative data and information about a process from the material handling standpoint.

33.5.3 Developing the Plan

Once the data are collected they must be used to develop a set of alternative solutions to the problem. For example, three different plant layouts, with the various process areas placed in different arrangements, represent alternative solutions to a problem of designing a facility in which to carry out a new manufactur-

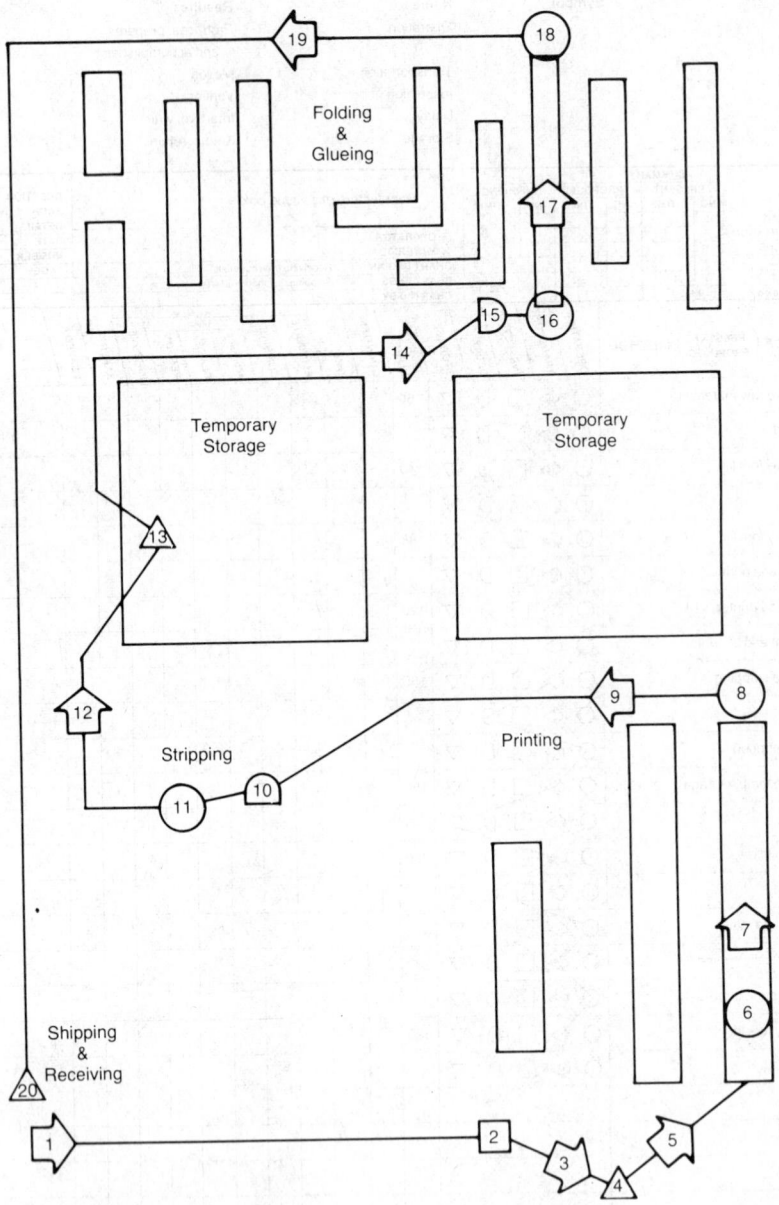

Fig. 33.24 Flow diagram.

ing process. These alternative layouts can be tested and compared using a combination of economic analysis, computer simulation, queueing analysis, linear programming, and routing analysis. A scoresheet can be constructed that gives each layout a numerical value for each of several predetermined criteria, weights the importance of the several criteria, and arrives at a total score which establishes the relative merit of the candidate layouts. The most meritorious design can be further tested and refined before implementation.

33.5.4 Implementing the Solution

After the best plan for solving the material handling problem has been identified, the final step is to implement the solution. Once total system costs—both initial costs and recurring costs—have been

To From	A	B	C	D	E	F	G
A		13	37	69	85	57	33
B	13		21	53	69	41	17
C	37	21		29	45	17	41
D	69	53	29		13	33	57
E	85	69	45	13		49	73
F	57	41	17	33	49		21
G	33	17	41	57	73	21	

A

Fig. 33.25 "From–to" chart.

calculated, approval of the project must be obtained from top management. Once approval is obtained, carefully prepared, written bid specifications typically are let to several vendors or contractors. Competing bids or proposals must then be evaluated closely to ascertain that all are quoting on the same type and grade of equipment and components.

Finally, each step of the acquisition process must be closely monitored to ensure that any construction is accomplished in a correct and timely manner and that equipment installation procedures are faithfully followed. Once the completed facility is available, it should be fully tested before final acceptance from the vendors and contractors. Then, personnel must be fully trained in the operation of the new facility.

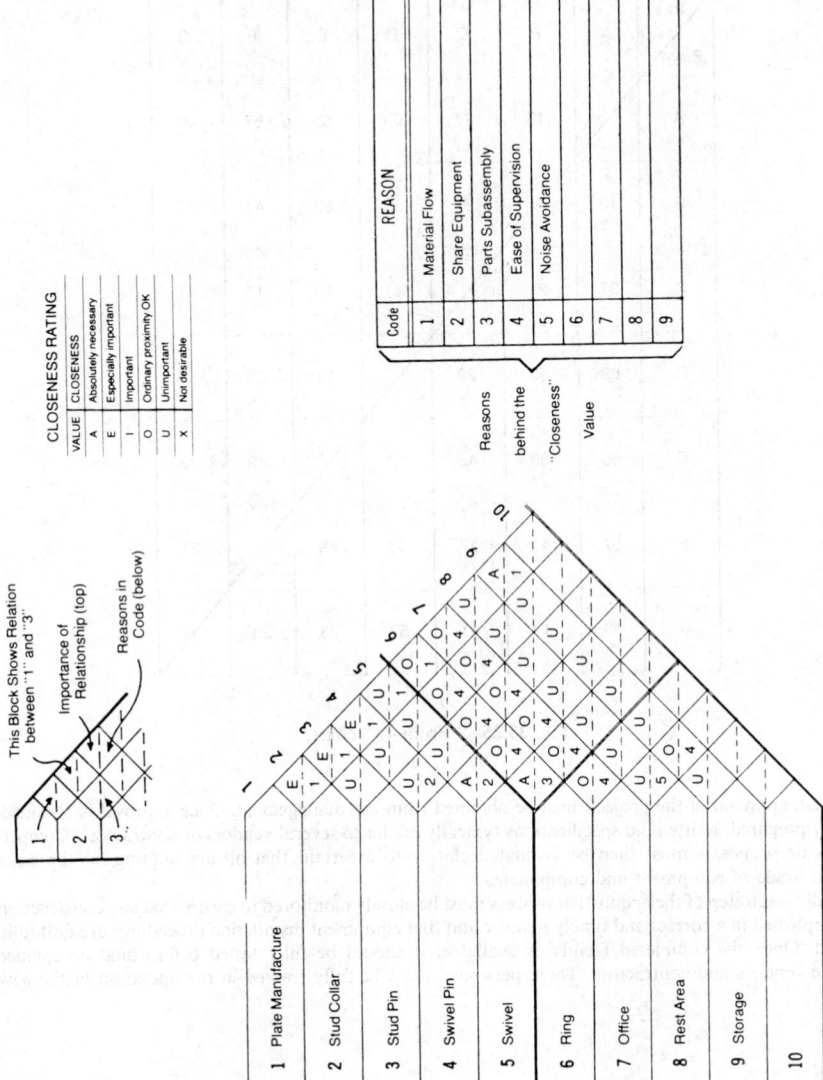

Fig. 33.26 Activity relationship chart.

CHAPTER 34

COMPUTER-AIDED MANUFACTURING

WILLIAM E. BILES
MAGD E. ZOHDI

Industrial Engineering Department
Louisiana State University
Baton Rouge, Louisiana

34.1 INTRODUCTION

Modern manufacturing systems and industrial robots are advanced automation systems that use computers as an integral part of their control. Computers are now a vital part of automated manufacturing. They control stand-alone manufacturing systems, such as various machine tools, welders, laser-beam cutters, robots, and automatic assembly machines. They control production lines and are beginning to take over control of the entire factory. The term "computer-integrated-manufacturing systems," or CIMS for short, is a reality in the modern industrial society. As illustrated in Fig. 34.1, CIMS combines computer-aided design (CAD), computer-aided manufacturing (CAM), computer-aided inspection (CAI), and computer-aided production planning (CAPP), along with automated material handling.

This chapter focuses on the more limited subject of computer-aided manufacturing, in both parts fabrication and assembly as shown in Fig. 34.1. It treats numerical-control (NC) machining, robotics, group technology, and shows how to fuse these into a CAM system. Actually, it is only necessary to then integrate these functions with automated material storage and handling to have a CIM system.

34.2 DEFINITIONS AND CLASSIFICATIONS

34.2.1 Automation

"Automation" is a relatively new word, having been coined in the 1930s as a substitute for the word "automatization," which referred to the introduction of automatic controls in manufacturing. Automation implies the performance of a task without human assistance. Manufacturing processes are classified as manual, semiautomatic, or automatic, depending on the extent of human involvement in the ongoing operation of the process.

The primary reasons for automating a manufacturing process are to:

1. Reduce the cost of the manufactured product, through savings in both material and labor.
2. Improve the quality of manufactured product by eliminating errors and reducing the variability in product quality.
3. Reduce the lead time for manufactured product, thus providing better service for customers.
4. Increase the rate of production.
5. Make the workplace safer.

The economic reality of the marketplace has provided the incentive for industry to automate its manufacturing processes. In Japan and West Germany, the shortage of citizen labor sparked the drive toward automation. In the United States, stern competition from Japanese and West German manufacturers in terms of both product cost and product quality has necessitated automation. Whatever the reasons, a strong movement toward automated manufacturing processes is being witnessed throughout the industrial nations of the world.

34.2.2 Production Operations

Production is a transformation process in which raw materials are converted into goods that are demanded in the marketplace. Labor, machines, tools, and energy are applied to materials at each of a sequence of steps that bring the materials closer to a marketable final state. These individual steps are called production operations.

There are three basic types of industries involved in transforming raw materials into marketable products:

1. *Basic Producers.* These transform natural resources into raw materials for use in manufacturing industry, for example, iron ore to steel ingot in a steel mill.
2. *Converters.* These take the output of basic producers and transform the raw materials into various industrial products, for example, steel ingot is converted into sheet metal.
3. *Fabricators.* These fabricate and assemble final products, for example, sheet metal is fabricated into body panels and assembled with other components into an automobile.

The concept of a computer-integrated-manufacturing system as depicted in Fig. 34.1 applies specifically to a "fabricator" type of industry. It is the "fabricator" industry that we focus on in this chapter.

The steps involved in creating a product is known as the "manufacturing cycle." In general, the following functions will be performed within a firm engaged in manufacturing a product:

1. *Sales and Marketing.* The order to produce an item stems either from customer orders or from production orders based on forecasted product demand.

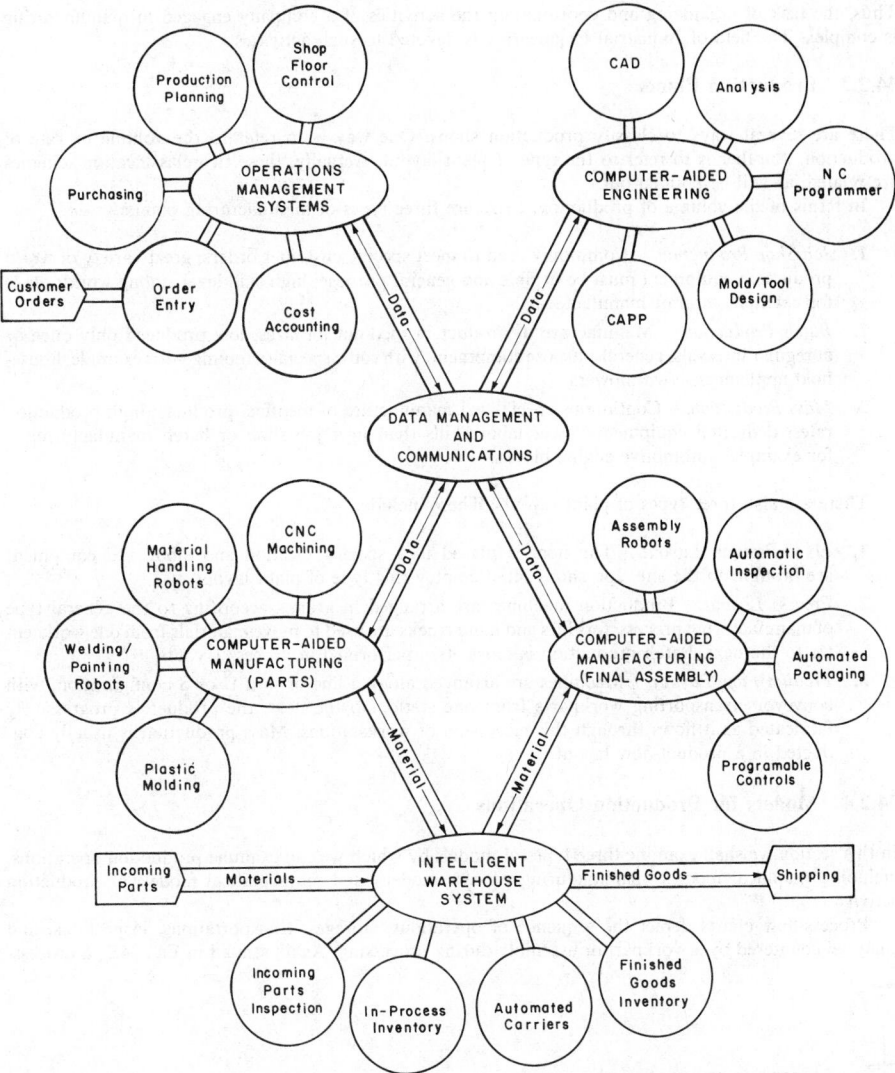

Fig. 34.1 Computer-integrated manufacturing system.

2. *Product Design and Engineering.* For proprietary products, the manufacturer is responsible for development and design, including component drawings, specifications, and bill of materials.

3. *Manufacturing Engineering.* Ensuring produceability of product designs, process planning, design of tools, jigs, and fixtures, and "troubleshooting" the manufacturing process.

4. *Industrial Engineering.* Determining work methods and time standards for each production operation.

5. *Production Planning and Control.* Determining the master production schedule, engaging in material requirements planning, operations scheduling, dispatching job orders, and expediting work schedules.

6. *Manufacturing.* Performing the operations that transform raw materials into finished goods.

7. *Material Handling.* Transporting raw materials, in-process components, and finished goods between operations.

8. *Quality Control.* Ensuring the quality of raw materials, in-process components, and finished goods.

9. *Shipping and Inventory Control.*

Thus, the task of organizing and coordinating the activities of a company engaged in manufacturing is complex. The field of Industrial Engineering is devoted to such activities.

34.2.3 Production Plants

There are several ways to classify production shops. One way is to refer to the volume or rate of production. Another is to refer to the type of plant layout. Actually, these two classification schemes are related, as will be pointed out.

In terms of the volume of production, there are three types of manufacturing plants:

1. *Job Shop Production.* Commonly used to meet specific customer orders; great variety of work; production equipment must be flexible and general purpose; high skill level among workforce—for example, aircraft manufacturing.

2. *Batch Production.* Manufacture of product in medium lot sizes; lots produced only once or at regular intervals; general-purpose equipment, with some specialty tooling—for example, household appliances, lawn mowers.

3. *Mass Production.* Continuous specialized manufacture of identical products; high production rates; dedicated equipment; lower labor skills than in a job shop or batch manufacturing—for example, automotive engine blocks.

There are also three types of plant layouts. These include:

1. *Fixed-Position Layout.* The item is placed in a specific location and labor and equipment are brought to the site. Job shops often employ this type of plant layout.

2. *Process Layout.* Production machines are arranged in groups according to the general type of manufacturing process; forklifts and hand trucks are used to move materials from one workcenter to the next. Batch production is most often performed in process layouts.

3. *Product-Flow Layout.* Machines are arranged along a line or in a U or S configuration, with conveyors transporting workparts from one station to the next; the product is progressively fabricated as it flows through the succession of workstations. Mass production is usually conducted in a product-flow layout.

34.2.4 Models for Production Operations

In this section we shall examine three types of models by which we can examine production operations, including graphical models, manufacturing process models, and mathematical models of production activity.

Process-flow charts depict the sequence of operations, storages, transportations, inspections, and delays encountered by a workpart or assembly during processing. As illustrated in Fig. 34.2, a process-

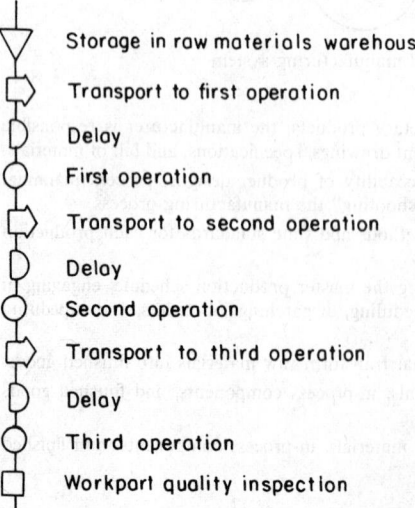

Storage in raw materials warehouse

Transport to first operation

Delay

First operation

Transport to second operation

Delay

Second operation

Transport to third operation

Delay

Third operation

Workpart quality inspection

Fig. 34.2 Flow process chart for a sample workpart.

flow chart gives no representation of the layout or physical dimensions of a process, but focuses on the succession of steps seen by the product. It is useful in analyzing the efficiency of the process, in terms of the proportion of time spent in transformation operations as opposed to transportations, storages, and delays.

The manufacturing-process model gives a graphical depiction of the relationship among the several entities that comprise the process. It is an input–output model. Its inputs are raw materials, equipment (machine tools), tooling and fixtures, energy, and labor. Its outputs are the completed workpiece, scrap, and waste. These are shown in Fig. 34.3. Also shown in this figure are the controls that are applied to the process to optimize the utilization of the inputs in producing completed workpieces, or in maximizing the production of completed workpieces at a given set of values describing the inputs.

Mathematical models of production activity quantify the elements incorporated into the process-flow chart. We distinguish between operation elements, which are involved whenever the workpart is on the machine and correspond to the circles in the process-flow chart, and nonoperation elements, which include storages, transportations, delays, and inspections. Letting T_o represent operation time per machine, T_{no} represent the nonoperation time associated with each operation, and n_m represent the number of machines or operations through which each part must be processed, then the total time required to process the part through the plant [called the manufacturing lead time (T_1)] is

$$T_1 = n_m(T_o + T_{no})$$

If there is a batch of p parts,

$$T_1 = n_m(pT_o + T_{no})$$

If a setup of duration T_{su} is required for each batch,

$$T_1 = n_m(T_{su} + pT_o + T_{no})$$

The total batch time per machine T_b is given by

$$T_b = T_{su} + pT$$

The average production time T_a per part is therefore

$$T_a = \frac{T_{su} + pT}{p}$$

The average production rate for each machine is

$$R_a = 1/T_a$$

As an example, a part requires six operations (machines) through the machine shop. The part is produced in batches of 100. A setup of 2.5 hr is needed. Average operation time per machine is 4.0 min. Average nonoperation time is 3.0 hr. Thus,

$$n = 6 \text{ machines}$$
$$p = 100 \text{ parts}$$
$$T_{su} = 2.5 \text{ hr}$$
$$T_o = 4/60 \text{ hr}$$
$$T_{no} = 3.0 \text{ hr}$$

Therefore, the total manufacturing lead time for this batch of parts is

$$T = 6[2.5 + 100(0.06667) + 3.0] = 73.0 \text{ hr}$$

If the shop operates on a 40-hr week, almost two weeks are needed to complete the order.

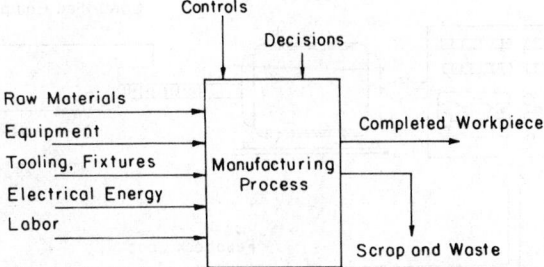

Fig. 34.3   General input–output model of the manufacturing process.

34.3 NUMERICAL-CONTROL MANUFACTURING SYSTEMS

34.3.1 Numerical Control

The most commonly accepted definition of numerical control (NC) is that given by the Electronic Industries Association (EIA): A system in which motions are controlled by the direct insertion of numerical data at some point. The system must automatically interpret at least some portion of these data.

The numerical control system consists of five basic, interrelated components as follows:

1. Data input devices.
2. Director (or machine control unit).
3. Machine tool or other controlled equipment.
4. Servo-drives for each axis of motion.
5. Feedback devices for each axis of motion.

The major components of a typical NC machine tool system are shown in Fig. 34.4.

The programmed codes that the machine control unit (MCU) can read may be perforated tape or punched tape, magnetic tape, tabulating cards, or signals directly from computer logic or some computer peripherals such as disk or drum storage. Direct computer control (DCC) is the most recent development, and one which affords the help of a computer in developing a part program.

34.3.2 The Coordinate System

The Cartesian coordinate system is the basic system used in NC control. The three primary linear motions for an NC machine are given as X, Y, and Z. Letters A, B, and C indicate the three rotational axes, as in Fig. 34.5.

NC machine tools are commonly classified as being either point-to-point or continuous path. The simplest form of NC is the point-to-point machine tool used for operations such as drilling, tapping, boring, punching, spot welding, or other operations that can be completed at a fixed coordinate position with respect to the workpiece. The tool does not contact the workpiece until the desired coordinate position has been reached; consequently, the exact path by which this position is reached is not important.

With continuous-path (contouring) NC systems, there is contact between the workpiece and the tool as the relative movements are made. Continuous-path NC systems are used primarily for milling and turning operations that can profile and sculpture workpieces. Other NC continuous-path operations include flame cutting, sawing, grinding, and welding, and even operations such as the application of adhesives. We should note that continuous-path systems can be programmed to perform point-to-point operations, although the reverse (while technically possible) is infrequently done.

34.3.3 Selection of Parts for NC Machining

Parts selection for NC should be based on an economic evaluation, including scheduling and machine availability. Economic considerations affecting NC part selection include alternative methods, tooling, machine loadings, manual versus computer-assisted part programming, and other applicable factors.

Thus, NC should be used only where it is more economical or does the work better or faster, or where it is more accurate than other methods. The selection of parts to be assigned to NC has a

Data Input Device   Director   Servo-Drive Device   Machine Tool Table or Other Controlled Equipment   Feedback Device or Transducer

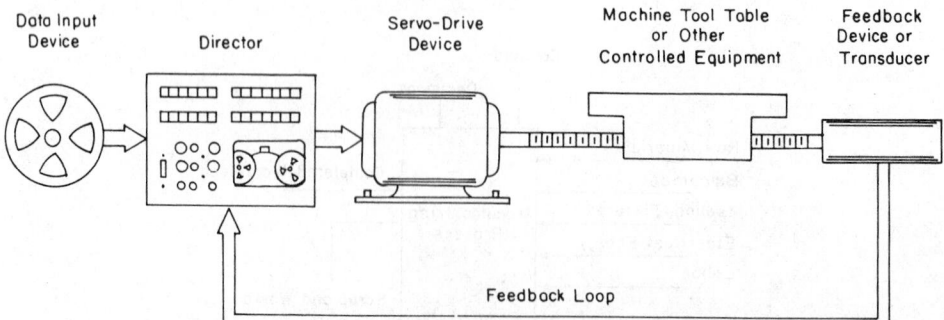

Feedback Loop

Fig. 34.4 Simplified numerical control system.

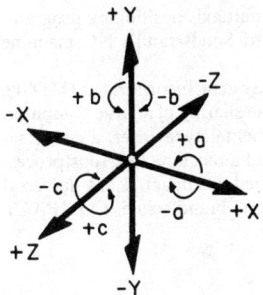

Fig. 34.5 An example of typical axis nomenclature for machine tools.

significant effect on its payoff. The following guidelines, which may be used for parts selection, describe those parts for which NC may be applicable:

1. Parts that require *substantial tooling costs* in relation to the total manufacturing costs by conventional methods.
2. Parts that require *long setup times* compared to the machine run time in conventional machining.
3. Parts which are machined in *small or variable lots.*
4. A *wide diversity of parts* requiring frequent changes of machine setup and a large tooling inventory if conventionally machined.
5. Parts that are *produced at intermittent times* because demand for them is cyclic.
6. Parts that have *complex configurations* requiring close tolerances and intricate relationships.
7. Parts that have *mathematically defined complex contours.*
8. Parts that require *repeatability* from part to part and lot to lot.
9. *Very expensive* parts where human error would be very costly and increasingly so as the part nears completion.
10. *High priority* parts where lead time and flow time are serious considerations.
11. Parts with *anticipated design changes.*
12. Parts that involve a *large number of operations* or *machine setups.*
13. Parts where *nonuniform cutting conditions* are required.
14. Parts which require 100% *inspection* or require measuring many checkpoints, resulting in high inspection costs.
15. *Family of parts.*
16. *Mirror-image parts.*
17. *New parts* for which conventional tooling does not already exist.
18. Parts that are suitable for *maximum machining* on NC machine tools.

34.3.4 Manual Part Programming

Manual part programming involves the detailed step-by-step listing of operations that the NC machine will perform. The steps are written in a definite order and form in what is known as manuscript. The person who writes the instructions is the part programmer.

34.3.5 Computer-Assisted Programming

The programming work is simplified by letting a computer do the tedious computations involved and make logical-type decisions. With accurate postprocessing, all NC units, even with different input specifications, can be programmed by using the same recognized problem-oriented language. There are several different problem-oriented languages in existence, some of which are listed below.

APT (Automatic Programmed Tools) is a three-dimensional language used to direct a tool in a well-defined path. APT requires a large-scale computer and is used extensively in complex contouring systems.

ADAPT (Air Material Command Developed APT) is an extension of APT and has been written specifically for small- to medium-sized computers.

AUTOSPOT (Automatic System for Positioning Tools) is a single-pass NC processor with three- or four-axis positioning and limited continuous-path capability.

SPLIT (Sundstrand Processing Language Internally Translated) is a multiaxis positioning programming system with some contouring capability. SPLIT is oriented toward Sundstrand's NC machine tools, but is available for some other NC machines.

COMPACT II is an NC language developed by Manufacturing Data System Incorporated (MDSI), a nationwide corporation based in Ann Arbor, MI. MDSI provides for time sharing of a large computer, use of the Compact II language, programming via remote terminal such as teletypewriter, a conversational system whereby instant corrections can be made to the program, and a machine link (postprocessor) that adapts the Compact II output for each machine to be programmed. Compact II has universal application to a wide variety of machines in both type and manufacturers' brand name. COMPACT II is a unique language for the following features:

1. Users are not required to have their own computer.

2. Instant return of the NC program is achieved.

3. The PLOT routine available in COMPACT II provides the user with a graphical display of the part program and debugging and modifications capability. The PLOT routine can be displayed on either a plotter or a cathode-ray tube.

4. MDSI utilizes a large nationwide network of computers. International services are available.

5. MDSI has developed stand alone units for COMPACT II at a reasonable cost and impressive capabilities.

34.3.6 Programming by Scanning and Digitizing

Programming may be done directly from a drawing, model, pattern, or template by digitizing or scanning. An optical reticle or other suitable viewing device connected to an arm is placed over the drawing. Transducers will identify the location and translate it either to a tape puncher or other suitable programming equipment. Digitizing is used in operations such as sheet-metal punching and hole drilling. A scanner enables an operator to program complex free-form shapes by manually moving a tracer over the contour of a model or premachined part. Data obtained through tracer movements are converted into tape by a minicomputer. Digitizing and scanning units have the capability of editing, modifying, or revising the basic data gathered.

34.3.7 Adaptive Control

Optimization processes have been developed to improve the operational characteristics of NC machine-tool systems. Two distinct methods of optimization are adaptive control and machinability data prediction. Although both techniques have been developed for metal-cutting operations, adaptive control finds application in other technological fields.

The adaptive control (AC) system is an evolutionary outgrowth of numerical control. AC optimizes an NC process by sensing and logically evaluating variables that are not controlled by position and velocity feedback loops. Essentially, an adaptive control system monitors process variables, such as cutting forces, tool temperatures, or motor torque, and alters the NC commands so that optimal metal removal or safety conditions are maintained.

A typical NC configuration (Fig. 34.6a) monitors position and velocity output of the servo system, using feedback data to compensate for errors between command and response. The AC feedback loop (Fig. 34.6b) provides sensory information on other process variables such as workpiece–tool air gaps, material property variations, wear, cutting depth variations, or tool deflection. This information is determined by techniques such as monitoring forces on the cutting tool, motor torque variations, or tool–workpiece temperatures. The data are processed by an adaptive controller that converts the process information into feedback data to be incorporated into the Machine Control Unit output.

34.3.8 Machinability Data Prediction

The specification of suitable feeds and speeds is essentially in conventional and NC cutting operations. Machinability data are used to aid in the selection of metal-cutting parameters based on the machining operation, the tool and workpiece material, and one or more production criteria. Techniques used to select machinability data for conventional machines have two important drawbacks in relation to NC applications; data are generally presented in a tabular form that requires manual interpolation, checkout, and subsequent revisions, and tests on the machine tool are required to find optimum conditions.

Specialized machinability data systems have been developed for NC applications to reduce the need for machinability data testing and to decrease expensive NC machining time. Part programming time is also reduced when machinability information is readily available.

A typical process schematic showing the relationship between machinability data and NC process flow is illustrated in Fig. 34.7.

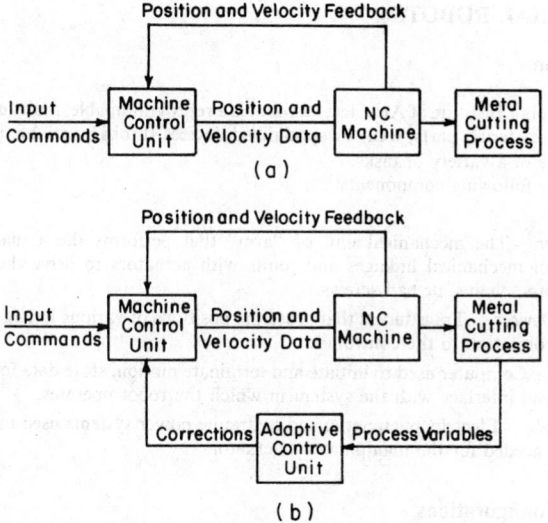

Fig. 34.6 Schematic diagrams for conventional and adaptive NC systems.

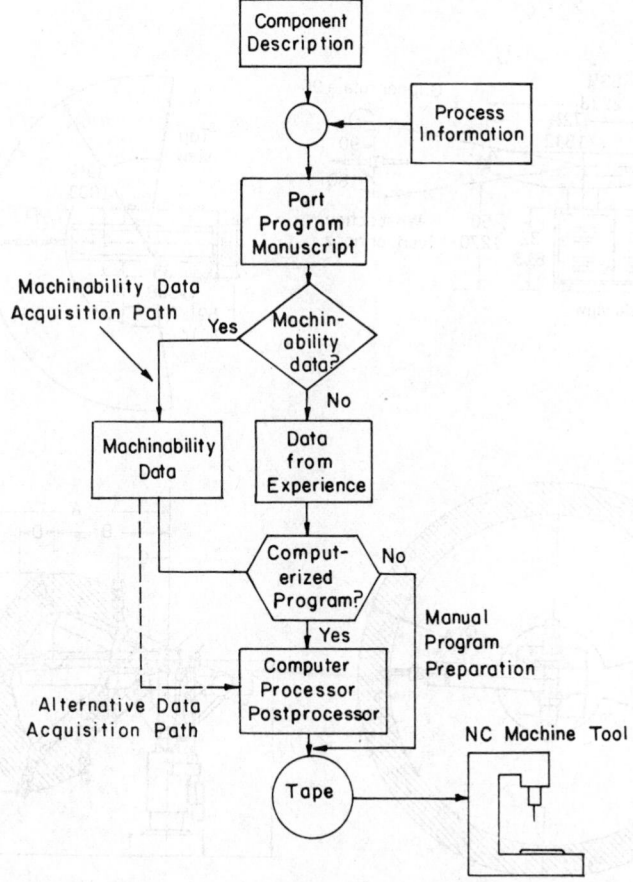

Fig. 34.7 Acquisition of machinability data in the NC process flow.

34.4 INDUSTRIAL ROBOTS

34.4.1 Definition

As defined by the Robot Institute of America, "a robot is a reprogrammable, multifunctional manipulator designed to handle material, parts, tools or specialized devices through variable programmed motions for the performance of a variety of tasks."

Robots have the following components:

1. *Manipulator.* The mechanical unit or "arm" that performs the actual work of the robot, consisting of mechanical linkages and joints with actuators to drive the mechanism directly through gears, chains, or ball screws.
2. *Feedback Devices.* Transducers that sense the positions of various linkages or joints and transmit this information to the controller.
3. *Controller.* Computer used to initiate and terminate motion, store data for position and motion sequence, and interface with the system in which the robot operates.
4. *Power Supply.* Electric, pneumatic, and hydraulic power systems used to provide and regulate the energy needed for the manipulator's actuators.

34.4.2 Robot Configurations

Industrial robots have one of three mechanical configurations, as illustrated in Fig. 34.8. Cylindrical coordinate robots have a work envelope that is composed of a portion of a cylinder. Spherical coordinate robots have a work envelope that is a portion of a sphere. Jointed-arm robots have a work envelope that approximates a portion of a sphere.

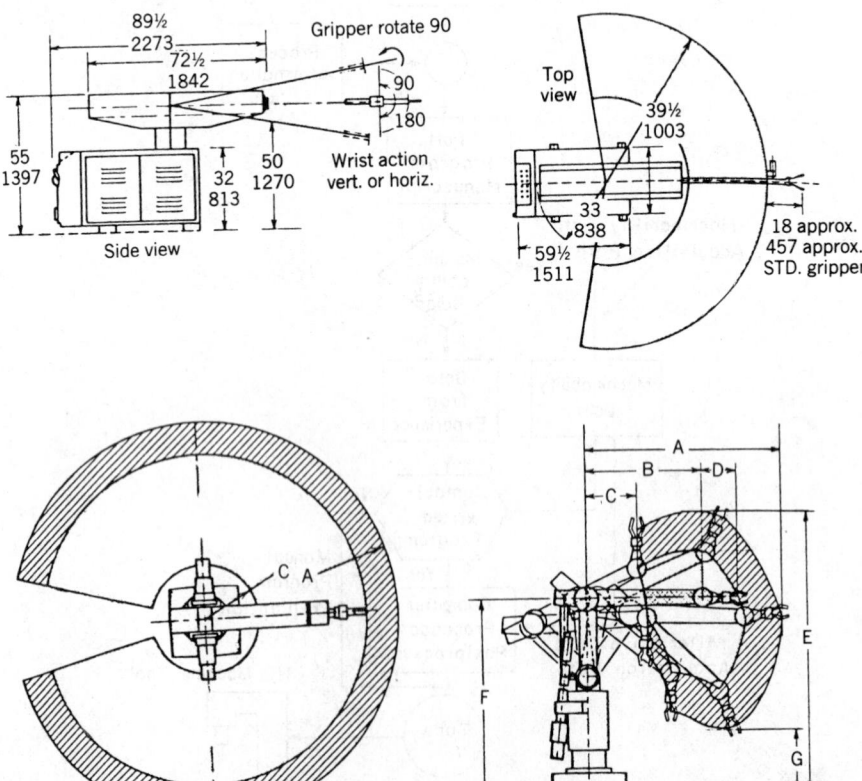

Fig. 34.8 Mechanical configurations of industrial robots.

There are six basic motions or degrees of freedom in the design of a robot—three arm and body motions and three wrist movements.

Arm and body motions:

1. Vertical traverse—an up-and-down motion of the arm.
2. Radial traverse—an in-and-out motion of the arm.
3. Rotational traverse—rotation about the vertical axis (right or left swivel of the robot body).

Wrist motions:

4. Wrist swivel—rotation of the wrist.
5. Wrist bend—up-and-down movement of the wrist.
6. Wrist yaw—right or left swivel of the wrist.

The mechanical hand movement, usually opening and closing, is not considered one of the basic degrees of freedom of the robot.

34.4.3 Robot Control and Programming

Robots can also be classified according to type of control. Point-to-point robot systems are controlled from one programmed point in the robot's control to the next point. These robots are characterized by high load capacity, large working range, and relatively ease of programming. They are suitable for pick-and-place, material handling, and machine loading tasks.

Contouring robots, on the other hand, possess the capacity to follow a closely spaced locus of points that describe a smooth, continuous path. The control of the path requires a large memory to store the locus of points. Continuous-path robots are therefore more expensive than point-to-point robots, but they can be used in such applications as seam welding, flame cutting, and adhesive beading.

There are three principal systems for programming robots:

1. *Manual Method.* Used in older, simpler robots, the program is set up by fixing stops, setting switches, and so on.
2. *Walkthrough.* The programmer "teaches" the robot by actually moving the hand through a sequence of motions or positions, which are recorded in the memory of the computer.
3. *Leadthrough.* The programmer drives the robot through a sequence of motions or positions using a console or teach pendant. Each move is recorded in the robot's memory.

34.4.4 Robot Applications

A current directory of robot applications in manufacturing include the following:

1. Material handling.
2. Machine loading and unloading.
3. Die casting.
4. Investment casting.
5. Forging and heat treating.
6. Plastic molding.
7. Spray painting and electroplating.
8. Welding (spot welding and seam welding).
9. Inspection.
10. Assembly.

Research and development efforts are under way to provide robots with sensory perception, including voice programming, vision and "feel." These capabilities will no doubt greatly expand the inventory of robot applications in manufacturing.

34.5 COMPUTERS IN MANUFACTURING

Flexible manufacturing systems combined with automatic assembly and product inspection, on the one hand, and integrated CAD/CAM systems on the other hand, are the basic components of the computer-integrated-manufacturing system. The overall control of such systems is predicated on hierarchical computer control such as illustrated in Fig. 34.9.

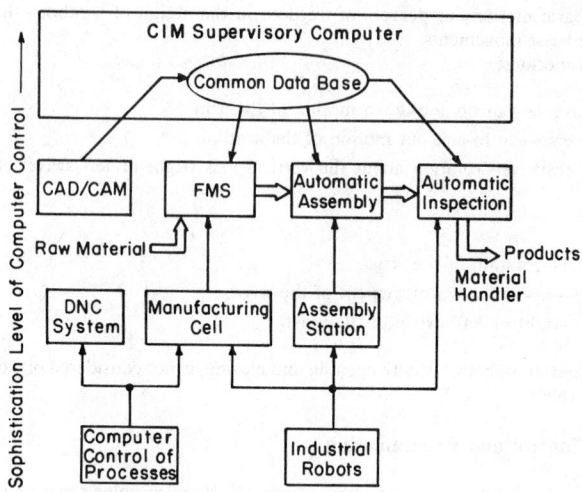

Fig. 34.9 Hierarchical Computer Control in Manufacturing.

34.5.1 Hierarchical Computer Control

The lowest level of the hierarchical computer control structure illustrated in Fig. 34.9 contains stand-alone computer control systems of manufacturing processes and industrial robots. The computer control of processes includes all types of CNC machine tools, welders, electrochemical machining (ECM), electrical discharge machining (EDM), and laser-cutting machines.

When a set of NC or CNC machine tools is placed under the direct control of a single computer, the resulting system is known as a direct-numerical-control (DNC) system. DNC systems can produce several different categories of parts or products, perhaps unrelated to one another. When several CNC machines and one or more robots are organized into a system for the production of a single part or family of parts, the resulting system is called a "manufacturing cell." The distinction between DNC systems and a manufacturing cell is that in DNC systems the same computer receives data from and issues instructions to several separate machines, whereas in manufacturing cells the computer coordinates the movements of several machines and robots working in concert. The computer receives "completion of job" signals from the machines and issues instructions to the robot to unload the machines and change their tools. The software includes strategies for handling machine breakdowns, tool wear, and other special situations.

The operation of several manufacturing cells can be coordinated by a central computer in conjunction with an automated material handling system. This is the next level of control in the hierarchical structure and is known as a "flexible manufacturing system" (FMS). The FMS receives incoming workpieces and processes them into finished parts, completely under computer control.

The parts fabricated in the FMS are then routed on a transfer system to automatic assembly stations where they are assembled into subassemblies or final product. These assembly stations can also incorporate robots for performing assembly operations. The subassemblies and final product may also be tested at automatic inspection stations.

As shown in Fig. 34.9, FMS, automatic assembly, and automatic inspection are integrated with CAD/CAM systems to minimize production lead time. These four functions are coordinated by means of the highest level of control in the hierarchical structure—computer-integrated-manufacturing (CIM) systems. This level of control is often called "supervisory computer control."

The increase in productivity associated with CIM systems will not come from a speedup of machining operations, but rather from minimizing the direct labor employed in the plant. Substantial savings will also be realized from reduced inventories, with reductions in the range of 80–90%.

34.5.2 CNC and DNC Systems

The distinguishing feature of a CNC system is a dedicated computer, usually a microcomputer, associated with a single machine tool such as a milling machine or a lathe. Programming the machine tool is managed through punched or magnetic tape, or directly from a keyboard.

DNC is another step beyond CNC, in that a number of CNC machines, ranging from a few to as many as 100, are connected directly to a remote computer. NC programs are downloaded directly to the CNC machine, which then processes a prescribed number of parts.

34.5.3 The Manufacturing Cell

The concept of a manufacturing cell is based on the notion of cellular manufacturing, wherein a group of machines served by one or more robots manufactures one part or one part family. Figure 34.10 depicts a typical manufacturing cell consisting of a CNC lathe, a CNC milling machine, a CNC drill, one conveyor to bring workparts into the cell, another to remove completed parts from the cell, and a robot to serve all these components.

Each manufacturing cell is self-contained and self-regulating. The cell is usually made up of 10 or fewer machines. Those cells that are not completely automated are usually staffed with fewer personnel than machines, with each operator trained to handle several machines or processes.

34.5.4 Flexible Manufacturing Systems

Flexible manufacturing systems (FMS) combine many different automation technologies into a single production system. These include NC and CNC machine tools, automatic material handling between machines, computer control over the operation of the material handling system and machine tools, and group technology principles. Unlike the manufacturing cell, which is typically dedicated to the production of a single parts family, the FMS is capable of processing a variety of part types simultaneously under NC control at the various workstations.

Human labor is used to perform the following functions to support the operations of the FMS:

Load raw workparts into the system.

Unload finished workparts from the system.

Change tools and tool settings.

Equipment maintenance and repair.

Robots can be used to replace human labor in certain of these functions, particularly those involving material or tool handling.

Figure 34.11 illustrates a sample FMS layout.

34.6 GROUP TECHNOLOGY

Group technology is a manufacturing philosophy in which similar parts are identified and grouped together to take advantage of similarities in design and/or manufacture. Similar parts are grouped into part families. For example, a factory that produces as many as 10,000 different part numbers can group most of these parts into as few as 50 distinct part families. Since the processing of each family would be similar, the production of part families in dedicated manufacturing cells facilitates work flow. Thus, group technology results in efficiencies in both product design and process design.

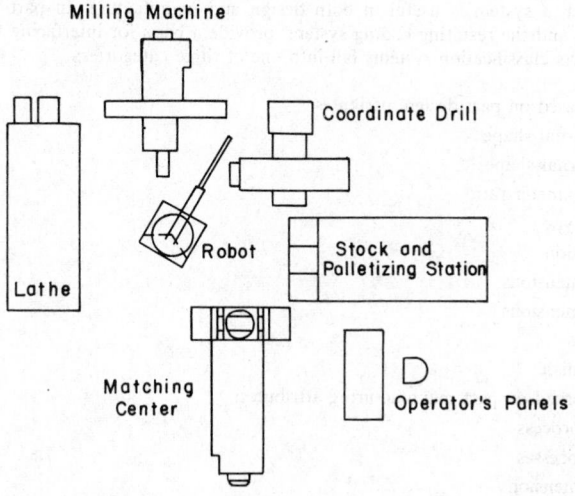

Fig. 34.10 A typical manufacturing cell.

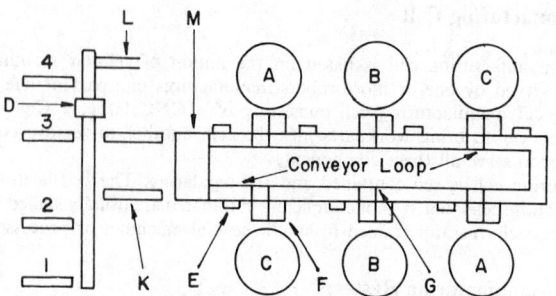

Fig. 34.11   A flexible manufacturing system.

34.6.1   Part Family Formation

The key to gaining efficiency in group-technology-based manufacturing is the formation of part families. A part family is a collection of parts that are similar either due to geometric features such as size and shape or because similar processing steps are required in their manufacture. Parts within a family are different, but are sufficiently similar in their design attributes (geometric size and shape) and/or manufacturing attributes (the sequence of processing steps required to make the part) to justify their identification as members of the same part family.

The biggest problem in initiating a group-technology-based manufacturing system is that of grouping parts into families. Three methods for accomplishing this grouping are:

1. *Visual Inspection.*   This method involves looking at the part, a photograph, or a drawing and placing the part in a group with similar parts. It is generally regarded as the most time-consuming and least accurate of the available methods.

2. *Parts Classification and Coding.*   This method involves examining the individual design and/or manufacturing attributes of each part, assigning a code number to the part on the basis of these attributes, and grouping similar code numbers into families. This is the most commonly used procedure for forming part families.

3. *Production Flow Analysis.*   This method makes use of the information contained on the routing sheets describing the sequence of processing steps involved in producing the part, rather than part drawings. Workparts with similar or identical processing sequences are grouped into a part family.

34.6.2   Parts Classification and Coding

As previously stated, parts classification and coding is the most frequently applied method for forming part families. Such a system is useful in both design and manufacture. In particular, parts coding and classification, and the resulting coding system, provide a basis for interfacing CAD and CAM in CIM systems. Parts classification systems fall into one of three categories:

1. Systems based on part design attributes:
   Basic external shape
   Basic internal shape
   Length/diameter ratio
   Material type
   Part function
   Major dimensions
   Minor dimensions
   Tolerances
   Surface finish
2. Systems based on part manufacturing attributes:
   Primary process
   Minor processes
   Major dimensions
   Length/diameter ratio

Surface finish
Machine tool
Operation sequence
Production time
Batch size
Annual production requirement
Fixtures needed
Cutting tools

3. Systems based on a combination of design and manufacturing attributes.

The part code consists of a sequence of numerical digits that identify the part's design and manufacturing attributes. There are two basic structures for organizing this sequence of digits:

1. Hierarchical structures in which the interpretation of each succeeding digit depends on the value of the immediately preceding digit.
2. Chain structures in which the interpretation of each digit in the sequence is position-wise fixed.

The Opitz system is perhaps the best known coding system used in parts classification and coding. The code structure is
12345 6789 ABCD
The first nine digits constitute the basic code that conveys both design and manufacturing data. The first five digits, 12345, are called the "form code" and give the primary design attributes of the part. The next four digits, 6789, constitute the "supplementary code" and indicate some of the manufacturing attributes of the part. The next four digits, ABCD, are called the "secondary code" and are used to indicate the production operations by type and sequence. Figure 34.12 gives the basic structure for the Opitz coding system. Note that digit 1 establishes two primary categories of parts, rotational and nonrotational, among nine separate part classes.

The MICLASS (Metal Institute Classification System) was developed by the Netherlands Organization for Applied Scientific Research to help automate and standardize a number of design, manufacturing, and management functions. MICLASS codes range from 12 to 30 digits, with the first 12 constituting a universal code that can be applied to any part. The remaining 18 digits can be made specific to any company or industry. The organization of the first 12 digits is as follows:

1st digit	Main shape
2nd and 3rd digits	Shape elements
4th digit	Position of shape elements
5th and 6th digits	Main dimensions
7th digit	Dimension ratio
8th digit	Auxiliary dimension
9th and 10th digits	Tolerance codes
11th and 12th digits	Material codes

MICLASS allows computer interactive parts coding, in which the user responds to a series of questions asked by the computer. The number of questions asked depends on the complexity of the part, and ranges from as few as 7 to more than 30, with an average of about 15.

34.6.3 Production Flow Analysis

Production flow analysis (PFA) is a method for identifying part families and associated grouping of machine tools. PFA is used to analyze the operations sequence and machine routing for the parts produced in a shop. It groups parts that have similar sequences and routings into a part family. PFA then establishes machine cells for producing part families.

The PFA procedure consists of the following steps:

1. Data collection is the gathering of part numbers and machine routings for each part produced in the shop.
2. Sorting process routings into "packs" according to similarity.
3. Constructing a PFA chart, such as depicted in Fig. 34.13, which shows the process sequence (in terms of machine code numbers) for each pack (denoted by a letter).

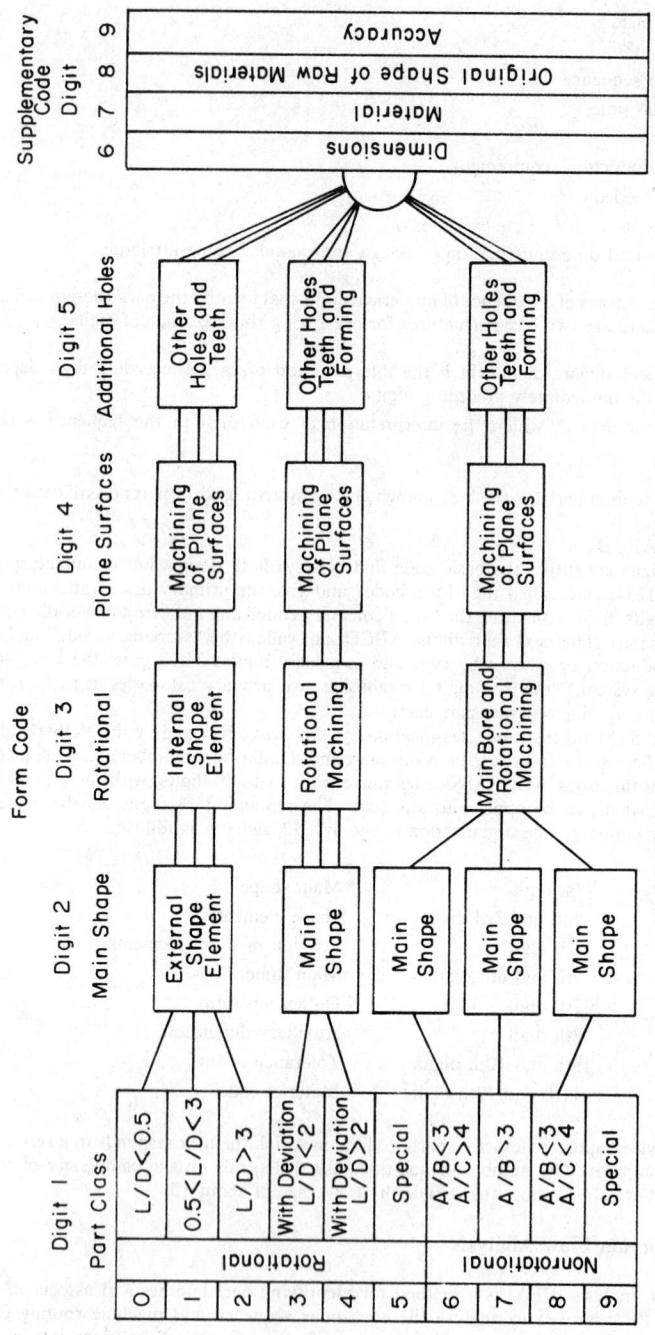

Fig. 34.12 Opitz parts classification and coding system.

Machine \ Part No.	1	2	3	4	5	6	7	8	9	10	11	12	13	14	15	16	17	18	19	20
Lathe	x	x		x	x		x	x	x		x	x		x	x		x	x	x	x
Milling Mach. I	x	x	x		x	x	x		x		x		x	x		x				x
Milling Mach. II			x	x				x		x		x	x		x		x	x	x	
Drilling Mach.	x	x	x	x		x	x	x		x	x	x	x	x		x	x	x		x
Grinding Mach.	x	x	x	x		x			x			x	x		x			x		x

Fig. 34.13   PFA chart.

4. Analysis of the PFA chart in an attempt to identify similar packs. This is done by rearranging the data on the original PFA chart into a new pattern that groups packs having similar sequences. Figure 34.14 shows the rearranged PFA chart. The machines grouped together within the blocks in this figure form logical machine cells for producing the resulting part family.

34.6.4 Types of Machine Cell Designs

The organization of machines into cells, whether based on parts classification and coding or PFA, follows one of three general patterns:

1. Single-machine cell.
2. Group-machine layout.
3. Flow line cell layout.

The single-machine pattern can be used for workparts whose attributes allow them to be produced using a single process. For example, a family composed of 40 different machine bolts can be produced on a single turret lathe.

The group-machine layout was illustrated in Fig. 34.13. The cell contains the necessary grouping of machine tools and fixtures for processing all parts in a given family, but material handling between machines is not fixed.

The flow line cell design likewise contains all machine tools and fixtures needed to produce a family of parts, but these are arranged in a fixed sequence with conveyors providing the flow of parts through the cell.

Machine \ Part No.	1	2	20	7	11	14	9	5	4	18	12	8	17	15	19	3	13	6	16	10
Lathe	x	x	x	x	x	x	x	x												
Milling Mach. I	x	x	x	x	x	x	x	x												
Drilling Mach.	x	x	x	x	x	x														
Grinding Mach.	x	x	x				x													
Lathe									x	x	x	x	x	x	x					
Milling Mach. II									x	x	x	x	x	x	x					
Drilling Mach.									x	x	x	x	x							
Grinding Mach.									x	x	x			x						
Milling Mach. I																x	x	x	x	
Milling Mach. II																x	x			x
Drilling Mach.																x	x	x	x	x
Grilling Mach.																x	x	x		

Fig. 34.14   Rearranged PFA chart.

34.6.5   Computer-Aided Process Planning

Computer-aided process planning (CAPP) involves the use of a computer to automatically generate the operation sequence (routing sheet) based on information about the workpart. CAPP systems require some form of classification and coding system, together with standard process plans for specific part families.

The flow of information in a CAPP system is initiated by having the user enter the part code for the workpart to be processed. The CAPP program then searches the part family matrix file to determine if a match exists. If so, the standard machine routing and the standard operation sequence are extracted from the computer file. If no such match exists, the user must then search the file for similar code numbers and manually prepare machine routings and operation sequences for dissimilar segments. Once this process has been completed, the new information becomes part of the master file so that the CAPP system generates an ever-growing data file.

BIBLIOGRAPHY

C. R. Asfahl, *Robots and Manufacturing Automation,* Wiley, New York, 1985.

CAD/CAM: Meeting Today's Productivity Challenge, Society of Manufacturing Engineers, Dearborn, MI, 1980.

J. J. Childs, *Principles of Numerical Control,* Industrial Press, New York, 1969.

Compact II, Programming Manual, Manufacturing Data Systems, Inc., Ann Arbor, MI, 1984.

Computer Integrated Manufacturing Systems: Selected 1985 Readings, Industrial Engineering and Management Press, Atlanta, GA, 1985.

M. P. Groover, *Automation, Production Systems, and Computer-Aided Manufacturing,* Prentice-Hall, Englewood Cliffs, NJ, 1980.

M. P. Groover and E. W. Zimmers, Jr., *CAD/CAM: Computer-Aided Design and Manufacturing,* Tech Tran Corporation, Naperville, IL, 1984.

V. D. Hunt, *Smart Robots,* Tech Tran Corporation, Naperville, IL, 1984.

Industrial Robots: A Summary and Forecast, Tech Tran Corporation, Naperville, IL, 1983.

E. Kafrissen and M. Stephens, *Industrial Robots and Robotics,* Tech Tran Corporation, Naperville, IL, 1984.

Y. Koren, *Computer Control of Manufacturing Systems,* McGraw-Hill, New York, 1983.

R. A. Lindberg, *Processes and Materials of Manufacturing,* 2nd ed., Allyn and Bacon, Boston, MA, 1977.

Machining Data Handbook, 3rd ed., Machinability Data Center, Cincinnati, OH, 1980.

Numerical Control, Vol. 1, *Fundamentals,* Society of Manufacturing Engineers, Dearborn, MI, 1981.

Numerical Control, Vol. 2, *Applications,* Society of Manufacturing Engineers, Dearborn, MI, 1981.

Y. C. Pao, *Elements of Computer-Aided Design and Manufacturing,* Wiley, New York, 1984.

R. S. Pressman and J. E. Williams, *Numerical Control and Computer-Aided Manufacturing,* Wiley, New York, 1977.

Roberts and Prentice, *Programming for Numerical Control Machines,* McGraw-Hill, New York, 1978.

C. Morgan, *Robots: Planning and Implementation,* Tech Tran Corporation, Naperville, IL, 1984.

Tool and Manufacturing Handbook, Society of Manufacturing Engineers, Dearborn, MI, 1984.

CHAPTER 35
QUALITY ASSURANCE

DUDLEY LESSER

RCA/Missile and Surface Radar Division
Moorestown, New Jersey

35.1 QUALITY ORGANIZATION

The organization of the quality function depends primarily on the nature of its charter. There are two basic charters that are usually used. Both have the common objectives of the formation of and responsibility for a company's overall Quality Program. The difference between them is in the area of implementation. While there are no doubt companies whose very product line dictates one or the other methods of implementation, the choice, when not founded on some natural basis, is significant. Each charter or implementation method is categorized either as an Assessment Charter or a Line Charter:

1. *Assessment Charter.* The basis for this charter is that the Quality Organization is made up of a relatively small group of highly experienced personnel whose primary function is to assess the implementation of the Quality Program. They do this mainly from observations, audits, and limited evaluations of the product being produced. They may be responsible for responding to customer complaints. They may also be responsible for the quality functions associated with material purchased and all field quality activities. The charter may include some highly specialized line functions, but the direct line functions associated with inspection, testing, calibration, and stockroom controls are not included. These functions are to be found in the production or engineering organizational charters. This form of organization is best suited to a company with a stable product line that has produced essentially the same product over a long period of time. Under these conditions most of the production problems associated with materials, parts, specifications, drawings, manufacturing processes, and other production tasks have been eliminated or controlled. With production problems at a tolerable level, the Quality Program will be concerned mainly with maintenance of an acceptable system. This is the ideal situation under which an Assessment Organization is most effective. The key element is stability and a minimum of periodic production innovations.

2. *Line Charter.* This charter is the most commonly used in companies engaged in relatively low-volume production of complex products. This organization is very much a part of the production process and usually has responsibility for inspection, testing, test equipment design and manufacture, purchased material and assembly quality acceptance, and other related line functions. The underlying basis for this type of organization is early detection of problems, followed by rapid corrective action and evaluation. It not only devises the Quality Program, but also has direct responsibility for its implementation in those areas directly related with the "goodness" of the product. The greatest problem this type of organization faces is its inherent tendency to become unduly influenced by meeting schedules and company billings. But there are many who would argue that this is exactly what is desired, for too often the Assessment Organization becomes too theoretical and attracts "purists" to its ranks.

The brief discussion of the two basic types of Quality Organization should provide some insights for the top management of any company faced with choices. Obviously there are compromises that can be devised whereby each organization can be seeded with functions normally found in the other. But no matter what form the end result takes, there is a necessity for every company to make clear which theory or philosophy its management embraces. Additionally, the choice must be communicated to all levels of supervisors. Uncertainty in this area leads to less than cost-effective operations and constant bickering among lower levels in the company.

Once the primary choice has been made and communicated, the details of fleshing out the Quality Organization should be left to the Quality Manager, since there are many acceptable and usable structures that will suit the specific requirements of any company.

One final comment is in order, and experience indicates it to be crucial. The easiest task any experienced Quality Manager faces is the generation of a Quality Program. Here the objective is clear and the actions to be taken are logical. The implementation, verification, evaluation, and corrective action that ensue after the adoption of a Quality Program becomes the challenging task. Technical as well as interpersonal skills have to be combined to achieve effective results. The question as to the reporting level of the Quality Manager, or whatever other title is contrived to meet the needs of the day, has at times been a troublesome one. Experience dictates that when the Quality Manager reports to the highest operating management, he or she becomes most effective. Not only are attempts to submerge deficiencies hindered, the psychological effect that flows from the very real possibility that a defect in an operational area will be given visibility at the highest level of operating management does more to solve problems than any other motivator. From a customer standpoint this level of reporting engenders a feeling of confidence that products will meet the requirements contracted for. Although the benefits that can accrue to top management are not always completely understood, this level of reporting makes it possible for executives to obtain a mostly unbiased assessment of the total company's effectiveness in its product activities.

It cannot be overstated that the choice of Quality Manager should be made with the same degree of circumspection used for any other key management assignment.

35.2 CONTRACT REVIEW

Fundamental to the quality function is the understanding of what job has to be done. Often, failure to adequately review and comprehend what has to be done is the root cause of uneconomical or flawed Quality Programs. On the surface it seems trivial to ask what has to be done. But the question can easily be deceptive, since what is to be done is commonly perceived differently by different groups in a company.

To describe the job to be done for a Quality Organization, normal biases must be put aside and reliance placed in source documents. A source document can take many forms. Obviously, the largest jobs are defined by a written contract agreed to by both buyer and seller. In many cases the Quality Organization has been involved before signing of any contract, having most likely participated in the proposing and estimating phase preceding contract signing. But not every job has a well documented contract. Often jobs that are of small value or represent work between entities within a large company are agreed to with an amazing lack of definitive information concerning what a customer or user might expect.

The Quality Organization must review a job regardless of its size and the amount of definitive contractual documentation. The review must result in an understanding of what is to be done and under what conditions the result will be acceptable. Such a review should result in a document that the Quality Organization will use to allocate the resources required to provide control adequate to ensure company management that there is a high probability that the deliverables will meet the requirements of the job contracted for.

Each review should identify the following:

1. The deliverables. Items should be identified. The quantity of each item and the delivery rate also should be specified.
2. Applicable drawings, specifications, or other engineering data to be used or delivered.
3. The method of control of the engineering data, if stipulated.
4. Special information that determines in any way how the deliverables are to be made, controlled, or delivered.
5. Inspection requirements.
6. Acceptance requirements. This should name the acceptor and specify the degree of formality associated with acceptance.
7. Packaging, packing, and shipping requirements.
8. Warranty requirements, if any.
9. Liaison requirements with the customer if specified.
10. Requirements for establishing the basis for closing or completing the effort, task, or contract.

In the event that any of the applicable items are unclear or ambiguous it is necessary that those items be clarified. Many times the marketing, sales, or contract organization is charged with the responsibility for interpreting a contract. Any interpretations of unclear or ambiguous items must be documented and become part of the contract. Later disagreements must be resolved using these documented interpretations. Care must be exercised to prevent curbstone opinions by unauthorized members of any organization from becoming the official position. When necessary, the Quality Organization should insist on obtaining clarifications from cognizant customer personnel. Most sales, marketing, or contracts departments are understandably reluctant to approach a customer on some subtle point when the ink is hardly dry on a hard-won contract. But procrastination in resolving crucial points can result in untimely negative impact. The Quality Organization must be diligent in resolving open questions as soon as they are detected.

35.3 QUALITY PLANNING

A key element in any Quality Program is planning. Flowing from it is the method of handling every aspect of the Quality Organization's involvement in production. Where multiple products are involved, there should be a documented Quality Program Plan for each product or product line. There may be many similarities between each of these plans, but it is not cost-effective to have one plan for all products.

The plan should be generated by personnel within the Quality Organization who have significant experience with Quality Programs previously used in the company. They must be knowledgeable about other operating organizations in the company. If the Quality Program is to be useful and effective, it must be practical and doable.

As a partial checklist the following items should be addressed in the Quality Program Plan:

1. Material procurement
    a. Incoming inspection and test
    b. Field quality utilization
    c. Statistical sampling
    d. Subcontractor control
2. Special stockroom control.
3. Manufacturing release inspection.
4. Process controls.
5. Operation sheet control.
6. Tooling design and inspection.
7. Test equipment design, manufacture, and test.
8. Calibration requirements.
9. Product inspection
    a. Parts
    b. Assembly
    c. Subsystem and system
10. Product test
    a. Test procedure and specification controls
    b. Subassembly
    c. Assembly
    d. System
11. Trouble, failure reporting.
12. Final product inspection.
13. Acceptance testing.
14. Packaging, packing, shipping inspection.
15. Scrap and rework reporting and controls.
16. Operator training and certification.
17. Corrective action.
18. Nonconforming material control.
19. Configuration controls.
20. Drawing and change control.
21. Quality records.
22. Reliability requirements and controls.
23. Qualification requirements and controls.
24. Work instructions.
25. Warranty controls.
26. Customer relations.
27. Quality skill levels required.
28. Quality acceptance stamps.
29. Engineering controls.
30. Technical data controls.
31. Customer-furnished material controls.
32. Computer program quality assurance.
33. Schedules and milestones.
34. Design reviews.
35. Safety requirements.
36. Quality audit schedules.
37. Special microwave requirements.
38. Spare parts and assembly requirements.

Most important, the scope of work spelled out in the Quality Program Plan should correspond to the cost estimate made during the proposal stage. If there is a discrepancy between estimates and budgets, there has to be a reconciliation between any differences. Care must be exercised to make certain the Plan agrees with accepted budgets.

Once approved and released the Plan provides the basis for all Quality Organization efforts. Adequate provisions must be established so that the Plan can be modified as the program changes over time.

The preceding listing is not meant to be followed without adequate analysis. Each program is different and requires a Plan suited to its needs. Most programs have a great deal in common and, accordingly Plans, will reflect this. From a cost standpoint not every plan need be a stand alone one and can have references to other plans.

One last point: the Plan has to support the Program effectively and economically. Personnel using it must be made aware that it must not be followed blindly.

35.4 PURCHASED MATERIAL

Of all the efforts required in a well-managed and -organized Quality Program those associated with purchased material are the most difficult to manage. In any company engaged in continually producing the same product, the control of purchased material is relatively simple. Once qualified sources are identified and the early problems detected and effective corrective action taken, the task becomes one of continual surveillance, ensuring a consistent and uninterrupted flow of good material. While sudden problems can arise, the probability is high that given adequate surveillance the flow of "good" material will require little management attention. The challenge is greater for companies engaged in producing a variety of products consisting of materials of high complexity, often approaching the state of the art, which are ordered at very low delivery rates.

Let us define what is meant by "purchased material" from a Quality Program viewpoint. The following distinct categories of Purchased Material exist:

1. Raw material.
2. Parts.
3. Assemblies.
4. Subsystems.
5. Systems.
6. Support services.
7. Computer programs.
8. Capital items.
9. Facilities.

Each of these categories has unique characteristics that must be dealt with in such a manner that a cost-effective program is maintained. Before launching into each category and discussing an effective way of dealing with it, several aspects common to all categories should be examined.

An effective Quality Program requires that a source selection system be used to identify suppliers that can provide required materials at the lowest possible prices. Source selection involves approvals by Purchasing Departments. Generally, approved suppliers must meet financial, business-practice, and experience standards. The purpose of source selection is to become satisfied that a proposed supplier has the expertise and facilities to produce the required material with reasonable assurance that the yield will be satisfactory. To have effective source selection a skilled group of design engineers, manufacturing engineers, production specialists, and Quality Assurance specialists must review prospective suppliers. Their evaluations provide the basis for placing orders. If purchasing decisions are made by the Purchasing Manager based solely on price, it is possible that product quality will be marginally acceptable. The Quality Manager's responsibility is to evaluate the findings of the source selection board of experts. If the evaluation leads to a decision different from the Purchasing Manager's, resolution of the difference may have to be made by higher management.

The Quality function should also maintain a vendor rating system. Simplistically, the rating system should distinguish "good" vendors from "bad" ones. In reality, vendor rating systems are very difficult to apply. A system must take into account categories of vendors, complexity of products, and circumstances of procurements. A vendor providing simple resistor networks cannot be compared to a vendor providing microwave devices operating in Ku band with a top military priority. Moreover, the information in a vendor rating system must be kept current.

All categories of purchased material should be subject to a Receiving Inspection. The Receiving Inspection verifies the acceptability of purchased materials. This verification is a vital link in the manufacturing chain. If a company does not inspect materials upon receipt, it may find that its finished products are unacceptable—and this can lead to costly schedule delays.

There are six distinct identifiable methods available to Receiving Inspection. These apply after routine paperwork correlations with Purchase Orders and inspection for obvious shipping damage at the receiving dock. The six methods are:

1. *Certificate of Compliance.* A certificate of compliance is a statement by a supplier that the material being furnished on a purchase order meets its requirements. The intention of this document is to provide confidence that the material can be put directly into stock without further inspection or test at its destination. The document is based on trust. But although it is practical for raw materials such as metals, chemicals, wood, rubber, and paints, a certificate of compliance is marginally unacceptable for other items that have direct performance-related impact. The method is considered completely unusable for sophisticated subsystem and system procurements.

2. *Sampling Inspection and Test.* Basic to this method are statistical techniques for determining the acceptability of lots of material. A sample of the material delivered from the supplier is inspected

and tested. Based on the results and previously agreed levels of defects, a whole lot is accepted, rejected, or subjected to more extensive inspection and test. The sampling technique normally follows industry standards. A common standard is contained in MIL-STD-105. Inherent in the method is the assumption that the sample is representative of the whole lot. Where large quantities of the same material are procured, this assumption is more likely to be realized. In very-low-volume procurements there exists the possibility of accepting nonconforming material or possibly rejecting good material. This method also presumes that a predetermined percentage of the material can be nonconforming. At issue is the cost effectiveness of accepting some fraction of nonconforming material at Receiving Inspection and detecting material defects at a later stage in the manufacturing sequence.

    3. *100% Inspection/Test on Every Item of Product for All Documented Parameters.* This method provides the highest level of verification. It is also by far the most costly, not only because of the effort required to perform the inspections and tests but also because of planning and facility requirements. This method should be restricted to instances where conformance with all requirements is an absolute must. Generally, this method is for items that have not been qualified previously, for very small quantities of material, and where procurements have been aperiodic. Imposition of this method should require the prior approval of the Quality Manager.

    4. *Inspection/Test on Every Item of Product for a Selected Group of Parameters.* This method is a variation of method 3, with a significant difference—less cost. Because the parameters being inspected and tested are less than the total possible, facilities and labor are reduced accordingly. This method is most often used when there are some vital characteristics that must be verified, and there is some apprehension as to the item's ability to consistently meet the performance requirement.

    The inherent shortcoming of these 100% methods has to do with the supplier's behavior. The possibility exists that most of the items for which these methods are used come from sole sources. A sole-source supplier aware of the Receiving Inspection technique being used may reduce his or her own testing and elect to replace rejects. All Quality Departments must be alert to this tactic, which raises their costs.

    5. *Inspection/Test Verification by Some Other Organization Within or External to the Company.* Consider specialized material purchased for use in-house only. Generally, such material is purchased in low quantity, and acceptability is determined by the ultimate user. Most often, the material is inspected somewhat superficially by Receiving Inspection and is delivered to the end user for any additional inspection or testing. Consider also a very specific parameter or group of parameters requiring inspection with test equipment not in the company's inventory. Radiographic testing could be a typical example. Such testing is often done at a commercial testing laboratory or a university. The results are always documented and certified. Care has to be exercised in shipping and receiving of the material being tested—frequent use of a very specialized inspection or test should be evaluated from the viewpoint of cost effectiveness. It may be cheaper to purchase the specialized equipment.

    6. *Acceptance at Source.* This method, which almost every company uses, requires close economic examination. But the advantages are significant. Most important, by detecting nonconforming material at the supplier's plant, it can cut out the waste in shipping, further inspecting and testing, and returning the material. Further gains accrue from firsthand knowledge of how the supplier manufactures, inspects, and tests the procured material. Material handled in this manner requires only shipping-damage inspection when it is received. Overall schedules are enhanced by this method. The disadvantages are not negligible, however. A highly skilled employee is required. He or she must have long experience with purchased material and good technical judgment. His or her travel costs can mount up. In-plant inspection and test capabilities may be lost. Finally, it is difficult to reject material accepted at source by a Quality Representative after it has been shipped and is being used. A supplier's normal reaction would be that the material was damaged by the user after shipment. The buyer would have to prove that a defect was latent.

    Let us now examine the various categories of Purchased Material:

    1. *Raw Material.* Raw material most often describes material used in manufacturing. The primary procurement requirements are composition, size, shape, and markings. Most often, raw material is purchased from recognized distributors, and shipments are accompanied by certificates of compliance. Since industrywide standards are directly applicable, the quality review of the purchase documents consists mainly in ascertaining that a supplier is a recognized source. Source selection deals less with manufacturers than with distributors, because considerations such as price and delivery are most important.

    2. *Parts.* The highest volume of purchased material falls into this category. Because items of almost every variety and degree of complexity are purchased as distinct entities to be used in manufacturing products, drawings are the primary source for describing physical and functional requirements. Occasionally there are additional specifications. Most parts can be used in many applications; a small percentage are uniquely designed for specific applications. Drawings and specifications protect needed characteristics from being changed unexpectedly by a supplier; a part is always purchased to a drawing

and/or specification controlled by the purchaser. Difficulties often arise because different customers need parts that are identical except for a few characteristics, such as different levels of testing or differences in the number of hours a part has been "burned in" before delivery. Any occurrence that threatens the supply of parts gets immediate management attention. Where lot sampling is used by Receiving Inspection, any reject of entire lots causes panic, and pressure is brought to bear to do whatever is necessary to maintain an adequate flow of parts. A company might have to change from lot sampling to 100% Inspection/Test. When such an operational change is overdone, Receiving Inspection operates in a less than cost-effective manner. Additionally, suppliers are not encouraged to take adequate corrective actions to prevent future problems.

Is parts qualification required? Not every product development and manufacturing effort needs parts qualification. This is an area of technical judgment based on the requirements of the product being manufactured. If an effective qualification program is not invoked and maintained, there is usually excessive inspection and testing performed at Receiving Inspection. For military parts procurement, maximum use of the QPL (Qualified Parts List) should be required. Parts standardization when applied to a product is another method of achieving a cost-effective parts procurement program and increasing the reliability of the manufactured product by reducing the total number of different parts to be purchased.

3. *Assemblies.*   There are two basically different types of assembly purchases. The first includes assemblies made by a subcontractor to a purchaser's in-house designs. The second includes assemblies procured only by specification and a Statement of Work. These assemblies are designed by a subcontractor. Each type has characteristics that require different responses from the Quality Organization.

The review of purchase documents for the manufacture of in-house designs by a subcontractor is an important area of concern for the Quality Organization. The review must ascertain that all necessary drawings and specifications are called out in the purchase order. Provisions must be made for change control and the disposition of nonconforming material at the subcontractor's facility. Requirements for acceptance of finished assemblies should be clear. Provision for in-process surveillance by the Quality Organization of the subcontractor's operations should be spelled out clearly. Any specialized controls or other process requirements should also be detailed.

For assemblies that are to be both designed and produced by a subcontractor there are a number of specific concerns that must be spelled out in the purchase order. One concern is the use of parts by the subcontractor. Questions about qualification, standardization, and reliability must be addressed. Another concern is design review: Who should attend, how are designs and changes approved, how are action items closed out? Selection of test equipment, approval of test requirements, and acceptance procedures also require special attention. In addition, the area of the management of change is of vital importance. The freedom to make changes, the restriction on certain kinds of changes, and approval requirements must all be detailed. The subject of requirements and standards for drawings must also be addressed. During the review of the purchase order the Quality Organization must be constantly aware of the need to transmit to the subcontractor requirements that are imposed on the total product by the prime customer. Details overlooked at this time become more painful to impose later.

4. *Subsystems.*   Subsystems may be viewed as assemblies of assemblies. The distinguishing difference between assemblies and subsystems deals with performance. Whereas an assembly has to function individually in accordance with a set of requirements, a subsystem has a number of assemblies that must function together to meet some higher level of performance. To the concerns associated with assemblies must now be added concerns of integration and interchangeability.

Integration involves the interconnection and functioning together of assemblies that make up a subsystem. Cabling, connectors, water lines, cooling ducts, grounding bases, waveguides, and data links are examples of interconnections between assemblies of a subsystem. They all must be defined, documented, and controlled.

Interchangeability requirements, important when more than one subsystem is being procured, must be stated in the purchase documents. Testing for interchangeability may also be a requirement that must be included.

A further consideration deals with packing and packaging. Since many times a subsystem is shipped directly to an end user from a subcontractor's facility, packing requirements, including holding devices, must be included in the purchase order. Most often a buyer's Quality Organization, after accepting the subsystem functionally, inspects and authorizes a shipment.

As a final consideration before the purchase order is released, warranty and required repair of early failures should be reviewed. This gets to be a controversial issue after delivery from a subcontractor.

5. *Systems.*   Considerations concerning procurement of entire systems are almost identical with those related to subsystems. The major difference is one of involvement resulting from increased complexity. Skills of personnel assigned to purchase order review, in-process surveillance, and final acceptance must be commensurate with the complexity of the system being procured.

Characteristically, system procurements involve many contributors. A by-product of this is many contacts with the subcontractor which many times manifest themselves as unofficial changes to the design of a system and its specifications. The Quality Organization must be alert to see that all these "changes" are documented and become part of the purchase order.

A further complication may involve Technical Manuals and Spares, important elements of any system procurement that require close attention. Acceptance criteria for individual spare items are many times troublesome if there is an attempt to prevent duplication of inspections and tests performed when the individual parts were received originally.

6. *Support Services.* Many items are procured to support the main effort of manufacturing a product. They are diverse and may be hardware, software, technical data, or services. Examples are calibration services, experimental parts, consulting services, testing laboratory services, computer programming, aircraft rental, X-ray filming, and specialized test equipment rentals.

Any purchase order for support services must deal with two questions: (1) Is the support service being purchased defined, and is there an agreed method of determining that it has been properly delivered? (2) How much does the support service affect the acceptability of the final product? It is the second question that must be verified by the Quality Organization. If there is a direct effect on acceptance of the final product, then the Quality Organization must be prepared to verify conformance with requirements for both the support service *and* the final product. Documentation of that verification must be maintained by the Quality Organization in order to permit final acceptance of the product. This last step is often poorly handled simply because the Quality Organization cannot readily recognize the applicability and accordingly handles the acceptance of support services differently from other acceptances.

7. *Computer Programs.* The purchase of computer programs is believed by many to be radically different from purchasing hardware. But aside from a new jargon there is fundamentally little difference between hardware and computer program purchasing. Each must be defined. Performance must be specified. Acceptance criteria must be set. There must be documentation. Change must be controlled. Because computer programs cannot be "inspected" visually with the same ease as hardware, certain variations of in-process verification have to be adopted. The emphasis has to be on control and verification. A great deal of emphasis has to be placed on basic engineering and design review by the equivalent in-house engineering organization responsible for the procurement. This emphasis is usually written in a statement or work carefully delineating the review process that the supplier must conform to. The process should contain verification milestones that must be identified. Appropriate pass–fail criteria must be spelled out.

After that process has been successfully completed the normal verification surveillance by the Quality Organization takes place. It includes the in-process surveillance (with documented audits) and test witnessing and acceptance verification sign-off prior to shipment. The identification and marking of computer programs is another essential aspect to be verified by the Quality Organization, no matter whether the delivery media are lists or reels of magnetic tape.

Many times it is expedient to have members of the Quality Organization specially trained to handle computer programs. Familiarization with the jargon permits more effective communication with the supplier and the in-house engineering organization.

8. *Capital Items.* These are purchases by a company with their own funds that become part of the inventory of equipment owned. These items do not include materials for manufacturing products. They range from machines to computing systems and test equipment. In many cases Receiving Inspection has little to do with such items, but since they are to last a relatively long period, there is every good reason that the Quality Organization should treat them in the same manner as any other equipment. The review of the purchase order should concentrate on performance requirements of the purchase and the method by which it is to be inspected and accepted. Particular attention should be given to a review of warranty requirements. If the item is large or complex, acceptance should be made by a qualified member of the Quality Organization. If the purchase is equipment that has to be installed and working in a plant, thought should be given to a conditional acceptance at the supplier's facility with final acceptance in-plant when the equipment is fully installed and operating in accordance with the purchase order requirements.

9. *Facilities.* The purchase of facilities represents a series of challenges for the Quality Organization. These items may be buildings, test sites, or similar complete entities. They may involve either the purchase of materials and the completion of the facility by in-plant construction people or the contracting with a general contractor for the total job of building a turnkey facility.

Often an architectural firm is engaged to design and specify a total facility which is subsequently subcontracted for. In any case, the Quality Organization has a rather complex acceptance function to perform. It is made somewhat easier if the architect is to supervise the erection of the facility and subsequently accept it.

If there is no supervising architect, an in-house engineering or plant engineering group will have supervision responsibility. In this case the Quality Organization should be responsible for adequately spelling out and seeing that acceptance criteria have been verified. Many times this means hiring building-trade consultants to test and certify various aspects of the facility. The key here is complete documentation of all inspections and tests so that a high confidence exists at acceptance that the facility meets all requirements.

35.5 WORK INSTRUCTIONS

One of the most contested aspects of a well-conceived Quality Program is the one dealing with work instructions. While everyone might agree that instructions are needed to ensure that unskilled or semi-skilled workers do their jobs, there is controversy over the amount of detail required. And there can be out and out disagreement with the need for instructions for skilled workers, such as engineers. This difference comes about primarily from the feeling that the more thought processes are involved in a job the less the need for following documented instructions. But reality does not seem to bear this out. Many errors traced to their origins can be attributed not only to poor judgment, but also to failure of highly skilled workers to know what to do at their jobs. It is a truth, rarely admitted, that all workers need work instructions. Differences between the work instructions involve degrees of skill. The unskilled require instructions geared to a step-by-step approach containing simple accept–reject criteria. The skilled may require less detail and more of the reason why something must be accomplished. While the unskilled require instructions with little judgment options, the skilled can be expected to exercise a wide range of judgment. The argument is not instructions versus no instructions, but rather the amount of detail of and options for exercise of judgment.

With the advent of more complex products the need for written work instructions has increased. While it is true that the work force is better educated and trained, it is also true that modern production technology has been moving toward greater complexity much more rapidly than the average skill of the work force.

Another reason work instructions have become more necessary is that the work force has higher mobility. Employees used to remain with one company for long periods of time, and it was not uncommon for many employees to spend their entire working careers with one company. What was usual in the past has now become the exception. Even an employee who remains at a single company for his or her entire career can experience several moves to various divisions, each dealing with some highly specialized area of the business. Management can no longer expect employees to be familiar with policies and methodology throughout a company, and if managers desire adherence to their way of doing things, it must be through the use of instructions to which compliance is required.

The challenge facing any Quality Organization is seeing to it that work instructions are documented, that they are in fact being followed, and that they are changed in response to changes made in the controlling documentation that dictates their content. The degree of detail in work instructions must be commensurate with the skill level of the work. The difference between general work instructions and those tailored to a specific job should be kept in mind. There is nothing wrong with using standard company instructions if following them will yield the results required.

Often the procedures or instructions for implementing a task will be called out in a contract. Examples are applying conformal coatings for printed circuit boards, providing certain data items, making repairs to defects in castings, preparing drawings, and performing certain types of environmental tests.

The Quality Organization uses periodic audits to provide management assurance that work instructions are in place and are being used. Because adherence to instructions directly affects the quality of products, early detection and corrective action for deficiencies is a mandatory Quality effort.

The adequate implementation of a work instruction program is one of the most difficult of all the tasks called for in a well-managed Quality Program. The benefits are significant for any company and therefore should be pursued diligently.

35.6 STOCKROOM CONTROL

An essential ingredient in a complete Quality Program involves control of materials in stockrooms. (In a given company there may exist many different stockrooms each containing specialized materials.) Control deals with materials' entering and leaving as well as their storage.

In any company engaged in making a finished product two primary stockrooms are likely to be found, one dealing in parts and assembly stock, the other containing raw material stock. Additional stockrooms may contain engineering material, tools and test equipment, spares, packing material, returned material (repairs), and finished product material. Depending on the type of product there may be other highly specialized stockrooms.

The essence of control for any stockroom is a set of procedures governing its use. These procedures are most often generated by an Industrial Engineering organization reporting to Production Management. The stockrooms are generally under the operational control of a Production Control organization with responsibility for seeing that materials required to support production schedules are on hand. The procedures should include:

1. The category of material stored in a given stockroom.
2. Stockroom supervision identification.
3. Stockroom security.

4. Authorized roster of who deposits and withdraws materials.
5. Documentation used for all stockroom transactions and the proper level of authority.
6. Method of storing the material.
7. Controls for age-sensitive material.
8. Centralized inventory methods.
9. Change procedures for material in the stockroom.
10. Inspections required for stockroom transactions.
11. Safe packing and transporting of material leaving the stockroom.
12. Environmental requirements.
13. Stocking sequence.

Some comments concerning the above:

1. The stockroom should easily be identified by the category of material it contains. A procedure should be in place that keeps out other categories of material.

2. There should be procedures for identifying a stockroom's supervisor plus his or her alternates. If necessary, a different supervisor should be identified for each shift. The supervisor need not be located in the stockroom; only his or her identification is necessary.

3. Each stockroom must be protected from unauthorized access. Only authorized personnel may have free access.

4. Coupled with physical security of the stockroom is the problem of the designation of personnel who can deliver and remove material. The same consideration mentioned in 3 above applies.

5. All documentation for delivery, storage, and distribution of stockroom material must be controlled. Records form an important part of the Quality Program. Dates of receipt, name of vendor, part number including revision level, level of change, and quantity and inspection status are all important information items used by the Quality Organization. This information, together with Receiving Inspection records, can be used effectively when spares are drawn and shipped directly from the stockroom. If records are verified, spares in most cases can be shipped safely without further inspection or test. A further consideration is stock balance, which is important for material ordering.

6. An important control element in any stockroom is the method of designation or identification of material. Depending on the type of material being stocked, part numbers are used. Where revision status is important, material is designated by both part number and revision level. There will be materials stored with the same part number and different revision levels. This can lead to confusion and even temptation when material at one revision level is consumed. Strict adherence to procedural control is required to prevent the wrong material from being released.

7. There has been a continuing problem in most stockrooms because a wide variety of materials are age sensitive. Strict adherence to procedural controls is required. Because variable quantities of the same age-sensitive material are being stocked at the same time, marking and detecting expired material is a function of adherence to procedure. The disposition of expired material is sometimes a problem; the most effective disposition is immediate scrapping of the material.

8. Centralized inventory is primarily an accounting function. It becomes a Quality concern when many different stockrooms are involved and many different products contain the same material. Accuracy and completeness of inventory are important. Consider, for example, the revision level reported for a given part number. If the revision level is not carried on the inventory, and there is a product requiring that level, the possibility of using the wrong material exists. Most companies use the Military Standard which says that if parts or materials are not interchangeable, part numbers must be different.

9. The need to execute a change on material in stock may occur when inspection and test problems encountered during manufacturing call for corrective action to parts or material still in stockrooms. The location for performing the necessary work is determined usually by Manufacturing Engineering. Wherever the changes are executed, adequate documentation and manufacturing support must be made readily available. One outcome of such changes may be a need to reidentify parts under new part numbers. Stockroom records must be updated to reflect the change status.

10. Inspections by the Quality Organization have to be implemented in stockrooms as at any other workstations. Stockroom procedures must clearly state what activities require inspections and how the completion of such inspections are recorded. Such direct inspections are in addition to periodic audits by the Quality Organization to determine that overall processes are being followed.

11. Packing and handling require attention. Almost all material needs some protection when it is moved around a factory. Every category of material has peculiarities. Some peculiarities are obvious, such as size, weight, or fragility. Other peculiarities may be hidden. The sensitive characteristics can be mechanical or electrical. Often, for purchased items, the original packing material is kept to facilitate internal transportation.

12. Specialized stockrooms many times have specific environmental requirements, normally temperature and humidity control over a specified range. If specified limits are a requirement, there must be adequate recording devices showing the actual temperature and relative humidity on a 24-hr, 7-day/week basis. These charts or records must be maintained with other stockroom documentation. Procedures should exist for actions to be taken when charts or records of temperature and relative humidity show out-of-limit recordings. By means of periodic audits the Quality Organization should verify that recording is taking place and appropriate records are being properly maintained.

13. Material placed in the stockroom must be procedurally controlled as to sequence of removal. If not specified procedurally, it is possible for new material to be stored in such a way that it becomes the material first released and old material stays in storage for extraordinarily long periods of time. After being stored for a given period of time, some materials must be processed in order to remain usable. An example: high vacuum tubes are required to be degassed if they remain in storage for a given length of time. Another consideration is warranty validity. Most materials purchased are warranted for a fixed period of time. If kept in stock for a longer period, it is extremely difficult to return them to the supplier if there is any latent defect discovered during the manufacturing cycle.

35.7 INSPECTION OPERATIONS

Considered the most important task in a Quality Program, inspection impinges on all activities associated with production. Fundamental to the inspection function is the basis for rating the "quality" of the effort being inspected. A commonly held belief is that after Inspection has accepted an item, it is completely error free. Only rarely is this a true statement. Because the costs of removing all errors are not always justified or required, the basis for determining the acceptability of inspected items is unique to each situation. Commonly, the basis for acceptability is a document such as a drawing, specification, process sheet, or procedure. In some inspections, however, the final decision on acceptability is made by the inspector. He or she may use a table associated with some statistical sampling plan. Some decisions are arrived at only when the item is mated with some other part; acceptability is then based on the degree of conformance of both items. Other decisions can only be made after certain physical data are taken and reduced by computer techniques.

Before discussing each of these decision methods some general comments regarding inspection and inspectors are in order. An often used cliche is "you can't inspect Quality into a product." The statement may be trite, but it is true. Inspectors do not design or build products, so if a product does not perform its function or must be repaired because of construction faults it is rarely due to an inspector's error. While inspectors can be used to separate good products from bad ones, it is certainly a poor production facility that depends on this method to deliver good products. It can be shown that several redundant inspections have to be performed to find most detectable errors. Design errors and latent manufacturing errors cannot be detected by inspection at all. The inspector performs his or her function to provide increased confidence that certain obvious and critical characteristics are in accordance with designated documentation. With simple products this method provides high confidence that a product will perform as specified and customers will be satisfied. Increasing complexity brings the usual method of inspection into contention. The subtleties cannot easily be inspected, and methods applicable to simple products are no longer sufficient for customer satisfaction. Process controls and testing methods have to be relied on. The inspector's role is shifting. In addition to being involved with physical and mechanical details, he or she now must see to it that all required processes, tests, and validations have been performed. A Quality Program currently has to view the entire spectrum of the production cycle to select, on a cost-effective basis, the critical contributors to meeting a product's requirements. Inspection can no longer be relied on in the same manner it once was to ensure customer satisfaction.

Worthy of some discussion are the qualifications for inspectors. The most common source of inspectors has been from inside factories. After many years' experience with various classes of products and factory methods, operators and assemblers have been promoted. If the nature of a factory's products remains the same, this method of promotion is adequate. These inspectors are invaluable in detecting errors and problems, and their decisions about product quality can be counted on heavily to ensure customer satisfaction. As the complexity and sophistication of new products replace the old and products continually change, the past source for inspection personnel is less adequate. Working with very sophisticated production equipment provides little knowledge about how the product actually should function. This is especially true of products made by computer-controlled machines. Additionally, in the past mechanical and electrical inspectors had great experience in their areas of concern. But now the line between what is mechanical and what is electrical is blurred. Microwave devices, lasers, integrated circuits, computer programs, memories, and fiber optics are all good examples where there are no sharp dividing lines. Another area of change has to do with the ability of inspectors to translate complicated engineering information into meaningful inspection criteria. In the past drawings provided the necessary criteria. In the future there will be a need for well-thought-out, written inspection procedures to supplement drawings.

The preceding discussion suggests a new breed of inspector relying less on factory experience and more on broad-based technical experience with only a minimum of hands-on production experience. Of continuing concern is what the source for this new breed will be. Since most inspectors are generated internally, there will be emphasis on internal training programs designed to supplement on-the-job experience.

Let us return to the methods for deciding whether a product is acceptable. Decisions are based on these methods:

1. Decisions using documents as a basis.
2. Decisions made by the inspector.
3. Decisions using statistical sampling.
4. Decisions made as a result of mating two items.
5. Decisions made by reducing physical data.

1. The first method relies on official documents. This is by far the most common method in use. The documents can be drawings, specifications, procedures, brochures, manuals, or similar items. The problem with this method is that a document may contain information somewhat different from that used to manufacture the product. This can result in unnecessary rejection because of inspection for many things beyond the normal range of inspection tasks. Because the documents contain more information than necessary, the inspection task becomes too costly and actually may not reveal all essential discrepancies. For simple mechanical products, this method can be effective; for complex products, it leads to less than cost-effective use of the available inspection force. It is much more effective if a stand-alone inspection document, which contains the minimum inspection requirement and the necessary accept/reject criteria, is generated by the Quality Organization and appropriately reviewed by the Engineering and Manufacturing organizations.

2. The second method, relying on the inspector's own judgment, has some great attractions and some obvious disadvantages. For highly skilled and intricate work, this method is especially effective. Considering that the source of inspectors usually is from the highly skilled work force, the inspector is most likely the best judge of these specialized tasks. In many cases inspection criteria cannot be properly specified without elaborate measurements. For example, to specify that an instrument gear train should be properly meshed and should run quietly would take a rather extensive document and involve a number of quantitative measurements. An experienced tool maker could determine the setting of the gear train by ear in considerably less time and with few measurements. The obvious disadvantages of this method are that not all inspectors are equally experienced and qualified; there are no data verifying the inspection; and some gear trains may require precise measurements. Care has to be exercised by Quality Management when this method is used.

3. The use of statistical sampling, although quite effective, has some significant drawbacks. For some products, quantities of the items being inspected are too small and schedules are too erratic. A need for uniformity can be a further consideration. No matter what level of sampling is used, other than 100%, there is a probability that some item will not meet its requirements. In cases where this is not tolerable, this method cannot be used. The selection of sampling tables and levels of acceptability must be carefully controlled by the Quality Organization.

4. Many times the acceptability of parts can only be determined after mating with other parts. Although mostly applicable to precision mechanical parts, this method can be used anywhere two parts are separately produced and subsequently mated to form one assembly, which is normally repairable only by factory personnel using proper equipment. Most often this method is used for high-cost, state-of-the-art parts when the cost is prohibitively high for applying tight individual tolerances to the individual parts. One drawback to this method involves yield: The yield of one part may not match that of the other part; therefore, there will be a reduced yield of the mated parts. This can cause schedule problems. Care must be exercised by Production Control to maintain significant data on the individual parts so they can be mated on paper and mates can easily be found. It should be kept in mind that this method can only be used when the two or more individual parts are formed into one assembly that is used and repaired as an inseparable assembly by the end user.

5. With the increasing availability of computers in manufacturing facilities, decisions concerning acceptability of some complex items are based on many data points being recorded and subsequently being reduced and analyzed. Data reduction facilities can be remote from factory areas. For this reason there is often a delay between taking measurements and a decision concerning acceptance. Consequently, the parts in question must be marked so they can easily be stored and found when the data have been reduced and a decision arrived at.

35.8 PRODUCT ACCEPTANCE

The role of the Quality Organization in product acceptance is a straightforward one. For clarification purposes product acceptance in this case means the process by which a product is accepted by a customer for subsequent delivery.

Usually a contract will contain a section devoted to the requirements for acceptance, which consists of successfully passing a set of tests described in a document delivered as part of the technical data required by the contract and approved by the customer. Many times a representative of the customer witnesses the tests and performs the act of acceptance. More often than not, these tests are selected demonstrations of the total performance of the product required by contract. This is done for two reasons: first, because it is an economic approach, considering the costs involved for demonstrating every item of performance for a large, complex product; second, because there is the inherent confidence that only a product that has satisfactorily passed its lower-level inspections and tests will be presented for customer acceptance at the complete product level. The confidence originates from the customer's periodic auditing of the Quality Program. If there is confidence that all the required inspections and tests have been satisfactorily performed, that is enough justification for accepting the product on the basis of selected tests.

Prior to production contracts for complex products, there has most likely been some sort of a Qualification Program during which performance has been assessed under expected end limit conditions. These conditions include environmental considerations. From this assessment production specifications and drawings are generated which, if complied with, will provide high confidence that performance will be as required. It is the Quality Program and its implementation that validates the product's conformance to its governing documentation. A further consideration is the use of qualified nonstandard parts. During production qualification, it is normal to qualify certain parts that have not been tested previously to demonstrate their uniquely required performance. After having qualified these parts it is necessary to make certain that the parts meet their requirements during the production phase. The material acceptance procedures usually found in the Quality Program are instrumental in providing the needed confidence. In addition, the Quality Organization provides final records for the major items which make up the product. These records, maintained by the Quality Organization, become useful as time goes by and possible questions arise as the product is used by the customer. The records are important for warranty considerations.

The Quality Organization provides the surveillance and inspection function during the packaging and packing of the product.

A final acceptance report is usually put together by the Quality Organization that documents all the findings during acceptance. All the tags usually found on the lower-level items are removed and logged to form the permanent record of the product as it finally is shipped.

A valuable fallout of the Quality Organization involvement is taking corrective action where indicated. This effort provides the feedback to other contributing organizations that improvements may be necessary to make the acceptance phase more cost-effective.

Generally, the customer's representative indicates his or her acceptance on the acceptance report as well as on the normal shipping documents. Sometimes, because of the required time for reduction of the acceptance test data, there is a lag between completion of the test demonstration and the literal acceptance by the customer. Usually, there is adequate indication of test compliance to permit the shipping sequence to commence while data are being reduced.

35.9 WARRANTY

Warranty considerations represent a somewhat troublesome element not only for the Quality Organization, but also for all managers. A company must be concerned not only with costs associated with warranty, but also with its image, which has a long-term effect on business prospects.

The first fact that must be realized when discussing warranty is that the customer pays for it. Another fact that should be borne in mind is that warranty considerations can and usually do apply to items other than equipment, such as technical data, computer programs, and services.

From the earliest stage of contract review, the Quality Organization should be aware of the warranty requirements and the obligations of the company. All other organizations within the company should also be familiar with warranty requirements and should treat them as design parameters. A most important consideration is the length of time covered by the warranty. Warranty needs should be addressed in all engineering designs before committing them to drawings and specifications. Early decisions have to be made regarding passing along warranty requirements to subcontractors, keeping in mind that costs will go up in proportion to the severity of requirements and the duration of liability. The Quality Organization representative who reviews purchase orders prior to their release should be aware of the prime contractor requirement and be alert to check applicability in all subcontract purchase orders. It is important to keep in mind that the absence of any explicit warranty requirement does not mean no warranty is applicable. Warranty implied by the Uniform Commercial Code may be applicable even though not expressly called out. What is costly is establishing very specific terms of the warranty and a specific period of time over which it is to be in force.

Intermingled with the concept of warranty is reliability. In contracts, reliability is stated in terms that can be equated to a given number of failures over a given period of time. Warranty refers to situations where the number of allowable failures is exceeded. The Quality Organization exercises judgment concerning warranty claims over time.

Items under warranty claims are usually returned for repair or replacement. Sometimes, because of the nature of the product, warranty claims are made by notifying the manufacturer, who must then send field repairmen out to investigate the claim. The field repairmen should be authorized to take the appropriate action. For sophisticated equipment, preventive maintenance instructions are provided to customers. There are occasions that may indicate warranty problems, but are really failures due to lack of maintenance. This may be overlooked officially, but the true nature of the problem should be reported to prevent unnecessary action from being taken.

Another problem that requires attention in handling warranty claims is associated with the extent of the repair to be made. Obviously, if an item is to be replaced totally, an item of the proper configuration level is required. But consider an item made sometime in the past that is returned for repair of a limited area. As the item is inspected, conditions other than the reported one may be discovered. They may not be such that they can cause imminent failure, but suppose they are below current governing standards. There is a tendency, normal for inspection personnel, to repair them. The management policy governing this activity should be quite clear.

The area of physical damage to products as a result of shipment should be a concern of the Quality Organization. Such damage can result from combinations of packaging, packing, and transportation methods. Each method should be called out in a specification, drawing, or procedure.

35.10   R & D AND THE QUALITY PROGRAM

It used to be believed that there was no place for a Quality Program during research and development (R & D) and that only during production should the Quality Organization get involved. Recent experiences have proved otherwise. In fact, it can be stated that any R & D effort that does not have a Quality Program in effect will most probably generate a product at higher than normal costs and with less than desirable operation characteristics. The time to introduce a disciplined and controlled environment in a company engaged in the development of complex products is during the R & D phase, when technical data are generated, design reviews are held, and a myriad of decisions are made concerning the product. Most R & D programs are well managed, and there are sincere attempts to control all activities. But because the overarching objective is a product that will outperform competitive products, some managers of R & D programs feel that the other necessary requirements of a new product will be supplied by other parts of the organization during the manufacturing phase. There is no denying that performance is probably the single most important aspect of any new product, but if quality and reliability are ignored there will be an economic impact that cannot be lightly dismissed. Once specifications have been decided on, drawings generated, qualification testing completed, and acceptance testing determined, changes become difficult and costly.

To minimize discoveries as a product moves from R & D to manufacturing, disciplined control must be invoked early in the process. The real challenge is not whether to have control, but how to have it and still maintain the flexibility so necessary during development. What is required is an early recognition of the need and the assignment of a qualified representative of the Quality Organization to generate a Quality Program that is appropriate to R & D. The program should describe the controls to be used and should provide for adequate flexibility. The program should be generated with the agreement and concurrence of the R & D manager. This need cannot be overemphasized. R & D managers have two primary reactions to a Quality Program. The hardest to cope with is the one that stems from not recognizing the need; the R & D manager will say, the program "has to happen and I'll make the best of it." The second reaction consists of concern over costs and the impact on schedules and flexibility. Obviously past experience of the R & D manager contributes most to his or her reaction. For this reason care must be exercised in selecting the Quality representative. His or her technical competence and experience are both important. If the technical community accepts the representative as a contributing team member, the implementation of the Quality Program will proceed smoothly.

The transition to production is enhanced by using the Quality representative who was responsible during the R & D phase. His or her ability to interpret requirements will provide for a smooth introduction. Many of the technical contributors during the R & D phase are usually reassigned as production starts up, and the only real bridge may be provided by the Quality representative who, because of his or her involvement, can maintain continuing contact with the R & D team.

With the advent of more sophisticated products that have extended R & D phases, the early Quality Program implementation may turn out to be the most significant event in making a new program successful and cost-effective.

One other observation should be made regarding very large R & D programs involving many engineers and several subcontractors. Such programs are almost large enough to become separate business entities. In these cases the Program Manager is most likely to be selected from the senior management of a company. Ultimately the R & D phase turns into the production phase with the whole R & D organization intact. In this case the Quality representative becomes in effect the Manager of a separate and distinct Quality Organization. The Quality Program is product oriented and is managed as part of the organizational structure dedicated to this single program.

35.11 DESIGN REVIEWS

Design reviews are effective in ensuring that a design developed by an Engineering Organization will perform to its controlling specification. Design review should also provide the assurance that a product is designed so that it can be manufactured and maintained through its intended life in a cost-effective manner. Usually design reviews are conducted on significant subassemblies as well as on the total product. Many times there is initially a preliminary design review and then a final one. The earlier design review is usually conducted after Engineering has made decisions concerning major concepts of product configuration and has allocated performance characteristics to elements of the configuration. The review is usually performed by a group of recognized experts representing many engineering disciplines and significant areas of fabrication and production expertise. During the presentations, especially the preliminary one, there is a review of alternatives investigated and the basis for design decisions. Trade-offs are presented when there are no conclusive results. Results of design reviews are documented and action items are assigned.

If the process is followed and the proper experts are made available, the resultant design should be a cost-effective one. The review is only as effective as the effort put into it and the skill of the reviewers.

The Quality Organization should have a representative at the reviews with a diverse background. He or she should be respected by both the engineering and the production organizations. He or she has a twofold responsibility. First, he or she must provide the necessary inputs concerning the design itself; second, he or she must provide the surveillance necessary to see that resultant action items are satisfactorily closed out.

In the first area of responsibility the Quality representative can provide guidance on the items provided in the checklist (Table 35.1). The second area of responsibility involves a key part of any design review: the timely and effective closeout of action items. Some of the items are crucial and

Table 35.1 Quality Program Design Review Checklist

Design Review Item: _____ Date Held: _____ Design Review Chairman: _____ Quality Representative: _____ Report Issued: No. ____ Date: _____	*Not Applicable*	*No Special Action Required*	*Action in Process*	*Further Analysis Required*	*Action Required*
1. Inspection and test skills needed					
2. Special processes required					
3. Safety considerations					
4. Ease of repair, rework, and maintenance					
5. Special inspection tools required					
6. Special test equipment required					
7. Calibration accuracies required					
8. New areas of operator certification required					
9. Levels of inspection and test required					
10. Introduction of accessible test points					
11. New part vendors and acceptance requirements					
12. Specialized computer controls and library requirements					
13. Special packaging, packing, and shipping requirements					
14. Special internal handling requirements					
15. Age-sensitive material requirements					
16. Nonstandard parts requirements					
17. Special qualification requirements					

normally get the close attention which leads to early closeout. But less critical items are sometimes neglected and accordingly fail to impact the design as they were expected to. Tracking and final closeout is normally either a prime responsibility or an area of auditing by the Quality representative.

35.12 COMPUTER PROGRAMS

Increasingly, computer programs are becoming integral elements of complex products. The engineering aspects associated with their development and qualification are complex and high technical skills are employed. Many of these skills are now being considered part of the production process. In the not too distant past, programming skills were used mostly in R & D activities where the only control being exercised was whatever the participants imposed on themselves.

To a large extent, as in any newly developed technology, a whole new language was developed, some of which can accurately be described as jargon. It is well understood by those engaged in the technology and almost unintelligible to others. Further complicating matters, each new application of computer technology seemed to adopt new jargon. As computer programs started to become part of the product itself, there was uncertainty as to how to identify, specify, and deliver computer programs. Computer programs could be furnished as a set of listings or put on punched cards, paper tape, magnetic tape, or transcribed onto disks. The methodology of incorporating change and controlling configuration became an evolving matter varying considerably from product to product and company to company.

Because of this background there has been some confusion concerning the application of a Quality Program. Many military customers have generated specialized Software Quality Assurance requirements that provide a unique method for the control of Computer Programs. Other groups within industry have also attempted to adopt some sort of Quality requirements. Before discussing the Quality involvement, it is necessary to make a clear distinction between the quality of the technical content of a computer program and the quality control of computer programs. Many times these two aspects of quality get confused. The difference is analogous to the distinction between the quality of a product in terms of performance and the quality of the physical entity incorporating the design. The quality of the content of the program is an evaluation that can best be made by the technical supervision of the group designing the program and the technical community reviewing the design before the program is incorporated in the product. After the program is incorporated in the product, the evaluation can only be made as part of the overall performance of the product. The Quality Assurance aspect of the control of the computer program is a continuous process mostly akin to the control exercised over a product's individual elements. Most effective control can be exercised over a computer program if it is treated as simply another element of the product. While there are some significant differences between software and hardware the essence of Quality control is identical. Some specialized aspects of control must be considered for computer programs, first because of the difficulty in easily inspecting a computer program to verify its conformance with a specification and second because of problems of change and configuration controls.

Once the computer program structure has been determined (in effect the design has been broken down to its building blocks), the Quality Organization can exercise control. For a hardware element the normal quality functions involve inspection, test, storage, and change control. These same control functions must be imposed in computer programs. These functions will be described and specialized aspects noted.

It is assumed that whatever hardware and software are generated in any programming organization are under the normal quality controls used elsewhere in the production facility. Each building block of a computer program is generated in accordance with the appropriate technical control. The building block is subjected to a documented performance test that is witnessed by a representative of the Quality Organization. During the testing phase changes can be made and compiled by the cognizant technical organization. Upon successful completion of the test, the building block is placed in a library, which is analogous to a tightly controlled stockroom. Copies of the program can be made for any purpose, but every time the building block is used to build up to a higher level of program, the latest controlled version of the building block must be used. The higher level is tested and its performance documented. The building process is repeated until there is a final controlled and tested computer program produced, its original resident in the library. The computer programs in the library are identified by version descriptors, which form the basis for configuration control. Each of the computer programs used from the library is marked by the Quality Organization and represents the only official computer program qualified for product use. There may be copies of that program being used by technical personnel, but they are considered unofficial.

Changes that are required as a result of official use are generated as patches to the original computer program and the official version descriptor now must reflect these changes or patches. When the computer program and its patches are compiled into a new version, it must pass new performance tests in order to become the official program. Each of these activities is under the direct control of the Quality Organization.

Many times a number of official computer programs are combined onto a magnetic disk, and the disk becomes the official set of computer programs. Marking and certification of the disk is under the control of the Quality Organization; appropriate marking and library controls are invoked.

The Quality Organization is also involved with test data recorded on data tapes requiring special processing to determine whether a particular test has successfully been completed. These data tapes are controlled directly by the Quality representative overseeing test operations. He or she furnishes a certified copy to the data reduction facility, which must operate under documented procedures with calibrated equipment. As with the computer programs, the data tapes must be marked and strict library controls must be implemented.

The management of any facility making a product that contains computer programs must constantly be sensitive to the tendency to impose very special controls that are expensive, but do not add any effectiveness. Only recently, computer programs have become part of deliverable products. This has led to the belief that because software technology is so different a whole new method of control is necessary. Experience has not borne this out, and before entering into new controls, management should challenge their need. Established methods of control will work with special attention to the differences that distinguish software from hardware.

35.13 CONFIGURATION CONTROL

There are a number of terms that have different meanings depending on the way they are used and who uses them. Configuration control, configuration management, baseline control, change control, drawing control, and design control are typical terms that are often confused.

Some brief nonrigorous definitions are in order:

1. Configuration control refers to all the methodology used to control production so that the product will conform to a structured configuration of material, computer programs, and technical data.

2. Configuration management is the most general term used to describe the coordinated efforts required to make a product conforming to specific performance and physical requirements. These requirements must be documented in specifications, drawings, and contracts.

3. Baseline control refers to the technique used to control production via specifications and drawings when multiple, distinctively different versions of the product are contracted for. This control is primarily an activity of the Engineering and Contracts departments, since it provides the basic information the rest of the operating organizations use.

4. Change control can describe two distinct activities. The first, which is simpler, is the activity associated with the methodology of controlling changes in documents located in files scattered throughout a company. Drawings and documents are charged out to production personnel. As these drawings and documents are changed, old ones must be retrieved and replaced with new ones. The second change control activity involves the actual change of technical content in drawings and documents. The method of authorization, initiation, approval, and execution of change is governed by change control.

5. Drawing control is the method by which information on drawings is changed to conform to contract and engineering design changes. One apparently minor change to a specified performance parameter may entail a large number of changes to manufacturing drawings and a large number of production changes which must be timed properly.

6. Design control usually refers to the methodology by which the Engineering Organization institutes and maintains control of the technical integrity of any product through its entire contractual history. The normal output of this effort is both whatever technical data are provided to the customer and any documentation required during the production phase.

The Quality Organization's role in configuration control covers a number of different functions:

1. Product inspection.
2. Product test.
3. Field quality.
4. Quality audits.
5. Quality records.

In each function the Quality Organization verifies, inspects, or tests to the latest released engineering information. In this manner, on an incremental basis there is a control of the configuration of a product as it is being manufactured. Through the audit function there is a periodic surveillance of the degree of compliance with all applicable procedures and instructions. This provides confidence that the procedures and policies of configuration management are being complied with. Through the retention of quality records there is continuous monitoring of product configuration. A record of all the subsidiary, serialized product subassemblies provides historical information for products already delivered.

Some contracts include so-called configuration audits, which call for developing a detailed record of the configuration of a product being readied for delivery. When compared to the mandated contractual configuration, this record usually produces a list of differences between as-designed and the as-built configurations. These differences generally must be adjudicated before customer acceptance. Differences can result from various sources. Managers of engineering, production, and configuration and quality control may become involved in the adjudication. Where adequate attention has been devoted to implementing configuration management policies and procedures the adjudication of any differences is a routine matter. Quality records usually are a vital part of any configuration audit.

35.14   CHANGE CONTROL

Change control is a key element in any well-managed production facility. Without it there is no way of knowing the true status of the product documentation or the product material. Almost any product except the most simple changes continually. Change originates in a number of ways:

1. Production deficiencies.
2. Changing customer requirements.
3. Vendor and subcontractor problems.
4. Warranty problems.

5. Safety requirements.
6. Economic changes.
7. State-of-the-art changes.
8. Application difficulties.

1. Production deficiencies relate to two separate problems. The first and most common has to do with defects. Because of lack of adequate design review by the production organization or because the work force is inadequate, a given design produces low yields. Under this condition the design must be changed to increase yield and lower production costs. The design change may be minor and may involve only a very small number of documents or parts and assemblies. Sometimes the changes are extensive and involve many items. The other problem has to do with unreliability of the basic technology inherent in the design. No matter how extensive design reviews are, they do not preclude this type of problem. A change must be introduced to prevent excessive customer complaints.

2. Customer needs may change: Previous performance may be inadequate; the customer may require additional functions. This can result in either minor or extensive changes to documentation and hardware.

3. Vendor and subcontractor problems are one of the most common causes of change. A vendor may perform poorly—or go out of business altogether. Such problems may force changes in documentation purchase orders, or, in severe cases, changes in manufacturing processes.

4. Warranty problems are a common cause for change. The costs resulting from warranty returns or repairs is a common driver of change. Most managements are extremely sensitive to warranty costs and usually demand timely changes to minimize these costs.

5. Safety problems demand immediate change. Although most safety problems are accounted for during design reviews, occasionally the problem is discovered during production, and until the change is incorporated, production will stop.

6. During the life of a product, especially if it lasts a reasonable length of time, many cost-saving innovations may be adopted. They can affect selection of materials, manufacturing processes, and design. In any well-managed company there is a steady stream of such changes.

7. As time goes on, technological change makes available new materials and ways to do things. Invariably this growth contributes not only to reduced production costs, but also to greatly enhanced performance. When technologically motivated improvements are incorporated into product design and production methodology, difficulties may arise because often new materials, parts, or processes are used before their reliability has been demonstrated.

8. Calls for changes can come from many sources. Because most people engaged in production are interested in improving the product or reducing cost, proposed changes flow in an endless stream. But not every change proposed is desirable or economically justifiable. The need for competent review becomes necessary. Every production organization must establish a review group that includes participants from all operating groups. The key person is its chairman. He or she must have the ability to understand the total engineering and production process, have a position in the organization that permits the review group to operate unimpeded, and have a strong enough personality to control the diverse group and reject, if necessary, very strong arguments for change. Above all, the chairman must be knowledgeable about the company's policies and objectives so that the review group has a unified purpose. The review group must recommend which changes are to be implemented, when they are to be broken in, and what to do, if anything, concerning items already shipped. The review group must also prepare or obtain cost estimates for the work to be done and schedules for its implementation. The review group passes approved changes on to a higher authority for final approval. After this authorization, the changes are then given to some version of a production control organization responsible for their implementation. At this point the control of documentation is of great concern. Special care must be taken to change applicable documentation in all areas affecting production. In many cases old information is physically removed and replaced with new, and accurate accounting of such transactions is maintained.

Two more aspects of change control should be discussed. The first concerns temporary change, which occurs when a performance defect is noted by the manufacturing or test organizations. Generally, engineers are asked to investigate the problem and provide corrective action. A typical result is a change in parts. If the change is implemented prior to the normal authorized method it is called a temporary change. Documentation is marked up by the engineer at the site of the problem, with provisions for the automatic expiration of the marked up documentation, generally in 30–45 days. This is enough time to process the change in the normal manner. The Quality Organization controls this activity in accordance with documented procedures, for if there is excessive "temporary" change, the implication is that the design is not adequate and a corrective action request should be sent to engineering management.

The other aspect of change control has to do with cumulative change, which is used for a prototype or first production model. The changes against a given part number are accumulated over a fixed period, usually until the item is through test for the first time.

35.15 RELIABILITY

Some experts believe that reliability is a sort of natural discipline that belongs in the Quality Organization. Others have argued that it belongs in the Engineering Organization. There is some justification for both views, so reliability can be found in either organization.

The proponents of the view favoring putting reliability in the Quality Organization believe that doing so guarantees that the discipline will be enforced and given management visibility. There is a concern over the tendency of Engineering to assign a lower priority to reliability than to meeting performance goals; trade-offs between performance and reliability will always favor performance. This will in turn result in not meeting contractual or internal goals set for reliability.

To those favoring putting reliability in Engineering, the argument revolves around the recognition that the reliability requirement is an item of performance equally as important as any other specified parameter. They argue that, if reliability performance is not a part of the ongoing design process, contractual requirements will not be met. Since the skills for meeting reliability requirements are the same as design skills, setting up another specialized design group in the Quality Organization is not very desirable. In the limit it could result in a competition for available skills. Another consideration involves the ability of a reliability group to keep current with the state of the art. A reliability group located in the Quality Organization might develop a propensity for "making tomorrow like yesterday" because of its limited exposure to new design innovations.

A cost-effective reliability program combines the strengths of both organizations. Reliability requirements must be incorporated in the overall design process. The engineering design organization must be charged with doing so at the outset. As for the Quality Organization, it must verify that the design incorporates appropriate reliability requirements and that finished goods conform to the requirements.

The Quality Organization develops special expertise in dealing with reliability requirements.

One parameter for specifying reliability is MTBF (mean time between failure). When the value assigned to MTBF is large, demonstration of compliance becomes time consuming unless the Quality Organization uses standardized procedures for such demonstrations or devises substitute tests that provide the necessary confidence that requirements are being verified. Of course, if requirements are contractual, liaison with the customer's technical and contractual representatives will be required.

Information gained from analysis of parts that fail during either production or subsequent field use provides insights into the behavior of the parts under various operating conditions or as a function of time. The Quality Organization remains concerned throughout production and field use.

As the Quality Organization inspects, tests, and accepts parts, it accumulates a great deal of experience and data. The Organization is the source for judgments concerning procurement risks when high reliability parts are required to meet product reliability specifications.

Over the life of a product many actions subsequent to the original design phase contribute to what is described as product reliability maturity. These actions include many production process changes as well as design enhancements. Over time, the effect of these actions should result in higher reliability. A reliability group in the Quality Organization with a broad view associated with many products is in a position to transfer the experience from one product to another. In this way reliability maturity of any given product can be achieved earlier and more efficiently.

Close association with the product through all phases of the production cycle provides a reliability group in the Quality Organization the opportunity to take timely corrective action. The result is an increased probability that reliability goals will be attained. This level of corrective action is different from the one normally associated with catastrophic failures in which corrective action prevents repetition. This level involves more subtle problems that deal with degradation of performance with time.

It has been said that it is possible to have a high quality product that is unreliable, but that it is impossible to have a highly reliable product of poor quality. Combining the skills of the Engineering and Quality Organizations may be the most cost-effective way to guarantee that products will be obtained of adequate quality and also meet reliability requirements.

35.16 STANDARDS

Many companies not only write standards for the materials and processes they use, they also maintain a standards group, usually part of the Engineering Organization, which adopts and promulgates standards on a corporate-wide basis. (Corporate standards usually have provision for adopting local standards for specific, authorized needs.)

Standards most commonly fall into the following major categories:

1. Parts and materials.   3. Processes.
2. Design.   4. Engineering practices.

The involvement of the Quality Program with these standards is an important one even though they are widely used with little or no special considerations. The involvement of the Quality Organization has to be an integral part of any standards program, built into the procedural methodology for selection and promulgation.

Recognizing that standards can cover an extremely large range of candidates for adoption, only a few will be discussed in each of the categories referred to previously to illustrate Quality Organization involvement.

1. *Parts and Materials. a. Parts.* Consider, first, parts selected by the Standards Group for R & D work. Even though they will not be used in deliverable products, the Quality Organization should require reviews of purchase orders for these parts and should monitor incoming inspections and tests. The obvious need for acceptable parts by those engaged in R & D efforts makes this a necessary area of control. The Standards Group also compiles lists of preferred parts for deliverable products. The process of standardization provides a number of benefits: cost economies, familiarity by designers, enhanced reliability of delivered products, and ease of ordering, accepting, and storing. In order to optimize this process, the Standards Group must analyze and test standard parts. Here the Quality Organization must exercise appropriate levels of control, from assistance in source selection, to surveillance of the qualification effort, and finally to the inspection and test of the parts purchased by the Standards Group.

b. Materials. The Standards Group usually provides information to engineering and design groups concerning the properties and characteristics of various materials. Design rules are also provided. The Standards Group provides instructions and selection guides for finishes and adhesives. Information as to the "dos" and "don'ts" of material combinations is also generally provided. The Quality Organization's involvement in this activity is mostly informational; the Organization is responsible for translating Quality instructions for use by inspection personnel.

2. *Design.* Standardized design practices and methodology are also provided by the Standards Group. These might involve tolerances, electrical grounding, sealing, weatherproofing, electromagnetic shielding, cabling techniques, drafting practices, including the various rules for serialization, nameplates, aperature cards, and change records. Again, the Quality Organization must bridge the gap between the Standards and Manufacturing Groups. Many times a Quality Organization representative is included in the Standards Group to assist in those areas which will find greatest application in production activities.

3. *Processes.* A typical production facility is engaged in many fabrication and assembly processes, including soldering, pinning, fastening, welding, brazing, annealing, sheet metal fabrication, applying corrosion protection, metal finishing, etching, and adhesive application. Each process can be performed in a number of ways, so if a process is uncontrolled, results will vary from operator to operator and from time to time. In order to obtain uniform results and prevent costly rejections, processes are specified and controlled by the Standards Group. The Quality Organization is normally an active participant with the Standards Group in this effort. In particular, a Quality representative is responsible for supplying the inputs regarding tolerances, accept–reject criteria, and repair and rework procedures.

4. *Engineering Practices.* The Standards Group is responsible for incorporating uniform practices in a company's engineering activities. Document control, change control, design reviews, engineering notebook control, and design configuration control are typical areas of concern. The Quality Organization's primary role is to be knowledgeable enough to perform meaningful audits.

A standard generated by the Quality Organization is a Quality Manual, which is primarily a collection of instructions and procedures that are used by Quality personnel. The instructions are independent of any particular product. Typical instructions are:

1. Inspection stamps—types and controls.
2. Instructions for completing tags and documents used by the Quality Organization.
3. Quality audit checklists.
4. Lot sampling plans.
5. Quality Organization record retention schedules.

6. Computer program controls.
7. Organization charts.
8. Quality Organization job descriptions.
9. Age-sensitive material controls.
10. Customer complaint procedures.
11. Vendor rating system.
12. Corrective action request procedures.
13. Specification review procedures.
14. Preparation of quality procedures.
15. Technical data review procedures.

35.17 TECHNICAL DATA

There is no widely accepted definition of the term technical data. In its broadest sense it includes all data associated with a product regardless of whether they are deliverable or not. Sometimes the term is defined as everything except drawings and specifications. On most military procurements by the U.S. Department of Defense, technical data include the items referenced in the DD1423 (Contract Data Items). The DD1423 describes all items of data that are deliverable and contains many items which may not normally be considered technical.

Regardless of how the term is defined the Quality Program must provide a level of control commensurate with technical data's importance to the product. There is a tendency in some Quality Organizations to concentrate all their resources on hardware and practically ignore the technical data, believing it to be an Engineering responsibility. An effective Quality Program plan accounts for all items that control the product and all deliverable items. Increasingly, these items include data.

Consider deliverable technical data. With some exceptions a Quality Program has to view technical data in the same manner as it views hardware. Requirements that define technical data are very similar to drawings that define hardware. In military procurement, the detailed requirement that defines the content of the technical data is called a DID (data item description). In nonmilitary procurements the basic contract generally defines the content of deliverable data. The Quality Organization is not responsible for the content, editing, and publication of technical data. Rather, the Organization is responsible for final inspection and acceptance.

There are several inspection points. There is an inspection of content. Only the technical organization generating data can adequately "inspect" content for technical accuracy and conformance with contractual requirements. The review and approval must be documented, usually on a sign-off sheet that accompanies a document as it proceeds from initiator to final delivery. A technical publication's organization inspects the document prior to printing to make sure all of it is there and that it conforms with the contractually required formats and all normal publication standards.

The printed document is reviewed and accepted by the Quality Organization. The challenge is to perform the review and acceptance effectively and economically.

After two previous inspections there should be nothing wrong with a document except for a printing or collating error. Experience has indicated otherwise, and further inspection is warranted. Sampling techniques have been cost-effective. After dividing possible errors into categories of severity and assigning levels of acceptability for each category, randomly selected pages are sampled, the number a function of a document's thickness. The document will not be completely free of all errors. Rather, it will meet a predetermined level of acceptability.

The question is, what should this level be? There is no single answer; the level should be determined by whatever use the document will be put to. Perhaps the best guide is experience. Because other documents delivered under different but similar contracts are usually available for error examination, they can provide a basis for establishing the standard which should be imposed.

35.18 TOOLS AND TEST EQUIPMENT

Tools and test equipment are devices that control the manufacture, inspection, and functional test of deliverable products. These devices control the quantitative conformance of a product to the drawings and specifications that define it—its physical aspects and performance. These devices fall into two main categories:

1. Standard, commercially available.
2. Specifically designed and manufactured.

Standard, commercially available devices have advertised and verifiable capabilities. If they are periodically calibrated and serviced their capabilities will be maintained over long periods of time.

Devices are specified in operation sheets, process sheets, test specifications, test procedures, and other documents. Problems can arise when operating personnel attempt to substitute other tools and test equipment that they believe have equivalent capabilities. But not all devices have the same capabilities under all environmental conditions, and this must be evaluated before equivalency is truly established. Either the equivalent should be listed in the applicable procedure or an evaluation should be made by a technically competent authority.

Specially designed and manufactured tools and test equipment present unique problems for all production organizations. To begin with, the need for a special design is a clear indication that a standard, commercially available instrument is not usable. Also, the special instrument must do what it has been designed for, and users must, through experience, establish requirements for calibrating the instrument and maintaining it in calibration.

Difficulties can start with the drawings that define a special instrument. Since the instrument usually is not deliverable, documentation may be incomplete. The degree of completeness will determine the amount of difficulty that will be experienced in proving out the design; the ease, stability, and periodicity of calibration; and the complexity of maintenance. The Production and Quality Organizations are responsible for establishing the standards that are to be adhered to by anyone designing, testing, or using special instruments. A distinction must be made between test equipment to be used in a factory and deliverable field test equipment, which is a product in its own right.

Another area of concern is the handling of instruments that a major subcontractor may design and manufacture for use solely on a product being procured. For maximum control, a unique instrument funded on a purchase order should be listed as a separate item with a delivery schedule called out. The instrument will have visibility to the Field Quality Group and, additionally, the special instrument can be delivered to the purchaser for other possible use. The costs for such instruments often are not included as part of product costs. If care is not exercised, that cost item will be repeated if separate purchase orders for the same product are issued over time.

35.19 CALIBRATION

In a Quality Program calibration refers to the calibration of instruments used in acceptance testing. Calibration has two major components:

1. Verifying the accuracy of a measuring tool or test instrument to a recognized standard (maintained, for example, by the National Bureau of Standards.)
2. The interval of time between such calibrations.

It is important to point out that there is a basic difference between precision and accuracy. A device can be precise but inaccurate. Precision means that a device will produce a repeatable reading in response to a repeated constant stimulus. Accuracy means that the reading the device produces in response to a given stimulus is the same as the reading a standard device would produce in response to the same stimulus. Calibration is meant to validate a device's accuracy.

Factory operation sheets and test procedures generally specify the instrument to be used for making a measurement, whether the instrument is manufactured in-house or purchased off-the-shelf. Included in the specification, either explicitly or implicitly, is the accuracy of the instrument. If an "equivalent" instrument is permitted, the specification should say so.

In a good Quality Program, there must be in place a documented calibration system. Elements of a proper system are:

1. An adequately equipped standards laboratory.
2. Ready identification of calibration status on all measuring instruments.
3. Calibration schedules for all measuring instruments.
4. A recall system that notifies personnel to return all measuring instruments in accordance with calibration schedules.
5. A crew of technicians trained to perform maintenance, troubleshooting, repair, and calibration.
6. Procedures for implementing the calibration system.
7. Records of failure rates of instruments and whether they were in or out of calibration when recalled.
8. A properly equipped library of instructional information, usually supplied by instrument manufacturers.
9. Stockroom of repair parts.
10. Schedule for calibrating the standard instruments used in the laboratory.
11. Procedures for using outside laboratories for calibrations which cannot be accomplished in-house.

All measuring instruments within the facility must carry some form of readily visible identification to indicate whether they are in calibration and the next date of calibration. Some facilities use color-coded stickers that can be read only by those who know the code. Others use stickers on which the information is written. Even instruments used for troubleshooting must be identified.

Calibration scheduling is one of the most important, and often most difficult, elements of the calibration system. The question of how long an instrument will retain its accuracy is not always easy to answer. Instrument manufacturers can help in determining initial calibration intervals, but the ultimate interval has to be established by experience during an instrument's normal service. Establishing intervals between calibrations and maintenance actions involves economic as well as accuracy considerations. Short intervals can produce excessive costs. Overly long intervals can undermine accuracy and increase the risk of equipment failure. Generally speaking, it is better that intervals be slightly on the short side. There is often a need to extend the interval length when the instrument is being used to take measurements during a test that cannot be interrupted. The ultimate validation of interval lengths is contained in any long-term records maintained by a calibration department. Whatever the result, interval lengths must be clearly spelled out in the procedures governing the calibration system.

The recall system ensures that instruments are in calibration. According to the calibration schedule, the calibration department collects the instruments, performs the required maintenance, and returns the instruments to stock or to service. The recall system must be documented and disseminated to everyone using measuring instruments.

The calibration department's personnel must be able to repair and maintain diverse instruments. Because these instruments require knowledge and skills to repair that are not always available in a production facility, personnel must be trained. Many instrumentmakers conduct training courses. Also required is a policy regarding the level of repair that will be done in-house. When it is no longer economically effective to do repairs in-house, arrangements with authorized outside sources should be made.

In order for the calibration system to be effective and economical, it must be governed by a set of clear and definitive procedures. Their use must be mandatory, and there must be periodic audits to make certain they are being carried out.

Current and accurate records keep a calibration system cost-effective. They can identify instruments that have high failure rates and, therefore, are uneconomical. The records also provide information as to which parts should be stored by the laboratory. Information about whether instruments were in calibration when recalled can be used to extend intervals between calibrations or to extend temporarily the recall of an instrument. The amount, extent, and format of records should be detailed in calibration system procedures.

Many instrument manufacturers make available service notes on an automatic basis if requested. Instructional material for instruments designed in-house should be available wherever they are to be calibrated, whether in a calibration laboratory or in some other area, or whether a combination of a special area and a calibration laboratory is needed to perform a complete calibration. Calibration system procedures should describe how this will be accomplished.

A stockroom containing replacement parts required for maintaining measuring instruments is necessary to minimize downtime. Calibration records and manufacturers' recommendations are key sources of information regarding the items that should be stocked. But it often is easy to obtain parts from manufacturers; therefore, because stocking of parts represents an additional inventory cost, it should not be done automatically. It should require approval of management.

Laboratory standard instruments require periodic maintenance and calibration. In the majority of cases, this cannot be done in-house. Because most standards are somewhat fragile and are affected by a variety of environmental conditions, packaging, packing, and shipping must be closely monitored if the accuracy of calibrations is to be preserved.

Some measuring instruments are maintained and calibrated at an outside facility. All aspects of shipping and handling require attention if accuracy is to be maintained. Because shipping and handling are done on a somewhat routine basis, there can be a tendency to delegate such tasks to personnel not completely aware of their importance. They must be documented and periodically audited. The facility doing the work should also be audited periodically to make certain that adequate controls are in effect.

Increasingly, there is a tendency to lease rather than own measurement instruments. This may mean that calibration is no longer directly the responsibility of a calibration department. Nevertheless, quality program management cannot ignore its responsibility to see that leased equipment used to make measurements for product acceptance has been calibrated properly. The Quality Organization must be able to provide the assurance that leased equipment is adequate to the job regardless of whether calibration is done in-house or purchased as a service.

Increasing product complexity leads to performance parameters on the frontiers of measurement capability. The measurements associated with such requirements are extremely difficult to perform within the capabilities of measurement instruments. Calibration to the order of accuracy of normal measurements becomes most illusive. Many times an ad hoc standard must be adopted and treated in special ways, but with procedural safeguards that do not diminish accuracy.

35.20   PERSONAL TOOLS

In almost every manufacturing company employees use hand tools that they own themselves. Many companies require assembly personnel to purchase some hand tools and often arrange to make them available at low cost. The most common examples are screwdrivers, cutters, long nose pliers, and small socket wrench sets. Tools and gages are inspected routinely either by the Quality Organization or by Manufacturing. In facilities engaged in extensive mechanical fabrication this is an important element of the Quality Program. The conditions of tools and gages directly affect the quality of workmanship. Nevertheless, the condition of personal hand tools cannot be overlooked. Employees often continue to use them after they should be replaced. Problems can arise:

1. A screwdriver that is the wrong size or is excessively worn can scratch a surface. It can burr screws, hindering easy removal and damaging their appearance. Rejections can result.

2. Side cutters that are nicked or whose cutting edges are not aligned can damage wires. Many such defects are not easily detected and become what are known as latent defects.

3. Long nose pliers that are nonaligned or burred can cause wiring or circuit damage. The need for rework or latent defects can result.

A Quality Program that is thorough and complete will include controls applicable to personal tools. The controls should depend on the nature of the product involved and product flow. An essential element in any control procedure is education. Employees should be informed of the necessity for periodic inspection of their own tools. Coupled with education should be a program for inexpensive replacement. There can be discontent when rejection of personal tools means an additional expenditure by an employee. The result of an inspection can be a disagreement regarding a tool's condition. Many manufacturing companies elect to provide all hand tools and bar the use of any personal tools.

There is another category of personal tools that is very hard to control: the special little aids that employees fashion for themselves to make their jobs easier. Often they jealously guard these aids so other employees do not find out about them. If they are hidden, it is difficult to identify and periodically inspect them. Gages and holding devices are the most common, but these aids are as diverse as human imagination. Here, again, making the employee aware of the consequences of faulty devices is an effective way to locate, identify, and inspect them. Many companies wisely try to reward employees who devise such aids. This not only provides recognition for the employees, but helps identify the aids, which then can be included in normal tool inventory and be easily controlled and inspected.

A Quality Organization must be alert to this problem and provide the impetus to prevent the use of personal tools from being detrimental to the product quality. Quality Management should institute what is often described as roving or patrol inspection to keep consistent surveillance over the use of personal tools.

35.21   TRAINING AND CERTIFICATION

Training and certification is an extremely broad subject that affects all activities in a company. It is directly related to the effectiveness of a Quality Program. This discussion will focus on the personnel directly concerned with manufacturing, inspection, and testing. Companies try to hire people with experience in particular lines of work. The problem is the experience was gained at another company. Whether by design or custom, each company devises its own methods and procedures, so experience is often unique. If the members of a company's work force are not trained to function in a like manner, the company's products may not all be the same. (There are certain classes of tasks where innovation and difference is desired, however, in order to enhance or improve existing ways of doing things and result in a more cost-effective operation. Recognizing this, management should be alert to these different ways of doing things so that they can be incorporated in the company's work instructions.)

Of concern here are those manufacturing tasks directly associated with deliverable products, ranging from blueprint reading to welding. Because of increasing complexity and miniaturization, many specialized manufacturing tasks have become necessary. The need to make certain that employees are qualified to perform their assigned operations has focused attention on training and certification. Many companies have instituted what are called vestibule training programs, which undertake to train every new manufacturing employee in basic skills such as use of small hand tools, soldering, blueprint reading, and basic harness techniques. This elementary training provides a common base for all manufacturing personnel. It may not deal with a particular skill to the depth required in a manufacturing task, however. Furthermore, frequent personnel shifts due to load changes may result in some operators being assigned to tasks for which they have not been trained adequately. Personnel shifts may also be dictated by labor agreements rather than by the particular skills of the operators. Job classification may not determine whether a particular employee has the skills required for performing a particular task. So personnel shifts can have a direct effect on product quality.

Periodic retraining has been used to reinforce the skills attained in vestibule training. But retraining may be unnecessary if there is a minimum of personnel shifts or if broad-based training is not required in the first place. By creating new occupations, specialized skills can be gathered together in such a way that certain tasks are immune to personnel shifts. This approach can create personnel problems when a company's business mix changes and the specialized skills are not needed at all. Creating new, highly skilled occupations often leads to higher manufacturing costs.

A recent innovation has been to adopt certification in certain skill areas, then adjusting labor contracts to recognize its impact. In the past, certification has been used primarily for tasks that cannot be evaluated by normal inspection techniques. It works this way: Workers are trained in depth for a specific task. Their output, usually test pieces, is evaluated by expert instructors with long experience and by sophisticated inspection methods. At the end of such training, workers are certified to perform the task for a specified period of time at the end of which the certification has to be renewed. Workers so certified are registered and carry objective evidence of certification such as a card. Welding is a prime example. With the advent of space hardware, soldering has become an area of certification.

More tasks are being brought under the umbrella of certification as the most effective method to ensure high-quality workmanship. The only caution that should be exercised is excessive use of the method. As more tasks require certified workers, this lowers the flexibility of the work force (unless many workers are certified in many areas). The expense of certification has to be economically justified. It has been used in some cases where it is not necessary, in order to enhance stability in a given work task. Each production facility has a unique set of circumstances that should dictate which areas need certification. The method is quite expensive and causes a loss of production flexibility that can only be justified by extreme need.

35.22 NONCONFORMING MATERIAL

Simply defined, nonconforming material is material that does not conform with either the documentation that describes its performance or its manufacturing drawings. Generally, everyone engaged in production proclaims that he or she will not condone the use of any material which is nonconforming. This is a desirable goal, but the realities of the production environment make achievement expensive. There is almost nothing produced that from time to time does not contain mistakes, errors, and substitutions. An arbitrary decision that eliminates any nonconforming material commits a total organization to rejection, repair, or rework of all nonconforming material regardless of its impact. It casts product design in a mold that is too rigid. While there are certain products that must be manufactured without any deviation from design, the overwhelming number of products do not fall into this category.

Many deviations from design have little effect on a product. The challenge for any well-integrated production organization is to identify defects early in the production cycle, then evaluate the impact and determine the proper response. Usually, the Quality Organization is assigned this task. This choice is based on the fact that the Quality Organization generally is the first to detect the defect. Since timeliness is an important ingredient in the total solution, it is logical to have the earliest detector of the discrepancy manage the total effort to dispose of it effectively.

Usually, a special group chaired by a member of the Quality Organization deals with nonconforming material. In U.S. Defense Department contracts this group is called a Material Review Board (MRB). It contains a member representing the government who has a right of veto over proposed actions.

It is convenient and expeditious to categorize defects by degrees of severity. This classification will permit the timely disposition of the defect in accordance with predetermined rules adopted by a material review group. One theory holds that the more classifications there are, the more routine the solutions to defects. But, experience indicates that a great deal of time will be expended on the discussion of which classification a defect should be put into, rather than on the necessary solution. Experience with military procurements has resulted in a three tier classification system: critical, major, and minor.

The question is, which classification should be applied to a given defect? The classification has to be based on the defect's impact on product performance. An illustration may be helpful. Consider a bracket with a dimension called out on a drawing to be x inches from the centerline of hole pattern A. Because of a manufacturing error, the dimension in question measures $0.125 + x$ inches. It is known that this defect does not interfere in any way with the bracket's use. For purposes of disposing of nonconforming material, the discrepant dimension should be treated as a minor defect.

The following should provide the necessary guidance for proper classification.

A defect is critical if it prevents a product from performing its intended function.

A major defect is one that has a significant impact on product performance or life, but does not make the product nonusable. Consider an integrated circuit that operates below specification at extreme temperatures. While the integrated circuit should perform equally well throughout its specified temperature range, degraded performance at the limits of the range is considered a major, not a critical, defect.

A minor defect has no important impact on the use of the product. An example: an extra hole in a machined part which causes no degradation of the part's performance.

The disposition of a defect first involves classifying it and second assessing the risk it poses to the user. If a defect in a product is removed by replacement or rework, the delivered product will no longer be nonconforming and no further action will be required. A customer may elect to take delivery of a nonconforming product with a critical or major defect. Financial considerations or agreement to replace the nonconforming material at a later date may influence the decision. Regardless of the decision and the reasons for it, the actions must be documented. Some companies refuse to deliver products with critical defects as a matter of policy. Minor defects, discovered after nonconforming materials have been incorporated into a part or assembly, are relatively easily disposed of since such defects carry no performance risk. Cost, convenience, and scheduling play decisive roles in the decision to accept a product with a minor defect. Minor defects can be fixed by standard repairs which, by prior agreements with other organizations in a company, can be made without specific approval in every case. Some examples are hole plugging, wire splicing, conformal coating, printed circuit board trace repair, and fastening. Standard repair can be used by material review groups without long discussion or government approval.

Care must be exercised that a good balance is arrived at between the cost of fixing minor defects and incorporating them into products in order to meet delivery schedules. Companies often are judged not only on performance, schedule, and cost, but on the number of defects found in their products, regardless of severity.

35.23 QUALITY COSTS

The items whose costs are considered most significant to a Quality Program are:

1. Rework.
2. Repair.
3. Scrap.
4. Warranty.
5. Field quality.
6. Incoming material inspection and test.

In most companies labor costs for rework are segregated and reported. Generally, if a previously established norm is not exceeded, there is no management action. If the norm is exceeded, there is an investigation to determine causes and corrective action is taken. Any drive to continually reduce rework costs has to be motivated by management initiative and requires highly skilled task forces to be successful. The problems contributing to rework have to be identified and actions taken to eliminate them. These problems can arise in any operating organization within a company, but, because it is the Quality Organization that detects the need for rework, the rework problem is often described as a Quality Problem and the costs associated with rework are called Quality Costs. Care has to be exercised that the Quality Organization is not identified as the sole contributor to rework costs. If it is, it will be expected that increased Quality actions will decrease rework costs and the expectations may go unsatisfied as the real causes are not investigated and corrected.

Repair costs usually are segregated and reported. Repairs generally must be made because of poor handling of products, poor equipment, poor processes, or simple carelessness. The remedies and cautions associated with rework apply here as well.

Scrap can result from many causes ranging from changes in the design to damage caused in troubleshooting. An example of a subtle cause is the purposeful destruction of a device or part because it works intermittently. Scrap costs are also segregated and reported separately in most companies. If costs do not exceed the historical norm, little if anything is done about it. As in the cases of rework and repair, the key to scrap cost reduction is analysis of causes and economics.

Costs associated with warranty are somewhat more difficult to assess. Since customers are involved and the intangible value of customer satisfaction plays an important role, analysis of warranty costs are sometimes clouded. Nonetheless, warranty costs contribute significantly to product price. Warranty costs are often thought of as Quality costs, simply because the concept of warranty conjures up thoughts of poor quality. Obviously, the causes of warranty claims are involved with every organization of the company and must be involved in cost-effective measures to reducing it.

The Field Quality Group of the Quality Organization is responsible for accepting purchased items at designated vendors. Items are accepted in this manner for economic reasons. Items accepted at their source are introduced into the product with usually only an inspection for shipping damage. Subsequent inspections and tests may reveal shortcomings with corrective measures indicated. Sometimes the resulting costs are considered Quality costs, perhaps because the Quality Organization identifies the causes of deficiencies and institutes the actions necessary to eliminate them. Usually, corrective action is undertaken at the vendor's facility. Occasionally, the action required involves purchase orders, statements of work, or specifications. In any case the Quality Organization has the primary responsibility.

The most common contributor to what can be labeled Quality Costs are the costs due to problems with purchased parts and materials. When a variety and large volume of items must be processed, the resources of inspection and test departments will be taxed. Since this function generally is a part of the Quality Organization, any difficulty with the parts or materials after they have passed into stock is considered a Quality problem and the costs associated with them are labeled Quality costs. Often incoming inspection and test departments resort to accepted, but not always effective, sampling methods. The motivation is economically based and often the results are less than desired. Corrective action is the responsibility of the Quality Organization.

Regardless of the myths and realities involved with Quality costs they have an economic impact and should get management attention. The only caution is their label. It is a mistake to look to the Quality Organization exclusively when corrective measures are in order. Costs for inefficient production are not created by a Quality Organization, and all operating organizations must contribute to their elimination or reduction. The way to label a problem is to determine which skills are required to solve it. Inspectors cannot fix design problems and all organizations must be mature enough to recognize that.

35.24 CORRECTIVE ACTION

Since the Quality Organization usually detects problems or defects in their earliest stages, it is the organization usually looked to for the evaluation of problems and institution of corrective action. Of all the elements of an effective quality program, corrective action is the one that potentially can provide the greatest benefit.

Undertaking large corrective action efforts whenever anything goes wrong is not an effective use of resources. More effective is a graduated response gauged to the severity of problems. Requests for corrective action range from simple verbal requests through action items identified at status reviews to formal, documented requests. The request method should be documented in an authorized Quality Organization procedure. The formal corrective action request should be a document generated and serialized by the Quality Organization. Provision should also be made for evaluation of the corrective action undertaken.

A request for corrective action must deal with the cause of a particular problem and the correction contemplated. The correction should be twofold: an immediate correction for the problem at hand and a correction that will eliminate the cause of the problem and prevent recurrence.

The Quality Organization reviewer must be on guard for responses that may appear to be effective, but in reality pay lip service and little else to long-term solutions. An example of such a response: The defect was caused inadvertently by an operator, and the corrective action will be to instruct the operator in the proper method of performing the task involved. Experience indicates that this sort of response will do little if anything to prevent future defects. To begin with, the operator should have been adequately trained before being assigned to the task, and if the operator was trained, the training should be reviewed to determine if it is adequate. Or training may be adequate, but the operator has a problem that makes it unlikely the job can be accomplished without error. This often happens when there is some disruption in the make-up of factory skills. Lay-offs, new contracts, union negotiations, or changes in product lines all can affect the labor force. Often, because of seniority or other considerations, personnel not normally used in certain operations are found to be temporarily engaged in those tasks.

Another item the reviewer must be on guard for is repeating corrective action for the same error or defect. Quality records must make it easy to detect repeat items. Repeated defects may indicate a fundamental problem such as poor design, defective machinery, or incomplete or wrong procedures. The corrective action that should be taken for repetitive errors is usually different from what is required for random errors.

Before finally closing out the corrective action request, the reviewer must be convinced that the actions taken have a high probability of preventing further occurrences.

The corrective action request represents the last official method for obtaining correction for problems encountered in implementing the Quality Program. Sometimes a company cannot respond with effective corrective action. When this happens, it becomes necessary for higher management to resolve the problem. No procedure can be written concerning how this can be accomplished. The problem must be turned over to the highest level of Quality Management for adjudication and developing possible solutions. It is to be hoped that in most well-managed companies the need for this method to obtain corrective action will be rare.

35.25 QUALITY REPORTS

What reports should be generated by a Quality Organization? Which reports should be promulgated, and to whom? A distinction should be made regarding those reports that are required contractually and those that are generated by the Quality Organization for internal consumption. Since the generation of reports is time consuming and therefore costly, and reports use up valuable storage space, Quality Management should authorize only those considered necessary.

Discretionary reports, that is, those not required contractually, generally can be segregated into four groupings:

1. Management reports.
2. Newspaper-type reports.
3. Magazine-type reports.
4. History-type reports.

Management reports provide information concerning Quality matters to the top management of a company. These reports should summarize the status of the quality of the company's activities. The report should use a graphic format that permits an unaided visual presentation. Highlights of achievements or of problems requiring top management assistance should be presented separately in a bullet format, not in a narrative. Back-up details should be available, but not presented until requested. There should be a graphic presentation of trends in selected aspects of the Quality Program. Care must be taken to prevent the presentation of subjective information without qualification. The objective in management reports should be a clear presentation of Quality Program status structured for the audience. Distribution of these reports should be kept at a minimum.

Newspaper-type reports cover information gathered by the Quality Organization that is important on a day-to-day basis. Examples of quality newspaper headlines are:

1. Low inspection yields.
2. High reject rate in test.
3. Repetitive defects for the same cause.
4. High failure rates for specific purchased items.
5. Low yield reported by Field Quality at subcontractor's facility.
6. Customer complaints about a common defect.
7. High technical data rejects.
8. High scrap rate.
9. High rework rate.

Stories under these headlines should have one common objective. They should deal with items that require immediate corrective action. The effort expended in these reports can yield dramatic results. It is essential for a report of this type to be distributed to the individual in the proper organization who can take the necessary corrective action.

Magazine-type reports provide information gathered over a longer period of time that can be analyzed to reveal trends and patterns. Examples of such information are inspection results by assembly area or shift; incoming material reject rate by vendor or commodity; test equipment failures; high test failures due to test procedure problems. The analysis of such reports can reveal problems more subtle than those causing very high and immediately recognizable defects.

These reports should be distributed to those charged with problem analysis in the various operating activities.

History-type reports, issued periodically, cover relatively long periods and reveal the long-term trends which are useful in determining systemic problems whose solution usually requires extensive analysis. These reports may confirm that problems do not exist or that corrective actions have been effective. Such reports might cover yearly scrap and rework results, rejects by product, rejects by plant areas, rejects due to plant facility failures, rejects by supplier. These reports can indicate corrective actions ranging from the need for personnel retraining to the need for identifying new vendors.

The analysis of the problems covered by these reports is generally performed by the Quality Organization. Suggested solutions to the problems are included with the reports.

35.26 MOTIVATIONAL PROGRAMS

There is little doubt that motivational programs can have a positive effect on product quality. But if motivational programs are sporadic and reactive, and they are allowed to slowly fade away as problems subside, little lasting impact is felt.

Motivational programs having a focus and maintained over a long period have an effect akin to long-term institutional advertising; they identify an area of concern and modify behavior in positive ways. Personal recognition is an excellent motivator and is many times used in conjunction with the motivational effort.

A motivation program in the Quality Program should focus on a specific area such as fastening or handling. For maximum impact, the focus should shift from time to time. Numerous posters and banners displayed prominently help reinforce the message. Well-managed motivational programs can be structured to teach as well as preach.

A more recent innovation in U.S. production facilities is the introduction of Quality Circles— basically groups of employees who meet together to discuss and suggest ways to improve the quality of the product they work on. Suggestions accepted by supervisors are put into effect and monitored. Improved product quality and increased productivity can result. Controversy in the application of Quality Circles has centered on the difficulty in linking results with economic rewards. The closer these two are related, the more positive the results. Establishing Quality Circles requires careful planning and continuing management attention. The potential benefits to quality and productivity are significant.

Mass production, which brings more products to more people at the lowest possible cost, has given rise to problems for those engaged in manufacturing. A lack of identification between an employee's job and a finished product contributes to a lack of job satisfaction that leads to early burnout of employees, accompanied by physical and mental disorders—and to lower quality of the product. The lower quality results from inattention and lack of pride.

One of the methods advocated for correcting this situation is Team Manufacturing, whereby a small group does all the work necessary to produce a product. The satisfaction gained from such a work experience is said to increase the pride of creating something useful and results in a higher quality product.

Can Team Manufacturing be used for all types of products? While it undoubtedly contributes to increased satisfaction, it is not certain that it contributes to higher product quality. Not everyone on the team is skilled and expert, and there will be some whose work will be marginal. Team Manufacturing requires very careful planning and should be applied only after thorough analysis.

35.27 RECORD RETENTION

Normally, a Quality Organization accumulates a significant quantity of records. At every point during the production process, starting with acceptance of materials and parts and ending with final acceptance of a product, a written record is produced. In addition, records are generated to document internal Quality Organization activities. Warranty records form another body of documents. Almost all these records are useful and some are invaluable. Past events can be reconstructed from written records. They may be needed to settle legal disputes. Records can constitute a data base for the analysis of engineering and production experiences. It is possible to use the analyzed data to establish norms and for later comparisons. Corrective action to effect changes in trends can be taken on a meaningful basis.

The challenge facing any Quality Organization is to determine what records should be retained and for how long. The mass of data accumulated in any large company is truly awesome. Indiscriminate retention is costly and in paper form can consume tremendous storage facilities. Additionally, easy retrieval, a normal requirement, can be quite difficult.

Since all the records can prove necessary for some purpose, it is generally not wise to keep some records and destroy others. Experience indicates that the key to rational record retention lies in the ease of retrieval.

What is needed is a scheme that leads to levels of retention associated with the desired record retrievability. Careful study by the Quality Organization with appropriate inputs from all the operating organizations and legal counsel is a necessity. The study should identify the various levels of retrievability. Typically this will reveal that levels vary with time. Some records, primarily the most recent, will require instant retrievability, while others will accept longer and longer retrieval times. The effectivity time for a given level, which can be changed with the passage of time, should also be a result of the study.

Only records that need to be retrieved immediately should be easily accessible, while other levels can eventually be relegated to long-term storage.

Periodic review of a company's retention scheme is a requirement. Without periodic review, the scheme will rarely be challenged. Most companies have retention schedules governing all kinds of company records that require yearly review. The records retained by a Quality Organization should be included in a company's record retention schedule.

35.28 QUALITY AUDITS

One of the challenges that every Quality Organization faces has to do with verification of the effectiveness of the Quality Program. Ways to accomplish this objective range from doing nothing until something goes wrong to establishing elaborate inspection procedures. While each method provides some degree of verification, the key question is: Is the method one that provides high confidence that all is going well? An effective Quality Program has to deal with all activities that have an impact on product quality. In reality, asking for verification of the effectiveness of a Quality Program is tantamount to asking if an entire company is working effectively.

Quality Audits are used extensively for this purpose. They are relatively inexpensive and easy to accomplish. If implemented over a long period of time, they provide a reliable indicator of the degree of compliance with an established Quality Program.

To be effective, a Quality Program must be implemented by every department in a company. A Quality Program is not the exclusive province of a company's Quality Organization. Whereas the Quality Organization may structure the Quality Program and be responsible for an assessing of its effectiveness, actual day-to-day implementation is allocated to all departments.

Once a Quality Program is adopted, each department is responsible for a set of operating instructions for implementation. The sum total of these instructions implements the company's approved and authorized Quality Program from day to day. It follows that an evaluation of compliance with these departmental instructions is a method for determining Quality Program compliance. The only difficulty is that the relatively large number of instructions makes the method time consuming if they are all verified in detail. The Quality Audit is the practical solution for providing compliance confidence.

There are two distinct categories of audit. The first is a series of questions on a checklist. The Quality Auditor can observe the results objectively. The second is a checklist that poses a series of questions to a designated member of the functional group being audited. The objective in either case is to assess the degree of compliance with procedures. On the basis of the completed audit checklists, the state of the Quality Program is determined. For major discrepancies noted in individual checklists, requests for corrective action are issued and pursued.

There are two vital ingredients in the Quality Audit method. One is the design of checklists. This is done either by experienced personnel in the Quality Organization or by experienced representatives of the functional organizations involved. The second vital ingredient is the frequency of the individual audits. There is no fixed number appropriate for each area. The frequency depends on the nature of business the company is engaged in and the significance of a particular department to product performance, reliability, or overall quality. Experience indicates audit frequency varies from monthly for critical areas to semiannually for areas such as packaging, packing, and shipping. The ultimate responsibility for Quality Audits rests with management of the Quality Organization.

The results of the audits must be documented. An overall evaluation by the Quality Organization's management and presented to the company's top management is an essential element. One word of caution must be added to enhance Quality Audits as a usable management tool. There must be flexibility: changing the checklist questions should be allowed. It must be remembered that the Audit is to ascertain the degree of compliance with procedures that functional personnel believe is necessary for *them* to know that desired product quality is being attained. Through these audits it is possible to know that the agreed-to Quality Program is being achieved on a continuing basis.

35.29 TRACEABILITY

Traceability requires an extremely high level of justification before implementation. It is a costly endeavor and in many cases requires drastic changes in internal procedures. These procedures impact purchasing, material control, storage, stockroom control, assembly, rework, testing, record keeping, and shipping.

Traceability is usually defined as the ability to trace the entire history of the material used to produce a deliverable item. It has inadvertently been specified when what was really required was good material controls. Traceability is most often used to analyze fully every product failure in order to prevent a repeat of the failure. Once the cause has been found, traceability provides the ability to segregate products suspected of having similar defects and to take appropriate corrective action. The need to accomplish this with absolute accuracy is the driving force for traceability. Products usually requiring such effort include space equipment, nuclear products, and certain types of medical products—products whose performance is critical.

Any contract containing a requirement for traceability must receive special attention by all operating departments. Records and material identification are essential elements. Traceability has a major impact on stockroom control. The entire method of stocking purchased and fabricated materials must be revised if they are stocked by part number with no recognition of revision level. With a traceability requirement, a stockroom has to store material, parts, and assemblies by revision level and received lot. Records of revision level and received lot must accompany material when it is released from the stockroom. All documents used during the production cycle, with all change, rework, or repair transactions, must accompany a product as it goes through its manufacturing cycle. Readily retrievable records must be retained after finished products have been shipped. Similar requirements apply to subcontractors.

When traceability is a requirement, the role of the Quality Organization involves frequent audits in every operating activity to demonstrate that proper procedures are in effect. It becomes extremely difficult, if not impossible, to reconstruct the detailed records once continuity is lost.

Because of the high cost involved in the implementation of traceability requirements top management approval prior to submitting proposals containing such requirements is considered prudent. After contract award an implementation committee should be established, chaired by a management representative of the Quality Organization.

CHAPTER 36
QUALITY CONTROL

PHILIP S. BRUMBAUGH

Quality Assurance, Inc.
St. Louis, Missouri

Reprinted from *Handbook of Industrial Engineering*, Wiley, New York, 1982, by permission of the publisher.

36.1 SPECIFICATIONS, TOLERANCES, AND ALLOWANCES

36.1.1 Specifications

Product specifications provide procedural instructions for the operator. These instructions should be clear and precise, but where they are not, or where there is conflict with, for example, past shop practice, resolution will have to be obtained at a higher level than that of the operator. Product specifications also provide standards for test and inspection. Again, clarity and precision are essential; where they are lacking, the inspection department will tend to "fill the void" and make decisions in this area, often with predictably unfortunate consequences for product reliability, safety, and fitness for use.

36.1.2 Tolerances

Tolerance is the maximum amount of variation that can be permitted from a nominal or stated specification. In general, close (or numerically small) tolerances must be maintained on parts that are to be assembled with other parts, and the closer the tolerances, the greater the ease of assembly. This is a particularly critical consideration in mass production situations, where interchangeability of parts is required. However, while close tolerances reduce assembly costs, they increase the cost of fabricating the individual parts. Hence a balance of the two types of costs must be achieved, with the objective of minimizing their sum. Expressed differently, it may at times be economically advantageous to incur the cost of some selective fitting of parts in order to realize the fabrication cost savings that result from a relaxation of tolerances.

36.1.3 Mandatory and Advisory Tolerances

Mandatory tolerances are tolerance requirements that must be met because they directly affect important quality characteristics of the end product. Advisory tolerances are provided for manufacturing convenience; they are designed to assist the operator in achieving the mandatory tolerance requirements. An example might be: "Add 10 oz $\pm$ 0.1 oz (283.49 g $\pm$ 283 g) of compound X to achieve a solution strength of 18% $\pm$ 0.5%." The weight tolerances are advisory; the strength tolerances are mandatory. In the absence of clear distinctions between the two types of tolerances, inspection departments will tend to interpret all tolerances as mandatory.

Unilateral and Bilateral Tolerances

Both unilateral and bilateral tolerances are encountered in practice. A unilateral tolerance permits variation in only one direction from a nominal value; a bilateral tolerance permits variation in both directions. For example, so-called 5% resistors carry ratings of X Ω $\pm$ 5%; this is a bilateral tolerance. Precision resistors, in contrast, carrying ratings of X Ω $^{+1\%}_{-0\%}$, this is a unilateral tolerance. Unilateral tolerances have the advantage of making it relatively simple to maintain a given allowance for mating parts while changing the tolerances. (See the example in the next part of this chapter.)

36.1.4 Allowances

An allowance is defined as the minimum specified clearance between mating parts.

36.1.5 Design Review Teams

Historically, the product design department has had the dominant influence in setting product specifications. With the evolution of increasingly complex products, and the ever-larger capital commitments required for their manufacture, it has become necessary to broaden the responsibility for determining specifications. Establishment of a design review team is a prevalent response to their requirement. In addition to representation from product design, the design review team may comprise, but need not be limited to, the following:

Market research specialist—what specifications do the customers really require?

Manufacturing specialist—what are the production cost implications of specification decisions?

Reliability engineer—what are the product reliability and maintainability implications of specification decisions?

Product safety specialist/lawyer—what are the safety and product liability implications of specification decisions?

Quality control engineer—what are the inspection and quality cost implications of specification decisions?

36.2 NATURAL AND ENGINEERING TOLERANCE LIMITS

36.2.1 Natural Tolerance Limits

True natural tolerance limits for a normally distributed variable are located three process standard deviations (SDs) from the process mean ($\mu \pm 3\sigma$). Numerical values for μ and σ, however, are rarely known with certainty. Even when calculations are based on a large amount of data, the resulting values cannot be considered perfectly accurate because of the likely presence of sampling error.

36.2.2 Engineering Tolerance Limits

In practice, engineering tolerances can be established through the use of statistical tolerance limits. The term "statistical" indicates that an estimation procedure is involved, utilizing sample data. If it can be assumed that the manufacturing process yields dimensions that are normally distributed, then several alternative approaches are available.

A common procedure is to take a random sample of n items from the process, measure each item for the dimension in question, and then compute the sample mean $\bar{x}$ and the sample SD s. The statistical tolerance limits are set at $\bar{x} \pm Ks$, where K is a factor obtained from a standard table. Tables of K are available from most standard references on quality control, for example, Bowker and Lieberman[1] and Juran et al.[2] The numerical value of K is governed by the sample size used, the confidence level chosen, and the proportion of the population to be included within the limits.

Two alternative techniques for establishing statistical tolerance limits under the assumption of normality make use of the sample range R and the mean of several sample ranges $\bar{R}$; these approaches avoid the need to calculate s, the sample SD. When the assumption of normality is not justified, a distribution-free method of establishing tolerances is available. For a detailed discussion of all of these techniques, see Section 22 of Juran et al.[2]

36.3 MERGER OF TOLERANCES

36.3.1 Additive Parts

Merger of tolerances occurs when parts are assembled in an additive manner. The conventional method of calculating the tolerance on the assembled dimension is to add the individual part tolerances. Thus, if an assembly A is formed by additively combining parts X, Y, and Z, then the tolerances T are conventionally related by

$$TA = TX + TY + TZ$$

The statistical approach recognizes that simultaneous occurrences of extreme values are very unlikely. The merged tolerance is then computed as

$$TA = \sqrt{(TX)^2 + (TY)^2 + (TZ)^2}$$

This method is based on the fact that, when random variables are added, their variances (which are squared quantities) are additive, but their SDs (which are first-power quantities) are not.

36.3.2 Mating Parts

Application of the principle of merger of tolerances is not limited to situations involving additive parts. In cases where assembled parts are mating, the concept is equally valid.

36.4 DEGREES OF SERIOUSNESS

When a product or service is evaluated in terms of customer usage, safety and economic considerations are vital. Any nonconformity that occurs with a severity sufficient to cause a product or service not to satisfy intended normal or reasonably foreseeable usage requirements is termed a "defect."[3] It is frequently useful in evaluating the product or service to employ the following classification of defects by degree of seriousness:

Class 1 (Very Serious).   Leads directly to severe injury or catastrophic economic loss.

Class 2 (Serious).   Leads directly to significant injury or significant economic loss.

Class 3 (Major).   Related to major problems with respect to intended normal or reasonably foreseeable use.

Class 4 (Minor). Related to minor problems with respect to intended normal or reasonably foreseeable use.

As part of such a rating system, it is sometimes advantageous to employ a modifier to describe the likelihood that the potential defect will be found in the operation of the product or service. The modifiers that are used form a likelihood continuum, with the terms "virtually certain" and "virtually no chance" at the extremes. The intermediate modifiers all relate to a 50% chance of occurrence and are termed "substantially above," "somewhat above," "around," "somewhat below," and "substantially below."

Other uses of classification systems are (1) to rank quality parameters according to their importance and (2) to determine an index of quality for purposes of process control. In conjunction with a rating or classification procedure, it is customary to use penalties or demerits, which are essentially arbitrary units that serve as weighting factors.

36.5 PROCESS CAPABILITY

36.5.1 Basic Concept

Process capability is the measured, inherent reproducibility of the product turned out by a process (Ref. 2, p. 16). This statement refers to a specific manufacturing process and is concerned with the degree of product homogeneity resulting from a state of statistical control. A process is in a state of statistical control when all variation in the output variable(s) can be attributed to chance. Expressed differently, statistical control exists in the absence of assignable (specific and identifiable) causes of variation. The rationale for studying process capability is that it is essential to be able (1) to predict whether satisfactory products will be produced by a given process and (2) to diagnose the problems when a process fails to produce satisfactory products.

36.5.2 Process Capability Measurement

Process capability can be measured in several ways; the most common technique is through the use of control charts. In this approach at least 10 subgroups, each customarily containing five consecutively produced items, are selected. Appropriate measurements are made on each item, and for each subgroup the values of the mean $\bar{x}$ and the range R are calculated and are plotted on $\bar{x}$ and R charts (this is discussed later in this chapter). If statistical control appears to be present, then the process capability is computed as six SDs, which can be estimated from the average range $\bar{R}$ as $6\bar{R}/d_2$. The value of d_2 is a function of n, the number of items in each subgroup; see Table 36.1. If statistical control does not appear to be present, then the assignable cause(s) of variation must be identified and eliminated and the process capability measured again. If the apparent lack of control is slight, it is often ignored.

36.5.3 Process Capability Analysis

Once process capability has been measured, it must be related to product tolerance; this procedure is known as "process capability analysis." One use of such analysis is predictive: Will process X be able to produce items that meet the required tolerance? The basic technique is to compare the computed six SD value to the tolerance. If this value is less than the tolerance, the process is capable of meeting the specifications; if not, the process is not capable (see Fig. 36.1).

The other use of process capability analysis is diagnostic: Why is process X not able to produce items that meet tolerance requirements? There are many potential reasons, such as improper setup, faulty measurements, commingling of products, and the occurrence of sudden or assignable causes of variation. Detailed discussion of the most widely used techniques of analysis is available in Juran et al.[2]: graphic methods, Section 16; evolutionary operation, Section 27A; and response surface methodology, Section 28. In addition, experimental design is a very valuable technique. General statistical concepts, along with the considerations involved in the planning of tests, are discussed in Juran and Gryna.[5]

36.6 PROCESS CONTROL—FRACTION NONCONFORMING

36.6.1 The p-Charts

The statistic p is the fraction of nonconforming units contained in a sample; the p-chart "tracks" its value through time in order to control the fraction nonconforming p' in the process output. The sampling distribution of p has a standard error $\sigma_p = \sqrt{p'(1-p')/n}$, where n is the size of the sample (or subgroup). In reality the value of p' is not likely to be known, and it is usually estimated by the statistic $\bar{p}$, which is the average value of p computed from a number of samples. The standard error σ_p is then estimated by $s_p = \sqrt{\bar{p}(1-\bar{p})/n}$. In setting up a control chart, a number k of

Table 36.1 Factors for Estimating σ' from $\overline{R}$ or $\overline{\sigma}$ (From Ref. 4)

Number of Observations in Subgroup n	d_2 Factor for Estimating σ' from $\overline{R}$ $(\sigma' = \overline{R}/d_2)$	c_2 Factor for Estimating σ' from $\overline{\sigma}$ $(\sigma' = \overline{\sigma}/c_2)$	Number of Observations in Subgroup n	d_2 Factor for Estimating σ' from $\overline{R}$ $(\sigma' = \overline{R}/d_2)$	c_2 Factor for Estimating σ' from $\overline{\sigma}$ $(\sigma' = \overline{\sigma}/c_2)$
2	1.128	0.5642	21	3.778	0.9638
3	1.693	0.7236	22	3.819	0.9655
4	2.059	0.7979	23	3.858	0.9670
5	2.326	0.8407	24	3.895	0.9684
			25	3.931	0.9697
6	2.534	0.8686	30	4.086	0.9748
7	2.704	0.8882	35	4.213	0.9784
8	2.847	0.9027	40	4.322	0.9811
9	2.970	0.9139	45	4.415	0.9832
10	3.078	0.9227	50	4.498	0.9849
11	3.173	0.9300	55	4.572	0.9863
12	3.258	0.9359	60	4.639	0.9874
13	3.336	0.9410	65	4.699	0.9884
14	3.407	0.9453	70	4.755	0.9892
15	3.472	0.9490	75	4.806	0.9900
16	3.532	0.9523	80	4.854	0.9906
17	3.588	0.9551	85	4.898	0.9912
18	3.640	0.9577	90	4.939	0.9916
19	3.689	0.9599	95	4.978	0.9921
20	3.735	0.9619	100	5.015	0.9925

Reproduced by permission from *ASTM Manual on Presentation of Data*, American Society for Testing and Materials, Philadelphia, 1945.

[a]Estimate of $\sigma' = \overline{R}/d_2$ or $\overline{\sigma}/c_2$ where $\overline{R}$ is the average of sample ranges and $\overline{\sigma}$ is the average of sample SDs.

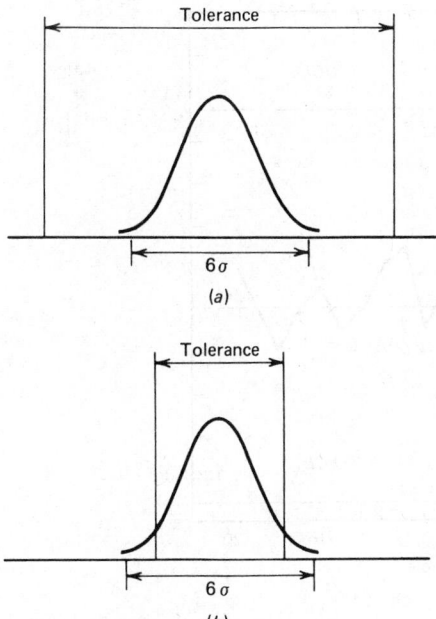

Fig. 36.1 Illustration of (a) adequate process capability and (b) inadequate capability.[4]

subgroups (usually $k = 20$), each containing n units, are selected in sequence and inspected for attributes. The values of $\bar{p}$ and s_p are calculated; the central line on the control chart is established at $\bar{p}$, and the trial control limits for p are set at $\bar{p} \pm 3s_p$. These "three-sigma" limits will not necessarily enclose 99.73% of the distribution of p, since p is not normally distributed. However, we may say that, if the process is in control with respect to fraction nonconforming, it is very unlikely that a computed value of p will fall outside of the limits. The principal reasons for a lower limit in controlling fraction nonconforming are to detect instances of careless inspection or the occurrence of a significant improvement in the manufacturing process. (See Fig. 36.2.)

The next step is to plot all k values of p in *proper sequence* on the chart. If any value of p falls outside a control limit, it is necessary to determine whether it can be explained by an assignable cause. If so, that value of p is discarded, $\bar{p}$ and s_p are recomputed on the basis of the remaining subgroups, and new trial control limits are established. If not, it is concluded that k is too small, and additional data are needed. If all plots of p are within the limits, we then check for randomness of the sequence. This can be done roughly by visual inspection of the pattern on the chart or statistically by means of a test for runs. If nonrandomness is present, we must identify and eliminate the causes and revise the control limits accordingly.

After the preceding steps have been accomplished, and assuming the average fraction $\bar{p}$ is satisfactory, the trial control limits can be tentatively accepted, and the chart can be used for controlling the process. Periodically, a subgroup of n items is inspected, and the value of p is computed and plotted on the chart. If a plot of p falls outside a control limit, or if there appears to be a drift toward a control limit, it is concluded that an assignable cause of variation has occurred; it is necessary to investigate and take appropriate corrective action. As long as the plots remain between the control limits and the sequence appears random, it is concluded that the process remains in a state of statistical control. As more data are accumulated, of course, the control limits can be further refined. The frequency of sampling is governed by the per unit cost of sampling and inspection, the cost of a "false alarm" (type 1 error), and the cost of operating out of control (type 2 error).[6]

36.6.2 Stabilized p-Charts

When the sample or subgroup size varies, the control limits can be made constant by standardizing the sample fraction nonconforming p as follows:

$$p^* = (p - p'')/\sigma_p''$$

where p^* = standardized sample fraction nonconforming
   p'' = specified or "goal" process fraction nonconforming
   $\sigma_p'' = \sqrt{p''\,(1 - p'')/n}$

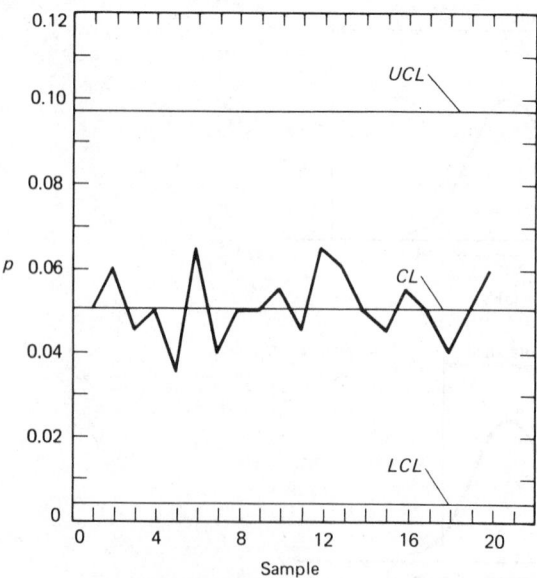

Fig. 36.2 A p-chart.[4]

The net effect of this procedure is that the variable being plotted is expressed in SD units. A stabilized p-chart always has its central line set at zero, and the upper and lower control limits are at $+3$ and -3, respectively.

36.6.3 The np-Charts

In situations where the subgroup size is constant, the use of an np-chart may be appropriate. The statistic np, which is plotted on the chart, is simply the observed number of nonconforming units in the subgroup; hence the computational step in the p-chart procedure of dividing the number of nonconforming units by n is avoided. The CL (control limit) and the UCL (upper control limit) and LCL (lower control limit) are computed as follows:

$$\text{CL} = n\bar{p}$$
$$\text{UCL} = n\bar{p} + 3s_{np}$$
$$\text{LCL} = n\bar{p} - 3s_{np}$$

where $s_{np} = \sqrt{n\bar{p}(1 - \bar{p})}$.

It is sometimes felt that the np-chart, in comparison with the p-chart, is more easily understood by shop personnel. However, this advantage may be more than offset by the added complexity of using both types of charts, since the np-chart has only limited applicability.

36.7 PROCESS CONTROL—NUMBER OF NONCONFORMITIES

36.7.1 The c-Charts

In some cases it may be difficult to identify discrete items for inspection. For example, electrical wire might be inspected for pinhole nonconformities in the insulation. It would be impossible to determine the number of locations on the insulation where a nonconformity might occur and to identify them. Therefore, instead of attempting to quantify the fraction of nonconforming locations, it is more logical to select some length of wire as an inspection unit and to count the number of nonconformities.

This variable, which is symbolized by c, tends to follow the form of the Poisson distribution (Ref. 7, p. 418). Theoretically, the central line of a c-chart will be at c', the process average number of nonconformities per inspection unit, and the control limits will be at $c' \pm 3\sigma_c$, where $\sigma_c = \sqrt{c'}$. In reality, of course, the value of c' is usually not known; its value is estimated by $\bar{c}$, the sample average number of nonconformities per inspection unit (generally computed from 20 inspection units). The trial control limits are then set at $\bar{c} \pm 3s_c$, where $s_c = \sqrt{\bar{c}}$. The principal reason for a lower limit in controlling the number of nonconformities is to detect instances of careless inspection or the occurrence of a significant improvement in the manufacturing process. (See Fig. 36.3.)

The 20 values of c are plotted *in proper sequence* on the chart. If any value of c falls outside a control limit, it is necessary to determine whether there is an assignable cause. If not, it is an indication that further data are needed. If so, the value of c outside the limit can be discarded, and c and the control limits are recomputed on the basis of the remaining data. When the control limits contain all the plots, it is necessary to determine whether the plots form a random sequence. This can be done roughly by visual examination of the pattern of the plots or statistically by a test for runs. After all of the preceding steps have been accomplished, the trial control limits may be tentatively accepted, and the chart may be used for controlling the process. As additional data are accumulated, of course, the control limits can be further refined.

36.7.2 The u-Charts

At times it may not be possible to fix the quantity of product inspected at a constant amount. This would be true, for example, where 100% inspection of output is required, since the volume of production will undoubtedly vary from day to day. It is advantageous to treat the sample as being composed of k inspection units, each of fixed size and hence containing equal opportunity for nonconformities to occur. The number of nonconformities per sample c can then be converted to the average number of nonconformities per inspection unit $u = c/k$. This procedure permits the establishment of a chart with a CL of constant value, $\bar{u}$. The UCL and LCL will vary from sample to sample as the number of inspection units varies.

$$\text{UCL} = \bar{u} + 3\sqrt{\bar{u}/k}$$
$$\text{LCL} = \bar{u} - 3\sqrt{\bar{u}/k}$$

In setting up a u-chart, the same procedural considerations are involved as with a c-chart.

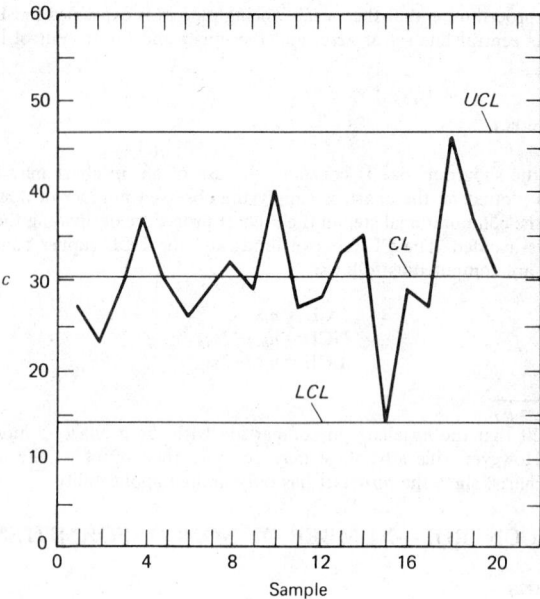

Fig. 36.3 A c-chart.[4]

36.8   PROCESS CONTROL—VARIABLES

36.8.1   The $\bar{x}$-Charts

In many cases it is economically advantageous to control a process on the basis of variables rather than of fraction nonconforming. The reason is that, since a set of measurements contains more information than the same number of counted observations, the same degree of control over a process can be maintained with fewer measurements than counts. Partially offsetting this advantage, however, is that the process of obtaining measurements and performing calculations with them may be more time consuming (and more costly) than the simple classification of items as conforming or nonconforming.

When a process is to be controlled on the basis of variables, it is necessary to control both the mean and the variability; the $\bar{x}$-chart is the most widely used technique for controlling the mean. The design of an $\bar{x}$-chart is based on the fact that the sample mean $\bar{x}$ is normally distributed and that it itself has a mean equal to the process mean μ and an SD (or standard error) $\sigma_{\bar{x}} = \sigma/\sqrt{n}$, where σ is the process standard deviation. When the process output is normally distributed, the normality of $\bar{x}$ is exact; when the process output is not normally distributed, the normality of $\bar{x}$ is approximate, with the degree of approximation improving as the sample size n increases. Ideally, if a process is in control with a specified mean μ'' and a specified SD σ'', a control chart utilizing the statistic $\bar{x}$ can be set up with CL at μ'' and three-sigma control limits at $\mu'' \pm 3\sigma_{\bar{x}}''$. This procedure will provide for a "false alarm" probability of 0.0027. (See Fig. 36.4.)

In reality, of course, when control over a process is yet to be established, the values of μ and σ are usually not known. It is necessary to take from the process a series of k samples or subgroups (usually $k = 20$), each containing n units. For each subgroup the mean $\bar{x}$ and the range R are calculated; the range is defined as the difference between the largest measurement and the smallest. The resulting k values of $\bar{x}$ are averaged to form $\bar{\bar{x}}$, and the k values of R are averaged to form $\bar{R}$. An $\bar{x}$-chart is then set up, with CL at $\bar{\bar{x}}$ and UCL and LCL as follows:

$$\text{UCL} = \bar{\bar{x}} + A_2\bar{R}$$
$$\text{LCL} = \bar{\bar{x}} - A_2\bar{R}$$

where A_2 has a value determined from Table 36.2 according to the size n of the subgroups. The product $A_2\bar{R}$ is an estimate of $3\sigma_{\bar{x}}$; alternatively, a computed value of $s_{\bar{x}}$ could be used to estimate $\sigma_{\bar{x}}$, but this is a computationally more cumbersome procedure.

The limits just calculated are trial control limits; the next step is to plot the k values of $\bar{x}$ on the chart *in proper sequence*. At this point it is necessary to stress that control over the process variability is equally as important as control over the process mean. Therefore, an R-chart should be developed concurrently with the $\bar{x}$-chart (see Section 36.8.2). If any plotted value of $\bar{x}$ falls outside a control

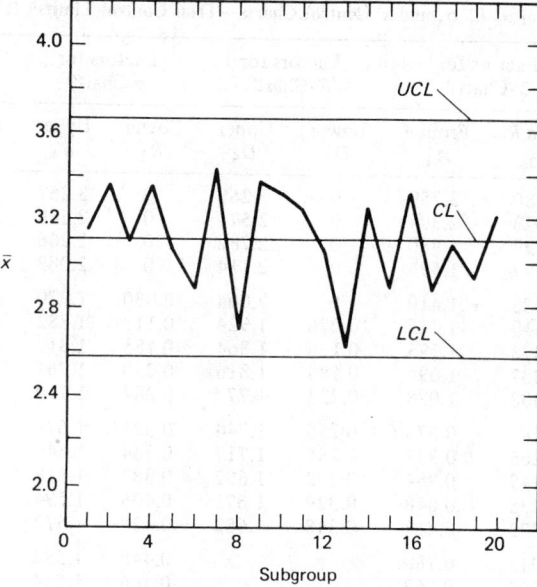

Fig. 36.4   An $\bar{x}$-chart.[4]

limit, it is necessary to determine whether there is an assignable cause. If not, it is an indication that further data are needed. If so, the value of $\bar{x}$ outside the limit can be discarded, and $\bar{\bar{x}}$, $\bar{R}$, and the control limits are recomputed on the basis of the remaining data. When the control limits contain all the plots, it is then necessary to determine whether the plots form a random sequence. This can be done roughly by visual examination of the pattern of the plots or statistically by a test for runs. After all of the preceding steps have been accomplished, the trial control limits may be tentatively accepted, and the chart used for controlling the process mean. As additional data are accumulated, of course, the control limits can be further refined. For a discussion of the economic factors to be considered in designing an $\bar{x}$-chart, see Saniga.[8]

36.8.2   The R-Charts

The purpose of an R-chart is to maintain control over the variability of a process. A series of k subgroups (usually $k = 20$), each containing n units, is selected from the process. For each subgroup the range R is calculated; the range is defined as the difference between the largest measurement and the smallest. The resulting k values of R are averaged to form $\bar{R}$. An R-chart is then set up, with CL at $\bar{R}$, UCL at $D_4\bar{R}$, and LCL at $D_3\bar{R}$. The factors D_3 and D_4 have values determined from Table 36.2 according to the size n of the subgroups. The product $D_4\bar{R}$ establishes a control limit that is approximately three standard errors of R above $\bar{R}$, and the product $D_3\bar{R}$ establishes a control limit that is approximately three standard errors of R below $\bar{R}$. The principal reason for an LCL on the process variability is to detect instances of careless inspection or the occurrence of a significant improvement in the manufacturing process. (See Fig. 36.5.)

These limits are trial control limits; the next step is to plot the k values of R on the chart *in proper sequence*. At this point it is necessary to stress that control over the process mean is equally as important as control over the process variability. Therefore an $\bar{x}$-chart should be developed concurrently with the R-chart (see Section 36.8.1). If any plotted value of R falls outside a control limit, it is necessary to determine whether there is an assignable cause. If not, it is an indication that further data are needed. If so, the value of R outside the limit can be discarded, and $\bar{R}$ and the control limits are recomputed on the basis of the remaining data. When the control limits contain all the plots, it is necessary to determine whether the plots form a random sequence. This can be done roughly by visual examination of the pattern of the plots or statistically by a test for runs. After all of the preceding steps have been accomplished, the trial control limits may be tentatively accepted, and the chart used for controlling the process variability. As additional data are accumulated, of course, the control limits can be further refined. For a discussion of the economic factors to be considered in designing an R-chart, see Saniga.[8]

Table 36.2   Factors for $\bar{x}$, R, σ, and x Control Charts—Trial Control Limits[a] (From Ref. 4)

Number of Observations in Subgroup	Factors for $\bar{x}$-Chart[b]		Factors for R-Chart[c]		Factors for σ-Chart[d]		Factors for x-Chart[e]	
	From $\bar{R}$ A_2	From $\bar{\sigma}$ A_1	Lower D_3	Upper D_4	Lower B_3	Upper B_4	From $\bar{R}$ E_2	From $\bar{\sigma}$ E_1
2	1.880	3.759	0	3.268	0	3.267	2.660	5.318
3	1.023	2.394	0	2.574	0	2.568	1.772	4.146
4	0.729	1.880	0	2.282	0	2.266	1.457	3.760
5	0.577	1.596	0	2.114	0	2.089	1.290	3.568
6	0.483	1.410	0	2.004	0.030	1.970	1.184	3.454
7	0.419	1.277	0.076	1.924	0.118	1.882	1.109	3.378
8	0.373	1.175	0.136	1.864	0.185	1.815	1.054	3.323
9	0.337	1.094	0.184	1.816	0.239	1.761	0.010	3.283
10	0.308	1.028	0.223	1.777	0.284	1.716	0.975	3.251
11	0.285	0.973	0.256	1.744	0.321	1.679	0.946	3.226
12	0.266	0.925	0.284	1.717	0.354	1.646	0.921	3.205
13	0.249	0.884	0.308	1.692	0.382	1.618	0.899	3.188
14	0.235	0.848	0.329	1.671	0.406	1.594	0.881	3.174
15	0.223	0.817	0.348	1.652	0.428	1.572	0.864	3.161
16	0.212	0.788			0.448	1.552		
17	0.203	0.762			0.466	1.534		
18	0.194	0.738			0.482	1.518		
19	0.187	0.717			0.497	1.503		
20	0.180	0.698			0.510	1.490		
21	0.173	0.680			0.523	1.477		
22	0.167	0.662			0.534	1.466		
23	0.162	0.647			0.545	1.455		
24	0.157	0.632			0.555	1.445		
25	0.153	0.619			0.565	1.435		
Over 25	†	†			†	†		

[a] Factors reproduced from 1950 *ASTM Manual on Quality Control of Materials* by permission of the American Society for Testing and Materials, Philadelphia. All factors in exhibit are based on a normal distribution.

[b] $UCL_{\bar{x}} = \bar{x} + A_2\bar{R}$ and $LCL_{\bar{x}} = \bar{x} - A_2\bar{R}$ or $UCL_{\bar{x}} = \bar{x} + A_1\bar{\sigma}$ and $LCL_{\bar{x}} = \bar{x} - A_1\bar{\sigma}$.

[c] $UCL_R = D_4\bar{R}$ and $LCL_R = D_3\bar{R}$.

[d] $UCL_\sigma = B_4\bar{\sigma}$ and $LCL_\sigma = B_3\bar{\sigma}$.

[e] $\bar{x} + E_2\bar{R}$ and $\bar{x} - E_2\bar{R}$ or $\bar{x} + E_1\bar{\sigma}$ and $\bar{x} - E_1\bar{\sigma}$.

† Values of these constants may be determined for larger sample sizes from formulas given in J. M. Juran, F. M. Gryna, and R. S. Bingham, Jr., *Quality Control Handbook*, 3rd ed., McGraw-Hill, New York, Appendix I, pp. 1–7.

36.8.3   Other Variables Charts

Additional chart techniques have been developed for special situations. For example, when the subgroup size n exceeds 12, the sample range R tends to lose efficiency. In such cases the subgroup can be subdivided, and the mean range $\bar{R}$ for the subdivisions can be plotted on a chart. Further techniques for controlling process variability include the s-chart for sample SDs and the s^2-chart for sample variances. Obtaining values for s and s^2 is computationally tedious, however.

Alternatives to the $\bar{x}$-chart for controlling the process mean include charts for the median and the midrange. These statistics are easy to calculate, but are less efficient than the mean. (Larger sample sizes are needed to attain given risk levels.) For details concerning all of these chart techniques, see Duncan (Ref. 7, pp. 453–455).

36.9   PROCESS CONTROL—OTHER TECHNIQUES

A criticism of the standard (or Shewhart) charts just described is that they provide decision rules based only on the most recent observations. The cumulative sum (cusum) chart permits decisions to

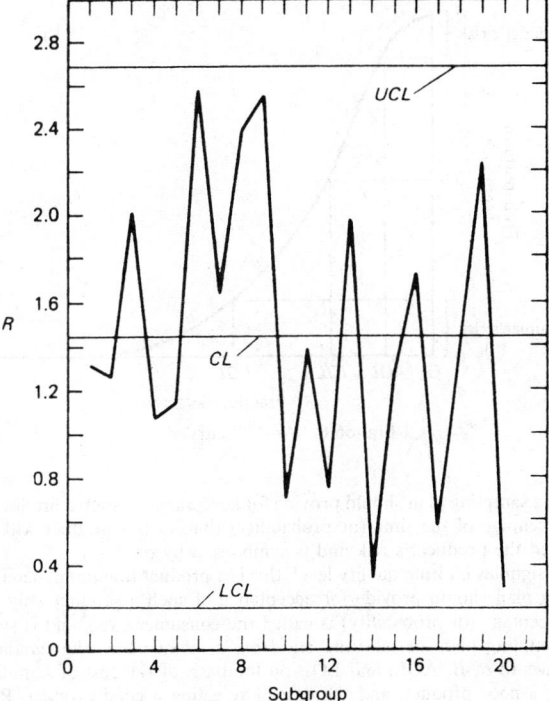

Fig. 36.5 An R-chart.[4]

be based upon more data. Cusum charts can be designed for attributes or for variables, for one-sided decision procedures or for two-sided procedures [see Duncan (Ref. 7, pp. 464–484)]. One study[9] has shown that, when both a cusum chart and a Shewhart chart are of minimum-cost design, there is little difference between them. However, when a Shewhart chart is nonoptimal, a comparable cusum chart will result in lower cost. A cusum chart, though, is more likely than a comparable Shewhart chart to react to minor process disturbances.

A technique that is useful when there is a natural "drift" in a process (e.g., due to tool wear), or when the inherent variability of a process is much narrower than the tolerance, is the acceptance control chart. Specific designs are treated in Duncan (Ref. 7, pp. 485–502). Adaptive control techniques are widely used in the chemical industry, where the type and amount of corrective action must be predicted; see Juran et al. (Ref. 2, pp. 29-41–29-48).

36.10 LOT ACCEPTANCE PLANS—ATTRIBUTES

36.10.1 Single-Sample Plans

In acceptance sampling a decision to accept or reject a collection of items—called a "lot"—is based upon the inspection of a sample of items taken from the lot. In many cases sampling is economically advantageous because inspection of the entire lot would be too costly. Sometimes, of course, sampling is necessary, as in those cases where testing is destructive.

A single-sample plan is defined by specifying values for n, the sample size, and c, the acceptance number. If the observed number of nonconforming units in the sample does not exceed c, the lot is deemed acceptable. If the observed number of nonconforming units does exceed c, then the lot is rejected. The properties of a given acceptance sampling plan are customarily shown by its operating characteristic (OC) curve. This is a graph in which the probability of accepting a lot is plotted against the lot or process fraction nonconforming. It is often useful to distinguish between a type-A OC curve, in which we are concerned with an isolated lot, and a type-B OC curve, in which we are concerned with a series of lots formed essentially at random from the output of some process (Ref. 7, pp. 157–161). No such distinction will be made in this chapter; all OC curves will be considered as type B. An OC curve is shown in Fig. 36.6.

In the figure AQL signifies the acceptable quality level; this is a product fraction nonconforming

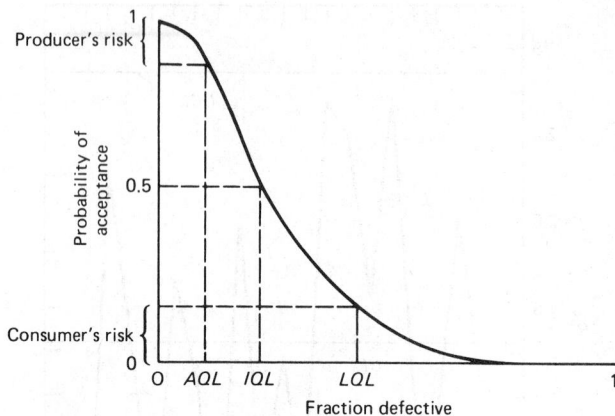

sufficiently small that a sampling plan should provide for acceptance of such a product a large percentage of the time. The percentage of the time (or probability) that such a product will be rejected by the sampling plan is called the producer's risk and is symbolized by α.

The symbol LQL signifies limiting quality level; this is a product fraction nonconforming sufficiently large that a sampling plan should provide for acceptance of such a product only a small percentage of the time. This percentage (or probability) is called the consumer's risk and is symbolized by β.

In designing a sampling plan, or selecting one from a set of plans, the standard procedure is to assign numerical values to α, β, AQL, and LQL on the basis of the cost of sampling and inspection, the cost of accepting a poor product, and the cost of rejecting a good product. Recent research has been directed toward optimization, that is, the design of sampling plans that will result in the best combinations of expected outgoing quality and expected cost of quality assurance.[10]

36.10.2 Military Standard 105D

Standard military procedures for acceptance sampling by attributes have been in existence for almost 40 years. The current version, designated MIL-STD-105D, was issued by the U.S. Department of Defense in 1963.[11] This standard is essentially a set of sampling plans indexed with respect to AQL. The selection of a specific plan from MIL-STD-105D depends on the lot size, the AQL that has been chosen, and the inspection level that has been chosen.

There are three general levels and four special levels of inspection; the level to be used in a given application depends on the degree of discrimination needed between good and bad lots, on the cost of the item, and on whether testing is destructive. There is a provision for a mandatory shift to tightened inspection when it appears that product quality has deteriorated significantly and a provision for an optional shift to reduced inspection when it appears that product quality is exceptionally good on a consistent basis. The standard recognizes the classification of defects into critical, major, and minor and also makes provision for double- and multiple-sample procedures if desired. Selected type-B OC curves are provided, as are average sample size curves for the double and multiple plans without curtailed inspection. In its mode of operation, MIL-STD-105D pushes the manufacturer to supply a product at least as good as the specified AQL.

36.10.3 Double-Sample Plans

In the area of sampling and inspection costs, a double-sample plan can provide significant savings over a comparable single-sample plan. Such a plan is defined by specifying values for the five numbers n_1, n_2, c_1, c_2, and c_3, with $c_1 < c_2$ and $c_2 < c_3$. If in a sample of n_1 units taken from a lot the number of nonconforming units is less than or equal to c_1, the lot is accepted. If the number of nonconforming units exceeds c_2, the lot is rejected. If the number of nonconforming units exceeds c_1, but does not exceed c_2, a second sample of size n_2 is taken. If there are c_3 or fewer nonconforming units in the combined samples, the lot is accepted; if there are more than c_3 nonconforming units, the lot is rejected. In many cases the value of c_2 is set equal to that of c_3.

The cost savings with respect to a comparable single-sample plan arise from two sources. (Comparable plans are those that have essentially equal values for AQL, LQL, α, and β.) First, n_1 for the double-sample plan is smaller than n for the single-sample plan, and therefore whenever the lot is accepted or rejected on the basis of the first sample, the amount of inspection will be less. Second, when inspecting the second sample, it is possible to utilize curtailed inspection when the lot is rejected. This means that inspection ceases as soon as c_3 has been exceeded. These relationships are shown graphically, without reference to any specific sampling plans, in the average sample size curves in Fig. 36.7. The average sample size is the weighted average of n_1 and $n_1 + n_2$, where the weights are the probabilities of only one sample being required and of a second sample being required. These probabilities are functions of the product fraction nonconforming.

Tables of double-sampling plans are given in MIL-STD-105D, discussed previously. For a detailed discussion of the design of double-sample plans, see Ref. 7, pp. 179–198.

36.10.4 Multiple-Sample Plans

A lot acceptance sampling plan containing more than two stages is known as a "multiple-sample plan" or, at times, as "group sequential sampling." There are typically seven or eight stages, at each of which a sample of n units is taken for inspection, unless an accept or reject decision was reached at the preceding stage. Thus at each stage the cumulative sample size is n, $2n$, $3n$, and so on.

In addition, at each stage the plan provides an acceptance number and a rejection number whose values, along with the value of n, are determined by the choice of AQL, LQL, α, and β. The rejection number is always larger than the acceptance number. The decision rule at each stage is to accept the lot if the total number of nonconforming units in the combined samples through that stage is less than or equal to the acceptance number. A decision to reject the lot is made if the number of nonconforming units is equal to or greater than the rejection number. If the number of nonconforming units is greater than the acceptance number and less than the rejection number, proceed to the next stage. A decision is forced at the last stage because the rejection number is equal to the acceptance number plus 1.

The inspection cost associated with a multiple-sample plan can be significantly less than that with a comparable single-sample plan because the average sample size of the former tends to be less than the fixed sample size for the latter. (See the discussion of the average sample size curve in the previous subsection.) This holds true even without curtailed inspection in the second and subsequent stages. The inspection cost advantage will be offset to some degree by the greater administrative complexity. Tables of multiple-sample plans are provided in MIL-STD-105D. For details concerning the construction of multiple-sample plans see Ref. 7, pp. 199–208.

36.10.5 Sequential Sampling Plans

When items are taken from a lot one at a time for inspection, the procedure is known as item-by-item sequential sampling. It was noted previously that, when the sample size n and the acceptance number c are fixed, it is not always possible to hold both α and β exactly at the desired levels because of the integer-value constraints on n and c. Sequential sampling permits α and β to be fixed very close to the desired values because the sample size n varies from one lot to another. At each sampling step there is an acceptance number and a rejection number whose values are determined by the values that have been assigned to AQL, LQL, α, and β. When the cumulative number of nonconforming units equals or falls below the acceptance number, the lot is accepted; when the cumulative number of nonconforming units equals or exceeds the rejection number, the lot is rejected. For values in between, the sampling is continued.

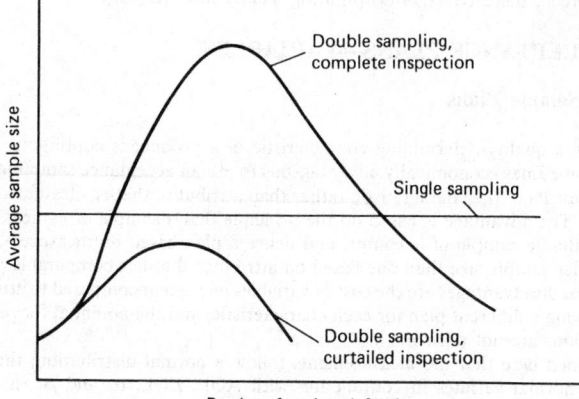

Fig. 36.7 Average sample size curves.[4]

A major advantage of sequential sampling is that inspection economies can often be realized. It has been estimated, for example, that the average sample size for a sequential sampling plan may be as low as one-half of the fixed sample size required in a comparable single-sample plan.[12] For details concerning the construction of sequential sampling plans, see Ref. 7, pp. 179–198.

36.11 LOT ACCEPTANCE PLANS—RECTIFICATION

36.11.1 Rectifying Inspection

In sampling inspection by attributes, a rejected lot may be subjected to screening or 100% inspection in order to remove all nonconforming units. This general procedure is called "rectifying inspection." The specific procedure used in any given situation may follow any one of several variations on the general theme: If a lot passes sampling inspection, the nonconforming units in the sample may or may not be removed, and, if removed, may or may not be replaced by conforming units; when a rejected lot is screened, the removed nonconforming units may or may not be replaced by conforming units. The purpose of rectification is to establish quality assurance in the sense of setting an upper limit on the outgoing fraction nonconforming. This fraction is called the "average outgoing quality" (AOQ) and is a weighted average of the fraction nonconforming contained in the rectified and the unrectified lots. The fraction nonconforming in rectified lots is taken to be zero, and that in the unrectified lots is the average fraction nonconforming turned out by the process from which the lots are formed. The weights are the probabilities of acceptance as given by the OC curve for the plan that is being used; these probabilities, of course, are a function of the product fraction nonconforming.

The equation for computing AOQ depends on which of the specific rectifying procedures described is being used. In all cases the AOQ function will have a maximum, called the "average outgoing quality limit" (AOQL), because when product quality is good, the unrectified lots will contain good quality, and when product quality is poor, most of the lots will be rectified. An AOQ curve, without reference to any specific sampling plan, is shown in Fig. 36.8.

Inspection cost can be taken into account in selecting a plan through the calculation of the average total inspection (ATI):

$$ATI = n + (1 - P_a)(N - n)$$

where n = sample size
N = lot size
P_a = probability of acceptance for a given incoming quality level (from the OC curve)

In addition to the single-sample application, rectifying inspection can also be employed with double-sample and multiple-sample attributes plans and with variables plans (see Ref. 7, pp. 333–357).

36.11.2 Dodge–Romig Plans

Tables have been developed that give sampling plans for rectifying inspection.[13] These are known as the Dodge–Romig tables and are indexed by AOQL. Both single-sample and double-sample plans are presented, and it is assumed that all rejected lots are screened. Unlike MIL-STD-105D, the Dodge–Romig tables do not provide for the classification of nonconformities and do not contain procedures for tightened and reduced inspection. The Dodge–Romig tables also provide sampling plans indexed by lot tolerance percent defective (LTPD) [limiting quality level (LQL)].

36.12 LOT ACCEPTANCE PLANS—VARIABLES

36.12.1 Single-Sample Plans

In situations where a quality-determining characteristic of a product is continuous and has a known distribution, it is sometimes economically advantageous to use an acceptance sampling plan that utilizes the actual measurements of the characteristic rather than attributes, that is, classification as conforming or nonconforming. The advantage is based on the principle that a sample of n measurements contains more information than a sample of n counts, and hence a plan based on measurements (or variables) will require a smaller sample size than one based on attributes that has comparable properties (similar OC curve). Principal disadvantages are the cost of variables inspection compared to attributes inspection, the necessity of having a different plan for each characteristic, and the potential for serious inaccuracies when the distributions are not normal.

It will be assumed here that the measurements follow a normal distribution; this assumption will permit the use of normal variates in conjunction with AQL, LQL, α, and β. In a situation where the SD σ of the process from which lots are formed is assumed to be constant, then the fraction nonconforming contained in such lots will be determined by the mean of the process. If there is a

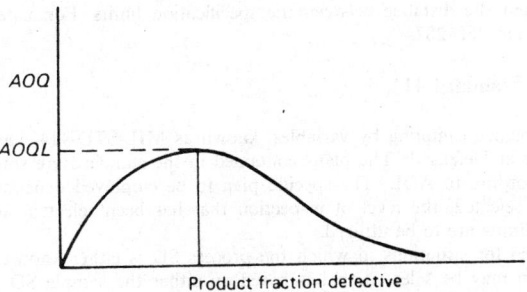

Fig. 36.8   An AOQ curve.[4]

single specification limit, for example, a lower limit L, then the process mean μ_1, which yields a lot fraction nonconforming equal to AQL, can be transformed into a standard normal variate z_1 as follows:

$$\frac{\mu_1 - L}{\sigma} = z_1$$

Similarly, the process mean μ_2, which yields a fraction nonconforming equal to LQL, can be transformed as follows:

$$\frac{\mu_2 - L}{\sigma} = z_2$$

The statistic used for acceptance or rejection of a lot will be the sample mean $\bar{x}$, and if k represents the critical value of the standardized form of $\bar{x}$, then

$$(k - z_1) \sqrt{n} = -z_\alpha$$

and

$$(k - z_2) \sqrt{n} = z_\beta$$

The symbol $-z_\alpha$ represents a value of the standard normal variate z that establishes the producer's risk α under the sampling distribution of $\bar{x}$ when the process mean $\bar{\mu} = \mu_1$. Similarly, the symbol z_β represents a value of z that establishes the consumer's risk β under the sampling distribution of $\bar{x}$ when the process mean $\mu = \mu_2$. The sampling plan can then be defined as

$$k = \frac{z_1 - z_\alpha}{\sqrt{n}} \quad\quad \text{or} \quad\quad k = \frac{z_2 + z_\beta}{\sqrt{n}}$$

and

$$n = \left(\frac{z_\alpha + z_\beta}{z_1 - z_2} \right)^2$$

The acceptance sampling procedure will be to take a random sample of n units from the lot, measure each for the appropriate characteristic, and compute the mean $\bar{x}$. If $(\bar{x} - L)/\sigma \geq k$, accept the lot; if $(\bar{x} - L)/\sigma < k$, reject the lot. This procedure is based on the assumption that σ is constant and that its value is known. If σ is constant, but its value is not known, then the computations must be modified as follows:

$$k = \frac{z_\alpha z_2 + z_\beta z_1}{z_\alpha + z_\beta}$$

and

$$n = \left(\frac{1 + k^2}{2} \right) \left(\frac{z_\alpha + z_\beta}{z_1 - z_2} \right)^2$$

Then, letting the symbol s represent the sample SD, the decision rule is: If $(\bar{x} - L)/s \geq k$, accept the lot; if $(\bar{x} - L)/s < k$, reject the lot. Because of the symmetry of the normal distribution, it is a relatively simple matter to modify the preceding techniques to accommodate a situation requiring a single upper specification limit U. The OC curve will resemble that for an attributes plan (see Fig. 36.6).

Variables acceptance sampling plans can be constructed for situations in which there are both upper and lower specification limits. The specific procedure to be used depends on the relative magnitudes of the process SD and the distance between the specification limits. For a detailed discussion, see Ref. 7, pp. 247–267 and 268–282.

36.12.2 Military Standard 414

Procedures for acceptance sampling by variables, known as MIL-STD-414, have been developed by the U.S. Department of Defense.[14] The plans contained in the standard are single-sample plans only and are indexed according to AQL. The specific plan to be employed depends on the lot size, the AQL that has been selected, the level of inspection that has been selected, and whether single or double specification limits are to be utilized.

Plans are provided for situations in which the process SD is either known or unknown, and in the latter case a plan may be selected that is based on either the sample SD or the sample range. Whatever specific plan is employed, the standard contains provisions for shifting to tightened or reduced inspection when appropriate. The rationale for the development of MIL-STD-414 was the possibility of a reduction in inspection costs through the use of smaller samples than would be required under MIL-STD-105D. However, the exercise of care in utilizing MIL-STD-414 is necessary for several reasons:

1. The risks associated with any given plan can be significantly altered if the assumption of normality is not valid.
2. The lot size classes are not the same as in MIL-STD-105D.
3. The rules for a shift to tightened inspection are not the same as in MIL-STD-105D.
4. The behavior of the producer's risk as a function of sample size is not the same as in MIL-STD-105D.

36.13 SPECIAL TECHNIQUES

36.13.1 Skip-Lot Sampling Plans

When an attributes-based lot acceptance sampling plan is being used, it is possible under certain conditions to reduce the total amount of inspection through a skip-lot technique.[15] The procedure is to impose sampling inspection upon only a certain fraction of the lots, provided that a predetermined number of consecutive lots have been accepted. The skip-lot procedure remains in effect until one lot is rejected, at which time all subsequent lots are subjected to the acceptance sampling plan. Reinstatement of the skip-lot procedure is permissible when the consecutive-lot criterion has been satisfied again. Rectifying inspection (see earlier parts of this chapter) can be utilized in conjunction with a skip-lot technique.

36.13.2 Chain Sampling Plans

In situations where the per unit cost of inspection is high, as in the case of destructive testing, the sample size n is of necessity very small. The only meaningful acceptance number is $c = 0$. Every such plan has an OC curve whose second derivative is positive throughout; that is, there is no inflection point. In such cases the plan has very limited ability to distinguish between good and bad lots; in particular, the producer has little protection against chance rejection of lots formed from a satisfactory process.

The chain sampling plan was devised to correct this situation.[16] The procedure is to select an appropriate attributes-based single-sample plan with a small value of n and with $c = 0$. At the same time, select a value for the parameter i to be used in conjunction with the plan ($i = 1, 2, 3, \ldots$). When i consecutive lots have each yielded a sample of n containing no nonconforming units, set $c = 1$ and continue. As soon as a sample is taken that contains one or more nonconforming units, set $c = 0$ and repeat the procedure.

There are certain conditions necessary for proper use of chain sampling:

1. The application must be a continuing one, not an isolated one.
2. The process quality must be essentially stable.
3. The consumer should have no particular reason to suspect poor quality in any given lot.
4. There must be confidence that the supplier will not on occasion deliberately submit a poor lot in the hope that it will pass acceptance sampling when $c = 1$.

36.13.3 Discovery Sampling

On a lot-by-lot basis, not all quality levels are equally likely to occur. Discovery sampling[17] requires the estimation of the fraction of partially nonconforming lots in the total number of lots in order to adjust the sample size periodically. An AOQL-type protection (see earlier part of this chapter) can be achieved, often with smaller sample sizes than would be necessary with conventional lot acceptance sampling plans.

36.13.4 Continuous Sampling Plans

When units are produced in a continuous production system, the formation of inspection lots creates both conceptual and practical difficulties. For this reason, continuous sampling plans have been developed that differ from the lot-by-lot plans described in Sections 36.10 through 36.12. The earliest techniques date from World War II,[18] and the original plan is known as CSP-1.

Values must be selected for two parameters, i (a positive integer) and f ($0 < f < 1$). The operation of the plan begins with 100% inspection until i consecutive units are found to be free of nonconformities. At this point the inspection rate drops from 100% to f until a nonconforming unit is found. The 100% inspection rate is then reinstated, and the cycle begins again. All nonconforming units found are corrected or replaced by conforming units. For any CSP-1 plan, the OC curve, the AOQL, and the average fraction of units inspected (AFI) can be computed.

In addition, a system of classification of nonconformities can be employed (see Section 36.4). Since a process that is in control can produce a random nonconforming unit, modifications known as CSP-2 and CSP-3 require that two out of k nonconforming units be found before reinstating 100% inspection. Multilevel continuous plans also have been developed that provide for decreasing values of f as long as no nonconforming units have been found. Others have developed continuous sampling plans that differ from Dodge's in that (1) units are inspected in groups rather than individually, (2) inspection begins on a sampling basis rather than at the 100% rate, and (3) an AOQL is assured regardless of whether the process is in control. For details, see Girschick[19] and Wald and Wolfowitz.[20]

36.13.5 Compressed Limit Gauging

In using an attributes-based acceptance sampling plan, it is sometimes advantageous to use tighter specification limits than those prescribed by engineering design. The purpose is to reduce the needed sample size and hence lower the total inspection cost. The usefulness of this technique is greatest when both AQL and LQL are small and when the cost of variables inspection is high. However, the risks associated with sampling can be precisely controlled only if there is no marked departure from normality and if the process standard deviation remains constant from lot to lot. The technique is sometimes known as narrow limit gauging or increased severity testing and is sometimes applicable to the construction of p-charts for process control. For further details, see Ref. 7, pp. 264–265.

36.13.6 Bulk Sampling

The most common objective of bulk sampling is to determine the mean quality of the material where the material may be a gas, a liquid, or a solid. If the material is contained in packages, and if it can be assumed that it is uniform within each package, then techniques for discrete-product quality control can be used (see Sections 36.10 through 36.12). However, a different approach is required if the contents of the packages are commingled, if there is reason to suspect that the within-package contents are not uniform, or if the material is not packaged at all, that is, if it is contained in a pipeline or it is stored on the ground in a pile.

For each situation it is necessary to develop a model that will determine the sampling procedure and the acceptance criteria. The model, in turn, requires the establishment of assumptions concerning the statistical properties of the materials. In general, models fall into two groups: (1) models for distinctly segmented bulk material and (2) models for bulk material moving in the stream. See Ref. 2, pp. 25A-1–25A-14, for a comprehensive overview of these models, along with an extensive list of references.

36.14 SYMBOLS

A_2 Factor for $\bar{x}$-chart trial control limits

α Producer's risk

β Consumer's risk

c Number of nonconformities per inspection unit (control charts); acceptance number (acceptance sampling)

a' Process average number of nonconformities per inspection unit

$\bar{c}$ Sample average number of nonconformities per inspection unit

d_2 Factor for estimating σ from $\bar{R}$

D_3 Factor for R-chart lower trial control limit

D_4 Factor for R-chart upper trial control limit

k	Number of subgroups (control charts); critical value of standardized $\bar{x}$ (acceptance sampling)
K	Statistical tolerance factor
L	Lower specification limit
μ	Process mean
n	Sample size
N	Lot size
p	Sample fraction nonconforming
p'	Process fraction nonconforming
p''	Specified value of process fraction nonconforming
$\bar{p}$	Mean of a number of values of p
p^*	Standardized p
P_a	Probability of acceptance
R	Sample range
$\bar{R}$	Mean of a number of values of R
s	Sample SD
s^2	Sample variance
s_c	Estimate of standard error of c
s_p	Estimate of standard error of p
$s_{\bar{x}}$	Estimate of standard error of $\bar{x}$
σ	Process SD
σ_c	Standard error of c
σ_p	Standard error of p
$\sigma_{\bar{x}}$	Standard error of $\bar{x}$
Σ	Summation
u	Average number of nonconformities per inspection unit
$\bar{u}$	Mean of a number of values of u
U	Upper specification limit
$\bar{x}$	Sample mean
$\bar{\bar{x}}$	Mean of a number of values of $\bar{x}$
z	Standardized normal variate

REFERENCES

1. A. H. Bowker and G. J. Lieberman, *Engineering Statistics,* 2nd ed., Prentice-Hall, Englewood Cliffs, NJ, 1972.
2. J. M. Juran, F. M. Gryna, Jr., and R. S. Bingham, Jr., *Quality Control Handbook,* 3rd ed., McGraw-Hill, New York, 1974.
3. American Society for Quality Control, "ANSI/ASQC Standard A2," Milwaukee, WI, 1978.
4. G. Salvendy (Ed.), *Handbook of Industrial Engineering,* Wiley, New York, 1982, pp. 8.3.1–8.3.23.
5. J. M. Juran and F. M. Gryna, Jr., "Statistical Aids for Planning and Analysis Tests," *Quality Planning and Analysis,* McGraw-Hill, New York, 1970.
6. D. C. Montgomery and R. G. Heikes, "Process Failure Mechanism and Optimal Design of Fraction Defective Control Charts," *AIIE Transactions,* 467–472 (Dec. 1976).
7. A. J. Duncan, *Quality Control and Industrial Statistics,* 4th ed., Irwin, Homewood, IL, 1974, p. 405.
8. E. M. Saniga, "Joint Economically Optimal Design of $\bar{x}$ and R Control Charts," *Management Science,* **24**(4), 420–431 (1977).
9. A. L. Goel, "A Comparative and Economic Investigation of $\bar{x}$ and Cumulative Sum Control Charts," unpublished doctoral thesis, University of Wisconsin–Madison, 1968.
10. C. W. Moreno, "A Performance Approach to Attribute Sampling and Multiple Action Decisions," *AIIE Transactions,* 183–197 (Sept. 1979).
11. U. S. Department of Defense, *Sampling Procedures and Tables for Inspection by Attributes,* MIL-STD-105D, Washington, DC, 1963.
12. A. Wald, *Sequential Analysis,* Wiley, New York, 1947, p. 57.
13. H. F. Dodge and H. G. Romig, *Sampling Inspection Tables, Single and Double Sampling,* 2nd ed., Wiley, New York, 1959.

14. U.S. Department of Defense, *Sampling Procedures and Tables for Inspection by Variables for Percent Defective,* MIL-STD-414, Washington, DC, 1957.

15. H. F. Dodge, "Skip-Lot Sampling Plans," *Industrial Quality Control,* 3–5 (Feb. 1955).

16. H. F. Dodge, "Chain Sampling Inspection Plan," *Industrial Quality Control,* 10–13 (Jan. 1955).

17. E. F. Taylor, "Discovery Sampling," *Proceedings of the Ninth Annual ASQC Convention,* ASQC, Milwaukee, WI, 1955.

18. H. F. Dodge, "A Sampling Inspection Plan for Continuous Production," *Annals of Mathematical Statistics,* **14,** 264–279 (1943).

19. M. A. Girschick, *A Sequential Inspection Plan for Quality Control,* Technical Report No. 16, Applied Mathematics and Statistics Laboratory, Stanford University, Stanford, CA, July 1954.

20. A. Wald and J. Wolfowitz, "Sampling Inspection Plans for Continuous Production Which Insure a Prescribed Limit on the Outgoing Quality," *Annals of Mathematical Statistics,* **16,** 30–49 (1945).

BIBLIOGRAPHY

A. V. Feigenbaum, *Total Quality Control—Engineering and Management,* McGraw-Hill, New York, 1961.

E. L. Grant and R. S. Leavenworth, *Statistical Quality Control,* 4th ed., McGraw-Hill, New York, 1972.

PART 4
SYSTEMS, CONTROLS, AND INSTRUMENTATION

CHAPTER 37

SYSTEMS ENGINEERING: ANALYSIS AND DESIGN

ANDREW P. SAGE

George Mason University
Fairfax, Virginia

37.1 INTRODUCTION

Systems engineering is a management technology. Technology involves the *organization* and *delivery* of *science* for the (presumed) betterment of humankind. Management involves the interaction of the *organization* with the *environment.* Here, we interpret environment in a very general sense to include the complete external milieu surrounding individuals and organizations. Hence, systems engineering as a management technology involves *science, organizations,* and their *environments.*

With respect to these, the process of systems engineering involves working with clients in order to assist them in the organization of knowledge to aid in judgment and choice activities. The activities result in the making of decisions and associated resource allocations through enhanced efficiency, effectiveness, equity, and explicability as a result of systems engineering efforts.

The result of a systems engineering study is a set of design, analysis, management, direction, and regulation activities relative to planning, development, production, and operation of total systems in order to maintain overall integrity and integration as related to performance and reliability.

This set of action alternatives is selected from a larger set, in accordance with a value system, in order to influence future conditions. Development of a set of rational policy or action alternatives must be based on formation and identification of candidate alternative policies and objectives against which to evaluate the impacts of these proposed activities, such as to enable selection of efficient, effective, and equitable alternatives for implementation.

Systems engineering is an appropriate combination of mathematical, behavioral, and management theories in a useful setting appropriate for the resolution of complex real world issues of large scale and scope. As such, systems engineering consists of the use of management, behavioral, and mathematical constructs to identify, structure, analyze, evaluate, and interpret generally incomplete, uncertain, imprecise, and otherwise imperfect information. When associated with a value system, this information leads to knowledge to permit decisions that have been evolved with maximum possible understanding of their impacts. A central problem of systems engineering is to select methods that are explicit, rational, and compatible with the implementation framework extant, and the perspectives and knowledge bases of those responsible for decision activities, such that decision making and the resulting policies become as efficient, effective, equitable, and explicable as possible.

Among the appropriate conditions for use of systems engineering are the following:

- There are many considerations and interrelations.
- There are far-reaching and controversial value judgments.
- There are multidisciplinary and interdisciplinary considerations.
- The available information is uncertain, imprecise, incomplete, or otherwise flawed.
- Future events are uncertain and difficult to predict.
- Institutional and organizational considerations play an important role.
- There is a need for explicit and explicable consideration of the efficiency, effectiveness, and equity of alternative courses of action.

There are a number of results potentially attainable from use of systems engineering approaches. These include:

- Identification of perceived needs in terms of identified objectives and values of a client group.
- Identification of a set of information requirements to enable judgment and choice.
- Enhanced identification of a wide range of proposed alternatives or policies that might satisfy needs and achieve objectives.
- Increased understanding of issues that led to the effort, and the impacts of alternative actions upon these issues.
- Proposed alternative actions will be ranked in terms of the utility (benefits and costs) in achieving objectives and satisfying needs.
- A set of alternatives is selected for implementation and this will lead to a plan for action.
- Ultimately the action plans result in a working system or policy.

To develop professionals capable of coping satisfactorily with diverse factors involved in wide-scope problem solving is a primary goal of systems engineering and systems engineering education. This does not imply that a single individual or even a small group can, despite their strong motivation, solve all of the problems involved in a systems study. Such a requirement would demand total and absolute intellectual maturity on the part of the systems engineer and such is surely not realistic. It is also unrealistic to believe that issues can be resolved without very close association with a number of people who have stakes, and who thereby become stakeholders, in problem solution efforts. Consequently, systems engineers must be capable of facilitation and communication of knowledge between the diverse group of professionals, and their publics, that are involved in wide-scope problem solving. This requires that systems engineers be knowledgeable and able to use not only the technical constructs of systems science and operations research, but the behavioral constructs and management abilities also needed for resolution of complex large-scale problems. Intelligence, imagination, and creativity are necessary but not sufficient for proper use of the procedures of systems engineering. In addition, facility in human relations and effectiveness as a broker of information among parties at interest in a systems study are very much needed as well.

It is this blending of the technical, managerial, and behavioral that is a normative goal of success for systems engineering education and for systems engineering professional practice. Thus, systems engineering involves: *science*, and various analysis perspectives associated with the sciences; *organizations*, and the understanding of organizational behavior; and the *environment*, and understanding of the diverse interactions of organizations of people, machines, and institutions with their environments.

Systems engineering must be taught and practiced at three levels: systems science and operations research, systems methodology and design, and systems management. Systems engineers must be aware of a wide variety of systems science and operations research methods for the formulation, analysis, and interpretation of contemporary issues. They must be familiar with systems methodology and design in order to be able to select eclectic approaches that are best suited to the task at hand. Finally, a knowledge of systems management is necessary in order to be able to evolve processes that are matched to behavioral and organizational concerns and realities.

All three of these levels are important. To neglect any of them in the practice of systems engineering is to invite failure. It is generally not fully meaningful to talk only of a method or algorithm or even a complete methodology as a useful design, planning, or decision making process. It is ultimately meaningful to talk of a particular process as being useful. A systemic process for design, planning, and decision making will depend on the methodology, the operational environment, and leadership associated with use of the methodology. Thus systems management, systems methodology and design, and systems science and operations research, separately and collectively, do play a fundamental role in systems engineering.

A fundamental difficulty in software systems engineering, for example, is that of determining proper human–computer interfaces. While it is possible and often desirable to delegate authority to a computer, responsibility cannot meaningfully be delegated to a machine. Behaviorally irrelevant delegation of authority to a machine generally means that managers will not accept responsibility for system use. Proper systems management can eliminate, or at least reduce, this authority–responsibility gap and leads to a reduction of one of the factors that acts to impede system implementation.

Each time that a person acquires new information, analyzes, and uses it for judgment and decision making, that person is engaged, to some extent, in the knowledge-based discipline that is systems engineering. There are great extremes of variation in what constitutes information as well as that which constitutes best methods of use of information. These are highly dependent on the contingency task structure of environment, task, and decision maker experiential familiarity with task and environment.

37.2 THE SYSTEM LIFE CYCLE AND FUNCTIONAL ELEMENTS OF SYSTEMS ENGINEERING

The systems engineering approach enables accomplishment of the purposes we have described through a structured approach to qualitative and quantitative *formulation, analysis,* and *interpretation* of the impacts of action alternatives upon the needs perspectives, the institutional perspectives, and the value perspectives of stakeholders to large-scale issues in the public and private sectors.

A successful approach to systems engineering as an adjuvant for increased innovation and productivity, and other contemporary challenges, must be capable of issue formulation, analysis, and interpretation at the level of institutions and values as well as at the level of symptoms. Systems engineering approaches must allow for the incorporation of need and value perspectives, as well as technology perspectives, into models and postulates used to evolve and evaluate policies or activities that may result in technological and other innovations.

In actual practice, the steps of the systems process (formulation, analysis, and interpretation) are applied iteratively, and there is much feedback from one step to the other. This occurs because of the learning that is accomplished in the process of problem solution. Underlying all of this is the need for a general understanding of the diversity of the many systemic algorithms available and their role in the systems engineering process. The knowledge taxonomy for systems engineering, which consists of the major intellectual categories into which systems efforts may be categorized, is of considerable importance. The categories include systems science and operations research, systems methodology and design, systems management, and the application of these to realistic and relevant problems.

Systems science and operations research involves the development and application of concepts that form the basis for problem formulation and solution in systems engineering. Numerous tools for mathematical systems theory have been developed, including operations research (linear programming, nonlinear programming, dynamic programming, graph theory, etc.), decision and control theory, statistical analysis, economic systems analysis, and modeling and simulation. Systems science is also concerned with psychology and human factors concepts, social interaction and human judgment research, nominal group processes, and other behavioral science efforts. Of very special significance for systems engineering is the interaction of the behavioral and the algorithmic components of systems science in the choice-making process. The combination of a set of systems science and operations research methods, and a set of relations among these methods and activities, constitute what is known as a *methodology*.

A *methodology* is an open set of procedures that provides the means for solving problems. The tools of systems engineering, or the content of systems engineering, consist of a variety of algorithms and concepts that use words, mathematics, and graphics. Words, mathematics, and graphics are also the elements of systems engineering methodology as well as the elements of communications. These are structured in ways that enable various problem solving activities within systems engineering. Particular sets of relations among tools and activities, which constitute the framework for systems engineering, are of special importance here. Existence of, and use of, an appropriate systems engineering methodology is of considerable utility with respect to dealing with the many considerations, interrelations, and controversial value judgments associated with contemporary problems. It is also of major use in the design of systems and processes that are appropriate for the knowledge perspectives of those who use these processes to aid in decision-making tasks. In addition, an appropriate methodology and framework for the systems approach should enable the design of knowledge-based systems that enhance problem resolution at institutional and value levels, as well as at symptomatic levels. It can be strongly argued that solutions offered only at the level of systems, as is often done, will likely not be solutions at all, especially over the long term.

Systems engineering can be and has been described in many ways. Of particular importance here is a *morphological* description of systems engineering. It is this morphological description, or description in terms of form, which leads to a specific methodology and design procedure that results in a systemic process that is useful for issue resolution. We can discuss the *knowledge dimension* of systems engineering. This would include the various disciplines and professions that may be needed in a systems team to allow it to accomplish intended purposes of the team, such as provision of the knowledge base. Alternatively, we may speak of the phases or *time dimension* of a systems effort, which would start with policy and program planning and proceed through other portions of the *systems life cycle*. These include system planning, development, operation, and finally modification or retirement and phase out. Of special interest are the steps of the logic structure or *logic dimension* of systems engineering which consists of:

> *Formulation* of issues, or identification of problems or issues in terms of needs and constraints, objectives, or values associated with issue resolution, and alternative policies, controls, hypotheses, or complete systems that might resolve or ameliorate issues.
>
> *Analysis* of impacts of alternative policies, courses of action, or complete systems.
>
> *Interpretation* or evaluation of the utility of alternatives and their impacts upon the affected stakeholder group, and selection of a set of action alternatives for implementation.

We could also associate *feedback and learning* steps to interconnect these steps one to another. The systems process is typically very iterative. We shall not explicitly show feedback and learning in our conceptual models of the systems process.

Here we have described a three-dimensional morphology of systems engineering. There are a number of systems engineering morphologies or frameworks. In many of these, the logic dimension is divided into a larger number of steps which are iterative in nature. A particular seven step framework involves:

1. *Problem definition,* in which a descriptive and normative scenario of needs, constraints, and alterables associated with an issue are developed. The purpose of problem definition is to clarify the issues under consideration such as to allow other steps of a systems engineering effort to be carried out.

2. *Value system design,* in which objectives and objectives measures or attributes, with which to determine success in achieving objectives, are determined. Also, the interrelationship between objectives and objectives measures, and the interaction between objectives and the elements in the problem definition step are determined. This establishes a measurement framework that is needed to establish the extent to which the impacts of proposed policies or decisions will achieve objectives.

3. *System synthesis,* in which candidate or alternative decisions, hypotheses, options, policies, or systems that might result in needs satisfaction and objective attainment are postulated. A key word here is "might," for it is not assumed, at this step, that the policies are desirable.

4. *Systems analysis and modeling,* in which models are constructed to allow determination of the consequences of pursuing policies. The purpose of systems analysis and modeling is to determine the behavior or subsequent conditions resulting from alternative policies and systems.

Forecasting and impact analysis are, therefore, the most important objectives of systems analysis and modeling.

5. *Optimization or refinement of each alternative,* in which the individual policies and/or systems are tuned, often by means of parameter adjustment methods, so that each individual policy or system is refined in some "best" fashion in accordance with the value system that has been identified earlier.

6. *Evaluation and decision making,* in which systems and/or policies and/or alternatives are evaluated in terms of the extent to which the impacts of the alternatives achieve objectives and satisfy needs. Needed to accomplish evaluation are the attributes of the impacts of proposed policies and associated objective and/or subjective measurement of attribute satisfaction for each proposed alternative. Often this results in a prioritization of alternatives, with one or more being selected for further planning and resource allocation.

7. *Planning for action,* in which implementation efforts, resource and management allocations, or plans for the next phase of a systems engineering effort are delineated.

More often than not, the information required to accomplish the seven steps of the systems engineering morphology presented is not perfect due to uncertainty, imprecision, or incompleteness effects. This presents a major challenge to the design of processes and to systems engineering practice.

The intensity of effort needed for the steps of systems engineering varies greatly with the type of problem being considered. Problems of large scale and scope will generally involve a number of perspectives. These interact and the intensity of their interaction and involvement with the issue under consideration determines the scope and type of effort needed in the various steps of the systems process. Figure 37.1 is a flow chart representation of these steps for the planning phase of a systems engineering effort. Selection of appropriate algorithms or approaches to enable completion of these steps and satisfactory transition to the next step are major systems engineering tasks.

It is here convenient and possible to aggregate problem definition, value system design, and system synthesis into a single step in the systems process called *formulation* or input description and specification. The systems analysis and modeling step and optimization or refinement of alternatives step can be aggregated into a single step called *analysis.* The evaluation, decision making, and planning for action steps can be combined into a single step called *interpretation.* Figure 37.2 illustrates the hierarchical structure of these steps. It does not, however, show the iterative feedback nature of systems engineering. Nor does it show the interaction with systems management considerations that invariably involve human judgment.

It is possible to cite a number of characterizations of morphological frameworks for systems engineering. For example, the application area of defense systems acquisition generally involves 10 phases of the time dimension. These are related to an often used seven-phase description of systems engineering, as indicated in Table 37.1. These phases constitute the life cycle of systems engineering.

A soundly based systems methodology is needed to evolve *design* procedures for the development of physical systems, as well as systemic process aids to human efforts in planning, design, and decision making. Generally, a systems engineering design morphology will involve five phases:

1. Requirement specifications.
2. Preliminary conceptual design.
3. Detailed design and testing.
4. Evaluation.
5. Operational deployment.

In a sense these phases of systems design might be regarded as phases of the time dimension of systems engineering. While this is not incorrect, such a view does not capture the general requirement for phases within phases in the overall systems life cycle. To design a systemic process for one of the time-dimension phases such as planning, for example, it will generally be desirable to proceed through the five phases we describe as the essential phases of the design process. For example, the human operator in a human–machine system will generally function at a cognitive level in various problem solving activities (involving detection, diagnosis, evaluation, and monitoring of system operation) and at a physiological level in controlling the system. The phases of design and the steps of the systems process will be accomplished by the human operator. Preceding this, there is the need to design the human–machine system itself. The phases of design occur here also. Section 37.5 will be devoted to the very important subject of systems design.

Each of these phases of systems design is very important for sound development of physical systems and systemic processes as aids for decision making and resource allocation activities such as planning and design. Relatively less attention appears to have been paid to the requirement specifications phase than to the other phases of the systems design process. In many ways the requirement specification phase of a systems engineering design effort is the most important. It is this phase that has as its

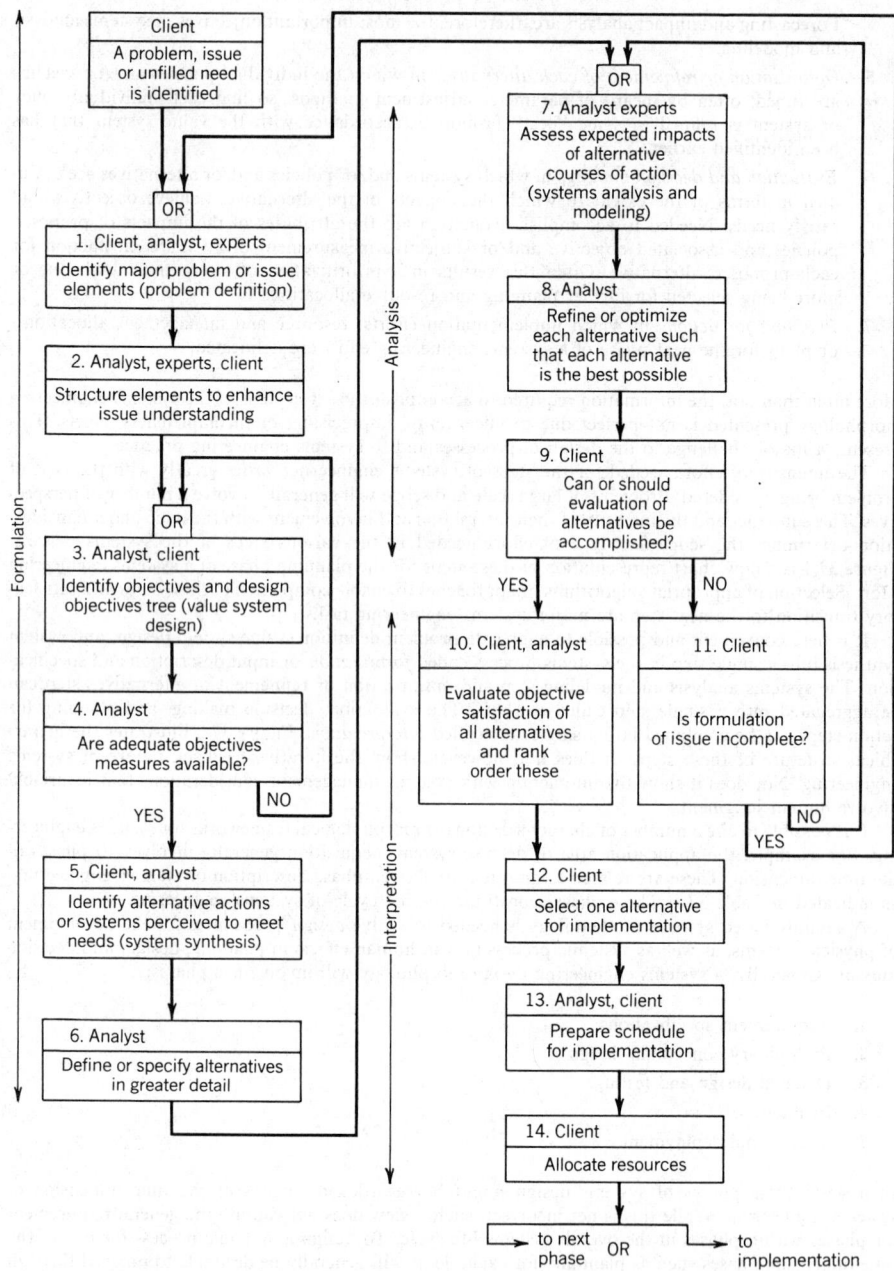

Fig. 37.1 Flow chart outlining steps of the systems engineering process.

goal the detailed definition of the needs, activities, and objectives to be fulfilled or achieved by the process to be ultimately developed. Thus this phase strongly influences all the phases that follow. It is this phase that describes preliminary design considerations that are needed to achieve successfully the fundamental goals underlying a systems engineering study. It is in this phase that the information requirements and the method of judgment and choice used for selection of alternatives are determined.

Systems engineering studies will often result in a knowledge-based system that enables organization of information to enhance decision making. The three logic steps of the systems engineering effort (formulation, analysis, and interpretation) will be needed to evolve requirements specifications. Of

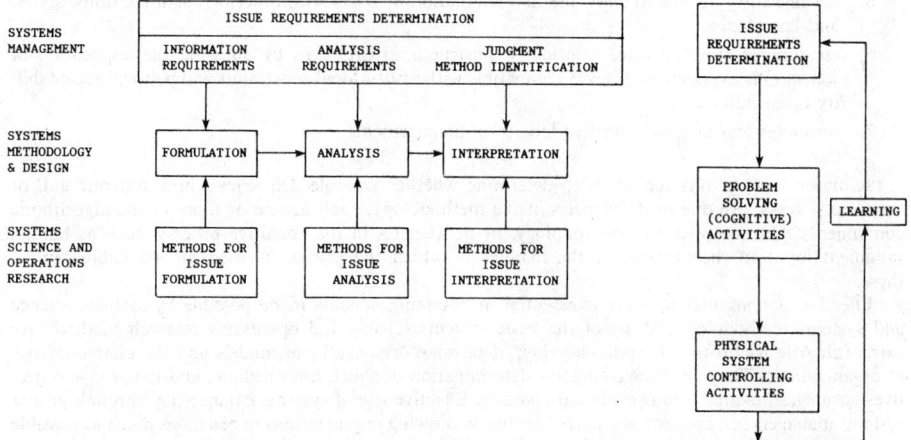

Fig. 37.2 Structure of the systems engineering process (with an illustration of human–machine interaction and learning).

course, they will be needed in other phases of a design effort as well as in the other phases of the time dimension or life cycle of the systems effort. We again see the need for phases within phases.

The result of a systems engineering effort is often a systemic process. A systemic process necessarily involves a methodology, which is the combination of a set of systems science algorithms, *and* human judgment, of the analyst and the client. Thus, a systemic process necessarily involves systems management considerations as the guiding force, especially at the cognitive meta-level where decisions concerning how to decide are made.

Effective design, including evaluation of a systemic aid, must include an operational evaluation component that will consider the strong interaction between a systemic methodology and the situational issues inherent in the incorporation of the methodology and system or the knowledge-based aid that results from a design effort, into a systemic support process. This operational evaluation of a systemic aid, for example, is needed in order to determine whether the systemic process

1. is logically sound;
2. is matched to the operational and organizational situation and environment extant;
3. supports a variety of cognitive skills, styles, and knowledge;
4. assists users or clients to develop and use their own cognitive skills, styles, and knowledge;

Table 37.1 The Time-Dimension Phases of the Systems Engineering Morphology Compared to the Phases of the Systems Acquisition Process

Systems Engineering Time-Dimension Phases	Systems Acquisition Life-Cycle Phases
Program planning	Mission analysis and development of MENS (mission element need statement)
Project planning	Conceptual
	Demonstration and validation
System development	Full-scale development
Production	Production
Distribution Operation	Deployment of phase in 1. Operation/support 2. Maintenance/repair 3. Modification/retrofit
Retirement	Retirement or phase out

5. is sufficiently flexible to allow use and adaptation by users with differing cognitive skills, styles, and knowledge;

6. encourages more effective solution of unstructured problems by allowing the application of job-specific experiences in a way compatible with institutional constraints and political acceptability constraints; and

7. promotes learning and effective long-term management.

A major task, in this regard, is to determine whether possible deficiencies in a systemic aid, or systemic process, are due to deficiencies in the methodology, such as one or more of the algorithmic components that comprise the methodology, or deficiencies in the cognitive process, such as human judgment input to the process, or the manner in which the process is used, or a combination of these.

Effective *systems management* is essential in obtaining benefits made possible by systems science and systems methodology. Many of the basic systems science and operations research methods are extraordinarily useful to this end. They help determine organizational models and the characteristics of organizations. They encourage effective determination of short, intermediate, and long-range objectives, strategies, tactics, and operational controls. Effective use of systems engineering, through proper systems management, appears singularly capable of allowing organizations to benefit as much as possible from combined efforts to the staff of an organization.

Even more importantly, however, appropriate systems management includes the interaction of the behavioral with the analytic. It allows for and encourages the use of multiple perspectives concerning issues in the systems process. This is very important, for it is the behavioral concerns that are often the dominant ones affecting effective systems management for innovation and productivity. The use of descriptive and normative studies of management and organization cognitive styles, and how cognitive styles depend on the contingency task structure of issue environment and decision maker experiential familiarity with these, can do much to assist in the development of systemic processes that can capture relevant expertise and use it as a part of the knowledge base used to determine the impact of identified alternatives. For it is wide scope processes that offer maximum support to the formulation, analysis, and interpretation of significant large scale issues. Knowledge-based systemic aids that encourage careful comprehensive formulation of issues, structuring of decision situations, and the subsequent analysis or determination of the impacts of proposed alternatives lead not only to identification of pertinent factors for planning and choice making, but also to illumination of these factors in a relatively easy to understand fashion. This explicability and communicability facilitates the interpretation necessary to select designs, plans, and decisions for implementation. Thus it is essential in the very important area of design of primarily cognitive systemic process aids, such as computer-aided-design systems, or physical systems or processes, such as integrated manufacturing systems.

It is generally not fully meaningful to talk only of a method or algorithm or even a complete systemic methodology as a useful planning or decision-making process. It is ultimately meaningful to talk of a particular systemic process as being useful. A systemic process for planning and decision making will depend on the systemic methodology, the operational environment, and leadership associated with use of the methodology. At the level of physical systems, such as an integrated manufacturing system, human operator and human supervisor concerns are very important. Thus systems management does indeed play a fundamental role in systems engineering. Its role in the iterative steps of the systems process is concerned with identification and explication of performance objectives for the process; determination of the contingency task structure which attempts a match between the cognitive and affective style of the decision maker and situational demands; and the resulting choice of an evaluation and decision rule for actual use in the existing decision task. Thus, we view systems management as a primarily cognitive meta-level activity that has the responsibility of resolving various "how-to-questions."

37.3 SYSTEMS ENGINEERING OBJECTIVES

Ten performance objectives appear to be of primary importance to those who desire to evolve quality plans, forecasts, decisions, or alternatives for action implementation. These are:

1. to identify needs, constraints, and alterables associated with the problem, issue, or requirement to be resolved (problem definition);

2. to identify a planning horizon or time interval for alternative action implementation, information flow, and objective satisfaction (planning horizon, identification);

3. to identify all significant objectives to be fulfilled, values implied by the choice of objectives, and objectives measures or attributes associated with various outcome states, with which to measure objective attainment (value system design);

4. to identify decisions, events, and event outcomes and the relations among them, such that a

structure of the possible paths among options, alternatives, or decisions, and the possible outcomes of these emerges (impact assessment);

5. to identify uncertainties and risks associated with the environmental influences affecting alternative decision outcomes (probability identification);

6. to identify measures associated with the costs and benefits or attributes of the various outcomes or impacts that result from judgment and choice (worth, value, or utility measurement);

7. to search for and evaluate new information, and the cost-effectiveness of obtaining this information relevant to improved knowledge of the time-varying nature of event outcomes that follow decisions or choice of alternatives (information acquisition and evaluation);

8. to enable selection of a best course of action in accordance with a rational procedure (choice making);

9. to reexamine the expected effectiveness of all feasible alternative courses of action, including those initially regarded as unacceptable, prior to making a final alternative selection (sensitivity analysis); and

10. to make detailed and explicit provisions for implementation of the selected action alternative, including contingency plans as needed (planning for implementation of action).

These objectives are, of course, very closely related to the aforementioned steps of the framework for systems engineering. To accomplish them requires attention to and knowledge of the methods of systems science and operations research such as to enable design of support systems, physical systems, and the many interface concerns that involve human interaction through use of the constructs of systems methodology and design. Also required is much effort at the level of systems management so that the resulting process is efficient, effective, equitable, and explicable. To ensure this, it is necessary to ensure that those involved in systems engineering efforts be concerned with technical knowledge of the issue under consideration, able to cope effectively with administrative concerns relative to the human elements of the issue, interested in and able to communicate across those actors involved in the issue, and capable of innovation and outscoping of relevant elements of the issue under consideration. These attributes (technical knowledge, human understanding and administrative ability, communicability, and innovativeness) are, of course, primary attributes of effective management.

37.4 SYSTEMS ENGINEERING METHODOLOGY AND METHODS

A variety of methods are suitable to accomplish the various steps of systems engineering. We shall briefly describe some of them here.

37.4.1 Issue Formulation

As has been indicated, issue formulation is the step of the systems engineering effort in which the problem or issue is defined (problem definition) in terms of the objectives of a client group (value system design) and where potential alternatives that might resolve needs are identified (system synthesis). Many studies have shown that the way in which an issue is resolved is critically dependent on the way in which the issue is formulated or framed. The issue formulation effort is concerned primarily with identification and description of the elements of the issue under consideration with, perhaps, some initial effort at structuring these in order to enhance understanding of the relations among these elements. Structural concerns are of importance in the analysis effort also. The systems process is iterative and interactive, and the results of preliminary analysis are used to refine the issue formulation effort. So the primary intent of issue formulation is to identify relevant elements that represent and are associated with issue definition, the objectives that should be achieved in order to satisfy needs, and potential action alternatives.

There are at least four ways to accomplish issue formulation:

1. *Asking* stakeholders to the issue under consideration.
2. *Descriptive identification* from a study of presently existing systems.
3. *Normative synthesis* from a study of documents describing what "should be," such as planning documents.
4. *Experimental discovery* based on experimentation with an evolving system.

These approaches are neither mutually exclusive nor exhaustive. Generally, the most appropriate efforts will use a combination of these approaches.

There are conflicting concerns with respect to which blend of these approaches is most appropriate for a specific task. The *asking* approach seems very appropriate when there is little uncertainty and imprecision associated with the issue under consideration, so that the issue is relatively well understood and may be easily structured, and where members of the client group possess much relevant expertise concerning the issue and the environment in which the issue is embedded. When these characteristics of the issue—lack of imprecision and presence of expert experiential knowledge—are present, then a direct declarative approach based on direct "asking" of "experts" is a simple and efficient approach. When there is considerable imprecision or when there is a lack of experiential familiarity with the issue under concern, then the other approaches take on greater significance. The asking approach is also prone to a number of cognitive information processing biases, as we shall discuss later. This is not as much of a problem in the other approaches.

Unfortunately, however, there are other difficulties with each of the other three approaches. *Descriptive identification,* from a study of existing systems of issue formulation elements, will very likely result in a new system or process that is based or anchored on an existing system, and tuned, adjusted, or perturbed from this existing system to yield incremental improvements. Thus, it is likely to result in incremental improvements to existing systems and procedures but not to result in major innovations or totally new systems and concepts.

Normative synthesis from a study of planning documents will result in an issue formulation effort that is based on what has been identified as desirable objectives and policies for the client group. A plan at any given phase may well not exist, or it may be flawed in any of several ways. Thus the information base may well not be present, or may be flawed. When these circumstances exist, it will not be a simple task to accomplish effective normative synthesis of issue formulation elements for the next phase of activity from a study of planning documents relative to the previous phase.

Often it is not possible to determine an appropriate set of issue formulation elements prior to use of the other steps of the systems effort. In some cases it will not be possible to define an appropriate set of issue formulation efforts prior to actual implementation of a preliminary system design. This often occurs. There are many important issues where there is an insufficient experiential basis to judge the effectiveness and completeness of a set of issue formulation efforts. Often, for example, clients will have difficulty in coping with very abstract formulation requirements and in visualizing the system or process that may ultimately evolve. Thus, it *may* be useful to identify an initial set of issue formulation elements and accomplish subsequent analysis and interpretation based on these, without extraordinary concern for completeness of the issue formulation efforts. A system or process, designed with ease of adaptation and change as a primary requirement, is implemented on a trial basis. As users become familiar with this new system or process, additions and modifications to the initially identified issue formulation elements result. This heuristic approach is used to iterate the system or process design requirements as experiential familiarity with the evolving prototype system grows.

From this discussion we could construct a balance sheet that profiles the concerns associated with these four approaches to issue formulation. Selection of an appropriate blend of approaches is, in itself, a systems engineering problem. There are multiple conflicting concerns involved in this selection that involve:

1. the overall system or plan;
2. the environment in which the overall system or plan is embedded;
3. the specific issue which is embedded in the overall system or plan;
4. the client or user group and their expertise and familiarity with respect to the overall system, the specific issue, and the environment surrounding these; and
5. the systems engineering team and its expertise and familiarity with respect to all of these.

The key parts of the problem definition step of issue formulation involve identification of needs, constraints, and alterables, and determination of the interactions among these elements and the group that they impact. Need is a condition requiring supply or relief or lack of something required, desired, or useful. In order to define a problem satisfactorily, we must determine the alterables—those items pertaining to the needs that can be changed. Alterables can be separated into those over which no control is possible and those over which control is possible. It is the controllable alterables that are of special concern in systems engineering since they can be changed or modified to assist in achieving particular outcomes. To define a problem adequately, we must also determine the limitations or constraints under which the needs can or must be satisfied and the range over which it is permissible to vary the controllable alterables. Finally, we must determine relevant groups of people who are affected by a given problem.

Value system design is concerned with defining objectives, determining their interactions, and ordering these into a hierarchical structure. Objectives and their attainment are, of course, related to the needs, alterables, and constraints associated with problem definition. Thus the objectives can, and should be, related to these problem definition elements. Finally, a set of measures is needed whereby to measure objective attainment. Generally, these are called attributes. It is necessary to ensure that all needs are satisfied by attainment of at least one objective.

The first step in system synthesis is to identify activities and alternatives for attaining each of the objectives, or the postulation of complete systems to this end. It is then desirable to determine interactions among the proposed activities and to illustrate relationships between the activities and the needs and objectives. Activities measures are needed to gauge the degree of accomplishment of proposed activities. Systemic methods useful for problem definition are generally useful for value system design and system synthesis as well. This is another reason that suggests the efficacy of aggregating these three steps under a single heading: *formulation.*

People generally encounter complexity in dealings with systems as well as with one another. This complexity comes about both because of the large number of elements involved in any complex issue and because of the large number of interactions. These will typically assume many forms, such as the elements of a large organization or the hardware components of a large technical system. Alternately, elements and interactions may exist only in the mind of a group of individuals, such as a set of multiple attributes and relationships among them that are important in a decision making situation.

Complex issues will have a structure associated with them. In some problem areas, structure is often well understood and well articulated. In other areas, it is not possible to articulate structure in such a clear fashion. There exists considerable motivation to develop techniques with which to enhance structure determination, as a system structure must always be dealt with by individuals or groups regardless of whether the structure is articulated or not. Furthermore, an individual or a group can deal much more effectively with systems and make better decisions when the structure of the underlying system is well defined and exposed *and* communicated clearly. One of the fundamental objectives of systems engineering is to structure knowledge elements such that they are capable of being better understood and communicated.

We now discuss formal methods appropriate for "asking" as a method of issue formulation. Then we shall contrast and compare some of these approaches. The methods associated with the other three generic approaches to issue formulation also involve approaches to analysis that will be discussed in the next subsection.

Several of the formal methods that are particularly helpful in the identification, through asking, of issue formulation elements are based on principles of collective inquiry in which a group of interested and motivated people is brought together to stimulate each other's creativity in generating issue formulation elements. We may distinguish two groups of collective inquiry modeling methods:

1. *Brainwriting, Brainstorming, Synectics, Nominal Group Technique, and Charette.* These approaches typically require a few hours of time, a group of knowledgeable people gathered in one place, and a group leader or facilitator. *Brainwriting* is typically better than *brainstorming* in reducing the influence of dominant individuals. Both methods can be very productive: 50–150 ideas or elements might be generated in less than 1 hr. *Synectics,* based on problem analogies, might be appropriate if there is a need for truly unconventional, innovative ideas. Considerable experience with the method is a requirement, however, particularly for the group leader. The *nominal group technique* is based on a sequence of idea generation, discussion, and prioritization. It can be very useful when an initial screening of a large number of ideas or elements is needed. *Charette* offers a conference or workshop type format for generation and discussion of ideas and/or elements.

2. *Questionnaires, Surveys, and DELPHI.* These three methods of collective inquiry modeling do not require the group of participants to gather at one place and time, but they typically take more time to achieve results than the first group of methods. In *questionnaires* and *surveys,* a usually large number of participants is asked, on an individual basis, for ideas or opinions, which are then processed to achieve an overall result. There is no interaction among participants. *DELPHI* usually provides for written interaction among participants in several rounds. Results of previous rounds are fed back to participants, and they are asked to comment, revise their views as desired, etc. A DELPHI can be very instructive, but usually takes several weeks or months to complete.

Use of most structuring methods, in addition to leading to greater clarity of the problem formulation elements, will also typically lead to identification of new elements and revision of element definitions. As we have indicated, most structuring methods contain an analytical component, and they may, therefore, be more properly labeled as analysis methods. The following element structuring aids are among the many modeling aids available:

- *Interaction Matrices.* These may be useful to identify clusters of closely related elements in a large set, in which case we have a self-interaction matrix; or to structure and identify the couplings between elements of different sets, for example, objectives and alternatives. In this case we produce *cross-interaction matrices,* such as shown in Fig. 37.3. Interaction matrices are useful for initial, comprehensive exploration of sets of elements. Learning about problem interrelationships during the process of constructing an interaction matrix is a major result of use of these matrices.

- *Trees.* Trees are graphical aids particularly useful to portray hierarchical or branching-type structures. They are excellent for communication, illustration, and clarification. Trees may be useful in all steps and phases of a systems effort. Figure 37.4 represents an attribute tree that represents those aspects of a proposal evaluation effort that will be formally considered in evaluation and prioritization of a set of proposals.

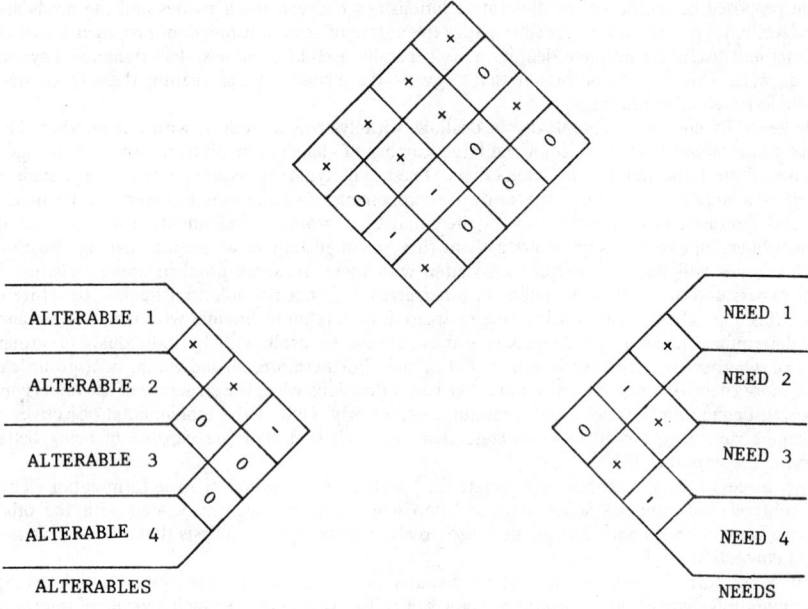

+ = enhancing interaction
- = inhibiting interaction
0 = no interaction

Fig. 37.3   Illustration of self- and cross-interaction matrices.

- *Causal Loop Diagrams.*   Causal loop diagrams, or influence diagrams, represent graphical pictures of causal interactions between sets of variables. They are particularly helpful to make explicit one's perception of the causes of change in a system, and can serve very well as communication aids. A causal loop diagram is also useful as the initial part of a detailed simulation model. Figure 37.5 represents a causal loop diagram of a belief structure.

Two other descriptive methods potentially useful for issue formulation are:

- *System Definition Matrix.*   The system definition matrix, options profile, decision balance sheet, or checklist provides a framework for specification of the essential aspects, options, or characteristics of an issue, a plan, a policy, or a proposed or existing system. It can be helpful for the design and specification of alternative policies, designs, or other options or alternatives. The system definition matrix is just a table that shows important aspects of the options that are important for judgment relative to selection of approaches to issue formulation.
- *Scenario Writing.*   This method is based on narrative and creative descriptions of existing or possible situations or developments. Scenario descriptions can be helpful for clarification and communication of ideas and obtaining feedback on those ideas. Scenarios may also be helpful in conjunction with various analysis and forecasting methods where they may represent alternative or opposing views.

Clearly, successful formulation of issues through "asking" requires creativity. Creativity may be much enhanced through use of a structured systems engineering framework. For example, group meetings for issue formulation involve idea *formulation,* idea *analysis,* and idea *interpretation.* The structure of a group meeting may be conceptualized within a systems engineering framework. This framework is especially useful for visualizing the tradeoffs that must be made among allocation of resources for formulation, analysis, and interpretation of ideas in the issue formulation step itself. If there is an emphasis on idea formulation, we shall likely generate too many ideas to cope with easily. This will lead to a lack of attention to detail. On the other hand, if there is a deeemphasis on idea formulation, we shall typically encourage defensive avoidance through undue efforts to support the present situation,

1.   Understanding of Problem
     1.1 Navy Cost Credentialing Process
     1.2 NAVELEX Cost Analysis Methodology
     1.3 DoD Procurement Procedures

2.   Technical Approach
     2.1 Establishment of a Standard Methodology
     2.2 Compatibility with Navy Acquisition Process

3.   Staff Experience
     3.1 Directly Related Experience in Cost Credentialing, Cost Analysis
          and Procurement Procedures
     3.2 Direct Experience with Navy R&D Programs

4.   Corporate Qualifications
     Relevant Experience in Cost Analysis for Navy R&D Programs

5.   Management Approach
     5.1 Quality/Relevance
     5.2 Organization and Control Effectiveness

6.   Cost
     6.1 Manner in which Elements of Cost Contribute Directly to Project
          Success
     6.2 Appropriate of Cost Mix to the Technical Effort

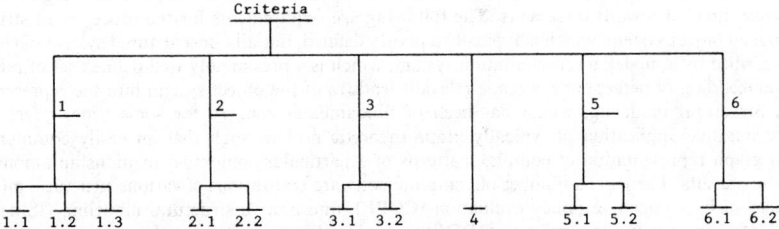

Fig. 37.4 Attribute tree and detailed listing of attributes for evaluation of proposals concerning cost credentialing.

or a rapid unconflicted change to a new situation. An overemphasis on analysis of ideas is usually time consuming and results in a meeting that seems to drown in details. There is merit inherent in encouraging a group to reach consensus. But the effort may be inappropriate also, since it may encourage arguments over concerns that are ineffective in influencing judgments.

Deemphasis on analysis of identified ideas will usually result in disorganized meetings in which hasty, poorly thought out ideas are accepted. Postmeeting disagreements concerning the results of the meeting are another disadvantage that often occurs. An emphasis on interpretation of ideas will

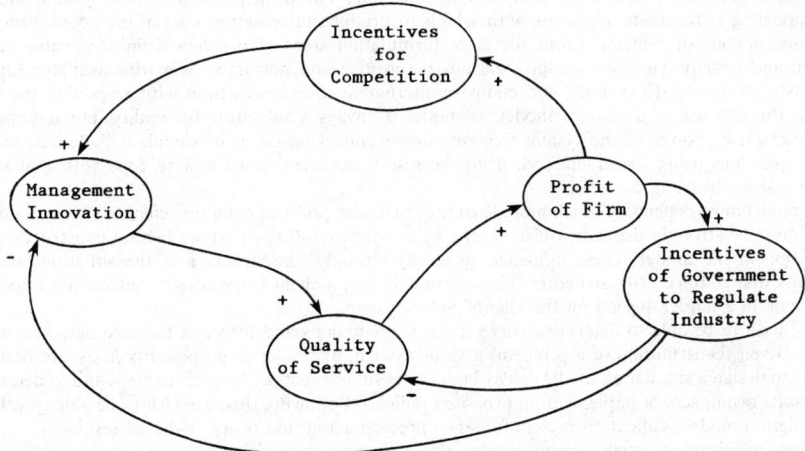

Fig. 37.5 Simple causal loop diagram.

produce a meeting that is emotional and people-centered. Misunderstandings will be frequent as issues become entrenched in an adversary, personality-centered process. On the other hand, deemphasis on interpretation of ideas results in meetings in which important information is not elicited. Consequently, the meeting is awkward and empty, and routine acceptance of ideas is a likely outcome.

37.4.2 Issue Analysis

Issue analysis in systems engineering involves forecasting and assessment of the impacts of proposed alternative courses of action. In turn, this suggests construction, testing, and validation of models. Impact assessment in systems engineering includes system analysis and modeling, and optimization and ranking or refinement of alternatives. First, the options or alternatives defined in issue formulation are structured, often as part of the issue formulation effort, and then analyzed in order to assess the anticipated impacts that may result from their implementation. Secondly, a refinement or optimization effort is often desirable. This is directed toward refinement or fine-tuning a viable alternative, and parameters within an alternative, so as to obtain maximum needs satisfaction, within given constraints, from a proposed policy.

To determine the structure of systems in the most effective manner requires the use of quantitative analysis to direct the structuring effort along the most important and productive paths. This is especially needed when time available to construct structural models of decision situations is limited. Cognitive maps, interaction matrices, intent structures, Delta charts, objective and attribute trees, causal loop diagrams, decision-outcome trees, signal flow graphs, etc., are all structural models that are very useful graphic aids to communications. There are also several approaches to knowledge structuring in artificial intelligence directed toward these ends. The following are requirements for the processes of structural modeling: an object system, which is typically a poorly defined, initially unstructured system of elements to be described by a model; a representation system, which is a presumably well defined set of relations; and an embedding of perceptions of some relevant features of the object system into the representation system. Structural modeling, which has been of fundamental concern for some time, refers to the systemic iterative application of typically graph-theoretic notions such that an easily communicable directed graph representation of complex patterns of a particular contextual relationship among a set of elements results. There are a number of computer software realizations of various structural modeling constructs such as cognitive policy evaluation (COPE), interpretive structural modeling (ISM), goal-directed decision structuring system (GODDESS), and multiattribute utility decomposition (MAUD). There are also several conceptual approaches to knowledge-based systems that have led to innovative approaches to knowledge structuring such as script applier mechanism (SAM), model of diagnostic reasoning (INTERNIST), and subjective understanding model of belief systems (POLITICS). The software associated with these systems is capable of comprehensive knowledge representation. Development of effective issue representation is essential as a necessary part of issue resolution. This can perhaps best be accomplished through development of procedures and processes that allow the structuring of uncertain, imprecise, and incomplete knowledge and that encourage relevant feedback to enhance cognitive abilities associated with issue perception. As suggested by the name, *structural modeling* is a transition activity that interrelates issue formulation with issue analysis.

Transformation of a number of identified issue formulation elements, which typically represent unclear, poorly articulated mental models of a system, into visible well-defined models useful for many purposes is the object of systems analysis and modeling. The principal objective of systems analysis and modeling is to create a process with which to produce information concerning consequences of proposed actions or policies. From the issue formulation steps of problem definition, value system design, and system synthesis, we have various descriptive and normative scenarios available for use. We wish, ultimately, to evaluate and compare alternative courses of action with respect to the value system through use of a systems model. A model is always a substitute for reality, but is hopefully one descriptive enough of the system elements under consideration to be useful. It is desired to pose policy questions using the model and, from the results obtained, learn how to cope with that subset of the real world being modeled.

A model must depend on much more than the particular problem definition elements being modeled; for it must be strongly dependent also on the value system and the purpose behind construction and utilization of the model. These influence, generally strongly, the structure of the situation, and the elements that comprise this structure. Those elements that a client believes to be sufficiently important to include in a model depend on the clients' value system.

We wish to be able to determine correctness of predictions and forecasts that are based on model usage. Given the definition of a problem, a value system, and a set of proposed policies, we desire to be able to design a simulation model consisting of relevant elements of these three steps and to determine the results or impacts of implementing proposed policies. Following this, we wish to be able to validate a simulation model. Validation must, of course, precede actual use of model-based results.

There are three essential steps in constructing a simulation model:

1. Determination of those problem definitions, value systems, and system synthesis elements that are most relevant to a particular problem.
2. Determination of the structural relationships among these elements.
3. Determination of parametric coefficients within the structure.

The use of methods for issue formulation and structural modeling may be especially rewarding with respect to satisfactory accomplishments of the first two steps of the systems engineering framework.

Usually we desire or need quantitative information for decision making. The results of many structural modeling efforts will yield only qualitative, at best ordinal, relations among the elements in a systems study. Generally there are a number of fundamental difficulties associated with the use of qualitative and ordinal, often binary, relations to infer preferences. Preference determination is a principal part of the interpretation step of systems engineering. While structural modeling methods cope well with causal and influential relations of a qualitative nature, quantitative dominance relations are needed to infer rational and transitive preferences. Therefore, we need more than a structural model of causes and influences. We need quantitative impacts that will be aggregated in the interpretation step, together with a value system, to evolve preferences. Accordingly, we proceed beyond determination of influential and causal structural models, and determine parametric relationships to enable construction of a quantitative model of the objective and subjective, quantitative and qualitative, constructs associated with an issue.

There are three uses to which models may normally be put. Model categories corresponding to these three uses are descriptive models, predictive or forecasting models, and policy or planning models. Representation and replication of important features of a given problem are the objects of a descriptive model. Good descriptive models are of considerable value in that they reveal much about the substance of complex issues and how, typically in a retrospective sense, change over time has occurred. One of the primary purposes behind constructing a descriptive model is to learn about the past. Often the past will be a good guide to the future.

In building a predictive or forecasting model, we must be especially concerned with determination of proper cause and effect, and various process relationships. If the future is to be predicted with integrity, we must have a method with which to determine exogenous variables accurately, and the model structure and parameters within the structure must be valid in order that the model be valid in a procedural as well as substantive sense. Often, it will not be possible to predict accurately all exogenous variables, and, in that case, conditional predictions can be made from particular assumed values of unknown exogenous variables.

The future is inherently uncertain. Consequently, predictive or forecasting models are often used to generate a variety of future scenarios, each a conditional prediction of the future based on some conditioning assumptions. In other words, we develop an "if–then" model.

Policy or planning models are much more than predictive or forecasting models, although any policy or planning model is also a predictive or forecasting model. The outcome from a policy or planning model must be evaluated in terms of a value system. Policy or planning efforts must not only predict outcomes from implementing alternative policies, but they must also present these outcomes in terms of the value system that is in a form useful and suitable for alternative ranking, evaluation, and decision making. Thus, a policy model must contain some provision for impact interpretation.

Model usefulness cannot be determined by objective truth criteria alone. Well-defined and well-stated functions and purposes for the simulation model are needed to determine simulation model usefulness. Fully objective criteria for model validity do not typically exist. There appears little likelihood of the development of a general-purpose context-free simulation model. The task is simply far too complicated. Thus, we must build models for specific purposes. Thus the question of model validity is context dependent.

Model credibility depends on the interaction between the model and model user. It appears very difficult to build a model that does not reflect the outlook and bias of the modeler. This activity is proscribed in effective systems engineering practice, since the *purpose of a model is to describe systematically the "view" of a situation held by the client, and not that held by the analyst.*

A great variety of approaches have been designed and used for the forecasting and assessment that are the primary goals of systems analysis. There are basically two classes of methods that we describe here: expert opinion methods and modeling and/or simulation methods.

Expert opinion methods are based on the assumption that knowledgeable people will be capable of saying sensible things about the impacts of alternative policies on the system, as a result of their experience with, or insight into, the issue or problem area. These methods are generally useful. They are particularly appropriate when there are no established theories or data concerning system operation, precluding the use of more precise analytical tools. Among the most prominent expert-opinion-based forecasting methods are surveys and DELPHI. There are, of course, many other ways of asking experts for their opinion—for example, hearings, meetings, and conferences. A particular problem with expert opinion (asking) methods is that cognitive bias is widespread, as are value incoherences; and incorporation of bias and incoherent values into the models that results from use of expert opinion methods often results in inconsistent and self-contradictory results. There exists a strong need in the forecasting and assessment community to recognize and ameliorate, by appropriate procedures, the effects of cognitive bias and value incoherencies in expert-opinion-modeling efforts. Expert opinion methods are often appropriate for the "asking" approach to issue formulation. They may be of considerably less value, especially when used as stand-alone approaches, for impact assessment and forecasting.

Simulation and *modeling methods* are based on the conceptualization and use of an abstraction or model of the real world which hopefully behaves in a similar way as the real system. Impacts of policy alternatives are studied in the model, which will hopefully lead to increased insight with respect to the actual situation.

Most simulation and modeling methods use the power of mathematical formulations and computers to keep track of many pieces of information at the same time. Two methods in which the power of the computer is combined with subjective expert judgments are *cross-impact analysis* and *workshop dynamic models*. Typically, experts provide subjective estimates of event probabilities and event interactions. These are processed by a computer to explore their consequences, and fed back to the analysts and thereafter to the experts for further study. The computer derives the resulting behavior of various model elements, over time, giving rise to renewed discussion and revision of assumptions.

Expert judgment is virtually always included in all modeling methods. Scenario writing can be an expert-opinion-modeling method, but typically this is done in a less direct and explicit way than in DELPHI, survey, ISM, cross-impact, or workshop dynamic models. As a result of this, internal inconsistency problems are reduced with those methods based on mathematical modeling. The following are additional forecasting methods based on mathematical modeling and simulation. In these methods, a structural model is generally formed on the basis of expert opinion and physical or social laws. Available data are then processed to determine parameters within the structure. Unfortunately, these methods are sometimes very data intensive and, therefore, expensive and time consuming to implement.

Trend extrapolation/time series forecasting is particularly useful when sufficient data about past and present developments are available, but there is little theory about underlying mechanisms causing change. The method is based on the identification of a mathematical description or structure that will be capable of reproducing the data into the future, typically over the short to medium term.

Continuous-time dynamic simulation is based on postulation and qualification of a causal structure underlying change over time. A computer is used to explore long-range behavior as it follows from the postulated causal structure. The method can be very useful as a learning and qualitative forecasting device, but its application may be rather costly and time consuming.

Discrete event digital simulation models are based on applications of queueing theory to determine the conditions under which system outputs or states will switch from one condition to another.

Input–output analysis has been specially designed for study of equilibrium situations and requirements in economic systems in which many industries are interdependent. Many economic data fit in directly to the method, which is, mathematically, relatively simple, and can handle many details.

Econometrics is another method mainly applied to economic description and forecasting problems. It is based on both theory and data with, usually, the main emphasis on specification of structural relations based on macroeconomic theory and the derivation of unknown parameters in behavioral equations from available economic data.

Microeconomic models represent an application of economic theories of firms and consumers who desire to maximize the profit and utility of their production and consumption alternatives.

Parameter estimation is a very important subject with respect to model construction and validation. Observation of basic data and estimation or identification of parameters within an assumed structure, often denoted as system identification, are essential steps in the construction and validation of system models. The simplest estimation procedure, in both concept and implementation, appears to be the least squares error estimator. Many estimation algorithms to accomplish this are available and are in actual use. The subjects of parameter estimation and system identification are being actively explored both in economics and systems engineering. There are numerous contemporary results, including algorithms for system identification and parameter estimation in very-large-scale systems representative of actual physical processes and organizations.

Verification of a model is necessary to ensure that the model behaves in a fashion intended by the model builder. If we can determine that the structure of the model corresponds to the structure of the elements obtained in the problem definition, value system design, and system synthesis steps, then the model is verified with respect to behaving in a gross, or structural, fashion as the model builder intends.

Even if a model is verified in a structural as well as parametric sense, there is still no assurance that the model is valid in the sense that predictions made from the model will occur. We can determine validity only with respect to the past. That is all that we can possibly have available at the present. Forecasts and predictions inherently involve the future. Since there may be structural and parametric changes as the future evolves, and since knowledge concerning results of policies not implemented may never be available, there is usually no way to completely validate a model. Nevertheless, there are several steps that can be used to validate a model. These include a reasonableness test in which we determine that the overall model, as well as model subsystems, respond to inputs in a reasonable way as determined by "knowledgeable" people. The model should also be valid according to statistical time series used to determine parameters within the model. Finally, the model should be epistemologically valid in that the policy interpretations of the various model parameters, structure, and recommendations are consistent with ethical, professional, and moral standards of the group affected by the model.

Once a model has been constructed it is often desirable to determine, in some best fashion, various policy parameters or controls that are subject to negotiation. The optimization or refinement of alternatives step is concerned with choosing parameters or controls to maximize or minimize a given performance index or criterion. Invariably there are constraints that must be respected in seeking this extremum. As previously noted, the *analysis* step of systems engineering consists of systems analysis and modeling, and optimization or refinement of alternatives and related methods that are appropriate in aiding effective judgment and choice.

There exist a number of methods for fine-tuning, refinement, or optimization of individual specific alternative policies or systems. These are useful to determine the best (in terms of needs satisfaction) control settings or rules of operation in a well-defined quantitatively describable system. A single scalar indicator of performance or desirability is typically needed. There are, however, approaches to multiple objective optimization that are based on welfare-type optimization concepts. It is these individually optimized policies or systems that are an input to the evaluation and decision-making effort in the interpretation step of systems engineering.

Among the many methods for optimization and refinement of alternatives are:

Mathematical programming, which is used extensively in operations research and analysis practice, for resource allocation under constraints, resolution of planning or scheduling problems, and similar applications. It is particularly useful when the best equilibrium or one-time setting has to be determined for a given policy or system. Figures 37.6 and 37.7 present some salient features of solutions to simple linear and nonlinear programming problems.

Optimum systems control addresses the problem of determining the best controls or actions when the system, the controls or actions, the constraints, and the performance index may change over time. A mathematical description of system change is necessary. Optimum systems control is particularly suitable for refining controls or parameters in systems in which trade-offs over time play an important part. Figure 37.8 presents salient features of the optimum systems control approach to a simple problem.

Application of the various refinement or optimization methods, like those described here, typically requires significant training and experience on the part of the systems analyst. Some of the many characteristics of analysis that are of importance for systemic efforts include the following:

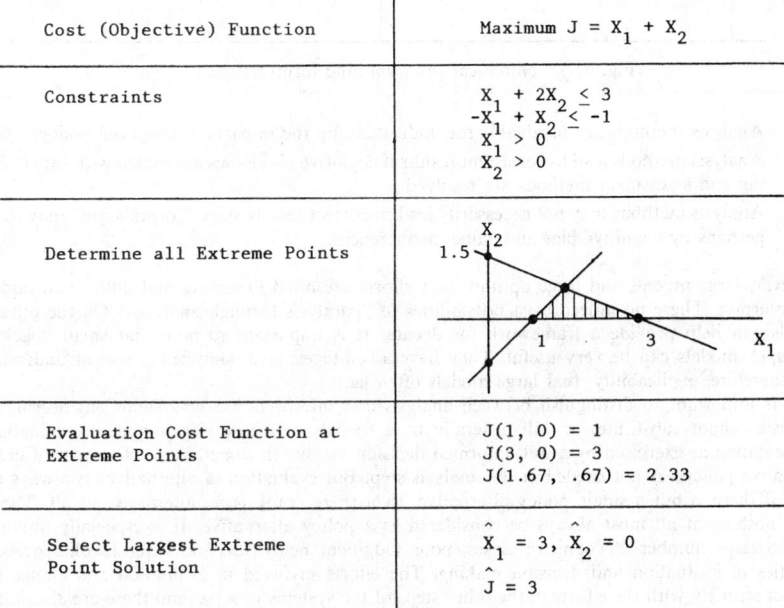

Cost (Objective) Function	Maximum $J = X_1 + X_2$
Constraints	$X_1 + 2X_2 \leq 3$ $-X_1 + X_2 \leq -1$ $X_1 > 0$ $X_2 > 0$
Determine all Extreme Points	
Evaluation Cost Function at Extreme Points	$J(1, 0) = 1$ $J(3, 0) = 3$ $J(1.67, .67) = 2.33$
Select Largest Extreme Point Solution	$\hat{X}_1 = 3, \ \hat{X}_2 = 0$ $\hat{J} = 3$

Fig. 37.6 Illustration of simple linear program solution. (There are many versions of a "simplex" method to systematize the search for a solution.)

	Cost (function) to be Maximized	$J = \phi(X)$
Problem to be Solved	Constraints	$C(X) \leq b$ $X \geq 0$
	Lagrangian	$L = \phi(X) + \lambda^T[b-C(X)]$
Necessary Conditions for Optimality	Necessary Conditions-- Kuhn Tucker Conditions	$\dfrac{\partial L}{\partial X} = \dfrac{\partial \phi(X)}{\partial X} - \dfrac{\partial C^T(X)}{\partial X}\lambda \leq 0$ $X^T \dfrac{\partial L}{\partial X} = X^T\left(\dfrac{\partial \phi}{\partial X} - \dfrac{\partial C^T(X)}{\partial X}\lambda\right) = 0$ $\dfrac{\partial L}{\partial \lambda} = b - C(X) \geq 0$ $\lambda^T \dfrac{\partial L}{\partial \lambda} = \lambda^T[b - C(X)] = 0$ $X \geq 0$ $\lambda \geq 0$

Fig. 37.7 Nonlinear programming formulation.

1. Analysis methods are invaluable for understanding the impacts of proposed policy.

2. Analysis methods lead to consistent results if cognitive bias issues associated with expert forecasting and assessment methods are resolved.

3. Analysis methods may not necessarily lead to correct results since "formulation" may be flawed, perhaps by cognitive bias and value incoherencies.

However, large models and large optimization efforts are often expensive and difficult to understand and interpret. There do indeed exist possibilities of "paralysis through analysis." On the other hand, models can help provide a framework for debate. It is important to note that small "back of the envelope" models can be very useful. They have advantages: cost, simplicity, ease of understanding, and, therefore, explicability, that large models often lack.

It is important to distinguish between analysis and interpretation in systems engineering efforts. Analysis cannot substitute, or will generally be a foolish substitute, for judgment, evaluation, and interpretation as exercised by a well-informed decision maker. In some cases, refinement of individual alternative policies is not needed in the analysis step. But evaluation of alternatives is always needed, since, if there is but a single policy alternative, then there really is no alternative at all. The option to do nothing at all must always be considered as a policy alternative. It is especially important to avoid a large number of cognitive biases, poor judgment heuristics, and value incoherencies in the activities of evaluation and decision making. The efforts involved in evaluation and choice making interact strongly with the efforts in the other steps of the systems process, and these are also influenced by cognitive bias, judgment heuristics, and value incoherencies. One of the fundamental tenets of the systems process is that by making the complete issue resolution process as explicit as possible, it is easier to detect and connect these deficiencies than it is in holistic intuitive processes.

Problem Specification	Cost (Objective) Function	$J = \theta[X(t_f)] + \int_{t_o}^{t_f} \phi[X(t), U(t), t]dt$
	System (Constraint) Equation	$\dot{X} = \dfrac{dX}{dt} = f[X(t), U(t), t]$
	Initial Condition	$X(t_o) = X_o$
Necessary Conditions for Optimality	Hamiltonian	$H = \phi + \lambda^T f$
	System Equation	$\dot{X} = \dfrac{\partial H}{\partial \lambda} = f$
	Adjoint Equation	$\dot{\lambda} = -\dfrac{\partial H}{\partial X} = -\dfrac{\partial \phi}{\partial X} - \dfrac{\partial f^T}{\partial X}\lambda$
	Initial Condition	$X(t_o) = X_o$
	Terminal Condition	$\lambda(t_f) = \dfrac{\partial \theta[X(t_f)]}{\partial X(t_f)}$

Fig. 37.8 Illustration of requirements for solution to optimum control problem.

37.4.3 Information Processing Biases

After completion of the *analysis* step, we begin the evaluation and decision-making effort of *interpretation*. Decisions must typically be made and policies formulated, evaluated, and applied in an atmosphere of uncertainty. The outcome from any proposed policy is seldom known with certainty. One of the purposes of analysis is to reduce, to the extent possible, uncertainties associated with the outcomes of proposed policies. Most planning, design, and resource allocation issues will involve a large number of decision makers who act according to their varied preferences. Often, these decision makers will have diverse and conflicting data available to them and the decision situation will be quite fragmented. Furthermore, outcomes resulting from actions can often only be adequately characterized by a large number of incommensurable attributes. Explicit informed comparison of alternatives across these attributes, by many stakeholders in an evaluation and choice-making process, is typically most difficult.

As a consequence of this, people will often search for and use some form of a dominance structure to enable rejection of alternatives that are perceived to be dominated by one or more other alternatives. An alternative is said to be "dominated" by another alternative when the other alternative has attribute scores at least as large as those that are associated with the dominated alternative with at least one attribute score that is larger. But, biases are systematic and prevalent in most unaided cognitive activities. Decisions and judgments are influenced by differential weights of information. They are influenced by a variety of human information processing deficiencies, such as base rates, representativeness, availability, adjustment, and anchoring. Often it is very difficult to disaggregate values of policy outcomes from causal relations determining these outcomes. Often correlation is used to infer causality. Wishful thinking and other forms of selective perception encourage us to not obtain potentially disconfirming information. The resulting confounding of values with facts can lead to great difficulties in discourse and related decision making.

It is especially important to avoid the large number of potential cognitive biases and flaws in the process of formulation, analysis, and interpretation for judgment and choice. These may well occur due to flaws in human information processing that are associated with the identification of problem elements, structuring of decision situations, and the probabilistic and utility assessment portions of the judgmental tasks of evaluation and decision making.

Among the cognitive biases and information processing flaws that have been identified are several which affect information formulation or acquisition, information analysis, and interpretation. Among these biases, which are not independent, are the following.

1. *Adjustment and Anchoring.*   Often a person finds that difficulty in problem solving is due, not to the lack of data and information, but rather to the existence of excess data and information. In such situations, the person often resorts to heuristics, which may reduce the mental efforts required to arrive at a solution. In using the anchoring and adjustment heuristic when confronted with a large amount of data, the person selects a particular datum, such as the mean, as an initial or starting point or anchor, and then adjusts that value improperly in order to incorporate the rest of that data, resulting in flawed information analysis.

2. *Availability.*   The decision maker uses only easily available information and ignores sources of significant but not easily available information. An event is believed to occur frequently, that is, with high probability, if it is easy to recall similar events.

3. *Base Rate.*   The likelihood of occurrence of two events is often compared by contrasting the number of times the two events occur and ignoring the rate of occurrence of each event. This bias often occurs when the decision maker has concrete experience with one event but only statistical or abstract information on the other. Generally abstract information will be ignored at the expense of concrete information. A base rate determined primarily from concrete information may be called a causal base rate, whereas that determined from abstract information is an incidental base rate. When information updates occur, this individuating information is often given much more weight than it deserves. It is much easier for the impact of individuating information to override incidental base rates than causal base rates.

4. *Conservatism.*   The failure to revise estimates as much as they should be revised, based on receipt of new significant information, is known as conservatism. This is related to data saturation and regression effects biases.

5. *Data Presentation Context.*   The impact of summarized data, for example, may be much greater than that of the same data presented in detailed, nonsummarized form. Also, different scales may be used to change the impact of the same data considerably.

6. *Data Saturation.*   People often reach premature conclusions on the basis of too small a sample of information while ignoring the rest of the data, which is received later, or stopping acquisition of data prematurely.

7. *Desire for Self-Fulfilling Prophecies.*   The decision maker values a certain outcome, interpretation, or conclusion and acquires and analyzes only information that supports this conclusion. This is another form of selective perception.

8. *Ease of Recall.*   Data that can easily be recalled or assessed will affect perception of the likelihood of similar events reoccurring. People typically weigh easily recalled data more in decision making than those data that cannot easily be recalled.

9. *Expectations.*   People often remember and attach higher validity to information that confirms their previously held beliefs and expectations than they do to disconfirming information. Thus, the presence of large amounts of information makes it easier for one to selectively ignore disconfirming information such as to reach any conclusion and thereby prove anything that one desires to prove.

10. *Fact–Value Confusion.*   Strongly held values may often be regarded and presented as facts. That type of information is sought which confirms or lends credibility to one's views and values. Information that contradicts one's views or values is ignored. This is related to wishful thinking in that both are forms of selective perception.

11. *Fundamental Attribution Error (Success/Failure Error).*   The decision maker associates success with personal inherent ability and associates failure with poor luck in chance events. This is related to availability and representativeness.

12. *Habit.*   Familiarity with a particular rule for solving a problem may result in reuse of the same procedure and selection of the same alternative when confronted with a similar type of problem and similar information. We choose an alternative because it has previously been acceptable for a perceived similar purpose or because of superstition.

13. *Hindsight.*   People are often unable to think objectively if they receive information that an outcome has occurred and they are told to ignore this information. With hindsight, outcomes that have occurred seem to have been inevitable. We see relationships much more easily in hindsight than in foresight and find it easy to change our predictions after the fact to correspond to what we know has occurred.

14. *Illusion of Control.*   A good outcome in a chance situation may well have resulted from a poor decision. The decision maker may assume a feeling of control over events that is not reasonable.

15. *Illusion of Correlation.*   A mistaken belief that two events covary when they do not covary is known as the illusion of correlation.

16.   *Law of Small Numbers.*   People are insufficiently sensitive to quality of evidence. They often express greater confidence in predictions based on small samples of data with nondisconfirming evidence than in much larger samples with minor disconfirming evidence. Sample size and reliability often have little influence on confidence.

17.   *Order Effects.*   The order in which information is presented affects information retention in memory. Typically the first piece of information presented (primacy effect) and the last presented (recency effect) assume undue importance in the mind of the decision maker.

18.   *Outcome Irrelevant Learning System.*   Use of an inferior processing or decision rule can lead to poor results and the decision maker can believe that these are good because of inability to evaluate the impacts of the choices not selected and the hypotheses not tested.

19.   *Representativeness.*   When making inference from data, too much weight is given to results of small samples. As sample size is increased, the results of small samples are taken to be representative of the larger population. The "laws" of representativeness differ considerably from the laws of probability and violations of the conjunction rule $P(A \cap B) \leq P(A)$ are often observed.

20.   *Selective Perceptions.*   People often seek only information that confirms their views and values. They disregard or ignore disconfirming evidence. Issues are structured on the basis of personal experience and wishful thinking. There are many illustrations of selective perception. One is "reading between the lines" such as, for example, to deny antecedent statements and, as a consequence, accept "if you don't promote me, I won't perform well" as following inferentially from "I will perform well if you promote me."

Of particular interest are circumstances under which these biases occur and their effects on activities such as planning and design. Through this, it may be possible to develop approaches that might result in debiasing or amelioration of the effects of cognitive bias. A number of studies have compared unaided expert performance with simple quantitative models for judgment and decision making. While there is controversy, most studies have shown that simple quantitative models perform better in human judgment and decision-making tasks, including information processing, than holistic expert performance in similar tasks. There are a number of prescriptions that might be given to encourage avoidance of possible cognitive biases and to debias those that do occur:

1.   Sample information from a broad data base and be especially careful to include data bases which might contain disconfirming information.

2.   Include sample size, confidence intervals, and other measures of information validity in addition to mean values.

3.   Encourage use of models and quantitative aids to improve upon information analysis through proper aggregation of acquired information.

4.   Avoid the hindsight bias by providing access to information at critical past times.

5.   Encourage people to distinguish good and bad decisions from good and bad outcomes.

6.   Encourage effective learning from experience. Encourage understanding of the decision situation and methods and rules used in practice to process information and make decisions such as to avoid outcome irrelevant learning systems.

In a sense, these results are disturbing. There is strong evidence that humans are very strongly motivated to understand, to cope with, and to improve themselves and the environment in which they function.

37.4.4   Interpretation

While there are a number of fundamental limitations to systemic efforts to assist in bettering the quality of human judgment, choice, decisions, and designs, there are also a number of desirable activities and these have resulted in several important holistic approaches that provide formal assistance in the evaluation and interpretation of the impacts of alternatives. These include the following.

Decision analysis, which is a very general approach to option evaluation and selection, involves identification of action alternatives and possible consequences, identification of the probabilities of these consequences, identification of the valuation placed by the decision maker on these consequences, computation of the expected utilities of the consequences, aggregating or summarizing these values for all consequences of each action. In doing this we obtain an expected utility evaluation of each alternative act, and the one with the highest value is the most preferred action or option. Figure 37.9 presents some of the salient features involved in the decision analysis of a simplified problem.

Multiattribute utility theory (MAUT) has been designed to facilitate comparison and ranking of alternatives with many attributes or characteristics. The relevant attributes are identified and structured, and a weight or relative utility is assigned by the decision maker to each basic attribute. The attribute measurements for each alternative are used to compute an overall worth or utility for each attribute.

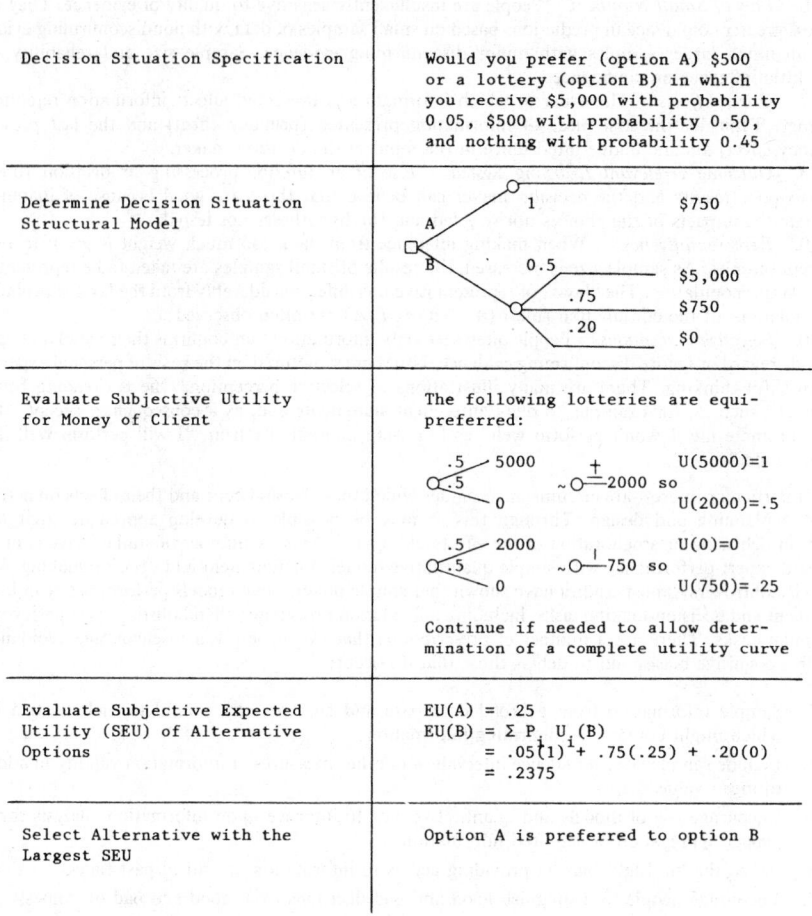

Decision Situation Specification	Would you prefer (option A) $500 or a lottery (option B) in which you receive $5,000 with probability 0.05, $500 with probability 0.50, and nothing with probability 0.45
Determine Decision Situation Structural Model	
Evaluate Subjective Utility for Money of Client	The following lotteries are equi-preferred:
Evaluate Subjective Expected Utility (SEU) of Alternative Options	$EU(A) = .25$ $EU(B) = \Sigma \, P_i \, U_i(B)$ $\quad = .05(1)^1 + .75(.25) + .20(0)$ $\quad = .2375$
Select Alternative with the Largest SEU	Option A is preferred to option B

Fig. 37.9 Illustration of simple single-attribute decision analysis effort.

Multiattribute utility theory allows for explicit recognition and incorporation of the decision maker's attitude toward risk in the utility computations. There are a number of variants of MAUT; many of them are simpler, more straightforward processes in which risk and uncertainty considerations are not taken into account. The method is very helpful to the decision maker in making values and preferences explicit, and making decisions that are consistent with those values. Figure 37.10 indicates some salient features of the MAUT approach for the particular case where there are no risks or uncertainties involved in the decision situation.

Policy capture (or *social judgment theory*) has also been designed to assist decision makers in making their values explicit and their decisions consistent with their values. In policy capture, the decision maker is asked to rank order a set of alternatives in a gestalt or holistic fashion. Then, alternative attributes and associated attribute measures are determined by elicitation from the decision maker. A mathematical procedure involving regression analysis is used to determine that relative importance weight of each attribute that will lead to a ranking as specified by the decision maker. The result is fed back to the decision maker, who, typically, will express the view that his or her values are different. In an iterative learning process, preference weights and/or overall rankings are modified until the decision maker is satisfied with both the weights and the overall alternative ranking.

There are many advantages to formal interpretation efforts in systems engineering. Among these are the following:

1. Developing decision situation models to aid in making the choice-making effort explicit helps one both to identify and to overcome the inadequacies of implicit mental models.

2. The decision situation model elements, especially the attributes of the outcomes of alternative actions, remind us of information we need to obtain about alternatives and their outcomes.

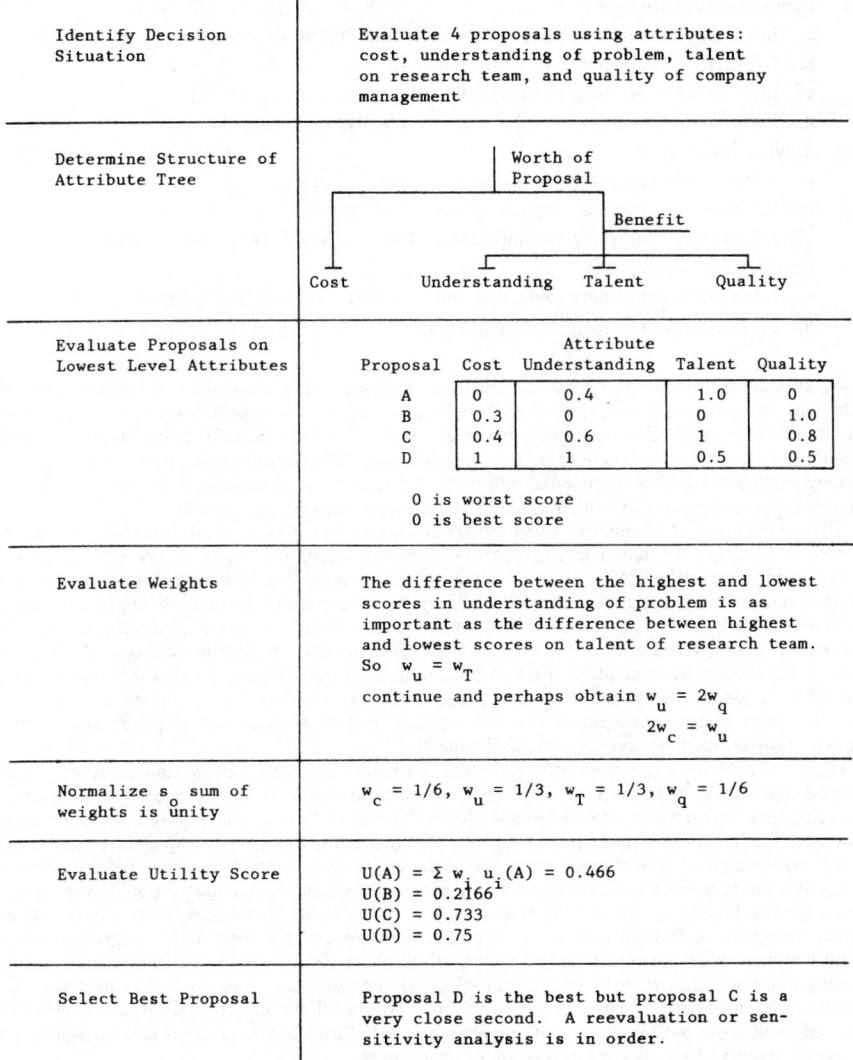

| Identify Decision Situation | Evaluate 4 proposals using attributes: cost, understanding of problem, talent on research team, and quality of company management |

Determine Structure of Attribute Tree

Evaluate Proposals on Lowest Level Attributes

		Attribute		
Proposal	Cost	Understanding	Talent	Quality
A	0	0.4	1.0	0
B	0.3	0	0	1.0
C	0.4	0.6	1	0.8
D	1	1	0.5	0.5

0 is worst score
0 is best score

Evaluate Weights

The difference between the highest and lowest scores in understanding of problem is as important as the difference between highest and lowest scores on talent of research team. So $w_u = w_T$

continue and perhaps obtain $w_u = 2w_q$
$$2w_c = w_u$$

Normalize s_o sum of weights is unity

$w_c = 1/6$, $w_u = 1/3$, $w_T = 1/3$, $w_q = 1/6$

Evaluate Utility Score

$U(A) = \Sigma w_i u_i(A) = 0.466$
$U(B) = 0.266$
$U(C) = 0.733$
$U(D) = 0.75$

Select Best Proposal

Proposal D is the best but proposal C is a very close second. A reevaluation or sensitivity analysis is in order.

Fig. 37.10 Illustration of simple multiple attribute utility theory evaluation under certainty.

3. We avoid such cognitive heuristics as evaluating one alternative on attribute *A* and another on attribute *B*.

4. We improve our ability to process information and, consequently, reduce the possibilities for cognitive bias.

5. We can aggregate facts and values in a prescribed systemic fashion rather than by adopting an agenda-dependent or intellect-limited approach.

6. We enhance brokerage, facilitation, and communication abilities among stakeholders to complex technological and social issues.

37.4.5 The Central Role of Information in Systems Engineering

Information is certainly a key ingredient supporting quality decisions and all of systems engineering efforts are based on appropriate acquisition and use of information. There are three basic types of information. These are fundamentally related to the three-step framework of systems engineering:

1. *Formulation Information*
   a. Information concerning the problem and associated needs, constraints, and alterables.
   b. Information concerning the value system.
   c. Information concerning possible option alternatives.
   d. Information concerning possible future alternative outcomes, states and scenarios.
2. *Analysis Information*
   a. Information concerning probabilities of future scenarios.
   b. Information concerning impacts of alternative options.
   c. Information concerning the importance of various value criterion or attributes.
3. *Interpretation Information*
   a. Information concerning evaluation and aggregation of facts and values.
   b. Information concerning implementation.

We see that useful and appropriate formulation, analysis, and interpretation of information is one of the most important and vital tasks in systems engineering efforts, since it is the efficient processing of information by the decision maker that produces effective decisions. A useful definition of information for our purposes is that it is data of value for decision making. The decision-making process is influenced by many contingency and environmental influences. A purpose of management technology is to provide systemic support processes to further enhance efficient decision-making activities.

After completion of evaluation and decision-making efforts it is generally necessary to become involved in planning for action to implement the chosen alternative option or the next phase of a systems engineering effort. More often than not, it will be necessary to iterate through the steps of systems engineering several times to obtain satisfactory closure upon one or more appropriate action alternatives. Planning for action also leads to questions concerning resource allocation, schedules, and management plans. There are, of course, a number of methods from systems science and operations research that support determination of schedules and implementation plans. Each of the steps is needed with different focus and emphasis at each phase of a systems effort. These phases depend on the particular effort under consideration but will typically include such phases as policy and program planning, project planning, and system development.

There are a number of complexities affecting "rational" planning, design, and decision making. We must cope with these in the design of effective systemic processes. The majority of these complexities involve systems management considerations. Herbert Simon and many others have indicated that the capacity of the human mind for formulating, analysis, and interpretation of complex large-scale issues is very small compared with the size and scope of the issues whose resolution is required for objective, substantive, and procedurally rational behavior. Anthony Downs has also indicated that decision quality is considerably limited by the human intellect. Among the limits to rationality are the fact that a planner, designer, or decision maker can formulate, analyze, and interpret only a restricted amount of information; can devote only a limited amount of time to decision making; and can become involved in many more activities than he or she can effectively consider and cope with simultaneously, and therefore must necessarily focus attention only on a portion of the major competing concerns. The direct effect of these is the presence of cognitive bias in information acquisition and processing and the use of cognitive heuristics for evaluation of alternatives.

In many cases, these cognitive heuristics will be flawed. This is not necessarily so, however. One of the hoped for results of the use of systems engineering approaches is the development of effective and efficient heuristics for enhanced judgment and choice.

There are many cognitive biases prevalent in most information-acquisition activities. The use of cognitive heuristics and decision rules is also prevalent and necessary to enable us to cope with the many demands on our time. One such heuristic is satisfying or searching for a solution that is "good enough." This may be quite appropriate if the stakes are small. In general, the quality of cognitive heuristics will be task dependent, and often the use of heuristics for evaluation will be both reasonable and appropriate. Rational decision making requires time, skill, wisdom, and other resources. It must, therefore, be reserved for the more important decisions. A goal of systems engineering is to enhance information acquisition, processing, and evaluation so that efficient and effective use of information is made in a process that is appropriate to the cognitive styles and time constraints of management.

A serious problem in practice is that we do get to use very simple heuristics that are appropriate for "concrete operational" situations in a familiar world in which we possess considerable intuitive skill, and we continue to use them in "formal operational" situations in an unfamiliar world in which they may be very inappropriate. Inadequate analysis and information acquisition and processing models resulting in a poor decision situation model may also occur because of poor experiential learning. Anthony Downs has indicated that the amount of information initially available is only a very small fraction of that potentially available; the "costs" of procurement, processing, and use of this additional information may be high; and important information, especially concerning future events, is often unavailable. Thus, uncertainty and imprecision are major factors inherent in any realistic situation in which information needs are present. One of the major goals in a systems design effort is to cope with these factors through eclectic use of the methods of systems science and operations research. A central purpose of all of this is to enable the efficient and effective acquisition, analysis, and interpretation of information. Another purpose is the development of effective and efficient heuristics. Finally, as our confidence and experiential familiarity with these heuristic rules develops, we hope to be able to develop holistic skills so that even the heuristic rules become unnecessary. In this way, we enable proper and appropriate knowledge-based, rule-based, or skill-based behavior as appropriate to the task at hand and the experiential familiarity of the problem solver with it.

37.5 SYSTEM DESIGN

This section discusses several topics relevant to the design and evaluation of systems. In order to develop our design methodology, we first discuss the purpose and objectives of system and systemic process design. Development of performance objectives for quality systems is important since evaluation of the logical soundness and performance of a system can be determined by measuring achievement of these objectives with and without the system. A discussion of general objectives for quality system design is followed by a presentation of a five-phase design methodology for system design. The section continues with leadership and training requirements for use of the resulting system and the impact of these requirements upon design considerations. While it is doubtlessly true that not every design process should, could, or would precisely follow each component in the detailed phases outlined here, we feel that this approach to systems design is sufficiently robust and generic that it can be used as a normative model of the design process and as a guide to the structuring and implementation of appropriate systems evaluation practices.

37.5.1 The Purposes of Systems Design

Contemporary issues that may result in the need for systems design are invariably complex. They typically involve a number of competing concerns, contain much uncertainty, and require expertise from a number of disparate disciplines for resolution. Thus it is not surprising that intuitive and affective judgments, often based on incomplete data, form the usual basis used for contemporary design and associated choice making. At the other extreme of the cognitive inquiry scale are the highly analytical, theoretical, and experimental approaches of the mathematical, physical, and engineering sciences. When intuitive judgment is skill based, it is generally effective and appropriate. One of the major challenges in system design is to develop processes that are appropriate for a variety of process users, some of whom may approach the design issue from a skill-based perspective, some from a rule-based perspective, and some from a knowledge-based perspective.

A central purpose of systems engineering and management science is to incorporate systems science and operations research methods into a methodology that, when it is associated with human judgment through systems management, results in a high-quality systems design procedure. By high-quality design, we mean one that will, with high probability, produce a system that is effective and efficient.

A systems design procedure must be specifically related to the operational environment for which the final system is intended. Control group testing and evaluation may serve many useful purposes with respect to determination of many aspects of algorithmic and behavioral efficacy of a system. Ultimate effectiveness involves user acceptability of the resulting system and evaluation of this process effectiveness will often involve testing and evaluation in the environment, or at least a closely simulated model of the environment, in which the system would be potentially deployed.

The potential benefits of systems engineering approaches to design can be interpreted as attributes or criteria for evaluation of the design approach itself. Achievement of many of these attributes may often not be experimentally measured except by inference, anecdotal, or testimonial and case study evidence taken in the operational environment for which the system is designed. Explicit evaluation of attribute achievement is a very important part of the overall systemic design process. This section describes the following:

1. A methodological framework for the design of systems and systemic processes such as planning and decision support systems.
2. An evaluation methodology which may be incorporated with or used independently of the design framework.

There exist a number of characteristics of effective systems efforts that can be identified. These form the basis for determining the attributes of systems and systemic design procedures. Some of these attributes will be more important for a given environment than others. Effective design must typically include an operational evaluation component that will consider the strong interaction between the system and the situational issues that led to the systems design requirement. This operational evaluation is needed in order to determine whether a system or a systemic process for incorporation into a physical system consisting of humans and machines

1. is logically sound;
2. is matched to the operational and organizational situation and environment extant;
3. supports a variety of cognitive skills, styles, and knowledge of the humans who must use the system;
4. assists users of the system to develop and use their own cognitive skills, styles, and knowledge;
5. is sufficiently flexible to allow use and adaptation by users with differing cognitive skills, styles, and knowledge;
6. encourages more effective solution of unstructured and unfamiliar issues allowing the application of job-specific experiences in a way compatible with various acceptability constraints; and
7. promotes effective long-term management.

37.5.2   The Operational Environment, the Decision Situation Model

A usable methodology that interrelates plans and/or designers, the values of those who will use the system, the design situation model, and inferences about states of nature or the environment that surrounds the design situation is needed as a methodological framework for systems design. To develop robust scenarios of planning and design situations in various operational environments, and specific instruments for evaluation, we first identify a mathematical and situational taxonomy of

* algorithmic constructs used in systemic design,
* performance objectives for quality planning and design, and
* operational environments for planning and design.

Planning, design, and decision support, as ultimate goals of many systemic processes, are basically efforts that involve obtaining, interpreting, forecasting the consequences of, evaluating, and prioritizing, and thereby organizing actions or action recommendations based on information, objectives, and existing situations or needs. A system or systemic process that is used to aid in the process of planning, design, or decision making should assist in or support the evaluation of alternatives relative to some criteria. It is necessary that planning, design, and decision-relevant information be condensed and described in ways that lead to effective analysis and problem structuring. Of equal importance is the need to be aware of the role of the affective in planning design and choice-making tasks such as to support different cognitive styles and needs, which vary from knowledge-based, to rule-based, to skill-based behavior. Again, we see the need for meta-level concerns in systems engineering. We desire to design efficient and effective physical control systems, problem solving (cognitive) systems, and interfaces between the two. This section is concerned with each of these.

Not all of the performance objectives for quality systems engineering will be, or need be, fully attained in all design instances, but it is generally true that the quality of a system or of a systems design process necessarily improves as more and more of these objectives are attained. Measures of quality of the resulting system, and therefore systemic process design quality, may be obtained by assessing the degree of achievement of these performance criteria by the resulting system, generally in an operational environment. In this way, an evaluation of the effectiveness of a planning, design, or decision support system may be conducted.

A taxonomy based on operational environments is necessary to describe particular situation models through which planning design and decision support are achieved. We are able to describe a large number of situations using elements or features of the three-component taxonomy described earlier. With these, we are able to evolve test instruments to estabish quantitative and qualitative evaluations of a system within an operational environment. The structural and functional properties of a system or a systemic process must be described in order that a purposeful evaluation can be accomplished. This purposeful evaluation of a systemic process is obtained by embedding the process into specific operational planning, design, or decision situations. Thus an evaluation effort also allows iteration and feedback to ultimately improve the overall systems design process. The evaluation methodology to be described is useful, therefore, as a part or phase of the design process. Also, it is useful, in and of itself, to evaluate and prioritize a set of systemic aids for planning, design, and decision support. It is also useful for evaluation of resulting system designs and operational systems providing a methodological framework both for the design and evaluation of physical systems and for systems that assist in the planning and design of systems.

37.5.3   The Design of Systemic Process Aids and Systems

The purpose of this section is to describe five important phases in the design of systems and systemic process aids. These phases serve as a guide not only for the sound design and development of systems and systemic aids for planning, design, and decision support processes, but for their evaluation and ultimate operational deployment as well. These five phases,

requirements specification,

preliminary conceptual design,

detailed design, testing, and implementation,

evaluation, and

operational deployment,

are applicable to design in general. Although the five phases will be described as if they are to be sequenced in a chronological fashion, sound design practice will generally necessitate iteration and feedback from a given phase to earlier phases.

Requirements Specification Phase

The requirements specification phase has as its goal the detailed definition of those needs, activities, and objectives to be fulfilled or achieved by the system or process, which is to result from the system design effort. Furthermore, the effort in this phase should result in a description of preliminary conceptual design considerations appropriate for the next phase. This must be accomplished in order to translate operational deployment needs, activities, and objectives into requirements specifications if, for example, that is the phase of the systems engineering design effort under consideration.

Among the many objectives of the requirements specifications phase of systems engineering are the following:

1. To define the problem to be solved, or range of problems to be solved, or issue to be resolved or ameliorated; including identification of needs, constraints, alterables, and stakeholder groups associated with operational deployment of the system or the systemic process.

2. To determine objectives for operational system or the operational aids for planning, design, and decision support.

3. To obtain commitment for prototype design of a system or systemic process aid from user group and management.

4. To search the literature and seek other expert opinions concerning the approach that is most appropriate for the particular situation extant.

5. To determine the estimated frequency and extent of need for the system or the systemic process.

6. To determine the possible need to modify the system or the systemic process to meet changed requirements.

7. To determine the degree and type of accuracy expected from the system or systemic process.

8. To estimate expected effectiveness improvement or benefits due to the use of the system or systemic process.

9. To estimate the expected costs of using the system or systemic process, including design and development costs, operational costs, and maintenance costs.

10. To determine typical planning horizons and periods to which the system or systemic process must be responsive.

11. To determine the extent of tolerable operational environment alteration due to use of the system or systemic process.

12. To determine what particular planning, design, or decision process appears best.

13. To determine the most appropriate roles for the system or systemic process to perform within the context of the planning, design, or decision situation and operational environment under consideration.

14. To estimate potential leadership requirements for use of the final system itself.

15. To estimate user group training requirements.

16. To estimate the qualifications required of the design team.

17. To determine preliminary operational evaluation plans and criteria.

18. To determine political acceptability and institutional constraints affecting use of an aided support process, and those of the system itself.

19. To document analytical and behavioral specifications to be satisfied by the support process and the system itself.

20. To determine the extent to which the user group can require changes during and after system development.

21. To determine potential requirements for contractor availability after completion of development and operational tests for additional needs determined by the user group, perhaps as a result of the evaluation effort.

22. To develop requirements specifications for prototype design of a support process and the operational system itself.

As a result of this phase, to which the four issue requirements identification approaches of Section 37.4.1 are fully applicable, there should exist a clear definition of typical planning, design, and decision issues, or problems requiring support, and other requirements specifications, so that it is possible to make a decision whether to undertake preliminary conceptual design. If the result of this phase indicates that the user group or client needs can potentially be satisfied in a cost effective manner, by a systemic process aid, for example, then documentation should be prepared concerning detailed specifications for the next phase, preliminary conceptual design, and initial specifications for the last three phases of effort. A design team is then selected to implement the next phase of the system life cycle. This discussion emphasizes the inherently coupled nature of these phases of the system life cycle, and illustrates why it is not reasonable to consider the phases as if they are uncoupled.

Preliminary Conceptual Design Phase

The preliminary conceptual design phase includes specification of the mathematical and behavioral content and associated algorithms for the system or process that should ultimately result from the effort, as well as the possible need for computer support to implement these. The primary goal of this phase is to develop conceptualization of a prototype system or process in response to the requirements specifications developed in the previous phase. Preliminary design according to the requirements specifications should be achieved. Objectives for preliminary conceptual design include the following:

1. To search the literature and seek other expert opinion concerning the particular approach to design and implementation that is likely to be most responsive to requirements specifications.

2. To determine the specific analytic algorithms to be implemented by the system or process.

3. To determine the specific behavioral situation and operational environment in which the system or process is to operate.

4. To determine the specific leadership requirements for use of the system in the operational environment extant.

5. To determine specific hardware and software implementation requirements, including type of computer programming language and input devices.

6. To determine specific information input requirements for the system or process.

7. To determine the specific type of output and interpretation of the output to be obtained from the system or process that will result from the design procedure.

8. To reevaluate objectives obtained in the previous phase, to provide documentation of minor changes, and to conduct an extensive reexamination of the effort if major changes are detected that could result in major modification and iteration through requirements specification or even termination of effort.

9. To develop a preliminary conceptual design of a prototype aid that is responsive to the requirements specification.

The expected product of this phase is a set of detailed design and testing specifications that, if followed, should result in a usable prototype system or process. User group confidence that an ultimately useful product should result from detailed design should be above some threshold, or the entire design effort should be redone. Another product of this phase is a refined set of specifications for the evaluation and operational deployment phases.

If the result of this phase is successful, the detailed design, testing, and implementation phase is begun. This phase is based on the products of the preliminary conceptual design phase, which should result in a common understanding among all interested parties about the planning and decision support design effort concerning the following:

1. Who the user group or responsive stakeholder is.

2. The structure of the operational environment in which plans, designs, and decisions are made.

3. What constitutes a plan, a design, or a decision.

4. How plans, designs, and decisions are made without the process or system and how they will be made with it.

5. What implementation, political acceptability, and institutional constraints affect the use of the system or process.

6. What specific analysis algorithms will be used in the system or process, and how these algorithms will be interconnected to form the methodological construction of the system or process.

Detailed Design, Testing, and Implementation Phase

In the third phase of design, a system or process that is presumably useful in the operational environment is produced. Among the objectives to be attained in this phase are the following:

1. To obtain and design appropriate physical facilities (physical hardware, computer hardware, output device, room, etc.).
2. To prepare computer software.
3. To document computer software.
4. To prepare a user's guide to the system and the process in which the system is embedded.
5. To prepare a leader's guide for the system and the associated process.
6. To conduct control group or operational (simulated operational) tests of the system and make minor changes in the aid as a result of the tests.
7. To complete detailed design and associated testing of a prototype system based on the results of the previous phase.
8. To implement the prototype system in the operational environment as a process.

The products of this phase are detailed guides to use of the system as well as, of course, the prototype system itself. It is very important that the user's guide and the leader's guide address, at levels appropriate for the parties interested in the effort, the way in which the performance objectives identified in Section 37.5.3 are satisfied. The description of system usage and leadership topics should be addressed in terms of the analytic and behavioral constructs of the system and the resulting process, as well as in terms of operational environment situation concerns. These concerns include:

1. Frequency of occurrence of need for the system or process.
2. Time available from recognition of need for a plan, design, or decision to identification of an appropriate plan, design, or decision.
3. Time available from determination of an appropriate plan, design, or decision to implementation of the plan, design, or decision.
4. Value of time.
5. Possible interactions with the plans, designs, or decisions of others.
6. Information base characteristics.
7. Organizational structure.
8. Top management support for the resulting system or process.

It is especially important that the portion of this phase which concerns implementation of the prototype system specifically address important questions concerning cognitive style and organizational differences among parties at interest and institutions associated with the design effort. Stakeholder understanding of environmental changes and side effects that will result from use of the system is critical for ultimate success. This need must be addressed. Evaluation specification and operational deployment specifications will be further refined as a result of this phase.

Evaluation Phase

Evaluation of the system in accordance with evaluation criteria, determined in the requirements specification phase and modified in the subsequent two design phases is accomplished in the fourth phase of systems development. This evaluation should always be assisted to the extent possible by all parties at interest to the systems design effort and the resultant systemic process. The evaluation effort must be adapted to other phases of the design effort so that it becomes an integral functional part of the overall design process. As noted, evaluation may well be an effort distinct from design that is used to determine usefulness or appropriateness for specified purposes of one or more previously designed systems. Among the objectives of system or process evaluation are the following:

1. To identify a methodology for evaluation.
2. To identify criteria on which the success of the system or process may be judged.
3. To determine effectiveness of the system in terms of success criteria.
4. To determine an appropriate balance between the operational environment evaluation and the control group evaluation.

5. To determine performance objective achievement of the system.
6. To determine behavioral or human factor effectiveness of the system.
7. To determine the most useful strategy for employment of the existing system.
8. To determine user group acceptance of the system.
9. To suggest refinements in existing systems for greater effectiveness of the process in which the new system has been embedded.
10. To evaluate the effectiveness of the system or process.

These objectives are obtained from a critical evaluation issue specification or evaluation need specification, which is the first or problem-definition step of the evaluation methodology. Generally, the critical issues for evaluation are minor adaptations of the elements that are present in the requirements specifications step of the design process outlined in the previous section. A set of specific evaluation test requirements and tests are evolved from these objectives and needs. These must be such that each objective measure and critical evaluation issue component can be determined from at least one evaluation test instrument.

If it is determined that the system and the resulting process support cannot meet user needs, the systems design process iterates to an earlier phase and development continues. An important by-product of evaluation is the determination of ultimate performance limitations and the establishment of a protocol and procedure for use of the system that results in maximum user group satisfaction. A report is written concerning results of the evaluation process, especially those factors relating to user group satisfaction with the designed system. The evaluation process should result in suggestions for improvement in design and in better methodologies for future evaluations.

Section 37.5.6 will present additional details of the methodologies framework for evaluation. These have applicability to cases where evaluation is a separate and independent effort as well as when it is one of the phases of the design process.

Operational Deployment Phase

The last phase of design concerns operational deployment and final implementation. This must be accomplished in such a way that all user groups obtain adequate instructions in use of the system and complete operating and maintenance documentation and instructions. Specific objectives for the operational deployment phase of the system design effort are:

1. To enhance operational deployment.
2. To accomplish final design of the system.
3. To provide for continuous monitoring of postimplementation effectiveness of the system and the process into which the system is embedded.
4. To provide for retrodesign of the system as indicated by effectiveness monitoring.
5. To provide proper training and leadership for successful continued operational use of the system.
6. To identify barriers to successful implementation of the final design product.
7. To provide for "maintenance" of the system.

37.5.4 Leadership and Training Requirements

The actual use of a system as contrasted with potential usefulness is directly dependent on the value that the user group of stakeholders associates with use of the system and the resulting process in an operational environment. This is dependent, in part, on how well the system satisfies performance objectives, and on how well it is able to cope with one or more of the pathologies or pitfalls of planning, design, and /or decision making under potentially stressful operational environment conditions.

Quality planning, design, and decision support is dependent on being able to obtain relatively complete identification of pertinent factors that influence plans, designs, and decisions. The careful comprehensive formulation of issues and associated requirements for issue resolution will lead to identification of pertinent critical factors for system design. These factors are ideally illuminated in a relatively easy to understand fashion that facilitates the interpretation necessary to evaluate and subsequently select plans, designs, and decisions for implementation. Success in this is, however, strongly dependent on adroitness in use of the system. It is generally not fully meaningful to talk only of an algorithm or even a complete system, which is, typically, a piece of hardware and software, but which may well be a carefully written set of protocols and procedures, as useful by itself. It is meaningful to talk of a particular systemic process as being useful. This process involves the interaction of a methodology with systems management at the cognitive process or human judgment level. A systemic process depends on the system, the operational environment, and leadership associated with use of the system.

A process involves design integration of a methodology with the behavioral concerns of human cognitive judgment in an operational environment.

Operational evaluation of a systemic process that involves human interaction, such as an integrated manufacturing complex, appears the only realistic way to extract truly meaningful information concerning process effectiveness of a given system design. This must necessarily include leadership and training requirements to use the system. There are necessary trade-offs associated with leadership and training for using a system and these are addressed in operational evaluation.

37.5.5 System Evaluation

Previous subsections have described a framework for a general system design procedure. They have indicated the role of evaluation in this process. Successful evaluation, especially operational evaluation, is strongly dependent on explicit development of a plan for evaluation developed prior to, and perhaps modified and improved during the course of, an actual evaluation. This section will concern itself with development of a methodological framework for system evaluation, especially for operational evaluation of systemic processes for planning, design, and decision support.

Evaluation Methodology and Evaluation Criteria

Objectives for evaluation of a system concern the following:

1. Identification of a methodology for operational evaluation.
2. Establishing criteria on which the success of the system may be judged.
3. Determining the effectiveness of the support in terms of these criteria.
4. Determining the most useful strategy for employment of an existing system and potential improvements such that effectiveness of the newly implemented system and the overall process might be improved.

Figure 37.11 illustrates a partial intent structure or objectives tree, which contributes to system evaluation. The lowest-level objectives contribute to satisfaction of the 10 performance objectives for systems engineering and systems design outlined in Section 37.3. These lowest-level elements form pertinent criteria for the operational system evaluation. They concern the algorithmic effectiveness or performance objective achievement of the system, the behavioral or human factor effectiveness of the system in the operational environment, and the system efficacy. Each of these three elements become top level criteria or attributes and each should be evaluated to determine evaluation of the system itself.

Subcriteria which support the three lowest level criteria of Fig. 37.11 may be identified. These are dependent on the requirements identified for the specific system that has been designed. Table 37.2 illustrates some generic criteria that may be used to identify specific criteria. Attainment of each of these criteria by the system may be measured by observation of the system within the operational environment, and by test instruments and surveys of user groups involved with the operational system and process.

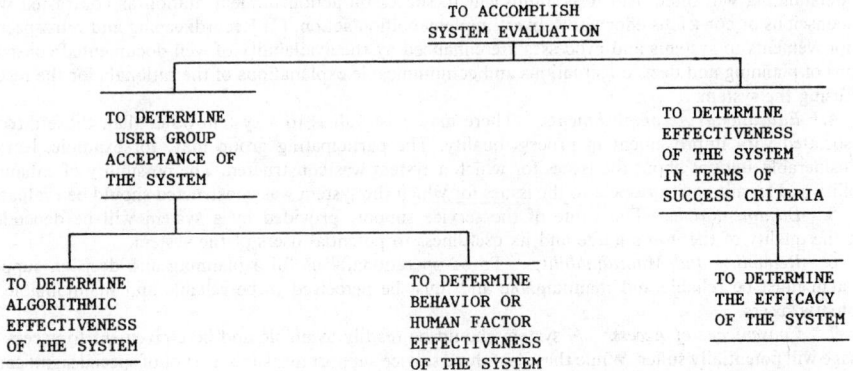

Fig. 37.11 Partial intent structure for system evaluation (criteria for evaluation are lowest-level elements).

Algorithmic Effectiveness of Performance Objectives Achievement Evaluation

A number of performance objectives can be cited, which, if achieved, should lead to a quality system. Achievement of these objectives is measured by logical soundness of the operational system and process; improved system quality as a result of using the system; and improvements in the way an overall process functions, compared to the way it typically functions without the system or with an alternative system.

Behavioral or Human Factors Evaluation

A system may be well structured algorithmically in the sense of achieving a high degree of satisfaction of the performance objectives, yet the process incorporating the system may seriously violate behavioral and human factor sensibilities. This will typically result in misuse or underuse. There are many cases where technically innovative systems have failed to achieve broad scope objectives because of human factor failures. Strongly influencing the acceptability of system implementation in operational settings are such factors as organizational slack; natural human resistance to change; and the present sophistication, attitude, and past experience of the user group and its management with similar systems and processes. Behavioral or human factor evaluation criteria used to evaluate performance include political acceptability, institutional constraint satisfaction, implementability evaluation, human workload evaluation, management procedural change evaluation, and side effect evaluation.

Efficacy Evaluation

Two of the three first-level evaluation criteria concern algorithmic effectiveness or performance objective achievement and behavioral or human factors effectiveness. It is necessary for a system to be effective in each of these in order that it be potentially capable of truly aiding in terms of improving process quality and being acceptable for implementation in the operational environment for which it was designed. There are a number of criteria or attributes related to usefulness, service support, or efficacy to which a system must be responsive. Thus evaluation of the efficacy of a system and the associated process is important in determining the service support value of the process. There are seven attributes of efficacy. Each of these will be discussed here.

    1. *Time Requirements.* The time requirements to use a system form an important service support criterion. Even if a system is potentially capable of excellent results, but the results can only be obtained after critical deadlines have passed, the overall process must be given a low rating with respect to a time responsiveness criterion.

    2. *Leadership and Training.* Leadership and training requirements for use of a system are important design considerations. It is important that there be an evaluation component directed at assessing leadership and training needs and trade-offs associated with the use of a system.

    3. *Communication Accomplishments.* Effective communication is important for two reasons. (1) Implementation action is often accomplished at a different hierarchical level, and therefore by a different set of actors, than the hierarchical level at which selection of alternative plans, designs, or decisions was made. Implementation action agents often behave poorly when an action alternative is selected that they regard as threatening or arbitrary, either personally or professionally, on an individual or a group basis. Widened perspectives of a situation are made possible by effective communication. Enhanced understanding will often lead to commitment to successful action implementation as contrasted with unconscious or conscious efforts to subvert implementation action. (2) Recordkeeping and retrospective improvements to systems and processes are enhanced by the availability of well-documented constructions of planning and decision situations and communicable explanations of the rationale for the results of using the system.

    4. *Educational Accomplishments.* There may exist values to a system other than those directly associated with improvement in process quality. The participating group may, for example, learn a considerable amount about the issues for which a system was constructed. The possibility of enhanced ability and learning with respect to the issues for which the system was constructed should be evaluated.

    5. *Documentation.* The value of the service support provided by a system will be dependent on the quality of the user's guide and its usefulness to potential users of the system.

    6. *Reliability and Maintainability.* To be operationally useful a planning and decision support system must be reliable and maintainable and must be perceived to be reliable and maintainable by potential users.

    7. *Convenience of Access.* A system should be readily available and be convenient to access, or usage will potentially suffer. While these last three service support measures are not of special significance with respect to justification of the need for a system, they may be important in determining operational usage, and, therefore, operational effectiveness of a system and the associated process.

37.5.6 Evaluation Test Instruments

Several special evaluation test instruments to satisfy test requirements and measure achievement of the evaluation criteria will need to be developed. These include investigations of effectiveness in terms of performance objective attainment; selection of appropriate scenarios that affect use of the system, use of the system subject to these scenarios by a test group, and completion of evaluation questionnaires; and questionnaires and interviews with operational users of the system.

Every effort must be made to ensure, to the extent possible, that evaluation test results will be credible and valid. Intentional redundancy should be provided to allow correlation of results obtained from the test instruments to ensure maximum supportability and reliability of the facts and opinions to be obtained from test procedures.

The evaluation team should take advantage of every opportunity to observe use of the system within the operational environment. Evaluation of personnel reactions to the aid should be based on observations, designed to be responsive to critical evaluation issues, and the response of operational environment personnel to test questionnaires. When any of a number of constraints make it difficult to obtain real time operational environment observation, experiential and anecdotal information becomes of increased value. Also, retrospective evaluation of use of a system is definitely possible and desirable if sufficiently documented records of past usage of an aided process are available.

Many other effectiveness questions will likely arise as an evaluation proceeds. Questions specific to a given evaluation are determined after study of the particular situation and the system being evaluated. It is, however, important to have an initial set of questions to guide the evaluation investigation and a purpose of this subsection to provide a framework for accomplishing this.

One of the important concerns in evaluation is that of those parts of the efficacy evaluation that deal with various "abilities" of a system. These include producibility, reliability, maintainability, and marketability. Table 37.2 presents a listing of attributes that may be used to "score" the performance of systems on relevant effectiveness criteria.

37.6 CONCLUSIONS

In this chapter we have discussed salient aspects concerning the systems engineering of large and complex systems. We have been concerned especially with development of a methodology for the design and evaluation of systems. There are a number of effectiveness attributes or aspects of effective systems. Design of an effective large-scale system necessarily involves integration of operational environ-

---

Table 37.2 Criteria for Evaluation of Systems and Systemic Processes

Algorithmic Effectiveness or Performance Objective Achievement Evaluation

    Logical soundness
    Improved performance quality
    Process changes

Behavioral or Human Factors Evaluation

    Political acceptability
    Institutional constraints
    Implementability
    Workload changes
    Procedural changes
    Side effects

Efficacy Evaluation

    Time requirements
    Leadership and training requirements
    Communication accomplishments
    Educational accomplishments
    Documentation
    Reliability (and other "abilities")
    Convenience of access

ment concerns involving human behavior and judgment with mechanistic and physical science concerns. An effective systemic design process should:

1. Allow a thorough and carefully conducted requirements specification effort to determine and specify needs of stakeholders prior to conceptual design of a system process to accomplish the desired task.
2. Be capable of dealing with both quantitative and qualitative criteria representing costs and effectiveness from their economic, social, environmental, and other perspectives.
3. Be capable of minimizing opportunities for cognitive bias, and provide debiasing procedures for those biases that occur.
4. Allow separation of opinions and facts from values, and separation of ends from means, or values from alternative acts.
5. Provide an objective communicable framework that allows identification, formulation, and display of the structure of the issue under consideration, as well as the rationale of the choice process.
6. Allow for considerations of trade-offs among conflicting and incommensurate criteria.
7. Provide flexibility and monitoring support to allow design process evaluation rule selection with due consideration to the task structure and operational environment constraints on the decision maker.
8. Provide an open process to allow consideration of new criteria and alternatives as values change and broadscope awareness of issues grows.

There are a number of potential benefits of the systems approach that should follow high achievement of each of the criteria for effective systemic processes. An appropriate systemic design process will:

1. Provide structure to relatively unstructured issues.
2. Facilitate conceptual formulation of issues.
3. Provide cognitive cues to search and discovery.
4. Encourage parsimonious collection, organization, and utilization of relevant data.
5. Extend and debias information processing abilities.
6. Encourage vigilant cognitive style.
7. Provide brokerage between parties at interest.

There are many imperfections and limits to processes designed using the methodologies from what we know as systems engineering and systems analysis. Some of these have been documented in this chapter. But what are the alternatives to appropriate systemic processes for the resolution of issues associated with the design of complex large-scale systems; and are not the fundamental limitations to these alternatives even greater?

37.7 APPLICATION AREAS AND GUIDE TO RELEVANT LITERATURE

It would be virtually impossible to detail an exhaustive listing of areas of potential application of systems engineering. Among the many areas in which systems engineering has been applied in order to organize a knowledge base of information for planning, design, and decision making are:

Computers	Telecommunications
Controls	Agricultural resources
Communications	Law enforcement
Large-scale networks	Resource management
Health service delivery	Military planning
Urban modeling	Product procurement
Transportation	System acquisition and procurement
Traffic control	Budgeting
Industrial automation	Portfolio analysis
Environmental protection	Marketing
Automated manufacturing	Management information systems
Energy	Communication, command, control, and intelligence system

REFERENCES

Our discussion of systems engineering is regrettably but necessarily incomplete. We shall conclude our effort with a presentation of 55 citations that complement and extend this presentation. A number of the personal references contain discussions that support and extend those contained here. References to the works of others are historical and/or seminal in nature. Appropriate references for a novice to systems engineering are 6, 8, 10, 12, 13, 21, 25, 27, 39, 41, 49, and 50.

There is considerable literature that concerns various facets of systems engineering. Some references that expand upon the general methodological discussion presented here are:

1. A. P. Sage, *Methodology for Large Scale Systems,* McGraw-Hill, New York, 1977.
2. A. P. Sage (Ed.), *Systems Engineering: Methodology and Applications,* IEEE Press and Wiley, New York, 1977.
3. A. P. Sage, "Methodological Considerations in the Design of Large Scale Systems Engineering Processes," in *Large Scale Systems,* Y. Haimes (Ed.), North Holland, Amsterdam, 1982, Chap. 5, pp. 99–141.

The last reference is to an edited work of specially written papers for the Case Western Reserve University Centennial that contains a number of contributions relevant to the related areas of systems engineering, management science, and operations research. Discussion of alternate viewpoints concerning systems engineering is contained in Refs. 2, 3, and

4. A. D. Hall, *A Methodology for Systems Engineering,* Van Nostrand, New York, 1962.
5. H. Chestnut, *Systems Engineering Methods,* Wiley, New York, 1967.
6. E. E. David, E. J. Piel, and J. G. Truxal (Eds.), *The Man Made World,* McGraw-Hill, New York, 1971.
7. H. H. Goode and R. E. Machol, *Systems Engineering,* McGraw-Hill, New York, 1957.
8. M. F. Rubinstein, *Patterns of Problem Solving,* Prentice-Hall, Englewood Cliffs, NJ, 1975.
9. N. Wiener, *Cybernetics,* Wiley, New York, 1948.
10. J. Beishon and G. Peters, *Systems Behavior,* Harper and Row, New York, 1976.
11. B. S. Blanchard and W. J. Fabrycky, *Systems Engineering and Analysis,* Prentice-Hall, Englewood Cliffs, NJ, 1981.
12. J. E. Robertshaw, S. J. Mecca, and M. N. Rerick, *Problem Solving: A Systems Approach,* Petrocelli, Princeton, NJ, 1978.
13. C. W. Churchman, *The Systems Approach,* Delacorte Press, New York, 1968.

References 4, 5, 7, and 9 are of interest from historical perspectives. References 6, 8, 10, and 12 are elementary works.

One of the principal concerns of systems engineering is planning and design. A very appropriate text which discusses this is

14. G. N. Nadler, *The Planning and Design Approach,* Wiley, New York, 1981.

Systems engineering involves fundamental issues that include the roles of reason, chance, cause, experience, discovery, logic, and argument. These support its role as a management technology in knowledge organization for decision making. A seminal text in this area is

15. A. W. Burks, *Chance, Cause, Reason,* University of Chicago Press, Chicago, IL, 1977.

The first step of the systems engineering framework is issue formulation. An appropriate source for discussion of a number of structured approaches to issue formulation is

16. A. B. Van Gundy, Jr., *Techniques of Structured Problem Solving,* Van Nostrand, New York, 1981.

There are many possible references that might be cited concerning systems analysis. Two excellent fundamental works that discuss the graph-theoretic basis for system analysis structural models are

17. F. Harary, R. Z. Norman, and D. Cartwright, *Structural Models: An Introduction to the Theory of Directed Graphs,* Wiley, New York, 1965.
18. F. Roberts, *Discrete Mathematical Models with Applications to Social, Biological, and Environmental Problems,* Prentice-Hall, Englewood Cliffs, NJ, 1976.

There have been many graph-theoretic applications in systems engineering. Many of these are discussed in

19. D. V. Steward, *Systems Analysis and Management: Structure, Strategy, Design,* Petrocelli, Princeton, NJ, 1981.

A discussion of the interpretive structural modeling approach, which is very useful for relating elements according to influential relations, may be found in Refs. 1, 2, 19, and

20. J. N. Warfield, *Societal Systems: Planning, Policy, Complexity,* Wiley, New York, 1976.

which also presents a number of methodological discussions.

Cognitive maps, which are very useful structural models of belief systems, are discussed in

21. C. Eden, S. Jones, and D. Sims, *Thinking in Organizations,* Macmillan, New York, 1979.

Discussions of artificial intelligence perspectives on knowledge representation may be found in

22. R. C. Schank and C. K. Riesbeck, *Inside Computer Understanding,* Erlbaum, Hillsdale, NJ, 1981.

23. R. S. Michalski et al. (Eds.), *Machine Learning, An Artificial Intelligence Approach,* Tioga, 1983.

24. A. Barr, P. R. Cohen, and E. A. Feigenbaum (Eds.), *Handbook of Artificial Intelligence,* Kaufman, Los Altos, CA, 1981 and 1982.

There are a variety of approaches to model building and simulation. Very useful introductory presentations are contained in

25. D. H. Burghes and A. D. Wood, *Mathematical Models in the Social, Management, and Life Sciences,* Wiley, New York, 1980.

26. C. L. Dym and E. J. Ivey, *Principles of Mathematical Modeling,* Academic Press, New York, 1980.

27. L. E. Alfeld and A. K. Graham, *Introduction to Urban Dynamics,* Wright Allen Press, 1976.

Especially recommended for its breadth of treatment is

28. S. I. Gass and R. Sisson (Eds.), *A Guide to Models in Government Planning and Operations,* Sauger Books, 1975.

An excellent perspective on the use and value of models is contained in the foregoing reference and particularly in

29. M. Greenberger, M. A. Crenson, and B. L. Crissey, *Models in the Policy Process,* Russell Sage Foundation, 1976.

Operations research models for analysis such as mathematical programming and queueing theory are presented in

30. F. S. Hillier and G. J. Lieberman, *Introduction to Operations Research,* Holden Day, San Francisco, CA, 1967.

31. H. M. Wagner, *Principles of Operations Research,* Prentice-Hall, Englewood Cliffs, NJ, 1975.

as well as a large number of other texts. Economic systems analysis approaches, including cost-benefit and cost-effectiveness analysis, are discussed in

32. R. De Neufville and J. H. Stafford, *Systems Analysis for Engineers and Managers,* McGraw-Hill, New York, 1971.

33. A. P. Sage, *Economic Systems Analysis,* North Holland, Amsterdam, 1983.

Forecasting the impacts of proposed policy is an especially important activity within systems engineering. Two definitive sources that contain a wealth of information relative to forecasting are

34. J. S. Armstrong, *Long Range Forecasting: From Crystal Ball to Computer,* Wiley, New York, 1978.

35. A. L. Porter et al., *A Guidebook for Technology Assessment and Impact Analysis,* North Holland, Amsterdam, 1980.

Many approaches in systems engineering are based on concepts from probability theory and statistics. References that discuss these important topics are

36. J. L. Melsa and A. P. Sage, *An Introduction to Probability and Stochastic Processes,* Prentice-Hall, Englewood Cliffs, NJ, 1973.

37. R. L. Winkler and W. L. Hays, *Statistics: Probability, Inference, and Decision,* Holt, Rinehart and Winston, New York, 1975.

Optimization of dynamic systems is another important topic in systems engineering. Among the many references that discuss advances in this area is

38. A. P. Sage and C. C. White, *Optimum Systems Control,* Prentice-Hall, New York, 1977.

The systems engineering step of interpretation is composed of decision making and planning for implementation action. Many of the references already cited contain sections on implementation. Four particularly good introductory discussions of decision analysis and decision making are contained in

39. H. Raiffa, *Decision Analysis—Introductory Lectures on Choices Under Uncertainty,* Addison Wesley, Reading, MA, 1968.

40. R. V. Brown, A. S. Kahr, and C. Peterson, *Decision Analysis for the Manager,* Holt, Rinehart and Winston, New York, 1974.

41. C. A. Holloway, *Decision Making Under Uncertainty, Models and Choice,* Prentice-Hall, Englewood Cliffs, NJ, 1979.

42. G. P. Huber, *Managerial Decision Making,* Scott, Foresman, Glenview, IL, 1980.

As we have indicated throughout our discussions, systems management concerns at the cognitive process level are especially important in systems engineering. References that support our discussions include those which deal with the fallibility of human judgment:

43. I. L. Janis and L. Mann, *Decision Making,* Free Press, 1977.

44. R. M. Hogarth, *Judgment and Choice,* Wiley, New York, 1980.

45. H. A. Simon, *Reason in Human Affairs,* Stanford University Press, Palo Alto, CA, 1983.

46. D. Kahneman, P. Slovic, and A. Tversky (Eds.), *Judgment Under Uncertainty: Heuristics and Biases,* Cambridge, England, 1982.

Closely related to this are studies involving risk and hazard such as

47. B. Fischhoff et al., *Acceptable Risk,* Cambridge University Press, England, 1981.

and the growing literature in knowledge-based systems for planning and decision support such as

48. P. G. W. Keen and M. S. Morton, *Decision Support Systems: An Organizational Perspective,* Addison Wesley, Reading, MA, 1978.

A long survey and interpretation of much of this literature is contained in

49. A. P. Sage, "Behavioral and Organizational Considerations in the Design of Information Systems and Processes for Planning and Decision Support," *IEEE Transactions on Systems, Man, and Cybernetics,* **SMC-11**(9), 640–678 (Sep. 1981).

Much of systems engineering is associated with measurement. Recent seminal works concerning this are

50. F. S. Roberts, *Measurement Theory,* Addison Wesley, Reading, MA, 1979.

51. R. O. Mason and E. B. Swanson (Eds.), *Measurement for Management Decision,* Addison Wesley, Reading, MA, 1981.

The Mason and Swanson reference contains a number of chapters relative to various perspectives on information and value measurements and, in particular, the use of these measurements in decision support systems.

The design of human–machine systems is but one of many contemporary areas in which systems engineering approaches are of much interest. Two recent works are

52. W. B. Rouse (Ed.), *Advances in Man Machine Systems Research,* JAI Press, 1983.

53. A. P. Sage (Ed.), Special Issue on "Knowledge Based Systems Design in Man-Machine Systems Research," *Automatica* (Dec. 1983).

Essential in the successful practice of systems engineering is an awareness of fundamental limitations and pitfalls associated with the use of the approaches of systems science and operations research, systems methodology and design, and systems management. Our final two references,

54. E. S. Quade and G. Majone, *Pitfalls of Analysis,* Wiley, New York, 1980.

55. A. P. Sage, "Systems Engineering: Fundamental Limits and Future Prospects," *IEEE Proceedings,* **69**(2), 158–166 (Feb. 1981).

discuss these important concerns as they relate to professional practice.

CHAPTER 38
HISTORICAL NOTE ON CONTROL

MICHAEL J. RABINS

College of Engineering
Wayne State University
Detroit, Michigan

The basic concept of feedback control goes back to antiquity. As long as humans have devised mechanisms and devices to control their environment, they have searched for ways to control these mechanisms and devices. Examples of early controls practice are the accurate regulation of waterflow in ancient Greek water clocks, overflow regulation in irrigation troughs to save precious water in arid lands, and in medieval times, early control of windmills to keep them pointing into the wind and turning at regulated speeds.

However, the concentrated start of control applications as an art is generally acknowledged to coincide with the use of the Watt Steam Engine Speed Governor and its continued improvement from the middle of the eighteenth century up to the present. The early beginnings of the theory of automatic control date back to a paper by J. C. Maxwell entitled "On Governors" presented to the Royal Society in England in 1868. In the second half of the nineteenth century and the early part of the twentieth century, regulators were developed that were related to power generation and transmission, and there was some early analyses of telescope control systems and autopilots for airplanes. A 1922 paper by Sikorsky on ship steering was seminal. However, these early analyses were rudimentary and very limited.

Early in the twentieth century there was significant progress in the practice of automatic controls via the development of instruments and regulators for the process or power industries. Following World War I, and on into the mid-1930s, there was additional development in the field of automatic controls to meet the needs of the communication industry. In particular, work done at the Bell Telephone Laboratories under Bode and Nyquist in the development of feedback amplifiers and telephone repeater systems laid the foundation for much of the later theoretical and design work. At the same time there was rapid development in the use of differential equations, frequency response, Laplace transforms, and the Routh–Hurwitz stability criteria. The Nyquist and Routh–Hurwitz stability criteria as well as Bode design techniques are still frequently used today in successful system development.

World War II saw the necessarily fast development of many control techniques and devices for weaponry and for advanced industrial controls. In particular the Ziegler–Nichols tuning method was developed late in the war in conjunction with the processing of fuels for nuclear weapons. This period also saw the development of many different kinds of electronic controllers that were used in communications and navigation. The extensive use of the frequency response method and the further development of Laplace domain techniques for system modeling and design, sometimes called "classical" control theory, were outcomes of these efforts.

Following World War II, particularly as secret information became declassified and available to the public, there was widespread industrial application of wartime control theory and techniques. Evans' root locus technique and z transform methods were developed. In the early 1960s, as digital computers started to become available, the use of the computer for direct digital control was developed. Automation in industry, based on computerization, began to find heavy application.

Also in the early 1960s, the state-space approach—sometimes called "modern" control theory— was developed by Kalman et al. based on turn-of-the-century Lyapunov concepts. Optimal control theory and the mathematical theory of controlled processes (e.g., system controllability and observability) made their appearances. Simultaneously, space projects, the biomedical field, manufacturing processes, and many other application areas began to make heavy use of control systems theory and practice. In the late 1960s and early 1970s there was detailed analysis of optimal controls and much progress toward dynamic optimization, particularly on U.S. Defense Department projects. This period saw the placement of a man on the moon, in which systems control and instrumentation techniques played a large part.

Through the late 1970s and early 1980s development in automatic controls systems theory and practice kept pace with the increasing availability of computing power at ever lower costs. In particular, the availability of stand-alone microcomputers saw the widespread development of decentralized control and the use of dedicated computer control systems for small segments of process and industrial manufacturing systems. This period witnessed the increased growth in the application of real-time, on-line control to manufacturing processes and in particular to the area of robotics. Presently, there is hardly a field of engineering and scientific endeavor that does not heavily draw on the techniques of systems control theory and instrumentation practice.

CHAPTER 39

MATHEMATICAL MODELS OF DYNAMIC PHYSICAL SYSTEMS

K. PRESTON WHITE, JR.

University of Virginia
Department of Systems Engineering
Charlottesville, Virginia

39.1 RATIONALE

The design of modern control systems is inextricably linked to the formulation and analysis of mathematical models of dynamic physical systems. This is simply because a model is more accessible to study than the physical system the model represents. Models typically are less costly and less time consuming to construct and test. Changes in the structure of a model are easier to implement, and changes in the behavior of a model are easier to isolate and understand. A model often can be used to achieve insight when the corresponding physical system cannot, because experimentation with the actual system is too dangerous or too demanding. Indeed, a model can be used to answer "what if" questions about a system that has not yet been realized or actually cannot be realized with current technologies.

The type of model used by the control engineer depends upon the nature of the system the model represents, the objectives of the engineer in developing the model, and the tools which the engineer has at his or her disposal for developing and analyzing the model. A mathematical model is a description of a system in terms of equations. Because the physical systems of primary interest to the control engineer are dynamic in nature, the mathematical models used to represent these systems most often incorporate difference or differential equations. Such equations, based on physical laws and observations, are statements of the fundamental relationships among the important variables that describe the system. Difference and differential equation models are expressions of the way in which the current values assumed by the variables combine to determine the future values of these variables.

Mathematical models are particularly useful because of the large body of mathematical theory that exists for the study and solution of equations. Based on this theory, a wide range of techniques has been developed specifically for the study of control systems. In recent years, computer programs have been written that implement virtually all of these techniques. Computer program packages are now commonly available for both simulation and computational assistance in the analysis and design of control systems (Refs. 3, 12, 13, 15, 17, 20, 25, 26, among others).

It is important to understand that a variety of models can be realized for any given physical system. The choice of a particular model always represents a tradeoff between the fidelity of the model and the effort required in model formulation and analysis. This tradeoff is reflected in the nature and extent of simplifying assumptions used to derive the model. In general, the more faithful the model is as a description of the physical system modeled, the more difficult it is to obtain general solutions. In the final analysis, the best engineering model is not necessarily the most accurate or precise. It is, instead, the simplest model that yields the information needed to support a decision. A classification of various types of models commonly encountered by control engineers is given in Section 39.8.

A large and complicated model is justified if the underlying physical system is itself complex, if the individual relationships among the system variables are well understood, if it is important to understand the system with a great deal of accuracy and precision, and if time and budget exist to support an extensive study. In this case, the assumptions necessary to formulate the model can be minimized. Such complex models cannot be solved analytically, however. The model itself must be studied experimentally, using the techniques of computer simulation. This approach to model analysis is treated in Section 39.7.

Simpler models frequently can be justified, particularly during the initial stages of a control system study. In particular, systems that can be described by linear difference or differential equations permit the use of powerful analysis and design techniques. These include the transform methods of classical control theory and the state-variable methods of modern control theory. Descriptions of these standard forms for linear systems analysis are presented in Sections 39.4, 39.5, and 39.6.

During the past several decades, a unified approach for developing lumped-parameter models of physical systems has emerged. This approach is based on the idea of idealized system elements, which store, dissipate, or transform energy. Ideal elements apply equally well to the many kinds of physical systems encountered by control engineers. Indeed, because control engineers most frequently deal with systems that are part mechanical, part electrical, part fluid, and/or part thermal, a unified approach to these various physical systems is especially useful and economic. The modeling of physical systems using ideal elements is discussed further in Sections 39.2, 39.3, and 39.4.

Frequently, more than one model is used in the course of a control system study. Simple models that can be solved analytically are used to gain insight into the behavior of the system and to suggest candidate designs for controllers. These designs are then verified and refined in more complex models, using computer simulation. If physical components are developed during the course of a study, it is often practical to incorporate these components directly into the simulation, replacing the corresponding model components. An iterative, evolutionary approach to control systems analysis and design is depicted in Fig. 39.1.

39.2 IDEAL ELEMENTS

Differential equations describing the dynamic behavior of a physical system are derived by applying the appropriate physical laws. These laws reflect the ways in which energy can be stored and transferred within the system. Because of the common physical basis provided by the concept of energy, a general

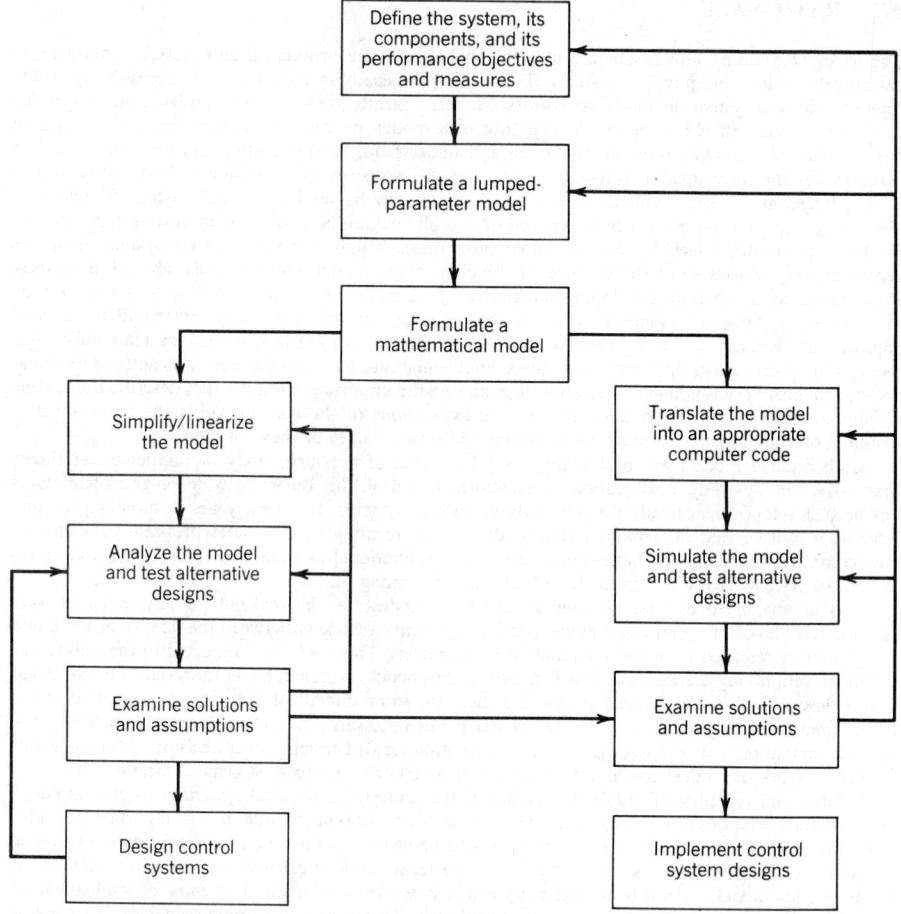

Fig. 39.1 An iterative approach to control system design, showing the use of mathematical analysis and computer simulation.

approach to deriving differential equation models is possible (Refs. 2, 8, 19, 23, 24, and 27). This approach applies equally well to mechanical, electrical, fluid, and thermal systems and is particularly useful for systems that are combinations of these physical types.

39.2.1 Physical Variables

An idealized *two-terminal* or *one-port* element is shown in Fig. 39.2. Two *primary physical variables* are associated with the element: a through variable $f(t)$ and an across variable $v(t)$. *Through variables* represent quantities that are transmitted through the element, such as the force transmitted through a spring, the current transmitted through a resistor, or the flow of fluid through a pipe. Through variables have the same value at both ends or terminals of the element. *Across variables* represent the difference in state between the terminals of the element, such as the velocity difference across the ends of a spring, the voltage drop across a resistor, or the pressure drop across the ends of a pipe. *Secondary physical variables* are the integrated through variable $h(t)$ and the integrated across variable $x(t)$. These represent the accumulation of quantities within an element as a result of the integration of the associated through and across variables. For example, the momentum of a mass is an integrated through variable, representing the effect of forces on the mass integrated or accumulated over time. Table 39.1 defines the primary and secondary physical variables for various physical systems.

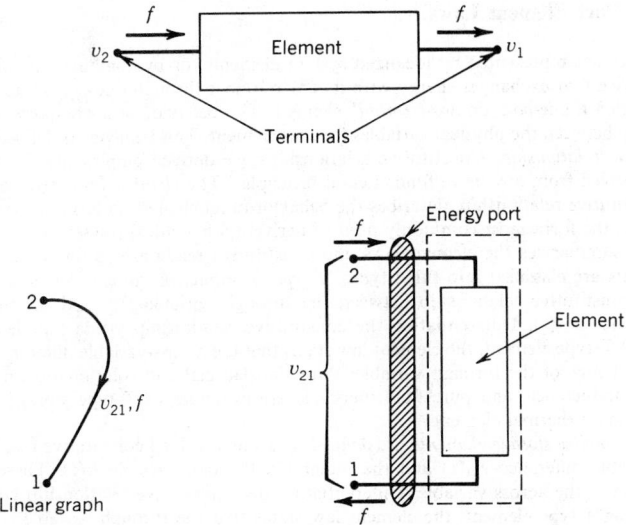

Fig. 39.2 A two-terminal or one-port element, showing through and across variables.[23]

39.2.2 Power and Energy

The flow of *power* $P(t)$ into an element through the terminals 1 and 2 is the product of the through variable $f(t)$ and the difference between the across variables $v_2(t)$ and $v_1(t)$. Suppressing the notation for time dependence, this may be written as

$$P = f(v_2 - v_1) = fv_{21}$$

A negative value of power indicates that power flows out of the element. The *energy* $E(t_a, t_b)$ transferred to the element during the time interval from t_a to t_b is the integral of power, that is,

$$E = \int_{t_a}^{t_b} P \, dt = \int_{t_a}^{t_b} fv_{21} \, dt$$

A negative value of energy indicates a net transfer of energy out of the element during the corresponding time interval.

Thermal systems are an exception to these generalized energy relationships. For a thermal system, power is identically the through variable $q(t)$, heat flow. Energy is the integrated through variable $H(t_a, t_b)$, the amount of heat transferred.

By the *first law of thermodynamics*, the net energy stored within a system at any given instant must equal the difference between all energy supplied to the system and all energy dissipated by the system. The generalized classification of elements given in the following sections is based on whether the element stores or dissipates energy within the system, supplies energy to the system, or transforms energy between parts of the system.

Table 39.1 Primary and Secondary Physical Variables for Various Systems[23]

System	Through Variable f	Integrated Through Variable h	Across Variable v	Integrated Across Variable x
Mechanical–translational	Force F	Translational momentum p	Velocity difference v_{21}	Displacement difference x_{21}
Mechanical–rotational	Torque T	Angular momentum h	Angular velocity difference Ω_{21}	Angular displacement difference Θ_{21}
Electrical	Current i	Charge q	Voltage difference v_{21}	Flux linkage λ_{21}
Fluid	Fluid flow Q	Volume V	Pressure difference P_{21}	Pressure–momentum Γ_{21}
Thermal	Heat flow q	Heat energy $\mathcal{H}$	Temperature difference θ_{21}	Not used in general

39.2.3 One-Port Element Laws

Physical devices are represented by idealized system elements, or by combinations of these elements. A physical device that exchanges energy with its environment through one pair of across and through variables is called a *one-port* or *two-terminal* element. The behavior of a one-port element expresses the relationship between the physical variables for that element. This behavior is defined mathematically by a *constitutive relationship*. Constitutive relationships are derived empirically, by experimentation, and are not derived from any more fundamental principles. The *element law*, derived from the corresponding constitutive relationship, describes the behavior of an element in terms of across and through variables and is the form most commonly used to derive mathematical models.

Table 39.2 summarizes the element laws and constitutive relationships for the one-port elements. Passive elements are classified into three types. *T-type* or *inductive storage* elements are defined by a single-valued constitutive relationship between the through variable $f(t)$ and the integrated across-variable difference $x_{21}(t)$. Differentiating the constitutive relationship yields the element law. For a linear (or ideal) T-type element, the element law states that the across-variable difference is proportional to the rate of change of the through variable. Pure translational and rotational compliance (springs), pure electrical inductance, and pure fluid inertance are examples of T-type storage elements. There is no corresponding thermal element.

A-type or *capacitive storage elements* are defined by a single-valued constitutive relationship between the across-variable difference $v_{21}(t)$ and the integrated through variable $h(t)$. These elements store energy by virtue of the across variable. Differentiating the constitutive relationship yields the element law. For a linear A-type element, the element law states that the through variable is proportional to the derivative of the across-variable difference. Pure translational and rotational inertia (masses), and pure electrical, fluid, and thermal capacitance are examples.

It is important to note that when a nonelectrical capacitance is represented by an A-type element, one terminal of the element must have a constant (reference) across variable, usually assumed to be zero. In a mechanical system, for example, this requirement expresses the fact that the velocity of a mass must be measured relative to a noninertial (nonaccelerating) reference frame. The constant velocity terminal of a pure mass may be thought of as being attached in this sense to the reference frame.

D-type or *resistive elements* are defined by a single-valued constitutive relationship between the across and the through variables. These elements dissipate energy, generally by converting energy into heat. For this reason, power always flows into a D-type element. The element law for a D-type energy dissipator is the same as the constitutive relationship. For a linear dissipator, the through variable is proportional to the across-variable difference. Pure translational and rotational friction (dampers or dashpots), and pure electrical, fluid, and thermal resistance are examples.

Energy-storage and energy-dissipating elements are called *passive* elements, because such elements do not supply outside energy to the system. The fourth set of one-port elements are *source elements,* which are examples of *active* or power-supplying elements. Ideal sources describe interactions between the system and its environment. A pure *A-type source* imposes an across-variable difference between its terminals, which is a prescribed function of time, regardless of the values assumed by the through variable. Similarly, a pure *T-type source* imposes a through-variable flow through the source element, which is a prescribed function of time, regardless of the corresponding across variable.

Pure system elements are used to represent physical devices. Such models are called *lumped-element models.* The derivation of lumped-element models typically requires some degree of approximation, since (1) there rarely is a one-to-one correspondence between a physical device and a set of pure elements and (2) there always is a desire to express an element law as simply as possible. For example, a coil spring has both mass and compliance. Depending on the context, the physical spring might be represented by a pure translational mass, or by a pure translational spring, or by some combination of pure springs and masses. In addition, the physical spring undoubtedly will have a nonlinear constitutive relationship over its full range of extension and compression. The compliance of the coil spring may well be represented by an ideal translational spring, however, if the physical spring is approximately linear over the range of extension and compression of concern.

39.2.4 Multiport Elements

A physical device that exchanges energy with its environment through two or more pairs of through and across variables is called a *multiport element*. The simplest of these, the idealized *four-terminal* or *two-port* element, is shown in Fig. 39.3. Two-port elements provide for transformations between the physical variables at different energy ports, while maintaining instantaneous continuity of power. In other words, net power flow into a two-port element is always identically zero:

$$P = f_a v_a + f_b v_b = 0$$

The particulars of the transformation between the variables define different categories of two-port elements.

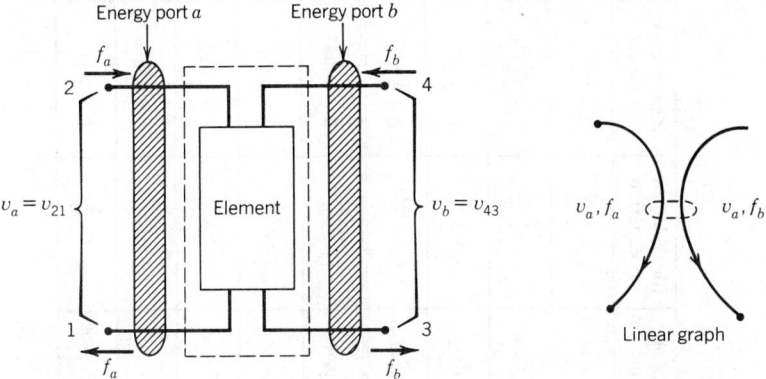

Fig. 39.3 A four-terminal or two-port element, showing through and across variables.

A *pure transformer* is defined by a single-valued constitutive relationship between the integrated across variables or between the integrated through variables at each port:

$$x_b = f(x_b) \quad \text{or} \quad h_b = f(h_a)$$

For a linear (or ideal) transformer, the relationship is proportional, implying the following relationships between the primary variables:

$$v_b = nv_a, \qquad f_b = -\frac{1}{n}f_a$$

where the constant of proportionality n is called the *transformation ratio*. Levers, mechanical linkages, pulleys, gear trains, electrical transformers, and differential-area fluid pistons are examples of physical devices that typically can be approximated by pure or ideal transformers. Figure 39.4 depicts some examples. *Pure transmitters*, which serve to transmit energy over a distance, frequently can be thought of as transformers with $n = 1$.

A *pure gyrator* is defined by a single-valued constitutive relationship between the across variable at one energy port and the through variable at the other energy port. For a linear gyrator, the following relations apply:

$$v_b = rf_a, \qquad f_b = \frac{-1}{r}v_a$$

where the constant of proportionality is called the *gyration ratio* or *gyrational resistance*. Physical devices that perform pure gyration are not as common as those performing pure transformation. A mechanical gyroscope is one example of a system that might be modeled as a gyrator.

In the preceding discussion of two-port elements, it has been assumed that the type of energy is the same at both energy ports. A *pure transducer*, on the other hand, changes energy from one physical medium to another. This change may be accomplished either as a transformation or a gyration. Examples of *transforming transducers* are gears with racks (mechanical rotation to mechanical translation), electric motors and electric generators (electrical to mechanical rotation and vice versa). Examples of *gyrating transducers* are the piston-and-cylinder (fluid to mechanical) and piezoelectric crystals (mechanical to electrical).

More complex systems may have a large number of energy ports. A common *six-terminal* or *three-port element* called a *modulator* is depicted in Fig. 39.5. The flow of energy between ports a and b is controlled by the energy input at the modulating port c. Such devices inherently dissipate energy, since

$$P_a + P_c \geq P_b$$

although most often the modulating power P_c is much smaller than the power input P_a or the power output P_b. When port a is connected to a pure source element, the combination of source and modulator is called a *pure dependent source*. When the modulating power P_c is considered the input and the modulated power P_b is considered the output, the modulator is called an *amplifier*. Physical devices that often can be modeled as modulators include clutches, fluid valves and couplings, switches, relays, transistors, and variable resistors.

Table 39.2 Element Laws and Constitutive Relationships for Various One-Port Elements[23]

Type of element	Physical element	Linear graph	Diagram	Constitutive relationship	Energy or power function	Ideal elemental equation	Ideal energy or power
T-type energy storage $\varepsilon > 0$ Pure: $x_{21}=f(f)$, $\varepsilon=\int_0^f f\,dx_{21}$ Ideal: $x_{21}=Lf$, $\varepsilon=\frac{1}{2}Lf^2$	Translational spring	2 — k — $\frac{v_{21}}{F}$ — 1	F; v_1,x_1; v_2,x_2	$x_{21}=f(F)$	$\varepsilon=\int_0^F F\,dx_{21}$	$v_{21}=\frac{1}{k}\frac{dF}{dt}$	$\varepsilon=\frac{1}{2}\frac{F^2}{k}$
	Rotational spring	2 — K — $\frac{\Omega_{21}}{T}$ — 1	T; Ω_1,θ_1; Ω_2,θ_2	$\theta_{21}=f(T)$	$\varepsilon=\int_0^T T\,d\theta_{21}$	$\Omega_{21}=\frac{1}{K}\frac{dT}{dt}$	$\varepsilon=\frac{1}{2}\frac{T^2}{K}$
	Inductance	2 — L — $\frac{v_{21}}{i}$ — 1	i; v_1,λ_1; v_2,λ_2	$\lambda_{21}=f(i)$	$\varepsilon=\int_0^i i\,d\lambda_{21}$	$v_{21}=L\frac{di}{dt}$	$\varepsilon=\frac{1}{2}Li^2$
	Fluid inertance	2 — I — $\frac{P_{21}}{Q}$ — 1	Q; P_1,Γ_1; P_2,Γ_2	$\Gamma_{21}=f(Q)$	$\varepsilon=\int_0^Q Q\,d\Gamma_{21}$	$P_{21}=I\frac{dQ}{dt}$	$\varepsilon=\frac{1}{2}IQ^2$
A-type energy storage $\varepsilon > 0$ Pure: $h=f(v_{21})$, $\varepsilon=\int_0^{v_{21}} v_{21}\,dh$ Ideal: $h=Cv_{21}$, $\varepsilon=\frac{1}{2}Cv_{21}^2$	Translational mass	2 — m — $\frac{v_{21}}{F}$ — 1	F,p; v_2; $v_1=$const	$p=f(v_2)$	$\varepsilon=\int_0^{v_2} v_2\,dp$	$F=m\frac{dv_2}{dt}$	$\varepsilon=\frac{1}{2}mv_2^2$
	Inertia	2 — J — $\frac{\Omega_{21}}{T}$ — 1	T,h; Ω_2; $\Omega_1=$const	$h=f(\Omega_2)$	$\varepsilon=\int_0^{\Omega_2}\Omega_2\,dh$	$T=J\frac{d\Omega_2}{dt}$	$\varepsilon=\frac{1}{2}J\Omega_2^2$
	Electrical capacitance	2 — C — $\frac{v_{21}}{i}$ — 1	i; q; v_2; v_1	$q=f(v_{21})$	$\varepsilon=\int_0^{v_{21}} v_{21}\,dq$	$i=C\frac{dv_{21}}{dt}$	$\varepsilon=\frac{1}{2}Cv_{21}^2$
	Fluid capacitance	2 — C_f — $\frac{P_{21}}{Q}$ — 1	Q,V; P_2; $P_1=$const	$V=f(P_2)$	$\varepsilon=\int_0^{P_2} P_2\,dV$	$Q=C_f\frac{dP_2}{dt}$	$\varepsilon=\frac{1}{2}C_f P_2^2$
	Thermal capacitance	2 — C_t — $\frac{\theta_{21}}{q}$ — 1	$q,\mathcal{H}$; θ_2; $\theta_1=$const	$\mathcal{H}=f(\theta_2)$	$\varepsilon=\int_0^{\theta_2} q\,dt=\mathcal{H}$	$q=C_t\frac{d\theta_2}{dt}$	$\varepsilon=C_t\theta_2$

			Pure		Ideal	
D-type energy dissipators $\mathscr{P} \geqslant 0$			$f = f(v_{21})$ $\mathscr{P} = v_{21}f(v_{21})$		$f = \frac{1}{R}v_{21}$ $\mathscr{P} = \frac{1}{R}v_{21}^2$ $= Rf^2$	
Translational damper			$F = f(v_{21})$	$\mathscr{P} = Fv_{21}$	$F = bv_{21}$	$\mathscr{P} = bv_{21}^2$
Rotational damper			$T = f(\Omega_{21})$	$\mathscr{P} = T\Omega_{21}$	$T = B\Omega_{21}$	$\mathscr{P} = B\Omega_{21}^2$
Electrical resistance			$i = f(v_{21})$	$\mathscr{P} = iv_{21}$	$i = \frac{1}{R}v_{21}$	$\mathscr{P} = \frac{1}{R}v_{21}^2$
Fluid resistance			$Q = f(P_{21})$	$\mathscr{P} = QP_{21}$	$Q = \frac{1}{R_f}P_{21}$	$\mathscr{P} = \frac{1}{R_f}P_{21}^2$
Thermal resistance			$q = f(\theta_{21})$	$\mathscr{P} = q$	$q = \frac{1}{R_t}\theta_{21}$	$\mathscr{P} = \frac{1}{R_t}\theta_{21}$
Energy sources A-type across-variable source $\mathscr{P} \gtrless 0$			$v_{21} = f(t)$	$\mathscr{P} = fv_{21}$		
T-type through-variable source $\mathscr{E} \gtrless 0$			$f = f(t)$	$\mathscr{P} = fv_{21}$		

Nomenclature

$\mathscr{E}$ = energy, $\mathscr{P}$ = power

f = generalized through-variable, F = force, T = torque, i = current, Q = fluid flow rate, q = **heat flow rate**

h = generalized integrated through-variable, p = translational momentum, h = angular momentum, q = charge, V = fluid volume displaced, $\mathscr{K}$ = heat

v = generalized across-variable, v = translational velocity, Ω = angular velocity, v = voltage, P = pressure, θ = temperature

x = generalized integrated across-variable, x = translational displacement, Θ = angular displacement, λ = flux linkage, Γ = pressure-momentum

$1/k$ = generalized ideal inductance, $1/K$ = reciprocal translational stiffness, $1/\mathcal{K}$ = reciprocal rotational stiffness, L = inductance, L = fluid inertance

C = generalized ideal capacitance, m = mass, J = moment of inertia, C = capacitance, C_f = fluid capacitance, C_t = thermal capacitance

R = generalized ideal resistance, $1/b$ = reciprocal translational damping, $1/B$ = reciprocal rotational damping, R = electrical resistance, R_f = fluid resistance, R_t = thermal resistance

System	Symbol	Pure transformer	Ideal transformer	Transformation ratio
Mechanical translation (lever)	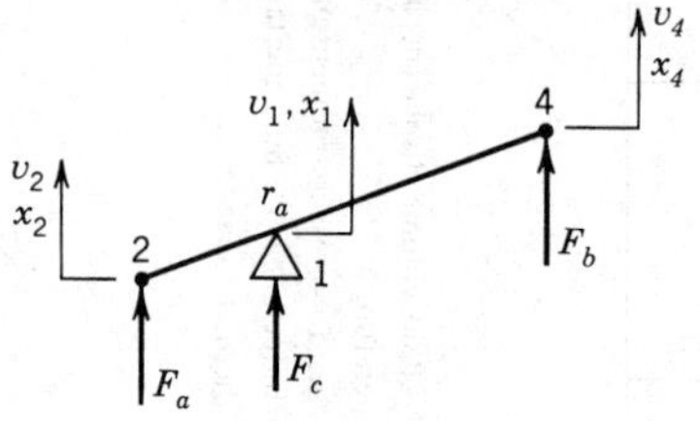	$x_{41} = f(x_{21})$	$v_{41} = n v_{21}$ $F_b = -\dfrac{1}{n} F_a$	$n = -\dfrac{r_b}{r_a}$ Lever ratio
Mechanical rotational (gears)	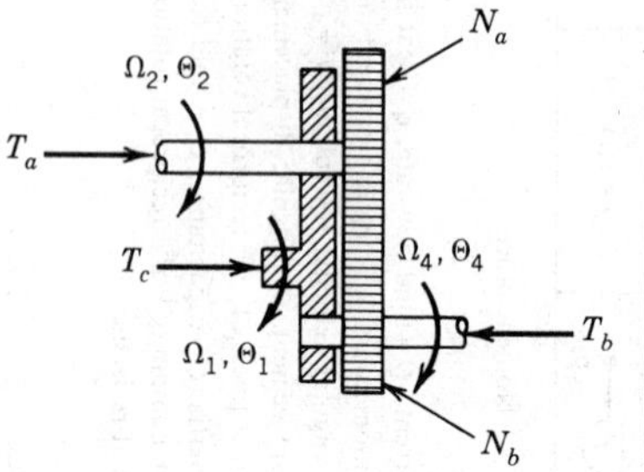	$\Theta_{41} = f(\Theta_2)$	$\Omega_{41} = n\Omega_{21}$ $T_b = -\dfrac{1}{n} T_a$	$n = -\dfrac{N_a}{N_b}$ Gear ratio
Electrical (magnetic)	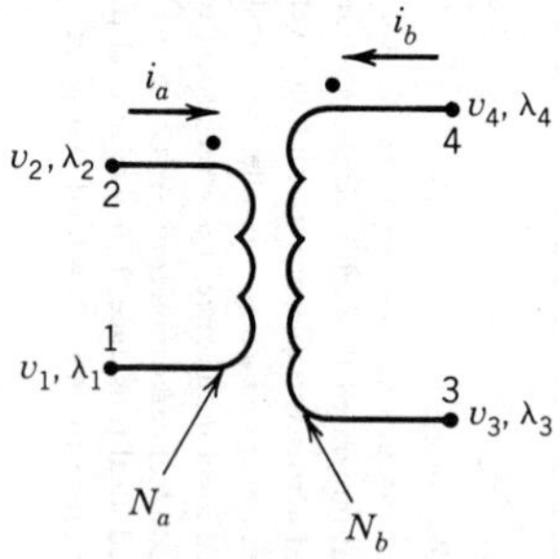	$\lambda_{43} = f(\lambda_{21})$	$v_{43} = n v_{21}$ $i_b = -\dfrac{1}{n} i_a$	$n = \dfrac{N_b}{N_a}$ Turns ratio
Fluid (differential piston)	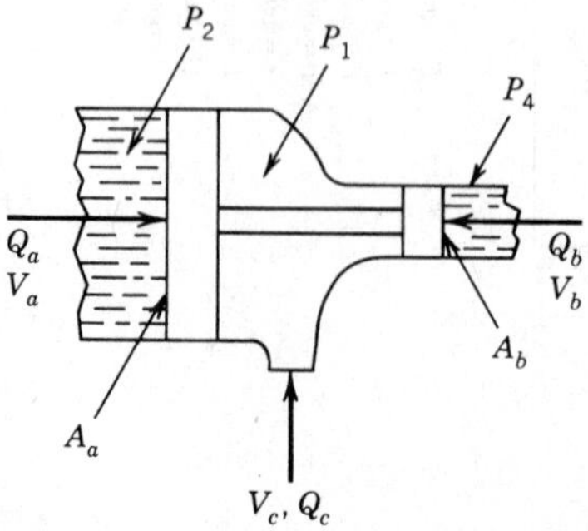	$V_b = f(V_a)$	$P_{41} = n P_{21}$ $Q_b = -\dfrac{1}{n} Q_a$	$n = \dfrac{A_a}{A_b}$ Area ratio

Fig. 39.4a   Examples of transforms and transducers: pure transformers.[23]

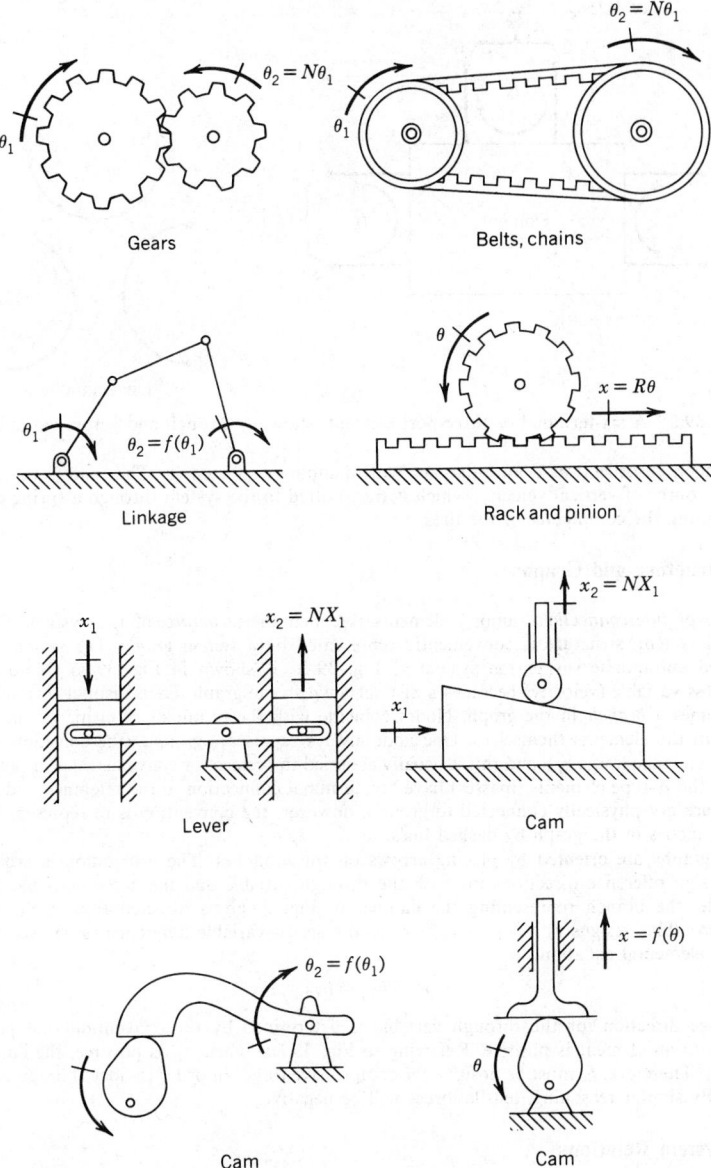

Fig. 39.4*b* Examples of transformers and transducers: pure mechanical transformers and transforming transducers.[7]

39.3 SYSTEM STRUCTURE AND INTERCONNECTION LAWS

39.3.1 A Simple Example

Physical systems are represented by connecting the terminals of pure elements in patterns that approximate the relationships among the properties of component devices. As an example, consider the mechanical-translational system depicted in Fig. 39.6*a*, which might represent an idealized automobile suspension system. The inertial properties associated with the masses of the chassis, passenger compartment, engine, and so on, all have been lumped together as the pure mass m_1. The inertial properties of the unsprung components (wheels, axles, etc.) have been lumped into the pure mass m_2. The compliance of the suspension is modeled as a pure spring with stiffness k_1 and the frictional effects (principally

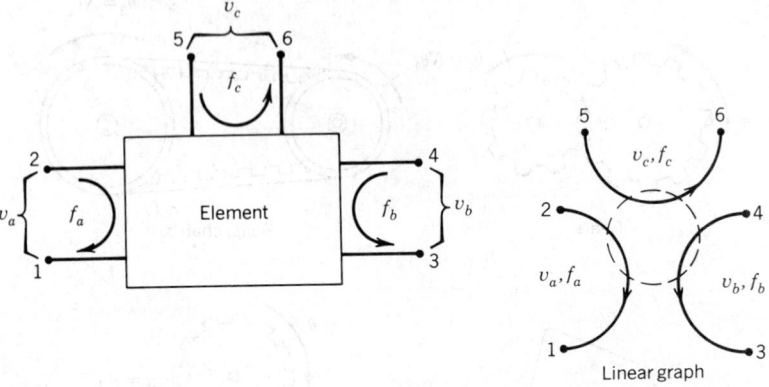

Fig. 39.5 A six-terminal or three-port element, showing through and across variables.

from the shock absorbers) as a pure damper with damping coefficient b. The road is represented as an input or source of vertical velocity, which is transmitted to the system through a spring of stiffness k_2, representing the compliance of the tires.

39.3.2 Structure and Graphs

The *pattern of interconnections* among elements is called the *structure* of the system. For a one-dimensional system, structure is conveniently represented by a *system graph*. The system graph for the idealized automobile suspension system of Fig. 39.6a is shown in Fig. 39.6b. Note that each distinct across variable (velocity) becomes a distinct *node* in the graph. Each distinct through variable (force) becomes a *branch* in the graph. Nodes coincide with the terminals of elements and branches coincide with the elements themselves. One node always represents *ground* (the constant velocity of the inertial reference frame v_g), and this is usually assumed to be zero for convenience. For nonelectrical systems, all the A-type elements (masses) have one terminal connection to the reference node. Because the masses are not physically connected to ground, however, the convention is to represent the corresponding branches in the graph by dashed lines.

System graphs are oriented by placing arrows on the branches. The orientation is arbitrary and serves to assign reference directions for both the through-variable and the across-variable difference. For example, the branch representing the damper in Fig. 39.6b is directed from node 2 (tail) to node 1 (head). This assigns $v_b = v_{21} = v_2 - v_1$ as the across-variable difference to be used in writing the damper elemental equation

$$f_b = bv_v = bv_{21}$$

The reference direction for the through variable is determined by the convention that power flow $P_b = f_b v_b$ into an element is positive. Referring to Fig. 39.6a, when v_{21} is positive, the damper is in compression. Therefore, f_b must be positive for compressive forces in order to obey the sign convention for power. By similar reasoning, tensile forces will be negative.

39.3.3 System Relations

The structure of a system gives rise to two sets of *interconnection laws* or *system relations*. Continuity relations apply to through variables and compatibility relations apply to across variables. The interpretation of system relations for various physical systems is given in Table 39.3.

Continuity is a general expression of dynamic equilibrium. In terms of the system graph, continuity states that the algebraic sum of all through variables entering a given node must be zero. Continuity applies at each node in the graph. For a graph with n nodes, continuity gives rise to n continuity equations, $n - 1$ of which are independent. For node i, the continuity equation is

$$\sum_j f_{ij} = 0$$

where the sum is taken over all branches (i, j) incident on i.

For the system graph depicted in Fig. 39.6b, the four continuity equations are

$$
\begin{aligned}
\text{node 1:} \quad & f_{k_1} + f_b - f_{m_1} = 0 \\
\text{node 2:} \quad & f_{k_2} - f_{k_1} - f_b - f_{m_2} = 0 \\
\text{node 3:} \quad & f_s - f_{k_2} = 0 \\
\text{node g:} \quad & f_{m_1} + f_{m_2} - f_s = 0
\end{aligned}
$$

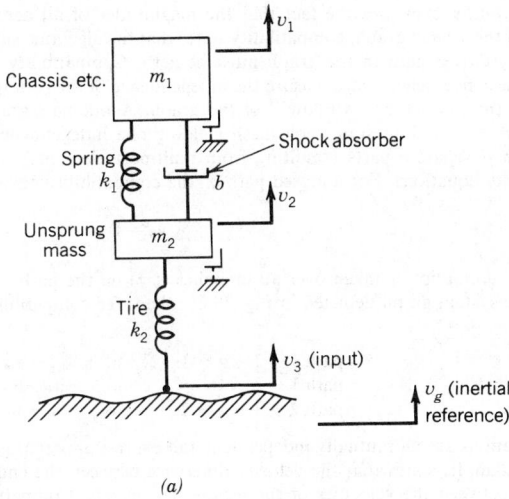

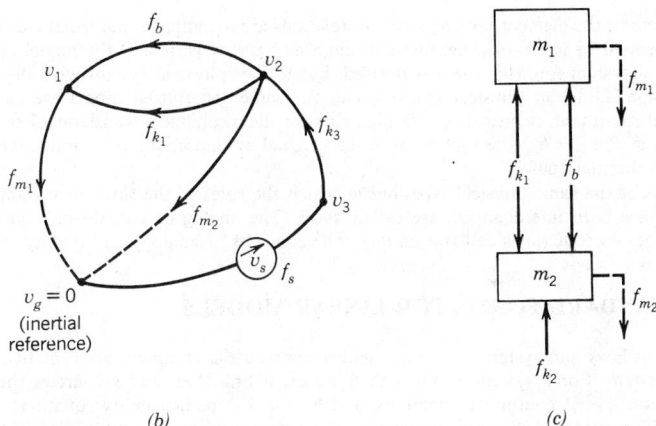

Fig. 39.6 An idealized model of an automobile suspension system: (a) lumped-element model, (b) system graph, (c) free-body diagram.

Only three of these four equations are independent. Note, also, that the equations for nodes 1 through 3 could have been obtained from the conventional *free-body diagrams* shown in Fig. 39.6c, where f_{m_1} and f_{m_2} are the *D'Alembert forces* associated with the pure masses. Continuity relations are also known as *vertex, node, flow,* and *equilibrium relations.*

Table 39.3 System Relations for Various Systems

System	Continuity	Compatibility
Mechanical	Newton's first and third laws (conservation of momentum)	Geometrical constraints (distance is a scalar)
Electrical	Kirchhoff's current law (conservation of charge)	Kirchhoff's voltage law (potential is a scalar)
Fluid	Conservation of matter	Pressure is a scalar
Thermal	Conservation of energy	Temperature is a scalar

Compatibility expresses the fact that the magnitudes of all across variables are scalar quantities. In terms of the system graph, compatibility states that the algebraic sum of the across-variable differences around any closed path in the graph must be zero. Compatibility applies to any closed path in the system. For convenience and to ensure the independence of the resulting equations, continuity is usually applied to the *meshes* or "windows" of the graph. A one-part graph with n nodes and b branches will have $b - n + 1$ meshes, each mesh yielding one independent compatibility equation. A planar graph with p separate parts (resulting from multiport elements) will have $b - n + p$ independent compatibility equations. For a closed path q, the compatibility equation is

$$\sum_q v_{ij} = 0$$

where the summation is taken over all branches (i, j) on the path.

For the system graph depicted in Fig. 39.6b, the three compatibility equations based on the meshes are

$$
\begin{aligned}
\text{path } 1 \to 2 \to g \to 1: & \qquad -v_b + v_{m_2} - v_{m_1} = 0 \\
\text{path } 1 \to 2 \to 1: & \qquad -v_{k_1} + v_b = 0 \\
\text{path } 2 \to 3 \to g \to 2: & \qquad -v_{k_2} - v_s - v_{m_2} = 0
\end{aligned}
$$

These equations are all mutually independent and express apparent geometric identities. The first equation, for example, states that the velocity difference between the ends of the damper is identically the difference between the velocities of the masses it connects. Compatibility relations are also known as *path, loop,* and *connectedness* relations.

39.3.4 Analogs and Duals

Taken together, the element laws and system relations are a complete mathematical model of a system. When expressed in terms of generalized through and across variables, the model applies not only to the physical system for which it was derived, but to any physical system with the same generalized system graph. Different physical systems with the same generalized model are called *analogs*. The mechanical rotational, electrical, and fluid analogs of the mechanical translational system of Fig. 39.6a are shown in Fig. 39.7. Note that because the original system contains an inductive storage element, there is no thermal analog.

Systems of the same physical type, but in which the roles of the through variables and the across variables have been interchanged, are called *duals*. The analog of a dual—or, equivalently, the dual of an analog—is sometimes called a *dualog*. The concepts of analogy and duality can be exploited in many different ways.

39.4 STANDARD FORMS FOR LINEAR MODELS

The element laws and system relations together constitute a complete mathematical description of a physical system. For a system graph with n nodes, b branches, and s sources, there will be $b - s$ element laws, $n - 1$ continuity equations, and $b - n + 1$ compatibility equations. This is a total of $2b - s$ differential and algebraic equations. For systems composed entirely of linear elements, it is always possible to reduce these $2b - s$ equations to either of two standard forms. The *input/output* or *I/O form* is the basis for *transform* or so-called *classical linear systems analysis*. The *state-variable form* is the basis for *state-variable* or so-called *modern linear systems analysis*.

39.4.1 I/O Form

The classical representation of a system is the "black box," depicted in Fig. 39.8. The system has a set of p inputs (also called *excitations* or *forcing functions*), $u_j(t)$, $j = 1, 2, \ldots, p$. The system also has a set of q outputs (also called *response variables*), $y_k(t)$, $k = 1, 2, \ldots, q$. Inputs correspond to sources and are assumed to be known functions of time. Outputs correspond to physical variables that are to be measured or calculated.

Linear systems represented in I/O form can be modeled mathematically by *I/O differential equations*. Denoting as $y_{kj}(t)$ that part of the kth output $y_k(t)$ that is attributable to the jth input $u_j(t)$, there are $(p \times q)$ I/O equations of the form

$$\frac{d^n y_{kj}}{dt^n} + a_{n-1}\frac{d^{n-1}y_{kj}}{dt^{n-1}} + \cdots + a_1\frac{dy_{kj}}{dt} + a_0 y_{kj}(t) = b_m\frac{d^m u_j}{dt^m} + b_{m-1}\frac{d^{m-1}u_j}{dt^{m-1}} + \cdots + b_1\frac{du_j}{dt} + b_0 u_j(t)$$

where $j = 1, 2, \ldots, p$ and $k = 1, 2, \ldots, q$. Each equation represents the dependence of one output and its derivatives on one input and its derivatives. By the *principle of superposition*, the kth output in response to all of the inputs acting simultaneously is

$$y_k(t) = \sum_{j=1}^{p} y_{kj}(t)$$

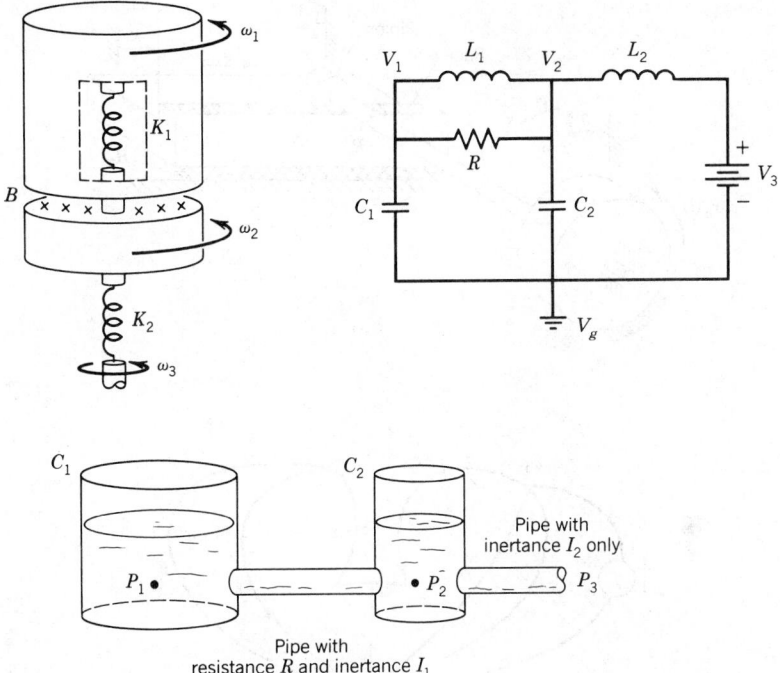

Fig. 39.7 Analogs of the idealized automobile suspension system depicted in Fig. 39.6.

A system represented by nth order I/O equations is called an nth *order system*. In general, the order of a system is determined by the number of *independent* energy-storage elements within the system, that is, by the number of T-type and A-type elements for which the initial energy stored can be independently specified.

The coefficients $a_0, a_1, \ldots, a_{n-1}$ and $b_0, b_1, \ldots, b_m$ are parameter groups made up of algebraic combinations of the system physical parameters. For a system with constant parameters, therefore, these coefficients are also constant. Systems with constant parameters are called *time-invariant* systems and are the basis for classical analysis.

39.4.2 Deriving the I/O Form—An Example

I/O differential equations are obtained by combining element laws and continuity and compatibility equations in order to eliminate all variables except the input and the output. As an example, consider the mechanical system depicted in Fig. 39.9a, which might represent an idealized milling machine. A rotational motor is used to position the table of the machine tool through a rack and pinion. The motor is represented as a torque source T with inertia J and internal friction B. A flexible shaft, represented as a torsional spring K, is connected to a pinion gear of radius R. The pinion meshes with a rack, which is rigidly attached to the table of mass m. Damper b represents the friction opposing the motion of the table. The problem is to determine the I/O equation that expresses the relationship between the input torque T and the position of the table x.

The corresponding system graph is depicted in Fig. 39.9b. Applying continuity at nodes 1, 2, and 3 yields

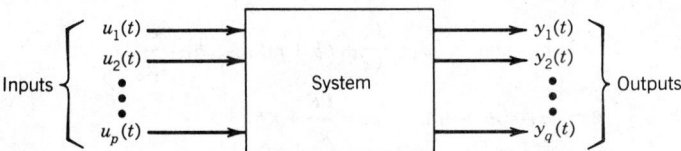

Fig. 39.8 Input/output (I/O) or "black box" representation of a dynamic system.

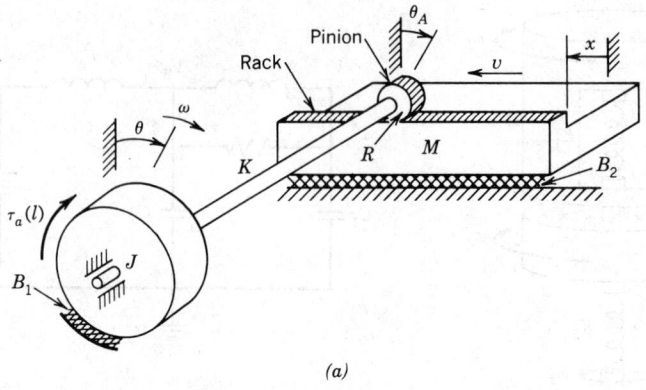

(a)

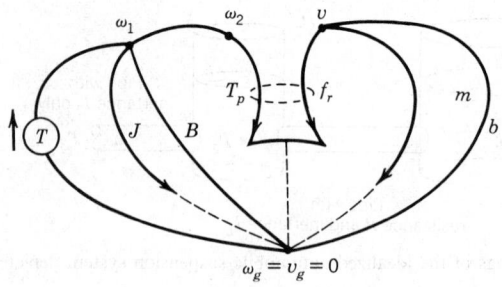

Fig. 39.9 An idealized model of a milling machine: (a) lumped-element model,[5] (b) system graph.

$$
\begin{aligned}
\text{node 1:} \quad & T - T_J - T_B - T_K = 0 \\
\text{node 2:} \quad & T_K - T_p = 0 \\
\text{node 3:} \quad & -f_r - f_m - f_b = 0
\end{aligned}
$$

Substituting the elemental equation for each of the one-port elements into the continuity equations and assuming zero ground velocities yields

$$
\begin{aligned}
\text{node 1:} \quad & T - J\dot{\omega}_1 - B\omega_1 - K \int (\omega_1 - \omega_2)dt = 0 \\
\text{node 2:} \quad & K \int (\omega_1 - \omega_2)dt - T_p = 0 \\
\text{node 3:} \quad & -f_r - m\dot{v} - bv = 0
\end{aligned}
$$

Note that the definition of the across variables for each element in terms of the node variables, as above, guarantees that the compatibility equations are satisfied. With the addition of the constitutive relationships for the rack and pinion

$$
\omega_2 = \frac{1}{R} v_2 \quad \text{and} \quad T_p = -Rf_r
$$

there are now five equations in the five unknowns ω_1, ω_2, v, T_p, and f_r. Combining these equations to eliminate all of the unknowns except v yields, after some manipulation,

$$
a_3 \frac{d^3 v}{dt^3} + a_2 \frac{d^2 v}{dt^2} + a_1 \frac{dv}{dt} + a_0 v = b_1 T
$$

where

$$
a_3 = Jm, \qquad a_1 = \frac{JK}{R^2} + Bb + mK, \qquad b_1 = \frac{K}{R}
$$

$$
a_2 = Jb + mB, \qquad a_0 = \frac{BK}{R^2} + Kb
$$

Differentiating yields the desired I/O equation

$$a_3 \frac{d^3 x}{dt^3} + a_2 \frac{d^2 x}{dt^2} + a_1 \frac{dx}{dt} + a_0 x = b_1 \frac{dT}{dt}$$

where the coefficients are unchanged.

For many systems, combining element laws and system relations can best be achieved by *ad hoc* procedures. For more complicated systems, formal methods are available for the orderly combination and reduction of equations. These are the so-called *loop method* and *node method* and correspond to procedures of the same names originally developed in connection with electrical networks. The interested reader should consult Ref. 23.

39.4.3 State-Variable Form

For systems with multiple inputs and outputs, the I/O model form can become unwieldy. In addition, important aspects of system behavior can be suppressed in deriving I/O equations. The "modern" representation of dynamic systems, called the *state-variable form*, largely eliminates these problems. A state-variable model is the maximum reduction of the original element laws and system relations that can be achieved without the loss of any information concerning the behavior of a system. State-variable models also provide a convenient representation for systems with multiple inputs and outputs and for systems analysis using computer simulation.

State variables are a set of variables $x_1(t)$, $x_2(t)$, . . . , $x_n(t)$ internal to the system from which any set of outputs can be derived, as depicted schematically in Fig. 39.10. A set of state variables is the minimum number of independent variables such that by knowing the values of these variables at any time t_0 and by knowing the values of the inputs for all time $t \geq t_0$, the values of the state variables for all future time $t \geq t_0$ can be calculated. For a given system, the number n of state variables is unique and is equal to the order of the system. The definition of the state variables is not unique, however, and various combinations of one set of state variables can be used to generate alternative sets of state variables. For a physical system, the state variables summarize the *energy state* of the system at any given time.

A complete state-variable model consists of two sets of equations, the *state* or *plant equations* and the *output equations*. For the most general case, the state equations have the form

$$\dot{x}_1(t) = f_1[x_1(t), x_2(t), \ldots, x_n(t), u_1(t), u_2(t), \ldots, u_p(t)]$$
$$\dot{x}_2(t) = f_2[x_1(t), x_2(t), \ldots, x_n(t), u_1(t), u_2(t), \ldots, u_p(t)]$$
$$\vdots$$
$$\dot{x}_n(t) = f_n[x_1(t), x_2(t), \ldots, x_n(t), u_1(t), u_2(t), \ldots, u_p(t)]$$

and the output equations have the form

$$y_1(t) = g_1[x_1(t), x_2(t), \ldots, x_n(t), u_1(t), u_2(t), \ldots, u_p(t)]$$
$$y_2(t) = g_2[x_1(t), x_2(t), \ldots, x_n(t), u_1(t), u_2(t), \ldots, u_p(t)]$$
$$\vdots$$
$$y_q(t) = g_q[x_1(t), x_2(t), \ldots, x_n(t), u_1(t), u_2(t), \ldots, u_p(t)]$$

These equations are expressed more compactly as the two vector equations

$$\dot{x}(t) = f[x(t), u(t)]$$
$$y(t) = g[x(t), u(t)]$$

where

$$\dot{x}(t) = \text{the } (n \times 1) \textit{ state vector}$$
$$u(t) = \text{the } (p \times 1) \textit{ input or control vector}$$
$$y(t) = \text{the } (q \times 1) \textit{ output or response vector}$$

and f and g are vector-valued functions.

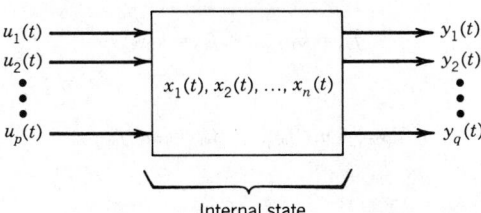

Internal state

Fig. 39.10 State-variable representation of a dynamic system.

For linear systems, the state equations have the form

$$\dot{x}_1(t) = a_{11}(t)x_1(t) + \cdots + a_{1n}(t)x_n(t) + b_{11}(t)u_1(t) + \cdots + b_{1p}(t)u_p(t)$$
$$\dot{x}_2(t) = a_{21}(t)x_1(t) + \cdots + a_{2n}(t)x_n(t) + b_{21}(t)u_1(t) + \cdots + b_{2p}(t)u_p(t)$$
$$\vdots$$
$$\dot{x}_n(t) = a_{n1}(t)x_1(t) + \cdots + a_{nn}(t)x_n(t) + b_{n1}(t)u_1(t) + \cdots + b_{np}(t)u_p(t)$$

and the output equations have the form

$$y_1(t) = c_{11}(t)x_1(t) + \cdots + c_{1n}(t)x_n(t) + d_{11}(t)u_1(t) + \cdots + d_{1p}(t)u_p(t)$$
$$y_2(t) = c_{21}(t)x_1(t) + \cdots + c_{2n}(t)x_n(t) + d_{21}(t)u_1(t) + \cdots + d_{2p}(t)u_p(t)$$
$$\vdots$$
$$y_n(t) = c_{q1}(t)x_1(t) + \cdots + c_{qn}(t)x_n(t) + d_{q1}(t)u_1(t) + \cdots + d_{qp}(t)u_p(t)$$

where the coefficients are groups of parameters. The linear model is expressed more compactly as the two linear vector equations.

$$\dot{x}(t) = A(t)x(t) + B(t)u(t)$$
$$y(t) = C(t)x(t) + D(t)u(t)$$

where the vectors x, u, and y are the same as the general case and the matrices are defined as

$$A = [a_{ij}] \text{ is the } (n \times n) \text{ system matrix}$$
$$B = [b_{jk}] \text{ is the } (n \times p) \text{ control, input, or}$$
$$\text{distribution matrix}$$
$$C = [c_{1j}] \text{ is the } (q \times n) \text{ output matrix}$$
$$D = [d_{1k}] \text{ is the } (q \times p) \text{ output distribution matrix}$$

For a time-invariant linear system, all of these matrices are constant.

39.4.4 Deriving the "Natural" State Variables—A Procedure

Because the state variables for a system are not unique, there are an unlimited number of alternative (but equivalent) state-variable models for the system. Since energy is stored only in generalized system storage elements, however, a natural choice for the state variables is the set of through and across variables corresponding to the independent T-type and A-type elements, respectively. This definition is sometimes called the set of *natural state variables* for the system.

For linear systems, the following procedure can be used to reduce the set of element laws and system relations to the natural state-variable model.

Step 1. For each independent T-type storage, write the element law with the derivative of the through variable isolated on the left-handed side, that is, $\dot{f} = L^{-1}v$.

Step 2. For each independent A-type storage, write the element law with the derivative of the across variable isolated on the left-hand side, that is, $\dot{v} = C^{-1}f$.

Step 3. Solve the compatibility equations, together with the element laws for the appropriate D-type and multiport elements, to obtain each of the across variables of the independent T-type elements in terms of the natural state variables and specified sources.

Step 4. Solve the continuity equations, together with the element laws for the appropriate D-type and multiport elements, to obtain the through variables of the A-type elements in terms of the natural state variables and specified sources.

Step 5. Substitute the results of step 3 into the results of step 1; substitute the results of step 4 into the results of step 2.

Step 6. Collect terms on the right-hand side and write in vector form.

39.4.5 Deriving the "Natural" State Variables—An Example

The six-step process for deriving a natural state-variable representation, outlined in the preceding section, is demonstrated for the idealized automobile suspension depicted in Fig. 39.6:

Step 1

$$\dot{f}_{k_1} = k_1 v_{k_1}, \qquad \dot{f}_{k_2} = k_2 v_{k_2}$$

Step 2

$$\dot{v}_{m_1} = m_1^{-1} f_{m_1}, \qquad \dot{v}_{m_2} = m^{-1} f_{m_2}$$

Step 3

$$v_{k_1} = v_b = v_{m_2} - v_{m_1}, \qquad v_{k_2} = -v_{m_2} - v_s$$

Step 4

$$f_{m_1} = f_{k_1} + f_b = f_{k_1} + b^{-1}(v_{m_2} - v_{m_1})$$
$$f_{m_2} = f_{k_2} - f_{k_1} - f_b = f_{k_2} - f_{k_1} - b^{-1}(v_{m_2} - v_{m_1})$$

Step 5

$$\dot{f}_{k_1} = k_1(v_{m_2} - v_{m_1}), \qquad \dot{v}_{m_1} = m_1^{-1}[f_{k_1} + b^{-1}(v_{m_2} - v_{m_1})]$$
$$\dot{f}_{k_2} = k_2(-v_{m_2} - v_3), \qquad \dot{v}_{m_2} = m_2^{-1}[f_{k_2} - f_{k_1} - b^{-1}(v_{m_2} - v_{m_1})]$$

Step 6

$$\frac{d}{dt}
\begin{bmatrix} f_{k_1} \\ f_{k_2} \\ v_{m_1} \\ v_{m_2} \end{bmatrix}
=
\begin{bmatrix}
0 & 0 & -k_1 & k_1 \\
0 & 0 & 0 & -k_2 \\
1/m_1 & 0 & -1/m_1 b & 1/m_1 b \\
-1/m_2 & 1/m_2 & 1/m_2 b & -1/m_2 b
\end{bmatrix}
\begin{bmatrix} f_{k_1} \\ f_{k_2} \\ v_{m_1} \\ v_{m_2} \end{bmatrix}
+
\begin{bmatrix} 0 \\ -k_2 \\ 0 \\ 0 \end{bmatrix} v_s$$

39.4.6 Converting from I/O to "Phase-Variable" Form

Frequently, it is desired to determine a state-variable model for a dynamic system for which the I/O equation is already known. Although an unlimited number of such models is possible, the easiest to determine uses a special set of state variables called the *phase variables*. The phase variables are defined in terms of the output and its derivatives as follows:

$$x_1(t) = y(t)$$
$$x_2(t) = \dot{x}_1(t) = \frac{d}{dt} y(t)$$
$$x_3(t) = \dot{x}_2(t) = \frac{d^2}{dt^2} y(t)$$
$$\vdots$$
$$x_n(t) = \dot{x}_{n-1}(t) = \frac{d^{n-1}}{dt^{n-1}} y(t)$$

This definition of the phase variables, together with the I/O equation of Section 39.4.1, can be shown to result in a state equation of the form

$$\frac{d}{dt}
\begin{bmatrix} x_1(t) \\ x_2(t) \\ \vdots \\ x_{n-1}(t) \\ x_n(t) \end{bmatrix}
=
\begin{bmatrix}
0 & 1 & 0 & \cdots & 0 \\
0 & 0 & 1 & \cdots & 0 \\
\vdots & \vdots & \vdots & \ddots & \vdots \\
0 & 0 & 0 & \cdots & 1 \\
-a_0 & -a_1 & -a_2 & \cdots & -a_{n-1}
\end{bmatrix}
\begin{bmatrix} x_1(t) \\ x_2(t) \\ \vdots \\ x_{n-1}(t) \\ x_n(t) \end{bmatrix}
+
\begin{bmatrix} 0 \\ 0 \\ \vdots \\ 0 \\ 1 \end{bmatrix} u(t)$$

and an output equation of the form

$$y(t) = \begin{bmatrix} b_0 & b_1 & \cdots & b_m \end{bmatrix}
\begin{bmatrix} x_1(t) \\ x_2(t) \\ \vdots \\ x_n(t) \end{bmatrix}$$

This special form of the system matrix, with ones along the upper off-diagonal and zeros elsewhere except for the bottom row, is called a *companion matrix*.

39.5 APPROACHES TO LINEAR SYSTEMS ANALYSIS

There are two fundamental approaches to the analysis of linear, time-invariant systems. *Transform methods* use rational functions obtained from the Laplace transformation of the system I/O equations. Transform methods provide a particularly convenient algebra for combining the component submodels of a system and form the basis of so-called *classical control theory*. *State-variable methods* use the vector state and output equations directly. State-variable methods permit the adaptation of important ideas from linear algebra and form the basis for so-called *modern control theory*. Despite the deceiving names of "classical" and "modern," the two approaches are complementary. Both approaches are widely used in current practice and the control engineer must be conversant with both.

39.5.1 Transform Methods

A *transformation* converts a given mathematical problem into an equivalent problem, according to some well-defined rule called a *transform*. Prudent selection of a transform frequently results in an equivalent problem that is easier to solve than the original. If the solution to the original problem can be recovered by an inverse transformation, the three-step process of (1) transformation, (2) solution in the *transform domain*, and (3) inverse transformation, may prove more attractive than direct solution of the problem in the original problem domain. This is true for fixed linear dynamic systems under the *Laplace transform*, which converts differential equations into equivalent algebraic equations.

Laplace Transforms: Definition

The one-sided Laplace transform is defined as

$$F(s) = \mathscr{L}[f(t)] = \int_0^\infty f(t)\, e^{-st}\, dt$$

and the inverse transform as

$$f(t) = \mathscr{L}^{-1}[F(s)] = \frac{1}{2\pi j} \int_{\sigma - j\omega}^{\sigma + j\omega} F(s)\, e^{-st}\, ds$$

The Laplace transform converts the function $f(t)$ into the transformed function $F(s)$; the inverse transform recovers $f(t)$ from $F(s)$. The symbol $\mathscr{L}$ stands for the "Laplace transform of"; the symbol $\mathscr{L}^{-1}$ stands for "the inverse Laplace transform of."

The Laplace transform takes a problem given in the *time domain*, where all physical variables are functions of the *real variable t*, into the *complex-frequency domain*, where all physical variables are functions of the complex frequency $s = \sigma + j\omega$, where $j = \sqrt{-1}$ is the imaginary operator. Laplace transform pairs consist of the function $f(t)$ and its transform $F(s)$. Transform pairs can be calculated by substituting $f(t)$ into the defining equation and then evaluating the integral with s held constant. For a transform pair to exist, the corresponding integral must converge, that is,

$$\int_0^\infty |f(t)| e^{-\sigma * t}\, dt < \infty$$

for some real $\sigma * > 0$. Signals that are physically realizable always have a Laplace transform.

Tables of Transform Pairs and Transform Properties

Transform pairs for functions commonly encountered in the analysis of dynamic systems rarely need to be calculated. Instead, pairs are determined by reference to a *table of transforms* such as that given in Table 39.4. In addition, the Laplace transform has a number of properties that are useful in determining the transforms and inverse transforms of functions in terms of the tabulated pairs. The most important of these are given in a *table of transform properties* such as that given in Table 39.5.

Poles and Zeros

The response of a dynamic system most often assumes the following form in the complex-frequency domain

$$F(s) = \frac{N(s)}{D(s)} = \frac{b_m s^m + b_{m-1} s^{m-1} + \cdots + b_1 s + b_0}{s^n + a_{n-1} s^{n-1} + \cdots + a_1 s + a_0} \tag{39.1}$$

Functions of this form are called *rational functions*, because these are the ratio of two polynomials $N(s)$ and $D(s)$. If $n \geq m$, then $F(s)$ is a *proper rational function*; if $n > m$, then $F(s)$ is a *strictly proper rational function*.

In factored form, the rational function $F(s)$ can be written as

$$F(s) = \frac{N(s)}{D(s)} = \frac{b_m (s - z_1)(s - z_2) \cdots (s - z_m)}{(s - p_1)(s - p_2) \cdots (s - p_n)} \tag{39.2}$$

The roots of the numerator polynomial $N(s)$ are denoted by z_j, $j = 1, 2, \ldots, m$. These numbers are called the *zeros* of $F(s)$, since $F(z_j) = 0$. The roots of the denominator polynomial are denoted by p_i, $i = 1, 2, \ldots, n$. These numbers are called the *poles* of $F(s)$, since $\lim_{s \to p_i} F(s) = \pm \infty$.

Table 39.4 Laplace Transform Pairs

$F(s)$	$f(t), t \geq 0$
1. 1	$\delta(t)$, the unit impulse at $t = 0$
2. $\dfrac{1}{s}$	1, the unit step
3. $\dfrac{n!}{s^{n+1}}$	t^n
4. $\dfrac{1}{s+a}$	e^{-at}
5. $\dfrac{1}{(s+a)^n}$	$\dfrac{1}{(n-1)!} t^{n-1} e^{-at}$
6. $\dfrac{a}{s(s+a)}$	$1 - e^{-at}$
7. $\dfrac{1}{(s+a)(s+b)}$	$\dfrac{1}{(b-a)}(e^{-at} - e^{-bt})$
8. $\dfrac{s+p}{(s+a)(s+b)}$	$\dfrac{1}{(b-a)}[(p-a)e^{-at} - (p-b)e^{-bt}]$
9. $\dfrac{1}{(s+a)(s+b)(s+c)}$	$\dfrac{e^{-at}}{(b-a)(c-a)} + \dfrac{e^{-bt}}{(c-b)(a-b)} + \dfrac{e^{-ct}}{(a-c)(b-c)}$
10. $\dfrac{s+p}{(s+a)(s+b)(s+c)}$	$\dfrac{(p-a)e^{-at}}{(b-a)(c-a)} + \dfrac{(p-b)e^{-bt}}{(c-b)(a-b)} + \dfrac{(p-c)e^{-ct}}{(a-c)(b-c)}$
11. $\dfrac{b}{s^2 + b^2}$	$\sin bt$
12. $\dfrac{s}{s^2 + b^2}$	$\cos bt$
13. $\dfrac{b}{(s+a)^2 + b^2}$	$e^{-at} \sin bt$
14. $\dfrac{s+a}{(s+a)^2 + b^2}$	$e^{-at} \cos bt$
15. $\dfrac{\omega_n^2}{s^2 + 2\zeta\omega_n s + \omega_n^2}$	$\dfrac{\omega_n}{\sqrt{1-\zeta^2}} e^{-\zeta\omega_n t} \sin \omega_n \sqrt{1-\zeta^2}\, t, \quad \zeta < 1$
16. $\dfrac{\omega_n^2}{s(s^2 + 2\zeta\omega_n s + \omega_n^2)}$	$1 + \dfrac{1}{\sqrt{1-\zeta^2}} e^{-\zeta\omega_n t} \sin(\omega_n \sqrt{1-\zeta^2}\, t + \phi)$

$$\phi = \tan^{-1} \frac{\sqrt{1-\zeta^2}}{\zeta} + \pi$$

(third quadrant)

Inversion by Partial-Fraction Expansion

The *partial-fraction expansion theorem* states that a strictly proper rational function $F(s)$ with *distinct* (*nonrepeated*) poles p_i, $i = 1, 2, \ldots, n$, can be written as the sum

$$F(s) = \frac{A_1}{s - p_1} + \frac{A_2}{s - p_2} + \cdots + \frac{A_n}{s - p_n} = \sum_{i=1}^{n} A_i \left(\frac{1}{s - p_i} \right) \qquad (39.3)$$

Table 39.5   Laplace Transform Properties

$f(t)$	$F(s) = \int_0^\infty f(t)e^{-st}\,dt$
1. $af_1(t) + bf_2(t)$	$aF_1(s) + bF_2(s)$
2. $\dfrac{df}{dt}$	$sF(s) - f(0)$
3. $\dfrac{d^2f}{dt^2}$	$s^2F(s) - sf(0) - \dfrac{df}{dt}\Big|_{t=0}$
4. $\dfrac{d^nf}{dt^n}$	$s^nF(s) - \displaystyle\sum_{k=1}^{n} s^{n-k}g_{k-1}$
	$g_{k-1} = \dfrac{d^{k-1}f}{dt^{k-1}}\Big|_{t=0}$
5. $\displaystyle\int_0^t f(t)\,dt$	$\dfrac{F(s)}{s} + \dfrac{h(0)}{s}$
	$h(0) = \displaystyle\int f(t)\,dt\,\Big|_{t=0}$
6. $\begin{cases} 0, & t < D \\ f(t-D), & t \geq D \end{cases}$	$e^{-sD}F(s)$
7. $e^{-at}f(t)$	$F(s+a)$
8. $f\left(\dfrac{t}{a}\right)$	$aF(as)$
9. $f(t) = \displaystyle\int_0^t x(t-\tau)y(\tau)\,d\tau$	$F(s) = X(s)Y(s)$
$\quad = \displaystyle\int_0^t y(t-\tau)x(\tau)\,d\tau$	
10.	$f(\infty) = \lim\limits_{s\to 0} sF(s)$
11.	$f(0+) = \lim\limits_{s\to\infty} sF(s)$

where the A_i, $i = 1, 2, \ldots, n$, are constants called *residues*. The inverse transform of $F(s)$ has the simple form

$$f(t) = A_1 e^{p_1 t} + A_2 e^{p_2 t} + \cdots + A_n e^{p_n t} = \sum_{i=1}^{n} A_i e^{p_i t}$$

The *Heaviside expansion theorem* gives the following expression for calculating the residue at the pole p_i,

$$A_i = (s - p_i)F(s)|_{s = p_i} \quad \text{for } i = 1, 2, \ldots, n$$

These values can be checked by substituting into Eq. (39.3), combining the terms on the right-hand side of Eq. (39.3), and showing the result yields the values for all the coefficients b_j, $j = 1, 2, \ldots, m$, originally specified in the form of Eq. (39.1).

Repeated Poles

When two or more poles of a strictly proper rational function are identical, the poles are said to be *repeated* or *nondistinct*. If a pole is repeated q times, that is, if $p_i = p_{i+1} = \cdots = p_{i+q-1}$, then the pole is said to be of *multiplicity* q. A strictly proper rational function with a pole of multiplicity q will contain q terms of the following form

$$\frac{A_{i1}}{(s - p_i)^q} + \frac{A_{i2}}{(s - p_i)^{q-1}} + \cdots + \frac{A_{iq}}{(s - p_i)}$$

in addition to the terms associated with the distinct poles. The corresponding terms in the inverse transform are

$$\left(\frac{1}{(q-1)!}A_{i\,1}t^{(q-1)}+\frac{1}{(q-2)!}A_{i\,2}t^{(q-2)}+\cdots+A_{iq}\right)e^{p_i t}$$

The corresponding residues are

$$A_{i\,1}=(s-p_i)^q F(s)|_{s=p_i}$$

$$A_{i\,2}=\left(\frac{d}{ds}\left[(s-p_i)^q F(s)\right]\right)\bigg|_{s=p_i}$$

$$\vdots$$

$$A_{iq}=\frac{1}{(q-1)!}\left(\frac{d^{(q-1)}}{ds^{(q-1)}}\left[(s-p_i)^q F(s)\right]\right)\bigg|_{s=p_i}$$

Complex Poles

A strictly proper rational function with complex conjugate poles can be inverted using partial-fraction expansion. Using a method called *completing the square*, however, is almost always easier. Consider the function

$$F(s)=\frac{B_1 s+B_2}{(s+\sigma-j\omega)(s+\sigma+j\omega)}$$

$$=\frac{B_1 s+B_2}{s^2+2\sigma s+\sigma^2+\omega^2}$$

$$=\frac{B_1 s+B_2}{(s+\sigma)^2+\omega^2}$$

From the transform tables the Laplace inverse is

$$f(t)=e^{-\sigma t}[B_1\cos\omega t+B_3\sin\omega t]$$

$$=Ke^{-\sigma t}\cos(\omega t+\phi)$$

where $B_3=(1/\omega)(B_2-aB_1)$
  $K=\sqrt{B_1^2+B_3^2}$
  $\phi=-\tan^{-1}(B_3/B_1)$

Proper and Improper Rational Functions

If $F(s)$ is not a strictly proper rational function, then $N(s)$ must be divided by $D(s)$ using *synthetic division*. The result is

$$F(s)=\frac{N(s)}{D(s)}=P(s)+\frac{N^*(s)}{D(s)}$$

where $P(s)$ is a polynomial of degree $m-n$ and $N^*(s)$ is a polynomial of degree $n-1$. Each term of $P(s)$ may be inverted directly using the transform tables. $N^*(s)/D(s)$ is a strictly proper rational function and may be inverted using partial-fraction expansion.

Initial-Value and Final-Value Theorems

The limits of $f(t)$ as time approaches zero or infinity frequently can be determined directly from the transform $F(s)$ without inverting. The *initial-value theorem* states that

$$f(0_+)=\lim_{s\to\infty} sF(s)$$

where the limit exists. If the limit does not exist (i.e., is infinite), the value of $f(0_+)$ is undefined. The *final-value theorem* states that

$$f(\infty)=\lim_{s\to 0} sF(s)$$

provided that (with the possible exception of a single pole at $s=0$) $F(s)$ has no poles with nonnegative real parts.

Transfer Functions

The Laplace transform of system I/O equation may be written in terms of the transform $Y(s)$ of the system response $y(t)$ as

$$Y(s) = \frac{G(s)N(s) + F(s)D(s)}{P(s)D(s)}$$

$$= \left(\frac{G(s)}{P(s)}\right)\left(\frac{N(s)}{D(s)}\right) + \frac{F(s)}{P(s)}$$

where

(a) $P(s) = a_n s^n + a_{n-1} + \cdots + a_1 s + a_0$ is the *characteristic polynomial* of the system,

(b) $G(s) = b_m s^m + b_{m-1}s^{m-1} + \cdots + b_1 s + b_0$ represents the *numerator dynamics* of the system,

(c) $U(s) = N(s)/D(s)$ is the transform of the input to the system, $u(t)$, assumed to be a rational

function, and

(d) $F(s) = a_n y(0)s^{n-1} + \left(a_n \dfrac{dy}{dt}(0) + a_{n-1}y(0)\right)s^{n-2} + \cdots$

$$+ \left(a_n \frac{d^{n-1}y}{dt^{n-1}}(0) + a_{n-1}\frac{d^{n-2}y}{dt}(0) + \cdots + a_1 y(0)\right)$$

reflects the initial system state [i.e., the initial conditions on $y(t)$ and its first $n-1$ derivatives].

The transformed response can be thought of as the sum of two components,

$$Y(s) = Y_{zs}(s) + Y_{zi}(s)$$

where

(e) $Y_{zs}(s) = [G(s)/P(s)][N(s)/D(s)] = H(s)U(s)$ is the transform of the *zero-state response*, that is, the response of the system to the input alone, and

(f) $Y_{zi}(s) = F(s)/P(s)$ is the transform of the *zero-input response*, that is, the response of the system to the initial state alone.

The rational function

(g) $H(s) = Y_{zs}(s)/U(s) = G(s)/P(s)$ is the *transfer function* of the system, defined as the Laplace transform of the ratio of the system response to the system input, assuming zero initial conditions.

The transfer function plays a crucial role in the analysis of fixed linear systems using transforms and can be written directly from knowledge of the system I/O equation as

$$H(s) = \frac{b_m s^m + \cdots + b_0}{a_n s^n + a_{n-1}s^{n-1} + \cdots + a_1 s + a_0}$$

Impulse Response

Since $U(s) = 1$ for a unit impulse function, the transform of the zero-state response to a unit impulse input is given by the relation (g) as

$$Y_{zs}(s) = H(s)$$

that is, the system transfer function. In the time domain, therefore, the unit *impulse response* is

$$h(t) = \begin{cases} 0 & \text{for } t \leq 0 \\ \mathcal{L}^{-1}[H(s)] & \text{for } t > 0 \end{cases}$$

This simple relationship is profound for several reasons. First, this provides for a direct characterization of time-domain response $h(t)$ in terms of the properties (poles and zeros) of the rational function $H(s)$ in the complex-frequency domain. Second, applying the convolution transform pair (Table 39.5) to relation (e) above yields

$$Y_{zs}(t) = \int_0^t h(\tau)u(t - \tau)d\tau$$

In words, the zero-state output corresponding to an arbitrary input $u(t)$ can be determined by convolution with the impulse response $h(t)$. In other words, the impulse response completely characterizes the system. The impulse response is also called the system *weighting function*.

Block Diagrams

Block diagrams are an important conceptual tool for the analysis and design of dynamic systems, because block diagrams provide a graphic means for depicting the relationships among system variables and components. A block diagram consists of unidirectional blocks representing specified system components or subsystems, interconnected by arrows representing system variables. Causality follows in the direction of the arrows, as in Fig. 39.11, indicating that the output is caused by the input acting on the system defined in the block.

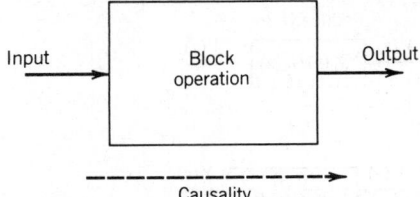

Fig. 39.11 Basic block diagram, showing assumed direction of causality or loading.

Combining transform variables, transfer functions, and block diagrams provides a powerful graphical means for determining the overall transfer function of a system, when the transfer functions of its component subsystems are known. The basic blocks in such diagrams are given in Fig. 39.12. A block diagram comprising many blocks and summers can be reduced to a single transfer function block by using the diagram transformations given in Fig. 39.13.

39.5.2 Transient Analysis Using Transform Methods

Basic to the study of dynamic systems are the concepts and terminology used to characterize system behavior or performance. These ideas are aids in *defining* behavior, in order to consider for a given context those features of behavior which are desirable and undesirable; in *describing* behavior, in order to communicate concisely and unambiguously various behavioral attributes of a given system; and in *specifying* behavior, in order to formulate desired behavioral norms for system design. Characterization of dynamic behavior in terms of standard concepts also leads in many cases to analytical shortcuts, since key features of the system response can frequently be determined without actually solving the system model.

Parts of the Complete Response

A variety of names is used to identify terms in the response of a fixed linear system. The complete response of a system may be thought of alternatively as the sum of:

	Input–Output Relations		
Type	Time Domain	Transform Domain	Symbol
(a) Multiplier	$y(t) = Kv(t)$	$Y(s) = KV(s)$	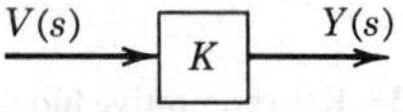 $V(s) \rightarrow \boxed{K} \rightarrow Y(s)$
(b) General transfer function	$y(t) = \mathscr{L}^{-1}[T(s)V(s)]$	$Y(s) = T(s)V(s)$	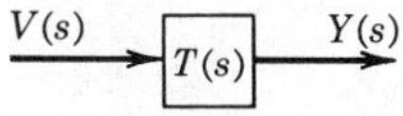 $V(s) \rightarrow \boxed{T(s)} \rightarrow Y(s)$
(c) Summer	$y(t) = v_1(t) + v_2(t)$	$Y(s) = V_1(s) + V_2(s)$	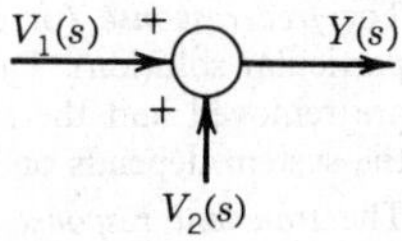 $V_1(s) \xrightarrow{+} \bigcirc \xrightarrow{} Y(s),\ +\ V_2(s)$
(d) Comparator	$y(t) = v_1(t) - v_2(t)$	$Y(s) = V_1(s) - V_2(s)$	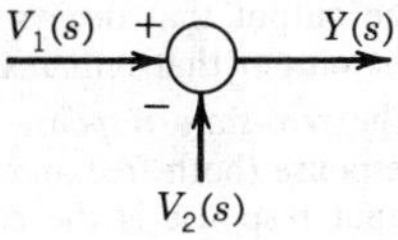 $V_1(s) \xrightarrow{+} \bigcirc \xrightarrow{} Y(s),\ -\ V_2(s)$
(e) Takeoff point	$y(t) = v(t)$	$Y(s) = V(s)$	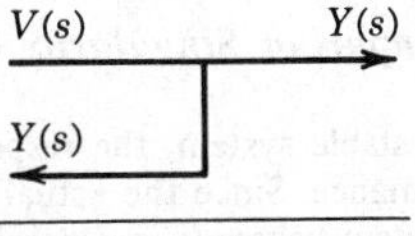 $V(s) \rightarrow Y(s),\ Y(s)$

Fig. 39.12 Basic block diagram elements.[19]

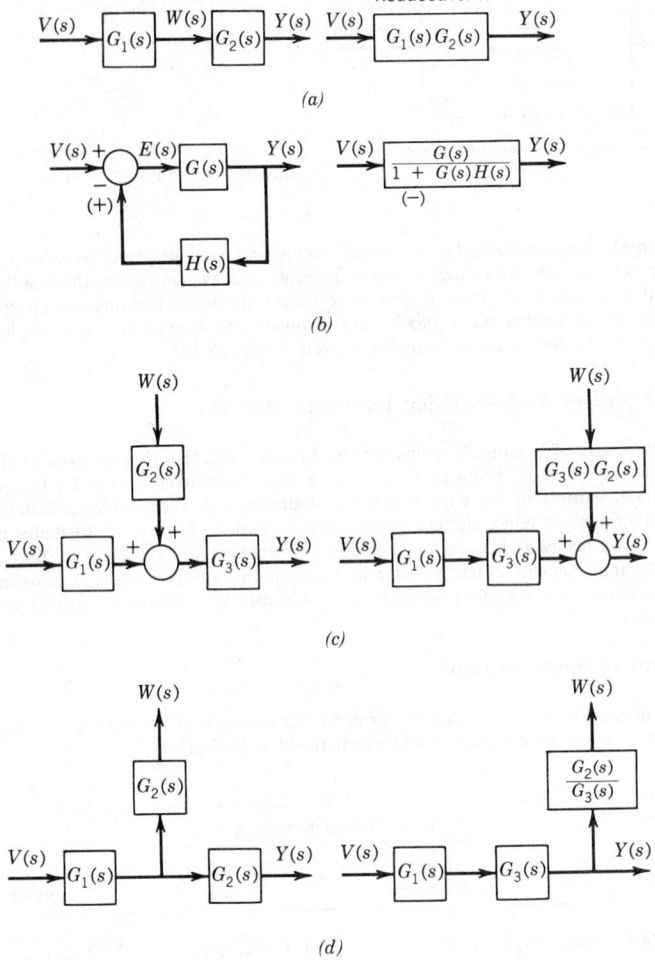

Fig. 39.13　Representative block diagram transformations: (*a*) series or cascaded elements, (*b*) feedback loop, (*c*) relocated summer, (*d*) relocated takeoff point.[19]

1. The *free response* (or complementary or homogeneous solution) and the *forced response* (or particular solution). The free response represents the natural response of a system when inputs are removed and the system responds to some initial stored energy. The forced response of the system depends on the form of the input only.

2. The *transient response* and the *steady-state response*. The transient response is that part of the output that decays to zero as time progresses. The steady-state response is that part of the output that remains after all the transients disappear.

3. The *zero-state response* and the *zero-input response*. The zero-state response is the complete response (both free and forced responses) to the input when the initial state is zero. The zero-input response is the complete response of the system to the initial state when the input is zero.

Test Inputs or Singularity Functions

For a stable system, the response to a specific input signal will provide several measures of system performance. Since the actual inputs to a system are not usually known *a priori*, characterization of the system behavior is generally given in terms of the response to one of a standard set of *test input signals*. This approach provides a common basis for the comparison of different systems. In addition, many inputs actually encountered can be approximated by some combination of standard inputs. The most commonly used test inputs are members of the family of *singularity functions*, depicted in Fig. 39.14.

First-Order Transient Response

The standard form of the I/O equation for a first-order system is

$$\frac{dy}{dt} + \frac{1}{\tau} y(t) = \frac{1}{\tau} u(t)$$

where the parameter τ is called the system *time constant*. The response of this standard first-order system to three test inputs is depicted in Fig. 39.15, assuming zero initial conditions on the output $y(t)$. For all inputs, it is clear that the response approaches its steady state monotonically (i.e., without oscillations) and that the *speed of response* is completely characterized by the time constant τ. The transfer function of the system is

$$H(s) = \frac{Y(s)}{U(s)} = \frac{1/\tau}{s + 1/\tau}$$

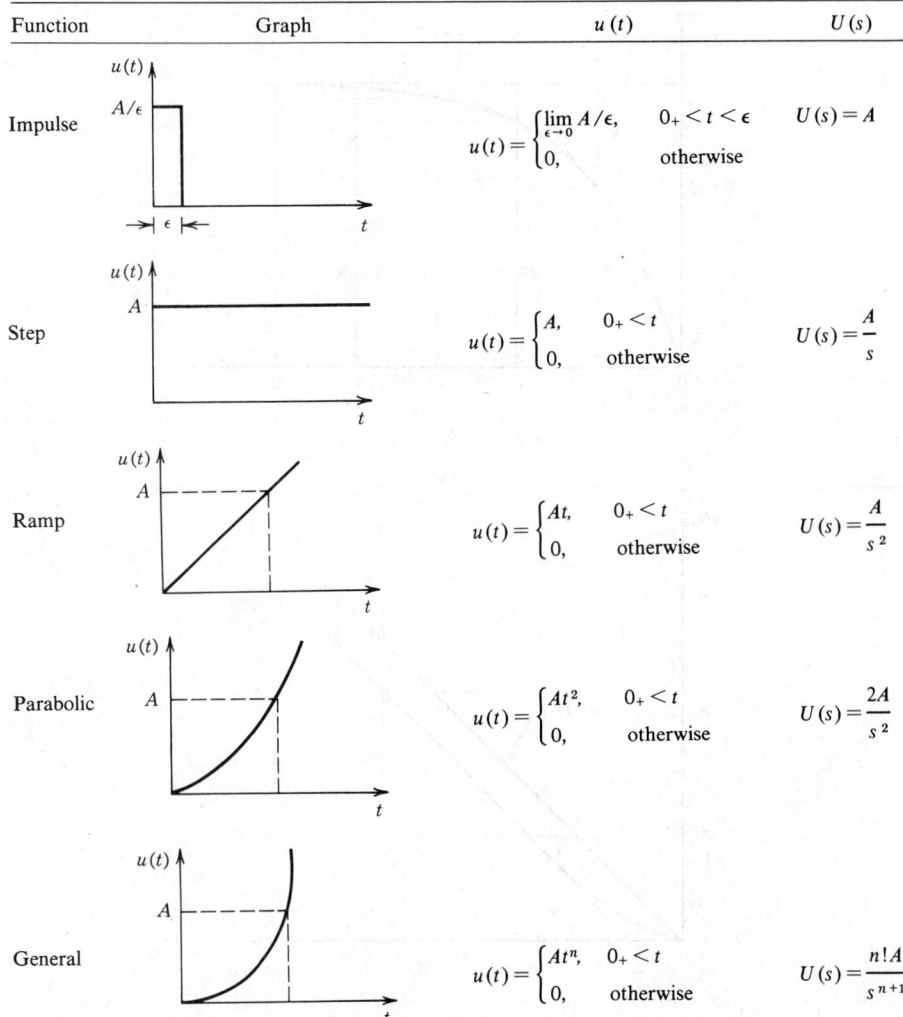

Function	Graph	$u(t)$	$U(s)$
Impulse		$u(t) = \begin{cases} \lim\limits_{\epsilon \to 0} A/\epsilon, & 0_+ < t < \epsilon \\ 0, & \text{otherwise} \end{cases}$	$U(s) = A$
Step		$u(t) = \begin{cases} A, & 0_+ < t \\ 0, & \text{otherwise} \end{cases}$	$U(s) = \dfrac{A}{s}$
Ramp		$u(t) = \begin{cases} At, & 0_+ < t \\ 0, & \text{otherwise} \end{cases}$	$U(s) = \dfrac{A}{s^2}$
Parabolic		$u(t) = \begin{cases} At^2, & 0_+ < t \\ 0, & \text{otherwise} \end{cases}$	$U(s) = \dfrac{2A}{s^2}$
General		$u(t) = \begin{cases} At^n, & 0_+ < t \\ 0, & \text{otherwise} \end{cases}$	$U(s) = \dfrac{n!A}{s^{n+1}}$

Fig. 39.14 Family of singularity functions commonly used as test inputs.

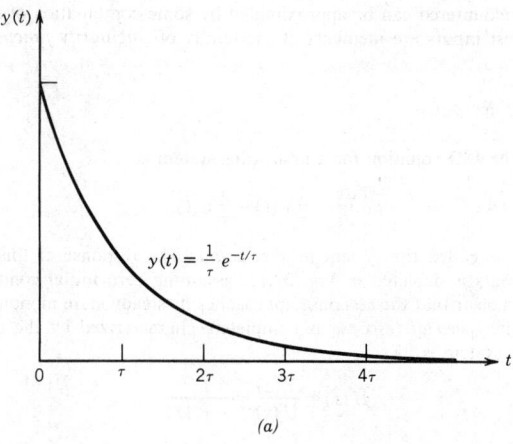

$$y(t) = \frac{1}{\tau} e^{-t/\tau}$$

(a)

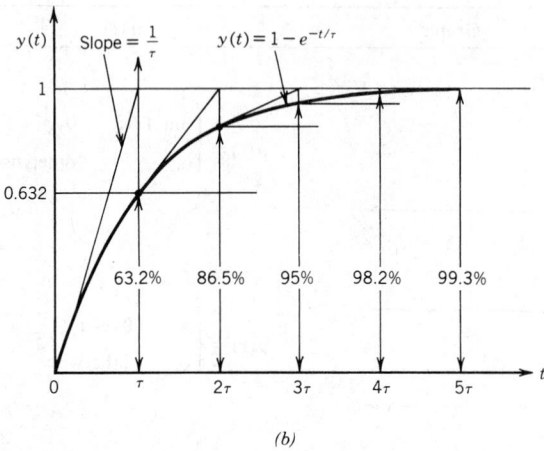

Slope $= \dfrac{1}{\tau}$

$y(t) = 1 - e^{-t/\tau}$

1

0.632

63.2% 86.5% 95% 98.2% 99.3%

(b)

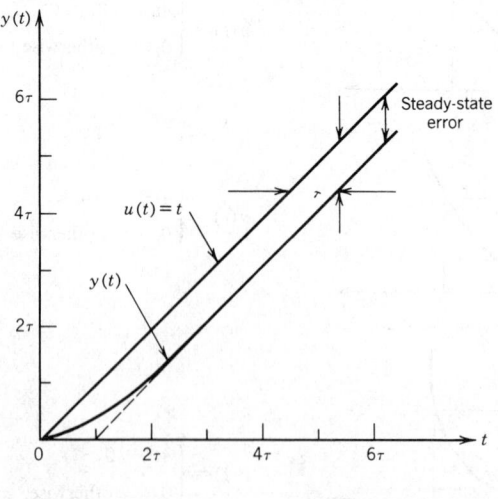

Steady-state error

$u(t) = t$

$y(t)$

(c)

Fig. 39.15 Response of a first-order system to (a) unit impulse, (b) unit step, and (c) unit ramp inputs.

and therefore $\tau = -p^{-1}$, where p is the system pole. As the absolute value of p increases, τ decreases and the response becomes faster.

The response of the standard first-order system to a step input of magnitude u for arbitrary initial condition $y(0) = y_0$ is

$$y(t) = y_{ss} - [y_{ss} - y_0]e^{-t/\tau}$$

where $y_{ss} = u$ is the steady-state response. Table 39.6 and Fig. 39.16 record the values of $y(t)$ and $\dot{y}(t)$ for $t = k\tau$, $k = 0, 1, \ldots, 6$. Note that over any time interval of duration τ, the response increases approximately 63% of the difference between the steady-state value and the value at the beginning of the time interval, that is,

$$y(t + \tau) - y(t) \approx 0.63212[y_{ss} - y(t)]$$

Note also that the slope of the response at the beginning of any time interval of duration τ intersects the steady-state value y_{ss} at the end of the interval, that is,

$$\frac{dy}{dt}(t) = \frac{y_{ss} - y(t)}{\tau}$$

Table 39.6 Tabulated Values of the Response of a First-Order System to a Unit Step Input

t	$y(t)$	$\dot{y}(t)$
0	0	τ^{-1}
τ	0.632	$0.368\tau^{-1}$
2τ	0.865	$0.135\tau^{-1}$
3τ	0.950	$0.050\tau^{-1}$
4τ	0.982	$0.018\tau^{-1}$
5τ	0.993	$0.007\tau^{-1}$
6τ	0.998	$0.002\tau^{-1}$

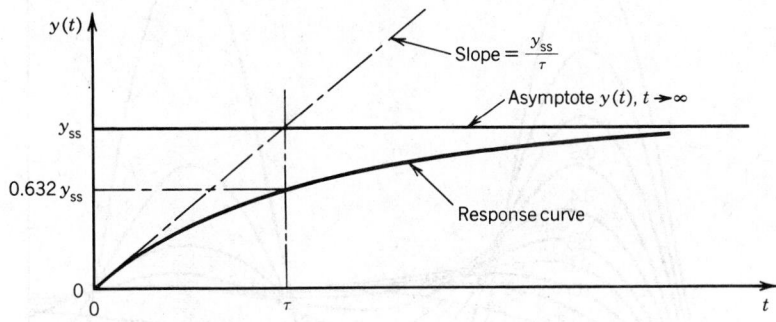

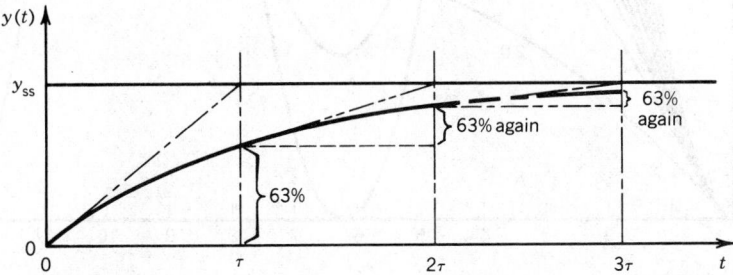

Fig. 39.16 Response of a first-order system to a unit step input, showing the relationship to the time constant.

Finally, note that after an interval of four time constants, the response is within 98% of the steady-state value, that is,

$$y(4\tau) \approx 0.98168(y_{ss} - y_0)$$

For this reason, $T_s = 4\tau$ is called the (2%) *setting time.*

Second-Order Transient Response

The standard form of the I/O equation for a second-order system is

$$\frac{d^2y}{dt^2} + 2\zeta\omega_n \frac{dy}{dt} + \omega_n^2 y(t) = \omega_n^2 u(t)$$

with transfer function

$$H(s) = \frac{Y(s)}{U(s)} = \frac{\omega_n^2}{s^2 + 2\zeta\omega_n s + \omega_n^2}$$

The system poles are obtained by applying the binomial theorem to the characteristic equation as

$$p_{1,2} = -\zeta\omega_n \pm j\omega_n\sqrt{1 - \zeta^2}$$

where the following parameters are defined: ζ is the *damping ratio*, ω_n is the *natural frequency*, and $\omega_d = \omega_n\sqrt{1 - \zeta^2}$ is the *damped natural frequency.*

The nature of the response of the standard second-order system to a step input depends on the value of the damping ratio, as depicted in Fig. 39.17. For a stable system, four classes of response are defined.

1. *Overdamped Response* ($\zeta > 1$). The system poles are real and distinct. The response of the second-order system can be decomposed into the response of two cascaded first-order systems, as shown in Fig. 39.18.

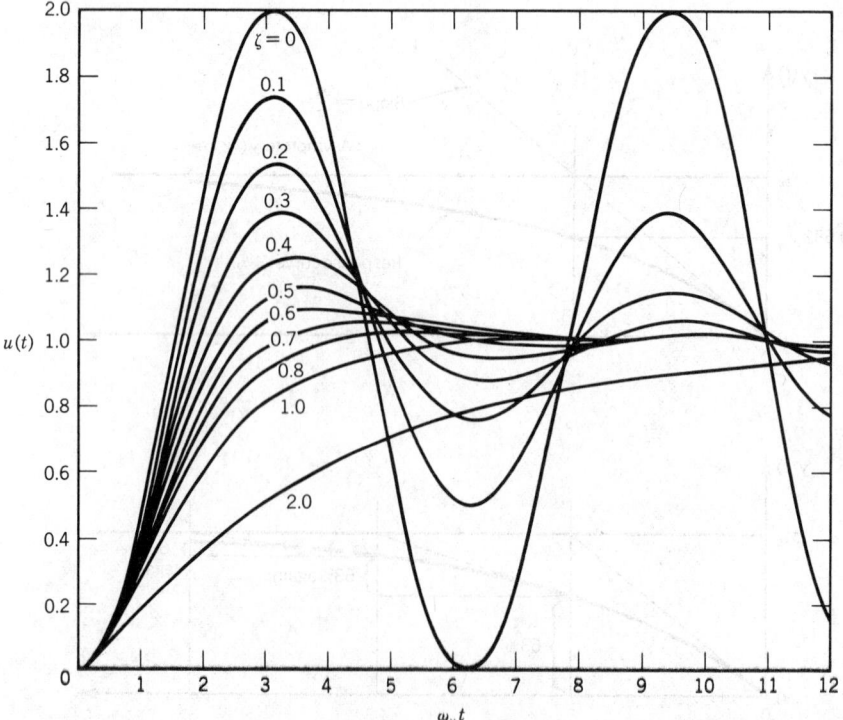

Fig. 39.17 Response of a second-order system to a unit step input for selected values of the damping ratio.

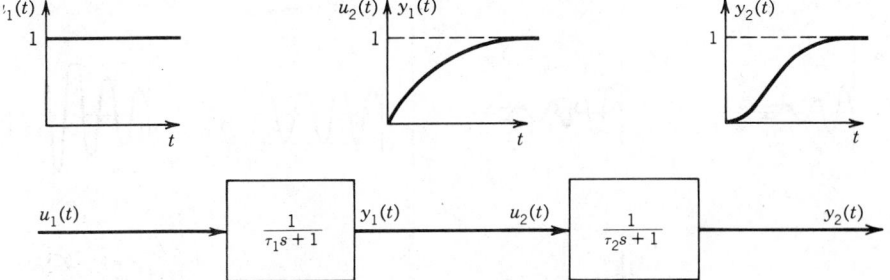

Fig. 39.18 Overdamped response of a second-order system decomposed into the responses of two first-order systems.

2. *Critically Damped Response* ($\zeta = 1$). The system poles are real and repeated. This is the limiting case of overdamped response, where the response is as fast as possible without overshoot.

3. *Underdamped Response* ($1 > \zeta > 0$). The system poles are complex conjugates. The response oscillates at the damped frequency ω_d. The magnitude of the oscillations and the speed with which the oscillations decay depend on the damping ratio ζ.

4. *Harmonic Oscillation* ($\zeta = 0$). The system poles are pure imaginary numbers. The response oscillates at the natural frequency ω_n and the oscillations are undamped (i.e., the oscillations are sustained and do not decay).

The Complex s-Plane

The location of the system poles (roots of the characteristic equation) in the *complex s-plane* reveals the nature of the system response to test inputs. Figure 39.19 shows the relationship between the location of the poles in the complex plane and the parameters of the standard second-order system. Figure 39.20 shows the unit impulse response of a second-order system corresponding to various pole locations in the complex plane.

Transient Response of Higher-Order Systems

The response of third- and higher-order systems to test inputs is simply the sum of terms representing component first- and second-order responses. This is because the system poles must either be real, resulting in first-order terms, or complex, resulting in second-order underdamped terms. Furthermore, because the transients associated with those system poles having the largest real part decay the most slowly, these transients tend to dominate the output. The response of higher-order systems therefore tends to have the same form as the response to the *dominant poles*, with the response to the *subdominant*

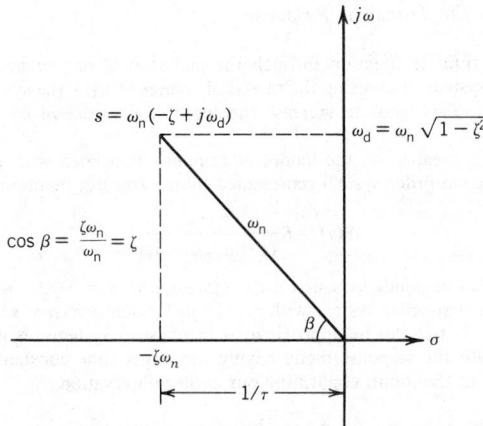

Fig. 39.19 Location of the upper complex pole in the s-plane in terms of the parameters of the standard second-order system.

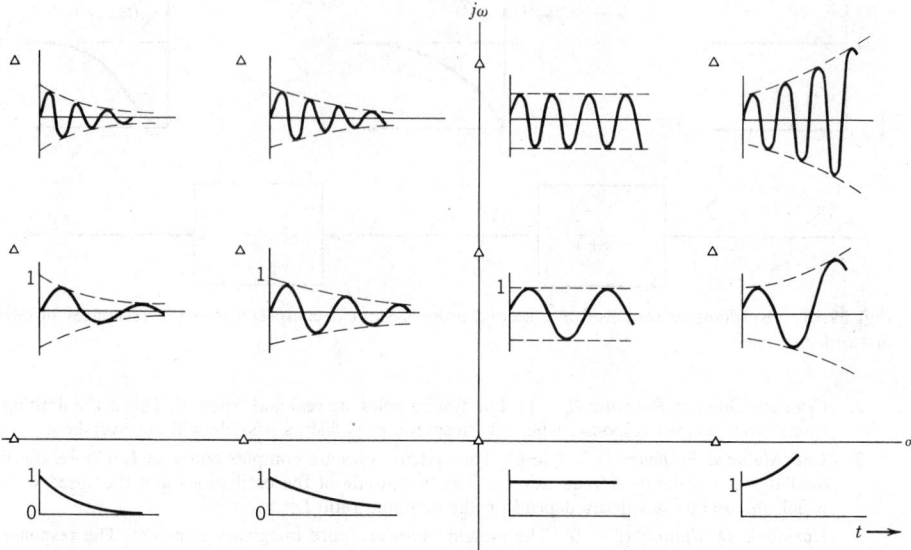

Fig. 39.20 Unit impulse response for selected upper complex pole locations in the s-plane.[8]

poles superimposed over it. Note that the larger the relative difference between the real parts of the dominant and subdominant poles, the more the output tends to resemble the dominant mode of response.

For example, consider a fixed linear third-order system. The system has three poles. The poles may either all be real, or one may be real while the other pair is complex conjugates. This leads to the three forms of step-response shown in Fig. 39.21, depending on the relative locations of the poles in the complex plane: (1) first-order dominant with two first-order subdominant modes; (2) first-order dominant with one second-order underdamped subdominant mode; or (3) second-order dominant with one first-order subdominant mode.

Transient Performance Measures

The transient response of a system is commonly described in terms of the measures defined in Table 39.7 and shown in Fig. 39.22. While these measures apply to any output, for a second-order system these can be calculated exactly in terms of the damping ratio and natural frequency, as shown in column three of the table. A common practice in control system design is to determine an initial design with dominant second-order poles that satisfy the performance specifications. Such a design can easily be calculated and then modified as necessary to achieve the desired performance.

The Effect of Zeros on the Transient Response

Zeros arise in a system transfer function through the inclusion of one or more derivatives of $u(t)$, among the inputs to the system. By sensing the rate(s) of change of $u(t)$, the system in effect *anticipates* the future values of $u(t)$. This tends to increase the speed of response of the system relative to the input $u(t)$.

The effect of a zero is greatest on the modes of response associated with neighboring poles. For example, consider the second-order system represented by the transfer function

$$H(s) = K\frac{s - z}{(s - p_1)(s - p_2)}$$

If $z = p_1$, then the system responds as a first-order system with $\tau = -p_2^{-1}$; whereas if $z = p_2$, then the system responds as a first-order system with $\tau = -p_1^{-1}$. Such *pole-zero cancellation* can only be achieved mathematically, but it can be approximated in physical systems. Note that by diminishing the residue associated with the response mode having the larger time constant, the system responds more quickly to changes in the input, confirming our earlier observation.

39.5.3   Response to Periodic Inputs Using Transform Methods

The response of a dynamic system to periodic inputs can be a critical concern to the control engineer. An input $u(t)$ is *periodic* if $u(t + T) = u(t)$ for all time t, where T is a constant called the period.

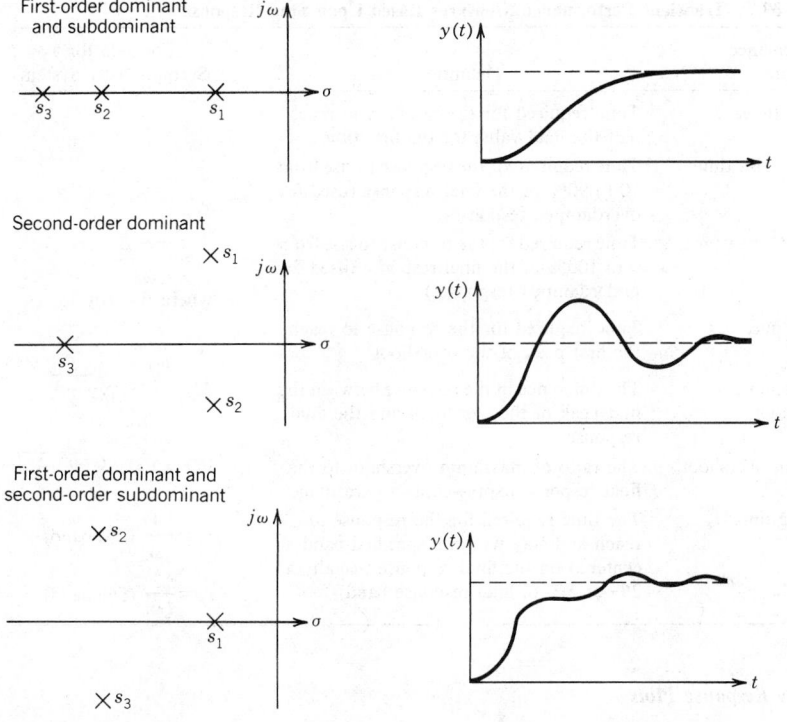

Fig. 39.21 Step response of a third-order system for alternative upper-complex pole locations in the s-plane.

Periodic inputs are important because these are ubiquitous: rotating unbalanced machinery, reciprocating pumps and engines, ac electrical power, and a legion of noise and disturbance inputs can be approximated by periodic inputs. Sinusoids are the most important category of periodic inputs, because these are frequently occurring, easily analyzed, and form the basis for analysis of general periodic inputs.

Frequency Response

The *frequency response* of a system is the steady-state response of the system to a sinusoidal input. For a linear system, the frequency response has the unique property that the response is a sinusoid of the same frequency as the input sinusoid, differing only in amplitude and phase. In addition, it is easy to show that the amplitude and phase of the response are functions of the input frequency, which are readily obtained from the system transfer function.

Consider a system defined by the transfer function $H(s)$. For an input

$$u(t) = A \sin \omega t$$

the corresponding steady-state output is

$$y_{ss}(t) = AM(\omega) \sin[\omega t + \phi(\omega)]$$

where $M(\omega) = |H(j\omega)|$ is called the *magnitude ratio*
$\phi(\omega) = \angle H(j\omega)$ is called the *phase angle*
$H(j\omega) = |H(s)|_{s=j\omega}$ is called the *frequency transfer function*

The frequency transfer function is obtained by substituting $j\omega$ for s in the transfer function $H(s)$. If the complex quantity $H(j\omega)$ is written in terms of its real and imaginary parts as $H(j\omega) = \text{Re}(\omega) + j\text{Im}(\omega)$, then

$$M(\omega) = [\text{Re}(\omega)^2 + \text{Im}(\omega)^2]^{1/2}$$
$$\phi(\omega) = \tan^{-1}[\text{Im}(\omega)/\text{Re}(\omega)]$$

and in polar form

$$H(j\omega) = M(\omega)e^{j\phi(\omega)}$$

Table 39.7   Transient Performance Measures Based Upon Step Response

Performance Measure	Definition	Formula for a Second-Order System
Delay time, t_d	Time required for the response to reach half the final value for the first time	
10–9% rise time, t_r	Time required for the response to rise from 10 to 90% of the final response (used for overdamped responses)	
0–100% rise time, t_r	Time required for the response to rise from 0 to 100% of the final response (used for underdamped responses)	$t_r = \dfrac{\pi - \beta}{\omega_d}$ where $\beta = \cos^{-1}\zeta$
Peak time, t_p	Time required for the response to reach the first peak of the overshoot	$t_p = \dfrac{\pi}{\omega_d}$
Maximum overshoot, M_p	The difference in the response between the first peak of the overshoot and the final response	$M_p = e^{-\zeta\pi/\sqrt{1-\zeta^2}}$
Percent overshoot, PO	The ratio of maximum overshoot to the final response expressed as a percentage	$PO = 100e^{-\zeta\pi/\sqrt{1-\zeta^2}}$
Setting time, t_s	The time required for the response to reach and stay within a specified band centered on the final response (usually a 2% or 5% of final response band)	$t_s = \dfrac{4}{\zeta\omega_n}$ (2% band) $t_s = \dfrac{3}{\zeta\omega_n}$ (5% band)

Frequency Response Plots

The frequency response of a fixed linear system is typically represented graphically, using one of three types of frequency response plots. A *polar plot* is simply a plot of the vector $H(j\omega)$ in the complex plane, where $\text{Re}(\omega)$ is the abscissa and $\text{Im}(\omega)$ is the ordinate. A *logarithmic plot* or *Bode diagram* consists of two displays: (1) the magnitude ratio in decibels $M_{db}(\omega)$ [where $M_{db}(\omega) = 20 \log M(\omega)$]

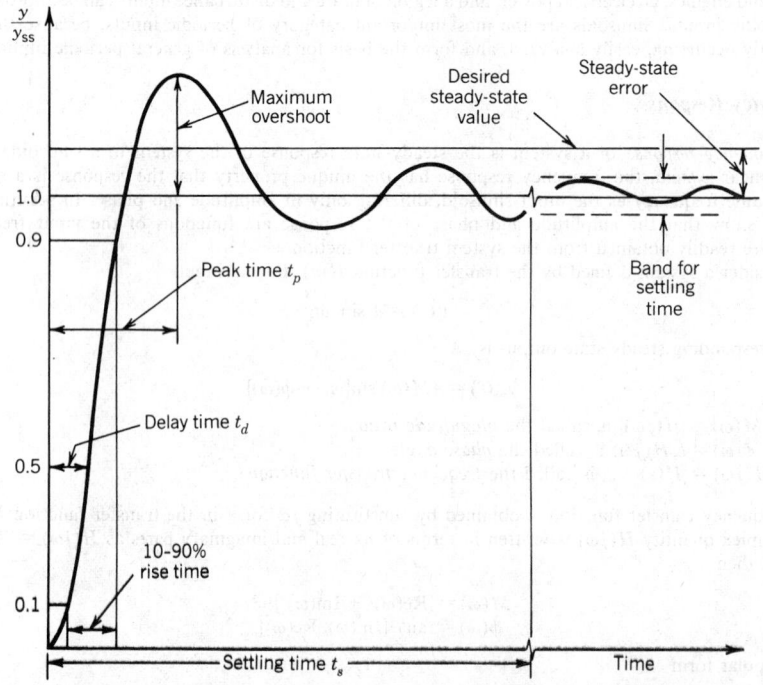

Fig. 39.22 Transient performance measures based on step response.

versus log ω, and (2) the phase angle in degrees $\phi(\omega)$ versus log ω. Bode diagrams for normalized first- and second-order systems are given in Fig. 39.23. Bode diagrams for higher-order systems are obtained by adding these first- and second-order terms, appropriately scaled. A *Nichols diagram* can be obtained by cross plotting the Bode magnitude and phase diagrams, eliminating log ω. Polar plots and Bode and Nichols diagrams for common transfer functions are given in Table 39.8.

Frequency Response Performance Measures

Frequency response plots show that dynamic systems tend to behave like *filters*, "passing" or even amplifying certain ranges of input frequencies, while blocking or attenuating other frequency ranges. The range of frequencies for which the amplitude ratio is no less than 3 db of its maximum value is called the *bandwidth* of the system. The bandwidth is defined by upper and lower *cutoff frequencies* ω_c, or by $\omega = 0$ and an upper cutoff frequency if $M(0)$ is the maximum amplitude ratio. Although the choice of "down 3 db" used to define the cutoff frequencies is somewhat arbitrary, the bandwidth is usually taken to be a measure of the range of frequencies for which a significant portion of the input is felt in the system output. The bandwidth is also taken to be a measure of the system speed of response, since attenuation of inputs in the higher-frequency ranges generally results from the inability of the system to "follow" rapid changes in amplitude. Thus, a narrow bandwidth generally indicates a sluggish system response.

Response to General Periodic Inputs

The *Fourier series* provides a means for representing a general periodic input as the sum of a constant and terms containing sine and cosine. For this reason the *Fourier series*, together with the superposition principle for linear systems, extends the results of frequency response analysis to the general case of arbitrary periodic inputs. The Fourier series representation of a periodic function $f(t)$ with period $2T$ on the interval $t^* + 2T \geq t \geq t^*$ is

$$f(t) = \frac{a_0}{2} + \sum_{n=1}^{\infty} \left(a_n \cos \frac{n \pi t}{T} + b_n \sin \frac{n \pi t}{T} \right)$$

where

$$a_n = \frac{1}{T} \int_{t^*}^{t^*+2T} f(t) \cos \frac{n \pi t}{T} \, dt$$

$$b_n = \frac{1}{T} \int_{t^*}^{t^*+2T} f(t) \sin \frac{n \pi t}{T} \, dt$$

If $f(t)$ is defined outside the specified interval by a periodic extension of period $2T$, and if $f(t)$ and its first derivative are piecewise continuous, then the series converges to $f(t)$ if t is a point of continuity, or to $\frac{1}{2} [f(t_+) + f(t_-)]$ if t is a point of discontinuity. Note that while the Fourier series in general is infinite, the notion of bandwidth can be used to reduce the number of terms required for a reasonable approximation.

39.6 STATE-VARIABLE METHODS

State-variable methods use the vector state and output equations introduced in Section 39.4 for analysis of dynamic systems directly in the time domain. These methods have several advantages over transform methods. First, state-variable methods are particularly advantageous for the study of multivariable (multiple input/multiple output) systems. Second, state-variable methods are more naturally extended for the study of linear time-varying and nonlinear systems. Finally, state-variable methods are readily adapted to computer simulation studies.

39.6.1 Solution of the State Equation

Consider the vector equation of state for a fixed linear system:

$$\dot{x}(t) = Ax(t) + Bu(t)$$

The solution to this system is

$$x(t) = \Phi(t)x(0) + \int_0^t \Phi(t - \tau)Bu(\tau) \, d\tau$$

where the matrix $\Phi(t)$ is called the *state-transition matrix*. The state-transition matrix represents the free response of the system and is defined by the matrix exponential series

$$\Phi(t) = e^{At} = I + At + \frac{1}{2!} A^2 t^2 + \cdots = \sum_{k=0}^{\infty} \frac{1}{k!} A^k t^k$$

where I is the identity matrix. The state transition matrix has the following useful properties:

$$\Phi(0) = I$$
$$\Phi^{-1}(t) = \Phi(-t)$$
$$\Phi^k(t) = \Phi(kt)$$
$$\Phi(t_1 + t_2) = \Phi(t_1)\Phi(t_2)$$
$$\Phi(t_2 - t_1)\Phi(t_1 - t_0) = \Phi(t_2 - t_0)$$
$$\dot{\Phi}(t) = A\,\Phi(t)$$

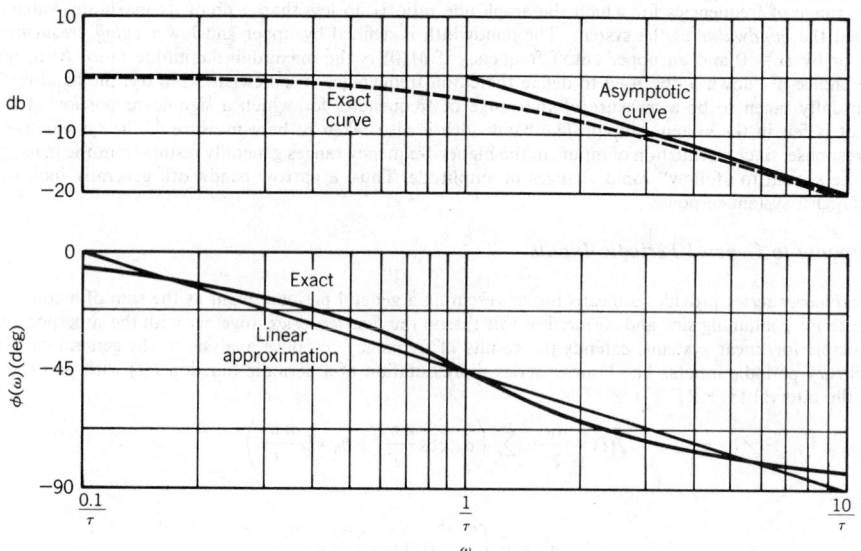

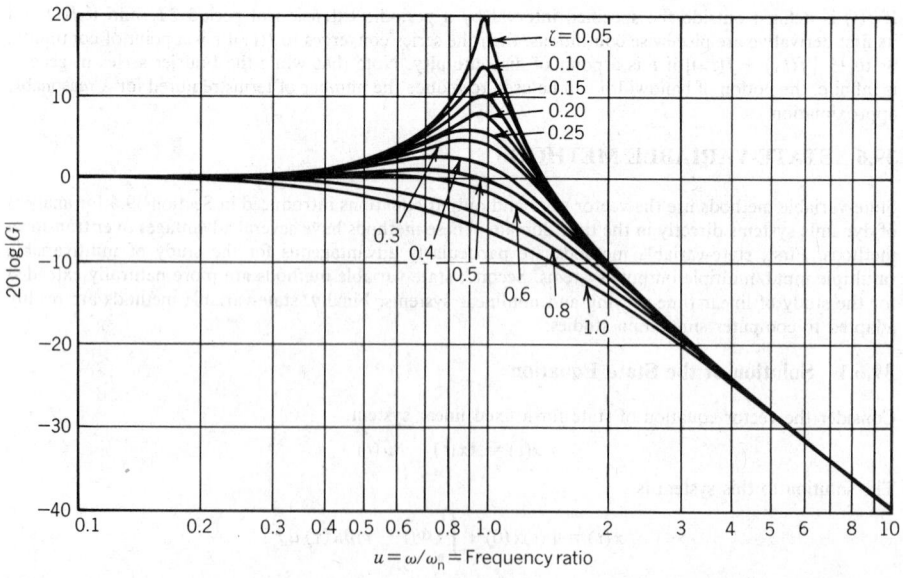

Fig. 39.23   Bode diagrams for normalized (a) first-order and (b) second-order systems.

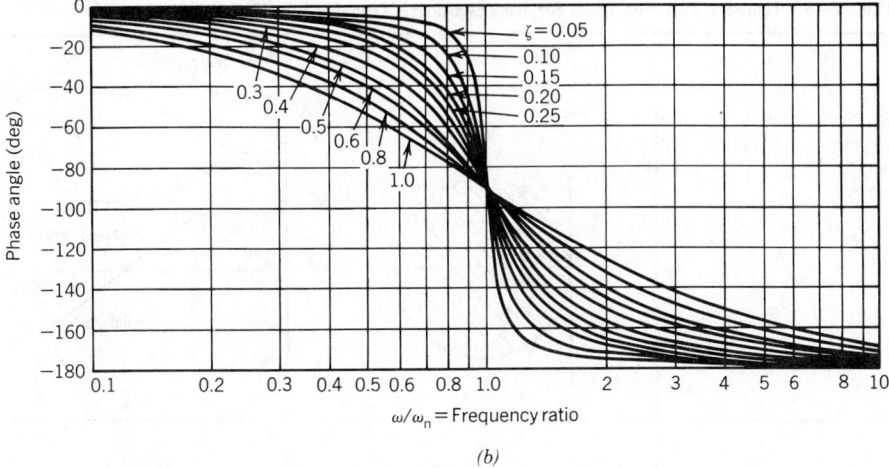

(b)

Fig. 39.23 *(Continued)*

The Laplace transform of the state equation is

$$sX(s) - x(0) = AX(s) + BU(s)$$

The solution to the fixed linear system therefore can be written as

$$x(t) = \mathcal{L}^{-1}[X(s)]$$
$$= \mathcal{L}^{-1}[\Phi(s)]x(0) + \mathcal{L}^{-1}[\Phi(s)BU(s)]$$

where

$$\Phi(t) = \mathcal{L}^{-1}[\Phi(s)] = \mathcal{L}^{-1}[sI - A]^{-1}$$

39.6.2 Eigenstructure

The internal structure of a system (and therefore its free response) is defined entirely by the system matrix A. The concept of matrix *eigenstructure,* as defined by the eigenvalues and eigenvectors of the system matrix, can provide a great deal of insight into the fundamental behavior of a system. In particular, the system eigenvectors can be shown to define a special set of first-order subsystems embedded within the system. These subsystems behave independently of one another, a fact that greatly simplifies analysis.

System Eigenvalues and Eigenvectors

For a system with system matrix A, the system *eigenvectors* v_i and associated *eigenvalues* λ_i are defined by the equation

$$Av_i = \lambda_i v_i$$

Note that the eigenvectors represent a set of special directions in the state space. If the state vector is aligned in one of these directions, then the homogeneous state equation becomes $\dot{v}_i = Av_i = \lambda_i v_i$, implying that each of the state variables changes at the *same* rate determined by the eigenvalue λ_i. This further implies that, in the absence of inputs to the system, a state vector that becomes aligned with a eigenvector will remain aligned with that eigenvector.

The system eigenvalues are calculated by solving the nth-order polynomial equation

$$|\lambda I - A| = \lambda^n + a_{n-1}\lambda^{n-1} + \cdots + a_1\lambda + a_0 = 0$$

This equation is called the *characteristic equation*. Thus the system eigenvalues are the roots of the characteristic equation, that is, the system eigenvalues are identically the system poles defined in transform analysis.

Each system eigenvector is determined by substituting the corresponding eigenvalue into the defining equation and then solving the resulting set of simultaneous linear equations. Only $n-1$ of the n components of any eigenvector are independently defined, however. In other words, the magnitude of an eigenvector is arbitrary, and the eigenvector describes a direction in the state space.

Table 39.8 Transfer Function Plots for Representative Transfer Functions[28]

$G(s)$	Polar plot	Bode diagram
1. $\dfrac{K}{s\tau_1 + 1}$		
2. $\dfrac{K}{(s\tau_1 + 1)(s\tau_2 + 1)}$		
3. $\dfrac{K}{(s\tau_1 + 1)(s\tau_2 + 1)(s\tau_3 + 1)}$		
4. $\dfrac{K}{s}$		

Table 39.8 (*Continued*)

Nichols diagram	Root locus	Comments
		Stable; gain margin $= \infty$
		Elementary regulator; stable; gain margin $= \infty$
		Regulator with additional energy-storage component; unstable, but can be made stable by reducing gain
		Ideal integrator; stable

Table 39.8 (Continued)

$G(s)$	Polar plot	Bode diagram

5.

$$\dfrac{K}{s(s\tau_1 + 1)}$$

Polar plot: $\omega \to 0$, $-\omega$, -1, $\omega = \infty$, $+\omega$, $\omega \to 0$

Bode diagram: -6 db/oct, $-90°$, ϕ M, Phase margin, $-180°$, 0 db, $\dfrac{1}{\tau_1}$, log ω, -12 db/oct

6.

$$\dfrac{K}{s(s\tau_1 + 1)(s\tau_2 + 1)}$$

Polar plot: $\omega \to 0$, $-\omega$, -1, $\omega = \infty$, $+\omega$, $\omega \to 0$

Bode diagram: $-90°$, ϕ M, -6, Phase margin, Gain margin, $-180°$, 0 db, $\dfrac{1}{\tau_1}$, -12, $1/\tau_2$, log ω, $-270°$, -18 db/oct

7.

$$\dfrac{K(s\tau_a + 1)}{s(s\tau_1 + 1)(s\tau_2 + 1)}$$

Polar plot: $\omega \to 0$, $-\omega$, -1, $\omega = \infty$, $+\omega$, $\omega \to 0$

Bode diagram: $-90°$, ϕ M, -6 db oct, Phase margin, -12, $1/\tau_2$, ϕ, $-180°$, 0 db, $\dfrac{1}{\tau_1}$, $\dfrac{1}{\tau_a}$, -6, log ω, -12 db/oct

8.

$$\dfrac{K}{s^2}$$

Polar plot: $\omega \to$, $-\omega$, -1, $\omega = \infty$, $+\omega$

Bode diagram: Gain margin = 0, Phase margin = 0, ϕ M, -12 db/oct, $-180°$, 0 db, ϕ, log $\omega \to$

1128

Table 39.8 (*Continued*)

Nichols diagram	Root locus	Comments
		Elementary instrument servo; inherently stable; gain margin = ∞
		Instrument servo with field-control motor or power servo with elementary Ward-Leonard drive; stable as shown, but may become unstable with increased gain
		Elementary instrument servo with phase-lead (derivative) compensator; stable
		Inherently unstable; must be compensated

Table 39.8 (*Continued*)

	$G(s)$	Polar plot	Bode diagram
9.	$\dfrac{K}{s^2(s\tau_1 + 1)}$		
10.	$\dfrac{K(s\tau_a + 1)}{s^2(s\tau_1 + 1)}$ $\tau_a > \tau_1$		
11.	$\dfrac{K}{s^3}$		
12.	$\dfrac{K(s\tau_a + 1)}{s^3}$		

Table 39.8 (*Continued*)

Nichols diagram	Root locus	Comments
		Inherently unstable; must be compensated
		Stable for all gains
		Inherently unstable
		Inherently unstable

Table 39.8 (*Continued*)

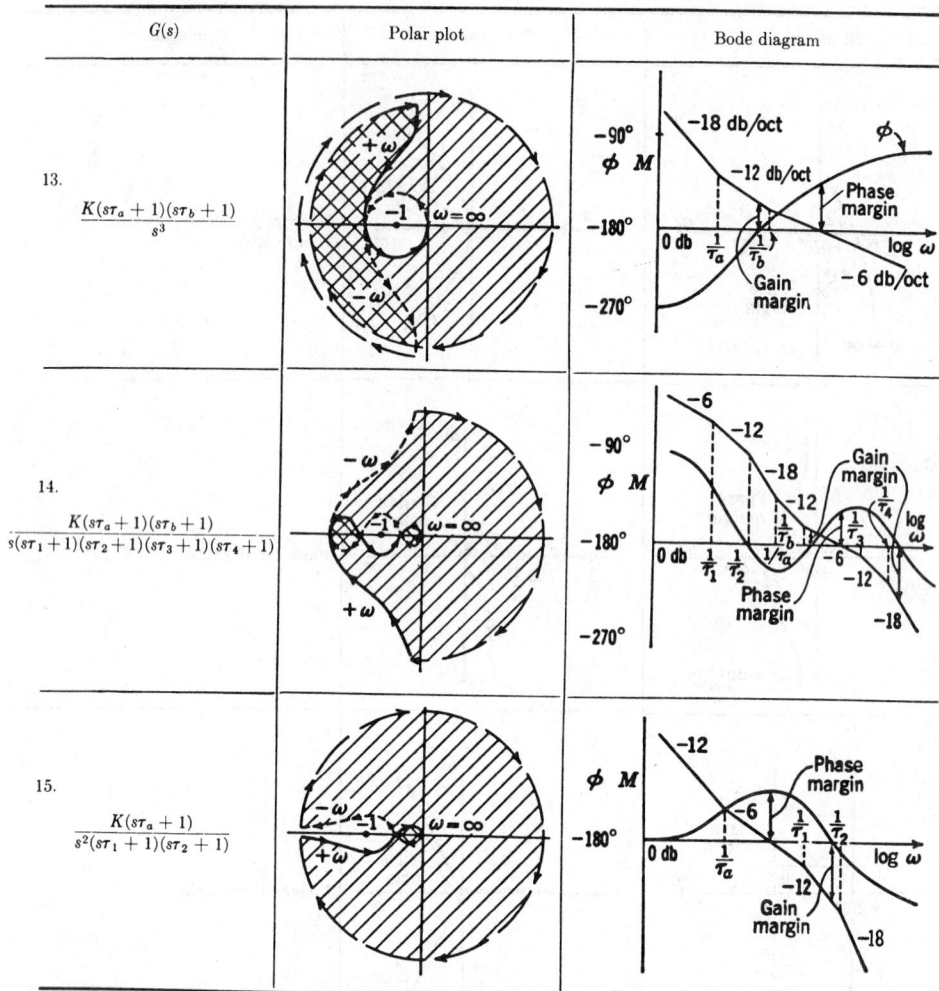

$G(s)$	Polar plot	Bode diagram
13. $\dfrac{K(s\tau_a + 1)(s\tau_b + 1)}{s^3}$		
14. $\dfrac{K(s\tau_a + 1)(s\tau_b + 1)}{s(s\tau_1 + 1)(s\tau_2 + 1)(s\tau_3 + 1)(s\tau_4 + 1)}$		
15. $\dfrac{K(s\tau_a + 1)}{s^2(s\tau_1 + 1)(s\tau_2 + 1)}$		

Diagonalized Canonical Form

There will be one linearly independent eigenvector for each distinct (nonrepeated) eigenvalue. If all of the eigenvalues of an nth-order system are distinct, then the n independent eigenvectors form a new basis for the state space. This basis represents new coordinate axes defining a set of state variables $z_i(t)$, $i=1,2,\ldots,n$, called the *diagonalized canonical variables*. In terms of the diagonalized variables, the homogeneous state equation is

$$\dot{z}(t) = \Lambda z$$

where Λ is a diagonal system matrix of the eigenvectors, that is,

$$\Lambda = \begin{bmatrix} \lambda_1 & 0 & \cdots & 0 \\ 0 & \lambda_2 & \cdots & 0 \\ \vdots & & \ddots & \vdots \\ 0 & 0 & \cdots & \lambda_n \end{bmatrix}$$

The solution to the diagonalized homogeneous system is

$$z(t) = e^{\Lambda t} z(0)$$

Table 39.8 *(Concluded)*

Nichols diagram	Root locus	Comments
		Conditionally stable; becomes unstable if gain is too low
		Conditionally stable; stable at low gain, becomes unstable as gain is raised, again becomes stable as gain is further increased, and becomes unstable for very high gains
		Conditionally stable; becomes unstable at high gain

where $e^{\Lambda t}$ is the diagonal state-transition matrix

$$e^{\Lambda t} = \begin{bmatrix} e^{\lambda_1 t} & 0 & \cdots & 0 \\ 0 & e^{\lambda_2 t} & \cdots & 0 \\ \vdots & \vdots & \ddots & \vdots \\ 0 & 0 & \cdots & e^{\lambda_n t} \end{bmatrix}$$

Modal Matrix

Consider the state equation of the nth-order system

$$\dot{x}(t) = Ax(t) + Bu(t)$$

which has real, distinct eigenvalues. Since the system has a full set of eigenvectors, the state vector $x(t)$ can be expressed in terms of the canonical state variables as

$$x(t) = v_1 z_1(t) + v_2 z_2(t) + \cdots + v_n z_n(t) = Mz(t)$$

where M is the $n \times n$ matrix whose columns are the eigenvectors of A, called the *modal matrix*. Using the modal matrix, the state-transition matrix for the original system can be written as

$$\Phi(t) = e^{At} = Me^{\Lambda t}M^{-1}$$

where $e^{\Lambda t}$ is the diagonal state-transition matrix. This frequently proves to be an attractive method for determining the state-transition matrix of a system with real, distinct eigenvalues.

Jordan Canonical Form

For a system with one or more repeated eigenvalues, there is not in general a full set of eigenvectors. In this case, it is not possible to determine a diagonal representation for the system. Instead, the simplest representation that can be achieved is block diagonal. Let $L_k(\lambda)$ be the $k \times k$ matrix

$$L_k(\lambda) = \begin{bmatrix} \lambda & 1 & 0 & \cdots & 0 \\ 0 & \lambda & 1 & \cdots & 0 \\ \vdots & \vdots & \lambda & \ddots & 0 \\ \vdots & \vdots & \vdots & \ddots & 1 \\ 0 & 0 & 0 & 0 & \lambda \end{bmatrix}$$

Then for any $n \times n$ system matrix A there is certain to exist a nonsingular matrix T such that

$$T^{-1}AT = \begin{bmatrix} L_{k_1}(\lambda_1) & & & \\ & L_{k_2}(\lambda_2) & & \\ & & \ddots & \\ & & & L_{k_r}(\lambda_r) \end{bmatrix}$$

where $k_1 + k_2 + \cdots + k_r = n$ and where the $\lambda_i, i = 1, 2, \ldots, r$, are the (not necessarily distinct) eigenvalues of A. The matrix $T^{-1}AT$ is called the *Jordan cononical form*.

39.7 SIMULATION

39.7.1 Simulation—Experimental Analysis of Model Behavior

Closed-form solutions for nonlinear or time-varying systems are rarely available. In addition, while explicit solutions for time-invariant linear systems can always be found, for high-order systems this is often impractical. In such cases it may be convenient to study the dynamic behavior of the system using *simulation*.

Simulation is the *experimental* analysis of model behavior. A *simulation run* is a controlled experiment in which a specific realization of the model is manipulated in order to determine the response associated with that realization. A *simulation study* comprises *multiple runs*, each run for a different combination of model parameter values and/or initial conditions. The generalized solution of the model must then be inferred from a finite number of simulated data points.

Simulation is almost always carried out with the assistance of computing equipment. *Digital simulation* involves the *numerical solution* of model equations using a digital computer. *Analog simulation* involves solving model equations by analogy with the behavior of a physical system using an analog computer. *Hybrid simulation* employs digital and analog simulation together using a hybrid (part digital and part analog) computer.

39.7.2 Digital Simulation

Digital continuous-system simulation involves the approximate solution of a state-variable model over successive time steps. Consider the general state-variable equation

$$\dot{x}(t) = f[x(t), u(t)]$$

to be simulated over the time interval $t_0 \le t \le t_K$. The solution to this problem is based on the repeated solution of the single-variable, single-step subproblem depicted in Fig. 39.24. The subproblem may be stated formally as follows:

Given:

1. $\Delta t(k) = t_k - t_{k-1}$, the length of the kth *time step*.
2. $x_i(t) = f_i[x(t), u(t)]$ for $t_{k-1} \le t \le t_k$, the ith equation of state defined for the state variable $x_i(t)$ over the kth time step.
3. $u(t)$ for $t_{k-1} \le t \le t_k$, the input vector defined for the kth time step.
4. $\tilde{x}(k-1) \simeq x(t_{k-1})$, an initial approximation for the state vector at the beginning of the time step.

Find:

5. $\tilde{x}_i(k) \simeq x_i(t_k)$, a final approximation for the state variable $x_i(t)$ at the end of the kth time step.

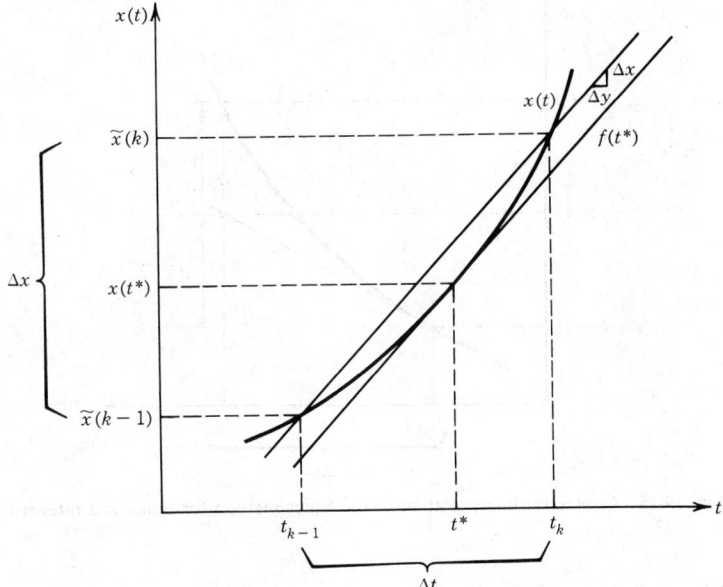

Fig. 39.24 Numerical approximation of a single variable over a single time step.

Solving this single-variable, single-step subproblem for each of the state variables $x_i(t)$, $i = 1, 2,$ ..., n, yields a final approximation for the state vector $\bar{x}(k) \simeq x(t_k)$ at the end of the kth time step. Solving the complete single-step problem K times over K time steps, beginning with the initial condition $\bar{x}(0) = x(t_0)$ and using the final value of $\bar{x}(t_k)$ from the kth time step as the initial value of the state for the $(k+1)$st time step, yields a discrete succession of approximations $\bar{x}(1) \simeq x(t_1)$, $\bar{x}(2) \simeq x(t_2)$, ..., $\bar{x}(K) \simeq x(t_K)$ spanning the solution time interval.

The basic procedure for completing the single-variable, single-step problem is the same regardless of the particular integration method chosen. It consists of two parts: (1) calculation of the average value of the ith derivative over the time step as

$$\dot{x}_i(t^*) = f_i[x(t^*), u(t^*)] = \frac{\Delta x_i(k)}{\Delta t(k)} \simeq \tilde{f}_i(k)$$

and (2) calculation of the final value of the simulated variable at the end of the time step as

$$\bar{x}_i(k) = \tilde{x}_i(k-1) + \Delta x_i(k)$$
$$\simeq \tilde{x}_i(k-1) + \Delta t(k)\tilde{f}_i(k)$$

If the function $f_i[x(t), u(t)]$ is continuous, then t^* is guaranteed to be on the time step, that is, $t_{k-1} \leq t^* \leq t_k$. Since the value of t^* is otherwise unknown, however, the value of $x(t^*)$ can only be approximated as $\tilde{f}(k)$.

Different *numerical integration* methods are distinguished by the means used to calculate the approximation $f_i(k)$. A wide variety of such methods is available for digital simulation of dynamic systems. The choice of a particular method depends on the nature of the model being simulated, the accuracy required in the simulated data, and the computing effort available for the simulation study. Several popular classes of integration methods are outlined in the following subsections.

Euler Method

The simplest procedure for numerical integration is the Euler method. The standard Euler method approximates the average value of the ith derivative over the kth time step using the derivative evaluated at the beginning of the time step, that is,

$$\tilde{f}_i(k) = f_i[\bar{x}(k-1), u(t_{k-1})] \simeq f_i(t_{k-1})$$

$i = 1, 2, \ldots, n$ and $k = 1, 2, \ldots, K$. This is shown geometrically in Fig. 39.25 for the scalar single-step case. A modification of this method uses the newly calculated state variables in the derivative calculation as these new values become available. Assuming the state variables are computed in numerical order according to the subscripts, this implies

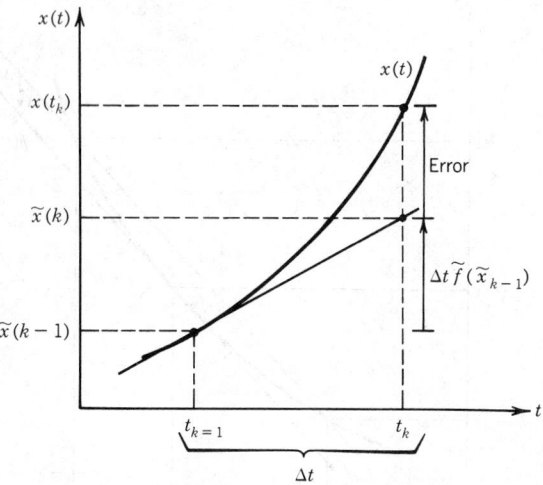

Fig. 39.25   Geometric interpretation of the Euler method for numerical integration.

$$\tilde{f}_i(k) = f_i[\tilde{x}_1(k), \ldots, \tilde{x}_{i-1}(k), \tilde{x}_i(k-1), \ldots, \tilde{x}_n(k-1), u(t_{k-1})]$$

The modified Euler method is modestly more efficient than the standard procedure and, frequently, is more accurate. In addition, since the input vector $u(t)$ is usually known for the entire time step, using an average value of the input, such as

$$u(k) = \frac{1}{\Delta t(k)} \int_{t_{k-1}}^{t_k} u(\tau) \, d\tau$$

frequently leads to a superior approximation of $\tilde{f}_i(k)$.

The Euler method requires the least amount of computational effort per time step of any numerical integration scheme. Local truncation error is proportional to Δt^2, however, which means that the error within each time step is highly sensitive to step size. Because the accuracy of the method demands very small time steps, the number of time steps required to implement the method successfully can be large relative to other methods. This can imply a large computational overhead and can lead to inaccuracies through the accumulation of roundoff error at each step.

Runge–Kutta Methods

Runge-Kutta methods precompute two or more values of $f_i[x(t),u(t)]$ in the time step $t_{k-1} \leq t \leq t_k$ and use some weighted average of these values to calculate $\tilde{f}_i(k)$. The *order* of a Runge–Kutta method refers to the number of derivative terms (or *derivative calls*) used in the scalar single-step calculation. A Runge–Kutta routine of order N therefore uses the approximation

$$\tilde{f}_i(k) = \sum_{j=1}^{N} w_j f_{ij}(k)$$

where the N approximations to the derivative are

$$f_{i1}(k) = f_i[\tilde{x}(k-1), u(t_{k-1})]$$

(the Euler approximation) and

$$f_{ij} = f_i\left[\tilde{x}(k-1) + \Delta t \sum_{l=1}^{j-1} Ib_{jl} f_{il}, u\left(t_{k-1} + \Delta t \sum_{l=1}^{j-1} b_{jl}\right)\right]$$

where I is the identity matrix. The weighting coefficients w_j and b_{jl} are not unique, but are selected such that the error in the approximation is zero when $x_i(t)$ is some specified Nth-order polynomial in t. Coefficients commonly used for Runge–Kutta integration are given in Table 39.9.

Among the most popular of the Runge–Kutta methods is fourth-order Runge–Kutta. Using the defining equations for $N = 4$ and the weighting coefficients from Table 39.9 yields the derivative approximation

$$\tilde{f}_i(k) = \frac{1}{6}[f_{i1}(k) + 2f_{i2}(k) + 2f_{i3}(k) + f_{i4}(k)]$$

Table 39.9 Coefficients Commonly Used for Runge–Kutta Numerical Integration[13]

Common Name	N	b_{jl}	w_j
Open or explicit Euler	1	All zero	$w_1 = 1$
Improved polygon	2	$b_{21} = \frac{1}{2}$	$w_1 = 0$
			$w_2 = 1$
Modified Euler or Heun's method	2	$b_{21} = 1$	$w_1 = \frac{1}{2}$
			$w_2 = \frac{1}{2}$
Third-order Runge–Kutta	3	$b_{21} = \frac{1}{2}$	$w_1 = \frac{1}{6}$
		$b_{31} = -1$	$w_2 = \frac{2}{3}$
		$b_{32} = 2$	$w_3 = \frac{1}{6}$
Fourth-order Runge–Kutta	4	$b_{21} = \frac{1}{2}$	$w_1 = \frac{1}{6}$
		$b_{31} = 0$	$w_2 = \frac{1}{3}$
		$b_{32} = \frac{1}{2}$	$w_3 = \frac{1}{3}$
		$b_{43} = 1$	$w_4 = \frac{1}{6}$

based on the four derivative calls

$$f_{i\,1}(k) = f_i\left[\bar{x}(k-1), u(t_{k-1})\right]$$
$$f_{i\,2}(k) = f_i\left[\bar{x}(k-1) + \frac{\Delta t}{2}\, I f_{i\,1}, u\left(t_{k-1} + \frac{\Delta t}{2}\right)\right]$$

$$f_{i\,3}(k) = f_i\left[\bar{x}(k-1) + \frac{\Delta t}{2}\, I f_{i\,2}, u\left(t_{k-1} + \frac{\Delta t}{2}\right)\right]$$

$$f_{i\,4}(k) = f_i\left[\bar{x}(k-1) + \Delta t\, I f_{i\,3}, u(t_k)\right]$$

where I is the identity matrix.

Because Runge–Kutta formulas are designed to be exact for a polynomial of order N, local truncation error is of the order Δt^{N+1}. This considerable improvement over the Euler method means that comparable accuracy can be achieved for larger step sizes. The penalty is that N derivative calls are required for each scalar evaluation within each time step.

Euler and Runge–Kutta methods are examples of *single-step methods* for numerical integration, so-called because the state $x(k)$ is calculated from knowledge of the state $x(k-1)$, without requiring knowledge of the state at any time prior to the beginning of the current time step. These methods are also referred to as *self-starting methods*, since calculations may proceed from any known state.

Multistep Methods

Multistep methods differ from the single-step methods previously described in that multistep methods use the stored values of two or more previously computed states and/or derivatives in order to compute the derivative approximation $\tilde{f}_i(k)$ for the current time step. The advantage of multistep methods over Runge–Kutta methods is that these require only one derivative call for each state variable at each time step for comparable accuracy. The disadvantage is that multistep methods are not self-starting, since calculations cannot proceed from the initial state alone. Multistep methods must be started, or restarted in the case of discontinuous derivatives, using a single-step method to calculate the first several steps.

The most popular of the multistep methods are the *Adams–Bashforth predictor methods* and the *Adams–Moulton corrector methods*. These methods use the derivative approximation

$$\tilde{f}_i(k) = \sum_{j=0}^{N} b_j f_i\left[\bar{x}(k-j), u(k-j)\right]$$

where the b_j are weighting coefficients. These coefficients are selected such that the error in the approximation is zero when $x_i(t)$ is a specified polynomial. Table 39.10 gives the values of the weighting coefficients for several Adams–Bashforth–Moulton rules. Note that the predictor methods employ an *open* or *explicit rule*, since for these methods $b_0 = 0$ and a prior estimate of $x_i(k)$ is not required. The corrector methods use a *closed* or *implicit rule*, since for these methods $b_0 \neq 0$ and a prior estimate of $x_i(k)$ is required. Note also that for all of these methods $\sum_{j=0}^{N} b_j = 1$, ensuring unity gain for the integration of a constant.

Predictor–Corrector Methods

Predictor–corrector methods use one of the multistep predictor equations to provide an initial estimate (or "prediction") of $x(k)$. This initial estimate is then used with one of the multistep corrector equations

Table 39.10 Coefficients Commonly Used for Adams–Bashforth–Moulton Numerical Integration[13]

Common Name	Predictor or Corrector	Points	b_{-1}	b_0	b_1	b_2	b_3
Open or explicit Euler	Predictor	1	0	1	0	0	0
Open trapezoidal	Predictor	2	0	$\frac{3}{2}$	$-\frac{1}{2}$	0	0
Adams three-point predictor	Predictor	3	0	$\frac{23}{12}$	$-\frac{16}{12}$	$\frac{5}{12}$	0
Adams four-point predictor	Predictor	4	0	$\frac{55}{24}$	$-\frac{59}{24}$	$\frac{37}{24}$	$-\frac{9}{24}$
Closed or implicit Euler	Corrector	1	1	0	0	0	0
Closed trapezoidal	Corrector	2	$\frac{1}{2}$	$\frac{1}{2}$	0	0	0
Adams three-point corrector	Corrector	3	$\frac{5}{12}$	$\frac{8}{12}$	$-\frac{1}{12}$	0	0
Adams four-point corrector	Corrector	4	$\frac{9}{24}$	$\frac{19}{24}$	$-\frac{5}{24}$	$\frac{1}{24}$	0

to provide a second and improved (or "corrected") estimate of $x(k)$, before proceeding to the next step. A popular choice is the four-point Adams–Bashforth predictor together with the four-point Adams–Moulton corrector, resulting in a prediction of

$$\tilde{x}_i(k) = \tilde{x}_i(k-1) + \frac{\Delta t}{24}[55\tilde{f}_i(k-1) - 59\tilde{f}_i(k-2) + 37\tilde{f}_i(k-3) - 9\tilde{f}_i(k-4)]$$

for $i = 1, 2, \ldots, n$, and a correction of

$$\tilde{x}_i(k) = \tilde{x}_i(k-1) + \frac{\Delta t}{24}\{9f_i[\tilde{x}(k), u(k)] + 19\tilde{f}_i(k-1) - 5\tilde{f}_i(k-2) + \tilde{f}_i(k-3)\}$$

Predictor–corrector methods generally incorporate a strategy for increasing or decreasing the size of the time step depending on the difference between the predicted and corrected $x(k)$ values. Such *variable time-step methods* are particularly useful if the simulated system possesses local time constants that differ by several orders of magnitude, or it there is little *a priori* knowledge about the system response.

Numerical Integration Errors

An inherent characteristic of digital simulation is that the discrete data points generated by the simulation $x(k)$ are only approximations to the exact solution $x(t_k)$ at the corresponding point in time. This results from two types of errors that are unavoidable in the numerical solutions. *Round-off errors* occur because numbers stored in a digital computer have finite word length (i.e., a finite number of bits per word) and therefore limited precision. Because the results of calculations cannot be stored exactly, round-off error tends to increase with the number of calculations performed. For a given total solution interval $t_0 \le t \le t_K$, therefore, round-off error tends to increase (1) with increasing integration-rule order (since more calculations must be performed at each time step) and (2) with decreasing step size Δt (since more time steps are required).

Truncation errors or *numerical approximation errors* occur because of the inherent limitations in the numerical integration methods themselves. Such errors would arise even if the digital computer had infinite precision. *Local* or *per-step truncation error* is defined as

$$e(k) = x(k) - x(t_k)$$

given that $x(k-1) = x(t_{k-1})$ and that the calculation at the kth time step is infinitely precise. For many integration methods, local truncation errors can be approximated at each step. *Global* or *total truncation error* is defined as

$$e(K) = x(K) - x(t_K)$$

given that $x(0) = x(t_0)$ and the calculations for all K time steps are infinitely precise. Global truncation error usually cannot be estimated, neither can efforts to reduce local truncation errors be guaranteed to yield acceptable global errors. In general, however, truncation errors can be decreased by using more sophisticated integration methods and by decreasing the step size Δt.

Time Constants and Time Steps

As a general rule, the step size Δt for simulation must be less than the smallest local time constant of the model simulated. This can be illustrated by considering the simple first-order system

$$\dot{x}(t) = \lambda x(t)$$

and the difference equation defining the corresponding Euler integration

$$x(k) = x(k-1) + \Delta t \lambda \, x(k-1)$$

The continuous system is stable for $\lambda < 0$, while the discrete approximation is stable for $|1 + \lambda \Delta t| < 1$. If the original system is stable, therefore, the simulated response will be stable for

$$\Delta t \leq 2 \, |1/\lambda|$$

where the equality defines the *critical step size*. For larger step sizes, the simulation will exhibit *numerical instability*. In general, while higher-order integration methods will provide greater per-step accuracy, the critical step size itself will not be greatly reduced.

A major problem arises when the simulated model has one or more time constants $|1/\lambda_i|$ that are small when compared to the total solution time interval $t_0 \leq t \leq t_K$. Numerical stability will then require very small Δt, even though the transient response associated with the higher-frequency (larger λ_i) subsystems may contribute little to the particular solution. Such problems can be addressed either by neglecting the higher-frequency components where appropriate, or by adopting special numerical integration methods for *stiff systems*.

Selecting an Integration Method

The best numerical integration method for a specific simulation is the method that yields an acceptable global approximation error with the minimum amount of round-off error and computing effort. No single method is best for all applications. The selection of an integration method depends on the model simulated, the purpose of the simulation study, and the availability of computing hardware and software.

In general, for well-behaved problems with continuous derivatives and no stiffness, a lower-order Adams predictor is often a good choice. Multistep methods also facilitate estimating local truncation error. Multistep methods should be avoided for systems with discontinuities, however, because of the need for frequent restarts. Runge–Kutta methods have the advantage that these are self-starting and provide fair stability. For stiff systems where high-frequency modes have little influence on the global response, special stiff-system methods enable the use of economically large step sizes. Variable-step rules are useful when little is known *a priori* about solutions. Variable-step rules often make a good choice as general-purpose integration methods.

Round-off error usually is not a major concern in the selection of an integration method, since the goal of minimizing computing effort typically obviates such problems. Double-precision simulation can be used where round off is a potential concern. An upper bound on step size often exists because of discontinuities in derivative functions or because of the need for response output at closely spaced time intervals.

Continuous System Simulation Languages

Digital simulation can be implemented for a specific model in any high-level language such as FORTRAN or PASCAL. The general process for implementing a simulation is shown in Fig. 39.26. In addition, many special-purpose continuous system simulation languages are commonly available. Such languages greatly simplify programming tasks and typically provide for good graphical output. Among the most popular simulation languages are CSMP (Continuous System Modeling Program, Ref. 26), CSSL (Continuous System Simulation Language, Ref. 3), DYNAMO (DYNAmic MOdeling, Refs. 9 and 20), and DARE (Differential Analyzer REplacement, Ref. 13). Microcomputer implementations of simulation packages also are becoming increasingly available.

39.7.3 Analog and Hybrid Simulation

Analog computers or *electronic differential analyzers* use continuously variable electric voltages to represent the variables of the model simulated. The computer comprises multiple, parallel, electronic components which can be set up to perform mathematical operations such as summing, multiplication, integration, and function generation. The computer is programmed by manipulating the individual components to establish an electrical circuit, such that the differential equations describing the circuit are identically those of the model to be simulated.

Computing Diagrams

The heart of an analog computer is the *operational amplifier* or *op-amp*. Op-amps are high gain (10^5–10^9) dc amplifiers. When wired as shown in Fig. 39.27, the relationship between the output voltage v_o and the input voltage v_i depends only on the impedance of the output and input circuit elements. Figure 39.28 shows the circuits corresponding to the basic linear operations, together with the corresponding analog diagram symbols.

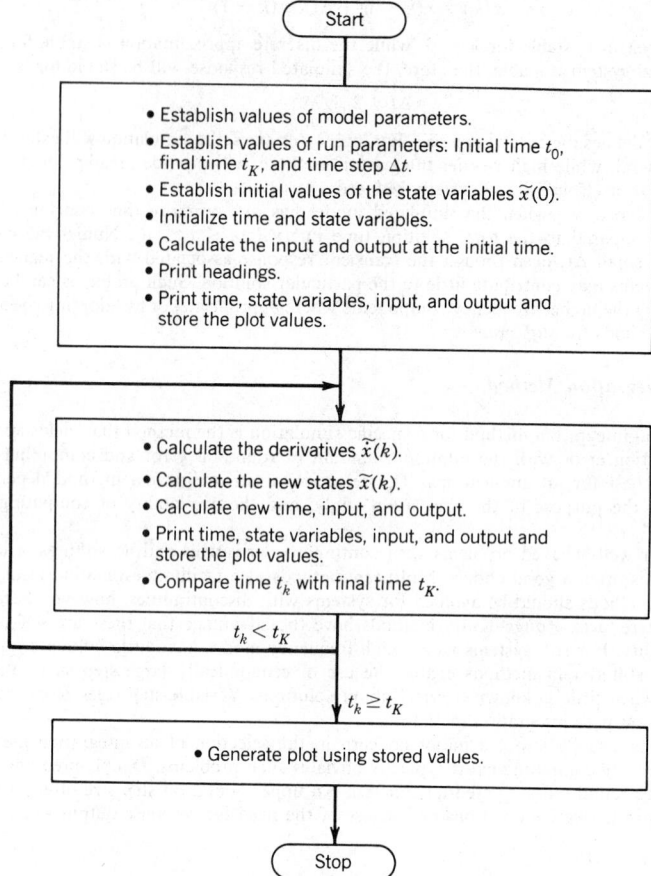

Fig. 39.26 General process for implementing digital simulation (adapted from Close and Frederick[5]).

Beginning with a state-variable model, an analog computer program is developed by first converting the model equation into a computing diagram. The diagram is constructed by assuming that signals corresponding to the derivatives $\dot{x}_i$ are available. Each of these signals is then integrated to yield the corresponding state variable x_i. For linear systems, signals representing the state variables are then multiplied by constants and summed according to each of the model equations. This results in signals representing the derivative terms $\dot{x}_i$, which were initially assumed to be available. The analog computing diagram for a second-order linear system is shown in Fig. 39.29.

Simulating nonlinear mathematical operations is achieved by using such elements as multipliers, diodes, function generators, and digital logic elements. Diodes are particularly useful for developing piecewise-linear approximations to complicated nonlinear relations. In general, however, nonlinear elements are less accurate, less stable, and more difficult to use than linear elements. For this reason, nonlinearities should be avoided in analog models to the greatest extent possible.

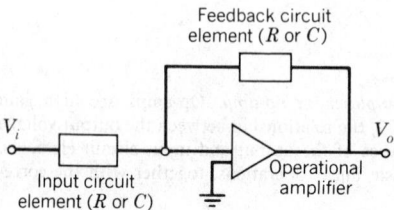

Fig. 39.27 Basic analog circuit.[1]

Fig. 39.28 Analog circuits and symbols for the basic linear operations.

Computer Operation	Input e_i	Output e_o
Constant multiplication	$f(t)$	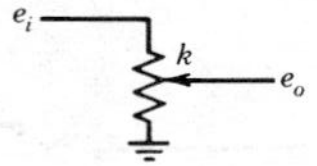 $kf(t), k \leq 1$ 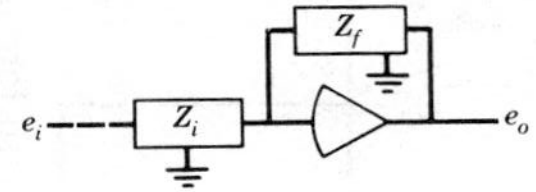
Addition	$f_i(t)$	 $-\sum_{i=1}^{n} k_i f_i(t)$ 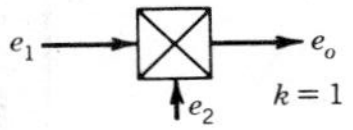
Integration	$f_i(t)$	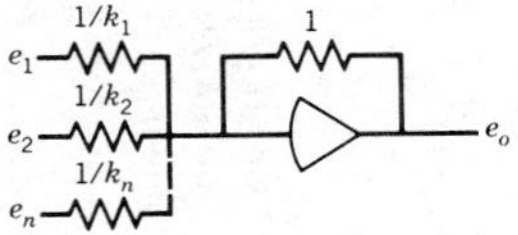 $-\int_0^1 \left(\sum_{i=1}^{n} k_i f_i(t)\right) dt + IC$ 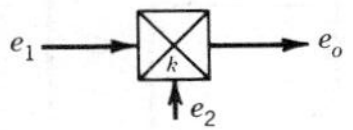
Complex transfer functions	$f(s)$	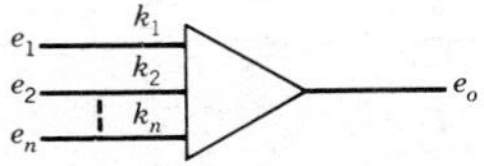 $-\dfrac{Z_f(s)}{Z_i(s)} f(s)$ 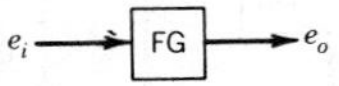
Multiplication	$f(t), g(t)$	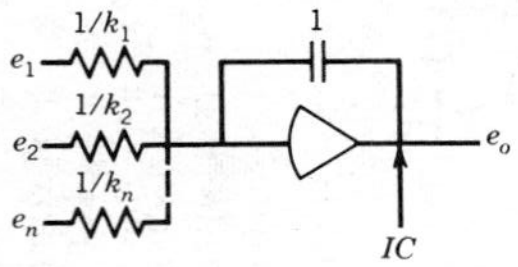 $kf(t)g(t)$ 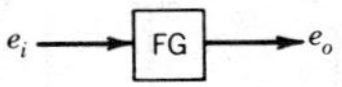
Function generation	$g(t)$	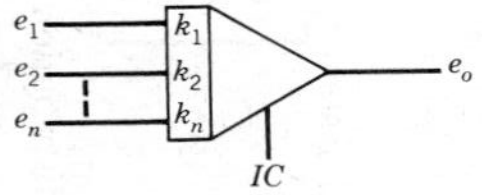 $f[g(t)]$ 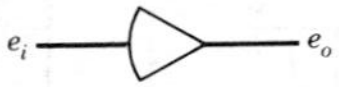
High-gain amplification	$f(t)$	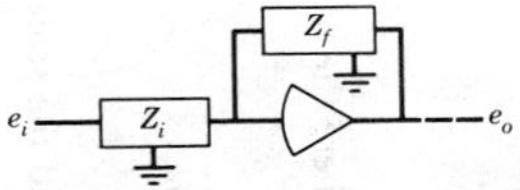 $-gf(t)$ 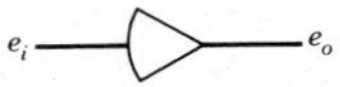

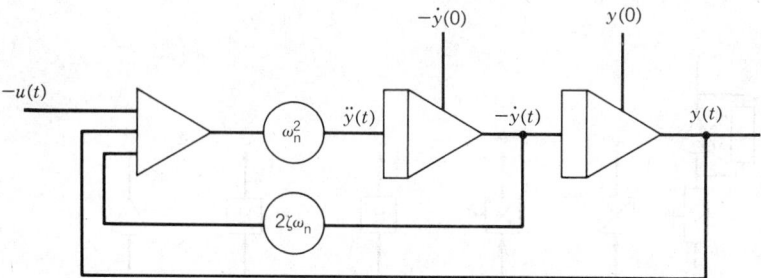

Fig. 39.29 Analog computing diagram for a second-order system in standard form.

Scaling

Two limitations are imposed on analog signals (*machine variables*) by the limited accuracy of the electronic components. First, analog voltages are necessarily restricted to some finite range, typically ±10 or ±100 V. Voltages out of range are distorted and frequently lead to saturation of the op-amps. Second, analog voltages are necessarily restricted to some minimum sensible level. Voltages below such levels can become corrupted with noise and are difficult to measure with accuracy. To account for these limitations, it is usually necessary to *scale* the model to be simulated.

Proper scaling transforms the problem variables x and u and real time t into machine variables v and v_u and machine time T, such that the range of each machine variable is as large as possible without exceeding the maximum.

For linear systems, scaling amounts to the following linear transformations on the problem variables

$$x_i = k_i v_i \qquad \text{for} \quad i = 1, 2, \ldots, n$$
$$u_j = k_{uj} v_{uj} \qquad \text{for} \quad j = 1, 2, \ldots, m$$
$$t = \alpha T$$

where the k_i, k_{uj}, and α are called *scale factors*. Substituting these expressions into the state-variable model

$$\frac{d}{dt} x(t) = Ax(t) + Bu(t)$$

yields the transformed state equation

$$\frac{d}{dT} v(T) = A^* v(T) + B^* v_u(T)$$

where the scaled system matrix is

$$A^* = \begin{bmatrix} k_t a_{11} & k_t a_{12}(k_2/k_1) & k_t a_{13}(k_3/k_1) & \cdots \\ k_t a_{21}(k_1/k_2) & k_t a_{22} & k_t a_{23}(k_3/k_2) & \cdots \\ k_t a_{31}(k_1/k_3) & k_t a_{32}(k_2/k_3) & k_t a_{33} & \cdots \\ \vdots & \vdots & \vdots & \end{bmatrix}$$

and the scaled distribution matrix is

$$B^* = \begin{bmatrix} k_t b_{11}(k_{u1}/k_1) & k_t b_{12}(k_{u2}/k_1) & k_t b_{13}(k_{u3}/k_1) & \cdots \\ k_t b_{21}(k_{u1}/k_2) & k_t b_{22}(k_{u2}/k_2) & k_t b_{23}(k_{u3}/k_2) & \cdots \\ k_t b_{31}(k_{u1}/k_3) & k_t b_{32}(k_{u2}/k_3) & k_t b_{33}(k_{u3}/k_3) & \cdots \\ \vdots & \vdots & \vdots & \end{bmatrix}$$

Scaling is completed by assigning values to each scale factor. This is usually accomplished as follows. First, select α such that all the diagonal elements a_{ii}^*, $i = 1, \ldots, n$, are within the range 0.1–10.0. Second, select the ratios (k_u/k_i) such that all the remaining elements a_{ij}^* are also within the 0.1–10.0 range. (Note that since all of the ratios are not independent, this selection is usually one of trial-and-error, involving a great many compromises.) Third, the ratios in the elements of B^* are selected in the same fashion. Finally, after all of the ratios have been fixed, the actual values of the k_i's and k_{uj}'s are selected. The objective of this final selection is to achieve scale factors that are as small as possible without violating the lower bounds:

$$k_i \geq \frac{\max|x_i|}{V} \qquad \text{for} \quad i = 1, 2, \ldots, n$$

$$k_{uj} \geq \frac{\max|u_j|}{V} \qquad \text{for} \quad j = 1, 2, \ldots, m$$

where $\max|x_i|$ and $\max|u_j|$ are the largest absolute values assumed by the problem variables during the course of a simulation run and V is the maximum absolute value permitted for any machine variable. Since good estimates of $\max|x_i|$ are not always available, *rescaling* after trial simulations is frequently required.

For stiff systems (systems with time constants differing by several orders of magnitude), the scaling procedure described will not always succeed. The situation is similar to the problem of selecting a suitable Δt for digital simulation of a stiff system. The usual remedy is also similar, requiring separate simulations of the slow and fast subsystems.

Hybrid Simulation

Practical analog simulation of even moderately large problems is most often carried out using a *hybrid analog–digital computer*. A hybrid computer consists of (1) an analog computer on which the bulk of the actual simulation is conducted; (2) a digital (usually mini- or micro-) computer, which is used to control the operation of the analog computer and to preprocess and postprocess simulation data; and (3) an *interface,* which permits communication between the computers through digital-to-analog and analog-to-digital signal conversions.

The digital computer accommodates the following options, which provide for more efficient operation of the analog simulation.

1. *Programming and Scaling.* Modern hybrid computers incorporating *autopatch* systems (electronic switching matrices instead of manual "patch" interconnections among the analog components) can use digital software routines to translate digital simulation language statements into the appropriate analog computer program. Scaling can also be accomplished automatically by digital software prior to patching.

2. *Setup and Static Check.* Coefficient values and initial conditions can be stored in digital memory. After retrieving these values, the digital machine can automatically set the parameter values of the analog components and check for correct propagation of initial conditions.

3. *Preprocessing and Postprocessing.* Raw data can be analyzed, reduced, and formatted using digital software in order to provide input data compatible with analog computer requirements. Output data from the analog computer can also be processed and stored on the digital machine for future analysis and report generation.

4. *Iterative Analog Computation.* Repeated analog simulations for *sensitivity analysis* can be set up, commanded, and controlled by the digital machine. *Time sharing* of the analog machine can be accomplished in a similar manner.

5. *Combined Simulation.* Computations that are difficult to implement within the analog computer, such as multivariate functions, can be performed on the digital computer during simulation.

Advantages and Disadvantages

Pure analog simulation is obsolete and hybrid simulation is (economically) justified only in select instances. The advantages of analog/hybrid simulation are (1) analog *component calculations are fast* (0.1% half-scale accuracies for input frequencies as high as 1000 Hz for most commercially available components) and (2) *parallel operation of components* makes overall computing speed virtually independent of problem size. These features are particularly useful for real-time simulation and control applications and for very large simulation studies requiring hundreds of op-amps and thousands of simulation runs (such as is true with larger process-plant and aerospace-vehicle studies).

The disadvantages of analog/hybrid simulation are (1) analog/hybrid computers require components that are precision-made for a limited market and therefore are relatively *expensive* (compared with digital machines of the same capacity using off-the-shelf components), (2) analog/hybrid computers have fixed and *limited component accuracy,* at best within 0.01% of half-scale (compared with digital machines that can improve accuracy indefinitely at the expense of computing speed and that can recover speed by paralleling inexpensive mini/microcomputers for large simulations), and (3) analog/hybrid computers are more complicated, more difficult, and more time-consuming to program and debug. As digital computers become increasingly less expensive and more common, analog/hybrid computer installations will continue to disappear.

39.8 MODEL CLASSIFICATIONS

Mathematical models of dynamic systems are distinguished by several criteria which describe fundamental properties of model variables and equations. These criteria in turn prescribe the theory and mathematical techniques that can be used to study different models. Table 39.11 summarizes these distinguishing criteria. In the following sections, the approaches adopted for the analysis of important classes of systems are briefly outlined.

Table 39.11   Classification of Mathematical Models of Dynamic Systems

Criterion	Classification	Description
Certainty	Deterministic	Model parameters and variables can be known with certainty. Common approximation when uncertainties are small.
	Stochastic	Uncertainty exists in the values of some parameters and/or variables. Model parameters and variables are expressed as random numbers or processes and are characterized by the parameters of probability distributions.
Spatial characteristics	Lumped	State of the system can be described by a finite set of state variables. Model is expressed as a discrete set of point functions described by ordinary differential or difference equations.
	Distributed	State depends on both time and spatial location. Model is usually described by variables that are continuous in time and space, resulting in partial differential equations. Frequently approximated by lumped elements. Typical in the study of structures and mass and heat transport.
Parameter variation	Fixed or time invariant	Model parameters are constant. Model described by differential of difference equations with constant coefficients. Model with same initial conditions and input delayed by t_d has the same response delayed by t_d.
	Time varying	Model parameters are time dependent.
Superposition property	Linear	Superposition applies. Model can be expressed as a system of linear difference or differential equations.
	Nonlinear	Superposition does not apply. Model is expressed as a system of nonlinear difference or differential equations. Frequently approximated by linear systems for analytical ease.
Continuity of independent variable (time)	Continuous	Dependent variables (input, output, state) are defined over a continuous range of the independent variable (time), even though the dependence is not necessarily described by a mathematically continuous function. Model is expressed as differential equations. Typical of physical systems.
	Discrete	Dependent variables are defined only at distinct instants of time. Model is expressed as difference equations. Typical of digital and nonphysical systems.
	Hybrid	System with continuous and discrete subsystems, most common in computer control and communication systems. Sampling and quantization typical in A/D (analog-to-digital) conversion; signal reconstruction for D/A conversion. Model frequently approximated as entirely continuous or entirely discrete.
Quantization of dependent variables	Nonquantized	Dependent variables are continuously variable over a range of values. Typical of physical systems at macroscopic resolution.
	Quantized	Dependent variables assume only a finite number of different values. Typical of computer control and communication systems (sample data systems).

39.8.1   Stochastic Systems

Systems in which some of the dependent variables (input, state, output) contain random components are called *stochastic systems*. Randomness may result from environmental factors, such as wind gusts or electrical noise, or simply from a lack of precise knowledge of the system model, such as when a human operator is included within a control system. If the randomness in the system can be described by some rule, then it is often possible to derive a model in terms of probability distributions involving, for example, the means and variances of model variables or parameters.

State-Variable Formulation

A common formulation is the fixed, linear model with additive noise

$$\dot{x}(t) = Ax(t) + Bu(t) + w(t)$$
$$y(t) = Cx(t) + v(t)$$

where $w(t)$ is a zero-mean Gaussian disturbance and $v(t)$ is a zero-mean Gaussian measurement noise. This formulation is the basis for many *estimation problems*, including the problem of *optimal filtering*. Estimation essentially involves the development of a rule or algorithm for determining the best estimate of the past, current, or future values of measured variables in the presence of disturbances or noise.

Random Variables

In the following, important concepts for characterizing random signals are developed. A *random variable* x is a variable that assumes values that cannot be precisely predicted *a priori*. The likelihood that a random variable will assume a particular value is measured as the *probability* of that value. The probability *distribution function* $F(x)$ of a continuous random variable x is defined as the probability that x assumes a value no greater than x, that is,

$$F(x) = \Pr(X \le x) = \int_{-\infty}^{x} f(x)\, dx$$

The probability *density function* $f(x)$ is defined as the derivative of $F(x)$.

The *mean* or *expected value* of a probability distribution is defined as

$$E(X) = \int_{-\infty}^{\infty} xf(x)\, dx = \overline{X}$$

The mean is the first moment of the distribution. The *nth moment* of the distribution is defined as

$$E(X^n) = \int_{-\infty}^{\infty} x^n f(x)\, dx$$

The mean square of the difference between the random variable and its mean is the *variance* or *second central moment* of the distribution,

$$\sigma^2(X) = E(X - \overline{X})^2 = \int_{-\infty}^{\infty} (x - \overline{X})^2 f(x)\, dx = E(X^2) - [E(X)]^2$$

The square root of the variance is the *standard deviation* of the distribution.

$$\sigma(X) = \sqrt{E(X^2) - [E(X)]^2}$$

The mean of the distribution therefore is a measure of the average magnitude of the random variable, while the variance and standard deviation are measures of the variability or dispersion of this magnitude.

The concepts of probability can be extended to more than one random variable. The *joint distribution* function of two random variables x and y is defined as

$$F(x,y) = \Pr(X < x \text{ and } Y < y) = \int_{-\infty}^{x} \int_{-\infty}^{y} f(x,y)\, dy\, dx$$

where $f(x,y)$ is the joint distribution. The *ij*th moment of the joint distribution is

$$E(x^i Y^j) = \int_{-\infty}^{\infty} x^i \int_{-\infty}^{\infty} y^j f(x,y)\, dy\, dx$$

The *covariance* of x and y is defined to be

$$E[(X - \overline{X})(Y - \overline{Y})]$$

and the normalized covariance or *correlation coefficient* as

$$\rho = \frac{E[(X - \overline{X})(Y - \overline{Y})]}{\sqrt{\sigma^2(X)\sigma^2(Y)}}$$

Although many distribution functions have proven useful in control engineering, far and away the most useful is the *Gaussian* or *normal distribution*

$$F(x) = \frac{1}{\sigma\sqrt{2\pi}} \exp[(-x - \mu)^2/2\sigma^2]$$

where μ is the mean of the distribution and σ is the standard deviation. The Gaussian distribution has a number of important properties. First, if the input to a linear system is Gaussian, the output will also be Gaussian. Second, if the input to a linear system is only approximately Gaussian, the output will tend to approximate a Gaussian distribution even more closely. Finally, a Gaussian distribution can be completely specified by two parameters, μ and σ, and therefore a zero-mean Gaussian variable is completely specified by its variance.

Random Processes

A *random process* is a set of random variables with time-dependent elements. If the statistical parameters of the process (such as σ for the zero-mean Gaussian process) do not vary with time, the process is *stationary*. The *autocorrelation function* of a stationary random variable $x(t)$ is defined by

$$\phi_{xx}(\tau) = \lim_{T \to \infty} \frac{1}{2T} \int_{-T}^{T} x(t)x(t + \tau)\, dt$$

a function of the fixed time interval τ. The autocorrelation function is a quantitative measure of the sequential dependence or time correlation of the random variable, that is, the relative effect of prior values of the variable on the present or future values of the variable. The autocorrelation function also gives information regarding how rapidly the variable is changing and about whether the signal is in part deterministic (specifically, periodic). The autocorrelation function of a zero-mean variable has the properties

$$\sigma^2 = \phi_{xx}(0) \geq \phi_{xx}(\tau), \qquad \phi_{xx}(\tau) = \phi_{xx}(-\tau)$$

In other words, the autocorrelation function for $\tau = 0$ is identically the variance and the variance is the maximum value of the autocorrelation function. From the definition of the function, it is clear that (1) for a purely random variable with zero mean, $\phi_{xx}(\tau) = 0$ for $\tau \neq 0$, and (2) for a deterministic variable, which is periodic with period T, $\phi_{xx}(k2\pi T) = \sigma^2$ for k integer. The concept of time correlation is readily extended to more than one random variable. The *cross-correlation function* between the random variables $x(t)$ and $y(t)$ is

$$\phi_{xy}(\tau) = \lim_{T \to \infty} \int_{-\infty}^{\infty} x(t)y(t + \tau)\, dt$$

For $\tau = 0$, the cross-correlation between two zero-mean variables is identically the covariance. A final characterization of a random variable is its *power spectrum,* defined as

$$G(\omega,x) = \lim_{T \to \infty} \frac{1}{2\pi T} \left| \int_{-T}^{T} x(t)e^{-j\omega t}\, dt \right|^2$$

For a stationary random process, the power spectrum function is identically the Fourier transform of the autocorrelation function

$$G(\omega,x) = \frac{1}{\pi} \int_{-\infty}^{\infty} \phi_{xx}(\tau)e^{-j\omega t}\, dt$$

with

$$\phi_{xx}(0) = \int_{-\infty}^{\infty} G(\omega,x)\, d\omega$$

39.8.2 Distributed-Parameter Models

There are many important applications in which the state of a system cannot be defined at a finite number of points in space. Instead, the system state is a continuously varying function of both time and location. When continuous spatial dependence is explicitly accounted for in a model, the independent variables must include spatial coordinates as well as time. The resulting *distributed-parameter model* is described in terms of *partial differential equations,* containing partial derivatives with respect to each of the independent variables.

Distributed-parameter models commonly arise in the study of mass and heat transport, the mechanics of structures and structural components, and electrical transmission. Consider as a simple example the unidirectional flow of heat through a wall, as depicted in Fig. 39.30. The temperature of the wall is not in general uniform, but depends on both the time t and position within the wall x, that is, $\theta = \theta(x,t)$. A distributed-parameter model for this case might be the first-order partial differential equation

$$\frac{d}{dt}\theta(x,t) = \frac{1}{C_t} \frac{\partial}{\partial x} \left[\frac{1}{R_t} \frac{\partial}{\partial x} \theta(x,t) \right]$$

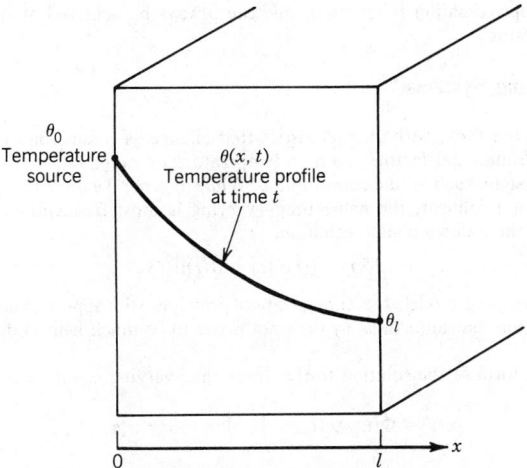

Fig. 39.30 Uniform heat transfer through a wall.

where C_t is the thermal capacitance and R_t is the thermal resistance of the wall (assumed uniform).

The complexity of distributed parameter models is typically such that these models are avoided in the analysis and design of control systems. Instead, distributed parameter systems are approximated by a finite number of spatial "lumps," each lump being characterized by some average value of the state. By eliminating the independent spatial variables, the result is a *lumped-parameter (or lumped-element) model* described by coupled ordinary differential equations. If a sufficiently fine-grained representation of the lumped microstructure can be achieved, a lumped model can be derived that will approximate the distributed model to any desired degree of accuracy. Consider, for example, the three temperature lumps shown in Fig. 39.31, used to approximate the wall of Fig. 39.30. The corresponding third-order lumped approximation is

$$
\frac{d}{dt}
\begin{bmatrix}
\theta_1(t) \\[2mm]
\theta_2(t) \\[2mm]
\theta_3(t)
\end{bmatrix}
=
\begin{bmatrix}
-\dfrac{9}{C_t R_t} & \dfrac{3}{C_t R_t} & 0 \\[3mm]
\dfrac{3}{C_t R_t} & -\dfrac{6}{C_t R_t} & \dfrac{3}{C_t R_t} \\[3mm]
0 & \dfrac{3}{C_t R_t} & -\dfrac{6}{C_t R_t}
\end{bmatrix}
\begin{bmatrix}
\theta_1(t) \\[2mm]
\theta_2(t) \\[2mm]
\theta_3(t)
\end{bmatrix}
+
\begin{bmatrix}
\dfrac{6}{C_t R_t} \\[3mm]
0 \\[3mm]
0
\end{bmatrix}
\theta_0(t)
$$

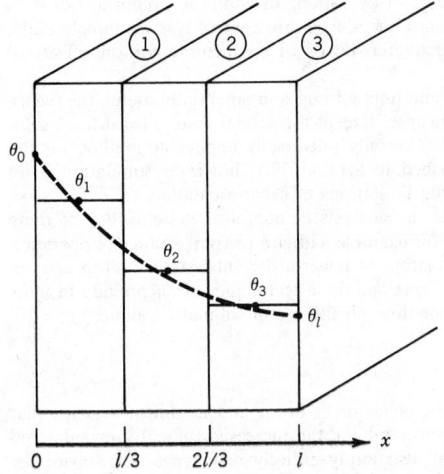

Fig. 39.31 Lumped-parameter model for uniform heat transfer through a wall.

If a more detailed approximation is required, this can always be achieved at the expense of adding additional, smaller lumps.

39.8.3   Time-Varying Systems

Time-varying systems are those with characteristics that change as a function of time. Such variation may result from environmental factors, such as temperature or radiation, or from factors related to the operation of the system, such as fuel consumption. While in general a model with variable parameters can be either linear or nonlinear, the name time—varying is most frequently associated with linear systems described by the following state equation:

$$\dot{x}(t) = A(t)x(t) + B(t)u(t)$$

For this linear time-varying model, the superposition principle still applies. Superposition is a great aid in model formulation, but unfortunately does not prove to be much help in determining the model solution.

Paradoxically, the form of the solution to the linear time-varying equation is well known[14]:

$$x(t) = \Phi(t,t_0)x(t_0) + \int_{t_0}^{t} \Phi(t,\tau)B(\tau)u(\tau)\,dt$$

where $\Phi(t,t_0)$ is the time-varying state-transition matrix. This knowledge is typically of little value, however, since it is not usually possible to determine the state-transition matrix by any straightforward method. By analogy with the first-order case, the relationship

$$\Phi(t,t_0) = \exp\left(\int_{t_0}^{t} A(\tau)\,d\tau\right)$$

can be proven valid *if and only if*

$$A(t)\int_{t_0}^{t} A(\tau)\,d\tau = \int_{t_0}^{t} A(\tau)\,d\tau A(t)$$

that is, if and only if $A(t)$ and its integral commute. This is a very stringent condition for all but a first-order system and, as a rule, it is usually easiest to obtain the solution using simulation.

Most of the properties of the fixed transition matrix extend to the time-varying case:

$$\Phi(t,t_0) = I$$
$$\Phi^{-1}(t,t_0) = \Phi(t_0,t)$$
$$\Phi(t_2,t_1)\Phi(t_1,t_0) = \Phi(t_2,t_0)$$
$$\dot{\Phi}(t,t_0) = A(t)\Phi(t,t_0)$$

39.8.4   Nonlinear Systems

The theory of fixed, linear, lumped-parameter systems is highly developed and provides a powerful set of techniques for control system analysis and design. In practice, however, all physical systems are nonlinear to some greater or lesser degree. The linearity of a physical system is usually only a convenient approximation, restricted to a certain range of operation. In addition, nonlinearities such as dead zones, saturation, or on–off action are sometimes introduced into control systems intentionally, either to obtain some advantageous performance characteristic or to compensate for the effects of other (undesirable) nonlinearities.

Unfortunately, while nonlinear systems are important, unbiquitous, and potentially useful, the theory of nonlinear differential equations is comparatively meager. Except for special cases, closed-form solutions to nonlinear systems are generally unavailable. The only universally applicable method for the study of nonlinear systems in *simulation*. As described in Section 39.7, however, simulation is an experimental approach, embodying all of the attending limitations of experimentation.

A number of special techniques are available for the analysis of nonlinear systems. All of these techniques are in some sense approximate, assuming, for example, either a restricted range of operation over which nonlinearities are mild or the relative isolation of lower-order subsystems. When used in conjunction with more complex simulation models, however, these techniques often provide insights and design concepts that would be difficult to discover through the use of simulation alone.[29]

Linear versus Nonlinear Behaviors

There are several fundamental differences between the behavior of linear and nonlinear systems that are especially important. These differences not only account for the increased difficulty encountered in the analysis and design of nonlinear systems, but also imply entirely new types of behavior for nonlinear systems that are not possible for linear systems.

The fundamental property of linear systems is *superposition*. This property states that if $y_1(t)$ is the response of the system to $u_1(t)$ and $y_2(t)$ is the response of the system to $u_2(t)$, then the response of the system to the linear combination $a_1u_1(t) + a_2u_2(t)$ is the linear combination $a_1y_1(t) + a_2y_2(t)$. An immediate consequence of superposition is that the responses of a linear system to inputs differing only in amplitude is qualitatively the same. Since superposition does not apply to nonlinear systems, the responses of a nonlinear system to large and small changes may be fundamentally different.

This fundamental difference in linear and nonlinear behaviors has a second consequence. For a linear system, interchanging two elements connected in series does not affect the overall system behavior. Clearly, this cannot be true in general for nonlinear systems.

A third property peculiar to nonlinear systems is the potential existence of *limit cycles*. A linear oscillator oscillates at an amplitude that depends on its initial state. A limit cycle is an oscillation of fixed amplitude and period, independent of the initial state, that is unique to the nonlinear system.

A fourth property concerns the response of nonlinear systems to sinusoidal inputs. For a linear system, the response to sinusoidal input is a sinusoid of the same frequency, potentially differing only in magnitude and phase. For a nonlinear system, the output will in general contain other frequency components, including possibly harmonics, subharmonics, and aperiodic terms. Indeed, the response need not contain the input frequency at all.

Linearizing Approximations

Perhaps the most useful technique for analyzing nonlinear systems is to approximate these with linear systems. While many linearizing approximations are possible, linearization can frequently be achieved by considering small excursions of the system state about a reference trajectory. Consider the nonlinear state equation

$$\dot{x}(t) = f[x(t), u(t)]$$

together with a reference trajectory $x^0(t)$ and reference input $u^0(t)$ that together satisfy the state equation

$$\dot{x}^0(t) = f[x^0(t), u^0(t)]$$

Note that the simplest case is to choose a static equilibrium or *operating point* $\bar{x}$ as the reference "trajectory," such that $0 = t(\bar{x}, 0)$. The actual trajectory is then related to the reference trajectory by the relationships

$$x(t) = x^0(t) + \delta x(t)$$
$$u(t) = u^0(t) + \delta u(t)$$

where $\delta x(t)$ is some small perturbation about the reference state and $\delta u(t)$ is some small perturbation about the reference input. If these perturbations are indeed small, then applying the Taylor's series expansion about the reference trajectory yields the linearized approximation

$$\delta \dot{x}(t) = A(t)\delta x(t) + B(t)\delta u(t)$$

where the state and distribution matrices are the *Jacobian matrices*

$$A(t) = \begin{bmatrix} \dfrac{\partial f_1}{\partial x_1} & \dfrac{\partial f_1}{\partial x_2} & \cdots & \dfrac{\partial f_1}{\partial x_n} \\[2mm] \dfrac{\partial f_2}{\partial x_1} & \dfrac{\partial f_2}{\partial x_2} & \cdots & \dfrac{\partial f_2}{\partial x_n} \\[2mm] \vdots & \vdots & & \vdots \\[2mm] \dfrac{\partial f_n}{\partial x_1} & \dfrac{\partial f_n}{\partial x_2} & \cdots & \dfrac{\partial f_n}{\partial x_n} \end{bmatrix}_{x(t)=x^0(t);\ u(t)=u^0(t)}$$

$$B(t) = \begin{bmatrix} \dfrac{\partial f_1}{\partial u_1} & \dfrac{\partial f_1}{\partial u_2} & \cdots & \dfrac{\partial f_1}{\partial u_m} \\[2mm] \dfrac{\partial f_2}{\partial u_1} & \dfrac{\partial f_2}{\partial u_2} & \cdots & \dfrac{\partial f_2}{\partial u_m} \\[2mm] \vdots & \vdots & & \vdots \\[2mm] \dfrac{\partial f_n}{\partial u_1} & \dfrac{\partial f_n}{\partial u_2} & \cdots & \dfrac{\partial f_n}{\partial u_m} \end{bmatrix}_{x(t)=x^0(t);\ u(t)=u^0(t)}$$

If the reference trajectory is a fixed operating point $\bar{x}$, then the resulting linearized system is time invariant and can be solved analytically. If the reference trajectory is a function of time, however, then the resulting system is linear, but time varying.

Describing Functions

The describing function method is an extension of the frequency transfer function approach of linear systems, most often used to determine the stability of limit cycles of systems containing nonlinearities. The approach is approximate and its usefulness depends on two major assumptions:

1. All the nonlinearities within the system can be aggregated mathematically into a single block, denoted as $N(M)$ in Fig. 39.32, such that the equivalent gain and phase associated with this block depend only on the amplitude M_d of the sinusoidal input $m(\omega t) = M \sin(\omega t)$ and are independent of the input frequency ω.

2. All the harmonics, subharmonics, and any dc component of the output of the nonlinear block are filtered out by the linear portion of the system, such that the effective output of the nonlinear block is well approximated by a periodic response having the same fundamental period as the input.

Although these assumptions appear to be rather limiting, the technique gives reasonable results for a large class of control systems. In particular, the second assumption is generally satisfied by higher-order control systems with symmetric nonlinearities, since (a) symmetric nonlinearities do not generate dc terms, (b) the amplitudes of harmonics are generally small when compared with the fundamental term and subharmonics are uncommon, and (c) feedback within a control system typically provides low-pass filtering to further attenuate harmonics, especially for higher-order systems. Because the method is relatively simple and can be used for systems of any order, describing functions have enjoyed wide practical application.

The describing function of a nonlinear block is defined as the ratio of the fundamental component of the output to the amplitude of a sinusoidal input. In general, the response of the nonlinearity to the input

$$m(\omega t) = M \sin \omega t$$

is the output

$$n(\omega t) = N_1 \sin(\omega t + \phi_1) + N_2 \sin(2\omega t + \phi_2) + N_3 \sin(3\omega t + \phi_3) + \cdots$$

and, hence, the describing function for the nonlinearity is defined as the complex quantity

$$N(M) = \frac{N_1}{M} e^{j\phi_1}$$

Derivation of the approximating function typically proceeds by representing the fundamental frequency by the Fourier series coefficients

$$A_1(M) = \frac{2}{T} \int_{-T/2}^{T/2} n(\omega t) \cos \omega t \, d(\omega t)$$

$$B_1(M) = \frac{2}{T} \int_{-T/2}^{T/2} n(\omega t) \sin \omega t \, d(\omega t)$$

The describing function is then written in terms of these coefficients as

$$N(M) = \frac{B_1(M)}{M} + j\frac{A_1(M)}{M} = \left[\left(\frac{B_1(M)}{M} \right)^2 + \left(\frac{A_1(M)}{M} \right)^2 \right]^{1/2} \exp\left[j \tan^{-1}\left(\frac{A_1(M)}{B_1(M)} \right) \right]$$

Note that if $n(\omega t) = -n(-\omega t)$, then the describing function is odd, $A_1(M) = 0$, and there is no phase shift between the input and output. If $n(\omega t) = n(-\omega t)$, then the function is even, $B_1(M) = 0$, and the phase shift is $\pi/2$.

The describing functions for a number of typical nonlinearities are given in Fig. 39.33. Reference 10 contains an extensive catalog. The following derivation for a dead zone nonlinearity demonstrates

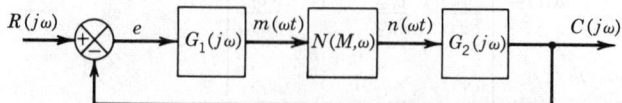

Fig. 39.32 General nonlinear system for describing function analysis.

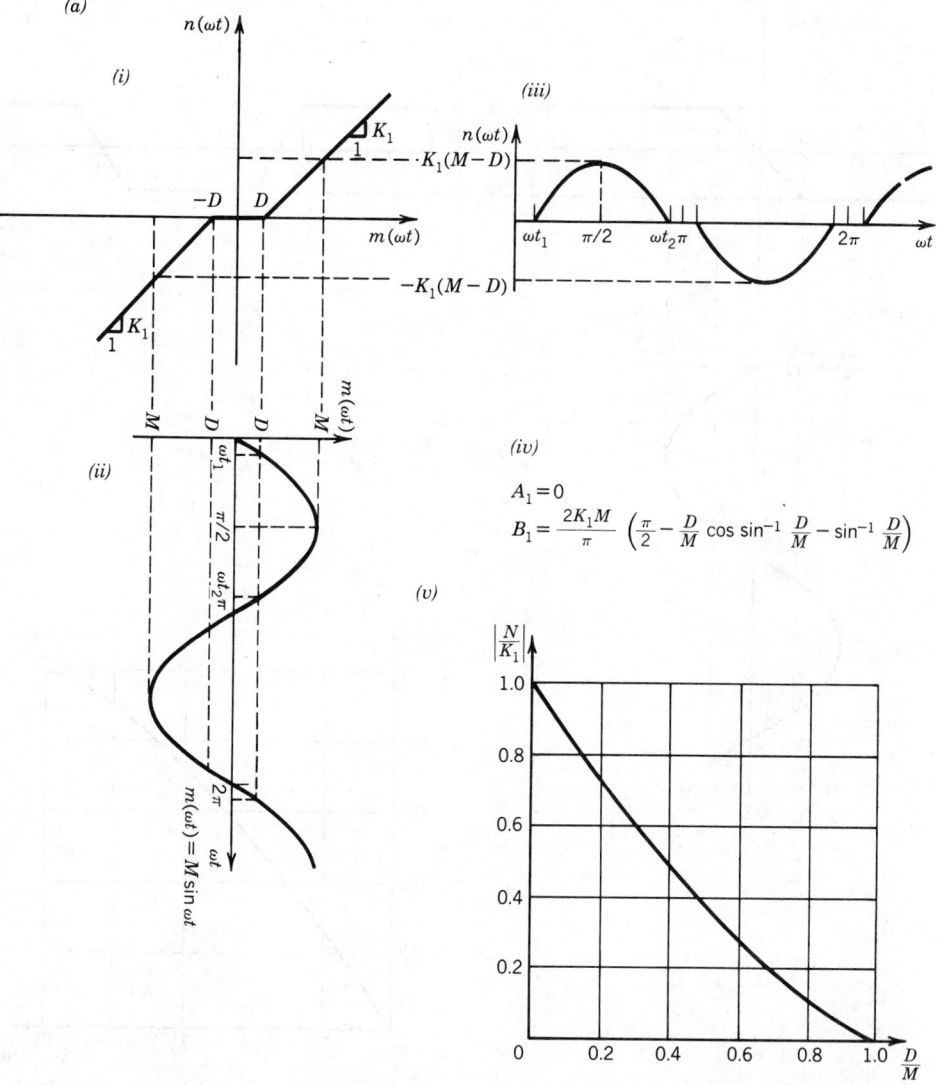

Fig. 39.33a Describing functions for typical nonlinearities (after Refs. 10 and 24). Dead zone nonlinearity: (*i*) nonlinear characteristic; (*ii*) sinusoidal input wave shape; (*iii*) output wave shape; (*iv*) describing-function coefficients; (*v*) normalized describing function.

the general procedure for deriving a describing function. For the saturation element depicted in Fig. 39.33a, the relationship between the input $m(\omega t)$ and output $n(\omega t)$ can be written as

$$n(\omega t) = \begin{cases} 0, & \text{for} \quad -D < m < D \\ K_1 M(\sin \omega t - \sin \omega_1 t), & \text{for} \quad m > D \\ K_1 M(\sin \omega t + \sin \omega_1 t), & \text{for} \quad m < -D \end{cases}$$

Since the function is odd, $A_1 = 0$. By the symmetry over the four quarters of the response period,

$$B_1 = 4\left[\frac{2}{\pi/2}\int_0^{\pi/2} n(\omega t)\sin \omega t \, d(\omega t)\right]$$

$$= \frac{4}{\pi}\left[\int_0^{\omega t_1} (0)\sin \omega t \, d(\omega t) + \int_{\omega t_1}^{\pi/2} K_1 M(\sin \omega t - \sin \omega_1 t)\sin \omega t \, d(\omega t)\right]$$

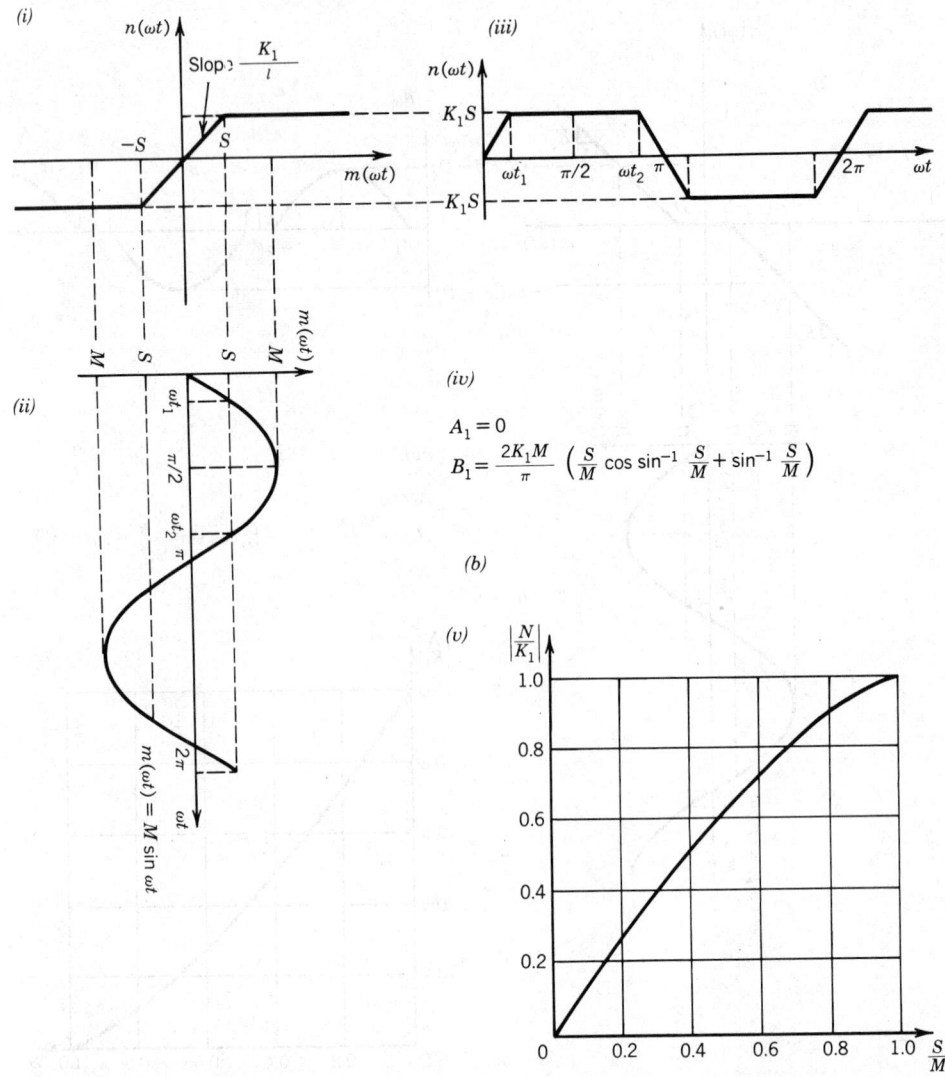

Fig. 39.33b Saturation nonlinearity: (*i*) nonlinear characteristic; (*ii*) sinusoidal input wave shape; (*iii*) output wave shape; (*iv*) describing-function coefficients; (*v*) normalized describing function.

where $\omega t_1 = \sin^{-1} (D/M)$. Evaluating the integrals and dividing by M yields the describing function listed in Fig. 39.33.

Phase-Plane Method

The *phase-plane method* is a graphical application of the state-space approach used to characterize the free-response of second-order nonlinear systems. While any convenient pair of state variables can be used, the *phase variables* originally were taken to be the displacement and velocity of the mass of a second-order mechanical system. Using the two state variables as the coordinate axis, the transient response of a system is captured on the *phase plane* as the plot of one variable against the other, with time implicit on the resulting curve. The curve for a specific initial condition is called a *trajectory* in the phase plane; a representative sample of trajectories is called the *phase portrait* of the system. The phase portrait is a compact and readily interpreted summary of the system response. Phase portraits for a sample of typical nonlinearities are shown in Fig. 39.34.

Four methods can be used to construct a phase portrait: (1) direct solution of the differential

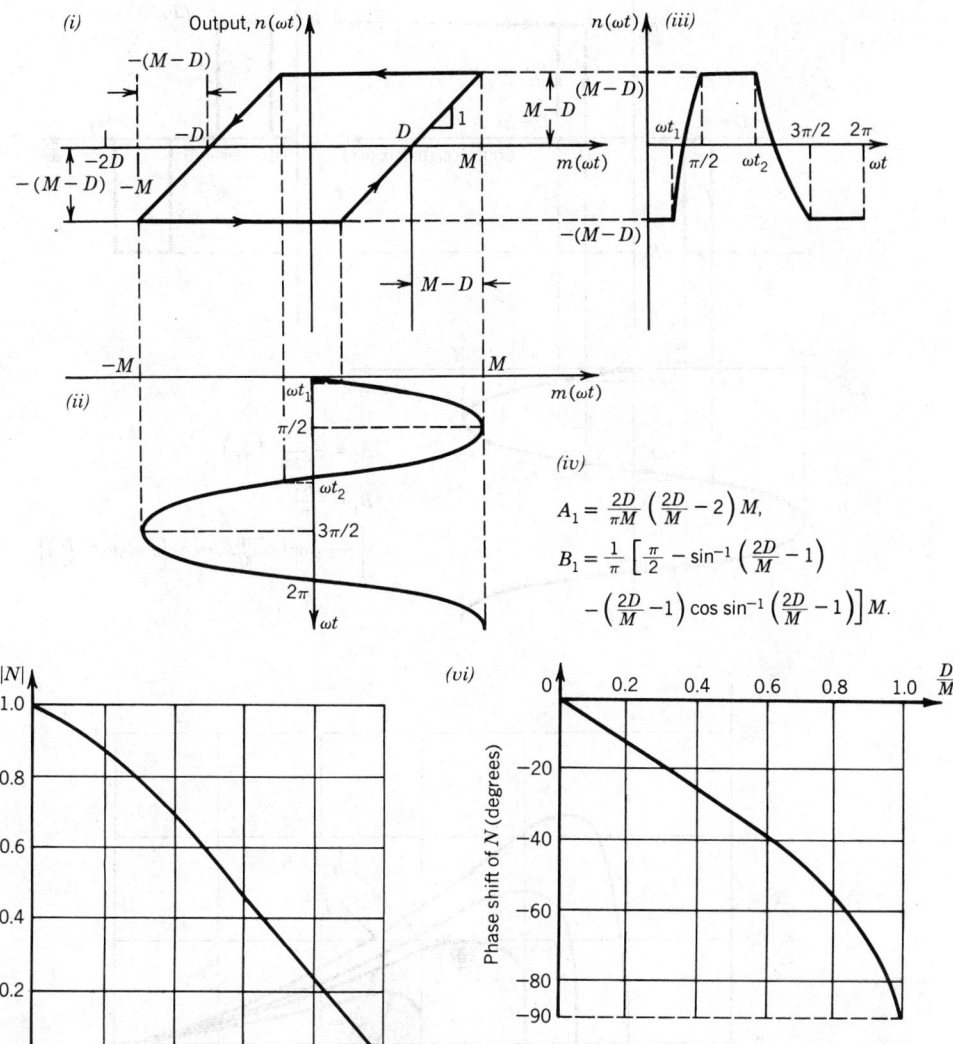

Fig. 39.33c Backlash nonlinearity: (*i*) nonlinear characteristic; (*ii*) sinusoidal input wave shape; (*iii*) output wave shape; (*iv*) describing-function coefficients; (*v*) normalized amplitude characteristics for the describing function; (*vi*) normalized phase characteristics for the describing function.

equation, (2) the graphical *method of isoclines,* (3) transformation of the second-order system (with time as the independent variable) into an equivalent first-order system (with one of the phase variables as the independent variable), and (4) numerical solution using simulation. The first and second methods are usually impractical; the third and fourth methods are frequently used in combination. For example, consider the second-order model

$$\frac{dx_1}{dt} = f_1(x_1, x_2), \qquad \frac{dx_2}{dt} = f_2(x_1, x_2)$$

Dividing the second equation by the first and eliminating the *dt* terms yields

$$\frac{dx_2}{dx_1} = \frac{f_2(x_1, x_2)}{f_1(x_1, x_2)}$$

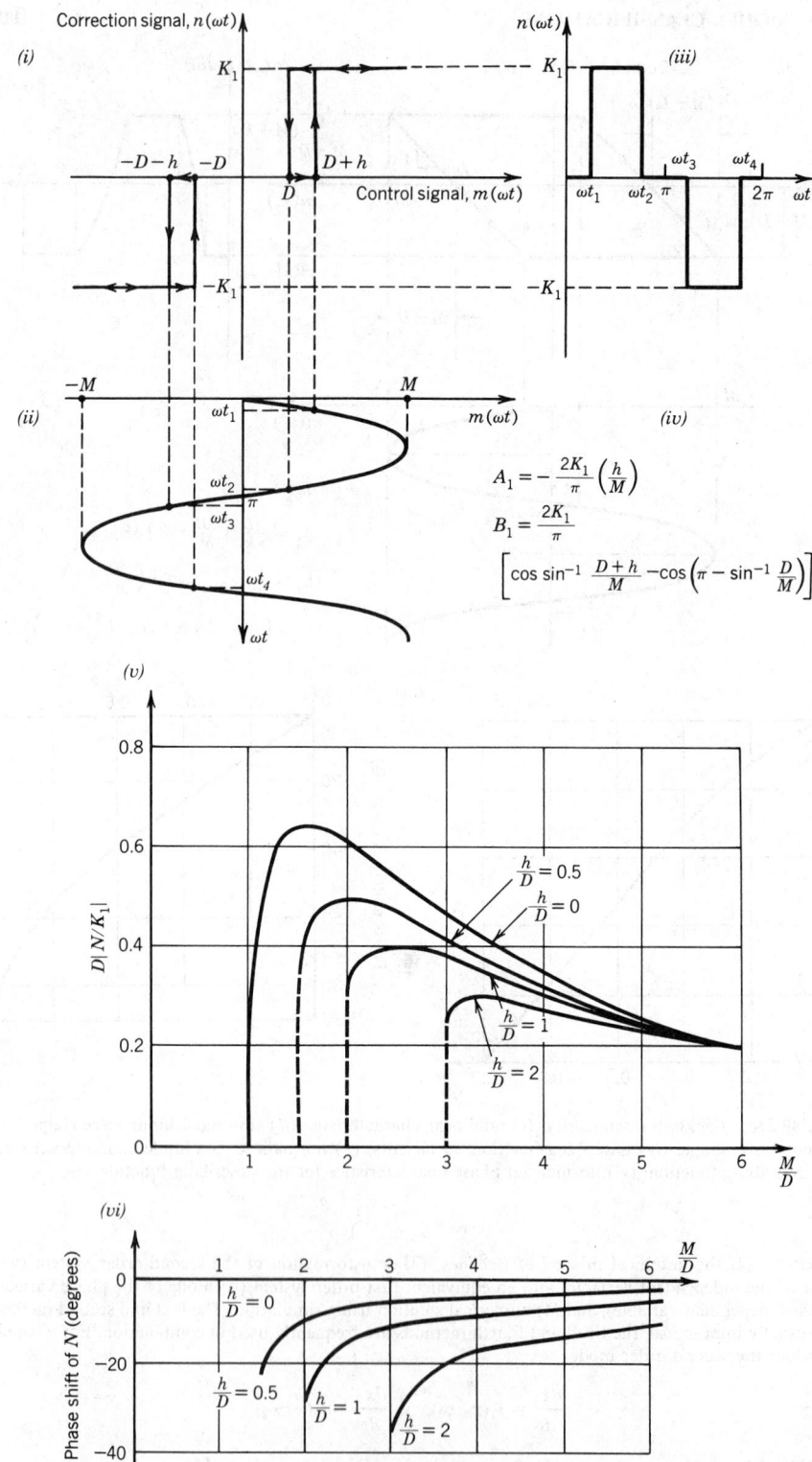

Fig. 39.33d Three-position on–off device with hysteresis: (*i*) nonlinear characteristic; (*ii*) sinusoidal input wave shape; (*ii*) output wave shape; (*iv*) describing-function coefficients; (*v*) normalized amplitude characteristics for the describing function; (*vi*) normalized phase characteristics for the describing function.

This first-order equation describes the phase-plane trajectories. In many cases it can be solved analytically. If not, it always can be simulated.

The phase-plane method complements the describing-function approach. A describing function is an approximate representation of the sinusoidal response for systems of any order, while the phase plane is an exact representation of the (free) transient response for first- and second-order systems. Of course, the phase-plane method can theoretically be extended for higher-order systems, but the difficulty of visualizing the nth order state space typically makes such a direct extension impractical. An approximate extension of the method has been used with some considerable success,[29] however, in order to explore and validate the relationships among pairs of variables in complex simulation models. The approximation is based on the assumptions that the paired variables define a second-order subsystem which, for the purposes of analysis, is weakly coupled to the remainder of the system.

39.8.5 Discrete and Hybrid Systems

A *discrete-time system* is one for which the dependent variables are defined only at distinct instants of time. Discrete-time models occur in the representation of systems that are inherently discrete, in the analysis and design of digital measurement and control systems, and in the numerical solution of differential equations (see Section 39.7). Because of the increasing use of digital computers in control engineering, discrete-time models are becoming increasingly important. The discrete-time nature of a computer's sampling of continuous physical signals also leads to the occurrence of *hybrid systems*, that is, systems that are in part discrete and in part continuous. Discrete-time models of hybrid systems are called *sampled-data systems.*

Difference Equations

Dynamic models of discrete-time systems most naturally take the form of *difference equations.* The input–output (I/O) form of an nth order difference equation model is

$$f[y(k+n), y(k+n-1), \ldots, y(k), u(k+n-1), \ldots, u(k)] = 0$$

which expresses the dependence of the $(k+n)$th value of the output, $y(k+n)$, on the n preceding values of the output y and input u. For a linear system, the I/O form can be written as

$$y(k+n) + a_{n-1}(k)y(k+n-1) + \cdots + a_1(k)y(k+1) + a_0(k)y(k)$$
$$= b_{n-1}(k)u(k+n-1) + \cdots + b_0(k)u(k)$$

In state-variable form, the discrete-time model is the vector difference equation

$$x(k+1) = f[x(k), u(k)]$$
$$y(k) = g[x(k), u(k)]$$

where x is the state-vector, u is the vector of inputs, and y is the vector of outputs. For a linear system, the discrete state-variable form can be written as

$$x(k+1) = A(k)x(k) + B(k)u(k)$$
$$y(k) = C(k)x(k) + D(k)u(k)$$

The mathematics of difference equations parallels that of differential equations in many important respects. In general, the concepts applied to differential equations have direct analogies for difference equations, although the mechanics of their implementation may vary (see Ref. 16 for a development of dynamic modeling based on difference equations). One important difference is that the general solution of nonlinear and time-varying difference equations can usually be obtained through *recursion.* For example, consider the discrete nonlinear model

$$y(k+1) = \frac{y(k)}{1 + y(k)}$$

Recursive evaluation of the equation beginning with the initial condition $y(0)$ yields

$$y(1) = \frac{y(0)}{1 + y(0)}$$

$$y(2) = \frac{y(1)}{1 + y(1)} = \left[\frac{y(0)}{1 + y(0)}\right] \bigg/ \left[1 + \frac{y(0)}{1 + y(0)}\right] = \frac{y(0)}{1 + 2y(0)}$$

$$y(3) = \frac{y(2)}{1 + y(2)} = \frac{y(0)}{1 + 3y(0)}$$

$$\vdots$$

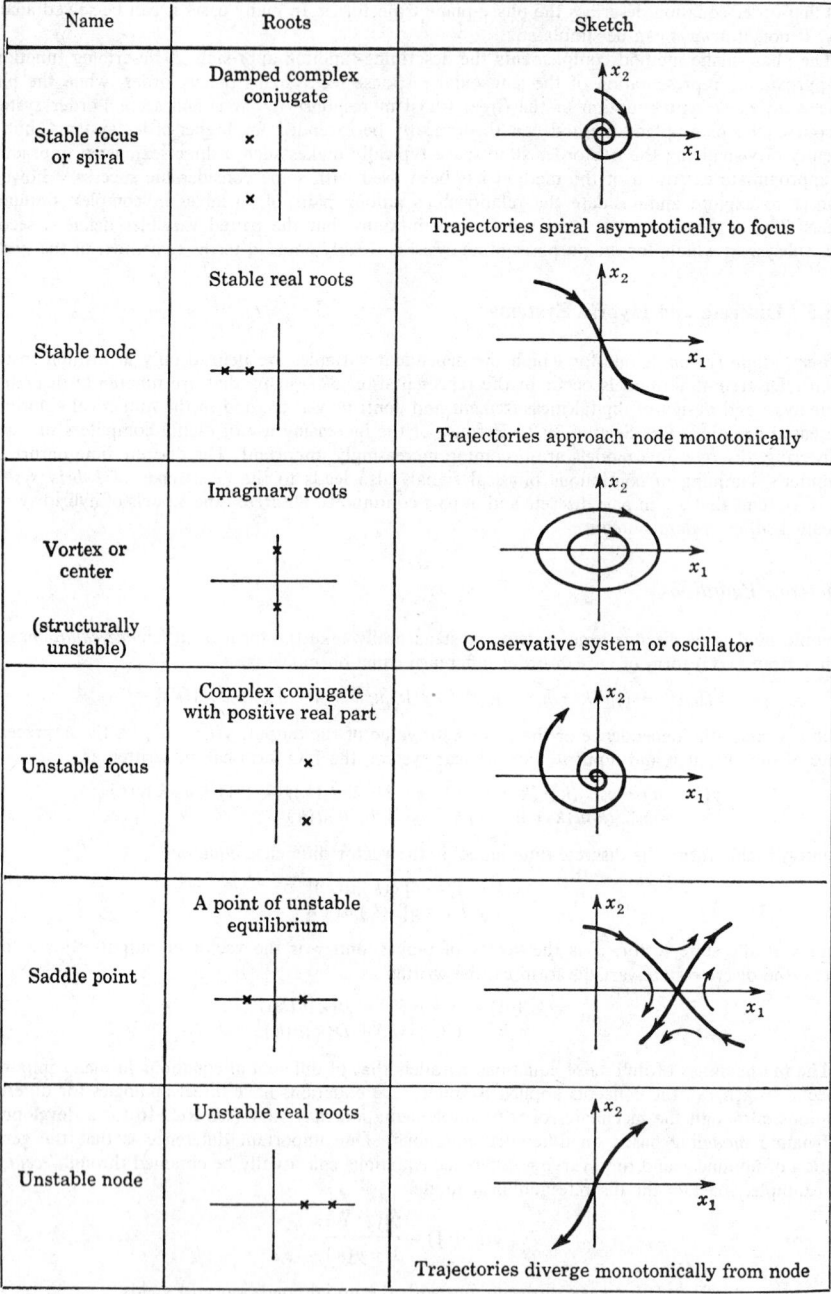

Name	Roots	Sketch
Stable focus or spiral	Damped complex conjugate	Trajectories spiral asymptotically to focus
Stable node	Stable real roots	Trajectories approach node monotonically
Vortex or center (structurally unstable)	Imaginary roots	Conservative system or oscillator
Unstable focus	Complex conjugate with positive real part	
Saddle point	A point of unstable equilibrium	
Unstable node	Unstable real roots	Trajectories diverge monotonically from node

Fig. 39.34 Typical phase-plane plots for second-order systems.[10]

the pattern of which reveals, by induction,

$$y(k) = \frac{y(0)}{1 + ky(0)}$$

as the general solution.

Uniform Sampling

Uniform sampling is the most common mathematical approach to *analog-to-digital (A/D) conversion*, that is, to extracting the discrete time approximation $y*(k)$ of the form

Name	Roots	Sketch
Spiral with nonlinearity viscous damping and coulomb friction		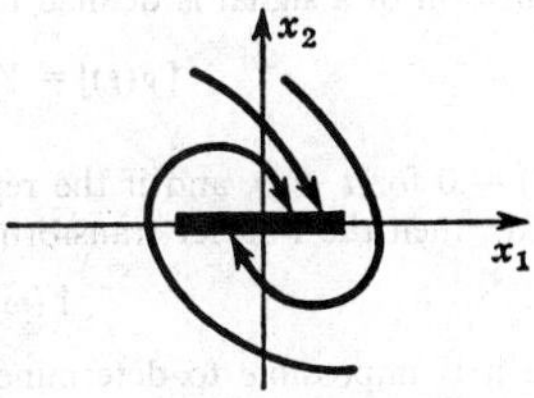 Spirals are warped by coulomb friction
Nonlinear system with coulomb damping and no viscous damping		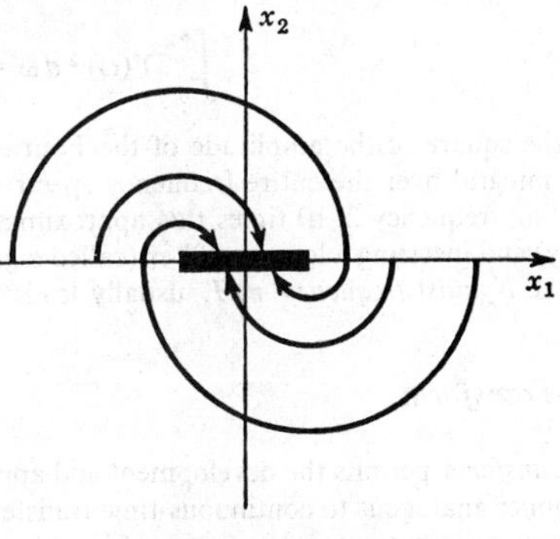 Semicircles center at ends of coulomb line
Spiral with backlash		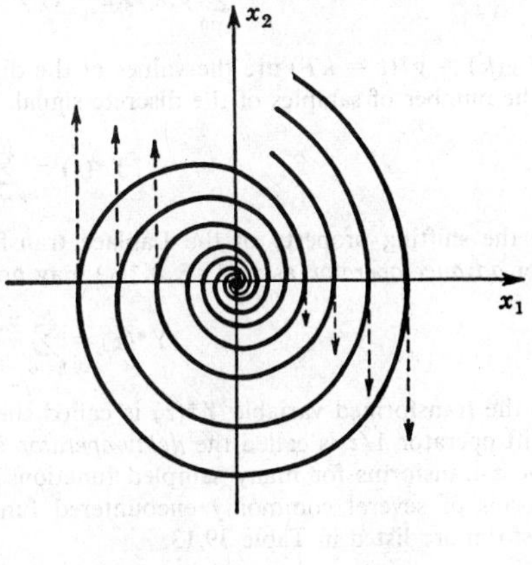

Fig. 39.34   (Continued)

$$y^*(k) = y(t = kT)$$

from the continuous-time signal $y(t)$, where T is a constant interval of time called the *sampling period*. If the sampling period is too large, however, it may not be possible to represent the continuous signal accurately. The *sampling theorem* guarantees that $y(t)$ can be reconstructed from the uniformly sampled values $y^*(k)$ if the sampling period satisfies the inequality

$$T \leq \frac{\pi}{\omega_u}$$

where ω_u is the highest frequency contained in the Fourier transform $Y(\omega)$ of $y(t)$, that is, if

$$Y(\omega) = 0 \qquad \text{for all} \quad \omega > \omega_u$$

The Fourier transform of a signal is defined to be

$$\mathscr{F}[y(t)] = Y(\omega) = \int_{-\infty}^{\infty} y(t)e^{-j\omega t}\, dt$$

Note that if $y(t) = 0$ for $t \geq 0$, and if the region of convergence for the Laplace transform includes the imaginary axis, then the Fourier transform can be obtained from the Laplace transform as

$$Y(\omega) = [Y(s)]_{s=j\omega}$$

For cases where it is impossible to determine the Fourier transform analytically, such as when the signal is described graphically or by a table, numerical solution based on the *fast Fourier transform* (*FFT*) *algorithm* is usually satisfactory.

In general, the condition $T \leq \pi/\omega_u$ cannot be satisfied exactly, since most physical signals have no finite upper frequency ω_u. A useful approximation is to define the upper frequency as the frequency for which 99% of the signal "energy" lies in the frequency spectrum $0 \leq \omega \leq \omega_u$. This approximation is found from the relation

$$\int_0^{\omega_u} |Y(\omega)|^2\, d\omega = 0.99 \int_0^{\infty} |Y(\omega)|^2\, d\omega$$

where the square of the amplitude of the Fourier transform $|Y(\omega)|^2$ is said to be the *power spectrum* and its integral over the entire frequency spectrum is referred to as the "energy" of the signal. Using a sampling frequency 2–10 times this approximate upper frequency (depending on the required factor of safety) and inserting a low pass filter (called a *guard filter*) before the sampler to eliminate frequencies above the *Nyquist frequency* π/T, usually leads to satisfactory results.[19]

The z-Transform

The *z-transform* permits the development and application of transfer functions for discrete-time systems, in a manner analogous to continuous-time transfer functions based on the Laplace transform. A discrete signal may be represented as a series of impulses

$$y^*(t) = y(0)\delta(t) + y(1)\delta(t - T) + y(2)\delta(t - 2T) + \cdots$$

$$= \sum_{k=0}^{N} y(k)\delta(t - kT)$$

where $y(k) = y^*(t = kT)$ are the values of the discrete signal, $\delta(t)$ is the unit impulse function, and N is the number of samples of the discrete signal. The Laplace transform of the series is

$$Y^*(s) = \sum_{k=0}^{N} y(k)e^{-ksT}$$

where the shifting property of the Laplace transform has been applied to the pulses. Defining the *shift* or *advance operator* as $z = e^{st}$, $Y^*(s)$ may now be written as a function of z

$$Y^*(z) = \sum_{k=0}^{N} \frac{y(k)}{z^k} = \mathscr{Z}[y(t)]$$

where the transformed variable $Y^*(z)$ is called the z-transform of the function $y^*(t)$. The inverse of the shift operator $1/z$ is called the *delay operator* and corresponds to a time delay of T.

The z-transforms for many sampled functions can be expressed in closed form. A listing of the transforms of several commonly encountered functions is given in Table 39.12. Properties of the z-transform are listed in Table 39.13.

Pulse Transfer Functions

The transfer function concept developed for continuous systems has a direct analog for sampled-data systems. For a continuous system with sampled output $u(t)$ and sampled input $y(t)$, the *pulse* or *discrete transfer function* $G(z)$ is defined as the ratio of the z-transformed output $Y(z)$ to the z-transformed input $U(z)$, that is, assuming zero initial conditions. In general, the pulse transfer function has the form

$$G(z) = \frac{Y(z)}{U(z)} = \frac{b_0 + b_1 z^{-1} + b_2 z^{-2} + \cdots + b_m z^{-m}}{1 + a_1 z^{-1} + a_2 z^{-1} + \cdots + a_n z^{-n}}$$

Table 39.12   z-Transform Pairs

	$X(s)$	$x(t)$ or $x(k)$	$X(z)$
1	1	$\delta(t)$	1
2	e^{-kTs}	$\delta(t-kT)$	z^{-k}
3	$\dfrac{1}{s}$	$1(t)$	$\dfrac{z}{z-1}$
4	$\dfrac{1}{s^2}$	t	$\dfrac{Tz}{(z-1)^2}$
5	$\dfrac{1}{s+a}$	e^{-at}	$\dfrac{z}{z-e^{-aT}}$
6	$\dfrac{a}{s(s+a)}$	$1-e^{-at}$	$\dfrac{(1-e^{-aT})z}{(z-1)(z-e^{-aT})}$
7	$\dfrac{\omega}{s^2+\omega^2}$	$\sin\omega t$	$\dfrac{z\sin\omega T}{z^2-2z\cos\omega T+1}$
8	$\dfrac{s}{s^2+\omega^2}$	$\cos\omega t$	$\dfrac{z(z-\cos\omega T)}{z^2-2z\cos\omega T+1}$
9	$\dfrac{1}{(s+a)^2}$	te^{-at}	$\dfrac{Tze^{-aT}}{(z-e^{-aT})^2}$
10	$\dfrac{\omega}{(s+a)^2+\omega^2}$	$e^{-at}\sin\omega t$	$\dfrac{ze^{-aT}\sin\omega T}{z^2-2ze^{-aT}\cos\omega T+e^{-2aT}}$
11	$\dfrac{s+a}{(s+a)^2+\omega^2}$	$e^{-at}\cos\omega t$	$\dfrac{z^2-ze^{-aT}\cos\omega T}{z^2-2ze^{-aT}\cos\omega T+e^{-2aT}}$
12	$\dfrac{2}{s^3}$	t^2	$\dfrac{T^2z(z+1)}{(z-1)^3}$
13		a	$\dfrac{z}{z-a}$
14		$a^k\cos k\pi$	$\dfrac{z}{z+a}$

Zero-Order Hold

The *zero-order data hold* is the most common mathematical approach to *digital-to-analog (D/A) conversion,* that is, to creating a piecewise continuous approximation $u(t)$ of the form

$$u(t)=u^*(k)\qquad \text{for}\quad kT\le t<(k+1)T$$

from the discrete time signal $u^*(k)$, where T is the period of the hold. The effect of the zero-order hold is to convert a sequence of discrete impulses into a staircase pattern, as shown in Fig. 39.35. The transfer function of the zero-order hold is

$$G(s)=\frac{1}{s}(1-e^{-Ts})=\frac{1-z^{-1}}{s}$$

Using this relationship, the pulse transfer function of the sampled-data system shown in Fig. 39.36 can be derived as

$$G(z)=(1-z^{-1})\,\mathscr{Z}\left[\mathscr{L}^{-1}\frac{G(s)}{s}\right]$$

The continuous system with transfer function $G(s)$ has a sampler and a zero-order hold at its input and a sampler at its output. This is a common configuration in many computer-control applications.

Table 39.13 z-Transform Properties

	$x(t)$ or $x(k)$	$\mathscr{Z}[x(t)]$ or $\mathscr{Z}[x(k)]$
1	$ax(t)$	$aX(z)$
2	$x_1(t) + x_2(t)$	$X_1(z) + X_2(z)$
3	$x(t+T)$ or $x(k+1)$	$zX(z) - zx(0)$
4	$x(t+2T)$	$z^2X(z) - z^2x(0) - zx(T)$
5	$x(k+2)$	$z^2X(z) - z^2x(0) - zx(1)$
6	$x(t+kT)$	$z^kX(z) - z^kx(0) - z^{k-1}x(T) - \cdots - zx(kT - T)$
7	$x(k+m)$	$z^mX(z) - z^mx(0) - z^{m-1}x(1) - \cdots - zx(m-1)$
8	$tx(t)$	$-Tz\dfrac{d}{dz}[X(z)]$
9	$kx(k)$	$-z\dfrac{d}{dz}[X(z)]$
10	$e^{-at}x(t)$	$X(ze^{aT})$
11	$e^{-ak}x(k)$	$X(ze^a)$
12	$a^kx(k)$	$X\left(\dfrac{z}{a}\right)$
13	$ka^kx(k)$	$-z\dfrac{d}{dz}\left[X\left(\dfrac{z}{a}\right)\right]$
14	$x(0)$	$\lim\limits_{z\to\infty} X(z)$ if the limit exists
15	$x(\infty)$	$\lim\limits_{z\to1} [(z-1)X(z)]$ if $\dfrac{z-1}{z}X(z)$ is analytic on and outside the unit circle
16	$\sum\limits_{k=0}^{\infty} x(k)$	$X(1)$
17	$\sum\limits_{k=0}^{n} x(kT)y(nT - kT)$	$X(z)Y(z)$

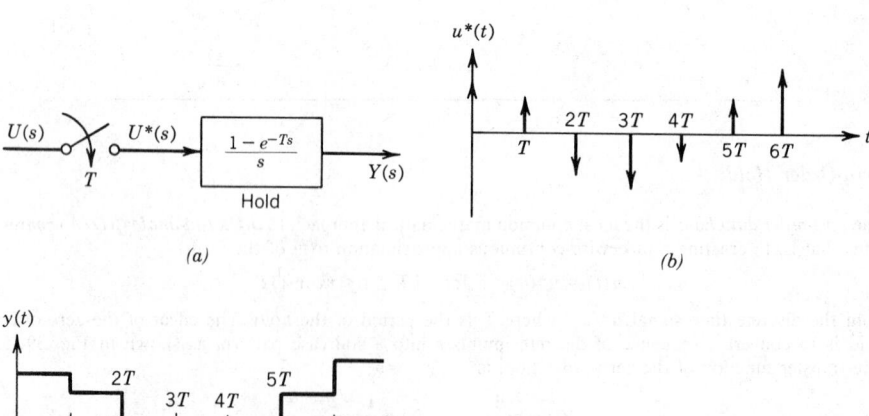

Fig. 39.35 Zero-order hold: (a) block diagram of hold with a sampler, (b) sampled input sequence, (c) analog output for the corresponding input sequence.[19]

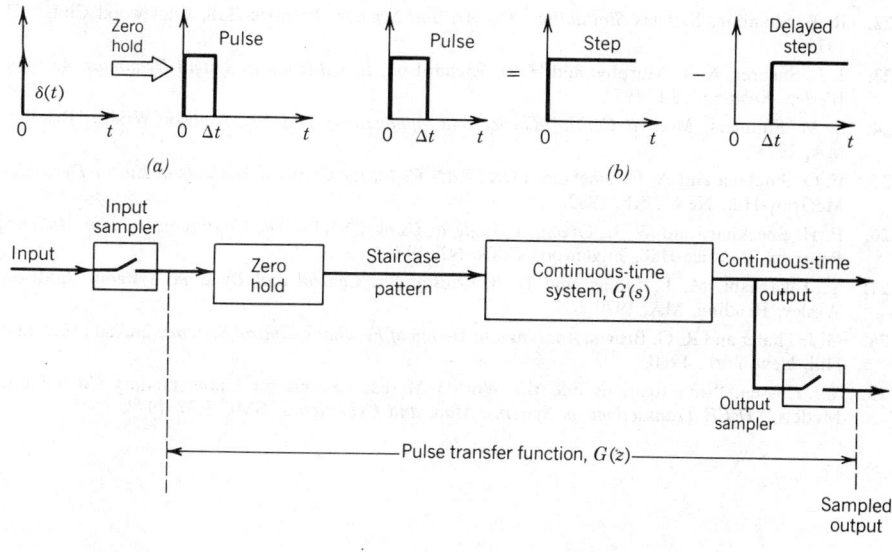

(c)

Fig. 39.36 Pulse transfer function of a continuous system with sampler and zero hold.[1]

REFERENCES

1. D. M. Auslander, Y. Takahashi, and M. J. Rabins, *Introducing Systems and Control,* McGraw-Hill, New York, 1974.
2. R. H. Cannon, Jr., *Dynamics of Physical Systems,* McGraw-Hill, New York, 1967.
3. CDC, Continuous System Simulation Language, Version 3, User's Guide, Control Data Corp., Sunnyvale, Calif., 1971.
4. Y. Chu, *Digital Simulation of Continuous Systems,* McGraw-Hill, New York, 1969.
5. C. M. Close and D. K. Frederick, *Modeling and Analysis of Dynamic Systems,* Houghton Mifflin, Boston, 1978.
6. W. J. Cunningham, *Introduction to Nonlinear Analysis,* McGraw-Hill, New York, 1958.
7. E. O. Doebelin, *System Dynamics: Modeling and Response,* Merrill, Columbus, OH, 1972.
8. R. C. Dorf, *Modern Control Systems,* 2nd ed., Addison-Wesley, Reading, MA, 1974.
9. J. W. Forrester, *Industrial Dynamics,* MIT Press, Cambridge, MA, 1961.
10. J. E. Gibson, *Nonlinear Automatic Control,* McGraw-Hill, New York, 1963.
11. G. Gordon, *System Simulation,* 2nd ed., Prentice-Hall, Englewood Cliffs, NJ, 1978.
12. M. Jamshidi and M. Malek-Zavarei, *Linear Control Systems: A Computer-Aided Approach,* Pergamon, Oxford, 1984.
13. G. A. Korn and J. V. Wait, *Digital Continuous System Simulation,* Prentice-Hall, Englewood Cliffs, NJ, 1978.
14. B. C. Kuo, *Automatic Control Systems,* 3rd ed., Prentice-Hall, Englewood Cliffs, NJ, 1975.
15. S. J. Larimer, *User's Manual for Total: An Interactive Computer-Aided Design Program for Digital and Continuous Control System Analysis and Synthesis,* Air Force Institute of Technology, 1978.
16. D. G. Luenberger, *Introduction to Dynamic Systems: Theory, Models, and Applications,* Wiley, New York, 1979.
17. J. L. Melsa and S. K. Jones, *Computer Programs for Computational Assistance in the Study of Linear Control Theory,* 2nd ed., McGraw-Hill, New York, 1973.
18. K. Ogata, *Modern Control Engineering,* Prentice-Hall, Englewood Cliffs, NJ, 1970.
19. W. J. Palm, III, *Modeling, Analysis, and Control of Dynamic Systems,* Wiley, New York, 1983.
20. A. Pugh, III, *DYNAMO II User's Manual,* MIT Press, Cambridge, MA, 1973.
21. D. G. Schultz and J. L. Melsa, *State Functions and Linear Control Systems,* McGraw-Hill, New York, 1967.

22. R. E. Shannon, *Systems Simulation: The Art and Science,* Prentice-Hall, Englewood Cliffs, NJ, 1975.

23. J. L. Shearer, A. T. Murphy, and H. H. Richardson, *Introduction to System Dynamics, Addison-Wesley, Reading, MA, 1971.*

24. S. M. Shinners, *Modern Control Theory and Application,* 2nd ed., Addison-Wesley, Reading, MA, 1978.

25. F. O. Smetana and A. O. Smetana, *FORTRAN Codes for Classical Methods in Linear Dynamics,* McGraw-Hill, New York, 1982.

26. F. H. Speckhart and W. L. Green, *A Guide to Using CSMP—The Continuous System Modeling Program,* Prentice-Hall, Englewood Cliffs, NJ, 1976.

27. Y. Takahashi, M. J. Rabins, and D. M. Auslander, *Control and Dynamic Systems,* Addison-Wesley, Reading, MA, 1970.

28. G. J. Thaler and R. G. Brown, *Analysis and Design of Feedback Control Systems,* 2nd ed., McGraw-Hill, New York, 1960.

29. W. Thissen, "Investigations into the World3 Model: Lessons for Understanding Complicated Models," *IEEE Transactions on Systems, Man, and Cybernetics,* **SMC-8**(3), 1978.

CHAPTER 40

BASIC CONTROL SYSTEMS DESIGN

WILLIAM J. PALM III

The University of Rhode Island
Kingston, Rhode Island

Revised from William J. Palm III, *Modeling, Analysis and Control of Dynamic Systems*, Wiley, 1983, by permission of the publisher.

40.1 INTRODUCTION

The purpose of a *control system* is to produce a desired *output.* This output is usually specified by a command *input,* and is often a function of time. For simple applications in well-structured situations, *sequencing* devices like timers can be used as the control system. But most systems are not that easy to control, and the controller must have the capability of reacting to disturbances, changes in its environment, and new input commands. The key element that allows a control system to do this is *feedback,* which is the process by which a system's output is used to influence its behavior. Feedback in the form of the room-temperature measurement is used to control the furnace in a thermostatically controlled heating system. Figure 40.1 shows the *feedback loop* in the system's *block diagram,* which is a graphical representation of the system's control structure and logic. Another commonly found control system is the pressure regulator shown in Fig. 40.2.

Feedback has several useful properties. A system whose individual elements are nonlinear can often be modeled as a linear one over a wider range of its variables with the proper use of feedback. This is because feedback tends to keep the system near its reference operating condition. Systems that can maintain the output near its desired value despite changes in the environment are said to have good *disturbance rejection.* Often we do not have accurate values for some system parameters, or these values might change with age. Feedback can be used to minimize the effects of parameter changes and uncertainties. A system that has both good disturbance rejection and low sensitivity to parameter variation is *robust.* The application that resulted in the general understanding of the properties of feedback is shown in Fig. 40.3. The electronic amplifier gain A is large, but we are uncertain of its exact value. We use the resistors R_1 and R_2 to create a feedback loop around the amplifier, and pick R_1 and R_2 so that $AR_2/R_1 \gg 1$. Then the input–output relation becomes $e_o \cong R_1 e_i/R_2$, which is independent of A as long as A remains large. If R_1 and R_2 are known accurately, then the system gain is now reliable.

Figure 40.4 shows the block diagram of a *closed-loop* system, which is a system with feedback. An *open-loop* system, such as a timer, has no feedback. Figure 40.4 serves as a focus for outlining the prerequisites for this chapter. The reader should be familiar with the *transfer-function* concept based on the Laplace transform, the *pulse-transfer function* based on the z-transform, for digital control, and the differential equation modeling techniques needed to obtain them. It is also necessary to understand block-diagram algebra, characteristic roots, the final-value theorem, and their use in evaluating system response for common inputs like the step function. Also required are stability analysis techniques, such as the Routh criterion, and transient performance specifications such as the damping ratio ζ, natural frequency ω_n, dominant time constant τ, maximum overshoot, settling time, and bandwidth. State vector notation is necessary for Section 40.12. The above material is reviewed in the previous chapter. Treatment in depth is given in Refs. 3, 4, 7, 9, 11, 13, and 14.

40.2 CONTROL SYSTEM STRUCTURE

The electromechanical position control system shown in Fig. 40.5 illustrates the structure of a typical control system. A load with an inertia I is to be positioned at some desired angle θ_r. A dc motor is provided for this purpose. The system contains viscous damping, and a disturbance torque T_d acts on the load, in addition to the motor torque T. Because of the disturbance the angular position θ of the load will not necessarily equal the desired value θ_r. For this reason a potentiometer is used to measure the displacement θ. The potentiometer voltage representing the controlled position θ is compared to the voltage generated by the command potentiometer. This device enables the operator to dial in the desired angle θ_r. The amplifier sees the difference e between the two potentiometer voltages. The basic function of the amplifier is to increase the small error voltage e up to the voltage level required by the motor and to supply enough current required by the motor to drive the load. In addition, the amplifier may shape the voltage signal in certain ways to improve the performance of the system.

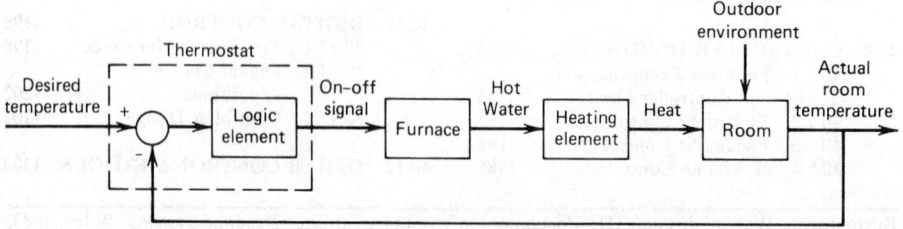

Fig. 40.1 Block diagram of the thermostat system for temperature control.[11]

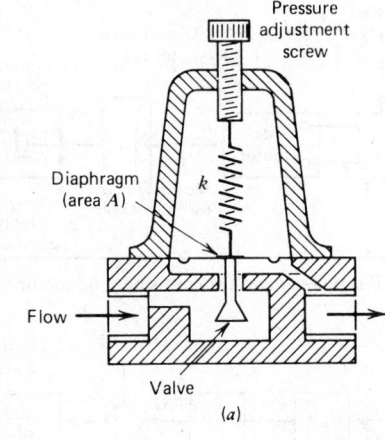

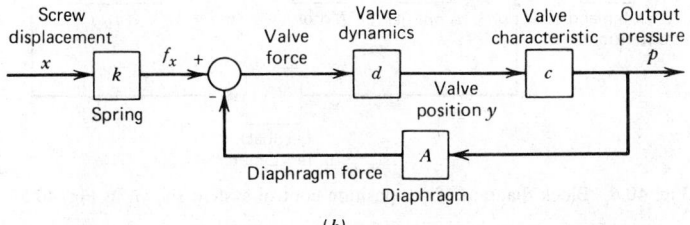

Fig. 40.2 Pressure regulator: (a) cutaway view; (b) block diagram.[11]

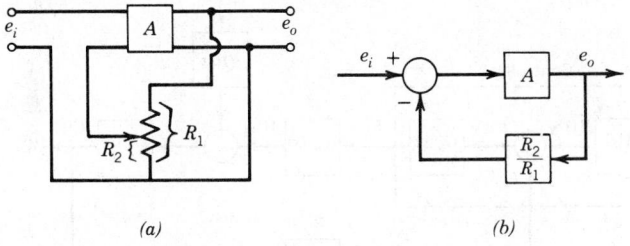

(a)   (b)

Fig. 40.3 A closed-loop system.

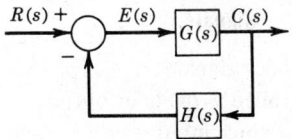

Fig. 40.4 Feedback compensation of an amplifier.

The control system is seen to provide two basic functions: (1) to respond to a command input that specifies a new desired value for the controlled variable, and (2) to keep the controlled variable near the desired value in spite of disturbances. The presence of the feedback loop is vital to both functions. A block diagram of this system is shown in Fig. 40.6.

The power supplies required for the potentiometers and the amplifier are not shown in block diagrams of control system logic because they do not contribute to the control logic.

40.2.1 A Standard Diagram

The electromechanical positioning system fits the general structure of a control system (Fig. 40.7). This figure also gives some standard terminology. Not all systems can be forced into this format, but it serves as a reference for discussion.

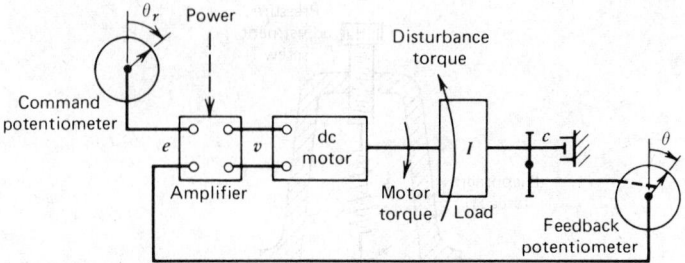

Fig. 40.5 Position-control system using a dc motor.[11]

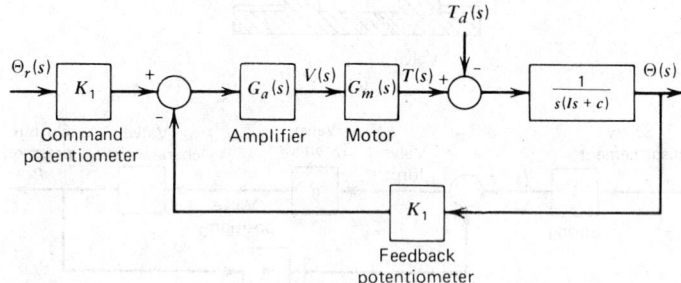

Fig. 40.6 Block diagram of the position-control system shown in Fig. 40.5.[11]

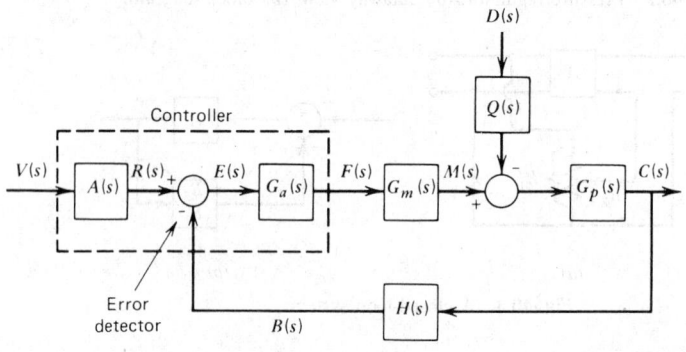

Elements		Signals	
$A(s)$	Input elements	$B(s)$	Feedback signal
$G_a(s)$	Control logic elements	$C(s)$	Controlled variable or output
$G_m(s)$	Final control elements	$D(s)$	Disturbance input
$G_p(s)$	Plant elements	$E(s)$	Error or actuating signal
$H(s)$	Feedback elements	$F(s)$	Control signal
$Q(s)$	Disturbance elements	$M(s)$	Manipulated variable
		$R(s)$	Reference input
		$V(s)$	Command input

Fig. 40.7 Terminology and basic structure of a feedback-control system.[11]

The controller is generally thought of as a logic element that compares the command signal with the measurement of the output and decides what should be done. The input and feedback elements are transducers for converting one type of signal into another type. This allows the error detector directly to compare two signals of the same type (e.g., two voltages). Not all functions show up as separate physical elements. The error detector in Fig. 40.5 is simply the input terminals of the amplifier.

The control logic elements produce the control signal, which is sent to the *final control elements*. These are the devices that develop enough torque, pressure, heat, etc., to influence the elements under control. Thus the final control elements are the "muscle" of the system, while the control logic elements are the "brain." Here we are primarily concerned with the design of the logic to be used by this brain.

The object to be controlled is the *plant*. The *manipulated variable* is generated by the final control elements for this purpose. The disturbance input also acts on the plant. This is an input over which the designer has no influence, and perhaps for which little information is available as to the magnitude, functional form, or time of occurrence. The disturbance can be a random input, such as wind gusts on a radar antenna, or deterministic, such as Coulomb friction effects. In the latter case, we can include the friction force in the system model by using a nominal value for the coefficient of friction. The disturbance input would then be the deviation of the friction force from this estimated value and would represent the uncertainty in our estimate.

Several control system classifications can be made with reference to Fig. 40.7. A *regulator* is a control system in which the controlled variable is to be kept constant in spite of disturbances. The command input for a regulator is its *set point*. A *follow-up system* is supposed to keep the controlled variable near a command value that is changing with time. An example of a follow-up system is a machine tool in which a cutting head must trace a specific path in order to shape the product properly. This is also an example of a *servomechanism*, which is a control system whose controlled variable is a mechanical position, velocity, or acceleration. A thermostat system is not a servomechanism, but is a *process-control system*, where the controlled variable describes a thermodynamic process. Typically such variables are temperature, pressure, flow rate, liquid level, chemical concentration, etc.

40.2.2 Transfer Functions

A transfer function is defined for each input–output pair of the system. A specific transfer function is found by setting all other inputs to zero, and reducing the block diagram. The *primary* or *command* transfer function for Fig. 40.7 is

$$\frac{C(s)}{V(s)} = \frac{A(s)G_d(s)G_m(s)G_p(s)}{1 + G_a(s)G_m(s)G_p(s)H(s)} \tag{40.1}$$

The *disturbance* transfer function is

$$\frac{C(s)}{D(s)} = \frac{-Q(s)G_p(s)}{1 + G_a(s)G_m(s)G_p(s)H(s)} \tag{40.2}$$

The transfer functions of a given system all have the same denominator.

40.2.3 System-Type Number and Error Coefficients

The error signal in Fig. 40.4 is related to the input as

$$E(s) = \frac{1}{1 + G(s)H(s)}R(s) \tag{40.3}$$

If the final value theorem can be applied, the steady-state error is

$$e_{ss} = \lim_{s \to 0} \frac{sR(s)}{1 + G(s)H(s)} \tag{40.4}$$

The *static error coefficient* c_i is defined as

$$c_i = \lim_{s \to 0} s^i G(s)H(s), \qquad i = 0, 1, 2 \ldots \tag{40.5}$$

A system is of *type n* if $G(s)H(s)$ can be written as $s^n F(s)$. Table 40.1 relates the steady-state error to the system type for three common inputs, and can be used to design systems for minimum error. The higher the system type, the better the system is able to follow a rapidly changing input. But higher-type systems are more difficult to stabilize, so a compromise must be made in the design. C_0, C_1, and C_2 are called the *position, velocity*, and *acceleration error coefficients*.

40.3 TYPICAL APPLICATIONS

Control systems can be used to influence just about any physical variable. We have already seen controllers for voltage, temperature, pressure, and position. Other commonly controlled variables include speed, current, flow, and liquid level. A liquid-level controller is shown in Fig. 40.8. The liquid is supplied under pressure p to the control valve. The valve motion is produced by a diaphragm with

Table 40.1 Steady-State Error e_{ss} for Different System Type Numbers

	System Type Number n			
$R(s)$	0	1	2	3
Step $1/s$	$\dfrac{1}{1+C_0}$	0	0	0
Ramp $1/s^2$	∞	$\dfrac{1}{C_1}$	0	0
Parabola $1/s^3$	∞	∞	$\dfrac{1}{C_2}$	0

pneumatic pressure on one side and a resisting spring on the other. The air pressure is produced by a pneumatic amplifier whose output is affected by the motion of a beam with a bellows at each end (this amplifier is discussed in Section 40.4). A bubbler tube filled with air indicates the liquid level by sensing the hydrostatic pressure in the tank. The air pressure in the tube is connected to the feedback bellows. The set point of the controller indicates the desired liquid level. This is set by changing the pressure in the set point bellows via an adjustable restriction in the supply line. The system is balanced when the level is at the desired value and the feedback pressure equals the set point pressure. If the level is below the desired value, the beam pivots counterclockwise. The amplifier and valve are designed to respond to this motion by increasing the flow rate q. The flow q is a function of p and the valve motion y. Two possible disturbances to this system are changes in the supply pressures of the air and the liquid.

Machine tools are one application of the hydraulic system shown in Fig. 40.9. The applied force f is supplied by the servomotor. The mass m represents that of a cutting tool and the power piston, while k represents the combined effects of the elasticity naturally present in the structure and that

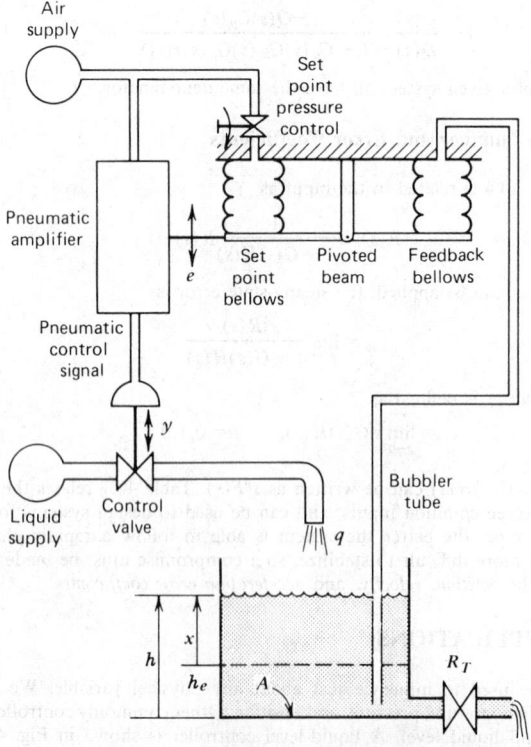

Fig. 40.8 Pneumatic liquid-level control system.[11]

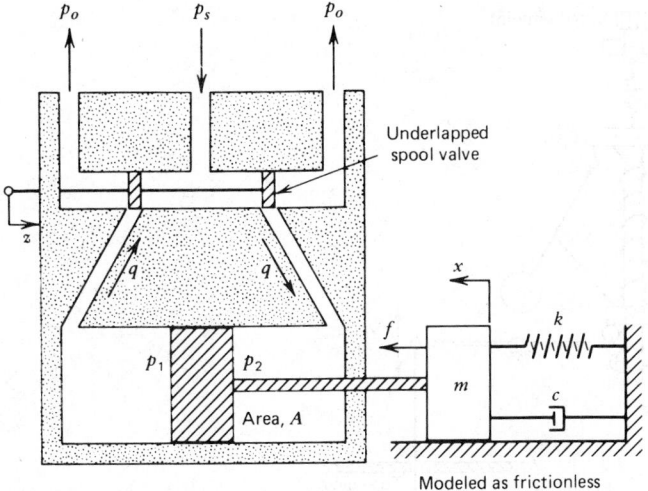

Fig. 40.9   Hydraulic servomotor with a load.[11]

introduced by the designer to achieve proper performance. A similar statement applies to the damping c. The valve displacement z is generated by another control system in order to move the tool through its prescribed motion. The spool valve shown in Fig. 40.9 has two *lands*. If the width of the land is greater than the port width, the valve is said to be *overlapped*. In this case a dead zone exists in which a slight change in the displacement z produces no power piston motion. Such dead zones create control difficulties and are avoided by designing the valve to be *underlapped* (the land width is less the port width). For such valves there will be a small flow opening even when the valve is in the neutral position ($z = 0$). This gives it a higher sensitivity than an overlapped valve.[9]

The variables z and $\Delta p = p_2 - p_1$ determine the volume flow rate, as

$$q = f(z, \Delta p)$$

For the reference equilibrium condition ($z = 0$, $\Delta p = 0$, $q = 0$), a linearization gives

$$q = C_1 z - C_2 \Delta p \tag{40.6}$$

The linearization constants are available from theoretical and experimental results.[8] The transfer function for the system is

$$T(s) = \frac{X(s)}{Z(s)} = \frac{C_1}{\dfrac{C_2 m}{A} s^2 + \left(\dfrac{cC_2}{A} + A\right)s + \dfrac{C_2 k}{A}} \tag{40.7}$$

The development of the steam engine led to the requirement for a speed-control device to maintain constant speed in the presence of changes in load torque or steam pressure. In 1788, James Watt of Glasgow developed his now famous flyball governor for this purpose (Fig. 40.10). Watt took the principle of sensing speed with the centrifugal pendulum of Thomas Mead and used it in a feedback loop on a steam engine. As the motor speed increases, the flyballs move outward and pull the slider upward. The upward motion of the slider closes the steam valve, thus causing the engine to slow down. If the engine speed is too slow, the spring force overcomes that due to the flyballs, and the slider moves down to open the steam valve. The desired speed can be set by moving the plate to change the compression in the spring. The principle of the flyball governor is still used for speed-control applications. Typically, the pilot valve of a hydraulic servomotor is connected to the slider to provide the high forces required to move large supply valves.

The challenge of modern control engineering is the integration of the multiple control systems required to control the numerous variables found in complex systems, such as the one shown in Fig. 40.11.

40.4   TRANSDUCERS AND ERROR DETECTORS

The control system structure shown in Fig. 40.7 indicates a need for physical devices to perform several types of functions. Here we present a brief overview of some available transducers and error detectors. Actuators and devices used to implement the control logic are discussed in Sections 5 and 6.

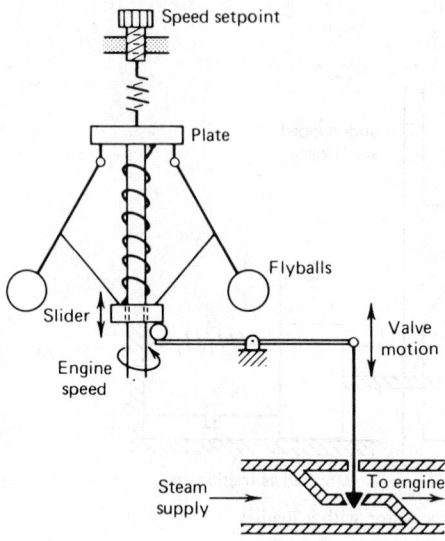

Fig. 40.10 James Watt's flyball governor for speed control of a steam engine.[11]

40.4.1 Displacement and Velocity Transducers

A *transducer* is a device that converts one type of signal into another type. An example is the potentiometer, which converts displacement into voltage, as in Fig. 40.12. In addition to this conversion, the transducer can be used to make measurements. In such applications the term *sensor* is more appropriate. Displacement can also be measured electrically with a *linear variable differential transformer* (LVDT) or a *synchro*. An LVDT measures the linear displacement of a movable magnetic core through a primary winding and two secondary windings (Fig. 40.13). An ac voltage is applied to the primary. The secondaries are connected together and also to a detector that measures the voltage and phase

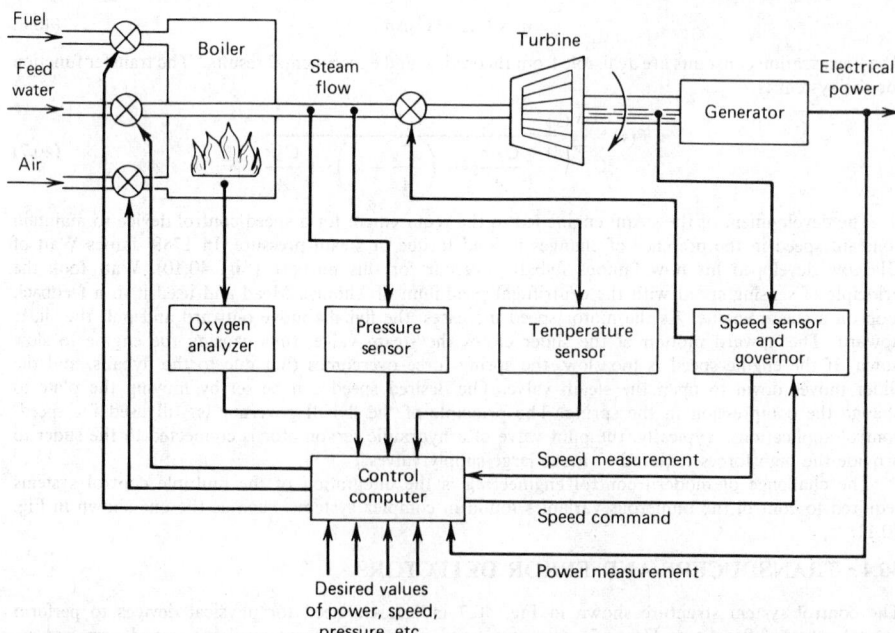

Fig. 40.11 Computer control system for a boiler–generator. Each important variable requires its own controller. The interaction between variables calls for coordinated control of all loops.[11]

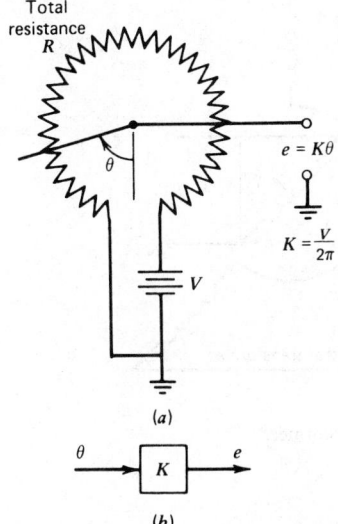

(a)

θ → [K] → e

(b)

Fig. 40.12 Rotary potentiometer.[11]

difference. A phase difference of 0° corresponds to a positive core displacement, while 180° indicates a negative displacement. The amount of displacement is indicated by the amplitude of the ac voltage in the secondary. The detector converts this information into a dc voltage e_o such that $e_o = Kx$. The LVDT is sensitive to small displacements. Two of them can be wired together to form an error detector.

A synchro is a rotary differential transformer, with angular displacement as either the input or output. They are often used in pairs (a *transmitter* and a *receiver*) where a remote indication of angular displacement is needed. When a transmitter is used with a synchro *control transformer,* two angular displacements can be measured and compared (Fig. 40.14). The output voltage e_o is approximately linear with angular difference within ± 70°, so that $e_o = K(\theta_1 - \theta_2)$.

Displacement measurements can be used to obtain forces and accelerations. For example, the displacement of a calibrated spring indicates the applied force. The accelerometer is another example. Still another is the *strain gage* used for force measurement. It is based on the fact that the resistance of a fine wire changes as it is stretched. The change in resistance is detected by a circuit that can be calibrated to indicate the applied force.

Velocity measurements in control systems are most commonly obtained with a *tachometer.* This is essentially a dc generator (the reverse of a dc motor). The input is mechanical (a velocity). The output is a generated voltage proportional to the velocity. Translational velocity is usually measured by converting it to angular velocity with gears, for example. Tachometers using ac signals are also available.

Other velocity transducers include a magnetic pickup that generates a pulse every time a gear

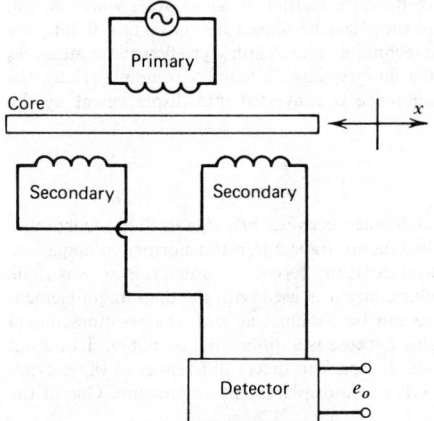

Fig. 40.13 Linear variable differential transformer (LVDT).[11]

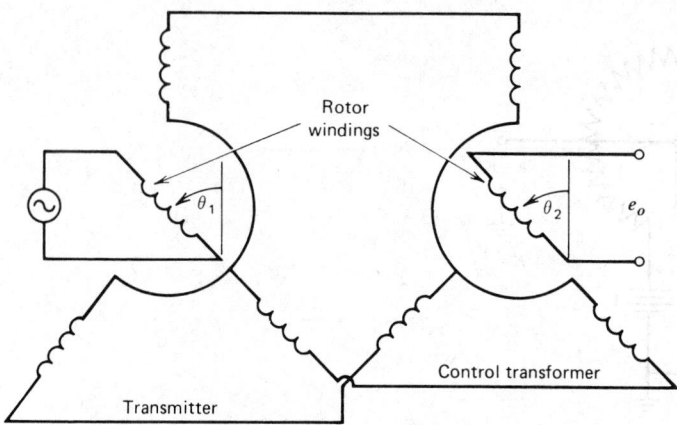

Fig. 40.14   Synchro transmitter-control transformer.[11]

tooth passes. If the number of gear teeth is known, a pulse counter and timer can be used to compute the angular velocity. A similar principle is employed by an *optical encoder*. A light beam is broken by a rotating slotted disk, and a photoelectric cell converts this information into pulses that can be analyzed as before. The outputs of these devices are especially suitable for digital control purposes.

40.4.2   Temperature Transducers

When two wires of dissimilar metals are joined together, a voltage is generated if the junctions are at different temperatures. If the reference junction is kept at a fixed, known temperature, the thermocouple can be calibrated to indicate the temperature at the other junction in terms of the voltage *v*. Electrical resistance changes with temperature. Platinum gives a linear relation between resistance and temperature, while nickel is less expensive and gives a large resistance change for a given temperature change. Semiconductors designed with this property are called *thermistors*. Different metals expand at different rates when the temperature is increased. This fact is used in the bimetallic strip transducer found in most home thermostats. Two dissimilar metals are bonded together to form the strip. As the temperature rises the strip curls, breaking contact and shutting off the furnace. The temperature gap can be adjusted by changing the distance between the contacts. The motion also moves a pointer on the temperature scale of the thermostat. Finally, the pressure of a fluid inside a bulb will change as its temperature changes. If the bulb fluid is air, the device is suitable for use in pneumatic temperature controllers.

40.4.3   Flow Transducers

Flow rates can be measured by introducing a flow restriction, such as an orifice plate, and measuring the pressure drop across the restriction. The flow-rate–pressure relation is $\Delta p = Rq^2$, where R can be found from calibration of the device. The pressure drop can be sensed by converting it into the motion of a diaphragm. Figure 40.15 illustrates a related technique. The Venturi-type flowmeter measures the static pressures in the constricted and unconstricted flow regions. Bernoulli's principle relates the pressure difference to the flow rate. This pressure difference is converted into displacement by the diaphragm.

40.4.4   Error Detectors

The error detector is simply a device for finding the difference between two signals. This function is sometimes an integral feature of sensors, such as with the synchro transmitter–transformer combination. A beam on a pivot provides a way of comparing displacements, forces, or pressures, as was done with the pneumatic level controller (Fig. 40.8). A similar concept is used with the diaphragm element shown in Fig. 40.15. A detector for voltage difference can be obtained as with the position-control system shown in Fig. 40.5. An amplifier intended for this purpose is a *differential amplifier*. Its output is proportional to the difference between the two inputs. In order to detect differences in other types of signals, such as temperature, they are usually converted to a displacement or pressure. One of the detectors mentioned previously can then be used.

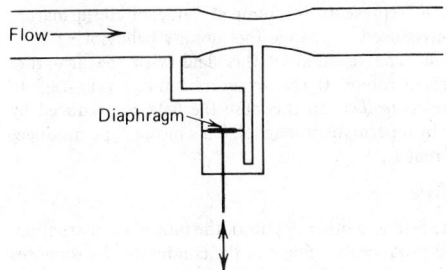

Fig. 40.15 Venturi-type flowmeter. The diaphragm displacement indicates the flow rate.[11]

40.4.5 Dynamic Response

The usual transducer and detector models are static models, and as such imply that the components respond instantaneously to the variable being sensed. Of course, any real component has a dynamic response of some sort, and this response time must be considered in relation to the controlled process when a sensor is selected. For example, the time constant of a thermocouple in air typically is between 10 and 100 sec. If the thermocouple is in a liquid, the thermal resistance is different, and so will be the time constant (it will be smaller). If the controlled process has a time constant at least 10 times greater, we probably would be justified in using a static sensor model.

40.5 ACTUATORS

An *actuator* is the final control element that operates on the low-level control signal to produce a signal containing enough power to drive the plant for the intended purpose. The armature-controlled dc motor, the hydraulic servomotor, and the pneumatic bellows are common examples of actuators.

40.5.1 Electromechanical Actuators

Figure 40.16 shows an electromechanical system consisting of an armature-controlled dc motor driving a load inertia. The rotating armature consists of a wire conductor wrapped around an iron core.

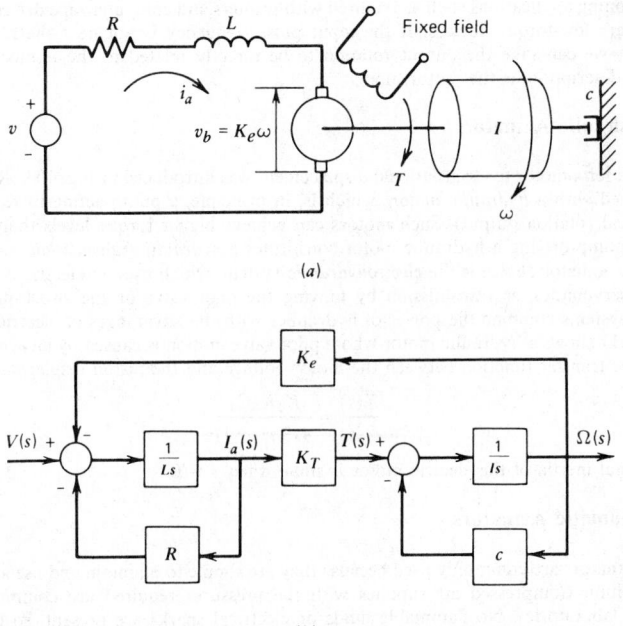

Fig. 40.16 Armature-controlled dc motor with a load, and the system's block diagram.[11]

This winding has an inductance L. The resistance R represents the lumped value of the armature resistance and any external resistance deliberately introduced to change the motor's behavior.

The armature is surrounded by a magnetic field. The reaction of this field with the armature current produces a torque that causes the armature to rotate. If the armature voltage v is used to control the motor, the motor is said to be *armature-controlled*. In this case the field is produced by an electromagnet supplied with a constant voltage or by a permanent magnet. This motor type produces a torque T that is proportional to the armature current i_a:

$$T = K_T i_a \qquad (40.8)$$

The torque constant K_T depends on the strength of the field and other details of the motor's construction. The motion of a current-carrying conductor in a field produces a voltage in the conductor that opposes the current. This voltage is called the *back emf* (electromotive force). Its magnitude is proportional to the speed and is given by

$$e_b = K_e \omega \qquad (40.9)$$

The transfer function for the armature-controlled dc motor is

$$\frac{\Omega(s)}{V(s)} = \frac{K_T}{LIs^2 + (RI + cL)s + cR + K_e K_T} \qquad (40.10)$$

Another motor configuration is the field-controlled dc motor. In this case the armature current is kept constant, and the field voltage v is used to control the motor. The transfer function is

$$\frac{\Omega(s)}{V(s)} = \frac{K_T}{(R + Ls)(Is + c)} \qquad (40.11)$$

where R and L are the resistance and inductance of the field circuit, and K_T is the torque constant. No back emf exists in this motor to act as a self-braking mechanism.

Two-phase ac motors can be used to provide a low-power, variable-speed actuator. This motor type can accept the ac signals directly from LVDTs and synchros without demodulation. However, it is difficult to design ac amplifier circuitry to do other than proportional action. For this reason the ac motor is not found in control systems as often as dc motors. The transfer function for this type is of the form of Eq. (40.11).

An actuator especially suitable for digital systems is the *stepper motor*, a special dc motor that takes a train of electrical input pulses and converts each pulse into an angular displacement of a fixed amount. Motors are available with resolutions ranging from about four steps per revolution to more than 800 steps per revolution. For 36 steps per revolution the motor will rotate by $10°$ for each pulse received. When not being pulsed, the motors lock in place. Thus they are excellent for precise positioning applications such as required with printers and computer tape drives. A disadvantage is that they are low-torque devices. If the input pulse frequency is not near the resonant frequency of the motor, we can take the output rotation to be directly related to the number of input pulses and use that description as the motor model.

40.5.2 Hydraulic Actuators

The hydraulic servomotor for translational displacement was introduced in Fig. 40.9. Rotational motion can be obtained with a *hydraulic motor*, which is, in principle, a pump acting in reverse (fluid input and mechanical rotation output). Such motors can achieve higher torque levels than electric motors. A hydraulic pump driving a hydraulic motor constitutes a *hydraulic transmission*.

A popular actuator choice is the *electrohydraulic* system, which uses an electric actuator to control a hydraulic servomotor or transmission by moving the pilot valve or the swash-plate angle of the pump. Such systems combine the power of hydraulics with the advantages of electrical systems.

Figure 40.17 shows a hydraulic motor whose pilot valve motion is caused by an armature-controlled dc motor. The transfer function between the motor voltage and the piston displacement is

$$\frac{X(s)}{V(s)} = \frac{K_1 K_2 C_1}{As^2(\tau s + 1)} \qquad (40.12)$$

If the rotational inertia of the electric motor is small, then $\tau \approx 0$.

40.5.3 Pneumatic Actuators

Pneumatic actuators are commonly used because they are simple to maintain and use a readily available working medium. Compressed air supplies with the pressures required are commonly available in factories and laboratories. No flammable fluids or electrical sparks are present, so these devices are considered the safest to use with chemical processes. Their power output is less than that of hydraulic systems, but greater than that of electric motors.

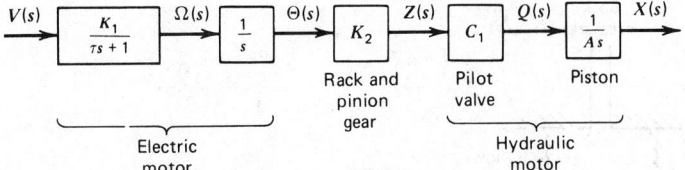

Fig. 40.17 Electrohydraulic system for translation.[11]

A device for converting pneumatic pressure into displacement is the bellows shown in Fig. 40.18. The transfer function for a linearized model is of the form

$$\frac{X(s)}{P(s)} = \frac{K}{\tau s + 1} \tag{40.13}$$

where x and p are deviations of the bellows displacement and input pressure from nominal values.

In many control applications a device is needed to convert small displacements into relatively large pressure changes. The *nozzle-flapper* serves this purpose (Fig. 40.19a). The input displacement y moves the flapper with little effort required. This changes the opening at the nozzle orifice. For a large enough opening the nozzle back pressure is approximately the same as atmospheric pressure p_a. At the other extreme position with the flapper completely blocking the orifice, the back pressure equals the supply pressure p_s. This variation is shown in Fig. 40.19b. Typical supply pressures are between 30 and 100 psia. The orifice diameter is approximately 0.01 in. Flapper displacement is usually less than one orifice diameter.

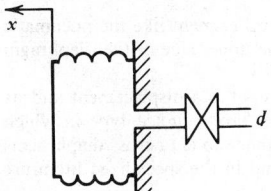

Fig. 40.18 Pneumatic bellows.[11]

The nozzle-flapper is operated in the linear portion of the back pressure curve. The linearized back pressure relation is

$$p = -K_f x \tag{40.14}$$

where $-K_f$ is the slope of the curve and is a very large number. From the geometry of similar triangles we have

$$p = -\frac{aK_f}{a+b} y \tag{40.15}$$

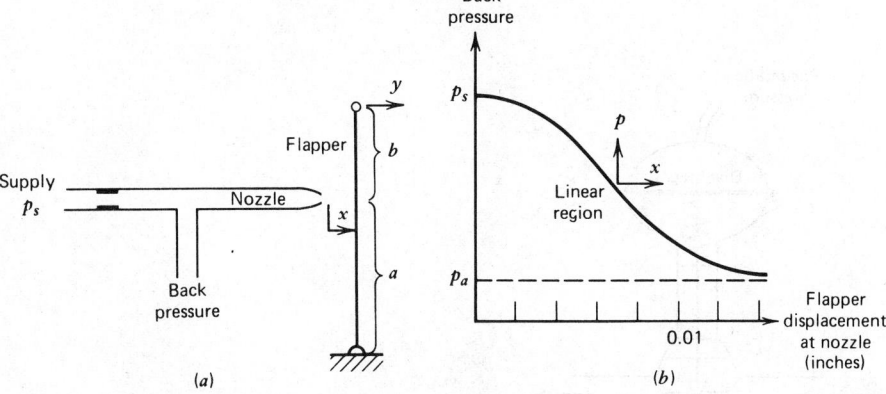

Fig. 40.19 Pneumatic nozzle-flapper amplifier and its characteristic curve.[11]

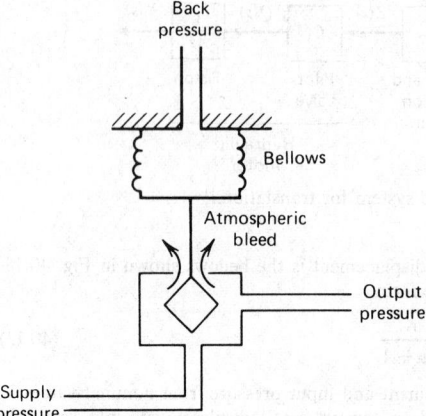

Fig. 40.20   Pneumatic relay.[11]

In its operating region the nozzle-flapper's back pressure is well below the supply pressure. If more output pressure is required, a pneumatic relay or amplifier can be used. Figure 40.20 illustrates this concept. As the back pressure increases, the relay closes off the supply line, and the output pressure approaches atmospheric pressure. As the back pressure decreases, the valve shuts off the atmospheric bleed, and the output pressure approaches the supply pressure. The relay is said to be reverse-acting because an increase in back pressure produces a decrease in output.

The output pressure from the relay can be used to drive a final control element like the pneumatic actuating valve shown in Fig. 40.21. The pneumatic pressure acts on the upper side of the diaphragm and is opposed by the return spring.

The nozzle-flapper is a *force–distance*-type of actuator because its input is a displacement and its output a force (pressure). The other type of pneumatic actuator is the *force–balance* type in which both the input and output are pressures (Fig. 40.22). The pressures are made to act across diaphragms whose resulting motion activates a pneumatic relay. Details can be found in the specialized literature (e.g., Ref. 15).

40.6   CONTROL LAWS

The control logic elements are designed to act on the actuating (error) signal to produce the control signal. The algorithm that is physically implemented for this purpose is the *control law* or *control action*. A nonzero error signal results from either a change in command or a disturbance. The general function of the controller is to keep the controlled variable near its desired value when these occur. More specifically, the control objectives might be stated as follows:

1. Minimize the steady-state error.
2. Minimize the settling time.
3. Achieve other transient specifications, such as minimizing the maximum overshoot.

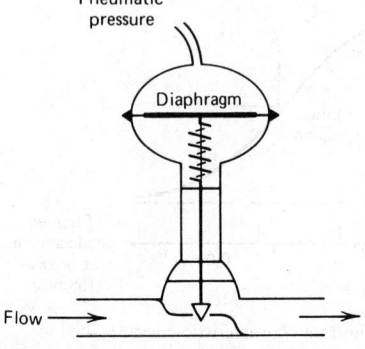

Fig. 40.21   Pneumatic flow-control valve.[11]

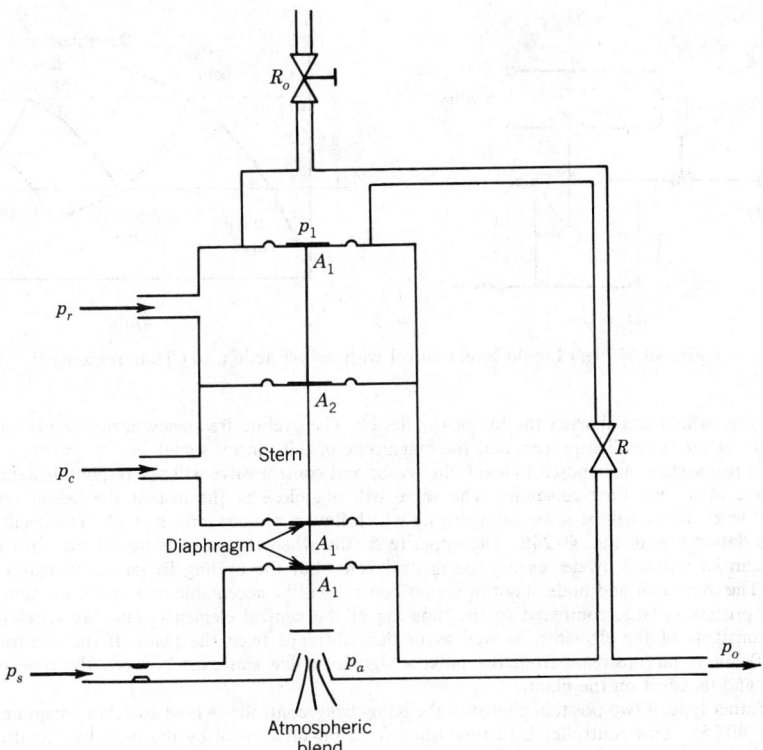

Fig. 40.22 Force-balance type pneumatic actuator.

In practice, the design specifications for a controller are more detailed. For example, the bandwidth might also be specified along with a safety margin for stability. We never know the numerical values of the system's parameters with true certainty, and some controller designs can be more sensitive to such parameter uncertainties than other designs. So a parameter sensitivity specification might also be included.

The following control laws form the basis of most control systems.

40.6.1 Two-Position Control

Two-position control is the most familiar type perhaps because of its use in home thermostats. The control output takes on one of two values. With the *on–off* controller, the controller output is either on or off (fully open or fully closed). The switching diagram for the on–off controller with hysteresis is shown in Fig. 40.23.

An example of an application of an on–off controller to a liquid-level system is shown in Fig. 40.24a. The time response shown in Fig. 40.24b with a solid line is for an ideal system in which the control valve acts instantaneously. The controlled variable cycles with an amplitude that depends on the width of the neutral zone or gap. This zone is provided to prevent frequent on–off switching, or

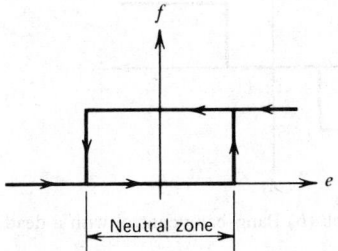

Fig. 40.23 Transfer characteristics of the on–off controller. The actuating error is $e = r - c$, where $r =$ set point, $c =$ controlled variable, and $f =$ control signal.[11]

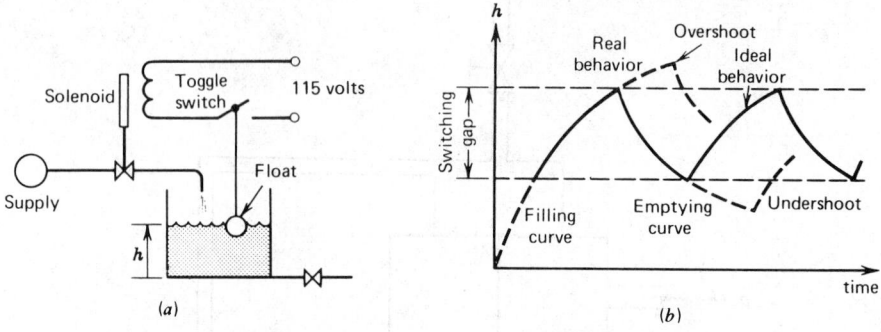

Fig. 40.24   (a) Liquid-level control with on–off action. (b) Time response.[11]

chattering, which can shorten the life of the device. The cycling frequency also depends on the time constant of the controlled process and the magnitude of the control signal.

In a real system, as opposed to ideal, the sensor and control valve will not respond instantaneously, but have their own time constants. The valve will not close at the instant the height reaches the desired level. There will be some delay during which flow continues into the tank. The result is shown by the dotted line in Fig. 40.24b. The opposite occurs when the valve is turned on. This unwanted effect can be reduced by decreasing the neutral zone, but the cycling frequency increases if this is done. The overshoot and undershoot in on–off control will be acceptable only when the time constant of the process is large compared to the time lag of the control elements. This lag is related to the time constants of the elements, as well as to their distance from the plant. If the control valve in Fig. 40.24a is far upstream from the tank, a significant lag can exist between the time of control action and its effect on the plant.

Another type of two-position control is the *bang–bang* controller whose switching diagram is shown in Fig. 40.25a. This controller is distinguished from on–off control by the fact that the direction or sign of the control signal can have two values. A motor with constant torque that can reverse quickly might be modeled as a bang–bang device. Because such perfect switching is impossible, a more accurate model would include a dead zone (Fig. 40.25b). When the error is within the zone, the controller output is zero.

40.6.2   Proportional Control

Two-position control is acceptable for many applications in which the requirements are not too severe. However, many situations require finer control.

Consider the tank system shown in Fig. 40.24a. To replace the two-position controller, we might try setting the control valve manually to achieve a flow rate that balances the system at the desired level. We might then add a controller that adjusts this setting in proportion to the deviation of the level from the desired value. This is *proportional control,* the algorithm in which the change in the control signal is proportional to the error. By convention block diagrams for controllers are drawn in terms of the deviations from a zero-error equilibrium condition. Applying this convention to the general terminology of Fig. 40.6, we see that proportional control is described by

$$F(s) = K_p E(s)$$

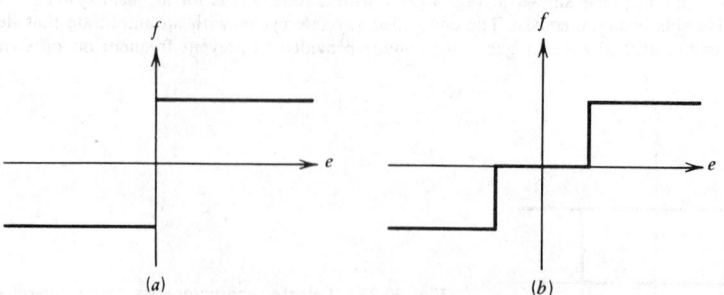

Fig. 40.25   Transfer characteristics: (a) Ideal bang–bang control. (b) Bang–bang control with a dead zone. The control signal is f; the error signal is e. [11]

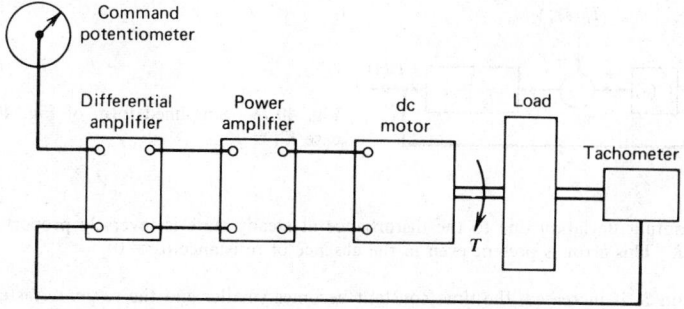

Fig. 40.26 Velocity-control system using a dc motor.[11]

where $F(s)$ is the deviation in the control signal and K_p is the *proportional gain*. If the total valve displacement is $y(t)$ and the manually created displacement is x, then

$$y(t) = K_p e(t) + x$$

The percent change in error needed to move the valve full scale is the *proportional band*. It is related to the gain as

$$K_p = \frac{100}{\text{band }\%}$$

The zero-error valve displacement x is the *manual reset.*

Proportional Control of a First-Order System

To investigate the behavior of proportional control, consider the speed-control system shown in Fig. 40.26; it is identical to the position controller shown in Fig. 40.6, except that a tachometer replaces the feedback potentiometer. We can combine the amplifier gains into one, denoted K_p. The system is thus seen to have proportional control. We assume the motor is field-controlled with a negligible inductance. The disturbance is a torque T_d, for example, resulting from friction. Choose the reference equilibrium condition to be $T_d = T = 0$ and $\omega_r = \omega = 0$. The block diagram is shown in Fig. 40.27. For a meaningful error signal to be generated, K_1 and K_2 should be chosen to be equal. With this simplification the diagram becomes that shown in Fig. 40.28, where $G(s) = K = K_1 K_p K_T / R$. A change in desired speed can be simulated by a unit step input for ω_r. For $\Omega_r(s) = 1/s$, the velocity approaches the steady-state value $\omega_{ss} = K/(c + K) < 1$. Thus the final value is less than the desired value of 1, but it might be close enough if the damping c is small. The time required to reach this value is approximately four time constants, or $4\tau = 4I/(c + K)$.

A sudden change in load torque can also be modeled by a unit step function $T_d(s) = 1/s$. The steady-state response due solely to the disturbance is $-1/(c + K)$. If $(c + K)$ is large, this error will be small.

The performance of the proportional control law thus far can be summarized as follows. For a first-order plant with step function inputs:

1. The output never reaches its desired value if resistance is present ($c \neq 0$), although it can be made arbitrarily close by choosing the gain K large enough. This is *offset* error.

2. The output approaches its final value without oscillation. The time to reach this value is inversely proportional to K.

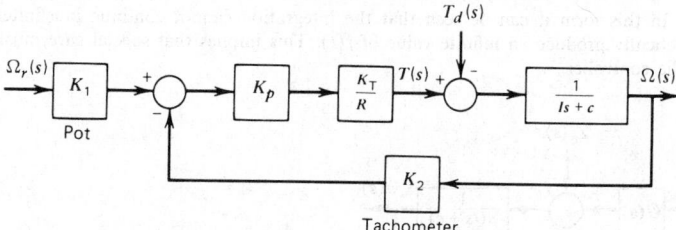

Fig. 40.27 Block diagram of the velocity-control system of Fig. 40.26.[11]

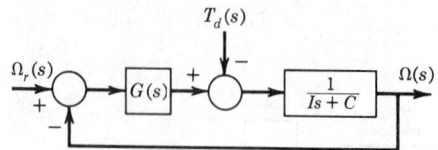

Fig. 40.28 Simplified form of Fig. 40.27 for the case $K_1 = K_2$.

3. The output deviation due to the disturbance at steady state is inversely proportional to the gain K. This error is present even in the absence of resistance ($c = 0$).

As the gain K is increased, the time constant becomes smaller and the response faster. Thus the chief disadvantage of proportional control is that it results in steady-state errors and can only be used when the gain can be selected large enough to reduce the effect of the largest expected disturbance. Since proportional control gives zero error only for one load condition (the reference equilibrium), the operator must change the manual reset by hand (hence the name).

An advantage to proportional control is that the control signal responds to the error instantaneously (in theory at least). It is used in applications requiring rapid action. Processes with time constants too small for the use of two-position control are likely candidates for proportional control.

The results of this analysis can be applied to any first-order system of the form in Fig. 40.28.

Proportional Control of a Second-Order System

Proportional control of a neutrally stable second-order plant is represented by the position controller of Fig. 40.6 if the amplifier transfer function is a constant $G_a(s) = K_a$. Let the motor transfer function be $G_m(s) = K_T/R$ as before. The modified block diagram is given in Fig. 40.29 with $G(s) = K = K_1K_aK_T/R$. The closed-loop system is stable if I, c, and K are positive. For no damping ($c = 0$), the closed-loop system is neutrally stable. With no disturbance and a unit step command, $\Theta_r(s) = 1/s$, the steady-state output is $\theta_{ss} = 1$. The offset error is thus zero if the system is stable ($c > 0$, $K > 0$). The steady-state output deviation due to a unit step disturbance is $-1/K$. This deviation can be reduced by choosing K large. The transient behavior is indicated by the damping ratio, $\zeta = c/2\sqrt{IK}$.

For slight damping the response to a step input will be very oscillatory and the overshoot large. The situation is aggravated if the gain K is made large to reduce the deviation due to the disturbance. We conclude therefore that proportional control of this type of second-order plant is not a good choice unless the damping constant c is large. We will see shortly how to improve the design.

40.6.3 Integral Control

The offset error that occurs with proportional control is a result of the system reaching an equilibrium in which the control signal no longer changes. This allows a constant error to exist. If the controller is modified to produce an increasing signal as long as the error is nonzero, the offset might be eliminated. This is the principle of *integral control*. In this mode the change in the control signal is proportional to the *integral* of the error. In the terminology of Fig. 40.7, this gives

$$F(s) = \frac{K_I}{s}E(s) \tag{40.16}$$

where $F(s)$ is the deviation in the control signal and K_I is the *integral gain*. In the time domain the relation is

$$f(t) = K_I \int_0^t e(t)\, dt \tag{40.17}$$

if $f(0) = 0$. In this form it can be seen that the integration cannot continue indefinitely because it would theoretically produce an infinite value of $f(t)$. This implies that special care must be taken to reinitialize the controller.

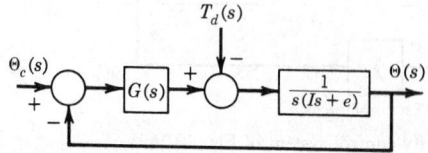

Fig. 40.29 Position servo.

Integral Control of a First-Order System

Integral control of the velocity in the system of Fig. 40.26 has the block diagram shown in Fig. 40.28, where $G(s) = K/s$, $K = K_1 K_I K_T / R$. The integrating action of the amplifier is physically obtained by the techniques to be presented in Section 40.8. The control system is stable for I, c, and K positive. For a unit step command input, $\omega_{ss} = 1$; so the offset error is zero. For a unit step disturbance, the steady-state deviation is zero if the system is stable. The steady-state performance using integral control is thus excellent, for this plant with step inputs.

The damping ratio is $\zeta = c/2\sqrt{IK}$. For slight damping the response will be oscillatory rather than exponential as with proportional control. Improved steady-state performance has thus been obtained at the expense of degraded transient performance. The conflict between steady-state and transient specifications is a common theme in control system design. As long as the system is underdamped, the time constant is $\tau = 2I/c$ and is not affected by the gain K, which only influences the oscillation frequency in this case. It might be physically possible to make K small enough so that $\zeta \geqslant 1$, and the nonoscillatory feature of proportional control recovered, but the response would tend to be sluggish. Transient specifications for fast response generally require $\zeta < 1$. The difficulty with $\zeta < 1$ is that τ is fixed by c and I. If c and I are such that $\zeta < 1$, then τ is large if $I \gg c$.

Integral Control of a Second-Order System

Proportional control of the position servomechanism in Fig. 40.29 gives a nonzero steady-state deviation due to the disturbance. Integral control $[G(s) = K/s]$ applied to this system results in the command transfer function

$$\frac{\Theta(s)}{\Theta_r(s)} = \frac{K}{Is^3 + cs^2 + K} \tag{40.18}$$

With the Routh criterion we immediately see that the system is not stable because of the missing s term. Integral control is useful in improving steady-state performance, but in general it does not improve and may even degrade transient performance. Improperly applied, it can produce an unstable control system. It is best used in conjunction with other control modes.

40.6.4 Proportional-Plus-Integral Control

Integral control raised the order of the system by one in the preceding examples, but did not give a characteristic equation with enough flexibility to achieve acceptable transient behavior. The instantaneous response of proportional control action might introduce enough variability into the coefficients of the characteristic equation to allow both steady-state and transient specifications to be satisfied. This is the basis for using *proportional-plus-integral control* (PI control). The algorithm for this two-mode control is

$$F(s) = K_p E(s) + \frac{K_I}{s} E(s) \tag{40.19}$$

The integral action provides an automatic, not manual, reset of the controller in the presence of a disturbance. For this reason it is often called *reset action*.

The algorithm is sometimes expressed as

$$F(s) = K_p \left(1 + \frac{1}{T_I s} \right) E(s) \tag{40.20}$$

where T_I is the *reset time*. The reset time is the time required for the integral action signal to equal that of the proportional term, if a constant error exists (a hypothetical situation). The reciprocal of reset time is expressed as repeats per minute and is the frequency with which the integral action repeats the proportional correction signal.

The proportional control gain must be reduced when used with integral action. The integral term does not react instantaneously to a zero-error signal but continues to correct, which tends to cause oscillations if the designer does not take this effect into account.

PI Control of a First-Order System

PI action applied to the speed controller of Fig. 40.26 gives the diagram shown in Fig. 40.28 with $G(s) = K_p + K_I/s$. The gains K_p and K_I are related to the component gains as before. The system is stable for positive values of K_p and K_I. For $\Omega_r(s) = 1/s$, $\omega_{ss} = 1$, and the offset error is zero as with only integral action. Similarly, the deviation due to a unit step disturbance is zero at steady state. The damping ratio is $\zeta = (c + K_p)/2\sqrt{IK_I}$. The presence of K_p allows the damping ratio to be selected without fixing the value of the dominant time constant. For example, if the system is underdamped, the time constant is $\tau = 2I/(c + K_p)$ $(\zeta < 1)$. The gain K_p can be picked to obtain the desired time constant, while K_I is used to set the damping ratio. A similar flexibility exists if $\zeta = 1$. Complete description of the transient response requires that the numerator dynamics present in the transfer functions be accounted for (Ref. 11, Section 5.7).

PI Control of a Second-Order System

Integral control for the position servomechanism of Fig. 40.29 resulted in a third-order system that is unstable. With a proportional term, the diagram becomes that of Fig. 40.29, with $G(s) = K_p + K_I/s$. The steady-state performance is acceptable as before if the system is assumed to be stable. This is true if the Routh criterion is satisfied; that is, if I, c, K_p, and K_I are positive and $cK_p - IK_I > 0$. The difficulty here occurs when the damping is slight. For small c, the gain K_p must be large in order to satisfy the last condition, and this can be difficult to implement physically. Such a condition can also result in an unsatisfactory time constant. The root-locus method of Section 40.10 provides the tools for analyzing this design further.

40.6.5 Derivative Control

Integral action tends to produce a control signal even after the error has vanished, which suggests that the controller be made aware that the error is approaching zero. One way to accomplish this is to design the controller to react to the derivative of the error with the *derivative control* law.

$$F(s) = K_D s E(s) \tag{40.21}$$

where K_D is the *derivative gain*. This algorithm is also called *rate action*. It is used to damp out oscillations. Since it depends only on the error rate, derivative control should never be used alone. When used with proportional action, the following PD-control algorithm results:

$$\begin{aligned} F(s) &= (K_p + K_D s)E(s) \\ &= K_p(1 + T_D s)E(s) \end{aligned} \tag{40.22}$$

where T_D is the *rate time* or *derivative time*. With integral action included, the proportional-plus-integral-plus-derivative (PID) control law is obtained.

$$F(s) = \left(K_p + \frac{K_I}{s} + K_D s \right) E(s) \tag{40.23}$$

This is a three-mode controller.

PD Control of a Second-Order System

Design of a controller with all three modes increases the cost of the system (except perhaps for digital systems, where the only change is a software modification). There are applications of the position servomechanism in which a nonzero deviation resulting from the disturbance can be tolerated, but an improvement in transient response over the proportional control result is desired. Integral action would not be required and rate action can be substituted to improve the transient response. Application of PD control to this system gives the block diagram of Fig. 40.29 with $G(s) = K_p + K_D s$.

The system is stable for positive values of K_D and K_p. The presence of rate action does not affect the steady-state response, and the steady-state results are identical to those with P control; namely, zero offset error and a deviation of $-1/K_p$ due to the disturbance. The damping ratio is $\zeta = (c + K_D)/2\sqrt{IK_p}$.

For P control, $\zeta = c/2\sqrt{IK_p}$. Introduction of rate action allows the proportional gain K_p to be selected large to reduce the steady-state deviation, while K_D can be used to achieve an acceptable damping ratio. The rate action also helps to stabilize the system by adding damping (if $c = 0$ the system with P control is not stable).

The equivalent of derivative action can be obtained by using a tachometer to measure the angular velocity of the load. The block diagram is shown in Fig. 40.30. The gain of the amplifier–motor–potentiometer combination is K_1, and K_2 is the tachometer gain. The advantage of this system is that it does not require signal differentiation, which is difficult to implement physically. The gains K_1 and K_2 can be chosen to yield the desired damping ratio and steady-state deviation as was done with K_p and K_D.

40.6.6 PID Control

The position servomechanism design with PI control is not completely satisfactory because of the difficulties encountered when the damping c is small. This problem can be solved by the use of the full PID-control law, as shown in Fig. 40.29 with $G(s) = K_p + K_D s + K_I/s$.

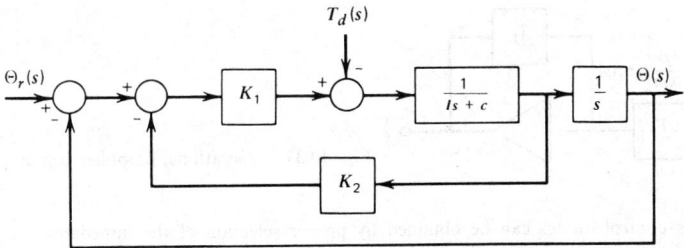

Fig. 40.30 Tachometer feedback arrangement to replace PD control for the position servo.[11]

A stable system results if all gains are positive and if $(c + K_D)K_p - IK_I > 0$. The presence of K_D relaxes somewhat the requirement that K_p be large to achieve stability. The steady-state errors are zero, and the transient response can be improved because three of the coefficients of the characteristic equation can be selected. To make further statements requires the root locus technique.

Proportional, integral, and derivative actions and their various combinations are not the only control laws possible, but they are the most common. It has been estimated that 90% of all controllers are of the PI type. This percentage will probably decrease as digital control with its great flexibility becomes more widely used. But the PI and PID controllers for some time will remain the standard against which any new designs must compete.

The conclusions reached concerning the performance of the various control laws are strictly true only for the plant model forms considered. These are the first-order model without numerator dynamics and the second-order model with a root at $s = 0$ and no numerator zeros. The analysis of a control law for any other linear system follows the preceding pattern. The overall system transfer functions are obtained, and all of the linear system analysis techniques can be applied to predict the system's performance. If the performance is unsatisfactory, a new control law is tried and the process repeated. When this process fails to achieve an acceptable design, more systematic methods of altering the system's structure are needed; they are discussed in later sections.

We have used step functions as the test signals because they are the most common and perhaps represent the severest test of system performance. Impulse, ramp, and sinusoidal test signals are also employed. The type to use should be made clear in the design specifications.

40.7 CONTROLLER HARDWARE

The control law must be implemented by a physical device before the control engineer's task is complete. The earliest devices were purely kinematic and were mechanical elements such as gears, levers, and diaphragms that usually obtained their power from the controlled variable. Most controllers now are analog electronic, hydraulic, pneumatic, or digital electronic devices. We now consider the analog electronic type. Digital control is taken up at the end of the chapter.

40.7.1 Feedback Compensation and Controller Design

Most controllers that implement versions of the PID algorithm are based on the following feedback principle. Consider the single loop system shown in Fig. 40.1. If the open-loop transfer function is large enough that $|G(s)H(s)| \gg 1$, the closed-loop transfer function is approximately given by

$$T(s) = \frac{G(s)}{1 + G(s)H(s)} \cong \frac{G(s)}{G(s)H(s)} = \frac{1}{H(s)}$$

The principle states that a power unit $G(s)$ can be used with a feedback element $H(s)$ to create any desired transfer function $T(s)$. The power unit must have a gain high enough that $|G(s)H(s)| \gg 1$, and the feedback elements must be selected so that $H(s) = 1/T(s)$. This principle was used in Section 40.1 to explain the design of a feedback amplifier.

40.7.2 Electronic Controllers

The op amp is a high-gain amplifier with a high input impedance. A diagram of an op amp with feedback and input elements with impedances $T_f(s)$ and $T_i(s)$ is shown in Fig. 40.31. An approximate relation is

$$\frac{E_o(s)}{E_i(s)} = -\frac{T_f(s)}{T_i(s)}$$

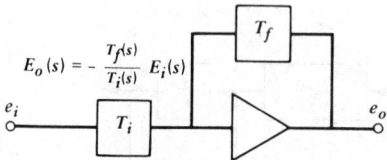

Fig. 40.31   Operational amplifier (op amp).[11]

The various control modes can be obtained by proper selection of the impedances. Because of the sign reversal property of the op amp, an *inverter* is often required.

A proportional controller can be obtained with two resistors, as shown in Fig. 40.32. An inverter is such a circuit with $R_f = R_i$. This multiplier circuit can be modified to act as an adder (Fig. 40.33).

PI control can be implemented with the circuit of Fig. 40.34. Many industrial controllers provide the operator with a choice of control modes, and the operator can switch from one mode to another when the process characteristics or control objectives change. When a switch occurs, it is necessary to provide any integrators with the proper initial voltages, or else undesirable transients will occur with the integrator is switched into the system. Commercially available controllers usually have built-in circuits for this purpose. A diagram of one such design for PI control is given in Ref. 12.

In practice the final control elements are always incapable of delivering energy to the controlled system above a certain rate. Also, integral control can suffer from a nonlinear effect called *reset windup*, which is in part caused by the finite capacity of the final control elements. This problem can be eliminated by limiting the output of the controller so that it cannot command the final control elements to deliver more power than they can. For circuits to do this, see Refs. 12 and 15.

In theory a differentiator can be created by interchanging the resistance and capacitance in the integrator. The difficulty with this design is that no electrical signal is "pure." Contamination always exists as a result of voltage spikes, ripple, and other transients generally categorized as "noise." These high-frequency signals have large slopes compared with the more slowly varying primary signal, and thus they will dominate the output of the differentiator. In practice this problem is solved by filtering out high-frequency signals either with a low-pass filter inserted in cascade with the differentiator, or by using a redesigned differentiator such as the one shown in Fig. 40.35. For the ideal PD controller, $R_1 = 0$. The attenuation curve for the ideal controller breaks upward at $\omega = 1/R_2C$ with a slope of 20 db/decade. The curve for the practical controller does the same but then becomes flat for $\omega > (R_1 + R_2)/R_1R_2C$. This provides the required limiting effect at high frequencies.

PID control can be implemented by joining the PI and PD controllers in parallel, but this is expensive because of the number of op amps and power supplies required. Instead, the usual implementation is that shown in Fig. 40.36. The circuit limits the effect of frequencies above $\omega = 1/\beta R_1C_1$. When $R_1 = 0$, ideal PID control results. This is sometimes called the *noninteractive* algorithm because the effect of each of the three modes is additive, and they do not interfere with one another. The form given for $R_1 \neq 0$ is the *real* or *interactive* algorithm. This name results from the fact that historically it was difficult to implement noninteractive PID control with mechanical or pneumatic devices.

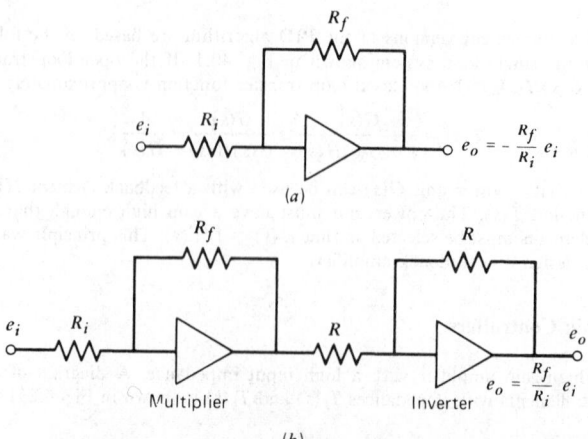

Fig. 40.32   Op-amp implementation of proportional control.[11]

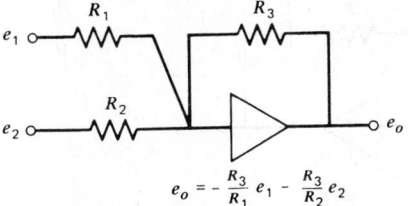

$$e_o = -\frac{R_3}{R_1}e_1 - \frac{R_3}{R_2}e_2$$

Fig. 40.33   Op-amp adder circuit.[11]

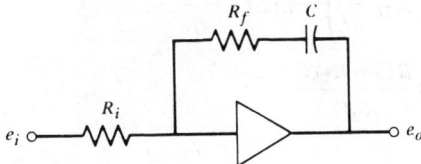

Fig. 40.34   Op-amp implementation of PI control.[11]

40.7.3   Pneumatic Controllers

The nozzle-flapper introduced in Section 40.5 is a high-gain device that is difficult to use without modification. The gain K_f is known only imprecisely and is sensitive to changes induced by temperature and other environmental factors. Also, the linear region over which Eq. (40.14) applies is very small. However, the device can be made useful by compensating it with feedback elements. PI and PID control can be obtained with two opposing bellows in the feedback path, as shown in Fig. 40.37. PI control is obtained if the inlet resistance to bellows 1 is zero ($R_D = 0$). The full PID-control law is obtainable when both resistances are nonzero and $R_I \gg R_D$.

40.7.4   Hydraulic Controllers

The basic unit for synthesis of hydraulic controllers is the hydraulic servomotor. The nozzle-flapper concept is also used in hydraulic controllers.[8] A PI controller is shown in Fig. 40.38. It can be modified for P-action. Derivative action has not seen much use in hydraulic controllers. This action supplies damping to the system, but hydraulic systems are usually highly damped intrinsically because of the viscous working fluid. PI control is the algorithm most commonly implemented with hydraulics.

40.8   FURTHER CRITERIA FOR GAIN SELECTION

Once the form of the control law has been selected, the gains must be computed in light of the performance specifications. In the examples of the PID family of control laws in Section 40.6, the damping ratio, dominant time constant, and steady-state error were taken to be the primary indicators of system performance in the interest of simplicity. In practice, the criteria are usually more detailed. For example, the rise time and maximum overshoot, as well as the other transient response specifications

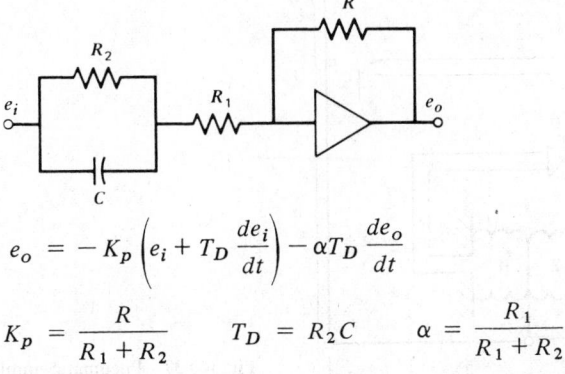

$$e_o = -K_p\left(e_i + T_D\frac{de_i}{dt}\right) - \alpha T_D\frac{de_o}{dt}$$

$$K_p = \frac{R}{R_1 + R_2} \qquad T_D = R_2 C \qquad \alpha = \frac{R_1}{R_1 + R_2}$$

Fig. 40.35   Practical op-amp implementation of PD control.[11]

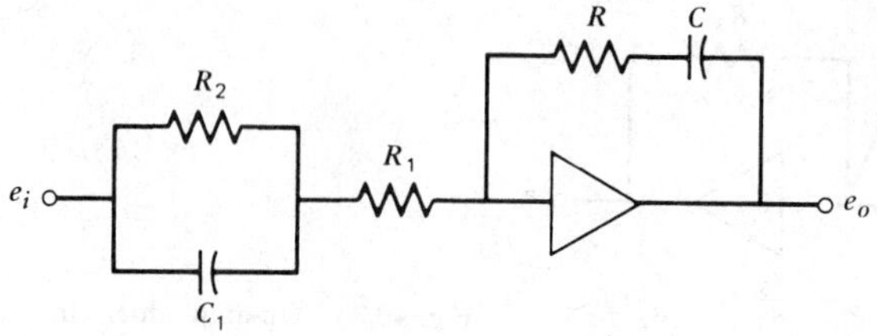

$$e_o = -\left(K_p e_i + K_I \int_0^t e_i\,dt + K_D \frac{de_i}{dt}\right) - \beta R_1 C_1 \frac{de_o}{dt}$$

$$\beta = \frac{R_2}{R_1 + R_2} \qquad K_p = \beta \frac{RC + R_2 C_1}{R_2 C}$$

$$K_I = \frac{\beta}{R_2 C} \qquad K_D = \beta R C_1$$

Fig. 40.36 Practical op-amp implementation of PID control.[11]

of the previous chapter, may be encountered. Requirements can also be stated in terms of frequency response characteristics, such as bandwidth, resonant frequency, and peak amplitude. Whatever specific form they take, a complete set of specifications for control system performance generally should include the following considerations, for given forms of the command and disturbance inputs:

1. Equilibrium specifications.
   (a) Stability.
   (b) Steady-state error.
2. Transient specifications.
   (a) Speed of response.
   (b) Form of response.
3. Sensitivity specifications.
   (a) Sensitivity to parameter variations.
   (b) Sensitivity to model inaccuracies.
   (c) Noise rejection (bandwidth, etc.).

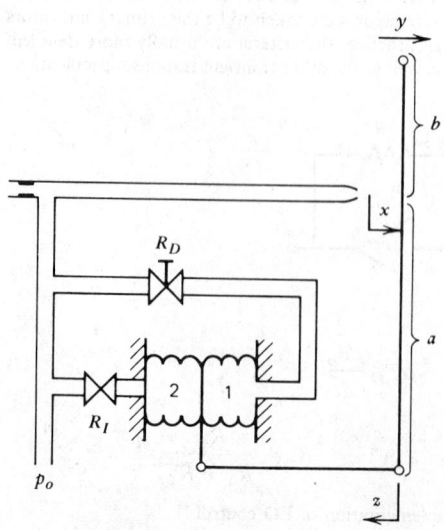

Fig. 40.37 Pneumatic implementation of PID control.[11]

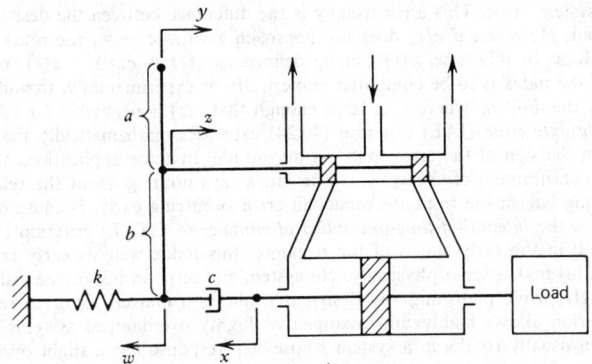

Fig. 40.38 Hydraulic implementation of PI control.[11]

4. Nonlinear effects.
   (a) Stability.
   (b) Final control element capabilities.

In addition to these performance stipulations, the usual engineering considerations of initial cost, weight, maintainability, etc., must be taken into account. The considerations are highly specific to the chosen hardware, and it is difficult to deal with such issues in a general way.

Two approaches exist for designing the controller. The proper one depends on the quality of the analytical description of the plant to be controlled. If an accurate model of the plant is easily developed, the approach is to design a specialized controller for the particular application. The range of adjustment of controller gains in this case can usually be made small because the accurate plant model allows the gains to be precomputed with confidence. This technique reduces the cost of the controller and can often be applied to electromechanical systems.

The second approach is used when the plant is relatively difficult to model, which is often the case in process control. A standard controller with several control modes and wide ranges of gains is used, and the proper mode and gain settings are obtained by testing the controller on the process in the field. This approach should be considered when the cost of developing an accurate plant model might exceed the cost of controller tuning in the field. Of course, the plant must be available for testing for this approach to be feasible.

40.8.1 Performance Indices

The performance criteria encountered thus far require a set of conditions to be specified—for example, one for steady-state error, one for damping ratio, and one for the dominant time constant. If there are many such conditions, and if the system is of high order with several gains to be selected, the design process can get quite complicated because transient and steady-state criteria tend to drive the design in different directions. An alternative approach is to specify the system's desired performance by means of one analytical expression called a *performance index*. Powerful analytical and numerical methods are available that allow the gains to be systematically computed by minimizing (or maximizing) this index.

To be useful a performance index must be selective. The index must have a sharply defined extremum in the vicinity of the gain values that give the desired performance. If the numerical value of the index does not change very much for large changes in the gains from their optimal values, the index will not be selective.

Any practical choice of a performance index must be easily computed, either analytically, numerically, or experimentally. Four common choices for an index are the following:

$$J = \int_0^\infty |e(t)|\, dt \qquad \text{(IAE)} \qquad\qquad (40.24)$$

$$J = \int_0^\infty t|e(t)|\, dt \qquad \text{(ITAE)} \qquad\qquad (40.25)$$

$$J = \int_0^\infty e^2(t)\, dt \qquad \text{(ISE)} \qquad\qquad (40.26)$$

$$J = \int_0^\infty te^2(t)\, dt \qquad \text{(ITSE)} \qquad\qquad (40.27)$$

1188

where $e(t)$ is the system error. This error usually is the difference between the desired and the actual values of the output. However, if $e(t)$ does not approach zero as $t \to \infty$, the preceding indices will not have finite values. In this case, $e(t)$ can be defined as $e(t) = c(\infty) - c(t)$, where $e(t)$ is the output variable. If the index is to be computed numerically or experimentally, the infinite upper limit can be replaced by the limit t_f, where t_f is large enough that $e(t)$ is negligible for $t > t_f$.

The *integral absolute-error* (IAE) criterion (40.24) expresses mathematically that the designer is not concerned with the sign of the error, only its magnitude. In some applications the IAE criterion describes the fuel consumption of the system. The index says nothing about the relative importance of an error occurring late in the response versus an error occurring early. Because of this, the index is not as selective as the *integral-of-time-multiplied absolute-error* (ITAE) criterion (40.25). Since the multiplier t is small in the early stages of the response, this index weights early errors less heavily than later errors. This makes sense physically. No system can respond instantaneously, and the index is lenient accordingly, while punishing any design that allows a nonzero error to remain for a long time. Neither criterion allows highly underdamped or highly overdamped systems to be optimum. The ITAE criterion usually results in a system whose step response has a slight overshoot and well-damped oscillations.

The *integral squared-error* (ISE) and *integral-of-time-multiplied squared-error* (ITSE) criteria are analogous to the IAE and ITAE criteria, except that the square of the error is employed, for three reasons: (1) in some applications, the squared error represents the system's power consumption; (2) squaring the error weights large errors much more heavily than small errors; (3) the squared error is much easier to handle analytically. The derivative of a squared term is easier to compute than that of an absolute value and does not have a discontinuity at $e = 0$. These differences are important when the system is of high order with multiple error terms.

The closed-form solution for the response is not required to evaluate a performance index. For a given set of parameter values, the response and the resulting index value can be computed numerically. Using systematic search procedures such as those given in Ref. 6 the optimum solution can be obtained; this makes the procedure suitable for use with nonlinear systems.

40.8.2 The Ziegler–Nichols Rules

The difficulty of obtaining accurate transfer function models for some processes has led to the development of empirically based rules of thumb for computing the optimum gain values for a controller. Commonly used guidelines are the Ziegler–Nichols rules, which have proved so helpful that they are still in use 40 years after their development. The rules actually consist of two separate methods. The first method requires the open-loop step response of the plant, while the second uses the results of experiments performed with the controller already installed. While primarily intended for use with systems for which no analytical model is available, the rules are also helpful even when a model can be developed.

Ziegler and Nichols developed their rules from experiments and analysis of various industrial processes. Using the IAE criterion with a unit step response, they found that controllers adjusted according to the following rules usually had a step response that was oscillatory but with enough damping so that the second overshoot was less than 25% of the first (peak) overshoot. This is the *quarter-decay* criterion and is sometimes used as a specification.

The first method is the *process-reaction method* and relies on the fact that many processes have an open-loop step response like that shown in Fig. 40.39. This is the process *signature* and is characterized by two parameters, R and L. R is the slope of a line tangent to the steepest part of the response curve, and L is the time at which this line intersects the time axis. First- and second-order linear systems do not yield positive values for L, and so the method cannot be applied to such systems.

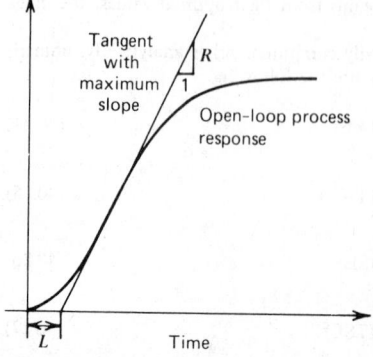

Fig. 40.39 Process signature for a unit step input.[11]

Table 40.2   The Ziegler–Nichols Rules

Controller transfer function $G(s) = K_p\left(1 + \dfrac{1}{T_I s} + T_D s\right)$

Control Mode	Process-Reaction Method	Ultimate-Cycle Method
P control	$K_p = \dfrac{1}{RL}$	$K_p = 0.5K_{pu}$
PI control	$K_p = \dfrac{0.9}{RL}$	$K_p = 0.45K_{pu}$
	$T_I = 3.3L$	$T_I = 0.83P_u$
PID control	$K_p = \dfrac{1.2}{RL}$	$K_p = 0.6K_{pu}$
	$T_I = 2L$	$T_I = 0.5P_u$
	$T_D = 0.5L$	$T_D = 0.125P_u$

However, third- and higher-order linear systems with sufficient damping do yield such a response. If so, the Ziegler–Nichols rules recommend the controller settings given in Table 40.2.

The *ultimate-cycle method* uses experiments with the controller in place. All control modes except proportional are turned off, and the process is started with the proportional gain K_p set at a low value. The gain is slowly increased until the process begins to exhibit sustained oscillations. Denote the period of this oscillation by P_u and the corresponding ultimate gain by K_{pu}. The Ziegler–Nichols recommendations are given in Table 40.2 in terms of these parameters. The proportional gain is lower for PI control than for P control and is higher for PID control because I action increases the order of the system and thus tends to destabilize it; thus a lower gain is needed. On the other hand, D action tends to stabilize the system; hence the proportional gain can be increased without degrading the stability characteristics. Because the rules were developed for a typical case out of many types of processes, final tuning of the gains in the field is usually necessary.

40.8.3   Nonlinearities and Controller Performance

All physical systems have nonlinear characteristics of some sort, although they can often be modeled as linear systems provided the deviations from the linearization reference condition are not too great. Under certain conditions, however, the nonlinearities have significant effects on the system's performance. One such situation can occur during the start-up of a controller if the initial conditions are much different from the reference condition for linearization. The linearized model is then not accurate, and nonlinearities govern the behavior. If the nonlinearities are mild, there might not be much of a problem. Where the nonlinearities are severe, such as in process control, special consideration must be given to start-up. Usually in such cases the control signal sent to the final control elements is manually adjusted until the system variables are within the linear range of the controller. Then the system is switched into automatic mode. Digital computers are often used to replace the manual adjustment process because they can be readily coded to produce complicated functions for the start-up signals. Care must also be taken when switching from manual to automatic. For example, the integrators in electronic controllers must be provided with the proper initial conditions.

40.8.4   Reset Windup

The saturation nonlinearity shown in Fig. 40.40 represents the practical behavior of many actuators and final control elements. For example, a motor–amplifier combination can produce a torque proportional to the input voltage over a limited range. However, no amplifier can supply an infinite current; there is a maximum current and thus a maximum torque that the system can produce. The final control elements are said to be overdriven when they are commanded by the controller to do something they cannot do. Since the limitations of the final control elements are ultimately due to the limited rate at which they can supply energy, it is important that all system performance specifications and controller designs be consistent with the energy-delivery capabilities of the elements to be used.

Any controller with integral action can exhibit the phenomenon called *reset windup* or *integrator buildup* when overdriven, if it is not properly designed. For a step change in set point, the proportional term responds instantly and saturates immediately if the set-point change is large enough. On the other hand, the integral term does not respond as fast. It integrates the error signal and saturates

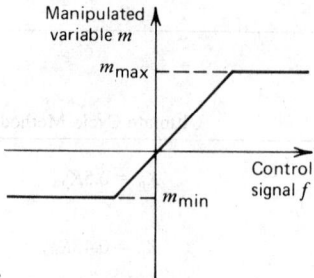

some time later if the error remains large for a long enough time. As the error decreases, the proportional term is no longer saturated. However, the integral term continues to increase as long as the error has not changed sign, and thus the manipulated variable remains saturated. Even though the output is very near its desired value, the manipulated variable remains saturated until after the error has reversed sign. The result can be a large overshoot in the response of the controlled variable.

40.8.5　Design with Power-Limited Elements

Reset windup can be avoided if we make sure the controller never allows the final control elements to saturate. One way of accomplishing this is to install limiters on the controller so that the voltages would not be allowed to exceed the value required to induce saturation of the actuator, and the overshoot in the response would be reduced. This technique is useful as a precaution in applications where the magnitudes of the inputs are difficult to predict in advance.

Another method is to select the gains so that saturation will never occur. This requires knowledge of the maximum input magnitude that the system will encounter. It can be generalized as follows (Ref. 11, Chap. 7):

1. For a model whose only nonlinearity results from saturation, assume the system is operating in the linear region, and compute the response for the most severe inputs the system is expected to deal with.

2. Compute the maximum gain values that can be allowed without inducing saturation. Steps 1 and 2 can be done analytically for first- and second-order systems. For higher-order systems, computer simulation can be used.

3. Use other criteria to decide whether the gains would be set to these maximum values or to lower values. Such criteria might be the desired values of time constants or damping ratio, or the performance indices (IAE, ISE, etc.).

40.9　COMPENSATION AND ALTERNATE STRUCTURES

If a control law can be found that satisfies most but not all of the specifications, a common design technique is to insert a *compensator* into the system. This is a device that alters the response of the controller so that the overall system will have satisfactory performance. The three categories of compensation techniques generally recognized are *series compensation, parallel* (or *feedback) compensation,* and *feed-forward compensation.* The three structures are loosely illustrated in Fig. 40.41, where we assume the final control elements have a unity transfer function. The transfer function of the controller is $G_1(s)$. The feedback elements are represented by $H(s)$ and the compensator by $G_c(s)$. We assume that the plant is unalterable, as is usually the case in control system design. The choice of compensation structure depends on what type of specifications must be satisfied. The physical devices used as compensators are similar to the pneumatic, hydraulic, and electrical devices treated previously.

40.9.1　Series Compensation

The most commonly used series compensators are the *lead,* the *lag,* and the *lead-lag* compensators. Electrical implementations of these are shown in Fig. 40.42. Other physical implementations are available. Generally, the lead compensator improves the speed of response; the lag decreases the steady-state error; and the lead-lag affects both. Graphical aids such as the root locus and frequency response plots are usually needed to design these compensators (Ref. 11, Chap. 8).

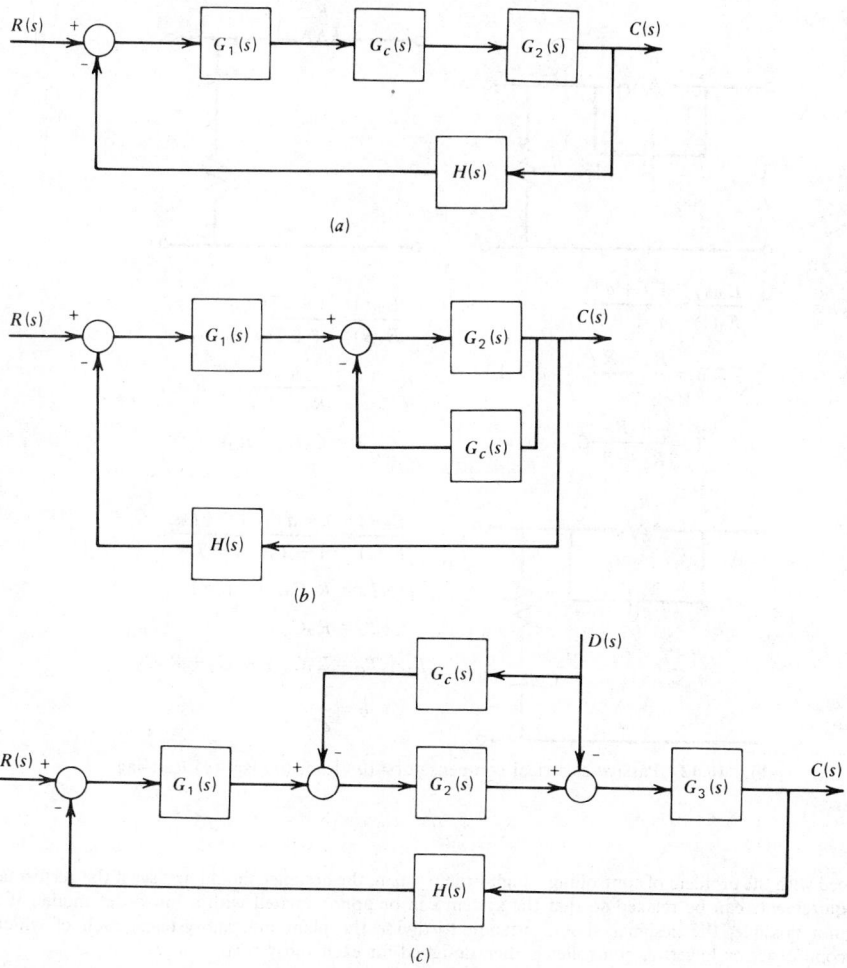

Fig. 40.41 General structures of the three compensation types: (*a*) series; (*b*) parallel (or feedback); (*c*) feed-forward. The compensator transfer function is $G_c(s)$.[11]

40.9.2 Feed-Forward Compensation

The control algorithms considered thus far have counteracted disturbances by using measurements of the output. One difficulty with this approach is that the effects of the disturbance must show up in the output of the plant before the controller can begin to take action. On the other hand, if we can measure the disturbance, the response of the controller can be improved by using the measurement to augment the control signal sent from the controller to the final control elements. This is the essence of feed-forward compensation.

40.9.3 Feedback Compensation and Cascade Control

The use of a tachometer to obtain velocity feedback, as in Fig. 40.30, is a case of feedback compensation. The feedback-compensation principle of Fig. 40.3 is another. Another form is *cascade control*, in which another controller is inserted within the loop of the original control system (Fig. 40.43). The new controller can be used to achieve better control of variables within the forward path of the system. Its set point is manipulated by the first controller.

Cascade control is frequently used when the plant cannot be satisfactorily approximated with a model of second order or lower. This is because the difficulty of analysis and control increases rapidly with system order. The characteristic roots of a second-order system can easily be expressed in analytical form. This is not so for third order or higher, and few general design rules are available. When

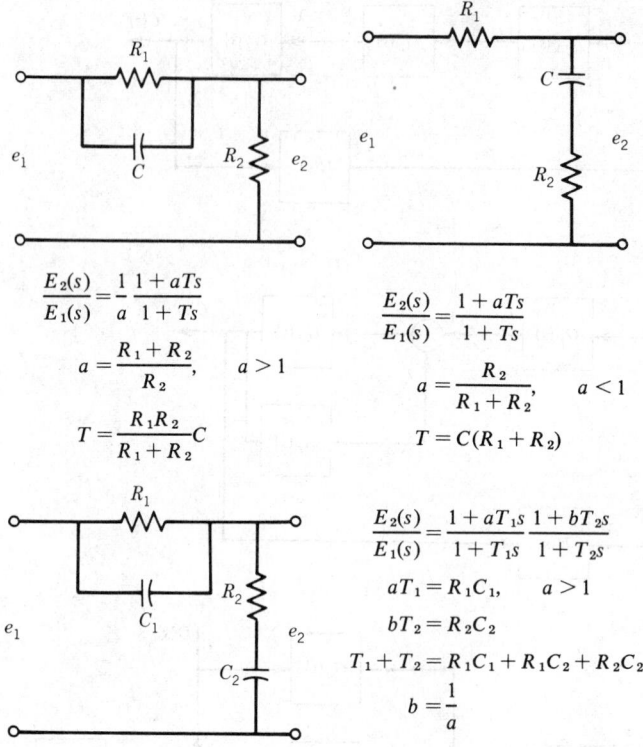

$$\frac{E_2(s)}{E_1(s)} = \frac{1}{a}\frac{1+aTs}{1+Ts}$$

$$a = \frac{R_1+R_2}{R_2}, \qquad a > 1$$

$$T = \frac{R_1 R_2}{R_1+R_2}C$$

$$\frac{E_2(s)}{E_1(s)} = \frac{1+aTs}{1+Ts}$$

$$a = \frac{R_2}{R_1+R_2}, \qquad a < 1$$

$$T = C(R_1+R_2)$$

$$\frac{E_2(s)}{E_1(s)} = \frac{1+aT_1s}{1+T_1s}\frac{1+bT_2s}{1+T_2s}$$

$$aT_1 = R_1C_1, \qquad a > 1$$

$$bT_2 = R_2C_2$$

$$T_1 + T_2 = R_1C_1 + R_1C_2 + R_2C_2$$

$$b = \frac{1}{a}$$

Fig. 40.42 Passive electrical compensators: (*a*) lead; (*b*) lag; (*c*) lead–lag.

faced with the problem of controlling a high-order system, the designer should first see if the performance requirements can be relaxed so that the system can be approximated with a low-order model. If this is not possible, the designer should attempt to divide the plant into subsystems, each of which is second order or lower. A controller is then designed for each subsystem.

40.9.4 State-Variable Feedback

There are techniques for improving system performance that do not fall entirely into one of the three compensation categories considered previously. In some forms these techniques can be viewed as a type of feedback compensation, while in other forms they constitute a modification of the control law. State-variable feedback (SVFB) is a technique that uses information about all the system's state variables to modify either the control signal or the actuating signal. These two forms are illustrated in Fig. 40.44. Both forms require that the state vector **x** be measurable or at least derivable from other information. Devices or algorithms used to obtain state variable information other than directly from measurements are variously termed *state reconstructors, estimators, observers,* or *filters* in the literature.

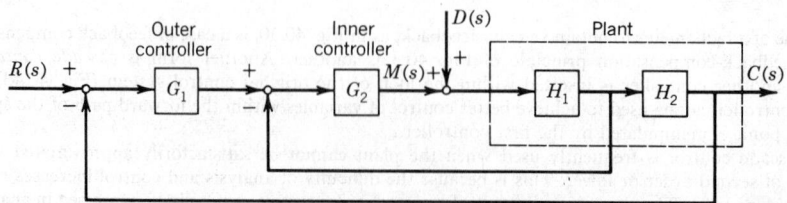

Fig. 40.43 Cascade control structure.

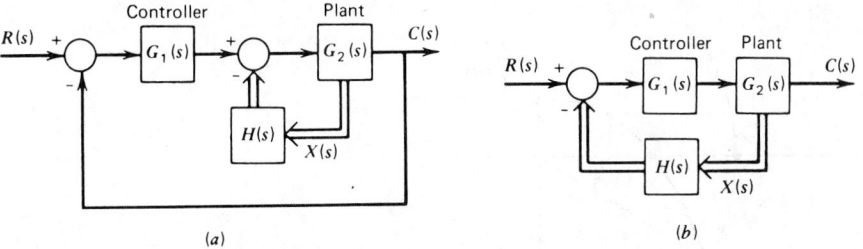

Fig. 40.44 Two forms of state-variable feedback: (a) internal compensation of the control signal; (b) modification of the actuating signal.[11]

40.9.5 Pseudoderivative Feedback (PDF)

Pseudoderivative feedback (PDF) is an extension of the velocity feedback compensation concept of Fig. 40.30.[11,12] It uses integral action in the forward path, plus an internal feedback loop whose operator $H(s)$ depends on the plant (Fig. 40.45). For $G(s) = 1/(Is + c)$, $H(s) = K_1$. For $G(s) = 1/Is^2$, $H(s) = K_1 + K_2s$. The primary advantage of PDF is that it does not need derivative action in the forward path to achieve the desired stability and damping characteristics.

40.10 GRAPHICAL DESIGN METHODS

Higher-order models commonly arise in control systems design. For example, integral action is often used with a second-order plant, and this produces a third-order system to be designed. Although algebraic solutions are available for third- and fourth-order polynomials, these solutions are cumbersome for design purposes. Fortunately there exist graphical techniques to aid the designer. Frequency response plots of both the open- and closed-loop transfer functions are useful. The *Bode plot*, the *Nyquist plot*, and the *Nichols chart* all present the frequency response information in different forms. Each form has its own advantages. The *root locus plot* shows the location of the characteristic roots for a range ·of values of some parameters, such as a controller gain. A tabulation of these plots for typical transfer functions is given in the previous chapter (Fig. 39.8). The design of two-position and other nonlinear control systems is facilitated by the *describing function*, which is a linearized approximation based on the frequency response of the controller (see Section 39.8.4). Graphical design methods are discussed in more detail in Refs. 1, 3, 7, 9, 11, 13, and 14.

40.10.1 The Nyquist Stability Theorem

The *Nyquist stability theorem* is a powerful tool for linear system analysis. If the open-loop system has no poles with positive real parts, we can concentrate our attention on the region around the point $-1 + i0$ on the polar plot of the *open-loop* transfer function.

Figure 40.46 shows the polar plot of the open-loop transfer function of an arbitrary system which is assumed to be open-loop stable. The Nyquist stability theorem is stated as follows:

> The system is closed-loop stable if and only if the point $-1 + i0$ lies to the left of the open-loop Nyquist plot relative to an observer traveling along the plot in the direction of increasing frequency ω.

Therefore, the system described by Fig. 40.46 is closed-loop stable.

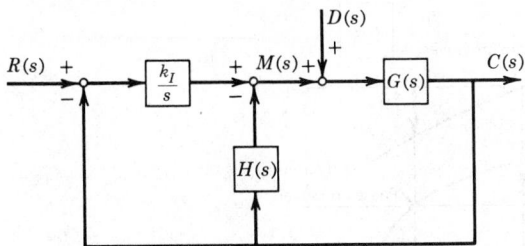

Fig. 40.45 Structure of pseudoderivative feedback (PDF).

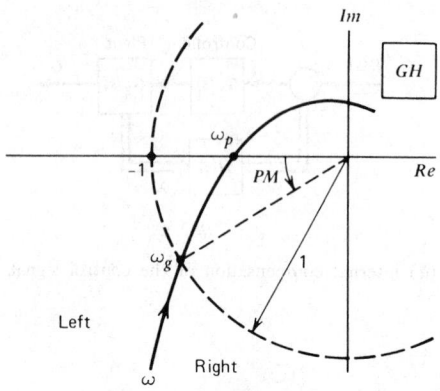

Fig. 40.46 Nyquist plot for a stable system.[11]

40.10.2 Phase and Gain Margins

The Nyquist theorem provides a convenient measure of the relative stability of a system. A measure of the proximity of the plot to the $-1 + i0$ point is given by the angle between the negative real axis and a line from the origin to the point where the plot crosses the unit circle (see Fig. 40.46). The frequency corresponding to this intersection is denoted ω_g. This angle is the *phase margin* (PM) and is positive when measured down from the negative real axis. The phase margin is the phase at the frequency ω_g where the magnitude ratio or "gain" of $G(i\omega)H(i\omega)$ is unity (0 db). The frequency ω_p, the phase crossover frequency, is the frequency at which the phase angle is $-180°$. The *gain margin* (GM) is the difference in decibels between the unity gain condition (0 db) and the value of $|GH|$ db at the phase crossover frequency. Thus

$$\text{gain margin} = -|G(i\omega_p)H(i\omega_p)| \text{ db} \tag{40.28}$$

A system is stable only if the phase and gain margins are both positive.

The phase and gain margins can be illustrated on the Bode plots shown in Fig. 40.47. The phase and gain margins can be stated as safety margins in the design specifications. A typical set of such specifications is as follows:

$$\text{gain margin} \geq 8 \text{ db} \quad \text{and} \quad \text{phase margin} \geq 30° \tag{40.29}$$

In common design situations, only one of these equalities can be met, and the other margin is allowed to be greater than its minimum value. It is not desirable to make the margins too large since this results in a low gain. The design will be sluggish and might have a large steady-state error. We note that another common set of specifications is

$$\text{gain margin} \geq 6 \text{ db} \quad \text{and} \quad \text{phase margin} \geq 40° \tag{40.30}$$

The 6 db limit corresponds to the quarter amplitude decay response obtained with the gain settings given by the Ziegler–Nichols ultimate-cycle method (Table 40.2).

40.10.3 Systems with Dead-Time Elements

The Nyquist theorem is particularly useful for systems with dead-time elements, especially when the plant is of an order high enough to make the root-locus method cumbersome. A delay D in either

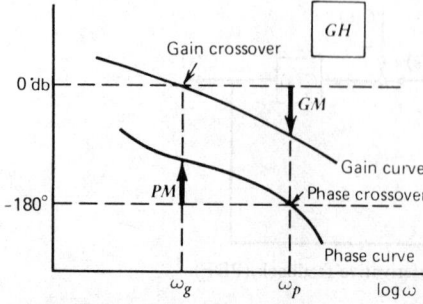

Fig. 40.47 Bode plot showing definitions of phase and gain margin.[11]

the manipulated variable or the measurement will result in an open-loop transfer function of the form

$$G(s)H(s) = e^{-Ds}P(s)$$

and

$$|G(i\omega)H(i\omega)| = |P(i\omega)||e^{-i\omega D}| = |P(i\omega)|$$
$$\angle G(i\omega)H(i\omega) = \angle P(i\omega) - \angle e^{-i\omega D}$$
$$= \angle P(i\omega) - \omega D$$

Thus the dead time decreases the phase proportionally to the frequency ω, but it does not change the gain curve. This makes the analysis of its effects easier to accomplish with the open-loop frequency response plot.

40.10.4 Open-Loop Design for PID Control

Some general comments can be made about the effects of proportional, integral, and derivative control actions on the phase and gain margins. P action does not affect the phase curve at all and thus can be used to raise or lower the open-loop gain curve until the specifications for the gain and phase margins are satisfied. If I action or D action is included, the proportional gain is selected last. Therefore, when using this approach to the design, it is best to write the PID algorithm with the proportional gain factored out, as

$$F(s) = K_p\left(1 + \frac{1}{T_I s} + T_D s\right)E(s) \tag{40.31}$$

D action affects both the phase and gain curves. Therefore, the selection of the derivative gain is more difficult than the proportional gain. The increase in phase margin due to the positive phase angle introduced by D action is partly negated by the derivative gain, which reduces the gain margin. Increasing the derivative gain increases the speed of response, makes the system more stable, and allows a larger proportional gain to be used to improve the system's accuracy. However, if the phase curve is too steep near $-180°$, it is difficult to use D action to improve the performance.

I action also affects both the gain and phase curves. It can be used to increase the open-loop gain at low frequencies. However, it lowers the phase crossover frequency ω_p and thus reduces some of the benefits provided by D action. If required, the D-action term is usually designed first, followed by I action and P action, respectively.

The classical design methods based on the Bode plots obviously have a large component of trial and error because usually both the phase and gain curves must be manipulated to achieve an acceptable design. Given the same set of specifications, two designers can use these methods and arrive at substantially different designs. Many rules of thumb and ad hoc procedures have been developed, but a general foolproof procedure does not exist. However, an experienced designer can often obtain a good design quickly with these techniques. The use of a computer plotting routine greatly speeds up the design process.

40.10.5 Closed-Loop Response and the Nichols Chart

Construction of the closed-loop response plots from the Nichols chart can be tedious. However, the chart is useful for determining the gain required to achieve a specified maximum (resonance) peak value m_{cp} in the closed-loop response. It can be shown with a Fourier series analysis that the frequency components of an input signal near the resonance peak dominate the system's response (Ref. 14, Section 9-4). Thus the specification of this peak value is an important design criterion. The control gain required to achieve the desired m_{cp} is found by plotting the closed-loop response line on the chart for a unity gain. The amount of gain that must be added (in decibels) is then found from the number of decibels by which the line must be translated to become tangent to the m_{cp} contour. A specification in common use is that M_{cp} should be less than 2 (m_{cp} less than 6 db). A typical value is $M_{cp} = 1.3$; that is, $m_{cp} = 2.28$ db. For many systems, this specification is approximately equivalent to the gain and phase margins specified by (40.29)—namely, $GM \geq 8$ db and $PM \geq 30°$.

40.10.6 Design with the Root Locus

The effect of D action as a series compensator can be seen with the root locus. The term $(1 + T_D s)$ in Fig. 40.31 can be considered as a series compensator to the proportional controller. The D action adds an open-loop zero at $s = -1/T_D$. For example, a plant with the transfer function $1/s(s + 1)(s + 2)$, when subjected to proportional control, has the root locus shown in Fig. 40.48a. If the proportional gain is high enough, the system will be unstable. If D action is used to put an open-loop zero at $s = -1.5$, the resulting root locus is given by Fig. 40.48b. The D action prevents the system from becoming unstable.

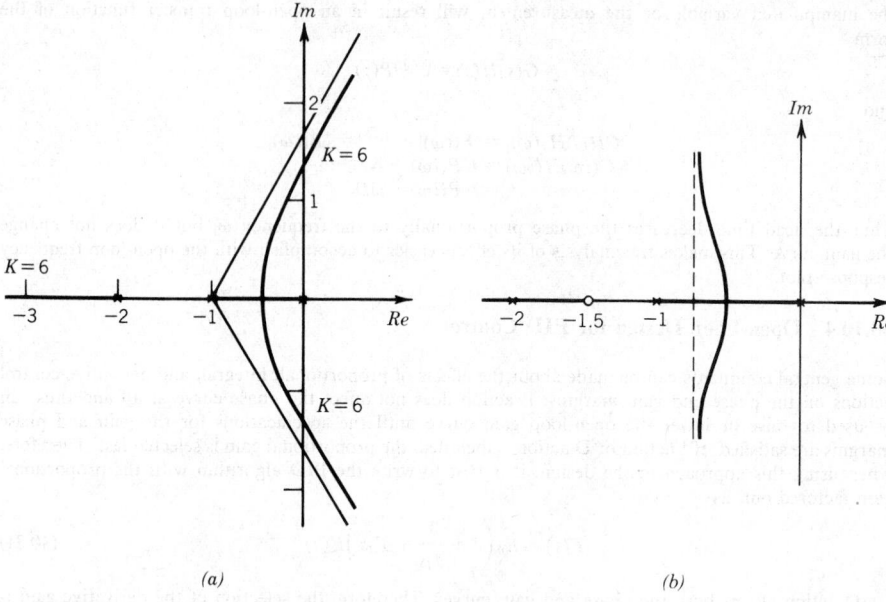

(a)   *(b)*

Fig. 40.48   (*a*) Root locus plot for $s(s + 1)(s + 2) + K = 0$, for $K \geq 0$. (*b*) The effect of PD control with $T_D = \frac{2}{3}$.

The integral action in PI control can be considered to add an open-loop pole at $s = 0$, and a zero at $s = -1/T_I$. Proportional control of the plant $1/(s + 1)(s + 2)$ gives a root locus like that shown in Fig. 40.49, with $a = 1$ and $b = 2$. A steady-state error will exist for a step input. With the PI compensator applied to this plant, the root locus is given by Fig. 40.48*b*, with $T_I = \frac{2}{3}$. The steady-state error is eliminated, but the response of the system has been slowed because the dominant paths of the root locus of the compensated system lie closer to the imaginary axis than for the uncompensated system.

As another example, let the plant-transfer function be

$$G_p(s) = \frac{1}{s^2 + a_2 s + a_1} \tag{40.32}$$

where $a_2 > 0$ and $a_1 > 0$. PI control applied to this plant gives the closed-loop command transfer function

$$T_1(s) = \frac{K_p s + K_I}{s^3 + a_2 s^2 + a_1 s + K_p s + K_I} \tag{40.33}$$

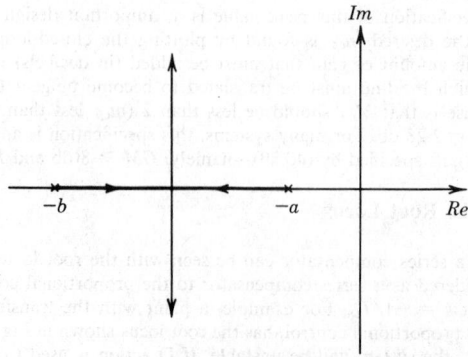

Fig. 40.49   Root-locus plot for $(s + a)(s + b) + K = 0$.

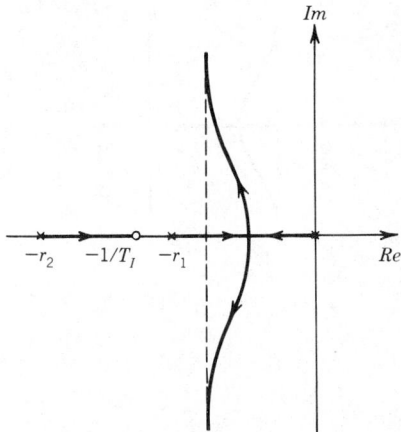

Fig. 40.50 Root-locus plot for PI control of a second-order plant.

Note that the Ziegler–Nichols rules cannot be used to set the gains K_p and K_I. The second-order plant, Eq. (40.32), does not have the S-shaped signature of Fig. 40.39, so the process-reaction method does not apply. The ultimate-cycle method requires K_I to be set to zero and the ultimate gain K_{pu} determined. With $K_I = 0$ in Eq. (40.33) the resulting system is stable for all $K_p > 0$, and thus a positive ultimate gain does not exist.

Take the form of the PI-control law given by Eq. (40.31) with $T_D = 0$, and assume that the characteristic roots of the plant (Fig. 40.50) are real values $-r_1$ and $-r_2$ such that $-r_2 < -r_1$. In this case the open-loop transfer function of the control system is

$$G(s)H(s) = \frac{K_p(s + 1/T_I)}{s(s + r_1)(s + r_2)} \qquad (40.34)$$

One design approach is to select T_I and plot the locus with K_p as the parameter. If the zero at $s = -1/T_I$ is located to the right of $s = -r_1$, the dominant time constant cannot be made as small as is possible with the zero located between the poles at $s = -r_1$ and $s = -r_2$ (Fig. 40.50). A large integral gain (small T_I and large K_p) is desirable for reducing the overshoot due to a disturbance, but the zero should not be placed to the left of $s = -r_2$ because the dominant time constant will be larger than that obtainable with the placement shown in Fig. 40.50 for large values of K_p. Sketch the root-locus plots to see this. A similar situation exists if the poles of the plant are complex.

The effects of the lead compensator in terms of time-domain specifications (characteristic roots) can be shown with the root-locus plot. Consider the second-order plant with the real distinct roots $s = -\alpha, -\beta$. The root locus for this system with proportional control is shown in Fig. 40.51a. The smallest dominant time constant obtainable is τ_1, marked in the figure. With lead compensation, the root locus becomes that shown in Fig. 40.51b. The pole and zero introduced by the compensator reshapes the locus so that a smaller dominant time constant can be obtained. This is done by choosing the proportional gain high enough to place the roots close to the asymptotes.

With reference to the system shown in Fig. 40.51a, suppose that the desired damping ratio ζ_1 and desired time constant τ_1 are obtainable with a proportional gain of K_{p1}, but the resulting steady-state error $\alpha\beta/(\alpha\beta + K_{p1})$ to a step input is too large. We need to increase the gain while preserving the desired damping ratio and time constant. With the lag compensator, the root locus is as shown in Fig. 40.51c. By considering specific numerical values, one can show that for the compensated system, roots with a damping ratio ζ_1 correspond to a high value of the proportional gain. Call this value K_{p2}. Thus $K_{p2} > K_{p1}$, and the steady-state error will be reduced. If the value of T is chosen large enough the pole at $s = -1/T$ is approximately canceled by the zero at $s = -1/aT$, and the open-loop transfer function is given approximately by

$$G(s)H(s) = \frac{aK_p}{(s + \alpha)(s + \beta)} \qquad (40.35)$$

Thus the system's response is governed approximately by the complex roots corresponding to the gain value K_{p2}. By comparing Fig. 40.51a with 40.51c, we see that the compensation leaves the time constant relatively unchanged. From Eq. (40.35) it can be seen that since $a < 1$, K_p can be selected as the larger value K_{p2}. The ratio of K_{p1} to K_{p2} is approximately given by the parameter a.

Design by pole–zero cancellation can be difficult to accomplish because a response pattern of the system is essentially ignored. The pattern corresponds to the behavior generated by the canceled pole

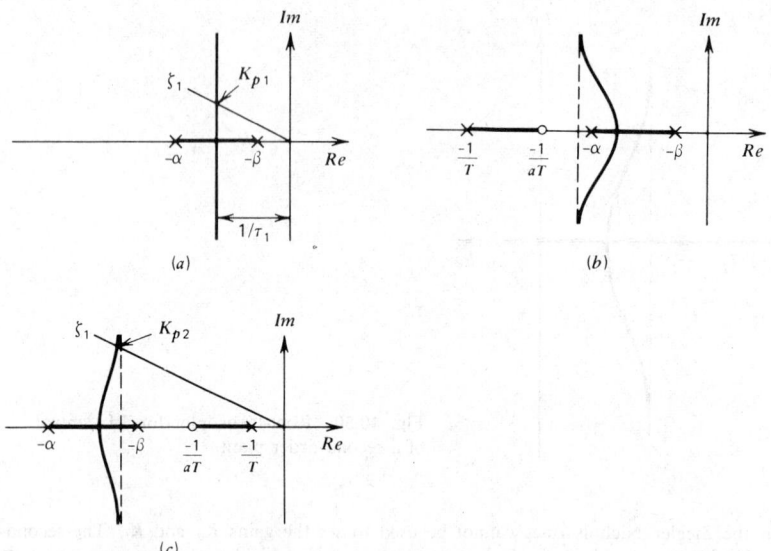

Fig. 40.51 Effects of series lead and lag compensators: (*a*) uncompensated system's root locus; (*b*) root locus with lead compensation; (*c*) root locus with lag compensation.

and zero, and this response can be shown to be beyond the influence of the controller. In this example, the canceled pole gives a stable response because it lies in the left-hand plane. However, another input not modeled here, such as a disturbance, might excite the response and cause unexpected behavior. The designer should therefore proceed with caution. None of the physical parameters of the system are known exactly, so exact pole-zero cancellation is not possible. A root-locus study of the effects of parameter uncertainty and a simulation study of the response are often advised before the design is accepted as final.

40.11 DIGITAL CONTROL

Digital control has several advantages over analog devices. A greater variety of control algorithms is possible, including nonlinear algorithms and ones with time-varying coefficients. Also, greater accuracy is possible with digital systems. Some disadvantages are that they require power (unlike a passive RC circuit), and their application is limited to signals whose time variation is slow enough to be handled by the samplers and the logic circuitry.

Sampling, discrete-time models, the z-transform, and pulse transfer functions were outlined in Section 39.8.5. In most digital control applications, the plant is an analog system, but the controller design is more easily done with discrete-time models. The discrete model of the plant is obtained as shown in Fig. 40.52, where $G(z) = \mathscr{Z}[g(t)]$; $g(t) = \mathscr{L}^{-1}[G(s)]$. Table 39.12 in the previous chapter facilitates this process. The cascaded-block rule of block diagram algebra must be modified when samplers are present. Figure 40.53 shows the required modification and the new notation: $G_1G_2(z) = \mathscr{Z}\{\mathscr{L}^{-1}[G_1(s)G_2(s)]\}$. Note that $G_1(z)G_2(z) \neq G_1G_2(z)$ in general. Figure 40.54 gives the input–output relations for several configurations. Note that it is not always possible to define a transfer function (Fig. 40.54*f*).

40.11.1 Transient Performance

The variable z is related to the Laplace variable s by

$$z = e^{sT} \tag{40.36}$$

A linear discrete system is unstable if $|z| > 1$, that is, if any of its characteristic roots lies outside a circle of unit radius centered on the origin of the complex plane. Stability tests originally developed to detect the occurrence of a root in the right half of the s plane can be converted for use with discrete models by using the transformation $z = (s + 1)/(s - 1)$, which maps the inside of the unit circle onto the left half of the s plane.

Formula (40.36) is useful in relating the behavior of a time function as specified by its roots in the s plane to the location of the corresponding roots in the z plane. Transient behavior can be

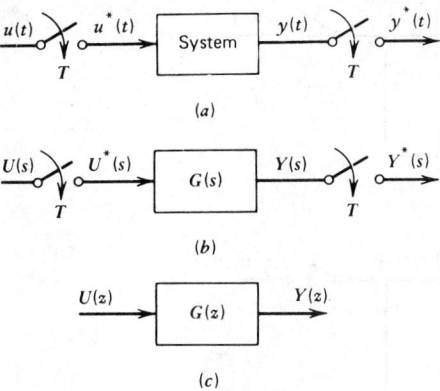

Fig. 40.52 Equivalent representations of a sampled-data system: (*a*) time-domain diagram; (*b*) *s*-domain diagram; (*c*) *z*-domain diagram (automatically implies the existence of an input sampler and of either a real or a fictitious output sampler).[11]

characterized by the damping ratio, natural frequency, and time constant of the dominant root in the *s* plane. Therefore Eq. (40.36) gives the root locations in the *z* plane required to produce the same transient behavior. The transformation (40.36) is not one to one. If $z = e^{s_1 T}$, then $z = e^{s_2 T}$ also, where

$$s_2 = s_1 + i\frac{2\pi n}{T}$$

and *n* is integer. However, if guard filters are used to prevent aliasing, we need not be concerned with the solutions for $n > 0$.

Consider a stable root pair $s = -a \pm ib$. The corresponding *z* values are

$$z = e^{-aT}e^{\pm ibT} = e^{-aT}(\cos bT \pm i \sin bT) \qquad (40.37)$$

Purely imaginary roots $s = \pm ib$ correspond to *z* roots on the unit circle $z = \cos bT \pm i \sin bT$. Horizontal lines of constant frequency and variable time constant in the *s* plane correspond to radial lines in the *z* plane (Fig. 40.55*a*). These make an angle of $\theta = \pm bT$ with the positive real axis. For small time constants, the *z* roots lie close to the origin (these roots are denoted by O in both plots).

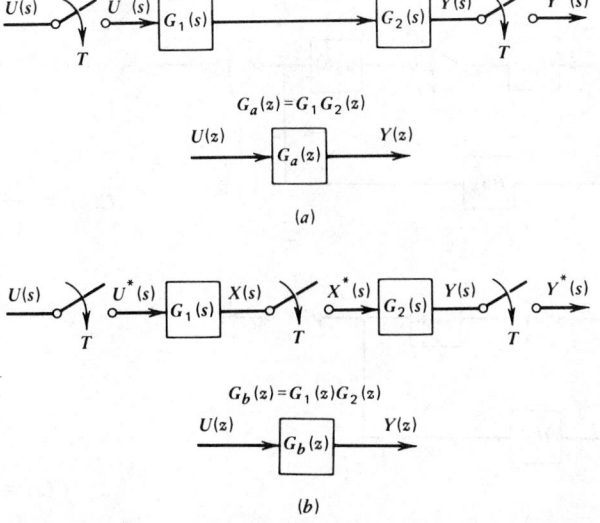

Fig. 40.53 Transfer functions for cascaded elements. (*a*) Analog elements not separated by a sampler, and their cascaded pulse-transfer function $G_a(z)$. (*b*) Analog elements separated by a sampler, and their cascaded pulse transfer function $G_b(z)$.[11]

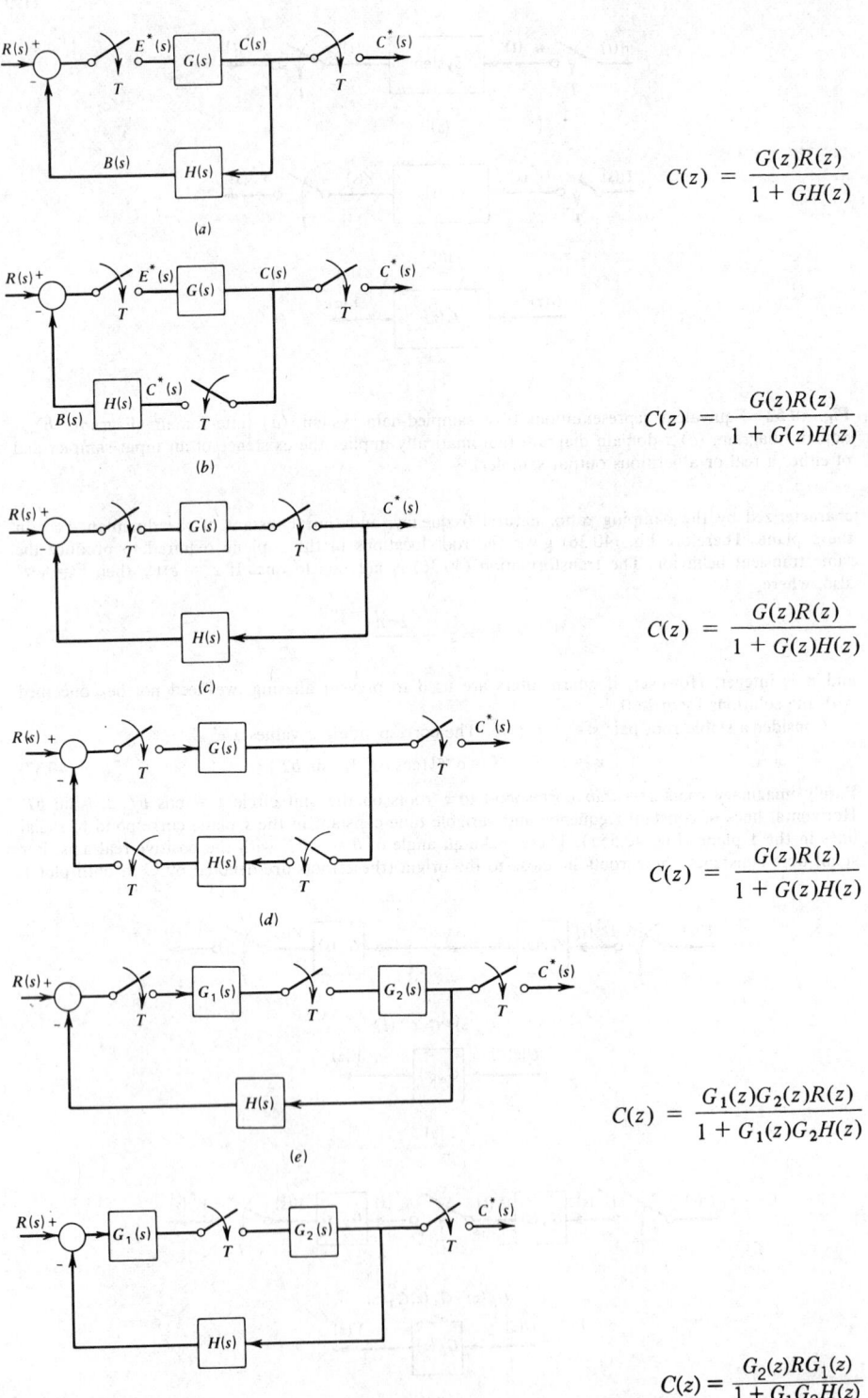

$$C(z) = \frac{G(z)R(z)}{1 + GH(z)}$$

(a)

$$C(z) = \frac{G(z)R(z)}{1 + G(z)H(z)}$$

(b)

$$C(z) = \frac{G(z)R(z)}{1 + G(z)H(z)}$$

(c)

$$C(z) = \frac{G(z)R(z)}{1 + G(z)H(z)}$$

(d)

$$C(z) = \frac{G_1(z)G_2(z)R(z)}{1 + G_1(z)G_2H(z)}$$

(e)

$$C(z) = \frac{G_2(z)RG_1(z)}{1 + G_1G_2H(z)}$$

(f)

Fig. 40.54 Output relations for various loop and sampler configurations.[11]

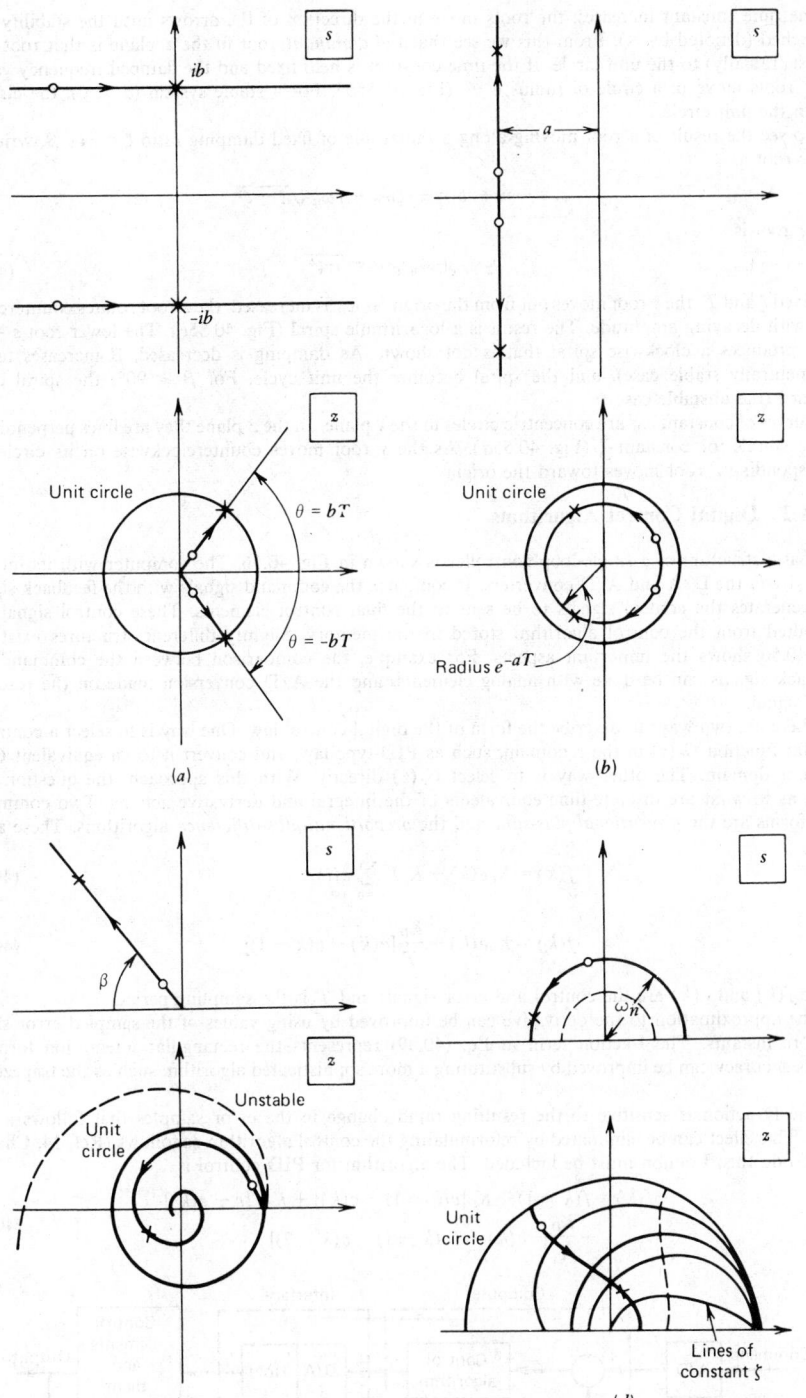

Fig. 40.55 Equivalent root paths in the s plane and z plane. (a) Roots with the same oscillation frequency b. (b) Roots with the same time constant $\tau = 1/a$. (c) Roots with the same damping ratio ζ. (d) Roots with the same natural frequency ω_n.[11]

As the time constant increases, the roots move in the direction of the arrows until the stability limit is reached (denoted by $\times$). From this we see that the *dominant* root in the z plane is that root lying closest (radially) to the unit circle. If the time constant is held fixed and the damped frequency varied, the z roots move in a circle of radius e^{-aT} (Fig. 40.55b). For a stable system ($a > 0$), the circle is within the unit circle.

To see the result of a root moving along a radial line of fixed damping ratio $\zeta = \cos\beta$, write the upper root as

$$s = -a + ib = -\zeta\omega_n + i\omega_n\sqrt{1-\zeta^2}$$

The z root is

$$z = e^{-\zeta\omega_n T}e^{i\omega_n T\sqrt{1-\zeta^2}} \tag{40.38}$$

For fixed ζ and T, the s root moves out from the origin as ω_n is increased; the z root rotates counterclockwise with decaying amplitude. The result is a logarithmic spiral (Fig. 40.55c). The lower root $s = -a - ib$ produces a clockwise spiral that is not shown. As damping is decreased, β increases to $90°$ (the neutrally stable case), and the spiral becomes the unit cycle. For $\beta \geq 90°$, the spiral opens outward (the unstable case).

Curves of constant ω_n are concentric circles in the s plane. In the z plane they are lines perpendicular to the spirals for constant ζ (Fig. 40.55d). As the s root moves counterclockwise on its circle, the corresponding z root moves toward the origin.

40.11.2 Digital Control Algorithms

The basic structure of a single-loop controller is shown in Fig. 40.56. The computer with its internal clock drives the D/A and A/D converters. It compares the command signals with the feedback signals and generates the control signals to be sent to the final control elements. These control signals are computed from the control algorithm stored in the memory. Slightly different structures exist, but Fig. 40.56 shows the important aspects. For example, the comparison between the command and feedback signals can be done with analog elements, and the A/D conversion made on the resulting error signal.

There are two ways to describe the form of the digital control law. One way is to select a controller transfer function $G_c(s)$ in the s domain, such as PID-type law, and convert it to an equivalent $G_c(z)$ in the z domain. The other way is to select $G_c(z)$ directly. With this approach, the question now arises as to what are discrete-time equivalents of the integral and derivative actions. Two commonly used forms are the *proportional-plus-sum* and the *proportional-plus-difference* algorithms. These are

$$f(k) = K_p e(k) + K_I T \sum_{i=0}^{k} e(i) \tag{40.39}$$

$$f(k) = K_p e(k) + \frac{K_D}{T}[e(k) - e(k-1)] \tag{40.40}$$

where $f(k)$ and $e(k)$ are the control and error signals and T is the sampling period.

The approximation to the derivative can be improved by using values of the sampled error signal at more instants. The I-action term in Eq. (40.39) represents the rectangular integration formula, and its accuracy can be improved by substituting a more sophisticated algorithm such as the trapezoidal rule.

The D action is sensitive to the resulting rapid change in the error samples that follows a step input. This effect can be eliminated by reformulating the control algorithm as follows (Ref. 14, Chapter 11). To do this, I action must be included. The algorithm for PID control is

$$f(k) = f(k-1) + K_p[c(k-1) - c(k)] + K_I T[r - c(k)]$$
$$+ \frac{K_D}{T}[-c(k) + 2c(k-1) - c(k-2)] \tag{40.41}$$

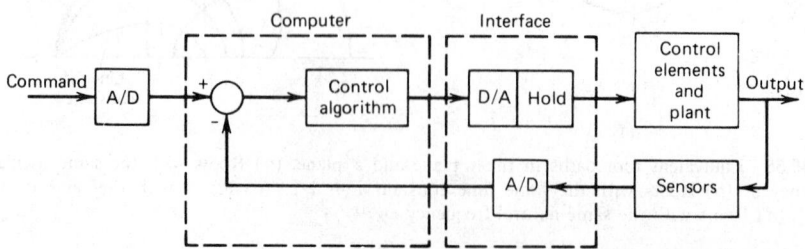

Fig. 40.56 Structure of a digital control system.[11]

40.11.3 Controller Design

Equations (40.39) through (40.41) are digital versions of the PID family of control laws that were obtained by mimicking the attributes of the analog versions. For example, the sum in the I-action term in (40.39) is supposed to represent the behavior of an analog integrator.

It is possible to obtain the digital control law by using other finite difference techniques to convert the continuous-time control law into discrete-time form. The most commonly used methods for this purpose are the Euler method and the Tustin method (Ref. 11, Section 4.9). The PID law (40.22) can be written as the differential equation

$$\frac{df}{dt} = K_p \frac{de}{dt} + K_I e + K_D \frac{d^2e}{dt^2} \tag{40.42}$$

Application of the Euler method with a step size T results in a difference equation that is equivalent to

$$\frac{F(z)}{E(z)} = \frac{a_1 z^2 + a_2 z + a_3}{z(z-1)} \tag{40.43}$$

But the Tustin method applied to (40.42) gives

$$\frac{F(z)}{E(z)} = \frac{a_1 z^2 + a_2 z + a_3}{z^2 - 1} \tag{40.44}$$

Both methods give the same form for PI control (40.19); namely,

$$\frac{F(z)}{E(z)} = \frac{a_1 z + a_2}{z - 1} \tag{40.45}$$

The relation of the a_i coefficients to K_p, K_I, and K_D is different for the Euler and Tustin methods.

Thus there are three possible approaches to designing a digital controller (Ref. 11, Section 6.12):

1. The performance is specified in terms of the desired continuous-time response, and the controller design is done entirely in the s-domain as with an analog controller. The resulting control law is then converted to discrete-time form by a finite-difference technique such as the Euler or Tustin methods, or by using forms (40.39), (40.40), or (40.41). If the sampling period is small, this method can be successfully applied. The technique is widely used for two reasons. When existing analog controllers are converted to digital control, the form of the control law and the values of its associated gains are known to have been satisfactory. Therefore, a digital version is designed. Second, because analog design methods are well established, many engineers prefer to take this route and then convert the design into a discrete-time equivalent. Other methods for developing discrete equivalents from analog transfer functions can be developed with frequency-response techniques (see Ref. 4).

2. The performance specifications are given in terms of the desired continuous-time response and/ or desired root locations in the s plane. From these the corresponding root locations in the z plane are found, and a discrete control law is designed. Any of the forms (40.39) through (40.45) can be used. This method is considered the most practical one to use. It avoids the approximation errors that are inherent in the first method.

3. The performance specifications are given in terms of the desired discrete-time response and/ or desired root locations in the z plane. The rest of the procedure follows as in the second method. Obviously, this approach is the most direct since the s plane is bypassed entirely. However, since most of our applications involve analog plants, it is difficult to state specifications in the z domain.

The graphical design methods of Section 40.10 can be applied even when the design is performed in the z domain. The root locus is especially useful, while the frequency response methods require some modification.[4]

40.12 OTHER CONTROL METHODS

Except for state-variable feedback, all of the design techniques presented here comprise "classical" control methods. Since the 1950s, a variety of new techniques have been developed that are loosely termed "modern" control. The term "modern" does not imply that these methods have replaced the classical methods. Indeed the classical methods remain predominant because when combined with some testing and computer simulation, an experienced engineer can rapidly achieve an acceptable design. The few significant mechanical engineering applications of modern control theory have been mainly in the control of aerospace vehicles. The main reason for this is that the high-order, multivariable systems that cannot be controlled with classical methods are also nonlinear, and the modern methods cannot yet deal with nonlinear systems in any general way. The current approach to multivariable systems like the one shown in Fig. 40.11 is to use classical methods to design a controller for each subsystem, since they can often be modeled with low-order linearized models. The coordination of the various low-level controllers is a nonlinear problem, and is usually done in an ad hoc fashion. Here lies the potential contribution of modern control theory, and research is active in this area.

Familiarity with the concepts and terminology of modern control theory will be helpful in keeping abreast of the latest developments. *Optimal control* includes a number of algorithms for systematic design of a control law to minimize a performance index, such as

$$J = \int_0^\infty (\mathbf{x}^T \mathbf{Q} \mathbf{x} + \mathbf{u}^T \mathbf{R} \mathbf{u}) \, dt \tag{40.46}$$

where $\mathbf{x}$ and $\mathbf{u}$ are the deviations of the state and control vectors from the desired reference values. The quadratic index (40.46) is a generalization of the ISE index. The matrices $\mathbf{Q}$ and $\mathbf{R}$ are chosen by the designer to provide relative weighting for the elements of $\mathbf{x}$ and $\mathbf{u}$. If the plant can be described by the linear state-variable model

$$\dot{\mathbf{x}} = \mathbf{A}\mathbf{x} + \mathbf{B}\mathbf{u} \tag{40.47}$$
$$\mathbf{y} = \mathbf{C}\mathbf{x} + \mathbf{D}\mathbf{u} \tag{40.48}$$

where $\mathbf{y}$ is the vector of outputs, then the solution of this *linear-quadratic* problem is a linear control law:

$$\mathbf{u} = \mathbf{K}\mathbf{y} \tag{40.49}$$

where $\mathbf{K}$ is a matrix of gains that can be found by several algorithms. A valid solution is guaranteed to yield a stable closed-loop system, a major benefit of this method.

Even if it is possible to formulate the control problem in this way, several practical difficulties arise. Some of the terms in (40.46) might be beyond the influence of the control vector $\mathbf{u}$; the system is then *uncontrollable*. Also, there might not be enough information in the output equation (40.48) to achieve control, and the system is then *unobservable*. Several tests are available to check *controllability* and *observability*. Not all of the necessary state variables might be available for feedback, or the feedback measurements might be noisy or biased. Algorithms known as *observers, state reconstructors, estimators,* and *filters* are available to compensate for the missing information. Another source of error is the uncertainty in the values of the coefficient matrices $\mathbf{A}$, $\mathbf{B}$, $\mathbf{C}$, and $\mathbf{D}$. *Identification* schemes can be used to compare the predicted and the actual system performance, and to adjust the coefficient values "on-line." Controllers capable of modifying their performance in response to new information or changing conditions are said to be *adaptive*.

For more information on modern control methods, see Refs. 1, 2, 4, 5, 10, 11, and 14.

REFERENCES

1. J. W. Brewer, *Control Systems: Analysis, Design, and Simulation,* Prentice-Hall, Englewood Cliffs, NJ, 1974.
2. A. E. Bryson and Y. C. Ho, *Applied Optimal Control,* Blaisdell, Waltham, MA, 1969.
3. J. J. D'Azzo and C. H. Houpis, *Linear Control System Analysis and Design,* McGraw-Hill, New York, 1981.
4. G. Franklin and J. Powell, *Digital Control of Dynamic Systems,* Addison-Wesley, Reading, MA, 1980.
5. A. Gelb (Ed.), *Applied Optimal Estimation,* MIT Press, Cambridge, 1974.
6. L. Hasdorff, *Gradient Optimization and Nonlinear Control,* Wiley, New York, 1976.
7. B. C. Kuo, *Automatic Control Systems,* 4th ed., Prentice-Hall, Englewood Cliffs, NJ, 1982.
8. D. McCloy and H. Martin, *The Control of Fluid Power,* Longman, London, 1973.
9. K. Ogata, *Modern Control Engineering,* Prentice-Hall, Englewood Cliffs, NJ, 1970.
10. K. Ogata, *System Dynamics,* Prentice-Hall, Englewood Cliffs, NJ, 1978.
11. W. J. Palm III, *Modeling, Analysis, and Control of Dynamic Systems,* Wiley, New York, 1983.
12. R. Phelan, *Automatic Control Systems,* Cornell University Press, Ithaca, NY, 1977.
13. S. M. Shinners, *Modern Control System Theory and Application,* Addison-Wesley, Reading, MA, 1978.
14. Y. Takahashi, M. Rabins, and D. Auslander, *Control,* Addison-Wesley, Reading, MA, 1970.
15. J. Truxal (Ed.), *Control Engineer's Handbook,* McGraw-Hill, New York, 1958.
16. J. Truxal, *Introductory Systems Engineering,* McGraw-Hill, New York, 1971.

CHAPTER 41

MEASUREMENTS

E. L. HIXSON
E. A. RIPPERGER

University of Texas
Austin, Texas

41.1 STANDARDS AND ACCURACY

41.1.1 Standards

Measurement is the process by which a quantitative comparison is made between a standard and a measurand. The measurand is the particular quantity of interest—the thing that is to be quantified. The standard of comparison is of the same character as the measurand, and, so far as mechanical engineering is concerned, the standards are defined by law and maintained by the National Bureau of Standards.[1] The four independent standards that have been defined are length, time, mass, and temperature. All other standards are derived from these four. Before 1960 the standard for length was the international prototype meter, kept at Sevres, France. In 1960 the meter was redefined as 1,650,763.73 wavelengths of krypton light. Then in 1983 the 17th General Conference on Weights and Measures adopted a new standard: a meter is the distance traveled in a vacuum by light in 1/299,792,458 seconds.[2] However, there is a copy of the international prototype meter, known as the National Prototype Meter, kept at the National Bureau of Standards. Below that level there are several bars known as National Reference Standards and below that there are the working standards. Interlaboratory standards in factories and laboratories are sent to the National Bureau of Standards for comparison with the working standards. These interlaboratory standards are the ones usually available to engineers.

Standards for the other three basic quantities have also been adopted by the Bureau of Standards, and accurate measuring devices for those quantities should be calibrated against those standards.

The standard mass is a cylinder of platinum–iridium, the international kilogram, also kept at Sevres, France. It is the only one of the basic standards that is still established by a prototype. In the United States, the basic unit of mass is the U.S. basic prototype kilogram No. 20. There are working copies of this standard that are used to determine the accuracy of interlaboratory standards. Force is not one of the fundamental quantities, but in the United States the standard unit of force is the pound, defined as the gravitational attraction for a certain platinum mass at sea level and 45° latitude.

Absolute time, or the time when some event occurred in history, is not of much interest to engineers; they are more likely to need to measure time intervals, that is, the time between two events. The second is the basic unit for time measurements. At one time the second was defined as 1/86,400 of the average period of rotation of the earth on its axis, but that is not a practical standard. The period varies and the earth is slowing down. Consequently, a new standard based on the oscillations associated with a certain transition within the cesium atom has been defined and adopted. The second is now "the duration of 9,192,631,770 periods of the radiation corresponding to the transition between the two hyperfine levels of the fundamental state of cesium 133."[3] Thus the Cesium "clock" is the basic frequency standard, but tuning forks, crystals, electronic oscillators, etc., may be used as secondary standards. For the convenience of anyone who requires a time signal of a high order of accuracy, the Bureau of Standards broadcasts continuously time signals of different frequencies from stations WWV, WWVB, and WWVL located in Fort Collins, Colorado, and WWVH located in Hawaii. Other nations also broadcast timing signals. For details on the time signal broadcasts, potential users should consult the Bureau of Standards.[4]

Temperature is one of four fundmental quantities in the international measuring system. Temperature is fundamentally different in nature from length, time, and mass. It is an intensive quantity, whereas the others are extensive. Join the two bodies that have the same temperature together and you will have a larger body at that same temperature. Join two bodies that have a certain mass and you will have one body of twice the mass of the original body. Two bodies are said to be at the same temperature if they are in thermal equilibrium. The international practical temperature scale, adopted in 1968 (IPTS-68) by the International Committee on Weights and Measurement,[5] is the one now in effect and the one with which engineers are primarily concerned. In this system the kelvin (K) is the basic unit of temperature. It is 1/273.16 of the temperature at the triple point of water—the temperature at which the solid, liquid, and vapor phases of water exist in equilibrium. Degrees celsius (°C) are related to degrees kelvin by the equation

$$t = T - 273.15$$

where t = degrees celsius
T = degrees kelvin

Zero celsius is the temperature established between pure ice and air-saturated pure water at normal atmospheric pressure. The IPTS-68 established six primary fixed reference temperatures and procedures for interpolating between them. These are the temperatures and procedures used for calibrating precise temperature measuring devices.

41.1.2 Accuracy

In measurement practice four terms are frequently used to describe an instrument. They are: accuracy, precision, sensitivity, and linearity. Accuracy as applied to an instrument is the closeness with which a reading approaches the true value. Since there is some error in every reading, the "true value" is never known. In the discussion of error analysis which follows, methods of estimating the "closeness" with which the determination of a measured value approaches the true value will be presented. Precision is the degree to which readings agree among themselves. If the same value is measured many times and all the measurements agree very closely, the instrument is said to have a high degree of precision. It may not, however, be a very accurate instrument. Accurate calibration is necessary for accurate measurement. Measuring instruments must, for accuracy, be from time to time compared to a standard. These will usually be laboratory or company standards, which are in turn compared from time to time with a working standard at the Bureau of Standards. This chain can be thought of as the pedigree of the instrument, and the calibration of the instrument is said to be traceable to NBS.

41.1.3 Sensitivity or Resolution

These two terms, as applied to a measuring instrument refer to the smallest change in the measured quantity to which the instrument responds. Obviously the accuracy of an instrument will depend to some extent on the sensitivity. If, for example, the sensitivity of a pressure transducer is one kilopascal, any particular reading of the transducer has a potential error of at least one kilopascal. If the readings expected are in the range of 100 kilopascals and a possible error of 1% is acceptable, then the transducer with a sensitivity of one kilopascal may be acceptable, depending on what other sources of error may be present in the measurement. A highly sensitive instrument is difficult to use. Therefore, a sensitivity significantly greater than that necessary to obtain the desired accuracy is no more desirable than one with insufficient sensitivity.

41.1.4 Linearity

The calibration curve for an instrument does not have to be a straight line. However, conversion from a scale reading to the corresponding measured value is most convenient if it can be done by multiplying by a constant rather than by referring to a nonlinear calibration curve or by computing from an equation. Consequently, instrument manufacturers generally try to produce instruments with a linear readout, and the degree to which an instrument approaches this ideal is indicated by its "linearity." Several definitions of "linearity" are used in instrument-specification practice. The so-called "independent linearity" is probably the most commonly used in specifications. For this definition the data for the instrument readout versus the input are plotted and then a "best straight line" fit is made using the method of least squares. Linearity is then a measure of the maximum deviation of any of the calibration points from this straight line. This deviation can be expressed as a percentage of the actual reading or a percentage of the full-scale reading. The latter is probably the most commonly used, but it may make an instrument appear to be much more linear than it actually is. A better specification is a combination of the two.[6] Thus, linearity = $\pm A\%$ of reading or $\pm B\%$ of full-scale, whichever is greater.

Sometimes the term "independent linearity" is used to describe linearity limits based on actual readings. Since both are given in terms of a fixed percentage, an instrument with $A\%$ proportional linearity is much more accurate at low reading values than an instrument with $A\%$ independent linearity.

It should be noted that although specifications may refer to an instrument as having $A\%$ linearity, what is really meant is $A\%$ nonlinearity. If the linearity is specified as independent linearity, the user of the instrument should try to minimize the error in readings by selecting a scale, if that option is available, such that the actual reading is close to full scale. Never take a reading near the low end of a scale if it can possibly be avoided.

41.2 IMPEDANCE CONCEPTS

A basic question that must be considered when any measurement is made is: How has the measured quantity been affected by the instrument used to measure it? Is the quantity the same as it would have been had the instrument not been there? If the answer to the question is no, the effect of the instrument is called "loading." To characterize the loading the concepts of "stiffness" and "input impedance" are used.

At the input of each component in a measuring system there exists a variable q_{i1}, which is the one we are primarily concerned with in the transmission of information. At the same point, however, there is associated with q_{i1} another variable q_{i2} such that the product $q_{i1}q_{i2}$ has the dimensions of power and represents the rate at which energy is being withdrawn from the system. When these two quantities are identified, the generalized input impedance Z_{gi} can be defined by

$$Z_{gi} = q_{i1}/q_{i2} \tag{41.1}$$

if q_{i1} is an "effort variable." The effort variable is also sometimes called the "across variable." The quantity q_{i2} is called the "flow variable" or "through variable."

The application of these concepts is illustrated by the example in Fig. 41.1. The output of the linear network in blackbox (a) is the open circuit voltage E_0 until the load Z_L is attached across the terminals A–B. If Thevenin's theorem is applied after the load Z_L is attached, the system in Fig. 41.1b is obtained. For that system the current is given by

$$i_m = E_0/(Z_{AB} + Z_L) \qquad (41.2)$$

and the voltage E_L across Z_L is

$$E_L = i_m Z_L = E_0 Z_L/(Z_{AB} + Z_L)$$

or

$$E_L = E_0/(1 + Z_{AB}/Z_L) \qquad (41.3)$$

In a measurement situation E_L would be voltage indicated by the voltmeter, Z_L would be the input impedance of the voltmeter, and Z_{AB} would be the output impedance of the linear network. The true output voltage E_0 has been reduced by the voltmeter, but it can be computed from the voltmeter reading if Z_{AB} and Z_L are known. From Eq. (41.3) it is seen that the effect of the voltmeter on the reading is minimized by making Z_L as large as possible.

If the generalized input and output impedances Z_{gi} and Z_{go} are defined for nonelectrical systems as well as electrical systems, Eq. (41.3) can be generalized to

$$q_{im} = q_{iu}/(1 + Z_{go}/Z_{gi}) \qquad (41.4)$$

where q_{im} is the measured value of the effort variable and q_{iu} is the undisturbed value of the effort variable. The output impedance Z_{go} is not always defined or easy to determine, consequently, Z_{gi} should be large. If it is large enough, knowing Z_{go} is unimportant.

If q_{i1} is a flow variable rather than an effort variable (current is a flow variable, voltage is an effort variable), it is better to define an input admittance,

$$Y_{gi} = q_{i1}/q_{i2} \qquad (41.5)$$

rather than the generalized input impedance

$$Z_{gi} = \text{effort variable/flow variable}$$

The power drain of the instrument is

$$P = q_{i1}q_{i2} = q_{i1}^2/Y_{gi} \qquad (41.6)$$

Hence to minimize power drain Y_{gi} must be large. For an electrical circuit

$$I_m = I_u/(1 + Y_o/Y_i) \qquad (41.7)$$

where I_m = measured current
I_u = actual current
Y_o = output admittance of the circuit
Y_i = input admittance of the meter

When the power drain is zero as in structures in equilibrium, as, for example, when deflection is to be measured, the concepts of impedance and admittance are replaced with the concepts of "static stiffness" and "static compliance." Consider the idealized structure in Fig. 41.2.

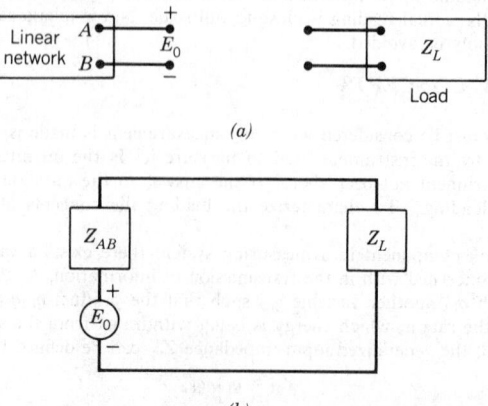

(a)

(b)

Fig. 41.1 Application of Thevenin's theorem.

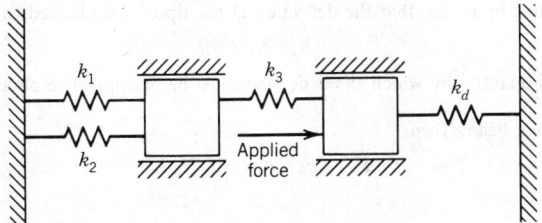

Fig. 41.2 Idealized elastic structure.

To measure the force in member K_2 an elastic link with a spring constant K_m is inserted in series with K_2. This link would undergo a deformation proportional to the force in K_2. If the link is very soft in comparison with K_1, no force can be transmitted to K_2. On the other hand, if the link is very stiff, it does not affect the force in K_2 but it will not provide a very good measure of the force. The measured variable is an effort variable and, in general, when it is measured, it is altered somewhat. To apply the impedance concept a flow variable whose product with the effort variable gives power is selected. Thus,

$$\text{flow variable} = \text{power/effort variable}$$

Mechanical impedance is then defined as force divided by velocity, or

$$Z = \text{force/velocity}$$

This is the equivalent of electrical impedance. However, if the static mechanical impedance is calculated for the application of a constant force, the impossible result

$$Z = \text{force/0} = \infty$$

is obtained.

This difficulty is overcome if energy rather than power is used in defining the variable associated with the measured variable. In that case the static mechanical impedance becomes the "stiffness" and

$$\text{stiffness} = S_g = \text{effort/}\!\int \text{flow } dt$$

In structures,

$$S_g = \text{effort variable/displacement}$$

When these changes are made, the same formulas used for calculating the error caused by the loading of an instrument in terms of impedances can be used for structures by inserting S for Z. Thus

$$q_{im} = q_{iu}/(1 + S_{go}/S_{gi}) \tag{41.8}$$

where q_{im} = measured value of the effort variable
q_{iu} = undisturbed value of the effort variable
S_{go} = static output stiffness of the measured system
S_{gi} = static input stiffness of the measuring system

For an elastic-force-measuring device such as a load cell, S_{gi} is the spring constant K_m. As an example consider the problem of measuring the reactive force at the end of a propped cantilever beam, as in Fig. 41.3.

According to Eq. (41.8) the force indicated by the load cell will be

$$F_m = F_u/(1 + S_{go}/S_{gi})$$

$$S_{gi} = K_m \quad \text{and} \quad S_{go} = 3EI/L^3$$

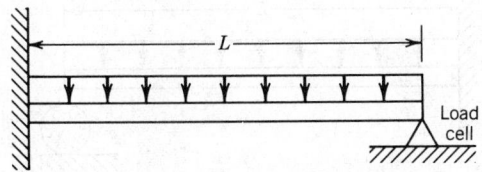

Fig. 41.3 Measuring the reactive force at the tip.

The latter is obtained by noting that the deflection at the tip of a tip loaded cantilever is given by

$$\delta = PL^3/3EI$$

The stiffness is the quantity by which the deflection must be multiplied to obtain the force producing the deflection.

For the cantilever beam, then,

$$F_m = F_u/(1 + 3EI/K_mL^3) \tag{41.9}$$

or

$$F_u = F_m(1 + 3EI/K_mL^3) \tag{41.10}$$

Clearly, if $K_m \gg 3EI/L^3$, the effect of the load cell on the measurement will be negligible.

To measure displacement rather than force, introduce the concept of compliance and define it as

$$C_g = \text{flow variable}/\int \text{effort variable } dt$$

Then

$$q_m = q_u/(1 + C_{go}/C_{gi}) \tag{41.11}$$

If displacements in an elastic structure are considered, the compliance becomes the reciprocal of stiffness, or the quantity by which the force must be multiplied to obtain the displacement caused by the force. The cantilever beam in Fig. 41.4 again provides a simple illustrative example.

If the deflection at the tip of this cantilever is to be measured using a dial gage with a spring constant K_m,

$$C_{gi} = 1/K_m \quad \text{and} \quad C_{go} = L^3/3EI$$

Thus

$$\delta_m = \delta_u(1 + K_mL^3/3EI) \tag{41.12}$$

Not all interactions between a system and a measuring device lend themselves to this type of analysis. A pitot tube, for example, inserted into a flow field distorts the flow field but does not extract energy from the field. Impedance concepts cannot be used to determine how the flow field will be affected.

There are also applications in which it is not desirable for a force-measuring system to have the highest possible stiffness. A subsoil pressure gage is an example. Such a gage, if it is much stiffer than the surrounding soil, will take a disproportionate share of the total load and will consequently indicate a higher pressure than would have existed in the soil if the gage had not been there.

41.3 ERROR ANALYSIS

41.3.1 Introduction

It may be accepted as axiomatic that there will always be errors in measured values. Thus, if a quantity X is measured, the correct value q and X will differ by some amount e. Hence

$$\pm(q - X) = e$$

or   (41.13)

$$q = X \pm e$$

It is essential, therefore, in all measurement work that a realistic estimate of e be made. Without such an estimate the measurement of X is of no value. There are two ways of estimating the error in a measurement. The first is the external estimate or ϵ_E, where $\epsilon = e/q$. This estimate is based on

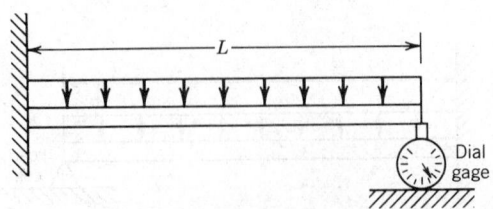

Fig. 41.4 Measuring the tip deflection.

knowledge of the experiment and measuring equipment, and to some extent on the internal estimate ϵ_I.

The internal estimate is based on an analysis of the data using statistical concepts.

41.3.2 Internal Estimates

If a measurement is repeated many times, the repeat values will not, in general, be the same. Engineers, it may be noted, do not usually have the luxury of repeating measurements many times. Nevertheless, the standardized means for treating results of repeated measurements are useful, even in the error analysis for a single measurement.

If some quantity is measured many times and it is assumed that the errors occur in a completely random manner, and that small errors are more likely to occur than large errors, and errors are just as likely to be positive as negative, the distribution of errors can be represented by the curve

$$F(X) = Y_0 e^{-(X-\bar{X})/2\sigma^2} \tag{41.14}$$

where $F(X)$ = number of measurements for a given value of $(X - \bar{X})$
   Y_0 = maximum height of the curve or the number of measurements for which $X = \bar{X}$
   $\bar{X}$ = value of X at the point where maximum height of the curve occurs

σ determines the lateral spread of the curve.

This curve is the normal, or Gaussian, frequency distribution. The area under the curve between X and δX represents the number of data points that fall between these limits, and the total area under the curve denotes the total number of measurements made. If the normal distribution is defined so that the area between X and δX is the probability that a data point will fall between those limits, the total area under the curve will be unity and

$$F(X) = \frac{\exp - (X - \bar{X})^2/2\sigma^2}{\sigma\sqrt{2\pi}} \tag{41.15}$$

and

$$P_x = \int_{-x}^{x} \frac{\exp - (X - \bar{X})^2/2\sigma^2}{\sigma\sqrt{2\pi}} \, dx \tag{41.16}$$

Now if $\bar{X}$ is defined as the average of all the measurements and σ as the standard deviation, then

$$\sigma = \left[\Sigma \frac{(X - \bar{X})^2}{N} \right]^{1/2} \tag{41.17}$$

where N is the total number of measurements. Actually this definition is used as the best estimate for a universe standard deviation, that is, for a very large number of measurements. For smaller subsets of measurements the best estimate of σ is given by

$$\sigma = \left[\Sigma \frac{(X - \bar{X})^2}{(n-1)} \right]^{1/2} \tag{41.18}$$

where n is the number of measurements in the subset. Obviously the difference between the two values of σ becomes negligible as n becomes very large (or as $n \to N$).

The probability curve based on these definitions is shown in Fig. 41.5.

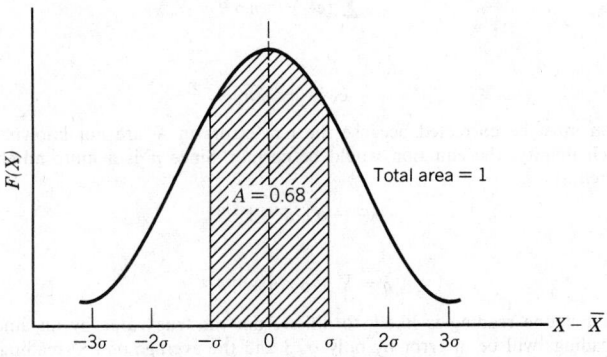

Fig. 41.5 Probability curve.

The area under this curve between $-\sigma$ and $+\sigma$ is 0.68. Hence 68% of the measurements can be expected to have errors that fall in the range of $\pm\sigma$. Thus the chances are 68/32, or better than 2 to 1 that the error in a measurement will fall in this range. For the range $\pm 2\sigma$ the area is 0.95. Hence 95% of all the measurement errors will fall in this range, and the odds are about 20:1 that a reading will be within this range. The odds are about 384:1 that any given error will be in the range of $\pm 3\sigma$.

Some other definitions related to the normal distribution curve are:

1. *Probable Error.* The error likely to be exceeded in half of all the measurements and not reached in the other half of the measurements. This error in Fig. 41.5 is about 0.67σ.

2. *Mean Error.* The arithmetic mean of all the errors regardless of sign. This is about 0.8σ.

3. *Limit of Error.* The error that is so large it is most unlikely ever to occur. It is usually taken as 4σ.

41.3.3 Use of Normal Distribution to Calculate the Probable Error in X

The foregoing statements apply strictly only if the number of measurements is very large. Suppose that n measurements have been made. That is a sample of n data points out of an infinite number. From that sample $\overline{X}$ and σ are calculated as above. How good are these numbers? To determine that proceed as follows:

Let

$$\overline{X} = F(X_1, X_2, X_3, \ldots, X_n) = (\Sigma X_i)/n \tag{41.19}$$

$$e_{\overline{x}} = \sum_{i=1}^{n} \frac{\partial F}{\partial X_i} e_{X_i} \tag{41.20}$$

where $e_{\overline{x}}$ = the error in $\overline{X}$
e_{X_i} = the error in X_i

$$e_{\overline{X}}^2 = \sum_{i=1}^{n} \left(\frac{\partial F}{\partial X_i} e_{X_i}\right)^2 + \sum_{i=1, j=1}^{n} \frac{\partial F}{\partial X_i} e_{X_i} \frac{\partial F}{\partial X_j} e_{X_j} \tag{41.21}$$

where $i \neq j$.

If the errors e_i to e_n are independent and symmetrical, the cross product terms will tend to disappear and

$$e_{\overline{X}}^2 = \sum_{i=1}^{n} \left(\frac{\partial F}{\partial X_i} e_{X_i}\right)^2 \tag{41.22}$$

Since $\partial F/\partial X_i = 1/n$

$$e_{\overline{x}} = \left[\sum_{i=1}^{n} \left(\frac{1}{n}\right)^2 e_{X_i}^2\right]^{1/2} \tag{41.23}$$

or

$$e_{\overline{x}} = \left[\left(\frac{1}{n}\right)^2 \sum_{i=1}^{n} (e_{X_i})^2\right]^{1/2} \tag{41.24}$$

from the definition of σ

$$\sum_{i=1}^{n} (e_{X_i})^2 = n\sigma^2 \tag{41.25}$$

and

$$e_{\overline{x}} = \sigma/\sqrt{n}$$

This equation must be corrected because the real errors in X are not known. If the number n were to approach infinity, the equation would be correct. Since n is a finite number, the corrected equation is written as

$$e_{\overline{x}} = \sigma/(n-1)^{1/2} \tag{41.26}$$

and

$$q = \overline{X} \pm \sigma/(n-1)^{1/2} \tag{41.27}$$

This says that if one reading is likely to differ from the true value by an amount σ, then the average of 10 readings will be in error by only $\sigma/3$ and the average of 100 readings will be in error by $\sigma/10$. To reduce the error by a factor of 2 the number of readings must be increased by a factor of 4.

41.3.4 External Estimates

In almost all experiments several steps are involved in making a measurement. It may be assumed that in each measurement there will be some error, and if the measuring devices are adequately calibrated, errors are as likely to be positive as negative. The worst condition insofar as accuracy of the experiment is concerned would be for all errors to have the same sign. In that case, assuming the errors are all much less than 1, the resultant error will be the sum of the individual errors, that is,

$$\epsilon_E = \epsilon_1 + \epsilon_2 + \epsilon_3 + \cdots \tag{41.28}$$

It would be very unusual for all errors to have the same sign. Likewise it would be very unusual for the errors to be distributed in such a way that

$$\epsilon_E = 0$$

A general method for treating problems that involve a combination of errors to determine what error is to be expected as a result of the combination follows:

Suppose that

$$V = F(a, b, c, d, e, \ldots, x, y, z) \tag{41.29}$$

where $a, b, c, \ldots, x, y, z$ represent quantities that must be individually measured to determine V.

$$\delta V = \sum_{n=a}^{z} \frac{\partial F}{\partial n} \delta n$$

$$e_E = \sum_{n=a}^{z} \frac{\partial F}{\partial n} e_n \tag{41.30}$$

The sum of the squares of the error contributions is given by

$$e_E^2 = \left(\sum_{a}^{z} \frac{\partial F}{\partial n} e_n \right)^2 \tag{41.31}$$

Now, as in the discussion of internal errors, assume that errors e_n are independent and symmetrical. This justifies taking the sum of the cross products as zero:

$$\sum_{n=a, m=a}^{z} \frac{\partial F}{\partial n} \frac{\partial F}{\partial m} e_n e_m = 0 \tag{41.32}$$

$$n \neq m$$

Hence

$$e_E^2 = \sum_{a}^{z} \left(\frac{\partial F}{\partial n} \right)^2 e_n^2$$

or

$$e_E = \left(\sum_{a}^{z} \left[\frac{\partial F}{\partial n} \right]^2 e_n^2 \right)^{1/2} \tag{41.33}$$

This is the "most probable value" of e_E. It is much less than the worst case

$$\epsilon_e = [|\epsilon_a| + |\epsilon_b| + |\epsilon_c| + \cdots + |\epsilon_z|] \tag{41.34}$$

As an application, the determination of g, the local acceleration of gravity, by use of a simple pendulum will be considered:

$$g = 4\pi^2 L/T^2 \tag{41.35}$$

where L = the length of the pendulum
T = the period of the pendulum

If an experiment is performed to determine g, the length L and the period T would be measured. To determine how the accuracy of g will be influenced by errors in measuring L and T write

$$\partial g/\partial L = 4\pi^2/T^2 \quad \text{and} \quad \partial g/\partial T = -8\pi^2 L/T^3 \tag{41.36}$$

The error in g is the variation in g written as follows:

$$\delta g = (\partial g/\partial L)\Delta L + (\partial g/\partial T)\Delta T \tag{41.37}$$

or

$$\delta g = (4\pi^2/T^2)\Delta L - (8\pi^2 L T/T^3)\Delta T \tag{41.38}$$

It is always better to write the errors in terms of percentages. Consequently, Eq. (41.38) is rewritten

$$\delta g = (4\pi^2 L/T^2)\Delta L/L - 2(4\pi^2 L/T^2)\Delta T/T \tag{41.39}$$

or

$$\delta g/g = \Delta L/L - 2\Delta T/T \tag{41.40}$$

then

$$e_g = [e_L^2 + (2e_T)^2]^{1/2} \tag{41.41}$$

where e_g is the "most probable error" in the measured value of g. That is to say

$$g = 4\pi^2 L/T^2 \pm e_g \tag{41.42}$$

where L and T are the measured values. Note that even though a positive error in T causes a negative error in the calculated value of g the contribution of the error in T to the "most probable error" is taken as positive. Note also that an error in T contributes four times as much to the "most probable error" as an error in L contributes. It is fundamental in measurements of this type that those quantities that appear in the functional relationship raised to some power greater than unity contribute more heavily to the "most probable error" than other quantities and must, therefore, be measured with greater care.

The determination of the "most probable error" is simple and straightforward. The question is how are the errors, such as $\Delta L/L$ and $\Delta T/T$ determined. If the measurements could be repeated often enough, the statistical methods discussed in the internal error evaluation could be used to arrive at a value. Even in that case it would be necessary to choose some representative error such as the standard deviation, or the mean error. Unfortunately, as was noted previously, in engineering experiments it usually is not possible to repeat measurements enough times to make statistical treatments meaningful. Engineers engaged in making measurements will have to use what knowledge they have of the measuring instruments and the conditions under which the measurements are made to make a reasonable estimate of the accuracy of each measurement. When all of this has been done and a "most probable error" has been calculated, it should be remembered that the result is not the actual error in the quantity being determined, but is, rather, the engineer's best estimate of the magnitude of the uncertainty in the final result.[8]

Consider again the problem of determining g. Suppose that the length L of the pendulum has been determined by means of a meter stick with 1 mm calibration marks and the error in the calibration is considered negligible in comparison with other errors. Suppose the value of L is determined to be 91.7 cm. Since the calibration marks are 1 mm apart, it can be assumed that ΔL is no greater than 0.5 mm. Hence maximum $\Delta L/L = 5.5 \times 10^{-4}$. Suppose T is determined with the pendulum swinging in a vacuum with an arc of $\pm 5°$ using a stopwatch that has an inherent accuracy of one part in 10,000. (If the arc is greater than $\pm 5°$, a nonisochronous swing error enters the picture.) This means that the error in the watch reading will be no more than 10^{-4} sec. However, errors are introduced in the period determination by human error in starting and stopping the watch as the pendulum passes a selected point in the arc. This error can be minimized by selecting the highest point in the arc because the pendulum has zero velocity at that point, and timing a large number of swings so as to spread the error out over that number of swings. Human reaction time may vary from as low as 0.2 sec to as high as 0.7 sec. A value of 0.5 sec will be assumed. Thus the estimated maximum error in starting and stopping the watch will be 1 sec (± 0.5 sec at the start and ± 0.5 sec at the stop). A total of 100 swings will be timed. Thus the estimated maximum error in the period will be 1/100 sec. If the period is determined to be 1.92 sec, the estimated maximum error will be $0.01/1.92 = 0.005$. Compared to this the error in the period due to the inherent accuracy of the watch is negligible. The nominal value of g calculated from the measured values of L and T is 982.03 cm/sec². The "most probable error" [Eq. (41.41)] is

$$[4(0.005)^2 + (5.5 \times 10^{-4})^2]^{1/2} = 0.01 \tag{41.43}$$

The uncertainty in the value of g is then ± 9.82 cm/sec², or, in other words, the value of g will be somewhere between 972.21 and 991.85 cm/sec².

Often it is necessary for the engineer to determine in advance how accurately the measurements must be made in order to achieve a given accuracy in the final calculated result. For example, in the pendulum problem it may be noted that the contribution of the error in T to the "most probable error" is more than 300 times the contribution of the error in the length measurement. This suggests, of course, that the uncertainty in the value of g could be greatly reduced if the error in T could be reduced. Two possibilities for doing this might be (1) find a way to do the timing that does not involve human reaction time, or, if that is not possible, (2) increase the number of cycles timed. If the latter alternative is selected and other factors remain the same the error in T, timed over 200 swings is 1/200 or 0.005 sec. As a percentage the error is $0.005/1.92 = 0.0026$. The "most probable error" in g then becomes

$$e_g = [4 \times (2.6 \times 10^{-3})^2 + (5.5 \times 10^{-4})^2]^{1/2} = 0.005 \tag{41.44}$$

This is approximately one-half of the "most probable error" in the result obtained by timing just 100 swings. With this new value of e_g the uncertainty in the value of g becomes ± 4.91 cm/sec² and g then can be said to be somewhere between 977.12 and 986.94 cm/sec^2. The procedure for reducing this uncertainty still further is now self-evident.

Clearly the value of this type of error analysis depends on the skill and objectivity of the engineer in estimating the errors in the individual measurements. Such skills are acquired only by practice and careful attention to all the details of the measurements.

41.4 APPENDIX

41.4.1 Vibration Measurement

For vibration measurements see Section 25.3.

41.4.2 Acceleration Measurement

For acceleration measurements see Section 25.4.

41.4.3 Shock Measurement

For shock measurement see Section 25.5.

41.4.4 Sound Measurement

Introduction

Sound[9] is an oscillation about the mean of pressure or stress in a fluid or solid medium. Sound is also an auditory sensation produced by those oscillations. Acoustics[1] is the science of sound, which includes its generation, transmission, and effects. Most appliances, machines, and vehicles generate unwanted sound called acoustic noise.[9] It is the purpose here to present methods for the accurate measurement of acoustic noise as a means for determining noise exposure of people in work spaces, as an aid in reducing noise, and for its use in diagnosing vibration problems.

Sound pressure is commonly determined by a microphone that converts pressure to an electrical signal, which is processed and presented on a visual display. Such devices are known as sound level meters. Three types of handheld meters, specified in an American National Standard,[10] are commercially available. The type required depends on the intended use. For example, the type 1 meter is required for precision laboratory measurements, while the type 3 is intended for sound surveys where high accuracy is unwarranted. Sound level meters are calibrated to give accurate measures of sound pressure and means are provided in some cases for a calibration check before making measurements.

Because of the extremely large range of pressures that can be sensed by the human ear, sound pressures are usually measured and expressed in logarithmic units. The *decibel notation* is used and the measured quantity is compared to a reference pressure. Logarithmic qualities are referred to as *levels*. Sound pressure level (SPL) is defined as

$$\text{SPL} = 20 \log_{10} \frac{P_x}{P_{\text{ref}}} \tag{41.45}$$

where P_x is the measured pressure and P_{ref} is 20×10^{-6} Pa (N/m). The units of SPL are dB reference to 20 μPa.

Since acoustic noise has a major impact on people, measurements are usually made in the frequency range of hearing. This is normally taken as 20 Hz to 20 kHz. The knowledge of the distribution of sound energy with frequency is important when dealing with noise impact on people and using radiated sound to identify machine vibrations. Thus, sound level meters have some means of determining the frequency distribution of the spectral components of sound pressure.

The basic configuration of a sound level meter is that shown in Fig. 41.6. The weighting networks are used to shape the noise spectrum so that a measure that relates to the hearing characteristic can be made. The frequency versus amplitude characteristics of the weighting networks are shown in Fig. 41.7. The A weighting that produces a level referred to as L_A in dBA is shaped as approximately the inverse of the loudness contour of the ear at the threshold of hearing. The B and C weightings are matched to successively higher loudness levels; however, most noise regulations require data to be presented in dBA. Some meters may have the D weighting, which is used as a measure of annoyance due to noise from aircraft flyovers. Some meters may have a "flat" weighting position, which eliminates the weighting networks. Then the frequency response is limited only by the characteristics of the microphone and amplifiers. In addition some meters provide for an external filter for a more complete frequency analysis.

Several other types of sound level meters are available that process the acoustic signal for special purposes. There are meters that will respond to instantaneous peak pressures and "hold" for convenient

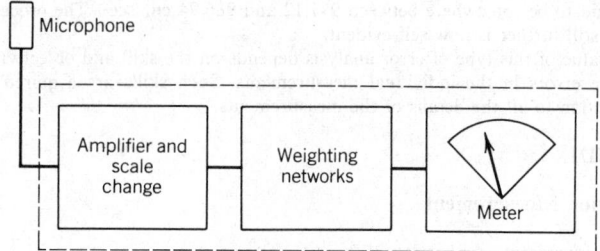

Fig. 41.6 Sound level meter.

readout. Some have a response function that gives increased accuracy for impulsive repetitive noise. Others will integrate over time to give an equivalent sound level (L_{eq}) for noise with large and random variations with time. A final type worn by a worker gives a continuous readout of the accumulated dose of excessive noise.

Measurements in Open Spaces[11]

In this situation the sound source is usually not inside a building and the acoustic waves propagate freely over appreciable distances. The measuring microphone is not in an enclosure. The noise impact may concern:

1. Community noise.
2. Transportation noise.
3. Industrial noise.

In open spaces the microphone will sense the pressure produced by a passing acoustic wave. At high frequencies when the microphone size is an appreciable fraction of the acoustic wavelength, waves are diffracted by the microphone and it becomes directional. For a 1-in.-diameter microphone these effects occur at frequencies about 2 kHz and higher. Under these conditions one should use a microphone designed for *free-field* conditions[12] and correct for angle of wave arrival.

An accurate measure of the sound pressure level of an acoustic wave cannot be made when there are reflecting surfaces nearby. An operator holding a sound level meter can cause reflections. Holding the microphone at arm's length to the side of the body minimizes this problem, however, a microphone on a remote cable is recommended.

Large plane surfaces near the propagation path or behind the operator can cause serious interference effects. At a particular frequency, a direct and reflected path difference of one half wavelength can

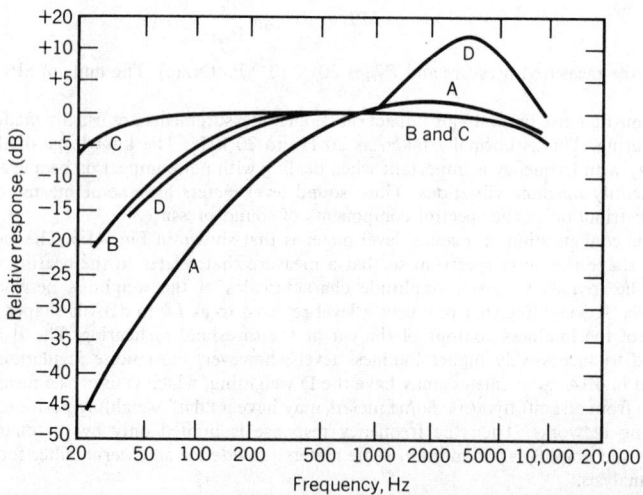

Fig. 41.7 Frequency characteristics for weighting functions of sound level meters.

cause almost complete cancellation. Thus, errors will depend on the frequency content of the sound wave.

The ground is a large reflecting surface and is usually present. However, if the microphone is placed directly on the ground, only the direct wave will be measured. A relatively open space without diffracting objects with the microphone placed a very few inches above the surface is recommended for accurate measurements.

Open space measurements usually imply exposure to all the elements of weather. Wind speeds of a few miles per hour can cause microphone noise that will mask the acoustic signal to be measured. *Wind screens* made of porous materials greatly reduce this effect. Ordinary sound level meters should not be exposed to temperature and moisture extremes. However, microphone and measuring systems can be obtained that operate over extreme weather conditions.

The sound radiated from a small sound source, one whose physical dimensions are small compared to the acoustic wavelength, spreads or diverges in all directions. This divergence is spherical, which results in the sound pressure decaying inversely with distance. The SPL will decrease 6 dB when the measuring distance is doubled. Acoustically large sources will not radiate equally well in all directions. However, in any radial direction the SPL will still decay 6 dB with a doubling of distance.

When acoustic measurements are to be made in an out-of-doors location, a sound pressure level will be present due to local and remote and sound sources. This level is referred to as the *ambient*, and it will provide a lower limit for other measurements. The energy adds in unrelated or uncorrelated sound fields. Thus, when the noise to be measured has a sound pressure equal to the ambient, the SPL will be 3 dB above the ambient.

When the ambient SPL and the addition of a sound source raises the SPL above the ambient, a correct source level can be obtained. Since the energy in an acoustic wave is proportional to sound pressure squared, the corrected pressure squared is

$$P_s^2 = P_m^2 - P_a^2 \tag{41.46}$$

where P_s is the correct source pressure, P_m is the measured pressure, and P_a is the ambient. P can be obtained from SPL as follows:

$$P = P_{ref}10^{(SPL/20)} \tag{41.47}$$

where P_{ref} is 20 μPa. The corrected SPL is

$$(SPL)_s = 20 \log_{10} \frac{P_s}{P_{ref}} \tag{41.48}$$

When it is necessary to do a significant analysis on a sound source, a simple level correction as above is not adequate. When the frequency distribution of energy and statistical amplitude distributions are desired, the sound signal level should be at least 20 dB above the ambient noise. Thus a measuring distance must be chosen to ensure this signal to noise ratio.

Measurement in Enclosed Spaces [11]

In many noisy situations a sound source radiates into a completely enclosed space, such as a room with closed doors and windows. Since sound waves travel at about 330 m/sec, it only takes a few miliseconds for a large number of acoustic rays to be traversing the paths between reflecting surfaces and filling the room with a *reverberant sound field*. The sound pressure level continues to increase until the absorbing surfaces in the room absorb the acoustic energy at the same rate as the source is supplying it. Thus, there will be a region near the source where the directly radiated sound will dominate over the reverberant sound, and the pressure will decay inversely with distance. At some distance, determined by the acoustic absorption in the room, the direct sound will equal the reverberant sound. In the rest of the room the reverberant sound will be relatively uniform and independent of the distance from the source. An exception to this uniformity occurs when large single-frequency components produce dominant standing waves in the room. More details on acoustic characteristics of rooms can be found in the acoustic literature.[13]

The sound pressure levels measured in enclosed spaces depend on many factors. The sound sources and the acoustical absorption in the room will determine the levels in the direct field close to the sources and in the reverberant field. All the sources in the room will contribute to the reverberant field. If the sound pressure level near a noisy machine is greater than the reverberant sound pressure level, the measurement is characteristic of that machine. If it is lower, no information is obtained. If other noise sources can be turned off, a single machine can be studied.

When the noise exposure of workers in a noisy environment is to be determined, it is essential to measure the sound pressure level at the work station at ear level. If the worker stays at one place, a single measurement may suffice. However, for a mobile worker an acoustic dosimeter worn by the individual is essential.

Acoustic Intensity—Diagnosis of Vibrations

The acoustic power radiated by a noisy machine and the frequency spectrum of that power are very useful quantities. When the directive nature of radiated power is known, the sound pressure level in the vicinity of such a machine placed in an open space can be determined. When such a machine is in an enclosed space, the reverberant sound field can be predicted. In this case the total absorption in the room controls the reverberant field and the directive nature is not needed.

When the spectral content of the sound power of a noisy machine is measured, sources of vibrations may be located. Rotating unbalanced parts, oscillating structures, and repetitive impacts will cause vibration of machine surfaces that radiate sound. The period of dominate frequency noise component can be related to the rotational, oscillating, or repetition frequency or a harmonic of that frequency. Such correlations can be used to locate problems due to excess wear in an old machine and locating the trouble areas in the design of quiet machines. For example, worn bearings and gears can be located by the characteristic noise they generate.

Vibrational diagnosis can be done with a single microphone used to measure sound pressure. The "ac" signal out of a sound level meter may be directed to a spectrum analyzer or filter set for this purpose. The determination of sound power, however, requires the quantitive measure of *acoustic intensity*. Intensity is the vector product of the real part of acoustic pressure times acoustic particle velocity:

$$I = \operatorname{Re}(p \ \mathbf{v}) \quad W/m^2 \tag{41.49}$$

Intensity is thus the power per unit area flowing through a surface surrounding a noisy machine. Since intensity is a directed quantity, the component in the direction away from the machine will be radiated as acoustic power. When the machine surface is subdivided into small areas as in Fig. 41.8, the average intensity times that area plus the products for the rest of the small areas give total power radiated:

$$w = \sum_i^n I_i A_i \quad W \tag{41.50}$$

Small areas should be selected that have small intensity variations; then the average value can be used to calculate power flow through the area.

Since measurements can be made near to a noise source and the product of pressure and velocity tend to discriminate against uncorrelated signals, sound power determination can be made in enclosed spaces. This is particularly advantageous for large machines that cannot be placed in anechoic rooms or reverberation chambers.

The acoustic variables are measured by two microphones as shown in Fig. 41.9. The average of the two measured pressures is the pressure between them. The acoustic partical velocity in the direction of the microphone axes is related to the pressure gradient. Then

$$V_x = \frac{1}{\rho_0 d} \int (P_2 - P_1) \, dt \quad m/sec \tag{41.51}$$

where ρ_0 is air density and d is the microphone spacing. Then the average intensity in the x direction becomes

$$I_x = \frac{1}{T} \int_0^T P_{av} \, v_x \, dt \tag{41.52}$$

where averaging time T is long compared to the time scale of P and v. Some commercial instruments measure intensity in this way.

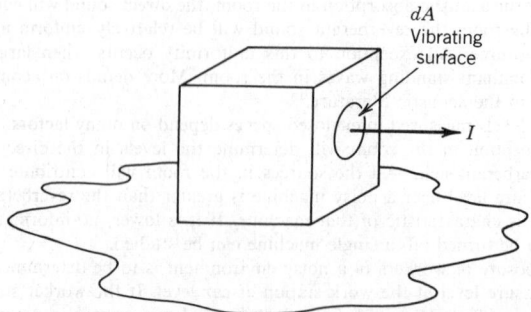

Fig. 41.8   Intensity and power measurements.

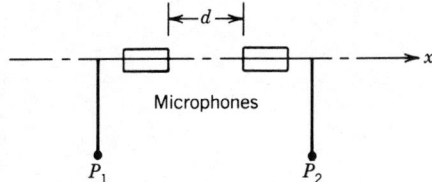

Fig. 41.9 Intensity measuring probe.

When a two-channel spectrum analyzer is available, intensity can be determined from the two microphone signals in another way. The frequency-domain representation of intensity can be obtained by the Fourier transform of the two pressure signals,[14,15]

$$I_x(f) = \frac{\text{Im}(S_{12})}{2\rho_0\, d\,\omega} \tag{41.53}$$

where $\text{Im}(S_{12})$ is the imaginary part of the cross spectral density $S_{12} = P_1(f)P_2^*(f)$ (* designates the complex conjugate). The angular frequency is ω. The frequency-domain representation is very informative since the frequency of predominate energy components are displayed. Vibrational diagnosis is then readily accomplished.

REFERENCES

1. W. A. Wildhack, "NBS Source of American Standards," *ISA Journal,* **8,** 2 (February, 1961).

2. P. Giacoma, "News from the BIPM," *Metrologia,* **20,** 1 (April, 1984).

3. *NBS Tech. News Bull.,* **52,** 1 (Jan., 1968).

4. G. Kamas and S. L. Howe (Eds.), *Time and Frequency User's Manual,* NBS Special Publication 559, U.S. Government Printing Office, Washington, D.C., 1979.

5. "The International Temperature Scale of 1968: A Committee Report," *Metrologia,* **5,** 2 (April, 1969).

6. Ernest O. Doebelin, *Measurement Systems—Applications and Design,* McGraw-Hill, New York, 1966, pp. 38–101.

7. N. H. Cook and E. Rabinowicz, *Physical Measurement and Analysis,* Addison Wesley, Reading, MA, 1963, pp. 29–68.

8. S. J. Kline and F. A. McClintock, "Describing Uncertainties in Single Sample Experiments," *Mech. Engr.,* **75,** 3 (Jan., 1953).

9. *Acoustical Terminology,* ANSI S1.1-1960 (R1971), American National Standards Institute, New York.

10. *Specification for Sound Level Meters,* ANSI S1.4-1971, American National Standards Institute, New York.

11. *Methods for the Measurement of Sound Pressure Level,* ANSI S1.13-1971, American National Standards Institute, New York.

12. *Specifications for Laboratory Standard Microphones,* ANSI S1.12-1967, American National Standards Institute, New York.

13. H. W. Lord, W. S. Gatley, and H. A. Evensen, *Noise Control for Engineers,* McGraw-Hill, New York, 1980.

14. F. J. Foley, "Measurement of Acoustic Intensity Using the Cross-Spectral Density of Two Microphone Signals," *J. Acous. Soc. Am.,* **62,** 1057–1059 (1977).

15. G. Krishnoppa, "Cross-Spectral Method of Measuring Acoustic Intensity by Correcting Phase and Gain Missmatch Errors by Microphone Calibration," *J. Acous. Soc. Am.,* **69,** 307–310 (1981).

PART 5
MANAGEMENT AND RESEARCH

PART 5

MANAGEMENT AND RESEARCH

CHAPTER 42

ENGINEERING MANAGEMENT

HANS THAMHAIN

Worcester Polytechnic Institute
Department of Management
Worcester, Massachusetts

42.1 ORGANIZING THE ENGINEERING FUNCTION

42.1.1 Organizing for Task Efficiency

Modern engineering practices are often too multidisciplinary to be structured strictly along functional lines. Fundamentally, an engineering department has two choices for organizing its operations:

1. *Functional/Matrix.* The engineering department can organize functionally, that is, its company has a single engineering department that does all of the company's engineering work. Each specialty group, headed by a department or unit manager, exists only once. Each resource is shared by all engineering activities that require it. When the engineering function has to perform several missions, tasks, or programs simultaneously, it automatically operates as a *matrix* as described in Table 42.1.

2. *Projectized.* The resources of the engineering department, such as the people, facilities, and office space, are partitioned into project units. Each unit has a manager or leader who is responsible for the proper management of that project. He or she also controls all the resources within that project unit without sharing them with other activities (Table 42.2).

For engineering departments of companies that are in one business, all products and technology developments go essentially through one engineering department, which most likely arranges its operations along matrix lines. On the other hand, companies that are in a number of unrelated businesses or execute projects for different clients, typical for the military and aerospace industry, are most likely organized along project lines. However, for many engineering departments the choice is not that simple. Their companies handle a mixture of large, small, short, and long-range projects for internal as well as external clients through their engineering functions, requiring the ability to adopt continuously to the changing multiproject requirements. Companies that fall into this category have often opted for hybrid structures that overlay the fundamental engineering organization with various organizational subsystems. These overlays may projectize certain engineering resources or establish miniature matrices within engineering, leading to so-called multidimensional matrices, quite common for complex businesses. They also accommodate the widest span of engineering activities with great flexibility while retaining most of the conventional functional stability and resource effectiveness.

These real-world project organizations are complex, both on paper and in practices. They must be carefully designed to accommodate the needed intrafunctional integration while maintaining effective use of engineering resources. The management tools available to define the workings of an engineering organization are: charters, management directives, policies and procedures, and job descriptions.

Table 42.1 Organization Design: Engineering Function—Matrix

An engineering function matrix is essentially an overlay of temporary project organizations to the functional engineering department. The engineering department retains traditional functional characteristics. A clear chain of command exists for all organizational components within the engineering function. Department and unit managers direct the functional engineering groups. Task managers or project engineers perform the engineering-internal coordination and interface with external operating groups along established matrix lines.

The task managers essentially have two bosses: the task managers are accountable (1) to the project manager who may report into engineering or another function for the implementation of the requirements, and (2) to the functional engineering manager for the quality of the work, the availability of necessary resources, and the use of these resources in accomplishing the desired results.

The advantages of this matrix overlay within engineering are a more rapid reaction time, better control, and more efficient integration of multidisciplinary activities. Yet it provides a high concentration of specialized resources and ensures their economical use.

The additional integrator points, via engineering task/project managers, provide an effective interface with engineering-external organizations while at the same time reducing the span of control needed by the program office. On the other hand, the additional organizational overlay is likely to increase cost overheads and requires a more sophisticated management style to operate smoothly and efficiently within an environment that is characterized by extensive sharing of resources, power, and responsibility.

In describing these multidimensional matrices, executives are mostly reaching for verbal tools such as charters, directives, and policies other than conventional charts, which focus on the primary command channels but often confuse the issues of dual accountability and multidimensional controls.

Table 42.2   Organizational Design: Engineering Function—Projectized

The engineering function is organized into project groups. In its purest form, an engineering project manager is assigned for each project directing a complete line organization with full authority over the personnel, facilities, and equipment with the charter to manage one specific project from start to finish. It represents the strongest form of project authority, which encourages performance, schedule, and cost trade-offs; usually represents the best interfaces to engineering-external organizations; and has the best reaction time. However, by comparison with the matrix, the projectized organization usually requires considerable start-up and phase-out efforts and offers less opportunity to share production elements, to use economics of scale, and to balance work loads.

Because of these limitations, companies seldom projectize their engineering departments unless these projects are large enough to use fully the dedicated resources. What is more common is a partially projectized organization. That is, the project manager may fully control *some* resources that are particularly critical to the project and/or can be fully used over the project life cycle, while other resources remain under the control of functional managers who allocate them to specific projects as needed, based on negotiated agreements with the project managers.

42.1.2 Describing and Communicating Organizational Structures

To work effectively, people must understand where they fit in the corporate structure and what their responsibilities are. Especially with the internally complex workings of a modern engineering department it is important to define the management process, the responsibilities, and reporting relations for all organizational components within engineering and its organizational interfaces.

Too often the engineering manager, under pressure to "get started" on a new activity or program, rushes into organizing the engineering team without establishing the proper organizational framework. While initially the prime focus is on staffing, the manager cannot effectively attract and hold quality people until certain organizational pillars are in place. At a minimum, the basic tasks and reporting relations must be defined before the recruiting effort can start.

Establishing the organizational foundation is necessary to communicate the project requirements, responsibilities, and relationships to new or prospective team members, but it is also needed to manage the anxiety that usually develops during the team formation.

Make Functional Ties Work for You

Strong ties of team members to other functional or project organizations may exist as part of the cultural network. These ties are desirable and often necessary for the successful management of multidisciplinary engineering efforts. An additional organizational dimension is introduced when the engineering activities are directed via a project office, which may be chartered within the engineering department or external to it. To build the proper organizational framework, the engineering manager must understand the global organizational structures, the cultures, and value systems. When working with such a program or project office, the two organizational axes are defined in principle as follows: The program office gives operational directions to the engineering personnel and is normally responsible for the budget and schedule. The functional/engineering organization provides technical guidance and is usually responsible for the quality of the work and for personnel administration. Both the engineering program manager and the functional engineering manager must understand this process and perform accordingly or severe jurisdictional conflicts can develop.

Structure Your Organization

The key to successfully building an organization for accommodating a new engineering program lies in clearly defined and communicated responsibility and reporting relationships. The tools come, in fact, from conventional management practices. They provide the basis for defining and communicating the organizational network and for directing the engineering activities effectively. The principal tools are:

- Organizational charters
- Organization charts
- Policies and procedures
- Management directives
- Operating guidelines
- Job descriptions

Examples of these management tools are shown in Tables 42.3 through 42.5. However, it should be emphasized that position titles, responsibilities, and matrix structures vary considerably in industry. Therefore, these examples provide primarily a typology to management for developing their own tools.

Table 42.3   Typical Charter of Engineering Manager

Position Title

Manager of Design Engineering

Authority

The Manager of Design Engineering has the delegated authority from the Director of Engineering to establish, develop, and deploy organizational resources according to established budgets and business needs, both long and short range. This authority includes the hiring, training, maintaining, and terminating of personnel; the development and maintenance of the physical plant, facilities, and equipment; and the direction of the engineering personnel regarding the execution and implementation of specific engineering programs.

Responsibility

The Manager of Design Engineering is accountable to the Director of Engineering for providing the necessary direction for planning, organizing, and executing the engineering efforts in his or her cognizant area. This includes (1) establishment of objectives, policies, technical requirements, schedules, and budgets as required for the effective execution of engineering programs; (2) assurance of engineering design quality and workmanship; and (3) cost-effective implementation of established engineering plans. He or she further seeks out and develops new methods and technology to prepare the company for future business opportunities.

Table 42.4   Typical Charter of Lead Engineer or Task Leader or Task Manager (Matrix Organization)

Position Title

Lead Engineer, Processor Development

Authority

Appointed jointly by the Director of Engineering and the Program Manager, the Lead Engineer has the authority to direct the implementation of the processor development within the functional organization according to the established plans and requirements.

Responsibility

Reporting to both the Director of Engineering and the program manager, the Lead Engineer is responsible: (1) to the Functional/Resource Manager(s) for the most cost-effective technical implementation of the assigned program task, including technical excellence and quality of workmanship; and (2) to the Program Manager for the most efficient implementation of the assigned program task according to the program plan, including cost, schedule, and specified performance.

Specifically, the lead engineer:

1. Represents the program manager within the functional organization.
2. Develops and implements detailed work package plans from the overall program plan.
3. Acts for the functional manager in directing personnel and other functional resources toward work package implementation.
4. Plans the functional resource requirements for his or her work package over the life cycle of the program and advises functional management accordingly.
5. Directs subfunctional task integration according to the work package plan.
6. Reports to program office and functional manager on the project status.

Table 42.5 Policy Directive: Engineering

Company Logo
ABC Division

Policy Directive No. 00.000.00   Effective Date: May 1, 1984
Engineering Management

1. *Purpose*

   To provide the basis for the management of engineering activities within the ABC Division. Specifically, the Policy establishes the relationship between program and functional operations and defines the responsibilities, authorities, and accountabilities necessary to ensure engineering quality and operational performance consistent with established business plans and company policies.

2. *Organizations Affected*

   All Engineering Departments of the ABC Division.

3. *Definitions*

   3.1 *Unit Manager.* Individual responsible for managing an engineering group. He or she reports to a Department Manager.

   3.2 *Department Manager.* Individual responsible for managing an Engineering Department, usually consisting of various engineering groups, each headed by a Unit Manager. The Department Manager reports to the Director of Engineering.

   3.3 *Lead Engineer or Task Manager.* Individual jointly appointed by the Program Manager and the Functional Superior. The Lead Engineer or Task Manager is responsible for the cross-functional integration of the assigned task.

   3.4 *Task Authorization.* A one-page document summarizing the program requirements, including tasks, results, schedules, budgets, and key personnel.

   3.5 *Annual Engineering Plan.* Document summarizing the specific goals and objectives of Engineering as well as specific programs to be funded for the next year.

   3.6 *Department Budget.* Document establishing the funding level for each Engineering Department based on activities and other organizational needs.

4. *Applicable Documents*

   4.1 Supervisor's Handbook.

   4.2 Program Management Guidelines.

   4.3 Procedures on Engineering Management, No. 00.11.00.

5. *Policy*

   5.1 All engineering departments are organized along functional lines with each individual reporting to one functional superior. The chain of command is established as: Unit Manager, Department Manager, Director of Engineering. The Director of Engineering is appointed by the Division General Managers.

   5.2 Specific Engineering Departments are chartered by the Director of Engineering who is also responsible for making the managerial appointments.

   5.3 The Engineering Organization, through its various departments, is chartered to accomplish the following functions:

   1. Execute engineering programs according to company-external or -internal requests.

   2. Support the Business Divisions in the identification pursuit, and acquisition of new contract business.

   3. Provide the necessary program direction and leadership within Engineering for planning, organizing, developing, and integrating technical efforts. This includes the establishment of program objectives, requirements, schedules, and budgets as defined by contract and customer requirements.

   4. Maintain close liaison with customer, vendor, and educational resources, keeping abreast of state-of-the-art and technological trends.

1227

Table 42.5   (*Continued*)

Company Logo
ABC Division

Policy Directive No. 00.000.00   Effective Date: May 1, 1984
Engineering Management

---

     5.   Plan, acquire, and direct outside and company-funded research that enhances the company's competitive position.

     6.   Plan, use, and develop all engineering resources to meet current and future requirements for quality, cost-effectiveness, and state-of-the-art technology.

5.4   The activity level is being defined by department budgets that are based on company-internal or -external requests for specific services or programs.

5.5   Resource requirements for new engineering programs can be introduced at any point of the engineering organization. However, these requests must be reported via the established chain of command to the Director of Engineering who coordinates and assigns the specific activities or programs to organizational units.

5.6   The basis for resource planning is the Annual Engineering Plan, which is to be coordinated with the Division's Annual Business Plan.

5.7   The Program Manager is responsible for defining a preliminary program plan as a basis for negotiating personnel resources with functional managers.

5.8   Each unit manager appoints a Lead Engineer or Task Manager for each program to be executed within his or her organization.

5.9   No engineering program shall be started without proper program definition, which includes at the minimum (1) the specific tasks to be performed, (2) the overall performance specifications, (3) a definition of the overall task configuration, (4) the deliverable items or results, (5) a master schedule, and (6) the program budget. Five percent of the total program budget estimate is a reasonable outlay for this program definition phase.

5.10   Responsibility for *defining* an engineering program rests with the Program Manager who must coordinate the requirements and capabilities between the sponsor and resource organizations, including engineering.

5.11   The principal instrument for summarizing contractional requirements of (1) technical requirements, (2) deliverable items, (3) schedules, (4) budgets, and (5) responsible personnel shall be the Task Authorization.

5.12   The Lead Engineer or Task Manager is responsible for the cross-functional integration of the assigned engineering task. He or she has dual accountability: (1) to his or her functional superior for the most cost-effective technical implementation, including technical excellency and quality of workmanship, and (2) to the Program Manager for the most efficient implementation of the assigned task, according to the established specifications, schedules, and budgets.

5.13   Functional management or their designated task leaders shall comply with the directions of the Program Manager regarding the contractional requirements, including performance, schedules, budgets, and changes, as well as reporting requirements and customer presentations.

5.14   The responsibility for (1) quality of workmanship and (2) method of task implementation rests explicitly with the engineering organization.

5.15   Conflicts over administrative or technical items between the functional and program organization must be dealt with in good faith and on a timely basis. If no resolution can be found, the problems must be reported to the next managerial level (if necessary up to the Division General Manager) who is responsible for resolving the conflict.

42.1.3   Organizing a New Engineering Department

The following 10-point process provides some guidelines for organizing a new engineering department or engineering task team.

Mission and Task Definition

1.   *Mission Statement.* Define the overall mission for the new organizational unit, including global business objectives, technical objectives, responsibilities, and timing. Write charter, agreed upon by senior management.

2. *Work Definition.* Define the type of work to be performed, both functions and projects. Use a functional organization chart or work breakdown structure as planning model.

3. *Organizational Scope and Budget.* Define the functional capabilities that should be included in the new department versus those that can be contracted from other support departments. The global organization chart and budgets for personnel, all facilities, and organizational development provide the basic tools. The organizational scope must be agreed upon by senior management before proceeding to the next step.

Organization Design

4. *Organizational Structure.* Define the principal functional units of the new department or project and its reporting structure. The organization chart and mission statements for each principal engineering unit are the basic tools.

5. *Reporting Relations and Control.* Define the specific reporting relations, responsibilities, and controls that make the new engineering department or project operational. These relations can be very intricate, especially in a matrix environment, and should be delineated in some basic policy statements, management directives, and organizational charters.

6. *Manpower and Facility Planning.* The specific staffing levels, facility, and equipment requirements should be delineated and summarized in an organization plan in a step-by-step fashion, showing: (1) the specific tasks necessary for implementing the organization, (2) the responsible individuals, (3) the timing, and (4) the budget.

Staffing

7. *Job Description.* Prepare job descriptions for all positions reporting directly to the head of the new department or project.

8. *Personnel Requisition, Hiring of Direct Reports.* Advertise your new positions, company internally and externally. Interview and staff those positions that directly report to you.

9. *Complete Staffing.* Each manager should be permitted to hire his or her own staff. Therefore, staffing is performed in organizational layers, repeating steps 7 through 9 until the organization is fully staffed, a process which is often *planned* over longer periods of time to give the new organization the flexibility to grow gradually with its operational capability.

Organization Development

10. *Team Building.* Building the personnel into a homogeneous team, unified and focused on the organizational goals, is a continuous process. It requires the whole spectrum of management skills and an understanding of the evolving culture and value system of the new engineering organization. For specific guidelines see Section 42.3.1.

Eliciting Commitment

Establishing firm commitment from engineering personnel is important in any situation. It is absolutely critical, however, in situations where managers have to step across functional lines and have to deal with personnel over whom they have little or no control. In such a matrix environment, engineering management personnel must sell each assignment primarily on the basis of professional interest, perceived career potential, and satisfaction of higher-order needs such as the need for recognition, visibility, praise, and accomplishment. Furthermore, the perception of how the task manager may *influence* the rewards administered by other functional organizations may influence the enthusiasm of personnel toward the new assignment. Therefore, it is critical for engineering task leaders to identify and satisfy the needs of their personnel, involve them during a project start-up, and clearly discuss new assignments, including requirements and career implications. The initial job review is an important element in the process to involve, assess needs, and elicit commitment. This interview should clearly address:

- Job requirements
- Project objectives, scope, and challenges
- Specific job content in terms of techniques, skills, experience, disciplines, and degree of difficulty
- Reporting relations and project organization
- Personnel transfer/assignment policy
- Risks and challenges
- Final project outcome and deliverables
- Specific measures for final success
- Reward system

Such a dialog will provide both parties with an insight into how compatible the requirements are with personal goals on one hand and capabilities on the other hand. In addition, there are many side benefits to a properly conducted interview prior to the assignment. It sets the tone of how the engineering task or project will be managed, recognizes the individual needs and wants, stimulates thinking, creates involvement and visibility at many levels, and establishes lines of communication for the future. If properly done, interviewing for engineering assignments is a highly pervasive process that provides a unifying force toward integrating and building a committed engineering project team.

42.2 COMPENSATION AND PERSONNEL DEVELOPMENT

42.2.1 Attracting and Holding Quality People

Since the mid 1950s there has been an increasing body of theory and practice concerned with attracting and holding of quality personnel. It stresses the importance of favorable overall organizational climate and the satisfaction of professional needs in addition to the crucial financial incentives.

Many factors contribute to an organization's ability to attract and hold quality people. Richardson and Lomicka[1] discuss eight "motivational dimensions": (1) work climate, *including pay, benefits,* security, and status; (2) performance goals; (3) performance feedback; (4) participative process; (5) job design; (6) organization design; (7) supervisory style; and (8) organization values and process. These factors must be considered when designing the engineering organization as well as during its operation.

Proper financial compensation and rewards are important to the morale and motivation of any organization. The engineering function is no exception. To be perceived fair and equitable these compensation practices must be based on job content, job performance, and the realities of the job market.

Four specific management systems are used to define the financial compensation and reward system:

1. Job classification
2. Job description
3. Base pay classification
4. Performance appraisal

Designing these systems requires more than writing just another procedural document. It requires a senior management perspective of understanding the organizational culture and motivational dimensions. In addition, there are several issues that often make it necessary to treat compensation practices of engineering personnel separately from the rest of the organization:

*   Job classifications and job descriptions for engineering personnel are usually not compatible with those existing for other professional jobs. It is often difficult to pick an existing classification and adopt it to engineering personnel. If not adjusted properly, the low formal authority and small number of direct reports may distort the position level of engineering managers in spite of their broad-range business responsibilities.
*   Dual accountability and dual reporting relationships, quite common in engineering organizations, raise the question of who should assess performance and control the rewards.
*   Bases for financial rewards are often difficult to establish, quantify, and administer. The criterion of "doing a good job" is difficult to quantify.
*   Special compensations for overtime, extensive travel, or living away from home should be considered. Bonus pay for preestablished results may be considered, but is a particularly difficult and delicate issue since often many people contribute to the incentivized results.

42.2.2 Performance Evaluation and Reward Systems

Some specific guidelines are provided in this section to help in establishing compensation systems for engineering organizations. The foundations of these compensation practices are based on the four previously listed systems: (1) job classification, (2) job description, (3) base pay classification, and (4) performance appraisal.

Job Classifications and Job Descriptions

Every effort should be made to fit the job classifications for the new department into the existing standard classification that has already been established for the organization.

The first step is to define job titles for various engineering personnel and their corresponding responsibilities. Since titles imply certain responsibilities, organizational status, and pay level, they should be carefully selected and each of them supported by a formal job description.

The job description provides the basic charter for the position. It should be written not just for one individual but more generally for all individuals who fit the respective job classification. A good job description is brief and concise, typically one page. It can be broken down into three sections: (1) overall responsibilities, (2) specific duties, and (3) qualifications. Table 42.6 provides a sample job description for a lead project engineer.

Base Pay Classifications

To be equitable, engineering pay scales must be carefully designed, considering the total spectrum of responsibilities, formal authorities over personnel and other resources, and accountabilities for business results. If done improperly over- or underpayment of engineering personnel may result, leading to dissatisfaction and morale problems throughout the organization.

Many companies that struggled with this problem found a solution by (1) working out compensation schemes as a team of senior managers and personnel specialists, (2) applying criteria of responsibility and business accountability, and (3) considering the supply and demand of factors of the local area. Once the proper range of compensation has been set, fine-tuning can be done during the hiring process. Managers can choose a salary from the established range based on judgment of actual position responsibilities, the candidate's qualifications, the available budget, and other considerations. Valuable guidance and perspective can be obtained from the personnel specialist.

Performance Appraisals

Performance appraisals are designed to accomplish one or several of the following objectives:

- Assess the employee's work performance, preferably against preestablished objectives
- Provide a justification for salary actions

Table 42.6 Sample Job Description

Job Description: Lead Project Engineer of Processor Development

Overall Responsibility

Responsible for directing the technical development of the new central processor including managing the technical personnel assigned to this development. The lead project engineer has dual responsibility (1) to his/her functional superior for the technical implementation and engineering quality and (2) to the project manager for managing the development within the established budget and schedule.

Specific Duties and Responsibilities

1. Provide necessary program direction for planning, organizing, developing, and integrating the engineering effort, including establishing the specific objectives, schedules, and budgets for the processor subsystem.
2. Provide technical leadership for analyzing and establishing requirements, preliminary designing, designing, prototyping, and testing of the processor subsystem.
3. Divide the work into discrete and clearly definable tasks. Assign tasks to technical personnel within the lead engineer's area of responsibility and other organizational units.
4. Define, negotiate, and allocate budgets and schedules according to the specific tasks and overall program requirements.
5. Measure and control cost, schedule, and technical performance against program plan.
6. Report deviations from program plan to program office.
7. Replan trade-off and redirect the development effort in case of contingencies such as to best utilize the available resources toward the overall program objectives.
8. Plan, maintain, and utilize engineering facilities to meet the long-range program requirements.

Qualifications

1. Strong technical background in state-of-the-art central processor development.
2. Prior task management experience with proven record for effective cost and schedule control of multidisciplinary technology-based task in excess of SIM.
3. Personal skills to lead, direct, and motivate senior engineering personnel.
4. Excellent oral and written communication skills.

- Establish new goals and objectives for the next review period
- Identify and deal with work-related problems
- Serve as a basis for career discussions

The realities are, however, that some of these objectives are in conflict. For example, a performance appraisal that focuses on salary actions is usually not conducive to future goal setting, problem-solving, or career planning.

In order to get around this dilemma, many companies have separated the salary discussion from other parts of the performance appraisal. Moreover, successful managers have carefully considered the complex issues involved and have built a performance appraisal system that is solid on content, measurability, and source of information.

The first challenge is on content, that is, to decide what to review and how to measure performance. Modern engineering management practices try to individualize accountability as much as possible. However, practices are often modified to ensure balance and equity for jointly performed responsibilities. In principle, engineering personnel are assessed on their ability to implement a specific project:

- Technical implementation measured against requirements, quality, schedules, and cost targets.
- Team performance as measured by ability to staff, to build an effective task group, to interface with other groups, and to integrate among various functions.

In addition, engineering managers are measured by the following:

- Business results as measured by contributions to profits, return on investment, new business, and income; also, on-time delivery, meeting of contractual requirements, and within-budget performance.
- Managerial performance as measured by overall engineering management effectiveness, organization, direction and leadership, and team performance; as well as technological advancement and readiness for future missions.

All of these areas are difficult to assess and even more difficult to quantify. Specific measures are shown in Tables 42.7 and 42.8 for both engineers and engineering managers.

Finally, one needs to decide who is to conduct the performance appraisal and to make the salary adjustment. Since engineering organizations often involve matrix structures with dual accountabilities, good practices call for inputs from "all bosses." Moreover, resource personnel may be shared among many activities and projects. Therefore, it should be the immediate functional superior, rather than the project manager, who integrates all performance feedback and makes the final appraisal.

42.2.3   Developing the Engineering Function

Developing the engineering organization and its personnel requires a disciplined approach and advanced planning. Because of the functional dynamics involved and the experiential skills required, many senior managers feel that engineering personnel must be trained and developed primarily on the job. However, there are sharply different approaches being taken by organizations regarding their methods and practices. Table 42.9 provides an overview of these methods, which break down into three categories: (1) on-the-job training, (2) schooling, and (3) organizational system developments.

A Planned Approach to Project Personnel Development

Successful personnel development programs should consider the total job cycle from need identification to placement. Four specific phases are defined. Each phase should be planned, regularly reviewed and updated, and its results used on a perpetuating basis as inputs for the next phase.

Phase 1: Identify Staffing Needs.   An important first step is to identify the number and types of engineering management personnel needed over the next few years. Three years may be a convenient planning horizon. This phase needs the involvement of all senior management personnel. Specific management tools to aid this phase are:

Business plans
Organization charts
Job descriptions
Manpower plans

Phase 2: Define the Work Environment.   The various management processes must be clearly spelled out, especially in a less structured environment that is often characterized by vertical and horizontal

Table 42.7 Performance Measures for Senior Engineers

Who Performs Appraisal

Functional superior of engineer

Source of Performance Data

Functional resource managers and project managers

Primary Measures

1. Success in directing the agreed-on task toward completion
   - Technical implementation according to requirements
   - Quality
   - Key milestones/schedules
   - Target costs, design-to-cost
   - Innovation
   - Trade-offs
2. Effectiveness as a team member or team leader
   - Building effective task team
   - Working together with others, participation, involvement
   - Interfacing with support organizations and subcontractors
   - Interfunctional coordination
   - Getting along with others
   - Change orientation
   - Making commitments

Secondary Measures

1. Success and effectiveness in performing special assignments in accordance with functional charter
   - Advancing technology
   - Developing organization
   - Resource planning
   - Functional direction and leadership
2. Administrative support services
   - Reports and reviews
   - Special task forces and committees
   - Project planning
   - Procedure development
3. New business development
   - Bid proposal support
   - Customer presentations
4. Professional development
   - Keeping abreast in professional field
   - Publications
   - Liaison with societies, vendors, customers, and educational institutions

Additional Considerations

1. Difficulty of tasks involved
   - Technical challenges
   - State-of-the-art considerations
   - Changes and contingencies
2. Managerial responsibilities
   - Task leader for number of project personnel
   - Multifunctional integration
   - Budget responsibility
   - Staffing responsibility
   - Specific accountabilities
3. Multiproject involvement
   - Number of different projects
   - Number and magnitude of functional tasks and duties
   - Overall work load

Table 42.8   Performance Measures for Engineering Managers

Who Performs Appraisal

Functional superior of employee

Source of Performance Data

Functional superior, resource managers, general manager, and project managers

Primary Measures

1. Engineering manager's success in using the company resources toward preestablished global objectives
    - Technology development
    - Personnel development
    - Resource/personnel utilization
    - Quality of engineering work
    - Participation in long-range business planning
2. Manager's effectiveness in leading specific missions and project, including establishing
    - Market measures, new business, follow-on contract
    - Accomplishing specific programs within preestablished budgets and schedules
    - Bid proposals
    - Ad hoc tasks

Secondary Measures

1. Ability to utilize organizational resources
    - Overhead cost reduction
    - Working with existing personnel
    - Cost-effective make–buy decisions
2. Ability to build effective task teams
    - Project staffing
    - Interfunctional communications
    - Low team conflict, complaints, and hassles
    - Professionally satisfied team members
    - Work with support groups
3. Effective work planning and plan implementation
    - Plan detail and measurability
    - Commitment by key personnel and management
    - Management involvement
    - Contingency provisions
    - Reports and reviews
4. Customer/client satisfaction
    - Perception of overall task performance by sponsor
    - Communications, liaison
    - Responsiveness to changes
5. Participation in business management
    - Keeping senior management informed of new project/product/business opportunities
    - Bid proposal work
    - Business planning, policy development

Additional Considerations

1. Difficulty of tasks involved
    - Technical tasks
    - Administrative and organizational complexity
    - Multidisciplinary nature
    - Staffing and start-up

Table 42.8 *(Continued)*

2. Scope of work
   - Total project budget
   - Number of personnel involved
   - Number of organizations and subcontractors involved
3. Changing work environment
   - Nature and degree of customer changes and redirections
   - Contingencies

lines of control. Furthermore, these processes become a crucial input to the design of any personnel development program. Three types of processes are used to define the work environment:

1. *Role Specifications.* These define the authority and reporting relations. They include job descriptions, policies, procedures, directives, and organization charts.
2. *Engineering Management Support System.* This provides the basis for integrated decision-making. It includes planning, bidding, reports, reviews, and controls.
3. *Reward System.* This establishes the accountability and assesses performance as a basis for rewards. It includes the appraisal system, salary and bonus structure, promotional policies, and career development.

The management tools to be developed and updated include:

Job description
Charter of specific program office
Project management policies and procedures
Project management guidelines
Appraisal and reward guidelines/policy

Table 42.9 Methods and Techniques for Developing Engineering Personnel, Including Managers

I. *Experiential Training/On-the-Job*

- Working with experienced professional leader
- Working with project team members
- Assigning a variety of project management responsibilities
- Bid proposal activities
- Job rotation
- Formal on-the-job training
- Supporting multifunctional activities
- Customer liaison activities

II. *Conceptional Training/Schooling*

- Courses, seminars, workshops
- Simulations, games, cases
- Group exercises
- Hands-on exercises in using project management techniques
- Professional meetings
- Conventions, symposia
- Readings, books, trade journals, professional magazines

III. *Organizational Development*

- Formally established and recognized engineering function
- Proper organization
- Support systems
- Engineering charter
- Engineering management directives, policies, and procedures

Phase 3: Establish the Personnel Development Plan. The development plan should include all aspects, from identifying candidates to placement. Along with training, companies have learned to pay more attention to the team approach of personnel development where personnel become intrinsically interested, involved, and motivated toward taking on responsibilities and grow with the engineering operations. The major facets of the development plan are:

Means of identifying and attracting candidates for various positions

Specific training and development methods

Appraisal and assessment of training effectiveness

Personnel placement and advancement

Phase 4: The Implement Plan. To be successful, the personnel development plan must have an approved budget and the total commitment of the organization, at all levels of management. This commitment will likely develop if management was actively involved in the generation of the plan. It seems to be important to assess the effectiveness of the personnel development activity from both sides, the management and the personnel in training. If people are not attracted to the development opportunity, it is doomed to fail. Regular interviews should be held with all participants of the program being involved. The specific management tools to aid this phase are:

Engineering personnel development plan

Formal and informal feedback from participants

Review meetings

Independent audit

42.3 DEVELOPING THE PROJECT TEAM

42.3.1 Team Building Is Crucial to Effective Engineering Management

Building effective task teams is one of the prime responsibilities of engineering managers. Team building involves a whole spectrum of management skills required to identify, commit, and integrate various task groups from traditional functional organizations into a multidisciplinary task/management system. This process has been known for centuries. However, it becomes more complex and requires more specialized management skills as bureaucratic hierarchies decline and horizontally oriented teams and work units evolve.[5]

Team building is an ongoing process that requires leadership skills and an understanding of the organization, its interfaces, authority and power structure, and motivational factors. It is a process particularly critical in certain engineering situations, such as:

- Establishing a new engineering program
- Improving project–client relations
- Organizing for a bid proposal
- Integrating new project personnel
- Resolving interfunctional problems
- Working toward a major milestone
- Transitioning the project into a new engineering phase

Characteristics of an Effective Engineering Team

To be effective the engineering manager must provide an atmosphere conducive to teamwork, that is, a climate with the following characteristics:

- Team members are committed to the program
- Good interpersonnel relations and team spirit exist
- The necessary resources are available
- Goals and program objectives are clearly defined
- Top management is involved and supportive
- Good program leadership exists
- Open communication among team members and the support organization
- A low degree of detrimental conflict

To build an effective and productive team certain needs and leadership expectations must be fulfilled as shown below:

Needs of Team Members	Expectations from Task Leader
Sense of belonging	Direction and leadership
Interest in work itself	Assistance in problem-solving
Professional achievement	Creation of stimulating environment
Encouragement: pride	Adaptation of new members
Recognition for accomplishment	Capacity to handle conflict
Protection from infighting	Resistance to change
Job security/continuity	Representation at higher management
Potential for career growth	Facilitation of career growth

Although they are often not apparent, these needs and expectations usually exist intrinsically in all teams. That is, team members may not always realize their needs and wants explicitly. However, members must feel comfortable in their work environment, have mutual trust, and be interested and involved in the work they are doing in order to form a coherent effective team.

42.3.2 Barriers to Project Team Development

The understanding of barriers to engineering team building can help in developing an environment conducive to effective teamwork. The barriers to team building discussed in Table 42.10 were identified and analyzed in a field study by Thamhain and Wilemon.[4] They are typical for many project environments.

Table 42.10   Barriers to Effective Team Building and Suggested Handling Approaches

Barrier	Suggestions for Effectively Managing Barriers (How to Minimize or Eliminate Barriers)
Differing outlooks, priorities, interests, and judgments of team members	Make an effort early in the project life cycle to discover these conflicting differences. Fully explain the scope of the project and the rewards that may be forthcoming upon successful project completion. Sell the "team" concept and explain responsibilities. Try to blend individual interests with the overall project objectives.
Role conflicts	As early in a project as feasible, ask team members where they see themselves fitting into the project. Determine how the overall project can best be divided into subsystems and subtasks (e.g., the work breakdown structure). Assign/negotiate roles. Conduct regular status review meetings to keep team informed on progress and watch for unanticipated role conflicts over the project's life.
Unclear project objectives/ outcomes	Ensure that all parties understand the overall and interdisciplinary project objectives. Clear and frequent communication with senior management and the client becomes critically important. Status review meetings can be used for feedback. Finally, a proper team name can help to reinforce the project objectives.
Dynamic project environments	The major challenge is to stabilize external influences. First, key project personnel must work out an agreement on the principal project direction and "sell" this direction to the total team. Also, educate senior management and the customer on the detrimental consequences of unwarranted change. It is critically important to forecast the environment within which the project will be developed. Develop contingency plans.

Table 42.10 *(Continued)*

Barrier	Suggestions for Effectively Managing Barriers (How to Minimize or Eliminate Barriers)
Competition over team leadership	Senior management must help establish the project manager's leadership role. On the other hand, the project manager needs to fulfill the leadership expectations of team members. Clear role and responsibility definition often minimizes competition over leadership.
Lack of team definition and structure	Project leaders need to sell the team concept to senior management as well as to their team members. Regular meetings with the team will reinforce the team notion as will clearly defined tasks, roles, and responsibilities. Also, visibility in memos and other forms of written media as well as senior management and client participation can unify the team.
Project personnel selection	Attempt to negotiate the project assignments with potential team members. Clearly discuss with potential team members the importance of the project, their role in it, what rewards might result upon completion, and the general "rules-of-the-road" of project management. Finally, if team members remain uninterested in the project, then replacements should be considered.
Credibility of project leader	Credibility of the project leader among team members is crucial. It grows with the image of a sound decision-maker in both general management and relevant technical expertise. Credibility can be enhanced by the project leader's relationship to other key managers who support the team's efforts.
Lack of team member commitment	Try to determine lack of team member commitment early in the life of the project and attempt to change possible negative views toward the project. Often, insecurity is a major reason for the lack of commitment: try to determine why insecurity exists, then work on reducing the team member's fears. Conflicts with other team members may be another reason for lack of commitment. It is important for the project leader to intervene and mediate the conflict quickly. Finally, if a team member's professional interests lie elsewhere, the project leader should examine ways to satisfy part of the team member's interests or consider replacement.
Communication problems	The project leader should devote considerable time communicating with individual team members about their needs and concerns. In addition, the leader should provide a vehicle for timely sessions to encourage communications among the individual team contributors. Tools for enhancing communications are status meetings, reviews, schedules, reporting system, and colocation. Similarly, the project leader should establish regular and thorough communications with the client and senior management. Emphasis is placed on written and oral communications with key issues and agreements in writing.
Lack of senior management support	Senior management support is an absolute necessity for dealing effectively with interface groups and proper resource commitment. Therefore, a major goal for project leaders is to maintain the continued interest and commitment of senior management in their projects. We suggest that senior management become an integral part of project reviews. Equally important, it is critical for senior management to provide the proper environment for the project to function effectively. Here the project leader needs to tell management at the onset of the program what resources are needed. The project manager's relationship with senior management and ability to develop senior management support are critically affected by his own credibility and the visibility and priority of his project.

How to Overcome Team Barriers

Managers who are successfully performing their role not only recognize these barriers but also know when in the life cycle of an engineering project they are most likely to occur. Moreover, these managers take preventive actions and are usually social architects who understand the interaction of organizational and behavioral variables and can foster a climate of active participation and minimal conflict. This requires carefully developed skills in leadership, administration, organization, and technical expertise. Specific suggestions are advanced in Table 42.10.

42.3.3 Team Building as an Ongoing Process

While proper attention to team building is critical during the early phases of an engineering project, it is a never-ending process. The manager is continually monitoring team functioning and performance to see what corrective action may be needed. Several barometers provide good clues to potential team dysfunctioning. First, noticeable changes in performance levels for the team and/or for individual team members should always be followed up. Such changes can be symptomatic of more serious problems, for example, conflict, lack of work integration, communication problems, and unclear objectives. Second, the project leader and team members want to be aware of the changing energy levels of team members. This, too, may signal more serious problems or that the team is tired and stressed. Sometimes changing the work pace, taking time off, or selling short-term targets can serve as a means to reenergize team members. More serious cases, however, can call for more drastic action, such as reappraising project objectives and/or the means to achieve them. Third, verbal and nonverbal clues from team members may be a source of information on team functioning. It is important to hear the needs and concerns of team members (verbal clues) and to observe how they act in carrying out their responsibilities (nonverbal clues). Finally, detrimental behavior of one team member toward another could signal a problem within the team. There are several indicators of effective and ineffective teams. At any point in the life of a team, the engineering manager should be aware of certain effectiveness/ineffectiveness indicators, which are summarized in Table 42.11.

It is highly recommended that project leaders hold regular meetings to evaluate overall team performance and deal with team functioning problems. The focus of these meetings can be directed toward "what are we doing well as a team" and "what areas need our team's attention." This approach often brings positive surprises in that the total team will be informed on progress in diverse project areas, for example, a breakthrough in technology development, a subsystem schedule met ahead of the original target, or a positive change in the client's behavior toward the project. After the positive issues have been discussed, attention should be devoted toward actual or potential problem areas. The meeting leader should ask each team member for observations and an open discussion should be held. Next, assignments should be agreed upon on how to best handle these problems. Finally, a plan for problem follow-up should be developed.

Table 42.11 Project Team Characteristics: Effective versus Ineffective

The Effective Team: Likely Characteristics	The Ineffective Team: Likely Characteristics
High performance and task efficiency	Low performance
Innovative/creative behavior	Resistance to change
Commitment	Low commitment to project objectives
Professional objectives of team members coinciding with project requirements	Need for excessive coaching
Team members highly interdependent, interface effectively	Unclear project objectives and fluid commitment levels from key participants
Capacity for conflict resolution, but conflict encouraged when it can lead to beneficial results	Unproductive gamesmanship, manipulation of others, hidden feelings, conflict avoided at all costs
Effective communication	Confusion, conflict, inefficiency
High trust levels	Subtle sabotage, fear, disinterest, or footdragging
Result-oriented	Lack of involvement
Interested in membership	Cliques, collusion, isolating members
High energy levels and enthusiasm	Lethargic/unresponsive
High morale	
Change-oriented	

42.4 MOTIVATION AND LEADERSHIP

42.4.1 New Realities in Engineering Management

More than in any other organizational form, an understanding of motivational forces and leadership skills are essential for effective management of multidisciplinary engineering activities. The ability to build project teams, motivate people, and create organizational structures conducive to innovative and effective work requires sophisticated interpersonal and organizational skills.

There is no single magic formula for successful engineering management. However, most senior managers agree that effective engineering managers need to understand the interaction of organizational and behavioral elements in order to build an environment conducive to their teams' motivational needs, a prerequisite for effective leadership of complex multidisciplinary engineering undertakings.

42.4.2 Motivational Forces in Engineering

Understanding people is important for the effective management of today's technical programs and engineering activities. The breed of managers that succeeds within these often unstructured work environments faces many challenges: Company-internally they must be able to deal effectively with a variety of interfaces and support personnel over whom they often have little or no control. Company-externally, managers have to cope with constant and rapid change regarding the technology, markets, regulations, and socioeconomic factors. Moreover, traditional methods of authority-based direction, performance measures, and control are virtually impractical in such contemporary environments.

Sixteen specific professional needs of engineering personnel are listed below. Research studies show that the fulfillment of these professional needs can drive project personnel to higher performance or, conversely, the inability to fulfill these needs may become a barrier to team work and high project performance. The rationale for this important correlation is found in the complex interaction of organizational and behavioral elements. Effective engineering management performance involves three primary issues: (1) people skills, (2) organizational structure, and (3) management style. All three issues are influenced by the specific task to be performed and the surrounding environment. The same three issues surface again in satisfying professional needs of engineering professionals. That is, the degree of satisfaction of any of the needs is a function of (1) having the right mix of people with appropriate skills and traits, (2) organizing the people and resources according to the tasks to be performed, and (3) adopting the right leadership style. For a detailed discussion of the field research study see Ref. 2.

1. *Interesting and Challenging Work.* Interesting and challenging work satisfies various professional esteem needs. It is oriented toward intrinsic motivation of the individual and helps to integrate personal goals with the objectives of the organization.

2. *Professionally Stimulating Work Environment.* This leads to professional involvement, creativity, and interdisciplinary support; it also fosters team building and is conducive to effective communication, conflict resolution, and commitment toward organizational goals. The quality of this work environment is defined through its organizational structure, facilities, and management style. It includes a very complex set of variables maybe closest resembled by Ouchi's Theory Z Organization.

3. *Professional Growth.* Professional growth is measured by promotional opportunities, salary advances, learning of new skills and techniques, and professional recognition. A particular challenge exists for management in limited-growth or zero-growth businesses to compensate for the lack in promotional opportunities by offering more intrinsic professional growth in terms of job satisfaction.

4. *Overall Leadership.* This involves dealing effectively with individual contributors, managers, and support personnel within a specific functional discipline as well as across organizational lines. It involves technical expertise, information-processing skills, effective communication, and decision-making skills. Taken together, leadership means satisfying the needs for clear direction and unified guidance toward established objectives.

5. *Tangible Rewards.* These include salary increases, bonuses, and incentives, as well as promotions, recognition, better offices, and educational opportunities. Although extrinsic, these financial rewards are necessary to sustain strong long-term efforts and motivation. Furthermore, they validate more intrinsic rewards such as recognition and praise, and reassure people that higher goals are attainable.

6. *Technical Expertise.* Personnel need to have all necessary interdisciplinary skills and expertise available within the engineering team to perform the required tasks. Technical expertise includes an understanding of the technicalities of the work, the technology and underlying concepts, theories and principles, design methods and techniques, and functioning and interrelationship of the various components that make up the total system.

7. *Assisting in Problem Solving.* Assisting in problem solving, such as facilitating solutions to technical, administrative, and personal problems, is a very important need. If not satisfied, it often leads to frustration, conflict, and poor quality engineering work.

8. *Clearly Defined Objectives.* Goals, objectives, and outcomes of an engineering effort must be clearly communicated to all affected personnel. Conflict can develop over ambiguities or missing information.

9. *Management Control.* Management control is important to engineering professionals for effective performance. The managers must understand the interaction of organizational and behavior variables in order to exert the direction, leadership, and control required to steer the engineering effort toward established organizational goals without stifling innovation and creativity.

10. *Job Security.* This is one of the very fundamental needs that must be satisfied before people consider higher-order growth needs.

11. *Senior Management Support.* Senior management support should be provided in four major areas: (1) financial resources, (2) effective operating charter, (3) cooperation from support departments, and (4) provision of necessary facilities and equipment. It is particularly crucial to larger, more complex undertaking.

12. *Good Interpersonal Relations.* These are required for effective engineering teamwork since they foster a stimulating work environment with involved, motivated personnel; low conflict; and high productivity.

13. *Proper Planning.* Proper planning is absolutely essential for successful management of complex engineering work. It requires communication and information-processing skills to define the actual resource requirements and administrative support necessary. It also requires the ability to negotiate resources and commitment from key personnel in various support groups across organizational lines.

14. *Clear Role Definition.* This helps to minimize role conflict and power struggles among team members and/or supporting organizations. Clear charters, plans, and good management direction are some of the powerful tools to facilitate clear role definitions.

15. *Open Communications.* This satisfies the need for a free flow of information both horizontally and vertically. It keeps personnel informed and functions as a pervasive integrator of the overall project effort.

16. *Minimizing Changes.* Although the engineering manager has to live with constant change, project personnel often see change as an unnecessary condition that impedes their creativity and productivity. Advanced planning and proper communications can help to minimize changes and lessen their negative impact.

42.4.3 The Power Spectrum in Engineering Management

Engineering managers must often cross functional lines to get the required support. This is especially true for managers who operate within a matrix structure. Almost invariably, the manager must build multidisciplinary teams into cohesive work groups and successfully deal with a variety of interfaces such as functional departments, staff groups, team members' clients, and senior management. This is a work environment where managerial power is shared by many individuals. In contrast to the traditional organization that provides position power largely in the form of legitimate authority, engineering managers derive their power partially from other sources such as earned authority: the power that comes from expertise and credibility, and the image of a sound decision maker.

In today's environment, most engineering management is characterized by:

- Authority patterns defined only in part by formal organization chart plans.
- Authority largely perceived by the members of the organization based on earned credibility, expertise, and perceived priorities.
- Dual accountability of most personnel, especially in project-oriented environments.
- Shared power between resource managers and project/task managers.
- Individual autonomy and participation greater than in traditional organizations.
- Weak superior–subordinate relationships in favor of stronger peer relationships.
- Subtle shifts of personnel loyalties from functional to project lines.
- Engineering performance depending on teamwork.
- Group decision-making tending to favor the strongest organizations.
- Reward and punishment power along both vertical and horizontal lines in a highly dynamic pattern.
- Influences to reward and punishment from many organizations and individuals.

- Multiproject involvement of support personnel and sharing of resources among many engineering activities.

Position power is a necessary prerequisite for effective engineering leadership. Like many other components of the management system, leadership style has also undergone changes over time. With increasing task complexity, increasing dynamics of the organizational environment, and the evolution of new organizational systems, such as the matrix, a more adaptive and skill-oriented management style evolved. This style complements the organizationally derived power bases such as authority, reward, and punishment with bases developed by the individual manager. Examples are technical and managerial expertise, friendship, work challenge, promotional ability, fund allocations, charisma, personal favors, project goal identification, recognition, and visibility. This so-called Style II management evolved particularly with the matrix. A descriptive summary is presented in Table 42.12. Effective engineering management combines both styles.

Various research studies by Gemmill, Thamhain, and Wilemon provide an insight into the power spectrum available to project managers.

Figure 42.1 indicates the importance of nine influence bases in gaining support from engineering subordinates and assigned personnel. Technical and managerial expertise, work challenge, and influence over salary are the most important influences that engineering managers seem to have, while authority, fund allocations, and penalty factors appear least important in gaining support from engineering subordinates.

More specifically, influence bases have been investigated[5] with regard to engineering management effectiveness. Managers who are perceived by their personnel as (1) *emphasizing* work challenge and expertise but (2) *deemphasizing* authority, penalty, and salary as influence base, foster a climate of good communications, high involvement, and strong support to the engineering tasks at hand. Ultimately, this style results in high performance ratings by upper management.

The relationship of managerial influence style and effectiveness was statistically measured. Kendall–Tau nonparametric correlation techniques were used to determine the association of each influence base to (1) perceived support, (2) openness of communication channels, (3) work involvement, and (4) managerial effectiveness rating. For specifics see Ref. 3.

One of the most interesting findings is the importance of work challenge as an influence method. Work challenge appears to integrate the personal goals and needs of personnel with organizational goals. That is, work challenge is primarily oriented toward the intrinsic motivation of engineering personnel, while other methods are oriented more toward extrinsic rewards with less regard to the personnel's professional needs. Therefore, to enrich the assignments of engineering personnel in a professionally challenging way may indeed have a beneficial effect on overall performance. Additionally, the assignment of challenging work is a variable over which engineering managers may have a great deal of control. Even if the total task structure is fixed, the method by which work is assigned and distributed is discretionary in most cases.

42.4.4 Skill Requirements for Engineering Managers

Engineering activities are invariably complex and multifaceted. Managing an engineering organization requires skills in technical, administrative, interpersonal, and leadership areas. To be effective, managers must consider all facets of the job. They must understand the people, the task, the tools, and the

Table 42.12 Bases of Influence

Authority	The ability to influence others because personnel perceive you as having the right to issue orders.	System I Style Organizationally derived bases of influence. May be sufficient in more closed, mechanistic organizational systems.
Reward Power	The ability to influence others because personnel value the rewards they believe you are capable of administering.	
Punishment Power	The ability to influence others because personnel want to avoid the "punishments."	
		System II Style
Expert Power	The ability to influence others because they respect your expertise and competence.	Power bases derived from the individual manager. Needed complementary to Style I in dynamic organizational systems.
Referent Power	The ability to influence others because they identify with you, your project/product, or your position in the organization.	

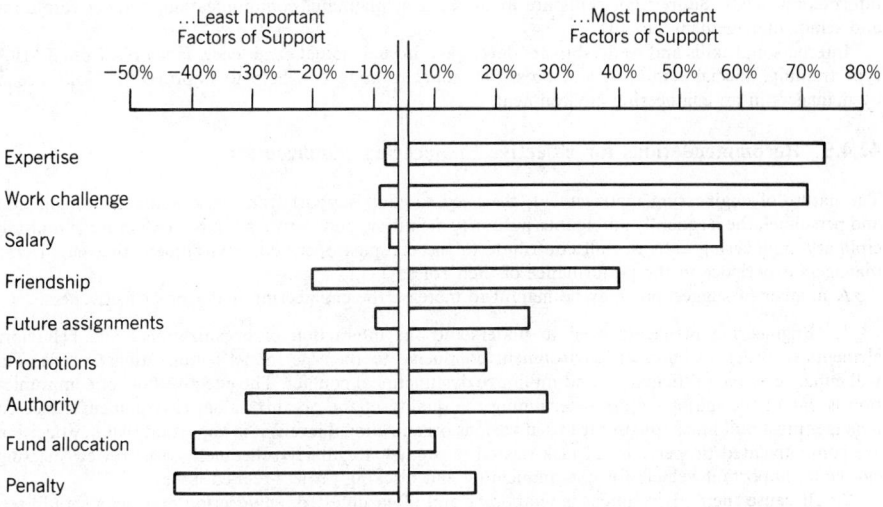

Fig. 42.1 Power spectrum of engineering managers.

organization. The days of the manager who gets by with technical expertise alone or pure administrative skills are gone.[6]

Technical Skills

The engineering manager rarely has all the technical expertise to direct the multidisciplinary activities handed. Nor is it necessary or desirable to do so. It is essential, however, that the engineering manager understands the technologies and their trends, the markets, and the business environment to participate effectively in the search for integrated solutions and technological innovations. Without this understanding the consequences of local decisions on the total program, the potential growth ramifications, and relationships to other business opportunities cannot be foreseen by the manager. Furthermore, technical expertise is necessary to communicate effectively with the project team, and to assess risks and make trade-offs between cost, schedule, and technical issues.

Administrative Skills

Administrative skills are essential. The engineering manager must be experienced in planning, staffing, budgeting, scheduling, and other control techniques.

While it is important that managers understand the company's operating procedures and the available tools, it is often necessary for the engineering manager to free himself or herself from the administrative details. Particularly in larger departments, engineering managers have to delegate considerable administrative tasks to support groups or to hire an administrator.

Interpersonal Skills and Leadership

Effective engineering leadership involves a whole spectrum of skills and abilities: clear direction and guidance; ability to integrate multidisciplinary efforts; ability to plan and elicit commitments; communication skills; assistance in problem solving; dealing effectively with managers and support personnel across functional lines often with little or no formal authority; information-processing skills, the ability to collect and filter relevant data valid for decision-making in a dynamic environment; and ability to integrate individual demands, requirements, and limitations into decisions that benefit the overall engineering task. It further involves the program manager's ability to resolve intergroup conflicts and team building—important factors to overall program performance.

In engineering, quality leadership depends heavily on the manager's personal experience, credibility, and understanding of the interaction of organizational and behavioral elements. The engineering manager must be a social architect, that is, understanding how the organization works and how to work with the organization. Organizational skills are particularly important during the start-up of a new engineering program when the manager forms the new team by integrating people from many different disciplines into an effective work group. It requires far more than simply constructing another organization chart. At a minimum, it requires defining the reporting relationships, responsibilities, lines of control, and information needs. Supporting skills are in the area of planning, communication, conflict resolution, and senior management support.

Interpersonal skills and leadership are developed through actual experience. However, formal MBA-type training, special seminars, and cross-functional training can help to improve the skills needed by managers in an engineering environment.

42.4.5 Recommendations for Effective Engineering Management

The nature of engineering management, the need to elicit support from various organizational units and personnel, the frequently ambiguous authority definition, and the often temporary nature of multidisciplinary engineering activities all contribute to the complex operating environment that engineering managers experience in the performance of their roles.

A number of suggestions may be helpful to increase the engineering manager's effectiveness.

1. Engineering managers need to understand the interaction of organizational and behavioral elements in order to build an environment conducive to their personnel's motivational needs. This will enhance active participation and minimize dysfunctional conflict. The effective flow of communication is one of the major factors determining the quality of the organizational environment. Since the manager must build task/project teams at various organizational levels, it is important that key decisions are communicated properly to all task-related personnel. Regularly scheduled status review meetings can be an important vehicle for communicating and tracking project-related issues.

2. Because their environment is temporary and often untested, engineering managers should seek a leadership style that allows them to adapt to the often conflicting demands existing within their organization, support departments, customers, and senior management. They must learn to "test" the expectations of others by observation and experimentation. Although difficult, they must be ready to alter their leadership style as demanded by both the specific tasks and their participants.

3. Engineering managers should try to accommodate the professional interests and desires of supporting personnel when negotiating their tasks. Task effectiveness depends on how well the manager provides work challenges to motivate those individuals who provide support. Work challenge further helps to unify personal goals with the goals and objectives of the organization. Although the total work of an engineering department may be fixed, the manager has the flexibility of allocating task assignments among various contributors.

4. Engineering managers should develop or maintain technical expertise in their field. Without an understanding of the technology to be managed, they are unable to win confidence among their team members to build credibility with the customer community, to participate in search for solutions, or to lead a unified engineering effort.

5. Effective planning early in the life cycle of a new engineering program is highly recommended. Planning is a pervasive activity that leads to personnel involvement, understanding, and commitment. It helps to unify the task team, provides visibility, and minimizes future dysfunctional conflict.

6. Finally, engineering managers can influence the work climate by their own actions. Their concern for personnel, their ability to integrate personal goals and needs of personnel with organizational goals, and their ability to create personal enthusiasm for the work itself can foster a climate that is high in motivation, work involvement, open communication, creativity, and engineering performance.

A situation approach to engineering management is presented in Fig. 42.2, which indicates that the intrinsic motivation of engineering personnel increases with the managers' emphasis on work challenge, their own expertise, and their ability to establish friendship ties. On the other hand, emphasis on penalty measures, authority, and inability to manage conflict lowers personnel motivation. Furthermore, managerial position power is determined by such variables as formal position within the organization, the scope and nature of work, earned authority, and ability to influence promotion and future work assignments.

Taken as a whole, engineering managers who have strong position power and can foster a climate of highly motivated personnel not only obtain higher support from their personnel, but also achieve high overall performance ratings from their superiors.

42.5 EFFECTIVE CONFLICT MANAGEMENT

42.5.1 Conflict in Engineering Organizations

Conflict is fundamental to complex task management and is often determined by the interplay of the engineering organization with its support functions. Complex organizational relationships, dual accountability, and shared managerial powers and resources are factors that contribute to a new frontier environment where conflict is inevitable. When conflict becomes dysfunctional, it often results in poor

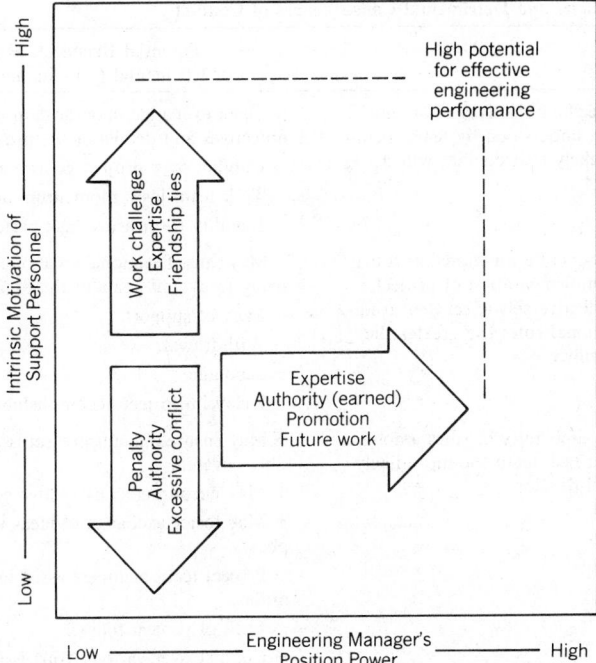

Fig. 42.2 Position power and motivational forces in engineering.

group decision-making, long delays over operational issues, and disruptions of the team's efforts. However, contrary to conventional wisdom, conflict can be beneficial. It produces involvement, new information, and competitive spirit.

42.5.2 Common Causes of Conflict in Engineering

Project managers frequently admit they are unprepared to deal effectively with conflict. Yet, understanding conflict and its determinants is a prerequisite for effectively dealing with the multitude of work and personality problems which come into play during the execution of modern engineering programs.

Conflict in engineering organizations has been investigated by Thamhain and Wilemon[4] who defined seven typical sources of conflict. Table 42.13 summarizes these conflict situations including potentially beneficial and detrimental consequences. Engineering managers can often prepare for and deal more effectively with operational problems if they can be anticipated.

42.5.3 Conflict in the Project Life Cycle

While it is important to examine some of the principal determinants of conflict from an aggregate perspective, more specific and useful insight can be gained by exploring the intensity of various conflict sources over the life cycle of an engineering project, namely, project formation, project build-up, main program phase, and phaseout.

What to Do about Conflict?

An important first step in effectively dealing with the inevitable conflict situations is to recognize the specific sources of conflict throughout the project life. Preventive measures are often simple and represent good management practices. Table 42.14 provides a summary of specific recommendations for minimizing unproductive conflict.

42.5.4 Conflict Resolution Approaches

When conflict situations arise, engineering managers often resolve the conflict at the project level rather than resorting to arbitration at higher management. Yet a formal hierarchical system is established in many organizations to arbitrate conflicts that cannot be resolved at the project level. However,

Table 42.13 Beneficial and Detrimental Consequences of Conflict

Conflict Situation	Potential Beneficial (+) and Detrimental (−) Characteristics
1. The less the specific objectives of an engineering task are understood by team members the more likely that conflict will develop.	+ Open exchange may develop which clarifies objectives and details of the project. − Conflict over project goals and priorities. − Task team loses momentum and energy. − Inability to measure task performance.
2. The more members of a functional area perceive that the implementation of project management will adversely affect their traditional organizational roles the greater the potential for conflict.	+ May cause functional area to relate more effectively to overall goals of the organization. − Lack of support. − Withdrawal. − Sabotage. − Delay in project accomplishment.
3. The greater the ambiguity of roles among participants of a task team the more likely that conflict will develop.	+ May ensure difficult project issues are responsibly covered. + May encourage constructive competition. + May foster exchange of ideas in early project phases. + Project team members assist in own role definition. − Lack of project focus. − Conflict over various "turf issues." − Confusion. − Avoidance of responsibility.
4. The greater the agreement on top management goals the lower the potential for detrimental conflict at the engineering project level.	+ Lowers potential for parochial conflict among departments. + Goal congruency. − Top management goals may not be applicable at project level. − Discourages dialogue of project issues with upper management.
5. The lower the task leader's authority and power of reward and punishment the greater the potential for conflict to develop.	+ Encourages open exchange between project manager and support groups. + Encourages participative decision-making. + Power varies according to situational needs, expertise, and credibility. + Good involvement and participation. − May delay decision-making process. − May not be able to reward satisfactorily key contributors.
6. The greater the diversity of expertise among the participants of a task team the greater the potential for conflict.	+ Enhances the decision-making process by providing high-quality informational inputs. + Team members can effectively search for solutions. + Improves overall communications and departmental interfaces. − May slow project decision-making process due to alternative problem-solving approaches suggested. − May lead to power struggle and grouping. − Search for excellency may impair engineering efficiency.

Table 42.14 Recommendations for Minimizing Unproductive Conflict

Engineering Project Life-Cycle Phase	Conflict Source	Recommendations
Project formation	Priorities	Clearly defined plans. Joint decision making and/or consultation with affected parties.
		Stress importance of project to goals of the organization.
	Procedures	Develop detailed administrative operating procedures to be followed in conduct of project. Secure approval from key administrators. Develop statement of understanding or charter.
	Schedules	Develop schedule commitments in advance of actual project commencement.
		Forecast other departmental priorities and possible impact on project.
Build-up phase	Priorities	Provide effective feedback to support areas on forecasted project plans and needs via status review sessions.
	Schedules	Carefully schedule work breakdown packages (project subunits) in cooperation with functional groups.
	Procedures	Contingency planning on key administrative issues.
Main program	Schedules	Continually monitor work in progress. Communicate results to affected parties.
		Forecast potential problems and consider alternatives.
		Identify potential "trouble spots" needing closer surveillance.
	Technical	Early resolution of technical problems.
		Communication of schedule and budget restraints to technical personnel.
		Emphasize adequate, early technical testing.
		Facilitate early agreement on final designs.
	Manpower	Forecast and communicate manpower requirements early.
		Establish manpower requirements and priorities with functional and staff groups.
Phaseout	Schedules	Close monitoring of schedules throughout project life cycle.
		Consider reallocation of available manpower to critical project areas prone to schedule slippages.
		Attain prompt resolution of technical issues which may impact schedules.
	Personality and manpower	Develop plans for reallocation of manpower upon project completion.
		Maintain harmonious working relationships with project team and support groups. Try to loosen up "high-stress" environment.

more typically, engineering personnel have straightforward meetings with other team members to address task-oriented conflicts. Such direct confrontation methods often can dispose of many problems before they become detrimental to the overall engineering task.

Five methods, defined by Blake and Mouton,[8] are frequently used to describe modes for handling conflict. These approaches are listed in Table 42.15.

What Conflict Resolution Works Best?

Engineering managers are quick to point out the highly situational effectiveness of each conflict resolution mode. However, in spite of their situational nature, certain modes are more important to engineering

Table 42.15   Five General Modes for Handling Conflict

Conflict Handling Modes	Characteristics
Withdrawing	Retreating from a conflict issue. Here, engineering managers do not deal with the disagreement. They may ignore it entirely, may withdraw out of fear, may feel inadequate to bring about an effective resolution, or may want to avoid rocking the boat. Withdrawal may intensify the conflict situation. On the other hand, withdrawal can be beneficial as either a temporary strategy to allow the other party to cool off or to buy time for further study of the issues.
Smoothing	Emphasizes common areas of agreement and deemphasizes areas of difference. Like withdrawing, smoothing may not address the real issues in a disagreement. Smoothing can be effective, however, in combination with other modes. Furthermore, it leads to identifying areas of agreement so one can more clearly focus on areas of disagreement, and project work can often continue in areas where there is agreement by the parties.
Compromising	Bargaining and searching for solutions that bring some degree of satisfaction to the parties involved in conflict. Since compromise yields less than optimum results, the project manager must weigh such actions against program goals. Compromise is always the outcome of a negotiation.
Forcing	Exerting one's viewpoint at the expense of another—characterized by competitiveness and win–lose behavior. To force, the leader must have the proper position power. Forcing is often used as a last resort by engineering managers since it may cause resentment and deterioration of the work climate. However, many organizational or final technical decisions are most effectively made via forcing.
Confronting or problem solving	Involves a rational problem-solving approach. Disputing parties solve differences by focusing on the issues, looking at alternative approaches, and selecting the best alternative. Confronting may contain elements of other modes such as compromising and smoothing.

managers, while others are less favored.[6] As indicated in Fig. 42.3, the problem-solving approaches of *confrontation* and *compromise* seem to be most important to engineering managers and most frequently used while *forcing* and *withdrawal* seem to be least favored.

The effectiveness of conflict resolution approaches is highly situational and depends on the type of conflict to be solved, the personnel and the organization involved, and the power relationship that exists among the personnel engaged over a particular problem. It appears, however, that confrontation, compromise, and the so-called problem-solving approach, not only are most frequently used but also

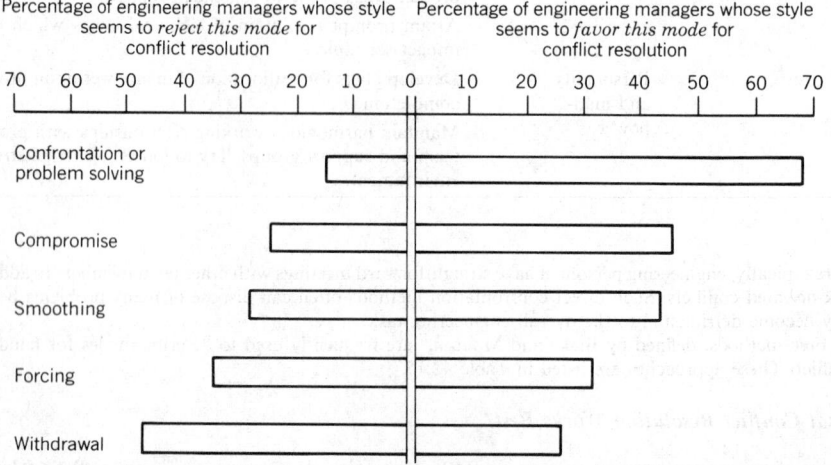

Fig. 42.3   Conflict resolution modes.

seem to result in a higher management performance as measured by general management personnel. This holds particularly in situations of complex, unstructured decision-making that does not follow traditional lines of authority or preestablished rationales.

Research in technologically oriented environments suggests that it is less important to search for a best mode of effective conflict management. It appears to be more significant that managers, in their capacity as integrators of diverse organizational resources, use the full range of conflict resolution modes. While confrontation seems to be the ideal approach under most circumstances, other approaches may be equally effective depending on the situational content of the disagreement. Withdrawal, for example, may be used effectively as a temporary measure until new information can be sought or to "cool off" a hostile reaction from a colleague. As a basic long-term strategy, however, withdrawal may actually escalate a disagreement if no resolution is eventually sought.

In other cases, compromise or smoothing might be considered an effective strategy by the engineering manager, if it is acceptable with regard to the overall project objectives. Forcing, on the other hand, often proves to be a win/lose mode. Nevertheless, some engineering managers find that forcing is the only viable mode in some situations, especially those dealing with organizational or personnel decisions. Confrontation, or the problem-solving mode, may encompass all conflict handling modes to some extent. For example, in solving a conflict a project manager may use confrontation in combination with withdrawal, compromise, forcing, and smoothing to eventually get an effective resolution to the issue in question whereby all affected parties can live with the eventual outcome. Finally, the manager can affect the situational effectiveness by fostering an environment that is high on work challenge, with people involved in professionally stimulating, interesting work, with proper forward planning, open communication channels, sufficient technical expertise, and good engineering leadership. This is a work environment that minimizes the detrimental conflict and maximizes its beneficial aspects.

REFERENCES

1. J. A. Richardson and R. F. Lomicka, "Non-Financial Motivation: Creating a Work Environment for Effective Human Performance," in *Handbook of Industrial Engineering*, Wiley-Interscience, New York, 1982.

2. H. J. Thamhain, "The Effective Engineering Manager," *Conference Record*, IEEE Engineering Management Conference, Dayton, Ohio, 1981.

3. G. R. Gemmill and H. J. Thamhain, "Influence Styles of Project Managers: Some Performance Correlates," *Academy of Management Journal*, 216–224 (June, 1974).

4. H. J. Thamhain and D. L. Wilemon "Diagnosing Conflict Determinants in Project Management," *IEEE Transactions of Engineering Management* (February, 1975).

5. H. J. Thamhain and D. L. Wileman, "Anatomy of a High-Performing New Product Team," in *Handbook of Technology Management*, Wiley, New York, 1986.

6. H. J. Thamhain, "Developing Engineering Management Skills," in *Management of R & D Engineering*, North Holland, Amsterdam, 1986.

7. R. R. Blake and J. S. Mouton, *The Managerial Grid*, Gulf Publishing Company, 1964; and R. J. Burke, "Methods of Resolving Interpersonal Conflict," *Personnel Administration*, 48–55 (July–August, 1969).

8. H. J. Thamhain and D. L. Wilemon, "Conflict Management in Project Lifecycles," *Sloan Management Review* (Spring, 1975).

BIBLIOGRAPHY

J. R. Adams and N. S. Kirchof, "A Training Technique for Developing Project Managers," *Project Management Quarterly* (March, 1983).

J. J. Aquilino, "Multi-skilled Work Teams: Productivity Benefits," *California Management Review* (Summer, 1977).

J. D. Aram and C. P. Morgan, "Role of Project Team Collaboration in R & D Performance," *Management Science* (June, 1976).

S. Atkins and A. Katcher, "Getting Your Team in Tune," *Nation's Business* (March, 1975).

J. Bachman et al., "Bases of Supervisory Power: A Comparative Study in Five Organizational Settings," in *Control in Organizations*, A. Tannenbaum (Ed.), McGraw-Hill, New York, 1968.

L. Benningson, "The Team Approach to Project Management," *Management Review*, **61**, 48–52 (January, 1972).

J. L. Bower, "Managing for Efficiency, Managing for Equity," *Harvard Business Review* (July–August, 1983).

R. J. Burke, "Methods of Managing Superior–Subordinate Conflict," *Canadian Journal of Behavioral Science*, **2**, 124–135 (1970).

R. Carzo, Jr., "Some Effects of Organization Structure on Group Effectiveness," *Admin. Science Quarterly* (March, 1963).

P. Chambers, "Matching Personalities to Create Effective Teams," *International Management* (November, 1973).

J. P. Cicero and D. L. Wilemon, "Project Authority: A Multidimensional View," *IEEE Transactions on Engineering Management* (May, 1970).

D. I. Cleland, "A Kaleidoscope of Matrix Management Systems," *AMA Management Review* (December, 1981).

D. I. Cleland and W. R. King, *Systems Analysis and Project Management,* McGraw-Hill, New York, 1983.

The Conference Board, *Matrix Organizations of Complex Businesses,* New York, 1979.

S. Dauzig, "How to Bridge the Technology Gap in Manpower Planning," *AMA Management Review* (April, 1982).

L. E. Davis, "Individuals and the Organization," *California Management Review* (Spring, 1980).

B. A. Diliddo, P. C. James, and H. J. Dietrich, "Managing R & D Creatively: B. F. Goodrich's Approach," *Management Review* (July, 1981).

D. D. Ely, "Team Building for Creativity," *Personnel Journal* (April, 1975).

W. M. Evan, "Superior–Subordinate Conflict in Research Organizations," *Administrative Science Quarterly,* **10,** 52–64 (1965).

W. M. Evan, "Conflict and Performance in R & D Organizations," *Industrial Management Review,* 37–46 (Fall, 1965).

W. L. Franch and R. W. Hollman, "Management by Objectives: The Team Approach," *Calif. Management Review* (September, 1975).

G. R. Gemmill and H. J. Thamhain, "The Effectiveness of Different Power Styles of Project Managers in Gaining Project Support," *IEEE Transactions on Engineering Management,* **20** (May, 1973); *Project Management Quarterly,* **5**(1) (Spring, 1974).

G. R. Gemmill and D. L. Wilemon, "The Power Spectrum in Project Management," *Sloan Management Review,* 15–25 (Fall, 1970).

W. C. Goggin, "How the Multidimensional Structure Works at Dow Corning," *Harvard Business Review* (January–February, 1974).

J. L. Hayes, "Teamwork," *Management Review* (November, 1977).

E. P. Hollander and J. W. Julian, "Contemporary Trends in the Analysis of Leadership Processes," *Psychological Bulletin,* **71,** 387–397 (1969).

D. S. Hopkins, "Roles of Project Teams and Venture Groups in New Product Development," *Research Management* (January, 1975).

R. J. Howe, "Building Teams for Increased Productivity," *Personnel Journal* (January, 1977).

S. A. Huesing, "Team Approach and Computer Development," *Journal Systems Management* (September, 1977).

L. V. Imundo, *Establishing Standards and Assessing Employee Performance,* AMA Publication, New York, 1980.

J. Ivancevich and J. Donnelly, "Leader Influence and Performance," *Personnel Psychology,* **23,** 539–549 (1970).

J. Kelly, "Make Conflict Work for You," *Harvard Business Review,* 103–113 (July–August, 1970).

J. T. Kidder, *The Soul of a New Machine,* Avon Books, Hearst Corporation, New York, 1982.

D. F. Kocaoglu (Ed.), *Management of R & D Engineering,* North Holland, Amsterdam, 1986.

D. F. Kocaoglu (Ed.), *Handbook of Technology Management,* Wiley, New York, 1986.

R. C. Linden and T. R. Mitchell, "The Effects of Group Interdependence on Supervisor Performance Evaluation," *Personnel Psychology* (Summer, 1983).

J. Mattill, "Critical 20% for Innovation," *Technical Review* (July–August, 1983).

C. J. Middleton, "How to Set Up a Project Organization," *Harvard Business Review* (March–April, 1967).

H. Mitzberg, "Organization Design: Fashion or Fit?," *Harvard Business Review* (January–February, 1981).

M. Moravec, "Performance Appraisals," *AMA Management Review* (June, 1981).

I. Mordlea, "A Comparison of an R & D Laboratory's Organization Structures," *IEEE Transactions on Engineering Management* (December, 1967).

D. H. Morton, "The Project Manager, Catalyst to Constant Change," *Project Management Quarterly* (January, 1965).

W. Ouchi, *Theory Z: How American Business Can Meet the Japanese Challenge,* Addison-Wesley, Reading, MA, 1981.

R. D. Paulson, "The Chief Executive as Change Agent," *Management Review* (February, 1982).

C. Pincus, "An Approach to Plan Development and Team Formation," *Project Management Quarterly* (December, 1982).

H. E. Pywell, "Engineering Management in a Multi-Matrix Organization," *IEEE Transactions on Engineering Management* (August, 1979).

H. E. Riggs, *Managing High-Technology Companies,* Lifetime Learning Publications, Belmont, CA, 1983.

W. P. Sullivan, "Do You Have What It Takes to Get to the Top?," *Management Review* (April, 1983).

R. Tannenbaum and W. H. Schmidt, "How to Choose a Leadership Pattern," *HBR Classic* (May–June, 1973).

H. J. Thamhain, "Managing Engineers Effectively," *IEEE Transactions on Engineering Management* (August, 1983).

H. J. Thamhain, "Production Within a Matrix Environment," Chapter 28 in D. I. Cleland (Ed.), *Matrix Management Systems Handbook,* Van Nostrand Reinhold, New York, 1983.

H. J. Thamhain, *Engineering Program Management,* Wiley, New York, 1985.

H. J. Thamhain and D. L. Wilemon, "Skill Requirements of Engineering Program Managers," *Convention, Record* and presentation at Twenty-Sixth Joint Engineering Management Conference, October, 1978.

H. J. Thamhain and D. L. Wilemon, "Leadership, Conflict, and Program Management Effectiveness," *Sloan Management Review* (Fall, 1977).

H. J. Thamhain, D. L. Wilemon, and H. S. Dugan, "Managing Change in Project Management," *Convention Record Ninth Annual Symposium of the Project Management Institute,* Chicago, IL, October, 1977.

H. J. Thamhain and D. L. Wilemon, "Leadership Effectiveness in Project Management," in Donald J. Reifer (Ed.), *Tutorial: Software Management,* IEEE Computer Society, Los Angeles, CA, 1983, pp. 235–241.

N. M. Tichy, C. J. Fobrun, and M. A. Devanna, "Strategic Human Resource Management," *Sloan Management Review* (Winter, 1982).

D. L. Wilemon, "Project Management and Its Conflicts: A View from Apollo," *Chemical Technology* (September, 1972).

J. H. Zenger and D. E. Miller, "Building Effective Teams," *Personnel* (March, 1974).

CHAPTER 43

PROGRAM PLANNING AND CONTROL

JOSEPH A. MACIARIELLO

Claremont Graduate School
and
Claremont McKenna College
Claremont, California

This chapter describes the organization behavior and planning and control processes that are appropriate for the management of many complex engineering and construction projects. Each of the techniques and processes must be adapted to specific projects in accordance with their complexity.

43.1 ORGANIZING FOR COMPLEX PROGRAMS

Complex programs, although requiring close coordination, may not be large enough or of a duration long enough to justify a separate project organization form, yet they often require far more coordination than is possible under a functional organization. In other words, neither extreme, a pure functional or a pure project organization, is ideally suited for such complex programs, and the trade-off implied in the choice between the two is often unacceptable.

As the programs of an organization have become more complex and numerous, the functional organization has been forced to recognize the limitations of its structure. Often this recognition has led to formation of ad hoc task forces of functional personnel to handle the "unique" coordination problems created by the high degree of interdependence required among functional disciplines on a given program. As these types of programs are recognized to be the very nature of the organization, task-force relationships often become less ad hoc and more formal.

Members of task forces are selected based on functional expertise from the relevant functional groups. Leadership of a task force is often assigned from the functional group that can make the largest contribution toward solving the problem for which the task force was formed. Once formed, it provides a device to achieve coordination across specialized functions and is an expedient means of getting complex tasks accomplished.

Although task forces are disbanded upon completion of a program, the number of programs requiring such an arrangement should be expected to increase for any organization engaged in complex programs. In these organizations the need for a more permanent structure that makes relationships among task-force members more formal soon becomes apparent. The new structure should allow the organization to coordinate these complex activities while retaining functional specialization and resulting scale economies. Neither of the two structures described allows an organization to achieve total coordination of complex activities and full-scale economies of specialization *simultaneously*.

A solution to this organizational dilemma was first provided approximately 30 years ago in the aerospace industry and has since been used in various forms by firms who are in businesses as divergent as food and produce and accounting services. This organization form has become known as the *matrix form* and is used today in many variants of project and product management. Regardless of the type of application, it embodies the assumption that it is not necessary to make an explicit trade-off in organizational design between advantages of scale economies and coordination. Rather it attempts to achieve high levels of each simultaneously and thus reach a higher level of effectiveness than either the functional or decentralized form.

With this as a background, we now turn to a general description of the matrix structure, followed by illustration of its application to project management.

43.1.1 The Matrix Organization

Regardless of the nature of the firm employing this hybrid organization form, it may be described using a matrix as shown in Table 43.1, with the various products, projects, or services identified in the columns of the matrix and the functions identified in the rows.

The matrix depicts an organization with n programs. These programs represent the primary output of the organization and therefore place demands upon the various functions. The functions, represented in the rows of the matrix, supply resources to various programs within the organization. The total output of a function for any particular period of time is found in the last column of the matrix and consists of the sum of all the contributions of a particular function to various programs. Each row of the last column represents the demand placed upon one function by multiple programs. This demand may be measured in physical units of input, such as man-hours, or in monetary units.

Functional contributions that are indirect, such as research and development, contract administration, personnel, business planning, public relations, and finance are included in the overhead row of the matrix.

The matrix itself, however, *does not* uniquely define the distinguishing characteristics of the organization form used in program control, although the term is widely used to describe the organization form. Any organization may be described as a matrix; it need not be a hybrid form. All organizations produce outputs that may be identified in the columns of a matrix and use inputs that may be identified in the rows of the matrix. That is, all organizations have a purpose and use inputs or processes to fulfill the purpose. The truly distinguishing characteristics of the matrix organization structure, in all its variations, lie in the dual dimensions of management embodied within it and the allocation of responsibility and authority resulting from the dual management dimensions.

The matrix organization uses two overlapping dimensions of management, each of which may be identified within the matrix. The dual dimensions of management may be identified by referring to

Table 43.1   The Structure of Matrix Organization

Functions	Programs				Total Functional Output
	Program 1	Program 2	Program 3	Program n	
Engineering					
Procurement					
Quality assurance					
Logistics support					
Manufacturing					
Program control					
Program management					
Overhead					
Total program requirements					

Table 43.1. Under the matrix organization structure full responsibility for the goals of a program is given to program managers, and this responsibility is identified by the column dimension of Table 43.1. However, functional personnel who perform the work on the programs of an organization receive direction from functional management under the matrix structure, thus providing the second and overlapping dimension of management for the programs of the organization. The functional dimension of management is identified by the rows of the matrix.

Therefore, although the program manager assumes full responsibility for delivery of a product that meets performance specifications on a timely basis and in accordance with contractual resource limits, he or she does not have direct authority over the functional organizations that actually perform the work. If he or she did have such authority, the organization would not be a hybrid at all; it would be simply a decentralized form. The distinguishing feature, therefore, of the matrix organization is the separation of the responsibility for the goals of a program from the authority to direct the work necessary to achieve those goals.

Furthermore, although there is a separation of responsibility and authority under the matrix structure, functional personnel actually do operate subject to dual sources of authority: the knowledge-based authority within the function and the resource-based authority of the program manager. Unity of command is thus broken.

Even though the hallmark of the matrix is the separation of responsibility and authority, in practice we find that rather than a clear separation of responsibility and authority, we have formal and informal relationships among program and functional personnel that lead to a distribution of responsibility and authority. Formal and informal relationships, however, vary depending on the application and personalities involved.

43.1.2   Project-Matrix Organizations

Project organizations are concerned with planning, coordination, and control of complex projects of an organization; projects require many activities proceeding both serially and simultaneously toward an *ultimate goal,* and continuous and intricate interaction among many different functional personnel of an organization. The goals for each project ordinarily include profit, either short term (i.e., the project only) or long term (i.e., future business). The goals are met by achieving agreed upon performance with respect to cost, quality, and time variables. Therefore, cost, quality, and schedule are ordinarily the key success variables for a project. Upon completion of all activities the life of the project ceases and the organization is *dissolved.*

Figure 43.1 provides an example of a typical organization for the management of complex programs with the matrix structure superimposed. The formal organization chart, however, does not illustrate the dual dimensions of management embodied within the matrix. These dual dimensions were illustrated in conjunction with Table 43.1.

The role of the project manager and his or her staff is to carry out the planning and control process of a project *without* getting involved in the actual direction of functional work. Under the matrix structure, the project manager plays the role of a planner, coordinator, and controller whose chief concern and responsibility is to produce a project on time, within cost constraints, and in accordance with quality specifications. The project manager of a medium- to large-scale project generally has a

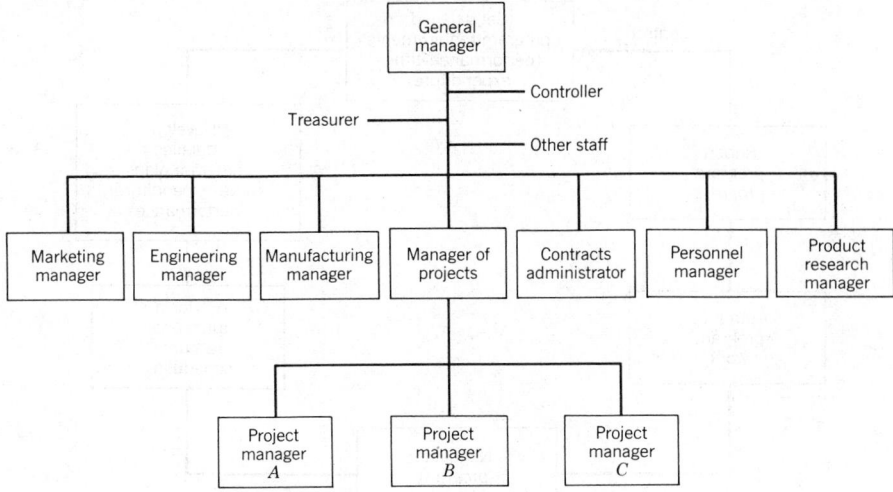

Fig. 43.1 Project-matrix organization structure.

small staff assigned directly to him or her. The staff is charged with responsibility for the planning, coordination, and control of subdivisions of the project.

Authority for directing functional work and accomplishing the technical tasks of a project lies with the managers and individual contributors of the various functions. The individual contributors to a project do not ordinarily report to the project manager, but rather report to their respective functional managers. Often in the project organization structure a connection is made between the project office and functional groups by the appointment of assistant project managers, project engineers, or project leaders for each of the major functions, and these assistants report either directly or indirectly to the project manager.

43.2 PROGRAM PLANNING AND CONTROL PROCESS: OVERVIEW

Figure 43.2 summarizes the program planning and control process. The process provides for planning according to goals and requirements and control by exception. As indicated by Fig. 43.2, the process is initiated by establishing detailed program requirements, and in meeting them we simultaneously achieve the profit goals of a program and the organization.

Detailed requirements are established by preparing a means–end (work) breakdown structure (WBS), which is a hierarchical subdivision of a program. The WBS provides the framework within which we may establish program requirements and prepare detailed plans for the time, expenditures, and performance variables of the program.

Once all end items (purposes and subpurposes) of the program have been established, the next step of the process requires logical, consistent, and coordinated plans to achieve the end items of the program. Network analysis provides a tool for identifying functional activities that must be performed to achieve a lowest-level end on the WBS. That is, in putting together the detailed plans for a complex program, we begin at a level of detail where we can identify functional activities with which we have had some prior experience and in this way break up the novel task into its known elements. This process tends to reduce the novelty of a complex program.

Network plans and the WBS provide the basis for estimating the expenditures of a program. Labor, material, and overhead costs assigned to each lower-level item of a WBS may be derived from estimates of activities contained on networks. By summarizing vertically (i.e., up the WBS) all expenditure estimates beginning at the lowest level of the WBS, we may arrive at expenditure estimates for any other level of the WBS. Standard costs do not exist for complex programs and must be established on an individual program basis.

From detailed network plans, constructed for each lowest-level end item of the WBS, detailed schedules are developed in *each* function with the goal of achieving the plans of the program. During the resource scheduling process, functional managers allocate their functional resources among competing programs to maximize compliance with all the program plans of the organization.

Financial planning must be done for the total program, yet the financial plan so derived cannot be utilized directly as a budget, since its construction assumes that activities will be accomplished in a manner considered optimum from the point of view of the program. The need to balance resources

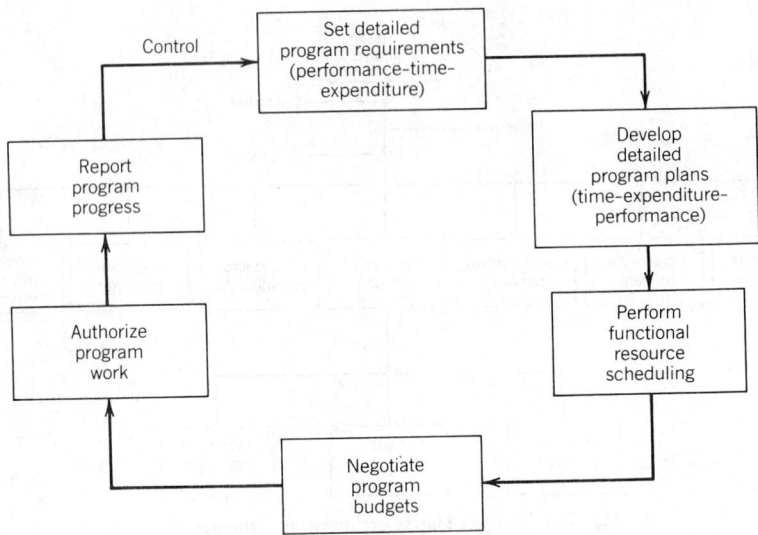

Fig. 43.2   Summary of program-control process.

among programs, observe institutional rules, and react to unexpected program change often requires us to accept less than optimum resource allocations. This means that *actual* resource allocation or scheduling decisions can be made only for those activities to be accomplished in a relatively short period of time, since long-range schedules would depend on long-range demands placed on a function by all programs, and these demands cannot be predicted with accuracy.

Once these allocation decisions are made, a block of work represented on the network, derived from the WBS, is authorized. The program-control process then turns to activities of control. Program office personnel are concerned with controlling actual performance to achieve a balance among expenditure, time, and quality variables of a program. Since we are required to achieve balance among these variables, our program-control system must contain and process progress information on all three variables.

It is necessary, therefore, both to calculate variances for expenditure, time, and quality goals and to derive measures of combined variable performance whenever possible. Techniques of variance analysis are available for combining the time and cost variables into planned and actual measures of *value of work performed*. Performance variables are usually introduced in a qualitative way, although in certain circumstances quantitative performance variances may be defined.

The reporting structure should be designed to conform to the means-end breakdown to the program contained in the WBS. It should be possible to retrieve actual versus planned data on each of the three key program variables for any level of the WBS. In addition, it should be possible to summarize information horizontally to obtain detailed planned and actual data for functional organizations.

The reporting system is part of the contribution made in the program planning and control process toward directing program effort to problem areas to resolve deviations that occur between program requirements and actual performance. It is not a substitute for a well-designed organizational structure, but it is intended to support the structure.

43.3   WORK BREAKDOWN STRUCTURE AND MEANS–END ANALYSIS

Work breakdown structures (WBSs) have been widely used in defense and aerospace activities, but as with the development of the matrix structure, the WBS has been a pragmatic response to the needs posed by new and complex programs and has not drawn guidance from a body of theory. A broad outline of a theory for the WBS does exist, however, and is described by March and Simon.[1] Some of the questions regarding construction and the use of WBSs for the elaboration of activities involved in new programs can be clarified by appealing to their work regarding means–end analysis. They state (Ref. 1, p. 190):

> In the elaboration of new programs, the principal technique of successive approximations is means–end analysis: (1) starting with the general goal to be achieved, (2) discovering a set of means, very generally specified, for accomplishing this goal, (3) taking each of these means, in turn, as a new subgoal and discovering a set of more detailed means for achieving it, etc.

How much detail should the WBS contain? Again referring to Ref. 1 (p. 191):

> *It proceeds until it reaches a level of concreteness where known, existing programs can be employed to carry out the remaining detail. Hence the process connects new general purposes with an appropriate subset of existing repertory of generalized means. When the new goal lies in a relatively novel area, this process may have to go quite far before it comes into contact with that which is already known and programmed; when the goal is of a familiar kind, only a few levels need to be constructed of the hierarchy before it can be fitted into available programmed sequences.*

The objective of the WBS, therefore, is to take innovative output requirements of a complex program and proceed through a hierarchical subdivision of the program down to a level of detail at which groups of familiar activities can be identified. Familiar activities are those for which the functional organizations have had some experience. What is familiar to one organization may not be familiar to another, depending on experience.

Program complexity is an organization-dependent variable, and the same program may require different levels of detail from different organizations. The *primary determinant* of complexity is organization-relevant technology. A program that is of relatively high technology for an organization requires more detailed analysis via the WBS than a program that is of relatively low technology. A program can be complex, however, even if the technology is low relative to what the organization is accustomed; that is, it may be ill structured with many design options available, organizationally interdependent with many interactions required among functional disciplines, or very large. Therefore, the degree of detail found in a WBS for a given program depends on the relative level of technology required, the number of design options available, the interdependence of functional activities, and its size.

43.3.1 WBS and Project Management

Figure 43.3 provides an example of a WBS for a construction project. The objective of the project is to construct a television transmission tower and an associated building for housing television transmission equipment. (This example is developed by permission of the publisher based upon case study "Peterson General Contractors," reproduced in R. A. Johnson, F. E. Kast, and J. E. Rosenzweig, *The Theory and Management of Systems,* 3rd ed., McGraw-Hill Book Company, Inc., New York, pp. 268–273.) As a contractor for the project, we are given specifications for both the tower and building by our customer. We set out to prepare a proposal for this task that will be evaluated by the management of the television station.

As we see from the WBS, the main purpose or end item of the project (i.e., level 0 of the WBS) is provision of the TV transmission system. The primary means for providing this system are shown in level 1 of the WBS. That is, to complete the system we must provide the TV tower, the equipment building, the cable connecting the two, and a service road between the building and tower. These level-1 items are means for constructing the TV transmission system, but are also ends unto themselves for the level-2 items. For example, in order to construct the tower, we must prepare the site, erect the structure, and install the electrical system. These level-2 items are means for accomplishing level-1 ends, which themselves were means for achieving the level-0 end.

Similarly, to provide an equipment building we must prepare the site, provide a structure, and install a fuel tank. These level-2 WBS ends are also means for constructing the structure of the equipment building. Furthermore, to provide a structure for the equipment building, we must provide a basement, main floor, roof, and interior. These level-3 WBS items are means for accomplishing the building, but also ends unto themselves.

For each level-1 WBS item we proceed to elaborate means and ends until we arrive at means that are very familiar tasks, at which point we cease factoring the project into more detailed means. The amount of factoring done on a given end item and project therefore depends on the relative novelty associated with the project. Note that for the service road we proceed *immediately* to final means (i.e., lay the base and grade) to achieve that end. Those two means are familiar activities to the organization and the factoring thus stops for that end item at level 1. Likewise for the level-1 WBS item "underground cable" we simply insert one activity ("install the cable") at level 2 and that ends the means–end chain for the cable.

Once we reach familiar means we identify these as activities rather than ends, simply because they are final means, and, although our detailed planning may separate each of these activities into two or more tasks, there is no utility in identifying more detailed means. All other WBS elements except at level 0 serve as both means and ends. Our detailed network planning begins at the level of the WBS where these final means or activities are identified. Network planning thus begins at different levels of the WBS for various level-1 ends. For example, network planning will begin at level 1 for the service road, but at level 2 for the equipment building.

The elements of Fig. 43.3 that remain to be explained are the level-1 ends "program management" and "overhead." Strictly speaking, we define our programs in terms of identifiable ends or outputs until we get down to the very last level, at which point we identify functions or activities; these

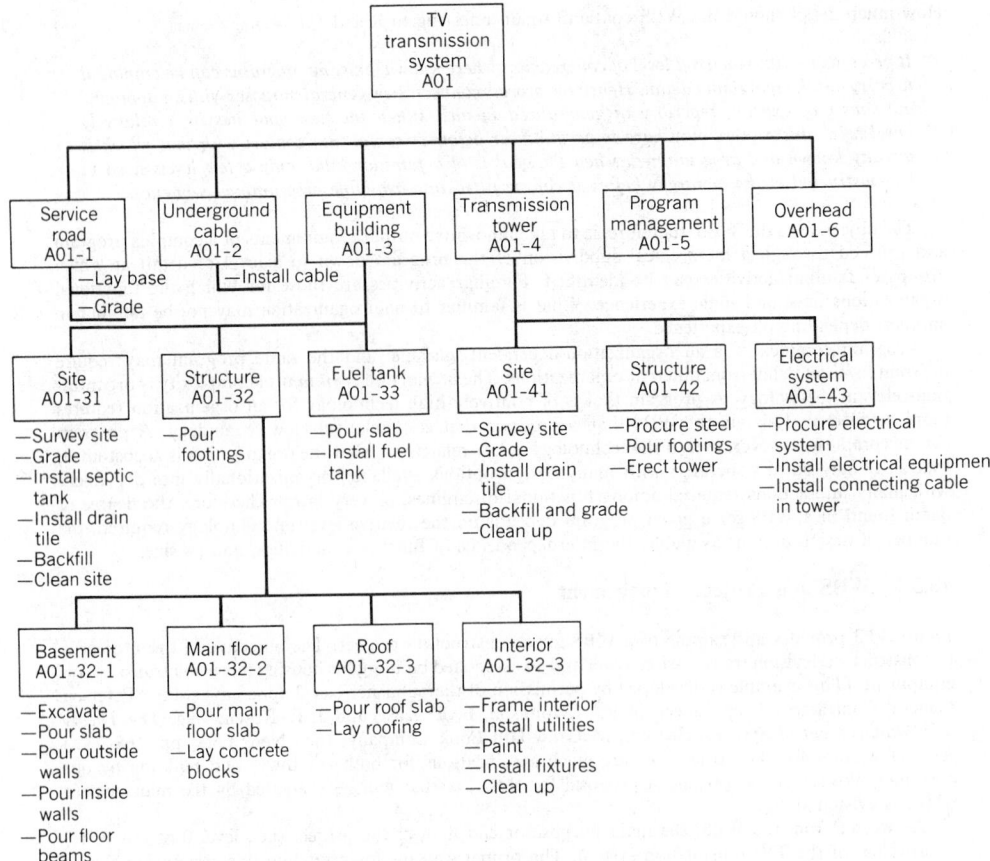

Fig. 43.3 WBS for a TV transmission system.

latter activities are inputs rather than identifiable outputs. Because the input of project management is primarily that of planning, decision-making, and control, it cannot be traced directly to any one WBS item, but rather must be assigned directly to the project itself. We accomplish this by making it a level-1 item so as to include within the WBS framework all the resource costs associated with the program. Similarly, when deriving the WBS, we initially trace only those means that are directly related to each end item. Yet we also want the WBS to provide an accounting framework for accumulating total project costs. Therefore, we assign all indirect resources to the level-1 item called "overhead."

Once the WBS is defined we can assign an account-code structure to it. The purpose of the account code is to provide unique identification for each end item of the WBS to serve as the basis for the cost accumulation and reporting system of the project.

Any combination of alphabetical and numerical characters may be used; the only real requirement for the identification system is that each end item contain in its identification the account letter and number of its parent. For example, the identification assigned to the TV transmission system is A01 at level 0 of the WBS. The equipment building is identified as A01-3, indicating that its parent is the TV transmission system and that it is the third, level-1 end item. The building structure is identified as A01-32, indicating that it is part of the equipment building (A01-3), which itself is part of the TV transmission system (A01). The account-code structure proceeds down to the last end item of the WBS. Functional activities below lowest level ends of the WBS are assigned resource code numbers or letters for purposes of estimating and reporting financial expenditures.

43.4 PLANNING TO ACHIEVE PROGRAM GOALS

The means–end analysis represented by the WBS is carried to a level of detail where relatively familiar activities may be identified to accomplish each of the lower-level items on the WBS. That is, our means–end analysis proceeds to a level of detail where the lowest level means for accomplishing ends are among the rather standard activities of functional organizations. Once these means have been classified in standard organizational terms, the ends that they serve can be accomplished with minimum difficulty. Each of the lowest level ends usually requires the completion of a number of interrelated tasks or activities by multiple functional disciplines. Often these different functional disciplines of an organization engaged in complex programs interact more or less continuously with one another.

To plan and control these interrelated activities to achieve program goals, we must define each activity and identify their interrelationships for each of the lowest-level ends. After accomplishing this detailed planning for each lowest-level end, we must identify properly the interrelationships *among these ends.* Once these two tasks are complete, we have completed the task of identifying the activities required to achieve program goals together with their interrelationships. After we have identified the means for achieving lowest-level (i.e., the most detailed) ends, it is only a matter of summarization to arrive at plans to achieve the higher-level ends.

In other words, the WBS and networks allow us to complete the definition of the means–end chain for our complex program. Without this detailed planning, program work would likely be characterized by random activity and lead to bottlenecks among functional resources. The resulting confusion is likely to occur regardless of the organization structure used. The WBS and networks bring to the program tools of planning, coordination, and communication that are of a complexity equal to that of program needs.

The basic approach to detailed planning in the program control process is provided by network analysis. A network is a diagrammatic model of the procedure in which detailed activities of an end item are to be accomplished. Once constructed, it becomes the basis for estimating time and resource requirements.

43.4.1 Network Construction

To begin, let us refer to Fig. 43.4, a typical network for an end item of a WBS. Note that this simple network consists of interrelated lines, some related serially to one another and others proceeding simultaneously or in parallel. The lines of the network are called *activities,* and they represent the tasks or subtasks identified at the lowest level of the WBS. These activities consume functional resources.

Each activity of a network is identified uniquely by two circles or *events.* Each event represents a point in time only and therefore does not involve the use of resources in any way. The purpose of an event is to indicate the *start* or *completion* of an activity; it merely represents an instant in time.

Each activity, to be identified uniquely, requires a "start event" and an "end event," formally known as a *preceding* and a *succeeding* event, respectively. An activity may share *either* a preceding or succeeding event with other activities, but not both.

Detailed network analysis begins at the lowest-level means on the WBS. We proceed to establish a diagrammatic plan for accomplishing each of these lowest-level end items. Small projects may require only one network. Large projects often require a network for each of the lowest-level end items of the WBS.

The initial "pass" at network construction should proceed under ideal assumptions regarding the order and interrelationships among activities of the network. That is, if a given activity may be accomplished either in series or in parallel with another, it is customary during the network construction phase to use the more conservative approach, which ordinarily implies a serial relationship among the activities.

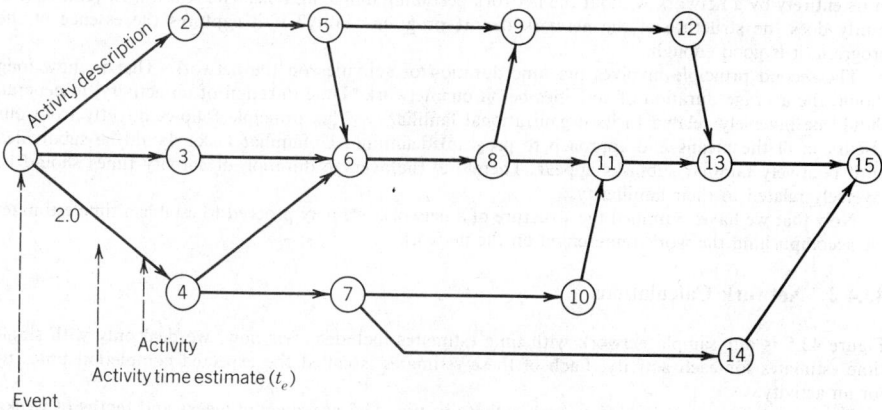

Fig. 43.4 A typical network with event numbers.

Each event on the network is numbered. Numbers are placed on the network consecutively beginning at the start of the network and proceeding toward the end. The order of the numbering system of the network is: left to right and top to bottom.

It is useful to distinguish two special types of events: *nodes* and *burst points*. A node is an event that serves as the succeeding event for two or more activities. Events 6, 9, 11, 13, and 15 of Fig. 43.4 are therefore nodes or nodal events. Events that serve as preceding events for two or more activities are called burst points. Events 1, 4, 5, 7, and 8 are burst points. Although none appear in Fig. 43.4, there are events that are both nodes and burst points.

A node represents the point of completion of multiple activities, any one of which may be complete without the node itself being complete. All activities of a node must be complete for the node to be complete and for the activity succeeding the node to begin. The node is considered complete, therefore, when the last activity of the node is complete.

A burst point represents the start event for multiple activities, which themselves proceed in parallel. Each activity for which the burst point serves as a start may proceed to completion while other parallel activities proceed at their own pace toward completion.

Events that are neither burst points nor nodes simply connect activities that must proceed serially or consecutively, one dependent on the other. Events 2, 3, 12, and 14 are examples of these serial events. One activity precedes such an event and one activity succeeds it.

If all the activities of an end item in an optimum arrangement should flow purely in series, there is hardly any need to consider network analysis as a planning tool. The task simply cannot be represented by a network. Therefore, burst points and nodes are the hallmark of networks and of any task for which accomplishment proceeds both in series and in parallel.

With this as background we can describe Fig. 43.4 completely. Event 1 is the start event for the network. Activities 1–2, 1–3, and 1–4 may begin immediately and simultaneously. Activity 2–5 is serially dependent on 1–2 and therefore cannot begin until event 2 is complete. Activity 5–6 is serially dependent on 2–5 but may be completed without event 6 being complete, for event 6 is complete only when activities 5–6, 3–6, and 4–6 are *all* complete. Similarly, activity paths 1–3, 3–6, and 1–4, 4–6 proceed in series to node event 6.

Activities 4–7, 7–10, and 10–11 are serial. Activities 7–14 and 14–15 represent a path, parallel with 7–10 and 10–11 from burst event 7, which itself proceeds serially. Activity 6–8 may begin only after the completion of node event 6. Event 9 is a node that cannot be complete until both activities 5–9 and 8–9 are complete. Similarly event 11, a node, requires the completion of activities 8–11 and 10–11 for its completion.

Event 13 is a node requiring the completion of activities 12–13 and 11–13 for its completion. Activity 12–13 is serially dependent for its start on 9–12. Finally, the network is complete when event 15 is complete, which, since it is a node, requires the completion of activities 13–15 and 14–15.

Note that activities of both nodes and burst points share a common event number. At burst points all activities have identical preceding event numbers; at nodes all activities share common succeeding event numbers.

A final consideration regarding the structure of networks is size. The amount of detail included in a network and the scope of the task modeled determine the size of the network. Although the question of size cannot be answered definitively for a given application, there are two principles that may be used to support judgment regarding the question. First the network is only a model of a complex program; as such it should capture the essence of the task while not distorting reality in a significant manner. We must realize that the reality of program activity is too complex to be captured in its entirety by a network without the network becoming unmanageable. The real test is: How significantly does the structure of the network distort program reality? If it captures the essence of the program, it is good enough.

The second principle involves the time duration of activities on the network. That is, how long should the average duration of activities be on our network. Time duration of an activity, in general, should be inversely related to its organizational familiarity. This principle follows directly from our discussion of the means–end approach to program definition. Unfamiliar tasks should be subdivided until relatively familiar subtasks appear. Therefore, the average duration of activity times should be inversely related to their familiarity.

Now that we have examined the structure of a network, we may proceed to establish time estimates for accomplishing the work represented on the network.

43.4.2 Network Calculations

Figure 43.5 is our sample network with time estimates included. For now, we deal only with single time estimates for each activity. Each of these estimates is called the expected completion time (t_e) for an activity.

Time estimates appearing below each activity in Fig. 43.5 are given in weeks and tenths of weeks. They are established on the assumption that the proper resources will be available when necessary to

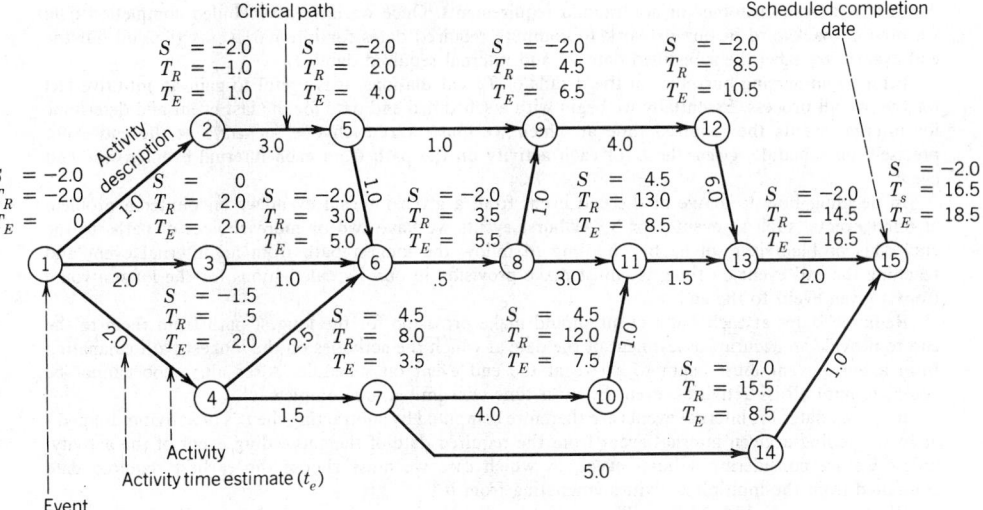

Fig. 43.5 A typical network, including calculations of slack.

perform the activity in the optimum manner. That is, in arriving at each time estimate we assume not only that the proper level of resources will be available, but that the task will be accomplished at the minimum total cost.

This planning premise is necessary to avoid confusion in the planning process. It requires that we estimate the time requirements for each activity under the assumption that all scale economies or efficiencies will be achieved. This means that just the most efficient amount of labor, material, and facilities will be used. Although we know that complexities of the real world often prevent us from performing each activity in the most efficient way, there is no alternative standard assumption that will facilitate the planning process as well as this one.

The objective of the calculation procedure of network analysis is to compute expected times, scheduled completion times, and slack for each event of the network. In making these calculations node and burst point events play the dominant role.

The expected completion time for an event is the sum of expected completion times (t_e) of individual activities on the longest path to that event. Therefore, at each node we take the longest path to that node as the expected completion time T_E for the node event. The expected completion time becomes the basis for calculations of T_E's on succeeding events that are connected to it. The expected completion time for an event is also the expected start time for succeeding activities. For example, event 6 is a node. There are three paths flowing into event 6 (i.e., 1–2, 2–5, 5–6; 1–3, 3–6; and 1–4, 4–6). The expected completion time for event 6 also is the *expected start time* for activity 6–8.

To find the T_E of each path leading to event 6, we simply sum t_e for all activities on that path. Note that even though we defined T_E as the expected completion time of an event, at a node T_E is associated with the longest path leading to the node, which implies that we must calculate all path T_E's and compare them. The path calculations to a node event that are shorter than the longest path are ignored in calculating the T_E of the event, *but* they are used later to make slack calculations. The lower path has a T_E of 4.5 weeks, the middle 3.0 weeks, and the upper 5.0 weeks. Therefore, the longest path to event 6 is the upper path; event 6 thus has T_E of 5.0. The events preceding 6 are all serial except event 1. All serial events have expected completion times equal to the sum of the T_E of the preceding event plus the t_e of the preceding activity.

Since we move forward in our calculations, burst points cause no difficulty in calculating T_E's; we simply take the T_E at the burst point and sum forward. The T_E at the initial event of a network is normally zero.

To summarize, expected completion times for each *event* are the sum of the longest path leading to that event. The only difficulty created in calculating T_E's is at node events, since there we must choose the longest among alternative T_E's, which then becomes the basis for future forward calculations in the network. The reader should be sure he or she can follow the calculations of T_E's to the end of our small sample network.

Once expected completion times have been calculated for each event of a network, we may begin computing required times for each event. To begin we either must be supplied with a scheduled completion time for the entire network or we must be willing to assume that the T_E for the end event is suitable as a scheduled completion date. Ordinarily we are supplied with a scheduled completion date derived from customer or contractual requirements. Once we have a scheduled completion date we proceed backward in our network to compute required dates for internal events (i.e., all but the end event). We label the scheduled date T_s and internal required dates T_R.

Prior to immersing ourselves in the details of T_R calculations, it is useful to gain an intuitive feel for the overall process. Essentially we begin with a scheduled end date for the last event and determine for internal events the required time at which we must start each one to arrive at the end event precisely on schedule, *given* the t_e of each activity on the path from each internal event to the end event.

In deciding how to arrive at the end event from a given internal event we encounter a problem at burst events, such as event 1 or 8. At burst events we have two or more choices of paths to the end! This problem is resolved by choosing to follow the longest path from an internal event and reaching the end event on time; we must make provision in our T_R calculations for the longest route from a given event to the end.

Required dates at each burst event should make provision for the longest path from them to the end to provide an accurate assessment of the date at which the activities on the longest path emanating from a burst event must begin to arrive at the end event on schedule. After all, a job cannot be complete until all its activities, even the most time consuming, are complete.

Required dates for internal events are therefore computed by subtracting the t_e's of activities immediately succeeding a given internal event from the required date of the succeeding event of the activity, unless we are considering a burst event, in which case we must choose the earliest required date computed from the multiple activities emanating from it.

We turn now to Fig. 43.5 to illustrate and gain practice in these calculations. Beginning as we must at the end event, we are given a T_s of 16.5 weeks for event 15. Therefore, the T_R of event 13 is 14.5. Similarly once the T_R is computed for event 13, we proceed to compute T_R for events 12 and 11 by merely subtracting the respective t_e. The process continues as simple subtraction until we arrive at burst events, such as events 1, 7, and 8.

Consider event 7. The required start time for activity 7–14 is found by subtracting from the required date of event 14 the t_e of activity 7–14, producing a required time of 10.5 weeks. The required start time for activity 7–10 is found to be 8.0 weeks, again by subtracting its t_e of 4.0 weeks from the required date of 12.0 weeks of its succeeding event (i.e., event 10). The T_R for burst event 7 is 8.0, the smaller of the two activity required times and is determined by activity 7–10. The logic of this is straightforward. Event 7 must be complete at 8.0 weeks in order to provide just the time required for the longest path from event 7 to the end of the network. The longest path from event 7 is 7–10, 10–11, 11–13, 13–15, which has a cumulative expected completion time of 8.5 weeks; when this is added to the required completion time of event 7 of 8.0 weeks, it yields 16.5 weeks, precisely the scheduled completion time of the entire network!

To summarize this section, required completion times for internal events are derived from the scheduled completion date of the end event by subtracting from the scheduled date the expected completion times of the longest path from that internal event to the end event. To implement this procedure, we simply subtract from the required date of an event the expected completion times of all activities flowing into it and assign the resultant required times to the *preceding* event. If a burst event is involved, we must choose the smallest required time of all activities for which the burst event serves as the predecessor as the required completion data for the burst event. The reader should trace through the calculations of all other required dates in the sample network.

43.4.3 Slack Calculations

Slack is the difference between the required T_R or scheduled T_s completion date of an event and its expected completion date. It is computed for each event on the network. Slack calculations provide a quantitative measure of how closely the plan represented by the network meets the time or schedule requirements of the program. Negative slack implies tardiness, positive slack earliness, and zero slack indicates the event is expected to be completed on time.

The longest path of activities from the beginning of a network to the end is the one with the least amount of slack algebraically. That is, the longest path has either the smallest amount of positive slack or the largest amount in terms of absolute value of negative slack. This makes sense intuitively, since the longest path of activities from the beginning to the end determines the expected completion time of the end event, and there can be no path of greater length in the network. Each event off the longest path has more slack algebraically than each event on the longest path.

To find the longest path, we simply trace back in the network those connected events that have identical slack beginning at the end event. Since the T_E of event 15 is 18.5 weeks and the T_s is 16.5 weeks, event 15 has a slack of −2.0 weeks. We find that event 13 also has a slack of −2.0 and is connected to event 15 by activity 13–15; it too is on the longest path of the network. Proceeding backward we find that events 12, 9, 8, 6, 5, 2, and 1 all have a slack of −2.0 and are connected to one another. Therefore, the activity path 1–2, 2–5, 5–6, 6–8, 8–9, 9–12, 12–13, and 13–15 is the longest path in the network. This path causes the plan to be 2.0 weeks late. It is therefore called the most limiting or *critical* path.

Slack on this most limiting path is constant because the t_e's that we accumulate to produce a T_E of 18.5 for event 15 are the same t_e's we subtract from the T_s of 16.5 to arrive at T_R's for each event on this longest path. The path that produced the greatest expected completion time on the network is also the one that produces the earliest required times. The slack on this path is therefore constant throughout.

Slack is only a meaningful measure given for a path, not a single event. For example, the events on the critical path have a slack of -2.0, indicating that the entire path is 2.0 weeks late. Given the T_s of event 15 and given the t_e's of each activity *on the path*, each event may be said to be 2.0 weeks late, but only because the whole path is 2.0 weeks late. This aspect of slack will be very important to us later.

Although we have identified the most limiting path, other less limiting paths also must be identified, because, if we are not careful, we may reduce the T_E on the most limiting path only to find that another path has become almost equally limiting, and that our reductions were only partially beneficial.

To isolate the second-most-limiting path, we must look for an event with the next smallest (algebraic) amount of slack. We note that event 4 has -1.5 weeks, and now must identify the activities of this path, since it is the second most limiting.

Clearly activity 1–4 is on the second-most-limiting path because it has a T_E of 2.0 and a T_R of 0.5, but does the path stop there? It does not. Note that the T_E at node event 6 is 5.0 weeks. Recall also that the T_E at node 6 was *caused* by the upper path (1–2, 2–5, 5–6). The T_E of the lower path (i.e., 1–4, 4–6) is 4.5 compared with a T_R at event 6 of 3.0, thus yielding slack of -1.5 for the entire path—1–4, 4–6. Therefore, to calculate slack at a node event for noncontrolling paths, we must compare the T_E of noncontrolling paths with the T_R at the node.

The importance of identifying secondary paths cannot be underestimated. For example, if any activity or series of activities on the critical path prior to node event 6 is expedited by 2.0 weeks in an attempt to eliminate all negative slack from the network, it will only reduce the T_E of the network by 0.5 weeks, simply because the path that was not reduced (i.e., 1–4, 4–6) now becomes dominant at node event 6. The critical path forward of node event 6 will remain the same, but now with slack of -1.5 instead of -2.0. We can only eliminate 2 weeks of slack at node event 6 by cutting both the upper and lower paths; the upper by 2.0 weeks and the lower by 1.5 weeks!

In a similar manner, we can find the slack associated with each activity and event of the network. The slack for activities emanating from burst points is found by comparing the T_E of each activity with the T_R of its succeeding event. Normally at nodes, the slack of all but one activity differs from the slack at the event. It is a simple matter, however, to compute the slack of the other activities.

Although it is not crucial for purposes of network planning, it is useful at this point to differentiate between slack and free slack. For purposes of resource scheduling, it is important to know if slack on an activity leading into a node is different (i.e., greater) from slack at the node. If the slack of the activity is greater than that at the node, the difference represents free slack or, in the terminology employed by the Critical Path Method (CPM), free float. Free slack is most useful for purposes of scheduling. It tells the functional manager by what amount he or she may extend an activity without affecting slack calculation on any activity or event that lies ahead of the node event.

43.4.4 Translating an Invalid Plan into a Valid One

Assuming the time estimates provided on a network are correct, we may proceed in three ways to produce a valid plan. First, we may consider taking more risk in the way we carry out our activities. That is, we might evaluate the effects of performing certain serial activities in parallel. Let us return to the sample network of Fig. 43.5, which contained slack of -2.0 weeks on its critical path. Figure 43.6 illustrates the effects of changing activities 12–13 and 13–15 from their present serial relationship to parallel construction. Since both of these activities are on the critical path, we know that performing them in parallel will reduce the expected completion time of the final event by the t_e of activity 13–15, exactly 2.0 weeks, thus producing a valid plan! We know this *a priori* because, although event 13 is a node event, activity 11–13 has positive slack.

What does this kind of change imply? It implies that we are not going to wait for activity 12–13 (perhaps a test of some kind) prior to undertaking 13–15. Yet it may turn out that when we perform 12–13 we might have to alter the work represented by 13–15, based on what we learn from 12–13. Therefore, if we alter the relationships of our otherwise optimum technological plan for performance, by definition we undertake additional risk. If no other acceptable way exists to achieve a valid plan, this course of action must be taken.

It is useful at this point to take a detour to provide a meaningful example of *free slack*. The critical path on our network passes through node event 6. At node event 6, there are two activities that have positive slack event although event 6 has zero slack. Activity 3–6 has a slack of 2.0 weeks and activity 4–6 has a slack of 0.5 weeks. The slack on both of those activities is free slack, since it may be used up for those activities *without* affecting the slack in any part of the network *forward* of

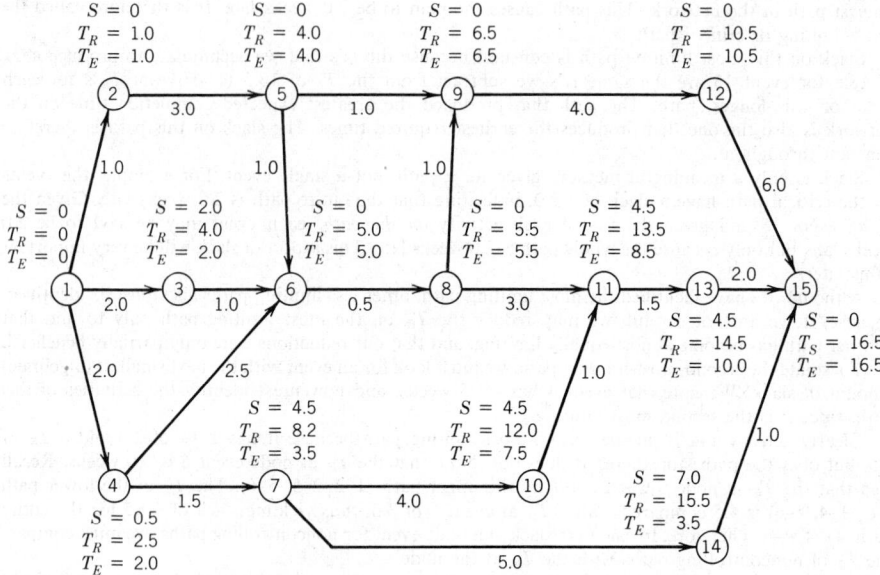

Fig. 43.6 Validating typical network.

the node event. Knowledge of free slack is useful to functional managers during the resource scheduling process, as we shall see later.

The second procedure for producing a valid plan attempts to expedite certain activities in the network to save time while maintaining the optimum performance plan.

If the first two procedures are impossible, we can only change the schedule date, with the concurrence of our customer or management, or redefine the program.

43.4.5   Precedence Networks versus *I, J* Networks

The method of network construction illustrated in the last section is not the only one possible. Many users prefer another network format, which is called the Precedent Format and is contrasted to the *I, J* format illustrated previously. In the *I, J* format, the activity takes place on the "line," whereas in a precedent network the activity takes place on the "node" or "event." In the *I, J* format each activity (or job) is described by a preceding event (*I*) and by a succeeding event (*J*). In a precedent network, each job is described by an event number alone.

We proceed to illustrate the difference between *I, J* and Precedent networks by an illustration of the construction of a custom built home.* Figure 43.7 presents a typical network in *I, J* notation for the custom home. This network is constructed using the principles discussed previously in this chapter.

To convert from *I, J* to precedent networks, each event and its notation is laid out as illustrated in Fig. 43.8a.

Figure 43.8b presents the same network logic for the custom built home but now in precedent notation.

Some users prefer precedent notation over *I, J* notation because networks are easier to construct and to understand. In the final analysis it appears to be a matter of taste as to which one is used. So long as the logic of a network is correct, the form is of lesser importance although it has been my experience that users hold strong views as to one format over another. Therefore, the format should be adapted to the particular users.

43.5   PROGRAM FINANCIAL PLANNING

A close relationship exists between the WBS, networks, and financial expenditure plans of a program. Since complex programs are characterized by the same uncertainty in the cost dimension, whatever forces extend the time dimension of a program ordinarily extend its expenditure dimension also.

---

* This example is adapted by permission from Construction Associates, Case No. 9-610-060, Harvard Business School, Cambridge, MA.

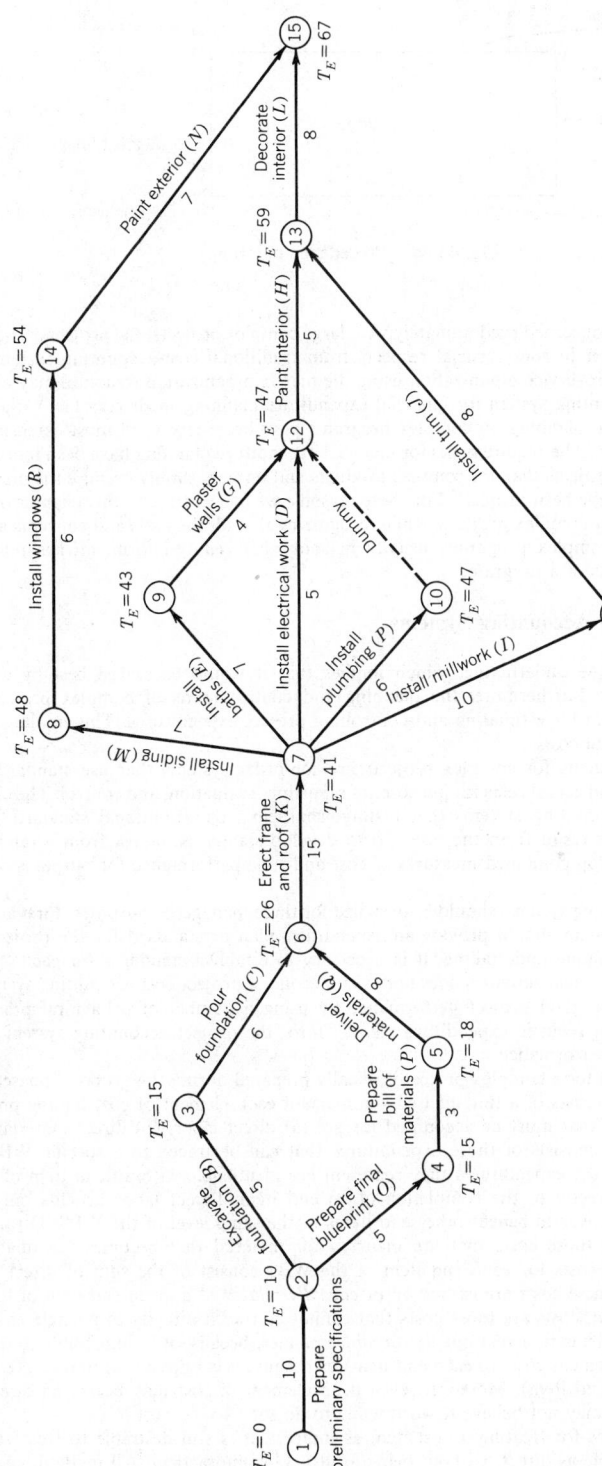

Fig. 43.7 Network for custom-built home—I, J notation.

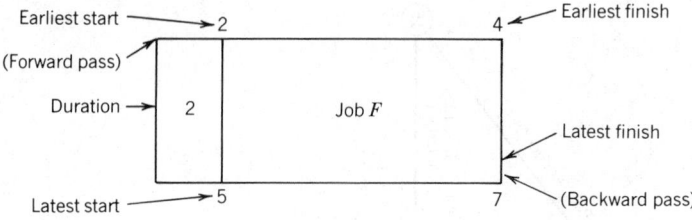

Fig. 43.8a   Precedence notation.

In an organization concerned predominately with large complex projects, the project cost-accounting system must be different in some crucial respects from traditional cost-accounting systems. On the other hand, a product or service organization using the matrix organization structure may often utilize a traditional cost-accounting system for financial expenditure planning, with very few variations.

Systems for financial planning on complex programs have been developed most extensively in the area of complex projects. The requirements for financial expenditure planning have been more stringent, as a rule, for complex projects than for complex products and services simply because the most complex programs of the past have been projects. For these reasons, we concentrate in this chapter on financial expenditure planning for complex projects, while recognizing that all the tools and concepts are directly transferable to other complex programs, usually in somewhat reduced form, always reflecting the managerial requirements of a program.

43.5.1   Project Cost-Accounting Systems

Each project is a unique undertaking, which implies that it would be served best by a job-order cost-accounting system. Furthermore, the planning and control needs of complex projects require that we provide a system for estimating and controlling project expenditures. This implies a need for both standard and actual costs.

Cost-accounting systems for complex projects are job-order systems that use standard costs for purposes of planning, and actual costs for purposes of reporting, evaluation, and control. The characteristics of project cost-accounting systems that distinguish them from traditional standard, job-order, cost-accounting systems result from the need (a) to develop standards almost from scratch for each project and (b) to develop combined measures of cost and time performance for purposes of program planning and control.

Project cost-accounting systems should be designed for three managerial purposes. First and perhaps most important, the system should provide an expenditure plan or standard for the project. Since a complex project is a unique undertaking, it is necessary to establish standards for each "job." Past experience regarding identical activities does not exist. Second, a project cost-accounting system should be designed to motivate good project performance by using a number of behavioral practices that produce challenging but realistic expenditure plans. Third, the project accounting system should be used for evaluation of performance.

An expenditure plan for a complex project is usually prepared during the proposal phase of project activity. Such a plan consists of a time-phased summary of each element of cost for the project. The three elements of costs that must be accounted for are (a) direct labor, (b) direct materials, and (c) overhead. Direct labor consists of those expenditures that can be traced to a specific WBS element of the project. That is, the expenditures they represent are identifiable with an end item of the WBS, and they contribute directly to the completion of an end item. Direct labor benefits only one end item and does not spill over to benefit other end items at the same level of the WBS. Direct material costs for a project are those costs that are incurred for material that becomes identifiable with a given end item. Direct costs for each end item of the WBS consist of the sum of direct labor and direct material costs. These costs are *caused by* or *causally related to* a given end item of the project.

Indirect, or overhead, costs are those costs that cannot be traced directly to a single end item but rather help many end items in a real but *untraceable* manner, because it is impossible to do so (e.g., tracing program management costs to each end item) or because it is impractical to do so (e.g., tracing office supplies to each end item). Moreover, even if an element of cost may be traced directly to an end item, management may not believe it worthwhile to do so.

Given all the reasons for treating a cost item as indirect, it is still desirable to treat as many of the total costs as possible as direct, without being arbitrary or impractical. All methods of allocating indirect costs tend to be imprecise. In addition to the three principal elements of costs, project proposals generally contain a catchall category of costs called "general and administrative" costs. These are overhead items that are considered to be even more remote from the project than are those items included in overhead. Examples include research and development activities, public relations, and

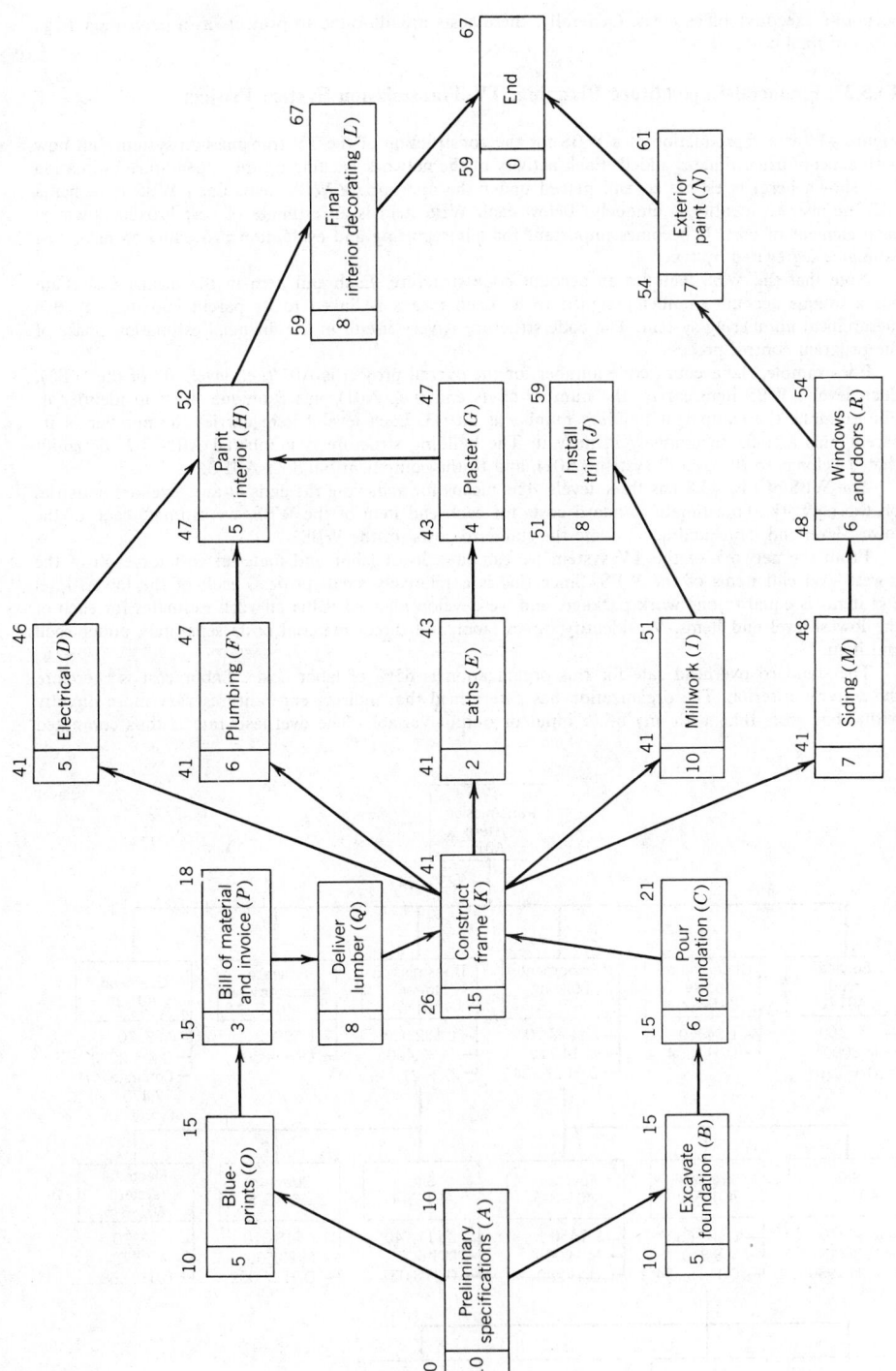

Fig. 43.8b Network for custom-built home, precedent notation.

corporate executive office costs. Generally, these costs are allocated to projects as a percentage (e.g., 15%) of total costs.

43.5.2 Financial-Expenditure Planning: TV Transmission System Project

Figure 43.9 is a reproduction of a WBS for the construction of the TV transmission system, but now with expenditure estimates added. Each activity of the network backing up the expenditure estimation (not shown here) is estimated and placed under the appropriate WBS item. Each WBS item has a code number to identify it uniquely. Below each WBS item is an estimate of cost broken down by each element of cost. It becomes important for our reporting and evaluation procedure to have cost estimates segregated by type.

Note that the WBS includes an account code structure. Each end item in the means–end chain has a unique account number assigned to it. Each means is linked to its parent end item by this hierarchical numbering system. The code structure is very useful in the financial estimation phase of the program control process.

For example, the account code number for the overall project is A01 (i.e., level "0" of the WBS). Each level-1 WBS item carries the number of its end (i.e., A01) plus a unique suffix to identify it. For example, the equipment building number is A01-3. Each level-2 item carries the number of its parent plus a suffix to uniquely identify it. The building structure is numbered A01-03-2 to signify that it belongs to the overall system (A01), and to the equipment building A01-3.

The WBS of Fig. 43.8 has three levels. The means for achieving the ends at any level are activities on the network. To estimate standard costs for each end item of the WBS, we estimate each of the lowest-level end items and accumulate the standard costs up the WBS.

From the network of the TV system we estimate direct labor and material cost for each of the lowest-level end items of the WBS. Since this is a relatively small project, each of the lowest-level end items is equal to one work package, and we develop planned value of work estimates for each of the lowest-level end items. We identify direct labor and direct material costs separately under each end item.

The standard overhead rate for this organization is 65% of labor costs. Labor cost is therefore the activity criterion. The organization has determined that indirect expenditures vary more directly with labor costs than with any other input or output variable. The overhead rate is thus computed

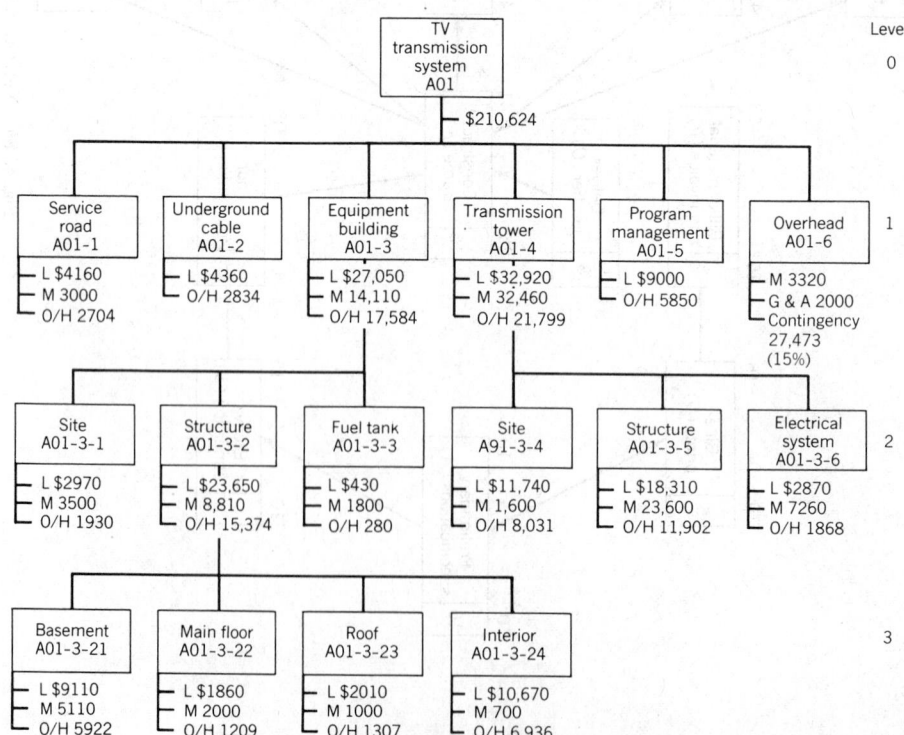

Fig. 43.9 WBS for TV transmission system.

by estimating overhead expenditures over the accounting period (normally a year) and dividing these expenditures by the expected or normal volume of labor costs for that same period.

Once we have arrived at the overhead rate, we simply apply it at each lowest-level end item to the standard assigned to the variable that serves as the activity criterion. This gives us the standard overhead charge for that end item. We then sum the three elements of cost to arrive at standard costs for an end item. Since we can relate a lowest-level end item to the network, we shall be in a position in the reporting phase to collect actual costs for work performed and compare them to the planned value of work performed. Finally, we sum standard costs for each end item to its parent to find successively higher levels of project costs until we arrive at the standard cost for the entire system (i.e., A01 on the WBS).

Note that there are costs for project management and certain other overhead items that we choose not to allocate to project end items, but rather we identify these separately at level 1 of the WBS. Of course, they too become part of our total estimated costs for the project. The estimated costs for the project may also be displayed by month as in Table 43.2. Table 43.2 becomes a control document, it does not contain profit or contingency thus displaying a total cost which is $44,579 lower than the costs appearing on the WBS in Fig. 43.8.

The work package, because it connects the WBS, the network, and the cost-accounting system for a meaningful segment of work, is the basic instrument for integrating the time and cost variables of a project. It is the lowest level of detail at which it is feasible to devise a combined measure of performance for time and cost.

The combined measure of performance is ordinarily called the *planned value of work* and it is arrived at simply by estimating the *budgeted value* of work represented on the network for each work package. Each work package thus contains estimates of its planned value, so that any major part of the work package is accorded a corresponding planned value.

Once work progresses, we collect data on actual expenditures and progress and assign *actual cost for work actually accomplished* for each work package. We then compare the *planned value for work actually accomplished* with the actual cost for work accomplished and compute the variance. The variance thus represents a measure of cost performance versus plan for the work actually accomplished. It integrates expenditures with schedule performance, thus achieving the joint measure of performance we seek. We shall discuss this integrated reporting measure further, later in Section 43.8.

43.6   RESOURCE ALLOCATION VERSUS PROGRAM PLANNING

Program plans represented by networks and financial plans provide functional management with the requirements, resources, and priorities for their function on each of the organization's programs. Although network plans provide a schedule for accomplishing the work, this schedule is not always practical or feasible when all other constraints placed on the function are considered. There are six specific requirements excluded during the planning process that must be considered during the resource allocation process. They are as follows:

1. Sufficient manpower to perform each activity in an optimum manner is assumed to be available when formulating and optimizing plans. Limited availability of manpower and the competition among programs for the same manpower must be taken into account during the resource-allocation process.

2. The pattern of resource demands from all of the program plans must be considered not only in the light of resources available but also in terms of the distribution of demand placed on resources over time. Functional management cannot be expected to increase continuously and reduce functional manpower in light of the fluctuating demands of each program. Functional manpower levels are determined based on long-term organizational demands and use must be relatively even from one period to the next.

3. Common facilities (e.g., computer time and test rigs) are often required simultaneously by activities of the same program or by activities of different programs. The allocation process must resolve these conflicts.

4. Cash flow requirements of the programs are not always feasible for the organization, and these limitations enter into the allocation function.

5. State work laws and regulations must be observed in allocation decisions when overtime is being considered.

6. The nature of the contract negotiated between contractor and customer with regard to the relative value of various programs to the organization, as well as the long-term objectives of the organization, affect the relative priority that should be accorded various programs by the organization. This is a consideration of the resource-allocation process.

Not only must we recognize scheduling as a distinct activity in the program-control process separate from, yet related to, planning, but we must also establish different time horizons for these two activities.

Table 43.2 Financial Expenditure Plan According to Expected Completion Dates

Element of Cost	1 (1–22)	2 (22–44)	3 (44–66)	4 (66–88)	5 (88–110)	6 (110–132)	7 (132–154)	8 (154–176)	9 (176–198)	10 (198–220)	Total
					Months (Days) Worked						
Labor	$17,200	$ 8,570	$ 5,220	$ 2,660	$ 15,100	$ 3,600	$ 14,110	$ 2,300			$ 68,760
Material expenditures	$30,860	$21,730									$ 52,590
Applied O/H (65% of labor)	$11,180	$ 5,571	$ 3,393	$ 1,729	$ 9,815	$ 2,340	$ 9,172	$ 1,495			$ 44,695
Project total cost	$59,240	$35,871	$ 8,613	$ 4,389	$ 24,915	$ 5,940	$ 23,282	$ 3,795			$166,045
Cumulative total cost	$59,240	$95,111	$103,724	$108,113	$133,028	$138,968	$162,250	$166,045			

Program planning must be carried out for the entire duration of the program. Scheduling, on the other hand, ordinarily may be done profitably only on a short-term time horizon.

Scheduling requires commitment of resources on the part of functional management to specific tasks of the many programs of the organization. As the network relationships indicate, however, activities of one functional organization are dependent on the completion of activities of other functional organizations. Because of the dynamic, constantly changing nature of complex programs, we cannot expect network relationships and time estimates to be very precise. Expected start and completion times of activities become more tenuous the greater the elapsed time from the present. Therefore, functional organizations cannot establish realistic long-term schedules for carrying out the work of multiple programs. It is usually futile to allocate resources to specific jobs unless they are to be performed in the near term. More accurate scheduling can be done for these near-term activities since most of the activities that limit their start are either in progress or complete.

For the examples presented later in this chapter, I have selected a time horizon of 10 weeks for scheduling and assume that the scheduling process takes place on a monthly basis. I believe this time horizon and scheduling interval to be realistic.

Start dates for activities that are scheduled by the functional organizations must find their way back to appropriate program plans. Scheduled start dates are *superimposed* on network calculations, and they supersede early expected start dates in calculation of the network so long as they are equal to or greater than expected start dates. Scheduled start dates that are later than expected start dates are invalid. Program office personnel must check the consistency of functional schedules and approve their implications. The portion of a program plan that has been scheduled is called a *scheduled plan*.

Although distant activities cannot be scheduled, it is important to preserve a valid plan for distant work since the time estimates and interrelationships of the entire plan determine the time requirements (required dates) of work that can be scheduled.

To summarize this section we may say that resource allocation or scheduling is a function with different purposes than planning. A network plan cannot ordinarily be used as a schedule for a project, yet it must serve as the basis for the schedule. Moreover, once activities are scheduled, these data must be incorporated into network plans. Thus, there is communication between these two important functions. If the plan alone is used as a schedule for performing the work, with slack used without considering other activities and competing programs, the ability to optimize performance in the organization is restricted, and the value of the program-control system is lessened.

The resource-allocation process consists of three distinct but interrelated tasks: *resource loading, resource leveling,* and *constrained resource scheduling.* Resource loading is concerned with deriving the total demands of all programs placed on the resources of a function during a specified period of time. Resource leveling attempts to "smooth out" the demands to eliminate major peaks and troughs. Constrained resource scheduling is concerned with achieving all demands of the programs of an organization within the resource constraints of the function at minimum disruption to the plans of each of the organization's programs.

43.6.1   Resource Loading

To understand the resource-loading process, it is convenient to view the problem in matrix form. The various programs of an organization place demands on resources during a particular period of time, and the functional organizations supply these resources. A matrix illustrating this process appears in Table 43.3. The matrix represents the total demands placed on each of five functions by each

Table 43.3   Resource Loading in Matrix Form[a]

	Programs					Total Functional Man-Hours
Functions	1	2	3	4	5	
A	20	40	8	5	0	73
B	30	30	15	0	0	75
C	25	25	20	8	20	98
D	10	15	20	14	30	89
E	40	10	12	10	0	72
Total program man-hours	125	120	75	37	50	407

[a]Resource allocation matrix: 3-month period based on earliest expected start dates.

program. These demands, however, are not time phased in this exhibit. The resource demands in the matrix are taken from the work packages that are expected to be performed during the scheduling period.

Each work package consists of a group of activities for a program. In addition, they contain time-phased estimates of the resources required to accomplish the work. These time-phased estimates for each function for the scheduling horizon must then be summed across all programs to produce a manpower loading report.

Information on demands placed on each functional group during the scheduling horizon is only part of the information required in the resource-loading process. In addition we require slack information from each of the program plans.

Table 43.4 is an example of a computer output report for one function, drafting, for one program for a 10-week period. This information is derived from program plans. The activities represented on the report have start dates, expected completion dates, required dates, and slack calculations.

Loading information from work packages, together with calculations of slack from program plans, are then combined into the manpower loading report for one functional organization. Figure 43.10 presents an example of a time-phased loading plan based on expected start and completion times for each of the activities. Where positive or negative slack exists, it is indicated by an extension of each bar to its right (for positive slack) or left (for negative slack). Within each bar we have placed the number of persons per week required to achieve each task and have summed the total demands placed on the function vertically by week. The row on the bottom of the chart therefore contains an estimate of the total demands in terms of man-weeks of effort placed on the function of drafting by all programs. Figure 43.11 presents the loading plan graphically.

From Fig. 43.11 we note that there is an uneven distribution of demand for drafting resources over the 10-week period, with very high demands occurring in weeks 7–8 and 8–9. Even if manpower is in good supply in the drafting organization, it is usually undesirable to have these large variations in the demand for manpower. The *resource-leveling process* attempts to remedy this situation by leveling or smoothing manpower demands within the constraints of required dates for the various activities.

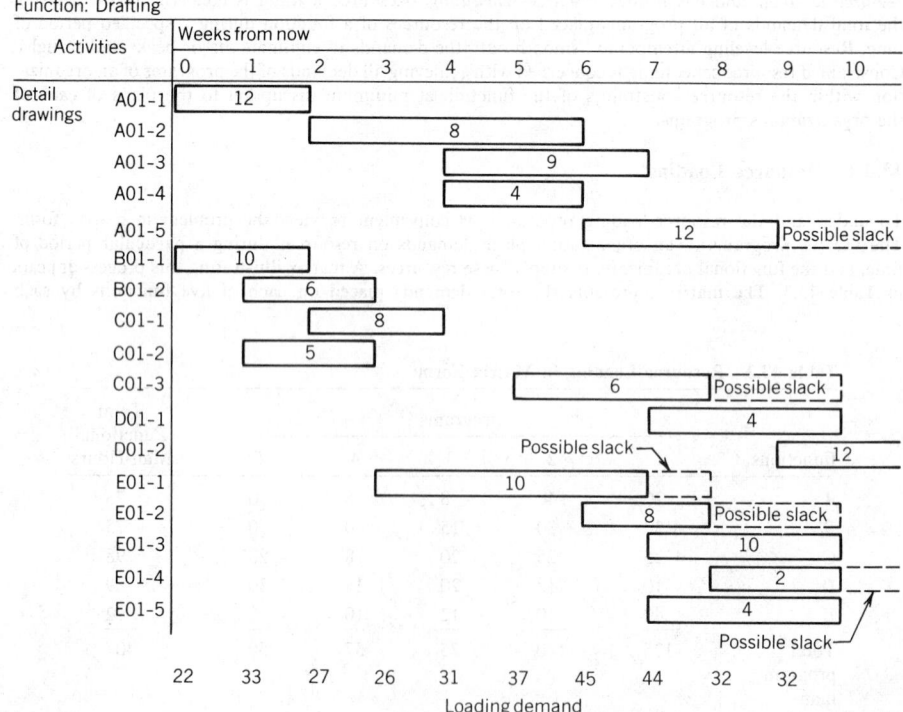

Fig. 43.10 Time-phased manpower loading plan.

Table 43.4 Planned Activities of Functional Organization

Function: Drafting
Program: XXX
Responsibility: John Smith

Preceding/Succeeding Event Numbers	Activity	Time Estimate	Start Date	Expected Completion Date	Required Completion Date	Slack
001–002	Prepare detail drawings—101	2.0	01/01	01/15	01/29	2.0
011–012	Prepare detail drawings—105	4.0	01/15	02/15	02/15	0.0
021–022	Prepare detail drawings—208	3.0	02/01	02/22	02/15	−1.0
031–032	Prepare detail drawings—304	1.5	02/01	02/11	02/04	−1.0
051–052	Prepare detail drawings—508	2.5	02/15	03/03	03/01	−0.4

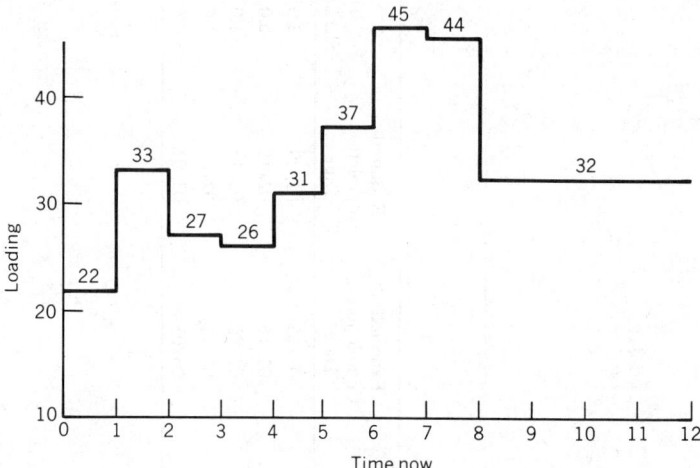

Fig. 43.11   Profile of demands for manpower.

43.6.2   Resource Leveling

The resource-leveling process begins with the resource-loading and slack calculations of Table 43.4 and the manpower profile of Fig. 43.10. It proceeds to *level* the demands for resources *without* exceeding the required dates of programs. The process is constrained by its leveling objectives and by the overlap time requirements of the programs, not by available resources.

By using slack calculations, some start times of activities may be adjusted to begin later than their earliest expected start date, thus shifting the demand for resources to a later point in time, *without* exceeding the original expected completion date of the network. Therefore, the resource-leveling process requires us to adjust performance times of activities according to slack calculations to produce a pattern of demand for resources that is as stable as possible over the scheduling horizon.

The resource requirements of the valid plan are taken as a beginning point for resource leveling. Required completion dates are treated as constraints, so that we maintain valid plans. Adjustments are made by each functional manager; these adjustments are coordinated with personnel assigned to the various program-management offices to ensure that each functional organization does not frustrate the schedules of the other. That is, the use of slack by the various functions must be coordinated by the various program offices.

Free slack (or float), which we defined to be that part of activity slack that if used does not affect slack calculations forward of the activity in the network, may be used immediately without approval of the program office, since its use cannot affect any other activity in the program. Normal slack, however, is identified with a path and, although we may use it during resource leveling, its use must be coordinated with program office personnel whose program is affected. Coordination is necessary since only one activity on a path may use its positive slack; if all functions represented by activities on a single path use the slack, the combined network calculations would produce a negative slack path!

The resource-leveling function may be carried out manually unless the networks are large and involve multiple resources. Computer programs are available commercially for more complex networks.

Resource leveling for our sample programs results in the revised loading plan of Fig. 43.12 and the manpower profile of Fig. 43.13. Note that in the leveling process we were able to reduce peak demands of weeks 7–8 and 8–9 significantly. This was accomplished by the selective use of positive slack that was available on the program. Assuming this use of slack meets the approval of the proper program office personnel, the leveling procedure results in a definite improvement in the distribution of resources of the drafting organization. The manpower profile produced by the resource-leveling procedure, however, may not be feasible in light of known manpower levels. The next problem therefore involves scheduling all these drafting tasks within the limits of available draftsmen. This is the task of scheduling subject to manpower constraints to which we turn now.

43.6.3   Resource Scheduling Subject to Manpower Constraints

If the resource-leveling process produces a loading profile within the manpower limitations of the function, all is well. If, however, the loading demands of the leveling process produce manpower

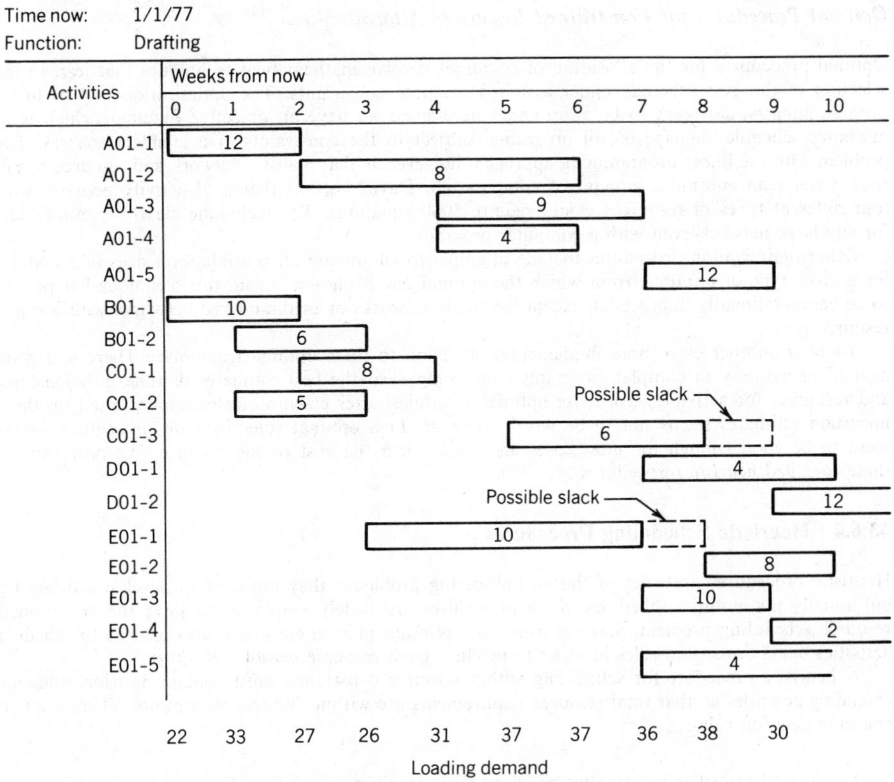

Time now:　　1/1/77
Function:　　Drafting

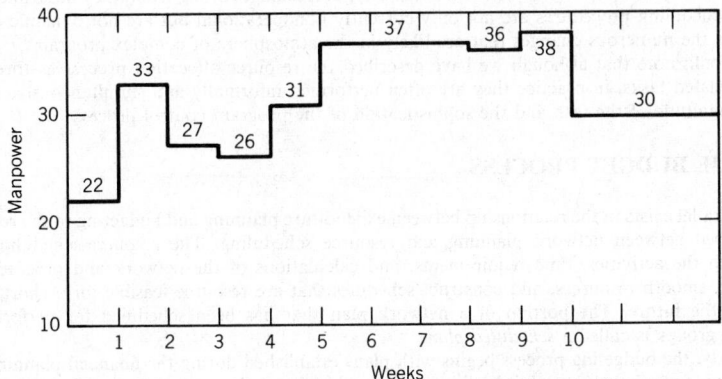

Fig. 43.12　Resource-leveled plan.

requirements that exceed manpower availability, including the use of overtime, then we must relax schedules until they "fit" within resource limits. This relaxation process must be done in such a way as to minimize the extensions required to the critical path while maintaining a reasonably smooth distribution of demand for resources.

When the resource-leveling procedure does not produce a feasible schedule under otherwise optimal planning conditions, we are forced to increase the duration of schedules of at least some programs of the organization. To determine which activities should be prolonged, we require a priority system. Two types of systems exist: optimal procedures and rule-of-thumb or heuristic procedures. We examine each category in turn.

Fig. 43.13　Revised profile of demand for manpower.

Optimal Procedures for Constrained Resources Allocation

Optimal procedures for the allocation of resources involve mathematical techniques that lead to the selection of the best schedule under assumed resource constraints. The optimization tool of linear programming would seem to be appropriate here since we have an objective function, which is to minimize schedule slippage on all programs, subject to the constraints of available resources. The problem with the linear programming approach, however, is that realistic network and resource conditions require an enormous amount of computation. Davis[2] reports that a 55-activity network with four different types of resources would require 5000 equations. The technique clearly is impractical for any large network even with a computer program.

Other optimization procedures include attempts to enumerate all possible sequences of schedules for a given type of resource, from which the optimal one is chosen. Again this technique has proved to be computationally impractical, except for small networks or medium-sized networks with a single resource type.

There is another even more fundamental objection to these optimal techniques. There is a good deal of uncertainty in complex programs concerning even the best estimates of time, relationships, and resources for activities. To divise optimal schedules, after elaborate calculations, based on these uncertain estimates seems not to be worth the cost. Less optimal rule-of-thumb procedures would seem to be good enough for most cases and well worth the cost of the exercise. We now turn to these so-called *heuristic* procedures.

43.6.4 Heuristic Scheduling Procedures

Heuristic procedures are rules of thumb for solving problems; they are used to develop satisfactory but usually not optimal schedules. Such procedures are widely employed to solve the constrained resource scheduling problem. Starting from the optimum plan, these procedures lead us to schedule activities based on certain rules in order to produce good resource feasible schedules.

A heuristic procedure for scheduling within resource constraints must contain decision rules for extending activities so that total resource requirements are within resource constraints. There are two common decision rules:

1. Accord priorities to activities based on their required completion dates, with activities having the earliest required completion dates scheduled ahead of those with later required completion dates.
2. Rank activities in order of duration and perform activities with the shortest duration first.

These two rules of thumb are given as examples of procedures that may be used to solve the constrained scheduling problem, given a leveled loading plan. All heuristic procedures proceed to extend activities that cannot be accommodated by available levels of manpower through the use of one of these rules. A heuristic procedure must also have secondary rules for breaking ties. For example, if two activities have identical required dates yet cannot be performed simultaneously because of resource constraints, we might decide to perform the one with the shortest duration first.

It is important to realize that these rules of thumb are not likely to produce optimal schedules. They are designed to produce satisfactory feasible schedules. When placed in the context of the uncertainties found in organizations engaged in complex programs, however, rules of thumb such as these are operational and flexible enough to respond to the inevitable changes brought about by these uncertainties. Optimal scheduling procedures are not only currently nonoperational but are not dynamic enough to respond to the numerous changes that are likely in the atmosphere of complex programs.

We should note that although we have described the resource allocation process as three distinct but interrelated tasks, in practice they are often performed informally and simultaneously, depending on the magnitude of the task and the sophistication of the program control process.

43.7   THE BUDGET PROCESS

A close parallel exists in the relationship between expenditure planning and budgeting to the relationship we described between network planning and resource scheduling. The resource-scheduling process begins with the activities, time requirements, and calculations of the network and proceeds to load manpower, smooth resources, and construct schedules that are resource feasible for a short period of time into the future. The portion of a network plan that has been scheduled for performance by functional groups is called a *scheduled plan*.

Similarly, the budgeting process begins with plans established during the financial planning process and proceeds to authorize expenditure limits within which, *on balance*, budgets are expected to adhere. The budget for a work package, however, is likely to differ in some important aspects from the financial plan, since the authorized work package must reflect decisions made in the resource-allocation process. The portion of the financial expenditure plan for which we have a budget is called a *budgeted plan.*

43.7.1 Work Package and Operating Budgets

As in all organizations, each responsibility center of an organization engaged in complex programs requires an operating budget. For organizations engaged in complex programs, however, it is almost impossible to prepare an annual operating budget for a functional organization with any degree of confidence that it will be followed closely. Yet each organizational unit must perform resource and expenditure planning over a longer horizon than that which it can forecast perfectly.

This apparent budgeting dilemma is resolved by requiring both work package and operating budgets. Work package budgets, covering a short period of time, serve as work authorization documents, whereas approved operating budgets serve to guide decisions regarding resource levels in each of the functional departments.

Financial data on work packages prepared during the expenditure planning process are far from ready to serve as budgets for the program. These financial plans were derived from estimates of network activities. As we saw previously network plans ordinarily cannot be converted directly into schedules, but rather must be considered in light of available resources and other competing demands. Therefore, the financial plans of a given work package cannot be converted into budgets until the activities included in the work package have been scheduled, for only then do we know precisely when activities will be done and by whom and what resources will be used. Only a small portion of the financial plan is eligible to serve as a budget, for only a small portion of the network plan upon which the expenditure plan is based has been scheduled. The budget for the portion of the network plan that has been scheduled is negotiated between program and functional personnel. The approved budget then serves as the document that authorizes functional work.

43.7.2 The Budget as an Authorization Document

We have seen that to achieve both scale economies and coordination, the matrix structure causes us to violate the classical principle of *unity of command.* The stresses produced by the dual sources of command to which functional personnel must respond are nowhere potentially more divisive than in the budgeting and authorization process. This process, however, if performed correctly also possesses opportunities to enhance identification with program goals, to improve performance, and to reduce or eliminate these natural tensions caused by dual lines of command.

Since the management function of directing program work formally lies with functional managers under the matrix structure, the program manager should use the budget and related authorizing documents to exert control over functional performance.

Once the schedule is prepared for functional work, the budget implications may be derived by applying rates for each cost element as established in the program cost-accounting system. The schedule for functional work and its supporting budget serve as authorizing documents for functional work. By reviewing and approving the schedules and budgets of these work packages, the program manager begins to assert control over his or her program. Thus a major portion of a program manager's time is spent negotiating budgets with functional managers in light of original financial planning for the work, current schedules, past performance by the various functions, and overall program status.

Program office personnel should have the following questions regarding proposed work package budgets:

1. Does the schedule as presented by the functional groups validate our program plans?
2. Will the scheduled work meet technical specifications?
3. Is the budget for the work consistent with planning estimates?

If each of these questions is answered in the affirmative, the program manager simply approves the budget and authorizes performance. He or she may authorize performance for the entire scheduling horizon, 10 weeks in our examples, for the duration of the work package, or for the scheduling period, assumed to be one month in our examples. The authorization period is controlled by specifying the time period during which the program manager will accept charges against the account number assigned to a given work package.

43.7.3 Authorization Document

The characteristics of complex programs require that we integrate time and expenditure plans into a measure of *planned value of work performed.* Once a schedule and budget are prepared and approved, *planned value* may be established for each work package, or it may be decided to integrate expenditure and time plans at a level somewhat higher on the WBS. The tightest control is achieved when planned values are established at the work package level.

If the integration is done at the work package level, the planned value of work becomes the approved budget for the work package. Later in the reporting process, actual costs for work performed on a work package can be compared with planned value for work to monitor in an integrated way time and cost performance.

The authorization document is a work package approved by the program manager or by his or her representative. It should contain detailed information on time, cost, planned value, and expected performance on that segment of work. When approved by both program and functional management, it becomes the agreement, or performance contract, between program and functional personnel. Figure 43.14 provides an illustration of an authorizing document.

Figure 43.14 contains a summary description of the work embodied in the work package, a milestone chart indicating scheduled completion dates for work package activities, a time-phased financial expenditure plan, and a time-phased work plan. Finally, it contains planned value of work calculations for major milestones.

If the process leading to the issuance of the authorizing document is properly organized, the standards established should induce sufficient motivation on the part of functional personnel to perform the task and achieve or exceed the standards established.

43.7.4 The Budget Procedure

Operating budgets are ordinarily prepared for each program and functional organization on an annual basis. Each program prepares an operating budget for each period, and the budget normally represents a portion of the overall budget for the entire program. Each program budget contains an estimate of revenue and expenses. Expenses are ordinarily derived from a combination of approved work package budgets and program expenditure plans for the period, and normally include some funds for contingency. Estimating annual revenue for programs that are expected to extend over many budgeting periods is a difficult problem.

Operating budgets are also prepared for each functional group for the budget period. If these functional budgets are tight and challenging, we must expect many functions to require some contingency funds for inevitable overexpenditures. These funds may be provided either through an appeal process to program managers, and a corresponding revision to the operating budget, or through a partial allocation of contingency funds from program budgets to functional organizations during the expenditure

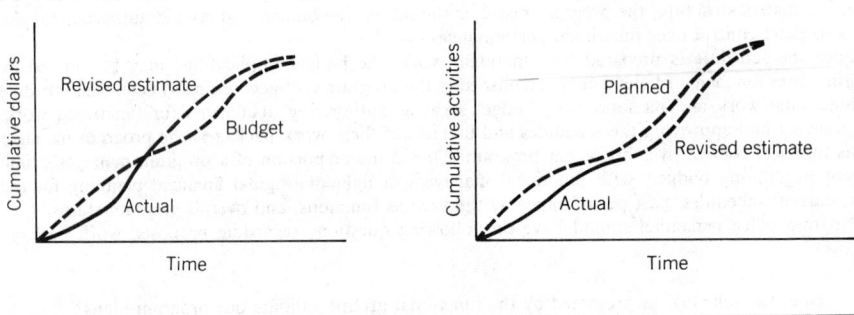

Planning value of work scheduled

Milestones:
1 Complete preliminary design
2 Complete final design
3 Prepare preliminary drawings
4 Prepare final detail drawings
5 Reliability summary
6 Manufacture
7 Reliability summary
8 Test

Work package description: The purpose of this task is to design and fabricate air inlet valve acording to specification xxx. WBS No. xxx-xxx-xxx

Fig. 43.14 Sample work package authorization document.

planning process. The latter procedure is likely to lead to goal-congruent behavior if functional organizations are treated as profit centers. The former procedure is likely to be most effective if functional organizations are treated as cost centers, which is the typical responsibility center designation for these organizational units.

Operating budgets for functional disciplines are prepared by summing approved work package budgets pertaining to a function with planned expenditures derived from approved program expenditure plans for the portion of the budget period not covered by authorized work packages.

Contingency funds are required in these operating budgets to allow for some overexpenditures that are expected with tight budgets as well as unanticipated but inevitable differences that occur between financial expenditure plans and approved work package plans. That is, annual operating budgets for each function must be approved based largely on expenditure plans of the organization. Later approved costs of work packages usually will deviate from the costs included in the original operating budget because of changes that occur between the planning process and the resource allocation process.

If differences between approved work packages and operating budgets are large, revisions are called for in operating budgets; if differences are small, they should be absorbed into a contingency account, which itself may be treated as an overhead account, along with all other nonprogram work. The responsibility center overhead account should also include idle time, company sponsored research and development, proposal effort, indirect functional supplies, and functional supervision.

43.7.5 Functional Overhead Budgets

Functional overhead budgets normally contain estimates for all functional expenditures that cannot be traced directly to a funded program. Functional organizations are normally held responsible for performance regarding these overhead budgets. Under these circumstances, functional managers will attempt to keep their personnel employed on either a contractual program or an approved company-funded project, such as a proposal or a research project. Otherwise, the time of functional personnel must be charged to a special account called "idle time," and although some charges are expected to this account because of resource-allocation problems and because of normal transitions from one program to another that often involve delays, these charges must be kept to a minimum for the functional manager to achieve good performance regarding his or her overhead budget.

Program managers, on the other hand, seek to remove functional manpower from their program as quickly as possible, since the performance of these managers is often evaluated based on profit. Functional managers must be able to use these personnel on other programs *for the organization as a whole* to achieve the cost reduction that is attributed to the program office. If the organization cannot use personnel so released, their time must be charged to idle time, which may cause functional managers to overexpend their overhead budgets! If functional overhead budgets are in jeopardy, the temptation is always present to mischarge functional time to contracts that appear to be able to absorb such charges and to prolong existing problem work longer than necessary.

Because many organizations treat their level of functional resources as essentially fixed during the short term, an unplanned underexpenditure on a program ordinarily shifts an equivalent amount of costs somewhere else in the organization and does not result in a comparable organizational saving *unless* these resources may be absorbed profitably on another contract or approved internally funded program.

These facts of organizational life lead us to place a premium upon planning and flexibility on the part of functional managers. Functional planning must always include provision for contingencies, whether this takes the form of preplanned effort on internally approved programs or plans to shift resources to new programs if they are released prematurely.

If this kind of contingency planning is not done within functional disciplines, pressures will build to mischarge contracts, overrun overhead budgets, and adjust the level of personnel in the organization at an undesirable rate.

Moreover, it is a mistake to place too much emphasis on performance regarding overhead budgets in the evaluation of functional managers to the exclusion of performance regarding cost, schedule, and quality on all programs that are supported by the function. In addition, the quality of planning for the use of functional resources should play an important role in performance evaluation.

43.8 SYSTEMS OF REPORTING FOR PROGRAM CONTROL

To put the general requirements of a reporting system for complex programs into perspective, it is necessary to remember that we are describing a system that replaces the cost-accounting system in the management control process. The program cost-accounting system, however, *does have similarities* with conventional cost-accounting systems (e.g., account code structure, standards, variances, and overhead allocations) as we have seen. Therefore, when it comes to designing program reporting systems, it is useful to begin by reviewing the reporting system established in conventional cost accounting for each element of cost. It turns out that each of the variances used in conventional cost accounting may be used in program cost accounting, but they must be supplemented with combined cost and schedule variances.

The labor variances of conventional cost accounting are subdivided into time and rate variances. The time variance is found for a task as follows:

$$\text{standard hours} - \text{actual hours} \times \text{standard rate} = \text{time variance} \qquad (43.1)$$

The rate variance for labor is found by

$$\text{standard rate} - \text{actual rate} \times \text{actual hours} = \text{rate variance} \qquad (43.2)$$

The total labor variance for a task is

$$\text{standard labor cost} - \text{actual labor cost} = \text{total labor variance} \qquad (43.3)$$

Material variances are similarly subdivided into quantity and price variances. The quantity variance is computed as follows:

$$\text{standard quantity} - \text{actual quantity} \times \text{standard price} = \text{quantity variance} \qquad (43.4)$$

The price variance is given by

$$\text{standard price} - \text{actual price} \times \text{actual quantity} = \text{price variance} \qquad (43.5)$$

The total material variance is

$$\text{standard material cost} - \text{actual material cost} = \text{total material variance} \qquad (43.6)$$

We omit the overhead variances, since our reporting system is normally concerned with controllable direct costs.

Now the difference between the requirements for program reporting and conventional cost reporting systems develops because the labor and material variances are only cost or spending variances. They essentially assume that the scheduled work was completed in the process of spending funds for labor and materials. This is a realistic assumption in most manufacturing operations. Not so, however, in the management of complex programs.

For any WBS item, we are interested in the relationship between *planned value of work* for a given time period and *actual cost of work* for the same time period. This will tell us our total variance for the task and we, by the computation of more detailed variances, seek to trace its causes. There are five potential causes for any total variance, and they are as follows:

1. We did more or less work than scheduled.
2. We used more or less labor than planned for the actual work we did.
3. We paid more or less than planned for the actual labor used.
4. We used more material than planned for the work we accomplished.
5. We paid more or less than planned for the material we actually used.

The portion of the total variance attributable to number 1 is called the *schedule variance*, and the portion attributable to numbers 2–5 is called the *spending variance*. Therefore, the total variance for a given task or end item of a program is

$$\text{budgeted value of work planned} - \text{actual cost of work accomplished} = \text{total variance} \qquad (43.7)$$

The schedule variance is given by

$$\text{budgeted value of work planned} - \text{budgeted value of work accomplished} = \text{schedule variance} \qquad (43.8)$$

The spending variance is

$$\text{budgeted value of work accomplished} - \text{actual cost of work accomplished} = \text{spending variance} \qquad (43.9)$$

The spending variance is then subdivided into labor and material variances according to Eqs. (43.1) and (43.6). Labor and material variances may be subdivided further into rate and quantity variances according to Eqs. (43.1), (43.2), (43.4), and (43.5).

These nine variances may be computed for each level of the WBS and for each functional organization at regular intervals throughout the life of a program. The total variances for a WBS end item tells the responsible manager whether there is a problem or not regarding cost and schedule performance. If a problem exists, he or she can request more detailed reporting information for the next level of the WBS and find exactly where the problem is and whether the problem is concerned with schedule slippage or with a labor or material spending variance.

Note the only variance that is new is the schedule variance. Each of the other four variances appears in conventional cost-accounting systems. The schedule variance requires for its computation data on planned and actual schedule performance together with normal cost data for each level of the WBS. The fact that there is only one new variance required should dispel any mystery surrounding the schedule and cost reporting requirements for complex programs.

Conceptually, the time-and-cost reporting system for complex programs may be represented by Fig. 43.15. We should be capable of calculating a schedule variance and a cost variance for any level of the WBS. We should be able to divide the cost variance into its labor and material elements. We should be able to trace the schedule variance to a scheduled plan.

The reporting system should also contain the capability to provide program information for each functional organization that is performing work on the program by WBS end item. This information should follow the same format as that for the program.

These then are the broad outlines that the formal reporting system should take while recognizing that the systems should be flexible and adaptable to each organization.

Let us look at an example of an application of our variance system. Let us assume that we are considering a work package for the structure (A01-3-2) of our TV transmission building. Moreover, let us assume we placed the material orders at the estimated price and that we have chosen not to include overhead in our project control reports since the project manager has little or no control over it. Therefore, our primary control variable is the estimated $23,650 of labor costs for this work package.

Approximately 6 weeks into the project we have an integrated progress chart drawn up for us. It is shown in Fig. 43.16. The chart shows that work on the structure is currently 3 weeks ahead of schedule, yet that is not the whole story. The schedule variance is positive, we have done more in the first 6 weeks than originally planned (i.e., BVWA > BVWP). Yet, we have spent more than budgeted for the work accomplished (i.e., ACWA > BVWA). Projecting these trends to completion, we will spend approximately $2000 more than estimated but we will finish 3 weeks early. Our conclusion at first look might be that the schedule gain was accomplished by spending more labor resources and further investigation might show that to be true. Nevertheless, unless there are some charges made, the work package will overrun by approximately $2000 at completion and that is bad news! The integrated report gives us a rather complete picture. It is much clearer than independent budget versus actual expenditure and schedule progress reports.

43.8.1 Reporting Delays and Bias

There always must be some delay between actual program progress and problems and their reporting, since the reporting process consumes time. All program status must be ascertained from the performing organizations. These data then must be processed. After processing these reports must be analyzed both to ensure that the processing was done correctly and to analyze progress and problems. It is not unusual for this processing and analysis work to consume 2 weeks or more on a complex program.

Moreover, the subsequent meetings and recommendations for action may take still another week or two. When action is finally taken, it may be to remedy a problem that existed a month ago! To further complicate the reporting problem, bias may creep into the reports.

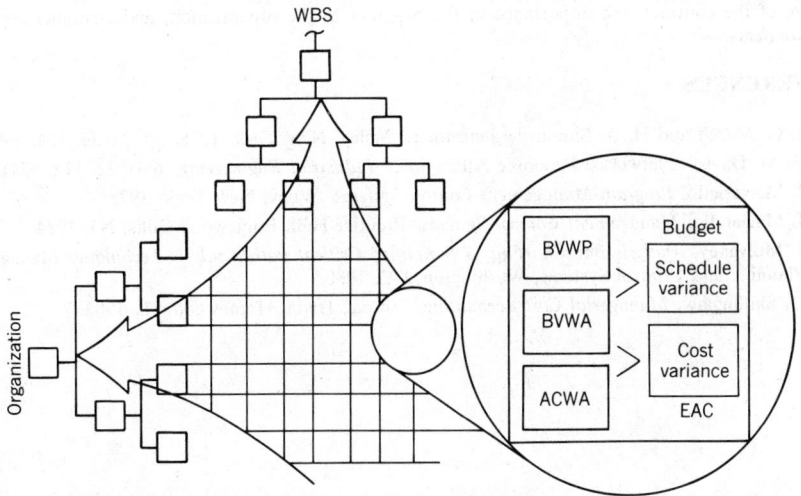

The variance terms in the diagram are defined as follows:
BVWP—budget value of work planned; BVWA—budget value of work
accomplished; ACWA—actual cost of work accomplished; EAC—estimate
of costs at completion.

Fig. 43.15 Integrated time–cost reporting system.

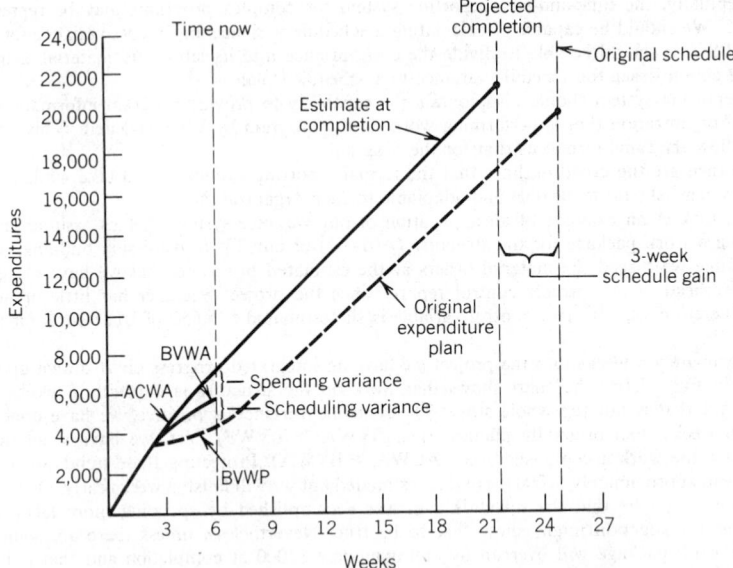

Fig. 43.16 Integrated progress chart for the structure of the TV transmission building A01-3-2. BVWA—budget value of work planned; ACWA—actual cost of work accomplished. (Adapted by permission from G. Schillinglaw, *Managerial Cost Accounting,* 4th ed., Richard D. Irwin, Homewood, IL, 1977, p. 763.)

If functional supervision is evaluated based on its rate of progress, then we should expect bias to enter into the reporting system. Bias can be prevented to some extent by explicit definition of activities that then become standard for the organization.

Often standardization of activity descriptions manifests itself in the preparation of a dictionary of terms and activities. The dictionary serves to accurately identify completed activities and improves communication within the program group while providing the basis for a historical file of time and cost date. This file becomes useful for estimating future programs containing similar activities.

Finally, the frequency of reporting should vary from one program to the next, depending on the nature of the contract, the importance of the program to the organization, and customer reporting requirements.

REFERENCES

1. J. G. March and H. A. Simon, *Organizations,* Wiley, New York, 1978, pp. 31–32, 154, 190–191.
2. E. M. Davis, "Networks: Resource Allocation," *Industrial Engineering,* **6**(4), 22–34 (1974).
3. J. Maciariello, *Program-Management Control Systems,* Wiley, New York, 1978.
4. J. Maciariello, *Management Control Systems,* Prentice-Hall, Englewood Cliffs, NJ, 1984.
5. J. Mulvaney, *Analysis Bar Charting, A Simplified Critical Path Analysis Technique,* Management Planning and Control Systems, Washington, DC, 1980.
6. G. Shillinglaw, *Managerial Cost Accounting,* 5th ed., Irwin, Homewood, IL, 1982.

CHAPTER 44

FINANCE AND THE ENGINEERING FUNCTION

WILLIAM BRETT

New York, New York

44.1 INTRODUCTION AND OUTLINE

Finance is fundamental; accounting is merely the set of procedures, techniques, and reports that make possible the effective execution of the finance function. Harold Geneen, the legendary chairman of International Telephone and Telegraph, included in his *Sayings of Chairman Hal,* "The worst thing a manager can do is run out of money." He meant it! The corporate function of Finance is that function which makes the decisions, or rather provides the recommendations to top management who really make the decisions, that prevent the enterprise from running out of money. Accounting gathers, organizes, and disseminates information that make it possible to make these decisions accurately and timely. In modern business, accounting performs many correlative functions, some in such detail and so esoteric as to appear to be an end in themselves.

The objectives of this chapter on finance and accounting are to describe:

- How accounting systems work to provide information for top managers and owners.
- How financial management is carried out.

Additionally, this chapter provides a concise description of how an accounting system is constructed to provide for the needs of middle management and staff groups such as engineers and marketers.

The purposes and uses of accounting systems, data, and reports are quite different for different people and functions in the business community. The engineer needs to understand accounting principles and processes as they apply to his or her function and also to understand the way in which others of the enterprise view business and what their information needs are. The following are five major groups that have distinctly differing points of view and objectives:

- Owners, investors, lenders, and boards of directors
- Top managers
- Middle managers of line functions
- Staff groups such as product planners, engineers, and market researchers
- Accountants

44.1.1 Needs of Owners, Investors, and Lenders

The first group—owners, investors, and lenders—have as their primary concern the preservation and protection of the capital or the assets of the business. The Board of Directors represents the interest of the owners and can be considered to be the agents of the owners (stockholders). The board members provide continuing review of the performance of top management as well as approval or disapproval of policies and key investment decisions.

This entire group wants to be assured that the property of the business—fixed plant and equipment, inventories, etc.—is being conserved. Next, they want to be assured that there will be sufficient liquidity, which means only that there will be enough cash available to pay all the bills as they come due. Finally, they want to see evidence of some combination of regular payout or growth in value—a financial return such as regular dividends or indications that the enterprise is increasing in value. Increase in value may be evidenced by growth in sales and profits, by increases in the market value of the stock, or by increased value of the assets owned. If the dividend payout is small, the growth expectations will be large.

The information available to the owners is, at a minimum, that which is published for public companies—the balance sheet, cash flow, and profit and loss statement. Special reports and analyses are also provided when indicated.

44.1.2 Needs of Top Managers

The top managers must be sensitive to the needs and desires of the owners as expressed by the Board of Directors and of the bankers and other lenders so that all of the purposes and objectives of owners and lenders are also the objectives of top managers. Additionally, top management has the sole responsibility for:

Developing long-range strategic plans and objectives
Approving short-range operating and financial plans
Ensuring that results achieved are measuring up to plan
Initiating broad gage corrective programs when results are not in conformance with objectives

Reports of financial results to this group must be in considerable detail and identified by major program, product, or operating unit in order to give insight sufficient to correct problems in time to prevent disasters. The degree of detail is determined by the management style of the top executive. Usually such reports are set up so that trouble points are automatically brought to the top executives' attention, and the detail is provided in order to make it possible to delve into the problems.

In addition to the basic financial reports to the owners, directors and top managers need:

- Long-term projections
- One-year budgets
- Periodic comparison of budget to actual
- Unit or facility results
- Product line results
- Performance compared to standard cost

44.1.3 Needs of Middle Managers of Line Functions

For our present purposes we will consider only managers of the sales and the manufacturing groups and their needs for financial, sales, and cost information. The degree to which the chief executive shares information down the line varies greatly among companies, ranging from a highly secretive handling of all information to a belief that sharing all the facts of the business improves performance and involvement through greater participation. In the great bulk of publicly held, large corporations, with modern management, most of the financial information provided to top management is available to staff and middle management, either on routine basis or on request. There are additional data that are needed by lower-level line managers where adequate operational control calls for much greater detail than that which is routinely supplied to top executives.

The fundamental assignment of the line manager in manufacturing and sales is to execute the policies of top management. In order to do this effectively, the manager needs to monitor actions and evaluate results. In an accounting context this means the manufacturing manager, either by formal rules or by setting his or her personal rules of thumb, needs to:

- Set production goals
- Set worker and machine productivity standards
- Set raw material consumption standards
- Set overhead cost goals
- Establish product cost standards
- Compare actual performance against goals
- Develop remedial action plans to correct deficiencies
- Monitor progress in correcting variances

The major accounting and control tools needed to carry out this mission include:

- Production standards
- Departmental budgets
- Standard costs
- Sales and production projections
- Variance reports
- Special reports

It is important that the line manager understands the profit and loss picture in his or her area of control and that job performance is not merely measured against preset standards but that he or she is considered to be an important contributor to the entire organization. It is, then, important that managers understand the total commercial environment in which they are working, so that full disclosure of product profits is desirable. Such a philosophy requires that accounting records and reports be clear and straightforward, with the objective of exposing operating issues rather than being designed for a tax accountant or lawyer.

The top marketing executive must have a key role in the establishment of prices and the determination of the condition of the market so that he or she is a full partner in managing the enterprise for profits. He or she therefore needs to participate with the manufacturing executive in the development of budgets and longer-range financial plans. Thus the budget becomes a joint document of marketing and manufacturing, with both committed to its successful execution.

The marketing executive needs to be furnished with all of the information indicated above as appropriate for the manufacturing manager.

44.1.4   Needs of Staff Groups (Product Planners, Engineers, Market Researchers)

The major requirement of accounting information for staff is that it provide a way to measure the economic effect of proposed changes to the enterprise. For the engineer this may mean changes in equipment or tooling or redesign of the product as a most frequent kind of change that must be evaluated before funds can be committed.

Accounting records that show actual and standard costs by individual product and discrete operation are invaluable in determining the effect of change in design or process. If changes in product or process can result in changes in total unit sales or in price, the engineer needs to know those projected effects. His or her final projections of improved profits will then incorporate the total effect of engineering changes.

The accounting records need to be in sufficient detail that new financial projections can be made reliably, with different assumptions of product features, sales volume, cost, and price.

44.1.5   Needs of Accountants

The accounting system must satisfy the strategic, operational, and control requirements of the organization as outlined above, but it has other external demands that must be satisfied. The accountants have the obligation to maintain records and prepare reports to shareholders that are "in conformity with generally accepted accounting principles consistently applied." Therefore, traditional approaches are essential so that the outside auditor as well as the tax collector will understand the reports and find them acceptable. There seems to be little need to sacrifice the development of good, effective control information for operating executives in order to satisfy the requirements of the tax collector or the auditor. The needs are compatible.

The key financial reporting and accounting systems typically used by each group are explained next.

44.2   SYSTEMS FOR OWNERS

A major concern of the owners or the Board of Directors and the lenders to the business must be to ensure the security of the assets of the business. The obvious way to do this in a small enterprise is occasionally to take a look. It is certainly appropriate for directors to visit facilities and places where inventories are housed to ensure that the assets really do exist, but this can only serve as a spot check and an activity comparable to a military inspection—everything looks very good when the troops know that the general is coming. The most useful and convenient way, as well as the most reliable way, to protect the assets is by careful study of financial records and a comparison with recent history to determine the trends in basic values within the business. A clear and consistent understanding of the condition of the assets of the business requires the existence of a uniform and acceptable system of accounting for them and for reporting their condition. The accounting balance sheet provides this.

In the remainder of this chapter, a set of examples based on the experience of one fictitious company is developed. The first element in the case study is the corporate balance sheet. From there the case moves back to the profit and loss and the cash flow statements. The case moves eventually back to the basic statements of expense and revenue to demonstrate how these records are used by the people managing the business—how these records enable them to make decisions concerning pricing, product mix, and investment in new plant and processes. The case will also show how these records help management to direct the business into growth patterns, a strengthened financial position, or increased payout to the owners.

The name of the fictitious company is the Commercial Construction Tool Company, Incorporated, and will be referred to as CCTCO throughout the remainder of this chapter. The company manufactures a precision hand tool, which is very useful in the positioning and nailing of various wooden structural members as well as sheathing in the construction of frame houses. The tool is a proprietary product on which the patents ran out some time ago; however, the company has had a reputation for quality and performance that has made it very difficult for competition to gain much headway. The tool has a reputation and prestige among users such that no apprentice carpenter would be without one. The product is sold through hardware distributors who supply lumber yards and independent retail hardware stores. About three years ago the company introduced a lighter weight and somewhat simplified model for use in the "do-it-yourself" market. Sales of the home-use model have been good and growing rapidly, and there is some concern that the HOMMODEL (home model) is cannibalizing sales of the COMMODEL (commercial model).

The company has one manufacturing facility and its general offices and sales offices are at the same location.

At the first directors' meeting after the year-end closing of the books the board is presented with the financial statements starting with the balance sheets for the beginning and end of the year. The principle of the balance sheet is that the enterprise has a net value to the owners (net worth) equal to the value of what is owned (the assets) less the amount owed to others (the liabilities).

44.3 BALANCE SHEET

When any business starts, the first financial statement is the balance sheet. In the case of CCTCO, the company was started many years ago to exploit the newly patented product. The beginning balance sheet was the result of setting up the initial financing. To get the enterprise started the original owners determined that $1000 (represents one million dollars, since in all of the exhibits and tables the last three zeros are deleted) was needed. The inventor and friends and associates put up $600 as the owners share—600,000 shares of common stock at a par value of $1 per share. Others, familiar with the product and the originators of the business, provided $400 represented by notes to be paid in 20 years—long-term debt. The original balance sheet was as shown below:

Assets		Liabilities and Net Worth	
		Liabilities	
Cash	1000	Long-term debt	400
		Net worth	
		Capital stock	600
Total assets	1000	Total liabilities and net worth	1000

The first financial steps of the company were to purchase equipment and machinery for $640 and raw materials for $120. The equipment was sent COD, but the raw material was to be paid for in 30 days. Immediately the balance sheet became more complex. There were now current assets—cash and inventory of raw materials—as well as fixed assets—machinery. Current liabilities showed up now in the form of accounts payable—the money owed for the raw material. All this before anything was produced. Now the balance sheet had become:

Assets		Liabilities and Net Worth	
		Liabilities	
Cash	360	Accounts payable	120
Inventories	120	Current liabilities	120
Current assets	480	Long-term debt	400
Fixed assets	640	Net worth	
		Capital stock	600
Total assets	1120	Total liabilities and net worth	1120

After a number of years of manufacturing and selling product the balance sheet became as shown below in Table 44.1. This important financial report requires explanation.

Assets are generally of three varieties:

- *Current.* Usually liquid and will probably be turned over at least once each year.
- *Fixed.* Usually real estate and the tools of production, frequently termed plant, property, and equipment.
- *Intangible.* Assets without an intrinsic value, such as good will or development costs which are not written off as a current expense but are declared an asset until the development has been commercialized.

44.3.1 Current Assets

In CCTCO's balance sheet the first item to occur is cash, which the company tries to keep relatively low, sufficient only to handle the flow of checks. Any excess over that amount the treasurer applies to pay off short-term debt, which has been arranged with local banks at one-half of one percent over the prime rate.

Accounts receivable are trade invoices not yet paid. The terms offered by CCTCO are typical—2% 10 days net 30, which means that if the bill is paid by the customer within 10 days after receipt, he or she can take a 2% discount, otherwise the total amount is due within 30 days. Distributors in the hardware field are usually hard pressed for cash and are frequently slow payers. As a result, receivables are the equivalent of two and a half month's sales, tying up a significant amount of the company's capital.

Table 44.1 Commercial Construction Tool Co., Inc.

Balance Sheet	Beginning
Assets	
Current assets	
Cash	52
Accounts receivable	475
Inventories	941
Total current assets	1468
Fixed assets	
Gross plant and equipment	2021
Less reserve for depreciation	471
Net plant and equipment	1550
Total assets	3018
Liabilities	
Current liabilities	
Accounts payable	457
Short-term debt	565
Long-term debt becoming current	130
Total current liabilities	1152
Long-term liabilities	
Interest-bearing debt	843
Total liabilities	1995
Net worth	
Capital stock	100
Earned surplus	923
Total net worth	1023
Total liabilities and net worth	3018

Inventories are the major element of current assets and consist of purchased raw materials, primarily steel, paint, and purchased parts; work in process, which includes all material that has left the raw material inventory point but has not yet reached the stage of completion where it is ready to be shipped; and finished goods. In order to provide quick delivery service to customers, CCTCO finds it necessary to maintain inventories at the equivalent of about three months' shipments—normally about 25% of the annual cost of goods sold.

44.3.2 Current Liabilities

Skipping to the liability section of the report, in order to look at all the elements of the liquid segment of the balance sheet, we next evaluate the condition of current liabilities. This section is composed of two parts: interest-bearing debt and debt that carries no interest charge. The noninterest-bearing part is primarily accounts payable, which is an account parallel but opposite to accounts receivable. It consists of the trade obligations not yet paid for steel, paint, and parts as well as office supplies and other material purchases. Sometimes included in this category are estimates of taxes that have been accrued during the period but not yet paid as well as other services used but not yet billed or paid for.

44.3.3 Accrual Accounting

At this point it is useful to define the term "accrued" or "accrual" as opposed to "cash" basis accounting. Almost all individual, personal accounting is done on a cash basis, that is, for individual tax accounting, no transaction exists until the money changes hands—by either writing a check or paying cash. In commercial and industrial accounting the accrual system is normally used, in which the transaction is deemed to occur at the time of some other overt act. For example, a sale takes place when the goods are shipped against a bona fide order, even though money will not change hands for another month. Taxes are charged based on the pro rata share for the year even though they may not be paid until the subsequent year. Thus costs and revenues are charged when it is clear that they are in fact obligated. This tends to anticipate and level out income and costs and to reduce the effect of fluctuations resulting only from the random effect of the time at which payments are made. Business managers wish to eliminate, as far as possible, wide swings in financial results and accrual accounting assists in this, as well as providing a more clearly cause-related set of financial statements. It also complicates the art of accounting quite considerably.

44.3.4   Interest-Bearing Current Liabilities

Interest-bearing current obligations are of two types: short-term bank borrowings and that portion of long-term debt that must be paid during the current year. Most businesses, and particularly those with a seasonal variation in sales, find it necessary to borrow from banks on a regular basis. The fashion clothing industry needs to produce three or four complete new lines each year and must borrow from the banks to provide the cash to pay for labor and materials to produce the fall, winter, and spring lines. When the shipments have been made to the distributors and large retail chains and their invoices have been paid, the manufacturer can "get out of the banks," only to come back to finance the next season's line. Because CCTCO's sales have a significant summer bulge at the retail level, they must have heavy inventories in the early spring, which drop to a fairly low level in the fall. Bank borrowings are usually required in February through May, but CCTCO is normally out of the banks by year end, so that the year-end balance sheet has a sounder look than it would have in April. The item "short-term borrowings" of $565 consists of bank loans that had not been paid back by the year's end.

The second part of interest-bearing current liabilities is that part of the long-term debt that matures within 12 months, and will have to be paid within the 12-month period. Such obligations are typically bonds or long-term notes. These current maturities represent an immediate drain on the cash of the business and are therefore classed as a current liability. As CCTCO has an important bond issue with maturities taking place uniformly over a long period, it has long-term debt maturing in practically every year.

44.3.5   Net Working Capital

The total of current assets less current liabilities is known as "net working capital." Although it is not usually defined in the balance sheet, it is important in the financial management of a business because it represents a large part of the capitalization of an enterprise and because, to some degree, it is controllable in the short run.

In times of high interest rates and cash shortages, companies tend to take immediate steps to collect their outstanding bills much more quickly. They will carefully "age" their receivables, which means that an analysis showing receivables ranked by the length of time they have been unpaid will be made and particular pressure will be brought to bear on those invoices that have been outstanding for a long time. On the other hand, steps will be taken to slow the payment of obligations; discounts may be passed up if the need for cash is sufficiently pressing and a general slowing of payments will occur.

Considerable pressure will be exerted to reduce inventories in the three major categories of raw material, work in process, and finished goods as well as stocks of supplies. Annual inventory turns can sometimes be significantly improved. There are, however, irreducible minimums for net working capital, and going beyond those points may result in real damage to the business through reducing service, increasing delivery times, damaging credit ratings, and otherwise upsetting customer and supplier relationships.

The effect of reducing net working capital, in a moderate and constructive way, spreads through the financial structure of the enterprise. The need for borrowing is reduced and interest expense is thereby reduced and profits are increased. Also, another effect on the balance sheet further improves the financial position. As the total debt level is reduced and the net worth is increased, the ratio of debt to equity is reduced, thus improving the financial community's assessment of strength. An improved rating for borrowing purposes may result, making the company eligible for lower interest rates. Other aspects of this factor will be covered in more detail in the discussion of net worth and long-term debt.

44.3.6   Current Ratio

The need to maintain the strength of another important analysis ratio puts additional resistance against the objective of holding net working capital to the minimum. Business owners feel the need to maintain a healthy "current ratio." In order to be in a position to pay current bills, the aggregate of cash, receivables, and inventories must be available in sufficient amount. One measure of the ability to pay current obligations is the ratio of current assets to current liabilities, the current ratio. In more conservative times and before the days of leverage, a ratio of 2.0 or even 3.0 was considered strong, an indication of financial stability. In times of high interest rates and with objectives of rapid growth, much lower ratios are acceptable and even desirable. CCTCO's ratio of 1.27 ($1468/$1152) is considered quite satisfactory.

44.3.7 Fixed Assets

Continuing the evaluation on the asset side of the balance sheet we find the three elements of fixed assets, that is, gross plant and equipment, reserve for depreciation, and net plant and equipment. Gross plant is the original cost of all the assets now owned and is a straightforward item. The concept of depreciation is one which is frequently misunderstood and partly because of the name "reserve for depreciation." The name seems to indicate that there is a reserve of cash, put away somewhere that can be used to replace the old equipment when necessary. This is not the case. Accountants have a very special meaning for the word reserve in this application. It means, to an accountant, the sum of the depreciation expense that has been applied over the life, up to now, of the asset.

When an asset, such as a machine, is purchased, it is assigned an estimated useful life in years. In a linear depreciation system, the value of the asset is reduced each year by the same percentage that one year is to its useful life. For example, an asset with a 12-year useful life would have an 8.33% annual depreciation rate (100 times the reciprocal of 12). The critical reason for reducing the value each year is to reduce the profit by an amount equivalent to the degree to which equipment is transformed into product. With high income taxes, the depreciation rate is critical to ensuring that taxes are held to the legal minimum. When the profit and loss statement is covered, the effect on profits and cash flow as a result of using nonlinear, accelerated depreciation rates will be covered. The important point to understand is that the reserve for depreciation does not represent a reserve of cash but only an accounting artifice to show how much depreciation expense has been taken (charged against profits) so far and, by difference, to show the amount of depreciation expense that may be taken in the future.

The difference between gross plant and reserve for depreciation, net plant and equipment, is not necessarily the remaining market value of the equipment at all, but is the amount of depreciation expense that may be charged against profits in future years. The understanding of this principle of depreciation is critical to the later understanding of profits and cash flow.

44.3.8 Total Capital

Together, the remaining items (long-term debt and net worth) on the liability side of the balance sheet make up the basic investment in the business. In the beginning, the entrepreneurs looked for money to get the business started. It came from two sources, equity investors and lenders. The equity investors were given an ownership share in the business, with the right to a portion of whatever profits might be made or a pro rata share of the proceeds of liquidation, if that became necessary. The lenders were given the right to regular and prescribed interest payments and were promised repayment of principal on a scheduled basis. They were not to share in the profits, if any. A third source of capital became available as the enterprise prospered. Profits not paid out in dividends were reinvested in plant and equipment and working capital. Each of these sources has an official name.

Lenders:	Long-Term Debt
Equity Investors:	Capital Stock
Profits Reinvested:	Earned Surplus

In many cases the cash from equity investors is divided into two parts, the par value of the common shares issued, traditionally $1 each and the difference between par and the actual proceeds from the sale of stock. For example, the sale of 1000 shares of par value $1 stock, for $8000 net of fees, would be expressed:

Capital Stock (1000 shares at $1 par):	$1000
Paid in Surplus:	$7000
Total Capitalization:	$8000

The final item on the balance sheet, earned surplus or retained earnings, represents the accumulated profits generated by the business which have not been paid out, but were reinvested.

Net worth is the total of capital stock and earned surplus and can also be defined as the difference, at the end of an accounting period, between the value of the assets, as stated on the corporate books, and the obligations of the business.

All of this is a simplified view of the balance sheet. In actual practice there are a number of other elements that may exist and take on great importance. These include preferred stocks, treasury stock, deferred income taxes, and goodwill. When any of these special situations occur, a particular review of the specific case is needed in order to understand the implications to the business and their effect on the financial condition of the enterprise.

Table 44.2 Commercial Construction Tool Co., Inc.—Costs and Revenues, Bad Year—Actual

Balance Sheet	Beginning	Ending	Change
Assets			
Current assets			
Cash	52	62	10
Accounts receivable	475	573	98
Inventories	941	1000	59
Total current assets	1468	1635	167
Fixed assets			
Gross plant and equipment	2021	2521	500
Less reserve for depreciation	471	744	273
Net plant and equipment	1550	1777	227
Total assets	3018	3413	395
Liabilities			
Current liabilities			
Accounts payable	457	503	46
Short-term debt	565	600	36
Long-term debt becoming current	130	130	0
Total current liabilities	1152	1233	82
Long-term liabilities			
Interest-bearing debt	843	1129	286
Total liabilities	1995	2362	368
Net worth			
Capital stock	100	100	0
Earned surplus	923	950	27
Total net worth	1023	1050	27
Total liabilities and net worth	3018	3413	395

44.3.9 Second Year Comparison

The balance sheet in Table 44.1 is a statement of condition. It tells the financial position of the company at the beginning of the period. At the end of the year the Board of Directors is presented two balance sheets—the condition of the business at the beginning and at the end of the period, as shown in Table 44.2. The Board is interested in the trends represented by the change in the balance sheet over a 1-year period.

Total assets have increased by $395 over the period—probably a good sign. Net worth or owners' equity has increased by $27, which is $368 less than the increase in assets. The money for the increase in assets comes from creating substantially more liabilities or obligations as well as the very small increase in the net worth. A look at the liabilities shows the following (note the errors from rounding that result from the use of computer models for financial statements):

	Increase
Accounts payable	$ 46
Short-term debt	36
Long-term debt	286
	$368

Changes in net working capital are evaluated to determine the efficiency in the use of cash and the soundness of the short-term position. No large changes that would raise significant questions have taken place. Current assets increased $167 and current liabilities by $82. These increases result from the fact that sales had increased, which had required higher inventories and receivables. The current ratio (current assets over current liabilities) had strengthened to 1.33 from 1.27 at the beginning of the period, indicating an improved ability to pay bills and probably increased borrowing power.

A major change in the left-hand (asset) side was the increase in fixed assets. Gross plant was up $500, nearly 25%, indicating an aggressive expansion or improvement program.

Table 44.3 Commercial Construction Tool Co., Inc.—Costs and Revenues, Bad Year—Actual

Proft and Loss Statement	
Sales	4772
Cost of production	4036
Underabsorbed burden	62
Fully absorbed cost	4097
Beginning inventory	941
Ending inventory	1000
Net change	59
Cost of sales	4038
Gross margin	734
Selling expense	177
Administrative	249
Operating profit	308
Interest	169
Profit before tax	138
Income tax	66
Net income	72

Net worth and earned surplus were up by $27, an important fact, sure to receive attention from the board.

In order to understand why the balance sheet had changed and to further evaluate the year's results, the directors needed a profit and loss statement and a cash flow statement.

44.4 PROFIT AND LOSS STATEMENT

The profit and loss statement (P&L) is probably the best understood and most used statement provided by accountants: It summarizes most of the important annual operating data and it acts as a bridge from one balance sheet to the next. It is a summary of transactions for the year—where the money came from and where most of it went. Table 44.3 is the P&L for CCTCO for the year.

For the sake of simplicity, net sales are shown as Sales. In many statements, particularly internal reports, gross sales are shown followed by returns and discounts to give a net sales figure. Cost of sales is a little more complex. Sales may be made from inventory or off the production line on special order. Stocks of finished goods or inventories are carried on the books at their cost of production. The formula for determining the cost of product shipped to customers is:

$$\text{beginning inventory} + \text{cost of production} - \text{ending inventory} = \text{cost of sales}$$

Additionally, CCTCO uses a standard burden rate system of applying overhead costs to production. The difference between the overhead charged to production at standard burden rates and the actual overhead costs for the period, in this case $62, is called unabsorbed burden and is added to the cost of production for the year, or it may be charged off as a period cost. The procedures for developing burden rates will be treated in more detail in a subsequent section.

Gross margin is the difference between sales dollars and the cost of manufacture. After deducting the costs of administrative overhead and selling expense, operating profit remains. Interest expense is part of the total cost of capital of the business and is therefore separate from operations. The last item, income tax, only occurs when there is a profit.

44.4.1 Financial Ratios

The combination of the P&L and the balance sheet makes it possible to calculate certain ratios that have great significance to investors. The ratios are shown in Table 44.4. The first and most commonly used as a measure of success is the return on sales. This is a valuable ratio to measure progress of a company from year to year, but is of less importance in comparing one company to another. A more useful ratio would be returns to value added. Value added is the difference between the cost of purchased

Table 44.4 Commercial Construction Tool
Co., Inc.—Costs and Revenues, Bad Year—
Actual

Financial Ratios	
Return on sales	1.51
Return on assets	2.24
Return on invested capital	3.56
Return on equity	6.94
Asset to sales ratio	0.67
Debt percent to debt plus equity	69
Average cost of capital	20.67

raw materials and net sales, and represents the economic contribution of the enterprise. It is a concept used more extensively in Europe than the United States and is the basis of the Value Added Tax (VAT), quite common in Europe and at this writing being considered in the United States.

Return on assets begins to get closer to the real interest of the investor. It represents the degree to which assets are profitable, and would indicate, from an overall economic point of view, whether the enterprise was an economic and competitive application of production facilities.

A ratio even more interesting to the investor is the return on invested capital. Total assets, as was described earlier, are financed by three sources:

- Equity—made up of stock, that is, owners' investment and profits retained in the business.
- Interest-bearing debt—composed of bonds, notes, and bank loans.
- Current liabilities—composed of operating debts such as accounts payable and taxes payable, which do not require interest payments.

Because the current liabilities are normally more than offset by current assets, the economic return is well described by the return on total or invested capital, which is net profit after taxes divided by the sum of equity plus interest-bearing debt.

A rate of return percentage of great interest to the owner is the return on equity. This rate of return compared to the return on total capital represents the degree to which the investment is or can be leveraged. It is to the interest of the investor to maximize the return on his or her dollars invested, so, to the degree that money can be borrowed at interest rates well below the capacity of the business to provide a return, the total profits to the owners will increase. Return on equity is a function of the ratio of debt to debt plus equity (total capital) and is a measure of the leverage percentage in the business. It is to the advantage of the owners to increase this ratio in order to increase the return on equity up to the point that the investment community, including bankers, concludes that the company is excessively leveraged and is in unsound financial condition. At that point it becomes more difficult to borrow money and interest rates of willing lenders increase significantly. Fashions in leverage change depending on the business cycle. In boom times with low interest rates, highly leveraged enterprises are popular, but tend to fall into disfavor when times are tough.

A more direct measure of leverage is "debt percent to debt plus equity" or debt to total capital. The 69% for CCTCO indicates that lenders really "own" 69% of the company and investors only 31%.

Another ratio of interest to investors is the asset turnover or asset to sales ratio. If sales from a given asset base can be very high, the opportunity to achieve high profits appears enhanced. On the other hand, it is very difficult to change the asset to sales ratio very much without changing the basic business. Certain industries or businesses are characterized as being capital intensive, which means they have a high asset to sales ratio or a low asset turnover. It is fundamental to the integrated forest products industries that they have a high asset to sales ratio, typically one to one. The opposite extreme, for example, the bakery industry, may have a ratio of 0.3–0.35 and turn over assets about three times per year. Good management and very effective use of facilities coupled with low inventories can make the best industry performer 10% better than the average, but there is no conceivable way that the fundamental level can be dramatically and permanently changed.

The final figure in Table 44.4, that of average cost of capital, cannot be calculated from only the P&L and balance sheet. One component of the total cost of capital is the dividend payout, which is not included in either report. It was stated previously that the P&L shows where most of the money went—it does not include dividends and payments for new equipment and other capital goods. For this we need the cash flow, also known as the source and application of funds, shown in Table 44.5.

Table 44.5 Commercial Construction Tool Co., Inc.—Costs and Revenues, Bad Year—Actual

Source and Application of Funds	
Net profit after tax	72
Depreciation expense	273
Cash generated	344
Increase in net working capital	
Change in cash	10
Change in receivables	98
Change in inventories	59
Change in payables	−46
Net change	121
Capital expenditures	500
Operating cash requirements	621
Operating cash flow	−276
Dividends	45
Net cash needs	−321
Increase in debt	321

44.5 CASH FLOW OR SOURCE AND APPLICATION OF FUNDS

There are two sources of operating cash for any business: the net profits after tax and noncash expenses. In Table 44.5, the cash generated by the business is shown as $344, the sum of net profit and depreciation. This is actually the operating cash generated and does not include financing cash sources, which are also very important. These sources include loans, capital contributions, and the sale of stock and are included in the cash flow statement as well as in the balance sheet where they have already been reviewed in a previous section of this chapter.

It seems clear and not requiring further explanation that the net profit after tax represents money remaining at the end of the period, but the treatment of noncash expenses as a source of operating funds is less self-evident. Included in the cost of production and sales in the previous section were materials and labor and many indirect expenses such as rent and depreciation, which were included in the P&L in order to achieve two objectives:

- Do not overstate annual earnings.
- Do not pay more income taxes than the law requires.

In the section on fixed assets, when discussing the balance sheet, it was pointed out that the reserve for depreciation is not an amount of money set aside and available for spending. It is the total of the depreciation expense charged so far against a still existing asset. The example was a piece of equipment with a useful life of 12 years, the total value of which was reduced by 8.33% (the reciprocal of 12 times 100) each year. This accounting action is taken to reduce profits to a level that takes into consideration the decreasing value of equipment over time, to reduce taxes, and to avoid overstating the value of assets. Depreciation expense is not a cash expense—no check is written—it is an accounting convention. The cash profit to the business is therefore understated in the P&L statement because less money was spent for expenses than indicated. The overstatement is the amount of depreciation and other noncash expenses included in costs for the year.

In the P&L in Table 44.4, included in the cost of sales of $4038, is $273 of depreciation expense. If this noncash item were not included as an expense of doing business, profit before tax would be increased from $138 to $411. Taxes were calculated at a 48% rate, so the revised net profit after tax would be $214. This new net profit would also be cash generated from operations instead of the $344 actually generated ($72 profit plus $273 depreciation) when noncash expenses are included as costs. The reduction in cash available to the business resulting from ignoring depreciation is exactly equal to the increase in taxes paid on profits. The anomaly is that the business has more money left at the end of the year when profits are lower!

44.5.1 Accelerated Depreciation

This is a logical place to examine various kinds of depreciation systems. So far, only a straight-line approach has been considered—the example used was a 12-year life resulting in an 8.33% annual expense or writedown rate. Philosophical arguments have been developed to support a larger writedown in the early years and reducing the depreciation rate in later years. Some of the reasons advanced include:

- A large loss in value is suffered when a machine becomes second hand.
- The usefulness and productivity of a machine is greater in the early years.
- Maintenance and repair costs of older machines are larger.
- The value of older machines does not change much from one year to the next.

The reason that accelerated systems have come into wide use is more practical than philosophical. With faster, early writedowns the business reduces its taxes now and defers them to a later date. Profits are reduced in the early years but cash flow is improved. There are two common methods of accelerating depreciation in the early years of a machine's life:

Sum of the digits
Double declining balance

Table 44.6 compares the annual depreciation expense for the two accelerated systems to the straight-line approach. For these examples a salvage value of zero is assumed at the end of the period of useful life. At the time of asset retirement and sale, a capital gain or loss would be realized as compared to the residual, undepreciated value of the asset, or zero, if fully depreciated.

The methods of calculation are represented by the following equations and examples where:

$$N = \text{number of years of useful life}$$
$$A = \text{year for which depreciation is calculated}$$
$$P = \text{original price of the asset}$$
$$D_a = \text{depreciation in year } A$$
$$B = \text{book value at year end}$$

The equation for straight-line depreciation is

$$D_a = \frac{1}{n} \times P$$
$$B = P - (D_1 + D_2 + \cdots + D_a)$$

In the example with an asset costing \$40,000 with an 8-year useful life:

$$D_a = \frac{1}{8} \times 40,000 = 0.125 \times 40,000 = 5000$$

To calculate depreciation by the sum of the years' digits method, use

$$D_a = [(N + 1 - A)/(N + N - 1 + N - 2 + \cdots + 1)] \times P$$

For the third year, for example,

$$D_3 = [(8 + 1 - 3)/(8 + 7 + 6 + 5 + 4 + 3 + 2 + 1)] \times 40,000$$
$$D_3 = [(6)/(36)] \times 40,000 = 0.1667 \times 40,000 = 6667$$

The depreciation rates shown in Table 44.6 under the double declining balance method are calculated to show a comparison of write-off rates between systems. The actual calculations are done quite differently:

$$D_a = \frac{2}{n} \times B_{a-1}$$

$$B_a = P - (D_1 + D_2 + \cdots D_{a-1})$$

In the third year, then,

and

$$D_3 = \frac{2}{8} \times 22,500 = 5625$$

$$B_3 = 40,000 - (10,000 + 7500 + 5625) = 16,875$$

Note in Table 44.6 that the double declining balance method, as should be expected, if allowed to continue forever, never succeeds in writing off the entire value. The residue is completely written off in the final year of the asset's life. The sum of the years' digits is a straight line and provides for a full write-off at the end of the period.

Figure 44.1 depicts, graphically, the annual depreciation expense using the three methods.

In many cases, a company will succeed in attaining both the advantages to cash flow and tax

Table 44.6   Accelerated Depreciation Methods[a]

Straight-Line Method[b]

Year	Rate	Depreciation Expense	Book Value, Year End
1	0.125	5000	35000
2	0.125	5000	30000
3	0.125	5000	25000
4	0.125	5000	20000
5	0.125	5000	15000
6	0.125	5000	10000
7	0.125	5000	5000
8	0.125	5000	0

Sum of the Years' Digits Method[c]

Year	Rate	Depreciation Expense	Book Value, Year End
1	0.2222222	8889	31111
2	0.1944444	7778	23333
3	0.1666667	6667	16667
4	0.1388889	5556	11111
5	0.1111111	4444	6667
6	0.0833333	3333	3333
7	0.0555556	2222	1111
8	0.0277778	1111	0

Double the Declining Balance Method[d]

Year	Rate Equivalent	Depreciation Expense	Book Value, Year End
1	0.25	10000	30000
2	0.1875	7500	22500
3	0.140625	5625	16875
4	0.1054688	4219	12656
5	0.0791016	3164	9492
6	0.0593262	2373	7119
7	0.0444946	1780	5339
8	0.1334839	5339	0

[a] Basic assumptions: equipment life, 8 years; original price, $40,000; estimated salvage value, $0.

[b] Annual rate equation: one divided by the number of years times the original price.

[c] Annual rate equation: sum of the number of years divided into the years of life remaining.

[d] Annual rate equation: twice the straight-line rate times the book value at the end of the preceding year.

minimization of accelerated depreciation as well as the maximizing of earnings by using straight-line depreciation. This is done by having one set of books for the tax collector and another for the shareholders and the investing public. This practice is an accepted approach and, where followed, is explained in the fine print of the annual report.

A number of special depreciation provisions and investment tax credit arrangements are available to companies from time to time. The provisions change as tax laws are revised either to encourage investment and growth or to plug tax loopholes, depending on which is politically popular at the

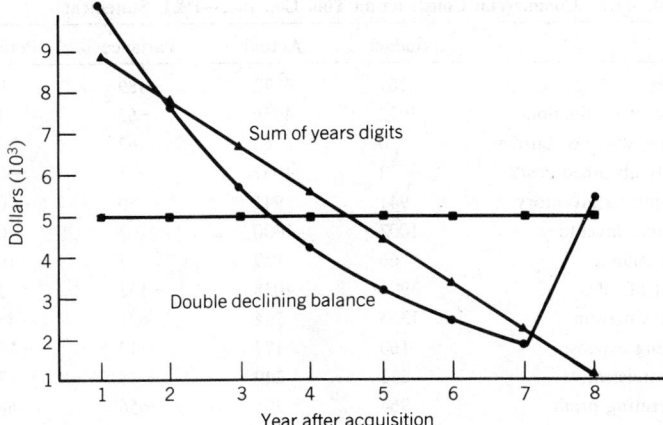

Fig. 44.1 Comparison of depreciation methods. ($40,000 original price; 8-year life; and no residual salvage value.)

time. The preceding explains the theory—applications vary considerably with changes in the law and differences in corporate objectives and philosophy.

The cash generated by the business has, as its first use, the satisfaction of the needs for working capital, that is, the needs for funds to finance increases in inventories, receivables, and cash in the bank. Each of these assets requires cash in order to provide them. Offsetting these uses of cash are the changes that may take place in the short-term debts of the enterprise and accounts payable. In Table 44.5, we see that $121 is required in increased net working capital, essentially all of which goes to provide for increased inventories and receivables needed to support sales increases.

The largest requirement for cash is the next item, that of capital expenditures, which has consumed $500 of the cash provided to the business. The total needs of the company for cash—the operating cash requirements—have risen to $621 compared to the cash generated of $344, and that is not the end of cash needs. The shareholders have become accustomed to a return on their investment—an annual cash dividend. The dividend is not considered part of operating cash flow nor is it a tax deductible expense as interest payments are. The dividend, added to the net operating cash flow of −$276, results in a borrowing requirement for the year of $321.

To summarize, the Board of Directors has been furnished a set of operating statements and financial ratios as shown in Tables 44.2–44.5. These ratios show a superficial picture of the economics of the enterprise from a financial viewpoint and present some issues and problems to the directors. The condition of the ratios and rates of return for CCTCO are of great concern to the directors and lead to some hard questions for management.

Why, when the total cost of capital, that is, interest plus dividends as a percentage of debt plus equity is 20.67%, is the return on total capital only 3.56%? Why does it take $0.67 worth of assets to provide $1 worth of sales in a year? Why is it that profit after tax is only 1.51% of sales? The board will not be pleased with performance and will want to know what can and will be done to improve. The banks will perhaps have concerns about further loans and shareholders or prospective shareholders will wonder about the price of the stock.

The answers to these questions require a level of cost and revenue information normally supplied to top management.

44.6 SYSTEMS FOR TOP MANAGERS

Corporate chief executives who allowed themselves to be as badly surprised by poor results at the end of the year as the chief executive of CCTCO would be unlikely to last long enough to take corrective action. However, the results at CCTCO can provide clear examples of the usefulness of accounting records in determining the cause of business problems and in pointing in the direction of practical solutions.

The first step of the chief executive at CCTCO was to compare actual results with those projected for the year. It had been the practice at CCTCO to prepare a comprehensive business plan and budget at the beginning of each year. Monthly and yearly, reports comparing actual with budget were made available to top officers of the company. Tables 44.7–44.10 show a comparison of the budgeted P&L, performance ratios, balance sheet, and cash flow for the year compared to the actual performance already reviewed by the board.

Table 44.7   Commercial Construction Tool Co., Inc.—P&L Statement

	Budget	Actual	Variance	Percent
Sales	5261	4772	−489	−9.29
Cost of production	3972	4036	−63	−1.59
Underabsorbed burden	0	62	62	NMF
Fully absorbed cost	3972	4097	−1	−0.03
Beginning inventory	941	941	0	0.00
Ending inventory	1007	1000	−7	−0.68
Net change	66	59	−7	−10.36
Cost of sales	3906	4038	−132	−3.37
Gross margin	1355	734	−621	−45.81
Selling expense	160	177	−17	−10.90
Administrative	231	249	−18	−7.88
Operating profit	964	308	−656	−68.08
Interest	154	169	−15	−9.93
Profit before tax	810	138	−671	−82.92
Income tax	389	66	322	82.92
Net income	421	72	−349	−82.92

Table 44.8   Commercial Construction Tool Co., Inc.—Financial Ratios

	Budget	Actual	Variance
Return on sales	8.00	1.51	−6.50
Return on assets	12.98	2.24	−10.74
Return on invested capital	20.47	3.56	−16.92
Return on equity	34.76	6.94	−27.82
Asset to sales ratio	0.62	0.67	−0.06
Debt percent to debt plus equity	60	69	−9.53
Average cost of capital	16.43	20.67	−4.24

An examination of the budget/actual comparisons revealed many serious deviations from plan. Net worth and long-term debt were trouble spots. Profits were far from expected results, and cash flow was far below plan.

The president searched the reports for the underlying causes in order to focus his attention and questions on those corporate functions and executives that appeared to be responsible for the failures. He concluded that there were seven critical variances from the budget, which when understood, should eventually lead to the underlying real causes. They included

Element	Variance	Percent
Sales	−489	−9.29
Underabsorbed burden	−62	—
Cost of sales	−132	−3.37
Selling expense	−17	−10.90
Administrative expense	−18	−7.88
Interest	−15	−9.93
Net working capital	−68	−14.50

The president asked the VP Sales and the VP Manufacturing to report to him as to what had happened to cause these variances from plan and what corrective action could be taken. He instructed the Controller to provide all the cost and revenue analyses needed to arrive at answers.

In two weeks the three executives made a presentation to the president that provided a comprehensive understanding of the problems, recommended solutions to them, and a timetable to implement the program. The following is a summary of that report.

Table 44.9 Commercial Construction Tool Co., Inc.—Variance Analysis, Balance Sheet

	Budget	Actual	Variance	Percent
Assets				
Current assets				
Cash	56	62	6	10.78
Accounts receivable	631	573	−59	−9.29
Inventories	1007	1000	−7	−0.68
Total current assets	1695	1635	−59	−3.51
Fixed assets				
Gross plant and equipment	2521	2521	0	0.00
Less reserve for depreciation	744	744	0	0.00
Net plant and equipment	1777	1777	0	0.00
Total assets	3472	3413	−59	−1.71
Liabilities				
Current liabilities				
Accounts payable	491	503	13	2.59
Short-term debt	604	600	−4	−0.68
Long-term debt becoming current	130	130	0	0.00
Total current liabilities	1225	1233	9	0.70
Long-term liabilities				
Interest-bearing debt	848	1129	281	33.17
Total liabilities	2073	2362	290	13.98
Net worth				
Capital stock	100	100	0	0.00
Earned surplus	1300	950	−349	−26.87
Total net worth	1400	1050	−349	−24.95
Total liabilities and net worth	3472	3413	−59	−1.71

Table 44.10 Commercial Construction Tool Co., Inc.—Variance Analysis, Source and Application of Funds

	Budget	Actual	Variance	Percent
Net profit after tax	419	72	−347	−82.84
Depreciation expense	273	273	0	0.00
Cash generated	692	344	−347	−50.20
Increase in net working capital				
Change in cash	4	10	6	150.90
Change in receivables	156	98	−59	−37.53
Change in inventories	66	59	−7	−10.36
Change in payables	−34	−46	−13	37.85
Net change	193	121	−72	−37.40
Capital expenditures	500	500	0	0.00
Operating cash requirements	693	621	−72	−10.42
Operating cash flow	−1	−276	−275	27500.00
Dividends	45	45	0	0.00
Net cash needs	−46	−321	−275	593.31
Increase in debt	46	321	275	593.31

44.6.1 Report to the President

Causes of Last Year's Results

The poor operating results of last year are caused almost entirely by a change in product mix from the previous year and not contemplated in the budget established 15 months ago. The introduction of the HOMMODEL nearly two years ago resulted in very few sales in the early months following its initial availability. However, early last year, sales accelerated dramatically, caught up with, and passed those of the COMMODEL. For a number of reasons this has had a poor effect on the financial structure of our company:

- Lack of experience on the new product has resulted in costs higher than standard.
- Standard margins are lower for the HOMMODEL.
- Travel and communications costs were high because of the new product introduction.
- Prices on the HOMMODEL were lower than standard because of special introductory dealer discounts and deals.
- Receivables increased because of providing initial stocking plans for new dealers handling the HOMMODEL.
- Higher interest expense resulted from higher debt—a direct result of cash flow shortfall.

The only significant variance unrelated to the new product was the fact that factory and office rents were raised during the year.

The following product mix table summarizes a number of accounting documents and shows the effect of product mix on profits. Product "A" is the COMMODEL and product "B" is the HOM-MODEL.

	Budget	Actual	Variance
Product "A"			
Sales—hundreds	740	530	−210
Sales—dollars	3130	2274	−857
Cost of sales	2303	1693	610
Margin	826	580	−246
Unit price	4.23	4.29	0.06
Unit cost	3.11	3.20	−0.08
Unit margin	1.12	1.09	−0.02
Product "B"			
Sales—hundreds	670	830	160
Sales—dollars	2131	2498	368
Cost of sales	1698	2334	−635
Margin	445	164	−281
Unit price	3.18	3.01	−0.17
Unit cost	2.54	2.81	−0.28
Unit margin	0.64	0.20	−0.45

Recommended Corrective Action

As the major problems are caused by the new product cannibalizing sales of the old COMMODEL, action is directed toward increasing margins on the HOMMODEL to nearly that of the COMMODEL and increasing the proportion of sales of the latter. This will be accomplished by simultaneously reducing unit cost and increasing selling price of the new product. The following program will be undertaken:

- Increase the unit price to 3.52 and eliminate deals and promotion pricing for a margin improvement of 0.34.
- Productivity improvements realized in the last two months of the year will reduce costs by 0.15 for the year.
- Proposed changes in material and finish will further reduce costs by 0.08.

These changes in price and cost will bring the standard margin of the HOMMODEL to 1.21, slightly more than that of the COMMODEL, thus eliminating any unfavorable effect of cannibalizing.

This report enabled the president to assure the board that the year to come would provide better results.

The following section describes the heart of the accounting system which made it possible for the VP Sales and the VP Manufacturing to develop a plan to solve the problems of the business.

44.7 SYSTEMS FOR MIDDLE MANAGEMENT

In the preceding sections, the need for cost and financial information and analyses has been demonstrated. In this section, the various elements of an accounting system will be developed to show how the needed management information is collected and organized. There is no one way or right way to do this—each company has its own approach and each practitioner of the accounting art has his or her favorite system. What is described here is only one of many possible approaches, but embodies most of the fundamental characteristics of a product accounting system.

In the CCTCO example there are six basic functions that generate costs; they are:

Administrative	Presses
Sales	Factory assembly
Maintenance	Pack

The classes of costs generated are

Fixed indirect	Direct
Variable indirect	

Fixed indirect costs are those that cannot readily be assigned directly to a product and do not change with short run fluctuations in the level of production. Important examples are rent expense or the salary of the superintendent of a production department. A long-term increase in production level would eventually require the addition of more factory floor space, so that fixed expense does in fact become variable in the long run, but for a normal budgeting period such costs are assumed to be nonvariable. Table 44.11 gives a tabulation of all of the expenses and costs of CCTCO for the year. Not included are capital expenditures and dividends which, as we have seen before, are not expenditures that affect the P&L. Interest expense is also excluded as a nonoperating expense comparable to dividends, both of which make up the total cost of capital. At the top of the table are shown fixed indirect costs generated by all six of the functions or cost centers inherent in CCTCO's business, and to which costs are distributed. Note that some cost elements are an integral part of the cost center, such as salaries and wages, and that others must be allocated among the various cost centers. Rent, for example, was $920 for the year and was allocated among the six cost centers on the basis of square feet occupied. The total of all fixed indirect costs, $1458 can then be applied to product costs or to General and Administrative (G&A), or Sales expense, which are separate items on the P&L.

In the next part of the cost statement: "Fixed Indirect Factory Cost," G&A, and Selling expense are removed from the total of all fixed expense, and maintenance costs are distributed to the three production departments. There are a number of ways in which the costs of maintenance could be distributed among departments receiving the service. They include proration based on square feet of floor space, man-hours as a percent of total or an engineering estimate. At CCTCO a maintenance order system that identifies the department and machine that each distinct maintenance job is applied to is in effect. This makes it possible to charge costs on an "as used" basis to the various production departments.

In a more complex, typical situation, there may be a large number of service cost centers for which the total costs must be determined and then distributed to the direct production departments. The total fixed indirect factory costs will then be applied to production through burden rates, the method of determining which will be covered in a subsequent paragraph, to ensure that all manufacturing costs are applied to the product.

Variable indirect costs are those which change in consumption with changing levels of production but are not consumed by, or applied directly to, the product. Variable indirect labor includes warehouse and materials handling labor. Variable indirect materials include lubricating oils and greases.

Direct costs include all labor and material that go into the product or operate on the product. They include the steel and plastic from which the product is made and the labor of machine operators and assemblers. At CCTCO, direct labor includes packing labor; also, direct materials include packing and display boxes. In many companies these items are considered indirect, since they do not add final utility value to the product.

Direct labor is charged to the product through the use of time reports identifying specific worker hours applied to a product or a production lot. Direct material costs can be applied through the use of a store's requisition drawn on a specific product. The amount of record keeping and tabulating necessary for such comprehensive cost identification is considerable and shortcuts are frequently used to reduce the clerical work required. Computer-based data entry systems at the workstation are feasible and can eliminate much of the cost of data collection.

Table 44.11 shows the total cost of production for the year:

Fixed indirect factory cost	$1032
Variable indirect factory cost	$ 113
Factory direct costs	$3004
Total cost of production	$4575

A comparison of these figures with the budget for the year shown in Table 44.12 points out where costs are different than projected for the year. The total variance analysis, Table 44.13, saves time in identifying problems. This report is of value in determining the reasons for differences or variances

Table 44.11 Commercial Construction Tool Co., Inc.—Costs and Revenues, Bad Year—Actual

Indirect Expense, Total Company	Total ($000)	Percent Total Cost	Administrative	Sales	Maintenance	Presses	Factory Assembly	Packing	Total
Indirect fixed									
Facilities cost									
Square feet	123753		17894	13421	2300	45069	27041	18028	123753
Rent	920	20.11	133	100	17	335	201	134	920
Insurance	40	0.87	6	4	1	15	9	6	40
Salaries									
Administration	60	1.31	60						60
Sales	34	0.74		34					34
Factory	40	0.87			5	19	11	5	40
Fringes	42	0.91	19	11	1	6	3	2	42
Travel	21	0.46	7	14					21
Communications	12	0.26	5	7					12
Wages	17	0.37			17				17
Depreciation	273	5.96	19	8	30	161	30	25	273
Total fixed indirect	1458	31.87	249	177	71	536	254	171	1458
Fixed indirect factory cost									
Maintenance					−71	36	21	14	0
General and administrative expense			−249						−249

	Total	%	A	B	C	D	E	F
Selling expense	1032	22.55	571	275	185	0	0	0
Indirect factory costs—fixed	1032							
Distributed indirect variable cost	−177							
Wages	82	1.79	24	35	23			
Supplies	31	0.68	23	0	8			
Indirect factory costs—variable	113	2.47	47	35	31			
Total indirect factory costs	1145	25.02	618	310	217			
Direct costs of manufacture								
Material	1772	38.74	1188	330	255			
Labor								
Direct labor hours	190494		58461	99359	32674			
Hourly rate			7.11	6.33	5.72			
Labor dollars	1231	26.92	416	629	187			
Total direct costs	3004	65.66	1604	959	441	0	0	0
General and administrative expense	249	5.45					249	
Sales expense	177	3.87						177
Grand total all costs	4575	100	2186	1248	644	71	249	177

Table 44.12 Commercial Construction Tool Co., Inc.—Budgeted Costs and Revenues

Indirect Expense, Total Company	Total ($000)	Percent Total Cost	Administrative	Sales	Maintenance	Presses	Factory Assembly	Packing	Total
Indirect fixed									
Facilities cost									
Square feet	123753		17894	13421	2300	45069	27041	18028	123753
Rent	820	18.39	119	89	15	299	179	119	820
Insurance	40	0.90	6	4	1	15	9	6	40
Salaries									
Administration	60	1.35	60						60
Sales	33	0.74		33					33
Factory	47	1.05			6	23	13	6	47
Fringes	43	0.97	19	10	2	7	4	2	43
Travel	18	0.40	6	12					18
Communications	6	0.13	3	3					6
Wages	14	0.31			14				14
Depreciation	273	6.11	19	8	30	161	30	25	273
Total fixed indirect	1354	30.36	231	160	67	504	235	158	1354
Fixed indirect factory cost									
Maintenance					−67	34	20	13	0

Account	Total	Per unit	−231	−160	0				Total
General and administrative expense			−231						−231
Selling expense				−160					−160
Indirect factory costs—fixed	963	21.60	0	0	0	537	255	171	963
Distributed indirect variable cost									
Wages	72	1.62				21	32	19	72
Supplies	24	0.54				18	0	6	24
Indirect factory costs—variable	96	2.16				39	32	25	96
Total indirect factory costs	1060	23.76				576	287	197	1060
Direct costs of manufacture									
Material	1805	40.48				1175	394	236	1805
Labor									
Direct labor hours	187125					55495	95425	36205	187125
Hourly rate						6.95	6.45	5.60	
Labor dollars	1204	27.00				386	615	203	1204
Total direct costs	3009	67.48	0	0	0	1560	1010	439	3009
General and administrative expense	231	5.18	231						231
Sales expense	160	3.58		160					160
Grand total all costs	4460	100	231	160	67	2103	1277	622	4460

Table 44.13   Commercial Construction Tool Co., Inc.—Costs and Revenues, Bad Year—Actual, Variance Analysis

Indirect Expense, Total Company	Total ($000)	Percent Total Cost	Administrative	Sales	Maintenance	Presses	Factory Assembly	Packing	Total
Indirect fixed									
Facilities cost									
Square feet									
Rent	-100		-14	-11	-2	-36	-22	-15	-100
Insurance	0		0	0	0	0	0	0	0
Salaries									
Administration	0		0	0	0	0	0	0	0
Sales	-1		0	-1	0	0	0	0	-1
Factory	7		0	0	1	3	2	1	7
Fringes	-3		-1	-2	0	0	0	0	-3
Travel	-6		-3	-3	0	0	0	0	-6
Communications	-3		0	0	-3	0	0	0	-3
Wages	0		0	0	0	0	0	0	0
Depreciation	0		0	0	0	0	0	0	0
Total fixed indirect	-104		-18	-17	-4	-32	-19	-13	-104
Fixed indirect factory cost									
Maintenance	0		0	0	4	-2	-1	-1	0

	(1)	(2)	(3)	(4)	(5)	(6)	(7)
General and administrative expense	18						18
Selling expense	17					17	
Indirect factory costs—fixed	−69	−14	−21	−34	0		
Distributed indirect variable cost							
Wages	−10	−4	−3	−3	0		
Supplies	−7	−2	0	−5	0		
Indirect factory costs—variable	−17	−6	−3	−8	0		
Total indirect factory costs	−85	−20	−24	−42	0		
Direct costs of manufacture							
Material	33	−18	65	−14			
Direct labor hours	−3369	3531	−3934	−2966			
Hourly rate		−0.12	0.12	−0.16			
Labor dollars	−28	16	−13	−30			
Total direct costs	5	−3	51	−43	0		
General and administrative expense	−18					−17	−18
Sales expense	−17					−17	−18
Grand total all costs	−115	−22	29	−83	−4	−17	

in fixed costs. Changes in operating levels, or production output as compared to plan are not shown, so over- or underspending of direct costs or variable indirect costs may be the result of greater or less output or efficiency variances. Additional facts are needed to identify the causes of variances.

These data are extremely useful and enable us to determine total cost of production, which when coupled with changes in inventory will lead to a figure of total profitability. This report does not, however, give a clue as to the relative costs or profitability or production levels of the two products that CCTCO manufactures. For this we need two additional reports; one is a burden analysis and the second a comparison of actual and standard product costs and revenues.

44.8   DETERMINING BURDEN RATES

The burden rate provides a method for applying indirect costs to each product in such a way as to absorb all indirect costs and to allow each product to carry its fair share of such costs. The application of burden costs to products assists in setting prices that will ensure that all of the costs of doing business will be charged to the customer. As in all of the accounting systems covered thus far, there are many ways to distribute overhead. At CCTCO, direct labor hour burden rates are used. At the time of budget preparation, the total standard man-hours for the budgeted production for the year is determined for each direct labor department. The indirect costs for each department are divided by the projected direct labor hours to produce an hourly expense rate. Table 44.14 shows the derivation of fixed, variable, and total burden rates for the three production departments of CCTCO.

44.8.1   Standard Sales and Costs

In the preceding paragraph, total standard man-hours for each department were used to derive the hourly burden rate. Table 44.15 contains all of the steps taken to develop a budgeted sales, production, inventory, and margin report for the two products produced by CCTCO. It was determined at budget time that 74,000 of Item A would be sold and production of 74,500 would be required. Production engineers determined that the labor hours required at a normal, achievable rate should be 153 per hundred units, of which 43 are to be required in the Press department, 79 in the Assembly department, and 31 in Packing. Multiplying in each department and adding across results in total standard hours of 113,985 projected to produce 74,500 units. The standard labor rates as presented in Table 44.12 are extended by the standard hours to produce the standard labor dollars for Item A of $732 (the 000 is omitted). The same process is followed for Item B; note that the total labor dollars tie in with the numbers in Table 44.12.

A similar process is followed for burden and for material to derive a total standard cost to be deducted from the standard sales in order to calculate the standard gross margin.

Comparing Table 44.15 budgeted sales and production with Table 44.16, actual sales and production, we find the essential causes for the poor results and the underlying support for the report to the President. In Table 44.17, the variance analysis shows clearly sales of Item B to be over budget and the sales of Item A to be under budget. Additionally, the selling price of Item B is well under expectations. The source of the conclusions and recommendations for the next year is now evident.

44.8.2   Unabsorbed Burden

The report to the president included, as an additional reduction in profits, $62 of unabsorbed burden. This resulted from a combination of the number of actual direct hours charged and overspending fixed and variable expense. Table 44.18 recaps the burden charged to product as shown in Table 44.17, compared to the actual indirect expense as tabulated in Table 44.11. When the amount of burden charged to product is insufficient to cover the actual period expense, the remainder must be absorbed as an additional cost of production, normally not allocated to product.

44.9   SYSTEMS FOR ENGINEERS AND OTHER STAFF

Essentially all of the various reports and analyses described so far are of use to the staff. An extension of the kinds of reports already discussed is necessary for staff to project the financial results of proposed changes in machinery and equipment, product, or marketing programs. A most important skill for the engineer is the ability to develop and use computer models that will demonstrate the effect of new programs under various assumptions of economic conditions and pricing situations. The financial statements presented in this chapter were all prepared on a personal computer using an electronic spread sheet. The capability of changing fundamental assumptions and running off long-term projections is essential to today's engineer in industry. An understanding of the basic relationships among all the elements of balance sheet, P&L, and cash flow statement is needed to build and exercise these financial models.

Table 44.14 Commercial Construction Tool Co., Inc.—Budgeted Costs and Revenues

	Total ($000)	Percent Total Cost	Administrative	Sales	Maintenance	Presses	Factory Assembly	Packing	Total
Burden rate determination									
Standard direct labor hours						55495	394	236	187125
Indirect factory costs—fixed	963					537	255	171	963
Indirect factory costs—variable	96					39	32	25	96
Fixed burden rate						9.68	2.67	4.73	
Variable hourly burden rate						0.70	0.34	0.70	
Total burden rate						10.38	3.01	5.43	

Table 44.15 Commercial Construction Tool Co., Inc.—Budgeted Costs and Revenues

Production and Sales

	Total ($000)	Percent Total Cost	Administrative	Sales	Maintenance	Presses	Factory Assembly	Packing	Total
Item A	Unit								
Production—hundreds	153					43	79	31	745
Standard hours per hundred									153
Total standard hours						32035	58855	23095	113985
Standard labor	0.98					223	380	129	732
Standard hourly rate						6.95	6.45	5.60	
Standard burden	0.85					333	177	125	635
Standard burden rate						10.38	3.01	5.43	
Standard material total	1.28					633	216	104	953
Cost per pound						1.25	0.65	0.87	
Pounds per hundred						6.80	4.46	1.60	
Total standard cost	3.11					1188	772	359	2319
Standard price	4.23								
Sales—hundreds									740
Sales—dollars									3130
Standard margin	1.12								826
Inventory change									16

Item B	Unit				
Production—hundreds					690
Standard hours per hundred	106	34	53	19	106
Total standard hours		23460	36570	13110	73140
Standard labor	0.68	163	236	73	472
Standard hourly rate		6.95	6.45	5.60	
Standard burden	0.62	244	110	71	425
Standard burden rate		10.38	3.01	5.43	
Standard material total	1.24	541	178	132	852
Cost per pound		1.18	0.58	1.20	
Pounds per hundred		6.65	4.46	1.60	
Total standard cost	2.54	948	524	277	1749
Standard price	3.18				2194
Sales—hundreds					670
Sales—dollars					2131
Standard margin	0.64				445
Inventory change					51

Table 44.16 Commercial Construction Tool Co., Inc.—Costs and Revenues, Bad Year—Actual

	Total ($000)	Percent Total Cost	Administrative	Sales	Maintenance	Presses	Factory Assembly	Packing	Total
Production and Sales									
Item A	Unit								
Production—hundreds	156								575
Actual—hours per hundred						48	81	27	156
Total actual hours						27477	46699	15357	89533
Actual labor	1.01					195	296	88	579
Actual hourly rate						7.11	6.33	5.72	
Actual burden	0.89					285	140	83	509
Standard burden rate						10.38	3.01	5.43	
Actual material—total	1.30					530	143	76	749
Cost per pound						1.33	0.58	0.93	
Pounds per hundred						6.93	4.30	1.42	
Total actual cost	3.20					1011	579	247	1837
Actual price	4.29								
Sales—hundreds									530
Sales—dollars									2274
Actual margin	1.09								580
Inventory change									144

Item B	Unit				
Production—hundreds		39	66	22	800
Actual—hours per hundred	126				126
Total actual hours		30984	52660	17317	100961
Actual labor	0.82	220	333	99	652
Actual hourly rate		7.11	6.33	5.72	
Actual burden	0.72	322	158	94	574
Standard burden rate		10.38	3.01	5.43	
Actual material—total	1.28	658	186	179	1023
Cost per pound		1.26	0.54	1.45	
Pounds per hundred		6.53	4.31	1.54	
Total actual cost	2.81	1200	678	372	2250
Actual price	3.01				2408
Sales—hundreds					830
Sales—dollars					2498
Actual margin	0.20				164
Inventory change					−84

Table 44.17 Commercial Construction Tool Co., Inc.—Costs and Revenues, Bad Year—Actual Variance Analysis

Item	Total ($000) Unit	Percent Total Cost	Administrative	Sales	Maintenance	Presses	Factory Assembly	Packing	Total
Production and Sales									
Item A									
Production—hundreds						0	0	0	−170
Actual—hours per hundred	−3					−5	−2	4	−3
Total actual hours						4558	12156	7738	24452
Actual labor						27	84	41	153
Actual hourly rate						−0.16	0.12	−0.12	
Actual burden						47	37	42	126
Standard burden rate	0.00					0.00	0.00	0.00	
Actual material—total	−0.02					103	73	28	204
Cost per pound						−0.08	0.07	−0.06	
Pounds per hundred						−0.13	0.16	0.18	
Total actual cost	−0.08					178	193	111	482
Actual price	0.06								
Sales—hundreds									−210
Sales—dollars									−857
Actual margin	−0.02								−246
Inventory change									−128

Item B	Unit				
Production—hundreds		0	0	0	110
Actual—hours per hundred	-20	-5	-13	-3	-20
Total actual hours		-7524	-16090	-4207	-27821
Actual labor		-57	-97	-26	-180
Actual hourly rate		-0.16	0.12	-0.12	
Actual burden		-78	-48	-23	-149
Standard burden rate	0.00	0.00	0.00	0.00	
Actual material—total		-117	-8	-46	-171
Cost per pound	-0.04	-0.08	0.04	-0.25	
Pounds per hundred		0.12	0.15	0.06	
Total actual cost	-0.28	-252	-154	-95	-500
Actual price	-0.17				160
Sales—hundreds					368
Sales—dollars					
Actual margin	-0.45				-281
Inventory change					135

Table 44.18 Commercial Construction Tool Co., Inc.—Costs and Revenues, Bad Year—Actual Burden Absorption Analysis

	Total ($000)	Percent Total Cost	Administrative	Sales	Maintenance	Presses	Factory Assembly	Packing	Total
Product "A" absorbed burden	509					285	140	83	509
Product "B" absorbed burden	574					322	158	94	574
Total absorbed burden	1083					607	299	178	1083
Total factory indirect cost	1145					618	310	217	1145
Underabsorbed burden	−62					−11	−12	−39	−62

44.10 SYSTEMS FOR ACCOUNTANTS

All of the reports that have been presented are prepared by the accounting staff, and it has been demonstrated that they are all useful in managing the business effectively. More reporting or analysis is not required to satisfy the auditors or the tax collector. What is needed for both is comprehensive and accurate records that will provide an audit trail, so that transactions can be traced from one account to the next. These are considerations not of interest to the engineer and manager.

44.11 CONCLUSIONS

This chapter is intended to portray the principles behind the financial function and the underlying cost accounting systems needed to manage a business. No accounting system will be found that operates in quite this way. All too frequently the systems in a company have become very complex and overlaid with subsystems designed to satisfy every need, even needs that no longer exist. Each system requires considerable study in order to understand its special idiosyncrasies.

CHAPTER 45

INVESTMENT ANALYSIS

BYRON W. JONES

Kansas State University
Manhattan, Kansas

45.1 ESSENTIALS OF FINANCIAL ANALYSIS

45.1.1 Sources of Funding for Capital Expenditures

Engineering projects typically require the expenditure of funds for implementation and in return provide a savings or increased income to the firm. In this sense an engineering project is an investment for the firm and must be analyzed as an investment. This is true whether the project is a major new plant, a minor modification of some existing equipment, or anything in between. The extent of the analysis of course must be commensurate with the financial importance of the project. Financial analysis of an investment has two parts: funding of the investment and evaluation of the economics of the investment. Except for very large projects, such as a major plant expansion or addition, these two aspects can be analyzed independently. All projects generally draw from a common pool of capital funds rather than each project being financed separately. The engineering function may require an in-depth evaluation of the economics of a project, while the financing aspect generally is not dealt with in detail if at all. The primary reason for being concerned with the funding of projects is that the economic evaluation often requires at least an awareness if not an understanding of this function.

The funds used for capital expenditures come from two sources: debt financing and equity financing. Debt financing refers to funds that are borrowed from outside the company. The two common sources are bank loans and the sale of bonds. Bank loans are typically used for short-term financing, and bonds are used for long-term financing. Debt financing is characterized by a contractual arrangement specifying interest payments and repayment. The lender does not share in the profits of the investments for which the funds are used nor does it share the associated risks except through the possibility of the company defaulting. Equity financing refers to funds owned by the company. These funds may come from profits earned by the company or from funds set aside for depreciation allowances. Or, the funds may come from the sale of new stock. Equity financing does not require any specified repayment; however, the owners of the company (stockholders) do expect to make a reasonable return on their investment.

The decisions of how much funding to secure and the relative amounts to secure from debt and equity sources are very complicated and require considerable subjective judgment. The current stock market, interest rates, projections of future market conditions, etc., must be addressed. Generally, a company will try to maintain approximately a constant ratio of funding from the different sources. This mix will be selected to maximize earnings without jeopardizing the company's financial well being. However, the ratio of debt to equity financing does vary considerably from company to company reflecting different business philosophies.

45.1.2 The Time Value of Money

The time value of money is frequently referred to as interest or interest rate in economic analyses. Actually, the two are not exactly the same thing. Interest is a fee paid for borrowed funds and is established when the loan is made. The time value of money is related to interest rates, but it includes other factors also. The time value of money must reflect the cost of money. That is, it must reflect the interest that is paid on loans and bonds, and it must also reflect the dividends paid to the stockholders. The cost of money is usually determined as a weighted average of the interest rates and dividend rates paid for the different sources of funds used for capital expenditures. The time value of money must also reflect opportunity costs. The opportunity cost is the return that can be earned on available, but unused, projects.

In principle, the time value of money is the greater of the cost of money and the opportunity cost. The determination of the time value of money is difficult, and the reader is referred to advanced texts on this topic for a complete discussion (see, for example, Bussey[1]). The determination is usually made at the corporate level and is not the responsibility of engineers. The time value of money is frequently referred to as the interest rate for economic evaluations, and one should be aware that the terms interest rate and time value of money are used interchangeably. The time value of money is also referred to as the required rate of return.

Another factor that may or may not be reflected in a given time value of money is inflation. Inflation results in a decreased buying power of the dollar. Consequently, the cost of money generally is higher during periods of high inflation, since the funds are repaid in dollars less valuable than those in which the funds were obtained. Opportunity costs are not necessarily directly affected by inflation except that inflation affects the cash flows used in evaluating the returns for the projects on which the opportunity costs are based. It is usually up to the engineer to verify that inflation has been included in a specified time value of money, since this information is not normally given. In some applications it is beneficial to use an inflation-adjusted time value of money. The relationship is

$$1 + i_r = \frac{1 + i_a}{1 + f} \tag{45.1}$$

where i_a is the time value of money, which reflects the higher cost of money and the higher opportunity cost due to inflation; f is the inflation rate; and i_r is the inflation-adjusted time value of money, which actually reflects the true cost of capital in terms of constant value. The variables i_r and i_a may be referred to as the real and apparent time values of money, respectively. i_r, i_a, and f are all expressed in fractional rather than percentage form in Eq. (45.1) and all must be expressed on the same time basis, generally an annual rate. See Section 45.3 for additional discussion on the use of i_r and i_a. Also see Jones[2] for a detailed discussion.

The time value of money may also reflect risks associated with a project. This is particularly true for projects where there is a significant probability of poor return or even failure (e.g., the development of a new product). In principle, risk can be evaluated in assessing the economics of a project by including the probabilities of various outcomes (see Riggs,[3] for example). However, these calculations are complicated and are often dependent on subjective judgment. The more common approach is to simply use a time value of money which is greater for projects that are more risky. This is why some companies will use different values of i for different types of investments (e.g., expansion versus cost reduction versus diversification, etc.). Such adjustments for risk are usually made at the corporate level and are based on experience and other subjective inputs as much as they are on formal calculations. The engineer usually is not concerned with such adjustments, at least for routine economic analyses. If the risks of a project are included in an economic analysis, then it is important that the time value of money also not be adjusted for risks, since this would represent an overcompensation and would distort the true economic picture.

45.1.3 Discounted Cash Flow and Interest Calculations

For the purpose of economic analysis, a project is represented as a group of cash flows showing the expenditures and the income or savings attributable to the project. The object of economic analysis is normally to determine the profitability of the project based on these cash flows. However, the profitability cannot be assessed simply by summing up the cash flows, owing to the effect of the time value of money. The time value of money results in the value of a cash flow depending not only on its magnitude but also on when it occurs according to the equation

$$P = F \frac{1}{(1+i)^n} \tag{45.2}$$

where F is a cash flow that occurs sometime in the future, P is the equivalent value of that cash flow now, i is the annual time value of money in fractional form, and n is the number of years from now when cash flow F occurs. Cash flow F is often referred to as the future value or future amount, while cash flow P is referred to as the present value or present amount. Equation (45.2) can be used to convert a set of cash flows for a project to a set of economically equivalent cash flows. These equivalent cash flows are referred to as discounted cash flows and reflect the reduced economic value of cash flows that occur in the future. Table 45.1 shows a set of cash flows that have been discounted.

Equation (45.2) is the basis for a more general principle referred to as economic equivalence. It shows the relative economic value of cash flows occurring at different points in time. It is not necessary that P refer to a cash flow that occurs at the present; rather, it simply refers to a cash flow that occurs n years before F. The equation works in either direction for computing equivalent cash flows. That is, it can be used to find a cash flow F that occurs n years after P and that is equivalent to P, or to find a cash flow P that is equivalent to F but which occurs n years before F. This principle of equivalence allows cash flows to be manipulated as needed to facilitate economic calculations.

The time value of money is usually specified as an annual rate. However, several other forms are

Table 45.1 Discounted Cash Flow Calculation

Year	Estimated Cash Flows for Project	Discounted Cash Flows[a]
0	−$120,000	−$120,000
1	− 75,000	− 68,200
2	+ 50,000	+ 41,300
3	+ 60,000	+ 45,100
4	+ 70,000	+ 47,800
5	+ 30,000	+ 18,000
6	+ 20,000	+ 11,300

[a] Based on $i = 10\%$.

sometimes encountered or may be required to solve a particular problem. An interest rate* as used in Eq. (45.2) is referred to as a discrete interest rate, since it specifies interest for a discrete time period of 1 year and allows calculations in multiples (n) of this time period. This time period is referred to as the compounding period. If it is necessary to change an interest rate stated for one compounding period to an equivalent interest rate for a different compounding period, it can be done by

$$i_1 = (1 + i_2)^{\Delta t_1/\Delta t_2} - 1 \qquad (45.3)$$

where Δt_1 and Δt_2 are compounding periods and i_1 and i_2 are the corresponding interest rates, respectively. Interest rates i_1 and i_2 are in fractional form. If an interest rate with a compounding period different than 1 year is used in Eq. (45.2) then n in that equation refers to the number of those compounding periods and not the number of years.

Interest may also be expressed in a nominal form. Nominal interest rates are frequently used to describe the interest associated with borrowing but are not used to express the time value of money. A nominal interest rate is stated as an annual interest rate but with a compounding period different from 1 year. A nominal rate must be converted to an equivalent compound interest rate before being used in calculations. The relationship between a nominal interest rate (i_n) and a compound interest rate (i_c) is $i_c = i_n/m$, where m is the number of compounding periods per year. The compounding period (Δt) for i_c is 1 year/m. For example, a 10% nominal interest rate compounded quarterly translates to a compound interest rate of 2.5% with a compounding period of $\frac{1}{4}$ year. Equation (45.3) may be used to convert the resulting interest rate to an equivalent interest rate with annual compounding. This later interest rate is referred to as the effective annual interest rate. For the 10% nominal interest above, the effective annual interest rate is 10.38%.

Interest may also be defined in continuous rather than discrete form. With continuous interest (sometimes referred to as continuous compounding), interest acrues continuously. Equation (45.2) can be rewritten for continuous interest as

$$P = F \times e^{-rt} \qquad (45.4)$$

where r is the continuous interest rate and has units of inverse time (but is normally expressed as a percentage per unit of time), t is the time between P and F, and e is the base of the natural logarithm. Note that the time units on r and t must be consistent with the year normally used. Discretely compounded interest may be converted to continuously compounded interest by

$$r = \frac{1}{\Delta t} \ln(1 + i)$$

and continuously compounded interest to discretely compounded interest by

$$i = e^{r \, \Delta t} - 1$$

Note that these are dimensional equations; the units for r and Δt must be consistent and the interest rates are in fractional form.

It is often desirable to manipulate groups of cash flows rather than just single cash flows. The principles used in Eq. (45.2) or (45.3) may be extended to multiple cash flows if they occur in a regular fashion or if they flow continuously over a period of time at some defined rate. The types of cash flows that can be readily manipulated are:

1. A uniform series in which cash flows occur in equal amounts on a regular periodic basis.
2. An exponentially increasing series in which cash flows occur on a regular periodic basis and increase by a constant percentage each year as compared to the previous year.
3. A gradient in which cash flows occur on a regular periodic basis and increase by a constant amount each year.
4. A uniform continuous cash flow where the cash flows at a constant rate over some period of time.
5. An exponentially increasing continuous cash flow where the cash flows continuously at an exponentially increasing rate.

These cash flows are illustrated in Figs. 45.1–45.5.

Any one of these groups of cash flows may be related to a single cash flow P as shown in these figures. The relationship between a group of cash flows and a single cash flow (or a single to single cash flow) may be reduced to an interest factor. The interest factors resulting in a single present amount are shown in Table 45.2. Derivations for most of these interest factors can be found in an

---

* The term interest rate is used here instead of the time value of money. The results apply to interest associated with borrowing and interest in the context of the time value of money.

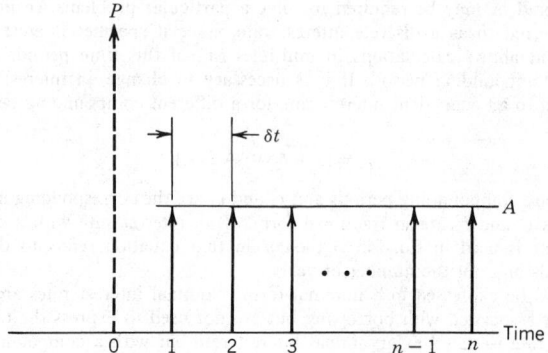

Fig. 45.1 Uniform series of cash flows.

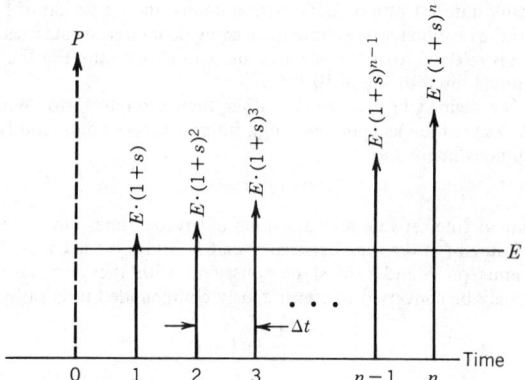

Fig. 45.2 Exponentially increasing series cash flows.

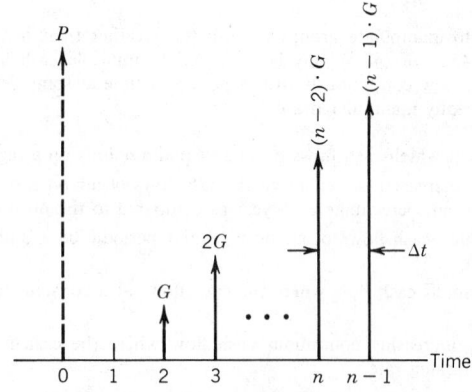

Fig. 45.3 Cash flow gradient.

introductory text on engineering economics (see, for example, Grant et al.[4]). The interest factor gives the relationship between the group of cash flows and the single present amount. For example,

$$P = A \cdot (P/A, i, n)$$

The term $(P/A, i, n)$ is referred to as the interest factor and gives the ratio of P to A and shows that it is a function of i and n. Other interest factors are used accordingly. The interest factors may be manipulated as if they are the mathematical ratio they represent. For example,

$$(A/P, i, n) = \frac{1}{(P/A, i, n)}$$

Thus for each interest factor represented in Table 45.2 a corresponding inverse interest factor may be generated as above. Two interest factors may be combined to generate a third interest factor in some cases. For example,

$$(F/A, i, n) = (P/A, i, n) \ (F/P, i, n)$$

or

$$(F/C, r, t) = (P/C, r, t) \ (F/P, r, t)$$

Again the interest factors are manipulated as the ratios they represent. In theory, any combination of interest factors may be used in this manner. However, it is wise not to mix interest factors using discrete interest with those using continuous interest, and it is usually best to proceed one step at a time to ensure that the end result is correct.

There are several important limitations when using the interest factors in Table 45.2. The time relationship between cash flows must be adhered to rigorously. Special attention must be paid to the time between P and the first cash flow. The time between the periodic cash flows must be equal to the compounding period when interest factors with discrete interest rates are used. If these do not match, Eq. (45.3) must be used to find an interest rate with the appropriate compounding period. It is also necessary to avoid dividing by zero. This is not usually a problem; however, it is possible that $i = s$ or $r = a$, resulting in division by zero. The interest factors reduce to simpler forms in these special cases:

$$(P/F, i, s, n) = n, \quad s = i$$
$$(P/C, r, a, t) = c \cdot t, \quad a = r$$

It is sometimes necessary to deal with groups of cash flows that extend over very long periods of time ($n \rightarrow \infty$ or $t \rightarrow \infty$). The interest factors in this case reduce to simpler forms; but, limitations may exist if a finite value of P is to result. These reduced forms and limitations are presented in Table 45.3.

Several of the more common interest factors may be referred to by name rather than the notation used here. These interest factors and the corresponding names are presented in Table 45.4.

45.2 INVESTMENT DECISIONS

45.2.1 Allocation of Capital Funds

In most companies there are far more projects available than there are funds to implement them. It is necessary, then, to allocate these funds to the projects that provide the maximum return on the funds invested. The question of how to allocate capital funds will generally be handled at several

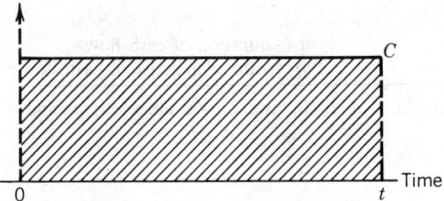

Fig. 45.4 Uniform continuous cash flow.

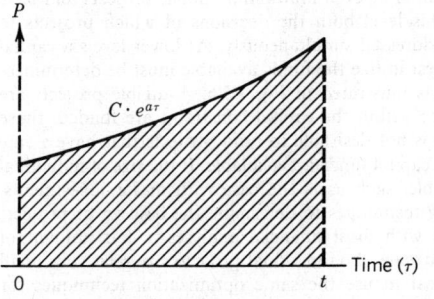

Fig. 45.5 Exponentially increasing continuous cash flow.

Table 45.2 Mathematical Expression of Interest Factors for Converting Cash Flows to Present Amounts[a]

Type of Cash Flows	Interest Factor	Mathematical Expression
Single	$(P/F, i, n)$	$= \dfrac{1}{(1+i)^n}$
Single	$(P/F, r, t)$	$n = \dfrac{t_2 - t_1}{\Delta t}$ $= e^{-rt}$ $t = t_2 - t_1$
Uniform series	$(P/A, i, n)$	$= \dfrac{1}{i}\left[1 - \dfrac{1}{(1+i)^n}\right]$ $n = $ number of cash flows
Uniform series	$(P/A, r, n, \Delta t)$	$= \dfrac{1}{e^{r\,\Delta t} - 1}(1 - e^{-rn\,\Delta t})$ $\Delta t = $ time between cash flows $n = $ number of cash flows
Exponentially increasing series	$(P/E, i, s, n)$	$= \dfrac{1+s}{i-s}\left[1 - \left(\dfrac{1+s}{1+i}\right)^n\right]$ $n = $ number of cash flows $s = $ escalation rate
Exponentially increasing series	$(P/E, r, s, \Delta t, n)$	$= \dfrac{1+s}{(e^{r\,\Delta t} - s - 1)}\left[1 - \left(\dfrac{1+s}{e^{r\,\Delta t}}\right)^n\right]$ $n = $ number of cash flows $\Delta t = $ time between cash flows $s = $ escalation rate
Gradient	$(P/G, i, n)$	$= \dfrac{1}{i}\left[\dfrac{(1+i)^n - 1}{i\,(1+i)^n} - \dfrac{n}{(1+i)^n}\right]$
Continuous	$(P/C, r, t)$	$= \dfrac{1 - e^{-rt}}{r}$ $t = $ duration of cash flow
Continuous increasing exponentially	$(P/C, r, a, t)$	$= \dfrac{1 - e^{-(r-a)t}}{r - a}$ $t = $ duration of cash flow $a = $ rate of increase of cash flow

[a] See Figs. 45.1–45.5 for definitions of variables n, Δt, s, a, E, G, and C. Variables i, r, a, and s are in fractional rather than percentage form.

different levels within the company, with the level at which the allocation is made depending on the size of the projects involved. At the top level, major projects such as plant additions or new product developments are considered. These may be multimillion or even multibillion dollar projects and have major impact on the future of the company. At this level both the decisions of which projects to fund and how much total capital to invest may be addressed simultaneously. At lower levels a capital budget may be established. Then, the projects that best utilize the funds available must be determined.

The basic principle utilized in capital rationing is illustrated in Fig. 45.6. Available projects are ranked in order of decreasing return. Those that are within the capital constraint are funded, those that are outside this constraint are not. However, it is not desirable to fund projects that have a rate of return less than the cost of money even if sufficient capital funds are available. If the size of individual projects is not small compared to the funds available, such as is the case with major investments, then it may be necessary to use linear programming techniques to determine the best set of projects to fund and the amount of funding to secure. As with most financial analyses, a fair amount of subjective judgment is also required. At the other extreme, the individual projects are relatively small compared to the capital available. It is not practical to use the same optimization techniques for such a large number of projects; and little benefit is likely to be gained anyway. Rather, a cutoff rate

Table 45.3 Interest Factors for $n \rightarrow \infty$ or $t \rightarrow \infty$

Interest Factor	Limitation
$(P/A, i, n) = \dfrac{1}{i}$	$i > 0$
$(P/A, r, n, \Delta t) = \dfrac{1}{e^{r\,\Delta t} - 1}$	$r > 0$
$(P/E, i, s, n) = \dfrac{1+s}{i-s}$	$s < i$
$(P/E, r, s, \Delta t, n) = \dfrac{1+s}{e^{r\,\Delta t} - s - 1}$	$s < e^{r\,\Delta t} - 1$
$(P/G, i, n) = \left(\dfrac{1}{i}\right)^2$	$i > 0$
$(P/C, r, t) = \dfrac{1}{r}$	$r > 0$
$(P/C, r, a, t) = \dfrac{1}{r-a}$	$a < r$

Table 45.4 Interest Factor Names

Factor	Name
$(F/P, i, n)$	Compound amount factor
$(P/F, i, n)$	Present worth factor
$(F/A, i, n)$	Series compound amount factor
$(P/A, i, n)$	Series present worth factor
$(A/P, i, n)$	Capital recovery factor
$(A/F, i, n)$	Sinking fund factor

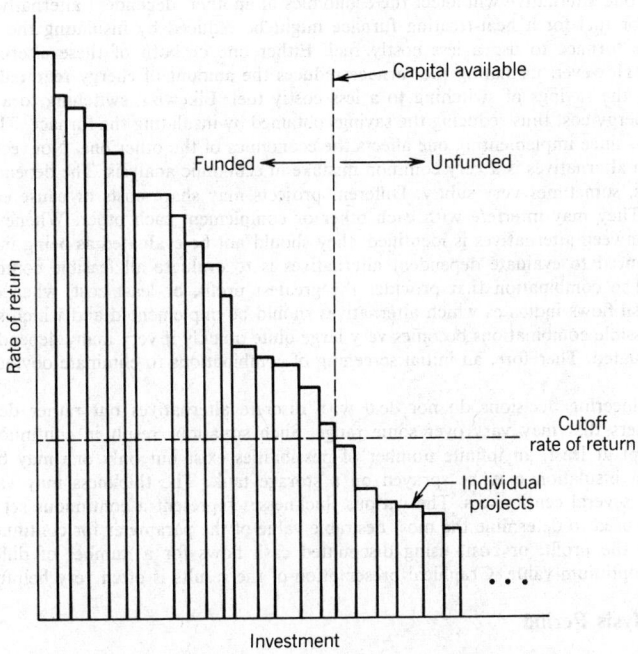

Fig. 45.6 Principle of capital allocation.

of return (also called required rate of return or minimum attractive rate of return) is established. Projects with a return greater than this amount are considered for funding; projects with a return less than this amount are not. If the cutoff rate of return is selected appropriately, then the total funding required for the projects considered will be approximately equal to the funds available. The use of a required rate of return allows the analysis of routine projects to be evaluated without using sophisticated techniques. The required rate of return may be thought of as an opportunity cost, since there are presumably unfunded projects available that earn approximately this return. The required rate of return may be used as the time value of money for routine economic analyses, assuming it is greater than the cost of money. For further discussion of the allocation of capital see Grant et al.[4] or other texts on engineering economics.

45.2.2   Classification of Alternatives

From the engineering point of view, economic analysis provides a means of selecting among alternatives. These alternatives can be divided into three categories for the purpose of economic analysis: independent, mutually exclusive, and dependent.

Independent alternatives do not affect one another. Any one or any combination may be implemented. The decision to implement one alternative has no effect on the economics of another alternative that is independent. For example, a company may wish to reduce the delivered cost of some heavy equipment it manufactures. Two possible alternatives might be to (1) add facilities for rail shipment of the equipment directly from the plant and (2) add facilities to manufacture some of the subassemblies in-house. Since these alternatives have no effect on one another, they are independent. Each independent alternative may be evaluated on its own merits. For routine purposes the necessary criterion for implementation is a net profit based on discounted cash flows or a return greater than the required rate of return.

Mutually exclusive alternatives are the opposite extreme from independent alternatives. Only one of a group of mutually exclusive alternatives may be selected, since implementing one eliminates the possibility of implementing any of the others. For example, some particular equipment in the field may be powered by a diesel engine, a gasoline engine, or an electric motor. These alternatives are mutually exclusive since once one is selected there is no reason to implement any of the others. Mutually exclusive alternatives cannot be evaluated separately but must be compared to each other. The single most profitable, or least costly, alternative as determined by using discounted cash flows is the most desirable from an economic point of view. The alternative with the highest rate of return is not necessarily the most desirable. The possibility of not implementing any of the alternatives should also be considered if it is a feasible alternative.

Dependent alternatives are like independent alternatives in that any one or any combination of a group of independent alternatives may be implemented. Unlike independent alternatives, the decision to implement one alternative will affect the economics of another, dependent alternative. For example, the expense for fuel for a heat-treating furnace might be reduced by insulating the furnace and by modifying the furnace to use a less costly fuel. Either one or both of these alternatives could be implemented. However, insulating the furnace reduces the amount of energy required by the furnace thus reducing the savings of switching to a less costly fuel. Likewise, switching to a less costly fuel reduces the energy cost thus reducing the savings obtained by insulating the furnace. These alternatives are dependent, since implementing one affects the economics of the other one. Not recognizing dependence between alternatives is a very common mistake in economic analysis. The dependence can occur in many ways, sometimes very subtly. Different projects may share costs or cause each other to be more costly. They may interfere with each other or complement each other. Whenever a significant dependence between alternatives is identified, they should not be evaluated as being independent. The approach required to evaluate dependent alternatives is to evaluate all feasible combinations of the alternatives. The combination that provides the greatest profit, or least cost, when evaluated using discounted cash flows indicates which alternatives should be implemented and which should not. The number of possible combinations becomes very large quite quickly if very many dependent alternatives are to be evaluated. Therefore, an initial screening of combinations to eliminate obviously undesirable ones is useful.

Many engineering decisions do not deal with discrete alternatives but rather deal with one or more parameters that may vary over some range. Such situations result in continuous alternatives, and, in concept at least, an infinite number of possibilities exist but only one may be selected. For example, foam insulation may be sprayed on a storage tank. The thickness may vary from a few millimeters to several centimeters. The various thicknesses represent a continuous set of alternatives. The approach used to determine the most desirable value of the parameter for continuous alternatives is to evaluate the profit, or cost, using discounted cash flows for a number of different values to determine an optimum value. Graphical presentation of the results is often very helpful.

45.2.3   Analysis Period

An important part of an economic analysis is the determination of the appropriate analysis period. The concept of life-cycle analysis is used to establish the analysis period. Life-cycle analysis refers to an analysis period that extends over the life of the entire project including the implementation (e.g., development and construction). In many situations an engineering project addresses a particular need (e.g., transporting fluid from a storage tank to the processing plant). When it is clear that this need exists only for a specific period of time, then this period of need may establish the analysis period. Otherwise the life of the equipment involved will establish the project life.

Decisions regarding the selection, replacement, and modification of particular equipment and machinery is often a necessary part of many engineering functions. The same principles used for establishing an analysis period for the larger projects apply in this situation as well. If the lives of the equipment are greater than the period of need, then that period of need establishes the analysis period. If the lives of the equipment are short compared to the analysis period, then the equipment lives establish the analysis period. The lives of various equipment alternatives are often significantly different. The concept of life-cycle analysis requires that each alternative be evaluated over its full life. Fairness in the comparison requires that the same analysis period be applied to all alternatives that serve the same function. In order to resolve these two requirements more than one life cycle for the equipment may be used to establish an analysis period that includes an integer number of life cycles of each alternative. For example, if equipment with a life of 6 years is compared to equipment with a life of 4 years, an appropriate analysis period is 12 years, two life cycles of the first alternative and three life cycles of the second.

Obsolescence must also be considered when selecting an analysis period. Much equipment becomes uneconomical long before it wears out. The life of equipment, then, is not set by how long it can function but rather by how long it will be before it is desirable to upgrade with a newer design. Unfortunately, there is no simple method to determine when this time will be since obsolescence is due to new technology and new designs. In a few cases, it is clear that changes will soon occur (e.g., a new technology that has been developed but that is not yet in the marketplace). Such cases are the exception rather than the rule and much subjective judgment is required to estimate when something will become obsolete.

The requirements to maintain acceptable liquidity in a firm also may affect the selection of an analysis period. An investment is expected to return a net profit. The return may be a number of years after the initial investment, however. If a company is experiencing cash flow difficulties or anticipates that they may, this delay in receiving income may not be acceptable. A long-term profit is of little value to a company that becomes insolvent. In order to maintain liquidity, an upper limit may be placed on the time allowed for an investment to show a profit. The analysis period must be shortened then to reflect this requirement if necessary.

45.3 EVALUATION METHODS

45.3.1 Establishing Cash Flows

The first part of any economic evaluation is necessarily to determine the cash flows that appropriately describe the project. It is important that these cash flows represent all economic aspects of the project. All hidden cost (e.g., increased maintenance) or hidden benefits (e.g., reduced downtime) must be included as well as the obvious expenses, incomes, or savings associated with the project. Wherever possible, nonmonetary factors (e.g., reduced hazzards) should be quantified and included in the analysis. Also, taxes associated with a project should not be ignored. (Some companies do allow a before-tax calculation for routine analysis.) Care should be taken that no factor be included twice. For example, high maintenance costs of an existing machine may be considered an expense in the alternative of keeping that machine or a savings in the alternative of replacing it with a new one, but it should not be considered both ways when comparing these two alternatives. Expenses or incomes that are irrelevant to the analysis should not be included. In particular, sunk costs, those expenses which have already been incurred, are not a factor in an economic analysis except for how they affect future cash flows (e.g., past equipment purchases may affect future taxes owing to depreciation allowances). The timing of cash flows over a project's life is also important, since cash flows in the near future will be discounted less than those that occur later. It is, however, customary to treat all of the cash flows that occur during a year as a single cash flow either at the end of the year (year-end convention) or at the middle of the year (mid-year convention).

Estimates of the cash flows for a project are generally determined by establishing what goods and services are going to be required to implement and sustain a project and by establishing the goods, services, benefits, savings, etc., that will result from the project. It is then necessary to estimate the associated prices, costs, or values. There are several sources of such information including historical data, projections, bids, or estimates by suppliers, etc. Care must be exercised when using any of these sources to be sure they accurately reflect the true price when the actual transaction will occur. Historical data are misleading, since they reflect past prices, not current prices, and may be badly in error owing to inflation that has occurred in recent years. Current prices may not accurately reflect future prices for the same reason. When historical data are used, they should be adjusted to reflect changes that have occurred in prices. This adjustment can be made by

$$p(t_0) = p(t_1)\frac{PI(t_0)}{PI(t_1)} \tag{45.5}$$

where $p(t)$ is the price, cost, or value of some item at time t; $PI(t)$ is the price index at time t, and t_0 is the present time. The price index reflects the change in prices for an item or group of items. Indexes for many categories of goods and services are available from the Bureau of Labor Statistics.[2] Current prices should also be adjusted when they refer to future transactions. Many companies have projections for prices for many of their more important products. Where such projections are not available, a relationship similar to Eq. (45.5) may be used except that price indexes for future years are not available. Estimates of future inflation rates may be substituted instead:

$$p(t_2) = p(t_0)(1+f)^n \tag{45.6}$$

where f is the annual inflation rate, n is the number of years from the present until time t_2 ($t_2 - t_0$ in years), and t_0 is the present time. The inflation rate in Eq. (45.6) is the overall inflation rate unless it is expected that the particular item in question will increase in price much faster or slower than prices in general. In this case, an inflation rate pertaining to the particular item should be used.

Changing prices often distort the interpretation of cash flows. This distortion may be minimized by expressing all cash flows in a reference year's dollars (e.g., 1980 dollars). This representation is referred to as constant dollar cash flows. Historic data may be converted to constant dollar representation by

$$Y^c = Y^d\frac{\overline{PI}(t_0)}{\overline{PI}(t)} \tag{45.7}$$

where Y^c is the constant dollar representation of a dollar cash flow Y^d, $\overline{PI}(t)$ is the value of the price index at time t, t_0 is the reference year, and t is the year in which Y^d occurred. An overall price index such as the Wholesale Price Index or the Gross National Product Implicit Price Deflator is used in this calculation, whereas a more specific price index is used in Eq. (45.5). Future cash flows may be expressed using constant dollar representation by

$$Y^c = Y^d\frac{1}{(1+f)^n} \tag{45.8}$$

where f is the projected annual inflation rate and n is the number of years after t_0 that Y^d occurs. It is usually convenient to let t_0 be the present. Then n is equal to the number of years from the present and, also, present prices may be used to make most constant dollar cash flow estimates. The use of constant dollar representation simplifies the economic analysis in many situations. However, it is important that the time value of money be adjusted as indicated in Eq. (45.2). Additional discussion on this topic may be found in Ref. 2.

Mutually exclusive alternatives and dependent alternatives often yield cash flows that are either all negative or predominantly negative, that is, they only deal with expenses. It is not possible to view each alternative in terms of an investment (initial expense) and a return (income). However, two alternatives may be used to create a set of cash flows that represent an investment and return as shown in Fig. 45.7. Alternative B is more expensive to implement than A but costs less to operate or sustain. Cash flow C is the difference between B and A. It shows the extra investment required for B and the savings it produces. Cash flow C may then be analyzed as an investment to determine if the extra investment required for C is worthwhile. This approach works well when only two alternatives are considered. If there are three or more alternatives, the comparison gets more complicated. Figure 45.8 shows the process required. Two alternatives are compared, the winner of that comparison is compared to a third, the winner of that comparison to a fourth, and so on until all alternatives have been considered and the single best alternative identified. When using this procedure, it is customary, but not necessary, to order the alternatives from least expensive to most expensive according to initial cost. This analysis of multiple alternatives may be referred to as incremental analysis, since only the difference, or incremental cash flow between alternatives, is considered. The same concept may be applied to continuous alternatives.

45.3.2 Present Worth

All of the cash flows for a project may be reduced to a single equivalent cash flow using the concepts of time value of money and cash flow equivalence. This single equivalent cash flow is usually calculated for the present time or the project initiation, hence the term present worth (also present value). However, this single cash flow can be calculated for any point in time if necessary. Occasionally, it is desired to calculate it for the end of a project rather than the beginning, the term future worth (also future value) is applied then. This single cash flow, either a present worth or future worth, is a measure of the net profit for the project and thus is an indication of whether or not the project is worthwhile economically. It may be calculated using interest factors and cash flow manipulations as described in

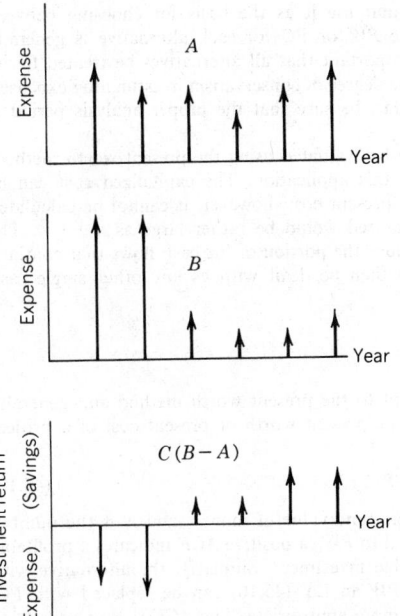

Fig. 45.7 Investment and return generated by comparing two alternatives.

preceding sections. Modern calculators and computers usually make it just as easy to calculate the present worth directly from the project cash flows. The present worth PW of a project is

$$PW = \sum_{j=0}^{n} Y_j \frac{1}{(1+i)^j} \qquad (45.9)$$

where Y_j is the project cash flow in year j, n is the length of the analysis period, and i is the time value of money. This equation uses the sign convention of income or savings being positive and expenses being negative.

In the case of independent alternatives, the PW of a project is a sufficient measure of the project's profitability. Thus, a positive present worth indicates an economically desirable project and a negative present worth indicates an economically undesirable project. In the case of mutually exclusive, dependent, or continuous alternatives the present worth of a given alternative means little. The alternative, or combination of dependent alternatives, that has the highest present worth is the most desirable economically. Often cost is predominant in these alternatives. It is customary then to reverse the sign convention in Eq. (45.9) and call the result the present cost. The alternative with the smallest present cost is then the most desirable economically. It is also valid to calculate the present worth of

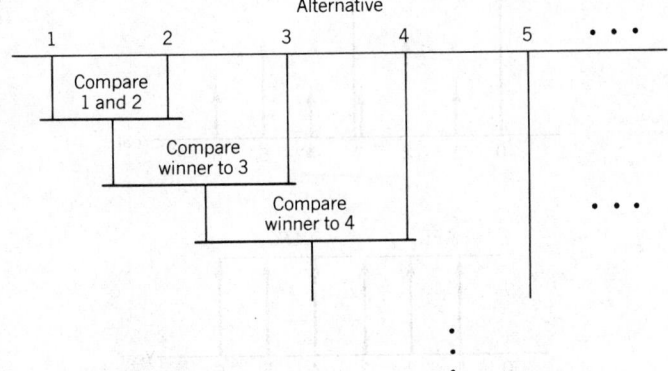

Fig. 45.8 Comparison of multiple alternatives using incremental cash flows.

the incremental cash flow (see Figs. 45.7 and 45.8) and use it as the basis for choosing between alternatives. However, the approach of calculating the *PW* or *PC* for each alternative is generally much easier. Regardless of the method chosen, it is important that all alternatives be treated fairly; use similar assumptions about future prices, use the same degree of conservatism in estimating expenses, make sure they all serve equally well, etc. In particular, be sure that the proper analysis period is selected when equipment lives differ.

Projects that have very long or indefinite lives may be evaluated using the present worth method. The present cost is referred to as capitalized cost in this application. The capitalized cost can be used for economic analysis in the same manner as the present cost; however, it cannot be calculated using Eq. (45.9), since the number of calculations required would be rather large as $n \rightarrow \infty$. The interest factors in Table 45.3 usually can be used to reduce the portion of the cash flows that continue indefinitely to a single equivalent amount, which can then be dealt with as any other single cash flow.

45.3.3   Annual Cash Flow

The annual cash flow method is very similar in concept to the present worth method and generally can be used whenever a present worth analysis can. The present worth or present cost of a project can be converted to an annual cash flow, *ACF*, by

$$ACF = PW \cdot (A/P, i, n) \tag{45.10}$$

where $(A/P, i, n)$ is the capital recovery factor, i is the time value of money, and n is the number of years in the analysis period. Since *ACF* is proportional to *PW*, a positive *ACF* indicates a profitable investment and a negative *ACF* indicates an unprofitable investment. Similarly, the alternative with the largest *ACF* will also have the largest *PW*. The *PW* in Eq. (45.10) can be replaced with *PC*, and *ACF* can be used to represent a cost when that is more appropriate. The *ACF* is thus equally as useful for economic analysis as *PW* or *PC*. It also has the advantage of having more intuitive meaning. *ACF* represents the equivalent annual income or cost over the life of a project.

Annual cash flow is particularly useful for analyses involving equipment with unequal lives. The n in Eq. (45.10) refers to the length of the analysis period and for unequal lives that means integer multiples of the life cycles. The annual cash flow for the analysis period will be the same as the analysis period for a single life cycle, as shown in Fig. 45.9, as long as the cash flows for the equipment repeat from one life cycle to the next. The annual cash flow for each equipment alternative can then be calculated for its own life cycle rather than for a number of life cycles. Unfortunately, prices generally increase from one life cycle to the next due to inflation and the cash flows from one life cycle to the next will be more like those shown in Fig. 45.10. The errors caused by this change from life cycle to life cycle usually will be acceptable if inflation is moderate (e.g., less than 5%) and the lives of various alternatives do not differ greatly (e.g., 7 versus 9 years). If inflation is high or if alternatives have lives that differ greatly, significant errors may result. The problem can often be circumvented by converting the cash flows to constant dollars using Eq. (45.8). With the inflationary price increases removed, the cash flows will usually repeat from one life cycle to the next.

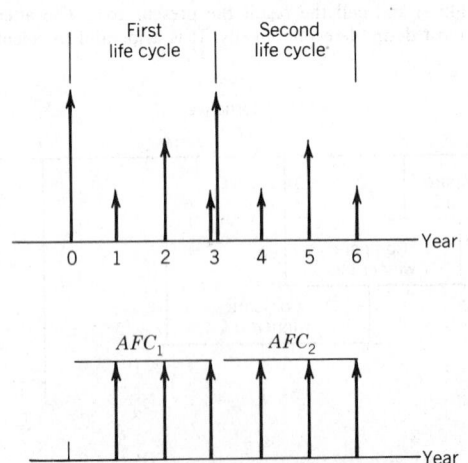

Fig. 45.9   Annual cash flow for equipment with repeating cash flows (3-year life).

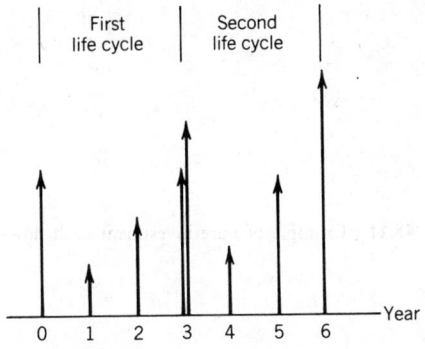

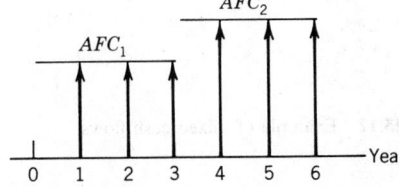

Fig. 45.10   Annual cash flow for equipment with nonrepeating cash flows (3-year life).

45.3.4   Rate of Return

The rate of return (also called the internal rate of return) method is the most frequently used technique for evaluating investments. The rate of return is based on Eq. (45.9) except that rather than solving for PW, the PW is set to zero and the equation is solved for i. The resulting interest rate is the rate of return of the investment:

$$0 = \sum_{j=0}^{n} Y_j \frac{1}{(1+i)^j} \qquad (45.11)$$

where Y_j is the cash flow for year j, i is the investment's rate of return (rather than the time value of money), and n is the length of the analysis period. It is usually necessary to solve Eq. (45.11) by trial and error, except for a few very simple situations. If constant dollar representation is used, the resulting rate of return is referred to as the real rate of return or the inflation-corrected rate of return. It may be converted to a dollar rate of return using Eq. (45.2).

A rate of return for an investment greater than the time value of money indicates that the investment is profitable and worthwhile economically. Likewise an investment with a rate of return less than the time value of money is unprofitable and is not worthwhile economically. The rate of return calculation is generally preferred over the present worth method or annual cash flow method by decision makers since it gives a readily understood economic measure. However, the rate of return method only allows a single investment to be evaluated or two projects compared using the incremental cash flow. When several mutually exclusive alternatives, dependent alternative combinations, or continuous alternatives exist, it is necessary to compare two investments at a time as shown in Fig. 45.8 using incremental cash flows. Present worth or annual cash flow methods are simpler to use in these instances. It is important to realize that with these types of decisions the alternative with the highest rate of return is not necessarily the preferable alternative.

The rate of return method is intended for use with classic investments as shown in Fig. 45.11. An expense (investment) is made initially and income (return) is generated in later years. If a particular set of cash flows, such as shown in Fig. 45.12, does not follow this pattern, it is possible that Eq. (45.11) will generate more than one solution. It is also very easy to misinterpret the results of such cash flows. Reference 4 explains how to proceed in evaluating cash flows of the nature shown in Fig. 45.12.

45.3.5   Benefit–Cost Ratio

The benefit–cost ratio (B/C) calculation is a form of the present worth method. With B/C, Eq. (45.9) is used to calculate the present worth of the income or savings (benefits) of the investment and the expenses (costs) separately. These two quantities are then combined to form the benefit–cost ratio:

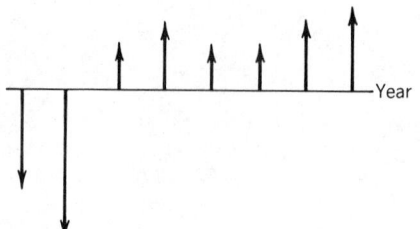

Fig. 45.11   Example of pure investment cash flows.

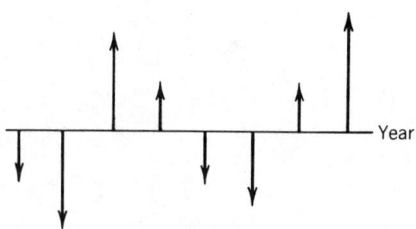

Fig. 45.12   Example of mixed cash flows.

$$B/C = \frac{PW_B}{PC_E}$$

where PW_B is the present worth of the benefits and PC_E is the present cost of the expenses. A B/C greater than 1 indicates that the benefits outweigh the costs and a B/C less than 1 indicates the opposite. The B/C also gives some indication as to how good an investment is. A B/C of about 1 indicates a marginal investment, whereas a B/C of 3 or 4 indicates a very good one. The PC_E usually refers to the initial investment expense. The PW_B includes the income and savings less operating cost and other expenses. There is some leeway in deciding whether a particular expense should be included in PC_E or subtracted from PW_B. The placement will change B/C some, but will never make a B/C which is less than 1 become greater than 1 or vice versa.

The benefit–cost calculation can only be applied to a single investment or used to compare two investments using the incremental cash flow. When evaluating several mutually exclusive alternatives, dependent alternative combinations, or continuous alternatives, the alternatives must be compared two at a time as shown in Fig. 45.8 using incremental cash flows. The single alternative with the largest B/C is not necessarily the preferred alternative in this case.

45.3.6   Payback Period

The payback calculation is not a theoretically valid measure of the profitability of an investment and is frequently criticized for this reason. However, it is widely used and does provide useful information. The payback period is defined as the period of time required for the cumulative net cash flow to be equal to zero; that is, the time required for the income or savings to offset the initial costs and other expenses. The payback period does not measure the profitability of an investment but rather its liquidity. It shows how fast the money invested is recovered. It is a useful measure for a company experiencing cash flow difficulties and which cannot afford to tie up capital funds for a long period of time. A maximum allowed payback period may be specified in some cases. A short payback period is generally indicative of a very profitable investment, but that is not ensured since there is no accounting for the cash flows that occur after the payback period. Most engineering economists agree that the payback period should not be used as the means of selecting among alternatives.

The payback period is sometimes calculated using discounted cash flows rather than ordinary cash flows. This modification does not eliminate the criticisms of the payback calculation although it does usually result in only profitable investments having a finite payback period. A maximum allowed payback period may also be used with this form. This requirement is equivalent to arbitrarily shortening the analysis period to the allowed payback period to reflect liquidity requirements. Since there are different forms of the payback calculation and the method is not theoretically sound, extreme care should be exercised in using the payback period in decision making.

REFERENCES

1. L. E. Bussey, *The Economic Analysis of Industrial Projects,* Prentice-Hall, Englewood Cliffs, NJ, 1978.
2. B. W. Jones, *Inflation in Engineering Economic Analysis,* Wiley, New York, 1982.
3. J. L. Riggs, *Engineering Economics,* 2nd ed., McGraw-Hill, New York, 1982.
4. E. L. Grant, W. G. Ireson, and R. S. Leavenworth, *Principles of Engineering Economy,* 7th ed., Wiley, New York, 1982.

CHAPTER 46

COST ESTIMATING

RODNEY STEWART

Mobile Data Services
Huntsville, Alabama

Reprinted from Rodney D. Stewart, *Cost Estimating,* Wiley-Interscience, New York, 1982, by permission of the publisher.

46.1 THE ANATOMY OF AN ESTIMATE

The cost estimating process, like the manufacture of a product, is comprised of parallel and sequential steps that flow together and interact to culminate in a completed estimate. Figure 46.1 shows the anatomy of an estimate. This figure depicts graphically how the various cost estimate ingredients are synthesized from the basic man-hour estimates and material quantity estimates. Man-hour estimates of each basic skill required to accomplish the job are combined with the labor rates for these basic skills to derive labor dollar estimates. In the meantime, material quantities are estimated in terms of the units by which they are measured or purchased, and these material quantities are combined with their costs per unit to develop direct material dollar estimates. Labor overhead or burden is applied to direct material costs. Then travel costs and other direct costs are added to produce total costs; general and administrative expenses and fee or profit are added to derive the "price" of the final estimate.

The labor rates that are applied to the basic man-hour estimates are usually "composite" labor rates; that is, they represent an average of the rates within a given skill category. For example, the engineering skill may include draftsmen, designers, engineering assistants, junior engineers, engineers, and senior engineers. The number and titles of engineering skills vary widely from company to company, but the use of a composite labor rate for the engineering skill category is common practice. The composite labor rate is derived by multiplying the labor rate for each skill by the percentage of man-hours of that skill required to do a given task and adding the results. For example, if each of the six skills have the following labor rates and percentages, the composite labor rate is computed as follows:

Skill	Labor Rate ($/hr)	Percentage in the Task
Draftsman	$ 6.00	7
Designer	$ 8.00	3
Engineering assistant	$10.00	10
Junior engineer	$13.00	20
Engineer	$15.00	50
Senior engineer	$18.00	10
. Total		100

Composite labor rate = $(0.07 \times \$6.00) + (0.03 \times \$8.00) + (0.10 \times \$10.00) + (0.20 \times \$13.00) + (0.50 \times \$15.00) + (0.10 \times \$18.00) = \$13.56$. Similar computations can be made to obtain the composite labor rate for skill within any of the other skill categories.

Another common practice is to establish separate overhead or burden pools for each skill category. These burden pools carry the peripheral costs that are related to and are a function of the man-

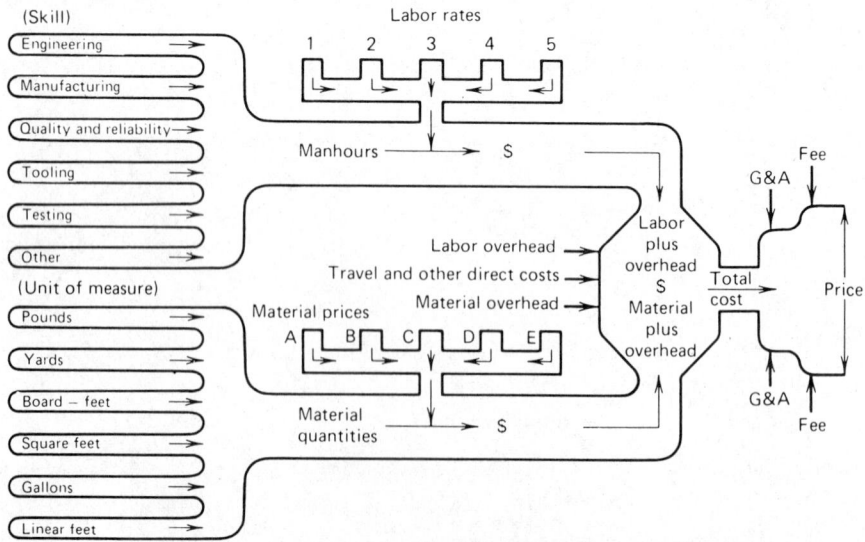

Fig. 46.1 Anatomy of an estimate.[1]

hours expended in that particular skill category. Assuming that the burden pool is established for each of the labor skills shown in Fig. 46.1, we can write an equation to depict the entire process. This equation is shown in Fig. 46.2. Although the equation appears to be complex because of the number of factors included, the mathematics consists only of multiplication and addition. So far we have only considered a one-element cost estimate. The addition of multielement work activities or work outputs will greatly increase the number of mathematical computations; and it becomes readily evident that the anatomy of an estimate is so complex that computer techniques for computation are highly desirable if not essential.

46.1.1 Time, Skills, and Labor Hours Required to Prepare an Estimate

The resources (skills, calendar time, and man-hours) required to prepare a cost estimate depend upon a number of factors. One factor is the estimating method used. Another factor is the level of technology or state of the art involved in the job or task being estimated. A rule of thumb can be used to develop a rough idea of the estimating time required. The calendar time required to develop an accurate and credible estimate is usually about 8% of the calendar time required to accomplish a task involving existing technology and 18% for a task involving a high technology (i.e., nuclear plant construction, aerospace projects). These percentages are divided approximately as shown in Table 46.1.

Note that the largest percentage of the estimating time required is for defining the output. This area is most important because it establishes a good basis for estimate credibility and accuracy as well as making it easier for the estimator to develop supportable man-hour and material estimates. These percentages also assume that the individuals who are going to perform the task or who have intimate working knowledge of the task are going to assist in estimate preparation. Hence the skill mix for estimating is very similar to the skill mix required for actually performing the task.

Man-hours required for preparation of a cost estimate can be derived from these percentages by multiplying the task's calendar period in years by 2000 man-hours per man-year, multiplying the result by the percentage in Table 46.1, and then multiplying the result by 0.1 and by the number of personnel on the estimating team. Estimating team size is a matter of judgment and depends on the complexity of the task, but it is generally proportional to the skills required to perform the task (as mentioned).

Examples of the application of these rules of thumb for determining the resources required to prepare a cost estimate follow:

1. A 3-year, high-technology project involving 10 basic skills or disciplines would require the following number of man-hours to estimate: $3 \times 2000 \times 0.18 \times 0.1 \times 10 = 1080$ man-hours
2. A 6-month "existing-technology" project requiring five skills or disciplines would require $0.6 \times 2000 \times 0.08 \times 0.1 \times 5 = 48$ man-hours to develop an estimate

These relationships are drawn from the author's experience in preparing and participating in a large number of cost estimates and can be relied on to give you a guideline in preparing for the estimating process. But remember that these are "rules of thumb," and exercise caution and discretion in their application.

$$
T = \Big\{ \; [(E_H \times E_R) \times (1 + E_0)] + [(M_H \times M_R) \times (1 + M_0)] + [(TO_H \times TO_R)
$$
$$
\times (1 + TO_0)] + [(Q_H \times Q_R) \times (1 + Q_0)] + [(TE_H + TE_R) \times (1 + TE_0)]
$$
$$
+ [(O_H \times O_R) \times (1 + O_0)] + S_D + S_O + M_D + [M_D \times (1 + M_{OH})]
$$
$$
+ T_D + C_D + OD_D \Big\} \times \Big\{ GA + 1.00 \Big\} \times \Big\{ F + 1.00 \Big\}
$$

(a)

$$
T = \Big\{ \; [(L1_H \times L1_R) \times (1 + L1_0)] + [(L2_H \times L2_R) \times (1 + L2_0) \cdots
$$
$$
+ [(LN_H \times LN_R \times (1 + LN_0)] + S_D + S_O + M_D + [M_D \times (1 + M_{OH})]
$$
$$
+ T_D + CD + OD_D \Big\} \times \Big\{ 1 + GA \Big\} \times \Big\{ 1 + F \Big\}
$$

Where: $L1, L2, \ldots LN$ are various labor rate categories

(b)

Fig. 46.2 Generalized equation for cost estimating: (a) specific labor rate categories; (b) general labor rate categories.[1]

Symbols:

T = total cost

E_H = engineering labor hours

E_R = engineering composite labor rate in dollars per hour

E_O = engineering overhead rate in decimal form (i.e.: 1.15 = 115%)

M_H = manufacturing labor hours

M_R = manufacturing composite labor rate in dollars per hour

M_O = manufacturing overhead rate in decimal form

TO_H = tooling labor hours

TO_R = tooling composite labor rate in dollars per hour

TO_O = tooling overhead in decimal form

Q_H = quality, reliability, and safety labor hours

Q_R = quality, reliability, and safety composite labor rate in dollars per hour

Q_O = quality, reliability, and safety overhead rate in decimal form

TE_H = testing labor hours

TE_R = testing composite labor rate in dollars per hour

TE_O = testing overhead rate in decimal form

O_H = other labor hours

O_R = labor rate for other hours category in dollars per hour

O_O = overhead rate for other hours category in decimal form

S_D = major subcontract dollar

S_O = other subcontract dollars

M_D = material dollars

M_{OH} = material overhead in decimal form (10% = 0.10)

T_D = travel dollars

C_D = computer dollars

OD_D = other direct dollars

GA = general and administrative expense in decimal form (25% = 0.25)

F = fee in decimal form (8% = 0.08)

Fig. 46.2 *(Continued)*

Table 46.1 Estimating Time as a Percentage of Total Job Time

	Existing Technology (%)	High Technology (%)
Defining the output	4.6	14.6
Formulating the schedule and ground rules	1.2	1.2
Estimating materials and labor-hours	1.2	1.2
Estimating overhead, burden, and G&A	0.3	0.3
Estimating fee, profit, and earnings	0.3	0.3
Publishing the estimate	0.4	0.4
Total	8.0	18.0

46.2   DISCUSSION OF TYPES OF COSTS

46.2.1   Initial Acquisition Cost

Businessmen, consumers, and government officials are becoming increasingly aware of the need to estimate accurately and to justify the initial acquisition cost of an item to be purchased, manufactured, or built. When we speak of initial acquisition costs, we are usually referring to the total costs to procure, install, and put into operation a piece of equipment, a product, or a structure. Initial acquisition costs do not consider costs associated with the use and possession of the item. Individuals or businesses who purchase products are beginning to give more serious consideration to maintenance, operation, depreciation, energy, insurance, storage, and disposal costs before purchasing an item, whether it be an automobile, home appliance, suit of clothes, or industrial equipment. Because of continuing inflation and the constant fluctuation in the rate of inflation, it is as important to estimate accurately the most probable initial acquisition cost as it is to consider the cost savings brought about by use of the item being acquired. Initial acquisition costs include planning, estimating, designing, and/or purchasing the components of the item; manufacturing, assembly, and inspection of the item; and installing and testing the item. Initial acquisition costs also include marketing, advertising, and markup of the price of the item as it flows through the distribution chain.

46.2.2   Fixed and Variable Costs

The costs of all four categories of productive outputs (processes, products, projects, and services) involve both fixed and variable costs. The relationship between fixed and variable costs depends on a number of factors, but it is principally related to the kind of output being estimated and the rate of output. Fixed cost is that group of costs involved in an ongoing activity whose total will remain relatively constant regardless of the quantity of output or the phase of the output cycle being estimated. Variable cost is the group of costs that vary in relationship to the rate of output. Therefore, where it is desirable to know the effect of output rate on costs, it is important to know the relationship between the two forms of cost as well as the magnitude of these costs. Fixed costs are meaningful only if they are considered at a given point in time, since inflation and escalation will provide a variable element to "fixed" costs. Fixed costs may only be truly fixed over a given range of outputs. Rental of floor space for a production machine is an example of a fixed cost, and its use of electrical power will be a variable cost.

46.2.3   Recurring and Nonrecurring Costs

Recurring costs are those that are repetitive in nature and depend on continued output of a like kind. They are similar to variable costs because they are dependent on the quantity or magnitude of output. Nonrecurring costs are those that are incurred to generate the very first item of output. It is important to separate recurring and nonrecurring costs if it is anticipated that the costs of continued or repeated production will be required at some future date.

46.2.4   Direct and Indirect Costs

As discussed earlier, direct costs are those that are attributable directly to the specific work activity or work output being estimated. Indirect costs are those that are spread across several projects and allocable on a percentage basis to each project.

Table 46.2 is a matrix giving examples of these costs for various work outputs.

46.3   COLLECTING THE INGREDIENTS OF THE ESTIMATE

Before discussing the finer points of estimating, it is important to define these ingredients and to give you a preview of the techniques and methods that will be used to collect these estimate ingredients.

46.3.1   Labor Hours

Since the expenditure of labor hours is the basic reason for the incurrence of costs, the estimating of labor hours is the most important aspect of cost estimating. Labor hours and manpower requirements are estimated by four basic techniques: (1) use of the methods, time, and motion (MTM) techniques; (2) the man-loading or staffing technique; (3) direct judgment of man-hours required; (4) and the use of man-hour/manpower estimating handbooks.

MTM methods are perhaps the most widespread methods of deriving man-hour and skill estimates for industrial processes. These methods are available from and taught by the MTM Association for Standards and Research, located in Fair Lawn, New Jersey. The association is international in scope

Table 46.2 Examples of Costs for Various Outputs

	Process	Product	Project	Service
Initial acquisition costs	Plant construction costs	Manufacturing costs, marketing costs, and profit	Planning costs, design costs, manufacturing costs, test and checkout costs, and delivery costs	
Fixed costs	Plant maintenance costs	Plant maintenance costs	Planning costs and design costs	Building rental
Variable costs	Raw material costs	Labor costs	Manufacturing costs, test and checkout costs, and delivery costs	Labor costs
Recurring costs	Raw material costs	Labor and material costs	Manufacturing costs, test and checkout costs, and delivery costs	Labor costs
Nonrecurring costs	Plant construction costs	Plant construction costs	Planning costs and design costs	Initial capital equipment investment
Direct costs	Raw materials	Manufacturing costs	Planning, design, manufacturing, test and checkout, and delivery costs	Labor and materials costs
Indirect costs	Energy costs	Marketing costs and profit	Energy costs	Energy costs

and has developed five generations of MTM systems for estimating all aspects of industrial, manufacturing, or machining operations. The MTM method subdivides operator motions into small increments that can be measured and provides a means for combining the proper manual operations in a sequence to develop man-hour requirements for accomplishing a job.

The man-loading or staffing technique is perhaps the simplest and most widely used method for estimating the man-hours required to accomplish a given job. In this method, the estimator envisions the job, the work location, and the equipment or machines required, and estimates the number of people and skills that would be needed to staff a particular operation. His or her estimate is usually in terms of a number of people for a given number of days, weeks, or months. From this staffing level, the estimated on-the-job man-hours required to accomplish a given task can be computed.

Another method that is closely related to this second method is the use of direct judgment of the number of man-hours required. This judgment is usually made by an individual who has had direct hands-on experience in either performing or supervising a like task.

Finally, the use of handbooks is a widely used and accepted method of developing man-hour estimates. Handbooks usually provide larger time increments than the MTM method and require a specific knowledge of the work content and operation being performed.

46.3.2 Materials and Subcontracts

Materials and subcontract dollars are estimated in three ways: (1) drawing "takeoffs" and handbooks; (2) dollar-per-pound relationships; and (3) direct quotations or bids. The most accurate way to estimate material costs is to calculate material quantities directly from a drawing or specification of the completed product. Using the quantities required for the number of items to be produced, the appropriate materials manufacturer's handbook, and an allowance for scrap or waste, one can accurately compute the material quantities and prices. Where detailed drawings of the item to be produced are not available, a dollar-per-pound relationship can be used to determine a rough order of magnitude cost. Firm quotations or bids for the materials or for the item to be subcontracted are better than any of the previously mentioned ways of developing a materials estimate because the supplier can be held to his or her bid.

46.3.3 Labor Rates and Factors

The labor rate, or number of dollars required per man-hour, is the quantity that turns a man-hour estimate into a cost estimate; therefore the labor rate and any direct cost factors that are added to it are key elements of the cost estimate. Labor rates vary by skill, geographical location, calendar date, and the time of day or week they are applied. Labor rates vary from over $3 per hour (the minimum wage in 1980) for hourly paid personnel to over $30 per hour for salaried personnel. Labor rates vary geographically by a factor of 2.61 to 1 in the continental United States and 3.61 to 1 if Alaska is included. The calendar date affects the labor rate because inflation must be added each year to adjust wages to the cost of living increases encountered during that year. For hourly paid personnel, overtime premiums of 150% are common for hours worked over 40 hours per week, and 250% for work accomplished on Sundays or holidays. Shift premiums and hazardous duty pay are also added to hourly wages to develop the actual labor rate to be used in developing a cost estimate. Wage rate structures vary considerably depending on union contract agreements. Once the labor rate is applied to the man-hour estimate to develop a labor cost figure, other factors are commonly used to develop other direct cost allowances such as travel costs and direct material costs.

46.3.4 Indirect Costs, Burden, and Overhead

Burden or overhead costs for engineering activities very often are as high as 100% of direct engineering labor costs, and manufacturing overheads go to 150% and beyond. A company that can keep its overhead from growing excessively, or a company that can successfully trim its overhead, can place itself in an advantageously competitive position. Since overhead more than doubles the cost of a work activity or work output, trimming the overhead has a significant effect on reducing overall costs.

46.3.5 Administrative Costs

Administrative costs range up to 20% of total direct and indirect costs for large companies. General and administrative costs are added to direct and overhead costs and are recognized as a legitimate business expense.

46.3.6 Fee, Profit, or Earnings

The fee, profit, or earnings will depend on the amount of risk the company is taking in marketing the product, the market demand for the item, and the required return on the company's investment.

This subject is one that deserves considerable attention by the cost estimator. Basically, the amount of profit depends on the astute business sense of the company's management. Few companies will settle for less than 10% profit, and many will not make an investment or enter into a venture unless they can see a 20–30% return on their investment.

46.3.7 Assembly of the Ingredients

Once resource estimates have been accumulated, the process of reviewing, compiling, organizing, and computing the estimate begins. This process is divided into two general subdivisions of work: (1) reviewing, compiling, and organizing the input resource data, and (2) computation of the costs based on desired or approved labor rates and factors. A common mistake made in developing cost estimates is the failure to perform properly the first of these work subdivisions. In the process of reviewing, compiling, and organizing the data, duplications in resource estimates are discovered and eliminated; omissions are located and remedied; overlapping or redundant effort is recognized and adjusted; and missing or improper rationale, backup data, or supporting data are identified, corrected, or supplied. A thorough review of the cost estimate input data by the estimator or estimating team, along with an adjustment and reconciliation process, will accomplish these objectives.

Computation of a cost estimate is mathematically simple since it involves only multiplication and addition. The number of computations can escalate rapidly, however, as the number of labor skills, fiscal years, and work breakdown structure elements are increased. One who works frequently in ground-up, industrial engineering man-hour and material cost estimating will quickly come to the conclusion that some form of computer assistance is required. Computer programs capable of performing the basic computations required to develop a cost estimate are available at a nominal cost from the federal government or from a number of computer software firms. Some programs can be obtained from the Computer Operations Services and Management Information Center (COSMIC) at the University of Georgia, Athens, Georgia. In addition, there are companies that specialize in automatic data processing equipment and services for accounting, economics, and estimating applications.

With the basic ingredients and basic tools available, we are now ready to follow the steps required to develop a good cost estimate. All steps are needed for any good cost estimate. The manner of accomplishing each step, and the depth of information needed and time expended on each step, will vary considerably depending on what work activity or work output is being estimated. These steps are as follows:

1. Define the product, process, project, service.
2. Develop a methodology and estimate schedule.
3. Formulate a project schedule.
4. Prepare estimate ground rules.
5. Estimate man-hours and materials.
6. Review, compile, organize, reconcile, and compute the estimate.
7. Publish the estimate.
8. Use the estimate.

46.4 THE FIRST QUESTIONS TO ASK (AND WHY)

Whether you are estimating the cost of a process, product, project, or service, there are some basic questions you must ask in order to get started on a cost estimate. These questions relate principally to the requirements, descriptions, location, and timing of the work.

46.4.1 What Is It?

A surprising number of cost estimates fail to be accurate or credible because of a lack of specificity in describing the work that is being estimated. The objectives, ground rules, constraints, and requirements of the work must be spelled out in detail to form the basis for a good cost estimate. First, it is necessary to determine which of the four generic work outputs (process, product, project, or service) or combination of work outputs best describes the work being estimated. Then it is necessary to describe the work in as much detail as possible.

46.4.2 What Does It Look Like?

Work descriptions usually take the form of detailed specifications, sketches, drawings, materials lists, and parts lists. Weight, size, shape, material type, power, accuracy, resistance to environmental hazards, and quality are typical factors that are described in detail in a specification. Processes and services are usually defined by the required quality, accuracy, speed, consistency, or responsiveness of the work. Products and projects, on the other hand, usually require a preliminary or detailed design of the item or group of items being estimated. In general, more detailed designs will produce more accurate cost estimates. The principal reason for this is that as a design proceeds, better definitions and descriptions of all facets of this design unfold. The design process is an interactive one in which component or subsystem designs proceed in parallel; component or subsystem characteristics reflect upon and affect one another to alter the configuration and perhaps even the performance of the end item. Another reason that a more detailed design results in a more accurate and credible cost estimate is that the amount of detail itself produces a greater awareness and visibility of potential inconsistencies, omissions, duplications, and overlaps.

46.4.3   When Is It to Be Available?

Production rate, production quantity, and timing of production initiation and completion are important ground rules to establish before starting a cost estimate. Factors such as raw material availability, labor skills required, and equipment utilization often force a work activity to conform to a specific time period. It is important to establish the optimum time schedule early in the estimating process, to establish key milestone dates, and to subdivide the overall work schedule into indentifiable increments that can be placed on a calendar time scale. A work output schedule placed on a calendar time scale will provide the basic inputs needed to compute start-up costs, fiscal-year funding, and inflationary effects.

46.4.4   Who Will Do It?

The organization or organizations that are to perform an activity, as well as the skill and salary levels within these organizations, must be known or assumed to formulate a credible cost estimate. Given a competent organization with competent employees, another important aspect of developing a competitive cost estimate is the determination of the make or buy structure and the skill mix needs throughout the time period of a work activity. Judicious selection of the performers, and wise time phasing of skill categories and skill levels can rapidly produce prosperity for any organization with a knowledge of its employees, its product, and its customer.

46.4.5   Where Will It Be Done?

Geographical factors have a strong influence on the credibility and competitive stature of a cost estimate. In addition to the wide variation in labor costs for various locations, material costs vary substantially from location to location, and transportation costs are entering even more heavily into the cost picture than in the past because of increasing energy costs. The cost estimator must develop detailed ground rules and assumptions concerning location of the work and then estimate costs accurately in keeping with all location-oriented factors.

46.5   THE ESTIMATE SKELETON: THE WORK ELEMENT STRUCTURE

The first step in developing a cost estimate of any type of work output is the development of a work element structure (sometimes called a work breakdown structure). The work element structure serves as a framework for collecting, accumulating, organizing, and computing the direct and directly related costs of a work activity or work output. It also can be and usually is used for managing and reporting resources and related costs throughout the lifetime of the work. There is considerable advantage in using the work element structure and its accompanying task descriptions as the basis for scheduling, reporting, tracking, and organizing as well as for initial costing. Hence, it is important to devote considerable attention to this phase of the overall estimating process. A work element structure is developed by subdividing a process, product, project, or service into its major work elements, then breaking the major work elements into subelements, and subelements into subsubelements, and so on. There are usually five to ten subelements under each major work element.

The purpose of developing the work element structure is fivefold:

1. To provide a lower-level breakout of small tasks that are easy to identify, man-load, schedule, and estimate.
2. To provide assurance that all required work elements are included in the work output.
3. To reduce the possibility of overlap, duplication, or redundancy of tasks.
4. To furnish a convenient hierarchical structure for the accumulation of resources estimates.
5. To give greater overall visibility as well as depth of penetration into the makeup of any work activity.

46.6   THE HIERARCHICAL RELATIONSHIP OF A WORK ELEMENT STRUCTURE

A typical work element structure is shown in Fig. 46.3. Note that the relationship resembles a hierarchy where each activity has a higher activity, parallel activities, and lower activities. A basic principle of work element structures is that the resources or content of each work element are made up of the sum of the resources or content of elements below it. No work element that has lower elements exceeds the sum of those lower elements in resource requirements. The bottommost elements are estimated at their own level and sum to higher levels. Many numbering systems are feasible and workable. The numbering system shown is one that has proven workable in a wide variety of situations.

One common mistake in using work element structures is to try to input or allocate effort to every block, even those at a higher level. Keep in mind that this should not be done because each block or work element contains only that effort included in those elements below it. If there are no blocks below it, then it can contain resources. If there is need to add work activities or resources not included in a higher level block, add an additional block below it to include the desired effort. Level one of a work element structure is usually the top level, with lower levels numbered sequentially as shown. The "Level" is usually equal to the number of digits in the work element block. For example, the block numbered 1.1.3.2 is in Level 4 because it contains four digits.

46.7   FUNCTIONAL ELEMENTS DESCRIBED

When subdividing a work activity or work output into its elements, the major subdivisions can be either functional or physical elements. The second level in a work element structure usually consists of a combination of functional and physical elements if a product or project is being estimated. For a process or service, all second-level activities could be functional. Functional elements of a production or project activity can include activities such as planning, project management, systems engineering and integration, testing, logistics, and operations. A process or service can include any of hundreds of functional elements. Typical examples of the widely dispersed functional elements that can be found in a work element structure for a service are advising, assembling, binding, cleaning, fabricating, inspecting, packaging, painting, programming, projecting, receiving, testing, and welding.

46.8   PHYSICAL ELEMENTS DESCRIBED

The physical elements of a work output are the physical structures, hardware, products, or end items that are supplied to the consumer. These physical elements represent resources because they take labor and materials to produce. Hence they can and should be a basis for the work element structure.

Figure 46.4 shows a typical work element structure of just the physical elements of a well-known consumer product, the automobile. The figure shows how just one automobile company chose to subdivide the components of an automobile. For any given product or project, the number of ways that a work element structure can be constructed are virtually unlimited. For example, the company could have included the carburetor and engine cooling system as part of the engine assembly (this might have been a more logical and workable arrangement since it is used in costing a mass production operation). Note that the structure shows a Level 3 breakout of the body and sheet metal element, and the door (a Level 3 element) is subdivided into its Level 4 components.

This physical element breakout demonstrates several important characteristics of a work element structure. First, note that Level 5 would be the individual component parts of each assembly or subassembly. It only took three subdivisions of the physical hardware to get down to a point where the next level breakout would be the individual parts. One can see rapidly that breaking down every Level 2 element three more levels (down to Level 5) would result in a very large work element structure. Second, to convert this physical hardware breakout into a true work element structure would require the addition of some functional activities. To provide the manpower as well as the materials required to procure, manufacture, assemble, test, and install the components of each block, it is necessary to add an "assembly," "fabrication," or "installation" activity block.

46.9   TREATMENT OF RECURRING AND NONRECURRING ACTIVITIES

Most work consists of both nonrecurring activities or "one-of-a-kind" activities needed to produce an item or to provide a service, and recurring or repetitive activities that must be performed to provide more than one output unit. The resources requirements (man-hours and materials) necessary to perform these nonrecurring and recurring activities reflect themselves in nonrecurring and recurring costs.

Although not all estimates require the separation of nonrecurring and recurring costs, it is often both convenient and necessary to separate costs because one may need to know what the costs are for an increased work output rate. Since work output rate principally affects the recurring costs, it is desirable to have these costs readily accessible and identifiable.

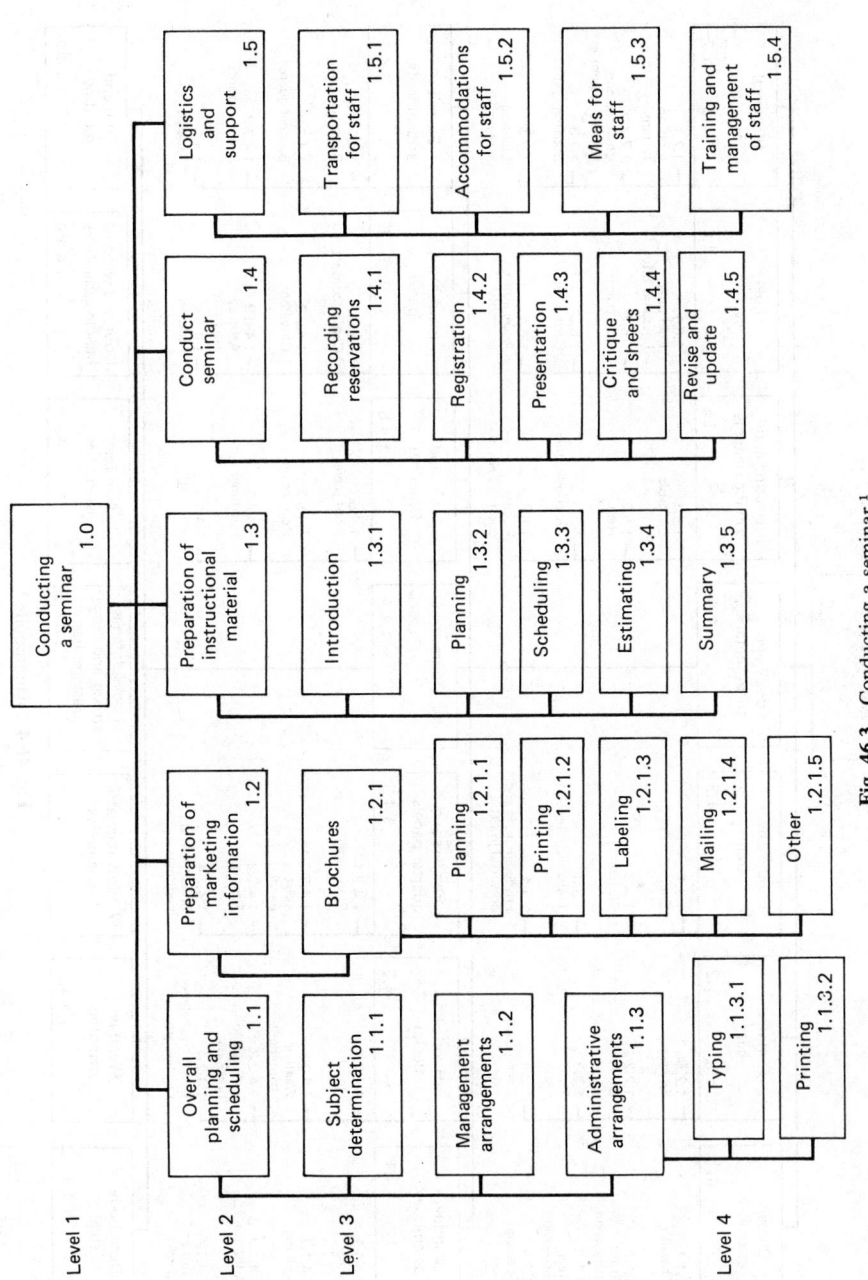

Fig. 46.3 Conducting a seminar.[1]

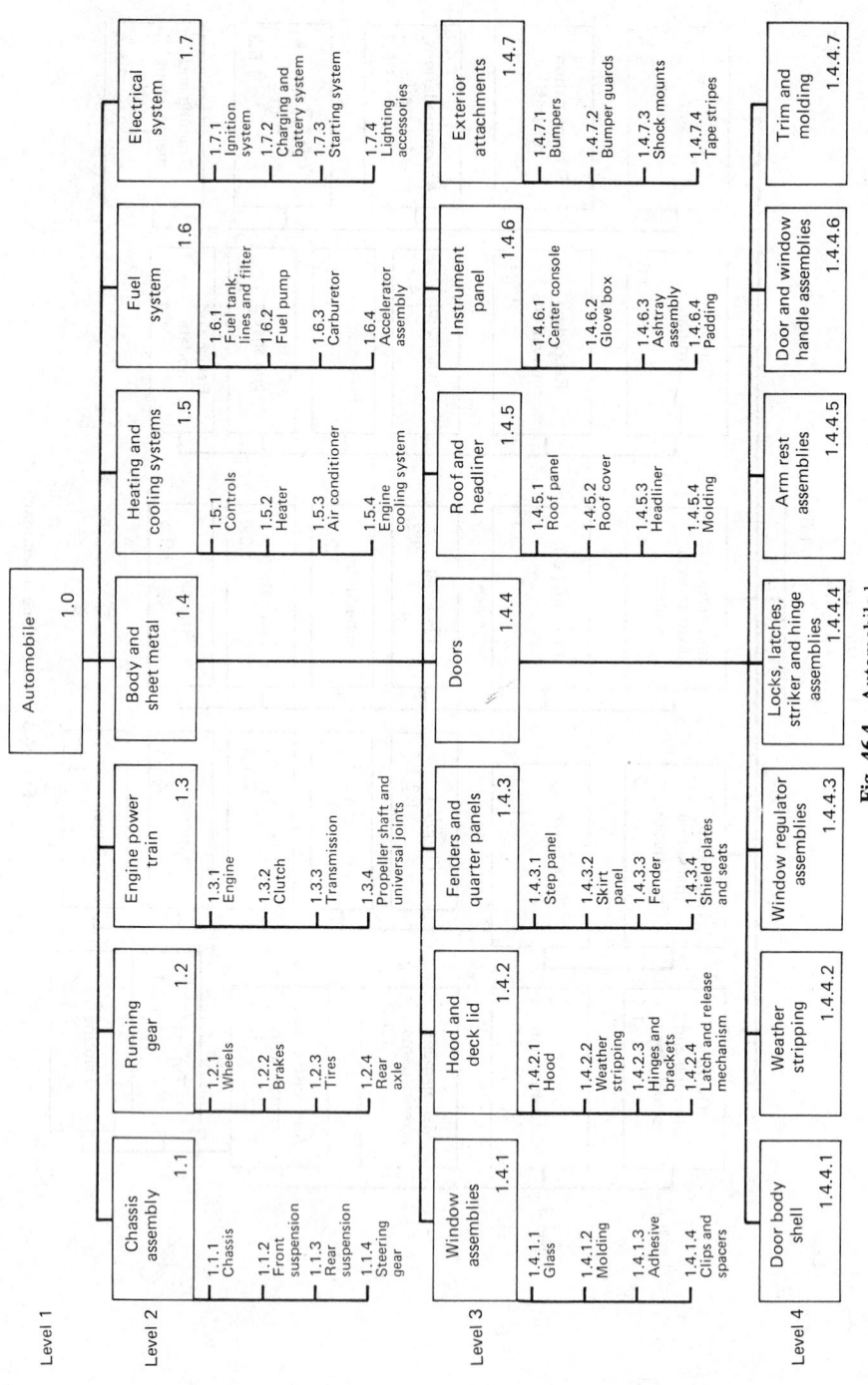

Fig. 46.4 Automobile.[1]

Separation of nonrecurring and recurring costs can be done in two ways through the use of the work element structure concept. First, the two costs can be identified, separated, and accounted for within each work element. Resources for each task block would, then, include three sets of resource estimates: (1) nonrecurring costs, (2) recurring costs, and (3) total costs for that block. The second convenient method of cost separation is to start with identical work element structures for both costs and develop two separate cost estimates. A third estimate, which sums the two cost estimates into a total can also use the same basic work element structure. If there are elements unique to each cost category, they can be added to the appropriate work element structure.

46.10 WORK ELEMENT STRUCTURE INTERRELATIONSHIPS

As shown in the automobile example, considerable flexibility exists concerning the placement of physical elements (the same is true with functional elements) in the work element structure. Because of this, and because it is necessary to define clearly where one element leaves off and the other takes over, it is necessary to provide a detailed definition of what is included in each work activity block. For example, in the automotive example, the rear axle unit could have been located and defined as part of the power train or as part of the chassis assembly rather than as part of the running gear. Where does the rear axle leave off and the power train begin? Is the differential or part of the differential included in the power train? These kinds of questions must be answered—and they usually are answered—before a detailed cost estimate is generated, in the form of a work element structure dictionary. The dictionary describes exactly what is included in each work element and what is excluded; it defines where the interface is located between two work elements; and it defines where the assembly effort is located to assemble or install two interfacing units.

A good work element structure dictionary will prevent many problems brought about by overlaps, duplications, and omissions because detailed thought has been given to the interfaces and content of each work activity.

46.10.1 Skill Matrix in a Work Element Structure

When constructing a work element structure, keep in mind that each work element will be performed by a person or group of people using one or more skills. There are two important facets of the labor or work activity for each work element: skill mix and skill level. The skill mix is the proportion of each of several skill categories that will be used in performing the work. Skill *categories* vary widely and depend on the type of work being estimated. For a residential construction project, for example, typical skills would be bricklayer, building laborer, carpenter, electrician, painter, plasterer, or plumber. Other typical construction skills are structural steelworker, cement finisher, glazier, roofer, sheet metal worker, pipefitter, excavation equipment operator, and general construction laborer. Professional skills such as lawyers, doctors, financial officers, administrators, project managers, engineers, printers, writers, and so forth are called on to do a wide variety of direct-labor activities. Occasionally, skills will be assembled into several broad categories (such as engineering, manufacturing, tooling, testing, and quality assurance) that correspond to overhead or burden pools.

Skill *level,* on the other hand, depicts the experience or salary level of an individual working within a given skill category. For example, engineers are often subdivided into various categories such as principal engineers, senior engineers, engineers, associate engineers, junior engineers, and engineering technicians. The skilled trades are often subdivided into skill level categories and given names that depict their skill level; for example, carpenters could be broken down into master carpenters, journeymen carpenters, apprentice carpenters, and carpenter helpers. Because skill categories and skill levels are designated for performing work within each work element, it is not necessary to establish separate work elements for performance of each skill. A work element structure for home construction would not have an element designated carpentry because carpentry is a skill needed to perform one or more of the work elements (i.e., roof construction, wall construction, etc.).

46.10.2 Organizational Relationships to a Work Element Structure

Frequently all or part of a work element structure will have a direct counterpart in the performing organization. Although it is not necessary for the work element structure to be directly correlatable to the organizational structure, it is often convenient to assign the responsibility for estimating and for performing a specific work element to a specific organizational segment. This practice helps to motivate the performer, since it gives him or her responsibility for an identifiable task; and it provides the manager greater assurance that each part of the work will be accomplished. In the planning and estimating process, early assignment of work elements to those who are going to be responsible for performing the work will motivate them to do a better job of estimating and will provide greater assurance of completion of the work within performance, schedule, and cost constraints because the functional organizations have set their own goals. Job performance and accounting for work accomplished versus funds spent can also be accomplished more easily if an organizational element is held responsible for a specific work element in the work element structure.

46.11   METHODS USED IN THE ESTIMATING PROCESS

46.11.1   Detailed Estimating

Detailed estimating involves the synthesis of a cost estimate from resource estimates made at the lowest possible level in the work element structure. Detailed estimating presumes that a detailed design of the product or project is available and that a detailed manufacturing, assembly, testing, and delivery schedule is available for the work. This type of estimating assumes that skills, man-hours, and materials can be identified for each work element through one or more of the methods that follow. A detailed estimate is usually developed through a synthesis of work element estimates developed by various methods.

46.11.2   Direct Estimating

A direct estimate is a judgmental estimate made in a "direct" method by an estimator or performer who is familiar with the task being estimated. The estimator will observe and study the task to be performed and then quote his or her estimate in terms of man-hours, materials, and/or dollars. For example, a direct estimate could be quoted as "so many dollars." Many expert estimators can size up and estimate a job with just a little familiarization. One estimator I know can take a fairly complex drawing and, within just a few hours, develop a rough order-of-magnitude estimate of the resources required to build the item. Direct estimating is a skill borne of experience in both estimating and in actually performing the "hands-on" work.

46.11.3   Estimating by Analogy (Rules of Thumb)

This method is similar to the direct estimating method in that considerable judgment is required, but an additional feature is the comparison with some existing or past task of similar description. The estimator collects resources information on a similar or analogous task and compares the task to be estimated with the similar or analogous one. The estimator would say that "this task should take about twice the time (man-hours, dollars, materials, etc.) as the one used as a reference." This judgmental factor (a factor of 2), would then be multiplied by the resources used for the reference task to develop the estimate for the new task. A significant pitfall in this method of estimating is the potential inability of the estimator to identify subtle differences in the two work activities and, hence, to be estimating the cost of a system based on one that is really not similar or analogous.

46.11.4   Firm Quotes

One of the best methods of estimating the resources required to complete a work element or to perform a work activity is the development of a firm quotation by the supplier or vendor. The two keys to the development of a realistic quotation are (1) the solicitation of bids from at least three sources and (2) the development of a detailed and well-planned request for quotation. Years of experience by many organizations in the field of procurement have indicated that three bids are the optimum from a standpoint of achieving the most realistic and reasonable price at a reasonable expenditure of effort. The acquisition of at least three bids provides sufficient check and balance and furnishes bid prices and conditions for comparison, evaluation, and selection. A good request for quotation (RFQ) is essential, however, to evaluate the bids effectively. The RFQ should contain ground rules, schedules, delivery locations and conditions, evaluation criteria, and specifications for the work. The RFQ should also state and specify the format required for cost information. A well-prepared RFQ will result in a quotation or proposal that will be easily evaluated, verified, and compared with independent estimates.

46.11.5   Handbook Estimating

Handbooks, catalogs, and reference books containing information on virtually every conceivable type of product, part, supplies, equipment, raw material, and finished material are available in libraries and bookstores, and directly from publishers. Many of these handbooks provide labor estimates for installation or operation as well as the purchase costs of the item. Some catalogs either do not provide price lists or provide price lists as a separate insert to permit periodic updates of prices without changing the basic catalog description. Information services provide microfilmed cassettes and viewing devices for access to the descriptions and costs of thousands and even tens of thousands of items.

  If you produce a large number of estimates, it may pay to subscribe to a microfilm catalog and handbook data access system, or, at least, to develop your own library of handbooks and catalogs.

46.11.6 The Learning Curve

The learning curve is a graphical representation of the reduction in time, resources, or costs either actually or theoretically encountered in the conduct of a repetitive production-motivated human activity. The theory behind the learning curve (and there are data available to confirm this theory for some specific applications) is that successive identical operations after the first one will take less time, use fewer resources, or cost less than preceding operations. The term "learning" is used because it relates primarily to the improvement of mental or manual skills observed when an operation is repeated, but "learning" can also be achieved by a shop or organization through the use of improved equipment, purchasing, production, or management techniques. When the "learning curve" is used in applications other than those involving the feedback loop and so bringing improvement of an individual's work activities, it is more properly named by one or more of the following terms:

Productivity improvement curve

Manufacturing progress function

Experience curve

Progress curve

Improvement curve

Production improvement curve

Production acceleration curve

Time reduction curve

Cost improvement curve

Learning curve theory is based on the concept that as the total quantity of units produced doubles, the hours required to produce the last unit of this doubled quantity will be reduced by a constant percentage. This means that the hours required to produce unit 2 will be a certain percent less than the hours required to produce unit 1; the hours required to produce unit 4 will be the same percent less than the hours required to produce unit 2; the hours required to produce unit 8 will be the same percent less than unit 4; and this constant percentage of reduction will continue for doubled quantities as long as uninterrupted production of the same item continues. The complement of this constant percentage of reduction is commonly referred to as the "slope." This means that if the constant percentage of reduction is 10 percent, the slope would be 90 percent. Table 46.3 gives an example of a learning curve with 90 percent slope when the number of hours required to produce the first unit is 100.

The reason for using the term "slope" in naming this reduction will be readily seen when the learning curve is plotted on coordinates with logarithmic scales on both the x and y axes (in this instance, the learning "curve" actually becomes a straight line). But first, let us plot the learning curve on conventional coordinates. You can see by the plot in Fig. 46.5 that it is truly a curve when plotted on conventional coordinates, and that the greater the production quantity, the smaller the incremental reduction in man-hours required from unit to unit.

When the learning curve is plotted on log-log coordinates, as shown in Fig. 46.6, it becomes a straight line. The higher the slope, the flatter the line; the lower the slope, the steeper the line.

The effects of plotting curves on different slopes can be seen on the chart shown in Fig. 46.7. This chart shows the effects on man-hour reductions of doubling the quantities produced twelve times.

Formulas for the unit curve and the cumulative average curve are shown in Table 46.4.

Care should be taken in the use of the learning curve to avoid an overly optimistic (low) learning curve slope and to avoid using the curve for too few units in production. Most learning curve textbooks will point out that this technique is credibly applicable only to operations that are done by hand (employ manual or physical operations) and that are highly repetitive.

Table 46.3 Learning Curve Values

Cumulative Units	Hours per Unit	Percent Reduction
1	100.00	
2	90.00	10
4	81.00	10
8	72.90	10
16	65.61	10
32	59.05	10

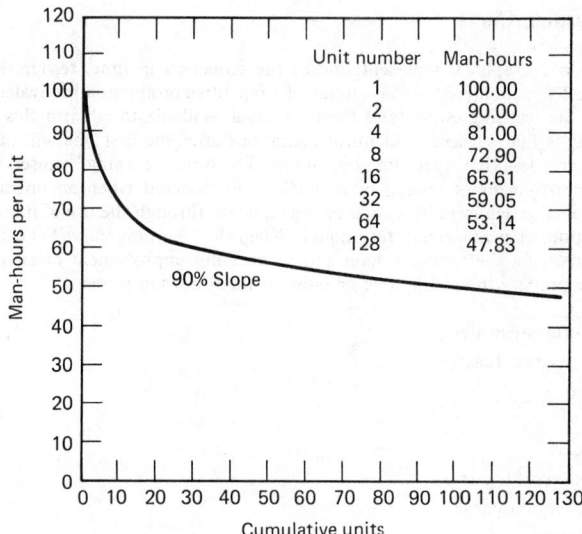

Fig. 46.5 Cumulative units versus man-hours per unit.[1]

46.11.7 Man-Loading Methods

One of the most straightforward methods of estimating resources or man-hours required to accomplish a task is the man-loading, or shop-loading method. This estimating technique is based on the fact that an experienced participant or manager of any activity can usually perceive, through judgment and knowledge of the activity being estimated, the number of individuals of various skills needed to accomplish a task. The shop-loading method is similar in that the estimator can usually predict what portion of an office or shop's capacity will be occupied by a given job. This percentage shop-loading factor can be used to compute man-hours or resources if the total shop manpower or total shop operation costs are known. Examples of the man-loading and shop-loading methods based on 1896 man-hours per man year are shown in Table 46.5.

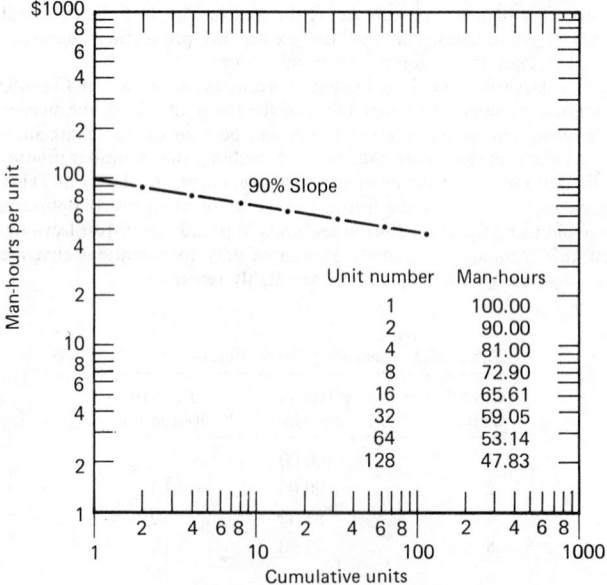

Fig. 46.6 Cumulative units versus man-hours per unit.[1]

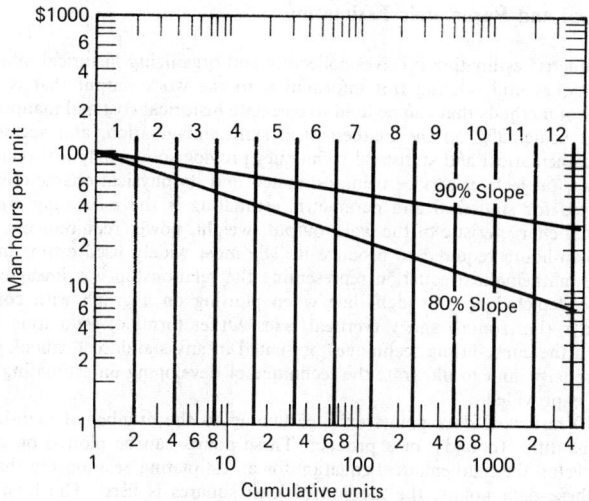

Fig. 46.7 Cumulative units versus man-hours per unit.[1]

Table 46.4 Learning Curve Formulas

Unit curve $Y_X = KX^N$

where Y_X = number of direct labor man-hours required to produce the Xth unit
K = number of direct labor man-hours required to produce the first unit
X = number of units produced
n = slope of curve expressed in positive hundredths (e.g., $n = 0.80$ for an 80% curve) and
$N = \dfrac{\log_{10}n}{\log_{10}2}$

$$\text{Cumulative average curve } V_X \approx \frac{K}{X(1+N)}[(X+0.5)(1+N)-(0.5)(1-N)]$$

where V_X = the cumulative average number of direct labor man-hours required to produce X units

Table 46.5 Man-Loading and Shop-Loading Methods

	Time Increment (Year)						
	1	2	3	4	5	6	7
Man-Loading Method							
Engineers	1	1	1	2	1	0	0
Hours	1,896	1,896	1,896	3,792	1,896	0	0
Technicians	3	4	4	6	2	1	0
Hours	5,688	7,584	7,584	11,376	3,792	1,896	0
Draftsmen	0	0	1	3	6	4	2
Hours	0	0	1,896	5,688	11,376	7,584	3,792
Shop-Loading Method							
Electrical shop (5 workers)	10%	15%	50%	50%	5%	0%	0%
Hours	948	1,422	4,740	4,740	474	0	0
Mechanical shop (10 workers)	5%	5%	10%	80%	60%	10%	5%
Hours	948	948	1,896	15,168	11,376	1,896	948

46.11.8 Statistical and Parametric Estimating

Statistical and parametric estimating involves collecting and organizing historical information through mathematical techniques and relating this information to the work output that is being estimated. There are a number of methods that can be used to correlate historical cost and manpower information; the choice depends principally on the mathematical skills, imagination, and access to data of the estimator. These mathematical and statistical techniques provide some analytical relationship between the process, product, project, or service being estimated and its physical characteristics. The format most commonly used for statistical and parametric estimating is the *estimating relationship*, which relates some physical characteristic of the work output (weight, power requirements, size, or volume) with the cost or man-hours required to produce it. The most widely used estimating relationship is linear. That is, the mathematical equation representing the relationship is a linear equation, and the relationship can be depicted by a straight line when plotting on a graph with conventional linear coordinates for the x (horizontal) and y (vertical) axes. Other forms of estimating relationships can be derived based on the curve-fitting techniques presented in any standard textbook on statistics. For the purpose of simplicity, and to illustrate the technique of developing an estimating relationship, the following example is provided.

Suppose that 10 historical data points exist pertaining to the number of man-hours required to produce various quantities (pounds) of a product. These points can be plotted on a chart as shown in Fig. 46.8. To develop the mathematical equation for an estimating relationship that best represents the aggregate of these data points, the method of least squares is used. The least squares method defines a straight line through the data points such that the sum of the squared deviations or distances from the line is smaller than it would be for any other straight line. The resulting straight line goes through the overall "mean" of the data. When the data points represent a random sample from a larger population of data, the least squares line is a best estimate of the relationship for the total population. The equation of the straight line to be defined is

$$y = mx + b$$

where m is the slope of the line, and b is the point where the line intercepts the y axis.

If n is the number of data points, and x_i and y_i are the coordinates of specific data points, the two equations shown in Table 46.6 can be solved simultaneously to define the line that best fits the data.

The data points in Fig. 46.8 are shown in Table 46.7 along with derived values of the expressions contained in the equations of Table 46.6.

Table 46.6 Sample Estimating Formulas

1. $\sum_{i=1}^{n} y_i = nb + m \sum_{i=1}^{n} x_i$

2. $\sum_{i=1}^{n} y_i x_i = b \sum_{i=1}^{n} x_i + m \sum_{i=1}^{n} (x_i)^2$

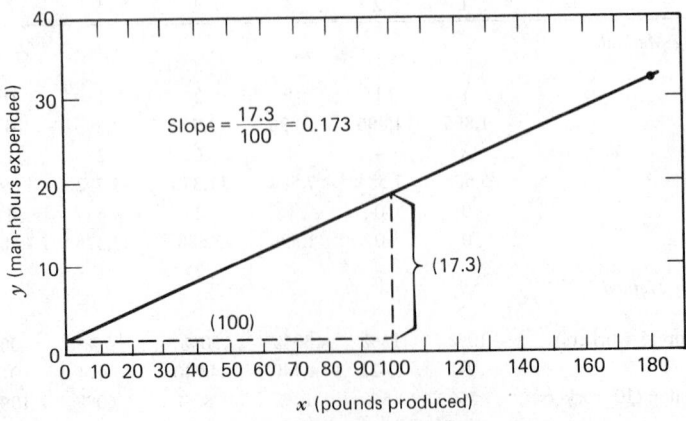

Slope = $\dfrac{17.3}{100}$ = 0.173

(17.3)

(100)

Fig. 46.8 x (pounds produced) versus y (man-hours expended).[1]

Table 46.7 Sample Estimating Chart

x_i (Pounds)	y_i (Man-hours)	$(x_i)^2$	$(x_i y_i)$
20	3.5	400	70
30	7.4	900	222
40	7.1	1,600	284
60	15.6	3,600	936
70	11.1	4,900	777
90	14.9	8,100	1,341
100	23.5	10,000	2,350
120	27.1	14,400	3,252
150	22.1	22,500	3,315
180	32.9	32,400	5,922
$\Sigma x_i = 860$	$\Sigma x_i = 165.2$	$\Sigma (x_i)^2 = 98,800$	$\Sigma (x_i y_i) = 18,469$

Substituting the appropriate values from Table 46.7 into the equations shown in Table 46.6 we get the following two equations:

$$165.2 = 10b + 860m$$
$$18,469 = 860b + 98,800m$$

By multiplying the first equation by -1, and dividing the second equation by 86, we can add the two equations as follows:

$$-165.2 = -10b - 860m$$
$$\underline{214.76 = 10b + 1148.84m}$$
$$49.56 = 288.84m$$

and

$$m = \frac{49.56}{288.84} = 0.172$$

Therefore $b = 1.728$.

The equation for the "best-fit line," then, can be expressed as $y = mx + b$ or man-hours = $(0.172x$ pounds$) + 1.728$.

Estimating relationships, whether based on a linear best-fit curve such as the one shown or on some other statistical averaging technique have certain advantages but certain distinct limitations. They have the advantage of providing a quick estimate even though very little is known about the work output except its physical characteristics. They correlate the present estimate with past history of resource utilization on similar items, and their use simplifies the estimating process. They require the use of statistical or mathematical skills rather than detailed estimating skills, which may be an advantage if detailed estimating skills are not available to the estimating organization.

On the other hand, because of their dependence on past (historical) data, they may erroneously indicate cost trends. Some products, such as mass-produced electronics, are providing more capability per pound, and lower costs per pound, volume, or component count every year. Basing electronics costs on past history may, therefore, result in noncompetitively high estimates. History should not be repeated if that history contains detrimental inefficiencies, duplications, unnecessary redundancies, rework, and overestimates. Often it is difficult to determine what part of historical data should be used to accurately reflect future resource requirements.

Finally, the parametric or statistical estimate, unless used at a very low level in the estimating process, does not provide in-depth visibility, and it does not permit determination of cost effects from subtle changes in schedule, performance, or design requirements. The way to use the statistical or parametric estimate most effectively is to subdivide the work into the smallest possible elements and then to use statistical or parametric methods to derive the resources required for these small elements.

One widely used "cost model" that uses statistical and parametric estimating techniques is the RCA "PRICE" (Programmed Review of Information for Costing and Evaluation) model. This system, designed primarily for high-technology aerospace or electronic products but adaptable to some commercial products, is marketed by RCA Corporation's Government and Commercial Systems Division in Moorestown, New Jersey and uses a centralized computer that can be accessed through commercial telephone lines. All that is required to become an expert user of PRICE is to attend a 2-week training course provided by RCA, to purchase an inexpensive computer terminal, and to pay a nominal user's fee for each access to the computer. The principal benefits of this system are quick access and the ability to make comparative cost assessments of alternative approaches quickly.

It would not be advisable to use this system for competitive cost estimates where no other method of cross-checking the estimate is available. The system uses a number of "complexity" factors that are subjective in nature, and which strongly affect the magnitude of the cost output. This fact, coupled with the previously mentioned limitations of statistical and parametric cost estimating, make the model useful only for rough order-of-magnitude estimates or for cost sensitivity studies.

46.12   DEVELOPING A SCHEDULE

Schedule elements are time-related groupings of work activities that are placed in sequence to accomplish an overall desired objective. Schedule elements for a process could be represented by very small (minutes, hours, or days) time periods. The scheduling of a process is represented by the time the raw material or raw materials take during each step to travel through the process. The schedule for manufacturing a product or delivery of a service is, likewise, a time flow of the various components or actions into a completed item or activity.

A project (the construction or development of a fairly large, complex, or multidisciplinary tangible work output) contains distinct schedule elements called milestones. These milestones are encountered in one form or another in almost all projects: (1) study and analysis; (2) design; (3) procurement of raw materials and purchased parts; (4) fabrication or manufacturing of components and subsystems; (5) assembly of the components and subsystems; (6) testing of the combined system to qualify the unit for operation in its intended environment; (7) acceptance testing, preparation, packaging, shipping, and delivery of the item; and (8) operation of the item.

46.13   TECHNIQUES USED IN SCHEDULE PLANNING

There are a number of analytical techniques used in developing an overall schedule of a work activity that help to ensure the correct allocation and sequencing of schedule elements: precedence and dependency networks, arrow diagrams, critical path bar charts, and program evaluation and review technique (PERT). These techniques use graphical and mathematical methods to develop the best schedule based on sequencing in such a way that each activity is performed only when the required predecessor activities are accomplished. A simple example of how these techniques work is shown in Fig. 46.9. Eight schedule elements have been chosen, the length of each schedule activity has been designated, and a relationship has been established between each activity and its predecessor activity as in Table 46.8.

Note several things about the precedence relationships: (1) some activities can be started before their predecessor activities are completed, (2) some activities must be fully completed before their follow-on activities can be started, and (3) some activities cannot be started until a given number of months after the 100% completion date of a predecessor activity. Once these schedule interrelationships are established, a total program schedule can be laid out by starting either from a selected beginning point and working forward in time until the completion date is reached, or by starting from a desired completion date and working backward in time to derive the required schedule starting date. In many

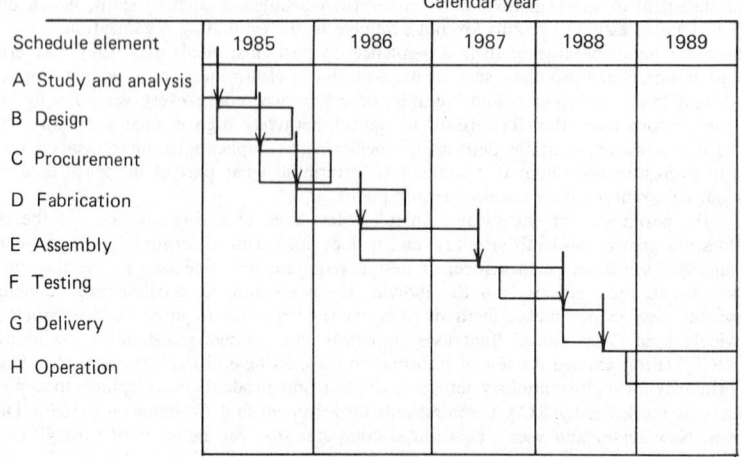

Fig. 46.9   Schedule element.[1]

Table 46.8 Schedule Relationships

Schedule Element	Title of Schedule Element	Time Required for Completion	Percent Completion Required[a]
A	Study and analysis	6 months	$33\frac{1}{3}$
B	Design	8 months	50
C	Procurement	8 months	50
D	Fabrication	12 months	$66\frac{2}{3}$
E	Assembly	12 months	100 plus 2 months
F	Testing	8 months	100
G	Delivery	4 months	100 plus 4 months
H	Operation	36 months	100

[a] Percent completion required before subsequent activity can be accomplished.

instances, you will find that both the start date and completion date are given. In that instance the length of schedule elements and their interrelationships must be established through an iterative process to develop a schedule that accomplishes a job in the required time. If all schedule activities are started as soon as their prerequisites are met, the result is the shortest possible time schedule to perform the work.

Most complex work activities have multiple paths of activity that must be accomplished parallel with each other. The longest of these paths is called a "critical path," and the schedule critical path is developed by connecting all of the schedule activity critical paths. Construction of a schedule such as that shown in Fig. 46.9 brings to light a number of other questions. The first of these is, "How do I establish the length of each activity?" This question strikes at the heart of the overall estimating process itself, since many costs are incurred by the passage of time. Costs of an existing work force, overhead (insurance, rental, and utilities), and material handling and storage continue to pile up in an organization whether there is a productive output or not. Hence it is important to develop the shortest possible overall schedule to accomplish a job and each schedule element in the shortest time and in the most efficient method possible. The length of each schedule activity is established by an analysis of that activity and the human and material resources available and required to accomplish it.

A second question is, "What do I do if there are other influences on the schedule such as availability of facilities, equipment, and manpower?" This is a factor that arises in most estimating situations. There are definite schedule interactions in any multiple-output organization that must be considered in planning a single work activity. Overall corporate planning must take into account these schedule interactions in its own critical path chart to ensure that facilities, manpower, and funds are available to accomplish all work activities in an effective and efficient manner. A final question is, "How do I establish a credible 'percent complete' figure for each predecessor work activity?" This is accomplished by breaking each activity into subactivities. For instance, "design" can be subdivided into conceptual design, preliminary design, and final design. If the start of the procurement activity is to be keyed to the completion of preliminary design, then the time that the preliminary design is complete determines the percentage of time and corresponding design activity that must be completed prior to the initiation of procurement.

46.14 ESTIMATING ENGINEERING ACTIVITIES

46.14.1 Engineering Skill Levels

The National Society of Professional Engineers has developed position descriptions and recommended annual salaries for nine levels of engineers. These skill levels are broad enough in description to cover a wide variety of engineering activities. The principal activities performed by engineers are described in the following paragraphs.

46.14.2 Design

The design activity for any enterprise includes conceptual design, preliminary design, final design, and design changes. The design engineer must design prototypes, components for development or preproduction testing, special test equipment used in development or preproduction testing, support equipment, and production hardware. Since design effort is highly dependent on the specific work output description, design hours must be estimated by a design professional experienced in the area being estimated.

46.14.3 Analysis

Analysis goes hand in hand with design and employs the same general skill level as design engineering. Categories of analysis that support, augment, or precede design are thermal analysis, stress analysis, failure analysis, dynamics analysis, manufacturing analysis, safety analysis, and maintainability analysis. Analysis is estimated by professionals skilled in analytical techniques. Analysis usually includes computer time as well as labor hours.

46.14.4 Drafting

Drafting, or engineering drawing, is one area in the engineering discipline where labor hours can be correlated to a product: the completed engineering drawing. Labor-hour estimates must still be quoted in ranges, however, because the labor hours required for an engineering drawing will vary considerably depending on the complexity of the item being drawn. The drafting times given in Table 46.9 are approximations for Class A drawings of nonelectronic (mechanical) parts where all the design information is available and where the numbers represent "board time," that is, actual time that the draftsman is working on the drawing. A Class A drawing is one that is fully dimensioned and has full supporting documentation. An additional 8 hr per drawing is usually required to obtain approval and sign-offs of stress, thermal, supervisors, and drawing release system personnel. If a "shop drawing" is all that is required (only sufficient information for manufacture of the part with some informal assistance from the designer and/or draftsman), the board time man-hours required would be approximately 50% of that listed in Table 46.9.

46.15 MANUFACTURING/PRODUCTION ENGINEERING

The manufacturing/production engineering activity required to support a work activity is preproduction planning and operations analysis. This differs from the general type of production engineering wherein overall manufacturing techniques, facilities, and processes are developed. Excluded from this categorization is the design time of production engineers who redesign a prototype unit to conform to manufacturing or consumer requirements, as well as time for designing special tooling and special test equipment. A listing of some typical functions of manufacturing engineering follows:

1. Fabrication planning.
   Prepare operations sheets for each part.
   List operational sequence for materials, machines, and functions.
   Recommend standard and special tooling.
   Make up tool order for design and construction of special tooling.
   Develop standard time data for operations sheets.
   Conduct liaison with production and design engineers.
2. Assembly planning.
   Develop operations sheets for each part.
   Build first sample unit.
   Itemize assembly sequence and location of parts.
   Order design and construction of special jigs and fixtures.
   Develop exact component dimensions.
   Build any special manufacturing aids, such as wiring harness jig boards.

Table 46.9 Engineering Drafting Times

Drawing Letter Designation	Size	Approximate Board-Time Hours for Drafting of Class A Drawings (hr)
A	8½ × 11	1–4
B	11 × 17	2–8
C	17 × 22	4–12
D	22 × 34	8–16
E and F	34 × 44 and 28 × 40	16–40
J	34 × 48 and larger	40–80

Apply standard time data to operations sheet.

Balance time cycles of final assembly line workstations.

Effect liaison with production and design engineers.

Set up material and layout of each workstation in accordance with operations sheet.

Instruct mechanics in construction of the first unit.

3. Test planning.

Determine overall test method to meet performance and acceptance specifications.

Break total test effort into positions by function and desired time cycle.

Prepare test equipment list and schematic for each position.

Prepare test equipment design order for design and construction of special-purpose test fixtures.

Prepare a step-by-step procedure for each position.

Effect liaison with production and design engineers.

Set up test positions and check out.

Instruct test operator on first unit.

4. Sustaining manufacturing engineering.

Debug, as required, engineering design data.

Debug, as required, manufacturing methods and processes.

Recommend more efficient manufacturing methods throughout the life of production.

The following formula may be helpful in deriving manufacturing engineering man-hour estimates for high production rates:

1. Total fabrication and assembly man-hours, divided by the number of units to be produced, multiplied by 20 gives manufacturing engineering start-up costs.

2. For sustaining manufacturing engineering, take the unit fabrication and assembly man-hours, multiply by 0.07. (These factors are suggested for quantities up to 100 units.)

46.15.1   Engineering Documentation

A large part of an engineer's time is spent in writing specifications, reports, manuals, handbooks, and engineering orders. The complexity of the engineering activity and the specific document requirements are important determining factors in estimating the engineering labor hours required to prepare engineering documentation.

The hours required for engineering documentation (technical reports, specifications, and technical manuals) will vary considerably depending on the complexity of the work output; however, average labor hours for origination and revision of engineering documentation have been derived based on experience, and these figures can be used as average labor hours per page of documentation. (See Tables 46.10 and 46.11.)

46.16   ESTIMATING MANUFACTURING/PRODUCTION AND ASSEMBLY ACTIVITIES

46.16.1   The Process Plan

A key to successful estimating of manufacturing activities is the process plan. A process plan is a listing of all operations that must be performed to manufacture a product or to complete a project, along with the labor hours required to perform each operation. The process plan is usually prepared by an experienced foreman, engineer, or technician who knows the company's equipment, personnel, and capabilities, or by a process planning department chartered to do all of the process planning. The process planner envisions the equipment, workstation, and environment; estimates the number of persons required, and estimates how long it will take to perform each step. From this information he or she derives the labor hours required. Process steps are numbered and space is left between operations listed to allow easy insertion or omission of operations or activities as the process is modified.

A typical process plan for a welded cylinder assembly is given in Table 46.12.

The process plan is used not only to plan and estimate a manufacturing or construction process, but also it is often used as part of the manufacturing or construction work order itself. As such, it shows the shop or construction personnel each step to take in the completion of the work activity. Fabrication of items from metals, plastics, or other materials in a shop is usually called "manufacturing," while fabrication of buildings, structures, bridges, dams, and public facilities on site is usually called construction. Different types of standards and estimating factors are used for each of these categories of work. First, we cover manufacturing activities.

Table 46.10   New Documentation

Function	Man-Hours per Page
Research, liaison, technical writing, editing, and supervision	5.7
Typing and proofreading	0.6
Illustrations	4.3
Engineering	0.7
Coordination	0.2
Total[a]	11.5

[a]A range of 8–12 man-hours per page can be used.

Table 46.11   Revised Documentation

Function	Man-Hours per Page
Research, liaison, technical writing, editing, and supervision	4.00
Typing and proofreading	0.60
Illustrations	0.75
Engineering	0.60
Coordination	0.20
Total[a]	6.15

[a]A range of 4–8 man-hours per page can be used.

46.16.2   Developing Labor-Hour Estimates Based on Time Standards

The process plan described can be based on the judgment of the estimator using man-loading or shop-loading techniques, or the process plan can be developed using time standards developed by industrial engineering time or job analysis techniques. There are several organizations that specialize in study, publication, and training in the development and use of time standards. Predominant among these is the MTM Association for Standards and Research in Fair Lawn, New Jersey.

Detailed time standards developed through job analysis usually involve the subdivision of a task down into each human body movement that is required to perform a job. These movements, such as "pick-up, place, or position," are accomplished in very small increments, usually 0.0001–0.001 minute. Because it is impractical for the estimator to break work activities into increments this small, aggregate time standards rather than detailed standards are used. These aggregate standards are developed through a synthesis of a number of small body movements into an overall body motion or work increment that ranges in time from 0.05 to 0.50 minute. The use of aggregate standards takes more estimating time to accomplish than the man-loading or shop-loading method, but there are certain advantages to using time standards. Some of these are (1) they can be used by persons that do not have an intimate familiarity with the job being estimated; (2) they promote consistency between estimates; and (3) resulting estimates are less likely to vary with the estimator because they are based on work content rather than judgment or opinion. Time standards are used extensively in estimating labor hours required for manufacturing activities.

46.17   MANUFACTURING ACTIVITIES

Manufacturing activities are broken into various categories of effort such as metal working and forming; welding, brazing, and soldering; application of fasteners; plating, printing, surface treating, and heat treating; and manufacturing of electronic components (a special category). The most common method of estimating the time and cost required for manufacturing activities is the industrial engineering approach whereby standards or target values are established for various operations. The term standards is used to indicate standard time data. All possible elements of work are measured, assigned a standard time for performance, and documented. When a particular job is to estimated, all of the applicable standards for all related operations are added together to determine the total time.

The use of standards produces more accurate and more easily justifiable estimates. Standards also promote consistency between estimates as well as among estimators. Where standards are used, personal experience is desirable or beneficial but not mandatory. Standards have been developed over a number of years through the use of time studies and synthesis of methods analysis. They are based on the level of efficiency that could be attained by a job shop producing up to 1000 units of any specific

Table 46.12 Process Plan[a]

Operation Number	Labor Hours	Description
010	—	Receive and inspect material (skins and forgings)
020	24	Roll form skin segments
030	60	Mask and chem-mill recessed pattern in skins
040	—	Inspect
050	36	Trim to design dimension and prepare in welding skin segments into cylinders (two)
060	16	Locate segments on automatic seam welder tooling fixture and weld per specification (longitudinal weld)
070	2	Remove from automatic welding fixture
080	18	Shave welds on inside diameter
090	16	Establish trim lines (surface plate)
100	18	Install in special fixture and trim to length
110	8	Remove from special fixture
120	56	Install center mandrel—center ring, forward and aft sections (cylinders)—forward and aft mandrel—forward and aft rings—and complete special feature setup
130	—	Inspect
140	24	Butt weld (four places)
150	8	Remove from special fixture and remove mandrels
160	59	Radiograph and dye penetrant inspect
170	—	Inspect dimensionally
180	6	Reinstall mandrels in preparation for final machining
190	14	Finish OD—aft
	10	Finish OD—center
	224	Finish OD—forward
200	40	Program for forward ring
220	30	Handwork (three rings)
230	2	Reinstall cylinder assembly with mandrels still in place or on the special fixture
240	16	Clock and drill index holes
250	—	Inspect
260	8	Remove cylinder from special fixture—remove mandrel
270	1	Install in holding cradle
280	70	Locate drill jig on forward end and hand drill leak check vein (drill and tap), and hand drill hole pattern
290	64	Locate drill jig on aft ring and hand drill hole pattern
300	—	Inspect forward and aft rings
310	8	Install protective covers on each end of cylinder
320	—	Transfer to surface treat
340	24	Remove covers and alodine
350	—	Inspect
360	8	Reinstall protective covers and return to assembly area

[a]Drawing No. D21216; Part No. 1D21254; Title: Cylinder Assembly (Welded).

work output. Standards are actually synoptical values of more detailed times. They are adaptations, extracts, or benchmark time values for each type of operation. The loss of accuracy occasioned by summarization and/or averaging is acceptable when the total time for a system is being developed. If standard values are used with judgment and interpolations for varying stock sizes, reasonably accurate results can be obtained.

Machining operations make up a large part of the manufacturing costs of many products and projects. Machining operations are usually divided into setup times and run times. Setup time is the time required to establish and adjust the tooling, to set speeds and feeds on the metal-removal machine, and to program for the manufacture of one or more identical or similar parts. Run time is the time required to complete each part. It consists of certain fixed positioning times for each item being machined as well as the actual metal-removal and cleanup time for each item. Values listed are for "soft" and "hard" materials. Soft values are for aluminum, magnesium, and plastics. Hard values are for stainless steel, tool steel, and beryllium. Between these two times would be standard values for brass, bronze, and medium steel.

46.18   IN-PROCESS INSPECTION

The amount of in-process inspection performed on any process, product, project, or service will depend on the cost of possible scrappage of the item as well as the degree of reliability required for the final work output. In high-rate production of relatively inexpensive items, it is often economically desirable to forego in-process inspection entirely in favor of scrapping any parts that fail a simple go, no-go inspection at the end of the production line. On the other hand, expensive and sophisticated precision-manufactured parts may require nearly 100% inspection. A good rule of thumb is to add 10% of the manufacturing and assembly hours for in-process inspection. This in-process inspection does not include the in-process testing that will be covered in the following paragraphs.

46.19   TESTING

Testing usually falls into three categories: (1) development testing, (2) qualification testing, and (3) production acceptance testing.

Rules of thumb are difficult to come by for estimating development testing because testing varies with the complexity, uncertainty, and technological content of the work activity. The best way to estimate the cost of development testing is to produce a detailed test plan for the specific project and to cost each element of this test plan separately, being careful to consider all skills, facilities, equipment, and material needed in the development test program.

Qualification testing is required in most commercial products and on all military or space projects to demonstrate adequately that the article will operate or serve its intended purpose in environments far more severe than those intended for its actual use. Automobile crash tests are an example. Military products must often undergo severe and prolonged tests under high shock, thermal, and vibration loads as well as heat, humidity, cold, and salt spray environments. These tests must be meticulously planned and scheduled before a reasonable estimate of their costs can be generated.

Receiving inspection, production testing, and acceptance testing can be estimated using experience factors and ratios available from previous like work activities. Receiving tests are tests performed on purchased components, parts, and/or subassemblies prior to acceptance by the receiving department. Production tests are tests of subassemblies, units, subsystems, and systems during and after assembly. Experience has shown, generally, that test labor varies directly with the amount of fabrication and assembly labor. The ratio of test labor to other production labor will depend on the complexity of the item being tested. Table 46.13 gives the test labor percent of direct fabrication and assembly labor for simple, average, and complex items.

Table 46.13   Test Estimating Ratios

	Percent of Direct Labor		
	Simple	Average	Complex
Fabrication and Assembly Labor Base			
Receiving test	1	2	4
Production test	9	18	36
Total	10	20	40
Assembly Labor Base			
Receiving test	2	3	7
Production test	15	32	63
Total	17	35	70

46.19.1   Special Tooling and Test Equipment

Special-purpose tooling and special-purpose test equipment are important items of cost because they are used only for a particular job; therefore, that job must bear the full cost of the tool or test fixture. In contrast to the special items, general-purpose tooling or test equipment is purchased as capital equipment and costs are spread over many jobs. Estimates for tooling and test equipment are included in overall manufacturing startup ratios shown in Table 46.14. Under "degree of precision and complexity," "high," means high-precision multidisciplinary systems, products, or subsystems; "medium" means moderately complex subsystems or components; and "low" means simple, straightforward designs of components or individual parts. Design hours required for test equipment are shown in Table 46.15.

46.20   COMPUTER SOFTWARE COST ESTIMATING

Because of the increasing number and types of computers and computer languages, it is difficult to generate overall ground rules or rules of thumb for computer software cost estimating. Productivity in computer programming is greatly affected by the skill and competence of the computer analyst or programmer. A general rule of thumb, however, is that it will take about 3 hr per program statement

Table 46.14   Manufacturing Startup Ratios

Cost Element	Degree of Precision and Complexity	Percent of Recurring Manufacturing Costs, Lot Quantity			
		10	100	1000	10,000
Production planning	High	20	6	1.7	0.5
	Medium	10	3	0.8	0.25
	Low	5	1.5	0.4	0.12
Special tooling	High	10	6	3.5	2
	Medium	5	3	2	1
	Low	3	1.5	1	—
Special test equipment	High	10	6	3.5	2
	Medium	6	3	2	1
	Low	3	1.5	1	0.5
Composite total	High	40	18	8.7	4.5
	Medium	21	9	4.8	2.25
	Low	11	4.5	2.4	1.12

Table 46.15   Design Hours for Test Equipment

Type Design	Hours/ Square Foot	Standard Drawing Size	Square Feet/ Drawing	Hours/ Drawing
Original concept	15	C	2.5	38
		D	5.0	75
		H	9.0	135
		J	11.0	165
Layout	10	B	1.0	10
		C	2.5	25
		D	5.0	50
		H	9.0	90
		J	11.0	110
Detail or copy	3	A	0.7	2.1
		B	1.0	3.0
		C	2.5	7.5
		D	5.0	15.0
		H	9.0	27.0
		J	11.0	33.0

or line of code to perform the requirements analysis, program design, coding, and software verification for a program written in machine language or assembly language. For a program written in a higher-order language (BASIC, FORTRAN, COBOL, or PASCAL), this figure is approximately 1.87 hr per program statement. In addition to these direct man-hours, approximately 10% of the direct hours should be added for program documentation and software program management. Complicated flight software for aircraft and space systems is subjected to design review and testing in simulations and on the actual flight computer hardware. A software critical design review is usually conducted about 43% the way through the program; an integrated systems test is performed at the 67% completion mark; prototype testing is done at 80% completion; installation with the hardware is started with about 7% of the time remaining (at the 93% completion point).

46.21 LABOR ALLOWANCES

"Standard times" assume that the workers are well trained and experienced in their jobs, that they apply themselves to the job 100% of the time, that they never make a mistake, take a break, lose efficiency, or deviate from the task for any reason. This, of course, is an unreasonable assumption because there are legitimate and numerous unplanned work interruptions that occur with regularity in any work activity. Therefore, labor allowances must be added to any estimate that is made up of an accumulation of standard times. These labor allowances can accumulate to a factor of 1.5–2.5. The total standard time for a given work activity, depending on the overall inherent efficiency of the shop, equipment, and personnel, will depend on the nature of the task. Labor allowances are made up of a number of factors that are described in the following section.

46.21.1 Variance from Measured Labor Hours

Standard hours vary from actual measured labor hours because workers often deviate from the standard method or technique used or planned for a given operation. This deviation can be caused by a number of factors ranging from the training, motivation, or disposition of the operator to the use of faulty tools, fixtures, or machines. Sometimes shortages of materials or lack of adequate supervision are causes of deviations from standard values. These variances can add 5–20% to standard time values.

46.21.2 Personal, Fatigue, and Delay (PFD) Time

Personal times are for personal activities such as coffee breaks, trips to the restroom or water fountain, unforeseen interruptions, or emergency telephone calls. Fatigue time is allocated because of the inability of a worker to produce at the same pace all day. Operator efficiency decreases as the job time increases. Delays include unavoidable delays caused by the need for obtaining supervisory instructions, equipment breakdown, power outages, or operator illness. PFD time can add 10–20% to standard time values.

46.21.3 Tooling and Equipment Maintenance

Although normal or routine equipment maintenance can be done during other than operating shifts, there is usually some operator-performed machine maintenance activity that must be performed during the machine duty cycle. These activities include adjusting tools, sharpening tools, and periodically cleaning and oiling machines. In electroplating and processing operations, the operator maintains solutions and compounds, and handles and maintains racks and fixtures. Tooling and equipment maintenance can account for 5–12% of standard time values.

46.21.4 Normal Rework and Repair

The overall direct labor hours derived from the application of the preceding three allowance factors to standard times must be increased by additional amounts to account for normal rework and repair. Labor values must be allocated for rework of defective purchased materials, rework of in-process rejects, final test rejects, and addition of minor engineering changes. Units damaged on receipt or during handling must also be repaired. This factor can add 10–20% direct labor hours to those previously estimated.

46.21.5 Engineering Change Allowance

For projects where design stability is poor, where production is initiated prior to final design release, and where field testing is being performed concurrently with production, an engineering change allowance should be added of up to 10% of direct labor hours. Change allowances vary widely for different types of work activities. Even fairly well-defined projects, however, should contain a change allowance.

46.21.6   Engineering Prototype Allowance

The labor hours required to produce an engineering prototype are greater than those required to produce the first production model. Reworks are more frequent, and work is performed from sketches or unreleased drawings rather than production drawings. An increase over first production unit labor of 15–25% should be included for each engineering prototype.

46.21.7   Design Growth Allowance

Where estimates are based on incomplete drawings, or where concepts only or early breadboards only are available prior to the development of a cost estimate, a design growth allowance is added to all other direct labor costs. This design growth allowance is calculated by subtracting the percentage of design completion from 100% as shown in the following tabulation:

Desirable Design Completion Percentage	Percentage of Design Completed	Design Growth Allowance
100%	50%	50%
100%	75%	25%
100%	80%	20%
100%	90%	10%
100%	100%	0%

46.21.8   Cost Growth Allowances

Occasionally a cost estimate will warrant the addition of allowances for cost growth. Cost growth allowances are best added at the lowest level of a cost estimate rather than at the top levels. These allowances include reserves for possible misfortunes, natural disasters, strikes, and other unforeseen circumstances. Reserves should not be used to account for normal design growth. Care should be taken in using reserves in a cost estimate because they are usually the first cost elements that come under attack for removal from the cost estimate or budget. Remember, cost growth with an incomplete design is a certainty, not a reserve or contingency! Defend your cost growth allowance, but be prepared to relinquish your reserve if necessary.

46.22   ESTIMATING SUPERVISION, MANAGEMENT, AND OTHER DIRECT CHARGES

Direct supervision costs will vary with the task and company organization. Management studies have shown that the span of control of a supervisor over a complex activity should not exceed 12 workers. For simple activities, the ratio of supervisors to employees can go down. But the 1 to 12 ratio (8.3%) will usually yield best results. Project management for a complex project can add an additional 10–14%. Other direct charges are those attributable to the project being accomplished but not included in direct labor or direct materials. Transportation, training, and reproduction costs, as well as special service or support contracts and consultants, are included in the category of "other direct costs."

Two cost elements of "other direct costs" that are becoming increasingly prominent are travel and transportation costs. Air, bus, and rail fares as well as private conveyance costs are increasing at a rate higher than the general inflation rate because rapidly increasing fuel costs comprise a large part of the cost per passenger mile. For this reason a frequent check on public and private conveyance rates and costs is mandatory. Most companies provide a private vehicle mileage allowance for employees who use their own vehicles in the conduct of company business. Rates differ and depend on whether the private conveyance is being used principally for the benefit of the company or principally for the convenience of the traveler. Regardless of which rate is used, the mileage allowance must be periodically updated to keep pace with actual costs. Many companies purchase or lease vehicles to be used by their employees on official business, and sometimes personal travel.

Per diem travel allowances or reimbursement for lodging, meals, and miscellaneous expenses must also be included in overall travel budgets. These reimbursable expenses include costs of a motel or hotel room; food, tips, and taxes; local transportation and communication; and other costs such as laundry, mailing costs, and on-site clerical services. Transportation costs include the transport of equipment, supplies, and products, as well as personnel, and can include packaging, handling, shipping, postage, and insurance charges.

46.23   THE USE OF "FACTORS" IN ESTIMATING

Although the use of "factors" in estimating is not recommended, the practice is becoming increasingly common, particularly in high-technology work activities and work outputs. One company uses an "allocation factor," which allocates miscellaneous labor-oriented functions to specific functions such as fabrication or assembly. This company adds 14.4% to fabrication hours and 4.1% to assembly hours to cover miscellaneous labor-hour expenditures associated with these two functions. It is also common to estimate hours for planning, tooling, quality and inspection, production support, and sustaining engineering based on percentages of manufacturing and/or assembly hours. Tooling materials and computer tapes are sometimes estimated based on so much cost per tooling hour, and miscellaneous shop hardware (units, bolts, fasteners, cleaning supplies, etc.), otherwise known as "panstock," is estimated at a cost per manufacturing hour.

The disadvantage of the use of such factors is that inefficiencies can become embedded in the factored allowances and eventually cause cost growth. A much better method of estimating the man-hours and materials required to accomplish these other direct activities is to determine the specific tasks and materials required to perform the job by man-loading, shop-loading, or process planning methods. When the materials, labor hours, and other direct costs have been estimated, the basic direct resources required to do the job have been identified. The next step in the estimating process is to determine labor rates, indirect costs, administrative expenses, and profit.

REFERENCE

1. R. D. Stewart, *Cost Estimating,* Wiley-Interscience, New York, 1982.

CHAPTER 47

SAFETY ENGINEERING

JACK B. RE VELLE

Hughes Aircraft Company
Fullerton, California

47.1 INTRODUCTION

47.1.1 Background

More than ever before, engineers are aware of and concerned with employee safety and health. The necessity for this involvement was accelerated with the passage of the OSHAct in 1970, but much of what has occurred since that time would have happened whether or not the OSHAct had become the law.

As workplace environments become more technologically complex, the necessity for protecting the work force from safety and health hazards continues to grow. Typical workplace operations from which workers should be protected are presented in Table 47.1. Whether they should be protected through the use of personal protective equipment, engineering controls, administrative controls, or a combination of these approaches, one fact is clear; it makes good sense to ensure that they receive the most cost-effective protection available. Arguments in support of engineering controls over personal protective equipment and vice versa are found everywhere in the current literature. Some of the most persuasive discussions are included in this chapter.

47.1.2 Employee Needs and Expectations

In 1981 Re Velle and Boulton asked the question, "Who cares about the safety of the worker on the job?" in their award-winning two-part article in *Professional Safety*, "Worker Attitudes and Perceptions of Safety." The purpose of their study was to learn about worker attitudes and perceptions of safety. To accomplish this objective, they established the following working definition:

WORKER ATTITUDES AND PERCEPTIONS As a result of continuing observation, an awareness is developed, as is a tendency to behave in a particular way regarding safety.

To learn about these beliefs and behaviors, they inquired to find out:

1. Do workers think about safety?
2. What do they think about safety in regard to:
   (a) Government involvement in their workplace safety.
   (b) Company practices in training and hazard prevention.
   (c) Management attitudes as perceived by the workers.
   (d) Coworkers' concern for themselves and others.
   (e) Their own safety on the job.
3. What do workers think should be done, and by whom, to improve safety in their workplace?

The major findings of the Re Velle–Boulton study are summarized here.*

Half the workers think that government involvement in workplace safety is about right; almost one-fourth think more intervention is needed in such areas as more frequent inspections, stricter regulations, monitoring, and control.

Workers in large companies expect more from their employers in providing a safe workplace than workers in small companies. Specifically, they want better safety programs, more safety training, better equipment and maintenance of equipment, more safety inspections and enforcement of safety regulations, and provision of more personal protective equipment.

Supervisors who talk to their employees about safety and are perceived by them to be serious are also seen as being alert for safety hazards and representative of their company's attitude.

Coworkers are perceived by other employees to care for their own safety and for the safety of others.

Only 20% of the surveyed workers consider themselves to have received adequate safety training. But more than three-fourths of them feel comfortable with their knowledge to protect themselves on the job.

Men are almost twice as likely to wear needed personal protective equipment as women.

Half the individuals responding said they would correct a hazardous condition if they saw it.

Employees who have had no safety training experience almost twice as many on-the-job accidents as their fellow workers who have received such training.

Workers who experienced accidents were generally candid and analytical in accepting responsibility for their part in the accident; and 85% said their accidents could have been prevented.

---

* Reprinted with permission from the January 1982 issue of *Professional Safety*, official publication of the American Society of Safety Engineers.

Table 47.1 Operations Requiring Engineering Controls and/or Personal Protective Equipment

Acidic/basic process and treatments	Grinding
Biological agent processes and treatments	Hoisting
Blasting	Jointing
Boiler/pressure vessel usage	Machinery (mills, lathes, presses)
Burning	Mixing
Casting	Painting
Chemical agent processes and treatments	Radioactive source processes and treatments
Climbing	Sanding
Compressed air/gas usage	Sawing
Cutting	Shearing
Digging	Soldering
Drilling	Spraying
Electrical/electronic assembly and fabrication	Toxic vapor, gas, and mists and dust exposure
Electrical tool usage	Welding
Flammable/combustible/toxic liquid usage	Woodworking

The remainder of this chapter addresses those topics and provides that information which engineering practitioners require to professionally perform their responsibilities with respect to the safety of the work force.

47.2 GOVERNMENT REGULATORY REQUIREMENTS*

Two relatively new agencies of the federal government enforce three laws that impact many of the operational and financial decisions of American businesses, large and small. The Environmental Protection Agency (EPA) has responsibility for administering the Toxic Substances Control Act (TSCA) and the Resource Conservation and Recovery Act (RCRA), both initially enforced in 1976. The Occupational Safety and Heath Act (OSHAct) of 1970 is enforced by the Occupational Safety and Health Administration (OSHA), a part of the Department of Labor. This section addresses the regulatory demands of these federal statutes from the perspective of whether to install engineering controls that would enable companies to meet these standards or simply to discontinue certain operations altogether, that is, can they justify the associated costs of regulatory compliance.

47.2.1 Environmental Protection Agency (EPA)

Toxic Substances Control Act (TSCA)

Until the TSCA, the federal government was not empowered to prevent chemical hazards to health and the environment by banning or limiting chemical substances at a germinal, premarket stage. Through the TSCA of 1975, production workers, consumers, indeed every American, would be protected by an equitably administered early warning system controlled by the EPA. This broad law authorizes the EPA Administrator to issue rules to prohibit or limit the manufacturing, processing, or distribution of any chemical substance or mixture that "may present an unreasonable risk of injury to health or the environment." The EPA Administrator may require testing—at a manufacturer's or processor's expense—of a substance after finding that:

- The substance may present an unreasonable risk to health or the environment.
- There may be a substantial human or environmental exposure to the substance.
- Insufficient data and experience exist for judging a substance's health and environmental effects.
- Testing is necessary to develop such data.

This legislation is designed to cope with hazardous chemicals like kepone, vinyl chloride, asbestos, fluorocarbon compounds (Freons), and polychlorinated biphenyls (PCBs).

* "Engineering Controls: A Comprehensive Overview," by Jack B. Re Velle. Used by permission of The Merritt Company, Publisher, from T. S. Ferry, *Safety Management Planning,* copyright © 1982, The Merritt Company, Santa Monica, CA 90406.

Resource Conservation and Recovery Act (RCRA)

Enacted in 1976 as an amendment to the Solid Waste Disposal Act, the RCRA sets up a "cradle-to-grave" regulatory mechanism, that is, a tracking system for such wastes from the moment they are generated to their final disposal in an environmentally safe manner. The act charges the EPA with the development of criteria for identifying hazardous wastes, creating a manifest system for tracking wastes through final disposal, and setting up a permit system based on performance and management standards for generators, transporters, owners, and operators of waste treatment, storage, and disposal facilities. It is expected that the RCRA will be a strong force for innovation and eventually lead to a broad rethinking of chemical processes, that is, to look at hazardous waste disposal *not* just in terms of immediate costs, but rather with respect to life-cycle costs.

47.2.2   Occupational Safety and Health Administration (OSHA)*

The Occupational Safety and Health Act (OSHAct), a federal law that became effective on April 28, 1971, is intended to pull together all federal and state occupational safety and health-enforcement efforts under a federal program designed to establish uniform codes, standards, and regulations. The expressed purpose of the act is "to assure, as far as possible, every working woman and man in the Nation safe and healthful working conditions, and to preserve our human resources." To accomplish this purpose, the promulgation and enforcement of safety and health standards is provided for, as well as research, information, education, and training in occupational safety and health.

Perhaps no single piece of federal legislation has been more praised and, conversely, more criticized than the OSHAct, which basically is a law requiring virtually all employers to ensure that their operations are free of hazards to workers.

Occupational Safety and Health Standards

When Congress passed the OSHAct of 1970, it authorized the promulgation, without further public comment or hearings, of groups of already codified standards. The initial set of standards of the act (Part 1910, published in the *Federal Register* on May 29, 1971) thus consisted in part of standards that already had the force of law, such as those issued by authority of the Walsh–Healey Act, the Construction Safety Act, and the 1958 amendments to the Longshoremen's and Harbor Workers' Compensation Act. A great number of the adopted standards, however, derived from voluntary national consensus standards previously prepared by groups such as the American National Standards Institute (ANSI) and the National Fire Protection Association (NFPA).

The OSHAct defines the term "occupational safety and health standard" as meaning "a standard which requires conditions or the adoption or use of one or more practices, means, methods, operations or processes, reasonably necessary or appropriate to provide safe or healthful employment and places of employment." Standards contained in Part 1910* are applicable to general industry. Those contained in Part 1926 are applicable to the construction industry; and standards applicable to ship repairing, shipbuilding, and longshoring are contained in Parts 1915–1918. These OSHA standards fall into the following four categories, with examples for each type:

1. *Specification Standards.*   Standards that give specific proportions, locations, and warning symbols for signs that must be displayed.
2. *Performance Standards.*   Standards that require achievement of, or within, specific minimum or maximum criteria.
3. *Particular Standards (Vertical).*   Standards that apply to particular industries, with specifications that relate to the individual operations.
4. *General Standards (Horizontal).*   Standards that can apply to any workplace and relate to broad areas (environmental control, walking surfaces, exits, illumination, etc.).

The Occupational Health and Safety Administration is authorized to promulgate, modify, or revoke occupational safety and health standards. It also has the authority to promulgate emergency temporary standards where it is found that employees are exposed to grave danger. Emergency temporary standards can take effect immediately on publication in the *Federal Register*. Such standards remain in effect until superseded by a standard promulgated under the procedures prescribed by the OSHAct—notice of proposed rule in the *Federal Register,* invitation to interested persons to submit their views, and a public hearing if required.

---

* R. De Reamer, *Modern Safety and Health Technology,* copyright © 1980. Reprinted by permission of Wiley, New York.
* The Occupation Safety and Health Standards, Title 29, CFR Chapter XVIII, Parts 1910, 1926, and 1915–1918 are available at all OSHA regional and area offices.

Required Notices and Records

During an inspection the compliance officer will ascertain whether the employer has:

- Posted notice informing employees of their rights under the OSHAct (Job Safety and Health Protection, OSHAct poster).
- Maintained log of recordable injuries and illnesses (OSHA Form No. 200, Log and Summary of Occupational Injuries and Illnesses).
- Maintained the Supplementary Record of Occupational Injuries and Illnesses (OSHA Form No. 101).
- Annually posted the Summary of Occupational Injuries and Illnesses (OSHA Form No. 200). This form must be posted no later than February 1 and must remain in place until March 1.
- Made a copy of the OSHAct and OSHA safety and health standards available to employees on request.
- Posted boiler inspection certificates, boiler licenses, elevator inspection certificates, and so on.

47.2.3 State-Operated Compliance Programs

The OSHAct encourages each state to assume the fullest responsibility for the administration and enforcement of occupational safety and health programs. For example, federal law permits any state to assert jurisdiction, under state law, over any occupational or health standard not covered by a federal standard.

In addition, any state may assume responsibility for the development and enforcement of its own occupational safety and health standards for those areas now covered by federal standards. However, the state must first submit a plan for approval by the Labor Department's Occupational Safety and Health Administration. Many states have done so.

Certain states are now operating under an approved state plan. These states may have adopted the existing federal standards or may have developed their own standards. Some states also have changed the required poster. You need to know whether you are covered by an OSHA-approved state plan operation, or are subject to the federal program, in order to determine which set of standards and regulations (federal or state) apply to you. The easiest way to determine this is to call the nearest OSHA Area Office.

If you are subject to state enforcement, the OSHA Area Office will explain this, explain whether the state is using the federal standards, and provide you with information on the poster and on the OSHA recordkeeping requirements. After that, the OSHA Area Office will refer you to the appropriate state government office for further assistance.

This assistance also may include free on-site consultation visits. If you are subject to state enforcement, you should take advantage of this service.

For your information, the following are operating under OSHA-approved state plans, as of May 1, 1985:

Alaska	New Mexico
Arizona	North Carolina
California	Oregon
Hawaii	Puerto Rico
Indiana	South Carolina
Iowa	Tennessee
Kentucky	Utah (no consultation)
Maryland	Vermont
Michigan	Virginia
Minnesota	Virgin Islands
Nevada	Washington
	Wyoming

47.3 SYSTEM SAFETY*

System safety is when situations having accident potential are examined in a step-by-step cause–effect manner, tracing a logical progression of events from start to finish. System safety techniques can

* R. De Reamer, *Modern Safety and Health Technology,* copyright © 1980. Reprinted by permission of Wiley, New York.

provide meaningful predictions of the frequency and severity of accidents. However, their greatest asset is the ability to identify many accident situations in the system that would have been missed if less detailed methods had been used.

47.3.1 Methods of Analysis

A system cannot be understood simply in terms of its individual elements or component parts. If an operation of a system is to be effective, all parts must interact in a predictable and a measurable manner, within specific performance limits and operational design constraints.

In analyzing any system, three basic components must be considered: (1) the equipment (or machines); (2) the operators and supporting personnel (maintenance technicians, material handlers, inspectors, etc.); and (3) the environment in which both workers and machines are performing their assigned functions. Several analysis methods are available:

- *Gross-Hazard Analysis.* Performed early in design; considers overall system as well as individual components; it is called "gross" because it is the initial safety study undertaken.
- *Classification of Hazards.* Identifies types of hazards disclosed in the gross-hazard analysis, and classifies them according to potential severity (Would defect or failure be catastrophic?); indicates actions and/or precautions necessary to reduce hazards. May involve preparation of manuals and training procedures.
- *Failure Modes and Effects.* Considers kinds of failures that might occur and their effect on the overall product or system. Example: effect on system that will result from failure of single component (e.g., a resistor or hydraulic valve).
- *Hazard-Criticality Ranking.* Determines statistical, or quantitative, probability of hazard occurrence; ranking of hazards in the order of "most critical" to "least critical."
- *Fault-Tree Analysis.* Traces probable hazard progression. Example: If failure occurs in one component or part of the system, will fire result? Will it cause a failure in some other component?
- *Energy-Transfer Analysis.* Determines interchange of energy that occurs during a catastrophic accident or failure. Analysis is based on the various energy inputs to the product or system and how these inputs will react in event of failure or catastrophic accident.
- *Catastrophe Analysis.* Identifies failure modes that would create a catastrophic accident.
- *System-Subsystem Integration.* Involves detailed analysis of interfaces, primarily between systems.
- *Maintenance-Hazard Analysis.* Evaluates performance of the system from a maintenance standpoint. Will it be hazardous to service and maintain? Will maintenance procedures be apt to create new hazards in the system?
- *Human-Error Analysis.* Defines skills required for operation and maintenance. Considers failure modes initiated by human error and how they would affect the system. The question of whether special training is necessary should be a major consideration in each step.
- *Transportation-Hazard Analysis.* Determines hazards to shippers, handlers, and bystanders. Also considers what hazards may be "created" in the system during shipping and handling.

There are other quantitative methods that have successfully been used to recommend a decision to adopt engineering controls, personal protective equipment, or some combination. Some of these methods are*:

- *Expected Outcome Approach.* Since safety alternatives involve accident costs that occur more or less randomly according to probabilities which might be estimated, a valuable way to perform needed economic analyses for such alternatives is to calculate expected outcomes.
- *Decision Analysis Approach.* A recent extension of systems analysis, this approach provides useful techniques for transforming complex decision problems into a sequentially oriented series of smaller, simpler problems. This means that a decision-maker can select reasoned choices that will be consistent with his or her perceptions about the uncertainties involved in a particular problem together with his or her fundamental attitudes toward risk-taking.
- *Mathematical Modeling.* Usually identified as an "operations research" approach, there are numerous mathematical models that have demonstrated potential for providing powerful analysis insights into safety problems. These include dynamic programming, inventory-type modeling, linear programming, queue-type modeling, and Monte Carlo simulation.

* J. B. Re Velle, *Engineering Controls: A Comprehensive Overview.* Used by permission of The Merritt Company, Publisher, from T. S. Ferry, *Safety Management Planning,* copyright © 1982, The Merritt Company, Santa Monica, CA 90406.

There is a growing body of literature about these formal analytical methods and others not mentioned in this chapter, including failure mode and effect (FME), technique for human error prediction (THERP), system safety hazard analysis, and management oversight and risk tree (MORT).

All have their place. Each to a greater or lesser extent provides a means of overcoming the limitations of intuitive, trial-and-error analysis.

Regardless of the method or methods used, the systems concept of hazard recognition and analysis makes available a powerful tool of proven effectiveness for decision making about the acceptability of risks. To cope with the complex safety problems of today and the future, engineers must make greater use of system safety techniques.

47.3.2 Fault Tree Technique*

When a problem can be stated quantitatively, management can assess the risk and determine the trade-off requirements between risk and capital outlay. Structuring key safety problems or vital decision-making in the form of fault paths can greatly increase communication of data and subjective reasoning. This technique is called fault-tree analysis. The transferability of data among management, engineering staff, and safety personnel is a vital step forward.

Another important aspect of this system safety technique is a phenomenon that engineers have long been aware of in electrical networks. That is, an end system formed by connecting several subsystems is likely to have entirely different characteristics from any of the subsystems considered alone. To fully evaluate and understand the entire system's performance with key paths of potential failure, the engineer must look at the entire system—only then can he or she look meaningfully at each of the subsystems.

Figure 47.1 introduces the most commonly used symbols used in fault-tree analysis.

47.3.3 Criteria for Preparation/Review of System Safety Procedures*

Correlation Between Procedure and Hardware

1. Statement of hardware configuration to which it was written?
2. Background descriptive or explanatory information where needed?
3. Reflect or reference latest revisions of drawings, manuals, or other procedures?

Adequacy of the Procedure

1. The best way to do the job?
2. Procedure easy to understand?
3. Detail appropriate—not too much, not too little?
4. Clear, concise, and free from ambiguity that could lead to wrong decisions?
5. Calibration requirements clearly defined?
6. Critical red-line parameters identified and clearly defined? Required values specified?
7. Corrective controls of above parameters clearly defined?
8. All values, switches, and other controls identified and defined?
9. Pressure limits, caution notes, safety distances, or hazards peculiar to this operation clearly defined?
10. Hard-to-locate components adequately defined and located?
11. Jigs and arrangements provided to minimize error?
12. Job safety requirements defined, for example, power off, pressure down, and tools checked for sufficiency?
13. System operative at end of job?
14. Hardware evaluated for human factors and behavioral stereotype problems? If not corrected, are any such clearly identified?
15. Monitoring points and methods of verifying adherence specified?
16. Maintenance and/or inspection to be verified? If so, is a log provided?
17. Safe placement of process personnel or equipment specified?

---

* R. De Reamer, *Modern Safety and Health Technology*. Copyright © 1980. Reprinted by permission of Wiley, New York.
* Reprinted from *MORT Safety Assurance Systems*, pp. 278–283, by courtesy of Marcel Dekker, Inc., New York.

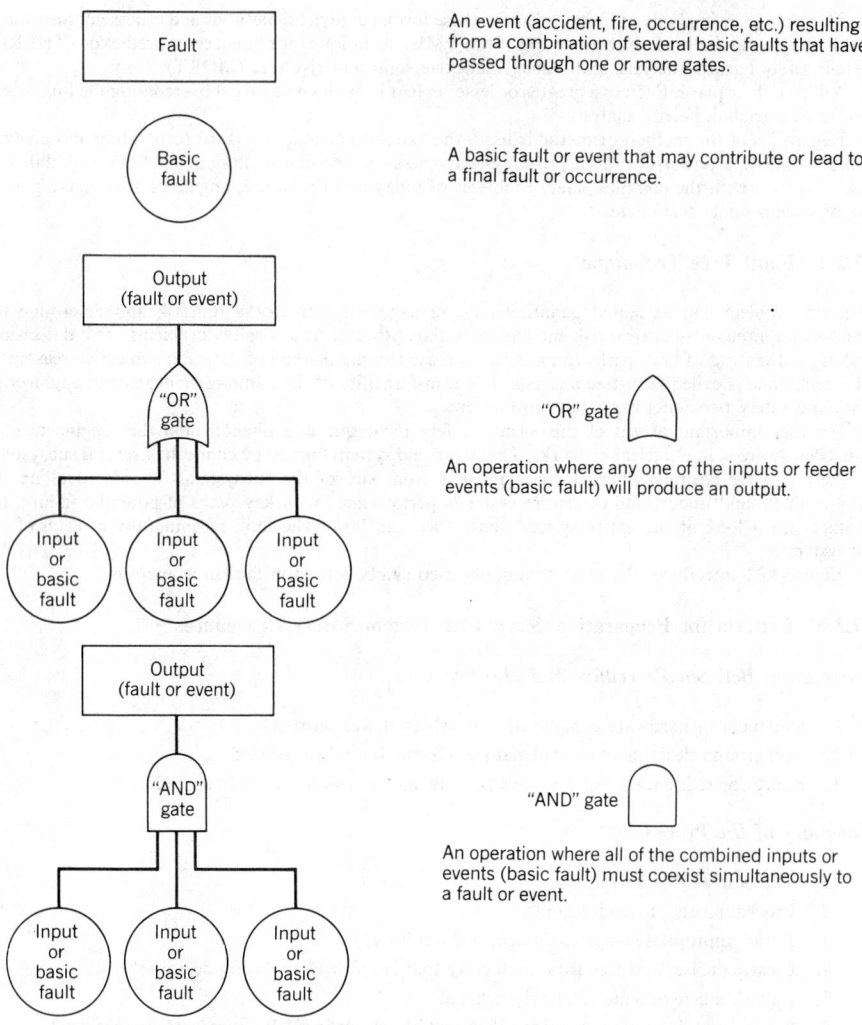

Fault — An event (accident, fire, occurrence, etc.) resulting from a combination of several basic faults that have passed through one or more gates.

Basic fault — A basic fault or event that may contribute or lead to a final fault or occurrence.

Output (fault or event)

"OR" gate — An operation where any one of the inputs or feeder events (basic fault) will produce an output.

Input or basic fault / **Input or basic fault** / **Input or basic fault**

Output (fault or event)

"AND" gate — An operation where all of the combined inputs or events (basic fault) must coexist simultaneously to a fault or event.

Input or basic fault / **Input or basic fault** / **Input or basic fault**

Fig. 47.1   Most common symbols used in fault-tree analysis.

18. Errors in previous, similar processes studied for cause? Does this procedure correct such causes?

Accuracy of the Procedure

1. Capacity to accomplish specified purpose verified by internal review?
2. All gauges, controls, valves, etc., called out, described, and labeled exactly as they actually are?
3. All setpoints or other critical controls, etc., compatible with values in control documents?
4. Safety limitations adequate for job to be performed?
5. All steps in the proper sequence?

Adequacy and Accuracy of Supporting Documentation

1. All necessary supporting drawings, manuals, data sheets, sketches, etc., either listed or attached?
2. All interfacing procedures listed?

Securing Provisions

1. Adequate instructions to return the facility or hardware to a safe operating or standby condition?
2. Securing instructions provide step-by-step operations?

Backout Provisions

1. Can procedure put any component or system in a condition which could be dangerous?
2. If so, does procedure contain emergency shutdown or backout procedures either in an appendix or as an integral part?
3. Backout procedure (or instructions for its use) included at proper place?

Emergency Measures

1. Procedures for action in case of emergency conditions?
2. Does procedure involve critical actions such that preperformance briefing on possible hazards is required?
3. Are adequate instructions either included or available for action to be taken under emergency conditions? Are they in the right place?
4. Are adequate shutdown procedures available? Cover all systems involved? Available for emergency reentry teams?
5. Specify requirements for emergency team for accident recovery, troubleshooting, or investigative purposes where necessary? Describe conditions under which emergency team will be used? Hazards they may encounter or must avoid?
6. Does procedure consider interfaces in shutdown procedures?
7. How will changes be handled? What are thresholds for changes requiring review?
8. Emergency procedures tested under range of conditions that may be encountered, for example, at night during power failure?

Caution and Warning Notes

1. Caution and warning notes included where appropriate?
2. Caution and warning notes precede operational steps containing potential hazards?
3. Adequate to describe the potential hazard?
4. Major cautions and warnings called out in general introduction, as well as prior to steps?
5. Separate entries with distinctive bold type or other emphatic display?
6. Do they include supporting safety control (health physics, safety engineer, etc.) if needed at specific required steps in procedure?

Requirements for Communications and Instrumentation

1. Adequate means of communication provided?
2. Will loss of communications create a hazard?
3. Course of action clearly defined for loss of required communications?
4. Verification of critical communication included prior to point of need?
5. Will loss of control or monitoring capability of critical functions create a hazard to people or hardware?
6. Alternate means, or a course of action to regain control or monitoring functions, clearly defined?
7. Above situations flagged by cautions and warnings?

Sequence-of-Events Considerations

1. Can any operation initiate an unscheduled or out-of-sequence event?
2. Could it induce a hazardous condition?
3. Identified by warnings or cautions?
4. Covered by emergency shutdown and backout procedures?
5. All steps sequenced properly? Sequence will not contribute to or create a hazard?
6. All steps which, if performed out-of-sequence, could cause a hazard identified and flagged?

7. Have all noncompatible simultaneous operations been identified and suitably restricted?
8. Have these been prohibited by positive callout or separation in step-by-step inclusion within the text of the procedure?

Environmental Considerations (Natural or Induced)

1. Environmental requirements specified that constrain the initiation of the procedure or require shutdown or evacuation, once in progress?
2. Induced environments (toxic or explosive atmospheres, etc.) considered?
3. All latent hazards (pressure, height, voltage, etc.) in adjacent environments considered?
4. Are there induced hazards from simultaneous performance of more than one procedure by personnel within a given space?

Personnel Qualification Statements

1. Requirement for certified personnel considered?
2. Required frequency of recheck of personnel qualifications considered?

Interfacing Hardware and Procedures Noted

1. All interfaces described by detailed callout?
2. Interfacing operating procedures identified, or written to provide ready equipment?
3. Where more than one organizational element is involved, are proper liaison and areas of responsibility established?

Procedure Sign-Off

1. Procedure to be used as an in-hand, literal checklist?
2. Step sign-off requirements considered and identified and appropriate spaces provided in the procedure?
3. Procedure completion sign-off requirements indicated (signature, authority, date, etc.)?
4. Supervisor verification of correct performance required?

General Requirements

1. Procedure discourages a shift change during performance or accommodates a shift change?
2. Where shift changes are necessary, include or reference shift overlap and briefing requirements?
3. Mandatory inspection, verification, and system validation required whenever procedure requires breaking into and reconnecting a system?
4. Safety prerequisites defined? All safety instructions spelled out in detail to all personnel?
5. Require prechecks of supporting equipment to ensure compatibility and availability?
6. Consideration for unique operations written in?
7. Procedures require walk-through or talk-through dry runs?
8. General supervision requirements, for example, what is protocol for transfer of supervisor responsibilities to a successor?
9. Responsibilities of higher supervision specified?

Reference Considerations

1. Applicable quality assurance and reliability standards considered?
2. Applicable codes, standards, and regulations considered?
3. Procedure complies with control documents?
4. Hazards and system safety degradations identified and considered against specific control manuals, standards, and procedures?
5. Specific prerequisite administrative and management approvals complied with?
6. Comments received from the people who will do the work?

Special Considerations

1. Has a documented safety analysis been considered for safety-related deviations from normal practices or for unusual or unpracticed maneuvers?
2. Have new restrictions or controls become effective that affect the procedure in such a manner that new safety analyses may be required?

47.4 HUMAN FACTORS ENGINEERING/ERGONOMICS*

47.4.1 Human–Machine Relationships

- Human factors engineering is defined as "the application of the principles, laws, and quantitative relationships which govern man's response to external stress to the analysis and design of machines and other engineering structures, so that the operator of such equipment will not be stressed beyond his/her proper limit or the machine forced to operate at less than its full capacity in order for the operator to stay within acceptable limits of human capabilities."†
- A principal objective of the supervisor and safety engineer in the development of safe working conditions is the elimination of bottlenecks, stresses and strains, and psychological booby traps that interfere with the free flow of work. It is an accepted concept that the less an operator has to fear from his or her job or machine, the more attention he or she can give to his or her work.
- In the development of safe working conditions, attention is given to many things, including machine design and machine guarding, personal protective equipment, plant layout, manufacturing methods, lighting, heating, ventilation, removal of air contaminants, and the reduction of noise. Adequate consideration of each of these areas will lead to a proper climate for accident prevention, increased productivity, and worker satisfaction.
- The human factors engineering approach to the solution of the accident problem is to build machines and working areas around the operator, rather than place him or her in a setting without regard to his or her requirements and capacities. Unless this is done, it is hardly fair to attribute so many accidents to human failure, as is usually the case.
- If this point of view is carried out in practice, fewer accidents should result, training costs should be reduced, and extensive redesign of equipment after it is put into use should be eliminated.
- All possible faults in equipment and in the working area, as well as the capacities of the operator, should be subjected to advance analysis. If defects are present, it is only a matter of time before some operator "fails" and has an accident.
- Obviously, the development of safe working conditions involves procedures that may go beyond the occasional safety appraisal or search for such obvious hazards as an oil spot on the floor, a pallet in the aisle, or an unguarded pinch point on a new lathe.

Human–machine relationships have improved considerably with increased mechanization and automation. Nevertheless, with the decrease in manual labor has come specialization, increased machine speeds, and monotonous repetition of a single task, which create work relationships involving several physiological and psychological stresses and strains. Unless this scheme of things is recognized and dealt with effectively, many real problems in the field of accident prevention may be ignored.

47.4.2 Human Factors Engineering Principles

- Human factors engineering or *ergonomics,* * as it is sometimes called, developed as a result of the experience in the use of highly sophisticated equipment in World War II. The ultimate potentialities of complex instruments of war could not be realized because the human operators lacked the necessary capabilities and endurance required to operate them. This discipline now has been extended to many areas. It is used extensively in the aircraft and aerospace industry and in many other industries to achieve more effective integration of humans and machines.

---

* R. De Reamer, *Modern Safety and Health Technology.* Copyright © 1980. Reprinted by permission of Wiley, New York.
† Theodore F. Hatch, Professor (retired), by permission.
* The term *ergonomics* was coined from the Greek roots *ergon* (work) and *nomos* (law, rule) and is now currently used to deal with the interactions between humans and such environmental elements as atmospheric contaminants, heat, light, sound, and all tools and equipment pertaining to the work-place.

- The analysis should consider all possible faults in the equipment, in the work area, and in the worker—including a survey of the nature of the task, the work surroundings, the location of controls and instruments, and the way the operator performs his or her duties. The questions of importance in the analysis of machines, equipment, processes, plant layout, and the worker will vary with the type and purpose of the operation, but usually will include the following (pertaining to the worker)*:

  1. What sense organs are used by the operator to receive information? Does he or she move into action at the sound of a buzzer, blink of a light, reading of a dial, verbal order? Does the sound of a starting motor act as a cue?
  2. What sort of discrimination is called for? Does the operator have to distinguish between lights of two different colors, tones of two different pitches, or compare two dial readings?
  3. What physical response is he or she required to make: Pull a handle? Turn a wheel? Step on a pedal? Push a button?
  4. What overall physical movements are required in the physical response? Do such movements interfere with his or her ability to continue receiving information through his or her sense organs? (For example, would pulling a handle obstruct his or her line of vision to a dial he or she is required to watch?) What forces are required (e.g., torque in turning a wheel)?
  5. What are the speed and accuracy requirements of the machine? Is the operator required to watch two pointers to a hairline accuracy in a split second? Or is fairly close approximation sufficient? If a compromise is necessary, which is more essential: speed or accuracy?
  6. What physiological and environmental conditions are likely to be encountered during normal operation of the machine? Are there any unusual temperatures, humidity conditions, crowded workspace, poor ventilation, high noise levels, toxic chemicals, and so on?

- Pertaining to the machine, equipment, and the surrounding area, these key questions should be asked:

  1. Can the hazard be eliminated or isolated by a guard, ventilating equipment, or other device?
  2. Should the hazard be identified by the use of color, warning signs, blinking lights, or alarms?
  3. Should interlocks be used to protect the worker when he or she forgets or makes the wrong move?
  4. Is it necessary to design the machine, the electrical circuit, or the pressure circuit so it will always be fail-safe?
  5. Is there need for standardization?
  6. Is there need for emergency controls, and are controls easily identified and accessible?
  7. What unsafe conditions would be created if the proper operating sequence were not followed?

47.4.3   General Population Expectations*

- The importance of standardization and normal behavior patterns has been recognized in business and industry for many years. A standard tool will more likely be used properly than will a nonstandard one, and standard procedures will more likely be followed.

- People expect things to operate in a certain way and certain conditions to conform with established standards. These general population "expectations"—the way in which the ordinary person will react to a condition or stimulus—must not be ignored or workers will be literally trapped into making mistakes. A list of "General Population Expectations" follows:

  1. Doors are expected to be at least 6 feet, 6 inches in height.
  2. The level of the floor at each side of a door is expected to be the same.
  3. Stair risers are expected to be of the same height.
  4. It is a normal pattern for persons to pass to the left on motorways (some countries excluded).
  5. People expect guardrails to be securely anchored.
  6. People expect the hot-water faucet to be on the left side of the sink, the cold-water faucet on the right, and the faucet to turn to the left (counterclockwise) to let the water run and to the right to turn the water off.
  7. People expect floors to be nonslippery.
  8. Flammable solvents are expected to be found in labeled, red containers.
  9. The force required to operate a lever, push a cart, or turn a crank is expected to go unchanged.
  10. Knobs on electrical equipment are expected to turn clockwise for "on," to increase current, and counterclockwise for "off."

---

* R. A. McFarland, "Application of Human Factors Engineering to Safety Engineering Problems," *National Safety Congress Transactions,* 1967, Vol. 12. Permission granted by the National Safety Council.
* R. De Reamer, *Modern Safety and Health Technology.* Copyright © 1980. Reprinted by permission of Wiley, New York.

11. For control of vehicles in which the operator is riding, the operator expects a control motion to the right or clockwise to result in a similar motion of his or her vehicle and vice versa.

12. Very large objects or dark objects imply "heaviness." Small objects or light-colored ones imply "lightness." Large heavy objects are expected to be "at the bottom." Small, light objects are expected to be "at the top."

13. Seat heights are expected to be at a certain level when a person sits down.

47.5 ENGINEERING CONTROLS FOR MACHINE TOOLS*

47.5.1 Basic Concerns

Machine tools (such as mills, lathes, shearers, punch presses, grinders, drills, and saws) provide an example of commonplace conditions where there is only a limited number of items of personal protective gear available for use. In such cases as these, the problem to be solved is not personal protective equipment versus engineering controls, but rather which engineering control(s) should be used to protect the machine operator. A summary of employee safeguards is contained in Table 47.2. The list of possible machinery-related injuries is presented in Section 47.10. There seem to be as many hazards created by moving machine parts as there are types of machines. Safeguards are essential for protecting workers from needless and preventable injuries.

A good rule to remember is: Any machine part, function, or process that may cause injury must be safeguarded. Where the operation of a machine or accidental contact with it can injure the operator or others in the vicinity, the hazard must be either controlled or eliminated.

Dangerous moving parts in these three basic areas need safeguarding:

- The point of operation: that point where work is performed on the material, such as cutting, shaping, boring, or forming of stock.
- Power transmission apparatus: all components of the mechanical system that transmit energy to the part of the machine performing the work. These components include flywheels, pulleys, belts, connecting rods, couplings, cams, spindles, chains, cranks, and gears.
- Other moving parts: all parts of the machine that move while the machine is working. These can include reciprocating, rotating, and transverse moving parts, as well as feed mechanisms and auxiliary parts of the machine.

A wide variety of mechanical motions and actions may present hazards to the worker. These can include the movement of rotating teeth, and any parts that impact or shear. These different types of hazardous mechanical motions and actions are basic to nearly all machines, and recognizing them is the first step toward protecting workers from the danger they present.

The basic types of hazardous mechanical motions and actions are:

Motions	Actions
Rotating (including in-running nip points)	Cutting
	Punching
Reciprocating	Shearing
Transverse	Bending

47.5.2 General Requirements

What must a safeguard do to protect workers against mechanical hazards? Engineering controls must meet these minimum general requirements:

- *Prevent Contact.* The safeguard must prevent hands, arms, or any other part of a worker's body from making contact with dangerous moving parts. A good safeguarding system eliminates the possibility of operators or workers placing their hands near hazardous moving parts.
- *Secure.* Workers should not be able to easily remove or tamper with the safeguard, because a safeguard that can easily be made ineffective is no safeguard at all. Guards and safety devices

* J. B. Re Velle, *Engineering Controls: A Comprehensive Overview.* Used by permission of the Merritt Company, Publisher, from T. S. Ferry, *Safety Management Planning,* copyright © 1982, The Merritt Company, Santa Monica, CA 90406.

Table 47.2 Summary of Employee Safeguards

To Protect	Personal Protective Equipment to Use	Engineering Controls to Use
Breathing	Self-contained breathing apparatus, gas masks, respirators, alarm systems	Ventilation, air-filtration systems, critical level warning systems, electrostatic precipitators
Eyes/face	Safety glasses, filtered lenses, safety goggles, face shield, welding goggles/helmets, hoods	Spark deflectors, machine guards
Feet/legs	Safety boots/shoes, leggings, shin guards	
Hands/arms/body	Gloves, finger cots, jackets, sleeves, aprons, barrier creams	Machine guards, lockout devices, feeding and ejection methods
Head/neck	Bump caps, hard hats, hair nets	Toe boards
Hearing	Ear muffs, ear plugs, ear valves	Noise reduction/isolation by equipment modification/substitution, equipment lubrication/maintenance programs, eliminate/dampen noise sources, reduce compressed air pressure, change operations[a]
Excessively high/ low temperatures	Reflective clothing, temperature controlled clothing	Fans, air conditioning, heating, ventilation, screens, shields, curtains
Overall	Safety belts, lifelines, grounding mats, slap bars	Electrical circuit grounding, polarized plugs/outlets, safety nets

[a]Examples of the types of changes that should be considered include:
- Grinding instead of chipping.
- Electric tools in place of pneumatic tools.
- Pressing instead of forging.
- Welding instead of riveting.
- Compression riveting over pneumatic riveting.
- Mechanical ejection in place of air-blast ejection.
- Wheels with rubber or composition tires on plant trucks and cars instead of all-metal wheels.
- Wood or plastic tote boxes in place of metal tote boxes.
- Use of an undercoating on machinery covers.
- Wood in place of all-metal workbenches.

should be made of durable material that will withstand the conditions of normal use. They must be firmly secured to the machine.

Machines often produce noise (unwanted sound), and this can result in a number of hazards to workers. Not only can it startle and disrupt concentration, but it can interfere with communications, thus hindering the worker's safe job performance. Research has linked noise to a whole range of harmful health effects, from hearing loss and aural pain to nausea, fatigue, reduced muscle control, and emotional disturbances. Engineering controls such as the use of sound-dampening materials, as well as less sophisticated hearing protection, such as ear plugs and muffs, have been suggested as ways of controlling the harmful effects of noise. Vibration, a related hazard that can cause noise and thus result in fatigue and illness for the worker, may be avoided if machines are properly aligned, supported, and, if necessary, anchored.

Because some machines require the use of cutting fluids, coolants, and other potentially harmful substances, operators, maintenance workers, and others in the vicinity may need protection. These substances can cause ailments ranging from dermatitis to serious illnesses and disease. Specially constructed safeguards, ventilation, and protective equipment and clothing are possible temporary solutions to the problem of machinery-related chemical hazards until these hazards can be better controlled or eliminated from the workplace.

- *Protect from Falling Objects.* The safeguard should ensure that no objects can fall into moving parts. A small tool that is dropped into a cycling machine could easily become a projectile that could strike and injure someone.
- *Create No New Hazards.* A safeguard defeats its own purpose if it creates a hazard of its own such as a shear point, a jagged edge, or an unfinished surface that can cause a laceration. The edges of guards, for instance, should be rolled or bolted in such a way that eliminates sharp edges.
- *Create No Interference.* Any safeguard which impedes a worker from performing the job quickly and comfortably might soon be overridden or disregarded. Proper safeguarding can actually enhance efficiency, since it can relieve the worker's apprehensions about injury.
- *Allow Safe Lubrication.* If possible, one should be able to lubricate the machine without removing the safeguards. Locating oil reservoirs outside the guard, with a line leading to the lubrication point, will reduce the need for the operator or maintenance worker to enter the hazardous area.

47.5.3 Danger Sources

All power sources for machinery are potential sources of danger. When using electrically powered or controlled machines, for instance, the equipment as well as the electrical system itself must be properly grounded. Replacing frayed, exposed, or old wiring will also help to protect the operator and others from electrical shocks or electrocution. High-pressure systems, too, need careful inspection and maintenance to prevent possible failure from pulsation, vibration, or leaks. Such a failure could cause explosions or flying objects.

47.6 MACHINE SAFEGUARDING METHODS*

47.6.1 General Classifications

There are many ways to safeguard machinery. The type of operation, the size or shape of stock, the method of handling the physical layout of the work area, the type of material, and production requirements or limitations all influence selection of the appropriate safeguarding method(s) for the individual machine.

As a general rule, power transmission apparatus is best protected by fixed guards that enclose the danger area. For hazards at the point of operation, where moving parts actually perform work on stock, several kinds of safeguarding are possible. One must always choose the most effective and practical means available.

1. Guards
   - (a) Fixed
   - (b) Interlocked
   - (c) Adjustable
   - (d) Self-adjusting
2. Devices
   - (a) Presence sensing
     Photoelectrical (optical)
     Radio frequency (capacitance)
     Electromechanical
   - (b) Pullback
   - (c) Restraint
   - (d) Safety controls
     Safety trip control
     Pressure-sensitive body bar
     Safety triprod
     Safety tripwire cable
     Two-hand control
     Two-hand trip

* J. B. Re Velle, *Engineering Controls: A Comprehensive Overview.* Used by permission of The Merritt Company, Publisher, from T. S. Ferry, *Safety Management Planning,* copyright © 1982, The Merritt Company, Santa Monica, CA 90406.

  (e) Gates
       Interlocked
       Other
  3. Location/distance
  4. Potential feeding and ejection methods to improve safety for the operator
   (a) Automatic feed
   (b) Semiautomatic feed
   (c) Automatic ejection
   (d) Semiautomatic ejection
   (e) Robot
  5. Miscellaneous aids
   (a) Awareness barriers
   (b) Miscellaneous protective shields
   (c) Hand-feeding tools and holding fixtures

47.6.2 Guards, Devices, and Feeding and Ejection Methods

Tables 47.3–47.5 provide the interested reader with specifics regarding machine safeguarding.

47.7 ALTERNATIVES TO ENGINEERING CONTROLS*

Engineering controls are an alternative to personal protective equipment, or is it the other way around? This chicken-and-egg situation has become an emotionally charged issue with exponents on both sides arguing their beliefs with little in the way of well-founded evidence to support their cases. The reason for this unfortunate situation is that there is no single solution to all the hazardous operations found in industry. The only realistic answer to the question concerning which of the two methods of abating personnel hazards is—it depends. Each and every situation requires an independent analysis considering all the known factors so that a truly unbiased decision can be reached.

   This section presents material useful to engineers in the selection and application of solutions to industrial safety and health problems. Safety and health engineering control principles are deceptively few: substitution; isolation; and ventilation, both general and localized. In a technological sense, an appropriate combination of these strategic principles can be brought to bear on any industrial safety or hygiene control problem to achieve a satisfactory quality of the work environment. It may not be, and usually is not, necessary or appropriate to apply all these principles to any specific potential hazard. A thorough analysis of the control problem must be made to ensure that a proper choice from among these methods will produce the proper control in a manner that is most compatible with the technical process, is acceptable to the workers in terms of day-to-day operation, and can be accomplished with optimal balance of installation and operating expenses.

47.7.1 Substitution

Although frequently one of the most simple engineering principles to apply, substitution is often overlooked as an appropriate solution to occupational safety and health problems. There is a tendency to analyze a particular problem from the standpoint of correcting rather than eliminating it. For example, the first inclination in considering a vapor-exposure problem in a degreasing operation is to provide ventilation of the operation rather than consider substituting a solvent having a much lower degree of hazard associated with its use. However, substitution of less hazardous substances, changing from one type of process equipment to another, or, in some cases, even changing the process itself, may provide an effective control of a hazard at minimal expense.

   This strategy is often used in conjunction with safety equipment: substituting safety glass for regular glass in some enclosures, replacing unguarded equipment with properly guarded machines, replacing safety gloves or aprons with garments made of material more impervious to the chemicals being handled. Since substitution of equipment frequently is done as an immediate response to an obvious problem, it is not always recognized as an engineering control, even though the end result is every bit as effective.

   Substituting one process or operation for another may not be considered except in major modifications. In general, a change in any process from a batch to a continuous type of operation carries with it an inherent reduction in potential hazard. This is true primarily because the frequency and duration of potential contact of workers with the process materials are reduced when the overall

* J. B. Re Velle, *Engineering Controls: A Comprehensive Overview.* Used by permission of The Merritt Company, Publisher, from T. S. Ferry, *Safety Management Planning,* copyright © 1982, The Merritt Company, Santa Monica, CA 90406.

Table 47.3 Machine Safeguarding: Guards

Method	Safeguarding Action	Advantages	Limitations
Fixed	Provides a barrier	Can be constructed to suit many specific applications In-plant construction is often possible Can provide maximum protection Usually requires minimum maintenance Can be suitable to high production, repetitive operations	May interfere with visibility Can be limited to specific operations Machine adjustment and repair often requires its removal, thereby necessitating other means of protection for maintenance personnel
Interlocked	Shuts off or disengages power and prevents starting of machine when guard is open; should require the machine to be stopped before the worker can reach into the danger area	Can provide maximum protection Allows access to machine for removing jams without time-consuming removal of fixed guards	Requires careful adjustment and maintenance May be easy to disengage
Adjustable	Provides a barrier that may be adjusted to facilitate a variety of production operations	Can be constructed to suit many specific applications Can be adjusted to admit varying sizes of stock	Hands may enter danger area—protection may not be complete at all times May require frequent maintenance and/or adjustment The guard may be made ineffective by the operator May interfere with visibility
Self-adjusting	Provides a barrier that moves according to the size of the stock entering danger area	Off-the-shelf guards are often commercially available	Does not always provide maximum protection May interfere with visibility May require frequent maintenance and adjustment

Table 47.4 Machine Safeguarding: Devices

Method	Safeguarding Action	Advantages	Limitations
Photoelectric	Machine will not start cycling when the light field is interrupted When the light field is broken by any part of the operator's body during the cycling process, immediate machine braking is activated	Can allow freer movement for operator	Does not protect against mechanical failure May require frequent alignment and calibration Excessive vibration may cause lamp filament damage and premature burnout Limited to machines that can be stopped
Radio frequency (capacitance)	Machine cycling will not start when the capacitance field is interrupted When the capacitance field is disturbed by any part of the operator's body during the cycling process, immediate machine braking is activated	Can allow freer movement for operator	Does not protect against mechanical failure Antennae sensitivity must be properly adjusted Limited to machines that can be stopped
Electromechanical	Contact bar or probe travels a predetermined distance between the operator and the danger area Interruption of this movement prevents the starting of machine cycle	Can allow access at the point of operation	Contact bar or probe must be properly adjusted for each application; this adjustment must be maintained properly
Pullback	As the machine begins to cycle, the operator's hands are pulled out of the danger area	Eliminates the need for auxiliary barriers or other interference at the danger area	Limits movement of operator May obstruct workspace around operator Adjustments must be made for specific operations and for each individual

Method	Safeguarding Action	Advantages	Limitations
Restraint (holdback)	Prevents the operator from reaching into the danger area	Little risk of mechanical failure	Requires frequent inspections and regular maintenance Requires close supervision of the operator's use of the equipment Limits movements of operator May obstruct workspace Adjustments must be made for specific operations and each individual Requires close supervision of the operator's use of the equipment
Safety trip controls Pressure-sensitive body bar Safety tripod Safety tripwire	Stops machine when tripped	Simplicity of use	All controls must be manually activated May be difficult to activate controls because of their location Only protects the operator May require special fixtures to hold work May require a machine brake
Two-hand control	Concurrent use of both hands is required, preventing the operator from entering the danger area	Operator's hands are at a pre-determined location Operator's hands are free to pick up a new part after first half of cycle is completed	Requires a partial cycle machine with a brake Some two-hand controls can be rendered unsafe by holding with arm or blocking, thereby permitting one-hand operation Protects only the operator

Table 47.4 (*Continued*)

Method	Safeguarding Action	Advantages	Limitations
Two-hand trip	Concurrent use of two hands on separate controls prevents hands from being in danger area when machine cycle starts	Operator's hands are away from danger area Can be adapted to multiple operations No obstruction to hand feeding Does not require adjustment for each operation	Operator may try to reach into danger area after tripping machine Some trips can be rendered unsafe by holding with arm or blocking, thereby permitting one-hand operation Protects only the operator May require special fixtures
Gate	Provides a barrier between danger area and operator or other personnel	Can prevent reaching into or walking into the danger area	May require frequent inspection and regular maintenance May interfere with operator's ability to see the work

Table 47.5 Machine Safeguarding: Feeding and Ejection Methods

Method	Safeguarding Action	Advantages	Limitations
Automatic feed	Stock is fed from rolls, indexed by machine mechanism, etc.	Eliminates the need for operator involvement in the danger area	Other guards are also required for operator protection—usually fixed barrier guards Requires frequent maintenance May not be adaptable to stock variation
Semiautomatic feed	Stock is fed by chutes, movable dies, dial feed, plungers, or sliding bolster		
Automatic ejection	Workpieces are ejected by air or mechanical means		May create a hazard of blowing chips or debris Size of stock limits the use of this method Air ejection may present a noise hazard
Semiautomatic ejection	Workpieces are ejected by mechanical means, which are initiated by the operator	Operator does not have to enter danger area to remove finished work	Other guards are required for operator protection May not be adaptable to stock variation
Robots	They perform work usually done by operator	Operator does not have to enter danger area Are suitable for operations where high stress factors are present, such as heat and noise	Can create hazards themselves Require maximum maintenance Are suitable only to specific operations

process approach becomes one of continuous operation. The substitution of processes can be applied on a fundamental basis, for example, substitution of airless spray for conventional spray equipment can reduce the exposure of a painter to solvent vapors. Substitution of a paint dipping operation for the paint spray operation can reduce the potential hazard even further. In any of these cases, the automation of the process can further reduce the potential hazard (Table 47.5).

47.7.2 Isolation

Application of the principle of isolation is frequently envisioned as consisting of the installation of a physical barrier (such as a machine guard or device—refer to Tables 47.3 and 47.4) between a hazardous operation and the workers. Fundamentally, however, this isolation can be provided *without* a physical barrier through the appropriate use of distance and, in some situations, time.

Perhaps the most common example of isolation as a control strategy is associated with storage and use of flammable solvents. The large tank farms with dikes around the tanks, underground storage of some solvents, the detached solvent sheds, and fireproof solvent storage rooms within buildings are all commonplace in American industry. Frequently, the application of the principle of isolation maximizes the benefits of additional engineering concepts such as excessive noise control, remote control materials handling (as with radioactive substances), and local exhaust ventilation.

47.7.3 Ventilation

Workplace air quality is affected directly by the design and performance of the exhaust system. An improperly designed hood or a hood evacuated with an insufficient volumetric rate of air will contaminate the occupational environment and affect workers in the vicinity of the hazard source. This is a simple, but powerful, symbolic representation of one form of the close relationship between atmospheric emissions (as regulated by the Environmental Protection Agency) and occupational exposure (as regulated by the Occupational Safety and Health Administration). What is done with gases generated as a result of industrial operations/processes? These emissions can be exhausted directly to the atmosphere, indirectly to the atmosphere (from the workplace through the general ventilation system), or recirculated to the workplace. The effectiveness of the ventilation system design and operation impacts directly on the necessity and type of respiratory gear needed to protect the work force.

47.8   DESIGN AND REDESIGN*

47.8.1   Hardware

Designers of machines must consider the performance characteristics of machine operators as a major constraint in the creation or modification of both mechanical and electrical equipment. To do less would be tantamount to ignoring the limitations of human capabilities. Equipment designers especially concerned with engineering controls to be incorporated into machines, whether at the time of initial conceptualization or later when alterations are to be made, must also be cognizant of the principles of human factors (ergonomics). Equipment designers are aware that there are selected tasks that people can perform with greater skill and dependability than machines, and vice versa. Some of these positive performance characteristics are noted in Table 47.6. In addition, designers of equipment and engineering controls are knowledgeable of human performance limitations, both physically and psychologically. They know that the interaction of forces between people and their operating environment presents a never-ending challenge in assessing the complex interrelationships that provide the basis for that often fine line between safety versus hazard or health versus contaminant. Table 47.7 identifies the six pertinent sciences most closely involved in the design of machines and engineering controls.

It is both rational and reasonable to expect that, when engineering controls are being considered to eliminate or reduce hazards or contaminants, designers make full use of the principles established by specialists in these human performance sciences.

47.8.2   Process

A stress (or stressor) is some physical or psychological feature of the environment that requires an operator to be unduly exerted to continue performing. Such exertion is termed strain as in "stress

* J. B. Re Velle, *Engineering Controls: A Comprehensive Overview.* Used by permission of The Merritt Company, Publisher, from T. S. Ferry, *Safety Management Planning,* copyright © 1982, the Merritt Company, Santa Monica, CA 90406.

Table 47.6 Positive Performance Characteristics—Some Things Done Better by

People	Machines
Detect signals in high noise fields	Respond quickly to signals
Recognize objects under widely different conditions	Sense energies outside human range
Perceive patterns	Consistently perform precise, routine, repetitive operations
Sensitive to a wide variety of stimuli	Recall and process enormous amounts of data
Long-term memory	Monitor people or other machines
Handle unexpected or low-probability events	Reason deductively
Reason inductively	Exert enormous power
Profit from experience	Relatively uniform performance
Exercise judgment	Rapid transmission of signals
Flexibility, improvision, and creativity	Perform several tasks simultaneously
Select and perform under overload conditions	Expendable
Adapt to changing environment	Resistance to many environmental stresses
Appreciate and create beauty	
Perform fine manipulations	
Perform when partially impaired	
Relatively maintenance-free	

and strain." Common physical stressors in industrial workplaces are poor illumination, excessive noise, vibration, heat, and the presence of excessive, harmful atmospheric contaminants.

Unfortunately, much less is known about their effects when they occur at the same time, in rapid sequence, or over extended periods of time. Research suggests that such effects are not simply additive, but synergistic, thus compounding their detrimental effects. In addition, when physical work environments are unfavorable to equipment operators, two or more stressors are generally present: high temperature and excessive noise, for example. The solution to process design and redesign is relatively easy to specify, but costly to implement—design the physical environment so that all physical characteristics are within an acceptable range.

Marketed in the United States since the early 1960s, industrial robots offer both hardware and process designers a technology that can be used when hazardous or uncomfortable working conditions are expected or already exist. Where a job situation poses potential dangers or the workplace is hot or in some other way unpleasant, a robot should be considered as a substitute for human operators. Hot forging, die casting, and spray painting fall into this category. If workparts or tools are awkward or heavy, an industrial robot may fill the job. Some robots are capable of lifting items weighing several hundred pounds.

An industrial robot is a general purpose, programmable machine that possesses certain humanlike capabilities. The most obvious characteristic is the robot's arm, which, when combined with the robot's capacity to be programmed, makes it ideally suited to a variety of uncomfortable/undesirable production tasks. Hardware and process designers now possess an additional capability for potential inclusion in their future designs and redesigns.

Table 47.7 People Performance Sciences

Anthropometry	Pertains to the measurement of physical features and characteristics of the static human body.
Biomechanics	A study of the range, strength, endurance, and accuracy of movements of the human body.
Ergonomics	Human factors engineering—especially biomechanics aspects.
Human factors engineering	Designing for human use.
Kinesiology	A study of the principles of mechanics and anatomy of human movement.
Systems safety engineering	Designing that considers the operator's qualities, the equipment, and the environment relative to successful task performance.

47.9 PERSONAL PROTECTIVE EQUIPMENT*

47.9.1 Background

Engineering controls, which eliminate the hazard at the source and do not rely on the worker's behavior for their effectiveness, offer the best and most reliable means of safeguarding. Therefore, engineering controls must be first choice for eliminating machinery hazards. But whenever an extra measure of protection is necessary, operators must wear protective clothing or personal protective equipment.

If it is to provide adequate protection, the protective clothing and equipment selected must always be:

- appropriate for the particular hazards
- maintained in good condition
- properly stored when not in use, to prevent damage or loss
- kept clean and sanitary

Protective clothing is available for every part of the human body. Hard hats can protect the head from falling objects when the worker is handling stock; caps and hair nets can help keep the worker's hair from being caught in machinery. If machine coolants could splash, or particles could fly into the operator's eyes or face, then face shields, safety goggles, glasses, or similar kinds of protection must be used. Hearing protection may be needed when workers operate noisy machinery. To guard the trunk of the body from cuts or impacts from heavy or rough-edged stock, there are certain protective coveralls, jackets, vests, aprons, and full-body suits. Workers can protect their hands and arms from the same kinds of injury with special sleeves and gloves. And safety shoes and boots, or other acceptable foot guards, can shield the feet against injury in case the worker needs to handle heavy stock which might drop.

It is important to note that protective clothing and equipment themselves can create hazards. A protective glove which can become caught between rotating parts, or a respirator facepiece which hinders the wearer's vision require alertness and careful supervision whenever they are used.

Other aspects of the worker's dress may present additional safety hazards. Loose-fitting clothing might possibly become entangled in rotating spindles or other kinds of moving machinery. Jewelry, such as bracelets and rings, can catch on machine parts or stock and lead to serious injury by pulling a hand into the danger area.

Naturally, each situation will vary. In some simple cases, respirators, chemical goggles, aprons, and gloves may be sufficient personal protective equipment to afford the necessary coverage. In more complicated situations, even the most sophisticated equipment may not be enough and engineering controls would become mandatory. Safety, industrial, and plant engineers should be expected to provide the necessary analyses to ascertain the extent of the hazard to employees whose work causes them to be exposed to the corrosive fumes.

47.9.2 Planning and Implementing the Use of Protective Equipment*

This section reviews ways to help plan, implement, and maintain personal protective equipment. This can be considered in terms of the following nine phases: (1) need analysis, (2) equipment selection, (3) program communication, (4) training, (5) fitting and adjustment, (6) target date setting, (7) break-in period, (8) enforcement, and (9) follow-through.

The first phase of promoting the use of personal protective equipment is called *need analysis*. Before selecting protective equipment, the hazards or conditions the equipment must protect the employee from must be determined. To accomplish this, questions such as the following must be asked:

- What standards does the law require for this type of work in this type of environment?
- What needs do our accident statistics point to?
- What hazards have we found in our safety and/or health inspections?
- What needs show up in our job analysis and job observation activities?
- Where is the potential for accidents, injuries, illnesses, and damage?
- Which hazards cannot be eliminated or segregated?

* J. B. Re Velle, *Engineering Controls: A Comprehensive Overview.* Used by permission of The Merritt Company, Publisher, from T. S. Ferry, *Safety Management Planning,* copyright © 1982, The Merritt Company, Santa Monica, CA 90406.
* J. B. Re Velle, *Safety Training Methods.* Copyright © 1980. Reprinted by permission of Wiley, New York.

The second phase of promoting the use of protective equipment is *equipment selection.* Once a need has been established, proper equipment must be selected. Basic consideration should include the following:

- Conformity to the standards.
- Degree of protection provided.
- Relative cost.
- Ease of use and maintenance.
- Relative comfort.

The third phase is *program communication.* It is not appropriate to simply announce a protective equipment program, put it into effect, and expect to get immediate cooperation. Employees tend to resist change unless they see it as necessary, comfortable, or reasonable. It is helpful to use various approaches to publicity and promotion to teach employees why the equipment is necessary. Various points can be covered in supervisor's meetings, in safety meetings, by posters, on bulletin boards, in special meetings, and in casual conversation. Gradually, employees will come to expect or to request protective equipment to be used on the job. The main points in program communication are to educate employees in why protective equipment is necessary and to encourage them to want it and to use it.

Training is an essential step in making sure protective equipment will be used properly. The employees should learn why the equipment is necessary, when it must be used, who must use it, where it is required, what the benefits are, and how to use it and take care of it. Do not forget that employee turnover will bring new employees into the work area. Therefore, you will continually need to train new employees in the use of the protective equipment they will handle.

After the training phase comes the *fitting and adjustment* phase. Unless the protective equipment fits the individual properly, it may not give the necessary protection. There are many ways to fit or to adjust protective equipment. For example, face masks have straps that hold them snug against the contours of the head and face and prevent leaks; rubberized garments have snaps or ties that can be drawn up snugly, to keep loose and floppy garments from getting caught in machinery.

The next phase is *target date setting.* After the other phases have been completed, set specific dates for completion of the various phases. For example, all employees shall be fitted with protective equipment before a certain date; all training shall be completed by a certain date; after a certain date, all employees must wear their protective equipment while in the production area.

After setting the target dates, expect a *break-in period.* There will usually be a period of psychological adjustment whenever a new personal protective program is established. Remember two things:

- Expect some gripes, grumbles, and problems.
- Appropriate consideration must be given to each individual problem; then strive toward a workable solution.

It might also be wise to post signs that indicate the type of equipment needed. For example, a sign might read, "Eye protection must be worn in this area."

After the break-in phase comes *enforcement.* If all the previous phases were successful, problems in terms of enforcement should be few. In case disciplinary action is required, sound judgment must be used and each case must be evaluated on an individual basis.

If employees fail to use protective equipment, they may be exposed to hazards. Do not forget, the employer can be penalized if employees do not use their protection.

The final phase is *follow-through.* Although disciplinary action may sometimes be necessary, positive motivation plays a more effective part in a successful protective equipment program. One type of positive motivation is a proper example set by management. Managers must wear their protective equipment, just as employees are expected to wear theirs.

47.9.3 Adequacy, Maintenance, and Sanitation

Before selling safety shoes and supplying safety goggles at a company store, the attendants must be guided by a well-structured program of equipment maintenance, preferably preventive maintenance.

Daily maintenance of different types of equipment might include: adjustment of the suspension system on a safety hat; cleaning of goggle lenses, glasses, or spectacles; scraping residue from the sole of a safety shoe; or proper adjustment of a face mask when donning an air-purifying respirator.

Performing these functions should be coupled with periodic inspections for weaknesses or defects in the equipment. How often this type of check is made, of course, depends on the particular type of equipment used. For example, sealed-canister gas masks should be weighed on receipt from the manufacturer, and the weight should be marked indelibly on each canister. Stored units should then be reweighed periodically, and those exceeding a recommended weight should be discarded even though the seal remains unbroken.

Sanitation, as spelled out in the OSHAct, is a key part of any operation, and it requires the use of personal protective equipment, not only to eliminate cross-infection among users of the same unit of equipment, but because unsanitary equipment is objectionable to the wearer.

Procedures and facilities that are necessary to sanitize or disinfect equipment can be an integral part of an equipment maintenance program. For example, the OSHAct says, "Respirators used routinely shall be inspected during cleaning." Without grime and dirt to hinder an inspection, gauges can be read better, rubber or elastomer parts can be checked for pliability and signs of deterioration, and valves can be checked.

47.10 MANAGING THE SAFETY FUNCTION

47.10.1 Supervisor's Role

The responsibilities of the first-line supervisor are many. Direction of the work force includes the following supervisory functions:

- Setting goals.
- Improving present work methods.
- Delegating work.
- Allocating manpower.
- Meeting deadlines.
- Controlling expenditures.
- Following progress of work.
- Evaluating employee performance.
- Forecasting manpower requirements.
- Supervising on-the-job training.
- Reviewing employee performance.
- Handling employee complaints.
- Enforcing rules.
- Conducting meetings.
- Increasing safety awareness.*

Supervisory understanding of the interrelationships of these responsibilities is a learned attribute. Organizations that expect their supervisors to offer a high quality of leadership to their employees must provide appropriate training and experiential opportunities to current supervisors and supervisory trainees alike.

47.10.2 Elements of Accident Prevention†

- Safety policy must be clearly defined and communicated to all employees.
- The safety record of a company is a barometer of its efficiency. An American Engineering Council study revealed "maximum productivity is ordinarily secured only when the accident rate tends toward the unreducible minimum."
- Unless line supervisors are accountable for the safety of all employees, no safety program will be effective. Top management must let all supervisors and managers know what is expected of them in safety.
- Periodic progress reports are required to let managers and employees know what they have accomplished in safety.
- Meetings with supervisors and managers to review accident reports, compensation costs, accident-cause analysis, and accident-prevention procedures are important elements of the overall safety program.
- The idea of putting on a big safety campaign with posters, slogans, and safety contests is wrong. The Madison Avenue approach does not work over the long run.
- Good housekeeping and the enforcement of safety rules show that management has a real concern for employee welfare. They are important elements in the development of good morale. (A

---

* B. D. Lewis, Jr., *The Supervisor in 1975*, copyright © September 1973. Reprinted with permission of *Personnel Journal*, Costa Mesa, CA; all rights reserved.

† R. De Reamer, *Modern Safety and Health Technology*. Copyright © 1980. Reprinted by permission of Wiley, New York.

Table 47.8 Requirements for Safety Files[a]

The following items are presented for your convenience as you review your administrative storage index to determine the adequacy of your safety-related files.

Number	Action Required	Action Completed Yes	No
1.	Is there a separate section for safety-related files?		
2.	Are the following subjects provided for in the safety section of the files:		
	a. Blank OSHA forms?		
	b. Completed OSHA forms?		
	c. Blank company safety forms?		
	d. Completed company safety forms?		
	e. Blank safety checklists?		
	f. Completed safety checklists?		
	g. Agendas of company safety meetings?		
	h. Minutes of company safety meetings?		
	i. Records of safety equipment purchases?		
	j. Records of safety equipment checkouts?		
	k. Incoming correspondence related to safety?		
	l. Outgoing correspondence related to safety?		
	m. Record of safety projects assigned?		
	n. Record of safety projects completed?		
	o. Record of fire drills (if applicable)?		
	p. Record of external assistance used to provide specialized safety expertise?		
	q. Record of inspections by fire department, insurance companies, state and city inspectors, and OSHA compliance officers?		
	r. National Safety Council catalogs and brochures for films, posters, and other safety-related materials?		
3.	Are the files listed in item 2 reviewed periodically:		
	a. To ensure that they are current?		
	b. To retire material over five years old?		
4.	Are safety-related files reviewed periodically to determine the need to eliminate selected files and to add new subjects?		
5.	Is the index to the file current, so that an outsider could easily understand the system?		

[a]J. B. Re Velle, *Safety Training Methods.* Copyright © 1980. Reprinted by permission of Wiley, New York.

U.S. Department of Labor study has revealed that workers are vitally concerned with safety and health conditions of the workplace. A surprisingly high percentage of workers ranked protection against work-related injuries and illness and pleasant working conditions as having a priority among their basic on-the-job needs. In fact, they rated safety higher than fringe benefits and steady employment.)

- The use of personal protective equipment (safety glasses, safety shoes, hard hats, etc.) must be a condition of employment in all sections of the plant where such protection is required.
- Safety files must be complete and up to date to satisfy internal information requirements as well as external inspections by OSHA Compliance Officers and similar officials (Table 47.8).

47.10.3 Management Principles*

- Regardless of the industry or the process, the role of supervisors and managers in any safety program takes precedence over any of the other elements. This is not to say that the managerial role is necessarily more important than the development of safe environments, but without manager and supervisor participation, the other elements have a lukewarm existence. There is a dynamic relationship between management and the development of safe working conditions,

* R. De Reamer, *Modern Safety and Health Technology.* Copyright © 1980. Reprinted by permission of Wiley, New York.

and management and the development of safety awareness, and the relationship must not be denied.

- Where responsibility for preventing accidents and providing a healthful work environment is sloughed off to the safety department or a safety committee, any reduction in the accident rate is minimal. To reduce the accident rate, and in particular, to make a good rate better, line managers must be held responsible and accountable for safety. Every member of the management team must have a role in the safety program. Admittedly, this idea is not new, but application of the concept still requires crystal-clear definition and vigorous promotion.
- Notwithstanding the many excellent examples of outstanding safety records that have been achieved because every member of management had assumed full responsibility for safety, there are still large numbers of companies, particularly the small establishments, using safety contests, posters, or safety committees as the focal point of their safety programs—but with disappointing results. Under such circumstances safety is perceived as an isolated aspect of the business operation with rather low ceiling possibilities at best. But there are some who feel that gimmicks must be used because foremen and the managers do not have time for safety.
- As an example of the case in point, a handbook on personnel contains the statement that "A major disadvantage of some company-sponsored safety programs is that the supervisor can't spare sufficient time from his regular duties for running the safety program." Significantly, this was not a casual comment in a chapter on safety. It was indented and in bold print for emphasis. Yet it is a firmly accepted fact that to achieve good results in safety, managers and supervisors *must* take the time to fulfill their safety responsibilities. Safety is one of their *regular* duties.
- The interrelationships of the many components of an effective industrial safety program are portrayed in Fig. 47.2.

47.10.4 Eliminating Unsafe Conditions*

The following steps should be taken to effectively and efficiently eliminate an unsafe condition:

- *Remove.* If at all possible, have the hazard eliminated.
- *Guard.* If danger point (i.e., high tension wires) cannot be removed, see to it that hazard is shielded by screens, enclosures, or other guarding devices.
- *Warn.* If guarding is impossible or impractical, warn of the unsafe condition. If a truck must back up across a sidewalk to a loading platform, the sidewalk cannot be removed or a fence built around the truck. All that can be done is to warn that an unsafe condition exists. This is done by posting a danger sign or making use of a bell, horn, whistle, signal light, painted striped lines, red flag, or other device.
- *Recommend.* If you cannot remove or guard an unsafe condition on your own, notify the proper authorities about it. Make specific recommendations as to how the unsafe condition can be eliminated.
- *Follow Up.* After a reasonable length of time, check to see whether the recommendation has been acted on, or whether the unsafe condition still exists. If it remains, the person or persons to whom the recommendations were made should be notified.

The following factors should be considered in organizing a plant that provides for maximum productivity and employee well-being:

- The general arrangement of the facility should be efficient, orderly, and neat.
- Workstations should be clearly identified so that employees can be assigned according to the most effective working arrangement.
- Material flow should be designed to prevent unnecessary employee movement for given work.
- Materials storage, distribution, and handling should be routinized for efficiency and safety.
- Decentralized tool storage should be used wherever possible. Where centralized storage is essential (e.g., general supply areas, locker areas, and project storage areas), care should be given to establish a management system that will avoid unnecessary crowding or congested traffic flow. (Certain procedures, such as time staggering, may reduce congestion.)

* J. B. Re Velle, *Safety Training Methods.* Copyright © 1980. Reprinted by permission of Wiley, New York.

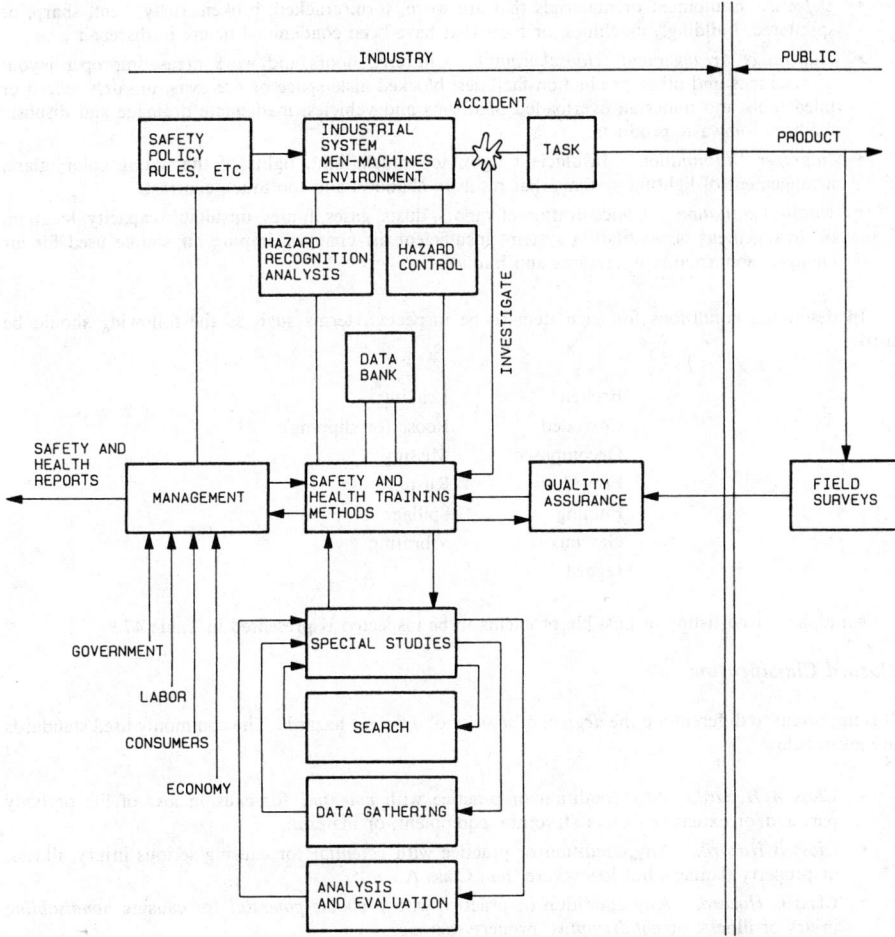

Fig. 47.2 Basic functions of an effective safety program. Reprinted with permission from *Industrial Engineering Magazine.* Copyright © American Institute of Industrial Engineers, Inc., 25 Technology Park/Atlanta, Norcross, GA 30092.

- Time-use plans should be established for frequently used facilities to avoid having workers wait for a particular apparatus.
- A warning system and communications network should be established for emergencies such as fire, explosion, storm, injuries, and other events that would affect the well-being of employees.

The following unsafe conditions checklist presents a variety of undesirable characteristics to which both employers and employees should be alert:

- *Unsafe Conditions—Mechanical Failure.* These are types of unsafe conditions that can lead to occupational accidents and injuries. *Note:* Keep in mind that unsafe conditions often come about as a result of unsafe acts.
- *Lack of Guards.* This applies to hazardous places like platforms, catwalks, or scaffolds where no guardrails are provided; power lines or explosive materials that are not fenced off or enclosed in some way; and machines or other equipment where moving parts or other danger points are not safeguarded.
- *Inadequate Guards.* Often a hazard that is partially guarded is more dangerous than it would be if there were no guards. The employee, seeing some sort of guard, may feel secure and fail to take precautions that would ordinarily be taken if there were no guards at all.

- *Defects.* Equipment or materials that are worn, torn, cracked, broken, rusty, bent, sharp, or splintered; buildings, machines, or tools that have been condemned or are in disrepair.
- *Hazardous Arrangement (Housekeeping).* Cluttered floors and work areas; improper layout of machines and other production facilities; blocked aisle space or fire exits; unsafely stored or piled tools and material; overloaded platforms and vehicles; inadequate drainage and disposal facilities for waste products.
- *Improper Illumination.* Insufficient light; too much light; lights of the wrong color; glare; arrangement of lighting systems that result in shadows and too much contrast.
- *Unsafe Ventilation.* Concentration of vapors, dusts, gases, fumes; unsuitable capacity, location, or arrangement of ventilation system; insufficient air changes, impure air source used for air changes; abnormal temperatures and humidity.

In describing conditions for each item to be inspected, terms such as the following should be used:

Broken	Leaking
Corroded	Loose (or slipping)
Decomposed	Missing
Frayed	Rusted
Fuming	Spillage
Gaseous	Vibrating
Jagged	

An alphabetized listing of possible problems to be inspected is presented in Table 47.9.

Hazard Classification

It is important to differentiate the *degrees of severity* of different hazards. The commonly used standards are given below.

- *Class A Hazard.* Any condition or practice with *potential* for causing *loss* of life or body part and/or extensive loss of structure, equipment, or material.
- *Class B Hazard.* Any condition or practice with *potential* for causing serious injury, illness, or property damage, but less severe than Class A.
- *Class C Hazard.* Any condition or practice with *probable potential* for causing *nondisabling* injury or illness, or *nondisruptive* property damage.

47.10.5 Unsafe Conditions Involving Mechanical or Physical Facilities*

The total working environment must be under constant scrutiny because of changing conditions, new employees, equipment additions and modifications, and so on. The following checklist is presented as a guide to identify potential problems:

1. **Building**
   Correct ceiling height.
   Correct floor type; in acceptable condition.
   Adequate illumination.
   Adequate plumbing and heating pipes and equipment.
   Windows with acceptable opening, closing, and holding devices; protection from breakage.
   Acceptable size doors with correct swing and operational quality.
   Adequate railing and nonslip treads on stairways and balconies.
   Adequate ventilation.
   Adequate storage facilities.
   Adequate electrical distribution system in good condition.
   Effective space allocation.
   Adequate personal facilities (restrooms, drinking fountains, washup facilities, etc.).

Table 47.9 List of Possible Problems to Be Inspected[a]

Acids	Dusts	Railroad cars
Aisles	Electric motors	Ramps
Alarms	Elevators	Raw materials
Atmosphere	Explosives	Respirators
Automobiles	Extinguishers	Roads
Barrels	Flammables	Roofs
Bins	Floors	Safety devices
Blinker lights	Forklifts	Safety glasses
Boilers	Fumes	Safety shoes
Borers	Gas cylinders	Scaffolds
Buggies	Gas engines	Shafts
Buildings	Gases	Shapers
Cabinets	Hand tools	Shelves
Cables	Hard hats	Sirens
Carboys	Hoists	Slings
Catwalks	Hoses	Solvents
Caustics	Hydrants	Sprays
Chemicals	Ladders	Sprinkler systems
Claxons	Lathes	Stairs
Closets	Lights	Steam engines
Connectors	Mills	Sumps
Containers	Mists	Switches
Controls	Motorized carts	Tanks
Conveyors	Piping	Trucks
Cranes	Pits	Vats
Crossing lights	Platforms	Walkways
Cutters	Power tools	Walls
Docks	Presses	Warning devices
Doors	Racks	

[a]*Principles and Practices of Occupational Safety and Health: A Programmed Instruction Course,* OSHA 2213, Student Manual Booklet 1, U.S. Department of Labor, Washington, DC, p. 40.

Efficient traffic flow.
Adequate functional emergency exits.
Effective alarms and communications systems.
Adequate fire prevention and extinguishing devices.
Acceptable interior color scheme.
Acceptable noise absorption factor.
Adequate maintenance and cleanliness.

2. **Machinery and Equipment**
Acceptable placement, securing, and clearance.
Clearly marked safety zones.
Adequate nonskid flooring around machines.
Adequate guard devices on all pulleys.
Sharp, secure knives and cutting edges.
Properly maintained and lubricated machines, in good working condition.
Functional, guarded, magnetic-type switches on all major machines.
Properly wired and grounded machines.
Functional hand and portable power tools, in good condition and grounded.
Quality machines adequate to handle the expected work load.
Conspicuously posted safety precautions and rules near each machine.
Guards for all pinch points within 7 ft of the floor.

47.11 SAFETY TRAINING

47.11.1 Specialized Courses*

First-Aid Training

First-aid courses pay big dividends in industry. This statement is based on clear evidence that people trained in first aid are more safety conscious and less likely to have an accident.

The importance of first-aid training from the safety standpoint is that it teaches much more than applying a bandage or a splint. According to the Red Cross, "The primary purpose of first aid training is the prevention of accidents." Each lesson teaches the student to analyze (1) how the accident happened, (2) how the accident could have been prevented, and (3) how to treat the injury. But the biggest dividend of first-aid training is the lives that have been saved because trainees were prepared to apply mouth-to-mouth resuscitation, to stop choking using the Heimlich maneuver (ejection of foreign object by forceful compression of diaphragm), or to stem the flow of blood.

Since the OSHAct, first-aid training has become a matter of federal law—the act stipulates that in absence of an infirmary, clinic, or hospital in proximity to the workplace, a person or persons shall be adequately trained to render first aid. The completion of the basic American National Red Cross first-aid course will be considered as having met this requirement. Just what constitutes *proximity* to a clinic or hospital? The OSH Review Commission recognizing that first aid must be given within 3 min of serious accidents concluded that an employer whose plant had no one trained in first aid present and was located 9 min from the nearest hospital violated the standard (1910.151, Medical and First Aid).

Driver Training

The number 1 accident killer of *employees* is the traffic accident. Each year more than 27,000 workers die in non-work-related motor-vehicle accidents, and an additional 3900 employees are killed in work-related accidents. The employer pays a heavy toll for these accidents. Those that are work related are compensable, but the others are, nonetheless, costly. The loss of a highly skilled worker, a key scientist, or a company executive could have a serious impact on the success of the business.

There is, fortunately, something constructive that employers can do to help protect their employees and their executives from the tragedy and waste of traffic accidents. Driver training for workers and executives can be provided either in-house or through community training agencies.

Companies that have conducted driver-training programs report that the benefits of such training were not limited to the area of improved traffic-accident performance. These companies also experienced lower on-the-job injury frequency rates (the training produced an increase in safety awareness) and improved employee–community relations.

Companies have taken several approaches to driver training:

- A course has been made available to employees on a volunteer basis, either on or off hours.
- Driver training has been made mandatory for employees who operate a motor vehicle on company business.
- The company has promoted employee attendance at community-agency-operated programs.
- Full-scale driver-training programs have been conducted for all employees and members of their families. This is done off-hours, and attendance is voluntary.

Fire Protection Training

All employees must know what to do when a fire alarm sounds. All employees must know something about the equipment provided for fire protection and what they can do toward preventing a fire. They must know:

- The plan established for evacuation of the building in case of emergency.
- How to use the first-aid fire appliances provided (extinguishers, hose, etc.).
- How to use other protective equipment. (Every employee should know that water to extinguish fires comes out of the pipes of the sprinkler systems and that stock must not be piled so close to sprinkler lines that it prevents good distribution of water from sprinkler heads on a fire in the piled material. They should know that fire doors must be kept operative and not obstructed by stock piles, tools, or other objects.)

* R. De Reamer, *Modern Safety and Health Technology.* Copyright © 1980. Reprinted by permission of Wiley, New York.

- How to give a fire alarm and how to operate plant fire alarm boxes and street boxes in the public alarm system.
- Where smoking in the plant is permitted and where, for fire-safety reasons, it is prohibited.
- The housekeeping routine (disposal of wiping rags and waste, handling of packing materials, and other measures for orderliness and cleanliness throughout the plant).
- Hazards of any special process in which the employee is engaged.

All these "what-to-do" items can appropriately be covered in training sessions and evacuation drills.

Other Specialized Courses

Some of the other specialized courses that can be given for safety training are:

- Accident investigation
- Accident report preparation
- Hazard inspection
- Personal protective equipment
- Powered equipment and vehicles
- Safety recordkeeping
- Specific disasters

47.11.2   Job Hazard Analysis Training*

Admittedly, the conventional mass approach to safety training takes little of the supervisor's time. Group training sessions, safety posters, films, and booklets are handled by the plant safety engineer or other staff people. On the other hand, where safety training is carried out on a personalized basis, the first-line supervisor must necessarily do the training. This will take more of his or her time and require more attention to detail, but this additional effort pays off because of the increased effectiveness of the training method.

In launching a personalized safety-training program, the first step is the preparation of a job-hazard analysis for each job in the plant. To make the job-hazard analysis in an organized manner, use of a form similar to the one shown in Table 47.10 is suggested. The key elements of the form are: (1) job description; (2) job location; (3) key job steps; (4) tool used; (5) potential health and injury hazards; and (6) safe practices, apparel, and equipment.

A review of the form will indicate the steps in making a job-hazard analysis. To start an analysis, the key steps of the job are listed in order in the first column of the form. Where pertinent, the tool used to perform the job step is listed in the second column. Then, in the third column opposite each job step, the hazards of the particular step are indicated. Finally, in the fourth column of the form are listed the safe practices the employee must be shown and discussed. Here the supervisor lists the safe work habits that must be stressed and the safety equipment and clothing required for the job.

In making the analysis, an organized approach is required so the less obvious accident hazards will not be missed. This means going out on the floor and actually watching the job being performed and jotting down key steps and hazards as they are observed. Supervisors who make such a job-hazard analysis are often surprised to find hazards in the job cycle that they had missed seeing in the past. Their original negative reaction to the thought of additional paperwork soon disappears. In the long run, supervisors realize that proper hazard analysis will help them do a better training job.

As previously stated, a job-hazard analysis is made for each job. In most cases, each supervisor will have to make from 5 to 10 different analyses. Of course, in maintenance and construction work, the variety of jobs covers a much wider range. Fortunately, these jobs can be grouped by the type of work performed and a job-hazard analysis can be made for each category of work, rather than for each job. For example, repair, installation, and relocation of equipment; cleaning motors; and unloading cars might be a few of the various categories of maintenance work to be analyzed.

47.11.3   Management's Overview of Training*

An effective accident prevention program requires proper job performance from everyone in the workplace. All employees must know about the materials and equipment they are working with, what

---

* R. De Reamer, *Modern Safety and Health Technology*. Copyright © 1980. Reprinted by permission of Wiley, New York.
* J. B. Re Velle, *Safety Training Methods*. Copyright © 1980. Reprinted by permission of Wiley, New York.

Table 47.10 Job Hazard Analysis

Job Description: __Three Spindle Drill Press—Impeller 34C6__

Job Location: __Bldg. 19-2, Pump Section__

Key Job Steps	Tool Used	Potential Health and Injury Hazard	Safe Practices, Apparel, and Equipment
Get material for operation	Tote box	Dropping tote box on foot.	Wear safety shoes. Have firm grip on box.
		Back strain from lifting.	Stress proper lifting methods.
		Picking up overloaded boxes.	Tell employee to get help or lighten load.
Inspect and set up drill press	Drill press	Check for defective machines.	Do not operate if defective. Attach red or yellow "do not operate" tag.
		Chuck wrench not removed.	Always remove chuck wrench immediately after use.
		Making adjustments when machine is running.	Always stop spindle before making adjustments.
Drilling		Hair, clothing, or jewelry catching on spindle.	Wear head covering, snug-fitting clothing. No loose sleeves. Avoid wearing rings, bracelets, or wristwatches.
		Spinning work or fixture.	Use proper blocks or clamps to hold work and fixture securely.
		Injury to hands—cuts, etc.	*Never wear gloves.* Use hook, brush, or other tool to remove chips. Use compressed air only when instructed.
		Drill sticks in work.	Stop spindle, free drill by hand.
		Flying chips.	Wear proper eye protection.
		Pinch points at belts.	Always stop press before adjusting belts.
		Broken drills.	Do not attempt to force drill, apply pressure.

James Black

Signature

4/22/86   1 of 3

Date   Page

hazards are known in the operation, and how these hazards have been controlled or eliminated. Each individual employee needs to know and understand the following points (especially if they have been included in the company policy and in a "code of safe practices"):

No employee is expected to undertake a job until he or she has received instruction on how to do it properly and has been authorized to perform that job.

No employee should undertake a job that appears to be unsafe.

Mechanical safeguards are in place and must be kept in place.

Each employee is expected to report all unsafe conditions encountered during work.

Even slight injury or illness suffered by an employee must be reported at once.

In addition to the points above, any safety rules that are a condition of employment, such as the use of safety shoes or eye protection, should be explained clearly and enforced at once.

The first-line supervisors must know how to train employees in the proper way of doing their jobs. Encourage and consider providing for supervisory training for these supervisors. (Many colleges offer appropriate introductory management training courses.)

Some specific training requirements in the OSHA standards must be met, such as those that pertain to first aid and powered industrial trucks (including forklifts). In general, they deal with situations where the use of untrained or improperly trained operators on skill machinery could cause hazardous situations to develop, not only for the operator, but possibly for nearby workers, too.

Particular attention must be given to new employees. Immediately on arriving at work, new employees begin to learn things and to form attitudes about the company, the job, their boss, and their fellow employees. Learning and attitude formation occur regardless of whether the employer makes a training effort. If the new employees are trained during those first few hours and days to do things the right way, considerable losses may be avoided later.

At the same time, attention must be paid to regular employees, including the old-timers. Old habits can be wrong habits. An employee who continues to repeat an unsafe procedure is not working safely, even if an "accident" has not resulted from this behavior.

Although every employee's attitude should be one of determination that "accidents" can be prevented, one thing more may be needed. It should be stressed that the responsibility assigned to the person in charge of the job—as well as to all other supervisors—is to be sure that there is a concerted effort under way at all times to follow every safe work procedure and health practice applicable to that job. It should be clearly explained to these supervisors that they should never silently condone unsafe or unhealthful activity in or around any workplace.

BIBLIOGRAPHY

"Accident Prevention: Your Key to Controlling Surging Workers' Compensation Costs," *Occupational Hazards,* 35 (November 1979).

"Accident Related Losses Make Cost Soar," *Industrial Engineering,* 26 (May 1979).

"Analyzing a Plant Energy-Management Program: Part I—Measuring Performance," *Plant Engineering,* 59 (October 30, 1980).

"Analyzing a Plant Energy-Management Program: Part II—Forecasting Consumption" *Plant Engineering,* 149 (November 13, 1980).

"Anatomy of a Vigorous In-Plant Program," *Occupational Hazards,* 32 (July 1979).

"A Shift Toward Protective Gear," *Business Week,* 56H (April 13, 1981).

"A Win for OSHA," *Business Week,* 62 (June 29, 1981).

"Buyers Should Get Set for Tougher Safety Rules," *Purchasing,* 34 (May 25, 1976).

"Complying with Toxic and Hazardous Substances Regulations—Part I," *Plant Engineering,* 283 (March 6, 1980).

"Complying with Toxic and Hazardous Substances Regulations—Part II," *Plant Engineering,* 157 (April 17, 1980).

"Computers Help Pinpoint Worker Exposure," *Chemecology,* 11 (May 1981).

"Conserving Energy by Recirculating Air from Dust Collection Systems," *Plant Engineering,* 151 (April 17, 1980).

"Control Charts Help Set Firm's Energy Management Goals," *Industrial Engineering,* 56 (December 1980).

"Controlling Noise and Reverberation with Acoustical Baffles," *Plant Engineering,* 131 (April 17, 1980).

"Controlling Plant Noise Levels," *Plant Engineering,* 127 (June 24, 1976).

"Cost-Benefit Decision Jars OSHA Reform," *Industry Week,* 18 (June 29, 1981).

"Cost Factors for Justifying Projects," *Plant Engineering,* 145 (October 16, 1980).

"Costs, Benefits, Effectiveness, and Safety: Setting the Record Straight," *Professional Safety,* 28 (August 1975).

"Costs Can Be Cut Through Safety," *Professional Safety,* 34 (October 1976).

"Cutting Your Energy Costs," *Industry Week,* 43 (February 23, 1981).

R. De Reamer, *Modern Safety and Health Technology,* Wiley, New York, 1980.

"Elements of Effective Hearing Protection," *Plant Engineering,* 203 (January 22, 1981).

"Energy Constraints and Computer Power Will Greatly Impact Automated Factories in the Year 2000," *Industrial Engineering,* 34 (November 1980).

"Energy Managers Gain Power," *Industry Week,* 62 (March 17, 1980).

"Energy Perspective for the Future," *Industry Week,* 67 (May 26, 1980).

"Engineering and Economic Considerations for Baling Plant Refuse, 34 (April 30, 1981).

Engineering Control Technology Assessment for the Plastics and Resins Industry, NIOSH Research Report Publication No. 78-159.

"Engineering Project Planner, A Way to Engineer Out Unsafe Conditions," *Professional Safety,* 16 (November 1976).

"EPA Gears Up to Control Toxic Substances," *Occupational Hazards,* 68 (May 1977).

T. S. Ferry, *Safety Management Planning,* The Merritt Company, Santa Monica, CA, 1982.

"Fume Incinerators for Air Pollution Control," *Plant Engineering,* 108 (November 13, 1980).

"Groping for a Scientific Assessment of Risk," *Business Week,* 120J (October 20, 1980).

"Hand and Body Protection: Vital to Safety Success," *Occupational Hazards,* 31 (February 1979).

"Hazardous Wastes: Coping with a National Health Menace," *Occupational Hazards,* 56 (October 1979).

"Hearing Conservation—Implementing an Effective Program," *Professional Safety,* 21 (October 1978).

"How Do You Know Your Hazard Control Program is Effective?," *Professional Safety,* 18 (June 1981).

"How to Control Noise," *Plant Engineering,* 90 (October 5, 1972).

"Human Factors Engineering—A Neglected Art," *Professional Safety,* 40 (March 1978).

"IE Practices Need Reevaluation Due to Energy Trends," *Industrial Engineering,* 52 (December 1980).

"Industrial Robots: A Primer on the Present Technology," *Industrial Engineering,* 54 (November 1980).

"Job-Safety Equipment Comes Under Fire, Are Hard Hats a Solution or a Problem?," *The Wall Street Journal,* 40 (November 18, 1977).

W. G. Johnson, *MORT Safety Assurance Systems,* Marcel Dekker, Inc., New York, 1979.

R. A. McFarland, "Application of Human Factors Engineering to Safety Engineering Problems," *National Safety Congress Transactions,* National Safety Council, Chicago, 1967, Vol. 12.

"New OSHA Focus Led to Noise-Rule Delay," *Industry Week,* 13 (June 15, 1981).

"OSHA Communique," *Occupational Hazards,* 27 (June 1981).

"OSHA Moves Health to Front Burner," *Purchasing,* 46 (September 26, 1979).

"OSHA to Analyze Costs, Benefits of Lead Standard," *Occupational Health & Safety,* 13 (June 1981).

Patty's Industrial Hygiene and Toxicology, 3rd revised ed., Wiley-Interscience, New York, 1978, Vol. 1.

"Practical Applications of Biomechanics in the Workplace," *Professional Safety,* 34 (July 1975).

"Private Sector Steps Up War on Welding Hazards," *Occupational Hazards,* 50 (June 1981).

"Putting Together a Cost Improvement Program," *Industrial Engineering,* 16 (December 1979).

"Reduce Waste Energy with Load Controls," *Industrial Engineering,* 23 (July 1979).

"Reducing Noise Protects Employee Hearing," *Chemecology,* 9 (May 1981).

"Regulatory Relief Has Its Pitfalls, Too," *Industry Week,* 31 (June 29, 1981).

J. B. Re Velle, *Safety Training Methods,* Wiley, New York, 1980.

"ROI Analysis for Cost-Reduction Projects," *Plant Engineering,* 109 (May 15, 1980).

"Safety & Profitability—Hand in Hand," *Professional Safety,* 36 (March 1978).

"Safety Managers Must Relate to Top Management on Their Terms," *Professional Safety,* 22 (November 1976).

"Superfund Law Spurs Cleanup of Abandoned Sites," *Occupational Hazards,* 67 (April 1981).

"Taming Coal Dust Emissions," *Plant Engineering,* 123 (May 15, 1980).

"The Cost-Benefit Argument—Is the Emphasis Shifting?," *Occupational Hazards,* 55 (February 1980).

"The Cost/Benefit Factor in Safety Decisions," *Professional Safety,* 17 (November 1978).

"The Design of Manual Handling Tasks," *Professional Safety,* 18 (March 1980).

"The Economics of Safety . . . A Review of the Literature and Perspective," *Professional Safety,* 31 (December 1977).

"The Hidden Cost of Accidents," *Professional Safety,* 36 (December 1975).

"The Human Element in Safe Man-Machine Systems," *Professional Safety,* 27 (March 1981).

"The Problem of Manual Materials Handling," *Professional Safety,* 28 (April 1976).

"Time for Decisions on Hazardous Waste," *Industry Week,* 51 (June 15, 1981).

"Tips for Gaining Acceptance of a Personal Protective Equipment Program," *Professional Safety,* 20 (March 1976).

"Toxic Substances Control Act," *Professional Safety,* 25 (December 1976).

"TSCA: Landmark Legislation for Control of Chemical Hazards," *Occupational Hazards,* 79 (May 1977).

"Were Engineering Controls 'Economically Feasible'?," *Occupational Hazards,* 27 (January 1981).

"Were Noise Controls 'Technologically Feasible'?," *Occupational Hazards*, 37 (January 1981).

"What Are Accidents Really Costing You?" *Occupational Hazards*, 41 (March 1979).

"What's Being Done About Hazardous Wastes?," *Occupational Hazards*, 63 (April 1981).

"Where OSHA Stands on Cost-Benefit Analysis," *Occupational Hazards*, 49 (November 1980).

"Worker Attitudes and Perceptions of Safety," *Professional Safety*, 28 (December 1981); 20 (January 1982).

CHAPTER 48

ENGINEERS' LIABILITY

VICTOR JOHN YANNACONE, JR.

Patchogue, New York

48.1 ENGINEERS' LIABILITY

The relationships between engineers and the products and projects they design and develop, or of which they supervise the construction, fabrication, or manufacture are as rich and varied as the profession itself. Legal responsibility or liability, however, is determined from the legal relationship that exists between engineers and their employers, their clients, consumers, and the public-at-large, all of whom may become parties to a lawsuit. Whether any individual engineer is a plaintiff or a defendant or perhaps both in any particular lawsuit depends on the circumstances of each incident when considered in the context of these relationships.

Engineers generally act in the eyes of the law as agents for others, the usual status being that of employee, although engineers may be partners or joint venturers or independent consultants.

48.2 THE ENGINEER AS AGENT

Most engineers in the United States today do not practice as independent professionals but are instead employed by corporate enterprises of some kind. The licensed professional engineer who is employed by an engineering consulting firm is still an employee. Engineers who are employed have the power to obligate their employers in matters within the known or apparent scope of the authority associated with their job title and function.

When a firm has given a particular engineer "sign-off authority" with respect to any defined aspect of a project, "signing off" is the equivalent of approval by the entire firm, no matter how large and complex a corporate entity it might be. Those who deal directly with such an employee, such as other engineers employed by other firms, vendors, and subcontractors, are, however, required to ascertain the scope of the individual engineer's authority. Although evidence about the scope of authority for an agent or employee must come from the alleged source of the authority, not the agent or employee, the legal doctrine concept of "apparent authority" complicates matters.

Where an agent or employee has sufficient appearance of authority as would lead a reasonable party dealing with the individual to believe actual authority existed, authority is "apparent" and the employer is bound by the actions of the employee or agent even though it could later be shown that the employee was not only acting outside the scope of authority granted, but had been expressly forbidden by the employer to act in that manner. When apparent authority can be inferred from the trappings and circumstances of the activity, the employer is prevented from raising "lack of authority" as a defense against claims based on the employee's acts or conduct.

Engineers who are employees or agents can be free of personal liability to third parties for the contractual and commercial obligations of their employer, even though they may be the person who spoke for the employer in a transaction, so long as it is clear from the circumstances of the transaction that they are acting as an employee or agent. This limitation on the liability of individual engineers is important and is easily protected by making sure that the employment or agency nature of the relation is always disclosed.

Nevertheless, even where employment or agency actually exists, unless it is unequivocally clear that the individual engineer is acting on behalf of an employer or other disclosed principal, a third party has the right to proceed against either the engineer or the employer/principal under the rule that an agent for an undisclosed principal is liable on the transaction together with his or her principal. Individual engineers acting as employees, or agents for a disclosed principal, should indicate as part of their "business" signature the status they hold, and the authority under which they execute any document, no matter how trivial or personal it may seem at the time. This is particularly important with respect to contracts, plans, specifications, purchase orders, and documents dealing with quality assurance and quality control.

Even though engineers acting with apparent authority may be protected from the direct claims of a third party, they are not protected from claims for contribution and indemnification by their employer or principal where they exceed the scope of their authority and their employer or principal becomes liable for the damage associated with their unauthorized acts. Because the errors, omissions, carelessness, and negligence of an engineer may be deemed "unauthorized" acts, permitting an employer to look to professional employees such as engineers for contribution and indemnity, some "hold harmless" agreement should be included in any contract for engineering services. The rules of agency exist as a snare for the unwary, and they should be considered in fashioning any employment agreement to which an engineer may be a party.

48.3 THE ENGINEER AS EMPLOYEE

As a condition of employment, an engineer may be requested to sign a variety of "employment" agreements—assignments of patents, confidentiality and "noncompetition" agreements. Noncompetition "clauses," which are agreements not to compete with the business efforts of the present employer, either as an independent or as the employee or agent of another, are usually only enforceable to the extent that they continue for a reasonable time after the termination of the present employment and are limited in occupational scope.

Occasionally an engineer may be asked to execute an "indemnity and hold harmless" agreement, in effect, agreeing to assume any responsibility befalling the employer as a result of the engineer employee's acts or omissions. While in practice it may seem difficult to enforce such agreements, the request is a strong sign of the employer's attitude toward personal accountability. Any engineer employee should ascertain that the employer is insured against professional errors and omissions, and that such insurance also covers the liability of individual employed engineers. Otherwise the engineer employee should obtain such insurance personally.

One aspect of the law of agency that arises often during projects involving many engineers employed by many different contractors and vendors is the right to assume that communication with an engineer involved in a particular aspect of a complex project constitutes communication to the employer of that engineer. The rule of law is essentially the rule of reason. If the individual receiving the communication has responsibility for the subject matter of the communication, then it is reasonable to assume that the communication has been properly received. Similarly, if it is known that the person to whom the communication is directed is obligated to report directly to the person affected by the communication, it may be said that the information has been properly transmitted. The obligation to transmit the information, however, must be direct and immediate.

With some exceptions, the rules of agency, developed primarily for commercial transactions, are generally respected in tort actions (actions claiming damages for bodily injury, and other injury to persons or property). An employer is generally liable for the negligence, carelessness, errors and omissions of its engineer employees, even if the employer may have forbidden the conduct of the employee before it took place and condemned the conduct afterward. On the other hand, an engineer employee cannot escape personal responsibility and liability as an engineer merely by showing that the professional malpractice was committed while acting as an employee or agent for a disclosed principal. This may entitle the employee to share responsibility with the employer, but not to evade it.

48.4   THE ENGINEER AS JOINT VENTURER

Among the forms of legal association available to modern engineering, construction, and technology none is more versatile than the joint venture. As the name implies, a joint venture is nothing more than the temporary consolidation of disparate talents and resources to accomplish a particular objective or attain a particular business or scientific goal. The *Research and Development Limited Partnership,* so common as a tax shelter, is but one particular form of "joint venture."

At its most formal, the joint venture will be organized as a discrete legal entity. Partnerships, whether among individuals, existing partnerships, corporations of varying size, or all of the business entities known to the law, are the common form under which joint ventures proceed. Limited partnerships permit some participants in the joint venture to share in the profits, but limit their liability for losses to the amount of their direct financial contribution to the venture, much like stockholders in a public corporation. The joint venture also may assume the corporate form with the participants becoming "true stockholders" in the venture. The usual rule is to structure the joint venture so as to gain the most favorable economic and tax status available at the time.

Although in a well-organized joint venture the roles and obligations of the various participants will be defined in a formal written agreement, and independent books and records (even a separate office) maintained, there are other joint ventures involving engineers where the organization is considerably less formal, often without any written agreement at all. In the absence of a written agreement, whether any relationship involving engineers (other than as employees or agents of a disclosed principal) is a joint venture requires answers to many questions of fact before the legal rights and responsibilities of the parties can be determined.

Once a joint venture has been established, whether formally by some written agreement or by determination of a court following some legal action, certain rules of agency apply.

One party to a joint venture generally may not legally bind other parties to the joint venture as individuals. Only the joint venture as an entity may be bound by one party to the joint venture acting with real or apparent authority. The rules of agency involving disclosed and undisclosed principals and real and apparent authority apply, but only with respect to the legal principal, the joint venture itself.

The liability of a joint venture may be affected by its legal form. While partnerships are often seen as a more flexible business organization and may hold certain tax advantages over corporations, they lack the limitation of liability inherent in the corporate entity. Any liability incurred by the joint venture may be passed directly through to the individual partners. Incorporating the joint venture permits the participants to limit the liability for the venture to the assets of the joint venture corporation provided only that the participants have respected the corporate independence of the joint venture. In general, establishing the corporate independence of a joint venture in order to limit liability depends on organizing the venture with adequate working capital; maintaining independent books, records, and bank accounts; avoiding any commingling of funds with those of the individual parties to the joint venture; and conducting the business of the joint venture separate and apart from the business activities of the participants.

Even though individuals employed by or acting for a corporation may be held liable in tort for their own acts and omissions, the corporation as a legal entity is respected and its stockholders are immune from liability even in tort actions.

Where the joint venture is organized as a general partnership, or where the "corporate veil is pierced" because of failure to establish and maintain corporate independence of the joint venture, the participants are jointly and severally responsible as partners for the liabilities of the joint venture. Joint and several liability means that even though there may be many partners in the venture, any one of them can be required to pay an entire judgment then *seek pro rata* contribution from the other venturers.

The laws of several states allow formation of a partnership that includes one general partner and any number of "limited partners" who are, essentially, investors, but who also may have active roles in the partnership. The only party with complete operating and management authority in a limited partnership is the general partner. Unless delegated specific authority by the general partner, the limited partners have no authority to bind the limited partnership at all. Liability follows authority to the extent that the limited partners have no further obligation to the creditors of the partnership beyond the stated terms of their investment; the general partner, alone, has unlimited liability for the obligations of the firm.

An engineer dealing with a joint venture would be well advised to make the effort to ascertain what the precise nature of the joint venture is. Friendly cooperation on an interesting project could turn into a legal and financial nightmare for the unprepared.

48.5 THE ENGINEER AS CONSULTANT

The engineer who acts as an independent consultant stands alone and is, therefore, totally unprotected from individual personal liability. In fact, there are a number of circumstances that will permit others to avoid liability themselves on the ground that they reasonably relied on the opinion and advice of an engineer—defined as one of a learned calling holding themselves out to the public as skilled in an engineering art.

Just as engineers who act as independent consultants are not able to use others as a shield from liability, they do not have to be concerned about vicarious liability as, for example, a joint venturer must be.

An independent engineering firm may take any form authorized by the laws of the state or states in which it will do business: sole proprietorship, partnership, limited partnership, corporation, and joint ventures. The relationship between such an engineering organization and the individual engineers with whom it contracts are governed by the terms of any contracts, whether written or otherwise, and the rules of agency.

Although there is a developing body of law that permits independent consulting engineers to have their relative responsibility determined separately from that of their clients, in a combined design/construction case (with a number of defendants including the landowner, the tenant, the architect, the contractor, the consulting engineer, and the municipality), the consulting engineer still faces the risk of bearing a disproportionate share of the liability for damages.

48.6 LIABILITY UNDER CONTRACT

The primary legal responsibility of the engineer as an independent professional is to the client who, under circumstances such as design defect claims, may be judicially determined to be the general public. As employed professionals, engineers are legally responsible under the terms of their employment agreements, whether written contracts or some other form of agreement, to their employer.

Engineers like any other individual or legal entity, are liable for failure to fulfill their obligations under the terms and conditions of any contract to which they are a party. Unfortunately, it is not always clear what obligations were assumed by a professional—one who exercises informed and skilled judgment—like an engineer. Some contracts are almost painfully precise as to what services are required of a professional, while other contracts speak of "performance" and leave most details up to the skill and judgment of the engineer.

While precise statements of the obligations of each party to a contract are preferred, there is an ancient rule of law derived from the Romans which provides that when an agreement seeks to spell out in detail all that is included in the scope of work, it will be presumed that anything not included is excluded. Attempting to "say it all" in an agreement increases the probability of errors (usually of omission) and increases the penalty for error by raising a presumption of intentional omission.

When an engineer makes an error in work performed for a client or an employer, the legal principles under which both responsibility and the measure of damages will be determined depend on the effects of the error—the legal "injury." If the error results in injury to a member of the general public, it will be governed by the principles of tort law—that body of legal principles concerned with imposing noncriminal liability for damages. On the other hand, if the error results in injury only to the client or employer, then the principles of contract law determine liability and provide the measure of damages.

Contract law generally permits claims to be made under a contract only by those parties who are "in privity"—those parties among whom a contractual relationship actually exists. Certain parties whom it is contended the nominal parties to the contract intended to benefit may also have certain legally enforceable rights as third party beneficiaries. Subcontractors generally are not permitted to seek redress against each other under concurrent but independent contracts with a general contractor.

Contract law also limits the amount of damages that can be recovered. Ordinarily, consequential damages may not be recovered in an action for breach of contract. A party who breaches a contract can be required to return all or a portion of their remuneration under the contract. They cannot, however, be required to reimburse the costs associated with the breach such as the additional expenses attributable to delay in completing a project. That requires a special undertaking in the original contract clearly stating that the parties intend such additional liability to apply.

Occasionally, where the parties foresee that it may be difficult to ascertain or prove the exact amount of damages upon breach of the terms and conditions of a contract, they may provide in advance for some determinate or determinable dollar amount that will be accepted as "liquidated damages" in the event of a failure in performance. So long as this device is not used by one party to impose an obviously excessive "penalty" on the defaulting party, it will be accepted in any litigation over the contract.

48.7 CONTRACT DEFENSES

Statutes of limitations are nothing more than legislative determinations of how much time should be allowed for a party to give legal notice and commence an action against another seeking damages for a legally cognizable injury. In an action for breach of contract, the time within which to bring the action begins to run from the time breach of the contract occurs, whether or not the party claiming damages discovers the facts establishing the breach of contract within the period of limitation.

One aspect of contract law that is perhaps unique to engineering is the defense of acceptance. Under the law of sales, which generally governs all transactions involving property other than real estate, once a purchaser or consumer has received the goods or services to be provided and has had a chance to examine the product or consider the effectiveness of the services and reject them if they do not conform to the terms and conditions of the contract, the goods or services are deemed accepted. Thereafter a purchaser or consumer will not be heard by a court to complain about any but latent defects. Since so many terms and conditions of the "engineering" contract depend on "acceptance," acceptance often becomes a formal and well-considered event. The general rule is that unconditional acceptance of even only a part of a project or a limited amount of professional service is acceptance "enough" to seriously limit the rights and remedies of the purchaser or consumer under the contract.

Acceptance does not always act to protect an engineer from further liability, however. Acceptance is a defense only as to those defects that could be discovered upon reasonable inspection. These are sometimes referred to as "patent" defects in contrast to "latent" defects that cannot be discovered upon reasonable inspection. The right of action against latent defects is not terminated by acceptance.

Derived as it is from the law of sales, the defense of acceptance tends to be more effective in litigation over work that results in a "project" or "product," like a building or machine, or a system component than it does in litigation over designs, plans, specifications, drawings, calculations, or some other exercise of engineering "art."

So long as the error or omission of an engineer does not result in personal injury and remains subject to the principles of contract law, the consequences can be considered determinate and determinable in the context of insurance underwriting. Relatively objective estimates can be made of the risk of loss and the extent of liability for breach of the terms and conditions of any particular contract. The availability of defenses such as the statute of limitations and acceptance also work to keep liability "under control."

A different set of rules pertains to liability for damages in tort, since engineers can be said to owe a duty to the general public to do their work "professionally."

48.8 LIABILITY IN TORT—NEGLIGENCE

How "tort" is actually defined is somewhat problematical. Historians suggest that, at the time the attempt was being made to end blood feuds and to get people to bring their quarrels off the "field of honor" and into court, the anglicized phonetic rendition of the Norman French word for injury or wrong, "tort," was adopted to characterize generically a basic claim of wrongdoing that was being prosecuted by an individual civilly rather than by the king or state criminally. Whatever the origins of the term, the Anglo-American jurisprudence proceeds from the assumption that every human being, and even legally constituted business entities such as corporations, partnerships, and joint ventures have the right to be secure in their persons and property. When that right to security of person and property is violated, the wrong is actionable at law.

Tort actions are by no means limited to those seeking recovery of damages for bodily injury or injury to property. Damages may be recovered for other "personal" injuries such as defamation and malicious interference with business relationships. Because of its rather "fundamentalist" approach to righting wrongs, tort law tends to take little cognizance of the rules of agency, corporate law, or even contract terms insofar as they may appear to shield a wrongdoer from responsibility. Thus, a corporate officer may be personally liable even if the tort was committed in the name of, and approved by, the corporation as a separate legal entity. Furthermore, while the doctrines of contract law may limit the number of persons who may enforce the terms and conditions of a contract, tort law ignores these strictures and places responsibility directly on the wrongdoer, whether "in privity" with the injured party or not.

While tort law may have been developed as an alternative to the blood feud and trial by combat as the means to redress intentional wrongs such as assault and battery; nevertheless, its most common applica tion today is as a remedy for unintentional wrongs—negligent and careless acts which cause injury and damage. Negligence is defined at law as failing to do that which a prudent person would do, or doing that which a prudent person would not do, under the circumstances. In the context of engineering practice, it is the failure to exercise that care which an engineer of ordinary prudence would have exercised. It is, at its most obvious, the failure to recheck calculations or to recheck material specifications.

The duty to exercise reasonable care and avoid negligence does not mean that engineers guarantee the results of their professional efforts. Quite the contrary, if an engineer can show that everything a reasonably prudent engineer might do was, in fact, done, and done correctly; then, by definition, there can be no liability for negligence. Indeed, numerous cases can be found where a defendant engineer convinced a court and jury that they had exercised due care consistent with their status as a professional notwithstanding the most disastrous consequences of their work. This contrasts sharply with actions based on failure to meet all the terms and conditions of a contract where neither good faith nor due care have any relevance and liability ensues upon the occurrence, and solely by reason of, the contract breach itself. In a negligence action, it is conduct that is on trial and if fault cannot be found, liability will not lie.

There is, however, a rule of evidence that often makes certain conduct itself proof of negligence: the doctrine of *res ipsa loquitur*. *Res ipsa loquitur* literally means "the thing speaks for itself." For example, if a bridge collapses and all the contractors, engineers, inspectors, fabricators, manufacturers, and others that might be responsible for the collapse are before the court, anyone injured by the collapse need merely prove that the persons before the court are the only persons who could have been responsible for the disaster. The law presumes that, by the very nature of the occurrence, someone had to be at fault. Since all possible "someones" are before the court, the only question is which one of them is the guilty party, a matter they can thrash out among themselves.

The doctrine of *res ipsa loquitur*, however, creates only an inference of negligence. It is still possible for any one of the parties accused of negligence to affirmatively prove that the disaster occurred for reasons that bespeak the negligence of no one and, if that position prevails, no liability will be imposed on any of the defendants.

One consideration in determining negligence is unique to engineers and other professionals. The more qualified the engineer for the task, the higher the standard of care to which that engineer will be held. While an expert with specialized skills or experience may be held liable under circumstances where others might be excused, fault, and therefore liability for damages may be predicated on attempting a job for which an engineer or engineering firm is not fully qualified or sufficiently experienced.

In design litigation, failure to meet industry standards, to comply with OSHA regulations, or similar criteria may serve to establish liability. However, since such criteria are regarded as minimum standards, mere compliance rarely succeeds as a defense to claims of negligence.

In contract actions an engineer is usually not responsible for consequential damages flowing from failure to perform all the terms and conditions of a contract. In negligence actions, however, the measure of damages includes all consequences naturally and proximately, albeit unexpectedly, flowing from any conduct not meeting professional standards.

It is impossible in this chapter to present a complete list of instances in which engineers have been held liable for failure to maintain professional standards. In some cases, as, for example, failure to recheck data and calculations, the breach of duty is obvious. In others, it may not be so clear. Whether production goes forward under a more expensive process less likely to yield a product contaminated with a toxic substance, or whether it goes forward with a less expensive process which, although standard throughout the industry, still, nevertheless yields a hazardous product, is a choice fraught with potentially catastrophic economic consequences. There is, as yet, no clearly defined rule of law for guidance.

One of the most common claims in liability litigation against engineers, particularly those involved in large and complex projects such as electrical power generating facilities, large buildings, and mass transit, is negligent failure to warn about hazards, risks, or latent defects known to the engineer, but unknown to his or her actual client or ultimate consumer. In one instance, a bearing manufacturer received an order from an airplane manufacturer. Upon ascertaining what the bearings were to be used for, however, the bearing company engineers concluded that the bearings ordered were inappropriate for the intended use. The airplane manufacturer was so advised; but replied that it was not interested in the bearing company's opinion. If it could not obtain the bearings ordered from that bearing company, it would get the same bearings elsewhere. The bearing company sold the bearings to the airplane manufacturer while simultaneously purchasing a large amount of additional liability insurance to cover the risk it perceived as associated with installation of the bearings.

Although the bearing company satisfied its duty to warn the airplane manufacturer and would be immune from any claims by the airplane manufacturer, the bearing company did not meet its obligation to warn others who might be at risk, about the danger it perceived. Because of negligent (perhaps even reckless) failure to warn, the bearing company would be independently liable to third parties, particularly passengers on those planes that were built using those inappropriate bearings.

48.9 NEGLIGENCE DEFENSES

For a negligence claim to be successful, the engineer must be at fault in some degree. There are several defenses important to engineers facing claims for damages arising out of negligence. If the state of the art were such at the time the engineer performed professional services that the event or occurrence which caused the injury and damages could not have been forseen at that time by a responsible and prudent engineer, liability may be avoided. Some states have adopted this "state-of-the-art defense," but it is by no means the general rule.

Another defense is premised on the fact that liability depends on a showing that the negligence of the engineer was the direct and proximate cause of the damage. If it can be shown that there are or were other superseding causes responsible for damages, the engineer can escape liability even though there was a deviation from professional standards. However, the law also recognizes that there may be more than one proximate cause of injury and damages, so that if an independent superseding cause is concurrent with other causes, all those responsible are liable for the damages. This rule of law is best summarized in an old English maxim, "in for a penny; in for a pound."

One special case of "intervening cause" is the contributory negligence of the party claiming damages. The rule of contributory negligence has harsh consequences. Often, a jury is told that if they find the plaintiff negligent "in the slightest degree," they can impose no liability upon a defendant no matter how great the negligence of the defendant might have been. To ameliorate this essential injustice to victims of negligent injury, most states have adopted some rule of comparative negligence, which permits the court to apportion responsibility for the damage according to the fault of the parties.

48.10 STATUTES OF LIMITATIONS

Interpreting statutes of limitations in tort cases has become a miasma of conflicting judicial interpretations concerning just when a cause of action for damages arising out of professional negligence can be said to accrue. In some states, the period of limitations begins to run not when the negligence occurs, but at the time damages attributable to that negligence become manifest to a particular person to whom some duty of care is owed. In actions claiming damages from medication and other products, for example, plaintiffs may be allowed to come into court many years after the product was placed on the market, and even some time after "some" damage may have occurred, because the connection between their damage and the product was only lately discovered. Such "discovery" rules mean that personal injury claims for damages attributable to concealed negligence or latent product defects may never go "away" in the absence of specific and explicit legislation.

Some state legislatures have attempted to quiet such claims forever by enacting "statutes of repose" which categorically bar all claims—regardless of whether they are based on latent defects or involve slow starting diseases such as cancer—after some definite number of years legislatively determined to be "reasonable."

48.11 LIABILITY WITHOUT FAULT

Negligence as a basis for professional liability is essentially fair because it offers engineers an opportunity to avoid damage awards by exercising care and prudence in the performance of their professional duties. To the extent that the spectre of liability for negligence encourages better engineering practices, it promotes the public interest as well. There are, however, certain theories of "strict liability" that impose an obligation to respond in damages regardless of the degree of care exercised by the engineers.

Requiring an injured layperson, whether aided by an attorney or not, to prove that an engineer failed to use due care and caution in accordance with professional standards, and that such failure caused injury to persons or property which resulted in actual damage, seemed more fair than it actually was. Injured plaintiffs often found themselves caught in the web of conflicting "expert" opinion. Often, too, the facts necessary to establish deviation from professional standards of care were within the exclusive possession and control of the defendants and were essentially undiscoverable. In practice, the obstacles to recovery by the "victims of technology" became, in the eyes of many jurists and legislators, too formidable to be fair. Accordingly, other and different forms of law to promote recovery of damages by victims were created.

48.12 WARRANTY

Initially, courts facing the difficulties in proof of negligent professional conduct sought refuge in the doctrines of contract law. It was argued that every manufacturer or seller warrants (guarantees) that their product meets a certain standard of safety. If a product did not meet that safety standard, it did not matter what care was taken in producing it, anyone injured by the product could recover damages under a theory of implied warranty, just as if the safety of the product or device had been expressly guaranteed.

The forms of warranty most commonly relied upon by the courts were the warranty of "merchantability," under which the manufacturer was deemed to have assured the consumer or user of the product that the product had at least the characteristics which enabled it to be accepted in the trade with goods of like description, and the warranty of "fitness for the intended use," under which anyone who purchased the product for a specific purpose known to the seller could assume that the seller warranted that the product or device was fit for that purpose.

When used in resolving personal injury claims, the courts tended to declare that such "warranty" actions, although derived from contract law, were really actions in tort, so that tort rules applied. Thus, the privity required in contract actions was overlooked, and any purchaser or later user was allowed to recover damages under the implied warranty. In some states, the statutes of limitation for these "implied warranty" actions were said to run from the time of discovery of the injury and damages.

In effect, by using the principle of implied warranty, the courts were able to impose liability for damages from defective products without requiring any proof of negligence in manufacture. The emphasis in litigation shifted from proving "fault" to establishing "defect."

There are circumstances in which engineers can become guarantors of the results of their professional efforts under the law of warranty, and that is by express agreement. One theory that is often used against physicians and surgeons in medical malpractice claims is that the doctor "guaranteed" that the patient would have a particular (good) result. If circumstances exist from which such a guaranty can be inferred, even if the "guaranty" is not in writing, a patient can recover damages from the doctor regardless of the doctor's good faith and skill, and regardless of whether the failure to achieve the promised result was due to forces beyond the doctor's control.

48.13 STRICT PRODUCT LIABILITY

During the 1960s the prestigious American Law Institute endorsed the proposition that the seller of a product that turned out to be defective in some respect should be liable to any user of the product who may be injured by such use (even an injured bystander to such use) without regard to the efforts of the manufacturer or seller to make the product safe. The rule was quickly extended to products that were far from obviously dangerous, and to other than bodily injuries. The philosophy underlying the new doctrine was stated by Justice Traynor of California who pointed out that while imposing liability for damages may promote better products and more careful manufacture, an even more important consideration is that the manufacturer and seller are in a position to assess and distribute the risk of injury among all the purchasers of the product. In effect, Justice Traynor argued that the manufacturer or seller should become insurers of the safety of their products.

It has been a short step from Justice Traynor's social insurance theory to the present day doctrine of strict product liability, which imposes liability on the manufacturers of a product even though the dangers to be avoided and the methods to reduce risk were not only unknown at the time of manufacture and design, but were actually unknowable.

There are a number of ways in which a product can be judicially determined to be "defective." The most widely accepted test is based on risk–utility analysis. If the utility of the product outweighs the risks of harm associated with its use, then the product is not "unreasonably dangerous" or "defective." For example, it is understood that chemotherapy for cancer patients involves high risk of untoward side effects which of themselves represent "injury." Nonetheless, the benefits of the procedures outweigh the harm.

In the evolving law of products liability, it has been suggested that a product cannot be "defective" as a matter of law if it fulfills "consumer expectations." This is a thinly veiled effort to restore assumption of risk as a defense. Whether the "consumer expectation test" will ever gain widespread judicial acceptance in products liability litigation is doubtful.

The liability that is associated with product design and manufacturing processes necessarily extends into the area of warnings. Generally, a product otherwise perfectly safe can become "defective" as a matter of law if it is not accompanied by appropriate instructions and suitable warnings. The most obvious example of this is a nonprescription medication use of which can lead to allergic reactions unless certain precautions are followed and premonitory symptoms heeded. The product without the warning is incomplete and, in the eyes of the law, defective. The duty to warn includes the duty, not only to inform the user of the risks inherent in the product so that an intelligent decision can be made whether or not to use the product at all, but also the duty to tell the user what actions can be taken to reduce the danger of harm and, if some danger is foreseeable and unavoidable (as in the case of cancer chemotherapy), the means that will reduce the consequences to a minimum.

While the duty to warn about what is known at the time of sale is well established, there is considerable controversy over the duty to warn about hazards that become known years after the time of sale or use. It is the thinking of the New Jersey courts (so far) that early lack of knowledge and the state of the art at the time of sale and use are irrelevant in determining liability for damages from the use of "now-known-to-be-defective" materials, products, or devices, such as asbestos. It appears that after acquired knowledge of a product defect imposes a duty and concomitant responsibility to reach out to those who can be reasonably expected to be harmed by the defect and warn them based upon the new information. To what extent liability will be imposed should depend on the age of the product or device, how "patent" or "latent" the defect was, and just what individual or entity has the most direct access to the consumer or ultimate user. Clearly, the engineer has some duty to warn. The questions the courts have not yet answered are, "Warn whom?" "Warn them how?" and "Just how much warning is required?"

Although strict products liability is most often imposed for bodily injury, it has been found in property damage cases as well, especially where the property damage has been associated with personal injury.

To date these "strict liability" or liability without fault theories have been applied most often to manufacturers and sellers of products. One can foresee, however, that manufacturers may seek to implead the professional consultants and engineers who helped bring the product into existence. At the least they will seek to obtain expert testimony without cost in support of the product. At that point in time, even though the strict liability applies most properly only to the manufacturer and seller, it is likely that the courts will impose the same standards to the design engineer on whose boards the product was born, or the production engineer who brought the product into being.

48.14   PRODUCT ABUSE AND ASSUMPTION OF RISK DEFENSES

Among the defenses to product liability claims, one of the most important is "abuse of product." Although this may be nothing more than contributory negligence in another guise, if a product or device is used for a purpose other than its intended purpose, or in a manner other than directed, liability may not be imposed if the misuse contributed to the occurrence of injury and damage.

The "assumption of risk" defense in claims for damage based on theories of negligence, warranty, and strict liability, seeks to avoid liability because the injured party was aware of an obvious and known danger associated with the product or device but used it anyway.

To the courts, however, the recklessness of the injured party must be flagrant and egregious. For example, in New York a printer who bypassed safety guards and other protective devices in order to remove ink specks from the rollers of the press he was operating recovered damages from the manufacturer of the printing press. The manufacturer was deemed sufficiently familiar with the practice of printers to bypass the kind of safety devices that had been installed to be held liable despite the deliberate defeat of the safety devices by the printer. The pressman's case is typical of so-called "third party liability cases" where the employers or insurance carriers responsible for worker's compensation benefits to injured employees seek to recover part of their losses by looking toward a defect in design or engineering of a product or machine.

48.15   INSURANCE FOR ENGINEERS

If practitioners of the engineering arts can be held liable for damages without fault no matter how great the skill they exercise or care they demonstrate, protection from liability lies with insurance. The concept of "strict" product liability was born of the lofty ideal of social insurance, but professional liability insurance is the inevitable practical result.

While the engineer as employee of a well-insured firm probably has limited practical liability exposure, and individual "errors and omissions" or professional liability insurance might be duplicative of that maintained by the employer, the professional engineering consultant or "moonlighting" engineer otherwise employed is in a different category. Professional liability or "errors and omissions" insurance is a necessity. The proper amount of coverage is something that should be worked out with a competent underwriter. Claims may be infrequent, but can be economically devastating when incurred. Perhaps the most important part of professional liability insurance coverage for engineers is not the assurance that the insurance carrier will pay the damages which may eventually be awarded an injured party, but that they will pay the cost of defending the claim, particularly the legal fees and expenses associated with expert testimony.

48.16   CONCLUSION

In part, the object and purpose of the law of torts is to spread the risks of injury and damages arising from devices, machines, materials, and products. However, it is equally true that the law of torts seeks to encourage the development of better engineering techniques and a safety consciousness among engineers. A jury which determines relative responsibility in the occurrence of an accident in court cases necessarily considers the actual fault of the participants. Second to maintaining proper insurance, then, the best defense to liability claims for the practicing engineer continues to be, as it has always been, good engineering.

CHAPTER 49
PATENTS

DAVID A. BURGE

David A. Burge Company, L.P.A.
Cleveland, Ohio

49.1 WHAT CAN BE PATENTED AND BY WHOM

For an invention to be patentable, it must meet several requirements set up to ensure that patents are not issued irresponsibly. Some of these standards are complex to understand and apply. Let us simplify and summarize the essence of these requirements.

49.1.1 Ideas, Inventions, and Patentable Inventions

"Invention" is a misleading term because it is used in so many different senses. In one, it refers to the act of inventing. In another, it refers to the product of the act of inventing. In still another, the term designates a patentable invention, the implication mistakenly being that if an invention is not patentable it is not an invention.

In the context of modern patent law, *invention* is the conception of a novel and useful contribution followed by its reduction to practice. *Conception* is the beginning of an invention; it is the creation in the mind of an inventor of a useful means for solving a particular problem. *Reduction to practice* can be either *actual,* as when an embodiment of the invention is tested to prove its successful operation under typical conditions of service, or *constructive,* as when a patent application is filed containing a complete description of the invention.

Ideas, per se, are not inventions and are not patentable. They are the tools of inventors, used in the development of inventions. Inventions are patentable only insofar as they meet certain criteria established by law. For an invention to be patentable, it must satisfy the following conditions:

1. Fit within one of the statutorily recognized classes of patentable subject matter.
2. Be the true and original product of the person seeking to patent the invention as its inventor.
3. Be new at the time of its invention by the person seeking to patent it.
4. Be useful in the sense of having some beneficial use in society.
5. Be nonobvious to one of ordinary skill in the art to which the subject matter of the invention pertains at the time of its invention.
6. Satisfy certain statutory bars that require the inventor to proceed with due diligence in pursuing efforts to file and prosecute a patent application.

49.1.2 The Requirement of Statutory Subject Matter

As stated in the Supreme Court decision of *Kewanee Oil* v. *Bicron Corp.,* 416 U.S. 470, 181 USPQ 673 (1974), no patent is available for any discovery, however useful, novel, and nonobvious, unless it falls within one of the categories of patentable subject matter prescribed by Section 101 of Title 35 of the United States Code. Section 101 makes this provision:

> *Whoever invents or discovers a new and useful process, machine, manufacture, or composition of matter, or any new and useful improvement thereof may obtain a patent therefor, subject to the conditions and requirements of this title.*

The effect of establishing a series of statutory classes of eligible subject matter has been to limit the pursuit of patent protection to the useful arts. Patents directed to processes, machines, articles of manufacture, and compositions of matter have come to be referred to as utility patents inasmuch as these statutorily recognized classes encompass the useful arts.

Three of the four statutorily recognized classes of eligible subject matter may be thought of as products, namely, machines, manufactures, and compositions of matter. "Machine" has been interpreted in a relatively broad manner to include a wide variety of mechanisms and mechanical elements. "Manufactures" is essentially a catch-all term covering products other than machines and compositions of matter. "Compositions of matter," another broad term, embraces such elements as new molecules, chemical compounds, mixtures, alloys, and the like. "Manufactures" and "compositions of matter" arguably include such genetically engineered life forms as are not products of nature. The fourth class, "processes," relates to procedures leading to useful results.

Subject matter held to be ineligible for patent protection includes printed matter, products of nature, ideas, and scientific principles. Alleged inventions of perpetual motion machines are refused patents. A mixture of ingredients such as foods and medicines cannot be patented unless there is more to the mixture than the mere cumulative effect of its components. So-called patent medicines are seldom patented.

While no patent can be issued on an old product despite the fact that it has been found to be derivable through a new process, the new process for producing the product may well be patentable. That a product has been reduced to a purer state than was previously available in the prior art does not render the product patentable, but the process of purification may be patentable. A new use for an old product does not entitle one to obtain *product* patent protection, but may entitle one to obtain *process* patent protection, assuming the process meets other statutory requirements.

A newly discovered law of nature, regardless of its importance, is not entitled to patent protection. Methods of conducting business and processes that either require a mental step to be performed or depend on aesthetic or emotional reactions have been held to not constitute statutory subject matter.

While the requirement of statutory subject matter falls principally within the bounds of 35 U.S.C. 101, other laws also operate to restrict the patenting of certain types of subject matter. For example, several statutes have been passed by Congress affecting patent rights in subject matter relating to atomic energy, aeronautics, and space. Still another statute empowers the Commissioner of Patents and Trademarks to issue secrecy orders regarding patent applications disclosing inventions that might be detrimental to the national security of the United States.

The foreign filing of patent applications on inventions made in the United States is prohibited for a brief period of time until a license has been granted by the Commissioner of Patents and Trademarks to permit foreign filing. This prohibition period enables the Patent and Trademark Office to review newly filed applications, locate any containing subject matter that may pose concerns to national security, and, after consulting with other appropriate agencies of government, issue secrecy orders preventing the contents of these applications from being publicly disclosed. If a secrecy order issues, an inventor may be barred from filing applications abroad on penalty of imprisonment for up to 2 years or a $10,000 fine or both. In the event a patent application is withheld under a secrecy order, the patent owner has a right to recover compensation from the government for damage caused by the secrecy order and/or for the use the government may have made of the invention.

Until only recently it was necessary for an inventor to wait for the expiration date of a six-month period measured from the date of filing of an application before proceeding with foreign filing unless a license permitting expedited foreign filing was specifically requested from and granted by the Patent and Trademark Office. Now, however, licenses permitting expedited foreign filing are almost always automatically granted by the Office at the time of issuing an official filing receipt, which advises the inventor of the filing date and serial number assigned to his or her application. Official filing receipts are now routinely being issued within a month of the date of filing, and bear a statement attesting to the grant of a foreign filing license.

49.1.3 The Requirement of Originality of Inventorship

Under United States patent law, only the true and original inventor or inventors may apply to obtain patent protection. If the inventor has derived an invention from any other source or person, he or she is not entitled to apply for or obtain a patent.

The laws of our country are strict regarding the naming of the proper inventor or joint inventors in a patent application. When one person acting alone conceives an invention, he or she is the sole inventor and he or she alone must be named as the inventor in a patent application filed on that invention. When a plurality of people contribute to the conception of an invention, these persons must be named as joint inventors if they have contributed to the inventive features that are claimed in a patent application filed on the invention.

The concept of joint inventorship is recognized by statute, but is undefined. Joint inventorship is generally deemed to have occurred if two or more persons have collaborated in some fashion, with each contributing to conception. It is not necessary that exactly the same idea should have occurred to each of the collaborators at the same time.

When a substantial number of patentable features relating to a single overall development have occurred as the result of different combinations of sole inventors acting independently and/or joint inventors collaborating at different times, the patent law places a burden on the inventors to sort out "who invented what." Patent protection on the overall development must be pursued in the form of a number of separate patent applications, each directed to such patentable aspects of the development as originated with a different inventor or group of inventors. In this respect, United States patent practice is unlike that of many foreign countries where the company for whom all the inventors work is often permitted to file a single patent application in its own name covering the overall development.

Misjoinder of inventors occurs when a person who is not a joint inventor has been named as such in a patent application. Nonjoinder of inventors occurs when there has been a failure to include a person who should have been named as a joint inventor. Misdesignation of inventorship occurs when none of the true inventors are named in an application. Only in recent years has correction of a misdesignation been permitted. If a problem of misjoinder, nonjoinder, or misdesignation has arisen without deceptive intent, provisions of the patent law permit correction of the error as long as such is pursued with diligence following the discovery.

49.1.4 The Requirement of Novelty

Section 101 of Title 35 of the United States Code requires that a patentable invention be new. What is meant by "new" is defined in Sections 102(a), 102(e), and 102(g). Section 102(a) bars the issuance of a patent on an invention "known or used by others in this country, or patented or described in a printed publication in this or a foreign country, before the invention thereof by the applicant for patent." Section 102(e) bars the issuance of a patent on an invention "described in a patent granted on an application for patent by another filed in the United States before the invention thereof by the applicant for patent." Section 102(g) bars the issuance of a patent on an invention that "before the applicant's invention thereof . . . was made in this country by another who had not abandoned, suppressed, or concealed it."

These novelty requirements amount to negative rules of invention, the effect of which is to prohibit the issuance of a patent on an invention if the invention is not new. The novelty requirements of 35 U.S.C. 102 should not be confused with the statutory bar requirements of 35 U.S.C. 102, which are discussed in the last section of this chapter. A comparison of the novelty and statutory bar requirements of 35 U.S.C. 102 is presented in Table 49.1. The statutory bar requirements are distinguishable from the novelty requirements in that they do not relate to the newness of the invention, but to ways an inventor, who would otherwise have been able to apply for patent protection, has lost that right by tardiness.

To understand the novelty requirements of 35 U.S.C. 102, one must understand the concept of anticipation. A claimed invention is anticipated if a single prior art reference contains all the essential elements of the claimed invention. If teachings from more than one reference must be combined to show that the claimed combination of elements exists, there is no anticipation, and novelty exists. Combining references to render a claimed invention unpatentable brings into play the nonobviousness requirements of 35 U.S.C. 103, not the novelty requirement of 35 U.S.C. 102. Novelty hinges on anticipation and is a much easier concept to understand and apply than that of nonobviousness.

35 U.S.C. 102(a) Known or Used by Others in This Country Prior to the Applicant's Invention

In interpreting whether an invention has been known or used in this country, it has been held that the knowledge must consist of a complete and adequate description of the claimed invention and that this knowledge must be available, in some form, to the public. Prior use of an invention in this country by another will only be disabling if the invention in question has actually been reduced to practice and its use has been accessible to the public in some minimal sense. For a prior use to be disabling under Section 102(a), the use must have been of a complete and operable product or process that has been reduced to practice.

35 U.S.C. 102(a) Described in a Printed Publication in This or a Foreign Country Prior to the Applicant's Invention

For a printed publication to constitute a full anticipation of a claimed invention, the printed publication must adequately describe the claimed invention. The description must be such that it enables a person

Table 49.1   Summary of the Novelty and Statutory Bar Requirements of 35 U.S.C. 102

Novelty Requirements

One may not patent an invention if, prior to its date of invention, the invention was any of the following:

1. Known or used by others in this country.
2. Patented or described in a printed publication in this or a foreign country.
3. Described in a patent granted on an application for patent by another filed in the United States.
4. Made in this country by another who had not abandoned, suppressed, or concealed it.

Statutory Bar Requirements

One may not patent an invention he or she has previously abandoned. One may not patent an invention if, more than one year prior to the time his or her patent application is filed, the invention was any of the following:

1. Patented or described in a printed publication in this or a foreign country.
2. In public use or on sale in this country.
3. Made the subject of an inventor's certificate in a foreign country.
4. Made the subject of a foreign patent application, which results in the issuance of a foreign patent before an application is filed in this country.

of ordinary skill in the art to which the invention pertains to understand and make the invention. The question of whether a *publication* has taken place is construed quite liberally by the courts to include almost any act that might legitimately constitute publication. The presence of a single thesis in a college library has been held to constitute publication. Similar liberality has been applied in construing the meaning of the term "printed."

35 U.S.C. 102(a) Patented in This or a Foreign Country

An invention is not deemed to be novel if it was patented in this country or any foreign country prior to the applicant's date of invention. For a patent to constitute a full anticipation and thereby render an invention unpatentable for lack of novelty, the patent must provide an adequate, operable description of the invention. The standard to be applied under Section 102(a) is whether the patent "describes" a claimed invention. A pending patent application is treated as constituting a "patent" for purposes of applying Section 102(a) as of the date of its issuance.

35 U.S.C. 102(e) Described in a Patent Filed in This Country Prior to the Applicant's Invention

Section 102(e) prescribes that if another inventor has applied to protect an invention before you invent the same invention, you cannot patent the invention. The effective date of a United States patent, for purposes of a Section 102(e) determination, is the filing date of its application rather than the date of patent issuance.

35 U.S.C. 102(g) Abandoned, Suppressed, or Concealed

For the prior invention of another person to stand as an obstacle to the novelty of one's invention under Section 102(g), the invention made by another must not have been abandoned, suppressed, or concealed. Abandonment, suppression, or concealment may be found when an inventor has been inactive for a significant period of time in pursuing reduction to practice of an invention. This is particularly true when the inventor's becoming active again has been spurred by knowledge of entry into the field of a second inventor.

49.1.5 The Requirement of Utility

To comply with the utility requirements of United States patent law, an invention must be capable of achieving some minimal useful purpose that is not illegal, immoral, or contrary to public policy. The invention must be operable and capable of being used for some beneficial purpose. The invention does not need to be a commercially successful product in order to satisfy the requirement of utility. While the requirement of utility is ordinarily a fairly easy one to meet, problems do occasionally arise with chemical compounds and processes, particularly in conjunction with various types of drugs. An invention incapable of being used to effect the proposed object of the invention may be held to fail the utility requirement.

49.1.6 The Requirement of Nonobviousness

The purpose of the novelty requirements of 35 U.S.C. 102 and the nonobviousness requirement of 35 U.S.C. 103 are the same—to limit the issuance of patents to those innovations that do, in fact, advance the state of the useful arts. While the requirements of novelty and nonobviousness may seem very much alike, the requirement of nonobviousness is a more sweeping one. This requirement maintains that if it would have been obvious (at the time an invention was made) to anyone ordinarily skilled in the art to produce the invention in the manner disclosed, then the invention does not rise to the dignity of a patentable invention and is therefore not entitled to patent protection.

The question of nonobviousness must be wrestled with by patent applicants in the event the Patent and Trademark Office rejects some or all their claims based on an assertion that the claimed invention is obvious in view of the teaching of one or a combination of two or more prior art references. When a combination of references is relied on in rejecting a claim, the argument the examiner is making is that it is obvious to combine the teachings of these references to produce the claimed invention. When such a rejection has been made, the burden is on the applicant to establish to the satisfaction of the examiner that the proposed combination of references would not have been obvious to one skilled in the art at the time the invention was made; and/or that, even if the proposed combination of references is appropriate, it still does not teach or suggest the claimed invention.

In an effort to ascertain whether a new development is nonobvious, the particular facts and circumstances surrounding the development must be considered and weighed as a whole. Moreover, the question of nonobviousness must be judged as of the time the invention was made and in light of the then-existing knowledge and state of the art. This test of nonobviousness has been found to be an extremely difficult one for courts to apply.

The statutory language prescribing the nonobviousness requirement appears at Title 35, Section 103, stating:

> A patent may not be obtained . . . if the differences between the subject matter sought to be patented and the prior art are such that the subject matter as a whole would have been obvious at the time the invention was made to a person having ordinary skill in the art to which said subject matter pertains.

In the landmark decision of *Graham* v. *John Deere*, 383 U.S. 1, 148 USPQ 459 (1966), the U.S. Supreme Court held that several basic factual inquiries should be made in determining nonobviousness. These inquiries prescribe a four-step procedure or approach for judging nonobviousness. First, the scope and content of the prior art in the relevant field or fields must be ascertained. Second, the level of ordinary skill in the pertinent art is determined. Third, the differences between the prior art and the claims at issue are examined. Fourth and finally, a determination is made as to whether these differences would have been obvious to one of ordinary skill in the applicable art at the time the invention was made. Since the *Graham* ruling in 1966, a great deal of conflicting verbiage has appeared in court decisions interpreting what was intended by the *Graham* decision and confounding the meaning of nonobviousness. However, the rulings that are now issuing from the newly established U.S. Court of Appeals for the Federal Circuit (CAFC), which now hears (among other types of cases) all appeals in patent cases from the various federal district courts, are doing much to fulfill the long-felt need for a crisp, clean, and clear understanding of the proper test for nonobviousness. The CAFC is intellectually the finest court in many years to hear patent cases. It has made it clear that evidence of what are referred to as "secondary considerations" should be considered as part of all of the evidence in determining nonobviousness, not just when the decision-maker remains in doubt after applying the four-step test of *Graham*. Secondary considerations include such things as commercial success of the invention in the marketplace, long-felt but unsatisfied need that is met by the invention, and prior failure by others who were trying to produce the same invention.

This brings us to the nonsubject of "synergism." In recent years, substantial consternation has arisen as the result of the supposed creation of a "synergism" test for nonobviousness. This test was said to have arisen from two Supreme Court decisions, one titled *Anderson's-Black Rock, Inc.* v. *Pavement Salvage Co., Inc.*, 396 U.S. 57, 163 USPQ 673 (1969), and the other titled *Sakraida* v. *Ag Pro, Inc.*, 425 U.S. 273, 189 USPQ 449 (1976). The concept of synergism has been said, by those mystics who have claimed to understand it, to arise when the combined effect of the several elements in a "combination patent"—a very poor term intended to refer to a patent that claims a combination-of-elements invention—amounts to more than the simple sum of the effects of the elements. The Court of Appeals of the Federal Circuit now has made it absolutely clear that such terms as "synergism" and "combination patent" are inappropriate, and that there is no support whatsoever for a "synergism" test in the patent statutes.

49.1.7 Statutory Bar Requirements

Despite the fact that an invention may be new, useful, and nonobvious and that it may satisfy the other requirements of the patent law, an inventor can still lose the right to pursue patent protection on the invention unless he or she complies with certain requirements of the law called statutory bars. The statutory bar requirements ensure that inventors will act with diligence in pursuing patent protection.

Since 35 U.S.C. 102 includes both the novelty and the statutory bar requirements of the law, the reader may wonder why the discussions of these two categories of requirements are not combined. Several reasons explain why these requirements should be considered separately. Section 102 is so easily misinterpreted that its content needs to be divided to be understood. Section 102 intertwines in a complex way the presentation of novelty and statutory bar requirements. The novelty requirements are basic to patentability in the same sense as are the requirements of statutory subject matter, originality, and nonobviousness. The statutory bar requirements are not basic to a determination of patentability, but rather operate to decline patent protection to an invention that may have been patentable at one time.

Section 102(b) bars the issuance of a patent if an invention was "in public use or on sale" in the United States more than one year prior to the date of the application for a patent. Section 102(c) bars the issuance of a patent if a patent applicant has previously abandoned the invention. Section 102(d) bars the issuance of a patent if the applicant has caused the invention to be first patented in a foreign country and has failed to file an application in the United States within one year after filing for a patent in a foreign country. Table 49.1 summarizes the statutory bar requirements of Section 102.

Once an invention has been made, the inventor is under no specific duty to file a patent application within any certain period of time. However, should one of the "triggering" events described in Section

102 occur, regardless of whether this occurrence may have been the result of action taken by the inventor or by actions of others, the inventor must apply for a patent within the prescribed period of time or be barred from obtaining a patent.

Some of the events that trigger statutory bar provisions are the patenting of an invention in this or a foreign country; the describing in a printed publication of the invention in this or a foreign country; public use of the invention in this country; or putting the invention on sale in this country. Some public uses and putting an invention on sale in this country will not trigger statutory bars if these activities were incidental to experimentation. Whether a particular activity amounts to experimental use has been the subject of much judicial dissension. The doctrine of experimental use is a difficult one to apply because of the conflicting decisions issued on this subject.

Certainly, the safest approach to take is to file for patent protection well within one year of the possibility of any statutory bar coming into play. If foreign patent protections are to be sought, the safest approach is to file an application in this country before any public disclosure is made of the invention.

49.2 PREPARING TO APPLY FOR A PATENT

Conducting a patentability search and preparing a patent application are two of the most important stages in efforts to pursue patent protection. This section points out pitfalls to avoid in both stages.

49.2.1 The Patentability Search

Conducting a patentability search prior to the preparation of a patent application can be extremely beneficial even when an inventor is convinced that no one has introduced a similar invention into the marketplace. A properly performed patentability study will guide not only the determination of the scope of patent protection to be sought, but also the claim-drafting approaches to be used. In almost every instance, a patent attorney who has at hand the results of a carefully conducted patentability study can do a better job of drafting a patent application, thereby helping to ensure that it will be prosecuted smoothly, at minimal expense, through the rigors of examination in the Patent and Trademark Office.

Occasionally, a patentability search will indicate that an invention is totally unpatentable. When this is the case, the search will have saved the inventor the cost of preparing and filing a patent application. At times a patentability search turns up one or more newly issued patents that pose infringement concerns. A patentability search is not, however, as extensive a search as is one conducted to locate possible infringement concerns when a great deal of money is being invested in a new product.

Some reasonable limitation is ordinarily imposed on the scope of a patentability search to keep search costs within a relatively small budget. The usual patentability search covers only United States patents and does not extend to foreign patents or to publications. Only the most pertinent Patent and Trademark Office subclasses are covered. However, despite the fact that patentability studies are not of exhaustive scope, a carefully conducted patentability search ordinarily can be relied on to give a decent indication of whether an invention is worthy of pursuing patent coverage to protect.

Searches do occasionally fail to turn up one or more pertinent references despite the best efforts of a competent searcher. Several reasons explain why a reference may be missed. One is that the files of the Public Search Room of the Patent and Trademark Office are incomplete. The Patent and Trademark Office estimates that as many as 7% of the Search Room references are missing or misfiled. Another reason is that the Public Search Room files do not contain some Patent Office subclasses. The searcher must review these missing subclasses in the "examiners' art," the files of patents used by examiners, where the examiners are free to remove references and take them to their offices as they see fit. Since most patents are cross-referenced in several subclasses, a careful searcher will try to ensure that the field encompassed by a search extends to enough subclasses that patents are located that should have been found in other subclasses, but were not.

49.2.2 Putting the Invention in Proper Perspective

It is vitally important that a client takes whatever time is needed to make certain that his or her patent attorney fully understands the character of an invention before the attorney undertakes the preparation of a patent application. The patent attorney should be given an opportunity to talk with those involved in the development effort from which an invention has emerged. He or she should be told what features these people believe are important to protect. Moreover, the basic history of the art to which the invention relates should be described, together with a discussion of the efforts made by others to address the problems solved by the present invention.

The client should also convey to his or her patent attorney how the present invention fits into the client's overall scheme of developmental activities. Much can be done in drafting a patent application to lay the groundwork for protection of future developments. Additionally, one's patent attorney needs to know how product liability concerns may arise with regard to the present invention so the statements he or she makes in the patent application will not be used to client's detriment in product liability litigation. Personal injury lawyers have been known to scrutinize the representations a manufacturer makes in his or her patents to find language that will assist in obtaining recoveries for persons injured by patented as well as nonpatented inventions.

Before preparation of an application is begun, careful consideration should be given to the scope and type of claims that will be included. In many instances, it is possible to pursue both process and product claims. Also, in many instances, it is possible to present claims approaching the invention from varying viewpoints so different combinations of features can be covered. Frequently, it is possible to couch at least two of the broadest claims in different language so efforts of competitors to design around the claim language will be frustrated.

Careful considerations must be given to approaches competitors may take in efforts to design around the claimed invention. The full range of invention equivalents must be determined to the best of the abilities of the inventor and his or her attorney so claims of appropriate scope will be presented in the resulting application.

49.2.3 Preparing the Application

A properly prepared patent application is a work of art. It should be a readable and understandable teaching document. If it is not, insist that your patent attorney rework the document. A patent application that accurately describes an invention without setting forth the requisite information in a clear and convincing format may be legally sufficient, but it does not represent the quality of work a client has the right to expect.

A well-drafted patent application should explain accurately, yet interestingly, the background of the invention and the character of the problems that are overcome. It should discuss the closest prior art known to the applicant and should indicate how the invention patentably differs from prior art proposals. It should present a summary of the invention that brings out the major advantages of the invention and explains how prior art drawbacks are overcome. These elements of a patent application may occupy several typed pages. They constitute an introduction to the remainder of the document.

Following this introductory section, the application should present a brief description of such drawings as may accompany the application. Then follows a detailed description of the best mode known to the inventor for carrying out the invention. In the detailed description, one or more preferred embodiments of the invention are described in sufficient detail to enable a person having ordinary skill in the art to which the invention relates to practice the invention. While some engineering details such as dimensions, materials of construction, circuit component values, and the like, may be omitted, all details critical to the practice of the invention must be included. If there is any question about the essential character of a detail, prudent practice would dictate its inclusion.

The written portion of the application concludes with a series of claims. The claims are the most difficult part of the application to prepare. While the claims tend to be the most confusing part of the application, the applicant should spend enough time wrestling with the claims and/or discussing this section with the patent attorney to make certain that the content of the claims is fully understood. Legal gibberish should be avoided, such as endless uses of the word "said." Elements unessential to the practice of the invention should be omitted from the claims. Essential elements should be described in the broadest possible terms in at least some of the claims so the equivalents of the preferred embodiment of the invention will be covered.

The patent application will usually include one or more sheets of drawings and will be accompanied by a suitable declaration or oath to be signed by the inventor or inventors. The drawings of a patent application should illustrate each feature essential to the practice of the invention and show every feature to which reference is made in the claims. The drawings must comply in size and format with several very technical rules promulgated by the Patent and Trademark Office. The preparation of patent drawings is ordinarily best left to an experienced patent draftsperson.

If a patent application is prepared properly, it should pave the way for smooth handling of the patent application during its prosecution. If a patent application properly tells the story of the invention, it should constitute a teaching document that will stand on its own and be capable of educating a court regarding the character of the art to which the invention pertains, as well as the import of this invention to that art. Since patent suits are tried before judges who rarely have technical backgrounds, it is important that a patent application make an effort to present the basic features of the invention in terms understandable by those having no technical training. It is unusual for an invention to be so impossibly complex that its basic thrust defies description in fairly simple terms. A patent application is suspect if it wholly fails to, at some point, set forth the pith of the invention in terms a grade school student can grasp.

49.2.4 Enablement, Best Mode, Description, and Distinctness Requirements

Once a patent application has been prepared and is in the hands of the inventor for review, it is important that the inventor keep in mind the enablement, best mode, description, and distinctness requirements of the patent law.

The enablement requirement calls for the patent application to present sufficient information to enable a person skilled in the relevant art to make and use the invention. The disclosure presented in the application must be such that it does not require one skilled in the art to experiment to any appreciable degree to practice the invention.

The best mode requirement mandates that an inventor disclose, at the time he or she files a patent application, the best mode he or she then knows about for carrying out or practicing the invention.

The description requirement also relates to the descriptive part of a patent application and the support it must provide for any claims that may need to be added after the application has been filed. Even though a patent application may adequately teach how to make and use the subject matter of the claimed invention, a problem can arise during the prosecution of a patent application where one determines it is desirable to add claims differing in language from those filed originally. If the claim language one wants to add does not find full support in the originally filed application, the benefit of the original filing date will be lost with regard to the subject matter of the claims to be added. Therefore, in reviewing a patent application prior to its being executed, an inventor should keep in mind that the description that forms a part of the application should include support for any language he or she may later want to incorporate in the claims of the application.

The distinctness requirement applies to the content of the claims. In reviewing the claims of a patent application, an inventor should endeavor to make certain the claims particularly point out and distinctly claim the subject matter that he or she regards as his or her invention. The claims must be definite in the sense that their language must clearly set forth the area over which an applicant seeks exclusive rights. The language used in the claims must find support in the earlier descriptive portion of the application. The claims must not include within their scope of coverage any prior art known to the inventor, and yet should present the invention in the broadest possible terms that patentably distinguish the invention over the prior art.

49.2.5 Functional Language in Claims

While functional language in claims may tend to draw objection, there is statutory support for using a particular type of functional claim language. Section 112 of Title 35 of the United States Code includes this statement:

> An element in a claim for a combination may be expressed as a means or step for performing a specified function without the recital of structure, material, or acts in support thereof, and such claim shall be construed to cover the corresponding structure, material, or acts described in the specification and equivalents thereof. [emphasis added]

Using a "means plus function" format to claim an invention can be one of the most effective avenues to take in an effort to achieve the broadest possible coverage of alternative approaches that competitors may explore. However, in drafting a claim in means-plus-function format, care must be taken to ensure that what is being claimed amounts to more than a single means (i.e., a single element defined in means-plus-function format), since the requirement of the patent law that means-plus-function language be used only in a claim for a combination is not met by such a claim. Such a claim is deemed to be of undue breadth for, in essence, it claims every conceivable means for achieving a stated result.

49.2.6 Product-by-Process Claims

In some instances it is possible to claim a product by describing the process or method of its manufacture. Traditionally, the Patent and Trademark Office has taken the position that product-by-process claims are permissible only when the product cannot be adequately defined in any other fashion. Since 1974, however, product-by-process claims have been deemed allowable by the Patent and Trademark Office even if the product can be described in purely structural terms, so long as other claim requirements, such as definiteness, are properly met. Some examiners still adhere to the more traditional view of the impropriety of product-by-process claims, and difficulties continue to be encountered in obtaining the allowance of these types of claims universally throughout the Patent and Trademark Office.

49.2.7 Claim Format

Each claim is a complete sentence. In many instances the first part of the sentence of each claim appears at the beginning of the claims section and reads "What is claimed is:." Each claim typically includes three parts: preamble, transition, and body. The preamble is the part that introduces the claim by summarizing the field of the invention, its relation to the prior art, and its intended use, or the like. The transition is a word or phrase connecting the preamble to the body. The terms "comprises" or "comprising" often perform this function. The body is the listing of elements and limitations that define the scope of what is being claimed.

Claims are either *independent* or *dependent*. An independent claim stands on its own and makes no reference to any other claim. A dependent claim refers to another claim that may be independent or dependent, and adds to the subject matter of the referenced claim. If a dependent claim depends from (makes reference to) more than one other claim, it is called a *multiple dependent claim*.

One type of claim format that can be used gained notoriety in a 1917 decision of the Commissioner of Patents in a case styled *Ex parte Jepson*, 1917 C.D. 62. In a claim of the Jepson format, the preamble recites all the elements deemed to be old, the body of the claim includes only such new elements as constitute improvements, and the transition separates the old from the new. The Patent and Trademark Office favors the use of Jepson-type claims since this type of claim is thought to assist in segregating what is old in the art from what the applicant claims as his or her invention.

In 1966, the Patent and Trademark Office sought to encourage the use of Jepson-type claims by prescribing the following Rule 75(e):

> Where the nature of the case admits, as in the case of an improvement, any independent claim should contain in the following order, (1) a preamble comprising a general description of the elements or steps of the claimed combination which are conventional or known, (2) a phrase such as "wherein the improvement comprises," and (3) those elements, steps and/or relationships which constitute that portion of the claimed combination which the applicant considers as the new or improved portion.

Thankfully, the use of the term "should" in Rule 75(e) makes use of Jepson-type claims permissive rather than mandatory. Many instances occur when it is desirable to include several distinctly old elements in the body of a claim. The preamble in a Jepson-type claim has been held to constitute a limitation for purposes of determining patentability and infringement, while the preambles of claims presented in other types of formats may not constitute limitations. A proper understanding of the consequences of presenting claims in various types of formats and the benefits thereby obtained will be taken into account by one's patent attorney.

49.2.8 Executing the Application

Once an inventor has satisfied himself or herself with the content of a proposed patent application, he or she should read carefully the oath or declaration accompanying the application. The required content of this formal document recently has been simplified. In it the inventor states that he or she:

1. Has reviewed and understands the content of the specification, including the claims, as amended by any amendment specifically referred to in the oath or declaration.
2. Believes the named inventor or inventors to be the original and first inventor or inventors of the subject matter which is claimed and for which a patent is sought.
3. Acknowledges the duty to disclose to the Patent and Trademark Office during examination of the application information which is material to the examination of the application.

If the application is being filed as a division, continuation, or continuation-in-part of one or more co-pending parent applications, the parent case or cases are identified in the oath or declaration. Additionally, if a claim to the benefit of a foreign-filed application is being made, it is recited in the oath or declaration.

Absolutely no changes should be made in any part of a patent application once it has been executed. If some change, no matter how ridiculously minor, is found to be required after an application has been signed, the executed oath or declaration must be destroyed and a new one signed after the application has been corrected. If an application is executed without having been inspected by the applicant or is altered after having been executed, it may be stricken from the files of the Patent and Trademark Office.

49.2.9 Patent and Trademark Office Fees

The Office charges a fee to file an application, a fee to issue a patent, and a host of other fees for such things as obtaining an extension of time to respond to an Office Action.

Effective October 1, 1982, a very dramatic increase in almost all fees was implemented. The most striking result of the October 1, 1982 fee increase was the dramatic increase in applications filed during September, 1982, as the date of the fee increase drew increasingly near. I, too, stood in the long line that snaked through the lobby of Crystal Plaza Two on September 30, 1982, to file the applications prepared for our clients during September, 1982, to take advantage of the then applicable filing fees. The sight of half a dozen Office clerks stamping in applications and tossing them into crates as fast as the papers could be sent airborne was an unprecedented picture I shall long remember.

As of this writing, the *basic fee* required to file a patent application has increased 680-fold to

$340 since it was first set at 50 cents for filing an application under the patent act of 1790. From 1965 through October 1, 1982, the basic fee for filing a utility application stood at $65, with additional charges being made for claims in excess of a total of 10 and independent claims in excess of one.

In addition to the basic fee of $340, $34 is charged for each independent claim in excess of a total of three; $12 is charged for each claim of any kind in excess of a total of 20; and $110 is charged for any application that includes one or more multiple dependent claims. However, if the applicant is entitled to claim the benefits of *small entity status,* the entire filing fee is halved, as are most other fees that are associated with the handling of a patent application.

The filing fee for a design application presently stands at $140 unless small entity status is established, whereupon this fee also may be halved.

New rules now permit the Office to assign a filing date before the filing fee and oath or declaration have been received. While the filing fee and an oath or declaration are still needed to complete an application, a filing date will now be assigned as of the date of receipt of the specification and any required drawings.

The *issue fee* charged by the Office for issuing a utility patent on an allowed application stands at $560. Small entity status reduces this fee to $280. The issue fee for a design application is $200, which also may be halved with the establishment of small entity status.

Maintenance fees have been enacted to keep an issued utility patent in force during its term. No maintenance fees are charged on design or plant patents or on utility patents that have issued from applications filed before December 12, 1980. Patents issuing on applications filed from December 12, 1980 through August 26, 1982 stand at $225, $445, and $670, to be paid, respectively, no later than 3½, 7½, and 11½ years from their dates of issuance, with no small entity status reductions being available. Patents issuing on applications filed after August 26, 1982, stand at $450, $890, and $1340 (with small entity "half price" reductions being available), to be paid, respectively, no later than 3½, 7½, and 11½ years from their issue dates.

Fee increases can be expected at 3-year intervals. The above schedules went into effect on October 1, 1985.

49.2.10   Small Entity Status

The practice of providing half price fees to individual inventors, nonprofit organizations, and small businesses came into existence concurrently with the implementation of the fee increase of October 1, 1982.

Qualification for small entity status requires only the filing of a verified statement prior to or with the first fee paid as a small entity. Statements as to qualification as a small entity must be filed by all entities having rights with respect to an application or patent in order to qualify. Once qualification has been achieved, there is a continuing duty to advise the Office before paying the next fee if qualification for small entity status has been lost.

Those who qualify for small entity status include:

1. A sole inventor who has not transferred his or her rights and is under no obligation to transfer his or her rights to an entity that fails to qualify.

2. Joint inventors where no one among them has transferred his or her rights and is under no obligation to transfer his or her rights to an entity that fails to qualify.

3. A nonprofit organization such as an institution of higher education or an IRS qualified and exempted nonprofit organization.

4. A small business that has not more than 500 employees after taking into account the average number of employees (including full time, part time, and temporary) during the fiscal year of the business entity in question and of its affiliates, with the term "affiliate" being defined by a broad-reaching "control" test.

Any attempt to fraudulently establish small entity status, or establishing such status improperly and through gross negligence, is considered a fraud on the Office. An application could be disallowed for such an act. Failure to establish small entity status on a timely basis forfeits the right to small entity status benefits with respect to a fee being paid. A good faith error made in establishing small entity status may be excused by paying any deficient fees. However, if the payment is made more than three months after the error occurred, a verified statement establishing good faith and explaining the error must be filed.

49.2.11   Express Mail Filing

During 1983 a new procedure was adopted by the Office that permits any paper or fee to be filed with the Office by using the "Express Mail Post Office to Addressee" service of the U.S. Postal Service.

When this is done, the filing date of the paper or fee will be that shown on the "Express Mail" mailing label.

To qualify for the filed-when-mailed advantage, each paper must bear the number of the "Express Mail" mailing label, must be addressed to the Commissioner of Patents and Trademarks, Washington, DC 20231, and must include a certificate of mailing by "Express Mail" signed by the person who makes the actual mail deposit.

The practical and very important effect of this new procedure is to eliminate the hassle that has long been associated with the last minute attempts to effect physical delivery of patent applications to the Office in time to meet a bar date or comply with a convention filing date.

49.3 PROSECUTING A PENDING PATENT APPLICATION

Once an executed patent application has been received by the Patent and Trademark Office, the patent application is said to be pending. The prosecution period of a patent application is the time during which an application is pending; it begins when a patent application is filed in the Patent and Trademark Office and continues until either a patent is granted or the application is abandoned. The activities that take place during this time are referred to as prosecution.

49.3.1 Patent Pending

Once an application for a patent has been received by the Patent and Trademark Office, the applicant may mark products embodying the invention with an indication of "Patent Pending" or "Patent Applied For." These expressions mean a patent application has been filed and has neither been abandoned nor issued as a patent. The terms do not mean the Patent and Trademark Office has taken up examination of the merits of an application, much less approved the application for issuance as a patent.

The fact that a patent application has been filed, or is pending or applied for, does not provide any protection against infringement by competitors. While pending patent applications are held in secrecy by the Patent and Trademark Office and therefore do not constitute a source of information available to competitors regarding the activities of an inventor, nothing prevents competitors from independently developing substantially the same invention and seeking to market it. Neither is there any legal basis for stopping a competitor from purchasing a product bearing a designation "Patent Pending" and copying the invention embodied in the purchased product.

As a practical matter, however, marking products with the designation "Patent Pending" often has the effect of discouraging competitors from copying an invention, whereby the term of the patent that eventually issues may effectively be extended to include the period during which the application is pending. In many instances, competitors will not risk a substantial investment in preparation for the manufacture and merchandising of a product marked with the designation "Patent Pending," for they know their efforts may be legally interrupted as soon as a patent issues.

49.3.2 Secrecy of Pending Applications

With the exception of applications filed to reissue existing patents, pending patent applications are maintained in strictest confidence by the Patent and Trademark Office. No information regarding a pending application will be given out by the Office without authority from the applicant or owner of the application. However, if an interested third party learns of the pendency of an application, he or she may file a protest to its issuance.

The file of a pending application can only be inspected as a matter of right by the named inventor, an assignee of record, an exclusive licensee, an attorney of record, or such persons as have received written authority from someone permitted by right to inspect the file. This provision of secrecy extends to abandoned applications as well as to pending applications. In the event an abandoned application is referred to in an issued patent, access to the file of the abandoned case will be granted to members of the public on request. Should a pending patent application be referred to in an issued patent, access may usually be obtained by petition. All reissue applications are open to inspection by the general public.

49.3.3 Duty of Candor

The Patent and Trademark Office has placed increased emphasis on the duty an applicant has to deal candidly with the Patent and Trademark Office.

In accordance with Patent Office guidelines, a patent applicant is urged to submit an Information Disclosure Statement either concurrently with the filing of an application or within three months of its filing. When these guidelines were implemented in 1977, what are now called Information Disclosure Statements were referred to as Prior Art Statements. An Information Disclosure Statement may be either separate from or incorporated in a patent application. It should include a listing of patents, publications, or other information that is believed to be "material," and a concise explanation of the relevance of each listed item. Items are deemed to be "material" where there is a "substantial likelihood that a reasonable examiner would consider it important in deciding whether to allow the application to issue as a patent."

The courts have held that those who participate in proceedings before the Office have the "highest duty of candor and good faith." While the courts differ markedly in their holding of the consequences of misconduct, fraud on the Patent and Trademark Office has been found to be a proper basis for taking a wide variety of punitive actions such as striking applications from the records of the Office, cancelling issued patents, denying enforcement of patents in infringement actions, awarding attorney's fees to defendants in infringement actions, and imposing criminal sanctions on those who were involved in fraudulently procuring patents. Inequitable conduct other than outright fraud has been recognized as a defense against enforcement of a patent, as a basis for awarding attorney's fees in an infringement action, and as a basis of antitrust liability.

In short, the duty of candor one has in dealings with the Office should be taken very seriously. Prudent practice would urge that if there is any question concerning whether a reference or other facts are "material," a citation should be made promptly to the Office so that the examiner can decide the issue.

49.3.4   Initial Review of Application

Promptly after an application is filed, it is examined to make certain it is complete and satisfies formal requirements sufficiently to permit its being assigned a filing date and serial number. Once a patent application has been received by the Patent and Trademark Office and assigned a filing date and serial number, the classification of the subject matter of the claimed invention is determined and the application is assigned to the appropriate examining group. In the group, the application is assigned to a particular examiner. Each examiner takes the applications assigned to him or her in the order of their filing.

The Patent and Trademark Office receives nearly 2000 patent applications per week. Although nearly 1400 examiners staff the Office, a backlog of several months of cases awaits action in most of the examining sections called "Group Art Units." This results in a delay of several months between the time an application is filed and when it receives its first thorough examination on the merits.

Once an examiner reaches an application and begins the initial review, he or she checks the application still further for compliance with formal requirements and conducts a search of the prior art to determine the novelty and nonobviousness of the claimed invention. The examiner prepares an Office Action in which he or she notifies the applicant of any objections to the application, or requirements regarding election of certain claims for present prosecution, and/or any rejections he or she believes should be made of the claims.

In the event the examiner deems all the claims in the application to be patentable, he or she notifies the applicant of this fact and issues a Notice of Allowance. It troubles me to receive a Notice of Allowance on the first Office Action. While I do not file applications with claims that are broader than I think a client is entitled to receive, if a Notice of Allowance issues on the first Office Action I am left with the nagging feeling that had I sought claims of even broader scope, perhaps they too might have been allowed.

In some instances, the examiner will find it necessary to object to the form of the application. One hopes that these formal objections are not debilitating and can be corrected by relatively minor amendments made in response to the Office Action.

In treating the merits of the claims, especially in a first Office Action, it is not uncommon for an examiner to reject a majority if not all of the claims. Some examiners feel strongly that they have a duty to cite the closest art they are able to find and to present rejections based on this art to encourage the inventor to put on record in the file of the application such arguments as are needed to illustrate to the public exactly how the claimed invention distinguishes itself patentably over the cited art.

49.3.5   Response to Office Action

ın the event the first Office Action issued by an examiner is adverse in any respect, the applicant may reply in almost any form that constitutes a bona fide attempt to advance the prosecution of the application. The applicant is entitled to at least one reexamination following the issuance of the first Office Action, even if the response filed by the applicant amounts to nothing more than a request for reconsideration.

Since the file of a patent application will become open to public inspection on the issuance of a patent and because an issued patent must be interpreted in view of the content of its file, the character of any arguments presented to the Patent and Trademark Office in support of a claimed invention are critical. In responding to an Office Action it is essential that care be taken in the drafting of arguments to ensure that no misrepresentations are made and that the arguments will not result in an unfavorable interpretation of allowed claims being made during the years when the resulting patent is in force.

It is important that every effort be made to advance the prosecution of the application in responding to an Office Action. While in previous years it was not unusual for half a dozen or more Office Actions to issue during the course of pendency of a patent application, in more recent years the Office has placed emphasis on "compacting" the prosecution of patent applications. Today it is not unusual for the prosecution of a patent application to be concluded on the issuance of the second or third Office Action. In an increasing number of cases, a final rejection is made as early as the second or third Office Action.

When an Office Action is mailed from the Patent and Trademark Office, a time period for filing a response begins. In the event a response is not filed within the time set by law, the application will automatically become abandoned, and rights to a patent may be lost forever. Ordinarily, a response must be filed within a three-month period from the mailing date of the Office Action. An extension of time of up to three months can be had so long as the statutory requirement of filing a response within six months of the issuance of an Office Action is met. In previous years, it was necessary to present reasons to justify the grant of an extension of time, and the grant of an extension request was discretionary. Now, however, extensions are granted automatically upon receipt of a petition for extension accompanied by a proper response to the Office Action and the required fee. The fee for obtaining an extension of time increases as the number of months covered by the extension is requested increases.

In responding to an Office Action, each objection and rejection made by the examiner must be treated. If the inventor agrees that certain of his or her claims should not be allowed in view of art cited by the examiner, these claims may be cancelled or amended to better distinguish the invention over the cited art.

Typical responses to Office Actions involve the addition, cancellation, substitution, or amendment of claims; the amendment of the descriptive portion of the application can be done if no "new matter" is added by the amendment; and the presentation of arguments regarding the allowability of the claimed invention in which explanations are provided about how the claims patentably distinguish themselves over the cited references. A response may also include the submission of an affidavit to overcome a cited reference either by establishing a date of invention before the effective date of the reference or by presenting factual evidence supporting patentability over the reference.

If the Office has objected to the drawings, corrections must now be made utilizing the services of a bonded patent drafting service, which has been approved by the Office, or substitute drawings including the required corrections supplied.

49.3.6 Reexamination Following a Response

Once the applicant has responded, the examiner reexamines the case and issues a second Office Action apprising the applicant of his or her findings. If the examiner agrees to allow all the claims that remain active in the application, prosecution on the merits is closed and the applicant may not present further amendments or add other claims as a matter of right. If the Office Action is adverse with regard to the merits of the claims, the prosecution of the case continues until such time as the examiner issues an Office Action including a final rejection.

The examiner makes a rejection *final* once a clear and unresolved issue has developed between the examiner and the applicant. After a final rejection has issued, the character of the responses that may be made by the applicant is limited. The applicant may appeal the final rejection to an intraagency Board of Appeals, cancel the rejected claims, comply with all of the requirements for allowance if any have been laid down by the examiner, or file a continuation application whereby the examination procedure is begun again.

If an initial appeal taken to the Board of Patent Appeals should result in an unfavorable decision, a further appeal may be taken either to the U.S. District Court for the District of Columbia or to the newly established U.S. Court of Appeals for the Federal Circuit (which, as of October 1, 1982, replaced what was previously known as the U.S. Court of Customs and Patent Appeals). In some instances, further appeals may be pursued to higher courts.

In the majority of instances during the period of prosecution, the application eventually reaches a form acceptable to the examiner handling the application, and the examiner will issue a Notice of Allowance. If it is impossible to reach accord with the examiner handling the application, the inventor can make use of the procedures for appeal.

One recently adopted practice of the Office that concerns many attorneys is worthy of mention. When the record of an application "does not otherwise reveal the reasons for allowance," an examiner may put a comment in the file to explain his or her reasons for allowing a case. While fine in theory, this practice has tended to result in examiners making very simplistic statements that focus on a single reason. If the reason stated by an examiner for allowing a patent is shown during litigation to be faulty, this can cause the patent to be held invalid. Accordingly, if a communication is received from the Office stating a reason for allowance, it should be reviewed with care. Often it is desirable to exercise one's right to promptly comment on the stated reason, for despite a statement in Office Rule 109 to the contrary, failure to comment may in real-world practice give rise to an implication that the applicant agrees.

49.3.7 Interviewing the Examiner

If, during the prosecution of a patent application, it appears that substantial differences of opinion or possible misunderstandings are being encountered in dealing with the examiner to whom the application has been assigned, it is advisable for the attorney to conduct a personal interview with the examiner.

Considering the relatively sterile and terse nature of many Office Actions, it is often impossible to determine accurately what the examiner's opinion may be regarding how the application should be further prosecuted. While the word processing equipment acquired by the Office has made it easier for examiners to expound the reasons underlying their rejections, situations still arise where it is quite clear that an examiner and an attorney are not communicating in the full sense of the word. At times, a personal interview will be found to provide valuable guidance for bringing the prosecution of the application to a successful conclusion. In other instances, an interview will be beneficial in ascertaining the true character of any difference of opinion between the applicant and the examiner, thereby enabling the exact nature of this issue to be addressed thoroughly in the next response filed by the applicant.

49.3.8 Restriction and Election Requirements

If a patent examiner determines that an application contains claims to more than one independent and distinct invention, the examiner may impose what is called a *restriction requirement*. In the event the examiner finds that the application claims alternative modes or forms of an invention, he or she may require the applicant to elect one of these species for present prosecution. This is called a *species election requirement*. Once a restriction or election requirement has been imposed, the applicant must elect one of the designated inventions or species for present prosecution in the original application. The applicant may file divisional applications on the nonelected inventions or species any time during the pendency of the original application.

When responding to an Office Action that includes a restriction and/or election requirement, it is usually desirable to present arguments in an effort to traverse the requirement and request its reconsideration. After traversing, the examiner is obliged to reconsider the requirement, but he or she may repeat it and make it final. Sometimes the examiner can be persuaded to modify or withdraw a restriction and/or election requirement, thereby permitting a larger number of claims to be considered during the prosecution of the pending application. Where a restriction and/or election requirement has been traversed (opposed by requesting reconsideration for stated reasons), upon allowance of elected claims, the Office will notify the applicant that the application is in condition for allowance except for the presence of such claims as remain in the case and are directed to a nonelected invention or species. The applicant will then be given a month to take appropriate action such as the filing of a divisional application or the filing of a petition for reconsideration of an election requirement. As a practical matter, unless the examiner has set out a restriction and/or election requirement that is utterly and completely absurd, a petition seeking reconsideration is usually a waste of effort. Thankfully, it is seldom that an examiner decides that a simple application defines two dozen separate inventions, whereby the need to petition is a rarity.

49.3.9 Double Patenting Rejections

Occasionally, one may receive a rejection based on the doctrine of double patenting. This doctrine precludes the issuance of second patent on the same invention already claimed in a previously issued patent.

One approach to overcoming a double patenting rejection is to establish a clear line of demarcation between the claimed subject matter of the second application and that of the earlier patent. If the line of demarcation is such that the claimed subject matter of the pending application is nonobvious in view of the invention claimed in the earlier patent, no double patenting rejection is proper. If the claimed subject matter of the pending application defines merely an obvious variation of the claimed invention of the earlier issued patent, the double patenting rejection may be overcome by the filing of a terminal disclaimer. The terminal portion of any patent issuing on the pending application is disclaimed so that any patent issuing on the pending application will expire on the same day the already existing patent expires. If the claimed subject matter of the pending application is identical to the claimed subject matter in the earlier issued patent, it is not possible to establish a line of demarcation between the two cases, and the pending application is not patentable even if a terminal disclaimer is filed.

The courts have held that double patenting problems can occur when a utility patent application and a design patent application have been filed on the same invention. The fact that both a utility patent and a design patent may have issued on various features of a common invention does not necessarily mean a double patenting problem exists. If it is possible to practice the ornamental appearance covered by the design patent without necessarily infringing any of the claims of the utility patent and if it is possible to practice the claimed invention of the utility patent without infringing the ornamental appearance covered by the design patent, no double patenting problem is present.

49.3.10 Patent Issuance

Once a Notice of Allowance has been mailed by the Office, the applicant has a period of three months to pay the issue fee. A patent will not issue unless this fee is paid. A few weeks before the patent issues, the Office mails a Notice of Issuance, which advises the applicant of the issue date and patent number.

Once a patent issues, its file history is no longer held secret. The several documents that form the complete file history of a patent are referred to collectively as the *file wrapper*.

Upon receipt of a newly issued patent, it should be reviewed with care to check for printing errors. If printing errors of misleading or otherwise significant nature are detected, it is desirable to petition for a Certificate of Correction. If errors of a clerical or typographical nature have been made by the applicant or by his or her attorney and if these errors are not the fault of the Patent and Trademark Office, a fee must be paid to obtain the issuance of a Certificate of Correction. If the errors are the fault of the Patent and Trademark Office, no such fee need be paid.

49.3.11 Safeguarding the Original Patent Document

The original patent document merits appropriate safeguarding. It is printed on heavy bond paper, its pages are fastened together by a silk ribbon, and it is sealed by the Official Seal of the United States Patent and Trademark Office. The patent owner should preserve this original document in a safe place as evidence of his or her proprietary interest in the invention. If an infringer must be sued, the patent owner may be called on to produce his or her original patent in court.

49.3.12 Continuation, Divisional, and Continuation-in-Part Applications

During the pendency of an application, it may be desirable to file either a continuation or a divisional application. A continuation application may be filed if the prosecution of a pending application has not proceeded as desired, whereby a further opportunity for reconsideration can be had before an appeal is taken. A divisional application may be filed when two or more inventions are disclosed in the original application, and claims to only one of these inventions are deemed to be proper for examination in the originally filed case. The division picks up the claims that were not permitted to be examined in the parent application.

It frequently occurs during the pendency of a patent application that a continuing research and development program being conducted by the inventor results in improvements in the original invention. Because of a prohibition in the patent law against amending the content of a pending patent application to include "new matter," any improvements made in the invention after the time an application is filed cannot be incorporated into the pending application. When improvements are made that are deemed to merit patent protection, a continuation-in-part application is filed. Such an application can be filed only during the pendency of an earlier-filed application commonly called the *parent* case. The continuation-in-part case receives the benefit of the filing date of the parent application with regard to such subject matter as is common to the parent application. Any subject matter uncommon to the parent application is entitled only to the benefit of the filing date of the continuation-in-part case.

In some instances when a continuation-in-part application has been filed, the improvement that forms the new subject matter of the continuation-in-part case is closely associated with the subject matter of the earlier-filed application, and the earlier application can be abandoned in favor of the continuation-in-part case. In other instances, the new matter that forms the basis of the continuation-in-part application clearly constitutes an invention in and of itself. In such a situation it may be desirable to continue the prosecution of the original application to obtain one patent that covers the invention claimed in the original application, and a second patent that covers the improvement invention.

49.3.13 Maintaining a Chain of Pending Applications

If a continuing development program is under way and produces a series of new improvements, it can be advantageous to maintain on file in the Patent and Trademark Office a continuing series of pending applications. If an original parent application is initially filed, and a series of continuation, division, and/or continuation-in-part applications are filed in such a manner that ensures the existence of an uninterrupted chain of pending cases, any patent or patents that may issue on the earlier cases cannot be used as references against later applications in the chain. This technique of maintaining a series or chain of pending applications is an especially important technique to use when the danger exists that the closest prior art the Office may be able to cite against the products of continuing research and development effort is the patent protection that issued on early aspects of the overall effort.

49.4 PATENT PROTECTIONS AVAILABLE ABROAD

United States patents provide no protection abroad and can be asserted against a foreigner only in the event the foreigner's activities infringe within the geographical bounds of our country. This section briefly outlines some of the factors one should consider if patent protection outside the United States is desired.

49.4.1 Canadian Filing

Many United States inventors file in Canada. Filing an application in Canada is relatively inexpensive compared with the cost of filing in other countries. With the exception of a stringently enforced unity requirement, which necessitates that all the claims in an application strictly define a single inventive concept, Canadian patent practice essentially parallels that of the United States. If one has success in prosecuting an application in the United States, it is not unusual for the Canadian Patent Office to agree to allow claims of substantially the same scope as those allowed in the United States.

49.4.2 Foreign Filing in Other Countries

Obtaining foreign patent protection in countries other than Canada, particularly in non-English-speaking countries, has long been an expensive undertaking. In almost all foreign countries, local agents or attorneys must be employed, and the requirements of the laws of each country must be met. Some countries exempt large areas of subject matter such as pharmaceuticals from what may be patented. It may be necessary to provide a certified copy of the United States case for filing in each foreign country selected. Translations are needed in most non-English-speaking countries. In such countries as Japan, even the retyping of a patent application to put it in proper form can be costly.

With the exception of a few English-speaking countries, it is not at all uncommon for the cost of filing an application in a single foreign country to equal, if not exceed, the total costs that have been incurred in the entire process of preparing and filing the original United States application. These seemingly unreasonably high costs prevail even though the United States application, from which a foreign application is prepared, already presents the essential elements of the foreign case.

49.4.3 Annual Maintenance Taxes and Working Requirements

In many foreign countries annual fees are charged to maintain the active status of a patent. In some countries, the fees escalate each year on the theory that the invention covered by the patent must be worth more as it is put into practice more extensively. These annual maintenance fees not only benefit foreign economies, but also cause many patent owners to dedicate their foreign invention rights to the public. Maintaining patents in force in several foreign countries is often unjustifiably expensive.

In many foreign countries there are requirements that an invention be "worked" or practiced within these countries if patents within these countries are to remain active. Licensing of a citizen of the country to practice an invention satisfies the working requirement in some countries.

49.4.4 Filing under International Convention

If applications are filed abroad within one year of the filing date of an earlier-filed United States case, the benefit of the filing date of the earlier-filed United States case may be attributed to the foreign applications. Filing within one year of the filing date of a United States case is known as filing under international convention. The convention referred to is the Paris Convention, which has been ratified by our country and by almost all other major countries.

Many foreign countries do not provide the one-year grace period afforded by United States statute to file an application. Instead, certain foreign countries require that an invention be "absolutely novel" at the time of filing of a patent application in these countries. If the United States application has been filed prior to any public disclosure of an invention, the absolute novelty requirements of certain foreign countries can be met by filing applications in these countries under international convention, whereby the effective filing date of the foreign cases is the same as that of the United States case.

49.4.5 Filing on a Country-by-Country Basis

If one decides to file abroad, one approach is to file separate applications in each selected country. Most United States patent attorneys have associates in foreign countries with whom they work in pursuing patent protections abroad. It is customary for the United States attorney to advise a foreign associate about how he or she believes the prosecution of an application should be handled, but to leave final decisions to the expertise of the foreign associate.

49.4.6 The Patent Cooperation Treaty

Since June, 1978, United States applicants have been able to file an application in the United States Patent and Trademark Office in accordance with the terms of the Patent Cooperation Treaty (PCT), which has been ratified by the United States. While PCT member countries are becoming quite numerous, the use which has been made to date of PCT filings has been relatively minimal. Only about 5000 PCT applications are being filed worldwide each year. Only about 1800 PCT cases are being filed in the United States each year.

PCT member countries include such major countries as Australia, Austria, Belgium, Brazil, Denmark, Finland, France, Hungary, Netherlands, Norway, Sweden, Soviet Union, Switzerland, United Kingdom, United States, and West Germany. In filing a PCT case, a United States applicant can designate the application for eventual filing in the national offices of such other countries as have ratified the treaty.

One advantage of PCT filing is that the applicant is afforded an additional eight months beyond the one year period he or she would otherwise have had under the Paris Convention to decide whether he or she wants to complete filings in the countries he or she has designated. Under the Patent Cooperation Treaty, an applicant has 20 months from the filing date of his or her U.S. application to make the final foreign filing decision.

Another advantage of PCT filing is that it can be carried out literally at the last minute of the one year convention period measured from the date of filing of a U.S. application. Thus, in situations where a decision to file abroad to effect filings has been postponed until it is impractical if not impossible to effect filings of separate applications in individual countries, a single PCT case can be filed on a timely basis in the United States Patent and Trademark Office designating the desired countries.

Still another feature of PCT filing is that, by the time the applicant must decide on whether to complete filings in designated countries, he or she has the benefit of the preliminary search report (a first Office Action) on which to base his or her decision. If the applicant had elected instead to file applications on a country-by-country basis under international convention, it is possible that he or she might not have received a first Office Action from the Patent and Trademark Office within the one year permitted for filing under international convention.

49.4.7 The European Patent Convention

Another option available to United States citizens since June, 1978, is to file a single patent application to obtain protection in one or more of the European countries that are parties to the so-called European Patent Convention (EPC). It should be noted that the EPC is a convention separate and apart from the Common Market, with EPC membership not including all Common Market countries, yet including some countries that do not belong to the Common Market. As of this writing, EPC member countries include Austria, Belgium, France, Italy, Liechtenstein, Luxembourg, Netherlands, Sweden, Switzerland, United Kingdom, and West Germany.

Two routes are available to United States citizens to effect EPC filing. One is to act directly through a European patent agent or attorney. The other is to use PCT filing through the United States Patent and Trademark Office and to designate EPC filing as a "selected country."

A European Patent Office (EPO) has been set up in Munich, West Germany. Before applications are examined by the EPO in Munich, a Receiving Section located at The Hague inspects newly filed applications for form. A novelty search report on the state of the art is provided by the International Patent Institute at The Hague. Within 18 months of filing The Hague will publish an application to seek views on patentability from interested parties. Once publication has been made and the examination fee paid by the applicant, examination moves to Munich where a determination is made of patentability and prosecution is carried out with the applicant responding to objections received from the examiner. The EPO decides whether a patent will issue, after which time a copy of the patent application is transferred to the individual patent offices of the countries designated by the applicant. The effect of EPC filing is that, while only a single initial application need be filed and prosecuted, in the end, separate and distinct patents issue in the designated countries. Any resulting patents have terms of 20 years measured from the date of filing of the original application.

49.4.8 Advantages and Disadvantages of International Filing

An advantage of both PCT and EPC filing is that the required applications can be prepared in exactly the same format. Their form and content will be accepted in all countries that have adhered to the EPC and/or PCT programs. Therefore, the expense of producing applications in several different formats and in different languages is eliminated. The fact that both PCT and EPC applications can, in their initial stages, be prepared and prosecuted in the English language is another important advantage for United States citizens.

A principal disadvantage of both of these types of international patent filings is their cost. Before savings over the country-by-country approach are achieved, filing must be anticipated in several countries, perhaps as many as four to six, depending on which countries are selected. A disadvantage of EPC filing is that a single examination takes place for all the designated countries, and patent protection in all these countries is determined through this single examination procedure.

49.4.9 Trends in International Patent Protection

With the advent of the PCT and EPC programs, a significant step forward has been taken that may someday lead to the development of a multinational patent system. For the predictable future, however, it seems clear that the major countries of the world intend to maintain intact their own patent systems. Certainly, the United States is reluctant to give full faith and credit to foreign patents in which foreign nationals would be permitted to enforce patents obtained in their countries in the United States. Similarly, other nations are reluctant to give United States citizens the right to exclude others from making, using, or selling inventions in their countries based solely on the issuance of patents in our country.

BIBLIOGRAPHY

T. Arnold and F. S. Vaden, III, *Invention Protection,* Barnes & Noble, New York, 1971.

R. A. Buckles, *Ideas, Inventions and Patents,* Wiley, New York, 1957.

D. A. Burge, *Patent and Trademark Tactics and Practice,* 2nd ed., Wiley, New York, 1984.

D. S. Chisum, *Patents,* Matthew Bender, New York, 1978, 1983.

E. W. Kitner and J. Lahr, *An Intellectual Property Law Primer,* 2nd ed., Clark Boardman, New York, 1982.

Manual of Patent Examining Procedure, 5th ed., U.S. Government Printing Office, August, 1983.

H. R. Mayers and B. G. Brunsvold, *Drafting Patent License Agreements,* 2nd ed., Bureau of National Affairs, Washington, DC, 1984.

P. D. Rosenberg, *Patent Law Fundamentals,* Clark Boardman, New York, 1975.

F. B. Schramm, *Handbook on Patent Disputes,* Harrison, Atlanta, 1974.

R. A. White, R. K. Caldwell, and J. F. Lynch, *Patent Litigation: Procedure and Tactics,* Matthew Bender, New York, 1978.

CHAPTER 50

SOURCES OF MECHANICAL ENGINEERING INFORMATION

FRITZ DUSOLD

Mid-Manhattan Library
Science and Business Department
New York, New York

50.1 INTRODUCTION

This chapter is designed to enable the engineer to find information efficiently and to take advantage of all available information. The emphasis is placed on publications and services designed to identify and obtain information. Because of space limitations references to individual works, which contain the required information, are limited to a few outstanding or unusual items.

50.2 THE PRIMARY LITERATURE

The most important source of information is the primary literature. It consists mainly of the articles published in periodicals and of papers presented at conferences. New discoveries are first reported in the primary literature. It is, therefore, a major source of current information. Peer review and editorial scrutiny, prior to publication of an article, are imposed to ensure that the article passes a standard of quality. Most engineers are familiar with a few publications, but are not aware of the extent of the total production of primary literature. *Engineering Index*[1] alone abstracted material from almost 4000 periodicals and conferences in 1982.

Handbooks and encyclopedias are part of the secondary literature. They are derived from primary sources and make frequent references to periodicals. Handbooks and encyclopedias are arranged to present related materials in an organized fashion and provide quick access to information in a condensed form.

While monographs—books written for professionals—are either primary or secondary sources of knowledge and information, textbooks are part of the tertiary literature. They are derived from primary and secondary sources. Textbooks provide extensive explanations and proofs for the material covered to provide the student with an opportunity to understand a subject thoroughly.

50.2.1 Periodicals

In most periodicals published by societies and commercial publishers, articles are identified usually by issue, and/or volume, date, and page number. Bibliographic control is excellent, and it is usually a routing matter to obtain a copy of a desired article. But some problems do exist. The two most common are periodicals that are known by more than one name, and the use of nonstandard abbreviations. Both of these problems could be solved by using the International Standard Serial Number (ISSN), which accurately identifies each publication. The increasing size and use of automated databases should provide an impetus to increased use of ISSN or some other standard.

The first scientific periodical, *Le Journal des Scavans,* was published January 4, 1665. The second, *Philosophical Transactions* of the Royal Society (London), appeared on March 6, 1665. The number of scientific periodicals has been increasing steadily with some setbacks caused by wars and natural catastrophies. The accumulated body of knowledge is tremendous. Much of this information can be retrieved by consulting indexes, abstracts, and bibliographies.

50.2.2 Conference Proceedings

The bibliographic control for papers presented at conferences is not nearly as good as for periodicals. The responsibility for publishing the papers usually falls upon the sponsoring agent or host group. For major conferences the sponsoring agency is frequently a professional society or a department of a university. In these cases an individual with some experience in publishing is usually found to act as an editor of the *Proceedings.* In other instances the papers are issued prior to the conference as *Preprints.* In still other situations the papers will be published in a periodical as a special issue or distributed over several issues of one or more periodicals. An additional, unknown, percentage of papers are never published and are only available in manuscript form from the author.

50.3 INDEXES AND ABSTRACTS

Toward the end of the last century the periodical literature had reached a volume that made it impossible for the "educated man" to review all publications. In order to retrieve the desired or needed information, indexes and abstracts were prepared by individual libraries and professional societies. In the 1960s computers became available for storing and manipulating information. This lead to the creation and marketing of automated data banks.

50.3.1 Manual Searching

The major abstracts typically provide the name of the author, a brief abstract of the article, and the title of the article, and identify where the article was published. Alphabetical author and subject indexes are usually provided, and a unique number is assigned to refer to the abstract. Many abstracts are published monthly or more frequently. Annual cumulations are available in many cases. The most important abstracts for engineers are:

Engineering Index[1]
Science Abstracts[2]
    Series A: Physics Abstracts
    Series B: Electrical and Electronics Abstracts
    Series V: Computer and Control Abstracts
Chemical Abstracts[3]
Metals Abstracts[4]

A comprehensive listing of abstracts and indexes can be found in *Ulrich's International Periodical Directory.*[5]

50.3.2 Online Searching

Most of the major indexes and abstracts are now available in machine-readable form. For a comprehensive list of databases and online vendors see *Information Industry Market Place.*[6] The names of online databases frequently differ from their paper counterparts. *Engineering Index,*[1] for example, offers COMPENDEX and EI Engineering Meetings online. Most of the professional societies producing online databases will undertake a literature search. A society member is frequently entitled to reduced charges for this service. In addition to indexes and abstracts, several encyclopedias and handbooks are, or will shortly be, available online. The *Kirk–Othmer Encyclopedia of Chemical Technology*[7] is one example. There seems to be virtually no limit to the information that can be made available online. The high demand for quick information retrieval ensures the expansion of this service. However, some problems exist in marketing, assessment of charges, legal aspects related to "fair use," and competition. These forces will undoubtedly cause some changes in the online industry.

    The most common method of obtaining access to these databases is through vendors. Leasing arrangements with database producers are also available in some cases. Leasing might be cost effective in instances where an individual database is frequently used, because of the elimination of online charges.

    In addition to the online indexes, several library networks and consortia produce online databases. These are essentially equivalent to the catalogs of member libraries and can be used to determine which library owns a particular book or subscribes to a particular periodical. Materials cataloged prior to the establishment of a machine-readable database are usually not included, since few libraries have the financial resources to do a retrospective conversion from print to machine-readable form. For a comprehensive listing of library networks and consortia offering online services, see *Information Industry Market Place 1983.*[6]

50.4 ENCYCLOPEDIAS AND HANDBOOKS

There are well over 300 encyclopedias and handbooks covering science and technology. A large number of these are more than ten years old. The date of publication should therefore be checked before using any of these works if the required information is likely to have been affected by recent progress. The following list represents only a sampling of available works of outstanding value.

    The *Kirk–Othmer Encyclopedia of Chemical Technology*[7] provides in 25 volumes, plus a separate index, a comprehensive and authoritative treatment of a wide range of subjects, with heaviest concentration on materials and processes. The basic set is updated by supplements. As previously mentioned, this work has become available online for ease of search and retrieval of information buried in the work's 9 million words.

    The *Encyclopedia of Computer Science and Technology*[8] is the only major reference tool covering this fast moving field. It provides a thorough treatment, but will have to be updated in some areas by using the primary literature.

    Metals Handbook[9] provides an encyclopedic treatment of metallurgy and related subjects. Each of the volumes is devoted to a separate topic such as mechanical testing, powder metallurgy, and heat treating. Each of the articles is written by a committee of experts on that particular topic. Volume 6 of the 9th edition was published in 1983; remaining volumes are expected to be published in yearly intervals.

    The *CRC Handbook of Chemistry and Physics,*[10] popularly known as the "Rubber Handbook," is probably the most widely available handbook. It is updated annually to include new materials and to provide more accurate information on previously published sections as soon as the information becomes available. The CRC Press is one of the major publishers of scientific and technical handbooks. A *Composite Index of CRC Handbooks*[11] provides access to the information covering the latest editions of 57 CRC handbooks, some of them multivolume sets, which were published prior to 1977.

The increasing concern with industrial health and safety has placed an additional responsibility on the engineer to see that materials are handled in a safe manner. *Dangerous Properties of Industrial Materials*[12] provides an authoritative treatment of this subject. This book also covers handling and shipping regulations for a large variety of materials.

Engineers have always been concerned with interaction between humans and machines. This area has become increasingly sophisticated and specialized. The *Human Factors Design Handbook*[13] is written for the design engineer rather than the human factor specialist. The book provides the engineer with guidelines for designing products for convenient use by people.

Engineering work frequently requires a variety of calculations. The *Standard Handbook of Engineering Calculations*[14] provides the answer to most problems. Although most of the information in this handbook is easily adaptable to computer programming, a new edition would probably take greater advantage of the increasing availability of computers.

Conservation of energy has become an important consideration, owing to energy's increasing cost. The *Energy Technology Handbook*[15] provides a large amount of information on energy-efficient design. In addition to conservation, the use of alternate sources of energy has been investigated intensively. The *Solar Energy Handbook*[16] covers this area in considerable detail.

The *Engineering and Industrial Graphics Handbook*[17] provides the latest information on methods for producing engineering graphics. Automated methods for the production of graphic materials are stressed.

Composite materials frequently offer advantages in properties and economy over conventional materials. A complete listing of composites, giving properties, typical applications, cost, and special fabrication requirements, can be found in the *Handbook of Composites*.[18] A theoretical treatment of these materials is also included.

Any required information on valves can be found in *Lyons' Valve Designers' Handbook*.[19] This book is heavily illustrated. Information is provided on construction, flow reduction, placement, and potential problems encountered with valves.

At the time when England converted to the metric system the British Standards Institution published *Metric Standards for Engineers*.[20] With the increasing worldwide distribution of products, metric units will gain in importance regardless of the official position of the U.S. government. This handbook offers the engineer an authoritative and detailed treatment of metrification.

The *Infrared Handbook*[21] was primarily prepared for use by the U.S. Navy. Although military uses of infrared radiation were the reason for writing this handbook, the coverage is extensive, as well as authoritative and detailed.

50.5 CODES, SPECIFICATIONS, AND STANDARDS

Codes, specifications, and standards are produced by government agencies, professional societies, businesses, and organizations devoted almost exclusively to the production of standards. In the United States the American National Standards Institute[22] (ANSI) acts as a clearinghouse for industrial standards. ANSI frequently represents the interests of U.S. industries at international meetings. Copies of standards from most industrial countries can be purchased from ANSI as well as from the issuer.

Copies of standards issued by government agencies are usually supplied by the agency along with the contract. They are also available from several centers maintained by the government for the distribution of publications. Most libraries do not collect government specifications.

Many of the major engineering societies issue specifications in areas related to their functions. These specifications are usually developed, and revised, by membership committees.

The American Society of Mechanical Engineers[23] (ASME) has been a pioneer in publishing codes concerned with areas in which mechanical engineers are active. In 1885 ASME formed a Standardization Committee on Pipe and Pipe Threads to provide for greater interchangeability. In 1911 the Boiler Code Committee was formed to enhance the safety of boiler operation. The 1983 ASME Boiler and Pressure Vessel Code was published in a metric (SI) edition, in addition to the edition using U.S. customary units, to reflect its increasing worldwide acceptance. The boiler code covers the design, materials, manufacture, installation, operation, and inspection of boilers and pressure vessels. The 1983 edition consists of 22 sections plus two code cases books. Revisions, additions, and deletions to the code are published twice yearly during the three year cycle of the code.

The most frequently used collection of specifications is the *Annual Book of Standards*[24] issued by the American Society for Testing and Materials (ASTM). These standards are prepared by committees drawn primarily from the industry most immediately concerned with the topic.

The standards written by individual companies are usually prepared by a member of the standards department. They are frequently almost identical to standards issued by societies and government agencies and make frequent references to these standards. The main reason for these "in-house" standards is to enable the company to revise a standard quickly in order to impose special requirements on a vendor.

The large number of standards issued by a variety of organizations has resulted in a number of identical or equivalent standards. The *Index and Directory of U.S. Industry Standards*[25] provides a listing of over 20,000 standards arranged by subject. Other features of this publication are a list of 400 standards-preparing organizations, a society/numeric index, and an ANSI concordance.

50.6 GOVERNMENT PUBLICATIONS

The U.S. government is probably the largest publisher in the world. Most of the publications are available from the Superintendent of Documents.[26] The *Monthly Catalog of Government Publications* serves as a subject index to these publications. These publications are provided, free of charge, to depository libraries throughout the country. Depository libraries are obligated to keep these publications for a minimum of 5 years and to make them readily available to the public.

The government agencies most likely to publish information of interest to engineers are probably the National Bureau of Standards, the Geological Survey, the National Oceanic and Atmospheric Administration, and the National Technical Information Service.

Since there are no copyright restrictions on government publications, they can readily be used by commercial publishers. The 1982–1983 edition of the *CRC Handbook of Chemistry and Physics*,[10] for example, updated the enthalpy, entropy, Gibbs free energy of formation, and heat capacity for many compounds and ions based on the recently revised National Bureau of Standards *Circular 500*.

50.7 ENGINEERING SOCIETIES

Engineering societies have exerted a strong influence on the development of the profession. The ASME publishes the following periodicals in order to keep individuals informed of new developments and to communicate other important information:

Applied Mechanics Reviews (monthly)

CIME (*Computers in Mechanical Engineering,* published by Springer Verlag, New York)

Mechanical Engineering (monthly)

Transactions (quarterly)

The *Transactions* cover the following fields: power, turbomachinery, industry, heat transfer, applied mechanics, bioengineering, energy resources technology, solar energy engineering, dynamic systems, measurement and control, fluids engineering, engineering materials and technology, pressure vessel technology, and tribology.

Many engineering societies have prepared a code of ethics in order to guide and protect engineers. Societies frequently represent the interests of the profession at government hearings and keep the public informed on important issues. They also provide an opportunity for continuing education, particularly for preparing for professional engineers examinations. The major societies and trade associations in the United States are

American Concrete Institute

American Institute of Chemical Engineers

American Institute of Steel Construction

American Society of Civil Engineers

American Society of Heating, Refrigerating, and Air-Conditioning Engineers

American Society of Mechanical Engineers

Institute of Electrical and Electronics Engineers

Instrument Society of America

National Association of Corrosion Engineers

National Electrical Manufacturers Association

National Fire Protection Association

Society of Automotive Engineers

Technical Association for the Pulp and Paper Industry

Underwriters Laboratories

50.8 LIBRARIES

The most comprehensive collections of engineering information can be found at large research libraries. Five of the largest in the United States are:

Engineering Societies Library
345 East 47th Street
New York, NY 10017

John Crerar Library
35 West 33rd Street
Chicago, IL 60616

Library of Congress
Washington, DC 20540

Linda Hall Library
5109 Cherry Street
Kansas City, MO 64110

New York Public Library
Fifth Avenue and 42nd Street
New York, NY 10016

These libraries are accessible to the public. They do provide duplicating services and will answer telephoned or written reference questions.

Substantial collections also exist at universities and engineering schools. These libraries are intended for use by faculty and students, but outsiders can frequently obtain permission to use these libraries by appointment, upon payment of a library fee, or through a cooperative arrangement with a public library.

Special libraries in business and industry frequently have excellent collections on the subjects most directly related to their activity. They are usually only available for use by employees and the company.

Public libraries vary considerably in size, and the collection will usually reflect the special interests of the community. Central libraries, particularly in large cities, may have a considerable collection of engineering books and periodicals. Online searching is becoming an increasingly frequent service provided by public libraries.

Regardless of the size of a library, the reference librarian should prove helpful in obtaining materials not locally available. These services include interlibrary loans from networks, issuing of courtesy cards to provide access to nonpublic libraries, and providing the location of the nearest library that owns needed materials.

50.9 LIBRARY NETWORKS

Not even the largest libraries have the resources to collect everything published. Library networks and consortia have been formed in order to share resources. There is, at the moment, no national network in the United States. Many networks exist on a state or regional level. There is enough interlocking to cover the whole country, but the connections are frequently cumbersome. Connections with libraries and library networks in other countries do exist so that copies of materials not available in the United States can be obtained. The Center for Research Libraries[27] provides one such link.

50.10 INFORMATION BROKERS

During the last decade a large number of information brokers have come into existence. For an international listing see *Information Industry Market Place.*[6] Information brokers can be of considerable use in researching the literature and retrieving information, particularly in situations where the engineer does not have the time and resources to do the searching. The larger brokers have a staff of trained information specialists skilled in online and manual searching. Retrieval of needed items is usually accomplished by sending a messenger to make copies at a library. It is therefore not surprising that most information brokers are located near research libraries or are part of an information center.

The larger information brokers usually cover all subjects and offer additional services, such as translating foreign language materials. Smaller brokers, and those associated with a specialized agency, frequently offer searching in a limited number of subjects. The selection of the most appropriate information broker should receive considerable attention if a large amount of work is required or a continuing relationship is expected.

REFERENCES

1. *Engineering Index,* Engineering Information Inc. (monthly).
2. *Science Abstracts,* INSPEC, Institution of Electrical Engineers (semimonthly).
3. *Chemical Abstracts,* American Chemical Society (weekly).
4. *Metals Abstracts,* Metals Information (monthly).
5. *Ulrich's International Periodicals Directory,* R. R. Bowker, New York (annual).
6. *Information Industry Market Place,* An International Directory of Information Products and Services, R. R. Bowker, New York.

7. *Kirk–Othmer Encyclopedia of Chemical Technology*, 3rd ed., Wiley-Interscience, New York, 1978–1984 (24 vols.).

8. *Encyclopedia of Computer Science and Technology*, Marcel Dekker, New York, 1975–1981 (15 vols.).

9. *Metals Handbook*, American Society for Metals, Metals Park, OH.

10. *CRC Handbook of Chemistry and Physics*, CRC Press Inc., Boca Raton, FL (annual).

11. *Composite Index of CRC Handbooks*, CRC Press Inc., Boca Raton, FL, 1977.

12. I. N. Sax, *Dangerous Properties of Industrial Materials*, Van Nostrand Reinhold, New York, 1984.

13. W. E. Woodson, *Human Factors Design Handbook*, McGraw-Hill, New York, 1981.

14. T. G. Hicks, *Standard Handbook of Engineering Calculations*, McGraw-Hill, New York, 1972.

15. *Energy Technology Handbook*, McGraw-Hill, New York, 1977.

16. H. C. Landa et al., *The Solar Energy Handbook*, Film Instruction Company of America, Milwaukee, WI, 1977.

17. *Engineering and Industrial Graphics Handbook*, McGraw-Hill, New York, 1982.

18. *Handbook of Composites*, Van Nostrand Reinhold, New York, 1982.

19. J. L. Lyons, *Lyons' Valve Designers' Handbook*, Van Nostrand Reinhold, New York, 1982.

20. British Standards Institution, *Metric Standards for Engineers*, BSI, London, 1967.

21. *Infrared Handbook*, Office of Naval Research, Washington, DC, 1978.

22. American National Standards Institute, 1430 Broadway, New York, NY 10018.

23. American Society of Mechanical Engineers, 345 East 47th Street, New York, NY 10017.

24. American Society for Testing and Materials, *Annual Book of Standards*, ASTM, 1916 Race Street, Philadelphia, PA.

25. *Index and Directory of U.S. Industry Standards*, Information Handling Services, 15 Inverness Way East, Englewood, CO 80150.

26. Superintendent of Documents, U.S. Government Printing Office, Washington, DC 20402.

27. Center for Research Libraries, 5721 Cottage Grove Avenue, Chicago, IL 60637.

CHAPTER 51

ONLINE DATABASES: THE NEW RESEARCH TOOLS

DAVID M. ABELS

JUDITH WANGER

Cuadra Associates, Inc.
Santa Monica, California

About 15 years ago, a new means for conducting research emerged in the form of online database services—computer-based services that provided for the retrieval and manipulation of stored collections of publicly available information. From a terminal located in the library, laboratory, or office, a user was able to dial a local telephone number and be connected to one of a handful of online service organizations that offered access to collections of scientific, technical, and medical bibliographic information, the full-text of legal materials, and economic and financial statistics.

For the most part, these collections of information—called databases—represented computer-readable versions of printed publications. For example, among the first online databases were a version of the *Chemical Abstracts Service Journal* (without abstracts) and COMPENDEX, which corresponded to the monthly *Engineering Index*. Using the system (computer program) provided by an online service, a user was able to specify the parameters of his or her research problem (e.g., by entering pertinent author names or subject terms) and, within seconds, could learn, say, how many articles represented in the database were potentially relevant to the user's research problem. Some or all of the citations (or citations with abstracts) could be printed immediately at the terminal or sent to the user through the mail. What might have taken hours or days of research with printed publications now could be done in a matter of minutes.

The online database services industry has grown considerably, in terms of the numbers of databases that are available and in the scope of their coverage. This growth, in turn, has stimulated considerable growth in the number of users of online database services. The purpose of this chapter is (1) to describe, in general, the types of databases that are now available for online users and (2) to illustrate some of the ways in which the online database services are used.

51.1 BACKGROUND AND DEFINITIONS

The online access of information stored in remotely located computers involves the following technical links:

1. The user's computer terminal or microcomputer must have a coupler into which the telephone handset is placed or a modem that is connected to a telephone outlet, which allows for interactive communications (i.e., to both send and receive data) with a remotely located computer.
2. Users achieve a connection with the remotely located computer by dialing a long-distance or, more likely, a local number that connects them to one of the telecommunications networks, which, in turn, is connected to the computers of the online services.
3. The user follows a set of procedures to identify the host computer to which the telecommunications network is to direct the communications, and to establish, through passwords, a right to access the computer. Once connected to the host computer—the online service—the user specifies the database that he or she wants to access.

By mid-1985, there were over 2750 online databases available through one or more of over 410 online services, primarily located in the United States, Canada, Europe, and Australia. Some online services provide for access throughout the world, whereas others limit access to users in their own countries or regions.

Databases are created by publishers, research organizations, government agencies, and many other types of commercial and nonprofit organizations. This group of suppliers is referred to as "database producers." In some cases, the online service and the database producer are the same organization, but in most cases they are not.

The types of databases that can be accessed online have been classified and described as follows by Cuadra Associates, Inc.[1]

1. *Reference Databases.* Refer or "point" users to another source (e.g., a document, an organization, or an individual) for additional details or for the complete text.
    a. *Bibliographic.* Contain citations, sometimes with abstracts, to the printed literature (e.g., journals).
    b. *Referral.* Contain references, sometimes with summaries, to nonpublished information sources (e.g., audiovisual materials or directories of companies).
2. *Source Databases.* Contain complete data or the full text of the original source information.
    a. *Numeric.* Contain original survey data and/or statistically manipulated representations of data (e.g., time series of economic or financial data, or stock and commodities data).
    b. *Textual–Numeric.* Are generally databases of records that contain some data elements of textual information and others of numeric data (e.g., handbook-type data of chemical and physical properties).
    c. *Full Text.* Contain records of the complete text of an item (e.g., a newspaper story, a magazine article, a specification, or a court decision).
    d. *Software.* Contain computer programs that can be downloaded for local use.

Most of the online services specialize in only one type of database or in certain combinations of databases. For example, several of the online services are referred to as "bibliographic database services," and they provide access to reference databases, as well as to other types of databases with *textual* information. The main reason for this specialization is that effective use of databases of different types requires appropriate systems. For example, the types of systems that are effective in the statistical manipulation of economic or financial data are likely to be totally inadequate in the retrieval of information from scientific/technical reference and full-text databases. Table 51.1 provides brief descriptions of database offerings for a few online database services, to illustrate the diversity among them.

51.2　SCOPE AND COVERAGE OF ONLINE DATABASES

Online databases cover a broad spectrum of disciplines and fields. The majority of databases are "source" databases—primarily numeric databases of economic, financial, and other business data. A significant percentage of databases are bibliographic databases, covering journal articles, conference reports, dissertations, research reports, and other types of published materials in various scientific and technical areas. Brief descriptions of several scientific and technical databases, as well as several that specifically relate to the field of engineering, are presented in Table 51.2. These examples are drawn from the Cuadra Associates' *Directory of Online Databases.*[1]

51.3　ONLINE SEARCH EXAMPLES

Use of the online database services involves some amount of preparation in order to select the databases that are appropriate for a particular research problem and to identify the way that the research problem should be expressed for retrieval in a given database, using a particular online system. Once online, the user can take advantage of the system interaction, to improve on that expression, as needed, to obtain the desired results.

Two search examples are provided here to highlight the characteristics of that system–user interaction. Both of these examples focus on databases containing textual information and on computer programs from the family of "information storage and retrieval" systems. These systems provide similar

Table 51.1　Illustrative List of Online Database Services

Online Service	Location	Types of Databases Offered
Battelle Columbus Laboratories	United States	Several scientific/technical bibliographic and textual–numeric databases (e.g., COPPER DATA CENTER; ELECTRONIC RIG STATS)
CISTI	Canada	Over 30 scientific/technical databases, primarily bibliographic (e.g., ALBERTA OIL SANDS INDEX)
Chase Econometrics/ Interactive Data	United States	Over 100 databases, primarily containing historical, current, and forecast data in various economic and financial areas (e.g., ALUMINUM FORECAST; WARD'S AUTOINFO-BANK)
DIALOG	United States	The supermarket of reference databases, with emphasis on business and chemical information, plus a growing number of source databases
ESA–IRS	Italy	Over 60 scientific/technical databases, primarily bibliographic, with an emphasis on those of interest to the European Space Agency community (e.g., SATELLITE DATA BANK; SPACECOMPS)
NewsNet, Inc.	United States	Over 230 databases of newsletters, corresponding to printed publications or written expressly for online use by various producers (e.g., ELECTRIC VEHICLE PROGRESS; HAZARDOUS WASTE NEWS)
Mead Data Central	United States	Two major full-text services: LEXIS, covering court cases and other legal information, and NEXIS, covering newspapers, magazines, and newsletters (e.g., *Chemical Engineering* and the *Engineering and Mining Journal*); and a source and reference database service: THE REFERENCE SERVICE
Telesystemes-Questel	France	Over 40, primarily scientific/technical bibliographic and textual–numeric databases (e.g., the MASS SPECTRAL SEARCH SYSTEM; NORIANE database on French and international industrial standards/regulations)

Table 51.2 Selected Databases of Interest to Mechanical Engineers

Name	Database Covers
CHEMICAL ENGINEERING ABSTRACTS	Chemical engineering worldwide, with an emphasis on practical applications in mechanical, civil, electrical, and instrumentational engineering (bibliographic)
COMPENDEX®	Most fields of engineering and technology, including mechanical engineering (bibliographic)
DELFT HYDRO	Civil and mechanical hydraulic engineering worldwide (bibliographic)
DOMA	Mechanical engineering worldwide, including technical aspects of materials, manufacturing processes, and environmental engineering (bibliographic)
EMIS (Electronic Materials Information Service)	Properties of solid-state materials; includes information related to use of materials in devices (referral; textual–numeric)
MANLABS–NPL MATERIALS DATA BANK	Thermochemical data on metals, inorganic compounds, alloys, and metals or gases in solutions (textual–numeric)
NTIS (National Technical Information Service)	Unrestricted reports from U.S. and non-U.S. government-sponsored research, development, and engineering analyses (bibliographic)
PATSEARCH™	U.S. utility patents in all areas of science and technology (bibliographic)
POLYPROBE	Information on currently manufactured plastics in the U.S. including name, generic type, manufacturer, and a number of physical properties (referral; textual–numeric)
SAE Abstracts	Technical papers on automotive technology and automotive-related industries that have been presented at Society of Automotive Engineers meetings and conferences (bibliographic)
WRCBASE	Detailed technical specifications for sludge and slurry pumps in the United Kingdom (referral; textual–numeric)

sets of functions, based on command-oriented interfaces that let users specify at any time, in any sequence, what it is that they want to do.

51.3.1 Retrieving Bibliographic Information

This example illustrates searching a bibliographic database to locate citations to the literature that would help to answer the following question: "What are the most effective acoustic methods for detecting abnormality in fast breeder reactor plants?"

The database selected for this search is ISMEC (Information Service in Mechanical Engineering), a bibliographic database produced by Cambridge Scientific Abstracts. It contains citations (with abstracts for the more recent years) to the worldwide literature published since 1973 in mechanical engineering, production engineering, and engineering management. Use of this database is illustrated with the DIALOG Information Services, Inc. system.

In planning a search strategy, the user has divided the research problem into two main "facets," which will be combined to produce the desired results. The first facet represents the concept of "acoustic methods," and the second facet represents the concept of the "fast breeder reactor." Search terms are selected to represent each of these two facets, as shown in Table 51.3 and explained further in the annotation that follows.

The bracketed numbers below refer to the numbered parts of the search example. The user's entries always allow a ?, which is the user's cue in the DIALOG system.

[1] The user specifies the database that is to be searched by entering its "file number," following the BEGIN command. The system responds with a confirmation that the user is connected to File 14, the ISMEC database.

Table 51.3　Retrieving Bibliographic Information

? BEGIN 14	[1]

File14:ISMEC: Mechanical Engineering—73–83/Jul
(Copr. Cambridge Sci. Abs.)

? SELECT acoustic? OR ultrasonic?	[2]

　　　　3610 ACOUSTIC?
　　　　1212 ULTRASONIC?

1　4503 ACOUSTIC? OR ULTRASONIC?	[3]
? SELECT fast(w)breeder(w)reactor?	[4]

　　2　　167 FAST(W)BREEDER(W)REACTOR?

? COMBINE 1 AND 2	[5]

　　3　　3 1 AND 2

? TYPE 3/5/1–2	[6]

3/5/1

　110606　　D80012599
　Acoustic measurement of penetration of liquid sodium into limestone concrete (nuclear reactor applications)
　Sutherland, H. J.; Smaardyk, J. E.; Kent, L. A.
　Sandia Labs., Albuquerque, NM 87185
　Nuclear Technology Vol. 49, No. 1 70–79 Jun 1980 Coden: NUTYBB
　Journal Addr: USA
　Treatment: P; T; X
　Document Type: 02
　(11 Refs)
　Descriptors: Acoustic variables measurement; Testing; Mathematical models; Absorption; Liquids; Concrete; Nuclear reactors; Chemical variables measurement
　Identifiers: pulse-echo method; penetration measurement; sodium absorption; sodium; limestone concrete; liquid-metal fast breeder reactors; interface reflectivity; interface conditions
　Class Codes: D2290; D8100

? SELECT S2 OR nuclear reactors	[7]

　　　　　1116 NUCLEAR REACTORS
　　4　1156 S2 OR nuclear reactors

? COMBINE 1 AND 4	[8]

　　5　　49　1 AND 4

? TYPE 5/2/1	[9]

5/2/1

　149763　　D83006808
　State-of-practice review of ultrasonic in-service inspection of class I system piping in commercial nuclear power plants
　Morris, C. J.; Becker, F. L.
　Battelle Pacific Northwest Lab., Richland, WA, USA
　1982
　Publ: NTIS, SPRINGFIELD, VA
　76 pp.
　Availability: SUMMARY LANGUAGE — ENGLISH; NUREG/CR-2468
　Languages: ENGLISH
　Descriptors: ultrasonics; nondestructive testing; pipes; nuclear reactors; inspection; process development
　Class Codes: D3450; D8100

---

[2]　The SELECT command is used to instruct the system to find the records in the database that contain the terms that follow. The Boolean operator OR in this expression indicates that the user wants the system to identify records that contain any one or more of the specified terms. In this first statement to the system, the user specifies the terms that represent the "acoustic methods" concept. To search a broad range of possibly relevant terms (e.g., acoustic equipment, acoustic devices, ultrasonic equipment) the user enters the word stems ACOUSTIC and ULTRASONIC with the truncation symbol "?." The truncation symbol is used to indicate that all single-word or multiword terms beginning with the stem are to be included in the search.

[3] In a matter of seconds, the system responds with the first level of feedback to the user. It first indicates for each of the truncated search terms the number of records that contain a form of the term. It then assigns a number to the set of retrieved records—Set 1—and reports the number of records (4503) that contain these terms.

[4] To search on the "fast breeder reactor" facet, the user enters the phrase FAST BREEDER REACTOR. The "with" proximity operator—(w)—is used to specify that the single words comprising this phrase should be adjacent to each other, in the order that is specified. The truncation symbol is used to specify that both the singular and plural form of the phrase is to be searched.

[5] The two search facets are COMBINEd using the Boolean AND, to have the system retrieve those records in which *both* facets are represented. The system reports that three records meet the search parameters.

[6] The user is now ready to "test" the quality of the search results—that is, to determine whether the selected search terms succeeded in retrieving relevant citations—by requesting that the system display several of the retrieved records. The user enters a TYPE command to receive an online display. The numbers that follow the TYPE indicate that the citations should be drawn from the results of Set 3; Format 5 should be displayed (i.e., the complete bibliographic citation, including abstract); and only the first two retrieved records (1–2) are to be displayed. (In Table 51.3 only the first retrieved record is shown.)

[7] The user spots the assigned subject vocabulary term NUCLEAR REACTORS in both records and decides to incorporate this term into the "fast breeder reactor" part of the search formulation. This is likely to increase the number of potentially relevant records that are retrieved. To accomplish this broadening of the strategy, the user OR's Set 2 (S2) with the new term NUCLEAR REACTORS.

[8] The two search facets are again COMBINEd using the Boolean AND. The system reports that 49 records now meet the search parameters.

[9] If the user is satisfied that the search has yielded satisfactory results, he or she decides whether to print the bibliography on a local printer or to specify that the online service is to produce the hardcopy listing and send it to the user at the end of the day. In this case the user decides to display the complete bibliographic citations online. (In Table 51.3, only the first citation is reproduced.)

To help ensure that a problem has been researched exhaustively, users can conduct similar searches on one or more other databases, available through the same or other online services. For example, among the numerous other databases potentially relevant to this particular research problem are DOE ENERGY (containing references to unclassified literature processed by the U.S. Department of Energy's Technical Information Center) and INTERNATIONAL NUCLEAR INFORMATION SYSTEM (produced by the International Atomic Energy Agency and covering the worldwide literature on peaceful applications of nuclear energy). Data on the characteristics of nuclear power stations and nuclear reactors worldwide are available in ELECNUC, a French database produced by the Commisariat à l'Energie Atomique (CEA). The EPRI DATABASE FOR ENVIRONMENTALLY ASSISTED CRACKING is produced by the Electric Power Research Institute and contains data on crack propagation in nuclear reactor environments that can be used for predicting the behavior of reactor components containing cracks or cracklike indications.

Although users will perform a similar set of system functions to retrieve information in these or other databases, the search terms they select are likely to vary from database to database. Knowing which databases are likely to yield the best results for a given problem and how best to express a research problem, for a given database on a particular system, is skill acquired through use.

51.3.2 Retrieving Factual Information

A second example is designed to illustrate searching a combined textual–numeric database to locate the answer to the following question: "What is the tensile strength of beryllium?"

The database selected for the search is KIRK-OTHMER/ONLINE, a database produced by John Wiley & Sons, Inc. The database contains all articles in the third edition of the *Kirk–Othmer Encyclopedia of Chemical Technology*. The *Encyclopedia* is a 25-volume set that was issued over a seven-year period, from 1978 to 1984, and covers chemical technology in such areas of energy, health, safety, and materials. Records in the database include data from the tables and figures that appear in the articles in the *Encyclopedia*. Use of this database is illustrated with the BRS system.

The facets of this search problem are represented by two sets of search terms: "tensile strength" and "beryllium." The system interaction is reproduced in Table 51.4.

Table 51.4 Retrieving Factual Information

ENTER DATA BASE NAME__: kirk   [1]
* KIRK-OTHMER ENCYCLOPEDIA OF CHEMICAL TECHNOLOGY.
* THIRD EDITION.
* COPYRIGHT JOHN WILEY & SONS, INC.
*SIGN-ON 18.57.43   08/27/83:
BRS/KIRK
BRS — SEARCH MODE — ENTER QUERY

1__:	tensile adj strength.ta.	[2]
RESULT	14	[3]
2__:	beryllium.ta.	[4]
RESULT	4	[5]
3__:	1 AND 2	[6]
RESULT	2	
4__:	..print 3 ti,ta/doc=1	[7]

- - - - - - - - - -

Table 3. Typical Tensile Properties**a of   [8]
          Beryllium at Room Temperature

COLUMN HEADINGS (5):
(1) Grades**b
(2) Orientation
(3) Ultimate strength, MPa
(4) Yield strength, MPa
(5) Elongation, %

(1)	(2)	(3)	(4)	(5)
vacuum hot pressed				
S-200	L**c	330.9	254.4	1.5
	T**d	371.6	262.7	2.6
S-65	L	337.8	237.2	3.4
	T	372.3	247.5	5.7
I-400	L	496.4	448.1	1.0
	T	517.1	468.8	2.0
cross-rolled sheet**e		565.3	420.5	22.0

a "Modulus of elasticity (Young's modulus) = 2.93 TIMES 105 MPa (to convert MPa to psi, multiply by 145).
**b "The tabulated properties are typical for the indicated Brush Wellman grades. Equivalent Kawecki Berylco grades are HP-20, CIP-HIP 1, HP-40, and PS-20, respectively.
**c "L = test in direction of pressing.
**d "T = test transverse to pressing direction.
**e "1.0–6.4 mm thick in plane of sheet.

[1] After logging into the system, the user enters the four-character database label for the KIRK-OTHMER ENCYCLOPEDIA. The system responds with an indication that the connection has been made and prompts the user to enter the first search query, after the "1__:."

[2] The first query that is entered specifies a search to be made of the two-word phrase—TENSILE STRENGTH. To ensure that the retrieval is limited to cases in which the two words are found together, in the specified order, the user links the two words with the word-adjacency operator—"adj." The entry ends with ".ta" to indicate that the words should be searched for only in the Table field (e.g., not in the Title or the Abstract fields of information).

[3] The system responds with a report that there are 14 records in the database that have TENSILE STRENGTH in their Table fields.

[4] The second query that is entered requests a search to be made of the term BERYLLIUM, also in the Table field.

[5] The system responds that there are four records in the database that have the word BERYLLIUM in their Table fields.

[6] In the third query, the user combines the results of the two previous searches, using the previous Set Numbers as search terms with the Boolean operator AND. This search expression

specifies that the system is to identify those records that have *both* BERYLLIUM and TENSILE STRENGTH in their Table fields.

[7] After learning that the system found two records that match the search specification, the user enters a print command, specifying that the title (TI) and table (TA) should be displayed for the first of the two documents.

[8] The relevant part of the resulting display—Table 3—is reproduced, to show the information that is provided.

At this point, the search may be completed, or the user may decide that additional information is needed. For example, the user may want to read parts or all of the article from which the table was extracted, and could have displayed the field that contains the volume and page number for the printed edition. Or, to get some context for the information displayed in the retrieved table, the user could request a display of the abstract.

With a printed publication, users work with back-of-the-book indexes and tables of contents and then work back and forth among the pages, skimming as needed, to locate the desired information. With online searching, users are able to specify with a great deal of precision what it is that they are looking for and, after the system identifies records that meet those specifications, users can tailor displays to include the number of records they want to see, with only as much information as they need.

51.4 USERS AND USES

The previous two examples highlighted uses that are fairly typical of bibliographic and other types of textual information databases that are searched through information retrieval systems. With certain other types of databases and systems that cover chemical and physical properties data, users can perform additional functions, for example, to identify materials with certain characteristics, to model the effects of temperature on a given compound, or to perform statistical analyses on the retrieved data.

Traditionally, users of bibliographic information databases have been librarians and information specialists, whereas users of databases with properties data and associated system capabilities for the analysis and manipulation of those data have been the end users themselves—engineers, scientists, researchers, economists, and businesspersons. This "division of labor" is changing, however, particularly as end users are being trained and encouraged to use the systems directly[2,3] and as microcomputers become more prevalent on the desks of professionals of all types.

Debates on the issue of whether the direct users of the online databases should be the system/database expert or the person with the information need are likely to continue for some time. However, there is little debate with regard to the importance of professionals in all fields becoming aware of the availability of online databases and understanding how they can be used to supplement already existing sources of information or, in some situations, be cost-effective and particularly timely replacements for traditional research methods.

REFERENCES

1. *Directory of Online Databases,* Cuadra Associates, Inc., Santa Monica, CA (published quarterly).
2. K. R. Walton and P. L. Dedert, "Experience at Exxon in Training End-Users to Search Technical Databases Online," *Online,* 42–50 (September 1983).
3. R. J. Richardson, "End User Online Searching in a High-Technology Engineering Environment," *Online,* 44–57 (October 1981).

BIBLIOGRAPHY

M. C. Berger and J. Wanger, "Retrieval, Analysis, and Display of Numerical Data," *Drexel Library Quarterly,* 11–26 (Summer–Fall 1982).

C.-C. Chen, *Online Bibliographic Searching: A Learning Manual,* Neal-Schuman, New York, 1981.

C. H. Fenichel and T. H. Hogan, *Online Searching: A Primer,* Learned Information, Marlton, NJ, 1981.

W. M. Henry et al., *Online Searching: An Introduction,* Butterworths, London, 1980.

C. T. Meadow and P. A. Cochrane, *Basics of Online Searching,* Wiley, New York, 1981.

R. C. Palmer, *Online Reference and Information Retrieval,* Littleton Co., Libraries Unlimited, 1983.

J. Wanger and R. N. Landau, "Nonbibliographic On-line Data Base Services," *Journal of the American Society for Information Science,* 171–180 (May 1980).

A number of journals cover the online database area. They include *Database,* Online, Inc., Weston, CT; *Online,* Online, Inc., Weston, CT; *Online Review,* Learned Information, Medford, NJ.

PART 6
ENERGY AND POWER

CHAPTER 52

THERMOPHYSICAL PROPERTIES OF FLUIDS

PETER E. LILEY

Purdue University
School of Mechanical Engineering
West Lafayette, Indiana

In this chapter, information is usually presented in the Système International des Unités, called in English the International System of Units and abbreviated SI. Various tables of conversion factors from other unit systems into the SI system and vice versa are available. The following table is only intended to enable rapid conversion to be made with moderate, that is, five significant figure, accuracy, usually acceptable in most engineering calculations. The references listed should be consulted for more exact conversions and definitions.

Table 52.1 Conversion Factors[a]

Density: 1 kg/m³ = 0.06243 lb_m/ft³ = 0.01002 lb_m/U.K. gallon = 8.3454 × 10^{-3} lb_m/U.S. gallon = 1.9403 × 10^{-3} slug/ft³ = 10^{-3} g/cm³

Energy: 1 kJ = 737.56 ft · lb_f = 239.01 cal_{th} = 0.94783 Btu = 3.7251 × 10^{-4} hp hr = 2.7778 × 10^{-4} kWhr

Specific energy: 1 kJ/kg = 334.54 ft · lb_f/lb_m = 0.4299 Btu/lb_m = 0.2388 cal/g

Specific energy per degree: 1 kJ/kg · K = 0.23901 Btu_{th}/lb · °F = 0.23901 cal_{th}/g · °C

Mass: 1 kg = 2.20462 lb_m = 0.06852 slug = 1.1023 × 10^{-3} U.S. ton = 10^{-3} tonne = 9.8421 × 10^{-4} U.K. ton

Pressure: 1 bar = 10^5 N/m² = 10^5 Pa = 750.06 mm Hg at 0°C = 401.47 in. H_2O at 32°F = 29.530 in. Hg at 0°C = 14.504 lb/in.² = 14.504 psia = 1.01972 kg/cm² = 0.98692 atm = 0.1 MPa

Temperature: T(K) = T(°C) + 273.15 = [T(°F) + 459.69]/1.8 = T(°R)/1.8

Temperature difference: ΔT(K) = ΔT(°C) = ΔT(°F)/1.8 = ΔT(°R)/1.8

Thermal conductivity: 1 W/m · K = 0.8604 kcal/m · hr · °C = 0.5782 Btu/ft · hr · °F = 0.01 W/cm · K = 2.390 × 10^{-3} cal/cm · sec · °C

Thermal diffusivity: 1 m²/sec = 38750 ft²/hr = 3600 m²/hr = 10.764 ft²/sec

Viscosity, dynamic: 1 N · sec/m² = 1 Pa · sec = 10^7 μP = 2419.1 lb_m/ft · hr = 10^3 cP = 75.188 slug/ft · hr = 10 P = 0.6720 lb_m/ft · sec = 0.02089 lb_f · sec/ft²

Viscosity, kinematic (*see* thermal diffusivity)

[a]E. Lange, L. F. Sokol, and V. Antoine, *Information on the Metric System and Related Fields,"* 6th ed., G. C. Marshall Space Flight Center, AL (exhaustive bibliography); C. H. Page and P. Vigoureux, *The International System of Units,* NBS S.P. 330, Washington, D.C., 1974; E. A. Mechtly, *The International System of Units, Physical Constants and Conversion Factors,* NASA S.P. 9012, 1973. Numerous revisions periodically appear; see, for example, *Pure Appl. Chem.,* **51**, 1–41 (1979) and later issues.

Table 52.2 Phase Transition Data for the Elements[a]

Name	Symbol	Formula Weight	T_m (K)	Δh_{fus} (kJ/kg)	T_b (K)	T_c (K)
Actinium	Ac	227.028	1323	63	3475	
Aluminum	Al	26.9815	933.5	398	2750	7850
Antimony	Sb	121.75	903.9	163	1905	5700
Argon	Ar	39.948	83	30	87.2	151
Arsenic	As	74.9216	885			2100
Barium	Ba	137.33	1002	55.8		4450
Beryllium	Be	9.01218	1560	1355	2750	6200
Bismuth	Bi	208.980	544.6	54.0	1838	4450
Boron	B	10.81	2320	1933	4000	3300
Bromine	Br	159.808	266	66.0	332	584
Cadmium	Cd	112.41	594	55.1	1040	2690
Calcium	Ca	40.08	1112	213.1	1763	4300
Carbon	C	12.011	3810		4275	7200
Cerium	Ce	140.12	1072	390		9750
Cesium	Cs	132.905	301.8	16.4	951	2015
Chlorine	Cl_2	70.906	172	180.7	239	417
Chromium	Cr	51.996	2133	325.6	2950	5500
Cobalt	Co	58.9332	1766	274.7	3185	6300

Table 52.2 *(Continued)*

Name	Symbol	Formula Weight	T_m (K)	Δh_{fus} (kJ/kg)	T_b (K)	T_c (K)
Copper	Cu	63.546	1357	206.8	2845	8280
Dysprosium	Dy	162.50	1670	68.1	2855	6925
Erbium	Er	167.26	1795	119.1	3135	7250
Europium	Eu	151.96	1092	60.6	1850	4350
Fluorine	F$_2$	37.997	53.5	13.4	85.0	144
Gadolinum	Gd	157.25	1585	63.8	3540	8670
Gallium	Ga	69.72	303	080.1	2500	7125
Germanium	Ge	72.59	1211	508.9	3110	8900
Gold	Au	196.967	1337	62.8	3130	7250
Hafnium	Hf	178.49	2485	134.8	4885	10400
Helium	He	4.00260	3.5	2.1	4.22	5.2
Holmium	Ho	164.930	1744	73.8	2968	7575
Hydrogen	H$_2$	2.0159	14.0		20.4	
Indium	In	114.82	430	28.5	2346	6150
Iodine	I$_2$	253.809	387	125.0	457	785
Iridium	Ir	192.22	2718	13.7	4740	7800
Iron	Fe	55.847	1811	247.3	3136	8500
Krypton	Kr	83.80	115.8	19.6	119.8	209.4
Lanthanum	La	138.906	1194	44.6	3715	10500
Lead	Pb	207.2	601	23.2	2025	5500
Lithium	Li	6.941	454	432.2	1607	3700
Lutetium	Lu	174.967	1937	106.6	3668	
Magnesium	Mg	24.305	922	368.4	1364	3850
Manganese	Mn	54.9380	1518	219.3	2334	4325
Mercury	Hg	200.59	234.6	11.4	630	1720
Molybdenum	Mo	95.94	2892	290.0	4900	1450
Neodymium	Nd	144.24	1290	49.6	3341	7900
Neon	Ne	20.179	24.5	16.4	27.1	44.5
Neptunium	Np	237.048	910		4160	12000
Nickel	Ni	58.70	1728	297.6	3190	8000
Niobium	Nb	92.9064	2740	283.7	5020	12500
Nitrogen	N$_2$	28.013	63.2	25.7	77.3	126.2
Osmium	Os	190.2	3310	150.0	5300	12700
Oxygen	O$_2$	31.9988	54.4	13.8	90.2	154.8
Palladium	Pd	106.4	1826	165.0	3240	7700
Phosphorus	P	30.9738	317		553	995
Platinum	Pt	195.09	2045	101	4100	10700
Plutonium	Pu	244	913	11.7	3505	10500
Potassium	K	39.0983	336.4	60.1	1032	2210
Praseodymium	Pr	140.908	1205	49	3785	8900
Promethium	Pm	145	1353		2730	
Protactinium	Pa	231	1500	64.8	4300	
Radium	Ra	226.025	973		1900	
Radon	Rn	222	202	12.3	211	377
Rhenium	Re	186.207	3453	177.8	5920	18900
Rhodium	Rh	102.906	2236	209.4	3980	7000
Rubidium	Rb	85.4678	312.6	26.4	964	2070
Ruthenium	Ru	101.07	2525	256.3	4430	9600
Samarium	Sm	150.4	1345	57.3	2064	5050
Scandium	Sc	44.9559	1813	313.6	3550	6410

Table 52.2 (*Continued*)

Name	Symbol	Formula Weight	T_m (K)	Δh_{fus} (kJ/kg)	T_b (K)	T_c (K)
Selenium	Se	78.96	494	66.2	958	1810
Silicon	Si	28.0855	1684	1802	3540	5160
Silver	Ag	107.868	1234	104.8	2435	6400
Sodium	Na	22.9898	371	113.1	1155	2500
Strontium	Sr	87.62	1043	1042	1650	4275
Sulfur	S	32.06	388	53.4	718	1210
Tantalum	Ta	180.948	3252	173.5	5640	16500
Technetium	Tc	98	2447	232	4550	11500
Tellurium	Te	127.60	723	137.1	1261	2330
Terbium	Tb	158.925	1631	67.9	3500	8470
Thallium	Tl	204.37	577	20.1	1745	4550
Thorium	Th	232.038	2028	69.4	5067	14400
Thulium	Tm	168.934	1819	99.6	2220	6450
Tin	Sn	118.69	505	58.9	2890	7700
Titanium	Ti	47.90	1943	323.6	3565	5850
Tungsten	W	183.85	3660	192.5	5890	15500
Uranium	U	238.029	1406	35.8	4422	12500
Vanadium	V	50.9415	2191	410.7	3680	11300
Xenon	Xe	131.30	161.3	17.5	164.9	290
Ytterbium	Yb	173.04	1098	44.2	1467	4080
Yttrium	Y	88.9059	1775	128.2	3610	8950
Zinc	Zn	65.38	692.7	113.0	1182	
Zirconium	Zr	91.22	2125	185.3	4681	10500

^{a} T_m = normal melting point; Δh_{fus} = enthalpy of fusion; T_b = normal boiling point; T_c = critical temperature.

Table 52.3 **Phase Transition Data for Compounds**^{a}

Substance	T_m (K)	Δh_m (kJ/kg)	T_b (K)	Δh_v (kJ/kg)	T_c (K)	P_c (bar)
Acetaldehyde	149.7	73.2	293.4	584	461	55.4
Acetic acid	289.9	195.3	391.7	405	594	57.9
Acetone	178.6	98	329.5	501	508	47
Acetylene		96.4	189.2	687	309	61.3
Air	60				133	37.7
Ammonia	195.4	331.9	239.7	1368	405.6	112.8
Aniline	267.2	113.3	457.6	485	699	53.1
Benzene	267.7	125.9	353.3	394	562	49
n-Butane	134.8	80.2	261.5	366	425.2	38
Butanol	188	125.2	391.2	593	563	44.1
Carbon dioxide	216.6	184	194.7	573	304.2	73.8
Carbon disulfide	161	57.7	319.6	352	552	79
Carbon monoxide	68.1	29.8	81.6	215	133	35
Carbon tetrachloride	250.3	173.9	349.9	195	556	45.6
Carbon tetrafluoride	89.5		145.2	138	227.9	37.4
Chlorobenzene	228		405	325	632.4	45.2
Chloroform	210	77.1	334.4	249	536.6	54.7
m-Cresol	285.1		475.9	421	705	45.5
Cyclohexane	279.6	31.7	356	357	554.2	40.7
Cyclopropane	145.5	129.4	240.3	477	397.8	54.9

Table 52.3 (*Continued*)

Substance	T_m (K)	Δh_m (kJ/kg)	T_b (K)	Δh_v (kJ/kg)	T_c (K)	P_c (bar)
n-Decane	243.2	202.1	447.3	276	617	21
Ethane	89.9	94.3	184.6	488	305.4	48.8
Ethanol	158.6	109	351.5	840	516	63.8
Ethyl acetate	190.8	119	350.2	366	523.3	38.3
Ethylene	104	119.5	169.5	480	283.1	51.2
Ethylene oxide	161.5	117.6	283.9	580	469	71.9
Formic acid	281.4	246.4	373.9	502	576	34.6
Heptane	182.6	140.2	371.6	316	540	27.4
Hexane	177.8	151.2	341.9	335	507	29.7
Hydrazine	274.7	395	386.7	1207	653	147
Hydrogen peroxide	271.2	310	431	1263		
iso-Butane	113.6	78.1	272.7	386	408.1	36.5
Methane	90.7	58.7	111.5	512	191.1	46.4
Methanol	175.4	99.2	337.7	1104	513.2	79.5
Methyl acetate	174.8		330.2	410	507	46.9
Methyl bromide	180	62.9	277.7	252	464	71.2
Methyl chloride	178.5	127.4	249.3	429	416.3	66.8
Methyl formate	173.4	125.5	305	481	487.2	60
Methylene chloride	176.4	54.4	312.7	328	510.2	60.8
Naphthalene	353.2	148.1	491	341	747	39.5
Nitric oxide	111	76.6	121.4	460	180.3	65.5
Octane	216.4	180.6	398.9	303	569.4	25
Pentane	143.7	116.6	309.2	357	469.8	33.7
Propane	86	80	231.1	426	370	42.5
Propanol	147	86.5	370.4	696	537	51.7
Propylene	87.9	71.4	225.5	438	365	46.2
Refrigerant 12	115	34.3	243.4	165	385	41.2
Refrigerant 13	92		191.8	148	302.1	38.7
Refrigerant 13B1	105.4		215.4	119	340	39.6
Refrigerant 21	138		282.1	242	451.7	51.7
Refrigerant 22	113	47.6	232.4	234	369	49.8
Steam/water	273.2	334	373.2	2257	647.3	221.2
Sulfuric acid	283.7	100.7	v	v	v	v
Sulfur dioxide	197.7	115.5	268.4	386	430.7	78.8
Toluene	178.2		383.8	339	594	41

[a] v = variable. T_m = normal melting point; Δh_m = enthalpy of fusion; T_b = normal boiling point; Δh_v = enthalpy of vaporization; T_c = critical temperature; P_c = critical pressure.

Table 52.4 Thermodynamic Properties of Liquid and Saturated Vapor Air[a]

T (K)	P_f (MPa)	P_g (MPa)	v_f (m³/kg)	v_g (m³/kg)	h_f (kJ/kg)	h_g (kJ/kg)	s_f (kJ/kg·K)	s_g (kJ/kg·K)
60	0.0066	0.0025	0.001027	6.876	−144.9	59.7	2.726	6.315
65	0.0159	0.0077	0.001078	2.415	−144.8	64.5	2.727	6.070
70	0.0340	0.0195	0.001103	1.021	−144.0	69.1	2.798	5.875
75	0.0658	0.0424	0.001130	0.4966	−132.8	73.5	2.896	5.714
80	0.1172	0.0826	0.001160	0.2685	−124.4	77.5	3.004	5.580
85	0.1954	0.1469	0.001193	0.1574	−115.2	81.0	3.114	5.464
90	0.3079	0.2434	0.001229	0.0983	−105.5	84.1	3.223	5.363
95	0.4629	0.3805	0.001269	0.0644	−95.4	86.5	3.331	5.272
100	0.6687	0.5675	0.001315	0.0439	−84.8	88.2	3.436	5.189
105	0.9334	0.8139	0.001368	0.0308	−73.8	89.1	3.540	5.110
110	1.2651	1.1293	0.001432	0.0220	−62.2	89.1	3.644	5.033
115	1.6714	1.5238	0.001512	0.0160	−49.7	87.7	3.750	4.955
120	2.1596	2.0081	0.001619	0.0116	−35.9	84.5	3.861	4.872
125	2.743	2.614	0.001760	0.0081	−19.7	78.0	3.985	4.770
132.5[b]	3.770	3.770	0.00309	0.0031	33.6	33.6	4.38	4.38

[a] v = specific volume; h = specific enthalpy; s = specific entropy; f = saturated liquid; g = saturated vapor. 1 MPa = 10 bar.
[b] Approximate critical point. Air is a multicomponent mixture.

Table 52.5 Ideal Gas Thermophysical Properties of Air[a]

T (K)	v (m³/kg)	h (kJ/kg)	s (kJ/kg · K)	c_p (kJ/kg · K)	γ	$\bar{v}_s$ (m/sec)	η (N · sec/m²)	λ (W/m · K)	Pr
200	0.5666	−103.0	6.4591	1.008	1.398	283.3	1.33.−5[b]	0.0183	0.734
210	0.5949	−92.9	6.5082	1.007	1.399	290.4	1.39.−5	0.0191	0.732
220	0.6232	−82.8	6.5550	1.006	1.399	297.3	1.44.−5	0.0199	0.730
230	0.6516	−72.8	6.5998	1.006	1.400	304.0	1.50.−5	0.0207	0.728
240	0.6799	−62.7	6.6425	1.005	1.400	310.5	1.55.−5	0.0215	0.726
250	0.7082	−52.7	6.6836	1.005	1.400	317.0	1.60.−5	0.0222	0.725
260	0.7366	−42.6	6.7230	1.005	1.400	323.3	1.65.−5	0.0230	0.723
270	0.7649	−32.6	6.7609	1.004	1.400	329.4	1.70.−5	0.0237	0.722
280	0.7932	−22.5	6.7974	1.004	1.400	335.5	1.75.−5	0.0245	0.721
290	0.8216	−12.5	6.8326	1.005	1.400	341.4	1.80.−5	0.0252	0.720
300	0.8499	−2.4	6.8667	1.005	1.400	347.2	1.85.−5	0.0259	0.719
310	0.8782	7.6	6.8997	1.005	1.400	352.9	1.90.−5	0.0265	0.719
320	0.9065	17.7	6.9316	1.006	1.399	358.5	1.94.−5	0.0272	0.719
330	0.9348	27.7	6.9625	1.006	1.399	364.0	1.99.−5	0.0279	0.719
340	0.9632	37.8	6.9926	1.007	1.399	369.5	2.04.−5	0.0285	0.719
350	0.9916	47.9	7.0218	1.008	1.398	374.8	2.08.−5	0.0292	0.719
360	1.0199	57.9	7.0502	1.009	1.398	380.0	2.12.−5	0.0298	0.719
370	1.0482	68.0	7.0778	1.010	1.397	385.2	2.17.−5	0.0304	0.719
380	1.0765	78.1	7.1048	1.011	1.397	390.3	2.21.−5	0.0311	0.719
390	1.1049	88.3	7.1311	1.012	1.396	395.3	2.25.−5	0.0317	0.719
400	1.1332	98.4	7.1567	1.013	1.395	400.3	2.29.−5	0.0323	0.719
410	1.1615	108.5	7.1817	1.015	1.395	405.1	2.34.−5	0.0330	0.719
420	1.1898	118.7	7.2062	1.016	1.394	409.9	2.38.−5	0.0336	0.719
430	1.2181	128.8	7.2301	1.018	1.393	414.6	2.42.−5	0.0342	0.718
440	1.2465	139.0	7.2535	1.019	1.392	419.3	2.46.−5	0.0348	0.718
450	1.2748	149.2	7.2765	1.021	1.391	423.9	2.50.−5	0.0355	0.718
460	1.3032	159.4	7.2989	1.022	1.390	428.3	2.53.−5	0.0361	0.718

470	1.3315	169.7	7.3209	1.024	1.389	433.0	2.57.−5	0.0367	0.718
480	1.3598	179.9	7.3425	1.026	1.389	437.4	2.61.−5	0.0373	0.718
490	1.3882	190.2	7.3637	1.028	1.388	441.8	2.65.−5	0.0379	0.718
500	1.4165	200.5	7.3845	1.030	1.387	446.1	2.69.−5	0.0385	0.718
520	1.473	221.1	7.4249	1.034	1.385	454.6	2.76.−5	0.0398	0.718
540	1.530	241.8	7.4640	1.038	1.382	462.9	2.83.−5	0.0410	0.718
560	1.586	262.6	7.5018	1.042	1.380	471.0	2.91.−5	0.0422	0.718
580	1.643	283.5	7.5385	1.047	1.378	479.0	2.98.−5	0.0434	0.718
600	1.700	304.5	7.5740	1.051	1.376	486.8	3.04.−5	0.0446	0.718
620	1.756	325.6	7.6086	1.056	1.374	494.4	3.11.−5	0.0458	0.718
640	1.813	346.7	7.6422	1.060	1.371	501.9	3.18.−5	0.0470	0.718
660	1.870	368.0	7.6749	1.065	1.369	509.3	3.25.−5	0.0482	0.717
680	1.926	389.3	7.7067	1.070	1.367	516.5	3.32.−5	0.0495	0.717
700	1.983	410.8	7.7378	1.075	1.364	523.6	3.38.−5	0.0507	0.717
720	2.040	432.3	7.7682	1.080	1.362	530.6	3.45.−5	0.0519	0.716
740	2.096	453.9	7.7978	1.084	1.360	537.5	3.51.−5	0.0531	0.716
760	2.153	475.7	7.8268	1.089	1.358	544.3	3.57.−5	0.0544	0.716
780	2.210	497.5	7.8551	1.094	1.356	551.0	3.64.−5	0.0556	0.716
800	2.266	519.4	7.8829	1.099	1.354	557.6	3.70.−5	0.0568	0.716
820	2.323	541.5	7.9101	1.103	1.352	564.1	3.76.−5	0.0580	0.715
840	2.380	563.6	7.9367	1.108	1.350	570.6	3.82.−5	0.0592	0.715
860	2.436	585.8	7.9628	1.112	1.348	576.8	3.88.−5	0.0603	0.715
880	2.493	608.1	7.9885	1.117	1.346	583.1	3.94.−5	0.0615	0.715
900	2.550	630.4	8.0136	1.121	1.344	589.3	4.00.−5	0.0627	0.715
920	2.606	652.9	8.0383	1.125	1.342	595.4	4.05.−5	0.0639	0.715
940	2.663	675.5	8.0625	1.129	1.341	601.5	4.11.−5	0.0650	0.714
960	2.720	698.1	8.0864	1.133	1.339	607.5	4.17.−5	0.0662	0.714
980	2.776	720.8	8.1098	1.137	1.338	613.4	4.23.−5	0.0673	0.714
1000	2.833	743.6	8.1328	1.141	1.336	619.3	4.28.−5	0.0684	0.714
1050	2.975	800.8	8.1887	1.150	1.333	633.8	4.42.−5	0.0711	0.714

Table 52.5 *(Continued)*

T (K)	v (m³/kg)	h (kJ/kg)	s (kJ/kg · K)	c_p (kJ/kg · K)	γ	$\bar{v}_s$ (m/sec)	η (N · sec/m²)	λ (W/m · K)	Pr
1100	3.116	858.5	8.2423	1.158	1.330	648.0	4.55.−5	0.0738	0.715
1150	3.258	916.6	8.2939	1.165	1.327	661.8	4.68.−5	0.0764	0.715
1200	3.400	975.0	8.3437	1.173	1.324	675.4	4.81.−5	0.0789	0.715
1250	3.541	1033.8	8.3917	1.180	1.322	688.6	4.94.−5	0.0814	0.716
1300	3.683	1093.0	8.4381	1.186	1.319	701.6	5.06.−5	0.0839	0.716
1350	3.825	1152.3	8.4830	1.193	1.317	714.4	5.19.−5	0.0863	0.717
1400	3.966	1212.2	8.5265	1.199	1.315	726.9	5.31.−5	0.0887	0.717
1450	4.108	1272.3	8.5686	1.204	1.313	739.2	5.42.−5	0.0911	0.717
1500	4.249	1332.7	8.6096	1.210	1.311	751.3	5.54.−5	0.0934	0.718
1550	4.391	1393.3	8.6493	1.215	1.309	763.2	5.66.−5	0.0958	0.718
1600	4.533	1454.2	8.6880	1.220	1.308	775.0	5.77.−5	0.0981	0.717
1650	4.674	1515.3	8.7256	1.225	1.306	786.5	5.88.−5	0.1004	0.717
1700	4.816	1576.7	8.7622	1.229	1.305	797.9	5.99.−5	0.1027	0.717
1750	4.958	1638.2	8.7979	1.233	1.303	809.1	6.10.−5	0.1050	0.717
1800	5.099	1700.0	8.8327	1.237	1.302	820.2	6.21.−5	0.1072	0.717
1850	5.241	1762.0	8.8667	1.241	1.301	831.1	6.32.−5	0.1094	0.717

1900	5.383	1824.1	8.8998	1.245	1.300	841.9	6.43.−5	0.1116	0.717
1950	5.524	1886.4	8.9322	1.248	1.299	852.6	6.53.−5	0.1138	0.717
2000	5.666	1948.9	8.9638	1.252	1.298	863.1	6.64.−5	0.1159	0.717
2050	5.808	2011.6	8.9948	1.255	1.297	873.5	6.74.−5	0.1180	0.717
2100	5.949	2074.4	9.0251	1.258	1.296	883.8	6.84.−5	0.1200	0.717
2150	6.091	2137.3	9.0547	1.260	1.295	894.0	6.95.−5	0.1220	0.717
2200	6.232	2200.4	9.0837	1.263	1.294	904.0	7.05.−5	0.1240	0.718
2250	6.374	2263.6	9.1121	1.265	1.293	914.0	7.15.−5	0.1260	0.718
2300	6.516	2327.0	9.1399	1.268	1.293	923.8	7.25.−5	0.1279	0.718
2350	6.657	2390.5	9.1672	1.270	1.292	933.5	7.35.−5	0.1298	0.719
2400	6.800	2454.0	9.1940	1.273	1.291	943.2	7.44.−5	0.1317	0.719
2450	6.940	2517.7	9.2203	1.275	1.291	952.7	7.54.−5	0.1336	0.720
2500	7.082	2581.5	9.2460	1.277	1.290	962.2	7.64.−5	0.1354	0.720

[a] v = specific volume; h = specific enthalpy; s = specific entropy; c_p = specific heat at constant pressure; γ = specific heat ratio, c_p/c_v (dimensionless); $\bar{v}_s$ = velocity of sound; η = dynamic viscosity; Pr = Prandtl number (dimensionless); λ = thermal conductivity; Condensed from S. Gordon, *Thermodynamic and Transport Combustion Properties of Hydrocarbons with Air*, NASA Technical Paper 1906, 1982, Vol. 1. These properties are based on constant gaseous composition. The reader is reminded that, at the higher temperatures, the influence of pressure can affect the composition and the thermodynamic properties.

[b] The notation 1.33.−5 signifies 1.33×10^{-5}.

Table 52.6 Thermophysical Properties of the U.S. Standard Atmosphere[a]

Z (m)	H (m)	T (K)	P (bar)	ρ (kg/m³)	g (m/sec²)	$\bar{v}_s$ (m/sec)
0	0	288.15	1.0133	1.2250	9.8067	340.3
1000	1000	281.65	0.8988	1.1117	9.8036	336.4
2000	1999	275.15	0.7950	1.0066	9.8005	332.5
3000	2999	268.66	0.7012	0.9093	9.7974	328.6
4000	3997	262.17	0.6166	0.8194	9.7943	324.6
5000	4996	255.68	0.5405	0.7364	9.7912	320.6
6000	5994	249.19	0.4722	0.6601	9.7882	316.5
7000	6992	242.70	0.4111	0.5900	9.7851	312.3
8000	7990	236.22	0.3565	0.5258	9.7820	308.1
9000	8987	229.73	0.3080	0.4671	9.7789	303.9
10000	9984	223.25	0.2650	0.4135	9.7759	299.5
11000	10981	216.77	0.2270	0.3648	9.7728	295.2
12000	11977	216.65	0.1940	0.3119	9.7697	295.1
13000	12973	216.65	0.1658	0.2667	9.7667	295.1
14000	13969	216.65	0.1417	0.2279	9.7636	295.1
15000	14965	216.65	0.1211	0.1948	9.7605	295.1
16000	15960	216.65	0.1035	0.1665	9.7575	295.1
17000	16954	216.65	0.0885	0.1423	9.7544	295.1
18000	17949	216.65	0.0756	0.1217	9.7513	295.1
19000	18943	216.65	0.0647	0.1040	9.7483	295.1
20000	19937	216.65	0.0553	0.0889	9.7452	295.1
22000	21924	218.57	0.0405	0.0645	9.7391	296.4
24000	23910	220.56	0.0297	0.0469	9.7330	297.7
26000	25894	222.54	0.0219	0.0343	9.7269	299.1
28000	27877	224.53	0.0162	0.0251	9.7208	300.4
30000	29859	226.51	0.0120	0.0184	9.7147	301.7
32000	31840	228.49	0.00889	0.01356	9.7087	303.0
34000	33819	233.74	0.00663	0.00989	9.7026	306.5
36000	35797	239.28	0.00499	0.00726	9.6965	310.1
38000	37774	244.82	0.00377	0.00537	9.6904	313.7
40000	39750	250.35	0.00287	0.00400	9.6844	317.2
42000	41724	255.88	0.00220	0.00299	9.6783	320.7
44000	43698	261.40	0.00169	0.00259	9.6723	324.1
46000	45669	266.93	0.00131	0.00171	9.6662	327.5
48000	47640	270.65	0.00102	0.00132	9.6602	329.8
50000	49610	270.65	0.00080	0.00103	9.6542	329.8

[a] Z = geometric attitude; H = geopotential attitude; ρ = density; g = acceleration of gravity; $\bar{v}_s$ = velocity of sound. Condensed and in some cases converted from *U.S. Standard Atmosphere 1976*, National Oceanic and Atmospheric Administration and National Aeronautics and Space Administration, Washington, DC. Also available as NOAA-S/T 76-1562 and Government Printing Office Stock No. 003-017-00323-0.

Table 52.7 Thermophysical Properties of n-Butane at Atmospheric Pressure[a]

	Temperature (K)									
	250	300	350	400	450	500	550	600	650	700
v	0.0016	0.411	0.485	0.558	0.630	0.701	0.773	0.844	0.915	0.986
h	236.6	718.9	810.7	913.1	1026	1149	1282	1403	1573	1731
s	3.564	5.334	5.616	5.889	6.155	6.414	6.667	6.913	7.152	7.386
c_p	2.21	1.73	1.94	2.15	2.36	2.56	2.74	2.92	3.08	3.22
Z	0.005	0.969	0.982	0.988	0.992	0.993	0.995	0.996	0.997	0.998
$\bar{v}_s$	1161	211	229	245	259	273	286	299	311	323
λ	0.0979	0.0161	0.0220	0.0270	0.0327	0.0390	0.046	0.054	0.063	0.074
η	2.545	0.076	0.088	0.101	0.111	0.124	0.135	0.147	0.160	0.174
Pr	5.75	0.84	0.82	0.81	0.80	0.80	0.80	0.79	0.78	0.77

[a] v = specific volume (m³/kg); h = specific enthalpy (kJ/kg); s = specific entropy (kJ/kg); h = specific enthalpy (kJ/kg); s = specific entropy (kJ/kg · K); c_p = specific heat at constant pressure (kJ/kg · K); Z = compressibility factor = Pv/RT; $\bar{v}_s$ = velocity of sound (m/sec); λ = thermal conductivity (W/m · K); η = viscosity (10^{-4}) (N · sec/m²) (thus, at 250 K the viscosity is 2.545 × 10^{-4} N · sec/m² = 0.0002545 Pa · sec); Pr = Prandtl number.

Table 52.8 Thermophysical Properties of Condensed and Saturated Vapor Carbon Dioxide from 200 K to the Critical Point[a]

T (K)	P (bar)	Specific Volume		Specific Enthalpy		Specific Entropy		Specific Heat (c_p)		Thermal Conductivity		Viscosity		Prandtl Number	
		Condensed[b]	Vapor	Condensed[b]	Vapor	Condensed[b]	Vapor	Condensed[b]	Vapor	Liquid	Vapor	Liquid	Vapor	Liquid	Vapor
200	1.544	0.000644	0.2362	164.8	728.3	1.620	4.439								
205	2.277	0.000649	0.1622	171.5	730.0	1.652	4.379								
210	3.280	0.000654	0.1135	178.2	730.9	1.682	4.319								
215	4.658	0.000659	0.0804	185.0	731.3	1.721	4.264								
216.6	5.180	0.000661	0.0718	187.2	731.5	1.736	4.250								
216.6	5.180	0.000848	0.0718	386.3	731.5	2.656	4.250	1.707	0.958	0.182	0.011	2.10	0.116	1.96	0.96
220	5.996	0.000857	0.0624	392.6	733.1	2.684	4.232	1.761	0.985	0.178	0.012	1.86	0.118	1.93	0.97
225	7.357	0.000871	0.0515	401.8	735.1	2.723	4.204	1.820	1.02	0.171	0.012	1.75	0.120	1.87	0.98
230	8.935	0.000886	0.0428	411.1	736.7	2.763	4.178	1.879	1.06	0.164	0.013	1.64	0.122	1.84	0.99
235	10.75	0.000901	0.0357	420.5	737.9	2.802	4.152	1.906	1.10	0.160	0.013	1.54	0.125	1.82	1.01
240	12.83	0.000918	0.0300	430.2	738.9	2.842	4.128	1.933	1.15	0.156	0.014	1.45	0.128	1.80	1.02
245	15.19	0.000936	0.0253	440.1	739.4	2.882	4.103	1.959	1.20	0.148	0.015	1.36	0.131	1.80	1.04
250	17.86	0.000955	0.0214	450.3	739.6	2.923	4.079	1.992	1.26	0.140	0.016	1.28	0.134	1.82	1.06
255	20.85	0.000977	0.0182	460.8	739.4	2.964	4.056	2.038	1.34	0.134	0.017	1.21	0.137	1.84	1.08
260	24.19	0.001000	0.0155	471.6	738.7	3.005	4.032	2.125	1.43	0.128	0.018	1.14	0.140	1.89	1.12
265	27.89	0.001026	0.0132	482.8	737.4	3.047	4.007	2.237	1.54	0.122	0.019	1.08	0.144	1.98	1.17
270	32.03	0.001056	0.0113	494.4	735.6	3.089	3.981	2.410	1.66	0.116	0.020	1.02	0.150	2.12	1.23
275	36.59	0.001091	0.0097	506.5	732.8	3.132	3.954	2.634	1.81	0.109	0.022	0.96	0.157	2.32	1.32
280	41.60	0.001130	0.0082	519.2	729.1	3.176	3.925	2.887	2.06	0.102	0.024	0.91	0.167	2.57	1.44
285	47.10	0.001176	0.0070	532.7	723.5	3.220	3.891	3.203	2.40	0.095	0.028	0.86	0.178	2.90	1.56
290	53.15	0.001241	0.0058	547.6	716.9	3.271	3.854	3.724	2.90	0.088	0.033	0.79	0.191	3.35	1.68
295	59.83	0.001322	0.0047	562.9	706.3	3.317	3.803	4.68		0.081	0.042	0.71	0.207	4.1	1.8
300	67.10	0.001470	0.0037	585.4	690.2	3.393	3.742			0.074	0.065	0.60	0.226		
304.2c	73.83	0.002145	0.0021	636.6	636.6	3.558	3.558								

[a] Specific volume, m³/kg; specific enthalpy, kJ/kg; specific entropy, kJ/kg · K; specific heat at constant pressure, kJ/kg · K; thermal conductivity, W/m · K; viscosity, 10^{-4} Pa · sec. Thus, at 250 K the viscosity of the saturated liquid is 1.28×10^{-4} N · sec/m² = 0.000128 N · sec/m² = 0.000128 Pa · sec. The Prandtl number is dimensionless.

[b] Above the solid line the condensed phase is solid; below the line, it is liquid.

[c] Critical point.

Table 52.9 Thermophysical Properties of Gaseous Carbon Dioxide at 1 bar Pressure[a]

								T (K)							
	300	350	400	450	500	550	600	650	700	750	800	850	900	950	1000
v (m³/kg)	0.5639	0.6595	0.7543	0.8494	0.9439	1.039	1.133	1.228	1.332	1.417	1.512	1.606	1.701	1.795	1.889
h (kJ/kg)	809.3	853.1	899.1	947.1	997.0	1049	1102	1156	1212	1269	1327	1386	1445	1506	1567
s (kJ/kg · K)	4.860	4.996	5.118	5.231	5.337	5.435	5.527	5.615	5.697	5.775	5.850	5.922	5.990	6.055	6.120
c_p (kJ/kg · K)	0.852	0.898	0.941	0.980	1.014	1.046	1.075	1.102	1.126	1.148	1.168	1.187	1.205	1.220	1.234
λ (W/m · K)	0.0166	0.0204	0.0243	0.0283	0.0325	0.0364	0.0407	0.0445	0.0481	0.0517	0.0551	0.0585	0.0618	0.0650	0.0682
μ (10⁻⁴ Pa · sec)	0.151	0.175	0.198	0.220	0.242	0.261	0.281	0.299	0.317	0.334	0.350	0.366	0.381	0.396	0.410
Pr	0.778	0.770	0.767	0.762	0.755	0.750	0.742	0.742	0.742	0.742	0.742	0.742	0.742	0.743	0.743

[a] v = specific volume; h = enthalpy; s = entropy; c_p = specific heat at constant pressure; λ = thermal conductivity; η = viscosity (at 300 K the gas viscosity is 0.0000151 N · sec/m² = 0.0000151 Pa · sec); Pr = Prandtl number.

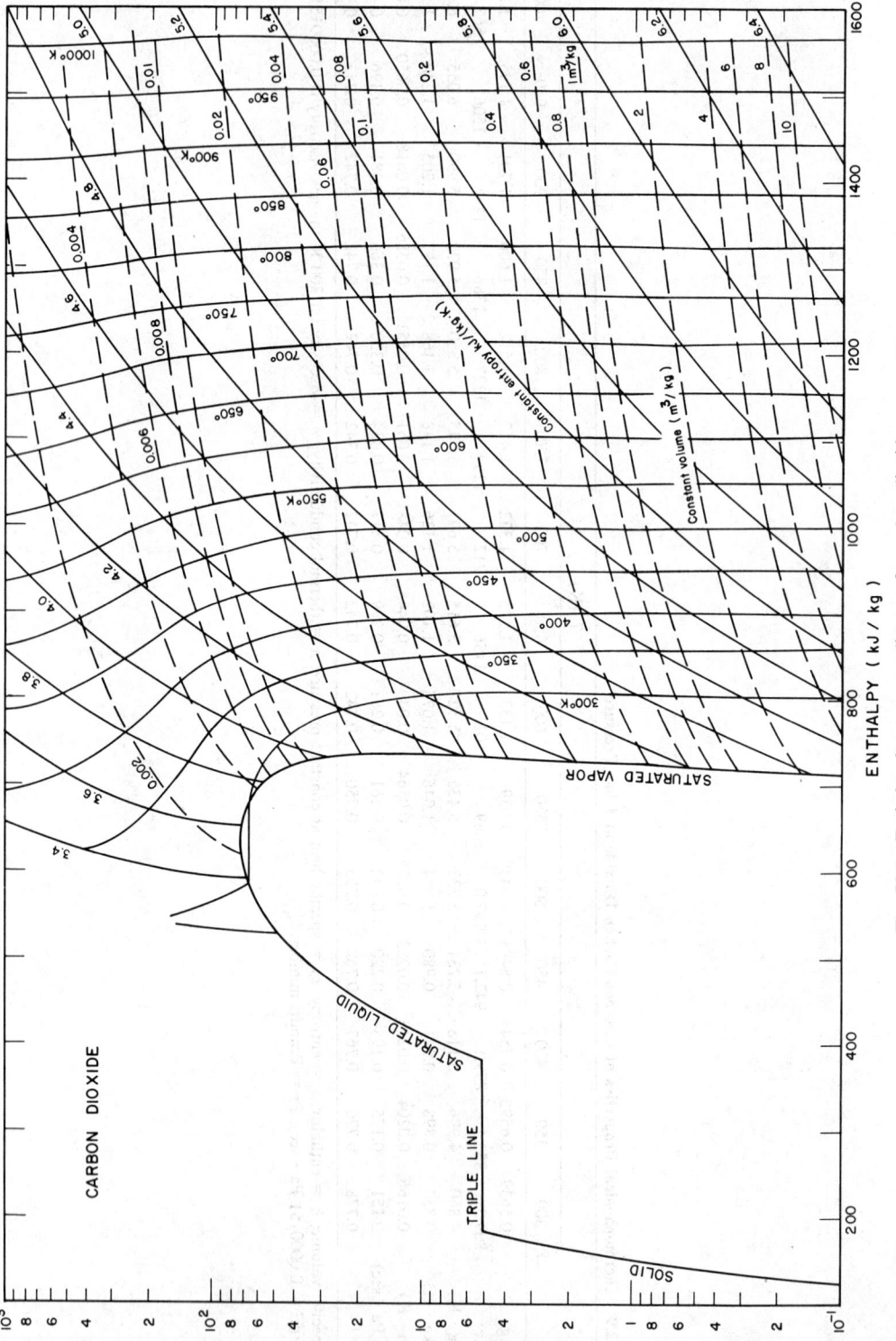

Fig. 52.1 Enthalpy–log pressure diagram for carbon dioxide.

Table 52.10 Thermophysical Properties of Ethane at Atmospheric Pressure[a]

	Temperature (K)									
	250	300	350	400	450	500	550	600	650	700
v	0.672	0.812	0.950	1.088	1.225	1.362	1.499	1.636	1.773	1.910
h	948.7	1068	1162	1265	1380	1505	1640	1783	1936	2097
s	7.330	7.634	7.854	8.198	8.467	8.730	8.986	9.237	9.482	9.720
c_p	1.58	1.76	1.97	2.18	2.39	2.60	2.79	2.97	3.14	3.30
Z	0.986	0.992	0.995	0.997	0.998	0.998	0.999	0.999	1.000	1.000
$\bar{v}_s$	287	312	334	355	374	392	410	427	443	459
λ	0.0103	0.0157	0.0219	0.0288	0.0361	0.0438	0.0518	0.0602	0.0688	0.0775
η	0.079	0.094	0.109	0.123	0.135	0.148	0.160	0.172	0.183	0.194
Pr	1.214	1.056	0.978	0.932	0.900	0.878	0.861	0.848	0.835	0.826

[a] v = specific volume (m³/kg); h = specific enthalpy (kJ/kg); s = specific entropy (kJ/kg · K); c_p = specific heat at constant pressure (kJ/kg · K); Z = compressibility factor = Pv/RT; $\bar{v}_s$ = velocity of sound (m/sec); λ = thermal conductivity (W/m · K); η = viscosity 10^{-4} N · sec/m² (thus, at 250 K the viscosity is 0.079×10^{-4} N · sec/m² = 0.0000079 Pa · sec); Pr = Prandtl number.

Table 52.11 Thermophysical Properties of Ethylene at Atmospheric Pressure[a]

	Temperature (K)					
	250	300	350	400	450	500
v	0.734	0.884	1.034	1.183	1.332	1.482
h	966.8	1039	1122	1215	1316	1415
s	7.556	7.832	8.085	8.331	8.568	8.800
c_p	1.40	1.57	1.75	1.93	2.10	2.26
Z	0.991	0.994	0.997	0.998	0.999	1.000
$\bar{v}_s$	306	330	353	374	394	403
λ	0.0149	0.0206	0.0271	0.0344	0.0425	0.0506
η	0.087	0.103	0.119	0.134	0.148	0.162
Pr	0.816	0.785	0.767	0.751	0.735	0.721

[a] v = specific volume (m³/kg); h = specific enthalpy (kJ/kg); s = specific entropy (kJ/kg · K); c_p = specific heat at constant pressure (kJ/kg · K); Z = compressibility factor = Pv/RT; $\bar{v}_s$ = velocity of sound (m/sec); λ = thermal conductivity (W/m · K); η = viscosity 10^{-4} N · sec/m² (thus, at 250 K the viscosity is 0.087 × 10^{-4} N · sec/m² = 0.0000087 Pa · sec); Pr = Prandtl number.

Table 52.12 Thermophysical Properties of n-Hydrogen at Atmospheric Pressure[a]

	Temperature (K)									
	250	300	350	400	450	500	550	600	650	700
v	10.183	12.218	14.253	16.289	18.324	20.359	22.39	24.48	26.47	28.50
h	3517	4227	4945	5669	6393	7118	7844	8571	9299	10029
s	67.98	70.58	72.79	74.72	76.43	77.96	79.34	80.60	81.76	82.85
c_p	14.04	14.31	14.43	14.48	14.50	14.51	14.53	14.55	14.57	14.60
Z	1.000	1.000	1.000	1.000	1.000	1.000	1.000	1.000	1.000	1.000
$\bar{v}_s$	1209	1319	1423	1520	1611	1698	1780	1859	1934	2006
λ	0.162	0.187	0.210	0.230	0.250	0.269	0.287	0.305	0.323	0.340
η	0.079	0.089	0.099	0.109	0.118	0.127	0.135	0.143	0.151	0.159
Pr	0.685	0.685	0.685	0.684	0.684	0.684	0.684	0.684	0.684	0.684

[a] v = specific volume (m^3/kg); h = specific enthalpy (kJ/kg); s = specific entropy (kJ/kg · K); c_p = specific heat at constant pressure (kJ/kg · K); Z = compressibility factor = Pv/RT; $\bar{v}_s$ = velocity of sound (m/sec); λ = thermal conductivity (W/m · K); η = viscosity 10^{-4} N · sec/m² (thus, at 250 K the viscosity is 0.079×10^{-4} N · sec/m² = 0.0000079 Pa · sec); Pr = Prandtl number.

Table 52.13 Thermodynamic Properties of Saturated Mercury[a]

T (K)	P (bar)	v (m³/kg)	h (kJ/kg)	s (kJ/kg · K)	c_p (kJ/kg · K)
0		6.873.−5[b]	0	0	0
20		6.875.−5	0.466	0.0380	0.0513
40		6.884.−5	1.918	0.0868	0.0894
60		6.897.−5	3.897	0.1267	0.1067
80		6.911.−5	6.129	0.1588	0.1156
100		6.926.−5	8.497	0.1852	0.1209
120		6.942.−5	10.956	0.2076	0.1248
140		6.958.−5	13.482	0.2270	0.1278
160		6.975.−5	16.063	0.2443	0.1304
180		6.993.−5	18.697	0.2598	0.1330
200		7.013.−5	21.386	0.2739	0.1360
220		7.034.−5	24.139	0.2870	0.1394
234.3	7.330.−10	7.050.−5	26.148	0.2959	0.1420
234.3	7.330.−10	7.304.−5	37.585	0.3447	0.1422
240	1.668.−9	7.311.−5	38.395	0.3481	0.1420
260	6.925.−8	7.339.−5	41.224	0.3595	0.1409
280	5.296.−7	7.365.−5	44.034	0.3699	0.1401
300	3.075.−6	7.387.−5	46.829	0.3795	0.1393
320	1.428.−5	7.413.−5	49.609	0.3885	0.1386
340	5.516.−5	7.439.−5	52.375	0.3969	0.1380
360	1.829.−4	7.472.−5	55.130	0.4048	0.1375
380	5.289.−4	7.499.−5	57.874	0.4122	0.1370
400	1.394.−3	7.526.−5	60.609	0.4192	0.1366
450	0.01053	7.595.−5	67.414	0.4352	0.1357
500	0.05261	7.664.−5	74.188	0.4495	0.1353
550	0.1949	7.735.−5	80.949	0.4624	0.1352
600	0.5776	7.807.−5	87.716	0.4742	0.1356
650	1.4425	7.881.−5	94.508	0.4850	0.1360
700	3.153	7.957.−5	101.343	0.4951	0.1372
750	6.197	8.036.−5	108.242	0.5046	0.1382
800	11.181	8.118.−5	115.23	0.5136	0.1398
850	18.816	8.203.−5	122.31	0.5221	0.1416
900	29.88	8.292.−5	129.53	0.5302	0.1439
950	45.23	8.385.−5	137.16	0.5381	0.1464
1000	65.74	8.482.−5	144.41	0.5456	0.1492

[a] v = specific volume; h = specific enthalpy; s = specific entropy, c_p = specific heat at constant pressure. Properties above the solid line are for the solid; below they are for the liquid. Condensed, converted, and interpolated from the tables of M. P. Vukalovich, A. I. Ivanov, L. R. Fokin, and A. T. Yakovlev, *Thermophysical Properties of Mercury,* Standartov, Moscow, USSR, 1971.
[b] The notation 6.873.−5 signifies 6.873×10^{-5}.

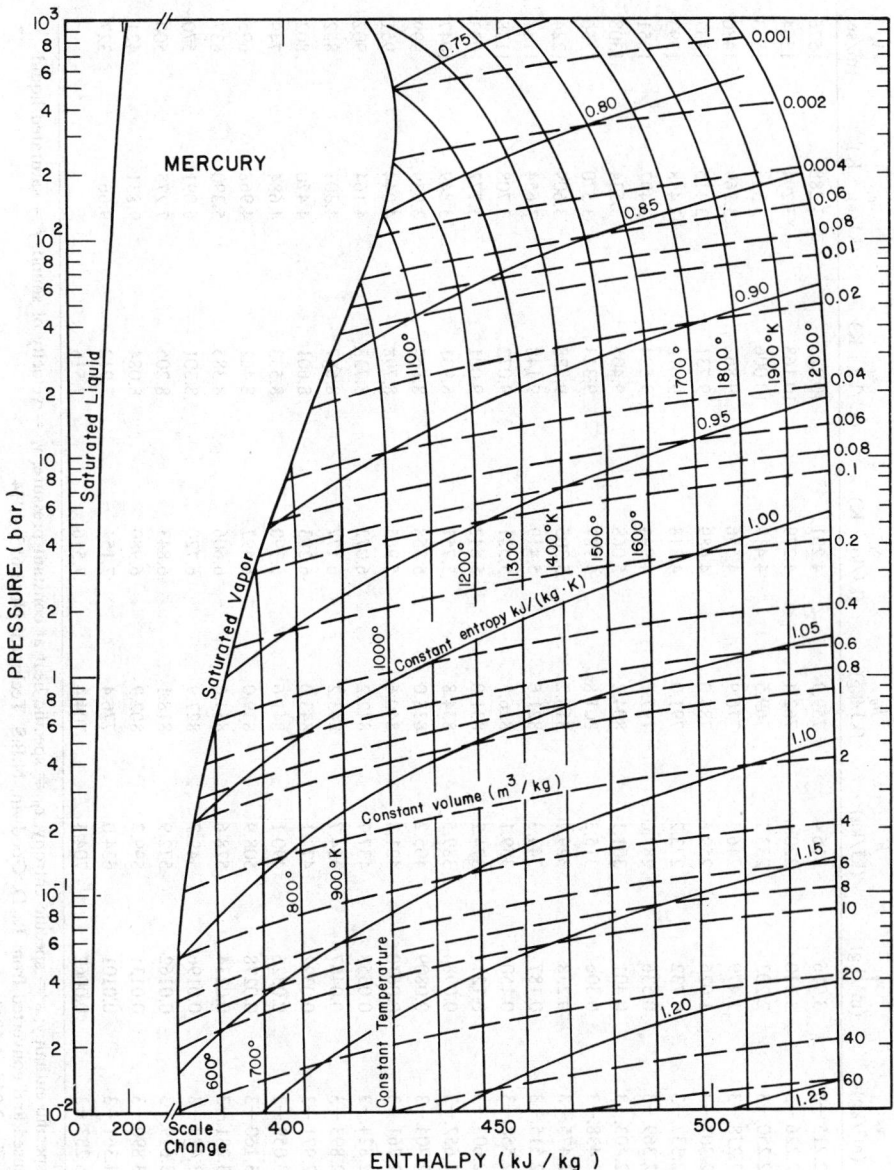

Fig. 52.2 Enthalpy–log pressure diagram for mercury.

Table 52.14 Thermodynamic Properties of Saturated Methane[a]

T (K)	P (bar)	v_f (m^3/kg)	v_g (m^3/kg)	h_f (kJ/kg)	h_g (kJ/kg)	s_f (kJ/kg · K)	s_g (kJ/kg · K)	c_{p_f} (kJ/kg · K)	$\bar{v}_s$ (m/sec)
90.68	0.117	2.215.−3[b]	3.976	216.4	759.9	4.231	10.225	3.288	1576
92	0.139	2.226.−3	3.410	220.6	762.4	4.279	10.168	3.294	1564
96	0.223	2.250.−3	2.203	233.2	769.5	4.419	10.006	3.326	1523
100	0.345	2.278.−3	1.479	246.3	776.9	4.556	9.862	3.369	1480
104	0.515	2.307.−3	1.026	259.6	784.0	4.689	9.731	3.415	1437
108	0.743	2.337.−3	0.732	273.2	791.0	4.818	9.612	3.458	1393
112	1.044	2.369.−3	0.536	287.0	797.7	4.944	9.504	3.497	1351
116	1.431	2.403.−3	0.401	301.1	804.2	5.068	9.405	3.534	1308
120	1.919	2.438.−3	0.306	315.3	801.8	5.187	9.313	3.570	1266
124	2.523	2.475.−3	0.238	329.7	816.2	5.305	9.228	3.609	1224
128	3.258	2.515.−3	0.187	344.3	821.6	5.419	9.148	3.654	1181
132	4.142	2.558.−3	0.150	359.1	826.5	5.531	9.072	3.708	1138
136	5.191	2.603.−3	0.121	374.2	831.0	5.642	9.001	3.772	1093
140	6.422	2.652.−3	0.0984	389.5	834.8	5.751	8.931	3.849	1047
144	7.853	2.704.−3	0.0809	405.2	838.0	5.858	8.864	3.939	999
148	9.502	2.761.−3	0.0670	421.3	840.6	5.965	8.798	4.044	951
152	11.387	2.824.−3	0.0558	437.7	842.2	6.072	8.733	4.164	902
156	13.526	2.893.−3	0.0467	454.7	843.2	6.177	8.667	4.303	852
160	15.939	2.971.−3	0.0392	472.1	843.0	6.283	8.601	4.470	802
164	18.647	3.059.−3	0.0326	490.1	841.6	6.390	8.533	4.684	749
168	21.671	3.160.−3	0.0278	508.9	839.0	6.497	8.462	4.968	695
172	25.034	3.281.−3	0.0234	528.6	834.6	6.606	8.385	5.390	637
176	28.761	3.428.−3	0.0196	549.7	827.9	6.720	8.301	6.091	570
180	32.863	3.619.−3	0.0162	572.9	818.1	6.843	8.205	7.275	500
184	37.435	3.890.−3	0.0131	599.7	802.9	6.980	8.084	9.831	421
188	42.471	4.361.−3	0.0101	634.0	776.4	7.154	7.912	19.66	327
190.56	45.988	6.233.−3	0.0062	704.4	704.4	7.516	7.516		

[a] v = specific volume; h = specific enthalpy; s = specific entropy; c_p = specific heat at constant pressure; $\bar{v}_s$ = velocity of sound; f = saturated liquid; g = saturated vapor. Condensed and converted from R. D. Goodwin, N.B.S. Technical Note 653, 1974.

[b] The notation 2.215.−3 signifies 2.215×10^{-3}.

Table 52.15 Thermophysical Properties of Methane at Atmospheric Pressure[a]

	Temperature (K)									
	250	300	350	400	450	500	550	600	650	700
v	1.275	1.532	1.789	2.045	2.301	2.557	2.813	3.068	3.324	3.580
h	1090	1200	1315	1437	1569	1709	1857	2016	2183	2359
s	11.22	11.62	11.98	12.30	12.61	12.91	13.19	13.46	13.73	14.00
c_p	2.04	2.13	2.26	2.43	2.60	2.78	2.96	3.16	3.35	3.51
Z	0.997	0.998	0.999	1.000	1.000	1.000	1.000	1.000	1.000	1.000
$\bar{v}_s$	413	450	482	511	537	562	585	607	629	650
λ	0.0276	0.0342	0.0417	0.0486	0.0571	0.0675	0.0768	0.0863	0.0956	0.1052
η	0.095	0.112	0.126	0.141	0.154	0.168	0.180	0.192	0.202	0.214
Pr	0.701	0.696	0.683	0.687	0.690	0.693	0.696	0.700	0.706	0.714

[a] v = specific volume (m³/kg); h = specific enthalpy (kJ/kg); s = specific entropy (kJ/kg); h = specific enthalpy (kJ/kg); s = specific entropy (kJ/kg · K); c_p = specific heat at constant pressure (kJ/kg · K); Z = compressibility factor = Pv/RT; $\bar{v}_s$ = velocity of sound (m/sec); λ = thermal conductivity (W/m · K); η = viscosity 10^{-4} N · sec/m² (thus, at 250 K the viscosity is 0.095×10^{-4} N · sec/m² = 0.0000095 Pa · sec); Pr = Prandtl number.

Table 52.16 Thermophysical Properties of Nitrogen at Atmospheric Pressure[a]

	Temperature (K)									
	250	300	350	400	450	500	550	600	650	700
v	0.7317	0.8786	1.025	1.171	1.319	1.465	1.612	1.758	1.905	2.051
h	259.1	311.2	363.3	415.4	467.8	520.4	573.4	626.9	681.0	735.6
s	6.650	6.840	7.001	7.140	7.263	7.374	7.475	7.568	7.655	7.736
c_p	1.042	1.041	1.042	1.045	1.050	1.056	1.065	1.075	1.086	1.098
Z	0.9992	0.9998	0.9998	0.9999	1.0000	1.0002	1.0002	1.0003	1.0003	1.0003
$\bar{v}_s$	322	353	382	407	432	455	476	496	515	534
λ	0.0223	0.0259	0.0292	0.0324	0.0366	0.0386	0.0414	0.0442	0.0470	0.0496
η	0.155	0.178	0.200	0.220	0.240	0.258	0.275	0.291	0.306	0.321
Pr	0.724	0.715	0.713	0.710	0.708	0.706	0.707	0.708	0.709	0.711

[a] v = specific volume (m³/kg); h = specific enthalpy (kJ/kg); s = specific entropy (kJ/kg · K); c_p = specific heat at constant pressure (kJ/kg · K); Z = compressibility factor = Pv/RT; $\bar{v}_s$ = velocity of sound (m/sec); λ = thermal conductivity (W/m · K); η = viscosity 10^{-4} N · sec/m² (thus, at 250 K the viscosity is 0.155×10^{-4} N · sec/m² = 0.0000155 Pa · sec); Pr = Prandtl number.

Table 52.17 Thermophysical Properties of Oxygen at Atmospheric Pressure[a]

	Temperature (K)									
	250	300	350	400	450	500	550	600	650	700
v	0.6402	0.7688	0.9790	1.025	1.154	1.282	1.410	1.539	1.667	1.795
h	226.9	272.7	318.9	365.7	413.1	461.3	510.3	560.1	610.6	661.9
s	6.247	6.414	6.557	6.682	6.793	6.895	6.988	7.075	7.156	7.232
c_p	0.915	0.920	0.929	0.942	0.956	0.972	0.988	1.003	1.018	1.031
Z	0.9987	0.9994	0.9996	0.9998	1.0000	1.0000	1.0001	1.0002	1.0002	1.0002
λ	0.0226	0.0266	0.0305	0.0343	0.0380	0.0416	0.0451	0.0487	0.0521	0.0554
η	0.179	0.207	0.234	0.258	0.281	0.303	0.324	0.344	0.363	0.381
Pr	0.725	0.716	0.713	0.710	0.708	0.707	0.708	0.708	0.709	0.710

[a] v = specific volume (m^3/kg); h = specific enthalpy (kJ/kg); s = specific entropy (kJ/kg · K); c_p = specific heat at constant pressure (kJ/kg · K); Z = compressibility factor = Pv/RT; $\bar{v}_s$ = velocity of sound (m/sec); λ = thermal conductivity (W/m · K); η = viscosity 10^{-4} N · sec/m^2 (thus, at 250 K the viscosity is 0.179 × 10^{-4} N · sec/m^2 = 0.0000179 Pa · sec); Pr = Prandtl number.

Table 52.18 Thermophysical Properties of Propane at Atmospheric Pressure[a]

	Temperature (K)									
	250	300	350	400	450	500	550	600	650	700
v	0.451	0.548	0.644	0.738	0.832	0.926	1.020	1.114	1.207	1.300
h	877.2	957.0	1048	1149	1261	1384	1517	1658	1808	1966
s	5.840	6.131	6.409	6.680	6.944	7.202	7.455	7.701	7.941	8.176
c_p	1.50	1.70	1.96	2.14	2.35	2.55	2.74	2.92	3.09	3.24
Z	0.970	0.982	0.988	0.992	0.994	0.996	0.997	0.997	0.998	0.998
$\bar{v}_s$	227	248	268	285	302	317	332	347	360	373
λ	0.0128	0.0182	0.0247	0.0296	0.0362	0.0423	0.047	0.053	0.060	0.069
η	0.070	0.082	0.096	0.108	0.119	0.131	0.143	0.155		
Pr	0.820	0.772	0.761	0.765	0.773	0.793	0.815	0.849		

[a] v = specific volume (m³/kg); h = specific enthalpy (kJ/kg); s = specific entropy (kJ/kg · K); c_p = specific heat at constant pressure (kJ/kg · K); Z = compressibility factor = Pv/RT; $\bar{v}_s$ = velocity of sound (m/sec); λ = thermal conductivity (W/m · K); η = viscosity 10^{-4} N · sec/m² (thus, at 250 K the viscosity is 0.070×10^{-4} N · sec/m² = 0.0000070 Pa · sec); Pr = Prandtl number.

Table 52.19 Thermophysical Properties of Propylene at Atmospheric Pressure[a]

	Temperature (K)					
	250	300	350	400	450	500
v	0.482	0.585	0.686	0.786	0.884	0.972
h	891.8	957.6	1040	1131	1235	1338
s	6.074	6.354	6.606	6.851	7.095	7.338
c_p	1.44	1.55	1.73	1.91	2.09	2.27
Z	0.976	0.987	0.992	0.994	0.995	0.996
$\bar{v}_s$	247	257	278	298	315	333
λ	0.0127	0.0177	0.0233	0.0296	0.0363	0.0438
η	0.072	0.087	0.101	0.115	0.128	0.141
Pr	0.814	0.769	0.754	0.742	0.731	0.728

[a] v = specific volume (m³/kg); h = specific enthalpy (kJ/kg); s = specific entropy (kJ/kg · K); c_p = specific heat at constant pressure (kJ/kg · K); Z = compressibility factor = Pv/RT; $\bar{v}_s$ = velocity of sound (m/sec); λ = thermal conductivity (W/m · K); η = viscosity 10^{-4} N · sec/m² (thus, at 250 K the viscosity is 0.072×10^{-4} N · sec/m² = 0.0000072 Pa · sec); Pr = Prandtl number.

Table 52.20 Thermophysical Properties of Saturated Refrigerant 22[a]

T (K)	P (bar)	v (m³/kg)		h (kJ/kg)		s (kJ/kg · K)		c_p (Liquid)	η (Liquid)	λ (Liquid)	τ (Liquid)
		Liquid	Vapor	Liquid	Vapor	Liquid	Vapor				
150	0.0017	6.209,−4[b]	83.40	268.2	547.3	3.355	5.215	1.059		0.161	
160	0.0054	6.293,−4	28.20	278.2	552.1	3.430	5.141	1.058		0.156	
170	0.0150	6.381,−4	10.85	288.3	557.0	3.494	5.075	1.057	0.770	0.151	
180	0.0369	6.474,−4	4.673	298.7	561.9	3.551	5.013	1.058	0.647	0.146	
190	0.0821	6.573,−4	2.225	308.6	566.8	3.605	4.963	1.060	0.554	0.141	
200	0.1662	6.680,−4	1.145	318.8	571.6	3.657	4.921	1.065	0.481	0.136	0.024
210	0.3116	6.794,−4	0.6370	329.1	576.5	3.707	4.885	1.071	0.424	0.131	0.022
220	0.5470	6.917,−4	0.3772	339.7	581.2	3.756	4.854	1.080	0.378	0.126	0.021
230	0.9076	7.050,−4	0.2352	350.6	585.9	3.804	4.828	1.091	0.340	0.121	0.019
240	1.4346	7.195,−4	0.1532	361.7	590.5	3.852	4.805	1.105	0.309	0.117	0.0172
250	2.174	7.351,−4	0.1037	373.0	594.9	3.898	4.785	1.122	0.282	0.112	0.0155
260	3.177	7.523,−4	0.07237	384.5	599.0	3.942	4.768	1.143	0.260	0.107	0.0138
270	4.497	7.733,−4	0.05187	396.3	603.0	3.986	4.752	1.169	0.241	0.102	0.0121
280	6.192	7.923,−4	0.03803	408.2	606.6	4.029	4.738	1.193	0.225	0.097	0.0104
290	8.324	8.158,−4	0.02838	420.4	610.0	4.071	4.725	1.220	0.211	0.092	0.0087
300	10.956	8.426,−4	0.02148	432.7	612.8	4.113	4.713	1.257	0.198	0.087	0.0071
310	14.17	8.734,−4	0.01643	445.5	615.1	4.153	4.701	1.305	0.186	0.082	0.0055
320	18.02	9.096,−4	0.01265	458.6	616.7	4.194	4.688	1.372	0.176	0.077	0.0040
330	22.61	9.535,−4	9.753,−3	472.4	617.3	4.235	4.674	1.460	0.167	0.072	0.0026
340	28.03	1.010,−3	7.479,−3	487.2	616.5	4.278	4.658	1.573	0.151	0.067	0.0014
350	34.41	1.086,−3	5.613,−3	503.7	613.3	4.324	4.637	1.718	0.130	0.062	0.0008
360	41.86	1.212,−3	4.036,−3	523.7	605.5	4.378	4.605	1.897	0.106		
369.3	49.89	2.015,−3	2.015,−3	570.0	570.0	4.501	4.501	∞		—	0

[a] c_p in units of kJ/kg · K; η = viscosity (10^{-4} Pa · sec); λ = thermal conductivity (W/m · K); τ = surface tension (N/m). Sources: P, v, T, h, s interpolated and extrapolated from I. I. Perelshteyn, *Tables and Diagrams of the Thermodynamic Properties of Freons 12, 13, 22*, Moscow, USSR, 1971. c_p, η, λ interpolated and converted from *Thermophysical Properties of Refrigerants*, ASHRAE, New York, 1976. τ calculated from V. A. Gruzdev et al., *Fluid Mech. Sov. Res.*, **3**, 172 (1974).
[b] The notation 6.209,−4 signifies 6.209×10^{-4}.

Table 52.21 Thermophysical Properties of Refrigerant 22 at Atmospheric Pressure[a]

	Temperature (K)					
	250	300	350	400	450	500
v	0.2315	0.2802	0.3289	0.3773	0.4252	0.4723
h	597.8	630.0	664.5	702.5	740.8	782.3
s	4.8671	4.9840	5.0905	5.1892	5.2782	5.3562
c_p	0.587	0.647	0.704	0.757	0.806	0.848
Z	0.976	0.984	0.990	0.994	0.995	0.996
$\bar{v}_s$	166.4	182.2	196.2	209.4	220.0	233.6
λ	0.0080	0.0110	0.0140	0.0170	0.0200	0.0230
η	0.109	0.130	0.151	0.171	0.190	0.209
Pr	0.800	0.765	0.759	0.761	0.766	0.771

[a] v = specific volume (m³/kg); h = specific enthalpy (kJ/kg); s = specific entropy (kJ/kg · K); c_p = specific heat at constant pressure (kJ/kg · K); Z = compressibility factor = Pv/RT; $\bar{v}_s$ = velocity of sound (m/sec); λ = thermal conductivity (W/m · K); η = viscosity 10^{-4} N · sec/m² (thus, at 250 K the viscosity is 0.109×10^{-4} N · sec/m² = 0.0000109 Pa · sec); Pr = Prandtl number.

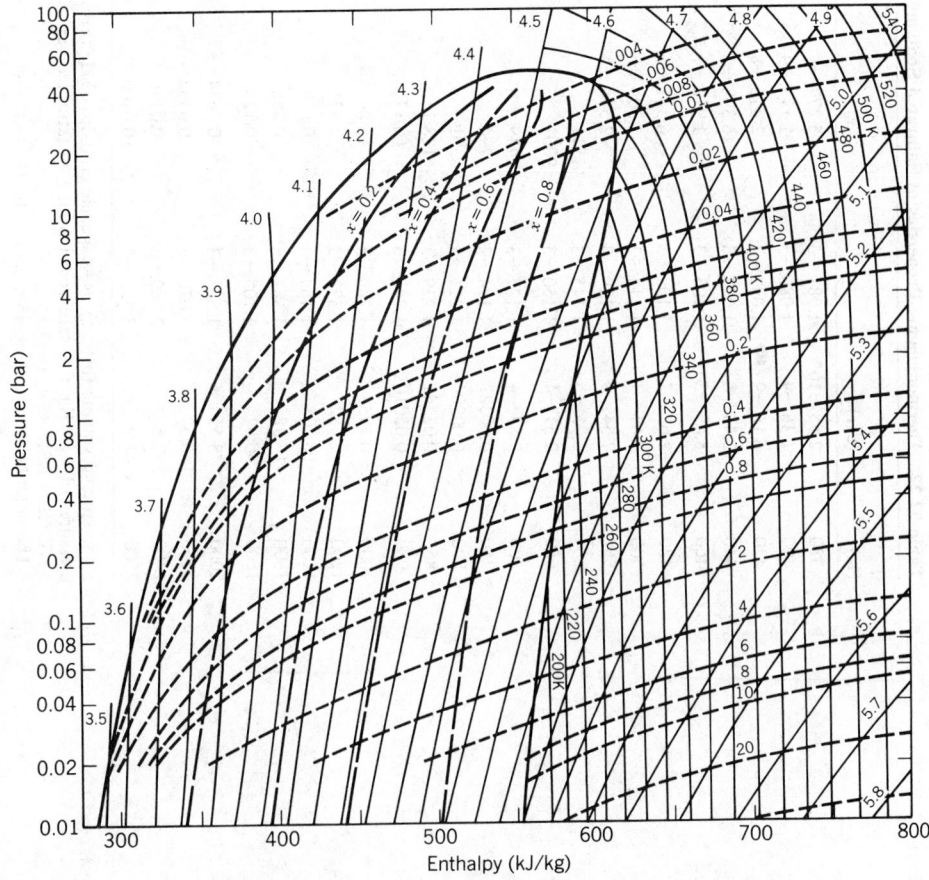

Fig. 52.3 Enthalpy–log pressure diagram for Refrigerant 22.

Table 52.22 Thermodynamic Properties of Saturated Sodium[a]

T (K)	P (bar)	v_f	v_g	h_f	h_g	s_f	s_g	c_{pf}	c_{pg}
380	2.55.−10[b]	1.081.−3	5.277.+9	0.219	4.723	2.853	14.705	1.384	0.988
400	1.36.−9	1.086.−3	2.173.+9	0.247	4.740	2.924	14.158	1.374	1.023
420	6.16.−9	1.092.−3	2.410.+8	0.274	4.757	2.991	13.665	1.364	1.066
440	2.43.−8	1.097.−3	6.398.+7	0.301	4.773	3.054	13.219	1.355	1.117
460	8.49.−8	1.103.−3	1.912.+7	0.328	4.790	3.114	12.814	1.346	1.176
480	2.67.−7	1.109.−3	6.341.+6	0.355	4.806	3.171	12.443	1.338	1.243
500	7.64.−7	1.114.−3	2.304.+6	0.382	4.820	3.226	12.104	1.330	1.317
550	7.54.−6	1.129.−3	2.558.+5	0.448	4.856	3.352	11.367	1.313	1.523
600	5.05.−5	1.145.−3	41511	0.513	4.887	3.465	10.756	1.299	1.745
650	2.51.−4	1.160.−3	9001	0.578	4.915	3.569	10.241	1.287	1.963
700	9.87.−4	1.177.−3	2449	0.642	4.939	3.664	9.802	1.278	2.160
750	0.00322	1.194.−3	794	0.705	4.959	3.752	9.424	1.270	2.325
800	0.00904	1.211.−3	301	0.769	4.978	3.834	9.095	1.264	2.452
850	0.02241	1.229.−3	128.1	0.832	4.995	3.910	8.808	1.260	2.542
900	0.05010	1.247.−3	60.17	0.895	5.011	3.982	8.556	1.258	2.597
1000	0.1955	1.289.−3	16.84	1.021	5.043	4.115	8.137	1.259	2.624
1200	1.482	1.372.−3	2.571	1.274	5.109	4.346	7.542	1.281	2.515
1400	6.203	1.469.−3	0.688	1.535	5.175	4.546	7.146	1.330	2.391
1600	17.98	1.581.−3	0.258	1.809	5.225	4.728	6.863	1.406	2.301
1800	40.87	1.709.−3	0.120	2.102	5.255	4.898	6.649	1.516	2.261
2000	78.51	1.864.−3	0.0634	2.422	5.256	5.064	6.480	1.702	2.482
2200	133.5	2.076.−3	0.0362	2.794	5.207	5.235	6.332	2.101	3.307
2400	207.6	2.480.−3	0.0196	3.299	5.025	5.447	6.166	3.686	8.476
2500	251.9	3.323.−3	0.0100	3.850	4.633	5.666	5.980		

[a] v = specific volume (m^3/kg); h = specific enthalpy (MJ/kg); s = specific entropy (kJ/kg · K); c_p = specific heat at constant pressure (kJ/kg · K); f = saturated liquid; g = saturated vapor. Converted from the tables of J. K. Fink, Argonne Nat. Lab. rept. ANL-CEN-RSD-82-4, 1982.

[b] The notation 2.55.−10 signifies 2.55×10^{-10}.

Table 52.23 Thermodynamic Properties of Ice/Water[a]

T (K)	P (bar)	v (m³/kg)	h (kJ/kg)	s (kJ/kg · K)	c_p (kJ/kg · K)
150	6.30.−11[b]	0.001073	94.7	1.328	1.224
160	7.72.−10	0.001074	107.3	1.409	1.291
170	7.29.−9	0.001076	120.6	1.489	1.357
180	5.38.−8	0.001077	134.5	1.569	1.426
190	3.23.−7	0.001078	149.1	1.648	1.495
200	1.62.−6	0.001079	164.4	1.726	1.566
210	7.01.−6	0.001081	180.4	1.805	1.638
220	2.65.−5	0.001082	197.1	1.882	1.711
230	8.91.−5	0.001084	214.6	1.960	1.785
240	2.73.−4	0.001085	232.8	2.038	1.860
250	7.59.−4	0.001087	251.8	2.115	1.936
260	0.00196	0.001088	271.5	2.192	2.013
270	0.00469	0.001090	292.0	2.270	2.091
273.15	0.00611	0.001091	298.7	2.294	2.116
273.15	0.00611	0.001000	632.2	3.515	4.217
280	0.00990	0.001000	661.0	3.619	4.198
290	0.01917	0.001001	702.9	3.766	4.184
300	0.03531	0.001003	744.7	3.908	4.179

[a] v = specific volume; h = specific enthalpy; s = specific entropy; c_p = specific heat at constant pressure. Ice values ($T \leq 273.15$ K) converted and rounded off from S. Gordon, NASA Tech. Paper 1906, 1982.
[b] The notation 6.30.−11 signifies 6.30×10^{-11}.

Table 52.24 Thermophysical Properties of Saturated Steam/Water[a]

P (bar)	T (K)	v_f	v_g	h_f	h_g	η_g	λ_f	λ_g	Pr_f	Pr_g
1.0	372.78	1.0434.−3[b]	1.6937	417.5	2675.4	0.1202	0.6805	0.0244	1.735	1.009
1.5	384.52	1.0530.−3	1.1590	467.1	2693.4	0.1247	0.6847	0.0259	1.538	1.000
2.0	393.38	1.0608.−3	0.8854	504.7	2706.3	0.1280	0.6866	0.0268	1.419	1.013
2.5	400.58	1.0676.−3	0.7184	535.3	2716.4	0.1307	0.6876	0.0275	1.335	1.027
3.0	406.69	1.0735.−3	0.6056	561.4	2724.7	0.1329	0.6879	0.0281	1.273	1.040
3.5	412.02	1.0789.−3	0.5240	584.3	2731.6	0.1349	0.6878	0.0287	1.224	1.050
4.0	416.77	1.0839.−3	0.4622	604.7	2737.6	0.1367	0.6875	0.0293	1.185	1.057
4.5	421.07	1.0885.−3	0.4138	623.2	2742.9	0.1382	0.6869	0.0298	1.152	1.066
5	424.99	1.0928.−3	0.3747	640.1	2747.5	0.1396	0.6863	0.0303	1.124	1.073
6	432.00	1.1009.−3	0.3155	670.4	2755.5	0.1421	0.6847	0.0311	1.079	1.091
7	438.11	1.1082.−3	0.2727	697.1	2762.0	0.1443	0.6828	0.0319	1.044	1.105
8	445.57	1.1150.−3	0.2403	720.9	2767.5	0.1462	0.6809	0.0327	1.016	1.115
9	448.51	1.1214.−3	0.2148	742.6	2772.1	0.1479	0.6788	0.0334	0.992	1.127
10	453.03	1.1274.−3	0.1943	762.6	2776.1	0.1495	0.6767	0.0341	0.973	1.137
12	461.11	1.1386.−3	0.1632	798.4	2782.7	0.1523	0.6723	0.0354	0.943	1.156
14	468.19	1.1489.−3	0.1407	830.1	2787.8	0.1548	0.6680	0.0366	0.920	1.175
16	474.52	1.1586.−3	0.1237	858.6	2791.8	0.1569	0.6636	0.0377	0.902	1.191
18	480.26	1.1678.−3	0.1103	884.6	2794.8	0.1589	0.6593	0.0388	0.889	1.206
20	485.53	1.1766.−3	0.0995	908.6	2797.2	0.1608	0.6550	0.0399	0.877	1.229
25	497.09	1.1972.−3	0.0799	962.0	2800.9	0.1648	0.6447	0.0424	0.859	1.251
30	506.99	1.2163.−3	0.0666	1008.4	2802.3	0.1684	0.6347	0.0449	0.849	1.278
35	515.69	1.2345.−3	0.0570	1049.8	2802.0	0.1716	0.6250	0.0472	0.845	1.306
40	523.48	1.2521.−3	0.0497	1087.4	2800.3	0.1746	0.6158	0.0496	0.845	1.331
45	530.56	1.2691.−3	0.0440	1122.1	2797.7	0.1775	0.6068	0.0519	0.849	1.358
50	537.06	1.2858.−3	0.0394	1154.5	2794.2	0.1802	0.5981	0.0542	0.855	1.386
60	548.70	1.3187.−3	0.0324	1213.7	2785.0	0.1854	0.5813	0.0589	0.874	1.442
70	558.94	1.3515.−3	0.0274	1267.4	2773.5	0.1904	0.5653	0.0638	0.901	1.503
80	568.12	1.3843.−3	0.0235	1317.1	2759.9	0.1954	0.5499	0.0688	0.936	1.573
90	576.46	1.4179.−3	0.0205	1363.7	2744.6	0.2005	0.5352	0.0741	0.978	1.651
100	584.11	1.4526.−3	0.0180	1408.0	2727.7	0.2057	0.5209	0.0798	1.029	1.737
110	591.20	1.4887.−3	0.0160	1450.6	2709.3	0.2110	0.5071	0.0859	1.090	1.837
120	597.80	1.5268.−3	0.0143	1491.8	2689.2	0.2166	0.4936	0.0925	1.163	1.963
130	603.98	1.5672.−3	0.0128	1532.0	2667.0	0.2224	0.4806	0.0998	1.252	2.126
140	609.79	1.6106.−3	0.0115	1571.6	2642.4	0.2286	0.4678	0.1080	1.362	2.343
150	615.28	1.6579.−3	0.0103	1611.0	2615.0	0.2373	0.4554	0.1307	1.502	2.571
160	620.48	1.7103.−3	0.0093	1650.5	2584.9	0.2497	0.4433	0.1280	1.688	3.041
170	625.41	1.7696.−3	0.0084	1691.7	2551.6	0.2627	0.4315	0.1404	2.098	3.344
180	630.11	1.8399.−3	0.0075	1734.8	2513.9	0.2766	0.4200	0.1557	2.360	3.807
190	634.58	1.9260.−3	0.0067	1778.7	2470.6	0.2920	0.4087	0.1749	2.951	8.021
200	638.85	2.0370.−3	0.0059	1826.5	2410.4	0.3094	0.3976	0.2007	4.202	12.16

Table 52.24 *(Continued)*

P (bar)	s_f	s_g	c_{p_f}	c_{p_g}	η_f	$\gamma_f{}'$	γ_g	$\bar{v}_{sf}{}'$	$\bar{v}_{sg}$	τ
1.0	1.3027	7.3598	4.222	2.048	2.801	1.136	1.321	438.74	472.98	0.0589
1.5	1.4336	7.2234	4.231	2.077	2.490	1.139	1.318	445.05	478.73	0.0566
2.0	1.5301	7.1268	4.245	2.121	2.295	1.141	1.316	449.51	482.78	0.0548
2.5	1.6071	7.0520	4.258	2.161	2.156	1.142	1.314	452.92	485.88	0.0534
3.0	1.6716	6.9909	4.271	2.198	2.051	1.143	1.313	455.65	488.36	0.0521
3.5	1.7273	6.9392	4.282	2.233	1.966	1.143	1.311	457.91	490.43	0.0510
4.0	1.7764	6.8943	4.294	2.266	1.897	1.144	1.310	459.82	492.18	0.0500
4.5	1.8204	6.8547	4.305	2.298	1.838	1.144	1.309	461.46	493.69	0.0491
5	1.8604	6.8192	4.315	2.329	1.787	1.144	1.308	462.88	495.01	0.0483
6	1.9308	6.7575	4.335	2.387	1.704	1.144	1.306	465.23	497.22	0.0468
7	1.9918	6.7052	4.354	2.442	1.637	1.143	1.304	467.08	498.99	0.0455
8	2.0457	6.6596	4.372	2.495	1.581	1.142	1.303	468.57	500.55	0.0444
9	2.0941	6.6192	4.390	2.546	1.534	1.142	1.302	469.78	501.64	0.0433
10	2.1382	6.5821	4.407	2.594	1.494	1.141	1.300	470.76	502.64	0.0423
12	2.2161	6.5194	4.440	2.688	1.427	1.139	1.298	472.23	504.21	0.0405
14	2.2837	6.4651	4.472	2.777	1.373	1.137	1.296	473.18	505.33	0.0389
16	2.3436	6.4175	4.504	2.862	1.329	1.134	1.294	473.78	506.12	0.0375
18	2.3976	6.3751	4.534	2.944	1.291	1.132	1.293	474.09	506.65	0.0362
20	2.4469	6.3367	4.564	3.025	1.259	1.129	1.291	474.18	506.98	0.0350
25	2.5543	6.2536	4.640	3.219	1.193	1.123	1.288	473.71	507.16	0.0323
30	2.6455	6.1837	4.716	3.407	1.143	1.117	1.284	472.51	506.65	0.0300
35	2.7253	6.1229	4.792	3.593	1.102	1.111	1.281	470.80	505.66	0.0280
40	2.7965	6.0685	4.870	3.781	1.069	1.104	1.278	468.72	504.29	0.0261
45	2.8612	6.0191	4.951	3.972	1.040	1.097	1.275	466.31	502.68	0.0244
50	2.9206	5.9735	5.034	4.168	1.016	1.091	1.272	463.67	500.73	0.0229
60	3.0273	5.8908	5.211	4.582	0.975	1.077	1.266	457.77	496.33	0.0201
70	3.1219	5.8162	5.405	5.035	0.942	1.063	1.260	451.21	491.31	0.0177
80	3.2076	5.7471	5.621	5.588	0.915	1.048	1.254	444.12	485.80	0.0156
90	3.2867	5.6820	5.865	6.100	0.892	1.033	1.249	436.50	479.90	0.0136
100	3.3606	5.6198	6.142	6.738	0.872	1.016	1.244	428.24	473.67	0.0119
110	3.4304	5.5595	6.463	7.480	0.855	0.998	1.239	419.20	467.13	0.0103
120	3.4972	5.5002	6.838	8.384	0.840	0.978	1.236	409.38	460.25	0.0089
130	3.5616	5.4408	7.286	9.539	0.826	0.956	1.234	398.90	453.00	0.0076
140	3.6243	5.3803	7.834	11.07	0.813	0.935	1.232	388.00	445.34	0.0064
150	3.6859	5.3178	8.529	13.06	0.802	0.916	1.233	377.00	437.29	0.0053
160	3.7471	5.2531	9.456	15.59	0.792	0.901	1.235	366.24	428.89	0.0043
170	3.8197	5.1855	11.30	17.87	0.782	0.867	1.240	351.19	420.07	0.0034
180	3.8765	5.1128	12.82	21.43	0.773	0.838	1.248	336.35	410.39	0.0026
190	3.9429	5.0332	15.76	27.47	0.765	0.808	1.260	320.20	399.87	0.0018
200	4.0149	4.9412	22.05	39.31	0.758	0.756	1.280	298.10	387.81	0.0011

[a] v = specific volume (m³/kg); h = specific enthalpy (kJ/kg); s = specific entropy (kJ/kg · K); c_p = specific heat at constant pressure (kJ/kg · K); η = viscosity (10^{-4} Pa · sec); λ = thermal conductivity (W/m · K); Pr = Prandtl number; $\gamma = c_p/c_v$ ratio; $\bar{v}_s$ = velocity of sound (m/sec); τ = surface tension (N/m); f' = wet saturated vapor; g = saturated vapor. Rounded off from values of C. M. Tseng, T. A. Hamp, and E. O. Moeck, Atomic Energy of Canada Report AECL-5910, 1977.
[b] The notation 1.0434.−3 signifies 1.0434×10^{-3}.

Table 52.25 Specific Heat (kJ/kg · K) at Constant Pressure of Saturated Liquids

Substance	Temperature (K)															
	250	260	270	280	290	300	310	320	330	340	350	360	370	380	390	400
Acetic acid	—[a]	—	—	—	2.03	2.06	2.09	2.12	2.16	2.19	2.23	2.26	2.29	2.33	2.36	2.39
Acetone	2.05	2.07	2.10	2.13	2.16	2.19	2.22	2.26	2.30	2.35	2.40					
Ammonia	4.48	4.54	4.60	4.66	4.73	4.82	4.91	5.02	5.17	5.37	5.64	6.04	6.68	7.80	10.3	21
Aniline	—	—	2.03	2.04	2.05	2.07	2.10	2.13	2.16	2.19	2.22	2.26	2.31	2.38	2.47	2.58
Benzene	—	—	—	1.69	1.71	1.73	1.75	1.78	1.81	1.84	1.87	1.91	1.94	1.98	2.03	2.09
n-Butane	2.19	2.23	2.27	2.32	2.37	2.43	2.50	2.58	2.67	2.76	2.86	2.97	3.08			
Butanol	2.13	2.17	2.22	2.27	2.33	2.38	2.44	2.51	2.58	2.65	2.73	2.82	2.93	3.06	3.20	3.36
Carbon tetrachloride	0.833	0.838	0.843	0.848	0.853	0.858	0.864	0.870	0.879	0.891	0.912	0.941	0.975			
Chlorobenzene	1.29	1.31	1.32	1.32	1.33	1.33	1.34	1.36	1.38	1.40	1.42	1.44	1.46	1.47	1.49	1.51
m-Cresol	—	—	—	—	2.04	2.07	2.11	2.14	2.18	2.21	2.24	2.27	2.30	2.32	2.35	2.38
Ethane	2.97	3.20	3.50	4.00	5.09	9.92	—									
Ethanol	—	2.24	2.28	2.33	2.38	2.45	2.54	2.64	2.75	2.86	2.99	3.12	3.26	3.41	3.56	3.72
Ethyl acetate	—	—	—	1.89	1.92	1.94	1.97	2.00	2.03	2.06	2.09	2.13	2.16	2.20	2.24	2.28
Ethyl sulfide	1.96	1.97	1.97	1.98	2.00	2.01	2.02	2.03								
Ethylene	3.25	3.78	5.0													
Formic acid	—	—	—	—	2.15	2.16	2.17	2.18	2.20	2.22	2.24	2.26	2.28	2.30	2.33	2.36
Heptane	2.08	2.10	2.13	2.17	2.20	2.24	2.28	2.32	2.36	2.41	2.45	2.49	2.54	2.59	2.64	2.70
Hexane	2.09	2.12	2.15	2.19	2.22	2.26	2.31	2.36	2.41	2.46	2.51	2.56	2.62	2.69	2.76	2.83
Methanol	2.31	2.34	2.37	2.41	2.46	2.52	2.58	2.65	2.73	2.82	2.91	3.01	3.12	3.24	3.36	3.49
Methyl formate					2.16	2.16										
Octane	2.07	2.10	2.13	2.16	2.19	2.22	2.26	2.31	2.35	2.39	2.43	2.47	2.52	2.57	2.62	2.69
Oil, linseed	1.58	1.61	1.65	1.69	1.73	1.78	1.82	1.87	1.91	1.95	1.99	2.03	2.08	2.13	2.17	2.21
Oil, olive	1.90	1.92	1.95	1.98	2.01	2.05	2.09	2.13	2.16	2.20	2.24	2.28	2.32	2.37	2.41	2.46
Pentane	1.96	2.02	2.08	2.14	2.21	2.28	2.35	2.42	2.49	2.56	2.63	2.70	2.77	2.84	2.91	2.98
Propane	2.35	2.41	2.48	2.56	2.65	2.76	2.89	3.06	3.28	3.62	4.23	5.98				
Propanol	2.04	2.10	2.16	2.24	2.32	2.41	2.51	2.62	2.74	2.86	2.99	3.12	3.26	3.40	3.55	3.71
Propylene	2.22	2.27	2.34	2.43	2.55	2.69	2.87	3.12	3.44	3.92	4.75	6.75				
Sulfuric acid	—	—	—	—	1.39	1.41	1.43	1.45	1.46	1.48	1.50	1.51	1.53	1.54	1.56	1.57
Sulfur dioxide	1.31	1.32	1.33	1.34	1.36	1.39	1.42	1.46	1.50	1.55	1.61	1.68	1.76	1.85	1.99	2.14
Turpentine	1.60	1.65	1.70	1.75	1.80	1.85	1.90	1.95	2.00	2.05	2.10	2.15	2.20	2.25	2.30	2.35

[a] Dashes indicate inaccessible states.

Table 52.26 Ratio of Principal Specific Heats, c_p/c_v, for Liquids and Gases at Atmospheric Pressure

Substance	Temperature (K)															
	200	220	240	260	280	300	320	340	360	380	400	420	440	460	480	500
Acetylene	1.313	1.294	1.277	1.261	1.247	1.234	1.222	1.211	1.200							
Air	1.399	1.399	1.399	1.399	1.399	1.399	1.399	1.398	1.397	1.396	1.395	1.394	1.392	1.390	1.388	1.386
Ammonia							1.307	1.299	1.291	1.284	1.278	1.271	1.265	1.260	1.255	1.249
Argon	1.663	1.665	1.666	1.666	1.666	1.666	1.666	1.666	1.666	1.666	1.666	1.666	1.666	1.666	1.666	1.666
i-Butane	1.357	1.356	1.357	1.362	1.113	1.103	1.096	1.089	1.084	1.079	1.075	1.071	1.068	1.065	1.063	1.060
n-Butane	1.418	1.413	1.409	1.406	1.112	1.103	1.096	1.089	1.084	1.079	1.075	1.072	1.069	1.066	1.063	1.061
Carbon dioxide		1.350	1.332	1.317	1.303	1.290	1.282	1.273	1.265	1.258	1.253	1.247	1.242	1.238	1.233	1.229
Carbon monoxide	1.405	1.404	1.403	1.402	1.402	1.401	1.401	1.400	1.399	1.398	1.396	1.395	1.393	1.391	1.389	1.387
Ethane		1.250	1.233	1.219	1.205	1.193	1.182	1.172	1.163	1.155	1.148	1.141	1.135	1.130	1.125	1.120
Ethylene				1.268	1.251	1.236	1.223	1.212	1.201	1.192	1.183	1.175	1.168	1.162	1.156	1.151
Fluorine	1.393	1.387	1.380	1.374	1.368	1.362										
Helium	1.667	1.667	1.667	1.667	1.667	1.667	1.667	1.667	1.667	1.667	1.667	1.667	1.667	1.667	1.667	1.667
n-Hydrogen	1.439	1.428	1.418	1.413	1.409	1.406	1.403	1.402	1.401	1.400	1.399	1.398	1.398	1.398	1.397	1.397
Krypton	1.649	1.654	1.657	1.659	1.661	1.662	1.662	1.662	1.662	1.662	1.662	1.663	1.663	1.664	1.666	1.667
Methane	1.337	1.333	1.328	1.322	1.314	1.306	1.296	1.287	1.278	1.268	1.258	1.249	1.241	1.233	1.226	1.219
Neon	1.667	1.667	1.667	1.667	1.667	1.667	1.667	1.667	1.667	1.667	1.667	1.667	1.667	1.667	1.667	1.667
Nitrogen	1.399	1.399	1.399	1.399	1.399	1.399	1.399	1.399	1.398	1.398	1.397	1.396	1.395	1.393	1.392	1.391
Oxygen	1.398	1.397	1.397	1.396	1.395	1.394	1.392	1.389	1.387	1.384	1.381	1.378	1.375	1.371	1.368	1.365
Propane	1.513	1.506	1.173	1.158	1.145	1.135	1.126	1.118	1.111	1.105	1.100	1.095	1.091	1.087	1.084	1.081
Propylene			1.133	1.122	1.111	1.101	1.091	1.082	1.072	1.063	1.055	1.046	1.038	1.030	1.023	1.017
R12				1.159	1.148	1.139	1.132	1.127	1.122	1.118	1.115	1.112	1.110	1.108	1.106	1.104
R21			1.171			1.179	1.165	1.156	1.148	1.142	1.137	1.133	1.129	1.126	1.123	1.120
R22					1.204	1.190	1.174	1.167	1.160	1.153	1.148	1.144	1.140	1.136	1.132	1.129
Steam									1.323	1.321	1.319	1.317	1.315	1.313	1.311	1.309
Xenon	1.623	1.632	1.639	1.643	1.651	1.655	1.658	1.661	1.662	1.662	1.662	1.662	1.662	1.662	1.662	1.662

Table 52.27 Surface Tension (N/m) of Liquids

Substance	Temperature (K)															
	250	260	270	280	290	300	310	320	330	340	350	360	370	380	390	400
Acetone	0.0291	0.0279	0.0266	0.0253	0.0240	0.0228	0.0214	0.0201	0.0187	0.0174	0.0162	0.0150	0.0139	0.0128	0.0117	0.0106
Ammonia	0.0317	0.0294	0.0271	0.0248	0.0226	0.0203	0.0181	0.0159	0.0138	0.0117	0.0099	0.0080	0.0059	0.0040	0.0021	0.0003
Benzene			0.0320	0.0306	0.0292	0.0278	0.0265	0.0252	0.0239	0.0227	0.0215	0.0203	0.0191	0.0179	0.0167	0.0155
Butane	0.0177	0.0165	0.0153	0.0141	0.0129	0.0116	0.0104	0.0092	0.0080	0.0069	0.0059	0.0049	0.0040	0.0031	0.0023	0.0016
CO_2	0.0092	0.0071	0.0051	0.0032	0.0016	0.0003	—									
Chlorine	0.0244	0.0228	0.0213	0.0198	0.0183	0.0168	0.0153	0.0138	0.0123	0.0108	0.0094	0.0080	0.0066	0.0052	0.0044	0.0037
Ethane	0.0059	0.0047	0.0035	0.0024	0.0013	0.0005	—									
Ethanol	0.0271	0.0261	0.0251	0.0242	0.0232	0.0223	0.0214	0.0205	0.0196	0.0186	0.0177	0.0167	0.0158	0.0148	0.0137	0.0126
Ethylene	0.0032	0.0019	0.0009	0.0002	—											
Heptane	0.0244	0.0234	0.0224	0.0214	0.0205	0.0195	0.0185	0.0176	0.0166	0.0156	0.0147	0.0137	0.0127	0.0118	0.0109	0.0099
Hexane	0.0229	0.0218	0.0207	0.0197	0.0186	0.0175	0.0165	0.0154	0.0145	0.0135	0.0125	0.0115	0.0106	0.0096	0.0086	0.0076
Mercury	0.474	0.472	0.470	0.468	0.466	0.464	0.462	0.460	0.458	0.456	0.454	0.452	0.450	0.448	0.446	0.444
Methanol						0.0223	0.0214	0.0205	0.0196	0.0187	0.0178	0.0168	0.0159	0.0149	0.0139	0.0128
Octane	0.0251	0.0243	0.0234	0.0225	0.0216	0.0207	0.0197	0.0188	0.0179	0.0170	0.0161	0.0152	0.0143	0.0135	0.0127	0.0120
Propane	0.0132	0.0118	0.0104	0.0091	0.0079	0.0067	0.0056	0.0046	0.0037	0.0028	0.0018	0.0009				
Propylene	0.0133	0.0120	0.0106	0.0091	0.0078	0.0065	0.0053	0.0042	0.0032	0.0023	0.0014	0.0006				
R12	0.0148	0.0135	0.0122	0.0109	0.0096	0.0083	0.0070	0.0058	0.0047	0.0036	0.0027	0.0018	0.0010	0.0003		
R13	0.0057	0.0042	0.0029	0.0018	0.0009	0.0003										
Toluene	0.0342	0.0327	0.0312	0.0298	0.0285	0.0272	0.0260	0.0249	0.0236	0.0225	0.0214	0.0203	0.0193	0.0183	0.0173	0.0163
Water			—	0.0746	0.0732	0.0716	0.0699	0.0684	0.0666	0.0650	0.0636	0.0614	0.0595	0.0575	0.0555	0.0535

Table 52.28 Thermal Conductivity (W/m · K) of Saturated Liquids

Substance	Temperature (K)															
	250	260	270	280	290	300	310	320	330	340	350	360	370	380	390	400
Acetic acid	—	—	—	—	0.165	0.164	0.162	0.161	0.160	0.158	0.157	0.156	0.154	0.153	0.152	0.150
Acetone	0.179	0.175	0.171	0.167	0.163	0.160	0.156	0.152	0.149	0.145	0.141	0.137	0.133	0.130	0.126	0.122
Ammonia	0.562	0.541	0.522	0.503	0.484	0.466	0.445	0.424	0.403	0.382	0.359	0.335	0.309	0.284	0.252	0.227
Aniline	—	—	0.175	0.175	0.174	0.173	0.172	0.172	0.171	0.170	0.169	0.168	0.167	0.166	0.165	0.165
Benzene	—	—	—	0.151	0.149	0.145	0.141	0.138	0.135	0.132	0.129	0.126	0.123	0.120	0.117	0.114
Butane	0.130	0.126	0.122	0.118	0.115	0.111	0.108	0.104	0.101	0.097	0.094	0.090	0.086	0.082	0.078	0.075
Butanol	0.161	0.159	0.157	0.155	0.153	0.151	0.149	0.147	0.145	0.143	0.141	0.139	0.138	0.136	0.135	0.133
Carbon tetrachloride	0.114	0.112	0.110	0.107	0.105	0.103	0.100	0.098	0.096	0.094	0.092	0.089	0.087	0.085	0.083	0.080
Chlorobenzene	0.137	0.135	0.133	0.131	0.129	0.127	0.125	0.123	0.121	0.119	0.117	0.115	0.113	0.111	0.109	0.107
m-Cresol	—	—	—	—	0.150	0.149	0.148	0.148	0.147	0.146	0.146	—	—	—	—	—
Ethane	0.105	0.098	0.091	0.084	0.074	0.065	—	—	—	—	—	—	—	—	—	—
Ethanol	—	0.178	0.176	0.173	0.169	0.166	0.163	0.160	0.158	0.155	0.153	0.151	0.149	0.147	0.145	0.143
Ethyl acetate	0.158	0.155	0.152	0.149	0.146	0.143	0.140	0.137	0.134	0.131	0.128	0.124	0.121	0.118	0.115	0.112
Ethyl sulfide	0.144	0.142	0.140	0.138	0.136	0.134	—	—	—	—	—	—	—	—	—	—
Ethylene	0.100	0.090	0.078	0.064	—	—	—	—	—	—	—	—	—	—	—	—
Formic acid	—	—	—	—	0.293	0.279	0.255	0.241	—	—	—	—	—	—	—	—
Heptane	0.138	0.136	0.133	0.130	0.127	0.125	0.122	0.119	0.117	0.114	0.111	0.108	0.106	0.104	0.102	0.100
Hexane	0.138	0.135	0.132	0.129	0.125	0.122	0.119	0.116	0.112	0.110	0.107	0.104	0.101	0.099	0.096	0.093
Methanol	0.218	0.215	0.211	0.208	0.205	0.202	0.199	0.196	0.194	0.191	0.188	0.186	0.183	0.180	0.178	0.176
Methyl formate	0.202	0.199	0.196	0.193	0.190	0.187	0.184	0.181	0.178	0.175	0.172	—	—	—	—	—
Octane	0.141	0.138	0.136	0.133	0.130	0.128	0.125	0.123	0.120	0.118	0.115	0.112	0.110	0.107	0.105	0.103
Oil, linseed	0.175	0.173	0.171	0.169	0.167	0.165	0.163	0.161	0.159	0.157	0.154	0.152	0.150	0.148	0.146	0.144
Oil, olive	0.170	0.170	0.169	0.169	0.168	0.168	0.167	0.167	0.166	0.165	0.165	0.164	0.163	0.163	0.162	0.161
Pentane	0.133	0.128	0.124	0.120	0.116	0.112	0.108	0.106	0.102	0.098	0.094	0.090	0.086	0.082	0.078	0.074
Propane	0.120	0.116	0.112	0.107	0.102	0.097	0.093	0.088	0.084	0.079	0.075	0.070	—	—	—	—
Propanol	0.166	0.163	0.161	0.158	0.156	0.154	0.152	0.150	0.148	0.146	0.144	0.142	0.141	0.139	0.138	0.136
Propylene	0.137	0.131	0.127	0.122	0.118	0.111	0.107	0.101	0.096	0.090	0.083	0.077	—	—	—	—
Sulfuric acid	—	—	—	—	0.324	0.329	0.332	0.336	0.339	0.342	0.345	0.349	0.352	0.355	0.358	0.361
Sulfur dioxide	0.227	0.221	0.214	0.208	0.202	0.196	0.190	0.184	0.172	0.158	—	—	—	—	—	—
Turpentine	—	—	—	0.128	0.127	0.127	0.126	—	—	—	—	—	—	—	—	—

Table 52.29 Viscosity (10^{-4} Pa · sec) of Saturated Liquids

Substance	Temperature (K)															
	250	260	270	280	290	300	310	320	330	340	350	360	370	380	390	400
Acetic acid	—	—	—	—	13.1	11.3	9.77	8.49	7.40	6.48	5.70	5.03	4.45	3.95	3.52	3.15
Acetone	5.27	4.63	4.11	3.68	3.32	3.01	2.73	2.49	2.28	2.10	1.94	—	—	—	—	—
Ammonia	2.20	1.94	1.74	1.58	1.44	1.31	1.20	1.08	0.98	0.88	0.78	0.70	0.64	0.58	0.53	0.48
Aniline	—	—	96.9	71.6	53.5	40.3	30.7	23.6	18.2	14.2	11.2	8.84	7.03	5.63	4.54	3.68
Benzene	—	—	—	8.08	6.80	5.84	5.13	4.52	4.05	3.60	3.29	2.98	2.70	2.45	2.24	2.04
Butane	2.63	2.37	2.15	1.95	1.77	1.61	1.56	1.33	1.21	1.10	1.00	0.90	0.81	0.73	0.66	0.60
Butanol	94.2	70.9	53.9	41.4	32.1	25.1	19.8	15.7	12.6	10.1	8.22	6.71	5.50	4.53	3.75	3.13
Carbon tetrachloride	—	—	14.0	11.9	10.2	8.78	7.70	6.81	6.06	5.43	4.89	4.41	4.00	3.65	3.34	3.07
Chlorobenzene	13.9	12.2	10.7	9.50	8.44	7.53	6.74	6.05	5.46	4.93	4.47	4.07	3.70	3.39	3.10	2.85
m-Cresol	—	—	—	—	261	136	72.0	39.0	21.5	12.1	6.90	—	—	—	—	—
Ethane	0.81	0.70	0.60	0.51	0.42	0.35	—	—	—	—	—	—	—	—	—	—
Ethanol	—	23.3	18.9	15.5	12.8	10.6	8.81	7.39	6.24	5.29	4.51	3.86	3.31	2.86	2.48	2.15
Ethyl acetate	7.31	6.53	5.86	5.28	4.78	4.33	3.95	3.60	3.30	3.03	2.79	2.57	2.38	2.20	2.04	1.90
Ethyl sulfide	7.69	6.68	5.86	5.19	4.64	4.18	3.78	3.45	3.16	2.92	2.70	2.51	2.34	2.20	2.06	1.95
Ethylene	0.582	0.505	0.418	0.31	—	—	—	—	—	—	—	—	—	—	—	—
Formic acid	48.3	37.4	29.5	23.6	19.2	15.9	13.3	11.2	9.56	8.24	7.16	6.27	5.53	4.91	4.39	3.94
Heptane	7.25	6.25	5.46	4.82	4.28	3.83	3.45	3.14	2.87	2.62	2.39	2.20	2.03	1.87	1.72	1.59
Hexane	5.03	4.46	3.97	3.57	3.22	2.93	2.68	2.45	2.24	2.07	1.92	1.78	1.63	1.50	1.38	1.26
Methanol	12.3	10.1	8.43	7.15	6.14	5.32	4.60	4.09	3.60	3.19	2.84	2.53	2.28	2.06	1.85	1.65
Methyl formate	5.71	5.04	4.50	4.04	3.66	3.34	3.06	2.82	2.61	2.43	2.27	2.13	2.01	1.89	1.79	1.70
Octane	10.5	8.80	7.49	6.45	5.66	5.01	4.48	4.03	3.65	3.33	3.05	2.80	2.58	2.38	2.18	2.02
Oil, linseed	2300	1500	1000	700	490	356	263	198	152	118	93.5	74.8	60.6	49.6	41.1	34.3
Oil, olive	8350	4600	2600	1600	1000	630	410	278	193	136	98.3	72.2	54.0	40.9	31.0	24.6
Pentane	3.50	3.16	2.87	2.62	2.39	2.20	2.03	1.87	1.73	1.59	1.46	1.34	1.22	1.11	1.01	0.92
Propane	1.71	1.53	1.37	1.23	1.10	0.98	0.87	0.76	0.66	0.58	0.51	0.45	—	—	—	—
Propanol	67.8	50.0	37.7	29.0	22.7	18.0	14.6	11.9	9.87	8.27	6.99	5.97	5.14	4.46	3.90	3.44
Propylene	1.41	1.27	1.16	1.05	0.94	0.84	0.75	0.67	0.60	0.53	0.47	0.42	—	—	—	—
Sulfuric acid	—	—	—	363	259	189	141	107	82.6	64.7	51.4	41.4	33.7	27.7	23.1	19.4
Sulfur dioxide	5.32	4.53	3.90	3.39	2.85	2.42	2.09	1.83	1.62	1.45	1.30	1.16	1.02	0.89	0.77	0.66
Turpentine	36.3	28.8	23.3	19.1	15.9	13.4	11.4	9.80	8.50	7.44	6.56	5.83	5.21	4.68	4.23	3.85

Table 52.30   Thermochemical Properties at 1.013 bar, 298.15 K

Substance	Formula	ΔH_f° (kJ/kg · mol)	ΔG_f° (kJ/kg · mol)	S° (kJ/kg · mol · K)
Acetaldehyde	$C_2H_4O(g)$	−166,000	−132,900	265.2
Acetic acid	$C_2H_4O_2(g)$	−436,200	+315,500	282.5
Acetone	$C_3H_6O(l)$	−248,000	−155,300	200.2
Acetylene	$C_2H_2(g)$	+226,740	+209,190	200.8
Ammonia	$NH_3(g)$	−46,190	−16,590	192.6
Aniline	$C_6H_7N(l)$	+35,300	+153,200	191.6
Benzene	$C_6H_6(l)$	+49,030	+117,000	172.8
Butanol	$C_4H_{10}O(l)$	−332,400	−168,300	227.6
n-Butane	$C_4H_{10}(l)$	−105,900		
n-Butane	$C_4H_{10}(g)$	−126,200	−17,100	310.1
i-Butane	$C_4H_{10}(g)$	−134,500	−20,900	294.6
Carbon dioxide	$CO_2(g)$	−393,510	−394,390	213.7
Carbon disulfide	$CS_2(g)$	−109,200		237.8
Carbon monoxide	$CO(g)$	−110,520	−137,160	197.6
Carbon tetrachloride	$CCl_4(g)$	−103,000	−66,100	311.3
Carbon tetrafluoride	$CF_4(g)$	−921,300	−878,200	261.5
Chloroform	$CHCl_3(g)$	−104,000	−70,500	295.6
Cyclohexane	$C_6H_{12}(g)$	−123,100	+31,800	298.2
Cyclopropane	$C_3H_6(g)$	+53,300	+104,300	237.7
n-Decane	$C_{10}H_{22}(l)$	−332,600	−17,500	425.5
Diphenyl	$C_{12}H_{10}(g)$	−172,800	−283,900	348.5
Ethane	$C_2H_6(g)$	−84,670	−32,900	229.5
Ethanol	$C_2H_6O(g)$	−235,200	−168,700	282.7
Ethanol	$C_2H_6O(l)$	−277,600	−174,600	160.7
Ethyl acetate	$C_4H_8O_2(g)$	−432,700	−325,800	376.8
Ethyl chloride	$C_2H_5Cl(g)$	−107,600	−55,500	274.8
Ethyl ether	$C_4H_{10}O(g)$	−250,800	−118,400	352.5
Ethylene	$C_2H_4(g)$	+52,280	+68,130	219.4
Ethylene oxide	$C_2H_4O(g)$	−38,500	−11,600	242.9
Heptane	$C_7H_{16}(g)$	−187,800	−427,800	166.0
Hexane	$C_8H_{18}(g)$	−208,400	16,500	270.7
Hydrazine	$N_2H_4(l)$	−50,600	−149,200	121.2
Hydrazine	$N_2H_4(g)$	+95,400	+159,300	238.4
Hydrogen peroxide	$H_2O_2(l)$	−187,500	−120,400	109.6
Methane	$CH_4(g)$	−74,840	−50,790	186.2
Methanol	$CH_4O(l)$	−238,600	−126,800	81.8
Methanol	$CH_4O(g)$	−200,900	−162,100	247.9
Methyl acetate	$C_3H_6O_2(l)$	−444,300		
Methyl bromide	$CH_3Br(g)$	−36,200	−25,900	246.1
Methyl chloride	$CH_3Cl(g)$	−86,300	−63,000	234.2
Methyl formate	$C_2H_4O_2(g)$	−335,100	−301,000	292.8
Methylene chloride	$CH_2Cl_2(g)$	−94,000	−67,000	270.2
Naphthalene	$C_{10}H_8(g)$	−151,500	−224,200	336.5
Nitric oxide	$NO(g)$	+90,300	−86,600	210.6
Nitrogen peroxide	$NO_2(g)$	+33,300		240.0
Nitrous oxide	$N_2O(g)$	+82,000	+104,000	219.9
Octane	$C_8H_{18}(l)$	−250,000	6,610	360.8

Table 52.30 (Continued)

Substance	Formula	ΔH_f° (kJ/kg · mol)	ΔG_f° (kJ/kg · mol)	S° (kJ/kg · mol · K)
Octane	$C_8H_{18}(g)$	$-208{,}400$	$16{,}500$	466.7
i-Pentane	$C_5H_{12}(g)$	$-154{,}500$	$-14{,}600$	343.6
n-Pentane	$C_5H_{12}(g)$	$-146{,}400$	$-8{,}370$	348.9
Propane	$C_3H_8(g)$	$-103{,}800$	$-107{,}200$	269.9
Propanol	$C_3H_8(g)$	$-258{,}800$	$-164{,}100$	322.6
Propylene	$C_3H_6(g)$	$+20{,}400$	$+62{,}700$	267.0
R11	$CFCl_3(g)$	$-284{,}500$	$-238{,}000$	309.8
R12	$CCl_2F_2(g)$	$-468{,}600$	$-439{,}300$	300.9
R13	$CClF_3(g)$	$-715{,}500$	$-674{,}900$	285.6
R13B1	$CF_3Br(g)$	$-642{,}700$	$-616{,}300$	297.6
R23	$CHF_3(g)$	$-682{,}000$	$-654{,}900$	259.6
Sulfur dioxide	$SO_2(g)$	$-296{,}900$	$-300{,}200$	248.1
Sulfur hexafluoride	$SF_6(g)$	$-1{,}207{,}900$	$-1{,}105{,}000$	291.7
Toluene	$C_7H_8(g)$	$-50{,}000$	$-122{,}300$	320.2
Water	$H_2O(l)$	$-285{,}830$	$-237{,}210$	70.0
Water	$H_2O(g)$	$-241{,}820$	$-228{,}600$	188.7

Table 52.31 Ideal Gas Sensible Enthalpies (kJ/kg · mol) of Common Products of Combustion[a,b]

T (K)	CO	CO_2	$H_2O(g)$	NO	NO_2	N_2	O_2	SO_2
200	-2858	-3414	-3280	-2950	-3494	-2855	-2860	-3736
220	-2276	-2757	-2613	-2345	-2803	-2270	-2279	-3008
240	-1692	-2080	-1945	-1749	-2102	-1685	-1698	-2266
260	-1110	-1384	-1277	-1148	-1390	-1105	-1113	-1508
280	-529	-667	-608	-547	-668	-525	-529	-719
300	54	67	63	54	67	55	60	75
320	638	822	736	652	813	636	650	882
340	1221	1594	1411	1248	1573	1217	1238	1702
360	1805	2383	2088	1847	2344	1800	1832	2541
380	2389	3187	2768	2444	3130	2383	2429	3387
400	2975	4008	3452	3042	3929	2967	3029	4251
420	3562	4841	4138	3641	4739	3554	3633	5123
440	4152	5689	4828	4247	5560	4147	4241	6014
460	4642	6550	5522	4849	6395	4737	4853	6923
480	5334	7424	6221	5054	7243	5329	5468	7827
500	5929	8314	6920	6058	8100	5921	6088	8749
550	7427	10580	8695	7589	10295	7401	7655	11180
600	8941	12915	10500	9146	12555	8902	9247	13545
650	10475	15310	12325	10720	14875	10415	10865	16015
700	12020	17760	14185	12310	17250	11945	12500	18550
750	13590	20270	16075	13890	19675	13490	14160	21150
800	15175	22820	17990	15550	22140	15050	15840	23720
850	16780	25410	19945	17200	24640	16630	17535	26390
900	18400	28040	21920	18860	27180	18220	19245	29020
950	20030	30700	23940	20540	29740	19835	20970	31700
1000	21690	33410	25980	22230	32340	21460	22710	34430

Table 52.31 *(Continued)*

T (K)	CO	CO$_2$	H$_2$O(g)	NO	NO$_2$	N$_2$	O$_2$	SO$_2$
1050	23350	36140	28060	23940	34950	23100	24460	37180
1100	25030	38890	30170	25650	37610	24750	26180	39920
1150	26720	41680	32310	27380	40260	26430	27990	42690
1200	28430	44480	34480	29120	42950	28110	29770	45460
1250	30140	47130	36680	30870	45650	29810	31560	48270
1300	31870	50160	38900	32630	48350	31510	33350	51070
1350	33600	53030	41130	34400	51090	33220	35160	53900
1400	35340	55910	43450	36170	53810	34940	36970	56720
1450	37090	58810	45770	37950	56560	36670	38790	59560
1500	38850	61710	48100	39730	59310	38410	40610	62400
1550	40610	64800	50460	41520	62070	40160	42440	65260
1600	42380	67580	52840	43320	64850	41910	44280	68120
1650	44156	70530	55240	45120	67640	43670	46120	71000
1700	45940	73490	57680	46930	70420	45440	47970	73870
1750	47727	76460	60130	48740	73220	47210	49830	76760
1800	49520	79440	62610	50560	76010	49000	51690	79640
1850	51320	82430	65100	52380	78810	50780	53560	82540
1900	53120	85430	67610	54200	81630	52570	55440	85440
1950	54930	88440	70140	56020	84450	54370	57310	88350
2000	56740	91450	72690	57860	87260	56160	59200	91250
2100	60380	97500	77830	61530	92910	59760	62990	97080
2200	64020	103570	83040	65220	98580	63380	66800	102930
2300	67680	109670	88290	68910	104260	67010	70630	108790
2400	71350	115790	93600	72610	109950	70660	74490	114670
2500	75020	121930	98960	76320	115650	74320	78370	120560
2600	78710	128080	104370	80040	121360	77990	82270	126460
2700	82410	134260	109810	83760	127080	81660	86200	132380
2800	86120	140440	115290	87490	132800	85360	90140	138300
2900	89830	146650	120810	91230	138540	89050	94110	144240
3000	93540	152860	126360	94980	144270	92750	98100	150180

^{a} Converted and usually rounded off from *JANAF Thermochemical Tables*, NSRDS-NBS-37, 1971.

^{b} To illustrate the term "sensible enthalpy," which is the difference between the actual enthalpy and the enthalpy at the reference temperature, 298.15 K (= 25°C = 77°F = 537°R), the magnitude of the heat transfer, in kJ/kg · mol fuel and in kJ/kg fuel, will be calculated for the steady-state combustion of acetylene in excess oxygen, the reactants entering at 298.15 K and the products leaving at 2000 K. All substances are in the gaseous phase.

The basic equation is

$$Q + W = \sum_P n_i \left(\Delta h_f^o + \Delta h_s\right)_i - \sum_R n_i \left(\Delta h_f^o + \Delta h_s\right)_i$$

where P signifies products and R reactants, s signifies sensible enthalpy, and the Δh_s are looked up in the table for the appropriate temperatures.

If the actual reaction was

$$C_2H_2 + 1\tfrac{1}{2} O_2 \rightarrow 2CO_2 + H_2O + 3O_2$$

then $W = 0$ and $Q = 2 (-393,510 + 91,450) + 1 (-241,810 + 72,690) + 3 (0 + 59,200) - (226,740 + 0) - 1\tfrac{1}{2} (0 + 0) = -604,120 + (-169,120) + 177,600 - 226,740 = -822,380$ kJ/kg mol. C$_2$H$_2$ $= -31,584$ kJ/kg C$_2$H$_2$. Had the fuel been burnt in air one would write the equation with an additional 3.76(5.5) N$_2$ on each side of the equation. In the above, the enthalpy of formation of the stable elements at 298.15 K has been set equal to zero. For further information, most undergraduate engineering thermodynamics texts may be consulted.

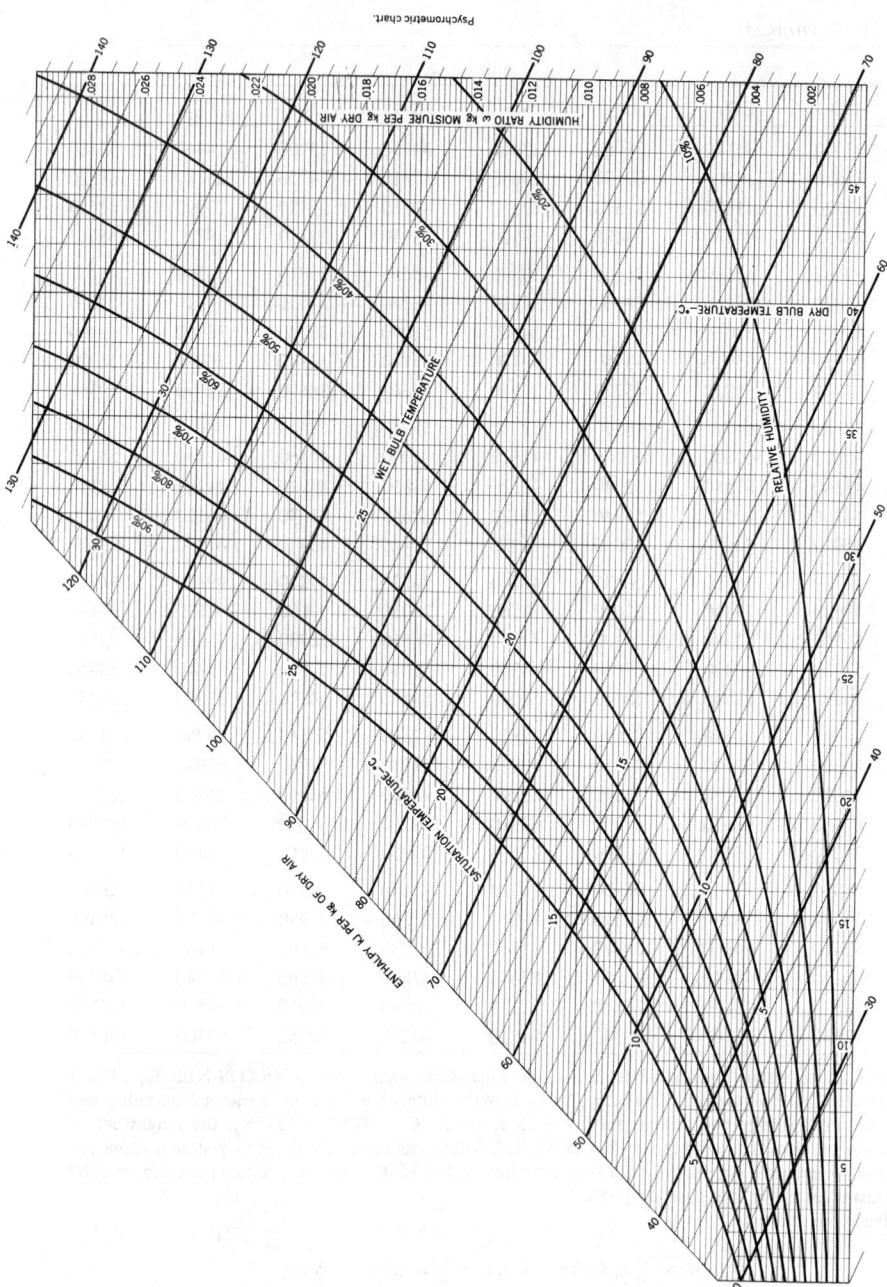

Fig. 52.4 Psychrometric chart. (ASHRAE chart No. 1, 1981. Prepared at the Center for Applied Thermodynamic Studies, University of Idaho, Moscow, ID. Copyright 1981 by the American Society of Heating, Refrigerating and Air Conditioning Engineers, Inc. and reproduced by permission of the copyright owner.)

CHAPTER 53

FLUID MECHANICS

REUBEN M. OLSON

Ohio University
College of Engineering and Technology
Athens, Ohio

All figures and tables produced, with permission, from *Essentials of Engineering Fluid Mechanics*, Fourth Edition, by Reuben M. Olson, copyright 1980, Harper & Row, Publishers.

53.1 DEFINITION OF A FLUID

A solid generally has a definite shape; a fluid has a shape determined by its container. Fluids include liquids, gases, and vapors, or mixtures of these. A fluid continuously deforms when shear stresses are present; it cannot sustain shear stresses at rest. This is characteristic of all real fluids, which are viscous. Ideal fluids are nonviscous (and nonexistent), but have been studied in great detail because in many instances viscous effects in real fluids are very small and the fluid acts essentially as a nonviscous fluid. Shear stresses are set up as a result of relative motion between a fluid and its boundaries or between adjacent layers of fluid.

53.2 IMPORTANT FLUID PROPERTIES

Density ρ and surface tension σ are the most important fluid properties for liquids at rest. Density and viscosity μ are significant for all fluids in motion; surface tension and vapor pressure are significant for cavitating liquids; and bulk elastic modulus K is significant for compressible gases at high subsonic, sonic, and supersonic speeds.

Sonic speed in fluids is $c = \sqrt{K/\rho}$. Thus, for water at 15°C, $c = \sqrt{2.18 \times 10^9/999} = 1480$ m/sec. For a mixture of a liquid and gas bubbles at nonresonant frequencies, $c_m = \sqrt{K_m/\rho_m}$, where m refers to the mixture. This becomes

$$c_m = \sqrt{\frac{p_g K_l}{[xK_l + (1-x)p_g][x\rho_g + (1-x)\rho_l]}}$$

where the subscript l is for the liquid phase and g is for the gas phase. Thus, for water at 20°C containing 0.1% gas nuclei by volume at atmospheric pressure, $c_m = 312$ m/sec. For a gas or a mixture of gases (such as air), $c = \sqrt{kRT}$, where $k = c_p/c_v$, R is the gas constant, and T is the absolute temperature. For air at 15°C, $c = \sqrt{(1.4)(287.1)(288)} = 340$ m/sec. This sonic property is thus a combination of two properties, density and elastic modulus.

Kinematic viscosity is the ratio of dynamic viscosity and density. In a Newtonian fluid, simple laminar flow in a direction x at a speed of u, the shearing stress parallel to x is $\tau_L = \mu(du/dy) = \rho\nu(du/dy)$, the product of dynamic viscosity and velocity gradient. In the more general case, $\tau_L = \mu(\partial u/\partial y + \partial v/\partial x)$ when there is also a y component of velocity v. In turbulent flows the shear stress resulting from lateral mixing is $\tau_T = -\rho u'v'$, a Reynolds stress, where u' and v' are instantaneous and simultaneous departures from mean values $\bar{u}$ and $\bar{v}$. This is also written as $\tau_T = \rho\epsilon(du/dy)$, where ϵ is called the turbulent eddy viscosity or diffusivity, an indirectly measurable flow parameter and not a fluid property. The eddy viscosity may be orders of magnitude larger than the kinematic viscosity. The total shear stress in a turbulent flow is the sum of that from laminar and from turbulent motion: $\tau = \tau_L + \tau_T = \rho(\nu + \epsilon)du/dy$ after Boussinesq.

53.3 FLUID STATICS

The differential equation relating pressure changes dp with elevation changes dz (positive upward parallel to gravity) is $dp = -\rho g\, dz$. For a constant-density liquid, this integrates to $p_2 - p_1 = -\rho g(z_2 - z_1)$ or $\Delta p = \gamma h$, where γ is in N/m³ and h is in m. Also $(p_1/\gamma) + z_1 = (p_2/\gamma) + z_2$; a constant piezometric head exists in a homogeneous liquid at rest, and since $p_1/\gamma - p_2/\gamma = z_2 - z_1$, a change in pressure head equals the change in potential head. Thus, horizontal planes are at constant pressure when body forces due to gravity act. If body forces are due to uniform linear accelerations or to centrifugal effects in rigid-body rotations, points equidistant below the free liquid surface are all at the same pressure. Dashed lines in Figs. 53.1 and 53.2 are lines of constant pressure.

Pressure differences are the same whether all pressures are expressed as gage pressure or as absolute pressure.

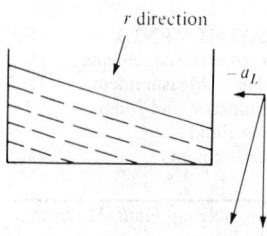

r direction

$-a_L$

a g **Fig. 53.1** Constant linear acceleration.

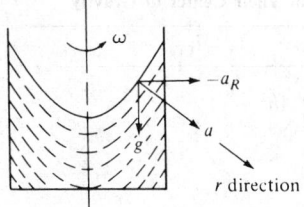

Fig. 53.2 Constant centrifugal acceleration.

53.3.1 Manometers

Pressure differences measured by barometers and manometers may be determined from the relation $\Delta p = \gamma h$. In a barometer, Fig. 53.3, $h_b = (p_a - p_v)/\gamma_b$ m.

An open manometer, Fig. 53.4, indicates the inlet pressure for a pump by $p_{\text{inlet}} = -\gamma_m h_m - \gamma y$ Pa gage. A differential manometer, Fig. 53.5, indicates the pressure drop across an orifice, for example, by $p_1 - p_2 = h_m(\gamma_m - \gamma_0)$ Pa.

Manometers shown in Figs. 53.3 and 53.4 are a type used to measure medium or large pressure differences with relatively small manometer deflections. Micromanometers can be designed to produce relatively large manometer deflections for very small pressure differences. The relation $\Delta p = \gamma \Delta h$ may be applied to the many commercial instruments available to obtain pressure differences from the manometer deflections.

53.3.2 Liquid Forces on Submerged Surfaces

The liquid force on any flat surface submerged in the liquid equals the product of the gage pressure at the centroid of the surface and the surface area, or $F = \bar{p}A$. The force F is not applied at the centroid for an inclined surface, but is always below it by an amount that diminishes with depth. Measured parallel to the inclined surface, $\bar{y}$ is the distance from 0 in Fig. 53.6 to the centroid and

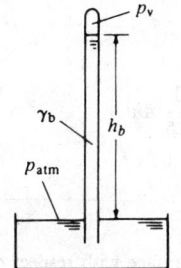

Fig. 53.3 Barometer.

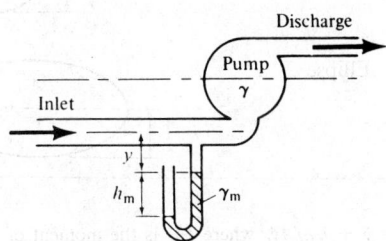

Fig. 53.4 Open manometer.

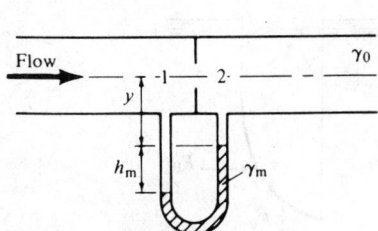

Fig. 53.5 Differential manometer.

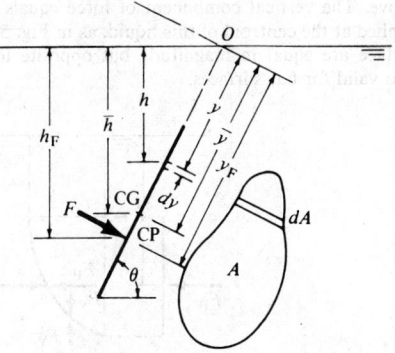

Fig. 53.6 Flat inclined surface submerged in a liquid.

Table 53.1 Moments of Inertia for Various Plane Surfaces about Their Center of Gravity

SURFACE		I_{CG}
Rectangle or square		$\dfrac{1}{12} Ah^2$
Triangle		$\dfrac{1}{18} Ah^2$
Quadrant of circle (or semicircle)		$\left(\dfrac{1}{4} - \dfrac{16}{9\pi^2}\right) Ar^2 = 0.0699\, Ar^2$
Quadrant of ellipse (or semiellipse)		$\left(\dfrac{1}{4} - \dfrac{16}{9\pi^2}\right) Aa^2 = 0.0699\, Aa^2$
Parabola		$\left(\dfrac{3}{7} - \dfrac{9}{25}\right) Ah^2 = 0.0686\, Ah^2$
Circle		$\dfrac{1}{16} Ad^2$
Ellipse		$\dfrac{1}{16} Ah^2$

$y_F = \bar{y} + I_{CG}/A\bar{y}$, where I_{CG} is the moment of inertia of the flat surface with respect to its centroid. Values for some surfaces are listed in Table 53.1.

For curved surfaces, the horizontal component of the force is equal in magnitude and point of application to the force on a projection of the curved surface on a vertical plane, determined as above. The vertical component of force equals the weight of liquid above the curved surface and is applied at the centroid of this liquid, as in Fig. 53.7. The liquid forces on opposite sides of a submerged surface are equal in magnitude but opposite in direction. These statements for curved surfaces are also valid for flat surfaces.

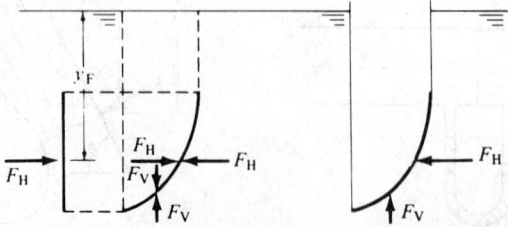

Fig. 53.7 Curved surfaces submerged in a liquid.

Buoyancy is the resultant of the surface forces on a submerged body and equals the weight of fluid (liquid or gas) displaced.

53.3.3 Aerostatics

The U.S. standard atmosphere is considered to be dry air and to be a perfect gas. It is defined in terms of the temperature variation with altitude (Fig. 53.8), and consists of isothermal regions and polytropic regions in which the polytropic exponent n depends on the lapse rate (temperature gradient).

Conditions at an upper altitude z_2 and at a lower one z_1 in an isothermal atmosphere are obtained by integrating the expression $dp = -\rho g\, dz$ to get

$$\frac{p_2}{p_1} = \exp \frac{-g(z_2 - z_1)}{RT}$$

In a polytropic atmosphere where $p/p_1 = (\rho/\rho_1)^n$,

$$\frac{p_2}{p_1} = \left[1 - g\frac{(n-1)}{n}\frac{(z_2 - z_1)}{RT_1}\right]^{n/(n-1)}$$

from which the lapse rate is $(T_2 - T_1)/(z_2 - z_1) = -g(n-1)/nR$ and thus n is obtained from $1/n = 1 + (R/g)(dt/dz)$. Defining properties of the U.S. standard atmosphere are listed in Table 53.2.

The U.S. standard atmosphere is used in measuring altitudes with altimeters (pressure gages) and, because the altimeters themselves do not account for variations in the air temperature beneath an aircraft, they read too high in cold weather and too low in warm weather.

53.3.4 Static Stability

For the *atmosphere* at rest, if an air mass moves very slowly vertically and remains there, the atmosphere is neutral. If vertical motion continues, it is unstable; if the air mass moves to return to its initial position, it is stable. It can be shown that atmospheric stability may be defined in terms of the polytropic exponent. If $n < k$, the atmosphere is stable (see Table 53.2); if $n = k$, it is neutral (adiabatic); and if $n > k$, it is unstable.

The stability of a body *submerged* in a fluid at rest depends on its response to forces which tend to tip it. If it returns to its original position, it is stable; if it continues to tip, it is unstable; and if it remains at rest in its tipped position, it is neutral. In Fig. 53.9 G is the center of gravity and B is the center of buoyancy. If the body in (a) is tipped to the position in (b), a couple Wd restores the

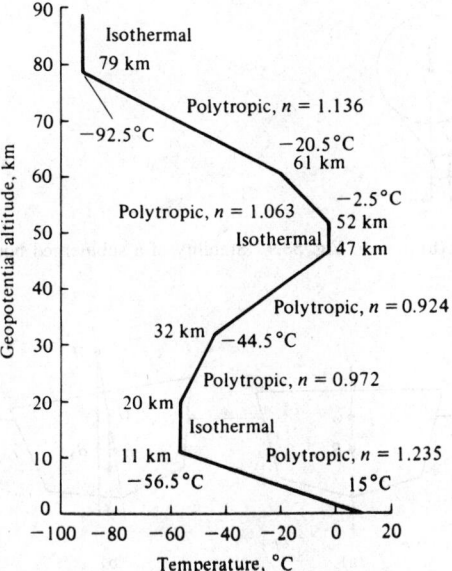

Fig. 53.8 U.S. standard atmosphere.

Table 53.2　Defining Properties of the U.S. Standard Atmosphere

ALTITUDE (m)	TEMPERATURE (°C)	TYPE OF ATMOSPHERE	LAPSE RATE (°C/km)	$\bar{g}$ (m/s²)	n	PRESSURE p(Pa)	DENSITY ρ(kg/m³)
0	15.0					1.013×10^5	1.225
		Polytropic	−6.5	9.790	1.235		
11,000	−56.5					2.263×10^4	3.639×10^{-1}
		Isothermal	0.0	9.759			
20,000	−56.5					5.475×10^3	8.804×10^{-2}
		Polytropic	+1.0	9.727	0.972		
32,000	−44.5					8.680×10^2	1.323×10^{-2}
		Polytropic	+2.8	9.685	0.924		
47,000	−2.5					1.109×10^2	1.427×10^{-3}
		Isothermal	0.0	9.654			
52,000	−2.5					5.900×10^1	7.594×10^{-4}
		Polytropic	−2.0	9.633	1.063		
61,000	−20.5					1.821×10^1	2.511×10^{-4}
		Polytropic	−4.0	9.592	1.136		
79,000	−92.5					1.038	2.001×10^{-5}
		Isothermal	0.0	9.549			
88,743	−92.5					1.644×10^{-1}	3.170×10^{-6}

body toward position (a) and thus the body is stable. If B were below G and the body displaced, it would move until B becomes above G. Thus stability requires that G is below B.

Floating bodies may be stable even though the center of buoyancy B is below the center of gravity G. The center of buoyancy generally changes position when a floating body tips because of the changing shape of the displaced liquid. The floating body is in equilibrium in Fig. 53.10a. In Fig. 53.10b the center of buoyancy is at B_1, and the restoring couple rotates the body toward its initial position in Fig. 53.10a. The intersection of BG is extended and a vertical line through B_1 is at M, the metacenter, and GM is the metacentric height. The body is stable if M is above G. Thus, the position of B relative to G determines stability of a submerged body, and the position of M relative to G determines the stability of floating bodies.

53.4　FLUID KINEMATICS

Fluid flows are classified in many ways. Flow is *steady* if conditions at a point do not vary with time, or for turbulent flow, if mean flow parameters do not vary with time. Otherwise the flow is *unsteady*. Flow is considered *one dimensional* if flow parameters are considered constant throughout

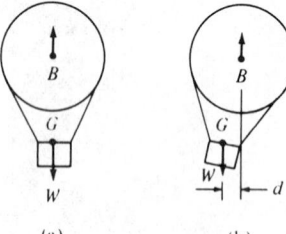

(a)　　　　　　(b)　　　　　Fig. 53.9　Stability of a submerged body.

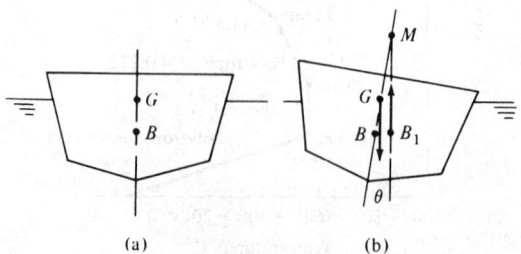

(a)　　　　　　　　　　　(b)

Fig. 53.10　Floating body.

a cross section, and variations occur only in the flow direction. *Two-dimensional* flow is the same in parallel planes and is not one dimensional. In *three-dimensional* flow gradients of flow parameters exist in three mutually perpendicular directions (*x, y,* and *z*). Flow may be *rotational* or *irrotational,* depending on whether the fluid particles rotate about their own centers or not. Flow is *uniform* if the velocity does not change in the direction of flow. If it does, the flow is *nonuniform. Laminar* flow exists when there are no lateral motions superimposed on the mean flow. When there are, the flow is *turbulent.* Flow may be intermittently laminar and turbulent; this is called flow in *transition.* Flow is considered *incompressible* if the density is constant, or in the case of gas flows, if the density variation is below a specified amount throughout the flow, 2–3%, for example. Low-speed gas flows may be considered essentially incompressible. Gas flows may be considered as *subsonic, transonic, sonic, supersonic,* or *hypersonic* depending on the gas speed compared with the speed of sound in the gas. Open-channel water flows may be designated as *subcritical, critical,* or *supercritical* depending on whether the flow is less than, equal to, or greater than the speed of an elementary surface wave.

53.4.1 Velocity and Acceleration

In Cartesian coordinates, velocity components are *u, v,* and *w* in the *x, y,* and *z* directions, respectively. These may vary with position and time, such that, for example, $u = dx/dt = u(x, y, z, t)$. Then

$$du = \frac{\partial u}{\partial x} dx + \frac{\partial u}{\partial y} dy + \frac{\partial u}{\partial z} dz + \frac{\partial u}{\partial t} dt$$

and

$$a_x = \frac{du}{dt} = \frac{\partial u}{\partial x}\frac{dx}{dt} + \frac{\partial u}{\partial y}\frac{dy}{dt} + \frac{\partial u}{\partial z}\frac{dz}{dt} + \frac{\partial u}{\partial t}$$

$$= \frac{Du}{Dt} = u\frac{\partial u}{\partial x} + v\frac{\partial u}{\partial y} + w\frac{\partial u}{\partial z} + \frac{\partial u}{\partial t}$$

The first three terms on the right-hand side are the *convective* acceleration, which is zero for uniform flow, and the last term is the *local* acceleration, which is zero for steady flow.

In natural coordinates (streamline direction *s,* normal direction *n,* and meridional direction *m* normal to the plane of *s* and *n*), the velocity *V* is always in the streamline direction. Thus, $V = V(s,t)$ and

$$dV = \frac{\partial V}{\partial s} ds + \frac{\partial V}{\partial t} dt$$

$$a_s = \frac{dV}{dt} = V\frac{\partial V}{\partial s} + \frac{\partial V}{\partial t}$$

where the first term on the right-hand side is the *convective* acceleration and the last is the *local* acceleration. Thus, if the fluid velocity changes as the fluid moves throughout space, there is a convective acceleration, and if the velocity at a point changes with time, there is a local acceleration.

53.4.2 Streamlines

A *streamline* is a line to which, at each instant, velocity vectors are tangent. A *pathline* is the path of a particle as it moves in the fluid, and for steady flow it coincides with a streamline.

The equations of streamlines are described by stream functions ψ, from which the velocity components in two-dimensional flow are $u = -\partial\psi/\partial y$ and $v = +\partial\psi/\partial x$. Streamlines are lines of constant stream function. In polar coordinates

$$v_r = -\frac{1}{r}\frac{\partial\psi}{\partial\theta} \quad \text{and} \quad v_\theta = +\frac{\partial\psi}{\partial r}$$

Some streamline patterns are shown in Figs. 53.11, 53.12, and 53.13. The lines at right angles to the streamlines are potential lines.

53.4.3 Deformation of a Fluid Element

Four types of deformation or movement may occur as a result of spatial variations of velocity: translation, linear deformation, angular deformation, and rotation. These may occur singly or in combination. Motion of the face (in the *x-y* plane) of an elemental cube of sides δx, δy, and δz in a time dt is shown in Fig. 53.14. Both translation and rotation involve motion or deformation without a change in shape of the fluid element. Linear and angular deformations, however, do involve a change in

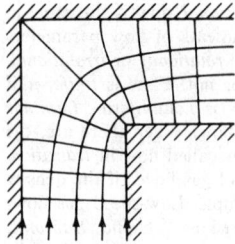

Fig. 53.11 Flow around a corner in a duct.

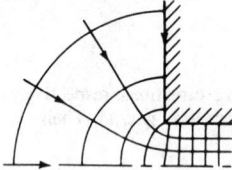

Fig. 53.12 Flow around a corner into a duct.

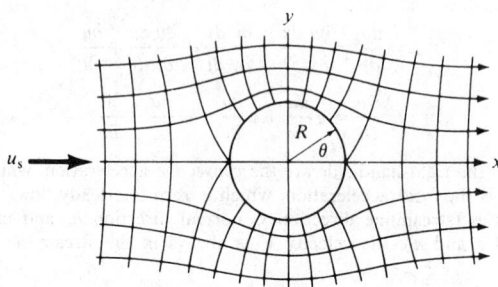

Fig. 53.13 Inviscid flow past a cylinder.

shape of the fluid element. Only through these linear and angular deformations are heat generated and mechanical energy dissipated as a result of viscous action in a fluid.

For linear deformation the relative change in volume is at a rate of

$$(\mathcal{V}_{dt} - \mathcal{V}_0)/\mathcal{V}_0 = \frac{\partial u}{\partial x} + \frac{\partial v}{\partial y} + \frac{\partial w}{\partial z} = \text{div } \mathbf{V}$$

which is zero for an incompressible fluid, and thus is an expression for the continuity equation. Rotation of the face of the cube shown in Fig. 53.14d is the average of the rotations of the bottom and left edges, which is

$$\frac{1}{2}\left(\frac{\partial v}{\partial x} - \frac{\partial u}{\partial y}\right) dt$$

The rate of rotation is the angular velocity and is

$$\omega_z = \frac{1}{2}\left(\frac{\partial v}{\partial x} - \frac{\partial u}{\partial y}\right) \quad \text{about the } z \text{ axis in the } x\text{-}y \text{ plane}$$

$$\omega_x = \frac{1}{2}\left(\frac{\partial w}{\partial y} - \frac{\partial v}{\partial z}\right) \quad \text{about the } x \text{ axis in the } y\text{-}z \text{ plane}$$

and

$$\omega_y = \frac{1}{2}\left(\frac{\partial u}{\partial z} - \frac{\partial w}{\partial x}\right) \quad \text{about the } y \text{ axis in the } x\text{-}z \text{ plane}$$

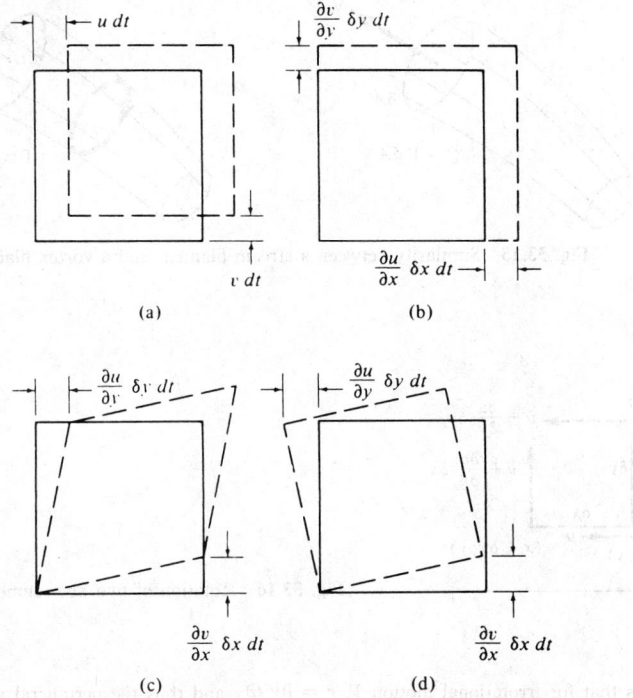

Fig. 53.14 Movements of the face of an elemental cube in the x-y plane: (a) translation; (b) linear deformation; (c) angular deformation; (d) rotation.

These are the components of the angular velocity vector Ω,

$$\Omega = \tfrac{1}{2}\ \text{curl}\ \mathbf{V} = \frac{1}{2} \begin{vmatrix} \mathbf{i} & \mathbf{j} & \mathbf{k} \\ \dfrac{\partial}{\partial x} & \dfrac{\partial}{\partial y} & \dfrac{\partial}{\partial z} \\ u & v & w \end{vmatrix} = \omega_x \mathbf{i} + \omega_y \mathbf{j} + \omega_z \mathbf{k}$$

If the flow is irrotational, these quantities are zero.

53.4.4 Vorticity and Circulation

Vorticity is defined as twice the angular velocity, and thus is also zero for irrotational flow. Circulation is defined as the line integral of the velocity component along a closed curve and equals the total strength of all vortex filaments that pass through the curve. Thus, the vorticity at a point within the curve is the circulation per unit area enclosed by the curve. These statements are expressed by

$$\Gamma = \oint \mathbf{V} \cdot d\mathbf{l} = \oint (u\,dx + v\,dy + w\,dz) \qquad \text{and} \qquad \zeta_A = \lim_{A \to 0} \frac{\Gamma}{A}$$

Circulation—the product of vorticity and area—is the counterpart of volumetric flow rate as the product of velocity and area. These are shown in Fig. 53.15.

Physically, fluid rotation at a point in a fluid is the instantaneous average rotation of two mutually perpendicular infinitesimal line segments. In Fig. 53.16 the line δx rotates positively and δy rotates negatively. Then $\omega_z = (\partial v/\partial x - \partial u/\partial y)/2$. In natural coordinates (the n direction is opposite to the radius of curvature r) the angular velocity in the s-n plane is

$$\omega = \frac{1}{2}\frac{\Gamma}{\delta A} = \frac{1}{2}\left(\frac{V}{r} - \frac{\partial V}{\partial n}\right) = \frac{1}{2}\left(\frac{V}{r} + \frac{\partial V}{\partial r}\right)$$

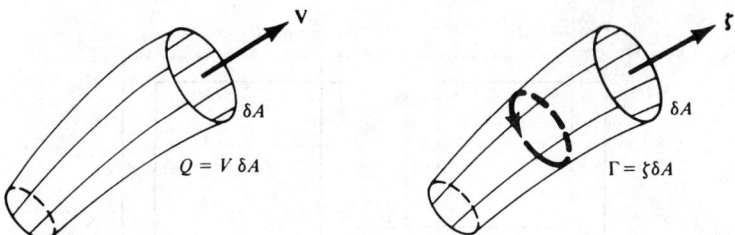

Fig. 53.15 Similarity between a stream filament and a vortex filament.

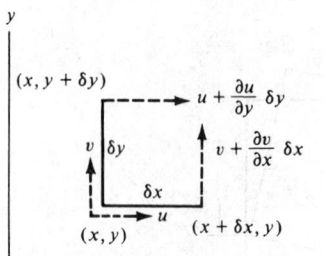

Fig. 53.16 Rotation of two line segments in a fluid.

This shows that for irrotational motion $V/r = \partial V/\partial n$ and thus the peripheral velocity V increases toward the center of curvature of streamlines. In an irrotational vortex, $V r = C$ and in a solid-body-type or rotational vortex, $V = \omega r$.

A combined vortex has a solid-body-type rotation at the core and an irrotational vortex beyond it. This is typical of a tornado (which has an inward sink flow superimposed on the vortex motion) and eddies in turbulent motion.

53.4.5 Continuity Equations

Conservation of mass for a fluid requires that in a *material* volume, the mass remains constant. In a *control* volume the net rate of influx of mass into the control volume is equal to the rate of change of mass in the control volume. Fluid may flow into a control volume either through the control surface or from internal sources. Likewise, fluid may flow out through the control surface or into internal sinks. The various forms of the continuity equations listed in Table 53.3 do not include sources and sinks; if they exist, they must be included.

The most commonly used forms for duct flow are $\dot{m} = V A \rho$ in kg/sec where V is the average flow velocity in m/sec, A is the duct area in m³, and ρ is the fluid density in kg/m³. In differential form this is $dV/V + dA/A + d\rho/\rho = 0$, which indicates that all three quantities may not increase nor all decrease in the direction of flow. For incompressible duct flow $Q = V A$ m³/sec where V and A are as above. When the velocity varies throughout a cross section, the average velocity is

$$V = \frac{1}{A}\int u \, dA = \frac{1}{n}\sum_{i=1}^{n} u_i$$

where u is a velocity at a point, and u_i are point velocities measured at the centroid of n equal areas. For example, if the velocity is u at a distance y from the wall of a pipe of radius R and the centerline velocity is u_m, $u = u_m (y/R)^{1/7}$ and the average velocity is $V = \frac{49}{60} u_m$.

53.5 FLUID MOMENTUM

The momentum theorem states that the net external force acting on the fluid within a control volume equals the time rate of change of momentum of the fluid plus the net rate of momentum flux or transport out of the control volume through its surface. This is one form of the Reynolds transport theorem, which expresses the conservation laws of physics for fixed mass systems to expressions for a control volume:

Table 53.3 Continuity Equations

GENERAL	$\dfrac{\partial \rho}{\partial t} + \nabla \cdot \rho \mathbf{V} = 0 \quad \text{or} \quad \dfrac{D\rho}{Dt} + \rho \nabla \cdot \mathbf{V} = 0$	VECTOR
Unsteady, compressible	$\dfrac{\partial \rho}{\partial t} + \dfrac{\partial(\rho u)}{\partial x} + \dfrac{\partial(\rho v)}{\partial y} + \dfrac{\partial(\rho w)}{\partial z} = 0$	Cartesian
	$\dfrac{\partial \rho}{\partial t} + \dfrac{\partial(\rho v_r)}{\partial r} + \dfrac{1}{r}\dfrac{\partial(\rho v_\theta)}{\partial \theta} + \dfrac{\partial(\rho v_z)}{\partial z} + \dfrac{\rho v_r}{r} = 0$	Cylindrical
	$\dfrac{\partial(\rho A)}{\partial t} + \dfrac{\partial}{\partial s}(\rho \mathbf{V} \cdot \mathbf{A}) = 0$	Duct
Steady, compressible	$\nabla \cdot \rho \mathbf{V} = 0$	Vector
	$\dfrac{\partial(\rho u)}{\partial x} + \dfrac{\partial(\rho v)}{\partial y} + \dfrac{\partial(\rho w)}{\partial z} = 0$	Cartesian
	$\dfrac{\partial(\rho v_r)}{\partial r} + \dfrac{1}{r}\dfrac{\partial(\rho v_\theta)}{\partial \theta} + \dfrac{\partial(\rho v_z)}{\partial z} + \dfrac{\rho v_r}{r} = 0$	Cylindrical
	$\rho \mathbf{V} \cdot \mathbf{A} = \dot{m}$	
Incompressible	$\nabla \cdot \mathbf{V} = 0$	Vector
Steady or unsteady	$\dfrac{\partial u}{\partial x} + \dfrac{\partial v}{\partial y} + \dfrac{\partial w}{\partial z} = 0$	Cartesian
	$\dfrac{\partial v_r}{\partial r} + \dfrac{1}{r}\dfrac{\partial v_\theta}{\partial \theta} + \dfrac{\partial v_z}{\partial z} + \dfrac{v_r}{r} = 0$	Cylindrical
	$\mathbf{V} \cdot \mathbf{A} = Q$	Duct

$$\Sigma \mathbf{F} = \frac{D}{Dt} \int_{\substack{\text{material} \\ \text{volume}}} \rho \mathbf{V}\, d\mathcal{V}$$

$$= \frac{\partial}{\partial t} \int_{\substack{\text{control} \\ \text{volume}}} \rho \mathbf{V}\, d\mathcal{V} + \int_{\substack{\text{control} \\ \text{surface}}} \rho \mathbf{V}(\mathbf{V} \cdot d\mathbf{s})$$

53.5.1 The Momentum Theorem

For steady flow the first term on the right-hand side of the preceding equation is zero. Forces include normal forces due to pressure and tangential forces due to viscous shear over the surface S of the control volume, and body forces due to gravity and centrifugal effects, for example. In scalar form the net force equals the total momentum flux leaving the control volume minus the total momentum flux entering the control volume. In the x direction

$$\Sigma F_x = (\dot{m} V_x)_{\text{leaving } S} - (\dot{m} V_x)_{\text{entering } S}$$

or when the same fluid enters and leaves,

$$\Sigma F_x = \dot{m}(V_{x \text{ leaving } S} - V_{x \text{ entering } S})$$

with similar expressions for the y and z directions.

For one-dimensional flow $\dot{m} V_x$ represents momentum flux passing a section and V_x is the average velocity. If the velocity varies across a duct section, the true momentum flux is $\int_A (u\rho\, dA)u$, and the ratio of this value to that based upon average velocity is the momentum correction factor β,

$$\beta = \frac{\displaystyle\int_A u^2\, dA}{V^2 A} \geq 1$$

$$\approx \frac{1}{V^2 n} \sum_{i=1}^{n} u_i^2$$

For laminar flow in a circular tube, $\beta = \frac{4}{3}$; for laminar flow between parallel plates, $\beta = 1.20$; and for turbulent flow in a circular tube, β is about 1.02–1.03.

53.5.2 Equations of Motion

For steady irrotational flow of an incompressible nonviscous fluid, Newton's second law gives the Euler equation of motion. Along a streamline it is

$$V\frac{\partial V}{\partial s} + \frac{1}{\rho}\frac{\partial p}{\partial s} + g\frac{\partial z}{\partial s} = 0$$

and normal to a streamline it is

$$\frac{V^2}{r} + \frac{1}{\rho}\frac{\partial p}{\partial n} + g\frac{\partial z}{\partial n} = 0$$

When integrated, these show that the sum of the kinetic, displacement, and potential energies is a constant along streamlines as well as across streamlines. The result is known as the Bernoulli equation:

$$\frac{V^2}{2} + \frac{p}{\rho} + gz = \text{constant energy per unit mass}$$

$$\frac{\rho V_1^2}{2} + p_1 + \rho gz_1 = \frac{\rho V_2^2}{2} + p_2 + \rho gz_2 = \text{constant total pressure}$$

and

$$\frac{V_1^2}{2g} + \frac{p_1}{g\rho} + z_1 = \frac{V_2^2}{2g} + \frac{p_2}{g\rho} + z_2 = \text{constant total head}$$

For a reversible adiabatic compressible gas flow with no external work, the Euler equation integrates to

$$\frac{V_1^2}{2} + \frac{k}{k-1}\left(\frac{p_1}{\rho_1}\right) + gz_1 = \frac{V_2^2}{2} + \frac{k}{k-1}\left(\frac{p_2}{\rho_2}\right) + gz_2$$

which is valid whether the flow is reversible or not, and corresponds to the steady-flow energy equation for adiabatic no-work gas flow.

Newton's second law written normal to streamlines shows that in horizontal planes $dp/dr = \rho V^2/r$, and thus dp/dr is positive for both rotational and irrotational flow. The pressure increases away from the center of curvature and decreases toward the center of curvature of curvilinear streamlines. The radius of curvature r of straight lines is infinite, and thus no pressure gradient occurs across these.

For a liquid rotating as a solid body

$$-\frac{V_1^2}{2g} + \frac{p_1}{\rho g} + z_1 = -\frac{V_2^2}{2g} + \frac{p_2}{\rho g} + z_2$$

The negative sign balances the increase in velocity and pressure with radius.

The differential equations of motion for a viscous fluid are known as the Navier–Stokes equations. For incompressible flow the x-component equation is

$$\frac{\partial u}{\partial t} + u\frac{\partial u}{\partial x} + v\frac{\partial u}{\partial y} + w\frac{\partial u}{\partial z} = X - \frac{1}{\rho}\frac{\partial p}{\partial x} + \nu\left(\frac{\partial^2 u}{\partial x^2} + \frac{\partial^2 u}{\partial y^2} + \frac{\partial^2 u}{\partial z^2}\right)$$

with similar expressions for the y and z directions. X is the body force per unit mass. Reynolds developed a modified form of these equations for turbulent flow by expressing each velocity as an average value plus a fluctuating component ($u = \bar{u} + u'$ and so on). These modified equations indicate shear stresses from turbulence ($\tau_T = -\rho\overline{u'v'}$, for example) known as the Reynolds stresses, which have been useful in the study of turbulent flow.

53.6 FLUID ENERGY

The Reynolds transport theorem for fluid passing through a control volume states that the heat added to the fluid less any work done by the fluid increases the energy content of the fluid in the control volume or changes the energy content of the fluid as it passes through the control surface. This is

$$Q - Wk_{\text{done}} = \frac{\partial}{\partial t}\int_{\substack{\text{control}\\\text{volume}}} (e\rho)\,d\mathcal{V} + \int_{\substack{\text{control}\\\text{surface}}} e\rho(\mathbf{V}\cdot d\mathbf{S})$$

and represents the first law of thermodynamics for a control volume. The energy content includes kinetic, internal, potential, and displacement energies. Thus, mechanical and thermal energies are included, and there are no restrictions on the direction of interchange from one form to the other implied in the first law. The second law of thermodynamics governs this.

53.6.1 Energy Equations

With reference to Fig. 53.17, the steady flow energy equation is

$$\alpha_1 \frac{V_1^2}{2} + p_1 v_1 + gz_1 + u_1 + q - w = \alpha_2 \frac{V_2^2}{2} + p_2 v_2 + gz_2 + u_2$$

in terms of energy per unit mass, and where α is the kinetic energy correction factor:

$$\alpha = \frac{\int_A u^3 \, dA}{V^3 A} \approx \frac{1}{V^3 n} \sum_{i=1}^{n} u_i^3 \geq 1$$

For laminar flow in a pipe, $\alpha = 2$; for turbulent flow in a pipe, $\alpha = 1.05$–1.06; and if one-dimensional flow is assumed, $\alpha = 1$.

For one-dimensional flow of compressible gases, the general expression is

$$\frac{V_1^2}{2} + h_1 + gz_1 + q - w = \frac{V_2^2}{2} + h_2 + gz_2$$

For adiabatic flow, $q = 0$; for no external work, $w = 0$; and in most instances changes in elevation z are very small compared with changes in other parameters and can be neglected. Then the equation becomes

$$\frac{V_1^2}{2} + h_1 = \frac{V_2^2}{2} + h_2 = h_0$$

where h_0 is the stagnation enthalpy. The stagnation temperature is then $T_0 = T_1 + V_1^2/2c_p$ in terms of the temperature and velocity at some point 1. The gas velocity in terms of the stagnation and static temperatures, respectively, is $V_1 = \sqrt{2c_p(T_0 - T_1)}$. An increase in velocity is accompanied by a decrease in temperature, and vice versa.

For one-dimensional flow of liquids and constant-density (low-velocity) gases, the energy equation generally is written in terms of energy per unit weight as

$$\frac{V_1^2}{2g} + \frac{p_1}{\gamma} + z_1 - w = \frac{V_2^2}{2g} + \frac{p_2}{\gamma} + z_2 + h_L$$

where the first three terms are velocity, pressure, and potential heads, respectively. The head loss $h_L = (u_2 - u_1 - q)/g$ and represents the mechanical energy dissipated into thermal energy irreversibly (the heat transfer q is assumed zero here). It is a positive quantity and increases in the direction of flow.

Irreversibility in compressible gas flows results in an entropy increase. In Fig. 53.18 reversible flow between pressures p' and p is from a to b or from b to a. Irreversible flow from p' to p is from b to d, and from p to p' it is from a to c. Thus, frictional duct flow from one pressure to another results in a higher final temperature, and a lower final velocity, in both instances. For frictional flow between given temperatures (T_a and T_b, for example), the resulting pressures are lower than for frictionless flow (p_c is lower than p_a and p_f is lower than p_b).

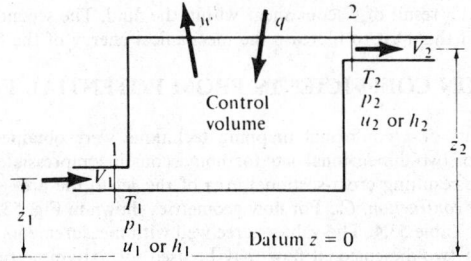

Fig. 53.17 Control volume for steady-flow energy equation.

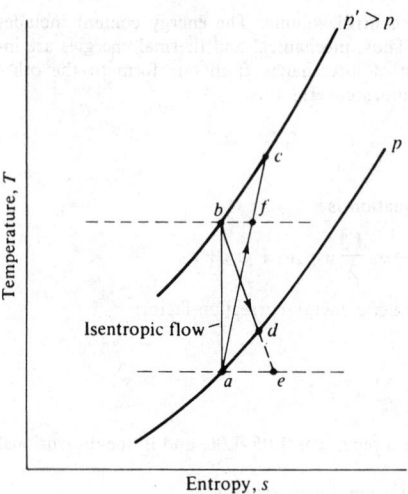

Fig. 53.18   Reversible and irreversible adiabatic flows.

53.6.2   Work and Power

Power is the rate at which work is done, and is the work done per unit mass times the mass flow rate, or the work done per unit weight times the weight flow rate.

Power represented by the work term in the energy equation is $P = w \, (V \, A \, \gamma) = w \, (V \, A \, \rho)$ W.

Power in a jet at a velocity V is $P = (V^2/2)(V \, A \, \rho) = (V^2/2g)(V \, A \, \gamma)$ W.

Power loss resulting from head loss is $P = h_L \, (V \, A \, \gamma)$ W.

Power to overcome a drag force is $P = F \, V$ W.

Power available in a hydroelectric power plant when water flows from a headwater elevation z_1 to a tailwater elevation z_2 is $P = (z_1 - z_2)(Q\gamma)$ W, where Q is the volumetric flow rate.

53.6.3   Viscous Dissipation

Dissipation effects resulting from viscosity account for entropy increases in adiabatic gas flows and the head loss term for flows of liquids. They can be expressed in terms of the rate at which work is done—the product of the viscous shear force on the surface of an elemental fluid volume and the corresponding component of velocity parallel to the force. Results for a cube of sides dx, dy, and dz give the dissipation function Φ:

$$\Phi = 2\mu \left[\left(\frac{\partial u}{\partial x} \right)^2 + \left(\frac{\partial v}{\partial y} \right)^2 + \left(\frac{\partial w}{\partial z} \right)^2 \right]$$

$$+ \mu \left[\left(\frac{\partial v}{\partial x} + \frac{\partial u}{\partial y} \right)^2 + \left(\frac{\partial w}{\partial y} + \frac{\partial v}{\partial z} \right)^2 + \left(\frac{\partial u}{\partial z} + \frac{\partial w}{\partial x} \right)^2 \right]$$

$$- \frac{2}{3} \mu \left(\frac{\partial u}{\partial x} + \frac{\partial v}{\partial y} + \frac{\partial w}{\partial z} \right)^2$$

The last term is zero for an incompressible fluid. The first term in brackets is the linear deformation, and the second term in brackets is the angular deformation and in only these two forms of deformation is there heat generated as a result of viscous shear within the fluid. The second law of thermodynamics precludes the recovery of this heat to increase the mechanical energy of the fluid.

53.7   CONTRACTION COEFFICIENTS FROM POTENTIAL FLOW THEORY

Useful engineering results of a conformal mapping technique were obtained by von Mises for the contraction coefficients of two-dimensional jets for nonviscous incompressible fluids in the absence of gravity. The ratio of the resulting cross-sectional area of the jet to the area of the boundary opening is called the *coefficient of contraction*, C_c. For flow geometries shown in Fig. 53.19, von Mises calculated the values of C_c listed in Table 53.4. The values agree well with measurements for low-viscosity liquids. The results tabulated for two-dimensional flow may be used for axisymmetric jets if C_c is defined by $C_c = b_{\text{jet}}/b = (d_{\text{jet}}/d)^2$ and if d and D are diameters equivalent to widths b and B, respectively.

Potential flow

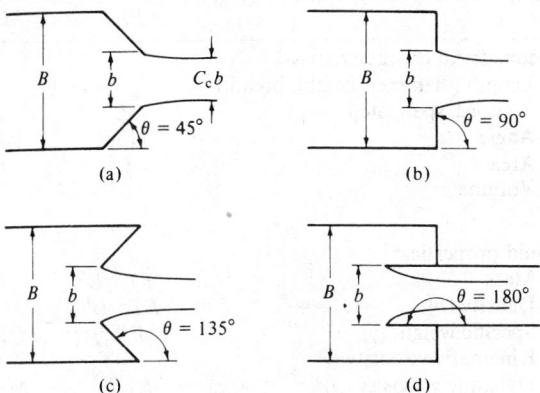

Fig. 53.19 Geometry of two-dimensional jets.

Thus, if a small round hole of diameter d in a large tank $(d/D \approx 0)$, the jet diameter would be $(0.611)^{1/2} = 0.782$ times the hole diameter, since $\theta = 90°$.

53.8 DIMENSIONLESS NUMBERS AND DYNAMIC SIMILARITY

Dimensionless numbers are commonly used to plot experimental data to make the results more universal. Some are also used in designing experiments to ensure dynamic similarity between the flow of interest and the flow being studied in the laboratory.

53.8.1 Dimensionless Numbers

Dimensionless numbers or groups may be obtained from force ratios, by a dimensional analysis using the Buckingham Pi theorem, for example, or by writing the differential equations of motion and energy in dimensionless form. Dynamic similarity between two geometrically similar systems exists when the appropriate dimensionless groups are the same for the two systems. This is the basis on which model studies are made, and results measured for one flow may be applied to similar flows.

The dimensions of some parameters used in fluid mechanics are listed in Table 53.5. The mass–length–time (MLT) and the force–length–time (FLT) systems are related by $F = M a = M L/T^2$ and $M = F T^2/L$.

Force ratios are expressed as

$$\frac{\text{inertia force}}{\text{viscous force}} = \frac{\rho L^2 V^2}{\mu V L} = \frac{\rho L V}{\mu}, \quad \text{the Reynolds number Re}$$

Table 53.4 Coefficients of Contraction for Two-Dimensional Jets

b/B	C_c $\theta = 45°$	C_c $\theta = 90°$	C_c $\theta = 135°$	C_c $\theta = 180°$
0.0	0.746	0.611	0.537	0.500
0.1	0.747	0.612	0.546	0.513
0.2	0.747	0.616	0.555	0.528
0.3	0.748	0.622	0.566	0.544
0.4	0.749	0.631	0.580	0.564
0.5	0.752	0.644	0.599	0.586
0.6	0.758	0.662	0.620	0.613
0.7	0.768	0.687	0.652	0.646
0.8	0.789	0.722	0.698	0.691
0.9	0.829	0.781	0.761	0.760
1.0	1.000	1.000	1.000	1.000

Table 53.5   Dimensions of Fluid and Flow Parameters

	FLT	MLT
Geometrical characteristics		
Length (diameter, height, breadth, chord, span, etc.)	L	L
Angle	None	None
Area	L^2	L^2
Volume	L^3	L^3
Fluid properties[a]		
Mass	FT^2/L	M
Density (ρ)	FT^2/L^4	M/L^3
Specific weight (γ)	F/L^3	M/L^2T^2
Kinematic viscosity (v)	L^2/T	L^2/T
Dynamic viscosity (μ)	FT/L^2	M/LT
Elastic modulus (K)	F/L^2	M/LT^2
Surface tension (σ)	F/L	M/T^2
Flow characteristics		
Velocity (V)	L/T	L/T
Angular velocity (ω)	$1/T$	$1/T$
Acceleration (a)	L/T^2	L/T^2
Pressure (Δp)	F/L^2	M/LT^2
Force (drag, lift, shear)	F	ML/T^2
Shear stress (τ)	F/L^2	M/LT^2
Pressure gradient ($\Delta p/L$)	F/L^3	M/L^2T^2
Flow rate (Q)	L^3/T	L^3/T
Mass flow rate ($\dot{m}$)	FT/L	M/T
Work or energy	FL	ML^2/T^2
Work or energy per unit weight	L	L
Torque and moment	FL	ML^2/T^2
Work or energy per unit mass	L^2/T^2	L^2/T^2

[a] Density, viscosity, elastic modulus, and surface tension depend upon temperature, and therefore temperature will not be considered a property in the sense used here.

$$\frac{\text{inertia force}}{\text{gravity force}} = \frac{\rho L^2 V^2}{\rho L^3 g} = \frac{V^2}{Lg} \quad \text{or} \quad \frac{V}{\sqrt{Lg}}, \quad \text{the Froude number Fr}$$

$$\frac{\text{pressure force}}{\text{inertia force}} = \frac{\Delta p L^2}{\rho L^2 V^2} = \frac{\Delta p}{\rho V^2} \quad \text{or} \quad \frac{\Delta p}{\rho V^2/2}, \quad \text{the pressure coefficient } C_p$$

$$\frac{\text{inertia force}}{\text{surface tension force}} = \frac{\rho L^2 V^2}{\sigma L} = \frac{V^2}{\sigma/\rho L} \quad \text{or} \quad \frac{V}{\sqrt{\sigma/\rho L}}, \quad \text{the Weber number We}$$

$$\frac{\text{inertia force}}{\text{compressibility force}} = \frac{\rho L^2 V^2}{K L^2} = \frac{V^2}{K/\rho} \quad \text{or} \quad \frac{V}{\sqrt{K/\rho}}, \quad \text{the Mach number M}$$

If a system includes n quantities with m dimensions, there will be at least $n - m$ independent dimensionless groups, each containing m repeating variables. Repeating variables (1) must include all the m dimensions, (2) should include a geometrical characteristic, a fluid property, and a flow characteristic and (3) should not include the dependent variable.

Thus, if the pressure gradient $\Delta p/L$ for flow in a pipe is judged to depend on the pipe diameter D and roughness k, the average flow velocity V, and the fluid density ρ, the fluid viscosity μ, and compressibility K (for gas flow), then $\Delta p/L = f(D, k, V, \rho, \mu, K)$ or in dimensions, $F/L^3 = f(L, L, L/T, FT^2/L^4, FT/L^2, F/L^2)$, where $n = 7$ and $m = 3$. Then there are $n - m = 4$ independent groups to be sought. If D, ρ, and V are the repeating variables, the results are

$$\frac{\Delta p}{\rho V^2/2} = f\left(\frac{D V \rho}{\mu}, \frac{k}{D}, \frac{V}{\sqrt{K/\rho}}\right)$$

or that the friction factor will depend on the Reynolds number of the flow, the relative roughness, and the Mach number. The actual relationship between them is determined experimentally. Results may be determined analytically for laminar flow. The seven original variables are thus expressed as four dimensionless variables, and the Moody diagram of Fig. 53.32 shows the result of analysis and experiment. Experiments show that the pressure gradient does depend on the Mach number, but the friction factor does not.

The Navier–Stokes equations are made dimensionless by dividing each length by a characteristic length L and each velocity by a characteristic velocity U. For a body force X due to gravity, $X = g_x = g(\partial z/\partial x)$. Then $x' = x/L$, etc., $t' = t(L/U)$, $u' = u/U$, etc., and $p' = p/\rho U^2$. Then the Navier–Stokes equation (x component) is

$$u'\frac{\partial u'}{\partial x'} + v'\frac{\partial u'}{\partial y'} + w'\frac{\partial u'}{\partial z'} + \frac{\partial u'}{\partial t'}$$

$$= \frac{gL}{U^2} - \frac{\partial p'}{\partial x'} + \frac{\mu}{\rho UL}\left(\frac{\partial^2 u'}{\partial x'^2} + \frac{\partial^2 u'}{\partial y'^2} + \frac{\partial^2 u'}{\partial z'^2}\right)$$

$$= \frac{1}{\mathrm{Fr}^2} - \frac{\partial p'}{\partial x'} + \frac{1}{\mathrm{Re}}\left(\frac{\partial^2 u'}{\partial x'^2} + \frac{\partial^2 u'}{\partial y'^2} + \frac{\partial^2 u'}{\partial z'^2}\right)$$

Thus for incompressible flow, similarity of flow in similar situations exists when the Reynolds and the Froude numbers are the same.

For compressible flow, normalizing the differential energy equation in terms of temperatures, pressure, and velocities gives the Reynolds, Mach, and Prandtl numbers as the governing parameters.

53.8.2 Dynamic Similitude

Flow systems are considered to be dynamically similar if the appropriate dimensionless numbers are the same. Model tests of aircraft, missiles, rivers, harbors, breakwaters, pumps, turbines, and so forth are made on this basis. Many practical problems exist, however, and it is not always possible to achieve complete dynamic similarity. When viscous forces govern the flow, the Reynolds number should be the same for model and prototype, the length in the Reynolds number being some characteristic length. When gravity forces govern the flow, the Froude number should be the same. When surface tension forces are significant, the Weber number is used. For compressible gas flow, the Mach number is used; different gases may be used for the model and prototype. The pressure coefficient $C_p = \Delta p/(\rho V^2/2)$, the drag coefficient $C_D = \mathrm{drag}/(\rho V^2/2)A$, and the lift coefficient $C_L = \mathrm{lift}/(\rho V^2/2)A$ will be the same for model and prototype when the appropriate Reynolds, Froude, or Mach number is the same. A cavitation number is used in cavitation studies, $\sigma_v = (p - p_v)/(\rho V^2/2)$ if vapor pressure p_v is the reference pressure or $\sigma_c = (p - p_c)/(\rho V^2/2)$ if a cavity pressure is the reference pressure.

Modeling ratios for conducting tests are listed in Table 53.6. Distorted models are often used for rivers in which the vertical scale ratio might be 1/40 and the horizontal scale ratio 1/100, for example, to avoid surface tension effects and laminar flow in models too shallow.

Incomplete similarity often exists in Froude–Reynolds models since both contain a length parameter. Ship models are tested with the Froude number parameter, and viscous effects are calculated for both model and prototype.

The specific speed of pumps and turbines results from combining groups in a dimensional analysis of rotary systems. That for pumps is $N_{s \text{ (pump)}} = N\sqrt{Q}/e^{3/4}$ and for turbines it is $N_{s \text{ (turbines)}} = N\sqrt{\mathrm{power}}/\rho^{1/2}e^{5/4}$, where N is the rotational speed in rad/sec, Q is the volumetric flow rate in m³/sec, and e is the energy in J/kg. North American practice uses N in rpm, Q in gal/min, e as energy per unit weight (head in ft), power as brake horsepower rather than watts, and omits the density term in the specific speed for turbines. The numerical value of specific speed indicates the type of pump or turbine for a given installation. These are shown for pumps in North America in Fig. 53.20. Typical values for North American turbines are about 5 for impulse turbines, about 20–100 for Francis turbines, and 100–200 for propeller turbines. Slight corrections in performance for higher efficiency of large pumps and turbines are made when testing small laboratory units.

53.9 VISCOUS FLOW AND INCOMPRESSIBLE BOUNDARY LAYERS

In viscous flows, adjacent layers of fluid transmit both normal forces and tangential shear forces, as a result of relative motion between the layers. There is no relative motion, however, between the fluid and a solid boundary along which it flows. The fluid velocity varies from zero at the boundary

Table 53.6 Modeling Ratios[a]

		MODELING PARAMETER			
RATIO	REYNOLDS NUMBER	FROUDE NUMBER, UNDISTORTED MODEL[b]	FROUDE NUMBER, DISTORTED MODEL[b]	MACH NUMBER, SAME GAS[d]	MACH NUMBER, DIFFERENT GAS[d]
Velocity $\dfrac{V_m}{V_p}$	$\dfrac{L_p}{L_m}\dfrac{\rho_p}{\rho_m}\dfrac{\mu_m}{\mu_p}$	$\left(\dfrac{L_m}{L_p}\right)^{1/2}$	$\left(\dfrac{L_m}{L_p}\right)_v^{1/2}$	$\left(\dfrac{\theta_m}{\theta_p}\right)^{1/2}$	$\left(\dfrac{k_m R_m \theta_m}{k_p R_p \theta_p}\right)^{1/2}$
Angular velocity $\dfrac{\omega_m}{\omega_p}$	$\left(\dfrac{L_p}{L_m}\right)^2\dfrac{\rho_p}{\rho_m}\dfrac{\mu_m}{\mu_p}$	$\left(\dfrac{L_p}{L_m}\right)^{1/2}$	—[c]	$\left(\dfrac{\theta_m}{\theta_p}\right)^{1/2}\dfrac{L_p}{L_m}$	$\left(\dfrac{k_m R_m \theta_m}{k_p R_p \theta_p}\right)^{1/2}\dfrac{L_p}{L_m}$
Volumetric flow rate $\dfrac{Q_m}{Q_p}$	$\dfrac{L_m}{L_p}\dfrac{\rho_p}{\rho_m}\dfrac{\mu_m}{\mu_p}$	$\left(\dfrac{L_m}{L_p}\right)^{5/2}$	$\left(\dfrac{L_m}{L_p}\right)_v^{3/2}\left(\dfrac{L_m}{L_p}\right)_H$	—[c]	—[c]
Time $\dfrac{t_m}{t_p}$	$\left(\dfrac{L_m}{L_p}\right)^2\dfrac{\rho_m}{\rho_p}\dfrac{\mu_p}{\mu_m}$	$\left(\dfrac{L_m}{L_p}\right)^{1/2}\left(\dfrac{g_p}{g_m}\right)^{1/2}$	$\left(\dfrac{L_m}{L_p}\right)_H\left(\dfrac{L_p}{L_m}\right)_v^{1/2}\left(\dfrac{g_p}{g_m}\right)^{1/2}$	$\left(\dfrac{\theta_p}{\theta_m}\right)^{1/2}\dfrac{L_m}{L_p}$	$\left(\dfrac{k_p R_p \theta_p}{k_m R_m \theta_m}\right)^{1/2}\dfrac{L_m}{L_p}$
Force $\dfrac{F_m}{F_p}$	$\left(\dfrac{\mu_m}{\mu_p}\right)^2\dfrac{\rho_p}{\rho_m}$	$\left(\dfrac{L_m}{L_p}\right)^3\dfrac{\rho_m}{\rho_p}$	$\dfrac{\rho_m}{\rho_p}\left(\dfrac{L_m}{L_p}\right)_H\left(\dfrac{L_m}{L_p}\right)_v^2$	$\dfrac{\rho_m}{\rho_p}\dfrac{\theta_m}{\theta_p}\left(\dfrac{L_m}{L_p}\right)^2$	$\dfrac{K_m}{K_p}\left(\dfrac{L_m}{L_p}\right)^2$

[a] Subscript m indicates model, subscript p indicates prototype.
[b] For the same value of gravitational acceleration for model and prototype.
[c] Of little importance.
[d] Here θ refers to temperature.

to a maximum or free stream value some distance away from it. This region of retarded flow is called the boundary layer.

53.9.1 Laminar and Turbulent Flow

Viscous fluids flow in a laminar or in a turbulent state. There are, however, transition regimes between them where the flow is intermittently laminar and turbulent. Laminar flow is smooth, quiet flow without lateral motions. Turbulent flow has lateral motions as a result of eddies superimposed on the main flow, which results in random or irregular fluctuations of velocity, pressure, and, possibly, temperature. Smoke rising from a cigarette held at rest in still air has a straight threadlike appearance for a few centimeters; this indicates a laminar flow. Above that the smoke is wavy and finally irregular lateral motions indicate a turbulent flow. Low velocities and high viscous forces are associated with

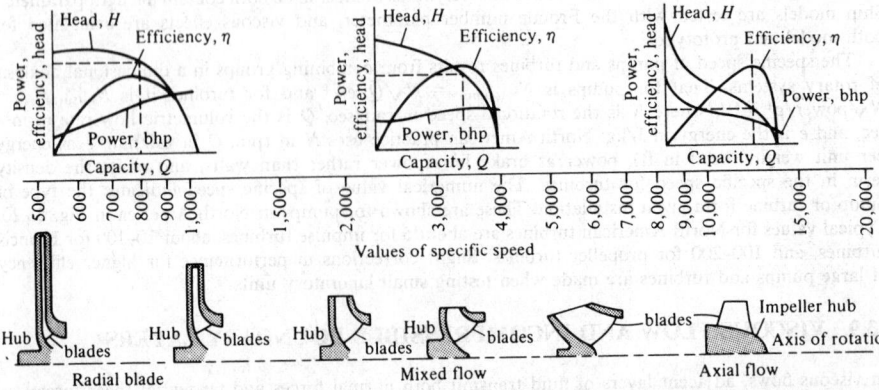

Fig. 53.20 Pump characteristics and specific speed for pump impellers. (Courtesy Worthington Corporation.)

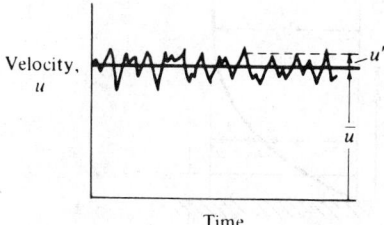

Fig. 53.21 Velocity at a point in steady turbulent flow.

laminar flow and low Reynolds numbers. High speeds and low viscous forces are associated with turbulent flow and high Reynolds numbers. Turbulence is a characteristic of flows, not of fluids. Typical fluctuations of velocity in a turbulent flow are shown in Fig. 53.21.

The axes of eddies in turbulent flow are generally distributed in all directions. In *isotropic* turbulence they are distributed equally. In flows of low turbulence, the fluctuations are small; in highly turbulent flows, they are large. The turbulence level may be defined as (as a percentage)

$$T = \frac{\sqrt{(\overline{u'^2} + \overline{v'^2} + \overline{w'^2})/3}}{\bar{u}} \times 100$$

where u', v', and w' are instantaneous fluctuations from mean values and $\bar{u}$ is the average velocity in the main flow direction (x, in this instance).

Shear stresses in turbulent flows are much greater than in laminar flows for the same velocity gradient and fluid.

53.9.2 Boundary Layers

The growth of a boundary layer along a flat plate in a uniform external flow is shown in Fig. 53.22. The region of retarded flow, δ, thickens in the direction of flow, and thus the velocity changes from zero at the plate surface to the free stream value u_s in an increasingly larger distance δ normal to the plate. Thus, the velocity gradient at the boundary, and hence the shear stress as well, decreases as the flow progresses downstream, as shown. As the laminar boundary thickens, instabilities set in and the boundary layer becomes turbulent. The transition from the laminar boundary layer to a turbulent boundary layer does not occur at a well-defined location; the flow is intermittently laminar and turbulent with a larger portion of the flow being turbulent as the flow passes downstream. Finally, the flow is completely turbulent, and the boundary layer is much thicker and the boundary shear greater in the turbulent region than if the flow were to continue laminar. A viscous sublayer exists within the turbulent boundary layer along the boundary surface. The shape of the velocity profile also changes when the boundary layer becomes turbulent, as shown in Fig. 53.22. Boundary surface roughness, high turbulence level in the outer flow, or a decelerating free stream causes transition to occur nearer the leading edge of the plate. A surface is considered rough if the roughness elements have an effect outside the viscous sublayer, and smooth if they do not. Whether a surface is rough or smooth depends not only on the surface itself but also on the character of the flow passing it.

A boundary layer will separate from a continuous boundary if the fluid within it is caused to

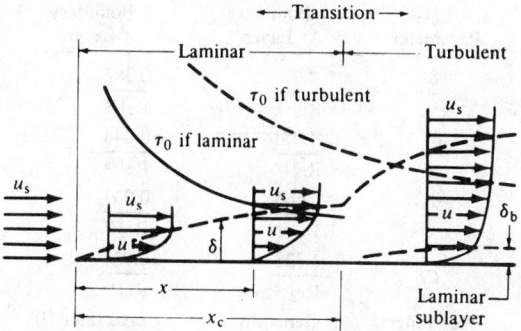

Fig. 53.22 Boundary layer development along a flat plate.

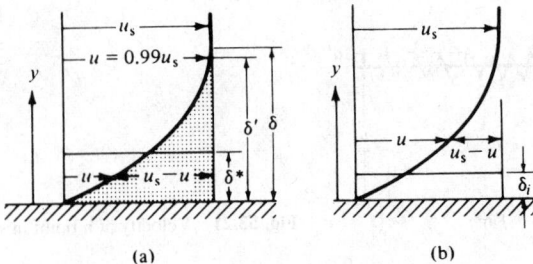

Fig. 53.23 Definition of boundary layer thickness: (*a*) displacement thickness; (*b*) momentum thickness.

slow down such that the velocity gradient du/dy becomes zero at the boundary. An adverse pressure gradient will cause this.

One parameter of interest is the boundary layer thickness δ, the distance from the boundary in which the flow is retarded, or the distance to the point where the velocity is 99% of the free stream velocity (Fig. 53.23). The displacement thickness is the distance the boundary is displaced such that the boundary layer flow is the same as one-dimensional flow past the displaced boundary. It is given by (see Fig. 53.23)

$$\delta_1 = \frac{1}{u_s} \int_0^\delta (u_s - u)\, dy = \int_0^\delta \left(1 - \frac{u}{u_s}\right) dy$$

A momentum thickness is the distance from the boundary such that the momentum flux of the free stream within this distance is the deficit of momentum of the boundary layer flow. It is given by (see Fig. 53.23)

$$\delta_2 = \int_0^\delta \left(1 - \frac{u}{u_s}\right)\frac{u}{u_s}\, dy$$

Also of interest is the viscous shear drag $D = C_f (\rho u_s^2/2)A$, where C_f is the average skin friction drag coefficient and A is the area sheared.

These parameters are listed in Table 53.7 as functions of the Reynolds number $Re_x = u_s \rho x/\mu$, where x is based on the distance from the leading edge. For Reynolds numbers between 1.8×10^5 and 4.5×10^7, $C_f = 0.045/Re_x^{1/6}$, and for Re_x between 2.9×10^7 and 5×10^8, $C_f = 0.0305/Re_x^{1/7}$. These results for turbulent boundary layers are obtained from pipe flow friction measurements for smooth pipes, by assuming the pipe radius equivalent to the boundary layer thickness, the centerline pipe velocity equivalent to the free stream boundary layer flow, and appropriate velocity profiles. Results agree with measurements.

When a turbulent boundary layer is preceded by a laminar boundary layer, the drag coefficient is given by the Prandtl–Schlichting equation:

$$C_f = \frac{0.455}{(\log_{10} Re_x)^{2.58}} - \frac{A}{Re_x}$$

Table 53.7   Boundary Layer Parameters

Parameter	Laminar Boundary Layer	Turbulent Boundary Layer
$\dfrac{\delta}{x}$	$\dfrac{4.91}{Re_x^{1/2}}$	$\dfrac{0.382}{Re_x^{1/5}}$
$\dfrac{\delta_1}{x}$	$\dfrac{1.73}{Re_x^{1/2}}$	$\dfrac{0.048}{Re_x^{1/5}}$
$\dfrac{\delta_2}{x}$	$\dfrac{0.664}{Re_x^{1/2}}$	$\dfrac{0.037}{Re_x^{1/5}}$
C_f	$\dfrac{1.328}{Re_x^{1/2}}$	$\dfrac{0.074}{Re_x^{1/5}}$
Re_x range	Generally not over 10^6	Less than 10^7

where A depends on the Reynolds number Re_c at which transition occurs. Values of A for various values of $\mathrm{Re}_c = u_s x_c / \nu$ are

Re_c	3×10^5	5×10^5	9×10^5	1.5×10^6
A	1035	1700	3000	4880

Some results are shown in Fig. 53.24 for transition at these Reynolds numbers for completely laminar boundary layers, for completely turbulent boundary layers, and for a typical ship hull. (The other curves are applicable for smooth model ship hulls.) Drag coefficients for flat plates may be used for other shapes that approximate flat plates.

The thickness of the viscous sublayer δ_b in terms of the boundary layer thickness is approximately

$$\frac{\delta_b}{\delta} = \frac{80}{(\mathrm{Re}_x)^{7/10}}$$

At $\mathrm{Re}_x = 10^6$, $\delta_b/\delta = 0.0050$ and when $\mathrm{Re}_x = 10^7$, $\delta_b/\delta = 0.001$, and thus the viscous sublayer is very thin.

Experiments show that the boundary layer thickness and local drag coefficient for a turbulent boundary layer preceded by a laminar boundary layer at a given location are the same as though the boundary layer were turbulent from the beginning of the plate or surface along which the boundary layer grows.

53.10 GAS DYNAMICS

In gas flows where density variations are appreciable, large variations in velocity and temperature may also occur and then thermodynamic effects are important.

53.10.1 Adiabatic and Isentropic Flow

In adiabatic flow of a gas with no external work and with changes in elevation negligible, the steady-flow energy equation is

$$\frac{V_1^2}{2} + h_1 = \frac{V_2^2}{2} + h_2 = h_0 = \text{constant}$$

for flow from point 1 to point 2, where V is velocity and h is enthalpy. Subscript 0 refers to a stagnation condition where the velocity is zero.

The speed of sound is $c = \sqrt{(\partial p/\partial s)_{\text{isentropic}}} \sqrt{K/\rho} = \sqrt{kp/\rho} = \sqrt{kRT}$. For air, $c = 20.04\sqrt{T}$ m/sec, where T is in degrees kelvin. A local Mach number is then $M = V/c = V/\sqrt{kRT}$.

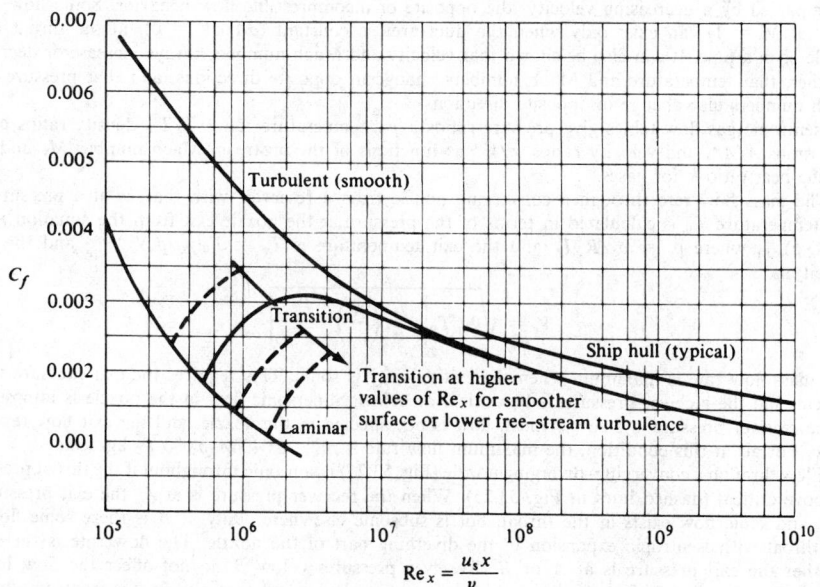

Fig. 53.24 Drag coefficients for smooth plane surfaces parallel to flow.

A gas at rest may be accelerated adiabatically to any speed, including sonic ($M = 1$) and theoretically to its maximum speed when the temperature reduces to absolute zero. Then,

$$c_p\,T_0 = c_p\,T + \frac{V^2}{2} = c_p\,T^* + \frac{V^{*2}}{2} = \frac{V_{\max}^2}{2}$$

where the asterisk (*) refers to a sonic state where the Mach number is unity.

The stagnation temperature T_0 is $T_0 = T + V^2/2c_p$, or in terms of the Mach number [$c_p = R\,k/(k - 1)$]

$$\frac{T_0}{T} = 1 + \frac{k - 1}{2}\,M^2 = 1 + 0.2M^2 \text{ for air}$$

The stagnation temperature is reached adiabatically from any velocity V where the Mach number is M and the temperature T. The temperature T^* in terms of the stagnation temperature T_0 is $T^*/T_0 = 2/(k + 1) = \frac{5}{6}$ for air.

The stagnation pressure is reached reversibly and is thus the isentropic stagnation pressure. It is also called the reservoir pressure, since for any flow a reservoir (stagnation) pressure may be imagined from which the flow proceeds isentropically to a pressure p at a Mach number M. The stagnation pressure p_0 is a constant in isentropic flow; if nonisentropic, but adiabatic, p_0 decreases:

$$\frac{p_0}{p} = \left(\frac{T_0}{T}\right)^{k/(k-1)} = \left(1 + \frac{k - 1}{2}\,M^2\right)^{k/(k-1)} = (1 + 0.2M^2)^{3.5} \text{ for air}$$

Expansion of this expression gives

$$p_0 = p + \frac{\rho V^2}{2}\left[1 + \frac{1}{4}\,M^2 + \frac{2 - k}{24}\,M^4 + \frac{(2 - k)(3 - 2k)}{192}\,M^6 + \cdots\right]$$

where the term in brackets is the compressibility factor. It ranges from 1 at very low Mach numbers to a maximum of 1.27 at $M = 1$, and shows the effect of increasing gas density as it is brought to a stagnation condition at increasingly higher initial Mach numbers. The equations are valid to or from a stagnation state for subsonic flow, and from a stagnation state for supersonic flow at M^2 less than $2/(k - 1)$, or M less than $\sqrt{5}$ for air.

53.10.2   Duct Flow

Adiabatic flow in short ducts may be considered reversible, and thus the relation between velocity and area changes is $dA/dV = (A/V)(M^2 - 1)$. For subsonic flow, dA/dV is negative and velocity changes relate to area changes in the same way as for incompressible flow. At supersonic speed, dA/dV is positive and an expanding area is accompanied by an increasing velocity; a contracting area is accompanied by a decreasing velocity, the opposite of incompressible flow behavior. Sonic flow in a duct (at $M = 1$) can exist only when the duct area is constant ($dA/dV = 0$), in the throat of a nozzle or in a pipe. It can also be shown that velocity and Mach numbers always increase or decrease together, that temperature and Mach numbers change in opposite directions, and that pressure and Mach numbers also change in opposite directions.

Isentropic gas flow tables give pressure ratios p/p_0, temperature ratios T/T_0, density ratios ρ/ρ_0, area ratios A/A^*, and velocity ratios V/V^* as functions of the upstream Mach number M_x and the specific heat ratio k for gases.

The mass flow rate through a converging nozzle from a reservoir with the gas at a pressure p_0 and temperature T_0 is calculated in terms of the pressure at the nozzle exit from the equation $\dot{m} = (V\,A\,\rho)_{\text{exit}}$, where $\rho_e = p_e/R\,T_e$ and the exit temperature is $T_e = T_0(p_e/p_0)^{(k-1)/k}$ and the exit velocity is

$$V_e = \sqrt{2c_p T_0\left[1 - \left(\frac{p_e}{p_0}\right)^{(k-1)/k}\right]}$$

The mass flow rate is maximum when the exit velocity is sonic. This requires the exit pressure to be critical, and the receiver pressure to be critical or below. Supersonic flow in the nozzle is impossible. If the receiver pressure is below critical, flow is not affected in the nozzle, and the exit flow remains sonic. For air at this condition, the maximum flow rate is $\dot{m} = 0.0404A_1p_0/\sqrt{T_0}$ kg/sec.

Flow through a converging–diverging nozzle (Fig. 53.25) is subsonic throughout if the throat pressure is above critical (dashed lines in Fig. 53.25). When the receiver pressure is at A, the exit pressure is also, and sonic flow exists at the throat, but is subsonic elsewhere. Only at B is there sonic flow in the throat with isentropic expansion in the diverging part of the nozzle. The flow rate is the same whether the exit pressure is at A or B. Receiver pressures below B do not affect the flow in the nozzle. Below A (at C, for example) a shock forms as shown and then the flow is isentropic to the shock, and beyond it, but not through it. When the throat flow is sonic, the mass flow rate is given

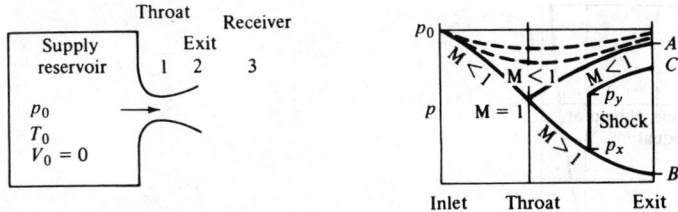

Fig. 53.25 Gas flow through converging–diverging nozzle.

by the same equation as for a converging nozzle with sonic exit flow. The pressures at A and B in terms of the reservoir pressure p_0 are given in isentropic flow tables as a function of the ratio of exit area to throat area, A_e/A^*.

53.10.3 Normal Shocks

The plane of a normal shock is at right angles to the flow streamlines. These shocks may occur in the diverging part of a nozzle, the diffuser of a supersonic wind tunnel, in pipes and forward of blunt-nosed bodies. In all instances the flow is supersonic upstream and subsonic downstream of the shock. Flow through a shock is not isentropic, although nearly so for very weak shocks. The abrupt changes in gas density across a shock allow for optical detection. The interferometer responds to density changes, the Schlieren method to density gradients, and the spark shadowgraph to the rate of change of density gradient. Density ratios across normal shocks in air are 2 at $M = 1.58$, 3 at $M = 2.24$, and 4 at $M = 3.16$ to a maximum value of 6.

Changes in fluid and flow parameters across normal shocks are obtained from the continuity, energy, and momentum equations for adiabatic flow. They are expressed in terms of upstream Mach numbers with upstream conditions designated with subscript x and downstream with subscript y. Mach numbers M_x and M_y are related by

$$\frac{1+kM_x^2}{M_x\left(1+\dfrac{k-1}{2}M_x^2\right)^{1/2}} = \frac{1+kM_y^2}{M_y\left(1+\dfrac{k-1}{2}M_y^2\right)^{1/2}} = f(M,k)$$

which is plotted in Fig. 53.26. The requirement for an entropy increase through the shock indicates M_x to be greater than M_y. Thus, the higher the upstream Mach number, the lower the downstream Mach number, and vice versa. For normal shocks, values of downstream Mach number M_y; temperature ratios T_y/T_x; pressure ratios p_y/p_x, p_{0y}/p_x, and p_{0y}/p_{0x}; and density ratios ρ_y/ρ_x depend only on the upstream Mach number M_x and the specific heat ratio k of the gas. These values are tabulated in books on gas dynamics and in books of gas tables.

The density ratio across the shock is given by the Rankine–Hugoniot equation

$$\frac{p_y}{\rho_x} = \left[\left(\frac{k+1}{k-1}\right)\frac{p_y}{p_x}+1\right] \Big/ \left[\frac{p_y}{p_x}+\left(\frac{k+1}{k-1}\right)\right]$$

and is plotted in Fig. 53.27, which shows that weak shocks are nearly isentropic, and that the density ratio approaches a limit of 6 for gases with $k = 1.4$.

Gas tables show that at an upstream Mach number of 2 for air, $M_y = 0.577$, the pressure ratio is $p_y/p_x = 4.50$, the density ratio is $\rho_y/\rho_x = 2.66$, the temperature ratio is $T_y/T_x = 1.68$, and the

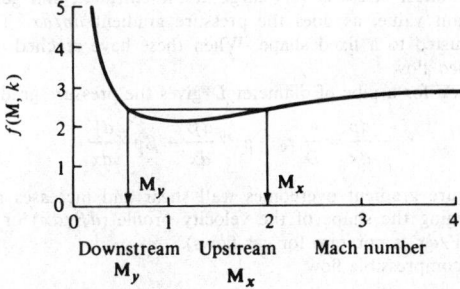

Fig. 53.26 Mach numbers across a normal shock, $k = 1.4$.

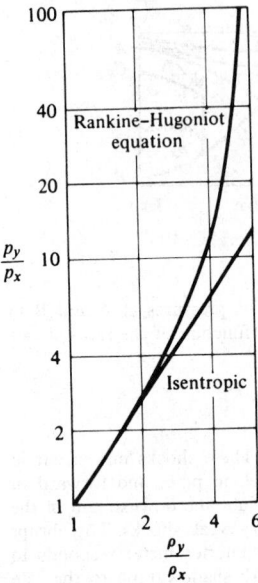

Fig. 53.27 Rankine–Hugoniot curve, $k = 1.4$.

stagnation pressure ratio is $p_{oy}/p_{ox} = 0.72$, which indicates an entropy increase of $s_y - s_x = -R \ln(p_{oy}/p_{ox}) = 94$ J/kg.

53.10.4 Oblique Shocks

Oblique shocks are inclined from a direction normal to the approaching streamlines. Figure 53.28 shows that the normal velocity components are related by the normal shock relations. From a momentum analysis, the tangential velocity components are unchanged through oblique shocks. The upstream Mach number M_1 is given in terms of the deflection angle θ, the shock angle β, and the specific heat ratio k for the gas as

$$\frac{1}{M_1^2} = \sin^2 \beta - \frac{(k+1)}{2} \frac{\sin \beta \sin \theta}{\cos(\beta - \theta)}$$

The geometry is shown in Fig. 53.29, and the variables in this equation are illustrated in Fig. 53.30. For each M_1 there is the possibility of two wave angles β for a given deflection angle θ. The larger wave angle is for strong shocks, with subsonic downstream flow. The smaller wave angle is for weak shocks, generally with supersonic downstream flow at a Mach number less than M_1.

Normal shock tables are used for oblique shocks if M_x is used for $M_1 \sin \beta$. Then $M_y = M_2 \sin(\beta - \theta)$ and other ratios of property values (pressure, temperature, and density) are the same as for normal shocks.

53.11 VISCOUS FLUID FLOW IN DUCTS

The development of flow in the entrance of a pipe with the development of the boundary layer is shown in Fig. 53.31. Wall shear stress is very large at the entrance, and generally decreases in the flow direction to a constant value, as does the pressure gradient dp/dx. The velocity profile also changes and becomes adjusted to a fixed shape. When these have reached constant conditions, the flow is called *fully developed* flow.

The momentum equation for a pipe of diameter D gives the pressure gradient as

$$-\frac{dp}{dx} = \frac{4}{D} \tau_0 + \rho V^2 \frac{d\beta}{dx} + \beta \rho V \frac{dV}{dx}$$

which shows that a pressure gradient overcomes wall shear and increases momentum of the fluid either as a result of changing the shape of the velocity profile ($d\beta/dx$) or by changing the mean velocity along the pipe (dV/dx is not zero for gas flows).

For fully developed incompressible flow

$$-\frac{dp}{dx} = \frac{\Delta p}{L} = \frac{4\tau_0}{D}$$

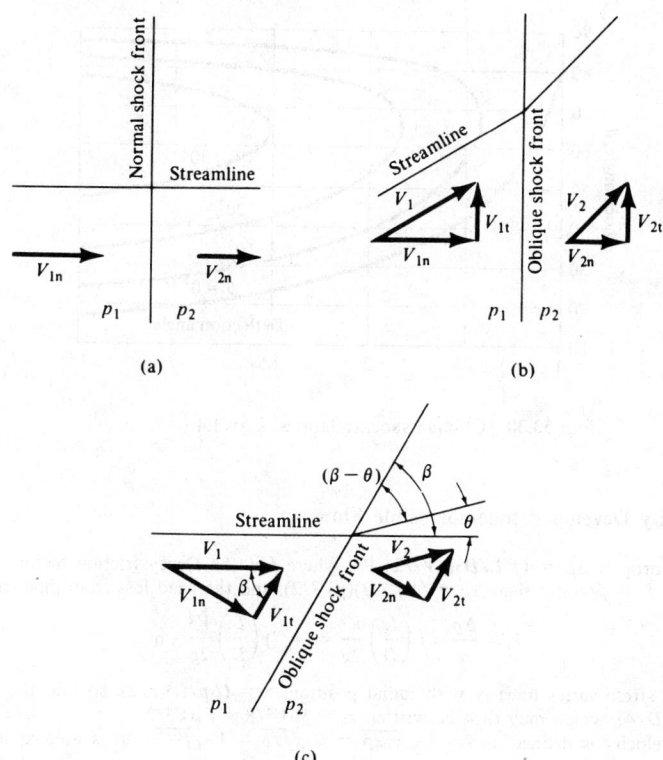

Fig. 53.28 Oblique shock relations from normal shock: (*a*) normal shock; (*b*) oblique shock; (*c*) oblique shock angles.

and a pressure drop simply overcomes wall shear.

For developing flow in the entrance, $\beta = 1$ initially and increases to a constant value downstream. Thus, the pressure gradient overcomes wall shear and also increases the flow momentum according to

$$-\frac{dp}{dx} = \frac{4\tau_0}{D} + \rho V^2 \frac{d\beta}{dx}$$

For fully developed flow, $\beta = \frac{4}{3}$ for laminar flow and $\beta \approx 1.03$ for turbulent flow in round pipes.

For compressible gas flow beyond the entrance, the velocity profile becomes essentially fixed in shape, but the velocity changes because of thermodynamic effects that change the density. Thus, the pressure gradient is

$$-\frac{dp}{dx} = \frac{4\tau_0}{D} + \beta \rho V \frac{dV}{dx}$$

Here β is essentially constant but dV/dx may be significant.

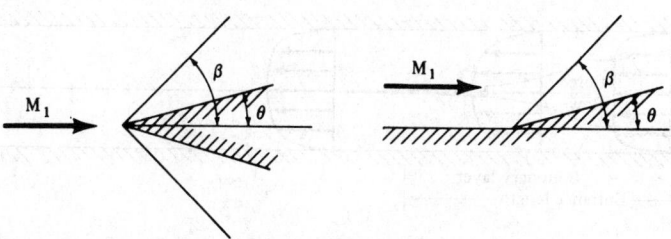

Fig. 53.29 Supersonic flow past a wedge and an inside corner.

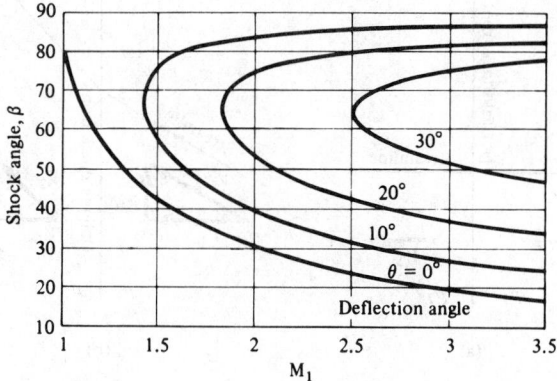

Fig. 53.30 Oblique shock relations, $k = 1.4$.

53.11.1 Fully Developed Incompressible Flow

The pressure drop is $\Delta p = (f\,L/D)(\rho V^2/2)$ Pa, where f is the Darcy friction factor. The Fanning friction factor $f' = f/4$ and then $\Delta p = (4\,f'/D)(\rho V^2/2)$, and the head loss from pipe friction is

$$h_f = \frac{\Delta p}{\gamma} = f\left(\frac{L}{D}\right)\frac{V^2}{2g} = (4f')\left(\frac{L}{D}\right)\frac{V^2}{2g} \quad \text{m}$$

The shear stress varies linearly with radial position, $\tau = (\Delta p/L)(r/2)$, so that the wall shear is $\tau_0 = (\Delta p/L)(D/4)$, which may then be written $\tau_0 = f\rho V^2/8 = f'\rho V^2/2$.

A shear velocity is defined as $v_* = \sqrt{\tau_0/\rho} = V\sqrt{f/8} = V\sqrt{f'/2}$ and is used as a normalizing parameter.

For noncircular ducts the diameter D is replaced by the hydraulic or equivalent diameter $D_h = 4A/P$, where A is the flow cross section and P is the wetted perimeter. Thus, an annulus between pipes of diameter D_1 and D_2, D_1 being larger, the hydraulic diameter is $D_2 - D_1$.

53.11.2 Fully Developed Laminar Flow in Ducts

The velocity profile in circular tubes is that of a parabola, and the centerline velocity is

$$u_{\max} = \frac{\Delta p}{L}\left(\frac{R^2}{4\mu}\right)$$

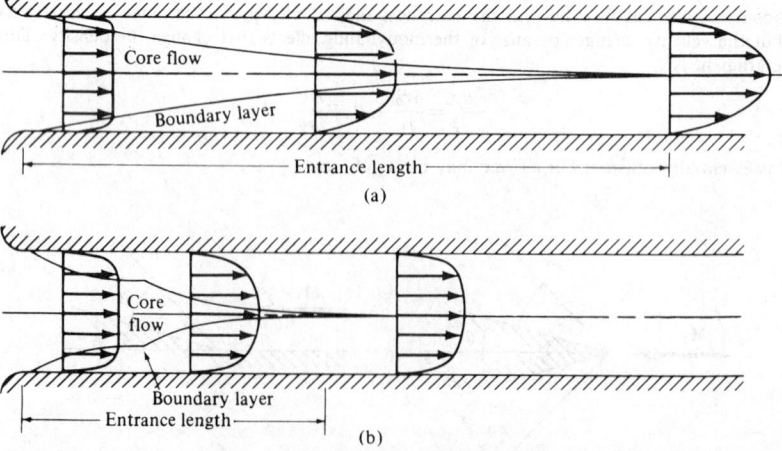

Fig. 53.31 Growth of boundary layers in a pipe: (a) laminar flow; (b turbulent flow.

and the velocity profile is

$$\frac{u}{u_{\max}} = 1 - \left(\frac{r}{R}\right)^2$$

where r is the radial location in a pipe of radius R. The average velocity is one-half the maximum velocity, $V = u_{\max}/2$.

The pressure gradient is

$$\frac{\Delta p}{L} = \frac{128\mu Q}{\pi D^4}$$

which indicates a linear increase with increasing velocity or flow rate. The friction factor for circular ducts is $f = 64/\mathrm{Re}_D$ or $f' = 16/\mathrm{Re}_D$ and applies to both smooth as well as rough pipes, for Reynolds numbers up to about 2000.

For *noncircular* ducts the value of the friction factor is $f = C/\mathrm{Re}$ and depends on the duct geometry. Values of $f\,\mathrm{Re} = C$ are listed in Table 53.8.

53.11.3 Fully Developed Turbulent Flow in Ducts

Knowledge of turbulent flow in ducts is based on physical models and experiments. Physical models describe lateral transport of fluid as a result of mixing due to eddies. Prandtl and von Kármán both derived expressions for shear stresses in turbulent flow based on the Reynolds stress ($\tau = -\rho \overline{u'v'}$) and obtained velocity defect equations for pipe flow. Prandtl's equation is

$$\frac{u_{\max} - u}{\sqrt{\tau_0/\rho}} = \frac{u_{\max} - u}{v_*} = 2.5 \ln \frac{R}{y}$$

where $u_{\max}$ is the centerline velocity and u is the velocity a distance y from the pipe wall. von Kármán's equation is

$$\frac{u_{\max} - u}{\sqrt{\tau_0/\rho}} = \frac{u_{\max} - u}{v_*}$$
$$= -\frac{1}{\kappa}\left[\ln\left(1 - \sqrt{1 - \frac{y}{R}}\right) + \sqrt{1 - \frac{y}{R}} \right]$$

In both, κ is an experimentally determined constant equal to 0.4 (some experiments show better agreement when $\kappa = 0.36$). Similar expressions apply to external boundary layer flow when the pipe radius R is replaced by the boundary layer thickness δ. Friction factors for smooth pipes have been developed from these results. One is the Blasius equation for $\mathrm{Re}_D = 10^5$ and is $f = 0.316/\mathrm{Re}_D^{1/4}$ obtained by using a power-law velocity profile $u/u_{\max} = (y/R)^{1/7}$. The value 7 here increases to 10 at higher Reynolds numbers. The use of a logarithmic form of velocity profile gives the Prandtl law of pipe friction for smooth pipes:

$$\frac{1}{\sqrt{f}} = 2\log(\mathrm{Re}_D \sqrt{f}) - 0.8$$

which agrees well with experimental values. A more explicit formula by Colebrook is $1/\sqrt{f} = 1.8\log(\mathrm{Re}_D/6.9)$, which is within 1% of the Prandtl equation over the entire range of turbulent Reynolds numbers.

The logarithmic velocity defect profiles apply for *rough* pipes as well as for smooth pipes, since the velocity defect ($u_{\max} - u$) decreases linearly with the shear velocity v_*, keeping the ratio of the two constant.

A relation between the centerline velocity and the average velocity is $u_{\max}/V = 1 + 1.33\sqrt{f}$, which may be used to estimate the average velocity from a single centerline measurement.

The Colebrook–White equation encompasses all turbulent flow regimes, for both smooth and rough pipes:

$$\frac{1}{\sqrt{f}} = 1.74 - 2\log\left(\frac{2k}{D} + \frac{18.7}{\mathrm{Re}_D \sqrt{f}}\right)$$

and this is plotted in Fig. 53.32, where k is the equivalent sand-grain roughness. A simpler equation by Haaland is

$$\frac{1}{\sqrt{f}} = -1.8\log\left[\frac{6.9}{\mathrm{Re}_D} + \left(\frac{k}{3.7D}\right)^{1.11}\right]$$

Table 53.8 Friction Factors for Laminar Flow

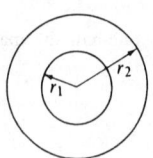
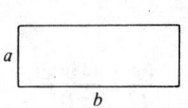
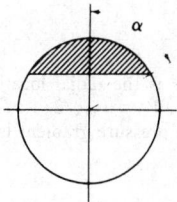

r_1/r_2	f Re	a/b	f Re	α	f Re
0.0001	71.78	0	96.00	0	62.2
0.001	74.68	1/20	89.91	10	62.2
0.01	80.11	1/10	84.68	20	62.3
0.05	86.27	1/8	82.34	30	62.4
0.10	89.37	1/6	78.81	40	62.5
0.20	92.35	1/4	72.93	60	62.8
0.40	94.71	2/5	65.47	90	63.1
0.60	95.59	1/2	62.19	120	63.3
0.80	95.92	3/4	57.89	150	63.7
1.00	96.00	1	56.91	180	64.0

	CIRCULAR SECTOR	ISOSCELES TRIANGLE	RIGHT TRIANGLE
α	f Re	f Re	f Re
0	48.0	48.0	48.0
10	51.8	51.6	49.9
20	54.5	52.9	51.2
30	56.7	53.3	52.0
40	58.4	52.9	52.4
50	59.7	52.0	52.4
60	60.8	51.1	52.0
70	61.7	49.5	51.2
80	62.5	48.3	49.9
90	63.1	48.0	48.0

which is explicit in f and is within 1.5% of the Colebrook–White equation in the range $4000 \leqq \mathrm{Re}_D \leqq 10^8$ and $0 \leqq k/D \leqq 0.05$.

Three types of problems may be solved:

1. *The Pressure Drop or Head Loss.* The Reynolds number and relative roughness are determined and calculations are made directly.

2. *The Flow Rate for Given Fluid and Pressure Drops or Head Loss.* Assume a friction factor, based on a high Re_D for a rough pipe, and determine the velocity from the Darcy equation. Calculate a Re_D, get a better f, and repeat until successive velocities are the same. A second method is to assume a flow rate and calculate the pressure drop or head loss. Repeat until results agree with the given pressure drop or head loss. A plot of Q versus h_L, for example, for a few trials may be used.

3. *A Pipe Size.* Assume a pipe size and calculate the pressure drop or head loss. Compare with given values. Repeat until agreement is reached. A plot of D versus h_L, for example, for a

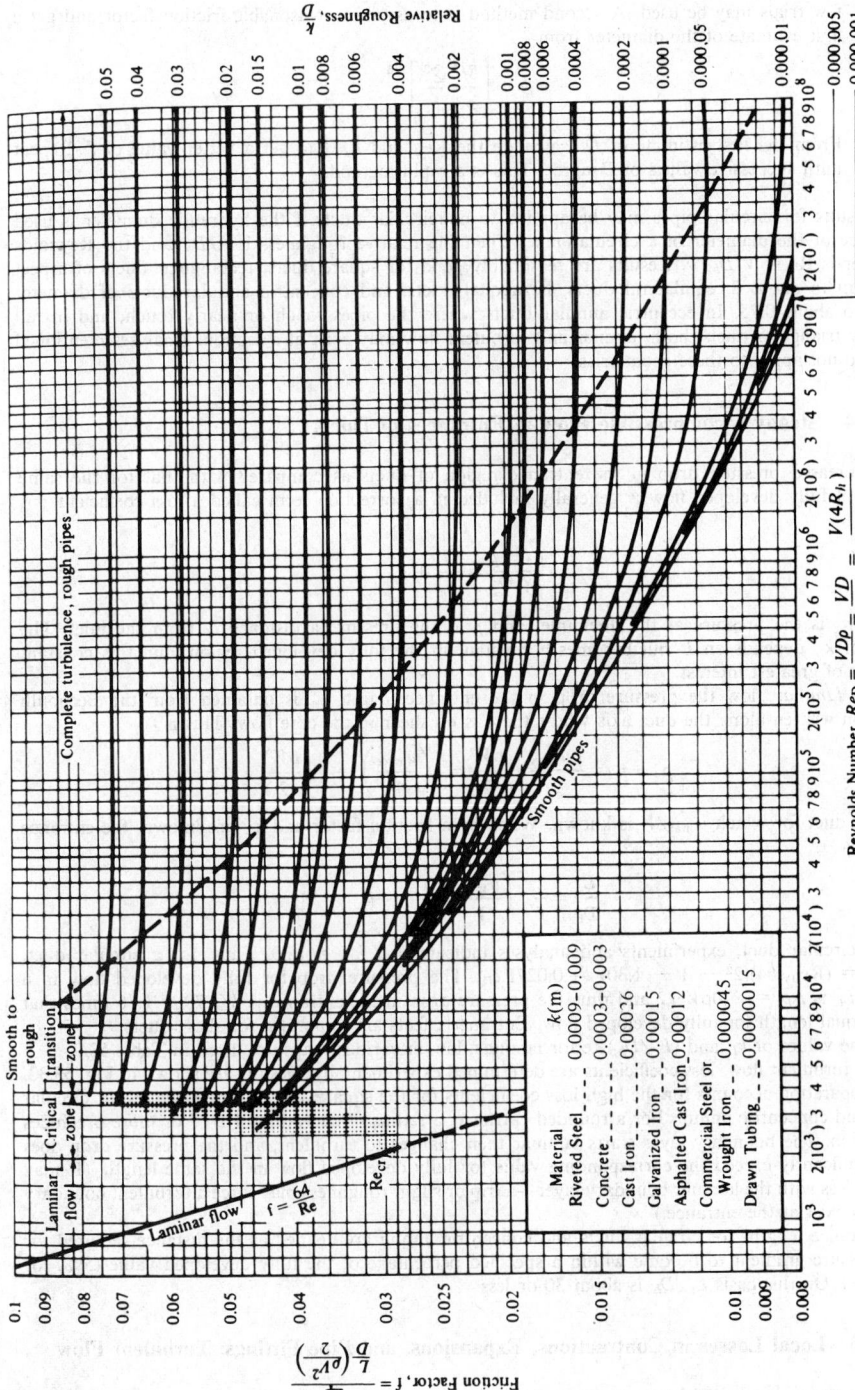

Fig. 53.32 Friction factors for commercial pipe. [From L. F. Moody, "Friction Factors for Pipe Flow," *Trans. ASME,* **66** (1944). Courtesy of The American Society of Mechanical Engineers.]

few trials may be used. A second method is to assume a reasonable friction factor and get a first estimate of the diameter from

$$D = \left[\frac{8fLQ^2}{\pi^2 g \, h_f} \right]^{1/5}$$

From the first estimate of D, calculate the Re_D and k/D to get a better value of f. Repeat until successive values of D agree. This is a rapid method.

Results for circular pipes may be applied to noncircular ducts if the hydraulic diameter is used in place of the diameter of a circular pipe. Then the relative roughness is k/D_h and the Reynolds number is $\mathrm{Re} = V \, D_h / \nu$. Results are reasonably good for square ducts, rectangular ducts of aspect ratio up to about 8, equilateral ducts, hexagonal ducts, and concentric annular ducts of diameter ratio to about 0.75. In eccentric annular ducts where the pipes touch or nearly touch, and in tall narrow triangular ducts, both laminar and turbulent flow may exist at a section. Analyses mentioned here do not apply to these geometries.

53.11.4 Steady Incompressible Flow in Entrances of Ducts

The increased pressure drop in the entrance region of ducts as compared with that for the same length of fully developed flow is generally included in a correction term called a loss coefficient, k_L. Then,

$$\frac{p_1 - p}{\rho V^2/2} = \frac{fL}{D_h} + k_L$$

where p_1 is the pressure at the duct inlet and p is the pressure a distance L from the inlet. The value of k_L depends on L but becomes a constant in the fully developed region, and this constant value is of greatest interest.

For *laminar* flow the pressure drop in the entrance length L_e is obtained from the Bernoulli equation written along the duct axis where there is no shear in the core flow. This is

$$p_1 - p_e = \frac{\rho u_{\max}^2}{2} - \frac{\rho V^2}{2} = \left[\left(\frac{u_{\max}}{V} \right)^2 - 1 \right] \frac{\rho V^2}{2}$$

for any duct for which $u_{\max}/V$ is known. When both friction factor and k_L are known, the entrance length is

$$\frac{L_e}{D_h} = \frac{1}{f} \left[\left(\frac{u_{\max}}{V} \right)^2 - 1 - k_L \right]$$

For a circular duct, experiments and analyses indicate that $k_L \approx 1.30$. Thus, for a circular duct, $L_e/D = (\mathrm{Re}_D/64)(2^2 - 1 - 1.30) = 0.027 \mathrm{Re}_D$. The pressure drop for fully developed flow in a length L_e is $\Delta p = 1.70 \rho V^2/2$ and thus the pressure drop in the entrance is $3/1.70 = 1.76$ times that in an equal length for fully developed flow. Entrance effects are important for short ducts.

Some values of k_L and $(L_e/D_h)\mathrm{Re}$ for laminar flow in various ducts are listed in Table 53.9.

For turbulent flow, loss coefficients are determined experimentally. Results are shown in Fig. 53.33. Flow separation accounts for the high loss coefficients for the square and reentrant shapes for circular tubes and concentric annuli. For a rounded entrance, a radius of curvature of $D/7$ or more precludes separation. The boundary layer starts laminar then changes to turbulent, and the pressure drop does not significantly exceed the corresponding value for fully developed flow in the same length. (It may even be less with the laminar boundary layer—a trip or slight roughness may force a turbulent boundary layer to exist at the entrance.)

Entrance lengths for circular ducts and concentric annuli are defined as the distance required for the pressure gradient to become within a specified percentage of the fully developed value (5%, for example). On this basis L_e/D_h is about 30 or less.

53.11.5 Local Losses in Contractions, Expansions, and Pipe Fittings; Turbulent Flow

Calculations of local head losses generally are approximate at best unless experimental data for given fittings are provided by the manufacturer.

Losses in *contractions* are given by $h_L = k_L V^2/2g$. Loss coefficients for a sudden contraction are shown in Fig. 53.34. For gradually contracting sections k_L may be as low as 0.03 for D_2/D_1 of 0.5 or less.

Losses in expansions are given by $h_L = k_L (V_1 - V_2)^2/2g$, section 1 being upstream. For a sudden expansion, $k_L = 1$, and for gradually expanding sections with divergence angles of 7° or 8°, k_L may

Table 53.9 Entrance Effects, Laminar Flow (See Table 53.8 for Symbols)

r_1/r_2	k_L		a/b	k_L	L_c/D_h, Re		α	k_L
0.0001	1.13		0	0.69	0.0059		0	1.74
0.001	1.07		1/8	0.88	0.0094		10	1.73
0.01	0.97		1/5	1.00	0.0123		20	1.72
0.05	0.86		1/4	1.08	0.0146		30	1.69
0.10	0.81		1/2	1.38	0.0254		40	1.65
0.20	0.75		3/4	1.52	0.0311		60	1.57
0.40	0.71		1	1.55	0.0324		90	1.46
0.60	0.69						120	1.39
0.80	0.69						150	1.34
1.00	0.69						180	1.33

α	CIRCULAR SECTOR k_L	ISOSCELES TRIANGLE k_L	RIGHT TRIANGLE k_L
0	2.97	2.97	2.97
10	2.06	2.14	2.40
20	1.71	1.85	2.09
30	1.58	1.79	1.94
40	1.53	1.83	1.88
50	1.50	1.95	1.88
60	1.49	2.14	1.94
70	1.48	2.38	2.09
80	1.47	2.72	2.40
90	1.46	2.97	2.97

be as low as 0.14 or even 0.06 for diffusers for low-speed wind tunnels or cavitation-testing water tunnels with curved inlets to avoid separation.

Losses in pipe fittings are given in the form $h_L = k_L V^2/2g$ or in terms of an equivalent pipe length by pipe-fitting manufacturers. Typical values for various fittings are given in Table 53.10.

53.11.6 Flow of Compressible Gases in Pipes with Friction

Subsonic gas flow in pipes involves a decrease in gas density and an increase in gas velocity in the direction of flow. The momentum equation for this flow may be written as

$$\frac{dp}{\rho V^2/2} + f\frac{dx}{D} + 2\frac{dV}{V} = 0$$

For *isothermal* flow the first term is $(2/\rho_1 V_1^2 p_1)p\, dp$, where the subscript 1 refers to an upstream section where all conditions are known. For $L = x_2 - x_1$, integration gives

$$p_1^2 - p_2^2 = \rho_1 V_1^2 p_1 \left(f\frac{L}{D} - 2\ln\frac{p_2}{p_1} \right)$$

or, in terms of the initial Mach number,

$$p_1^2 - p_2^2 = kM_1^2 p_1^2 \left(f\frac{L}{D} - 2\ln\frac{p_2}{p_1} \right)$$

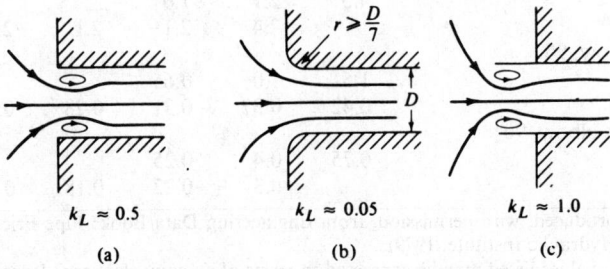

Fig. 53.33 Pipe entrance flows: (a) square entrance; (b) round entrance; (c) reentrant inlet.

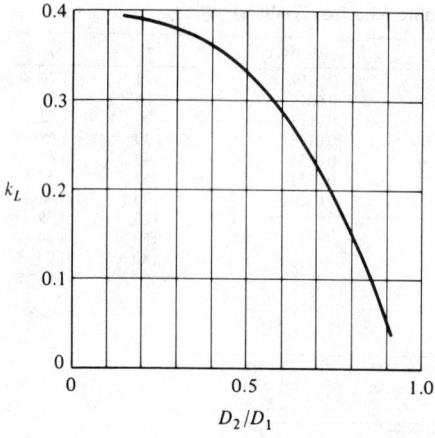

Fig. 53.34 Loss coefficients for abrupt contraction in pipes.

The downstream pressure p_2 at a distance L from section 1 may be obtained by trial by neglecting the term $2 \ln(p_2/p_1)$ initially to get a p_2, then including it for an improved value. The distance L to a section where the pressure is p_2 is obtained from

$$f\frac{L}{D} = \frac{1}{kM_1^2}\left[1 - \left(\frac{p_2}{p_1}\right)^2\right] - 2\ln\frac{p_1}{p_2}$$

A limiting condition (designated by an asterisk) at a length L^* is obtained from an expression of dp/dx to get

$$\frac{dp}{dx} = \frac{pf/2D}{1 - p/\rho V^2} = \frac{(f/D)(\rho V^2/2)}{kM^2 - 1}$$

Table 53.10   Typical Loss Coefficients for Valves and Fittings

VALVE OR FITTING	NOMINAL DIAMETER, CM					
	2.5	5	10	15	20	25
Globe valve, wide open:						
Screwed	9	7	5.5			
Flanged	12	9	6	6	5.5	5.5
Gate valve, wide open:						
Screwed	0.24	0.18	0.13			
Flanged		0.35	0.16	0.11	0.08	0.06
Foot valve, wide open			0.80 for all sizes			
Swing check valve, wide open						
Screwed	3.0	2.3	2.1			
Flanged			2.0 for all sizes			
Angle valve, wide open:						
Screwed	4.5	2.1	1.0			
Flanged		2.4	2.1	2.1	2.1	2.1
Regular elbow, 90°						
Screwed	1.5	1.0	0.65			
Flanged	0.42	0.37	0.31	0.28	0.26	0.25
Long-radius elbow, 90°						
Screwed	0.75	0.4	0.25			
Flanged		0.3	0.22	0.18	0.15	0.14

SOURCE: Reproduced, with permission, from Engineering Data Book: Pipe Friction Manual (Cleveland: Hydraulic Institute, 1979).

Note: The k_L values listed may be expressed in terms of an equivalent pipe length for a given installation and flow by equating $k_L = fL_e/D$ so that $L_e = k_L D/f$.

For a low subsonic flow at an upstream section (as from a compressor discharge) the pressure gradient increases in the flow direction with an infinite value when $M^* = 1/\sqrt{k} = 0.845$ for $k = 1.4$ (air, for example). For M approaching zero, this equation is the Darcy equation for incompressible flow. The limiting pressure is $p^* = p_1 M_1 \sqrt{k}$, and the limiting length is given by

$$\frac{fL^*}{D} = \frac{1}{kM_1^2} - 1 - \ln\frac{1}{kM_1^2}$$

Since the gas at any two locations 1 and 2 in a long pipe has the same limiting condition, the distance L between them is

$$\frac{fL}{D} = \left(\frac{fL^*}{D}\right)_{M_1} - \left(\frac{fL^*}{D}\right)_{M_2}$$

Conditions along a pipe for various initial Mach numbers are shown in Fig. 53.35.

For *adiabatic* flow the limiting Mach number is $M^* = 1$. This is from an expression for dp/dx for adiabatic flow:

$$\frac{dp}{dx} = -\frac{fkp}{2D}M^2\left[\frac{1 + (k-1)M^2}{1 - M^2}\right] = -\frac{f}{D}\frac{\rho V^2}{2}\left[\frac{1 + (k-1)M^2}{1 - M^2}\right]$$

The limiting pressure is

$$\frac{p^*}{p_1} = M_1\sqrt{\frac{2[1 + \frac{1}{2}(k-1)M_1^2]}{k+1}}$$

and the limiting length is

$$\frac{\bar{f}L^*}{D} = \frac{1 - M_1^2}{kM_1^2} + \frac{k+1}{2k}\ln\frac{(k+1)M_1^2}{2[1 + \frac{1}{2}(k-1)M_1^2]}$$

Except for subsonic flow at high Mach numbers, isothermal and adiabatic flow do not differ appreciably. Thus, since flow near the limiting condition is not recommended in gas transmission pipelines because of the excessive pressure drop, and since purely isothermal or purely adiabatic flow is unlikely, either adiabatic or isothermal flow may be assumed in making engineering calculations. For example, for methane from a compressor at 2000 kPa absolute pressure, 60°C temperature and 15 m/sec velocity ($M_1 = 0.032$) in a 30-cm commercial steel pipe, the limiting pressure is 72 kPa absolute at $L^* = 16.9$ km for isothermal flow, and 59 kPa at $L^* = 17.0$ km for adiabatic flow. A pressure of 500 kPa absolute would exist at 16.0 km for either type of flow.

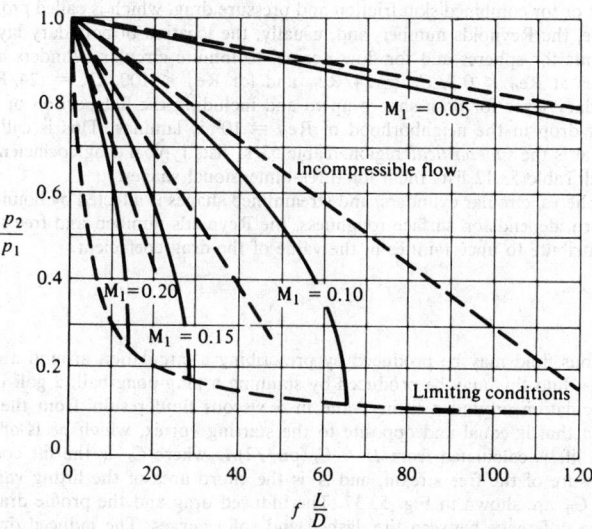

Fig. 53.35 Isothermal gas flow in a pipe for various initial Mach numbers, $k = 1.4$.

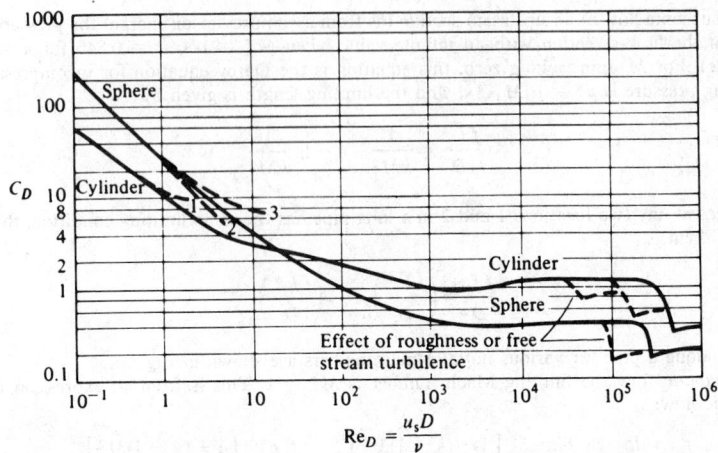

Fig. 53.36 Drag coefficients for infinite circular cylinders and spheres: (1) Lamb's solution for cylinder; (2) Stokes' solution for sphere; (3) Oseen's solution for sphere.

53.12   DYNAMIC DRAG AND LIFT

Two types of forces act on a body past which a fluid flows: a pressure force normal to any infinitesimal area of the body and a shear force tangential to this area. The components of these two forces integrated over the entire body in a direction parallel to the approaching flow is the *drag* force, and in a direction normal to it is the *lift* force. *Induced* drag is associated with a lift force on finite airfoils or blade elements as a result of downwash from tip vortices. Surface waves set up by ships or hydrofoils, and compression waves in gases such as Mach cones are the source of *wave* drag.

53.12.1   Drag

A drag force is $D = C(\rho u_s^2/2)A$, where C is the drag coefficient, $\rho u_s^2/2$ is the dynamic pressure of the free stream, and A is an appropriate area. For pure viscous shear drag C is C_f, the skin friction drag coefficient of Section 53.9.2 and A is the area sheared. In general, C is designated C_D, the drag coefficient for drag other than that from viscous shear only, and A is the chord area for lifting vanes or the projected frontal area for other shapes.

The drag coefficient for incompressible flow with pure pressure drag (a flat plate normal to a flow, for example) or for combined skin friction and pressure drag, which is called *profile* drag, depends on the body shape, the Reynolds number, and, usually, the location of boundary layer transition.

Drag coefficients for spheres and for flow normal to infinite circular cylinders are shown in Fig. 53.36. For spheres at $Re_D < 0.1$, $C_D = 24/Re_D$ and for $Re_D < 100$, $C_D = (24/Re_D)(1 + 3 Re_D/16)^{1/2}$. The boundary layer for both shapes up to and including the flat portion of the curves before the rather abrupt drop in the neighborhood of $Re_D = 10^5$ is laminar. This is called the *subcritical* region; beyond that is the *supercritical* region. Table 53.11 lists typical drag coefficients for two-dimensional shapes, and Table 53.12 lists them for three-dimensional shapes.

The drag of spheres, circular cylinders, and streamlined shapes is affected by boundary layer separation, which, in turn, depends on surface roughness, the Reynolds number, and free stream turbulence. These factors contribute to uncertainties in the value of the drag coefficient.

53.12.2   Lift

Lift in a nonviscous fluid may be produced by prescribing a circulation around a cylinder or lifting vane. In a viscous fluid this may be produced by spinning a ping-pong ball, a golf ball, or a baseball, for example. Circulation around a lifting vane in a viscous fluid results from the bound vortex or countercirculation that is equal and opposite to the starting vortex, which peels off the trailing edge of the vane. The lift is calculated from $L = C_L(\rho u_s^2/2)A$, where C_L is the lift coefficient, $\rho u_s^2/2$ is the dynamic pressure of the free stream, and A is the chord area of the lifting vane. Typical values of C_L as well as C_D are shown in Fig. 53.37. The induced drag and the profile drag are shown. The profile drag is the difference between the dashed and solid curves. The induced drag is zero at zero lift.

Table 53.11   Drag Coefficients for Two-Dimensional Shapes at Re = 10^5 Based on Frontal Projected Area (Flow Is from Left to Right)

Shape		C_D	Shape	C_D		
Plate	│	2.0	Rectangle			
			1:1	1.18		
Open tube	⊂	1.2	5:1	1.2		
	⊃	2.3	10:1	1.3		
			20:1	1.5		
Half cylinder	◖	1.16	Elliptical	Below	Above	
	◗	1.7	Cylinder	Re_c	Re_c	
			2:1	0.6	0.20	
Square cylinder	□	2.05	4:1	0.36	0.10	
	◇	1.55	8:1	0.26	0.10	
Equilateral triangle	▷	2.0				
	◁	1.6				

Table 53.12   Drag Coefficients for Three-Dimensional Shapes Re between 10^4 and 10^6 (Flow Is from Left to Right)

Shape		C_D
Disk	│	1.17
Open hemisphere	⊂	0.38
	⊃	1.42
Solid hemisphere	◖	0.42
	◗	1.17
Cube	⬚	1.05[a]
	◇	0.80[a]
Cone, 60°	◁	0.50

[a] Mounted on a boundary wall.

53.13   FLOW MEASUREMENTS

Fluid flow measurements generally involve determining static pressures, local and average velocities, and volumetric or mass flow rates.

53.13.1   Pressure Measurements

Static pressures are measured by means of a small hole in a boundary surface connected to a sensor—a manometer, a mechanical pressure gage, or an electrical transducer. The surface may be a duct wall or the outer surface of a tube, such as those shown in Fig. 53.38. In any case, the surface past which the fluid flows must be smooth, and the tapped holes must be at right angles to the surface.

Total or stagnation pressures are easily measured accurately with an open-ended tube facing into the flow, as shown in Fig. 53.38.

53.13.2   Velocity Measurements

A combined pitot tube (Fig. 53.38) measures or detects the difference between the total or stagnation pressure p_0 and the static pressure p. For an incompressible fluid the velocity being measured is V

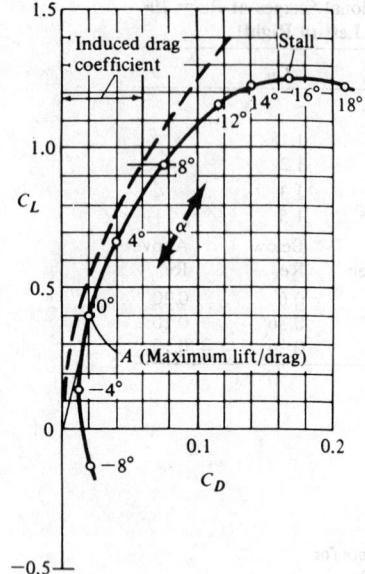

Fig. 53.37 Typical polar diagram showing lift–drag characteristics for an airfoil of finite span.

$= \sqrt{2(p_0 - p)/\rho}$. For subsonic gas flow the velocity of a stream at a temperature T and pressure p is

$$V = \sqrt{\frac{2kRT}{k-1}\left[\left(\frac{p_0}{p}\right)^{(k-1)/k} - 1\right]}$$

and the corresponding Mach number is

$$M = \sqrt{\frac{2}{k-1}\left[\left(\frac{p_0}{p}\right)^{(k-1)/k} - 1\right]}$$

For supersonic flow the stagnation pressure p_{0y} is downstream of a shock, which is detached and ahead of the open stagnation tube, and the static pressure p_x is upstream of the shock. In a wind tunnel the static pressure could be measured with a pressure tap in the tunnel wall. The Mach number M of the flow is

$$\frac{p_{0y}}{p} = \left(\frac{k+1}{2}M^2\right)^{k/(k-1)} \left(\frac{2k}{k+1}M^2 - \frac{k-1}{k+1}\right)^{1/(1-k)}$$

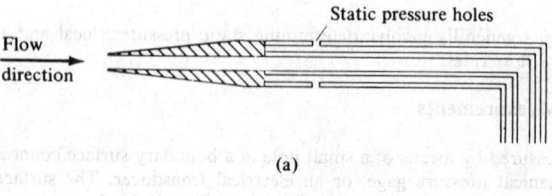

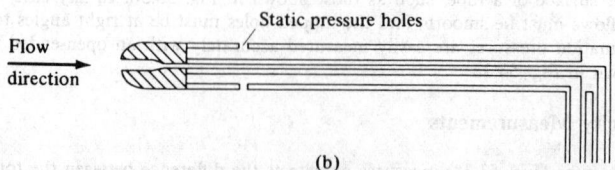

Fig. 53.38 Combined pitot tubes: (a) Brabbee's design; (b) Prandtl's design—accurate over a greater range of yaw angles.

which is tabulated in gas tables.

In a mixture of gas bubbles and a liquid for gas concentrations C no more than 0.6 by volume, the velocity of the mixture with the pitot tube and manometer free of bubbles is

$$V_{mixture} = \sqrt{\frac{2(p_0 - p_1)}{(1 - C)\rho_{liquid}}} = \sqrt{\frac{2gh_m}{(1 - C)}\left(\frac{\gamma_m}{\gamma_{liquid}} - 1\right)}$$

where h_m is the manometer deflection in meters for a manometer liquid of specific weight γ_m. The error in this equation from neglecting compressible effects for the gas bubbles is shown in Fig. 53.39. A more correct equation based on the gas–liquid mixture reaching a stagnation pressure isentropically is

$$\frac{V_1^2}{2} = \frac{p_0 - p_1}{\rho_w(1 - C)} + \frac{C}{1 - C}\left(\frac{p_1}{\rho_w}\right)\left[\frac{k}{k - 1}\left(\frac{p_0}{p_1}\right)^{(k-1)/k} - \frac{1}{k - 1} - \left(\frac{p_0}{p_1}\right)\right]$$

but is cumbersome to use. As indicated in Fig. 53.39 the error in using the first equation is very small for high concentrations of gas bubbles at low speeds and for low concentrations at high speeds.

If n velocity readings are taken at the centroid of n subareas in a duct, the average velocity V from the point velocity readings u_i is

$$V = \frac{1}{n}\sum_{i=1}^{n} u_i$$

In a circular duct, readings should be taken at $(r/R)^2 = 0.05, 0.15, 0.25, \ldots, 0.95$. Velocities measured at other radial positions may be plotted versus $(r/R)^2$, and the area under the curve may be integrated numerically to obtain the average velocity.

Other methods of measuring fluid velocities include length–time measurements with floats or neutral-buoyancy particles, rotating instruments such as anemometers and current meters, hot-wire and hot-film anemometers, and laser-doppler anemometers.

53.13.3 Volumetric and Mass Flow Fluid Measurements

Liquid flow rates in pipes are commonly measured with commercial water meters; with rotameters; and with venturi, nozzle, and orifice meters. These latter types provide an obstruction in the flow and make use of the resulting pressure change to indicate the flow rate.

The continuity and Bernoulli equations for liquid flow applied between sections 1 and 2 in Fig. 53.40 give the ideal volumetric flow rate as

$$Q_{ideal} = \frac{A_2\sqrt{2g\,\Delta h}}{\sqrt{1 - (A_2/A_1)^2}}$$

where Δh is the change in piezometric head. A form of this equation generally used is

$$Q = K\left(\frac{\pi d^2}{4}\right)\sqrt{2g\,\Delta h}$$

where K is the flow coefficient, which depends on the type of meter, the diameter ratio d/D, and the viscous effects given in terms of the Reynolds number. This is based on the length parameter d and the velocity V through the hole of diameter d. Approximate flow coefficients are given in Fig. 53.41. The relation between the flow coefficient K and this Reynolds number is

$$Re_d = \frac{Vd}{\nu} = \frac{Qd}{\frac{1}{4}\pi d^2\nu} = K\frac{d\sqrt{2g\,\Delta h}}{\nu}$$

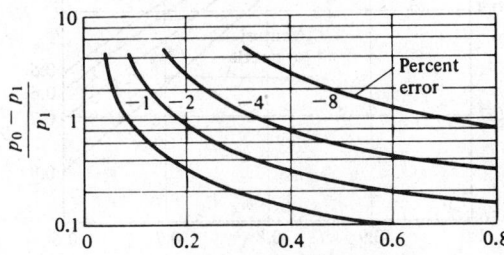

Air concentration C, by volume

Fig. 53.39 Error in neglecting compressibility of air in measuring velocity of air–water mixture with a combined pitot tube.

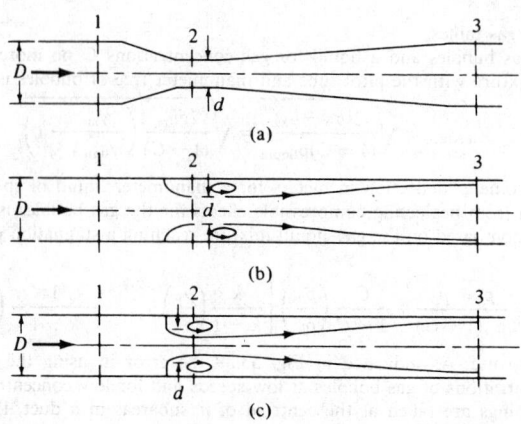

Fig. 53.40　Pipe flow meters: (a) venturi; (b) nozzle; (c) concentric orifice.

The dimensionless parameter $d\sqrt{2g\,\Delta h}/\nu$ can be calculated, and the intersection of the appropriate line for this parameter and the appropriate meter curve gives an approximation to the flow coefficient K. The lower values of K for the orifice result from the contraction of the jet beyond the orifice, where pressure taps may be located. Meter throat pressures should not be so low as to create cavitation. Meters should be calibrated in place or purchased from a manufacturer and installed according to instructions.

Elbow meters may be calibrated in place to serve as metering devices, by measuring the difference in pressure between the inner and outer radii of the elbow as a function of flow rate.

For compressible gas flows, isentropic flow is assumed for flow between sections 1 and 2 in Fig. 53.40. The mass flow rate is $\dot{m} = K\,Y\,A_2\sqrt{2\rho_1(p_1-p_2)}$, where K is as shown in Fig. 53.41 and $Y = Y(k, p_2/p_1, d/D)$ and is the expansion factor shown in Fig. 53.42. For nozzles and venturi tubes

$$Y = \sqrt{\dfrac{\left(\dfrac{k}{k-1}\right)\left(\dfrac{p_2}{p_1}\right)^{2/k}\left[1-\left(\dfrac{p_2}{p_1}\right)^{(k-1)/k}\right]\left[1-\left(\dfrac{d}{D}\right)^4\right]}{\left[1-\left(\dfrac{p_2}{p_1}\right)\right]\left[1-\left(\dfrac{d}{D}\right)^4\left(\dfrac{p_2}{p_1}\right)^{2/k}\right]}}$$

and for orifice meters

$$Y = 1 - \dfrac{1}{k}\left[0.41 + 0.35\left(\dfrac{d}{D}\right)^4\right]\left(1 - \dfrac{p_2}{p_1}\right)$$

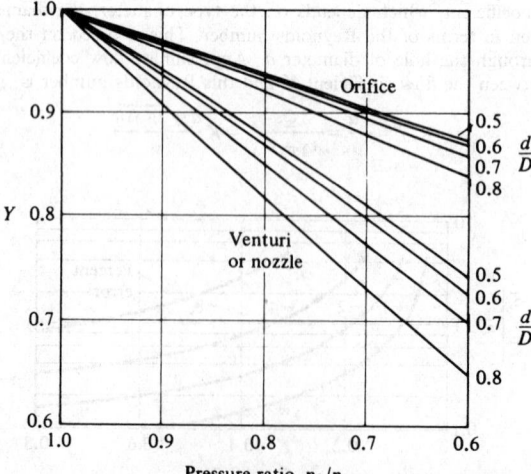

Fig. 53.41　Approximate flow coefficients for pipe meters.

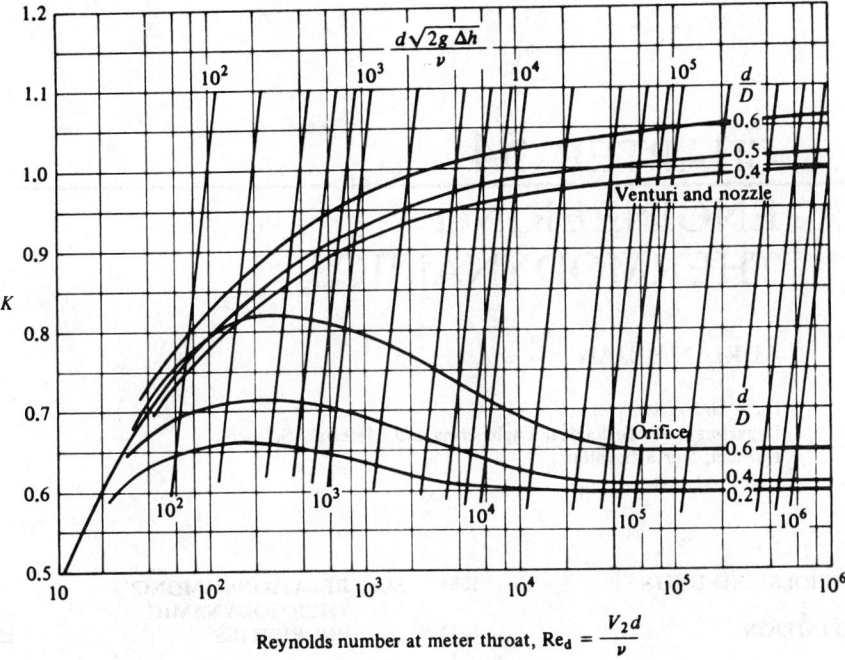

Fig. 53.42 Expansion factors for pipe meters, $k = 1.4$.

These are the basic principles of fluid flow measurements. Utmost care must be taken when accurate measurements are necessary, and reference to meter manufacturers' pamphlets or measurements handbooks should be made.

BIBLIOGRAPHY

General

R. M. Olson, *Essentials of Engineering Fluid Mechanics,* 4th ed., Harper and Row, New York, 1980.
V. L. Streeter (Ed.), *Handbook of Fluid Dynamics,* McGraw-Hill, New York, 1961.
V. L. Streeter and E. B. Wylie, *Fluid Mechanics,* McGraw-Hill, New York, 1979.

Section 53.9

H. Schlichting, *Boundary Layer Theory* (translated by J. Kestin), 7th ed., McGraw-Hill, New York, 1979.

Section 53.10

A. H. Shapiro, *The Dynamics and Thermodynamics of Compressible Fluid Flow,* Ronald Press, New York, 1953, Vol. I.

Section 53.12

S. F. Hoerner, *Fluid-Dynamic Drag,* S. F. Hoerner, Midland Park, NJ, 1958.

Section 53.13

R. W. Miller, *Flow Measurement Engineering Handbook,* McGraw-Hill, New York, 1983.
E. Ower and R. C. Pankhurst, *Measurement of Air Flow,* Pergamon Press, Elmsford, NY, 1977.

CHAPTER 54

ENGINEERING THERMODYNAMICS

ADRIAN BEJAN

Duke University
Department of Mechanical Engineering and Materials Science
Durham, North Carolina

The engineering field of thermodynamics has historically grown out of man's determination—as Sadi Carnot put it—to capture "the motive power of fire." Relative to mechanical engineering, thermodynamics describes the relationship between mechanical work and other forms of available energy. There are two facets to contemporary thermodynamics that must be stressed in a review such as this. The first is the equivalence of *work* and *heat* as two possible forms of energy exchange: This first facet is encapsulated in the first law of thermodynamics. The second aspect is the irreversibility of all processes (changes) that occur in nature. As summarized by the second law of thermodynamics, irreversibility or entropy generation is what prevents us from extracting the most possible work from various fuels; it is also what prevents us from doing the most with the work that is already at our disposal. The objective of this chapter is to review the first and second laws of thermodynamics and their implications in mechanical engineering, particularly with respect to such issues as energy conversion and energy conservation. The analytical aspects (the formulas) of engineering thermodynamics are reviewed primarily in terms of the behavior of a pure substance, as would be the case of the working fluid in a heat engine or in a refrigeration machine.

SYMBOLS AND UNITS

c	specific heat of incompressible substance, $J/kg \cdot K$
c_P	specific heat at constant pressure, $J/kg \cdot K$
c_T	constant temperature coefficient, m^3/kg
c_v	specific heat at constant volume, $J/kg \cdot K$
COP	coefficient of performance
E	energy, J
f	specific Helmholtz free energy $(u - Ts)$, J/kg
$\bar{F}$	force vector, N
g	gravitational acceleration, m/sec^2
g	specific Gibbs free energy $(h - Ts)$, J/kg
h	specific enthalpy $(u + Pv)$, J/kg
K	isothermal compressibility, m^2/N
m	mass of closed system, kg
$\dot{m}$	mass flow rate, kg/sec
m_i	mass of component in a mixture, kg
M	mass inventory of control volume, kg
M	molar mass, g/mol or kg/kmol
n	number of moles, mol
N_0	Avogadro's number
P	pressure
δQ	infinitesimal heat transfer interaction, J
$\dot{Q}$	heat transfer rate, W
$\bar{r}$	position vector, m
R	ideal gas constant, $J/kg \cdot K$
s	specific entropy, $J/kg \cdot K$
S	entropy, J/K
S_{gen}	entropy generation, J/K
$\dot{S}_{gen}$	entropy generation rate, W/K
T	absolute temperature, K
u	specific internal energy, J/kg
U	internal energy, J
v	specific volume, m^3/kg
$\bar{v}$	specific volume of incompressible substance, m^3/kg
V	volume, m^3
V	velocity, m/sec
δW	infinitesimal work transfer interaction, J
$\dot{W}_{lost}$	rate of lost available work, W
$\dot{W}_{sh}$	rate of shaft (shear) work transfer, W
x	linear coordinate, m

x	quality of liquid and vapor mixture
Z	vertical coordinate, m
β	coefficient of thermal expansion, $1/K$
γ	ratio of specific heats, c_P/c_v
η	"efficiency" ratio
η_I	first-law efficiency
η_{II}	second-law efficiency
θ	relative temperature, °C

Subscripts

( )$_{in}$	=	inlet port
( )$_{out}$	=	outlet port
( )$_{rev}$	=	reversible path
( )$_H$	=	high-temperature reservoir
( )$_L$	=	low-temperature reservoir
( )$_{max}$	=	maximum
( )$_T$	=	turbine
( )$_C$	=	compressor
( )$_N$	=	nozzle
( )$_D$	=	diffuser
( )$_1$	=	initial state
( )$_2$	=	final state
( )$_0$	=	reference state
( )$_f$	=	saturated liquid state (f = "fluid")
( )$_g$	=	saturated vapor state (g = "gas")
( )$_s$	=	saturated solid state (s = "solid")
( )$_*$	=	moderately compressed liquid state
( )$_+$	=	slightly superheated vapor state

DEFINITIONS

THERMODYNAMIC SYSTEM is the region or the collection of matter in space selected for analysis.

ENVIRONMENT is the thermodynamic system external to the system of interest, that is, external to the region selected for analysis or for discussion.

BOUNDARY is the real or imaginary surface delineating the system of interest. The boundary separates the system from its environment. The boundary is an unambiguously defined surface.

CLOSED SYSTEM is the thermodynamic system whose boundary is not penetrated (crossed) by the flow of mass.

OPEN SYSTEM, or flow system, is the thermodynamic system whose boundary is permeable to mass flow. Open systems have their own nomenclature, so that the thermodynamic system itself is usually referred to as the *control volume,* the boundary of the open system is the *control surface,* and the particular regions of the boundary that are crossed by mass flows are the *inlet* or *outlet* ports.

STATE is the condition (the being) of a thermodynamic system at a particular point in time, as described by an ensemble of quantities called *thermodynamic properties* (e.g., pressure, volume, temperature). Thermodynamic properties are only those quantities that depend solely on the instantaneous state of the system. Thermodynamic properties do not depend on the "history" of the system between two different states. Quantities that depend on the system history between states are not thermodynamic properties (examples of nonproperties are the work, heat, and mass transfer interactions; the entropy transfer interactions; the entropy generation; and the lost available work—see also the definition of *process*).

EXTENSIVE PROPERTIES are properties whose values depend on the size of the system (e.g., mass, volume, energy, enthalpy, entropy).

INTENSIVE PROPERTIES are properties whose values do not depend on the system size (e.g., pressure, temperature, specific energy, specific enthalpy, specific entropy). The collection of all intensive properties (or the properties of an infinitesimally small element of the system) constitutes the *intensive state*.

PHASE   is the collection of all system elements that have the same intensive state (e.g., the liquid droplets dispersed in a liquid–vapor mixture have the same intensive state, that is, the same pressure, temperature, specific volume, specific entropy, etc.).

PROCESS   is the one-word reference to the change of state from one initial state to a final state. In addition to the end states, knowledge of the process implies knowledge of the *interactions* experienced by the system while in communication with its environment (e.g., the work transfer, the heat transfer, the mass transfer, and the entropy transfer). To know the process also means to know the *path* (the history, or the succession of states) followed by the system from the initial to the final state. Note that the quantities that are not thermodynamic properties are quantities whose values depend on the path of the process as well as on the end states. Conversely, the change in a thermodynamic property depends solely on the end states of the process.

CYCLE   is a special process whose final state coincides with the initial state.

54.1   THE FIRST LAW OF THERMODYNAMICS FOR CLOSED SYSTEMS

The first law of thermodynamics is a statement that brings together three new concepts in thermodynamics: work transfer, heat transfer, and energy change. Out of these three concepts only energy change or, simply, energy, is, in general, a thermodynamic property. Before stating the first law and before writing down the *equation* that accounts for this statement, it is necessary to review the concepts of work transfer, heat transfer, and energy change.

Consider the force F_x experienced by a certain system at a point on its boundary. Relative to this system, the infinitesimal *work transfer* interaction between system and environment is

$$\delta W = -F_x \, dx$$

where the boundary displacement dx is defined as positive in the direction of the force F_x. When the force $\bar{F}$ and the displacement of its point of application $d\bar{r}$ are not collinear, the general definition of infinitesimal work transfer is

$$\delta W = -\bar{F} \cdot d\bar{r}$$

The work transfer interaction is considered positive when the system does work on its environment, in other words, when $\bar{F}$ and $d\bar{r}$ point in opposite directions. This sign convention has its origin in heat engine engineering, since the purpose of heat engines as thermodynamic systems is to do work on the environment.

In order for a system to experience work transfer, two things must happen: (1) a force must be present on the boundary, and (2) the point of application of this force (hence, the boundary) must move. The mere presence of forces on the boundary, without the displacement or the deformation of the boundary, does not mean work transfer. Likewise, the mere presence of boundary displacement without a force opposing or driving this motion does not mean work transfer. (For example, in the free expansion of a gas into an evacuated space, the gas system does not experience work transfer because throughout the expansion the pressure at the imaginary system–environment interface is zero.)

If a closed system can interact with its environment only via work transfer (i.e., in the absence of heat transfer, δQ, discussed later), then it is observed that the work transfer during a change of state from state 1 to state 2 is the same for all processes linking states 1 and 2,

$$-\left(\int_1^2 \delta W \right)_{\delta Q = 0} = E_2 - E_1$$

In this special case the work transfer interaction $(W_{1-2})_{\delta Q = 0}$ is a property of the system, since its value depends solely on the end states. This thermodynamic property is the *energy change* of the system, $E_2 - E_1$. The statement that preceded the last equation is the first law of thermodynamics for closed systems that do not experience heat transfer.[1]

Heat transfer is, like work transfer, an energy interaction that can take place between a system and its environment. An important distinction between δQ and δW is made by the second law of thermodynamics discussed in the next section: Heat transfer is the energy interaction accompanied by entropy transfer, whereas work transfer is the energy interaction taking place in the absence of entropy transfer. The transfer of heat is driven by the *temperature difference* established between the system and its environment. The temperature is the thermodynamic property accounting for the system's potential to experience heat transfer with another system. The system temperature is measured by placing the system in thermal communication with a test system called thermometer. The result of this measurement is the *relative temperature* θ expressed in degrees Celsius, $\theta(°C)$, or Fahrenheit, $\theta(°F)$; these alternative temperature readings are related through the conversion formulas

$$\theta(°C) = \tfrac{5}{9}[\theta(°F) - 32]$$
$$\theta(°F) = \tfrac{9}{5}\theta(°C) + 32$$

and

$$1°F = \tfrac{5}{9} °C$$

The boundary that prevents the transfer of heat, regardless of the magnitude of the system–environment temperature difference, is termed *adiabatic*. Conversely, the boundary that is the locus of heat transfer even in the limit of vanishingly small system–environment temperature difference is termed *diathermal*.

It is observed that a closed system undergoing a change of state $1 \to 2$ in the absence of work transfer experiences a heat-transfer interaction whose magnitude depends solely on the end states:

$$\left(\int_1^2 \delta Q \right)_{\delta W = 0} = E_2 - E_1$$

In the special case of zero work transfer, the heat-transfer interaction is a thermodynamic property of the system, which is by definition equal to the energy change experienced by the system in going from state 1 to state 2. The last equation is the first law of thermodynamics for closed systems incapable of experiencing work transfer.[1] Note that, unlike work transfer, the heat transfer is considered positive when it increases the energy of the system.

Most thermodynamic systems do not manifest the purely mechanical ($\delta Q = 0$) or purely thermal ($\delta W = 0$) behaviors exhibited by the simple systems for which the first law of thermodynamics was stated above. Most systems manifest a *coupled* mechanical and thermal behavior.[2] The preceding first-law statements can be used to show that the first law of thermodynamics for a process executed by a closed system experiencing both work and heat transfer is

$$\underbrace{\int_1^2 \delta Q}_{\substack{\text{heat} \\ \text{transfer}}} - \underbrace{\int_1^2 \delta W}_{\substack{\text{work} \\ \text{transfer}}} = \underbrace{E_2 - E_1}_{\substack{\text{energy} \\ \text{change} \\ \text{(property)}}}$$

$$\underbrace{}_{\substack{\text{energy interactions} \\ \text{(nonproperties)}}}$$

In words, the first law means that the net heat transfer into the system equals the work done by the system on the environment, plus the increase in the energy of the system. The first law of thermodynamics for a cycle or for an integral number of cycles executed by a closed system is

$$\oint \delta Q = \oint \delta W = 0$$

Note that the net change in the thermodynamic property energy is zero during a cycle or an integral number of cycles.

The energy change term $E_2 - E_1$ appearing on the right-hand side of the first law can be replaced by a more general notation that distinguishes between macroscopically identifiable forms of energy storage (kinetic, gravitational) and energy stored internally,

$$\underbrace{E_2 - E_1}_{\substack{\text{energy} \\ \text{change}}} = \underbrace{U_2 - U_1}_{\substack{\text{internal} \\ \text{energy} \\ \text{change}}} + \underbrace{\frac{mV_2^2}{2} - \frac{mV_1^2}{2}}_{\substack{\text{kinetic} \\ \text{energy} \\ \text{change}}} + \underbrace{mgZ_2 - mgZ_1}_{\substack{\text{gravitational} \\ \text{energy} \\ \text{change}}}$$

If the closed system expands or contracts *quasi-statically* (i.e., slowly enough, in mechanical equilibrium internally and with the environment) so that at every point in time the pressure P is uniform throughout the system, then the work transfer term can be calculated as the work done by all the boundary pressure forces as they move with their respective points of application,

$$\int_1^2 \delta W = \int_1^2 P \, dV$$

The work-transfer integral can be evaluated provided the path of the quasi-static process, $P(V)$, is known; this illustrates the fact that the work transfer is path dependent (i.e., not a thermodynamic property).

54.2 THE SECOND LAW OF THERMODYNAMICS FOR CLOSED SYSTEMS

A *heat reservoir* is a thermodynamic system that experiences only heat-transfer interactions and whose temperature remains constant during such interactions. Consider first a closed system executing a cycle or an integral number of cycles *while in thermal communication with no more than one heat reservoir.* To state the second law for this case is to observe that the net work transfer during each cycle cannot be positive,

$$\oint \delta W \le 0$$

in other words, a closed system cannot deliver work during one cycle, while in communication with one heat reservoir or with no heat reservoir at all. Examples of such cyclic operation are the vibration of a spring–mass system, or the bouncing of a basketball on the pavement: in order for these systems to return to their respective initial heights, that is, in order for them to execute cycles, the environment (humans) must perform work on them. The limiting case of frictionless cyclic operation is termed *reversible,* because in this limit the system returns to its initial state without intervention (work transfer) from the environment. Therefore, the distinction between reversible and irreversible cycles executed by closed systems in communication with no more than one heat reservoir is

$$\oint \delta W = 0 \text{ (reversible)}$$

$$\oint \delta W < 0 \text{ (irreversible)}$$

To summarize, the first and second laws for closed systems operating cyclically in contact with no more than one heat reservoir are (Fig. 54.1)

$$\oint \delta W = \oint \delta Q \le 0$$

The above statement of the second law can be used to show that in the case of a closed system executing one or an integral number of cycles *while in communication with two heat reservoirs,* the following inequality holds (Fig. 54.1)

$$\frac{Q_H}{T_H} + \frac{Q_L}{T_L} \le 0$$

where H and L denote the high-temperature and the low-temperature heat reservoirs, respectively. Symbols Q_H and Q_L stand for the value of the cyclic integral $\oint \delta Q$, where δQ is in one case exchanged

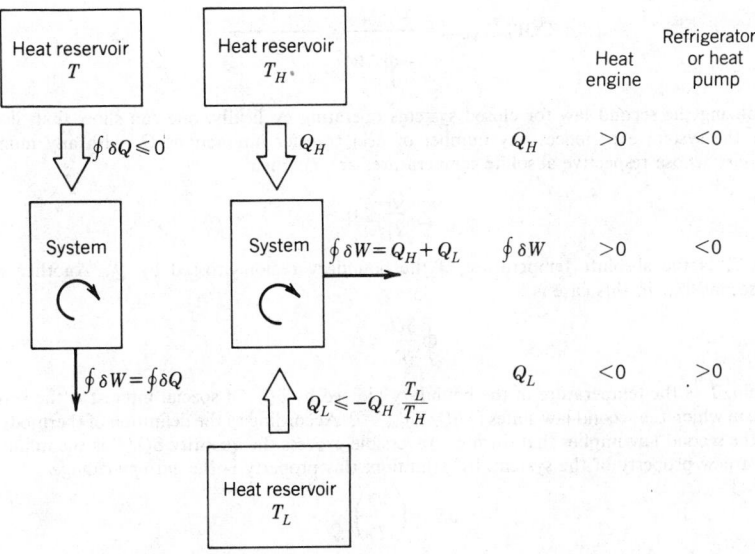

One heat reservoir   Two heat reservoirs

Fig. 54.1 The first and second laws of thermodynamics for a closed system operating cyclically while in communication with one or two heat reservoirs.

only with the H reservoir, and in the other with the L reservoir. In the reversible limit, the second law reduces to $T_H/T_L = -Q_H/Q_L$, which serves as definition for the absolute *thermodynamic temperature* scale denoted by symbol T. Absolute or "ideal gas" temperatures are expressed either in degrees Kelvin, $T(K)$, or in degrees Rankine, $T(R)$; the relationships between absolute and relative temperatures are

$$T(K) = \theta(°C) + 273.15 \text{ K}; \qquad T(R) = \theta(°F) + 459.67 \text{ R}$$
$$1 \text{ K} = 1°C \qquad\qquad\qquad 1 \text{ R} = 1°F$$

A *heat engine* is a special case of a closed system operating cyclically while in thermal communication with two heat reservoirs, a system that during each cycle intakes heat and delivers work, that is,

$$\oint \delta W = \oint \delta Q = Q_H + Q_L > 0$$

The goodness of the heat engine can be described in terms of the heat engine efficiency or the first-law efficiency

$$\eta_I = \frac{\oint \delta W}{Q_H} \leq 1 - \frac{T_L}{T_H}$$

Alternatively, the second-law efficiency of the heat engine is defined as[3]

$$\eta_{II} = \frac{\oint \delta W}{\left(\oint \delta W\right)_{\text{maximum (reversible case)}}} = \frac{\eta_I}{1 - T_L/T_H}$$

A *refrigerating machine* or a *heat pump* operates cyclically between two heat reservoirs in such a way that during each cycle it intakes work and delivers net heat to the environment,

$$\oint \delta W = \oint \delta Q = Q_H + Q_L < 0$$

The goodness of such machines can be expressed in terms of a coefficient of performance (Fig. 54.1)

$$\text{COP}_{\text{refrigerator}} = \frac{Q_L}{-\oint \delta W} \leq \frac{1}{T_H/T_L - 1}$$

$$\text{COP}_{\text{heat pump}} = \frac{-Q_H}{-\oint \delta W} \leq \frac{1}{1 - T_L/T_H}$$

Generalizing the second law for closed systems operating cyclically, one can show that, if during each cycle the system experiences any number of heat transfer interactions Q_i with any number of *heat reservoirs* whose respective absolute temperatures are T_i, then

$$\sum_i \frac{Q_i}{T_i} \leq 0$$

Note that T_i is the absolute temperature of the boundary region crossed by Q_i. Another way to write the second law in this case is

$$\oint \frac{\delta Q}{T} \leq 0$$

where, again, T is the temperature of the boundary pierced by δQ. Of special interest is the reversible cycle limit, in which the second law states ($\oint \delta Q/T)_{\text{rev}} = 0$. According to the definition of thermodynamic property, the second law implies that during a reversible process the quantity $\delta Q/T$ is the infinitesimal change in a new property of the system: by definition, this property is the *entropy change*

$$dS = \left(\frac{\delta Q}{T}\right)_{\text{rev}}$$

or

$$S_2 - S_1 = \left(\int_1^2 \frac{\delta Q}{T}\right)_{\text{rev}}$$

Combining this definition with the second-law statement for a cycle, $\oint \delta Q/T \le 0$, yields the second law of thermodynamics for *any process* executed by a closed system,

$$\underbrace{S_2 - S_1}_{\substack{\text{entropy} \\ \text{change} \\ \text{(property)}}} - \underbrace{\int_1^2 \frac{\delta Q}{T}}_{\substack{\text{entropy} \\ \text{transfer} \\ \text{(nonproperty)}}} \ge 0$$

The left-hand side in this inequality is the *entropy generation* associated with the process,

$$S_{\text{gen}} = S_2 - S_1 - \int_1^2 \frac{\delta Q}{T}$$

The entropy generation is a measure of the inequality sign in the second law, hence, a measure of the irreversibility of the process. As shown in the next section, the entropy generation is proportional to the useful work destroyed during the process.[3] Note also that any heat-transfer interaction (δQ) is accompanied by entropy transfer ($\delta Q/T$), whereas the work transfer δW is not. Thus, the concept of entropy transfer distinguishes between heat and work as two possible forms of energy interaction.

54.3 THE LAWS OF THERMODYNAMICS FOR OPEN SYSTEMS

If $\dot{m}$ represents the mass flow rate of working fluid through a port in the control surface, then the principle of *mass conservation* in the control volume reads

$$\underbrace{\sum_{\text{in}} \dot{m} - \sum_{\text{out}} \dot{m}}_{\text{mass transfer}} = \underbrace{\frac{\partial M}{\partial t}}_{\text{mass change}}$$

Subscripts in and out refer to summation over all the inlet and outlet ports, respectively, while M stands for the instantaneous mass inventory of the control volume.

The *first law of thermodynamics* is more general than the statement encountered earlier for closed systems, because this time we must account for the flow of energy associated with the $\dot{m}$ streams:

$$\underbrace{\sum_{\text{in}} \dot{m}\left(h + \frac{V^2}{2} + gZ\right) - \sum_{\text{out}} \dot{m}\left(h + \frac{V^2}{2} + gZ\right) + \sum_i \dot{Q}_i - \dot{W}_{\text{sh}}}_{\text{energy transfer}} = \underbrace{\frac{\partial E}{\partial t}}_{\substack{\text{energy} \\ \text{change}}}$$

On the left-hand side of the equal sign we have the energy transfer interactions: heat, shaft or shear work, and the energy transfer associated with mass flow across the control surface. The specific enthalpy h, fluid velocity V, and height Z are evaluated right at the boundary. On the right-hand side, E is the instantaneous system energy integrated over the control volume.

The *second law of thermodynamics* for an open system assumes the form

$$\underbrace{\sum_{\text{in}} \dot{m}s - \sum_{\text{out}} \dot{m}s + \sum_i \frac{\dot{Q}_i}{T_i}}_{\text{entropy transfer}} \le \underbrace{\frac{\partial S}{\partial t}}_{\text{entropy change}}$$

The specific entropy s is representative of the thermodynamic state of each stream right at the system boundary. The *entropy generation rate*

$$\dot{S}_{\text{gen}} = \frac{\partial S}{\partial t} + \sum_{\text{out}} \dot{m}s - \sum_{\text{in}} \dot{m}s - \sum_i \frac{\dot{Q}_i}{T_i}$$

is a measure of the irreversibility of open system operation. The engineering importance of $\dot{S}_{\text{gen}}$ stems from its proportionality to the rate of one-way destruction of available work. If the following parameters are fixed—all the mass flows ($\dot{m}$), the peripheral conditions (h, s, V, Z), and the heat-transfer interactions (Q_i, T_i) except (Q_o, T_o)—then one can use the first law and the second law to show that the work transfer rate $\dot{W}_{\text{sh}}$ cannot exceed a theoretical maximum:

$$\dot{W}_{\text{sh}} \le \sum_{\text{in}} \dot{m}\left(h + \frac{V^2}{2} + gZ - T_o s\right)$$

$$-\sum_{\text{out}} \dot{m}\left(h + \frac{V^2}{2} + gZ - T_o s\right) - \frac{\partial}{\partial t}(E - T_o s)$$

The right-hand side in this inequality is the maximum work transfer rate $\dot{W}_{\text{sh,max}}$, which exists only in the ideal limit of reversible operation. The rate of *lost work*, or available work destruction, is defined as

$$\dot{W}_{\text{lost}} = \dot{W}_{\text{sh,max}} - \dot{W}_{\text{sh}}$$

Again, using both laws, one can show that lost work is directly proportional to entropy generation,

$$\dot{W}_{\text{lost}} = T_o \dot{S}_{\text{gen}}$$

Conservation of useful (available) work in thermodynamic systems can only be achieved based on the systematic minimization of entropy generation in all the components of the system. Engineering applications of entropy generation minimization as a system design philosophy may be found in Ref. 3.

54.4 RELATIONS AMONG THERMODYNAMIC PROPERTIES

The analytical forms of the first and second laws of thermodynamics contain properties such as internal energy, enthalpy, and entropy, which cannot be measured directly. The values of these properties are derived from measurements that can be carried out in the laboratory (e.g., pressure, volume, temperature, specific heat); the formulas connecting the derived properties to the measurable properties are reviewed in this section. Consider an infinitesimal change of state experienced by a closed system. If kinetic and gravitational energy changes can be neglected, the first law reads

$$\delta Q_{\text{any path}} - \delta W_{\text{any path}} = dU$$

which emphasizes that dU is path independent. In particular, for a reversible path rev, the same dU is given by

$$\delta Q_{\text{rev}} - \delta W_{\text{rev}} = dU$$

Note that from the second law for closed systems we have $\delta Q_{\text{rev}} = T\, dS$. Reversibility (or zero entropy generation) also requires internal mechanical equilibrium at each stage during the process, hence, $\delta W_{\text{rev}} = P\, dV$, as for a quasi-static change in volume. The infinitesimal change in U is therefore

$$T\, dS - P\, dV = dU$$

Note that this formula holds for an infinitesimal change of state along any path (because dU is path independent), however, $T\, dS$ matches δQ and $P\, dV$ matches δW only if the path is free of entropy generation (reversible). In general, $\delta Q < T\, dS$ and $\delta W < P\, dV$, as shown in Fig. 54.2. The formula derived above for dU is called the *canonical relation;* for a closed system of unit mass it reads

$$T\, ds - P\, dv = du$$

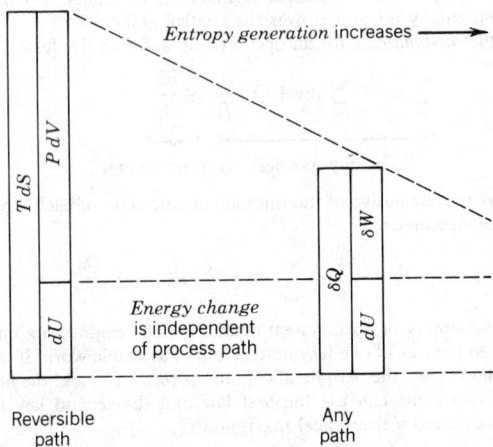

Fig. 54.2 The applicability of the canonical relation $dU = T\, dS - P\, dV$ to any infinitesimal process. (In this drawing all the quantities are assumed positive.)

Additional identities implied by this relation are

$$T = \left(\frac{\partial u}{\partial s}\right)_v \qquad -P = \left(\frac{\partial u}{\partial v}\right)_s$$

$$\frac{\partial^2 u}{\partial s \, \partial v} = \left(\frac{\partial T}{\partial v}\right)_s = -\left(\frac{\partial P}{\partial s}\right)_v$$

where the subscript indicates which variable is held constant during partial differentiation. Similar canonical relations and partial derivative identities exist in conjunction with other derived functions such as enthalpy, Gibbs free energy, and Helmholtz free energy:

1. Enthalpy (defined as $h = u + Pv$)

$$dh = T \, ds + v \, dP$$

$$T = \left(\frac{\partial h}{\partial s}\right)_P \qquad v = \left(\frac{\partial h}{\partial P}\right)_s$$

$$\frac{\partial^2 h}{\partial s \, \partial P} = \left(\frac{\partial T}{\partial P}\right)_s = \left(\frac{\partial v}{\partial s}\right)_P$$

2. Gibbs free energy (defined as $g = h - Ts$)

$$dg = -s \, dT + v \, dP$$

$$-s = \left(\frac{\partial g}{\partial T}\right)_P \qquad v = \left(\frac{\partial g}{\partial P}\right)_T$$

$$\frac{\partial^2 g}{\partial T \, \partial P} = -\left(\frac{\partial s}{\partial P}\right)_T = \left(\frac{\partial v}{\partial T}\right)_P$$

3. Helmholtz free energy (defined as $f = u - Ts$)

$$df = -s \, dT - P \, dv$$

$$-s = \left(\frac{\partial f}{\partial T}\right)_v \qquad -P = \left(\frac{\partial f}{\partial v}\right)_T$$

$$\frac{\partial^2 f}{\partial T \, \partial v} = -\left(\frac{\partial s}{\partial v}\right)_T = -\left(\frac{\partial P}{\partial T}\right)_v$$

In addition to the (P,v,T) surface, which can be determined based on measurements, Fig. 54.3, the following partial derivatives are furnished by special experiments:

1. The specific heat at constant volume, $c_v = (\partial u/\partial T)_v$, follows directly from the constant volume ($\delta W = 0$) heating of a unit mass of pure substance.
2. The specific heat at constant pressure, $c_P = (\partial h/\partial T)_P$, is determined during the constant-pressure heating of a unit mass of pure substance.
3. The Joule–Thomson coefficient, $\mu = (\partial T/\partial P)_h$, is measured during a throttling process, that is, during the flow of a stream through an adiabatic duct with friction (see the first law for an open system in the steady state).
4. The coefficient of thermal expansion, $\beta = (1/v)(\partial v/\partial T)_P$.
5. The isothermal compressibility, $K = (-1/v)(\partial v/\partial P)_T$.
6. The constant temperature coefficient, $c_T = (\partial h/\partial P)_T$.

Two noteworthy relationships between some of the partial-derivative measurements are

$$c_P - c_v = \frac{Tv\beta^2}{K}$$

$$\mu = \frac{1}{c_P}\left[T\left(\frac{\partial v}{\partial T}\right)_P - v\right]$$

Finally, the general equations relating the derived properties (u,h,s) to measurable quantities are

$$du = c_v \, dT + \left[T\left(\frac{\partial P}{\partial T}\right)_v - P\right] dv$$

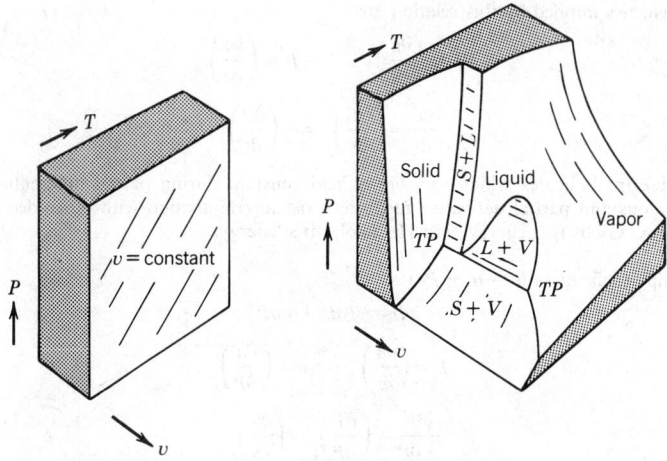

Incompressible substance   Pure substance

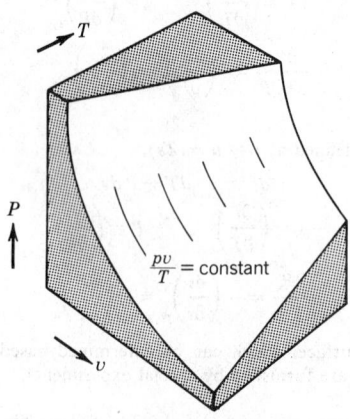

Ideal gas

Fig. 54.3 The (P,v,T) surface for a pure substance that contracts upon freezing, showing regions of ideal gas and incompressible fluid behavior. (In this figure, S = solid, V = vapor, L = liquid, TP = triple point.)

$$dh = c_P\, dT + \left[-T\left(\frac{\partial v}{\partial T}\right)_P + v \right] dP$$

$$ds = \frac{c_v}{T}\, dT + \left(\frac{\partial v}{\partial T}\right)_v dv \quad \text{or} \quad ds = \frac{c_P}{T}\, dT - \left(\frac{\partial v}{\partial T}\right)_P dP$$

These relations also suggest the following identities:

$$\left(\frac{\partial u}{\partial T}\right)_v = T\left(\frac{\partial s}{\partial T}\right)_v = c_v \qquad \left(\frac{\partial h}{\partial T}\right)_P = T\left(\frac{\partial s}{\partial T}\right)_P = c_P$$

54.5 IDEAL GASES

The relationships between thermodynamic properties and the analysis associated with applying the laws of thermodynamics are simplified considerably in cases where the pure substance exhibits ideal-gas or perfect-gas behavior. As shown in Fig. 54.3, this behavior sets in at relatively high temperatures and low pressures; in this limit, the (P, v, T) surface is fitted closely by the simple expression

$$\frac{Pv}{T} = R \quad \text{(constant)}$$

where R is the ideal gas constant of the substance of interest (Table 54.1). The formulas for internal energy, enthalpy, and entropy, which concluded the preceding section, assume the following form in the ideal-gas limit:

$$du = c_v \, dT; \; c_v = c_v(T)$$
$$dh = c_P \, dT; \; c_P = c_P(T) = c_v + R$$

$$ds = \frac{c_v}{T} dT + \frac{R}{v} dv \quad \text{or} \quad ds = \frac{c_P}{T} dT - \frac{R}{P} dP \quad \text{or} \quad ds = \frac{c_v}{P} dP + \frac{c_P}{v} dv$$

If the coefficients c_v and c_P are constant in the temperature domain of interest, then the *changes* in specific internal energy, enthalpy, and entropy relative to a reference state $(\;)_0$ are given by the formulas

$$u - u_0 = c_v(T - T_0)$$
$$h - h_0 = c_P(T - T_0)$$

where

$$h_0 = u_0 + RT_0;$$

$$s - s_0 = c_v \ln \frac{T}{T_0} + R \ln \frac{v}{v_0}$$

or

$$s - s_0 = c_P \ln \frac{T}{T_0} - R \ln \frac{P}{P_0}$$

or

$$s - s_0 = c_v \ln \frac{P}{P_0} + c_P \ln \frac{v}{v_0}$$

The ideal-gas model rests on two empirical constants, c_v and c_P, or c_v and R, or c_P and R. The ideal-gas limit is also characterized by

Table 54.1 Values of Ideal-Gas Constant and Specific Heat at Constant Volume for Gases Encountered in Mechanical Engineering[4]

Ideal Gas	R $\left(\dfrac{J}{kg \cdot K}\right)$	c_v $\left(\dfrac{J}{kg \cdot K}\right)$
Air	286.8	715.9
Argon, Ar	208.1	316.5
Butane, C_4H_{10}	143.2	1595.2
Carbon dioxide, CO_2	188.8	661.5
Carbon monoxide, CO	296.8	745.3
Ethane, C_2H_6	276.3	1511.4
Ethylene, C_2H_4	296.4	1423.5
Helium, He_2	2076.7	3152.7
Hydrogen, H	4123.6	10216.0
Methane, CH_4	518.3	1687.3
Neon, Ne	412.0	618.4
Nitrogen, N_2	296.8	741.1
Octane, C_8H_{18}	72.85	1641.2
Oxygen, O_2	259.6	657.3
Propane, C_3H_8	188.4	1515.6
Steam, H_2O	461.4	1402.6

$$\mu = 0, \qquad \beta = \frac{1}{T}, \qquad K = \frac{1}{P}, \qquad c_T = 0$$

The extent to which a thermodynamic system destroys available work is intimately tied to the system's entropy generation, that is, to the system's departure from the theoretical limit of reversible operation. Idealized processes that can be modeled as reversible occupy a central role in engineering thermodynamics, because they can serve as standards in assessing the goodness of real processes. Two benchmark reversible processes executed by closed ideal-gas systems are particularly simple and useful. A *quasi-static adiabatic process* $1 \rightarrow 2$ executed by a closed ideal-gas system has the following characteristics:

1. Energy interactions

$$\int_1^2 \delta Q = 0$$

$$\int_1^2 \delta W = \frac{P_2 V_2}{\gamma - 1} \left[\left(\frac{V_2}{V_1} \right)^{\gamma - 1} - 1 \right]$$

   where $\gamma = c_P / c_v$.
2. Path: $PV^\gamma = P_1 V_1^\gamma = P_2 V_2^\gamma$, constant.
3. Entropy change: $S_2 - S_1 = 0$, hence the name isoentropic or isentropic for this process.
4. Entropy generation

$$S_{\text{gen}_{1 \rightarrow 2}} = S_2 - S_1 - \int_1^2 \frac{\delta Q}{T} = 0$$

hence, the process is reversible.

A *quasi-static isothermal process* $1 \rightarrow 2$ executed by an ideal-gas closed system in communication with a single heat reservoir of temperature T is characterized by

1. Energy interactions

$$\int_1^2 \delta Q = \int_1^2 \delta W = mRT \ln \frac{V_2}{V_1}$$

2. Path: $T = T_1 = T_2$, constant, or $PV = P_1 V_1 = P_2 V_2$, constant.
3. Entropy change

$$S_2 - S_1 = mR \ln \frac{V_2}{V_1}$$

4. Entropy generation

$$S_{\text{gen}_{1 \rightarrow 2}} = S_2 - S_1 - \int_1^2 \frac{\delta Q}{T} = 0$$

hence, the process is reversible.

Mixtures of ideal gases also behave as ideal gases in the high-temperature, low-pressure limit. If a certain mixture of mass m contains ideal gases mixed in mass proportions m_i, and if the ideal-gas constants of each component are (c_{v_i}, c_{P_i}, R_i), then the equivalent ideal gas constants of the mixture are

$$c_v = \frac{1}{m} \sum_i m_i c_{v_i}$$

$$c_P = \frac{1}{m} \sum_i m_i c_{P_i}$$

$$R = \frac{1}{m} \sum_i m_i R_i$$

where $m = \sum_i m_i$.

One *mole* is the amount of substance of a system that contains as many elementary entities (e.g., molecules) as there are in 12 g of carbon 12; the number of such entities is Avogadro's number, $N_0 \cong 6.022 \times 10^{23}$. The mole is not a mass unit, since the mass of 1 mole is not the same for all substances.[1] The *molar mass M* of a given molecular species is the mass of 1 mole of that species, so that the total mass m is equal to the molar mass M times the *number of moles n*,

$$m = nM$$

Thus, the ideal-gas equation of state can be written as

$$PV = nMRT$$

where the product MR is the *universal gas constant*

$$\bar{R} = MR = 8.314 \frac{J}{mol \cdot K}$$

The equivalent molar mass of a mixture of ideal gases with individual molar masses M_i is

$$M = \frac{1}{n} \sum n_i M_i$$

where $n = \Sigma n_i$. The molar mass of air, as a mixture of nitrogen, oxygen, and traces of other gases, is 28.966 g/mol (or 28.966 kg/kmol). A more useful model of the air gas mixture relies on only nitrogen and oxygen as constituents, in the proportion 3.76 moles of nitrogen to every mole of oxygen; this simple model is used frequently in the field of combustion.

54.6 INCOMPRESSIBLE SUBSTANCES

At the opposite end of the spectrum, that is, at sufficiently high pressures and low temperatures in Fig. 54.3, solids and liquids behave so that their density or specific volume is practically constant. In this limit the (P, v, T) surface is adequately represented by the equation

$$v = \bar{v} \quad \text{(constant)}$$

The formulas for calculating changes in internal energy, enthalpy, and entropy become (see the end of the section on relations among thermodynamic properties)

$$du = c\, dT$$
$$dh = c\, dT + \bar{v}\, dP$$

$$ds = \frac{c}{T} dT$$

where c is the sole specific heat of the incompressible substance,

$$c = c_v = c_P$$

The specific heat c is a function of temperature only. In a sufficiently narrow temperature range where c can be regarded as constant, the finite changes in internal energy, enthalpy, and entropy relative to a reference state denoted by $(\ )_0$ are

$$u - u_0 = c(T - T_0)$$
$$h - h_0 = c(T - T_0) + \bar{v}(P - P_0)$$

where

$$h_0 = u_0 + P_0 \bar{v}$$
$$s - s_0 = c \ln \frac{T}{T_0}$$

The incompressible substance model rests on two empirical constants, c and $\bar{v}$.

54.7 TWO-PHASE STATES

As shown in Fig. 54.3, the domains in which the pure substance behaves either as an ideal gas or as an incompressible substance are interrupted by regions where the substance exists as a mixture of two phases, liquid and vapor, solid and liquid, or solid and vapor. The two-phase regions themselves intersect along the *triple point* line labeled *TP–TP* on the middle sketch of Fig. 54.3. In engineering cycle calculations, more useful are the projections of the (P, v, T) surface on the P–v plane or, through the relations reviewed earlier, on the T–s plane. The terminology associated with two-phase equilibrium states is easier to understand by focusing on the P–v diagram of Fig. 54.4a and by imagining the isothermal compression of a unit mass of substance (a closed system). Decreasing steadily the specific volume v, the substance ceases to be a pure vapor at state g, where the first droplets of liquid are formed. State g is a *saturated vapor state*. It is observed that isothermal compression beyond g proceeds at constant pressure up to state f where the last bubble (immersed in liquid) is suppressed. State f is a *saturated liquid state*. Isothermal compression beyond f is accompanied by a steep rise in pressure,

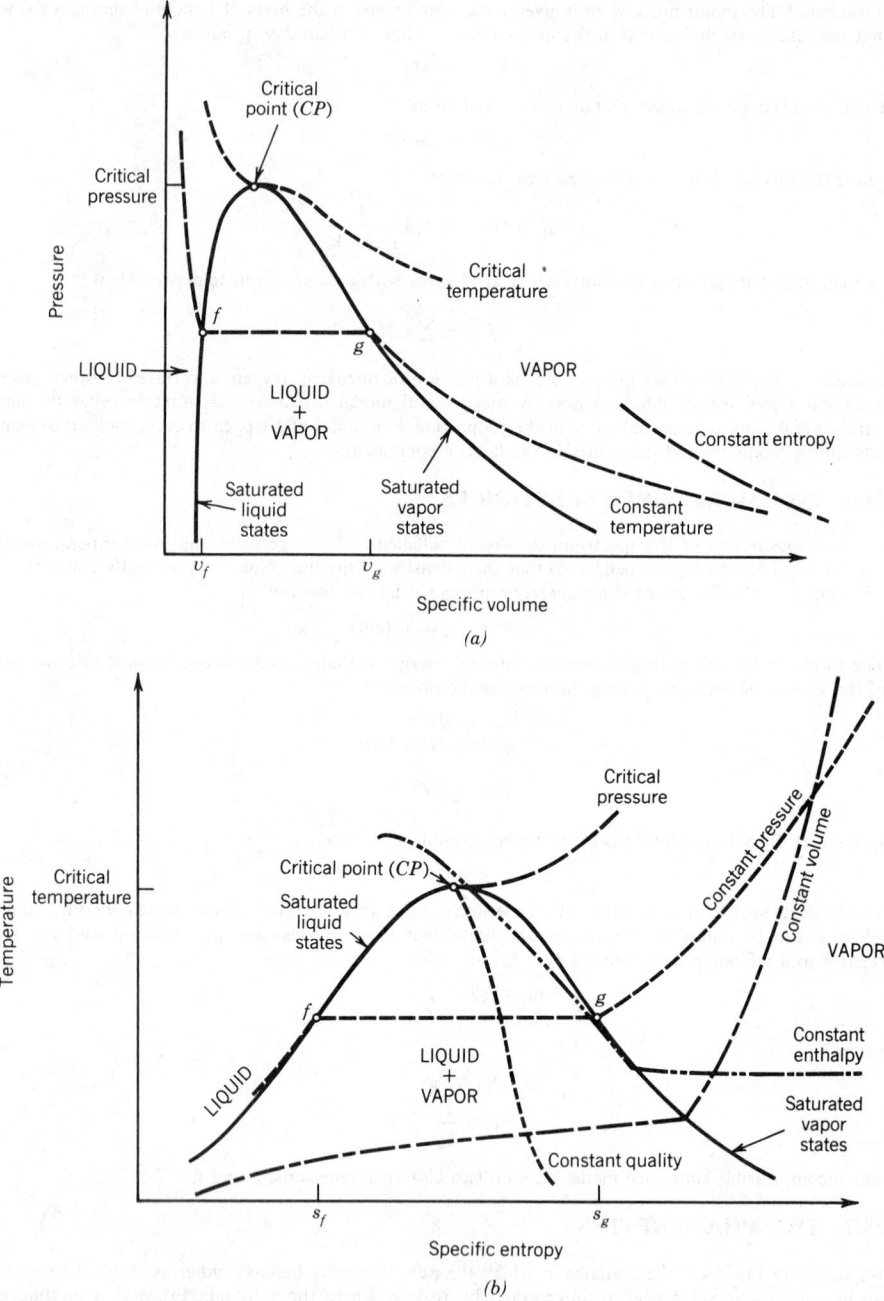

Fig. 54.4 The locus of two-phase (liquid and vapor) states, projected on the (*a*) *P–v* and (*b*) *T–s* planes.

depending on the compressibility of the liquid phase. The *critical state* is the intersection of the locus of saturated vapor states with the locus of saturated liquid states (Fig. 54.4*a*). The temperature and pressure corresponding to the critical state are termed *critical temperature* and *critical pressure*. Table 54.2 contains a compilation of critical-state properties of some of the more common substances.

Figure 54.4*b* shows the projection of the liquid and vapor domain on the *T–s* plane. On the same drawing is shown the relative positioning (the relative slope) of the traces of various constant-property cuts through the three-dimensional surface on which all the equilibrium states are positioned.

In the two-phase region, the temperature is a unique function of pressure: This one-to-one relationship is indicated also by the *Clapeyron* relation

$$\left(\frac{dP}{dT}\right)_{sat} = \frac{h_g - h_f}{T(v_g - v_f)} = \frac{s_g - s_f}{v_g - v_f}$$

where the subscript sat is a reminder that the relation holds for saturated states (such as g and f) and for mixtures of two saturated phases. Subscripts g and f indicate properties corresponding to the saturated vapor and liquid states found at temperature T_{sat} (and pressure P_{sat}). Built into the last equation is the identity

$$h_g - h_f = T(s_g - s_f)$$

which is equivalent to the statement that the Gibbs free energy is the same for the saturated states and their mixtures found at the same temperature,

$$g_g = g_f$$

The properties of a two-phase mixture depend on the proportion in which saturated vapor, m_g, and saturated liquid, m_f, enter the mixture. A thermodynamic property of the mixture is the *quality*,

$$x = \frac{m_g}{m_f + m_g}$$

whose value varies between 0 at state f and 1 at state g. Other properties of the mixture can be calculated in terms of the properties of the saturated states found at the same temperature,

$$u = u_f + x u_{fg}$$
$$h = h_f + x h_{fg}$$
$$s = s_f + x s_{fg}$$
$$v = v_f + x v_{fg}$$

Table 54.2 Critical-State Properties (after a Compilation in Ref. 5)

Fluid	Critical Temperature [K (°C)]		Critical Pressure [MPa (atm)]		Critical Specific Volume (cm³/g)
Air	133.2	(−140)	3.77	(37.2)	2.9
Alcohol (methyl)	513.2	(240)	7.98	(78.7)	3.7
Alcohol (ethyl)	516.5	(243.3)	6.39	(63.1)	3.6
Ammonia	405.4	(132.2)	11.3	(111.6)	4.25
Argon	150.9	(−122.2)	4.86	(48)	1.88
Butane	425.9	(152.8)	3.65	(36)	4.4
Carbon dioxide	304.3	(31.1)	7.4	(73)	2.2
Carbon monoxide	134.3	(−138.9)	3.54	(35)	3.2
Carbon tetrachloride	555.9	(282.8)	4.56	(45)	1.81
Chlorine	417	(143.9)	7.72	(76.14)	1.75
Ethane	305.4	(32.2)	4.94	(48.8)	4.75
Ethylene	282.6	(9.4)	5.85	(57.7)	4.6
Helium	5.2	(−268)	0.228	(2.25)	14.4
Hexane	508.2	(235)	2.99	(29.5)	4.25
Hydrogen	33.2	(−240)	1.30	(12.79)	32.3
Methane	190.9	(−82.2)	4.64	(45.8)	6.2
Methyl chloride	416.5	(143.3)	6.67	(65.8)	2.7
Neon	44.2	(−288.9)	2.7	(26.6)	2.1
Nitric oxide	179.3	(−93.9)	6.58	(65)	1.94
Nitrogen	125.9	(−147.2)	3.39	(33.5)	3.25
Octane	569.3	(296.1)	2.5	(24.63)	4.25
Oxygen	154.3	(−118.9)	5.03	(49.7)	2.3
Propane	368.7	(95.6)	4.36	(43)	4.4
Sulfur dioxide	430.4	(157.2)	7.87	(77.7)	1.94
Water	647	(373.9)	22.1	(218.2)	3.1

with the notation $( )_{fg} = ( )_g - ( )_f$.

Similar relations can be used to calculate the properties of two-phase states other than liquid and vapor, namely, solid and vapor or solid and liquid. For example, the enthalpy of a solid and liquid mixture is given by $h = h_s + xh_{sf}$, where subscript s stands for the *saturated solid state* found at the same temperature as for the two-phase state, and where h_{sf} is the latent heat of melting or solidification.

In general, the states situated immediately outside the two-phase dome sketched in Figs. 54.3 and 54.4 do not follow very well the limiting models discussed already (ideal gas, incompressible substance). Since the properties of closely neighboring states are usually not available in tabular form, the following approximate calculation proves useful. For a *moderately compressed liquid state,* that is, for a state situated immediately to the left of the dome in Fig. 54.4, the properties may be calculated as slight deviations from those of the saturated liquid state found at the same temperature as the compressed liquid state of interest,

$$h_* \cong (h_f)_{T*} + (v_f)_{T*}[P_* - (P_f)_{T*}]$$
$$s_* \cong (s_f)_{T*}$$

The subscript $( )_*$ denotes properties that belong to the moderately compressed liquid state.

For a *slightly superheated vapor state,* that is, a state situated immediately to the right of the dome in Fig. 54.4, the properties may be estimated in terms of those of the saturated vapor state found at the same temperature,

$$h_+ \cong (h_g)_{T+}$$
$$s_+ \cong (s_g)_{T+} + \left(\frac{P_g v_g}{T_g}\right)_{T+} \ln \frac{(P_g)_{T+}}{P_+}$$

In these expressions subscript $( )_+$ indicates the properties of the slightly superheated vapor state.

54.8 ANALYSIS OF ENGINEERING SYSTEM COMPONENTS

This section contains a summary of the equations obtained by applying the first and second laws of thermodynamics to the components encountered in most engineering systems, such as power cycles and refrigeration cycles. It is assumed that each component operates in *steady flow*.

Valve (throttle) or adiabatic duct with friction (Fig. 54.5a):

First law $h_1 = h_2$
Second law $\dot{S}_{gen} = \dot{m}(s_2 - s_1) > 0$

Expander or *turbine* with negligible heat transfer to the ambient (Fig. 54.5b):

First law $\dot{W}_T = \dot{m}(h_1 - h_2)$
Second law $\dot{S}_{gen} = \dot{m}(s_2 - s_1) \geqslant 0$

Efficiency $\eta_T = \dfrac{h_1 - h_2}{h_1 - h_{2,rev}} \leqslant 1$

Compressor or *pump* with negligible heat transfer to the ambient (Fig. 54.5c):

First law $\dot{W}_C = \dot{m}(h_2 - h_1)$
Second law $\dot{S}_{gen} = \dot{m}(s_2 - s_1) \geqslant 0$

Efficiency $\eta_C = \dfrac{h_{2,rev} - h_1}{h_2 - h_1} \leqslant 1$

Nozzle with negligible heat transfer to the ambient (Fig. 54.5d):

First law $\frac{1}{2}(V_2^2 - V_1^2) = h_1 - h_2$
Second law $\dot{S}_{gen} = \dot{m}(s_2 - s_1) \geqslant 0$

Efficiency $\eta_N = \dfrac{V_2^2 - V_1^2}{V_{2,rev}^2 - V_1^2} \leqslant 1$

Diffuser with negligible heat transfer to the ambient (Fig. 54.5e):

First law $h_2 - h_1 = \frac{1}{2}(V_1^2 - V_2^2)$
Second law $\dot{S}_{gen} = \dot{m}(s_2 - s_1) \geqslant 0$

Efficiency $\eta_D = \dfrac{h_{2,rev} - h_1}{h_2 - h_1} \leqslant 1$

Heat exchangers with negligible heat transfer to the ambient (Figs. 54.5f and 54.5g)

First law $\dot{m}_{hot}(h_1 - h_2) = \dot{m}_{cold}(h_4 - h_3)$
Second law $\dot{S}_{gen} = \dot{m}_{hot}(s_2 - s_1) + \dot{m}_{cold}(s_4 - s_3) \geqslant 0$

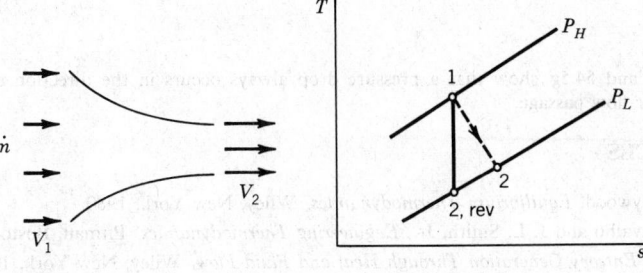

Fig. 54.5 Engineering system components, and their inlet and outlet states on the T–s plane. (P_H = high pressure; P_L = low pressure.)

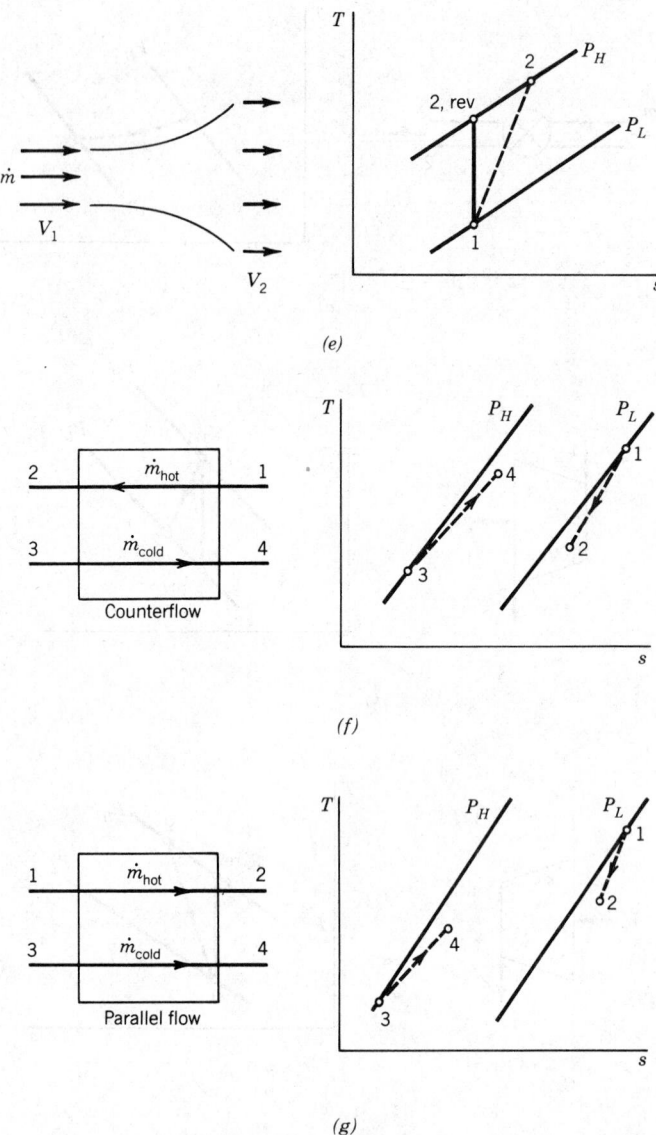

(e)

(f)

(g)

Fig. 54.5 (*Continued*)

Figures 54.5f and 54.5g show that a pressure drop always occurs in the direction of flow, in any heat exchanger flow passage.

REFERENCES

1. R. W. Haywood, *Equilibrium Thermodynamics*, Wiley, New York, 1980.
2. E. G. Cravalho and J. L. Smith, Jr., *Engineering Thermodynamics*, Pitman, Boston, 1981.
3. A. Bejan, *Entropy Generation Through Heat and Fluid Flow*, Wiley, New York, 1982.
4. A. Bejan and H. M. Paynter, *Solved Problems in Thermodynamics*, 1976 Fall Edition, Mechanical Engineering Department, Massachusetts Institute of Technology, Cambridge, Massachusetts.
5. C. O. Mackey, "Engineering Thermodynamics," *Kent's Mechanical Engineer's Handbook*, 12th ed., J. K. Salisbury (Ed.), Wiley, New York, 1967.

CHAPTER 55

THE EXERGY METHOD OF SYSTEMS ANALYSIS

JOHN E. AHERN

Glendora, California

55.1 FUNDAMENTALS OF EXERGY

One means of applying the second law of thermodynamics to systems analysis is the exergy method, which has been extensively applied in Europe and the Soviet Union. It is now being used to some extent in Japan and the United States. The exergy method will readily account for the irreversible losses in the system that must be considered for a realistic performance evaluation, especially for an energy conservation analysis. When the exergy method is used, the quality of the energy at stations throughout the system is developed relative to the surrounding condition, which is considered to be in a "dead state" having no further potential to do work. Exergy can be defined as the maximum work that is available at any station in a system relative to the surrounding condition. It is the goal of a good system design to obtain the maximum useful work from that which is available. Work is lost in system processes due to temperature differential losses and pressure differential losses. Heat and mass leakage between the atmosphere and the system and leakage within the system can cause losses. These work losses are minimized throughout a system to improve the system energy efficiency. In order to effectively minimize these losses, their magnitude and location must be determined as well as the cause of the losses and means of reducing them. It is for this reason that an exergy analysis is made of the design of a new system or an evaluation of an existing system for possible modification. The work in a process is made up of two components:

$$W_{total} = W_{reversible} + W_{irreversible}$$

The fundamentals of the exergy method involve analysis of the irreversible production of entropy associated with the irreversible work and then using the reference system to determine the distribution of the work or exergy losses.

55.2 THE IRREVERSIBLE PRODUCTION OF ENTROPY

Entropy is a property that can be defined as

$$ds = dQ/T \tag{55.1}$$

when the process is reversible. In an irreversible process the value of ds is always greater than the term on the right-hand side of Eq. (55.1). In a reversible adiabatic process, there is no heat transfer and the process is isentropic ($ds = 0$). Isentropic processes are ideal and do not occur in real-life systems. All energy systems have mechanical or thermal losses to some degree as an inherent feature. Entropy values are tabulated for many engineering working fluids including refrigerants. Of primary interest is the change in entropy through a system, so the absolute value of entropy is of secondary importance as long as the unit system used is consistent throughout the analysis.

Consider the heat flowing through a rod shown in Fig. 55.1. The rod is perfectly insulated except at each end where it is in intimate contact with an infinite heat source and an infinite heat sink. The amount of heat leaving the rod at point 2 is the same as that entering at point 1 by virtue of the first-law energy conservation. For heat to flow from point 1 to point 2 a temperature difference must exist and from Eq. (55.1) the equivalent entropy must be higher at point 2 than at point 1 because of its lower temperature. For example, if 100 W (342 Btu/hr) of heat flows through the rod from an infinite heat source at 811 K (1000°F) to an infinite heat sink at 533 K (500°F), then the equivalent entropy at the inlet is

$$s_1 = Q/T_1 = 100/811 = 0.123 \text{ W/K}$$

and at the outlet

$$s_2 = Q/T_2 = 100/533 = 0.188 \text{ W/K}$$

The increase in entropy in the rod because of the irreversible heat flow is

$$ds = s_2 - s_1 = 0.188 - 0.123 = 0.065 \text{ W/K}$$

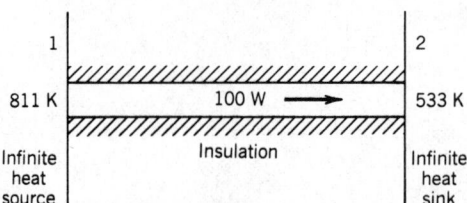

Fig. 55.1 Heat flow in a rod.

The irreversibly produced entropy calculated above is indicative of a loss of available work and is to be minimized where practical for an efficiently designed system. The determination of the irreversibly produced entropy values throughout a system will identify the losses in the system. However, for practical evaluation of the magnitude of these losses, the entropy changes are converted to work terms through the use of the reference system.

55.3 THE REFERENCE SYSTEM

The magnitude of the losses in the different processes in a system can be evaluated and directly compared by relating all values of work—input, output, or loss—to a common reference. The most commonly used reference is the sea-level ambient condition, which is the surrounding "dead state" medium for most terrestrial systems. The reference system used in exergy analysis is a tool that makes the second-law analysis of systems realistic and relatively simple. Small variations in the atmospheric conditions and the small exergy losses associated with atmospheric mixing can be neglected for conventional systems analysis. For example, when exhausting the flue gas products from a power plant, the exergy losses in the flue gas momentum and mixing of the gas with the atmosphere are usually negligibly small relative to other exergy losses in most systems.

The work available from a system referenced to its surroundings is defined by the second-law general energy equation

$$\text{work} = \underset{\substack{\text{energy}\\\text{transfer}\\\text{with surroundings}}}{(E - T_0 ds)} + \underset{\substack{\text{heat}\\\text{transfer}\\\text{with}\\\text{surroundings}}}{Q_i(1 - T_0/T_i)} - \underset{\substack{\text{internal}\\\text{system}\\\text{irreversible}\\\text{losses}}}{T_0 ds_{\text{irreversible}}} \tag{55.2}$$

where T_0 is the temperature of the surroundings, E is the difference in energy from the surroundings $(h_i - h_0)$, and ds is the difference in entropy from the surroundings $(s_i - s_0)$. The subscript i refers to any steady-state point in the system. This equation also represents the exergy or maximum available work at a given station in the system by accounting for all the work transferred with the surroundings or lost internally in the system from the point under consideration.

55.4 METHOD OF ANALYSIS

The exergy analysis of a system is simplified when the first-law heat and energy balance data are used as the basis for determining the exergy values at stations throughout the system. Generally, only the entropy values at each station are not available from the first-law analysis. When the entropy values are obtained from tables or charts, the exergy value at each station is calculated referenced to the surrounding "dead-state" condition where exergy is zero.

The exergy values at the state points in the system are calculated by the equation

$$\text{Ex} = (h - h_0) - T_0(s - s_0) \tag{55.3}$$

where T_0, h_0, and s_0 are the temperature, enthalpy, and entropy values of the system working fluid at the reference zero-exergy condition. The data from an exergy analysis are most conveniently developed into two tables. One table shows the properties at each station in the system including exergy, and the other table displays the magnitude, location, and type of the exergy losses as well as the total exergy change and the heat transferred for each process or component. These tables are shown as Tables 55.1 and 55.2 in the following example.

Example 55.1

As an example to demonstrate the procedure for performing an exergy analysis, the simple steam power plant shown schematically in Fig. 55.2 is used. The fuel has a heat content of 30,000 J/g and 12% of the input energy is assumed lost from the furnace as leaks and in the flue gas. The steam leaves the steam generator at 100 atm pressure and 520°C temperature and is condensed at 1 atm pressure. A 20-atm pressure drop is assumed in the steam line from the steam generator to the turbine. The steam turbine has an 80% efficiency, and the pump is assumed isentropic. The reference condition for the exergy calculations is taken as water at 22°C. The exergy values are given as J/g of fuel.

Pressurized water at 100 atm enters the steam generator at station 6 in Fig. 55.2 along with the fuel and air for combustion. Steam at 100 atm and 520°C leaves the steam generator at station 2. The properties of the water and steam at these stations are given in Table 55.1.

The heat loss in the steam generator furnace is

$$\text{Ex}_a = 30,000 \times 0.12 = 3600 \text{ J/g fuel}$$

and the heat transferred to the steam is

$$30,000 - 3600 = 26,400 \text{ J/g fuel}$$

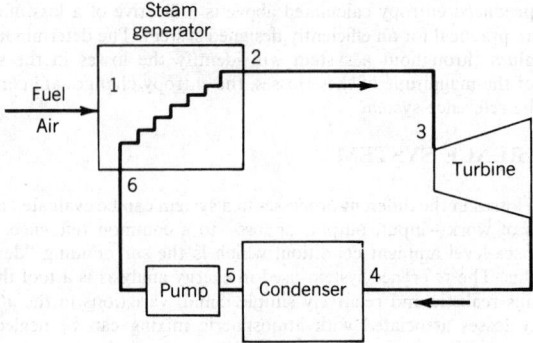

Fig. 55.2 Basic steam power plant.

A heat balance in the steam generator gives the steam flow/fuel flow ratio, r,

$$r = \frac{26{,}400 \text{ J/g fuel}}{(3425 - 428) \text{ J/g steam}} = 8.8$$

The exergy of the water entering the steam generator is

$$Ex_6 = 8.8[(428 - 105) - 295(1.303 - 0.367)] = 410 \text{ J/g fuel}$$

To this is added the exergy in the fuel of 30,000 J/g so that

$$Ex_1 = 410 + 30{,}000 = 30{,}410 \text{ J/g fuel}$$

The exergy of the steam leaving the steam generator is

$$Ex_2 = 8.8[(3425 - 105) - 295(6.662 - 0.367)] = 12{,}870 \text{ J/g fuel}$$

The exergy increase from heating the water is

$$12{,}870 - 410 = 12{,}460 \text{ J/g fuel}$$

The total exergy loss in the furnace from combustion and heat transfer is

$$Ex_b = 26{,}400 - 12{,}460 = 13{,}900 \text{ J/g fuel}$$

To separate this total into its components requires combustion calculations too detailed for this example. The isenthalpic pressure drop in the steam line from the steam generator to the turbine results in the fluid properties at the turbine entrance that are shown in Table 55.1 as station 3. The exergy at station 3 is

$$Ex_3 = 8.8[(3425 - 105) - 295(6.758 - 0.367)] = 12{,}630 \text{ J/g fuel}$$

The exergy loss from the flow pressure drop is then

$$Ex_c = 12{,}870 - 12{,}630 = 240 \text{ J/g fuel}$$

The properties of the fluid leaving the turbine at 1 atm saturated vapor are given in Table 55.1 for station 4 and the exergy value is

$$Ex_4 = 8.8[(2676 - 105) - 295(7.355 - 0.367)] = 4490 \text{ J/g fuel}$$

Table 55.1 Station State Properties (Steamside)

Station	Pressure (atm)	Temperature (°C)	Enthalpy (J/g)	Entropy (J/g · K)	Unit Exergy (J/g Steam)	Steam/Fuel Flow Rate	Total Exergy (J/g Fuel)
2	100	520	3425	6.662	1463	8.8	12,870
3	80	511	3425	6.758	1435	8.8	12,630
4	1	100	2676	7.355	510	8.8	4,490
5	1	100	418	1.303	36	8.8	320
6	100	—	428	1.303	46	8.8	410
Reference	1	22	105	0.367	0		

Table 55.2 Distribution of Exergy Losses (in Units of J/g Fuel)

Station or Component	Exergy	Change in Exergy	Work	Exergy Losses					
				Combustion and Heat Transfer	Heat Transfer	Heat Rejection	Friction	Machine Inefficiency	Total
1	30,410								
Steam generator									
2	12,870	17,540		13,940					13,940 (Ex$_b$)
Piping									
3	12,630	240					240		240 (Ex$_c$)
Turbine									
4	4,490	8,140	6,590 out					1,550	1,550 (Ex$_d$)
Condenser									
5	320	4,170				4,170			4,170 (Ex$_e$)
Pump									
6	410	90	90 in						
Furnace									
1	30,410	30,000	30,000 in			3,600			3,600 (Ex$_d$)
									23,500

The total exergy change in the turbine is then

$$\text{Ex} = 12{,}630 - 4490 = 8140 \text{ J/g fuel}$$

This total exergy includes the useful work removed from the turbine and the exergy loss from machine inefficiency. The exergy loss is calculated to be

$$\text{Ex}_d = 8.8[295(7.355 - 6.758)] = 1550 \text{ J/g fuel}$$

and the useful work is

$$\text{work} = 8140 - 1550 = 6590 \text{ J/g fuel}$$

The useful work can also be obtained from the enthalpy difference

$$\text{work} = h_3 - h_4 = 8.8(3425 - 2676) = 6590 \text{ J/g fuel}$$

The working fluid leaves the condenser, station 5, as a saturated liquid at 1 atm pressure. The calculation of exergy using the fluid properties in Table 55.1 gives

$$\text{Ex}_5 = 8.8[(418 - 105) - 295(1.303 - 0.367)] = 320 \text{ J/g fuel}$$

and assuming that the heat of condensation is not used, the exergy loss in the condenser is

$$\text{Ex}_e = \text{Ex}_4 - \text{Ex}_5 = 4490 - 320 = 4170 \text{ J/g fuel}$$

Assuming isentropic liquid pumping, the pump work is

$$\text{pump work} = \text{Ex}_6 - \text{Ex}_5 = 410 - 320 = 90 \text{ J/g fuel}$$

The data calculated are entered into Table 55.2, which shows the magnitude, location, and type of exergy losses in the system. This table is useful for systems analysis with the goals of improving the system efficiency and energy conservation. Table 55.3 gives a comparison of the energy and exergy balances for this system. The first-law energy balance shows the largest system loss to be in the condenser. When the quality of the energy is considered in the exergy balance, the condenser loss is small relative to the irreversible losses from combustion and heat transfer in the steam generator unit.

The low-temperature cryogenic region is especially suitable for an exergy analysis because of the high irreversible losses encountered at low temperatures. Processes such as the isenthalpic Joule–Thomson expansion are associated with high exergy losses at the low temperatures. Heat and mass leaks within the system and heat leaks into the system from the surroundings become more important in determining the system operating efficiency as the temperature of operation decreases. Techniques to reduce these losses involve cascading cycles, heat regeneration, and staged heat leak interception. The effectiveness of the latter technique is examined in the next example.

Example 55.2

The basis for this example is a typical helium liquefaction system shown in Fig. 55.3. For this analysis the system is divided into three temperature zones with average temperatures of 140, 40, and 5 K. The outside casing and the surrounding ambient temperatures are 295 K. The power dissipation caused by heat leaks can be evaluated by referring to the schematics in Fig. 55.4, where Fig. 55.4a shows direct heat leaks from the surroundings to each temperature zone, and Fig. 55.4b shows the heat leaks intercepted by staged temperature shields. Heat transfer with the surroundings is involved in this problem, and the second term on the right-hand side of Eq. (55.2) applies

$$\text{exergy (work) required} = q(1 - T_0/T_i)$$

Table 55.3 Comparison of Energy and Exergy Balances (in Units of J/g Fuel)

	Energy Balance	Exergy Balance
Steam generator	3,600	17,540
Piping	0	240
Turbine work	6,590	6,590
Turbine loss	0	1,550
Condenser	19,900	4,170
Pump work	−90	−90
	30,000	30,000

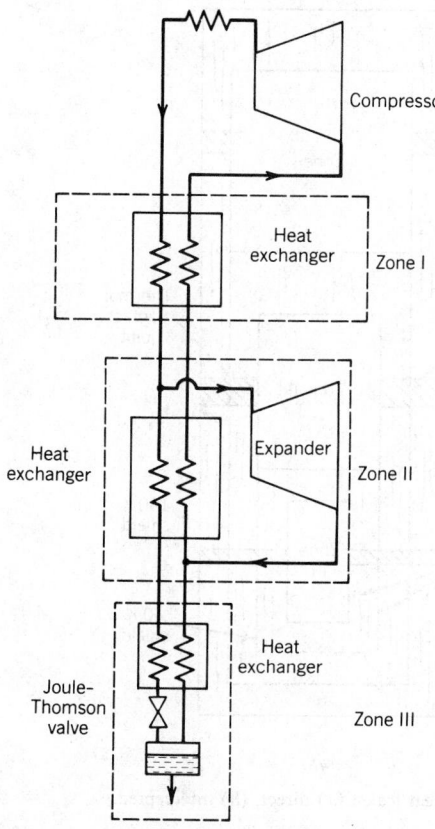

Fig. 55.3 Helium liquefaction system.

Since the internal system temperatures are lower than the surrounding ambient temperature T_0 for this cryogenic system, the exergy will be negative, indicating that this is additional specific work required to be put into the system to account for the heat leakage.

Table 55.4 gives the results of the heat leak calculations. An equivalent conductance of 0.01 W/K is assumed for the parasitic heat transfer through the supports and insulation. When the heat leaks are directly into the three zones without staged heat leak interception, the additional power

Table 55.4 Comparison of Direct and Intercepted Heat Leaks[a]

Zone	Temperature Difference (K)	Heat Flow q (W)	$(1 - T_0/T_i)$	Equivalent Work Input Required (W)
Direct Heat Leak				
I	$295 - 140 = 155$	1.55	-1.1	-1.8
II	$295 - \ \ 40 = 255$	2.55	-6.4	-16.6
III	$295 - \ \ \ 5 = 290$	2.9	-58.0	-168.2
			Total	-186.6
Intercepted Heat Leak				
I	$295 - 140 = 155$	$1.55 \times 3 = 4.65$	-1.1	-5.1
II	$140 - \ \ 40 = 100$	$1.0 \times 2 = 2.0$	-6.4	-12.4
III	$40 - \ \ \ 5 = \ \ 35$	$0.35 \times 1 = 0.35$	-58.0	-20.3
			Total	-37.8

[a] Assumed total insulation and support conductance is 0.01 W/K.

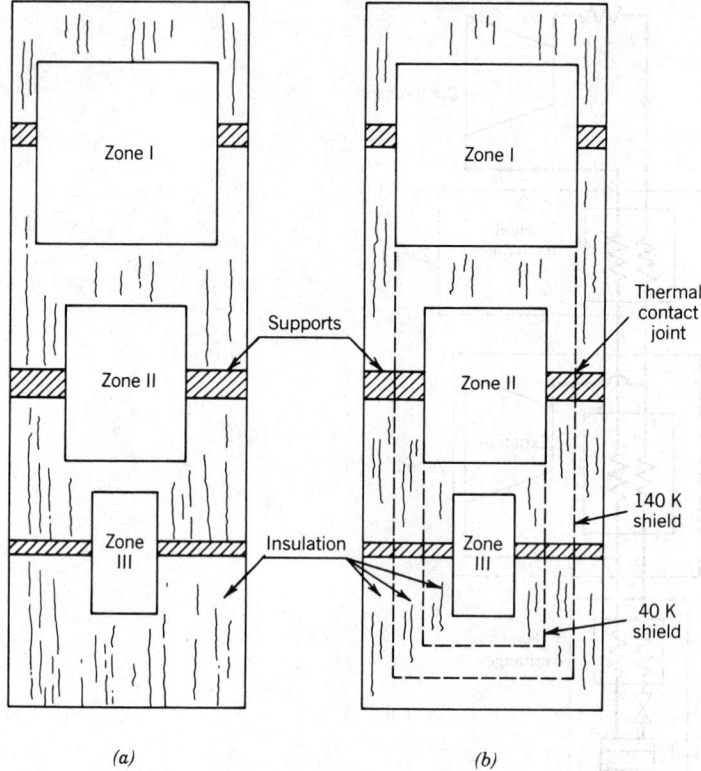

Fig. 55.4 Direct and intercepted heat leaks: (*a*) direct; (*b*) intercepted.

required by the system compressor is 186.6 W. This is reduced to 37.8 W when the heat leaks to the lower-temperature zones are intercepted by the higher-temperature shields. The cost of the additional operating power without staged heat leak interception must be traded off with the additional initial cost of shield construction to establish the most effective approach. However, the increasing importance of conserving energy resources for terrestrial systems and minimizing power requirements for cryogenic spacecraft applications dictates that second-law considerations be given a prominent role in refrigeration system design.

BIBLIOGRAPHY

J. E. Ahern, *The Exergy Method of Energy Systems Analysis,* Wiley-Interscience, New York, 1980.

A. Bejan, "Discrete Cooling of Low Heat Leak Supports to 4.2 K," *Cryogenics,* 290–293 (May 1975).

F. Bosnjakovic, *Technical Thermodynamics,* 3rd ed., P. L. Blackshear, Jr. (Ed.), Holt, Rinehart & Winston, New York, 1965.

D. E. Daney, "Low Temperature Losses in Supercritical Helium Refrigerators," in *Advances in Cryogenic Engineering,* Plenum, New York, 1976, Vol. 21, pp. 205–212.

J. H. Keenan, *Thermodynamics,* Wiley, New York, 1941.

A. Keller, "The Evaluation of Steam Power Plant Losses by Means of the Entropy Balance Diagram," *Transact. ASME,* 949–953 (October 1950).

R. Petela, "Exergy of Heat Radiation," *ASME J. Heat Transf.,* 187–192 (May 1964).

R. C. Tolman and P. C. Fine, "On the Irreversible Production of Entropy," *Rev. Mod. Phys.,* **10**(1) (1948).

C. Trepp, "Refrigeration Systems for Temperatures Below 25 K with Turboexpanders," in *Advances in Cryogenic Engineering,* Plenum, New York, 1961.

G. J. Van Wylen and R. E. Sonntag, *Fundamentals of Classical Thermodynamics,* Wiley, New York, 1965.

CHAPTER 56

HEAT TRANSFER FUNDAMENTALS

SHUN CHEN

Fellow Engineer
Westinghouse Electric Corporation
Steam Turbine Division
Orlando, Florida

SYMBOLS AND UNITS

A	area of heat transfer surface, m²
Bi	Biot number, hL/k, dimensionless
C	circumference, m
C_p	specific heat under constant pressure, J/kg · K
D	diameter, m
e	emissive power, W/m²
F	cross flow correction factor, dimensionless
F_{i-j}	configuration factor from surface i to surface j, dimensionless
Fo	Fourier number, $\alpha t A^2/V^2$, dimensionless
$F_{0-\lambda T}$	radiation function, dimensionless
G	irradiation, W/m²; mass velocity, kg/m² · sec
g	local gravitational acceleration, 9.8 m/sec²
g_c	proportionality constant, 1 kg · m/N · sec²
Gr	Grashof number, $gL^3\beta\Delta T/\nu^2$, dimensionless
h	convection heat transfer coefficient, equals $q/A\Delta T$, W/m² · K
h_{fg}	heat of vaporization, J/kg
J	radiocity, W/m²
k	thermal conductivity, W/m · K
L	length, m
Nu	Nusselt number, $\mathrm{Nu}_L = hL/k$, $\mathrm{Nu}_D = hD/k$, dimensionless
$\overline{\mathrm{Nu}}$	Nusselt number averaged over length, dimensionless
Pe	Peclet number, RePr, dimensionless
Pr	Prandtl number, $C_p\mu/k$, dimensionless
q	rate of heat transfer, W
q''	rate of heat transfer per unit area, W/m²
R	distance, m; thermal resistance, K/W
r	radial coordinate, m; recovery factor, dimensionless
Ra	Rayleigh number, GrPr; $\mathrm{Ra}_L = \mathrm{Gr}_L\mathrm{Pr}$, dimensionless
Re	Reynolds number, $\mathrm{Re}_L = \rho VL/\mu$, $\mathrm{Re}_D = \rho VD/\mu$, dimensionless
S	conduction shape factor, m
T	temperature, K or °C
t	time, sec
T_{as}	adiabatic surface temperature, K
T_{sat}	saturation temperature, K
T_b	fluid bulk temperature or base temperature of fins, K
T_e	excessive temperature, $T_s - T_{sat}$, K or °C
T_f	film temperature, $(T_\infty + T_s)/2$, K
T_i	initial temperature; at $t = 0$, K
T_0	stagnation temperature, K
T_s	surface temperature, K
T_∞	free stream fluid temperature, K
U	overall heat transfer coefficient, W/m² · K
V	fluid velocity, m/sec; volume, m³
x	one of the axes of Cartesian reference frame, m
α	thermal diffusivity, $k/\rho C_p$, m²/sec; absorptivity, dimensionless
β	coefficient of volume expansion, 1/K
Γ	mass flow rate of condensate per unit width, kg/m · sec
ΔT	temperature difference, K
δ	thickness of cavity space, m
ϵ	emissivity, dimensionless
η_f	fin efficiency, dimensionless
λ	wavelength, μm

μ	viscosity, kg/m · sec
ν	kinematic viscosity, m²/sec
ρ	reflectivity, dimensionless; density, kg/m³
σ	surface tension, N/m; Stefan–Boltzmann constant, 5.729×10^{-8} W/m² · K⁴
τ	transmissivity, dimensionless

Subscripts

b	black body; base condition of fin
c	convection
l	liquid
m	mean
0	stagnation condition
r	radiation
s	surface
v	vapor
λ	spectral
∞	free stream

Heat transfer encompasses all phenomena occurring in the transport of energy from one region to another as a result of a temperature difference between them. This transport can take place in three distinct modes: *conduction, convection,* and *radiation.* A particular convective heat transfer occurs when a change of phase takes place known as *boiling* and *condensation.*

56.1 CONDUCTION HEAT TRANSFER

CONDUCTION   is the mode of heat transfer by which energy exchange takes place from a region of high temperature to a region of low temperature by the kinetic or direct impact of molecules.

FOURIER'S LAW OF HEAT CONDUCTION   states that the rate of heat flow by conduction in a given direction is proportional to the temperature gradient in that direction and to the area normal to the heat flow direction. Namely, the heat flow in x direction, q_x, is given by

$$q_x = -kA \frac{\partial T}{\partial x}$$

where A is the area normal to the heat flow, $\partial T/\partial x$ is the temperature gradient, and k is the thermal conductivity of the material. For a three-dimensional body of constant thermal property without heat generation and under steady heat conduction the temperature satisfies

$$\nabla^2 T = 0$$

56.1.1 Thermal Conductivity

In general, a solid is a better heat conductor than a liquid, and a liquid is a better heat conductor than a gas. Also, metals have higher conductivities than nonmetals. Pure metals have thermal conductivities that decrease with increasing temperature. The trend may be reversed owing to the presence of impurities or alloy elements. The addition of another metal, or the presence of an impurity, usually decreases the thermal conductivity of a pure metal. The conductivities of nonmetallic solids generally increase with increases in temperature and density.

Liquids, in general, have thermal conductivities decreasing with increases in temperature; water, which has the highest thermal conductivity of all the nonmetallic liquids, however, is an exception. The thermal conductivities of gases increase with increasing temperature and decrease with increasing molecular weight. The thermal conductivities of some metals and nonmetals are given, respectively, in Tables 56.1 and 56.2. Insulating materials are used to obstruct the heat flow between enclosure and its surroundings. Examples of such materials together with their thermal properties are shown in Tables 56.3 and 56.4. The conductivities for some liquids and molten metals are given, respectively, in Tables 56.5 and 56.6. The thermal conductivities of gases are given in Table 56.7.

Table 56.1 Thermal Properties of Metallic Solids[a]

Composition	Melting Point (K)	Properties at 300 K				Properties at Various Temperatures (K) k (W/m·K); C_p (J/kg·K)		
		ρ (kg/m³)	C_p (J/kg·K)	k (W/m·K)	$\alpha \times 10^6$ (m²/sec)	100	600	1200
Aluminum	933	2702	903	237	97.1	302; 482	231; 1033	
Copper	1358	8933	385	401	117	482; 252	379; 417	339; 480
Gold	1336	19300	129	317	127	327; 109	298; 135	255; 155
Iron	1810	7870	447	80.2	23.1	134; 216	54.7; 574	28.3; 609
Lead	601	11340	129	35.3	24.1	39.7; 118	31.4; 142	
Magnesium	923	1740	1024	156	87.6	169; 649	149; 1170	
Molybdenum	2894	10240	251	138	53.7	179; 141	126; 275	105; 308
Nickel	1728	8900	444	90.7	23.0	164; 232	65.6; 592	76.2; 594
Platinum	2045	21450	133	71.6	25.1	77.5; 100	73.2; 141	82.6; 157
Silicon	1685	2330	712	148	89.2	884; 259	61.9; 867	25.7; 967
Silver	1235	10500	235	429	174	444; 187	412; 250	361; 292
Tin	505	7310	227	66.6	40.1	85.2; 188		
Titanium	1953	4500	522	21.9	9.32	30.5; 300	19.4; 591	22.0; 620
Tungsten	3660	19300	132	174	68.3	208; 87	137; 142	113; 152
Zinc	693	7140	389	116	41.8	117; 297	103; 436	

[a] Adapted from Ref. 1.

Table 56.2 Thermal Properties of Nonmetals

Description/Composition	Temperature (K)	Density, ρ (kg/m³)	Thermal Conductivity, k (W/m · K)	Specific Heat, C_p (J/kg · K)	$\alpha \times 10^6$ (m²/sec)
Bakelite	300	1300	0.232	1465	0.122
Brick, refractory					
Carborundum	872	—	18.5	—	—
Chrome-brick	473	3010	2.32	835	0.915
Fire clay brick	478	2645	1.0	960	0.394
Clay	300	1460	1.3	880	1.01
Coal, anthracite	300	1350	0.26	1260	0.153
Concrete (stone mix)	300	2300	1.4	880	0.692
Cotton	300	80	0.059	1300	0.567
Glass, window	300	2700	0.78	840	0.344
Rock, limestone	300	2320	2.15	810	1.14
Rubber, hard	300	1190	0.160	—	—
Soil, dry	300	2050	0.52	1840	0.138
Teflon	300	2200	0.35	—	—
	400	—	0.45	—	—

Table 56.3 Thermal Properties of Building and Insulating Materials (at 300 K)[a]

Description/Composition	Density ρ (kg/m³)	Thermal Conductivity, k (W/m · K)	Specific Heat, C_p (J/kg · K)	$\alpha \times 10^6$ (m²/sec)
Building boards				
Plywood	545	0.12	1215	0.181
Acoustic tile	290	0.058	1340	0.149
Hardboard, siding	640	0.094	1170	0.126
Woods				
Hardwoods (oak, maple)	720	0.16	1255	0.177
Softwoods (fir, pine)	510	0.12	1380	0.171
Masonry materials				
Cement mortar	1860	0.72	780	0.496
Brick, common	1920	0.72	835	0.449
Plastering materials				
Cement plaster, sand aggregate	1860	0.72	—	—
Gypsum plaster, sand aggregate	1680	0.22	1085	0.121
Blanket and batt				
Glass fiber, paper faced	16	0.046	—	—
Glass fiber, coated; duct liner	32	0.038	835	1.422
Board and slab				
Cellular glass	145	0.058	1000	0.400
Wood, shredded/cemented	350	0.087	1590	0.156
Cork	120	0.039	1800	0.181
Loose fill				
Glass fiber, poured or blown	16	0.043	835	3.219
Vermiculite, flakes	80	0.068	835	1.018

[a] Adapted from Ref. 1.

56.1.2 One-Dimensional Steady Heat Conduction

For the steady conduction of heat through a homogeneous material the rate of heat conduction can be given as $q = \Delta T / R$, where ΔT is the temperature difference and R is the *thermal resistance*. The reciprocal of R is called *thermal conductance* ($C = 1/R$). The rate of heat transfer, temperature distribution, and thermal resistance for plane wall, cylinder, and sphere are shown in Table 56.8.

Table 56.4 Thermal Conductivities for Some Industrial Insulating Materials[a]

Description/Composition	Maximum Service Temperature (K)	Typical Density (kg/m³)	Typical Thermal Conductivity, k (W/m · K), at Various Temperatures (K)			
			200	300	420	645
Blankets						
Blanket, mineral fiber, glass; fine fiber organic bonded	450	10		0.048		
		48		0.033		
Blanket, alumina-silica fiber	1530	48				0.105
Felt, semirigid; organic bonded	480	50–125		0.038	0.063	0.087
Felt, laminated; no binder	920	120			0.051	
Blocks, boards, and pipe insulations						
Asbestos paper, laminated and corrugated, 4-ply	420	190		0.078		
Calcium silicate	920	190			0.063	0.089
Polystyrene, rigid						
Extruded (R-12)	350	56	0.023	0.027		
Molded beads	350	16	0.026	0.040		
Rubber, rigid foamed	340	70		0.032		
Insulating cement						
Mineral fiber (rock, slag, or glass)						
With clay binder	1255	430			0.088	0.123
With hydraulic setting binder	922	560			0.123	
Loose fill						
Cellulose, wood or paper pulp	—	45	0.036	0.039		
Perlite, expanded	—	105		0.053		
Vermiculite, expanded	—	122		0.068		

[a] Adapted from Ref. 1.

Table 56.5 Thermal Properties of Saturated Liquids[a]

T (K)	ρ (kg/m³)	c_p (kJ/kg · K)	$\nu \times 10^6$ (m²/sec)	$k \times 10^3$ (W/m · K)	$\alpha \times 10^7$ (m²/sec)	Pr	$\beta \times 10^3$ (K⁻¹)
Ammonia, NH₃							
223	703.7	4.463	0.435	547	1.742	2.60	2.45
323	564.3	5.116	0.330	476	1.654	1.99	2.45
Carbon Dioxide, CO₂							
223	1,156.3	1.84	0.119	85.5	0.402	2.96	14.0
303	597.8	36.4	0.080	70.3	0.028	28.7	14.0
Engine Oil (Unused)							
273	899.1	1.796	4,280	147	0.910	47,000	0.70
430	806.5	2.471	5.83	132	0.662	88	0.70
Ethylene Glycol, C₂H₄(OH)₂							
273	1,130.8	2.294	57.6	242	0.933	617.0	0.65
373	1,058.5	2.742	2.03	263	0.906	22.4	0.65
Glycerin, C₃H₅(OH)₃							
273	1,276.0	2.261	8,310	282	0.977	85,000	0.47
320	1,247.2	2.564	168	287	0.897	1,870	0.50
Freon (Refrigerant-12), CCl₂F₂							
230	1,528.4	0.8816	0.299	68	0.505	5.9	1.85
320	1,228.6	1.0155	0.190	68	0.545	3.5	3.50

[a] Adapted from Ref. 2. See Table 56.22 for H_2O.

For the plane wall, heat is assumed to conduct only in the x direction; for cylinder and sphere, heat conduction is only in the radial direction.

For heat conduction through N *layers of plane walls* (no thermal resistance at the contact surface), with each layer having thickness Δx_n and thermal conductivity k_n ($n = 1, 2, \ldots, N$), the thermal resistance is

$$R = \sum_{n=1}^{N} \frac{\Delta x_n}{k_n A}$$

For heat conduction in radial direction through N *concentric, hollow cylinders* with zero thermal contact resistance,

$$R = \sum_{n=1}^{N} \frac{\ln(r_{n+1}/r_n)}{2\pi k_n L}$$

where r_1 = inner radius, r_{N+1} = outer radius.

For N *concentric, hollow spheres* with zero thermal contact resistance,

$$R = \sum_{n=1}^{N} \left(\frac{1}{r_n} - \frac{1}{r_{n+1}} \right) \Big/ 4\pi k_n$$

where r_1 = inner radius, r_{N+1} = outer radius.

56.1.3 Two-Dimensional Steady Heat Conduction

For an isotropic, homogeneous material without a heat source, solution of the *Laplace equation*, $\partial^2 T/\partial X^2 + \partial^2 T/\partial y^2 = 0$, is required. Exact solutions may be obtained only for simple geometries

Table 56.6 Thermal Properties of Liquid Metals[a]

Composition	Melting Point (K)	T (K)	ρ (kg/m³)	C_p (kJ/kg·K)	$\nu \times 10^7$ (m²/sec)	k (W/m·K)	$\alpha \times 10^5$ (m²/sec)	Pr
Bismuth	544	589	10,011	0.1444	1.617	16.4	0.138	0.0142
		1033	9,467	0.1645	0.8343	15.6	1.001	0.0083
Lead	600	644	10,540	0.159	2.276	16.1	1.084	0.024
		755	10,412	0.155	1.849	15.6	1.223	0.017
Mercury	234	273	13,595	0.140	1.240	8.180	0.429	0.0290
		600	12,809	0.136	0.711	11.95	0.688	0.0103
Potassium	337	422	807.3	0.80	4.608	45.0	6.99	0.0066
		977	674.4	0.75	1.905	33.1	6.55	0.0029
Sodium	371	366	929.1	1.38	7.516	86.2	6.71	0.011
		977	778.5	1.26	2.285	59.7	6.12	0.0037
NaK (56%/44%)	292	366	887.4	1.130	6.522	25.6	2.55	0.026
		977	740.1	1.043	2.174	28.9	3.74	0.0058
PbBi (44.5%/55.5%)	398	422	10,524	0.147	—	9.05	0.586	—
		644	10,236	0.147	1.496	11.86	0.790	0.189

[a] Adapted from *Liquid Materials Handbook*, The Atomic Energy Commission, Department of the Navy, Washington, DC, 1952.

Table 56.7   Thermal Properties of Gases at Atmospheric Pressure[a]

T (K)	ρ (kg/m³)	C_p (kJ/kg $\cdot$ K)	$\nu \times 10^6$ (m²/sec)	k (W/m $\cdot$ K)	$\alpha \times 10^4$ (m²/sec)	Pr
Air						
100	3.6010	1.0266	1.923	0.009246	0.0250	0.768
300	1.1774	1.0057	16.84	0.02624	0.2216	0.708
2500	0.1394	1.688	543.0	0.175	7.437	0.730
Ammonia, NH₃						
220	0.3828	2.198	19.0	0.0171	0.2054	0.93
473	0.4405	2.395	37.4	0.0467	0.4421	0.84
Carbon Dioxide						
220	2.4733	0.783	4.490	0.01081	0.0592	0.818
600	0.8938	1.076	30.02	0.04311	0.4483	0.668
Carbon Monoxide						
220	1.5536	1.0429	8.903	0.01906	0.1176	0.758
600	0.5685	1.0877	52.06	0.04446	0.7190	0.724
Helium						
33	1.4657	5.200	3.42	0.0353	0.04625	0.74
900	0.05286	5.200	781.3	0.298	10.834	0.72
Hydrogen						
30	0.8472	10.840	1.895	0.0228	0.02493	0.759
300	0.0819	14.314	109.5	0.182	1.554	0.706
1000	0.0245	14.968	822	0.440	11.997	0.686
Nitrogen						
100	3.4808	1.0722	1.971	0.009450	0.02531	0.786
300	1.1421	1.0408	15.63	0.0262	0.2204	0.713
1200	0.2851	1.2037	156.1	0.07184	2.0932	0.748
Oxygen						
100	3.9918	0.9479	1.946	0.00903	0.02388	0.815
300	1.3007	0.9203	15.86	0.02676	0.2235	0.709
600	0.6504	1.0044	52.15	0.04832	0.7399	0.704
Steam (H₂O Vapor)						
380	0.5863	2.060	21.6	0.0246	0.2036	1.060
850	0.2579	2.186	115.2	0.0637	1.130	1.019

[a] Adapted from Ref. 2.

and special boundary conditions. Methods of solving the equation include the use of *analytical, graphical, electrical analog,* and *numerical approaches.* The rate of heat transfer between two isotherms, T_1 and T_2, can be expressed in terms of the *conduction shape factor,* defined by

$$q = kS(T_1 - T_2)$$

Table 56.9 shows the values of S for some simple geometries.

Table 56.8 One-Dimensional Heat Conduction

Geometry	Heat-Transfer Rate and Temperature Distribution	Heat-Transfer Rate and Overall Heat-Transfer Coefficient with Convection at the Boundaries
Plane wall	$q = \dfrac{T_1 - T_2}{(x_2 - x_1)/kA}$ $T = T_1 + \dfrac{T_2 - T_1}{x_2 - x_1}(x - x_1)$ $R = (x_2 - x_1)/kA$	$q = UA(T_{\infty,1} - T_{\infty,2})$ $U = \dfrac{1}{\dfrac{1}{h_1} + \dfrac{x_2 - x_1}{k} + \dfrac{1}{h_2}}$
Hollow cylinder	$q = \dfrac{T_1 - T_2}{[\ln(r_2/r_1)]/2\pi kL}$ $T = T_1 + \dfrac{T_2 - T_1}{\ln(r_2/r_1)}\ln\dfrac{r}{r_1}$ $R = \dfrac{\ln(r_2/r_1)}{2\pi kL}$	$q = 2\pi r_1 L U_1 (T_{\infty,1} - T_{\infty,2})$ $\;\;= 2\pi r_2 L U_2 (T_{\infty,1} - T_{\infty,2})$ $U_1 = \dfrac{1}{\dfrac{1}{h_1} + \dfrac{r_1 \ln(r_2/r_1)}{k} + \dfrac{r_1}{r_2}\dfrac{1}{h_2}}$ $U_2 = \dfrac{1}{\left(\dfrac{r_2}{r_1}\right)\dfrac{1}{h_1} + \dfrac{r_2 \ln(r_2/r_1)}{k} + \dfrac{1}{h_2}}$
Hollow sphere	$q = \dfrac{T_1 - T_2}{\left(\dfrac{1}{r_1} - \dfrac{1}{r_2}\right)\Big/ 4\pi k}$ $T = \dfrac{1}{\left(1 - \dfrac{r_1}{r_2}\right)}\left[\dfrac{r_1}{r}(T_1 - T_2) + \left(T_2 - T_1\dfrac{r_1}{r_2}\right)\right]$ $R = \left(\dfrac{1}{r_1} - \dfrac{1}{r_2}\right)\Big/ 4\pi k$	$q = 4\pi r_1^2 U_1 (T_{\infty,1} - T_{\infty,2})$ $\;\;= 4\pi r_2^2 U_2 (T_{\infty,1} - T_{\infty,2})$ $U_1 = \dfrac{1}{\dfrac{1}{h_1} + r_1^2\left(\dfrac{1}{r_1} - \dfrac{1}{r_2}\right)\Big/k + \left(\dfrac{r_1}{r_2}\right)^2\dfrac{1}{h_2}}$ $U_2 = \dfrac{1}{\left(\dfrac{r_2}{r_1}\right)^2\dfrac{1}{h_1} + r_2^2\left(\dfrac{1}{r_1} - \dfrac{1}{r_2}\right)\Big/k + \dfrac{1}{h_2}}$

56.1.4 Heat Conduction with Convection Heat Transfer on the Boundaries

When a solid is surrounded by fluid (or gas), heat transfer takes place by convection (and/or radiation) between the solid surface and the fluid. Heat transfer due to convection is given by $q = hA\,\Delta T$, where h is the *convection heat transfer coefficient* (Section 56.2) and ΔT is the temperature difference between the solid surface and the fluid. The surface area in contact with the fluid is denoted by A. The *thermal resistance* for the convection heat transfer is given by $1/hA$ (*convection resistance*). The overall heat transfer by combined conduction and convection is frequently expressed in terms of an *overall heat transfer coefficient* U defined by $q = UA\,\Delta T$.

Table 56.8 shows the overall heat transfer coefficients for some simple geometries. Note that U may be based either on the inner surface (U_1) or on the outer surface (U_2) for the cylinders and spheres.

Critical Radius of Insulation for Cylinders.   Insulation materials are often used on the outside surfaces of cylindrical pipes to reduce heat loss. There are, however, situations in which the addition of insulation does not reduce heat loss, depending on whether the outside radius of the insulation is smaller or larger than the critical radius defined by $r_{cr} = k/h$. If the outer radius is less than r_{cr}, then the addition of insulation will increase heat loss. For outer radii greater than r_{cr} an increase in insulation thickness will cause a decrease in heat loss.

Extended Surfaces.   The rate of heat transfer between a structure and the surrounding ambient fluid may be increased by attaching extended surfaces to the primary surface. Examples may be found in the cooling fins of air-cooled engines—the radiator fins.

For a constant base temperature T_b, constant cross-sectional area A, circumference $C = 2W + 2t$, and length L which is much larger than the thickness t, the temperature distribution of the fin satisfies (Fig. 56.1)

$$\frac{d^2T}{dx^2} - \frac{hC}{kA}\,(T - T_\infty) = 0$$

The solutions of this equation depend on the boundary conditions at the tip $x = L$. Table 56.10 shows the temperature distribution and heat transfer rate for fins of uniform cross section subjected to different boundary conditions, assuming a constant h.

The *fin effectiveness* is defined as the ratio of fin heat transfer rate to the heat transfer rate that would exist without the fin. The use of fin is justified only if the fin effectiveness is much greater than 2.

The *fin efficiency* η_f is the ratio of the heat transfer rate from a fin to the heat transfer rate that would be obtained if the entire fin surface is maintained at the same temperature as the fin base temperature. Therefore,

$$q = \eta_f\, hA_f\,(T_b - T_\infty)$$

where A_f is the total surface area of the fin and T_b is the temperature of the fin at the base. Fins are usually applied to natural and forced convection of gases, and are generally not effective for two-phase flow and for application involving boiling and condensation.

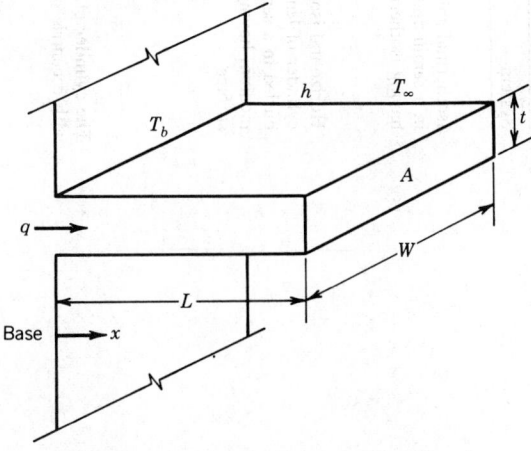

Fig. 56.1 Heat transfer by extended surfaces.

Table 56.9 Conduction Shape Factors

System	Schematic	Restrictions	Shape Factor
Isothermal sphere buried in a semiinfinite medium having isothermal surface		$z > D/2$	$\dfrac{2\pi D}{1 - D/4z}$
Horizontal isothermal cylinder of length L buried in a semiinfinite medium having isothermal surface		$L \gg D$	$\dfrac{2\pi L}{\cosh^{-1}(2z/D)}$
		$\left.\begin{array}{l} L \gg D \\ z > 3D/2 \end{array}\right\}$	$\dfrac{2\pi L}{\ln(4z/D)}$
The cylinder of length L with eccentric bore		$L \gg D_1, D_2$	$\dfrac{2\pi L}{\cosh^{-1}\left(\dfrac{D_1^2 + D_2^2 - 4\epsilon^2}{2D_1 D_2}\right)}$
Conduction between two cylinders of length L in infinite medium		$L \gg W$	$\dfrac{2\pi L}{\cosh^{-1}\left(\dfrac{4W^2 - D_1^2 - D_2^2}{2D_1 D_2}\right)}$

Circular cylinder of length L in a square solid	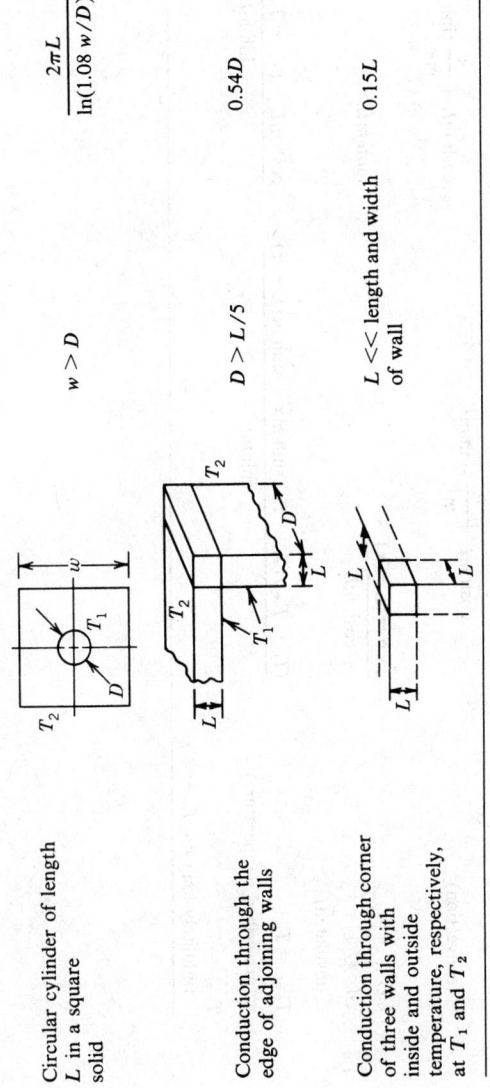	$w > D$	$\dfrac{2\pi L}{\ln(1.08\, w/D)}$
Conduction through the edge of adjoining walls		$D > L/5$	$0.54D$
Conduction through corner of three walls with inside and outside temperature, respectively, at T_1 and T_2		$L \ll$ length and width of wall	$0.15L$

Table 56.10 Temperature Distribution and Heat Transfer Rate at the Fin Base ($m = \sqrt{hC/kA}$)

Condition at $x = L$	$\dfrac{T - T_\infty}{T_b - T_\infty}$	Heat Transfer Rate $q/mkA\,(T_b - T_\infty)$
$h(T_{x=L} - T_\infty) = -k\left(\dfrac{dT}{dx}\right)_{x=L}$ (convection)	$\dfrac{\cosh m(L-x) + \dfrac{h}{mk}\sinh m(L-x)}{\cosh mL + \dfrac{h}{mk}\sinh mL}$	$\dfrac{\sinh mL + \dfrac{h}{mk}\cosh mL}{\cosh mL + \dfrac{h}{mk}\sinh mL}$
$\left(\dfrac{dT}{dx}\right)_{x=L} = 0$ (insulated)	$\dfrac{\cosh m(L-x)}{\cosh mL}$	$\tanh mL$
$T_{x=L} = T_L$ (prescribed temperature)	$\dfrac{(T_L - T_\infty)/(T_b - T_\infty)\sinh mx + \sinh m(L-x)}{\sinh ml}$	$\dfrac{\cosh mL - (T_L - T_\infty)/(T_b - T_\infty)}{\sinh mL}$
$T_{x=L} = T_\infty$ (infinitely long fin, $L \to \infty$)	e^{-mx}	1

56.1.5 Transient Heat Conduction

When a solid body at uniform temperature T_i is immersed in a fluid of different temperature T_∞, the solid body surface is subject to convection heat transfer, whereas conduction takes place inside the body. If the *Biot number* Bi $= hL/k$ ($L = V/A$ = volume of solid/surface area of solid) is less than 0.1, the temperature of the solid may be assumed uniform; that is, it will depend only on time. Analysis based on such assumption is called the *lumped heat-capacity* method. The temperature of the solid is given by

$$\frac{T - T_\infty}{T_i - T_\infty} = \exp\left(\frac{-t}{\tau_t}\right) = \exp(-\text{Bi Fo})$$

where $\tau_t = \rho C_p V/hA$ is called the *time constant.* Increasing the value of τ_t will cause a solid to respond more slowly to its environment and will increase the time required to reach thermal equilibrium ($T = T_\infty$).

The notation Fo $= \alpha t A^2/V^2$ is a nondimensional time called *Fourier number,* which, with the Biot number, characterizes transient heat-conduction problems. The total heat flow through the surface up to time t is

$$Q = \rho V C_p (T_i - T_\infty) [1 - \exp(-t/\tau_t)]$$

Transient Heat Transfer for Infinite Plate, Infinite Cylinder, and Sphere Subjected to Surface Convection

Analytical solutions have been worked out and presented in graphical forms by Heisler[3]; these are often referred to as the *Heisler charts.* The results for plane wall, cylinder, and sphere are presented in Figs. 56.2–56.10. At time zero each solid is assumed to have uniform temperature T_i, and is suddenly immersed in a fluid at temperature T_∞. Assuming that the convection heat-transfer coefficient h is constant, the resulting time history of temperature at any point within the solid can be found by combining Figs. 56.2 and 56.3 for plane walls; Figs. 56.5 and 56.6 for cylinders; Figs. 56.8 and 56.9 for spheres. The total amount of energy Q transferred across the solid surface up to any time t can be found from Figs. 56.4, 56.7, and 56.10.

56.2 CONVECTION HEAT TRANSFER

Convection is a mode of energy transport in which the energy is transferred by means of fluid motion. If the fluid motion is caused by external forces, the energy transfer is called *forced convection.* If the fluid motion arises from a buoyance effect caused by density difference, the energy transfer is called *free convection* or *natural convection.* The heat-transfer rate q due to overall effect of convection is expressed by *Newton's law of cooling:*

$$q = hA\,\Delta T$$

where A is the surface area and ΔT is the temperature difference defined depending on the problems. The quantity h is called the *convection heat-transfer coefficient* or *film coefficient,* which is governed by the velocity and physical properties of the fluid, and the shape and nature of the surface. The nondimensional heat-transfer coefficient Nu $= hL/k$, where L is a characteristic length and k is the thermal conductivity of the fluid, is called the *Nusselt number.*

56.2.1 Forced Convection—Internal Flow

Unless specified otherwise, the convection heat-transfer coefficient is based on $\Delta T = T_s - T_b$, so that $q = hA(T_s - T_b)$. The *bulk temperature,* or *mixing-cup temperature,* T_b, is the energy-average fluid temperature across the tube, and is defined as

$$T_b = \frac{\int C_p T\, d\dot{m}}{\int C_p\, d\dot{m}}$$

where $\dot{m}$ is the axial flow rate.

The flow at the entrance region is quite different from that far away from the entrance. The rate of heat transfer differs significantly depending on whether the flow is *laminar* or *turbulent.* It is generally observed that the flow becomes turbulent when $\text{Re}_D = V_m D/v > 2300$ for a smooth tube. The transition from laminar to turbulent also depends on the roughness of tube wall and other factors. The generally accepted range for transition is $200 < \text{Re}_D < 4000$.

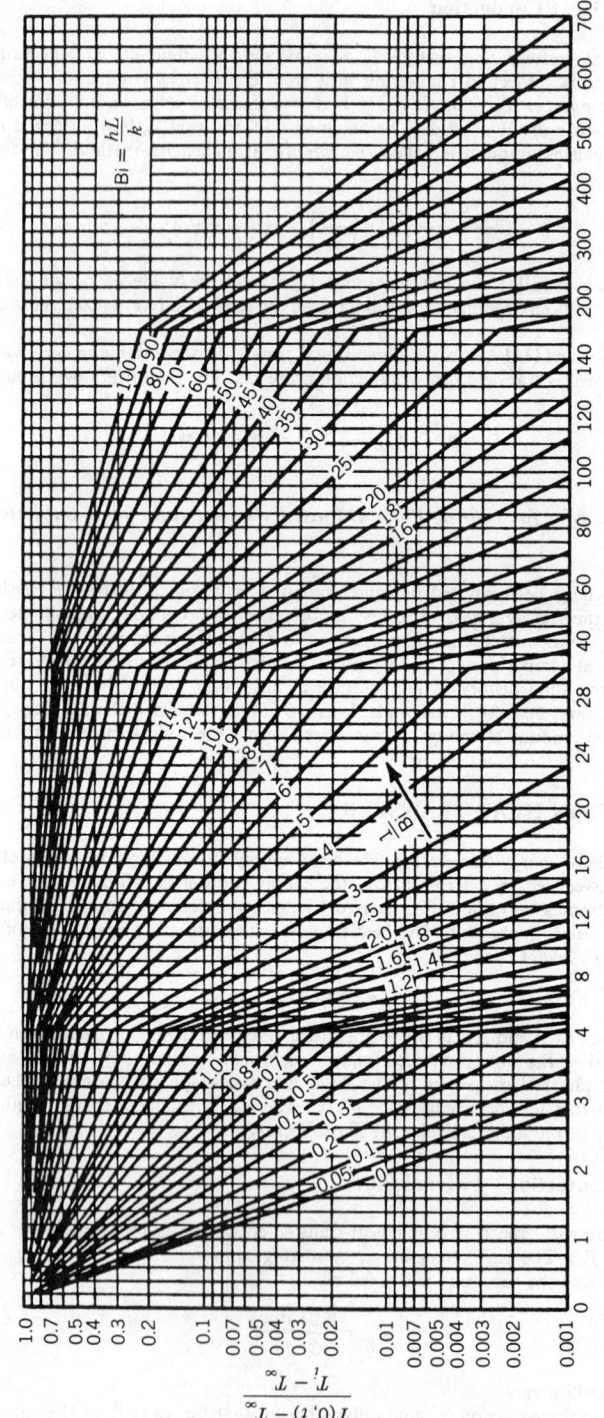

Fig. 56.2 Midplane temperature as a function of time for a plane wall of thickness 2L. Adapted from Heisler.[3]

$$Bi = \frac{hL}{k}$$

$$\frac{\alpha t}{L^2}$$

$$\frac{T(0,t) - T_\infty}{T_i - T_\infty}$$

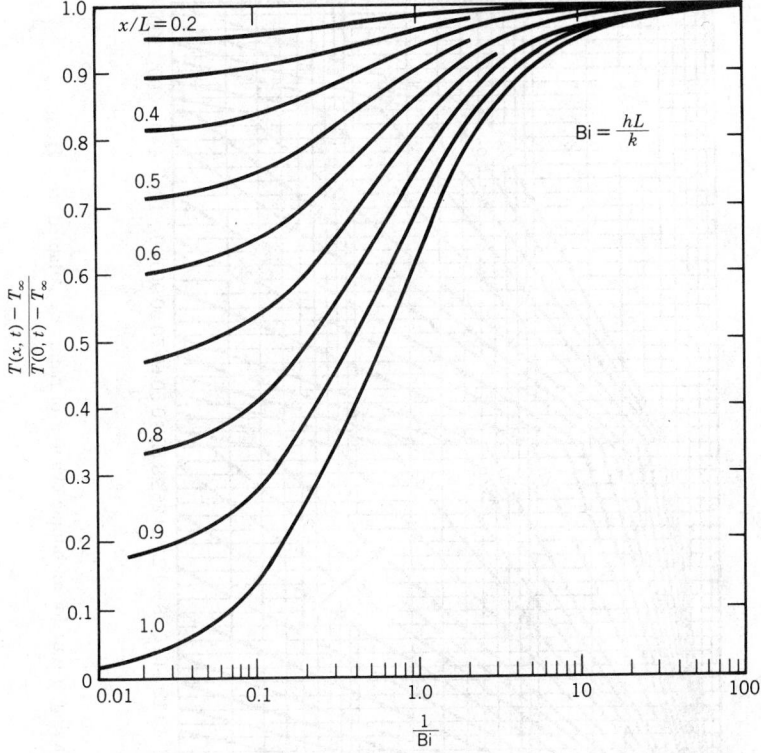

Fig. 56.3 Temperature distribution in a plane wall of thickness $2L$. Adapted from Heisler.[3]

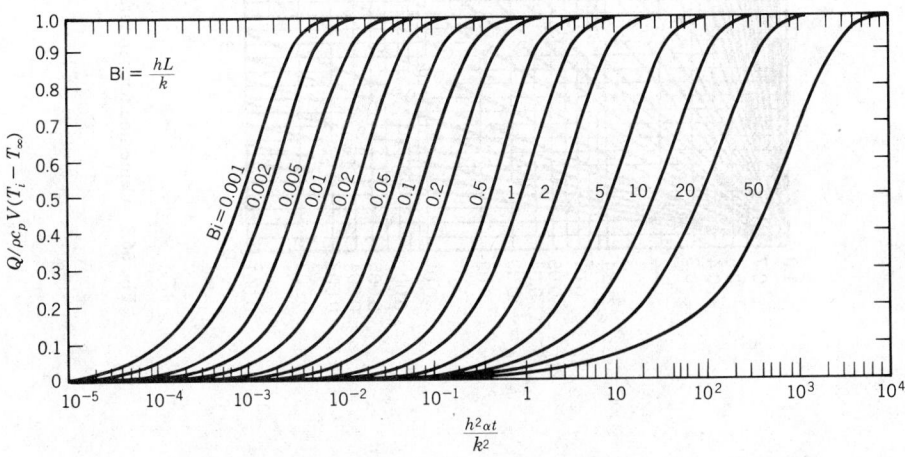

Fig. 56.4 Internal energy change as a function of time for a plane wall of thickness $2L$.[4] Used with the permission of McGraw-Hill Book Company.

Laminar Fully Developed Flow

The Nusselt numbers are constant when both temperature and velocity profiles are fully developed. The values depend on the thermal boundary conditions. For *circular tubes* and for $Pr \geq 0.6$, $x/D\,Re_D\,Pr > 0.05$, $Nu_D = 3.66$ and 4.36, respectively, for the constant temperature and the constant heat flux conditions on the tube walls. The fluid properties are based on the mean bulk temperature.

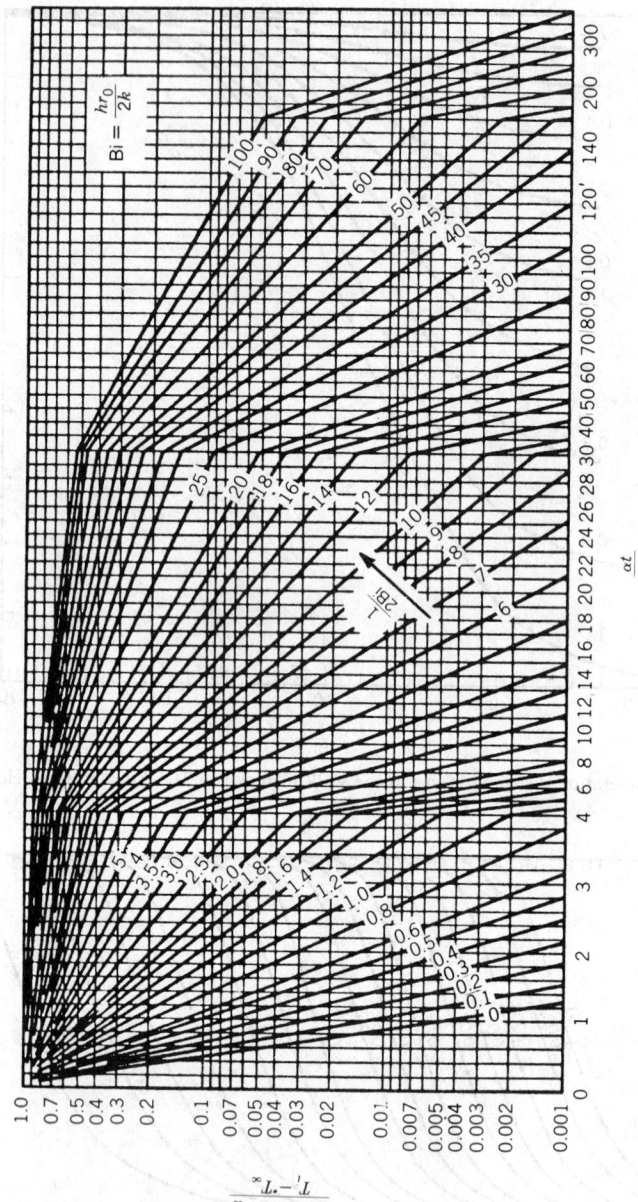

Fig. 56.5 Centerline temperature as a function of time for an infinite cylinder of radius r_0. Adapted from Heisler.[3]

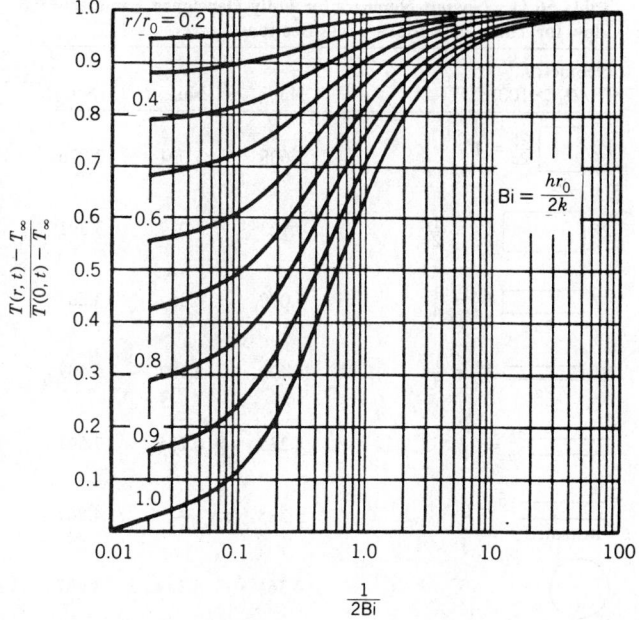

Fig. 56.6 Temperature distribution in an infinite cylinder of radius r_0. Adapted from Heisler.[3]

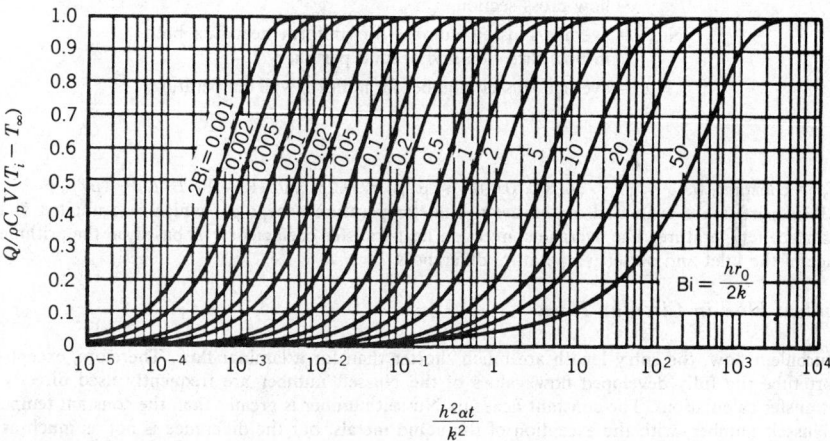

Fig. 56.7 Internal energy change as a function of time for an infinite cylinder of radius r_0.[4] Used with the permission of McGraw-Hill Book Company.

For *noncircular tubes*, the hydraulic diameter, $D_h = 4 \times$ flow cross-sectional area/wetted perimeter, must be used to define the Nusselt number Nu_D and the Reynolds number Re_D. Table 56.11 shows the Nusselt numbers based on hydraulic diameter for various cross sections.

Laminar Flow for Short Tubes

The Nusselt numbers are, in principle, infinite at the entrance and decrease to their asymptotic values of fully developed flow as the distance from the entrance increases. The Sieder–Tate equation[5] gives good correlation for the *combined entry length* (both thermal and velocity profiles are developing) and short tubes:

$$\overline{Nu_D} = \frac{\bar{h}D}{k} = 1.86 \,(Re_D \; Pr)^{1/3} \left(\frac{D}{L}\right)^{1/3} \left(\frac{\mu}{\mu_s}\right)^{0.14}$$

Table 56.11 Nusselt Numbers for Fully Developed Laminar Flow for Tubes of Various Cross Sections[a]

Geometry ($L/D_h > 100$)		Nu_{H1}	Nu_{H2}	Nu_r
$2b$ ☐ $2a$	$\dfrac{2b}{2a} = 1$	3.608	3.091	2.976
$2b$ ☐ $2a$	$\dfrac{2b}{2a} = \dfrac{1}{2}$	4.123	3.017	3.391
$2b$ ☐ $2a$	$\dfrac{2b}{2a} = \dfrac{1}{4}$	5.099	4.35	3.66
$2b$ ☐ $2a$	$\dfrac{2b}{2a} = \dfrac{1}{8}$	6.490	2.904	5.597
———	$\dfrac{2b}{2a} = 0$	8.235	8.235	7.541
——— Insulated	$\dfrac{b}{a} = 0$	5.385	—	4.861
◯		4.364	4.364	3.657

[a] Nu_{H1} = average Nusselt number for uniform heat flux in flow direction and uniform wall temperature at particular flow cross section.

Nu_{H2} = average Nusselt number for uniform heat flux both in flow direction and around periphery.

Nu_r = average Nusselt number for uniform wall temperature.

for T_s = constant, $0.48 < Pr < 16{,}700$, $0.0044 < \mu/\mu_s < 9.75$, and $(Re_D\ Pr\ D/L)^{1/3}\ (\mu/\mu_s)^{0.14} \geq 2$.

All properties are evaluated at the mean bulk temperature except μ_s, which is evaluated at the wall surface temperature. The average convection heat-transfer coefficient $\bar{h}$ is based on the arithmetic average of the inlet and outlet temperature differences.

Turbulent Flow in Circular Tubes

For turbulent flow, the entry length are much shorter than for a laminar flow. Therefore, except for a short tube the fully developed flow values of the Nusselt number are frequently used directly in heat-transfer calculations. The constant heat flux Nusselt number is greater than the constant temperature Nusselt number, with the exception of the liquid metals, but the difference is not as much as in the laminar flow and becomes negligible for $Pr > 1.0$. The Dittus–Boelter equation[6] in the following form is most often used if the difference between the pipe surface temperature and the bulk fluid temperature is less than 6°C (10°F) for liquids or 56°C (100°F) for gases:

$$Nu_D = 0.023\ Re_D^{0.8}\ Pr^n$$

for $0.7 \leqslant Pr \leqslant 160$, $Re_D \geq 10{,}000$, and $L/D \geq 60$, where

$$n = 0.4 \text{ for heating, } T_s > T_b$$
$$= 0.3 \text{ for cooling, } T_s < T_b$$

For temperature differences greater than specified above, use[5]

$$Nu_D = 0.027\ Re_D^{0.8}\ Pr^{1/3}\left(\frac{\mu}{\mu_s}\right)^{0.14}$$

for $0.7 \leqslant Pr \leqslant 16{,}700$, $Re_D \geq 10{,}000$, and $L/D \geq 60$.

The properties are evaluated at the mean bulk fluid temperature except μ_s, which is evaluated at the tube surface temperature.

For *concentric tube annulus*, the hydraulic diameter $D_h = D_o - D_i$ (outer diameter − inner

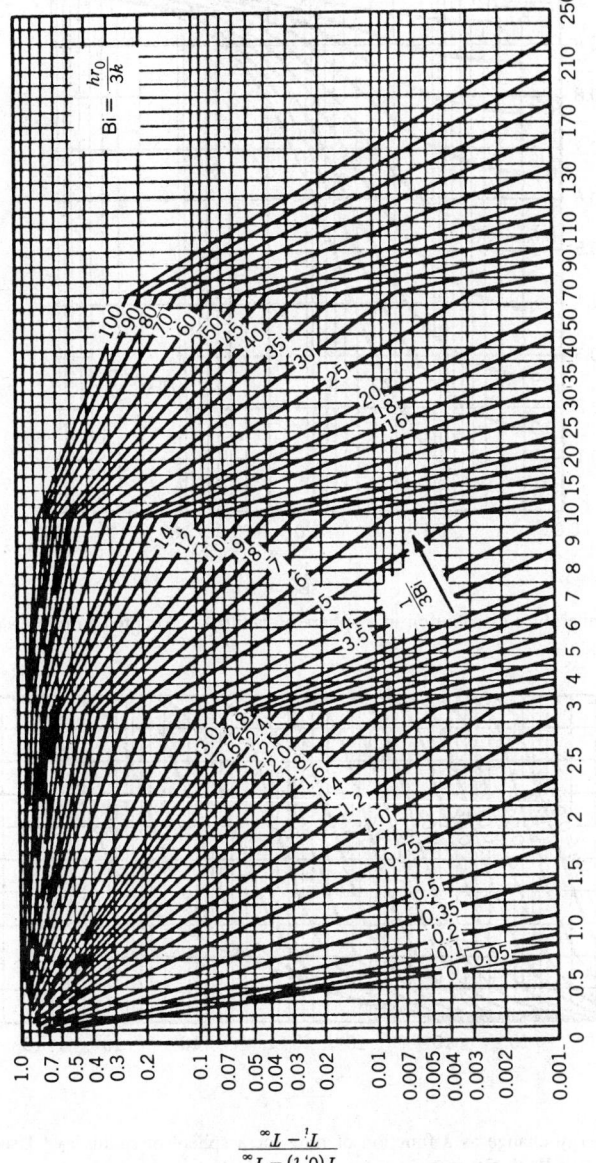

Fig. 56.8 Center temperature as a function of time in a sphere of radius r_0. Adapted from Heisler.[3]

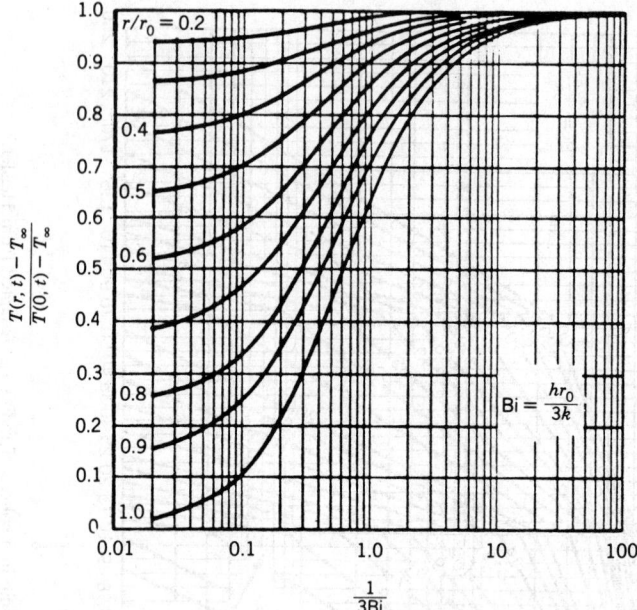

Fig. 56.9 Temperature distribution in a sphere of radius r_0. Adapted from Heisler.[3]

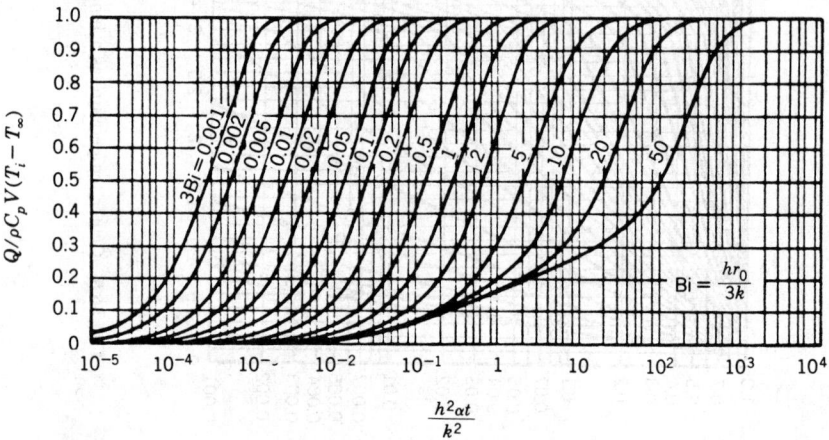

Fig. 56.10 Internal energy change as a function of time for a sphere of radius r_0.[4] Used with the permission of McGraw-Hill Book Company.

diameter) must be used for Nu_D and Re_D, and the coefficient h at either surface of the annulus must be evaluated from the Dittus–Boelter equation.

It should be noted that the foregoing equations apply for smooth surfaces. The heat-transfer rate will be larger for rough surfaces. Also, they are not applicable to the liquid metals.

Fully Developed Turbulent Flow of Liquid Metals in Circular Tubes

The Prandtl numbers for liquid metals are on the order of magnitude of 0.01, and the Nusselt number depends on the *Peclet number* defined by Pe = RePr:

$$Nu_D = 5.0 + 0.025 \, Pe_D^{0.8}$$

for T_s = constant and valid for $Pe_D > 100$ and $L/D > 60$;

$$\mathrm{Nu}_D = 4.82 + 0.0185\,\mathrm{Pe}_D^{0.827}$$

for $q'' = $ constant and valid for $3.6 \times 10^3 < \mathrm{Re}_D < 9.05 \times 10^5$, $10^2 < \mathrm{Pe}_D < 10^4$, and $L/D > 60$.

56.2.2 Forced Convection—External Flow

The heat transfer coefficient h is based on the temperature difference between the wall surface temperature and the fluid temperature in the free stream outside the thermal boundary layer. The total heat-transfer rate from the wall to the fluid is given by $q = hA(T_s - T_\infty)$. The Reynolds numbers are based on the free stream velocity. The fluid properties are evaluated either at the free stream temperature T_∞ or at the film temperature $T_f = (T_s + T_\infty)/2$.

Laminar Flow on a Flat Plate

For uniform velocity flow along a constant temperature semiinfinite plate the boundary layer originates from the leading edge and is laminar. The flow remains laminar until the local Reynolds number $\mathrm{Re}_x = U_\infty x/\nu$ reaches Re_c, the *critical Reynolds number*. For a smooth surface, it is generally assumed that $\mathrm{Re}_c = 5 \times 10^5$, although its value depends on surface roughness and many other factors.

The *local Nusselt number* at distance x from the leading edge and the *average Nusselt number* between $x = 0$ and $x = L$ are given below (Re_x and $\mathrm{Re}_L \leq 5 \times 10^5$):

$$\left.\begin{array}{l} \mathrm{Nu}_x = hx/k = 0.332\mathrm{Re}_x^{0.5}\,\mathrm{Pr}^{1/3} \\[2mm] \overline{\mathrm{Nu}}_L = \bar{h}L/k = 0.664\mathrm{Re}_L^{0.5}\,\mathrm{Pr}^{1/3} \end{array}\right\} \quad \text{for } \mathrm{Pr} \geq 0.6$$

$$\left.\begin{array}{l} \mathrm{Nu}_x = 0.565(\mathrm{Re}_x\,\mathrm{Pr})^{0.5} \\[2mm] \overline{\mathrm{Nu}}_L = 1.13(\mathrm{Re}_L\,\mathrm{Pr})^{0.5} \end{array}\right\} \quad \text{for } \mathrm{Pr} \leq 0.6$$

In the above equations all fluid properties are evaluated at the mean film temperature.

Turbulent Flow on a Flat Plate

For flows which are turbulent starting from the leading edges and all fluid properties based on the mean film temperature,

$$\mathrm{Nu}_x = 0.0292\mathrm{Re}_x^{0.8}\,\mathrm{Pr}^{1/3}$$

$$\overline{\mathrm{Nu}}_L = 0.036\mathrm{Re}_L^{0.8}\,\mathrm{Pr}^{1/3}$$

for $5 \times 10^5 \leq \mathrm{Re}_x$ and $\mathrm{Re}_L \leq 10^8$ and $0.6 \leq \mathrm{Pr} \leq 60$.

The Average Nusselt Number Between x = 0 and x = L with Transition

If the transition occurs instantaneously when the critical Reynolds number Re_c is reached, then[7]

$$\overline{\mathrm{Nu}}_L = 0.036\mathrm{Pr}^{1/3}[\mathrm{Re}_L^{0.8} - \mathrm{Re}_c^{0.8} + 18.44\mathrm{Re}_c^{0.5}]$$

for $5 \times 10^5 \leq \mathrm{Re}_L \leq 10^8$ and $0.6 \leq \mathrm{Pr} \leq 60$. In particular,

$$\overline{\mathrm{Nu}}_L = 0.036\mathrm{Pr}^{1/3}(\mathrm{Re}_L^{0.8} - 18,700)$$

For $\mathrm{Re}_c = 4 \times 10^5$,

$$\overline{\mathrm{Nu}}_L = 0.036\mathrm{Pr}^{1/3}(\mathrm{Re}_L^{0.8} - 23,100)$$

for $\mathrm{Re}_c = 5 \times 10^5$. All the fluid properties are evaluated at the mean film temperature for the above equations.

Circular Cylinder in Cross Flow

$$\overline{\mathrm{Nu}}_D = (0.4\mathrm{Re}_D^{0.5} + 0.06\mathrm{Re}_D^{2/3})\,\mathrm{Pr}^{0.4}\,(\mu_\infty/\mu_s)^{0.25}$$

for $0.67 < \mathrm{Pr} < 300$, $10 < \mathrm{Re}_D < 10^5$, and $0.25 < \mu_\infty/\mu_s < 5.2$. The fluid properties are evaluated at the free stream temperature except μ_s, which is evaluated at the surface temperature.[8]

Cylinders of Noncircular Cross Section in Cross Flow of Gases

$$\overline{\mathrm{Nu}}_D = C(\mathrm{Re}_D)^m\,\mathrm{Pr}^{1/3}$$

where C and m are listed in Table 56.12, and the fluid properties are evaluated at the mean film temperature.[9]

Table 56.12   Constants C and m for Noncircular Cylinders in Cross Flow

Geometry		Re_D	C	m
Square				
$V \rightarrow \diamondsuit \quad \updownarrow D$		$5 \times 10^3 - 10^5$	0.246	0.588
$V \rightarrow \square \quad \updownarrow D$		$5 \times 10^3 - 10^5$	0.102	0.675
Hexagon				
$V \rightarrow \hexagon \quad \updownarrow D$		$5 \times 10^3 - 1.95 \times 10^4$	0.160	0.638
		$1.95 \times 10^4 - 10^5$	0.0385	0.782
$V \rightarrow \hexagon \quad \updownarrow D$		$5 \times 10^3 - 10^5$	0.153	0.638
Vertical Plate				
$V \rightarrow \vert \quad \updownarrow D$		$4 \times 10^3 - 1.5 \times 10^4$	0.228	0.731

Flow Past a Sphere

$$\overline{Nu}_D = 2 + (0.4 Re_D^{0.5} + 0.06 Re_D^{2/3})\, Pr^{0.4} \, (\mu_\infty / \mu_s)^{0.25}$$

which is valid for $3.5 < Re_D < 8 \times 10^4$, $0.7 < Pr < 380$, and $1.0 < \mu_\infty / \mu_s < 3.2$. The fluid properties are evaluated at the free stream temperature except μ_s, which is evaluated at the surface temperature.[8]

Flow across Banks of Tubes

The tube arrangement may be either *staggered* or *aligned* (Fig. 56.11). The heat-transfer coefficient for the first row is approximately equal to that for a single tube. In turbulent flow, the heat-transfer coefficients of tubes in the first row are smaller than that of the subsequent rows. However, the coefficient becomes approximately constant beyond the fourth or fifth row. The average Nusselt number for the entire tube bundle, valid for number of rows ≥ 20, $0.7 < Pr < 500$, and $1000 < Re_{D.max} < 2 \times 10^6$, is[10]

$$\overline{Nu}_D = C (Re_{D.max})^m \, Pr^{0.36} \, (Pr_\infty / Pr_s)^{0.25}$$

where all fluid properties are evaluated at T_∞ except Pr_s, which is evaluated at the surface temperature. The constants C and m are listed in Table 56.13. The Reynolds number is based on the maximum fluid velocity occurring at the minimum free area available for the fluid flow. The maximum fluid velocity is determined by (see Fig. 56.11)

$$V_{max} = \frac{S_T}{S_T - D} V$$

aligned or staggered if

$$\sqrt{S_L^2 + (S_T/2)^2} > (S_T + D)/2$$

and

$$V_{max} = \frac{S_T}{2\sqrt{S_L^2 + (S_T/2)^2} - D} V$$

staggered if

$$\sqrt{S_L^2 + (S_T/2)^2} < (S_T + D)/2$$

Liquid Metals in Cross Flow over Banks of Tubes

For tubes in the inner rows, valid for $2 \times 10^4 < Re_{D.max} < 8 \times 10^4$ and $Pr < 0.03$,

$$\overline{Nu}_D = 4.03 + 0.228 (Re_{D.max} Pr)^{0.67}$$

in which fluid properties are evaluated at the mean film temperature.[11]

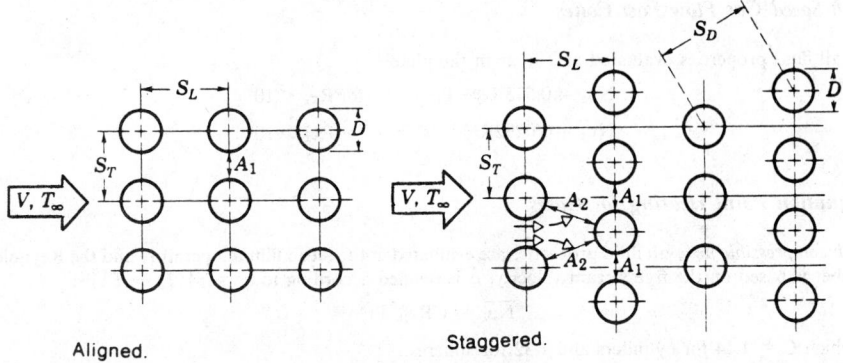

Fig. 56.11 Tube arrangement.

High-Speed Flow over a Flat Plate

For high free stream velocity the effects due to viscous dissipation and fluid compressibility must be taken into account in the heat-transfer calculation. The convection heat-transfer coefficient is defined according to $q = hA(T_s - T_{as})$, where T_{as} is the *adiabatic surface temperature* or *recovery temperature*, which is related to the *recovery factor* by $r = (T_{as} - T_\infty)/(T_0 - T_\infty)$. The *stagnation temperature* T_0 is related to the free stream static temperature T_∞ by

$$\frac{T_0}{T_\infty} = 1 + \frac{\gamma - 1}{2} M_\infty^2$$

where γ is the specific heat ratio of the fluid and $M_\infty =$ free stream velocity/acoustic velocity. For $0.6 < \mathrm{Pr} < 15$,

$$r = \mathrm{Pr}^{1/2} \qquad \text{for laminar flow } (\mathrm{Re}_x < 5 \times 10^5)$$
$$r = \mathrm{Pr}^{1/3} \qquad \text{for turbulent flow } (\mathrm{Re}_x > 5 \times 10^5)$$

With all fluid properties evaluated at the reference temperature $T_{\mathrm{ref}} = T_\infty + 0.5(T_s - T_\infty) + 0.22(T_{as} - T_\infty)$, the local heat-transfer coefficients at a distance x from the leading edge are given below[2]:

$$\mathrm{Nu}_x = 0.332 \mathrm{Re}_x^{0.5}\, \mathrm{Pr}^{1/3} \qquad \text{for } \mathrm{Re}_x < 5 \times 10^5$$
$$\mathrm{Nu}_x = 0.0292 \mathrm{Re}_x^{0.8}\, \mathrm{Pr}^{1/3} \qquad \text{for } 5 \times 10^5 < \mathrm{Re}_x < 10^7$$
$$\mathrm{Nu}_x = 0.185 \mathrm{Re}_x (\log_{10} \mathrm{Re}_x)^{-2.584} \qquad \text{for } 10^7 < \mathrm{Re}_x < 10^9$$

For very high free stream velocities, dissociation of gas may occur and there will be a large variation of properties in the boundary layer. The heat-transfer coefficient is then defined in terms of enthalpy difference: namely, $q = hA(i_s - i_{as})$, and the recovery factor is given by $r = (i_{as} - i_\infty)/(i_0 - i_\infty)$ (i_{as} is the enthalpy at the adiabatic wall condition). The same equations for Nu_x as shown above can be used except that all properties are evaluated at a reference enthalpy $i_{\mathrm{ref}} = i_\infty + 0.5(i_s - i_\infty) + 0.22(i_{as} - i_\infty)$.

Table 56.13 Constants C and m of Heat-Transfer Coefficient for the Tube Banks in Cross Flow

Configuration	$\mathrm{Re}_{D,\mathrm{max}}$	C	m
Aligned	10^3–2×10^5	0.27	0.63
Staggered $(S_T/S_L < 2)$	10^3–2×10^5	$0.35(S_T/S_L)^{1/5}$	0.60
Staggered $(S_T/S_L > 2)$	10^3–2×10^5	0.40	0.60
Aligned	2×10^5–2×10^6	0.021	0.84
Staggered	2×10^5–2×10^6	0.022	0.84

High-Speed Gas Flow Past Cones

For all fluid properties evaluated at T_{ref} as in the plate[12]:

$$\mathrm{Nu}_x = 0.575\mathrm{Re}_x^{0.5}\,\mathrm{Pr}^{1/3} \qquad \text{for } \mathrm{Re}_x < 10^5$$

$$\mathrm{Nu}_x = 0.0292\mathrm{Re}_x^{0.8}\,\mathrm{Pr}^{1/3} \qquad \text{for } \mathrm{Re}_x > 10^5$$

Stagnation Point Heating for Gases

For *incompressible flow,* all fluid properties are evaluated at the mean film temperature and the Reynolds number is based on the free stream velocity. $\bar{h}$ is defined according to $q = \bar{h}A(T_s - T_\infty)$[13]:

$$\overline{\mathrm{Nu}}_D = C\mathrm{Re}_D^{0.5}\mathrm{Pr}^{0.4}$$

in which $C = 1.14$ for cylinders and 1.32 for spheres.

For *supersonic flow,* a bow shock will stand in front of the body. For all fluid properties evaluated at the stagnation state of flow behind the bow shock:

$$\overline{\mathrm{Nu}}_D = C\mathrm{Re}_D^{0.5}\mathrm{Pr}^{0.4}(\rho_\infty/\rho_0)^{0.25}$$

in which $C = 0.95$ for cylinders and 1.28 for spheres; ρ_∞ is the free stream gas density and ρ_0 is the stagnation density of stream behind the bow shock. The heat-transfer rate is given by $q = \bar{h}A(T_s - T_0)$.

56.2.3 Free Convection

In free convection the fluid motion is caused by the buoyant force resulting from the density difference near the body surface, which is at a temperature different from that of the free fluid far removed from the surface where velocity is zero. In all free convection correlations, except for the enclosed cavities, the fluid properties are usually evaluated at the mean film temperature $T_f = (T_s + T_\infty)/2$. The thermal expansion coefficient β, however, is evaluated at the free fluid temperature T_∞. The convection heat transfer coefficient h is based on the temperature difference between the surface and the free fluid.

Free Convection from Flat Plates and Cylinders

The average Nusselt number $\overline{\mathrm{Nu}}_L$ is given by[4]

$$\overline{\mathrm{Nu}}_L = C(\mathrm{Gr}_L\,\mathrm{Pr})^m$$

The constants C and m are listed in Table 56.14. The product $\mathrm{Gr}_L\,\mathrm{Pr}$ is called the *Rayleigh number* (Ra_L). For certain ranges of the Rayleigh number, Figs. 56.12 and 56.13 are used instead of the above equation. For *three-dimensional shapes,* such as short cylinders and blocks, approximate results can be obtained with $C = 0.6$ and $m = \frac{1}{4}$ for $10^4 < \mathrm{Ra}_L < 10^9$ provided that the characteristic

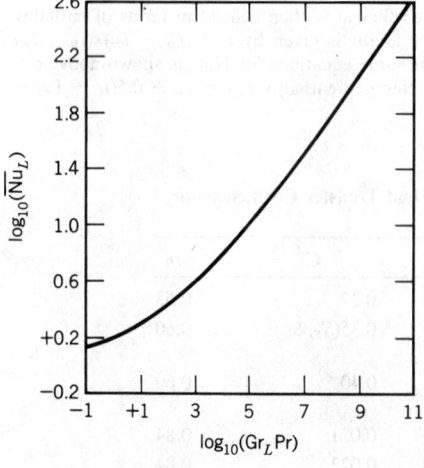

Fig. 56.12 Free convection heat-transfer correlation for heated vertical plates and cylinders. Adapted from Ref. 14. Used with the permission of McGraw-Hill Book Company.

Table 56.14 Constants for Free Convection from Flat Plates and Cylinders

Geometry	$Gr_L Pr$	C	m	L
Vertical flat plates and cylinders	10^{-1}–10^4	Use Fig. 56.12	Use Fig. 56.12	Height of plates and cylinders; restricted to $D/L \geq 35/Gr_L^{1/4}$ for cylinders
	10^4–10^9	0.59	$\tfrac{1}{4}$	
	10^9–10^{13}	0.10	$\tfrac{1}{3}$	
Horizontal cylinders	0–10^{-5}	0.4	0	Diameter D
	10^{-5}–10^4	Use Fig. 56.13	Use Fig. 56.13	
	10^4–10^9	0.53	$\tfrac{1}{4}$	
	10^9–10^{12}	0.13	$\tfrac{1}{3}$	
Upper surface of heated plates or lower surface of cooled plates	2×10^4–8×10^6	0.54	$\tfrac{1}{4}$	Length of a side for square plates, the average length of the two sides for rectangular plates
	8×10^6–10^{11}	0.15	$\tfrac{1}{3}$	
Lower surface of heated plates or upper surface of cooled plates	10^5–10^{11}	0.58	$\tfrac{1}{5}$	$0.9D$ for circular disks

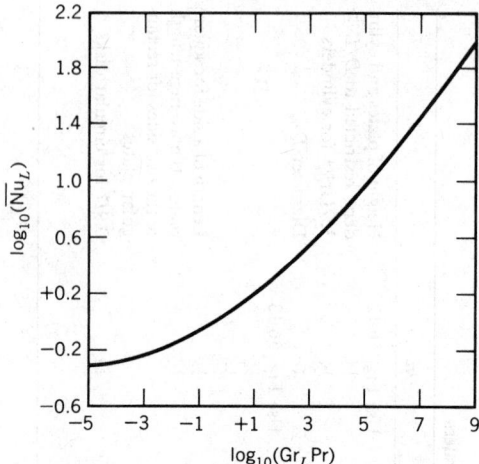

Fig. 56.13 Free convection heat-transfer correlation from heated horizontal cylinders. Adapted from Ref. 14. Used with the permission of McGraw-Hill Book Company.

length L is determined by $1/L = 1/L_{hor} + 1/L_{ver}$, where L_{ver} is the height and L_{hor} the horizontal dimension of the body. For *unsymmetrical horizontal surfaces,* other than square, rectangle, or circle, the characteristic length L can be calculated from $L = A/P$, where A is the area and P is the wetted perimeter of the surface.

Free Convection from Spheres

The following correlation is primarily for gases, $Pr \approx 1$, but may be used for liquid as an approximation [15]:

$$\overline{Nu}_D = 2 + 0.43(Gr_D\,Pr)^{0.25} \qquad \text{for } 1 < Gr_D < 10^5$$

Free Convection in Enclosed Spaces

When a temperature difference is imposed on the two walls that enclose a space filled with fluid, heat transfer will be experienced. For a small value of Rayleigh number the heat transfer may be due to conduction only. As the Rayleigh number is increased the free convection will be encountered. For the following correlations, the fluid properties are evaluated at the average temperature of the two walls.

Cavities between Two Horizontal Walls at Temperatures T_1 and T_2 Separated by Distance δ (T_1 for Lower Wall, $T_1 > T_2$)

$$q'' = \bar{h}(T_1 - T_2)$$

$$\overline{Nu}_\delta = 0.069 Ra_\delta^{1/3}\,Pr^{0.074} \qquad \text{for } 3 \times 10^5 < Ra_\delta < 7 \times 10^9$$
$$= 1.0 \qquad\qquad\quad\ \text{for } Ra_\delta < 1700$$

where $Ra_\delta = g\beta(T_1 - T_2)\,\delta^3/\alpha\nu$; δ is the thickness of the space.[16]

Cavities between Two Vertical Walls of Height H at Temperatures T_1 and T_2 Separated by Distance δ[17,18]

$$q'' = \bar{h}(T_1 - T_2)$$

$$\overline{Nu}_\delta = 0.22\left(\frac{Pr}{0.2 + Pr}\,Ra_\delta\right)^{0.28}\left(\frac{\delta}{H}\right)^{0.25}$$

for $2 < H/\delta < 10$, $Pr < 10^5$, and $Ra_\delta < 10^{10}$;

$$\overline{Nu}_\delta = 0.18\left(\frac{Pr}{0.2 + Pr}\,Ra_\delta\right)^{0.29}$$

for $1 < H/\delta < 2$, $10^{-3} < \mathrm{Pr} < 10^5$, and $10^3 < \mathrm{Ra}_\delta \mathrm{Pr}/(0.2 + \mathrm{Pr})$;

$$\overline{\mathrm{Nu}}_\delta = 0.42 \mathrm{Ra}_\delta^{0.25} \mathrm{Pr}^{0.012}(\delta/H)^{0.3}$$

for $10 < H/\delta < 40$, $1 < \mathrm{Pr} < 2 \times 10^4$, and $10^4 < \mathrm{Ra}_\delta < 10^7$.

56.2.4 The Log Mean Temperature Difference

In a double-pipe heat exchanger (Fig. 56.14) the heat transfer between two fluids for constant overall heat-transfer coefficient (Table 56.8) and constant fluid specific heat is given by $q = UA\,\Delta T_m$, where

$$\Delta T_m = \frac{\Delta T_2 - \Delta T_1}{\ln(\Delta T_2/\Delta T_1)}$$

The temperature difference ΔT_m is referred to as the *log mean temperature difference* (LMTD); ΔT_1 is the temperature difference of the two fluids at one end and ΔT_2 is that at the other end. If the ratio $\Delta T_2/\Delta T_1$ is less than 2, the *arithmetic mean temperature difference*, $(\Delta T_2 + \Delta T_1)/2$, may be used to calculate heat-transfer rate without much error. Note that (see Fig. 56.14)

$$\Delta T_1 = T_{h,i} - T_{c,i}, \qquad \Delta T_2 = T_{h,o} - T_{c,o} \qquad \text{for parallel flow}$$
$$\Delta T_1 = T_{h,i} - T_{c,o}, \qquad \Delta T_2 = T_{h,o} - T_{c,i} \qquad \text{for counterflow}$$

Cross Flow Coefficient

For a heat exchanger, other than the double-pipe type, in which there may be considerable variation in the values of U over the entire surface, the LMTD may not be the true mean temperature difference. For such applications a correction factor is used, so that q is determined by

$$q = UAF\,\Delta T_m$$

where ΔT_m is computed under the assumption of counterflow conditions, that is, $\Delta T_1 = T_{h,i} - T_{c,i}$ and $\Delta T_2 = T_{h,o} - T_{c,o}$. Some examples for the *correction factor F* are shown in Figs. 56.15 and 56.16.

56.3 RADIATION HEAT TRANSFER

Radiation is the transmission of energy by electromagnetic waves, which are characterized by their wavelength (λ) or frequency (ν). The transmitted energy is called *radiant energy*. However, the term *radiation* is also used to describe the radiant energy itself. The radiant energy travels in space on a straight path with the velocity of light (or wave velocity) and requires no intervening medium for its

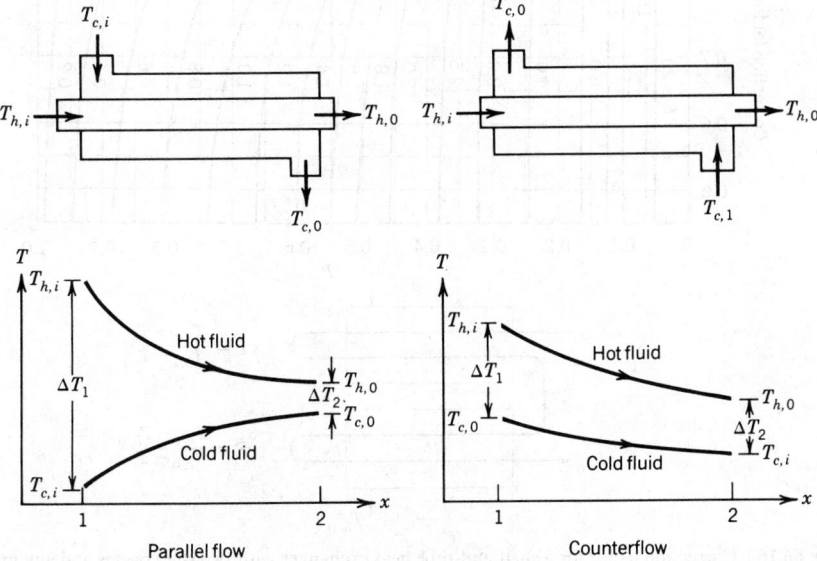

Fig. 56.14 Temperature profiles for parallel flow and counterflow in double-pipe heat exchanger.

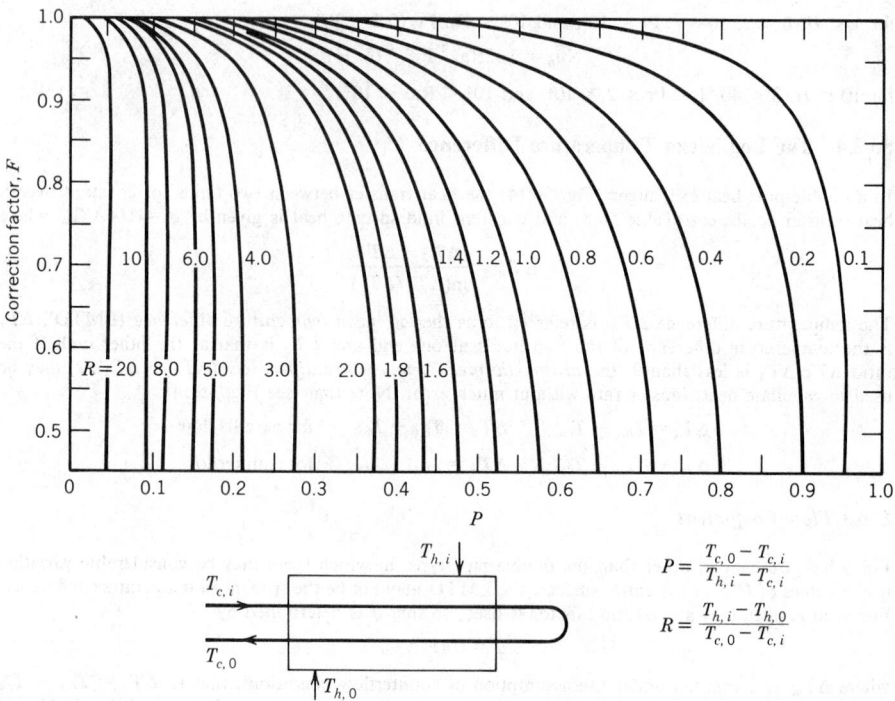

Fig. 56.15 Correction factor for a shell-and-tube heat exchanger with one shell and any multiple of two tube passes (two, four, etc., tube passes).

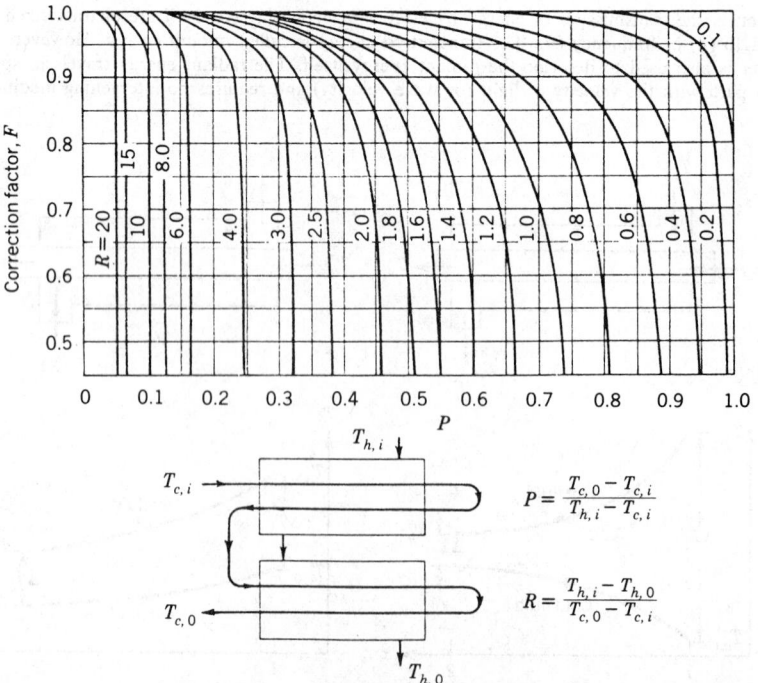

Fig. 56.16 Correction factor for a shell-and-tube heat exchanger with two shell passes and any multiple of four tube passes (four, eight, etc., tube passes).

propagation. Radiation can also be viewed as the transport of energy by photons. The energy of a photon is given by $h\nu$, with h representing Planck's constant. The wavelength λ and the frequency ν are related by $c = \lambda\nu$, with c representing the velocity of light. The velocity of light in vacuo has the value $c_0 = 2.9979 \times 10^8$ m/sec. All media in nature have wave velocities smaller than that in vacuo, and are related to c_0 by $c = c_0/n$, with n indicating the refractive index. The value of n for air is approximately equal to 1. The wavelength of radiation depends on the nature of the sources. Radiation in the wavelength range 0.1–100 μm (micrometer or micron; equals 10^{-6} m) is classified as *thermal radiation,* for which the radiant energy is governed by the temperature of the emitting body. A heated body emits and loses energy continuously by thermal radiation. When the radiant energy strikes a body, the energy may be absorbed, reflected, or transmitted. Figure 56.17 shows the spectrum of electromagnetic radiation. Note that the radiation is visible only in the range $\lambda = 0.4$–0.7 μm.

56.3.1 Black-Body Radiation

A *black body* is an ideal body that absorbs all the radiant energy falling upon it, regardless of the wavelength and direction. Such a body emits maximum energy for a prescribed temperature and wavelength. Black-body radiation is independent of direction; that is, the black body is a *diffuse emitter.*

The Stefan–Boltzmann Law

The rate at which the energy is radiated per unit area from a black body is proportional to the fourth power of its absolute temperature:

$$e_b = \sigma T^4$$

where e_b is the *total emissive power* and σ is the Stefan–Boltzmann constant, which has the value 5.729×10^{-8} W/m^2 · K^4 (0.173×10^{-8} Btu/hr · ft^2 · °R^4).

Planck's Distribution Law

At any given temperature the *spectral emissive power* of a black body, $e_{\lambda b}$, is a function of wavelength. The function, which Planck derived from the quantum theory, is

$$e_{\lambda b} = 2\pi \, C_1/\lambda^5[\exp(C_2/\lambda T) - 1]$$

where $e_{\lambda b}$ has a unit W/m^2 · μm (Btu/hr · ft^2 · μm).

The constants C_1 and C_2 are, respectively, 0.59544×10^{-16} W · m^2 (0.18892×10^8 Btu · μm^4/ hr · ft^2) and 14,388 μm · K (25,898 μm · °R). Figure 56.18 shows the distribution of spectral emissive power for a black body at different temperatures. The energy emitted at all wavelengths is shown to increase as the temperature increases. The peaks of constant temperature curves in Fig. 56.18 shift to the shorter wavelengths for the higher temperatures. These maximum points are related by *Wien's displacement law,* $\lambda_{max} T = 2897.8$ μm · K (5216.0 μm · °R).

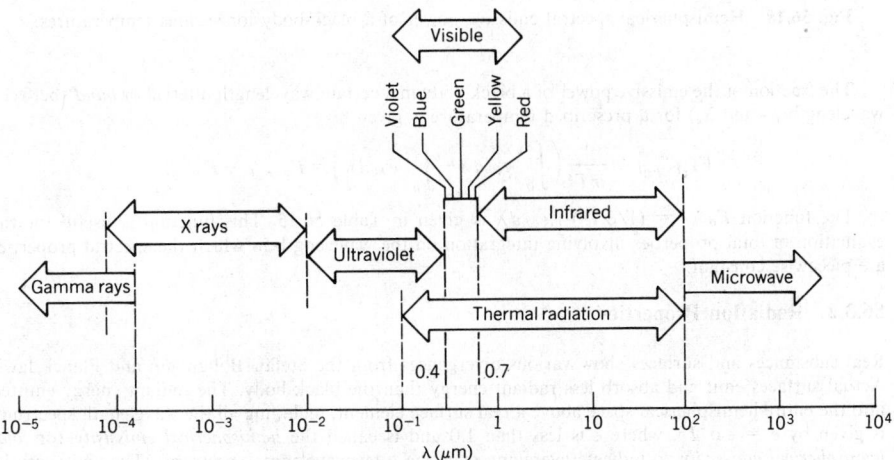

Fig. 56.17 Electromagnetic radiation spectrum.

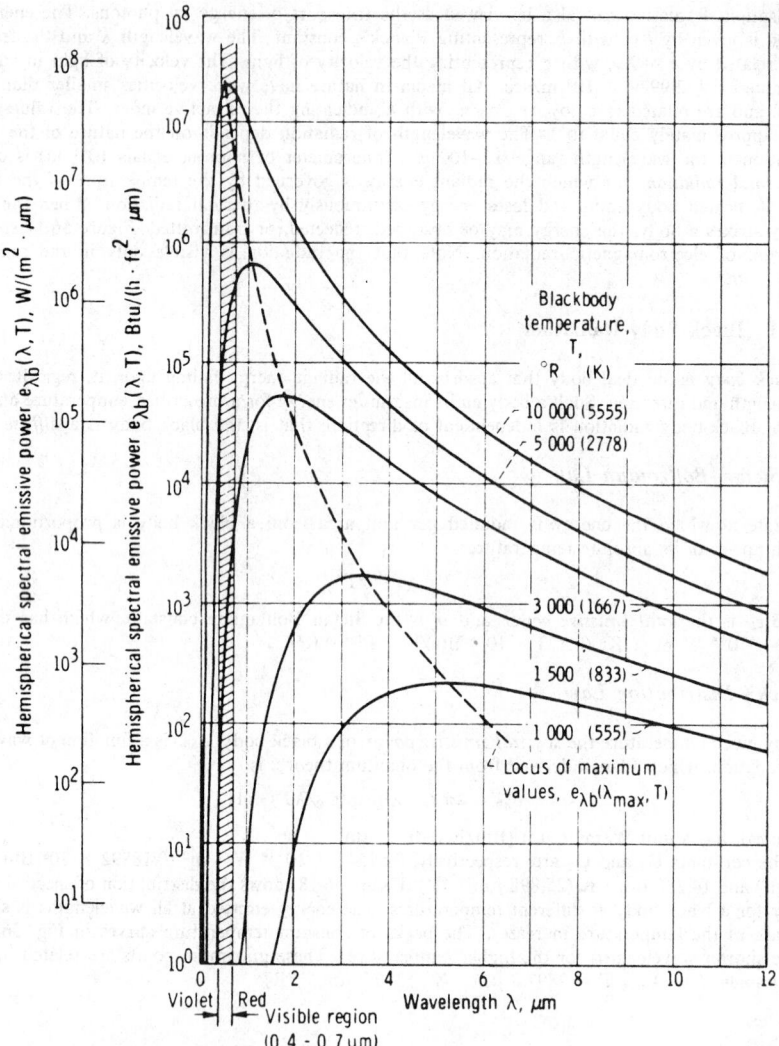

Fig. 56.18 Hemispherical spectral emissive power of a black-body for various temperatures.

The fraction of the emissive power of a black body in a certain wavelength interval or *band* (between wavelength λ_1 and λ_2) for a prescribed temperature is given by

$$F_{\lambda_1 T - \lambda_2 T} = \frac{1}{\sigma T^4} \left(\int_0^{\lambda_1} e_{\lambda b} d\lambda - \int_0^{\lambda_2} e_{\lambda b} d\lambda \right) = F_{0 - \lambda_1 T} - F_{0 - \lambda_2 T}$$

The function $F_{0-\lambda T} = (1/\sigma T^4) \int_0^\lambda e_{\lambda b} d\lambda$ is given in Table 56.15. This function is useful for the evaluation of total properties involving integration on the wavelength in which the spectral properties are piecewise constant.

56.3.2 Radiation Properties

Real substances and surfaces show various divergences from the Stefan–Boltzmann and Planck laws. Actual surfaces emit and absorb less radiant energy than the black body. The radiant energy emitted into the entire hemispherical space above a real surface element, including all the wavelength spectrum, is given by $e = \epsilon \sigma T^4$, where ϵ is less than 1.0 and is called the *hemispherical emissivity* (or *total hemispherical emissivity* to indicate averaging over the total wavelength spectrum). The emissivity is, in general, a function of the material, temperature, and surface conditions such as roughness, oxide

Table 56.15 Radiation Function $F_{0-\lambda T}$

λT			λT			λT		
μm · K	μm · °R	$F_{0-\lambda T}$	μm · K	μm · °R	$F_{0-\lambda T}$	μm · K	μm · °R	$F_{0-\lambda T}$
400	720	0.1864×10^{-11}	3400	6120	0.3617	6400	11,520	0.7692
500	900	0.1298×10^{-8}	3500	6300	0.3829	6500	11,700	0.7763
600	1080	0.9290×10^{-7}	3600	6480	0.4036	6600	11,880	0.7832
700	1260	0.1838×10^{-5}	3700	6660	0.4238	6800	12,240	0.7961
800	1440	0.1643×10^{-4}	3800	6840	0.4434	7000	12,600	0.8081
900	1620	0.8701×10^{-4}	3900	7020	0.4624	7200	12,960	0.8192
1000	1800	0.3207×10^{-3}	4000	7200	0.4809	7400	13,320	0.8295
1100	1980	0.9111×10^{-3}	4100	7380	0.4987	7600	13,680	0.8391
1200	2160	0.2134×10^{-2}	4200	7560	0.5160	7800	14,040	0.8480
1300	2340	0.4316×10^{-2}	4300	7740	0.5327	8000	14,400	0.8562
1400	2520	0.7789×10^{-2}	4400	7920	0.5488	8200	14,760	0.8640
1500	2700	0.1285×10^{-1}	4500	8100	0.5643	8400	15,120	0.8712
1600	2880	0.1972×10^{-1}	4600	8280	0.5793	8600	15,480	0.8779
1700	3060	0.2853×10^{-1}	4700	8460	0.5937	8800	15,840	0.8841
1800	3240	0.3934×10^{-1}	4800	8640	0.6075	9000	16,200	0.8900
1900	3420	0.5210×10^{-1}	4900	8820	0.6209	10,000	18,000	0.9142
2000	3600	0.6673×10^{-1}	5000	9000	0.6337	11,000	19,800	0.9318
2100	3780	0.8305×10^{-1}	5100	9180	0.6461	12,000	21,600	0.9451
2200	3960	0.1009	5200	9360	0.6579	13,000	23,400	0.9551
2300	4140	0.1200	5300	9540	0.6694	14,000	25,200	0.9628
2400	4320	0.1402	5400	9720	0.6803	15,000	27,000	0.9689
2500	4500	0.1613	5500	9900	0.6909	20,000	36,000	0.9856
2600	4680	0.1831	5600	10,080	0.7010	25,000	45,000	0.9922
2700	4860	0.2053	5700	10,260	0.7108	30,000	54,000	0.9953
2800	5040	0.2279	5800	10,440	0.7201	35,000	63,000	0.9970
2900	5220	0.2505	5900	10,620	0.7291	40,000	72,000	0.9979
3000	5400	0.2732	6000	10,800	0.7378	45,000	81,000	0.9985
3100	5580	0.2958	6100	10,980	0.7461	50,000	90,000	0.9989
3200	5760	0.3181	6200	11,160	0.7541	55,000	99,000	0.9992
3300	5940	0.3401	6300	11,340	0.7618	60,000	108,000	0.9994

layer thickness, and chemical contamination. The emissivity is a measure of how well a real body can radiate energy as compared with a black body of the same temperature. For a given wavelength the *spectral hemispherical emissivity* ϵ_λ of a real surface is defined as

$$\epsilon_\lambda = e_\lambda / e_{\lambda b}$$

in which e_λ is the hemispherical emissive power of the real surface and $e_{\lambda b}$ is that of a black body, both at the same temperature.

The *spectral irradiation* G_λ (W/m² · μm), which encompasses radiation incident from all directions, is defined as the rate at which radiation is incident upon a surface per unit area of the surface, per unit wavelength about the wavelength λ.

The *spectral hemispherical reflectivity* ρ_λ is defined as

$$\rho_\lambda = \text{(radiant energy reflected per unit time, area, and wavelength)}/G_\lambda$$

The *spectral hemispherical absorptivity* α_λ is defined as

$$\alpha_\lambda = \text{(radiant energy absorbed per unit time, area, and wavelength)}/G_\lambda$$

The *spectral hemispherical transmissivity* is defined as

$$\tau_\lambda = \text{(radiant energy transmitted per unit time, area, and wavelength)}/G_\lambda$$

The term total is used to designate the averaging over the entire wavelength from $\lambda = 0$ to ∞. Thus the *total hemispherical reflectivity, total hemispherical absorptivity,* and *total hemispherical transmissivity* are defined as follows:

$$\rho = \int_0^\infty \rho_\lambda G_\lambda \, d\lambda / G$$

$$\alpha = \int_0^\infty \alpha_\lambda G_\lambda \, d\lambda / G$$

and

$$\tau = \int_0^\infty \tau_\lambda G_\lambda \, d\lambda / G$$

where

$$G = \int_0^\infty G_\lambda \, d\lambda$$

For an *opaque* surface, $\tau_\lambda = 0$, the incident energy will either be reflected or absorbed without any transmission. For such a surface,

$$\alpha_\lambda + \rho_\lambda = 1$$

and

$$\alpha + \rho = 1$$

When the incident and the reflected rays lie symmetrically with respect to the surface normal at the point of incidence, the reflection is called *specular*. The reflection is called *diffuse* if the intensity of the reflected radiation is constant for all angles of reflection and is independent of the incident direction.

The terms *reflectance, absorptance,* and *transmittance* are used by some authors for the real surfaces and the terms reflectivity, absorptivity, and transmissivity are reserved for the properties of the ideal surfaces (i.e., those optically smooth and pure substances perfectly uncontaminated).

While the emissivity is a function of the material, temperature, and surface conditions, the absorptivity and reflectivity depend on both the surface characteristics and the nature of the incident radiation.

If the radiation properties are independent of the direction, the surface is called a *diffuse surface*. If they are independent of the wavelength, the surface is called a *gray surface*. A *diffuse-gray surface* absorbs a fixed fraction of incident radiation from any direction and at any wavelength, and $\alpha_\lambda = \epsilon_\lambda = \alpha = \epsilon$.

Kirchhoff's Law of Radiation

For the radiation in a particular direction the directional spectral emissivity is equal to the directional spectral absorptivity for the surface irradiated by a black body at the same temperature. The most general form stated that $\alpha'_\lambda = \epsilon'_\lambda$. If the incident radiation is independent of angle or if the surface is diffuse, then $\alpha_\lambda = \epsilon_\lambda$ for the hemispherical properties. The law can have various conditions imposed on it, depending on whether spectral, total, directional, or semispherical quantities are being considered.[19]

Emissivity of Metallic Surfaces

Pure smooth metal surfaces are often characterized by low values of emissivity and absorptivity, and high values of reflectivity. The spectral emissivity of metals tends to increase with decreasing wavelength, and exhibits a peak near the visible region. At wavelengths $\lambda > \sim 5$ μm the spectral emissivity increases with increasing temperature. This trend is reversed in the shorter wavelengths ($\lambda < \sim 1.27$ μm). The surface roughness has a pronounced effect on the hemispherical emissivity and absorptivity. Large *optical roughness* (mean square roughness of the surface/wavelength) will increase the hemispherical emissivity. If the optical roughness is small, the directional properties approach those for smooth surfaces. Impurities, such as an oxide or any other nonmetallic contaminants, will change the properties significantly and increase the emissivity of an otherwise pure metallic body. Table 56.16 gives a limited tabulation of the normal total emissivities for metals. It is to be noted that the hemispherical emissivity for metals may be 10–30% higher than the normal emissivity.

Emissivity of Nonmetallic Materials

Nonmetals are characterized by large values of total hemispherical emissivity and absorptivity at moderate temperatures. The temperature dependencies of the properties of nonmetals are, in general, small

Table 56.16 Normal Total Emissivity of Metals[a]

Materials	Surface Temperature (K)	Normal Total Emissivity
Aluminum		
Highly polished plate	480–870	0.038–0.06
Polished plate	373	0.095
Heavily oxidized	370–810	0.20–0.33
Bismuth, bright	350	0.34
Chromium, polished	310–1370	0.08–0.40
Copper		
Highly polished	310	0.02
Slightly polished	310	0.15
Black oxidized	310	0.78
Gold, highly polished	370–870	0.018–0.035
Iron		
Highly polished, electrolytic	310–530	0.05–0.07
Polished	700–760	0.14–0.38
Wrought iron, polished	310–530	0.28
Cast iron, rough, strongly oxidized	310–530	0.95
Lead		
Polished	310–530	0.06–0.08
Rough unoxidized	310	0.43
Mercury, unoxidized	280–370	0.09–0.12
Molybdenum, polished	310–3030	0.05–0.29
Nickel		
Electrolytic	310–530	0.04–0.06
Electroplated on iron, not polished	293	0.11
Nickel oxide	920–1530	0.59–0.86
Platinum, electrolytic	530–810	0.06–0.10
Silver, polished	310–810	0.01–0.03
Steel		
Polished sheet	90–420	0.07–0.14
Mild steel, polished	530–920	0.27–0.31
Sheet with rough oxide layer	295	0.81
Tin, polished sheet	310	0.05
Tungsten, clean	310–810	0.03–0.08
Zinc		
Polished	310–810	0.02–0.05
Gray oxidized	295	0.23–0.28

[a] Adapted from Ref. 19.

based on the limited number of measured data. Table 56.17 shows the normal total emissivity of some nonmetals.

Absorptivity for Solar Incident Radiation

The solar radiation is a form of thermal radiation with a particular spectral distribution, which can be approximated by a black-body radiation at an equivalent temperature of about 5800 K (10,000° R). At the outer limit of the atmosphere the average solar irradiation is 1353 W/m² (429 Btu/ft² · hr). This value is called the *solar constant* and is subject to modification whenever more precise measured data become available. The solar irradiation received at the surface of the earth will be smaller than the solar constant owing to scattering by air molecules, water vapor, and dust, and also owing to the absorption by O_3, H_2O, and CO_2 in the air. The absorptivity depends not only on the surface properties but also on the sources of incident radiation. Since solar radiation is concentrated at a shorter wavelength (high temperature), the absorptivities for certain materials under solar radiation may be entirely different from that for the low-temperature radiation for which the radiation is mostly in a longer-wavelength range. Table 56.18 gives a brief comparison of absorptivities of materials for both solar and low-temperature radiation.

56.3.3 Configuration Factor

To compute the magnitude of radiant energy exchange between two surfaces, the configuration factor (sometimes called *radiation shape factor, view factor, angle factor,* or *interception factor*) must be determined. The configuration factor F_{i-j} is defined as the fraction of radiation leaving a black surface i, which is intercepted by surface j (see Fig. 56.19). The configuration factor is a geometrical quantity determined from the relative position and shapes of the two surfaces. The configuration factor dF_{i-dj} indicates a differential fraction of energy from finite area A_i is intercepted by an infinitesimal area dA_j. Mathematical expressions for various cases are given below:

Infinitesimal area dA_i *to infinitesimal area* dA_j

$$dF_{di-dj} = \frac{\cos \theta_i \cos \theta_j}{\pi R^2} dA_j$$

Table 56.17 Normal Total Emissivity of Nonmetals[a]

Materials	Surface Temperature (K)	Normal Total Emissivity
Asbestos, board	310	0.96
Brick		
White refractory	1370	0.29
Rough red	310	0.93
Carbon, lampsoot	310	0.95
Concrete, rough	310	0.94
Ice, smooth	273	0.966
Magnesium oxide, refractory	420–760	0.69–0.55
Paint		
Oil, all colors	373	0.92–0.96
Lacquer, flat black	310–370	0.96–0.98
Paper, white	310	0.95
Plaster	310	0.91
Porcelain, glazed	295	0.92
Rubber, hard	293	0.92
Sandstone	310–530	0.83–0.90
Silicon carbide	420–920	0.83–0.96
Snow	270	0.82
Water, deep	273–373	0.96
Wood, sawdust	310	0.75

[a] Adapted from Ref. 19.

Table 56.18 Comparison of Absorptivities of Various Surfaces to Solar and Low-Temperature Thermal Radiation[a]

Surface	Absorptivity	
	For Solar Radiation	For Low-Temperature Radiation (~ 300 K)
Aluminum, highly polished	0.15	0.04
Copper, highly polished	0.18	0.03
Tarnished	0.65	0.75
Cast iron	0.94	0.21
Stainless steel, No. 301, polished	0.37	0.60
White marble	0.46	0.95
Asphalt	0.90	0.90
Brick, red	0.75	0.93
Gravel	0.29	0.85
Flat black lacquer	0.96	0.95
White paints, various types of pigments	0.12–0.16	0.90–0.95

[a] Adapted from Ref. 20 after J. P. Holman, *Heat Transfer*, McGraw-Hill, New York, 1981.

Infinitesimal area dA_i *to finite area* A_j

$$F_{di\text{-}j} = \int_{A_j} \frac{\cos \theta_i \, \cos \theta_j}{\pi R^2} \, dA_j$$

Finite area A_i *to finite area* A_j

$$F_{i\text{-}j} = \frac{1}{A_i} \int_{A_i} \int_{A_j} \frac{\cos \theta_i \, \cos \theta_j}{\pi R^2} \, dA_i \, dA_j$$

The analytical expressions of configuration factors can be found only for simple geometries. Some examples are presented in Figs. 56.20–56.23 for surfaces that emit and reflect diffusely.

Reciprocity Relations

The following reciprocal properties between configuration factors determine one configuration factor from the knowledge of the others:

$$dA_i \, dF_{di\text{-}dj} = dA_j \, dF_{dj\text{-}di}$$

$$dA_i F_{di\text{-}j} = A_j \, dF_{j\text{-}di}$$

$$A_i F_{i\text{-}j} = A_j F_{j\text{-}i}$$

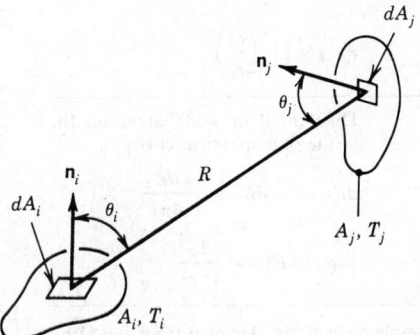

Fig. 56.19 Configuration factor for radiation exchange between surfaces of area dA_i and dA_j.

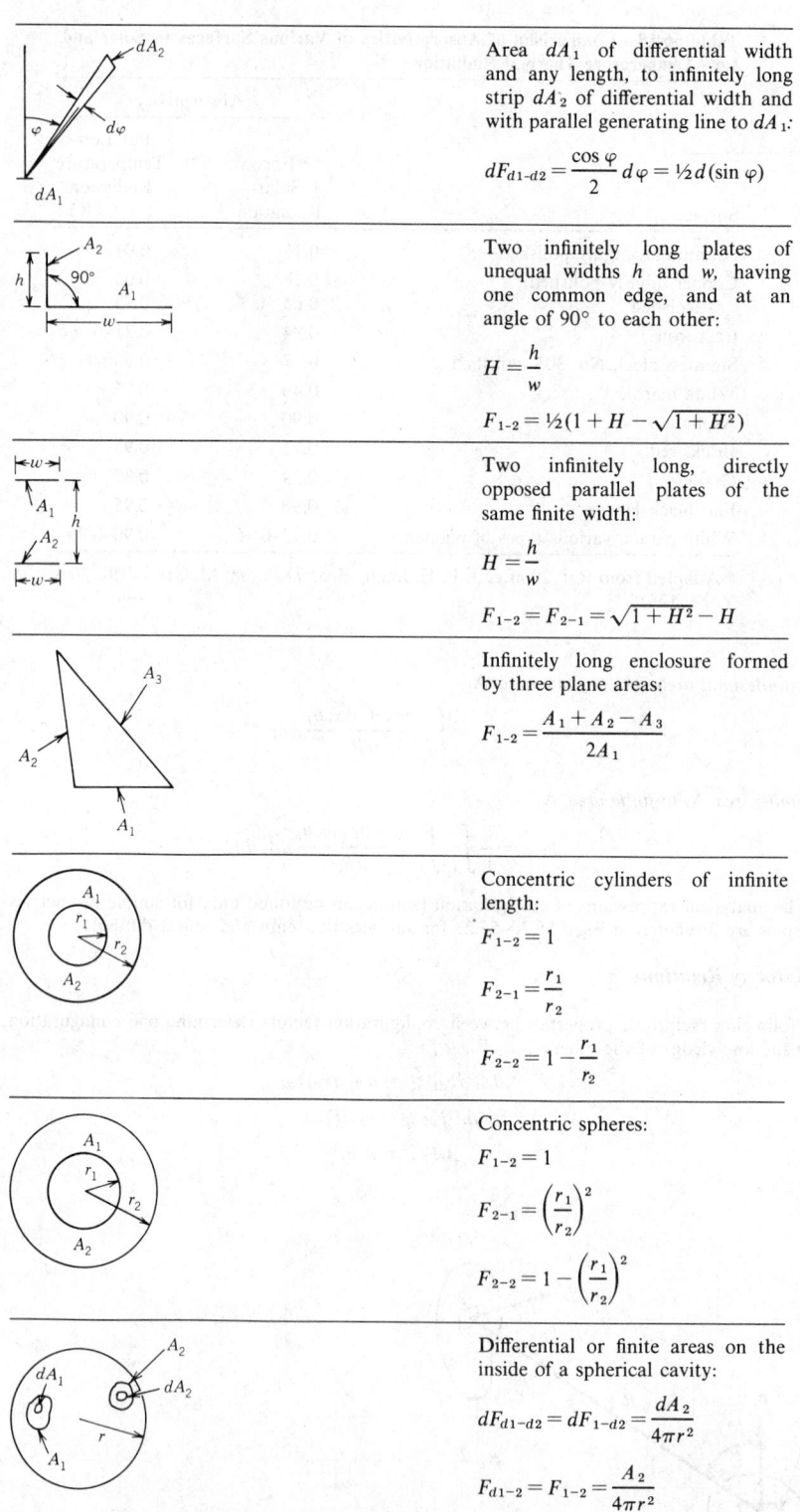

Area dA_1 of differential width and any length, to infinitely long strip dA_2 of differential width and with parallel generating line to dA_1:

$$dF_{d1\text{-}d2} = \frac{\cos \varphi}{2} \, d\varphi = \tfrac{1}{2} d(\sin \varphi)$$

Two infinitely long plates of unequal widths h and w, having one common edge, and at an angle of 90° to each other:

$$H = \frac{h}{w}$$

$$F_{1\text{-}2} = \tfrac{1}{2}(1 + H - \sqrt{1 + H^2})$$

Two infinitely long, directly opposed parallel plates of the same finite width:

$$H = \frac{h}{w}$$

$$F_{1\text{-}2} = F_{2\text{-}1} = \sqrt{1 + H^2} - H$$

Infinitely long enclosure formed by three plane areas:

$$F_{1\text{-}2} = \frac{A_1 + A_2 - A_3}{2A_1}$$

Concentric cylinders of infinite length:

$$F_{1\text{-}2} = 1$$

$$F_{2\text{-}1} = \frac{r_1}{r_2}$$

$$F_{2\text{-}2} = 1 - \frac{r_1}{r_2}$$

Concentric spheres:

$$F_{1\text{-}2} = 1$$

$$F_{2\text{-}1} = \left(\frac{r_1}{r_2}\right)^2$$

$$F_{2\text{-}2} = 1 - \left(\frac{r_1}{r_2}\right)^2$$

Differential or finite areas on the inside of a spherical cavity:

$$dF_{d1\text{-}d2} = dF_{1\text{-}d2} = \frac{dA_2}{4\pi r^2}$$

$$F_{d1\text{-}2} = F_{1\text{-}2} = \frac{A_2}{4\pi r^2}$$

Fig. 56.20 Configuration factors for some simple geometries. Adapted from Ref. 19.

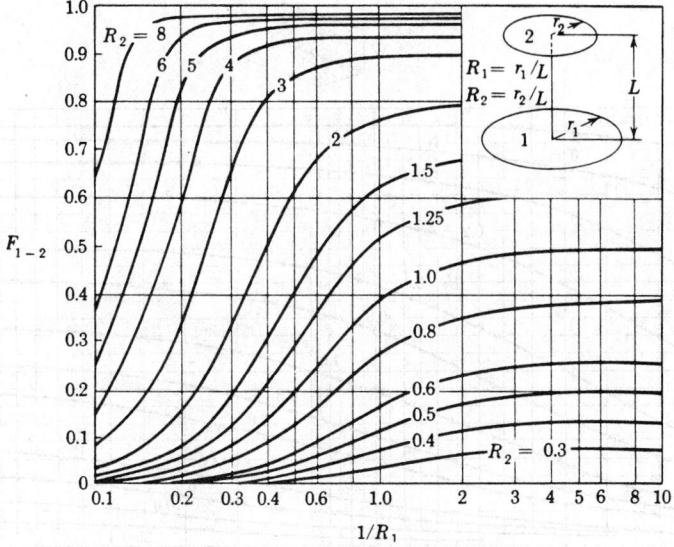

Fig. 56.21 Configuration factor for coaxial parallel circular disks.

The Additive Property

If surface A_i is subdivided into N parts $(A_{i_1}, A_{i_2}, \ldots, A_{i_N})$ and surface A_j is subdivided into M parts $(A_{j_1}, A_{j_2}, \ldots, A_{j_M})$, then

$$A_i F_{i-j} = \sum_{n=1}^{N} \sum_{m=1}^{M} A_{i_n} F_{i_n - j_m}$$

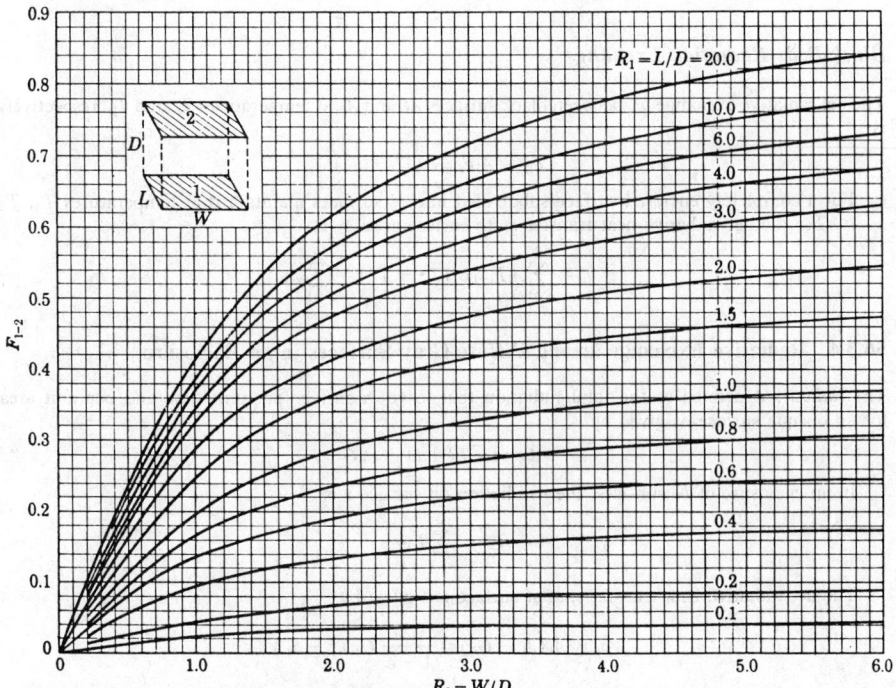

Fig. 56.22 Configuration factor for aligned parallel rectangles.

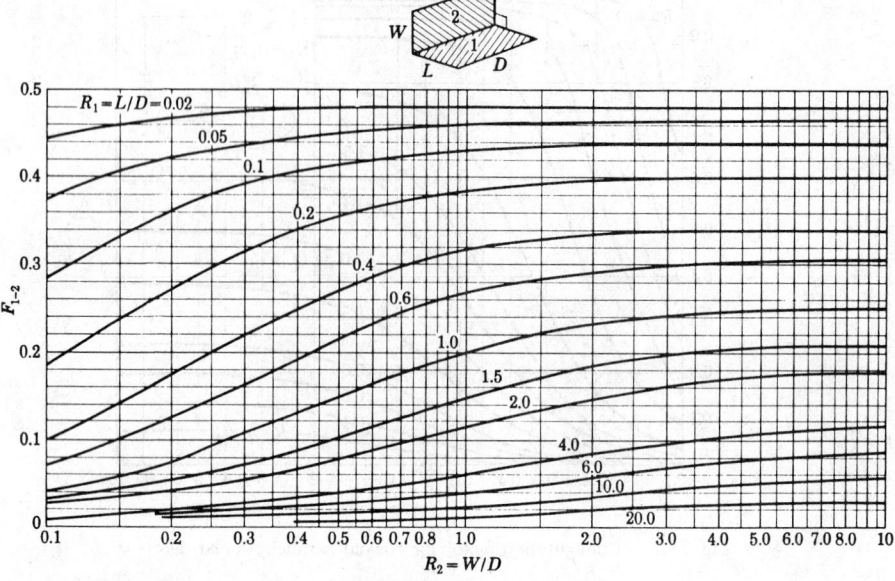

Fig. 56.23 Configuration factor for rectangles with common edge.

Relation in an Enclosure

For a completely enclosed surface, if the surface is subdivided into N parts having areas A_1, A_2, $\ldots$, A_N, then

$$\sum_{j=1}^{N} F_{i-j} = 1$$

Black-Body Radiation Exchange

The net radiative exchange q_{ij} between black surfaces A_i and A_j at temperatures T_i and T_j, respectively, is

$$q_{ij} = A_i F_{i-j}\, \sigma(\mathrm{T}_i^4 - T_j^4)$$

For a completely enclosed surface subdivided into N surfaces maintained at temperatures T_1, T_2, $\ldots$, T_N, the net radiative heat transfer q_i to surface area A_i is

$$q_i = \sum_{j=1}^{N} A_i F_{i-j}\, \sigma\, (T_i^4 - T_j^4) = \sum_{j=1}^{N} q_{ij}$$

56.3.4 Radiative Exchange among Diffuse-Gray Surfaces in an Enclosure

The *radiosity* J defined as the total radiation that leaves a surface per unit time and per unit area, for an opaque surface, is given by

$$J = \epsilon\sigma\, T^4 + (1 - \epsilon)\, G$$

In an enclosure of N surfaces, the irradiation on surface i is

$$G_i = \sum_{j=1}^{N} J_j F_{i-j}$$

The net radiative heat-transfer rate at surface i is given by

$$q_i = A_i\, (J_i - G_i\,) = \frac{\epsilon_i A_i}{1 - \epsilon_i}\, (\sigma T_i^4 - J_i\,)$$

On each surface either a uniform temperature or a constant heat-transfer rate can be specified; if the temperature is given on a surface, then the heat-transfer rate will be found for that surface and vice versa.

Case I: Temperatures T_i $(i = 1, 2, \ldots, N)$ for each of the N surfaces are prescribed. The values of J_i are solved from

$$\sum_{j=1}^{N} \{\delta_{ij} - (1 - \epsilon_i)F_{i-j}\} J_i = \epsilon_i \sigma T_i^4, \qquad 1 \leq i \leq N$$

and the net heat-transfer rate to surface i is

$$q_i = A_i \frac{\epsilon_i}{1 - \epsilon_i} (\sigma T_i^4 - J_i), \qquad 1 \leq i \leq N$$

where $\delta_{ij} = 0$ for $i \neq j$ and $\delta_{ij} = 1$ for $i = j$.

Case II: Heat-transfer rates q_i $(i = 1, 2, \ldots, N)$ to each of the N surfaces are prescribed. The values of J_i are solved from

$$\sum_{j=1}^{N} \{\delta_{ij} - F_{i-j}\} J_j = q_i / A_i, \qquad 1 \leq i \leq N$$

and the surface temperature is determined by

$$T_i = \left[\frac{1}{\sigma} \left(\frac{1 - \epsilon_i}{\epsilon_i} \frac{q_i}{A_i} + J_i \right) \right]^{1/4}, \qquad 1 \leq i \leq N$$

Case III: Temperatures T_i $(i = 1, \ldots, N_1)$ for N_1 surfaces and heat-transfer rates q_i $(i = N_1 + 1, \ldots, N)$ for $(N - N_1)$ surfaces are prescribed. The radiosities are determined by

$$\sum_{j=1}^{N} \{\delta_{ij} - (1 - \epsilon_i) F_{i-j}\} J_j = \epsilon_i \alpha T_i^4, \qquad 1 \leq i \leq N_1$$

$$\sum_{j=1}^{N} \{\delta_{ij} - F_{i-j}\} J_j = \frac{q_i}{A_i}, \qquad N_1 + 1 \leq i \leq N$$

and the net heat-transfer rate and temperature are found by

$$q_i = A_i \frac{\epsilon_i}{1 - \epsilon_i} (\sigma T_i^4 - J_i), \qquad 1 \leq i \leq N_1$$

$$T_i = \left[\frac{1}{\sigma} \left(\frac{1 - \epsilon_i}{\epsilon_i} \frac{q_i}{A_i} + J_i \right) \right]^{1/4}, \qquad N_1 + 1 \leq i \leq N$$

Two Diffuse-Gray Surfaces Forming an Enclosure

The net radiative exchange q_{12} between two surfaces that form an enclosure for some simple geometries are presented in Table 56.19.

Radiation Shield

The radiation heat transfer between two surfaces may be reduced by putting a highly reflective surface between them. The ratio of net radiative exchange with a shield to that without a shield is given by

$$\frac{q_{12 \text{ with shield}}}{q_{12 \text{ without shield}}} = \frac{1}{1 + \chi}$$

Table 56.20 shows the values of χ for shields of parallel plates, concentric cylinders, and concentric spheres. For a special case of parallel plates involving N shields, for which all the emissivities are equal, the value of χ becomes N.

Radiation Heat-Transfer Coefficient

The radiation heat-transfer rate is often expressed in terms of the temperature difference $T_1 - T_2$ by $q = h_r A (T_1 - T_2)$, where h_r is called the radiation heat-transfer coefficient or *radiation film coefficient*. For the case of radiation between two large parallel plates with emissivities, respectively, of ϵ_1 and ϵ_2,

$$h_r = \frac{\sigma(T_1^4 - T_2^4)}{(T_1 - T_2)\left(\dfrac{1}{\epsilon_1} + \dfrac{1}{\epsilon_2} - 1\right)}$$

Table 56.19   Net Radiative Exchange between Two Surfaces Forming an Enclosure

Large (Infinite) Parallel Planes

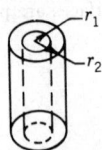

$$A_1 = A_2 = A$$

$$q_{12} = \frac{A\sigma(T_1^4 - T_2^4)}{\dfrac{1}{\epsilon_1} + \dfrac{1}{\epsilon_2} - 1}$$

Long (Infinite) Concentric Cylinders

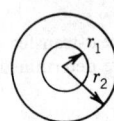

$$\frac{A_1}{A_2} = \frac{r_1}{r_2}$$

$$q_{12} = \frac{\sigma A_1(T_1^4 - T_2^4)}{\dfrac{1}{\epsilon_1} + \dfrac{1 - \epsilon_2}{\epsilon_2}\left(\dfrac{r_1}{r_2}\right)}$$

Concentric Spheres

$$\frac{A_1}{A_2} = \frac{r_1^2}{r_2^2}$$

$$q_{12} = \frac{\sigma A_1(T_1^4 - T_2^4)}{\dfrac{1}{\epsilon_1} + \dfrac{1 - \epsilon_2}{\epsilon_2}\left(\dfrac{r_1}{r_2}\right)^2}$$

Small Convex Object in a Large Cavity

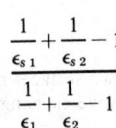

$$\frac{A_1}{A_2} \approx 0$$

$$q_{12} = \sigma A_1 \epsilon_1 (T_1^4 - T_2^4)$$

Table 56.20   Values of χ for Radiation Shields

Geometry	χ	
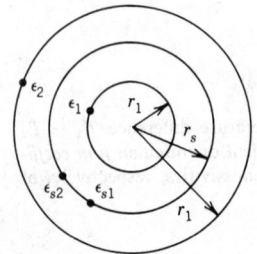	$\dfrac{\dfrac{1}{\epsilon_{s1}} + \dfrac{1}{\epsilon_{s2}} - 1}{\dfrac{1}{\epsilon_1} + \dfrac{1}{\epsilon_2} - 1}$	Infinitely long parallel plates
	$\dfrac{\left(\dfrac{r_1}{r_s}\right)^n\left(\dfrac{1}{\epsilon_{s1}} + \dfrac{1}{\epsilon_{s2}} - 1\right)}{\dfrac{1}{\epsilon_1} + \left(\dfrac{1}{\epsilon_2} - 1\right)\left(\dfrac{r_1}{r_2}\right)^n}$	$n = 1$ for infinitely long concentric cylinders $n = 2$ for concentric spheres

56.3.5 Thermal Radiation Properties of Gases

Gases such as air, oxygen (O_2), hydrogen (H_2), and nitrogen (N_2) have a symmetrical molecular structure and neither emit nor absorb radiation at low to moderate temperatures. Hence, for most engineering applications, such *nonparticipating gases* can be ignored. On the other hand, polyatomic gases such as water vapor (H_2O), carbon dioxide (CO_2), carbon monoxide (CO), sulfur dioxide (SO_2), and various hydrocarbons, emit and absorb significant amounts of radiation. These *participating gases* absorb and emit radiation in limited spectral ranges, called *bands*. In calculating the emitted or absorbed radiation for a gas layer, its thickness, shape, surface area, pressure, and temperature distribution must be taken into account. Precise method of calculation is quite complex. For engineering applications, the approximate method of Hottel[21] yields results of satisfactory accuracy. Hottel evaluated the effective total emissivities of carbon dioxide and water vapor with the temperature and the product of the partial pressure and the mean beam length as parameters (Figs. 56.24 and 56.25). The *mean beam length* L_e is a characteristic length corresponding to the radius of a gas hemisphere such that it radiates energy flux to the center of its base equal to the average flux radiated to the area of interest by the actual gas volume. Table 56.21 lists the mean beam lengths of several simple shapes. For a geometry for which L_e has not been determined, it is generally approximated by $L_e = 3.6V/A$ for an entire gas volume V radiating to its entire boundary surface A. The data in Figs. 56.24 and 56.25 were obtained for a total pressure of 1 atm and zero partial pressure of the water vapor. For other total and partial pressures the emissivities are corrected by multiplying C_{CO_2} (Fig. 56.26) and C_{H_2O} (Fig. 56.27), respectively, to ϵ_{CO_2} and ϵ_{H_2O}, which are found from Figs. 56.24 and 56.25.

The foregoing results apply when water vapor or carbon dioxide appears separately in a mixture with other nonparticipating gases. For a mixture of CO_2 and water vapor in a nonparticipating gas, the total emissivity of the mixture ϵ_g is found from

$$\epsilon_g = C_{CO_2}\epsilon_{CO_2} + C_{H_2O}\epsilon_{H_2O} - \Delta\epsilon$$

where $\Delta\epsilon$ is a correction factor given in Fig. 56.28.

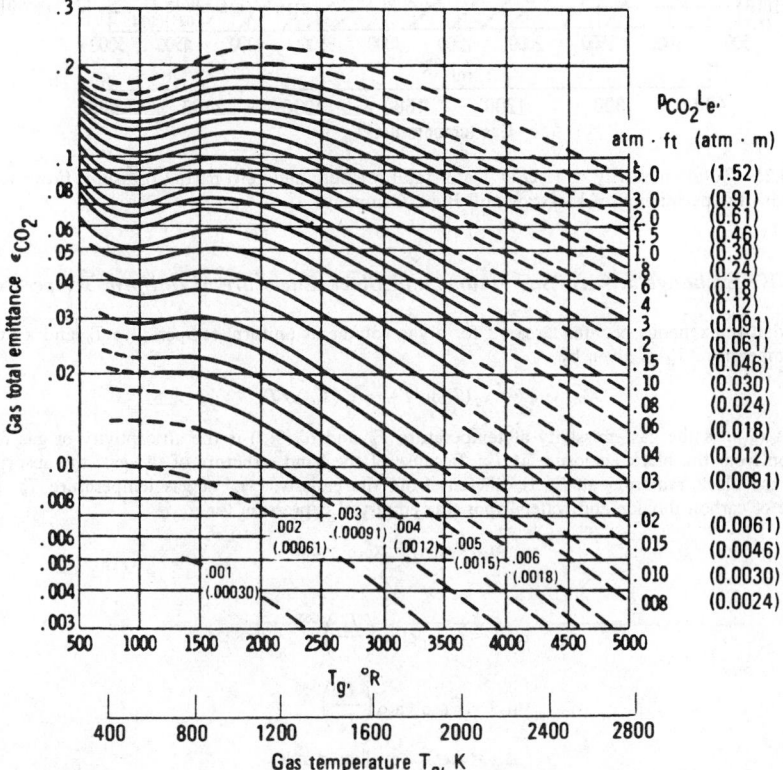

Fig. 56.24 Total emissivity of CO_2 in a mixture having a total pressure of 1 atm. From Ref. 21. Used with the permission of McGraw-Hill Book Company.

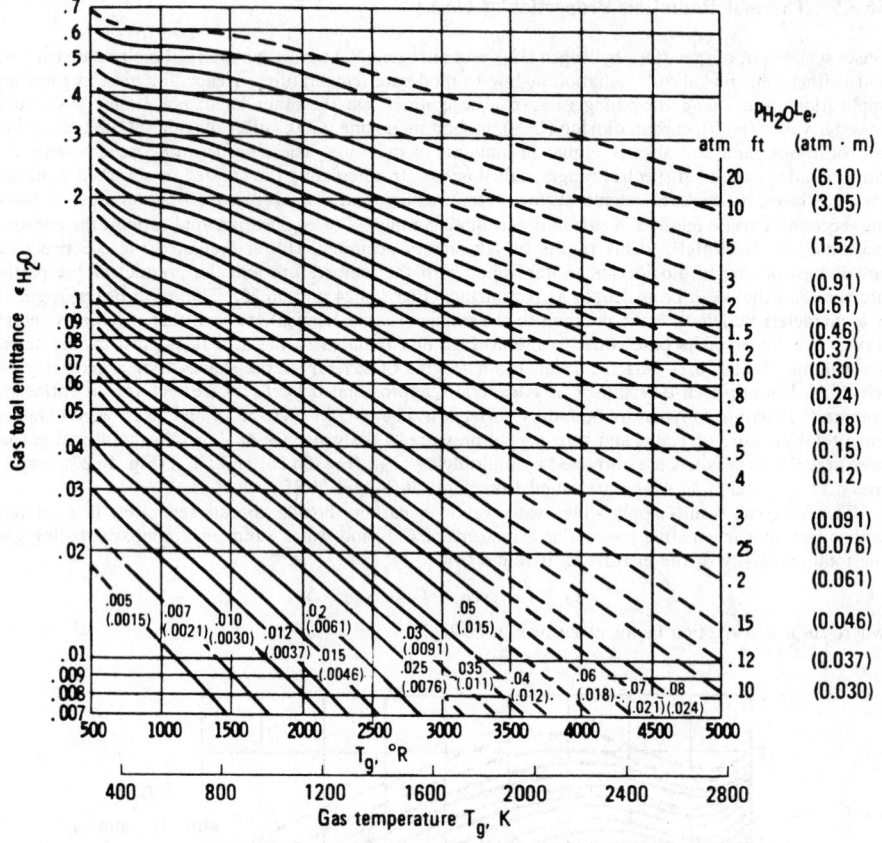

Fig. 56.25 Total emissivity of H_2O at 1 atm total pressure and zero partial pressure. From Ref. 21. Used with the permission of McGraw-Hill Book Company.

Radiative Exchange between Gas Volume and Black Enclosure of Uniform Temperature

The radiative exchange per unit area q'' for a gas volume at uniform temperature T_g and a uniform wall temperature T_w is given by

$$q'' = \epsilon_g(T_g)\sigma T_g^4 - \alpha_g(T_w)\sigma T_w^4$$

where $\epsilon_g(T_g)$ is the gas emissivity at temperature T_g and $\alpha_g(T_w)$ is the absorptivity of gas for the radiation from the black enclosure at T_w. Because of the band structure of the gas, the absorptivity α_g for the black radiation at T_w is different from the emissivity ϵ_g at gas temperature T_g. For a mixture of carbon dioxide and water vapor, the empirical expression for α_g is

$$\alpha_g = \alpha_{CO_2} + \alpha_{H_2O} - \Delta\alpha$$

where

$$\alpha_{CO_2} = C_{CO_2}\,\epsilon'_{CO_2}\left(\frac{T_g}{T_w}\right)^{0.65}$$

$$\alpha_{H_2O} = C_{H_2O}\,\epsilon_{H_2O}\left(\frac{T_g}{T_w}\right)^{0.45}$$

$$\Delta\alpha = \Delta\epsilon \qquad \text{evaluated at } T_w$$

The values of ϵ'_{CO_2} and ϵ_{H_2O} are found from Figs. 56.24 and 56.25 with an abscissa of T_w, but with the parameters $p_{CO_2}L_e$ and $p_{H_2O}L_e$ replaced, respectively, by $p_{CO_2}L_eT_w/T_g$ and $p_{H_2O}L_eT_w/T_g$.

Table 56.21 Mean Beam Length[a]

Geometry of Gas Volume	Characteristic Length	L_e
Hemisphere radiating to element at center of base	Radius R	R
Sphere radiating to its surface	Diameter D	$0.65D$
Circular cylinder of infinite height radiating to concave bounding surface	Diameter D	$0.95D$
Circular cylinder of semi-infinite height radiating to:		
Element at center of base	Diameter D	$0.90D$
Entire base	Diameter D	$0.65D$
Circular cylinder of height equal to diameter radiating to:		
Element at center of base	Diameter D	$0.71D$
Entire surface	Diameter D	$0.60D$
Circular cylinder of height equal to two diameters radiating to:		
Plane end	Diameter D	$0.60D$
Concave surface	Diameter D	$0.76D$
Entire surface	Diameter D	$0.73D$
Infinite slab of gas radiating to:		
Element on one face	Slab thickness D	$1.8D$
Both bounding planes	Slab thickness D	$1.8D$
Cube radiating to a face	Edge X	$0.6X$
Gas volume surrounding an infinite tube bundle and radiating to a single tube:		
Equilateral triangular array:	Tube diameter D,	
$S = 2D$	and spacing between	$3.0(S - D)$
$S = 3D$	tube centers, S	$3.8(S - D)$
Square array:		$3.5(S - D)$
$S = 2D$		

[a] Adapted from Ref. 19.

Radiative Exchange between Gas Volume and Gray Enclosure

When the emissivity of the enclosure, ϵ_w, is larger than 0.8, the rate of heat transfer may be approximated by

$$q_{gray} = \left(\frac{\epsilon_w + 1}{2}\right) q_{black}$$

where q_{gray} is the heat-transfer rate for gray enclosure and q_{black} is that for black enclosure. For values of $\epsilon_w < 0.8$, the band structures of the participating gas must be taken into account for heat-transfer calculations.

56.4 CONDENSATION AND BOILING

Condensation and boiling are forms of convection in which the fluid medium is undergoing a change of phase. The process of condensation is usually accomplished by allowing the vapor to come into contact with a surface at a temperature below the saturation temperature of the vapor. When a liquid is in contact with a solid surface maintained at a temperature above the saturation temperature of the liquid, boiling may occur. Condensation is associated with the change of phase from the vapor to the liquid state, resulting in heat transfer to the solid; boiling is associated with the phase change from the liquid to the vapor state, resulting in heat transfer from the solid. Heat-transfer coefficients for condensation and boiling are generally larger than that for convection without phase change. Application of boiling and condensation heat transfer may be seen in a closed-loop power cycle. In a power

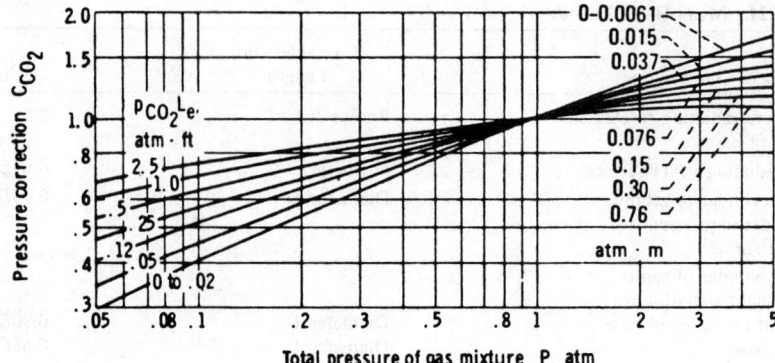

Fig. 56.26 Pressure correction for CO_2 total emissivity for values of P other than 1 atm. Adapted from Ref. 21. Used with the permission of McGraw-Hill Book Company.

cycle, the liquid is vaporized in a boiler at high pressure and temperature. After delivering work by expansion in a turbine, the vapor is condensed to its liquid state in a condenser, whereupon it is returned to the boiler to repeat the cycle.

56.4.1 Condensation

Depending on the surface conditions, the condensation may be a *film condensation* or a *dropwise condensation*. Film condensation usually occurs when a vapor, relatively free of impurities, is allowed to condense on a clean, uncontaminated surface. Dropwise condensation occurs on highly polished surfaces or on surfaces coated with substances that inhibit wetting. The condensate provides a resistance to heat transfer between the vapor and the surface. Therefore, it is desirable to use short vertical surfaces or horizontal cylinders to prevent the condensate from growing too thick. The heat-transfer rate for dropwise condensation is usually an order of magnitude larger than that for film condensation under similar conditions. Silicones, Teflon, and certain fatty acids can be used to coat the surfaces to promote dropwise condensation. However, such coatings may lose their effectiveness owing to oxidation or outright removal. Thus, except under carefully controlled conditions, film condensation may be expected to occur in most instances, and the condenser design calculations are often based on the assumption of film condensation.

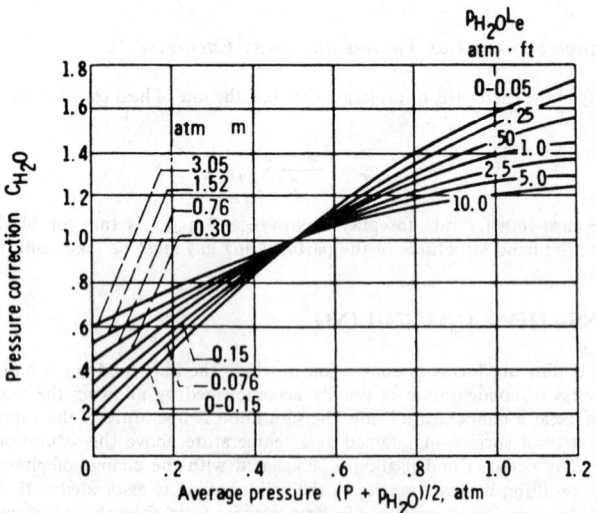

Fig. 56.27 Pressure correction for water vapor total emissivity for values of P_{H_2O} and P other than 0 and 1 atm, respectively. Adapted from Ref. 21. Used with the permission of McGraw-Hill Book Company.

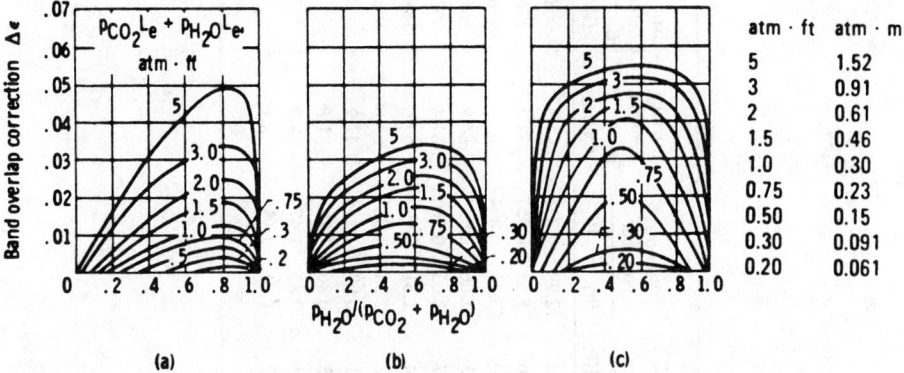

Fig. 56.28 Correction on total emissivity for band overlap when both CO_2 and water vapor are present: (a) gas temperature $T_g = 400$ K (720°R); (b) gas temperature $T_g = 810$ K (1460°R); (c) gas temperature, $T_g = 1200$ K (2160°R). Adapted from Ref. 21. Used with the permission of McGraw-Hill Book Company.

For condensation on a surface at temperature T_s the total heat-transfer rate to the surface is given by $q = \bar{h}_L A\ (T_{sat} - T_s)$, where T_{sat} is the saturation temperature of the vapor. The mass flow rate is determined by $\dot{m} = q/h_{fg}$; h_{fg} is the latent heat of vaporization of the fluid (see Table 56.22 for saturated water). Correlations are based on the evaluation of liquid properties at $T_f = (T_s + T_{sat})/2$ except h_{fg}, which is to be taken at T_{sat}.

Film Condensation on a Vertical Plate

The *Reynolds number for condensate flow* is defined by $Re_\Gamma = \rho_l\, V_m\, D_h/\mu_l$, where ρ_l and μ_l are the density and viscosity of the liquid, V_m is the average velocity of condensate, and D_h is the hydraulic diameter defined by $D_h = 4 \times$ condensate film cross-sectional area/wetted perimeter. For the condensation on a vertical plate $Re_\Gamma = 4\Gamma/\mu_l$, where Γ is the mass flow rate of condensate per unit width evaluated at the lowest point on the condensing surface. The condensate flow is generally considered to be laminar for $Re_\Gamma < 1800$, and turbulent for $Re_\Gamma > 1800$. The average Nusselt number is given by[22]

$$\overline{Nu}_L = 1.13 \left[\frac{g\rho_l\ (\rho_l - \rho_v)\ h_{fg}L^3}{\mu_l\, k_l\ (T_{sat} - T_s)} \right]^{0.25} \quad \text{for } Re_\Gamma < 1800$$

$$\overline{Nu}_L = 0.0077 \left[\frac{g\rho_l\ (\rho_l - \rho_v)\ L^3}{\mu_l^2} \right]^{1/3} Re_\Gamma^{0.4} \quad \text{for } Re_\Gamma > 1800$$

Film Condensation on the Outside of Horizontal Tubes and Tube Banks

$$\overline{Nu}_D = 0.725 \left[\frac{g\rho_l\ (\rho_l - \rho_v)\ h_{fg}D^3}{N\mu_l\ (T_{sat} - T_s)\ k_l} \right]^{0.25}$$

where N is the number of horizontal tubes placed one above the other; $N = 1$ for a single tube[23].

Film Condensation Inside Horizontal Tubes

For low vapor velocities such that Re_D based on the vapor velocities at the pipe inlet is less than 3500[24]

$$\overline{Nu}_D = 0.555 \left[\frac{g\rho_l\ (\rho_l - \rho_v)\ h'_{fg}D^3}{\mu_l\, k_l\ (T_{sat} - T_s)} \right]^{0.25}$$

where $h'_{fg} = h_{fg} + \frac{3}{8}\, c_{p,l}\ (T_{sat} - T_s)$.

For higher flow rate,[25] $Re_G > 5 \times 10^4$,

$$\overline{Nu}_D = 0.0265\, Re_G^{0.8}\ Pr^{1/3}$$

where the Reynolds number $Re_G = GD/\mu_l$ is based on the equivalent mass velocity $G = G_l + G_v\ (\rho_l/\rho_v)^{0.5}$. The mass velocity for the liquid G_l and for vapor G_v are calculated as if each occupied the entire flow area.

Table 56.22 Thermophysical Properties of Saturated Water

Temperature, T (K)	Pressure, P (bar)[a]	Specific Volume (m³/kg) $v_f \times 10^3$	Specific Volume (m³/kg) v_v	Heat of Vaporization, h_{fg} (kJ/kg)	Specific Heat (kJ/kg·K) $C_{p,l}$	Specific Heat (kJ/kg·K) $C_{p,v}$	Viscosity (N·sec/m²) $\mu_l \times 10^6$	Viscosity (N·sec/m²) $\mu_v \times 10^6$	Thermal Conductivity (W/m·K) $k_l \times 10^3$	Thermal Conductivity (W/m·K) $k_v \times 10^3$	Prandtl Number Pr_l	Prandtl Number Pr_v	Surface Tension, $\sigma_l \times 10^3$ (N/m)	Expansion Coefficient, $\beta_l \times 10^6$ (K⁻¹)
273.15	0.00611	1.000	206.3	2502	4.217	1.854	1750	8.02	659	18.2	12.99	0.815	75.5	−68.05
300	0.03531	1.003	39.13	2438	4.179	1.872	855	9.09	613	19.6	5.83	0.857	71.7	276.1
320	0.1053	1.011	13.98	2390	4.180	1.895	577	9.89	640	21.0	3.77	0.894	68.3	436.7
340	0.2713	1.021	5.74	2342	4.188	1.930	420	10.69	660	22.3	2.66	0.925	64.9	566.0
360	0.6209	1.034	2.645	2291	4.203	1.983	324	11.49	674	23.7	2.02	0.960	61.4	697.9
380	1.2869	1.049	1.337	2239	4.226	2.057	260	12.29	683	25.4	1.61	0.999	57.6	788
400	2.455	1.067	0.731	2183	4.256	2.158	217	13.05	688	27.2	1.34	1.033	53.6	896
450	9.319	1.123	0.208	2024	4.40	2.56	152	14.85	678	33.1	0.99	1.14	42.9	
500	26.40	1.203	0.0766	1825	4.66	3.27	118	16.59	642	42.3	0.86	1.28	31.6	
550	61.19	1.323	0.0317	1564	5.24	4.64	97	18.6	580	58.3	0.87	1.47	19.7	
600	123.5	1.541	0.0137	1176	7.00	8.75	81	22.7	497	92.9	1.14	2.15	8.4	
647.3	221.2	3.170	0.0032	0	∞	∞	45	45.0	238	238	∞	∞	0.0	

[a] 1 bar = 10⁵ N/m².

The Effect of Noncondensable Gases

If noncondensable gas such as air is present in a vapor, even in a small amount, the heat-transfer coefficient for condensation may be greatly reduced. It has been found that the presence of a few percent of air by volume in steam reduces the coefficient by 50% or more. Therefore, it is desirable in the condenser design to vent the noncondensable gases as much as possible.

56.4.2 Boiling

When boiling occurs in a quiescent liquid without external agitation, it is called *pool boiling*. In contrast, for *forced-convection boiling*, forced convection occurs simultaneously with boiling in which fluid motion is induced by external means. If the temperature of the liquid is below the saturation temperature, the process is called *subcooled*, or *local boiling*. If the liquid temperature is maintained or exceeds the saturation temperature, the process is referred to as *saturated*, or *bulk, boiling*. Figure 56.29 depicts the surface heat flux q'' as a function of excess temperature $\Delta T_e = T_s - T_{sat}$ for a typical pool boiling of water using an electrically heated wire. In the region $0 < \Delta T_e < \Delta T_{e,A}$ bubbles occur only on selected spots of the heating surface, and the heat transfer is mainly by free convection (*free convection boiling*). For $\Delta T_{e,A} < \Delta T_e < \Delta T_{e,C}$, the heating surface is densely populated with bubbles. The bubble separation induces considerable stirring action in the fluid near the surface, which increases heat transfer substantially. This region is referred to as *nucleate boiling*. When the excess temperature is raised to $\Delta T_{e,C}$, the heat flux reaches a maximum, and further increase of the temperature causes a decrease in the heat flux. The maximum heat flux is called the *critical heat flux*. In the *film boiling* regime, $\Delta T_e > \Delta T_{e,D}$, the entire heating surface is covered by a vapor film blanket, and the heat transfer to the liquid is caused by conduction and radiation through the vapor. Between points C and D, the heat flux decreases with increasing ΔT_e. In this region, part of the surface is covered by bubbles and part by a film. Evaporation in this region is called *transition boiling* or *partial film boiling*. The point of maximum heat flux, point C, is called the *burnout point* or the *Lindenfrost point*. Many engineering applications involve controlling the heat flux. When the heat flux is increased gradually, the temperature rises steadily until point C is reached. A slight increase of heat flux beyond the value q_C'', however, changes abruptly the surface temperature to $T_s = T_{sat} + \Delta T_{e,E}$, which generally exceeds the solid melting point. Although it is desirable to operate the vapor-producing apparatus with the heat flux close to q_C'', since this would permit the maximum use of the surface area, complications as described above will be encountered.

Nucleate Pool Boiling

The heat flux data are best correlated by[26]

$$q'' = \mu_l h_{fg} \left(\frac{g(\rho_l - \rho_v)}{g_c \, \sigma} \right)^{1/2} \left(\frac{c_{p,l} \, \Delta T_e}{C h_{fg} \, \mathrm{Pr}_l^{1.7}} \right)^3$$

where the subscripts l and v denote saturated liquid and vapor, respectively. The surface tension of the liquid is σ (N/m). The quantity g_c is the proportionality constant equal to $1 \text{ kg} \cdot \text{m/N} \cdot \text{sec}^2$. The quantity g is the local gravitational acceleration in m/sec². The values of C are given in Table 56.23. The above equation may be applied to different geometries such as plates, wire, or cylinders.

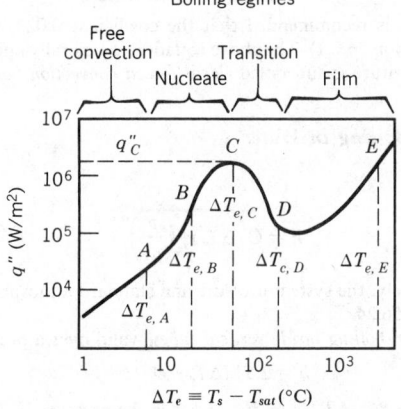

Fig. 56.29 Typical boiling curve for a wire in a pool of water at atmospheric pressure.

Table 56.23 Values of the Constant C for Various Liquid–Surface Combinations

Fluid–Heating-Surface Combinations	C
Water with polished copper, platinum, or mechanically polished stainless steel	0.0130
Water with brass or nickel	0.006
Water with ground and polished stainless steel	0.008
Water with Teflon-plated stainless steel	0.0058

[a] Adapted from Ref. 27.

The *critical heat flux* (point C of Fig. 56.29) is given by[28]

$$q_c'' = \frac{\pi}{24} h_{fg} \, \rho_v \left(\frac{\sigma \, g g_c \, (\rho_l - \rho_v)}{\rho_v^2} \right)^{0.25} \left(1 + \frac{\rho_v}{\rho_l} \right)^{0.5}$$

For a water–steel combination, $q_c'' \approx 1290$ KW/m² and $\Delta T_{e,c} \approx 30°C$. For water–chrome-plated copper, $q_c'' \approx 940$–1260 KW/m² and $\Delta T_{e,c} = 23$–28°C.

Film Pool Boiling

The heat transfer from the surface to the liquid is due to both convection and radiation. A total heat-transfer coefficient is defined by the combination of convection and radiation heat-transfer coefficients of the following form[29] for the outside surfaces of horizontal tubes:

$$h^{4/3} = h_c^{4/3} + h_r h^{1/3}$$

where

$$h_c = 0.62 \left(\frac{k_v^3 \, \rho_v \, (\rho_l - \rho_v) \, g \, (h_{fg} + 0.4 \, c_{p,v} \, \Delta T_e)}{\mu_v \, D \, \Delta T_e} \right)^{1/4}$$

and

$$h_r = \frac{5.73 \times 10^{-8} \, \epsilon (T_s^4 - T_{sat}^4)}{T_s - T_{sat}}$$

The vapor properties are evaluated at the film temperature $T_f = (T_s + T_{sat})/2$. The temperatures T_s and T_{sat} are in kelvins for the evaluation of h_r. The emissivity of the metallic solids can be found from Table 56.16. Note that $q = hA(T_s - T_{sat})$.

Nucleate Boiling in Forced Convection

The total heat-transfer rate can be obtained by simply superimposing the heat transfer due to nucleate boiling and forced convection:

$$q'' = q''_{boiling} + q''_{forced\,convection}$$

For forced convection, it is recommended that the coefficient 0.023 be replaced by 0.014 in the Dittus–Boelter equation (Section 56.2.1). The above equation is generally applicable to forced convection where the bulk liquid temperature is subcooled (*local forced convection boiling*).

Simplified Relations for Boiling in Water

For *nucleate boiling*,[30]

$$h = C \, (\Delta T_e)^n \left(\frac{p}{p_a} \right)^{0.4}$$

where p and p_a are, respectively, the system pressure and standard atmospheric pressure. The constants C and n are listed in Table 56.24.

For *local forced convection boiling inside vertical tubes*, valid over a pressure range of 5–170 atm,[31]

$$h = 2.54 \, (\Delta T_e)^3 \, e^{p/1.551}$$

where h has the unit W/m² · °C, ΔT_e is in °C, and p is the pressure in 10⁶ N/m³.

Table 56.24 Values of C and n for Simplified Relations for Boiling in Water[a]

Surface	q'' (KW/m²)	C	n
Horizontal	$q'' < 16$	1042	$\frac{1}{3}$
	$16 < q'' < 240$	5.56	3
Vertical	$q'' < 3$	537	$\frac{1}{7}$
	$3 < q'' < 63$	7.96	3

[a] Adapted from Ref. 30.

REFERENCES

1. F. P. Incropera and D. P. Dewitt, *Fundamentals of Heat Transfer,* Wiley, New York, 1981.

2. E. R. G. Eckert and R. M. Drake, Jr., *Analysis of Heat and Mass Transfer,* McGraw-Hill, New York, 1972.

3. M. P. Heisler, "Temperature Charts for Induction and Constant Temperature Heating," *Trans. ASME,* **69,** 227 (1947).

4. H. Grober and S. Erk, *Fundamentals of Heat Transfer,* McGraw-Hill, New York, 1961.

5. E. N. Sieder and C. E. Tate, "Heat Transfer and Pressure Drop of Liquids in Tubes," *Ind. Eng. Chem.,* **28,** 1429 (1936).

6. F. W. Dittus and L. M. K. Baelter, *Univ. Calif., Berkeley, Pub. Eng.,* **2,** 443 (1930).

7. A. J. Chapman, *Heat Transfer,* Macmillan, New York, 1974.

8. S. Whitaker, *AICHE J.,* **18,** 361 (1972).

9. M. Jakob, *Heat Transfer,* Wiley, New York, 1949, Vol. 1.

10. A. Zhukauska, "Heat Transfer from Tubes in Cross Flow," in *Advances in Heat Transfer,* J. P. Hartnett and T. F. Irvine, Jr. (eds.), Academic, New York, 1972, Vol. 8.

11. F. Kreith, *Principles of Heat Transfer,* Harper and Row, New York, 1973.

12. H. A. Johnson and M. W. Rubesin, "Aerodynamic Heating and Convective Heat Transfer," *Trans. ASME,* **71,** 447 (1949).

13. C. C. Lin (Ed.), *Turbulent Flows and Heat Transfer, High Speed Aerodynamics and Jet Propulsion,* Princeton University Press, Princeton, NJ, 1959, Vol. V.

14. W. H. McAdams, *Heat Transmission,* McGraw-Hill, New York, 1954.

15. T. Yuge, "Experiments on Heat Transfer from Spheres Including Combined Natural and Forced Convection," *J. Heat Transf.,* **82,** 214 (1960).

16. S. Globe and D. Dropkin, "Natural Convection Heat Transfer in Liquids Confined between Two Horizontal Plates," *J. Heat Transfer,* **81C,** 24 (1959).

17. I. Catton, "Natural Convection in Enclosures," *Proc. 6th International Heat Transfer Conference,* 6, Toronto, Canada, 1978.

18. R. K. MacGregor and A. P. Emery, "Free Convection through Vertical Plane Layers: Moderate and High Prandtl Number Fluids," *J. Heat Transfer,* **91,** 391 (1969).

19. R. Siegel and J. R. Howell, *Thermal Radiation Heat Transfer,* McGraw-Hill, New York, 1981.

20. G. G. Gubareff, J. E. Janssen, and R. H. Torborg, *Thermal Radiation Properties Survey,* 2nd ed., Minneapolis Honeywell Regulator Co., Minneapolis, MN, 1960.

21. H. C. Hottel, *Heat Transmission,* W. C. McAdams (ed.), McGraw-Hill, New York, 1954, Chap. 2.

22. W. H. McAdams, *Heat Transmission,* 3rd ed., McGraw-Hill, New York, 1954.

23. W. M. Rohsenow, "Film Condensation" in W. M. Rohsenow and J. P. Hartnett (eds.), *Handbook of Heat Transfer,* McGraw-Hill, New York, 1973.

24. J. C. Chato, "Laminar Condensation Inside Horizontal and Inclined Tubes," *J. Am. Soc. Heating Refrig. Aircond. Engrs.,* **4,** 52 (1962).

25. W. W. Akers, H. A. Deans, and O. K. Crosser, "Condensing Heat Transfer Within Horizontal Tubes," *Chem. Eng. Prog., Sym. Ser.,* **55**(29), 171 (1958).

26. W. M. Rohsenow, "A Method of Correlating Heat Transfer Data for Surface Boiling Liquids," *Trans. ASME,* **74,** 969 (1952).

27. J. P. Holman, *Heat Transfer,* McGraw-Hill, New York, 1981.

28. N. Zuber, "On the Stability of Boiling Heat Transfer," *Trans. ASME*, **80**, 711 (1958).
29. L. A. Bromley, "Heat Transfer in Stable Film Boiling," *Chem. Eng. Prog.*, **46**, 221 (1950).
30. M. Jacob and G. A. Hawkins, *Elements of Heat Transfer*, Wiley, New York, 1957.
31. M. Jacob, *Heat Transfer*, Wiley, New York, 1957, Vol. 2, p. 584.

BIBLIOGRAPHY

American Society of Heating, Refrigerating and Air Conditioning Engineering, *ASHRAE Handbook of Fundamentals*, 1972.

V. S. Arpaci, *Conduction Heat Transfer*, Addison-Wesley, Reading, MA, 1966.

H. S. Carslaw and J. C. Jager, *Conduction of Heat in Solid*, Oxford University Press, London, 1959.

J. A. Duffie and W. A. Beckman, *Solar Engineering of Thermal Process*, Wiley, New York, 1980.

B. Gebhart, *Heat Transfer*, McGraw-Hill, New York, 1971.

H. C. Hottel and A. F. Saroffin, *Radiative Transfer*, McGraw-Hill, New York, 1967.

W. M. Kays, *Convective Heat and Mass Transfer*, McGraw-Hill, New York, 1966.

J. G. Knudsen and D. L. Katz, *Fluid Dynamics and Heat Transfer*, McGraw-Hill, New York, 1958.

M. N. Ozisik, *Radiative Transfer and Interaction with Conduction and Convection*, Wiley, New York, 1973.

M. N. Ozisik, *Heat Conduction*, Wiley, New York, 1980.

M. Planck, *The Theory of Heat Radiation*, Dover, New York, 1959.

W. M. Rohsenow and H. Y. Choi, *Heat, Mass, and Momentum Transfer*, Prentice-Hall, Englewood Cliffs, NJ, 1961.

W. M. Rohsenow and J. P. Hartnett, *Handbook of Heat Transfer*, McGraw-Hill, New York, 1973.

H. Schlichting, *Boundary-Layer Theory*, McGraw-Hill, New York, 1979.

P. J. Schneider, *Conduction Heat Transfer*, Addison-Wesley, Reading, MA, 1955.

E. M. Sparrow and R. D. Cess, *Radiation Heat Transfer*, Wadsworth Publishing Co., Inc., Belmont, CA, 1966.

W. C. Turner and J. F. Malloy, *Thermal Insulation Handbook*, McGraw-Hill, New York, 1981.

N. B. Vargafik, *Table of Thermophysical Properties of Liquids and Gases*, Hemisphere Publishing Corp., Washington, DC, 1975.

J. A. Wiebelt, *Engineering Radiation Heat Transfer*, Holt, Rinehart and Winston, New York, 1966.

CHAPTER 57
COMBUSTION

RICHARD J. REED

North American Manufacturing Company
Cleveland, Ohio

57.1 FUNDAMENTALS OF COMBUSTION

57.1.1 Air–Fuel Ratios

Combustion is rapid oxidation, usually for the purpose of changing chemical energy into thermal energy—heat. This energy usually comes from oxidation of carbon, hydrogen, sulfur, or compounds containing C, H, and/or S. The oxidant is usually O_2—molecular oxygen from the air.

The stoichiometry of basic chemical equation balancing permits determination of the air required to burn a fuel. For example,

$$1CH_4 + 2O_2 \rightarrow 1CO_2 + 2H_2O$$

where the units are moles or volumes; therefore, 1 ft³ of methane (CH_4) produces 1 ft³ of CO_2; or 1000 m³ CH_4 requires 2000 m³ O_2 and produces 2000 m³ H_2O. Knowing that the atomic weight of C is 12, H is 1, N is 14, O is 16, and S is 32, it is possible to use the balanced chemical equation to predict weight flow rates: 16 lb/hr CH_4 requires 64 lb/hr O_2 to burn to 44 lb/hr CO_2 and 36 lb/hr H_2O.

If the oxygen for combustion comes from air, it is necessary to know that air is 20.99% O_2 by volume and 23.20% O_2 by weight, most of the remainder being nitrogen.

It is convenient to remember the following ratios:

$$air/O_2 = 100/20.99 = 4.76 \text{ by volume}$$
$$N_2/O_2 = 3.76 \text{ by volume}$$
$$air/O_2 = 100/23.20 = 4.31 \text{ by weight}$$
$$N_2/O_2 = 3.31 \text{ by weight}$$

Rewriting the previous formula for combustion of methane,

$$1CH_4 + 2O_2 + 2(3.76)N_2 \rightarrow 1CO_2 + 2H_2O + 2(3.76)N_2$$

or

$$1CH_4 + 2(4.76)air \rightarrow 1CO_2 + 2H_2O + 2(3.76)N_2$$

Table 57.1 lists the amounts of air required for stoichiometric (quantitatively and chemically correct) combustion of a number of pure fuels, calculated by the above method. (Table 59.1c lists similar information for typical fuels that are mixtures of compounds, calculated by the above method, but weighted for the percentages of the various compounds in the fuels.)

The stoichiometrically correct (perfect, ideal) air/fuel ratio from the above formula is therefore $2 + 2(3.76) = 9.52$ volumes of air per volume of the fuel gas. More than that is called a "lean" ratio, and includes excess air and produces an oxidizing atmosphere. For example, if the actual air/fuel ratio were 10:1, the %excess air would be

$$\frac{10 - 9.52}{9.52} \times 100 = 5.04\%$$

Communications problems sometimes occur because some people think in terms of air/fuel ratios, others in fuel/air ratios; some in weight ratios, others in volume ratios; and some in mixed metric units (such as normal cubic meters of air per metric tonne of coal), others in mixed American units (such as ft³ air/gal of oil). To avoid such confusions, the following method from Ref. 1 is recommended.

It is more convenient to specify air/fuel ratio in unitless terms such as %air (%aeration), %excess air, %deficiency of air, or equivalence ratio. Those experienced in this field prefer to converse in terms of %excess air. The scientific community favors equivalence ratio. The %air is easiest to use and explain to newcomers to the field: "*100% air*" is the correct (stoichiometric) amount; 200% air is twice as much as necessary, or 100% excess air. *Equivalence ratio*, widely used in combustion research, is the actual amount of fuel expressed as a fraction percent of the stoichiometrically correct amount of fuel. The Greek letter phi, ϕ, is usually used: $\phi = 0.9$ is lean; $\phi = 1.1$ is rich; and $\phi = 1.0$ is "on-ratio."

Formulas relating %air, ϕ, %excess air (%XS), and %deficiency of air (%def) are

$$\%air = 100/\phi = \%XS + 100 = 100 - \%def$$

$$\phi = \frac{100}{\%XS + 100} = \frac{1}{1 - (\%def/100)}$$

$$\%XS = \%air - 100 = \frac{1 - \phi}{\phi} \times 100$$

$$\%def = 100 - \%air = \frac{\phi - 1}{\phi} \times 100$$

Table 57.1 Proper Combining Proportions for Perfect Combustion[a]

Fuel	$\dfrac{\text{vol } O_2}{\text{vol fuel}}$	$\dfrac{\text{vol air}}{\text{vol fuel}}$	$\dfrac{\text{wt } O_2}{\text{wt fuel}}$	$\dfrac{\text{wt air}}{\text{wt fuel}}$	$\dfrac{\text{ft}^3 O_2}{\text{lb fuel}}$	$\dfrac{\text{ft}^3 \text{ air}}{\text{lb fuel}}$	$\dfrac{\text{m}^3 O_2}{\text{kg fuel}}$	$\dfrac{\text{m}^3 \text{ air}}{\text{kg fuel}}$
Acetylene, C_2H_2	2.50	11.9	3.08	13.3	36.5	174	2.28	10.8
Benzene, C_6H_6	7.50	35.7	3.08	13.3	36.5	174	2.28	10.8
Butane, C_4H_{10}	6.50	31.0	3.59	15.5	42.5	203	2.65	12.6
Carbon, C	—	—	2.67	11.5	31.6	150	1.97	9.39
Carbon monoxide, CO	0.50	2.38	0.571	2.46	6.76	32.2	0.422	2.01
Ethane, C_2H_6	3.50	16.7	3.73	16.1	44.2	210	2.76	13.1
Hydrogen, H_2	0.50	2.38	8.00	34.5	94.7	451	5.92	28.2
Hydrogen sulfide, H_2S	1.50	7.15	1.41	6.08	16.7	79.5	1.04	4.97
Methane, CH_4	2.00	9.53	4.00	17.2	47.4	226	2.96	14.1
Naphthalene, $C_{10}H_8$	—	—	3.00	12.9	35.5	169	2.22	10.6
Octane, C_8H_{18}	—	—	3.51	15.1	41.6	198	2.60	12.4
Propane, C_3H_8	5.00	23.8	3.64	15.7	43.1	205	2.69	12.8
Propylene, C_3H_6	4.50	21.4	3.43	14.8	40.6	193	2.54	12.1
Sulfur, S	—	—	1.00	4.31	11.8	56.4	0.74	3.52

[a] Reproduced with permission from *Combustion Handbook*.[1] (See Ref. 1)

Table 57.2 lists a number of equivalent terms for convenience in converting values from one "language" to another.

Excess air is undesirable, because, like N_2, it passes through the combustion process without chemical reaction; yet it absorbs heat, which it carries out the flue. The percent available heat (best possible fuel efficiency) is highest with zero excess air. (See Fig. 57.1.)

Excess fuel is even more undesirable because it means there is a deficiency of air and some of the fuel cannot be burned. This results in formation of soot and smoke. The accumulation of unburned fuel or partially burned fuel can represent an explosion hazard.

Enriching the oxygen content of the combustion "air" above the normal 20.9% reduces the nitrogen and thereby reduces the loss due to heat carried up the stack. This also raises the flame temperature, improving heat transfer, especially that by radiation.

Vitiated air (containing less than the normal 20.9% oxygen) results in less fuel efficiency, and may result in flame instability. Vitiated air is sometimes encountered in incineration of fume streams or in staged combustion.

57.1.2 Fuels

Fuels used in practical industrial combustion processes have such a major effect on the combustion that they must be studied simultaneously with combustion. Fuels are covered in detail in later chapters, so the treatment here is brief, relating only to the aspects having direct bearing on the combustion process.

Gaseous fuels are generally easier to burn, handle, and control than are liquid or solid fuels. Molecular mixing of a gaseous fuel with oxygen need not wait for vaporization nor mass transport within a solid. Burning rates are limited only by mixing rates and the kinetics of the combustion reactions; therefore, combustion can be compact and intense. Reaction times as short as 0.001 sec and combustion volumes from 10^4 to 10^7 Btu/hr · ft^3 are possible at atmospheric pressure.[2] Gases of low calorific value may require such large volumes of air that their combustion rates will be limited by the mixing time.

Combustion stability (ability of a flame to burn steadily and reliably without a pilot or spark) depends on burner nozzle configuration, plus the following fuel properties: minimum ignition temperature, upper and lower limits of flammability, and flame velocity. These properties are discussed in Chapter 59. The burner geometry must maintain conditions within these limits at the point where flame is initiated throughout the burner's full range of firing rates.

Liquid fuels are usually not as easily burned, handled, or controlled as are gaseous fuels. Mixing with oxygen can occur only after the liquid fuel is evaporated; therefore, burning rates are limited

Table 57.2 Equivalent Ways to Express Fuel-to-Air or Air-to-Fuel Ratios[1]

	ϕ	%air	%def	%XS
Fuel rich	2.50	40	60	
(air lean)	1.67	60	40	
	1.25	80	20	
	1.11	90	10	
	1.05	95	5	
Stoichiometric	1.00	100	0	0
Fuel lean	0.95	105		5
(air rich)	0.91	110		10
	0.83	120		20
	0.78	130		30
	0.71	140		40
	0.62	160		60
	0.56	180		80
	0.50	200		100
	0.40	250		150
	0.33	300		200
	0.25	400		300
	0.20	500		400
	0.167	600		500
	0.091	1100		1000
	0.048	2100		2000

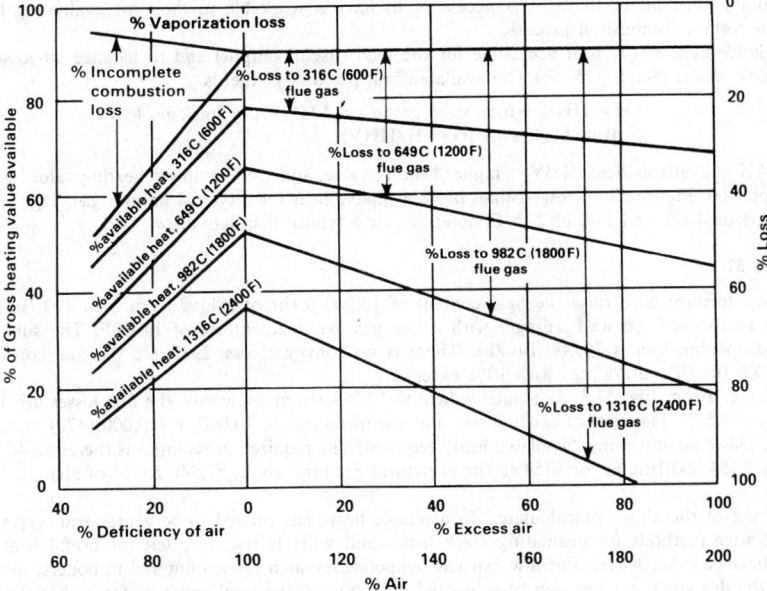

Fig. 57.1 Percent available heat (best possible efficiency) peaks at stoichiometric air/fuel ratio.[1]

by vaporization rates. In practice, combustion intensities are usually less with liquid fuels than with high calorific gaseous fuels such as natural gas.

Because vaporization is such an integral part of most liquid fuel burning processes, much of the emphasis in evaluating liquid fuel properties is on factors that relate to vaporization, including viscosity, which hinders good atomization, the primary method for enhancing vaporization. Much concern is also devoted to properties that affect storage and handling because, unlike gaseous fuels that usually come through a public utility's mains, liquid fuels must be stored and distributed by the user.

The stability properties (ignition temperature, flammability limits, and flame velocity) are not readily available for liquid fuels, but flame stability is often less critical with liquid fuels.

Solid fuels are frequently more difficult to burn, handle, and control than liquid or gaseous fuels. After initial volatilization, the combustion reaction rate depends on diffusion of oxygen into the remaining char particle, and the diffusion of carbon monoxide back to its surface, where it burns as a gas. Reaction rates are usually low and required combustion volumes high, even with pulverized solid fuels burned in suspension. Some fluidized bed and cyclone combustors have been reported to reach the intensities of gas and oil flames.[2]

Most commonly measured solid fuel properties apply to handling in stokers or pulverizers. See Chapter 61.

Wastes, by-product fuels, and gasified solids are being used more as fuel costs rise. Operations that produce such materials should attempt to consume them as energy sources. Handling problems, the lack of a steady supply, and pollution problems often complicate such fuel usage.

For the precise temperature control and uniformity required in many industrial heating processes, the burning of solids, especially the variable quality solids found in wastes, presents a critical problem. Such fuels are better left to very large combustion chambers, particularly boilers. When solids and wastes must be used as heat sources in small and accurate heating processes, a better approach is to convert them to low-Btu (producer) gas, which can be cleaned and then controlled more precisely.

57.2 PURPOSES OF COMBUSTION

The purposes of combustion, for the most part, center around elevating the temperature of something. This includes the first step in all successive combustion processes—the pilot flame—and, similarly, the initiation of incineration. Elevating the temperature of something can also make it capable of transmitting light or thermal energy (radiation and convection heat transfer), or it can cause chemical dissociation of molecules in the products of combustion to generate a special atmosphere gas for protection of materials in industrial heat processing.

All of the above functions of combustion are minor in comparison to the heating of air, water and steam, metals, nonmetallic minerals, and organics for industrial processing, and for space comfort conditioning. For all of these, it is necessary to have a workable method for evaluating the heat available from a combustion process.

Available heat is the heat accessible for the load (useful output) and to balance all losses other than stack losses. (See Fig. 57.2.) The available heat per unit of fuel is

$$AH = HHV - \text{total stack gas loss} = LHV - \text{dry stack gas loss}$$
$$\% \text{ available heat} = 100(AH/HHV)$$

where AH = available heat, HHV = higher heating value, and LHV = lower heating value, as defined in Chapter 60. Figure 57.3 shows values of % available heat for a typical natural gas; Fig. 57.4 for a typical residual oil; and Fig. 68.2 in Chapter 68, for a typical distillate oil.

Example 57.1

A process furnace is to raise the heat content of 10,000 lb/hr of a load from 0 to 470 Btu/lb in a continuous furnace (no wall storage) with a flue gas exit temperature of 1400°F. The sum of wall loss and opening loss is 70,000 Btu/hr. There is no conveyor loss. Estimate the fuel consumption using 1000 Btu/ft³ natural gas with 10% excess air.

Solution: From Fig. 57.3, % available heat = 58.5%. In other words, the flue losses are 100% − 58.5% = 41.5%. The sum of other losses and useful output = 70,000 + (10,000)(470) = 4,770,000 Btu/hr. This constitutes the "available heat" required. The required gross input is therefore 4,770,000/0.585 = 8,154,000 Btu/hr, or 8154 ft³/hr of natural gas (and about 81,540 ft³/hr of air).

The use of the above precalculated % available heats has proved to be a practical way to avoid long iterative methods for evaluating stack losses and what is therefore left for useful heat output and to balance other losses. For low exit gas temperatures such as encountered in boilers, ovens, and dryers, the dry stack gas loss can be estimated by assuming the total exit gas stream has the specific heat of nitrogen, which is usually a major component of the poc (products of combustion).

$$\frac{\text{dry stack loss}}{\text{unit of fuel}} = \left(\frac{\text{lb dry poc}}{\text{unit fuel}}\right)\left(\frac{0.253 \text{ Btu}}{\text{lb poc (°F)}}\right)(T_{\text{exit}} - T_{\text{in}})$$

or

$$\left(\frac{\text{scf dry poc}}{\text{unit fuel}}\right)\left(\frac{0.0187 \text{ Btu}}{\text{scf poc (°F)}}\right)(T_{\text{exit}} - T_{\text{in}})$$

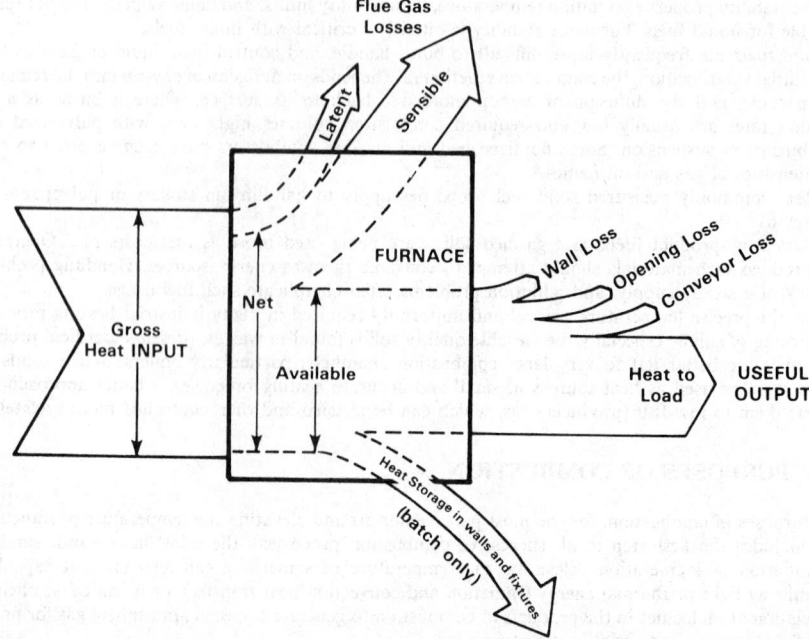

Fig. 57.2 Sankey diagram for a furnace, oven, kiln, incinerator, boiler, or heater—a qualitative and roughly quantitative method for analyzing efficiency of fuel-fired heating equipment.

All curves for hot air are based on 10% excess air.
All curves for excess air are based on 60 F temperature.

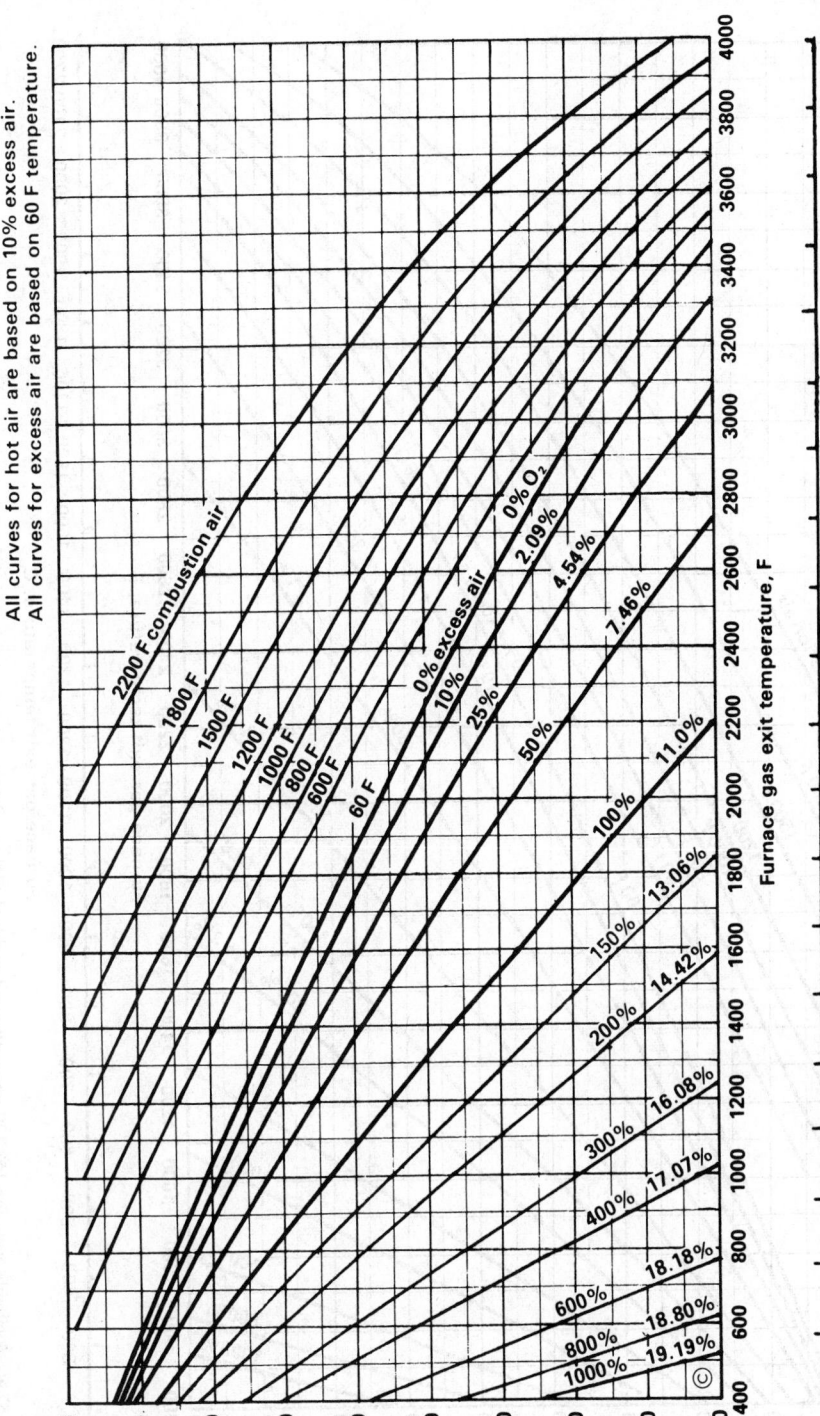

Furnace gas exit temperature, F

Furnace gas exit temperature. C

Fig. 57.3 Available heat for 1000 Btu/ft³ natural gas. In a furnace with 1600°F flue temperature and 10% excess air, 54% of the gross heat input is available for heating the load and balancing the losses other than stack losses.

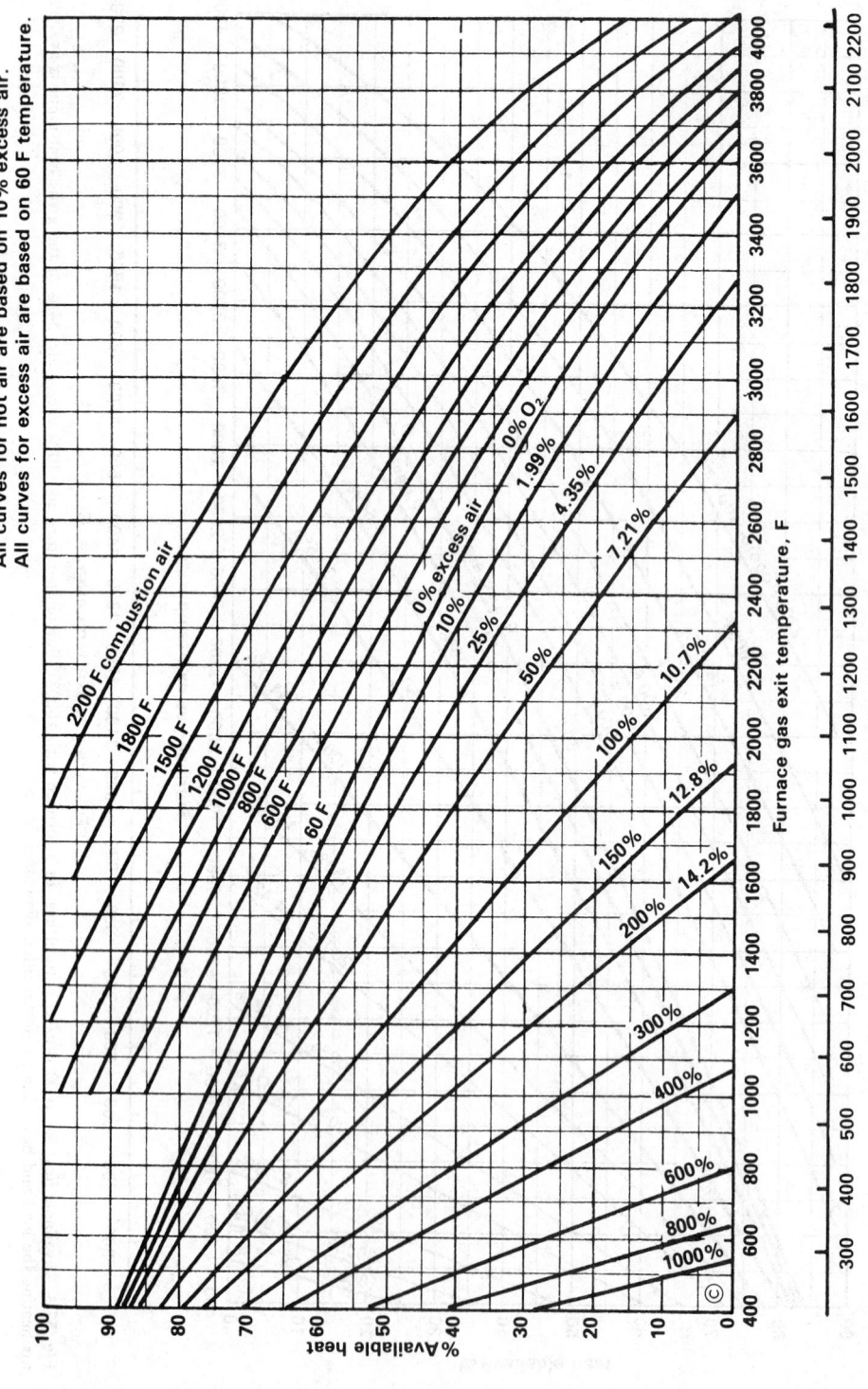

All curves for hot air are based on 10% excess air.
All curves for excess air are based on 60 F temperature.

Fig. 57.4 Available heat for 153,120 gross Btu/gal residual fuel oil (heavy, No. 6). With 2200°F gases leaving a furnace, 1000°F air entering the burners,

For a gaseous fuel, the "unit fuel" is usually scf (standard cubic foot), where "standard" is at 29.92 in. Hg and 60°F or nm³ (normal cubic meter), where "normal" is at 1.013 bar and 15°C.

Heat transferred from combustion takes two forms: radiation and convection. Both phenomena involve transfer to a surface.

Flame radiation comes from particle radiation and gas radiation. The visible yellow–orange light normally associated with a flame is actually from solid soot or char particles in the flame, and the "working" portion of this form of heat transfer is in the infrared wavelength range. Because oils have higher C/H ratios than gaseous fuels, oil flames are usually more yellow than gas flames (although oil flames can be made blue). Gas flames can be made yellow, by a delayed-mixing burner design, for the purpose of increasing their radiating capability.

Particulate radiation follows the Stefan–Boltzmann law for solids, but depends on the concentration of particles within the flame. Estimating or measuring the particle temperature and concentration is difficult.

Gas radiation and blue flame radiation contain more ultraviolet radiation and tend to be less intense. Triatomic gases (CO_2, H_2O, and SO_2) emit radiation that is largely invisible. Gases beyond the tips of both luminous and nonluminous flames continue to emit this gas radiation. As a very broad generalization, blue or nonluminous flames tend to be hotter, smaller, and less intense radiators than luminous flames. Gas radiation depends on the concentrations (or partial pressures) of the triatomic molecules and the beam thickness of their "cloud." Their temperatures are very transient.

Convection from combustion products beyond the flame tip follows conventional convection formulas—largely a function of velocity. This is the reason for recent emphasis on high-velocity burners. Flame convection by actual flame impingement is more difficult to evaluate because (a) flame temperatures change so rapidly and are so difficult to measure or predict, and (b) this involves extrapolating many convection formulas into ranges where good data are lacking.

Refractory radiation is a second stage of heat transfer. The refractory must first be heated by flame radiation and/or convection. A gas mantle, so-called "infrared" burners, and "radiation burners" use flame convection to heat some solid (refractory or metal) to incandescence so that they become good radiators.

57.3 BURNERS

In some cases, a burner may be nothing more than a nozzle. Some would say it includes a mixing device, windbox, fan, and controls. In some configurations, it is difficult to say where the burner ends and the combustion chamber or furnace begins. In this section, the broadest sense of the terms will generally be used.

A combustion system provides (1) fuel, (2) air, (3) mixing, (4) proportioning, (5) ignition, and (6) flame holding. In the strictest sense, a burner does only function 6; in the broadest sense, it may do any or all of these functions.

57.3.1 Burners for Gaseous Fuels

Open and natural draft-type burners rely on a negative pressure in the combustion chamber to pull in the air required for combustion, usually through adjustable shutters around the fuel nozzles. The suction in the chamber may be natural draft (chimney effect) or induced draft fans. A crude "burner" may be nothing more than a gas gun and/or atomizer inserted through a hole in the furnace wall. Fuel–air mixing may be poor, and fuel–air ratio control may be nonexistent. Retrofitting for addition of preheated combustion air is difficult. (See Fig. 57.5.)

Sealed-in and power burners have no intentional "free" air inlets around the burner, nor are there air inlets in the form of louvers in the combustion chamber wall. All air in-flow is controlled, usually by a forced draft blower or fan pushing the air through pipes or a windbox.

These burners usually have a higher air pressure drop at the burner, so air velocities are higher, enabling more thorough mixing and better control of flame geometry. Air flow can be measured, so automatic air–fuel ratio control is easy. (See Fig. 57.6.)

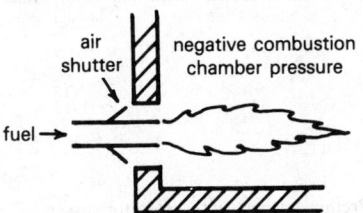

Fig. 57.5 Open, natural draft-type burner.

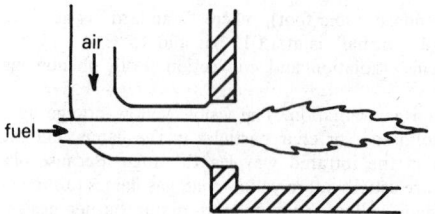

Fig. 57.6   Sealed-in, power burner.

Windbox burners often consist of little more than a long atomizer and a gas gun or gas ring. There are popular for boilers and air heaters where economic reasons have dictated that the required large volumes of air be supplied at very low pressure (2–10 in. wc) (in. wc = inches of water column). Precautions are necessary to avoid fuel flowback into the windbox. (See Fig. 57.7.)

Packaged burners usually consist of bolt-on arrangements with an integral fan and perhaps integral controls. These are widely used for new and retrofit installations from very small up to about 50×10^6 Btu/hr. (See Fig. 57.8.)

Premix burner systems may be found in any of the above configurations. Gas and air are thoroughly mixed upstream of the flame-holding nozzle. Most domestic appliances incorporate premixing, using some form of gas injector or inspirator (gas pressure inducing air through a venturi). Small industrial multiport burners of this type facilitate spreading a small amount of heat over a large area, as for heating kettles, vats, rolls, small boilers, moving webs, and low-temperature processing of conveyorized products. (See Fig. 57.9.) Large single port premix burners have been replaced by nozzle-mix burners. Better fuel–air ratio control is possible by use of aspirator mixers. (Air injection provides the energy to draw in the proper proportion of gas.) (See Fig. 57.10.) Many small units have undersized blowers,

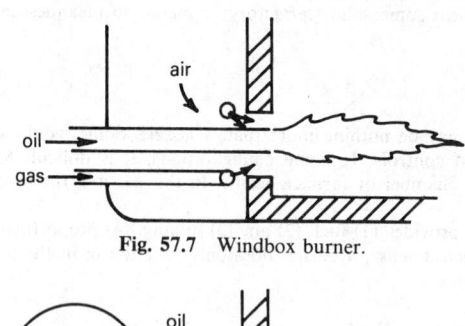

Fig. 57.7   Windbox burner.

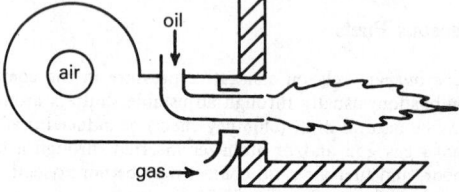

Fig. 57.8   Integral fan burner.

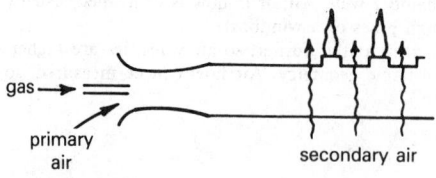

Fig. 57.9   Premix burners with inspirator mixer.

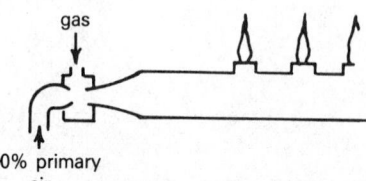

Fig. 57.10   Premix burners with aspirator mixer.

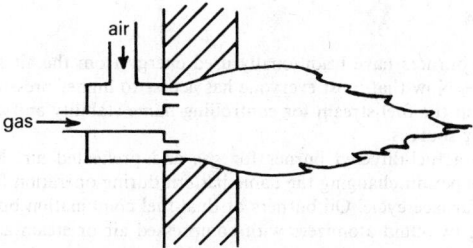

Fig. 57.11 Air-directed nozzle-mix burner.

relying on furnace draft to provide secondary air. As fuel costs rise, the unwarranted excess air involved in such arrangements makes them uneconomical.

Larger than a 4-in. (100-mm) inside-diameter mixture manifold is usually considered too great an explosion risk. For this reason, mixing in a fan inlet is rarely used.

Nozzle-mix burner systems constitute the most common industrial gas burner arrangement today. Gas and air are mixed as they enter the combustion chamber through the flame holder. (See Fig. 57.11.) They permit a broad range of fuel–air ratios, a wide variety of flame shapes, and multifuel firing. A very wide range of operating conditions are now possible with stable flames, using nozzle-mix burners. For processes requiring special atmospheres, they can even operate with very rich (50% excess fuel) or lean (1500% excess air). They can be built to allow very high velocities (420,000 scfh/in.^{2} of refractory nozzle opening) for emphasizing convection heat transfer. (See Fig. 57.12.) Others use centrifugal and coanda effects to cause the flame to scrub an adjacent refractory wall contour, thus enhancing wall radiation. (See Fig. 57.13.) By engineering the mixing configuration, nozzle-mix burner designers are able to provide a wide range of mixing rates, from a fast, intense ball of flame ($L/D = 1$) to conventional feather-shaped flame ($L/D = 5$–10) to long flames ($L/D = 20$–50). Changeable flame patterns are also possible.

Delayed-mix burners are a special form of nozzle mix, in which mixing is intentionally slow. (A raw gas torch is an unintentional form of delayed mixing.) Ignition of a fuel with a shortage of air results in polymerization or thermal cracking that forms soot particles only a few microns in diameter. These solids in the flame absorb heat and glow immediately, causing a delayed mix flame to be yellow or orange. The added luminosity enhances flame radiation heat transfer, which is one of the reasons for using delayed-mix flames. The other reason is that delayed mixing permits stretching the heat release over a great distance for uniform heating down the length of a radiant tube or a long kiln or furnace that can only be fired from one end.

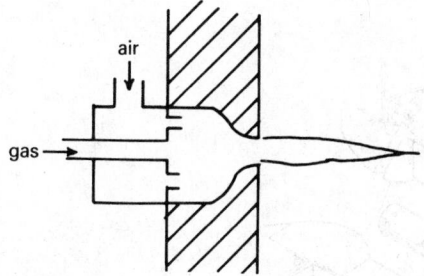

Fig. 57.12 High-velocity burner.

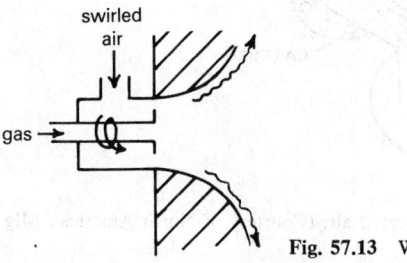

Fig. 57.13 Wall radiating burner (flat flame or coanda type).

Fuel-Directed Burners

Most industrial process burners have traditionally used energy from the air stream to maintain flame stability and flame shape. Now that most everyone has access to higher-pressure fuel supplies, it makes sense to use the energy in the fuel stream for controlling flame stability and shape, thereby permitting use of lower pressure air sources.

Figure 57.14 shows a fuel-directed burner for gas and preheated air. Multiple supply passages and outlet port positions permit changing the flame pattern during operation for optimum heat transfer during the course of a furnace cycle. Oil burners or dual-fuel combination burners can be constructed in similar manner using two-fluid atomizers with compressed air or steam as the atomizing medium.

57.3.2   Burners for Liquid Fuels

Much of what has been said above for gas burners applies as well for oil burning. Liquids do not burn; therefore, they must be vaporized first. Kettle boiling or hot air can be used to produce a hot vapor stream that is directly substitutable for gas in premix burners. Unless there are many burners or they are very small, it is generally more practical (less maintenance) to convert to combination (dual-fuel) burners of the nozzle-mix type.

Vaporization by Atomization

Almost all industrial liquid fuel burners use atomization to aid vaporization by exposing the large surface area (relative to volume) of millions of droplets in the size range of 100–400 μm. Mass transfer then occurs at a rapid rate even if the droplets are not exposed to furnace radiation or hot air.

Pressure atomization (as with a garden hose) uses the pressure energy in the liquid stream to cause the kinetic energy to overcome viscous and surface tension forces. If input is turned down by reducing fuel pressure, however, atomizing quality suffers; therefore, this method of atomization is limited to on–off units or cases where more than 250 psi fuel pressure is available.

Two-fluid atomization is the method most commonly used in industrial burners. Viscous friction by a high-velocity second fluid surrounding the liquid fuel stream literally tears it into droplets. The second fluid may be low-pressure air (<2 psi, or <13.8 kPa), compressed air, gaseous fuel, or steam. Many patented atomizer designs exist—for a variety of spray angles, sizes, turndown ranges, droplet sizes. Emulsion mixing usually gives superior atomization (uniformly small drops with relatively small consumption of atomizing medium) but control is complicated by interaction of the pressures and flows of the two streams. External mixing is just the opposite. A compromise called tip-emulsion atomization is the current state of the art.

Rotary-cup atomization delivers the liquid fuel to the center of a fast spinning cup surrounded by an air stream. Rotational speed and air pressure determine the spray angle. This is still used in some large boilers, but the moving parts near the furnace heat have proved to be too much of a maintenance problem in higher-temperature process furnaces and on smaller installations where a strict preventive maintenance program could not be effected.

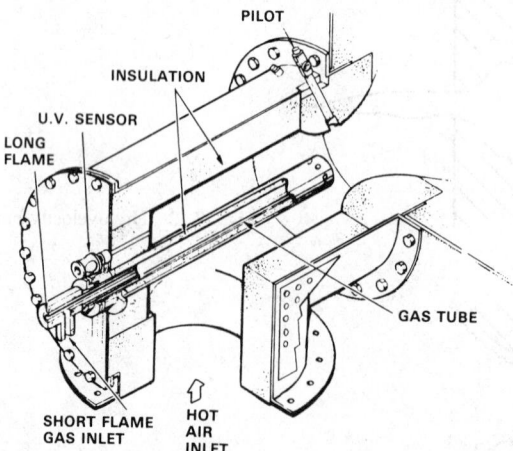

Fig. 57.14   Fuel-directed gas burner for use with preheated air. (Courtesy of North American Mfg. Company.)

Sonic and ultrasonic atomization systems create very fine drops, but impart very little motion to them. For this reason they do not work well with conventional burner configurations, but require an all new design.

Liquid Fuel Conditioning

A variety of additives can be used to reduce fuel degeneration in storage, minimize slagging, lessen surface tension, reduce pollution, and lower the dew point. Regular tank draining and cleaning and the use of filters are recommended.

Residual oils must be heated to reduce their viscosity for pumping, usually to 500 SSU (110 cSt). For effective atomization, burner manufacturers specify viscosities in the range of 100–150 SSU (22–32 cSt). In all but tropic climates, blended oils (Nos. 4 and 5) also require heating. In Arctic situations, distillate oils need heating. Figure 57.15 enables one to predict the oil temperature necessary for a specified viscosity. It is best, however, to install extra heating capacity because delivered oil quality may change.

Oil heaters can be steam or electric. If oil flow stops, the oil may vaporize or char. Either reduces heat transfer from the heater surfaces, which can lead to catastrophic failure in electric heaters. Oil must be circulated through heaters, and the system must be fitted with protective limit controls.

Hot oil lines must be insulated and traced with steam, induction, or resistance heating. The purpose of tracing is to balance heat loss to the environment. Rarely will a tracing system have enough capacity to heat up an oil line from cold. When systems are shut down, arrangements must be made to purge

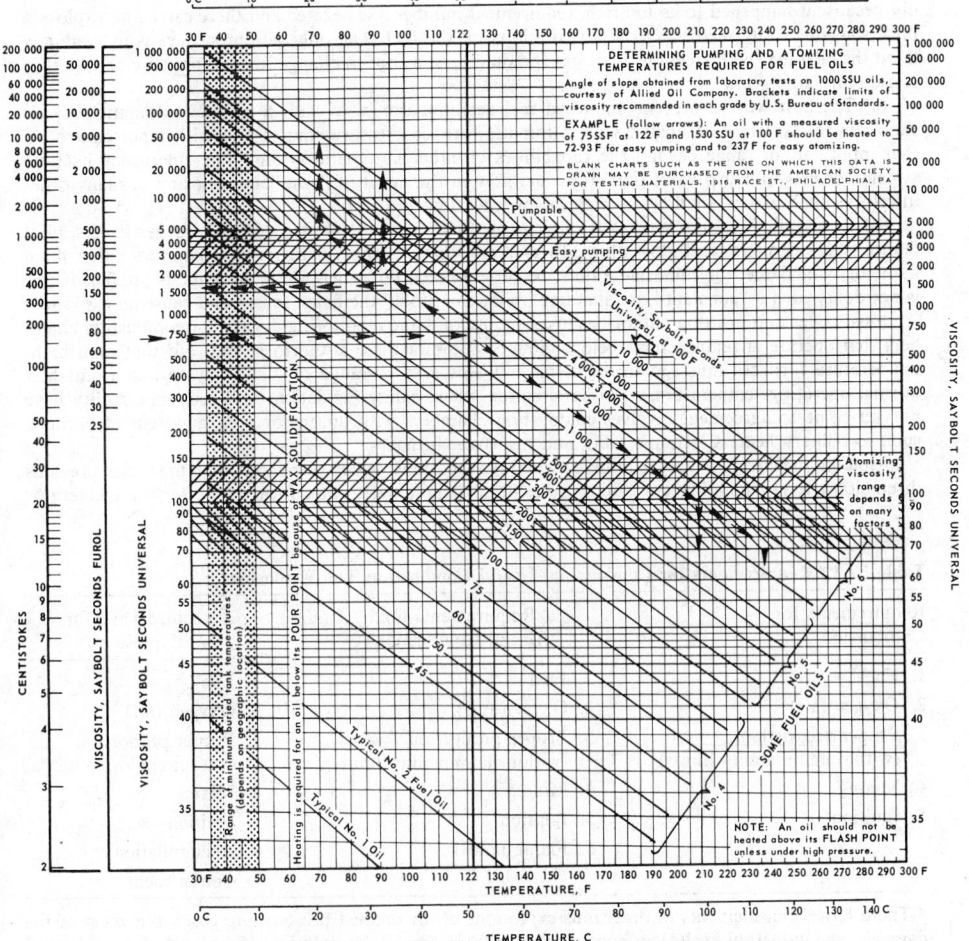

Fig. 57.15 Viscosity–temperature relations for typical fuel oils.

the heavy oil from the lines with steam, air, or (preferably) distillate oil. Oil standby systems should be operated regularly, whether needed or not. In cold climates, they should be started before the onset of cold weather and kept circulating all winter.

57.4 SAFETY CONSIDERATIONS

Operations involving combustion must be concerned about all the usual safety hazards of industrial machinery, plus explosions, fires, burns from hot surfaces, and asphyxiation. Less immediately severe, but long-range health problems related to combustion result from overexposure to noise and pollutants.

Preventing explosions should be *the* primary operating and design concern of *every* person in any way associated with combustion operations, because an explosion can be so devastating as to eliminate *all* other goals of anyone involved. The requirements for an explosion include the first five requirements for combustion (Table 57.3); therefore, striving to do a good job of combustion may set you up for an explosion. The statistical probability of having all seven explosion requirements at the same time and place is so small that people become careless, and therein lies the problem. Continuing training and retraining is the only answer.

The lower and upper limits of flammability are the same as the lower and upper explosive limits for any combustible gas or vapor. Table 59.3 lists these values for gases. Table 57.4 lists similar information for some common liquids. References 3, 4, and 5 list explosion-related data for many industrial solvents and off-gases.

Electronic safety control programs for most industrial combustion systems are generally designed (a) to prevent accumulation of unburned fuel when any source of ignition is present or (b) to immediately remove any source of ignition when something goes wrong, causing fuel accumulation. Of course, this is impossible in a furnace operating above 1400°F. If a burner in such a furnace should snuff out because it happened to go too rich, requirement number 3 is negated and there can be no explosion until someone (untrained) opens a port or shuts off the fuel. The only safe procedure is to gradually flood the chamber with steam or inert gas (gradually, so as not to change furnace pressure and thereby cause more air in-flow).

(a) The best way to prevent unburned fuel accumulation is to have a reliable automatic fuel–air ratio control system coordinated with automatic furnace pressure control and with input control so that input cannot range beyond the capabilities of either automatic system. The emergency back-up system consists of a trip valve that stops fuel flow in the event of flame failure or any of many other interlocks such as low air flow, or high or low fuel flow.

(b) Removal of ignition sources is implemented by automatic shutoff of other burner flames, pilot flames, spark igniters, and glow plugs. In systems where a single flame sensor monitors either main flame or pilot flame, the pilot flame must be programmed out when the main flame is proven. If this is not done, such a "constant" or "standing" pilot can "fool" the flame sensor and cause an explosion.

Most codes and insuring authorities insist on use of flame monitoring devices for combustion chambers that operate at temperatures below 1400°F. Some of these authorities point out that even high-temperature furnaces must go through this low-temperature range on their way to and from their normal operating temperature. Another situation where safety regulations and economic reality have not yet come to agreement involves combustion chambers with dozens or even hundreds of burners, such as refinery heaters, ceramic kilns, and heat-treat furnaces.

Avoiding fuel-fed fires first requires preventing explosions, which often start such fires. (See previous discussion.) Every building containing a fuel-fired boiler, oven, kiln, furnace, heater, or incinerator

Table 57.3 Requirements for Combustion, Useful Combustion, and Explosion[a]

Requirements for Combustion	Requirements for Useful Combustion	Requirements for Explosion
1. Fuel	1. Fuel	1. Fuel
2. Oxygen (air)	2. Oxygen (air)	3. Oxygen (air)
3. Proper proportion (within flammability limits)	3. Proper proportion (within flammability limits)	3. Proper proportion (within explosive limits)
4. Mixing	4. Mixing	4. Mixing
5. Ignition	5. Ignition	5. Ignition
	6. Flame holder	6. Accumulation
		7. Confinement

[a] There have been incidents of disastrous explosions of unconfined fast-burning gases, but most of the damage from industrial explosions comes from the fragments of the containing furnace that are propelled like shrapnel. Lightup explosions are often only "puffs" if large doors are kept open during startup.

Table 57.4 Flammability Data for Liquid Fuels[a]

Liquid Fuel	Flash Point, °F (°C) (Closed Cup Method)	Flammability Limits (%) Volume in Air		Autoignition Temperature, °F (°C)	Vapor density, G(air = 1)	Boiling Temperature, °F (°C)
		Lower	Upper			
Butane, -n	−76 (−60)	1.9	8.5	761 (405)	2.06	31 (−1)
Butane, -iso	−117 (−83)	1.8	8.4	864 (462)	2.06	11 (−12)
Ethyl alcohol (ethanol)	55 (13)	3.5	19	737 (392)	1.59	173 (78)
Ethyl alcohol, 30% in water	85 (29)	3.6	10	—	—	203 (95)
Fuel oil, #1	114–185 (46–85)	0.6	5.6	445–560 (229–293)	—	340–555 (171–291)
Fuel oil (diesel), #1-D	>100 (>38)	1.3	6.0	350–625 (177–329)	—	<590 (<310)
Fuel oil, #2	126–230 (52–110)	—	—	500–705 (260–374)	—	340–640 (171–338)
Fuel oil (diesel), #2-D	>100 (>38)	1.3	6.0	490–545 (254–285)	—	380–650 (193–343)
Fuel oil, #4	154–240 (68–116)	1	5	505 (263)	—	425–760 (218–404)
Fuel oil, #5	130–310 (54–154)	1	5	—	—	—
Fuel oil, #6	150–430 (66–221)	1	5	765 (407)	—	—
Gasoline, automotive	−50± (−46±)	1.3–1.4	6.0–7.6	700 (371)	3–4	91–403 (33–206)
Gasoline, aviation	−50± (−46±)	1	6.0–7.6	800–880 (427–471)	3–4	107–319 (42–159)
Jet fuel, JP-4	−2 (−19)	0.8	6.2	468 (242)	—	140–490 (60–254)
Jet fuel, JP-5	105 (41)	0.6	4.6	400 (204)	—	370–530 (188–277)
Jet fuel, JP-6	127 (53)	—	—	500 (260)	—	250–500 (121–260)
Kerosene	110–130 (43–54)	0.6	5.6	440–560 (227–293)	4.5	350–550 (177–288)
Methyl alcohol (methanol)	54 (12)	5.5	36.5	878 (470)	1.11	147 (64)
Methyl alcohol, 30% in water	75 (24)	—	—	—	—	167 (75)
Naphtha, dryclean	100–110 (38–43)	0.8	5.0	440–500 (227–260)	—	300–400 (149–204)
Naphtha, 76%, vm&p	20–45 (−7–+7)	0.9	6.0	450–500 (232–260)	3.75	200–300 (93–149)
Nonane, -n	88 (31)	0.74	2.9	403 (206)	4.41	303 (151)
Octane, -iso	10 (−12)	1.0	6.0	784 (418)	3.93	190–250 (88–121)
Propane	−156 (−104)	2.2	9.6	871 (466)	1.56	−44 (−42)
Propylene	−162 (−108)	2.0	11.1	927–952 (497–511)	1.49	−54 (−48)

[a] Reproduced with permission from Combustion Handbook.[1]

should have a spring-operated manual reset fuel shutoff valve outside the building with panic buttons at the guardhouse or at exits to allow shutting off fuel as one leaves the burning building. Gas-fuel lines should be overhead where crane and truck operators cannot rupture them, or underground. If underground, keep and use records of their locations to avoid digging accidents. Overhead fuel lines must have well-marked manual shutoff valves where they can be reached without a ladder. Liquid-fuel lines should be underground; otherwise, a rupture will pour or spray fuel on a fire.

The greatest contributor to fuel-fed fires is the fuel shutoff valve that will not work. All such valves, manual or automatic, must be tested on a regular maintenance schedule. Such testing may cause nuisance shutdowns of related equipment; therefore, a practical procedure is to have the maintenance crew do the day's end (or week's end) shutdown about once each month. The same test can check for leaking. If it is a fully automatic or manual-reset automatic valve, shutdown should be accomplished by simulating flame failure and, in succession, each of the interlocks.

Because so much depends on automatic fuel shutoff valves, it makes sense (a) to have a backup valve (blocking valve), and (b) to replace it before it hangs up—at least every 5 years, more often in adverse environments.

Maintenance and management people must be ever alert for open side panels and covers on safety switches and fuel shutoff valves. Remove wedges, wires, or blocks that are holding circuits closed or valves open. Remove jumper wires unless the manufacturer's wiring diagram specifies that they be left in. Eliminate valve by-passes unless they contain a similar valve. Then, get to the root of the problem by finding the cause of the nuisance that caused someone to try to bypass the safety system.

Storage of LP gas, oils, or solid fuels requires careful attention to applicable codes. If the point of use, an open line, or a large leak is below the oil storage elevation, large quantities may siphon out and flow into an area where there is a water heater or other source of ignition. Steam heaters in heavy oil tanks need regular inspections, or leaks can emulsify the oil, causing an overflow. It is advisable to make provision for withdrawing the heater for repair without having to drain the whole tank.

LP gas is heavier than air. Workers have been suffocated by this invisible gas when it leaked into access pits below equipment.

Codes and regulations are proliferating, many by local authorities or insuring groups. Most refer, as a base, to publications of the NFPA, the National Fire Protection Association, Batterymarch Park, Quincy, MA 02269. Their publications usually represent the consensus of technically competent volunteer committees from industries involved with the topic.

REFERENCES

1. R. J. Reed (ed.), *Combustion Handbook,* North American Mfg. Co., Cleveland, OH, 1978, pp. 5, 68, 387.

2. R. H. Essenhigh, "An Introduction to Stirred Reactor Theory Applied to Design of Combustion Chambers," in H. B. Palmer and J. M. Beer (eds.), *Combustion Technology,* Academic, New York, 1974, pp. 389–391.

3. National Fire Protection Association, NFPA No. 86B, *Industrial Furnaces,* National Fire Protection Association, Quincy, MA, 1982.

4. National Fire Protection Association, *Flash Point Index of Trade Name Liquids,* National Fire Protection Association, Quincy, MA, 1978.

5. National Fire Protection Association, NFPA No. 325M, *Fire Hazard Properties of Flammable Liquids, Gases, Volatile Solids,* National Fire Protection Association, Quincy, MA, 1977.

6. Factory Mutual Engineering Corp., *Handbook of Industrial Loss Prevention,* McGraw-Hill, New York, 1967.

CHAPTER 58
FURNACES

CARROLL CONE

Toledo, Ohio

58.1 SCOPE AND INTENT

This chapter has been prepared for the use of engineers with access to an electronic calculator and to standard engineering reference books, but not necessarily to a computer terminal. The intent is to provide information needed for the solution of furnace engineering problems in areas of design, performance analysis, construction and operating cost estimates, and improvement programs.

In selecting charts and formulas for problem solutions, some allowance has been made for probable error, where errors in calculations will be minor compared with errors in the assumptions on which calculations are based. Conscientious engineers are inclined to carry calculations to a far greater degree of accuracy than can be justified by probable errors in data assumed. Approximations have accordingly been allowed to save time and effort without adding to probable margins for error. The symbols and abbreviations used in this chapter are given in Table 58.1.

58.2 STANDARD CONDITIONS

Assuming that the user will be using English rather than metric units, calculations have been based on pounds, feet, Btu's, and degrees Fahrenheit, with conversion to metric units provided in the following text (see Table 58.2).

Assumed standard conditions include: ambient temperature for initial temperature of loads, for heat losses from furnace walls or open cooling of furnace loads—70°F.

Condition of air entering system for combustion or convection cooling: temperature, 70°F; absolute pressure, 14.7 psia; relative humidity, 60% at 70°F, for a water vapor content of about 1.4% by volume.

58.2.1 Probable Errors

Conscientious furnace engineers are inclined to carry calculations to a far greater degree of accuracy than can be justified by uncertainties in basic assumptions such as thermal properties of materials, system temperatures and pressures, radiation view factors and convection coefficients. Calculation procedures recommended in this chapter will, accordingly, include some approximations, identified in the text, that will result in probable errors much smaller than those introduced by basic assumptions, where such approximations will expedite problem solutions.

58.3 FURNACE TYPES

Furnaces may be grouped into two general types:

1. As a source of energy to be used elsewhere, as in firing steam boilers to supply process steam, or steam for electric power generation, or for space heating of buildings or open space.
2. As a source of energy for industrial processes, other than for electric power.

The primary concern of this chapter will be the design, operation, and economics of industrial furnaces, which may be classified in several ways:
By function:

Heating for forming in solid state (rolling, forging)
Melting metals or glass
Heat treatment to improve physical properties
Preheating for high-temperature coating processes, galvanizing, vitreous enameling, other coatings
Smelting for reduction of metallic ores
Firing of ceramic materials
Incineration

By method of load handling:

Batch furnaces for cyclic heating, including forge furnaces arranged to heat one end of a bar or billet inserted through a wall opening, side door, stationary-hearth-type car bottom designs
Continuous furnaces with loads pushed through or carried by a conveyor
Tilting-type furnace

To avoid the problems of door warpage or leakage in large batch-type furnaces, the furnace can be a refractory-lined box with an associated firing system, mounted above a stationary hearth, and

Table 58.1 **Symbols and Abbreviations**

A	area in ft^2
a	absorptivity for radiation, as fraction of black body factor for receiver temperature:
	$\quad a_g \quad$ combustion gases
	$\quad a_w \quad$ furnace walls
	$\quad a_s \quad$ load surface
	$\quad a_m \quad$ combined emissivity–absorptivity factor for source and receiver
C	specific heat in Btu/lb · °F or cal/g · °C
cfm	cubic feet per minute
D	diameter in ft or thermal diffusivity (k/dC)
d	density in lb/ft^3
e	emissivity for radiation as fraction of black-body factor for source temperature, with subscripts as for a above
F	factor in equations as defined in text
fpm	velocity in ft/min
G	mass velocity in lb/ft^2 · hr
g	acceleration by gravity (32.16 ft/sec^2)
H	heat-transfer coefficient (Btu/hr · ft^2 · °F)
	$\quad H_r \quad$ for radiation
	$\quad H_c \quad$ for convection
	$\quad H_t \quad$ for combined $H_r + H_c$
HHV	higher heating value of fuel
h	pressure head in units as defined
k	thermal conductivity (Btu/hr · ft · °F)
L	length in ft, as in effective beam length for radiation, decimal rather than feet and inches
LHV	lower heating value of fuel
ln	logarithm to base e
MTD	log mean temperature difference
N	a constant as defined in text
psi	pressure in lb/in^2
	psig, pressure above atmospheric
	psia, absolute pressure
Pr	Prandtl number $(\mu C/k)$
Q	heat flux in Btu/hr
R	thermal resistance (r/k) or ratio of external to internal thermal resistance (k/rH)
Re	Reynolds number (DG/μ)
r	radius or depth of heat penetration in ft
T	temperature in °F, except for radiation calculations where °S = (°F + 460)/100
	$\quad T_g$, combustion gas temperature
	$\quad T_w$, furnace wall temperature
	$\quad T_s$, heated load surface
	$\quad T_c$, core or unheated surface of load
t	time in hr
μ	viscosity in lb/hr · ft
wc	inches of water column as a measure of pressure
V	volume in ft^3
v	velocity in ft/sec
W	weight in lb
X	time factor for nonsteady heat transfer (tD/r^2)
x	horizontal coordinate
y	vertical coordinate
z	coordinate perpendicular to plane xy

Table 58.2 Conversion of Metric to English Units

Length	1 m = 3.281 ft
	1 cm = 0.394 in
Area	1 m² = 10.765 ft²
Volume	1 m³ = 35.32 ft³
Weight	1 kg = 2.205 lb
Density	1 g/cm³ = 62.43 lb/ft²
Pressure	1 g/cm² = 2.048 lb/ft² = 0.0142 psi
Heat	1 kcal = 3.968 Btu
	1 kwh = 3413 Btu
Heat content	1 cal/g = 1.8 Btu/lb
	1 kcal/m² = 0.1123 Btu/ft³
Heat flux	1 W/cm² = 3170 Btu/hr · ft²
Thermal conductivity	$\dfrac{1\ cal}{sec\ cm\ °C} = \dfrac{242\ Btu}{hr\ ft\ °F}$
Heat transfer	$\dfrac{1\ cal}{sec\ cm^2\ °C} = \dfrac{7373\ Btu}{hr\ ft^2\ °F}$
Thermal diffusivity	$\dfrac{1\ cal/sec\ ·\ cm\ ·\ °C}{C\ ·\ g/cm^3} = \dfrac{3.874\ Btu/hr\ ·\ ft\ ·\ °F}{C\ ·\ lb/ft^3}$

arranged to be tilted around one edge of the hearth for loading and unloading by manual handling, forklift trucks, or overhead crane manipulators.

For handling heavy loads by overhead crane, without door problems, the furnace can be a portable cover unit with integral firing and temperature control. Consider a cover-type furnace for annealing steel strip coils in a controlled atmosphere. The load is a stack of coils with a common vertical axis, surrounded by a protective inner cover and an external heating cover. To improve heat transfer parallel to coil laminations, they are loaded with open coil separators between them, with heat transferred from the inner cover to coil ends by a recirculating fan. To start the cooling cycle, the heating cover is removed by an overhead crane, while atmosphere circulation by the base fan continues. Cooling may be enhanced by air-blast cooling of the inner cover surface.

For heating heavy loads of other types, such as weldments, castings, or forgings, car bottom furnaces may be used with some associated door maintenance problems. The furnace hearth is a movable car, to allow load handling by an overhead traveling crane. In one type of furnace, the door is suspended from a lifting mechanism. To avoid interference with an overhead crane, and to achieve some economy in construction, the door may be mounted on one end of the car and opened as the car is withdrawn. This arrangement may impose some handicaps in access for loading and unloading.

Loads such as steel ingots can be heated in pit-type furnaces, preferably with units of load separated to allow radiating heating from all sides except the bottom. Such a furnace would have a cover displaced by a mechanical carriage and would have a compound metal and refractory recuperator arrangement. Loads are handled by overhead crane equipped with suitable gripping tongs.

Continuous-Type Furnaces

The simplest type of continuous furnace is the hearth-type pusher furnace. Pieces of rectangular cross section are loaded side by side on a charge table and pushed through the furnace by an external mechanism. In the design shown, the furnace is fired from one end, counterflow to load travel, and is discharged through a side door by an auxiliary pusher lined up by the operator.

Furnace length is limited by thickness of the load and alignment of abutting edges, to avoid buckling up from the hearth.

A more complex design would provide multiple zone firing above and below the hearth, with recuperative air preheating.

Long loads can be conveyed in the direction of their length in a roller-hearth-type furnace. Loads can be bars, tubes, or plates of limited width, heated by direct firing, by radiant tubes, or by electric-resistor-controlled atmosphere, and conveyed at uniform speed or at alternating high and low speeds for quenching in line.

Sequential heat treatment can be accomplished with a series of chain or belt conveyors. Small parts can be loaded through an atmosphere seal, heated in a controlled atmosphere on a chain belt conveyor, discharged into an oil quench, and conveyed through a washer and tempering furnace by a series of mesh belts without intermediate handling.

Except for pusher-type furnaces, continuous furnaces can be self-emptying. To secure the same advantage in heating slabs or billets for rolling and to avoid scale loss during interrupted operation, loads can be conveyed by a walking-beam mechanism. Such a walking-beam-type slab heating furnace would have loads supported on water-cooled rails for over- and underfiring, and would have an overhead recuperator.

Thin strip materials, joined in continuous strand form, can be conveyed horizontally or the strands can be conveyed in a series of vertical passes by driven support rolls. Furnaces of this type can be incorporated in continuous galvanizing lines.

Unit loads can be individually suspended from an overhead conveyor, through a slot in the furnace roof, and can be quenched in line by lowering a section of the conveyor.

Small parts or bulk materials can be conveyed by a moving hearth, as in the rotary-hearth-type or tunnel kiln furnace. For roasting or incineration of bulk materials, the shaft-type furnace provides a simple and efficient system. Loads are charged through the open top of the shaft and descend by gravity to a discharge feeder at the bottom. Combustion air can be introduced at the bottom of the furnace and preheated by contact with the descending load before entering the combustion zone, where fuel is introduced through sidewalls. Combustion gases are then cooled by contact with the descending load, above the combustion zone, to preheat the charge and reduce flue gas temperature.

With loads that tend to agglomerate under heat and pressure, as in some ore-roasting operations, the rotary kiln may be preferable to the shaft-type furnace. The load is advanced by rolling inside an inclined cylinder. Rotary kilns are in general use for sintering ceramic materials.

Classification by Source of Heat

The classification of furnaces by source of heat is as follows:

Direct-firing with gas or oil fuels.

Combustion of material in process, as by incineration with or without supplemental fuel.

Internal heating by electrical resistance or induction in conductors, or dielectric heating of nonconductors.

Radiation from electric resistors or radiant tubes, in controlled atmospheres or under vacuum.

58.4 FURNACE CONSTRUCTION

The modern industrial furnace design has evolved from a rectangular or cylindrical enclosure, built up of refractory shapes and held together by a structural steel binding. Combustion air was drawn in through wall openings by furnace draft, and fuel was introduced through the same openings without control of fuel/air ratios except by the judgment of the furnace operator. Flue gases were exhausted through an adjacent stack to provide the required furnace draft.

To reduce air infiltration or outward leakage of combustion gases, steel plate casings have been added. Fuel economy has been improved by burner designs providing some control of fuel/air ratios, and automatic controls have been added for furnace temperature and furnace pressure. Completely sealed furnace enclosures may be required for controlled atmosphere operation, or where outward leakage of carbon monoxide could be an operating hazard.

With the steadily increasing costs of heat energy, wall structures are being improved to reduce heat losses and heat demands for cyclic heating. The selection of furnace designs and materials should be aimed at a minimum overall cost of construction, maintenance, and fuel or power over a projected service life. Heat losses in existing furnaces can be reduced by adding external insulation or rebuilding walls with materials of lower thermal conductivity. To reduce losses from intermittent operation, the existing wall structure can be lined with a material of low heat storage and low conductivity, to substantially reduce mean wall temperatures for steady operation and cooling rates after interrupted firing.

Thermal expansion of furnace structures must be considered in design. Furnace walls have been traditionally built up of prefired refractory shapes with bonded mortar joints. Except for small furnaces, expansion joints will be required to accommodate thermal expansion. In sprung arches, lateral expansion can be accommodated by vertical displacement, with longitudinal expansion taken care of by lateral slots at intervals in the length of the furnace. Where expansion slots in furnace floors could be filled by scale, slag, or other debris, they can be packed with a ceramic fiber that will remain resilient after repeated heating.

Differential expansion of hotter and colder wall surfaces can cause an inward-bulging effect. For stability in self-supporting walls, thickness must not be less than a critical fraction of height.

Because of these and economic factors, cast or rammed refractories are replacing prefired shapes for lining many types of large, high-temperature furnaces. Walls can be retained by spaced refractory shapes anchored to the furnace casing, permitting reduced thickness as compared to brick construction. Furnace roofs can be suspended by hanger tile at closer spacing, allowing unlimited widths.

Cast or rammed refractories, fired in place, will develop discontinuities during initial shrinkage that can provide for expansion from subsequent heating, to eliminate the need for expansion joints.

As an alternate to cast or rammed construction, insulating refractory linings can be gunned in place by jets of compressed air and retained by spaced metal anchors, a construction increasingly popular for stacks and flues.

Thermal expansion of steel furnace casings and bindings must also be considered. Where the furnace casing is constructed in sections, with overlapping expansion joints, individual sections can be separately anchored to building floors or foundations. For gas-tight casings, as required for controlled atmosphere heating, the steel structure can be anchored at one point and left free to expand elsewhere. In a continuous galvanizing line, for example, the atmosphere furnace and cooling zone can be anchored to the foundation near the casting pot, and allowed to expand toward the charge end.

58.5   FUELS AND COMBUSTION

Heat is supplied to industrial furnaces by combustion of fuels or by electrical power. Fuels now used are principally fuel oil and fuel gas. Because possible savings through improved design and operation are much greater for these fuels than for electric heating or solid fuel firing, they will be given primary consideration in this section.

Heat supply and demand may be expressed in units of *Btu* or *kcal* or as gallons or barrels of fuel oil, tons of coal or *kwh* of electric power. For the large quantities considered for national or world energy loads, a preferred unit is the "quad," one quadrillion or 10^{15} Btu. Conversion factors are

$$
\begin{aligned}
1 \text{ quad} &= 10^{15} \text{ Btu} \\
&= 172 \times 10^6 \text{ barrels of fuel oil} \\
&= 44.34 \times 10^6 \text{ tons of coal} \\
&= 10^{12} \text{ cubic feet of natural gas} \\
&= 2.93 \times 10^{11} \text{ kwh electric power}
\end{aligned}
$$

At 30% generating efficiency, the fuel required to produce 1 quad of electrical energy is 3.33 quads. One quad fuel is accordingly equivalent to 0.879×10^{11} kwh net power.

Fuel demand, in the United States during recent years, has been about 75 quads per year from the following sources:

Coal	15 quads
Fuel oil	
Domestic	18 quads
Imported	16 quads
Natural gas	23 quads
Other, including nuclear	3 quads

Hydroelectric power contributes about 1 quad net additional. Combustion of waste products has not been included, but will be an increasing fraction of the total in the future.

Distribution of fuel demand by use is estimated at:

Power generation	20 quads
Space heating	11 quads
Transportation	16 quads
Industrial, other than power	25 quads
Other	4 quads

Net demand for industrial furnace heating has been about 6%, or 4.56 quads, primarily from gas and oil fuels.

The rate at which we are consuming our fossil fuel assets may be calculated as (annual demand)/ (estimated reserves). This rate is presently highest for natural gas, because, besides being available at wellhead for immediate use, it can be transported readily by pipeline and burned with the simplest type of combustion system and without air pollution problems. It has also been delivered at bargain prices, under federal rate controls.

As reserves of natural gas and fuel oil decrease, with a corresponding increase in market prices, there will be an increasing demand for alternative fuels such as synthetic fuel gas and fuel oil, waste materials, lignite, and coal.

Synthetic fuel gas and fuel oil are now available from operating pilot plants, but at costs not yet competitive.

As an industrial fuel, coal is primarily used for electric power generation. In the form of metallurgical coke, it is the source of heat and the reductant in the blast furnace process for iron ore reduction, and as fuel for cupola furnaces used to melt foundry iron. Powdered coal is also being used as fuel and reductant in some new processes for solid-state reduction of iron ore pellets to make synthetic scrap for steel production.

Since the estimated life of coal reserves, particularly in North America, is so much greater than for other fossil fuels, processes for conversion of coal to fuel gas and fuel oil have been developed almost to the commercial cost level, and will be available whenever they become economical. Processes for coal gasification, now being tried in pilot plants, include:

1. *Producer Gas.* Bituminous coal has been commercially converted to fuel gas of low heating value, around 110 Btu/scf LHV, by reacting with insufficient air for combustion and steam as a source of hydrogen. Old producers delivered a gas containing sulfur, tar volatiles, and suspended ash, and have been replaced by cheap natural gas. By reacting coal with a mixture of oxygen and steam, and removing excess carbon dioxide, sulfur gases, and tar, a clean fuel gas of about 300 Btu/scf LHV can be supplied. Burned with air preheated to 1000°F and with a flue gas temperature of 2000°F, the available heat is about 0.69 HHV, about the same as for natural gas.

2. *Synthetic Natural Gas.* As a supplement to dwindling natural gas supplies, a synthetic fuel gas of similar burning characteristics can be manufactured by adding a fraction of hydrogen to the product of the steam–oxygen gas producer and reacting with carbon monoxide at high temperature and pressure to produce methane. Several processes are operating successfully on a pilot plant scale, but with a product costing much more than market prices for natural gas. The process may yet be practical for extending available natural gas supplies by a fraction, to maintain present market demands. For gas mixtures or synthetic gas supplies to be interchangeable with present gas fuels, without readjustment of fuel/air ratio controls, they must fit the Wobbe Index:

$$\frac{\text{HHV Btu/scf}}{(\text{specific gravity})^{0.5}}$$

The fuel gas industry was originally developed to supply fuel gas for municipal and commercial lighting systems. Steam was passed through incandescent coal or coke, and fuel oil vapors were added to provide a luminous flame. The product had a heating value of around 500 HHV, and a high carbon monoxide content, and was replaced as natural gas or coke oven gas became available. Coke oven gas is a by-product of the manufacture of metallurgical coke that can be treated to remove sulfur compounds and volatile tar compounds to provide a fuel suitable for pipeline distribution. Blast furnace gas can be used as an industrial or steam-generating fuel, usually after enrichment with coke oven gas. Gas will be made from replaceable sources such as agricultural and municipal wastes, cereal grains, and wood, as market economics for such products improve.

Heating values for fuels containing hydrogen can be calculated in two ways:

1. Higher heating value (HHV) is the total heat developed by burning with standard air in a ratio to supply 110% of net combustion air, cooling products to ambient temperature, and condensing all water vapor from the combustion of hydrogen.

2. Lower heating value (LHV) is equal to HHV less heat from the condensation of water vapor. It provides a more realistic comparison between different fuels, since flue gases leave most industrial processes well above condensation temperatures.

HHV factors are in more general use in the United States, while LHV values are more popular in most foreign countries.

For example, the HHV value for hydrogen as fuel is 319.4 Btu/scf, compared to a LHV of 270.2.

The combustion characteristics for common fuels are tabulated in Table 58.3, for combustion with 110% standard air. Weights in pounds per 10^6 Btu HHV are shown, rather than corresponding volumes, to expedite calculations based on mass flow. Corrections for flue gas and air temperatures other than ambient are given in charts to follow.

The heat released in a combustion reaction is:

total heats of formation of combustion products − total heats of formation of reactants

Heats of formation can be conveniently expressed in terms of Btu per pound mol, with the pound mol for any substance equal to a weight in pounds equal to its molecular weight. The heat of formation for elemental materials is zero. For compounds involved in common combustion reactions, values are shown in Table 58.4.

Data in Table 58.4 can be used to calculate the higher and lower heating values of fuels. For methane:

$$CH_4 + 2O_2 = CO_2 + 2H_2O$$

Table 58.3 Combustion Characteristics of Common Fuels

Fuel	Btu/scf	Fuel	Air	Flue Gas
		Weight in lb/10^6 Btu		
Natural gas (SW U.S.)	1073	42	795	837
Coke oven gas	539	57	740	707
Blast furnace gas	92	821	625	1446
Mixed blast furnace and coke oven gas:				
Ratio CO/BF 1/1	316	439	683	1122
1/3	204	630	654	1284
1/10	133	752	635	1387
Hydrogen	319	16	626	642
	Btu/lb			
No. 2 fuel oil	19,500	51	810	861
No. 6 fuel oil	18,300	55	814	
With air atomization				869
With steam atomization at 3 lb/gal				889
Carbon	14,107	71	910	981

HHV

$$169,290 + (2 \times 122,976) - 32,200 = 383,042 \text{ Btu/lb} \cdot \text{mol}$$
$$383,042/385 = 995 \text{ Btu/scf}$$

LHV

$$169,290 + (2 \times 104,040) - 32,200 = 345,170 \text{ Btu/lb} \cdot \text{mol}$$
$$345,170/385 = 897 \text{ Btu/scf}$$

Available heats from combustion of fuels, as a function of flue gas and preheated air temperatures, can be calculated as a fraction of the HHV. The net ratio is one plus the fraction added by preheated air less the fraction lost as sensible heat and latent heat of water vapor, from combustion of hydrogen, in flue gas leaving the system.

Available heats can be shown in chart form, as in the following figures for common fuels. On each chart, the curve on the right is the fraction of HHV available for combustion with 110% cold air, while the curve on the left is the fraction added by preheated air, as functions of air or flue gas temperatures. For example, the available heat fraction for methane burned with 110% air preheated to 1000°F, and with flue gas out at 2000°F, is shown in Fig. 58.1: 0.41 + 0.18 − 0.59 HHV.

Values for other fuels are shown in charts that follow:

Fig. 58.2, fuel oils with air or steam atomization

Fig. 58.3, by-product coke oven gas

Table 58.4 Heats of Formation

Material	Formula	Molecular Weight	Heats of Formation (Btu/lb · mol[a])
Methane	CH_4	16	32,200
Ethane	C_2H_6	30	36,425
Propane	C_3H_8	44	44,676
Butane	C_4H_{10}	58	53,662
Carbon monoxide	CO	28	47,556
Carbon dioxide	CO_2	44	169,290
Water vapor	H_2O	18	104,040
Liquid water			122,976

[a] The volume of 1 lb mol, for any gas, is 385 scf.

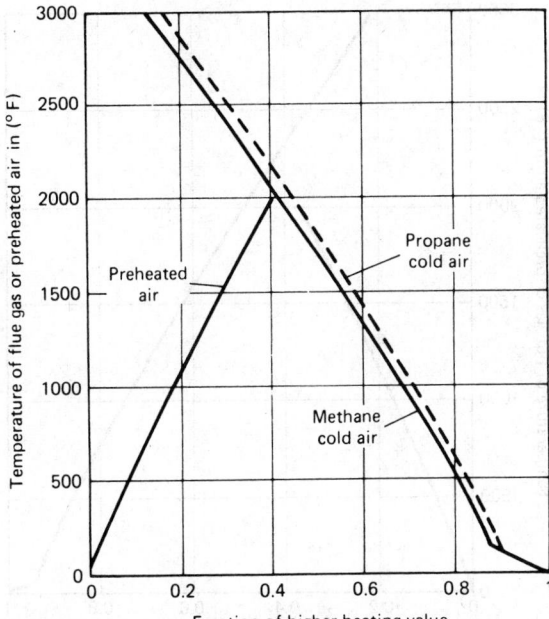

Fig. 58.1 Available heat for methane and propane combustion. Approximate high and low limits for commercial natural gas.[1]

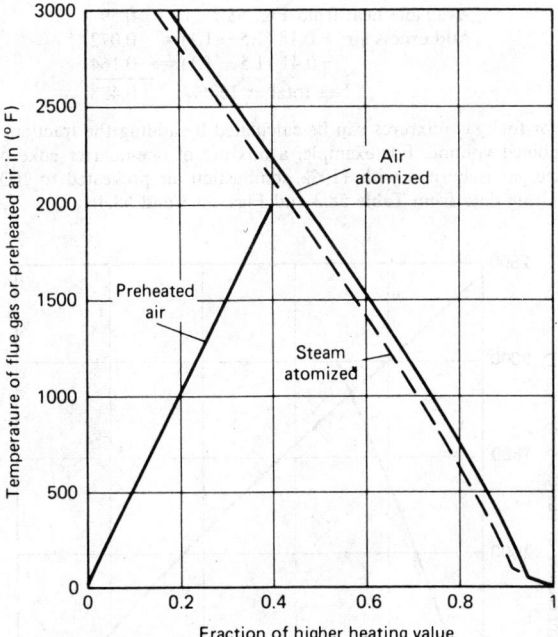

Fig. 58.2 Available heat ratios for fuel oils with air or steam atomization.[1]

Fig. 58.4, blast furnace gas

Fig. 58.5, methane

For combustion with other than 110% of net air demand, the corrected available heat can be calculated as follows. For methane with preheated air at 1000°F and flue gas out at 2000°F and 150% net air supply:

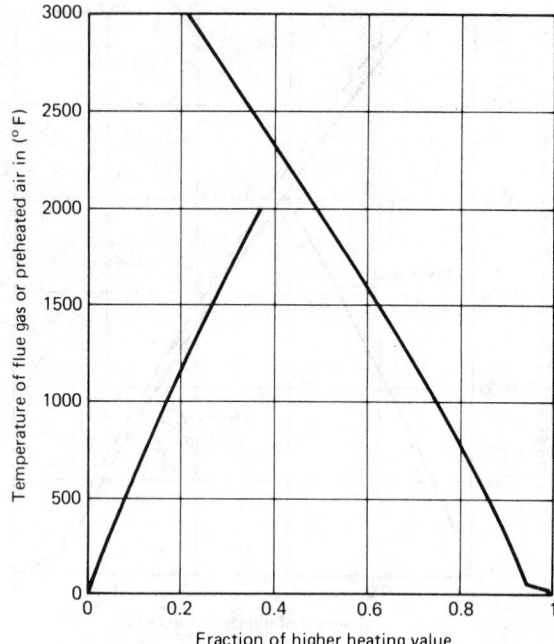

Fig. 58.3 Available heat ratios for by-product coke oven gas.[1]

Available heat from Fig. 58.1	0.59
Add excess air $+0.18\ (1.5-1.1)=$	0.072
$-0.41\ (1.5-1.1)=$	-0.164
Net total at 150%	0.498

Available heats for fuel gas mixtures can be calculated by adding the fractions for either fuel and dividing by the combined volume. For example, a mixture of one-quarter coke oven gas and three-quarters blast furnace gas is burned with 110% combustion air preheated to 1000°F, and with flue gas out at 2000°F. Using data from Table 58.3 and Figs. 58.3 and 58.4:

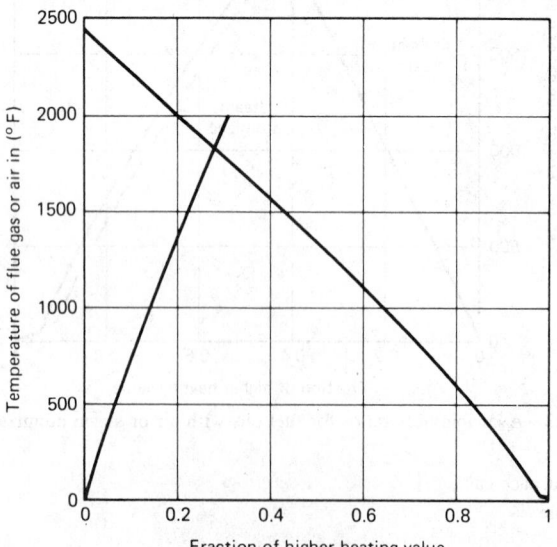

Fig. 58.4 Available heat ratios for blast furnace gas.[1]

$$CO\ (539 \times 0.25 = 134.75)\ (0.49 + 0.17)\ =\ 88.93$$
$$BF\ (92 \times 0.75 = \underline{69.00})\ (0.21 + 0.144) = \underline{24.43}$$
$$HHV\ \overline{203.75} \qquad Available = \overline{113.36}$$
$$Net:\ 113.36/203.75 = 0.556\ combined\ HHV$$

58.6 OXYGEN ENRICHMENT OF COMBUSTION AIR

The available heats of furnace fuels can be improved by adding oxygen to combustion air. Some studies have been based on a total oxygen content of 35%, which can be obtained by adding 21.5 scf pure oxygen or 25.45 scf of 90% oxygen per 100 scf of dry air. The available heat ratios are shown in the chart in Fig. 58.5.

At present market prices, the power needed to concentrate pure oxygen for enrichment to 35% will cost more than the fuel saved, even with metallurgical oxygen from an in-plant source. As plants are developed for economical concentration of oxygen to around 90%, the cost balance may become favorable for very-high-temperature furnaces.

In addition to fuel savings by improvement of available heat ratios, there will be additional savings in recuperative furnaces by increasing preheated air temperature at the same net heat demand, depending on the ratio of heat transfer by convection to that by gas radiation in the furnace and recuperator.

58.7 THERMAL PROPERTIES OF MATERIALS

The heat content of some materials heated in furnaces or used in furnace construction is shown in the chart in Fig. 58.6, in units of Btu/lb. Vertical lines in curves represent latent heats of melting or other phase transformations. The latent heat of evaporation for water in flue gas has been omitted from the chart. The specific heat of liquid water is, of course, about 1.

Thermal conductivities in English units are given in reference publications as: (Btu/(ft² · hr))/ (°F/in.) or as (Btu/(ft² · hr))/(°F/ft). To keep dimensions consistent, the latter term, abbreviated to $k = $ Btu/ft · hr · °F will be used here. Values will be $\frac{1}{12}$th of those in terms of °F/in.

Thermal conductivities vary with temperature, usually inversely for iron, steel, and some alloys, and conversely for common refractories. At usual temperatures of use, average values of k in Btu/ (ft · hr · °F) are in Table 58.5.

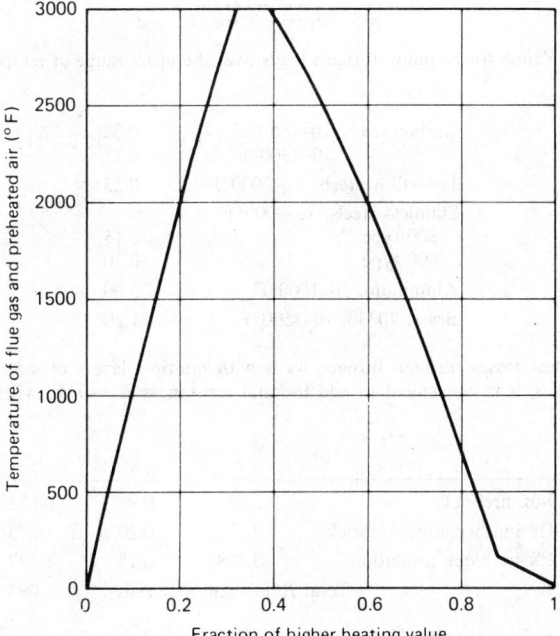

Fig. 58.5 Available heat ratios for combustion of methane with 110% air containing 35% O_2.[1]

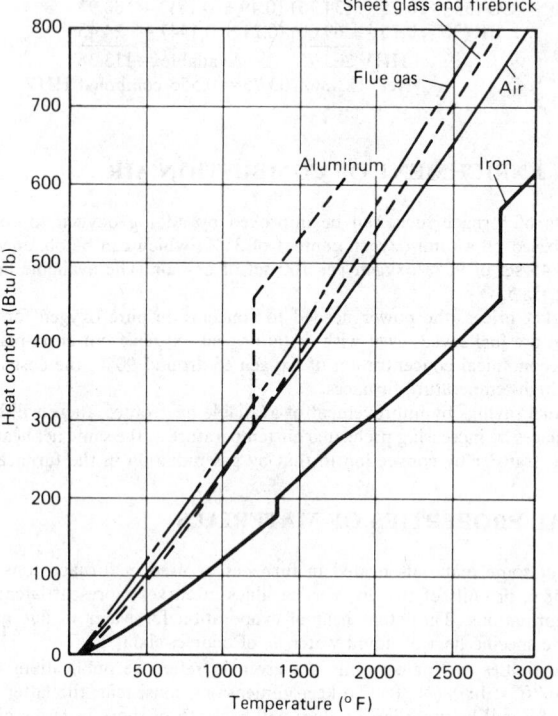

Fig. 58.6 Heat content of materials at temperature.[1]

To expedite calculations for nonsteady conduction of heat, it is convenient to use the factor for "thermal diffusivity," defined as

$$D = \frac{k}{dC} = \frac{\text{thermal conductivity}}{\text{density} \times \text{specific heat}}$$

in consistent units. Values for common furnace loads over the usual range of temperatures for heating are:

Carbon steels, 70–1650°F	0.32
70–2300°F	0.25
Low-alloy steels, 70–2000°F	0.23
Stainless steels, 70–2000°F	
300 type	0.15
400 type	0.20
Aluminum, 70–1000°F	3.00
Brass, 70/30, 70–1500°F	1.20

In calculating heat losses through furnace walls with multiple layers of materials with different thermal conductivities, it is convenient to add thermal resistances $R = r/k$, where r is thickness in ft. For example,

	r	k	r/k
9-in. firebrick	0.75	0.9	0.833
4½-in. insulating firebrick	0.375	0.20	1.875
2¼-in. block insulation	0.208	0.15	1.387
Total R for wall materials			4.095

Overall thermal resistance will include the factor for combined radiation and convection from the outside of the furnace wall to ambient temperature. Wall losses as a function of wall surface temperature,

Table 58.5 Average Values of k (Btu/ft · hr · ° F)

	Mean Temperature (°F)				
	100	1000	1500	2000	2500
Steel, SAE 1010	33	23	17	17	
Type HH HRA	8	11	14	16	
Aluminum	127	133			
Copper	220	207	200		
Brass, 70/30	61	70			
Firebrick	0.81	0.82	0.85	0.89	0.93
Silicon carbide	11	10	9	8	6
Insulating firebrick	0.12	0.17	0.20	0.24	

for vertical surfaces in still air, are shown in Fig. 58.7, and are included in the overall heat loss data for furnace walls shown in the chart in Fig. 58.8.

The chart in Fig. 58.9 shows the thermodynamic properties of air and flue gas, over the usual range of temperatures, for use in heat-transfer and fluid flow problems. Data for other gases, in formula form, are available in standard references.

Linear coefficients of thermal expansion are the fractional changes in length per °F change in temperature. Coefficients in terms of $10^6 \times$ net values are listed below for materials used in furnace construction and for the usual range of temperatures:

Carbon steel	9
Cast HRA	10.5
Aluminum	15.6
Brass	11.5
Firebrick, silicon carbide	3.4
Silica brick	4.7

Coefficients for cubical expansion of solids are about $3 \times$ linear coefficients. The cubical coefficient for liquid water is about 185×10^{-6}.

58.8 HEAT TRANSFER

Heat may be transmitted in industrial furnaces by radiation—gas radiation from combustion gases to furnace walls or direct to load, and solid-state radiation from walls, radiant tubes, or electric heating

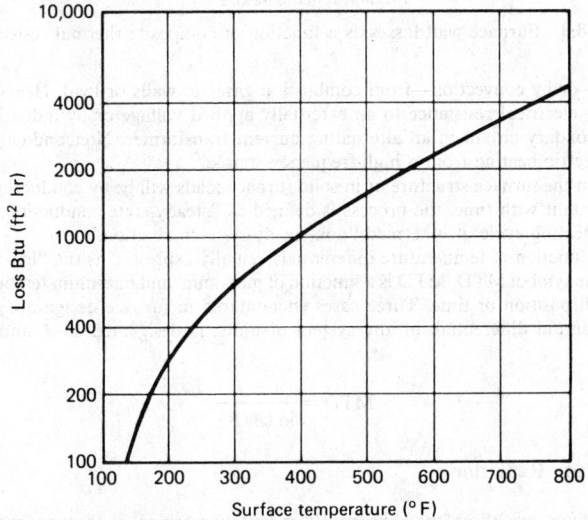

Fig. 58.7 Furnace wall losses as a function of surface temperature.[1]

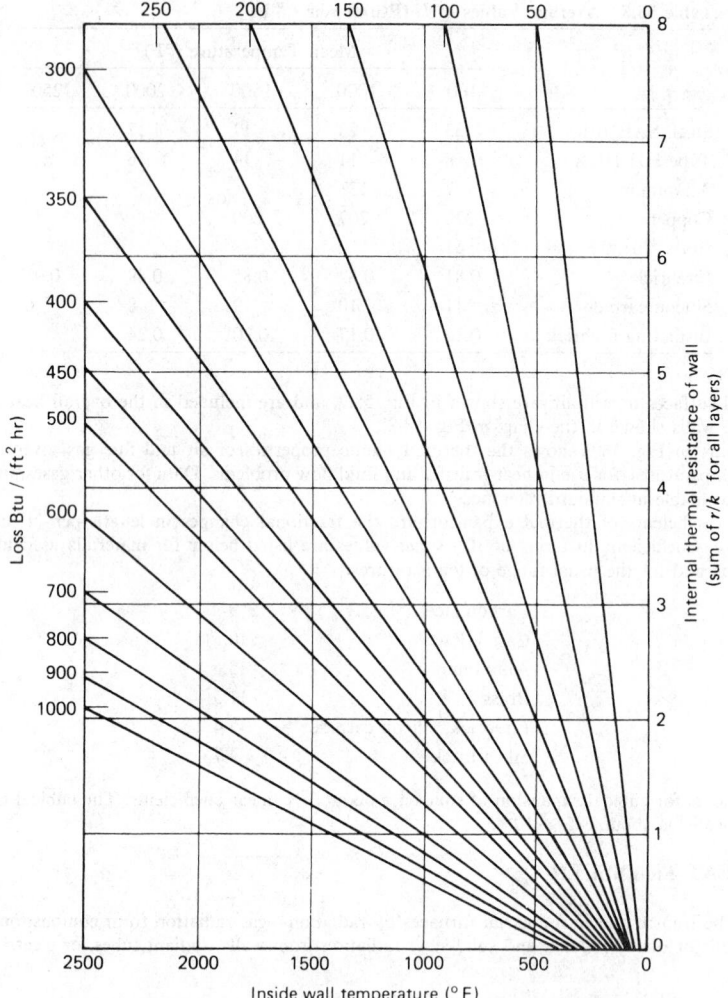

Fig. 58.8 Furnace wall losses as a function of composite thermal resistance.[1]

elements to load—or by convection—from combustion gases to walls or load. Heat may be generated inside the load by electrical resistance to an externally applied voltage or by induction, with the load serving as the secondary circuit in an alternating current transformer. Nonconducting materials may be heated by dielectric heating from a high-frequency source.

Heat transfer in the furnace structure or in solid furnace loads will be by conduction. If the temperature profile is constant with time, the process is defined as "steady-state conduction." If temperatures change during a heating cycle, it is termed "non-steady-state conduction."

Heat flow is a function of temperature differentials, usually expressed as the "log mean temperature difference" with the symbol MTD. MTD is a function of maximum and minimum temperature differences that can vary with position or time. Three cases encountered in furnace design are illustrated in Fig. 58.10. If the maximum differential, in any system of units, is designated as A and the minimum is designated by B:

$$\text{MTD} = \frac{A - B}{\ln (A/B)}$$

58.8.1 Solid-State Radiation

"Black-body" surfaces are those that absorb all radiation received, with zero reflection, and exist only as limits approached by actual sources or receivers of solid radiation. Radiation between black bodies is expressed by the Stefan–Boltzmann equation:

$$Q/A = N(T^4 - T_0^4) \quad \text{Btu/hr} \cdot \text{ft}^2$$

where N is the Stefan–Boltzmann constant, now set at about 0.1713×10^{-8} for T and T_0, source and receiver temperatures, in °R. Because the fourth powers of numbers representing temperatures in °R are large and unwieldy, it is more convenient to express temperatures in °S, equivalent to (°F + 460)/100. The constant N is then reduced to 0.1713.

With source and receiver temperatures identified as T_s and T_r in °S, and with allowance for emissivity and view factors, the complete equation becomes:

$$Q/A = 0.1713 \times em \times Fr(T_s^4 - T_r^4) \quad \text{Btu/hr} \cdot \text{ft}^2$$

at the receiving surface, where

em = combined emissivity and absorptivity
      factors for source and receiving surfaces
Fr = net radiation view factor for receiving
      surface
T_s and T_r = source and receiving temperature in °S

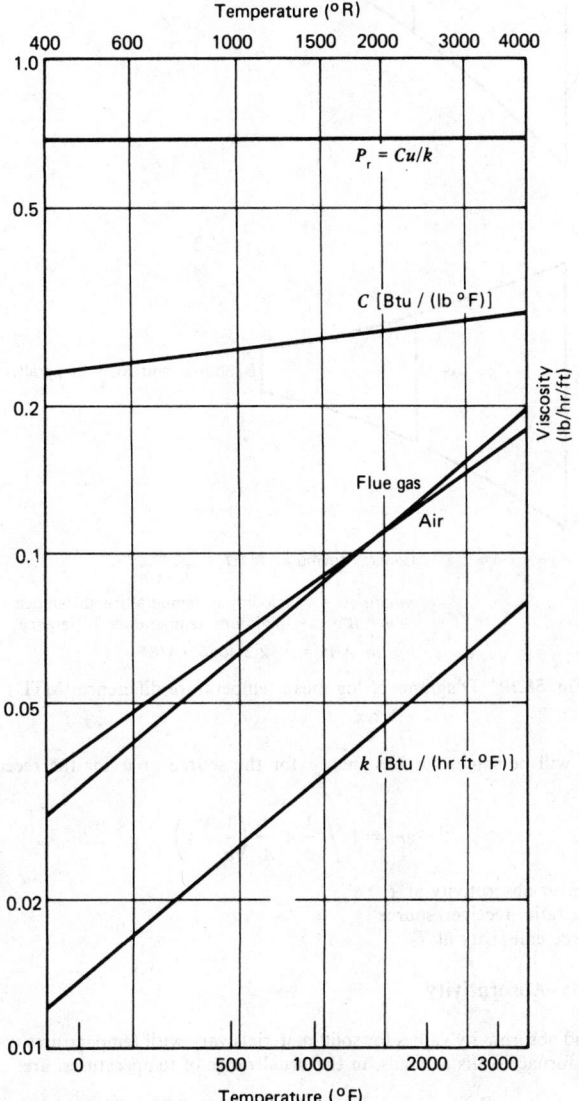

Fig. 58.9 Thermodynamic properties of air and flue gas.[1]

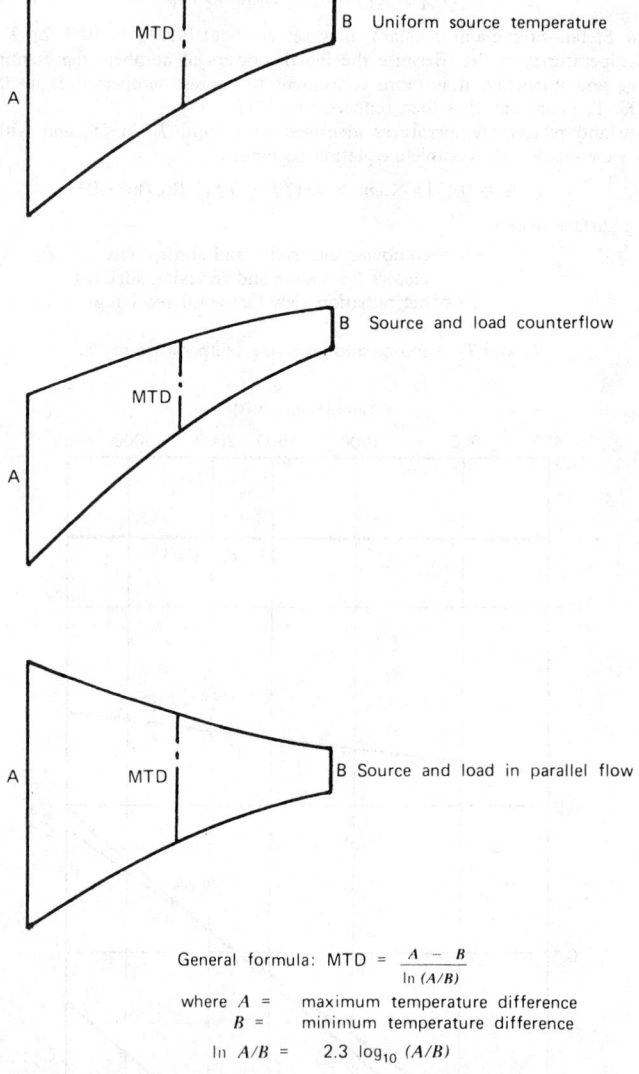

$$\text{General formula: MTD} = \frac{A - B}{\ln (A/B)}$$

where A =   maximum temperature difference
B =   minimum temperature difference

$\ln A/B$ = $2.3 \log_{10} (A/B)$

Fig. 58.10   Diagrams of log mean temperature difference (MTD).[1]

The factor em will be somewhat less than e for the source or a for the receiving surface, and can be calculated:

$$em = 1 \left/ \frac{1}{a} + \frac{A_r}{A_s} \left(\frac{1}{e} - 1 \right) \right.$$

where a = receiver absorptivity at T_r
A_r/A_s = area ratio, receiver/source
e = source emissivity at T_s

58.8.2   Emissivity–Absorptivity

While emissivity and absorptivity values for solid materials vary with temperatures, values for materials commonly used as furnace walls or loads, in the usual range of temperatures, are:

Refractory walls	0.80–0.90
Heavily oxidized steel	0.85–0.95

Bright steel strip	0.25–0.35
Brass cake	0.55–0.60
Bright aluminum strip	0.05–0.10
Hot-rolled aluminum plate	0.10–0.20
Cast heat-resisting alloy	0.75–0.85

For materials such as sheet glass, transparent in the visible light range, radiation is reflected at both surfaces at about 4% of incident value, with the balance absorbed or transmitted. Absorptivity decreases with temperature, as shown in Fig. 58.11.

The absorptivity of liquid water is about 0.96.

58.8.3 Radiation Charts

For convenience in preliminary calculations, black-body radiation, as a function of temperature in °F, is given in chart form in Fig. 58.12. The value for the receiver surface is subtracted from that of the source to find net interchange for black-body conditions, and the result is corrected for emissivity and view factors. Where heat is transmitted by a combination of solid-state radiation and convection, a black-body coefficient, in Btu/hr · °F, is shown in the chart in Fig. 58.13. This can be added to the convection coefficient for the same temperature interval, after correcting for emissivity and view factor, to provide an overall coefficient (H) for use in the formula

$$Q/A = H(T - T_r)$$

58.8.4 View Factors for Solid-State Radiation

For a receiving surface completely enclosed by the source of radiation, or for a flat surface under a hemispherical radiating surface, the view factor is unity. Factors for a wide range of geometrical

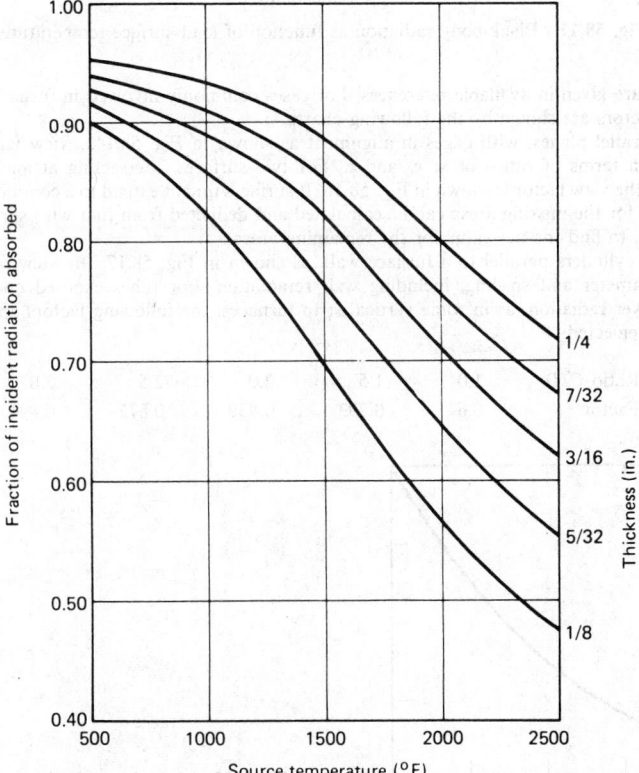

Fig. 58.11 Radiation absorptivity of sheet glass with surface reflection deducted.[1]

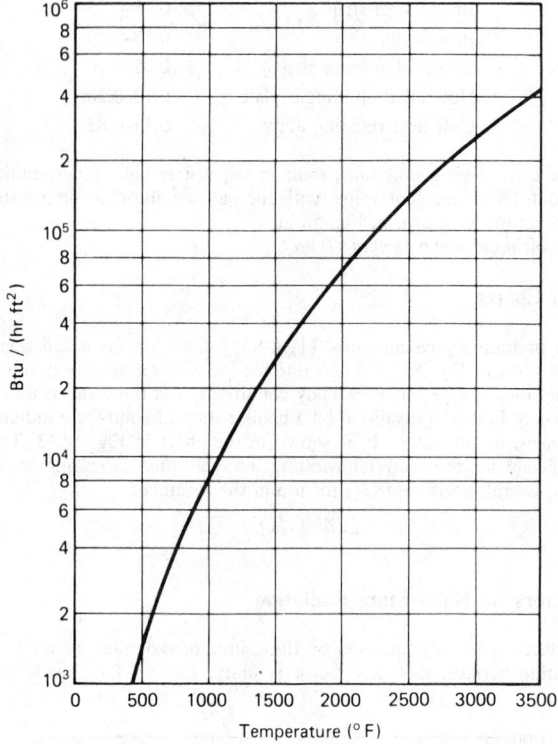

Fig. 58.12 Black-body radiation as function of load surface temperature.

configurations are given in available references. For cases commonly involved in furnace heat-transfer calculations, factors are shown by the following charts.

For two parallel planes, with edges in alignment as shown in Fig. 58.14a, view factors are given in Fig. 58.15 in terms of ratios of x, y, and z. For two surfaces intersecting at angle of 90° at a common edge, the view factor is shown in Fig. 58.16. If surfaces do not extend to a common intersection, the view factor for the missing areas can be calculated and deducted from that with surfaces extended as in the figure, to find the net value for the remaining areas.

For spaced cylinders parallel to a furnace wall, as shown in Fig. 58.17, the view factor is shown in terms of diameter and spacing, including wall reradiation. For tubes exposed on both sides to source or receiver radiation, as in some vertical strip furnaces, the following factors apply if sidewall reradiation is neglected:

Ratio C/D	1.0	1.5	2.0	2.5	3.0
Factor	0.67	0.793	0.839	0.872	0.894

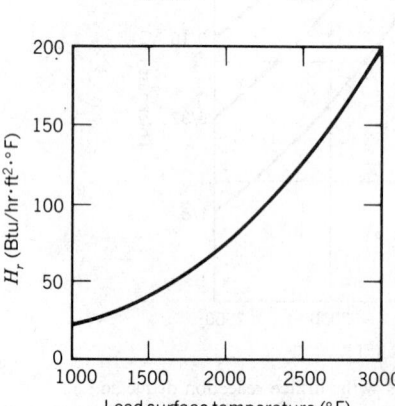

Fig. 58.13 Black-body radiation coefficient for source temperature uniform at 50–105° above final load surface temperature.

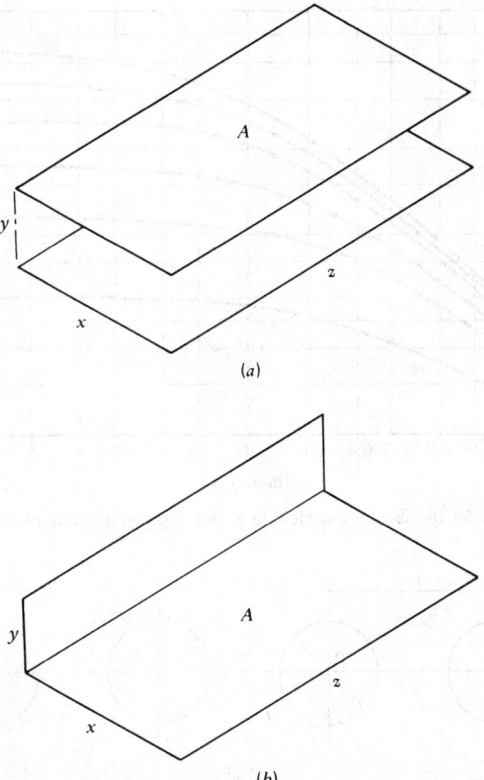

Fig. 58.14 Diagram of radiation view factors for parallel and perpendicular planes.[1]

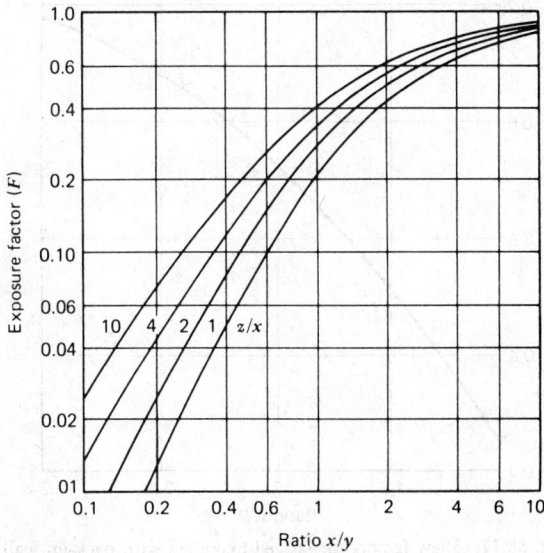

Fig. 58.15 Radiation view factors for parallel planes.[1]

For ribbon-type electric heating elements, mounted on a back-up wall as shown in Fig. 58.18, exposure factors for projected wall area and for total element surface area are shown as a function of the (element spacing)/(element width) ratio. Wall reradiation is included, but heat loss through the back-up wall is not considered. The emission rate from resistor surface will be W/in.² = $Q/491A$, where

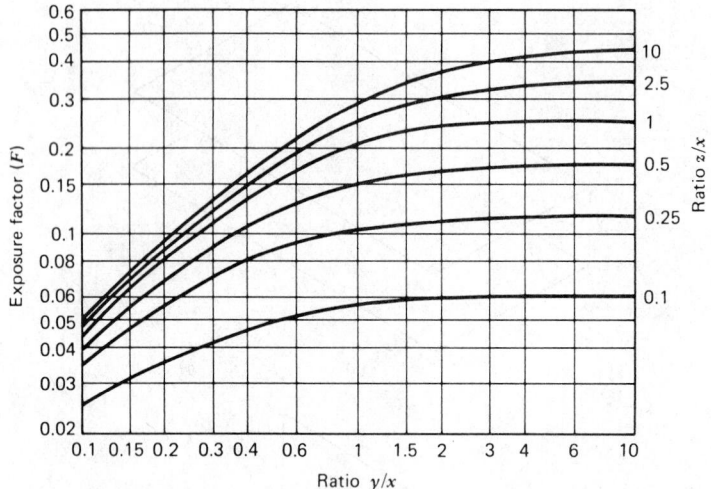

Fig. 58.16 Radiation view factors for perpendicular planes.[1]

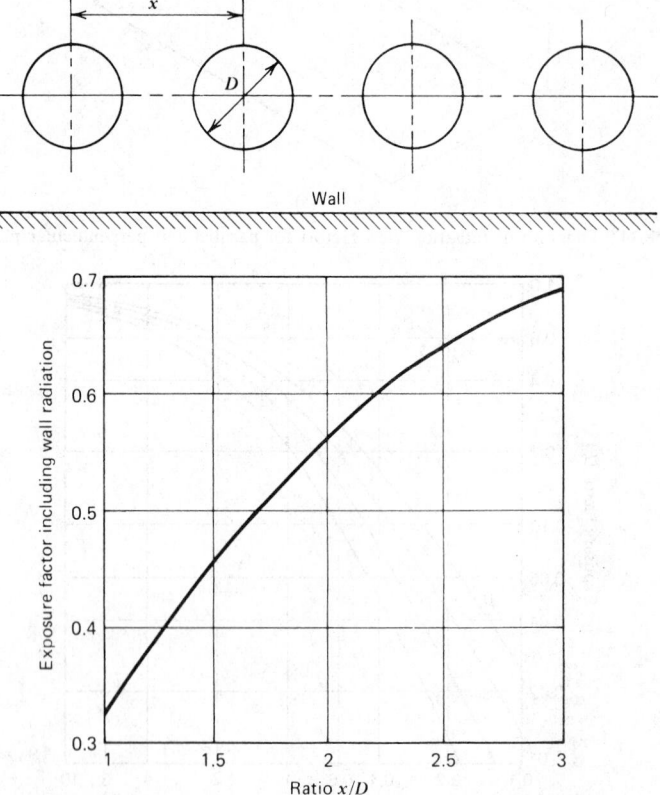

Fig. 58.17 View factors for spaced cylinders with back-up wall.[1]

$$\frac{Q}{A} = \frac{Btu/hr}{ft^2}$$

For parallel planes of equal area, as shown in Fig. 58.14, connected by reradiating walls on four sides, the exposure factor is increased as shown in Fig. 58.19. Only two curves, for $z/x = 1$ and $z/x = 10$ have been plotted for comparison with Fig. 58.13.

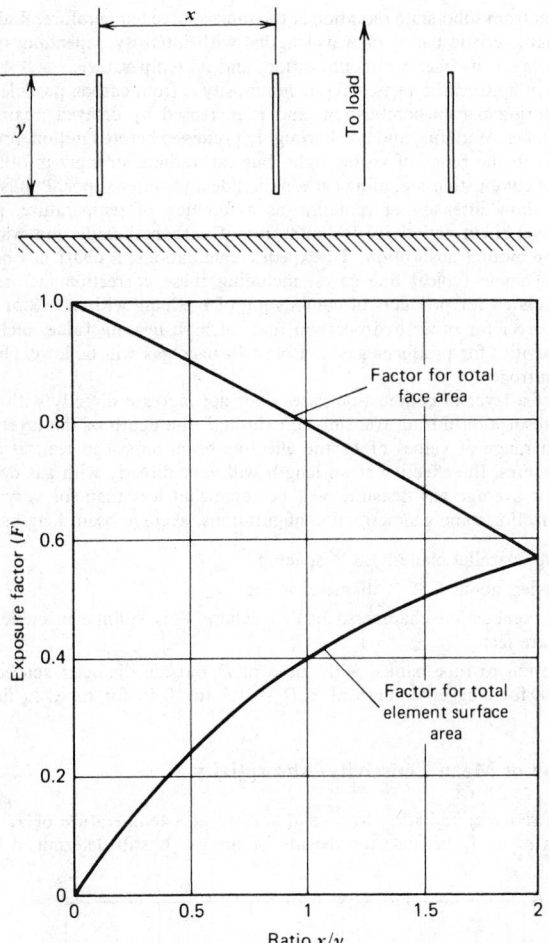

Fig. 58.18 View factors for ribbon-type electric heating elements mounted on back-up wall.[1]

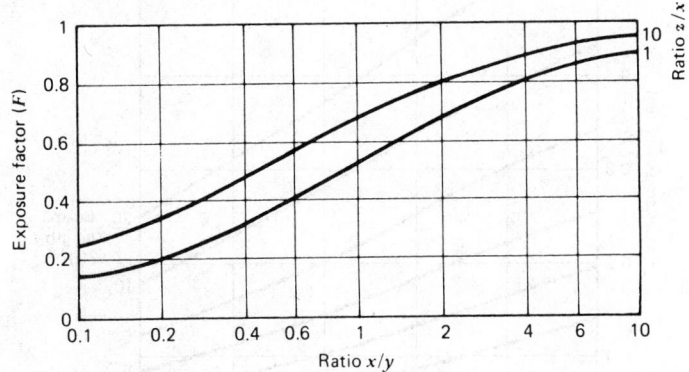

Fig. 58.19 View factors for parallel planes connected by reradiating sidewalls.[1]

58.8.5 Gas Radiation

Radiation from combustion gases to walls and load can be from luminous flames or from nonluminous products of combustion. Flame luminosity results from suspended solids in combustion gases, either incandescent carbon particles or ash residues, and the resulting radiation is in a continuous spectrum corresponding to that from solid-state radiation at the same source temperature. Radiation from nonlumi-nous gases is in characteristic bands of wavelengths, with intensity depending on depth and density of the radiating gas layer, its chemical composition, and its temperature.

For combustion of hydrocarbon gases, flame luminosity is from carbon particles formed by cracking of unburned fuel during partial combustion, and is increased by delayed mixing of fuel and air in the combustion chamber. With fuel and air thoroughly premixed before ignition, products of combustion will be nonluminous in the range of visible light, but can radiate strongly in other wavelength bands for some products of combustion including carbon dioxide and water vapor. Published data on emissivi-ties of these gases show intensity of radiation as a function of temperature, partial pressure, and beam length. The combined emissivity for mixtures of carbon dioxide and water vapor requires a correction factor for mutual absorption. To expedite calculations, a chart has been prepared for the overall emissivity of some typical flue gases, including these correction factors. The chart in Fig. 58.20 has been calculated for products of combustion of methane with 110% of net air demand, and is approximately correct for other hydrocarbon fuels of high heating value, including coke oven gas and fuel oils. Emissivities for producer gas and blast furnace gas will be lower, because of dilution of radiating gases by nitrogen.

The emissivity of a layer of combustion gases does not increase directly with thickness or density, because of partial absorption during transmission through the depth of the layer. The chart provides several curves for a range of values of L, the effective beam length in feet, at a total pressure of 1 atm. For other pressures, the effective beam length will vary directly with gas density.

Beam lengths for average gas densities will be somewhat less than for very low density because of partial absorption. For some geometrical configurations, average beam lengths are:

Between two large parallel planes, $1.8 \times$ spacing.

Inside long cylinder, about $0.85 \times$ diameter in feet.

For rectangular combustion chambers, $3.4V/A$ where V is volume in cubic feet and A is total wall area in square feet.

Transverse radiation to tube banks, with tubes of D outside diameter spaced at x centers; L/D ranges from 1.48 for staggered tubes at $x/D = 1.5$ to 10.46 for tubes in line and $x/D = 3$ in both directions.

58.8.6   Evaluation of Mean Emissivity–Absorptivity

For a gas with emissivity e_g radiating to a solid surface at a temperature of T_s °F, the absorptivity a_g will be less than e_g at T_s because the density of the gas is still determined by T_g. The effective

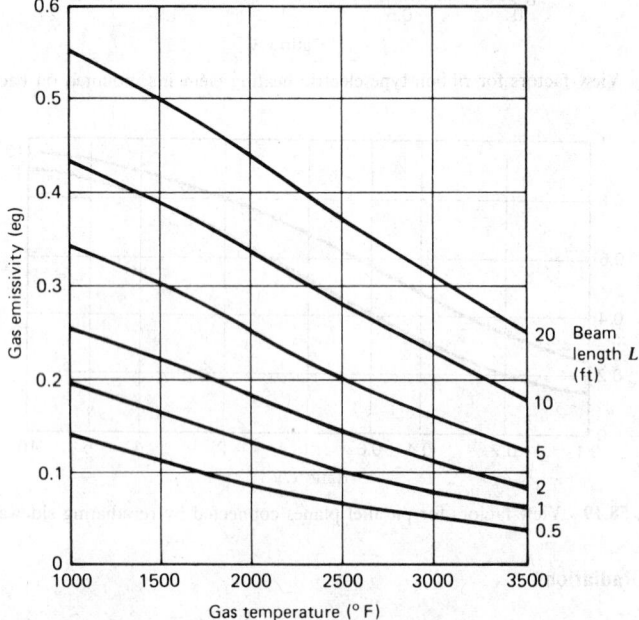

Fig. 58.20   Gas emissivity for products of combustion of methane burned with 110% air. Approximate for fuel oils and coke oven gas.[1]

PL becomes $T_s/T_g \times PL$ at T_s. Accurate calculation of the combined absorptivity for carbon dioxide and water vapor requires a determination of a_g for either gas and a correction factor for the total. For the range of temperatures and PL factors encountered in industrial heat transfer, the net heat transfer can be approximated by using a factor e_{gm} somewhat less than e_g at T_g in the formula:

$$Q/A = 0.1713 e_{gm} F(T_g^4 - T_s^4)$$

where T_g is an average of gas temperatures in various parts of the combustion chamber; the effective emissivity will be about $e_{gm} = 0.9 e_g$ at T_g and can be used with the chart in Fig. 58.20 to approximate net values.

58.8.7 Combined Radiation Factors

For a complete calculation of heat transfer from combustion gases to furnace loads, the following factors will need to be evaluated in terms of the equivalent fraction of black-body radiation per unit area of the exposed receiving surface:

$F_{gs} = $ Coefficient for gas direct to load, plus radiation reflected from walls to load.

$F_{gw} = $ Coefficient for gas radiation absorbed by walls.

$F_{ws} = $ Coefficient for solid-state radiation from walls to load.

Convection heat transfer from gases to walls and load is also involved, but can be eliminated from calculations by assuming that gas to wall convection is balanced by wall losses, and that gas to load convection is equivalent to a slight increase in load surface absorptivity. Mean effective gas temperature is usually difficult to measure, but can be calculated if other factors are known. For example, carbon steel slabs are being heated to rolling temperature in a fuel-fired continuous furnace. At any point in the furnace, neglecting convection,

$$F_{gw} (T_g^4 - T_w^4) = F_{ws} (T_w^4 - T_s^4)$$

where T_g, T_w, and T_s are gas, wall, and load surface temperatures in °S.

For a ratio of 2.5 for exposed wall and load surfaces, and a value of 0.17 for gas-to-wall emissivity, $F_{gw} = 2.5 \times 0.17 = 0.425$. With wall to load emissivity equal to $F_{ws} = 0.89$, wall temperature constant at 2350°F (28.1°S), and load temperature increasing from 70 to 2300°F at the heated surface ($T_s = 5.3$–27.6°S), the mean value of gas temperature (T_g) can be determined:

$$\text{MTD, walls to load} = \frac{2280 - 50}{\ln(2280/50)} = 584°F$$

Mean load surface temperature $T_{sm} = 2350 - 584 = 1766°F$ (22.26°S)

Q/A per unit of load surface, for reradiation:

$$0.425 \times 0.1713(T_g^4 - 28.1^2) = 0.89 \times 0.1713(28.1^4 - 22.26^4) = 57,622 \text{ Btu/hr} \cdot \text{ft}^2$$

$$T_g = 34.49°S \ (2989°F)$$

With a net wall emissivity of 0.85, 15% of gas radiation will be reflected to the load, with the balance being absorbed and reradiated. Direct radiation from gas to load is then

$$1.15 \times 0.17 \times 0.1713(34.49^4 - 22.26^4) = 47,389 \text{ Btu/hr} \cdot \text{ft}^2$$

$$\text{Total radiation: } 57,622 + 47,389 = 105,011 \text{ Btu/hr} \cdot \text{ft}^2$$

For comparison, black-body radiation from walls to load, without gas radiation, would be 64,743 Btu/hr · ft² or 62% of the combined total.

With practical furnace temperature profiles, in a counterflow, direct-fired continuous furnace, gas and wall temperatures will be depressed at the load entry end to reduce flue gas temperature and stack loss. The resulting net heating rates will be considered in Section 58.8.12.

Overall heat-transfer coefficients have been calculated for constant wall temperature, in the upper chart in Fig. 58.21, or for constant gas temperature in the lower chart. Coefficients vary with mean gas emissivity and with A_w/A_s, the ratio of exposed surface for walls and load, and are always less than one for overall radiation from gas to load, or greater than one for wall to load radiation. Curves can be used to find gas, wall, or mean load temperatures when the other two are known.

58.8.8 Steady-State Conduction

Heat transfer through opaque solids and motionless layers of liquids or gases is by conduction. For constant temperature conditions, heat flow is by "steady-state" conduction and does not vary with

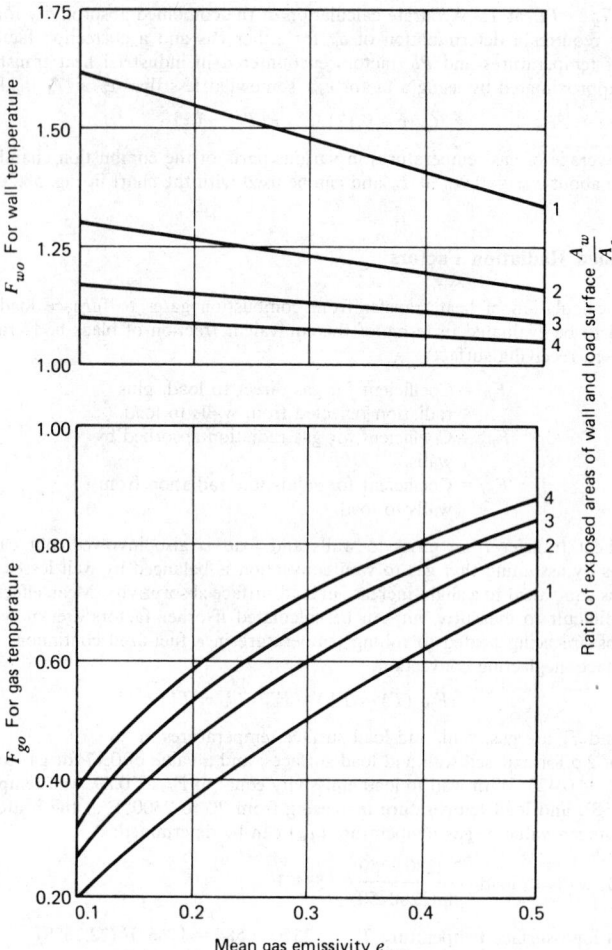

Fig. 58.21 Overall heat-transfer coefficients for gas and solid radiation, as function of gas emissivity and wall-to-load area ratio, for uniform gas or wall temperature, compared to black-body radiation.[1]

time. For objects being heated or cooled, with a continuous change in internal temperature gradients, conduction is termed "non-steady-state."

Thermal conduction in some solid materials is a combination of heat flow through the material, radiation across internal space resulting from porosity, and convection within individual pores or through the thickness of porous layers.

Conductivities of refractory and insulating materials tend to increase with temperature, because of porosity effects. Values for most metals decrease with temperature, partly because of reduced density. Conductivity coefficients for some materials used in furnace construction or heated in furnaces are listed in Table 58.5.

A familiar problem in steady-state conduction is the calculation of heat losses through furnace walls made up of multiple layers of materials of different thermal conductivities. A convenient method of finding overall conductance is to find the thermal resistance (r/k = thickness/conductivity in consistent units) and add the total for all layers. Because conductivities vary with temperature, mean temperatures for each layer can be estimated from a preliminary temperature profile for the composite wall. Overall resistance will include the effects of radiation and conduction between the outer wall surface and its surroundings.

A chart showing heat loss from walls to ambient surrounding at 70°F, combining radiation and convection for vertical walls, is shown in Fig. 58.7. The corresponding thermal resistance is included in the overall heat-transfer coefficient shown in Fig. 58.8 as a function of net thermal resistance of the wall structure and inside face temperature.

As an example of application, assume a furnace wall constructed as follows:

Material	r	k	r/k
9 in. firebrick	0.75	0.83	0.90
4½ in. 2000°F insulation	0.375	0.13	2.88
2½ in. ceramic fiber block	0.208	0.067	3.10
Total R for solid wall			6.88

With an inside surface temperature of 2000°F, the heat loss from Fig. 58.7 is about 265 Btu/ft · hr². The corresponding surface temperature from Fig. 58.8 is about 200°F, assuming an ambient temperature of 70°F.

Although not a factor affecting wall heat transfer, the possibility of vapor condensation in the wall structure must be considered by the furnace designer, particularly if the furnace is fired with a sulfur-bearing fuel. As the sulfur dioxide content of fuel gases is increased, condensation temperatures increase to what may exceed the temperature of the steel furnace casing in normal operation. Resulting condensation at the outer wall can result in rapid corrosion of the steel structure.

Condensation problems can be avoided by providing a continuous membrane of aluminum or stainless steel between layers of the wall structure, at a point where operating temperatures will always exceed condensation temperatures.

58.8.9 Non-Steady-State Conduction

Heat transfer in furnace loads during heating or cooling is by transient or non-steady-state conduction, with temperature profiles within loads varying with time. With loads of low internal thermal resistance, heating time can be calculated for the desired load surface temperature and a selected time–temperature profile for furnace temperature. With loads of appreciable thermal resistance from surface to center, or from hot to colder sides, heating time will usually be determined by a specified final load temperature differential, and a selected furnace temperature profile for the heating cycle.

For the case of a slab-type load being heated on a furnace hearth, with only one side exposed, and with the load entering the furnace at ambient temperature, the initial gradient from the heated to the unheated surface will be zero. The heated surface will heat more rapidly until the opposite surface starts to heat, after which the temperature differential between surfaces will taper off with time until the desired final differential is achieved.

In Fig. 58.22 the temperatures of heated and unheated surface or core temperature are shown as a function of time. In the lower chart temperatures are plotted directly as a function of time. In the upper chart the logarithm of the temperature ratio (Y = load temperature/source temperature) is plotted as a function of time for a constant source temperature. After a short initial heating time, during which the unheated surface or core temperature reaches its maximum rate of increase, the two curves in the upper diagram become parallel straight lines.

Factors considered in non-steady-state conduction and their identifying symbols are listed in Table 58.6.

Table 58.6 Non-Steady-State Conduction Factors and Symbols

T_f = Furnace temperature, gas or wall as defined

T_s = Load surface temperature

T_c = Temperature at core or unheated side of load

T_0 = Initial load temperature with all temperatures in units of (°F − 460)/100 or °S

$$Y_s = \frac{T_f - T_s}{T_f - T_0}$$

$$Y_c = \frac{T_f - T_c}{T_f - T_0}$$

R = External/internal thermal resistance ratio = k/rH

X = Time factor = tD/r^2

D = Diffusivity as defined in Section 58.7

r = Depth of heat penetration in feet

k = Thermal conductivity of load (Btu/ft · hr · °F)

H = External heat transfer coefficient (Btu/ft² · hr · °F)

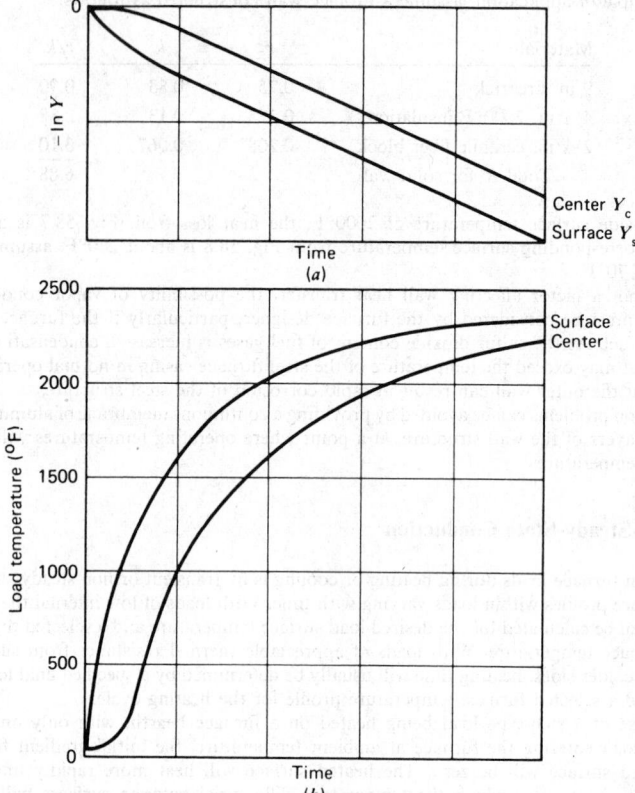

Fig. 58.22 Maximum and minimum load temperatures, and $-\ln Y_s$ or $-\ln Y_c$ as a function of heating time with constant source temperature.[1]

Charts have been prepared by Gurney-Lurie, Heisler, Hottel, and others showing values for Y_s and Y_c for various R factors as a function of X. Separate charts are provided for Y_s and Y_c, with a series of curves representing a series of values of R. These curves are straight lines for most of their length, curving to intersect at $Y = 1$ and $X = 0$. If straight lines are extended to $Y = 1$, the curves for Y_c at all values of R converge at a point near $X = 0.1$ on the line for $Y_c = 1$. It is accordingly possible to prepare a single line chart for $-\ln Y_c/(X - 0.1)$ to fit selected geometrical shapes. This has been done in Fig. 58.23 for slabs, long cylinders, and spheres. Values of Y_c determined with this chart correspond closely with those from conventional charts for $X - 0.1$ greater than 0.2.

Because the ratio Y_s/Y_c remains constant as a function of R after initial heating, it can be shown in chart form, as in Fig. 58.24, to allow Y_s to be determined after Y_c has been found.

By way of illustration, a carbon steel slab 8 in. thick is being heated from cold to $T_s = 2350°F$ in a furnace with a constant wall temperature of 2400°F, with a view factor of 1 and a mean emissivity–absorptivity factor of 0.80. The desired final temperature of the unheated surface is 2300°F, making the Y_c factor

$$Y_c = \frac{2400 - 2300}{2400 - 70} = 0.0429$$

From Fig. 58.23 $H_r = 114 \times 0.80 = 91$; $r = 8/12 = 0.67$; R is assumed at 17. The required heating time is determined from Fig. 58.24:

$$R = \frac{17}{0.67 \times 91} = 0.279$$

$$\frac{-\ln Y_c}{X - 0.1} = 1.7$$

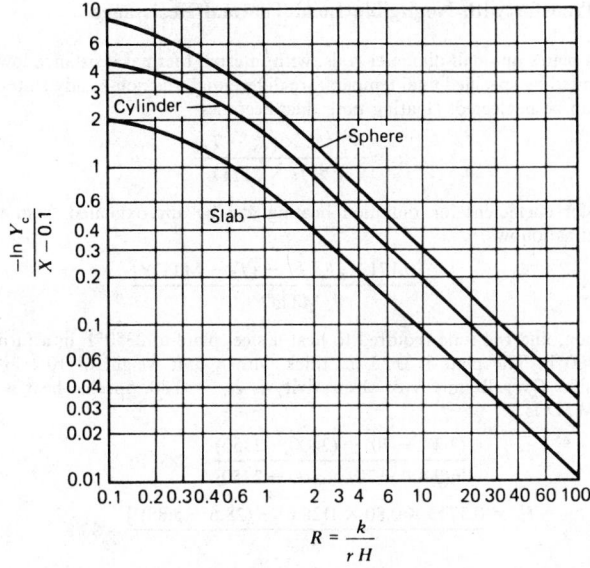

Fig. 58.23 A plot of $-\ln Y_c/(X - 0.1)$ as a function of R. [1]

and

$$X = \frac{-\ln 0.0429}{1.7} + 0.1 = 1.95 = tD/r^2$$

With $D = 0.25$, from Section 58.7,

$$t = \frac{Xr^2}{D} = \frac{1.95 \times 0.67^2}{0.25} = 3.50 \text{ hr}$$

Slabs or plates heated from two sides are usually supported in the furnace in a horizontal position on spaced conveyor rolls or rails. Support members may be uncooled, in which case radiation to the bottom surface will be reduced by the net view factor. If supports are water cooled, the additional heat input needed to balance heat loss from load to supports can be balanced by a higher furnace temperature on the bottom side. In either case, heating times will be greater than for a uniform input from both sides.

Furnace temperatures are normally limited to a fraction above final load temperatures to avoid local overheating during operating delays. Without losses to water cooling, top and bottom furnace temperature will accordingly be about equal.

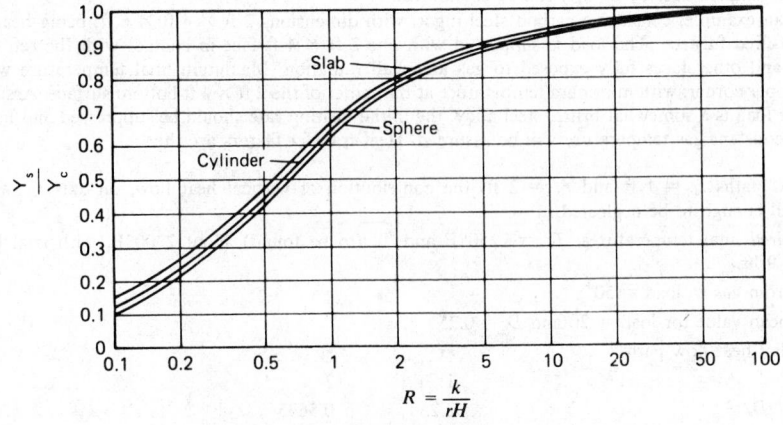

Fig. 58.24 The ratio Y_s/Y_c plotted as a function of R. [1]

58.8.10   Heat Transfer with Negligible Load Thermal Resistance

When heating thin plates or small-diameter rods, with internal thermal resistance low enough to allow heating rates unlimited by specified final temperature differential, the non-steady-state-conduction limits on heating rates can be neglected. Heating time then becomes

$$t = \frac{W \times C \times (T_s - T_0)}{A \times H \times \text{MTD}}$$

The heat-transfer coefficient for radiation heating can be approximated from the chart in Fig. 58.13 or calculated as follows:

$$H_r = \frac{0.1713 e_m F_s [T_f^4 - (T_f - \text{MTD})^4]}{\text{MTD} \times A_s}$$

As an illustration, find the time required to heat a steel plate to 2350°F in a furnace at a uniform temperature of 2400°F. The plate is 0.25 in. thick with a unit weight of 10.2 lb/ft² and is to be heated from one side. Overall emissivity–absorptivity is $e_m = 0.80$. Specific heat is 0.165. The view factor is $F_s = 1$. MTD is

$$\frac{(2400 - 70) - (2400 - 2350)}{\ln(2400 - 70)/(2400 - 2350)} = 588°F$$

$$H_r = \frac{0.1713 \times 0.80 \times 1[28.6^4 - (28.6 - 5.88)^4]}{588} = 93.8$$

$$t = \frac{10.2 \times 0.165(2350 - 70)}{1 \times 93.8 \times 588} = 0.069 \text{ hr}$$

58.8.11   Newman Method

For loads heated from two or more perpendicular sides, final maximum temperatures will be at exposed corners, with minimum temperatures at the center of mass for heating from all sides, or at the center of the face in contact with the hearth for hearth-supported loads heated equally from the remaining sides. For surfaces not fully exposed to radiation, the corrected H factor must be used.

The Newman method can be used to determine final load temperatures with a given heating time t. To find time required to reach specified maximum and minimum final load temperatures, trial calculations with several values of t will be needed.

For a selected heating time t, the factors Y_s and Y_c can be found from charts in Figs. 58.23 and 58.24 for the appropriate values of the other variables—T_s, T_c, H, k, and r—for each of the heat flow paths involved—r_x, r_y, and r_z. If one of these paths is much longer than the others, it can be omitted from calculations:

$$Y_c = Y_{cx} \times Y_{cy} \times Y_{cz}$$
$$Y_s = Y_{sx} \times Y_{sy} \times Y_{sz}$$

For two opposite sides with equal exposure, only one is considered. With T_c known, T_s and T_f (furnace temperature, T_g or T_w) can be calculated.

As an example, consider a carbon steel ingot, with dimensions 2 ft × 4 ft × 6 ft, being heated in a direct-fired furnace. The load is supported with one 2 ft × 4 ft face in contact with the refractory hearth and other faces fully exposed to gas and wall radiation. Maximum final temperature will be at an upper corner, with minimum temperature at the center of the 2 ft × 4 ft bottom surface. Assuming that the load is a somewhat brittle steel alloy, the initial heating rate should be suppressed and heating with a constant gas temperature will be assumed. Heat-transfer factors are then

Flow paths $r_s = 1$ ft and $r_y = 2$ ft, the contribution of vertical heat flow, on axis r_z, will be small enough to be neglected.

Desired final temperatures: $T_c = 2250°F$ and T_s (to be found) about 2300°F, with trial factor $t = 9$ hr.

H from gas to load = 50

k mean value for load = 20 and $D = 0.25$

Radial heat flow path	rx	ry
r	1	2
$X = tD/r^2$	2.25	0.5625
$R = k/H_r$	0.4	0.2

$-\ln Y_c/(X-0.1)$ from Fig. 58.23   1.3   1.7
Y_s/Y_c from Fig. 58.24   0.41   0.26
Y_c 0.0611   0.455
Y_s 0.025   0.119

Combined factors:

$$Y_c = 0.0611 \times 0.455 = 0.0278 = \frac{T_g - T_c}{T_g - 70}$$

$$Y_s = 0.025 \times 0.119 = 0.003 = \frac{T_g - T_s}{T_g - 70}$$

For $T_c = 2250°F$, $T_g = 2316°F$
$T_s = 2309°F$

This is close enough to the desired $T_s = 2300°F$.

The time required to heat steel slabs to rolling temperature, as a function of the thickness heated from one side and the final load temperature differential, is shown in Fig. 58.25. Relative heating times for various hearth loading arrangements, for square billets, are shown in Fig. 58.26. These have been calculated by the Newman method, which can also be used to evaluate other loading patterns and cross sections.

58.8.12 Furnace Temperature Profiles

To predict heating rates and final load temperatures in either batch or continuous furnaces, it is convenient to assume that source temperatures, gas (T_g) or furnace wall (T_w), will be constant in time. Neither condition is achieved with contemporary furnace and control system designs. With constant gas temperature, effective heating rates are unnecessarily limited, and the furnace temperature control system is dependent on measurement and control of gas temperatures, a difficult requirement. With uniform wall temperatures, the discharge temperature of flue gases at the beginning of the heating cycle will be higher than desirable. Three types of furnace temperature profiles, constant T_g, constant T_w, and an arbitrary pattern with both variables, are shown in Fig. 58.27.

Contemporary designs of continuous furnaces provide for furnace temperature profiles of the third type illustrated, to secure improved capacity without sacrificing fuel efficiency. The firing system comprises three zones of length: a preheat zone that can be operated to maintain minimum flue gas temperatures in a counterflow firing arrangement, a firing zone with a maximum temperature and firing rate consistent with furnace maintenance requirements and limits imposed by the need to avoid overheating

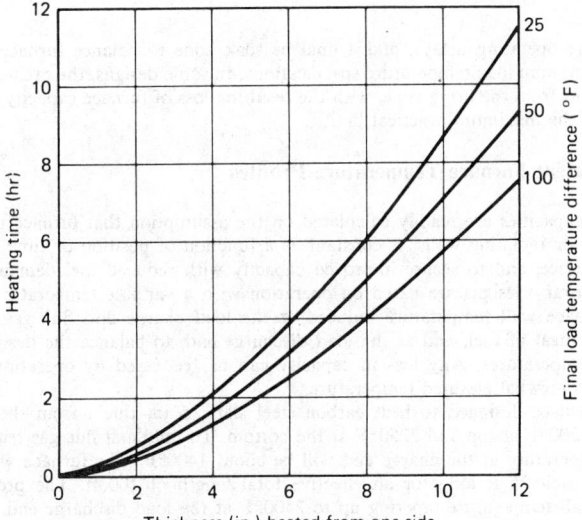

Fig. 58.25 Relative heating time for square billets as a function of loading pattern.[1]

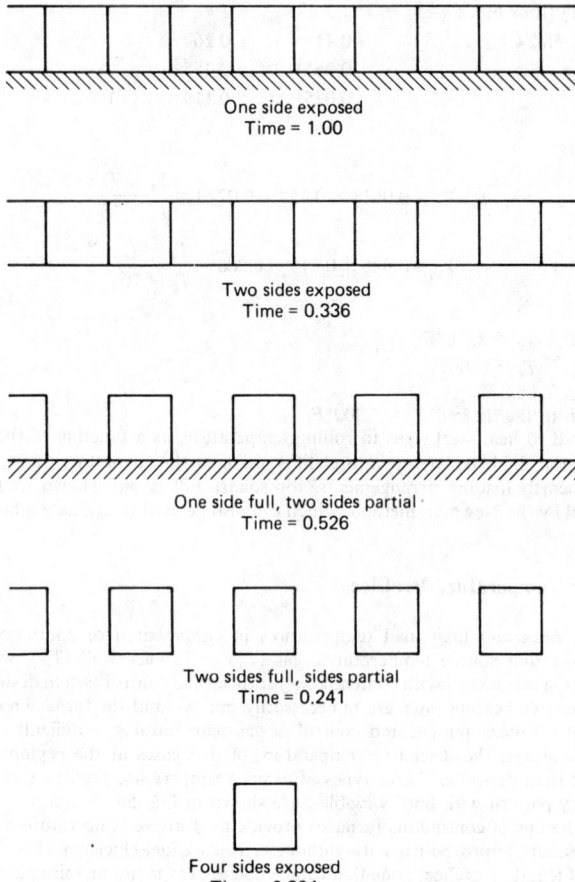

One side exposed
Time = 1.00

Two sides exposed
Time = 0.336

One side full, two sides partial
Time = 0.526

Two sides full, sides partial
Time = 0.241

Four sides exposed
Time = 0.204

Fig. 58.26 Heating time for carbon steel slabs to final surface temperature of 2300°F, as a function of thickness and final load temperature differential.[1]

of the load during operating delays, and a final or soak zone to balance furnace temperature with maximum and minimum load temperature specifications. In some designs, the preheat zone is unheated except by flue gases from the firing zone, with the resulting loss of furnace capacity offset by operating the firing zone at the maximum practical limit.

58.8.13 Equivalent Furnace Temperature Profiles

Furnace heating capacities are readily calculated on the assumption that furnace temperature, either combustion gases or radiating walls, is constant as a function of position or time. Neither condition is realized in practice; and to secure improved capacity with reduced fuel demand in a continuous furnace, contemporary designs are based on operation with a variable temperature profile from end to end, with furnace wall temperature reduced at the load charge and flue gas discharge end, to improve available heat of fuel, and at the load discharge end, to balance the desired maximum and minimum load temperatures. Any loss in capacity can be recovered by operating the intermediate firing zones at a somewhat elevated temperature.

Consider a furnace designed to heat carbon steel slabs, 6 in. thick, from the top only to final temperatures of 2300°F at top and 2250°F at the bottom. To hold exit flue gas temperature to about 2000°F, wall temperature at the charge end will be about 1400°F. The furnace will be fired in four zones of length, each 25 ft long for an effective total length of 100 ft. The preheat zone will be unfired, with a wall temperature tapering up to 2400°F at the load discharge end. That temperature will be held through the next two firing zones and dropped to 2333°F to balance final load temperatures in the fourth or soak zone. With overall heating capacity equal to the integral of units of length

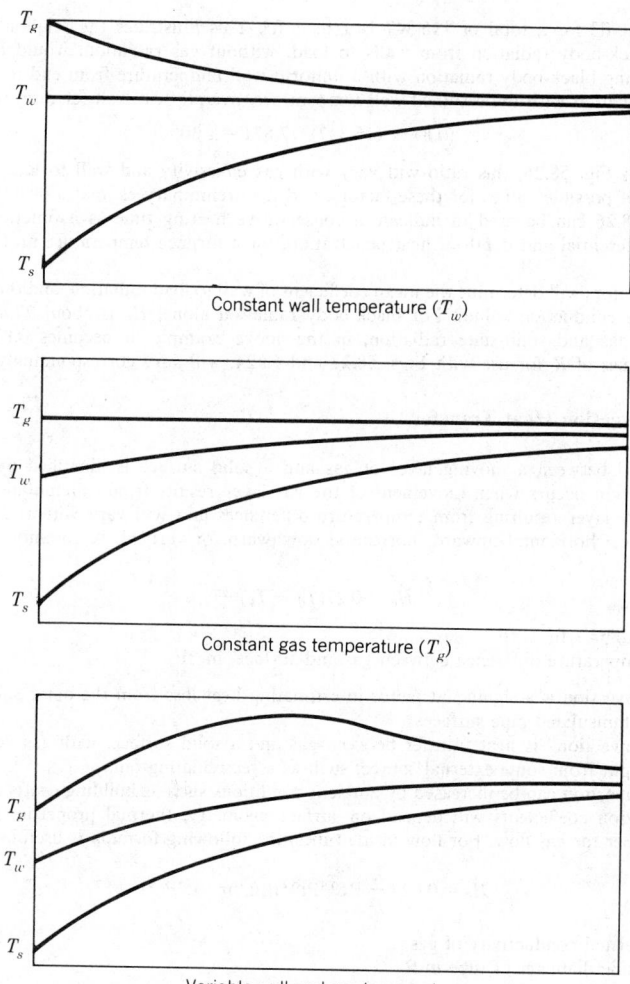

Constant wall temperature (T_w)

Constant gas temperature (T_g)

Variable wall and gas temperatures

Fig. 58.27 Furnace temperature profiles.

times their absolute temperatures, effective heat input will be about 87% of that for a uniform temperature of 2400°F for the entire length.

Heat transfer from combustion gases to load will be by direct radiation from gas to load, including reflection of incident radiation from walls, and by radiation from gas to walls, absorbed and reradiated from walls to load. Assuming that wall losses will be balanced by convection heat transfer from gases, gas radiation to walls will equal solid-state radiation from walls to load:

$$A_w/A_s \times 0.1713 \times e_{gm}(T_g^4 - T_w^4) = e_{ws} \times 0.1713(T_w^4 - T_s^4)$$

where A_w/A_s = exposed area ratio for walls and load
e_{gm} = emissivity–absorptivity, gas to walls
e_{ws} = emissivity–absorptivity, walls to load

At the midpoint in the heating cycle, MTD = 708°F and mean load surface temperature = $T_{sm} = 1698$°F.

With $a_s = 0.85$ for refractory walls, 15% of gas radiation will be reflected to load, and total gas to load radiation will be:

$$1.15 \times e_{gm} \times 0.1713(T_g^4 - T_s^4)$$

For $A_w/A_s = 2.5$, $e_{gm} = 0.17$, and $e_{ws} = 0.89$ from walls to load, the mean gas temperature = $T_g = 3108$°F, net radiation, gas to load = 47,042 Btu/hr · ft² and gas to walls = walls to load =

69,305 Btu/hr · ft² for a total of 116,347 Btu/hr · ft². This illustrates the relation shown in Fig. 58.21, since black-body radiation from walls to load, without gas radiation, would be 77,871 Btu/hr · ft². Assuming black-body radiation with a uniform wall temperature from end to end, compared to combined radiation with the assumed wall temperature, overall heat transfer ratio will be

$$(0.87 \times 116,347)/77,871 = 1.30$$

As shown in Fig. 58.26, this ratio will vary with gas emissivity and wall to load areas exposed. For the range of possible values for these factors, and for preliminary estimates of heating times, the chart in Fig. 58.26 can be used to indicate a conservative heating time as a function of final load temperature differential and depth of heat penetration, for a furnace temperature profile depressed at either end.

Radiation factors will determine the mean coefficient of wall to load radiation, and the corresponding non-steady-state conduction values. For black-body radiation alone, H_r is about $77,871/708 = 110$. For combined gas and solid-state radiation, in the above example, it becomes $0.87 \times 116,347/708 = 143$. Values of R for use with Figs. 58.23 and 58.24, will vary correspondingly ($R = k/4H$).

58.8.14 Convection Heat Transfer

Heat transferred between a moving layer of gas and a solid surface is identified by "convection." Natural convection occurs when movement of the gas layer results from differentials in gas density of the boundary layer resulting from temperature differences and will vary with the position of the boundary surface: horizontal upward, horizontal downward, or vertical. A commonly used formula is:

$$H_c = 0.27(T_g - T_s)^{0.25}$$

where H_c = Btu/hr · ft² · °F
$T_g - T_s$ = temperature difference between gas and surface, in °F

Natural convection is a significant factor in estimating heat loss from the outer surface of furnace walls or from uninsulated pipe surfaces.

"Forced convection" is heat transfer between gas and a solid surface, with gas velocity resulting from energy input from some external source, such as a recirculating fan.

Natural convection can be increased by ambient conditions such as building drafts and gas density. Forced convection coefficients will depend on surface geometry, thermal properties of the gas, and Reynolds number for gas flow. For flow inside tubes, the following formula is useful:

$$H_c = 0.023 \frac{k}{D} \text{Re}^{0.8}\text{Pr}^{0.4} \text{Btu/hr} \cdot \text{ft}^2 \cdot °\text{F}$$

where k = thermal conductivity of gas
D = inside diameter of tube in ft
Re = Reynolds number
Pr = Prandtl number

Forced convection coefficients are given in chart form in Fig. 58.28 for a Prandtl number assumed at 0.70.

For forced convection over plane surfaces, it can be assumed that the preceding formula will apply for a rectangular duct of infinitely large cross section, but only for a length sufficient to establish uniform velocity over the cross section and a velocity high enough to reach the Re value needed to promote turbulent flow.

In most industrial applications, the rate of heat transfer by forced convection as a function of power demand will be better for perpendicular jet impingement from spaced nozzles than for parallel flow. For a range of dimensions common in furnace design, the heat-transfer coefficient for jet impingement of air or flue gas is shown in Fig. 58.29, calculated for impingement from slots 0.375 in. wide spaced at 18–24 in. centers and with a gap of 8 in. from nozzle to load.

Forced convection factors for gas flow through banks of circular tubes are shown in the chart in Fig. 58.30 and for tubes spaced as follows:

A: staggered tubes with lateral spacing equal to diagonal spacing.
B: tubes in line, with equal spacing across and parallel to direction of flow.
C: tubes in line with lateral spacing less than half longitudinal spacing.
D: tubes in line with lateral spacing over twice longitudinal spacing.

With F the configuration factor from Fig. 58.30, heat-transfer coefficients are

$$H_c = Fk\,\text{Re}^{0.6}/D$$

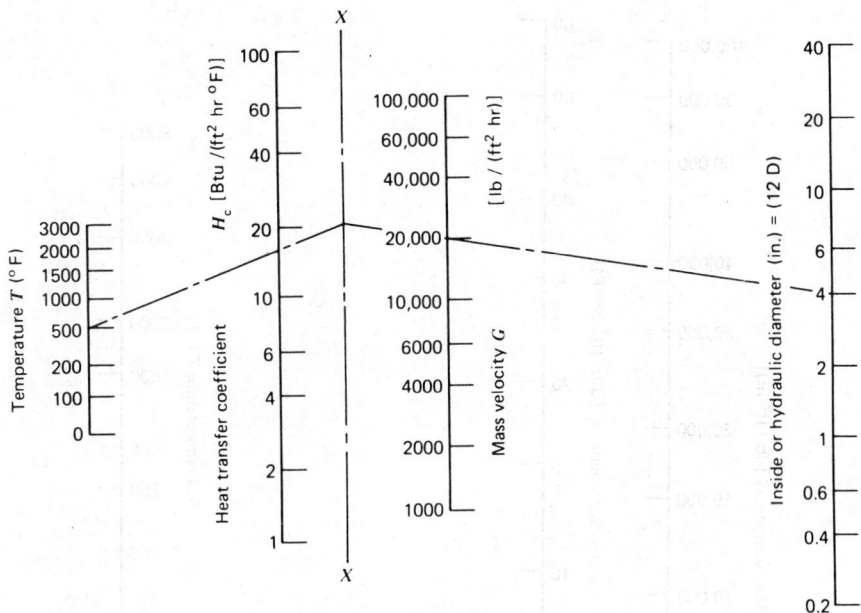

Fig. 58.28 Convection coefficient (H_c) for forced convection inside tubes to air or flue gas.[1]

Convection coefficients from this formula are approximately valid for 10 rows of tubes or more, but are progressively reduced to a factor of 0.65 for a single row.

For gas to gas convection in a cross-flow tubular heat exchanger, overall resistance will be the sum of factors for gas to the outer diameter of tubes, tube wall conduction, and inside diameter of tubes to gas. Factors for the outer diameter of tubes may include gas radiation as calculated in Section 58.7.5.

58.8.15 Fluidized-Bed Heat Transfer

For gas flowing upward through a particulate bed, there is a critical velocity when pressure drop equals the weight of bed material per unit area. Above that velocity, bed material will be suspended in the gas stream in a turbulent flow condition. With the total surface area of suspended particles on the order of a million times the inside surface area of the container, convection heat transfer from gas to bed material is correspondingly large. Heat transfer from suspended particles to load is by conduction during repeated impact. The combination can provide overall coefficients upward of 10 times those available with open convection, permitting the heating of thick and thin load sections to nearly uniform temperatures by allowing a low gas to load thermal head.

58.8.16 Combined Heat-Transfer Coefficients

Many furnace heat-transfer problems will combine two or more methods of heat transfer, with thermal resistances in series or in parallel. In a combustion chamber, the resistance to radiation from gas to load will be parallel to the resistance from gas to walls to load, which is two resistances in series. Heat flow through furnace walls combines a series of resistances in series, combustion gases to inside wall surface, consecutive layers of the wall structure, and outside wall surface to surroundings, the last a combination of radiation and convection in parallel.

As an example, consider an insulated, water-cooled tube inside a furnace enclosure. With a tube outside diameter of 0.5 ft and a cylindrical insulation enclosure with an outside diameter of 0.75 ft, the net thickness will be 0.125 ft. The mean area at midthickness is $\pi(0.5 + 0.75)/2$, or 1.964 ft^2 per ft of length. Outer surface area of insulation is 0.75π, or 2.36 ft^2 per linear foot. Conductivity of insulation is $k = 0.20$. The effective radiation factor from gas to surface is assumed at 0.5 including reradiation from walls. For the two resistances in series,

$$0.1713 \times 0.5 \times 2.36(29.6^4 - T_s^4) = 1.964(T_s - 150) \times \frac{0.20}{0.125}$$

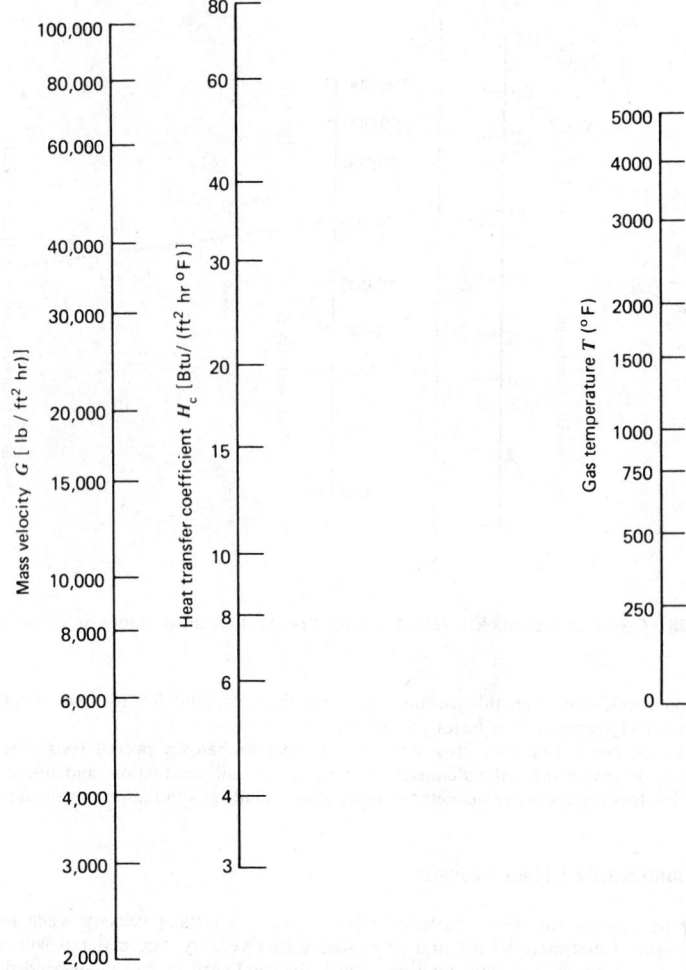

Fig. 58.29 Convection coefficient (H_c) for jet impingement of air or flue gas on plane surfaces, for spaced slots, 0.375 in. wide at 18–24 in. centers, 8 in. from load.[1]

By trial, the receiver surface temperature is found to be about 2465°F. Heat transfer is about 7250 Btu/hr · linear ft or 9063 Btu/hr · ft² water-cooled tube surface.

If the insulated tube in the preceding example is heated primarily by convection, a similar treatment can be used to find receiver surface temperature and overall heat transfer.

For radiation through furnace wall openings, heat transfer in Btu/hr · ft² · °F is reduced by wall thickness, and the result can be calculated similarly to the problem of two parallel planes of equal size connected by reradiating walls, as shown in Fig. 58.19.

Heat transfer in internally fired combustion tubes ("radiant tubes") is a combination of convection and gas radiation from combustion gases to tube wall. External heat transfer from tubes to load will be direct radiation and reradiation from furnace walls, as illustrated in Fig. 58.19. The overall factor for internal heat transfer can be estimated from Fig. 58.31, calculated for 6 in. and 8 in. inside diameter tubes. The convection coefficient increases with firing rate and to some extent with temperature. The gas radiation factor depends on temperature and inside diameter. The effect of flame luminosity has not been considered.

58.9   FLUID FLOW

Fluid flow problems of interest to the furnace engineer include the resistance to flow of air or flue gas, over a range of temperatures and densities through furnace ductwork, stacks and flues, or recupera-

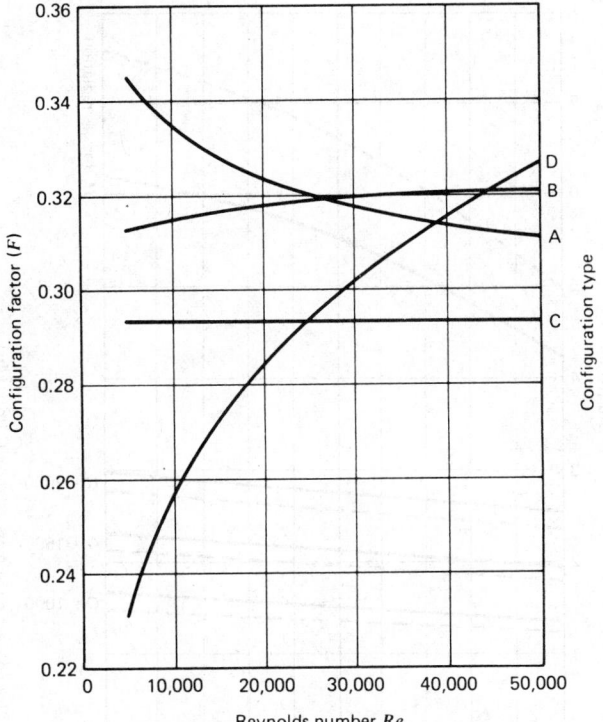

Fig. 58.30 Configuration factors for convection heat transfer, air or flue gas through tube banks.[1]

tors and regenerators. Flow of combustion air and fuel gas through distribution piping and burners will also be considered. Liquid flow, of water and fuel oil, must also be evaluated in some furnace designs but will not be treated in this chapter.

To avoid errors resulting from gas density at temperature, velocities will be expressed as mass velocities in units of $G = $ lb/hr $\cdot$ ft². Because the low pressure differentials in systems for flow of air or flue gas are usually measured with a manometer, in units of inches of water column (in. H_2O), that will be the unit used in the following discussion.

The relation of velocity head h_v in in. H_2O to mass velocity G is shown for a range of temperatures in Fig. 58.32. Pressure drops as multiples of h_v are shown, for some configurations used in furnace design, in Figs. 58.33 and 58.34. The loss for flow across tube banks, in multiples of the velocity head, is shown in Fig. 58.35 as a function of the Reynolds number.

The Reynolds number Re is a dimensionless factor in fluid flow defined as $\text{Re} = DG/\mu$, where D is inside diameter or equivalent dimension in feet, G is mass velocity as defined above, and μ is viscosity as shown in Fig. 58.9. Values of Re for air or flue gas, in the range of interest, are shown in Fig. 58.36. Pressure drop for flow through long tubes is shown in Fig. 58.37 for a range of Reynolds numbers and equivalent diameters.

58.9.1 Preferred Velocities

Mass velocities used in contemporary furnace design are intended to provide an optimum balance between construction costs and operating costs for power and fuel; some values are listed below:

Medium	Mass Velocity G	Velocity Head (in. H_2O)
Cold air	15,000	0.7
800°F air	10,000	0.3
2200°F flue gas	1,750	0.05
1500°F flue gas	2,000	0.05

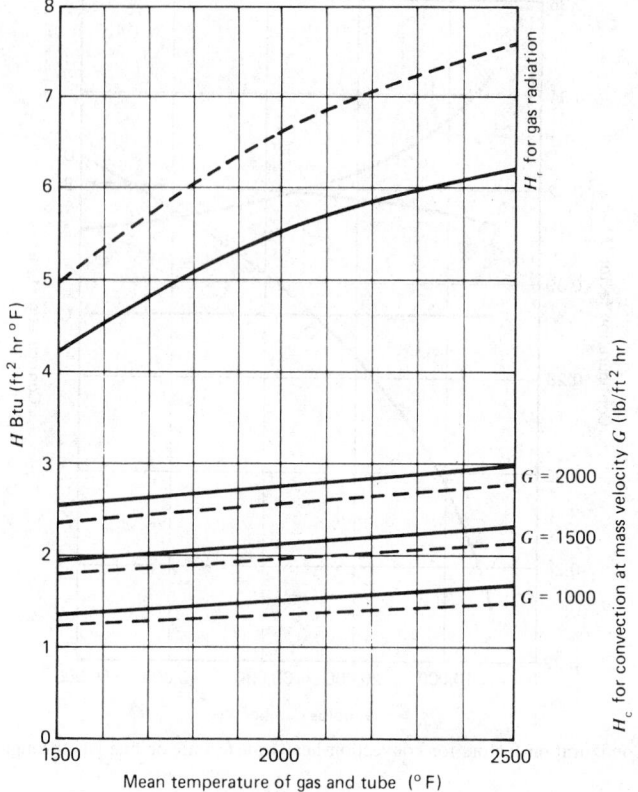

Fig. 58.31 Gas radiation (H_r) and convection (H_c) coefficients for flue gas inside radiant tubes.[1]

The use of these factors will not necessarily provide an optimum cost balance. Consider a furnace stack of self-supporting steel construction, lined with 6 in. of gunned insulation. For $G = 2000$ and $h_v = 0.05$ at 1500°F, an inside diameter of 12 ft will provide a flow of 226,195 lb/hr. To provide a net draft of 1 in. H_2O with stack losses of about 1.75 h_v or 0.0875 in., the effective height from Fig. 58.38 is about 102 ft. By doubling the velocity head to 0.10 in. H_2O, G at 1500°F becomes 3000. For the same mass flow, the inside diameter is reduced to 9.8 ft. The pressure drop through the stack increases to about 0.175 in., and the height required to provide a net draft of 1 in. increases to about 110 ft. The outside diameter area of the stack is reduced from 4166 ft² to $11 \times 3.1416 \times 110 = 3801$ ft². If the cost per square foot of outside surface is the same for both cases, the use of a higher stack velocity will save construction costs. It is accordingly recommended that specific furnace designs receive a more careful analysis before selecting optimum mass velocities.

Stack draft, at ambient atmospheric temperature of 70°F, is shown in Fig. 58.38 as a function of flue gas temperature. Where greater drafts are desirable with a limited height of stack, a jet-type stack can be used to convert the momentum of a cold air jet into stack draft. Performance data are available from manufacturers.

58.9.2 Centrifugal Fan Characteristics

Performance characteristics for three types of centrifugal fans are shown in Fig. 58.39. More exact data are available from fan manufacturers. Note that the backward curved blade has the advantage of limited horsepower demand with reduced back pressure and increasing volume, and can be used where system resistance is unpredictable. The operating point on the pressure–volume curve is determined by the increase of duct resistance with flow, matched against the reduced outlet pressure, as shown in the upper curve.

58.9.3 Laminar and Turbulent Flows

The laminar flow of a fluid over a boundary surface is a shearing process, with velocity varying from zero at the wall to a maximum at the center of cross section or the center of the top surface for

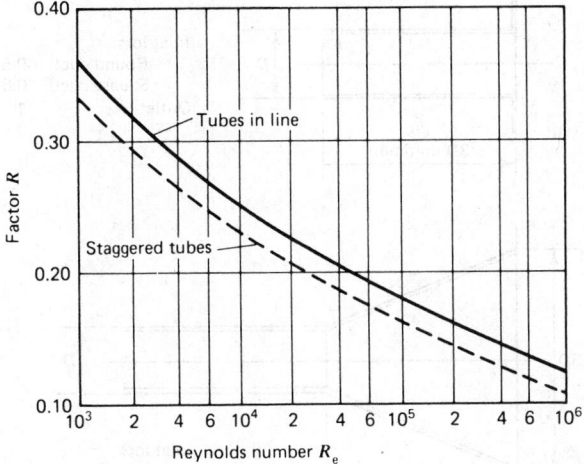

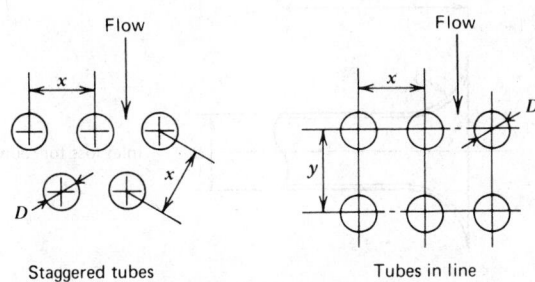

Fig. 58.32 Heat loss for flow of air or flue gas across tube banks at atmospheric pressure (velocity head) $\times F \times R$.

Staggered Tubes		Tubes in Line		Factor F for x/D		
x/D	Factor F	y/D	1.5	2	3	4
1.5	2.00	1.25	1.184	0.576	0.334	0.268
2	1.47	1.5	1.266	0.656	0.387	0.307
3	1.22	2	1.452	0.816	0.497	0.390
4	1.14	3	1.855	1.136	0.725	0.572
		4	2.273	1.456	0.957	0.761

liquids in an open channel. Above a critical Reynolds number, between 2000 and 3000 in most cases, flow becomes a rolling action with a uniform velocity extending almost to the walls of the duct, and is identified as turbulent flow.

With turbulent flow the pressure drop is proportional to D; the flow in a large duct can be converted from turbulent to laminar by dividing the cross-sectional area into a number of parallel channels. If flow extends beyond the termination of these channels, the conversion from laminar to turbulent flow will occur over some distance in the direction of flow.

Radial mixing with laminar flow is by the process of diffusion, which is the mixing effect that occurs in a chamber filled with two different gases separated by a partition after the partition is removed. Delayed mixing and high luminosity in the combustion of hydrocarbon gases can be accomplished by "diffusion combustion," in which air and fuel enter the combustion chamber in parallel streams at equal and low velocity.

58.10 BURNER AND CONTROL EQUIPMENT

With increasing costs of fuel and power, the fraction of furnace construction and maintenance costs represented by burner and control equipment can be correspondingly increased. Burner designs should

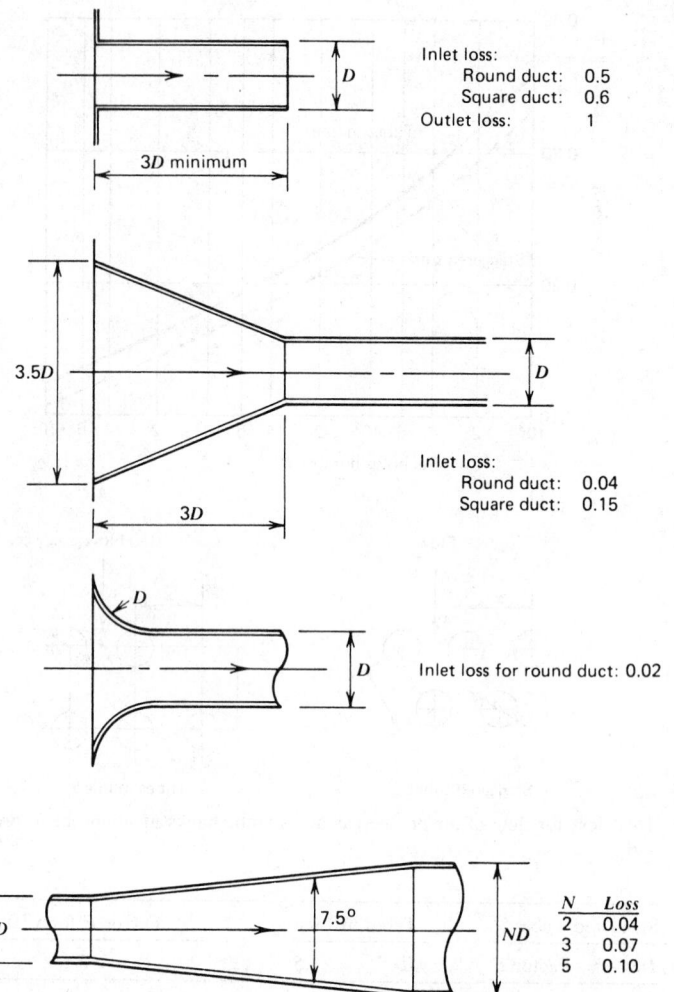

Fig. 58.33 Pressure drop in velocity heads for flow of air or flue gas through entrance configurations or expansion sections.[1]

be selected for better control of flame pattern over a wider range of turndown and for complete combustion with a minimum excess air ratio over that range.

Furnace functions to be controlled, manually or automatically, include temperature, internal pressure, fuel/air ratio, and adjustment of firing rate to anticipated load changes. For intermittent operation, or for a wide variation in required heating capacity, computer control may be justified to anticipate required changes in temperature setting and firing rates, particularly in consecutive zones of continuous furnaces.

58.10.1 Burner Types

Burners for gas fuels will be selected for the desired degree of premixing of air and fuel, to control flame pattern, and for the type of flame pattern, compact and directional, diffuse or flat flame coverage of adjacent wall area. Burners for oil fuels, in addition, will need provision for atomization of fuel oil over the desired range of firing rates.

The simplest type of gas burner comprises an opening in a furnace wall, through which combustion air is drawn by furnace draft, and a pipe nozzle to introduce fuel gas through that opening. Flame pattern will be controlled by gas velocity at the nozzle and by excess air ratio. Fuel/air ratio will be manually controlled for flame appearance by the judgment of the operator, possibly supplemented by

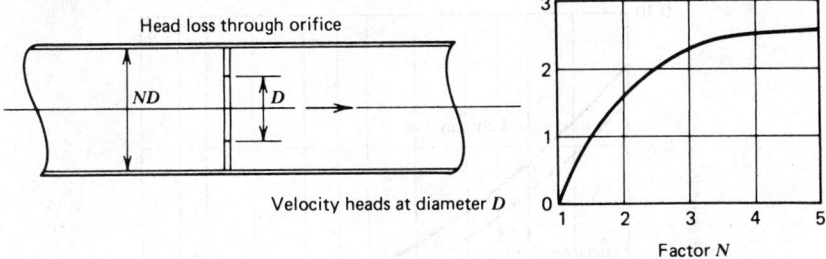

Head loss through orifice

Velocity heads at diameter D

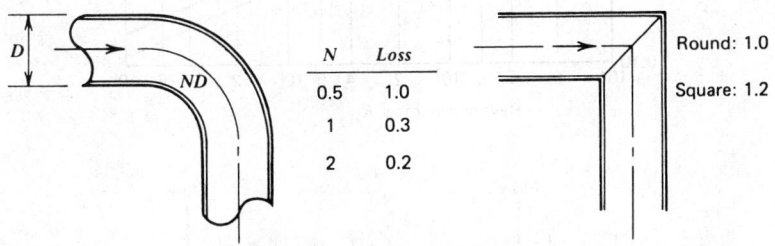

Head loss in pipe or duct elbows

N	$Loss$
0.5	1.0
1	0.3
2	0.2

Round: 1.0

Square: 1.2

Proportioning Piping for uniform distribution

Total pressure = static pressure + velocity head

Area at D should exceed 2.5 × combined areas of A, B, and C

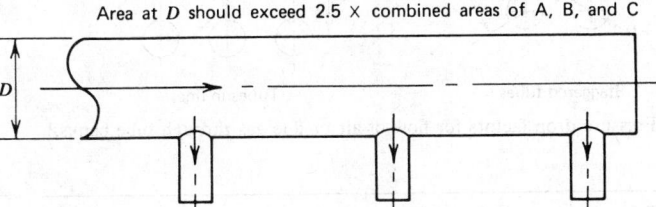

Fig. 58.34 Pressure drop in velocity heads for flow of air or flue gas through orifices, elbows, and lateral outlets.[1]

continuous or periodic flue gas analysis. In regenerative furnaces, with firing ports serving alternately as exhaust flues, the open pipe burner may be the only practical arrangement.

For one-way fired furnaces, with burner port areas and combustion air velocities subject to control, fuel/air ratio control can be made automatic over a limited range of turndown with several systems, including:

Mixing in venturi tube, with energy supplied by gas supply inducing atmospheric air. Allows simplest piping system with gas available at high pressure, as from some natural gas supplies.

Venturi mixer with energy from combustion air at intermediate pressure. Requires air supply piping and distribution piping from mixing to burners.

With both combustion air and fuel gas available at intermediate pressures, pressure drops through adjustable orifices can be matched or proportioned to hold desired flow ratios. For more accurate control, operation of flow control valves can be by an external source of energy.

Proportioning in venturi mixers depends on the conservation of momentum—the product of flow rate and velocity or of orifice area and pressure drop. With increased back pressure in the combustion chamber, fuel/air ratio will be increased for the high pressure gas inspirator, or decreased with air pressure as the source of energy, unless the pressure of the induced fluid is adjusted to the pressure in the combustion chamber.

The arrangement of a high-pressure gas inspirator system is illustrated in Fig. 58.40. Gas enters the throat of the venturi mixer through a jet on the axis of the opening. Air is induced through the

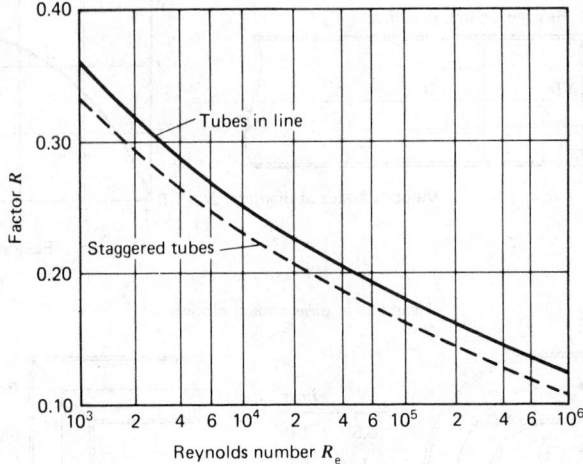

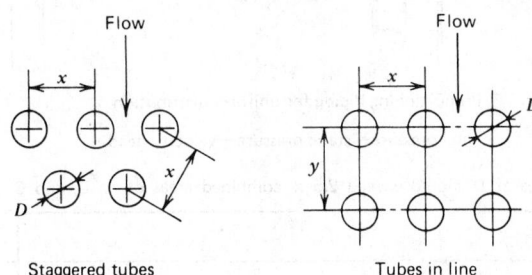

Fig. 58.35 Pressure drop factors for flow of air or flue gas through tube banks.[1]

Staggered Tubes		Tubes in Line		Factor F for x/D		
x/D	Factor F	y/D	1.5	2	3	4
1.5	2.00	1.25	1.184	0.576	0.334	0.268
2	1.47	1.5	1.266	0.656	0.387	0.307
3	1.22	2	1.452	0.816	0.497	0.390
4	1.14	3	1.855	1.136	0.725	0.572
		4	2.273	1.456	0.957	0.761

surrounding area of the opening, and ratio control can be adjusted by varying the air inlet opening by a movable shutter disk. A single inspirator can supply a number of burners in one firing zone, or a single burner.

For the air primary mixing system, a representative arrangement is shown in Fig. 58.41. The gas supply is regulated to atmospheric, or to furnace gas pressure, by a diaphragm-controlled valve. Ratio control is by adjustment of an orifice in the gas supply line. With air flow the only source of energy, errors in proportioning can be introduced by friction in the gas-pressure control valve. Each mixer can supply one or more burners, representing a control zone.

With more than one burner per zone, the supply manifold will contain a combustible mixture that can be ignited below a critical port velocity to produce a backfire that can extinguish burners and possibly damage the combustion system. This hazard has made the single burner per mixer combination desirable, and many contemporary designs combine mixer and burner in a single structure.

With complete premixing of fuel and air, the flame will be of minimum luminosity, with combustion complete near the burner port. With delayed mixing, secured by introducing fuel and air in separate streams, through adjacent openings in the burner, or by providing a partial premix of fuel with a fraction of combustion air, flame luminosity can be controlled to increase flame radiation.

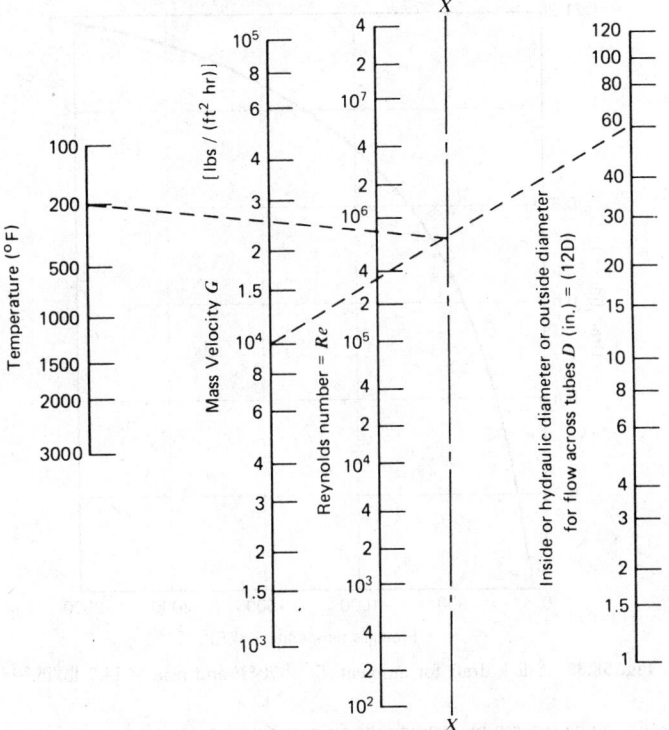

Fig. 58.36 Reynolds number (Re) for flow of air or flue gas through tubes or across tube banks.[1]

In a burner providing no premix ahead of the combustion chamber, flame pattern is determined by velocity differentials between air and fuel streams, and by the subdivision of air flow into several parallel streams. This type of burner is popular for firing with preheated combustion air, and can be insulated for that application.

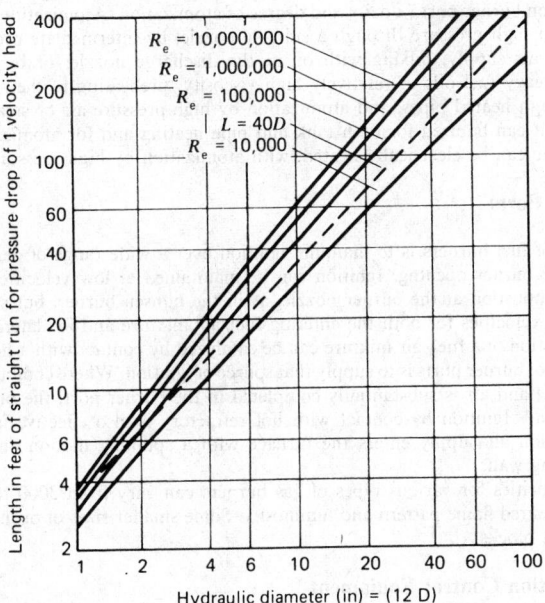

Fig. 58.37 Length in feet for pressure drop of one velocity head, for flow of air or flue gas, as a function of Re and D.[1]

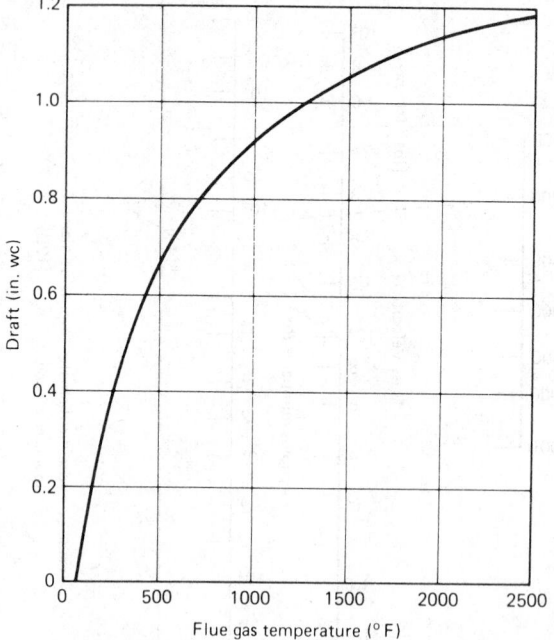

Fig. 58.38 Stack draft for ambient $T_g = 70°F$ and psia $= 14.7$ lb/in.^{2}.[1]

Partial premix can be secured by dividing the air flow between a mixing venturi tube and a parallel open passage.

With the uncertainty of availability of contemporary fuel supplies, dual fuel burners, optionally fired with fuel gas or fuel oil, can be used. Figure 58.42 illustrates the design of a large burner for firing gas or oil fuel with preheated air. For oil firing, an oil-atomizing nozzle is inserted through the gas tube. To avoid carbon buildup in the oil tube from cracking of residual oil during gas firing, the oil tube assembly is removable.

Oil should be atomized before combustion in order to provide a compact flame pattern. Flame length will depend on burner port velocity and degree of atomization. Atomization can be accomplished by delivery of oil at high pressure through a suitable nozzle; by intermediate pressure air, part or all of the combustion air supply, mixing with oil at the discharge nozzle; or by high-pressure air or steam. For firing heavy fuel oils of relatively high viscosity, preheating in the storage tank, delivery to the burner through heated pipes, and atomization by high-pressure air or steam will be needed. If steam is available, it can be used for both tank and pipe heating and for atomization. Otherwise, the tank and supply line can be electrically heated, with atomization by high-pressure air.

58.10.2 Burner Ports

A major function of fuel burners is to maintain ignition over a wide range of demand and in spite of lateral drafts at the burner opening. Ignition can be maintained at low velocities by recirculation of hot products of combustion at the burner nozzle, as in the bunsen burner, but stability of ignition is limited to low port velocities for both the entering fuel/air mixture and for lateral drafts at the point of ignition. Combustion of a fuel/air mixture can be catalyzed by contact with a hot refractory surface. A primary function of burner ports is to supply that source of ignition. Where combustion of a completely mixed source of fuel and air is substantially completed in the burner port, the process is identified as "surface combustion." Ignition by contact with hot refractory is also effective in flat flame burners, where the combustion air supply enters the furnace with a spinning motion and maintains contact with the surrounding wall.

Burner port velocities for various types of gas burners can vary from 3000 to 13,000 lb/hr · ft², depending on the desired flame pattern and luminosity. Some smaller sizes of burners are preassembled with refractory port blocks.

58.10.3 Combustion Control Equipment

Furnace temperature can be measured by a bimetallic thermocouple inserted through the wall or by an optical sensing of radiation from furnace walls and products of combustion. In either case, an

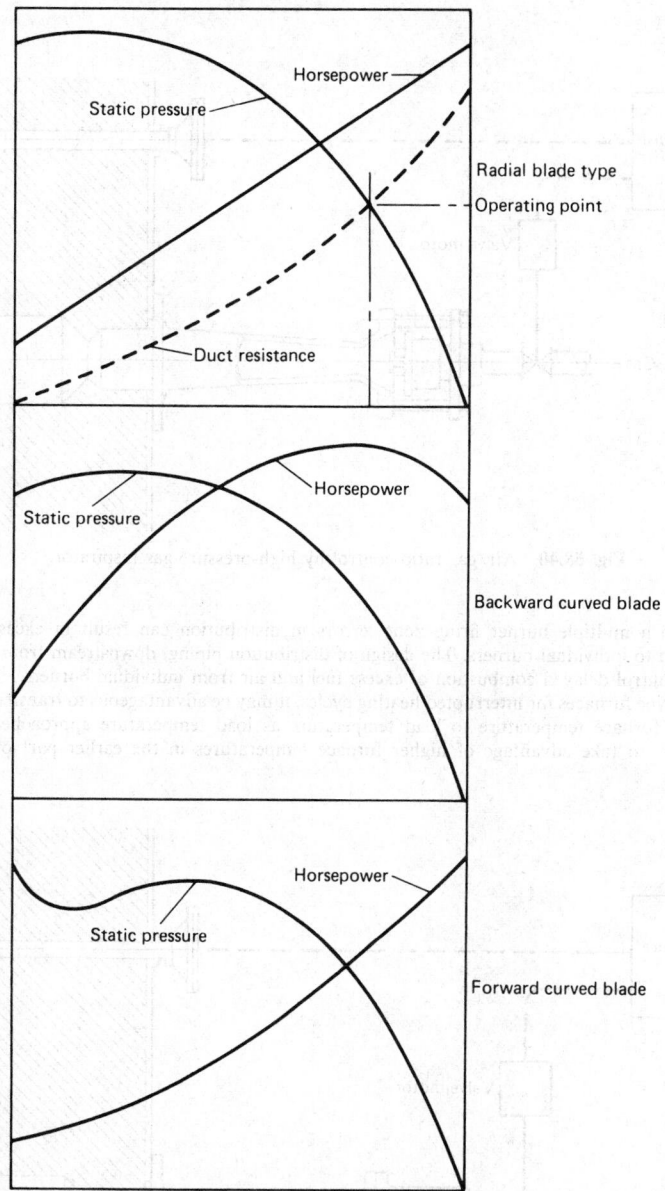

Fig. 58.39 Centrifugal fan characteristics.[1]

electrical impulse is translated into a temperature measurement by a suitable instrument and the result indicated by a visible signal and optionally recorded on a moving chart. For automatic temperature control, the instrument reading is compared to a preset target temperature, and the fuel and air supply adjusted to match through a power-operated valve system.

Control may be on–off, between high and low limits; three position, with high, normal, and off valve openings; or proportional with input varying with demand over the full range of control. The complexity and cost of the system will, in general, vary in the same sequence. Because combustion systems have a lower limit of input for proper burner operation or fuel/air ratio control, the proportioning temperature control system may cut off fuel input when it drops to that limit.

Fuel/air ratios may be controlled at individual burners by venturi mixers or in multiple burner firing zones by similar mixing stations. To avoid back firing in burner manifolds, the pressures of air and·gas supplies can be proportioned to provide the proper ratio of fuel and air delivered to individual burners through separate piping. Even though the desired fuel/air ratio can be maintained for the

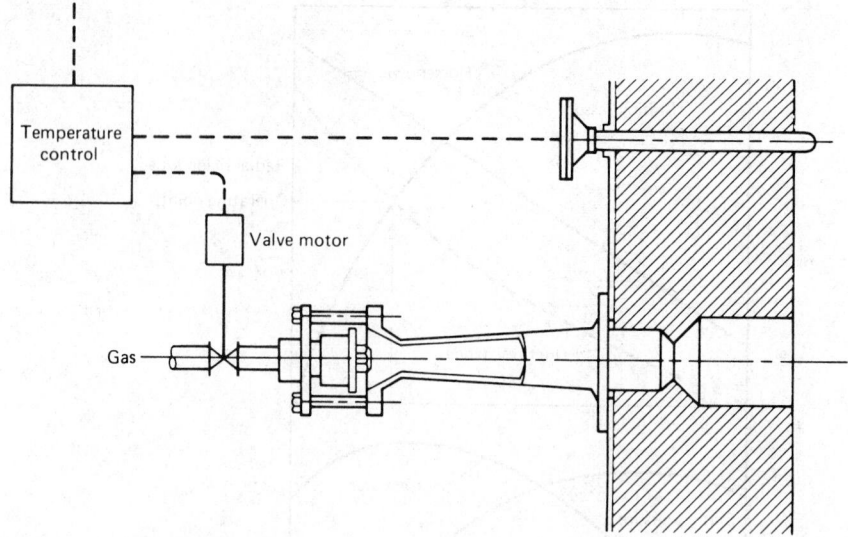

Fig. 58.40   Air/gas ratio control by high-pressure gas inspirator.[1]

total input to a multiple burner firing zone, errors in distribution can result in excess air or fuel being supplied to individual burners. The design of distribution piping, downstream from ratio control valves, will control delayed combustion of excess fuel and air from individual burners.

In batch-type furnaces for interrupted heating cycles, it may be advantageous to transfer temperature control from furnace temperature to load temperature as load temperature approaches the desired level, in order to take advantage of higher furnace temperatures in the earlier part of the heating

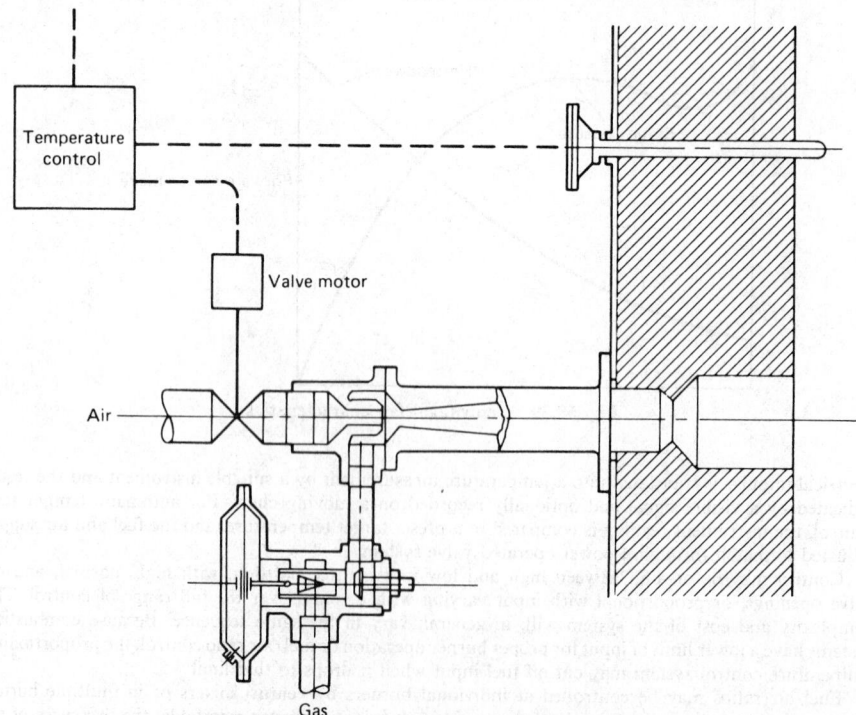

Fig. 58.41   Air/gas ratio control by air inspirator.[1]

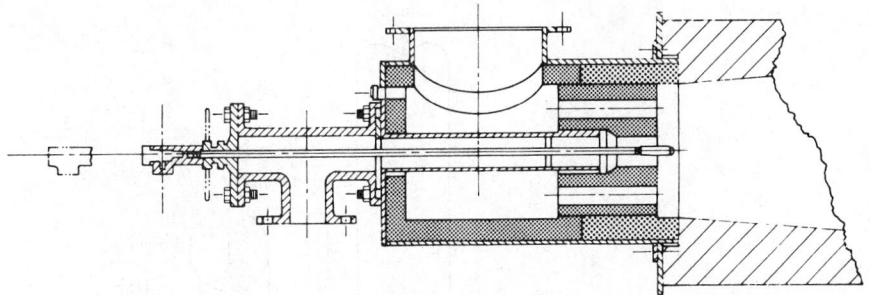

Fig. 58.42 Dual fuel burner with removable oil nozzle.[1] (Courtesy Bloom Engineering Company.)

cycle. An example is a furnace for annealing steel strip coils. Because heat flow through coil laminations is a fraction of that parallel to the axis of the coil, coils may be stacked vertically with open coil separators between them, to provide for heat transfer from recirculated furnace atmosphere to the end surfaces of coils. For bright annealing, the furnace atmosphere will be nonoxidizing, and the load will be enclosed in an inner cover during heating and cooling, with the atmosphere recirculated by a centrifugal fan in the load support base, to transfer heat from the inner cover to end faces of coils. There will also be some radiation heat transfer from the inner cover to the cylindrical surface of the coil stack.

Inner covers are usually constructed of heat-resisting alloy, with permissible operating temperatures well above the desired final load temperature. A preferred design provides for initial control of furnace inside wall temperature from a thermocouple inserted through the furnace wall, with control switched to a couple in the support base, in contact with the bottom of the coil stack, after load temperature reaches a present level below the desired final temperature.

To avoid leakage of combustion gases outward through furnace walls, with possible overheating of the steel enclosure, or infiltration of cold air that could cause nonuniform wall temperatures, control of internal furnace pressure to slightly above ambient is desirable. This can be accomplished by an automatic damper in the outlet flue, adjusted to hold the desired pressure at the selected point in the furnace enclosure. In furnaces with door openings at either end, the point of measurement should be close to hearth level near the discharge end. A practical furnace pressure will be 0.01–0.05 in. H_2O.

With recuperative or regenerative firing systems, the preferred location of the control damper will be between the waste-heat recovery system and the stack, to operate at minimum temperature. In high-temperature furnaces without waste-heat recovery, a water-cooled damper may be needed.

With combusiton air preheated before distribution to several firing zones, the ratio control system for each zone will need adjustment to entering air temperature. However, if each firing zone has a separate waste-heat recovery system, the zone air supply can be measured before preheating to maintain the balance with fuel input.

The diagram of a combustion control system in Fig. 58.43 shows how these control functions can be interlocked with the required instrumentation.

For automatic furnace pressure control to be effective, it should be used in combination with proportioning-type temperature control. With on–off control, for example, the control of furnace pressure at zero firing rate cannot be accomplished by damper adjustment, and with a continuous variation in firing rate between maximum and minimum limits, or between maximum and off, the adjustment of damper position to sudden changes in firing rate will involve a time-lag factor that can make control ineffective.

An important function of a furnace control system is to guard against safety hazards, such as explosions, fires, and personal injury. Requirements have been well defined in codes issued by industrial insurers, and include provision for continuous ignition of burners in low-temperature furnaces, purging of atmosphere furnaces and combustion of hydrogen or carbon monoxide in effluent atmospheres, and protection of operating personnel from injury by burning, mechanical contact, electrical shock, poisoning by inhalation of toxic gases, or asphyxiation. Plants with extensive furnace operation should have a safety engineering staff to supervise selection, installation, and maintenance of safety hazard controls and to coordinate the instruction of operating personnel in their use.

58.10.4 Air Pollution Control

A new and increasing responsibility of furnace designers and operators is to provide controls for toxic, combustible, or particulate materials in furnace flue gases, to meet federal or local standards

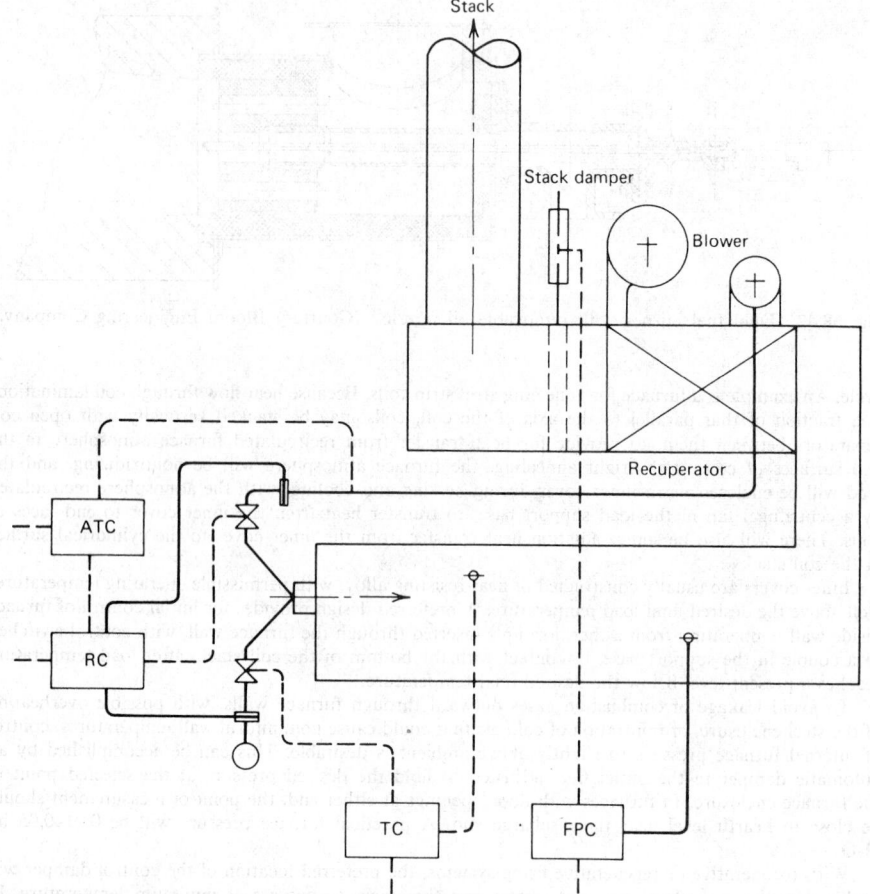

Fig. 58.43 Combustion control diagram for recuperative furnace.[1]

for air quality. Designs for furnaces to be built in the immediate future should anticipate a probable increase in restrictions of air pollution in coming years.

Toxic contaminants include sulfur and chlorine compounds, nitrogen oxides, carbon monoxide, and radioactive wastes. The epidemic of "acid rain" in areas downwind from large coal-burning facilities is an example.

Combustible contaminants include unburned fuel, soot, and organic particulates from incinerators, and the visible constituents of smoke, except for steam. Other particulates include suspended ash and suspended solids from calcination processes.

Types of control equipment include:

1. Bag filters or ceramic fiber filters to remove suspended solids. Filters require periodic cleaning or replacement, and add to the pressure drop in flue gases leaving the system.

2. Electrostatic filters, in which suspended particles pass through a grid to be electrically charged, and are collected on another grid or on spaced plates with the opposite potential. Smaller units are cleaned periodically by removal and washing. Large industrial units are cleaned in place. A possible objection to their use is a slight increase in the ozone content of treated air.

3. Wet scrubbers are particularly effective for removing water-soluble contaminants such as sulfur and chlorine compounds. They can be used in place of filters for handling heavy loads of solid particulates such as from foundry cupola furnaces, metal-refining processes, and lime kilns. Waste material is collected as a mud or slurry, requiring proper disposal to avoid solid-waste problems.

4. Combustible wastes, such as the solvent vapors from organic coating ovens, may be burned in incinerator units by adding combustion air and additional fuel as required. Fuel economy

may be improved by using waste heat from combustion to preheat incoming gases through a recuperator. The same system may be used for combustible solid particulates suspended in flue gases.

5. Radioactive wastes from nuclear power plants will usually be in the form of suspended solids that can be treated accordingly if suitable facilities for disposal of collected material are available, or as radioactive cooling water for which a suitable dumping area will be needed.

58.11 WASTE HEAT RECOVERY SYSTEMS

In fuel-fired furnaces, a fraction of the energy from combustion leaves the combustion chamber as sensible heat in waste gases, and the latent heat of evaporation for any water vapor content resulting from the combustion of hydrogen. Losses increase with flue gas temperature and excess air, and can reach 100% of input when furnace temperatures equal theoretical flame temperatures.

Waste heat can be recovered in several ways:

1. Preheating incoming loads in a separate enclosure ahead of the furnace.
2. Generating process steam, or steam for electric power generation. Standby facilities will be needed for continuous demand, to cover interruptions of furnace operation.
3. Preheating combustion air, or low-Btu fuels, with regenerative or recuperative firing systems.

58.11.1 Regenerative Air Preheating

For the high flue gas temperatures associated with glass- and metal-melting processes, for which metallic recuperators are impractical, air may be preheated by periodical reversal of the direction of firing, with air passing consecutively through a hot refractory bed or checker chamber, the furnace combustion chamber, and another heat-storage chamber in the waste-gas flue. The necessary use of the furnace firing port as an exhaust port after reversal limits the degree of control of flame patterns and the accuracy of fuel/air control in multiple port furnaces. Regenerative firing is still preferred, however, for open hearth furnaces used to convert blast furnace iron to steel, for large glass-melting furnaces, and for some forging operations.

A functional diagram of a regenerative furnace is shown in Fig. 58.44. The direction of flow of

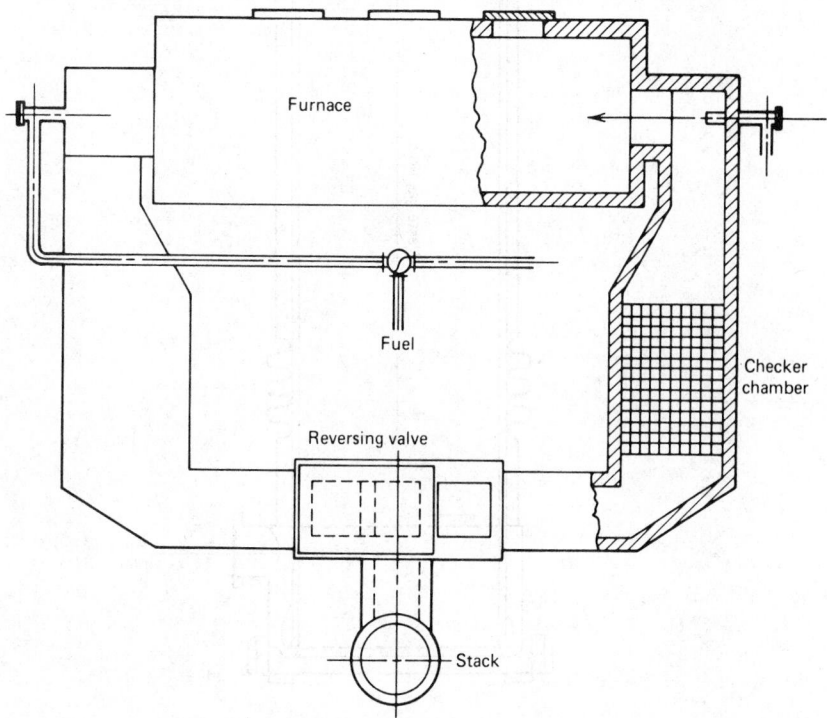

Fig. 58.44 Regenerative furnace diagram.[1]

combustion air and flue gas is reversed by a valve arrangement, connecting the low-temperature end of the regenerator chamber to either the combustion air supply or the exhaust stack. Fuel input is reversed simultaneously, usually by an interlocked control. Reversal can be in cycles of from 10 to 30 min duration, depending primarily on furnace size.

58.11.2 Recuperator Systems

Recuperative furnaces are equipped with a heat exchanger arranged to transfer heat continuously from outgoing flue gas to incoming combustion air. Ceramic heat exchangers, built up of refractory tubes or refractory block units arranged for cross flow of air and flue gas, have the advantage of higher temperature limits for incoming flue gas, and the disadvantage of leakage of air or flue gas between passages, with leakage usually increasing with service life and pressure differentials. With the improvement in heat-resistant alloys to provide useful life at higher temperatures, and with better control of incoming flue gas temperatures, metallic recuperators are steadily replacing ceramic types.

Metal recuperators can be successfully used with very high flue gas temperatures if entering temperatures are reduced by air dilution or by passing through a high-temperature waste-heat boiler.

Familiar types of recuperators are shown in the accompanying figures:

Figure 58.45: radiation or stack type. Flue gases pass through an open cylinder, usually upward,

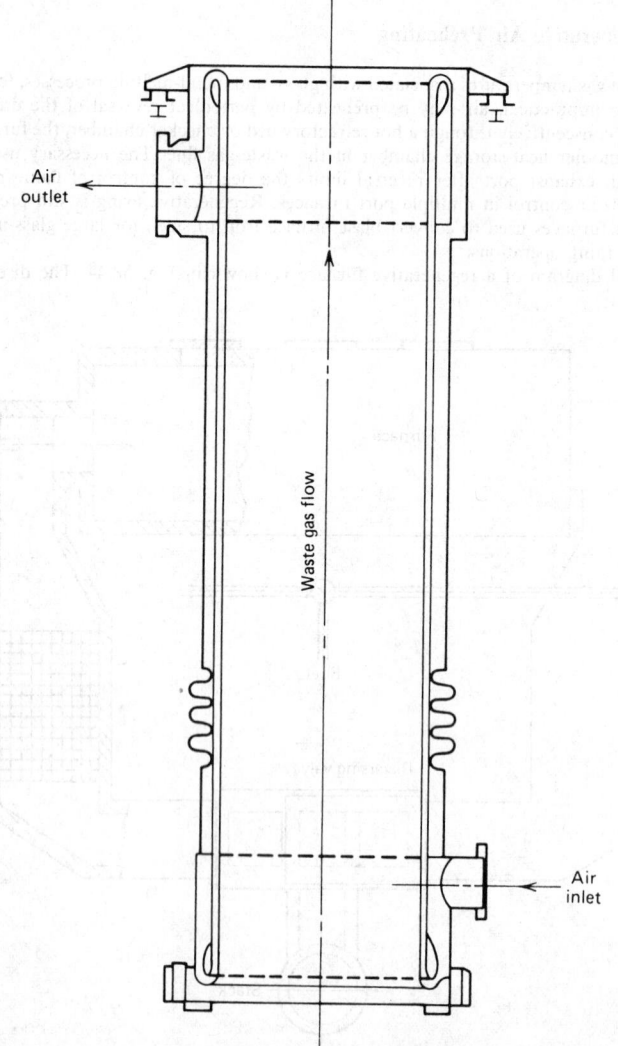

Fig. 58.45 Stack-type recuperator.[1] (Courtesy Morgan Engineering Company.)

with heat transfer primarily by gas radiation to the surrounding wall. An annular air passage is provided between inner and outer cylinders, in which heat is transferred to air at high velocity by gas radiation and convection, or by solid-state radiation from inner to outer cylinders and convection. The radiation recuperator has the advantage of acting as a portion of an exhaust stack, usually with flue gas and air counterflow. Disadvantages are distortion and resulting uneven distribution of air flow, resulting from differential thermal expansion of the inner tube, and the liability of damage from secondary combustion in the inner chamber.

Figure 58.46: cross-flow tubular type. By passing air through a series of parallel passes, as in a tube assembly, with flue gas flowing across tubes, relatively high heat-transfer rates can be achieved. It will ordinarily be more practical to use higher velocities on the air side, and use an open structure on the flue gas side to take some advantage of gas radiation. Figure 58.46 shows a basic arrangement, with air tubes in parallel between hot and cold air chambers at either end. Some problems may be introduced by differential thermal expansion of parallel tubes, and tubes may be curved to accommodate variations in length by lateral distortion.

A popular design avoids the problems of thermal expansion by providing heat-exchange tubes with concentric passages and with connections to inlet and outlet manifolds at the same end. Heat transfer from flue gas to air is by gas radiation and convection to the outer tube surface, by convection

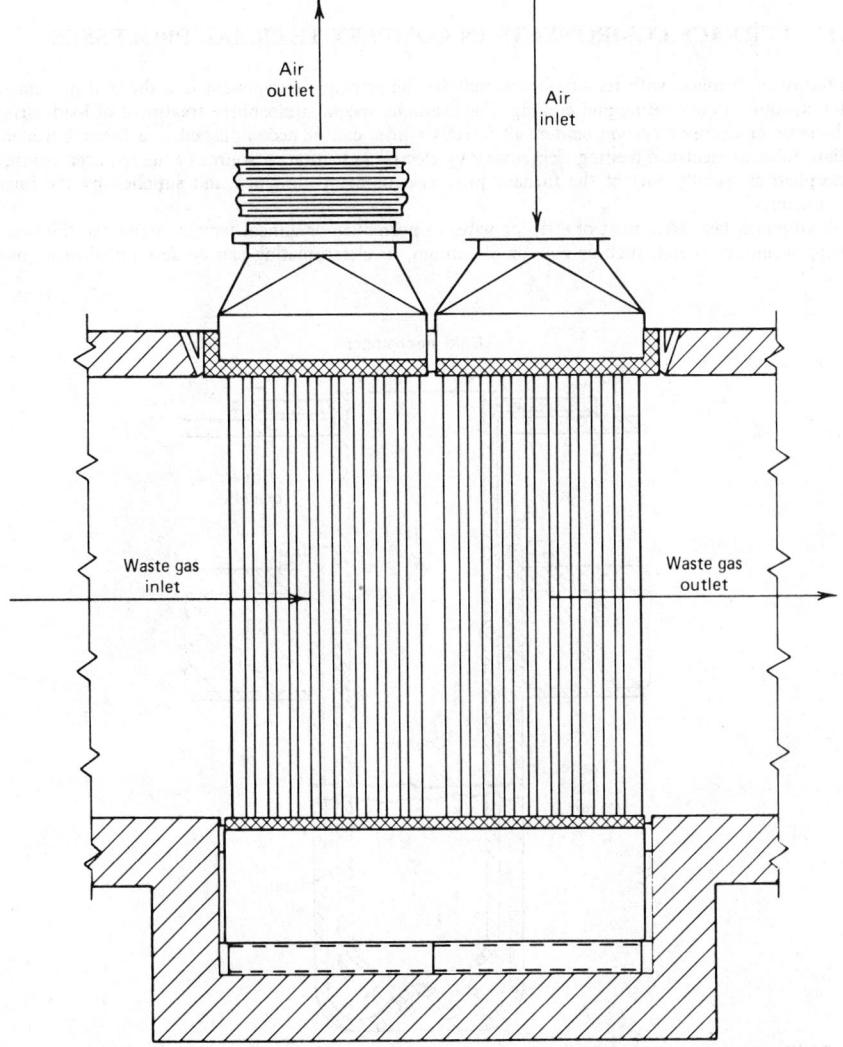

Fig. 58.46 Cross-flow-type recuperator in waste-gas flue.[1] (Courtesy Thermal Transfer Corporation.)

from the inner surface to high-velocity air, and by solid-state radiation between outer and inner tubes in series with convection from inner tubes to air. Concentric tube recuperators are usually designed for replacement of individual tube units without a complete shutdown and cooling of the enclosing flue. The design is illustrated in Fig. 58.47.

58.11.3 Recuperator Combinations

To provide preheated air at pressure required for efficient combustion, without excessive air leakage from the air to the flue gas side in refractory recuperators, the air pressure can be increased between the recuperator and burner by a booster fan. Top air temperatures will be limited by fan materials. As an alternate, air temperatures can be boosted by a jet pump with tolerance for much higher temperatures.

In a popular design for recuperator firing of soaking pits, flue gases pass through the refractory recuperator at low pressure, with air flowing counterflow at almost the same pressure. Air flow is induced by a jet pump, and, to increase the jet pump efficiency, the jet air can be preheated in a metal recuperator between the refractory recuperator and the stack. Because the metal recuperator can handle air preheated to the limit of the metal structure, power demand can be lowered substantially below that for a cold air jet.

Radiant tubes can be equipped with individual recuperators, as shown in Fig. 58.48. Some direct-firing burners are available with integral recuperators.

58.12 FURNACE COMPONENTS IN COMPLEX THERMAL PROCESSES

An industrial furnace, with its auxiliaries, may be the principal component in a thermal process with functions other than heating and cooling. For example, special atmosphere treatment of load surfaces, to increase or decrease carbon content of ferrous alloys, can be accomplished in a furnace heated by radiant tubes or electrical heating elements or by electric induction. A source of the required controlled atmosphere is usually part of the furnace process equipment, designed and supplied by the furnace manufacturer.

Continuous heat treatment of strip or wire, to normalize or anneal ferrous materials, followed by coating in molten metal, such as zinc or aluminum, or electroplating can be accomplished by one of

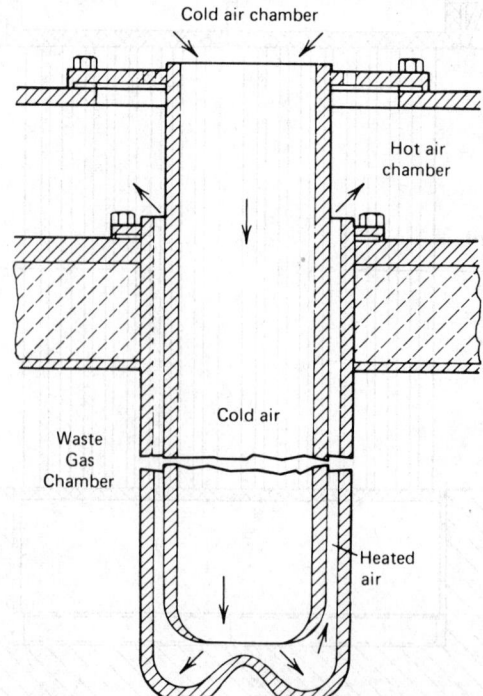

Fig. 58.47 Concentric tube recuperator, Hazen type.[1] (Courtesy C-E Air Preheater Division, Combustion Engineering, Inc.)

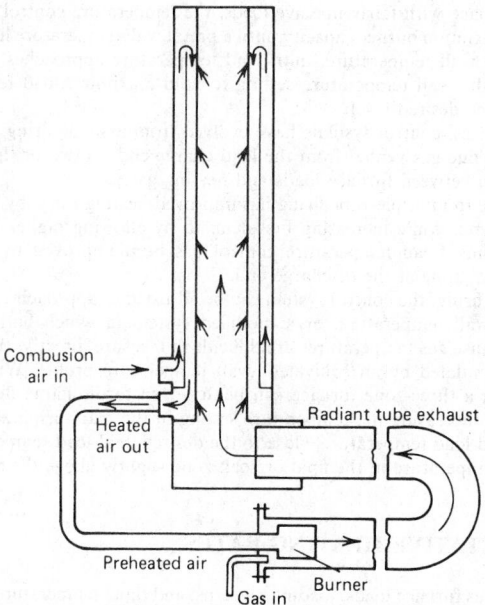

Combusion air in →

Heated air out

Radiant tube exhaust

Preheated air

Gas in

Burner

Fig. 58.48 Radiant tube recuperator.[1] (Courtesy Holcroft Division, Thermo-Electron Corp.)

two arrangements for furnace coating lines. One arrangement has a sequence of horizontal passes, with a final cooling zone to regulate strip temperature to the approximate temperature of the coating bath, and an integral molten-metal container. Strip is heat treated in a controlled atmosphere to avoid oxidation, with the same atmosphere maintained to the point of immersion in molten metal. The second arrangement is for higher velocities and longer strands in heating and cooling passes. In this arrangement, strip may be processed in a series of vertical strands, supported by conveyor rolls.

Furnace lines designed for either galvanizing or aluminum coating may be designed with two molten-metal pots, with the entry strand arranged to be diverted to either one, and with the cooling zone adjustable to discharge the strand to either pot at the required temperature.

Thermal processing lines may include furnace equipment for heating the load to the temperature required for annealing, normalizing, or hardening, a quench tank for oil or water cooling to develop hardness, a cleaning station to remove quench oil residues, and a separate tempering furnace to develop the desired combination of hardness and toughness. Loads may be in continuous strand form, or in units carried by trays or fixtures that go through the entire process or carried on a series of conveyors. The required atmosphere generator will be part of the system.

Where exposure to hydrogen or nitrogen in furnace atmospheres may be undesirable, as in heat treatment of some ferrous alloys, heating and cooling can be done in a partial vacuum, usually with heat supplied by electrical resistors. Quenching can be done in a separate chamber with a controlled atmosphere suitable for brief exposure.

Systems for collecting operating data from one or more furnaces, and transmitting the data to a central recording or controlling station, may also be part of the responsibility of the furnace supplier.

58.13 FURNACE CAPACITY

Factors limiting the heating capacity of industrial furnaces include building space limitations, available fuel supplies, limited temperature of heat sources such as electric resistors or metal radiant tubes, and limits on final load temperature differentials. Other factors under more direct control by furnace designers are the choice between batch and continuous heating cycles; time–temperature cycles to reach specified final load temperatures; fuel firing arrangements; and control systems for furnace temperature, furnace pressure, and fuel/air ratios. In addition, the skills and motivation of furnace operating personnel, as the result of training, experience, and incentive policies, will directly affect furnace efficiency.

58.14 FURNACE TEMPERATURE PROFILES

Time–temperature patterns can be classified as uniform wall temperature (T_w), uniform combustion gas temperature (T_g), or variable T_w and T_g designed to secure the best combination of heating capacity and fuel efficiency.

In a batch-type furnace with fairly massive loads, the temperature control system can be arranged to allow firing at the maximum burner capacity until a preset wall temperature limit is reached, adjusting firing rate to hold that wall temperature, until load temperature approaches the limit for the heated surface, and reducing the wall temperature setting to hold maximum load temperature T_s while the minimum T_c reaches the desired level.

In continuous furnaces, control systems have evolved from a single firing zone, usually fired from the discharge end with flue gas vented from the load charge end, to two or three zone firing arranged for counterflow relation between furnace loads and heating gases.

Progress from single to multiple zone firing has improved heating rates, by raising furnace temperatures near the charge end, while increasing fuel demand by allowing higher temperatures in flue gas leaving the preheat zone. Load temperature control has been improved by allowing lower control temperatures in the final zone at the discharge end.

With multiple zone firing, the control system can be adjusted to approach the constant-gas-temperature model, constant wall temperature, or a modified system in which both T_g and T_w vary with time and position. Because gas temperatures are difficult to measure directly, the constant-gas-temperature pattern can be simulated by an equivalent wall temperature profile. With increasing fuel costs, temperature settings in a three-zone furnace can be arranged to discharge flue gases at or below the final load temperature, increasing the temperature setting in the main firing zone to a level to provide an equilibrium wall and load temperature, close to the desired final load temperature, during operating delays, and setting a temperature in the final or soak zone slightly above the desired final load surface temperature.

58.15   REPRESENTATIVE HEATING RATES

Heating times for various furnace loads, loading patterns, and time–temperature cycles can be calculated from data on radiation and non-steady-state conduction. For preliminary estimates, heating times for steel slabs to rolling temperatures, with a furnace temperature profile depressed at the entry end, have been estimated on a conservative basis as a function of thickness heated from one side and final load temperature differential and are shown in Fig. 58.26. The ratios for heating time required for square steel billets, in various loading patterns, are shown in Fig. 58.25. For other rectangular cross sections and loading patterns, heating times can be calculated by the Newman method.

Examples of heating times required to reach final load temperatures of $T_s = 2300°F$ and $T_c = 2350°F$, with constant furnace wall temperatures, are:

1. 12-in.-thick carbon steel slab on refractory hearth with open firing: 9 hr at 54.4 lb/hr · ft².
2. 4-in.-thick slab, same conditions as 1: 1.5 hr at 109 lb/hr · ft².
3. 4 in. square carbon steel billets loaded at 8 in. centers on a refractory hearth: 0.79 hr at 103 lb/hr · ft².
4. 4 in. square billets loaded as in 3, but heated to $T_s = 1650°F$ and $T_c = 1600°F$ for normalizing: 0.875 hr at 93 lb/hr · ft².
5. Thin steel strip, heated from both sides to 1350°F by radiant tubes with a wall temperature of 1700°F, total heating rate for both sides: 70.4 lb/hr · ft².
6. Long aluminum billets, 6 in. diameter, are to be heated to 1050°F. Billets will be loaded in multiple layers separated by spacer bars, with wind flow parallel to their length. With billets in lateral contact and with wind at a mean temperature of 1500°F, estimated heating time is 0.55 hr.
7. Small aluminum castings are to be heated to 1000°F on a conveyor belt, by jet impingement of heated air. Assuming that the load will have thick and thin sections, wind temperature will be limited to 1100°F to avoid overheating thinner sections. With suitable nozzle spacing and wind velocity, the convection heat-transfer coefficient can be $H_c = 15$ Btu/hr · ft² and the heating rate 27 lb/hr · ft².

58.16   SELECTING NUMBER OF FURNACE MODULES

For a given heating capacity and with no limits on furnace size, one large furnace will cost less to build and operate than a number of smaller units with the same total hearth area. However, furnace economy may be better with multiple units. For example, where reheating furnaces are an integral part of a continuous hot strip mill, the time required for furnace repairs can reduce mill capacity unless normal heating loads can be handled with one of several furnaces down for repairs. For contemporary hot strip mills, the minimum number of furnaces is usually three, with any two capable of supplying normal mill demand.

Rolling mills designed for operation 24 hr per day may be supplied by batch-type furnaces. For example, soaking-pit-type furnaces are used to heat steel ingots for rolling into slabs. The mill rolling rate is 10 slabs/hr. Heating time for ingots with residual heat from casting averages 4 hr, and the time allowed for reloading an empty pit is 2 hr, requiring an average turnover time of 6 hr. The required number of ingots in pits and spaces for loading is accordingly 60, requiring six holes loaded 10 ingots per hole.

If ingots are poured after a continuous steelmaking process, such as open hearth furnaces or oxygen retorts, and are rolled on a schedule of 18 turns per week, it may be economical at present fuel costs to provide pit capacity for hot storage of ingots cast over weekends, rather than reheating them from cold during the following week.

With over- and underfired slab reheating furnaces, with slabs carried on insulated, water-cooled supports, normal practice has been to repair pipe insulation during the annual shutdown for furnace maintenance, by which time some 50% of insulation may have been lost. By more frequent repair, for example, after 10% loss of insulation, the added cost of lost furnace time, material, and labor may be more than offset by fuel savings, even though total furnace capacity may be increased to offset idle time.

58.17 FURNACE ECONOMICS

The furnace engineer may be called on to make decisions, or submit recommendations for the design of new furnace equipment or the improvement of existing furnaces. New furnaces may be required for new plant capacity or addition to existing capacity, in which case the return on investment will not determine the decision to proceed. Projected furnace efficiency will, however, influence the choice of design.

If new furnace equipment is being considered to replace obsolete facilities, or if the improvement of existing furnaces is being considered to save fuel or power, or to reduce maintenance costs, return on investment will be the determining factor. Estimating that return will require evaluation of these factors:

Projected service life of equipment to be improved.

Future costs of fuel, power, labor for maintenance, or operating supervision and repairs, for the period assumed.

Cost of production lost during operating interruptions for furnace improvement or strikes by construction trades.

Cost of money during the improvement program and interest available from alternative investments.

Cost of retraining operating personnel to take full advantage of furnace improvements.

58.17.1 Operating Schedule

For a planned annual capacity, furnace size will depend on the planned hours per year of operation, and fuel demand will increase with the ratio of idle time to operating time, particularly in furnaces with water-cooled load supports. If furnace operation will require only a two- or three-man crew, and if furnace operation need not be coordinated with other manufacturing functions, operating costs may be reduced by operating a smaller furnace two or three turns per day, with the cost of overtime labor offset by fuel savings.

On the other hand, where furnace treatment is an integral part of a continuous manufacturing process, the provision of standby furnace capacity to avoid plant shutdowns for furnace maintenance or repairs may be indicated.

If furnace efficiency deteriorates rapidly between repairs, as with loss of insulation from water-cooled load supports, the provision of enough standby capacity to allow more frequent repairs may reduce overall costs.

58.17.2 Investment in Fuel-Saving Improvements

At present and projected future costs of gas and oil fuels, the added cost of building more efficient furnaces or modifying existing furnaces to improve efficiency can usually be justified. Possible improvements include better insulation of the furnace structure, modified firing arrangements to reduce flue gas temperatures or provide better control of fuel/air ratios, programmed temperature control to anticipate load changes, more durable insulation of water-cooled load supports and better maintenance of insulation, proportioning temperature control rather than the two position type, and higher preheated air temperatures. For intermittent furnace operation, the use of a low-density insulation to line furnace walls and roofs can result in substantial savings in fuel demand for reheating to operating temperature after idle periods.

The relative costs and availability of gas and oil fuels may make a switch from one fuel to another desirable at any future time, preferably without interrupting operations. Burner equipment and control systems are available, at some additional cost, to allow such changeovers.

The replacement of existing furnaces with more fuel-efficient designs, or the improvement of existing furnaces to save fuel, need not be justified in all cases by direct return on investment. Where present plant capacity may be reduced by future fuel shortages, or where provision should be made for increasing capacity with fuel supplies limited to present levels, cost savings by better fuel efficiency may be incidental.

Government policies on investment tax credits or other incentives to invest in fuel-saving improvements can influence the return on investment for future operation.

REFERENCE

1. C. Cone, *Energy Management for Industrial Furnaces,* Wiley, New York, 1980.

CHAPTER 59

GASEOUS FUELS

RICHARD J. REED

North American Manufacturing Company
Cleveland, Ohio

59.1 INTRODUCTION

Gaseous fuels are generally easier to handle and burn than are liquid or solid fuels. Gaseous fossil fuels include natural gas (primarily methane and ethane) and liquefied petroleum gases (LPG; primarily propane and butane). Gaseous man-made or artificial fuels are mostly derived from liquid or solid fossil fuels. Liquid fossil fuels have evolved from animal remains through eons of deep underground reaction under temperature and pressure, while solid fuels evolved from vegetable remains. Figure 59.1, adapted from Ref. 1, shows the ranges of hydrogen/carbon ratios for most fuels.

59.2 NATURAL GAS

59.2.1 Uses and Distribution

Although primarily used for heating, natural gas is also frequently used for power generation (via steam turbines, gas turbines, diesel engines, and Otto cycle engines) and as feedstock for making chemicals, fertilizers, carbon-black, and plastics. It is distributed through intra- and intercontinental pipe lines in a high-pressure gaseous state and via special cryogenic cargo ships in a low-temperature, high-pressure liquid phase (LNG).

Final street-main distribution for domestic space heating, cooking, water heating, and steam generation is at regulated pressures on the order of a few inches of water column to a few pounds per square inch, gage, depending on local facilities and codes. Delivery to commercial establishments and institutions for the same purposes, plus industrial process heating, power generation, and feedstock, may be at pressures as high as 100 or 200 psig (800 or 1500 kPa absolute). A mercaptan odorant is usually added so that people will be aware of leaks.

Before the construction of cross-country natural gas pipe lines, artificial gases were distributed through city pipe networks, but gas generators are now usually located adjacent to the point of use.

The American Gas Association[2] reported that natural gas provided 20.5 of the 76.2 quadrillion Btu (6.006×10^{12} of the 22.33×10^{12} kWh) of energy consumed in the United States in 1980. Its utilization was approximately as follows:

Residential	31.3%
Commercial	15.8%
Industrial	51.0%

59.2.2 Environmental Impact

The environmental impact of natural gas combustion is minimal—particulates, if burners are poorly adjusted (too rich, poor mixing, quenching); nitrogen oxides in a few cases of intense combustion with preheated air. Because of its easy mixing and relative lack of unsaturated hydrocarbons, natural gas has a reputation as the clean fuel—to burn, to store, and to transport.

59.2.3 Sources, Supply, and Storage

Natural gas is found with oil deposits (animal fossils) and coal deposits (plant fossils). As-yet untapped supplies are known to exist (1) near the coast of the Gulf of Mexico in very deep geopressured/geothermal aquifers and (2) in difficult-to-separate Appalachian shale formations.

Except for these hard-to-extract supplies, U.S. natural gas supplies have been variously predicted to last 10–20 years, but such predictions are questionable because of the effects of economic and regulatory variations on consumption, production, and exploration. Except for transoceanic LNG vessels, distribution is by pipe line, using a small fraction of the fuel in diesel-driven compressors to provide pumping power.

Storage facilities are maintained by many local gas utilities as a cushion for changing demand. These may be low-pressure gas holders with floating bell-covers, old wells or mines (medium pressure), or cryogenic vessels for high-pressure liquefied gas.

59.2.4 Types and Composition

Natural gases are classified as "sweet" or "sour," depending on their content of sulfur compounds. Most such compounds are removed before distribution. Odorants added (so that leaks can be detected) are usually sulfur compounds, but the amount is so minute that it has no effect on performance or pollution.

Various geographic sources yield natural gases that may be described as "high methane," "high Btu," or "high inert."

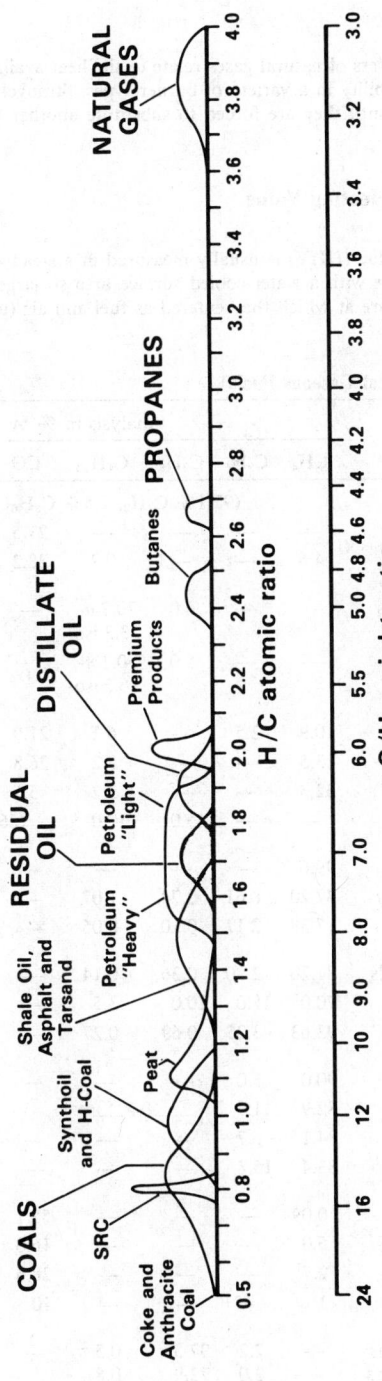

Fig. 59.1 Hydrogen/carbon ratios of fossil and synthetic fuels. (Adapted from Ref. 1.)

59.2.5 Properties

Properties that concern most users of natural gases relate to the heat available from their combustion, flow characteristics, and burnability in a variety of burner types. Strangely, few people pay attention to the properties of their gas until they are forced to substitute another fuel for it. Some properties are listed in Table 59.1.[3]

59.2.6 Calorific Value or Heating Value

The gross or higher heating value (HHV) is usually measured in a steady-state calorimeter, which is a small fire-tube heat exchanger with a water-cooled surface area so large that it cools the products of combustion to the temperature at which they entered as fuel and air (usually 60°F). HHV can be

Table 59.1a Analyses of Typical Gaseous Fuels[3]

Type of Gas	Analysis in % by Volume								
	CH_4	C_2H_6	C_3H_8	C_4H_{10}	CO	H_2	CO_2	O_2	N_2
Acetylene, commercial	(97.1% C_2H_2, 2.5% C_3H_6O)							0.084	0.28
Blast furnace	—	—	—	—	27.5	1.0	11.5	—	60.0
Blue (water), bituminous	4.6	—	—	0.7	28.2	32.5	5.5	0.9	27.6
Butane, commercial, natural gas	—	—	6.0	70.7 n-, 23.3 iso-	—	—	—	—	—
Butane, commercial, refinery gas	—	—	5.0	50.1 n-, 16.5 iso-	(28.3% C_4H_6)				
Carbureted blue, low gravity	10.9	2.5	—	6.1	21.9	49.6	3.6	0.4	5.0
Carbureted blue, heavy oil	13.5	—	—	8.2	26.8	32.2	6.0	0.9	12.4
Coke oven, by-product	32.3	—	—	3.2	5.5	51.9	2.0	0.3	4.8
Mapp	—	—	15.0	10.0	(66.0% C_3H_4, 9.0% C_3H_6)				
Natural, Alaska	99.6	—	—	—	—	—	—	—	0.4
Natural, Algerian LNG, Canvey	87.20	8.61	2.74	1.07	—	—	—	—	0.36
Natural, Gaz de Lacq	97.38	2.17	0.10	0.05	—	—	—	—	0.30
Natural, Groningen, Netherlands	81.20	2.90	0.36	0.14	—	—	0.87	—	14.40
Natural, Libyan LNG	70.0	15.0	10.0	3.5	—	—	—	—	0.90
Natural, North Sea, Bacton	93.63	3.25	0.69	0.27	—	—	0.13	—	1.78
Natural, Birmingham, AL	90.0	5.0	—	—	—	—	—	—	5.0
Natural, Cleveland, OH	82.9	11.9	0.3		—	—	0.2	0.3	4.4
Natural, Kansas City, MO	84.1	6.7	—	—	—	—	0.8	—	8.4
Natural, Pittsburgh, PA	83.4	15.8	—	—	—	—	—	—	0.8
Producer, Koppers–Totzek[a]	0.09	—	—	—	55.1	33.7	9.8	—	1.3
Producer, Lurgi[b]	5.0	—	—	—	16.0	25.0	14.0	—	40.0
Producer, W-G, bituminous[b]	2.7	—	—	—	28.6	15.0	3.4	0	50.3
Producer, Winkler[b]	1	—	—	—	10	12	22	—	55
Propane, commercial, natural gas	—	2.2	97.3	0.5	—	—	—	—	—
Propane, commercial, refinery gas	—	2.0	72.9	0.8	(24.3% C_3H_6)				
Sasol, South Africa	28.0	—	—	—	22.0	48.0	1.0		
Sewage, Decatur	68.0	—	—	—	—	2.0	22.0	—	6.0
SNG, no methanation	79.3	—	—	—	1.2	19.0	0.5	—	—

[a] O_2-blown.
[b] Air-blown.

calculated from a volumetric analysis and the calorific values of the pure compounds in the gas (Table 59.2). For example, for a natural gas having the analysis shown in column 2 below, the tabulation shows how a weighted average method can be used to determine the calorific value of the mixture:

Col. 1, Constituent	Col. 2, % Volume	Col. 3, HHV from Table 59.2 (Btu/ft³)	Col. 4 = (Col. 3 × Col. 2)/100
Methane, CH_4	90	1013	912
Ethane, C_2H_6	6	1763	106
Nitrogen, N_2	4	0	0
Total	100%		1018 Btu/ft³

Table 59.1b Properties of Typical Gaseous Fuels[3]

Type of Gas	Gas Gravity	Calorific Value Btu/ft³ Gross	Net	kcal/m³ Gross	Net	Gross Btu/ft³ of Standard Air	Gross kcal/m³ of Standard Air
Acetylene, commercial	0.94	1410	1360	12548	12105	115.4	1027
Blast furnace	1.02	92	91	819	819	135.3	1204
Blue (water), bituminous	0.70	260	239	2314	2127	126.2	1121
Butane, commercial, natural gas	2.04	3210	2961	28566	26350	104.9	932.6
Butane, commercial, refinery gas	2.00	3184	2935	28334	26119	106.1	944.2
Carbureted blue, low gravity	0.54	536	461	4770	4102	106.1	944.2
Carbureted blue, heavy oil	0.66	530	451	4716	4013	101.7	905.0
Coke oven, by-product	0.40	569	509	5064	4530	105.0	934
Mapp	1.48	2406	2282	21411	20308	113.7	1011.86
Natural, Alaska	0.55	998	906	8879	8063	104.8	932.6
Natural, Algerian LNG, Canvey	0.64	1122	1014	9985	9024	104.3	928.2
Natural, Gaz de Lacq	0.57	1011	911	8997	8107	104.1	927.3
Natural, Groningen, Netherlands	0.64	875	789	7787	7021	104.4	927.3
Natural, Libyan LNG	0.79	1345	1223	11969	10883	106.1	928.2
Natural, North Sea, Bacton	0.59	1023	922	9104	8205	105.0	934.4
Natural, Birmingham, AL	0.60	1002	904	8917	8045	106.1	945.1
Natural, Cleveland, OH	0.635	1059	959	9424	8534	106.2	942.4
Natural, Kansas City, MO	0.63	974	879	8668	7822	106.3	946.0
Natural, Pittsburgh, PA	0.61	1129	1021	10047	9086	106.3	945.1
Producer, Koppers–Totzek[a]	0.78	288	271	2563	2412	135.2	1203
Producer, Lurgi[b]	0.80	183	167	1629	1486	125.3	1115
Producer, W-G, bituminous[b]	0.84	168	158	1495	1406	129.2	1150
Producer, Winkler[b]	0.98	117	111	1041	988	188.7	1679
Propane, commercial, natural gas	1.55	2558	2358	22764	20984	107.5	956.6
Propane, commercial, refinery gas	1.77	2504	2316	22283	20610	108.0	961.1
Sasol, South Africa	0.55	500	448	4450	3986	114.9	1022
Sewage, Decatur	0.79	690	621	6140	5526	105.3	936.2
SNG, no methanation	0.47	853	765	7591	6808	105.8	943.3

[a] O_2-blown.

[b] Air-blown.

Table 59.1c Combustion Characteristics of Typical Gaseous Fuels[3]

Type of Gas	Wobbe Index	Vol. Air Req'd per Vol. Fuel	Stoichiometric Products of Combustion					Flame Temperature (°F)[b]
			%CO$_2$ Dry[a]	%H$_2$O Wet	%N$_2$ Wet	Total Vol. Vol. Fuel		
Acetylene, commercial	1559	12.14	17.4	8.3	75.8	12.66		3966
Blast furnace	91.0	0.68	25.5	0.7	74.0	1.54		2559
Blue (water), bituminous	310.8	2.06	17.7	16.3	68.9	2.77		3399
Butane, commercial, natural gas	2287	30.6	14.0	14.9	73.2	33.10		3543
Butane, commercial, refinery gas	2261	30.0	14.3	14.4	73.4	32.34		3565
Carbureted blue, low gravity	729.4	5.05	14.0	18.9	69.8	5.79		3258
Carbureted blue, heavy oil	430.6	5.21	15.7	16.6	70.3	6.03		3116
Coke oven, by-product	961.2	5.44	10.8	21.4	70.1	6.20		3525
Mapp	1947	21.25	15.6	11.9	74.4	22.59		3722
Natural, Alaska	1352	9.52	11.7	18.9	71.6	10.52		3472
Natural, Algeria LNG, Canvey	1423	10.76	12.1	18.3	71.9	11.85		3483
Natural, Gaz de Lacq	1365	9.71	11.7	18.8	71.6	10.72		3474
Natural, Groningen, Netherlands	1107	8.38	11.7	18.4	72.0	9.40		3446
Natural, Kuwait, Burgan	1364	10.33	12.2	18.3	71.7	10.40		3476
Natural, Libya LNG	1520	12.68	12.5	17.4	72.2	13.90		3497
Natural, North Sea, Bacton	1345	9.74	11.8	18.7	71.7	10.77		3473

Fuel							
Natural, Birmingham, AL	1291	9.44	11.7	18.6	71.8	10.47	3468
Natural, East Ohio	1336	9.70	11.9	18.7	71.7	10.72	3472
Natural, Kansas City, MO	1222	9.16	11.8	18.5	71.9	10.19	3461
Natural, Pittsburgh, PA	1446	10.62	12.0	18.3	71.9	11.70	3474
Producer, BCR, W. Kentucky	444	3.23	22.3	14.7	66.0	3.88	3514
Producer, IGT, Lignite	562	4.43	18.7	17.5	67.0	5.24	3406
Producer, Koppers–Totzek	326.1	2.13	27.7	12.6	63.2	2.69	3615
Producer, Lurgi	204.6	1.46	18.4	15.5	68.9	2.25	3074
Producer, Lurgi, subbituminous	465	2.49	23.4	19.6	61.5	3.20	3347
Producer, W-G, bituminous	183.6	1.30	18.5	9.8	73.5	2.08	3167
Producer, Winkler	118.2	0.62	24.1	9.3	68.9	1.51	3016
Propane, commercial, natural gas	2029	23.8	13.7	15.5	73.0	25.77	3532
Propane, commercial, refinery gas	2008	23.2	14.0	14.9	73.2	25.10	3560
Sasol, South Africa	794.4	4.30	12.8	21.0	68.8	4.94	3584
Sewage, Decatur	791.5	6.55	14.7	18.4	69.7	7.52	3368
SNG, no methanation	1264	8.06	11.3	19.8	71.1	8.96	3485

[a] Ultimate.

[b] Theoretical (calculated) flame temperatures, dissociation considered, with stoichiometrically correct air/fuel ratio. Although these temperatures are lower than those reported in the literature, they are all computed on the same basis; so they offer a comparison of the relative flame temperatures of various fuels.

Table 59.2   Calorific Properties of Some Compounds Found in Gaseous Fuels

Compound	Wobbe Index	Gross Heating Value (Btu/ft³)	Net Heating Value (Btu/ft³)	Pounds, Dry poc[a] per std ft³ of Fuel	Pounds H_2O per std ft³ of Fuel	Air Volume per Fuel Volume
Methane, CH_4	1360	1013	921	0.672	0.0950	9.56
Ethane, C_2H_6	1729	1763	1625	1.204	0.1425	16.7
Propane, C_3H_8	2034	2512	2328	1.437	0.1900	23.9
Butane, C_4H_{10}	2302	3264	3034	2.267	0.2375	31.1
Carbon Monoxide, CO	328	323	323	0.255	0	2.39
Hydrogen, H_2	1228	325	279	0	0.0474	2.39
Hydrogen Sulfide, H_2S	588	640	594	0.5855	0.0474	7.17
N_2, O_2, H_2O, CO_2, SO_2	0	0	0	[b]	[c]	0

[a] poc = products of combustion.

[b] Weight of N_2, O_2, CO_2, and SO_2 in fuel.

[c] Weight of H_2O in fuel.

It is a convenient coincidence that most solid fossil fuels release about 96–99 gross Btu/ft³ of standard *air;* liquid fossil fuels release about 101–104 Btu/ft³; gaseous fossil fuels about 104–108 Btu/ft³.

This would say that the natural gas in the example above should require about 1017 Btu/ft³ gas divided by 106 Btu/ft³ air = 9.6 ft³ air/ft³ gas. Precise stoichiometric calculations would say 0.909(9.53) + 0.06(16.7) = 9.58 ft³ air/ft³ gas.

59.2.7   Net Heating Value

Because a calorimeter cools the exit gases below their dew point, it retrieves the latent heat of condensation of any water vapor therein. But that latent heat is not recapturable in most practical heating equipment because of concern about corrosion; therefore, it is more realistic to subtract the latent heat from HHV, yielding a net or lower heating value, LHV. This is approximately

$$\frac{\text{LHV}}{\text{unit of fuel}} = \frac{\text{HHV}}{\text{unit of fuel}} - \left(\frac{970 \text{ Btu}}{\text{lb } H_2O} \times \frac{\text{lb } H_2O}{\text{unit of fuel}} \right)$$

Values for the latter term are listed in Table 59.2. LHV can also be calculated using data from Table 59.2 in the same manner as was illustrated for HHV. (Note that available heat was discussed in Chapter 57.)

59.2.8   Flame Stability

Flame stability is influenced by burner and combustion chamber configuration (aerodynamic and heat transfer characteristics) and by the fuel properties tabulated in Table 59.3.

59.2.9   Gas Gravity

Gas gravity, G (Table 59.1), is the ratio of the actual gas density relative to the density of dry air at standard temperature and pressure (0.0765 lb/ft³). This should not be confused with "specific gravity," which is the ratio of actual density relative to that of water. Gas gravity for natural gases typically ranges from 0.58 to 0.64, and is used in determination of flow rates and pressure drops through pipe lines, orifices, burners, and regulators:

$$\text{flow} = \text{flow coefficient} \times \text{area (ft}^2) \times \sqrt{2g\,(\text{psf pressure drop})/\rho}$$

where g = 32.2 ft/sec² and ρ = gas gravity × 0.0765. Unless otherwise emphasized, gas gravity is measured and specified at standard temperature and pressure (60°F and 29.92 in. Hg).

59.2.10   Wobbe Index

Wobbe index or Wobbe number (Table 59.2) is a convenient indicator of heat input considering the flow resistance of a gas-handling system. Wobbe index is equal to gross heating value divided by the square root of gas gravity; $W = \text{HHV}/\sqrt{G}$.

Table 59.3 Fuel Properties That Influence Flame Stability[3,a]

Fuel	Minimum Ignition Temperature, °F(°C)	Calculated Flame Temperature, °F(°C)[b]		Flammability Limits, % Fuel Gas by Volume[c]		Laminar Flame Velocity, fps(m/sec)		Percent Theoretical Air for Maximum Flame Velocity
		In Air	In O₂	Lower	Upper	In Air	In O₂	
Acetylene, C_2H_2	581(305)	4770(2632)	5630(3110)	2.5	81.0	8.75(2.67)	—	83
Blast furnace gas	—	2650(1454)	—	35.0	73.5	—	—	—
Butane, commercial	896(480)	3583(1973)	—	1.86	8.41	2.85(0.87)	—	—
Butane, n-C_4H_{10}	761(405)	3583(1973)	—	1.86	8.41	1.3(0.40)	—	97
Carbon monoxide, CO	1128(609)	3542(1950)	—	12.5	74.2	1.7(0.52)	—	55
Carbureted water gas	—	3700(2038)	5050(2788)	6.4	37.7	2.15(0.66)	—	90
Coke oven gas	—	3610(1988)	—	4.4	34.0	2.30(0.70)	—	90
Ethane, C_2H_6	882(472)	3540(1949)	—	3.0	12.5	1.56(0.48)	—	98
Gasoline	536(280)	—	—	1.4	7.6	—	—	—
Hydrogen, H_2	1062(572)	4010(2045)	5385(2974)	4.0	74.2	9.3(2.83)	—	57
Hydrogen sulfide, H_2S	558(292)	—	—	4.3	45.5	—	—	—
Mapp gas, C_3H_4	850(455)	—	—	3.4	10.8	—	15.4(4.69)	—
Methane, CH_4	1170(632)	3484(1918)	5301(2927)	5.0	15.0	1.48(0.45)	14.76(4.50)	90
Methanol, CH_3OH	725(385)	3460(1904)	—	6.7	36.0	1.6(0.49)	—	—
Natural gas	—	3525(1941)	4790(2643)	4.3	15.0	1.00(0.30)	15.2(4.63)	100
Producer gas	—	3010(1654)	—	17.0	73.7	0.85(0.26)	—	90
Propane, C_3H_8	871(466)	3573(1967)	5130(2832)	2.1	10.1	1.52(0.46)	12.2(3.72)	94
Propane, commercial	932(500)	3573(1967)	—	2.37	9.50	2.78(0.85)	—	—
Propylene, C_3H_6	—	—	5240(2893)	—	—	—	—	—
Town gas (Br. coal)	700(370)	3710(2045)	—	4.8	31.0	—	—	—

[a] For combustion with air at standard temperature and pressure.

[b] Flame temperatures are theoretical—calculated for stoichiometric ratio, dissociation considered.

[c] In a fuel-air mix. Example for methane: the lower flammability limit or lower explosive limit, LEL = 5% or 95 volumes air/5 volumes gas = 19.1 air/gas ratio. From Table 59.2, stoichiometric ratio is 9.56:1. Therefore excess air is $19 - 9.56 = 9.44$ ft³ air/ft³ gas or $9.44/9.56 \times 100 = 99.4\%$ excess air.

If air can be mixed with a substitute gas to give it the same Wobbe index as the previous gas, the existing burner system will pass the same gross Btu/hr input. This is often invoked when propane–air mixtures are used as standby fuels during natural gas curtailments. To be precise, the amount of air mixed with the propane should then be subtracted from the air supplied through the burner.

The Wobbe index is also used to maintain a steady input despite changing calorific value and gas gravity. Because most process-heating systems have automatic input control (temperature control), maintaining steady input may not be as much of a problem as maintaining a constant furnace atmosphere (oxygen or combustibles).

59.2.11 Flame Temperature

Flame temperature depends on burner mixing aerodynamics, fuel/air ratio, and heat loss to surroundings. It is very difficult to measure with repeatability. Calculated adiabatic flame temperatures, corrected for dissociation of CO_2 and H_2O, are listed in Tables 59.1 and 59.3 for 60°F air; in Chapter 68 it is listed for elevated air temperatures. Obviously, higher flame temperatures produce better heat-transfer rates from flame to load.

59.2.12 Minimum Ignition Temperature

Minimum ignition temperature, Table 59.3, relates to safety in handling, ease of light-up, and ease of continuous self-sustained ignition (without pilot or igniter, which is preferred). In mixtures of gaseous

Table 59.4a   Physical Properties[a] of LP Gases[b,6]

	Pro-pane	iso-Butane	Butane
Molecular weight	44.09	58.12	58.12
Boiling point, °F	−43.7	+10.9	+31.1
Boiling point, °C	−42.1	−11.7	−0.5
Freezing point, °F	−305.8	−255.0	−216.9
Density of liquid			
Specific gravity, 60°F/60°F	0.508	0.563	0.584
Degrees, API	147.2	119.8	110.6
Lb/gal	4.23	4.69	4.87
Density of vapor (ideal gas)			
Specific gravity (air = 1)	1.522	2.006	2.006
Ft³ gas/lb	8.607	6.53	6.53
Ft³ gas/gal of liquid	36.45	30.65	31.8
Lb gas/1000 ft³	116.2	153.1	153.1
Total heating value (after vaporization)			
Btu/ft³	2,563	3,369	3,390
Btu/lb	21,663	21,258	21,308
Btu/gal of liquid	91,740	99,790	103,830
Critical constants			
Pressure, psia	617.4	537.0	550.1
Temperature, °F	206.2	272.7	306.0
Specific heat, Btu/lb, °F			
c_p, vapor	0.388	0.387	0.397
c_v, vapor	0.343	0.348	0.361
c_p/c_v	1.13	1.11	1.10
c_p, liquid 60°F	0.58	0.56	0.55
Latent heat of vaporization at boiling			
point, Btu/lb	183.3	157.5	165.6
Vapor pressure, psia			
0°F	37.8	11.5	7.3
70°F	124.3	45.0	31.3
100°F	188.7	71.8	51.6
100°F (ASTM), psig max	210		70
130°F	274.5	109.5	80.8

[a] Properties are for commercial products and vary with composition.

[b] All values at 60°F and 14.696 psia unless otherwise stated.

Table 59.4b Combustion Data[a] for LP Gases[b,6]

	Pro-pane	iso-Butane	Butane
Flash temperature, °F (calculated)	−156	−117	−101
Ignition temperature, °F	932	950	896
Maximum flame temperature in air, °F			
Observed	3497	3452	3443
Calculated	3573	3583	3583
Flammability limits, % gas in air			
Lower	2.37	1.80	1.86
Higher	9.50	8.44	8.41
Maximum rate flame propagation in 1 in. tube			
Inches per second	32	33	33
Percentage gas in air	4.6–4.8	3.6–3.8	3.6–3.8
Required for complete combustion (ideal gas)			
Air, ft³ per ft³ gas	23.9	31.1	31.1
lb per lb gas	15.7	15.5	15.5
Oxygen, ft³ per ft³ gas	5.0	6.5	6.5
lb per lb gas	3.63	3.58	3.58
Products of combustion (ideal gas)			
Carbon dioxide, ft³ per ft³ gas	3.0	4.0	4.0
lb per lb gas	2.99	3.03	3.03
Water vapor, ft³ per ft³ gas	4.0	5.0	5.0
lb per lb gas	1.63	1.55	1.55
Nitrogen, ft³ per ft³ gas	18.9	24.6	24.6
lb per lb gas	12.0	11.8	11.8

[a] Properties are for commercial products and vary with composition.

[b] All values at 60°F and 14.696 psia unless otherwise stated.

compounds, such as natural gas, the minimum ignition temperature of the mixture is that of the compound with the lowest ignition temperature.

59.2.13 Flammability Limits

Flammability limits (Table 59.3, formerly termed "limits of inflammability") spell out the range of air-to-fuel proportions that will burn with continuous self-sustained ignition. "Lower" and "upper" flammability limits [also termed lower explosive limit (LEL) and upper explosive limit (UEL)] are designated in % gas in a gas–air mixture. For example, the flammability limits of a natural gas are 4.3% and 15%. The 4.3% gas in a gas–air mixture means 95.7% must be air; therefore, the "lean limit" or "lower limit" air/fuel ratio is 95.7/4.3 = 22.3∶1 (volume ratio), which means that more than 22.3∶1 (volume ratio) will be too lean to burn. Similarly, less than (100 − 15)/15 = 5.67∶1 is too rich to burn.

For the flammability limits of fuel mixtures other than those listed in Table 59.3, the Le Chatelier equation[4] and U.S. Bureau of Mines data[5] can be used.

59.3 LIQUEFIED PETROLEUM GASES

LP gases (LPG) are by-products of natural gas production and of refineries. They consist mainly of propane (C_3H_8), with some butane, propylene, and butylene. They are stored and shipped in liquefied form under high pressure; therefore, their flow rates are usually measured in gallons per hour or pounds per hour. When expanded and evaporated, LPG are heavier than air. Workmen have been asphixiated by LPG in pits beneath leaking LPG equipment.

The rate of LPG consumption is much less than that of natural gas or fuel oils. Practical economics usually limit use to (a) small installations inaccessible to pipe lines, (b) transportation, or (c) standby for industrial processes where oil burning is difficult or impossible.

LPG can usually be burned in existing natural gas burners, provided the air/gas ratio is properly readjusted. On large multiple burner installations an automatic propane–air mixing station is usually installed to facilitate quick changeover without changing air–gas ratios. (See the discussion of Wobbe index, Section 59.2.10.) Some fuel must be consumed to produce steam or hot water to operate a vaporizer for most industrial installations.

Table 59.4 lists some properties of commercial LPG, but it is suggested that more specific information be obtained from the local supplier.

REFERENCES

1. M. G. Fryback, "Synthetic Fuels—Promises and Problems," *Chemical Engineering Progress* (May, 1981).
2. American Gas Association, *Gas Facts, 1980 Data,* American Gas Association, Arlington, VA, 1981, p. 81.
3. R. J. Reed (ed.), *Combustion Handbook,* North American Mfg. Co., Cleveland, OH, 1978, pp. 12, 36–38.
4. F. E. Vandeveer and C. G. Segeler, "Combustion," in C. G. Segeler (ed.), *Gas Engineers Handbook,* The Industrial Press, New York, 1965, pp. 2/75–2/76.
5. H. F. Coward and G. W. Jones, *Limits of Flammability of Gases and Vapors* (U.S. Bureau of Mines Bulletin 503), U.S. Government Printing Office, Washington, DC, 1952, pp. 20–81.
6. E. W. Evans and R. W. Miller, "Testing and Properties of LP-Gases," in C. G. Segeler (ed.), *Gas Engineers Handbook,* The Industrial Press, New York, 1965, p. 5/11.

CHAPTER 60

LIQUID FOSSIL FUELS
FROM PETROLEUM

RICHARD J. REED

North American Manufacturing Company
Cleveland, Ohio

For most of the information in this chapter, the author is deeply indebted to John W. Thomas, retired
Chief Mechanical Engineer of the Standard Oil Company (Ohio).

60.1 INTRODUCTION

The major source of liquid fuel is crude petroleum; other sources are shale and tar sands. Synthetic hydrocarbon fuels—gasoline and methanol—can be made from coal and natural gas. Ethanol, some of which is used as an automotive fuel, is derived from vegetable matter.

Crude petroleum and refined products are a mix of a wide variety of hydrocarbons—aliphatics (straight- or branched-chained paraffins and olefins), aromatics (closed rings, six carbons per ring with alternate double bonds joining the ring carbons, with or without aliphatic side chains), and naphthenic or cycloparaffins (closed single-bonded carbon rings, five to six carbons).

Very little crude petroleum is used in its natural state. Refining is required to yield marketable products that are separated by distillation into fractions including a specific boiling range. Further processing (such as cracking, reforming, and alkylation) alters molecular structure of some of the hydrocarbons and enhances the yield and properties of the refined products.

Crude petroleum is the major source of liquid fuels in the United States now and for the immediate future. Although the oil embargo of 1973–1974 intensified development of facilities for extraction of oil from shale and of hydrocarbon liquids from coal, the economics do not favor early commercialization of these processes. Their development has been slowed by an apparently adequate supply of crude oil. Tar sands are being processed in small amounts in Canada, but no commercial facility exists in the United States. (See Table 60.1.)

In 1981,[1] the United States processed (refined, reformed, separated) 12.58×10^6 barrels of crude per day. Of this, 8.58×10^6 barrels were produced (pumped from wells) in the United States. Crude imports peaked in 1977 at 6.594×10^6 barrels per day. This was 47% of the total crude run. Of the total product demand for 1982, 55%, or 15.53×10^6 barrels per day, was for transportation. (See Table 60.2.)

Except for commercial propane and butane, fuels for heating and power generation are generally heavier and less volatile than fuels used in transportation. The higher the "flash point," the less hazardous is handling of the fuel. (Flash point is the minimum temperature at which the fuel oil will catch fire if exposed to naked flame. Minimum flash points are stipulated by law for safe storage and handling of various grades of oils.)

Properties of fuels reflect the characteristics of the crude. Paraffinic crudes have a high concentration of straight-chain hydrocarbons, which may leave a wax residue with distillation. Aromatic and naphthenic crudes have concentrations of ring hydrocarbons. Asphaltic crudes have a preponderance of heavier ring hydrocarbons and leave a residue after distillation. (See Table 60.3.)

Table 60.1 Principal Uses of Liquid Fuels

Heat and Power

Fuel oil	Space heating (residential, commercial, industrial)
	Steam generation for electric power
	Industrial process heating
	Refinery and chemical feedstock
Kerosene	Supplemental space heating
Turbine fuel	Stationary power generation
Diesel fuel	Stationary power generation
Liquid propane[a]	Isolated residential space heating
	Standby industrial process heating

Transportation

Jet fuel	Aviation turbines
Diesel fuel	Automotive engines
	Marine engines
	Truck engines
Gasoline	Automotive
	Aviation
Liquid propane and butane[a]	Limited automotive use

[a] See Chapter 59 on gaseous fossil fuels.

Table 60.2 Petroleum Product Demand and Supply in 1981[1]

Product	Millions of Barrels per Day		
	Supplied in 1981		Supplied Peak (and Year)
Petroleum distillates 1–4	5.34	6.4 total	7.412 (1978)
Gasoline petroleum distillates 5	1.060		
Jet fuel, kerosene	0.77	0.97 total	
Jet fuel, naphtha	0.20		
Distillates—diesel railroad, truck, marine	1.15	2.76 total	3.432 (1977)
Distillates—commercial and industrial heating	1.17		
Distillates—other	0.44		
Residuals—for utilities	0.93	1.85 total	
Residuals—commercial and industrial heating	0.92		
Liquefied petroleum gas + ethane	1.44		

Table 60.3 Ultimate Chemical Analyses of Various Crudes[a,6]

Crude Petroleum Source	% wt of					Specific Gravity (at temperature, °F)	Base
	C	H	N	O	S		
Baku, USSR	86.5	12.0		1.5		0.897	
California	86.4	11.7	1.14		0.60	0.951 (at 59°F)	Naphthene
Colombia, South America	85.62	11.91	0.54				
Kansas	85.6	12.4			0.37	0.912	Mixed
Mexico	83.0	11.0	___1.7___		4.30	0.97 (at 59°F)	Naphthene
Oklahoma	85.0	12.9			0.76		Mixed
Pennsylvania	85.5	14.2				0.862 (at 59°F)	Paraffin
Texas	85.7	11.0	___2.61___		0.70	0.91	Naphthene
West Virginia	83.6	12.9		3.6		0.897 (at 32°F)	Paraffin

[a] See, also, Table 60.7.

60.2 FUEL OILS

Liquid fuels in common use are broadly classified as follows:

1. Distillate fuel oils derived directly or indirectly from crude petroleum.
2. Residual fuel oils that result after crude petroleum is topped; or viscous residuums from refining operations.
3. Blended fuel oils, mixtures of the above.

The distillate fuels have lower specific gravity and are less viscous than residual fuel oils. Petroleum refiners burn a varying mix of crude residue and distilled oils in their process heaters. The changing gravity and viscosity require maximum oil preheat for atomization good enough to assure complete combustion. Tables 60.4–60.9 describe oils in current use. Some terms used in those tables are defined below.

Aniline point is the lowest Fahrenheit temperature at which an oil is completely miscible with an equal volume of freshly distilled aniline.

API gravity is a scale of specific gravity for hydrocarbon mixtures referred to in "degrees API"

Table 60.4   Some Properties of Liquid Fuels[2]

Property	Gasoline	Kerosene	Diesel Fuel	Light Fuel Oil	Heavy Fuel Oil	Coal Tar Fuel	Bituminous Coal (for Comparison)
Analysis, % wt							
C	85.5	86.3	86.3	86.2	86.2	90.0	80.0
H	14.4	13.6	12.7	12.3	11.8	6.0	5.5
N						1.2	1.5
O						2.5	7
S	0.1	0.1	1.0	1.5	2.0	0.4	1
Boiling range, °F	104–365	284–536	356 up	392 up	482 up	392 up	
Flash point, °F	−40	102	167	176	230	149	
Gravity specific at 59°F	0.73	0.79	0.87	0.89	0.95	1.1	1.25
Heat value, net							
cal/g	10,450	10,400	10,300	10,100	9,900	9,000	7,750
Btu/lb	18,810	18,720	18,540	18,180	17,820	16,200	13,950
Btu/US gal	114,929	131,108	129,800	131,215	141,325		
Residue, % wt at 662°F			15	50	60	60	
Viscosity, kinematic							
Centistokes at 59°F	0.75	1.6	5.0	50	1,200	1,500	
Centistokes at 212°F		0.6	1.2	3.5	20	18	

(for American Petroleum Institute). The relationships between API gravity, specific gravity, and density are:

$$\text{sp gr } 60/60°F = \frac{141.5}{°API + 131.5}$$

where °API is measured at 60°F (15.6°C).

$$\text{sp gr } 60/60°F = \frac{lb/ft^3}{62.3}$$

where lb/ft^3 is measured at 60°F (15.6°C).

SSU (or SUS) is seconds, Saybolt Universal, a measure of kinematic viscosity determined by measuring the time required for a specified quantity of the sample oil to flow by gravity through a specified orifice at a specified temperature. For heavier, more viscous oils, a larger (Furol) orifice is used, and the results are reported as SSF (seconds, Saybolt Furol).

kin visc in centistokes = 0.226 × SSU − 195/SSU, for SSU 32–100
kin visc in centistokes = 0.220 × SSU − 135/SSU, for SSU > 100
kin visc in centistokes = 2.24 × SSF − 184/SSF, for SSF 25–40
kin visc in centistokes = 2.16 × SSF − 60/SSF, for SSF > 40
1 centistoke (cSt) = 0.000001 m^2/sec

Unlike distillates, residual oils contain noticeable amounts of inorganic matter, ash content ranging from 0.01% to 0.1%. Ash often contains vanadium, which causes serious corrosion in boilers and heaters. (A common specification for refinery process heaters requires 50% nickel–50% chromium alloy for tube supports and hangers when the vanadium exceeds 150 ppm.) V_2O_5 also lowers the eutectic of many refractories, causing rapid disintegration. Crudes that often contain high vanadium are

Venezuela, Bachaqoro	350 ppm
Iran	350–400 ppm
Alaska, North Slope	80 ppm

60.2.1   Kerosene

Kerosene is a refined petroleum distillate consisting of a homogeneous mixture of hydrocarbons. It is used mainly in wick-fed illuminating lamps and kerosene burners. Oil for illumination and for domestic

Table 60.5 Gravities and Related Properties of Liquid Petroleum Products

Typical Ranges for (left-margin brackets):
- **Diesel Fuels:** 1D, 2D
- **Aviation Turbine Fuels:** JET A (48)(47), JP5 (48), JP4 (48)(56)
- **Fuel Oils:** #6, #5, #4, #2, #1 (48)

$°API_o$	Specific Gravity 60°F/60°F (15.6°C/15.6°C)	lb/gal	kg/m³	Gross Btu/gal[a]	Gross kcal/liter[a]	% H, wt[a]	Net Btu/gal[a]	Net kcal/liter[a]	Specific Heat @ 40°F	Specific Heat @ 300°F	Temperature Correction °API/°F[a]	ft³ 60°F air/gal	Ultimate % CO_2
0	1.076	8.969	1075	160,426	10,681	8.359	153,664	10,231	0.391	0.504	0.045	1581	—
2	1.060	8.834	1059	159,038	10,589	8.601	152,183	10,133	0.394	0.508	—	—	—
4	1.044	8.704	1043	157,692	10,499	8.836	150,752	10,037	0.397	0.512	—	—	18.0
6	1.029	8.577	1028	156,384	10,412	9.064	149,368	9,945	0.400	0.516	0.048	1529	17.6
8	1.014	8.454	1013	155,115	10,328	9.285	148,028	9,856	0.403	0.519	0.050	1513	17.1
10[b]	1.000[b]	8.335[b]	1000[b]	153,881	10,246	10.00	146,351	9,744	0.406	0.523	0.051	1509	16.7
12	0.986	8.219	985.0	152,681	10,166	10.21	145,100	9,661	0.409	0.527	0.052	1494	16.4
14	0.973	8.106	971.5	151,515	10,088	10.41	143,888	9,580	0.412	0.530	0.054	1478	16.1
16	0.959	7.996	958.3	150,380	10,013	10.61	142,712	9,502	0.415	0.534	0.056	1463	15.8
18	0.946	7.889	945.5	149,275	9,939	10.80	141,572	9,426	0.417	0.538	0.058	1448	15.5
20	0.934	7.785	933.0	148,200	9,867	10.99	140,466	9,353	0.420	0.541	0.060	1433	15.2
22	0.922	7.683	920.9	147,153	9,798	11.37	139,251	9,272	0.423	0.545	0.061	1423	14.9
24	0.910	7.585	909.0	146,132	9,730	11.55	138,210	9,202	0.426	0.548	0.063	1409	14.7
26	0.898	7.488	897.5	145,138	9,664	11.72	137,198	9,135	0.428	0.552	0.065	1395	14.5
28	0.887	7.394	886.2	144,168	9,599	11.89	136,214	9,069	0.431	0.555	0.067	1381	14.3
30	0.876	7.303	875.2	143,223	9,536	12.06	135,258	9,006	0.434	0.559	0.069	1368	14.0
32	0.865	7.213	864.5	142,300	9,475	12.47	134,163	8,933	0.436	0.562	0.072	1360	13.8
34	0.855	7.126	854.1	141,400	9,415	12.63	133,259	8,873	0.439	0.566	0.074	1347	13.6
36	0.845	7.041	843.9	140,521	9,356	12.78	132,380	8,814	0.442	0.569	0.076	1334	13.4
38	0.835	6.958	833.9	139,664	9,299	12.93	131,524	8,757	0.444	0.572	0.079	1321	13.3
40	0.825	6.887	824.2	138,826	9,243	13.07	130,689	8,702	0.447	0.576	0.082	1309	13.1
42	0.816	6.798	814.7	138,007	9,189	—	—	—	0.450	0.579	0.085	—	13.0
44	0.806	6.720	805.4	137,207	9,136	—	—	—	0.452	0.582	0.088	—	12.8

[a] For gravity measured at 60°F (15.6°C) only.
[b] Same as H_2O.

Table 60.6 Heating Requirements for Products Derived from Petroleum[3]

Commercial Fuels	Specific Gravity at 60°F/60°F (15.6°C)	Distillation Range, °F(°C)	Vapor Pressure,[a] psia(mm Hg)	Latent Btu/gal[b] to Vaporize	Btu/gal[b] to Heat from 32°F (0°C) to		
					Pumping Temperature	Atomizing Temperature	Vapor
No. 6 oil	0.965	600–1000(300–500)	0.054 (2.8)	764	371	996	3619[c]
No. 5 oil	0.945	600–1000(300–500)	0.004 (0.2)	749	133	635	3559[c]
No. 4 oil	0.902	325–1000(150–500)	0.232 (12)	737	—	313	2725[c]
No. 2 oil	0.849	375– 750(150–400)	0.019 (1)	743	—	—	2704[c]
Kerosene	0.780	256– 481(160–285)	0.039 (2)	750	—	—	1303[c]
Gasoline	0.733	35– 300(37–185)	0.135 (7)	772	—	—	1215[c]
Methanol	0.796	148 (64)	4.62 (239)	3140	—	—	3400[d]
Butane	0.582	31 (0)	31(1604)	808	—	—	976[d]
Propane	0.509	–44 (–42)	124(6415)	785	—	—	963[d]

[a] At the atomizing temperature or 60°F, whichever is lower. Based on a sample with the lowest boiling point from column 3.
[b] To convert Btu/US gallon to kcal/liter, multiply by 0.666. To convert Btu/US gallon to Btu/lb, divide by 8.335 × sp gr, from column 2. To convert Btu/US gallon to kcal/kg, divide by 15.00 × sp gr, from column 2.
[c] Calculated for boiling at midpoint of distillation range, from column 3.
[d] Includes latent heat plus sensible heat of the vapor heated from boiling point to 60°F (15.6°C).

Table 60.7 Analyses and Characteristics of Selected Fuel Oils[3]

	Source	Ultimate Analysis (% Weight)						ppm if > 50	% wt Asphaltine	% wt C Residue	°API at 60°F	Flash Point, °F	HV, Btu/lb		Pour Point, °F	Viscosity, SSU	
		C	H	N	S	Ash	O[a]						Gross	Net		At 140°F	At 210°F
Distillates	Alaska	86.99	12.07	0.007	0.31	<0.001	0.62	—	—	—	33.1	—	—	—	—	33.0	29.5
	California	86.8	12.52	0.053	0.27	<0.001	0.36	—	—	—	32.6	—	19,330	—	—	30.8	29.5
	West Texas	88.09	9.76	0.026	1.88	<0.001	0.24	—	—	—	18.3	—	—	—	—	32.0	28.8
Residuals	Alaska	86.04	11.18	0.51	1.63	0.034	0.61	50 Ni 67 V	5.6	12.9	15.6	215	18,470	17,580	38	1071	194
	California	86.66	10.44	0.86	0.99	0.20	0.85	—	8.62	15.2	12.6	180	18,230	17,280	42	720	200
	DFM (shale)	86.18	13.00	0.24	0.51	0.003	1.07	[b]	0.036	4.1	33.1	182	19,430	18,240	40	36.1	30.7
	Gulf of Mexico	84.62	10.77	0.36	2.44	0.027	1.78	—	7.02	14.8	13.2	155	18,240	17,260	40	835	181
	Indo/Malaysia	86.53	11.93	0.24	0.22	0.036	1.04	101 V	0.74	3.98	21.8	210	19,070	17,980	61	199	65
	Middle East[c]	86.78	11.95	0.18	0.67	0.012	0.41	—	3.24	6.0	19.8	350	19,070	17,980	48	490	131.8
	Pennsylvania[d]	84.82	11.21	0.34	2.26	0.067	1.3	65 Na 82 V	4.04	12.4	15.4	275	18,520	17,500	66	1049	240
	Venezuela	85.24	10.96	0.40	2.22	0.081	1.10	52 Ni 226 V	8.4	6.8	14.1	210	18,400	17,400	58	742	196.7
	Venezuela desulfurized	85.92	12.05	0.24	0.93	0.033	0.83	101 V	2.59	5.1	23.3	176	18,400	17,300	48	113.2	50.5

a By difference.
b 91 Ca, 77 Fe, 88 Ni, 66 V.
c Exxon.
d Amerada Hess.

Table 60.8 Fuel Oil Specifications[8]

Grade of Fuel Oil[a]	Flash Point, °C (°F) Min	Pour Point, °C (°F) Max	Water and Sediment, Vol % Max	Carbon Residue on 10% Bottoms, % Max	Ash, Weight % Max	Distillation Temperatures, °C (°F) 10% Point Max	90% Point Min	90% Point Max	Saybolt Viscosity, s[d] Universal at 38°C (100°F) Min	Universal Max	Furol at 50°C (122°F) Min	Furol Max	Kinematic Viscosity, cSt[d] At 38°C (100°F) Min	At 38°C Max	At 40°C (104°F) Min	At 40°C Max	At 50°C (122°F) Min	At 50°C Max	Specific Gravity, 60/60°F (deg API) Max	Copper Strip Corrosion Max	Sulfur, % Max
No. 1 A distillate oil intended for vaporizing pot-type burners and other burners requiring this grade of fuel	38 (100)	−18[c] (0)	0.05	0.15	—	215 (420)	—	288 (550)	—	—	—	—	1.4	2.2	1.3	2.1	—	—	0.8499 (35 min)	No. 3	0.5
No. 2 A distillate oil for general purpose heating for use in burners not requiring No. 1 fuel oil	38 (100)	−6[c] (20)	0.05	0.35	—	—	282[c] (540)	338 (640)	(32.6)	(37.9)	—	—	2.0[c]	3.6	1.9[c]	3.4	—	—	0.8762 (30 min)	No. 3	0.5[b]
No. 4 (Light) Preheating not usually required for handling or burning	38 (100)	−6[c] (20)	0.50	—	0.05	—	—	—	(32.6)	(45)	—	—	2.0	5.8	—	—	—	—	0.876[g] (30 max)	—	—
No. 4 Preheating not usually required for handling or burning	55 (130)	−6[c] (20)	0.50	—	0.10	—	—	—	(45)	(125)	—	—	5.8	26.4[h]	5.5	24.0[f]	—	—	—	—	—

Grade	Flash Point °C (°F)	Pour Point	Water and sediment, vol %	Carbon residue on 10%	Ash, wt %	Distillation temperature	Kinematic viscosity at 40°C (cSt) min	Kinematic viscosity at 40°C (cSt) max	Kinematic viscosity at 100°C (cSt) min	Kinematic viscosity at 100°C (cSt) max	Saybolt viscosity Universal at 38°C (100°F) min	Saybolt viscosity Universal at 38°C (100°F) max	Saybolt viscosity Furol at 50°C (122°F) min	Saybolt viscosity Furol at 50°C (122°F) max
No. 5 (Light) Preheating may be required depending on climate and equipment	55 (130)	—	1.00	—	—	—	(>125)	(300)	—	—	>26.4	65^f	>24.0	58^f	—	—	—
No. 5 (Heavy) Preheating may be required for burning and, in cold climates, may be required for handling	55 (130)	—	1.00	—	—	—	(>300)	(900)	(23)	(40)	>65	194^f	>58	168^f	(42)	(81)	—
No. 6 Preheating required for burning and handling	60 (140)	g	2.00^e	—	—	—	(>900)	(9000)	(>45)	(300)	—	—	—	—	—	>92	638^f

a It is the intent of these classifications that failure to meet any requirement of a given grade does not automatically place an oil in the next lower grade unless in fact it meets all requirements of the lower grade.

b In countries outside the United States other sulfur limits may apply.

c Lower or higher pour points may be specified whenever required by conditions of storage or use. When pour point less than −18°C (0°F) is specified, the minimum viscosity for grade No. 2 shall be 1.7 cSt (31.5 SUS) and the minimum 90% point shall be waived.

d Viscosity values in parentheses are for information only and not necessarily limiting.

e The amount of water by distillation plus the sediment by extraction shall not exceed 2.00%. The amount of sediment by extraction shall not exceed 0.50%. A deduction in quantity shall be made for all water and sediment in excess of 1.0%.

f Where low-sulfur fuel oil is required, fuel oil falling in the viscosity range of a lower numbered grade down to and including No. 4 may be supplied by agreement between purchaser and supplier. The viscosity range of the initial shipment shall be identified and advance notice shall be required when changing from one viscosity range to another. This notice shall be in sufficient time to permit the user to make the necessary adjustments.

g This limit guarantees a minimum heating value and also prevents misrepresentation and misapplication of this product as Grade No. 2.

h Where low-sulfur fuel oil is required, Grade 6 fuel oil will be classified as low pour +15°C (60°F) max or high pour (no max). Low-pour fuel oil should be used unless all tanks and lines are heated.

Table 60.9 Application of ASTM Fuel Oil Grades, as Described by One Burner Manufacturer

Fuel Oil	Description
No. 1	Distillate oil for vaporizing-type burners
No. 2	Distillate oil for general purpose use, and for burners not requiring No. 1 fuel oil
No. 4	Blended oil intended for use without preheating
No. 5	Blended residual oil for use with preheating; usual preheat temperature is 120–220°F
No. 6	Residual oil for use with preheaters permitting a high-viscosity fuel; usual preheat temperature is 180–260°F
Bunker C	Heavy residual oil, originally intended for ocean-going ships

stoves must be high in paraffins to give low smoke. The presence of naphthenic and especially aromatic hydrocarbons increases the smoking tendency. A "smoke point" specification is a measure of flame height at which the tip becomes smoky. The "smoke point" is about 73 mm for paraffins, 34 mm for naphthalenes, and 7.5 mm for aromatics and mixtures.

Low sulfur content is necessary in kerosenes because:

1. Sulfur forms a bloom on glass lamp chimneys and promotes carbon formation on wicks.
2. Sulfur forms oxides in heating stoves. These swell, are corrosive and toxic, creating a health hazard, particularly in nonvented stoves.

Kerosene grades[9] (see Table 60.10) in the United States are:

No. 1 K: A special low-sulfur grade kerosene suitable for critical kerosene burner applications.

No. 2 K: A regular-grade kerosene suitable for use in flue-connected burner applications and for use in wick-fed illuminating lamps.

60.2.2 Aviation Turbine Fuels

The most important requirements of aircraft jet fuel relate to freezing point, distillation range, and level of aromatics. Fluidity at low temperature is important to ensure atomization. A typical upper viscosity limit is 7–10 cSt at 0°F, with the freezing point as low as −60°F.

Aromatics are objectionable because (1) coking deposits from the flame are most pronounced with aromatics of high C/H ratio and less pronounced with short-chain compounds, and (2) they must be controlled to keep the combustor liner at an acceptable temperature.

Jet fuels for civil aviation are identified as Jet A and A1 (high-flash-point, kerosene-type distillates), and Jet B (a relatively wide boiling range, volatile distillate).

Jet fuels for military aviation are identified as JP4 and JP5. The JP4 has a low flash point and a wide boiling range. The JP5 has a high flash point and a narrow boiling range. (See Table 60.11.)

Table 60.10 Chemical and Physical Requirements for Kerosene[9]

Property	Limit
Distillation temperature	
10% recovered	401°F (205°C)
Final boiling point	572°F (300°C)
Flash point	100°F (38°C)
Freezing point	−22°F (−30°C)
Sulfur, % weight	
No. 1 K	0.04 maximum
No. 2 K	0.30 maximum
Viscosity, kinematic	
at 104°F (40°C), centistokes	1.0 min/1.9 max

Gas turbine fuel oils for other than use in aircraft must be free of inorganic acid and low in solid or fibrous materials. (See Tables 60.12 and 60.13.) All such oils must be homogeneous mixtures that do not separate by gravity into light and heavy components.

60.2.3 Diesel Fuels

Diesel engines, developed by Rudolf Diesel, rely on the heat of compression to achieve ignition of the fuel. Fuel is injected into the combustion chamber in an atomized spray at the end of the compression stroke, after air has been compressed to 450–650 psi and has reached a temperature, due to compression, of at least 932°F (500°C). This temperature ignites the fuel and initiates the piston's power stroke. The fuel is injected at about 2000 psi to ensure good mixing.

Diesels are extensively used in truck transport, rail trains, and marine engines. They are being used more in automobiles. In addition, they are employed in industrial and commercial stationary power plants.

Fuels for diesels vary from kerosene to medium residual oils. The choice is dictated by engine characteristics, namely, cylinder diameter, engine speed, and combustion wall temperature. High-speed

Table 60.11 Specifications[10] and Typical Properties[7] of Aviation Turbine Fuels

Property	Specifications			Typical, 1979		
	Jet A	Jet A1	Jet B	26 Samples JP4	7 Samples JP5	60 Samples Jet A
Aromatics, % vol	20	20	20	13.0	16.4	17.9
Boiling point, final, °F	572	572	—	—	—	—
Distillation, max temperature, °F						
For 10% recovered	400	400	—	208	387	375
For 20% recovered	—	—	290	—	—	—
For 50% recovered	—	—	370	293	423	416
For 90% recovered	—	—	470	388	470	473
Flash point, min, °F	100	100	—	—	—	—
Freezing point, max, °F	−40	−53	−58	−110	−71	−56
Gravity, API, max	51	51	57	53.5	41.2	42.7
Gravity, API, min	37	37	45	—	—	—
Gravity, specific 60°F min	0.7753	0.7753	0.7507	0.765	0.819	0.812
Gravity, specific 60°F max	0.8398	0.8398	0.8017	—	—	—
Heating value, gross Btu/lb	—	—	—	18,700	18,530	18,598
Heating value, gross Btu/lb min	18,400	18,400	18,400	—	—	—
Mercaptan, % wt	0.003	0.003	0.003	0.0004	0.0003	0.0008
Sulfur, max % wt	0.3	0.3	0.3	0.030	0.044	0.050
Vapor pressure, Reid, psi	—	—	3	2.5	—	0.2
Viscosity, max SSU						
At −4°F	52	—	—	—	—	—
At −30°F	—	—	—	34–37	60.5	54.8

Table 60.12 Nonaviation Gas Turbine Fuel Grades per ASTM[11]

Grade	Description
No. 0-GT	A naphtha or low-flash-point hydrocarbon liquid
No. 1-GT	A distillate for gas turbines requiring cleaner burning than No. 2-GT
No. 2-GT	A distillate fuel of low ash suitable for gas turbines not requiring No. 1-GT
No. 3-GT	A low ash fuel that may contain residual components
No. 4-GT	A fuel containing residual components and having higher vanadium content than No. 3-GT

Table 60.13 Specifications[11] for Nonaviation Gas Turbine Fuels

Property	Specifications				
	0-GT	1-GT	2-GT	3-GT	4-GT
Ash, max % wt	0.01	0.01	0.01	0.03	—
Carbon residue, max % wt	0.15	0.15	0.35	—	—
Distillation, 90% point, max °F	—	(550)[a]	(640)	—	—
Distillation, 90% point, min °F	—	—	(540)	—	—
Flash point, min °F	—	(100)	(100)	(130)	(150)
Gravity, API min	—	(35)	(30)	—	—
Gravity, spec 60°F max	—	0.850	0.876	—	—
Pour point, max °F	—	(0)	(20)	—	—
Viscosity, kinematic					
Min SSU at 100°F	—	—	(32.6)	(45)	(45)
Max SSU at 100°F	—	(34.4)	(40.2)	—	—
Max SSF at 122°F	—	—	—	(300)	(300)
Water and sediment, max % vol	0.05	0.05	0.05	1.0	1.0

[a] Values in parentheses are approximate.

small engines require lighter fuels and are more sensitive to fuel quality variations. Slow-speed, larger industrial and marine engines use heavier grades of diesel fuel oil.

Ignition qualities and viscosity are important characteristics that determine performance. The ignition qualities of diesel fuels may be assessed in terms of their cetane numbers or diesel indices. Although the diesel index is a useful indication of ignition quality, it is not as reliable as the cetane number, which is based on an engine test:

$$\text{diesel index} = (\text{aniline point, °F}) \times (\text{API gravity}/100)$$

The diesel index is an arbitrary figure having a significance similar to cetane number, but having a value 1–5 numbers higher.

The cetane number is the percentage by volume of cetane in a mixture of cetane with an ethylnaphthalene that has the same ignition characteristics as the fuel. The comparison is made in a diesel engine equipped either with means for measuring the delay period between injection and ignition or with a surge chamber, separated from the engine intake port by a throttle in which the critical measure below which ignition does not occur can be measured. Secondary reference fuels with specific cetane numbers are available. Cetane number is a measure of ignition quality and influences combustion roughness.

The use of a fuel with too low a cetane number results in accumulation of fuel in the cylinder before combustion, causing "diesel knock." Too high a cetane number will cause rapid ignition and high fuel consumption.

The higher the engine speed, the higher the required fuel cetane number. Suggested rpm values for various fuel cetane numbers are shown in Table 60.14.[5] Engine size and operating conditions are important factors in establishing approximate ignition qualities of a fuel.

Too viscous an oil will cause large spray droplets and incomplete combustion. Too low a viscosity may cause fuel leakage from high-pressure pumps and injection needle valves. Preheating permits use of higher viscosity oils.

Table 60.14 Fuel Cetane Numbers for Various Engine Speeds[5]

Engine Speed (rpm)	Cetane Number
Above 1500	50–60
500–1500	45–55
400– 800	35–50
200– 400	30–45
100– 200	15–40
Below 200	15–30

To minimize injection system wear, fuels are filtered to remove grit. Fine gage filters are considered adequate for engines up to 8 Hz, but high-speed engines usually have fabric or felt filters. It is possible for wax to crystallize from diesel fuels in cold weather, therefore, preheating before filtering is essential.

To minimize engine corrosion from combustion products, control of fuel sulfur level is required. (See Tables 60.15 and 60.16.)

60.2.4 Summary

Aviation jet fuels, gas turbine fuels, kerosenes, and diesel fuels are very similar. The following note from Table 1 of Reference 11 highlights this:

No. 0-GT includes naphtha, Jet B fuel, and other volatile hydrocarbon liquids. No. 1-GT corresponds in general to Spec D396 Grade No. 1 fuel and Classification D975 Grade No. 1-D Diesel fuel in physical properties. No. 2-GT corresponds in general to Spec D396 Grade No. 2 fuel and Classification D975 Grade No. 2 Diesel fuel in physical properties. No. 3-GT and No. 4-GT viscosity range brackets Spec D396 and Grade No. 4, No. 5 (light), No. 5 (heavy), No. 6, and Classification D975 Grade No. 4-D Diesel fuel in physical properties.

60.3 SHALE OILS

As this is written, there is no commercial producing shale oil plant in the United States. Predictions are that the output products will be close in characteristics and performance to those made from petroleum crudes.

Table 60.17 lists properties of a residual fuel oil (DMF) from one shale pilot operation and of a shale crude oil.[13] Table 60.18 lists ultimate analyses of oils derived from shales from a number of locations.[14] Properties will vary with the process used for extraction from the shale. The objective of all such processes is only to provide feedstock for refineries. In turn, the refineries' subsequent processing will also affect the properties.

If petroleum shortages occur, they will probably provide the economic impetus for completion of developments already begun for the mining, processing, and refining of oils from shale.

60.4 OILS FROM TAR SANDS

At the time that this is written, the only commercially practical operation for extracting oil from tar sands is at Athabaska, Alberta, Canada, using surface mining techniques. When petroleum supplies become short, economic impetus therefrom will push completion of developments already well under way for mining, processing, and refining of oils from tar sands.

Table 60.19 lists chemical and physical properties of several tar sand bitumens.[15] Further refining will be necessary because of the high density, viscosity, and sulfur content of these oils.

Extensive deposits of tar sands are to be found around the globe, but most will have to be recovered by some *in situ* technique, fireflooding, or steam flooding. Yields tend to be small and properties vary with the recovery method, as illustrated in Table 60.20.[15]

Table 60.15 Diesel Fuel Descriptions[12]

Grade	Description
No. 1D	A volatile distillate fuel oil for engines in service requiring frequent speed and load changes
No. 2D	Distillate fuel oil of lower volatility for engines in industrial and heavy mobile service
No. 4D	A fuel oil for low and medium speed diesel engines
Type CB	For buses, essentially 1D
Type TT	For trucks, essentially 2D
Type RR	For railroads, essentially 2D
Type SM	For stationary and marine use, essentially 2D or heavier

Table 60.16a Detailed Requirements for Diesel Fuel Oils[a,12]

Grade of Diesel Fuel Oil	Flash Point, °C (°F) Min	Cloud Point, °C (°F) Max	Water and Sediment, Vol % Max	Carbon Residue on 10% Residuum, % Max	Ash, Weight % Max	Distillation Temperatures, °C (°F), 90% Point		Viscosity				Sulfur,[d] Weight % Max	Copper Strip Corrosion Max	Cetane Number[e] Min
								Kinematic, cSt[g] at 40°C		Saybolt, SUS at 100°F				
						Min	Max	Min	Max	Min	Max			
No. 1-D A volatile distillate fuel oil for engines in service requiring frequent speed and load changes	38 (100)	b	0.05	0.15	0.01	—	288 (550)	1.3	2.4	—	34.4	0.50	No. 3	40[f]
No. 2-D A distillate fuel oil of lower volatility for engines in industrial and heavy mobile service	52 (125)	b	0.05	0.35	0.01	282[c] (540)	338 (640)	1.9	4.1	32.6	40.1	0.50	No. 3	40[f]
No. 4-D A fuel oil for low and medium speed engines	55 (130)	b	0.50	—	0.10	—	—	5.5	24.0	45.0	125.0	2.0	—	30[f]

[a] To meet special operating conditions, modifications of individual limiting requirements may be agreed upon between purchaser, seller, and manufacturer.

[b] It is unrealistic to specify low-temperature properties that will ensure satisfactory operation on a broad basis. Satisfactory operation should be achieved in most cases if the cloud point (or wax appearance point) is specified at 6°C above the tenth percentile minimum ambient temperature for the area in which the fuel will be used. This guidance is of a general nature; some equipment designs, using improved additives, fuel properties, and/or operations, may allow higher or require lower cloud point fuels. Appropriate low-temperature operability properties should be agreed on between the fuel supplier and purchaser for the intended use and expected ambient temperatures.

[c] When cloud point less than −12°C (10°F) is specified, the minimum viscosity shall be 1.7 cSt (or mm²/sec) and the 90% point shall be waived.

[d] In countries outside the United States, other sulfur limits may apply.

[e] Where cetane number by Method D613 is not available, ASTM Method D976, Calculated Cetane Index of Distillate Fuels may be used as an approximation. Where there is disagreement, Method D613 shall be the referee method.

[f] Low-atmospheric temperatures as well as engine operation at high altitudes may require use of fuels with higher cetane ratings.

[g] cSt = 1 mm²/sec.

[h] The values stated in SI units are to be regarded as the standard. The values in U.S. customary units are for information only.

1704

Table 60.16b Typical Properties of Diesel Fuels[7]

| | All United States, 1981 | | | | | | Eastern United States, 1981 | | | | | | | | | | | |
| | 48 Samples No. 1D | | | 112 Samples No. 2D | | | 24 Samples Type CB | | | 44 Samples Type TT | | | 13 Samples Type RR | | | 4 Samples Type SM | | |
Property	Min	Avg	Max	Min	Avg	Max	Min	Avg	Max	Min	Avg	Max	Min	Avg	Max	Min	Avg	Max
Ash, % wt	0.000	0.001	0.005	0.000	0.002	0.020	—	0.001	0.005	—	0.002	0.015	—	0.000	0.001	—	0.001	0.001
Carbon residue, % wt	0.000	0.059	0.067	0.000	0.101	0.300	—	0.67	0.21	0.101	—	0.25	—	0.121	0.23	—	0.148	0.21
Cetane number	36	46.7	53.0	29.0	45.6	52.4	—	49.8	—	—	45.6	—	—	44.8	—	—	—	—
Distillation, 90% point, °F	445	448	560	493	587	640	451	512	640	451	571	640	540	590	640	482	577	640
Flash point, °F	104	138	176	132	166	240	120	140	240	120	162	240	156	164	192	136	162	180
Gravity, API	37.8	42.4	47.9	22.8	34.9	43.1	—	41.5	—	—	36.3	—	—	33.8	—	—	35.3	—
spec, 60/60°F	0.836	0.814	0.789	0.917	0.850	0.810	—	0.818	—	—	0.843	—	—	0.856	—	—	0.848	—
Sulfur, % wt	0.000	0.070	0.25	0.010	0.283	0.950	—	0.086	0.24	—	0.198	0.46	—	0.283	0.580	—	0.155	0.28
Viscosity, SSU, at 100°F	32.6	33.3	35.7	33.8	36.0	40.3	32.9	34.3	40.2	32.9	35.7	40.2	34.2	36.0	37.8	36.0	—	37.8

Table 60.17   Properties of Shale Oils[13]

Property	DMF Residual	Crude
Ultimate analysis		
Carbon, % wt	86.18	84.6
Hydrogen, % wt	13.00	11.3
Nitrogen, % wt	0.24	2.08
Sulfur, % wt	0.51	0.63
Ash, % wt	0.003	0.026
Oxygen, % wt by difference	1.07	1.36
Conradson carbon residue, %	4.1	2.9
Asphaltene, %	0.036	1.33
Calcium, ppm	0.13	1.5
Iron, ppm	6.3	47.9
Manganese, ppm	0.06	0.17
Magnesium, ppm	—	5.40
Nickel, ppm	0.43	5.00
Sodium, ppm	0.09	11.71
Vanadium, ppm	0.1	0.3
Flash point, °F	182	250
Pour point, °F	40	80
API gravity at 60°F	33.1	20.3
Viscosity, SSU at 140°F	36.1	97
SSU at 210°F	30.7	44.1
Gross heating value, Btu/lb	19,430	18,290
Net heating value, Btu/lb	18,240	17,260

Table 60.18   Elemental Content of Shale Oils, % wt[14]

Source	Carbon, C			Hydrogen, H			Nitrogen, N			Sulfur, S			Oxygen, O		
	Min	Avg	Max	Min	Avg	Max	Min	Avg	Max	Min	Avg	Max	Min	Avg	Max
Colorado	83.5	84.2	84.9	10.9	11.3	11.7	1.6	1.8	1.9	0.7	1.2	1.7	1.3	1.7	2.1
Utah	84.1	84.7	85.2	10.9	11.5	12.0	1.6	1.8	2.0	0.5	0.7	0.8	1.2	1.6	2.0
Wyoming	83.1	81.3	84.4	11.2	11.7	12.2	1.4	1.8	2.2	0.4	1.0	1.5	1.7	2.0	2.3
Kentucky	83.6	84.4	85.2	9.6	10.2	10.7	1.0	1.3	1.6	1.4	1.9	2.4	1.8	2.3	2.7
Queensland, Australia (four locations)	80.0	82.2	85.5	10.0	11.1	12.8	1.0	1.2	1.6	0.3	1.9	6.0	1.1	3.0	6.6
Brazil		85.3			11.2			0.9			1.1			1.5	
Karak, Jordan	77.6	78.3	79.0	9.4	9.7	9.9	0.5	0.7	0.8	9.3	10.0	10.6	0.9	1.4	1.9
Timahdit Morocco	79.5	80.0	80.4	9.7	9.8	9.9	1.2	1.4	1.6	6.7	7.1	7.4	1.8	2.0	2.2
Sweden	86.5	86.5	86.5	9.0	9.4	9.8	0.6	0.7	0.7	1.7	1.9	2.1	1.4	1.6	1.7

REFERENCES

1. "Journal Forecast Supply & Demand," *Oil and Gas Journal,* 131 (Jan. 25, 1982).
2. J. D. Gilchrist, *Fuels and Refractories,* Macmillan, New York, 1963.
3. R. J. Reed, *Combustion Handbook,* 2nd ed., North American Manufacturing Co., Cleveland, Ohio, 1978.
4. Braine and King, *Fuels—Solid, Liquid, Gaseous,* St. Martin's Press, New York, 1967.
5. *Kempe's, Engineering Yearbook,* Morgan Grompium, London.
6. W. L. Nelson, *Petroleum Refinery Engineering,* McGraw-Hill, New York, 1968.
7a. E. M. Shelton, *Diesel Oils, DOE/BETC/PPS—81/5,* U.S. Department of Energy, Washington, DC, 1981.

Table 60.19 Chemical and Physical Properties of Several Tar Sand Bitumens[15]

	Uinta Basin, Utah	Southeast Utah	Athabasca, Alberta	Trapper Canyon, WY[a]	South TX	Santa Rosa, NM[a]	Big Clifty, KY	Bellamy, MO
Carbon, % wt	85.3	84.3	82.5	82.4	—	85.6	82.4	86.7
Hydrogen, % wt	11.2	10.2	10.6	10.3	—	10.1	10.8	10.3
Nitrogen, % wt	0.96	0.51	0.44	0.54	0.36	0.22	0.64	0.10
Sulfur, % wt	0.49	4.46	4.86	5.52	~10	2.30	1.55	0.75
H/C ratio	1.56	1.44	1.53	1.49	1.34	1.41	1.56	1.42
Vanadium, ppm	23	151	196	91	85	25	198	—
Nickel, ppm	96	62	82	53	24	23	80	—
Carbon residue, % wt	10.9	19.6	13.7	14.8	24.5	22.1	16.7	—
Pour point, °F	125	95	75	125	180	—	85	—
API gravity	11.6	9.2	9.5	5.4	−2.0	5	8.7	10

Viscosities range from 50,000 to 600,000 SSF (100,000 to 1,300,000 cSt).

[a] Outcrop samples.

Table 60.20 Elemental Composition of Bitumen and Oils Recovered from Tar Sands by Methods C and S[a,15]

	Bitumen	Light Oil C[b]	Heavy Oil C 1–4 Mo.	Heavy Oil C 5–6 Mo.	Product Oil C	Product Oil S[c]
Carbon, % wt	86.0	86.7	86.1	86.7	86.6	85.9
Hydrogen, % wt	11.2	12.2	11.8	11.3	11.6	11.3
Nitrogen, % wt	0.93	0.16	0.82	0.66	0.82	1.17
Sulfur, % wt	0.45	0.30	0.39	0.33	0.43	0.42
Oxygen, % wt[d]	1.42	0.64	0.89	1.01	0.55	1.21

[a] These percentages are site and project specific.

[b] C = reverse-forward combustion.

[c] S = steamflood.

[d] By difference.

7b. E. M. Shelton, *Heating Oils, DOE/BETC/PPS—80/4,* U.S. Department of Energy, Washington, DC, 1980.

7c. E. M. Shelton, *Aviation Turbine Fuel, DOE/BETC/PPS—80/2,* Department of Energy, Washington, DC, 1979.

8. ANSI/ASTM D396, *Standard Specification for Fuel Oils,* American Society for Testing and Materials, Philadelphia, PA, 1980.

9. ANSI/ASTM D3699, *Standard Specification for Kerosene,* American Society for Testing and Materials, Philadelphia, PA, 1982.

10. ANSI/ASTM D1655, *Standard Specification for Aviation Turbine Fuels,* American Society for Testing and Materials, Philadelphia, PA, 1982.

11. ANSI/ASTM D2880, *Standard Specification for Gas Turbine Fuel Oils,* American Society for Testing and Materials, Philadelphia, PA, 1982.

12. ANSI/ASTM D975, *Standard Specification for Diesel Fuel Oils,* American Society for Testing and Materials, Philadelphia, PA, 1981.

13. M. Heap et al., *The Influence of Fuel Characteristics on Nitrogen Oxide Formation—Bench Scale Studies,* Energy and Environmental Research Corp., Irvine, CA, 1979.

14. H. Tokairin and S. Morita, "Properties and Characterizations of Fischer-Assay-Retorted Oils from Major World Deposits," *Synthetic Fuels from Oil Shale and Tar Sands,* Institute of Gas Technology, Chicago, IL, 1983.

15. K. P. Thomas et al., "Chemical and Physical Properties of Tar Sand Bitumens and Thermally Recovered Oils," *Synthetic Fuels from Oil Shale and Tar Sands,* Institute of Gas Technology, Chicago, IL, 1983.

CHAPTER 61
COALS, LIGNITE, PEAT

JAMES G. KEPPELER

Manager of Coal Operations
Electric Fuels Corporation

61.1 INTRODUCTION

61.1.1 Nature

Coal is a dark brown to black sedimentary rock derived primarily from the unoxidized remains of carbon-bearing plant tissues. It is a complex, combustible mixture of organic, chemical, and mineral materials found in strata, or "seams," in the earth, consisting of a wide variety of physical and chemical properties.

The principal types of coal, in order of metamorphic development, are lignite, subbituminous, bituminous, and anthracite. While not generally considered a coal, peat is the first development stage in the "coalification" process, in which there is a gradual increase in the carbon content of the fossil organic material, and a concomitant reduction in oxygen.

Coal substance is composed primarily of carbon, hydrogen, and oxygen, with minor amounts of nitrogen and sulfur, and varying amounts of moisture and mineral impurities.

61.1.2 Reserves—Worldwide and United States

According to the World Coal Study (see Ref. 3), the total geological resources of the world in "millions of tons of coal equivalent" (mtce) is 10,750,212, of which 662,932, or 6%, is submitted as "Technically and Economically Recoverable Resources."

Millions of tons of coal equivalent is based on the metric ton (2205 lb) with a heat content of 12,600 Btu/lb (7000 kcal/kg).

A summary of the percentage of technically and economically recoverable reserves and the percentage of total recoverable by country is shown in Table 61.1.

As indicated in Table 61.1, the United States possesses over a quarter of the total recoverable reserves despite the low percentage of recovery compared to other countries.

It is noted that the interpretation of "technical and economic" recovery is subject to considerable variation and also to modification, as technical development and changing economic conditions dictate. It should also be noted that there are significant differences in density and heating values in various coals and, therefore, the mtce definition should be kept in perspective.

In 1977, the world coal production was approximately 2450 mtce,[3] or about $\frac{1}{270}$th of the recoverable reserves.

According to the U.S. Geological Survey, the remaining U.S. Coal Reserves total almost 4000 billion tons,[4] with overburden to 6000 ft in seams of 14 in. or more for bituminous and anthracite and in seams of 2½ ft or more for subbituminous coal and lignite. The U.S. Bureau of Mines and U.S. Geological Survey have further defined "Reserve Base" to provide a better indication of the technically and economically mineable reserves, where a higher degree of identification and engineering evaluation is available.

Table 61.1

	Percentage of Recoverable[a] of Geological Resources	Percentage of Total Recoverable Reserves
Australia	5.5	4.9
Canada	1.3	0.6
Peoples Republic of China	6.8	14.9
Federal Republic of Germany	13.9	5.2
India	15.3	1.9
Poland	42.6	9.0
Republic of South Africa	59.7	6.5
United Kingdom	23.7	6.8
United States	6.5	25.2
Soviet Union	2.2	16.6
Other Countries	24.3	8.4
		100.0

[a] Technically and economically recoverable reserves. Percentage indicated is based on total geological resources reported by country.

Source: World Coal Study, *Coal—Bridge to the Future,* 1980.

A summary of the reserve base of U.S. coal is provided in Table 61.2.[5]

61.1.3 Classifications

Coals are classified by "rank," according to their degree of metamorphism, or progressive alteration, in the natural series from lignite to anthracite. Perhaps the most widely accepted standard for classification of coals is ASTM D388, which ranks coals according to fixed carbon and calorific value (expressed in Btu/lb) calculated to the mineral-matter-free basis. Higher-rank coals are classified according to fixed carbon on the dry basis; the lower-rank coals are classed according to calorific value on the moist basis. Agglomerating character is used to differentiate between certain adjacent groups. Table 61.3 shows the classification requirements.

Table 61.2 Demonstrated Reserve Base[a] of Coal in the United States on January, 1980, by Rank (Millions of Short Tons)

State[b]	Anthracite	Bituminous	Subbituminous	Lignite	Total[c]
Alabama[d]	—	3,916.8	—	1,083.0	4,999.8
Alaska	—	697.5	5,443.0	14.0	6,154.5
Arizona	—	410.0	—	—	410.0
Arkansas	96.4	288.7	—	25.7	410.7
Colorado[d]	25.5	9,086.1	3,979.9	4,189.9	17,281.3
Georgia	—	3.6	—	—	3.6
Idaho	—	4.4	—	—	4.4
Illinois[d]	—	67,606.0	—	—	67,606.0
Indiana	—	10,586.1	—	—	10,586.1
Iowa	—	2,197.1	—	—	2,197.7
Kansas	—	993.8	—	—	993.8
Kentucky					
Eastern[d]	—	12,927.5	—	—	12,927.5
Western	—	21,074.4	—	—	21,074.4
Maryland	—	822.4	—	—	822.4
Michigan[d]	—	127.7	—	—	127.7
Missouri	—	6,069.1	—	—	6,069.1
Montana	—	1,385.4	103,277.4	15,765.2	120,428.0
New Mexico[d]	2.3	1,835.7	2,683.4	—	4,521.4
North Carolina	—	10.7	—	—	10.7
North Dakota	—	—	—	9,952.3	9,952.3
Ohio[d]	—	19,056.1	—	—	19,056.1
Oklahoma	—	1,637.8	—	—	1,637.8
Oregon	—	—	17.5	—	17.5
Pennsylvania	7,092.0	23,188.8	—	—	30,280.8
South Dakota	—	—	—	366.1	366.1
Tennessee[d]	—	983.7	—	—	983.7
Texas[d]	—	—	—	12,659.7	12,659.7
Utah[d]	—	6,476.5	1.1	—	6,477.6
Virginia	125.5	3,345.9	—	—	3,471.4
Washington[d]	—	303.7	1,169.4	8.1	1,481.3
West Virginia	—	39,776.2	—	—	39,776.2
Wyoming[d]	—	4,460.5	65,463.5	—	69,924.0
Total[c]	7,341.7	239,272.9	182,035.0	44,063.9	472,713.6

[a] Includes measured and indicated resource categories defined by USBM and USGS and represents 100% of the coal in place.

[b] Some coal-bearing states where data are not sufficiently detailed or where reserves are not currently economically recoverable.

[c] Data may not add to totals due to rounding.

[d] Data not completely reconciled with demonstrated reserve base data.

Table 61.3 ASTM (D388) Classification of Coals by Rank[a]

Class	Group	Fixed Carbon Limits, Percent (Dry, Mineral-Matter-Free Basis)		Volatile Matter Limits, Percent (Dry, Mineral-Matter-Free Basis)		Calorific Value Limits, Btu/lb (Moist,[b] Mineral-Matter-Free Basis)		Agglomerating Character
		Equal to or Greater Than	Less Than	Greater Than	Equal to or Less Than	Equal to or Greater Than	Less Than	
I Anthracitic	1. Metaanthracite	98	—	—	2	—	—	Nonagglomerating
	2. Anthracite	92	98	2	8	—	—	
	3. Semianthracite[c]	86	92	8	14	—	—	
II Bituminous	1. Low-volatile bituminous	78	86	14	22	—	—	
	2. Medium-volatile bituminous	69	78	22	31	—	—	
	3. High-volatile A bituminous	—	69	31	—	14,000[d]	—	Commonly agglomerating[e]
	4. High-volatile B bituminous	—	—	—	—	13,000[d]	14,000	
	5. High-volatile C bituminous	—	—	—	—	11,500	13,000	
						10,500	11,500	Agglomerating
III Subbituminous	1. Subbituminous A	—	—	—	—	10,500	11,500	
	2. Subbituminous B	—	—	—	—	9,500	10,500	
	3. Subbituminous C	—	—	—	—	8,300	9,500	
IV Lignitic	1. Lignite A	—	—	—	—	6,300	8,300	
	2. Lignite B	—	—	—	—	—	6,300	

[a] This classification does not include a few coals, principally nonbanded varieties, that have unusual physical and chemical properties and that come within the limits of fixed carbon or calorific value of the high-volatile bituminous and subbituminous ranks. All of these coals either contain less than 48% dry, mineral-matter-free fixed carbon or have more than 15,500 moist, mineral-matter-free British thermal units per pound.

[b] Moist refers to coal containing its natural inherent moisture but not including visible water on the surface of the coal.

[c] If agglomerating, classify in the low-volatile group of the bituminous class.

[d] Coals having 69% or more fixed carbon on the dry, mineral-matter-free basis shall be classified according to fixed carbon, regardless of calorific value.

[e] It is recognized that there may be nonagglomerating varieties in these groups of the bituminous class and that there are notable exceptions in the high-volatile C bituminous group.

Agglomerating character is determined by examination of the residue left after the volatile determination. If the residue supports a 500-g weight without pulverizing or shows a swelling or cell structure, it is said to be "agglomerating."

The mineral-matter-free basis is used for ASTM rankings, and formulas to convert Btu, fixed carbon, and volatile matter from "as-received" bases are provided. Parr formulas—Eqs. (61.1)–(61.3) are appropriate in case of litigation. Approximation formulas—Eqs. (61.4)–(61.6) are otherwise acceptable.

Parr formulas

$$\text{Dry, MM-Free } FC = \frac{FC - 0.15S}{100 - (M + 1.08A + 0.55S)} \times 100 \tag{61.1}$$

$$\text{Dry, MM-Free } VM = 100 - \text{Dry, MM-Free } FC \tag{61.2}$$

$$\text{Moist, MM-Free Btu} = \frac{\text{Btu} - 50S}{100 - (1.08A + 0.55S)} \times 100 \tag{61.3}$$

Approximation formulas

$$\text{Dry, MM-Free } FC = \frac{FC}{100 - (M + 1.1A + 0.1S)} \times 100 \tag{61.4}$$

$$\text{Dry, MM-Free } VM = 100 - \text{Dry, MM-Free } FC \tag{61.5}$$

$$\text{Moist, MM-Free Btu} = \frac{\text{Btu}}{100 - (1.1A + 0.1S)} \times 100 \tag{61.6}$$

where MM = mineral matter
Btu = British thermal unit
FC = percentage of fixed carbon
VM = percentage of volatile matter
A = percentage of ash
S = percentage of sulfur

Other classifications of coal include the International Classification of Hard Coals, the International Classification of Brown Coals, the "Lord" value based on heating value with ash, sulfur, and moisture removed, and the Perch and Russell Ratio, based on the ratio of Moist, MM-Free Btu to Dry, MM-Free VM.

61.2 CURRENT USES—HEAT, POWER, STEELMAKING, OTHER

According to statistics compiled by the U.S. Department of Energy, the primary use of coals produced in the United States in recent years has been for electric utilities; comprising over ⅘ths of the almost 700 million short tons consumed domestically in 1980. Moreover, utility use of coal has accounted for over 50% of the domestic consumption in every year since 1962.

The second largest use, at 9.5% in 1980, was the manufacture of coke. This demand for coal, greater than that for utility use in the 1950s, has since declined owing to more efficient steelmaking, greater use of scrap metal, increased use of substitute fuels in blast furnaces, and other factors. The production of coke from coal is accomplished by heating certain coals in the absence of air to drive off volatile matter and moisture. To provide a suitable by-product coke, and parent coal must possess certain properties including low ash, low sulfur, low coking pressure, and high coke strength. By-product coking ovens, the most predominant, are so named for their ability to recapture otherwise wasted by-products driven off by heating the coal, such as coke oven gas, coal-tar, ammonia, oil, and useful chemicals. Beehive ovens, named for their shape and configuration, are also used, albeit much less extensively, in the production of coke.

Industrial use of coal, other than coking plants, comprised 8.5% of the domestic consumption in 1980. This group consumes coal for process steam primarily and also uses coal in open-fired applications, such as kilns, and in process heaters.

61.3 TYPES

Anthracite is the least abundant of U.S. coal forms. Sometimes referred to as "hard" coal, it is shiny black or dark silver-gray and relatively compact. Inasmuch as it is the most advanced form in the coalification process, it is sometimes found deeper in the earth than bituminous. As indicated earlier, the ASTM definition puts upper and lower bounds of dry, mineral-matter-free fixed carbon percent at 98% and 86%, respectively, which limits volatile matter to not more than 14%. Combustion in turn is characterized by higher ignition temperatures and longer burnout times than bituminous coals.

Excepting some semianthracites that have a granular appearance, they have a consolidated appearance, unlike the layers seen in many bituminous coals. Typical Hardgrove Grindability Index ranges from 20 to 60 and specific gravity typically ranges 1.55 ± 0.10.

Anthracitic coals can be found in Arkansas, Colorado, Pennsylvania, Massachusetts, New Mexico, Rhode Island, Virginia, and Washington, although by far the most abundant reserves are found in Pennsylvania.

Bituminous coal is by far the most plentiful and utilized coal form, and within the ASTM definitions includes low-, medium-, and high-volatile subgroups. Sometimes referred to as "soft" coal, it is named after the word bitumen, based on general tendency toward forming a sticky mass on heating.

At a lower stage of development in the coalification process, carbon content is less than the anthracites, from a maximum of 86% to less than 69% on a dry, mineral-matter-free basis. Volatile matter, at a minimum of 14% on this basis, is greater than the anthracites, and, as a result, combustion in pulverized form is somewhat easier for bituminous coals. Production of gas is also enhanced by their higher volatility.

The tendency of bituminous coals to produce a cohesive mass on heating lends them to coke applications. Dry, mineral-matter-free oxygen content generally ranges from 5% to 10%, compared to a value as low as 1% for anthracite. They are commonly banded with layers differing in luster.

The low-volatile bituminous coals are grainier and more subject to size reduction in handling.

The medium-volatile bituminous coals are sometimes distinctly layered, and sometimes only faintly layered and appearing homogeneous. Handling may or may not have a significant impact on size reduction.

The high-volatile coals (A, B, and C) are relatively hard and less sensitive to size reduction from handling than low- or medium-volatile bituminous.

Subbituminous coals, like anthracite and lignite, are generally noncaking. "Caking" refers to fusion of coal particles after heating in a furnace, as opposed to "coking," which refers to the ability of a coal to make a good coke, suitable for metallurgical purposes.

Oxygen content, on a dry, mineral-matter-free basis, is typically 10–20%.

Brownish black to black in color, this type coal is typically smooth in appearance with an absence of layers.

High in inherent moisture, it is ironic that these fuels are often dusty in handling and appear much like drying mud as they disintegrate on sufficiently long exposure to air.

The Healy coal bed in Wyoming has the thickest seam of coal in the United States at 220 ft. It is subbituminous, with an average heating value of 7884 Btu/lb, 28.5% moisture, 30% volatile matter, 33.9% fixed carbon, and 0.6% sulfur. Reported strippable reserves of this seam are approximately 11 billion tons.[4]

Lignites, often referred to as "brown coal," often retain a woodlike or laminar structure in which wood fiber remnants may be visible. Like subbituminous coals, they are high in seam moisture, up to 50% or more, and also disintegrate on sufficiently long exposure to air.

Both subbituminous coals and lignites are more susceptible than higher rank coals to storage, shipping, and handling problems, owing to their tendency for slacking (disintegration) and spontaneous ignition. During the slacking, a higher rate of moisture loss at the surface than at the interior may cause higher rates and stresses at the outside of the particles, and cracks may occur with an audible noise.

Peat is decaying vegetable matter formed in wetlands; it is the first stage of metamorphosis in the coalification process. Development can be generally described as anaerobic, often in poorly drained flatlands or former lake beds. In the seam, peat moisture may be 90% or higher, and, therefore, the peat is typically "mined" and stacked for drainage or otherwise dewatered prior to consideration as a fuel. Because of its low bulk density at about 15 lb/ft³ and low heating value at about 6000 Btu/lb (both values at 35% moisture), transportation distances must be short to make peat an attractive energy option.

In addition, it can be a very difficult material to handle, as it can arch in bins, forming internal friction angles in excess of 70°.

Chemically, peat is very reactive and ignites easily. It may be easily ground, and unconsolidated peat may create dusting problems.

61.4 PHYSICAL AND CHEMICAL PROPERTIES—DESCRIPTION AND TABLES OF SELECTED VALUES

There are a number of tests, qualitative and quantitative, used to provide information on coals; these tests will be of help to the user and/or equipment designer. Among the more common tests are the following, with reference to the applicable ASTM test procedure.

A. *"Proximate" analysis* (D3172) includes moisture, "volatile matter," "fixed carbon," and ash as its components.

Percent moisture (D3173) is determined by measuring the weight loss of a prepared sample (D2013) when heated to between 219°F (104°C) and 230°F (110°C) under rigidly controlled conditions. The results of this test can be used to calculate other analytical results to a dry basis. This moisture is referred to as "residual," and must be added to moisture losses incurred in sample preparation, called "air-dry losses" in order to calculate other analytical results to an "as-received" basis. The method which combines both residual and air dry moisture is D3302.

Percent volatile matter (D3175) is determined by establishing the weight loss of a prepared sample (D2013) resulting from heating to 1740°F (950°C) in the absence of air under controlled conditions. This weight loss is corrected for residual moisture, and is used for an indication of burning properties, coke yield, and classification by rank.

Percent ash (D3174) is determined by weighing the residue remaining after burning a prepared sample under rigidly controlled conditions. Combustion is in an oxidizing atmosphere and is completed for coal samples at 1290–1380°F (700–750°C).

Fixed carbon is a calculated value making up the fourth and final component of a proximate analysis. It is determined by subtracting the volatile, moisture, and ash percentages from 100.

Also generally included with a proximate analysis are calorific value and sulfur determinations.

B. *Calorific value*, Btu/lb (J/g, cal/g), is most commonly determined (D2015) in an "adiabatic bomb calorimeter," but is also covered by another method (D3286), which uses an "isothermal jacket bomb calorimeter." The values determined by this method are called gross or high heating values and include the latent heat of water vapor in the products of combustion.

C. *Sulfur* is determined by one of three methods provided by ASTM, all covered by D3177: the Eschka method, the bomb washing method, and a high-temperature combustion method.

The Eschka method requires that a sample be ignited with an "Eschka mixture" and sulfur be precipitated from the resulting solution as $BaSO_4$ and filtered, ashed, and weighed.

The bomb wash method requires use of the oxygen-bomb calorimeter residue; sulfur is precipitated as $BaSO_4$ and processed as in the Eschka method.

The high-temperature combustion method produces sulfur oxides from burning of a sample at 2460°F (1350°C), which are absorbed in a hydrogen peroxide solution for analysis. This is the most rapid of the three types of analysis.

D. *Sulfur forms* include sulfate, organic, and pyritic, and, rarely, elemental sulfur. A method used to quantify sulfate, pyritic sulfur, and organic sulfur is D2492. The resulting data are sometimes used to provide a first indication of the maximum amount of sulfur potentially removable by mechanical cleaning.

E. *Ultimate analysis* (D3176) includes total carbon, hydrogen, nitrogen, oxygen, sulfur, and ash. These data are commonly used to perform combustion calculations to estimate combustion air requirements, products of combustion, and heat losses such as incurred by formation of water vapor by hydrogen in the coal.

Chlorine (D2361) and phosphorus (D2795) are sometimes requested with ultimate analyses, but are not technically a part of D3176.

F. *Ash mineral analysis* (D2795) includes the oxides of silica (SiO_2), alumina (Al_2O_3), iron (Fe_2O_3), titanium (TiO_2), phosphorous (P_2O_5), calcium (CaO), magnesium (MgO), sodium (Na_2O), and potassium (K_2O).

These data are used to provide several indications concerning ash slagging or fouling tendencies, abrasion potential, electrostatic precipitator operation, and sulfur absorption potention.

See Section 61.8 for further details.

G. *Grindability* (D409) is determined most commonly by the Hardgrove method to provide an indication of the relative ease of pulverization or grindability, compared to "standard" coals having grindability indexes of 40, 60, 80, and 110. As the index increases, pulverization becomes easier, that is, an index of 40 indicates a relatively hard coal; an index of 100 indicates a relatively soft coal.

Standard coals may be obtained from the U.S. Bureau of Mines.

A word of caution is given: grindability may change with ash content, moisture content, temperature, and other properties.

H. *Free swelling index* (D720) also referred to as a "coke-button" test, provides a relative index (1–9) of the swelling properties of a coal. A sample is burned in a covered crucible, and the resulting index increases as the swelling increases, determined by comparison of the button formed with standard profiles.

I. *Ash fusion temperatures* (D1857) are determined from triangular-core-shaped ash samples, in a reducing atmosphere and/or in an oxidizing atmosphere. Visual observations are recorded of temperatures at which the core begins to deform, called "initial deformation"; where height equals width, called "softening"; where height equals one-half width, called "hemispherical"; and where the ash is fluid.

The hemispherical temperature is often referred to as the "ash fusion temperature."

While not definitive, these tests provide a rough indication of the slagging tendency of coal ash.

Analysis of petrographic constituents in coals has been used to some extent in qualitative and semiquantitative analysis of some coals, most importantly in the coking coal industry. It is the application of macroscopic and microscopic techniques to identify maceral components related to the plant origins of the coal. The macerals of interest are vitrinite, exinite, resinite, micrinite, semifusinite, and fusinite. A technique to measure reflectance of a prepared sample of coal and calculate the volume percentages of macerals is included in ASTM Standard D2799.

Table 61.4 shows selected analyses of coal seams for reference.

61.5 BURNING CHARACTERISTICS

The ultimate analysis, described in the previous section, provides the data required to conduct fundamental studies of the air required for stoichiometric combustion, the volumetric and weight amounts of combustion gases produced, and the theoretical boiler efficiencies. These data assist the designer in such matters as furnace and auxiliary equipment sizing. Among the items of concern are draft equipment for supplying combustion air requirements, drying and transporting coal to the burners and exhausting the products of combustion, mass flow and velocity in convection passes for heat transfer and erosion considerations, and pollution control equipment sizing.

The addition of excess air must be considered for complete combustion and perhaps minimization of ash slagging in some cases. It is not uncommon to apply 25% excess air or more to allow operational flexibility.

As rank decreases, there is generally an increase in oxygen content in the fuel, which will provide a significant portion of the combustion air requirements.

The theoretical weight, in pounds, of combustion air required per pound of fuel for a stoichiometric condition is given by

$$11.53C + 34.34 [H_2 - \tfrac{1}{8}O_2] + 4.29S \tag{61.7}$$

where C, H_2, O_2, and S are percentage weight constituents in the ultimate analysis.

The resulting products of combustion, again at a stoichiometric condition and complete combustion, are

$$CO_2 = 3.66C \tag{61.8}$$

$$H_2O = 8.94H_2 + H_2O \text{ (wt\% } H_2O \text{ in fuel)} \tag{61.9}$$

$$SO_2 = 2.00S \tag{61.10}$$

$$N_2 = 8.86C + 26.41 (H_2 - \tfrac{1}{8}O_2) + 3.29S + N_2 \text{ (wt\% nitrogen in fuel)} \tag{61.11}$$

The combustion characteristics of various ranks of coal can be seen in Fig. 61.1, showing "burning profiles" obtained by thermal gravimetric analysis. As is apparent from this figure, ignition of lower rank coals occurs at a lower temperature and combustion proceeds at a more rapid rate than higher rank coals. This information is, of course, highly useful to the design engineer in determination of the size and configuration of combustion equipment.

The predominant firing technique for combustion of coal is in a pulverized form. To enhance ignition, promote complete combustion, and, in some cases, mitigate the effects of large particles on slagging and particulate capture, guidelines are generally given by the boiler manufacturer for pulverizer output (burner input).

Typical guidelines are as follows:

Coal Class/Group	Percentage Passing a 200 Mesh Sieve	Percentage Retained on a 50 Mesh Sieve	Allowable Coal/Air Temperature (°F)
Anthracite	80	2.0	200
Low-volatile bituminous	70–75	2.0	180
High-volatile bituminous A	70–75	2.0	170
High-volatile bituminous C	65–72	2.0	150–160
Lignite	60–70	2.0	110–140

It is noted that these guidelines may vary for different manufacturers, ash contents, and equipment applications and, of course, the manufacturer should be consulted for fineness and temperature recommendations.

The sieve designations of 200 and 50 refer to U.S. Standard sieves. The 200 mesh sieve has 200 openings per linear inch, or 40,000 per square inch. The ASTM designations for these sieves are 75 and 300 micron, respectively.

Finally, agglomerating character may also have an influence on the fineness requirements, since this property might inhibit complete combustion.

Table 61.4 Selected Values—Coal and Peat Quality

Parameter	East Kentucky, Skyline Seam (Washed)	Pennsylvania, Pittsburgh #8 Seam (Washed)	Illinois, Harrisburg 5 (Washed)	Wyoming, Powder River Basin (Raw)	Florida Peat, Sumter County In situ	Florida Peat, Sumter County Dry
Moisture % (total)	8.00	6.5	13.2	25.92	86.70	—
Ash %	6.48	6.5	7.1	6.00	0.54	4.08
Sulfur %	0.82	1.62	1.28	0.25	0.10	0.77
Volatile %	36.69	34.40	30.6	31.27	8.74	65.73
Grindability (HGI)	45	55	54	57	36	69[a]
Calorific value (Btu/lb, as received)	12,500	13,100	11,700	8,500	1,503	11,297
Fixed carbon	48.83	52.60	49.1	37.23	4.02	30.19
Ash minerals						
SiO_2	50.87	50.10	48.90	32.02	58.29	
Al_2O_3	33.10	24.60	25.50	15.88	19.50	
TiO_2	2.56	1.20	1.10	1.13	1.05	
CaO	2.57	2.2	2.90	23.80	1.95	
K_2O	1.60	1.59	3.13	0.45	1.11	
MgO	0.80	0.70	1.60	5.73	0.94	
Na_2O	0.53	0.35	1.02	1.27	0.40	
P_2O_5	0.53	0.38	0.67	1.41	0.09	
Fe_2O_3	5.18	16.20	12.20	5.84	14.32	
SO_3	1.42	1.31	1.96	11.35	2.19	
Undetermined	0.84	1.37	1.02	1.12	0.16	

Ultimate Analysis (Dry)

Ash	7.04	7.0	8.23	7.53	4.08
Hydrogen	5.31	5.03	4.95	4.80	4.59
Carbon	75.38	78.40	76.57	69.11	69.26
Nitrogen	1.38	1.39	1.35	0.97	1.67
Sulfur	0.89	1.73	1.47	0.34	0.77
Oxygen (by difference)	9.95	6.35	7.03	17.24	19.33
Chlorine	0.05	0.10	0.40	0.01	0.30

Ash Fusion Temperatures (°F)

Initial deformation (reducing)	2800+	2350	2240	2204	1950
Softening ($H = W$) (reducing)	2800+	2460	2450	2226	2010
Hemispherical ($H = \frac{1}{2} W$) (reducing)	2800+	2520	2500	2250	2060
Fluid (reducing)	2800+	2580	2700+	2302	2100

a At 9% H_2O.

Rate of weight loss,

Mg/minute

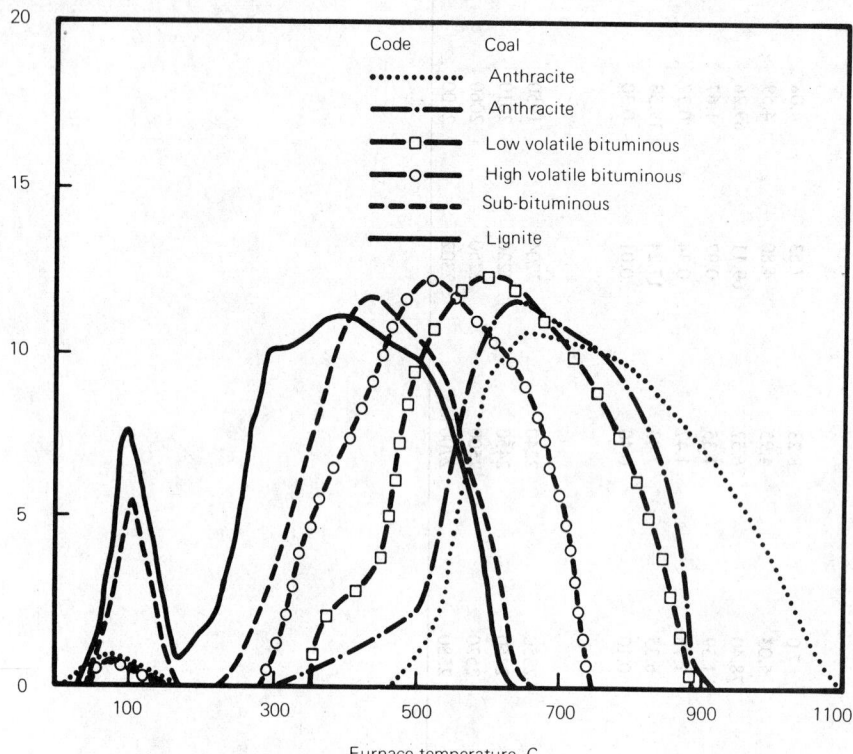

Furnace temperature, C

Fig. 61.1 Comparison of burning profiles for coals of different rank.

61.6 ASH CHARACTERISTICS

Ash is an inert residue remaining after the combustion of coal and can result in significant challenges for designers and operators of the combustion, ash handling, and pollution control equipment. The quantity of ash in the coal varies widely from as little as 6% or less to more than 30% by weight. Additionally, diverse physical and chemical properties of ashes can pose substantial problems with slagging, abrasion, and fouling of boilers. Electrostatic precipitators, used for pollution control, can experience material changes in collection efficiency depending on the mineral constituents of the ash.

"Slagging" is a term that generally refers to the formation of high-temperature fused ash deposits on furnace walls and other surfaces primarily exposed to radiant heat.

"Fouling" generally refers to high-temperature bonded ash deposits forming on convection tube banks, particularly superheat and reheat tubes.

Indication of ash-slagging tendencies can be measured by tests such as viscosity–temperature tests or by ash-softening tests. In addition, there are many empirical equations that are used to provide information as to the likelihood of slagging and fouling problems.

ASTM Standard number D1857 is the most common test used for slagging indication. In this test, ash samples are prepared as triangular cones and then are heated at a specified rate. Observations are then made and recorded of temperatures at prescribed stages of ash deformation, called initial deformation, softening temperature, hemispherical temperature, and fluid temperature. These tests are conducted in reducing and/or oxidizing atmospheres.

Another method used, although far more costly, involves measurement of the torque required to rotate a platinum bob suspended in molten slag. A viscosity–temperature relationship is established as a result of this test, which is also conducted in reducing and/or oxidizing atmospheres. A slag is generally considered liquid when its viscosity is below 250 poise, although tapping from a boiler may require a viscosity of 50–100 poise. It is plastic when its viscosity is between 250 and 10,000 poise. It is in this region where removal of the slag is most troublesome.

Ash mineral analyses are used to calculate empirical indicators of slagging problems. In these analyses are included metals reported as equivalent oxide weight percentages of silica, alumina, iron, calcium, magnesium, sodium, potassium, titania, phosphorous, and sulfur, as follows:

$$SiO_2 + Al_2O_3 + Fe_2O_3 + CaO + MgO + Na_2O + K_2O + TiO_2 + P_2O_5 + SO_3 = 100\%$$

Some ratios calculated using these data are

Base: Acid Ratio, B/A

$$\frac{B}{A} = \frac{base}{acid} = \frac{Fe_2O_3 + CaO + MgO + Na_2O + K_2O}{SiO_2 + Al_2O_3 + TiO_2}$$

It has been reported[1] that a base/acid ratio in the range of 0.4 to 0.7 results typically in low ash fusibility temperatures and, hence, more slagging problems.

Slagging Factor, R_s

$$R_s = B/A \times \% \text{ sulfur, dry coal basis}$$

It has been reported[15] that coals with bituminous-type ashes exhibit a high slagging potential with a slagging factor above 2 and severe slagging potential with a slagging factor of more than 2.6. Bituminous-type ash refers to those ashes where iron oxide percentage is greater than calcium *plus* magnesium oxide.

Silica/Alumina Ratio

$$\frac{silica}{alumina} = \frac{SiO_2}{Al_2O_3}$$

It has been reported[1] that the silica in ash is more likely to form lower-melting-point compounds than is alumina and for two coals having the same base/acid ratio, the coal with a higher silica/alumina ratio should result in lower fusibility temperatures. However, it has also been reported[2] that for low base/acid ratios the opposite is true.

Iron/Calcium Ratio

$$\frac{iron}{dolomite} = \frac{Fe_2O_3}{CaO + MgO}$$

This ratio and its use are essentially the same as the iron/calcium ratio.

Silica Percentage (SP)

$$SP = \frac{SiO_2 \times 100}{SiO_2 + Fe_2O_3 + CaO + MgO}$$

This parameter has been correlated with ash viscosity. As silica ratio increases, the viscosity of slag increases. Graphical methods[2] are used in conjunction with this parameter to estimate the T_{250} temperature—the temperature where the ash would have a viscosity of 250 poise. Where the acidic content is less than 60% and the ash is lignitic, the *dolomite percentage (DP)* is used in preference to the silica percentage, along with graphs to estimate the T_{250}:

$$DP = \frac{(CaO + MgO) \times 100}{Fe_2O_3 + CaO + MgO + Na_2O + K_2O}$$

where the sum of the basic and acidic components are adjusted, if necessary, to equal 100%. For bituminous ash or lignitic-type ash having acidic content above 60%, the base/acid ratio is used in conjunction with yet another graph.

Fouling Factor (R_F)

$$R_F = \text{acid base} \times \% \text{ Na}_2 \text{ (bituminous ash)}$$

or

$$R_F = \% \text{ Na}_2\text{O (lignitic ash)}$$

For bituminous ash, the fouling factor[17] is "low" for values less than 0.1, "medium" for values between 0.1 and 0.25, "high" for values between 0.25 and 0.7, and "severe" values above 0.7. For lignitic-type ash, the percentage of sodium is used, and low, medium, and high values are < 3.0, 3.0–6.0, and > 6.0, respectively.

The basis for these factors is that sodium is the most important single factor in ash fouling, volatilizing in the furnace and subsequently condensing and sintering ash deposits in cooler sections.

Chlorine has also been used as an indicator of fouling tendency of eastern-type coals. If chlorine, from the ultimate analysis, is less than 0.15%, the fouling potential is low; if between 0.15 and 0.3, it has a medium fouling potential; and if above 0.3, its fouling potential is high.[1]

Ash resistivity can be predicted from ash mineral and coal ultimate analyses, according to a method described by Bickelhaupt.[14] Electrostatic precipitator sizing and/or performance can be estimated using the calculated resistivity. For further information, the reader is referred to Ref. 14.

61.7 SAMPLING

Coals are by nature heterogeneous and, as a result, obtaining a representative sample can be a formidable task. Its quality parameters such as ash, moisture, and calorific value can vary considerably from seam to seam and even within the same seam. Acquisition of accurate data to define the nature and ranges of these values adequately is further compounded by the effects of size gradation, sample preparation, and analysis accuracy.

Inasmuch as these data are used for such purposes as pricing, control of the operations in mines, and preparation plants, determination of power plant efficiency, estimation of material handling and storage requirements for the coal and its by-products, and in some cases for determination of compliance with environmental limitations, it is important that samples be taken, prepared, and analyzed in accordance with good practice.

To attempt to minimize significant errors in sampling, ASTM D2234 was developed as a standard method for the collection of a gross sample of coal and D2013 for preparation of the samples collected for analysis. It applies to lot sizes up to 10,000 tons (9080 mg) per gross sample and is intended to provide an accuracy of $\pm\frac{1}{10}$th of the average ash content in 95 of 100 determinations.

The number and weight of increments of sample comprising the gross sample to represent the lot, or consignment, is specified for nominal top sizes of ⅝ in. (16 mm), 2 in. (50 mm), and 6 in. (150 mm), for raw coal or mechanically cleaned coal. Conditions of collection include samples taken from a stopped conveyor belt (the most desirable), full and partial stream cuts from moving coal consignments, and stationary samples.

One recommendation made in this procedure and worth special emphasis is that the samples be collected under the direct supervision of a person qualified by training and experience for this responsibility.

This method does not apply to the sampling of reserves in the ground, which is done by core drilling methods or channel sampling along outcrops as recommended by a geologist or mining engineer. It also does not apply to the sampling of coal slurries.

A special method (D197) was developed for the collection of samples of pulverized coals to measure size consist or fineness, which is controlled to maintain proper combustion efficiency.

Sieve analyses using this method may be conducted on No. 16 (1.18 mm) through No. 200 (750 mm) sieves, although the sieves most often referred to for pulverizer and classifier performance are the No. 50, the No. 100, and the No. 200 sieves. The number refers to the quantity of openings per linear inch.

Results of these tests should plot as a straight line on a sieve distribution chart. A typical fineness objective, depending primarily on the combustion characteristics might be for 70% of the fines to pass through a 200-mesh sieve and not more than 2% to be retained on the 50-mesh sieve.

D2013 covers the preparation of coal samples for analysis, or more specifically, the reduction and division of samples collected in accordance with D2234. A "referee" and "nonreferee" method are delineated, although the "nonreferee" method is the most commonly used. "Referee" method is used to evaluate equipment and the nonreferee method.

Depending on the amount of moisture and, therefore, the ability of the coal to pass freely through reduction equipment, samples are either predried in air drying ovens or on drying floors or are processed directly by reduction, air drying, and division. Weight losses are computed for each stage of air drying to provide data for determination of total moisture, in combination with D3173 or 3302 (see Section 61.6).

Samples must ultimately pass the No. 60 (250 mm) sieve (D3173) or the No. 8 (2.36 mm) sieve (D3302) prior to total moisture determination.

Care must be taken to adhere to this procedure, including the avoidance of moisture losses while awaiting preparation, excessive time in air drying, proper use of riffling or mechanical division equipment, and verification and maintenance of crushing equipment size consist.

61.8 COAL CLEANING

Partial removal of impurities in coal such as ash and pyritic sulfur has been conducted since before 1900, although application and development has intensified during recent years owing to a number of factors, including the tightening of emissions standards, increasing use of lower quality seams, and increasing use of continuous mining machinery. Blending of two or more fuels to meet tight emissions standards, or other reasons, often requires that each of the fuels is of a consistent grade, which in turn may indicate some degree of coal cleaning.

Coal cleaning may be accomplished by physical or chemical means, although physical coal cleaning is by far the most predominant.

Primarily, physical processes rely on differences between the specific gravity of the coal and its impurities. Ash, clay, and pyritic sulfur have a higher specific gravity than that of coal. For example, bituminous coal typically has a specific gravity in the range 1.12–1.35, while pyrite's specific gravity is between 4.8 and 5.2.

One physical process that does not benefit from specific gravity differences is froth flotation. It is only used for cleaning coal size fractions smaller than 28 mesh (½ mm). Basically the process requires coal fines to be agitated in a chamber with set amounts of air, water, and chemical reagents, which creates a froth. The coal particles are selectively attached to the froth bubbles, which rise to the surface and are skimmed off, dewatered, and added to other clean coal fractions.

A second stage of froth flotation has been tested successfully at the pilot scale for some U.S. coals. This process returns the froth concentrate from the first stage to a bank of cells where a depressant is used to sink the coal and a xanthate flotation collector is used to selectively float pyrite.

The predominant commercial methods of coal cleaning use gravity separation by static and/or dynamic means. The extent and cost of cleaning naturally depends on the degree of end product quality desired, the controlling factors of which are primarily sulfur, heating value, and ash content.

Although dry means may be used for gravity separation, wet means are by far the more accepted and used techniques.

The first step in designing a preparation plant involves a careful study of the washability of the coal. "Float and sink" tests are run in a laboratory to provide data to be used for judging application and performance of cleaning equipment. In these tests the weight percentages and composition of materials are determined after subjecting the test coal to liquid baths of different specific gravities.

Pyritic sulfur and/or total sulfur percent, and heating value are typically determined for both the float (called "yield") and sink (called "reject") fractions.

Commonly, the tests are conducted on three or more size meshes, such as 1½ in. × 0 mesh, ⅜ in. × 100 mesh, and minus 14 mesh, and at three or more gravities of such as 1.30, 1.40, and 1.60. Percentage recovery of weight and heating value are reported along with other data on a cumulative basis for the float fractions. An example of this, taken from a Bureau of Mines study[11] of 455 coals is shown in Table 61.5.

Many coals have pyrite particles less than one micron in size (0.00004 in.), which cannot be removed

Table 61.5   Cumulative Washability Data[a]

Product	Recovery Weight	% Btu	Btu/lb	Ash %	Sulfur (%) Pyritic	Sulfur (%) Total	lb SO₂/ MBtu
Sample Crushed to Pass 1½ in.							
Float—1.30	55.8	58.8	14,447	3.3	0.26	1.09	1.5
Float—1.40	90.3	93.8	14,239	4.7	0.46	1.33	1.9
Float—1.60	94.3	97.3	14,134	5.4	0.53	1.41	2.0
Total	100.0	100.0	13,703	8.3	0.80	1.67	2.4
Sample Crushed to Pass ⅜ in.							
Float—1.30	58.9	63.0	14,492	3.0	0.20	1.06	1.5
Float—1.40	88.6	93.2	14,253	4.6	0.37	1.26	1.8
Float—1.60	92.8	96.8	14,134	5.4	0.47	1.36	1.9
Total	100.0	100.0	13,554	9.3	0.77	1.64	2.4
Sample Crushed to Pass 14 Mesh							
Float—1.30	60.9	65.1	14,566	2.5	0.16	0.99	1.4
Float—1.40	88.7	93.1	14,298	4.3	0.24	1.19	1.7
Float—1.60	93.7	97.1	14,134	5.4	0.40	1.29	1.8
Total	100.0	100.0	13,628	8.8	0.83	1.72	2.5

[a] State: Pennsylvania (bituminous); coal bed: Pittsburgh; county: Washington; raw coal moisture: 2.0%.

practically by mechanical means. Moreover, the cost of coal cleaning increases as the particle size decreases, as a general rule, and drying and handling problems become more difficult. Generally, coal is cleaned using particle sizes as large as practical to meet quality requirements.

It is interesting to note that a 50-micron pyrite particle (0.002 in.) inside a 14-mesh coal particle (0.06 in.) does not materially affect the specific gravity of a pure coal particle of the same size.

Washability data are usually organized and plotted as a series of curves, including:

1. Cumulative float–ash, sulfur %.
2. Cumulative sink ash %.
3. Elementary ash % (not cumulative).
4. % recovery (weight) versus specific gravity.

Types of gravity separation equipment include jigs, concentrating tables, water-only cyclones, dense-media vessels, and dense-media cyclones.

In jigs, the coal enters the vessel in sizes to 8 in. and larger and stratification of the coal and heavier particles occurs in a pulsating fluid. The bottom layer, primarily rock, ash, and pyrite, is stripped from the mixture and rejected. Coal, the top layer, is saved. A middle layer may also be collected and saved or rejected depending on quality.

Concentrating tables typically handle coals in the ⅜ in. × 0 or ¼ in. × 0 range and use water cascading over a vibrating table tilted such that heavier particles travel to one end of the table while lighter particles, traveling more rapidly with the water, fall over the adjacent edge.

In "water-only" and dense-media cyclones, centrifugal force is used to separate the heavier particles from lighter particles.

In heavy-media vessels, the specific gravity of the media is controlled typically in the range of 1.45 to 1.65. Particles floating are saved as clean coal, while those sinking are reject. Specific gravity of the media is generally maintained by the amount of finely ground magnetite suspended in the water, as in heavy-media cyclones. Magnetite is recaptured in the preparation plant circuitry by means of magnetic separators.

Drying of cleaned coals depends on size. The larger sizes, ¼ in. or ⅜ in. and larger, typically require little drying and might only be passed over vibrating screens prior to stockpiling. Smaller sizes, down to, say, 28 mesh, are commonly dried on stationary screens followed by centrifugal driers. Minus 28 mesh particles, the most difficult to dry, are processed in vibrating centrifuges, high-speed centrifugal driers, high-speed screens, or vacuum filters.

REFERENCES

1. J. G. Singer, *Combustion—Fossil Power Systems,* Combustion Engineering, Inc., Windsor, CT, 1981.
2. *Steam—Its Generation and Use,* Babcock & Wilcox, New York, 1978.
3. C. L. Wilson, *Coal—Bridge to the Future,* Vol. 1, Report of the World Coal Study, Ballinger, Cambridge, MA, 1980.
4. *Keystone Coal Industry Manual,* McGraw-Hill, New York, 1981.
5. *1979/1980 Coal Data,* National Coal Association and U.S. DOE, Washington, DC, 1982.
6. R. A. Meyers, *Coal Handbook,* Dekker, New York, 1981.
7. *ASTM Standards, Part 26: Gaseous Fuels; Coal and Coke, Atmospheric Analysis,* Philadelphia, PA, 1982.
8. F. M. Kennedy, J. G. Patterson, and T. W. Tarkington, *Evaluation of Physical and Chemical Coal Cleaning and Flue Gas Desulfurization,* EPA 600/7-79-250, U.S. EPA, Washington, DC, 1979.
9. Gibbs and Hill, Inc., *Coal Preparation for Combustion and Conversion,* Electric Power Research Institute, Palo Alto, CA, 1978.
10. *Coal Data Book,* President's Commission on Coal, U.S. Gov't. Ptg. Office, Washington, DC, 1980.
11. J. A. Cavallaro and A. W. Deurbrouch, *Sulfur Reduction Potential of the Coals of the United States,* RI8118, U.S. Dept. of Interior, Washington, DC, 1976.
12. R. A. Schmidt, *Coal in America,* 1979.
13. Auth and Johnson, *Fuels and Combustion Handbook,* McGraw-Hill, New York, 1951.
14. R. E. Bickelhaupt, *A Technique for Predicting Fly Ash Resistivity,* EPA 600/7-79-204, Southern Research Institute, Birmingham, AL, August, 1979.
15. R. C. Attig and A. F. Duzy, *Coal Ash Deposition Studies and Application to Boiler Design,* Babcock and Wilcox, Atlanta, GA, 1969.

16. A. F. Duzy and C. L. Wagoner, *Burning Profiles for Solid Fuels,* Babcock and Wilcox, Atlanta, GA, 1975.
17. E. C. Winegartner, *Coal Fouling and Slagging Parameters,* Exxon Research and Engineering, Baytown, TX, 1974.
18. J. B. McIlroy and W. L. Sage, Relationship of Coal-Ash Viscosity to Chemical Composition, *The Journal of Engineering for Power,* Babcock and Wilcox, New York, 1960.
19. R. P. Hensel, *The Effects of Agglomerating Characteristics of Coals on Combustion in Pulverized Fuel Boilers,* Combustion Engineering, Windsor, CT, 1975.
20. W. T. Reed, *External Corrosion and Deposits in Boilers and Gas Turbines,* Elsevier, New York, 1971.
21. O. W. Durrant, *Pulverized Coal—New Requirements and Challenges,* Babcock and Wilcox, Atlanta, GA, 1975.

CHAPTER 62

SOLAR ENERGY APPLICATIONS

JAN F. KREIDER

Jan F. Kreider and Associates, Inc.
Boulder, Colorado

62.1 SOLAR ENERGY AVAILABILITY

Solar energy is defined as that radiant energy transmitted by the sun and intercepted by earth. It is transmitted through space to earth by electromagnetic radiation with wavelengths ranging between 0.20 and 15 microns. The availability of solar flux for terrestrial applications varies with season, time of day, location, and collecting surface orientation. In this chapter we shall treat these matters analytically.

62.1.1 SOLAR GEOMETRY

Two motions of the earth relative to the sun are important in determining the intensity of solar flux at any time—the earth's rotation about its axis and the annual motion of the earth and its axis about the sun. The earth rotates about its axis once each day. A solar day is defined as the time that elapses between two successive crossings of the local meridian by the sun. The local meridian at any point is the plane formed by projecting a north–south longitude line through the point out into space from the center of the earth. The length of a solar day on the average is slightly less than 24 hr, owing to the forward motion of the earth in its solar orbit. Any given day will also differ from the average day owing to orbital eccentricity, axis precession, and other secondary effects embodied in the equation of time described below.

Declination and Hour Angle

The earth's orbit about the sun is elliptical with eccentricity of 0.0167. This results in variation of solar flux on the outer atmosphere of about 7% over the course of a year. Of more importance is the variation of solar intensity caused by the inclination of the earth's axis relative to the ecliptic plane of the earth's orbit. The angle between the ecliptic plane and the earth's equatorial plane is 23.45°. Figure 62.1 shows this inclination schematically.

The earth's motion is quantified by two angles varying with season and time of day. The angle varying on a seasonal basis that is used to characterize the earth's location in its orbit is called the solar "declination." It is the angle between the earth–sun line and the equatorial plane as shown in Fig. 62.2. The declination δ_s is taken to be positive when the earth–sun line is north of the equator and negative otherwise. The declination varies between $+23.45°$ on the summer solstice (June 21 or 22) and $-23.45°$ on the winter solstice (December 21 or 22). The declination is given by

$$\sin \delta_s = 0.398 \cos [0.986(N - 173)] \tag{62.1}$$

in which N is the day number.

The second angle used to locate the sun is the solar-hour angle. Its value is based on the nominal 360° rotation of the earth occurring in 24 hr. Therefore, 1 hr is equivalent to an angle of 15°. The hour angle is measured from zero at solar noon. It is denoted by h_s and is positive before solar noon and negative after noon in accordance with the right-hand rule. For example 2:00 PM corresponds to $h_s = -30°$ and 7:00 AM corresponds to $h_s = +75°$.

Solar time, as determined by the position of the sun, and clock time differ for two reasons. First, the length of a day varies because of the ellipticity of the earth's orbit; and second, standard time is determined by the standard meridian passing through the approximate center of each time zone. Any position away from the standard meridian has a difference between solar and clock time given by [(local longitude − standard meridian longitude)/15] in units of hours. Therefore, solar time and local standard time (LST) are related by

$$\text{solar time} = \text{LST} - \text{EoT} - (\text{local longitude} - \text{standard meridian longitude})/15 \tag{62.2}$$

in units of hours. EoT is the equation of time which accounts for difference in day length through a year and is given by

$$\text{EoT} = 12 + 0.1236 \sin x - 0.0043 \cos x + 0.1538 \sin 2x + 0.0608 \cos 2x \tag{62.3}$$

in units of hours. The parameter x is

$$x = \frac{360(N - 1)}{365.24} \tag{62.4}$$

where N is the day number counted from January 1 as $N = 1$.

Solar Position

The sun is imagined to move on the celestial sphere, an imaginary surface centered at the earth's center and having a large but unspecified radius. Of course, it is the earth that moves, not the sun, but the analysis is simplified if one uses this Ptolemaic approach. No error is introduced by the moving

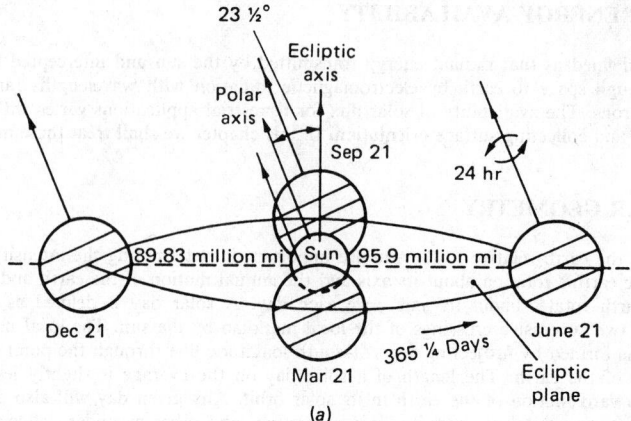

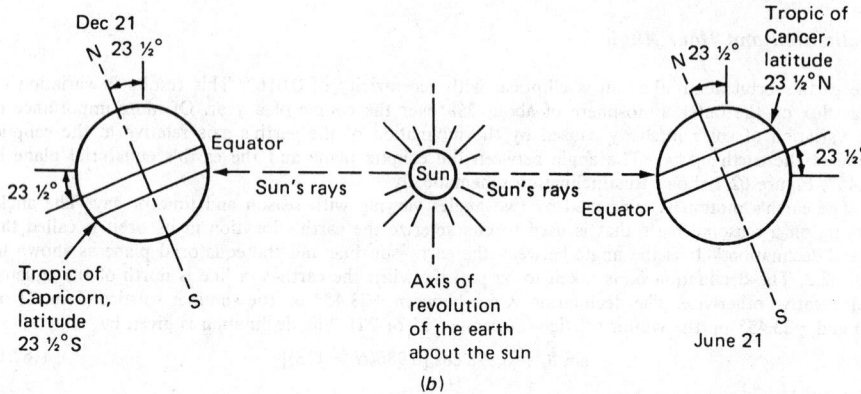

Fig. 62.1 (*a*) Motion of the earth about the sun. (*b*) Location of tropics. Note that the sun is so far from the earth that all the rays of the sun may be considered as parallel to one another when they reach the earth.

sun assumption, since the relative motion is the only motion of interest. Since the sun moves on a spherical surface, two angles are sufficient to locate the sun at any instant. The two most commonly used angles are the solar-altitude and azimuth angles (see Fig. 62.3) denoted by α and a_s, respectively. Occasionally, the solar-zenith angle, defined as the complement of the altitude angle, is used instead of the altitude angle.

The solar-altitude angle is related to the previously defined declination and hour angles by

$$\sin \alpha = \cos L \cos \delta_s \cos h_s + \sin L \sin \delta_s \qquad (62.5)$$

in which L is the latitude, taken positive for sites north of the equator and negative for sites south of the equator. The altitude angle is found by taking the inverse sine function of Eq. (62.5).

The solar-azimuth angle is given by[1]

$$\sin a_s = \frac{\cos \delta_s \sin h_s}{\cos \alpha} \qquad (62.6)$$

To find the value of a_s, the location of the sun relative to the east–west line through the site must be known. This is accounted for by the following two expressions for the azimuth angle:

$$a_s = \sin^{-1}\left(\frac{\cos \delta_s \sin h_s}{\cos \alpha}\right), \qquad \cos h_s > \frac{\tan \delta_s}{\tan L} \qquad (62.7)$$

$$a_s = 180° - \sin^{-1}\left(\frac{\cos \delta_s \sin h_s}{\cos \alpha}\right), \qquad \cos h_s < \frac{\tan \delta_s}{\tan L} \qquad (62.8)$$

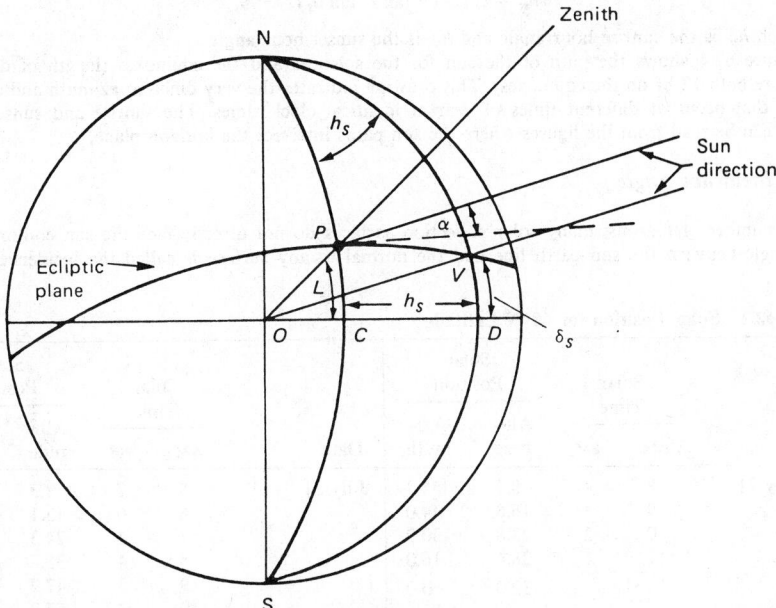

Fig. 62.2 Definition of solar-hour angle h_s (*CND*), solar declination δ_s (*VOD*), and latitude L (*POC*); P, site of interest. Modified from J. F. Kreider and F. Kreith, *Solar Heating and Cooling*, revised 1st ed., Hemisphere, Washington, DC, 1977.

Table 62.1 lists typical values of altitude and azimuth angles for latitude $L = 40°$. Complete tables are contained in Refs. 1 and 2.

62.1.2 Sunrise and Sunset

Sunrise and sunset occur when the altitude angle $\alpha = 0$. As indicated Fig. 62.4, this occurs when the center of the sun intersects the horizon plane. The hour angle for sunrise and sunset can be found from Eq. (62.5) by equating α to zero. If this is done, the hour angles for sunrise and sunset are found to be

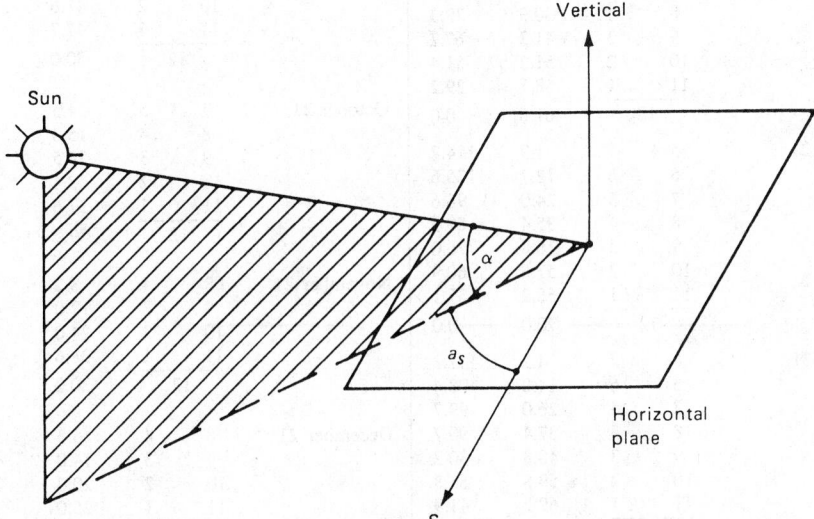

Fig. 62.3 Diagram showing solar-altitude angle α and solar-azimuth angle a_s.

$$h_{sr} = \cos^{-1}(-\tan L \tan \delta_s) = -h_{ss} \qquad (62.9)$$

in which h_{sr} is the sunrise hour angle and h_{ss} is the sunset hour angle.

Figure 62.4 shows the path of the sun for the solstices and the equinoxes (length of day and night are both 12 hr on the equinoxes). This drawing indicates the very different azimuth and altitude angles that occur at different times of year at identical clock times. The sunrise and sunset hour angles can be read from the figures where the sun paths intersect the horizon plane.

Solar Incidence Angle

For a number of reasons, many solar collection surfaces do not directly face the sun continuously. The angle between the sun–earth line and the normal to any surface is called the incidence angle.

Table 62.1   Solar Position for 40°N Latitude

Date	Solar Time AM	Solar Time PM	Solar Position Altitude	Solar Position Azimuth	Date	Solar Time AM	Solar Time PM	Solar Position Altitude	Solar Position Azimuth
January 21	8	4	8.1	55.3	July 21	5	7	2.3	115.2
	9	3	16.8	44.0		6	6	13.1	106.1
	10	2	23.8	30.9		7	5	24.3	97.2
	11	1	28.4	16.0		8	4	35.8	87.8
	12		30.0	0.0		9	3	47.2	76.7
						10	2	57.9	61.7
February 21	7	5	4.8	72.7		11	1	66.7	37.9
	8	4	15.4	62.2		12		70.6	0.0
	9	3	25.0	50.2					
	10	2	32.8	35.9					
	11	1	38.1	18.9	August 21	6	6	7.9	99.5
	12		40.0	0.0		7	5	19.3	90.9
						8	4	30.7	79.9
March 21	7	5	11.4	80.2		9	3	41.8	67.9
	8	4	22.5	69.6		10	2	51.7	52.1
	9	3	32.8	57.3		11	1	59.3	29.7
	10	2	41.6	41.9		12		62.3	0.0
	11	1	47.7	22.6					
	12		50.0	0.0	September 21	7	5	11.4	80.2
						8	4	22.5	69.6
April 21	6	6	7.4	98.9		9	3	32.8	57.3
	7	5	18.9	89.5		10	2	41.6	41.9
	8	4	30.3	79.3		11	1	47.7	22.6
	9	3	41.3	67.2		12		50.0	0.0
	10	2	51.2	51.4					
	11	1	58.7	29.2	October 21	7	5	4.5	72.3
	12		61.6	0.0		8	4	15.0	61.9
May 21	5	7	1.9	114.7		9	3	24.5	49.8
	6	6	12.7	105.6		10	2	32.4	35.6
	7	5	24.0	96.6		11	1	37.6	18.7
	8	4	35.4	87.2		12		39.5	0.0
	9	3	46.8	76.0					
	10	2	57.5	60.9	November 21	8	4	8.2	55.4
	11	1	66.2	37.1		9	3	17.0	44.1
	12		70.0	0.0		10	2	24.0	31.0
June 21	5	7	4.2	117.3		11	1	28.6	16.1
	6	6	14.8	108.4		12		30.2	0.0
	7	5	26.0	99.7					
	8	4	37.4	90.7	December 21	8	4	5.5	53.0
	9	3	48.8	80.2		9	3	14.0	41.9
	10	2	59.8	65.8		10	2	20.0	29.4
	11	1	69.2	41.9		11	1	25.0	15.2
	12		73.5	0.0		12		26.6	0.0

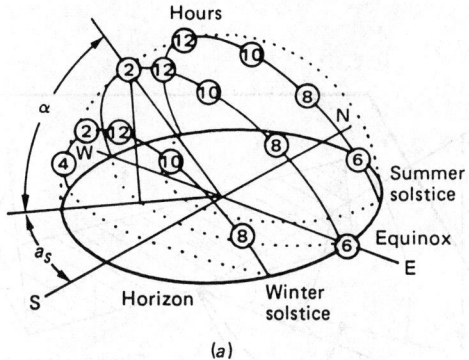

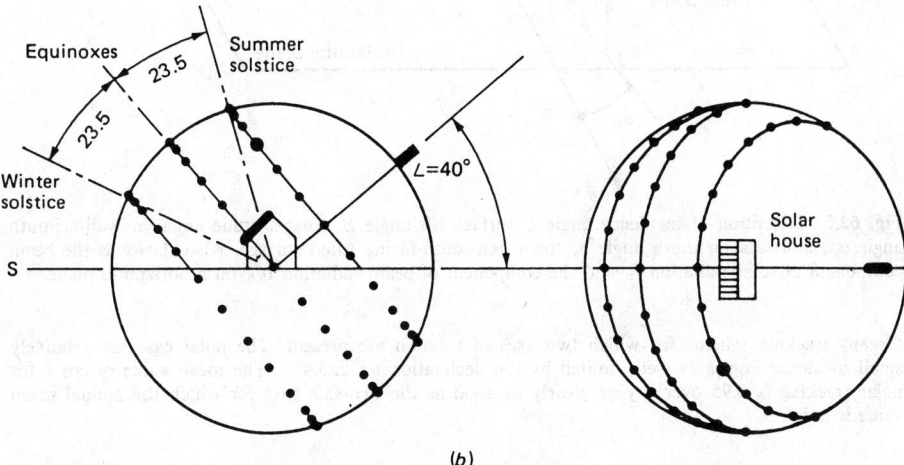

Fig. 62.4 Sun paths for the summer solstice (6/21), the equinoxes (3/21 and 9/21), and the winter solstice (12/21) for a site at 40°N: (*a*) isometric view; (*b*) elevation and plan views.

The intensity of off-normal solar radiation is proportional to the cosine of the incidence angle. For example, Fig. 62.5 shows a fixed planar surface with solar radiation intersecting the plane at the incidence angle i measured relative to the surface normal. The intensity of flux at the surface is $I_b \times \cos i$, where I_b is the beam radiation along the sun–earth line; I_b is called the direct, normal radiation. For a fixed surface such as that in Fig. 62.5 facing the equator, the incidence angle is given by

$$\cos i = \sin \delta_s (\sin L \cos \beta - \cos L \sin \beta \cos a_w)$$
$$+ \cos \delta_s \cos h_s (\cos L \cos \beta + \sin L \sin \beta \cos a_w) \quad (62.10)$$
$$+ \cos \delta_s \sin \beta \sin a_w \sin h_s$$

in which a_w is the "wall" azimuth angle and β is the surface tilt angle relative to the horizontal plane, both as shown in Fig. 62.5.

For fixed surfaces that face due south, the incidence angle expression simplifies to

$$\cos i = \sin(L - \beta)\sin \delta_s + \cos(L - \beta)\cos \delta_s \cos h_s \quad (62.11)$$

A large class of solar collectors move in some fashion to track the sun's diurnal motion, thereby improving the capture of solar energy. This is accomplished by reduced incidence angles for properly tracking surfaces vis-à-vis a fixed surface for which large incidence angles occur in the early morning and late afternoon (for generally equator-facing surfaces). Table 62.2 lists incidence angle expressions for nine different types of tracking surfaces. The term "polar axis" in this table refers to an axis of rotation directed at the north or south pole. This axis of rotation is tilted up from the horizontal at an angle equal to the local latitude. It is seen that normal incidence can be achieved (i.e., $\cos i = 1$)

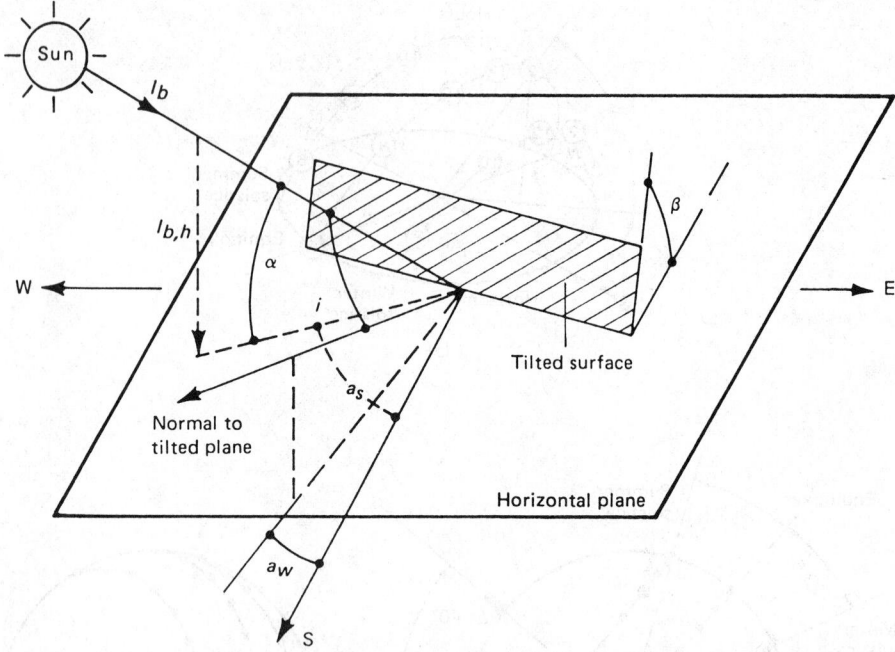

Fig. 62.5 Definition of incidence angle i, surface tilt angle β, solar-altitude angle α, wall-azimuth angle a_w, and solar-azimuth angle a_s for a non-south-facing tilted surface. Also shown is the beam component of solar radiation I_b and the component of beam radiation $I_{b,h}$ on a horizontal plane.

for any tracking scheme for which two axes of rotation are present. The polar case has relatively small incidence angles as well, limited by the declination to $\pm23.45°$. The mean value of $\cos i$ for polar tracking is 0.95 over a year, nearly as good as the two-axis case for which the annual mean value is unity.

62.1.3 Quantitative Solar Flux Availability

The previous section has indicated how variations in solar flux produced by seasonal and diurnal effects can be quantified. However, the effect of weather on solar energy availability cannot be analyzed theoretically; it is necessary to rely on historical weather reports and empirical correlations for calculations of actual solar flux. In this section this subject is described along with the availability of solar energy at the edge of the atmosphere—a useful correlating parameter, as seen shortly.

Extraterrestrial Solar Flux

The flux intensity at the edge of the atmosphere can be calculated strictly from geometric considerations if the direct-normal intensity is known. Solar flux incident on a terrestrial surface, which has traveled from sun to earth with negligible change in direction, is called beam radiation and is denoted by I_b. The extraterrestrial value of I_b averaged over a year is called the solar constant, denoted by I_{sc}. Its value is 429 Btu/hr · ft² or 1353 W/m². Owing to the eccentricity of the earth's orbit, however, the extraterrestrial beam radiation intensity varies from this mean solar constant value. The variation of I_b over the year is given by

$$I_{b,0}(N) = \left[1 + 0.034 \cos\left(\frac{360\,N}{365}\right)\right] \times I_{sc} \qquad (62.12)$$

in which N is the day number as before.

In subsequent sections the total daily, extraterrestrial flux will be particularly useful as a nondimensionalizing parameter for terrestrial solar flux data. The instantaneous solar flux on a horizontal, extraterrestrial surface is given by

$$I_{b,h0} = I_{b,0}(N) \sin \alpha \qquad (62.13)$$

Table 62.2 Solar Incidence Angle Equations for Tracking Collectors

Description	Axis (Axes)	Cosine of Incidence Angle (cos i)
Movements in altitude and azimuth	Horizontal axis and vertical axis	1
Rotation about a polar axis and adjustment in declination	Polar axis and declination axis	1
Uniform rotation about a polar axis	Polar axis	$\cos \delta_s$
East–west horizontal	Horizontal, east–west axis	$\sqrt{1 - \cos^2 \alpha \sin^2 a_s}$
North–south horizontal	Horizontal, north–south axis	$\sqrt{1 - \cos^2 \alpha \cos^2 a_s}$
Rotation about a vertical axis of a surface tilted upward L (latitude) degrees	Vertical axis	$\sin (\alpha + L)$
Rotation of a horizontal collector about a vertical axis	Vertical axis	$\sin \alpha$
Rotation of a vertical surface about a vertical axis	Vertical axis	$\cos \alpha$
Fixed "tubular" collector	North–south tiled up at angle β	$\sqrt{1 - [\sin (\beta - L) \cos \delta_s \cos h_s + \cos (\beta - L) \sin \delta_s]^2}$

as shown in Fig. 62.5. The daily total, horizontal radiation is denoted by I_0 and is given by

$$I_0(N) = \int_{t_{sr}}^{t_{ss}} I_{b,0}(N) \sin \alpha \, dt \tag{62.14}$$

$$I_0(N) = \frac{24}{\pi} I_{sc} \left[1 + 0.034 \cos \left(\frac{360\,N}{365} \right) \right] \times (\cos L \cos \delta_s \sin h_{sr} + h_{sr} \sin L \sin \delta_s) \tag{62.15}$$

in which I_{sc} is the solar constant. The extraterrestrial flux varies with time of year via the variations of δ_s and h_{sr} with time of year. Table 62.3 lists the values of extraterrestrial, horizontal flux for various latitudes averaged over each month. The monthly averaged, horizontal, extraterrestrial solar flux is denoted by $\bar{H}_0$.

Terrestrial Solar Flux

Values of instantaneous or average terrestrial solar flux cannot be predicted accurately owing to the complexity of atmospheric processes that alter solar flux magnitudes and directions relative to their extraterrestrial values. Air pollution, clouds of many types, precipitation, and humidity all affect the values of solar flux incident on earth. Rather than attempting to predict solar availability accounting for these complex effects, one uses long-term historical records of terrestrial solar flux for design purposes.

The U.S. National Weather Service (NWS) records solar flux data at a network of stations in the United States. The pyranometer instrument, as shown in Fig. 62.6, is used to measure the intensity of horizontal flux. Various data sets are available from the National Climatic Center (NCC) of the NWS. Prior to 1975, the solar network was not well maintained; therefore, the pre-1975 data were rehabilitated in the late 1970s and are now available from the NCC on magnetic media. Also, for the period 1950–1975, synthetic solar data have been generated for approximately 250 U.S. sites where solar flux data were not recorded. The predictive scheme used is based on other widely available meteorological data. Finally, since 1977 the NWS has recorded hourly solar flux data at a 38-station network with improved instrument maintenance. In addition to horizontal flux, direct-normal data are recorded and archived at the NCC. Figure 62.7 is a contour map of annual, horizontal flux for the United States based on recent data. The appendix to this chapter contains tabulations of average, monthly solar flux data for approximately 250 U.S. sites.

The principal difficulty with using NWS solar data is that they are available for horizontal surfaces only. Solar-collecting surfaces normally face the general direction of the sun and are, therefore, rarely horizontal. It is necessary to convert measured horizontal radiation to radiation on arbitrarily oriented collection surfaces. This is done using empirical approaches to be described.

Hourly Solar Flux Conversions

Measured, horizontal solar flux consists of both beam and diffuse radiation components. Diffuse radiation is that scattered by atmospheric processes; it intersects surfaces from the entire sky dome, not just from the direction of the sun. Separating the beam and diffuse components of measured, horizontal radiation is the key difficulty in using NWS measurements.

The recommended method for finding the beam component of total (i.e., beam plus diffuse) radiation is described in Ref. 1. It makes use of the parameter k_T called the clearness index and defined as the ratio of terrestrial to extraterrestrial hourly flux on a horizontal surface. In equation form k_T is

$$k_T \equiv \frac{I_h}{I_{b,h\,0}} = \frac{I_h}{I_{b,0}(N) \sin \alpha} \tag{62.16}$$

in which I_h is the measured, total horizontal flux. The beam component of the terrestrial flux is then given by the empirical equation

$$I_b = (ak_T + b)I_{b,0}(N) \tag{62.17}$$

in which the empirical constants a and b are given in Table 62.4. Having found the beam radiation, the horizontal diffuse component $I_{d,h}$ is found by the simple difference

$$I_{d,h} = I_h - I_b \sin \alpha \tag{62.18}$$

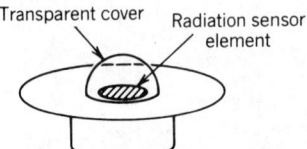

Transparent cover   Radiation sensor
                      element

Fig. 62.6 Schematic drawing of a pyranometer used for measuring the intensity of total (direct plus diffuse) solar radiation.

Table 62.3 Average Extraterrestrial Radiation on a Horizontal Surface $\overline{H}_0$ in SI Units and in English Units Based on a Solar Constant of 429 Btu/hr · ft² or 1.353 kW/m²

Latitude, Degrees	January	February	March	April	May	June	July	August	September	October	November	December
SI Units, W · hr/m² · Day												
20	7415	8397	9552	10,422	10,801	10,868	10,794	10,499	9791	8686	7598	7076
25	6656	7769	9153	10,312	10,936	11,119	10,988	10,484	9494	8129	6871	6284
30	5861	7087	8686	10,127	11,001	11,303	11,114	10,395	9125	7513	6103	5463
35	5039	6359	8153	9869	10,995	11,422	11,172	10,233	8687	6845	5304	4621
40	4200	5591	7559	9540	10,922	11,478	11,165	10,002	8184	6129	4483	3771
45	3355	4791	6909	9145	10,786	11,477	11,099	9705	7620	5373	3648	2925
50	2519	3967	6207	8686	10,594	11,430	10,981	9347	6998	4583	2815	2100
55	1711	3132	5460	8171	10,358	11,352	10,825	8935	6325	3770	1999	1320
60	963	2299	4673	7608	10,097	11,276	10,657	8480	5605	2942	1227	623
65	334	1491	3855	7008	9852	11,279	10,531	8001	4846	2116	544	97
English Units, Btu/ft² · Day												
20	2346	2656	3021	3297	3417	3438	3414	3321	3097	2748	2404	2238
25	2105	2458	2896	3262	3460	3517	3476	3316	3003	2571	2173	1988
30	1854	2242	2748	3204	3480	3576	3516	3288	2887	2377	1931	1728
35	1594	2012	2579	3122	3478	3613	3534	3237	2748	2165	1678	1462
40	1329	1769	2391	3018	3455	3631	3532	3164	2589	1939	1418	1193
45	1061	1515	2185	2893	3412	3631	3511	3070	2410	1700	1154	925
50	797	1255	1963	2748	3351	3616	3474	2957	2214	1450	890	664
55	541	991	1727	2585	3277	3591	3424	2826	2001	1192	632	417
60	305	727	1478	2407	3194	3567	3371	2683	1773	931	388	197
65	106	472	1219	2217	3116	3568	3331	2531	1533	670	172	31

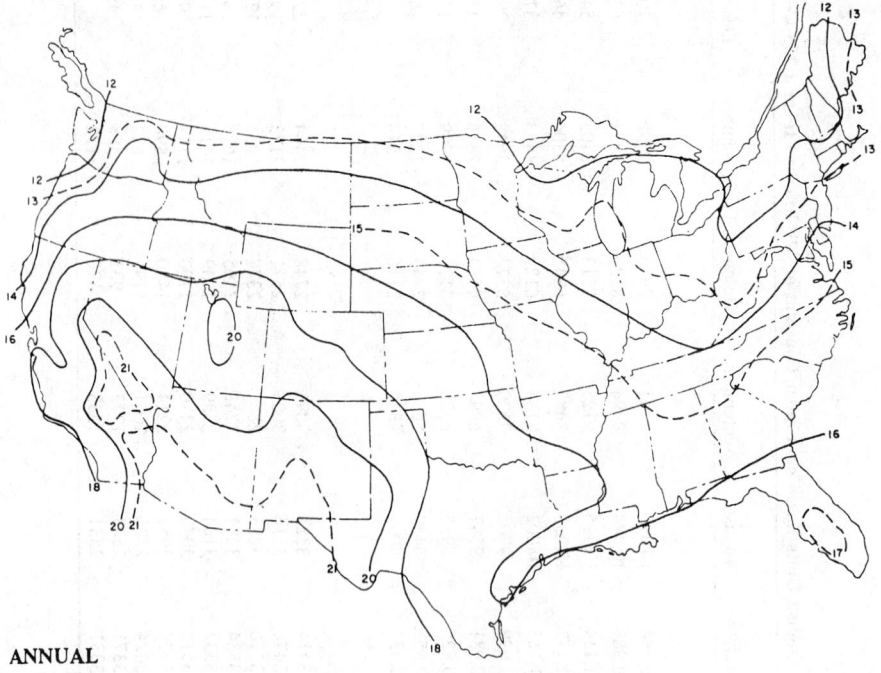

ANNUAL

*1 mJ/m² = 88.1 Btu/ft².

Fig. 62.7 Mean daily solar radiation on a horizontal surface in megajoules per square meter for the continental United States.

The separate values of horizontal beam and diffuse radiation can be used to find radiation on any surface by applying appropriate geometric "tilt factors" to each component and forming the sum accounting for any radiation reflected from the foreground. The beam radiation incident on any surface is simply $I_b \cos i$. If one assumes that the diffuse component is isotropically distributed over the sky dome, the amount intercepted by any surface tilted at an angle β is $I_{d,h} \cos^2(\beta/2)$. The total beam and diffuse radiation intercepted by a surface I_c is then

$$I_c = I_b \cos i + I_{d,h} \cos^2(\beta/2) + \rho I_h \sin^2(\beta/2) \qquad (62.19)$$

The third term in this expression accounts for flux reflected from the foreground with reflectance ρ.[1]

Monthly Averaged, Daily Solar Flux Conversions

Most performance prediction methods make use of monthly averaged solar flux values. Horizontal flux data are readily available (see the appendix), but monthly values on arbitrarily positioned surfaces

Table 62.4 Empirical Coefficients for Eq. (62.17)

Interval for k_T	a	b
0.00, 0.05	0.04	0.00
0.05, 0.15	0.01	0.002
0.15, 0.25	0.06	−0.006
0.25, 0.35	0.32	−0.071
0.35, 0.45	0.82	−0.246
0.45, 0.55	1.56	−0.579
0.55, 0.65	1.69	−0.651
0.65, 0.75	1.49	−0.521
0.75, 0.85	0.27	0.395

must be calculated using a method similar to that previously described for hourly tilted surface calculations. The monthly averaged flux on a tilted surface $\bar{I}_c$ is given by

$$\bar{I}_c = \bar{R}\,\bar{H}_h \tag{62.20}$$

in which $\bar{H}_h$ is the monthly averaged, daily total of horizontal solar flux and $\bar{R}$ is the overall tilt factor given by Eq. (62.21) for a fixed, equator-facing surface:

$$\bar{R} = \left(1 - \frac{\bar{D}_h}{\bar{H}_h}\right)\bar{R}_b + \frac{\bar{D}_h}{\bar{H}_h}\cos^2\frac{\beta}{2} + \rho\sin^2\frac{\beta}{2} \tag{62.21}$$

The ratio of monthly averaged diffuse to total flux, $\bar{D}_h/\bar{H}_h$ is given by

$$\frac{\bar{D}_h}{\bar{H}_h} = 0.775 + 0.347\left[h_{sr} - \frac{\pi}{2}\right] - \left[0.505 + 0.261\left(h_{sr} - \frac{\pi}{2}\right)\right]\cos\left[(\bar{K}_T - 0.9)\frac{360}{\pi}\right] \tag{62.22}$$

in which $\bar{K}_T$ is the monthly averaged clearness index analogous to the hourly clearness index. $\bar{K}_T$ is given by

$$\bar{K}_T \equiv \bar{H}_h/\bar{H}_0$$

where H_0 is the monthly averaged, extraterrestrial radiation on a horizontal surface at the same latitude at which the terrestrial radiation $\bar{H}_h$ was recorded. The monthly averaged beam radiation tilt factor $\bar{R}_b$ is

$$\bar{R}_b = \frac{\cos(L - \beta)\cos\delta_s\sin h_{sr}' + h_{sr}'\sin(L - \beta)\sin\delta_s}{\cos L \cos\delta_s \sin h_{sr} + h_{sr}\sin L \sin\delta_s} \tag{62.23}$$

The sunrise hour angle is found from Eq. (62.9) and the value of h_{sr}' is the smaller of (1) the sunrise hour angle h_{sr} and (2) the collection surface sunrise hour angle found by setting $i = 90°$ in Eq. (62.11). That is, h_{sr}' is given by

$$h_{sr}' = \min\{\cos^{-1}[-\tan L \tan\delta_s], \cos^{-1}[-\tan(L - \beta)\tan\delta_s]\} \tag{62.24}$$

Expressions for solar flux on a tracking surface on a monthly averaged basis are of the form

$$\bar{I}_c = \left[r_T - r_d\left(\frac{\bar{D}_h}{\bar{H}_h}\right)\right]\bar{H}_h \tag{62.25}$$

in which the tilt factors r_T and r_d are given in Table 62.5. Equation (62.22) is to be used for the diffuse to total flux ratio $\bar{D}_h/\bar{H}_h$.

62.2 SOLAR THERMAL COLLECTORS

The principal use of solar energy is in the production of heat at a wide range of temperatures matched to a specific task to be performed. The temperature at which heat can be produced from solar radiation is limited to about 6000°F by thermodynamic, optical, and manufacturing constraints. Between temperatures near ambient and this upper limit very many thermal collector designs are employed to produce heat at a specified temperature. This section describes the common thermal collectors.

62.2.1 Flat-Plate Collectors

From a production volume standpoint, the majority of installed solar collectors are of the flate-plate design; these collectors are capable of producing heat at temperatures up to 100°C. Flat-plate collectors are so named since all components are planar. Figure 62.8a is a partial isometric sketch of a liquid-cooled flat-plate collector. From the top down it contains a glazing system—normally one pane of glass, a dark colored metal absorbing plate, insulation to the rear of the absorber, and, finally, a metal or plastic weatherproof housing. The glazing system is sealed to the housing to prohibit the ingress of water, moisture, and dust. The piping shown is thermally bonded to the absorber plate and contains the working fluid by which the heat produced is transferred to its end use. The pipes shown are manifolded together so that one inlet and one outlet connection, only, are present. Figure 62.8b shows a number of other collector designs in common use.

The energy produced by flat-plate collectors is the difference between the solar flux absorbed by the absorber plate and that lost from it by convection and radiation from the upper (or "front") surface and that lost by conduction from the lower (or "back") surface. The solar flux absorbed is the incident flux I_c multiplied by the glazing system transmittance τ and by the absorber plate absorptance α. The heat lost from the absorber in steady state is given by an overall thermal conductance U_c multiplied by the difference in temperature between the collector absorber temperature T_c and the surrounding, ambient temperature T_a. In equation form the net heat produced q_u is then

$$q_u = (\tau\alpha)I_c - U_c(T_c - T_a) \tag{62.26}$$

Table 62.5 Concentrator Tilt Factors

Collector Type	r_T [a,b,c,d]	r_d [e]
Fixed aperture concentrators that do not view the foreground	$[\cos(L - \beta)/(d \cos L)][-ah_{coll} \cos h_{sr} (i = 90°)]$ $+ [a - b \cos h_{sr} (i = 90°)] \sin h_{coll}$ $+ (b/2)(\sin h_{coll} \cos h_{coll} + h_{coll})\}$	$(\sin h_{coll}/d)[\cos(L + \beta)/\cos L] - [1/(CR)]\}$ $+ (h_{coll}/d)[\cos h_{sr}/(CR)]$ $- [\cos(L - \beta)/\cos L] \cos h_{sr} (i = 90°)\}$
East–west axis tracking [f]	$(1/d)\int_0^{h_{coll}} \{[(a + b \cos x)/\cos L]$ $\times \sqrt{\cos^2 x + \tan^2 \delta_s}\, \} \, dx$	$(1/d)\int_0^{h_{coll}} \{(1/\cos L)\sqrt{\cos^2 x + \tan^2 \delta_s}$ $- [1/(CR)][\cos x - \cos h_{sr}]\} \, dx$
Polar tracking	$(ah_{coll} + b \sin h_{coll})/(d \cos L)$	$(h_{coll}/d)[(1/\cos L) + [\cos h_{sr}/(CR)]$ $- \sin h_{coll}/[d(CR)]$
Two-axis tracking	$(ah_{coll} + b \sin h_{coll})/(d \cos \delta_s \cos L)$	$(h_{coll}/d)[1/\cos \delta_s \cos L) + [\cos h_{sr}/(CR)]$ $- \sin h_{coll}/[d(CR)]$

[a] The collection hour angle value h_{coll} not used as the argument of trigonometric functions is expressed in radians; note that the total collection interval, $2h_{coll}$, is assumed to be centered about solar noon.

[b] $a = 0.409 + 0.5016 \sin(h_{sr} - 60°)$.

[b] $c = 0.6609 - 0.4767 \sin(h_{sr} - 60°)$.

[d] $d = \sin h_{sr} - h_{sr} \cos h_{sr}$; $\cos h_{sr}$ $(i = 90°) = -\tan \delta_s \tan(L - \beta)$.

[e] CR is the collector concentration ratio.

[f] Use elliptic integral tables to evaluate terms of the form of $\int_0^h \sqrt{\cos^2 x + \tan^2 \delta_s}\, dx$ contained in r_T and r_d.

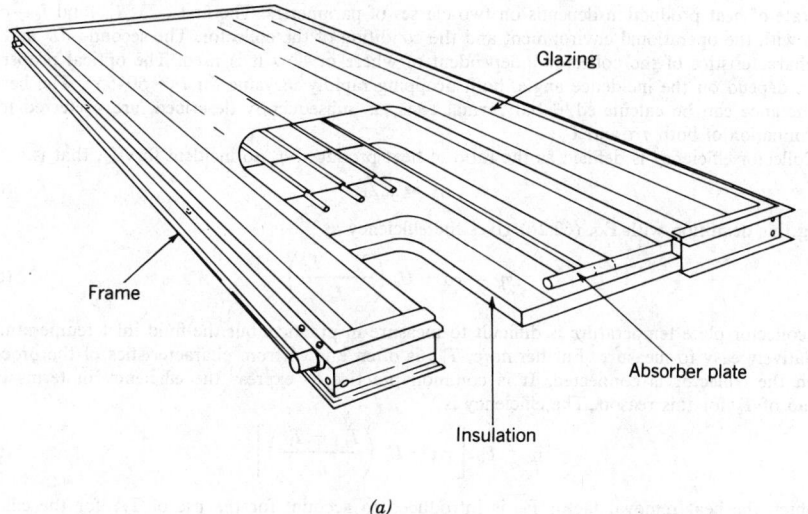

Glazing

Frame

Absorber plate

Insulation

(a)

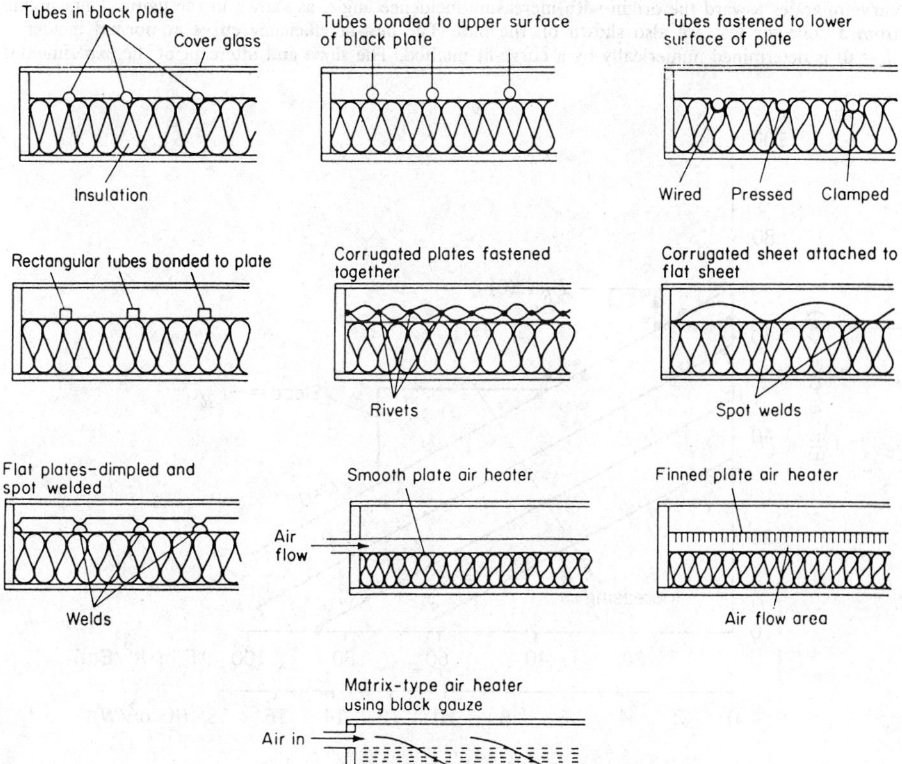

Tubes in black plate

Cover glass

Insulation

Tubes bonded to upper surface of black plate

Tubes fastened to lower surface of plate

Wired   Pressed   Clamped

Rectangular tubes bonded to plate

Corrugated plates fastened together

Rivets

Corrugated sheet attached to flat sheet

Spot welds

Flat plates–dimpled and spot welded

Welds

Smooth plate air heater

Air flow

Finned plate air heater

Air flow area

Matrix-type air heater using black gauze

Air in

Side view

Fig. 62.8 (*a*) Schematic diagram of solar collector with one cover. (*b*) Cross sections of various liquid- and air-based flat-plate collectors in common use.

The rate of heat production depends on two classes of parameters. The first—T_c, T_a, and I_c—having to do with the operational environment and the condition of the collector. The second—U_c and $\tau\alpha$—are characteristics of the collector independent of where or how it is used. The optical properties τ and α depend on the incidence angle, both dropping rapidly in value for $i > 50$–$55°$. The heat loss conductance can be calculated,[1,2] but formal tests, as subsequently described, are preferred for the determination of both $\tau\alpha$ and U_c.

Collector efficiency is defined as the ratio of heat produced q_u to incident flux I_c, that is,

$$\eta_c \equiv q_u/I_c \qquad (62.27)$$

Using this definition with Eq. (62.26) gives the efficiency as

$$\eta_c = \tau\alpha - U_c \left(\frac{T_c - T_a}{I_c} \right) \qquad (62.28)$$

The collector plate temperature is difficult to measure in practice, but the fluid inlet temperature $T_{f,i}$ is relatively easy to measure. Furthermore, $T_{f,i}$ is often known from characteristics of the process to which the collector is connected. It is common practice to express the efficiency in terms of $T_{f,i}$ instead of T_c for this reason. The efficiency is

$$\eta_c = F_R \left[\tau\alpha - U_c \left(\frac{T_{f,i} - T_a}{I_c} \right) \right] \qquad (62.29)$$

in which the heat removal factor F_R is introduced to account for the use of $T_{f,i}$ for the efficiency basis. F_R depends on the absorber plate thermal characteristics and heat loss conductance.[2]

Equation (62.29) can be plotted with the group of operational characteristics $(T_{f,i} - T_a)/I_c$ as the independent variable as shown in Fig. 62.9. The efficiency decreases linearly with the abscissa value. The intercept of the efficiency curve is the optical efficiency $\tau\alpha$ and the slope is $-F_R U_c$. Since the glazing transmittance and absorber absorptance decrease with solar incidence angle, the efficiency curve migrates toward the origin with increasing incidence angle, as shown in the figure. Data points from a collector test are also shown on the plot. The best-fit efficiency curve at normal incidence ($i = 0$) is determined numerically by a curve-fit method. The slope and intercept of the experimental

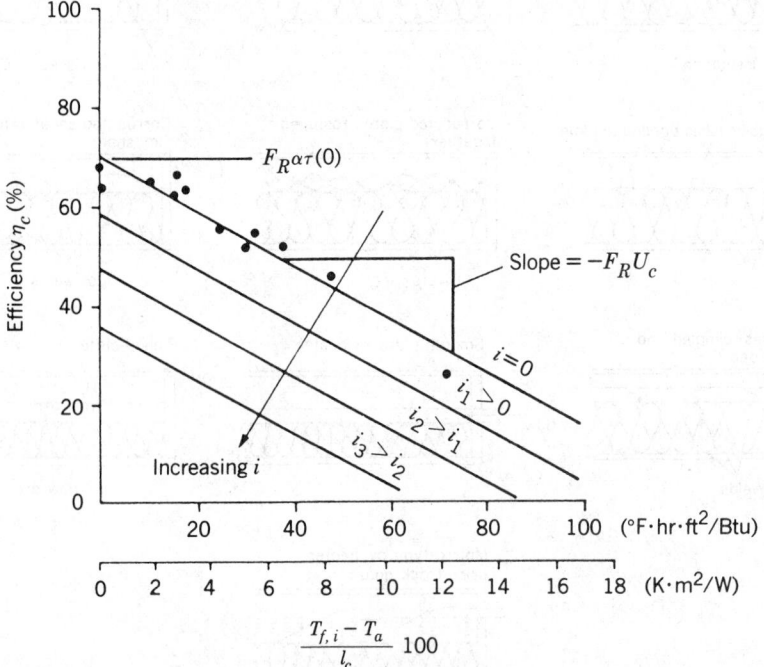

Fig. 62.9 Typical collector performance with 0° incident beam flux angle. Also shown qualitatively is the effect of incidence angle i, which may be quantified by $\overline{\tau\alpha}(i)/\overline{\tau\alpha}(0) = 1.0 + b_0(1/\cos i - 1.0)$, where b_0 is the incidence angle modifier determined experimentally (ASHRAE 93-77) or from the Stokes and Fresnel equations.

curve, so determined, are the preferred values of the collector parameters as opposed to those calculated theoretically.

Selective Surfaces

One method of improving efficiency is to reduce radiative heat loss from the absorber surface. This is commonly done by using a low emittance (in the infrared region) surface having high absorptance for solar flux. Such surfaces are called (wavelength) selective surface and are used on very many flat-plate collectors to improve efficiency at elevated temperature. Table 62.6 lists emittance and absorptance values for a number of common selective surfaces. Black chrome is very reliable and cost effective.

62.2.2 Concentrating Collectors

Another method of improvng the efficiency of solar collectors is to reduce the parasitic heat loss embodied in the second term of Eq. (62.29). This can be done by reducing the size of the absorber relative to the aperture area. Relatively speaking, the area from which heat is lost is smaller than the heat collection area and efficiency increases. Collectors that focus sunlight onto a relatively small absorber can achieve excellent efficiency at temperatures above which flat-plate collectors produce no net heat output. In this section a number of concentrators are described.

Trough Collectors

Figure 62.10 shows cross sections of five concentrators used for producing heat at temperatures up to 650°F at good efficiency. Figure 62.10a shows the parabolic "trough" collector representing the most common concentrator design available commercially. Sunlight is focused onto a circular pipe absorber located along the focal line. The trough rotates about the absorber centerline in order to maintain a sharp focus of incident beam radiation on the absorber. Selective surfaces and glass enclosures are used to minimize heat losses from the absorber tube.

Figures 62.10c and 62.10d show Fresnel-type concentrators in which the large reflector surface is subdivided into several smaller, more easily fabricated and shipped segments. The smaller reflector elements are easier to track and offer less wind resistance at windy sites; futhermore, the smaller reflectors are less costly. Figure 62.10e shows a Fresnel lens concentrator. No reflection is used with this approach; reflection is replaced by refraction to achieve the focusing effect. This device has the advantage that optical precision requirements can be relaxed somewhat relative to reflective methods.

Figure 62.10b shows schematically a concentrating method in which the mirror is fixed, thereby avoiding all problems associated with moving large mirrors to track the sun as in the case of concentrators described above. Only the absorber pipe is required to move to maintain a focus on the focal line.

The useful heat produced Q_u by any concentrator is given by

$$Q_u = A_a \eta_0 I_c - A_r U_c'(T_c - T_a) \tag{62.30}$$

in which the concentrator optical efficiency (analogous to $\tau\alpha$ for flat-plate collectors) is η_0, the aperture area is A_a, the receiver or absorber area is A_r, and the absorber heat loss conductance is U_c'. Collector efficiency can be found from Eq. (62.27) and is given by

Table 62.6 Selective Surface Properties

Material	Absorptance[a] α	Emittance ϵ	Comments
Black chrome	0.87–0.93	0.1	
Black zinc	0.9	0.1	
Copper oxide over aluminum	0.93	0.11	
Black copper over copper	0.85–0.90	0.08–0.12	Patinates with moisture
Black chrome over nickel	0.92–0.94	0.07–0.12	Stable at high temperatures
Black nickel over nickel	0.93	0.06	May be influenced by moisture
Black iron over steel	0.90	0.10	

[a] Dependent on thickness.

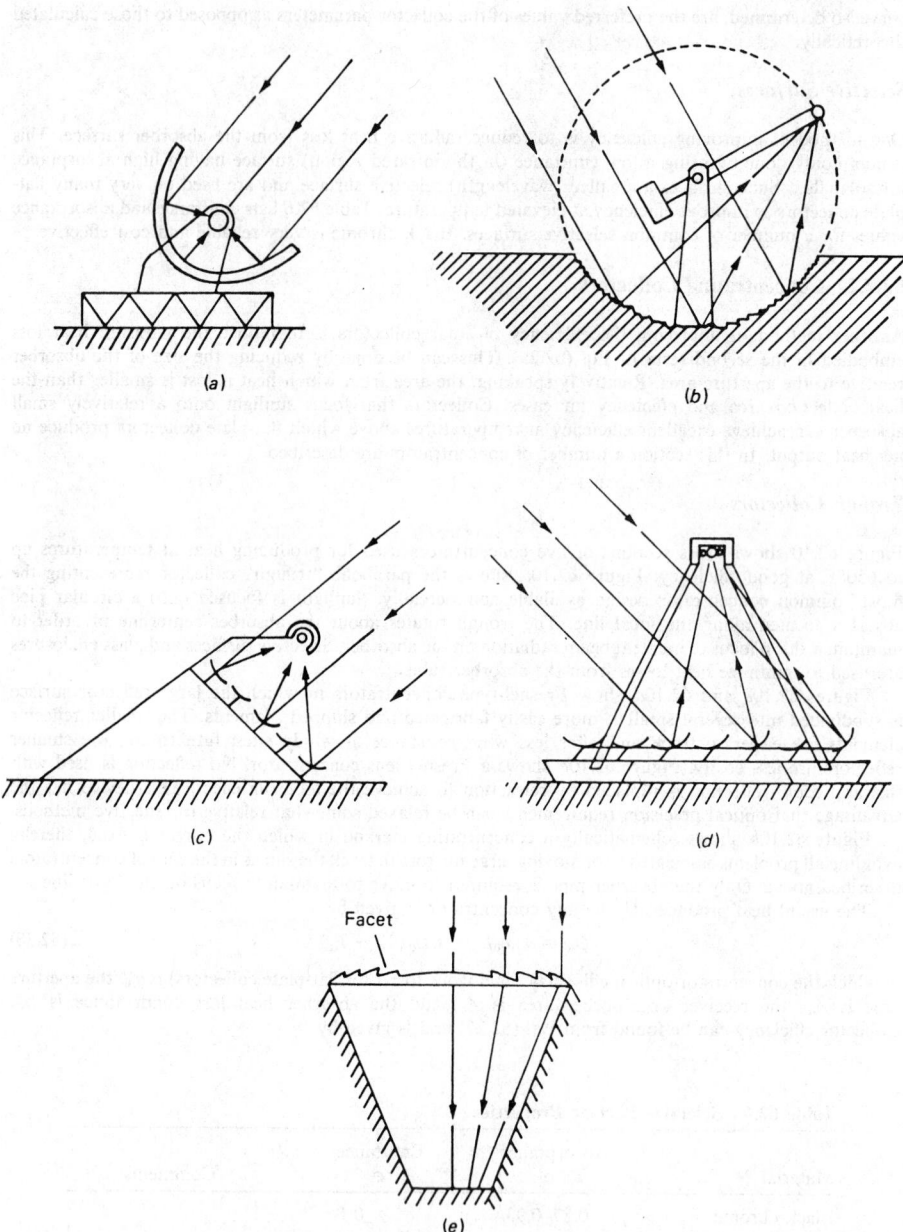

Fig. 62.10 Single-curvature solar concentrators: (a) parabolic trough; (b) fixed circular trough with tracking absorber; (c) and (d) Fresnel mirror designs; and (e) Fresnel lens.

$$\eta_c = \eta_0 - \frac{A_r}{A_a} U_c' \left(\frac{T_c - T_a}{I_c} \right) \tag{62.31a}$$

The aperture area–receiver area ratio $A_a/A_r > 1$ is called the geometric concentration ratio CR. It is the factor by which absorber heat losses are reduced relative to the aperture area:

$$\eta_c = \eta_0 - \frac{U_c'}{CR} \left(\frac{T_c - T_a}{I_c} \right) \tag{62.31b}$$

As with flat-plate collectors, efficiency is most often based on collector fluid inlet temperature $T_{f,i}$. On this basis, efficiency is expressed as

$$\eta_c = F_R \left[\eta_0 - U_c \left(\frac{T_{f,i} - T_a}{I_c} \right) \right] \tag{62.32}$$

in which the heat loss conductance U_c on an aperture area basis is used ($U_c = U_c'/CR$).

The optical efficiency of concentrators must account for a number of factors not present in flat-plate collectors including mirror reflectance, shading of aperture by receiver and its supports, spillage of flux beyond receiver tube ends at off-normal incidence conditions, and random surface, tracking, and construction errors that affect the precision of focus. In equation form the general optical efficiency is given by

$$\eta_0 = \rho_m \tau_c \alpha_r f_t \delta F(i) \tag{62.33}$$

where ρ_m is the mirror reflectance (0.8–0.9), τ_c is the receiver cover transmittance (0.85–0.92), α_r is the receiver surface absorptance (0.9–0.92), f_t is the fraction of aperture area not shaded by receiver and its supports (0.95–0.97), δ is the intercept factor accounting for mirror surface and tracking errors (0.90–0.95), and $F(i)$ is the fraction of reflected solar flux intercepted by the receiver for perfect optics and perfect tracking. Values for these parameters are given in Refs. 2 and 4.

Compound Curvature Concentrators

Further increases in concentration and concomitant reductions in heat loss are achievable if "dish-type" concentrators are used. This family of concentrators is exemplified by the paraboloidal dish concentrator, which focuses solar flux at a point instead of along a line as with trough collectors. As a result the achievable concentration ratios are approximately the square of what can be realized with single curvature, trough collectors. Figures 62.11 and 62.12 show a paraboloidal dish concentrator assembly. These devices are of most interest for power production and some elevated industrial process heat applications.

For very large aperture areas it is impractical to construct paraboloidal dishes consisting of a single reflector. Instead the mirror is segmented as shown in Fig. 62.13. This collector system called the central receiver has been used in several solar thermal power plants in the 1–15 MW range. This power production method is discussed in the next section.

The efficiency of compound curvature dish collectors is given by Eq. (62.32), where the parameters involved are defined in the context of compound curvature optics.[4] The heat loss term at high temperatures achieved by dish concentrators is dominated by radiation; therefore, the second term of the efficiency equation is represented as

$$\eta_c = \eta_0 - \frac{\epsilon_r \sigma(T_c^4 - T_a'^4)}{CR} \tag{62.34}$$

where ϵ_r is the infrared emittance of the receiver, σ is the Stefan–Boltzmann constant, and T_a' is the equivalent ambient temperature for radiation depending on ambient humidity and cloud cover.

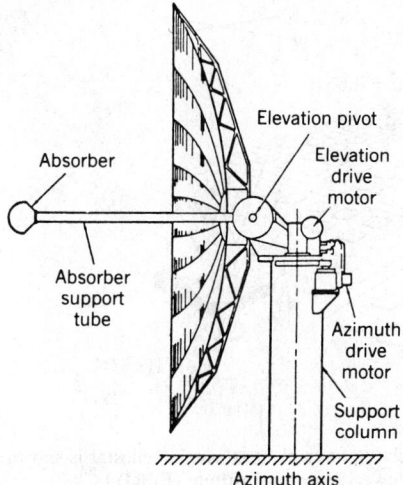

Elevation pivot

Elevation drive motor

Absorber

Absorber support tube

Azimuth drive motor

Support column

Azimuth axis

Fig. 62.11 Segmented mirror approximation to paraboloidal dish designed by Raytheon, Inc. Paraboloid is approximated by about 200 spherical mirrors. Average CR is 118, while maximum local CR is 350.

Fig. 62.12 Commercial paraboloidal solar concentrator. The receiver assembly has been removed from the focal zone for this photograph. (Courtesy of Omnium-G Corp., Anaheim, CA.)

For clear, dry conditions T_a' is about 15–20°F below the ambient dry bulb temperature. As humidity decreases, T_a approaches the dry bulb temperature.

The optical efficiency for the central receiver is expressed in somewhat different terms than those used in Eq. (62.33). It is referenced to solar flux on a horizontal surface and therefore includes the geometric tilt factor. For the central receiver, the optical efficiency is given by

$$\eta_0 = \Phi \mathcal{P} \rho_m \, \alpha_r f_t \, \delta \qquad (62.35)$$

in which the last four parameters are defined as in Eq. (62.33). The ratio of redirected flux to horizontal flux is $\mathcal{P}$ and is given approximately by

$$\mathcal{P} = 0.78 + 1.5(1 - \alpha/90)^2 \qquad (62.36)$$

from Ref. 4. The ratio of mirror area to ground area Φ depends on the size and economic factors applicable to a specific installation. Values for Φ have been in the range 0.4–0.5 for installations made through 1985.

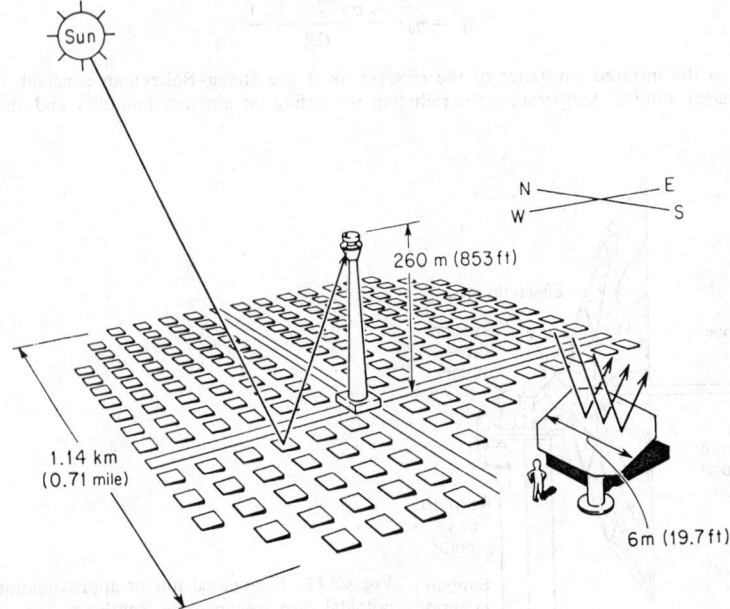

Fig. 62.13 Schematic diagram of a 50-MW$_e$ central receiver power plant. A single heliostat is shown in the inset to indicate its human scale. [From Electric Power Research Institute (EPRI).]

62.2.3 Collector Testing

In order to determine the optical efficiency and heat loss characteristics of flat-plate and concentrating collectors (other than the central receiver, which is difficult to test because of its size), testing under controlled conditions is preferred to theoretical calculations. Such test data are required if comparisons among collectors are to be made objectively. As of the mid-1980s very few consensus standards had been adopted by the U.S. solar industry. The ASHRAE Standard Number 93-77 applies to flat-plate collectors that contain either a liquid or a gaseous working fluid.[5] Collectors in which a phase change occurs are not included. In addition, the standards do not apply well to concentrators, since additional procedures are needed to find the optical efficiency and aging effects. Testing of concentrators uses sections of the above standard where applicable plus additional procedures as needed; however, no industry standard exists. (The ASTM has promulgated standard E905 as the first proposed standard for concentrator tests.) ASHRAE Standard Number 96-80 applies to very-low-temperature collectors manufactured without any glazing system.

Figure 62.14 shows the test loop used for liquid-cooled flat-plate collectors. Tests are conducted with solar flux at near-normal incidence to find the normal incidence optical efficiency $(\tau\alpha)_n$ along with the heat loss conductance U_c. Off-normal optical efficiency is determined in a separate test by orienting the collector such that several substantially off-normal values of $\tau\alpha$ or η_0 can be measured. The fluid used in the test is preferably that to be used in the installed application, although this is not always possible. If operational and test fluids differ, an analytical correction in the heat removal factor F_R is to be made.[2] An additional test is made after a period of time (nominal one month) to determine the effect of aging, if any, on the collector parameters listed above. A similar test loop and procedure apply to air-cooled collectors.[5]

The development of full system tests has only begun. Of course, it is the entire solar system (see next section) not just the collector that ultimately must be rated in order to compare solar and other energy-conversion systems. Testing of full-size solar systems is very difficult owing to their large size and cost. Hence, it is unlikely that full system tests will ever be practical except for the smallest systems such as residential water heating systems. For this one group of systems a standard test procedure (ASHRAE 95-81) exists. Larger-system performance is often predicted, based on component tests, rather than measured.

62.3 SOLAR THERMAL APPLICATIONS

One of the unique features of solar heat is that it can be produced over a very broad range of temperatures—the specific temperature being selected to match the thermal task to be performed. In this

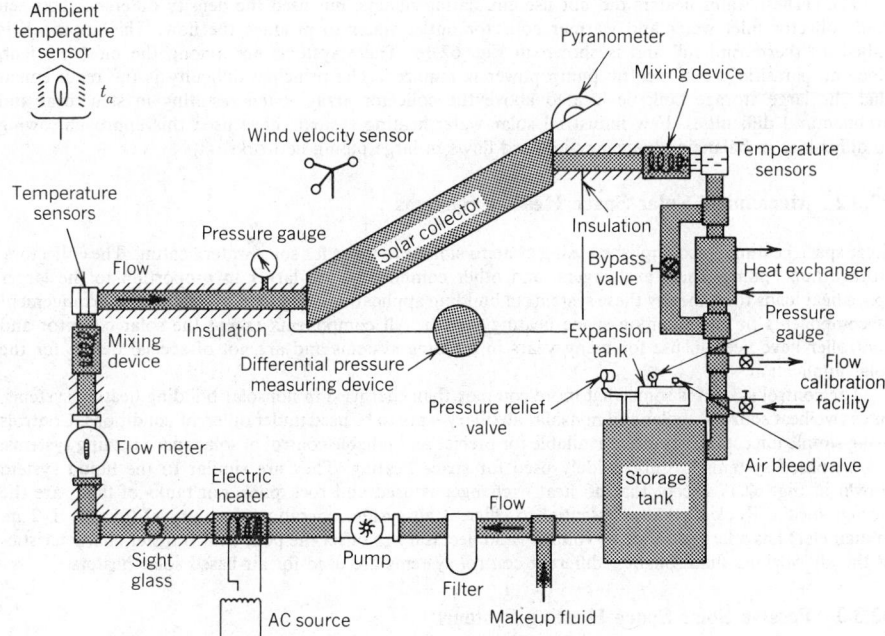

Fig. 62.14 Closed-loop testing configuration for the solar collector when the transfer fluid is a liquid.

section the most common thermal applications will be described in summary form. These include low-temperature uses such as water and space heating (30–100°C), intermediate temperature industrial processes (100–300°C), and high-temperature thermal power applications (500–850°C and above). Methods for predicting performance, where available, will also be summarized. Nonthermal solar applications are described in the next section.

62.3.1 Solar Water Heating

The most often used solar thermal application is for the heating of water for either domestic or industrial purposes. Relatively simple systems are used, and the load exists relatively uniformly through a year resulting in a good system load factor. Figure 62.15a shows a single-tank water heater schematically. The key components are the collector (0.5–1.0 ft²/gal day load), the storage tank (1.0–2.0 gal/ft² of collector), a circulating pump, and controller. The check valve is essential to prevent backflow of collector fluid, which can occur at night when the pump is off if the collectors are located some distance above the storage tank. The controller actuates the pump whenever the collector is 15–30°F warmer than storage. Operation continues until the collector is only 1.5–5°F warmer than the tank, at which point it is no longer worthwhile to operate the pump to collect the relatively small amounts of solar heat available.

The water-heating system shown in Fig. 62.15a uses an electrical coil located near the top of the tank to ensure a hot water supply during periods of solar outage. This approach is only useful in small residential systems and where nonsolar energy resources other than electricity are not available. Most commercial systems are arranged as shown in Fig. 62.15b, where a separate preheat tank, heated only by solar heat, is connected upstream of the nonsolar, auxiliary water heater tank or plant steam-to-water heat exchanger. This approach is more versatile in that any source of backup energy whatever can be used when solar heat is not available. Additional parasitic heat loss is encountered, since total tank surface area is larger than for the single tank design.

The water-heating systems shown in Fig. 62.15 are of the indirect type, that is, a separate fluid is heated in the collector and heat thus collected is transferred to the end use via a heat exchanger. This approach is needed in locations where freezing occurs in winter and antifreeze solutions are required. The heat exchanger can be eliminated, thereby reducing cost and eliminating the unavoidable fluid temperature decrement between collector and storage fluid streams, if freezing will never occur at the application site. The exchanger can also be eliminated if the "drain-back" approach is used. In this system design the collectors are filled with water only when the circulating pump is on, that is, only when the collectors are warm. If the pump is not operating, the collectors and associated piping all drain back into the storage tank. This approach has the further advantage that heated water otherwise left to cool overnight in the collectors is returned to storage for useful purposes.

The earliest water heaters did not use circulating pumps, but used the density difference between cold collector inlet water and warmer collector outlet water to produce the flow. This approach is called a "thermosiphon" and is shown in Fig. 62.16. These systems are among the most efficient, since no parasitic use of electric pump power is required. The principal difficulty is the requirement that the large storage tank be located above the collector array, often resulting in structural and architectural difficulties. Few industrial solar water-heating systems have used this approach, owing to difficulties in balancing buoyancy-induced flows in large piping networks.

62.3.2 Mechanical Solar Space Heating Systems

Solar space heating is accomplished using systems similar to those for solar water heating. The collectors, storage tank, pumps, heat exchangers, and other components are larger in proportion to the larger space heat loads to be met by these systems in building applications. Figure 62.17 shows the arrangement of components in one common space heating system. All components except the solar collector and controller have been in use for many years in building systems and are not of special design for the solar application.

The control system is somewhat more complex than that used in nonsolar building heating systems, since two heat sources—solar and nonsolar auxiliary—are to be used under different conditions. Controls using simple microprocessors are available for precise and reliable control of solar space heating systems.

Air-based systems are also widely used for space heating. They are similar to the liquid system shown in Fig. 62.17 except that no heat exchanger is used and rock piles, not tanks of fluid, are the storage media. Rock storage is essential to efficient air-system operation since gravel (usually 1–2 in. in diameter) has a large surface-to-volume ratio necessary to offset the poor heat transfer characteristics of the air working fluid. Slightly different control systems are used for air-based solar heaters.

62.3.3 Passive Solar Space Heating Systems

A very effective way of heating residences and small commercial buildings with solar energy and without significant nonsolar operating energy is the "passive" heating approach. Solar flux is admitted

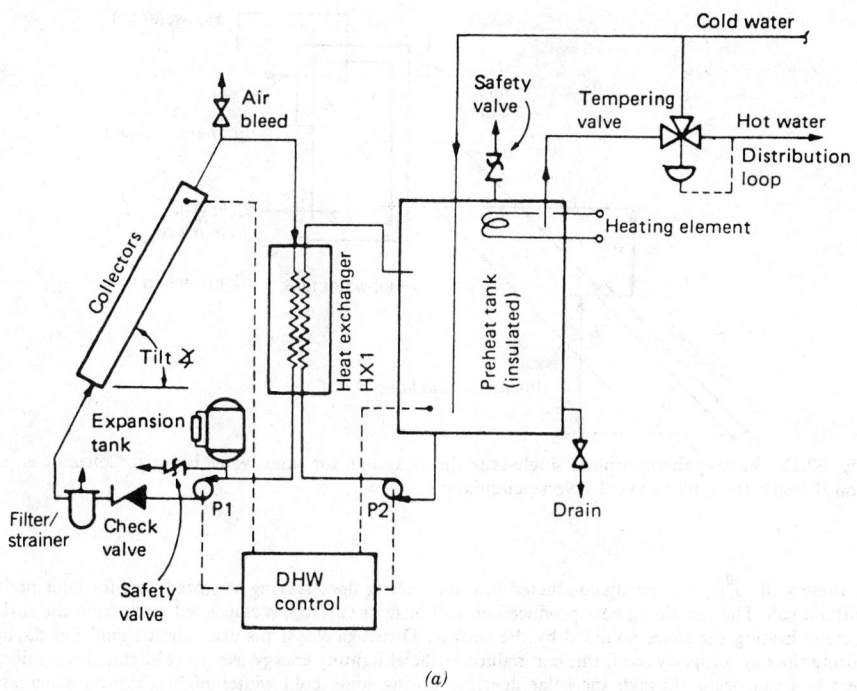

(a)

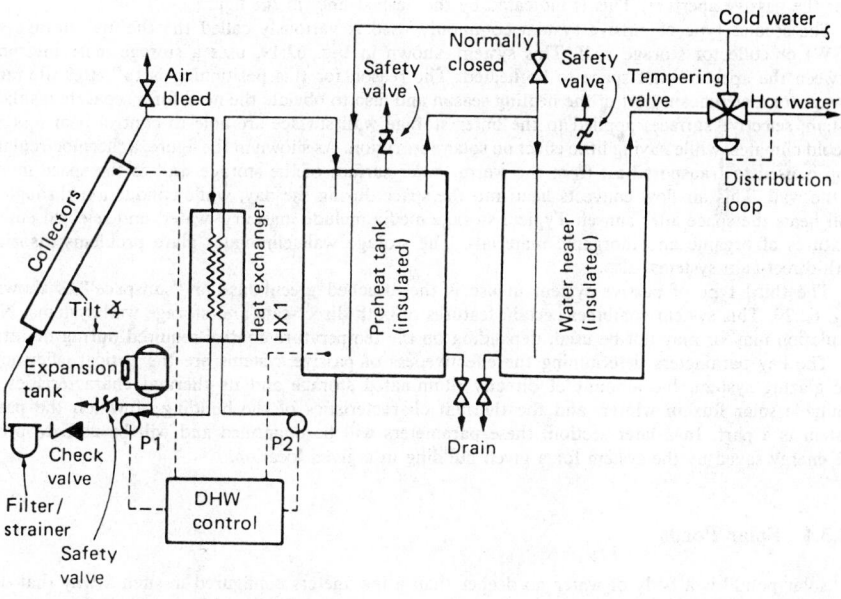

(b)

Fig. 62.15 (a) Single-tank indirect solar water-heating system. (b) Double-tank indirect solar water-heating system. Instrumentation and miscellaneous fittings are not shown.

into the space to be heated by large sun-facing apertures. In order that overheating not occur during sunny periods, large amounts of thermal storage are used, often also serving a structural purpose. A number of classes of passive heating systems have been identified and are described in this section.

Figure 62.18 shows the simplest type of passive system known as "direct gain." Solar flux enters a large aperture and is converted to heat by absorption on dark colored floors or walls. Heat produced

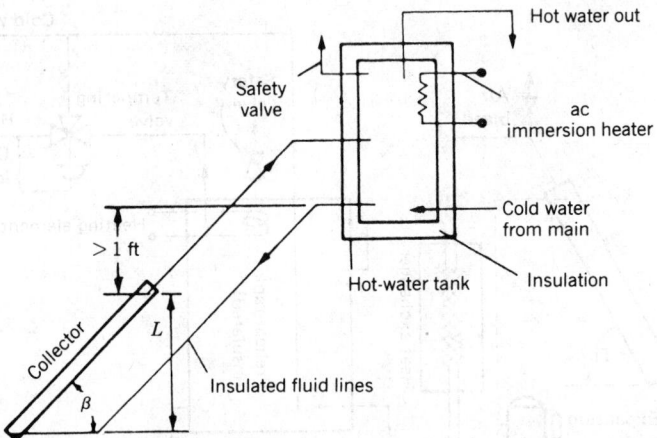

Fig. 62.16 Passive thermosiphon single-tank direct system for solar water heating. Collector is positioned below the tank to avoid reverse circulation.

at these wall surfaces is partly conducted into the wall or floor serving as stored heat for later periods without sun. The remaining heat produced at wall or floor surfaces is convected away from the surface thereby heating the space bounded by the surface. Direct-gain systems also admit significant daylight during the day; properly used, this can reduce artificial lighting energy use. In cold climates significant heat loss can occur through the solar aperture during long, cold winter nights. Hence, a necessary component of efficient direct-gain systems is some type of insulation system put in place at night over the passive aperture. This is indicated by the dashed lines in the figure.

The second type of passive system commonly used is variously called the thermal storage wall (TSW) or collector storage wall. This system, shown in Fig. 62.19, uses a storage mass interposed between the aperture and space to be heated. The reason for this positioning is to better illuminate storage for a significant part of the heating season and also to obviate the need for a separate insulation system; selective surfaces applied to the outer storage wall surface are able to control heat loss well in cold climates, while having little effect on solar absorption. As shown in the figure, a thermocirculation loop is used to transport heat from the warm, outer surface of the storage wall to the space interior to the wall. This air flow convects heat into the space during the day, while conduction through the wall heats the space after sunset. Typical storage media include masonry, water, and selected eutectic mixtures of organic and inorganic materials. The storage wall eliminates glare problems associated with direct-gain systems, also.

The third type of passive system in use is the attached greenhouse or "sunspace" as shown in Fig. 62.20. This system combines certain features of both direct-gain and storage wall systems. Night insulation may or may not be used, depending on the temperature control required during nighttime.

The key parameters determining the effectiveness of passive systems are the optical efficiency of the glazing system, the amount of directly illuminated storage and its thermal characteristics, the available solar flux in winter, and the thermal characteristics of the building of which the passive system is a part. In a later section, these parameters will be quantified and will be used to predict the energy saved by the system for a given building in a given location.

62.3.4 Solar Ponds

A "solar pond" is a body of water no deeper than a few meters configured in such a way that usual convection currents induced by solar absorption are suppressed. The oldest method for convection suppression is the use of high concentrations of soluble salts in layers near the bottom of the pond with progressively smaller concentrations near the surface. The surface layer itself is usually fresh water. Incident solar flux is absorbed by three mechanisms. Within a few millimeters of the surface the infrared component (about one-third of the total solar flux energy content) is completely absorbed. Another third is absorbed as the visible and ultraviolet components traverse a pond of nominal 2-m depth. The remaining one-third is absorbed at the bottom of the pond. It is this component that would induce convection currents in a freshwater pond thereby causing warm water to rise to the top where convection and evaporation would cause substantial heat loss. With proper concentration

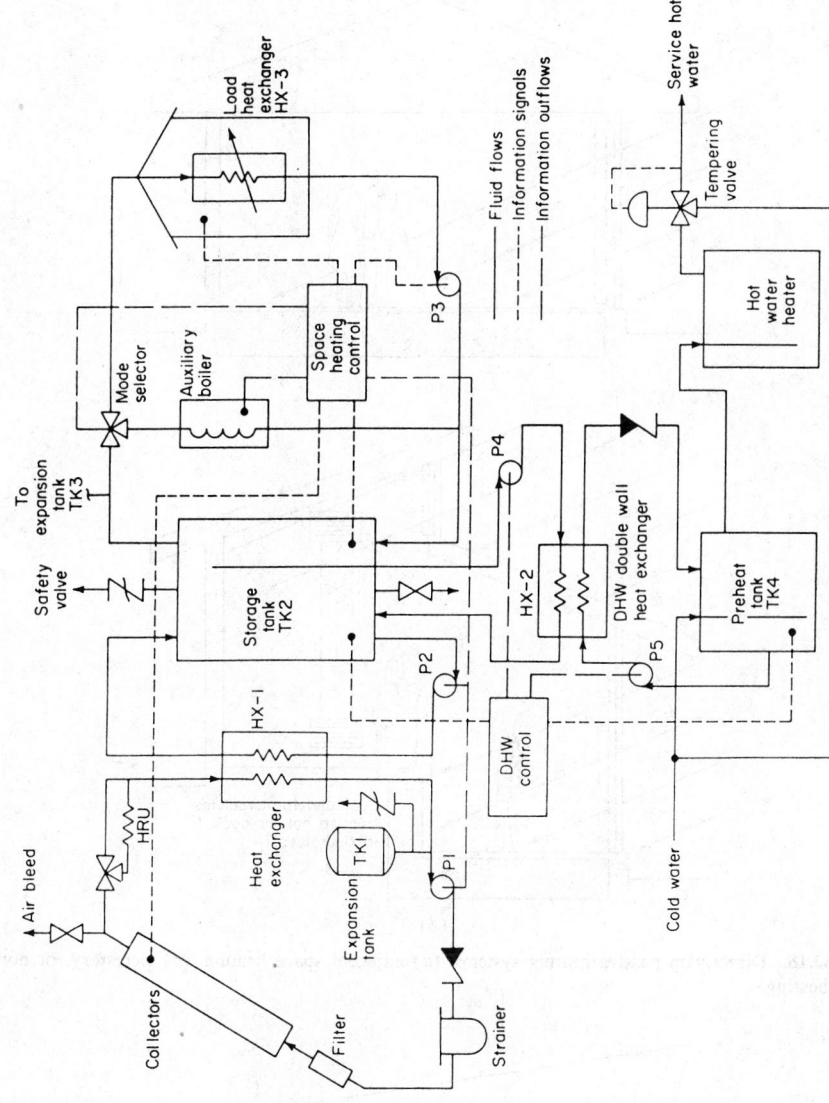

Fig. 62.17 Schematic diagram of a typical liquid-based space heating system with domestic water preheat.

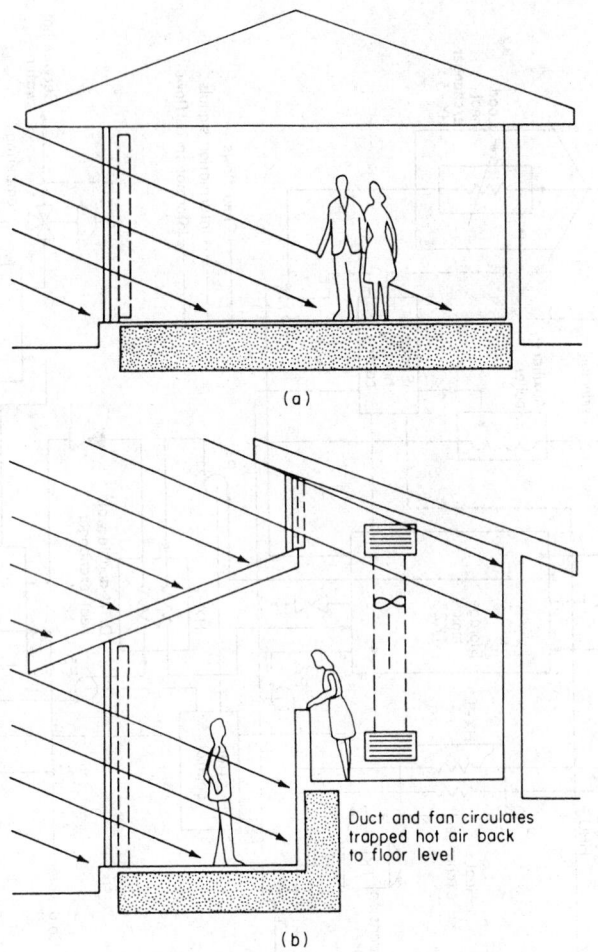

Fig. 62.18 Direct-gain passive heating systems: (*a*) adjacent space heating; (*b*) clerestory for north zone heating.

Duct and fan circulates
trapped hot air back
to floor level

(a)

(b)

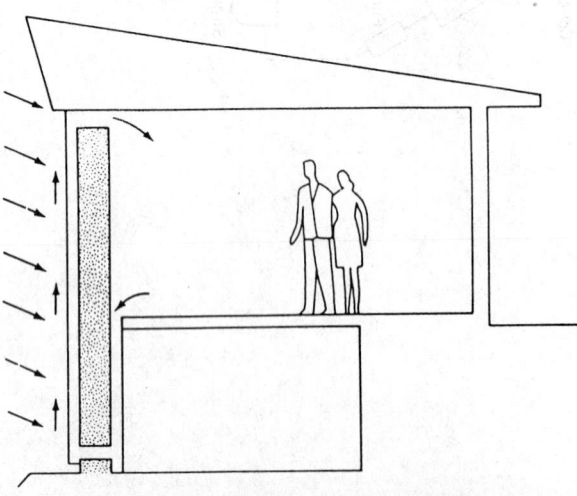

Fig. 62.19 Indirect-gain passive system—TSW system.

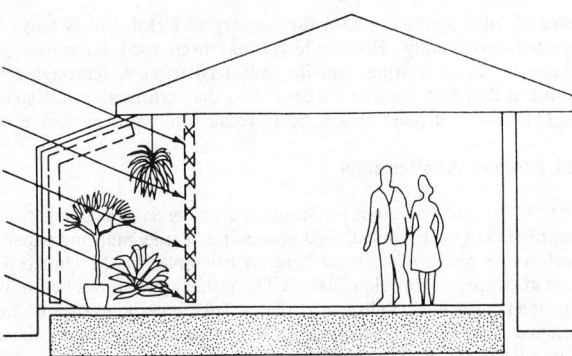

Fig. 62.20 Greenhouse or attached sun-space passive heating system using a combination of direct gain into the greenhouse and indirect gain through the thermal storage wall, shown by cross-hatching, between the greenhouse and the living space.

gradient, convection can be completely suppressed and significant heat collection at the bottom layer is possible. Salt gradient ponds are hydrodynamically stable if the following criterion is satisfied:

$$\frac{d\rho}{dz} = \frac{\partial\rho}{\partial s}\frac{ds}{dz} + \frac{\partial\rho}{\partial T}\frac{dT}{dz} > 0 \tag{62.37}$$

where s is the salt concentration, ρ is the density, T is the temperature, and z is the vertical coordinate measured positive downward from the pond surface. The inequality requires that the density must decrease upward.

Useful heat produced is stored in and removed from the lowest layer as shown in Fig. 62.21. This can be done by removing the bottom layer of fluid, passing it through a heat exchanger, and returning the cooled fluid to another point in the bottom layer. Alternatively, a network of heat-removal pipes can be placed on the bottom of the bond and the working fluid passed through for heat collection. Depending on the design, solar ponds also may contain substantial heat storage capability if the lower convective zone is relatively thick. This approach is used when uniform heat supply is necessary over a 24-hr period but solar flux is available for only a fraction of the period. Other convection-suppression techniques and heat-removal methods have been proposed but not used in more than one installation at most.

The requirements for an effective solar pond installation include the following. Large amounts of nearly free water and salt must be available. The subsoil must be stable in order to minimize changes in pond shape that could fracture the waterproof liner. Adequate solar flux is required year around; therefore, pond usage is confined to latitudes within 40° of the equator. Freshwater aquifers used for potable water should not be nearby in the event of a major leak of saline water into the groundwater. Other factors include low winds to avoid surface waves and windblown dust collection within the pond (at the neutral buoyancy point), low soil conductivity (i.e., low water content) to minimize conduction heat loss, and durable liner materials capable of remaining leakproof for many years.

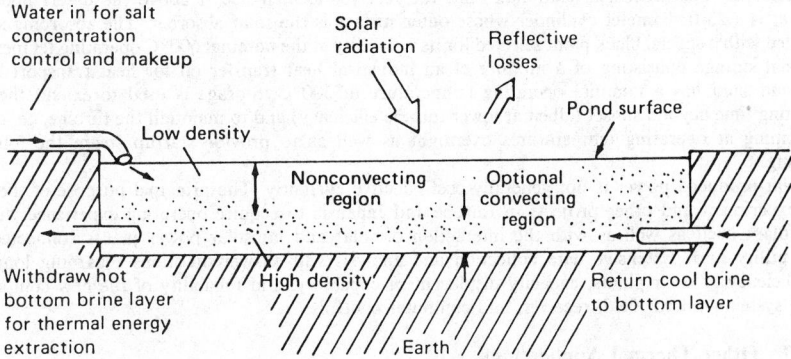

Fig. 62.21 Schematic diagram of a nonconvecting solar pond showing conduits for heat withdrawal, surface washing, and an optional convecting zone near the bottom.

The principal user of solar ponds has been the country of Israel. Ponds tens of acres in size have been built and operated successfully. Heat collected has been used for power production with an organic Rankine cycle, for space heating, and for industrial uses. A thorough review of solar pond technology is contained in Ref. 6. A method for predicting the performance of a solar pond is presented in the next section. The theory of pond optics, heat production, and heat loss is contained in Ref. 1.

62.3.5 Industrial Process Applications

Process heat up to 300°C for industry can be produced by commercially available trough-type concentrators. Most of the heat needed in the chemical, food processing, textile, pulp and paper, and petrochemical industries can, therefore, be provided by solar heat, in principle. In the United States about half of industrial heat is used at temperatures below 300°C. The viable applications below 300°C use collectors ranked in increasing temperature capability—flat-plates, solar ponds, evacuated tubes, and parabolic or other trough designs. Above 300°C, solar applications have been few in the major industries— primary metals, stone–clay–glass, electric power production.

Since many industrial processes operate around the clock, it may appear prudent at first glance to use substantial storage to permit high daily and annual load factors. This is appropriate from a thermal viewpoint but economic constraints have dictated that the most cost effective systems built to date have only sufficient storage to carry through solar transients of no longer than 30 min. The difficulty is the unavailability of an inexpensive, high-temperature, storage medium with proper heat-transport properties.

Solar industrial process heat was not yet a mature technology in the mid-1980s. Less than 100 systems existed worldwide and many of the earliest systems performed well below expectations. The key reasons were poor control function, lower than expected collector efficiency, parasitic heat losses in extensive piping networks, and poor durability of important components. However, the early experiments were very valuable in significantly improving this new technology. Later generation systems worked well, and the promise of solar heat applications is good under certain conditions of available land area for large arrays, adequate solar flux, and favorable economic conditions—advantageous tax consideration and expensive, nonsolar fuels. Significant reductions in system cost are needed for widespread application.

62.3.6 Solar Thermal Power Production

Solar energy has very high thermodynamic availability owing to the high effective temperature of the source. Therefore, production of shaft power and electric power therefrom is thermodynamically possible. Two fundamentally different types of systems can be used for power production: (1) a large array of concentrating collectors of several tens of meters in area connected by a fluid or electrical network and (2) a single, central receiver using mirrors distributed over a large area but producing heat and power only at one location. The determination of which approach is preferred depends on required plant capacity. For systems smaller than 10 MW the distributed approach appears more economical with existing steam turbines. For systems greater than 10 MW, the central receiver appears more economical.[7] However, if highly efficient Brayton or Stirling engines were available in the 10–20 kW range, the distributed approach would have lowest cost for any plant size. Such systems are expected to be available by the year 2000.

The first U.S. central receiver began operating in the fall of 1982. Located in the Mojave Desert this 10-MW plant (called "Solar One") is connected to the southern California electrical grid. The collection system consists of 1818 heliostats totaling 782,000 ft² in area. Each 430 ft² mirror is computer controlled to focus reflected solar flux onto the receiver located 300 ft above the desert floor. The receiver is a 23-ft diameter cyclinder whose outer surface is the solar absorber. The absorbing surface is coated with a special black paint selected for its reliability at the nominal 600°C operating temperature. Thermal storage consisting of a mixture of an industrial heat transfer oil for heat transport and of rock and sand has a nominal operating temperature of 300°C. Storage is used to extend the plant operating time beyond sunset (albeit at lower turbine efficiency) and to maintain the turbine, condenser, and piping at operating temperatures overnight as well as to provide startup steam the following morning.

Solar-produced power is not generally cost effective currently. The principal purpose of the Solar One experiment and other projects in Europe and Japan is to acquire operating experience with the solar plant itself as well as with the interaction of solar and nonsolar power plants connected in a large utility grid. Extensive data collection and analysis will answer questions regarding long-term net efficiency of solar plants, capacity displacement capability, and reliability of the new components of the system—mirror field, receiver, and computer controls.

62.3.7 Other Thermal Applications

The previous sections have discussed the principal thermal applications of solar energy that have been reduced to practice in at least five different installations and that show significant promise for economic displacement of fossil or fissile energies. In this section two other solar-conversion technologies are summarized.

Solar-powered cooling has been demonstrated in many installations in the United States, Europe, and Japan. Chemical absorption, organic Rankine cycle, and desiccant dehumidifaction processes have all been shown to be functional. Most systems have used flat-plate collectors, but higher coefficients of performance are achievable with mildly concentrating collectors. References 1, 2, and 8 describe solar-cooling technologies. To date, economic viability has not been generally demonstrated, but further research resulting in reduced cost and improved efficiency is expected to continue.

Thermal energy stored in the surface layers of the tropical oceans has been used to produce electrical power on a small scale. A heat engine is operated between the warmest layer at the surface and colder layers several thousand feet beneath. The available temperature difference is of the order of 20°C, therefore, the cycle efficiency is very low—only a few percent. However, this type of power plant does not require collectors or storage. Only a turbine capable of operating efficiently at low temperature is needed. Some cycle designs also require very large heat exchangers, but new cycle concepts without heat exchangers and their unavoidable thermodynamic penalties show promise.

62.3.8 Performance Prediction for Solar Thermal Processes

In a rational economy the single imperative for use of solar heat for any of the myriad applications outlined heretofore must be cost competitiveness with other energy sources—fossil and fissile. The amount of useful solar energy produced by a solar-conversion system must therefore be known along with the cost of the system. In this section the methods usable for predicting the performance of widely deployed solar systems are summarized. Special systems such as the central receiver, the ocean thermal power plant, and solar cooling are not included. The methods described here require a minimum of computational effort, yet embody all important parameters determining performance.

Solar systems are connected to end uses characterized by an energy requirement or "load" L and by operating temperature that must be achievable by the solar-heat-producing system. The amount of solar-produced heat delivered to the end use is the useful energy Q_u. This is the net heat delivery accounting for parasitic losses in the solar subsystem. The ratio of useful heat delivered to the requirement L is called the "solar fraction" denoted by f_s. In equation form the solar fraction is

$$f_s \equiv \frac{Q_u}{L} \tag{62.38}$$

Empirical equations have been developed relating the solar fraction to other dimensionless groups characterizing a given solar process. These are summarized shortly.

A fundamental concept used in many predictive methods is the solar "utilizability" defined as that portion of solar flux absorbed by a collector that is capable of providing heat to the specified end use. The key characteristic of the end use is its temperature. The collector must produce at least enough heat to offset losses when the collector is at the minimum temperature $T_{\min}$ usable by the given process. Figure 62.22 illustrates this idea schematically. The curve represents the flux absorbed over a day by a hypothetical collector. The horizontal line intersecting this curve represents the threshold flux that must be exceeded for a net energy collection to take place. In the context of the efficiency equation [Eq. (62.32)], this critical flux I_{cr} is that which results in a collector efficiency of exactly zero when the collector is at the minimum usable process temperature $T_{\min}$. Any greater flux will result in net heat production. From Eq. (62.32) the critical intensity is

$$I_{cr} = \frac{U_c(T_{\min} - T_a)}{\tau \alpha} \tag{62.39}$$

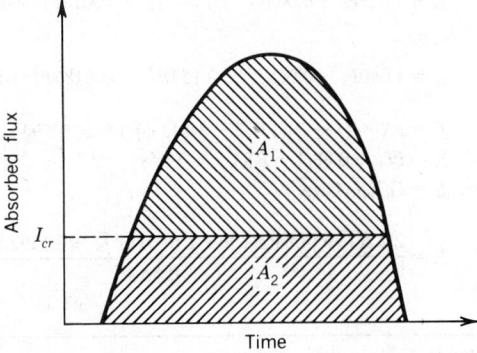

Fig. 62.22 Daily absorbed solar flux ($A_1 + A_2$) and useful solar flux (A_1) at intensities above I_{cr}.

The solar utilizability is the ratio of the useful daily flux (area above I_{cr} line in Fig. 62.22) to the total absorbed flux (area $A_1 + A_2$) beneath the curve. The utilizability denoted by ϕ is

$$\phi = \frac{A_1}{A_1 + A_2} \tag{62.40}$$

This quantity is a solar radiation statistic depending on I_{cr}, characteristics of the incident solar flux and characteristics of the collection system. It is a very useful parameter in predicting the performance of solar thermal systems.

Table 62.7 summarizes empirical equations used for predicting the performance of the most common solar-thermal systems. These expressions are given in terms of the solar fraction defined above and dimensionless parameters containing all important system characteristics. The symbols used in this table are defined in Table 62.8. In the brief space available in this chapter, all details of these prediction methodologies cannot be included. The reader is referred to Refs. 1, 4, 9, 10, and 11 for details.

62.4 NONTHERMAL SOLAR ENERGY APPLICATIONS

In this section the principal nonthermal solar conversion technology is described. Photovoltaic cells are capable of converting solar flux directly into electric power. This process, first demonstrated in the 1950s, holds considerable promise for significant use in the future. Major cost reductions have been accomplished. In this section the important features of solar cells are described.

Photovoltaic conversion of sunlight to electricity occurs in a thin layer of semiconductor material exposed to solar flux. Photons free electric charges, which flow through an external circuit to produce useful work. The semiconductor materials used for solar cells are tailored to be able to convert the majority of terrestrial solar flux; however, low-energy photons in the infrared region are usually not usable. Figure 62.23 shows the maximum theoretical conversion efficiency of seven common materials used in the application. Each material has its own threshold band-gap energy, which is a weak function of temperature. The energy contained in a photon is $E = h\nu$. If E is greater than the band-gap energy shown in this figure, conversion can occur.

Figure 62.23 also shows the very strong effect of temperature on efficiency. For practical systems it is essential that the cell be maintained as near to ambient temperature as possible.

Solar cells produce current proportional to the solar flux intensity with wavelengths below the band-gap threshold. Figure 62.24 shows the equivalent circuit of a solar cell. Both internal shunt and series resistances must be included. These result in unavoidable parasitic loss of part of the power produced by the equivalent circuit current source of strength I_s. Solving the equivalent circuit for the power P produced and using an expression from Ref. 2 for the junction leakage I_J results in

$$P = [I_s - I_0 (e^{e_0 V/kT} - 1)]V \tag{62.41}$$

in which e_0 is the electron charge, k is the Boltzmann constant, and T is the temperature. The current source I_s is given by

$$I_s = \eta_0 (1 - \rho_c) \alpha e_0 n_p \tag{62.42}$$

in which η_0 is the collector carrier efficiency, ρ_c is the cell surface reflectance, α is the absorptance of photons, and n_p is the flux density of sufficiently energetic photons.

Table 62.7 Empirical Solar Fraction Equations[a]

System Type	f_s Expression	Time Scale
Water heating and liquid-based space heating	$f_s = 1.029 P_s - 0.065 P_L - 0.245 P_s^2 + 0.0018 P_L^2 + 0.0215 P_s^3$	Monthly
Space heating— air-based systems	$f_s = 1.040 P_s - 0.065 P_L - 0.159 P_s^2 + 0.00187 P_L^2 + 0.0095 P_s^3$	Monthly
Passive direct gain	$f_s = PX + (1 - P)(3.082 - 3.142\,\overline{\phi})(1 - e^{-0.329x})$	Monthly
Passive storage wall	$f_s = Pf_\infty + 0.88(1 - P)(1 - e^{-1.26f_\infty})$	Monthly
Concentrating collector systems	$f_s = F_R \overline{\eta}_0 \overline{I}_c A_c N \overline{\phi}' / L$	Monthly
Solar ponds (pond radius R to provide annual pond temperature $\overline{T}_p$)	$R = \dfrac{2.2\,\overline{\Delta T} + [4.84(\Delta T)^2 + \overline{L}(0.3181\,\overline{I}_p - 0.1592\Delta T)]^{1/2}}{\overline{I}_p - 0.5\Delta T}$	Annual

[a] See Table 62.8 for symbol definitions.

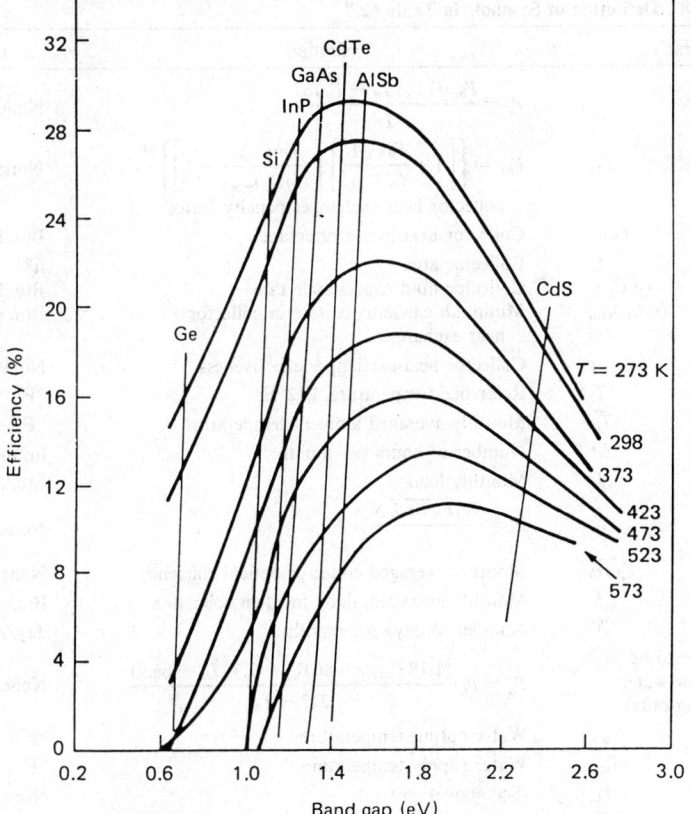

Fig. 62.23 Maximum theoretical efficiency of photovoltaic converters as a function of band-gap energy for several materials.

In addition to the solar cell, complete photovoltaic systems also must contain electrical storage and a control system. The cost of storage presents another substantial cost problem in the widespread application of photovoltaic power production. The costs of the entire conversion system must be reduced by an order of magnitude in order to be competitive with other power sources. Vigorous research in the United States, Europe, and Japan has made significant gains in the past decade. The installed capacity of photovoltaic systems is expected to be at least 10,000 MW by the year 2000.

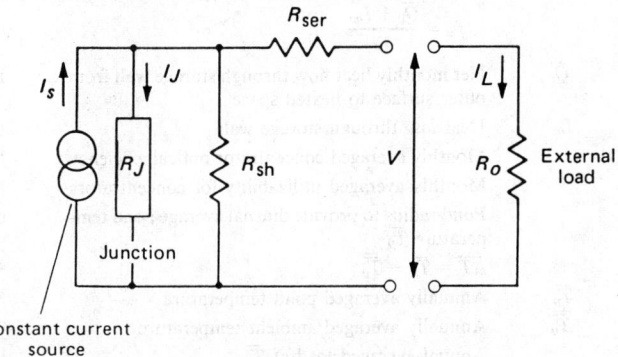

Fig. 62.24 Equivalent circuit of an illuminated p-n photocell with internal series and shunt resistances and nonlinear junction impedance R_J.

Table 62.8 Definition of Symbols in Table 62.7

Parameters		Definition	Units[a]
P_L		$P_s = \dfrac{F_{hx} F_R U_c (T_r - \overline{T}_a)\Delta t}{L}$	None
	F_{hx}	$F_{hx} = \left\{ \left[1 + \dfrac{F_R U_c A_c}{(\dot{m}C_p)_c} \right] \left[\dfrac{(\dot{m}C_p)_c}{(\dot{m}C_p)_{\min}\epsilon} - 1 \right] \right\}^{-1},$ collector heat exchanger penalty factor	None
	$F_R U_c$	Collector heat-loss conductance	Btu/hr · ft² · °F
	A_c	Collector area	ft²
	$(\dot{m}C_p)_c$	Collector fluid capacitance rate	Btu/hr · °F
	$(\dot{m}C_p)_{\min}$	Minimum capacitance rate in collector heat exchanger	Btu/hr · °F
	ϵ	Collector heat-exchanger effectiveness	None
	T_r	Reference temperature, 212°F	°F
	$\overline{T}_a$	Monthly averaged ambient temperature	°F
	Δt	Number of hours per month	hr/month
	L	Monthly load	Btu/month
P_s		$P_s = \dfrac{F_{hx} F_R \overline{\tau\alpha}\, \overline{I}_c N}{L}$	None
	$F_R \overline{\tau\alpha}$	Monthly averaged collector optical efficiency	None
	$\overline{I}_c$	Monthly averaged, daily incident solar flux	Btu/day · ft²
	N	Number of days per month	day/month
$(P_L'$ — to be used for water heating only)		$P_L' = P_L \dfrac{(1.18 T_{wo} + 3.86 T_{wi} - 2.32 \overline{T}_a - 66.2)}{212 - \overline{T}_a}$	None
	T_{wo}	Water output temperature	°F
	T_{wi}	Water supply temperature	°F
	P_L	(See above)	None
P		$P = (1 - e^{-0.294 Y})^{0.652}$	None
	Y	Storage–vent ratio, $Y = \dfrac{C \Delta T}{\overline{\phi}\, \overline{I}_c\, \overline{\tau\alpha}\, A_c}$	None
	C	Passive storage capacity	Btu/°F
	ΔT	Allowable diurnal temperature saving in heated space	°F
$\overline{\phi}$		Monthly averaged utilizability (see below)	None
X		Solar-load ratio, $X = \dfrac{\overline{I}_c \overline{\tau\alpha} A_c N}{L}$	None
	L	Monthly space heat load	Btu/month
f_∞		Solar fraction with hypothetically infinite storage, $f_\infty = \dfrac{\overline{Q}_i + L_w}{L}$	None
	$\overline{Q}_i$	Net monthly heat flow through storage wall from outer surface to heated space	Btu/month
	L_w	Heat *loss* through storage wall	Btu/month
$F_R \overline{\eta}_0$		Monthly averaged concentrator optical efficiency	None
$\overline{\phi}'$		Monthly averaged utilizability for concentrators	None
R		Pond radius to provide diurnal average pond temperature $\overline{T}_p$	m
ΔT		$\overline{\Delta T} = \overline{T}_p - \overline{\overline{T}}_a$	°C
	$\overline{T}_p$	Annually averaged pond temperature	°C
	$\overline{\overline{T}}_a$	Annually averaged ambient temperature	°C
$\overline{L}$		Annual averaged load at $\overline{T}_p$	W
$\overline{I}_p$		Annual averaged insolation absorbed at pond bottom	W/m²

Table 62.8 *Continued*

Parameters	Definition	Units[a]
$\bar{\phi}$	Monthly flat-plate utilizability (equator facing collectors), $$\bar{\phi} = \exp\{[A + B(\bar{R}_N/\bar{R})](\bar{X}_c + C\bar{X}_c^2)\}$$	None
A	$A = 7.476 - 20.0\bar{K}_T + 11.188\bar{K}_T^2$	None
B	$B = -8.562 + 18.679\bar{K}_T - 9.948\bar{K}_T^2$	None
C	$C = -0.722 + 2.426\bar{K}_T + 0.439\bar{K}_T^2$	None
$\bar{R}$	Tilt factor, see Eq. (62.21)	None
$\bar{R}_N$	Monthly averaged tilt factor for hour centered about noon (see Ref. 9)	None
$\bar{X}_c$	Critical intensity ratio, $\bar{X}_c = \dfrac{I_{cr}}{r_{T,N}\bar{R}_N\bar{H}_h}$	None
$r_{T,N}$	Fraction of daily total radiation contained in hour about noon, $$r_{T,N} = r_{d,n}[1.07 + 0.025\sin(h_{sr} - 60)]$$	day/hr
	$r_{d,n} = \dfrac{\pi}{24}\dfrac{1 - \cos h_{sr}}{\sin h_{sr} - h_{sr}\cos h_{sr}}$	day/hr
I_{cr}	Critical intensity [see Eq. (62.39)]	Btu/hr · ft²
$\bar{\phi}'$	Monthly concentrator utilizability, $$\bar{\phi}' = 1.0 - (0.049 + 1.49\bar{K}_T)\bar{X} + 0.341\bar{K}_T\bar{X}^2$$ $$0.0 < \bar{K}_T < 0.75,\ 0 < \bar{X} < 1.2)$$ $$\bar{\phi}' = 1.0 - \bar{X}$$ $$(\bar{K}_T > 0.75,\ 0 < \bar{X} < 1.0)$$	None
$\bar{X}$	Concentrator critical intensity ratio, $$\bar{X} = \dfrac{U_c(T_{f,i} - \bar{T}_a)\Delta t_c}{\bar{\eta}_0\bar{I}_c}$$	None
$T_{f,i}$	Collector fluid inlet temperature—assumed constant	°F
Δt_c	Monthly averaged solar system operating time	hr/day

[a] USCS units shown except for solar ponds; SI units may also be used for all parameters shown in USCS units.

REFERENCES

1. J. F. Kreider and F. Kreith, *Solar Energy Handbook,* McGraw-Hill, New York, 1981.

2. F. Kreith and J. F. Kreider, *Principles of Solar Engineering,* Hemisphere/McGraw-Hill, New York, 1978.

3. M. Collares-Pereira and A. Rabl, "The Average Distribution of Solar Energy," *Solar Energy,* **22,** 155–164 (1979).

4. J. F. Kreider, *Medium and High Temperature Solar Processes,* Academic Press, New York, 1979.

5. ASHRAE Standard 93-77, *Methods of Testing to Determine the Thermal Performance of Solar Collectors,* ASHRAE, Atlanta, GA, 1977.

6. H. Tabor, "Solar Ponds—Review Article," *Solar Energy,* **27,** 181–194 (1981).

7. T. Fujita et al., *Projection of Distributed Collector Solar Thermal Electric Power Plant Economics to Years 1990–2000,* JPL Report No. DOE/JPL-1060-77/1, 1977.

8. J. F. Kreider and F. Kreith, *Solar Heating and Cooling,* Hemisphere/McGraw-Hill, New York, 1982.

9. W. A. Monsen, Master's Thesis, University of Wisconsin, 1980.

10. M. Collares-Pereira and A. Rabl, "Simple Procedure for Predicting Long Term Average Performance of Nonconcentrating and of Concentrating Solar Collectors," *Solar Energy,* **23,** 235–253 (1979).

11. J. F. Kreider, *The Solar Heating Design Process,* McGraw-Hill, New York, 1982.

APPENDIX

MONTHLY AVERAGED INSOLATION DATA

Notes:
1. Climatic data and horizontal solar flux taken from V. Cinquemani et al., *Input Data for Solar Systems*, National Climatic Center, U.S. Department of Commerce, Asheville, NC, 1978. Blank table entries indicate data unavailable from this source.

2. Solar flux on tilted surfaces (tilts equal to latitude, latitude +15°, and vertical are used) calculated using the method described in the text [Eq. (62.20) ff].

3. Tabulated solar flux values on tilted surfaces *do not include radiation reflected from the fore-ground*. To include this effect add the term $\overline{H}_h \rho \sin^2 (\beta/2)$ where $\overline{H}_h$ is the horizontal flux, ρ is the diffuse foreground reflectance, and β is the surface tilt angle.

4. Units used in the table are Btu/(day · ft²) for solar flux, °F for temperatures, and °F-days for degree days. $\overline{K}_r$ has no units. The degree-day basis is 65°F.

STA NO: 3133
YUCCA FLATS
NV LAT = 36.95

	Jan	Feb	Mar	Apr	May	Jun	Jul	Aug	Sep	Oct	Nov	Dec	Ann
HORIZ INSOL:	953.	1273.	1764.	2246.	2577.	2734.	2653.	2382.	2022.	1516.	1041.	853.	1835.
TILT = LAT:	1635.	1851.	2147.	2265.	2259.	2247.	2247.	2266.	2306.	2126.	1722.	1534.	
TILT = LAT+15:	1748.	1897.	2078.	2036.	1908.	1841.	1867.	1986.	2174.	2151.	1824.	1657.	
TILT = 90:	1541.	1499.	1363.	983.	680.	552.	607.	850.	1281.	1629.	1572.	1499.	
KT:	.62	.64	.68	.71	.72	.73	.73	.72	.72	.70	.64	.61	
AMB TEMP:	0	0	0	0	0	0	0	0	0	0	0	0	0
HTG DEG DAYS:	0	0	0	0	0	0	0	0	0	0	0	0	0

STA NO: 3812
ASHEVILLE
NC LAT = 35.43

	Jan	Feb	Mar	Apr	May	Jun	Jul	Aug	Sep	Oct	Nov	Dec	Ann
HORIZ INSOL:	722.	971.	1306.	1664.	1804.	1854.	1776.	1627.	1361.	1147.	849.	658.	1312.
TILT = LAT:	1086.	1285.	1492.	1642.	1594.	1564.	1532.	1527.	1460.	1471.	1266.	1032.	
TILT = LAT+15:	1139.	1297.	1429.	1478.	1368.	1314.	1301.	1348.	1465.	1468.	1323.	1094.	
TILT = 90:	961.	983.	905.	718.	541.	479.	498.	608.	787.	1065.	1099.	950.	
KT:	.45	.47	.49	.52	.50	.50	.49	.49	.48	.51	.49	.44	
AMB TEMP:	37.9	39.4	45.9	55.9	63.7	70.8	73.5	72.8	66.7	56.8	46.3	38.7	55.7
HTG DEG DAYS:	840	717	592	279	100	14	0	0	50	269	561	815	4237

STA NO: 3813
MACON
GA LAT = 32.70

	Jan	Feb	Mar	Apr	May	Jun	Jul	Aug	Sep	Oct	Nov	Dec	Ann
HORIZ INSOL:	769.	1020.	1363.	1735.	1895.	1919.	1785.	1718.	1439.	1247.	940.	729.	1379.
TILT = LAT:	1087.	1295.	1526.	1702.	1671.	1630.	1550.	1613.	1526.	1552.	1335.	1077.	
TILT = LAT+15:	1139.	1307.	1464.	1535.	1436.	1371.	1321.	1427.	1430.	1551.	1396.	1143.	
TILT = 90:	932.	954.	681.	688.	509.	444.	459.	586.	777.	1084.	1129.	967.	
KT:	.44	.46	.50	.53	.52	.52	.49	.51	.49	.53	.50	.45	
AMB TEMP:	47.8	50.4	56.5	65.8	73.5	79.6	81.4	80.9	75.8	65.7	55.2	48.3	65.1
HTG DEG DAYS:	543	423	298	66	6	0	0	0	0	82	304	518	2240

STA NO: 3820
AUGUSTA
GA LAT = 33.37

	Jan	Feb	Mar	Apr	May	Jun	Jul	Aug	Sep	Oct	Nov	Dec	Ann
HORIZ INSOL:	751.	1015.	1338.	1728.	1865.	1904.	1803.	1667.	1410.	1220.	917.	721.	1362.
TILT = LAT:	1075.	1303.	1504.	1696.	1652.	1614.	1563.	1565.	1498.	1528.	1317.	1084.	
TILT = LAT+15:	1125.	1316.	1442.	1529.	1419.	1357.	1329.	1384.	1403.	1526.	1377.	1151.	
TILT = 90:	927.	970.	679.	698.	517.	453.	471.	585.	774.	1076.	1121.	981.	
KT:	.43	.47	.49	.53	.52	.51	.49	.51	.49	.52	.50	.45	
AMB TEMP:	45.8	48.3	54.6	63.8	71.7	78.2	80.4	79.6	74.2	64.1	53.7	46.4	63.4
HTG DEG DAYS:	601	475	346	90	10	0	0	0	0	104	344	577	2547

STA NO: 3822
SAVANNAH
GA LAT = 32.13

	Jan	Feb	Mar	Apr	May	Jun	Jul	Aug	Sep	Oct	Nov	Dec	Ann
HORIZ INSOL:	795.	1044.	1399.	1761.	1852.	1844.	1783.	1621.	1364.	1217.	941.	754.	1365.
TILT = LAT:	1117.	1318.	1562.	1725.	1645.	1572.	1551.	1551.	1435.	1496.	1319.	1105.	
TILT = LAT+15:	1170.	1331.	1500.	1557.	1415.	1327.	1322.	1347.	1344.	1493.	1378.	1173.	
TILT = 90:	954.	965.	893.	684.	495.	431.	451.	554.	723.	1031.	1106.	988.	
KT:	.44	.47	.51	.54	.51	.50	.49	.48	.46	.51	.50	.45	
AMB TEMP:	49.9	52.1	58.0	66.1	73.3	79.1	81.1	80.6	76.2	67.1	57.1	50.4	65.9
HTG DEG DAYS:	483	379	256	63	0	0	0	0	0	60	253	458	1952

STA NO: 3860 — HUNTINGTON, WV — LAT = 38.37

	Jan	Feb	Mar	Apr	May	Jun	Jul	Aug	Sep	Oct	Nov	Dec	Annual
HORIZ INSOL:	526.	757.	1067.	1448.	1710.	1844.	1769.	1580.	1306.	1004.	638.	467.	1176.
TILT = LAT:	787.	1002.	1219.	1425.	1505.	1542.	1515.	1386.	1424.	1322.	958.	726.	0
TILT = LAT+15:	815.	1002.	1160.	1279.	1289.	1292.	1282.	1307.	1315.	1329.	990.	760.	0
TILT = 90:	691.	775.	770.	680.	569.	521.	540.	641.	813.	989.	832.	660.	0
KT:	.36	.39	.42	.46	.48	.49	.48	.48	.48	.48	.41	.35	
AMB TEMP:	34.3	36.1	44.3	55.7	64.5	72.4	75.3	73.9	67.7	57.1	45.5	36.0	55.2
HTG DEG DAYS:	952	809	649	293	115	11	0	0	46	265	585	899	4624

STA NO: 3870 — GREENVILLE-SPARTANBURG, SC — LAT = 34.90

	Jan	Feb	Mar	Apr	May	Jun	Jul	Aug	Sep	Oct	Nov	Dec	Annual
HORIZ INSOL:	730.	982.	1329.	1697.	1839.	1918.	1830.	1699.	1406.	1180.	680.	670.	1347.
TILT = LAT:	1083.	1288.	1513.	1671.	1625.	1618.	1579.	1597.	1508.	1507.	1305.	1030.	0
TILT = LAT+15:	1136.	1301.	1450.	1504.	1394.	1358.	1340.	1409.	1411.	1505.	1366.	1101.	0
TILT = 90:	953.	978.	910.	718.	538.	477.	497.	619.	804.	1035.	1130.	951.	0
KT:	.44	.47	.50	.53	.51	.51	.50	.51	.49	.52	.50	.44	
AMB TEMP:	42.3	44.4	50.9	61.0	69.1	75.9	78.3	77.5	71.7	61.7	51.0	42.9	60.6
HTG DEG DAYS:	704	577	450	144	29	0	0	0	9	145	420	685	3163

STA NO: 3927 — FORT WORTH, TX — LAT = 32.83

	Jan	Feb	Mar	Apr	May	Jun	Jul	Aug	Sep	Oct	Nov	Dec	Annual
HORIZ INSOL:	805.	1069.	1409.	1616.	1890.	2153.	2155.	1983.	1621.	1293.	938.	766.	1475.
TILT = LAT:	1157.	1373.	1585.	1580.	1576.	1816.	1856.	1865.	1739.	1622.	1336.	1151.	0
TILT = LAT+15:	1215.	1390.	1522.	1425.	1440.	1567.	1567.	1646.	1633.	1624.	1398.	1225.	0
TILT = 90:	1001.	1020.	919.	650.	512.	456.	491.	649.	884.	1139.	1133.	1044.	0
KT:	.46	.49	.51	.50	.53	.58	.59	.59	.56	.55	.51	.47	
AMB TEMP:	44.8	48.7	55.0	65.2	72.5	80.6	84.8	84.9	77.7	67.6	55.8	47.9	65.5
HTG DEG DAYS:	626	456	335	88	0	0	0	0	0	60	287	530	2332

STA NO: 3928 — WICHITA, KS — LAT = 37.65

	Jan	Feb	Mar	Apr	May	Jun	Jul	Aug	Sep	Oct	Nov	Dec	Annual
HORIZ INSOL:	784.	1058.	1405.	1783.	2036.	2264.	2239.	2032.	1616.	1250.	871.	690.	1502.
TILT = LAT:	1302.	1497.	1663.	1774.	1790.	1880.	1907.	1924.	1798.	1699.	1401.	1196.	0
TILT = LAT+15:	1379.	1522.	1598.	1593.	1526.	1559.	1599.	1690.	1705.	1687.	1472.	1279.	0
TILT = 90:	1207.	1198.	1055.	812.	622.	556.	597.	773.	1014.	1287.	1261.	1147.	0
KT:	.52	.54	.55	.55	.57	.60	.61	.61	.58	.59	.51	.51	
AMB TEMP:	31.3	36.3	43.6	58.6	66.1	75.8	80.7	79.7	70.6	59.6	44.8	34.5	56.6
HTG DEG DAYS:	1045	804	671	275	90	7	0	0	32	211	606	946	4587

STA NO: 3937 — LAKE CHARLES, LA — LAT = 30.12

	Jan	Feb	Mar	Apr	May	Jun	Jul	Aug	Sep	Oct	Nov	Dec	Annual
HORIZ INSOL:	728.	1010.	1313.	1570.	1849.	1970.	1788.	1657.	1435.	1381.	917.	706.	1365.
TILT = LAT:	954.	1228.	1434.	1528.	1649.	1684.	1563.	1556.	1558.	1682.	1220.	962.	0
TILT = LAT+15:	990.	1236.	1375.	1382.	1422.	1419.	1335.	1381.	1462.	1686.	1269.	1011.	0
TILT = 90:	775.	866.	788.	589.	463.	408.	424.	530.	748.	1135.	987.	822.	0
KT:	.38	.43	.46	.47	.51	.53	.49	.49	.50	.56	.46	.40	
AMB TEMP:	52.3	55.1	60.3	68.9	75.2	80.7	82.4	82.2	78.4	70.0	60.2	54.3	68.3
HTG DEG DAYS:	415	306	200	26	0	0	0	0	0	36	177	338	1498

STA NO: 3940 — JACKSON, MS — LAT = 32.32

	Jan	Feb	Mar	Apr	May	Jun	Jul	Aug	Sep	Oct	Nov	Dec	Annual
HORIZ INSOL:	754.	1026.	1369.	1709.	1941.	2024.	1909.	1780.	1509.	1271.	902.	709.	1409.
TILT = LAT:	1049.	1296.	1528.	1672.	1721.	1716.	1655.	1672.	1604.	1578.	1257.	1028.	0
TILT = LAT+15:	1096.	1308.	1466.	1509.	1478.	1440.	1406.	1479.	1504.	1578.	1311.	1087.	0
TILT = 90:	890.	950.	876.	671.	509.	443.	465.	594.	808.	1097.	1050.	912.	0
KT:	.42	.46	.50	.52	.54	.54	.52	.53	.51	.54	.48	.43	0
AMB TEMP:	47.1	49.8	56.1	65.7	72.7	79.4	81.7	81.2	76.0	65.8	55.3	48.9	55.0
HTG DEG DAYS:	569	442	313	74	6	0	0	0	0	91	301	504	2300

STA NO: 3945 — COLUMBIA, MO — LAT = 38.82

	Jan	Feb	Mar	Apr	May	Jun	Jul	Aug	Sep	Oct	Nov	Dec	Annual
HORIZ INSOL:	612.	875.	1179.	1526.	1880.	2090.	2116.	1878.	1450.	1101.	703.	523.	1328.
TILT = LAT:	981.	1215.	1376.	1510.	1652.	1726.	1802.	1778.	1609.	1492.	1105.	865.	0
TILT = LAT+15:	1028.	1226.	1314.	1354.	1410.	1444.	1513.	1561.	1505.	1490.	1150.	913.	0
TILT = 90:	892.	967.	882.	723.	613.	561.	604.	751.	928.	1135.	982.	808.	0
KT:	.43	.45	.47	.49	.53	.56	.58	.57	.53	.53	.46	.40	0
AMB TEMP:	29.3	33.6	41.7	55.0	64.4	73.0	77.3	76.0	68.3	58.0	43.9	32.8	54.4
HTG DEG DAYS:	1107	879	730	314	117	11	0	5	42	247	633	998	5033

STA NO: 3947 — KANSAS CITY, MO — LAT = 39.30

	Jan	Feb	Mar	Apr	May	Jun	Jul	Aug	Sep	Oct	Nov	Dec	Annual
HORIZ INSOL:	648.	895.	1203.	1575.	1873.	2080.	2102.	1862.	1452.	1092.	737.	562.	1340.
TILT = LAT:	1080.	1265.	1417.	1564.	1645.	1726.	1789.	1764.	1619.	1493.	1196.	975.	0
TILT = LAT+15:	1136.	1278.	1355.	1403.	1403.	1436.	1502.	1548.	1514.	1492.	1250.	1035.	0
TILT = 90:	998.	1017.	917.	755.	620.	568.	612.	756.	942.	1144.	1078.	929.	0
KT:	.46	.47	.48	.50	.52	.55	.58	.57	.54	.53	.49	.44	0
AMB TEMP:	27.1	32.3	40.7	54.2	64.1	73.0	77.5	76.5	68.0	57.6	42.3	31.3	53.7
HTG DEG DAYS:	1175	916	753	336	127	15	0	0	50	259	681	1045	5357

STA NO: 4725 — BINGHAMTON, NY — LAT = 42.22

	Jan	Feb	Mar	Apr	May	Jun	Jul	Aug	Sep	Oct	Nov	Dec	Annual
HORIZ INSOL:	386.	576.	861.	1242.	1496.	1681.	1659.	1425.	1131.	779.	414.	297.	996.
TILT = LAT:	591.	768.	986.	1221.	1309.	1396.	1411.	1339.	1249.	1044.	604.	445.	0
TILT = LAT+15:	606.	760.	932.	1091.	1120.	1168.	1191.	1174.	1160.	1031.	613.	456.	0
TILT = 90:	524.	605.	653.	639.	568.	548.	573.	642.	763.	805.	517.	394.	0
KT:	.31	.33	.36	.41	.42	.45	.46	.44	.43	.41	.31	.27	0
AMB TEMP:	22.0	22.8	31.3	44.7	55.1	64.8	69.1	67.3	60.2	50.3	38.2	25.4	46.0
HTG DEG DAYS:	1333	1182	1045	609	320	75	21	40	172	456	804	1228	7285

STA NO: 12832 — APALACHICOLA, FL — LAT = 29.73

	Jan	Feb	Mar	Apr	May	Jun	Jul	Aug	Sep	Oct	Nov	Dec	Annual
HORIZ INSOL:	853.	1126.	1474.	1873.	2091.	1998.	1814.	1688.	1535.	1371.	1040.	818.	1474.
TILT = LAT:	1150.	1387.	1624.	1835.	1858.	1709.	1586.	1585.	1611.	1660.	1409.	1145.	0
TILT = LAT+15:	1205.	1403.	1552.	1659.	1595.	1439.	1355.	1407.	1514.	1663.	1476.	1215.	0
TILT = 90:	957.	986.	886.	668.	475.	403.	420.	530.	765.	1112.	1156.	1000.	0
KT:	.44	.48	.52	.57	.58	.54	.50	.50	.51	.55	.52	.45	0
AMB TEMP:	53.7	55.8	60.7	68.3	74.9	80.0	81.4	81.5	78.6	70.8	61.1	55.2	68.5
HTG DEG DAYS:	368	290	175	30	0	0	0	0	0	22	158	318	1361

STA NO:12834
DAYTONA BEACH
FL
LAT = 29.18

	JAN	FEB	MAR	APR	MAY	JUN	JUL	AUG	SEP	OCT	NOV	DEC	YEAR
HORIZ INSOL:	958.	1213.	1548.	1884.	1968.	1826.	1784.	1682.	1478.	1251.	1036.	870.	1458.
TILT = LAT:	1308.	1500.	1706.	1839.	1755.	1573.	1564.	1580.	1542.	1485.	1385.	1219.	
TILT = LAT+15:	1380.	1522.	1643.	1663.	1511.	1332.	1338.	1403.	1449.	1482.	1449.	1298.	
TILT = 90:	1101.	1067.	921.	657.	457.	392.	411.	520.	726.	976.	1126.	1068.	
KT:	.49	.51	.54	.57	.55	.50	.49	.48	.49	.50	.51	.48	.48
AMB TEMP:	58.4	59.6	63.9	69.7	75.0	79.4	81.0	81.1	79.5	73.3	65.1	59.6	70.5
HTG DEG DAYS:	241	210	120	17	0	0	0	0	0	5	97	212	902

STA NO:12839
MIAMI
FL
LAT = 25.80

	JAN	FEB	MAR	APR	MAY	JUN	JUL	AUG	SEP	OCT	NOV	DEC	YEAR
HORIZ INSOL:	1057.	1314.	1603.	1859.	1844.	1708.	1763.	1630.	1456.	1303.	1119.	1019.	1473.
TILT = LAT:	1369.	1568.	1730.	1807.	1660.	1495.	1563.	1534.	1498.	1495.	1421.	1361.	
TILT = LAT+15:	1445.	1594.	1670.	1639.	1438.	1275.	1342.	1369.	1410.	1493.	1487.	1455.	
TILT = 90:	1111.	1063.	869.	581.	399.	354.	369.	461.	653.	930.	1109.	1163.	
KT:	.50	.52	.54	.55	.51	.47	.49	.48	.47	.49	.51	.51	.50
AMB TEMP:	67.2	67.8	71.3	75.0	78.0	81.0	82.3	82.9	81.7	77.8	72.2	68.3	75.5
HTG DEG DAYS:	53	67	17	0	0	0	0	0	0	0	13	56	205

STA NO:12841
ORLANDO
FL
LAT = 28.55

	JAN	FEB	MAR	APR	MAY	JUN	JUL	AUG	SEP	OCT	NOV	DEC	YEAR
HORIZ INSOL:	999.	1244.	1582.	1899.	1989.	1831.	1801.	1673.	1497.	1304.	1096.	926.	1487.
TILT = LAT:	1356.	1529.	1739.	1851.	1775.	1581.	1581.	1572.	1559.	1544.	1463.	1294.	
TILT = LAT+15:	1432.	1553.	1676.	1675.	1529.	1339.	1353.	1397.	1465.	1543.	1534.	1382.	
TILT = 90:	1138.	1079.	927.	647.	448.	384.	404.	509.	722.	1008.	1188.	1135.	
KT:	.50	.52	.55	.57	.55	.50	.49	.49	.49	.52	.53	.50	.51
AMB TEMP:	60.3	61.5	65.9	71.3	76.4	80.2	81.4	81.8	80.1	74.3	66.6	61.5	71.8
HTG DEG DAYS:	197	184	94	13	0	0	0	0	0	0	75	170	733

STA NO:12842
TAMPA
FL
LAT = 27.97

	JAN	FEB	MAR	APR	MAY	JUN	JUL	AUG	SEP	OCT	NOV	DEC	YEAR
HORIZ INSOL:	1011.	1259.	1594.	1908.	1998.	1847.	1753.	1653.	1492.	1346.	1108.	935.	1492.
TILT = LAT:	1357.	1538.	1745.	1860.	1785.	1597.	1543.	1553.	1550.	1590.	1464.	1291.	
TILT = LAT+15:	1433.	1563.	1682.	1684.	1538.	1353.	1323.	1382.	1457.	1590.	1535.	1378.	
TILT = 90:	1131.	1077.	918.	637.	438.	377.	394.	496.	709.	1030.	1180.	1124.	
KT:	.50	.52	.55	.57	.56	.50	.49	.49	.49	.53	.53	.49	.49
AMB TEMP:	60.4	61.8	66.0	72.0	77.2	81.0	81.9	82.2	80.8	74.7	66.8	61.6	72.2
HTG DEG DAYS:	203	176	90	9	0	0	0	0	0	0	71	169	718

STA NO:12844
WEST PALM BEACH
FL
LAT = 26.68

	JAN	FEB	MAR	APR	MAY	JUN	JUL	AUG	SEP	OCT	NOV	DEC	YEAR
HORIZ INSOL:	1000.	1233.	1556.	1814.	1845.	1706.	1779.	1663.	1419.	1224.	1060.	958.	1438.
TILT = LAT:	1304.	1475.	1685.	1764.	1657.	1488.	1571.	1564.	1462.	1407.	1356.	1290.	
TILT = LAT+15:	1374.	1496.	1625.	1594.	1435.	1269.	1348.	1394.	1375.	1402.	1416.	1376.	
TILT = 90:	1063.	1008.	863.	590.	411.	362.	379.	479.	653.	864.	1064.	1105.	
KT:	.48	.50	.53	.54	.51	.47	.49	.49	.46	.47	.49	.49	.49
AMB TEMP:	65.5	66.1	69.8	73.9	77.5	80.5	81.9	82.3	81.5	77.0	71.0	66.8	74.5
HTG DEG DAYS:	83	91	25	0	0	0	0	0	0	0	22	78	299

STA NO:12907 LAREDO TX LAT = 27.53

	Jan	Feb	Mar	Apr	May	Jun	Jul	Aug	Sep	Oct	Nov	Dec	Year
HORIZ INSOL:	959.	1196.	1516.	1727.	1952.	2073.	2131.	2009.	1705.	1406.	1041.	890.	1555.
TILT = LAT:	1263.	1440.	1647.	1679.	1747.	1782.	1862.	1888.	1782.	1663.	1348.	1203.	0
TILT = LAT+15:	1328.	1459.	1586.	1522.	1500.	1500.	1580.	1673.	1679.	1667.	1408.	1278.	0
TILT = 90:	1036.	995.	859.	586.	429.	371.	394.	541.	796.	1074.	1069.	1030.	0
KT:	.47	.49	.52	.52	.54	.57	.59	.59	.56	.55	.49	.46	
AMB TEMP:	56.5	60.9	67.6	76.3	81.3	86.0	87.9	87.7	82.9	75.5	65.2	58.6	73.9
HTG DEG DAYS:	299	177	87	0	0	0	0	0	0	8	74	231	876

STA NO:12916 NEW ORLEANS LA LAT = 29.98

	Jan	Feb	Mar	Apr	May	Jun	Jul	Aug	Sep	Oct	Nov	Dec	Year
HORIZ INSOL:	835.	1112.	1415.	1780.	1968.	2004.	1814.	1717.	1514.	1335.	973.	779.	1437.
TILT = LAT:	1126.	1372.	1555.	1738.	1751.	1712.	1585.	1612.	1588.	1615.	1307.	1086.	0
TILT = LAT+15:	1179.	1388.	1494.	1571.	1507.	1441.	1354.	1430.	1492.	1617.	1364.	1150.	0
TILT = 90:	937.	978.	853.	647.	471.	406.	424.	540.	760.	1084.	1065.	944.	0
KT:	.44	.48	.50	.54	.55	.54	.50	.51	.51	.54	.49	.44	
AMB TEMP:	52.9	55.6	60.7	68.6	75.1	80.4	81.9	81.9	78.2	69.8	60.1	54.8	63.3
HTG DEG DAYS:	403	299	188	29	0	0	0	0	0	40	179	327	1465

STA NO:12917 PORT ARTHUR TX LAT = 29.95

	Jan	Feb	Mar	Apr	May	Jun	Jul	Aug	Sep	Oct	Nov	Dec	Year
HORIZ INSOL:	800.	1071.	1353.	1610.	1871.	2011.	1846.	1736.	1527.	1321.	953.	754.	1404.
TILT = LAT:	1067.	1312.	1480.	1566.	1668.	1718.	1613.	1630.	1603.	1596.	1274.	1042.	0
TILT = LAT+15:	1114.	1324.	1421.	1417.	1438.	1446.	1376.	1446.	1506.	1596.	1328.	1100.	0
TILT = 90:	881.	930.	811.	598.	463.	406.	425.	543.	766.	1069.	1035.	899.	0
KT:	.42	.46	.48	.49	.52	.54	.51	.51	.51	.53	.48	.42	
AMB TEMP:	52.0	55.1	60.1	68.9	75.0	80.8	83.0	83.1	78.9	69.9	60.2	54.2	68.5
HTG DEG DAYS:	420	302	202	33	0	0	0	0	0	35	184	342	1518

STA NO:12919 BROWNSVILLE TX LAT = 25.90

	Jan	Feb	Mar	Apr	May	Jun	Jul	Aug	Sep	Oct	Nov	Dec	Year
HORIZ INSOL:	913.	1135.	1458.	1737.	1927.	2115.	2213.	2027.	1694.	1439.	1055.	862.	1548.
TILT = LAT:	1153.	1343.	1563.	1687.	1732.	1827.	1938.	1905.	1757.	1672.	1329.	1118.	0
TILT = LAT+15:	1206.	1343.	1505.	1531.	1497.	1537.	1643.	1691.	1657.	1676.	1386.	1182.	0
TILT = 90:	913.	888.	788.	559.	403.	348.	364.	509.	753.	1050.	1029.	927.	0
KT:	.43	.45	.49	.52	.54	.58	.61	.59	.55	.55	.48	.43	
AMB TEMP:	60.3	63.4	67.7	74.9	79.3	82.8	84.4	84.4	81.6	75.7	68.1	62.8	73.8
HTG DEG DAYS:	225	151	89	0	0	0	0	0	0	5	35	145	650

STA NO:12921 SAN ANTONIO TX LAT = 29.53

	Jan	Feb	Mar	Apr	May	Jun	Jul	Aug	Sep	Oct	Nov	Dec	Year
HORIZ INSOL:	895.	1154.	1450.	1612.	1895.	2069.	2121.	1947.	1638.	1350.	1009.	847.	1499.
TILT = LAT:	1214.	1423.	1592.	1568.	1690.	1767.	1843.	1829.	1725.	1626.	1352.	1190.	0
TILT = LAT+15:	1276.	1441.	1531.	1420.	1457.	1485.	1562.	1619.	1622.	1628.	1414.	1265.	0
TILT = 90:	1016.	1012.	865.	592.	458.	400.	428.	573.	812.	1084.	1101.	1043.	0
KT:	.46	.49	.51	.49	.53	.56	.58	.57	.54	.54	.50	.47	
AMB TEMP:	50.7	54.5	60.8	69.6	76.0	82.2	84.7	84.7	79.3	70.5	59.7	53.2	68.8
HTG DEG DAYS:	451	310	194	31	0	0	0	0	0	32	179	373	1570

STA NO:12924 CORPUS CHRISTI TX — LAT = 27.77

	JAN	FEB	MAR	APR	MAY	JUN	JUL	AUG	SEP	OCT	NOV	DEC	ANN
HORIZ INSOL:	898.	1147.	1430.	1642.	1866.	2094.	2186.	1991.	1687.	1416.	1043.	845.	1521.
TILT = LAT:	1173.	1379.	1549.	1596.	1672.	1798.	1906.	1870.	1764.	1679.	1357.	1136.	0
TILT = LAT+15:	1229.	1394.	1489.	1446.	1445.	1511.	1615.	1657.	1661.	1683.	1418.	1203.	0
TILT = 90:	954.	951.	812.	570.	429.	374.	397.	543.	661.	1089.	1080.	967.	0
KT:	.44	.47	.49	.49	.52	.57	.60	.58	.55	.55	.49	.44	.44
AMB TEMP:	56.3	59.6	64.9	72.8	77.9	82.4	84.8	85.1	81.0	73.9	64.9	59.1	71.9
HTG DEG DAYS:	304	199	120	0	0	0	0	0	0	7	81	219	930

STA NO:12928 KINGSVILLE TX — LAT = 27.52

	JAN	FEB	MAR	APR	MAY	JUN	JUL	AUG	SEP	OCT	NOV	DEC	ANN
HORIZ INSOL:	912.	1161.	1435.	1663.	1864.	2036.	2111.	1922.	1625.	1390.	1034.	849.	1500.
TILT = LAT:	1189.	1393.	1552.	1615.	1671.	1752.	1845.	1805.	1693.	1639.	1338.	1137.	0
TILT = LAT+15:	1247.	1409.	1493.	1465.	1444.	1476.	1567.	1602.	1594.	1641.	1397.	1204.	0
TILT = 90:	966.	958.	810.	571.	425.	371.	394.	529.	760.	1057.	1059.	965.	0
KT:	.45	.47	.49	.50	.52	.56	.58	.56	.53	.54	.49	.44	0
AMB TEMP:													0
HTG DEG DAYS	0	0	0	0	0	0							0

STA NO:12960 HOUSTON TX — LAT = 29.98

	JAN	FEB	MAR	APR	MAY	JUN	JUL	AUG	SEP	OCT	NOV	DEC	ANN
HORIZ INSOL:	772.	1034.	1297.	1527.	1775.	1898.	1826.	1506.	1471.	1276.	924.	730.	1351.
TILT = LAT:	1023.	1261.	1414.	1479.	1585.	1627.	1597.	1583.	1541.	1533.	1278.	1000.	0
TILT = LAT+15:	1066.	1270.	1356.	1339.	1369.	1373.	1363.	1406.	1446.	1532.	1278.	1054.	0
TILT = 90:	839.	890.	774.	573.	454.	404.	424.	534.	738.	1024.	993.	858.	0
KT:	.40	.44	.46	.46	.49	.51	.50	.50	.49	.52	.46	.41	.41
AMB TEMP:	52.1	55.3	60.8	69.4	75.8	81.1	83.3	83.4	79.2	70.9	61.1	54.6	68.9
HTG DEG DAYS:	416	294	189	23	0	0	0	0	0	24	155	333	1434

STA NO:13721 PATUXENT RIVER MD — LAT = 38.28

	JAN	FEB	MAR	APR	MAY	JUN	JUL	AUG	SEP	OCT	NOV	DEC	ANN
HORIZ INSOL:	608.	862.	1181.	1538.	1763.	1893.	1817.	1627.	1357.	1021.	707.	537.	1243.
TILT = LAT:	955.	1178.	1371.	1520.	1551.	1582.	1555.	1531.	1487.	1347.	1093.	878.	0
TILT = LAT+15:	998.	1187.	1309.	1364.	1327.	1323.	1315.	1347.	1388.	1340.	1138.	926.	0
TILT = 90:	860.	928.	871.	719.	579.	526.	547.	656.	848.	1008.	965.	817.	0
KT:	.42	.45	.47	.49	.49	.51	.50	.49	.49	.49	.45	.40	0
AMB TEMP:													0
HTG DEG DAYS:													0

STA NO:13722 RALEIGH-DURHAM NC — LAT = 35.87

	JAN	FEB	MAR	APR	MAY	JUN	JUL	AUG	SEP	OCT	NOV	DEC	ANN
HORIZ INSOL:	694.	943.	1276.	1644.	1808.	1864.	1776.	1611.	1377.	1105.	812.	636.	1295.
TILT = LAT:	1046.	1250.	1459.	1620.	1596.	1570.	1530.	1513.	1484.	1418.	1212.	1002.	0
TILT = LAT+15:	1096.	1261.	1397.	1457.	1369.	1318.	1298.	1335.	1387.	1413.	1265.	1062.	0
TILT = 90:	926.	959.	891.	717.	549.	486.	504.	610.	807.	1029.	1053.	923.	0
KT:	.43	.46	.47	.51	.50	.50	.49	.48	.50	.50	.48	.43	0
AMB TEMP:	40.5	42.2	49.2	59.5	67.4	74.4	77.5	76.5	70.6	60.2	50.0	41.2	59.1
HTG DEG DAYS:	760	638	502	180	48	0	0	0	12	186	450	738	3514

STA NO:13723 GREENSBORO NC LAT = 35.08

	Jan	Feb	Mar	Apr	May	Jun	Jul	Aug	Sep	Oct	Nov	Dec	Year
HORIZ INSOL:	715.	970.	1313.	1683.	1868.	1953.	1864.	1697.	1418.	1141.	839.	659.	1343.
TILT = LAT:	1096.	1300.	1511.	1661.	1648.	1640.	1602.	1595.	1535.	1478.	1272.	1058.	
TILT = LAT+15:	1151.	1313.	1448.	1494.	1411.	1373.	1357.	1406.	1435.	1476.	1330.	1125.	
TILT = 90:	979.	1004.	928.	737.	563.	499.	519.	639.	838.	1081.	1114.	985.	
KT:	.45	.47	.50	.53	.52	.52	.51	.51	.50	.52	.50	.45	
AMB TEMP:	38.7	40.6	47.8	58.6	67.1	74.4	77.7	76.0	69.7	59.2	48.3	39.6	58.1
HTG DEG DAYS:	815	683	544	203	59	0	0	0	24	209	501	787	3825

STA NO:13737 NORFOLK VA LAT = 36.90

	Jan	Feb	Mar	Apr	May	Jun	Jul	Aug	Sep	Oct	Nov	Dec	Year
HORIZ INSOL:	678.	932.	1281.	1677.	1888.	2000.	1853.	1680.	1396.	1083.	811.	624.	1325.
TILT = LAT:	1051.	1259.	1482.	1659.	1663.	1674.	1590.	1581.	1517.	1409.	1249.	1016.	
TILT = LAT+15:	1102.	1271.	1419.	1490.	1422.	1399.	1346.	1392.	1418.	1404.	1306.	1079.	
TILT = 90:	942.	980.	923.	751.	581.	517.	531.	649.	842.	1038.	1102.	949.	
KT:	.44	.46	.49	.53	.53	.53	.51	.51	.50	.50	.49	.44	
AMB TEMP:	40.5	41.4	48.1	57.8	66.7	74.5	78.3	76.9	71.8	61.7	51.6	42.3	59.3
HTG DEG DAYS:	760	661	532	226	53	0	0	0	9	141	402	704	3488

STA NO:13739 PHILADELPHIA PA LAT = 39.88

	Jan	Feb	Mar	Apr	May	Jun	Jul	Aug	Sep	Oct	Nov	Dec	Year
HORIZ INSOL:	555.	795.	1108.	1434.	1660.	1811.	1758.	1575.	1281.	959.	619.	470.	1169.
TILT = LAT:	898.	1104.	1297.	1417.	1458.	1510.	1501.	1503.	1412.	1208.	970.	783.	
TILT = LAT+15:	938.	1110.	1236.	1270.	1247.	1263.	1268.	1302.	1316.	1280.	1005.	824.	
TILT = 90:	817.	882.	843.	699.	580.	539.	560.	664.	829.	981.	860.	732.	
KT:	.40	.43	.45	.46	.46	.48	.48	.48	.48	.47	.42	.38	
AMB TEMP:	32.3	33.9	41.9	52.9	63.2	72.3	76.8	74.8	68.1	57.4	46.2	35.2	54.6
HTG DEG DAYS:	1014	871	716	367	122	0	0	0	38	249	564	924	4865

STA NO:13740 RICHMOND VA LAT = 37.50

	Jan	Feb	Mar	Apr	May	Jun	Jul	Aug	Sep	Oct	Nov	Dec	Year
HORIZ INSOL:	632.	877.	1210.	1565.	1762.	1872.	1774.	1601.	1348.	1033.	733.	567.	1248.
TILT = LAT:	977.	1183.	1398.	1546.	1552.	1569.	1523.	1504.	1466.	1345.	1118.	915.	
TILT = LAT+15:	1022.	1192.	1336.	1388.	1329.	1315.	1290.	1325.	1369.	1338.	1163.	967.	
TILT = 90:	874.	922.	877.	716.	566.	512.	528.	634.	823.	995.	980.	849.	
KT:	.42	.44	.47	.49	.49	.50	.49	.49	.49	.48	.46	.41	
AMB TEMP:	37.5	39.4	46.9	57.8	66.5	74.2	77.9	76.3	70.0	59.3	49.0	39.0	57.8
HTG DEG DAYS:	853	717	569	226	64	0	0	0	21	203	480	806	3939

STA NO:13741 ROANOKE VA LAT = 37.32

	Jan	Feb	Mar	Apr	May	Jun	Jul	Aug	Sep	Oct	Nov	Dec	Year
HORIZ INSOL:	661.	899.	1236.	1581.	1764.	1882.	1796.	1620.	1358.	1080.	765.	591.	1270.
TILT = LAT:	1029.	1216.	1429.	1561.	1554.	1578.	1542.	1523.	1477.	1416.	1173.	960.	
TILT = LAT+15:	1079.	1226.	1367.	1402.	1331.	1322.	1305.	1341.	1379.	1411.	1224.	1017.	
TILT = 90:	925.	948.	895.	719.	564.	510.	528.	637.	826.	1050.	1032.	895.	
KT:	.44	.45	.48	.50	.49	.50	.49	.49	.49	.50	.47	.43	
AMB TEMP:	36.4	38.1	45.3	55.9	64.4	71.7	75.2	74.1	68.0	57.8	46.7	37.4	55.9
HTG DEG DAYS:	887	753	611	283	101	0	0	0	32	235	549	856	4307

STA NO:13754
CHERRY POINT
NC
LAT = 34.90

	JAN	FEB	MAR	APR	MAY	JUN	JUL	AUG	SEP	OCT	NOV	DEC	YR
HORIZ INSOL:	757.	1025.	1387.	1794.	1925.	1939.	1830.	1634.	1427.	1170.	907.	718.	1376.
TILT = LAT:	1135.	1357.	1598.	1770.	1700.	1634.	1578.	1534.	1534.	1491.	1354.	1135.	
TILT = LAT+15:	1193.	1337.	1524.	1593.	1456.	1371.	1340.	1355.	1435.	1489.	1419.	1209.	
TILT = 90:	1004.	1037.	957.	754.	552.	478.	497.	601.	817.	1072.	1178.	1051.	
KT:	.46	.49	.52	.56	.54	.52	.50	.49	.50	.52	.52	.47	.50
AMB TEMP:													
HTG DEG DAYS:													

STA NO:13781
WILMINGTON
DE
LAT = 39.67

	JAN	FEB	MAR	APR	MAY	JUN	JUL	AUG	SEP	OCT	NOV	DEC	YR
HORIZ INSOL:	571.	827.	1149.	1480.	1710.	1883.	1823.	1615.	1318.	984.	645.	489.	1206.
TILT = LAT:	925.	1155.	1349.	1465.	1502.	1568.	1556.	1521.	1455.	1324.	1016.	818.	
TILT = LAT+15:	968.	1164.	1238.	1313.	1284.	1309.	1313.	1337.	1357.	1317.	1054.	862.	
TILT = 90:	843.	925.	876.	717.	589.	546.	569.	675.	851.	1008.	903.	767.	
KT:	.41	.44	.46	.47	.48	.50	.50	.49	.49	.43	.43	.39	.47
AMB TEMP:	32.0	33.6	41.6	52.3	62.4	71.4	75.8	74.1	67.9	57.2	45.7	34.7	54.0
HTG DEG DAYS:	1023	879	725	381	128	10	0	0	32	254	579	939	4940

STA NO:13865
MERIDIAN
MS
LAT = 32.33

	JAN	FEB	MAR	APR	MAY	JUN	JUL	AUG	SEP	OCT	NOV	DEC	YR
HORIZ INSOL:	744.	1012.	1328.	1662.	1860.	1963.	1823.	1739.	1454.	1258.	897.	699.	1370.
TILT = LAT:	1033.	1276.	1477.	1625.	1650.	1667.	1583.	1633.	1540.	1559.	1249.	1011.	
TILT = LAT+15:	1079.	1286.	1416.	1466.	1420.	1401.	1348.	1445.	1444.	1558.	1302.	1068.	
TILT = 90:	875.	933.	846.	656.	500.	440.	458.	584.	777.	1083.	1043.	895.	
KT:	.42	.45	.48	.51	.52	.53	.50	.51	.50	.53	.48	.42	.50
AMB TEMP:	46.9	49.8	56.1	65.4	72.4	79.2	81.2	80.7	75.3	64.8	54.2	47.9	64.5
HTG DEG DAYS:	575	443	312	79	7	0	0	0	0	111	331	530	2388

STA NO:13866
CHARLESTON
WV
LAT = 38.37

	JAN	FEB	MAR	APR	MAY	JUN	JUL	AUG	SEP	OCT	NOV	DEC	YR
HORIZ INSOL:	498.	707.	1010.	1356.	1639.	1776.	1682.	1514.	1272.	972.	613.	440.	1123.
TILT = LAT:	730.	917.	1142.	1329.	1413.	1488.	1443.	1422.	1303.	1271.	909.	668.	
TILT = LAT+15:	753.	913.	1085.	1192.	1238.	1248.	1224.	1252.	1290.	1262.	938.	695.	
TILT = 90:	634.	702.	719.	638.	553.	511.	524.	619.	789.	947.	785.	598.	
KT:	.34	.37	.40	.43	.46	.47	.46	.46	.46	.46	.39	.33	.43
AMB TEMP:	34.5	36.5	44.5	55.9	64.5	72.0	75.0	73.6	67.5	57.0	45.4	36.2	55.2
HTG DEG DAYS:	946	798	642	287	113	10	0	0	46	267	588	893	4590

STA NO:13874
ATLANTA
GA
LAT = 33.65

	JAN	FEB	MAR	APR	MAY	JUN	JUL	AUG	SEP	OCT	NOV	DEC	YR
HORIZ INSOL:	718.	969.	1304.	1686.	1854.	1914.	1812.	1709.	1422.	1200.	883.	674.	1345.
TILT = LAT:	1022.	1238.	1464.	1654.	1642.	1621.	1569.	1605.	1515.	1506.	1266.	1004.	
TILT = LAT+15:	1068.	1248.	1402.	1491.	1410.	1362.	1334.	1418.	1419.	1503.	1322.	1061.	
TILT = 90:	879.	920.	859.	689.	520.	457.	476.	600.	787.	1064.	1077.	902.	
KT:	.42	.45	.48	.52	.52	.51	.50	.51	.49	.52	.49	.43	.50
AMB TEMP:	42.4	45.0	51.1	61.1	69.1	75.6	78.0	77.5	72.3	62.4	51.4	43.5	60.8
HTG DEG DAYS:	701	560	443	144	27	0	0	0	8	137	408	667	3095

STA NO:13876 BIRMINGHAM AL LAT = 33.57

HORIZ INSOL:	707.	967.	1296.	1674.	1857.	1919.	1810.	1724.	1455.	1211.	858.	661.	1345.
TILT = LAT:	1000.	1234.	1453.	1641.	1645.	1625.	1567.	1619.	1552.	1520.	1219.	976.	0
TILT = LAT+15:	1043.	1243.	1392.	1479.	1413.	1366.	1333.	1431.	1454.	1573.	1271.	1031.	0
TILT = 90:	856.	915.	851.	683.	519.	456.	474.	602.	804.	1073.	1271.	873.	0
KT:	.41	.45	.48	.51	.52	.52	.50	.51	.50	.52	.47	.42	
AMB TEMP:	44.2	46.9	53.3	63.2	70.5	77.4	79.9	79.2	73.9	63.3	52.1	45.2	62.4
HTG DEG DAYS:	654	517	389	116	20	0	0	0	6	137	391	614	2844

STA NO:13880 CHARLESTON SC LAT = 32.90

HORIZ INSOL:	744.	995.	1339.	1732.	1860.	1844.	1799.	1585.	1394.	1193.	934.	721.	1345.
TILT = LAT:	1049.	1262.	1498.	1695.	1649.	1568.	1561.	1487.	1476.	1478.	1332.	1068.	0
TILT = LAT+15:	1097.	1272.	1436.	1532.	1418.	1322.	1329.	1317.	1382.	1475.	1393.	1133.	0
TILT = 90:	897.	930.	868.	690.	509.	442.	464.	557.	755.	1030.	1129.	960.	0
KT:	.42	.45	.49	.53	.52	.50	.49	.47	.48	.51	.50	.44	
AMB TEMP:	48.6	50.5	56.5	64.6	72.1	77.9	80.2	79.6	75.2	66.1	56.3	49.3	64.7
HTG DEG DAYS:	521	419	300	69	5	0	0	0	0	74	271	487	2146

STA NO:13881 CHARLOTTE NC LAT = 35.22

HORIZ INSOL:	719.	971.	1318.	1695.	1856.	1921.	1831.	1695.	1416.	1173.	866.	672.	1344.
TILT = LAT:	1073.	1279.	1503.	1670.	1639.	1619.	1578.	1593.	1523.	1505.	1289.	1054.	0
TILT = LAT+15:	1125.	1291.	1441.	1503.	1405.	1358.	1339.	1405.	1425.	1503.	1348.	1119.	0
TILT = 90:	946.	975.	909.	724.	546.	482.	502.	624.	817.	1088.	1119.	971.	0
KT:	.44	.46	.50	.53	.52	.51	.50	.51	.50	.52	.50	.45	
AMB TEMP:	42.1	44.0	50.6	60.8	68.8	75.9	78.5	77.7	72.0	61.7	51.0	42.5	60.5
HTG DEG DAYS:	710	588	461	145	34	0	0	0	10	152	420	698	3218

STA NO:13882 CHATTANOOGA TN LAT = 35.03

HORIZ INSOL:	631.	859.	1176.	1550.	1732.	1831.	1735.	1630.	1336.	1108.	773.	580.	1245.
TILT = LAT:	900.	1096.	1319.	1519.	1532.	1548.	1499.	1530.	1426.	1402.	1112.	863.	0
TILT = LAT+15:	935.	1099.	1260.	1368.	1317.	1302.	1275.	1351.	1333.	1396.	1155.	906.	0
TILT = 90:	772.	819.	791.	665.	523.	471.	487.	602.	763.	1005.	946.	771.	0
KT:	.38	.41	.44	.48	.48	.49	.48	.49	.47	.49	.44	.38	
AMB TEMP:	40.2	42.9	49.8	60.5	68.5	76.0	78.8	78.0	71.9	60.8	48.9	41.2	59.8
HTG DEG DAYS:	769	625	483	165	51	0	0	0	9	182	483	738	3505

STA NO:13883 COLUMBIA SC LAT = 33.95

HORIZ INSOL:	762.	1021.	1355.	1747.	1895.	1947.	1842.	1703.	1439.	1211.	921.	722.	1380.
TILT = LAT:	1113.	1325.	1533.	1717.	1676.	1646.	1592.	1599.	1538.	1530.	1346.	1107.	0
TILT = LAT+15:	1167.	1339.	1470.	1547.	1438.	1381.	1352.	1413.	1440.	1529.	1409.	1177.	0
TILT = 90:	971.	996.	906.	717.	531.	464.	484.	604.	804.	1087.	1157.	1012.	0
KT:	.45	.47	.50	.54	.53	.52	.50	.51	.50	.53	.51	.46	
AMB TEMP:	45.4	47.6	54.2	64.1	72.1	78.8	81.2	80.2	74.5	64.2	53.8	46.0	63.5
HTG DEG DAYS:	608	493	360	83	12	0	0	0	0	112	341	589	2598

STA NO:13883 JACKSONVILLE FL LAT = 30.50	Jan	Feb	Mar	Apr	May	Jun	Jul	Aug	Sep	Oct	Nov	Dec	Ann
HORIZ INSOL:	900.	1164.	1522.	1856.	1956.	1885.	1802.	1694.	1442.	1223.	996.	818.	1438.
TILT = LAT:	1250.	1460.	1692.	1815.	1740.	1614.	1573.	1591.	1512.	1471.	1360.	1168.	0
TILT = LAT+15:	1316.	1480.	1629.	1640.	1496.	1362.	1343.	1411.	1419.	1467.	1423.	1242.	0
TILT = 90:	1063.	1055.	938.	678.	479.	411.	430.	544.	734.	987.	1122.	1033.	0
KT:	.48	.50	.54	.56	.54	.51	.50	.48	.50	.50	.47	.47	
AMB TEMP:	54.6	56.3	61.2	68.1	74.3	79.2	81.0	81.0	78.2	70.5	61.2	55.4	62.4
HTG DEG DAYS:	348	282	176	24	0	0	0	0	0	19	161	317	1327

STA NO:13891 KNOXVILLE TN LAT = 35.82	Jan	Feb	Mar	Apr	May	Jun	Jul	Aug	Sep	Oct	Nov	Dec	Ann
HORIZ INSOL:	621.	863.	1191.	1599.	1803.	1902.	1804.	1666.	1383.	1121.	759.	569.	1273.
TILT = LAT:	903.	1120.	1348.	1573.	1592.	1600.	1554.	1566.	1491.	1440.	1110.	864.	0
TILT = LAT+15:	939.	1125.	1288.	1415.	1365.	1342.	1318.	1381.	1394.	1436.	1154.	909.	0
TILT = 90:	783.	848.	820.	698.	547.	489.	507.	626.	810.	1047.	953.	780.	0
KT:	.39	.42	.45	.50	.51	.51	.49	.50	.49	.51	.45	.39	
AMB TEMP:	40.6	42.8	49.9	60.3	68.4	75.5	78.2	77.3	71.6	60.9	49.2	41.5	59.7
HTG DEG DAYS:	756	630	484	173	47	0	0	0	10	175	474	729	3478

STA NO:13893 MEMPHIS TN LAT = 35.05	Jan	Feb	Mar	Apr	May	Jun	Jul	Aug	Sep	Oct	Nov	Dec	Ann
HORIZ INSOL:	683.	945.	1278.	1633.	1885.	2045.	1972.	1824.	1471.	1205.	817.	629.	1366.
TILT = LAT:	999.	1233.	1450.	1611.	1665.	1718.	1696.	1717.	1587.	1548.	1193.	959.	0
TILT = LAT+15:	1043.	1242.	1386.	1450.	1427.	1437.	1434.	1513.	1486.	1547.	1243.	1014.	0
TILT = 90:	870.	934.	873.	699.	548.	489.	516.	656.	848.	1120.	1024.	871.	0
KT:	.42	.45	.48	.51	.52	.55	.54	.54	.52	.54	.47	.42	
AMB TEMP:	40.5	43.8	51.0	62.5	70.9	78.6	81.6	80.4	73.6	63.0	50.9	42.7	51.6
HTG DEG DAYS:	760	594	457	131	22	0	0	0	7	142	423	691	3227

STA NO:13894 MOBILE AL LAT = 30.68	Jan	Feb	Mar	Apr	May	Jun	Jul	Aug	Sep	Oct	Nov	Dec	Ann
HORIZ INSOL:	828.	1100.	1408.	1722.	1972.	1869.	1715.	1642.	1449.	1290.	955.	759.	1385.
TILT = LAT:	1134.	1369.	1555.	1681.	1666.	1599.	1500.	1541.	1522.	1580.	1298.	1070.	0
TILT = LAT+15:	1188.	1385.	1493.	1519.	1436.	1350.	1283.	1367.	1427.	1580.	1355.	1133.	0
TILT = 90:	953.	986.	864.	644.	474.	413.	426.	536.	741.	1070.	1067.	937.	0
KT:	.44	.48	.50	.52	.52	.50	.47	.48	.49	.53	.49	.43	
AMB TEMP:	51.2	54.0	59.4	67.9	74.8	80.3	81.6	81.5	77.5	68.9	58.5	52.9	67.4
HTG DEG DAYS:	451	337	221	40	0	0	0	0	0	39	211	385	1684

STA NO:13895 MONTGOMERY AL LAT = 32.30	Jan	Feb	Mar	Apr	May	Jun	Jul	Aug	Sep	Oct	Nov	Dec	Ann
HORIZ INSOL:	752.	1013.	1341.	1729.	1897.	1972.	1841.	1746.	1468.	1262.	915.	719.	1388.
TILT = LAT:	1046.	1276.	1492.	1693.	1683.	1674.	1598.	1639.	1556.	1564.	1280.	1047.	0
TILT = LAT+15:	1092.	1287.	1431.	1527.	1447.	1407.	1360.	1450.	1459.	1563.	1336.	1108.	0
TILT = 90:	887.	933.	855.	677.	503.	440.	459.	585.	784.	1086.	1072.	931.	0
KT:	.42	.45	.49	.53	.53	.53	.51	.50	.52	.53	.49	.43	
AMB TEMP:	47.5	50.6	56.5	65.2	72.4	78.9	81.0	80.7	76.0	65.8	55.0	48.5	64.8
HTG DEG DAYS:	556	419	299	76	8	0	0	0	0	93	306	512	2269

STA NO:13897 NASHVILLE TN LAT = 36.12

	Jan	Feb	Mar	Apr	May	Jun	Jul	Aug	Sep	Oct	Nov	Dec	Annual
HORIZ INSOL:	580.	824.	1130.	1544.	1825.	1963.	1891.	1737.	1398.	1114.	711.	521.	1270.
TILT = LAT:	832.	1063.	1273.	1517.	1610.	1648.	1625.	1634.	1511.	1437.	1031.	774.	.
TILT = LAT+15:	862.	1065.	1214.	1364.	1380.	1379.	1375.	1440.	1413.	1433.	1068.	809.	.
TILT = 90:	716.	803.	776.	682.	556.	500.	523.	652.	826.	1048.	880.	689.	.
KT:	.37	.40	.43	.48	.51	.52	.50	.51	.50	.51	.42	.36	.
AMB TEMP:	38.3	41.0	48.7	60.1	68.5	76.6	79.6	78.5	72.0	60.9	48.4	40.4	59.4
HTG DEG DAYS:	828	672	524	176	45	0	0	0	10	180	498	763	3696

STA NO:13923 SHERMAN TX LAT = 33.72

	Jan	Feb	Mar	Apr	May	Jun	Jul	Aug	Sep	Oct	Nov	Dec	Annual
HORIZ INSOL:	794.	1037.	1366.	1610.	1852.	2114.	2077.	1932.	1580.	1268.	919.	744.	1441.
TILT = LAT:	1164.	1346.	1543.	1577.	1640.	1780.	1788.	1818.	1701.	1603.	1333.	1141.	.
TILT = LAT+15:	1224.	1361.	1481.	1421.	1408.	1488.	1511.	1603.	1596.	1610.	1395.	1215.	.
TILT = 90:	1020.	1010.	909.	664.	521.	471.	502.	656.	882.	1144.	1141.	1044.	.
KT:	.46	.48	.50	.50	.51	.57	.57	.57	.55	.55	.51	.47	.
AMB TEMP:	41.7	45.9	52.3	63.7	71.2	79.4	83.6	83.7	76.0	65.8	53.4	44.8	63.5
HTG DEG DAYS:	722	535	411	114	13	0	0	0	0	90	353	626	2864

STA NO:13957 SHREVEPORT LA LAT = 32.47

	Jan	Feb	Mar	Apr	May	Jun	Jul	Aug	Sep	Oct	Nov	Dec	Annual
HORIZ INSOL:	762.	1038.	1342.	1613.	1886.	2065.	2014.	1877.	1554.	1303.	929.	731.	1426.
TILT = LAT:	1069.	1318.	1496.	1576.	1673.	1748.	1741.	1764.	1657.	1628.	1308.	1073.	.
TILT = LAT+15:	1118.	1331.	1434.	1422.	1438.	1464.	1475.	1559.	1555.	1631.	1367.	1137.	.
TILT = 90:	911.	970.	860.	642.	505.	447.	475.	618.	836.	1138.	1100.	959.	.
KT:	.43	.47	.49	.49	.52	.56	.55	.56	.53	.55	.50	.44	.
AMB TEMP:	47.2	50.5	56.8	66.4	73.4	80.2	83.2	83.2	77.4	67.5	56.2	49.2	65.9
HTG DEG DAYS:	552	416	291	65	5	0	0	0	0	70	278	490	2167

STA NO:13958 AUSTIN TX LAT = 30.30

	Jan	Feb	Mar	Apr	May	Jun	Jul	Aug	Sep	Oct	Nov	Dec	Annual
HORIZ INSOL:	865.	1125.	1429.	1605.	1834.	2072.	2106.	1931.	1606.	1333.	987.	825.	1476.
TILT = LAT:	1184.	1397.	1576.	1563.	1634.	1765.	1826.	1814.	1696.	1620.	1339.	1175.	.
TILT = LAT+15:	1243.	1414.	1515.	1413.	1410.	1482.	1547.	1605.	1594.	1621.	1399.	1250.	.
TILT = 90:	997.	1003.	870.	603.	465.	412.	442.	586.	814.	1093.	1100.	1038.	.
KT:	.46	.48	.49	.49	.51	.56	.58	.57	.54	.54	.50	.47	.
AMB TEMP:	49.7	53.3	59.5	68.6	75.2	81.6	84.6	84.7	78.9	70.1	59.1	52.3	68.1
HTG DEG DAYS:	483	344	223	44	5	0	0	0	0	39	205	399	1737

STA NO:13959 WACO TX LAT = 31.62

	Jan	Feb	Mar	Apr	May	Jun	Jul	Aug	Sep	Oct	Nov	Dec	Annual
HORIZ INSOL:	833.	1096.	1428.	1612.	1774.	2112.	2130.	1958.	1601.	1301.	957.	803.	1467.
TILT = LAT:	1168.	1385.	1591.	1573.	1578.	1790.	1841.	1841.	1702.	1605.	1329.	1178.	.
TILT = LAT+15:	1226.	1402.	1529.	1420.	1361.	1499.	1557.	1626.	1599.	1606.	1389.	1254.	.
TILT = 90:	997.	1012.	901.	627.	478.	434.	466.	618.	842.	1105.	1109.	1057.	.
KT:	.46	.48	.51	.49	.49	.57	.58	.58	.54	.54	.50	.47	.
AMB TEMP:	47.0	50.9	57.2	67.3	74.5	81.9	85.6	85.7	78.9	69.1	57.5	49.8	67.1
HTG DEG DAYS:	558	401	280	56	0	0	0	0	0	51	241	471	2058

STA NO:13960 DALLAS TX LAT = 32.85

	Jan	Feb	Mar	Apr	May	Jun	Jul	Aug	Sep	Oct	Nov	Dec	Year
HORIZ INSOL:	822.	1071.	1422.	1627.	1889.	2135.	2122.	1950.	1587.	1276.	936.	780.	1468.
TILT = LAT:	1187.	1376.	1601.	1591.	1674.	1801.	1829.	1834.	1699.	1598.	1334.	1180.	
TILT = LAT+15:	1248.	1393.	1538.	1435.	1438.	1506.	1545.	1619.	1595.	1598.	1396.	1257.	
TILT = 90:	1031.	1023.	929.	654.	512.	456.	489.	642.	864.	1121.	1131.	1073.	
KT:	.47	.49	.52	.50	.52	.57	.58	.58	.54	.54	.51	.48	
AMB TEMP:	45.4	49.4	55.8	66.4	73.8	81.6	85.7	85.0	78.2	68.0	55.9	48.2	66.2
HTG DEG DAYS:	608	437	314	71	0	0	0	0	0	55	284	521	2290

STA NO:13962 ABILENE TX LAT = 32.43

	Jan	Feb	Mar	Apr	May	Jun	Jul	Aug	Sep	Oct	Nov	Dec	Year
HORIZ INSOL:	924.	1183.	1576.	1843.	2037.	2209.	2139.	1956.	1598.	1315.	1008.	863.	1554.
TILT = LAT:	1358.	1537.	1790.	1810.	1804.	1862.	1845.	1840.	1707.	1645.	1445.	1323.	
TILT = LAT+15:	1436.	1562.	1724.	1632.	1546.	1554.	1558.	1624.	1603.	1648.	1517.	1417.	
TILT = 90:	1194.	1149.	1033.	717.	521.	451.	482.	635.	860.	1150.	1230.	1216.	
KT:	.52	.53	.57	.56	.57	.59	.59	.58	.55	.56	.54	.52	
AMB TEMP:	43.7	47.9	54.5	65.2	72.4	80.3	83.9	83.6	76.1	66.1	54.1	46.4	64.5
HTG DEG DAYS:	660	479	354	104	11	0	0	0	0	89	336	577	2610

STA NO:13963 LITTLE ROCK AR LAT = 34.73

	Jan	Feb	Mar	Apr	May	Jun	Jul	Aug	Sep	Oct	Nov	Dec	Year
HORIZ INSOL:	731.	1003.	1313.	1611.	1929.	2107.	2032.	1861.	1518.	1228.	847.	674.	1404.
TILT = LAT:	1081.	1317.	1490.	1581.	1704.	1769.	1747.	1751.	1639.	1576.	1238.	1039.	
TILT = LAT+15:	1133.	1331.	1428.	1423.	1460.	1477.	1476.	1544.	1536.	1576.	1292.	1102.	
TILT = 90:	948.	1000.	893.	683.	550.	488.	516.	660.	869.	1136.	1064.	950.	
KT:	.44	.47	.49	.50	.54	.56	.55	.55	.53	.54	.48	.44	
AMB TEMP:	39.5	42.9	50.3	61.7	69.8	78.1	81.4	80.6	73.3	62.4	50.3	41.6	61.0
HTG DEG DAYS:	791	619	470	139	21	0	0	0	5	143	441	725	3354

STA NO:13964 FORT SMITH AR LAT = 35.33

	Jan	Feb	Mar	Apr	May	Jun	Jul	Aug	Sep	Oct	Nov	Dec	Year
HORIZ INSOL:	744.	999.	1312.	1616.	1912.	2069.	2065.	1877.	1502.	1201.	851.	682.	1404.
TILT = LAT:	1125.	1326.	1498.	1588.	1688.	1752.	1772.	1768.	1626.	1550.	1267.	1077.	
TILT = LAT+15:	1182.	1341.	1435.	1427.	1445.	1463.	1495.	1558.	1524.	1549.	1324.	1145.	
TILT = 90:	999.	1017.	907.	696.	557.	498.	531.	676.	874.	1126.	1099.	996.	
KT:	.46	.48	.49	.50	.53	.56	.57	.56	.53	.54	.49	.46	
AMB TEMP:	39.0	43.3	50.3	62.2	70.1	78.0	82.2	81.4	74.0	63.2	50.4	41.5	61.3
HTG DEG DAYS:	806	608	471	132	17	0	0	0	5	135	438	729	3336

STA NO:13966 WICHITA FALLS TX LAT = 33.97

	Jan	Feb	Mar	Apr	May	Jun	Jul	Aug	Sep	Oct	Nov	Dec	Year
HORIZ INSOL:	862.	1123.	1472.	1763.	2017.	2221.	2166.	1969.	1602.	1291.	957.	799.	1520.
TILT = LAT:	1301.	1487.	1683.	1734.	1782.	1864.	1861.	1854.	1729.	1650.	1412.	1260.	
TILT = LAT+15:	1375.	1510.	1618.	1562.	1525.	1553.	1569.	1634.	1623.	1653.	1482.	1348.	
TILT = 90:	1159.	1132.	999.	723.	548.	480.	513.	671.	901.	1181.	1221.	1171.	
KT:	.51	.52	.54	.54	.56	.60	.59	.59	.56	.53	.53	.51	
AMB TEMP:	41.5	45.9	52.5	64.3	72.3	81.3	85.8	85.5	77.0	66.0	52.9	44.2	64.1
HTG DEG DAYS:	729	535	409	112	13	0	0	0	0	92	369	645	2904

STA NO:13967 — OKLAHOMA CITY, OK — LAT = 35.40

	Jan	Feb	Mar	Apr	May	Jun	Jul	Aug	Sep	Oct	Nov	Dec	Year
HORIZ INSOL:	801.	1055.	1400.	1725.	1918.	2144.	2128.	1950.	1554.	1233.	901.	725.	1461.
TILT = LAT:	1239.	1419.	1614.	1702.	1693.	1795.	1824.	1839.	1690.	1601.	1363.	1171.	0
TILT = LAT+15:	1308.	1439.	1550.	1531.	1449.	1497.	1537.	1619.	1585.	1602.	1429.	1249.	0
TILT = 90:	1115.	1097.	982.	739.	560.	503.	538.	698.	909.	1167.	1194.	1094.	0
KT:	.49	.51	.53	.54	.53	.57	.58	.58	.55	.52	.52	.49	
AMB TEMP:	36.8	41.3	48.2	60.4	69.3	76.8	81.5	81.1	73.0	62.4	49.2	40.0	59.9
HTG DEG DAYS:	874	664	532	180	36	0	0	0	12	148	474	775	3695

STA NO:13968 — TULSA, OK — LAT = 36.20

	Jan	Feb	Mar	Apr	May	Jun	Jul	Aug	Sep	Oct	Nov	Dec	Year
HORIZ INSOL:	732.	978.	1306.	1603.	1822.	2021.	2030.	1965.	1473.	1164.	827.	659.	1373.
TILT = LAT:	1133.	1316.	1503.	1578.	1608.	1693.	1740.	1759.	1602.	1516.	1254.	1064.	0
TILT = LAT+15:	1191.	1331.	1440.	1419.	1377.	1415.	1468.	1548.	1500.	1515.	1310.	1132.	0
TILT = 90:	1017.	1020.	925.	707.	557.	507.	543.	691.	877.	1113.	1097.	992.	0
KT:	.46	.48	.50	.50	.51	.54	.56	.56	.56	.53	.52	.49	
AMB TEMP:	36.6	41.2	48.3	60.8	68.8	77.3	82.1	81.4	73.3	62.9	49.4	39.8	60.2
HTG DEG DAYS:	880	668	528	176	28	0	0	0	10	143	468	781	3680

STA NO:13970 — BATON ROUGE, LA — LAT = 30.53

	Jan	Feb	Mar	Apr	May	Jun	Jul	Aug	Sep	Oct	Nov	Dec	Year
HORIZ INSOL:	785.	1054.	1379.	1681.	1871.	1926.	1746.	1677.	1464.	1301.	920.	737.	1379.
TILT = LAT:	1057.	1300.	1519.	1640.	1665.	1647.	1526.	1574.	1537.	1580.	1237.	1027.	0
TILT = LAT+15:	1104.	1312.	1458.	1482.	1436.	1388.	1305.	1396.	1443.	1580.	1288.	1084.	0
TILT = 90:	878.	929.	841.	629.	472.	413.	427.	541.	746.	1068.	1008.	891.	0
KT:	.42	.46	.51	.51	.52	.52	.48	.49	.49	.53	.47	.42	
AMB TEMP:	51.0	53.9	59.7	68.4	74.8	80.3	82.0	81.6	77.5	68.5	58.6	52.9	67.4
HTG DEG DAYS:	451	335	208	33	0	0	0	0	0	54	208	381	1670

STA NO:13985 — DODGE CITY, KS — LAT = 37.77

	Jan	Feb	Mar	Apr	May	Jun	Jul	Aug	Sep	Oct	Nov	Dec	Year
HORIZ INSOL:	827.	1122.	1476.	1886.	2090.	2358.	2295.	2055.	1687.	1301.	894.	732.	1560.
TILT = LAT:	1401.	1614.	1763.	1885.	1837.	1953.	1953.	1948.	1889.	1788.	1454.	1299.	0
TILT = LAT+15:	1488.	1646.	1696.	1693.	1564.	1616.	1636.	1710.	1773.	1798.	1530.	1394.	0
TILT = 90:	1309.	1303.	1124.	859.	634.	566.	606.	783.	1068.	1363.	1315.	1257.	0
KT:	.55	.57	.58	.60	.58	.63	.63	.63	.61	.61	.56	.54	
AMB TEMP:	30.8	35.2	41.2	54.0	64.0	73.7	79.2	78.1	68.9	57.9	42.6	33.4	54.9
HTG DEG DAYS:	1060	834	738	344	115	21	0	0	41	247	666	980	5046

STA NO:13994 — ST. LOUIS, MO — LAT = 38.75

	Jan	Feb	Mar	Apr	May	Jun	Jul	Aug	Sep	Oct	Nov	Dec	Year
HORIZ INSOL:	627.	886.	1205.	1564.	1871.	2092.	2049.	1817.	1459.	1100.	718.	531.	1327.
TILT = LAT:	1013.	1232.	1410.	1550.	1645.	1739.	1747.	1717.	1619.	1488.	1134.	881.	0
TILT = LAT+15:	1062.	1244.	1348.	1390.	1404.	1447.	1469.	1508.	1514.	1487.	1182.	930.	0
TILT = 90:	924.	981.	904.	739.	610.	560.	593.	729.	933.	1131.	1010.	825.	0
KT:	.44	.46	.48	.50	.52	.56	.55	.53	.53	.53	.47	.41	
AMB TEMP:	31.3	35.1	43.3	56.5	65.8	74.9	78.6	77.2	69.6	59.1	45.0	34.6	55.9
HTG DEG DAYS:	1045	837	682	272	103	10	0	0	35	224	600	942	4750

STA NO:13995 — SPRINGFIELD, MO — LAT = 37.23

	Jan	Feb	Mar	Apr	May	Jun	Jul	Aug	Sep	Oct	Nov	Dec	Ann
HORIZ INSOL	684.	926.	1235.	1604.	1882.	2075.	2063.	1873.	1481.	1144.	775.	603.	1362.
TILT = LAT	1074.	1258.	1426.	1585.	1657.	1732.	1763.	1769.	1625.	1514.	1191.	983.	0
TILT = LAT+15	1128.	1270.	1364.	1424.	1417.	1444.	1485.	1555.	1521.	1513.	1243.	1042.	0
TILT = 90	969.	983.	892.	728.	585.	530.	566.	715.	908.	1128.	1049.	918.	0
KT	.45	.47	.48	.50	.52	.56	.55	.56	.53	.48	.43		
AMB TEMP	32.9	37.0	44.0	56.5	65.1	73.6	77.8	77.0	69.3	59.0	45.5	36.0	56.1
HTG DEG DAYS	995	784	660	275	94	10	6	35	227	585	899		4570

STA NO:13996 — TOPEKA, KS — LAT = 39.07

	Jan	Feb	Mar	Apr	May	Jun	Jul	Aug	Sep	Oct	Nov	Dec	Ann
HORIZ INSOL	681.	941.	1257.	1642.	1915.	2128.	2126.	1910.	1516.	1147.	772.	584.	1385.
TILT = LAT	1143.	1340.	1487.	1633.	1683.	1811.	1764.	1811.	1696.	1576.	1259.	1016.	0
TILT = LAT+15	1205.	1357.	1424.	1465.	1435.	1520.	1466.	1588.	1587.	1578.	1318.	1081.	0
TILT = 90	1060.	1080.	961.	781.	625.	611.	570.	767.	983.	1210.	1138.	971.	0
KT	.48	.50	.54	.54	.54	.58	.57	.58	.56	.56	.51	.45	
AMB TEMP	28.0	33.4	41.2	54.5	64.5	73.5	78.2	77.2	68.2	57.6	42.9	31.8	54.3
HTG DEG DAYS	1147	885	745	329	118	13	0	0	55	259	663	1029	5243

STA NO:14601 — BANGOR, ME — LAT = 44.80

	Jan	Feb	Mar	Apr	May	Jun	Jul	Aug	Sep	Oct	Nov	Dec	Ann
HORIZ INSOL	455.	725.	1094.	1440.	1729.	1859.	1857.	1611.	1255.	839.	471.	379.	1143.
TILT = LAT	856.	1131.	1370.	1452.	1511.	1571.	1526.	1533.	1449.	1225.	817.	749.	0
TILT = LAT+15	897.	1142.	1308.	1296.	1283.	1316.	1267.	1340.	1349.	1218.	844.	792.	0
TILT = 90	817.	962.	966.	794.	676.	662.	623.	768.	930.	991.	751.	735.	0
KT	.41	.45	.48	.48	.49	.51	.49	.51	.50	.50	.47	.39	
AMB TEMP	0	0	0	0	0	0	0	0	0	0	0	0	0
HTG DEG DAYS	0	0	0	0	0	0	0	0	0	0	0	0	0

STA NO:14607 — CARIBOU, ME — LAT = 46.87

	Jan	Feb	Mar	Apr	May	Jun	Jul	Aug	Sep	Oct	Nov	Dec	Ann
HORIZ INSOL	419.	724.	1133.	1414.	1578.	1762.	1757.	1501.	1103.	688.	366.	311.	1063.
TILT = LAT	861.	1214.	1485.	1433.	1374.	1485.	1438.	1428.	1275.	1004.	629.	642.	0
TILT = LAT+15	905.	1231.	1421.	1282.	1166.	1242.	1193.	1246.	1183.	991.	644.	677.	0
TILT = 90	839.	1064.	1086.	820.	657.	669.	630.	753.	843.	819.	575.	633.	0
KT	.42	.49	.52	.48	.45	.48	.47	.49	.45	.41	.36	.36	
AMB TEMP	10.7	12.9	23.6	36.7	49.7	59.6	64.9	62.3	54.1	43.8	31.4	16.1	38.8
HTG DEG DAYS	1683	1459	1283	849	474	170	84	122	327	657	1008	1516	9632

STA NO:14732 — NEW YORK CITY (LA GUARDIA), NY — LAT = 40.77

	Jan	Feb	Mar	Apr	May	Jun	Jul	Aug	Sep	Oct	Nov	Dec	Ann
HORIZ INSOL	548.	795.	1118.	1457.	1690.	1784.	1802.	1583.	1280.	951.	593.	457.	1171.
TILT = LAT	914.	1130.	1324.	1445.	1483.	1520.	1498.	1493.	1421.	1299.	946.	784.	0
TILT = LAT+15	956.	1137.	1263.	1294.	1256.	1282.	1252.	1310.	1325.	1292.	980.	826.	0
TILT = 90	842.	914.	875.	726.	600.	579.	550.	682.	849.	1003.	844.	740.	0
KT	.41	.44	.46	.47	.48	.48	.49	.48	.48	.48	.41	.38	
AMB TEMP	32.1	33.1	40.6	51.7	61.8	71.5	76.7	74.9	68.1	58.1	47.3	35.6	54.3
HTG DEG DAYS	1020	893	756	399	145	0	0	0	30	224	531	911	4909

STA NO:14733
BUFFALO
NY
LAT = 42.93

													Year
HORIZ INSOL:	349.	546.	889.	1315.	1596.	1804.	1776.	1513.	1152.	784.	403.	283.	1034.
TILT = LAT:	526.	729.	1034.	1303.	1397.	1491.	1507.	1428.	1285.	1071.	599.	429.	0
TILT = LAT+15:	536.	720.	978.	1164.	1191.	1242.	1267.	1251.	1194.	1058.	608.	439.	0
TILT = 90:	462.	576.	695.	689.	606.	583.	611.	691.	795.	836.	518.	382.	0
KT:	.29	.32	.38	.43	.45	.48	.49	.47	.45	.42	.31	.26	
AMB TEMP:	23.7	24.4	32.1	44.9	55.1	65.7	70.1	68.4	61.6	51.5	39.8	27.9	47.1
HTG DEG DAYS:	1280	1137	1020	603	321	58	12	33	138	419	756	1150	6927

STA NO:14734
NEWARK
NJ
LAT = 40.70

													Year
HORIZ INSOL:	552.	793.	1109.	1449.	1687.	1795.	1760.	1565.	1273.	951.	596.	454.	1165.
TILT = LAT:	920.	1125.	1311.	1437.	1480.	1493.	1500.	1475.	1411.	1298.	950.	775.	0
TILT = LAT+15:	963.	1132.	1250.	1286.	1264.	1248.	1266.	1294.	1315.	1290.	984.	816.	0
TILT = 90:	848.	909.	864.	721.	599.	548.	573.	674.	842.	1000.	848.	731.	0
KT:	.42	.44	.45	.47	.47	.48	.48	.48	.48	.48	.42	.38	
AMB TEMP:	31.4	32.6	40.6	51.7	61.9	71.4	76.4	74.6	67.8	57.5	46.2	34.5	53.9
HTG DEG DAYS:	1042	907	756	399	143	0	0	0	34	243	564	946	5034

STA NO:14735
ALBANY
NY
LAT = 42.75

													Year
HORIZ INSOL:	457.	688.	986.	1335.	1570.	1730.	1725.	1499.	1170.	817.	457.	356.	1066.
TILT = LAT:	775.	990.	1171.	1325.	1373.	1433.	1464.	1414.	1306.	1124.	713.	604.	0
TILT = LAT+15:	806.	992.	1112.	1183.	1172.	1196.	1233.	1238.	1214.	1113.	730.	630.	0
TILT = 90:	716.	810.	792.	697.	596.	565.	596.	682.	807.	879.	629.	564.	0
KT:	.38	.40	.42	.44	.44	.46	.47	.46	.45	.43	.35	.33	
AMB TEMP:	21.5	23.5	33.4	46.9	57.7	67.5	72.0	69.6	61.9	51.4	39.6	25.9	47.6
HTG DEG DAYS:	1349	1162	980	543	253	39	9	22	135	422	762	1212	6888

STA NO:14737
ALLENTOWN
PA
LAT = 40.65

													Year
HORIZ INSOL:	528.	764.	1078.	1410.	1637.	1777.	1765.	1546.	1238.	926.	568.	430.	1139.
TILT = LAT:	853.	1070.	1267.	1395.	1436.	1479.	1504.	1456.	1367.	1254.	888.	716.	0
TILT = LAT+15:	901.	1075.	1207.	1249.	1228.	1237.	1269.	1278.	1273.	1246.	918.	751.	0
TILT = 90:	789.	860.	833.	700.	585.	544.	573.	666.	814.	963.	786.	667.	0
KT:	.40	.42	.44	.45	.46	.47	.48	.47	.46	.47	.40	.36	
AMB TEMP:	27.8	29.4	38.1	49.9	60.1	69.5	74.1	71.7	64.7	54.1	42.3	30.7	51.0
HTG DEG DAYS:	1153	997	834	453	190	21	0	6	85	344	681	1063	5827

STA NO:14740
HARTFORD
CT
LAT = 41.93

													Year
HORIZ INSOL:	478.	715.	979.	1315.	1568.	1686.	1649.	1422.	1154.	853.	497.	385.	1058.
TILT = LAT:	795.	1016.	1147.	1299.	1374.	1400.	1403.	1335.	1276.	1164.	777.	650.	0
TILT = LAT+15:	827.	1019.	1089.	1161.	1174.	1172.	1185.	1171.	1186.	1154.	798.	679.	0
TILT = 90:	730.	826.	765.	673.	585.	545.	566.	637.	776.	903.	686.	607.	0
KT:	.38	.41	.41	.43	.44	.45	.45	.44	.44	.44	.36	.34	
AMB TEMP:	24.8	26.8	35.6	47.7	58.3	67.8	72.7	70.4	62.8	52.6	41.3	28.2	49.1
HTG DEG DAYS:	1246	1070	911	519	226	24	0	12	106	384	711	1141	6350

STA NO:14742 BURLINGTON VT LAT = 44.47

	Jan	Feb	Mar	Apr	May	Jun	Jul	Aug	Sep	Oct	Nov	Dec	YEAR
HORIZ INSOL:	385.	607.	940.	1296.	1574.	1729.	1721.	1475.	1122.	741.	375.	283.	1021.
TILT = LAT:	660.	880.	1133.	1291.	1374.	1425.	1456.	1394.	1266.	1033.	575.	470.	0
TILT = LAT+15:	684.	879.	1075.	1151.	1170.	1187.	1223.	1219.	1175.	1020.	584.	485.	0
TILT = 90:	611.	727.	785.	704.	621.	589.	620.	699.	804.	820.	504.	434.	0
KT:	.34	.38	.41	.43	.45	.46	.47	.46	.44	.41	.36	.29	
AMB TEMP:	16.8	18.6	29.1	43.0	54.8	65.2	69.8	67.4	59.3	48.8	37.0	22.6	44.4
HTG DEG DAYS:	1494	1299	1113	660	331	63	20	49	191	502	840	1314	7876

STA NO:14745 CONCORD NH LAT = 43.20

	Jan	Feb	Mar	Apr	May	Jun	Jul	Aug	Sep	Oct	Nov	Dec	YEAR
HORIZ INSOL:	460.	686.	974.	1317.	1582.	1705.	1675.	1455.	1140.	817.	463.	362.	1053.
TILT = LAT:	799.	999.	1160.	1307.	1384.	1410.	1421.	1371.	1273.	1136.	739.	636.	0
TILT = LAT+15:	834.	1002.	1102.	1167.	1180.	1178.	1197.	1201.	1182.	1125.	759.	665.	0
TILT = 90:	745.	823.	790.	695.	606.	566.	590.	670.	792.	895.	659.	601.	0
KT:	.39	.41	.42	.43	.45	.45	.46	.45	.44	.44	.36	.34	
AMB TEMP:	20.6	22.6	32.3	44.2	55.1	64.7	69.6	67.2	59.5	49.3	38.0	24.8	45.6
HTG DEG DAYS:	1376	1187	1014	624	315	58	16	45	182	487	810	1246	7360

STA NO:14751 HARRISBURG PA LAT = 40.22

	Jan	Feb	Mar	Apr	May	Jun	Jul	Aug	Sep	Oct	Nov	Dec	YEAR
HORIZ INSOL:	536.	771.	1083.	1411.	1652.	1805.	1764.	1551.	1267.	934.	579.	447.	1150.
TILT = LAT:	866.	1072.	1267.	1394.	1451.	1503.	1505.	1460.	1397.	1256.	896.	741.	0
TILT = LAT+15:	903.	1076.	1207.	1248.	1240.	1257.	1270.	1282.	1302.	1247.	925.	778.	0
TILT = 90:	788.	856.	827.	694.	583.	542.	566.	661.	826.	959.	790.	690.	0
KT:	.40	.42	.44	.45	.46	.48	.48	.47	.47	.47	.40	.37	
AMB TEMP:	30.1	32.3	41.0	52.8	63.1	72.0	76.1	73.9	67.0	55.8	43.8	32.6	53.4
HTG DEG DAYS:	1082	916	744	370	128	0	0	0	51	293	636	1004	5224

STA NO:14764 PORTLAND ME LAT = 43.65

	Jan	Feb	Mar	Apr	May	Jun	Jul	Aug	Sep	Oct	Nov	Dec	YEAR
HORIZ INSOL:	450.	682.	970.	1304.	1567.	1712.	1659.	1461.	1158.	822.	459.	363.	1051.
TILT = LAT:	794.	1004.	1162.	1295.	1370.	1414.	1407.	1378.	1302.	1159.	746.	655.	0
TILT = LAT+15:	829.	1008.	1104.	1156.	1167.	1180.	1185.	1206.	1210.	1149.	767.	687.	0
TILT = 90:	744.	833.	797.	695.	608.	573.	592.	680.	816.	919.	670.	625.	0
KT:	.39	.41	.42	.43	.44	.46	.46	.46	.45	.45	.36	.35	
AMB TEMP:	21.5	22.9	31.8	42.7	52.7	62.2	68.0	66.4	58.7	49.1	38.6	25.7	45.0
HTG DEG DAYS:	1349	1179	1029	669	381	106	27	55	200	493	792	1218	7498

STA NO:14765 PROVIDENCE RI LAT = 41.73

	Jan	Feb	Mar	Apr	May	Jun	Jul	Aug	Sep	Oct	Nov	Dec	YEAR
HORIZ INSOL:	506.	739.	1032.	1374.	1655.	1775.	1695.	1499.	1209.	907.	538.	419.	1112.
TILT = LAT:	855.	1055.	1220.	1361.	1457.	1473.	1443.	1411.	1343.	1252.	858.	726.	0
TILT = LAT+15:	893.	1060.	1160.	1218.	1237.	1230.	1217.	1238.	1250.	1243.	886.	763.	0
TILT = 90:	790.	858.	814.	701.	606.	559.	574.	665.	815.	975.	766.	686.	0
KT:	.40	.42	.43	.45	.47	.47	.47	.46	.46	.47	.39	.37	
AMB TEMP:	28.4	29.4	36.9	47.3	56.9	66.4	72.1	70.4	63.4	53.7	43.3	31.5	50.0
HTG DEG DAYS:	1135	997	871	531	259	36	0	10	93	350	651	1039	5972

STA NO:14768 ROCHESTER NY LAT = 43.12

	JAN	FEB	MAR	APR	MAY	JUN	JUL	AUG	SEP	OCT	NOV	DEC	YEAR
HORIZ INSOL:	364.	560.	903.	1339.	1606.	1817.	1781.	1519.	1160.	782.	404.	281.	1043.
TILT = LAT:	566.	757.	1058.	1331.	1405.	1501.	1510.	1435.	1297.	1072.	605.	428.	
TILT = LAT+15:	580.	749.	1002.	1188.	1198.	1250.	1270.	1256.	1206.	1059.	615.	438.	
TILT = 90:	505.	603.	714.	706.	612.	588.	615.	696.	806.	838.	525.	382.	
KT:	.30	.33	.39	.44	.45	.48	.49	.47	.45	.42	.31	.26	
AMB TEMP:	24.0	24.8	33.0	46.1	56.5	66.9	71.2	69.3	62.3	52.3	40.5	28.3	47.9
HTG DEG DAYS:	1271	1126	992	567	285	46	9	26	126	398	735	1138	6719

STA NO:14771 SYRACUSE NY LAT = 43.12

	JAN	FEB	MAR	APR	MAY	JUN	JUL	AUG	SEP	OCT	NOV	DEC	YEAR
HORIZ INSOL:	385.	571.	890.	1324.	1578.	1778.	1758.	1504.	1165.	777.	399.	285.	1034.
TILT = LAT:	615.	779.	1039.	1314.	1380.	1469.	1491.	1419.	1305.	1064.	594.	439.	
TILT = LAT+15:	633.	772.	984.	1174.	1177.	1225.	1254.	1243.	1213.	1051.	602.	449.	
TILT = 90:	555.	623.	701.	697.	604.	609.	609.	690.	811.	831.	513.	393.	
KT:	.32	.34	.38	.44	.45	.48	.47	.47	.45	.42	.31	.27	
AMB TEMP:	23.6	24.6	33.2	46.5	56.8	66.9	71.5	69.7	62.8	52.5	41.0	28.1	48.1
HTG DEG DAYS:	1283	1131	986	555	272	46	11	18	120	392	720	1144	6678

STA NO:14777 WILKES-BARRE-SCRANTO PA LAT = 41.33

	JAN	FEB	MAR	APR	MAY	JUN	JUL	AUG	SEP	OCT	NOV	DEC	YEAR
HORIZ INSOL:	455.	689.	991.	1339.	1591.	1746.	1760.	1513.	1199.	897.	490.	368.	1086.
TILT = LAT:	723.	952.	1156.	1322.	1394.	1466.	1463.	1425.	1326.	1223.	742.	589.	
TILT = LAT+15:	748.	952.	1098.	1183.	1192.	1253.	1223.	1250.	1233.	1214.	760.	611.	
TILT = 90:	651.	762.	764.	676.	582.	579.	551.	665.	798.	946.	647.	538.	
KT:	.35	.39	.41	.43	.45	.48	.47	.47	.45	.46	.35	.32	
AMB TEMP:	26.0	27.3	36.0	48.5	58.9	67.9	72.2	70.0	62.9	52.6	40.8	29.1	49.4
HTG DEG DAYS:	1209	1056	899	495	219	28	7	18	116	391	726	1113	6277

STA NO:14780 LAKEHURST NJ LAT = 40.03

	JAN	FEB	MAR	APR	MAY	JUN	JUL	AUG	SEP	OCT	NOV	DEC	YEAR
HORIZ INSOL:	560.	797.	1109.	1456.	1672.	1775.	1703.	1533.	1261.	956.	621.	475.	1160.
TILT = LAT:	913.	1113.	1300.	1441.	1468.	1480.	1455.	1442.	1388.	1287.	980.	799.	
TILT = LAT+15:	954.	1119.	1239.	1291.	1255.	1239.	1230.	1267.	1293.	1279.	1016.	842.	
TILT = 90:	834.	891.	847.	712.	585.	535.	552.	651.	817.	983.	871.	750.	
KT:	.41	.43	.45	.47	.47	.47	.47	.47	.47	.47	.42	.39	
AMB TEMP:	0	0	0	0	0	0	0	0	0	0	0	0	0
HTG DEG DAYS:	0	0	0	0	0	0	0	0	0	0	0	0	0

STA NO:14819 CHICAGO IL LAT = 41.78

	JAN	FEB	MAR	APR	MAY	JUN	JUL	AUG	SEP	OCT	NOV	DEC	YEAR
HORIZ INSOL:	507.	760.	1107.	1459.	1789.	2007.	1944.	1719.	1354.	969.	566.	402.	1215.
TILT = LAT:	859.	1095.	1327.	1453.	1568.	1658.	1650.	1631.	1531.	1362.	923.	685.	
TILT = LAT+15:	897.	1102.	1266.	1300.	1335.	1377.	1388.	1429.	1428.	1357.	950.	718.	
TILT = 90:	794.	898.	891.	745.	643.	602.	629.	755.	933.	1070.	831.	644.	
KT:	.40	.43	.46	.48	.50	.53	.53	.53	.52	.50	.41	.35	
AMB TEMP:	24.3	27.4	36.8	49.9	60.0	70.5	74.7	73.7	65.9	55.4	40.4	28.5	50.6
HTG DEG DAYS:	1262	1053	874	453	208	26	0	8	57	316	738	1132	6127

STA NO:14820
CLEVELAND
OH
LAT = 41.40

	JAN	FEB	MAR	APR	MAY	JUN	JUL	AUG	SEP	OCT	NOV	DEC	YEAR
HORIZ INSOL:	388.	601.	922.	1350.	1681.	1843.	1828.	1583.	1240.	867.	466.	318.	1591.
TILT = LAT:	576.	797.	1061.	1334.	1474.	1529.	1554.	1494.	1378.	1174.	694.	474.	0
TILT = LAT+15:	589.	790.	1005.	1193.	1258.	1275.	1309.	1310.	1283.	1164.	709.	486.	0
TILT = 90:	502.	624.	698.	683.	608.	567.	598.	693.	832.	905.	600.	419.	0
KT:	.30	.34	.38	.44	.47	.49	.50	.49	.47	.44	.33	.28	
AMB TEMP:	26.9	27.9	36.1	48.3	58.3	67.9	71.4	70.0	63.9	53.8	41.6	30.3	49.7
HTG DEG DAYS:	1181	1039	896	501	244	40	9	17	95	354	702	1076	6154

STA NO:14821
COLUMBUS
OH
LAT = 40.00

	JAN	FEB	MAR	APR	MAY	JUN	JUL	AUG	SEP	OCT	NOV	DEC	YEAR
HORIZ INSOL:	459.	677.	980.	1353.	1647.	1813.	1755.	1641.	1282.	945.	538.	387.	1123.
TILT = LAT:	693.	900.	1122.	1332.	1446.	1510.	1498.	1548.	1414.	1268.	805.	596.	0
TILT = LAT+15:	714.	897.	1065.	1193.	1237.	1263.	1265.	1359.	1318.	1260.	826.	617.	0
TILT = 90:	610.	703.	724.	662.	578.	541.	561.	691.	832.	967.	698.	536.	0
KT:	.34	.37	.40	.43	.46	.48	.48	.50	.48	.47	.37	.31	
AMB TEMP:	28.4	30.3	39.2	51.2	61.1	70.4	73.6	71.9	65.2	54.2	41.7	30.7	51.5
HTG DEG DAYS:	1135	972	800	418	176	13	0	8	76	342	699	1063	5702

STA NO:14822
DETROIT
MI
LAT = 42.42

	JAN	FEB	MAR	APR	MAY	JUN	JUL	AUG	SEP	OCT	NOV	DEC	YEAR
HORIZ INSOL:	417.	680.	1000.	1399.	1716.	1866.	1835.	1576.	1253.	876.	478.	344.	1120.
TILT = LAT:	670.	966.	1186.	1392.	1503.	1543.	1558.	1489.	1409.	1218.	749.	562.	0
TILT = LAT+15:	692.	967.	1127.	1244.	1280.	1285.	1309.	1305.	1313.	1209.	768.	583.	0
TILT = 90:	605.	785.	798.	726.	633.	587.	616.	708.	867.	955.	662.	518.	0
KT:	.34	.39	.42	.46	.48	.50	.50	.49	.48	.46	.36	.31	
AMB TEMP:	25.5	26.9	35.4	48.1	58.4	69.1	73.3	71.9	64.5	54.3	41.1	29.6	49.9
HTG DEG DAYS:	1225	1067	918	507	238	26	0	11	80	342	717	1097	6228

STA NO:14826
FLINT
MI
LAT = 42.97

	JAN	FEB	MAR	APR	MAY	JUN	JUL	AUG	SEP	OCT	NOV	DEC	YEAR
HORIZ INSOL:	383.	636.	957.	1339.	1658.	1813.	1797.	1555.	1196.	829.	429.	309.	1075.
TILT = LAT:	606.	897.	1132.	1330.	1451.	1498.	1524.	1470.	1342.	1150.	657.	492.	0
TILT = LAT+15:	623.	896.	1075.	1186.	1236.	1248.	1281.	1287.	1248.	1140.	670.	508.	0
TILT = 90:	544.	728.	767.	703.	625.	585.	616.	709.	832.	904.	575.	448.	0
KT:	.32	.38	.41	.44	.47	.48	.49	.48	.46	.44	.33	.29	
AMB TEMP:	22.3	23.8	32.6	45.9	55.8	65.8	69.7	68.2	61.0	51.2	38.3	26.8	46.8
HTG DEG DAYS:	1324	1154	1004	573	306	65	14	36	147	433	801	1184	7041

STA NO:14827
FORT WAYNE
IN
LAT = 41.00

	JAN	FEB	MAR	APR	MAY	JUN	JUL	AUG	SEP	OCT	NOV	DEC	YEAR
HORIZ INSOL:	455.	698.	982.	1361.	1672.	1842.	1787.	1594.	1274.	924.	516.	370.	1123.
TILT = LAT:	713.	960.	1139.	1344.	1466.	1529.	1522.	1504.	1416.	1261.	789.	583.	0
TILT = LAT+15:	737.	960.	1081.	1203.	1252.	1276.	1283.	1320.	1320.	1253.	810.	605.	0
TILT = 90:	636.	766.	748.	682.	599.	560.	583.	691.	849.	973.	691.	530.	0
KT:	.35	.39	.40	.44	.47	.49	.49	.49	.48	.48	.36	.31	
AMB TEMP:	25.3	27.6	36.5	49.3	59.6	69.5	73.0	71.3	64.5	53.6	40.2	28.6	49.9
HTG DEG DAYS:	1231	1047	884	471	216	23	0	12	90	363	744	1128	6209

STA NO:14837 — MADISON, WI — LAT = 43.13

	Jan	Feb	Mar	Apr	May	Jun	Jul	Aug	Sep	Oct	Nov	Dec	Annual
HORIZ INSOL:	515.	804.	1136.	1398.	1743.	1948.	1934.	1708.	1299.	911.	504.	389.	1191.
TILT = LAT:	936.	1227.	1396.	1395.	1526.	1605.	1638.	1624.	1481.	1302.	832.	704.	0
TILT = LAT+15:	983.	1241.	1334.	1247.	1298.	1333.	1374.	1421.	1380.	1297.	860.	741.	0
TILT = 90:	887.	1031.	961.	738.	652.	615.	652.	778.	924.	1037.	753.	674.	0
KT:	.43	.48	.49	.46	.49	.52	.53	.53	.50	.49	.39	.37	
AMB TEMP:	16.8	20.3	30.2	45.3	56.0	65.8	70.1	68.7	59.7	49.9	34.7	21.9	44.9
HTG DEG DAYS:	1494	1252	1079	591	297	72	14	39	173	474	909	1336	7730

STA NO:14839 — MILWAUKEE, WI — LAT = 42.95

	Jan	Feb	Mar	Apr	May	Jun	Jul	Aug	Sep	Oct	Nov	Dec	Annual
HORIZ INSOL:	479.	737.	1089.	1443.	1768.	1977.	1962.	1719.	1310.	908.	525.	378.	1191.
TILT = LAT:	838.	1088.	1323.	1443.	1548.	1629.	1662.	1634.	1492.	1291.	873.	669.	0
TILT = LAT+15:	876.	1096.	1262.	1289.	1317.	1352.	1393.	1430.	1391.	1285.	903.	702.	0
TILT = 90:	784.	902.	905.	759.	657.	617.	655.	779.	928.	1025.	792.	635.	0
KT:	.40	.43	.46	.47	.50	.53	.54	.53	.51	.48	.40	.35	
AMB TEMP:	19.4	22.5	31.4	44.7	54.2	64.5	69.9	69.2	61.1	51.0	36.5	24.2	45.7
HTG DEG DAYS:	1414	1190	1042	609	348	90	15	36	140	440	855	1265	7444

STA NO:14847 — SAULT STE. MARIE, MI — LAT = 46.47

	Jan	Feb	Mar	Apr	May	Jun	Jul	Aug	Sep	Oct	Nov	Dec	Annual
HORIZ INSOL:	325.	603.	1029.	1383.	1683.	1811.	1835.	1523.	1049.	673.	332.	253.	1042.
TILT = LAT:	572.	935.	1308.	1401.	1473.	1483.	1547.	1449.	1196.	962.	528.	449.	0
TILT = LAT+15:	591.	938.	1248.	1248.	1229.	1229.	1293.	1264.	1108.	949.	536.	466.	0
TILT = 90:	533.	796.	944.	793.	689.	638.	684.	757.	783.	778.	470.	424.	0
KT:	.32	.40	.47	.47	.48	.48	.51	.48	.43	.40	.30	.29	
AMB TEMP:	14.2	15.2	24.0	38.2	49.0	58.7	63.8	63.2	55.3	46.2	32.8	20.1	40.0
HTG DEG DAYS:	1575	1394	1271	804	496	200	96	125	291	583	966	1392	9193

STA NO:14848 — SOUTH BEND, IN — LAT = 41.70

	Jan	Feb	Mar	Apr	May	Jun	Jul	Aug	Sep	Oct	Nov	Dec	Annual
HORIZ INSOL:	416.	660.	993.	1387.	1723.	1922.	1912.	1852.	1666.	1291.	909.	497.	1138.
TILT = LAT:	645.	909.	1164.	1376.	1510.	1590.	1571.	1574.	1578.	1255.	769.	533.	0
TILT = LAT+15:	663.	907.	1105.	1231.	1287.	1324.	1302.	1324.	1382.	1247.	790.	551.	0
TILT = 90:	574.	727.	774.	707.	623.	566.	634.	608.	732.	978.	677.	483.	0
KT:	.33	.37	.41	.45	.48	.51	.51	.51	.51	.47	.36	.30	
AMB TEMP:	24.0	26.3	35.3	48.1	58.4	68.6	72.3	71.0	63.8	53.4	39.6	28.2	49.1
HTG DEG DAYS:	1271	1084	921	507	245	35	6	24	98	368	762	1141	6462

STA NO:14850 — TRAVERSE CITY, MI — LAT = 44.73

	Jan	Feb	Mar	Apr	May	Jun	Jul	Aug	Sep	Oct	Nov	Dec	Annual
HORIZ INSOL:	311.	568.	1001.	1405.	1729.	1912.	1910.	1609.	1165.	754.	377.	257.	1083.
TILT = LAT:	483.	810.	1228.	1413.	1511.	1571.	1614.	1531.	1327.	1065.	587.	409.	0
TILT = LAT+15:	492.	806.	1169.	1261.	1283.	1302.	1351.	1338.	1233.	1053.	597.	419.	0
TILT = 90:	429.	665.	860.	772.	675.	634.	674.	766.	849.	850.	518.	371.	0
KT:	.28	.35	.44	.47	.49	.51	.53	.51	.46	.42	.31	.26	
AMB TEMP:	20.8	20.7	28.7	42.7	52.8	63.7	68.7	67.5	59.4	49.8	36.9	25.9	44.8
HTG DEG DAYS:	1370	1240	1125	669	367	104	33	66	178	471	843	1212	7698

STA NO:14852 YOUNGSTOWN OH LAT = 41.27

Parameter	1	2	3	4	5	6	7	8	9	10	11	12	Year
HORIZ INSOL:	385.	587.	890.	1278.	1586.	1759.	1734.	1506.	1194.	851.	457.	315.	1045.
TILT = LAT:	566.	768.	1015.	1257.	1390.	1462.	1476.	1418.	1318.	1144.	670.	465.	0
TILT = LAT+15:	577.	760.	960.	1124.	1189.	1222.	1246.	1244.	1226.	1133.	683.	476.	0
TILT = 90:	491.	598.	664.	644.	580.	550.	576.	661.	793.	878.	576.	409.	0
KT:	.30	.33	.37	.41	.45	.47	.48	.46	.45	.44	.33	.27	0
AMB TEMP:	25.7	26.7	35.3	47.7	57.6	67.0	70.7	69.2	62.7	52.6	40.3	28.8	48.7
HTG DEG DAYS:	1218	1072	921	519	258	42	9	22	118	384	741	1122	6426

STA NO:14858 HOUGHTON MI LAT = 47.17

Parameter	1	2	3	4	5	6	7	8	9	10	11	12	Year
HORIZ INSOL:	244.	484.	933.	1366.	1660.	1838.	1838.	1521.	1010.	671.	291.	192.	1004.
TILT = LAT:	379.	704.	1174.	1387.	1447.	1501.	1547.	1451.	1152.	978.	447.	297.	0
TILT = LAT+15:	383.	698.	1116.	1234.	1226.	1243.	1293.	1265.	1066.	965.	450.	301.	0
TILT = 90:	337.	587.	850.	796.	691.	656.	697.	769.	762.	799.	393.	266.	0
KT:	.25	.33	.43	.47	.47	.49	.51	.49	.42	.40	.27	.23	0
AMB TEMP:	0	0	0	0	0	0	0	0	0	0	0	0	0
HTG DEG DAYS:	0	0	0	0	0	0	0	0	0	0	0	0	0

STA NO:14860 ERIE PA LAT = 42.08

Parameter	1	2	3	4	5	6	7	8	9	10	11	12	Year
HORIZ INSOL:	346.	577.	920.	1359.	1646.	1847.	1833.	1455.	1201.	827.	416.	278.	1059.
TILT = LAT:	499.	767.	1068.	1347.	1442.	1529.	1557.	1369.	1338.	1124.	606.	397.	0
TILT = LAT+15:	506.	759.	1011.	1204.	1230.	1274.	1309.	1200.	1245.	1112.	614.	403.	0
TILT = 90:	429.	603.	710.	699.	608.	578.	610.	653.	817.	871.	518.	344.	0
KT:	.28	.33	.39	.44	.46	.49	.50	.45	.46	.43	.31	.25	0
AMB TEMP:	25.1	25.2	32.9	44.8	54.6	64.6	68.7	67.5	61.4	51.6	40.1	29.1	47.1
HTG DEG DAYS:	1237	1114	995	606	336	80	24	43	141	415	747	1113	6851

STA NO:14895 AKRON-CANTON OH LAT = 40.92

Parameter	1	2	3	4	5	6	7	8	9	10	11	12	Year
HORIZ INSOL:	428.	650.	964.	1357.	1668.	1839.	1787.	1596.	1272.	908.	505.	353.	1111.
TILT = LAT:	650.	872.	1113.	1340.	1483.	1528.	1522.	1505.	1413.	1230.	762.	543.	0
TILT = LAT+15:	669.	868.	1056.	1199.	1249.	1275.	1283.	1321.	1317.	1221.	781.	560.	0
TILT = 90:	574.	687.	729.	679.	597.	559.	582.	690.	846.	947.	663.	487.	0
KT:	.33	.36	.40	.44	.47	.49	.49	.49	.48	.46	.35	.30	0
AMB TEMP:	26.3	27.7	36.2	48.5	58.7	68.3	71.7	70.3	63.7	53.3	40.7	29.4	49.6
HTG DEG DAYS:	1200	1044	893	495	231	33	9	16	101	369	729	1104	6224

STA NO:14898 GREEN BAY WI LAT = 44.48

Parameter	1	2	3	4	5	6	7	8	9	10	11	12	Year
HORIZ INSOL:	451.	725.	1104.	1439.	1719.	1908.	1889.	1622.	1218.	821.	465.	350.	1143.
TILT = LAT:	832.	1118.	1378.	1448.	1503.	1568.	1597.	1542.	1394.	1181.	790.	652.	0
TILT = LAT+15:	871.	1128.	1317.	1293.	1276.	1301.	1337.	1348.	1297.	1172.	815.	685.	0
TILT = 90:	790.	947.	968.	786.	667.	629.	664.	767.	889.	943.	721.	628.	0
KT:	.40	.45	.48	.48	.49	.51	.52	.51	.48	.46	.38	.35	0
AMB TEMP:	15.4	18.0	28.6	43.8	54.5	64.5	69.2	67.7	58.9	49.2	34.1	20.9	43.7
HTG DEG DAYS:	1538	1316	1128	636	338	91	22	54	191	490	927	1367	8098

STA NO:14913 DULUTH MN LAT = 46.83

HORIZ INSOL:	389.	673.	1034.	1373.	1643.	1767.	1854.	1547.	1095.	725.	381.	292.	1064.
TILT = LAT:	769.	1099.	1326.	1392.	1432.	1446.	1562.	1476.	1264.	1074.	665.	581.	0
TILT = LAT+15:	805.	1110.	1266.	1239.	1214.	1199.	1305.	1287.	1173.	1064.	665.	610.	0
TILT = 90:	742.	955.	963.	793.	679.	632.	696.	776.	835.	881.	612.	567.	0
KT:	.39	.45	.48	.47	.47	.47	.51	.49	.45	.43	.35	.34	
AMB TEMP:	8.5	12.1	23.5	38.6	49.4	59.0	65.6	64.1	54.4	45.3	28.4	14.4	38.6
HTG DEG DAYS:	1751	1481	1287	792	484	194	67	104	318	611	1098	1569	9756

STA NO:14914 FARGO ND LAT = 46.90

HORIZ INSOL:	415.	706.	1098.	1476.	1835.	1994.	2120.	1825.	1304.	874.	457.	337.	1203.
TILT = LAT:	850.	1175.	1429.	1509.	1604.	1628.	1786.	1764.	1557.	1374.	873.	730.	0
TILT = LAT+15:	893.	1189.	1367.	1345.	1356.	1343.	1486.	1538.	1451.	1373.	908.	774.	0
TILT = 90:	828.	1027.	1043.	860.	748.	691.	774.	918.	1039.	1152.	826.	729.	0
KT:	.42	.47	.51	.50	.52	.53	.59	.58	.54	.52	.42	.40	
AMB TEMP:	5.9	10.7	24.2	42.3	54.6	64.7	70.7	69.2	57.9	47.0	28.6	13.0	40.8
HTG DEG DAYS:	1832	1520	1265	681	334	97	13	33	234	558	1092	1612	9271

STA NO:14918 INTERNATIONAL FALLS MN LAT = 48.57

HORIZ INSOL:	356.	663.	1046.	1444.	1716.	1853.	1921.	1618.	1121.	704.	346.	272.	1088.
TILT = LAT:	758.	1153.	1391.	1488.	1497.	1509.	1615.	1558.	1331.	1092.	636.	598.	0
TILT = LAT+15:	795.	1168.	1330.	1324.	1265.	1246.	1345.	1357.	1236.	1083.	655.	631.	0
TILT = 90:	743.	1025.	1039.	877.	735.	683.	748.	843.	907.	916.	594.	596.	0
KT:	.40	.47	.50	.50	.49	.50	.53	.52	.47	.44	.34	.36	
AMB TEMP:	1.9	7.0	20.6	38.2	50.1	60.4	65.8	63.2	53.0	43.5	24.9	8.7	36.5
HTG DEG DAYS:	1956	1624	1376	804	462	168	66	112	364	667	1203	1745	10547

STA NO:14920 LA CROSSE WI LAT = 43.87

HORIZ INSOL:	481.	765.	1101.	1426.	1713.	1905.	1900.	1666.	1242.	864.	494.	370.	1161.
TILT = LAT:	883.	1176.	1360.	1430.	1498.	1568.	1608.	1585.	1416.	1240.	836.	682.	0
TILT = LAT+15:	926.	1188.	1298.	1277.	1273.	1302.	1348.	1386.	1318.	1233.	864.	717.	0
TILT = 90:	838.	993.	945.	767.	655.	618.	657.	774.	894.	993.	762.	655.	0
KT:	.42	.46	.48	.48	.48	.51	.52	.52	.49	.47	.39	.36	
AMB TEMP:	16.1	20.0	31.1	47.6	59.0	68.5	72.8	71.4	61.8	51.8	35.4	21.8	45.4
HTG DEG DAYS:	1516	1260	1051	522	224	39	10	17	130	421	888	1339	7417

STA NO:14922 MINNEAPOLIS-ST. PAUL MN LAT = 44.88

HORIZ INSOL:	464.	764.	1104.	1442.	1737.	1928.	1970.	1687.	1255.	860.	480.	353.	1170.
TILT = LAT:	885.	1214.	1387.	1455.	1518.	1582.	1664.	1610.	1450.	1267.	843.	679.	0
TILT = LAT+15:	930.	1229.	1325.	1298.	1289.	1311.	1391.	1407.	1350.	1261.	873.	716.	0
TILT = 90:	849.	1040.	980.	796.	680.	640.	693.	805.	933.	1030.	778.	660.	0
KT:	.42	.48	.49	.48	.49	.51	.54	.53	.50	.48	.40	.37	
AMB TEMP:	12.2	16.5	28.3	45.1	57.1	66.9	71.2	70.2	60.0	50.0	32.4	18.6	44.1
HTG DEG DAYS:	1637	1358	1138	597	271	65	11	21	173	472	978	1438	8159

STA NO:14923
MOLINE
IL
LAT = 41.45

	JAN	FEB	MAR	APR	MAY	JUN	JUL	AUG	SEP	OCT	NOV	DEC	YEAR
HORIZ INSOL:	535.	812.	1119.	1452.	1754.	1969.	1939.	1715.	1357.	996.	595.	433.	1224.
TILT = LAT:	912.	1183.	1338.	1452.	1537.	1629.	1647.	1625.	1529.	1398.	974.	751.	0.
TILT = LAT+15:	955.	1194.	1276.	1299.	1310.	1355.	1383.	1424.	1427.	1394.	1012.	790.	0.
TILT = 90:	846.	971.	894.	740.	628.	590.	622.	746.	926.	1096.	880.	711.	0.
KT:	.42	.46	.46	.47	.49	.52	.53	.53	.51	.51	.43	.38	
AMB TEMP:	21.5	25.7	35.7	50.6	61.1	70.8	74.5	72.9	64.6	54.4	39.2	26.6	49.8
HTG DEG DAYS:	1349	1100	908	436	184	20	0	11	79	344	774	1190	6395

STA NO:14925
ROCHESTER
MN
LAT = 43.92

	JAN	FEB	MAR	APR	MAY	JUN	JUL	AUG	SEP	OCT	NOV	DEC	YEAR
HORIZ INSOL:	477.	753.	1082.	1410.	1696.	1902.	1909.	1662.	1250.	870.	494.	370.	1156.
TILT = LAT:	874.	1154.	1333.	1413.	1483.	1565.	1615.	1581.	1428.	1253.	838.	685.	0.
TILT = LAT+15:	916.	1165.	1271.	1261.	1261.	1300.	1353.	1383.	1329.	1246.	867.	721.	0.
TILT = 90:	829.	973.	926.	759.	651.	618.	660.	774.	902.	1005.	765.	660.	0.
KT:	.41	.46	.47	.47	.48	.51	.52	.52	.49	.48	.39	.36	
AMB TEMP:	12.9	16.9	27.8	44.5	56.2	66.0	70.1	68.6	59.3	49.6	32.6	18.9	43.6
HTG DEG DAYS:	1615	1347	1153	615	292	78	21	35	185	485	972	1429	8227

STA NO:14931
BURLINGTON
IA
LAT = 40.78

	JAN	FEB	MAR	APR	MAY	JUN	JUL	AUG	SEP	OCT	NOV	DEC	YEAR
HORIZ INSOL:	579.	859.	1165.	1538.	1876.	2121.	2085.	1828.	1416.	1061.	664.	481.	1306.
TILT = LAT:	987.	1248.	1391.	1533.	1645.	1753.	1770.	1735.	1596.	1488.	1100.	843.	0.
TILT = LAT+15:	1036.	1261.	1329.	1373.	1401.	1454.	1484.	1521.	1491.	1488.	1147.	891.	0.
TILT = 90:	918.	1021.	922.	767.	647.	602.	639.	776.	956.	1163.	999.	803.	0.
KT:	.44	.47	.48	.50	.53	.57	.57	.56	.53	.54	.46	.40	
AMB TEMP:	22.9	27.3	36.9	51.3	61.8	71.4	75.4	73.9	65.4	55.3	39.8	27.6	50.8
HTG DEG DAYS:	1305	1056	871	416	172	16	0	8	70	320	756	1159	6149

STA NO:14933
DES MOINES
IA
LAT = 41.53

	JAN	FEB	MAR	APR	MAY	JUN	JUL	AUG	SEP	OCT	NOV	DEC	YEAR
HORIZ INSOL:	581.	861.	1181.	1557.	1868.	2125.	2097.	1828.	1434.	1068.	658.	487.	1312.
TILT = LAT:	1024.	1278.	1428.	1558.	1637.	1753.	1778.	1738.	1630.	1528.	1121.	891.	0.
TILT = LAT+15:	1077.	1293.	1365.	1394.	1393.	1452.	1489.	1522.	1524.	1528.	1171.	945.	0.
TILT = 90:	963.	1057.	959.	791.	659.	617.	656.	793.	991.	1208.	1028.	861.	0.
KT:	.45	.47	.49	.51	.52	.57	.57	.56	.54	.55	.47	.42	
AMB TEMP:	19.4	24.2	33.9	49.5	60.9	70.5	75.1	73.3	64.3	54.3	37.8	25.0	49.0
HTG DEG DAYS:	1414	1142	964	465	186	26	0	13	94	350	816	1240	6710

STA NO:14935
GRAND ISLAND
NE
LAT = 40.97

	JAN	FEB	MAR	APR	MAY	JUN	JUL	AUG	SEP	OCT	NOV	DEC	YEAR
HORIZ INSOL:	661.	917.	1265.	1692.	1972.	2242.	2216.	1939.	1509.	1138.	739.	569.	1405.
TILT = LAT:	1189.	1363.	1537.	1701.	1729.	1848.	1878.	1847.	1719.	1630.	1276.	1075.	0.
TILT = LAT+15:	1258.	1383.	1473.	1524.	1470.	1528.	1570.	1618.	1608.	1635.	1339.	1148.	0.
TILT = 90:	1129.	1127.	1028.	848.	674.	623.	666.	822.	1035.	1287.	1178.	1052.	0.
KT:	.50	.51	.52	.55	.55	.60	.61	.59	.57	.58	.52	.48	
AMB TEMP:	22.3	27.7	35.5	49.9	60.7	70.7	76.3	75.0	64.4	53.7	38.2	27.0	50.1
HTG DEG DAYS:	1324	1044	915	461	184	35	6	5	107	362	804	1178	6425

1778

STA NO:14936 HURON SD LAT = 44.38

HORIZ INSOL:	488.	745.	1114.	1530.	1872.	2101.	2183.	1892.	1418.	988.	577.	405.	1276.
TILT = LAT:	926.	1155.	1391.	1549.	1638.	1722.	1843.	1817.	1663.	1433.	1063.	806.	0
TILT = LAT+15:	973.	1166.	1328.	1383.	1388.	1423.	1536.	1588.	1553.	1494.	1110.	855.	0
TILT = 90:	887.	979.	976.	837.	713.	668.	735.	887.	1065.	1222.	998.	793.	0
KT:	.43	.46	.49	.51	.53	.56	.60	.59	.56	.55	.47	.41	
AMB TEMP:	12.5	17.9	29.0	45.8	57.0	67.1	73.7	72.1	60.7	49.6	32.4	19.2	44.8
HTG DEG DAYS:	1627	1319	1116	576	273	72	9	13	169	482	978	1420	8054

STA NO:14940 MASON CITY IA LAT = 43.15

HORIZ INSOL:	554.	836.	1168.	1519.	1895.	2114.	2084.	1833.	1405.	1010.	600.	443.	1288.
TILT = LAT:	1035.	1291.	1444.	1527.	1660.	1737.	1764.	1750.	1622.	1485.	1059.	852.	0
TILT = LAT+15:	1092.	1309.	1381.	1365.	1409.	1437.	1474.	1531.	1514.	1485.	1105.	904.	0
TILT = 90:	990.	1091.	996.	804.	696.	647.	686.	831.	1015.	1195.	982.	832.	0
KT:	.46	.50	.50	.50	.53	.56	.57	.57	.55	.54	.46	.42	
AMB TEMP:	14.2	18.5	29.0	45.7	57.4	67.2	71.3	69.9	60.2	50.5	33.6	20.1	44.8
HTG DEG DAYS:	1575	1302	1116	579	265	64	13	31	165	457	942	1392	7901

STA NO:14943 SIOUX CITY IA LAT = 42.40

HORIZ INSOL:	569.	842.	1170.	1578.	1901.	2124.	2122.	1845.	1421.	1038.	643.	469.	1310.
TILT = LAT:	1035.	1273.	1431.	1587.	1666.	1748.	1797.	1759.	1629.	1507.	1125.	686.	0
TILT = LAT+15:	1091.	1289.	1368.	1419.	1415.	1447.	1503.	1539.	1522.	1507.	1176.	941.	0
TILT = 90:	983.	1064.	975.	821.	684.	634.	679.	819.	1006.	1204.	1041.	863.	0
KT:	.46	.49	.49	.52	.54	.57	.58	.57	.55	.55	.48	.43	
AMB TEMP:	18.0	23.4	33.2	49.4	60.9	70.3	75.3	73.5	63.4	53.1	36.3	23.5	48.4
HTG DEG DAYS:	1457	1165	986	474	189	33	0	10	113	378	861	1287	6953

STA NO:14944 SIOUX FALLS SD LAT = 43.57

HORIZ INSOL:	533.	802.	1152.	1543.	1894.	2100.	2150.	1845.	1410.	1005.	608.	441.	1290.
TILT = LAT:	1002.	1240.	1430.	1557.	1658.	1725.	1817.	1764.	1635.	1492.	1098.	867.	0
TILT = LAT+15:	1056.	1255.	1367.	1391.	1407.	1426.	1517.	1542.	1527.	1493.	1147.	921.	0
TILT = 90:	960.	1048.	992.	827.	704.	652.	709.	846.	1032.	1209.	1025.	852.	0
KT:	.45	.48	.50	.51	.54	.56	.59	.57	.55	.55	.48	.43	
AMB TEMP:	14.2	19.4	30.0	46.1	57.7	67.6	73.3	71.8	60.9	50.2	33.1	20.0	45.4
HTG DEG DAYS:	1575	1277	1085	567	259	65	10	18	165	465	957	1395	7838

STA NO:14991 EAU CLAIRE WI LAT = 44.87

HORIZ INSOL:	452.	746.	1090.	1426.	1681.	1872.	1886.	1621.	1196.	826.	451.	341.	1132.
TILT = LAT:	851.	1177.	1366.	1437.	1468.	1538.	1594.	1543.	1371.	1204.	768.	643.	0
TILT = LAT+15:	892.	1190.	1304.	1282.	1247.	1276.	1334.	1348.	1275.	1196.	792.	675.	0
TILT = 90:	813.	1006.	965.	787.	661.	627.	671.	774.	879.	973.	702.	621.	0
KT:	.41	.47	.48	.48	.48	.50	.52	.51	.48	.46	.37	.35	
AMB TEMP:	11.7	15.4	27.3	44.5	56.2	66.1	70.5	68.4	58.7	48.7	32.0	18.0	43.1
HTG DEG DAYS:	1652	1389	1169	615	293	65	14	37	202	505	990	1457	8388

STA NO:22010 DEL RIO TX LAT = 29.37

	JAN	FEB	MAR	APR	MAY	JUN	JUL	AUG	SEP	OCT	NOV	DEC	YEAR
HORIZ INSOL:	958.	1206.	1580.	1700.	1827.	2024.	2054.	1936.	1584.	1360.	1060.	903.	1516.
TILT = LAT:	1314.	1494.	1747.	1655.	1632.	1732.	1789.	1819.	1663.	1636.	1429.	1281.	
TILT = LAT+15:	1386.	1516.	1684.	1498.	1409.	1457.	1518.	1611.	1563.	1638.	1498.	1367.	
TILT = 90:	1109.	1065.	947.	613.	450.	398.	424.	568.	781.	1088.	1170.	1132.	
KT:	.49	.51	.55	.51	.51	.55	.57	.57	.53	.54	.52	.50	
AMB TEMP:	50.8	55.7	62.6	72.0	78.2	84.3	86.7	86.1	80.2	71.2	59.6	52.3	70.0
HTG DEG DAYS:	449	283	163	16	0	0	0	0	0	34	184	394	1523

STA NO:23023 MIDLAND-ODESSA TX LAT = 31.93

	JAN	FEB	MAR	APR	MAY	JUN	JUL	AUG	SEP	OCT	NOV	DEC	YEAR
HORIZ INSOL:	1081.	1383.	1839.	2192.	2430.	2562.	2389.	2210.	1844.	1522.	1176.	1000.	1802.
TILT = LAT:	1627.	1833.	2110.	2165.	2142.	2142.	2052.	2081.	1989.	1935.	1724.	1565.	
TILT = LAT+15:	1735.	1876.	2050.	1951.	1822.	1770.	1723.	1833.	1872.	1950.	1824.	1689.	
TILT = 90:	1455.	1386.	1216.	809.	537.	431.	478.	672.	984.	1362.	1491.	1462.	
KT:	.60	.62	.66	.67	.68	.69	.65	.63	.54	.64	.62	.60	
AMB TEMP:	43.6	47.8	54.3	64.3	72.3	79.9	82.3	81.8	75.4	65.8	53.3	45.9	63.9
HTG DEG DAYS:	663	482	349	98	0	0	0	0	0	81	356	592	2621

STA NO:23034 SAN ANGELO TX LAT = 31.37

	JAN	FEB	MAR	APR	MAY	JUN	JUL	AUG	SEP	OCT	NOV	DEC	YEAR
HORIZ INSOL:	962.	1208.	1606.	1851.	2031.	2186.	2123.	1966.	1607.	1337.	1044.	895.	1566.
TILT = LAT:	1386.	1547.	1810.	1813.	1801.	1850.	1836.	1848.	1707.	1649.	1542.	1339.	
TILT = LAT+15:	1467.	1572.	1744.	1637.	1545.	1547.	1553.	1632.	1604.	1652.	1542.	1434.	
TILT = 90:	1207.	1140.	1023.	696.	501.	431.	461.	614.	839.	1134.	1237.	1219.	
KT:	.52	.53	.57	.56	.56	.59	.58	.58	.54	.55	.54	.52	
AMB TEMP:	46.4	50.4	57.1	67.2	74.5	81.6	84.7	84.5	76.8	67.2	55.5	48.3	66.2
HTG DEG DAYS:	577	413	287	74	0	0	0	0	0	73	298	518	2240

STA NO:23042 LUBBOCK TX LAT = 33.65

	JAN	FEB	MAR	APR	MAY	JUN	JUL	AUG	SEP	OCT	NOV	DEC	YEAR
HORIZ INSOL:	1031.	1332.	1762.	2168.	2396.	2544.	2412.	2208.	1820.	1468.	1116.	935.	1766.
TILT = LAT:	1613.	1813.	2056.	2151.	2108.	2119.	2063.	2083.	1986.	1910.	1694.	1522.	
TILT = LAT+15:	1721.	1855.	1988.	1937.	1792.	1750.	1729.	1833.	1868.	1924.	1792.	1641.	
TILT = 90:	1468.	1403.	1221.	852.	578.	473.	518.	717.	1024.	1379.	1491.	1441.	
KT:	.60	.62	.65	.67	.67	.68	.66	.66	.63	.64	.62	.59	
AMB TEMP:	39.1	42.7	48.9	60.0	68.5	77.0	79.7	78.4	71.0	61.0	48.8	41.3	59.7
HTG DEG DAYS:	803	624	508	190	29	0	0	0	8	162	486	735	3545

STA NO:23043 ROSWELL NM LAT = 33.40

	JAN	FEB	MAR	APR	MAY	JUN	JUL	AUG	SEP	OCT	NOV	DEC	YEAR
HORIZ INSOL:	1047.	1373.	1807.	2218.	2459.	2610.	2441.	2242.	1913.	1527.	1131.	952.	1810.
TILT = LAT:	1631.	1871.	2111.	2201.	2163.	2171.	2087.	2115.	2094.	1994.	1811.	1545.	
TILT = LAT+15:	1741.	1917.	2042.	1982.	1837.	1789.	1749.	1860.	1972.	2012.	1810.	1667.	
TILT = 90:	1483.	1448.	1249.	860.	575.	463.	512.	717.	1071.	1440.	1503.	1462.	
KT:	.61	.63	.66	.68	.68	.70	.67	.66	.66	.66	.62	.60	
AMB TEMP:	38.1	42.9	49.3	59.7	68.5	77.0	79.2	77.9	70.4	59.6	46.9	39.3	59.1
HTG DEG DAYS:	834	619	487	185	20	0	0	0	17	195	543	797	3697

STA NO:23044 — EL PASO, TX — LAT = 31.80

	JAN	FEB	MAR	APR	MAY	JUN	JUL	AUG	SEP	OCT	NOV	DEC	YEAR
HORIZ INSOL:	1125.	1480.	1909.	2364.	2601.	2682.	2450.	2284.	1987.	1639.	1244.	1031.	1900.
TILT = LAT:	1703.	1984.	2208.	2341.	2287.	2235.	2102.	2151.	2157.	2110.	1842.	1620.	0
TILT = LAT+15:	1820.	2037.	2140.	2109.	1938.	1639.	1763.	1894.	2034.	2134.	1955.	1751.	0
TILT = 90:	1529.	1510.	1265.	850.	532.	416.	474.	681.	1059.	1494.	1603.	1517.	0
KT:	.62	.66	.69	.72	.72	.72	.67	.67	.67	.68	.65	.61	
AMB TEMP:	43.6	48.4	54.6	63.9	72.2	80.3	82.3	80.5	74.2	64.0	51.6	44.4	63.4
HTG DEG DAYS:	663	465	328	89	0	0	0	0	0	92	402	639	2678

STA NO:23047 — AMARILLO, TX — LAT = 35.23

	JAN	FEB	MAR	APR	MAY	JUN	JUL	AUG	SEP	OCT	NOV	DEC	YEAR
HORIZ INSOL:	960.	1244.	1631.	2019.	2212.	2393.	2280.	2103.	1760.	1404.	1033.	872.	1659.
TILT = LAT:	1552.	1726.	1916.	2007.	1947.	1993.	1950.	1985.	1938.	1864.	1611.	1474.	0
TILT = LAT+15:	1654.	1763.	1849.	1806.	1659.	1651.	1637.	1747.	1821.	1876.	1701.	1588.	0
TILT = 90:	1430.	1358.	1171.	845.	598.	511.	548.	733.	1037.	1375.	1434.	1412.	0
KT:	.59	.60	.61	.63	.62	.64	.62	.63	.62	.63	.60	.58	
AMB TEMP:	36.0	39.7	45.6	56.5	65.6	74.6	78.7	77.6	69.8	59.5	46.3	38.5	57.4
HTG DEG DAYS:	899	708	601	275	81	10	0	0	20	206	561	822	4183

STA NO:23048 — TUCUMCARI, NM — LAT = 35.18

	JAN	FEB	MAR	APR	MAY	JUN	JUL	AUG	SEP	OCT	NOV	DEC	YEAR
HORIZ INSOL:	1009.	1297.	1712.	2096.	2314.	2484.	2349.	2164.	1829.	1443.	1073.	910.	1724.
TILT = LAT:	1650.	1815.	2026.	2090.	2035.	2064.	2006.	2045.	2022.	1926.	1687.	1555.	0
TILT = LAT+15:	1763.	1858.	1957.	1880.	1730.	1706.	1682.	1797.	1902.	1941.	1785.	1680.	0
TILT = 90:	1530.	1434.	1240.	873.	609.	511.	551.	747.	1080.	1424.	1509.	1499.	0
KT:	.62	.62	.64	.65	.64	.67	.64	.65	.64	.64	.62	.61	
AMB TEMP:	37.0	41.1	46.7	56.9	65.6	75.1	78.4	76.7	69.6	58.7	46.2	38.6	57.6
HTG DEG DAYS:	868	669	567	260	57	7	0	0	20	217	564	818	4047

STA NO:23050 — ALBUQUERQUE, NM — LAT = 35.05

	JAN	FEB	MAR	APR	MAY	JUN	JUL	AUG	SEP	OCT	NOV	DEC	YEAR
HORIZ INSOL:	1017.	1342.	1768.	2228.	2538.	2679.	2489.	2290.	1972.	1547.	1134.	928.	1827.
TILT = LAT:	1659.	1887.	2098.	2227.	2227.	2214.	2120.	2167.	2198.	2091.	1804.	1587.	0
TILT = LAT+15:	1773.	1935.	2030.	2003.	1886.	1819.	1772.	1903.	2071.	2114.	1913.	1715.	0
TILT = 90:	1537.	1495.	1263.	914.	623.	502.	553.	772.	1168.	1556.	1623.	1530.	0
KT:	.62	.64	.66	.69	.71	.72	.68	.68	.69	.69	.65	.62	
AMB TEMP:	35.2	40.0	45.8	55.8	65.3	74.6	78.7	76.6	70.1	58.2	44.5	36.2	56.8
HTG DEG DAYS:	924	700	595	282	58	0	0	0	7	218	615	893	4292

STA NO:23051 — CLAYTON, NM — LAT = 36.45

	JAN	FEB	MAR	APR	MAY	JUN	JUL	AUG	SEP	OCT	NOV	DEC	YEAR
HORIZ INSOL:	962.	1241.	1652.	2040.	2222.	2418.	2284.	2097.	1802.	1434.	1028.	861.	1670.
TILT = LAT:	1623.	1770.	1974.	2038.	1953.	2006.	1948.	1984.	2010.	1960.	1564.	1520.	0
TILT = LAT+15:	1734.	1811.	1906.	1832.	1662.	1659.	1634.	1743.	1890.	1977.	1760.	1641.	0
TILT = 90:	1521.	1418.	1236.	887.	627.	539.	575.	762.	1105.	1479.	1506.	1478.	0
KT:	.62	.61	.63	.64	.62	.65	.62	.63	.64	.66	.62	.60	
AMB TEMP:	33.1	36.1	40.4	50.8	60.0	69.2	73.6	72.4	65.0	54.8	42.3	35.1	52.7
HTG DEG DAYS:	989	809	763	431	172	38	0	0	5	324	681	927	5212

STA NO:23062
DENVER
CO
LAT = 39.75

	JAN	FEB	MAR	APR	MAY	JUN	JUL	AUG	SEP	OCT	NOV	DEC	YR
HORIZ INSOL:	840.	1127.	1530.	1879.	2135.	2351.	2273.	2044.	1727.	1301.	884.	732.	1568.
TILT = LAT:	1550.	1709.	1890.	1894.	1873.	1938.	1929.	1946.	1980.	1871.	1539.	1420.	0
TILT = LAT+15:	1655.	1748.	1821.	1698.	1591.	1601.	1613.	1705.	1859.	1884.	1625.	1531.	0
TILT = 90:	1491.	1422.	1252.	908.	686.	610.	648.	830.	1167.	1471.	1429.	1411.	
KT:	.61	.62	.62	.60	.57	.63	.64	.62	.64	.64	.59	.59	
AMB TEMP:	29.9	32.8	37.0	47.5	57.0	66.0	73.0	71.6	62.8	52.0	39.4	32.6	50.1
HTG DEG DAYS:	1089	902	868	525	253	80	0	0	120	408	768	1004	6016

STA NO:23063
EAGLE
CO
LAT = 39.65

	JAN	FEB	MAR	APR	MAY	JUN	JUL	AUG	SEP	OCT	NOV	DEC	YR
HORIZ INSOL:	754.	1078.	1502.	1933.	2255.	2509.	2385.	2084.	1767.	1307.	869.	691.	1594.
TILT = LAT:	1338.	1612.	1845.	1951.	1979.	2063.	2021.	1985.	2030.	1878.	1500.	1308.	0
TILT = LAT+15:	1421.	1644.	1777.	1750.	1677.	1697.	1686.	1739.	1907.	1892.	1582.	1407.	0
TILT = 90:	1269.	1332.	1219.	931.	707.	621.	660.	841.	1195.	1476.	1387.	1290.	
KT:	.54	.58	.60	.62	.63	.67	.65	.63	.65	.64	.58	.55	
AMB TEMP:	18.0	23.3	31.1	41.9	51.3	58.9	65.9	63.7	55.6	44.8	30.9	20.3	42.2
HTG DEG DAYS:	1457	1168	1051	693	425	190	43	79	285	626	1023	1386	8426

STA NO:23065
GOODLAND
KS
LAT = 39.37

	JAN	FEB	MAR	APR	MAY	JUN	JUL	AUG	SEP	OCT	NOV	DEC	YR
HORIZ INSOL:	789.	1056.	1424.	1829.	2062.	2357.	2319.	2046.	1643.	1268.	857.	695.	1529.
TILT = LAT:	1405.	1558.	1726.	1836.	1810.	1945.	1968.	1945.	1862.	1796.	1457.	1301.	0
TILT = LAT+15:	1494.	1588.	1660.	1647.	1539.	1607.	1645.	1705.	1746.	1806.	1535.	1398.	0
TILT = 90:	1334.	1280.	1131.	875.	662.	602.	645.	821.	1088.	1403.	1340.	1278.	
KT:	.56	.56	.57	.58	.58	.63	.64	.62	.61	.62	.57	.55	
AMB TEMP:	27.6	31.5	36.3	48.7	58.9	69.1	75.8	74.1	64.3	52.8	38.5	30.1	50.6
HTG DEG DAYS:	1159	938	890	489	216	55	0	0	108	387	795	1082	6119

STA NO:23066
GRAND JUNCTION
CO
LAT = 39.12

	JAN	FEB	MAR	APR	MAY	JUN	JUL	AUG	SEP	OCT	NOV	DEC	YR
HORIZ INSOL:	791.	1119.	1553.	1986.	2380.	2599.	2465.	2182.	1834.	1345.	918.	731.	1659.
TILT = LAT:	1395.	1665.	1906.	2004.	2087.	2134.	2089.	2080.	2108.	1922.	1580.	1377.	0
TILT = LAT+15:	1483.	1700.	1837.	1799.	1767.	1753.	1741.	1822.	1983.	1938.	1669.	1482.	0
TILT = 90:	1321.	1372.	1250.	940.	716.	612.	656.	859.	1227.	1502.	1461.	1357.	
KT:	.56	.59	.62	.63	.67	.69	.67	.66	.67	.65	.60	.57	
AMB TEMP:	26.6	33.6	41.2	51.7	62.2	71.3	78.7	75.4	67.2	54.9	39.8	29.5	52.7
HTG DEG DAYS:	1190	879	738	404	133	20	0	0	60	324	756	1101	5605

STA NO:23090
FARMINGTON
NM
LAT = 36.75

	JAN	FEB	MAR	APR	MAY	JUN	JUL	AUG	SEP	OCT	NOV	DEC	YR
HORIZ INSOL:	945.	1281.	1693.	2133.	2452.	2666.	2478.	2252.	1934.	1479.	1047.	837.	1766.
TILT = LAT:	1603.	1855.	2040.	2140.	2151.	2196.	2106.	2137.	2186.	2050.	1722.	1484.	0
TILT = LAT+15:	1712.	1901.	1971.	1924.	1822.	1803.	1758.	1875.	2059.	2072.	1824.	1601.	0
TILT = 90:	1505.	1499.	1287.	932.	663.	549.	596.	812.	1209.	1560.	1569.	1444.	
KT:	.61	.64	.65	.67	.68	.71	.68	.68	.69	.68	.64	.59	
AMB TEMP:	28.6	35.0	40.6	49.7	59.5	67.9	75.0	72.6	64.6	52.9	39.2	30.1	51.3
HTG DEG DAYS:	1128	840	756	455	184	36	6	67	67	375	774	1082	5713

STA NO:23129
LONG BEACH
CA
LAT = 33.82

	JAN	FEB	MAR	APR	MAY	JUN	JUL	AUG	SEP	OCT	NOV	DEC	YEAR
HORIZ INSOL:	928.	1215.	1610.	1938.	2064.	2140.	2300.	2100.	1701.	1326.	1004.	847.	1598.
TILT = LAT:	1420.	1629.	1859.	1913.	1823.	1800.	1971.	1979.	1845.	1699.	1491.	1350.	0
TILT = LAT+15:	1507.	1660.	1792.	1723.	1560.	1504.	1657.	1743.	1734.	1704.	1568.	1449.	0
TILT = 90:	1276.	1250.	1104.	780.	551.	474.	518.	698.	957.	1216.	1295.	1263.	0
KT:	.54	.56	.59	.60	.57	.57	.63	.62	.59	.58	.56	.54	
AMB TEMP:	54.2	55.5	57.2	60.6	64.1	67.3	72.2	73.3	71.8	66.6	60.6	55.5	63.3
HTG DEG DAYS:	339	273	247	148	71	23	0	0	7	48	155	295	1606

STA NO:23153
TONOPAH
NV
LAT = 38.07

	JAN	FEB	MAR	APR	MAY	JUN	JUL	AUG	SEP	OCT	NOV	DEC	YEAR
HORIZ INSOL:	918.	1274.	1777.	2251.	2577.	2788.	2703.	2438.	2043.	1521.	1031.	827.	1846.
TILT = LAT:	1625.	1907.	2202.	2283.	2259.	2283.	2284.	2330.	2364.	2191.	1769.	1545.	0
TILT = LAT+15:	1737.	1957.	2132.	2051.	1907.	1866.	1894.	2039.	2230.	2221.	1876.	1671.	0
TILT = 90:	1546.	1571.	1429.	1023.	713.	582.	641.	901.	1347.	1711.	1638.	1527.	0
KT:	.62	.65	.70	.71	.72	.74	.74	.74	.74	.72	.65	.62	
AMB TEMP:	30.2	34.6	39.6	48.1	56.9	65.3	73.0	70.7	63.5	52.1	39.8	31.9	50.5
HTG DEG DAYS:	1079	851	787	512	269	92	0	13	108	407	756	1026	5900

STA NO:23154
ELY
NV
LAT = 39.28

	JAN	FEB	MAR	APR	MAY	JUN	JUL	AUG	SEP	OCT	NOV	DEC	YEAR
HORIZ INSOL:	820.	1141.	1606.	2009.	2311.	2513.	2447.	2230.	1935.	1408.	926.	723.	1672.
TILT = LAT:	1470.	1714.	1987.	2030.	2027.	2067.	2073.	2129.	2249.	2045.	1609.	1365.	0
TILT = LAT+15:	1567.	1752.	1918.	1822.	1717.	1701.	1728.	1865.	2118.	2067.	1701.	1470.	0
TILT = 90:	1402.	1419.	1310.	955.	709.	612.	659.	879.	1316.	1611.	1493.	1347.	0
KT:	.58	.61	.64	.64	.65	.67	.67	.68	.71	.69	.61	.57	
AMB TEMP:	23.8	27.9	32.8	41.3	50.0	57.7	67.2	65.5	56.7	46.0	34.0	26.2	44.1
HTG DEG DAYS:	1283	1039	998	711	470	241	23	62	265	589	930	1203	7814

STA NO:23155
BAKERSFIELD
CA
LAT = 35.42

	JAN	FEB	MAR	APR	MAY	JUN	JUL	AUG	SEP	OCT	NOV	DEC	YEAR
HORIZ INSOL:	766.	1102.	1595.	2095.	2509.	2749.	2684.	2421.	1992.	1458.	942.	677.	1749.
TILT = LAT:	1172.	1497.	1872.	2088.	2202.	2266.	2276.	2296.	2231.	1959.	1443.	1072.	0
TILT = LAT+15:	1234.	1521.	1805.	1878.	1865.	1857.	1892.	2014.	2102.	1976.	1516.	1139.	0
TILT = 90:	1048.	1163.	1147.	878.	631.	507.	562.	811.	1196.	1456.	1272.	991.	0
KT:	.47	.53	.60	.65	.70	.74	.73	.72	.70	.65	.55	.46	
AMB TEMP:	47.5	52.4	56.6	62.7	69.8	76.9	83.9	81.6	76.6	66.6	56.0	47.9	64.9
HTG DEG DAYS:	543	353	266	140	22	0	0	0	0	55	276	530	2185

STA NO:23159
BRYCE CANYON
UT
LAT = 37.70

	JAN	FEB	MAR	APR	MAY	JUN	JUL	AUG	SEP	OCT	NOV	DEC	YEAR
HORIZ INSOL:	914.	1236.	1685.	2133.	2454.	2655.	2424.	2157.	1920.	1465.	1015.	818.	1740.
TILT = LAT:	1592.	1817.	2055.	2150.	2152.	2184.	2059.	2048.	2189.	2070.	1711.	1499.	0
TILT = LAT+15:	1700.	1861.	1986.	1931.	1821.	1792.	1719.	1796.	2061.	2093.	1812.	1619.	0
TILT = 90:	1507.	1483.	1319.	962.	689.	575.	616.	812.	1236.	1599.	1573.	1472.	0
KT:	.61	.63	.66	.67	.68	.71	.66	.65	.69	.64	.64	.60	
AMB TEMP:	19.8	23.2	28.7	37.7	46.2	54.1	61.6	59.9	52.9	42.8	30.7	22.4	40.0
HTG DEG DAYS:	1401	1170	1125	819	583	330	128	176	363	688	1029	1321	9133

STA NO:23160 TUCSON AZ LAT = 32.12

	Jan	Feb	Mar	Apr	May	Jun	Jul	Aug	Sep	Oct	Nov	Dec	Ann
HORIZ INSOL:	1099.	1432.	1864.	2363.	2671.	2730.	2341.	2183.	1979.	1602.	1208.	996.	1872.
TILT = LAT:	1669.	1918.	2156.	2343.	2345.	2269.	2012.	2055.	2152.	2064.	1792.	1567.	0
TILT = LAT+15:	1782.	1967.	2088.	2111.	1983.	1864.	1692.	1811.	2029.	2085.	1900.	1691.	0
TILT = 90:	1500.	1462.	1243.	860.	538.	418.	482.	672.	1066.	1465.	1560.	1466.	0
KT:	.61	.64	.67	.72	.74	.73	.64	.64	.67	.67	.64	.60	
AMB TEMP:	50.9	53.5	57.6	65.5	73.6	82.1	86.3	83.8	80.1	70.1	58.5	52.0	67.8
HTG DEG DAYS:	442	333	243	61	0	0	0	0	0	29	221	403	1752

STA NO:23161 DAGGETT CA LAT = 34.87

	Jan	Feb	Mar	Apr	May	Jun	Jul	Aug	Sep	Oct	Nov	Dec	Ann
HORIZ INSOL:	958.	1281.	1772.	2274.	2591.	2766.	2603.	2383.	2008.	1516.	1085.	876.	1843.
TILT = LAT:	1529.	1775.	2100.	2274.	2273.	2281.	2213.	2256.	2240.	2032.	1695.	1463.	0
TILT = LAT+15:	1628.	1815.	2031.	2046.	1922.	1869.	1845.	1980.	2111.	2052.	1794.	1576.	0
TILT = 90:	1401.	1393.	1279.	924.	620.	490.	549.	786.	1185.	1503.	1512.	1396.	0
KT:	.58	.61	.66	.70	.72	.74	.71	.71	.70	.67	.62	.58	
AMB TEMP:	47.3	52.0	56.7	64.3	72.3	80.1	87.3	85.5	79.2	68.1	55.5	48.0	66.4
HTG DEG DAYS:	549	371	271	118	14	0	0	0	0	57	296	527	2203

STA NO:23169 LAS VEGAS NV LAT = 36.08

	Jan	Feb	Mar	Apr	May	Jun	Jul	Aug	Sep	Oct	Nov	Dec	Ann
HORIZ INSOL:	978.	1340.	1823.	2319.	2646.	2778.	2588.	2355.	2037.	1540.	1086.	881.	1864.
TILT = LAT:	1637.	1930.	2206.	2335.	2319.	2284.	2197.	2235.	2305.	2126.	1764.	1544.	0
TILT = LAT+15:	1749.	1981.	2137.	2099.	1957.	1870.	1830.	1960.	2174.	2151.	1870.	1667.	0
TILT = 90:	1530.	1553.	1380.	981.	658.	524.	582.	817.	1255.	1608.	1601.	1498.	0
KT:	.62	.65	.69	.72	.74	.74	.71	.71	.70	.70	.64	.61	
AMB TEMP:	44.2	49.1	54.8	63.8	73.3	82.3	89.6	87.4	80.1	67.1	53.3	45.2	65.8
HTG DEG DAYS:	645	451	324	126	10	0	0	0	0	74	357	614	2601

STA NO:23174 LOS ANGELES CA LAT = 33.93

	Jan	Feb	Mar	Apr	May	Jun	Jul	Aug	Sep	Oct	Nov	Dec	Ann
HORIZ INSOL:	926.	1214.	1619.	1951.	2060.	2119.	2307.	2080.	1681.	1317.	1004.	849.	1594.
TILT = LAT:	1422.	1631.	1873.	1928.	1819.	1783.	1977.	1960.	1824.	1688.	1496.	1359.	0
TILT = LAT+15:	1509.	1663.	1806.	1736.	1556.	1490.	1661.	1726.	1713.	1692.	1574.	1459.	0
TILT = 90:	1280.	1253.	1115.	788.	552.	475.	521.	696.	948.	1210.	1302.	1274.	0
KT:	.54	.56	.60	.60	.57	.57	.63	.62	.58	.57	.56	.54	
AMB TEMP:	54.5	55.6	56.5	58.8	61.9	64.5	68.5	69.5	68.7	65.2	60.5	56.9	61.7
HTG DEG DAYS:	331	270	267	195	114	71	19	15	23	77	158	279	1819

STA NO:23179 NEEDLES CA LAT = 34.77

	Jan	Feb	Mar	Apr	May	Jun	Jul	Aug	Sep	Oct	Nov	Dec	Ann
HORIZ INSOL:	985.	1353.	1825.	2317.	2652.	2791.	2541.	2278.	2015.	1537.	1124.	913.	1861.
TILT = LAT:	1578.	1895.	2170.	2319.	2324.	2301.	2163.	2154.	2246.	2064.	1767.	1538.	0
TILT = LAT+15:	1683.	1943.	2102.	2086.	1964.	1883.	1806.	1892.	2118.	2085.	1873.	1660.	0
TILT = 90:	1450.	1495.	1321.	935.	618.	484.	547.	762.	1185.	1527.	1582.	1474.	0
KT:	.60	.64	.68	.72	.74	.75	.70	.68	.70	.68	.64	.60	
AMB TEMP:	51.6	56.5	61.6	70.4	79.6	88.3	95.4	93.3	86.9	74.3	60.7	52.7	72.6
HTG DEG DAYS:	421	261	150	42	0	0	0	0	0	10	163	381	1428

STA NO:23183 PHOENIX AZ LAT = 33.43

	JAN	FEB	MAR	APR	MAY	JUN	JUL	AUG	SEP	OCT	NOV	DEC	YR
HORIZ INSOL:	1021.	1374.	1814.	2355.	2677.	2739.	2486.	2293.	2015.	1576.	1150.	932.	1869.
TILT = LAT:	1584.	1875.	2121.	2346.	2347.	2269.	2124.	2164.	2220.	2073.	1749.	1505.	0
TILT = LAT+15:	1688.	1921.	2052.	2112.	1903.	1862.	1902.	1902.	2094.	2095.	1852.	1622.	0
TILT = 90:	1435.	1451.	1256.	902.	577.	453.	513.	727.	1135.	1504.	1541.	1421.	0
KT:	.59	.63	.67	.72	.74	.74	.68	.68	.70	.68	.63	.58	
AMB TEMP:	51.2	55.1	59.7	67.7	76.3	84.6	91.2	89.1	83.0	72.2	59.8	52.5	70.3
HTG DEG DAYS:	428	292	185	60	0	0	0	0	0	17	182	388	1552

STA NO:23184 PRESCOTT AZ LAT = 34.65

	JAN	FEB	MAR	APR	MAY	JUN	JUL	AUG	SEP	OCT	NOV	DEC	YR
HORIZ INSOL:	1016.	1335.	1777.	2275.	2629.	2762.	2309.	2092.	1955.	1543.	1140.	927.	1813.
TILT = LAT:	1636.	1858.	2101.	2273.	2305.	2279.	1976.	1973.	2168.	2068.	1794.	1561.	0
TILT = LAT+15:	1747.	1904.	2032.	2045.	1949.	1868.	1659.	1736.	2042.	2090.	1902.	1696.	0
TILT = 90:	1507.	1461.	1274.	917.	614.	484.	537.	716.	1142.	1528.	1606.	1497.	0
KT:	.61	.63	.66	.70	.73	.74	.63	.62	.68	.68	.65	.61	
AMB TEMP:	37.1	40.5	44.3	52.0	60.4	69.1	75.5	73.0	68.1	57.2	45.8	38.6	55.1
HTG DEG DAYS:	865	686	642	394	165	33	0	0	23	254	576	818	4456

STA NO:23185 RENO NV LAT = 39.50

	JAN	FEB	MAR	APR	MAY	JUN	JUL	AUG	SEP	OCT	NOV	DEC	YR
HORIZ INSOL:	800.	1150.	1649.	2159.	2523.	2701.	2692.	2406.	1998.	1431.	912.	706.	1761.
TILT = LAT:	1438.	1741.	2058.	2192.	2212.	2212.	2272.	2308.	2343.	2100.	1590.	1336.	0
TILT = LAT+15:	1531.	1781.	1989.	1974.	1867.	1811.	1883.	2019.	2209.	2125.	1680.	1437.	0
TILT = 90:	1371.	1447.	1365.	1032.	743.	625.	684.	940.	1378.	1665.	1477.	1318.	0
KT:	.57	.61	.66	.69	.71	.72	.74	.73	.74	.70	.61	.56	
AMB TEMP:	31.9	37.1	40.3	46.8	54.6	61.5	69.3	66.9	60.2	50.3	40.1	33.0	49.4
HTG DEG DAYS:	1026	781	766	546	328	145	17	50	168	456	747	992	6022

STA NO:23188 SAN DIEGO CA LAT = 32.73

	JAN	FEB	MAR	APR	MAY	JUN	JUL	AUG	SEP	OCT	NOV	DEC	YR
HORIZ INSOL:	976.	1266.	1632.	1937.	2003.	2062.	2186.	2057.	1717.	1373.	1063.	904.	1598.
TILT = LAT:	1465.	1676.	1866.	1907.	1773.	1744.	1802.	1936.	1851.	1738.	1554.	1415.	0
TILT = LAT+15:	1556.	1709.	1800.	1719.	1520.	1461.	1588.	1707.	1739.	1744.	1637.	1520.	0
TILT = 90:	1305.	1270.	1085.	753.	523.	451.	490.	663.	936.	1226.	1339.	1316.	0
KT:	.55	.57	.59	.59	.56	.55	.60	.61	.59	.58	.57	.55	
AMB TEMP:	55.2	56.7	58.0	60.7	63.3	65.5	69.6	71.4	69.9	66.1	60.8	56.7	62.9
HTG DEG DAYS:	314	237	219	144	79	52	6	0	16	43	140	257	1507

STA NO:23194 WINSLOW AZ LAT = 35.02

	JAN	FEB	MAR	APR	MAY	JUN	JUL	AUG	SEP	OCT	NOV	DEC	YR
HORIZ INSOL:	985.	1327.	1780.	2283.	2595.	2712.	2347.	2141.	1928.	1513.	1119.	894.	1802.
TILT = LAT:	1591.	1860.	2115.	2285.	2276.	2239.	2005.	2022.	2142.	2053.	1772.	1511.	0
TILT = LAT+15:	1697.	1906.	2046.	2056.	1924.	1838.	1681.	1778.	2017.	2053.	1879.	1630.	0
TILT = 90:	1466.	1470.	1292.	932.	624.	499.	547.	737.	1138.	1507.	1591.	1449.	0
KT:	.60	.63	.67	.71	.72	.73	.64	.64	.68	.67	.64	.59	
AMB TEMP:	32.6	39.1	44.8	53.7	62.7	71.8	78.3	76.1	69.5	57.3	43.2	33.8	55.3
HTG DEG DAYS:	1004	725	626	348	124	14	0	0	19	252	654	967	4733

STA NO:23195 — YUMA, AZ — LAT = 32.67

	Jan	Feb	Mar	Apr	May	Jun	Jul	Aug	Sep	Oct	Nov	Dec	Year
HORIZ INSOL:	1096.	1443.	1919.	2413.	2728.	2814.	2453.	2329.	2051.	1623.	1215.	1000.	1924.
TILT = LAT:	1691.	1958.	2243.	2401.	2392.	2329.	2101.	2196.	2250.	2117.	1832.	1604.	0
TILT = LAT+15:	1807.	2010.	2175.	2162.	2019.	1906.	1761.	1931.	2123.	2142.	1944.	1734.	0
TILT = 90:	1532.	1507.	1310.	891.	551.	420.	495.	712.	1127.	1521.	1609.	1514.	0
KT:	.62	.65	.70	.74	.76	.76	.67	.69	.70	.69	.65	.61	
AMB TEMP:	55.4	59.4	63.9	71.2	78.7	85.8	93.7	92.8	87.1	75.9	63.5	56.3	73.7
HTG DEG DAYS:	308	192	97	24	0	0	0	0	0	5	108	276	1010

STA NO:23230 — OAKLAND, CA — LAT = 37.73

	Jan	Feb	Mar	Apr	May	Jun	Jul	Aug	Sep	Oct	Nov	Dec	Year
HORIZ INSOL:	708.	1018.	1456.	1922.	2211.	2350.	2322.	2053.	1701.	1212.	822.	647.	1535.
TILT = LAT:	1144.	1429.	1735.	1923.	1942.	1947.	1976.	1945.	1907.	1640.	1305.	1102.	0
TILT = LAT+15:	1205.	1451.	1668.	1727.	1651.	1611.	1654.	1708.	1790.	1644.	1367.	1175.	0
TILT = 90:	1046.	1140.	1105.	874.	655.	564.	608.	782.	1077.	1240.	1167.	1049.	0
KT:	.47	.52	.57	.61	.62	.63	.64	.62	.57	.57	.50	.48	
AMB TEMP:	48.6	51.9	53.7	56.1	58.9	61.9	63.1	63.5	63.5	61.1	55.3	49.9	57.4
HTG DEG DAYS:	508	367	350	270	193	114	80	74	59	135	291	468	2909

STA NO:23232 — SACRAMENTO, CA — LAT = 38.52

	Jan	Feb	Mar	Apr	May	Jun	Jul	Aug	Sep	Oct	Nov	Dec	Year
HORIZ INSOL:	597.	939.	1458.	2004.	2435.	2684.	2688.	2368.	1907.	1315.	782.	538.	1643.
TILT = LAT:	939.	1319.	1755.	2017.	2135.	2271.	2263.	2263.	2191.	1843.	1256.	890.	0
TILT = LAT+15:	982.	1335.	1688.	1811.	1807.	1805.	1884.	1981.	2062.	1855.	1315.	940.	0
TILT = 90:	846.	1054.	1134.	931.	708.	597.	654.	897.	1260.	1423.	1128.	832.	0
KT:	.41	.49	.58	.64	.68	.72	.74	.72	.70	.63	.50	.41	
AMB TEMP:	45.1	49.8	53.0	58.3	64.3	70.5	75.2	74.1	71.5	63.3	53.0	45.8	60.3
HTG DEG DAYS:	617	426	372	227	120	20	0	0	5	101	360	595	2843

STA NO:23234 — SAN FRANCISCO, CA — LAT = 37.62

	Jan	Feb	Mar	Apr	May	Jun	Jul	Aug	Sep	Oct	Nov	Dec	Year
HORIZ INSOL:	708.	1009.	1455.	1920.	2226.	2377.	2392.	2116.	1742.	1226.	821.	642.	1553.
TILT = LAT:	1139.	1411.	1730.	1920.	1955.	1969.	2033.	2008.	1956.	1659.	1299.	1087.	0
TILT = LAT+15:	1199.	1432.	1664.	1724.	1661.	1628.	1699.	1762.	1837.	1663.	1360.	1158.	0
TILT = 90:	1039.	1123.	1099.	870.	654.	564.	612.	798.	1102.	1253.	1159.	1031.	0
KT:	.47	.51	.57	.61	.62	.63	.65	.64	.63	.58	.51	.47	
AMB TEMP:	48.3	51.2	53.0	55.3	58.3	61.6	62.5	63.0	64.1	61.0	55.3	49.7	56.9
HTG DEG DAYS:	518	386	372	291	210	120	93	84	66	137	291	474	3042

STA NO:23244 — SUNNYVALE, CA — LAT = 37.42

	Jan	Feb	Mar	Apr	May	Jun	Jul	Aug	Sep	Oct	Nov	Dec	Year
HORIZ INSOL:	738.	1037.	1485.	1944.	2277.	2453.	2441.	2167.	1760.	1248.	843.	660.	1588.
TILT = LAT:	1193.	1453.	1767.	1946.	1999.	2029.	2074.	2057.	1975.	1689.	1334.	1118.	0
TILT = LAT+15:	1259.	1476.	1701.	1746.	1698.	1675.	1732.	1804.	1855.	1694.	1399.	1192.	0
TILT = 90:	1093.	1156.	1120.	875.	658.	563.	611.	807.	1108.	1274.	1192.	1062.	0
KT:	.49	.52	.58	.61	.64	.65	.67	.65	.63	.58	.52	.48	
AMB TEMP:	0	0	0	0	0	0	0	0	0	0	0	0	0
HTG DEG DAYS:	0	0	0	0	0	0	0	0	0	0	0	0	0

STA NO:23273 — SANTA MARIA, CA — LAT = 34.90

	Jan	Feb	Mar	Apr	May	Jun	Jul	Aug	Sep	Oct	Nov	Dec	Year
HORIZ INSOL:	854.	1141.	1582.	1921.	2141.	2349.	2341.	2106.	1730.	1353.	974.	804.	1608.
TILT = LAT:	1323.	1544.	1844.	1902.	1886.	1960.	2000.	1988.	1896.	1773.	1481.	1312.	0
TILT = LAT+15:	1400.	1571.	1777.	1712.	1609.	1626.	1678.	1748.	1781.	1781.	1557.	1407.	0
TILT = 90:	1193.	1195.	1118.	801.	582.	503.	544.	726.	1007.	1295.	1301.	1237.	0
KT:	.52	.54	.59	.60	.60	.63	.64	.63	.61	.60	.56	.53	
AMB TEMP:	50.5	52.0	52.8	54.9	57.1	59.6	62.1	62.3	62.6	60.4	56.1	51.8	56.9
HTG DEG DAYS:	450	364	378	303	245	167	112	102	94	159	270	409	3053

STA NO:24011 — BISMARCK, ND — LAT = 46.77

	Jan	Feb	Mar	Apr	May	Jun	Jul	Aug	Sep	Oct	Nov	Dec	Year
HORIZ INSOL:	467.	776.	1168.	1459.	1848.	2060.	2184.	1877.	1354.	908.	507.	373.	1248.
TILT = LAT:	998.	1326.	1538.	1489.	1616.	1681.	1840.	1817.	1626.	1438.	1005.	839.	0
TILT = LAT+15:	1055.	1348.	1474.	1327.	1366.	1386.	1530.	1584.	1518.	1439.	1050.	894.	0
TILT = 90:	984.	1168.	1126.	847.	750.	705.	789.	940.	1085.	1207.	960.	846.	0
KT:	.47	.52	.54	.50	.53	.55	.60	.60	.56	.54	.46	.43	
AMB TEMP:	8.2	13.5	25.1	43.0	54.4	63.8	70.8	69.2	57.5	46.8	28.9	15.6	41.4
HTG DEG DAYS:	1761	1442	1237	660	339	122	18	35	252	564	1083	1531	9044

STA NO:24013 — MINOT, ND — LAT = 48.27

	Jan	Feb	Mar	Apr	May	Jun	Jul	Aug	Sep	Oct	Nov	Dec	Year
HORIZ INSOL:	384.	656.	1044.	1461.	1846.	1975.	2098.	1800.	1277.	850.	439.	310.	1176.
TILT = LAT:	830.	1123.	1380.	1505.	1614.	1607.	1765.	1748.	1549.	1386.	891.	715.	0
TILT = LAT+15:	874.	1137.	1319.	1340.	1362.	1325.	1467.	1522.	1444.	1386.	928.	760.	0
TILT = 90:	818.	993.	1026.	881.	778.	711.	797.	939.	1058.	1181.	855.	722.	0
KT:	.42	.46	.50	.51	.53	.53	.58	.58	.54	.53	.43	.40	
AMB TEMP:	7.9	12.8	23.6	41.1	52.8	62.0	68.8	67.2	56.2	46.1	27.9	14.7	40.1
HTG DEG DAYS:	1770	1462	1283	717	384	150	27	70	286	586	1113	1559	9407

STA NO:24018 — CHEYENNE, WY — LAT = 41.15

	Jan	Feb	Mar	Apr	May	Jun	Jul	Aug	Sep	Oct	Nov	Dec	Year
HORIZ INSOL:	766.	1068.	1433.	1771.	1995.	2258.	2230.	1966.	1667.	1242.	823.	671.	1491.
TILT = LAT:	1458.	1662.	1786.	1787.	1749.	1860.	1890.	1874.	1931.	1828.	1482.	1355.	0
TILT = LAT+15:	1555.	1698.	1718.	1601.	1486.	1537.	1579.	1641.	1812.	1840.	1564.	1460.	0
TILT = 90:	1412.	1402.	1208.	891.	683.	629.	672.	837.	1170.	1461.	1390.	1357.	0
KT:	.59	.60	.59	.57	.56	.60	.61	.60	.63	.63	.58	.57	
AMB TEMP:	26.6	29.0	31.6	42.7	52.4	61.3	69.1	67.6	58.2	47.9	35.5	29.2	45.9
HTG DEG DAYS:	1190	1008	1035	669	394	156	22	31	225	530	885	1110	7255

STA NO:24023 — NORTH PLATTE, NE — LAT = 41.13

	Jan	Feb	Mar	Apr	May	Jun	Jul	Aug	Sep	Oct	Nov	Dec	Year
HORIZ INSOL:	692.	958.	1333.	1724.	1968.	2267.	2277.	1990.	1565.	1177.	759.	605.	1445.
TILT = LAT:	1274.	1448.	1639.	1735.	1743.	1842.	1929.	1898.	1795.	1708.	1333.	1178.	0
TILT = LAT+15:	1351.	1472.	1573.	1555.	1481.	1542.	1611.	1662.	1681.	1715.	1400.	1262.	0
TILT = 90:	1219.	1207.	1103.	868.	681.	630.	680.	845.	1085.	1357.	1237.	1165.	0
KT:	.53	.55	.56	.56	.56	.60	.62	.61	.59	.60	.54	.52	
AMB TEMP:	23.4	28.1	34.3	47.8	58.3	68.0	74.3	73.0	62.3	51.0	36.2	26.8	48.6
HTG DEG DAYS:	1290	1033	952	522	238	65	7	8	141	439	864	1184	6743

STA NO:24025 PIERRE SD LAT = 44.38

	Jan	Feb	Mar	Apr	May	Jun	Jul	Aug	Sep	Oct	Nov	Dec	Ann
HORIZ INSOL:	530.	795.	1206.	1614.	1966.	2195.	2278.	1993.	1496.	1052.	623.	442.	1349.
TILT = LAT:	1039.	1258.	1532.	1643.	1722.	1797.	1923.	1920.	1770.	1616.	1179.	914.	
TILT = LAT+15:	1097.	1275.	1467.	1467.	1458.	1482.	1600.	1677.	1656.	1622.	1236.	974.	
TILT = 90:	1006.	1076.	1081.	886.	741.	687.	757.	932.	1137.	1331.	1117.	909.	
KT:	.47	.49	.53	.54	.56	.58	.63	.62	.51	.59	.51	.45	
AMB TEMP:	15.6	20.4	29.8	46.3	57.4	67.4	75.2	73.9	62.1	44.7	33.8	21.5	45.2
HTG DEG DAYS:	1531	1249	1091	561	267	74	6	10	152	451	936	1349	7677

STA NO:24027 ROCK SPRINGS WY LAT = 41.60

	Jan	Feb	Mar	Apr	May	Jun	Jul	Aug	Sep	Oct	Nov	Dec	Ann
HORIZ INSOL:	735.	1089.	1530.	1941.	2344.	2574.	2547.	2240.	1833.	1306.	826.	651.	1635.
TILT = LAT:	1409.	1729.	1945.	1983.	2056.	2106.	2151.	2155.	2170.	1973.	1519.	1330.	
TILT = LAT+15:	1501.	1769.	1876.	1777.	1737.	1727.	1785.	1684.	2041.	1993.	1604.	1433.	
TILT = 90:	1366.	1472.	1333.	992.	775.	677.	732.	952.	1331.	1597.	1433.	1335.	
KT:	.57	.62	.64	.63	.66	.69	.70	.69	.70	.67	.60	.57	
AMB TEMP:	19.2	23.4	28.9	40.1	50.4	58.9	68.2	66.1	56.4	44.7	30.7	22.6	42.5
HTG DEG DAYS:	1420	1165	1119	747	453	198	18	49	269	629	1029	1314	8410

STA NO:24028 SCOTTSBLUFF NE LAT = 41.87

	Jan	Feb	Mar	Apr	May	Jun	Jul	Aug	Sep	Oct	Nov	Dec	Ann
HORIZ INSOL:	676.	951.	1307.	1668.	1933.	2237.	2284.	2000.	1599.	1145.	723.	575.	1425.
TILT = LAT:	1275.	1464.	1619.	1681.	1694.	1840.	1932.	1912.	1855.	1681.	1288.	1141.	
TILT = LAT+15:	1353.	1489.	1554.	1505.	1440.	1520.	1612.	1673.	1738.	1687.	1352.	1222.	
TILT = 90:	1228.	1232.	1102.	856.	682.	641.	698.	866.	1138.	1346.	1200.	1132.	
KT:	.53	.54	.55	.54	.54	.60	.63	.62	.61	.59	.53	.51	
AMB TEMP:	24.9	29.5	34.3	46.2	56.5	65.9	73.7	71.6	61.2	50.2	36.2	27.6	48.2
HTG DEG DAYS:	1243	994	952	564	280	31	0	8	160	459	864	1159	6774

STA NO:24029 SHERIDAN WY LAT = 44.77

	Jan	Feb	Mar	Apr	May	Jun	Jul	Aug	Sep	Oct	Nov	Dec	Ann
HORIZ INSOL:	518.	788.	1205.	1537.	1883.	2156.	2329.	2006.	1502.	1005.	591.	441.	1330.
TILT = LAT:	1026.	1260.	1539.	1560.	1647.	1765.	1965.	1937.	1787.	1543.	1118.	934.	
TILT = LAT+15:	1083.	1277.	1475.	1393.	1396.	1456.	1633.	1691.	1672.	1546.	1171.	996.	
TILT = 90:	996.	1082.	1094.	850.	723.	687.	777.	948.	1156.	1272.	1058.	934.	
KT:	.47	.49	.53	.51	.53	.57	.64	.63	.60	.56	.49	.46	
AMB TEMP:	21.0	25.9	31.0	43.6	53.1	61.1	70.4	69.2	57.9	47.8	33.4	25.5	45.0
HTG DEG DAYS:	1364	1095	1054	642	375	168	28	31	245	533	948	1225	7708

STA NO:24033 BILLINGS MT LAT = 45.80

	Jan	Feb	Mar	Apr	May	Jun	Jul	Aug	Sep	Oct	Nov	Dec	Ann
HORIZ INSOL:	486.	763.	1189.	1526.	1913.	2174.	2384.	2022.	1470.	987.	561.	421.	1325.
TILT = LAT:	996.	1252.	1544.	1557.	1674.	1775.	2009.	1962.	1768.	1554.	1098.	935.	
TILT = LAT+15:	1052.	1269.	1480.	1389.	1416.	1462.	1667.	1712.	1653.	1558.	1150.	998.	
TILT = 90:	974.	1087.	1115.	867.	753.	712.	816.	985.	1164.	1298.	1048.	944.	
KT:	.46	.49	.54	.52	.54	.58	.66	.64	.59	.57	.49	.46	
AMB TEMP:	21.9	27.4	32.6	44.5	54.5	62.6	71.8	70.1	58.9	49.3	35.7	26.8	46.3
HTG DEG DAYS:	1336	1053	1004	612	333	131	10	15	221	487	879	1184	7265

STA NO:24036 LEWISTOWN MT LAT = 47.05

	Jan	Feb	Mar	Apr	May	Jun	Jul	Aug	Sep	Oct	Nov	Dec	Ann
HORIZ INSOL:	420.	692.	1128.	1444.	1807.	2059.	2288.	1901.	1372.	905.	502.	363.	1240.
TILT = LAT:	873.	1151.	1482.	1475.	1579.	1679.	1927.	1845.	1658.	1445.	1007.	823.	0
TILT = LAT+15:	919.	1165.	1419.	1314.	1335.	1384.	1599.	1609.	1548.	1446.	1052.	877.	0
TILT = 90:	854.	1006.	1087.	843.	741.	710.	824.	960.	1113.	1218.	965.	832.	0
KT:	.43	.47	.52	.49	.52	.55	.63	.61	.57	.54	.46	.43	
AMB TEMP:	19.1	23.8	27.5	40.1	49.6	56.6	65.5	64.4	54.0	45.5	32.2	24.5	41.9
HTG DEG DAYS:	1423	1154	1163	747	477	265	70	94	348	605	984	1256	8536

STA NO:24037 MILES CITY MT LAT = 46.43

	Jan	Feb	Mar	Apr	May	Jun	Jul	Aug	Sep	Oct	Nov	Dec	Ann
HORIZ INSOL:	457.	745.	1185.	1542.	1896.	2146.	2293.	1977.	1444.	961.	551.	399.	1300.
TILT = LAT:	949.	1241.	1555.	1580.	1659.	1751.	1932.	1920.	1746.	1531.	1107.	905.	0
TILT = LAT+15:	1000.	1259.	1491.	1410.	1402.	1442.	1604.	1674.	1632.	1535.	1160.	966.	0
TILT = 90:	929.	1085.	1134.	891.	760.	719.	810.	981.	1162.	1287.	1063.	916.	0
KT:	.45	.49	.54	.52	.54	.57	.63	.63	.59	.57	.49	.45	
AMB TEMP:	15.4	21.6	30.2	45.3	56.3	64.9	74.4	72.5	59.9	48.8	32.4	22.0	45.3
HTG DEG DAYS:	1538	1215	1079	591	288	117	9	16	217	508	978	1333	7889

STA NO:24089 CASPER WY LAT = 42.92

	Jan	Feb	Mar	Apr	May	Jun	Jul	Aug	Sep	Oct	Nov	Dec	Ann
HORIZ INSOL:	683.	1014.	1441.	1847.	2204.	2501.	2535.	2225.	1750.	1219.	765.	594.	1565.
TILT = LAT:	1362.	1642.	1850.	1883.	1932.	2044.	2138.	2150.	2089.	1872.	1453.	1364.	0
TILT = LAT+15:	1450.	1678.	1782.	1691.	1634.	1678.	1774.	1879.	1963.	1888.	1533.	1360.	0
TILT = 90:	1331.	1412.	1291.	979.	775.	703.	767.	988.	1313.	1533.	1382.	1276.	0
KT:	.57	.60	.61	.61	.62	.67	.70	.69	.68	.65	.58	.55	
AMB TEMP:	23.2	26.8	31.0	42.7	52.7	61.9	71.0	69.6	58.7	47.7	33.9	26.2	45.4
HTG DEG DAYS:	1296	1070	1054	669	388	147	13	17	229	536	933	1203	7555

STA NO:24090 RAPID CITY SD LAT = 44.05

	Jan	Feb	Mar	Apr	May	Jun	Jul	Aug	Sep	Oct	Nov	Dec	Ann
HORIZ INSOL:	542.	827.	1229.	1587.	1887.	2131.	2223.	1963.	1518.	1064.	647.	476.	1341.
TILT = LAT:	1053.	1309.	1557.	1612.	1652.	1748.	1877.	1888.	1793.	1624.	1222.	995.	0
TILT = LAT+15:	1112.	1328.	1492.	1440.	1411.	1444.	1564.	1649.	1678.	1630.	1282.	1063.	0
TILT = 90:	1018.	1118.	1095.	864.	711.	668.	737.	909.	1145.	1333.	1157.	994.	0
KT:	.47	.50	.53	.53	.53	.57	.61	.61	.60	.59	.52	.47	
AMB TEMP:	21.9	25.8	31.2	44.6	55.2	64.2	72.6	71.6	60.5	50.0	35.4	26.5	46.6
HTG DEG DAYS:	1336	1098	1048	612	319	134	13	17	191	474	888	1194	7324

STA NO:24121 ELKO NV LAT = 40.83

	Jan	Feb	Mar	Apr	May	Jun	Jul	Aug	Sep	Oct	Nov	Dec	Ann
HORIZ INSOL:	689.	1034.	1463.	1900.	2303.	2534.	2623.	2316.	1893.	1322.	812.	617.	1626.
TILT = LAT:	1249.	1581.	1821.	1926.	2020.	2077.	2214.	2227.	2233.	1964.	1438.	1190.	0
TILT = LAT+15:	1323.	1613.	1753.	1726.	1709.	1706.	1836.	1947.	2102.	1983.	1515.	1276.	0
TILT = 90:	1189.	1324.	1227.	948.	746.	653.	719.	955.	1349.	1574.	1341.	1175.	0
KT:	.52	.57	.60	.61	.65	.68	.72	.71	.71	.67	.57	.52	
AMB TEMP:	23.2	29.2	35.0	43.5	51.9	59.6	69.5	67.0	57.6	46.9	34.8	25.9	45.4
HTG DEG DAYS:	1296	1002	930	645	406	190	27	60	248	561	906	1212	7483

STA NO:24127 SALT LAKE CITY, UT — LAT = 40.77

	Jan	Feb	Mar	Apr	May	Jun	Jul	Aug	Sep	Oct	Nov	Dec	Year
HORIZ INSOL:	639.	989.	1454.	1894.	2362.	2561.	2590.	2254.	1843.	1293.	788.	570.	1603.
TILT = LAT:	1126.	1491.	1806.	1919.	2072.	2099.	2187.	2163.	2162.	1906.	1379.	1065.	
TILT = LAT+15:	1189.	1517.	1739.	1721.	1751.	1723.	1815.	1892.	2034.	1921.	1450.	1137.	
TILT = 90:	1061.	1240.	1215.	943.	756.	653.	713.	931.	1303.	1521.	1279.	1040.	
KT:	.48	.55	.60	.61	.66	.68	.71	.69	.69	.65	.59	.48	
AMB TEMP:	28.0	33.4	39.6	49.2	58.3	66.2	76.7	74.5	64.8	52.4	39.1	30.3	51.0
HTG DEG DAYS:	1147	885	787	474	237	88	0	5	105	402	777	1076	5983

STA NO:24128 WINNEMUCCA, NV — LAT = 40.90

	Jan	Feb	Mar	Apr	May	Jun	Jul	Aug	Sep	Oct	Nov	Dec	Year
HORIZ INSOL:	691.	1028.	1472.	1967.	2362.	2569.	2678.	2348.	1907.	1322.	810.	618.	1648.
TILT = LAT:	1256.	1571.	1836.	2002.	2071.	2105.	2258.	2260.	2254.	1967.	1436.	1198.	
TILT = LAT+15:	1332.	1602.	1768.	1794.	1751.	1727.	1871.	1976.	2124.	1986.	1513.	1285.	
TILT = 90:	1198.	1315.	1239.	984.	760.	658.	725.	968.	1365.	1578.	1339.	1184.	
KT:	.52	.57	.61	.64	.66	.68	.73	.72	.72	.67	.57	.52	
AMB TEMP:	28.2	34.1	37.6	45.1	53.8	61.7	71.0	67.8	59.2	48.3	37.3	30.4	47.9
HTG DEG DAYS:	1141	865	849	597	359	149	6	42	199	518	831	1073	6629

STA NO:24131 BOISE, ID — LAT = 43.57

	Jan	Feb	Mar	Apr	May	Jun	Jul	Aug	Sep	Oct	Nov	Dec	Year
HORIZ INSOL:	435.	840.	1304.	1827.	2277.	2463.	2613.	2197.	1737.	1138.	628.	437.	1495.
TILT = LAT:	880.	1315.	1659.	1874.	1997.	2012.	2202.	2126.	2091.	1746.	1149.	856.	
TILT = LAT+15:	922.	1334.	1593.	1676.	1666.	1652.	1823.	1857.	1964.	1756.	1203.	909.	
TILT = 90:	832.	1118.	1162.	986.	810.	714.	798.	996.	1330.	1433.	1078.	840.	
KT:	.41	.50	.56	.60	.64	.66	.72	.68	.68	.62	.49	.42	
AMB TEMP:	29.0	35.5	41.1	49.0	57.4	64.8	74.5	72.2	63.1	52.1	39.8	32.1	50.9
HTG DEG DAYS:	1116	826	741	480	252	97	0	12	127	406	756	1020	5933

STA NO:24134 BURNS, OR — LAT = 43.58

	Jan	Feb	Mar	Apr	May	Jun	Jul	Aug	Sep	Oct	Nov	Dec	Year
HORIZ INSOL:	490.	792.	1187.	1649.	2052.	2280.	2460.	2083.	1620.	1043.	594.	431.	1390.
TILT = LAT:	893.	1220.	1482.	1674.	1798.	1867.	2076.	2008.	1924.	1563.	1064.	838.	
TILT = LAT+15:	936.	1234.	1419.	1497.	1522.	1539.	1723.	1754.	1804.	1566.	1111.	889.	
TILT = 90:	845.	1030.	1031.	886.	749.	685.	773.	949.	1221.	1271.	991.	821.	
KT:	.42	.48	.51	.55	.58	.61	.68	.65	.63	.57	.46	.42	
AMB TEMP:	25.2	31.0	36.1	44.2	52.2	59.0	68.4	66.1	58.2	47.3	35.8	27.9	46.0
HTG DEG DAYS:	1234	952	896	624	402	205	30	68	226	549	876	1150	7212

STA NO:24137 CUT BANK, MT — LAT = 48.60

	Jan	Feb	Mar	Apr	May	Jun	Jul	Aug	Sep	Oct	Nov	Dec	Year
HORIZ INSOL:	402.	688.	1128.	1485.	1883.	2045.	2287.	1897.	1352.	871.	480.	334.	1238.
TILT = LAT:	911.	1215.	1530.	1536.	1647.	1663.	1925.	1853.	1667.	1449.	1036.	825.	
TILT = LAT+15:	962.	1233.	1466.	1368.	1389.	1368.	1599.	1614.	1556.	1452.	1085.	880.	
TILT = 90:	906.	1084.	1150.	905.	799.	736.	862.	1000.	1148.	1243.	1009.	844.	
KT:	.45	.49	.54	.52	.54	.55	.63	.61	.57	.55	.48	.44	
AMB TEMP:	16.2	22.4	26.8	39.5	49.6	56.5	64.4	62.2	53.2	44.1	29.7	21.4	40.5
HTG DEG DAYS:	1513	1193	1184	765	477	267	82	125	368	648	1059	1352	9033

STA NO:24138 DILLON MT LAT = 45.25

	Jan	Feb	Mar	Apr	May	Jun	Jul	Aug	Sep	Oct	Nov	Dec	Year
HORIZ INSOL:	527.	846.	1279.	1639.	1989.	2143.	2392.	2023.	1521.	1023.	602.	450.	:370.
TILT = LAT:	1080.	1406.	1670.	1679.	1742.	1753.	2017.	1959.	1927.	1601.	1175.	990.	0
TILT = LAT+15:	1142.	1431.	1604.	1499.	1473.	1446.	1674.	1710.	1710.	1607.	1233.	1059.	0
TILT = 90:	1057.	1225.	1201.	922.	766.	694.	804.	970.	1193.	1332.	1122.	1000.	0
KT:	.49	.54	.57	.55	.56	.57	.66	.64	.61	.58	.51	.48	
AMB TEMP:	20.2	25.5	29.6	41.1	50.4	57.5	66.4	64.6	54.7	45.0	31.8	23.9	42.6
HTG DEG DAYS:	1389	1106	1097	717	453	238	54	85	325	620	996	1274	8354

STA NO:24143 GREAT FALLS MT LAT = 47.48

	Jan	Feb	Mar	Apr	May	Jun	Jul	Aug	Sep	Oct	Nov	Dec	Year
HORIZ INSOL:	421.	720.	1170.	1489.	1848.	2101.	2329.	1933.	1379.	925.	498.	336.	1262.
TILT = LAT:	900.	1235.	1564.	1529.	1615.	1712.	1962.	1882.	1678.	1507.	1018.	759.	0
TILT = LAT+15:	948.	1253.	1500.	1363.	1365.	1409.	1626.	1640.	1567.	1511.	1065.	807.	0
TILT = 90:	885.	1091.	1158.	882.	764.	729.	846.	988.	1135.	1281.	981.	765.	0
KT:	.44	.50	.55	.51	.53	.56	.64	.62	.57	.56	.47	.41	
AMB TEMP:	20.5	26.6	30.5	43.4	53.3	60.8	69.3	67.4	57.3	48.3	34.6	26.5	44.9
HTG DEG DAYS:	1380	1075	1070	648	367	162	18	42	260	524	912	1194	7652

STA NO:24144 HELENA MT LAT = 46.60

	Jan	Feb	Mar	Apr	May	Jun	Jul	Aug	Sep	Oct	Nov	Dec	Year
HORIZ INSOL:	419.	709.	1146.	1487.	1850.	2040.	2334.	1930.	1412.	926.	521.	364.	1262.
TILT = LAT:	847.	1168.	1497.	1519.	1527.	1665.	1967.	1872.	1705.	1468.	1034.	801.	0
TILT = LAT+15:	890.	1182.	1433.	1354.	1376.	1374.	1631.	1632.	1593.	1470.	1081.	852.	0
TILT = 90:	823.	1018.	1091.	860.	751.	697.	824.	962.	1137.	1232.	989.	804.	0
KT:	.42	.47	.52	.51	.53	.54	.64	.61	.58	.55	.47	.42	
AMB TEMP:	18.1	25.4	30.6	42.7	52.2	59.2	67.9	66.2	55.5	45.3	31.7	23.3	43.2
HTG DEG DAYS:	1454	1109	1066	669	401	194	33	57	304	611	999	1293	8190

STA NO:24149 LEWISTON ID LAT = 46.38

	Jan	Feb	Mar	Apr	May	Jun	Jul	Aug	Sep	Oct	Nov	Dec	Year
HORIZ INSOL:	340.	609.	1020.	1435.	1842.	2015.	2336.	1931.	1435.	860.	413.	286.	1210.
TILT = LAT:	610.	944.	1292.	1458.	1611.	1646.	1968.	1871.	1731.	1324.	731.	545.	0
TILT = LAT+15:	632.	947.	1232.	1300.	1363.	1359.	1633.	1632.	1618.	1321.	754.	570.	0
TILT = 90:	572.	804.	930.	823.	741.	687.	819.	957.	1151.	1100.	676.	525.	0
KT:	.33	.40	.46	.49	.53	.54	.65	.61	.59	.51	.37	.32	
AMB TEMP:	31.2	38.1	42.9	50.3	58.1	65.0	73.4	71.5	63.3	51.8	40.5	34.8	51.7
HTG DEG DAYS:	1048	753	685	441	232	84	0	17	124	439	735	936	5464

STA NO:24153 MISSOULA MT LAT = 46.92

	Jan	Feb	Mar	Apr	May	Jun	Jul	Aug	Sep	Oct	Nov	Dec	Year
HORIZ INSOL:	312.	574.	982.	1382.	1783.	1933.	2327.	1881.	1358.	813.	410.	267.	1169.
TILT = LAT:	553.	888.	1245.	1403.	1557.	1579.	1961.	1823.	1634.	1251.	746.	508.	0
TILT = LAT+15:	570.	889.	1185.	1249.	1317.	1305.	1626.	1589.	1525.	1246.	771.	531.	0
TILT = 90:	515.	756.	901.	801.	730.	676.	831.	946.	1093.	1041.	696.	490.	0
KT:	.32	.39	.45	.47	.51	.52	.64	.60	.56	.49	.37	.31	
AMB TEMP:	20.8	27.2	33.3	43.9	52.2	58.9	66.6	65.0	55.3	44.1	32.3	24.7	43.7
HTG DEG DAYS:	1370	1058	983	633	397	201	39	71	301	648	981	1249	7931

STA NO:24155 — PENDLETON, OR — LAT = 45.68

	JAN	FEB	MAR	APR	MAY	JUN	JUL	AUG	SEP	OCT	NOV	DEC	YR
HORIZ INSOL	348.	614.	1044.	1503.	1926.	2144.	2396.	1994.	1502.	908.	438.	293.	1259.
TILT = LAT	607.	931.	1313.	1529.	1685.	1752.	2019.	1932.	1811.	1391.	770.	538.	0
TILT = LAT+15	627.	933.	1253.	1364.	1425.	1444.	1675.	1686.	1694.	1390.	795.	561.	
TILT = 90	564.	785.	937.	850.	755.	703.	816.	968.	1191.	1150.	710.	514.	
KT	.33	.40	.47	.51	.55	.57	.63	.61	.52	.38	.32		
AMB TEMP	32.0	38.9	43.8	50.9	58.5	65.6	73.5	71.5	64.0	52.6	41.4	35.7	52.4
HTG DEG DAYS	1023	731	657	423	220	70	6	13	97	384	708	908	5240

STA NO:24156 — FOCATELLO, ID — LAT = 42.92

	JAN	FEB	MAR	APR	MAY	JUN	JUL	AUG	SEP	OCT	NOV	DEC	YR
HCRIZ INSOL	539.	882.	1371.	1820.	2280.	2480.	2600.	2239.	1769.	1203.	689.	477.	1529.
TILT = LAT	987.	1373.	1743.	1859.	2000.	2027.	2192.	2165.	2118.	1841.	1362.	933.	0
TILT = LAT+15	1038.	1395.	1676.	1664.	1689.	1665.	1816.	1892.	1990.	1855.	1325.	993.	
TILT = 90	938.	1163.	1212.	964.	794.	700.	777.	993.	1332.	1506.	1186.	918.	
KT	.45	.52	.58	.60	.64	.66	.71	.69	.68	.64	.52	.45	
AMB TEMP	23.2	29.4	35.4	45.3	54.4	61.8	71.5	69.5	59.4	48.4	35.7	26.9	46.7
HTG DEG DAYS	1296	997	918	591	336	138	0	20	192	515	879	1181	7063

STA NO:24157 — SPOKANE, WA — LAT = 47.63

	JAN	FEB	MAR	APR	MAY	JUN	JUL	AUG	SEP	OCT	NOV	DEC	YR
HORIZ INSOL	315.	606.	1041.	1495.	1918.	2083.	2357.	1942.	1435.	841.	398.	255.	1224.
TILT = LAT	589.	983.	1357.	1538.	1679.	1696.	1986.	1893.	1765.	1339.	743.	500.	0
TILT = LAT+15	610.	989.	1296.	1370.	1417.	1396.	1645.	1649.	1650.	1337.	768.	522.	
TILT = 90	558.	853.	998.	889.	792.	727.	857.	997.	1199.	1129.	698.	485.	
KT	.33	.42	.49	.51	.55	.56	.65	.62	.60	.51	.38	.31	
AMB TEMP	25.4	32.2	37.5	46.1	54.7	61.5	69.7	68.0	59.6	47.8	35.5	29.0	47.3
HTG DEG DAYS	1228	918	853	567	327	144	21	47	196	533	885	1116	6835

STA NO:24172 — LOVELOCK, NV — LAT = 40.07

	JAN	FEB	MAR	APR	MAY	JUN	JUL	AUG	SEP	OCT	NOV	DEC	YR
HORIZ INSOL	804.	1165.	1657.	2165.	2555.	2750.	2784.	2484.	2027.	1451.	929.	714.	1791.
TILT = LAT	1482.	1800.	2087.	2213.	2240.	2246.	2346.	2393.	2403.	2172.	1666.	1396.	0
TILT = LAT+15	1580.	1844.	2018.	1985.	1888.	1836.	1939.	2092.	2267.	2201.	1765.	1505.	
TILT = 90	1424.	1511.	1399.	1054.	769.	643.	705.	983.	1431.	1740.	1563.	1389.	
KT	.59	.63	.67	.70	.72	.73	.76	.76	.75	.72	.63	.58	
AMB TEMP	28.9	35.2	40.1	48.5	57.5	65.6	71.3	71.3	62.7	51.2	38.4	30.8	50.4
HTG DEG DAYS	1119	834	772	495	255	86	0	17	126	428	798	1060	5990

STA NO:24215 — MOUNT SHASTA, CA — LAT = 41.32

	JAN	FEB	MAR	APR	MAY	JUN	JUL	AUG	SEP	OCT	NOV	DEC	YR
HORIZ INSOL	561.	857.	1250.	1756.	2186.	2436.	2577.	2213.	1735.	1155.	659.	505.	1491.
TILT = LAT	967.	1264.	1523.	1773.	1917.	1999.	2176.	2125.	2027.	1676.	1114.	927.	0
TILT = LAT+15	1015.	1278.	1459.	1588.	1624.	1645.	1806.	1859.	1904.	1682.	1162.	985.	
TILT = 90	902.	1042.	1024.	888.	732.	655.	728.	933.	1234.	1332.	1018.	897.	
KT	.43	.48	.52	.57	.61	.65	.71	.68	.66	.59	.47	.44	
AMB TEMP	33.6	37.8	40.4	46.3	53.3	60.0	67.8	66.0	61.2	51.4	41.7	35.5	49.6
HTG DEG DAYS	973	762	763	561	371	178	37	64	145	422	699	915	5890

STA NO:24216 RED BLUFF, CA — LAT = 40.15

	JAN	FEB	MAR	APR	MAY	JUN	JUL	AUG	SEP	OCT	NOV	DEC	YR
HORIZ INSOL:	570.	893.	1354.	1910.	2375.	2600.	2672.	2311.	1845.	1228.	706.	511.	1581.
TILT = LAT:	941.	1289.	1646.	1931.	2083.	2131.	2255.	2216.	2148.	1757.	1167.	890.	0
TILT = LAT+15:	985.	1304.	1580.	1731.	1761.	1749.	1869.	1939.	2021.	1766.	1219.	942.	0
TILT = 90:	864.	1049.	1090.	934.	742.	640.	702.	931.	1279.	1381.	1058.	847.	0
KT:	.42	.48	.55	.61	.67	.69	.73	.71	.69	.61	.48	.42	
AMB TEMP:	45.2	50.0	53.2	59.5	67.4	75.5	82.3	79.9	75.3	65.0	53.7	46.4	62.8
HTG DEG DAYS:	614	420	366	218	64	8	0	0	0	82	339	577	2688

STA NO:24225 MEDFORD, OR — LAT = 42.37

	JAN	FEB	MAR	APR	MAY	JUN	JUL	AUG	SEP	OCT	NOV	DEC	YR
HORIZ INSOL:	407.	737.	1133.	1639.	2034.	2278.	2475.	2121.	1589.	982.	504.	337.	1353.
TILT = LAT:	644.	1072.	1375.	1653.	1782.	1870.	2090.	2038.	1852.	1404.	806.	544.	0
TILT = LAT+15:	664.	1078.	1313.	1479.	1511.	1543.	1737.	1782.	1736.	1401.	830.	563.	0
TILT = 90:	578.	880.	934.	852.	718.	658.	743.	929.	1148.	1115.	719.	498.	0
KT:	.33	.43	.48	.54	.57	.61	.68	.66	.61	.52	.38	.31	
AMB TEMP:	36.6	41.3	44.8	50.2	57.3	64.3	71.7	70.4	64.4	53.4	43.5	37.7	53.0
HTG DEG DAYS:	880	664	626	444	250	94	11	21	89	360	645	846	4930

STA NO:24227 OLYMPIA, WA — LAT = 46.97

	JAN	FEB	MAR	APR	MAY	JUN	JUL	AUG	SEP	OCT	NOV	DEC	YR
HORIZ INSOL:	269.	503.	845.	1255.	1632.	1693.	1913.	1549.	1157.	636.	339.	222.	1001.
TILT = LAT:	438.	738.	1033.	1260.	1422.	1386.	1611.	1478.	1352.	906.	562.	375.	0
TILT = LAT+15:	446.	733.	978.	1121.	1206.	1151.	1345.	1289.	1256.	891.	573.	385.	0
TILT = 90:	396.	617.	738.	721.	678.	615.	716.	779.	898.	733.	508.	347.	0
KT:	.27	.34	.39	.43	.47	.45	.53	.49	.48	.38	.26		
AMB TEMP:	37.2	41.0	43.2	48.2	54.0	58.9	63.6	62.8	58.6	50.6	43.3	39.5	50.1
HTG DEG DAYS:	862	672	676	504	341	197	89	103	198	446	651	791	5530

STA NO:24229 PORTLAND, OR — LAT = 45.60

	JAN	FEB	MAR	APR	MAY	JUN	JUL	AUG	SEP	OCT	NOV	DEC	YR
HORIZ INSOL:	310.	554.	895.	1304.	1663.	1772.	2037.	1674.	1217.	724.	388.	260.	1067.
TILT = LAT:	505.	806.	1085.	1310.	1452.	1455.	1719.	1600.	1411.	1034.	639.	441.	0
TILT = LAT+15:	517.	803.	1026.	1167.	1232.	1209.	1434.	1397.	1313.	1021.	654.	455.	0
TILT = 90:	458.	669.	762.	730.	667.	615.	724.	813.	918.	832.	577.	410.	0
KT:	.29	.36	.40	.44	.47	.47	.56	.53	.49	.42	.33	.28	
AMB TEMP:	38.1	42.8	45.7	50.5	56.7	62.0	67.1	66.6	62.2	53.8	45.3	40.7	52.6
HTG DEG DAYS:	834	622	598	432	264	128	48	56	119	347	591	753	4792

STA NO:24230 REDMOND, OR — LAT = 44.27

	JAN	FEB	MAR	APR	MAY	JUN	JUL	AUG	SEP	OCT	NOV	DEC	YR
HORIZ INSOL:	491.	775.	1190.	1683.	2080.	2287.	2446.	2069.	1584.	999.	572.	425.	1383.
TILT = LAT:	928.	1212.	1504.	1714.	1822.	1871.	2063.	1999.	1891.	1510.	1044.	856.	0
TILT = LAT+15:	975.	1226.	1440.	1536.	1541.	1540.	1712.	1745.	1772.	1511.	1090.	909.	0
TILT = 90:	888.	1031.	1059.	923.	772.	702.	789.	962.	1214.	1235.	978.	846.	0
KT:	.43	.48	.52	.56	.59	.61	.67	.65	.63	.55	.46	.43	
AMB TEMP:	30.2	35.8	38.6	38.6	51.3	58.2	65.7	63.8	57.7	48.4	39.0	33.4	47.2
HTG DEG DAYS:	1079	818	818	618	425	220	55	102	233	515	780	980	6643

STA NO:24232 — SALEM, OR — LAT = 44.92

	JAN	FEB	MAR	APR	MAY	JUN	JUL	AUG	SEP	OCT	NOV	DEC	YR
HORIZ INSOL:	332.	588.	947.	1370.	1738.	1849.	2142.	1775.	1328.	769.	410.	277.	1127.
TILT = LAT:	540.	856.	1151.	1375.	1519.	1519.	1808.	1700.	1551.	1099.	672.	467.	0
TILT = LAT+15:	554.	854.	1093.	1227.	1289.	1261.	1507.	1485.	1446.	1088.	689.	483.	0
TILT = 90:	490.	709.	804.	755.	680.	623.	737.	846.	1000.	882.	605.	434.	0
KT:	.30	.37	.42	.46	.49	.49	.59	.56	.53	.43	.34	.29	
AMB TEMP:	38.8	42.9	45.2	49.8	55.7	61.2	66.6	66.1	61.9	53.2	45.2	40.9	52.3
HTG DEG DAYS:	812	619	614	456	295	133	43	53	120	366	594	747	4352

STA NO:24233 — SEATTLE-TACOMA, WA — LAT = 47.45

	JAN	FEB	MAR	APR	MAY	JUN	JUL	AUG	SEP	OCT	NOV	DEC	YR
HORIZ INSOL:	262.	495.	849.	1294.	1714.	1802.	2248.	1616.	1148.	656.	337.	211.	1053.
TILT = LAT:	433.	735.	1040.	1306.	1495.	1472.	1893.	1550.	1348.	958.	573.	359.	0
TILT = LAT+15:	441.	730.	993.	1162.	1265.	1218.	1572.	1351.	1252.	945.	585.	368.	0
TILT = 90:	394.	618.	755.	754.	715.	651.	822.	823.	902.	784.	522.	333.	0
KT:	.27	.34	.40	.44	.49	.48	.62	.52	.48	.40	.32	.26	
AMB TEMP:	38.2	42.3	44.1	48.7	54.9	59.8	64.5	63.8	59.6	52.2	44.6	40.5	51.1
HTG DEG DAYS:	831	636	648	489	313	167	80	82	170	397	612	760	5165

STA NO:24243 — YAKIMA, WA — LAT = 46.57

	JAN	FEB	MAR	APR	MAY	JUN	JUL	AUG	SEP	OCT	NOV	DEC	YR
HORIZ INSOL:	365.	666.	1122.	1598.	2009.	2169.	2358.	1975.	1483.	891.	444.	295.	1281.
TILT = LAT:	689.	1074.	1459.	1645.	1750.	1769.	1987.	1919.	1806.	1394.	823.	579.	0
TILT = LAT+15:	717.	1083.	1396.	1468.	1485.	1456.	1648.	1673.	1690.	1393.	853.	607.	0
TILT = 90:	655.	928.	1061.	929.	801.	727.	830.	983.	1206.	1165.	771.	563.	0
KT:	.36	.44	.51	.54	.57	.58	.65	.63	.61	.53	.40	.34	
AMB TEMP:	27.5	35.7	41.8	49.5	57.9	64.5	70.7	68.6	61.3	50.1	38.4	31.3	49.8
HTG DEG DAYS:	1163	820	719	465	239	94	20	37	147	462	798	1045	6009

STA NO:24255 — WHIDBEY ISLAND, WA — LAT = 48.35

	JAN	FEB	MAR	APR	MAY	JUN	JUL	AUG	SEP	OCT	NOV	DEC	YR
HORIZ INSOL:	283.	532.	918.	1345.	1760.	1820.	1981.	1593.	1173.	655.	357.	233.	1054.
TILT = LAT:	523.	843.	1176.	1372.	1537.	1483.	1666.	1530.	1401.	984.	659.	457.	0
TILT = LAT+15:	540.	844.	1118.	1220.	1298.	1226.	1387.	1333.	1303.	972.	679.	477.	0
TILT = 90:	493.	727.	866.	806.	748.	670.	763.	830.	954.	816.	616.	443.	0
KT:	.31	.38	.44	.47	.51	.49	.55	.51	.50	.41	.35	.30	
AMB TEMP:	0	0	0	0	0	0	0	0	0	0	0	0	0
HTG DEG DAYS:	0	0	0	0	0	0	0	0	0	0	0	0	0

STA NO:24283 — ARCATA, CA — LAT = 40.98

	JAN	FEB	MAR	APR	MAY	JUN	JUL	AUG	SEP	OCT	NOV	DEC	YR
HORIZ INSOL:	529.	793.	1133.	1587.	1843.	1962.	1808.	1579.	1342.	936.	593.	470.	1214.
TILT = LAT:	879.	1133.	1349.	1586.	1616.	1625.	1539.	1489.	1503.	1280.	953.	824.	0
TILT = LAT+15:	918.	1141.	1288.	1421.	1376.	1353.	1297.	1307.	1403.	1272.	988.	870.	0
TILT = 90:	808.	920.	895.	795.	643.	581.	587.	685.	903.	989.	853.	784.	0
KT:	.40	.44	.47	.51	.52	.52	.50	.48	.51	.47	.42	.40	
AMB TEMP:	0	0	0	0	0	0	0	0	0	0	0	0	0
HTG DEG DAYS:	0	0	0	0	0	0	0	0	0	0	0	0	0

STA NO:24284 — NORTH BEND, OR — LAT = 43.42

	JAN	FEB	MAR	APR	MAY	JUN	JUL	AUG	SEP	OCT	NOV	DEC	ANN
HORIZ INSOL	439.	705.	1058.	1510.	1857.	1994.	2108.	1786.	1377.	893.	525.	381.	1219.
TILT = LAT	755.	1041.	1287.	1519.	1626.	1641.	1782.	1704.	1589.	1279.	891.	694.	0
TILT = LAT+15	786.	1046.	1227.	1358.	1380.	1360.	1489.	1490.	1483.	1272.	923.	730.	0
TILT = 90	701.	864.	885.	805.	690.	629.	697.	817.	999.	1020.	814.	665.	0
KT	.37	.42	.45	.50	.52	.53	.58	.56	.54	.48	.41	.36	
AMB TEMP	44.6	46.6	46.9	49.1	53.1	56.9	59.0	59.7	58.4	54.9	50.1	46.5	52.2
HTG DEG DAYS	632	515	561	477	369	243	188	168	201	313	447	574	4688

STA NO:25704 — ADAK, AK — LAT = 51.88

	JAN	FEB	MAR	APR	MAY	JUN	JUL	AUG	SEP	OCT	NOV	DEC	ANN
HORIZ INSOL	231.	433.	716.	1033.	1180.	1182.	1120.	949.	759.	528.	308.	187.	719.
TILT = LAT	507.	734.	920.	1034.	1010.	955.	927.	870.	860.	830.	677.	454.	0
TILT = LAT+15	527.	734.	868.	914.	854.	797.	779.	754.	788.	818.	702.	477.	0
TILT = 90	496.	651.	699.	646.	557.	508.	504.	517.	603.	708.	657.	457.	0
KT	.32	.36	.37	.37	.34	.32	.31	.31	.34	.37	.37	.32	
AMB TEMP	0	0	0	0	0	0	0	0	0	0	0	0	0
HTG DEG DAYS	0	0	0	0	0	0	0	0	0	0	0	0	0

STA NO:93037 — COLORADO SPRINGS, CO — LAT = 38.82

	JAN	FEB	MAR	APR	MAY	JUN	JUL	AUG	SEP	OCT	NOV	DEC	ANN
HORIZ INSOL	891.	1178.	1550.	1931.	2129.	2369.	2212.	2025.	1759.	1359.	944.	782.	1594.
TILT = LAT	1610.	1762.	1892.	1942.	1869.	1957.	1881.	1923.	2002.	1933.	1621.	1484.	0
TILT = LAT+15	1721.	1803.	1824.	1743.	1589.	1617.	1576.	1687.	1881.	1949.	1714.	1603.	0
TILT = 90	1542.	1454.	1234.	907.	664.	590.	618.	800.	1157.	1505.	1498.	1470.	0
KT	.62	.62	.62	.61	.61	.63	.61	.61	.64	.65	.61	.60	
AMB TEMP	28.6	31.3	35.3	46.2	55.5	64.6	70.7	69.1	60.9	50.5	37.5	31.0	48.4
HTG DEG DAYS	1128	944	921	564	301	103	9	13	155	456	825	1054	6473

STA NO:93044 — ZUNI, NM — LAT = 35.10

	JAN	FEB	MAR	APR	MAY	JUN	JUL	AUG	SEP	OCT	NOV	DEC	ANN
HORIZ INSOL	986.	1297.	1688.	2167.	2473.	2602.	2264.	2078.	1895.	1496.	1088.	893.	1744.
TILT = LAT	1599.	1812.	1990.	2162.	2172.	2155.	1937.	1962.	2102.	2009.	1714.	1512.	0
TILT = LAT+15	1706.	1855.	1922.	1945.	1841.	1775.	1627.	1726.	1979.	2028.	1814.	1631.	0
TILT = 90	1476.	1430.	1215.	895.	620.	507.	544.	724.	1120.	1490.	1534.	1451.	0
KT	.60	.62	.63	.67	.69	.70	.62	.62	.67	.67	.63	.59	
AMB TEMP	30.3	34.6	39.6	48.1	56.6	65.4	71.4	69.4	63.3	52.5	40.1	32.0	50.3
HTG DEG DAYS	1076	851	787	507	264	68	0	13	91	388	747	1023	5815

STA NO:93045 — TRUTH OR CONSEQUENCE, NM — LAT = 33.23

	JAN	FEB	MAR	APR	MAY	JUN	JUL	AUG	SEP	OCT	NOV	DEC	ANN
HORIZ INSOL	1118.	1451.	1886.	2338.	2557.	2650.	2365.	2216.	1940.	1579.	1217.	1003.	1860.
TILT = LAT	1764.	1996.	2213.	2326.	2246.	2202.	2026.	2089.	2124.	2069.	1867.	1641.	0
TILT = LAT+15	1890.	2051.	2145.	2095.	1904.	1813.	1701.	1838.	2001.	2091.	1982.	1776.	0
TILT = 90	1616.	1552.	1307.	890.	573.	456.	507.	707.	1082.	1496.	1654.	1562.	0
KT	.64	.66	.69	.72	.71	.71	.65	.66	.68	.67	.62	.62	
AMB TEMP	40.0	44.9	50.2	59.5	68.2	76.9	79.3	77.4	71.6	61.3	48.7	40.8	59.9
HTG DEG DAYS	775	563	459	188	19	0	0	0	5	91	489	750	3392

STA NO:93058 PUEBLO, CO — LAT = 38.28

	Jan	Feb	Mar	Apr	May	Jun	Jul	Aug	Sep	Oct	Nov	Dec	Ann
HORIZ INSOL:	894.	1172.	1564.	1956.	2162.	2434.	2312.	2102.	1779.	1361.	954.	782.	1623.
TILT = LAT:	1584.	1725.	1898.	1964.	1899.	2010.	1965.	1996.	2018.	1913.	1610.	1449.	0
TILT = LAT+15:	1691.	1764.	1830.	1763.	1614.	1659.	1644.	1751.	1896.	1928.	1702.	1563.	0
TILT = 90:	1506.	1412.	1226.	903.	659.	583.	619.	811.	1153.	1478.	1480.	1426.	0
KT:	.61	.60	.62	.62	.60	.65	.63	.64	.65	.65	.61	.59	0
AMB TEMP:	30.1	34.7	40.0	51.7	61.1	70.7	76.4	74.5	66.2	54.5	40.8	33.0	52.8
HTG DEG DAYS:	1082	848	775	405	148	28	0	0	55	335	726	992	5394

STA NO:93101 EL TORO, CA — LAT = 33.67

	Jan	Feb	Mar	Apr	May	Jun	Jul	Aug	Sep	Oct	Nov	Dec	Ann
HORIZ INSOL:	947.	1236.	1610.	1928.	2070.	2194.	2363.	2155.	1737.	1357.	1026.	869.	1625.
TILT = LAT:	1451.	1658.	1856.	1903.	1829.	1844.	2024.	2032.	1887.	1747.	1527.	1389.	0
TILT = LAT+15:	1541.	1691.	1790.	1714.	1564.	1538.	1698.	1788.	1773.	1607.	1492.	1301.	0
TILT = 90:	1304.	1271.	1099.	773.	548.	473.	517.	706.	974.	1246.	1327.	1301.	0
KT:	.55	.57	.59	.59	.58	.59	.65	.64	.60	.59	.57	.55	0
AMB TEMP:	0	0	0	0	0	0	0	0	0	0	0	0	0
HTG DEG DAYS:	0	0	0	0	0	0	0	0	0	0	0	0	0

STA NO:93104 CHINA LAKE, CA — LAT = 35.68

	Jan	Feb	Mar	Apr	May	Jun	Jul	Aug	Sep	Oct	Nov	Dec	Ann
HORIZ INSOL:	909.	1229.	1735.	2233.	2549.	2747.	2612.	2616.	1980.	1473.	1034.	841.	1830.
TILT = LAT:	1470.	1720.	2070.	2238.	2235.	2263.	2218.	2491.	2221.	1994.	1635.	1431.	0
TILT = LAT+15:	1564.	1757.	2001.	2013.	1891.	1854.	1847.	2180.	2093.	2012.	1728.	1541.	0
TILT = 90:	1353.	1360.	1280.	936.	641.	515.	571.	855.	1198.	1490.	1465.	1373.	0
KT:	.57	.60	.66	.70	.71	.73	.71	.78	.70	.66	.61	.57	0
AMB TEMP:	0	0	0	0	0	0	0	0	0	0	0	0	0
HTG DEG DAYS:	0	0	0	0	0	0	0	0	0	0	0	0	0

STA NO:93111 POINT MUGU, CA — LAT = 34.12

	Jan	Feb	Mar	Apr	May	Jun	Jul	Aug	Sep	Oct	Nov	Dec	Ann
HORIZ INSOL:	927.	1220.	1636.	1951.	2018.	2055.	2118.	1935.	1608.	1296.	1006.	856.	1552.
TILT = LAT:	1432.	1647.	1899.	1929.	1782.	1731.	1820.	1822.	1738.	1661.	1509.	1384.	0
TILT = LAT+15:	1521.	1679.	1832.	1737.	1525.	1448.	1536.	1606.	1631.	1665.	1588.	1486.	0
TILT = 90:	1292.	1270.	1135.	792.	551.	474.	513.	666.	909.	1192.	1317.	1302.	0
KT:	.55	.57	.61	.60	.56	.55	.58	.58	.56	.57	.57	.55	0
AMB TEMP:	0	0	0	0	0	0	0	0	0	0	0	0	0
HTG DEG DAYS:	0	0	0	0	0	0	0	0	0	0	0	0	0

STA NO:93129 CEDAR CITY, UT — LAT = 37.70

	Jan	Feb	Mar	Apr	May	Jun	Jul	Aug	Sep	Oct	Nov	Dec	Ann
HORIZ INSOL:	682.	1180.	1636.	2092.	2467.	2706.	2503.	2241.	1968.	1460.	992.	786.	1743.
TILT = LAT:	1521.	1715.	1984.	2106.	2235.	2222.	2123.	2131.	2253.	2061.	1660.	1421.	0
TILT = LAT+15:	1621.	1752.	1916.	1892.	1831.	1822.	1770.	1869.	2123.	2083.	1757.	1531.	0
TILT = 90:	1433.	1392.	1272.	945.	690.	574.	622.	836.	1273.	1591.	1522.	1388.	0
KT:	.59	.60	.64	.66	.69	.72	.68	.68	.71	.69	.62	.58	0
AMB TEMP:	28.7	33.1	38.4	47.1	56.2	65.0	73.2	71.3	63.2	51.5	38.8	30.8	49.8
HTG DEG DAYS:	1125	893	825	537	281	86	0	0	114	424	786	1060	6137

STA NO:93193 — FRESNO, CA — LAT = 36.77

	JAN	FEB	MAR	APR	MAY	JUN	JUL	AUG	SEP	OCT	NOV	DEC	ANN
HORIZ INSOL:	657.	1012.	1566.	2093.	2484.	2733.	2665.	2423.	1985.	1429.	889.	574.	1711.
TILT = LAT:	1003.	1390.	1863.	2097.	2179.	2247.	2273.	2306.	2253.	1966.	1397.	905.	0
TILT = LAT+15:	1049.	1408.	1796.	1885.	1844.	1841.	1888.	2021.	2123.	1903.	1467.	955.	0
TILT = 90:	892.	1091.	1170.	917.	667.	547.	602.	854.	1246.	1490.	1245.	831.	0
KT:	.42	.50	.60	.66	.69	.73	.73	.73	.71	.66	.54	.41	
AMB TEMP:	45.3	49.9	53.9	60.3	67.4	73.9	80.6	78.3	73.8	64.2	53.5	45.8	62.3
HTG DEG DAYS:	611	423	344	182	51	9	0	0	0	90	345	595	2650

STA NO:93721 — BALTIMORE, MD — LAT = 39.18

	JAN	FEB	MAR	APR	MAY	JUN	JUL	AUG	SEP	OCT	NOV	DEC	ANN
HORIZ INSOL:	587.	840.	1162.	1488.	1714.	1879.	1823.	1600.	1330.	998.	660.	499.	1215.
TILT = LAT:	942.	1165.	1359.	1471.	1507.	1567.	1558.	1506.	1464.	1333.	1031.	825.	0
TILT = LAT+15:	985.	1173.	1298.	1319.	1289.	1310.	1315.	1324.	1367.	1326.	1071.	870.	0
TILT = 90:	855.	927.	876.	712.	582.	538.	562.	661.	849.	1009.	913.	770.	0
KT:	.42	.44	.46	.47	.48	.50	.50	.49	.49	.49	.44	.39	
AMB TEMP:	33.4	34.8	42.8	53.8	63.7	72.4	76.6	74.9	68.5	57.4	46.1	35.3	55.0
HTG DEG DAYS:	980	846	688	340	110	0	0	0	27	250	567	921	4729

STA NO:93729 — CAPE HATTERAS, NC — LAT = 35.27

	JAN	FEB	MAR	APR	MAY	JUN	JUL	AUG	SEP	OCT	NOV	DEC	ANN
HORIZ INSOL:	686.	952.	1326.	1774.	1962.	2036.	1921.	1705.	1470.	1137.	873.	659.	1375.
TILT = LAT:	1011.	1250.	1516.	1751.	1731.	1710.	1652.	1603.	1589.	1450.	1305.	1028.	0
TILT = LAT+15:	1057.	1260.	1453.	1576.	1481.	1430.	1399.	1414.	1488.	1446.	1366.	1090.	0
TILT = 90:	885.	951.	918.	755.	564.	492.	513.	627.	853.	1046.	1135.	944.	0
KT:	.42	.46	.50	.55	.55	.55	.53	.51	.52	.51	.51	.44	
AMB TEMP:	45.3	45.8	50.6	58.9	67.0	74.3	78.0	77.5	73.7	65.2	56.0	47.7	61.7
HTG DEG DAYS:	611	538	458	188	47	0	0	0	0	76	277	536	2731

STA NO:93734 — WASHINGTON-STERLING, DC — LAT = 38.95

	JAN	FEB	MAR	APR	MAY	JUN	JUL	AUG	SEP	OCT	NOV	DEC	ANN
HORIZ INSOL:	572.	815.	1125.	1459.	1718.	1901.	1818.	1617.	1340.	1004.	651.	481.	1208.
TILT = LAT:	901.	1116.	1305.	1439.	1511.	1585.	1554.	1523.	1474.	1337.	1003.	776.	0
TILT = LAT+15:	940.	1122.	1245.	1291.	1293.	1325.	1312.	1339.	1375.	1330.	1041.	814.	0
TILT = 90:	812.	881.	836.	695.	579.	538.	557.	664.	851.	1009.	883.	716.	0
KT:	.40	.43	.45	.46	.48	.51	.50	.49	.49	.49	.43	.37	
AMB TEMP:	32.1	33.8	41.8	53.1	62.6	71.1	75.3	73.6	66.9	55.9	44.7	34.0	53.7
HTG DEG DAYS:	1020	874	719	357	131	5	0	0	43	291	609	961	5010

STA NO:93805 — TALLAHASSEE, FL — LAT = 30.38

	JAN	FEB	MAR	APR	MAY	JUN	JUL	AUG	SEP	OCT	NOV	DEC	ANN
HORIZ INSOL:	877.	1138.	1479.	1823.	1936.	1883.	1748.	1675.	1493.	1318.	1009.	813.	1433.
TILT = LAT:	1207.	1418.	1639.	1782.	1723.	1612.	1529.	1573.	1568.	1600.	1377.	1156.	0
TILT = LAT+15:	1269.	1436.	1576.	1610.	1482.	1361.	1307.	1395.	1472.	1601.	1441.	1229.	0
TILT = 90:	1020.	1020.	956.	667.	475.	409.	425.	538.	758.	1080.	1136.	1020.	0
KT:	.46	.49	.52	.55	.54	.51	.48	.49	.50	.54	.51	.46	
AMB TEMP:	52.6	54.8	60.3	67.9	74.8	80.0	81.1	81.1	78.1	69.3	58.9	53.2	67.7
HTG DEG DAYS:	408	323	187	34	0	0	0	0	0	31	204	376	1563

STA NO:93814
CINCINNATI (COVINGTO)
OH
LAT = 39.07

	Jan	Feb	Mar	Apr	May	Jun	Jul	Aug	Sep	Oct	Nov	Dec	Year
HORIZ INSOL:	501.	738.	1027.	1399.	1672.	1837.	1771.	1634.	1312.	990.	589.	433.	1158.
TILT = LAT:	754.	986.	1175.	1376.	1471.	1534.	1514.	1540.	1440.	1317.	881.	671.	
TILT = LAT+15:	779.	986.	1117.	1234.	1259.	1284.	1280.	1353.	1343.	1310.	908.	699.	
TILT = 90:	664.	769.	749.	669.	571.	530.	551.	672.	833.	994.	764.	607.	
KT:	.35	.39	.41	.45	.47	.49	.48	.50	.48	.48	.39	.34	
AMB TEMP:	31.1	33.3	41.7	53.9	63.2	72.1	75.6	74.4	67.8	56.8	43.8	33.7	54.0
HTG DEG DAYS:	1051	888	722	341	138	9	0	0	44	271	636	970	5070

STA NO:93815
DAYTON
OH
LAT = 39.90

	Jan	Feb	Mar	Apr	May	Jun	Jul	Aug	Sep	Oct	Nov	Dec	Year
HORIZ INSOL:	489.	725.	1025.	1403.	1699.	1874.	1810.	1646.	1319.	969.	564.	408.	1161.
TILT = LAT:	754.	983.	1183.	1384.	1493.	1560.	1544.	1552.	1458.	1305.	856.	639.	
TILT = LAT+15:	780.	983.	1125.	1240.	1276.	1302.	1303.	1363.	1360.	1298.	882.	665.	
TILT = 90:	671.	775.	765.	685.	590.	549.	571.	691.	857.	996.	747.	580.	
KT:	.36	.39	.41	.45	.48	.50	.50	.50	.49	.48	.38	.33	
AMB TEMP:	28.1	30.4	39.0	51.4	61.6	71.3	74.6	73.0	66.3	55.5	41.8	30.9	52.0
HTG DEG DAYS:	1144	969	806	413	166	13	0	7	63	307	696	1057	5641

STA NO:93817
EVANSVILLE
IN
LAT = 38.05

	Jan	Feb	Mar	Apr	May	Jun	Jul	Aug	Sep	Oct	Nov	Dec	Year
HORIZ INSOL:	574.	823.	1151.	1501.	1783.	1983.	1920.	1735.	1403.	1087.	683.	499.	1262.
TILT = LAT:	876.	1106.	1327.	1480.	1569.	1655.	1642.	1636.	1540.	1447.	1037.	785.	
TILT = LAT+15:	912.	1111.	1266.	1328.	1342.	1381.	1386.	1439.	1440.	1443.	1076.	824.	
TILT = 90:	778.	862.	838.	698.	579.	534.	560.	688.	875.	1086.	906.	718.	
KT:	.39	.42	.45	.47	.50	.53	.53	.52	.51	.52	.43	.37	
AMB TEMP:	32.6	35.9	44.3	56.7	65.7	74.7	77.8	76.2	69.1	58.2	44.9	35.3	56.0
HTG DEG DAYS:	1004	815	653	263	95	5	0	0	34	236	603	921	4629

STA NO:93819
INDIANAPOLIS
IN
LAT = 39.73

	Jan	Feb	Mar	Apr	May	Jun	Jul	Aug	Sep	Oct	Nov	Dec	Year
HORIZ INSOL:	496.	747.	1037.	1399.	1688.	1868.	1806.	1643.	1324.	977.	579.	417.	1165.
TILT = LAT:	763.	1017.	1197.	1377.	1433.	1556.	1542.	1550.	1463.	1314.	882.	654.	
TILT = LAT+15:	790.	1018.	1139.	1235.	1268.	1300.	1301.	1361.	1365.	1307.	909.	682.	
TILT = 90:	678.	802.	773.	680.	584.	545.	567.	687.	858.	1001.	771.	595.	
KT:	.36	.40	.42	.45	.47	.50	.49	.50	.49	.48	.39	.33	
AMB TEMP:	27.9	30.7	39.7	52.3	62.2	71.7	75.0	73.2	66.3	55.7	41.7	30.9	52.3
HTG DEG DAYS:	1150	960	784	387	159	11	5	5	63	302	699	1057	5577

STA NO:93820
LEXINGTON
KY
LAT = 38.03

	Jan	Feb	Mar	Apr	May	Jun	Jul	Aug	Sep	Oct	Nov	Dec	Year
HORIZ INSOL:	546.	780.	1100.	1479.	1747.	1897.	1850.	1685.	1362.	1044.	657.	486.	1219.
TILT = LAT:	818.	1032.	1258.	1457.	1538.	1586.	1584.	1587.	1490.	1377.	986.	756.	
TILT = LAT+15:	848.	1034.	1198.	1308.	1317.	1327.	1339.	1396.	1391.	1372.	1021.	792.	
TILT = 90:	719.	798.	792.	688.	572.	523.	548.	671.	845.	1029.	857.	688.	
KT:	.37	.40	.43	.47	.50	.51	.51	.51	.49	.49	.42	.36	
AMB TEMP:	32.9	35.3	43.6	55.3	64.7	73.0	76.2	75.0	68.6	57.8	44.6	35.5	55.2
HTG DEG DAYS:	995	832	673	302	106	8	0	0	40	246	612	915	4729

STA NO:93821
LOUISVILLE
KY
LAT = 38.18

	JAN	FEB	MAR	APR	MAY	JUN	JUL	AUG	SEP	OCT	NOV	DEC	YEAR
HORIZ INSOL:	546.	789.	1102.	1467.	1720.	1904.	1838.	1680.	1361.	1042.	653.	488.	1216.
TILT = LAT:	822.	1052.	1263.	1444.	1514.	1591.	1573.	1583.	1490.	1378.	982.	766.	0
TILT = LAT+15:	852.	1055.	1204.	1296.	1296.	1331.	1329.	1392.	1392.	1373.	1017.	803.	0
TILT = 90:	724.	817.	797.	685.	568.	526.	549.	672.	848.	1032.	854.	699.	0
KT:	.37	.41	.43	.46	.48	.51	.50	.51	.49	.50	.42	.37	
AMB TEMP:	33.3	35.8	44.0	55.9	64.8	73.3	76.9	75.9	69.1	58.1	45.0	35.6	55.6
HTG DEG DAYS:	983	818	661	286	105	5	0	0	35	241	600	911	4645

STA NO:93822
SPRINGFIELD
IL
LAT = 39.83

	JAN	FEB	MAR	APR	MAY	JUN	JUL	AUG	SEP	OCT	NOV	DEC	YEAR
HORIZ INSOL:	585.	861.	1143.	1515.	1865.	2097.	2058.	1806.	1454.	1068.	677.	490.	1302.
TILT = LAT:	961.	1221.	1344.	1503.	1638.	1737.	1751.	1710.	1629.	1470.	1089.	828.	0
TILT = LAT+15:	1007.	1233.	1283.	1347.	1396.	1443.	1470.	1500.	1523.	1468.	1135.	873.	0
TILT = 90:	882.	985.	875.	737.	628.	581.	615.	748.	958.	1133.	978.	779.	0
KT:	.42	.46	.46	.49	.52	.56	.56	.55	.54	.53	.46	.39	
AMB TEMP:	26.7	30.4	39.4	53.1	63.4	72.9	76.1	74.4	67.2	56.6	41.9	30.5	52.7
HTG DEG DAYS:	1187	959	794	363	132	12	0	8	48	282	693	1070	5558

STA NO:93987
LUFKIN
TX
LAT = 31.23

	JAN	FEB	MAR	APR	MAY	JUN	JUL	AUG	SEP	OCT	NOV	DEC	YEAR
HORIZ INSOL:	794.	1069.	1376.	1624.	1867.	2055.	2006.	1864.	1531.	1349.	963.	768.	1439.
TILT = LAT:	1090.	1337.	1523.	1584.	1660.	1747.	1740.	1751.	1618.	1663.	1328.	1102.	0
TILT = LAT+15:	1140.	1350.	1462.	1431.	1429.	1466.	1476.	1549.	1519.	1666.	1388.	1169.	0
TILT = 90:	917.	967.	855.	624.	483.	427.	454.	591.	795.	1142.	1103.	975.	0
KT:	.48	.49	.49	.49	.52	.55	.55	.55	.52	.56	.56	.45	
AMB TEMP:	48.8	52.2	58.0	67.3	74.1	80.3	83.0	83.1	77.5	68.2	57.2	50.8	66.7
HTG DEG DAYS:	509	371	256	56	0	0	0	0	0	52	256	440	1940

STA NO:94008
GLASGOW
MT
LAT = 48.22

	JAN	FEB	MAR	APR	MAY	JUN	JUL	AUG	SEP	OCT	NOV	DEC	YEAR
HORIZ INSOL:	388.	671.	1105.	1488.	1828.	2047.	2193.	1863.	1340.	877.	479.	334.	1218.
TILT = LAT:	841.	1157.	1479.	1536.	1597.	1666.	1846.	1814.	1640.	1443.	1007.	799.	0
TILT = LAT+15:	885.	1172.	1416.	1368.	1349.	1371.	1532.	1580.	1531.	1445.	1054.	851.	0
TILT = 90:	829.	1024.	1103.	898.	770.	729.	825.	971.	1122.	1232.	975.	813.	0
KT:	.42	.47	.52	.52	.52	.55	.61	.60	.56	.55	.47	.43	
AMB TEMP:	9.2	15.2	25.2	42.8	54.2	62.0	70.5	69.0	57.2	46.4	29.0	17.1	41.5
HTG DEG DAYS:	1730	1394	1234	666	344	151	15	30	263	577	1080	1485	8959

STA NO:94224
ASTORIA
OR
LAT = 46.15

	JAN	FEB	MAR	APR	MAY	JUN	JUL	AUG	SEP	OCT	NOV	DEC	YEAR
HORIZ INSOL:	315.	545.	866.	1253.	1608.	1626.	1746.	1499.	1183.	713.	387.	261.	1000.
TILT = LAT:	534.	804.	1050.	1253.	1402.	1335.	1473.	1424.	1374.	1030.	657.	461.	0
TILT = LAT+15:	549.	801.	995.	1115.	1190.	1112.	1234.	1242.	1278.	1018.	674.	478.	0
TILT = 90:	491.	671.	743.	706.	657.	586.	653.	740.	901.	835.	599.	434.	0
KT:	.31	.36	.39	.46	.43	.43	.48	.47	.48	.42	.34	.29	
AMB TEMP:	40.6	43.6	44.4	47.8	52.3	56.5	60.0	60.3	58.4	52.8	46.5	42.8	50.5
HTG DEG DAYS:	756	599	639	516	394	255	163	151	201	378	555	688	5295

STA NO:94701
BOSTON
MA
LAT = 42.37

	JAN	FEB	MAR	APR	MAY	JUN	JUL	AUG	SEP	OCT	NOV	DEC	YEAR
HORIZ INSOL:	476.	710.	1016.	1326.	1620.	1817.	1749.	1486.	1260.	890.	503.	403.	1105.
TILT = LAT:	806.	1019.	1208.	1313.	1419.	1504.	1486.	1401.	1417.	1240.	803.	711.	0
TILT = LAT+15:	840.	1023.	1149.	1173.	1210.	1254.	1251.	1227.	1320.	1232.	827.	747.	0
TILT = 90:	745.	833.	814.	686.	605.	577.	596.	670.	871.	974.	716.	675.	0
KT:	.38	.41	.43	.43	.46	.48	.48	.46	.48	.47	.37	.37	
AMB TEMP:	29.2	30.4	38.1	48.6	58.6	68.0	73.3	71.3	64.5	55.4	45.2	33.0	51.3
HTG DEG DAYS:	1110	969	834	492	218	27	0	8	76	301	594	992	5621

STA NO:94725
MASSENA
NY
LAT = 44.93

	JAN	FEB	MAR	APR	MAY	JUN	JUL	AUG	SEP	OCT	NOV	DEC	YEAR
HORIZ INSOL:	391.	620.	978.	1343.	1613.	1779.	1751.	1484.	1124.	736.	388.	294.	1042.
TILT = LAT:	693.	921.	1197.	1345.	1408.	1463.	1480.	1405.	1275.	1038.	619.	514.	0
TILT = LAT+15:	720.	922.	1138.	1199.	1197.	1216.	1242.	1227.	1183.	1026.	632.	535.	0
TILT = 90:	648.	769.	839.	739.	641.	607.	636.	711.	816.	829.	552.	484.	0
KT:	.36	.39	.43	.45	.46	.48	.47	.45	.45	.42	.32	.31	
AMB TEMP:	14.5	16.7	27.6	42.2	54.1	64.3	66.7	64.3	59.2	48.5	35.9	20.1	43.2
HTG DEG DAYS:	1566	1352	1159	684	350	78	22	57	192	512	873	1392	8237

STA NO:94728
NEW YORK CITY (CENTRAL)
NY
LAT = 40.78

	JAN	FEB	MAR	APR	MAY	JUN	JUL	AUG	SEP	OCT	NOV	DEC	YEAR
HORIZ INSOL:	500.	721.	1037.	1364.	1636.	1688.	1710.	1483.	1214.	895.	533.	404.	1039.
TILT = LAT:	807.	997.	1212.	1346.	1435.	1439.	1425.	1394.	1338.	1206.	818.	657.	0
TILT = LAT+15:	839.	998.	1153.	1205.	1227.	1216.	1194.	1224.	1245.	1196.	841.	686.	0
TILT = 90:	732.	796.	796.	680.	587.	559.	534.	644.	798.	924.	717.	607.	0
KT:	.38	.40	.43	.44	.46	.45	.46	.45	.46	.45	.37	.34	
AMB TEMP:	32.2	33.4	41.1	52.1	62.3	71.6	76.6	74.9	68.4	58.7	47.4	35.5	54.5
HTG DEG DAYS:	1017	885	741	397	137	0	0	0	29	209	528	915	4848

STA NO:94823
PITTSBURGH
PA
LAT = 40.50

	JAN	FEB	MAR	APR	MAY	JUN	JUL	AUG	SEP	OCT	NOV	DEC	YEAR
HORIZ INSOL:	424.	625.	943.	1317.	1602.	1689.	1762.	1510.	1209.	895.	505.	347.	1069.
TILT = LAT:	631.	821.	1078.	1295.	1406.	1442.	1467.	1421.	1328.	1199.	750.	518.	0
TILT = LAT+15:	648.	815.	1022.	1159.	1203.	1219.	1228.	1248.	1236.	1188.	768.	533.	0
TILT = 90:	552.	638.	699.	652.	574.	555.	539.	650.	788.	914.	648.	458.	0
KT:	.32	.34	.38	.42	.45	.46	.47	.46	.45	.45	.35	.29	
AMB TEMP:	28.1	29.3	38.1	50.2	59.8	68.6	71.9	70.2	63.8	53.2	41.3	30.5	50.4
HTG DEG DAYS:	1144	1000	834	444	208	26	7	16	98	372	711	1070	5930

STA NO:94830
TOLEDO
OH
LAT = 41.60

	JAN	FEB	MAR	APR	MAY	JUN	JUL	AUG	SEP	OCT	NOV	DEC	YEAR
HORIZ INSOL:	435.	680.	997.	1384.	1717.	1849.	1878.	1616.	1276.	911.	498.	355.	1133.
TILT = LAT:	685.	944.	1168.	1372.	1505.	1572.	1556.	1527.	1427.	1255.	767.	567.	0
TILT = LAT+15:	707.	944.	1110.	1227.	1293.	1322.	1296.	1339.	1330.	1247.	788.	588.	0
TILT = 90:	614.	757.	776.	704.	620.	606.	576.	710.	865.	977.	674.	517.	0
KT:	.34	.39	.41	.45	.48	.51	.50	.48	.47	.47	.36	.31	
AMB TEMP:	24.8	27.1	35.8	48.4	58.8	68.9	72.3	70.8	63.8	53.0	39.6	28.0	49.3
HTG DEG DAYS:	1246	1061	905	498	229	32	5	18	99	379	762	1147	6381

STA NO:94849
ALPENA
MI
LAT = 45.07

	JAN	FEB	MAR	APR	MAY	JUN	JUL	AUG	SEP	OCT	NOV	DEC	YR
HORIZ INSOL:	362.	617.	1028.	1407.	1720.	1879.	1885.	1583.	1156.	743.	382.	270.	1086.
TILT = LAT:	622.	917.	1276.	1418.	1503.	1542.	1592.	1506.	1320.	1055.	609.	454.	0
TILT = LAT+15:	643.	919.	1216.	1264.	1276.	1279.	1332.	1315.	1226.	1043.	621.	468.	0
TILT = 90:	575.	767.	900.	780.	677.	632.	674.	760.	848.	845.	543.	420.	0
KT:	.33	.39	.46	.47	.49	.50	.52	.50	.46	.42	.32	.28	
AMB TEMP:	17.8	18.3	26.2	40.1	50.5	60.9	65.5	64.2	56.3	47.3	34.9	23.4	42.1
HTG DEG DAYS:	1463	1308	1203	747	455	150	75	110	265	549	903	1290	6518

STA NO:94860
GRAND RAPIDS
MI
LAT = 42.88

	JAN	FEB	MAR	APR	MAY	JUN	JUL	AUG	SEP	OCT	NOV	DEC	YR
HORIZ INSOL:	370.	648.	1014.	1412.	1755.	1956.	1914.	1676.	1262.	858.	446.	311.	1135.
TILT = LAT:	572.	918.	1214.	1409.	1537.	1613.	1622.	1591.	1428.	1199.	691.	494.	0
TILT = LAT+15:	586.	917.	1155.	1259.	1307.	1339.	1361.	1393.	1330.	1190.	707.	510.	0
TILT = 90:	509.	746.	825.	741.	652.	612.	643.	759.	886.	945.	609.	450.	0
KT:	.31	.38	.43	.46	.50	.52	.53	.52	.49	.46	.34	.29	
AMB TEMP:	23.2	24.5	33.1	46.5	57.1	67.4	71.5	70.0	62.4	52.0	38.7	27.4	47.8
HTG DEG DAYS:	1296	1134	989	555	270	44	8	27	114	409	789	1166	6801

STA NO:94918
NORTH OMAHA
NE
LAT = 41.37

	JAN	FEB	MAR	APR	MAY	JUN	JUL	AUG	SEP	OCT	NOV	DEC	YR
HORIZ INSOL:	634.	892.	1223.	1558.	1873.	2123.	2107.	1858.	1373.	1050.	644.	511.	1321.
TILT = LAT:	1144.	1331.	1484.	1558.	1642.	1751.	1787.	1768.	1549.	1490.	1082.	946.	0
TILT = LAT+15:	1209.	1349.	1421.	1395.	1397.	1452.	1496.	1548.	1446.	1489.	1128.	1005.	0
TILT = 90:	1086.	1104.	997.	789.	657.	613.	654.	801.	937.	1173.	986.	918.	0
KT:	.49	.50	.51	.51	.53	.57	.58	.57	.52	.54	.46	.44	
AMB TEMP:	20.2	25.5	34.6	50.0	60.9	70.2	75.1	73.7	64.4	54.4	37.9	25.7	49.4
HTG DEG DAYS:	1389	1106	942	456	186	33	7	10	99	342	813	1218	6601

CHAPTER 63
GEOTHERMAL RESOURCES

PETER D. BLAIR

U.S. Congress
Office of Technology Assessment
Washington, DC

63.1 GEOTHERMAL RESOURCES: AN INTRODUCTION

Geothermal resources are traditionally divided into three basic classes: (1) hydrothermal convection systems, including both vapor-dominated and liquid-dominated systems; (2) geopressured resources; and (3) hot dry rock and molten magma systems. The three basic resource categories are distinguished by geologic characteristics and the manner in which heat is transferred to the earth's surface (see Table 63.1). In the following we briefly discuss the characteristics and location of these resource categories in the United States.* Only the first of these resource types is commercially exploited today in the United States.

In 1975, the U.S. Geological Survey completed a national assessment of geothermal resources in the United States and published the results in USGS Circular 726 (subsequently updated in 1978 as Circular 790). This assessment defined a "geothermal resource base" for the United States based on geologic estimates of all stored heat in the earth above 15°C and within 6 miles of the surface, ignoring recoverability. In addition, these resources were catalogued according to the classes given above. The end result is a set of 108 known geothermal resource areas (KGRAs) encompassing over 3 million acres in the 11 western states.

63.1.1 Hydrothermal Resources

Hydrothermal convection systems are formed when underground reservoirs carry the earth's heat toward the surface by convective circulation of water (liquid-dominated resources) or steam (vapor-dominated resources). Only three vapor-dominated systems have been identified in the United States, that is, The Geysers and Mount Lassen in California and the Mud Volcano system in Yellowstone National Park. The remaining U.S. resources are liquid-dominated (see Fig. 63.1).

Vapor-Dominated Resources

In a vapor-dominated hydrothermal system, boiling of deep subsurface water produces water vapor, which is also often superheated by the hot surrounding rock. Geologists speculate that as the vapor moves toward the surface, a level of cooler, near-surface rock may induce condensation, which, along with cooler groundwater from the margins of the reservoir, serves to recharge the reservoir.[4] Since

Table 63.1 Geothermal Resource Classification[2]

Resource Type	Temperature Characteristics
1. Hydrothermal convection resource (heat carried upward from depth by convection of water or steam)	
a. Vapor-dominated	About 240°C (464°F)
b. Hot-water dominated	
1. High temperature	150–350°C+ (300–660°F)
2. Intermediate temperature	90–150°C (190–300°F)
3. Low temperature	Less than 90°C
2. Hot igneous resources (rock intruded in molten form from depth)	
a. Part still molten	Higher than 650°C (1200°F)
b. Not molten—"hot dry rock"	90–650°C (190–1200°F)
3. Conduction-dominated resources (heat carried upward by conduction through rock)	
a. Radiogenic (heat generated by radioactive decay)	30–150°C (86–300°F)
b. Sedimentary basins (hot fluid in sedimentary rocks)	30–150°C
c. Geopressured (hot fluid under high pressure)	150–200°C (300–390°F)

* Many expositions of the definition and classification of geothermal resources exist; some of the more recent ones include Refs. 1 and 4.

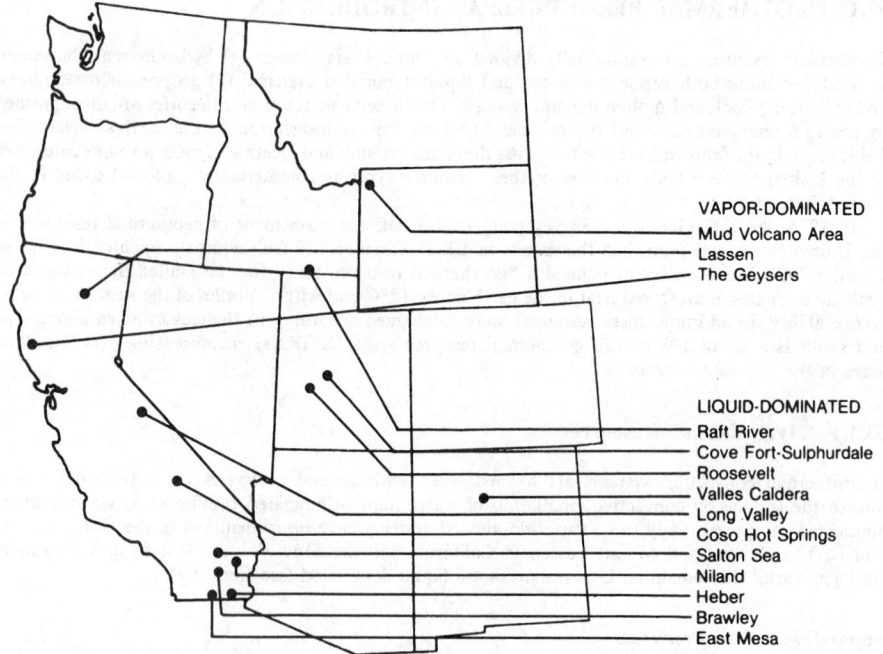

VAPOR-DOMINATED
Mud Volcano Area
Lassen
The Geysers

LIQUID-DOMINATED
Raft River
Cove Fort-Sulphurdale
Roosevelt
Valles Caldera
Long Valley
Coso Hot Springs
Salton Sea
Niland
Heber
Brawley
East Mesa

Fig. 63.1   Major U.S. hydrothermal resources.[3]

fluid convection is taking place constantly, the temperature in the vapor-filled area of the reservoir is relatively uniform and a well drilled into this region will yield high-quality superheated steam.

As mentioned earlier, the most developed geothermal resource in the world is The Geysers in northern California, which is a high-quality, vapor-dominated hydrothermal convection system. Currently, steam is produced from this resource at a depth of 5000–10,000 ft and piped directly into turbine–generators to produce electricity. Over 900 MW of electric generating capacity have been installed at The Geysers, and several capacity additions are scheduled over the next few years (see Fig. 63.2).

White and Williams[2] report that the commercially produced vapor-dominated systems at The Geysers, Lardarello (Italy), and Matsukawa (Japan) are all characterized by reservoir temperatures in excess of 450°F. Accompanying the water vapor are small concentrations, that is, less than 5%, of noncondensible gases (mostly carbon dioxide, hydrogen sulfide, and ammonia).

The Monte Amiata field (Italy) is a different type of vapor-dominated resource, which is characterized by lower temperatures than The Geysers-type resource and by much higher gas content (carbon dioxide and hydrogen sulfide). The geology of this category of vapor-dominated resource is not yet well understood, but may turn out to be more common than The Geysers-type resource because its existence is much more difficult to detect.

Liquid-Dominated Resources

Hot-water or wet-steam hydrothermal deposits are more commonly found than dry-steam deposits. Hot-water systems are often associated with a hot spring that discharges at the surface. When wet-steam deposits occur at considerable depths, the resource temperature is often well above the normal boiling point of water at atmospheric pressures. These temperatures are known to range from 100 to 700°F at pressures of 50–150 psig. When such resources penetrate to the surface, either through wells or through natural geologic anomalies, the water often flashes into steam.

The types of impurities found in wet-steam deposits vary dramatically. Commonly found dissolved salts and minerals include sodium, potassium, lithium, chlorides, sulfates, borates, bicarbonates, and silica. Salinity concentrations can vary from thousands to hundreds of thousands of parts per million. The Wairakei (New Zealand) and Cerro Prieto (Mexico) fields are examples of currently exploited liquid-dominated fields. In the United States many liquid-dominated systems have been identified and are being considered for development (see Fig. 63.1).

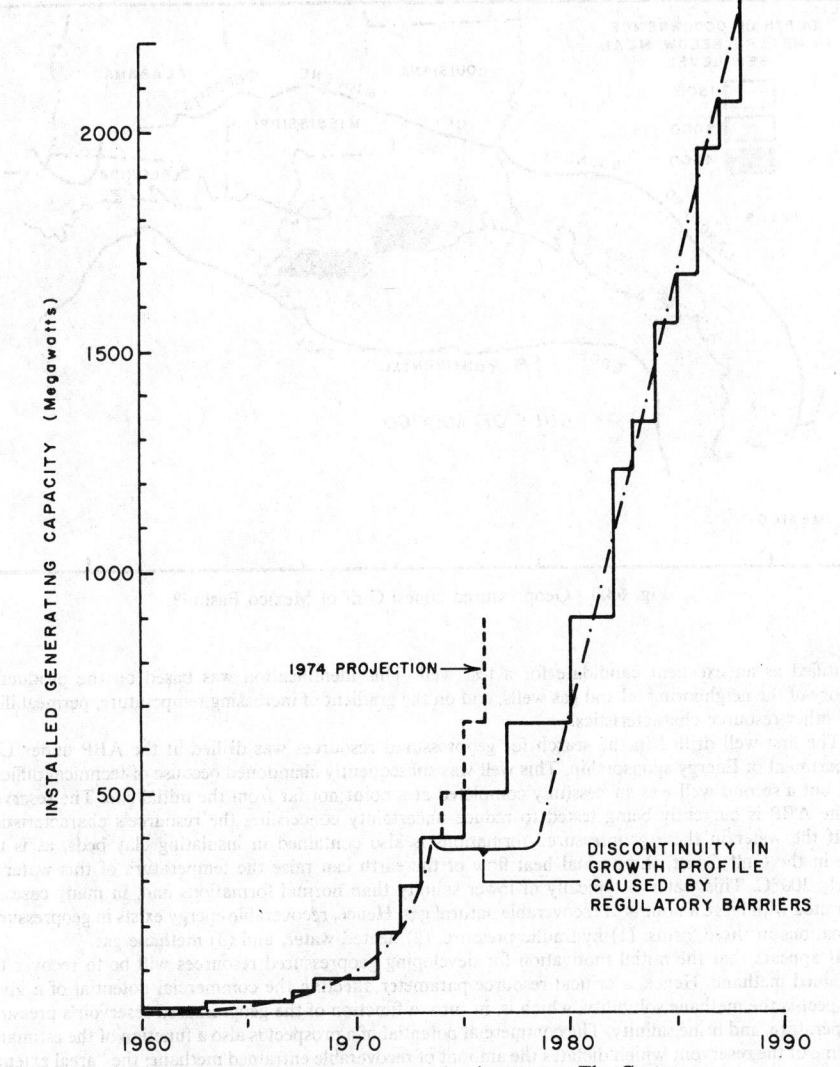

Fig. 63.2 Geothermal power development at The Geysers.

63.1.2 Geopressured Resources

Near the Gulf Coast of the United States there are a number of deep sedimentary basins that are geologically very young, that is, less than 60 million years. In such regions, fluid located in subsurface rock formation carries a part of the overburden load, thereby increasing the pressure within the formation. Such formations are referred to as geopressured and are judged by some geologists to be quite promising sources of energy over the next several decades. (Papadopulus[5] discusses the nature of geopressured resources in detail.)

Geopressured basins exist in several areas within the United States, but those of current interest are located along the Texas–Louisiana coast. These are of particular interest because they are very large in terms of both areal extent and thickness, and the geopressured liquids appear to have a great deal of dissolved methane. In a recent investigation of the Gulf Coast,[6] a number of "geopressured fairways" were identified; these are thick sandstone bodies expected to contain geopressured fluids of at least 300°F. Detailed studies of the fairways of the Frio Formation in East Texas were carried out in 1979; although only one, Brazoria (see Fig. 63.3), met the requirements for further well testing. Within this fairway, a particularly promising site known as the Austin Bayou Prospect (ABP) was

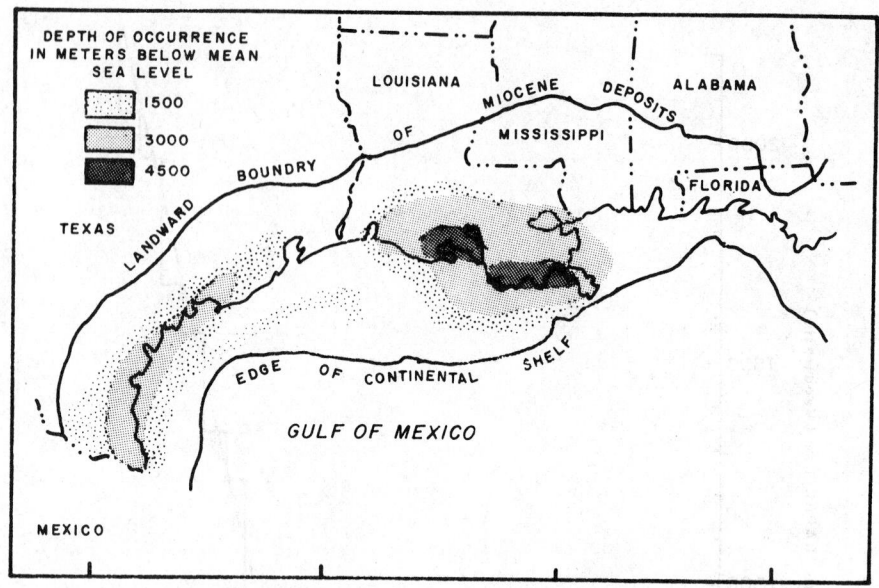

Fig. 63.3   Geopressured zones: Gulf of Mexico Basin.[10]

identified as an excellent candidate for a test well. This identification was based on the productive history of the neighboring oil and gas wells, and on the gradient of increasing temperature, permeability, and other resource characteristics.

The first well drilled in the search for geopressured resources was drilled at the ABP under U.S. Department of Energy sponsorship. This well was subsequently abandoned because of technical difficulties, but a second well was successfully completed at a point not far from the initial site. The reservoir at the ABP is currently being tested to reduce uncertainty concerning the resource's characteristics.

If the water in these geopressured formations is also contained in insulating clay beds, as is the case in the Gulf Coast, the normal heat flow of the earth can raise the temperature of this water to nearly 300°C. This water is typically of lower salinity than normal formations and, in many cases, is saturated with large amounts of recoverable natural gas. Hence, recoverable energy exists in geopressured formations in three forms: (1) hydraulic pressure, (2) heated water, and (3) methane gas.

It appears that the initial motivation for developing geopressured resources will be to recover the entrained methane. Hence, a critical resource parameter affecting the commercial potential of a given prospect is the methane solubility, which is, in turn, a function of the geopressured reservoir's pressure, temperature, and brine salinity. The commercial potential of a prospect is also a function of the estimated volume of the reservoir, which dictates the amount of recoverable entrained methane; the "areal extent," which dictates how much methane can ultimately be recovered from the prospect site; and the "pay thickness," which dictates the initial production rate and the rate of production decline over time.

63.1.3   Hot Dry Rock Resources

In some areas of the western United States, geologic anomalies such as tectonic plate movement and volcanic activity have created pockets of impermeable rocks covering a magma chamber. The temperature in these pockets increases with depth and proximity to the magma chamber, but, because of their impermeable nature, they lack a water aquifer. They are often referred to as hot dry rock (HDR) deposits. Several schemes for useful energy production from hot dry rock resources have been proposed, but all basically involve creation of an artificial aquifer that is used to bring heat to the surface. The concept is being tested by the U.S. Department of Energy at Fenton Hill near Los Alamos, New Mexico.

A typical two-well hot dry rock system is shown in Fig. 63.4. Water is injected at high pressure through the first well to the reservoir and returns to the surface through the second well at approximately the temperature of the reservoir. The water (steam) is used to generate electric power and is then recirculated through the first well.

The critical parameters affecting the commercial feasibility of HDR resources are the geothermal gradient and the achievable well flow rate.

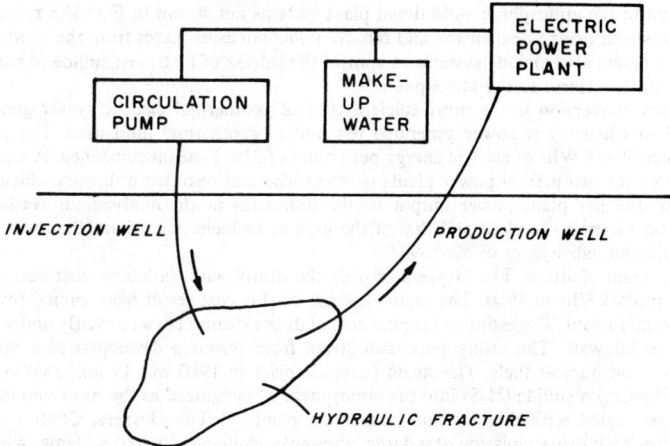

Fig. 63.4 Hot dry rock power plant configuration.

Development of HDR resources is still experimental and will not be considered here. An investigation of the commercial potential of HDR electric power has been carried out by Cummings et al.[7]

63.2 GEOTHERMAL ENERGY CONVERSION

63.2.1 Direct Steam Conversion

The geothermal resources of central Italy and The Geysers are characterized as "vapor-dominated" resources; that is, the fluids they produce are predominantly dry or superheated steam, usually with small amounts of noncondensible gases such as carbon dioxide (CO_2), hydrogen sulfide (H_2S), methane (CH_4), nitrogen (N_2), oxygen (O_2), hydrogen (H_2), and ammonia (NH_3).

The conversion of geothermal energy into electric energy at a vapor-dominated resource is a straight-forward process. The naturally pressurized steam is piped from wells to a power plant where it expands through a turbine–generator to produce electric energy. The geothermal steam is supplied to the turbine directly, save for the relatively simple removal of entrained solids in gravity separators or the removal of noncondensible gases in degassing vessels. From the turbine, steam is exhausted to a condenser, condensed to its liquid state, and pumped from the plant. Usually this condensate is reinjected to the subterranean aquifer. Unfortunately, vapor-dominated geothermal resources occur infrequently in nature. To date, electric power generation from natural dry steam occurs at only one area, Matuskawa in Japan, other than central Italy and The Geysers.

A simplified flow diagram illustrating the direct steam conversion process is shown in Fig. 63.5. The major components of the systems are the steam turbine–generator, condenser, and cooling towers. Dry steam from the geothermal production well is expanded through the turbine, which drives an electric generator. The wet steam exhausting from the turbine is condensed and the condensate is piped from the plant for reinjection or other disposal. The cooling towers reject the waste heat released

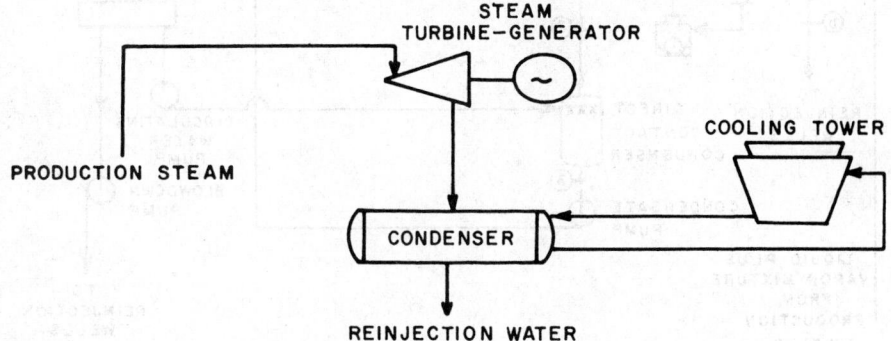

Fig. 63.5 Direct steam conversion.

by condensation to the atmosphere. Additional plant systems not shown in Fig. 63.5 remove entrained solids from the steam prior to expansion and remove noncondensible gases from the condenser. Recent plants at The Geysers also include systems to control the release of hydrogen sulfide (a noncondensible gas contained in the steam) to the atmosphere.

Direct steam conversion is the most efficient type of geothermal electric power generation. One measure of plant efficiency is power generated per unit of geothermal fluid used. The plants at The Geysers produce 50–55 Whr of electric energy per pound of 350°F steam consumed. A second measure of efficiency used for geothermal power plants is the geothermal resource utilization efficiency, defined as the ratio of the net plant power output to the difference in thermodynamic availability of the geothermal fluid entering the plant and that of the fluid at ambient conditions. Plants at The Geysers operate at utilization efficiencies of 50–56%.[8]

The direct steam plants at The Geysers provide the utility with its lowest cost base load thermal power, 20–30 mills/kWhr in 1980. The main elements of this cost result from capital investment and costs of geothermal steam. The estimated capital cost of direct steam plants currently under construction is $600–650 per kilowatt. The utility purchases steam from resource developers at a price based on its cost of fossil and nuclear fuels. The steam purchase price in 1980 was 17 mills/kWhr.

Release of hydrogen sulfide (H_2S) into the atmosphere is recognized as the most important environmental issue associated with direct steam conversion plants at The Geysers. Control measures are required to meet California emission standards. Presently available control systems, which treat the steam after it has passed through the turbine, result in significant penalties in capital and operating cost. These systems include the iron/caustic/peroxide process, which has been installed on six Geysers units, and the Stretford process, which is used on the two newest plants. Other more economical processes, which treat the steam before it reaches the turbine, are currently under development.

63.2.2 Flashed Steam Conversion

As was mentioned earlier, most geothermal resources produce not dry steam, but a pressurized two-phase mixture of steam and water. The majority of plants currently operating at these liquid-dominated resources use a flashed steam energy conversion process. Flashed steam plants are currently operating in New Zealand, Japan, Iceland, and Mexico.

Figure 63.6 is a simplified schematic of a flashed steam plant. In addition to the turbine, condenser, and cooling towers found in the direct steam process, the flashed steam plant contains a separator or flash vessel. The geothermal fluid from the production wells first enters this vessel where saturated steam is flashed from the liquid brine. This steam enters the turbine, while the unflashed brine is piped from the plant for reinjection or disposal. The remainder of the process is similar to the direct steam process.

Multiple stages of flash vessels are often used in the flashed steam systems to improve the plant efficiency and increase power output. Figure 63.7 shows a flow diagram of a two-stage flash plant. In this case, the unflashed brine leaving the initial flash vessel enters a second flash vessel that operates

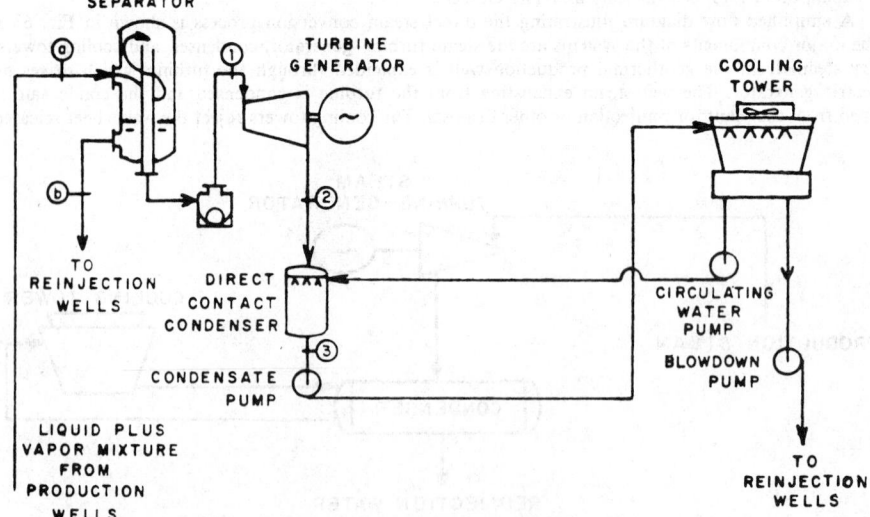

Fig. 63.6   Flashed steam conversion.

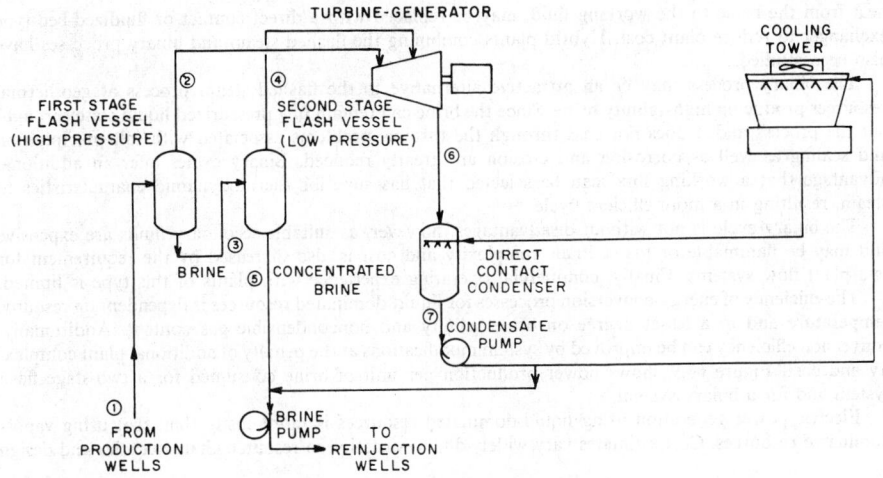

Fig. 63.7 Two-stage flash conversion.

at a lower pressure, causing additional steam to be flashed. This lower-pressure steam is admitted to the low-pressure section of the turbine, recovering energy that would have been lost if a single-stage flash process had been used. In a design study for a geothermal plant to be located near Heber, California, the two-stage flash process resulted in a 37% improvement in plant performance over a single-stage flash process. Addition of a third flash stage showed an incremental improvement of 6% and was determined not to be cost-effective.[9]

63.2.3 Binary Cycle Conversion

Binary cycle conversion plants are an alternative approach to flashed steam plants for electric power generation at liquid-dominated geothermal resources. In this type of plant a secondary fluid, usually a fluorocarbon or hydrocarbon, is used as the working fluid in a Rankine cycle, and the geothermal brine is used to heat this working fluid. Binary plants are not as well developed or commercially proven as flashed steam plants. However, test plants have been built and operated in Japan and at East Mesa in California's Imperial Valley.

Figure 63.8 shows the main components and flow steams in a binary conversion process. Geothermal brine from production wells passes through a heat exchanger, where it transfers sensible heat to the secondary working fluid. The cooled brine is then reinjected into the well field. The secondary working fluid is vaporized and superheated in the heat exchanger and expanded through a turbine, which drives an electric generator. The turbine exhaust is condensed in a surface condenser, and the condensate is pressurized and returned to the heat exchanger to complete the cycle. A cooling tower and circulating water system rejects the heat of condensation to the atmosphere.

Several variations of this cycle have been considered for geothermal power generation. A regenerator may be added between the turbine and condenser to recover energy from the turbine exhaust for condensate heating and to improve plant efficiency. The surface-type heat exchanger, which passes

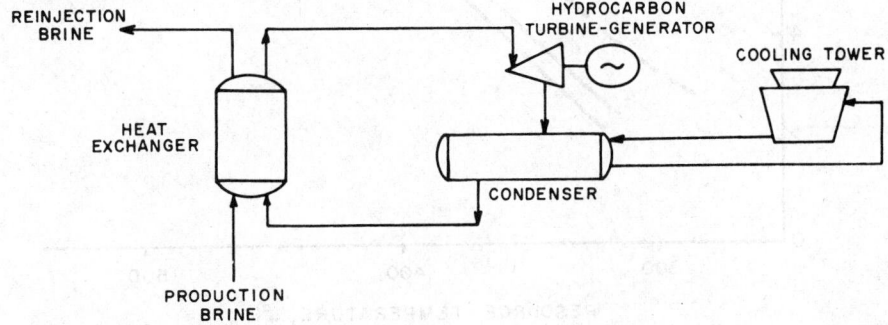

Fig. 63.8 Binary cycle conversion.

heat from the brine to the working fluid, may be replaced with a direct contact or fluidized-bed-type exchanger to reduce plant cost. Hybrid plants combining the flashed steam and binary processes have also been studied.

The binary process may be an attractive alternative to the flashed steam process at geothermal resources producing high-salinity brine. Since the brine can remain in a pressurized liquid state throughout the process and it does not pass through the turbine, problems associated with salt precipitation and scaling as well as corrosion and erosion are greatly reduced. Binary cycles offer an additional advantage that a working fluid can be selected that has superior thermodynamic characteristics to steam, resulting in a more efficient cycle.

The binary cycle is not without disadvantages, however, as suitable secondary fluids are expensive and may be flammable or toxic. Plant complexity and cost is also increased by the requirement for two plant flow systems. Finally, commercial operating experience with plants of this type is limited.

The efficiency of energy-conversion processes for liquid-dominated resources is dependent on resource temperature and to a lesser degree on brine salinity and noncondensible gas content. Additionally, conversion efficiency can be improved by system modifications at the penalty of additional plant complexity and cost. Figure 63.9 shows power production per unit of brine consumed for a two-stage flash system and for a binary system.

Electric power generation using liquid-dominated resources is more costly than that using vapor-dominated resources. Cost estimates vary widely due to variation in resource characteristics and design

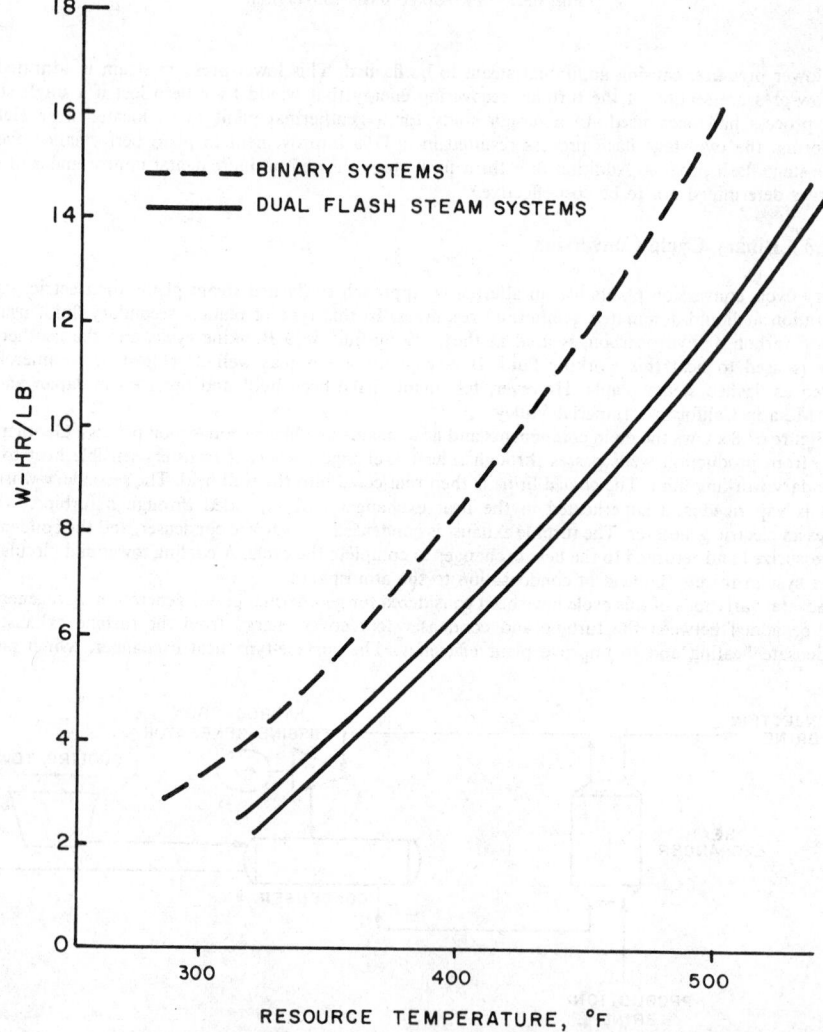

Fig. 63.9 Net geothermal brine effectiveness.

uncertainties. A summary of cost of energy estimates by several public and private organizations showed costs ranging from 40 to 130 mills/kWhr in 1980 dollars.[10]

Emissions of hydrogen sulfide (H_2S) at liquid-dominated geothermal plants are lower than for direct steam processes. Flashed steam plants emit 30–50% less H_2S than direct steam plants. Binary plants would not routinely emit H_2S because the brine would remain contained and pressurized throughout the process.

However, there are other environmental considerations inherent in liquid-dominated systems. A major question is the possibility of land surface subsidence caused by the withdrawal of the brine from the geothermal resource. Although reinjection of the brine after use in the plant may reduce or eliminate land subsidence, faulty reinjection could contaminate local fresh groundwater. Also, if all brine is reinjected, an external source of water is required for plant-cooling-water makeup.

63.2.4 Hybrid Fossil/Geothermal Plants

The hybrid fossil/geothermal power plant uses both fossil energy and geothermal heat to produce electric power. Several candidate systems have been proposed and analyzed by researchers at Brown University.[8] These include the "geothermal preheat" system in which geothermal brine is used for the initial feedwater heating in an otherwise conventional fossil-fired plant. Also proposed is a "fossil superheat" concept that incorporates a fossil-fired heater to superheat geothermal steam prior to expansion in a turbine.

Preliminary studies of these systems have indicated that they may offer the potential for reducing fossil fuel consumption while providing power generation at competitive costs. In a study for Roosevelt Hot Springs in Utah, geothermal preheating reduced the coal consumption of a coal-fired power plant by approximately 15%. Cost of power from hybrid plant configurations for several sites was reported to range from 20 to 30 mills/kWhr in 1977 dollars. However, the remoteness of most geothermal sites and resulting logistic and economic problems of coal supply and power transmission may preclude serious consideration of such hybrid designs at many areas. Furthermore, current uncertainties pertaining to geothermal fluid production and reinjection may pose unacceptable risks in consideration of the high capital cost of a coal-fired power plant.

REFERENCES

1. H. C. Armstead, *Geothermal Energy*, Wiley, New York, 1978.

2. D. E. White and D. L. Williams, "Assessment of Geothermal Resources in the U.S.: 1975," U.S. Geological Survey Circular 726, 1975.

3. R. L. Smith and H. R. Shaw, U.S. Geological Survey Circular 726, 1975.

4. P. M. Wright, "Nature and Occurrence of Geothermal Resources," Earth Science Laboratory, University of Utah Research Institute, 1980.

5. S. Papadopulus, "The Energy Potential of Geopressured Reservoirs: Hydrogeologic Factory," in H. Dorman and R. Deller (eds.), *Proceedings of the First Geopressured Energy Conference*, 1975.

6. D. Bebout et al., "Frio Sandstone Reservoirs in the Deep Subsurface Along the Texas Gulf Coast," U.S. Department of Energy, AT-ET40-1-4891, 1978.

7. R. Cummings et al., "Mining Earth's Heat: Hot Dry Rock Geothermal Energy," *Technology Review* (Feb. 1979).

8. R. DiPippo, *Geothermal Energy as a Source of Electricity*, U.S. Government Printing Office, Washington, DC, 1980.

9. Bechtel Corporation, "Conceptual Design of Commercial Geothermal Power Plants at Heber and Niland, California," U.S. Department of Energy, San Francisco, 1976.

10. P. Blair, T. A. V. Cassel, and R. H. Edelstein, *Geothermal Energy: Investment Decisions and Commercial Development*, Wiley, New York, 1982.

CHAPTER 64

ENERGY AUDITING

CARL BLUMSTEIN

Universitywide Energy Research Group
University of California
Berkeley, California

PETER KUHN

Kuhn and Kuhn,
Industrial Energy Consultants
Golden Gate Energy Center
Sausalito, California

64.1 ENERGY MANAGEMENT AND THE ENERGY AUDIT

Energy auditing is the practice of surveying a facility to identify opportunities for increasing the efficiency of energy use. A facility may be a residence, a commercial building, an industrial plant, or other installation where energy is consumed for any purpose. Energy management is the practice of organizing financial and technical resources and personnel to increase the efficiency with which energy is used in a facility. Energy management typically involves the keeping of records on energy consumption and equipment performance, optimization of operating practices, regular adjustment of equipment, and replacement or modification of inefficient equipment and systems.

Energy auditing is a part of an energy management program. The auditor, usually someone not regularly associated with the facility, reviews operating practices and evaluates energy using equipment in the facility in order to develop recommendations for improvement. An energy audit can be, and often is, undertaken when no formal energy management program exists. In simple facilities, particularly residences, a formal program is impractical and informal procedures are sufficient to alter operating practices and make simple improvements such as the addition of insulation. In more complex facilities, the absence of a formal energy management program is usually a serious deficiency. In such cases a major recommendation of the energy audit will be to establish an energy management program.

There can be great variation in the degree of thoroughness with which an audit is conducted, but the basic procedure is universal. The first step is to collect data with which to determine the facility's major energy uses. These data always include utility bills, nameplate data from the largest energy-using equipment, and operating schedules. The auditor then makes a survey of the facility. Based on the results of this survey, he or she chooses a set of energy conservation measures that could be applied in the facility and estimates their installed cost and the net annual savings that they would provide. Finally, the auditor presents his or her results to the facility's management or operators. The audit process can be as simple as a walkthrough visit followed by a verbal report or as complex as a complete analysis of all of a facility's energy using equipment that is documented by a lengthy written report.

The success of an energy audit is ultimately judged by the resulting net financial return (value of energy saved less costs of energy saving measures). Since the auditor is rarely in a position to exercise direct control over operating and maintenance practices or investment decisions, his or her work can come to naught because of the actions or inaction of others. Often the auditor's skills in communication and interpersonal relations are as critical to obtaining a successful outcome from an energy audit as his or her engineering skills. The auditor should stress from the outset of his or her work that energy management requires a sustained effort and that in complex facilities a formal energy management program is usually needed to obtain the best results. Most of the auditor's visits to a facility will be spent in the company of maintenance personnel. These personnel are usually conscientious and can frequently provide much useful information about the workings of a facility. They will also be critical to the success of energy conservation measures that involve changes in operating and maintenance practices. The auditor should treat maintenance personnel with respect and consideration and should avoid the appearance of "knowing it all." The auditor must also often deal with nontechnical managers. These managers are frequently involved in the decision to establish a formal energy management program and in the allocation of capital for energy saving investments. The auditor should make an effort to provide clear explanations of his or her work and recommendations to nontechnical managers and should be careful to avoid the use of engineering jargon when communicating with them.

While the success of an energy audit may depend in some measure on factors outside the auditor's control, a good audit can lead to significant energy savings. Table 64.1 shows the percentage of energy saved as a result of implementing energy audit recommendations in 172 nonresidential buildings. The average savings is more than 20%. The results are especially impressive in light of the fact that most of the energy-saving measures undertaken in these buildings were relatively inexpensive. The median value for the payback on energy-saving investments was in the 1–2 year range (i.e., the value of the energy savings exceeded the costs in 1–2 years). An auditor can feel confident in stating that an energy saving of 20% or more is usually possible in facilities where systematic efforts to conserve energy have not been undertaken.

64.2 PERFORMING AN ENERGY AUDIT—ANALYZING ENERGY USE

A systematic approach to energy auditing requires that an analysis of existing energy-using systems and operating practices be undertaken before efforts are made to identify opportunities for saving energy. In practice, the auditor may shift back and forth from the analysis of existing energy-use patterns to the identification of energy-saving opportunities several times in the course of an audit—first doing the most simple analysis and identifying the most obvious energy-saving opportunities, then performing more complex analyses, and so on. This strategy may be particularly useful if the audit is to be conducted over a period of time that is long enough for some of the early audit recommendations to be implemented. The resultant savings can greatly increase the auditor's credibility with the

Table 64.1   The Percentage of Energy Saved as a Result of Implementing Energy
Audit Recommendations in 172 Nonresidential Buildings[a,4]

	Site		Source	
Building Category	Savings (%)	Sample Size	Savings (%)	Sample Size
Elementary school	24	72	21	72
Secondary school	30	38	28	37
Large office	23	37	21	24
Hospital	21	13	17	10
Community center	56	3	23	18
Hotel	25	4	24	4
Corrections	7	4	5	4
Small office	33	1	30	1
Shopping center	11	1	11	1
Multifamily apartment	44	1	43	1

[a] Electricity is counted at 3413 Btu/kWhr for site energy and 11,500 Btu/kWhr
for source energy (i.e., including generation and transmission losses).

facility's operators and management, so that he or she will receive more assistance in completing his or her work and his or her later recommendations will be attended to more carefully.

The amount of time devoted to analyzing energy use will vary, but, even in a walkthrough audit, the auditor will want to examine records of past energy consumption. These records can be used to compare the performance of a facility with the performance of similar facilities. Examination of the seasonal variation in energy consumption can give an indication of the fractions of a facility's use that are due to space heating and cooling. Records of energy consumption are also useful in determining the efficacy of past efforts to conserve energy.

In a surprising number of facilities the records of energy consumption are incomplete. Often records will be maintained on the costs of energy consumed but not on the quantities. In periods of rapidly escalating prices, it is difficult to evaluate energy performance with such records. Before visiting a facility to make an audit, the auditor should ask that complete records be assembled and, if the records are not on hand, suggest that they be obtained from the facility's suppliers. Good record keeping is an essential part of an energy management program. The records are especially important if changes in operation and maintenance are to be made, since these changes are easily reversed and often require careful monitoring to prevent backsliding.

In analyzing the energy use of a facility, the auditor will want to focus his or her attention on the systems that use the most energy. In industrial facilities these will typically involve production processes such as drying, distillation, or forging. Performing a good audit in an industrial facility requires considerable knowledge about the processes being used. Although some general principles apply across plant types, industrial energy auditing is generally quite specialized. Residential energy auditing is at the other extreme of specialization. Because a single residence uses relatively little energy, highly standardized auditing procedures must be used to keep the cost of performing an audit below the value of potential energy savings. Standardized procedures make it possible for audits to be performed quickly by technicians with relatively limited training.

Commercial buildings lie between these extremes of specialization. The term "commercial building" as used here refers to those nonresidential buildings that are not used for the production of goods and includes office buildings, schools, hospitals, and retail stores. The largest energy-using systems in commercial buildings are usually lighting and HVAC (heating, ventilating, and air conditioning). Refrigeration consumes a large share of the energy used in some facilities (e.g., food stores) and other loads may be important in particular cases (e.g., research equipment in laboratory buildings). Table 64.2 shows the results of a calculation of the amount of energy consumed in a relatively energy-efficient office building for lighting and HVAC in different climates. Office buildings (and other commercial buildings) are quite variable in their design and use. So, while the proportions of energy devoted to various uses shown in Table 64.2 are not unusual, it would be unwise to treat them (or any other proportions) as "typical." Because of the variety and complexity of energy-using systems in commercial buildings and because commercial buildings frequently use quite substantial amounts of energy in their operation, an energy audit in a commercial building often warrants the effort of a highly trained professional. In the remainder of this section commercial buildings will be used to illustrate energy auditing practice.

Lighting systems are often a good starting point for an analysis of energy in commercial buildings.

Table 64.2   Results of a Calculation of the Amount of Energy Consumed in a Relatively Energy-Efficient Office Building for Lighting and HVAC[5]

	Energy Use (kBtu/ft²/yr)			
	Miami	Los Angeles	Washington	Chicago
Lights	34.0	34.0	34.0	34.0
HVAC auxiliaries	8.5	7.7	8.8	8.8
Cooling	24.4	9.3	10.2	7.6
Heating	0.2	2.9	17.7	28.4
Total	67.1	53.9	70.7	78.8

They are the most obvious energy consumers, are usually easily accessible, and can provide good opportunities for energy saving. As a first step the auditor should determine the hours of operation of the lighting systems and the watts per square foot of floorspace that they use. These data, together with the building area, are sufficient to compute the energy consumption for lighting and can be used to compare the building's systems with efficient lighting practice. Next, lighting system maintenance practices should be examined. As shown in Fig. 64.1, the accumulation of dirt on lighting fixtures can significantly reduce light output. Fixtures should be examined for cleanliness and the auditor should determine whether or not a regular cleaning schedule is maintained. As lamps near the end of their rated life, they lose efficiency. Efficiency can be maintained by replacing lamps in groups before they reach the end of their rated life. This practice also reduces the higher maintenance costs associated with spot relamping. Fixtures should be checked for lamps that are burned out or show signs of excessive wear, and the auditor should determine whether or not a group-relamping program is in effect.

After investigating lighting operation and maintenance practices, the auditor should measure the levels of illumination being provided by the lighting systems. These measurements can be made with a relatively inexpensive photometer. Table 64.3 gives recommended levels of illumination for a variety of activities. A level much in excess of these guidelines usually indicates an opportunity for saving energy. However, the auditor should recognize that good seeing also depends on other factors such as glare and contrast and that the esthetic aspects of lighting systems (i.e., their appearance and the effect they create) can also be important. More information about the design of lighting systems can be found in Ref. 1.

Analysis of HVAC systems in a commercial building is generally more complicated and requires more time and effort than lighting systems. However, the approach is similar in that the auditor will usually begin by examining operating and maintenance practices and then proceed to measure system performance.

Determining the fraction of a building's energy consumption that is devoted to the operation of its HVAC systems can be difficult. The approaches to this problem can be classified as either deterministic or statistical. In the deterministic approaches an effort is made to calculate HVAC energy consumption from engineering principles and data. First, the building's heating and cooling loads are calculated. These depend on the operating schedule and thermostat settings, the climate, heat gains and losses from radiation and conduction, the rate of air exchange, and heat gains from internal sources. Then energy use is calculated by taking account of the efficiency with which the HVAC systems meet these loads. The efficiency of the HVAC systems depends on the efficiency of equipment such as boilers and chillers and losses in distribution through pipes and ducts; equipment efficiency and distribution losses are usually dependent on load. In all but the simplest buildings, the calculation of HVAC energy consumption is sufficiently complex to require the use of computer programs; a number of such programs are available (see, for example, Ref. 2). The auditor will usually make some investigation of all of the factors necessary to calculate HVAC energy consumption. However, the effort involved in obtaining data that are sufficiently accurate and preparing them in suitable form for input to a computer program is quite considerable. For this reason, the deterministic approach is not recommended for energy auditing unless the calculation of savings from energy conservation measures requires detailed information on building heating and cooling loads.

Statistical approaches to the calculation of HVAC energy consumption involve the analysis of records of past energy consumption. In one common statistical method, energy consumption is analyzed as a function of climate. Regression analysis with energy consumption as the dependent variable and some function of outdoor temperature as the independent variable is used to separate "climate-dependent" energy consumption from "base" consumption. The climate-dependent fraction is considered to be the energy consumption for heating and cooling, and the remainder is assumed to be due to other uses. This method can work well in residences and in some small commercial buildings where heating and cooling loads are due primarily to the climate. It does not work as well in large commercial buildings because much of the cooling load in these buildings is due to internal heat gains and because

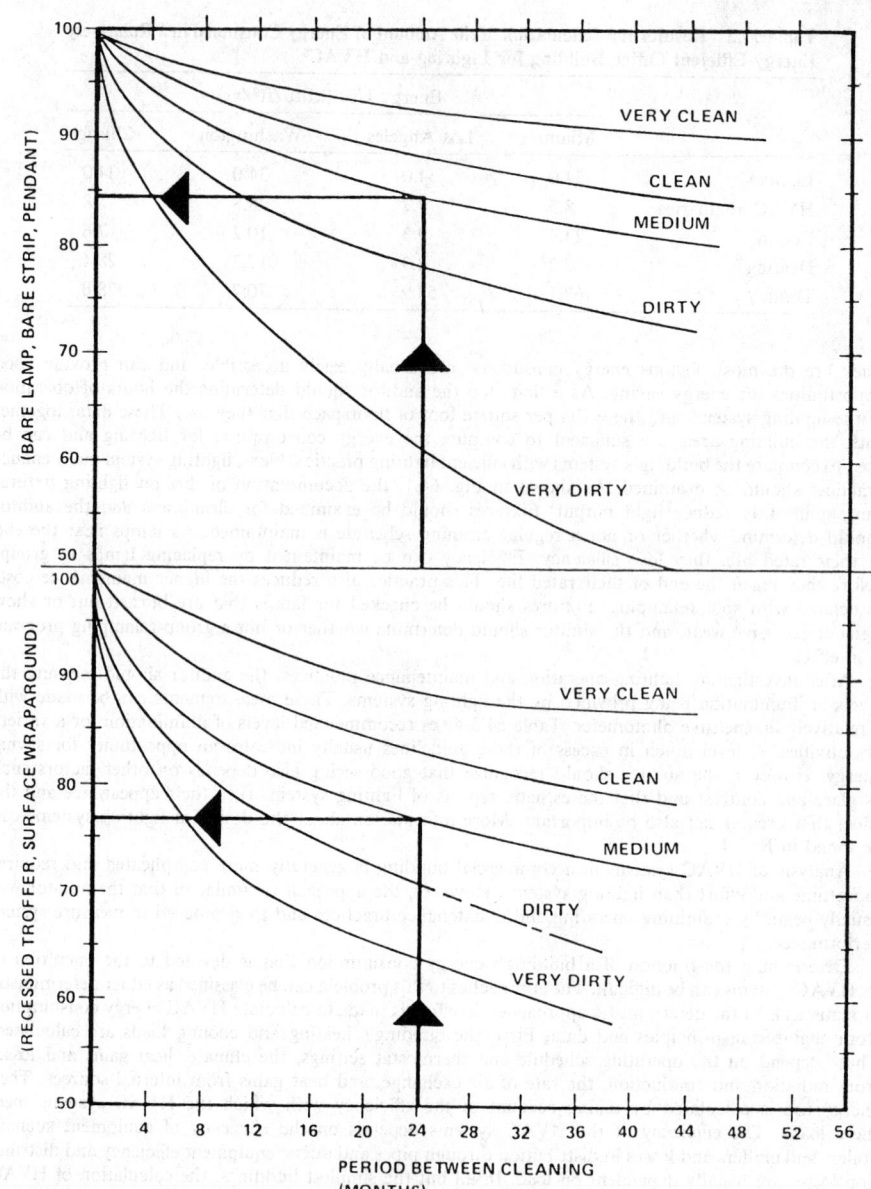

Fig. 64.1 Reduction in light output from fluorescent fixtures as a function of fixture cleaning frequency and the cleanliness of the fixture's surroundings.[3]

a significant part of the heating load may be for reheat (i.e., air that is precooled to the temperature required for the warmest space in the building may have to be reheated in other spaces). The easiest statistical method to apply, and the one that should probably be attempted first, is to calculate the energy consumption for all other end uses (lighting, domestic hot water, office equipment, etc.) and subtract this from the total consumption; the remainder will be HVAC energy consumption. If different fuel types are used for heating and cooling, it will be easy to separate consumption for these uses; if not, some further analysis of the climate dependence of consumption will be required. Energy consumption for ventilation can be calculated easily if the operating hours and power requirements for the supply and exhaust fans are known.

Whatever approach is to be taken in determining the fraction of energy consumption that is used for HVAC systems, the auditor should begin his or her work on these systems by determining their

Table 64.3 Range of Illuminances Appropriate for Various Types of Activities and Weighting Factors for Choosing the Footcandle Level[a] within a Range of Illuminance[6]

Category	Range of Illuminances (Footcandles)	Type of Activity
A	2–3–5	Public areas with dark surroundings
B	5–7.5–10	Simple orientation for short temporary visits
C	10–15–20	Working spaces where visual tasks are only occasionally performed
D	20–30–50	Performance of visual tasks of high contrast or large size: for example, reading printed material, typed originals, handwriting in ink and good xerography; rough bench and machine work; ordinary inspection; rough assembly
E	50–75–100	Performance of visual tasks of medium contrast or small size: for example, reading medium-pencil handwriting, poorly printed or reproduced material; medium bench and machine work; difficult inspection; medium assembly
F	100–150–200	Performance of visual tasks of low contrast or very small size: for example, reading handwriting in hard pencil or very poorly reproduced material; very difficult inspection
G	200–300–500	Performance of visual tasks of low contrast and very small size over a prolonged period: for example, fine assembly; very difficult inspection; fine bench and machine work
H	500–750–1000	Performance of very prolonged and exacting visual tasks: for example, the most difficult inspection; extra-fine bench and machine work; extra-fine assembly
I	1000–1500–2000	Performance of very special visual tasks of extremely low contrast and small size: for example, surgical procedures

Weighting Factors

Worker or task characteristics	−1	0	+1
Workers' age	Under 40	40–65	Over 65
Speed and/or accuracy	Not important	Important	Critical
Reflectance of task background	Greater than 70%	30–70%	Less than 30%

[a] To determine a footcandle level within a range of illuminance, find the weighting factor for each worker or task characteristic and sum the weighting factors to obtain a score. If the score is −3 or −2, use the lowest footcandle level; if −1, 0, or 1, use the middle footcandle level; if 2 or 3, use the highest level.

operating hours and control settings. These can often be changed to save energy with no adverse effects on a building's occupants. Next, maintenance practices should be examined. This examination will usually be initiated by determining whether or not a preventive maintenance (PM) program is being conducted. If there is a PM program, much can be learned about the adequacy of maintenance practices by examining the PM records. Often only a few spot checks of the HVAC systems will be required to verify that the records are consistent with actual practice. If there is no PM program, the auditor will usually find that the HVAC systems are in poor condition and should be prepared to make extensive checks for energy-wasting maintenance problems. Establishment of a PM program as part of the energy management program is a frequent recommendation from an energy audit.

Areas for HVAC maintenance that are important to check include heat exchanger surfaces, fuel–air mixture controls in combustors, steam traps, and temperature controllers. Scale on the water side of boiler tubes and chiller condenser tubes reduces the efficiency of heat transfer. Losses of efficiency can also be caused by the buildup of dirt on finned-tube air-cooled condensers. Improper control of fuel–air mixtures can cause significant losses in combustors. Leaky steam traps are a common cause of energy losses. Figure 64.2 shows the annual rate of heat loss through a leaky trap as a function of

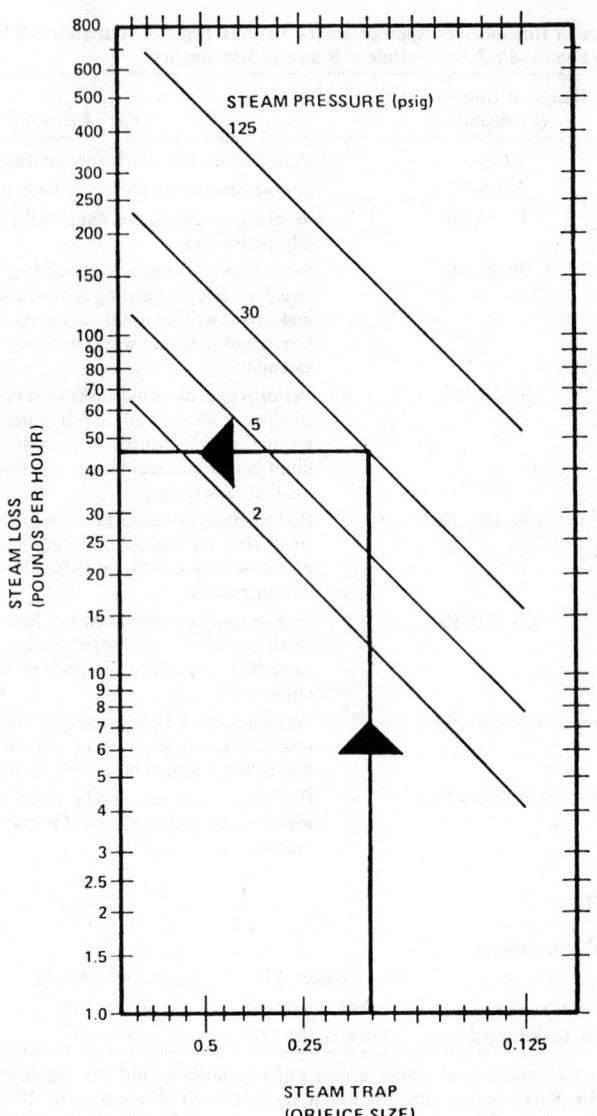

Fig. 64.2 Steam loss through leaking steam traps as a function of steam pressure and trap orifice size.[3]

the size of the trap orifice and steam pressure. Poorly maintained room thermostats and other controls such as temperature reset controllers can also cause energy waste. While major failures of thermostats can usually be detected as a result of occupant complaints or behavior (e.g., leaving windows open on cold days), drifts in these controls that are too small to cause complaints can still lead to substantial waste. Other controls, especially reset controls, can sometimes fail completely and cause an increase in energy consumption without affecting occupant comfort.

After investigating HVAC operation and maintenance practices, the auditor should make measurements of system performance. Typical measurements will include air temperature in rooms and ducts, water temperatures, air flow rates, pressure drops in air ducts, excess air in stack gases, and current drawn by electric motors operating fans and pumps. Instruments required include a thermometer, a pitot tube or anemometer, a manometer, a strobe light, a combustion test kit, and an ammeter. The importance of making measurements instead of relying on design data cannot be emphasized too strongly. Many, if not most, buildings operate far from their design points. Measurements may point to needed adjustments in temperature settings or air flow rates. Table 64.4 gives recommended air flow rates

Table 64.4 Recommended Rates of Outside-Air Flow for Various Applications[3]

1. Office Buildings

Work space	5 cfm/person
Heavy smoking areas	15 cfm/person
Lounges	5 cfm/person
Cafeteria	5 cfm/person
Conference rooms	15 cfm/person
Doctors' offices	5 cfm/person
Toilet rooms	10 air changes/hr
Lobbies	0
Unoccupied spaces	0

2. Retail Stores

Trade areas	6 cfm/customer
Street level with heavy use (less than 5,000 ft.² with single or double outside door)	0
Unoccupied spaces	0

3. Religious Buildings

Halls of worship	5 cfm/person
Meeting rooms	10 cfm/person
Unoccupied spaces	0

for various applications. Detailed analysis of the measured data requires a knowledge of HVAC system principles.

After measuring HVAC system performance, the auditor should make rough calculations of the relative importance of the different sources of HVAC system loads. These are primarily radiative and conductive heat gains and losses through the building's exterior surfaces, gains and losses from air exchange, and gains from internal heat sources. Rough calculations are usually sufficient to guide the auditor in selecting conservation measures for consideration. More detailed analyses can await the selection of specific measures.

While lighting and HVAC systems will usually occupy most of the auditor's time in a commercial building, other systems such as domestic hot water may warrant attention. The approach of first investigating operation and maintenance practices and then measuring system performance is usually appropriate for these systems.

64.3 PERFORMING AN ENERGY AUDIT—IDENTIFYING OPPORTUNITIES FOR SAVING ENERGY

In almost every facility one can discover a surprisingly large number of opportunities to save energy. These opportunities range from the obvious such as use of light switches to exotic approaches involving advanced energy conversion technologies. Identification of ways to save energy requires imagination and resourcefulness as well as a sound knowledge of engineering principles.

The auditor's job is to find ways to *eliminate unnecessary energy-using tasks* and ways to *minimize the work required to perform necessary tasks*. Some strategies that can be used to eliminate unnecessary tasks are improved controls, "leak plugging," and various system modifications. Taking space conditioning as an example, it is necessary to provide a comfortable interior climate for building occupants, but it is usually not necessary to condition a building when it is unoccupied, it is not necessary to heat and cool the outdoors, and it is not necessary to cool air from inside the building if air outside the building is colder. Controls such as time clocks can turn space-conditioning equipment off when a building is unoccupied, heat leaks into or out of a building can be plugged using insulation, and modification of the HVAC system to add an air-conditioner economizer can eliminate the need to cool inside air when outside air is colder.

Chapter 55, The Exergy Method of Energy Systems Analysis, discusses methods of analyzing the minimum amount of work required to perform tasks. While the theoretical minimum cannot be achieved in practice, analysis from this perspective can reveal inefficient operations and indicate where there may be opportunities for large improvements. Strategies for minimizing the work required to perform necessary tasks include heat recovery, improved efficiency of energy conversion, and various system modifications. Heat recovery strategies range from complex systems to cogenerate electrical and thermal energy to simple heat exchangers that can be used to heat water with waste heat from equipment. Examples of improved conversion efficiency are more efficient motors for converting electrical energy to mechanical work and more efficient light sources for converting electrical energy to light. Some system modifications that can reduce the work required to perform tasks are the replacement of resistance heaters with heat pumps and the replacement of dual duct HVAC systems with variable air volume systems.

There is no certain method for discovering all of the energy-saving opportunities in a facility. The most common approach is to review lists of energy conservation measures that have been applied elsewhere to see if they are applicable at the facility being audited. A number of such lists have been compiled (see, for example, Ref. 3). However, while lists of measures are useful, they cannot substitute for intelligent and creative engineering. The energy auditor's recommendations need to be tailored to the facility, and the best energy conservation measures often involve novel elements.

In the process of identifying energy saving opportunities, the auditor should concentrate first on low-cost conservation measures. The savings potential of these measures should be estimated before more expensive measures are evaluated. Estimates of the savings potential of the more expensive measures can then be made from the reduced level of energy consumption that would result from implementing the low-cost measures. While this seems obvious, there have been numerous occasions on which costly measures have been used but simpler, less expensive alternatives have been ignored.

64.3.1 Low-Cost Conservation

Low-cost conservation measures include turning off energy-using equipment when it is not needed, reducing lighting and HVAC services to recommended levels, rescheduling of electricity-intensive operations to off-peak hours, proper adjustment of equipment controls, and regular equipment maintenance. These measures can be initiated quickly, but their benefits usually depend on a sustained effort. An energy management program that assigns responsibility for maintaining these low-cost measures and monitors their performance is necessary to ensure good results.

In commercial buildings it is often possible to achieve very large energy savings simply by shutting down lighting and HVAC systems during nonworking hours. This can be done manually or, for HVAC systems, by inexpensive time clocks. If time clocks are already installed, they should be maintained in good working order and set properly. During working hours lights should be turned off in unoccupied areas. Frequent switching of lamps does cause some decrease in lamp life, but this decrease is generally not significant in comparison to energy savings. As a rule of thumb, lights should be turned out in a space that will be unoccupied for more than 5 min.

Measurements of light levels, temperatures, and air flow rates taken during the auditor's survey will indicate if lighting or HVAC services exceed recommended levels. Light levels can be decreased by relamping with lower-wattage lamps or by removing lamps from fixtures. In fluorescent fixtures, except for instant-start lamps, ballasts should also be disconnected because they use some energy when the power is on even when the lamps are removed.

If the supply of outside air is found to be excessive, reducing the supply can save heating and cooling energy (but see below on air-conditioner economizers). If possible, the reduction in air supply should be accomplished by reducing fan speed rather than by restricting air flow by the use of dampers, since the former procedure is more energy efficient. Also, too much air flow restriction can cause unstable operation in some fans.

Because most utilities charge more for electricity during their peak demand periods, rescheduling the operation of some equipment can save considerable amounts of money. It is not always easy to reschedule activities to suit the utility's peak demand schedule, since the peak demand occurs when most facilities are engaging in activities requiring electricity. However, a careful examination of major electrical equipment will frequently reveal some opportunities for rescheduling. Examples of activities that have been rescheduled to save electricity costs are firing of electric ceramic kilns, operation of swimming pool pumps, finish grinding at cement plants, and pumping of water from wells to storage tanks.

Proper adjustment of temperature and pressure controls in HVAC distribution systems can cut losses in these systems significantly. Correct temperature settings in air supply ducts can greatly reduce the energy required for reheat. Temperature settings in hot water distribution systems can usually be adjusted to reduce heat loss from the pipes. Temperatures are often set higher than necessary to provide enough heating during the coldest periods; during milder weather, the distribution temperature can be reduced to a lower setting. This can be done manually or automatically using a reset control. Reset controls are generally to be preferred, since they can adjust the temperature continuously. In steam distribution systems, lowering the distribution pressure will reduce heat loss from the flashing of condensate (unless the condensate return system is unvented) and also reduce losses from the surface of the pipes. Figure 64.3 shows the percentage of the heat in steam that is lost due to condensate flashing at various pressures. Raising temperatures in chilled-water distribution systems also saves energy in two ways. Heat gain through pipe surfaces is reduced, and the chiller's efficiency increases due to the higher suction head on the compressor (see Fig. 64.4).

A PM program is needed to ensure that energy-using systems are operating efficiently. Among the activities that should be conducted regularly in such a program are cleaning of heat exchange surfaces, surveillance of steam traps so that leaky traps can be found and repaired, combustion efficiency testing, and cleaning of light fixtures. Control equipment such as thermostats, time clocks, and reset controllers need special attention. This equipment should be checked and adjusted frequently.

64.3.2 Capital-Intensive Energy Conservation Measures

Major additions, modifications, or replacement of energy-using equipment usually require significant amounts of capital. These measures consequently undergo a more detailed scrutiny before a facility's management will decide to proceed with them. While the fundamental approach of eliminating unnecessary tasks and minimizing the work required for necessary tasks is unchanged, the auditor must pay much more attention to the tasks of estimating costs and savings when considering capital-intensive conservation measures.

This subsection will describe only a few of the many possible capital-intensive measures. These measures have been chosen because they illustrate some of the more common approaches to energy saving. However, they are not appropriate in all facilities and they will not encompass the majority of savings in many facilities.

Energy Management Systems

An energy management system (EMS) is a centralized computer control system for building services, especially HVAC. Depending on the complexity of the EMS, it can function as a simple time clock to turn on equipment when necessary, it can automatically cycle the operation of large electrical equipment to reduce peak demand, and it can program HVAC system operation in response to outdoor and indoor temperature trends so that, for example, the "warm-up" heating time before a building is occupied in the morning is minimized. While such a system can be a valuable component of complex building energy service systems, the energy auditor should recognize that the functions of an EMS often duplicate the services of less costly equipment such as time clocks, temperature controls, and manual switches.

Air-Conditioner Economizers

In many areas, outdoor temperatures are lower than return air temperatures during a large part of the cooling season. An air-conditioner economizer uses outside air for cooling during these periods so that the load on the compressor is reduced or eliminated. The economizer is a system of automatic dampers on the return air duct that are controlled by return-air and outside-air temperature sensors. When the outside air is cooler than the return air, the dampers divert the return air to the outdoors and let in enough fresh outside air to supply all the air to the building. In humid climates, economizers must be fitted with "enthalpy" sensors that measure wet-bulb as well as dry-bulb temperature so that the economizer will not let in outside air when it is too humid for use in the building.

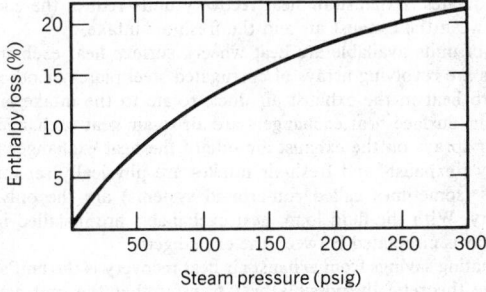

Fig. 64.3 Percentage of heat that is lost due to condensate flashing at various pressures.

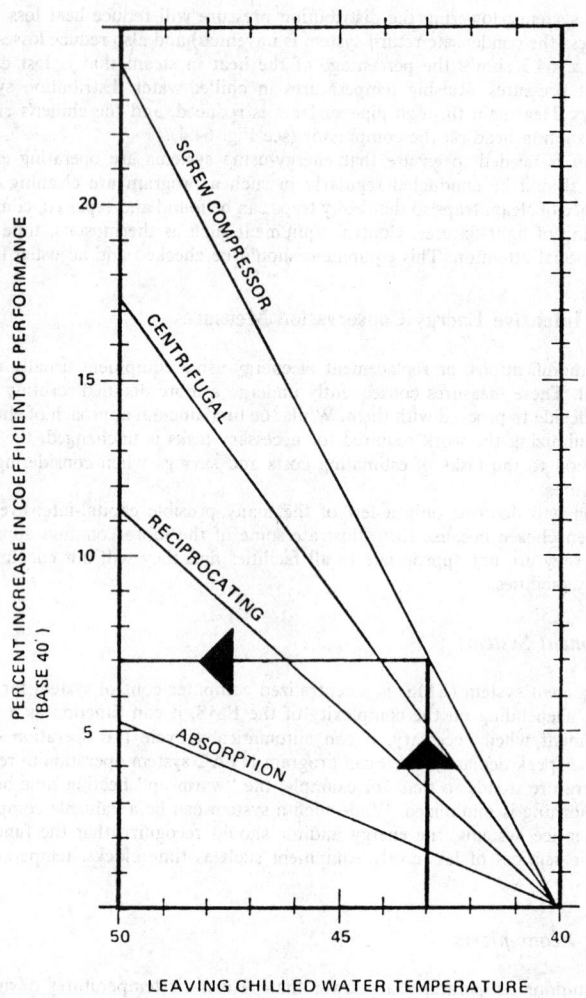

Fig. 64.4 Adjusting air-conditioner controls to provide higher chilled-water temperatures improves chiller efficiency.[3]

Building Exhaust-Air Heat Recovery Units

Exhaust-air heat recovery can be practical for facilities with large outside-air flow rates in relatively extreme climates. Hospitals and other facilities that are required to have once-through ventilation are especially good candidates. Exhaust-air heat recovery units reduce the energy loss in exhaust air by transferring heat between the exhaust air and the fresh air intake.

The common types of units available are heat wheels, surface heat exchangers, and heat-transfer-fluid loops. Heat wheels are revolving arrays of corrugated steel plates or other media. In the heating season, the plates absorb heat in the exhaust air duct, rotate to the intake air duct, and reject heat to the incoming fresh air. Surface heat exchangers are air-to-air heat exchangers. Some of these units are equipped with water sprays on the exhaust air side of the heat exchanger for indirect evaporative cooling. When a facility's exhaust- and fresh-air intakes are physically separated by large distances, heat-transfer-fluid loops (sometimes called run-around systems) are the only practical approach to exhaust-air heat recovery. With the fluid loop, heat exchangers are installed in both the exhaust and intake ducts and the fluid is circulated between the exchangers.

A key factor in estimating savings from exhaust air heat recovery is the unit's effectiveness, expressed as the percentage of the theoretically possible heat transfer that the unit actually achieves. With a 40°F temperature difference between the exhaust and intake air in the heating mode, a 60% effective unit will raise the intake air temperature by 24°F. In units with indirect evaporative cooling, the effectiveness indicates the extent to which the unit can reduce the difference between the intake air dry-bulb temperature and the exhaust air wet-bulb temperature. The effectiveness of commercially available exhaust air heat recovery units ranges from 50% to 80%; greater effectiveness is usually obtained at a higher price per unit of heat recovery capacity.

Refrigeration Heat Recovery

Heat recovery from refrigerators and air conditioners can replace fuel that would otherwise be consumed for low-temperature heating needs. Heat recovery units that generate hot water consist of water storage tanks with an integral refrigerant condenser that supplements or replaces the existing condenser on the refrigerator or air conditioner. These units reduce the facility's fuel or electricity consumption for water heating, and also increase the refrigeration or air conditioning system's efficiency due to the resulting cooler operating temperature of the condenser.

The most efficient condensing temperature will vary, depending on the compressor design and refrigerant, but in most cases it will be below 100°F. In facilities requiring water at higher temperatures, the refrigeration heat recovery unit can preheat water for the existing water heater, which will then heat the water to the final temperature.

Boiler Heat Recovery Devices

Part of the energy conversion losses in a boiler room can be reduced by installing a boiler economizer, air preheater, or a blowdown heat recovery unit. Both the economizer and the air preheater recover heat from the stack gases. The economizer preheats boiler feedwater and the air preheater heats combustion air. The energy savings from these devices are typically 5–10% of the boiler's fuel consumption. The savings depend primarily on the boiler's stack gas temperature. Blowdown heat recovery units are used with continuous blowdown systems and can either supply low-pressure steam to the deaerator or preheat makeup water for the boiler. Their energy savings are typically 1–2% of boiler fuel consumption. The actual savings will depend on the flow rate of the boiler blowdown and the boiler's steam pressure or hot-water temperature.

More Efficient Electric Motors

Replacement of integral-horsepower conventional electric motors with high-efficiency motors will typically yield an efficiency improvement of 2–5% at full load (see Table 64.5). While this saving is relatively small, replacement of fully loaded motors can still be economical for motors that operate continuously in areas where electricity costs are high. Motors that are seriously underloaded are better candidates for replacement. The efficiency of conventional motors begins to fall sharply at less than 50% load, and replacement with a smaller high-efficiency motor can yield a quick return. Motors that must run at part load for a significant part of their operating cycle are also good candidates for replacement, since high-efficiency motors typically have better part-load performance than conventional motors.

High-efficiency motors typically run faster than conventional motors with the same speed rating because high-efficiency motors operate with less slip. The installation of a high-efficiency motor to drive a fan or pump may actually increase energy consumption due to the increase in speed, since power consumption for fans and pumps increases as the cube of the speed. The sheaves in the fan or pump drive should be adjusted or changed to avoid this problem.

More Efficient Lighting Systems

Conversion of lighting fixtures to more efficient light sources is often practical when the lights are used for a significant portion of the year. Table 64.6 lists some of the more common conversions and the difference in power consumption. Installation of energy-saving ballasts in fluorescent lights provides a small (5–12%) percentage reduction in fixture power consumption, but the cost can be justified by energy cost savings if the lights are on most of the time. Additional lighting controls such as automatic dimmers can reduce energy consumption by making better use of daylight. Attention should also be given to the efficiency of the luminaire and (for indoor lighting) interior wall surfaces in directing light to the areas where it is needed. Reference 1 provides data for estimating savings from more efficient luminaires and more reflective wall and ceiling surfaces.

64.4 EVALUATING ENERGY CONSERVATION OPPORTUNITIES

The auditor's evaluation of energy conservation opportunities should begin with a careful consideration of the possible effects of energy conservation measures on safety, health, comfort, and productivity

Table 64.5 Comparative Efficiencies and Power Factors (%) for U-Frame, T-Frame, and Energy-Efficient Motors[7]

For Smaller Motors

| Horsepower Range: | 3–30 hp 3600 rpm | | | 3–30 hp 1800 rpm | | | 1.5–20 hp 1200 rpm | | |
| Speed: | | | | | | | | | |
Type:	U	T	EEM	U	T	EEM	U	T	EEM
Efficiency									
4/4 load	84.0	84.7	86.9	86.0	86.2	89.2	84.1	82.9	86.1
3/4 load	82.6	84.0	87.4	85.3	85.8	91.1	83.5	82.3	86.1
1/2 load	79.5	81.4	85.9	82.8	83.3	83.3	81.0	79.6	83.7
Power factor									
4/4 load	90.8	90.3	86.6	85.3	83.5	85.8	78.1	77.0	73.7
3/4 load	88.7	87.8	84.1	81.5	79.2	81.9	72.9	70.6	67.3
1/2 load	83.5	81.8	77.3	72.8	70.1	73.7	60.7	59.6	56.7

For Larger Motors

| Horsepower Range: | 40–100 hp 3600 rpm | | | 40–100 hp 1800 rpm | | | 25–75 hp 1200 rpm | | |
| Speed: | | | | | | | | | |
Type:	U	T	EEM	U	T	EEM	U	T	EEM
Efficiency									
4/4 load	89.7	89.6	91.6	90.8	90.9	92.9	90.4	90.1	92.1
3/4 load	88.6	89.0	92.1	90.2	90.7	93.2	90.3	90.3	92.8
1/2 load	85.9	87.2	91.3	88.1	89.2	92.5	89.2	89.3	92.7
Power factor									
4/4 load	91.7	91.5	89.1	88.7	87.4	87.6	88.3	88.5	86.0
3/4 load	89.9	89.8	88.8	87.1	85.4	86.3	86.6	86.4	83.8
1/2 load	84.7	85.0	85.2	82.0	79.2	81.1	80.9	80.3	77.8

within a facility. A conscientious effort should be made to solicit information from knowledgeable personnel and those who have experience with conservation measures in similar facilities. For energy conservation measures that do not interfere with the main business of a facility and the health and safety of its occupants, the determinant of action is the financial merit of a given measure.

Most decisions regarding the implementation of an energy conservation measure are based on the auditor's evaluation of the annual dollar savings and the initial capital cost, if any, associated with the measure. Estimation of the cost and savings from energy conservation measures is thus a critically important part of the analytical work involved in energy auditing.

When an energy conservation opportunity is first identified, the auditor should make a rough estimate of costs and savings in order to assess the value of further investigation. A rough estimate of the installed cost of a measure can often be obtained by consulting a local contractor or vendor who has experience with the type of equipment that the measure would involve. For commercial building energy conservation measures, a good guide to costs can be obtained from one of the annually published building construction cost estimating guides. The most valuable guides provide costs for individual mechanical, electrical, and structural components in a range of sizes or capacities. Rough estimates of the annual dollar savings from a measure can use simplified approaches to estimating energy savings such as assuming that a motor operates at its full nameplate rating for a specified percentage of the time.

If further analysis of a measure is warranted, a more accurate estimate of installed cost can be developed by preparing a clear and complete specification for the measure and obtaining quotations from experienced contractors or vendors. In estimating savings, one should be careful to calculate the measure's effect on energy use using accurate data for operating schedules, temperatures, flow rates, and other parameters. One should also give careful consideration to the measure's effect on maintenance requirements and equipment lifetimes, and include a dollar figure for the change in labor or depreciation costs in the savings estimate.

64.5 PRESENTING THE RESULTS OF AN ENERGY AUDIT

Effective presentation of the energy audit's results is crucial to achieving energy savings. The presentation may be an informal conversation with maintenance personnel, or it may be a formal presentation to

Table 64.6 Some Common Lighting Conversions

	Present Fixture			Replacement Fixture			
Type	Power Consumption (W)	Light Output (lumens)	Lifetime (hr)	Type	Power Consumption (W)	Light Output (lumens)	Lifetime (hr)
100-W incandescent	100	1,740	750	8-in. 22-W circline adapter in same fixture	34	980	7,500
500-W incandescent	500	10,600	1,000	175-W metal halide fixture	205	12,000	7,500
Four 40-W rapid-start warm-white 48-in. fluorescent tubes in two-ballast fixture	184	12,800	18,000	Four 34-W energy saver rapid-start tubes in same fixture	160	11,600	18,000
Two 75-W slimline warm-white 96-in. fluorescent tubes in one-ballast fixture	172	12,800	12,000	Two 60-W energy saver slim-line tubes in same fixture	142	10,680	12,000
150-W incandescent reflector floodlight (R-40)	150	1,200 (beam candlepower)	5,000	75-W incandescent projector floodlight (PAR-38)	75	1,430 (beam candlepower)	5,000
250-W mercury vapor streetlamp	285	10,400	24,000	150-W high-pressure sodium streetlamp	188	14,400	24,000

management with a detailed financial analysis. In some cases the auditor may also need to make a written application to an outside funding source such as a government agency.

The basic topics that should be covered in most presentations are the following:

1. The facility's historical energy use, in physical and dollar amounts broken down by end use.
2. A review of the existing energy management program (if any) and recommendations for improvement.
3. A description of the energy conservation measures being proposed and the means by which they will save energy.
4. The cost of undertaking the measures and the net benefits the facility will receive each year.
5. Any other effects the measure will have on the facility's operation, such as changes in maintenance requirements or comfort levels.

The auditor should be prepared to address these topics with clear explanations geared to the interests and expertise of the audience. A financial officer, for example, may want considerable detail on cash flow analysis. A maintenance foreman, however, will want information on the equipment's record for reliability under conditions similar to those in his or her facility. Charts, graphs, and pictures may help to explain some topics, but they should be used sparingly to avoid inundating the audience with information that is of secondary importance.

The financial analysis will be the most important part of a presentation that involves recommendations of measures requiring capital expenditures. The complexity of the analysis will vary, depending on the type of presentation, from a simple estimate of the installed cost and annual savings to an internal rate of return or discounted cash flow calculation.

The more complex types of calculations involve assumptions regarding future fuel and electricity price increases, interest rates, and other factors. Because these assumptions are judgmental and may critically affect the results of the analysis, the more complex analyses should not be used in presentations to the exclusion of simpler indices such as simple payback time or after-tax return on investment. These methods do not involve numerous projections about the future.

REFERENCES

1. J. E. Kaufman (ed.), *IES Lighting Handbook,* Illuminating Engineers Society of North America, New York, 1981.
2. M. Lokmanhekim et al., *DOE-2: A New State-of-the-Art Computer Program for the Energy Utilization Analysis of Buildings,* Lawrence Berkeley Laboratory Report, LBL-8974, Berkeley, CA, 1979.
3. U.S. Department of Energy, *Architects and Engineers Guide to Energy Conservation in Existing Buildings,* Federal Energy Management Program Manual, U.S. Department of Energy, Federal Programs Office, Conservation and Solar Energy, NTIS Report DOE/CS-1302, February 1, 1980.
4. L. W. Wall and J. Flaherty, *A Summary Review of Building Energy Use Compilation and Analysis (BECA) Part C: Conservation Progress in Retrofitted Commercial Buildings,* Lawrence Berkeley Laboratory Report, LBL-15375, Berkeley, CA, 1982.
5. F. C. Winkelmann and M. Lokmanhekim, *Life-Cycle Cost and Energy-Use Analysis of Sun Control and Daylighting Options in a High-Rise Office Building,* Lawrence Berkeley Laboratory Report, LBL-12298, Berkeley, CA, 1981.
6. California Energy Commission, *Institutional Conservation Program Energy Audit Report: Minimum Energy Audit Guidelines,* California Energy Commission, Publication No. P400-82-022, Sacramento, CA, 1982.
7. W. C. Turner (ed.), *Energy Management Handbook,* Wiley-Interscience, New York, 1982.

CHAPTER 65
COGENERATION

J. L. BOYEN

Consulting Engineer

65.1 DEFINITION

A cogeneration facility is defined as a system used to produce electrical or mechanical energy and forms of useful thermal energy for industrial, commercial heating, or cooling purposes through the sequential use of fuel. Sequential use of fuel means that heat rejected from a power production or heating process is used to produce additional power or process heat. It is this cascading use of energy in sequential processes that gives rise to the energy-conserving characteristic of cogeneration.

65.2 COMMERCIAL

Examples of commercial cogeneration are the following:

1. Reciprocating engine generator with low-pressure heat recovery from its exhaust and/or water jacket in the form of hot water or low-pressure steam, used in heating or cooling applications, as in Fig. 65.1.
2. A gas turbine generator with low-pressure recovery in the form of water or steam for use in heating, cooling, or process applications, as in Fig. 65.2.
3. Waste incineration with the production of steam for power generation, process heat or both.

65.3 PROCESS INDUSTRIES

65.3.1 Food Processing

Gas turbine plus heat recovery in the form of steam generation at 125–150 psi for process use: Since most food processing plants operate on a seasonal basis, the system can be adapted so that the steam generated by turbine heat recovery can be reinjected to the gas turbine combustor to produce added electric power for sale during off season periods. Figure 65.3 illustrates this system. Two examples are:

1. Boiler, gas or oil fired, generating high-pressure steam and fueled by conventional fossil fuels with the high-pressure steam driving an extraction steam turbine generator, with extraction steam used for process and the portion not used for process passes through the turbine to a condenser.

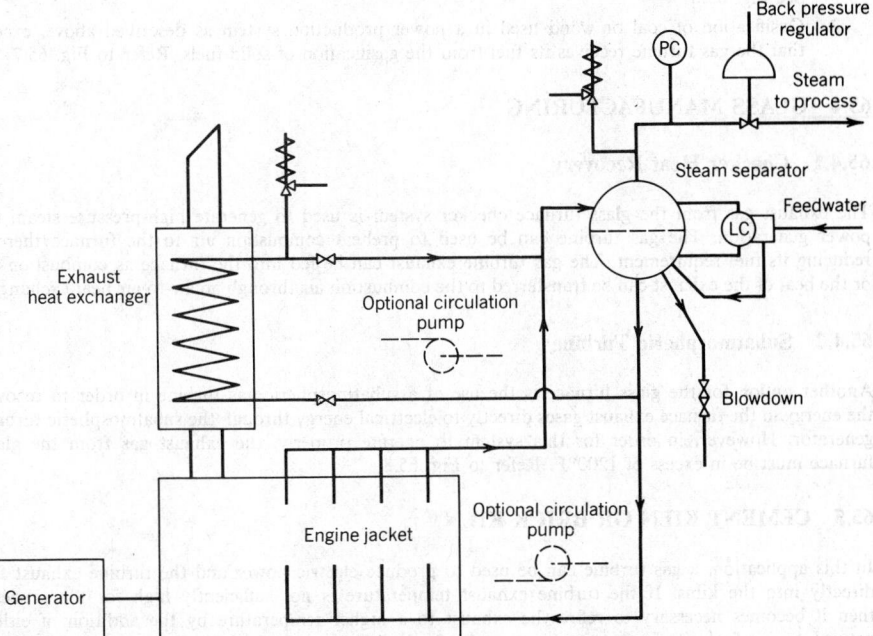

Fig. 65.1 Reciprocating engine jacket and exhaust heat recovery, low-pressure steam. PC, pressure controller; LC, level controller.

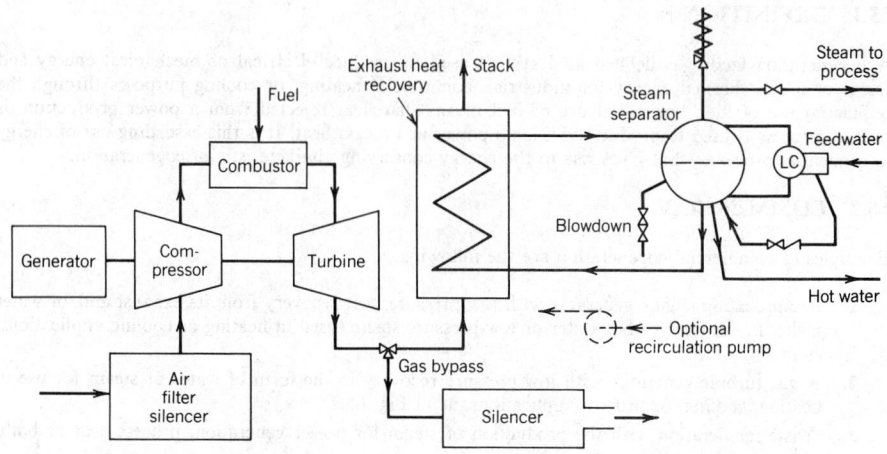

Fig. 65.2 Gas turbine exhaust heat recovery, low-pressure steam or hot water.

2. Boiler (biomass-fired) generating high-pressure steam for steam turbine operation; the steam turbine operating in an extraction mode for process. Steam not required for process passes through the turbine to the condenser for additional power generation. Refer to Fig. 65.4. Another version of the biomass-fired system is the fluidized-bed boiler, which can then operate a steam turbine in the same mode as the above. Refer to Fig. 65.5.

65.3.2 Paper and Pulp

Two examples of cogeneration in the paper and pulp industry are:

1. Gas- or oil-fired combustion turbine with heat recovery and supplementary firing, if necessary, generating high-pressure steam for power production. Extraction steam goes to the process. See Fig. 65.6.
2. Gasification of coal or wood used in a power production system as described above, except that the gas turbine receives its fuel from the gasification of solid fuels. Refer to Fig. 65.7.

65.4 GLASS MANUFACTURING

65.4.1 Checker Heat Recovery

The exhaust gas from the glass furnace checker system is used to generate high-pressure steam for power generation. The gas turbine can be used to preheat combustion air to the furnace thereby reducing its fuel requirement. The gas turbine exhaust can be fed into the furnace as combustion air or the heat of the exhaust can be transferred to the combustion air through an air-to-air heat exchanger.

65.4.2 Subatmospheric Turbine

Another option for the glass furnace is the use of a subatmospheric gas turbine in order to recover the energy in the furnace exhaust gases directly to electrical energy through the subatmospheric turbine generator. However, in order for this system to operate properly, the exhaust gas from the glass furnace must be in excess of 1200°F. Refer to Fig. 65.8.

65.5 CEMENT KILN OR BRICK KILN

In this application, a gas turbine can be used to produce electric power and the turbine exhaust fed directly into the kilns. If the turbine exhaust temperature is not sufficiently high for the process, then it becomes necessary to refire the exhaust to a higher temperature by the addition of either natural gas or fuel oil. The gas turbine can be fired with natural gas, fuel oil, or gas derived from the gasification of coal, wood, or refuse. Refer to Fig. 65.9.

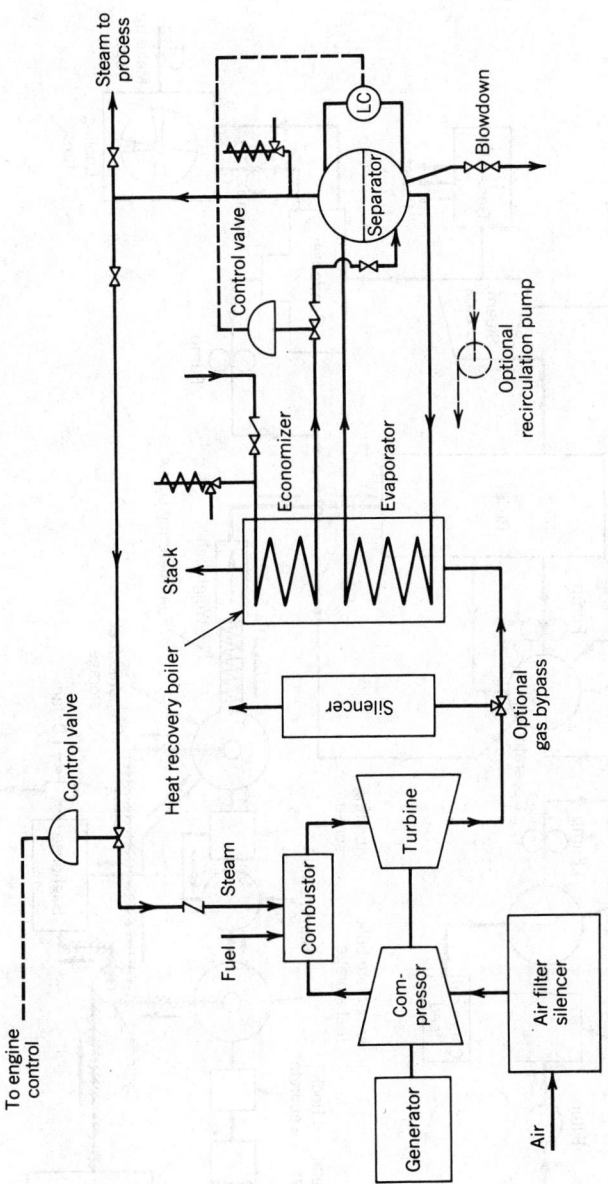

Fig. 65.3 Gas turbine exhaust heat recovery, high-pressure steam for process and injection to turbine combustor.

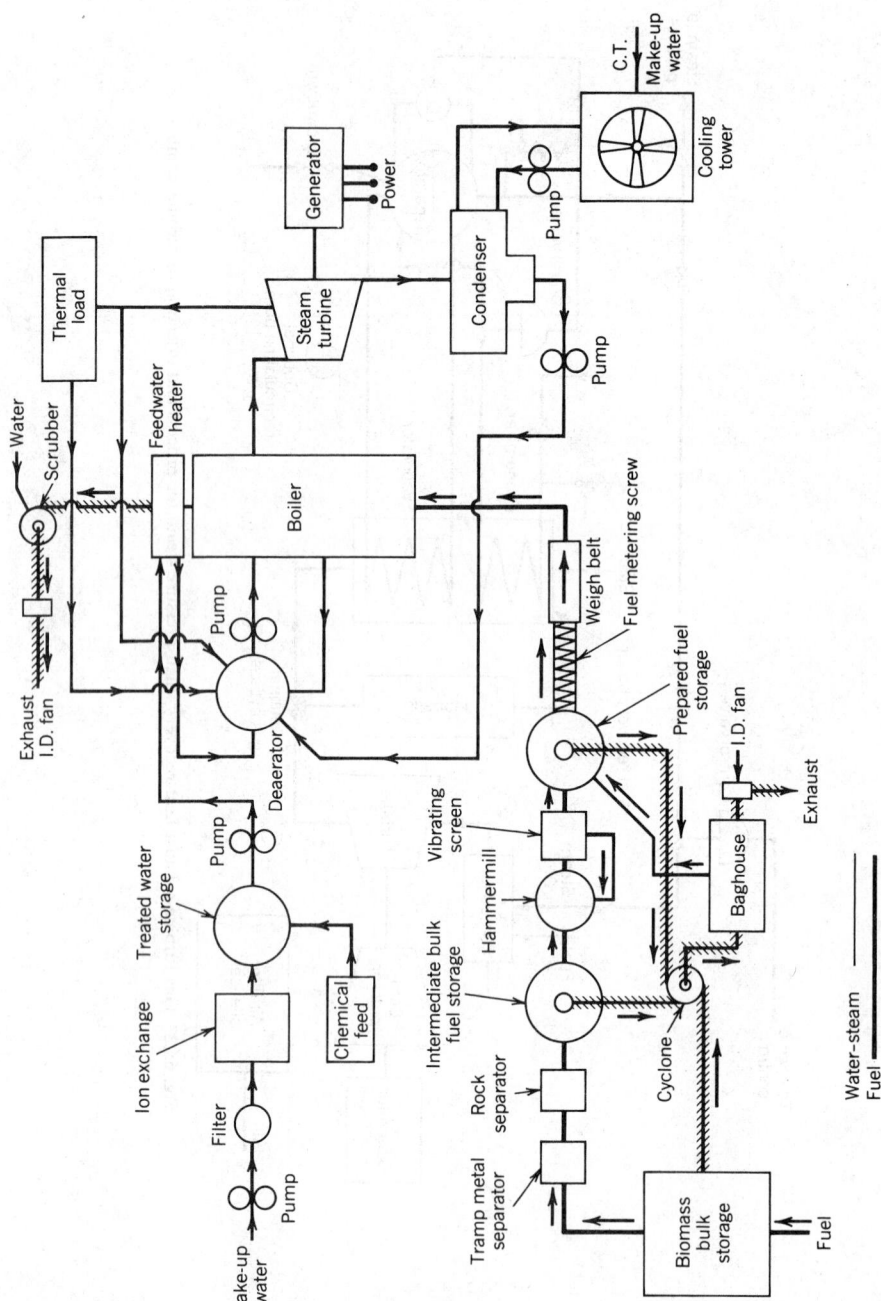

Fig. 65.4 Biomass-fired boiler, steam turbine, steam extraction for process, with condensing. I.D., induced draft; C.T., cooling tower.

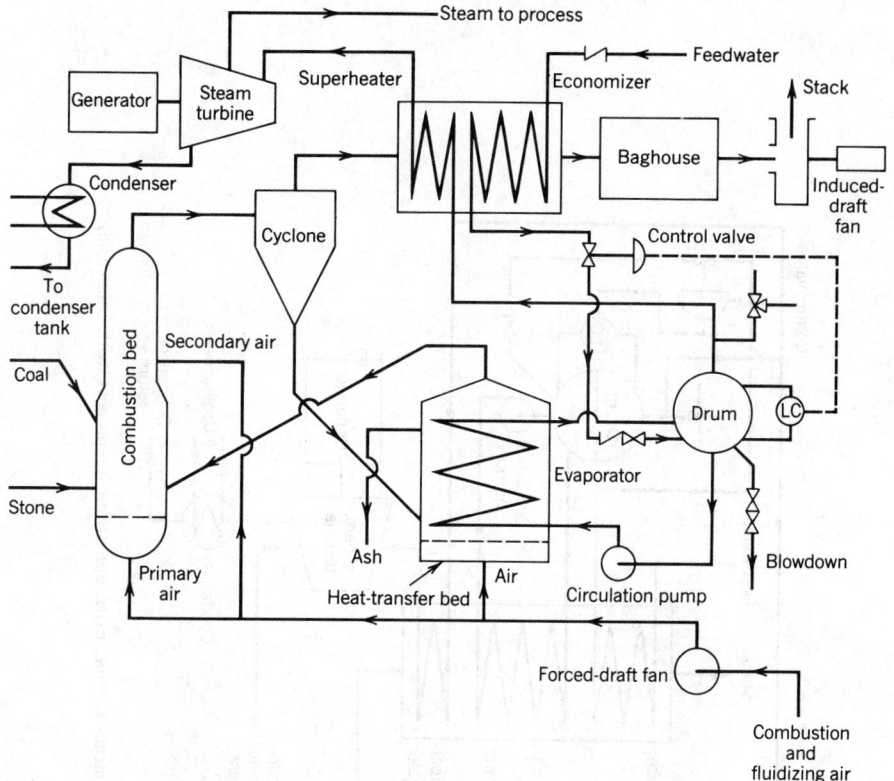

Fig. 65.5 Fluidized-bed boiler, steam turbine, steam extraction for process, with condensing.

65.6 AGRICULTURAL PRODUCE DEHYDRATION

65.6.1 Direct Fired

Direct-fired dehydrators using natural gas as a fuel can be converted to the use of gas turbine exhaust plus supplementary firing thereby producing electrical power and thermal energy for the dehydrating operation.

65.6.2 Indirect Heating

Indirect heating of the dehydration process by steam to air coils.

65.6.3 Biomass Fired

Biomass fuel properly prepared can be fired into a high-pressure boiler that produces steam for an extraction steam turbine to produce the electrical energy. The steam can be extracted from the turbine at the pressure required for dehydration. During off season with no extraction, all of the steam generated is used for the generation of electrical power. The steam generator used to burn biomass fuel must be selected to suit the type of fuel being burned and the manner in which it is prepared for combustion. If the fuel is available only in large pieces, it must be fired on a grate. If the fuel can be prepared and ground to the consistency of fine sawdust, it can be burned in suspension directly in the boiler combustion chamber. This results in a smaller boiler and, consequently, lower cost. For orchard prunings, vineyard prunings, etc., which can be ground up and dried, the boiler furnace should be sized for a heat release rate not in excess of 25,000 Btu/ft³ of furnace volume. Table 65.1 shows high heating values of some solid fuels. Table 65.2 shows high heating values of some wastes.

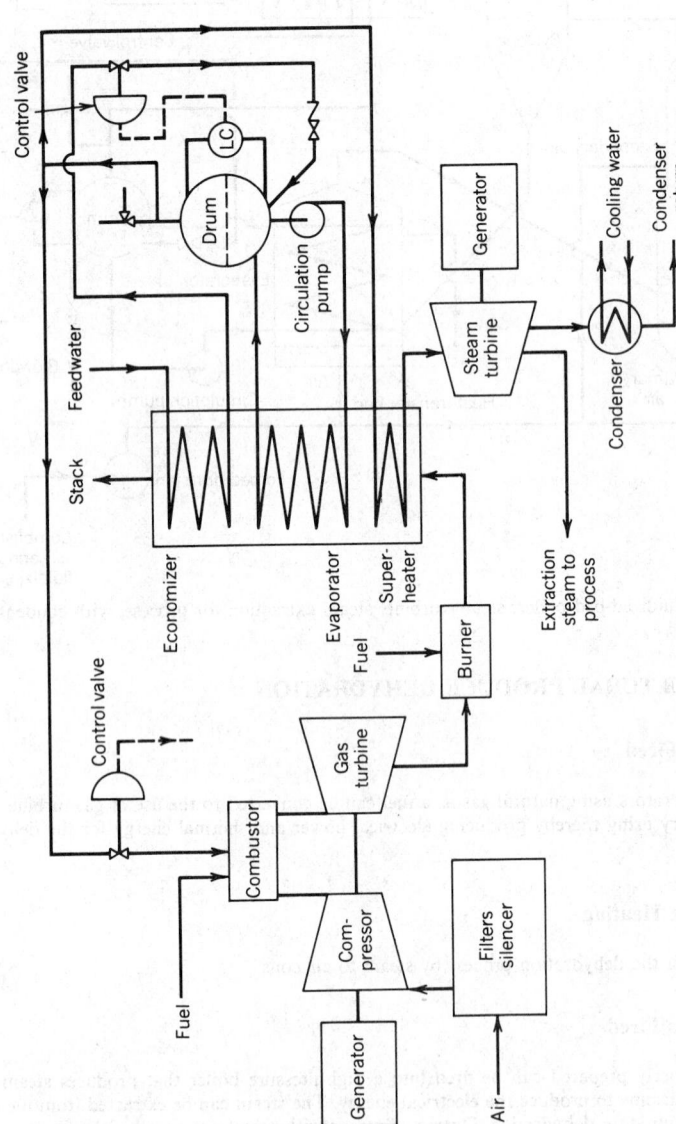

Fig. 65.6 Gas turbine exhaust heat recovery system with steam turbine, extraction steam for process, with condensing.

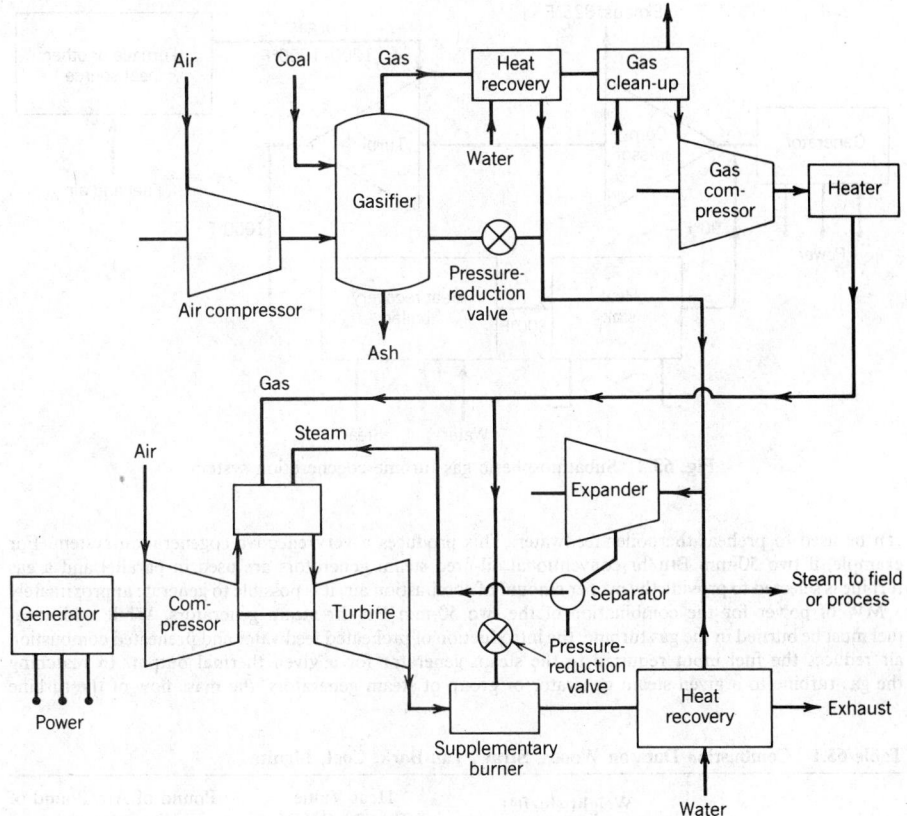

Fig. 65.7 Air-blown coal gasification cogeneration system.

65.7 ENHANCED OIL RECOVERY

Cogeneration to be used in the oil field in enhanced oil recovery operations can take several forms. Some of these are discussed in the following subsections.

65.7.1 Topping Steam Turbine

Conventional oil field steam generators can be used to produce dry saturated steam for use in a topping steam turbine. This mode of operation can be fully effective only if the wellhead pressure is low compared to the steam pressure generated. For example, if the steam generator is capable of operating at 1500 psig and the wellhead pressure is less than 600 psig, then a topping steam turbine system can be cost-effective. Since the great majority of oil field steam generators are producing wet steam of the order of 80% quality or less, it is necessary to use a steam separator to produce the dry saturated steam required for the topping turbine. The water component separated out can then be flashed down to the wellhead pressure, generating some additional steam, but producing wet saturated steam to the field as before except at a slightly reduced quality. Figure 65.10 illustrates such a system.

65.7.2 Topping Gas Turbine

A gas turbine plus heat recovery, plus a conventional steam generator, properly sized, is used to preheat combustion air to the steam generator and provide preheated water to the boiler. Since most gas turbines operate with exhaust temperature in excess of 800°F, a temperature reduction is required to reduce the exhaust temperature down to approximately 500°F before the exhaust can be directly injected as a combustion air. This is to protect standard burner components that may be damaged if the turbine exhaust were to be directly injected. In order to reduce the exhaust temperature, a convection water heater is placed in the turbine exhaust ahead of the injection point. The convection water heater

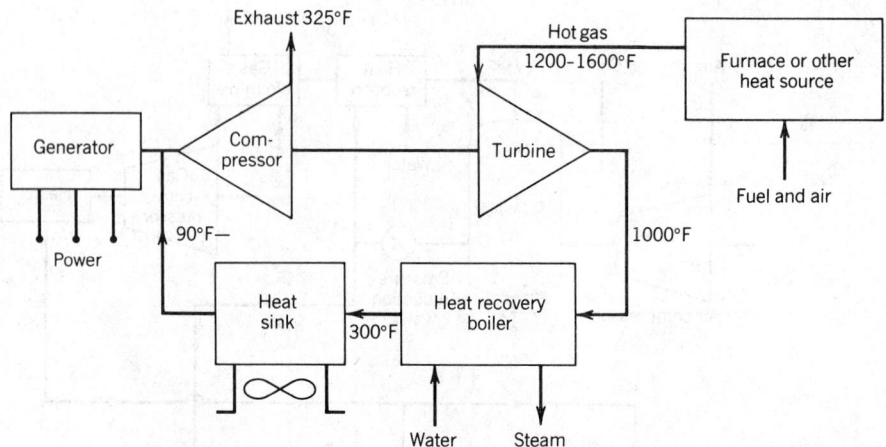

Fig. 65.8 Subatmospheric gas turbine cogeneration system.

can be used to preheat the boiler feedwater. This produces a very effective cogeneration system. For example, if two 50-mm Btu/hr conventional oil-fired steam generators are used in parallel and a gas turbine is selected to provide the proper amount of combustion air, it is possible to generate approximately 3 MW of power for the combination of the two 50-mm Btu/hr steam generators. While additional fuel must be burned in the gas turbine, the introduction of preheated feedwater and preheated combustion air reduces the fuel input required to the steam generator for a given thermal output. In matching the gas turbine to a given steam generator or group of steam generators, the mass flow of the turbine

Table 65.1 Combustion Data on Woods, Straw, Tan Bark, Coal, Lignite

Material	Weight (lb/ft³)		Heat Value (Btu/lb) (HHV) Dry Basis	Pound of Air/Pound of Dry Material (Stoichiometric)
	Air-Dried	Green		
Ash, white	42	47	8,220	6.70
Beech	43	54	8,210	6.32
Birch, white	38	51	7,920	6.10
Cedar, white	21	28	7,730	5.91
Cypress	29	47	9,120	6.94
Elm	44	53	8,130	6.23
Fir	27	52	8,410	6.45
Hemlock	25	49	8,095	6.29
Hickory	57	65	7,925	6.10
Maple	44	58	8,435	6.44
Oak, black	42	61	7,412	5.78
Oak, red	45	65	7,935	6.10
Oak, white	48	59	8,145	6.24
Pine, pitch	36	54	10,540	7.87
Pine, white	27	39	8,245	6.37
Pine, yellow	29	49	8,930	6.75
Poplar	29	49	8,200	6.32
Willow	25		7,925	6.10
Straw			6,500	4.25
Tan bark			9,500	6.21
Coal (anthracite)			14,600	9.90
Coal (bituminous B)			13,500	8.92
Lignite			7,500	7.09

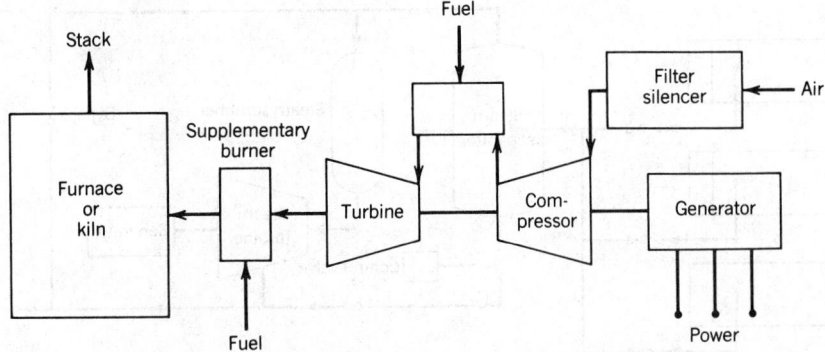

Fig. 65.9 Gas-turbine-exhaust-fired kiln cogeneration system.

exhaust must be approximately equal to or slightly greater than the combustion air requirement for the steam generator under normal operating conditions. This system is illustrated by Fig. 65.11.

65.7.3 Steam Turbine and Gas Turbine

If a topping steam turbine is used in combination with the gas turbine as in Section 65.7.2, then increased electric power can be generated, thereby greatly enhancing the economics of the cogeneration system. As in Section 65.7.1, the same criteria in determining whether a topping steam turbine can be efficiently applied must be observed. The combination gas turbine and steam turbine system is illustrated by Fig. 65.12.

65.7.4 Wet Steam Turbine

Some development work has been in progress on a steam turbine capable of accepting wet saturated steam, such as 80% quality steam normally produced by a conventional oil field steam generator.

Table 65.2 Heating Values of Some Wastes

Material	Percent Weight				Low Heating Value (Btu/lb)
	Ash	H_2O	Cl	S	
Car tires	6.3			1.2	15,570
Used fats	Varies			0.7	13,700
Refinery residue	3.5	20–30		1–7.0	9,880–10,800
Oil sludge	40.5	30		0.5–1.5	3,860
Oil carbon	27.0			4.5	11,700
Tank residues	1.0	30–50		0.5–1.5	7,190–9,880
Tar residues	3.0			1.1	13,700
Acid sludge	Varies			12–20	7,190
Oil filter residues	53	3.5		0.60	6,920
PVC residues	0.5		48.5		8,080
Polyethylene residues	Traces				18,000
Municipal solid waste (type 1)[a]	10.0				8,500
Municipal solid waste (type 2)[b]	50.0				4,500
Municipal solid waste (type 3)[c]	70.0				2,500
Sewage sludge (dewatered)	45–55				10,000 (avg.)

[a] Type 1: Municipal waste contains paper, cardboard, wood, some plastics, laminated paper, oily rags, rubber soap.

[b] Type 2: Municipal waste contains approximately 50% of type 1, plus garbage.

[c] Type 3: Municipal waste contains animal and vegetable wastes from restaurants, hotels, hospitals, etc.

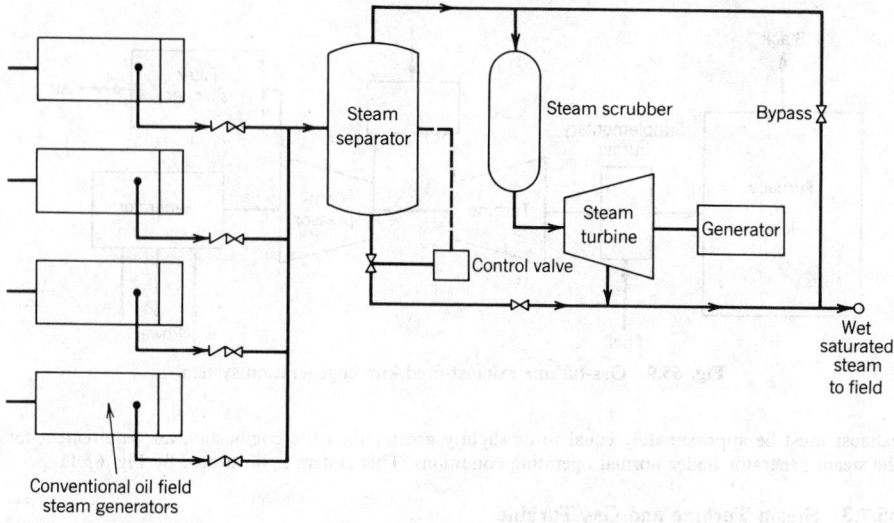

Fig. 65.10 Topping steam turbine cogeneration system for enhanced oil recovery steam injection.

Such a system will not require the steam separator, however, it will require a sufficiently high pressure difference between the generated pressure and the wellhead pressure in order for it to become effective. Whereas the efficiency of the dry steam turbine can be as high as 70% or possibly higher, depending on the number of stages used, the efficiency of the wet steam turbine in its present configuration is 50–55%, or less.

65.7.5 Gas Turbine with Heat Recovery

In cases where the field is new and there are very few or no existing steam generators, or if the wellhead pressure is so high as to preclude the use of a topping steam turbine, the concept discussed in Section 65.7.2 can be used, or a separate gas turbine and heat recovery system can be developed. In most cases, supplementary firing is required in order to generate the proper amount of steam at the correct pressure. In some cases, in order to maximize the electric power generation, two gas turbines can be used in parallel, exhausting into a single heat recovery steam generator to produce the steam generation rate required. Generally, if the steam generation pressure is 600 psi or less, it will not be necessary to supplementary fire the system in order to produce the desired efficiency. In this case the heat recovery steam generator must be designed so as to operate on a pinch point temperature difference as low as possible, preferably down to approximately 40°F.

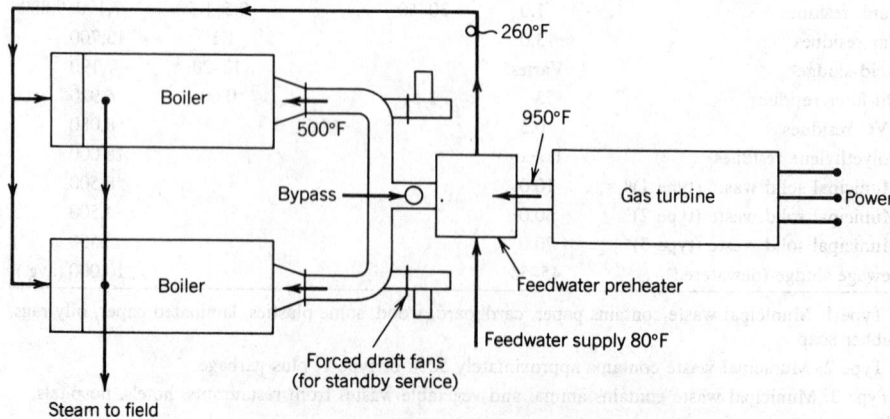

Fig. 65.11 Topping gas turbine cogeneration system for enhanced oil recovery steam injection.

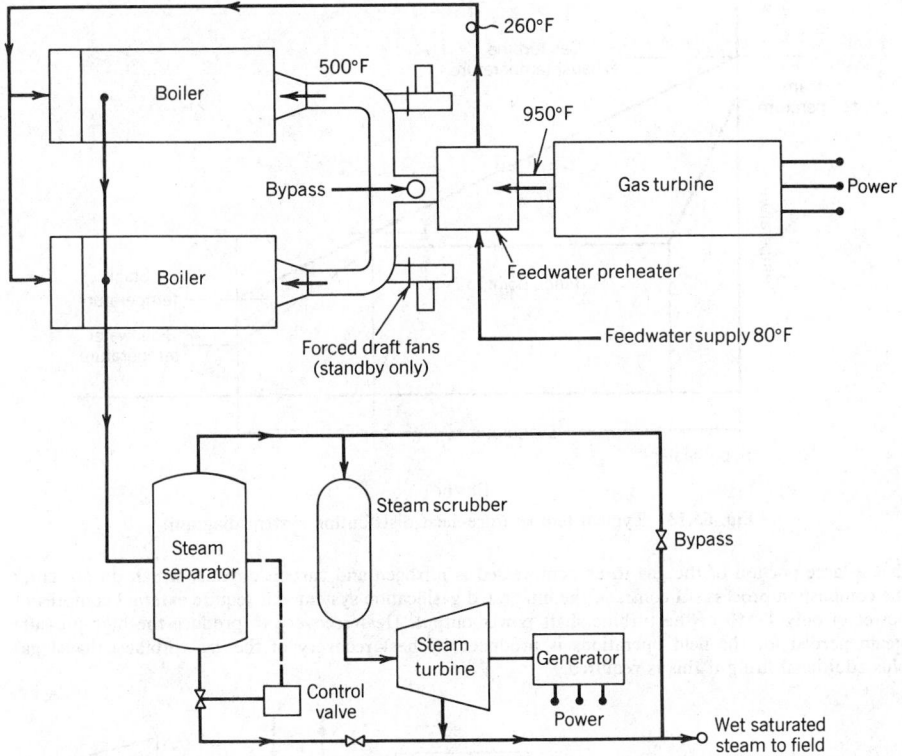

Fig. 65.12 Combination topping gas turbine and steam turbine cogeneration system for enhanced oil recovery.

65.7.6 Supplementary Firing of Turbine Exhaust

If the turbine exhaust temperature is under 900°F, an unfired system operating at 900 psi or higher will have a substantially high boiler stack temperature. In order to improve the efficiency of the system, supplementary firing is used to increase the gas temperature going to the boiler; at the same pinch point conditions this will result in a reduced stack temperature, thereby increasing the overall efficiency. The supplementary firing of natural gas or No. 2 diesel fuel is very efficient when used in this mode. The efficiency of the supplementary burner itself is approximately 95%; therefore, the incremental fuel that is burned in the supplementary burner is used more effectively than the fuel that is burned in a conventional boiler. Figure 65.13 illustrates a typical temperature–heat diagram for a gas turbine high-pressure heat recovery system. The gas turbine plus heat recovery system can be used in individual units of one turbine, one boiler or can be clustered into central plants using modules sized to provide the necessary total steam output.

If steam conditions are met, the system of Section 65.7.5 can be combined with a topping steam turbine to further increase the electrical power output.

65.7.7 Coal Gasification (Integrated)

In areas where it is economical to consider the use of coal as the fuel for enhanced oil recovery, an integrated gasification system can be used such as illustrated by Fig. 65.14. The integrated gas turbine system, which includes a high-pressure gasifier, is economic even for an air-blown gasification system because the compressor load has been materially reduced, since air for gasification is drawn from the turbine compressor and the same mass flow plus the gasification products are returned to the gas turbine combustor after being raised in pressure by a relatively small compressor, which only compensates for the pressure drop through the gasification system.

65.7.8 Coal Gasification (Independent)

The independent gasification system can require up to 14% of the turbine shaft power in order to compress the gas to the proper pressure for entry into the combustor. This is aggravated by the fact

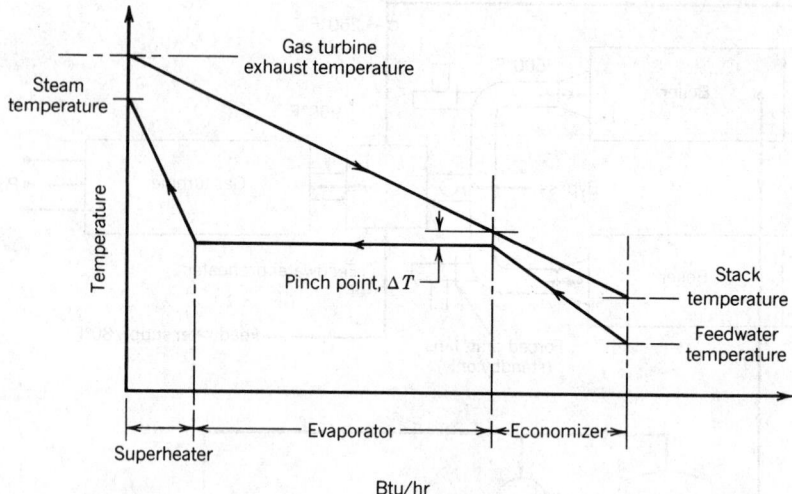

Fig. 65.13 Typical temperature–heat distribution system diagram.

that a large portion of the gas to be compressed is nitrogen and carbon dioxide, which do not enter the combustion process. In contrast, the integrated gasification system will require external compressor power of only 4–5% of the turbine shaft power output. Heat recovery to produce the high-pressure steam needed for the field operations is produced by heat recovery of the gas turbine exhaust gas plus additional firing if this is required.

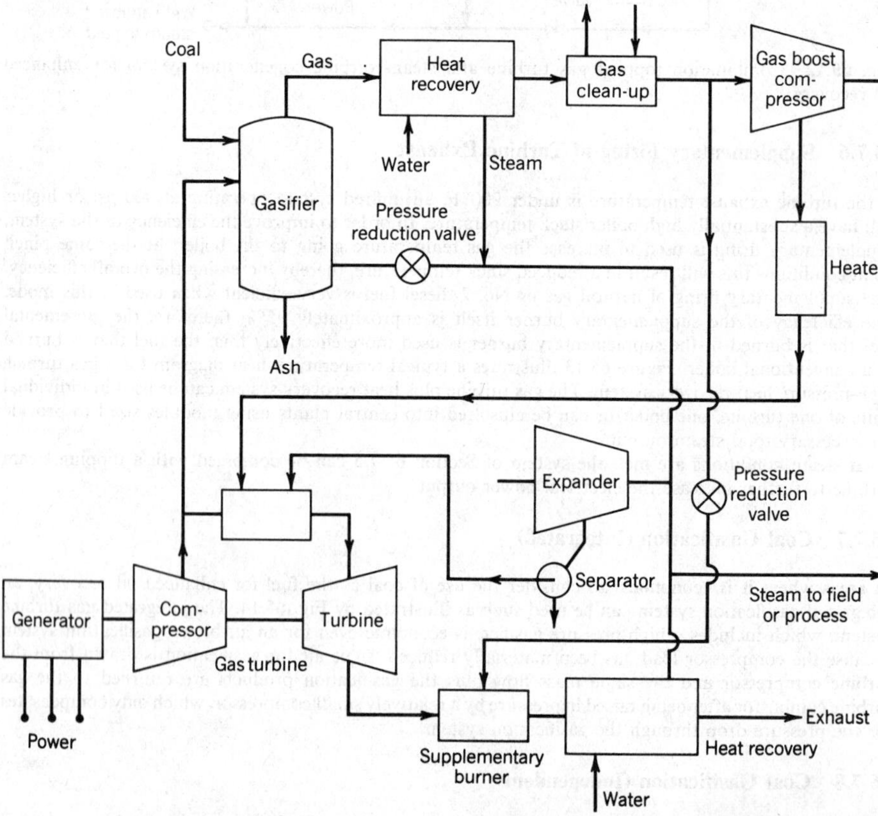

Fig. 65.14 Integrated gas-turbine–coal-gasification cogeneration system.

65.7.9   Fluidized-Bed Combustion

Coal or petroleum coke can be burned in a fluid-bed combustor system to generate the high-pressure steam and to generate hot air for the turbine cycle. The recirculating type of fluid-bed system is preferred because the heat-transfer tubes are not exposed to the combustion process, since combustion takes place in a separate vessel. The air-heating tubes can be immersed in the heat-transfer bed to produce hot air, up to approximately 1500°F, for the turbine. The heat-transfer tubes for the high-pressure steam generation can be convection tubes above the bed and in the gas turbine exhaust. Such a system is illustrated schematically by Fig. 65.15. Most modern gas turbines are designed to operate at turbine inlet temperatures in excess of 1500°F. Therefore, a turbine used in this mode of operation must be derated accordingly. The gas turbine power output can be increased by one or both of the following options:

1. A fossil-fueled topping combustor added after the immersed air-heater tubes in order to raise the air temperature from 1500°F to a higher temperature depending on the turbine limitations.
2. Steam generated by the turbine exhaust can be used in part as injection steam at the inlet of the immersed air-heater tubes in order to increase the mass flow to the turbine. See Fig. 65.15.

Figure 65.16 is a series of curves of theoretical steam rates for a topping steam turbine operating under various steam conditions. The theoretical steam rate must be corrected for the steam turbine efficiency, gear box efficiency, and generator efficiency in order to determine the actual steam rate for a given condition. The actual steam rate is

$$ASR = \frac{ESR}{E_t E_g E_r}$$

The gear box efficiency is usually 0.98 and the generator efficiency is 0.95. The turbine efficiency will vary depending on the type of turbine selected and must be determined from the supplier. Figure 65.17 is a series of curves showing the value of the specific heat of wet exhaust gas for various temperature ranges. Table 65.3 shows high heating values of gas and liquid fuels.

65.8   HEAT RECOVERY MODES FOR COGENERATION SYSTEMS

65.8.1   Steam Generation

Low Pressure

Steam generation at low pressures—up to 15 psig—is used for space heating or absorption air conditioning. The heat source may be the exhaust of gas turbines or engines or the effluent gas from an incinerator. In some cases heat in the reciprocating engine exhaust and jacket is recovered by means of a dual heat recovery unit. However, because of the temperature limitation in the engine cooling system, steam generation by water-jacket heat recovery is limited to a maximum pressure of about 12 psig depending on the limitations of the engine. The heating surface in a reciprocating engine water-jacket heat recovery system is the actual surface of the engine water jacket.

Most systems operate in a natural circulation mode. The steam separator vessel is mounted over the engine in order to create a sufficient head of water on the inlet side of the engine jacket to promote natural circulation. In some cases a pump is used to provide forced circulation. The engine-jacket heat recovery mode of operation must be carefully designed and the temperature must be accurately controlled in order to prevent the formation of steam bubbles on the engine cylinder walls. This steam bubble formation creates hot spots resulting in the breakdown of lubrication, thereby shortening the life of the engine. The outlet end of the jacket is connected to the steam separator inlet nozzle. The differential head created by the all-water downcomer and the steam–water riser induces water circulation through the jackets. A portion of the circulated water evaporates. The resulting steam is released from the steam separator steam nozzle. Unevaporated water is recirculated. The evaporated water is made up in the steam separator by the addition of water in response to a level control system. The system should also incorporate an auxiliary radiator for water-temperature control in cases where all the steam generated by the engine may not be required. This also permits engine operation in the event that the steam-generating portion is isolated for maintenance.

When exhaust heat from the engine is also recovered, a heat exchanger is added to the separator vessel for heat transfer from exhaust gas to water at the same temperature as the water produced by the engine jacket.

Some points to be observed in the design of reciprocating engine heat recovery systems are:

1. A large volume of water and an ample steam-liberation area are necessary to avoid the return of superheated water to the engine jacket. This avoids the previously stated possibility of steam

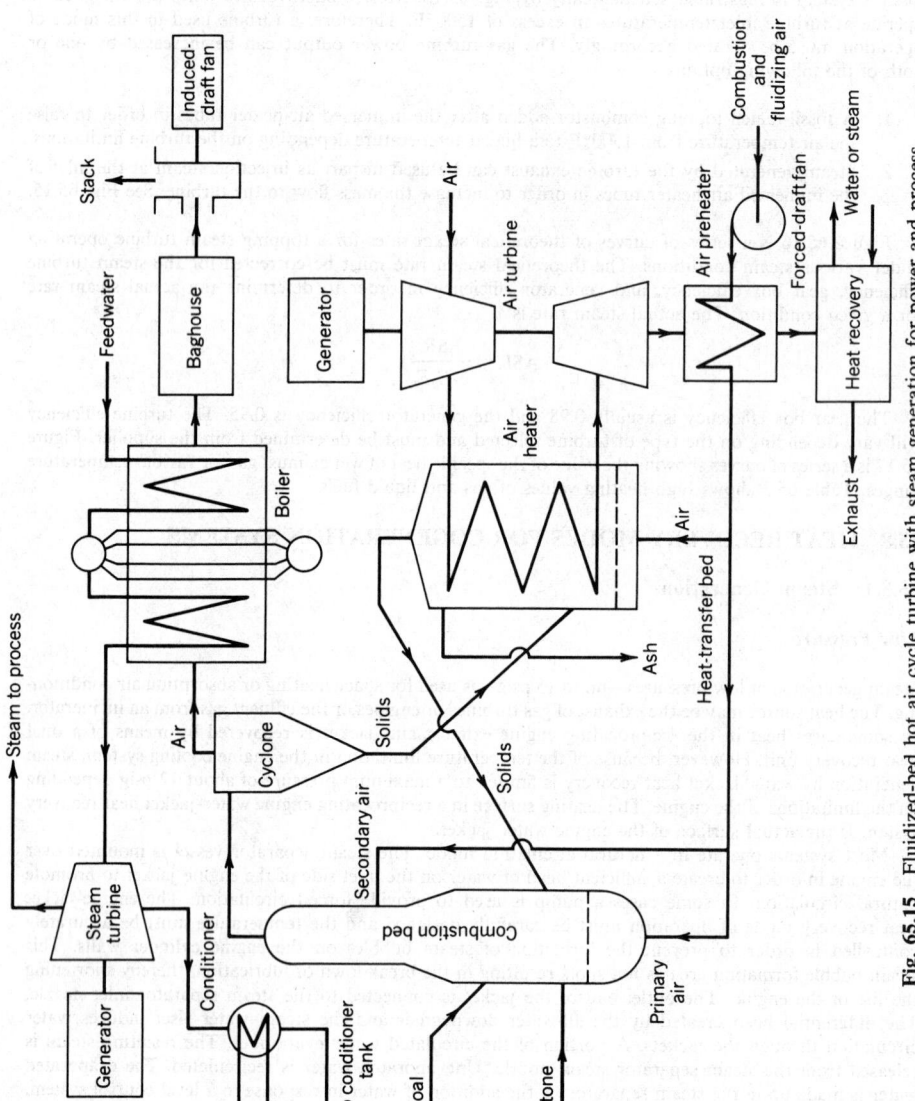

Fig. 65.15 Fluidized-bed hot air cycle turbine with steam generation for power and process.

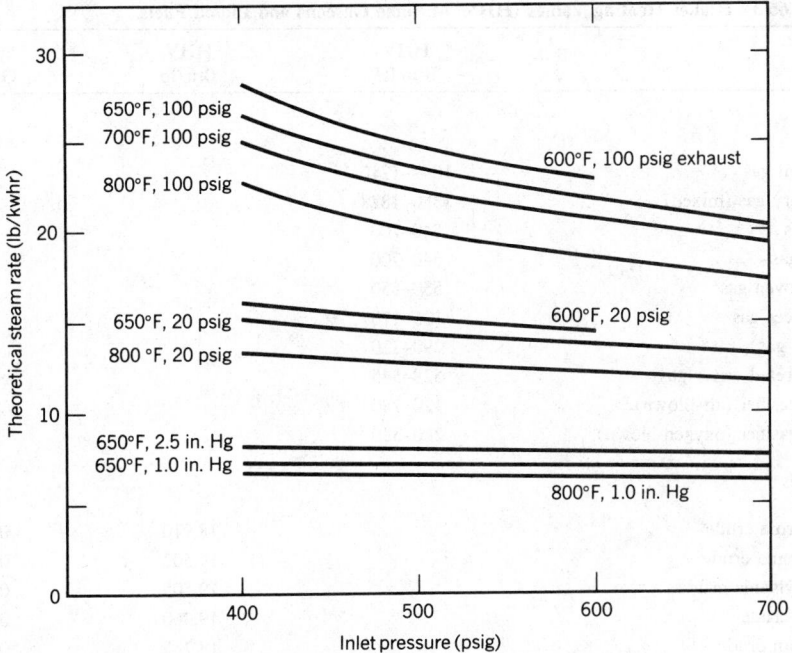

Fig. 65.16 Theoretical steam turbine steam rate curves.

bubbles forming in the engine jacket. There must be a good sedimentation chamber away from the engine to remove contaminants either brought in by the water or by chemicals added to the water for removing corrosive gases and dissolved solids.

2. Gas to water-heating surfaces must be accessible for easy inspection and repair.

3. Engine exhaust turnaround chambers within the boiler should be designed for a minimum pressure drop and to act as pulsation dampeners for exhaust noise suppression.

4. The heat recovery boiler must be equipped with a high-level cutoff, high-level alarm, low-level pump start, low-water alarm, separate mechanical water feeder for emergency service,

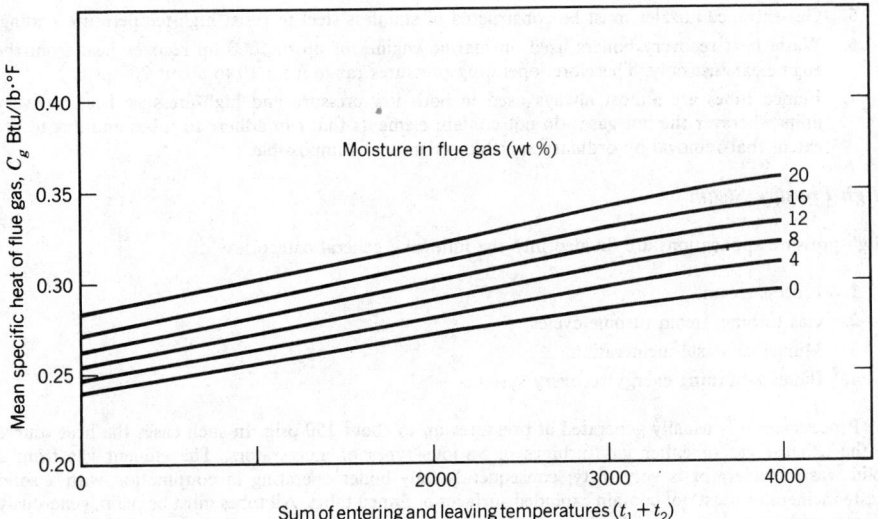

Fig. 65.17 Values of specific heat for wet exhaust air versus temperature.

Table 65.3 Higher Heating Values (HHV) of Some Gaseous and Liquid Fuels

Fuel	HHV Btu/ft³	HHV Btu/lb	Specific Gravity
Gases			
Natural gas	1047–1210		
Refinery gas (mixed)	1380–1828		
Oil gas	540–700		
Coal gas	540–700		
Coke-oven gas	550–650		
Producer gas	135–170		
Water gas	290–320		
Carbureted water gas	528–545		
Coal gasifier (air-blown)	120–140		
Coal gasifier (oxygen-blown)	280–330		
Liquids			
California crude		18,910	0.917
Oklahoma crude		19,502	0.869
Pennsylvania crude		19,505	0.813
Texas crude		19,460	0.875
Mexican crude		18,755	0.975
Gasoline		20,750	0.739
Kerosene		19,810	0.819
Gas oil		19,200	0.863
Diesel fuel No. 2		19,600	0.857
Fuel oil (midcontinent)		19,376	0.892
Fuel oil (California)		18,835	0.955

low-water cutoff, vent valve, ASME safety valve, pressure gauge, and blowdown valve, and also the required interconnecting piping. The boiler must be properly insulated—covered with a metal jacket.

5. Gas-entrance nozzles must be constructed of stainless steel to resist high-temperature scaling.
6. Waste heat recovery boilers used on marine engines of up to 5000 hp recover heat from the engine exhaust only. Therefore, operating pressures range from 10 to about 200 psig.
7. Finned tubes are almost always used in both low-pressure and high-pressure heat recovery units wherever the hot gases do not contain elements that can adhere to tubes and fins to the extent that removal by ordinary soot blower means is impossible.

High-Pressure Steam

High-pressure applications are divided into the following general categories:

1. Process steam.
2. Gas turbine, steam turbine cycles.
3. Municipal waste incineration.
4. Biomass-burning energy recovery systems.

Process steam is usually generated at pressures up to about 150 psig. In such cases the heat source is the exhaust gas of either gas turbines or various types of incinerators. The effluent gas from a solid waste incinerator is very dirty; consequently, any boiler operating in conjunction with a solid waste incinerator must not contain extended surfaces or finned tubes. All tubes must be plain, generously spaced, with pneumatic or preferably steam-operated soot blowers arranged to cover all surfaces effectively.

If the system is operating at steam pressures in excess of 125 psig, it is usually economical to equip the heat recovery boiler with an economizer. In cases where condensate makeup is high, it may also be advantageous to use a feedwater preheater section before the economizer. Since the temperature differences and the terminal temperatures are comparatively low, the feedwater heating section must be designed with dew point temperature limitations in mind. Figure 65.18 is a chart showing minimum metal temperatures for various fuels and sulfur contents. Since the sulfur content of the fuel has a tendency to raise the dew point temperature of the exhaust, there are severe limitations on the extent to which heat recovery can be achieved at these low temperatures without resorting to corrosion-resistant materials.

High-pressure steam generation for power generation and process is in excess of 150 psig and preferably in the 600-psig range with some superheat to drive a topping steam turbine operating at an exhaust pressure that can be effectively used in a process.

65.8.2 Water Heating

Heat recovery by water heating in cogeneration systems can be listed as follows:

1. Feedwater preheater for a fired water-heating system.
2. Engine jacket and engine exhaust heat recovery by water heating.
3. Combustion turbine exhaust heat recovery by water heating.
4. Heat recovery water heating for space heating, air conditioning, and domestic hot water.

A typical water-heater heat recovery system is illustrated by Fig. 65.19. This shows a forced circulation water heater utilizing a pressurized expansion tank at the circulating pump inlet and a system for the control of water flow to the process whereby water not required by the process is returned to the heat recovery loop by means of a bypass valve. This diagram shows a typical gas turbine exhaust heat recovery system using a gas bypass system that is actuated by the temperature of the water leaving the heat recovery unit. Obviously such a system is not economic at part-load conditions owing to the large amounts of turbine exhaust gas that may have to be bypassed in order to satisfy the temperature requirements. It is always desirable to use the recovered heat to its fullest extent possible. Since this type of system is a closed loop, there will be little or no makeup. However, the makeup water must be deionized in order to protect the system. The process load can be a large domestic hot water system, an absorption chilled water system for air conditioning, or a process using hot water. The temperature at which the system operates can vary from 180°F to over 400°F; however, at temperatures in excess of 212°F, the entire system must be pressurized.

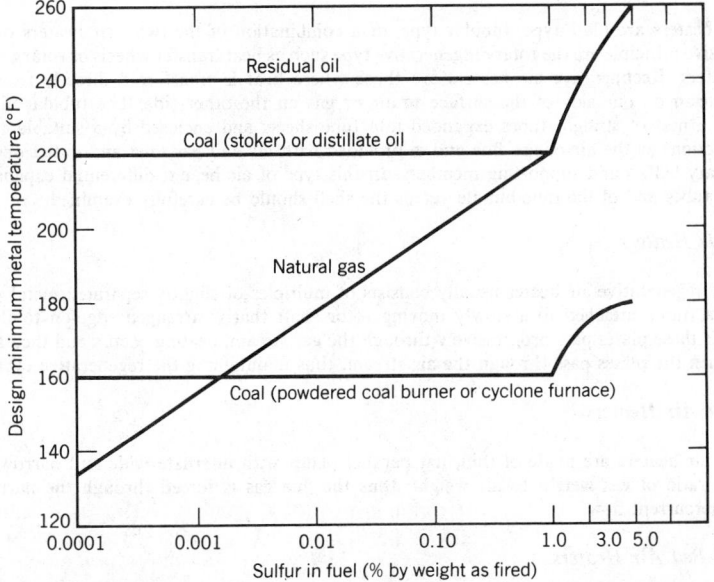

Fig. 65.18 Chart of minimum metal temperatures versus sulfur for various fuels. From *Steam: Its Generation and Use,* 38th ed., Babcock and Wilcox Co., New York, p. 13-4.

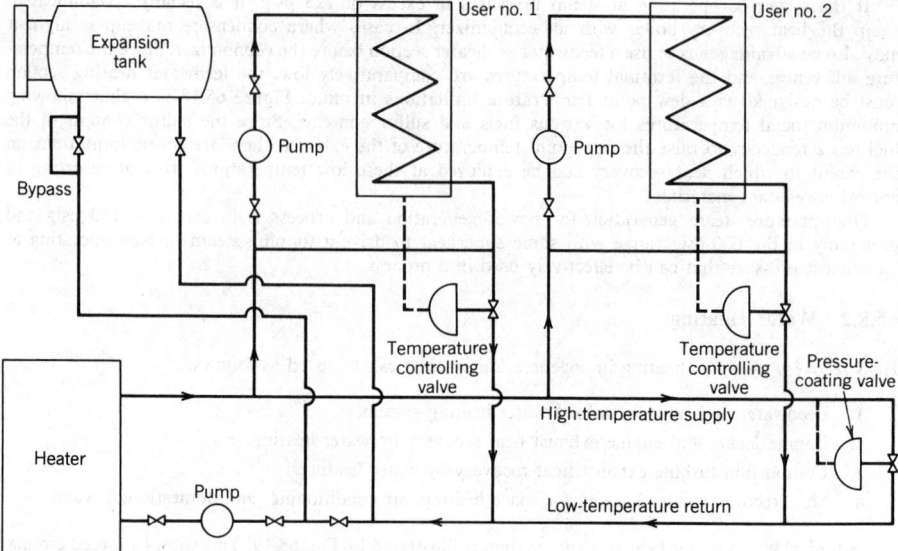

Fig. 65.19 Water-heating system with several loads.

65.8.3 Air Heating

Air heating in a cogeneration system can take the following forms:

1. Combustion air heating by waste gas or steam.
2. Compressed air heating for an air turbine cycle such as a fluidized-bed air heating cycle or a direct-fired air cycle.
3. Low-temperature air heating for space heating.
4. Regenerative air heaters.

Recuperative or Regenerative

These air heaters are plate type, tubular type, or a combination of the two. Air heaters operating in a regenerative principle are the rotary regenerative type such as heat transfer wheels or rotary combustion air preheaters. Recuperative air heaters are those where heat is transferred directly from the heat gases or steam on one side of the surface to air or gas on the other side. The tubular air heater is essentially a nest of straight tubes expanded into tube sheets and enclosed by a suitable casing. The casing functions as the air or gas flue and is provided with air and gas inlet and outlet openings and the necessary baffles and supporting members. In this type of air heater, differential expansion of the individual tubes and of the tube bundle versus the shell should be carefully examined.

Rotary Air Heaters

The rotary regenerative air heater usually consists of multiples of slightly separated metal plates supported in a frame attached to a slowly moving rotor shaft that is arranged edge-on to the gas and air flow. As these plates pass progressively through the gas stream, heating occurs and then subsequent cooling when the plates pass through the air stream, thus maintaining the regenerative cycle.

Plate-Type Air Heaters

Plate-type air heaters are made of thin, flat parallel plates with alternate wide and narrow spaces to match the ratio of gas weight to air weight; thus the flue gas is forced through the narrow spaces with countercurrent flow.

Fluidized-Bed Air Heaters

These are classified as indirect combustion-type air heaters. Tubes can be arranged both over the bed and in the bed. In this mode of operation the air heating can be done to provide the heat energy required for an air cycle turbine. Coefficients of heat transfer on the tubeside internally are over 20 Btu/hr/ft²/°F. Coefficients of heat transfer of tubes immersed in the bed can be as high as 90 Btu/hr/ft²/°F. If extended surface tubes are immersed in the bed, the fin effectiveness factor must be applied. Theoretically, the effectiveness factor for an unfinned tube is 1. This factor can be reduced to as low as 0.2 for tubes having up to 5 fins/in. and at fluidizing air mass velocity ratio of approximately 2. However, the effectiveness factor can be as high as 0.6 for tubes having 2 fins/in. and an air mass velocity ratio of approximately 1.7. The geometry of the immersed tubes must be carefully considered in computing the outside heat-transfer coefficients. The operating tube metal temperature is limited by the material selection. Table 65.4 shows the maximum allowable metal temperature for various materials that can be used as immersed air heater tubes in a fluidized-bed system. The sulfur content of the fuel burned in the fluid bed is also a factor in determining tube material.

65.8.4 Direct Heating

If the exhaust gas from a combustion turbine is applied directly into a dehydrator, an oven, or a kiln, this constitutes an acceptable load by the cogeneration definition. Sometimes the exhaust gas can be used directly in the kiln. In other cases it must be refired in order to achieve the temperature necessary for the process.

Supplementary firing can be by an oil- or gas-fired system, usually in the form of a duct burner. Since the exhaust gas from a combustion turbine contains from 15% to 17% oxygen, a substantial amount of burning can be done with the oxygen content of the exhaust gas alone. If a combustion turbine is used for this purpose, the pressure drop or backpressure on the turbine must be kept to as low a value as possible, since high backpressure results in reduced output. Generally, this exhaust pressure is limited to approximately 6 in. of water column. If very high exhaust pressures are encountered, like 10 in. or more, then the combustion turbine compressor air inlet should be supercharged to a pressure at least compensating for the exhaust pressure loss.

Table 65.4   Characteristics of Tube Materials for Fluid-Bed Air Heaters

Material	Temperature (°F)	Ultimate Strength (psi)	Yield Strength (psi)	Stress to Rupture (psi, 1000 hr)	Corrosion Resistance to
Hasteloy 25[a]	1700	37,000	28,000	8400	Wet Cl, HCl, oxidation or reducing atmosphere
Multi-Met[a]	1700	26,000	—	7000	H_2NO_3, HCl, oxidation or reducing atmosphere
Hasteloy C[a]	1600	55,000	37,000	6000	Wet Cl, HCl, H_2NO_3, H_2SO_4, oxidation or reducing atmosphere
Hasteloy "X"[a]	1700 1850	25,000	25,000	2500	Wet Cl, HCl, oxidation or reducing atmosphere
Incoloy 825[b]	1500	32,000	19,700	—	H_2SO_4, SO_2, SO_3
Inconel 601[b]	1800	8,000	7,000	—	H_2S, oxidizing atmosphere
Inconel 600[b]	1600	20,000	11,000	1850	SO_2, H_2S, oxidizing atmosphere
TP.310[c] Stmls.	1600	27,000	15,500	2000	S, SO_2
TP.347[c] Stmls.	1600	20,000	10,000	—	H_2NO_3, S, SO_2
Ceramic	2200+	—	—	—	—

[a] Union Carbide Corporation trademark.

[b] International Nickel Company trademark.

[c] Carpenter Steel Company trademark.

65.9   THERMAL FLUID HEATERS

Organic heat-transfer fluids have been used where high temperatures are required for process. The greatest advantage of using these fluids lies in their high-temperature, low-pressure characteristics. The major economy in using organic heat-transfer fluids over steam or other vapor systems is the lower cost of installation and operation. Unpressurized or low-pressure systems can significantly reduce capital cost. Since the heat is being transmitted by a liquid rather than a vapor, the piping and control devices can be in smaller sizes. The following are the major advantages of an organic fluid system over a common oil system:

1. Greater versatility in operating temperature.
2. Greater chemical stability.
3. Reduced makeup.
4. Closer control and faster response to changes in temperature.

Various organic heat-transfer fluids are available. The fluid selected depends on the temperature requirements of its operation. The fluids are ideally suited for systems that must deliver uniform heat at temperatures up to 800°F with precise temperature control and fast response. The following are the types of fluids available:

1. Polyalkylene glycols have a good thermal stability up to 500°F.
2. Ethylene glycols have an effective temperature range of −50–350°F and can be used in water solution, in air conditioning equipment, gas compressor engines, snow-melting systems, skating rinks, and other closed heating and cooling systems.
3. Fluorocarbons are primarily refrigerants having very low boiling temperatures.
4. Silicate fluids offer a wide range of operating temperatures from −40 to 600°F.
5. Inorganic salt mixtures are usable to temperatures up to 1000°F. Solidification occurs below 300°F, therefore, inorganic salt mixtures must be kept in continuous operation and drained after shutdown.
6. Diphenyl or diphenyl oxide eutectic mixtures have a usable range of 50–750°F, the boiling point being at 496°F. A closed system is required. The freezing point is at 54°F. Therefore, system lines must be steam traced or electrically traced in order to prevent freezing at the lower temperatures.
7. Chlorinated biphenyls are fire-resistant fluids with a useful range of 50–600°F. These fluids liberate hydrogen chloride at high temperature. A closed corrosion-resistant system is required.

Table 65.5 lists the physical characteristics of various common heat-transfer fluids. A typical system schematic is illustrated by Fig. 65.20. Three user zones are shown. More than three may be used,

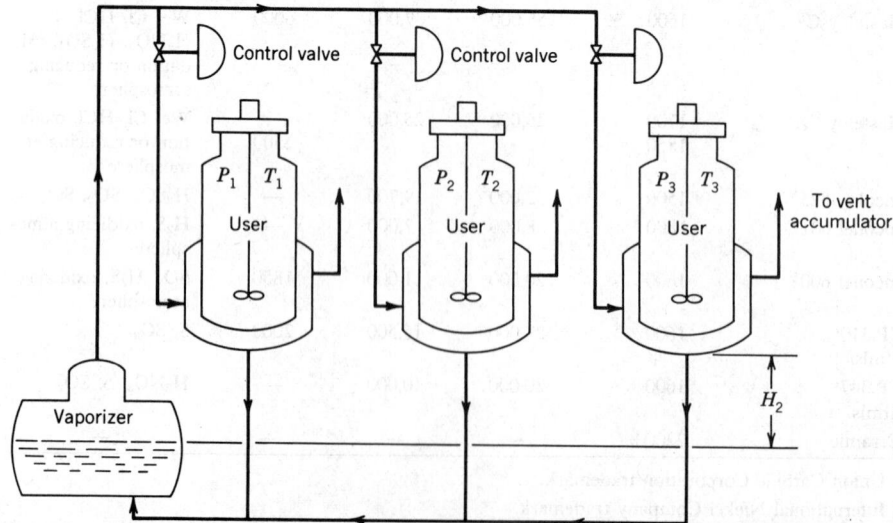

Fig. 65.20   Organic heat-transfer fluid system with several loads. From J. L. Boyen, *Thermal Energy Recovery*, Wiley, New York, 1979.

Table 65.5 Characteristics of Some Organic Heat Transfer Fluids

Property	Ethylene Glycol	Dowtherm "A"a	Dowtherm "G"a	Therminol 44b	Therminol 88b
Chemical stability	Stable	Stable (water dangerous)	Stable	Stable	Stable
Oxidative	Low	Stable	To 150°F in air	To 150°F in air	To 150°F
Thermal stability	Stable to 400°F	Stable to 650°F	Stable to 700°F	Stable to 425°F	Stable to 800°F
Maximum temperature					
Film	400°F	805°F	750°F	475°F	850°F
Bulk	325°F	700°F	700°F	425°F	800°F
Safety					
Fire resistant	Burns	Burns	Burns	Burns	Burns
Explosive	None	Possible in mist	Possible in mist	Negligible	Negligible
Density (lb/gal)					
70°F	9.3	8.82	9.16	7.75	9.40
300°F	Not used	7.98	8.36	6.85	8.42
500°F		6.60	7.62	Not used	7.67
600°F		6.03	7.22	Not used	6.58
Specific heat					
70°F	0.625	0.379	0.38	0.46	—
300°F	0.710	0.458	0.45	0.54	0.467
500°F	Not used	0.560	0.51	Not used	0.525
600°F		0.591	0.54	Not used	0.613
Thermal conductivity (Btu/ft²·hr·°F·ft)					
70°F	0.167	0.081	0.08	0.083	—
300°F	Not used	0.072	0.073	0.071	0.0712
500°F		0.061	0.070	Not used	0.0686
600°F		0.057	0.069	Not used	0.0608
Viscosity (C_p)					
70°F	20.0	4.50	30.0	6.20	1.57
300°F	1.0	0.60	1.0	0.44	0.55
500°F	Not used	0.35	0.22	Not used	0.23
600°F		0.30	0.21	Not used	

a DOW Chemical Company trademark.
b MONSANTO Chemical Company trademark.

each with its own temperature control. This is particularly useful in processes requiring the close temperature control and various stages operating at different temperatures.

The principal components of an organic fluid system are described in the following subsections.

65.9.1 The Heater

Two basic types of fluid heaters are liquid tube type and fire tube type. Either type may be fired or heated by a hot gas source such as a gas turbine or an incinerator. In liquid tube heaters, fluid is pumped through the tubes as it is heated. In fire tube heaters, fluid flows through the shell of the heaters surrounding the fire tubes. Velocities can be more accurately controlled in the liquid tube type. Since organic fluids break down chemically at elevated temperatures, it is important to design the heater for maximum heat flux rates, tube temperatures, and mass flow rates in order to control the maximum liquid film temperature. Fluid velocities in the tubes are typically 4–12 ft/sec. Equations for determining heat flux rates and tube temperatures are given in Section 65.14.

65.9.2 The Pump

Pumps used must have sufficient capacity and pressure head to circulate the fluid at the proper rate. Pumps usually are centrifugal type with bearings and seals suitable for the high temperatures involved and resistant to attack by the heat-transfer fluid. Mechanical seals are preferred to stuffing boxes. Water-cooled bearings and seals should be used for temperatures over 450°F.

65.9.3 Filters

Filters should be installed in a new system. The filter is preferably located at the suction side of the circulating pump. These filters may be wire mesh strainers or edge-type filters.

65.9.4 Materials

Most materials normally used in high-temperature systems can be used in thermal fluid heaters except for the heater coils. Mild steel is widely used, and copper, aluminum, bronze, brass alloys, and similar metals should be used minimally, mainly because they lose mechanical strength at higher temperatures. Tubes for the heater should be seamless steel tubes selected in accordance with the applicable rules of the ASME code.

65.9.5 Expansion Tank

Expansion tank must be installed at the highest point in the system and connected to the suction side of the pump. See Fig. 65.20. The tank is the main venting point of the system. The double drop leg expansion tank provides the greatest flexibility. Oxidation of the fluid in the expansion tank should be prevented and contact with ambient air at temperatures in excess of 100°F must be avoided. The tank can be blanketed with an inert gas such as nitrogen. The expansion tank is sized so that it is one-quarter full when the system is at room temperature and three-fourths full when the system is at operating temperature. It should be fitted with a sight glass at the full range and with a float-operated minimum level switch to shut off the heater in the event of accidental fluid loss.

65.9.6 Pipe Work

External piping can be a schedule 40 seamless carbon steel pipe. In order to prevent leakage at joints, with fittings and valves, it is desirable to use welded construction throughout. Piping must be designed for minimum pressure drop.

65.9.7 Packing

Flexible metal or solid graphite packings give best service with organic fluids. Graphite- or Teflon-impregnated short fiber asbestos gasketing is used for flanges, etc. Generally, five rings of packing are specified on valve stems to ensure seal. A metal bellows is the ideal sealing device for valve stems.

65.9.8 Valves

Cast steel valves with deep stuffing boxes are best for organic fluid systems. Outside screw valves should be used throughout the system to improve sealing.

65.9.9 Controls

Controls must be installed on both the heater and the heat-using units. Heater controls should be installed to regulate the firing mechanism or waste-gas bypass as in a heat recovery unit in direct proportion to the required output. Modulating control is preferable, although smaller systems may be operated as on and on/off types of systems. User controls must be installed to regulate the flow of heat-transfer fluid in proportion to the heat absorption of the user. In multiple-user unit systems, separate controls must be installed on each unit.

65.9.10 Fire Protection

The commonly used method of fire prevention in the event of the tube rupture in fired heaters is to pipe steam or CO_2 into the area as a snuffer. This can be either an automatic or a manual system.

65.9.11 Safety Controls

Safety controls should include the following:

1. High-temperature cutoff at heater outlet.
2. Flow-sensing switch to shut off burner or bypass heat source in the event of loss of flow; low-level switch in expansion bank to shut off heater or bypass hot gas in the event of loss of fluid.
3. Organic heat-transfer fluids are commonly used in chemical process operations where accurate temperature control is required for platen heating, crude oil heating in oil-production facilities, glycol heating for glycol regeneration in natural gas drying systems, and crew-quarter heating in offshore platforms.
4. A cogeneration system utilizing thermal fluid heaters can typically be a combustion turbine generating electric power with the turbine exhaust being the heat source for the fluid-heating system. Alternately, the reciprocating engine exhaust heat recovery can be used, usually in smaller applications.

65.10 RANKINE BOTTOMING CYCLES

Organic heat-transfer fluids having temperature–entropy characteristics that make them suitable for Rankine cycle power systems can be used in cogeneration systems where suitable waste heat is available. Organic bottoming cycles can be used on the exhaust gas of gas turbines, reciprocating engines, low-temperature process-heating energy recovery, and other sources of low- and medium-temperature gas or fluid. The fluid used is selected for the characteristics of the available waste heat source; for example, on low-temperature gas sources, typically less than 300°F, low-boiling-point fluids such as the fluorocarbons can be used. Table 65.6 shows several Rankine cycle fluids and their general characteristics.

In the organic Rankine cycle, superheated vapor is expanded through an expander. The low-pressure vapor from the expander exhaust, which contains substantial superheat, passes through the gas side of a regenerator transferring heat to the boiler feed liquid. The vapor from the regenerator is condensed and pumped to boiler pressure, preheated in the regenerator, and then passed through the boiler. Turbine expanders are preferred in cases where constant speed can be maintained.

Table 65.6 Some Rankine Cycle Fluids

	Use Temperature (°F)		
Fluid	Maximum	Minimum	Safety
Benzene	800	42	Flammable
Toluene	750	−139	Flammable
Chlorobenzene	630	−50	Flammable
Pyridine[a]	670	−43	Flammable[b]
Hexafluorobenzene	750	41	Nonflammable
Isobutane	450	−255	Flammable
Freon 113[c]	300	−31	Nonflammable

[a] Union Carbide Corporation trademark.

[b] Pyrene in concentrated form.

[c] Dupont Chemical Corporation trademark.

Not all fluids will require the use of a regenerator. This is determined by the condition of the vapor emerging from the expander discharge. If the vapor is substantially superheated at this point, a regenerator must be used to improve cycle efficiency. If the vapor at the expander exhaust is at or near saturation, then a regenerator is not required.

65.10.1 The Cycle

Figure 65.21 is a temperature–entropy diagram of a typical organic Rankine cycle. Conditions marked on Fig. 65.21 are 1–2, expander; 2–3, regenerator; 4–5, condenser; 5–6, pump to pressure; 6–7, regenerators; 7–8, liquid-phase heating; 8–9, latent heat addition; 9–1, superheating. In the diagram, the vapor is at 600°F entering the expander. At point 2 the expander exhaust temperature is 300°F, but still in the superheat region. The vapor enters one side of the regenerator at 2 and emerges at 3, still slightly superheated, where it enters the condenser at 1. The condenser latent heat removal is shown from 4 to 5. The system feedpump raises the pressure to 6, where it enters the other side of the regenerator. The fluid is heated to approximately 220°F to 7 where it emerges from the regenerator and enters the liquid-phase heater section of the vapor generator. It is heated in liquid form to 8 at approximately 440°F, at which point evaporation begins and proceeds to 9 at constant temperature. At 9 a vapor enters the superheater where it is heated to 600°F; it then again enters the expander.

65.10.2 Fluid Power Cycles

Organic Rankine cycle fluids can be broadly classified according to the type of power cycle produced:

1. Wetting cycle, where expansion through the expander results in a wet condition at the expander exhaust.

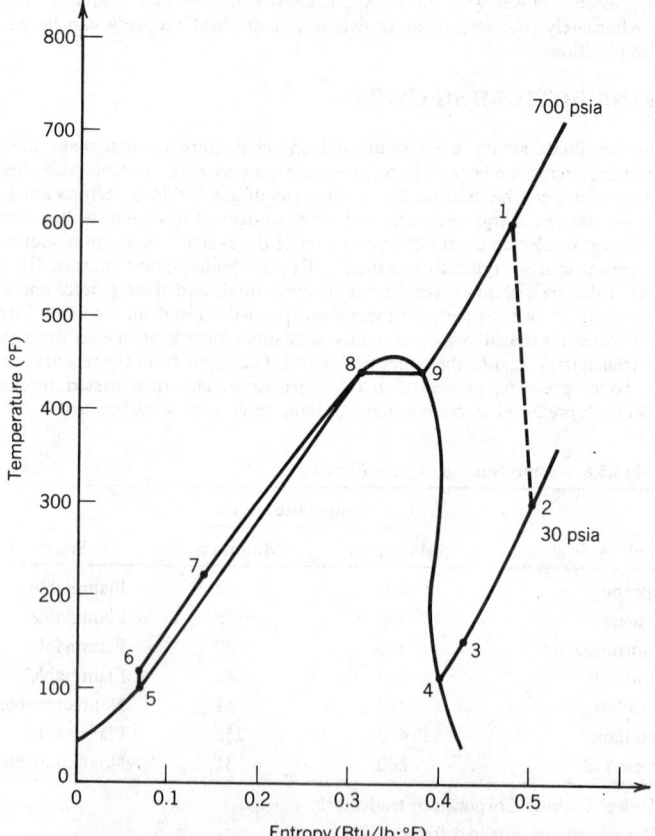

Fig. 65.21 Typical *T–S* diagram for an organic Rankine cycle power generation system.

2. A drying cycle, where the expander exhaust is in superheat and requires a regenerator.

3. Isentropic cycle, where the expansion line is at constant entropy.

This condition can be modified in some fluids by the addition of water. This is particularly true of the pyridines. Figure 65.22 illustrates this effect. Table 65.7 shows typical power performance of three working fluids. Table 65.8 shows typical cycle efficiencies for combustion turbine and organic Rankine cycle combinations as compared to a typical steam turbine, gas turbine combined cycle system. Organic Rankine cycle systems can best be used in subfreezing conditions where steam may become a problem.

65.11 STEAM INJECTION CYCLES

Steam injection into a gas turbine combustor as a source of incremental power has been the subject of investigation for some time. Usually the steam for injection is generated by the recovery of gas turbine exhaust heat. The steam may be injected as saturated steam or with some superheat. If all the steam generated is used for injection, the use of superheated steam is preferable, since this will reduce the incremental fuel increase required to superheat the injected steam from its entry condition to the turbine inlet temperature condition. The amount of steam that can be injected into a particular combustion turbine is limited by the turbine compressor characteristics. Usually injection up to approximately 6% of turbine air flow can be tolerated without compressor modifications. Massive injection of steam can result in the compressor operating at or near the surge condition, which must be avoided. The increase in power resulting from injection is typically 3% for each 1% of air flow injected as steam. This varies with the characteristics of the turbine.

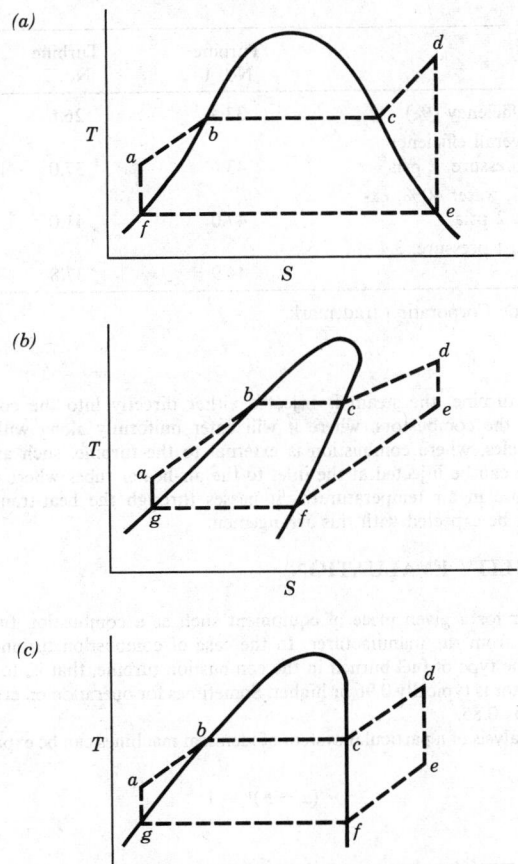

Fig. 65.22 Three types of organic Rankine cycle power *T–S* diagrams: (*a*) wetting cycle; (*b*) drying cycle; (*c*) isentropic cycle.

Table 65.7   Power Performance of Three Working Fluids[a]

	Steam	Pyridine 74.5–25.5	Pyridine 60–40	Toluene
Expander inlet pressure (psia)	300	900	900	270
Vapor temperature (°F)	775	700	700	600
Exhaust temperature (°F)	125	200	200	150
Exhaust pressure (psia)	2	14.7	14.7	3.4
Stack temperature (boiler)	326	331	328	354
Pinchpoint ΔT (°F)	35	35	35	35
Vapor generation (lb/h)	71,540	166,200	132,360	358,750
Recovered power (hp)	8479	8883	9315	9979
Prime power (hp) (gas turbine)	28,923	28,923	28,923	28,923
Total power (hp)	37,420	37,806	38,238	38,902
Overall efficiency (%)	44.70	45.15	45.67	46.46
Overall heat rate (Btu/hph)	5693	5635	5571	5476
Regenerator	No	Yes	No	Yes

[a] Fuel input to gas turbine prime power generator is 213.03×10^6 Btu/h for all cases.

Table 65.8   Typical Cycle Efficiencies of Several Organic Rankine Cycle Fluids and Steam

	Turbine No. 1	Turbine No. 2	Turbine No. 3
Simple cycle efficiency (%)	33.4	26.6	34.5
Steam cycle overall efficiency (%); exhaust pressure, 2 psia	43.6	37.0	44.7
Pyridine[a] 60%, water 40%; exhaust pressure, 2 psia	47.0	41.0	49.9
Toluene; exhaust pressure, 3.4 psia	44.9	37.8	46.5

[a] Union Carbide Corporation trademark.

In a combustion turbine, the steam is injected either directly into the combustors or into the airspace surrounding the combustors, where it will enter uniformly along with the turbine process air. In air turbine cycles, where combustion is external to the turbine, such as in fluidized-bed air-heater systems, steam can be injected at the inlet to the air-heater tubes where it will be superheated along with the increase in air temperature as it passes through the heat-transfer tubes. The same increase in power can be expected with this arrangement.

65.12   AVAILABILITY EVALUATION

The availability factor for a given piece of equipment such as a combustion turbine or a pump can generally be obtained from the manufacturer. In the case of combustion turbines, this factor is very much dependent on the type of fuel burned in the combustion turbine, that is, for operation of natural gas the availability factor is typically 0.96 or higher. Sometimes for operation on crude oil the availability factor can be as low as 0.85.

The availability analysis of a particular system of identical machines can be expressed by the binomial theorem as follows:

$$(a + b)^n = 1$$

where

$a + b = 1$
a = probability of being available for operation
b = probability of being unavailable for operation
n = number of identical units

For a system of three identical units:

$$(a + b)^3 = a^3 + 3a^2b + 3ab^2 + b^3$$

Substituting $a = 0.96$, $b = 1 - 0.96 = 0.04$, $n = 3$:

$$\underset{(1)}{0.8848} + \underset{(2)}{0.1105} + \underset{(3)}{0.0046} + \underset{(4)}{0.001} = 1.0$$

where

Term 1 is the probability that three units will be operating simultaneously.

Term 2 is the probability that one will be off and two on.

Term 3 is the probability that two will be off and one on.

Term 4 is the probability that all three units will be off simultaneously.

This analysis may be used for any number of identical machines, all operating under the same conditions and having the same performance characteristics.

65.13   ECONOMIC EVALUATION

In order to calculate the revenues or savings from cogeneration, the user must choose one of three possible power purchase options.

65.13.1   The Net Option

The cogenerator sells all the electricity generated to the utility. The cogenerator may also elect to sell capacity in kilowatts. Therefore, the revenues from cogeneration under this option are equal to energy sales plus any capacity sales.

65.13.2   All Power Consumed Internally

The cogenerator consumes all the electricity internally. As a result, the cogenerator does not receive revenues but rather savings from displaced purchase power.

65.13.3   The Surplus Option

The cogenerator sells excess energy to the electrical utility. The cogenerator may also choose to sell excess capacity to the utility. Under this option the cogenerator receives revenues from energy and capacity sales and savings from displaced purchased power.

65.13.4   Financing the Investment

Corporate Financing Option

With standard corporate financing, the investment is funded out of the firm's capital budget, which consists of internal cash, debt, and equity. The cost of these funds is a firm's hurdle rate, the minimum required return on the entire investment.

Project Financing Option

With project financing, a separate entity such as a partnership is often set up specifically for making the investment. The financing costs are equal to the interest on the debt portion plus the partners desired return on the equity portion on the investment. The program allows the debt portion to be financed in two ways: (1) a short-term construction loan that is refinanced with long-term financing, and (2) permanent long-term financing only.

 The above assumes that all thermal energy generated is consumed internally. If the thermal energy so generated is also sold, then the value of this energy must be included in the product output of the cogeneration plant. In any event, it should be considered as a saving over a conventional means of generating the thermal energy by the use of a separate fuel source.

Financial Analysis

In making the financial analysis, the following items must be considered:

1. *Investments.* Capital cost of plant, including site development, utility connections, spare parts, etc.

2. *Owning Costs.* Interest on investment, inflation rate, income taxes, property taxes, insurance.

3. *Operating Costs.* Fuel, water, chemicals, operating labor, maintenance labor, general and adminis- trative costs, administrative support labor, maintenance materials.

4. *Income.* Investment tax credit, sale of electric power, sale of thermal output such as process steam.

Maintenance labor, maintenance material, administrative support labor, general and administrative expenses, taxes, and insurance are usually stated in terms of percentage of capital investment.

65.14 APPLICABLE EQUATIONS

The same general equations are used throughout the preceding sections, therefore, all of the equations are grouped here.

65.14.1 Log Mean Temperature Difference

$$\Delta t_{Lm} = \frac{\Delta t_1 - \Delta t_2}{\log_e (\Delta t_1/\Delta t_2)}$$

65.14.2 General Heat Exchange

$$q = U_o A \, \Delta t_{Lm}$$

65.14.3 Overall Coefficient of Heat Transfer

$$U_{oa} = \frac{1}{\dfrac{1}{h_o} + \dfrac{1}{h_i} + \dfrac{1}{h_w} + \dfrac{1}{h_f} + \dfrac{1}{h_x}}$$

65.14.4 Coefficient of Heat Transfer for Gases Inside of Tubes

$$\frac{h_i D_i}{k} = 0.0255 \left(\frac{D_i V \rho}{\mu}\right)^{0.8} \left(\frac{C_p \mu}{k}\right)^{0.645}$$

The preceding equation is for either heating or cooling, provided that $G = V\rho$ is greater than $1650\rho^{0.645}$, and values of $G_{min} = (V\rho)_{min} = (G_1)_{min}$ are not less than

Pressure (psig)	$(G_1)_{min} = (V_1\rho)_{min}$
0	0.46
50	1.19
100	1.73
150	2.18
200	2.59
250	2.97
300	3.32

65.14.5 Coefficient of Heat Transfer for Gases Outside of Tubes

$$\frac{h_o D_o}{k} = 0.385 \left(\frac{C_p \mu}{k}\right)^{0.3} \left(\frac{DV\rho}{\mu}\right)^{0.56}$$

All physical properties are evaluated at the film temperature t_f. For estimates with air as the working fluid, the value of $C_p/k = 0.73$, and the preceding equation becomes

$$\frac{h_o D_o}{k} = 0.350 \left(\frac{DV\rho}{\mu}\right)^{0.56}$$

Another physical equation for air or flue gas estimates is

$$h_o = \frac{(0.80)C \, T_a^{1/3} \, G_1^{(0.60 + 0.08 \log d_o)}}{d_o^{0.53}}$$

$C = 1.25$ for air in staggered tube arrangement
$C = 1.00$ for in-line arrangement

65.14.6 Tube Wall Temperature

$$\frac{q}{A} = \Delta t_w k_w \frac{2\pi l}{\log_e (d_o/d_i)}$$

$$t_w = \frac{h_i t_i + h_o t_o}{h_i + h_o}$$

$$\Delta t_w = \frac{(q/A)(\log_e d_o/d_i)}{2\pi l k_w}$$

65.14.7 Gas Pressure Drop, inside Smooth Tubes, Turbulent Flow

$$\Delta P/N = 4f\rho V^2/2gDi$$

$$f = \frac{0.653}{(DV\rho/\mu)^{0.228}}$$

65.14.8 Coefficient of Heat Transfer, Liquids inside Tubes

$$\frac{hD_i}{k} = 0.022 \left(\frac{DG}{\mu}\right)_b^{0.8} \left(\frac{C_p \mu}{k}\right)_b^{0.4} \left(\frac{\mu_b}{\mu_w}\right)^{0.16}$$

65.14.9 Overall Coefficient of Heat Transfer with Externally Finned Tubes (Estimate)

$$U_{oa} = \frac{1}{(R_o + R_i + R_w + R_{fo} + R_{fi})}$$

$$R_o = 1/h_o E$$

$$R_i = \frac{A_o/A_i}{h_i}$$

$$R_w = \frac{1}{12(k_w/t_w)}$$

R_{fo} = external fouling factor (assumed)
R_{fi} = internal fouling factor (assumed)

65.14.10 Pressure Drop in Steam Piping

$$p_i = 0.8NF^{1.85}/\rho d_i^{4.97} \text{ psi}$$

NOMENCLATURE

A	total effective heat transfer surface, ft^2
a_r	ratio of external surface to internal surface
C	constant (dimensionless)
C_p	specific heat at constant pressure, Btu/lb·°F
D_i	tube inside diameter, ft
D_o	tube outside diameter, ft
d_i	tube inside diameter, in.
d_o	tube outside diameter, in.
E	fin efficiency (dimensionless)
F	flow, lb/sec
f	friction factor (dimensionless)
G	mass velocity, lb/hr/ft^2 of cross section
G_1	mass velocity, lb/sec/ft^2 of cross section
h_o	outside coefficient of heat transfer, Btu/hr·ft^2·°F
h_i	inside coefficient of heat transfer, Btu/hr·ft^2·°F
h_w	tube wall coefficient, Btu/hr·ft^2·°F
h_f	fouling coefficient, Btu/hr·ft^2·°F

k	thermal conductivity, Btu/ft·hr·°F
k_w	thermal conductivity of tube wall, Btu/ft·hr·°F
l	length of pipe or tube equal to 1 ft² of surface
N	length of tube or pipe, ft
ΔP	pressure drop, psi
q	heat-transfer rate, Btu/hr
R	resistance
T_a	average temperature, °R
t_i	temperature of fluid inside tubes, °F
t_o	temperature of fluid outside tubes, °F
Δt_1	greater terminal temperature difference, °F
Δt_2	lesser terminal temperature difference, °F
Δt_w	temperature difference across tube wall, °F
Δt_{Lm}	log mean temperature difference, °F
U_{oa}	overall coefficient of heat transfer, Btu/hr·ft²·°F
V	linear velocity, ft/hr
ΔP	pressure difference, psi
ΔT	temperature difference, °F
π	3.1416
ρ	density, lb/ft³
μ	viscosity (absolute), lb/ft·hr

Subscripts

i	inside diameter or coefficient
o	outside diameter or coefficient
a	average conditions
f	fouling coefficients
w	pipe or tube wall
oa	overall
Lm	log mean
e	Naperian logarithm

BIBLIOGRAPHY

J. L. Boyen, *Practical Heat Recovery,* Wiley, New York, 1975.

J. L. Boyen, *Thermal Energy Recovery,* Wiley, New York, 1979.

R. W. Foster-Pegg, *Coal Fired Air Turbine Cogenerations,* Westinghouse Electrical Corporation.

R. W. Foster-Pegg and R. V. Garland, *Atmospheric Fluid Bed Air Turbine Cogenerations System,* Westinghouse Electric Corporation, for American Power Conference, April 1982, pp. 6–10.

J. S. Davis, G. C. Dupontiel, and L. J. Sweet, *Utilization of Circulating Gluid Beds in Air Turbine Cogenerations.* Struthers Wells Corp., Struthers Wells, United Kingdom, 1981, pp. 7–11.

J. S. Davis, W. W. Young, and Carl J. Lyons, "Use of Solid Fuel Possible for Field Steam Generation," Batelle Columbus Laboratories, *Oil & Gas Journal* (June 8, 1981).

C. L. Marksberry and B. C. Lindahl, "The Applications of Indirectly Fired Open Cycle Gas Turbine Systems Utilizing Atmospheric Fluidized Bed Combustions to Industrial Cogenerations Situations," ASME Gas Turbine Division, Paper No. 79-GT-16.

C. L. Marksberry and B. C. Lindahl, "Industrial AFB-Fired Gas Turbine Systems with Topping Combustors," ASME Gas Turbine Division, Paper No. 80-GT-163.

CHAPTER 66
HEAT-RECOVERY BOILERS

J. L. BOYEN

Consulting Engineer

This chapter on industrial heat-recovery boilers will be confined to boilers that are suitable for heat recovery from the exhaust of gas turbines, diesel engine hot gas, or air sources such as furnaces. Typically, five types will be covered: fire tube, natural circulation, forced recirculation, forced circulation spillover cycle, and forced circulation monotube.

66.1   FIRE TUBE

Fire-tube boilers are the simplest and least expensive when compared to corresponding water-tube boilers; however, their inherent design limits their size and operating pressure. They are seldom built for pressures in excess of 150 psig. Their overload capability is limited, and exit gas temperature rises rapidly with increased output. One advantage is their large water-storage capacity. Because of this feature, wide and sudden fluctuations in steam demand are met with very little change in pressure. There are several types of fire-tube boilers in use at this time. The most popular one is the horizontal return tubular type as illustrated by Fig. 66.1.

66.1.1   Types of Fire-Tube Boilers

Fire-tube boilers may be generally classified as follows: (1) external furnace—this classification includes the horizontal return tubular type (HRT), short firebox type, or compact type; and (2) internal furnace type—this classification includes the horizontal tubular such as the locomotive type, short firebox compact, scotch, and vertical tubular. The most common fire-tube boiler in use on cogeneration systems at this time is the HRT with external firebox. Essentially, the HRT consists of a cylindrical shell with flat-tube sheets at both ends supporting a large number of 3 or 4 in. fire tubes. The hot gases pass through the immersed tubes generating steam in the cylindrical shell. The control of an HRT heat-recovery boiler is comparatively simple, being simply a water-level controller that in turn controls the flow of feedwater to the boiler shell as a function of water level.

The fire-tube boiler has been used to some extent as a heat-recovery boiler; however, its use is limited owing to the fact that the tubes are generally not finned on the inside. Some, however, do have spiral turbulence promoters. Therefore, a given rating in the fired condition is considerably reduced when used as a heat-recovery boiler because the gas temperatures entering the tubes will be as low as 25% or 30% of the temperature that would exist if it were directly fired. For this reason it is very limited in application as an industrial heat-recovery device.

66.2   NATURAL CIRCULATION—BENT-TUBE TYPE

There are basically two bent-tube-type boilers (Fig. 66.2) in use in cogeneration systems: three-drum low-head boilers as illustrated by Fig. 66.3 and two-drum natural circulation boilers.

66.2.1   Three-Drum Low-Head Boiler

Three-drum low-head boilers were designed for use in small spaces, especially in areas where head room is limited. However, like other water-tube boilers, they are seldom designed for pressures under 150 psig. Their low overall height limits their capacity because the hydraulic head differential between the bottom drum and the upper drum is reduced, which reduces the capability of providing circulation

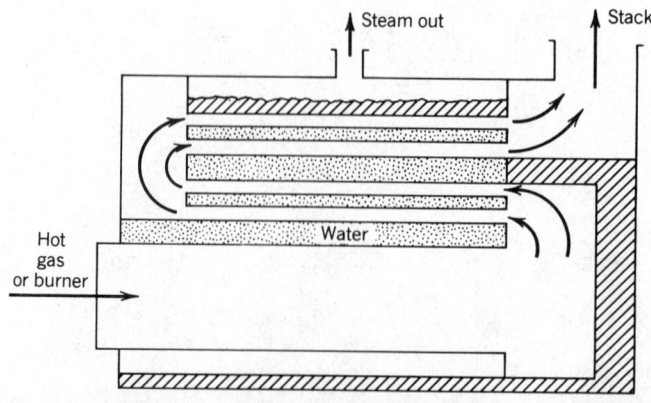

Fig. 66.1   Outline of typical fire-tube boiler.

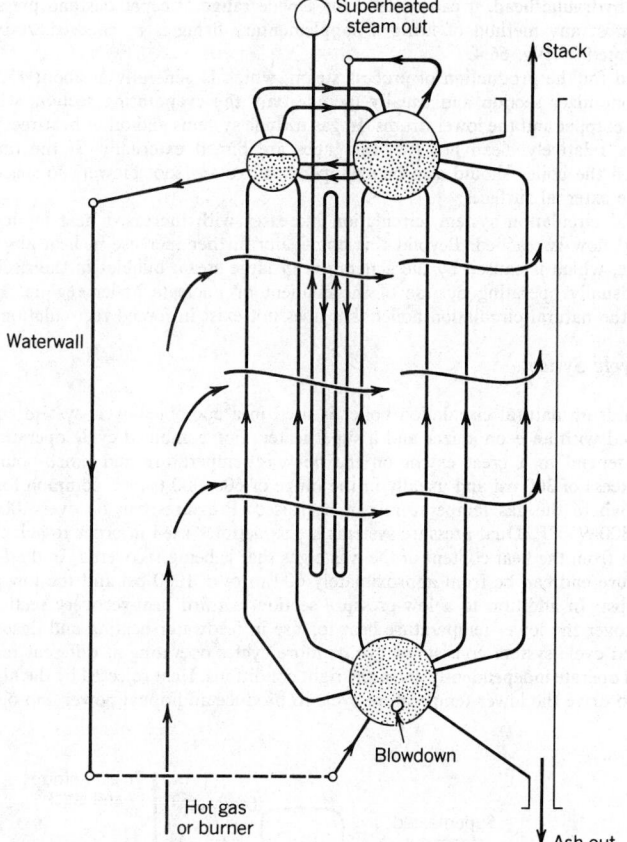

Fig. 66.2 Natural circulation bent-tube boiler.

for higher flow rates. Baffles are usually arranged for natural draft operation with a normal stack height.

66.2.2 Two-Drum Boiler

The two-drum natural circulation boiler is the most widely used boiler in heat-recovery applications. It is the simplest of the bent-tube types and is available in many designs. Because of its design flexibility

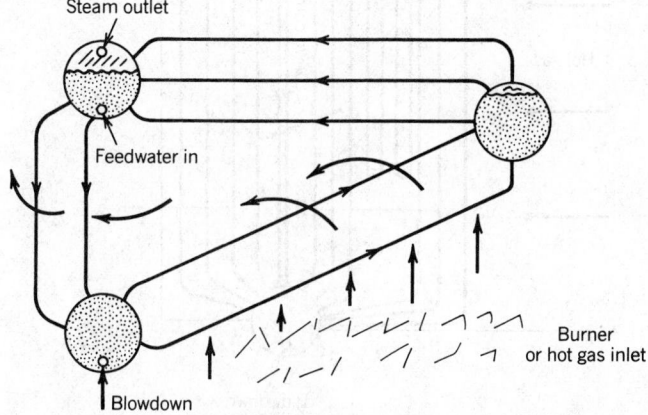

Fig. 66.3 Natural circulation three-drum low-head boiler.

and increased hydraulic head, it can be used for a wide range of capacities and pressures and can be adapted to almost any method of firing if supplementary firing is to be used. A typical two-drum boiler is illustrated by Fig. 66.4.

Boilers used for the production of process steam, which is generally at about 150 psi, very rarely include an economizer section and usually include only the evaporating section, which is the tubes that connect the upper and the lower drums. In gas turbine systems and other heat-recovery applications that produce a relatively clean hot gas, the tubes are finned externally. If the turbine is fired by liquid fuel, then the boiler should also be equipped with rotary soot blowers so spaced that they will cover all of the external surfaces.

In a natural circulation system, circulation increases with increased heat input until a point of maximum fluid flow is reached. Beyond this point, any further increase in heat absorption results in a flow decrease, which is caused by the generation of large steam bubbles in the risers. At this point, the boiler is usually operating outside of the ebulient or nucleate boiler regime. This is a distinct limitation for the natural circulation boiler that does not exist in forced recirculation systems.

Combined Cycle System

When the two-drum natural circulation boiler is used in a combined-cycle-system configuration, it is always equipped with an economizer and a superheater. For combined cycle operation, the operating pressure will depend to a great extent on the hot gas temperature and pinch-point considerations. Pressures in excess of 300 psi and usually in the range of 600–900 psi are common for supplementary-fired systems where the gas temperature can be raised. Pressures can be over 1000 psi and steam temperatures 800–950°F. Dual-pressure systems are sometimes used in order to achieve the maximum possible benefit from the heat content of the waste gas that is being recovered. In dual-pressure systems the high-pressure end can be from approximately 600 to over 1000 psi and the low-pressure end can be at 50–100 psi. In addition to a low-pressure section, a third heat-recovery section can be added, which will recover the lower-temperature heat for use in feedwater heating and deaeration.

A combined cycle system consists of two or more cycles operating at different temperatures, each of which could operate independently given the right conditions. Heat rejected by the higher-temperature cycle is used to drive the lower-temperature cycle to produce additional power and operate at a higher

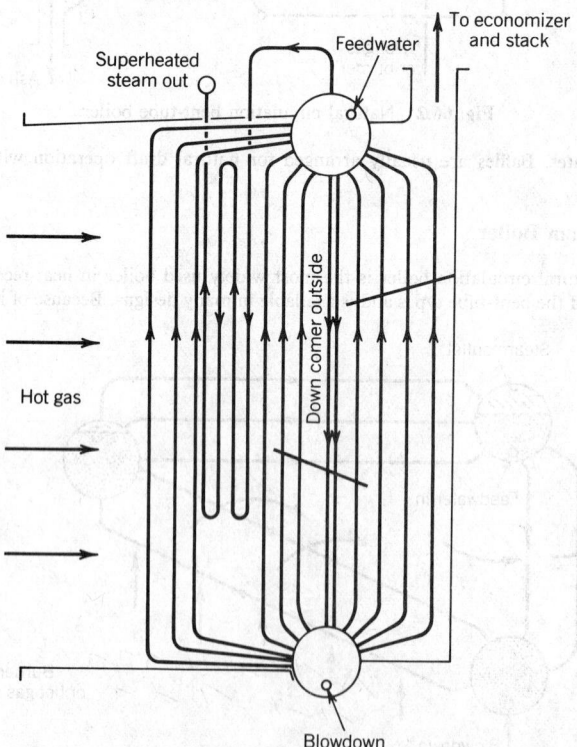

66.4 Natural circulation two-drum heat-recovery boiler.

overall efficiency than either cycle by itself could achieve. To be a combined cycle, the different cycles must operate on separate fluids. Combined cycles can consist of a diesel engine with steam power by heat recovery, a diesel-engine–organic-fluid power cycle, gas-turbine–steam power system, gas-turbine–organic-fluid cycle system, liquid-metal–steam system, MHD–steam power, gas-turbine–dual-fluid air to steam power.

The higher-temperature cycle is commonly called the topping cycle. The lower-temperature cycle is called the bottoming cycle. Topping cycles have operated as Otto, Brayton, and Rankine cycles. All bottoming cycles have been Rankine cycles; they operate on the heat rejected by an open cycle system such as a diesel engine exhaust, gas turbine exhaust, and the exhaust or waste heat resulting from the production of power by the prime power source whatever that power might be.

Most of the practical combined cycle systems now in use are gas turbine, steam turbine cycles. These require the least amount of innovation and can as a rule take advantage of existing hardware. Therefore, steam still remains as a dominant medium because of the low cost, availability, stability, high specific heat, high heat-transfer rates, and relative inertness of water. On the other hand, the organic fluids suitable for Rankine cycle systems are generally expensive and are not readily available, particularly in remote areas. Furthermore, they are unstable at higher temperatures, have low specific heat, low thermal conductivity, and lower heat-transfer coefficients.

All heat-transfer sections in a waste heat boiler are generally made of externally finned tubes. This takes advantage of the lower heat-transfer coefficients existing on the air side as compared to the higher coefficients existing on the fluid side.

Heat recovery boilers for straight unfired heat recovery systems usually operate at pressures up to 600 psig and steam temperatures as high as possible depending on the temperature of the waste gas; 50–75°F below the hot gas inlet temperature is usually a good compromise. Higher steam pressures and temperatures result in low theoretical water rate values for the steam turbine, which then produce more power at a given steam flow rate. However, the higher steam conditions usually also mean less steam is being generated. Therefore, there is a trade-off between low water rates and the steam flow available. The effect of less steam at higher steam conditions can be offset to a great degree by operating at two steam pressures, one high—300–600 psig—and one lower—35–60 psig—to capture the energy available at the low end. An analysis must be made in each case. Pinch-point conditions are also a factor. A good practical value of pinch point ΔT is 40°F; however, some systems have been designed having pinch point ΔT of 15–30°F. Here again is a trade-off that must be made because pinch point ΔT directly affects log mean ΔT and, hence, the required effective heating surface of the economizer and the evaporater sections. The cost of the additional surface must be compared to the benefits in the form of additional steam production and power realized. In this regard, the value of the additional power over the write-off period for the plant must be compared to the additional investment for greater surface. An example of the temperature–heat distribution chart for a two-pressure steam system is illustrated in Section 66.3, Fig. 66.8. This is the same regardless of whether a natural circulation or a forced circulation system is being considered. Note that it is possible to end up with a rather low final stack temperature. Care must be taken to select the correct materials for the economizer section if there is any chance that condensation will take place.

Overall cycle efficiencies of unfired combined cycle plants can range from 38% to 45% and sometimes higher depending on the gas turbine efficiency, the steam turbine efficiency, and the utilization of the heat available in the turbine exhaust. Gas turbines having simple cycle efficiencies of 30–34% are now available. Steam turbines in the range of 75% to over 80% are also available depending on the size of the turbine and the operating conditions in the system. Therefore, the careful selection of the steam cycle containing as much steam as possible under the selected conditions is the other factor involved in the final result.

Steam turbine exhaust pressure is an important factor in obtaining overall low water rates. The exhaust pressure selected is largely a function of the condenser cooling medium conditions and atmospheric ambient conditions. If a water-cooled condenser is used and the cooling water temperature is under 70°F, it is possible to attain a turbine exhaust pressure of less than 2 psia.

If an air-cooled condenser must be used and if the dry bulb temperature is at 100°F, the condenser pressure will be about 4 psia. Cooling water or a water-cooled condenser can also be obtained by using a cooling tower system. However, such a system is also subject to atmospheric conditions that limit condensing pressure. The steam condenser system is an important part of any combined cycle system and must receive careful consideration.

The air-cooled condenser is logical in arctic areas where water in suitable quantities is not available, such as on pipe-line applications. In those regions, ambient temperatures are low enough most of the time to permit low-pressure operation with an air-cooled condenser. However, even though ambient temperatures are low, air-cooled condensers can get quite large because of the low air-side heat-transfer coefficients. Finned tubing is always used, however, in order to limit size. The power required to drive the cooling fans is several times the power required to drive the water pumps in a water-cooled system. Therefore, the parasitic load on the combined cycle system is increased accordingly and net power output is reduced.

Freezing is also a problem in arctic areas. This is particularly true during periods of shutdown.

For normal operation, the condenser can be enclosed with controllable louvres in the top and sides to permit the partial recirculation of cooling air thereby keeping its temperature above ambient. The tubes must have sufficient slope to permit the rapid draining of condensate as it forms. The condenser hot well should be insulated and kept above freezing by the use of electrical immersion heaters. The condenser hot well should be of sufficient size to permit the draining of the entire system including the boiler into the heated and insulated hot well to prevent periods of freezing during prolonged shutdown. Both the boiler and the condenser should be designed to minimize the system-contained water.

Supplementary-fired and exhaust-fired heat-recovery boilers for combined cycle systems can operate up to 1450 psig, 900°F. Supplementary firing usually consists of a duct-type gas burner located directly in the air duct between the gas turbine exhaust and the heat-recovery boiler inlet. Exhaust-fired systems have a burner that is integral with the boiler as in a conventionally fired boiler except that the exhaust gas from the turbine is fed directly into the boiler furnace, with some of the exhaust gas serving as combustion air. In an unfired heat-recovery system the steam portion produces 30–35% of the total plant capability. In supplementary-fired systems the steam turbine produces up to 60% of the total power, and with exhaust-fired systems the steam plant produces up to 80% of the total plant output.

The unfired combined cycle is the simplest of the combined cycle systems, using only the energy in the gas turbine exhaust for steam generation. It eliminates the requirement for burner controls and safety devices associated with industrial gas burners. It is therefore less expensive and simpler to operate and maintain. However, the supplemental-fired heat-recovery combined cycle appears to be the most popular among utility users because the supplementary-firing feature provides a peaking possibility at high efficiency. The supplemental-fired heat-recovery combined cycle system can be designed for excellent part load heat rate. A good part load performance requires maximum utilization of gas turbine power with minimum of no firing of the boiler unless the gas turbines are fully loaded. Also, the control of air flow to the boiler by multiple-shaft turbines or the modulation of inlet guide vanes on single-shaft gas turbine compressors is required for maximum part and load efficiency.

One major disadvantage of the supplemental-fired system is the sharp variation in steam temperature with load, because air flow is not varied and the superheater is the element nearest the burners. Some systems have been installed with a superheater upstream of the burner, alleviating this situation. Such an arrangement is particularly applicable to cycles using low-temperature steam, approximately 75–100°F below gas turbine exhaust temperature. Steam pressure is controlled in this system by a wide range initial pressure governor on the steam turbine, regulated on the high end by a conventional combustion control system that limits maximum pressure by modulating the firing rate. Some systems now have a computer control on the steam turbine governor that varies steam pressure with flow. This type of control minimizes part load fuel rate and reduces moisture in the steam in the low-pressure turbine stages during light load operation.

Cogeneration System with Combined Cycle

The conventional combined cycle turbine system can be a true cogeneration system, according to the official definition of such a system, if the steam turbine is either an extraction turbine or a back pressure turbine that delivers low-pressure steam to some process. This is a particularly desirable type of system since the steam not diverted to process generates electric power, because it passes through the turbine to the condenser and can be modulated over a wide range of demand. In such a case the heat-recovery boiler is identical to any boiler that would be placed in a conventional combined cycle system. The feature that makes it cogenerating is the bleed steam from the steam turbine to process.

Control

The heat-recovery boiler water control is of the conventional type using a flow-sensing device on the steam drum that admits feedwater through the economizer to the drum as conditions demand. There is a point, however, concerning the economizer that must be taken into consideration. If the economizer is slightly overdesigned, it is possible that boiling can take place in the economizer at part load. This is due to the large heating surface as compared to the thermal demand. One way to avoid boiling in the economizer at part load is to provide a temperature control that senses water temperature at the outlet end of the economizer. This temperature control is set at approximately 5–8°F subcooled and activates a water bypass valve that serves to bypass water back to the deaerator if the temperature reaches a set point. This ensures continuously high flow through the economizer without reaching a boiling condition. Another means for accomplishing this is to provide a temperature-controlled gas bypass through or around the economizer section to unload the economizer at part load. If the heat-recovery boiler is supplementary fired, steam pressure is controlled by controlling the supplementary firing rate. If the heat-recovery boiler is unfired, steam pressure will be determined by the turbine throttle. Excess pressure would be relieved through a bypass valve to the condenser. Another method used is to divert the turbine exhaust gas; this may not be desirable and under most conditions is wasteful.

66.2.3 Adaptation of Natural Circulation to Ebulient Cooling for Diesel Engine Applications

This type of system is commonly used in conjunction with diesel engine heat recovery and refers principally to the heat recovery from the cooling jacket. Obviously, such heat recovery is limited to low-pressure applications not over 15 psig. Figure 66.5 illustrates such a system. This system must be used with care, otherwise engine damage can result due to insufficient cylinder wall cooling. Some details regarding the design of reciprocating engine heat recovery must be observed that are peculiar to that application:

1. A large volume of water and ample steam liberation area are necessary to avoid the return of superheated water to the engine jacket, thus avoiding the possibility of steam bubbles forming in the engine jacket, which results in hot spots and possible engine deterioration.

2. There should be a good sedimentation chamber away from the engine to remove contaminants either brought in by the water or by chemicals added to the water for removing corrosive gases and dissolved solids.

3. Gas to water-heating surfaces must be accessible for easy inspection and repair.

4. Engine exhaust turnaround chambers within the boiler should be designed for minimum pressure drop and to act as pulsation dampers for exhaust noise suppression.

5. The heat-recovery boiler must be furnished with a high-level cutoff, high-level alarm, low-level pump start, low-water alarm, separate mechanical water feeder for emergency service, low-water cutoff, vent valve, ASME safety valve, pressure gauge, blowdown valve, and the required interconnecting piping.

6. The boiler must be properly insulated and covered by a metal jacket. Turnaround chambers are internally insulated for pulsation dampening and noise suppression.

7. Gas entrance nozzles must be constructed of stainless steel to resist high-temperature scaling. This assumes that the boiler water circulates through the engine jacket and the engine exhaust passes through immersed tubes in the boiler drum.

It is also possible to have two separate systems on one engine. The water jacket recovery can drive a hot water or low-pressure steam activated absorption chiller, while the exhaust can generate

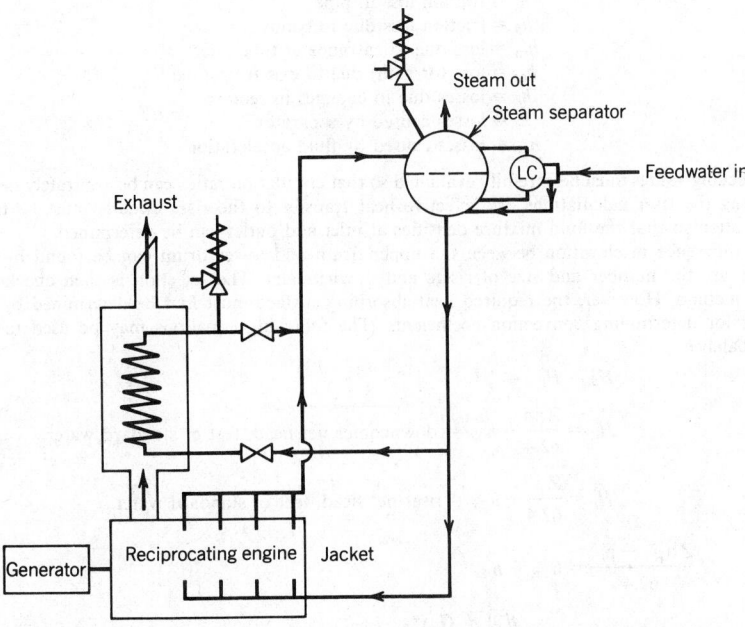

Fig. 66.5 Natural circulation diesel engine heat recovery. LC, level controller.

process steam at high pressure, medium-temperature hot water for process or heating, steam for operating a bottoming cycle, or heat an organic fluid for process or power. Practically all of the jacket heat can be recovered and approximately two-thirds of the exhaust heat can be recovered, making this cycle quite efficient.

66.2.4 Design Consideration and Analysis of Natural Circulation Boiler Systems

In a natural circulation system, circulation increases with increased heat input until a point of maximum fluid flow is reached. Beyond this point any further increase in heat absorption results in a flow decrease. In the natural circulation system, an increased flow rate can result only from the increasing difference in the densities of the fluids in the downcomers and risers. When the rate of increase of these losses becomes greater than the gain from increasing density difference, flow rate begins to drop. Therefore, it is important to design all circuits to operate in the region of the rising portion of the curve.

When design conditions are limited to the rising portion of the circulation curve, the natural circulation boiler tends to be self-compensating for the numerous variations in heat-absorption conditions encountered in an operating unit. These include sudden overloads, changes in cleanliness of heat absorbing surfaces, nonuniform heat transfer condition, or changes in burner operations if it is a supplementary-fired boiler.

The method of producing flow, whether natural or forced, in boiler circuits has practically no bearing on the effectiveness of heat-absorbing surfaces as long as the inside surface is wetted at all times by the water in a steam–water mixture of suitable quality to maintain nucleate boiling. If this requirement is met, the water film resistance to heat flow is negligibly small and the overall heat conductance depends on the gas-side conditions. Natural circulation is best used when large changes in density occur as a result of heat absorption. Therefore, natural circulation is restricted to subcritical applications where there is a substantial difference in densities between the downcomer fluid and the riser fluid.

In the design of a natural circulation boiler, the downcomer and riser circuits must be so proportioned that the downcomer net head is equal to the riser net head. This evaluation is usually made at standard water conditions, density at 62.4 lb/ft³. The sum of the downcomer losses is equal to the sum of the riser losses.

The sum of the downcomer losses is

$$h_f + h_b = h_{en} + h_{ex} + h_k$$

The sum of the riser losses is

$$h_f + h_b + h_{en} + h_{ex} + h_k + h_s + h_{ac}$$

where

$$
\begin{aligned}
h_f &= \text{friction loss in pipe} \\
h_b &= \text{friction loss due to bends} \\
h_{en} &= \text{loss due to entrance at tube inlet} \\
h_{ex} &= \text{loss (or gain) due to exit from tube} \\
h_k &= \text{losses due to changes in section} \\
h_s &= \text{losses caused by separator} \\
h_{ac} &= \text{losses caused by fluid acceleration}
\end{aligned}
$$

The preceding values must be carefully evaluated so that circulation ratios can be accurately determined. In making the riser calculations, the effective heat transfer to the riser circuits must be taken into consideration so that the fluid mixture densities at inlet and outlet can be determined.

The difference in elevation between the upper drum and lower drum can be found by trial and error, as are the number and size of risers and downcomers. The selection is then checked by the balance method. However, the required heat-absorbing surface must first be determined by standard methods for determining convection coefficients. The following equations may be used to calculate system balance.

$$H_d = H_r \tag{66.1}$$

$$H_d = \frac{Z\rho d}{62.4} - h_{\Sigma d} = \text{downcomer net head, feet of standard water} \tag{66.2}$$

$$H_r = \frac{Z\rho r}{62.4} + h_{\Sigma r} = \text{riser net head, feet of standard water} \tag{66.3}$$

$$\frac{Z(\rho_d - \rho_r)}{62.4} = h_{\Sigma d} + h_{\Sigma r} \tag{66.4}$$

$$h_f = 2.31 \frac{fl}{de} \frac{I}{\rho} \left(\frac{G}{10^5}\right)^2 \tag{66.5}$$

where

$$
\begin{aligned}
Z &= \text{difference in elevation, ft} \\
\rho_d &= \text{downcomer fluid density, lb/ft}^3 \\
\rho_{ex} &= \text{mixture density of riser fluid, lb/ft}^3 \\
\rho_r &= \text{mean density of riser fluid, lb/ft}^3
\end{aligned}
$$

$$\frac{Z\rho r}{62.4} = \text{riser gravity head, feet of standard water}$$

$$
\begin{aligned}
h_{\Sigma d} &= \text{sum of downcomer losses} \\
h_{\Sigma r} &= \text{sum of riser losses} \\
f &= \text{moody friction factor} \\
L &= \text{tube length, ft} \\
de &= \text{tube equivalent diameter, in.} \\
\rho &= \text{fluid density, lb/ft}^3 \\
G &= \text{mass velocity, lb/ft}^2 \cdot \text{hr} \\
N &= \text{velocity head loss factor, dimensionless}
\end{aligned}
$$

High-pressure heat-recovery boilers sometimes have a low-pressure section after the high-pressure economizer. The low-pressure section can be either a natural circulation unit or a forced recirculation unit having its own feedwater system. Low-pressure steam from this section is used for deaeration and feedwater heating and provides a distinct advantage in overall economy.

66.3 FORCED RECIRCULATION SYSTEMS

Figure 66.6 illustrates a typical forced recirculation system schematic. This form of steam generation for heat-recovery applications is in a class by itself because there are instances that demand forced recirculation systems that prevent vapor separation within the steam-generating tubes. The rate of recirculation is usually 5–10 times the maximum steaming rate of the boiler. Forced recirculation makes possible the use of smaller-diameter tubes, which reduce boiler size and weight and reduce the time required for steam up. Only one drum is required, which is actually a steam separator and not a steam drum as is common in natural circulation boilers. Some forced recirculation boilers, particularly smaller ones, have only a cyclone steam separator with an enlarged lower section for water-level control instead of the conventional drum. Boilers having only the cyclone separator are particularly adaptable to shipboard use because they are not particularly sensitive to pitch and roll.

The primary advantage of forced recirculation is probably the assurance that there is always forced water circulation in all boiler tubes. Since heat-recovery boilers very often have limited space, there may not be sufficient distance between the upper and lower drums to promote effective natural circulation. Forced recirculation is usually the solution in such circumstances.

A forced recirculation boiler is nearly always more compact than a comparable natural circulation boiler, since the former does not require a fixed relationship between the drum and the heating surfaces.

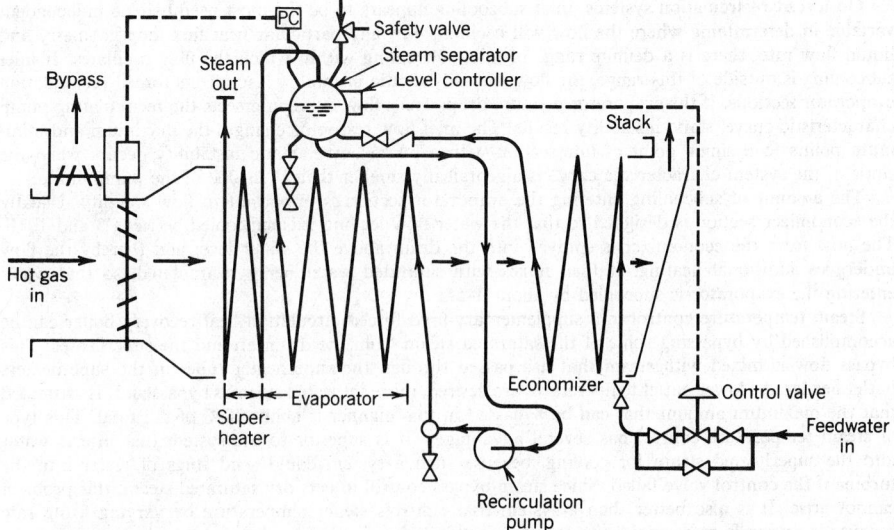

Fig. 66.6 Typical forced recirculation heat-recovery boiler schematic. PC, pressure controller.

The drum or separator can be located almost anywhere adjacent to the heating-surface assembly and does not necessarily have to be elevated above the heating-surface assembly.

Heating-surface design can be quite flexible. Vertical or inclined tubes are not required. The heating surfaces can be arranged in a horizontal return bend configuration or a circular shape consisting of flat spirally formed coils or other shapes. However, the heating surface must be arranged in such a way that the tubes are always completely filled, particularly at low flow rates, because if a steam–water interface exists, corrosion can often begin at this interface. Furthermore, with a completely filled tube, the tube wall temperature is uniform around the circumference of the tube. This would not be the case in a horizontal tube in which a water–steam interface existed, resulting in the steam portion of the tube running hotter than the water portion thereby causing thermal stresses.

External cleanout plugs permit internal cleaning of individual tube circuits either chemically or mechanically without disassembling the boiler casing. Cleaning of external surfaces of the tubes is by conventional soot blowers operated by either air or steam. A typical forced recirculation boiler schematic is illustrated by Fig. 66.6. This shows a boiler having a superheater, an evaporator, and an economizer and is typical of a single-pressure boiler such as would be used in a combined cycle system. In some single-pressure systems, a second economizer or feedwater heater is added after the primary economizer to heat makeup water from the feed treatment equipment to the deaerator. If there is no makeup, the second economizer can be used to heat condensate flowing from the condensate return tank to the deaerator.

No internal baffles are required in forced recirculation. Internal baffles are difficult to install and maintain and can cause ash accumulation unless they are steeply inclined or vertical. Even inclined baffles require some blowing to keep them clear of soot and ash.

Forced recirculation boilers are not used as much in the United States as they are in Europe, where they are almost exclusively used on high-pressure combined cycle systems and for power generation. The versatility of forced recirculation systems has not yet been fully appreciated, and it will probably require the importation of European designs creating competition to encourage boiler manufacturers in the United States to consider more seriously this alternative. Probably the most criticized part of a forced recirculation boiler is the recirculating pump. Obviously, if the pump fails, all water circulation ceases in the boiler tubes and the boiler must be immediately removed from service and isolated from its source of hot gas. However, a standby pump is usually provided for such an emergency. If the recirculating pump is property selected, particularly as to its mechanical shaft seal, it will provide long trouble-free service. The shaft seal and its environment comprise one of the most critical parts of the system. If operating temperatures are in excess of 250°F, the seal should be water cooled. A strainer should be in the piping ahead of the pump, and the seal faces must be continuously flushed with clean water. Another important item in the proper selection of a pump is NPSH. This could be important on close-coupled systems with only a few feet of head on the pump suction. Failure to consider this factor could result in severe cavitation and consequent recirculation failure.

An increase in total dissolved solids in the drum has a destabilizing effect based on experience with natural circulation designs. Physiochemical hydrodynamic effects resulting in increased pressure drop in the separator or drum located at the exit end of the circulation loop are responsible for destabilization. The solids level must therefore be kept at a reasonable level (not over 2000 ppm to prevent carryover).

On forced recirculation systems, inlet subcooling appears to be the most useful single independent variable in determining where the flow will oscillate. For any particular heat flux loop geometry and liquid flow rate, there is a definite range of inlet subcooling within which the flow oscillates. If inlet subcooling is outside of this range, the flow is steady. Static instability also affects forced recirculation evaporator sections. If the evaporator pressure drop versus flow curve intersects the recirculating pump characteristic curve, static instability results. The inlet flow restrictor changes the maximum and minimum points to a single point of intersection with a pump curve. Static instability occurs when the slope of the system characteristic curve is algebraically smaller than the slope of the pump curve.

The amount of subcooling entering the evaporator section is important to flow stability. Usually the economizer section is designed so that the water flow leaving it is subcooled between 5 and 10°F. The flow from the economizer is sprayed into the drum above the water level and thereby the flow undergoes additional heating. It then mixes with saturated water being recirculated, so that water entering the evaporator is subcooled by about 3–5°F.

Steam temperature control in a supplementary-fired forced circulation heat-recovery boiler can be accomplished by bypassing some of the saturated steam from the drum around the superheater. This bypass flow is mixed with steam that has passed through the superheater tubes in the superheaters outlet header. As lower outlet temperatures are desired, more saturated steam is bypassed. It is estimated that the maximum amount that can be bypassed in this manner is about 20% of the total. This type of steam temperature control has several advantages. It is superior to the system that injects water into the superheated steam for cooling, because such a system could send slugs of water into the turbine if the control valve failed. Since steam bypass control injects dry saturated steam, this problem cannot arise. It is also better than a system that controls steam temperature by varying firing rate because it responds more rapidly to load swings. Stability in the superheater and economizer can be accomplished by designing these sections for a reasonably high pressure drop, 10–20 psi for the super-heater and approximately 15 psi for the economizer. The economizer outlet temperature is controlled by constant flow to the economizer, bypassing the portion not required to satisfy the demand, diverted back to the deaerator. Steaming in the economizer must be avoided because it causes stagnation in some tubes that begin to steam with an accompanying increase in pressure drop, which in turn leads to less flow, more steaming, and, ultimately, stagnation. With a high-pressure drop across the entire economizer in the beginning, the increase in pressure drop caused by steaming in some tubes is small compared to the total, resulting in significant changes in flow and thereby preventing stagnation.

66.3.1 Multiple-Pressure Systems

Figure 66.7 illustrates the schematic of a dual-pressure forced recirculation system. The heat-recovery boiler essentially consists of five separate heat-transfer sections, namely, economizer, low-pressure evaporator, high-pressure evaporator, low-pressure superheater, and high-pressure superheater. All of the water for the system, including both the high-pressure and low-pressure systems, flows through the economizer to drum No. 1, which is the low-pressure drum. Water is circulated from the drum by means of a recirculation pump through the low-pressure evaporator and then back to the drum. Steam from the low-pressure drum passes to the low-pressure superheater and then out to the low-pressure steam turbine. Drum No. 1 gets its water directly from the common economizer section. Drum No. 2, which is the high-pressure drum, receives its water by means of a feedwater pump from drum No. 1. Therefore, it is receiving water at a temperature corresponding to the saturation temperature of the steam in the low-pressure section. A recirculating pump circulates water from drum No. 2 to the high-pressure evaporator and then back to drum No. 2. Saturated steam from the high-pressure drum No. 2 then passes to the high-pressure superheater and out to the high-pressure end of the steam turbine. The steam coming from the exhaust of the high-pressure turbine is at approximately the same pressure as the low-pressure steam coming from the low-pressure superheater and mixes with the steam from the low-pressure superheater before entering the first stages of the low-pressure turbine. Then all of the steam generated passes through the low-pressure turbine and to the condenser. This is considered to be a very efficient system in that a minimum of losses exist and the heat content of the waste gas stream is utilized to the greatest extent possible.

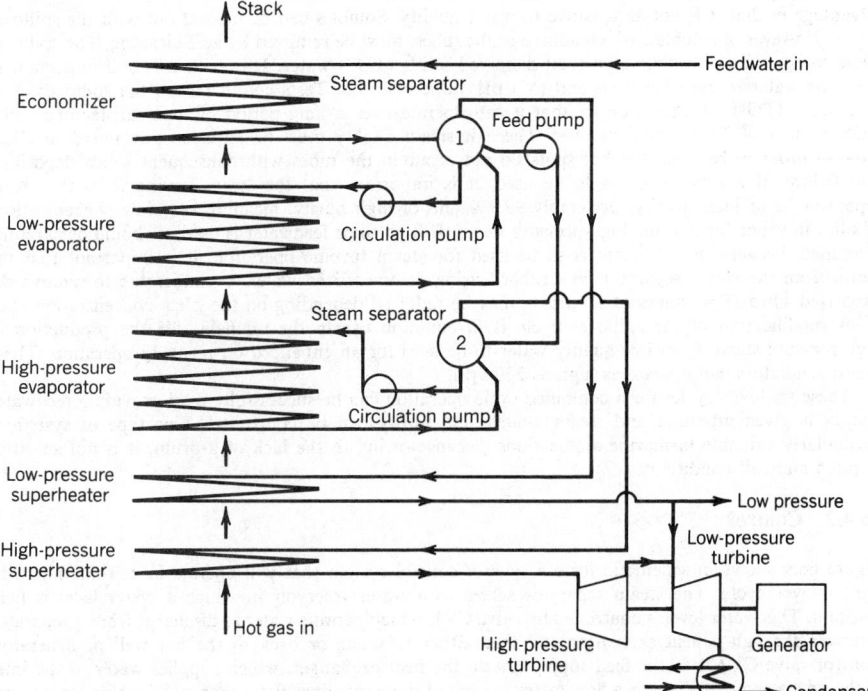

Fig. 66.7 Typical multiple-pressure forced recirculation heat-recovery boiler schematic.

Pinch-Point Effects

Figure 66.8 is a typical enthalpy–temperature chart showing the relation of the waste gas temperature and the temperature existing in the heat-recovery steam generator. This diagram is intended to illustrate typical conditions as they might exist in a dual-pressure forced recirculation steam generator as illustrated by Fig. 66.7. The pinch-point ΔT actually determines the amount of surface required for the evaporator and the economizer sections and should be selected taking into consideration the cost of additional surface versus the value of the additional steam generated by closing the pinch point. A good average pinch-point ΔT is 30–40°F, although units have been constructed with pinch-point ΔT values of less than this.

66.3.2 Supplementary Firing

The forced recirculation heat-recovery boiler, regardless of whether it is a single-pressure unit or a multiple-pressure unit, can be supplementary fired as was explained in the natural circulation section. The primary difference is that a forced recirculation system will respond faster to changes in firing rate than will a natural circulation system, because the forced recirculation system has smaller diameter tubes of less mass. There is less water contained in the tubes and, consequently, response to changes in temperature are more rapid. This can be valuable. However, it also means that more sensitive controls must be used, preferably electronic controls, which can respond sufficiently fast to effect the necessary automatic adjustments.

66.4 ONCE-THROUGH SPILLOVER CYCLE

66.4.1 Description

The forced circulation once-through spillover cycle is a modification of the once-through monotube cycle. In the spillover cycle, the controls are set to produce wet steam approximately 90%+ quality at the exit end of the evaporator zone. At this point, a cyclone separator is used to separate the 10%+ water from the wet steam. The water is tapped off through a feedwater heater and is then either recovered or discarded, depending on water availability and quality. Saturated steam from the separator then goes to the superheater section for further heating. This cycle is shown schematically by Fig. 66.9.

In the spillover cycle, the advantages of the monotube boiler are present, but it has a further advantage in that it is not as sensitive to water quality. Solubles can be flushed out with the spillover water. However, insolubles, which adhere to the tubes, must be removed by acid cleaning. The spillover cycle can operate on rather high total dissolved solids (TDS) water. However, it is still important to treat the water to zero hardness and to a pH value of 8–9.5. Tests conducted on the spillover cycle with high TDS feedwater indicate that it can operate over a long period of time satisfactorily with TDS content of 25,000 ppm or over. The exit steam quality must be adjusted downward to about 75% in order to be sure that hot spots do not occur in the tubes with consequent solids deposition and failure. If a superheater is to be used, it is important that the steam issuing from the steam separator be of high quality, preferably 99.9% and of high purity, meaning very low concentrations of silica in vapor form in the high-pressure steam. If high silica feedwater is used, it should be carefully examined, because, if the steam is to be used for steam turbine operation, it is important that the steam from the steam separator be scrubbed using demineralized water or condensate to remove the vaporized silica. Two stages of scrubbing may be required depending on the silica concentration.

A modification of the spillover cycle is in common use in the oil fields for the production of high-pressure steam from low-quality water to be used for an enhanced oil recovery operation. These steam generators can operate as high as 2500 psi.

These spillover cycles for a combined cycle operation can be successfully used providing feedwater quality is given attention and steam quality and purity can be controlled. This type of system is particularly valuable in marine applications, because owing to the lack of a drum, it is not sensitive to pitch and roll conditions.

66.4.2 Control

Figure 66.9 shows in schematic form a type of control system that will operate in conjunction with the spillover cycle. The steam separator serves as a water reservoir in which a water level is held constant. This water level is controlled by valve CV1, which permits water to discharge from a separator reservoir through a heat exchanger and then either to waste or back to the hot well or deaerator. Control valve CV2 is in the feed supply line to the heat exchanger, which supplies water to the inlet of the tube assembly. There is a flow meter in each of these two lines that sends out signals proportional to the respective flow rates. These signals are impressed on pulse converters, which convert the electrical

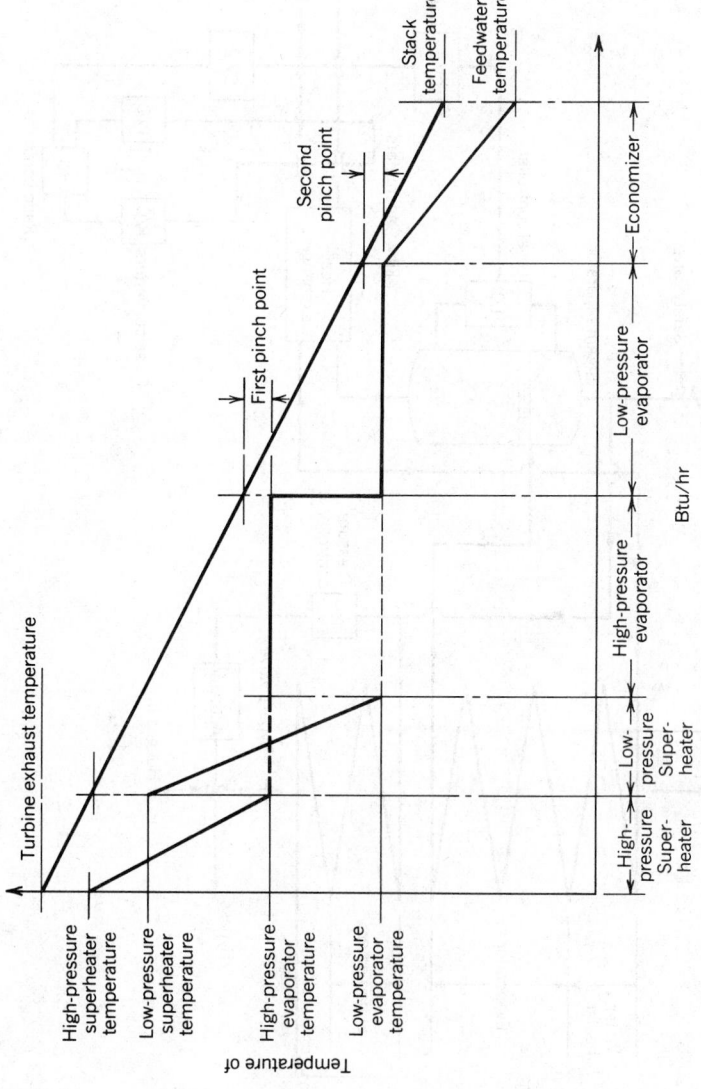

Fig. 66.8 Diagram showing pinch-point effects.

Turbine exhaust temperature

Temperature of

High-pressure superheater temperature

Low-pressure superheater temperature

High-pressure evaporator temperature

Low-pressure evaporator temperature

First pinch point

Second pinch point

Stack temperature

Feedwater temperature

High-pressure Super-heater

Low-pressure Super-heater

High-pressure evaporator

Low-pressure evaporator

Economizer

Btu/hr

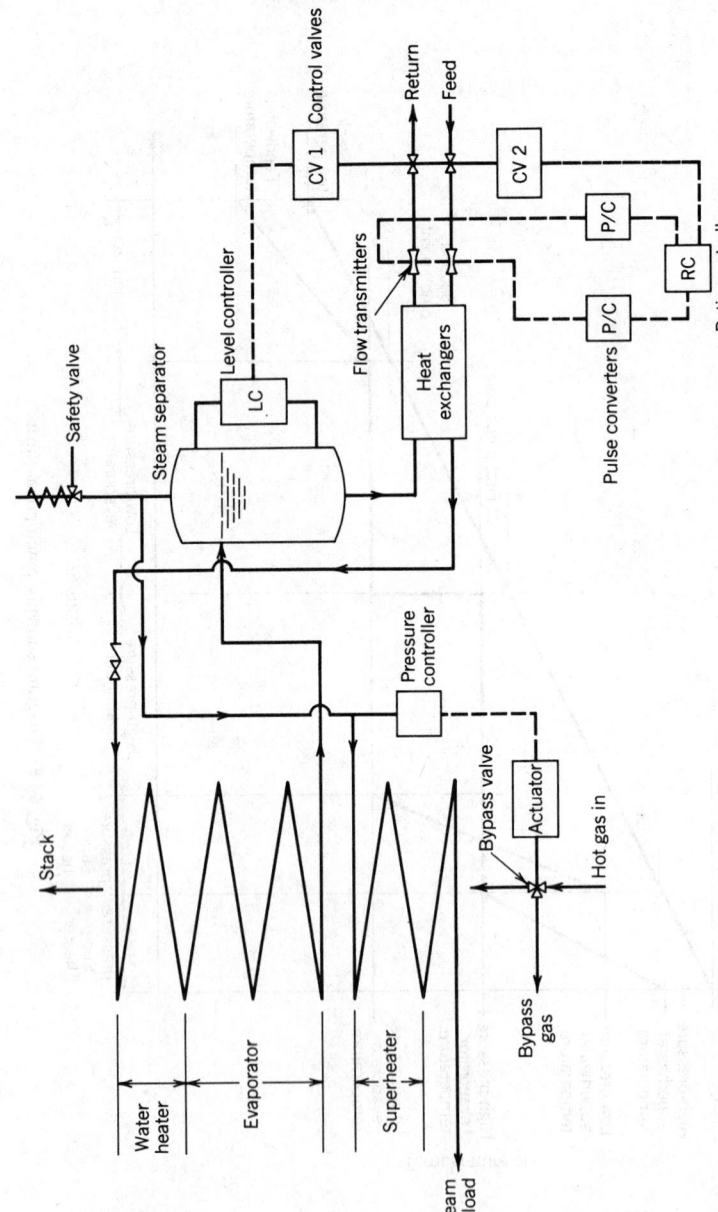

Fig. 66.9 Schematic of once-through spillover cycle heat-recovery system.

pulses from the flow meter to a direct current 3–20 mA. These two outputs are then impressed on a ratio controller, which is set at a predetermined point to proportion the feed to the spillover. For example, if the spillover rate is 20% of the total feed, then the ratio controller is set at 5. In this diagram, pressure is controlled by controlling the hot gas bypass. In most cases this would be an overriding control, since the system pressure would be set by a steam turbine or other load on the system. Steam temperature is controlled by bypassing saturated steam to the superheater outlet as was explained previously.

The drawing shows a distinct break between the liquid heater and the evaporator. However, since this is one continuous tube, this point is constant only for a given flow and pressure condition. At all other flow conditions the point at which evaporation begins moves up or down as the case might be in the length of the circuit.

A single-tube system can be used on flow rates up to approximately 50,000 lb/hr. However, the tubes become quite large at this point, usually in the range of 2½–3½ in. OD. Therefore, for larger installations, parallel tubes must be used. This brings up an important control problem, namely, the equal distribution of water among the several parallel circuits. A common method of accomplishing this is by orifices at the inlet of the liquid-phase heater. Inlet restrictions increase stability. Therefore, each tube in a multicircuit unit must have a flow restrictor. The restrictor must be located where no boiling occurs and where there is true single-phase flow. Stable flow can be achieved with a pressure drop in the flow restrictor equal to 50–150% of the two-phase pressure drop of the entire liquid-phase heater and evaporator circuit.

Hydrodynamic flow instability is a potential problem in forced recirculation evaporators. Instability can result in periodic oscillations in the steam drum water level (slugging steam flow and steam pressure). Instability may be either static or dynamic. Density wave instability is the most common type of dynamic instability. The effect of various parameters have been investigated, and the more pertinent ones are:

1. An inlet restriction increases stability.
2. An exit restriction decreases stability.
3. An increase in two-phase-flow pressure drop decreases stability.
4. An increase in steam pressure decreases stability.

In summary, the potential problem of hydrodynamic instability has been given thorough consideration and features such as inlet flow restrictors, large-diameter tubes, and subcooling at the evaporator section inlet have been used successfully to produce stable operation.

In addition to providing a pressure drop, a flow restrictor must fulfill two other requirements:

1. There can be no accumulation of suspended solids at the restrictor, or it will plug.
2. The restrictor must be fully drainable, which is important during acid cleaning operations.

Therefore, the restrictor should not be a simple thin plate or sharp-edged orifice, but a larger diameter section that produces a required drop in pressure, but is not susceptible to plugging.

As was previously stated an exit restriction decreases stability. Therefore, the steam separator, which is at the exit end of the evaporation section, should have a minimum pressure drop consistent with efficient separation. It is also important to keep the evaporator section at a low-pressure drop. For this reason, a larger tube or multiple parallel circuits are used.

Balanced circuits are vitally important in the spillover cycle. However, since this cycle operates with 10–25% excess water, it is still possible for the tube circuits to be slightly unequal and yet not be in superheat at the entrance to the steam separator. It is important that wet steam and not superheated steam enters the separator, because the solubles in the feedwater appear in concentrated form at the end of the steam-generating section.

It is important to locate the inlet restriction in a straight section of tube and not in a section with a bend or elbow close to the restrictor, since turbulence and impingement on the tube will cause excessive erosion of the tube wall. In some cases, the thicker tube wall is provided in areas of possible erosion. Since the inlet restrictor must necessarily operate and be exposed to high water velocity, it should be constructed of a hard material such as stellite, carboloy, or a ceramic material, in order to reduce the possibility of erosion. If excessive erosion occurs, then the orifice will lose its flow pressure drop characteristic and be much less effective as a stabilizer.

Another, more sophisticated means for securing equal flow in parallel circuits is the use of flow meters at each entrance and control valves ahead of the flow meter so that the impulses from these flow meters can be processed by a microprocessor, with a corrective signal going to the respective control valves in order to make all of the parallel circuits equal to a standard which must be used in order that the various circuits can be compared.

66.5   ONCE-THROUGH MONOTUBE CYCLE

66.5.1   Description

The forced circulation once-through monotube cycle consists essentially of a single-tube boiler with no intervening drums or separators. See Fig. 66.10 for a typical schematic of the system. All the water entering the inlet is heated, evaporated, and superheated within the tube. In other words, water enters in one end of the tube and superheated steam leaves the other end. More than one tube may be used in parallel. The number of tubes is governed by the steam requirement. The same criteria for parallel flow operation and control apply here as with the forced circulation spillover cycle explained previously. This cycle depends on having steam in superheat for control. A once-through boiler for production of saturated steam only will not control, because the degree of wetness cannot be accurately controlled. Water feed must be accurately proportional to heat input, otherwise the system will cycle from wet to superheat. A version of this cycle is known as the normalizer system. In the normalizer system, approximately 90% of the total water is introduced into the inlet. The remaining 10% is introduced into the superheater zone, and the proportion is then controlled by steam temperature. The normalizer system is more stable than a straight monotube, although it has been known to flood and starve when loads change abruptly. Tube circuit lengths must be kept as short as possible to help reduce the time factor for response to the control signal. For the same reasons, water and steam velocities must be high (30–50 ft/sec).

66.5.2   Applications

The monotube design is used in modified form by most supercritical utility boilers. Many smaller boilers of this type have been built for prototype plants, testing, vehicles, etc. At this time, at least one company is manufacturing a multiple-circuit monotube-type boiler for use in a combined cycle plant; while the factory test results appear to be satisfactory, it has yet to be approved in actual practice. This type of system is the most responsive of all of the steam-generating systems discussed here. Accordingly, its control is extremely critical.

66.5.3   Water-Quality Requirements

Owing to the fact that it is truly a monotube boiler in that all evaporation takes place in a single tube and that the superheating also takes place within this same tube, any water solids present in

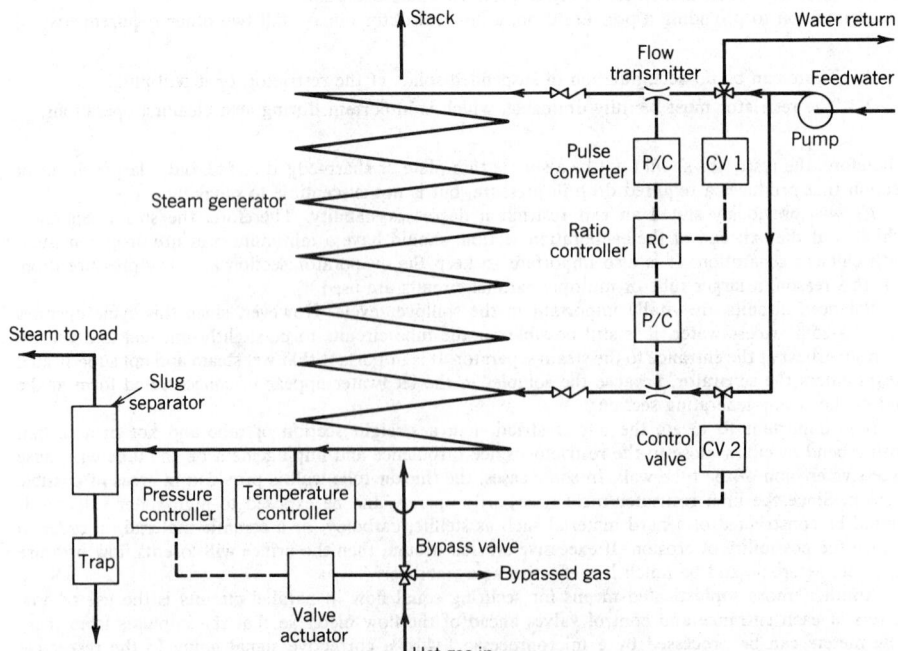

Fig. 66.10 Schematic of once-through monotube heat-recovery boiler system.

the feedwater will deposit out on the tube. Accordingly, feedwater for this type of boiler must be of the highest quality, preferably polished, demineralized water or condensate filtered to remove suspended particles before it is used as boiler feedwater. Preferably, all piping valves, tanks, etc., that come in contact with the feedwater should be of stainless steel, including the boiler tubes, in order to minimize the possibility of introducing rust or scale into the system that will plug the inlet orifices.

66.5.4　Control

Figure 66.10 illustrates a typical control system for a normalizer monotube cycle. In this system flow meters are placed at the boiler entrance and at the normalizer entrance. Signals from these flow meters are impressed on pulse converters which in turn activate a ratio controller. The ratio controller in turn sends a control signal to CV1 which is the control valve for the feedwater to the boiler. CV1 is actually a three-way valve that bypasses water not required by the steam demand. The second stream of water is controlled and goes to flow meter 2, which meters the water going into the superheater as normalizer water. CV2 controls this flow. CV2 is controlled by the outlet steam temperature. Steam pressure is controlled by bypassing the hot gas exhaust. Note that there is a separator at the outlet of the superheater. This is placed there to guard equipment such as turbines receiving the steam from water slugs. If there is an upset in the water-control system and for a momentary period the water feed exceeds the requirement, a water slug can pass through the entire system and emerge at the outlet. The separator at the outlet is intended to catch this slug and divert the water back to the condenser in order to protect the equipment receiving the steam.

　　The control of multiple-circuit monotube boilers is especially critical and complex. Care must be taken to ensure that all circuits are equal in heat absorption and waterflow. The use of orifices at the tube inlets as explained in Section 66.4 is recommended to promote circuit stability. The same pressure drop considerations apply.

　　The monotube system operates better at higher pressures than at lower pressures. For this reason it can operate successfully in the supercritical region. It is actually more stable at the higher pressures.

BIBLIOGRAPHY

J. L. Boyen, *Practical Heat Recovery,* Wiley, New York, 1975.

J. L. Boyen, *Thermal Energy Recovery,* Wiley, New York, 1979.

O. W. Eshback and M. Sonders, *Handbook of Engineering Fundamentals,* Wiley, New York, 1975.

CHAPTER 67

HEAT EXCHANGERS, VAPORIZERS, CONDENSERS

JOSEPH W. PALEN

Heat Transfer Research, Inc.
Alhambra, California

67.1 HEAT EXCHANGER TYPES AND CONSTRUCTION

Heat exchangers permit exchange of energy from one fluid to another, usually without permitting physical contact between the fluids. The following configurations are commonly used in the power and process industries.

67.1.1 Shell and Tube Heat Exchangers

Shell and tube heat exchangers normally consist of a bundle of tubes fastened into holes drilled in metal plates called tubesheets. The tubes may be rolled into grooves in the tubesheet, welded to the tubesheet, or both to ensure against leakage. When possible, U-tubes are used, requiring only one tubesheet. The tube bundle is placed inside a large pipe called a shell, see Fig. 67.1. Heat is exchanged between a fluid flowing inside the tubes and a fluid flowing outside the tubes in the shell.

When the tubeside heat-transfer coefficient is as high as three times the shellside heat-transfer coefficient, it may be advantageous to use low integral finned tubes. These tubes can have outside heat-transfer coefficients as high as plain tubes, or even higher, but increase the outside heat-transfer area by a factor of about 2.5–4. For design methods using finned tubes, see Ref. 11 for single-phase heat exchangers and Ref. 14 for condensers. Details of construction practices are described by Saunders.[58]

The Tubular Exchanger Manufacturers Association (TEMA) provides a manual of standards for construction of shell and tube heat exchangers,[1] which contains designations for various types of shell and tube heat exchanger configurations. The most common types are summarized below.

E-Type

The E-type shell and tube heat exchanger, illustrated in Figs. 67.2a and 67.2b, is the workhorse of the process industries, providing economical rugged construction and a wide range of capabilities.

Baffles support the tubes and increase shellside velocity to improve heat transfer. More than one pass is usually provided for tubeside flow to increase the velocity, Fig. 67.2a. However, for some cases, notably vertical thermosiphon vaporizers, a single tubepass is used, as shown in Fig. 67.2b.

The E-type shell is usually the first choice of shell types because of lowest cost, but sometimes requires more than the allowable pressure drop, or produces a temperature "pinch" (see Section 67.4.4), so other, more complicated types are used.

F-Type Shell

If the exit temperature of the cold fluid is greater than the exit temperature of the hot fluid, a temperature cross is said to exist. A slight temperature cross can be tolerated in a multitubepass E-type shell (see below), but if the cross is appreciable, either units in series or complete countercurrent flow is required. A solution sometimes used is the F-type or two-pass shell, as shown in Fig. 67.3.

The F-type shell has a number of potential disadvantages, such as thermal and fluid leakage around the longitudinal baffle and high pressure drop, but it can be effective in some cases if well designed.

J-Type

When an E-type shell cannot be used because of high pressure drop, a J-type or divided flow exchanger, shown in Fig. 67.4, is considered. Since the flow is divided and the flow length is also cut in half,

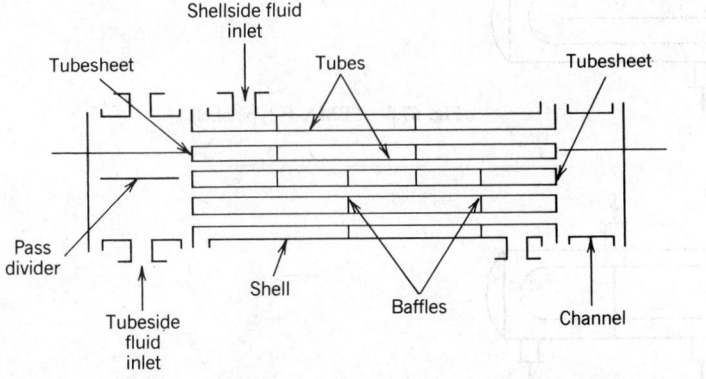

Fig. 67.1 Schematic illustration of shell and tube heat exchanger construction.

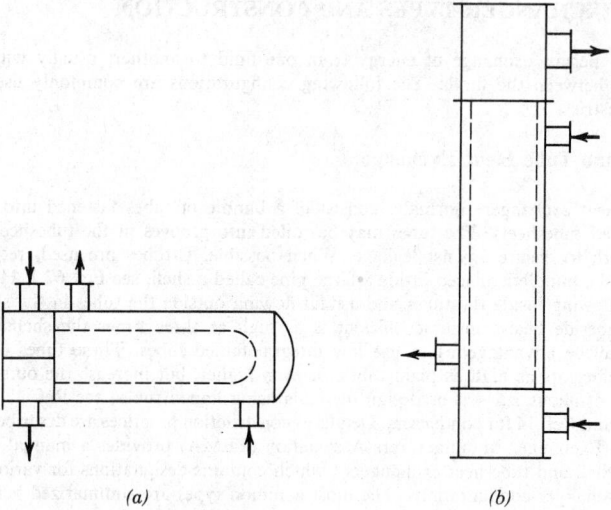

Fig. 67.2   TEMA E-type shell: (*a*) horizontal multitubepass; (*b*) vertical single tubepass.

the shellside pressure drop is only about one-eighth to one-fifth that of an E-type shell of the same dimensions.

X-Type

When a J-type shell would still produce too high a pressure drop, an X-type shell, shown in Fig. 67.5, may be used. This type is especially applicable for vacuum condensers, and can be equipped with integral finned tubes to counteract the effect of low shellside velocity on heat transfer. It is usually necessary to provide a flow distribution device under the inlet nozzle.

G-Type

This shell type, shown in Fig. 67.6, is sometimes used for horizontal thermosiphon shellside vaporizers. The horizontal baffle is used especially for boiling range mixtures and provides better flow distribution than would be the case with the X-type shell. The G-type shell also permits a larger temperature cross than the E-type shell with about the same pressure drop.

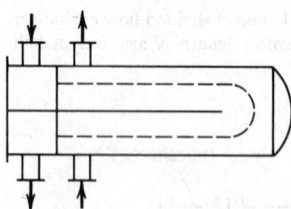

Fig. 67.3   TEMA F-type shell.

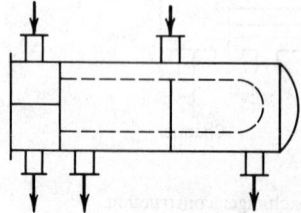

Fig. 67.4   TEMA J-type shell.

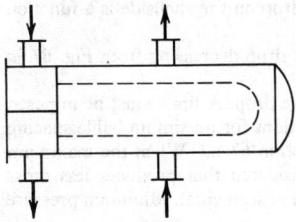

Fig. 67.5 TEMA X-type shell.

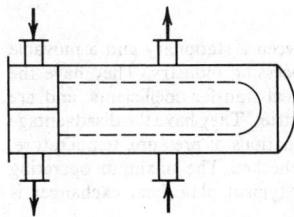

Fig. 67.6 TEMA G-type shell.

H-Type

If a G-type is required but pressure drop would be too high, an H-type may be used. This configuration is essentially just two G-types in parallel, as shown in Fig. 67.7.

K-Type

This type is used exclusively for kettle reboilers and vaporizers, and is characterized by the oversized shell intended to separate vapor and liquid phases, Fig. 67.8. Shell-sizing relationships are given in Ref. 25. Usually, the shell diameter is about 1.6–2.0 times the bundle diameter.

Baffle Types

Baffles are used to increase velocity of the fluid flowing outside the tubes ("shellside" fluid) and to support the tubes. Higher velocities have the advantage of increasing heat transfer and decreasing fouling (material deposit on the tubes), but have the disadvantage of increasing pressure drop (more

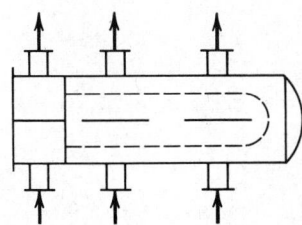

Fig. 67.7 TEMA H-type shell.

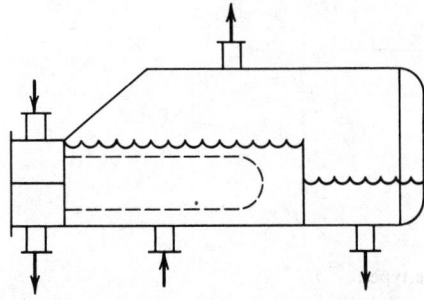

Fig. 67.8 TEMA K-type shell.

energy consumption per unit of fluid flow). The amount of pressure drop on the shellside is a function of baffle spacing, baffle cut, and baffle type.

Baffle types commonly used are shown in Fig. 67.9, with pressure drop decreasing from Fig. 67.9a to Fig. 67.9c.

Baffle spacing is increased when it is necessary to decrease pressure drop. A limit must be imposed to prevent tube sagging or flow-induced tube vibration. Recommendations for maximum baffle spacing are given in Ref. 1. Tube vibration is discussed in more detail in Section 67.4.2. When the maximum spacing still produces too much pressure drop, a baffle type is considered that produces less cross flow and more longitudinal flow, for example, double segmental instead of segmental. Minimum pressure drop is obtained if baffles are replaced by rod-type tube supports.[52]

67.1.2 Plate-Type Heat Exchangers

Composed of a series of corrugated or embossed plates clamped between a stationary and a movable support plate, these exchangers were originally used in the food-processing industry. They have the advantages of low fouling rates, easy cleaning, and generally high heat-transfer coefficients, and are becoming more frequently used in the chemical process and power industries. They have the disadvantage that available gaskets for the plates are not compatible with all combinations of pressure, temperature, and chemical composition. Suitability for specific applications must be checked. The maximum operating pressure is usually considered to be about 1.5 MPa (220 psia).[3] A typical plate heat exchanger is shown in Fig. 67.10.

67.1.3 Spiral Plate Heat Exchangers

These exchangers are also becoming more widely used, despite limitations on maximum size and maximum operating pressure. They are made by wrapping two parallel metal plates, separated by spacers, into a spiral to form two concentric spiral passages. A schematic example is shown in Fig. 67.11.

Spiral plate heat exchangers can provide completely countercurrent flow, permitting temperature crosses and close approaches, while maintaining high velocity and high heat-transfer coefficients. Since all flow for each fluid is in a single channel, the channel tends to be flushed of particles by the flow, and the exchanger can handle sludges and slurries more effectively than can shell and tube heat exchangers. The most common uses are for difficult-to-handle fluids with no phase change. However, the low-pressure-drop characteristics are beginning to promote some use in two-phase flow as condensers and reboilers.

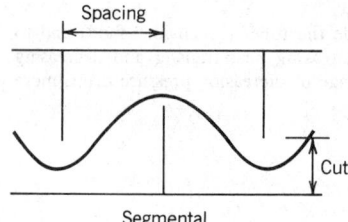

Segmental

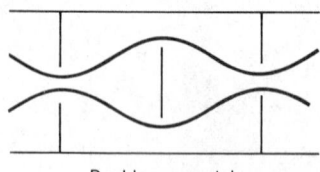

Double segmental

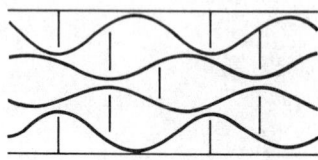

Triple segmental   **Fig. 67.9**   Baffle types.

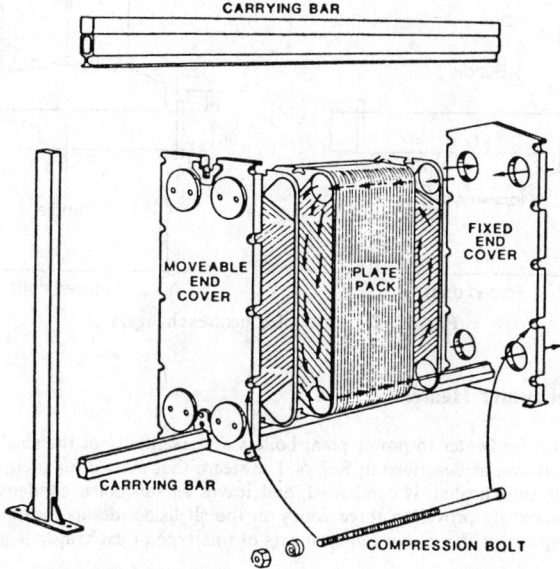

Fig. 67.10 Typical plate-type heat exchanger.

67.1.4 Air-Cooled Heat Exchangers

It is sometimes economical to condense or cool hot streams inside tubes by blowing air across the tubes rather than using water or other cooling liquid. They usually consist of a horizontal bank of finned tubes with a fan at the bottom (forced draft) or top (induced draft) of the bank, as illustrated schematically in Fig. 67.12.

Tubes in air-cooled heat exchangers (Fig. 67.12) are often 1 in. (25.4 mm) in outside diameter with $\frac{5}{8}$ in. (15.9 mm) high annular fins, 0.4–0.5 mm thick. The fins are usually aluminum and may be attached in a number of ways, ranging from tension wrapped to integrally extruded (requiring a steel or alloy insert), depending on the severity of service. Tension wrapped fins have an upper temperature limit (~300°F) above which the fin may no longer be in good contact with the tube, greatly decreasing the heat-transfer effectiveness. Various types of fins and attachments are illustrated in Fig. 67.13.

A more detailed description of air-cooled heat exchanger geometries is given Refs. 2 and 3.

67.1.5 Compact Heat Exchangers

The term compact heat exchanger normally refers to one of the many types of plate fin exchangers used extensively in the aerospace and cryogenics industries. The fluids flow alternately between parallel plates separated by corrugated metal strips that act as fins and that may be perforated or interrupted to increase turbulence. Although relatively expensive to construct, these units pack a very large amount of heat-transfer surface into a small volume, and are therefore used when exchanger volume or weight must be minimized. A detailed description with design methods is given in Ref. 4.

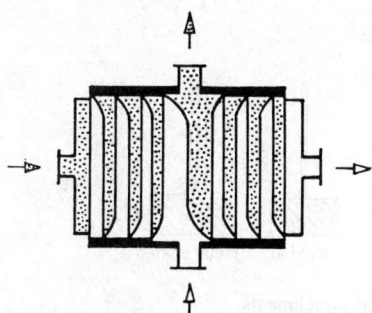

Fig. 67.11 Spiral plate heat exchanger.

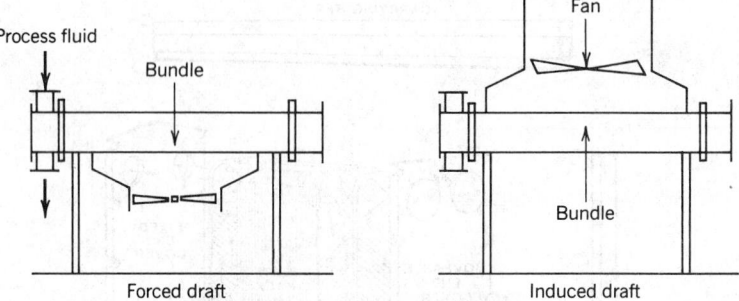

Fig. 67.12   Air-cooled heat exchangers.

67.1.6   Boiler Feedwater Heaters

Exchangers to preheat feedwater to power plant boilers are essentially of the shell and tube type but have some special features, as described in Ref. 5. The steam that is used for preheating the feedwater enters the exchanger superheated, is condensed, and leaves as subcooled condensate. More effective heat transfer is achieved by providing three zones on the shellside: desuperheating, condensing, and subcooling. A description of the design requirements of this type of exchanger is given in Ref. 5.

67.1.7   Recuperators and Regenerators

These heat exchangers are used typically to conserve heat from furnace off-gas by exchanging it against the inlet air to the furnace. A recuperator does this in the same manner as any other heat exchanger except the construction may be different to comply with requirements for low pressure drop and handling of the high-temperature, often dirty, off-gas stream.

The regenerator is a transient batch-type exchanger in which packed beds are alternately switched from the hot stream to the cold stream. A description of the operating characteristics and design of recuperators and regenerators is given in Refs. 6 and 59.

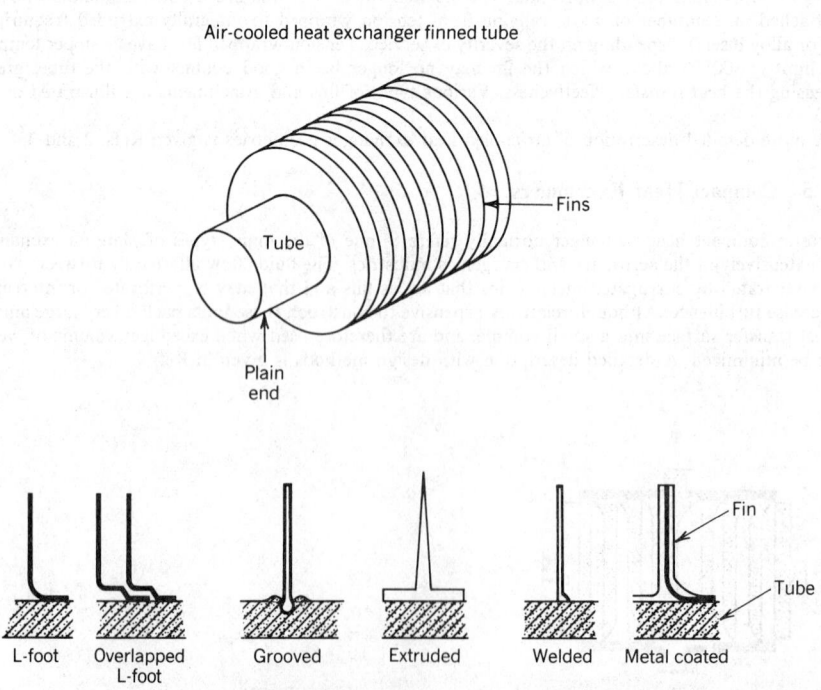

Fig. 67.13   Typical finned tube and attachments.

67.2 ESTIMATION OF SIZE AND COST

In determining the overall cost of a proposed process plant or power plant, the cost of heat exchangers is of significant importance. Since cost is roughly proportional to the amount of heat-transfer surface required, some method of obtaining an estimate of performance is necessary, which can then be translated into required surface. The term "surface" refers to the total area across which the heat is transferred. For example, with shell and tube heat exchangers "surface" is the tube outside circumference times the tube length times the total number of tubes. Well-known basic equations taken from Newton's law of cooling relate the required surface to the available temperature difference and the required heat duty.

67.2.1 Basic Equations for Required Surface

The following well-known equation is used:

$$A_o = \frac{Q}{U_o \times \text{MTD}} \tag{67.1}$$

The required duty (Q) is related to the energy change of the fluids:

(a) *Sensible Heat Transfer*

$$Q = W_1 C_{p\,1}(T_2 - T_1) \tag{67.2a}$$

$$= W_2 C_{p\,2}(t_1 - t_2) \tag{67.2b}$$

where W_1 = flow rate of "hot" fluid
  W_2 = flow rate of "cold" fluid
  $C_{p\,1}$ = heat capacity of hot fluid
  $C_{p\,2}$ = heat capacity of cold fluid
  T_1, T_2 = inlet and outlet temperatures of hot fluid
  t_1, t_2 = inlet and outlet temperatures of cold fluid

(b) *Latent Heat Transfer*

$$Q = W\lambda \tag{67.3}$$

where W = flow rate of boiling or condensing fluid
  λ = latent heat of respective fluid

The mean temperature difference (MTD) and the overall heat transfer coefficient (U_o) in Eq. (67.1) are discussed in Sections 67.2.2 and 67.2.3, respectively. Once the required surface, or area, (A_o) is obtained, heat exchanger cost can be estimated. A comprehensive discussion on cost estimation for several types of exchangers is given in Ref. 7. Cost charts for small- to medium-sized shell and tube exchangers, developed in 1982, are given in Ref. 8.

67.2.2 Mean Temperature Difference

The mean temperature difference (MTD) in Eq. (67.1) is given by the equation

$$\text{MTD} = \frac{F(T_A - T_B)}{\ln(T_A/T_B)} \tag{67.4}$$

where

$$T_A = T_1 - t_2 \tag{67.5}$$

$$T_B = T_2 - t_1 \tag{67.6}$$

The temperatures (T_1, T_2, t_1, t_2) are illustrated for the base case of countercurrent flow in Fig. 67.14.

The factor F in Eq. (67.4) is the multitubepass correction factor. It accounts for the fact that heat exchangers with more than one tubepass can have some portions in concurrent flow or cross flow, which produce less effective heat transfer than countercurrent flow. Therefore, the factor F is less than 1.0 for multitubepass exchangers, except for the special case of isothermal boiling or condensing streams for which F is always 1.0. Charts for calculating F are available in most heat-transfer textbooks. A comprehensive compilation for various types of exchangers is given by Taborek.[9]

In a properly designed heat exchanger, it is unusual for F to be less than 0.7, and if there is no temperature cross ($T_2 > t_2$), F will be 0.8 or greater. As a first approximation for preliminary sizing and cost estimation, F may be taken as 0.9 for multitubepass exchangers with temperature change of both streams and 1.0 for other cases.

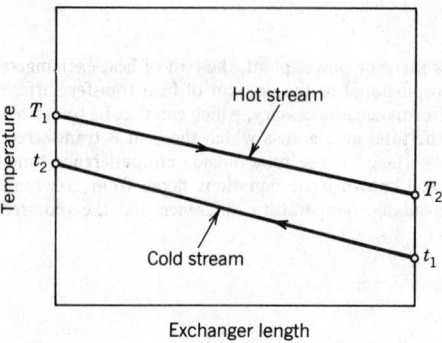

Fig. 67.14 Temperature profiles illustrated for countercurrent flow.

67.2.3 Overall Heat-Transfer Coefficient

The factor (U_o) in Eq. (67.1) is the overall heat-transfer coefficient. It may be calculated by procedures described in Section 67.3, and is the reciprocal of the sum of all heat-transfer resistances, as shown in the equation

$$U_o = 1/(R_{h_o} + R_{f_o} + R_w + R_{h_i} + R_{f_i}) \qquad (67.7)$$

where

$$R_{h_o} = 1/h_o \qquad (67.8)$$

$$R_{h_i} = (A_o/A_i h_i) \qquad (67.9)$$

$$R_w = \frac{A_o x_w}{A_m k_w} \qquad (67.10)$$

and h_o, h_i = outside, inside heat transfer coefficients
    R_{f_o}, R_{f_i} = outside, inside fouling resistances
    A_o, A_i, A_m = outside, inside, mean surface areas
        x_w = wall thickness
        k_w = wall thermal conductivity

Calculation of the heat-transfer coefficients h_o and h_i can be time consuming, since they depend on the fluid velocities, which, in turn, depend on the exchanger geometry. This is usually done now by computer programs that guess correct exchanger size, calculate heat-transfer coefficients, check size, adjust, and reiterate until satisfactory agreement between guessed and calculated size is obtained. For first estimates by hand before size is known, values of h_o and h_i, as well as values of the fouling resistances, R_{f_o} and R_{f_i}, are recommended by Bell for shell and tube heat exchangers.[10]

Very rough, first approximation values for the overall heat-transfer coefficient are given in Table 67.1.

Table 67.1 Approximate Values for Overall Heat Transfer Coefficient of Shell and Tube Heat Exchangers (Including Allowance for Fouling)

	U_o	
Fluids	Btu/hr·ft²·°F	W/m²·K
Water–water	250	1400
Oil–water	75	425
Oil–oil	45	250
Gas–oil	15	85
Gas–water	20	115
Gas–gas	10	60

67.2.4 Pressure Drop

In addition to calculation of the heat-transfer surface required, it is usually necessary to consider the pressure drop consumed by the heat exchanger, since this enters into the overall cost picture. Pressure drop is roughly related to the individual heat-transfer coefficients by an equation of the form,

$$\Delta P = Ch^m + EX \tag{67.11}$$

where ΔP = shellside or tubeside pressure drop
$\quad h$ = heat-transfer coefficient
$\quad C$ = coefficient depending on geometry
$\quad m$ = exponent depending on geometry—always greater than 1.0, and usually about 3.0
$\quad EX$ = extra pressure drop from inlet, exit, and pass turnaround momentum losses

Pressure drop is sensitive to the type of exchanger selected. In the final design it is attempted, where possible, to define the exchanger geometry so as to use all available pressure drop and thus maximize the heat-transfer coefficient. This procedure is subject to some constraints, however, as follows. The product of density times velocity squared ρv^2 is limited to minimize the possibility of erosion or tube vibration. A limit often used is $\rho v^2 < 4000$ lbm/ft·sec². This results in a velocity for liquids in the range of 7–10 ft/sec. For flow entering the shellside of an exchanger and impacting the tubes, an impingement plate is recommended to prevent erosion if $\rho v^2 > 1500$. Other useful design recommendations may be found in Ref. 1.

For condensing vapors, pressure drop should be limited to a fraction of the operating pressure for cases with close temperature approach to prevent severe decrease of the MTD owing to lowered equilibrium condensing temperature. As a safe "rule of thumb," the pressure drop for condensing is limited to about 10% of the operating pressure. For other cases, "reasonable" design pressure drops for heat exchangers roughly range from about 5 psi for gases and boiling liquids to as high as 20 psi for pumped nonboiling liquids.

67.3 RATING METHODS

After the size and basic geometry of a heat exchanger has been proposed, the individual heat-transfer coefficients h_o and h_i may be calculated based on actual velocities, and the required surface may be checked, based on these updated values. The pressure drops are also checked at this stage. Any inadequacies are adjusted and the exchanger is rechecked. This process is known as "rating." Different rating methods are used depending on exchanger geometry and process type, as covered in the following sections.

67.3.1 Shell and Tube Single-Phase Exchangers

Before the individual heat-transfer coefficients can be calculated, the heat exchanger tube geometry, shell diameter, shell type, baffle type, baffle spacing, baffle cut, and number of tubepasses must be decided. As stated above, lacking other insight, the simplest exchanger—E-type with segmental baffles—is tried first.

Tube Length and Shell Diameter

For shell and tube exchangers the tube length is normally about 5–8 times the shell diameter. Tube lengths are usually 8–20 ft long in increments of 2 ft. However, very large size exchangers with tube lengths up to 40 ft are more frequently used as economics dictate smaller MTD and larger plants. A reasonable trial tube length is chosen and the number of tubes (NT) required for surface A_o, Section 67.2, is calculated as follows:

$$NT = \frac{A_o}{a_o L} \tag{67.12}$$

where a_o = the surface/unit length of tube.
For plain tubes (as opposed to finned tubes),

$$a_o = \pi D_o \tag{67.13}$$

where D_o = the tube outside diameter
$\quad L$ = the tube length

The tube bundle diameter (D_b) can be determined from the number of tubes, but also depends on the number of tubepasses, tube layout, and bundle construction. Tube count tables providing this information are available from several sources. Accurate estimation equations are given by Taborek.[11]

A simple basic equation that gives reasonable first approximation results for typical geometries is the following:

$$D_b = P_t \left(\frac{NT}{\pi/4} \right)^{0.5} \tag{67.14}$$

where P_t = tube pitch (spacing between tube diameters). Normally, P_t/D_o = 1.25, 1.33, or 1.5.

The shell diameter D_s is larger than the bundle diameter D_b by the amount of clearance necessary for the type of bundle construction. Roughly, this clearance ranges from about 0.5 in. for U-tube or fixed tubesheet construction to 3–4 in. for pull-through floating heads, depending on the design pressure and bundle diameter. (For large clearances, sealing strips are used to prevent flow bypassing the bundles.) After the bundle diameter is calculated, the ratio of length to diameter is checked to see if it is in an acceptable range, and the length is adjusted if necessary.

Baffle Spacing and Cut

Baffle spacing L_{bc} and cut B_c (see Fig. 67.9) cannot be decided exactly until pressure drop is evaluated. However, a reasonable first guess ratio of baffle spacing to shell diameter (L_{bc}/D_s) is about 0.45. The baffle cut (B_c, a percentage of D_s) required to give good shellside distribution may be estimated by the following equation:

$$B_c = 16.25 + 18.75 \left(\frac{L_{bc}}{D_s} \right) \tag{67.15}$$

For more detail, see the recommendations of Taborek.[11]

Cross-Sectional Flow Areas and Flow Velocities

The cross-sectional flow areas for tubeside flow S_t and for shellside flow S_s are calculated as follows:

$$S_t = \left(\frac{\pi}{4} D_i^2 \right) \left(\frac{NT}{NP} \right) \tag{67.16}$$

where D_i = tube inside diameter
NP = number of tube passes

$$S_s = 0.785(D_b)(L_{bc})(P_t - D_o)/P_t \tag{67.17}$$

where L_{bc} = baffle spacing.

Equation (67.17) is approximate in that it neglects pass partition gaps in the tube field, it approximates the bundle average chord, and it assumes an equilateral triangular layout. For more accurate equations see Ref. 11.

The tubeside velocity V_t and the shellside velocity V_s are calculated as follows:

$$V_t = \frac{W_t}{S_t \rho_t} \tag{67.18}$$

$$V_s = \frac{W_s}{S_s \rho_s} \tag{67.19}$$

where ρ_t = density of the tubeside fluid
ρ_s = density of the shellside fluid
W_t, W_s = tubeside, shellside flow rates

Heat-Transfer Coefficients

The individual heat-transfer coefficients, h_o and h_i, in Eq. (67.1) can be calculated with reasonably good accuracy (±20–30%) by semiempirical equations found in several design-oriented textbooks.[11,12] Simplified approximate equations are the following:

(a) Tubeside Flow

$$Re = \frac{D_o V_t \rho_t}{\mu_t} \tag{67.20}$$

where μ_t = tubeside fluid viscosity.
If Re < 2000, laminar flow,

$$h_i = 1.86 \left(\frac{k_t}{D_i} \right) \left(Re\, Pr\, \frac{D_i}{L} \right)^{0.33} \tag{67.21}$$

If Re > 10,000, turbulent flow,

$$h_i = 0.024 \left(\frac{k_t}{D_i}\right) \mathrm{Re}^{0.8} \, \mathrm{Pr}^{0.4} \tag{67.22}$$

If 2000 < Re < 10,000, prorate linearly.

(b) *Shellside Flow*

$$\mathrm{Re} = \frac{D_o V_s \, \rho_s}{\mu_s} \tag{67.23}$$

If Re < 500, see Refs. 11 and 12.
If Re > 500,

$$h_o = 0.38 \, C_b \left(\frac{k_s}{D_o}\right) \mathrm{Re}^{0.6} \, \mathrm{Pr}^{0.33} \tag{67.24}$$

The term Pr is the Prandtl number and is calculated as $C_p \, \mu / k$.

The constant (C_b) in Eq. (67.24) depends on the amount of bypassing or leakage around the tube bundle.[13] As a first approximation, the values in Table 67.2 may be used.

Pressure Drop

Pressure drop is much more sensitive to exchanger geometry, and, therefore, more difficult to accurately estimate than heat transfer, especially for the shellside. The so-called Bell–Delaware method[11] is considered the most accurate method in open literature, which can be calculated by hand. The following very simplified equations are provided for a rough idea of the range of pressure drop, in order to minimize preliminary specification of unrealistic geometries.

(a) *Tubeside (contains about 30% excess for nozzles)*

$$\Delta P_t = \left[\frac{0.025(L)(\mathrm{NP})}{D_i} + 2(\mathrm{NP} - 1)\right] \frac{\rho_t V_t^2}{g_c} \tag{67.25}$$

where NP = number of tubepasses.

(b) *Shellside (contains about 30% excess for nozzles)*

$$\Delta P_s = \frac{0.24(L)(D_b)(\rho_s)(0.6V_s)^2}{g_c L_{bc} P_t} \tag{67.26}$$

where g_c = gravitational constant (4.17×10^8 for velocity in ft/hr and density in lb/ft^3).

67.3.2   Shell and Tube Condensers

The condensing vapor can be on either the shellside or tubeside depending on process constraints. The "cold" fluid is often cooling tower water, but can also be another process fluid, which is sensibly heated or boiled. In this section, the condensing-side heat-transfer coefficient and pressure drop are discussed. Single-phase coolants are handled, as explained in the last section. Boiling fluids will be discussed in the next section.

Selection of Condenser Type

The first task in designing a condenser, before rating can proceed, is to select the condenser configuration. Mueller[14] presents detailed charts for selection based on the criteria of system pressure, pressure drop, temperature, fouling tendency of the coolant, fouling tendency of the vapor, corrosiveness of the vapor, and freezing potential of the vapor. Table 67.3 is an abstract of the recommendations of Mueller.

Table 67.2   Approximate Bypass Coefficient for Heat Transfer, C_b

Bundle Type	C_b
Fixed tubesheet or U-tube	0.70
Split ring floating head, seal strips	0.65
Pull-through floating head, seal strips	0.55

Table 67.3   Condenser Selection Chart

Process Condition	Suggested Condenser Type[a]
Potential coolant fouling	HS/E, J, X
High condensing pressure	VT/E
Low condensing pressure drop	HS/J, X
Corrosive or very-high-temperature vapors	VT/E
Potential condensate freezing	HS/E
Boiling coolant	VS/E or HT/K, G, H

[a] V, vertical; H, horizontal; S, shellside condensation; T, tubeside condensation; /E, J, H, K, X, TEMA shell styles.

The suggestions in Table 67.3 may, of course, be ambiguous in case of more than one important criterion, for example, corrosive vapor together with a fouling coolant. In these cases, the most critical constraint must be respected, as determined by experience and engineering judgment. Corrosive vapors are usually put on the tubeside, and chemical cleaning used for the shellside coolant, if necessary. Since most process vapors are relatively clean (not always the case!), the coolant is usually the dirtier of the two fluids and the tendency is to put it on the tubeside for easier cleaning. Therefore, the most common shell and tube condenser is the shellside condenser using TEMA types E, J, or X, depending on allowable pressure drop; see Section 67.1. An F-type shell is sometimes specified if there is a large condensing range and a temperature cross (see below), but, owing to problems with the F-type, E-type units in series are often preferred in this case.

In addition to the above condenser types the vertical E-type tubeside condenser is sometimes used in a "reflux" configuration with vapor flowing up and condensate flowing back down inside the tubes. This configuration may be useful in special cases, such as when it is required to strip out condensable components from a vent gas that is to be rejected to the atmosphere. The disadvantage of this type of condenser is that the vapor velocity must be very low to prevent carryover of the condensate (flooding), so the heat-transfer coefficient is correspondingly low, and the condenser rather inefficient. Methods used to predict the limiting vapor velocity are given in Ref. 14.

Temperature Profiles

For a condensing pure component, if the pressure drop is less than about 10% of the operating pressure, the condensing temperature is essentially constant and the LMTD applies ($F = 1.0$) for the condensing section. If there are desuperheating and subcooling sections,[5] the MTD and surface for these sections must be calculated separately. For a condensing mixture, with or without noncondensables, the temperature profile of the condensing fluid with respect to fraction condensed should be calculated according to vapor–liquid equilibrium (VLE) relationships.[15] A number of computer programs are available to solve VLE relationships; a version suitable for programmable calculator is given in Ref. 16.

Calculations of the condensing temperature profile may be performed either integrally, which assumes vapor and liquid phases are well mixed throughout the condenser, or differentially, which assumes separation of the liquid phase from the vapor phase. In most actual condensers the phases are mixed near the entrance where the vapor velocity is high and separated near the exit where the vapor velocity is lower. The "differential" curve produces a lower MTD than the "integral" curve and is safer to use where separation is expected.

For most accuracy, condensers are rated incrementally by stepwise procedures such as those explained by Mueller.[14] These calculations are usually performed by computers.[17] As a first approximation, to get an initial size, a straight-line temperature profile is often assumed for the condensing section (not including desuperheating or subcooling sections!). As illustrated in Fig. 67.15, the true condensing curve is usually more like curve I, which gives a larger MTD than the straight line, curve II, making the straight-line approximation conservative. However, a curve such as curve III is certainly possible, especially with immiscible condensates, for which the VLE should always be calculated. For the straight-line approximation, the condensing heat-transfer coefficient is calculated at average conditions, as shown below.

Heat-Transfer Coefficients, Pure Components

For condensers, it is particularly important to be able to estimate the two-phase flow regime in order to predict the heat-transfer coefficient accurately. This is because completely different types of correlations are required for the two major flow regimes.

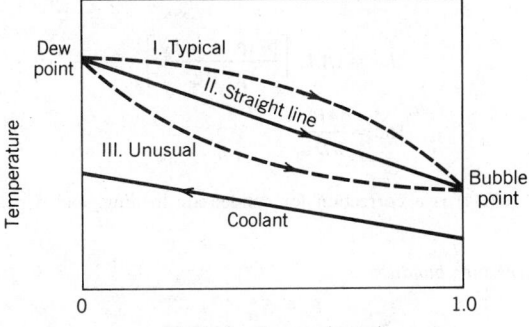

Fig. 67.15 Condensation profiles illustrated.

Shear Controlled Flow. The vapor shear force on the condensate is much greater than the gravity force. This condition can be estimated, according to Ref. 18, as,

$$J_g > 1.5 \tag{67.27}$$

where

$$J_g = \left[\frac{Gy}{gD_j \rho_v (\rho_l - \rho_v)} \right]^{0.5} \tag{67.28}$$

$D_j = D_i$ for tubeside condensation
  $= P_t - D_o$ for shellside condensation
G = total flow mass velocity
y = weight fraction vapor
ρ_v = vapor density
ρ_l = liquid density
g = acceleration of gravity

For shear-controlled flow, the condensate film heat-transfer coefficient (h_{cf}) is a function of the convective heat-transfer coefficient for liquid flowing alone and the two-phase pressure drop.[18]

$$h_{cf} = h_l (\phi_l^2)^{0.45} \tag{67.29}$$

$$h_l = h_i (1 - y)^{0.8} \tag{67.30}$$

or

$$h_l = h_o (1 - y)^{0.6} \tag{67.31}$$

$$\phi_l^2 = 1 + \frac{C}{X_{tt}} + \frac{1}{X_{tt}^2} \tag{67.32}$$

$C = 20$ (tubeside flow), $C = 9$ (shellside flow)

$$X_{tt} = \left[\frac{1-y}{y} \right]^{0.9} \left[\frac{\rho_v}{\rho_l} \right]^{0.5} \left[\frac{\mu_l}{\mu_v} \right]^{0.1} \tag{67.33}$$

μ_l = liquid viscosity, μ_v = vapor viscosity

Gravity Controlled Flow. The vapor shear force on the condensate is small compared to the gravity force, so condensate drains by gravity. This condition can be estimated, according to Ref. 18, when $J_g < 0.5$. Under gravity-controlled conditions, the condensate film heat-transfer coefficient is calculated as follows:

$$h_{cf} = F_g h_N \tag{67.34}$$

The term h_N is the heat-transfer coefficient from the well-known Nusselt derivation, given in Ref. 14 as

Horizontal tubes

$$h_N = 0.725 \left[\frac{k_l^3 \rho_l (\rho_l - \rho_v) g \lambda}{\mu_l (T_s - T_w) D} \right]^{0.25} \tag{67.35}$$

where λ = latent heat.

Vertical tubes

$$h_N = 1.1 \, k_l \left[\frac{\rho_l \, (\rho_l - \rho_v) \, g}{\mu_l^2 \, \mathrm{Re}_c} \right]^{0.33} \tag{67.36}$$

$$\mathrm{Re}_c = \frac{4W_c}{\pi D \, \mu_l} \tag{67.37}$$

The term F_g in Eq. (67.34) is a correction for condensate loading, and depends on the exchanger geometry.[14]

On horizontal X-type tube bundles:

$$F_g = N_{rv}^{-1/6} \tag{67.38}$$

(Ref. 12), where N_{rv} = number of tubes in a vertical row.

On baffled tube bundles (*owing to turbulence*):

$$F_g = 1.0 \quad \text{(frequent practice)} \tag{67.39}$$

In horizontal tubes:

$$F_g = \left[\frac{1}{1 + (1/(y - 1)) \, (\rho_v/\rho_l)^{0.667}} \right]^{0.75} \quad \text{(from Ref. 14)} \tag{67.40}$$

or

$$F_g = 0.8 \quad \text{(from Ref. 18)} \tag{67.41}$$

Inside or outside vertical tubes:

$$F_g = 0.73 \, \mathrm{Re}_c^{0.11} \quad \text{(rippled film region)} \tag{67.42}$$

or

$$F_g = 0.021 \, \mathrm{Re}_c^{0.58} \, \rho_r^{0.33} \quad \text{(turbulent film region)} \tag{67.43}$$

Use higher value of Eq. (67.42) or (67.43).

For quick hand calculations, the gravity-controlled flow equations may be used for h_{cf}, and will usually give conservative results.

Correction for Mixture Effects

The above heat-transfer coefficients apply only to the condensate film. For mixtures with a significant difference between the dew-point and bubble-point temperatures (condensing range), the vapor-phase heat-transfer coefficient must also be considered as follows:

$$h_c = \frac{1}{(1/h_{cf} + 1/h_v)} \tag{67.44}$$

where h_c = condensation heat-transfer coefficient
h_{cf} = condensate-film heat-transfer coefficient
h_v = vapor-phase heat-transfer coefficient

The vapor-phase heat-transfer rate depends on mass diffusion rates in the vapor. The well-known Colburn–Hougen method and other more recent approaches are summarized by Butterworth.[19] Methods for mixtures forming immiscible condensates are discussed in Ref. 20.

Diffusion-type methods require physical properties not usually available to the designer except for simple systems. Therefore, the vapor-phase heat-transfer coefficient is often estimated in practice by a "resistance-proration"-type method such as the Bell–Ghaly method.[21] In these methods the vapor-phase resistance is prorated with respect to the relative amount of duty required for sensible cooling of the vapor, resulting in the following expression:

$$h_v = (q_t/q_{sv}) h_{sv} \tag{67.44a}$$

where q_{sv} = heat flux to sensibly cool the vapor
q_t = total heat flux
h_{sv} = convective heat-transfer coefficient for sensible heat-transfer to the vapor phase

For more detail in application of the resistance proration method for mixtures, see Refs. 14 or 21.

Pressure Drop

For the condensing vapor, pressure drop is composed of three components—friction, momentum, and static head—as covered in Ref. 14. An approximate estimate on the conservative side can be obtained in terms of the friction component, using the Martinelli separated flow approach:

$$\Delta P_f = \Delta P_l \; \phi_l^2 \tag{67.45}$$

where ΔP_f = two-phase friction pressure drop
ΔP_l = friction loss for liquid phase alone

The Martinelli factor ϕ_l^2 may be calculated as shown in Eq. (67.32). Alternative methods for shellside pressure drop are presented by Diehl[22] and by Grant and Chisholm.[23] These methods were reviewed by Ishihara[24] and found reasonably representative of the available data. However, Eq. (67.32), also evaluated in Ref. 24 for shellside flow, should give about equivalent results.

67.3.3 Shell and Tube Reboilers and Vaporizers

Heat exchangers are used to boil liquids in both the process and power industries. In the process industry they are often used to supply vapors to distillation columns and are called reboilers. The same types of exchangers are used in many applications in the power industry, for example, to generate vapors for turbines. For simplicity these exchangers will all be called "reboilers" in this section. Often the heating medium is steam, but it can also be any hot process fluid from which heat is to be recovered, ranging from chemical reactor effluent to geothermal hot brine.

Selection of Reboiler Type

A number of different shell and tube configurations are in common use, and the first step in design of a reboiler is to select a configuration appropriate to the required job. Basically, the type of reboiler should depend on expected amount of fouling, operating pressure, mean temperature difference (MTD), and difference between temperatures of the bubble point and the dew point (boiling range).

The main considerations are as follows: (1) fouling fluids should be boiled on the tubeside at high velocity; (2) boiling either under deep vacuum or near the critical pressure should be in a kettle to minimize hydrodynamic problems unless means are available for very careful design; (3) at low MTD, especially at low pressure, the amount of static head must be minimized; (4) for wide boiling range mixtures, it is important to maximize the amount of countercurrent flow.

These and other criteria are discussed in more detail in Ref. 25, and summarized in a selection guide, which is abstracted in Table 67.4.

In addition to the above types covered in Ref. 25, falling film evaporators[26] may be preferred in cases with very low MTD, viscous liquids, or very deep vacuum for which even a kettle provides too much static head.

Temperature Profiles

For pure components or narrow boiling mixtures, the boiling temperature is nearly constant and the LMTD applies with $F = 1.0$. Temperature profiles for boiling range mixtures are very complicated, and although the LMTD is often used, it is not a recommended practice, and may result in underdesigned

Table 67.4   Reboiler Selection Guide

Process Conditions	Suggested Reboiler Type[a]
Moderate pressure, MTD, and fouling	VT/E
Very high pressure, near critical	HS/K or (F)HT/E
Deep vacuum	HS/K
High or very low MTD	HS/K, G, H
Moderate to heavy fouling	VT/E
Very heavy fouling	(F)HT/E
Wide boiling range mixture	VT/E, HS/G or /H
Very wide boiling range, viscous liquid	(F)HT/E

[a] V, vertical; H, horizontal; S, shellside boiling; T, tubeside boiling; (F), forced flow, else natural convection; /E, G, H, K, TEMA shell styles.

Table 67.5 Reboiler MTD Estimation

Reboiler Type[a]	T_A	T_B	MTD
HS/K	$T_1 - t_2$	$T_2 - t_2$	Eq. (67.7), $F = 1$
HS/X, G, H	$T_1 - t_1$	$T_2 - t_2$	Eq. (67.7), $F = 0.9$
VT/E	$T_1 - t_2$	$T_2 - t_1$	Eq. (67.7), $F = 1$
(F)HT/E or (F)HS/E	$T_1 - t_2$	$T_2 - t_1$	Eq. (67.7), $F = 0.9$
All types	Isothermal	$T_A = T_B$	T_A

[a] V, vertical; H, horizontal; S, shellside boiling; T, tubeside boiling; (F), forced flow, else natural convection; /E, G, H, K, TEMA shell styles.

reboilers unless compensated by excessive design fouling factors. Contrary to the case for condensers, using a straight-line profile approximation always tends to give too high MTD for reboilers, and can be tolerated only if the temperature rise across the reboiler is kept low through a high circulation rate.

Table 67.5 gives suggested procedures to determine an approximate MTD to use for initial size estimation, based on temperature profiles illustrated in Fig. 67.16.

It should be noted that the MTD values in Table 67.5 are intended to be on the safe side and that excessive fouling factors are not necessary as additional safety factors if these values are used. See Section 67.4.1 for suggested fouling factor ranges.

Heat-Transfer Coefficients

The two basic types of boiling mechanisms that must be taken into account in determining boiling heat-transfer coefficients are nucleate boiling and convective boiling. A detailed description of both types is given by Collier.[27] For all reboilers, the nucleate and convective boiling contributions are additive, as follows:

$$h_b = h_{nb} + h_{cb} \qquad (67.46)$$

where h_b = boiling heat-transfer coefficient
h_{nb} = nucleate boiling heat-transfer coefficient
h_{cb} = convective boiling heat-transfer coefficient

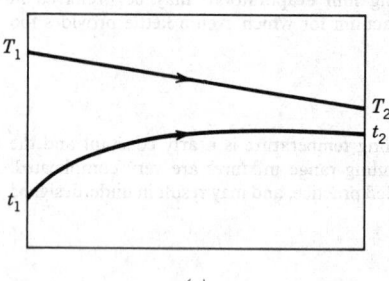

(a)

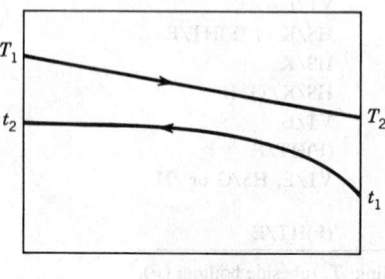

(b)

Fig. 67.16 Reboiler temperature profiles illustrated: (a) use for kettle and horizontal thermosiphon; (b) use for tubeside boiling vertical thermosiphon.

The convective boiling coefficient h_{cb} depends on the liquid-phase convective heat-transfer coefficient h_l, according to the same relationship, Eq. (67.29), given for shear-controlled condensation. For all reboiler types, except forced flow, the flow velocities required to calculate h_l depend on complex pressure balances for which computers are necessary for practical solution. Therefore, the convective component is sometimes approximated as a multiplier to the nucleate boiling component for quick estimations,[25] as in the following equation:

$$h_b = h_{nb}F_b \tag{67.47}$$

$$F_b = \frac{h_{nb} + h_{cb}}{h_{nb}} \tag{67.48}$$

where F_b is approximated as follows:

For tubeside reboilers (VT/E thermosiphon)

$$F_b = 1.5 \tag{67.49}$$

For shellside reboilers (HS/X, G, H, K)

$$F_b = 2.0 \tag{67.50}$$

Equations (67.49) and (67.50) are intended to give conservative results for first approximations. For more detailed calculations see Refs. 28–30.

The nucleate boiling heat-transfer coefficient (h_{nb}) is dependent not only on physical properties, but also on the temperature profile at the wall and the microscopic topography of the surface. For a practical design, many simplifications must be made, and the approximate nature of the resulting coefficients should be recognized. A reasonable design value is given by the following simple equation[25]:

$$h_{nb} = 0.025 \, F_c P_c^{0.69} q^{0.70} (P/P_c)^{0.17} \tag{67.51}$$

where h_{nb}, Btu/hr·ft^2·°F
P = system pressure, psi
P_c = critical pressure, psi
q = heat flux, Btu/hr·ft^2·°F

The term F_c is a correction for the effect of mixture composition on the boiling heat-transfer coefficient. The heat-transfer coefficient for boiling mixtures is lower than that of any of the pure components if boiled alone, as summarized in Ref. 27. This effect can be explained in terms of the change in temperature profile at the wall caused by the composition gradient at the wall, as illustrated in Ref. 31. Since the liquid–phase diffusional methods necessary to predict this effect theoretically are still under development and require data not usually available to the designer, an empirical relationship in terms of mixture boiling range (BR) is recommended in Ref. 25:

$$F_c = [1 + 0.018q^{0.15}BR^{0.75}]^{-1} \tag{67.52}$$

(BR = difference between dew-point and bubble-point temperatures, °F.)

Maximum Heat Flux

Above a certain heat flux, the boiling heat-transfer coefficient can decrease severely, owing to vapor blanketing, or the boiling process can become very unstable, as described in Refs. 27, 31, and 32. Therefore, the design heat flux must be limited to a practical maximum value. For many years the limit used by industry was in the range of 10,000–20,000 Btu/hr·ft^2 for hydrocarbons and about 30,000 Btu/hr·ft^2 for water. These rules of thumb are still considered reasonable at moderate pressures, although the limits, especially for water, are considerably conservative for good designs. However, at both very high and very low pressures the maximum heat fluxes can be severely decreased. Also, the maximum heat fluxes must be a function of geometry to be realistic. Empirical equations are presented in Ref. 25; the equations give much more accurate estimates over wide ranges of pressure and reboiler geometry.

(*a*) *For kettle (HS/K) and horizontal thermosiphon (HS/X,G,H)*

$$q_{max} = 803 \, P_c \left(\frac{P}{P_c}\right)^{0.35} \left(1 - \frac{P}{P_c}\right)^{0.9} \phi_b \tag{67.53}$$

$$\phi_b = 9.7 \left[\frac{D_b L}{A_o}\right] \tag{67.54}$$

In the limit, for $\phi_b > 1.0$, let $\phi_b = 1.0$. For $\phi_b < 0.1$, consider larger tube pitch or vapor relief channels.[25] Design heat flux should be limited to less than 0.7 q_{max}.

(b) For vertical thermosiphon (VT/E)

$$q_{max} = 16,080 \left(\frac{D_i^2}{L}\right)^{0.35} P_c^{0.61} \left(\frac{P}{P_c}\right)^{0.25} \left(1 - \frac{P}{P_c}\right)$$

(67.55)

In addition to the preceding check, the vertical tubeside thermosiphon should be checked to insure against mist flow (dryout). The method by Fair[28] was further confirmed in Ref. 33 for hydrocarbons. For water, extensive data and empirical correlations are available as described by Collier.[27] In order to determine the flow regime by these methods it is necessary to determine the flow rate, as described, for example, in Ref. 28. However, for preliminary specification, it may be assumed that the exit vapor weight fraction will be limited to less than 0.35 for hydrocarbons and less than 0.10 for aqueous solutions and that under these conditions dryout is unlikely.

67.3.4 Air-Cooled Heat Exchangers

Detailed rating of air-cooled heat exchangers requires selection of numerous geometrical parameters, such as tube type, number of tube rows, length, width, number and size of fans, etc., all of which involve economic and experience considerations beyond the scope of this chapter. Air-cooled heat exchangers are still designed primarily by the manufacturers using proprietary methods. However, recommendations for initial specifications and rating are given by Paikert[2] and by Mueller.[3] A preliminary rating method proposed by Brown[34] is also sometimes used for first estimates owing to its simplicity.

Heat-Transfer Coefficients

For a first approximation of the surface required, the bare-surface-based overall heat-transfer coefficients recommended by Smith[35] may be used. A list of these values from Ref. 3 is abstracted in Table 67.6. The values in Table 67.6 were based on performance of finned tubes, having a 1 in. outside diameter base tube on $2\frac{3}{8}$ in. triangular pitch, $\frac{5}{8}$ in. high aluminum fins ($\frac{1}{8}$ in. spacing between fin tips), with eight fins per inch. However, the values may be used as first approximations for other finned types.

As stated by Mueller, air-cooled heat exchanger tubes have had approximately the preceding dimensions in the past, but fin densities have tended to increase and now more typically range from 10 to 12 fins/in. For a more detailed estimate of the overall heat-transfer coefficient, the tubeside coefficients are calculated by methods given in the preceding sections and the airside coefficients are obtained as functions of fin geometry and air velocity from empirical relationships such as given by Gnielinski et al.[36] Rating at this level of sophistication is now done mostly by computer.

Temperature Difference

Air-cooled heat exchangers are normally "cross-flow" arrangements with respect to the type of temperature profile calculation. Charts for determination of the F-factor for such arrangements are presented by Taborek.[9] Charts for a number of arrangements are also given by Paikert[2] based on the "NTU

Table 67.6 Typical Overall Heat-Transfer Coefficients (U_0), Based on
Bare Tube Surface, for Air-Cooled Heat Exchangers

Service	U_o	
	Btu/hr·ft²·°F	W/m²·K
Sensible Cooling		
Process water	105–120	600–680
Light hydrocarbons	75–95	425–540
Fuel oil	20–30	114–170
Flue gas, 10 psig	10	57
Condensation		
Steam, 0–20 psig	130–140	740–795
Ammonia	100–200	570–680
Light hydrocarbons	80–95	455–540
Refrigerant 12	60–80	340–455
Mixed hydrocarbons, steam, and noncondensables	60–70	340–397

method." According to Paikert, optimum design normally requires NTU to be in the range of 0.8–1.5, where,

$$\text{NTU} = \frac{t_2 - t_1}{\text{MTD}} \qquad (67.56)$$

For first approximations, a reasonable air-temperature rise $(t_2 - t_1)$ may be assumed, MTD calculated from Eq. (67.4) using $F = 0.9$–1.0, and NTU checked from Eq. (67.56). It is assumed that if the air-temperature rise is adjusted so that NTU is about 1, the resulting preliminary size estimation will be reasonable. Another design criterion often used is that the face velocity V_f should be in the range of 300–700 ft/min (1.5–3.5 m/sec):

$$V_f = \frac{W_s}{L\, W_d\, \rho_v} \qquad (67.57)$$

where W_s = air rate, lb/min
L = tube length, ft
W_d = bundle width, ft
ρ_v = air density, lb/ft^3

Fan Power Requirement

One or more fans may be used per bundle. Good practice requires that not less than 40–50% of the bundle face area be covered by the fan diameter. The bundle aspect ratio per fan should approach 1 for best performance. Fan diameters range from about 4 to 12 ft (1.2 to 3.7 m), with tip speeds usually limited to less than 12,000 ft/min (60 m/sec) to minimize noise. Pressure drops that can be handled are in the range of only 1–2 in. water (0.035–0.07 psi, 250–500 Pa). However, for typical bundle designs and typical air rates, actual bundle pressure drops may be in the range of only $\frac{1}{4}$–1 in. water.

Paikert[2] gives the expression for fan power as follows:

$$P_f = \frac{V(\Delta p_s + \Delta p_d)}{E_f} \qquad (67.58)$$

where V = volumetric air rate, m^3/sec
Δp_s = static pressure drop, Pa
Δp_d = dynamic pressure loss, often 40–60 Pa
E_f = fan efficiency, often 0.6–0.7
P_f = fan power, W

67.3.5 Other Exchangers

For spiral, plate, and compact heat exchangers the heat-transfer coefficients and friction factors are sensitive to specific proprietary designs and such units are best sized by the manufacturer. However, preliminary correlations have been published. For spiral heat exchangers, see Mueller[3] and Minton.[37] For plate-type heat exchangers, Figs. 67.9 and 67.10, recommendations are given by Cooper[38] and Marriott.[39] For plate-fin and other compact heat exchangers, a comprehensive treatment is given by Webb.[4] For recuperators and regenerators the methods of Hausen are recommended.[6] Heat pipes are extensively covered by Chisholm.[40] Design methods for furnaces and combustion chambers are presented by Truelove.[41] Heat transfer in agitated vessels is discussed by Penney.[42] Double-pipe heat exchangers are described by Guy.[43]

67.4 COMMON OPERATIONAL PROBLEMS

When heat exchangers fail to operate properly in practice, the entire process is often affected, and sometimes must be shut down. Usually, the losses incurred by an unplanned shutdown are many times more costly than the heat exchanger at fault. Poor heat-exchanger performance is usually due to factors having nothing to do with the heat-transfer coefficient. More often the designer has overlooked the seriousness of some peripheral condition not even addressed in most texts on heat-exchanger design. Although only long experience, and numerous "experiences," can come close to uncovering all possible problems waiting to plague the heat-exchanger designer, the following subsections relating the more obvious problems are included to help make the learning curve less eventful.

67.4.1 Fouling

The deposit of solid insulating material from process streams on the heat-transfer surface is known as fouling, and has been called "the major unresolved problem in heat transfer."[44] Although this problem is recognized to be important (see Ref. 45) and is even being seriously researched,[45,46] the nature of the fouling process makes it almost impossible to generalize. As discussed by Mueller,[3] fouling can be caused by (1) precipitation of dissolved substances, (2) deposit of particulate matter, (3) solidification of material through chemical reaction, (4) corrosion of the surface, (5) attachment and growth of biological organisms, and (6) solidification by freezing. The most important variables affecting fouling (besides concentration of the fouling material) are velocity, which affects types 1, 2, and 5, and surface temperature, which affects types 3-6. For boiling fluids, fouling is also affected by the fraction vaporized. As stated in Ref. 25, it is usually impossible to know ahead of time what fouling mechanism will be most important in a particular case. Fouling is sometimes catalyzed by trace elements unknown to the designer. However, most types of fouling are retarded if the flow velocity is as high as possible, the surface temperature is as low as possible (exception is biological fouling[48]), the amount of vaporization is as low as possible, and the flow distribution is as uniform as possible.

The expected occurrence of fouling is usually accounted for in practice by assignment of fouling factors, which are additional heat-transfer resistances, Eq. (67.7). The fouling factors are assigned for the purpose of oversizing the heat exchanger sufficiently to permit adequate on-stream time before cleaning is necessary. Often in the past the fouling factor has also served as a general purpose "safety factor" expected to make up for other uncertainties in the design. However, assignment of overly large fouling factors can produce poor operation caused by excessive overdesign.[49,50]

For shell and tube heat exchangers it has been common practice to rely on the fouling factors suggested by TEMA.[1] Fouling in plate heat exchangers is usually less, and is discussed in Ref. 38. The TEMA fouling factors have been used for over 30 years and, as Mueller states, must represent some practical validity or else complaints would have forced their revision. As of this writing, a joint committee of TEMA and HTRI members has been formed to review the TEMA fouling recommendations and see if they require updating in view of new knowledge. In addition to TEMA, fouling resistances are presented more recently by Bell[10] and values recommended for reboiler design are given in Ref. 25. In general, the minimum value commonly used for design is 0.0005 °F·hr·ft²/Btu for condensing steam or light hydrocarbons. Typical values for process streams or treated cooling water are around 0.001-0.002 °F·hr·ft²/Btu, and for heavily fouling streams values in the range of 0.003-0.01 °F·hr·ft²/Btu are used. For reboilers (which have been properly designed) a design value of 0.001 °F·hr·ft²/Btu is usually adequate, although for wide boiling mixtures other effects in addition to fouling tend to limit performance.

67.4.2 Vibration

A problem with shell and tube heat exchangers that is becoming more frequent as heat exchangers tend to become larger and design velocities tend to become higher is tube failure due to flow-induced tube vibration. Summaries including recommended methods of analysis are given by Chenoweth[51] and by Mueller.[3] In general, tube vibration problems tend to occur when the distance between baffles or tube-support plates is too great. Maximum baffle spacings recommended by TEMA were based on the maximum unsupported length of tube that will not sag significantly. Experience has shown that flow-induced vibration can still occur at TEMA maximum baffle spacing, but at about 0.7–0.8 times this spacing most vibration can be eliminated at normal design velocities (see Section 67.2.4). Taborek[11] gives the following equations for TEMA maximum unsupported tube lengths (L_{su}), inches.

Steel and steel alloy tubes

For $D_o = \frac{3}{4}-2$ in,

$$L_{su} = 52\,D_o + 21$$

(67.59)

For $D_o = \frac{1}{4}-\frac{3}{4}$ in,

$$L_{su} = 68\,D_o + 9$$

(67.60)

Aluminum and copper alloy tubes

For $D_o = \frac{3}{4}-2$ in,

$$L_{su} = 46\,D_o + 17$$

(67.61)

For $D_o = \frac{1}{4}-\frac{3}{4}$ in,

$$L_{su} = 60\,D_o + 7$$

(67.62)

For segmental baffles with tubes in the windows, Fig. 67.9, the maximum baffle spacing is one-half the maximum unsupported tube length.

For very large bundle diameters, segmental or even double segmental baffles may not be suitable, since the spacing required to prevent vibration may produce too high pressure drops. (In addition, flow distribution considerations require that the ratio of baffle spacing to shell diameter not be less than about 0.2.) In such cases, one commonly used solution is to eliminate tubes in the baffle windows so that intermediate support plates can be used and baffle spacing can be increased; see Fig. 67.17. Another solution, with many advantages is the rod-type tube support in which the flow is essentially longitudinal and the tubes are supported by a cage of rods. A proprietary design of this type exchanger (RODbaffle) is licensed by Phillips Petroleum Co. Calculation methods are published in Ref. 52.

67.4.3 Flow Maldistribution

Several types of problems can occur when the flow velocities or fluid phases become distributed in a way not anticipated by the designer. This occurs in all types of exchangers, but the following discussion is limited to shell and tube and air-cooled exchangers, in which maldistribution can occur on either shellside or tubeside.

Shellside Flow

Single-phase flow can be maldistributed on the shellside owing to bypassing around the tube bundle and leakage between tubes and baffle and between baffle and shell. Even for typical well-designed heat exchangers, these ineffective streams can comprise as much as 40% of the flow in the turbulent regime and as much as 60% of the flow in the laminar regime. It is especially important for laminar flow to minimize these bypass and leakage streams, which cause both lower heat-transfer coefficients and lower effective MTD.[13] This can, of course, be done by minimizing clearances, but economics dictate that more practical methods include use of bypass sealing strips, increasing tube pitch, increasing baffle spacing, and using an optimum baffle cut to provide more bundle penetration. Methods for calculating the effects of these parameters are described by Taborek.[11]

Another type of shellside maldistribution occurs in gas–liquid two-phase flow in horizontal shells when the flow velocity is low enough that the vapor and liquid phases separate, with the liquid flowing along the bottom of the shell. For condensers this is expected and taken into account. However, for some other types of exchangers, such as vapor–liquid contactors or two-phase reactor feed-effluent exchangers, separation may cause unacceptable performance. An estimation of separation can be obtained by calculating the flow regime parameter J_g, Eq. (67.28). For $J_g < {\sim}1.5$, some phase separation would be expected; while for $J_g < {\sim}0.5$ complete separation would be expected. For such cases, if it is important to keep the phases mixed, a vertical heat exchanger is recommended. Improvement in mixing is obtained for horizontal exchangers if horizontal rather than vertical baffle cut is used.

Tubeside Flow

Several types of tubeside maldistribution have been experienced. For single-phase flow with axial nozzles into a single-tubepass exchanger, the dynamic head of the entering fluid can cause higher flow in the central tubes, sometimes even producing backflow in the peripheral tubes. This effect can be prevented by using an impingement plate on the centerline of the axial nozzle.

Another type of tubeside maldistribution occurs in cooling viscous liquids. Cooler tubes in parallel flow will tend to completely plug up in this situation, unless a certain minimum pressure drop is obtained, as explained by Mueller.[53]

For air-cooled single pass condensers, a backflow can occur owing to the difference in temperature driving force between bottom and top tube rows, as described by Berg and Berg.[54] This can cause an accumulation of noncondensables in air-cooled condensers, which can significantly affect performance, as described by Breber et al.[55] In fact, in severe cases, this effect can promote freezeup of tubes, or even destruction of tubes by water hammer. Backflow effects are eliminated if a small amount of excess vapor is taken through the main condenser to a backup condenser or if the number of fins per inch on bottom rows is less than on top rows to counteract the difference in temperature driving force.

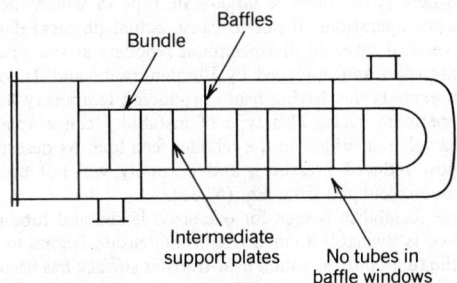

Fig. 67.17 Segmental baffles with no tubes in window.

For multipass tubeside condensers, or tubeside condensers in series, the vapor and liquid tend to separate in the headers with liquid running in the lower tubes. The fraction of tubes filled with liquid tends to be greater at higher pressures. In most cases the effect of this separation on the overall condenser heat-transfer coefficient is not serious. However, for multicomponent mixtures the effect on the temperature profile will be such as to decrease the MTD. For such cases, the temperature profile should be calculated by the differential flash procedure, Section 67.3.2. In general, because of unpredictable effects, entering a pass header with two phases should be avoided when possible.

67.4.4 Temperature Pinch

When the hot and cold streams reach approximately the same temperature in a heat exchanger, heat transfer stops. This condition is referred to as a temperature pinch. For shellside single-phase flow, unexpected temperature pinches can be the result of excessive bypassing and leakage combined with a low MTD and possibly a temperature cross. An additional factor, "temperature profile distortion factor," is needed as a correction to the normal F factor to account for this effect.[11,13] However, if good design practices are followed with respect to shellside geometry, this effect normally can be avoided.

In condensation of multicomponent mixtures, unexpected temperature pinches can occur in cases where the condensation curve is not properly calculated, especially when the true curve happens to be of type III in Fig. 67.15. This can happen when separation of liquid containing heavy components occurs, as mentioned above, and also when the condensing mixture has immiscible liquid phases with more than one dew point.[20] In addition, condensing mixtures with large desuperheating and subcooling zones can produce temperature pinches and must be carefully analyzed. In critical cases it is safer and may even be more effective to do desuperheating, condensing, and subcooling in separate heat exchangers. This is especially true of subcooling.[3]

Reboilers can also suffer from temperature-pinch problems in cases of wide boiling mixtures and inadequate liquid recirculation. Especially for thermosiphon reboilers, if poorly designed and the circulation rate is not as high as expected, the temperature rise across the reboiler will be greater than expected and a temperature pinch may result. This happens most often when the reboiler exit piping is too small and consumes an unexpectedly large amount of pressure drop. This problem normally can be avoided if the friction and momentum pressure drop in the exit piping is limited to less than 30% of the total driving head and the exit vapor fraction is limited to less than 0.25 for wide boiling range mixtures. For other recommendations, see Ref. 25.

67.4.5 Critical Heat Flux in Vaporizers

Owing to a general tendency to use lower temperature differences for energy conservation, critical heat flux problems are not now frequently seen in the process industries. However, for waste heat boilers, where the heating medium is usually a very hot fluid, surpassing the critical heat flux is a major cause of tube failure. The critical heat flux is that flux (Q/A_0) above which the boiling process departs from the nucleate or convective boiling regimes and a vapor film begins to blanket the surface, causing a severe rise in surface temperature, approaching the temperature of the heating medium. This effect can be caused by either of two mechanisms: (1) flow of liquid to the hot surface is impeded and is insufficient to supply the vaporization process or (2) the local temperature exceeds that for which a liquid phase can exist.[32] Methods of estimating the maximum design heat flux are given in Section 67.3.3, and the subject of critical heat flux is covered in great detail in Ref. 27. However, in most cases where failures have occurred, especially for shellside vaporizers, the problem has been caused by local liquid deficiency, owing to lack of attention to flow distribution considerations.

67.4.6 Instability

The instability referred to here is the massive large-scale type in which the fluid surging is of such violence as to at least disrupt operations, if not to cause actual physical damage. One version is the boiling instability seen in vertical tubeside thermosiphon reboilers at low operating pressure and high heat flux. This effect is discussed and analyzed by Blumenkrantz and Taborek.[56] It is caused when the vapor acceleration loss exceeds the driving head, producing temporary flow stoppage or backflow, followed by surging in a periodic cycle. This type of instability can always be eliminated by using more frictional resistance, a valve or orifice, in the reboiler feed line. As described in Ref. 32, instability normally only occurs at low reduced pressures, and normally will not occur if design heat flux is less than the maximum value calculated from Eq. (67.55).

Another type of massive instability is seen for oversized horizontal tubeside pure component condensers. When more surface is available than needed, condensate begins to subcool and accumulate in the downstream end of the tubes until so much heat-transfer surface has been blanketed by condensate that there is not enough remaining to condense the incoming vapor. At this point the condensate is blown out of the tube by the increasing pressure and the process is repeated. This effect does not occur in vertical condensers since the condensate can drain out of the tubes by gravity. This problem can sometimes be controlled by plugging tubes or injecting inert gas, and can always be eliminated by taking a small amount of excess vapor out of the main condenser to a small vertical backup condenser.

67.4.7 Inadequate Venting, Drainage, or Blowdown

For proper operation of condensers it is always necessary to provide for venting of noncondensables. Even so-called pure components will contain trace amounts of noncondensables that will eventually build up sufficiently to severely limit performance unless vented. Vents should always be in the vapor space near the condensate exit nozzle. If the noncondensable vent is on the accumulator after the condenser, it is important to ensure that the condensate nozzle and piping are large enough to provide unrestricted flow of noncondensables to the accumulator. In general, it is safer to provide vent nozzles directly on the condenser.

If condensate nozzles are too small, condensate can accumulate in the condenser. It is recommended that these nozzles be large enough to permit weir-type drainage (with a gas core in the center of the pipe) rather than to have a full pipe of liquid. Standard weir formulas[57] can be used to size the condensate nozzle. A rule of thumb used in industry is that the liquid velocity in the condensate piping, based on total pipe cross section, should not exceed 3 ft/sec (0.9 m/sec).

The problem of inadequate blowdown in vaporizers is similar to the problem of inadequate venting for condensers. Especially with kettle-type units, trace amounts of heavy, high-boiling, or nonboiling components can accumulate, not only promoting fouling but also increasing the effective boiling range of the mixture, thereby decreasing the effective heat-transfer coefficient. Therefore, means of continuous or at least periodic removal of liquid from the reboiler (blowdown) should be provided to ensure good operation. Even for thermosiphon reboilers, if designed for low heat fluxes (below about 2000 BTU/hr/ft², 6300 W/m²), the circulation through the reboiler may not be high enough to prevent heavy components from building up, and some provision for blowdown may be advisable in the bottom header.

REFERENCES

Note: Many of the following references are taken from the *Heat Exchanger Design Handbook* (HEDH), Hemisphere, Washington, DC, 1982, which will be referred to for simplicity as HEDH.

1. *Standards of Tubular Heat Exchanger Manufacturers Association*, 6th ed., TEMA, New York, 1978.

2. P. Paikert, "Air-Cooled Heat Exchangers," Section 3.8, HEDH.

3. A. C. Mueller, in *Handbook of Heat Transfer*, Rohsenow and Hartnet (eds.), McGraw-Hill, New York, 1983, Chap. 18.

4. R. L. Webb, "Compact Heat Exchangers," Section 3.9, HEDH.

5. F. L. Rubin, "Multizone Condensers, Desuperheating, Condensing, Subcooling," *Heat Transfer Eng.*, 3(1), 49–59 (1981).

6. H. Hausen, *Heat Transfer in Counterflow, Parallel Flow, and Crossflow*, McGraw-Hill, New York, 1983.

7. D. Chisholm et al., "Costing of Heat Exchangers," Section 4.8, HEDH.

8. R. S. Hall, J. Matley, and K. J. McNaughton, "Current Costs of Process Equipment," *Chem. Eng.*, 89(7), 80–116 (Apr. 5, 1982).

9. J. Taborek, "Charts for Mean Temperature Difference in Industrial Heat Exchanger Configurations," Section 1.5, HEDH.

10. K. J. Bell, "Approximate Sizing of Shell-and-Tube Heat Exchangers," Section 3.1.4, HEDH.

11. J. Taborek, "Shell and Tube Heat Exchangers, Single-Phase Flow," Section 3.3, HEDH.

12. D. Q. Kern, *Process Heat Transfer*, McGraw-Hill, New York, 1950.

13. J. W. Palen and J. Taborek. "Solution of Shellside Heat Transfer and Pressure Drop by Stream Analysis Method," *Chem. Eng. Prog. Symp. Series*, 65(92) (1969).

14. A. C. Mueller, "Condensers," Section 3.4, HEDH.

15. B. D. Smith, *Design of Equilibrium Stage Processes*, McGraw-Hill, New York, 1963.

16. V. L. Rice, "Program Performs Vapor-Liquid Equilibrium Calculations," *Chem. Eng.*, 77–86 (June 28, 1982).

17. R. S. Kistler and A. E. Kassem, "Stepwise Rating of Condensers," *Chem. Eng. Prog.*, 77(7), 55–59 (1981).

18. G. Breber, J. Palen, and J. Taborek, "Prediction of Horizontal Tubeside Condensation of Pure Components Using Flow Regime Criteria," *Heat Transfer Eng.*, 1(2), 72–79 (1979).

19. D. Butterworth, "Condensation of Vapor Mixtures," Section 2.6.3, HEDH.

20. R. G. Sardesai, "Condensation of Mixtures Forming Immiscible Liquids," Section 2.5.4, HEDH.

21. K. J. Bell and A. M. Ghaly, "An Approximate Generalized Design Method for Multicomponent/Partial Condensers," *AIChE Symp. Ser.,* No. 131, 72–79 (1972).

22. J. E. Diehl, "Calculate Condenser Pressure Drop," *Pet. Refiner,* **36**(10), 147–153 (1957).

23. I. D. R. Grant and D. Chisholm, "Two-Phase Flow on the Shell-side of a Segmentally Baffled Shell-and-Tube Heat Exchanger," *Trans. ASME J. Heat Transfer,* **101**(1), 38–42 (1979).

24. K. Ishihara, J. W. Palen, and J. Taborek, "Critical Review of Correlations for Predicting Two-Phase Flow Pressure Drops Across Tube Banks," *Heat Transfer Eng.,* **1**(3) (1979).

25. J. W. Palen, "Shell and Tube Reboilers," Section 3.6, HEDH.

26. R. A. Smith, "Evaporaters," Section 3.5, HEDH.

27. J. G. Collier, "Boiling and Evaporation," Section 2.7, HEDH.

28. J. R. Fair, "What You Need to Design Thermosiphon Reboilers," *Pet. Refiner,* **39**(2), 105 (1960).

29. J. R. Fair and A. M. Klip, "Thermal Design of Horizontal Type Reboilers," *Chem. Eng. Prog.,* **79**(3) (1983).

30. J. W. Palen and C. C. Yang, "Circulation Boiling Model of Kettle and Internal Reboiler Performance," Paper presented at the 21st National Heat Transfer Conference, Seattle, WA, 1983.

31. J. W. Palen, A. Yarden, and J. Taborek, "Characteristics of Boiling Outside Large Scale Multitube Bundles," *Chem. Eng. Prog. Symp. Ser.,* **68**(118), 50–61 (1972).

32. J. W. Palen, C. C. Shih, and J. Taborek, "Performance Limitations in a Large Scale Thermosiphon Reboiler," *Proceedings of the 5th International Heat Transfer Conference,* Tokyo, 1974, Vol. 5, pp. 204–208.

33. J. W. Palen, C. C. Shih, and J. Taborek, "Mist Flow in Thermosiphon Reboilers," *Chem. Eng. Prog.,* **78**(7), 59–61 (1982).

34. R. Brown, "A Procedure for Preliminary Estimate of Air-Cooled Heat Exchangers," *Chem. Eng.,* **85**(8), 108–111 (Mar. 27, 1978).

35. E. C. Smith, "Air-Cooled Heat Exchangers," *Chem. Eng.* (Nov. 17, 1958).

36. V. Gnielinski, A. Zukauskas, and A. Skrinska, "Banks of Plain and Finned Tubes," Section 2.5.3, HEDH.

37. P. Minton, "Designing Spiral-Plate Heat Exchangers," *Chem. Eng.,* **77**(9) (May 4, 1970).

38. A. Cooper and J. D. Usher, "Plate Heat Exchangers," Section 3.7, HEDH.

39. J. Marriott, "Performance of an Alfaflex Plate Heat Exchanger," *Chem. Eng. Prog.,* **73**(2), 73–78 (1977).

40. D. Chisholm, "Heat Pipes," Section 3.10, HEDH.

41. J. S. Truelove, "Furnaces and Combustion Chambers," Section 3.11, HEDH.

42. W. R. Penney, "Agitated Vessels," Section 3.14, HEDH.

43. A. R. Guy, "Double-Pipe Heat Exchangers," Section 3.2, HEDH.

44. J. Taborek et al., "Fouling—The Major Unresolved Problem in Heat Transfer," *Chem. Eng. Prog.,* **65**(92), 53–67 (1972).

45. *Proceedings of the Conference on Progress in the Prevention of Fouling in Process Plants,* sponsored by the Institute of Corrosion Science Technology and the Institute of Chemical Engineers, London, 1981.

46. J. W. Suitor, W. J. Marner, and R. B. Ritter, "The History and Status of Research in Fouling of Heat Exchangers in Cooling Water Service," *Canad. J. Chem. Eng.,* **55** (Aug., 1977).

47. A. Cooper, J. W. Suitor, and J. D. Usher, "Cooling Water Fouling in Plate Exchangers," *Heat Transfer Eng.,* **1**(3) (1979).

48. R. B. Ritter and J. W. Suitor, "Seawater Fouling of Heat Exchanger Tubes," *Proceedings of the 2nd National Conference on Complete Water Reuse,* Chicago, 1975.

49. C. H. Gilmour, "No Fooling–No Fouling," *Chem. Eng. Prog.,* **61**(7), 49–54 (1965).

50. J. V. Smith, "Improving the Performance of Vertical Thermosiphon Reboilers," *Chem. Eng. Prog.,* **70**(7), 68–70 (1974).

51. J. C. Chenoweth, "Flow-Induced Vibration," Section 4.6, HEDH.

52. C. C. Gentry, R. K. Young, and W. M. Small, "RODbaffle Heat Exchanger Thermal-Hydraulic Predictive Methods," *Proceedings of the 7th International Heat Transfer Conference,* Munich, 1982.

53. A. C. Mueller, "Criteria for Maldistribution in Viscous Flow Coolers," *Proceedings of the 5th International Heat Transfer Conference,* HE 1.4, Tokyo, Vol. 5, pp. 170–174.

54. W. F. Berg and J. L. Berg, "Flow Patterns for Isothermal Condensation in One-Pass Air-Cooled Heat Exchangers," *Heat Transfer Eng.*, **1**(4), 21–31 (1980).

55. G. Breber, J. W. Palen, and J. Taborek, "Study on Non-Condensable Vapor Accumulation in Air-Cooled Condensers," *Proceedings of the 7th International Heat Transfer Conference*, Munich, 1982.

56. A. Blumenkrantz and J. Taborek, "Application of Stability Analysis for Design of Natural Circulation Boiling Systems and Comparison with Experimental Data," *AIChE Symp. Ser.*, **68**(118) (1971).

57. V. L. Streeter, *Fluid Mechanics*, McGraw-Hill, New York, 1958.

58. E. A. D. Saunders, "Shell and Tube Heat Exchangers, Elements of Construction," Section 4.2, HEDH.

59. F. W. Schmidt, "Thermal Energy Storage and Regeneration," *Heat Exchangers Theory and Practice*, J. Taborek et al. (Eds.), Hemisphere, McGraw-Hill, New York.

CHAPTER 68

AIR HEATING

RICHARD J. REED

North American Manufacturing Company
Cleveland, Ohio

68.1 AIR-HEATING PROCESSES

To recover heat from a heating process, the hot waste gases (usually products of combustion) can be used to heat air for space heating, for ovens and drying, or for preheating combustion air. In a few cases, such air heating can be "direct," that is, by mixing the waste gas stream with cold fresh air. Such cases are limited by local codes and by the presence of harmful or undesirable substances (such as odors) in the waste stream. It is often impractical to heat space or preheat combustion air by direct-fired air heating because of the lack of adequate oxygen in the resulting mixture.

If the waste stream is largely nitrogen, and if the temperatures of both streams are between 0 and 800°F, where specific heats are about 0.24, a simplified heat balance can be used to evaluate the mixing conditions:

heat content of the waste stream + heat content of the fresh air = heat content of the mixture

or

$$W_w T_w + W_f T_f = W_m T_m = (W_w + W_f) T_m$$

where W = weight and T = temperature of waste gas, fresh air, and mixture (subscripts w, f, and m).

Example 68.1

If a 600°F waste gas stream flowing at 100 lb/hr is available to mix with 10°F fresh air and fuel, how many pounds per hour of 110°F makeup air can be produced?

Solution:

$$(100 \times 600) + 10 W_f = (100 + W_f) \times (110)$$

Solving, we find $W_f = 490$ lb/hr of fresh air can be heated to 110°F, but the 100 lb/hr of waste gas will be mixed with it; so the delivered stream, W_m will be 100 + 490 = 590 lb/hr.

If "indirect" air heating is necessary, a heat exchanger (recuperator or regenerator) must be used. These may take many forms such as plate-type heat exchangers, shell and tube heat exchangers, double-pipe heat exchangers, heat-pipe exchangers, heat wheels, pebble heater recuperators, and refractory checkerworks. The supplier of the heat exchanger should be able to predict the air preheat temperature and the final waste gas temperature. The amount of heat recovered Q is then $Q = W c_p (T_2 - T_1)$, where W is the weight of air heated, c_p is the specific heat of air (0.24 when below 800°F), T_2 is the delivered hot air temperature, and T_1 is the cold air temperature entering the heat exchanger. Tables and graphs later in this chapter permit estimation of fuel savings and efficiencies for cases involving preheating of combustion air.

If a waste gas stream is only a few hundred degrees Fahrenheit hotter than the air stream temperature required for heating space, an oven, or a dryer, such uses of recovered heat are highly desirable. For higher waste gas stream temperatures, however, the second law of thermodynamics would say that we can make better use of the energy by stepping it down in smaller temperature increments, and preheating combustion air usually makes more sense. This also simplifies accounting, since it returns the recovered heat to the process that generated the hot waste stream.

Preheating combustion air is a very logical method for recycling waste energy from flue gases in direct-fired industrial heating processes such as melting, forming, ceramic firing, heat treating, chemical and petroprocess heaters, and boilers. (It is always wise, however, to check the economics of using flue gases to preheat the load or to make steam in a waste heat boiler.)

68.2 COSTS

In addition to the cost of the heat exchanger for preheating the combustion air, there are many other costs that have to be weighed. Retrofit or add-on recuperators or regenerators may have to be installed overhead to keep the length of heat-losing duct and pipe to a minimum; therefore, extra foundations and structural work may be needed. If the waste gas or air is hotter than about 800°F, carbon steel pipe and duct should be insulated on the inside. For small pipes or ducts where this would be impractical, it is necessary to use an alloy with strength and oxidation resistance at the higher temperature and to insulate on the outside.

High-temperature air is much less dense; therefore, the flow passages of burners, valves, and pipe must be greater for the same input rate and pressure drop. Burners, valves, and piping must be constructed of better materials to withstand the hot air stream. The front face of the burner is exposed to more intense radiation because of the higher flame temperature resulting from preheated combustion air.

If the system is to be operated at a variety of firing rates, the output air temperature will vary; so temperature-compensating fuel/air ratio controls are essential to avoid wasting fuel. Also, to protect the investment in the heat exchanger, it is only logical that it be protected with high-limit temperature controls.

68.3 WARNINGS

Changing temperatures from end to end of high-temperature heat exchangers and from time to time during high-temperature furnace cycles cause great thermal stress, often resulting in leaks and shortened heat-exchanger life. Heat-transfer surfaces fixed at both ends (welded or rolled in) can force something to be overstressed. Recent developments in the form of high-temperature slip seal methods, combined with sensible location of such seals in cool air entrance sections, are opening a whole new era in recuperator reliability.

Corrosion, fouling, and condensation problems continue to limit the applications of heat-recovery equipment of all kinds. Heat-transfer surfaces in air heaters are never as well cooled as those in water heaters and waste heat boilers; therefore, they must exist in a more hostile environment. However, they may experience fewer problems from acid-dew-point condensation. If corrosives, particulates, or condensables are emitted by the heating process at limited times, perhaps some temporary bypassing arrangement can be instituted. High waste gas design velocities may be used to keep particulates in suspension until they reach an area where they can be safely dropped out.

Figure 68.1 shows recommended minimum temperatures to avoid "acid rain" in the heat exchanger.[1] Although a low final waste gas temperature is desirable from an efficiency standpoint, the shortened equipment life seldom warrants it. Acid forms from combination of water vapor with SO_3, SO_2, or CO_2 in the flue gases.

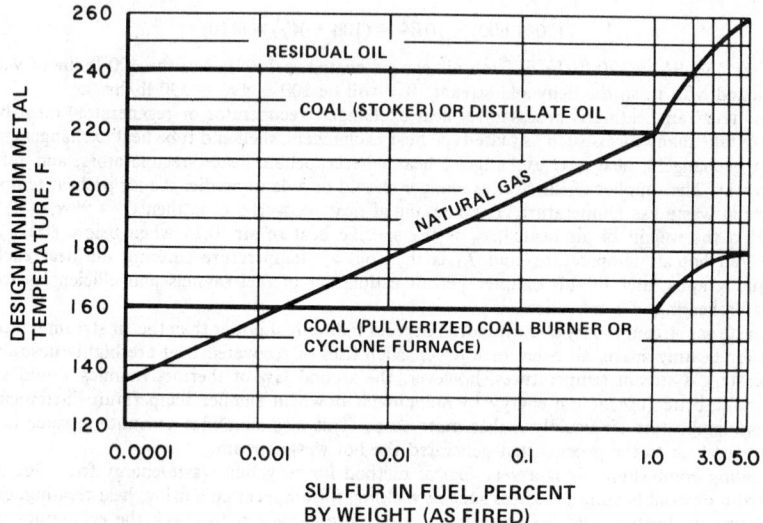

Fig. 68.1 Recommended minimum temperatures to avoid "acid rain" in heat exchangers.

68.4 BENEFITS

Despite all the costs and warnings listed above, combustion air preheating systems do pay. As fuel costs rise, the payback is more rewarding, even for small installations. Figure 68.2 shows percent available heat[2] (best possible efficiency) with various amounts of air preheat and a variety of furnace exit (flue) temperatures. All curves for hot air are based on 10% excess air.* The percentage of fuel saved by addition of combustion air preheating equipment can be calculated by the formula

$$\% \text{ fuel saved} = 100 \times \left(1 - \frac{\% \text{ available heat before}}{\% \text{ available heat after}} \right)$$

Table 68.1 lists fuel savings calculated by this method.[2]

---

* It is advisable to tune a combustion system for closer to stoichiometric air/fuel ratio *before* attempting to preheat combustion air. This is not only a quicker and less costly fuel conservation measure, but it then allows use of smaller heat-exchange equipment.

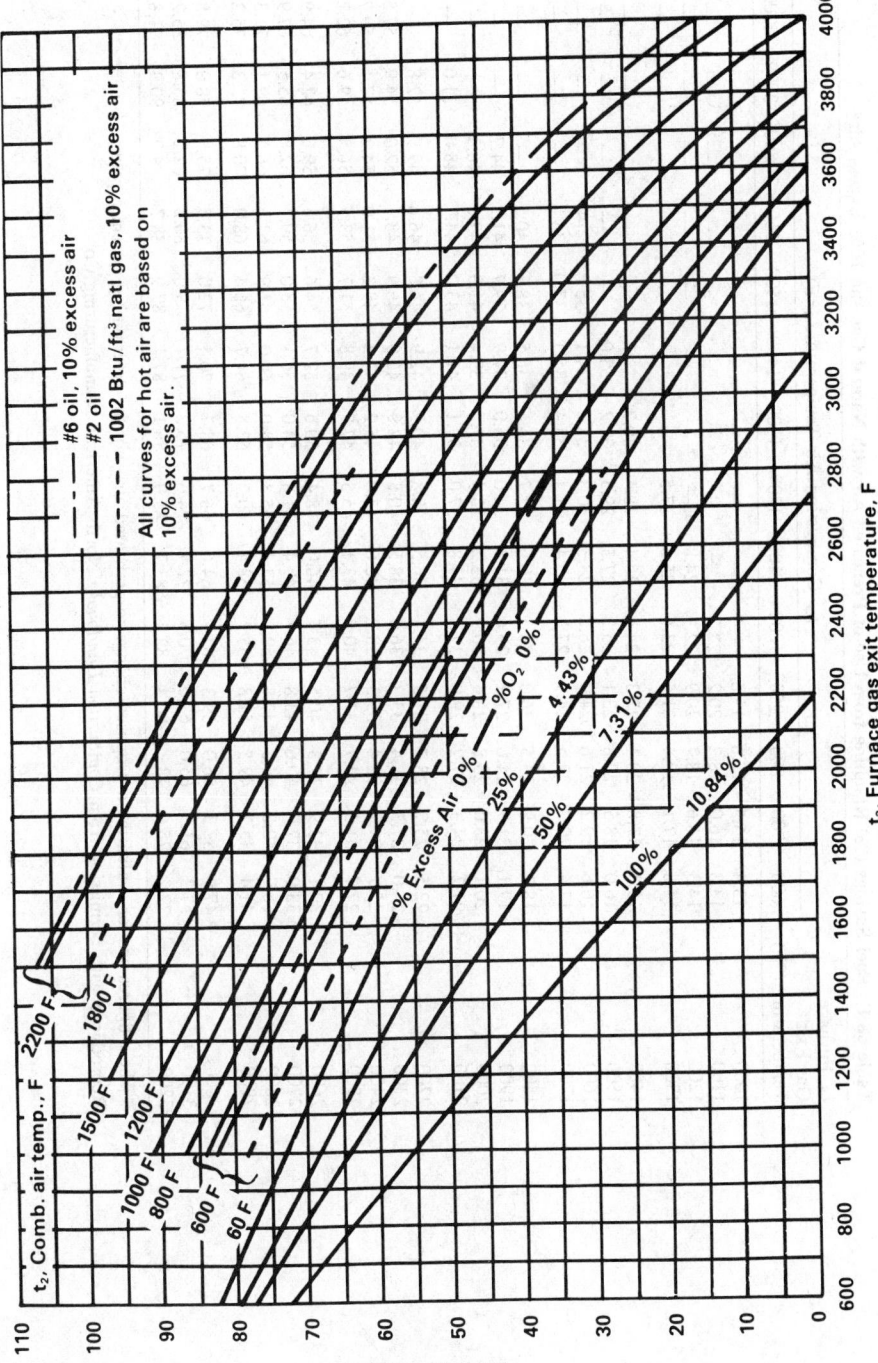

Fig. 68.2 Available heat with preheated combustion air at 10% excess air. Applicable only if there is no unburned fuel in the products of combustion. Corrected for dissociation. Reproduced with permission from *Combustion Handbook.*[2]

Table 68.1 Fuel Savings (%) Resulting from Use of Preheated Air with Natural Gas and 10% Excess Air[a]

t_3, Furnace Gas Exit Temperature (°F)	t_2, Combustion Air Temperature (°F)													
	600	700	800	900	1000	1100	1200	1300	1400	1500	1600	1800	2000	2200
1000	13.4	15.5	17.6	19.6	—	—	—	—	—	—	—	—	—	—
1100	13.8	16.0	18.2	20.2	22.2	—	—	—	—	—	—	—	—	—
1200	14.3	16.6	18.7	20.9	22.9	24.8	—	—	—	—	—	—	—	—
1300	14.8	17.1	19.4	21.5	23.6	25.6	27.5	—	—	—	—	—	—	—
1400	15.3	17.8	20.1	22.3	24.4	26.4	28.4	30.2	—	—	—	—	—	—
1500	16.0	18.5	20.8	23.1	25.3	27.3	29.3	31.2	33.0	—	—	—	—	—
1600	16.6	19.2	21.6	24.0	26.2	28.3	30.3	32.2	34.1	35.8	—	—	—	—
1700	17.4	20.0	22.5	24.9	27.2	29.4	31.4	33.4	35.3	37.0	38.7	—	—	—
1800	18.2	20.9	23.5	26.0	28.3	30.6	32.7	34.6	36.5	38.3	40.1	—	—	—
1900	19.1	21.9	24.6	27.1	29.6	31.8	34.0	36.0	37.9	39.7	41.5	44.7	—	—
2000	20.1	23.0	25.8	28.4	30.9	33.2	35.4	37.5	39.4	41.3	43.0	46.3	—	—
2100	21.2	24.3	27.2	29.9	32.4	34.8	37.0	39.1	41.1	43.0	44.7	48.0	51.0	—
2200	22.5	25.7	28.7	31.5	34.1	36.5	38.8	40.9	42.9	44.8	46.6	49.9	52.8	—
2300	24.0	27.3	30.4	33.3	36.0	38.5	40.8	42.9	45.0	46.9	48.7	52.0	54.9	57.5
2400	25.7	29.2	32.4	35.3	38.1	40.6	43.0	45.2	47.2	49.2	51.0	54.2	57.1	59.7
2500	27.7	31.3	34.7	37.7	40.5	43.1	45.5	47.7	49.8	51.7	53.5	56.8	59.6	62.2
2600	30.1	33.9	37.3	40.5	43.4	46.0	48.4	50.6	52.7	54.6	56.4	59.6	62.4	64.9
2700	33.0	37.0	40.6	43.8	46.7	49.4	51.8	54.0	56.1	58.0	59.7	62.8	65.5	67.9
2800	36.7	40.8	44.5	47.8	50.8	53.4	55.8	58.0	60.0	61.9	63.5	66.5	69.1	71.3
2900	41.4	45.7	49.5	52.8	55.7	58.4	60.7	62.8	64.7	66.4	68.0	70.8	73.2	75.2
3000	47.9	52.3	56.0	59.3	62.1	64.6	66.7	68.7	70.4	72.0	73.5	75.9	78.0	79.8
3100	57.3	61.5	65.0	68.0	70.5	72.7	74.6	76.2	77.7	79.0	80.2	82.2	83.8	85.2
3200	72.2	75.6	78.3	80.4	82.2	83.7	85.0	86.1	87.1	87.9	88.7	89.9	90.9	91.8

[a] These figures are for evaluating a proposed change to preheated air—not for determining system capacity.
Reproduced with permission from Combustion Handbook, North American Manufacturing Co.

Table 68.2 Effect of Combustion Air Preheat on Flame Temperature

Excess Air (%)	Preheated Combustion Air Temperature (°F)	Adiabatic Flame Temperature (°F)		
		With 1000 Btu/scf Natural Gas	With 137,010 Btu/gal Distillate Fuel Oil	With 153,120 Btu/gal Residual Fuel Oil
0	60	3468	3532	3627
10	60	3314	3374	3475
10	600	3542	3604	3690
10	700	3581	3643	3727
10	800	3619	3681	3763
10	900	3656	3718	3798
10	1000	3692	3754	3831
10	1100	3727	3789	3864
10	1200	3761	3823	3896
10	1300	3794	3855	3927
10	1400	3826	3887	3957
10	1500	3857	3918	3986
10	1600	3887	3948	4014
10	1700	3917	3978	4042
10	1800	3945	4006	4069
10	1900	3973	4034	4095
10	2000	4000	4060	4121
0	2000	4051	4112	4171

Preheating combustion air raises the flame temperature and thereby enhances radiation heat transfer in the furnace, which should lower the exit gas temperature and further improve fuel efficiency. Table 68.2 shows the magnitude of flame temperature increase when operating with 10% excess air,* but it is difficult to quantify the resultant saving.

Preheating combustion air has some lesser benefits. Flame stability is enhanced by the faster flame velocity and broader flammability limits. If downstream pollution control equipment is required (scrubber, baghouse), such equipment can be smaller and of less costly materials because the heat exchanger will have cooled the waste gas stream before it reaches such equipment.

REFERENCES

1. *Steam—Its Generation and Use,* Babcock and Wilcox Company, New York, 1978.
2. R. J. Reed, *Combustion Handbook,* 2nd ed., North American Manufacturing Co., Cleveland, OH, 1978.

---

* Although 0% excess air (stoichiometric air/fuel ratio) is ideal, practical considerations usually dictate operation with 5–10% excess air. During changes in firing rate, time lag in valve operation may result in smoke formation if some excess air is not available prior to the change. Heat exchangers made of 300 series stainless steels may be damaged by alternate oxidation and reduction (particularly in the presence of sulfur). For these reasons, it is wise to have an accurate air/fuel ratio controller with very limited time-delay deviation from air/fuel ratio setpoint.

CHAPTER 69

COOLING ELECTRONIC EQUIPMENT

ALLAN KRAUS

Naval Postgraduate School
Monterey, California

69.1 INTRODUCTION

69.1.1 Thermal Environments

The thermal environmental requirements for military electronic equipment are best indicated by the classifications provided by the MIL-E-5400R series of specifications summarized in Table 69.1. However, the permitted localized environment depends on the reliability requirement of the system in which the component is to be installed and the reliability of the component varies inversely with component temperature (see Section 69.1.3).

The term "environment" is preferred to the term "ambient," and the words "ambient temperature" have been so widely misused that they have become ambiguous and meaningless. The environment is generally called the ambient. However, ambient is frequently used to mean the average temperature of the cooling medium or the temperature of the coolant close to a component. Ambient has also been used to designate the temperature surrounding the carrier vehicle (aircraft, spacecraft, sea-going vehicle, or ground vehicle). When it is understood that ambient temperature refers to the temperature surrounding the component, it is rarely precisely clear which temperature is meant.

The only true way to specify an allowable component temperature is to use a temperature that will exist on or within the component. Then, if the resistance to heat flow between the point selected and some other point is known or can be calculated, allowable heat dissipations can be calculated on a realistic basis.

For example, if the heat-transfer mechanism between the case of a semiconductor device and the surroundings can be predicted by analysis or firm experimental data, then a thermal resistance from case to surroundings, R_{CS}, may be proposed. This resistance, coupled with the junction to case resistance, R_{JC}, permits a calculation for the allowable heat dissipation at a particular surrounding temperature T_S, via an electrothermal analog (Fig. 69.1)

$$q = \frac{T_J - T_S}{R_{CS} + R_{JC}} \tag{69.1}$$

where T_J is the allowable junction temperature.

69.1.2 Component Characteristics

In the selection of electronic components, the environmental requirements are dictated by the characteristics of the individual components. The components themselves are required to meet certain standards that are codified in detailed component specifications. While these specifications apply to both commercial and military equipment, Table 69.2 provides a list of military specifications for some of the more common components.

69.1.3 Component Reliability

Reliability is the probability of a device or system performing without failure for the period intended under the operating conditions encountered. The component failure rate $\lambda(t)$ is a weak function of time over the useful life of the component and may be treated as a constant, λ. The reliability is

$$R = e^{-\lambda t} \tag{69.2}$$

Table 69.1 Environmental Specifications for Airborne Electronic Components (MIL-E-5400R)

Environment	Classification[a]
50,000 ft altitude and continuous sea-level operation over the temperature range −54 to +55°C (+71°C intermittent operation)	Class 1
30,000 ft altitude and continuous sea-level operation over the temperature range −54 to +55°C (+71°C intermittent operation)	Class 1A
15,000 ft altitude and continuous sea-level operation over the temperature range −40 to +55°C (+71°C intermittent operation)	Class 1B
70,000 ft altitude and continuous sea-level operation over the temperature range −54 to +71°C (+95°C intermittent operation)	Class 2
100,000 ft altitude and continuous sea-level operation over the temperature range −54 to +95°C (+125°C intermittent operation)	Class 3
100,000 ft altitude and continuous sea-level operation over the temperature range −54 to +125°C (+150°C intermittent operation)	Class 4

[a] The addition of the letter x (e.g., Class 2x) identifies the equipment as operating in the ambient environment of the class but requiring cooling from a source external to the equipment.

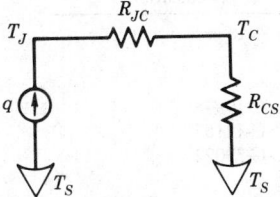

Fig. 69.1 Electrothermal analog circuit.

and the mean time between failures (MTBF) is the reciprocal of the failure rate:

$$\text{MTBF} = \frac{1}{\lambda} \tag{69.3}$$

Failure rates vary directly with component operating temperature and typical failure rate data are available.[1-4] Some failure rates for typical components are provided in Table 69.3[3] where the marked impact of operating temperature may be observed.

System Reliability

The reliability of a system composed of many components can be determined from the reliability of the individual components.

Series Reliability. For the system composed of n components in series in Fig. 69.2a, the total system reliability is the product of the reliabilities of the individual components:

$$R_T = \prod_{k=1}^{n} R_k \tag{69.4}$$

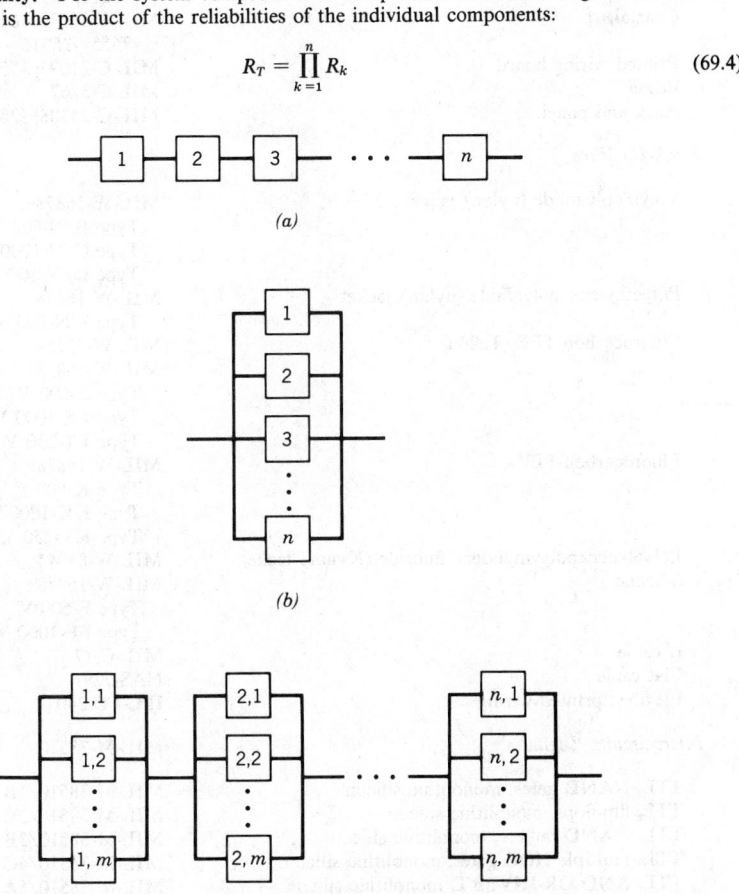

Fig. 69.2 (a) n subsystems or components in series; (b) n subsystems in parallel; and (c) n subsystems in series, each containing m subsystems.

Table 69.2   Governing Specifications for Some Electronic Components

Component	Specification
Capacitors	
Paper/plastic	
Fixed, plastic	MIL-C-19978
Fixed, paper–plastic	MIL-C-14157
Fixed, metalized	MIL-C-39022
Mica	
Fixed	MIL-C-5
Fixed, button style	MIL-C-10950
Fixed, established reliability	MIL-C-39001
Glass	MIL-C-23269
Ceramic	
Fixed, temperature-compensating	MIL-C-20
Fixed, general-purpose	MIL-C-11015
Fixed, established reliability	MIL-C-39014
Electrolytic	
Fixed (DC, aluminum, dry, polarized)	MIL-C-62
Fixed, nonsolid, tantalum	MIL-C-3965
Connectors	
Circular	MIL-C-5015; -26482; -38999; -81511; -83723
Coaxial, rf	MIL-C-3607; -3643; -3650; -3655; -25516; -39012
Printed wiring board	MIL-C-21097; -55302
Power	MIL-C-3767
Rack and panel	MIL-C-24308; -28748; -83733
Hook-Up Wire	
Vinyl/polyamide (nylon) jacket	MIL-W-16878
	Type B/N-600 V
	Type C/N-1000 V
	Type D/N-300 V
Polyethylene/polyamide (nylon) jacket	MIL-W-16878
	Type J/N-600 V
Fluorocarbon-TFE (Teflon)	MIL-W-22759
	MIL-W-16878
	Type E-600 V
	Type EE-1000 V
	Type ET-250 V
Fluorocarbon-FEP	MIL-W-16878
	Type K-600 V
	Type KK-1000 V
	Type KT-250 V
Polyalkene/polyvinylidene fluoride (Kynar) jacket	MIL-W-81044
Silicone	MIL-W-16878
	Type F-600 V
	Type FF-1000 V
rf cable	MIL-C-17
Flat cable	NAS-729
Flexible, printed wiring	IPC-FC-240
Microcircuits, Digital	MIL-M-38510
TTL, NAND gates, monolithic silicon	MIL-M-38510/1B
TTL, flip-flops, monolithic silicon	MIL-M-38510/2E
TTL, NAND buffers, monolithic silicon	MIL-M-38510/3B
TTL, multiple NOR gates, monolithic silicon	MIL-M-38510/4C
TTL, AND-OR-INVERT, monolithic silicon	MIL-M-38510/5A
TTL, binary full address, monolithic silicon	MIL-M-38510/6C
TTL, exclusive-OR gates, monolithic silicon	MIL-M-38510/7B
TTL, buffers/drivers, monolithic silicon	MIL-M-38510/8B
TTL, shift registers, monolithic silicon	MIL-M-38510/9C
TTL, decoders, monolithic silicon	MIL-M-38510/10B

Table **69.2** (*Continued*)

Component	Specification
TTL, arithmetic logic units, monolithic silicone	MIL-M-38510/11C
TTL, monostable multivibrators, monolithic silicone	MIL-M-38510/12E
TTL, counters, monolithic silicone	MIL-M-38510/13C
TTL, data selectors/multiplexers, monolithic silicone	MIL-M-38510/14B
TTL, bistable latches, monolithic silicone	MIL-M-38510/15
DTL, NAND gates, monolithic silicone	MIL-M-38510/30A
DTL, flip-flops, monolithic silicone	MIL-M-38510/33
CMOS, NAND gates, monolithic silicone	MIL-M-38510/50C
CMOS, flip-flops, monolithic silicone	MIL-M-38510/51C
CMOS, NOR gates, monolithic silicone	MIL-M-38510/52B
ECL, flip-flops, monolithic silicone	MIL-M-38510/61A
ECL, AND/NAND gates, monolithic silicone	MIL-M-38510/62
ECL, QUAD translator, monolithic silicone	MIL-M-38510/63
Schottky TTL, NAND gates, monolithic silicone	MIL-M-38510/70
Schottky TTL, flip-flops, monolithic silicone	MIL-M-38510/71A
Schottky TTL, shift registers, monolithic silicone	MIL-M-38510/76
Linear, voltage regulator, monolithic silicone	MIL-M-38510/102A
Linear, transistor arrays, monolithic silicone	MIL-M-38510/108
Linear, precision timers, monolithic silicone	MIL-M-38510/109
Prom, 512-bit bipolar programmable, read-only memory (P-ROM), monolithic silicone	MIL-M-38510/201A

Resistors

Composition, fixed	
Insulated	MIL-R-11
Established reliability	MIL-R-39008
Film, fixed	
High stability	MIL-R-10509
Insulated	MIL-R-22684
Established reliability	MIL-R-39017
Power type	MIL-R-11804
Wire-wound, fixed	
Power type	MIL-R-26
Accurate	MIL-R-93
Power type, chasis mounted	MIL-R-18546
Thermistor	
Insulated	MIL-T-23648
Composition, variable	
Standard	MIL-R-94
Lead screw actuated	MIL-R-22097
Wire-wound, variable	
Low operating temperature	MIL-R-19
Power type	MIL-R-22
Precision	MIL-R-12934
Lead screw actuated	MIL-R-27208
Established reliability	MIL-R-39015

Relays, Electrical, Hermetically Sealed — MIL-R-5757

Component	Specification
Six PDT contacts, low level and 2 A	MIL-R-5757/1
Three stud mounting, DPSTNO & DPDT contacts	MIL-R-5757/3
DPDT, 2 A, 40 mW	MIL-R-5757/13
SPDT, 15 A rms, 12 kV	MIL-R-5757/35
Current responsive, 1.5 A, inductive	MIL-R-5757/46
AC operated, DPDT contacts, 5 A	MIL-R-5757/49
Subminiature, DPDT contacts, 5 A	MIL-R-5757/77
Time-delay, fixed (300–500 msec)	MIL-R-5757/81

Transformers and Inductors — MIL-T-27

Coils, Radio Frequency — MIL-C-15305

Transformers, Pulse, Low Power — MIL-T-21038

Table 69.3   Some Typical Failure Rates[3]

Part Description	λ_b, Failures per Million Hours; Base Failure Rate		ΔT (°C)	Ratio of High to Low Failure Rate
	High Temperature	Low Temperature		
PNP silicon transistors	0.063 at 130°C and 0.3 stress	0.0096 at 25°C and 0.3 stress	105	7:1
NPN silicon transistors	0.033 at 130°C and 0.3 stress	0.0064 at 25°C and 0.3 stress	105	5:1
Glass capacitors	0.047 at 120°C and 0.5 stress	0.001 at 25°C and 0.5 stress	95	47:1
Transformers and coils, MIL-T-27, class Q	0.0267 at 85°C	0.0008 at 25°C	60	33:1
Resistors, carbon composition	0.0065 at 100°C and 0.5 stress	0.0003 at 25°C and 0.5 stress	75	22:1

Parallel Reliability. For the system composed of n components in parallel in Fig. 69.2b, the total system reliability is given by

$$R_T = 1 - \prod_{k=1}^{n} (1 - R_k) \tag{69.5}$$

Series–Parallel Reliability. For the system composed of n groups of components in series where each group of components contains m parallel paths, as shown in Fig. 69.2c where $n = 3$ and $m = 2$, the system reliability will be

$$R_T = 1 - (1 - R^n)^m \tag{69.6}$$

69.1.4 Thermal Control Options

The determination of which heat-transfer process or combination of processes is to be used in a particular situation is linked to Eq. (69.1). If the dissipation, the component, and the environmental temperatures and the component interior (junction to case) resistance are known, then from Eq. (69.1)

$$R_{CS} = \frac{T_J - T_S}{q} - R_{JC} \tag{69.7}$$

The value of R_{CS} is strongly governed by the heat-transfer mode and by the fluid used in the cooling system. These two facts are clearly shown in Fig. 69.3,[5] which presents the temperature difference required to drive a particular heat flux from the surface of a component to the environment. This figure reveals that for typical allowable temperature differences, such as 50°C, natural convection cooling in the presence of radiation in air cannot be relied upon for heat fluxes above 0.06 W/cm². However, if the air is set in motion so that forced convection prevails, the heat flux capability increases about seven times to 0.4 W/cm².

While the designer rarely has control over the selection of a coolant fluid, he or she can augment the heat-transfer surface to either enhance the mode of heat transfer or increase the dissipating area of the component surface.

69.2 HEAT-TRANSFER CORRELATIONS FOR ELECTRONIC EQUIPMENT COOLING

The reader should use the material in this section which pertains to heat-transfer correlations in geometries peculiar to electronic equipment in conjunction with the correlations provided in Chapter 56.

69.2.1 Forced Convection

External Flow on a Plane Surface

For an unheated starting length of the plane surface, x_0, in laminar flow, the local Nusselt number can be expressed by

$$Nu_x = \frac{0.332 Re^{1/2} Pr^{1/3}}{[1 - (x_0/x)^{3/4}]^{1/3}} \qquad (69.8)$$

where Re is the Reynolds number,

$$Re = \frac{\rho V x}{\mu} \qquad (69.9a)$$

Pr is the Prandtl number,

$$Pr = \frac{c_p \mu}{k} \qquad (69.9b)$$

and Nu is the Nusselt number,

$$Nu = \frac{hx}{k} \qquad (69.9c)$$

In the turbulent range ($Re > 3 \times 10^5$) for airflow[6] over Length L

$$Nu = \frac{hL}{k_f} = 0.0280 Re^{4/5} \left[1 + 0.4 \left(\frac{x_0}{x} \right)^{2.75} \right] \qquad (69.10)$$

where the subscript f indicates that the air property should be evaluated at the film temperature, which is the arithmetic mean of the average fluid bulk temperature and the average plate surface temperature (the average boundary layer temperature).

Cylinders in Crossflow

For airflow around single cylinders at all but very low Reynolds numbers, Hilpert[7] has proposed

$$Nu = \frac{hd}{k_f} = B \left(\frac{\rho V_\infty d}{\mu_f} \right)^n \qquad (69.11)$$

where V_∞ is the free stream velocity and where the constants B and n depend on the Reynolds number as indicated in Table 69.4.

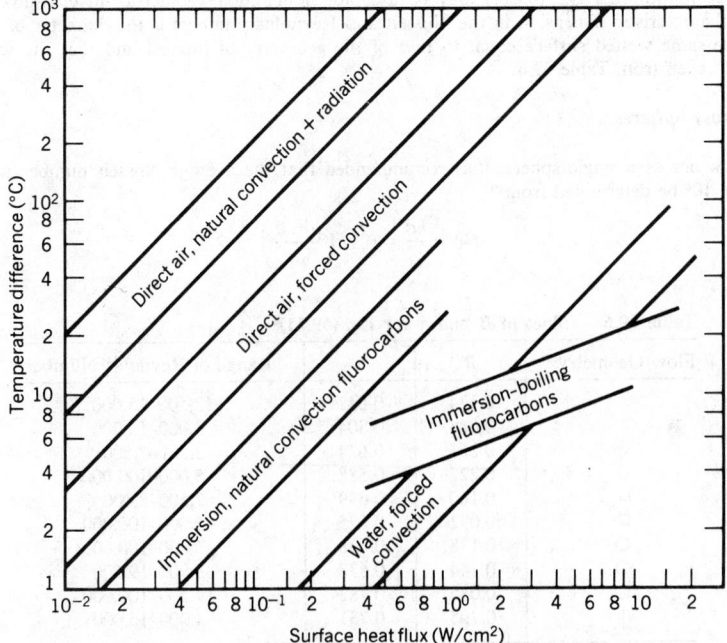

Fig. 69.3 Temperature differences attainable as a function of heat flux for various heat-transfer modes and various coolant fluids.

Table 69.4 Constants for Eq. (69.11)

Reynolds Number Range	B	n
1–4	0.891	0.330
4–40	0.821	0.385
40–4000	0.615	0.466
4000–40,000	0.174	0.618
40,000–400,000	0.0239	0.805

It has been pointed out[8] that Eq. (69.11) assumes a natural turbulence level in the oncoming air stream and that the presence of augmentative devices can increase n by as much as 50%. The modifications to B and n due to some of these devices are displayed in Table 69.5.

Table 69.5 Flow Disturbance Effects on B and n in Eq. (69.11)

Disturbance	Re Range	B	n
1. Longitudinal fin, 0.1d thick on front of tube	1000–4000	0.248	0.603
2. 12 longitudinal grooves, 0.7d wide	3500–7000	0.082	0.747
3. Same as 2 with burrs	3000–6000	0.0368	0.86

Equation (69.11) can be extended to other fluids[9] spanning a range of $1 < \text{Re} < 10^5$ and $0.67 < \text{Pr} < 300$:

$$\text{Nu} = \frac{hd}{k} = (0.4\text{Re}^{0.5} + 0.06\text{Re}^{0.67})\text{Pr}^{0.4}\left(\frac{\mu}{\mu_w}\right)^{0.25} \tag{69.12}$$

where all fluid properties are evaluated at the free stream temperature except μ_w, which is the fluid viscosity at the wall temperature.

Noncircular Cylinders in Crossflow

It has been found[8] that Eq. (69.11) may be used for noncircular geometries in crossflow provided that the characteristic dimension in the Nusselt and Reynolds numbers is the diameter of a cylinder having the same wetted surface equal to that of the geometry of interest and that the values of B and n are taken from Table 69.6.

Flow across Spheres

For airflow across a single sphere, it is recommended that the average Nusselt number when $17 < \text{Re} < 7 \times 10^4$ be determined from[10]

$$\text{Nu} = \frac{hd}{k_f} = 0.37\left(\frac{\rho V_\infty d}{\mu_f}\right)^{0.6} \tag{69.13}$$

Table 69.6 Values of B and n for Eq. (69.11)[a]

Flow Geometry	B	n	Range of Reynolds Number
	0.224	0.612	2,500–15,000
	0.085	0.804	3,000–15,000
◇	0.261	0.624	2,500–7,500
◇	0.222	0.588	5,000–100,000
☐	0.160	0.699	2,500–8,000
☐	0.092	0.675	5,000–100,000
◯	0.138	0.638	5,000–100,000
◯	0.144	0.638	5,000–19,500
◯	0.035	0.782	19,500–100,000
|	0.205	0.731	4,000–15,000

[a] From M. Jakob, *Heat Transfer* Wiley, New York, 1949, by permission of John Wiley & Sons, Inc.

and for $1 < \text{Re} < 25$ (Ref. 11),

$$\text{Nu} = \frac{hd}{k} = 2.2\text{Pr} + 0.48\text{Pr}(\text{Re})^{0.5} \tag{69.14}$$

For both gases and liquids in the range $3.5 < \text{Re} < 7.6 \times 10^4$ and $0.7 < \text{Pr} < 380$ (Ref. 9),

$$\text{Nu} = \frac{hd}{k} = 2 + (4.0\text{Re}^{0.5} + 0.06\text{Re}^{0.67})\text{Pr}^{0.4}\left(\frac{\mu}{\mu_w}\right)^{0.25} \tag{69.15}$$

Flow across Tube Banks

For the flow of fluids flowing normal to banks of tubes,[12]

$$\text{Nu} = \frac{hd}{k_f} = C\left(\frac{\rho V_\infty d}{\mu_f}\right)^{0.6}\left(\frac{c_p \mu}{k}\right)_f^{0.33} \phi \tag{69.16}$$

which is valid in the range $2000 < \text{Re} < 32,000$.

For in-line tubes, $C = 0.26$, whereas for staggered tubes, $C = 0.33$. The factor ϕ is a correction factor for sparse tube banks, and values of ϕ are provided in Table 69.7.

For air in the range where Pr is nearly constant ($\text{Pr} \approx 0.7$ over the range 25–200°C), Eq. (69.16) can be reduced to

$$\text{Nu} = \frac{hd}{k_f} = C'\left(\frac{\rho V_\infty d}{\mu_f}\right)^{n'} \tag{69.17}$$

where C' and n' may be determined from values listed in Table 69.8. This equation is valid in the range $2000 < \text{Re} < 40,000$ and the ratios x_L and x_T denote the ratio of centerline diameter to tube spacing in the longitudinal and transverse directions, respectively.

For fluids other than air, the curve shown in Fig. 69.4 should be used for staggered tubes.[10] For in-line tubes, the values of

$$j = \left(\frac{hd_0}{k}\right)\left(\frac{c_p \mu}{k}\right)^{-1/3}\left(\frac{\mu}{\mu_w}\right)^{-0.14}$$

should be reduced by 10%.[13]

Flow across Arrays of Pin Fins

For air flowing normal to banks of staggered cylindrical pin fins or spines,[14]

$$\text{Nu} = \frac{hd}{k} = 1.40\left(\frac{\rho V_\infty d}{\mu}\right)^{0.8}\left(\frac{c_p \mu}{k}\right)^{1/3} \tag{69.18}$$

Flow of Air over Electronic Components

For single prismatic electronic components, either normal or parallel to the sides of the component in a duct,[15] for $2.5 \times 10^3 < \text{Re} < 8 \times 10^3$,

$$\text{Nu} = 0.446\left[\frac{\text{Re}}{(1/6) + (5A_n/6A_0)}\right]^{0.57} \tag{69.19}$$

Table 69.7   Correlation Factor ϕ for Sparse Tube Banks

Number of rows, N	In Line	Staggered
1	0.64	0.68
2	0.80	0.75
3	0.87	0.83
4	0.90	0.89
5	0.92	0.92
6	0.94	0.95
7	0.96	0.97
8	0.98	0.98
9	0.99	0.99
10	1.00	1.00

Table 69.8 Values of the Constants C' and n' in Eq. (69.17)

$x_L = \dfrac{S_L}{d_0}$	$x_T = \dfrac{S_T}{d_0} = 1.25$		$x_T = \dfrac{S_T}{d_0} = 1.50$		$x_T = \dfrac{S_T}{d_0} = 2.00$		$x_T = \dfrac{S_T}{d_0} = 3.00$	
	C'	n'	C'	n'	C'	n'	C'	n'
Staggered								
0.600							0.213	0.636
0.900					0.446	0.571	0.401	0.581
1.000			0.497	0.558				
1.125					0.478	0.565	0.518	0.560
1.250	0.518	0.556	0.505	0.554	0.519	0.556	0.522	0.562
1.500	0.451	0.568	0.460	0.562	0.452	0.568	0.488	0.568
2.000	0.404	0.572	0.416	0.568	0.482	0.556	0.449	0.570
3.000	0.310	0.592	0.356	0.580	0.440	0.562	0.421	0.574
In line								
1.250	0.348	0.592	0.275	0.608	0.100	0.704	0.0633	0.752
1.500	0.367	0.586	0.250	0.620	0.101	0.702	0.0678	0.744
2.000	0.418	0.570	0.299	0.602	0.229	0.632	0.198	0.648
3.000	0.290	0.601	0.357	0.584	0.374	0.581	0.286	0.608

where the Nusselt and Reynolds numbers are based on the prism side dimension and where A_0 and A_n are the gross and net flow areas, respectively.

For staggered prismatic components, Eq. (69.19) may be modified to[15]

$$\mathrm{Nu} = 0.446 \left[\frac{\mathrm{Re}}{(1/6) + (5A_n/A_0)} \right]^{0.57} \left[1 + 0.639 \left(\frac{S_T}{S_{T,\,\mathrm{max}}} \right) \left(\frac{d}{S_L} \right)^{0.172} \right] \tag{69.20}$$

where d is the prism side dimension, S_L is the longitudinal separation, S_T is the transverse separation, and $S_{T,\,\mathrm{max}}$ is the maximum transverse spacing if different spacings exist.

When cylindrical heat sources are encountered in electronic equipment, a modification of Eq. (69.11) has been proposed[16]:

$$\mathrm{Nu} = \frac{hd}{k_f} = FB \left(\frac{\rho V_\infty d}{\mu} \right)^n \tag{69.21}$$

where F is an arrangement factor depending on the cylinder geometry (see Table 69.9) and where the constants B and n are given in Table 69.10.

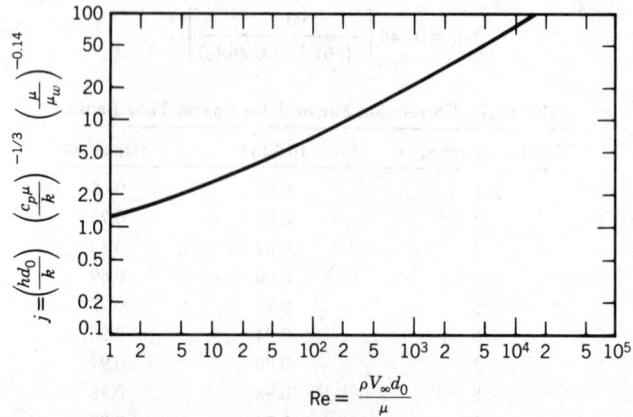

Fig. 69.4 Recommended curve for estimation of heat transfer coefficient for fluids flowing normal to staggered tubes 10 rows deep.[10]

Table 69.9 Values of F to Be Used in Eq. (69.21)[a]

Single cylinder in free stream: $F = 1.0$
Single cylinder in duct: $F = 1 + d/w$
In-line cylinders in duct:

$$F = \left(1 + \sqrt{\frac{1}{S_T}}\right)\left\{1 + \left(\frac{1}{S_L} - \frac{0.872}{S_L^2}\right)\left(\frac{1.81}{S_T^2} - \frac{1.46}{S_T} + 0.318\right)\left[Re^{0.526-(0.354/S_T)}\right]\right\}$$

Staggered cylinders in duct:

$$F = \left(1 + \sqrt{\frac{1}{S_T}}\right)\left\{1 + \left[\frac{1}{S_L}\left(\frac{15.50}{S_T^2} - \frac{16.80}{S_T} + 4.15\right) - \frac{1}{S_L}\left(\frac{14.15}{S_T^2} - \frac{15.33}{S_T} + 3.69\right)\right]Re^{0.13}\right\}$$

[a] Re to be evaluated at film temperature. S_L = ratio of longitudinal spacing to cylinder diameter. S_T = ratio of transverse spacing to cylinder diameter.

Table 69.10 Values of B and n for Use in Eq. (69.21)

Reynolds Number Range	B	n
1000–6000	0.409	0.531
6000–30,000	0.212	0.606
30,000–100,000	0.139	0.806

Forced Convection in Tubes, Pipes, Ducts, and Annuli

For heat transfer in tubes, pipes, ducts, and annuli, use is made of the equivalent diameter

$$d_e = \frac{4A}{WP} \tag{69.22}$$

in the Reynolds and Nusselt numbers unless the cross section is circular, in which case d_e and $d_i = d$.
  In the laminar regime[17] where Re < 2100,

$$Nu = hd_e/k = 1.86\,[RePr(d_e/L)]^{1/3}\,(\mu/\mu_w)^{0.14} \tag{69.23}$$

with all fluid properties except μ_w evaluated at the bulk temperature of the fluid.
  For Reynolds numbers above transition, Re > 2100,

$$Nu = 0.023(Re)^{0.8}\,(Pr)^{1/3}\,(\mu/\mu_w)^{0.14} \tag{69.24}$$

and in the transition region, 2100 < Re < 10,000,[18]

$$Nu = 0.116\,[(Re)^{2/3} - 125]\,(Pr)^{1/3}\,(\mu/\mu_w)^{0.14}\,[1 + (d_e/L)^{2/3}] \tag{69.25}$$

  London[19] has proposed a correlation for the flow of air in rectangular ducts. It is shown in Fig. 69.5. This correlation may be used for air flowing between longitudinal fins.

69.2.2 Natural Convection in Confined Spaces

For natural convection in confined horizontal spaces the recommended correlations for air are[8]

$$\begin{array}{ll} Nu = 0.195(Gr)^{1/4}, & 10^4 < Gr < \quad 4 \times 10^5 \\ Nu = 0.068(Gr)^{1/3}, & \quad\quad Gr > \quad 10^5 \end{array} \tag{69.26}$$

where Gr is the Grashof number,

$$Gr = \frac{g\rho^2\beta L^3\Delta T}{\mu^2} \tag{69.27}$$

and where, in this case, the significant dimension L is the gap spacing in both the Nusselt and Grashof numbers.
  For liquids[20]

$$Nu = 0.069(Gr)^{1/3}Pr^{0.407}, \quad 3 \times 10^5 < Ra < 7 \times 10^9 \tag{69.28}$$

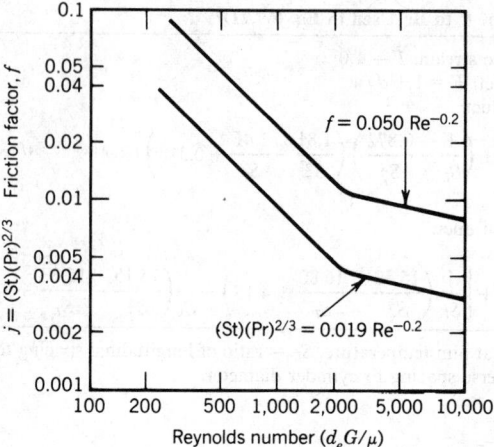

Fig. 69.5 Heat transfer and friction data for forced air through rectangular ducts. St is the Stanton number, $St = hG/c_p$.

where Ra is the Rayleigh number,

$$Ra = GrPr \tag{69.29}$$

For horizontal gaps with $Gr < 1700$, the conduction mode predominates and

$$h = \frac{k}{b} \tag{69.30}$$

where b is the gap spacing. For $1700 < Gr < 10,000$, use may be made of the Nusselt–Grashof relationship given in Fig. 69.6.[21,22]

For natural convection in confined vertical spaces containing air, the heat-transfer coefficient depends on whether the plates forming the space are operating under isoflux or isothermal conditions.[23]

For the symmetric isoflux case, a case that closely approximates the heat transfer in an array of printed circuit boards, the correlation for Nu is formed by using the method of Churchhill and Usagi[24] by considering the isolated plate case[25-27] and the fully developed limit[28]:

$$Nu = \left[\frac{12}{Ra''} + \frac{1.88}{(Ra'')^{2/5}}\right]^{-1/2} \tag{69.31}$$

where Ra'' is the modified channel Rayleigh number,

$$Ra'' = \frac{g\beta\rho^2 q'' c_p b^5}{\mu k^2 L} \tag{69.32}$$

The optimum spacing for the symmetrical isoflux case is

$$b_{opt} = 1.472R^{-0.2} \tag{69.33}$$

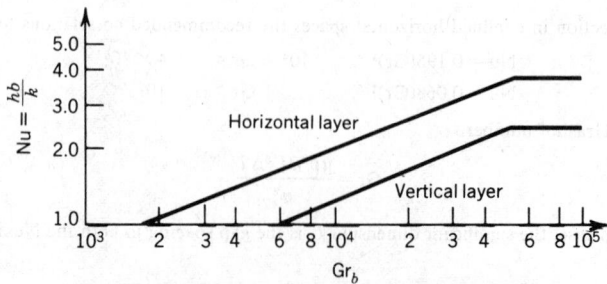

Fig. 69.6 Heat transfer through enclosed air layers.[21,22]

where

$$R = \frac{g\beta\rho^2 c_p q''}{\mu k^2 L} \tag{69.34}$$

For the symmetric isothermal case, a case that closely approximates the heat transfer in a vertical array of extended surface or fins, the correlation is again formed using the Churchhill and Usagi[24] method by considering the isolated plate case[10] and the fully developed limit[28-30]:

$$Nu = \left[\frac{576}{(Ra')^2} + \frac{2.873}{(Ra')^{1/2}} \right]^{-1/2} \tag{69.35}$$

where Ra' is the channel Rayleigh number

$$Ra' = \frac{g\beta\rho^2 c_p b^4}{\mu k L} \tag{69.36}$$

The optimum spacing for the symmetrical isothermal case is

$$b_{opt} = \frac{2.714}{P^{1/4}} \tag{69.37}$$

where

$$P = \frac{g\beta\rho^2 c_p \Delta T}{\mu k L} \tag{69.38}$$

69.2.3 Contact Resistance

In a package of electronic components, many components are "buried" within the package enclosure and there are many metal-to-metal interfaces or contacts between the component and the ultimate heat sink. Each of these contacts provides a thermal contact resistance and exhibits a temperature differential. It is necessary to account for each of these temperature differentials in calculating the component operating temperature. In Fig. 69.7, ΔT_c is the temperature drop across the contact and a contact conductance may be defined as

$$h_c = \frac{q}{A_c \Delta T_c} \tag{69.39}$$

Factors Influencing Contact Resistance

The heat transfer across an interface of two materials in contact is a very complex phenomenon and is a function of many parameters. The following appear to be of the greatest importance:

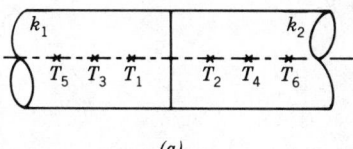

(a)

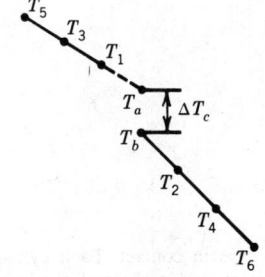

(b)

Fig. 69.7 Two solid metal bars in contact: (*a*) temperature designations for simulated probe; (*b*) linear temperature profile for this real interface condition.

1. The number of contact spots.
2. The shape of the contact spots: circular, elliptic, band, or rectangular.
3. The size of the contact spots.
4. The disposition or arrangement of the contact spots.
5. The geometry of the contacting surfaces with regard to roughness and waviness.
6. The average thickness of the void space (the noncontact region).
7. The fluid in the void space: gas, liquid, grease, vacuum.
8. The pressure of the fluid in the void space.
9. Thermal conductivities of the contacting solids and of the fluid in the void space.
10. The hardness of the contacting asperities.
11. The moduli of elasticity of the contacting asperities.
12. The average temperature of the interface.
13. The past history of the contact with regard to the number of previous matings.
14. The contact pressure.
15. The duration of the contact with regard to relaxation effects.
16. Vibration effects.
17. Directional effects.
18. Contact cleanliness.

Correlations for the Contact Coefficient

For smooth, wavy surfaces in a vacuum, a study[31] of two circular cylinders, each with radius b, the same Poisson ratio, and smooth hemispherical caps (see Fig. 69.8), provided a correlation

$$\frac{hb}{k} = \frac{2}{\pi} \frac{\epsilon}{f(\epsilon)} \qquad (\epsilon < 0.65) \qquad\qquad (69.40)$$

where with P_a, the apparent contact pressure,

$$\epsilon = 1.285 \left(\frac{P_a b}{Ed} \right)^{1/3}$$

$$f(\epsilon) = 1 - 1.40925\epsilon + 0.29591\epsilon^3 + 0.05204\epsilon^5$$

$$P_a = \frac{F}{\pi b^2}$$

where F is the load and for the two materials (material 1 and material 2)

$$k = \frac{2k_1 k_2}{k_1 + k_2}, \qquad E = \frac{2E_1 E_2}{E_1 + E_2}$$

and for the values of d shown in Fig. 69.8

$$d = d_1 + d_2$$

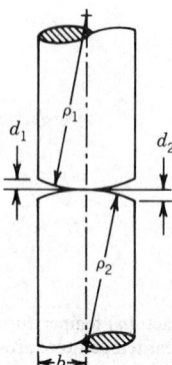

Fig. 69.8 Two cylinders with hemispherical caps in contact. Each cylinder has a modulus of elasticity E, a Poisson ratio ν, and a thermal conductivity k.

For nominally flat, rough surfaces in vacuum[32]

$$h = 1.45 \frac{k(P_a/H)^{0.985}}{\sigma} |\tan \bar{\theta}| \qquad (69.41)$$

where again

$$k = \frac{2k_1 k_2}{k_1 + k_2}$$

and P_a is the apparent contact pressure and H is the hardness of the softer material.
The rms roughness is

$$\sigma = (\sigma_1^2 + \sigma_2^2)^{1/2}$$

and $|\tan \bar{\theta}|$ is the average absolute asperity angle as indicated in Fig. 69.9.
For rough, wavy surfaces in vacuum[33]

$$h_c = 1 \left/ \left[\frac{\delta \psi(\epsilon)}{\sqrt{\pi} n k \epsilon} + \frac{\delta \psi(\lambda_e)}{k \lambda_e} \right] \right. \qquad (69.42)$$

where P_a, k, E, and H are as previously defined, n is the number of contacts, δ is the void space thickness, and

$$\epsilon^2 \lambda_e^2 = \frac{P_a}{H}$$

$$\lambda_e = 1.285 \left(\frac{P_a b}{2Ed} \right)^{1/3}$$

with b the apparent contact radius and d the "out of flatness" of solids 1 and 2 and[34]

$$\psi(x) = 1 - \frac{4}{\pi} x^2 - x^2$$

where x can take on values of either λ_e or ϵ.

69.3 THERMAL CONTROL TECHNIQUES

69.3.1 Extended Surface and Heat Sinks

The heat flux from a surface, q/A, can be reduced if the surface area A is increased. The use of extended surface or fins in a common method of achieving this reduction. Another way of looking at this is through the use of Newton's law of cooling:

$$q = hA \Delta T \qquad (69.43)$$

and considering that ΔT can be reduced for a given heat flow q by increasing h, which is difficult for a specified coolant, or by increasing the surface area A.

The common extended surface shapes are the longitudinal fin of rectangular profile, the radial fin of rectangular profile, and the cylindrical spine shown, respectively, in Figs. 69.10a, 69.10b, and 69.10c.

Assumptions in Extended Surface Analysis

The analysis of extended surface is subject to the following simplifying assumptions[35,36]:

1. The heat flow is steady; that is, the temperature at any point does not vary with time.
2. The fin material is homogeneous, and the thermal conductivity is constant and uniform.
3. The coefficient of heat transfer is constant and uniform over the entire face surface of the fin.
4. The temperature of the surrounding fluid is constant and uniform.

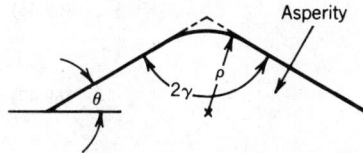

Asperity

Fig. 69.9 Nomenclature for an asperity. The slope of the asperity is the tangent of the angle θ. The radius of curvature of the asperity is ρ. Half of the vertex angle is the angle γ.

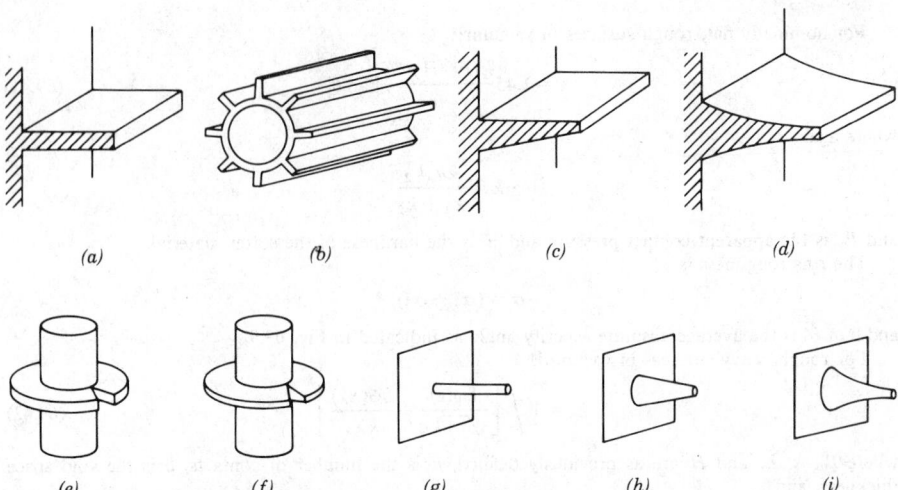

Fig. 69.10 Some typical examples of extended surfaces: (a) longitudinal fin of rectangular profile; (b) cylindrical tube equipped with longitudinal fins; (c) longitudinal fin of trapezoidal profile; (d) longitudinal fin of truncated concave parabolic profile; (e) cylindrical tube equipped with radial fin of rectangular profile; (f) cylindrical tube equipped with radial fin of truncated triangular profile; (g) cylindrical spine; (h) truncated conical spine; (i) truncated concave parabolic spine.

5. There are no temperature gradients within fin other than along the fin height.
6. There is no bond resistance to the flow of heat at the base of the fin.
7. The temperature at the base of the fin is uniform and constant.
8. There are no heat sources within the fin itself.
9. There is a negligible flow of heat from the tip and sides of the fin.
10. The heat flow from the fin is proportioned to the temperature difference or temperature excess, $\theta(x) = T(x) - T_s$, at any point on the face of the fin.

The Fin Efficiency

Because a temperature gradient always exists along the height of a fin when heat is being transferred to the surrounding environment by the fin, there is a question regarding the temperature to be used in Eq. (69.43). If the base temperature T_b (and the base temperature excess, $\theta_b = T_b - T_s$) is to be used, then the surface area of the fin must be modified by the computational artifice known as the fin efficiency, defined as the ratio of the heat actually transferred by the fin to the ideal heat transferred if the fin were operating over its entirety at the base temperature excess. In this case, the surface area A in Eq. (69.43) becomes

$$A = A_b + \eta_f A_f \tag{69.44}$$

The Longitudinal Fin of Rectangular Profile

With the origin of the height coordinate x taken at the fin tip, which is presumed to be adiabatic, the temperature excess at any point on the fin is

$$\theta(x) = \theta_b \frac{\cosh mx}{\cosh mb} \tag{69.45}$$

where

$$m = \left(\frac{2h}{k\delta}\right)^{1/2} \tag{69.46}$$

The heat dissipated by the fin is

$$q_b = Y_0 \theta_b \tanh mb \tag{69.47}$$

where Y_0 is called the characteristic admittance

$$Y_0 = (2hk\delta)^{1/2} L \tag{69.48}$$

and the fin efficiency is

$$\eta_f = \frac{\tanh mb}{mb} \tag{69.49}$$

The heat-transfer coefficient in natural convection may be determined from the symmetric isothermal case pertaining to vertical plates in Section 69.2.2. For forced convection, the London correlation described in Section 69.2.1 applies.

The Radial Fin of Rectangular Profile

With the origin of the radial height coordinate taken at the center of curvature and with the fin tip at $r = r_a$ presumed to be adiabatic, the temperature excess at any point on the fin is

$$\theta(r) = \theta_b \left[\frac{K_1(mr_a)I_0(mr) + I_1(mr_a)K_0(mr)}{I_0(mr_b)K_1(mr_a) + I_1(mr_a)K_0(mr_b)} \right] \tag{69.50}$$

where m is given by Eq. (69.46). The heat dissipated by the fin is

$$q_b = 2\pi r_b km\,\theta_b \left[\frac{I_1(mr_a)K_1(mr_b) - K_1(mr_a)I_1(mr_b)}{I_0(mr_b)K_1(mr_a) + I_1(mr_a)K_0(mr_b)} \right] \tag{69.51}$$

and the fin efficiency is

$$\eta_f = \frac{2r_b}{m(r_a^2 - r_b^2)} \left[\frac{I_1(mr_a)K_1(mr_b) - K_1(mr_a)I_1(mr_b)}{I_0(mr_b)K_1(mr_a) + I_1(mr_a)K_0(mr_b)} \right] \tag{69.52}$$

Tables of the fin efficiency are available,[37] and they are organized in terms of two parameters, the radius ratio

$$\rho = \frac{r_b}{r_a} \tag{69.53a}$$

and a parameter ϕ

$$\phi = (r_a - r_b)\left(\frac{2h}{kA_p}\right)^{1/2} \tag{69.53b}$$

where A_p is the profile area of the fin:

$$A_p = \delta(r_a - r_b) \tag{69.53c}$$

For air under forced convection conditions, the correlation for the heat-transfer coefficient developed by Briggs and Young[38] is applicable:

$$\frac{h}{2r_b k} = \left(\frac{2\rho Vr_b}{\mu}\right)^{0.681}\left(\frac{c_p\mu}{k}\right)^{1/3}\left(\frac{s}{r_a - r_b}\right)^{0.200}\left(\frac{s}{\delta}\right)^{0.1134} \tag{69.54}$$

where all thermal properties are evaluated at the bulk air temperature, s is the space between the fins, and r_a and r_b pertain to the fins.

The Cylindrical Spine

With the origin of the height coordinate x taken at the spine tip, which is presumed to be adiabatic, the temperature excess at any point on the spine is given by Eq. (69.45), but for the cylindrical spine

$$m = \left(\frac{4h}{kd}\right)^{1/2} \tag{69.55}$$

where d is the spine diameter. The heat dissipated by the spine is given by Eq. (69.47), but in this case

$$Y_0 = (\pi^2 hkd^3)^{1/2}/2 \tag{69.56}$$

and the spine efficiency is given by Eq. (69.49).

Algorithms for Combining Single Fins into Arrays

The differential equation for temperature excess that can be developed for any fin shape can be solved to yield a particular solution, based on prescribed initial conditions of fin base temperature excess and fin base heat flow, that can be written in matrix form[39,40] as

$$\begin{bmatrix} \theta_a \\ q_a \end{bmatrix} = [\Gamma] \begin{bmatrix} \theta_b \\ q_b \end{bmatrix} = \begin{bmatrix} \gamma_{11} & \gamma_{12} \\ \gamma_{21} & \gamma_{22} \end{bmatrix} \begin{bmatrix} \theta_b \\ q_b \end{bmatrix} \tag{69.57}$$

The matrix $[\Gamma]$ is called the thermal transmission matrix and provides a linear transformation from tip to base conditions. It has been cataloged for all of the common fin shapes.[39-41] For the longitudinal fin of rectangular profile

$$[\Gamma] = \begin{bmatrix} \cosh mb & -\dfrac{1}{Y_0} \sinh mb \\ -Y_0 \sinh mb & \cosh mb \end{bmatrix} \tag{69.58}$$

and this matrix possesses an inverse called the inverse thermal transmission matrix

$$[\Lambda] = [\Gamma]^{-1} = \begin{bmatrix} \cosh mb & \dfrac{1}{Y_0} \sinh mb \\ Y_0 \sinh mb & \cosh mb \end{bmatrix} \tag{69.59}$$

The assembly of fins into an array may require the use of any or all of three algorithms.[39-41] The objective is to determine the input admittance of the entire array

$$Y_{\text{in}} = \frac{q_b}{\theta_b} \bigg|_A \tag{69.60}$$

which can be related to the array (fin) efficiency by

$$\eta_f = \frac{Y_{\text{in}}}{hA_f} \tag{69.61}$$

The determination of Y_{in} can involve as many as three algorithms for the combination of individual fins into an array.

The Cascade Algorithm: For n fins in cascade as shown in Fig. 69.11a, an equivalent inverse thermal transmission matrix can be obtained by a simple matrix multiplication, with the individual fins closest to the base of the array acting as permultipliers:

$$\{\Lambda\}_e = \{\Lambda\}_n \{\Lambda\}_{n-1} \{\Lambda\}_{n-2} \cdots \{\Lambda\}_2 \{\Lambda\}_1 \tag{69.62}$$

For the case of the tip of the most remote fin adiabatic, the array input admittance will be

$$Y_{\text{in}} = \frac{\lambda_{21,e}}{\lambda_{11,e}} \tag{69.63}$$

If the tip of the most remote fin is not adiabatic, the heat flow to temperature excess ratio at the tip which is designated as μ

$$\mu = \frac{q_a}{\theta_a} \tag{69.64}$$

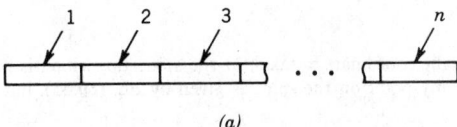

(a)

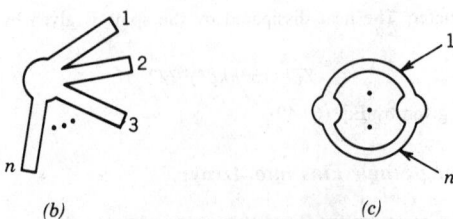

(b)   *(c)*

Fig. 69.11 (a) n fins in cascade, (b) n fins in cluster, and (c) n fins in parallel.

will be known. For example, for a fin dissipating to the environment through its tip designated by the subscript a:

$$\mu = hA_a \tag{69.65}$$

In this case, Y_{in} may be obtained through successive use of what is termed the reflection relationship (actually a bilinear transformation):

$$Y_{in,k-1} = \frac{\lambda_{21,k-1} + \lambda_{22,k-1}(q_a/\theta_a)}{\lambda_{11,k-1} + \lambda_{12,k-1}(q_a/\theta_a)} \tag{69.66}$$

The Cluster Algorithm. For n fins in cluster, as shown in Fig 69.11b, the equivalent thermal transmission ratio will be the sum of the individual fin input admittances:

$$\mu_e = \sum_{k=1}^{n} Y_{in,k} = \sum_{k=1}^{n} \frac{q_b}{\theta_b}\bigg|_k \tag{69.67}$$

Here, $Y_{in,k}$ can be determined for each individual fin via Eq. (69.64) if the fin has an adiabatic tip or via Eq. (69.66) if the tip is not adiabatic. It is obvious that this holds if subarrays containing more than one fin are in cluster.

The Parallel Algorithm. For n fins in parallel, as shown in Fig. 69.11c, an equivalent thermal admittance matrix $[Y]_e$ can be obtained from the sum of the individual thermal admittance matrices:

$$[Y]_e = \sum_{k=1}^{n} [Y]_k \tag{69.68}$$

where the individual thermal admittance matrices can be obtained from

$$[Y] = \begin{bmatrix} y_{11} & y_{12} \\ y_{21} & y_{22} \end{bmatrix} = \begin{bmatrix} -\dfrac{\gamma_{11}}{\gamma_{12}} & \dfrac{1}{\gamma_{12}} \\ -\dfrac{1}{\gamma_{12}} & \dfrac{\gamma_{22}}{\gamma_{12}} \end{bmatrix} = \begin{bmatrix} \dfrac{\lambda_{22}}{\lambda_{12}} & -\dfrac{1}{\lambda_{12}} \\ \dfrac{1}{\lambda_{12}} & -\dfrac{\lambda_{21}}{\lambda_{22}} \end{bmatrix} \tag{69.69}$$

If necessary, $[\Lambda]$ may be obtained from $[Y]$ using

$$[\Lambda] = \begin{bmatrix} \lambda_{11} & \lambda_{12} \\ \lambda_{21} & \lambda_{22} \end{bmatrix} = \begin{bmatrix} -\dfrac{y_{22}}{y_{21}} & \dfrac{1}{y_{21}} \\ -\dfrac{\Delta_Y}{y_{21}} & \dfrac{y_{11}}{y_{21}} \end{bmatrix} \tag{69.70}$$

where $\Delta_Y = y_{11}y_{22} - y_{12}y_{21}$

Singular Fins. There will be occasions when a singular fin, one whose tip comes to a point, will be used as the most remote fin in an array. In this case the $[\Gamma]$ and $[\Lambda]$ matrices do not exist and the fin is characterized by its input admittance.[39-41] Such a fin is the longitudinal fin of triangular profile where

$$Y_{in} = \frac{q_b}{\theta_b} = \frac{2hI_1(2mb)}{mI_0(2mb)} \tag{69.71}$$

where

$$m = \left(\frac{2h}{k\delta_b}\right)^{1/2} \tag{69.72}$$

69.3.2 The Cold Plate

The cold plate heat exchanger or forced cooled electronic chassis is used to provide a "cold wall" to which individual components and, for that matter, entire packages of equipment may be mounted. Its design and performance evaluation follows a certain detailed procedure that depends on the type of heat loading and whether the heat loading is on one or two sides of the cold plate. These configurations are displayed in Fig 69.12.

The design procedure is based on matching the available heat-transfer effectiveness ϵ to the required effectiveness ϵ determined from the design specifications. These effectivenesses are for the isothermal case in Fig. 69.12a

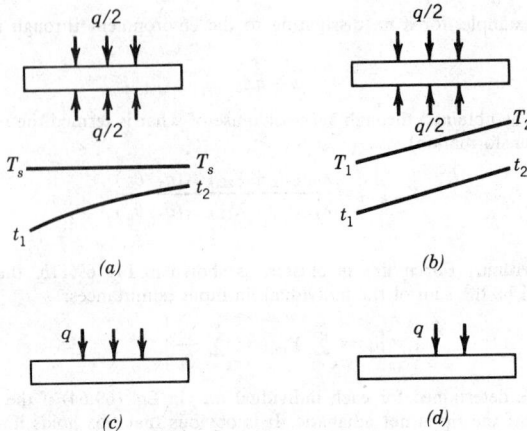

Fig. 69.12 (a) Double-sided, evenly loaded cold plate—isothermal case; (b) double-sided, evenly loaded cold plate—isoflux case; (c) single-sided, evenly loaded cold plate—isothermal case; and (d) single-sided, evenly loaded cold plate—isoflux case.

$$\epsilon = \frac{t_2 - t_1}{T_s - t_1} = 1 - e^{-N_{tu}} \tag{69.73}$$

and for the isoflux case in Fig. 69.12b

$$\epsilon = \frac{t_2 - t_1}{T_2 - t_1} \tag{69.74}$$

where the "number of transfer units" is

$$N_{tu} = \frac{h \eta_0 A}{W c_p} \tag{69.75}$$

and the overall passage efficiency is

$$\eta_0 = 1 - \frac{A_f}{A}(1 - \eta_f) \tag{69.76}$$

The surfaces to be used in the cold plate are those described by Kays and London[42] where physical, heat-transfer, and friction data are provided.

The detailed design procedure for the double-side-loaded isothermal case is as follows:

1. Design specification
   - (a) Heat load, q, W
   - (b) Inlet air temperature, t_1, °C
   - (c) Airflow, W, kg/sec
   - (d) Allowable pressure loss, cm H_2O
   - (e) Overall envelope, H, W, D
   - (f) Cold plate material thermal conductivity, k_m, W/m·°C
   - (g) Allowable surface temperature, T_s, °C
2. Select surface[42]
   - (a) Type
   - (b) Plate spacing, b, m
   - (c) Fins per meter, fpm
   - (d) Hydraulic diameter, d_e, m
   - (e) Fin thickness, δ, m
   - (f) Heat transfer area/volume, β, m²/m³
   - (g) Fin surface area/total surface area, A_f/A, m²/m³
3. Plot j and f data from data[42]

$$j = (St)(Pr)^{2/3} = f_1(Re) = f_1\left(\frac{d_e G}{\mu}\right)$$

where St is the Stanton number

$$\text{St} = \frac{hG}{c_p} \tag{69.77}$$

and f is the friction factor

$$f = f_2(\text{Re}) = f_2\left(\frac{d_e G}{\mu}\right)$$

4. Establish physical data
   (a) $\alpha = (b/H)\beta$, m²/m³
   (b) $r_h = d_e/4$, m
   (c) $\sigma = \alpha r_h$
   (d) $A_{fr} = WH$, m²   (frontal area)
   (e) $A_c = \sigma A_{fr}$, m²   (flow areas)
   (f) $V = DWH$ (volume)
   (g) $A = \alpha V$, m²   (total surface)

5. Heat balance
   (a) Assume average fluid specific heat, c_p, J/kg·°C
   (b) $\Delta t = t_2 - t_1 = q/Wc_p$, °C
   (c) $t_2 = t_1 + \Delta t$, °C
   (d) $t_{av} = \frac{1}{2}(t_1 + t_2)$
   (e) Check assumed value of c_p. Make another assumption if necessary

6. Fluid properties at t_{av}
   (a) c_p (already known), J/kg·°C
   (b) μ, N/sec·m²
   (c) k, W/m·°C
   (d) $(\text{Pr})^{2/3} = (c_p \mu/k)^{2/3}$

7. Heat-transfer coefficient
   (a) $G = W/A_c$, kg/sec·m²
   (b) $\text{Re} = d_e G/\mu$
   (c) Obtain j from curve (see item 3)
   (d) Obtain f from curve (see item 3)
   (e) $h = jGc_p/(\text{Pr})^{2/3}$, W/m²·°C

8. Fin efficiency
   (a) $m = (2h/k\delta)^{1/2}$, m⁻¹
   (b) $mb/2$ is a computation
   (c) $\eta_f = (\tanh mb/2)/mb/2$

9. Overall passage efficiency
   (a) Use Eq. (69.76)

10. Effectiveness
    (a) Required $\epsilon = (t_2 - t_1)/(T_s - T_1)$
    (b) Form N_{tu} from Eq. (69.75)
    (c) Actual available $\epsilon = 1 - e^{-N_{tu}}$
    (d) Compare required ϵ and actual ϵ and begin again with step 1 if comparison fails. If comparison is satisfactory go on to pressure loss calculation

11. Pressure loss
    (a) Establish v_1 (specific volume), m³/kg
    (b) Establish v_2, m³/kg
    (c) $v_m = \frac{1}{2}(v_1 + v_2)$, m³/kg
    (d) Form v_m/v_1
    (e) Form v_2/v_1
    (f) Obtain K_c and K_e (Ref. 42)
    (g) Determine ΔP, cm

$$\Delta P = 0.489 \frac{G^2 v_1}{2g} \left[(1 + K_c - \sigma^2) + f \frac{A}{A_c} \frac{v_m}{v_1} + 2 \left(\frac{v_2}{v_1} - 1 \right) - (1 - \sigma^2 - K_e) \frac{v_2}{v_1} \right] \qquad (69.78)$$

(h) Compare ΔP with specified ΔP. If comparison fails select a different surface or adjust the dimensions and begin again with step 1

If the cold plate is loaded on one side only, an identical procedure is followed except in steps 8 and 9. For single-side loading and for double and triple stacks, use must be made of the cascade and cluster algorithms for the combination of fins described in Section 69.3.1. Detailed examples of both of the foregoing cases may be found in Ref. 43.

69.3.3 Thermoelectric Coolers

Two thermoelectric effects are traditionally considered in the design and performance evaluation of a thermoelectric cooler:

The Seebeck effect concerns the net conversion of thermal energy into electrical energy under zero current conditions when two dissimilar materials are brought into contact. When the junction temperature differs from a reference temperature, the effect is measured as a voltage called the Seebeck voltage E_s.

The Peltier effect concerns the reversible evolution or absorption of heat that occurs when an electric current traverses the junction between two dissimilar materials. The Peltier heat absorbed or rejected depends on and is proportional to the current flow. There is an additional thermoelectric effect known as the Thomson effect, which concerns the reversible evolution or absorption of heat that occurs when an electric current traverses a single homogeneous material in the presence of a temperature gradient. This effect, however, is a negligible one and is neglected in considerations of thermoelectric coolers operating over moderate temperature differentials.

Equations for the Thermoelectric Effects

Given a pair of thermoelectric materials, A and B, with each having a thermoelectric power α_A and α_B,[44] the Seebeck coefficient is

$$\alpha = |\alpha_A| + |\alpha_B| \qquad (69.79)$$

The Seebeck coefficient is the proportionality constant between the Seebeck voltage and the junction temperature with respect to some reference temperature

$$dE_s = \pm \alpha \, dT$$

and it is seen that

$$\alpha = \frac{dE_s}{dt}$$

The Peltier heat is proportional to the current flow and the proportionality constant is Π, the Peltier voltage

$$q_p = \pm \Pi I \qquad (69.80)$$

The Thomson heat is proportional to a temperature difference dT and the proportionality constant is σ, the Thomson coefficient. With $dq_T = \pm \sigma I \, dt$, it is observed that $\sigma \, dT$ is a voltage and the Thomson voltage is defined by

$$E_T = \pm \int_{T_1}^{T_2} \sigma \, dT$$

Considerations of the second law of thermodynamics and the Kirchhoff voltage law show that the Peltier voltage is related to the Seebeck coefficient[43]

$$\Pi = \alpha T \qquad (69.81)$$

and if the Seebeck coefficient is represented as a polynomial[44]

$$\alpha = a + bT + \cdots$$

then

$$\Pi = aT + bT^2 + \cdots$$

Design Equations

In Fig. 69.13, which shows a pair of materials arranged as a thermoelectric cooler, there is a cold junction at T_c and a hot junction at T_h. The materials possess a thermal conductivity k and an electrical resistivity ρ. A voltage is provided so that a current I flows through the cold junction from B to A and through the hot junction from A to B. This current direction is selected to guarantee that $T_c < T_h$.

The net heat absorbed at the cold junction is the Peltier heat

$$q_p = \Pi T_c = \alpha I T_c \qquad (69.82a)$$

minus one-half of the I^2R loss (known as the Joule heat or Joule effect)

$$q_J = \tfrac{1}{2} I^2 R \qquad (69.82b)$$

and minus the heat regained at the cold junction (known as the Fourier heat or Fourier effect) due to the temperature difference $\Delta T = T_h - T_c$

$$q_F = K \, \Delta T = K(T_h - T_c) \qquad (69.82c)$$

Thus the net heat absorbed at the cold junction is

$$q = \alpha I T_c - \tfrac{1}{2} I^2 R - K \, \Delta T \qquad (69.83)$$

where the total resistance of the couple is the series resistance of material A and material B having areas A_A and A_B, respectively (both have length L),

$$R = \left(\frac{\rho_A}{A_A} + \frac{\rho_B}{A_B} \right) L \qquad (69.84)$$

and where the overall conductance K is the parallel conductance of the elements A and B:

$$K = \frac{1}{L} (k_A A_A + k_B A_B) \qquad (69.85)$$

In order to power the device, a voltage equal to the sum of the Seebeck voltages at the hot and cold junctions plus the voltage necessary to overcome the resistance drop must be provided:

$$V = \alpha T_h - \alpha T_c + RI = \alpha \, \Delta T + RI$$

and the power is

$$P = VI = (\alpha \, \Delta T + RI)I \qquad (69.86)$$

The coefficient of performance (COP) is the ratio of the net cooling effect to the power provided:

$$\text{COP} = \frac{q}{P} = \frac{\alpha T_c I - \tfrac{1}{2} I^2 R - K \, \Delta T}{\alpha \Delta T I + I^2 R} \qquad (69.87)$$

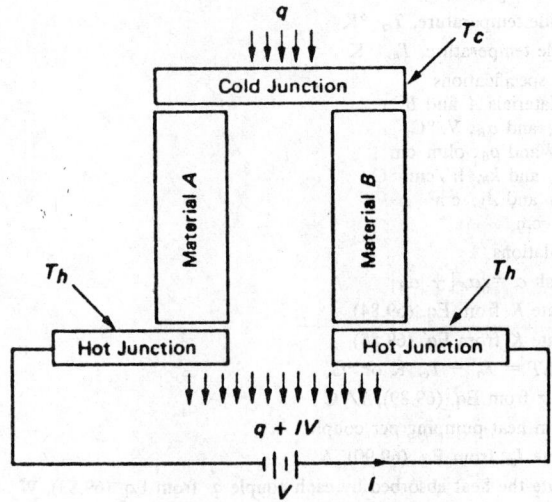

Fig. 69.13   Thermoelectric cooler.

Optimizations

The maximum possible temperature differential $\Delta T = T_h - T_c$ will occur when there is no net heat absorbed at the cold junction:

$$\Delta T_m = \tfrac{1}{2} z T_c^2 \tag{69.88}$$

where z is the figure of merit of the material

$$z = \frac{\alpha^2}{KR} \tag{69.89}$$

The current that yields the maximum amount of heat absorbed at the cold junction can be shown to be[43]

$$I = I_m = \frac{\alpha T_c}{R} \tag{69.90}$$

and the coefficient of performance in this case will be

$$\mathrm{COP}_m = \frac{1 - \Delta T/\Delta T_m}{2(1 + \Delta T/T_c)} \tag{69.91}$$

The current that optimizes or maximizes the coefficient of performance can be shown to be

$$I_0 = \frac{\alpha\,\Delta T}{R[(1 + zT_a)^{1/2} - 1]} \tag{69.92}$$

where $T_a = \tfrac{1}{2}(T_h + T_c)$. In this case, the optimum coefficient of performance will be

$$\mathrm{COP}_0 = \frac{T_c}{\Delta T}\left[\frac{\gamma - (T_h/T_c)}{\gamma + 1}\right] \tag{69.93}$$

where

$$\gamma = [1 + \tfrac{1}{2}z\,(T_h + T_c)]^{1/2} \tag{69.94}$$

Analysis of Thermoelectric Coolers

In the event that a manufactured thermoelectric cooling module is being considered for a particular application, the designer will need to specify the number of junctions required. A detailed procedure for the selection of the number of junctions is as follows:

1. Design specifications
   - (a) Total cooling load, q_T, W
   - (b) Cold-side temperature, T_c, °K
   - (c) Hot-side temperature, T_h, °K
   - (d) Cooler specifications
       - i. Materials A and B
       - ii. α_A and α_B, V/°C
       - iii. ρ_A and ρ_B, ohm·cm
       - iv. k_A and k_B, W/cm·°C
       - v. A_A and A_B, cm^2
       - vi. L, cm
2. Cooler calculations
   - (a) Establish $\alpha = |\alpha_A| + |\alpha_B|$
   - (b) Calculate R from Eq. (69.84)
   - (c) Calculate K from Eq. (69.85)
   - (d) Form $\Delta T = T_h - T_c$, K or °C
   - (e) Obtain z from Eq. (69.89), 1/°C
3. For maximum heat pumping per couple
   - (a) Calculate I_m from Eq. (69.90), A
   - (b) Calculate the heat absorbed by each couple q, from Eq. (69.83), W
   - (c) Calculate ΔT_m from Eq. (69.88), K or °C
   - (d) Determine COP_m from Eq. (69.91)

    (e) The power required per couple will be $p = q/\text{COP}_m$, W

    (f) The heat rejected per couple will be $p + q$, W

    (g) The required number of couples will be $n = q_T/q$

    (h) The total power required will be $p_T = nP$, W

    (i) The total heat rejected will be $q_{RT} = nq_R$, W

3A. For optimum coefficient of performance

    (a) Determine $T_a = \frac{1}{2}(T_h + T_c)$, K

    (b) Calculate I_0 from Eq. (69.92), A

    (c) Calculate the heat absorbed by each couple, q, from Eq. (69.83), W

    (d) Determine γ from Eq. (69.94)

    (e) Determine COP_0 from Eq. (69.93)

    (f) The power required per couple will be $P = q/\text{COP}_0$, W

    (g) The heat rejected per couple will be $q_R = P + q$, W

    (h) The required number of couples will be $n = q_T/q$

    (i) The total power required will be $P_T = nP$, W

    (j) The total heat rejected will be $q_{RT} = nq_R$, W

REFERENCES

1. J. W. Thornell, W. A. Fahley, and W. L. Alexander, *Hybrid Microcircuit Design and Procurement Guide*, Boeing Company, document available from National Technical Information Service, Springfield, VA, DOC AD 705974, 1972.

2. C. A. Harper, *Handbook of Thick Film Microelectronics*, McGraw-Hill, New York, 1974.

3. *Reliability Prediction of Electronic Equipment*, U.S. Dept. of Defense, MIL-HDBK-217B, NTIS, Springfield, VA, 1974.

4. W. F. Hilbert and F. H. Kube, *Effects on Electronic Equipment Reliability of Temperature Cycling in Equipment*, Final Report, Grumman Aircraft Engineering Corporation, Report No. EC-69-400, Bethpage, NY, Feb. 1969.

5. A. D. Kraus and A. Bar-Cohen, *Thermal Analysis and Control of Electronic Equipment*, Hemisphere Publishing Co., New York, 1983.

6. M. Jakob and W. M. Dow, "Heat Transfer from a Cylindrical Surface to Air in Parallel Flow with and without Unheated Starting Sections," *Trans. ASME*, **68** (1946).

7. R. Hilpert, Warmeabgue von Geheizten Drähten und Rohren in Lufstrom, *Forsch, Ing-Wes.*, **4**, 215–224 (1933).

8. M. Jakob, *Heat Transfer*, Wiley, New York, 1949.

9. S. Whitaker, "Forced Convection Heat Transfer Correlations for Flow in Pipes, Past Flat Plates, Single Cylinders, Single Spheres and for Flow in Packed Beds and Tube Bundles," *AIChE Journal*, **18**, 361–371 (1972).

10. W. H. McAdams, *Heat Transmission*, 3rd ed., McGraw-Hill, New York, 1954.

11. F. Kreith, *Principles of Heat Transfer*, International Textbook Co., Scranton, PA, 1959.

12. A. P. Colburn, "A Method of Correlating Forced Convection Heat Transfer Data and a Comparison of Fluid Friction," *Trans. AIChE*, **29**, 174–210 (1933).

13. "Standards of the Tubular Exchanger Manufacturer's Association," New York, 1949.

14. W. Drexel, "Convection Cooling," *Sperry Engineering Review*, **14**, 25–30 (December, 1961).

15. W. Robinson and C. D. Jones, *The Design of Arrangements of Prismatic Components for Crossflow Forced Air Cooling*, Ohio State University Research Foundation Report No. 47, Columbus, OH, 1955.

16. W. Robinson, L. S. Han, R. H. Essig, and C. F. Heddleson, *Heat Transfer and Pressure Drop Data for Circular Cylinders in Ducts and Various Arrangements*, Ohio State University Research Foundation Report No. 41, Columbus, OH, 1951.

17. E. N. Sieder and G. E. Tate, "Heat Transfer and Pressure Drop of Liquids in Tubes," *Ind. Eng. Chem.*, **28**, 1429–1436 (1936).

18. H. Hausen, *Z. VDI, Beih. Verfahrenstech.*, **4**, 91–98 (1943).

19. A. L. London, "Air Coolers for High Power Vacuum Tubes," *Trans. IRE*, **ED-1**, 9–26 (April, 1954).

20. S. Globe and D. Dropkin, "Natural Convection Heat Transfer in Liquids Confined by Two Horizontal Plates and Heated from Below," *J. Heat Transfer, Series C*, **81**, 24–28 (1959).

21. W. Mull and H. Rieher, "Der Warmeschutz von Luftschichten," *Gesundh-Ing. Beihefte,* **28** (1930).

22. J. G. A. DeGraaf and E. F. M. von der Held, "The Relation Between the Heat Transfer and the Convection Phenomena in Enclosed Plane Air Layers," *Appl. Sci. Res., Sec. A,* **3,** 393–410 (1953).

23. A. Bar-Cohen and W. M. Rohsenow, "Thermally Optimum Spacing of Vertical, Natural Convection Cooled, Parallel Plates," *J. Heat Transfer,* **106,** 116–123 (1984).

24. S. W. Churchhill and R. A. Usagi, "A General Expression for the Correlation of Rates of Heat Transfer and Other Phenomena," *AIChE J.,* **18**(6), 1121–1138 (1972).

25. N. Sobel, F. Landis, and W. K. Mueller, "Natural Convection Heat Transfer in Short Vertical Channels Including the Effect of Stagger," *Proceedings of the Third International Heat Transfer Conference,* Chicago, IL, 1966, Vol. 2, pp. 121–125.

26. W. Aung, L. S. Fletcher, and V. Sernas, "Developing Laminar Free Convection Between Vertical Flat Plates with Asymmetric Heating," *Int. J. Heat Mass Transfer,* **15,** 2293–2308 (1972).

27. O. Miyatake, T. Fujii, M. Fujii, and H. Tanaka, "Natural Convection Heat Transfer Between Vertical Parallel Plates—One Plate with a Uniform Heat Flux and the Other Thermally Insulated," *Heat Transfer Japan Research,* **4,** 25–33 (1973).

28. W. Aung, "Fully Developed Laminar Free Convection Between Vertical Flat Plates Heated Asymmetrically," *Int. J. Heat Mass Transfer,* **15,** 1577–1580 (1972).

29. W. Elenbaas, "Heat Dissipation of Parallel Plates by Free Convection," *Physica,* **9**(1), 665–671 (1942).

30. J. R. Bodoia and J. F. Osterle, "The Development of Free Convection Between Heated Vertical Plates," *J. Heat Transfer,* **84,** 40–44 (1964).

31. A. M. Clausing and B. T. Chao, *Thermal Contact Resistance in a Vacuum Environment,* NASA Report, ME-TN-242-1, University of Illinois, Champaign-Urbana, IL, 1963.

32. M. G. Cooper, B. B. Mikic, and M. M. Yovanovich, "Thermal Contact Resistance," *Int. J. Heat Mass Transfer,* **12,** 279–300 (1969).

33. B. B. Mikic, M. M. Yovanovich, and W. M. Rohsenow, *The Effect of Surface Roughness and Waviness upon the Overall Thermal Contact Resistance,* EPL Report No. 79361-43, Massachusetts Institute of Technology, Cambridge, MA, 1966.

34. M. M. Yovanovich, "Thermal Contact Conductance in a Vacuum," ScD Dissertation, Massachusetts Institute of Technology, Cambridge, MA, 1967.

35. K. A. Gardner, "Efficiency of Extended Surfaces," *Trans. ASME,* **67,** 621–631 (1945).

36. W. M. Murray, "Heat Transfer Through an Annular Disc or Fin of Uniform Thickness," *J. Appl. Mech.,* **5,** A78–A80 (1938).

37. D. Q. Kern and A. D. Kraus, *Extended Surface Heat Transfer,* McGraw-Hill, New York, 1972.

38. D. E. Briggs and E. H. Young, "Convection Heat Transfer and Pressure Drop of Air Flowing across Triangular Pitch Banks of Finned Tubes," *Chem. Eng. Prog. Symp. Ser.,* **41**(59), 1–10 (1963).

39. A. D. Kraus, A. D. Snider, and L. F. Doty, "An Efficient Algorithm for Evaluating Arrays of Extended Surface," *J. Heat Transfer,* **100,** 288–293 (1978).

40. A. D. Kraus, *Analysis and Evaluation of Extended Surface Thermal Systems,* Hemisphere Publishing Corp., 1982.

41. A. D. Kraus and A. D. Snider, "New Parametrizations for Heat Transfer in Fins and Spines," *J. Heat Transfer,* **102,** 415–419 (1980).

42. W. M. Kays and A. L. London, *Compact Heat Exchangers,* 3rd ed., McGraw-Hill, New York, 1984.

43. A. D. Kraus and A. Bar-Cohen, *Thermal Analysis and Control of Electronic Equipment,* Hemisphere Publishing Corp., New York, 1983.

44. *Handbook of Chemistry and Physics (CRC),* Chemical Rubber Co., Cleveland, OH, 1954.

CHAPTER 70

PUMPS AND FANS

WILLIAM A. SMITH

College of Engineering
University of South Florida
Tampa, Florida

70.1 PUMP AND FAN SIMILARITY

The performance characteristics of centrifugal pumps and fans (i.e., rotating fluid machines) are described by the same basic laws and derived equations and, therefore, should be treated together and not separately. Both fluid machines provide the input energy to create flow and a pressure rise in their respective fluid systems and both use the principle of fluid acceleration as the mechanism to add this energy. If the pressure rise across a fan is small (5000 Pa), then the gas can be considered as an incompressible fluid, and the equations developed to describe the process will be the same as for pumps.

Compressors are used to obtain large pressure increases in a gaseous fluid system. With such devices the compressibility of the gas must be considered, and a new set of derived equations must be developed to describe the compressor's performance. Because of this, the subject of gas compressors will be included in a separate chapter.

70.2 SYSTEM DESIGN: THE FIRST STEP IN PUMP OR FAN SELECTION

70.2.1 Fluid System Data Required

The first step in selecting a pump or fan is to finalize the design of the piping or duct system (i.e., the "fluid system") into which the fluid machine is to be placed. The fluid machine will be selected to meet the flow and developed head requirements of the fluid system. The developed head is the energy that must be added to the fluid by the fluid machine, expressed as the potential energy of a column of fluid having a height H_p (meters). H_p is the "developed head." Consequently, the following data must be collected before the pump or fan can be selected:

1. Maximum flow rate required and variations expected.
2. Detailed design (including layout and sizing) of the pipe or duct system, including all elbows, valves, dampers, heat exchangers, filters, etc.
3. Exact location of the pump or fan in the fluid system, including its elevation.
4. Fluid pressure and temperature available at start of system (suction).
5. Fluid pressure and temperature required at end of system (discharge).
6. Fluid characteristics (density, viscosity, corrosiveness, and erosiveness).

70.2.2 Determination of Fluid Head Required

The fluid head required is calculated using both the Bernoulli and D'Arcy equations from fluid mechanics. The Bernoulli equation represents the total mechanical (nonthermal) energy content of the fluid at any location in the system:

$$E_{T\,(1)} = P_1 v_1 + Z_1 g + V_1^2/2 \qquad (70.1)$$

where $E_{T\,(1)}$ = total energy content of the fluid at location (1), J/kg
$\quad P_1$ = absolute pressure of fluid at (1), Pa
$\quad v_1$ = specific volume of fluid at (1), m³/kg
$\quad Z_1$ = elevation of fluid at (1), m
$\quad g$ = gravity constant, m/sec²
$\quad V_1$ = velocity of fluid at (1), m/sec

The D'Arcy equation expresses the loss of mechanical energy from a fluid through friction heating between any two locations in the system:

$$\bar{v}\Delta P_f(i,j) = f L_e(i-j)\, V^2/2D \quad \text{J/kg·m} \qquad (70.2)$$

where $\quad \bar{v}$ = average fluid specific volume between two locations (i and j) in the system, m³/kg
$\quad \Delta P_f(i,j)$ = pressure loss due to friction between two locations (i and j) in the system, Pa
$\quad f$ = Moody's friction factor, an empirical function of the Reynolds number and the pipe roughness, nondimensional
$\quad L_e(i-j)$ = equivalent length of pipe, valves, and fittings between two locations i and j in the system, m
$\quad D$ = pipe internal diameter (i.d.), m

An example best illustrates the method.

Example 70.1

A piping system is designed to provide 2.0 m³/sec of water (Q) to a discharge header at a pressure of 200 kPa. Water temperature is 20°C. Water viscosity is 0.0013 N·sec/m². Pipe roughness is 0.05

mm. The gravity constant (g) is 9.81 m/sec². Water suction is from a reservoir at atmospheric pressure (101.3 kPa). The level of the water in the reservoir is assumed to be at elevation 0.0 m. The pump will be located at elevation 1.0 m. The discharge header is at elevation 50.0 m. Piping from the reservoir to the pump suction flange consists of the following:

1   20 m length of 1.07 m i.d. steel pipe
3   90° elbows, standard radius
2   gate valves
1   check valve
1   strainer

Piping from the pump discharge flange to the discharge header inlet flange consists of the following:

1   100 m length of 1.07 m i.d. steel pipe
4   90° elbows, standard radius
1   gate valve
1   check valve

Determine the "total developed head," H_p (m), required of the pump.
Solution:

Let location (1) be the surface of the reservoir, the system "suction location."
Let location (2) be the inlet flange of the pump.
Let location (3) be the outlet flange of the pump.
Let location (4) be the inlet flange to the discharge header, the system "discharge location."

By energy balances

$$E_{T\,(1)} - \bar{v}\Delta P_f(1-2) = E_{T\,(2)}$$

$$E_{T\,(2)} + E_p = E_{T\,(3)}$$

$$E_{T\,(3)} - \bar{v}\Delta P_f(3-4) = E_{T\,(4)}$$

where E_p is the energy input required by the pump. When E_p is described as the potential energy equivalent of a height of liquid, this liquid height is the "total developed head" required of the pump:

$$H_p = E_p/g \text{ m}$$

where H_p = total developed head, m.
   For the data given, assuming incompressible flow:

$P_1 = 101.3$ kPa	$Z_2 = +1.0$ m
$v_1 = 0.001$ m³/kg = constant	$Z_3 = +1.0$ m
$Z_1 = 0.0$ m	$Z_4 = +50.0$ m
$V_1 = 0.0$ m/sec	$P_4 = 200$ kPa

A_p = internal cross sectional area of the pipe, m²
$V_2 = Q/A = (2.0)(4)/\pi(1.07)^2 = 2.22$ m/sec
Assume $V_3 = V_4 = V_2 = 2.22$ m/sec
Viscosity (μ) = 0.0013 N·sec/m²
Reynolds number = $D\,V/v\mu$
   $= (1.07)(2.22)/(0.001)(0.0013) = 1.82 \times 10$
Pipe roughness (ϵ) = 0.05 mm
$\epsilon/D = 0.05/(1000)(1.07) = 0.000047$
From Moody's chart, $f = 0.009$ (see references on fluid mechanics)

From tables of equivalent lengths (see references on fluid mechanics):

Fitting	Equivalent Length, L_e (m)
Elbow	1.6
Gate valve (open)	0.3
Check valve	0.3
Strainer	1.8

$$L_e(1\text{--}2) = 20 + (3)(1.6) + 2(0.3) + 0.3 + 1.8 = 27.5 \text{ m}$$

$$L_e(3\text{--}4) = 100 + (4)(1.6) + 0.3 + 0.3 = 107.0 \text{ m}$$

$$\bar{v}\Delta P(1\text{--}2) = (0.009)(27.5)(2.22)^2/(2)(1.07) = 0.57 \text{ J/kg}$$

$$\bar{v}\Delta P(3\text{--}4) = (0.009)(107.0)(2.22)^2/(2)(1.07) = 2.21 \text{ J/kg}$$

$$E_{T(1)} = P_1 v_1 + Z_1 g + V_1^2/2$$

$$= (101,300)(0.001) + 0 + 0 = 101.30 \text{ J/kg}$$

$$E_{T(2)} = E_{T(1)} - \bar{v}\Delta P_f(1\text{--}2)$$

$$= 101.3 - 0.57 = 100.7 \text{ J/kg}$$

$$E_{T(4)} = P_4 v_4 + Z_4 g + V_4^2/2$$

$$= (200,000)(0.001) + (50.0)(9.81) + (2.22)^2/2$$

$$= 692.9 \text{ J/kg}$$

$$E_{T(3)} = E_{T(4)} + \bar{v}\Delta P_f(3\text{--}4)$$

$$= 692.9 + 2.21 = 695.1 \text{ J/kg}$$

$$E_p = E_{T(3)} - E_{T(2)}$$

$$= 695.1 - 100.7 = 594.4 \text{ J/kg}$$

$$H_p = E_p/g = 594.4/9.81 = 60.6 \text{ m of water}$$

It is seen that a pump capable of providing 2.0 m³/sec flow with a developed head of 60.6 m of water is required to meet the demands of this fluid system.

70.2.3 Total Developed Head of a Fan

The procedure for finding the total developed head of a fan is identical to that described for a pump. However, the fan head is commonly expressed in terms of a height of water instead of a height of the gas being moved, since water manometers are used to measure gas pressures at the inlet and outlet of a fan. Consequently,

$$H_{fw} = (\rho_g/\rho_w)H_{fg}$$

where H_{fw} = developed head of the fan, expressed as a head of water, m
$\quad H_{fg}$ = developed head of the fan, expressed as a head of the gas being moved, m
$\quad \rho_g$ = density of gas, kg/m³
$\quad \rho_w$ = density of water in manometer, kg/m³

As an example, if the head required of a fan is found to be 100 m of air by the method described in Section 70.2.2, the air density is 1.21 kg/m³, and the water density in the manometer is 1000 kg/m³, then the developed head, in terms of the column of water, is

$$H_{fw} = (1.21/1000)(100) = 0.121 \text{ m of water}$$

In this example the air is assumed to be incompressible, since the pressure rise across the fan was small (only 0.12 m of water, or 1177 Pa).

70.2.4 Engineering Data for Pressure Loss in Fluid Systems

In practice, only rarely will an engineer have to apply the D'Arcy equation to determine pressure losses in fluid systems. Tables and figures for pressure losses of water, steam, and air in pipe and duct systems are readily available from a number of references. (See Figs. 70.1 and 70.2.)

70.2.5 Systems Head Curves

A systems head curve is a plot of the head required by the system for various flow rates through the system. This plot is necessary for analyzing system performance for variable flow application and is desirable for pump and fan selection and system analysis for constant flow applications.

The curve to be plotted is H versus Q, where

$$H = [E_{T(3)} - E_{T(2)}]/g \tag{70.3}$$

Assuming that $V_1 = 0$ and $V = V_4$ in Eqs. (70.1) and (70.2), and letting $V = Q/A$, then Eq. (70.3) reduces to

$$H = K_1 + K_2 Q^2 \tag{70.4}$$

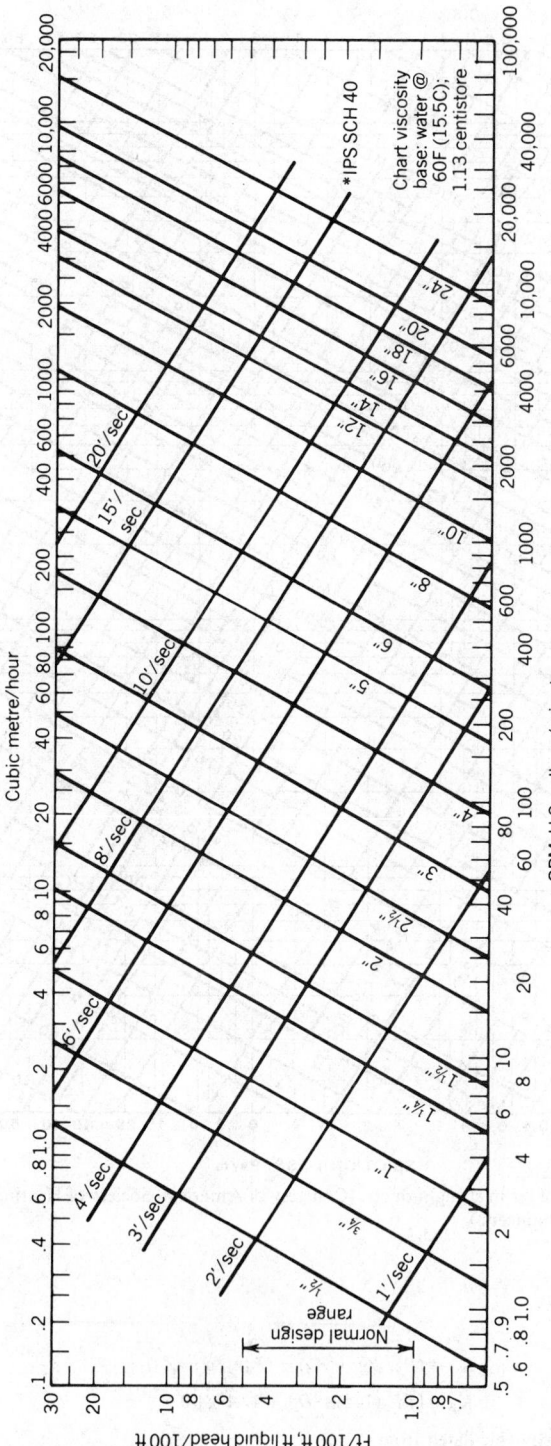

Fig. 70.1 Friction loss for water in commercial steel pipe (schedule 40). (Courtesy of American Society of Heating, Refrigerating and Air Conditioning Engineers.)

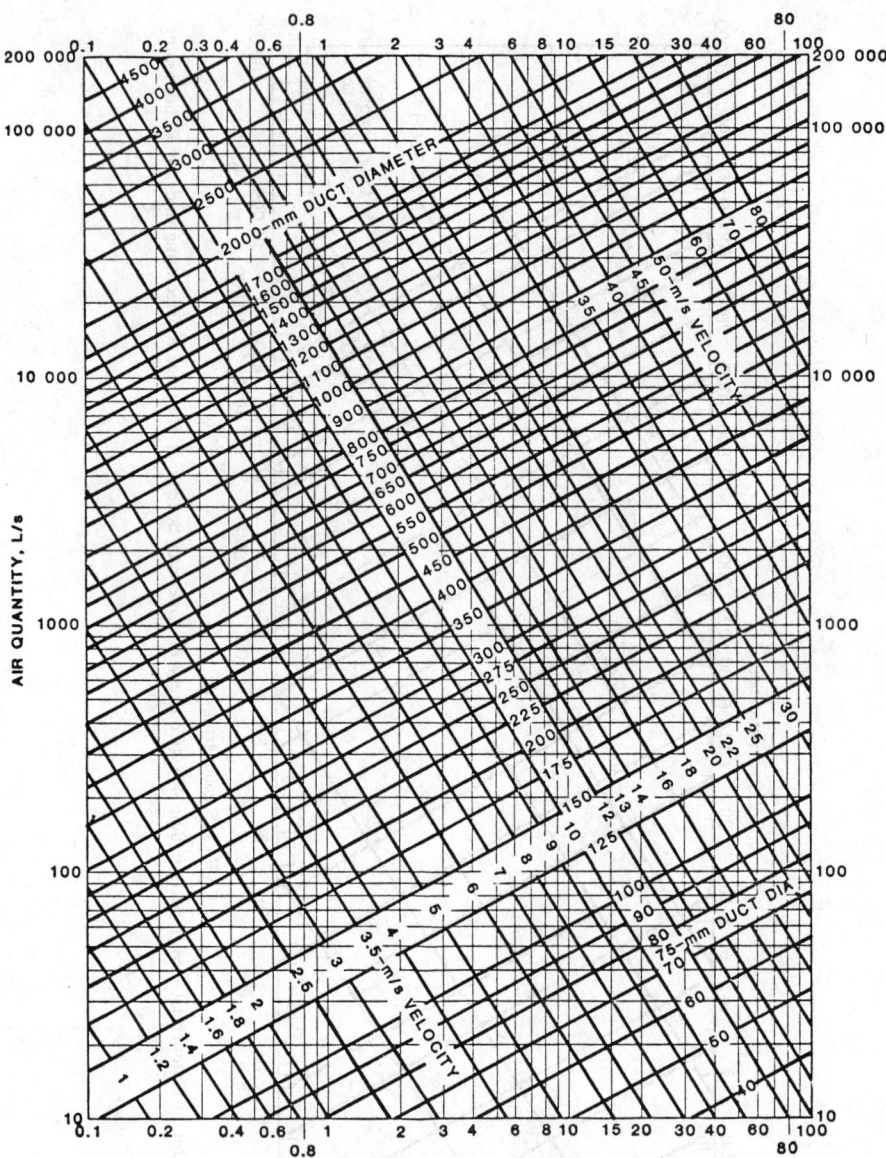

Fig. 70.2 Friction loss of air in straight ducts. (Courtesy of American Society of Heating, Refrigerating and Air Conditioning Engineers.)

where

$$K_1 = (P_4 v_4/g + Z_4) - (P_1 v_1/g + Z_1)$$

$$K_2 = [f L_e (1\text{-}4) A^2 Dg + 1/A^2 g](0.5)$$

However, K_2 is more easily calculated from

$$K_2 = (H - K_1)/Q^2$$

since both H and Q are known from previous calculations.

For Example 70.1:

$$K_1 = (200{,}000)(0.001)/9.81 + 50 - (101{,}300)(0.001)/9.81 + 0$$

$$= 60.0 \text{ m}$$

$$K_2 = (60.6 - 60.0)/(7200)^2 = 0.012 \times 10^{-6} \text{ hr}^2/\text{m}$$

A plot of this curve [Eq. (70.4)] would show a shallow parabola displaced from the origin by 60.0 m. (This will be shown in Fig. 70.10. Its usefulness will be discussed in Sections 70.6 and 70.7.)

70.3   CHARACTERISTICS OF ROTATING FLUID MACHINES

70.3.1   Energy Transfer in Rotating Fluid Machines

Most pumps and fans are of the rotating type. In a centrifugal machine the fluid enters a rotor at its eye and is accelerated radially by centrifugal force until it leaves at high velocity. The high velocity is then reduced by an area increase (either a volute or diffuser ring of a pump, or scroll of a fan) in which, by Bernoulli's law, the pressure is increased. This pressure rise causes only negligible density changes, since liquids (in pumps) are nearly incompressible and gases (in fans) are not compressed significantly by the small pressure rise (up to 0.5 m of water, or 5000 Pa, or 0.05 bar) usually encountered. For fan pressure rises exceeding 0.5 m of water, compressibility effects should be considered, especially if the fan is a large one (above 50 kW).

The principle of increasing a fluid's velocity, and then slowing it down to get the pressure rise, is also used in mixed flow and axial flow fluid machines. A mixed flow machine is one where the fluid acceleration is in both the radial and axial directions. In an axial machine, the fluid acceleration is intended to be axial but, in practice, is also partly radial, especially in those fans (or propellors) without any constraint (shroud) to prevent flow in the radial direction.

The classical equation for the developed head of a centrifugal machine is that given by Euler:

$$H = (C_{t_2}U_2 - C_{t_1}U_1)/g \quad \text{m} \tag{70.5}$$

where H is the developed head, m, of fluid in the machine; C_t is the tangential component of the fluid velocity C in the rotor; subscript 2 stands for the outer radius of the blade, r_2, and subscript 1 for the inner radius, r_1, m/sec; U is the tangential velocity of the blade, subscript 2 for outer tip and subscript 1 for the inner radius; and U_2 is the "tip speed," m/sec. The velocity vector relationships are shown in Fig. 70.3.

The assumptions made in the development of the theory are:

1. Fluid is incompressible.
2. Angular velocity is constant.

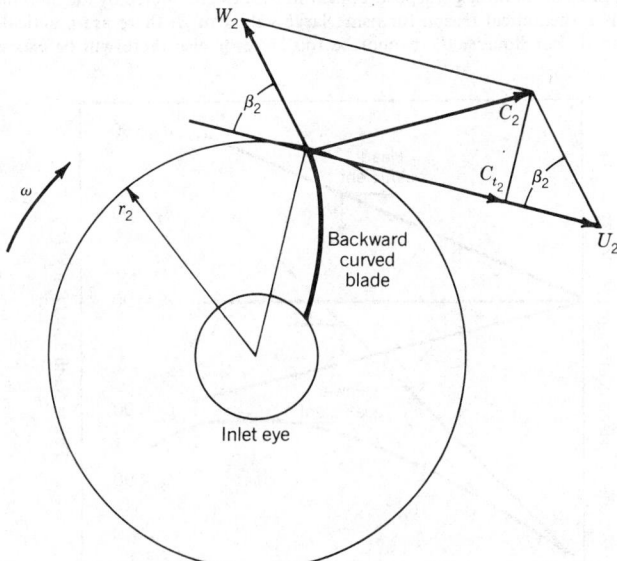

Fig. 70.3   Relationships of velocity vectors used in Euler's theory for the developed head in a centrifugal fluid machine; W is the fluid's velocity with respect to the blade; β is the blade angle; ω is the angular velocity, 1/sec.

3. There is no rotational component of fluid velocity while the fluid is between the blades, that is, the velocity vector W exactly follows the curvature of the blade.

4. No fluid friction.

The weakness of the third assumption is such that the model is not good enough to be used for design purposes. However, it does provide a guidepost to designers on the direction to take to design rotors for various head requirements.

If it is assumed that C_{t_1} is negligible (and this is reasonable if there is no deliberate effort made to cause prerotation of the fluid entering the rotor eye), then Eq. (70.5) reduces to

$$gH = \pi^2 N^2 D^2 - NQ \cot(\beta/b) \qquad (70.6)$$

where Q = the flow rate, m³/sec
$\quad\quad D$ = the outer diameter of the rotor, m
$\quad\quad b$ = the rotor width, m
$\quad\quad N$ = the rotational frequency, Hz

70.3.2 Nondimensional Performance Characteristics of Rotating Fluid Machines

Equation (70.6) can also be written as

$$(H/N^2D^2) = \pi^2/g - [D \cot(\beta/gb)](Q/ND^3) \qquad (70.7)$$

In Eq. (70.7) H/N^2D^2 is called the "head coefficient" and Q/ND^3 is the "flow coefficient." The theoretical power, P (W), to drive the unit is given by $P = QgH$, and this reduces to

$$(P/\rho N^3D^5) = (\pi^2)(Q/ND^3) - [D \cot(\beta/b)](Q/ND^3)^2 \qquad (70.8)$$

where $P/\rho N^3D^5$ is called the "power coefficient." Plots of Eqs. (70.7) and (70.8) for a given D/b ratio are shown in Fig. 70.4.

Analysis of Fig. 70.4 reveals that:

1. For a given Q, N, and D, the developed head increases as β gets larger, that is, as the blade tips are curved more into the direction of rotation.

2. For a given N and D, the head either rises, stays the same, or drops as Q increases, depending on the value of β.

3. For a given N and D, the power required continuously increases as Q increases for β's of 90° or larger, but has a peak value if β is less than 90°.

The practical applications of these guideposts appear in the designs offered by the fluid machine industry. Although there is a theoretical reason for using large values of β, there are practical reasons why β must be constrained. For liquids, β's cannot be too large or else there will be excessive turbulence,

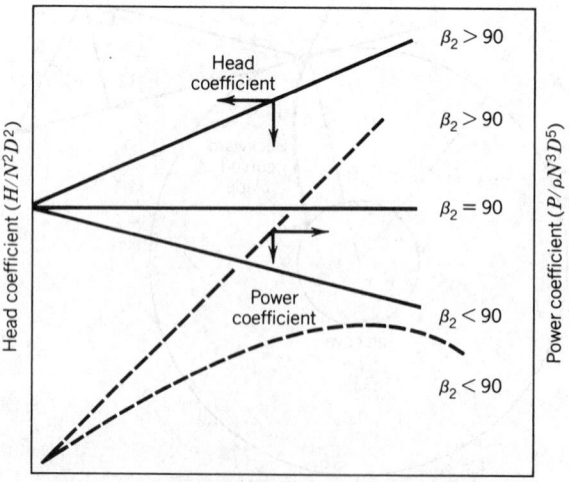

Fig. 70.4 Theoretical (Euler's) head and power coefficients plotted against the flow coefficient for constant D/b ratio and for values of $\beta < 90°$, equal to 90°, and $>90°$.

vibration, and erosion. Blades in pumps are always backward curved ($\beta < 90°$). For gases, however, β's can be quite large before severe turbulence sets in. Blade angles are constrained for fans not only by the turbulence but also by the decreasing efficiency of the fan and the negative economic effects of this decreasing efficiency. Many fan sizes utilize β's $> 90°$.

One important characteristic of fluid machines with blade angles less than $90°$ is that they are "limit load"; that is, there is a definite maximum power they will draw regardless of flow rate. This is an advantage when sizing a motor for them. For fans with radial ($90°$) or forward curved blades, the motor size selected for one flow rate will be undersized if the fan is operated at a higher flow rate. The result of undersizing a motor is overheating, deterioration of the insulation, and, if badly undersized, cutoff due to overcurrent.

70.3.3 Importance of the Blade Inlet Angle

While the outlet angle, β_2, sets the head characteristic the inlet angle, β_1, sets the flow characteristic, and by setting the flow characteristic, β_1 also sets the efficiency characteristic.

The inlet vector geometry is shown in Fig. 70.5.

If the rotor width is b at the inlet and there is no prerotation of the fluid prior to its entering the eye (i.e., $C_{t_1} = 0$), then the flow rate into the rotor is given by $Q = D_1 b_1 c_1$ and β_1 is given by:

$$\beta_1 = \arctan(C_1/U_1) = \tan^{-1}(Q/ND_1^3)(D_1/b_1)(1/\pi^2) \tag{70.9}$$

It is seen that β_1 is fixed by any choice of Q, N, D, and b_1. Also, a machine of fixed dimensions ($D_1 b_1$, β_1) and operated at one angular frequency (N) is properly designed for only one flow rate, Q. For flow rates other than its design value, the inlet geometry is incorrect, turbulence is created, and efficiency is reduced. A typical efficiency curve for a machine of fixed dimensions and constant angular velocity is shown in Fig. 70.6.

A truism of all fluid machines is that they operate at peak efficiency only in a narrow range of flow conditions (H and Q). It is the task of the system designer to select a fluid machine that operates at peak efficiency for the range of heads and flows expected in the operation of the fluid system.

70.3.4 Specific Speed

Besides the flow, head, and power coefficients, there is one other nondimensional coefficient that has been found particularly useful in describing the characteristics of rotating fluid machines, namely the specific speed N_s. Specific speed is defined as $NQ^{0.5}/H^{0.75}$ at peak efficiency. It is calculated by using the Q and H that a machine develops at its peak efficiency (i.e., when operated at a condition where its internal geometry is exactly right for the flow conditions required). The specific speed coefficient has usefulness when applying a fluid machine to a particular fluid system. Once the flow and head requirements of the system are known, the best selection of a fluid machine is that which has a

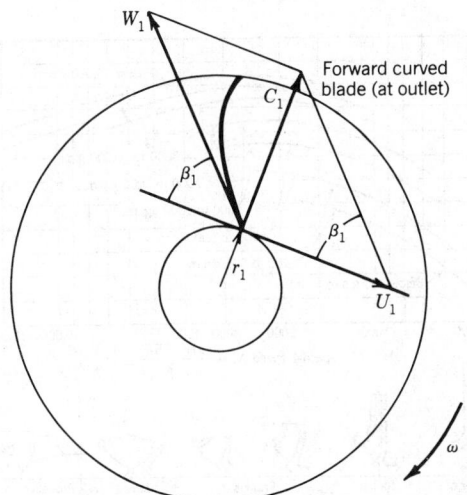

Fig. 70.5 Relationship of velocity vectors at the inlet to the rotor. Symbols are defined in Section 70.3.1.

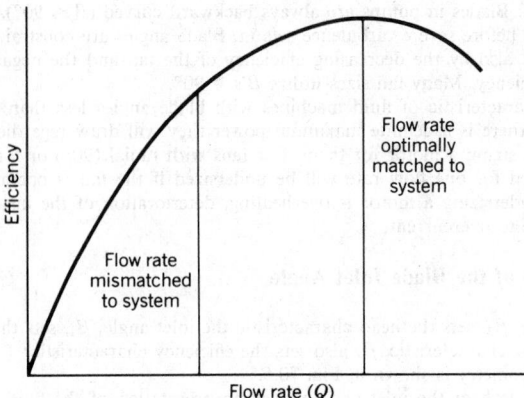

Fig. 70.6 Typical efficiency curve for fluid machines of fixed geometry and constant angular frequency.

specific speed equal to $NQ^{0.5}/H^{0.75}$, where the N, Q, and H are the actual operating parameters of the machine.

Since the specific speed of a machine is dependent on its structural geometry, the physical appearance of the machine as well as its application can be associated with the numerical value of its specific speed. Figure 70.7 illustrates this for a variety of pump geometries. The figure also gives approximate efficiencies to be expected from these designs for a variety of system flow rates (and pump sizes).

It is observed that centrifugal machines with large D/b ratios have low specific speeds and are suitable for high-head and low-flow applications. At the other extreme, the axial flow machines are suitable for low-head and large-flow applications. This statement holds for fans as well as pumps.

As an example of the use of specific speed, consider the pump application of Example 70.1. The head required was found to be 60.6 m. The flow was 7200 m³/hr. If a pump selected for this service is to have a rotational frequency of 14.75 Hz, then it should have a (nondimensional) specific speed of

$$N_s = (14.75)(2\pi)(2.0)^{0.5}/(9.81)^{0.75} (60.6)^{0.75} = 1.089$$

Its dimensional equivalent in the English system of units (rpm, gpm, ft) is 2972. Looking at Fig. 70.7 it is seen that a pump for this service would be of the centrifugal type, with an impeller that is wide and not very large in diameter. It is a large pump (31,700 gpm) and its efficiency is expected to be high (90%).

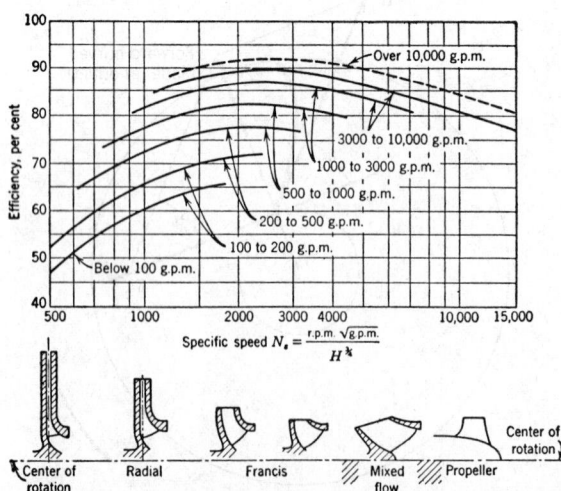

Fig. 70.7 Variation of physical appearance and expected efficiency with specific speed for a variety of pump designs and sizes. (Courtesy of Worthington Corporation.)

Assuming an efficiency of 90%, then the power requirement (P) would be

$$P = \rho QgH/\text{eff} = (1000)(2.0)(9.81)(60.6/(0.9)(1000)$$
$$= 1321 \text{ kW (or 1770 hp)}$$

70.3.5 Modeling of Rotating Fluid Machines

A "family" of fluid machines is one in which each member has the same geometric proportions (and physical appearance) as every other member, except for overall size. The largest member is merely a blown-up version of the smallest member.

Since the geometric proportions of each are the same, all members of a family have the same specific speed. They also have (theoretically) the same performance characteristics (Q, H, P, N) when the performance characteristics are expressed nondimensionally. Practically, the performance character- istics between members of a family differ slightly owing to changes in clearance distances, relative roughness, and Reynolds number that occur between sizes. These differences are called "secondary effects."

Ignoring secondary effects (and structural effects such as vibrations) the performance of an as yet unbuilt, large prototype can be predicted from tests on a small-scale model. Assume that the test data on a pump model, expressed nondimensionally, are as given in Fig. 70.8. It can be assumed that these results will be identical to those obtained on the prototype. If the prototype is to have a diameter of 0.81 m and a rotational frequency of 14.75 Hz, then, at peak efficiency, it can be predicted that the prototype will have the following flow, head, and power characteristics:

$$(Q/ND^3)_{\text{prototype}} = (Q/ND^3)_{\text{model}}$$
$$Q_p = (Q/ND^3)_{\text{model}} (ND^3)_{\text{prototype}}$$
$$= (0.0406)(2\pi)(14.75)(0.81)^3 = 2.0 \text{ m}^3/\text{sec}$$
$$(H/N^2D^2)_{\text{prototype}} = (H/N^2D^2)_{\text{model}}$$
$$H_p = (H/N^2D^2)_{\text{model}} (N^2D^2)_{\text{prototype}}$$
$$= (0.1054)(2\pi)^2(14.75)^2(0.81)^2/9.81 = 60.6 \text{ m}$$
$$P = \rho QgH/\text{eff} = 1321 \text{ kW } \textit{(from the previous section)}$$

If the model had a diameter of 0.1 m and a rotational frequency of 29 Hz, then, at peak efficiency, its flow, head, and power were:

$$Q_m = (Q/ND^3)(ND^3) = (2.0)(29/14.75)(0.1/0.81)^3$$
$$= 0.0074 \text{ m}^3/\text{sec}$$
$$H_m = (H/N^2D^2)p(N^2D^2)m = 60.6 \ (29/14.75)^2(0.1/0.81)^2$$
$$= 3.57 \text{ m}$$
$$P_m = (1000)(0.0074)(9.81)(3.57)/(0.9)(1000) = 0.29 \text{ kW}$$

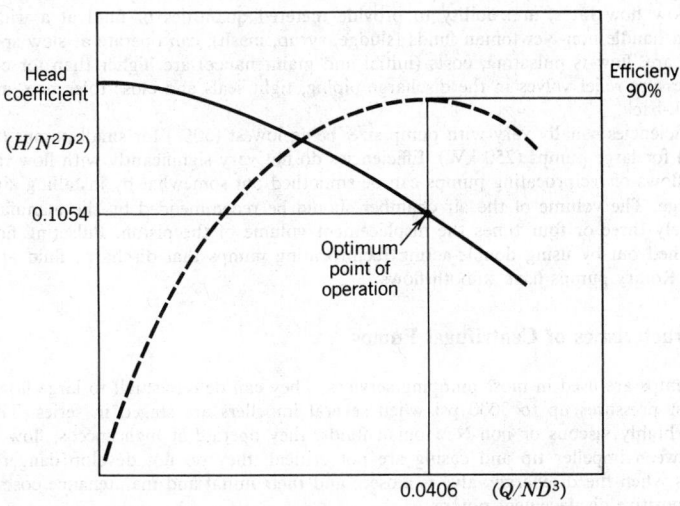

Fig. 70.8 Performance characteristics of a model pump, expressed nondimensionally.

Manufacturers of fluid machines often do not have facilities large enough (fluid quantities and power) to test their largest products. Consequently, the performance of such large machines is estimated from model tests.

70.3.6 Summary of Modeling Laws

Neglecting secondary effects (changes in Reynolds number, size, and clearance distances) the nondimensional performance relationships between a model and a prototype, for any single point of operation (i.e., one point on a common nondimensional curve of performance characteristics), can be summarized as follows:

Nondimensional Characteristic	Model	Prototype
Flow coefficient	Q/ND^3	$= Q/ND^3$
Head coefficient	H/N^2D^2	$= H/N^2D^2$
Power coefficient	$P/\rho N^3D^5$	$= P/\rho N^3D^5$
Efficiency	eff	$=$ eff
Specific speed	$NQ^{0.5}/H^{0.75}$	$= NQ^{0.5}/H^{0.75}$

The same relationships can be used to determine the changes in flow, head, or power of a single-fluid machine whenever its diameter or angular frequency is changed in an unchanging fluid system. For a machine of constant diameter, the flow, head, and power will vary with angular frequency as follows:

$$Q \propto N$$
$$H \propto N^2$$
$$P \propto N^3$$

70.4 PUMP SELECTION

70.4.1 Basic Types: Positive Displacement and Centrifugal (Kinetic)

Positive displacement pumps are best suited for systems requiring high heads and low flow rates, or for use with very viscous fluids. The common types are reciprocating (piston and cylinder) and rotary (gears, lobes, vanes, screws). Centrifugal pumps are well suited for the majority of pumping services. The common types are radial, centrifugal, mixed flow, and propellor (axial flow).

70.4.2 Characteristics of Positive Displacement Pumps

Some advantages of positive displacement pumps, besides their inherent ability to provide high discharge pressures at low flow rates, are: ability to provide metered quantities of fluid at a wide range of viscosities; can handle non-Newtonian fluids (sludge, syrup, mash); can operate at slow speeds. Some disadvantages are: flow is pulsating; costs (initial and maintenance) are higher than for centrifugals; must have pressure relief valves in the discharge piping; tight seals and close tolerances are essential to prevent leak-back.

Overall efficiencies usually vary with pump size, being lowest (50%) for small pumps (2 kW) and highest (90%) for large pumps (250 kW). Efficiencies do not vary significantly with flow rate.

Pulsating flows of reciprocating pumps can be smoothed out somewhat by installing air chambers in the discharge. The volume of the air chamber should be recommended by the manufacturer, but is approximately three or four times the displacement volume of the piston. Pulsating flows can be further smoothed out by using double-acting reciprocating pumps that discharge fluid at both ends of the stroke. Rotary pumps have smooth flows.

70.4.3 Characteristics of Centrifugal Pumps

Centrifugal pumps are used in most pumping services. They can deliver small to large flow rates and operate against pressures up to 3000 psi when several impellers are staged in series. They do not work well on highly viscous or non-Newtonian fluids; they operate at high speeds; flow is smooth; clearances between impeller tip and casing are not critical; they do not develop dangerously high head pressures when the discharge valve is closed; and their initial and maintenance costs are lower than that for positive displacement pumps.

Efficiencies of centrifugal pumps are about the same as their corresponding-sized positive displacement pumps if they are carefully matched to their systems. However, their efficiencies vary significantly with flow rate when operated at constant speed, and their efficiencies can be very poor if mismatched to their system, as seen in Figs. 70.6 and 70.8.

70.4.4 Net Positive Suction Head (NPSH)

The liquid static pressure at the suction of both positive displacement and centrifugal pumps must be higher than the liquid's vapor pressure to prevent vaporization at the inlet. Vaporization at the inlet, called "cavitation," causes a drop in developed head and, in severe cases, a complete loss of flow. Cavitation also causes pitting of the impeller that, in time and if severe enough, can destroy the impeller.

Net positive suction head (NPSH) is the difference between the static pressure and the vapor pressure of the liquid at the pump inlet flange, expressed in meters:

$$\text{NPSH} = (P_s - P_v)/\rho g \quad \text{m} \tag{70.10}$$

where P_s is the static pressure at the pump inlet flange, Pa; P_v is the liquid's vapor pressure, Pa; and ρ is the liquid density, kg/m³.

There are two NPSHs that a system designer must consider. One is the NPSH available (NPSHA), which is dependent on the design of the piping system (most importantly the relative elevations of the pump and the source of liquid being pumped). The second is the NPSH required (NPSHR) by the pump selected for the service. There is a static pressure loss within the pump as the liquid passes through the inlet casing and enters the blades. The severity of this loss is dependent on the design of the casing and the amount of acceleration (and turbulence) that the liquid experiences as it enters the blading. Manufacturers test for the NPSHR for each model of pump and report these requirements on the engineering performance specification sheets for the model. The task of the system designer is to ensure that the NPSHA exceeds the NPSHR. Using the data in Example 70.1, the NPSHA is calculated as follows:

P_v at 20°C = 2237 Pa
P_s is found from the calculation for the total energy at the pump suction flange, $E_{t(2)}$
$E_{t(2)} = 100.7 \text{ J/kg} = (P_s/\rho + Zg + V^2/2) \text{ at (2)}$
$P_s = (100.7)(1000) - (1)(9.81)(1000) - (2.21)^2(1000)/2 = 88,400 \text{ Pa}$
$\text{NPSHA} = (P_s - P_v)/\rho g = (88,400 - 2,237)/(9.81)(1000) = 8.8 \text{ m}$

This NPSHA (8.8 m) is considered large and quite adequate for most pump models. However, if, after a survey of available pumps, it is found that none can operate with this net positive suction head, then the design of the piping system will have to be changed: the pump will have to be placed at a lower elevation to ensure adequate suction static pressure.

70.4.5 Selection of Centrifugal Pumps

The pump selected for a fluid system must deliver the specified flow and required head at or near the pump's maximum efficiency, and have a NPSHR less than the NPSHA. However, only rarely will one find a pump model, even from a survey of several manufacturers, that exactly matches the system; that is, a pump whose flow and head at maximum efficiency exactly match the flow and head required.

The first step in pump selection is to contact several pump manufacturers and obtain the performance curves of the pumps they recommend for the specified service. A typical pump curve is shown in Fig. 70.9.

It is seen that on this one curve data are presented giving flow, head, efficiency, power, and NPSHR for a variety of impeller sizes (diameters). The curves for impeller sizes in between those shown can be estimated by extrapolation. Since clearance distances between the impeller tip and the pump casing are not critical, it is possible to install any of several different size impellers in one casing. It is also possible to cut down an existing impeller to a smaller size if it is found advantageous to do so after delivery of a pump and installation in its system.

An important selection parameter is the motor size. Note that there is a maximum power that the pump will require regardless of flow. It is advisable to specify a motor with a power rating at least equal to this maximum power required, since, in most applications, there will be times when the pump is called upon to deliver higher flow rates than originally expected.

For the pump in Example 70.1, the purchase specifications would be:

Flow	7200 m³/hr (2.0 m³/sec)
Head	60.6 m
NPSHA	8.8 m (28.9 ft)

It is seen, in Fig. 70.9, that the 32 in.-diameter impeller in the model 3420 would be adequate for the head and flow. However, the NPSHR is 10.29 m (33 ft), which is more than is available, and

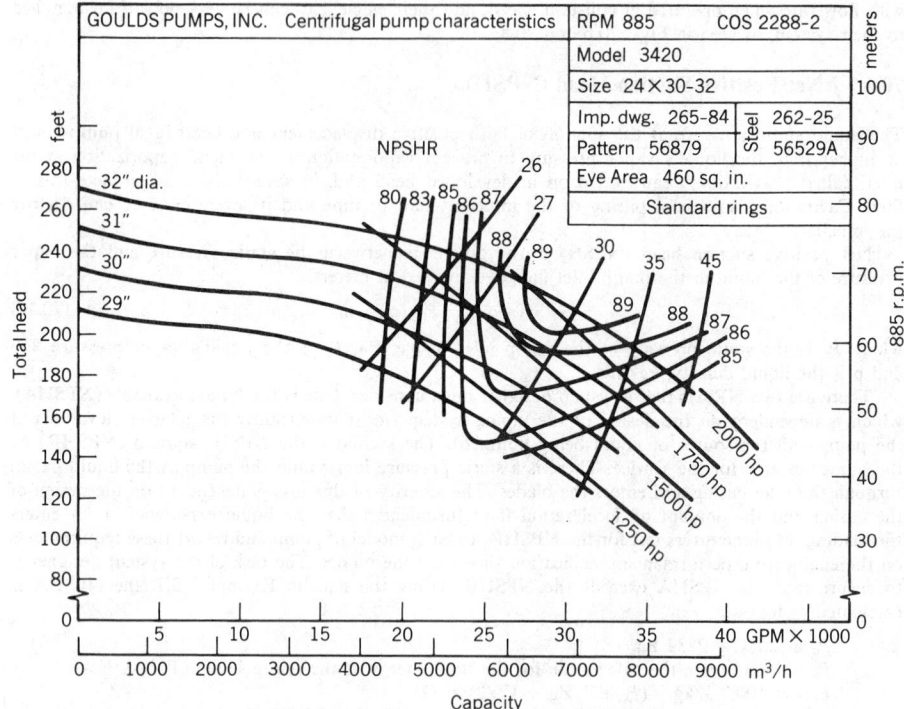

Fig. 70.9 Example of pump curve provided by manufacturer. (Courtesy Goulds Pumps, Inc., Seneca Falls, NY.)

therefore not acceptable. The efficiency is 89%, which is close to what was expected. The usual procedure is to survey more manufacturers in hopes of finding a better match, a higher efficiency, and one requiring less NPSH. If the model 3420 were finally selected, it is recommended that the 2000 hp (1492 kW) motor be specified. Also, the piping system would have to be altered to lower the pump elevation and provide more NPSHA.

Referring again to Fig. 70.9, it is seen that the size is given by the numbers 24 × 30–32. It is standard practice in the industry to use a size designation number that gives, in order, the diameter of the discharge flange, the diameter of the suction flange, and then the impeller diameter, all in inches.

70.4.6 Operating Performance of Pumps in a System

The actual point of operation (head and flow) of a pump in a piping system is found from the intersection of the pump curve (Fig. 70.9) and the system head curve [Eq. (70.4)]. Both curves represent the energy required to cause a specified flow rate. By the law of conservation of energy the energy input to the fluid by the pump must equal the energy required by the piping system for a specified flow. Figure 70.10 shows both curves plotted on the same coordinate system and establishes the point of operation of the pump in the system.

The actual point of operation of the model 3420 pump and the system of Example 70.1 is 7450 m³/hr at 60.7 m. If this flow is too large, the system will have to be throttled (by closing in a valve) until the flow is reduced to the desired value. If the system is throttled to 7200 m³/hr, the actual developed head will be 63.1 m. This means the head loss created in the valve, at 7200 m³/hr, is 63.1–60.6 or 2.5 m. The power wasted in this throttling process is $\rho QHg/\text{eff} = (1000)(2)(2.5)(9.81)/(0.89)(1000) = 55.1$ kW, which is converted into heat.

In large pumps (100 kW) even small differences in operating power (2%) can make large differences in operating economy. For this reason it is important for the purchaser of a pump to seek the best possible match of the pump to the system to optimize efficiency and avoid having to throttle the flow. For example, if the difference in operating power between two pumps capable of meeting a specified service (head and flow) is as small as 2 kW but the pump is operated continuously (8760 hours per year), then the energy difference is 17,520 kWhr which, at $0.05/kWhr, has a value of

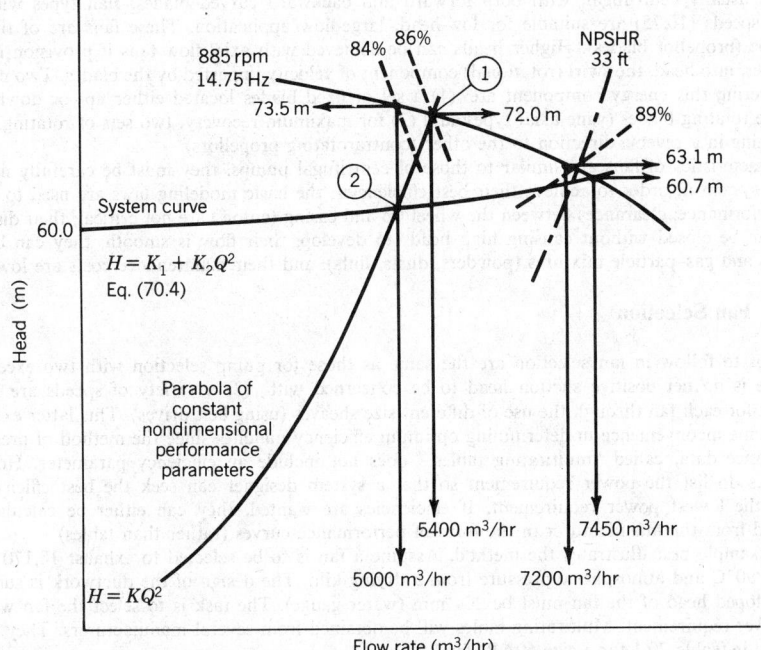

Fig. 70.10 Point of operation of a pump in a system.

$876 per year. If the additional cost (if any) of the more economical pump can be amortized over its financial lifetime for less than $876 per year, then the better pump should be purchased.

70.4.7 Throttling versus Variable Speed Drive

If the pump is to be operated at reduced flow rates for extended periods of time, it may be economically justifiable to use a variable speed drive.

As an example, assume that the system in Example 70.1 is operated at 5000 m³/hr for 2500 hours per year. If throttled to 5000 m³/hr, the pump head (from Fig. 70.10) would be 73.5 m and the efficiency would be (about) 84%. The energy consumed at this point of operation would be $\rho QHgh/$ eff $= (1000)(5000/3600)(93.5)(9.81)(2500)/(0.84)(1000) = 2.98 \times 10^6$ kWhr per year.

The operating points for the variable speed drive are determined by using the modeling laws (Section 70.3.5). If the diameter is constant, the H/N^2 and Q/N are constant and, for variable N, $H = KQ^2$, which is a parabola through the origin, as shown in Fig. 70.10. The operating points (1) and (2) on this parabola are related by the equations $H_1/N_1^2 = H_2/N_2^2$ and $Q_1/N_1 = Q_2/N_2$, where $H_2 = 60.0 + 0.012 \times 10^{-6} (5000)^2 = 60.3$ m. The K of the parabola is $H_2/Q_2^2 = 60.3/(5000)^2$. The intersection of this parabola with the original pump curve, point (1), is $H_1 = 72.0$ m and $Q_1 = 5400$ m³/hr. The reduced speed $N_1 = N_2 Q_2/Q_1 = (14.75)(5000)/5400 = 13.66$ Hz. The efficiency at (2), 86%, equals the efficiency at (1) since all nondimensional parameters at (1) and (2) are the same.

The energy consumed at the reduced pump speed (13.66 Hz) to provide 5000 m³/hr for 2500 hours per year is $(1000)(5000)/(3600)(60.3)(9.81)(2500)/(0.86)(1000) = 2.38 \times 10^6$ kWhr per year. The saving of 600,000 kWhr, at $0.05/kWhr, is worth $30,000 per year. If the cost of a variable speed drive in this example can be amortized over its financial lifetime for less than $30,000 per year, it should be purchased.

70.5 FAN SELECTION

70.5.1 Types of Fans; Their Characteristics

Fans, the same as pumps, are made in a large variety of types in order to serve a large variety of applications. There are also options in both cost and efficiency for applications requiring low power (5 kW). High-power applications require high efficiencies.

Fan types with low specific speeds (0.17) are suitable for high-head, low-flow application. These fans are usually centrifugal, with both forward and backward curved blades. Fan types with high specific speeds (16.75) are suitable for low-head, large-flow application. These fans are of the axial flow type (propellor blades). Higher heads can be achieved with axial flow fans if provision is made to recover, into head, the swirl (rotational) component of velocity imparted by the blades. Two methods of recovering this energy component are: (1) a set of fixed blades located either up- or downstream from the rotating blades (vane axial type); and (2) for maximum recovery, two sets of rotating blades, one turning in a reverse direction to the other (contrarotating propellors).

Characteristics of fans are similar to those of centrifugal pumps: they must be carefully matched to their system in order to achieve their best efficiencies; the basic modeling laws are used to predict their performance; clearances between the wheel tip and casing (cutoff) are not critical; their discharge ducts can be closed without causing high heads to develop; their flow is smooth; they can be used on gases and gas–particle mixtures (powders, dusts, lints); and their maintenance costs are low.

70.5.2 Fan Selection

The steps to follow in fan selection are the same as those for pump selection with two exceptions: (1) there is no net positive suction head to be concerned with; (2) a variety of speeds are usually available for each fan through the use of different size sheaves (using belt drives). This latter exception causes some inconvenience in determining optimum efficiency matches since the method of presenting performance data, called "multirating tables," does not include an efficiency parameter. However, the tables do list the power requirement so that a system designer can seek the best efficiency by seeking the lowest power requirement. If efficiencies are wanted, they can either be calculated or requested from the manufacturer in the form of performance curves (rather than tables).

An example best illustrates the method. Assume a fan is to be selected to exhaust 18,170 m^3/hr of air at 90°C and atmospheric pressure from a drying kiln. The design of the ductwork is such that the developed head of the fan must be 204 mm (water gauge). The task is to select the fan with the least power requirement. Multirating tables will be obtained from several manufacturers. They appear as shown in Table 70.1 for a size 60AW fan.

Table 70.1 Example of Multirating Table for Fans[a]

Capacity (cfm)	Outlet Velocity (fpm)	$5\frac{1}{2}''$ S.P.		6″ S.P.		$6\frac{1}{2}''$ S.P.	
		rpm	bhp	rpm	bhp	rpm	bhp
3000	921						
3700	1136	715	5.83				
4400	1351	715	6.37	746	7.10	777	7.84
5100	1566	715	6.98	747	7.73	777	8.48
5800	1781	717	7.63	748	8.41	778	9.24
6500	1996	722	8.40	752	9.21	781	10.04
7200	2211	730	9.26	759	10.11	787	10.98
7900	2426	738	10.17	766	11.07	794	12.00
8600	2641	748	11.12	776	12.12	803	13.08
9300	2856	759	12.18	786	13.21	812	14.21
10000	3071	771	13.29	798	14.37	823	15.46
10700	3286	783	14.45	809	15.60	835	16.76
11400	3501	796	15.69	822	16.87	847	18.11
12100	3716	810	16.99	836	18.28	860	19.52
12800	3931	825	18.40	850	19.71	874	21.07
13500	4146	841	19.91	865	21.27	888	22.64
14200	4361	857	21.45	881	22.92	904	24.35
14900	4576	873	23.03	896	24.60	919	26.16
15600	4791	890	24.75	913	26.34	935	27.98
16300	5006	907	26.48	930	28.22	952	29.88
17000	5221	924	28.22	947	30.10	969	31.92
17700	5436	941	30.18	963	32.00	985	33.96
18400	5651	960	32.30	981	34.13	1002	36.02
19100	5866	978	34.43	999	36.41	1020	38.31
19800	6081	997	36.87	1017	38.73	1038	40.77

[a] Courtesy of Buffalo Forge Co., Buffalo, N.Y.

The data presented in multirating tables are based on an air density of 1.201 kg/m³ (air at 760 mm mercury pressure and 21.11°C). It is easiest to adjust the system head required (at 1 atm and 90°C for the example) to an equivalent head based on the standard density (at STP) used in the tables. The relationship is: $H_{STP} = (H_{req})(21.11 + 273)(mmHG)/(°C + 273)(760 mm)$. Therefore $H_{STP} = (204)(294.11)(760)/(363)(760) = 165.1$ mm (6.5 in.) water gauge. The flow rate is unaffected by density changes since fans are constant volume devices. However, the power, as well as the head, is affected by density changes so that the power listed in Table 70.1 must be adjusted by the same factor that was used to adjust the head (0.81). Efficiency is independent of density, and 18,170 m³/hr is 10,000 cfm (cubic feet per minute). From the data in Table 70.1, one selection of fan would be the size 60AW operated at 823 rpm (13.72 Hz). The power required would be (15.46)(0.81) hp (9.34 kW). The efficiency can be calculated from the STP data by eff = $\rho QHg/P$ = $(1000)(18,170)(0.1651)(9.81)/(11.5)(1000)(3600) = 0.71$, where the density of water in the gauge is assumed to be 1000 kg/m³.

The multirating tables of different size fans of the same manufacturer as well as those of other manufacturers should be surveyed to find the one with the lowest power requirement. As an example, for the manufacturer of the fan in Table 70.1, the power requirement and efficiency of other sizes, all of which meet the head and flow requirements, are as follows:

Model	Size Wheel (m)	Power (kW)	Efficiency
45AW	0.83	11.48	0.58
50AW	0.92	10.43	0.64
55AW	1.01	9.67	0.69
60AW	1.10	9.34	0.71
70AW	1.29	8.57	0.77
80AW	1.47	9.00	0.74
90AW	1.66	9.76	0.68

The 70AW model is seen to be the best choice for this service.

70.5.3 Control of Fans for Variable Volume Service

Two common applications of fans requiring variable volume operation are combustion air to boilers and conditioned air to rooms in a building. Common methods of controlling air volume are outlet dampers; inlet dampers; inlet vanes (which impart a prerotation or swirl velocity to the air entering the wheel); variable pitch blades (on axial fans); and variable speed drives on the fan motor.

Figure 70.11 illustrates the effectiveness of these methods by comparing power ratios with flow ratios at reduced flows. The variable speed drive is the most effective method and, with the commercializa-

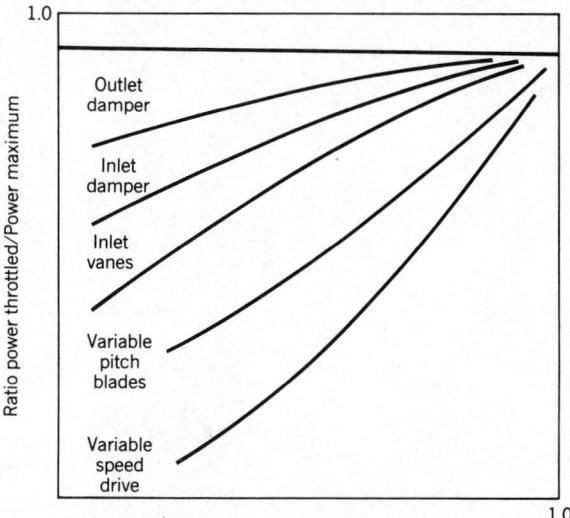

Fig. 70.11 Effectiveness of various methods of controlling fans in variable volume service.

tion of solid-state motor controls (providing variable frequency and variable voltage electrical service to standard induction motors), is becoming the most popular method for fan speed control.

BIBLIOGRAPHY

ASHRAE Handbook of Fundamentals, American Society of Heating, Refrigerating and Air Conditioning Engineers, Atlanta, GA, 1980.

Cameron Hydraulic Data, Ingersoll Rand Co., Woodcliff Lake, NJ, 1977.

G. T. Csanady, *Theory of Turbo Machines,* McGraw-Hill, New York, 1964.

Fans and Systems, Publication 201; *Troubleshooting,* Publication 202; *Field Performance Measurements,* Publication 203; Air Moving and Conditioning Association, Arlington Heights, IL.

Fans in Air Conditioning, The Trane Co., La Crosse, WI.

Flow of Fluids Through Valves, Fittings and Pipe, Technical Paper No. 410, Crane Co., Chicago, IL, 1976.

T. G. Hicks and T. W. Edwards, *Pump Application Engineering,* McGraw-Hill, New York, 1971.

Hydraulic Institute Standards, Hydraulic Institute, Cleveland, OH, 1975.

I. J. Karassick, *Centrifugal Pump Clinic,* Marcel Dekker, New York, 1981.

Laboratory Methods of Testing Fans for Ratings, Standard 210-74, Air Moving and Conditioning Association, Arlington Heights, IL, 1974.

CHAPTER 71

NUCLEAR POWER

WILLIAM KERR

Department of Nuclear Engineering
University of Michigan
Ann Arbor, Michigan

71.1 HISTORICAL PERSPECTIVE

71.1.1 The Birth of Nuclear Energy

The first large-scale application of nuclear energy was in a weapon. The second use was in submarine propulsion systems. Subsequent development of fission reactors for electric power production has been profoundly influenced by these early military associations, both technically and politically. It appears likely that the military connection, tenuous though it may be, will continue to have a strong political influence on applications of nuclear energy.

Fusion, looked on by many as a supplement to, or possibly as an alternate to fission for producing electric power, was also applied first as a weapon. Most of the fusion systems now being investigated for civilian applications are far removed from weapons technology. A very few are related closely enough that further civilian development could be inhibited by this association.

71.1.2 Military Propulsion Units

The possibilities inherent in an extremely compact source of fuel, the consumption of which requires no oxygen, and produces a small volume of waste products, was recognized almost immediately after World War II by those responsible for the improvement of submarine propulsion units. Significant resources were soon committed to the development of a compact, easily controlled, quiet, and highly reliable propulsion reactor. As a result, a unit was produced which revolutionized submarine capabilities.

The decisions that led to a compact, light-water-cooled and -moderated submarine reactor unit, using enriched uranium for fuel, were undoubtedly valid for this application. They have been adopted by other countries as well. However, the technological background and experience gained by U.S. manufacturers in submarine reactor development was a principal factor in the eventual decision to build commercial reactors that were cooled with light water and that used enriched uranium in oxide form as fuel. Whether this was the best approach for commercial reactors is still uncertain.

71.1.3 Early Enthusiasm for Nuclear Power

Until the passage, in 1954, of an amendment to the Atomic Energy Act of 1946, almost all of the technology that was to be used in developing commercial nuclear power was classified. The 1954 Amendment made it possible for U.S. industry to gain access to much of the available technology, and to own and operate nuclear power plants. Under the amendment the Atomic Energy Commission (AEC), originally set up for the purpose of placing nuclear weapons under civilian control, was given responsibility for licensing and for regulating the operation of these plants.

In December of 1953 President Eisenhower, in a speech before the General Assembly of the United Nations, extolled the virtues of peaceful uses of nuclear energy and promised the assistance of the United States in making this potential new source of energy available to the rest of the world. Enthusiasm over what was then viewed as a potentially inexpensive and almost inexhaustible new source of energy was a strong force which led, along with the hope that a system of international inspection and control could inhibit proliferation of nuclear weapons, to formation of the International Atomic Energy Agency (IAEA) as an arm of the United Nations. The IAEA, with headquarters in Vienna, continues to play a dual role of assisting in the development of peaceful uses of nuclear energy, and in the development of a system of inspections and controls aimed at making it possible to detect any diversion of special nuclear materials, being used in or produced by civilian power reactors, to military purposes.

71.1.4 U.S. Development of Nuclear Power

Beginning in the early 1950s the AEC, in its national laboratories, and with the participation of a number of industrial organizations, carried on an extensive program of reactor development. A variety of reactor systems and types were investigated analytically and several prototypes were built and operated.

In addition to the light water reactor (LWR), gas-cooled graphite-moderated reactors, liquid-fueled reactors with fuel incorporated in a molten salt, liquid-fueled reactors with fuel in the form of a uranium nitrate solution, liquid-sodium-cooled graphite-moderated reactors, solid-fueled reactors with organic coolant, and liquid-metal solid-fueled fast spectrum reactors have been developed and operated, at least in pilot plant form in the United States. All of these have had enthusiastic advocates. Most, for various reasons, have not gone beyond the pilot plant stage. Two of these, the high-temperature gas-cooled reactor (HTGR) and the liquid-metal-cooled fast breeder reactor (LMFBR), have been built and operated as prototype power plants.

Some of these have features associated either with normal operation, or with possible accident situations, which seem to make them attractive alternatives to the LWR. The HTGR, for example, operates at much higher outlet coolant temperature than the LWR and thus makes possible a significantly more efficient thermodynamic cycle as well as permitting use of a physically smaller steam turbine. The reactor core, primarily graphite, operates at a much lower power density than that of LWRs. This lower power density and the high-temperature capability of graphite make the HTGR's core much more tolerant of a loss-of-coolant accident than the LWR core.

The long, difficult, and expensive process needed to take a conceptual reactor system to reliable commercial operation has unquestionably inhibited the development of a number of alternative systems.

71.2 CURRENT POWER REACTORS, AND FUTURE PROJECTIONS

Although a large number of reactor types have been studied for possible use in power production, the number now receiving serious consideration is rather small.

71.2.1 Light-Water-Moderated Enriched-Uranium-Fueled Reactor

The only commercially viable power reactor systems operating in the United States today use LWRs. This is likely to be the case for the next decade or so. France has embarked on a construction program that will eventually lead to productions of about 90% of its electric power by LWR units. Great Britain has under consideration the construction of a number of LWRs. The Federal Republic of Germany has a number of LWRs in operation with additional units under construction. Russia and a number of other Eastern European countries are operating LWRs, and are constructing additional plants. Russia is also building a number of smaller, specially designed LWRs near several population centers. It is planned to use these units to generate steam for district heating. The first one of these reactors is scheduled to go into operation soon near Gorki.

71.2.2 Gas-Cooled Reactor

Several designs exist for gas-cooled reactors. In the United States the one that has been most seriously considered uses helium for cooling. Fuel elements are large graphite blocks containing a number of vertical channels. Some of the channels are filled with enriched uranium fuel. Some, left open, provide a passage for the cooling gas. One small power reactor of this type is in operation in the United States. Carbon dioxide is used for cooling in some European designs. Both metal fuels and graphite-coated fuels are used. A few gas-cooled reactors are being used for electric power production both in England and in France.

71.2.3 Heavy-Water-Moderated Natural-Uranium-Fueled Reactor

The goal of developing a reactor system that does not require enriched uranium led Canada to a natural-uranium-fueled, heavy-water-moderated, light-water-cooled reactor design dubbed Candu. A number of these are operating successfully in Canada. Argentina and India each uses a reactor power plant of this type, purchased from Canada, for electric power production.

71.2.4 Liquid-Metal-Cooled Fast Breeder Reactor

France, England, Russia, and the United States all have prototype liquid-metal-cooled fast breeder reactors (LMFBRs) in operation. Experience and analysis provide evidence that the plutonium-fueled LMFBR is the most likely, of the various breeding cycles investigated, to provide a commercially viable breeder. The breeder is attractive because it permits as much as 80% of the available energy in natural uranium to be converted to useful energy. The LWR system, by contrast, converts at most 3–4%.

Because plutonium is an important constituent of nuclear weapons, there has been concern that development of breeder reactors will produce nuclear weapons proliferation. This is a legitimate concern, and must be dealt with in the design of the fuel cycle facilities that make up the breeder fuel cycle.

71.2.5 Fusion

It may be possible to use the fusion reaction, already successfully harnessed to produce a powerful explosive, for power production. Considerable effort in the United States and in a number of other countries is being devoted to development of a system that would use a controlled fusion reaction to produce useful energy. At the present stage of development the fusion of tritium and deuterium nuclei appears to be the most promising reaction of those that have been investigated. Problems in the design, construction, and operation of a reactor system that will produce useful amounts of economical power appear formidable. However, potential fuel resources are enormous, and are readily available to any country that can develop the technology.

71.3 CATALOG AND PERFORMANCE OF OPERATING REACTORS, WORLDWIDE

Worldwide, the operation of nuclear power plants in 1982 produced more than 10% of all the electrical energy used. Table 71.1 contains a listing of reactors in operation in the United States and in the rest of the world.

71.4 U.S. COMMERCIAL REACTORS

As indicated earlier, the approach to fuel type and core design used in LWRs in the United States comes from the reactors developed for marine propulsion by the military.

71.4.1 Pressurized-Water Reactors

Of the two types developed in the United States, the pressurized water reactor (PWR) and the boiling water reactor (BWR), the PWR is a more direct adaptation of marine propulsion reactors. PWRs are operated at pressures in the pressure vessel (typically about 2250 psi) and temperatures (primary inlet coolant temperature is about 564°F with an outlet temperature about 64°F higher) such that bulk boiling does not occur in the core during normal operation. Water in the primary system flows through the core as a liquid, and proceeds through one side of a heat exchanger. Steam is generated on the other side at a temperature slightly less than that of the water that emerges from the reactor vessel outlet. Figure 71.1 shows a typical PWR vessel and core arrangement. Figure 71.2 shows a steam generator.

The reactor pressure vessel is an especially crucial component. Current U.S. design and operational philosophy assumes that systems provided to ensure maintenance of the reactor core integrity under both normal and emergency conditions will be able to deliver cooling water to a pressure vessel whose integrity is virtually intact after even the most serious accident considered in the safety analysis of hypothesized accidents required by U.S. licensing. A special section of the ASME Pressure Vessel Code, Section III, has been developed to specify acceptable vessel design, construction, and operating practices. Section XI of the code specifies acceptable inspection practices.

Practical considerations in pressure vessel construction and operation determine an upper limit to the primary operating pressure. This in turn prescribes a maximum temperature for water in the primary. The resulting steam temperature in the secondary is considerably lower than that typical of modern fossil-fueled plants. (Typical steam temperatures and pressures are about 1100 psi and 556°F at the steam generator outlet.) This lower steam temperature has required development of massive steam turbines to handle the enormous steam flow of the low-temperature steam produced by the large PWRs of current design.

71.4.2 Boiling-Water Reactors

As the name implies, steam is generated in the BWR by boiling, which takes place in the reactor core. Early concerns about nuclear and hydraulic instabilities led to a decision to operate military propulsion reactors under conditions such that the moderator–coolant in the core remains liquid. In the course of developing the BWR system for commercial use, solutions have been found for the instability problems.

Although some early BWRs used a design that separates the core coolant from the steam which flows to the turbine, all modern BWRs send steam generated in the core directly to the turbine. This arrangement eliminates the need for a separate steam generator. It does, however, provide direct communication between the reactor core and the steam turbine and condenser, which are located outside the containment. This leads to some problems not found in PWRs. For example, the turbine–condenser system must be designed to deal with radioactive nitrogen-16 generated by an (n,p) reaction of fast neutrons in the reactor core with oxygen-16 in the cooling water. Decay of the short-lived nitrogen-16 (half-life 7.1 sec) produces high-energy (6.13-MeV) highly penetrating gamma rays. As a result, the radiation level around an operating BWR turbine requires special precautions not needed for the PWR turbine. The direct pathway from core to turbine provided by the steam pipes also affords a possible avenue of escape and direct release outside of containment for fission products that might be released from the fuel in a core-damaging accident. Rapid-closing valves in the steam lines are provided to block this path in case of such an accident.

The selection of pressure and temperature for the steam entering the turbine that are not markedly different from those typical of PWRs leads to an operating pressure for the BWR pressure vessel that is typically less than half that for PWRs. (Typical operating pressure at vessel outlet is about 1050 psi with a corresponding steam temperature of about 551°F.)

Because it is necessary to provide for two-phase flow through the core, the core volume is larger than that of a PWR of the same power. The core power density is correspondingly smaller. Figure 71.3 is a cutaway of a BWR vessel and core arrangement. The in-vessel steam separator for removing

Table 71.1 Operating Power Reactors (1983)

Country	Reactor Type[a]	Number in Operation	Net MWe
Argentina	PHWR	1	335
Belgium	PWR	5	3450
Bulgaria	PWR	4	1760
Canada	PHWR	12	6622
Czechoslovakia	PWR	2	880
Federal Republic of Germany	PWR	7	6609
	BWR	4	3145
	PHWR	1	52
Finland	PWR	2	890
	BWR	2	1320
France	GCR	7	2205
	GCHWR	1	70
	PWR	27	23,720
	LMFBR	1	233
German Democratic Republic	PWR	5	1830
Hungary	PWR	1	440
India	BWR	2	400
	PHWR	2	404
Italy	GCR	1	150
	PWR	1	260
	BWR	1	875
Japan	PWR	11	7504
	BWR	12	8789
	GCR	1	150
	HWLWR	1	148
	LMFBR	3	280
Korea	PWR	3	1790
Netherlands	PWR	1	445
	BWR	1	50
Pakistan	PHWR	1	125
Spain	BWR	1	440
	PWR	3	1970
	GCR	1	480
Sweden	BWR	7	4725
	PWR	3	2600
Switzerland	BWR	1	320
	PWR	3	1620
Taiwan	BWR	4	3110
UK	GCR	26	5206
	AGR	5	3100
	HWLWR	1	92
	LMFBR	1	250
United States	BWR	26	20,518
	PWR	49	39,951
	HTGR	1	330
	LMFBR	1	18
USSR	LGR	18	10,900
	PWR	12	6895
	LMFBR	2	950
	BWR	1	50
Yugoslavia	PWR	1	615

[a] PWR = pressurized water reactor; BWR = boiling water reactor; AGR = advanced gas-cooled reactor; GCR = gas-cooled reactor; HTGR = high-temperature gas-cooled reactor; LMFBR = liquid-metal fast-breeder reactor; LGR = light-water-cooled graphite-moderated reactor; HWLWR = heavy-water-moderated light-water-cooled reactor; PHWR = pressurized heavy-water-moderated-and-cooled reactor; GCHWR = gas-cooled heavy-water-moderated reactor.

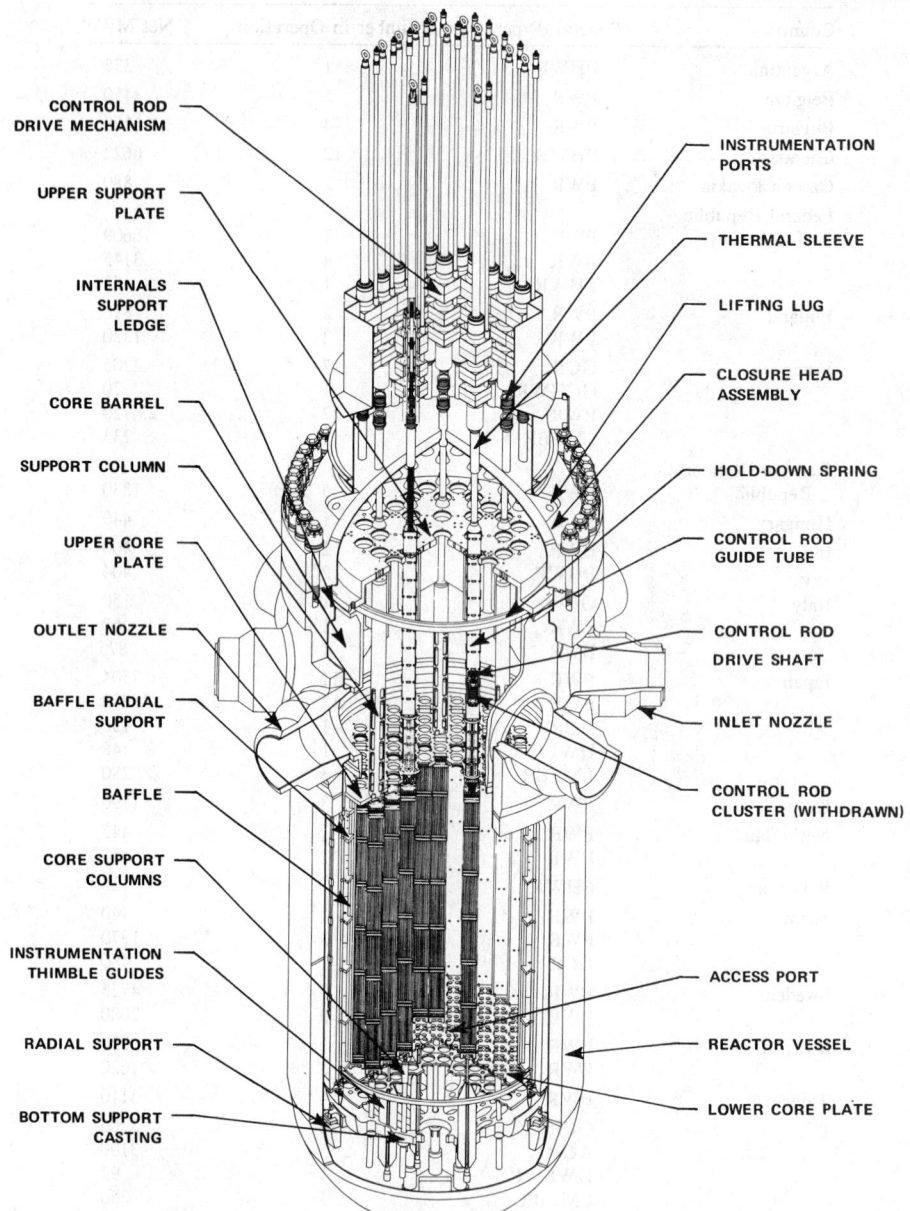

CONTROL ROD
DRIVE MECHANISM

UPPER SUPPORT
PLATE

INTERNALS
SUPPORT
LEDGE

CORE BARREL

SUPPORT COLUMN

UPPER CORE
PLATE

OUTLET NOZZLE

BAFFLE RADIAL
SUPPORT

BAFFLE

CORE SUPPORT
COLUMNS

INSTRUMENTATION
THIMBLE GUIDES

RADIAL SUPPORT

BOTTOM SUPPORT
CASTING

INSTRUMENTATION
PORTS

THERMAL SLEEVE

LIFTING LUG

CLOSURE HEAD
ASSEMBLY

HOLD-DOWN SPRING

CONTROL ROD
GUIDE TUBE

CONTROL ROD
DRIVE SHAFT

INLET NOZZLE

CONTROL ROD
CLUSTER (WITHDRAWN)

ACCESS PORT

REACTOR VESSEL

LOWER CORE PLATE

Fig. 71.1 Typical vessel and core configuration for PWR. (Courtesy Westinghouse.)

moisture from the steam is located above the core assembly. Figure 71.4 is a BWR fuel assembly. The assembly is contained in a channel box, which directs the two-phase flow. Fuel pins and fuel pellets are not very different in either size or shape from those for PWRs, although the cladding thickness for the BWR pin is somewhat larger than that of PWRs.

71.4.3 High-Temperature Gas-Cooled Reactors

Experience with the high-temperature gas-cooled reactor (HTGR) in the United States is limited. A 40-MWe plant was operated from 1967 to 1974. A 330-MWe plant has been in operation since 1976. A detailed design was developed for a 1000-MWe plant, but plans for its construction were abandoned.

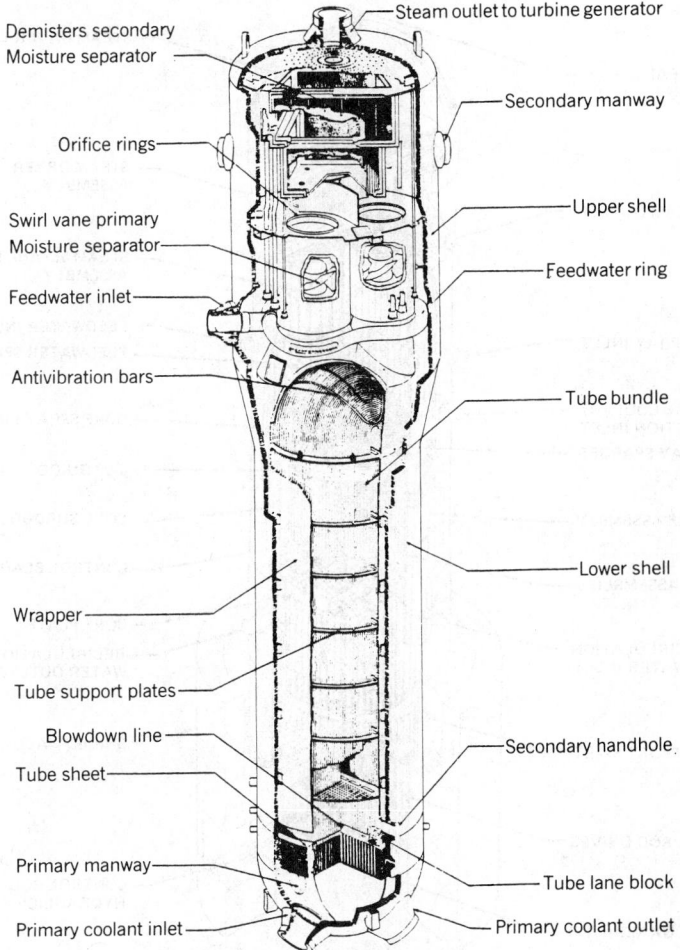

Fig. 71.2   Typical PWR steam generator.

Fuel elements for the plant in operation are hexagonal prisms of graphite about 31 in. tall and 5.5 in. across flats. Vertical holes in these blocks allow for passage of the helium coolant. Fuel elements for the larger proposed plant were similar. Figure 71.5 shows core and vessel arrangement. Typical helium-coolant outlet temperature for the reactor now in operation is about 1300°F. Typical steam temperature is 1000°F. The large plant was also designed to produce 1000°F steam.

The fuel cycle for the HTGR was originally designed to use fuel that combined highly enriched uranium with thorium. This cycle would convert thorium to uranium-233, which is also a fissile material, thereby extending fuel lifetime significantly. This mode of operation also produces uranium-233, which can be chemically separated from the spent fuel for further use. Recent work has resulted in the development of a fuel using low-enriched uranium in a once-through cycle similar to that used in LWRs.

The use of graphite as a moderator and helium as coolant allows operation at temperatures significantly higher than those typical of LWRs, resulting in higher thermal efficiencies. The large thermal capacity of the graphite core and the large negative temperature coefficient of reactivity make the HTGR insensitive to inadvertent reactivity insertions and to loss-of-coolant accidents. Operating experience to date gives some indication that the HTGR has advantages in increased safety and in lower radiation exposure to operating personnel. These possible advantages plus the higher thermal efficiency that can be achieved make further development attractive. However, the high cost of developing a large commercial unit, plus the uncertainties that exist because of the limited operating experience with this type reactor have so far outweighed the perceived advantages.

As the data in Table 71.1 indicate, there is significant successful operating experience with several types of gas-cooled reactors in a number of European countries.

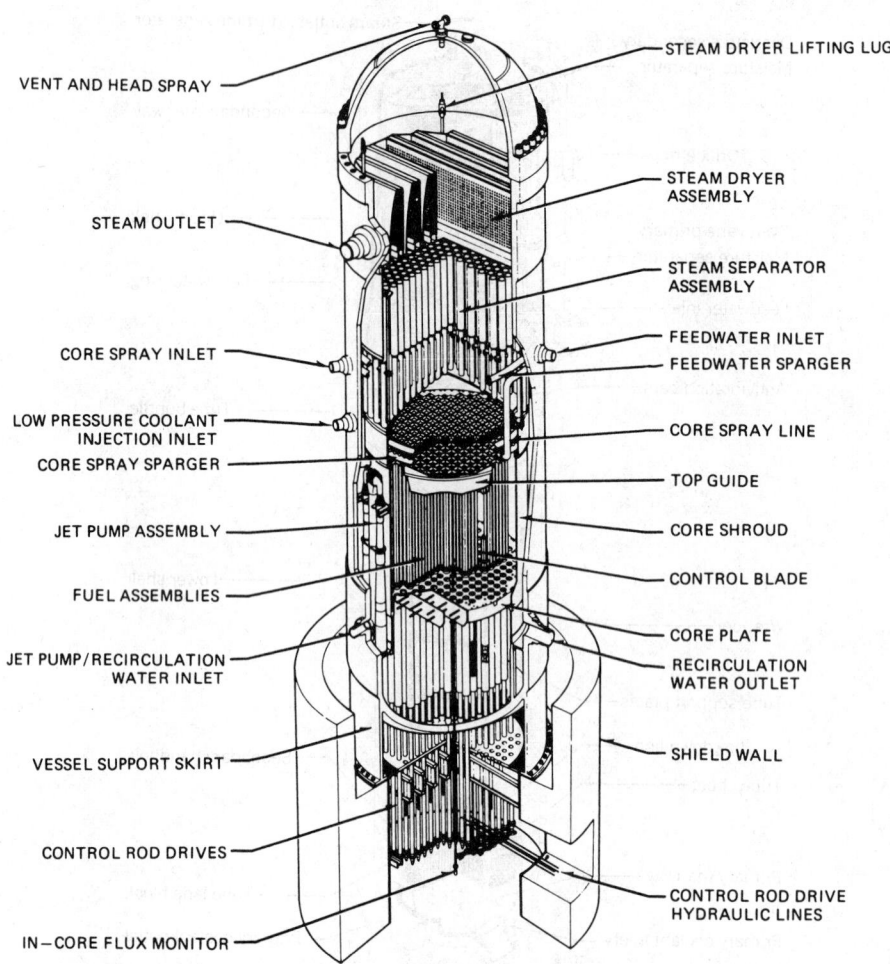

STEAM DRYER LIFTING LUG

VENT AND HEAD SPRAY

STEAM DRYER ASSEMBLY

STEAM OUTLET

STEAM SEPARATOR ASSEMBLY

FEEDWATER INLET

CORE SPRAY INLET

FEEDWATER SPARGER

LOW PRESSURE COOLANT INJECTION INLET

CORE SPRAY LINE

CORE SPRAY SPARGER

TOP GUIDE

JET PUMP ASSEMBLY

CORE SHROUD

FUEL ASSEMBLIES

CONTROL BLADE

CORE PLATE

JET PUMP/RECIRCULATION WATER INLET

RECIRCULATION WATER OUTLET

VESSEL SUPPORT SKIRT

SHIELD WALL

CONTROL ROD DRIVES

CONTROL ROD DRIVE HYDRAULIC LINES

IN–CORE FLUX MONITOR

Fig. 71.3 Typical BWR vessel and core configuration. (Courtesy General Electric.)

71.4.4 Constraints

Reactors being put into operation today are based on designs that were originally conceived as much as 20 years earlier. The incredible time lag between the beginning of the design process and the operation of the plant is one of the unfortunate products of a system of industrial production and federal regulation that moves ponderously and uncertainly toward producing a power plant that may be technically obsolescent by the time it begins operation. The combination of the large capital investment required for plant construction, the long period during which this investment remains unproductive for a variety of reasons, and the high interest rates charged for borrowed money have recently led to plant capital costs some 5–10 times larger than those for plants that came on line in the early to mid 1970s. Added to the above constraints is a widespread concern about dangers of nuclear power. These concerns span a spectrum that encompasses fear of contribution to nuclear weapons proliferation, on the one hand, to a strong aversion to high technology, on the other hand. This combination of technical, economic, and political constraints places a severe burden on those working to develop this important alternate source of energy.

71.4.5 Availability

A significant determinant in the cost of electrical energy produced by nuclear power plants is the plant capacity factor. The capacity factor is defined as a fraction calculated by dividing actual energy

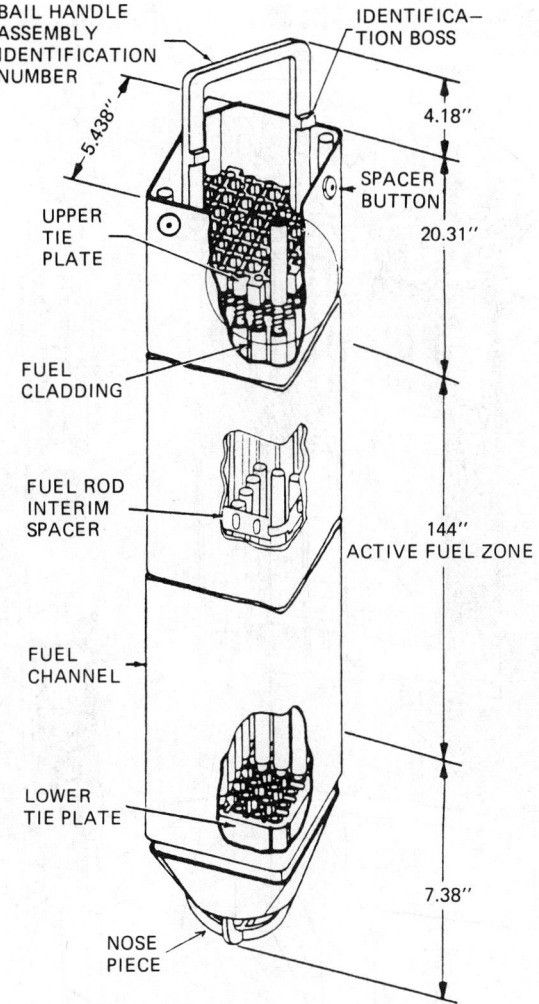

BAIL HANDLE
ASSEMBLY
IDENTIFICATION
NUMBER

IDENTIFICA-
TION BOSS

5.438"

4.18"

SPACER
BUTTON

UPPER
TIE
PLATE

20.31"

FUEL
CLADDING

FUEL ROD
INTERIM
SPACER

144"
ACTIVE FUEL ZONE

FUEL
CHANNEL

LOWER
TIE PLATE

7.38"

NOSE
PIECE

Fig. 71.4 BWR fuel assembly.

production during some specified time period by the amount that would have been produced by continuous power production at 100% of plant capacity. Many of the early estimates of power cost for nuclear plants were made with the assumption of a capacity factor of 0.80. Experience indicates an average for U.S. power plants of about 0.60. The contribution of capital costs to energy production has thus been more than 30% higher than the early estimates. Since capital costs typically represent anywhere between about 40–80% (depending on when the plant was constructed) of the total energy cost, this difference in goal and achievement is a significant factor in some of the recently observed cost increases for electricity produced by nuclear power. Examination of the experience of individual plants reveals a wide range of capacity factors. A few U.S. plants have achieved a cumulative capacity factor near 0.80. Some have capacity factors as low as 0.40. There is reason to believe that improvements can be made in many of those with low capacity factors. It should also be possible to go beyond 0.80. Capacity factor improvement is a fruitful area for better resource utilization and realization of lower energy costs.

71.5 POLICY

The Congress, in the 1954 amendment to the Atomic Energy Act, made the development of nuclear power national policy. Responsibility for ensuring safe operation of nuclear power plants was originally given to the Atomic Energy Commission. In 1975 this responsibility was turned over to a Nuclear

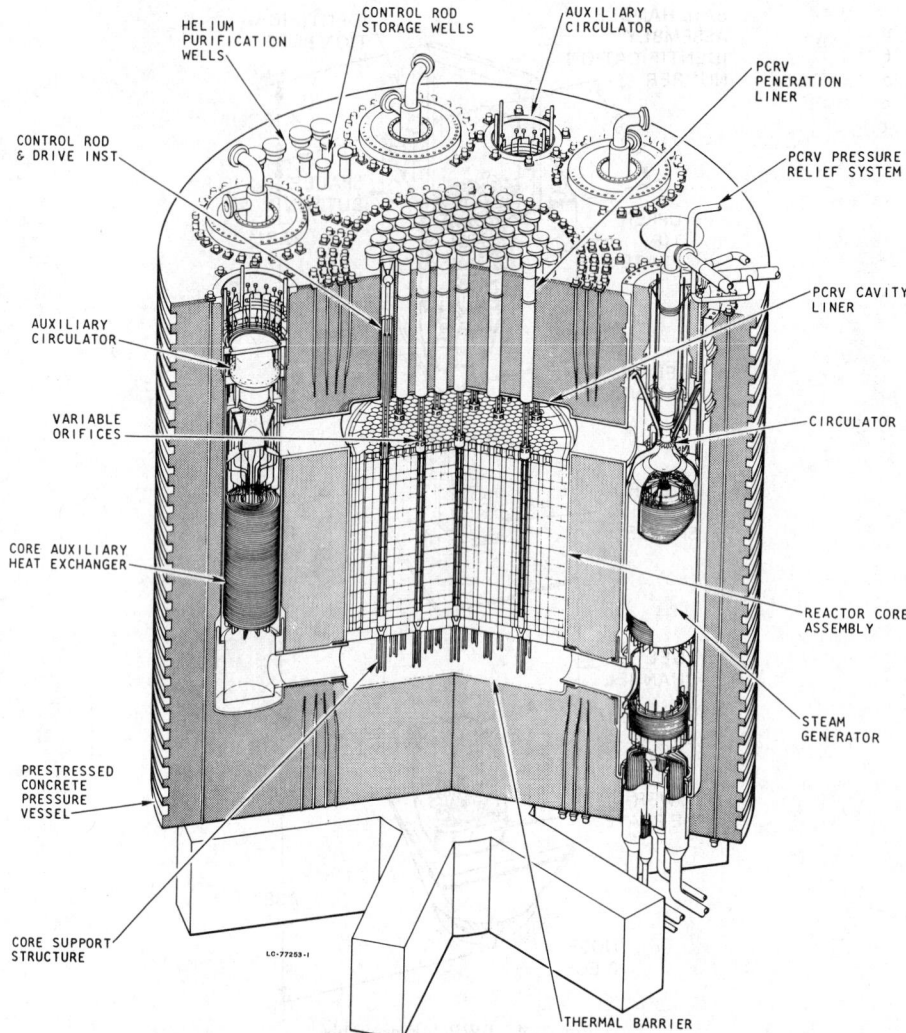

CONTROL ROD
STORAGE WELLS

HELIUM
PURIFICATION
WELLS

AUXILIARY
CIRCULATOR

PCRV
PENERATION
LINER

CONTROL ROD
& DRIVE INST

PCRV PRESSURE
RELIEF SYSTEM

AUXILIARY
CIRCULATOR

PCRV CAVITY
LINER

VARIABLE
ORIFICES

CIRCULATOR

CORE AUXILIARY
HEAT EXCHANGER

REACTOR CORE
ASSEMBLY

STEAM
GENERATOR

PRESTRESSED
CONCRETE
PRESSURE
VESSEL

CORE SUPPORT
STRUCTURE

LC-77253-1

THERMAL BARRIER

Fig. 71.5 HTGR pressure vessel and core arrangement. (Used by permission of Marcel Dekker, Inc., New York.)

Regulatory Commission (NRC), set up for this purpose as an independent federal agency. Nuclear power is the most highly regulated of all the existing sources of energy. Much of the regulation is at the federal level. However, nuclear power plants and their operators are subject to a variety of state and local regulations as well. Under these circumstances nuclear power is of necessity highly responsive to any energy policy that is pursued by the federal government, or of local branches of government, including one of bewilderment and uncertainty.

71.5.1 Safety

The principal safety concern is the possibility of exposure of people to the radiation produced by the large (in terms of radioactivity) quantity of radioactive material produced by the fissioning of the reactor fuel. In normal operation of a nuclear power plant all but a minuscule fraction of this material is retained within the reactor fuel and the pressure vessel. Significant exposure of people outside the plant can occur only if a catastrophic and extremely unlikely accident should release a large fraction of the radioactive fission products from the pressure vessel and from the surrounding containment system, and if these radioactive materials are then transported to locations where people are exposed to their radiation.

The uranium eventually used in reactor fuel is itself radioactive. The radioactive decay process, which begins with uranium, proceeds to produce several radioactive elements. One of these, radon-226, is a gas and can thus be inhaled by uranium miners. Hence, those who work in the mines are exposed to some hazard. Waste products of the mining and milling of uranium are also radioactive. When stored or discarded above ground, these wastes subject those in the vicinity to radon-226 exposure. These wastes or mill tailings must be dealt with to protect against this hazard. One method of control involves covering the wastes with a layer of some impermeable material such as asphalt.

The fresh fuel elements are also radioactive because of the contained uranium. However, the level of radioactivity is sufficiently low that the unused fuel assemblies can be handled safely without shielding.

71.5.2 Disposal of Radioactive Wastes

The used fuel from a power reactor is highly radioactive, although small in volume. The spent fuel produced by a year's operation of a 1000-MWe plant typically weighs about 40 tons and could be stored in a cube less than 5 ft on a side. It must be kept from coming in contact with people or other living organisms for long periods of time. (After 1000 years of storage the residual radioactivity of the spent fuel is about that of the original fresh fuel.) This spent fuel, or the radioactive residue that remains if most of the unused uranium and the plutonium generated during operation are chemically separated, is called high-level radioactive waste. Up to the present a variety of considerations, many of them political, have led to postponement of a decision on the choice of a permanent storage method for this material. The problem of safe storage has several solutions that are both technically and economically feasible. Technical solutions that currently exist include aboveground storage in air-cooled metal cannisters (for an indefinite period if desirable, with no decrease of safety over the period), as well as permanent disposal in deep strata of salt or of various impermeable rock formations. There have also been proposals to place the radioactive materials in deep ocean caverns. This method, although probably technically possible, is not yet developed. It would require international agreements not now in place. As indicated earlier, an operating plant also generates radioactive material in addition to fission products. Some of this becomes part of the various process streams that are part of the plant's auxiliary systems. These materials are typically removed by filters or ion-exchange systems, leaving filters or ion-exchange resins that contain radioactive materials. Tools, gloves, clothing, paper, and other materials may become slightly contaminated during plant operation. If the radioactive contamination has a half-life of more than a few weeks, these materials, described as low-level radioactive waste, must be stored or disposed of. The currently used disposal method involves burial in comparatively shallow trenches. Because of insufficient attention having been given to design and operation of some of the earlier burial sites, small releases of radioactive material have been observed. Several early burial sites are no longer in operation. Current federal legislation provides for compacts among several states that could lead to cooperative operation, by these states, of burial sites for low-level waste.

71.5.3 Economics

Nuclear power plants that began operation in the 1970s produce power at a cost considerably less than coal-burning plants of the same era. The current cost of power produced by oil-burning plants is two to three times as great as that produced by these nuclear plants. Nuclear power plants coming on line in the 1980s are much more expensive in capital cost (in some cases by a factor of 10!). The cost of the power they produce will be correspondingly greater. The two major contributors to the cost increase are high interest rates and the long construction period that has been required for most of these plants. Average construction time for plants now coming on line is about 11 years! It is likely that construction times can be decreased for new plants. The changes that were required as a result of the TMI accident have now been incorporated into regulations, into existing plants, and into new designs, eliminating the costly and time-consuming back fits that were required for plants under construction when the accident occurred. In Japan the average construction time for nuclear power plants is about 54 months. In Russia it is said to be about 77 months.

Standard plants are being designed and licensed that should make the licensing of an individual plant much faster and less involved. Concern over the pollution of the ecosphere caused by fossil-fueled plants (acid rain, CO_2) will call for additional pollution control, which will drive up costs of construction and operation of these plants. It is reasonable to expect nuclear power to be economically competitive with alternative methods of electric power generation in both the near and longer term.

71.5.4 Environmental Considerations

The environmental pollution produced by an operating nuclear power plant is far less than that caused by any other currently available method of producing electric power. The efficiency of the thermodynamic cycle for water reactors is lower than that of modern fossil-fuel plants because current design of reactor pressure vessels limits the steam temperature. Thus, the amount of waste heat rejected is greater for a nuclear plant than for a modern fossil-fuel plant of the same rated power. However, current methods of waste heat rejection (typically cooling towers) handle this with no particular environmental degradation. Nuclear power plants emit no carbon dioxide, no sulfur, no nitrous oxides. No large coal storage area is required. The tremendous volumes of sulfur compounds removed during coal combustion and the enormous quantities of ash produced by coal plants are problems with which those who operate nuclear plants do not have to deal.

Table 71.2 provides a comparison of emissions and wastes from a large coal-burning plant and from a nuclear power plant of the same rated power. Although there is a small release of radioactive material to the biosphere from the nuclear power plant, the resulting increase in exposure to a member of the population in the immediate vicinity of the plant is typically about 1% of that produced by naturally occurring background radiation.

71.5.5 Proliferation

Nuclear power plants are thought by some to increase the probability of nuclear weapons proliferation. It is true that a country with the trained engineers and scientists, the facilities, and the resources required to produce nuclear power can develop a weapons capability more rapidly than one without this background. However, for a country starting from scratch, the development of nuclear power is a detour that would consume needless time and resources. None of the countries that now possess nuclear weapons capability has used the development of civil nuclear power as a route to weapons development.

Nevertheless, it must be recognized that plutonium, an important constituent of weapons, is produced in light-water nuclear power plants. Plutonium is the preferred fuel for breeder reactors. The development of any significant number of breeder reactors would thus involve the production and handling of large quantities of plutonium.

As will be discussed in a later section, plutonium-239 can be produced by the absorption of a neutron in uranium-238. Since most of the uranium in the core of an LWR is uranium-238, plutonium is produced during operation of the reactor. However, if the plutonium-239 is left in a power reactor

Table 71.2 Waste Material from Different Types of 1000-MWe Power Plants (Capacity Factor = 0.8)

	Coal Fired	Water Reactor
Typical thermal efficiency, %	39	32
Thermal wastes (in thermal megawatts)		
To cooling water	1,170	1,970
To atmosphere	400	150
Total	1,570	2,120
Solid wastes		
Fly ash or slag, tons/year	330,000	0
cubic feet/year	7,350,000	0
railroad carloads/year	3,300	0
Radioactive wastes		
Fuel to reprocessing plant, assemblies/year	0	160
railroad carloads/year	0	5
Solid waste storage		
From reprocessing plant, cubic feet/year	0	100
From power plant, cubic feet/year	0	5,000
Gaseous and liquid wastes[a]		
(tons per day/10^6 cubic feet per day)		
Carbon monoxide	2/8	0
Carbon dioxide	21,000/53,200	0
Sulfur dioxide: 1% sulfur fuel	140/325	0
2.5% sulfur fuel	350/812	0
Nitrogen oxides	82/305	0
Particulates to atmosphere (tons/day)	0.4	0
Radioactive gases or liquids, equivalent dose mrem/year at plant boundary	Minor	5

[a] For 3,000,000 tons/year coal total ash content of 11%, fly ash precipitator efficiency of 99.5%, and 15% of sulfur remaining in ash.

core for the length of time typical of the fuel cycle used for LWRs or for breeders, neutrons are absorbed by some fraction of the plutonium to produce plutonium-240. This isotope also absorbs neutrons to produce plutonium-241. These heavier isotopes make the plutonium undesirable as weapons material. Thus, although the plutonium produced in power reactors can be separated chemically from the other materials in a used fuel element, it is not what would be considered weapons-grade material. A nation with the goal of developing weapons would almost certainly design and use a reactor and a fuel cycle designed specifically for producing weapons-grade material. On the other hand, if a drastic change in government produced a correspondingly drastic change in political objectives in a country that had a civil nuclear power program in operation, it would probably be possible to make use of power reactor plutonium to produce some sort of low-grade weapon.

71.6 BASIC ENERGY PRODUCTION PROCESSES

Energy can be produced by nuclear reactions that involve either fission (the splitting of a nucleus) or fusion (the fusing of two light nuclei to produce a heavier one). If energy is to result from fission, the resultant nuclei must have a smaller mass per nucleon (which means they are more tightly bound) than the original nucleus. If the fusion process is to produce energy, the fused nucleus must have a smaller mass per nucleon (i.e., be more tightly bound) than the original nuclei. Figure 71.6 is a curve of nuclear binding energies. Observe that only the heavy nuclei are expected to produce energy on fission, and that only the light nuclei yield energy in fusion. The differences in mass per nucleon before and after fission or fusion are available as energy.

71.6.1 Fission

In the fission process this energy is available primarily as kinetic energy of the fission fragments. Gamma rays are also produced as well as a few free neutrons, carrying a small amount of kinetic energy. The radioactive fission products decay (in most cases there is a succession of decays) to a stable nucleus. Gamma and beta rays are produced in the decay process. Most of the energy of these radiations is also recoverable as fission energy. Table 71.3 lists typical energy production due to fission of uranium by thermal neutrons, and indicates the form in which the energy appears. The quantity of energy available is of course related to the nuclear mass change by

$$\Delta E = \Delta mc^2$$

Fission in reactors is produced by the absorption of a neutron in the nucleus of a fissionable atom. In order to produce significant quantities of power, fission must occur as part of a sustained chain reaction, that is, enough neutrons must be produced in the average fission event to cause at

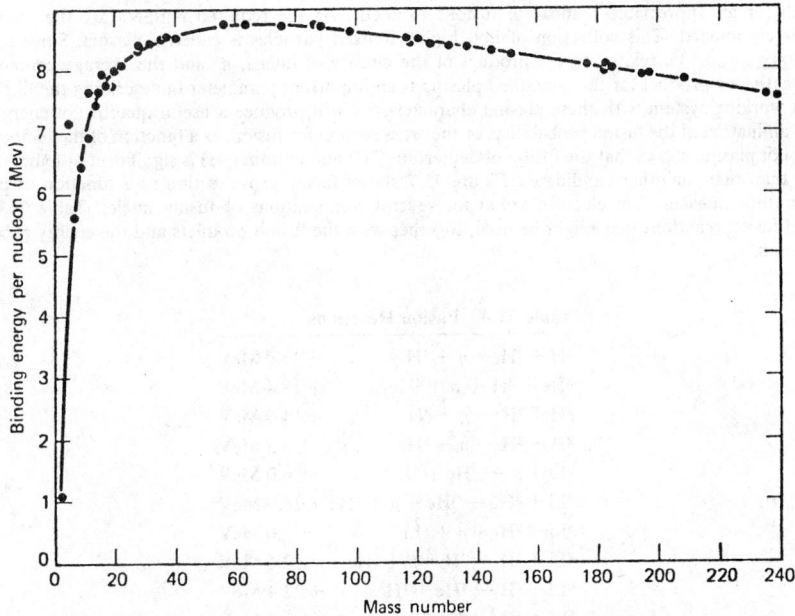

Fig. 71.6 Binding energy per nucleon versus mass number.

Table 71.3 Emitted and Recoverable Energies from Fission of ^{235}U

Form	Emitted Energy (MeV)	Recoverable Energy (MeV)
Fission fragments	168	168
Fission product decay		
β rays	8	8
γ rays	7	7
Neutrinos	12	—
Prompt γ rays	7	7
Fission neutrons (kinetic energy)	5	5
Capture γ rays	—	3–12
Total	207	198–207

least one new fission event to occur when absorbed in fuel material. The number of nuclei that are available and that have the required characteristics to sustain a chain reaction is limited to uranium-235, plutonium-239, and uranium-233. Only uranium-235 occurs in nature in quantities sufficient to be useful. (And it occurs as only 0.71% of natural uranium.) The other two can be manufactured in reactors. The reactions are indicated below:

$$^{238}U + n \longrightarrow {}^{239}U \longrightarrow {}^{239}Np \longrightarrow {}^{239}Pu$$

Uranium-239 has a half-life of 23.5 min. It decays to produce neptunium-239, which has a half-life of 2.35 days. The neptunium-239 decays to plutonium, which has a half-life of about 24,400 years.

$$^{232}Th + n \longrightarrow {}^{233}Th \longrightarrow {}^{233}Pa \longrightarrow {}^{233}U$$

Thorium-233 has a half-life of 22.1 min. It decays to protactinium-233, which has a half-life of 27.4 days. The protactinium decays to produce uranium-233 with a half-life of about 160,000 years.

71.6.2 Fusion

Fusion requires that two colliding nuclei have enough kinetic energy to overcome the Coulomb repulsion of the positively charged nuclei. If the fusion rate is to be useful in a power-producing system, there must also be a significant probability that fusion-producing collisions occur. These conditions can be satisfied for several combinations of nuclei if a collection of atoms can be heated to a temperature typically in the neighborhood of hundreds of millions of degrees and held together for a time long enough for an appreciable number of fusions to occur. At the required temperature the atoms are completely ionized. This collection of hot, highly ionized particles is called a plasma. Since average collision rate can be related to the product of the density of nuclei, n, and the average containment time, τ, the $n\tau$ product for the contained plasma is an important parameter in describing the likelihood that a working system with these plasma characteristics will produce a useful quantity of energy.

Examination of the fusion probability, or the cross section for fusion, as a function of the temperature of the hot plasma shows that the fusion of deuterium (^{2}H) and tritium (^{3}H) is significant at temperatures lower than that for other candidates. Figure 71.7 shows fusion cross section as a function of plasma temperature (measured in electron volts) for several combinations of fusing nuclei. Table 71.4 lists several fusion reactions that might be used, together with the fusion products and the energy produced per fusion.

Table 71.4 Fusion Reactions

$^3H + {}^2H \rightarrow n + {}^4He$	$+ 17.6$ MeV
$^3He + {}^2H \rightarrow p + {}^4He$	$+ 18.4$ MeV
$^2H + {}^2H \rightarrow p + {}^3H$	$+ 4.0$ MeV
$^2H + {}^2H \rightarrow n + {}^3He$	$+ 3.3$ MeV
$^6Li + p \rightarrow {}^3He + {}^4He$	$+ 4.0$ MeV
$^6Li + {}^3He \rightarrow {}^4He + p + {}^4He$	$+ 16.9$ MeV
$^6Li + {}^2H \rightarrow p + {}^7Li$	$+ 5.0$ MeV
$^6Li + {}^2H \rightarrow {}^3H + p + {}^4He$	$+ 2.6$ MeV
$^6Li + {}^2H \rightarrow {}^4He + {}^4He$	$+ 22.4$ MeV
$^6Li + {}^2H \rightarrow n + {}^7Be$	$+ 3.4$ MeV
$^6Li + {}^2H \rightarrow n + {}^3He + {}^4He$	$+ 1.8$ MeV

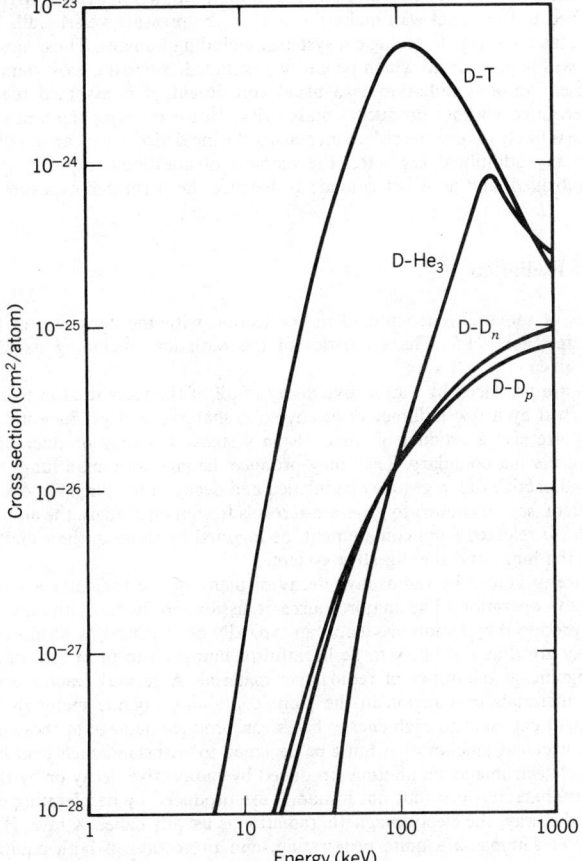

Fig. 71.7 Fusion cross section versus plasma temperature.

One of the problems with using the D–T reaction is the large quantity of fast neutrons that results, and the fact that a large fraction of the energy produced appears as kinetic energy of these neutrons. Some of the neutrons are absorbed in and activate the plasma-containment-system walls, making it highly radioactive. They also produce significant damage in most of the candidate materials for the containment walls. For these reasons there are some who advocate that work with the D–T reaction be abandoned in favor of the development of a system that depends on a set of reactions that is neutron-free.

Another problem with using the D–T reaction is that tritium does not occur in nature in sufficient quantity to be used for fuel. It must be manufactured. Typical systems propose to produce tritium by the absorption in lithium of neutrons resulting from the fusion process. Natural lithium consists of ^6Li (7.5%) and ^7Li (92.5%). The reactions are

$$^6Li + n \rightarrow {}^4He + {}^3H + 4.8 \text{ MeV (thermal neutrons)}$$

and

$$^7Li + n \rightarrow {}^4He + {}^3H + n + 2.47 \text{ MeV (threshold reaction)}$$

Considerations of neutron economy dictate that most of the neutrons produced in the fusion process be absorbed in lithium in order to breed the needed quantities of tritium. The reactions shown produce not only tritium, but also additional energy. The $(^6Li, n)$ reaction, for example, produces 4.7 MeV per reaction. If this energy can be recovered, it effectively increases the average available energy per fusion by about 27%.

71.7 CHARACTERISTICS OF THE RADIATION PRODUCED BY NUCLEAR SYSTEMS

An important by-product of the processes used to generate nuclear power is a variety of radiations in the form of either particles or electromagnetic photons. These radiations can produce damage in the materials that make up the systems and structures of the power reactors. High-energy neutrons, for example, absorbed in the vessel wall make the steel in the pressure vessel walls less ductile.

Radiation also causes damage to biological systems, including humans. Thus, most of the radiations must be contained within areas from which people are excluded. Since the ecosystem to which humans are normally exposed contains radiation as a usual constituent, it is assumed that some additional exposure can be permitted without producing undue risk. However, since the best scientific judgment concludes that there is likely to be some risk of increasing the incidence of cancer and of other undesirable consequences with any additional exposure, the amount of additional exposure permitted is small and is carefully controlled, and an effort is made to balance the permitted exposure against perceived benefits.

71.7.1 Types of Radiation

The principal types of radiation encountered in connection with the operation of fission and fusion systems are listed in Table 71.5. Characteristics of the radiation, including its charge and energy spectrum, are also given.

Alpha particles are produced by radioactive decay of all of the fuels used in fission reactors. They are, however, absorbed by a few millimeters of any solid material and produce no damage in typical fuel material. They are also a product of some fusion systems and may produce damage to the first wall that provides a plasma boundary. They may produce damage to human lungs during the mining of uranium when radioactive radon gas may be inhaled and decay in the lungs. In case of a catastrophic fission reactor accident, severe enough to generate aerosols from melted fuels, the alpha-emitting materials in the fuel might, if released from containment, be ingested by those in the vicinity of the accident, thus entering both the lungs and the digestive system.

Beta particles are produced by radioactive decay of many of the radioactive substances produced during fission reactor operation. The major source is fission products. Although more penetrating than alphas, betas produced by fission products can typically be absorbed by at most a few centimeters of most solids. They are thus not likely to be harmful to humans except in case of accidental release and ingestion of significant quantities of radioactive material. A serious reactor accident might also release radioactive materials to a region in the plant containing organic materials such as electrical insulation. A sufficient exposure to high-energy betas can produce damage to these materials. Reactor systems needed for accident amelioration must be designed to withstand such beta irradiations.

Gamma rays are electromagnetic photons produced by radioactive decay or by the fission process. Photons identical in characteristics (but not in name) are produced by decelerating electrons or betas. When produced in this way, the electromagnetic radiation is usually called X rays. High-energy (above several hundred keV) gammas are quite penetrating, and protection of both equipment and people requires extensive (perhaps several meters of concrete) shielding to prevent penetration of significant quantities into the ecosystem or into reactor components or systems that may be subject to damage from gamma absorption.

Neutrons are particles having about the same mass as that of the hydrogen nucleus or proton, but with no charge. They are produced in large quantities by fission and by some fusion interactions including the D–T fusion referred to earlier. High-energy (several MeV) neutrons are highly penetrating. They can produce significant biological damage. Absorption of fast neutrons can induce a decrease in the ductility of steel structures such as the pressure vessel in fission reactors or the inner wall of fusion reactors. Fast-neutron absorption also produces swelling in certain steel alloys.

71.8 BIOLOGICAL EFFECTS OF RADIATION

Observations have indicated that the radiations previously discussed can cause biological damage to a variety of living organisms, including humans. The damage that can be done to human organisms

Table 71.5 Radiation Encountered in Nuclear Power Systems

Name	Description	Charge (in Units of Electron Charge)	Energy Spectrum (MeV)
Alpha	Helium nucleus	+2	0 to about 5
Beta	Electron	+1, −1	0 to several
Gamma	Electromagnetic radiation	0	0 to about 10
Neutron		0	0 to about 20

includes death within minutes or weeks if the exposure is sufficiently large, and if it occurs during an interval of minutes or at most a few hours.

Radiation exposure has also been found to increase the probability that cancer will develop. It is considered prudent to assume that the increase in probability is directly proportional to exposure. However, there is evidence to suggest that at very low levels of exposure, say an exposure comparable to that produced by natural background, the linear hypothesis is not a good representation. Radiation exposure has also been found to induce mutations in a number of biological organisms. Studies of the survivors of the two nuclear weapons exploded in Japan have provided the largest body of data available for examining the question of whether harmful mutations are produced in humans by exposure of their forebears to radiation. Analyses of these data have led those responsible for the studies to conclude that the existence of an increase in harmful mutations has not been demonstrated unequivocally. However, current regulations of radiation exposure, in order to be conservative, assume that increased exposure will produce an increase in harmful mutations. There is also evidence to suggest that radiation exposure produces life shortening.

The Nuclear Regulatory Commission has the responsibility for regulating exposure due to radiation produced by reactors and by radioactive material produced by reactors. The standards used in the regulatory process are designed to restrict exposures to a level such that the added risk is not greater than that from other risks in the workplace or in the normal environment. In addition, effort is made to see that radiation exposure is maintained as "low as reasonably achievable."

71.9   THE CHAIN REACTION

Setting up and controlling a chain reaction is fundamental to achieving and controlling a significant energy release in a fission system. The chain reaction can be produced and controlled if a fission event, produced by the absorption of a neutron, produces more than one additional neutron. If the system is arranged such that one of these fission-produced neutrons produces, on the average, another fission, there exists a steady-state chain reaction. Competing with fission for the available neutrons are leakage out of the fuel region and absorptions that do not produce fission.

We observe that if only one of these fission-neutrons produces another fission, the average fission rate will be constant. If more than one produces fission, the average fission rate will increase at a rate that depends on the average number of new fissions produced for each preceding fission and the average time between fissions.

Suppose, for example, each fission produced two new fissions. One gram of uranium-235 contains 2.56×10^{21} nuclei. It would therefore require about 71 generations ($2^{71} \sim 2.4 \times 10^{21}$) to fission 1 g of uranium-235. Since fission of each nucleus produces about 200 MeV, this would result in an energy release of about 5.12×10^{23} MeV or 5.12×10^{10} J. The time interval during which this release takes place depends on the average generation time. Note, however, that in this hypothesized situation only about the last 10 generations contribute any significant fraction of the total energy. Thus, for example, if a generation could be made as short as 10^{-8} sec, the energy production rate could be nearly 5.12×10^{17} J/sec/g.

In power reactors the generation time is typically much larger than 10^{-8} sec by perhaps four or five orders of magnitude. Furthermore, the maximum number of new fissions produced per old fission is much less than two. Power reactors (in contrast to explosive devices) cannot achieve the rapid energy release hypothesized in the above example, for the very good reasons that the generation time and the multiplication inherent in these machines make it impossible.

71.9.1   Reactor Behavior

As indicated, it is neutron absorption in the nuclei of fissile material in the reactor core that produces fission. Furthermore, the fission process produces neutrons that can generate new fissions. This process sustains a chain reaction at a fixed level, if the relationship between neutrons produced by fission and neutrons absorbed in fission-producing material can be maintained at an appropriate level.

One can define neutron multiplication k as

$$k = \frac{\text{neutrons produced in a generation}}{\text{neutrons produced in the preceding generation}}$$

A reactor is said to be critical when k is 1. We examine the process in more detail by following neutron histories. The probability of interaction of neutrons with the nuclei of some designated material can be described in terms of a mean free path for interaction. The inverse, which is the interaction probability per unit path length, is also called macroscopic cross section. It has dimensions of inverse length.

We designate a cross section for absorption, Σ_a, a cross section for fission, Σ_f, and a cross section for scattering, Σ_s. If, then, we know the number of path lengths per unit time, per unit volume, traversed by neutrons in the reactor (for monoenergetic neutrons this will be nv, where n is neutron density and v is neutron speed), usually called the neutron flux, we can calculate the various interaction rates associated with these cross sections and with a prescribed neutron flux, as a product of the flux and the cross section.

A diagrammatic representation of neutron history, with the various possibilities that are open to the neutrons produced in the fission process, is shown below:

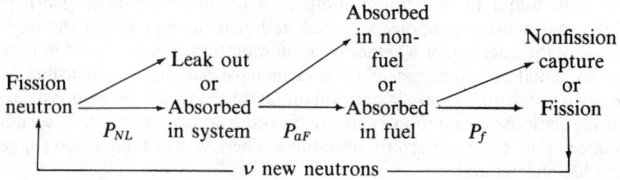

where P_{NL} = probability that neutron will not leak out of system before being absorbed
P_{aF} = probability that a neutron absorbed is absorbed in fuel
P_f = probability that a neutron absorbed in fuel produces a fission

In terms of the cross sections for absorption in fuel, Σ_a^F and for absorption, Σ_a

$$P_{aF} = f = \Sigma_a^F / \Sigma_a$$

where f is called the utilization factor. We can describe P_f as

$$P_f = \Sigma_f^F / \Sigma_a^F$$

Making use of the average number of neutrons produced per fission v, we calculate the quantity η, the average number of neutrons produced per neutron absorbed in fuel, as

$$\eta = v\, \Sigma_f^F / \Sigma_a^F$$

With these definitions, and guided by the preceding diagram, we conclude that the number of offspring neutrons produced by a designated fission neutron can be calculated as

$$N = \eta f P_{NL}$$

We conclude that the multiplication factor k is thus equal to $N/1$ and write

$$k = \eta f P_{NL}$$

Alternatively, making use of the earlier definitions we write

$$k = (v\, \Sigma_f^F / \Sigma_a^F)(\Sigma_a^F / \Sigma_a)$$

and if we describe Σ_a as

$$\Sigma_a = \Sigma_a^F + \Sigma_a^{nF}$$

where Σ_a^{nF} is absorption in the nonfuel constituents of the core, we have

$$k = v\, \Sigma_f^F P_{NL} / (\Sigma_a^F + \Sigma_a^{nF})$$

Observe that from this discussion one can also define a neutron generation time l as

$$l = N(t)/L(t)$$

where $N(t)$ and $L(t)$ represent, respectively, the neutron population and the rate of neutron loss (through absorption and leakage) at a time t.

For large reactors, the size of those now in commercial power production, the nonleakage probability is high, typically about 97%. For many purposes it can be neglected. For example, small changes in multiplication, produced by small changes in concentration of fissile or nonfissile material in the core, can be assumed to have no significant effect on the nonleakage probability, P_{NL}. Under these circumstances, and assuming that appropriate cross-sectional averaging can be done, the following relationships can be shown to hold. If we rewrite an earlier equation for k as

$$k = \eta f \left(n_f \sigma_f \middle/ \sum_i n_i \sigma_i \right)$$

where n_f, η_i represent, respectively, the concentration of fissile and nonfissile materials, and

$$\eta_f \sigma_f = \Sigma_f$$
$$n_i \sigma_i = \Sigma_{a_i}$$

where the last equation in the macroscopic cross section of the ith nonfissile isotope.

Variation of k with the variation in concentration of the fissile material (i.e., n_f) is given by

$$\frac{\delta k}{k} = \frac{\delta n_f}{n_f}\left(\frac{\Sigma_i n_i\,\sigma_i}{\eta_f\,\sigma_f + \Sigma_i\,\eta_i\,\sigma_i}\right)$$

This says that the fractional change in multiplication is equal to the fractional change in concentration of the fissile isotope times the ratio of the neutrons absorbed in all of the nonfissile isotopes to the total neutrons absorbed in the core.

Variation of k with variation in concentration of the jth nonfissile isotope is given by

$$\frac{\delta k}{k} = -\frac{\delta m_j}{m_j}\left(\frac{n_j\,\sigma_j}{n_f\,\sigma_f + \Sigma_i n_i\,\sigma_i}\right), \qquad i \neq j$$

This says that the fractional increase in multiplication is equal to the fractional decrease in concentration of the jth nonfissile isotope times the ratio of neutrons absorbed in that isotope to total neutrons absorbed. Although these are approximate expressions, they provide useful guidance in estimating effects of small changes in the material in the core on neutron multiplication.

71.9.2 Time Behavior of Reactor Power Level

We assume that the fission rate, and hence the reactor power, is proportional to neutron population. We express the rate of change of neutron population $N(t)$ as

$$l\left(\frac{dN}{dt}\right) = (k - 1)N$$

that is, in one generation the change in neutron population should be just the excess over the previous generation, times the multiplication k. The preceding equation has as a solution

$$N = N_0 \exp\left[\left(\frac{k-1}{l}\right)t\right]$$

where N_0 is the neutron population at time zero. One observes an exponential increase or decrease, depending on whether k is larger or smaller than unity. The associated time constant or e-folding time is

$$\tau = l/(k - 1)$$

For a $k - 1$ of 0.001 and l of 10^{-4} sec, the e-folding time is 0.10 sec. Thus in 1 sec the power level increases by e^{10} or about 10^4.

71.9.3 Effect of Delayed Neutrons on Reactor Behavior

Dynamic behavior as rapid as that described by the previous equations would make a reactor almost impossible to control. Fortunately there is a mode of operation in which the time constant is significantly greater than that predicted by these oversimplified equations. A small fraction of neutrons produced by fission, typically about 0.7%, come from radioactive decay of fission products. Six such fission products are identified for uranium fission. The mean time for decay varies from about 0.3 to about 79 sec. For an approximate representation, it is reasonable to assume a weighted mean time to decay for the six of about 17 sec. Thus, about 99.3% of the neutrons (prompt neutrons) may have a generation time of, say, 10^{-4} sec, while 0.7% have an effective generation time of 17 sec plus that of the prompt neutrons. An effective mean lifetime can be estimated as

$$\bar{l} = (0.993)l + 0.007(l + \bar{\lambda}^{-1})$$

For an l of 10^{-4} sec and a $\bar{\lambda}^{-1}$ of 17 sec we calculate

$$l \approx 0.993 \times 10^{-4} + 0.007(10^{-4} + 17) \approx 0.12 \text{ sec}$$

This suggests, given a value for $k - 1$ of 10^{-3}, an e-folding time of 120 sec. Observe that with this model the delayed neutrons are a dominant factor in determining time behavior of the reactor power level. A more detailed examination of the situation reveals that for a reactor slightly subcritical on prompt neutrons alone, but supercritical when delayed neutrons are considered (such a reactor is said to be "delayed supercritical"), the delayed neutrons almost alone determine time behavior. If, however, the multiplication is increased to the point that the reactor is critical on prompt neutrons alone (i.e., "prompt critical"), the time behavior is determined by the prompt neutrons, and changes in power level may be too rapid to be controlled by any external control system. Reactors that are meant to be controlled are designed to be operated in a delayed critical mode. Fortunately, if the reactor should inadvertently be put in a prompt critical mode, there are inherent physical phenomena that decrease the multiplication to a controllable level when a power increase occurs.

71.10 POWER PRODUCTION BY REACTORS

Most of the nuclear-reactor-produced electric power in the United States, and in the rest of the world, comes from light-water-moderated reactors (LWRs). Nuclear power reactors produce heat that is converted, in a thermodynamic cycle, to electrical energy. The two types now in use, the pressurized water reactor (PWR) and the boiling water reactor (BWR), use fuel that is very similar, and produce steam having about the same temperature and pressure. In both types water serves both as a coolant and a moderator. We will examine some of the salient features of each system and identify some of the differences.

71.10.1 The Pressurized-Water Reactor

The arrangement of fuel in the reactor core, and of the core in the pressure vessel, are shown in Fig. 71.1. As indicated earlier, bulk boiling is avoided by operation at high pressures. Liquid water is circulated through the core by large electric-motor-driven pumps located outside the pressure vessel in the cold leg of the piping that connects the vessel to a steam generator. Current designs use from two to four separate loops, each containing a steam generator. Each loop contains at least one pump. One current design uses two pumps in the cold leg of each loop. A schematic of the arrangement is shown in Fig. 71.8. Reactor pressure vessel and primary coolant loops, including the steam generator, are located inside a large containment vessel. A typical containment structure is shown in Fig. 71.9.

The containment, typically a massive 3–4-ft-thick structure of reinforced concrete, with a steel inner-liner, has two principal functions: protection of pressure vessel and primary loop from external damage (e.g., tornadoes, aircraft crashes) and containment of fission products that might be released outside the primary pressure boundary in case of serious damage to the reactor in an accident.

The steam generator is markedly different from the boiler in a fossil-fueled plant. It is essentially a heat exchanger containing several thousand metal tubes that carry the hot water coming from the reactor vessel outlet. Water surrounding the outside of these tubes is converted to steam. The rest of the energy-conversion cycle is similar in principle to that found in a fossil-fueled plant.

Experience with reactor operation has indicated that very careful control of water chemistry is necessary to preclude erosion and corrosion of the steam generator tubes (SGT). An important contribu-

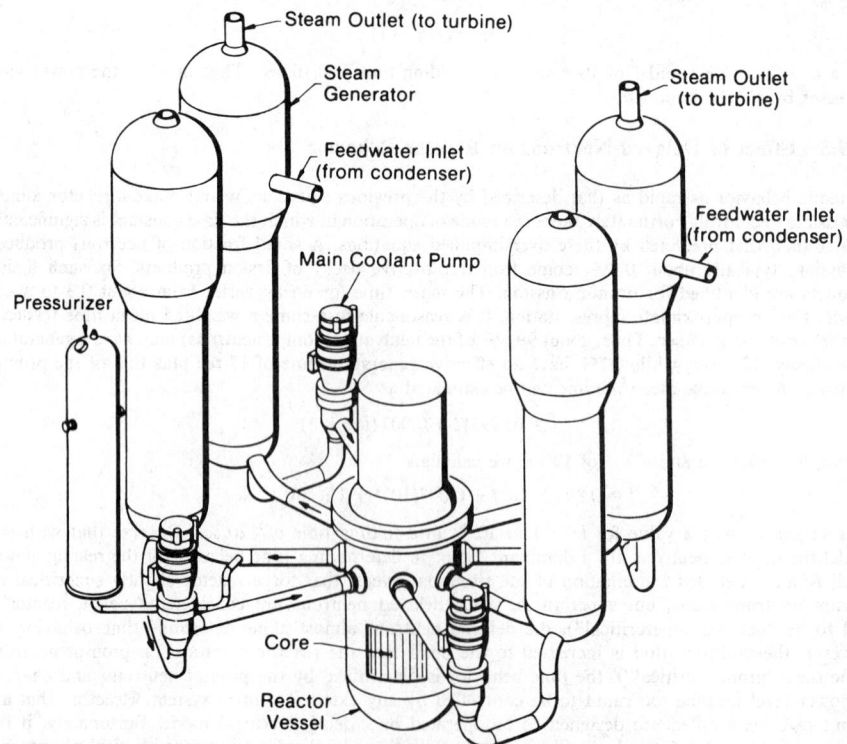

Fig. 71.8 Typical arrangement of PWR primary system.

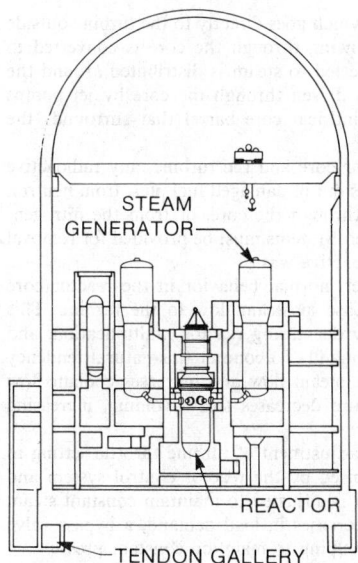

STEAM
GENERATOR

REACTOR

TENDON GALLERY   **Fig. 71.9** Typical large dry PWR containment.

tor to SGT damage has been leakage in the main condenser, which introduces impurities into the secondary water system. A number of early PWRs have retubed or otherwise modified their original condensers to reduce contamination caused by in-leakage of condenser cooling water.

The performance of steam generator tubes is of crucial importance because: (1) These tubes are part of the primary pressure boundary. SGT rupture can initiate a loss of coolant accident. (2) Leakage or rupture of SGTs usually leads to opening of the steam system safety valves because of the high primary system pressure. Since these valves are located outside containment, this accident sequence can provide an uncontrolled path for release of any radioactive material in the primary system directly to the atmosphere outside containment.

The reactor control system controls power level by a combination of solid control rods, containing neutron-absorbing materials that can be moved into and out of the core region, and by changing the concentration of a neutron-absorbing boron compound (typically boric acid) in the primary coolant. Control rod motion is typically used to achieve rapid changes in power. Slower changes, as well as compensation for burnup of uranium-235 in the core, are accomplished by boron-concentration changes. In the PWR the control rods are inserted from the top of the core. In operation enough of the absorber rods are held out of the core to produce rapid shutdown, or scram, when inserted. In an emergency, if rod drive power should be lost, the rods automatically drop into the core, driven by gravity.

The PWR is to some extent load following. Thus, for example, an increase in turbine steam flow caused by an increase in load produces a decrease in reactor coolant–moderator temperature. In the usual mode of operation a decrease in moderator temperature produces an increase in multiplication (the size of the effect depends on the boron concentration in the coolant–moderator), leading to an increase in power. The increase continues (accompanied by a corresponding increase in moderator–coolant temperature) until the resulting decrease in reactor multiplication leads to a return to criticality at an increased power level. Since the size of the effect changes significantly during the operating cycle (as fuel burnup increases the boron concentration is decreased), the inherent load-following characteristic must be supplemented by externally controlled changes in reactor multiplication.

A number of auxiliaries are associated with the primary. These include a water purification and makeup system, which also permits varying the boron concentration for control purposes, and an emergency cooling system to supply water for decay heat removal from the core in case of an accident that causes loss of the primary coolant.

Pressure in the primary is controlled by a pressurizer, which is a vertical cylindrical vessel connected to the hot leg of the primary system. In normal operation the bottom 60% or so of the pressurizer tank contains liquid water. The top 40% contains a steam bubble. System pressure can be decreased by water sprays located in the top of the tank. A pressure increase can be achieved by turning on electric heaters in the bottom of the tank.

71.10.2 The Boiling-Water Reactor

Fuel and core arrangement in the pressure vessel are shown in Fig. 71.3. Boiling in the core produces a two-phase mixture of steam and water, which flows out of the top of the core. Steam separators above the vessel water level remove moisture from the steam, which goes directly to the turbine outside of containment. Typically about one-seventh of the water flowing through the core is converted to steam during each pass. Feedwater to replace the water converted to steam is distributed around the inside near the top of the vessel from a spray ring. Water is driven through the core by jet pumps located in the annulus between the vessel wall and the cylindrical core barrel that surrounds the core and defines the upward flow path for coolant.

Because there is direct communication between the reactor core and the turbine, any radioactive material resulting, for example, from leakage of fission products out of damaged fuel pins, from neutron activation of materials carried along with the flow of water through the core, or from the nitrogen-16 referred to earlier, has direct access to turbine and condenser. Systems must be provided for removal from the coolant and for dealing with these materials as radioactive waste.

Unlike the PWR, the BWR is not load following. In fact, normal behavior in the reactor core produces an increase in the core void volume with an increase in steam flow to the turbine. This increased core voiding will increase neutron leakage, thereby decreasing reactor multiplication, and leading to a decrease in reactor power in case of increased demand. To counter this natural tendency of the reactor, a control system senses an increase in turbine steam flow and increases coolant flow through the core. The accompanying increase in core pressure decreases steam voiding, increasing multiplication and producing an increase in reactor power.

Pressure regulation in the BWR is achieved primarily by adjustment of turbine throttle setting to achieve constant pressure. An increase in load demand is sensed by the reactor control system and produces an increase in reactor power. Turbine valve position is adjusted to maintain constant steam pressure at the throttle. In rapid transients, which involve decreases in load demand, a bypass valve can be opened to send steam directly to the condenser, thus helping to maintain constant pressure.

BWRs in the United States make use of a pressure-suppression containment system. The hot water and steam released during a loss-of-coolant accident are forced to pass through a pool of water, condensing the steam. Pressure buildup is markedly less than if the two-phase mixture is released directly into containment. Figure 71.10 shows a Mark III containment structure of the type being used with the latest BWRs. Passing the fission products and the hot water and steam from the primary containment through water also results in significant removal of some of the fission products. The designers of this containment claim decontamination factors of 10,000 for some of the fission products that are usually considered important in producing radiation exposure following an accident.

As previously indicated, the control system handles normal load changes by adjusting coolant flow in the core. For rapid shutdown and for compensating for core burnup, the movable control rods are used. In a normal operating cycle several groups of control rods will be in the core at the beginning of core life, but will be completely out of the core at the end of the cycle. At any stage in core life some absorber rods are outside the core. These can be inserted to produce rapid shutdown.

Because of the steam-separator structure above the core, control rods in the BWR core must be inserted from the bottom. Insertion is thus not gravity assisted. Control rod drive is hydraulic. Compressed gas cylinders provide for emergency insertion if needed.

No neutron absorber is dissolved in the coolant, hence, control of absorber concentration is not required. However, cleanup of coolant containments, both solid and gaseous, is continuous. Maintaining a low oxygen concentration is especially important for the inhibition of stress-assisted corrosion cracking that has occurred in the primary system piping of a number of BWRs.

71.11   REACTOR SAFETY ANALYSIS

Under existing law the Nuclear Regulatory Commission has the responsibility for licensing power reactor construction and operation. (Those who operate the controls of the reactor and those who exercise immediate supervision of the operation must also be licensed by the Commission.) Commission policy provides for the granting of an operating license only after it has been formally determined by Commission review that the reactor power plant can be operated without undue risk to the health and safety of the public.

The current review process includes a detailed analysis of reactor system behavior under both normal and accident conditions. The existing approach involves the postulating of a set of design basis accidents (DBAs) and carrying out a deterministic analysis, which must demonstrate that the consequences of the hypothesized accidents are within a defined acceptable region. A number of the accident scenarios used for this purpose are of sufficiently low probability that they have not been observed in operating reactors. It is not practical to simulate the accidents using full-scale models or existing reactors. Analysis of reactor system behavior under the hypothesized situations must depend on analytical modeling. A number of large and complicated computer codes have been developed for this purpose.

Although the existing approach to licensing involves analysis of DBAs that can cause significant damage to the reactor power plant, none of the DBAs produces any calculable damage to personnel. Indeed core damage severe enough to involve melting of the core is not included in any of the sequences that are considered. However, in the design of the plant allowance is made, on a nonmechanistic

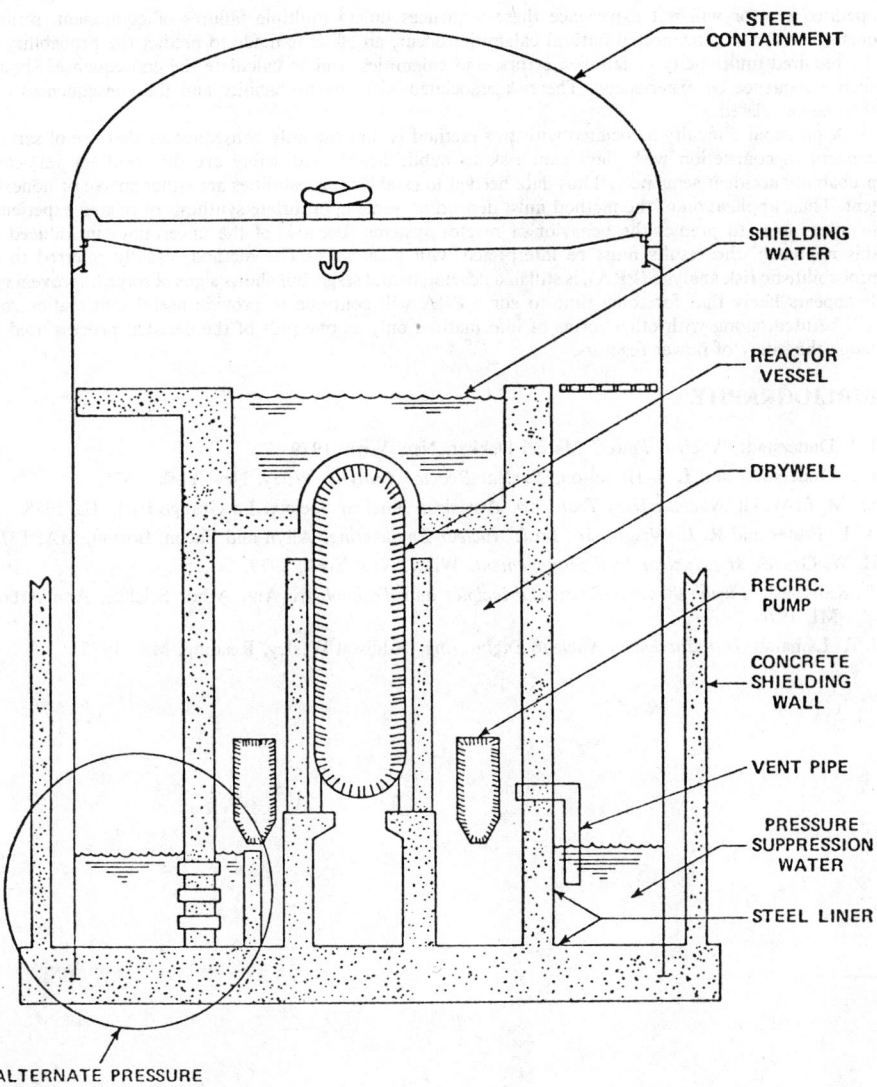

STEEL
CONTAINMENT

SHIELDING
WATER

REACTOR
VESSEL

DRYWELL

RECIRC.
PUMP

CONCRETE
SHIELDING
WALL

VENT PIPE

PRESSURE
SUPPRESSION
WATER

STEEL LINER

ALTERNATE PRESSURE
SUPPRESSION SYSTEM

Fig. 71.10   Mark III containment for BWR. (Used by permission of Pergamon Press, Inc., New York.)

basis, for consequences beyond those calculated for the DBA. This part of the design is not based on the results of an analytical description of a specific serious accident, but rather on nonmechanistic assumptions meant to encompass a bounding event.

This method of analysis, developed over a period of about two decades, has been used in the licensing and in the regulation of the reactors now in operation. It is likely to be a principal component of the licensing and regulatory processes for at least the next decade. However, the accident at Three Mile Island in 1979 convinced most of those responsible for reactor analysis, reactor operation, and reactor licensing that a spectrum of accidents broader than that under the umbrella of the DBA should be considered.

In the early 1970s, under the auspices of the Atomic Energy Commission, an alternative approach to dealing with the analysis of severe accidents was developed. The result of an application of the method to two operating reactor power plants was published in 1975 in an AEC report designated as WASH-1400 or the Reactor Safety Study. This method postulates accident sequences that may lead to undesirable consequences such as melting of the reactor core, breach of the reactor containment, or exposure of members of the public to significant radiation doses. Since a properly designed and operated reactor will not experience these sequences unless multiple failures of equipment, serious operator error, or unexpected natural calamities occur, an effort is made to predict the probability of the required multiplicity of failures, errors, and calamities, and to calculate the consequences should such a sequence be experienced. The risk associated with the probability and the consequences can then be calculated.

A principal difficulty associated with this method is that the only consequences that are of serious concern in connection with significant risk to public health and safety are the result of very-low-probability accident sequences. Thus data needed to establish probabilities are either sparse or nonexistent. Thus, application of the method must depend on some appropriate synthesis of related experience in other areas to predict the behavior of reactor systems. Because of the uncertainty introduced in this approach, the results must be interpreted with great care. The method, usually referred to as probabilitistic risk analysis (PRA), is still in a developmental stage, but shows signs of some improvement. It appears likely that for some time to come PRA will continue to provide useful information, but will be used, along with other forms of information, only as one part of the decision process used to judge the safety of power reactors.

BIBLIOGRAPHY

J. J. Duderstadt, *Nuclear Power,* Marcel Dekker, New York, 1979.

J. J. Duderstadt and L. J. Hamilton, *Nuclear Reactor Analysis,* Wiley, New York, 1975.

M. M. El-Wakil, *Nuclear Heat Transport,* American Nuclear Society, La Grange Park, IL, 1978.

A. L. Foster and R. L. Wright, Jr., *Basic Nuclear Engineering,* Allyn and Bacon, Boston, MA, 1973.

H. W. Graves, Jr., *Nuclear Fuel Management,* Wiley, New York, 1979.

T. Kammash, *Fusion Reactor Physics, Principles and Technology,* Ann Arbor Science, Ann Arbor, MI, 1976.

J. R. Lamarsh, *Introduction to Nuclear Engineering,* Addison-Wesley, Reading, MA, 1975.

APPENDIX TABLES OF NUCLEAR DATA

Table 71.6 Nuclear Data for Several Isotopes[a]

Atomic Number Z	Element or Isotope	Abundance (%)	Atomic Mass Weight (u)	Density (g/cm³)	Half-Life	Absorption at 0.025 eV	Scattering Thermal	Scattering Epithermal
1	H		1.00797			0.332	38	20.4
	¹H	~100	1.007825			0.332	38	20.4
	²H (D)	0.0151	2.01410			0.00046	7	3.4
	³H (T)	—	3.01605		12.6 yr			
2	He		4.0026			0.007	0.8	0.83
	³He	0.00013	3.01603			5500	0.8	
	⁴He	~100	4.00260			0	0.8	
3	Li		6.939	0.53		70	1.4	0.9
	⁶Li	7.52	6.01513			945(n, α)		
	⁷Li	92.48	7.01601			0.033		
	⁸Li	—			0.845 sec			
4	⁸Be	—	8.00531		~3×10^{-16} sec			
	⁹Be	100	9.01219	1.82		0.010	7	6.11
5	B		10.811	2.54		755	4	3.7
	¹⁰B	19.8	10.01294			3813(n, α)		
	¹¹B	80.2	11.00931			$<50 \times 10^{-3}$		
	¹²B	—			0.019 sec			
6	C		12.01115	2.22		0.0034	4.8	4.66
	¹²C	98.89	12.00000					
	¹³C	1.11	13.00335			0.0005		
	¹⁴C	—			5570 yr			
7	N		14.0067			1.88	10	9.9
	¹³N	—			10 mo			
	¹⁴N	99.63	14.00307			1.75		
	¹⁵N	0.37	15.00011					
	¹⁶N	—			7.4 sec			
8	O		15.9994			0.00019	4.2	3.75

Neutron Cross Sections (barns). Absorption at 0.025 eV; Scattering: Thermal, Epithermal.

Table 71.6. *(Continued)*

Atomic Number Z	Element or Isotope	Abundance (%)	Atomic Mass Weight (u)	Density (g/cm³)	Half-Life	Absorption at 0.025 eV	Scattering Thermal	Scattering Epithermal
	^{16}O	99.759	15.99491					
	^{17}O	0.037	16.99914					
	^{18}O	0.204	17.99916					
	^{19}O	—			29 sec			
9	^{19}F	100	18.99840			0.010	3.9	3.6
11	^{23}Na	100	22.98977	0.971		0.53	4.0	3.1
	^{24}Na	—			15.05 hr			
12	Mg		24.312	1.74		0.063	3.6	3.4
	^{24}Mg	78.60	23.98504			0.034		
	^{25}Mg	10.11	24.98584			0.280		
	^{26}Mg	11.29	25.98259			0.060		
	^{27}Mg	—			9.45 mo			
13	^{27}Al	100	26.9815	2.70		0.230	1.4	1.4
	^{28}Al	—			2.3 mo			
14	Si		28.086	2.4		0.16	1.7	2.2
	^{28}Si	92.27	27.97693			0.080		
	^{29}Si	4.68	28.97649			0.280		
	^{30}Si	3.05	29.97376			0.40		
	^{31}Si	—			2.62 yr			
15	^{31}P	100	30.97376	2.34		0.20	5	3.4
17	^{37}Cl	24.47	36.96590			56		
18	^{37}Ar	—	36.96678		35.1 days			
19	K		39.102	0.87		2.07	2.5	2.1
22	Ti		47.90	4.5		5.8	4	4.2
23	V		50.942	6.0		4.5	5.0	
24	Cr		51.996	6.92		3.1	3.0	3.9
	^{53}Cr	9.55	52.9407			1.82		
25	^{55}Mn	100	54.9381	7.2		13.2	2.3	1.9

26	56Mn				2.58 hr		11	11.4
	Fe	—	55.847	7.87		2.62		
	54Fe	5.84	53.9396			2.3		
	55Fe				2.6 yr			
	56Fe	91.68	55.9349			2.7		
	57Fe	2.17	56.9354			2.5		
	58Fe	0.31	57.9333			1.2		
	59Fe				45 days			
27	59Co	100	58.9332	8.71		18 & 19	7	5.8
	60Co	—			10.5 mo and 5.24 yr			
28	Ni		58.71	8.9		4.8	17.5	17.4
29	Cu		63.54	8.96		3.77	7.2	7.7
	63Cu	69.1	62.9298			4.6		
	64Cu				12.9 hr	2.2		
	65Cu	30.9	64.9278					
	66Cu				5.1 mo			
37	87Rb	27.85			4.7×10^{10} yr	0.12		
40	Zr		91.22	6.44		0.180	8	6.2
41	93Nb	100	92.90638	8.57		1.15	5	6.5
	94Nb	—			6.6 mo			
42	Mo		95.911	10.22		2.7	7	6
47	Ag		107.870	10.50		63	6	6.4
	107Ag	51.35	106.9051			31		
	108Ag				2.3 mo			
	109Ag	48.65	108.9047					
	110Ag	—			24 sec	87		
48	Cd		112.40	8.64		2450	7	
49	In		114.82	7.28		196	2.2	
	113In	4.2	112.9043					
	114In				72 sec and 49 days			
	115In	95.77	114.9039		6×10^{14} yr	50 and 150		
	116In				14 sec and 54 mo			
54	Xe		131.30			35	4.3	
	135Xe				9.2 hr	2.72×10^{6}		
	136Xe	8.87	135.9072			0.15		
62	Sm		150.35	7.7		5600		
	149Sm	13.84	148.9169			40,800		

Table 71.6. (*Continued*)

Atomic Number Z	Element or Isotope	Abundance (%)	Atomic Mass Weight (u)	Density (g/cm³)	Half-Life	Absorption at 0.025 eV	Neutron Cross Sections (barns) Scattering	
							Thermal	Epithermal
64	Gd		157.25	7.94		46,000		
	155Gd	14.73	154.9226			61,000		
	157Gd	15.68	156.9239			240,000		
79	197Au	100	196.9666	19.3		98.8	9.3	
	198Au	—			2.7 days	26,000		
82	Pb		207.19	11.35		0.170	11	11.3
	204Pb	1.5	203.9731			0.8		
	205Pb	—			1.4×10^{17} yr			
					3×10^7 yr			
	206Pb	23.6	205.9745			0.025		
	207Pb	22.6	206.9759			0.70		
	208Pb	52.3	207.9766			0.03		
	209Pb		208.9810		3.3 hr			
83	209Bi	100	208.9804	9.8		0.019 and 0.015	9	9.28
	210Bi	—			2.6×10^6 yr and 50 days			
					2.15 mo			
	211Bi	—						
84	210Po	—	209.9829		138 days			
	213Po	—	212.9928		4 μsec			
88	226Ra	—	226.0254		1620 yr	20		
90	230Th	—	230.0331		7.6×10^4 yr			
	232Th	100	232.0382	11.5	1.45×10^{10} yr	7.56	12.5	12.5
	233Th	—			22.1 mo			
91	233Pa	—			27.4 days		43	
92	U		238.03	19.0		7.68	8.3	
	233U		233.0395		1.62×10^5 yr	$527(n,f)$ $54(n,\alpha)$		
	234U	0.0057	234.0409		2.5×10^5 yr	105		
	235U	0.714	235.0439		7.1×10^8 yr	$582(n,f)$ $112(n,\gamma)$	10	
	236U	—	236.0457		2.39×10^7 yr	7		

Z	Isotope	Abundance (%)	Atomic mass		Half-life	Cross sections (barns)	
	^{237}U	—			6.75 days		
	^{238}U	99.28	238.0508		4.51×10^9 yr	2.71 / 14(n,f)	8.3
	^{239}U	—			23.5 mo		
93	^{239}Np	—			2.35 days		
94	^{239}Pu	—	239.0522	19.6	24,360 yr	746(n,f) / 280(n,γ)	9.6
	^{240}Pu	—	240.0540		6760 yr	<0.1(n,f) / 295(n,γ)	
	^{241}Pu	—			13.2 yr	1025(n,f) / 375(n,γ)	
	^{242}Pu	—	242.0587		3.79×10^5 yr	<0.2(n,f) / 30(n,γ)	
95	^{241}Am	—	241.0567		433 yr	710(n,γ) / 3(n,f)	
96	^{242}Cm	—	242.0588		163 days	25(n,γ) / <5(n,f)	
	^{244}Cm	—	244.0628		18.1 yr	~13(n,γ) / 1(n,f)	
98	^{252}Cf	—	252.08		2.65 yr(α) / 85.5 yr(s. f.)[b]	20(n,γ)	

[a] This table is from the second edition of *Basic Nuclear Engineering* by Foster and Wright, and is reproduced with permission of the publisher, Allyn and Bacon, Boston, MA.

[b] s. f. = spontaneous fission.

Table 71.7 Table of Radioisotopes[a]

Isotope	Half-Life	Type Decay	Most Predominant Energy (ies) (MeV)
$^{1}_{0}n$	12.8 mo	β^-	0.78
$^{3}_{1}H$	12.6 yr	β^-	0.18
$^{8}_{3}Li$	0.845 sec	β^-	13
		2α	3.2(total)
$^{8}_{4}Be$	$<1.4 \times 10^{-16}$ sec	2α	0.047(each)
$^{14}_{6}C$	5568 yr	β^-	0.155
$^{13}_{7}N$	10 mo	β^+	1.24
$^{22}_{11}Na$	2.6 yr	β^+, γ	0.542, 1.28
$^{24}_{11}Na$	15 hr	$\beta^-, \gamma_1 - \gamma_2$	1.39, 1.37–2.75
$^{28}_{13}Al$	2.27 mo	β^-, γ	2.86, 1.78
$^{32}_{15}P$	14.3 days	β^-	1.707
$^{40}_{19}K$	1.2×10^9 yr	β^-	1.33(89%)
		EC, γ	1.46(11%)
$^{47}_{20}Ca$	4.8 days	β^-, γ	0.66, 1.3
$^{47}_{21}Sc$	3.43 days	β^-, γ	0.439, 0.160(60%)
		β^-	0.60 (40%)
$^{48}_{23}V$	16.0 days	EC, $\beta^+, \gamma_1, \gamma_2$	0.69, 0.986, 1.314
$^{53}_{23}V$	2.0 mo	β^-, γ	2.50, 1.00
$^{55}_{26}Fe$	2.6 yr	EC	
$^{59}_{26}Fe$	45 days	β^-, γ	0.460, 1.10(54%)
			0.270, 1.29(46%)
$^{60}_{27}Co$	5.24 yr	$\beta^-, \gamma_1, \gamma_2$	0.302, 1.33, 1.17
$^{64}_{29}Cu$	12.9 hr	EC, γ	1.34 (42%)
		β^-	0.571 (39%)
		β^+	0.657 (19%)
$^{66}_{29}Cu$	5.1 mo	β^-	2.63 (91%)
		β^-, γ	1.5, 1.04(9%)
$^{65}_{30}Zn$	245 days	EC	(55%)
		EC, γ	1.12(45%)
$^{85}_{36}Kr$	10.3 yr	β^-	0.695 (98.5%)
		β^-, γ	0.15, 0.54(0.65%)
$^{90}_{38}Sr$	27.7 yr	β^-	0.545
$^{90}_{39}Y$	64.2 hr	β^-	2.26
$^{108}_{47}Ag$	2.3 mo	β^-	1.77(97%)
		EC, γ	0.45(1.5%)
$^{110}_{47}Ag$	24 sec	β^-	2.82 (40%)
		β^-, γ	2.24, 0.66(60%)
$^{116}_{49}In$	13 sec	β^-	3.29
$^{131}_{53}I$	8.08 days	β^-, γ	0.608, 0.364(87.2%)
		β^-, γ	0.335, 0.638(9.3%)
$^{137}_{53}I$	22 sec	β^-	(94%)
		β^-, n	0.56(6%)
$^{135}_{54}Xe$	9.23 hr	β^-, γ	0.51, 0.250(97%)
$^{137}_{54}Xe$	3.9 mo	β^-	3.5
$^{132}_{55}Cs$	7.1 days	EC, γ	0.67
$^{137}_{55}Cs$	30 yr	β^-	0.51 (92%)
		β^-	1.17 (8%)
$^{133}_{56}Ba$	7.2 yr	EC, γ	0.320, 0.081
$^{140}_{56}Ba$	12.8 days	β^-, γ	1.021, 0.16–0.03(60%)
		β^-, γ	0.48, 0.54–0.03 (30%)
$^{140}_{57}La$	40 hr	$\beta^-, \gamma_1-\gamma_2-\gamma_3$	1.32, 0.33–0.49–1.60(70%)
		$\beta^-, \gamma_1-\gamma_2$	1.67, 0.49–1.60 (20%)

Table 71.7 *(Continued)*

Isotope	Half-Life	Type Decay	Most Predominant Energy (ies) (MeV)
$^{144}_{58}$Ce	285 days	β^-	0.309
$^{147}_{61}$Pm	2.6 yr	β^-	0.223
$^{192}_{77}$Ir	74.4 days	EC, β^-, γ	0.66, 11 at different energies
$^{197}_{80}$Hg	65 hr	EC, γ	0.077
$^{214}_{82}$Pb	26.8 mo	β^-, γ	0.67, 0.295–0.352
$^{213}_{83}$Bi	47 mo	β^-	1.39　　(98%)
		β^-, γ	0.959, 0.434(2%)
		α	5.86
$^{208}_{84}$Po	2.93 yr	α	5.108
$^{210}_{84}$Po	138 days	α	5.30
$^{213}_{84}$Po	4.2×10^{-6} sec	α	8.34
$^{226}_{88}$Ra	1622 yr	α	4.78　　(94%)
		α, γ	4.77, 0.186(5.7%)
$^{231}_{90}$Th	25.6 hr	β^-, γ	0.308, 0.084　　(44%)
		β^-, γ	0.094, 0.058, 0.026(45%)
$^{233}_{92}$U	1.62×10^5 yr	α	4.823
$^{234}_{92}$U	2.48×10^5 yr	α	4.76 (73%)
	s. f.[b] 2×10^{16} yr	α, γ	4.72, 0.05(27%)
$^{235}_{92}$U	7.1×10^8 yr	α	4.40 (83%)
	s. f.[b] 1.9×10^{17} yr		
$^{238}_{92}$U	4.51×10^9 yr	α	4.195(77%)
	s. f.[b] 8×10^{18} yr	α, γ	4.18, 0.048(23%)
$^{239}_{92}$U	23.5 mo	β^-, γ	1.2, 0.73
$^{238}_{93}$Np	2.1 days	β^-, γ	1.272, 0.044(47%)
		β^-, γ	0.258, 1.03　(53%)
$^{238}_{94}$Pu	89.6 yr	α	5.49　　(72%)
	s. f.[b] 3.8×10^{10} yr	α, γ	5.45, 0.044(28%)
$^{241}_{95}$Am	458 yr	α, γ	5.477, 0.060
$^{242}_{96}$Cm	35 yr	α	6.11　　(73.7%)
		α, γ	6.07, 0.044 (26.3%)
$^{244}_{96}$Cm	17.9 yr	α	5.801　　(76.7%)
	s. f.[b] 1.4×10^7 yr	α, γ	5.759, 0.043 (23.3%)
$^{252}_{98}$Cf	2.65 yr(97%)	α	6.12(82%)
	s. f.[b] 85 yr(3%)		6.08(15%)
$^{254}_{98}$Cf	s. f.[b] 60.5 days	n	
		γ	
$^{247}_{97}$Bk	1.4×10^3 yr	α, γ	5.68(37%)
			5.52(58%)
$^{253}_{99}$Es	20.5 days	α, β	6.64, 0.017, 0.027
$^{257}_{100}$Fm	80 days	α	6.53(94%)
	s. f.[b] 100 yr		
$^{257}_{101}$Md	3 hr	EC, α	7.25(97%), 7.08(3%)

[a] This table is from the second edition of *Basic Nuclear Engineering* by Foster and Wright, and is reproduced with permission of the publisher, Allyn and Bacon, Boston, MA.

[b] s. f. = spontaneous fission.[c] Beta energies are the maximum energies.

[c] Beta energies are the maximum energies.

CHAPTER 72

GAS TURBINES

RICHARD HARMAN

University of Canterbury
Christchurch, New Zealand

Reprinted with additions from Richard Harman, *Gas Turbine Engineering*, Halsted Press, 1981, by permission of the author.

72.1 INTRODUCTION

72.1.1 Basic Operating Principles

Most types of engine are designed to extract as much as possible of the energy released by the combustion of a fuel and to deliver it in usable form at a rotating output shaft.

Burning the fuel in an open fire would simply produce waste heat. The transfer of energy from the fire to the hardware is therefore achieved by the appropriate use of a gaseous *working fluid,* often air, which is made to flow through the engine. The normal way to handle the working fluid is by the *thermodynamic cycle* of induction, compression, heating, expansion, and exhaust. In a *reciprocating engine,* these processes are performed in sequence in the same closed space, formed by a piston and cylinder which operate on the working fluid one mass at a time. In contrast, the working fluid flows without interruption through a *gas turbine engine,* passing continuously from one single purpose device to the next.

The arrangement of a simple gas turbine engine is shown in Fig. 72.1*a.* The rotating compressor acts as a fan to drive the working fluid into the heating system. The fluid is heated by internal combustion (burning a controlled supply of fuel in the airflow) or by heat exchange from an external source (passing the compressed fluid through heated passages). The turbine acts as a windmill, being blown round by the flow to drive the compressor and an output load via rotating shafts.

The loss of *pressure* as the fluid releases its energy to the turbine is akin to the voltage drop across an electric motor or the water-level difference required to create a useful water flow. While an incompressible fluid (for example, water) may be pressurized and depressurized, its negligible volume change means that little energy interchange is involved. In contrast, the use of a *compressible fluid* (gas) permits the absorption or release of considerable amounts of energy. For instance, in an expanding cylinder, the gas continues to push on the retreating piston and releases energy equivalent to the mean force times the distance traveled.

Similarly, in a turbine, the energy is released as the gas presses against retreating blades while compressor blades add energy by approaching the flow. The energy content of the gas at any time is basically the kinetic energy (half the mass times velocity squared) of its molecules, which is proportional to the gas *temperature.* The action and motion of the blades, in changing the *velocity* of the flow, therefore also change the mean molecular velocity and energy and the gas temperature and pressure. While the pressure changes are essential to the working of the engine, its performance is assessed by the temperature changes.

The type of *pressure and temperature measurements* used is very important. Usually they are absolute, stagnation (total) values, measured by instruments that face into the approaching flow to give an indication of the energy in the fluid flow at any point. The work done in compression or expansion is proportional to the change of stagnation temperature in the working fluid, in the form of heating during a compression process or cooling during an expansion process. The *temperature ratio,* between the temperatures before and after the process, is related to the *pressure ratio* across the process by the expression $T_b/T_a = (P_b/P_a)^{(\gamma-1)/\gamma}$, where γ is the ratio of specific heats at constant pressure and volume (as listed in the Appendix for different gases) and the temperature and pressure are stagnation values. It is the interaction between the temperature change and ratio, at different starting temperature levels, which permits the engine to generate a useful work output.

This can be demonstrated by a simple numerical example using the Kelvin scale for temperature. For a starting temperature of 300 K (27°C), a temperature ratio of 1.5 in compression yields a final temperature of 450 K and a change of 150°C. Starting instead at 400 K, the same ratio would yield a change of 200°C and a final temperature of 600 K. Note that absolute temperature levels are expressed in Kelvins, while changes are in degrees Celsius. The equivalent pressure ratio would ideally be 4.13, as calculated from $1.5^{1.4/0.4}$. These numbers show that, working over the same temperature ratio, the

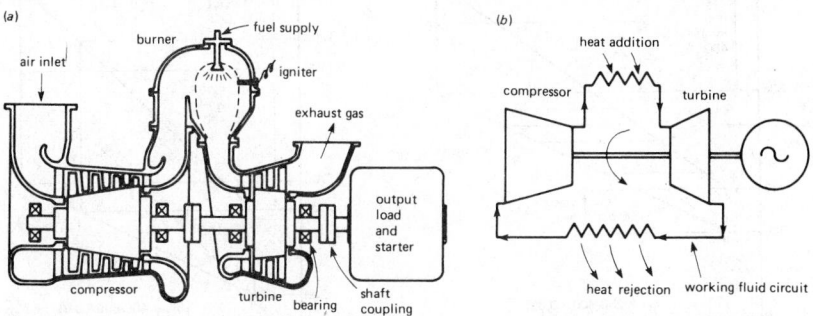

Fig. 72.1 Simple engine types: (*a*) open cycle; (*b*) closed cycle (diagrammatic).[30]

temperature change and, therefore, the work involved in the process vary in proportion to the starting temperature level.

This conclusion can be depicted graphically in a very significant way. If the temperature changes are drawn as vertical lines *ab* and *cd,* and are separated horizontally to avoid overlap, the resultant Fig. 72.2*a* forms the basis for a standard type of graph used to assess engine cycles. Assuming the starting and finishing pressures to be the same for the two processes, the thin lines through *ac* and *bd* depict two of a family of lines of constant pressure, which diverge as shown. In this ideal case, expansion processes could be represented by the same diagram, simply by proceeding down the lines *ba* and *dc.* Alternatively, if *ab* is taken as a compression process, *bd* as heat addition, *dc* as an expansion process, and *ca* as a heat rejection process, then the figure *abdc* represents the ideal cycle to which the working fluid of the engine is subjected.

In practice, the 600 K level at point *d* is too low a temperature from which to start the expansion. Figure 72.2*b* is more realistic, with line *ef* going from 800 to 1200 K and, clearly, much longer than line *ab.* Now, *ab* represents the work input required by the compressor. Of the expansion work capacity *fe,* only the fraction *fg* is required to drive the compressor, where *ag* is parallel to *bf.* The remaining 250°C, line *ge,* is energy which can be used to perform useful external work, by further expansion through additional turbine capacity or by blowing through a nozzle to provide jet thrust. Thus, by expanding from a high temperature, the required temperature drop of 150°C to drive the compressor can be obtained with a reduced temperature ratio of only 1200/1050 = 1.143 in this case, leaving a residual exhaust gas pressure of 2.59 bar for further expansion.

Now consider line *bf.* Its length is proportional to the heat added, and the limit to the maximum turbine entry temperature level at *f* is set by the turbine technology. The ability of the engine to generate a useful output arises from its use of the energy in the input fuel flow, but not all of the fuel energy can be recovered usefully. In this example, the heat input of 1200 − 450 = 750°C compares with the excess output of 250°C (line *ge*) to represent an *efficiency* of 0.33 (250/750), or 33%. If more fuel could be used, raising the maximum temperature level at the same pressure, then more useful work could be obtained at the same efficiency. This calculation of efficiency is consistent with the ideal expression: $\eta = 1 - (T_a/T_b)$.

The line *ea* represents heat rejection. This could involve passing the exhaust gas through a cooler, before returning it to the compressor, and this would be a *closed cycle* (Fig. 72.1*b*). Figure 72.1*a*

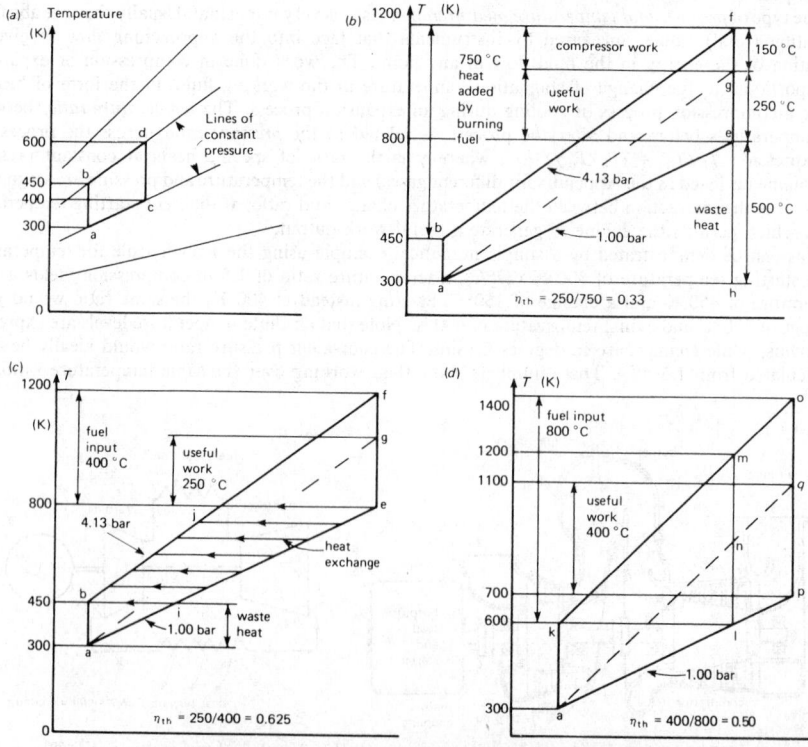

Fig. 72.2 Temperature changes in various ideal engine cycles.[30]

shows an *open-cycle* engine, which takes air from the atmosphere and exhausts back to the atmosphere. In this case, line *ea* still represents heat rejection, but the path from *e* to *a* involves the whole atmosphere and very little of the gas finds its way immediately from *e* to *a*. It is fundamental to the cycle shown that the remaining 500°C, line *eh*, is wasted heat because point *e* is at atmospheric pressure. The gas is therefore unable to expand further, so can do no more work.

An engine modification to improve the efficiency of the cycle is possible at this stage. The use of a *regenerator* or heat exchanger to transfer as much as possible of the heat rejected along *ea* into the compressed fluid can reduce the quantity of heat to be supplied between *b* and *f*. In Fig. 72.2c, the heat rejected along *ei* reappears in the compressed fluid along *bj*. The fuel added is then equivalent to 1200 − 800 = 400°C, which, with the same work output of 250°C, yields an overall efficiency of 0.625. In practice, the efficiency is much lower.

An alternative modification is the use of a higher pressure ratio, which also improves the overall efficiency. In Fig. 72.2d, the temperature ratio of 2 corresponds to a pressure ratio of 11.3. The efficiency has risen to 0.5 at the same turbine entry temperature (1200 K), because the raised compressor delivery temperature at *k* reduces the possible fuel input along *km* to 600°C and the lowered temperature level after expansion (*l*) reduces the heat wastage to 300°C. However, the useful output of 300°C is not sufficient alone to justify the increased complexity and cost of raising the pressure ratio: it needs to be matched by a raised turbine entry temperature as shown, for example, at *o*. In turn, this calls for a raised level of turbine technology in terms of materials and means of blade cooling.

The efficiency levels quoted are very high because the explanation so far has ignored many factors. *Inefficiency* of the compressor and turbine increases the compressor work demand while reducing the turbine work output, thereby drastically reducing the useful work output and efficiency. The effect of inefficiency is that, for a given temperature change, the compressor generates less than the ideal pressure level while the turbine drops the pressure further than in the ideal case: both leave less pressure for the next process. There are also pressure losses in the heat addition and heat rejection processes, and the regenerator transfers less heat than the ideal case quoted. There may be variations in the fluid mass flow rate and its specific heat (energy input divided by consequent temperature rise) around the cycle. These factors can easily combine to reduce the overall efficiency to zero, giving an unworkable engine.

72.1.2 A Brief History of Gas Turbine Development and Use

The use of a turbine driven by the rising flue gases above a fire dates back to Hero of Alexandria in 150 BC, and the Chinese were operating windmills at about the same period. It was not until AD 1791 that John Barber patented the forerunner of the gas turbine, proposing the use of a reciprocating compressor, a combustion system, and an impulse turbine. Even then, he foresaw the need to cool the turbine blades, for which he proposed water injection.

The year 1808 saw the introduction of the first *explosion type* of gas turbine, which in later forms used valves at entry and exit from the combustion chamber to provide intermittent combustion in a closed space. The pressure thus generated blew the gas through a nozzle to drive an impulse turbine. These operated successfully but inefficiently for Karavodine and Holzwarth from 1906 onward, and the type died out after a Brown, Boveri model designed in 1939.[1]

Developments of the continuous flow machine suffered from lack of knowledge, as different configurations were tried. Stolze in 1872 designed an engine with a seven-stage axial flow compressor, heat addition through a heat exchanger by external combustion, and a 10-stage reaction turbine. It was tested from 1900 to 1904 but did not work because of its very inefficient compressor.[1] Parsons was equally unsuccessful in 1884, when he tried to run a reaction turbine in reverse as a compressor. These failures resulted from the lack of understanding of aerodynamics prior to the advent of aircraft. As a comparison, in typical modern practice, a single-stage turbine drives about six or seven stages of axial compressor with the same mass flow.

The first successful dynamic compressor was Rateau's centrifugal type in 1905. Three assemblies of these, with a total of 25 impellers in series giving an overall pressure ratio of 4, were made by Brown, Boveri and used in the first working gas turbine engine, built by Armengaud and Lemale in the same year. The exhaust gas heated a boiler behind the turbine to generate low-pressure steam, which was directed through turbines to cool the blades and augment the power. Low component efficiencies and flame temperature (828 K) resulted in low work output and an overall efficiency of 0.03. By 1939, the use of *industrial* gas turbines had become well established: experience with the Velox boiler led Brown, Boveri into diverging applications; a Hungarian engine (Jendrassik) with axial flow compressor and turbine used regeneration to achieve an efficiency of 0.21; and the Sun Oil Co. (USA) was using a gas turbine engine to improve a chemical process.

The use of gas turbine engines for *aircraft propulsion* dates from 1930, when Whittle saw that its exhaust gas conditions ideally matched the requirements for jet propulsion and took out a patent.[2] His first model was built by British Thomson-Houston and ran as the Power Jets Type U in 1937, with a double-sided centrifugal compressor, a long combustion chamber which was curled round the outside of the turbine and an exhaust nozzle just behind the turbine. Problems of low compressor and turbine efficiency were matched by hardware problems and the struggle to control the combustion in a very small space. Ten reverse flow combustion chambers were introduced in 1938, the aim still being to keep the compressor and turbine as close together as possible to avoid shaft whirl problems (Fig. 72.3). Whittle's first flying engine was the W1, with 850 lb thrust, in 1941. It was made by Rover, whose gas turbine establishment was taken over by Rolls-Royce in 1943.

By this time there was a wide variety of configurations made by numerous British, American, and German companies. A Heinkel experimental engine flew in 1939 and a General Electric version of the W1 flew in 1941. Several jet engines were operational by the end of the Second World War, but the first commercial engine did not enter service until 1953, the Rolls-Royce Dart turboprop in the *Viscount*, followed by the turbojet de Havilland Ghost in the *Comet* of 1954. The subsequent growth of the use of jet engines has been visible and audible to most of the world, and has forced the growth of design and manufacturing technology.[3] By 1970 a range of standard configurations for different tasks had become established, and some aircraft engines were established in industrial applications and in ships.

Gas turbines entered the *surface transportation* fields also during their early stages of development. The first railway locomotive application was in Switzerland in 1941, with a Brown, Boveri engine driving an electric generator and electric motors driving the wheels. The engine efficiency approached 0.19, using regeneration. Many other turboelectric systems and engines with other transmission systems have entered service since 1950, some involving high-speed, lightweight passenger trains, but only in very small numbers of each type. Road use started with a Rover car of 1950, followed by Chrysler and other companies, but commercial use has been limited to trucks, particularly by Ford. Automotive gas turbine development has been largely independent of other types, and has forced the pace of development of regenerators; engines are being tested in most vehicle manufacturing countries. Hovercraft entered operation in the United Kingdom in 1963, using aircraft-derived gas turbine engines for their high power-to-weight ratio.

72.1.3   Component Characteristics and Capabilities

The compressors used are of the dynamic type as distinct from the positive displacement, reciprocating, and rotary types. Each stage of a dynamic compressor works to increase the flow velocity, and then lets it slow down again (diffuse) to gain pressure. It thereby causes the working fluid to flow continuously into a region of high pressure, but this flow pattern may break down if the outlet pressure becomes excessive. The compressor may then enter a stalled condition, in which the flow surges backward and forward through the compressor, unless provision is made in the engine design to control the loading of the blades.

Dynamic compressors are efficient, compact, relatively cheap, and handle vast quantities of working fluid. Considering air at 1.013 bar, 288 K, the inlet flow is typically 150 kg/sec (122 m³/sec or 259,000 ft³/min) per square meter of inlet area at Mach 0.4. As the intake diameters of the various types range from about 0.1 m (4 in.) to 2.44 m (8 ft), the range of mass flows varies from about 1.2 to 700 kg/sec (about 1 to 570 m³/sec or 2000 to 1.2 million ft³/min). They operate without lubrication, so the delivery is not contaminated by oil. This may be especially useful when the compressor is involved in a chemical process. The isentropic efficiency typically ranges from 0.7 to 0.9 depending on the type and quality of compressor.

The dynamic compressor is available in two forms, both represented diagrammatically as shown in Fig. 72.4*a*. The *centrifugal compressor* provides small to moderate airflows (up to about 50 kg/sec) at a pressure ratio of up to about 5 with subsonic flow, or higher flows at lower pressure ratio,

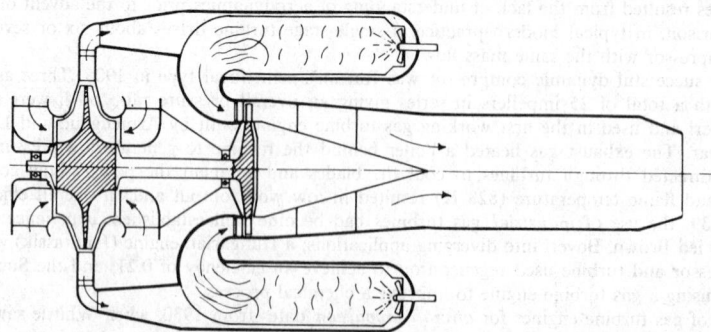

Fig. 72.3 Simplified arrangement of an early Whittle jet engine, with double-sided centrifugal compressor and reverse-flow combustion chambers. (Redrawn from Ref. 2 by permission of the Council of the Institution of Mechanical Engineers.)

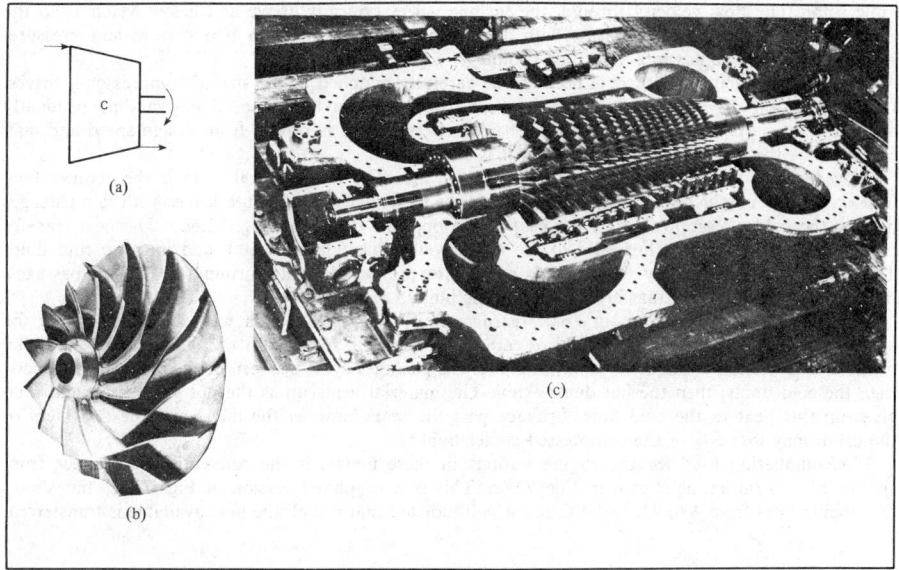

Fig. 72.4 Dynamic compressors: (a) diagrammatic representation; (b) an advanced centrifugal compressor impeller (by courtesy of Dr. A. Whitfield, University of Bath); (c) an axial-flow compressor with the top half casing removed [by courtesy of Sulzer Bros. (U.K.) Ltd.].

using a single rotor which inhales axially and exhausts radially (Fig. 72.4b). Two or more rotors are sometimes mounted on the same shaft, handling the fluid in series to boost its final pressure. A pressure ratio of over 10 may be generated in a single-stage centrifugal compressor with supersonic flow conditions, but this is only used in small machines.

The *axial compressor* is built up from a number of stages, like a series of fans of reducing flow area (Fig. 72.4c). It can handle much larger flows than the centrifugal compressor because its inlet occupies most of its frontal area, and it can generate a pressure ratio of 10 or more if sufficient stages are used. In practice, high pressure ratios are better achieved by running two or more compressor rotors in series, with a net reduction in the overall number of stages and reduced susceptibility to stall and other problems. Shorter compressors offer easier operation and also have higher critical whirl speeds, which can significantly ease the structural design process. In small engines, the optimal means of achieving a high pressure ratio sometimes involves the use of a single rotor with a few axial stages followed by a final centrifugal stage.

Gas turbine *internal combustion* is a continuous process taking place at constant pressure. A steady supply of fuel and air mixes and burns as it flows through a flame zone, which is located in the manner of a cloud in the lee of a mountain. The flame does not touch its container, being stabilized by the inlet airflow pattern, which also cools the container walls. Mixtures with a very wide range of fuel/air ratios can be burned, as the mixture in the flame zone is maintained at normal strength in the vicinity of the fuel inlet while any excess air is mixed in downstream of the flame. The combustion process involves very highly developed control of flame stability and can be tuned to emit very low levels of smoke, carbon monoxide, hydrocarbons and oxides of nitrogen. The combustion chamber volume is very small for the rate of heat release, partly because combustion takes place at high pressure: in aircraft engines the volume may be only 5% of that used in a boiler for the same fuel flow. It is conveniently labeled with the symbol B (for burning) in diagrams. Some engines use heat addition by heat exchange from an *external combustion* process, which permits the use of poorer fuels.

The *turbine* is the almost universal means of extracting energy from a high-pressure source of gas. It is represented diagrammatically in Fig. 72.5a, indicating that the gas expands as its pressure falls and it loses energy. Two types are used, the *radial turbine* (Fig. 72.5b), which is similar to a centrifugal compressor used in reverse, and the *axial flow turbine* (Fig. 72.5c). Typical isentropic efficiency is 0.75–0.9.

The axial turbine is akin to an axial compressor, but its blading is more complicated than a simple reversal of compressor blading. It extracts considerable energy, and with one stage can drive a compressor of six to seven stages passing the same mass flow. A single-stage turbine drops the gas stagnation pressure by about 30–50% from its absolute value upstream, with a corresponding drop in stagnation temperature of about 8–16%. For increased work output, several stages are often combined in the same rotor. The flow velocity through the turbine varies typically above and below Mach 1, so the gas imparts its energy very rapidly, with a corresponding rapid fall in temperature and pressure, which can be used to advantage for fast cooling of a gas.

The work capacity of a turbine needs to be carefully matched to the size of compressor it drives. If the turbine is to drive a load other than the engine compressor, so that its speed may vary independently of the rate of supply of gas, then the torque it exerts rises as it is slowed from design speed and may exceed twice the design speed torque when it is stalled.

Regeneration is achieved by two types of hardware. The conventional form is the counter flow *heat exchanger*, in which thin passages containing the hot exhaust gas are interleaved with thin passages passing the colder, compressed air (or gas) in the opposite direction (Fig. 72.6a). The heat transfer occurs in stages: from the hot fluid to the metal walls, through the metal, and into the cold fluid. The metal temperature may not be high enough to prevent deposits forming in the hot passages, reducing the rate of heat transfer, and the unit is bulky.

The other type is the *rotating drum regenerator* (Fig. 72.6b), which was developed mainly for automotive use. It is typically of metal or ceramic honeycomb form, with thin axial passages parallel to the axis. Rotation at about 20 rev/min makes the passages cycle between connecting the hot ducts, then the cold ducts, then the hot ducts again. The material heats up as the hot gas passes, and then gives up this heat to the cold flow. Leakage past the seals between the ducts and the end faces of the drum may lose 5% of the compressed cooler fluid.[4]

The numbering used for the engine stations in these figures is the conventional sequence from engine inlet to outlet, as shown in Fig. 72.6c. This is a simplified version of Fig. 72.2c, but shows the sloping lines from 4 to 2X and 4X to 2 which indicate that not all the heat available is transferred.

72.1.4   Engine Types and Characteristics

Open-cycle engines take a number of different forms. The simple engine of Fig. 72.7a generates a high-pressure source of gas, which may be accelerated through a nozzle to provide jet propulsion as a *turbojet* engine. In this form, the combustion system is squeezed between the compressor and turbine. Alternatively, a larger turbine may extract more of the exhaust energy, permitting it to drive an external load as well as the compressor as a *turboshaft* engine (Figs. 72.1a and 72.7b). Both of these engines are simple-cycle, *single-shaft* engines.

A variety of terms is used to describe the configurations of *multishaft* engines. The turbojet engine may be coupled to an industrial load by placing an independent turbine in its exhaust stream, rather than making the engine turbine larger. The second turbine is then a free *power turbine,* coupled to the load and running at a different speed from that of the engine. The core engine itself is then called the *gas generator* and the combination of the gas generator with the free power turbine makes a *two-shaft engine* (Fig. 72.7c).

The gas generator may itself have more than one shaft, in split-shaft or compound form. This is usually done to obtain satisfactory compressor operation in high-pressure-ratio engines. Figure 72.7d shows such a configuration, in which the load could be coupled to either shaft or to a separate power turbine. There are no unique terms to describe these configurations, but this chapter will refer to *low pressure* (l-p) and *high pressure* (h-p) components. Thus, the l-p compressor is driven by the l-p turbine and the h-p compressor by the h-p turbine, with the working fluid passing in the sequence shown. The shafts would be located concentrically in the case of an aircraft engine: Fig. 72.7e shows a two-shaft or *twin-spool* turbojet. Some engines have three shafts or spools in which the middle rotor is the *intermediate pressure* (i-p) system. It is convenient here to refer to the highest-pressure or core engine rotor as the h-p shaft, even in single-shaft engines, when discussing engine operation.

The shafts of two- and three-shaft engines run at their own equilibrium speeds, with no mechanical interconnection. The speed of rotation may range from 2000 to 80,000 rev/min but has little intrinsic significance as regards performance: it varies inversely with rotor diameter to keep the blade tangential velocity at a design speed of about 300–450 m/sec and to suit the flow Mach number. The direction of rotation is also of little significance, engines being designed to run in whichever direction suits the blade shape, but a few twin-spool aircraft engines use *contrarotating shafts* to minimize gyroscopic effects in turns and other attitude changes. However, the speed is significant in the case of a turboshaft engine because, depending on the engine size, a gear ratio of up to 20 to 1 may be required after the final turbine to reduce the speed of the output shaft to a conventional speed level for the driven machinery.

In addition to these variations of shaft configuration, there are several means of varying the thermodynamic cycle of industrial engines. The example given in Section 72.1.1 shows that, in compressing gas through a chosen pressure ratio, the work input required varies with the starting temperature. The use of an *intercooler* between compression stages usefully reduces the work required for the h-p compressor of some industrial engines (Fig. 72.8a). Conversely, it is desirable for expansion through the turbine to start at a high temperature, to reduce the pressure drop relative to the work extracted. For this reason, a second combustion process (*reheat*) may be used before the l-p turbine (Fig. 72.8b).

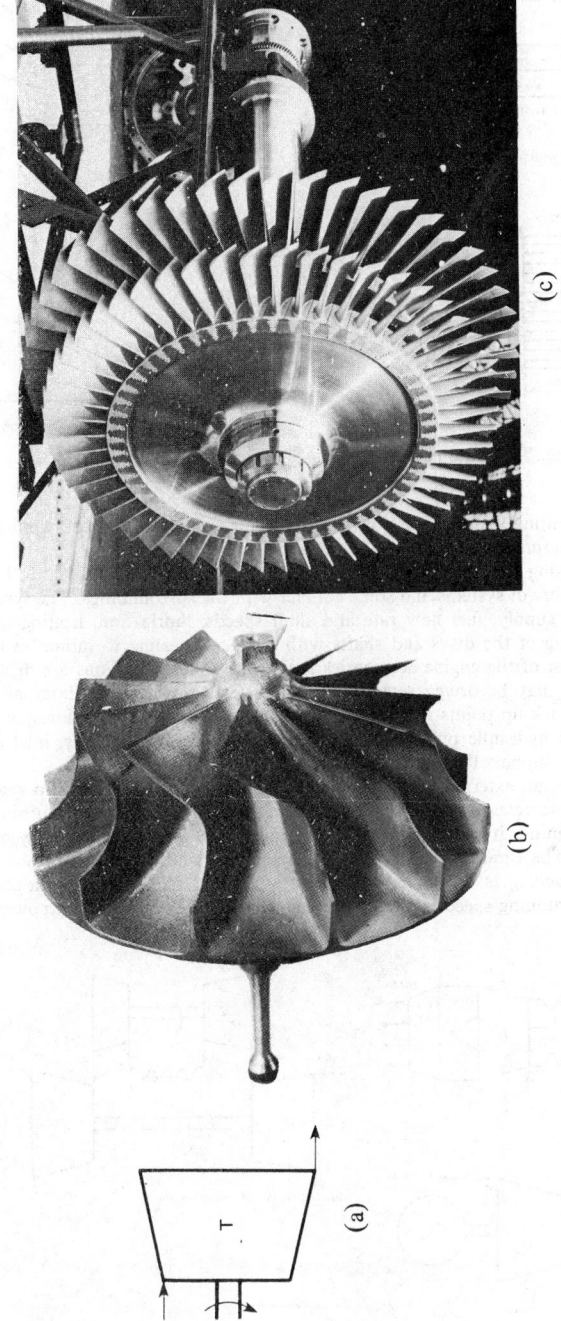

Fig. 72.5 Types of turbine: (*a*) diagrammatic representation; (*b*) a radial turbine rotor (by courtesy of Noel Penny Turbines Ltd.); (*c*) an axial turbine rotor (by courtesy of GEC Gas Turbines Ltd.).

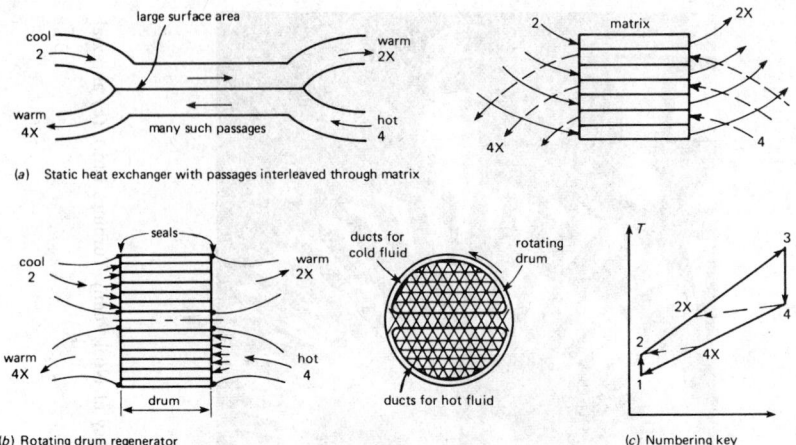

(a) Static heat exchanger with passages interleaved through matrix

(b) Rotating drum regenerator

(c) Numbering key

Fig. 72.6   Static and rotating types of regenerator.[30]

Both intercooling and reheating increase the work capacity of a given size of plant, while its efficiency may be increased by *regeneration,* as shown diagrammatically in Fig. 72.8c.

So far, only the operating cycle and the shaft arrangement have been considered. The complete engine also requires a number of systems and some contact with the surroundings. The systems include means to control the fuel supply, fuel flow rate and shaft speeds, lubrication, heating or cooling of the fuel and oil, air cooling of the disks and shafts with means of sealing to minimize leakage, and compressor operation. Most of the engine accessories that handle these functions are driven from the h-p shaft, although many may be driven separately in industrial applications. External connections may include a fuel inlet; pick-up points for the operator's controls; attachments for slave accessories such as the generator and hydraulic pump on an aircraft engine, and the starter; inlet and exhaust ducting; and mountings to support the engine.

Engine *starting* requires an external power source to rotate and accelerate the h-p shaft, until the combustion system is able to release enough heat to overcome the compressor and turbine losses and make the engine self-sustaining. It calls for a particular sequence of actions, which may be controlled by an automatic system. The time taken to start, and reach full load, ranges from about 15 sec to several minutes. Engine *stopping* is accomplished by shutting off the fuel supply, but a cooling period of several minutes at low running speed is advisable, since shutting down from high power may result

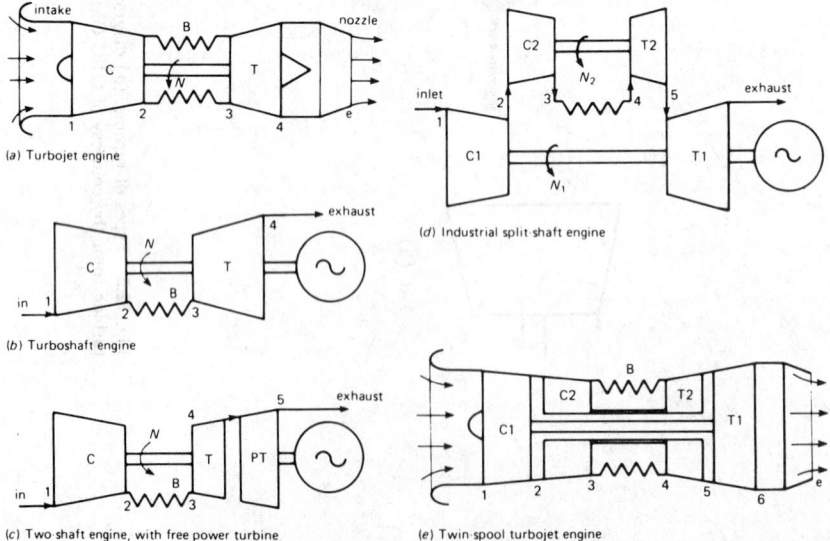

(a) Turbojet engine

(b) Turboshaft engine

(c) Two-shaft engine, with free power turbine

(d) Industrial split-shaft engine

(e) Twin-spool turbojet engine

Fig. 72.7   A selection of basic engine arrangements.[30]

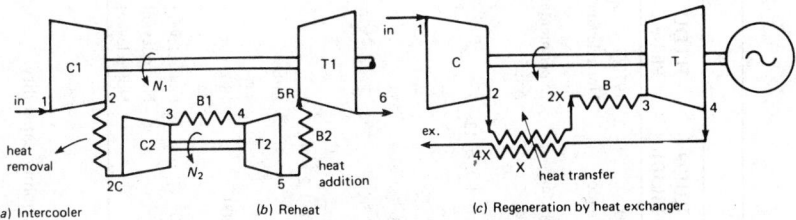

Fig. 72.8 Basic cycle modifications for industrial engines.[30]

in internal seizure if a casing cools faster and shrinks onto the rotor it contains. In large industrial engines, the rotor may need to be barred (rotated) as it cools after shutdown, to prevent it from setting with a permanent sag causing subsequent clearance or vibration problems.

Engine operation may call for variable speed or load (turndown). When the speed is fixed by, for instance, the coupling of an alternator to a power grid, a single-shaft engine permits rapid changes of load simply by changing the input fuel flow within the stable operating range but has low efficiency at part load, when the compressor and turbine losses remain high, permitting less useful output. A multishaft engine is more efficient at low load, when the h-p shaft slows down while the power turbine runs at the speed required by the load, but incurs a time delay in response to changed load requirements because of the time taken for the h-p shaft to change speed.

The gas turbine only delivers useful quantities of power in the upper 10–15% of its speed range, and its maximum speed is controlled by an operating limit such as maximum safe casing pressure or disk stress or turbine blade temperature. At any constant speed, the power demands of the compressor and load balance the turbine's power supply. Changes of speed are caused by increasing or reducing the fuel supply at a controlled rate, which directly alters the power output from the turbine and upsets the equilibrium. The torque excess or deficit and the inertia of the rotors then determine the rate of acceleration or deceleration, which is limited to avoid operational problems, which may include rich or weak extinction of the combustion, turbine blade thermal fatigue cracking, or the risk of increasing the compressor outlet pressure to the level at which it stalls.

The foregoing may apply to gas turbines in any application. The variants used specifically in aircraft are discussed more fully in Section 72.2.1. Industrial engines are considerably more variable, to suit the applications described later in Section 72.2. Their *origins* started with heavily built engines drawing to some extent from steam turbine practice. It was found feasible in the late 1950s to use aircraft engines as gas generators. These have evolved to the stage where the blading and combustion remain as developed for flight, but the structure may be made heavier and simpler. One engine alone, or a group of several engines side by side, may supply gas to an industrial power turbine. There are now also the medium-weight industrial engines, which, like the heavy engines, may use regeneration to provide efficiency levels comparable with the better *aeroderivative* engines. Some comparisons between these types, and between gas turbines and other types of power plant, are made in Sections 72.3.1 and 72.3.2.

72.1.5 Gas Turbine Engine Trends

The power output of a gas turbine engine increases relative to its cost by the use of higher turbine inlet temperatures and higher component efficiencies, but the higher temperature demands a correspondingly raised compression ratio to maximize the power output.

The temperature and pressure limits are set by the best available materials. The pressure ratio is limited by rotor tip velocity, which implies material stress, and the turbine entry temperature (TET) limit is set by turbine blade creep with the loss of strength at elevated temperature. The efficiency is very dependent on the aerodynamic design, and on the quality of sealing to minimize leakage flows in the engine and at the ends of the blades.

The trend of design progress with time is illustrated in Table 72.1 by data from a series of engines from one manufacturer. The engine listed as HTDU is the high-temperature demonstrator unit described in Ref. 5. Pressure ratios have risen as aerodynamic understanding has led to transonic and supersonic flow at the compressor inlet. The temperature limit has been avoided by artificial cooling of the blades, which keeps the metal temperature below about 1250 K in all engines, and may in time be met by the introduction of ceramic materials. Turbine efficiency has been increased by shrouding the ends of the blades to improve the sealing, and by detailed improvements to the blade profile and geometry: discussion in Ref. 6 foresees polytropic efficiencies reaching 0.93. In conjunction with other progress[3] engine efficiencies of 0.4 to 0.5 are foreseen.

While this progress is led by the aircraft industry, it steadily filters into industrial engine designs. A relative reduction in their cost will continue their encroachment into the industrial power field,

Table 72.1 Progress and Development of Aircraft Engine Design (Data Courtesy of Rolls-Royce Ltd.)

						Engine					
	Welland	Derwent V	Avon RA7	Avon 531	Conway RCo12	Spey 506	RB211-22B	RB211-524D4	RB211-535E4	RB199 Mk101	HTDU Phase 2
Entry into service	1944	1946	1951	1959	1960	1965	1972	1981	1984	1980	1970
Aircraft	Meteor I	Meteor IV	Canberra	Caravelle	Boeing 707	BAC 1-11	Lockheed L1011	Boeing 747	Boeing 757	Tornado	Not relevant
Engine type	Turbojet	Turbojet	Turbojet	Turbojet	Turbofan	Turbofan	Turbofan	Turbofan	Turbofan	Turbofan	Experimental[c]
Compressor type	Centrif	Centrif	Axial	Axial	Axial	Axial	Axial	Axial	Axial	Axial	Axial
Number of shafts	1	1	1	1	2	2	3	3	3	3	1
Number of compressor stages	1	1	12	17	$7 + 9 = 16$	$4 + 12 = 16$	$1 + 7 + 6 = 14$	$1 + 7 + 6 = 14$	$1 + 6 + 6 = 13$	$3 + 3 + 6^c = 12$	7
Mass flow (lb/sec)	31	62	119	184	280	203	1380	1550	1150	154	—
Bypass ratio					0.29	1	4.7	4.4	4.1	>1	—
Pressure ratio—total	3.9	4.2	6.4	10.3	14.8	17.1	25.5	29.3	28.5	>23	24
—split	—	—	—	—	3.66×4.03	2.6×6.6	$1.36 \times 4.6 \times 4.1$	$1.57 \times 4.37 \times 4.27$	$1.6 \times 3.5 \times 5.0$	—	—
Number of turbine stages	1	1	2	3	$1 + 2 = 3$	$2 + 2 = 4$	$1 + 1 + 3 = 5$	$1 + 1 + 3 = 5$	$1 + 1 + 3 = 5$	$1 + 1 + 2 = 4$	1
Shrouded blades	No	No	Yes	Yes	Yes	Yes	Yes	Yes	Yes	Yes	Either
Cooled blades	No	No	No	Yes	Yes	Yes	Yes	Yes	Yes	Yes	Yes
TET—max TO (K)	1060	1100	1170	1262	1310	1350	1500	1600	1520	>1600	1800
—cruise (K)	950	1000	1040	1115	1030	1190	1350	1360	1260	—	—
Thrust (lbf)	1600	3600	7150	12600	17960	10400	42000	53000	40100	7900[b]	Not relevant
SFC cruise (lb/lbf/hr)	1.12[a]	1.05[a]	0.92	0.95	0.68	0.76	0.63	0.61	0.58	—	Not relevant
Spec wt (lb/lbf)	0.53	0.31	0.34	0.29	0.26	0.22	0.22	0.19	0.18	—	Not relevant

[a] TO sfc (take-off specific fuel consumption).

[b] Dry value given: max afterburning = 16,000 lbf thrust.

[c] HTDU is a demonstrator used primarily to develop high-temperature technology. Later phases will continue to increase turbine blade and NGV cooling capability.

displacing some traditional power plants (Section 72.3.1). Further success is assured by current serious research work into the use of coal-derived fuel, and heat transfer from external combustion to reduce turbine erosion when using low-grade fuels.

72.2 APPLICATIONS

72.2.1 Engines for Aircraft

Aircraft engines exert a forward thrust on the airframe in reaction to the rearward acceleration imparted to part of the passing airflow. This is achieved by means of a propeller, or by generating a high-pressure jet which emerges through a nozzle. The propeller may be turned by any type of engine, but the gas turbine is by far the most effective means of providing jet propulsion.

The choice of a propeller or jet propulsion is determined by the required *flight speed*. Propulsive efficiency is lost if the outlet flow from the propeller or jet has too great a rearward velocity, which represents wasted kinetic energy. The efflux velocity from a propeller is low, which suits low flight velocities: medium and high jet velocities suit medium and high-speed flight.

The jet engine thus overcomes the speed limitation of the propeller, and is also well suited to operation at different *altitudes*. Its thrust varies with atmospheric pressure, being high at low altitudes where the aircraft drag is high, and falling at high altitude where the lower air density also reduces the drag. Its thrust falls only slightly as the flight speed increases, permitting very high speeds. Nevertheless, jet aircraft fly at a range of speeds which calls for a range of engine configurations, as discussed later.

Given the required thrust in an efficient engine configuration, the next most important criterion for a heavier-than-air machine is its *weight*. A heavier engine requires a heavier supporting structure, and more fuel to carry it throughout the aircraft's life. The extra fuel needs a bigger, heavier tank, which, in turn, needs more fuel and a more powerful engine to lift it.

Its high power-to-weight ratio and its jet favor the gas turbine over the piston engine in all but the smallest aircraft. It also has a longer life and is more reliable. The low weight calls for very sophisticated design, and the lightness of the rotating parts confers good handling properties by permitting low acceleration times. The high power is achieved by high internal pressures and temperatures.

Aircraft gas turbine engines are made by about 50 companies; the engines are made mostly in a few, well-known *configurations*. They may be considered as variations of the single or multispool turbojet (Figs. 72.7a and 72.7e), in which the combustion system is squeezed between the compressor and turbine to minimize weight and frontal area (cf. Fig. 72.1a). In turboshaft form (Section 72.1.4), its shaft may drive the rotors of a helicopter. If the reduction gear is built into the engine to drive a propeller, it becomes a *turboprop* or *prop-jet* engine (Fig. 72.9a), from which the exhaust gas exits through a nozzle to give a small residual jet thrust.

Enlargement of the l-p compressor (C1 in Fig. 72.7e) makes a *bypass-jet* engine if the excess l-p air is ducted to mix with the exhaust (Fig. 72.9b), or a *fan-jet* or *turbofan* if it emerges separately (Fig. 72.9c). This flow bypasses the h-p compressor, combustion system, and turbines. The *bypass ratio* is the ratio of this flow to the flow through the h-p system, and was typically 0.3–1.5 for the

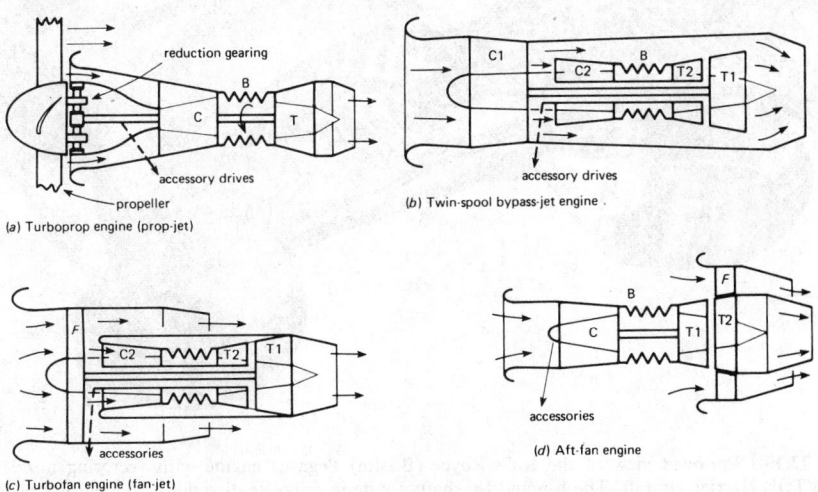

Fig. 72.9 Various aircraft engine configurations.[30]

bypass engine, or up to 6 for the fan engine. This rule is broken by the last entry in Table 72.1, however, to obtain better efficiency and other benefits. The higher bypass ratio engine provides better propulsive and overall efficiency for subsonic flight, because it generates its thrust by accelerating a large mass flow to a low velocity.

The fan or l-p compressor may have one or several stages of blading, and is driven by the l-p turbine in a two or three-spool engine. There is little balance of advantage for either number of shafts: the three-shaft engine[5] may be a little more expensive and heavy, but may offer higher efficiency and better handling response. The fan could instead be mounted as an *aft-fan* (Fig. 72.9*d*) behind an existing engine, to improve its efficiency, thrust, and noise. This avoids the need for a shaft: the turbine blades or buckets are extended to form fan blades protruding into a large duct outside the nacelle, the one-piece combined blades being called "bluckets."

Fan engines also make good *lift engines* for vertical (VTOL) or short take-off and landing (STOL). They may be installed vertically for direct lift, possibly rotating to horizontal for forward flight. The l-p air may be ducted to a lifting wing[7] for STOL, or may be taken directly through vectoring nozzles (Fig. 72.10) for VTOL.

Supersonic flight requires a considerably increased jet velocity. The *afterburner* reheats the exhaust gas after the turbines, permitting it to accelerate to an appropriate level above the flight velocity and boosting the thrust to overcome the increased drag.

The last aircraft engine type is the *auxiliary power unit* (APU). It is a small turboshaft engine that provides air-conditioning and electric and hydraulic power while the main engines are stopped, but its main function is to start the main engines without external assistance. The APU is usually started from batteries by an electric motor. When it reaches its operating speed, a substantial flow of air is bled from its compressor outlet and is ducted to drive the air turbine starters on the main engines. The fuel flow is increased when the turbine air supply is reduced by the air bleed, to provide the energy required for compression. These engines are also found on ground carts, which may be temporarily connected to an aircraft to service it. They may also have uses in industrial plants requiring air at 3 or 4 bar.

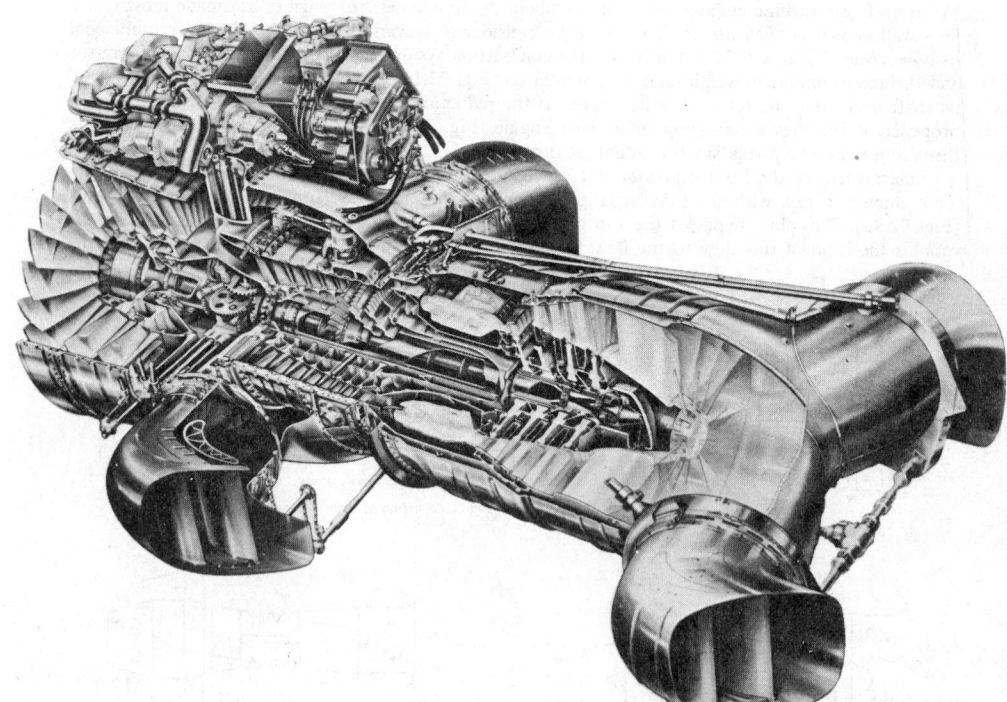

Fig. 72.10 Sectioned view of the Rolls-Royce (Bristol) Pegasus engine with vectoring nozzles for the VTOL Harrier aircraft. The h-p and l-p shafts rotate in opposite directions (by courtesy of Rolls-Royce Ltd.).

72.2.2 Engines for Surface Transportation

This category includes engines for rail, road, and off-road transport, some of which may be for military purposes, and marine use. The low weight and volume are assets in all cases, but the relatively high fuel consumption is more acceptable in military use. The engines are of turboshaft type with a free power turbine: they can burn a variety of liquid or gaseous fuels and need no liquid cooling system.

Railway applications are rare, because of the poor efficiency relative to diesel—electric or all-electric systems. Regeneration is essential to make the fuel consumption acceptable. The transmission may be mechanical, electrical, or hydraulic, and the task is made easier by the use of gentle gradients and long journeys.

Road use uncovers two more drawbacks. The engine has slow torque response to accelerator pedal movement, and inadequate torque to cope with gradients if geared for level running. The slow torque response stems from the loss of gas generator rotor speed at part throttle, and its long acceleration time. Some makers use variable-angle nozzle guide vanes (VNGVs) to maintain a high gas generator speed. Variable power turbine NGVs may also be used, even to the extent of providing reverse rotation when required. Volvo uses a turbine to drive differentially both the compressor shaft and the output shaft: at low road speeds the compressor speed remains high while, at higher speeds, the power is passed mainly to the output shaft. This also overcomes the second drawback: generally, the modest rate of torque gain with reduced speed necessitates the use of a multiratio gearbox or automatic transmission. Considerable reduction gearing is needed between the high turbine speed and low wheel speed, and regeneration is also essential.

Despite all these disadvantages, the automotive gas turbine still has good prospects for the future because of its low levels of pollution. With the Stirling cycle reciprocating engine it may offer the solution to the very tight emission control requirements of some countries. Work is in progress to introduce ceramic turbines, permitting higher temperatures and improving the cycle efficiency. It is compared with alternative engines in Ref. 8.

Off-road applications are usually military and may include landing craft, armored cars, tanks, hovercraft, and trucks. They may also use regeneration, and inlet air filtration is essential.

The gas turbine is well suited to *marine* use, since it matches the fluid nature of the supporting and propelling medium. The shaft drives the propeller or water jet unit or, in the case of hovercraft, the lifting fan and air propeller.[3] The engines are usually aeroderivatives and are the almost universal power plant for nonnuclear, new naval ships. Regeneration is not used because of its bulk and weight: it also adds to the complexity of changing to a spare engine if problems arise. The more efficient later engines are ousting the diesel and steam plant, which were sometimes used for economical cruise and supplemented by gas turbines for "sprint" use. Gas turbines are also finding favor in liquefied natural gas (LNG) tankers, in which they use as fuel the small "boil-off" quantity that evaporates to keep the main bulk cool. The boil-off can be increased by exhaust heat if necessary.

The main environmental problem is *corrosion*. It may affect all parts—disks, shafts, casings, blades and blade roots, and external parts. This can be controlled on steel components by the use of sacrificial coatings, often containing aluminium, iron, and chromium, or by using nickel or cadmium (on cold parts only). A more severe problem is "hot corrosion" of nickel and cobalt turbine blade alloys, resulting from combination of sodium from the salt atmosphere with sulfur from the fuel. This problem concerns operation at sea, and also land-based operation near the sea. A second problem is compressor fouling or erosion, by salt or sand or dirt deposits, causing a loss of performance. Both of these problems call for extensive filtration plants.

72.2.3 Applications in Electricity Generation

A turboshaft engine is ideal for driving an alternator, and most manufacturers will supply the engine matched to the electrical equipment as a turbogenerator set. Such units come from all origins and may have one or more shafts. The single-shaft engine in this application may be designed for a single operating speed and the shaft speed may be governed by the alternator or may be controlled to provide very accurate *frequency control* for specialist tasks. An example of this is the power supply for computers.

These units are suitable for use by national electricity generating authorities. Many are used on *base load*, where their efficiency is little less than that of other plants. The advantage of such units is that they can be distributed throughout the country in medium sizes, being located near the demand with no need for supplies of cooling water, and thereby minimizing transmission losses.

Local electricity boards and companies use turbogenerators for *peak lopping*. It is common practice for all the electricity supplied in a given period to be priced in proportion to the maximum rate of usage at any time during that period. If the peaks are lopped by, for instance, using additional local generating capacity to absorb the evening meal cooking load, then the savings to the local board, which buys power from the main grid, may pay for the peak lopping engine within a few days. However, the engine life is reduced by such duty and may be shorter still on *reserve peak* duty: each engine run may last only a few minutes, which gives a high proportion of starts to hours. Aeroderivative engines are often chosen for this duty.

72.2.4 The Oil and Gas Industry

Turboshaft engines are extensively used by the oil and gas industry, which can afford to burn its own fuel at, sometimes, relatively low efficiency.

A major application is *pumping oil and gas* through pipe lines over long distances, through deserts, arctic lands, sea, and other remote places. As the pressure falls owing to frictional losses in the pipe line, pumping stations are built at intervals along the line to boost the pressure. Gas is transmitted at a typical pressure of 68 bar (1000 lb/in.²), but the pressure varies considerably as the pipe line serves the secondary function of storage. Two-shaft engines are suitable for this duty, as the gas generator may then change speed with only moderate loss of efficiency to cater for the load variations. The engine is usually fueled by tapping off a small amount of the pumped oil or gas, and may operate on a remote site for weeks or months, without supervision. It is controlled from a main center, started, stopped, and monitored as required.

Gas turbines are also used extensively on offshore *oil drilling rigs.* There, they generate electricity, supply shaft power for pumping duties, and may heat the rig with their exhaust gas. The low weight and bulk, and supply as a prepackaged and tested unit, are of benefit. The danger to personnel in this environment is often reduced by the provision of considerable excess engine capacity.

The oil industry also uses gas turbine components separately, as discussed in Section 72.2.9.

72.2.5 Combined Cycles and Cogeneration

The relatively low efficiency of the turboshaft engine results from the high temperature remaining in the exhaust gas when the pressure is too low to provide useful work. The heat may be used to raise the efficiency by regeneration—heating the compressed air before the combustion process takes place. In the appropriate plant, the heat may be used more effectively to raise steam by feeding the engine exhaust gas into a boiler, to drive a steam turbine and generate more power. A suitably designed plant may permit the gas and steam sections to be operated independently as well as together,[9] while an old steam plant may be repowered in this form.[10] An unusual variant is discussed in Section 72.2.10.

Cogeneration gives higher efficiency than either a gas turbine or steam plant by itself. Typical gas turbine efficiency is 0.2–0.3, but at the higher pressure ratio, aeroderivative engines may exceed 0.35. Steam plants used for electricity generation alone can achieve peak levels of 0.35–0.38 in very large plant sizes (over 200 MW), while smaller units (under 20 MW) may peak at only 0.3.

In an *unfired* cogeneration plant, where no additional fuel is burned in the boiler, the overall plant efficiency may lie between 0.4 and 0.5. In a *fired* plant, the additional fuel may be of poorer quality and will add to the power output but, in the limit, as the gas turbine fuel input falls to only a small proportion of the plant total consumption, the efficiency falls back toward that of the steam plant alone. The plant then begins to resemble an energy topping application.

Energy topping is a method of improving the cycle efficiency of existing or proposed steam heating plants. It involves mounting a small gas turbine engine to use the boiler's air intake flow before discharging it into the boiler (Fig. 72.11). As the gas turbine burns more fuel, the boiler burns less fuel. While the engine runs at idle, supplying no output power, all the fuel energy input appears as exhaust heat, which reduces the boiler fuel input correspondingly. When shaft power is drawn off, perhaps 30% of the fuel energy input may be used. The total plant fuel input must be increased to supply that power in addition to the normal quantity of steam. In fact, the total fuel input may be increased by 8–10%, as also is the air mass flow: all of that 8–10% emerges as shaft power, an incremental efficiency of 1.0.

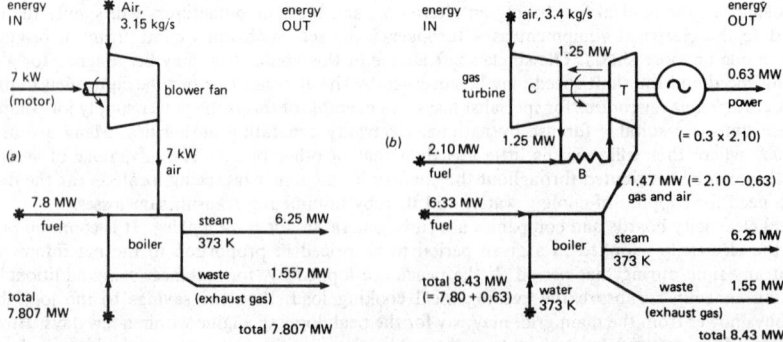

Fig. 72.11 Gas turbine topping of a process steam plant: (*a*) the standard boiler generating 10,000 kg/hr of steam; (*b*) the changed energy flows with the engine using the air first.[30]

An example of the energy balance at full load is shown in Fig. 72.11*b*. It assumes a boiler capacity of 10,000 kg/hr and an engine with efficiency of 0.3 and compression ratio of 12.5 to 1. Note that some of the energy in the gas passing through the combustor is recycled through the turbine, shaft, and compressor to reenter the combustor. This application offers a uniquely efficient means of obtaining power for the lighting or other loads in a plant, with two qualifications. The engine is likely to need higher quality and more expensive fuel than is normally used in the boiler. Second, the engine taking only 3.4 kg/sec of air in this example is a very small engine and its power output is quite low: the smallest engine available would probably suit a boiler of 4000 kg/hr but its compression ratio may be only 4 to 1, limiting the power output to 170 kW. The principle could be applicable to completely different types of fuel.

The most efficient combined cycle is the *total energy* package. Efficiencies of up to 0.85 are feasible, where the steam exhaust from the turbine is used for process work or heating. The steam turbine may then be run to a chosen backpressure, which provides the process temperature required. The residual exhaust gas heat may also be usable, but the cost of recovering all the waste heat may be prohibitive. A number of factories use these units for all their power and heat requirements, controlling this balance by varying the amount of supplementary firing.[11]

A pioneering combination of the gas turbine with steam was the Brown, Boveri *Velox boiler* of the 1930s (Fig. 72.12). The combustion chamber of the engine is the flame tube of the boiler, which, by virtue of the high gas pressure, is considerably more compact than a boiler having its combustion at atmospheric pressure. This made it suitable for rail and marine use as well as stationary use, but its efficiency could not match the use of exhaust heat for steam generation. It is described in Ref. 12.

Fifty years later, cogeneration is the most active growth area for the gas turbine plant. The search for increased efficiency and reduced NO_x formation sometimes involves mixing water with the engine airflow. Water injection directly into the combustor increases power output in proportion to the extra mass flow but reduces efficiency. When injected for direct evaporative intercooling before the h-p compressor, water increases power output and efficiency. In both cases, and when process steam is injected into the combustor, the pollution is reduced because the additional inert flow reduces the peak temperature. Research also concerns two-stage, rich–lean combustion systems and other ways to minimize pollution.[13]

72.2.6 Uses for the Exhaust Gases

The exhaust gas heat may have other uses than steam raising. It may be used for drying processes, with cold air mixed in to reduce the temperature as required. Applications in industry include kiln drying of cement, or tobacco, or blood from a slaughterhouse.

The combination of heat and power is used extensively for the desalination of seawater, particularly in the Arabian Gulf and other desert regions.

The partial use of the oxygen from the intake air permits the exhaust gas to form the basis of an *inert gas.* More fuel must be burned in it to use up the remaining oxygen and it will then need cooling before it can be used. A significant use for it is *purging* empty oil tanks, removing the oxygen to eliminate the risk of explosion. With the large mass flows involved, even a small engine provides plenty of gas.

In arctic or other regions, the gas from a portable gas turbine may be used to *thaw* material and equipment. It serves the purpose of a blow torch, but the use of the compressor and turbine provides considerably greater mass flow at more moderate temperatures.

A more powerful version uses the blast from the jet in addition to the heat, for *snow clearing* from airfield taxiways and runways. Old turbojet engines are used for this: the high fuel consumption

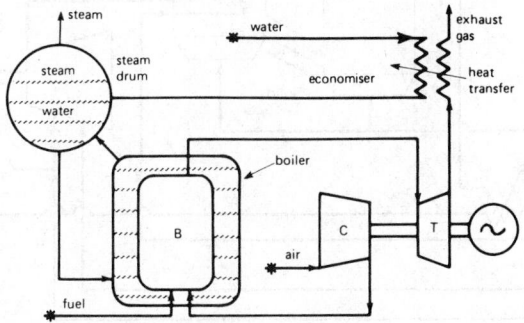

Fig. 72.12 Simplified diagram of Brown, Boveri Velox boiler.[30]

is offset by the time saving, or even the ability to open the airfield at all on any given day. It could also be used on roads and railways. A fishtail nozzle is used to provide a wide but shallow jet, with its area trimmed to retain the correct engine matching. The jet needs to be located near the ground, pointing slightly downward, and the engine is mounted on a heavy vehicle with adequate brakes and traction. An air mass flow of 10–20 kg/sec at a low power setting is very effective. People should be kept clear of the blast and the blown snow and ice.

72.2.7   External Combustion and Other Heat Sources

Many well-known fuels are unsuitable for gas turbine engines, because they are solid, slow burning, corrosive, or form ash, which damages the turbine. Research work is in hand to modify or gasify coal, to make it suitable for internal combustion,[5,14] but many more fuels could be used with external combustion.

The traditional means of getting their heat into the working fluid would be via a heat exchanger. Not all of the heat can be exchanged, and the working fluid could not reach the original temperature of the hot gas, but its cleanliness could provide indefinite turbine blade life. The undesirable combustion products could instead attack or deposit on the heat-exchanger tubes, reducing their effectiveness and requiring frequent cleaning. The problems are thus transferred but not eliminated.

The solution may be to link the gas turbine engine to the circulating fluidized bed type of combustor, as pioneered by Westinghouse and Struthers Wells in the United States and by GEC in the UK.[15,16] This type of combustor is derived from process plants developed since 1940 for cracking petroleum products and for calcining alumina, and further developed since 1970 for coal-fired boilers. The system used by GEC is shown in Fig. 72.13. The combustion takes place in a fluidized bed of solid material and the outlet gas carries over the lighter solids into a cyclone separator. The hot, separated solids replenish a second fluidized bed in which are immersed high-temperature alloy heat-exchange tubes containing the gas turbine's compressed air. After heating the air, the solids return to the first bed, while the hot gas escapes to give up further heat to the steam system.

By thus separating the combustion and heat-transfer processes, the combustion can be completed safely in a refractory lined tower, while the fluidizing air for the heat-transfer bed provides a mildly oxidizing, scrubbing environment that avoids corrosion and deposition problems. Further advantages may be gained from the greatly reduced heat exchanger size, resulting from taking heat from a solid

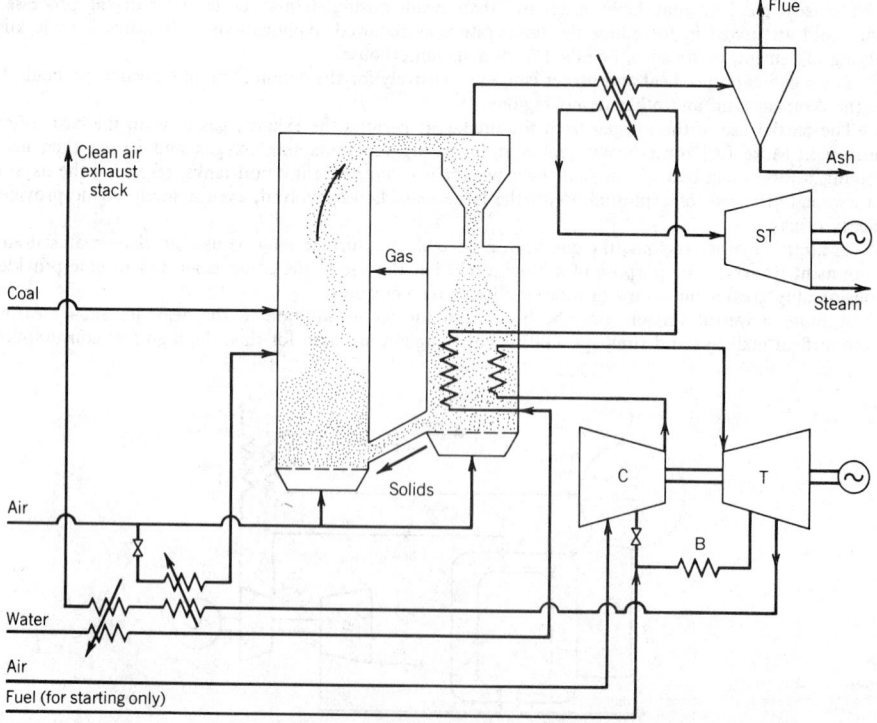

Fig. 72.13   External combustion using a circulating fluidized bed type of combustor.[30]

rather than gaseous source. The absence of condensing, corrosive products permits the turbine gas to be expanded and cooled to a low temperature to enhance the cycle efficiency, although the normal temperature limitations still apply to the gas exhaust. The possibility of differential variation of the heat inputs to the steam and air circuits makes it a flexible topping cycle, in which the proportions of electrical and heat energy are widely variable.

There are other sources of heat apart from combustion products. Two options are solar heat and the heat of a chemical reaction. There is also nuclear heat, which is capable of being handled by closed cycle gas turbines (Section 72.2.8).

72.2.8 Closed-Cycle Gas Turbines

The closed-cycle gas turbine uses the same working fluid repeatedly. The compressed gas is heated by external combustion or other heat source, and the expanded gas is cooled by a process which can make use of its heat. Given a need for this heat, which is often used for district heating in Europe, efficiencies of 0.85 are feasible.

Closed-cycle engines appear to compete well with steam plants in sizes from 6 MW upward.[17] Their capital cost is greater for size below 100 MW, but their efficiency, unlike that of steam plants, does not fall with size. Their efficiency also remains high at part load, and a wide variation of power can be achieved at constant speed by varying the total mass (and pressure) of working fluid in the system.

There is also the option of using different working fluids. Air works well, and incurs no cost of replenishment to compensate for leakage losses. Other gases are used in specific applications where their properties are advantageous. For instance, *helium* is considered better than air in generation plants over 30 to 50 MW, where the expense of the fluid is more than offset by the reduced size of the turbomachinery. *Carbon dioxide* is another relatively inert gas which is a viable working fluid, and might even be replenished from the combustion gases. *Nitrogen* is used in the regasification of LNG, which provides the cooling necessary before recompression: in this application, the regenerative cycle with external combustion also provides useful shaft power for electricity generation. The use of *hydrogen* is another possibility, as discussed in Section 72.2.10.

External combustion experience is expected to lead to the use of helium closed-cycle gas turbines as the means of extracting power from advanced gas-cooled *nuclear reactors*. The helium would be heated in its passage between spheres of active material, and would then flow through the turbomachinery, as shown in Fig. 72.14. The avoidance of the need for intermediate fluids, with heat exchange from the reactor coolant to the turbomachinery circuit, yields the highest possible temperature at turbine entry. The field is covered in Ref. 18 and in those references listed in Ref. 17.

72.2.9 Chemical and Process Industry Applications

The ability to compress or expand large quantities of air or gas is used in many plants to adjust between the different temperature and pressure levels of the process. This may be done by compressors and turbines acting separately, or together, or as part of an engine. Normally, a dynamic compressor handles so much air and absorbs so much power that the turbine is the essential driver: it is fortunate that a compression stage is usually accompanied by an expansion process, making the units compatible. The turbine may alternatively be coupled to an external load if the source of pressure arises by other means.

In all cases, the turbomachinery works without contact between its lubricant and the working fluid. The chemical process may therefore remain *uncontaminated* by oil, an advantage relative to most other types of compressor.

The earliest chemical industry application was the *Houdry process* of 1937, which converted a wide range of hydrocarbons by catalytic cracking to relatively uniform, high-grade gasoline.[19] Three

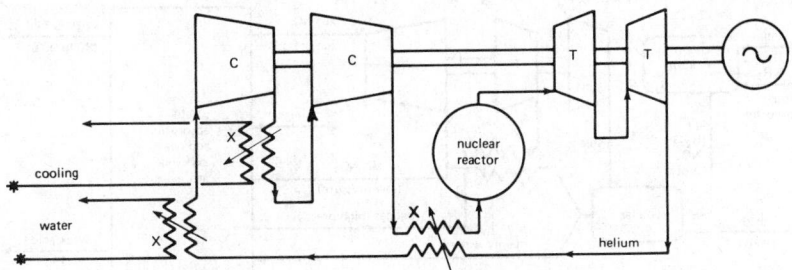

Fig. 72.14 Gas turbine in a helium-cooled reactor.[30]

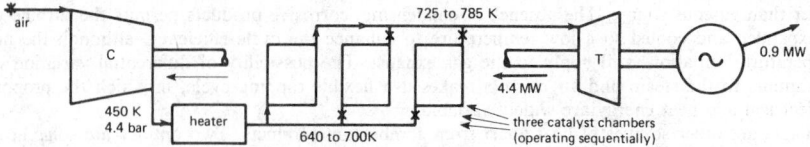

Fig. 72.15 The Houdry process for catalytic cracking of gasoline.[30]

beds of catalyst (clay and other materials) were used sequentially for 10 min each, by which time they were coated with carbon. Each bed was then purged for 5 min to recover the oil and vapor content, regenerated for 10 min with hot compressed air from the gas turbine, and purged for 5 min more before its next working period. The gas turbine section of the plant is shown in Fig. 72.15. This process has been superseded by catalyst in fluidized beds, partly because the turbine blades suffered erosion from catalyst dust.

The turbomachinery is an integral part of the SABAR process for making *nitric acid* (HNO_3).[20] Figure 72.16 shows the simplified flow diagram. It starts with liquid ammonia being heated for evaporation, mixed with air and compressed by a slave power source to enter the reactor, where it is converted catalytically to water vapor and various oxides of nitrogen (NO_x). The heat of reaction is taken for steam generation by a boiler and other heat exchange processes (not all shown), and the steam turbine provides power to compress the nitrous gases for the absorption processes. The tail gas, with very little NO_x remaining, is expanded to help the compression process.

Energy topping could be applied to industrial *combustion processes,* which burn waste products or materials not normally regarded as fuels. Waste products include *methane* gas, produced from sewage or manure or the digestion of offal, or *blast furnace gas* in which the carbon monoxide content warrants further combustion. There may be a requirement for such fuels to be filtered, decontaminated to protect the turbine, and dry.

A less usual fuel is *sulfur,* which is burned in large quantities to make sulfuric acid via sulfur dioxide (SO_2) and trioxide (SO_3) stages. The heat released is sometimes dissipated unused, sometimes partially recovered in a steam plant to generate electricity for the plant and the local community. It is shown in Ref. 18 that topping such a plant could increase its electricity output by over 60% at very moderate capital cost. Over 50% of the fuel may be burned in the engine, compared with 25% in Fig. 72.11, at zero fuel cost since the fuel would be burned anyway. The problems of handling the fuel in liquid form are well known, and the corrosion problem may well be negligible even though sulfur is undesirable in normal fuels. Many other unusual materials, which are converted by combustion processes, could similarly be burned in gas turbine engines. Corrosion, erosion, and combustion intensity would need to be satisfactory, along with numerous other factors. If in doubt, the use of an external combustion engine would increase the operational feasibility including, perhaps, that of burning sulfur.

The *turboexpander* shown in Fig. 72.17, which was derived from Ref. 22, is used variously to separate carbon dioxide, hydrogen sulfide, or nitrogen from natural gas. The gas often comes at high temperature and pressure from the ground, thermally degraded (CO_2 or N_2) or soured (H_2S). A turbine extracts work, cooling the gas to permit demethanization after partial condensation. The turbine drives a compressor to repressurize and heat the gas for delivery, and a steam turbine may additionally be required. Older plants often used refrigeration.

A turbine may also be used for *ultrafast cooling* processes, and may cool a flow of gas at a rate as high as 500,000°C/sec and with a temperature drop of up to 500°C. This capability could be of use where the high-temperature equilibrium condition of a gaseous mixture is required at lower temperature, where the equilibrium would normally be different. The gases involved would need to be checked

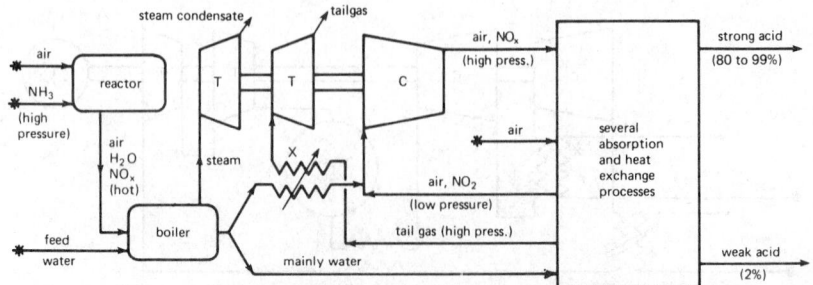

Fig. 72.16 The SABAR process for making nitric acid (HNO_3).[30]

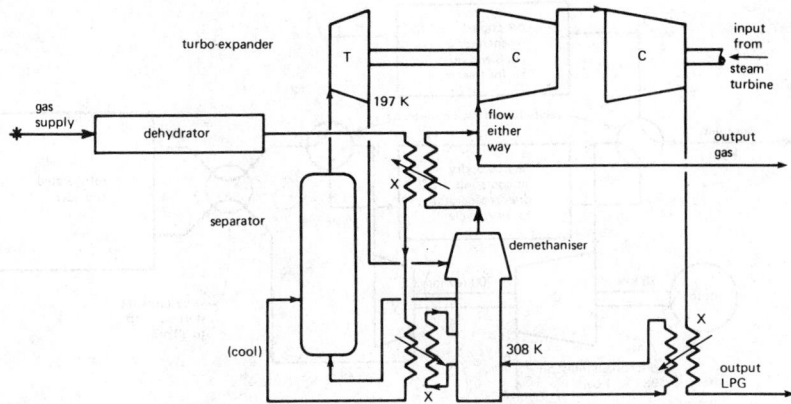

Fig. 72.17 Turboexpander used to produce LPG.[30]

for compatibility with the turbine material; a reducing or corrosive atmosphere may not be suitable and the temperature at turbine entry could not normally exceed 1200 K.

72.2.10 Unusual Applications

There are several miscellaneous configurations which are considered feasible, and some are now in operation.

Electrical load variations during the day may be smoothed by the *air-storage* gas turbine, which absorbs energy from the grid at low demand periods and feeds the grid at times of peak demand. It involves an electric motor driving a compressor, to pressurize overnight a large underground rock cavern.[23] When required, the air is released through a combustion system to drive a turbine and alternator. Further heat may be recovered from the exhaust gas.

Another possibility is *air-cycle refrigeration*. The simple arrangement of Fig. 72.18 cools the compressed air, which is then expanded to a very low temperature. The turbine extracts less energy than the compressor requires, so another engine or electric motor is required to supply the deficit. This unit would require predried air: the cycle of a unit handling humid air (Fig. 72.19) is discussed in Ref. 24. Neither has a very high coefficient of performance, but they offer a portable and very powerful source of cold air for blast freezing, available almost at the instant of startup, and without the need for freon or ammonia. The subject is covered more fully in Ref. 25.

Waste heat at low pressure, which accounts for the inefficiency of many thermal cycles, may be partially recovered by *subatmospheric pressure cogeneration*. This could involve a turbine expanding the hot exhaust gas from a boiler down to a low pressure, cooling in a heat exchanger, and then recompression to exhaust to the atmosphere. The turbomachinery would be quite small, although bigger than the energy topping engine which works at above atmospheric pressure. A high boiler exhaust temperature and high expansion and recompression ratios are required to obtain a worthwhile output.

A study[26] has shown the feasibility of a high-efficiency power cycle which minimizes the usage of high-grade energy for compression. It involves a closed cycle in which the hydrogen working fluid is compressed by an *absorption/desorption* cycle, using metal hydride ($FeTiH_x$) beds, as shown in Fig. 72.20. It is based on the fact that, as the gas is absorbed, it gives out heat, while, to desorb, heat must be fed in. Two or more hydride beds are used, with inlet or outlet valving sequenced to let one absorb as the other desorbs. Absorption takes place at low pressure, with the heat removed by water. To desorb into the high-pressure ducting, heat is added at about 373 K from a solar or geothermal source. After regeneration, the high-pressure gas is further heated with high-grade energy from a source at about 973 K. The turbine then drives only the electrical load. It is anticipated that about

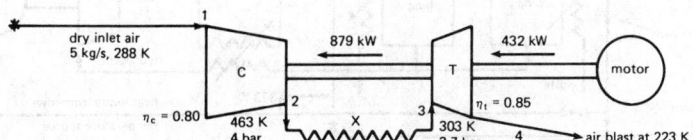

Fig. 72.18 Air-cycle refrigeration system for blast freezing (simple).[30]

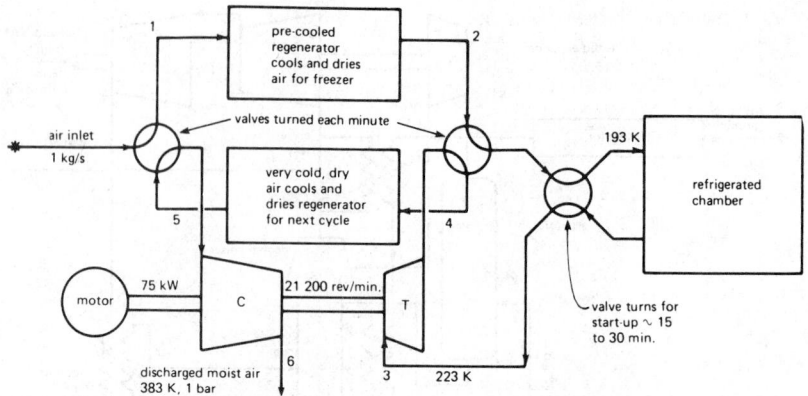

Fig. 72.19 Air-cycle refrigeration system (practical).[30]

90% of the high-grade energy is recovered as electricity, while about three times as much low-grade heat is also required. The x in $FeTiH_x$ varies during the cycle from about 0.2 to 0.6.

72.3 SELECTION

72.3.1 Diesel Engine, Steam Turbine Plant, or Gas Turbine?

This question has already been answered for many industries, each of these power plants having the clear balance of advantage in certain applications. The gas turbine has found its niche in aircraft and naval ship propulsion, many oil industry applications, intermittent electricity generation, and total energy systems (Section 72.2). To some extent, the nature of the power plant selects it for a given duty; for instance, diesel engines to drive reciprocating pumps and compressors; gas turbines to drive high-speed machinery such as centrifugal or axial flow pumps or compressors; and steam plants to provide process steam from the turbine exhaust. Not all are so distinct: in some applications the diesel engine or steam turbine may compete strongly with the gas turbine: the choice then depends on several factors, some of which are listed in very general terms in Table 72.2. Several items listed may be contentious and merit further discussion.

The *capital cost* data are very approximate, and apply to engine/generator sets in 1978/79. For gas turbines, the price depends very much on the maker or supplier, the country of origin, the size range, the duty or rating, and the point of delivery: the capital cost data are FOB, while the installed prices are for a turnkey contract, ready to switch on and start. Combined cycle plants cost about $ US 150–300/kW FOB.

The *installation* requirements depend on the power range. Smaller diesel and gas turbine generator sets are often preassembled as a transportable package. The gas turbine sets in particular may then just rest conveniently on the ground, or be attached by three-point mounting within the plant or on the roof. Very large gas turbine units tend to be installed as separate units, individually mounted on the floor and connected by ducting as a permanent installation.

The *efficiency* figures are not strictly comparable between the three power plants. The diesel engine figure applies to most engines and operating conditions. The gas turbine values cover a wide range of makes and models: the highest values apply to very-high-pressure-ratio engines or those with regeneration. A normal operating efficiency is about 0.25–0.3, and this will fall at part load. The steam plant

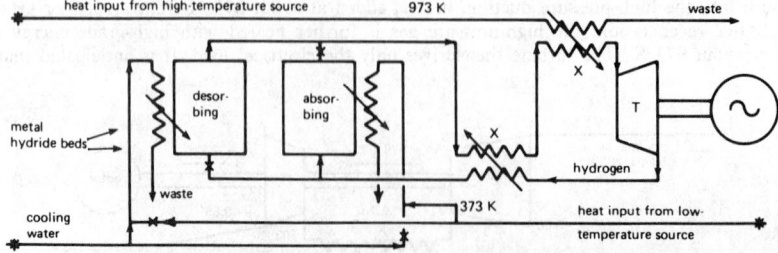

Fig. 72.20 Gas turbine with absorption/desorption compressor.[30]

Table 72.2 Comparison between Different Types of Power Plants[30]

Parameter	Diesel engine	Gas turbine	Steam turbine plant
Usual power range	10 kW to 10 MW	100 kW to 100 MW	1 MW to 1000 MW
Capital cost (approx)	$U.S. 110–180/kW	$U.S. 80–150/kW (aero) –280/kW (heavy)	$U.S. 140–240/kW (turbine only)
Installed cost (approx)		$U.S. 150–300/kW	$U.S. 800–1200/kW
Size and weight/kW	Medium	Small	Large
Specific Installation requirements	Heavy foundations Vibration isolation Quietening enclosure	Intake air filters Intake and exhaust mufflers	Ash handling facility Copious supplies of water Exhaust de-pollution
Plant life (typical) Inspection/overhaul periods	10 to 20 years Annual	15 to 30 years Several years	25 to 35 years Annual
Efficiency (typical) η factor at half load	0.35 to 0.40 1.00	0.20 to 0.40 0.8 to 0.85	0.30 to 0.40 0.91 to 0.96
Fuel grade and type	High Liquid (diesel grade)	High to medium Liquid or gas (see section 2.7)	Medium to low Gas, liquid or solid
Staffing requirements Lubrication needs Maintenance periods	A few Frequent oil and filter changes Weekly and monthly	None (extra to plant) Oil topping up Monthly and annual	Many Negligible Daily

works at only slightly better efficiency: 0.35–0.38 is the limit set by the typical pressure of 170 bar (2500 lb/in.²) and temperature of 810 K (1000°F).[27] In practice this only applies to very large plants, which operate long term as low as 0.3. Smaller plants have lower peak efficiency, even below 0.3. Combined-cycle installation is not tabulated here, but its efficiency reaches toward 0.5 peak, while total energy systems lie between 0.6 and 0.85.

No mention is made of *variable costs*. These may be inferred from the efficiency, fuel cost, extra staff, maintenance, and life data. The *staffing* requirement for the gas turbine is usually so low on a day-to-day basis that, if it is associated with other plants or a factory, no extra personnel are required. The main feature of a steam plant which offsets its high capital cost is the low cost of its relatively low-grade fuel, but it is suggested[6] that, when environmentalists demand the removal of smoke and sulfur dioxide from its exhaust gas to make it as clean as the gas turbine, there is little operating cost difference (per kWhr) between the two types. The maintenance costs for a gas turbine depend on many factors, which will be discussed separately in Section 72.3.2: a typical low cost may be $15 per kW year.

The *maintenance tasks* for diesel and steam plants are well known. Some of the tasks to be expected with gas turbines are cleaning of the inlet filter (Section 72.3.4) and compressor to maintain as new performance at the governed turbine entry temperature limit, and monthly topping-up of the oil system. The oil in many engines is never changed, and the filter life is long. The control system requires little servicing unless operational problems are encountered. The turbine blades may be inspected and a few replaced after several years, or they may all be recoated for another life. After such an overhaul, the performance should be restored to its original level and no running-in period is required.

72.3.2 Gas Turbine Type and Duty

This choice is influenced by numerous factors. The competitive engines which could perform the required function (Section 72.3.3), and are of comparable capital cost, may offer substantially different operating costs in terms of fuel usage and maintenance. Table 72.3 summarizes the parameters which affect the long-term maintenance and repair costs: some of these are discussed below.

Engine Rating and Life

The main influence on maintenance costs and engine life is the condition of the turbine blades, which is strongly dependent on the engine rating. The rating basically refers to turbine entry temperature, but may also involve rotational speed. These affect the blade creep life: typically the blade creep life is halved for every 20°C increase in temperature near its rated operating point. An engine rated for severe short-term operation may have a turbine blade life of only 2000 or 3000 hr (Section 72.2.3).

**Table 72.3   Typical Maintenance Cost Factors—Currency
Units/kWhr[30]**

Parameter	Operating Condition	Factor[1]
Rating	Base load	1.0
	Mid-range	1.5
	Peak load	2.5
	Reserve peak	5.0
Fired hours per start[2]	1000	1.0
	100	1.25
	10	2.0
	1	3.0
Fuel	Dry clean gas	1.0
	Class A distillate oil	1.5
	Treated heavy oil[3]	3.0
Installation space	Large building	1.0
	Close fitting enclosure	1.2
Experience at rating offered	Proven	1.0
	Limited development	2.0
	Prototype	4+
Design philosophy[4]	Heavy industrial	1.0
	Medium industrial	2.0
	Aero-derivative	2.5
Standard of operation and maintenance	Good	1.0
	Average	2.0
	Poor	4+

1 Multiply all factors.
2 Double for fast start.
3 Low sodium and vanadium.
4 Factors reduce with experience.
(Adapted from reference 4D by courtesy of the Council of the Institution
of Mechanical Engineers).

The turbine life is also strongly dependent on *starting time*. Although the gas turbine is nominally a quick starting engine, the full use of this capability reduces its life and, in peak lopping duty, will introduce thermal fatigue as the limiting factor. *Fuel quality* also has a marked effect, even a relatively clean, heavy oil being more damaging to the turbine than gas. Poorer oil and gaseous fuels are several times more destructive, and basically require the use of relatively low turbine entry temperatures with reduced efficiency and increased capital cost per kW. The use of external combustion for a closed-cycle engine permits the use of cheaper fuels. Operating experience in connection with different fuels is given in Refs. 14 and 28.

Aeroderivative or Industrial Origin?

Table 72.3 may be initially misleading, in that the aeroderivative engine has an extensive development background from which it is de-rated for gas generation duty in industrial use, while the industrial engine and power turbine often have little development. Internal technicalities, such as the number of shafts or turbine blade cooling, have little effect: for instance, the gas temperature in an engine with blade cooling is raised to bring the metal temperature up to that of the uncooled blade. The aeroderivative engine is likely to cost a little less to buy, but more to maintain. The aeroderivative engine is likely to be mechanically safe because of its passenger-carrying background, while the heavy industrial engine is safe by conservative design. The former may have an overhaul life of 10,000–40,000 hr, the latter 25,000–100,000 hr. The aeroderivative engine's higher efficiency may be matched by regeneration in the industrial engine.

The main functional difference is in the *repair procedure*. In the event of an internal failure, or scheduled internal inspection, the aeroderivative engine is usually removed and shipped to the maker or an approved overhaul facility. This is often necessitated by the special tooling, the delicate components, and critical settings during reassembly. A replacement engine may be installed within a few hours, either from the user's spares or as a rental unit. Plant downtime is minimized but the cost may be high.

The industrial engine is usually subject to *on-site repair*, but is likely to need such attention less often than the aeroderivative. The features and construction permit handling by normal maintenance personnel. A half casing may be lifted for inspection, or a component such as the turbine may be removed for reblading. The time involved may be less than a day, or several weeks, which may involve plant downtime. Smaller industrial gas generators may also be exchanged if necessary, and returned to the factory for repair. The remoteness of the site, personnel accommodation, cost of downtime, and standby arrangements may well determine the acceptable repair procedure and, therefore, the engine origin. This is discussed further in Ref. 3.

Also of great importance to this choice is the *reputation* of the makers and engines on the short list. Reputations are variable among engines from each origin, and the outstanding reputation of one of the engines offered may determine the choice above all other considerations.

Finally, these two origins may be further subdivided. The aeroderivative engine may stem from a *civil* or *military* background. The former may be more reliable, with fewer random failures, because long life and passenger safety are more important to the civil role. The industrial engine may be of *heavy* or *medium* design, the former probably being longer lived and more reliable, costing more to build and install but less to maintain, and having a less efficient operating cycle.

Sources of Information

It may be a problem for the newcomer to determine the makers' names, let alone their reputation. There are several useful journals, of which some are (alphabetically).

Diesel and Gas Turbine Progress

Gas Turbine World

International Power Generation

Offshore

Oil and Gas Journal

Turbomachinery International

These journals have up-to-date information and advertisements, and some issue an annual catalogue which lists the makers and their products. *Gas Turbine World* regularly lists competitive bids and orders, which indicate the range of current prices, as in Table 72.2. Some of these cost data were also obtained from Ref. 27.

The makers' and engines' reputations may be gauged from other users, whose names the makers should be proud to supply, and at meetings and conferences such as the Annual International Gas Turbine Conference. There is also the Diesel Engineers' and Users' Association (London), which publishes information on operation and costs. For a serious study, and any subsequent engineering or contract work, there are many engineering consultants who specialize in gas turbine work, and who are listed in some of the catalogues.

It has been assumed above that engine comparisons relate to the same *power level*. Until recently, powers have been quoted on different bases, but ISO 3977 rationalizes ratings into classes of the operating hours per year, and into ranges of the number of starts required. These are listed in Table 72.4 and two common examples are included. The performance ratings are determined by acceptance test procedures laid down in ISO 2314.

ISO 3977, *Gas Turbines—Procurement*, is also a very useful source of general information on definitions, fuels, control, and environmental pollution. It is of assistance with the contract side, specifying the technical information to be supplied by both the purchaser when inquiring and the manufacturer when tendering. Many countries issue their own version of it.

72.3.3 Planning a Gas Turbine Installation

The first requirement is to satisfy the installation's *functional* and *performance* needs. The relevant parameters to be chosen or specified may include power (or thrust), shaft speed, fuel consumption and fuel type, air mass flow, and exhaust gas temperature: the relative importance of these may differ

Table 72.4   Standard Ratings Specified in ISO 3977–1978 (E)[30]

Class	Annual running hours	Name
A	up to 500	Reserve peak
B	up to 2000	Peak-load
C	up to 6000	Mid-range
D	up to 8760	Base-load

Range	Annual average number of starts
I	over 500
II	up to 500
III	up to 100
IV	up to 25
V	Continuous operation without planned shut-down for inspection and/or maintenance within a specified period

Examples
ISO standard peak-load rating (2000 h, 500 starts) Class B: Range II.
ISO standard base-load rating (8760 h, 25 starts) Class D: Range IV.

widely between applications. The parameters may be determined by preliminary cycle calculations. It is advisable to err on the conservative side as regards losses and to allow a generous performance margin for plant tuning or uprating. Excessive generosity may reduce the plant efficiency, but considerably increases the engine life and reduces maintenance costs. A margin which is too small and is lost, so that the engine finally proves too small, is a far worse penalty.

The nominal installation power may range from considerable excess of *standby capacity* to zero. Adequate standby could avoid the loss of production time involved in engine repair. Where personal safety is concerned, such as on offshore drilling platforms, the standby could be 100%, involving the use of two engines at half power where one could suffice. Lower standby levels, which may keep a plant at full production, may be obtained by using several engines as gas generators, all running to supply the power turbine. If one should fail, or be shut down for scheduled maintenance, the others may be raised from part load to base-load level, or perhaps from continuous base-load to a higher, peak rating.

Where the standby capacity is inadequate to maintain full operation with one engine out, the concept of *availability* is used. It also allows for the loss of more than one engine; it is defined in several ways, a typical one being

$$\text{availability} = \frac{\text{period time} - \text{planned and unplanned downtime}}{\text{period time (time when operation is required)}}$$

Typical availability for a gas turbine plant ranges from 0.9 to 0.99. Analysis in Ref. 29 showed that, with the likely failure rate of the engines, planning for 0.97 availability would result in 1% loss of gas transmission delivery, while 0.91 availability would lose 7%. It would be worthwhile to run an engine at a very high rating to avoid such a loss. Such analysis also shows the importance of being able to plan the downtime, and avoid unexpected problems.

Plants designed for *cogeneration* typically extract about 10% less of the exhaust gas energy than could feasibly be extracted. This permits a reduction of cost in the steam plant section of about 40%, and forms a reasonable compromise.

There are many further considerations, including contracts and finance, which are beyond the scope of this chapter. The engineering analysis should, however, be balanced by liaison with the local *power authority*. Some plants generate surplus electricity, for which the authority may pay a reasonable rate and which may help it at peak times. A plant which generates most of its power needs in-house, but which may call upon the local authority for a large emergency supply in the event of an engine failure, may find it pays a heavy penal rate. This again emphasizes the desirability of spare capacity.

72.3.4   Ancillary Requirements

A range of ancillary equipment is available, and some items are essential to ensure a satisfactory plant operation and life. Engine manufacturers can supply or arrange for these units, and some supply them as plug-in modules, which are shipped separately as complete systems. Examples may include the air filters, exhaust silencers, fuel treatment plant, starting system, control room, and oil-cooling plant.

Air filters are used for silencing, for atmospheric dirt and sand, and for airborne salt. Salt is important in marine environments but may also be present in air hundreds of kilometers from the sea,[3] and is more serious in hotter engine types. Inertial separators remove the heavier particles, but an oil-wetted and washed nylon mesh coalescer is often used to remove sand and salt after an inertial stage. Paper and cloth bag filters are also used, but need considerable area for the high mass flow. Filter systems may use several stages of different types to ensure safety from the corrosive elements, which are only permitted in minute quantities. Some filters are self-cleaning but may cost more, offsetting the saving in maintenance.

Fuel treatment includes filtering and temperature control for liquid fuels and gas. Wet gas, with drops or larger slugs of liquid fuel, can result in turbine burnout from a sudden, excessive rate of heat release. Traps, demisters, and filters are essential to avoid slugs and to reduce droplets to a very low, finely dispersed level. Heating the gas to 20°C above its dew point may prevent condensation. Other aspects of fuel control, and the control system design itself, also need careful planning.

Oil cooling is likely to be the only need for cooling water, unless compressor intercooling is used. Only small quantities are required,[27] and air cooling may be used if the environment is cool.

The location of working areas for personnel, and the *control room*, should be out of the area where the very rare case of a major shaft, disk, or casing failure could result in injury.

APPENDIX   CHEMICAL AND THERMODYNAMIC PROPERTIES OF WORKING FLUIDS AND FUELS

The data in Tables 72.5–72.8 are adapted and calculated from the following sources (cited in the tables by superscripts *a–g*):

[a] R. C. Weast (Ed.), *Handbook of Chemistry and Physics,* 56th ed., CRC Press, 1975–1976.

[b] J. W. Rose and J. R. Cooper, *Technical Data on Fuel,* British National Committee, World Energy Conference, 7th ed., 1977.

[c] K. K. Kelly, U.S. Bureau of Mines, Bulletin 584, 1960.

[d] H. M. Spencer, *Ind. Engng Chem.,* **40** 2152–2154 (1948).

[e] G. J. Van Wylen and R. E. Sonntag, *Fundamentals of Classical Thermodynamics,* 2nd ed., Wiley, New York, 1976.

[f] Mobil Technical Bulletin, *Heavy Fuel Oil.*

[g] I. Gilmour, private communication.

Table 72.5 Properties of Typical Working Fluids and Process Gases[30]

	Chemical formula	Molecular[a] weight M.W.	Boiling[a] point @ 1 atm (K)	Density[a,b] @ISA (kg/m³)	Gas const R(=pv/T) 8314.3/M.W.	(Cp/Cv) @ISA γ	Specific heat capacity[c,d] $C_p = a + bT + cT^n$ where T = temperature/100 (J/kg K)				Dynamic[a] viscosity @ISA, μ (cp × 10³)	Thermal[a] conductivity @ISA (J/s m K)
							a	b	c	n		
Acetylene	C_2H_2	26.04	189	1.1019	319.3	1.112	2188	376.7	−11.6	+2	9.90	0.0200
Air	–	28.964	83	1.2256	287.1	1.399	947	21.3	−0.31	+2	18.00	0.0252
Ammonia	NH_3	17.03	240	0.7308	488.2	1.310	1746	147.4	−909	−2	9.65	0.0233
Argon	A	39.93	88	1.6908	208.2	1.666	521	0	0	–	21.87	0.0173
n-Butane	C_4H_{10}	58.17	273	2.4580	143.0	1.094	277	527.9	−16.3	+2	–	0.0149
Carbon dioxide	CO_2	44.01	195	1.8761	189.0	1.296	1005	20.0	−1959	−2	14.57	0.0157
Carbon disulphide	CS_2	76.13	319	–	109.2	1.227	684	8.8	−989	−2	9.67	0.0074
Carbon monoxide	CO	28.01	82	1.1853	296.9	1.400	1015	14.6	−164	−2	17.20	0.0242
Chlorine	Cl_2	70.91	236	3.0466	117.2	1.326	522	0.9	−401.2	−2	13.07	0.0083
Ethane	C_2H_6	30.09	184	1.2720	276.4	1.192	313	531.5	−15.4	+2	8.94	0.0201
Ethylene	C_2H_4	28.07	169	1.1860	296.3	1.238	422	426.6	−13.0	+2	9.60	0.0194
Helium	He	4.00	4	0.1692	2078.6	1.666	5199	0	0	–	19.20	0.1473
Hydrogen	H_2	2.016	20	0.0853	4124.1	1.405	13532	161.9	+2491	−2	8.65	0.1816
Hydrogen bromide	HBr	80.92	206	3.4241	102.7	1.401	331	6.4	+77.6	−2	17.95	0.0083
Hydrogen chloride	HCl	36.46	188	1.5428	228.0	1.397	720	14.2	+344.3	−2	13.98	0.0135
Hydrogen cyanide	HCN	27.03	299	–	307.6	1.307	1457	41.8	−2229	−2	–	0.0117
Hydrogen sulphide	H_2S	34.08	212	1.4410	244.0	1.325	959	36.3	−565	−2	12.33	0.0140
Methane	CH_4	16.05	109	0.6785	518.2	1.318	882	470.7	−11.2	+2	10.71	0.0327
Neon	Ne	20.18	27	0.8535	411.8	1.666	1030	0	0	–	30.76	0.0472
Nitric oxide	NO	30.01	120	1.2704	277.1	1.387	980	12.8	−195	−2	18.50	0.0250
Nitrogen	N_2	28.02	77	1.1854	296.7	1.401	1020	13.4	−179	−2	17.25	0.0252
Nitrogen dioxide	NO_2	46.01	294	1.9469	180.7	1.295	916	20.7	−1519	−2	14.25	0.0164
Nitrous oxide	N_2O	44.01	185	1.8748	188.9	1.281	1038	19.6	−1939	−2	19.90	0.0258
Oxygen	O_2	32.00	90	1.3530	259.8	1.399	936	13.1	−523	−2	7.90	0.0166
Propane	C_3H_8	44.14	231	1.8650	188.4	1.129	229	542.5	−16.6	+2	12.10	0.0166
Steam	H_2O	18.03	373	0.7620	461.1	–	1695	57.1	0	–	–	–
Sulphur dioxide	SO_2	64.06	263	2.7107	129.8	1.269	721	12.3	−1202	−2	12.10	–

Table 72.6 The Properties of Some Gaseous Fuels[30]

	Chemical formula	Molecular weight	Density @ ISA (kg/m^3)	Lower calorific value @ 288 K		
				(MJ/kg)	(MJ/m^3)[b]	(Btu/scf)
n-Butane	C_4H_{10}	58.17	2.458	45.71	112.41	2917
Carbon monoxide	CO	28.01	1.185	10.10	11.97	311
Ethane	C_2H_6	30.09	1.272	47.48	60.43	1584
Hydrogen	H_2	2.016	0.085	119.95	10.22	265
Methane	CH_4	16.04	0.678	50.01	33.95	881
Propane	C_3H_8	44.14	1.865	46.35	86.42	2290

Table 72.7 The Properties of Some Liquid Fuels[30]

Pure liquids	Chemical formula	Molecular weight M.W.	Boiling[b] point @ 1 atm (K)	Density[b] @ 228 K (kg/litre)	Lower (net) calorific value @ 298 K	
					(MJ/kg)[b, e]	(Btu/lb)
Benzene	C_6H_6	78.11	353	0.879	40.14	17255
n-Butane (L.P.G.)	C_4H_{10}	58.17	273	0.575	45.34	19490
Ethanol	CH_5OH	34.06	352	0.794	27.23	17705
n-Heptane	C_7H_{16}	100.21	371	0.684	44.56	19155
Methanol	CH_3OH	32.04	338	0.796	19.94	8570
n-Octane	C_8H_{18}	114.23	399	0.703	44.43	19100
Propane (L.P.G.)	C_3H_8	44.14	231	0.505	45.98	19765

Miscellaneous fuels	Density ρ @ 288 K (kg/litre)	Boiling range (K)	Kinematic viscosity @ temperature		Lower (net) calorific value @ 288 K	
			(cSt)	(K)	(MJ/kg)	(Btu/lb)
Diesel[b]	0.84	453 to 613	5.0	288	42.9	18440
Fuel oil[b, f]	0.80 to 0.97	453 to ?	1.4 to 70	355	$53.1-12.3\rho$	—
Gasoline[b]	0.72 to 0.76	301 to 473	0.5 to 0.65	288	41.9 to 44.0	18010 18915
Kerosine [36]	0.73 to 0.83	—	2.1 to 3.0	288	$51.7-10.5\rho$	—
Sulphur [23]	1.789 (410 K)	718	4.58	410	9.20	3955
Tar (coke oven)[b]	1.138 to 1.18	372 to 645+	—	—	36.0 to 37.7	15475 16205

Table 72.8 The Properties of Some Solid Fuels[30]

	Density[b] (no pores) (kg/m^3)	Bulk density[b] (with pores) (kg/m^3)	Higher (gross) calorific value*	
			(MJ/kg)	(Btu/lb)
Carbon (to CO)	1800 to 2300	—	9.20[e]	3955
(to CO_2)	—	—	32.76[e]	14080
Charcoal	1400 to 1700	—	34.30[e]	14745
Coal	1250 to 1700	580 to 830	32 to 36[b]	{ 13755 15475
Coke	1750 to 2000	300 to 510	33.60[b]	14445
Peat	1150 to 1250	—	19.70[b]	8470
Pine bark	—	220 to 320	19.77[g]	8500
Wood	500 to 1100	—	17.30[b]	7435

* These calorific values apply to the combustible content only. They may be degraded by content of ash and water approximately in proportion to their fraction of the weight.

REFERENCES

1. A. Meyer, "The Combustion Gas Turbine, Its History and Development," *Proc. Instn. Mech. Engrs.*, **141**, 197–222 (1939).

2. F. Whittle, "The Early History of the Whittle Jet Propulsion Gas Turbine," *Proc. Instn. Mech. Engrs.*, **152**, 419–435 (1945).

3. *Gas Turbines—Status and Prospects,* Institute of Mechanical Engineers Conference Publications C.P. 1976–1.

4. R. N. Penny, "Regenerators for High Temperature Gas Turbine Engines," in "Technical Advances in Gas Turbine Design," *Proc. Instn. Mech. Engrs.*, **183** (part 3N) (1969).

5. E. M. Eltis and G. L. Wilde, "The Rolls-Royce RB211 Turbofan Engine," *Proc. Instn. Mech. Engrs.*, **188**, 549–575 (1974).

6. H. C. Hottel and J. B. Howard, *New Energy Technology—Some Facts and Assessments,* MIT Press, Cambridge, MA, 1971.

7. D. C. Whittley, "The Augmentor Wing: Powered Life STOL a Proven Concept, *Interavia,* **2**, 143–145 (1974).

8. J. P. O'Brien, Ed., *Gas Turbines for Automotive Use,* Noyes Data Corporation, Park Ridge, NJ, 1980.

9. H. Czermak and K. Goebel, "Operating Experience with a Combined Cycle Gas Turbine and Steam Turbine Plant at the Hohe Wand Power Station of the Niederösterreichische Electrizitäts-werke Aktiengesellschaft," *Modern Steam Plant Practice,* Institution of Mechanical Engineers, London, 1971, pp. 107–113.

10. R. Coats, "The Efficient Use of Gas Turbines for Production," *Chart. Mech. Engr.,* **25**, 51–58 (1978).

11. H. J. Perkins and W. J. R. Stocks, "Gas Turbine Total Energy—The Key to High Thermal Utilization, *Proceedings of the Conference on Energy Recovery in Process Plants,* CP2–1975, Institution of Mechanical Engineers, London, pp. 81–88, 1975.

12. A. Meyer, "The Velox Steam Generator," *Trans. Am. Soc. Mech. Engrs.,* **57**, 469–478 (1935).

13. I. Stambler, "Strict NO_x Codes Call for Advanced Control Technology," *Gas Turbine World,* 57–60 (Sept.–Oct., 1983).

14. *Joint Conference on Combustion,* Proc. Am. Soc. Mech. Engrs. (Boston, June 1955) and Instn. Mech. Engrs. (London, Oct. 1955).

15. R. W. Foster-Pegg and J. S. Davis, "New Coal-Fired Air Turbine Cogeneration Design Ready for Marketing," *Modern Power Systems* (May, 1983).

16. A. R. Eves, *Gas Turbine Cogeneration and Combined Cycle Plant for Enhanced Fuel Efficiency,* GEC Gas Turbines Ltd., Technical Publication No. 151683, June 1983.

17. R. A. Harmon, C. A. Kinney, and W. M. Crim, Jr., "Cogeneration in Europe and the Combined Cycle Gas Turbine," *Turbomachinery International,* **20**, 29–34 (March 1979).

18. K. Bammert, *A General Review of Closed Cycle Gas Turbines Using Fossil, Nuclear and Solar Energy,* Verlag Karl Thiemig, Munich, 1975.

19. W. L. Nelson, *Petroleum Refinery Engineering,* 2nd ed., McGraw-Hill, New York, 1941.

20. "Nitric Acid (SABAR) Process," *Hydrocarbon Processing,* **54**, 164 (November, 1975).

21. R. T. C. Harman and A. G. Williamson, "Gas Turbine Topping for Increased Energy Recovery in Sulphuric Acid Manufacture," *Applied Energy,* **3**, 23–40 (January, 1977).

22. A. R. Valdes, "Use Expander Cycles for LPG Recovery," *Hydrocarbon Processing,* **53**, 89–91 (December, 1974).

23. M. Bergman (ed.), "Storage in Excavated Rock Caverns," *Proceedings of the First International Symposium in Stockholm, Rockstore 77,* Pergamon, Oxford, 1977.

24. *Turborefrigerating Machine,* TXM-300, Techmashexport, Moscow.

25. W. F. Rush, S. A. Weil, R. J. Dufour, and M. J. Lassota, "Brayton Refrigeration Systems," in *An Assessment of Selected Heat Pump Systems,* Project HC-4-20, American Gas Association, pp. 67–137.

26. J. R. Powell, F. J. Solzano, W. S. Yu, and J. Milan, *A Closed Brayton Cycle Using Hydrogen as a Work Fluid,* Brookhaven National Laboratory, Upton, New York.

27. A. B. Shearer, "On-Site Generation: Turbine Generators," in series *Power Supply Basics, International Power Generation,* **1**(2), 23–26 (December/January 1977–78).

28. W. P. Auer, "Operating Experience of Installed Gas Turbines," in "Operating Experience of High Duty Prime Movers," *Proc. Instn. Mech. Engrs.,* **178** (Part 3K), 100–115 (1963–64).

29. T. C. Heard and R. P. Lang, "The Concept of Availability for Gas Turbine Application Evaluation," *J. Engng. Pwr.,* **100,** 452–456 (July 1978).

30. R. T. C. Harman, *Gas Turbine Engineering,* Macmillan Press, Great Britain, 1981.

CHAPTER 73
STEAM TURBINES

GEORGE SILVESTRI, JR.

Westinghouse Electric Corporation
Orlando, Florida

The basic function of a steam turbine is to convert the stored thermal energy of steam into mechanical work. The stored thermal energy its first converted into kinetic energy or velocity by expanding the steam from a higher pressure to a lower pressure. The generation of the steam jet and its conversion into work occurs in a combination of stationary passages called nozzles, stationary vanes, or stationary blades and rotating passages called rotating blades or buckets. The combination is called a stage. The passages are contoured to produce both the desired velocity and direction of the jet. Shaft work is produced by directing the steam jet from the nozzles against the curved rotating blades, which are mounted on the rotating rotor and from the reaction of the jet as it leaves the rotating blades.

Figure 73.1 illustrates typical shapes for the stationary and rotating rows of impulse and reaction stages. In the impulse stage the work is produced by the change in direction of the velocity in the rotating row, while in a reaction stage, work is produced by a change in both the magnitude and direction of the steam velocity in the rotating row.

The turbine can drive a variety of equipment such as electric generators, ship propellers, pumps, fans, compressors, and other variable speed devices. The variable speed capability can greatly enhance the efficiency of the driver–driven combinations in sizes up to 50 MW. Modern central station electric power plants have unit capacities up to 1300 MW.

73.1 STEAM TURBINE CLASSIFICATIONS

There are a number of classifications of steam turbines, some of which are identified in the subsequent subsections.

73.1.1 Stage Design

The classical descriptions of turbines are impulse and reaction. Under the classical impulse design criteria all of the stage pressure drop occurs in the stationary blade. The classical reaction design takes 50% of the stage pressure drop in the rotating blade.

The impulse designs can be further divided into either velocity-compounded (Curtis) or pressure-compounded (Rateau) stages. With velocity compounding, the velocity resulting from the pressure drop in the nozzle is so high that the velocity is reduced in stages using two or more rows of rotating blades. Reversing vanes, which change the direction of the steam jet, are interspersed between the rotating blades. No pressure drop occurs in the rotating blades and reversing vanes. The original Curtis stage contained four to six rotating rows of blading. Only two rotating row Curtis stages are

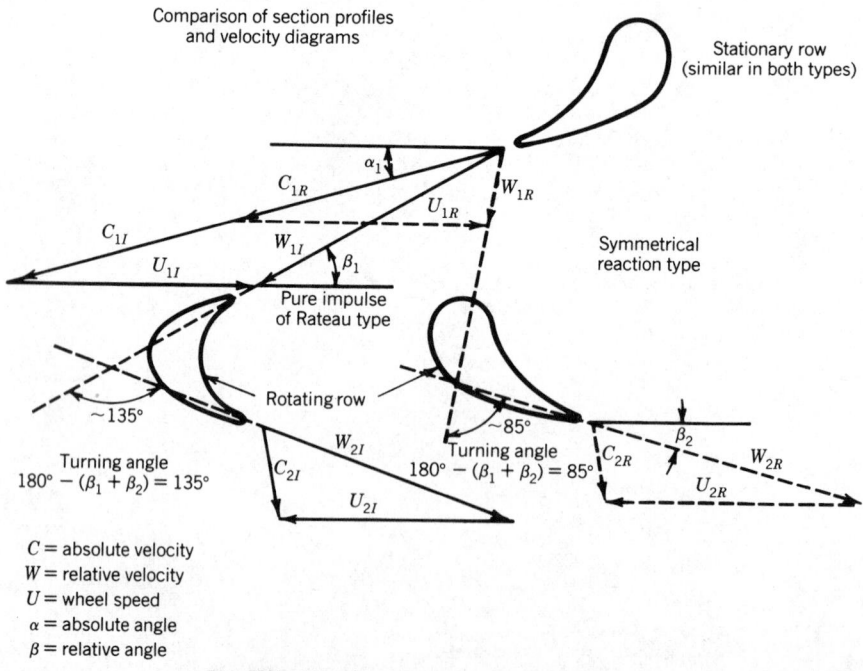

C = absolute velocity
W = relative velocity
U = wheel speed
α = absolute angle
β = relative angle

Fig. 73.1 Reaction-type versus impulse-type stage.

used in present day designs. Figure 73.2 is a representation of a classical Curtis stage. The symbols p, C, W, and U relate to pressure, absolute steam velocity, relative steam velocity, and rotating blade velocity, respectively.

In the case of pressure compounding, the pressure drop is taken in two or more Rateau stages, each of which consists of a row of nozzles and a row of rotating blades. The velocity resulting from the pressure drop in the nozzle is reduced in the rotating blade as shown in Fig. 73.3, similar to Fig. 73.2. Figure 73.4 is a representation of pressure compounding on a Mollier chart with two Rateau stages.

The Mollier chart is an enthalpy (h)–entropy (s) chart with lines of constant pressure and temperature. Figure 73.4 is a simplified representation, and the line joining the actual total state points (work done in each stage) is called the expansion or condition line. The slope of the line is a measure of the efficiency, being vertical for 100% and horizontal for 0% (in throttling, for example).

The available energy is the maximum amount of energy that is capable of being converted into velocity and then into shaft work when the steam is expanded from the inlet total pressure to the exit static pressure of the stage, Δh_s. The inlet total or stagnation conditions relate to the case where either the inlet velocity to the stage is zero (infinitely large plenum chamber) or the inlet velocity is converted into pressure isentropically (without any loss or heat transfer). The work Δh_w is less than Δh_s because of inefficiencies in the process. The work is the difference between the inlet and exit total enthalpies of the stage.

The steam leaving this stage and entering the next stage has velocity, carry-in, which is capable of producing work. This energy, in conjunction with the energy from expanding between the static inlet pressure and the static exit pressure, establishes the available energy of the second stage.

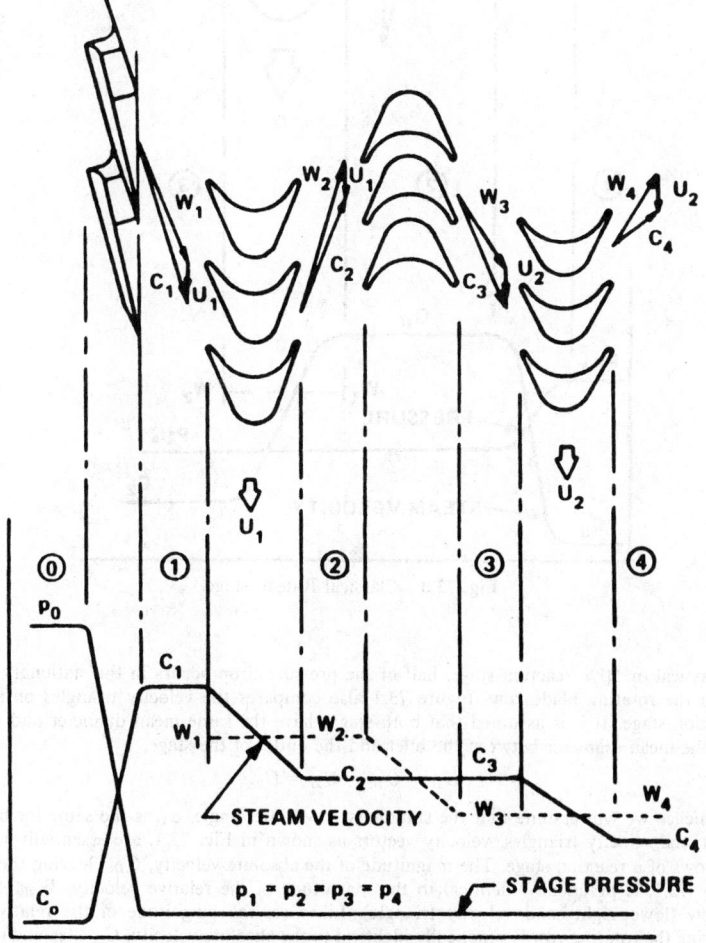

Fig. 73.2 Classical Curtis stage.

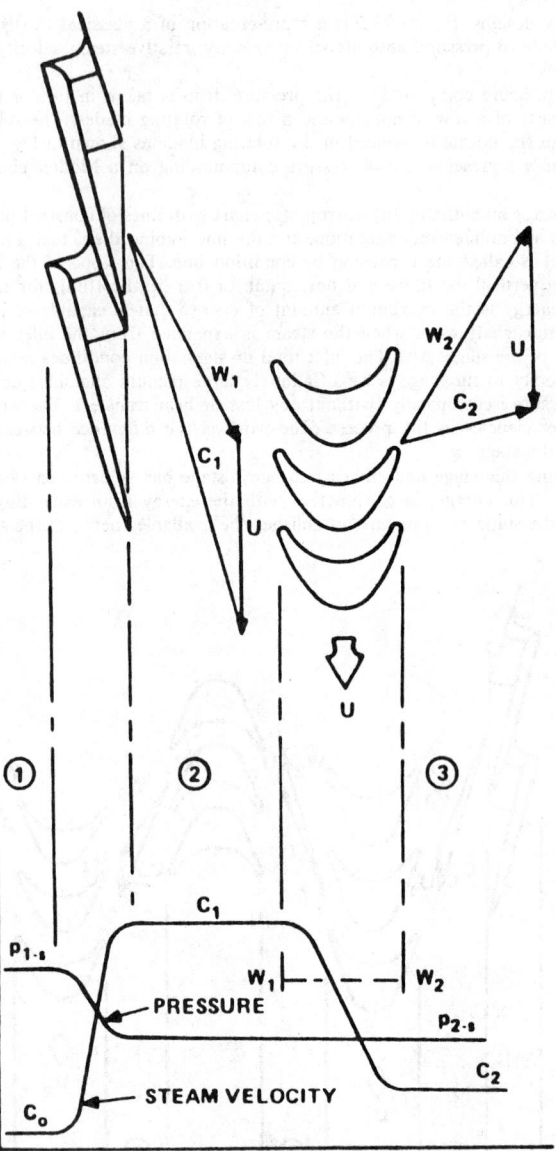

Fig. 73.3   Classical Rateau stage.

In a classical or 50% reaction stage, half of the pressure drop occurs in the stationary blade row and half in the rotating blade row. Figure 73.1 also compares the velocity triangles on an impulse and a reaction stage. If it is assumed that both stages have the same mean diameter and there is no change in the mean diameter between the inlet and the outlet of the stage,

$$U_{1I} = U_{1R} = U_{2I} = U_{2R}$$

For convenience we will assume that the stationary row outlet angle, α_1, is the same for both stages. As a result, the velocity triangles, velocity vectors as shown in Fig. 73.1, are essentially identical on the blade rows of a reaction stage. The magnitude of the absolute velocity, C_{1R}, leaving the stationary row of Fig. 73.1, is practically identical to the magnitude of the relative velocity, W_{2R}, leaving the rotating row (lower-right-hand velocity triangle). Likewise, the magnitude of the relative velocity, W_{1R}, entering the rotating row is practically identical to the absolute velocity, C_{2R}, leaving the rotating row. C_{2R} is also the absolute velocity entering the stationary row of the next reaction stage. W_{2R} is

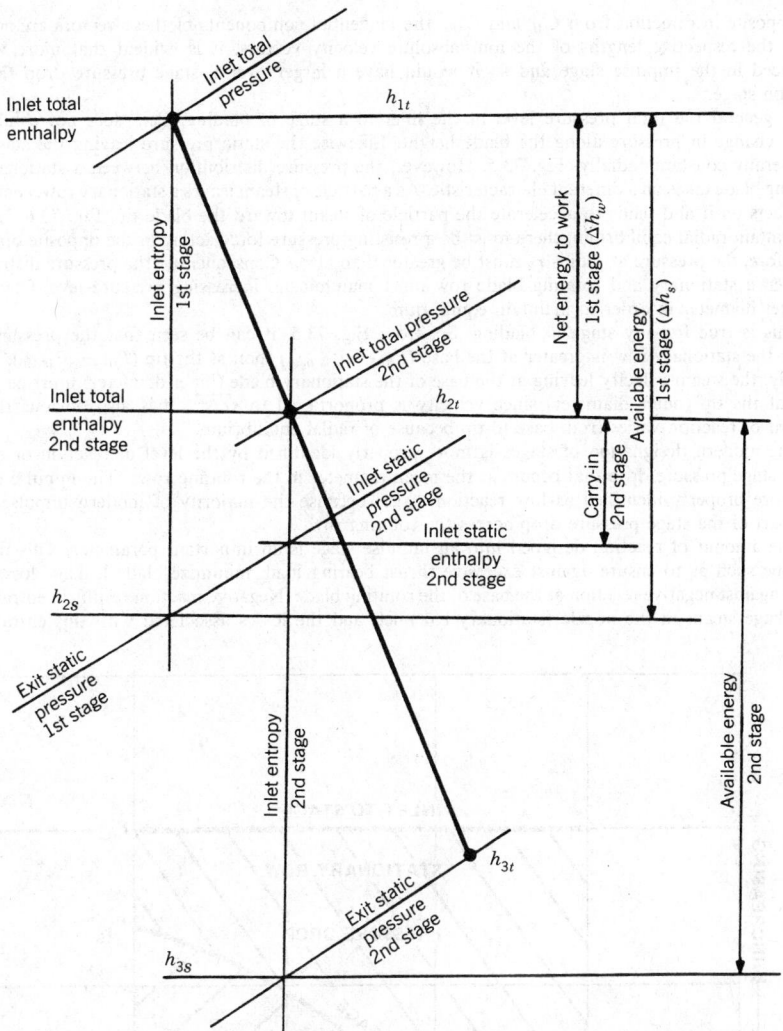

Fig. 73.4 Complete expansion lines for two Rateau stages in a group (pressure compounding).

greater than W_{1R} because of the pressure drop across the rotating row and its conversion into velocity. The length of C_{1R} is indicated by the intersection of the U_{1R} vector with the absolute velocity vector leaving the stationary row at α_1.

Note that the velocity triangle at inlet to the rotating row is similar to the velocity triangle leaving the rotating row. In such a situation the stage is said to be symmetric, and the rotating and stationary blades have identical shapes as shown by the rightmost blades in Fig. 73.1. In a symmetric stage α_1 equals β_2.

In the impulse stage the absolute velocity, C_{1I}, leaving the stationary row is larger than the absolute velocity leaving the stationary row of the reaction stage, C_{1R}, Fig. 73.1. Since the velocity of the rotating row, U_{1I}, is the same as U_{1R}, the relative inlet velocity of the rotating row, W_{1I}, is larger than W_{1R}. If there is no pressure drop across the rotating row, the relative outlet velocity, W_{2I}, is equal to W_{1I}. C_{2I} is the absolute velocity leaving the rotating blade of the impulse stage. Note that the velocity triangles at inlet and exit of the rotating blade of the impulse stage differ substantially from one another. This is reflected in the great difference in the blade shapes of the stationary and rotating rows.

The impulse stage turns the steam through a much greater angle, about 135°, than the reaction stage, about 85°. The work of a stage is equal to $U \times \Delta C_u$ where U is the wheel speed and C_u is the tangential component (plane of the wheel rotation) of the absolute velocity. Since C_{2I} and C_{2R}

are opposite in direction from C_{1I} and C_{1R}, the tangential components of these vectors are additive. From the respective lengths of the four absolute velocity vectors, it is evident that more work is produced in the impulse stage and so it would have a larger overall stage pressure drop than the reaction stage.

In general the total pressure level at the inlet to a stage of blading, P_{0t}, does not exhibit any radial change in pressure along the blade height. Likewise the static pressure leaving the stage, P_{2s}, is generally constant radially, Fig. 73.5. However, the pressure distribution between a stationary and rotating blade takes on a different characteristic. As a particle of steam leaves a stationary row, centrifugal force acts on it and tends to accelerate the particle of steam toward the blade tip, Fig. 73.6. In order to maintain radial equilibrium, there must be a resisting pressure force acting in the opposite direction. Therefore, the pressure at radius r_2 must be greater than at r_1. Consequently, the pressure distribution between a stationary and rotating blade row must maintain an increasing pressure level from inner to outer diameter in order to maintain equilibrium.

This is true for any stage of blading. So, from Fig. 73.5, it can be seen that the pressure drop across the stationary row is greater at the base ($P_{0t} - P_{1s\ base}$) than at the tip ($P_{0t} - P_{1s\ tip}$). Consequently, the steam velocity leaving at the base of the stationary blade (inner diameter) must be greater than at the tip (outer diameter) since velocity is proportional to $\sqrt{\Delta p}$. It is also obvious that the amount of reaction varies from base to tip because of radial equilibrium.

The modern designation of stages is more properly identified by the level of reaction or portion of the stage pressure drop that occurs at the mean diameter in the rotating rows. The impulse designs are more properly identified as low reaction stages because the majority of modern impulse stages take part of the stage pressure drop across the rotating row.

The amount of reaction designed into an impulse stage is an important parameter. This reaction must be such as to ensure against excessive thrust bearing load, minimize blade leakage losses, and guard against negative reaction at the base of the rotating blade. Negative reaction results in entrainment of leakage steam in the nozzle (stationary row) jet, and the losses associated with this entrainment

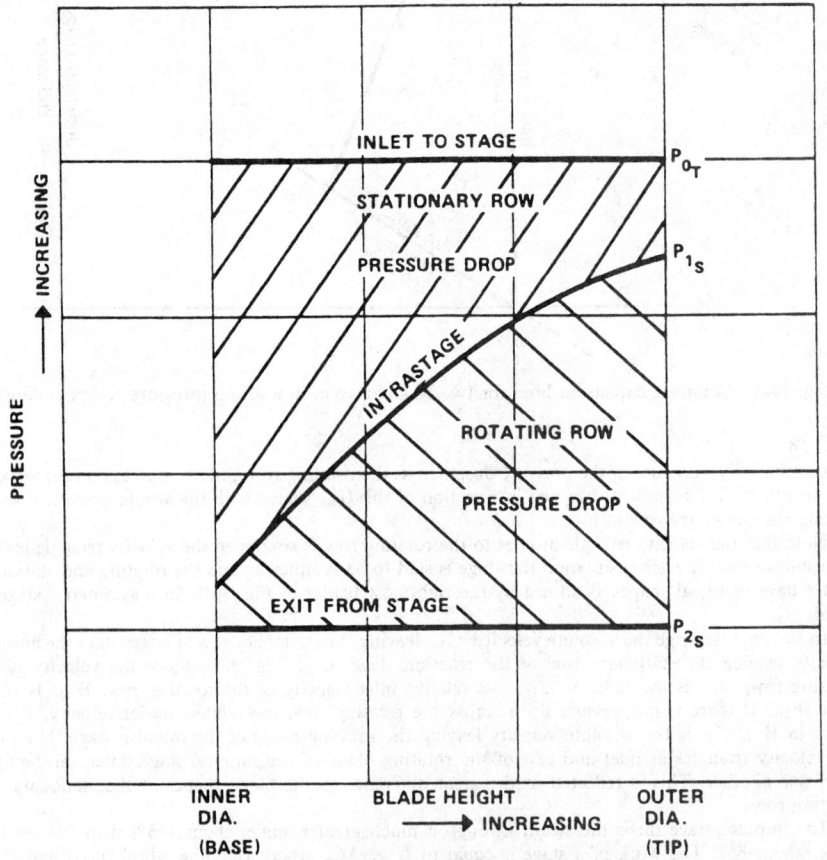

Fig. 73.5 Pressure distribution of a turbine stage.

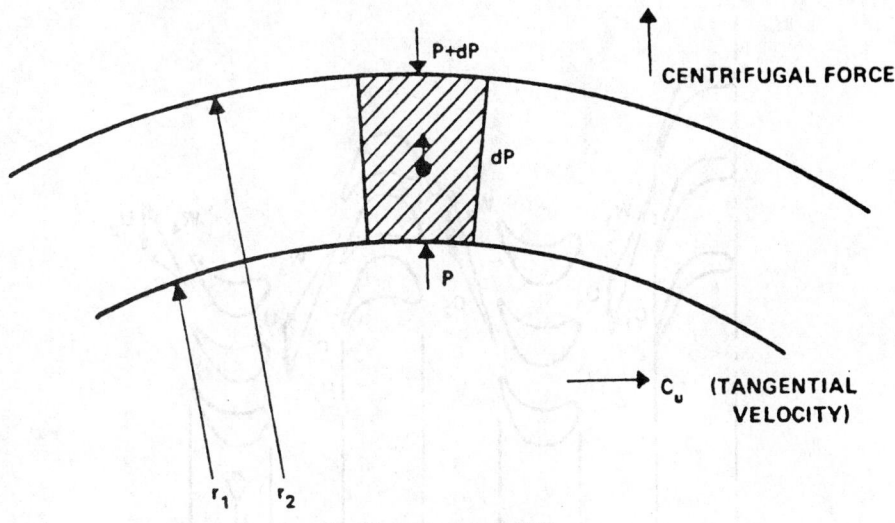

Fig. 73.6 Radial equilibrium.

are much greater than the losses associated with leakage steam bypassing the rotating blade row. Thus, it is necessary to determine the reaction required at the mean diameter to prevent negative reaction at the base diameter. In addition, the adverse pressure gradient that occurs with negative reaction in conjunction with the high turning in an impulse stage can result in flow separation and its attendant flow losses. Figures 73.7 and 73.8 are representations of modern Curtis and Rateau stage performance, respectively, with some design reaction.

The high reaction stages include the classical reaction stages. The reaction is typically defined as the percentage of the stage pressure drop that occurs at the stage mean diameter. While the definition of high reaction is imprecise, stages with mean diameter reactions of 25% or greater can be considered as high reaction designs.

Because of radial equilibrium and the corresponding increase in reaction from base to tip, the longer the blade, the larger the variation in reaction between base and tip. So, in the case of long blades, the differences between impulse and reaction blades are academic. The base section must of necessity have low reaction and the tip section must have high reaction.

73.1.2 Shaft or Shell Arrangements

The turbines can be classified according to the number and arrangement of the turbine shafts and shells (casings, cylinders, or elements). The shells are the pressure vessels that confine the steam to the blade path.

Currently, single-shell or -casing designs that expand steam from the boiler supply pressure down to the unit exhaust pressure are built for lower steam pressures and temperatures (up to 1250 psig and 950°F) and generally for nonreheat applications with low plant power ratings.

As unit power ratings increase, sufficient exhaust annulus flow area cannot be provided in single-case designs because of bearing span limitations or last-row-blade-length limitations.* So, the low-pressure expansion is taken in a separate casing. In current practice, anywhere from two to eight low-pressure flow paths are provided with a given high-pressure casing. The division of the turbine into high-pressure and low-pressure elements also results in the selection of rotating and stationary part materials that are more suited to the pressures and temperatures present in these elements.

When the multiple casings are on the same shaft, the design is tandem compound (TC). When some of the casings are on different shafts, it is cross compound (CC). With a cross-compound design, it is possible to use different shaft speeds and achieve a more beneficial selection in regard to the constraints of exhaust annulus area and thermal stress. This is discussed in more detail in Section 73.4.5. Use of low-speed, 1800-rpm, low-pressure (LP) elements makes it easier to provide large exhaust annulus areas because of low blade and rotor stress levels. Element efficiency is higher, but so is cost.

* The exhaust annulus area is the area between the base diameter and the tip diameter of the last row.

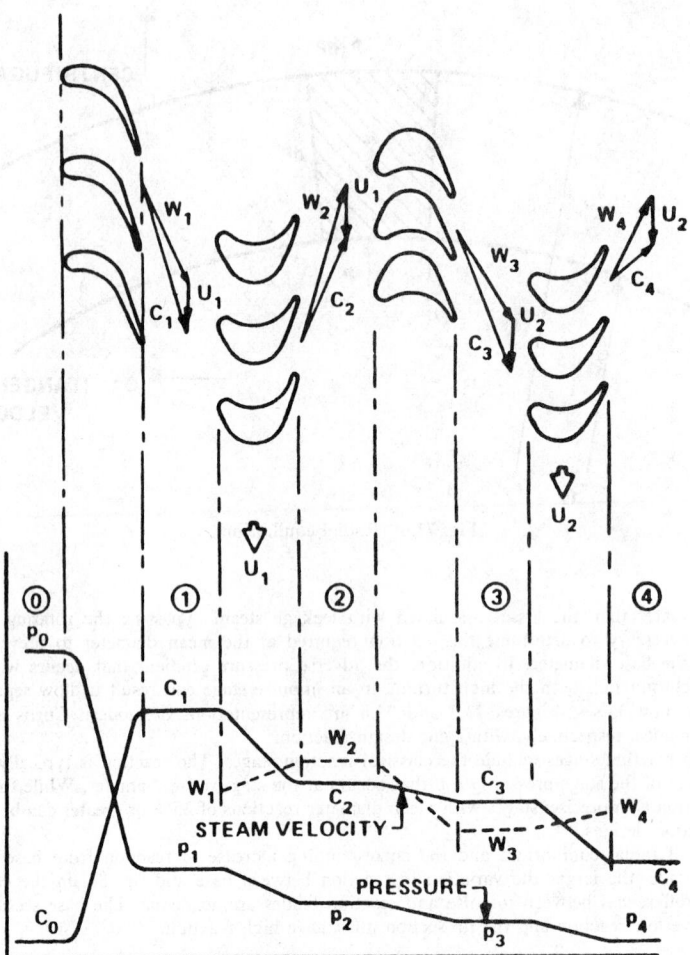

Fig. 73.7   Modern Curtis stage.

The number of stages required to expand the steam from a given inlet pressure to a given exhaust pressure depends on the blade lengths, the mean diameters, and the rpm or, more properly, the blade rotative speeds. The lower the rotative speed, the larger the number of stages required to expand the steam. Since 1800-rpm LP elements have lower blade rotative speeds, they have more stages than 3600-rpm LP elements, or they must have large rotor diameters to compensate for the lower rpm.

The larger rotor disks, rotor forgings (unfinished rotor bodies that are forged rather than cast by the steel suppliers), and shells increase the temperature differences between the inner and outer diameters of these components during unit startups and load changes. These temperature differences and the larger masses of the 1800-rpm components result in larger thermal stresses than 3600-rpm designs. So, for large-rating, high-pressure, high-temperature units, it is not uncommon to use high-speed, 3600-rpm, high-pressure (HP) elements and low-speed, 1800-rpm, LP elements.

Since most fossil-fueled units are reheat designs, the steam expansion is taken in three stages, HP, IP (intermediate pressure), and LP elements. For unit ratings up to about 600 MW, most U.S. reheat units use combined HP–IP elements. In this instance, the HP and IP expansions occur in a single outer casing or shell as shown in Fig. 73.9. An inner shell and/or separate blade rings separate the HP and IP expansions. Combined elements result in plant cost savings.

Referring to Fig. 73.9, the left pipe of the two small pipes in the upper half and the companion pipe in the lower half bring throttle (main) steam to the HP section. The flow passes through the partial-arc admission Rateau-type impulse stage, flowing to the right. At the exit from the impulse stage, the flow does a 180° reversal, flowing along the inner surface of the inner shell and around the main steam inlet pipes. The flow then passes through eight stages of reaction blading from right

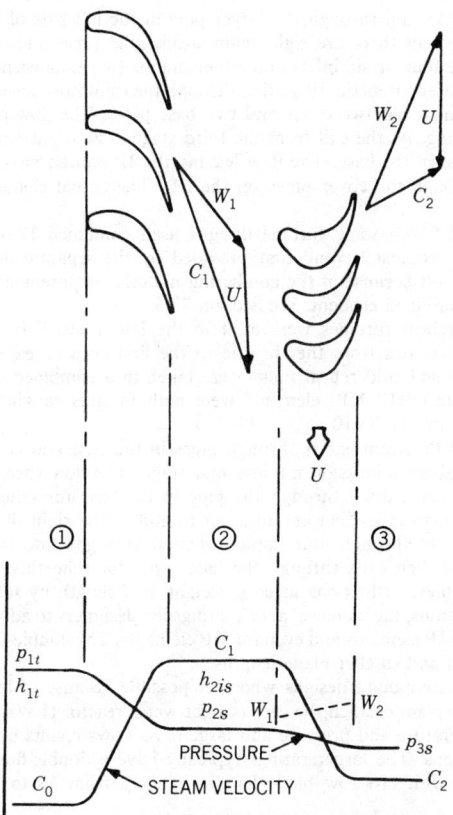

Fig. 73.8 Modern Rateau stage.

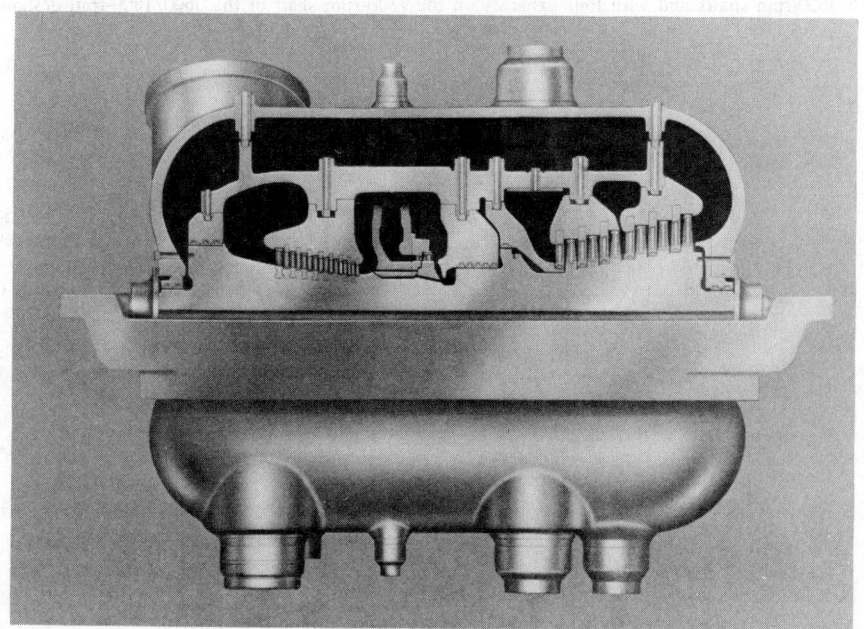

Fig. 73.9 Combined HP–IP element (single reheat).

to left, exits from the HP section through the larger pipe in the left side of the base and goes to the reheater. In the actual element there are eight main steam inlet pipes and two exhaust pipes. Other similar elements may have four or six inlets and either one or two exhaust pipes.

The flow from the reheater enters the IP section through the rightmost cover pipe and the companion base pipe. The actual element has two cover and two base pipes. The flow goes to the right through six stages of reaction blading. At the exit from the third stage, flow is extracted for feedwater heating through the rightmost pipe in the base. The flow leaving the IP section passes between the inner and outer shells and exits through the cover pipes on the left. The actual element has two pipes at this location.

During the 1950s and 1960s some turbine designs used combined IP–LP elements in order to provide an additional LP exhaust beyond that provided by the separate double flow LP elements. This design is no longer built because of the conflicting material requirements for the rotor in the IP and LP portions of the combined element. See Section 73.6.2.

A number of double-reheat turbines were built in the late 1960s and early 1970s. In many of these applications the expansions from the throttle to the first cold reheat point and from the first hot reheat point to the second cold reheat point were taken in a combined element. Combined very-high-pressure–high-pressure (VHP–HP) elements were built in sizes ranging from 300 to 765 MW. A typical design is shown in Fig. 73.10.

In Fig. 73.10, the throttle steam enters through pipes in the base and cover and flows to the left, passing through the partial-arc admission impulse first stage. The flow then passes through five full-arc admission impulse stages, exiting through the pipe in the base for reheating the first time. The reheated steam enters through pipes that are adjacent to and to the right of the throttle steam pipes. The steam flows to the right through four impulse stages, arranged into two groups for extraction for feedwater heating, and then exits through the base pipes for reheating the second time. Just as designers were able to achieve reductions in cost, weight, and length by incorporating LP elements into double-flow configurations, the increase in unit ratings led designers to adopt a double-flow arrangement for high-temperature IP elements and even for HP elements. The double-flow arrangement resulted in lower forging diameters and smaller blade lengths.

Designers use tandem-compound designs wherever possible because of lower space requirements, less complexity, and lower plant cost. In the case of light-water reactor (LWR) plants, the combination of low initial steam temperature and pressure and large mass flows results in the selection of tandem-compound 1800-rpm designs. The larger ratings typically have a double-flow HP element and three double-flow LP elements with last row blade lengths ranging from 38 to 54 in. as shown in Fig. 73.11 and Table 73.1.

The number of exhausts supplied with currently active fossil designs ranges from two to six on tandem-compound 3600-rpm designs with last row blade lengths from 20 to 34 in. as shown in Table 73.2. Cross-compound designs up to 1300 MW have been supplied with a total of eight exhausts on both 3600-rpm shafts and with four exhausts on the 1800-rpm shaft of the 3600/1800-rpm designs.

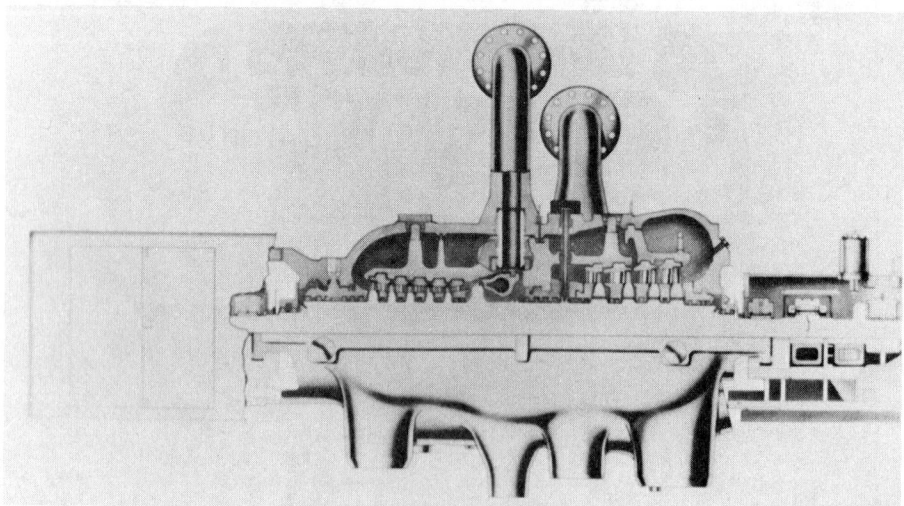

Fig. 73.10 Combined VHP–HP element (double reheat).

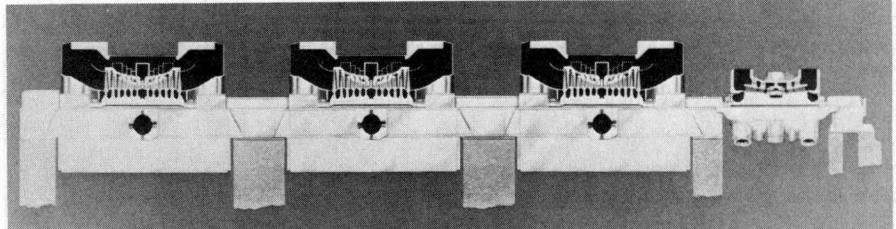

Fig. 73.11   LWR turbine.

73.1.3   Steam Conditions and Cycles

Steam turbines can be further classified by steam conditions and cycle. Nonreheat cycles are rarely purchased today except for combined steam and gas turbine cycles in which a gas turbine exhaust heat recovery boiler supplies steam to a steam turbine operating on a straight condensing cycle. The reheat cycles use regenerative feedheating for improved cycle efficiency. Several plants with two reheats have also been built.

In regard to the LWR turbines, the early plants incorporated moisture separators to remove the moisture formed during the initial expansion in the turbine. The steam, essentially dry and saturated, was returned to the LP element of the unit. While there was no temperature increase in the separation process, the removal of moisture increased the energy (enthalpy) level of the steam, and is thermodynamically equivalent to reheating. Later, plants followed the moisture separation with steam to steam reheating using throttle steam as the reheating steam source. Still later, plants used two stages of reheat with partially expanded steam supplying the low-pressure reheat stage and throttle steam supplying the high-pressure reheat stage. The reheaters and separator section are incorporated into one vessel, a moisture separator–reheater (MSR), as shown in Fig. 73.12.

The HP exhaust steam enters the MSR shown in Fig. 73.12 through the opening on the right side of the longitudinal view. It passes through two distribution manifolds, circular pipes on either side, shown in the transverse view. The flow leaving the manifolds passes through chevron-type moisture separators (some designs use wire mesh pads) to remove the water droplets. The dried steam passes over two bundles of tubes—two-stage reheater. The lower bundle uses partially expanded steam to do the heating, while the upper bundle uses throttle steam as the heat source.

In each tube bundle the heating steam flows inside the tubes, and the steam to be heated passes over the outside of the tubes. The outside of the tubes are finned to improve the heat-transfer capability. The regenerative cycle, in which partially expanded steam is used to preheat the condensed steam (feedwater) before it enters the steam generator, is applied extensively because of the thermodynamic effects and resulting improvements in turbine energy conversion (as noted in Section 73.7.2 and Section 73.7.3).

Table 73.1   1800 rpm LP Elements

Last-Row Blade Length (in.)	Exhaust Annulus Area (ft²)
37–40	91.2–105.7
42–45	117.3–131.8
52–54	168–174

Table 73.2   3600 rpm LP Elements

Last-Row Blade Length (in.)	Exhaust Annulus Area (ft²)
22–23	31.8–34
25–28	41.1–45.7
28.5–31	53.7–58.3
31–34	65.8–67.4

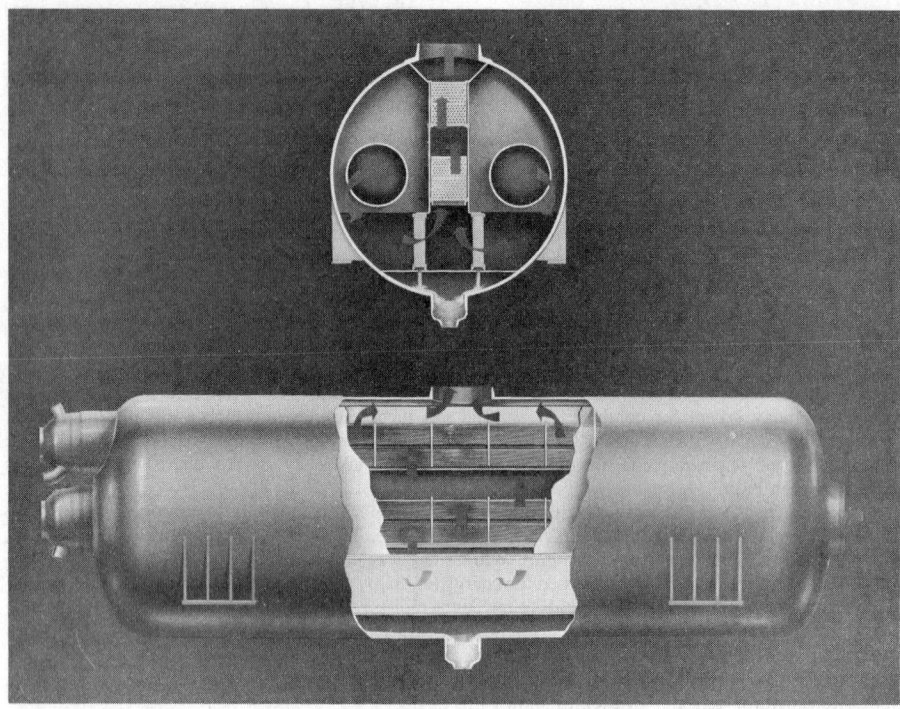

Fig. 73.12 Moisture separator–reheater.

The turbine cycles can also be classified as either condensing or noncondensing. In the condensing cycle, the turbine exhaust discharges directly into a heat exchanger, condenser, where heat is removed from the steam to convert it to liquid water. In a noncondensing cycle, the exhaust steam is discharged at elevated pressure to other equipment that may not be part of the cycle.

In many instances, some of the steam is extracted from the turbine at one or more intermediate locations for processes that may or may not be an integral part of the cycle. When control valves are incorporated into the turbine design just after the extraction point in order to regulate the extraction pressure, it is called automatic or controlled extraction. When the extraction pressure is allowed to vary in response to mass flow changes in the turbine and the extraction piping, it is an uncontrolled extraction cycle. Uncontrolled extraction is used in regenerative feedheating cycles (see Section 73.7.2). In some instances, when a source of waste heat is available, it can be used to generate steam and is introduced at a lower-pressure location in the turbine. This has been termed induction. Finally, turbines can be characterized by the type of equipment that they drive. Turbine speed is kept constant when it drives an electrical generator. In most cases, when steam turbines drive nonelectrical equipment, the turbine operates at variable speed. For practically all applications with throttle pressures of 2400 psig and higher and/or turbine ratings of 300 MW or larger in fossil-fueled plants, variable-speed auxiliary turbines are used to drive the main feedwater pumps. In a few instances, variable-speed auxiliary turbines are used to drive the boiler fans in electric power plants. Turbines are also used to drive pumps, fans, and compressors in industrial plants.

Most large naval ships and many merchant ships used variable-speed steam turbines to drive the ships' propellers by using reduction gears. The variable-speed feature results in higher efficiency over a wide operating range as compared to constant-speed drives.

73.2 UNIT SIZES

73.2.1 Single-Casing Designs

Single-casing designs, as illustrated by Fig. 73.13, are available in sizes up to about 110 MW, with the upper end of this size range being applied on combined-cycle plants with heat-recovery steam generators.

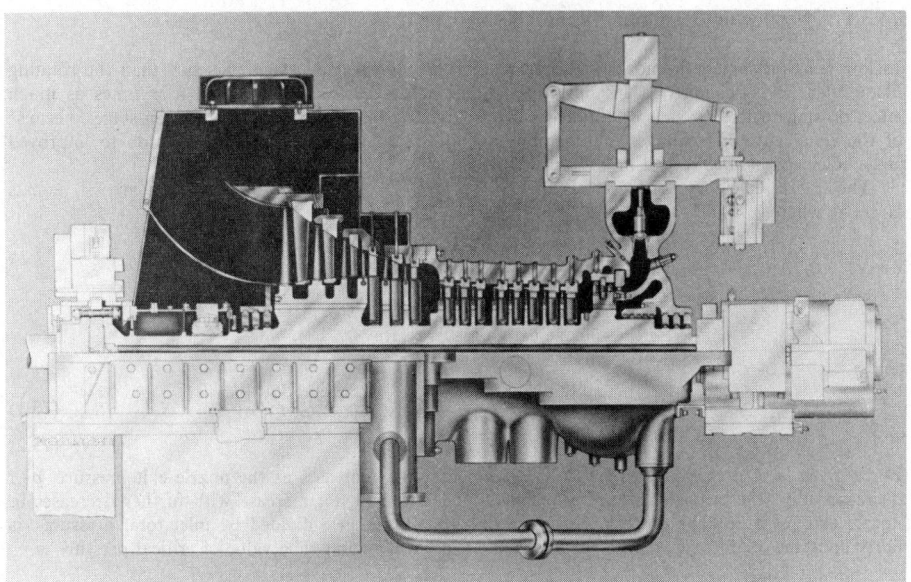

Fig. 73.13 Single-casing design.

73.2.2 Multiple-Casing Unit

Most multiple-casing reheat units for ratings up to 425 MW are tandem-compound 3600-rpm designs with a single double-flow LP element, TC2F (tandem-compound, two-flow LP). The HP and IP expansions are generally accomplished in a combined HP–IP element resulting in a two-casing design. In this size range, ratings have clustered about 180, 250, 325, and 425 MW. For sizes from 425 to 825 MW, most units use two double-flow 3600-rpm LP elements, TC4F. For ratings up to 550 MW, combined HP–IP elements are used almost exclusively and are often used up to 600 MW. For sizes above 600 MW, separate HP and IP elements are used. With sizes above 825 MW, tandem-compound 3600-rpm designs with three double-flow LP elements, TC6F, are used. In the 1000–1300-MW range, cross-compound designs, both 3600/3600-rpm and 3600/1800-rpm arrangements, are used.

LWR ratings have clustered around 650, 1000, 1150, and 1300 MW, being tandem-compound 1800-rpm designs. Many of the 650 and 1000 MW units use a double-flow HP element and two double-flow LP elements with appropriate last row blade lengths. Units of 1000 MW and higher include an additional LP element.

As was mentioned in the previous section, condensing steam turbines are often used as the feedwater pump drives on both fossil and LWR units. The drive turbines are generally supplied with steam in the 125–175 psia range, which is taken from a location just preceding the LP elements. In a number of instances, a single turbine drive is used for main unit ratings up to 1300 MW, resulting in drive turbine sizes up to 50 MW. It should be noted that a 50 MW main unit rating was considered to be large in the early 1950s.

73.3 STAGE DESIGN

As was stated in Section 73.1.1, a turbine stage consists of nozzles or stationary blades and one or more rows of rotating blades or buckets. If we confine our discussion to modern Rateau or typical reaction stages, there is only one row of rotating blades associated with a given row of nozzles. Considering the more general case where the approaching working fluid to the stage has velocity, the inlet fluid is denoted by static and total state points, both pressure and enthalpy. The total or stagnation point relates to an isentropic conversion of the velocity into pressure. See Section 73.7 for more detailed discussion of total and static state points.

The velocity and, consequently, the theoretical work that results from an expansion from the stage inlet total pressure to the stage exit static pressure is the isentropic enthalpy difference, Δh_s, between these two state points. The heat drop, Δh_s, results in an equivalent velocity of $223.8\sqrt{\Delta h_s}$ ft/sec or $44.7\sqrt{\Delta h_s}$ m/sec. The actual process is not isentropic and the stage work, Δh_w, is somewhat lower.

73.3.1 Nozzles or Stationary Vanes

In low reaction stages, the nozzle efficiency has a greater effect on stage efficiency than the rotating blade efficiency. Generally, a 1% improvement in nozzle efficiency has about four times as much effect on stage efficiency as a 1% improvement in rotating blade efficiency. In reaction stages, because of the equal enthalpy changes in the rotating and stationary blades, the contribution to improved stage efficiency is about equal in both parts of the stage.

The nozzle efficiency, Fig. 73.8, is $(h_{1t} - h_{2s})/(h_{1t} - h_{2is})$. The nozzle throat area A_t equals $G v_t / C_1$, where

$$p_{1t} = \text{nozzle inlet total pressure}$$
$$h_{1t} = \text{nozzle inlet total enthalpy}$$
$$h_{2is} = \text{nozzle exit isentropic state point}$$
$$h_{2s} = \text{actual nozzle static enthalpy}$$
$$p_{2s} = \text{nozzle exit static pressure}$$
$$G = \text{mass flow}$$
$$v_t = \text{throat specific volume at } p_{2s} \text{ and } h_{2s}$$
$$C_1 = \text{throat velocity} = 44.7 \sqrt{h_{1t} - h_{2s}} \text{ m/sec} \tag{73.1}$$
$$= 223.8 \sqrt{h_{1t} - h_{2s}} \text{ ft/sec} \tag{73.2}$$

For a given inlet nozzle pressure, p_{1t}, the steam flow, G, increases as the nozzle exit pressure, p_{2s}, decreases until the critical pressure is reached. The flow will not increase with further decreases in nozzle exit pressure. The critical pressure ratio—throat pressure divided by inlet total pressure—is fairly constant (0.55) over a large portion of the superheat region. The value of critical pressure is

$$\text{critical pressure, } p_c = p_{1t} \left(\frac{2}{\gamma + 1} \right)^{\gamma/(\gamma - 1)} \tag{73.3}$$

where γ is the isentropic exponent for steam. Until the critical pressure ratio is reached, the rate of increase of velocity exceeds the rate of increase in volume, so a convergent or nonexpanding passage is most suitable.

At pressure ratios above the critical value, the throat conditions and the nozzle exit or mouth conditions are identical. When the pressure ratio decreases below the critical value, the throat pressure remains at the value corresponding to critical pressure and further expansion occurs between the throat and the mouth, causing the jet to expand in all directions. Inclusion of an additional expanding section results in a controlled expansion to higher velocity and volume. Since the rate of increase in volume is greater than the rate of increase of velocity, the passage area must be increased. The term convergent–divergent or expanding nozzle has been applied to this type of design.

The design of the expanding section of a convergent–divergent nozzle is more exacting because of the supersonic nature of the flow and the shock losses that are present under some conditions of operation. Nozzles with a curved expanding section, Prandtl–Meyer nozzles or cusp nozzles, are more efficient and expensive than those with straight-wall expanding sections.

The velocity coefficient of a nozzle or rotating blade is the ratio of the average exit velocity leaving the foil divided by the theoretical velocity produced by an isentropic expansion from the inlet total pressure to the exit static pressure. Test values of velocity coefficients and the test procedures have been reported by Keenan, Kraft, Berry, and New.[1-4] Some designers prefer to use energy coefficients (the square of velocity coefficient) in their evaluations rather than velocity coefficients.

The greater the amount of turning in the blade passages, the lower the blade efficiency, as illustrated in Fig. 73.14.[5] However, the stage efficiencies calculated from blade passage velocity or energy coefficients are always higher than those obtained from turbine tests. The losses associated with step-ups or overlap, leakage, entrainment, and moisture effects are not accounted for in the blade row or cascade tests. See discussion by New.[4] In addition, the effects of turbulence, losses in upstream blade rows, and realistic boundary layers are difficult to simulate in the cascade tests. As a consequence, designers rely increasingly on multistage tests for verification of both high-pressure and low-pressure blading performance.[6-10] Cascade tests are still used to verify relative performance of blade passage configurations as well as to evaluate losses at various passage sections.

Operation with supersonic passage velocities results in lower passage velocity coefficients.

The nozzle profiles on both low- and high-reaction stages are aerodynamic or airfoil shapes in converging passages and are often referred to as curved foil nozzles. The shapes appear similar on both types of designs. However, the ratio of blade throat opening to blade pitch or gauging is lower on low-reaction stages. The gauging or sine of the outlet angle generally ranges from 0.17 to 0.20 on low-reaction stages and from 0.25 to 0.40 on high-reaction stages.

73.3.2 Rotating Blades or Buckets

In contrast to the similar nozzle shapes on low- and high-reaction stages, the profiles of low- and high-reaction rotating blades are markedly different from one another. In high-reaction stages with

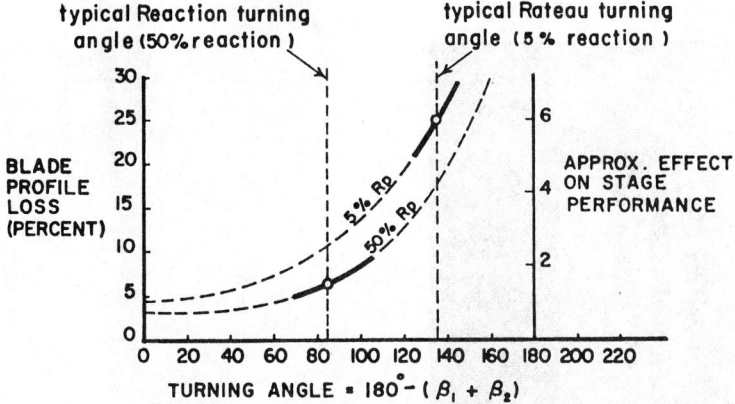

typical Reaction turning
angle (50% reaction)

typical Rateau turning
angle (5 % reaction)

BLADE
PROFILE
LOSS
(PERCENT)

APPROX. EFFECT
ON STAGE
PERFORMANCE

TURNING ANGLE = 180° - (β_1 + β_2)

Fig. 73.14   Profile losses versus turning and reaction.

constant width and profile (with height), the stationary and rotating blades have identical profiles and efficiency levels. When longer blades are used, the rotating blades are twisted and tapered. In this instance, the stationary and rotating blade profiles differ appreciably from one another.

In low-reaction stages, the turning in the nozzles is much less than that of the rotating blades. In addition, the leading edge of the rotating blades is very sharp and the inlet angle is very selective. The upstream and downstream halves of the rotating blade are practically mirror images. As the blades become longer, tapered and twisted designs are used so that the blade cross sections of the low- and high-reaction designs become more alike.

In most instances, the stationary blades are shrouded at the base as are the tips of the rotating blades. However, as the rotating blades become larger, centrifugal stress may make it undesirable to incorporate a shroud. The shrouds can either be integral to or mechanically joined to the blades. The shroud has a dual function. It makes it possible to achieve more effective sealing, thereby, reducing leakage losses. In addition, it enables the designers to tune the frequency of the blades by changing the number of blades that are joined together and by changing the mass of the shroud–blade combination. The shroud also limits the vibration amplitude of the individual blades, thereby, reducing the stress level.

Lashing wires are another means of tuning blades by joining them together at an intermediate blade height. Blades can incorporate lashing wires or shrouds or a combination of the two. Most large 3600-rpm blades have a maximum total of two lashing wires and/or shrouds while some 1800-rpm designs have as many as three. In some designs, freestanding blades without lashing wires or shrouds are used. The freestanding blades are characterized by much larger chords and a greater rate of taper than designs with lashing wires or shrouds.

Because of the greater turning in low-reaction rotating blades, the velocity and energy coefficients are lower than high-reaction designs.

73.3.3   Drum Construction

High-reaction designs have rotor configurations in which the sealing surfaces under the stationary blade are at approximately the same diameter as the base of the adjacent rotating blades. The rotor envelope is approximately conical or cylindrical with only minor steps in the rotor profile as shown in Fig. 73.15. For high-reaction designs that incorporate separate disks because of rotor forging limitations, the stationary blades seal at radial locations that approximate the base of adjacent rotating blades like the monoblock drum designs.

73.3.4   Diaphragm and Wheel Construction

The low-reaction designs use a construction in which the rotating blades are mounted on a series of disks. The stationary blades are mounted on diaphragms that project into the spaces between disks and seal at much lower diameters than the base sections of the adjacent rotating blades. Special sealing arrangements and balance holes in the disks are used to achieve zero static pressure drop across the disk and, therefore, prevent large rotor thrusts from being developed, Fig. 73.16. The disks are integral on large high-pressure, intermediate-pressure, and many low-pressure designs. As was the case with high-reaction designs, large 1800-rpm low-pressure elements may require shrunk-on disks. However, in this case the disk and diaphragm construction and sealing configurations of the integral disk rotors are retained.

Fig. 73.15 Springback seal (reaction stage).

73.3.5 Losses

There are a number of losses that occur in a turbine stage. The relative level of the particular loss or even its presence depends on the amount of reaction in the stage, the blade length, and the fluid state.

Profile (Aerodynamic) Losses

Flow losses that are related to the shape or profile of the blade foil are called profile or aerodynamic losses. The profile loss is evaluated at the midheight of the flow passage. The profile losses include the losses associated with the trailing-edge wakes at a reference-edge thickness. The loss portion of the velocity and energy coefficients that were discussed in Section 73.3.1 are a measure of the profile loss.

End-Wall Losses

Among the other factors affecting blade passage efficiency are end effects. The end effects relate to the losses associated with the friction at the top and bottom surfaces of the passage, which reduces the centrifugal force and results in a local pressure rise. Radial (secondary) flow then occurs along the partition, both inward and outward, with the formation of vortices and distortion of flow. The larger the ratio of passage height to passage throat width (aspect ratio), the lower the end effects.[5] End-wall losses are also increased by operation at nonoptimum velocity ratios, which changes the blade inlet angle.

Leakage Losses

Leakage losses relate to the flow that bypasses the blade passages and that passes through the clearance space between the rotating and stationary parts. The leakage loss is not only related to the fact that this leakage flow does not do work but also results in turbulence losses associated with the reentry and mixing of the leakage steam with the flow that passes through the blade path.

This reentry loss is less severe on high-reaction stages because the pressure drop in the blades minimizes the radial penetration of the low-velocity leakage steam and accelerates it when it passes

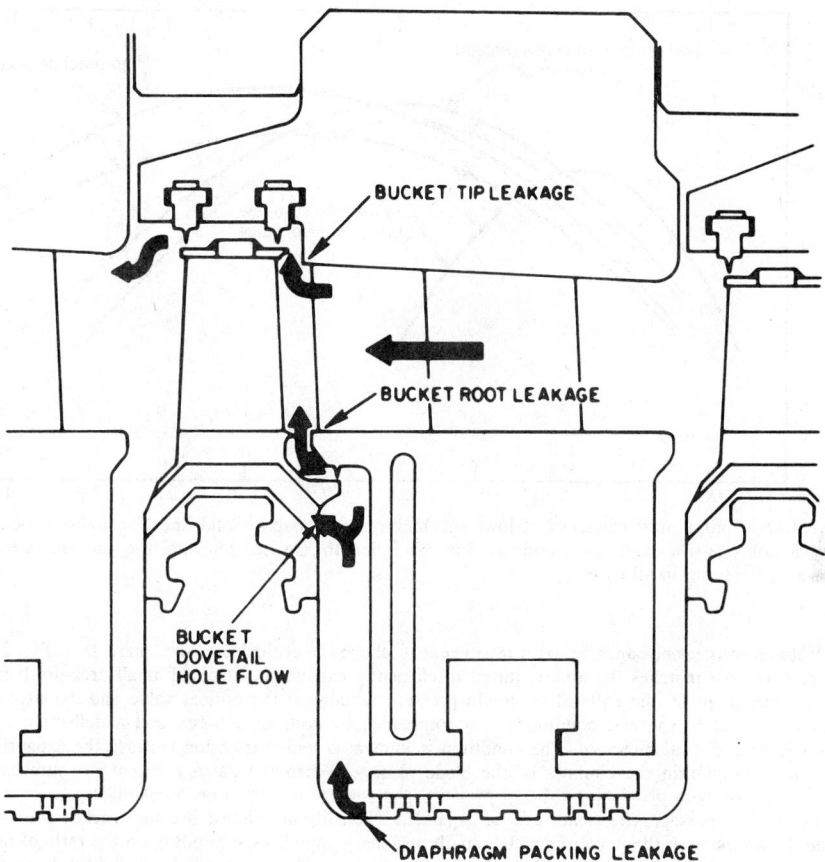

Fig. 73.16 Impulse wheel and diaphragm construction.

through the blades. Because of the low base-section reaction of impulse blades and the sharp rotating blade inlet profiles, the radial penetration of the leakage steam is larger and its effect is more pronounced. In addition, there is only a small pressure drop to accelerate this reentry steam. As a consequence, some manufacturers of low-reaction designs attempt to prevent the leakage steam from reentering the blade path as shown in Fig. 73.16.

Because of the larger pressure drops in the rotating blades of high-reaction designs, the tip leakage is greater as compared to low-reaction designs. In addition, the leakage on the stationary blades is lower on a low-reaction design because the sealing is done at a smaller diameter and a greater number of seals are used. Since the high-reaction stages have lower aerodynamic losses and the low-reaction designs have lower leakage losses, the comparative performance of the two designs with short blade heights depends on the relative importance of the aerodynamic and leakage losses. Figure 73.17 shows the effect of blade height on stage efficiency.

Secondary Flow Losses

Secondary flow losses result from the fact that this flow is at right angles to the main flow resulting a partial loss of fluid contact with the passage wall and attendant destruction of useful energy. The severity of secondary flow affects the movement of moisture in the boundary layer. In Ref. 12, it was noted that low-pressure turbine stationary vanes with less severe secondary flow reduced the erosion associated with moisture.

Over- and Underexpansion Losses

In Section 73.3.1, the concept of critical pressure ratio was presented. In addition, two types of passages, convergent and convergent–divergent (C–D) or expanding passages were described.

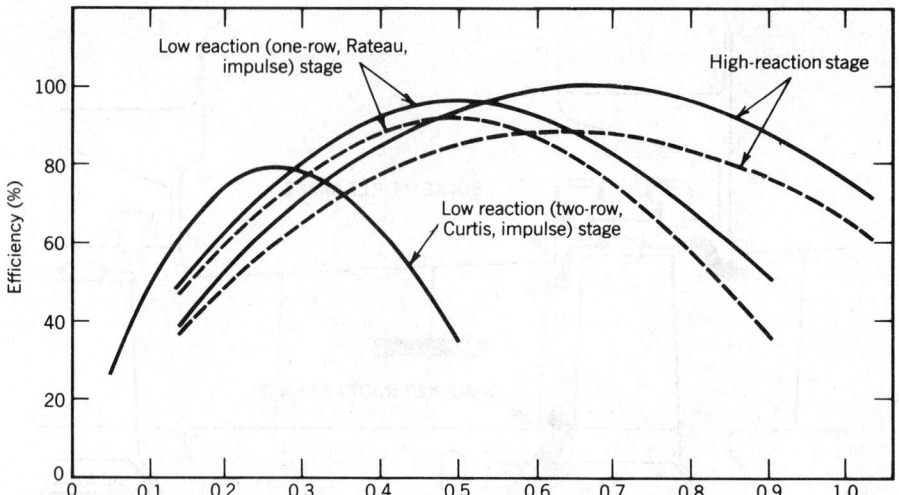

Fig. 73.17 Comparative efficiency of low- and high-reaction stages. Solid lines for high-reaction and low-reaction one-row stage correspond to 5-in. blade length. Dashed lines are for 1-in. blade length. Clearance is $\frac{1}{32}$ in. in all cases.

With convergent passages, at exit pressures at or above the critical pressure, point D of Fig. 73.18, the passage area matches the area required to efficiently expand the steam. For all pressure between point E and point D, the exhaust or mouth pressure remains at the critical value and the expansion from the throat to the exit conditions is accompanied by additional losses and a deflection of the steam jet in the axial direction. The condition is known as underexpansion because the expansion is not completed within the confines of the blade passage. When the passage throat pressure can no longer decrease with decreasing exhaust pressure, the passage is said to be "choked."

For a C–D passage, the steam can be expanded efficiently at exhaust pressures below the critical value. However, even this type of passage finally reaches a condition, depending on the ratio of mouth area to throat area, where further decreases in exhaust pressure are not accompanied by decreases in pressure at the exit plane of the passage. Point E of Fig. 73.18 of the C–D passage is equivalent to point D of the convergent passage. Further decreases of the exhaust pressure below E result in an expansion from the passage outlet plane to the exit pressure with additional losses and deflection of the steam jet.

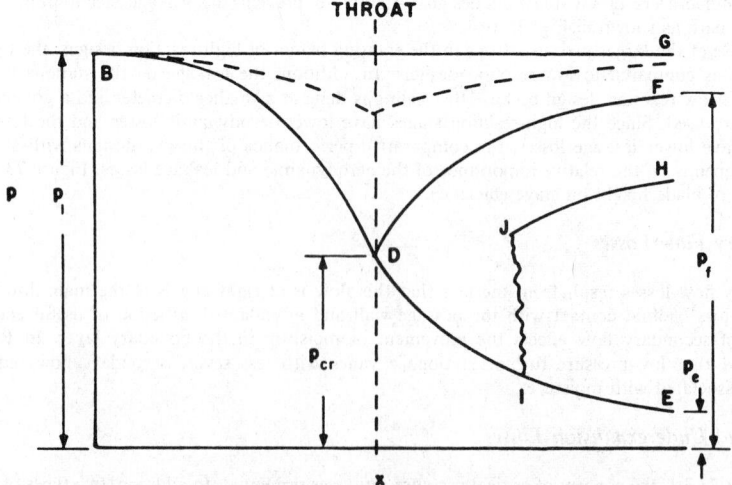

Fig. 73.18 Pressure distribution across a converging–diverging nozzle.

If the exit pressure of the C–D passage is very high (point *G*) relative to the design exit pressure, the velocity at the throat never reaches the critical value and the fluid diffuses up to the exhaust pressure. Because of the comparatively poor performance of the diverging portion of the passage as a diffuser (because of the high rate of passage divergence), the passage losses increase. When the exhaust pressure is low enough and critical pressure occurs at the throat, the fluid continues to expand in the divergent section to point *I*, for example, which is below the exit pressure, point *H*. A normal shock occurs in the passage, point *J*, and the fluid then diffuses to achieve the exit pressure. This process is known as overexpansion. The losses associated with overexpansion are much greater than those associated with underexpansion. Because of this, designers use converging passages wherever possible.

Edge Thickness Losses

Because of structural and manufacturing constraints, very thin trailing edges are not practical. Consequently, the basic aerodynamic efficiency of a blade is related to a given level of edge thickness. Deviations from this reference thickness result in variations in the losses associated with the trailing-edge wakes. Not only do the wakes reduce the blade velocity coefficients but they also increase losses in the downstream blades.

Incidence Losses

All blades have optimum inlet angles. The increased losses associated with inlet flow angles at nonoptimum values (incidence) are not accounted for in the aerodynamic losses and the velocity coefficients. The magnitude of the incidence loss, Fig. 73.19, is greater on blades with sharp or selective inlets. Since incidence can result in local flow separation, a higher degree of reaction in a stage will minimize the size of the separated zone on the rotating blades. If the incidence angle is large enough, it can result in stall flutter (flow-induced vibration) of the blade. This is a concern at the tip section of last row LP blades when operating at low mass flow and high exhaust pressure.

Windage Losses

On turbines in which there is no steam admission over a part of the blade periphery, there is a pumping action of blades on the stagnant steam in the unadmitted areas. This loss is primarily related to partial-arc admission first stages. This loss is not present in high-reaction designs.

Disk and Shroud Losses

Because of the small pressure drop on the rotating blades of low-reaction designs, there are friction losses on both sides of the blade disk and the outer surface of the blade shroud. These losses are not present on drum-type high-reaction designs.

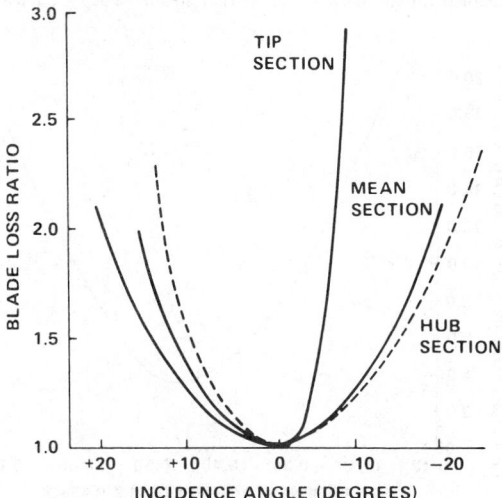

Fig. 73.19 Blade incidence loss.

Other Partial-Arc Admission Losses

On partial-arc admission designs, displacement losses occur when the blading enters and leaves the steam jet issuing from the nozzle groups. These losses are encountered with each interruption of the steam admission and disappear at full admission.

Moisture and Supersaturation Losses

Additional losses result from the presence of moisture in the steam. The losses are related to:

1. Nonequilibrium or supersaturation effects that result in less change in energy level for a given expansion as compared to thermodynamic equilibrium.
2. Losses associated with momentum interchange and friction between the moisture droplets and the steam.
3. Retardation or braking losses associated with moisture striking the backs on the blades because of the low-moisture velocity.

Typical moisture losses vary between 1.0% and 1.2% for each percent average moisture in the stage. In addition, over a period of time moisture erodes the blades and increases the losses associated with the blade profile.

Leaving and Hood Loss

If the velocity of the steam leaving a stage is dissipated partially or totally, a leaving loss is incurred. With properly designed intermediate stages, a substantial fraction of the velocity energy leaving a stage is available for conversion into work in the following stage. The leaving velocity is a minimum at a specific value of velocity ratio or stage exit volumetric flow and has a characteristic shape as shown in Fig. 73.20.

   If the turbine stage is the last one in a casing, especially the condensing section, there may also be a loss in static pressure in the blade passage or hood between the blade exhaust annulus and the exhaust opening in the shell. This is designated as hood loss. In modern units it is possible to design a diffusing exhaust hood in which the blade exit static pressure is lower than the casing exit static pressure, Fig. 73.21. In this instance, part of the leaving energy of the last stage is converted into static pressure resulting in increased work.[13] Designers combine the hood and leaving losses into a single quantity, exhaust loss, that represents the total energy dissipation between the blade exit and the casing exhaust as illustrated in Fig. 73.22.[14]

73.3.6 Basic Stage Efficiency

The efficiency of a blade row or stage varies with the change in a dimensionless quantity known as velocity ratio v. Velocity ratio is defined as the blade speed, U, divided by the steam speed, C. The velocity ratio can be defined on the basis of the actual steam velocity C_1 leaving the blade passage

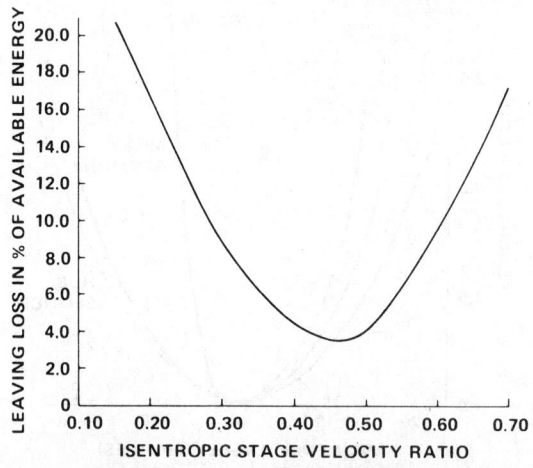

Fig. 73.20 Turbine stage leaving loss.

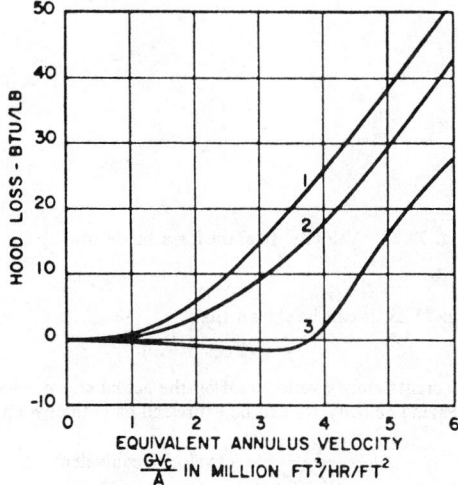

Fig. 73.21 Effect of hood design on hood loss. 1 and 2, conventional exhaust hoods; 3, new design exhaust hood.

(actual velocity ratio, ν_a). It can also be defined on the basis of the ideal velocity C_1 that would result from an isentropic expansion from the stage or blade row inlet total pressure to the exit static pressure. For a given exit flow angle from the blade row, there is a value of velocity ratio at which the stage efficiency is maximum. In Fig. 73.23, for a stationary row, C_1 is the absolute velocity leaving the nozzle and approaching the rotating blade, U is the velocity of the rotating blade, and α is the steam outlet angle from the nozzle, while C_2 is the inlet relative velocity to the rotating blade. In the case of a rotating blade, C_1 is the relative velocity leaving the blade, U is the velocity of the rotating blade, α is the relative steam outlet angle, and C_2 is the absolute steam velocity at the rotating blade exit.

In analyzing the performance of a blade row, the performance index, diagram ratio, ϵ, is defined by the equation

$$\frac{(C_2)^2}{(C_1)^2} = 1 - \epsilon \tag{73.4}$$

$$\epsilon = 1 - \frac{(C_2)^2}{(C_1)^2} \tag{73.5}$$

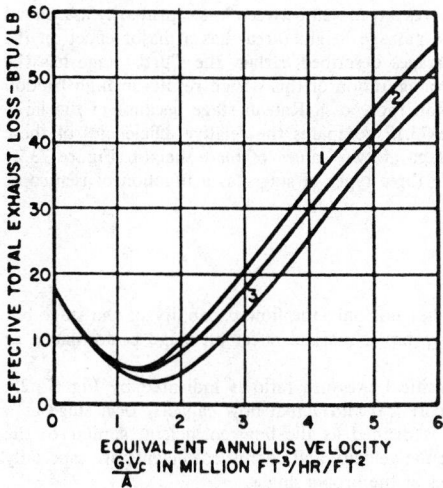

Fig. 73.22 Effect of hood design on total exhaust loss. 1 and 2, conventional exhaust hoods; 3, new design exhaust hood.

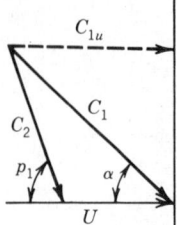

Fig. 73.23   Velocity diagram for a blade row.

By vector addition in Fig. 73.23, it can be shown that

$$\epsilon = \nu \, (2 \cos \alpha - \nu) \tag{73.6}$$

In this instance, ν is the actual velocity ratio based on the actual steam velocity. Moreover, the steam jet velocity C_1 and the leaving velocity C_2 can be expressed as enthalpy equivalents:

$$\Delta h_j = \frac{(C_1)^2}{(223.8)^2} - \text{jet velocity equivalent} \tag{73.7}$$

$$\Delta h_{LL} = \frac{(C_2)^2}{(223.8)^2} - \text{leaving velocity equivalent} \tag{73.8}$$

Inserting these quantities in Eq. (73.4):

$$\frac{\Delta h_{LL}}{\Delta h_j} = 1 - \epsilon \tag{73.9}$$

$$\Delta h_{LL} = (1 - \epsilon)(\Delta h_j) \tag{73.10}$$

If the exit state point, steam flow, passage area, and passage orientation are known, the jet velocity can be calculated from the continuity equation. Then the velocity ratio and diagram ratio can be calculated. Applying either a velocity coefficient to C_1 or an energy coefficient to Δh_c enables the user to calculate the stage inlet total pressure and total enthalpy. From the calculated leaving velocity from the upstream row, the row inlet static pressure (outlet static pressure of upstream row) can be calculated. The procedure is repeated for each blade row until the entire blade path is traversed. A somewhat analogous but simpler procedure is used for calculating performance for a stage or group of stages. In this instance, experimental data are used to develop curves or empirical relationships between isentropic velocity ratio and a quantity known as diagram efficiency. This efficiency curve is then adjusted for variations in blade height, aspect ratio, leakage, and other losses from some reference level.

73.3.7   Comparative Stage Efficiencies

The selection of low- versus high-reaction stages involves a trade-off between losses primarily associated with profile and those related to leakage. The blade passage height (area) has a major effect on the comparative leakage losses. Of the three types of stages described earlier, the Curtis stage has the lowest efficiency. The high heat drop and blading configuration of this design results in high friction losses. In addition, the leakage losses are higher than those of a Rateau stage because of the large sealing diameter under the stationary vane. Figure 73.17 compares the relative efficiencies of these two types of stages and a conventional reaction stage at two values of blade height. Figure 73.24 shows the relative efficiencies and work levels of the three types of stages as a function of isentropic velocity ratio.

73.3.8   Passage Sizing

Passage sizing in a stage is critical, since it establishes not only the flow capability of the stage but the velocity distribution as well. The latter quantity has a marked effect on stage performance as indicated by Eq. (73.6) and Figs. 73.17 and 73.24.

The theoretical flow capacity of a stage with critical pressure ratio is indicated by Fig. 73.25. Test data establish the relationship between the theoretical and actual flow capacity of a stage as a function of stage pressure ratio. This relationship is defined as the flow coefficient, similar to the discharge coefficient used in hydraulics. In addition, the velocity and energy coefficients are especially important to ensure that the fluid enters the passages at the proper angle.

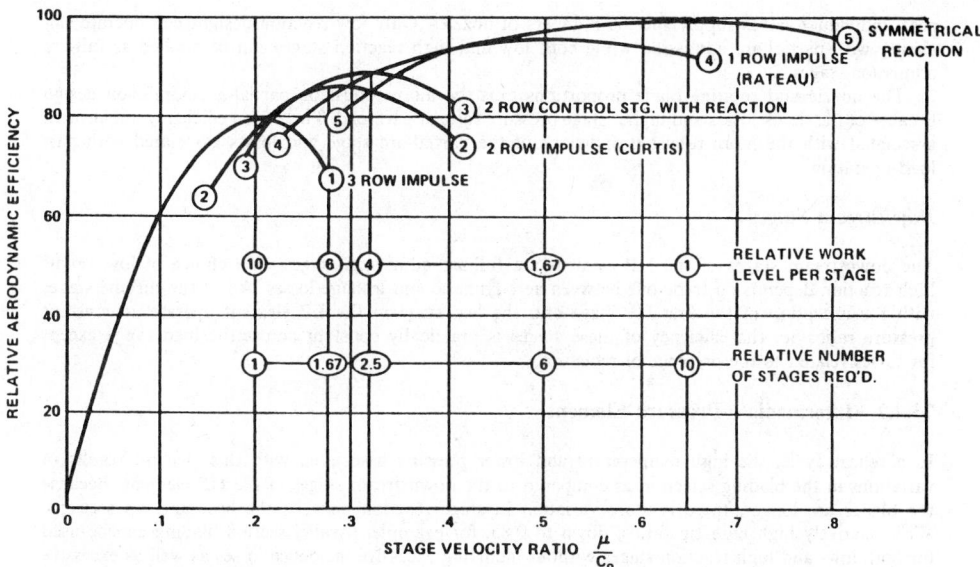

Fig. 73.24 Effect of stage type and work level on aerodynamic efficiency.

73.4 ELEMENT DESIGN

The large change in volumetric flow from the turbine throttle to the LP element exhaust results in the selection of considerably different blading at the two locations. The high temperatures, small blade heights, and modest rotor diameters of the initial stage or stages result in the selection of parallel section blading, often of low reaction. The low-temperature, large blade heights, and large rotor diameters of the exhaust stages require tapered, twisted blades in which the reaction varies considerably from the base to the tip.

73.4.1 High-Pressure Element

Flow or load control is achieved by valves that regulate the steam flow into the first-stage nozzles. The first stage can be one in which all of the nozzles are active under all conditions of load, full-arc admission, and all of the control valves move simultaneously. In contrast, the number of active nozzles could be varied as load is varied by sequential movement of the valves, partial-arc admission. In this

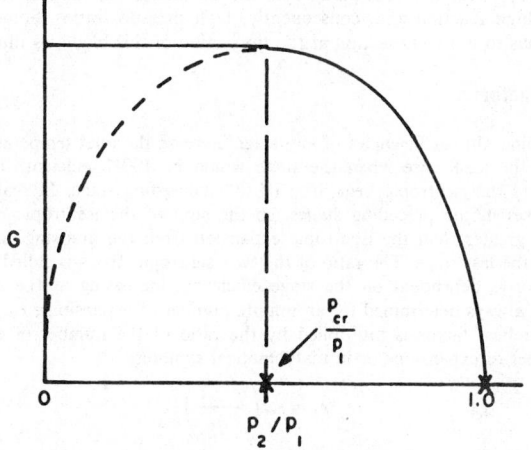

Fig. 73.25 Flow through a converging nozzle.

case, individual valves supply specific numbers of nozzles. Only low reaction designs can operate efficiently with partial-arc admission while both low and high reaction stages can be applied as full-arc admission stages.

The nozzle and rotating blade proportions must be more robust on partial-arc admission design because of the shock losses and higher stage pressure ratio at part load. There is an efficiency impairment associated with the more robust proportions of the partial-arc stage and losses associated with part load operation.

Downstream Stages

The downstream stages of the HP element are full-arc admission stages. The choice of low versus high reaction depends on trade-offs between aerodynamic and leakage losses. All of the turbine stages with the exception of the first HP stage and the last stage in the LP element operate at constant pressure ratio. So, the efficiency of these stages is practically constant across the load range except for losses related to the presence of moisture.

73.4.2 Intermediate-Pressure Element

In a reheat cycle, the high temperature and lower pressure associated with this element results in variations in the blading selection as compared to the downstream stages of the HP element. Because the blades are longer, there is more variation in stage reaction between the base and tip sections. With relatively high base–tip ratios, down to 0.85, for example, parallel section blading can be used for both low- and high-reaction stages without incurring excessive incidence losses as well as excessive blade pitch. Tapered, twisted blades reduce the incidence losses on stages with low base–tip ratios.

The blade incidence is related to the variation in blade speed between base and tip and radial equilibrium that causes the stage reaction to vary between base and tip.

Suppliers of low-reaction designs intentionally increase the mean diameter reaction of the stages as the blade length increases to avoid negative reaction. Mean diameter reaction values ranging from 10% on the initial HP stage to 35% on the last IP stage are not uncommon.

73.4.3 Low-Pressure Element

The initial stages of the LP element have high base–tip ratios, so designers use blading typical of the HP and IP element. Because of the large increases in specific volume at low pressure, the blade heights of the latter stages increase rapidly. The base–tip ratios are low, and there is considerable radial flow.

The end result is stages with appreciable changes in reaction from base to tip, ranging from slightly positive to beyond 50%. Conventional two-dimensional flow analysis is inadequate, especially at the blade tip. The large radial flow component of velocity can only be accounted for with flow-field analysis on high-speed computers. In this instance, a number of stages are analyzed as part of the same design problem and loss mechanisms are included in the modeling.

Convergent–divergent nozzles have generally been applied with mechanical turbines to achieve high power at low cost. In recent years, C–D passages have been applied under vastly different circumstances. As was noted in the previous paragraph, the tip sections of the last stage rotating blades of LP elements have high reaction and, consequently, high pressure ratios. Some manufacturers have found it advantageous to use C–D section at the tip section of this blade, as illustrated by Fig. 73.26.

73.4.4 Reheat Factor

In a multistage turbine, the inefficiencies of any stage increase the inlet temperature of the succeeding stage compared to the ideal case when the stage would be 100% efficient. The increase in steam temperature increases the isentropic heat drop of the succeeding stage. In reality, there is a partial recovery of the losses in the preceding stages, so the sum of the isentropic enthalpy drops of the individual stages is greater than the isentropic expansion from the first-stage inlet conditions to the exhaust pressure of the last stage. The ratio of the two isentropic drops is called reheat factor, $1 + r$.

The reheat factor is dependent on the stage efficiency, increasing as the stage efficiency drops. The reheat factor is always determined for an infinite number of expansions, r_∞. With a finite number of expansions, the reheat factor is multiplied by the ratio of the number of expansions minus one over the total number of expansions or in mathematical symbols:

$$r_1 = r_\infty \left(\frac{q - 1}{q} \right)$$

where q is the number of expansions.

The reheat factor also depends on the isentropic exponent γ of the fluid. If γ is constant, as in an ideal gas, the reheat factor is a function of the pressure ratio over the group of expansions only.

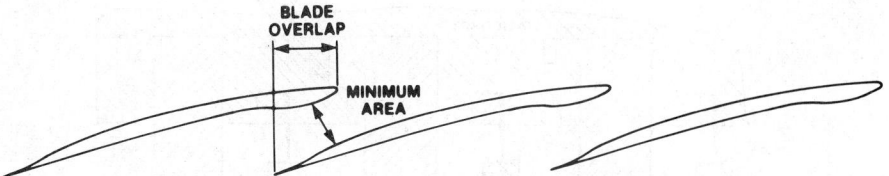

Fig. 73.26   Last stage convergent–divergent passage.

Since superheated steam closely approaches an ideal gas, the reheat factor in this region can be determined as a function of the pressure ratio. In the wet region, the reheat factor can be plotted as a function of the isentropic heat drop.[15,16]

73.4.5 Speed Selection

The speed, rpm, of a turbine element will affect its weight, cost, thermal performance, and cyclic duty capability. Two geometrically similar turbines with linear dimensions varying by a factor of 2 and with rpm varying inversely as the relative linear dimensions would have dynamically similar operation and would have the following relationships:

Parameters	High-Speed Turbine *A*	Low-Speed Turbine *B*
Speed, rpm	3600	1800
Relative linear dimensions	1	2
Relative flow area and power	1	4
Relative metal volume and weight	1	8
Relative power to weight ratio	2	1

In actual practice, the relative power to weight ratio is about 1.8 to 1. Since untuned blades of modest heights are used in the HP and IP elements, blade heights can be increased to obtain equivalent capacities for the high-speed turbine with lower rotor diameters and rotor lengths and, consequently, much lower rotor volume. The reduced diameters further result in smaller shells with thinner walls. Because of the reduced rotor mass (particularly the rotor diameter), the potential thermal stress is reduced, resulting in a superior design for high-temperature and cyclic-duty applications. The reduced rotor and shell weights result in lower material cost. However, where large flow areas are required and the steam temperatures are modest, slow speed elements, 1800 rpm, are advantageous. Examples of this are 1800-rpm LP exhausts on large capacity fossil-fired plants and 1800-rpm HP and LP elements for LWR plants.

73.4.6 Nonblading Losses

While blading accounts for a major portion of the losses in a turbine cycle, a sizable portion of the total is related to other parasitic losses.[17]

Pressure Losses

The transport of steam into and out of turbine shells means losses in the associated valves and piping. These losses vary as the square of the steam velocity. Typical main steam pressure losses of about 4% occur in the valves (steam chest) and piping at the HP element inlet. At the IP element inlet, valve and associated piping losses range from 2% to 2.5%. Whenever any of the HP inlet governor valves are throttling, the steam chest pressure drop can increase considerably.

Depending on the size of the extraction slots and chambers, the steam extracted for feedheating can incur not only a loss in the total pressure but also a loss in static pressure as well.

Shaft Leakages

Shaft leakage occurs where the turbine shafts pass through the shells confining the steam. The shaft leakages may occur at the large diameters associated with dummy or balance pistons that affect the thrust resulting from the pressure drop in the rotating blades. Leakages may also occur at the relatively modest diameters such as the glands, where the shafts penetrate the outer shells. In each instance, a multiplicity of seals is used to dissipate the pressure and velocity of the leakage steam. While axial seals have been used on nonreheat units in the past (Fig. 73.27), the large differential expansions and

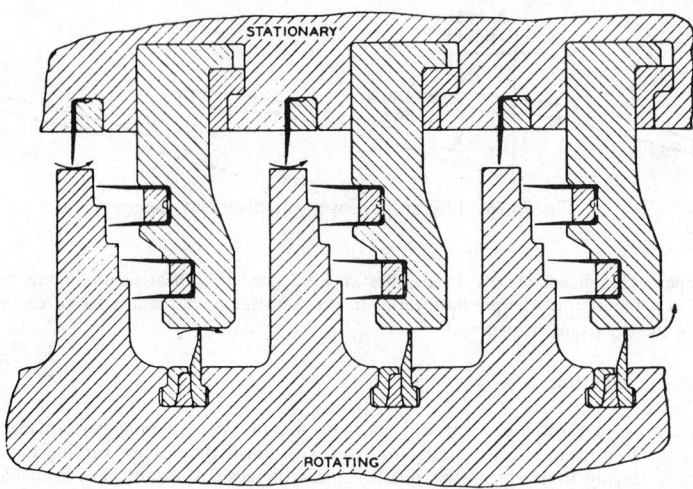

Fig. 73.27   Axial seals.

shaft lengths associated with reheat turbines require axially insensitive seal designs or radial seals (Figs. 73.28 and 73.29). To more effectively dissipate the kinetic energy in the individual seal throttlings and reduce the leakage flow, staggered-type seals are used (Fig. 73.28). At the LP elements, because of the low pressure, the availability loss associated with the shaft leakage is small and the differential expansion is large, so straight through radial seals are often used, Fig. 73.29.

The amount of seal leakage is proportional to the seal leakage area, the seal configuration (staggered or straight through), the number of throttlings or effective seals, the pressure ratio of the seal group, and the seal group inlet pressure and specific volume. Martin's formula is a satisfactory representation of seal leakage and is as follows:[19]

$$G = \frac{(C_1)(C_2)(K)(A)(P_1)}{\sqrt{P_1 V_1}} \sqrt{\frac{1 - 1/\rho^2}{n + \ln \rho}} \tag{73.11}$$

where G = leakage mass flow per unit time

$\quad C_1$ = constant adjusting for units

$\quad C_2$ = seal flow coefficient (varies between 0.6 and 0.8)

$\quad K$ = kinetic energy annihilation coefficient depending upon seal configuration (equals 1.0 for step type or staggered seals)

$\quad A$ = seal clearance area

$\quad P_1$ = inlet pressure to seal groups

$\quad V_1$ = inlet specific volume to seal group

$\quad \rho$ = ratio of inlet pressure to outlet pressure of seal group

$\quad n$ = number of throttling

$\quad \ln$ = natural log

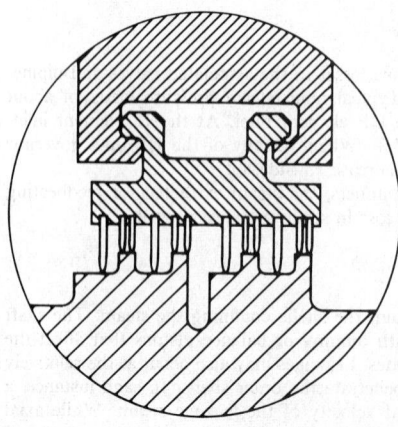

Fig. 73.28   Staggered radial seals.

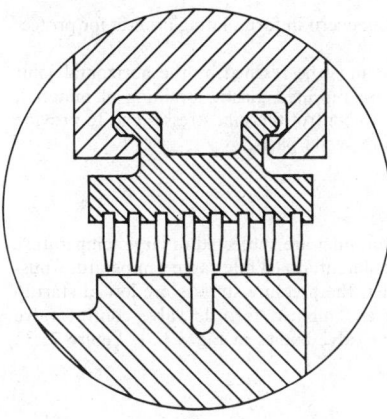

Fig. 73.29 Straight through radial seals.

Valve Stem Leakages and Bushing Leakages

The leakage associated with pressure-vessel penetrations for the activating mechanisms or stems of the control valves is small. However, the pressure levels and pressure ratios are large, while the length of sealing surface is small. Close-fitting bushings in conjunction with grooved valve stems (to increase the losses by producing alternate sudden enlargements and contractions) are effective in minimizing the amount of leakage.[20] The leakage losses can be calculated by using Martin's formula and replacing the number of throttling with the product of the resistive coefficient of the valve stem and the length of the bushing divided by the radial clearance between the bushing and the valve stem.

When both inner and outer shells are used on high-temperature units, the steam inlet pipes use either stacked rings or piston rings on the inlet pipe to allow relative movement between the inlet pipe and the inner shell while controlling the steam leakage. This is commonly referred to as bushing loss, and the amount of leakage ranges from 0.5% to 1.0% of the pipe flow.

Bearing and Lubrication System Losses

The combined losses of the oil pump, thrust bearing, and journal bearings equal about 1% of the turbine shaft output on central station units. The thrust bearing is used to locate the turbine rotor axially with respect to the stationary parts and at the same time carry any axial steam loads, thrust, which may be imposed on the rotor. The journal bearings are used to locate the rotor radially with respect to the stationary parts and, at the same time, support the weight of the rotor as well as vertical or transverse loads imposed on the rotor.

Dummy or balance pistons on high-reaction turbine elements reduce the unbalanced thrust to a magnitude comparable to that of low-reaction designs. The thrust bearing must provide sufficient margin to account for thrust variations resulting from unsymmetrical extraction arrangements (on double-flow elements), off-design operation and equipment deterioration (feedheaters out of service, seal degradation, and blading deposits, for example), and transients such as tripouts.

Journal bearings are of the plain or sleeve type and the tilting pad type. Journal bearings must also be designed to run at speed and overspeed without oil whip. Oil whip is an instability inherent in certain bearing configurations. Tilting pad bearings, while more expensive, have greater resistance to oil whip and can reduce the magnitude of steam whirl.

The oil pump losses are related to the quantity of oil required for the bearings and controls as well as the pump efficiency. Many oil systems incorporate a relatively low-capacity, high-pressure oil pump that supplies motive energy to an ejector which delivers a relatively large quantity of low-pressure oil.

Generator Losses

The generator losses total less than 1.5% of the turbine shaft output. They include the electrical, rotor windage, and generator cooling losses.

73.5 MECHANICAL DESIGN

Steam turbine pressure vessels are subjected to pressure and thermal loads. Pressure loads generally give rise to primary stress systems needed to satisfy equilibrium and, to some extent, secondary systems needed to satisfy compatibility. These are the designer's chief concern in formulating criteria for protection against tensile or creep–rupture failures.

The thermal stress systems create bending moments that must be resisted by the horizontal joint bolting. In addition, transient temperature changes can cause repeated plastic straining of material, leading to fatigue damage and cracking. It is often necessary to control thermal stresses and to provide flexibility for cyclic differential thermal expansions in pressure vessel parts.

73.5.1 Steam Chests and Valve Bodies

The steam chest and/or control valve bodies at the main steam inlet are subjected to large temperature mismatches during unit startup as well as large pressure differentials. While large temperature mismatches also occur on the valve bodies at the hot reheat inlet, the pressure stresses are low at startup and increase as load is increased. Figure 73.30 relates to a combined throttle valve–control valve assembly, while the throttle and control valves have separate valve bodies in Fig. 73.31. Figure 73.32 is a reheat stop-interceptor valve assembly.

73.5.2 Shells and Bolting

Turbine shells have widely varying zone temperatures and, hence, metal temperature gradients between zones. In the case of an inner cylinder, there is also a radial temperature gradient across the shell wall, since the internal and external steam temperatures are usually different. In designing a turbine cylinder, these temperature gradients can often be attenuated by judiciously locating the zones and/or flow passages of the steam.

The total pressure and temperature drops from inlet to exhaust in a turbine can be as much as 3300 psi and several hundred degrees. The double-casing design concept, Fig. 73.33, divides these large pressure and temperature drops into two or more steps, using nozzle chambers, inner cylinders, and outer cylinders as required. In the case of the high-pressure element, the outer shell is a relatively thin-walled structure containing a pressure of only about a quarter of that of the turbine inlet. In the case of IP element shells, Fig. 73.34, since the steam temperature at the exit of the inner shell does not vary much over the load range, the outer shell is essentially a constant-temperature device.

It is important to have a permanently tight horizontal joint. In order to avoid gasket yielding and creep, turbine cylinder joints are designed with metal-to-metal contact. The bolts are prestressed when cold in order to seal the horizontal joint surface and maintain a clamping force on the joint. Since a long-lasting clamping force is necessary to keep a metal joint permanently sealed, bolting materials are selected primarily on the basis of their relaxation strength. Relaxation involves creep at successively lower stresses. It is measured by determining the stress time curve for a constant total strain. The rate of relaxation decreases with time and design allowances are based on this.

High-alloy, high-strength materials are usually less ductile than low-strength steel. In addition, the bolting materials should not be notch sensitive. Tapered threads are used to reduce the stress concentration and spherical seated washers are often used to reduce any bending stress in the bolt in case of misalignment.

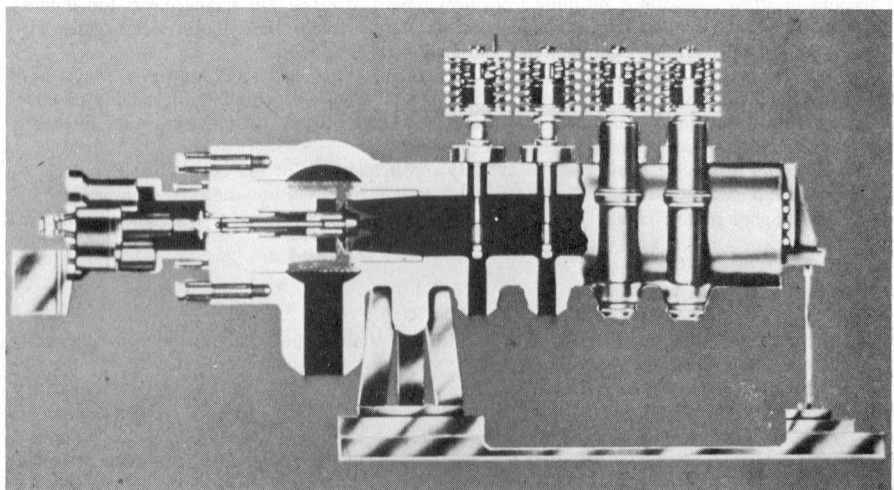

Fig. 73.30 Combined throttle valve–control valve assembly.

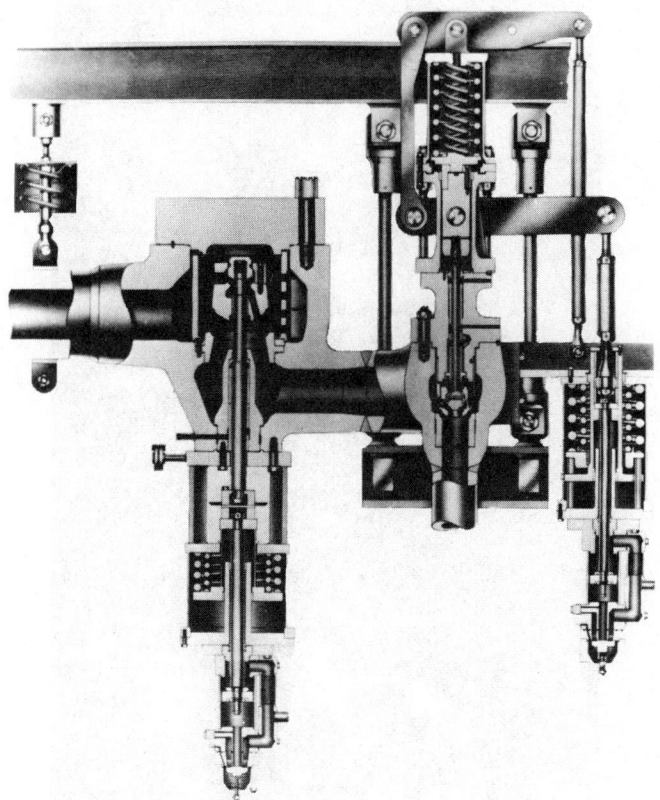

Fig. 73.31 Separate throttle valve–control valve bodies.

As steam temperatures increased above 800°F, the phenomenon of creep became increasingly important and pressure-vessel materials evolved accordingly. The progression of steps to improve high-temperature creep resistance began with the addition of Mo to carbon steel and was followed by additions of Cr to prevent graphitization embrittlement. The Cr–Mo steels of various compositions are used for steam temperatures up to 1050°F. The Mo contributes to high-temperature strength and the Cr stabilizes the carbon and enhances the oxidation resistance.

While austenitic steels have higher creep strengths than ferritic steels, their low thermal conductivity and higher thermal expansion increase susceptibility to thermal distortion and cracking. Current practice is to use ferritic steels as pressure-vessel materials for steam temperatures up to 1050°F. Steam turbines for LWR applications are able to use carbon steel for the high-pressure shells because of the modest steam temperatures.

73.5.3 Nozzle Chambers

Partial-arc admission, while improving turbine efficiency, results in sizable temperature differences between the nozzles of the active and inactive arcs of admission. If the chamber in which the nozzles are mounted includes both active and inactive nozzles, thermal stresses are induced in the walls separating the two groups of nozzles and in the adjoining structure. If the nozzle chamber is integral with the inner cylinder, the walls will be thicker and large thermal stresses can result under partial-arc operation.

Separating the nozzles from the inner cylinder and mounting them in a separate structure or chamber (nozzle channels or nozzle boxes) reduces the thermal stress considerably. The separated chambers may be divided into subchambers activated by individual valves, Fig. 73.35, or into individual chambers without subdivisions, Fig. 73.36.

73.5.4 Blade Rings

Attaching stationary blades to the turbine casing requires cutting grooves in the wall and may make it undesirably thick, thus creating a potential source for high thermal stress. A solution to this problem

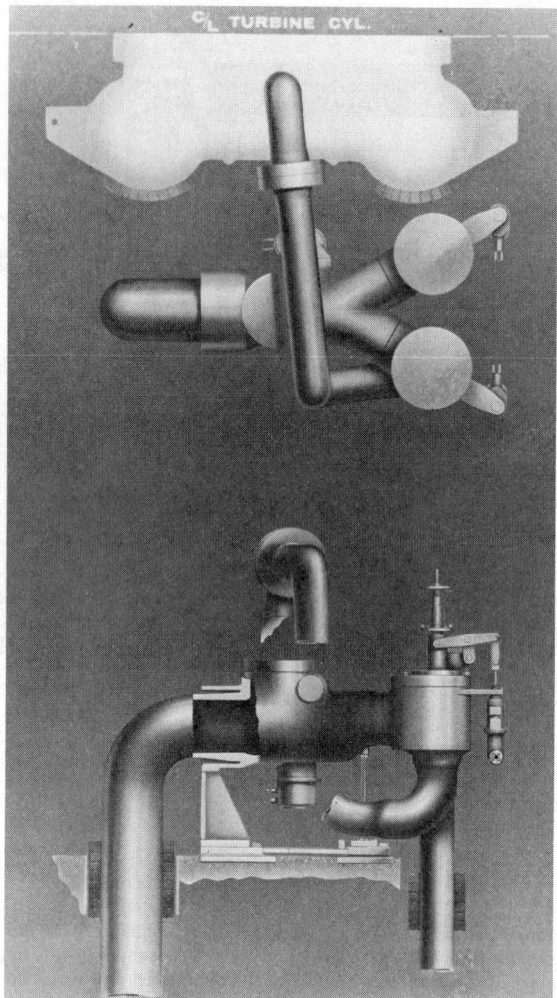

Fig. 73.32 Reheat stop-interceptor valve assembly.

in the high-pressure, high-temperature sections is to separate the blade rings from the turbine shells. The blade rings usually have a different metal temperature from that of the shell. Since they are separate parts from the turbine shell, they can be supported and guided by keys and dowels to allow unrestricted thermal expansions.

Current U.S. practice is to use both inner and outer shells as well as separate blade rings (on high-reaction designs) or separate diaphragms (on low-reaction disk and diaphragm designs) for the HP element and for at least the first blade group of the IP element. In contrast, European designs use inner and outer cylinders or outer cylinders and blade rings for a blade group but not all three.

73.5.5 Valves

Flow and pressure regulation are the primary tasks of the steam valves in conjunction with quick shutoff of flow during emergency conditions. Some of the emergency conditions are overspeeding of the unit, loss of vacuum, low oil pressure, excessive axial expansion of the rotor, and other equipment malfunctions. Valve closing times are between 0.15 and 0.20 sec.

The valves are opened and positioned by hydraulically actuated mechanisms and are closed by mechanical spring force. Throttle steam is controlled by the valves in the steam chest, consisting of the throttle valves and the governor valves.

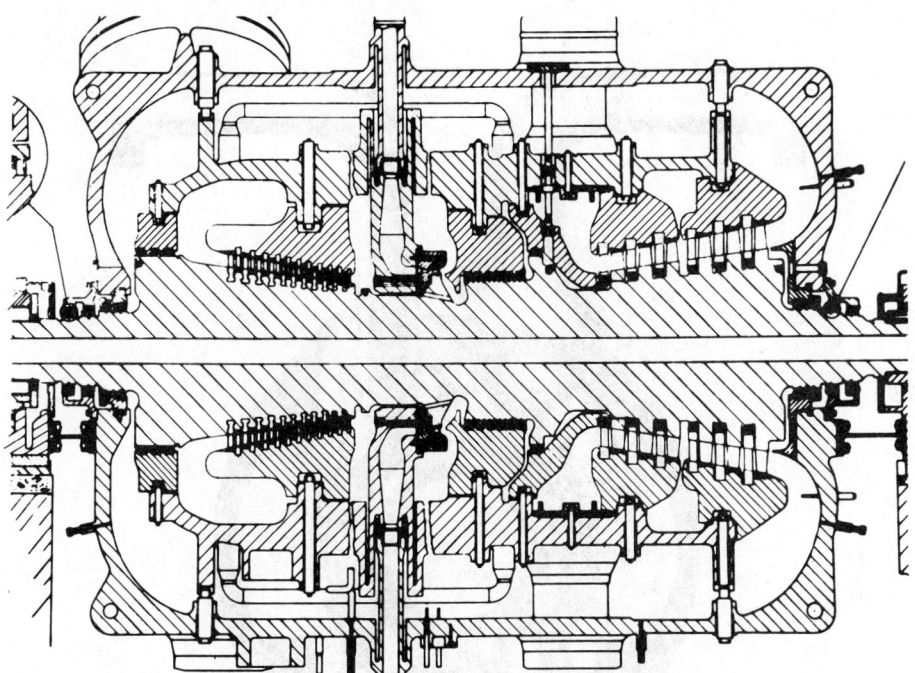

Fig. 73.33 Combined element stationary components.

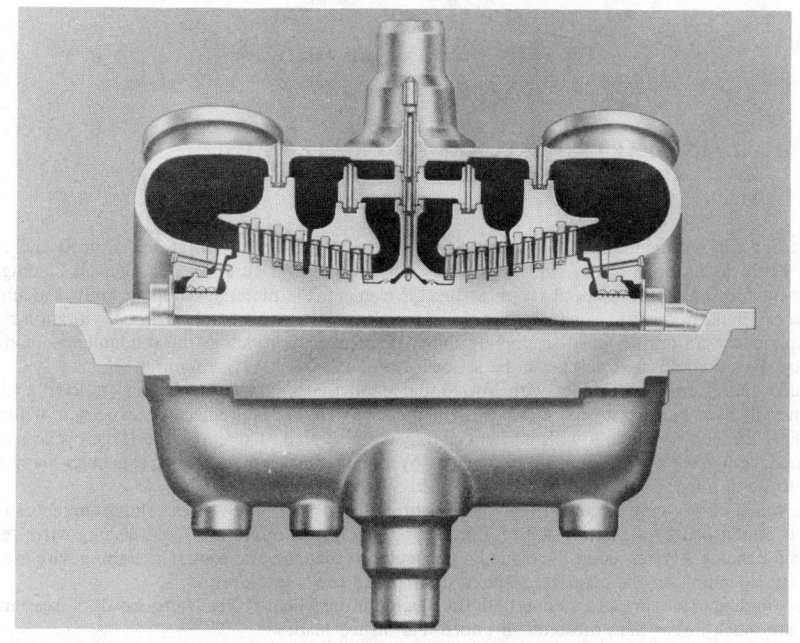

Fig. 73.34 Separate IP element.

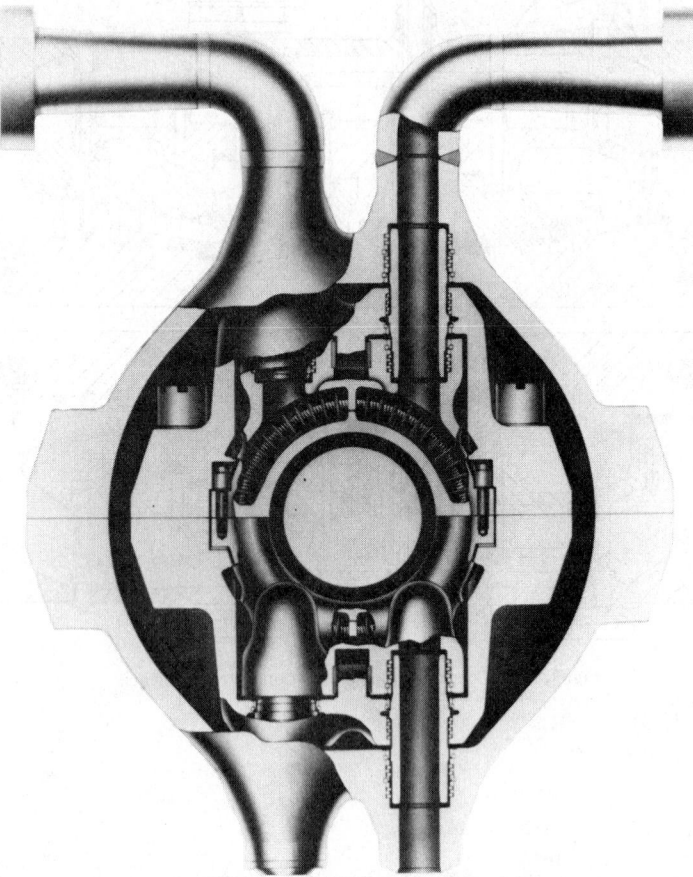

Fig. 73.35 Nozzle box with subchambers.

73.6 ROTATING PARTS (FEATURES AND MATERIALS)

73.6.1 Blading

The blading can be divided into two categories from the standpoint of vibration: untuned and tuned blades. With the untuned blades, which generally comprise almost all of the HP element blading, the IP element blading, and the initial stages of the LP element, the material damping limits the amount of vibratory buildup due to resonance. The blades are designed using a combination of vibration theory and experience and are designed to operate under resonant conditions in the fundamental modes of vibration. Required strength levels can be set based on analysis of service failures.

Tuned blades are those blades with low fundamental frequencies and are long, tapered, twisted, and generally used in the last three rows of LP elements. Strength requirements, necessary to withstand resonance, are not compatible with efficient aerodynamic design. Therefore, the large cyclic stresses associated with resonant buildup are eliminated by tuning the lower natural frequencies away from resonance with integral multiples of running speed.

The tuned blades may be joined together by shrouds and/or lashing wires. The grouping of tuned blades is established by a rotating test of a prototype row of blades. Shrouds and lashing wires reduce the steam bending stresses along the blade, in particular at the base and root. The lashing wire location is chosen depending on the blade frequencies desired and the stress levels.

Freestanding blades are also used as both tuned and untuned blades. The blade chords of freestanding blades are much larger than those on shrouded and lashed blades.

The blading material most widely applied in modern turbines is type 403 (12% Cr) steel, which

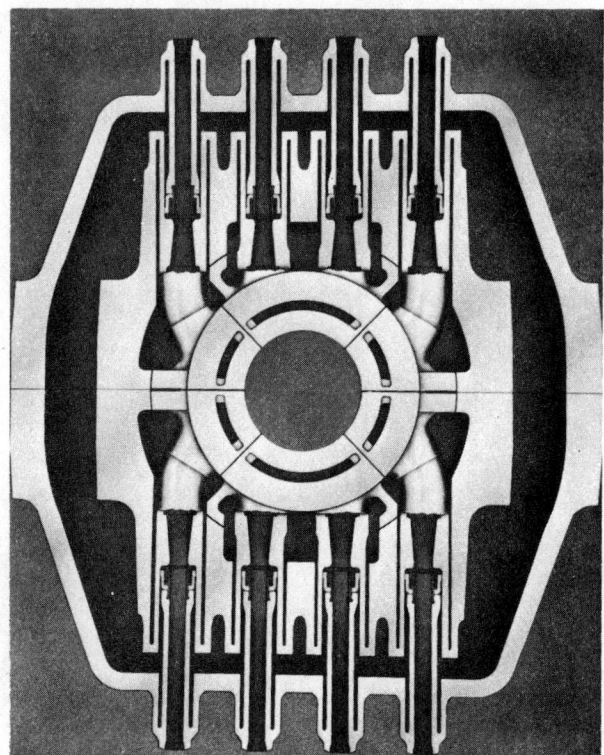

Fig. 73.36 Individual nozzle chambers.

is used in both the high- and low-temperature sections of the turbine. A modified version of the 12% Cr material, type 422, is used in the highest-temperature blade rows, especially for the partial-arc admission first stage. The longer last row blades may use 17-4 PH, while titanium is being more widely applied in the next to last rotating blades because of its resistance to corrosion resulting from deposition of contaminants in the steam. Ferritic steels are used for service at 1050°F and lower. Higher-temperature applications have used austenitic superalloys like K42B in the Eddystone #1 superpressure element.

A variety of blade root fastenings are used and include rationalized single-tee roots, the double-tee root, the side entry root (axial, angled, and curved) and a variety of pinned types, Figs. 73.37 and 73.38.

The blades are subjected to steady forces (centrifugal and bending). Superimposed on this are vibratory stresses associated with blade wakes and other sources of excitation. Partial-arc admission stages are also subject to shock loading as the rotating blades successively pass through active and inactive arcs of steam.

The general force systems acting on partial-arc first stage blades are shown diagrammatically in Fig. 73.39. As with full-arc admission blades, the forces are both steady and alternating. In contrast to the long blades used in the LP turbines, the first-stage blades are short and operate at much smaller mean diameters. Centrifugal forces are a minor factor in partial-arc stages except for the rotor body directly under the stage when tangential stresses are high and in the rotor side entry steeple grooves which receive the blade root. Steam forces on the rotating blades are those associated with the torque on the rotor and with the pressure drop across the row. The two forces determine the steady steam stresses in the blade. The centrifugal and bending stresses are used to evaluate the available endurance strength from a Goodman diagram.

During partial-arc operation, the blading is subjected to shock loading as a result of entering and leaving an active arc of admission every revolution. The shock force pattern and the resulting vibrations are illustrated in Fig. 73.40. The shock force induces free vibrations, which occur at the blade natural frequency. Since blade loading depends on the overall stage pressure ratio and is somewhat inversely proportional to the active admission arc length, blade vibrations resulting from partial-arc operation are most severe when the first valve point is reached.

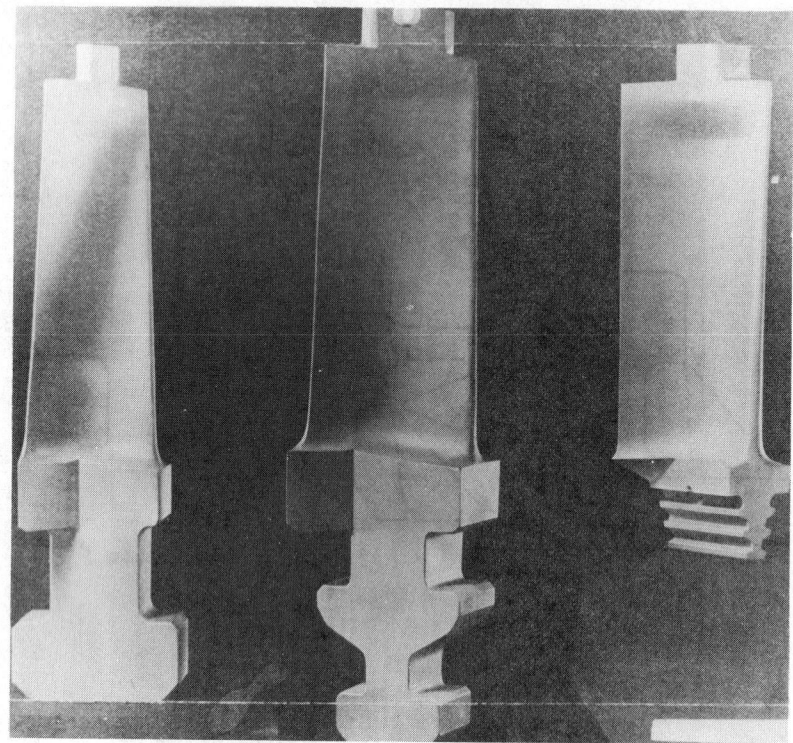

Fig. 73.37   Single-tee, double-tee, and side entry roots.

In addition to the partial-arc shocks, other higher-frequency alternating forces are present. Since steam velocities exiting from the nozzles are not uniform and the nozzle vanes are of finite thickness, steam forces acting on the rotating blades have vibrations over the nozzle pitch that are repeatable for every active nozzle. The diagram in Fig. 73.41 illustrates the exciting forces occurring at a frequency equal to the number of nozzles passed per second. Resonance occurs when the nozzle wake frequency coincides with the natural frequency of one of the blade group modes.

73.6.2   Rotors

For steam temperatures up to about 900°F, rotor alloys contain Ni with some Cr and Mo. As temperatures increase further, the Mo level is increased along with the addition of V. Many of the rotors used at 1000 and 1050°F have the familiar Cr–Mo–V composition.[21] Some vendors use a 12% Cr rotor alloy at 1000°F. While the high-temperature strength of this alloy is somewhat higher than that of Cr–Mo–V, it has considerably better oxidation resistance, especially at 1100°F.

The alloying elements used in the high-temperature rotors increase their FATT (fracture appearance transition temperature). Below the FATT, the materials are notch sensitive and brittle fracture is a concern. As a result, the LP rotors use an Ni–Cr–Mo–V material that has a low transition temperature. However, as temperatures increase beyond 700°F, this material becomes increasingly sensitive to temper

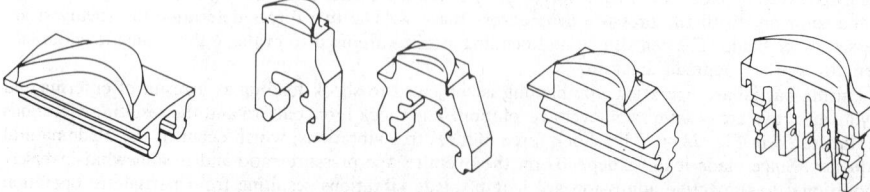

Fig. 73.38   Finger roots.

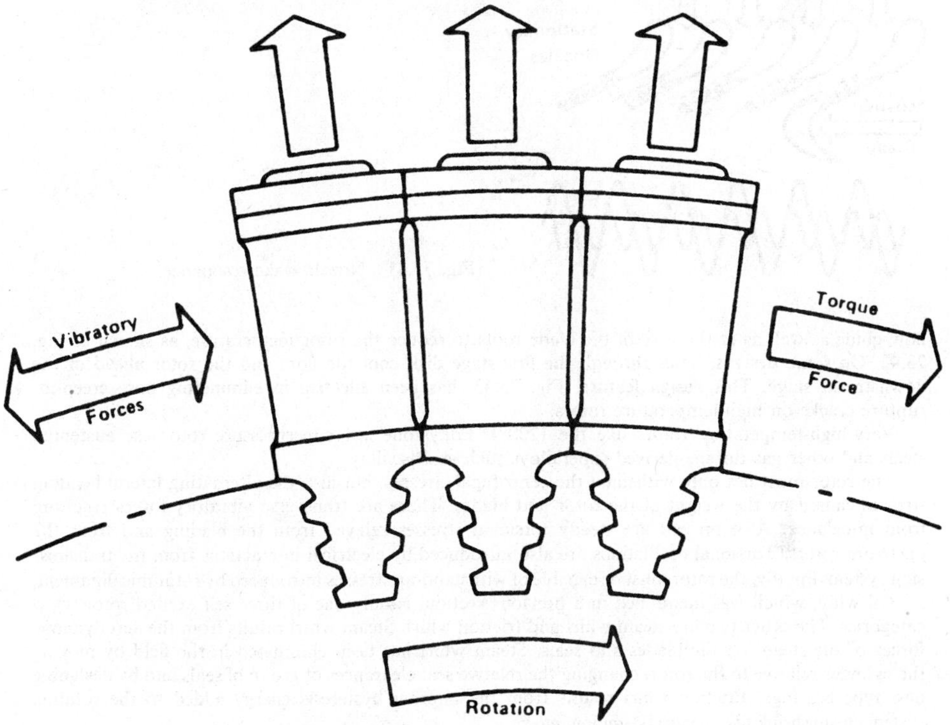

Centrifugul Force

Vibratory

Forces

Torque

Force

Rotation

Fig. 73.39 Forces acting on partial-arc blades.

embrittlement.[22] Because of the low throttle temperatures on LWR plants, LP rotor materials are an appropriate choice for the HP element.

Temper embrittlement relates to the segregation of tramp elements, such as phosphorous, tin, arsenic, and antimony, to grain boundaries and consequent embrittlement. The grain boundary segregation of tramp elements is exacerbated by the presence of Mn and Si in the steel.

The high-temperature rotor materials are judged primarily by their resistance to rupture and elongation (creep–rupture) as well as by their high-temperature notch sensitivity (toughness). Modern forgings increase toughness by using techniques such as vacuum carbon deoxidation, vacuum arc remelt, electroslag remelt, and low-sulfur melting.[23]

To increase the rotor groove wall strength at the HP and IP inlets, cooling schemes are often implemented. In the IP inlet, cool steam from the HP element exhaust is introduced under the inlet

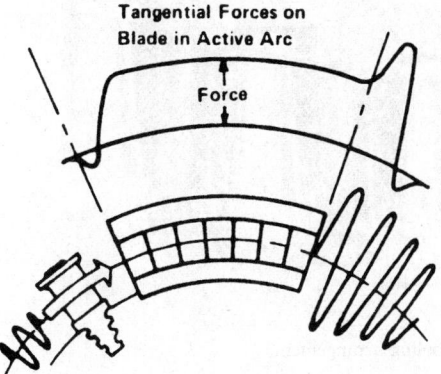

**Tangential Forces on
Blade in Active Arc**

Force

Fig. 73.40 Partial admission shock.

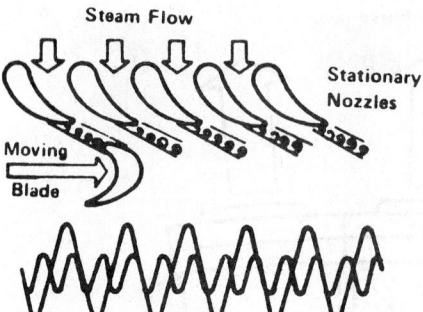

Fig. 73.41   Nozzle wake resonance.

flow guide as well as at the base of the blade roots to reduce the rotor temperature, as shown in Fig. 73.42. On some designs, holes through the first-stage disk cool the bore and the rotor ahead of the downstream stage. This design feature, Fig. 73.43, has been effective in eliminating bore creep-to-rupture cracks on high-temperature rotors.

Very-high-temperature rotors like the 1200°F Eddystone #1 superpressure rotor use austenitic steels and other gas-turbine-derived superalloys such as Discalloy.

The rotor must not only withstand the centrifugal stresses, but also the alternating lateral bending stresses caused by the weight of the rotor and blades. There are transverse vibratory forces resulting from imbalance. Also present are steady torsional stresses relayed from the blading and from the upstream rotors. Torsional oscillations are also introduced by electrical interactions from the transmission system. Finally, the rotor must be capable of withstanding stresses introduced by rotor misalignment.

Oil whip, which was mentioned in a previous section, is only one of three self-excited rotor whirl categories. The other two are steam whirl and friction whirl. Steam whirl results from the aerodynamic forces of the steam on the blades and seals. Steam whirl has been eliminated in the field by moving the cylinder relative to the rotor, changing the relative seal clearances of group of seals and by designing new type bearings. Friction whirl results from the internal hysteresis energy added to the rotating system due to shrink fits, material damping, etc.[24]

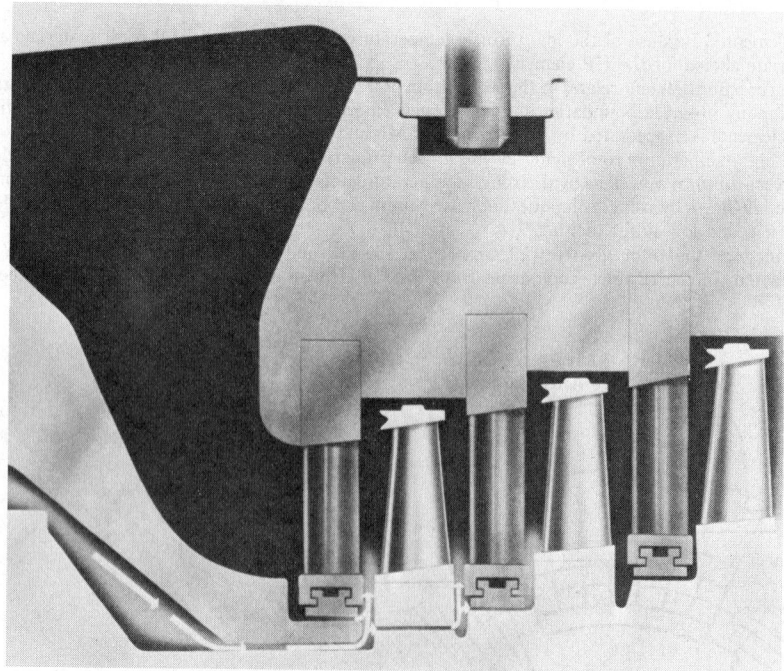

Fig. 73.42   IP rotor cooling arrangement.

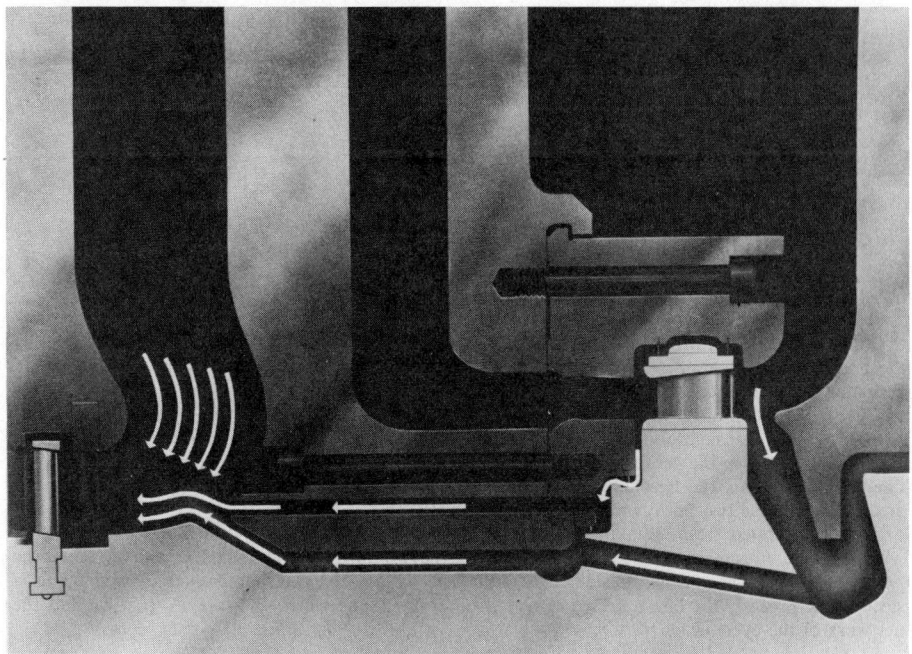

Fig. 73.43 HP rotor cooling arrangement.

73.7 CYCLE PERFORMANCE

The evolution of the steam cycle resulted in successive applications of the Rankine cycle, the regenerative Rankine cycle, and the reheat cycle. The Rankine cycle is the simplest and is applied today for mechanical drives, heat recovery cycles, and applications where maximum power to weight cycle applications are important.

73.7.1 Rankine Cycle

The Rankine cycle, which was used in the early steam plants, consists of compression of liquid water, heating and evaporation in the heat source or boiler (heat addition), expansion of the steam in the prime mover, and condensation (heat rejection) of the exhaust steam. Rankine cycles involve a straight expansion of the steam with no internal heat transfer. There is only one stage of heat addition which distinguishes it from other steam cycles.

A major portion of the heat added to the working fluid of a saturated steam Rankine cycle occurs during evaporation. For example, consider a cycle that generates 2.76 MPa (400 psia) steam and exhausts at 21.2°C (70°F). The enthalpy of the saturated water is 986.7 kJ/kg (424.2 Btu/lb) and the enthalpy of the saturated steam is 2801.9 kJ/kg (1204.6 Btu/lb). The enthalpy of liquid water at 21.1°C (70°F) is 88.4 kJ/kg (38.0 Btu/lb). So the change in enthalpy during boiling is 1815.2 kJ/kg (780.4 Btu/lb) compared to a total enthalpy change of 2713.1 kJ/kg (1166.6 Btu/lb) or 67% of the total heat added.

However, since 33% of the heat added is required to heat the water to its saturation temperature, the Rankine cycle effective heat-addition temperature is lower than a Carnot cycle with a heat added temperature of 299.4°C (445°F), the saturation temperature at 2.76 MPa (400 psia). The effective heat-addition temperature of the Rankine steam cycle is 215.6°C (420°F), so it would have a lower thermal efficiency than a Carnot cycle.

The effect of pressure on the relative magnitudes of the sensible and the evaporative (latent) heat during heat addition with a saturated steam cycle, as well as the level of exhaust moisture after an isentropic expansion, is illustrated in the following table for an ideal Rankine cycle. The data are for a condensing temperature of 21.2°C (70°F):

Pressure (psia)	Sensible Heat (Btu/lb)	Latent Heat (Btu/lb)	Exhaust Moisture (%)
400	388.2	780.4	29.5
800	471.8	689.6	32.9
1200	533.9	613.0	35.3
1600	586.2	540.3	37.4
2000	634.1	466.2	39.3
2400	681.0	384.8	41.4

Figure 73.44 is a representation of the idealized 2.76 MPa (400 psia) saturated cycle on a temperature–entropy, T–s, diagram. Idealized T–s cycle diagrams are a simple and convenient method of illustrating the effects of cycle parameters and cycle variations. While the actual cycles do not exploit the full potential of the ideal counterparts, the underlying principles are still valid.

In Fig. 73.44, line AB represents the liquid heating, line BC represents the evaporation, line CD represents the expansion, and line DA represents the heat rejection of a theoretical saturated steam Rankine cycle. Since the compression power is so small, it was ignored in the figure. The area below line ABC represents the heat added, the area below line DA represents the heat rejected, and the area between these two lines represents the work.

In the saturated steam Rankine cycle diagram of Fig. 73.44, the heat added is $Q_A = h_c - h_a$, where h_c and h_a are the respective enthalpies of the working fluid at points C and A, since the process in the boiler can be considered as either steady flow or constant pressure. Since DA is a steady flow process, the heat rejected is $Q_R = h_d - h_a$. Using work, $W = Q_A - Q_R$, we find the net work of the cycle to be $W = h_c - h_a - (h_d - h_a) = h_c - h_a + h_a - h_d = h_c - h_d$.

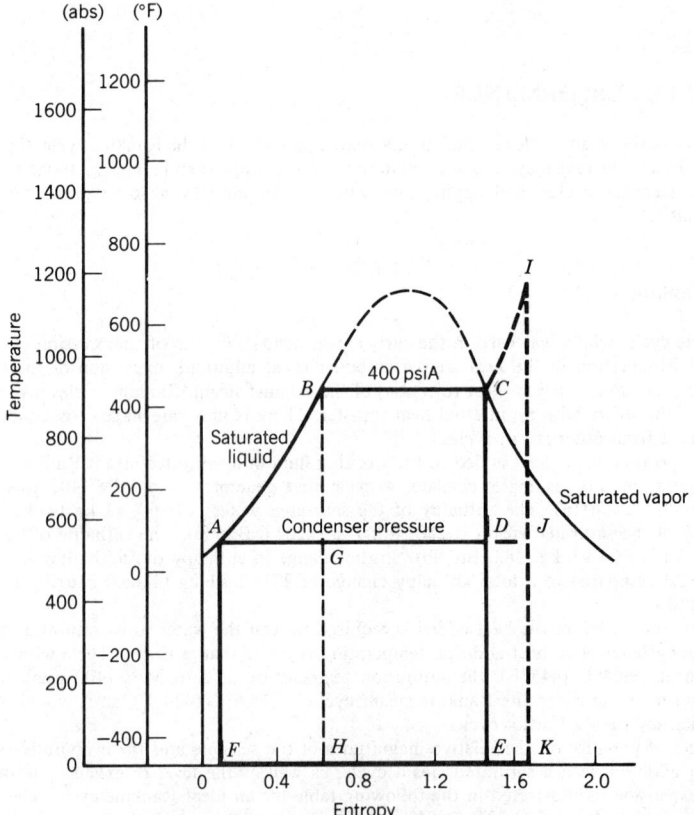

Fig. 73.44 Theoretical Rankine cycle.

The Rankine cycle thermal efficiency (neglecting pump work) is

$$\eta = \frac{h_c - h_d}{h_c - h_a} \tag{73.12}$$

where the subscripts refer to Fig. 73.44 and when, in general, h_c is the enthalpy of the steam (whether wet, dry, or superheated) as it leaves the steam generating unit, h_d is the enthalpy of the steam at the end of the isentropic expansion or at the turbine exhaust, and h_a is the enthalpy of the saturated liquid at the exhaust or condenser pressure.

In actual practice, the presence of moisture reduces the prime mover efficiency during the expansion process. Since the average moisture of the expansion of the superheated steam cycle of Fig. 73.44 (dashed line) is lower than that of the saturated steam cycle, its turbine efficiency is higher. All of the expansion and compression processes (shown on the $T\text{–}s$ diagrams) represent 100% prime mover efficiency.

73.7.2 Regenerative Rankine Cycle

Suppose that the liquid heating line AB of Fig. 73.44 could be eliminated. Since the ratio of the area above line AG to the area below line AB is poorer than the ratio of the area above line DG to the area below line BC, the thermal efficiency would improve if the heat addition from A to B were eliminated. The regenerative Rankine cycle is the means of eliminating all or part of the external heating of the liquid water to its boiling point.

In this cycle, a small amount of expanded steam is extracted from a series of pressure zones during the expansion process of the cycle, line CD of Fig. 73.45, to heat liquid water in a multiplicity of heat exchangers to a higher temperature, point B. The resulting effect is the removal of the cross-hatched area of Fig. 73.45 from the Rankine diagram of Fig. 73.44. The theoretical and the practical regenerative cycles reduce both the heat added to the cycle (because the external heating of the liquid is either reduced or eliminated completely) and the heat rejected from the cycle. The cycle work is

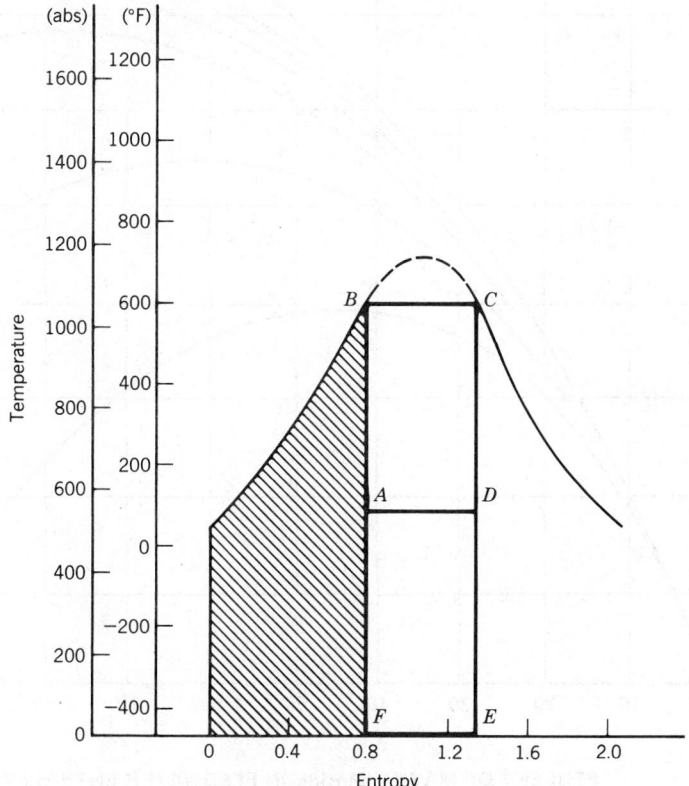

Fig. 73.45 Theoretical regenerative cycle.

reduced as flow is continuously removed during the expansion to preheat the liquid. However, the ratio of work to heat added is improved.

In actual plants, only a finite number of stages of feedheating are practical. In addition, the decreasing incremental efficiency improvements and the increasing incremental cost additions associated with higher and higher final feedwater temperatures place a practical limit on the maximum final feedwater temperature. The optimum final feedwater temperature increases as main steam pressure increases.

73.7.3 Superheat and Reheat Cycles

Superheat Cycles

The efficiency of the steam cycle can be improved if the steam were superheated, that is, heated beyond its saturation temperature. This is illustrated by the heavy dashed line in Fig. 73.44 where the final steam temperature is 371.1°C (700°F), rather than 238.7°C (460°F) with the saturated steam cycle. The line CI represents the superheating. The incorporation of CI has improved the ratio of the work (area above the heat rejection line) to the heat added (area below the heat added line) in the cycle, so the thermal efficiency is improved.

In addition to improving the ideal or theoretical cycle efficiency, superheating of steam improves the turbine efficiency by reducing the moisture level in the steam during the expansion process. The presence of moisture in the steam results in efficiency losses in the turbine and can also cause erosion

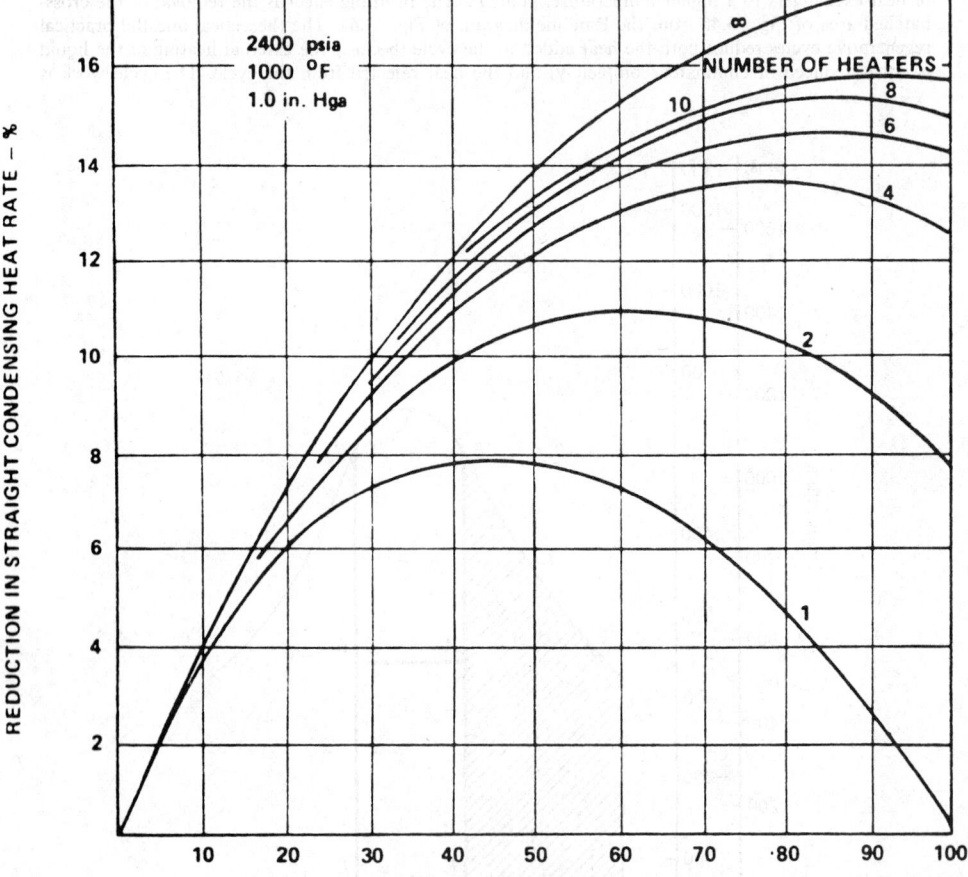

Fig. 73.46   Regenerative cycle heat rate effect.

of the blades. The latter condition can result in blade failure and further reduces blading efficiency. The amount of moisture in the steam is proportional to the distance the expansion line (*CD* for the saturated cycle and *IJ* for the superheat cycle) is inside the steam dome.

Superheat is incorporated not only in the Rankine cycle but also in the regenerative Rankine cycle and the reheat cycle. The incorporation of superheat increases the losses that occur during the internal heat-transfer processes of a regenerative cycle. However, the beneficial effects of superheat, as shown in Fig. 73.44, more than offset the increased losses during regenerative feedheating with superheated steam.

As was noted in the previous section, the optimum final feed temperature increases as main steam (throttle) pressure is increased. For example, many plants with throttle steam conditions of 58.6 MPa, 482°C (850 psig, 900°F) had final feedwater temperatures up to 210°C (410°F), while plants with throttle steam conditions of 10 MPa, 538°C (1450 psig, 1000°F) had final feedwater temperatures up to 232°C (450°F). In addition, the optimum final feedwater temperature is related to the number of stages of feedheating. Figure 73.46 illustrates the effect of the number of stages of feedheating on both the optimum temperature rise and the cycle efficiency.

While the Carnot cycle indicates that increasing the initial (throttle) temperature increases cycle efficiency is maintained. To be specific, an increase in throttle pressure for a given plant capacity will the area above the condenser pressure level of Fig. 73.47 is proportionately larger than the area under the heat addition line for the 10.0 MPa (1450 psia) cycles as compared to the 2.76 MPa (400 psia) cycle.

The potential for increased cycle efficiency as shown in Fig. 73.47 can only be realized if component efficiency is maintained. To be specific, an increase in throttle pressure for a given plan capacity will reduce the required blade heights and flow area of the initial turbine stages, resulting in an increase in leakage losses and a turbine efficiency impairment. The reduction in blade heights can also increase other blading losses. To restore the turbine efficiency or minimize the decrease if efficiency, an increase in plant capacity might be required. So, increases in throttle steam pressure must be judicious in order to preserve the theoretical increases in thermal efficiency.

In addition, an increase in initial pressure without any change in initial temperature on a Rankine

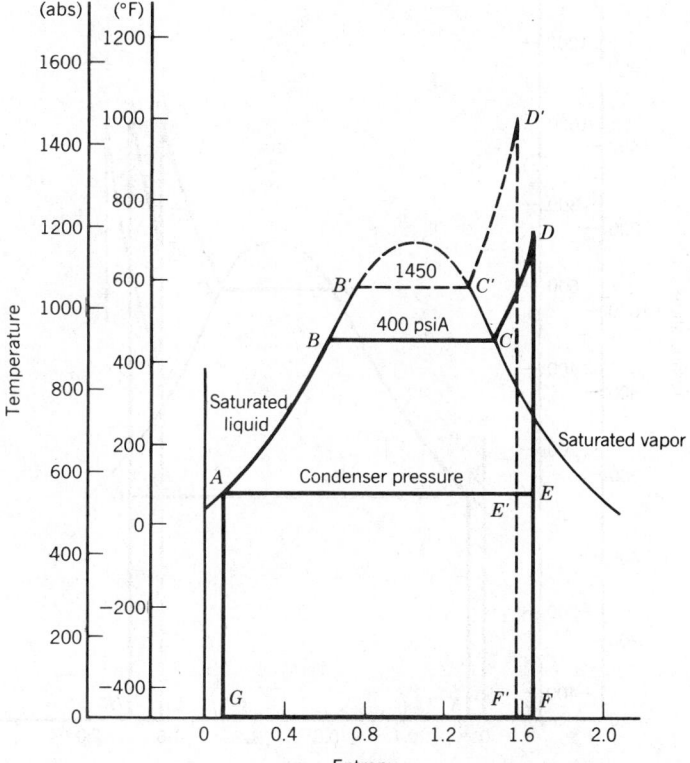

Fig. 73.47 Effects of pressure and temperature on cycle efficiency.

cycle would increase the moisture in the latter stages of the turbine. This would decrease turbine efficiency and offset some of the theoretical gain attributable to increased initial pressure.

The incorporation of regenerative feedheating reduces plant output for a given throttle steam flow because both the cycle heat addition and the cycle heat rejection are decreased. In order to maintain plant output, the throttle flow must be increased. This has the desirable effect of increasing the heights of the short blades in the initial turbine stages while the exhaust stages of the turbine, which are already quite long, receive less flow. The overall result is an increase in turbine efficiency.

Reheat Cycle

In the reheat cycle, the steam, after partially expanding through the turbine, is returned to the reheater section of the boiler where additional heat is added to the steam to increase its temperature. After leaving the reheater, the steam completes its expansion in the turbine.

Figure 73.48 is a representation of a theoretical reheat cycle on a *T–s* diagram. If the throttle steam at point *D* expands only a slight amount to point *E* and is reheated to *F*, the heat addition temperature of the reheat energy is very high. However, a very small amount of heat is added, so it has a very small effect in raising the combined heat-addition temperature of the total boiler heat input, line *ABCD* and line *EF*. In contrast, an initial expansion from point *D* to point *I* (the expansion indicated by the dashed line) has a lower effective heat-addition temperature, line *IJ*, than *EF*. However, much more heat is added during *IJ* and so it has greater weight in the total heat addition to the cycle.

The optimum reheat pressure in the case of infinite stages of feedwater heating and full regeneration is about one-third of the throttle pressure for the steam conditions given.[25] For plants with practical final feedwater temperatures (highest-temperature heater being supplied with cold reheat steam) and number of heaters, the optimum reheat pressure is 20–25% of the throttle pressure. When an additional heater is added (supplied with steam at a point between the cold reheat line and the throttle), the optimum reheat pressure is somewhat lower. The incorporation of reheat in a practical regenerative

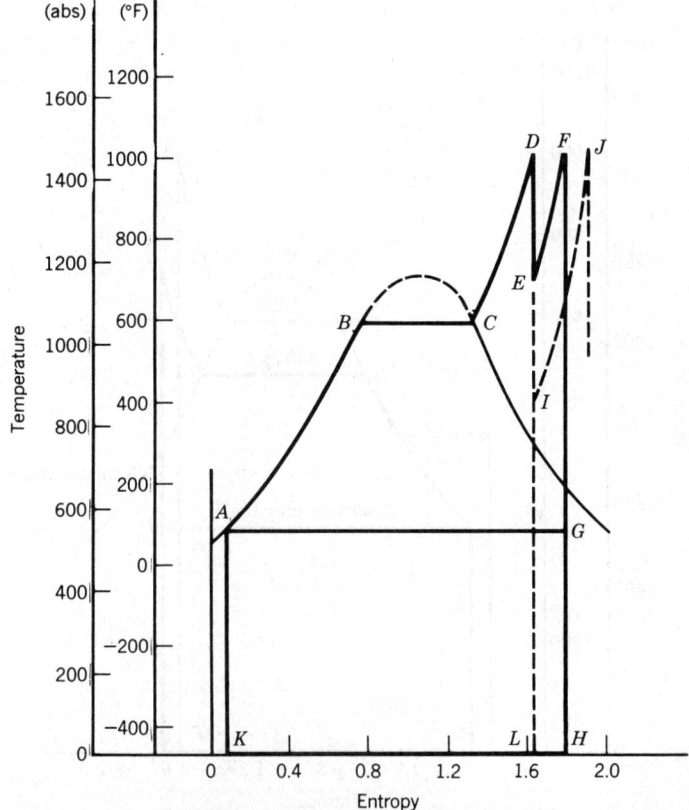

Fig. 73.48 Theoretical reheat cycle.

reheat cycle improves heat rate or efficiency by 4.0–5.0%, the higher values corresponding to higher throttle pressures.[26,27]

An evaluation of reheat cycles noted that a significant portion of the theoretical gain from reheating is overcome by pressure losses in the reheat piping and the reheater.[17] However, the accompanying reduction in turbine moisture losses recoups much of the theoretical improvement.

The widespread adoption of reheat in the early 1950s resulted in relatively rapid increases in throttle pressure from 10 MPa (1450 psig), to 12.4 MPa (1800 psig), to 12.8 MPa (2000 psig), and finally to 16.55 MPa (2400 psig).* Throttle temperature/reheat temperature combinations were 538°C/538°C (1000°F/1000°F), 565°C/538°C (1050°F/1000°F), and 565°C/565°C (1050°F/1050°F). A number of 16.55 MPa, 593°C/565°C (2400 psig, 1100°F/1050°F) units were also built.

A second stage of reheat improves cycle efficiency by about one-half the value realized from adopting single reheat. Moreover, to avoid the occurrence of superheated steam at the turbine exhaust, double reheat has been used only on plants with supercritical throttle pressures [above the critical pressure of 22.12 MPa (3208 psia)].

The first supercritical double reheat units had steam pressures considerably above 23.13 MPa (3500 psig) and steam temperatures well above 538°C (1000°F). The 120 MW, 31.03 MPa, 621°C (4500 psig, 1150°F) Philo #6 unit with reheats to 565°C (1050°F) and 538°C (1000°F) and the 325 MW, 34.47 MPa 649°C (5000 psig, 1200°F) Eddystone unit with two reheats to 565°C (1050°F) were the pioneering supercritical units.[28,29] Subsequently, utilities purchased 21 additional supercritical double-reheat units and over 150 supercritical single-heat units in sizes ranging from 250 to 1300 MW.

When a second reheat is incorporated in a 23.13 MPa, 538°C/538°C (3500 psig, 1000°F/1000°F) single-reheat cycle, plant efficiency increases by 1.6–2.0%. In each case, the highest pressure feedheater extracts steam from the cold reheat line of the first reheat.

The majority of double reheat units have steam conditions of 23.13 MPa, 538°C/552°C/565°C (3500 psig, 1000°F/1025°F/1050°F), while most of the supercritical single-reheat units operated at 23.13 MPa, 538°C/538°C (3500 psig, 1000°F/1000°F).

73.7.4 Moisture Separation and Steam Reheat

Until the 1950s, essentially all steam plants used fossil fuels as the energy source. The introduction of nuclear energy produced a radical change in the cycle and related hardware as compared to the advanced technology of the fossil applications. While the cycles and generic steam conditions of the nuclear plants were similar to those of the early days of power generation, the large equipment sizes and plant complexity required cycle innovation. Thermodynamic concepts similar to fossil units, but with special variations, were developed to meet reliability and efficiency concerns.[30]

The current generation of nuclear plants are practically all LWR plants. Two types of LWR plants have been developed: pressurized water reactors (PWRs) and boiling water reactors (BWRs). The PWRs can be either recirculating types, which produce slightly moist steam (less than 1% moisture), or once-through designs, which produce somewhat superheated steam [less than 56°C (100°F) of superheat]. The BWRs produce slightly wet steam like the recirculating PWRs. Initial steam pressures on the early PWR plants were about 4.14 MPa (600 psig) and were 6.55 MPa (950 psig) on BWRs. Current PWR pressures are in the 6.21–7.58 MPa (900–1100 psig) range, while BWR pressures have changed only slightly.

Cycle with External Moisture Removal

Because of the high moisture, 20–25%, that would be formed in a direct expansion to condenser pressure, the early plants used an external moisture separator at pressure levels ranging from 10% to 25% of the inlet throttle pressure. This arangement reduced the turbine exhaust moisture to a satisfactory level of 13% from the standpoint of blade erosion and resulted in a 2.5% improvement in thermal efficiency over a simple regenerative cycle. Figure 73.49 illustrates an expansion line with external moisture removal as indicated by the dotted line from point A to point B.

The expansion line is plotted on an enthalpy–entropy, h–s, diagram or Mollier chart. The discontinuities on the expansion line between point B and point H, as shown at points C, D, E, F, and G, indicate the use of internal (blade path) moisture-removal devices. Similar discontinuities are shown on the expansion lines of Figs. 73.50 and 73.51. While the cycle of Fig. 73.49 is often referred to as a nonreheat cycle, the removal of moisture (dotted line from point A to point B) is thermodynamically equivalent to reheating.

---

* There was a trend toward reheat in the 1920s. The cycle fell from favor in the 1930s because of operational difficulties, control complexity, and advancement in regenerative Rankine cycle technology.

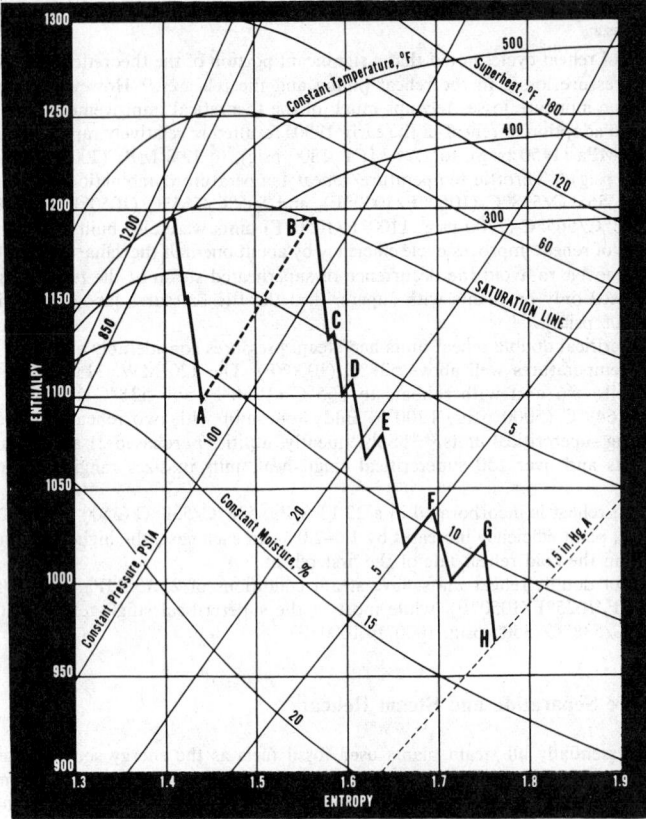

Fig. 73.49 Nonreheat expansion line.

Steam-to-Steam Reheat Cycle

The later generation plants (and practically all current applications) made use of steam-to-steam reheating immediately after external moisture separation. Steam from either the throttle (single-stage reheat) or a combination of throttle and extraction steam (two-stage reheat) is used to reheat the separator exhaust steam to a temperature that is about 14°C (25°F) below the saturation temperature of the heating steam.

The moisture separator is denoted by the dotted line between points B and C of Fig. 73.50, while the single-stage reheat is denoted by the dotted line between points C and D. For the two-stage reheat cycle, Fig. 73.51, the dotted line up to point A relates to the moisture separation, the portion between points A and B relates to the first-stage (extraction steam) reheat and the portion between B and C relates to the second-stage (throttle steam) reheat.

Based solely on ideal considerations, the use of steam-to-steam reheat results in a loss in theoretical cycle efficiency or very little gain. However, the reduction in downstream stage moisture results in improved turbine efficiency. As a result, the application of single-stage reheat improves the cycle efficiency by 1.5% (depending on the throttle pressure) over a regenerative cycle with external moisture separation. The gain from single reheat can be increased by an additional 0.3–0.5% by adopting two-stage reheat.

73.7.5 Heat Rate and Heat Balances

Rather than using the conventional measure of performance, thermal efficiency η, power plant designers use a quantity known as heat rate. The heat rate is the number of heat units that are required to generate 1 kWhr of electricity. The heat rate is most often defined in kilojoules per kilowatt hour, kJ/kWhr, or Btu per kilowatt hour, Btu/kWhr,

$$\text{heat rate} = \frac{Q_A}{KW} \tag{73.13}$$

since

$$\eta = \frac{W}{Q_A} \tag{73.14}$$

It can be seen that heat rate is the reciprocal of thermal efficiency with the work and heat added being defined in the proper units.

An essential analysis tool in the design and evaluation of a power plant is an accounting of the mass flow and energy in the various components of the plant system. Tabular and graphical representations that exhibit this information are described as energy balances. A more familiar term that has been applied to the graphical representation is "heat balance."

In power practice, heat balances may relate to the complete plant system or to the turbine generator and its related equipment. The turbine cycle heat balances may be categorized according to the way in which the feedwater pump power is treated. Prior to the 1950s main steam pressures were comparatively low by today's standards and the pump power was a relatively minor quantity. So, the power output on the turbine cycle heat balances did not account for the power required to drive the feed pumps. As a result the turbine cycle output and heat rate were gross values. With the increase in main stream pressures since 1950, which increased pump power considerably, and the increasing use of steam turbine drives for the pumps, a more realistic approach was adopted. In this case the turbine cycle output and the heat rate accounted for the power required by the feedwater pumps. The heat rates and turbine outputs are defined as net values in this instance. Figure 73.52 illustrates the influence of throttle pressure and main unit rating on pump power.

Since the cycle heat input of a turbine heat rate includes only the heat that is transferred to the steam and not the heat released by the fuel, the plant heat rate will be higher than turbine heat rate. In addition, there is power required to drive plant auxiliaries such as pulverizers, soot blowers, fans,

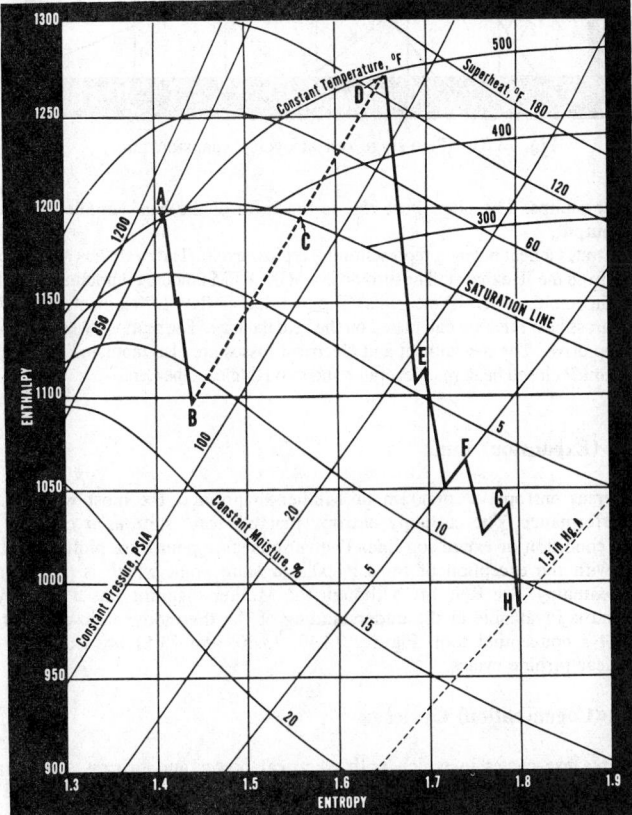

Fig. 73.50 Reheat cycle expansion line.

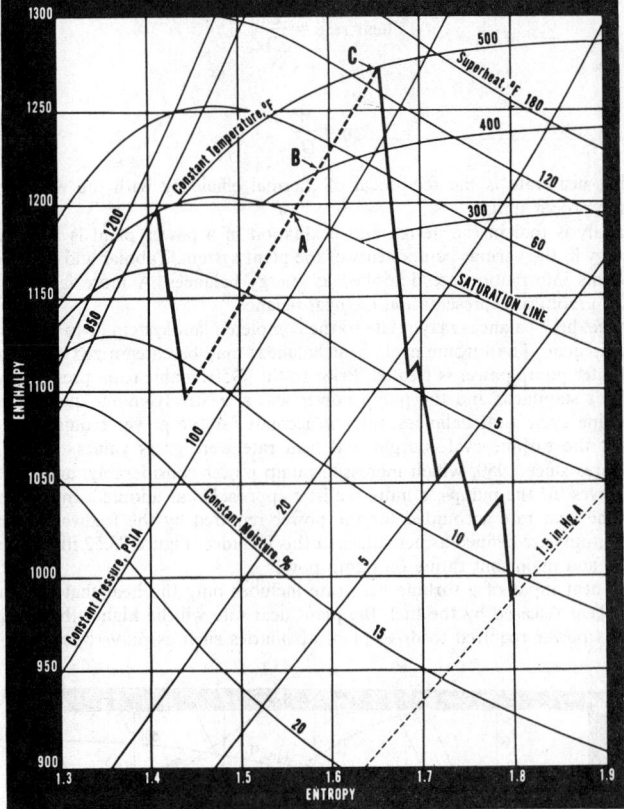

Fig. 73.51 Two-stage reheat cycle expansion line.

and circulating water pumps. The plant heat rate is the ratio of the heat supplied by the fuel to the net plant electrical output.

Figure 73.53 illustrates a heat balance representing a typical cycle. The cycle has a turbine driven pump which extracts steam from the IP exhaust. The turbine is a 3600 RPM tandem compound design. The HP and IP expansions occur in combined HP-IP elements. The two double-flow LP elements have 26-in. last-row blades. Both net and gross heat rates are calculated on the heat balance. The pump power, 11455KW, includes the losses in the pump drive. The mechanical and electrical losses are also tabulated. The cycle has seven feedwater heaters: four LP closed heaters, adeaerator and two HP closed heaters.

73.7.6 Condition (Expansion) Lines

The enthalpy, h, versus entropy, s, diagram or Mollier diagram is the most widely used chart in established cycle performance. The enthalpy–entropy plot of steam state as it expands through the turbine is called the condition or expansion line. The various state points are plotted at static pressure and total enthalpy with the exception of the LP exhaust state point which is plotted at condenser pressure and static enthalpy. See Ref. 14. Although the Mollier diagram was a primary tool in the past for engineers and is invaluable in the understanding of the thermodynamic cycles, the computer has made it more of a conceptual tool. Figures 73.49, 73.50, and 73.51 are examples of condition lines for typical nuclear turbine cycles.

73.7.7 Industrial (Cogeneration) Cycles

Many industrial plants use cycles in which both electrical power and process steam are produced; this is termed cogeneration. In most instances, steam is generated at a higher pressure and temperature than is required for the process. The steam is partially expanded in a turbine before being discharged to the process at the desired pressure and temperature. The turbines can be noncondensing or condensing

FEED PUMP POWER REQUIREMENTS VS. MAIN UNIT RATING

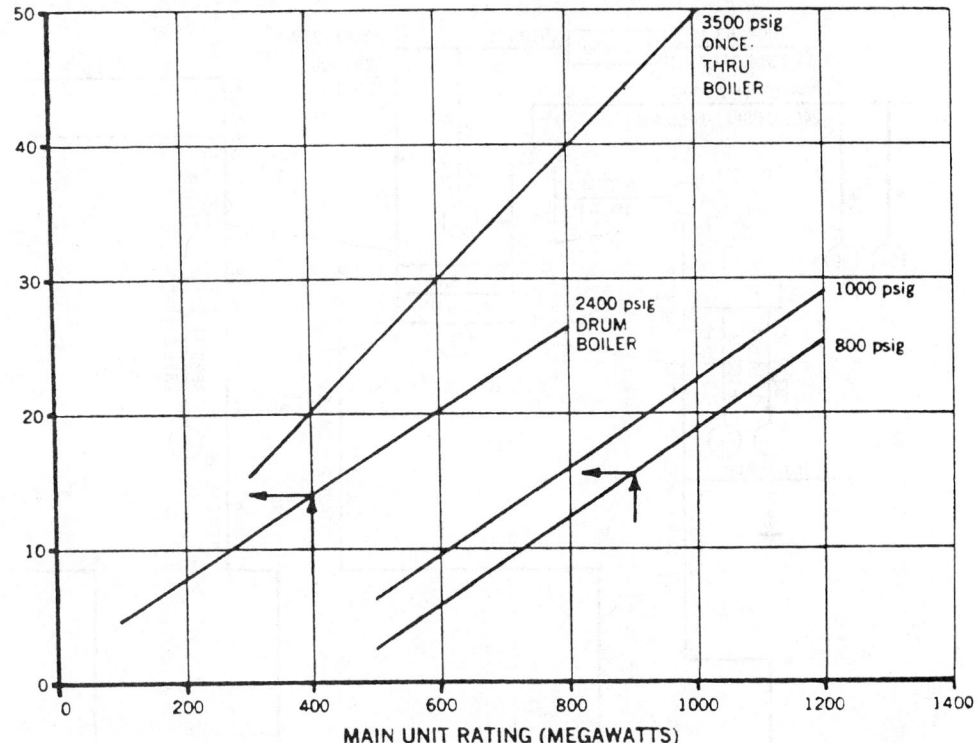

Fig. 73.52 Feed pump power requirements versus main unit rating.

designs. When the process steam pressure is higher than the turbine exhaust pressure, valving is incorporated at the pertinent locations in the turbine blading to maintain the pressure at the desired value. The turbines are typified by large mass flows and comparatively modest outputs. Cycle makeup is large and only a few stages of regenerative feedheating are used.

Prior to the Arab oil embargo in 1973, typical throttle steam pressures were 6.21 MPa (900 psig) and below, while steam temperatures were 441°C (825°F) or lower to simplify water treatment and reduce boiler cost. Current fuel prices have made it desirable to increase steam conditions to 8.62 MPa, 510°C (1250 psig, 950°F) and 10.0 MPa, 538°C (1450 psig, 1000°F).[31]

The operating range in which both process steam requirements (desired total energy, desired pressure, and desired temperature) and electrical power requirements are exactly satisfied is achieved by the use of turbine valves at the extraction locations. However, the range is limited and so either process requirements or electrical requirements must be unsatisfied at times. Since the processes are often continuous and should not be interrupted, the usual procedure has been to meet process demands and to supplement the electrical demand with purchases from an electric utility. In addition, pressure- and temperature-reducing stations are incorporated into the plant equipment so that boiler steam can be used to supplement and replace the steam supplied by the turbine to the process.

Figure 73.54 is a throttle steam–extraction flow–power output diagram for a single automatic extraction turbine (pressure level is controlled to meet process requirements). Some noteworthy, large, HP, high-temperature, congeneration plants have been or are being built on utility systems. For example, Linden 1, Public Service Electric & Gas Co. (PSE&G), rated at 225,000 kW, has steam conditions of 13.79 MPa, 538°C (2000 psig, 1050°F) and supplies almost 340,000 kg/hr (750,000 lb/hr) of steam to the Bayway Refinery of Exxon. Linden 2, also rated at 225,000 kW, has steam conditions of 16.20 MPa, 593°C/565°C (2350 psig, 1100°F/1050°F) and supplies 181,400 kg/hr (400,000 lb/hr) of process steam to the same refinery.[32,33] The two nuclear units at the Midland Nuclear Power Station, Consumers Power Co., will supply up to 1,840,000 kg/hr (4,050,000 lb/hr) of process steam to Dow Chemical Co., Michigan Division, Midland, Michigan, while generating 1,300,000 kW of electrical power.[34]

In some cogeneration applications additional steam is inducted into a lower-pressure zone of the turbine. The energy source for this low-pressure induction steam is generally waste heat from a process.

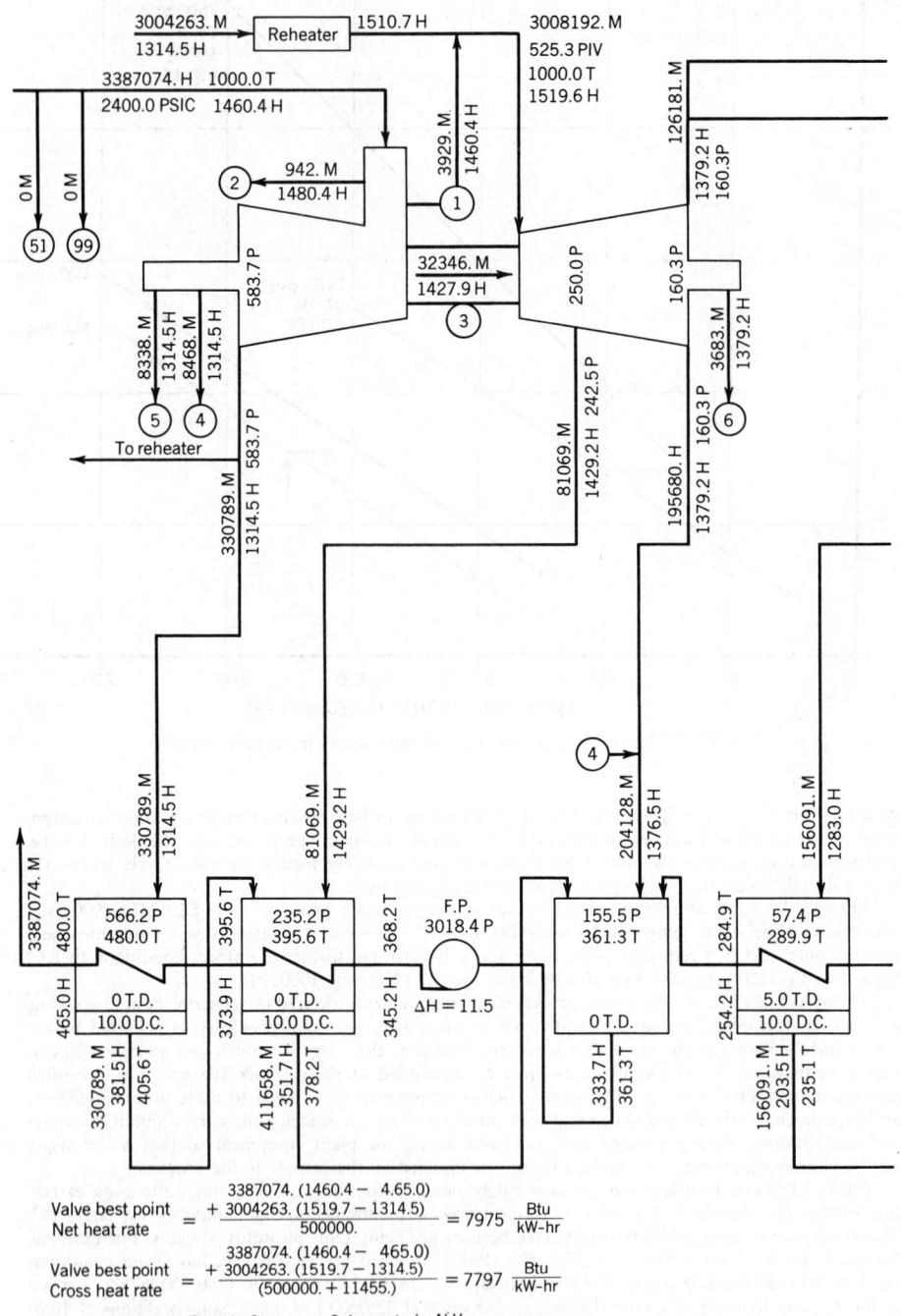

Fig. 73.53 Reheat turbine heat balance diagram.

Turbine and extraction arrangement is schematic only
For study purposes only

The value of the generator output shown on this heat balance is after all power for
excitation and other turbine-generator auxiliaries has been deducted.

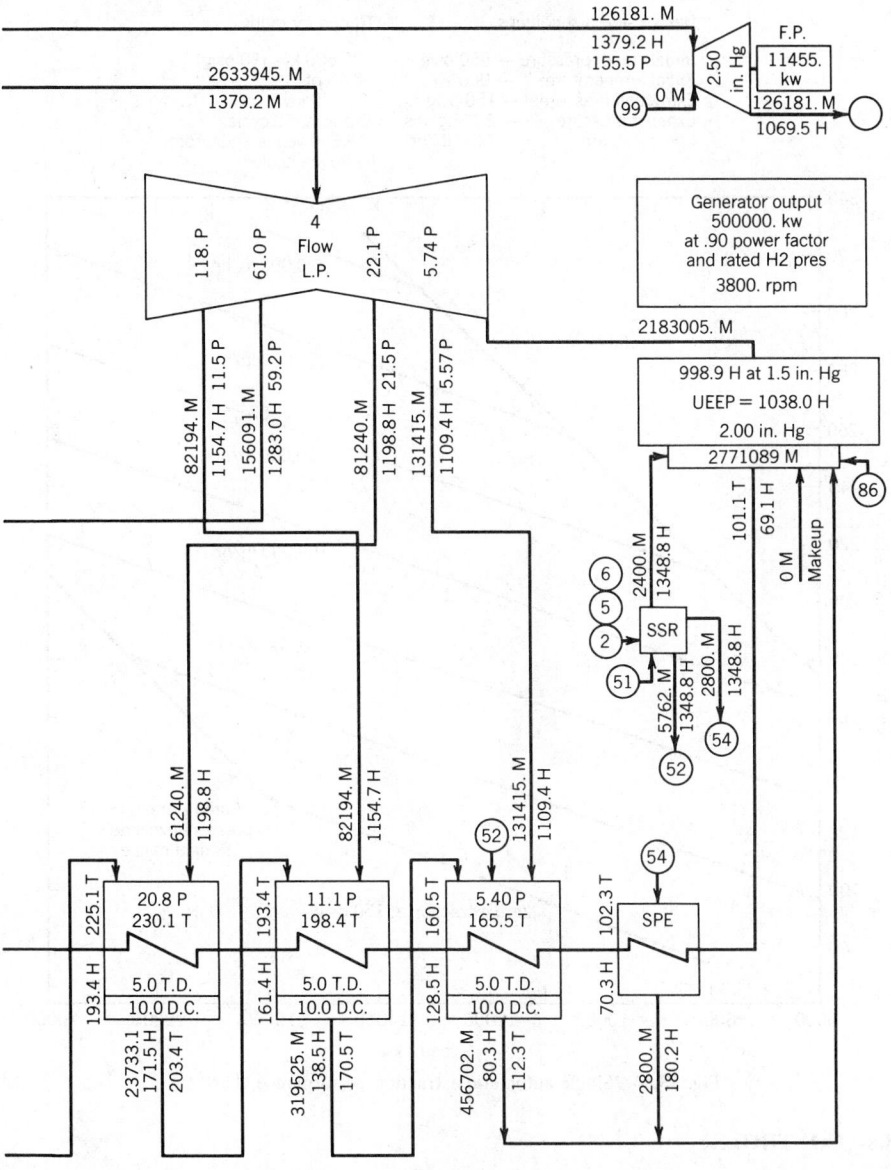

Generator output
500000. kw
at .90 power factor
and rated H2 pres
3800. rpm

Legend—calculations based
on 1967 Asme steam tables
M—Flow-lb/hr
P—Pressure-psia
H—Enthalpy-Btu/lb
T—Temperature-F degrees

500000. kW 2.00 in. Hg abs.   0 PCT MU
TC4F 26.0 in. LSB   3600 rpm
2400.0 psig   1000. / 1000. T
GEN-600000. KVA   .90 PF LIQ

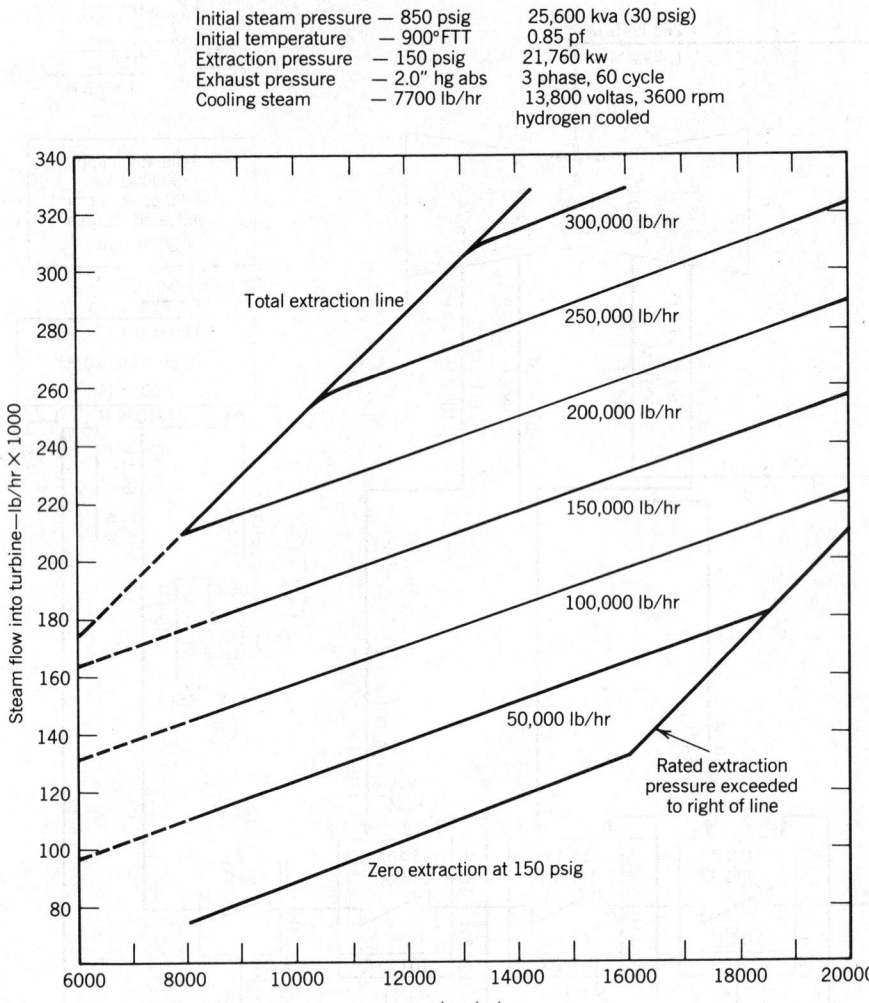

Fig. 73.54 Single automatic extraction performance chart.

73.8 CONTROLS

The controls in steam power plants can be grouped into two classes: emergency or protective systems and regulating systems that control load or flow or speed or some plant parameter or a combination thereof. The controls are supplemented with dedicated instrumentation or sensors that provide a basis for control action, equipment shutdown, or activation of alarms. The distinction between the two types of controls is not rigid in that some systems perform both a protective and a regulating function.

Emergency or Protective Controls

Major plant components have their own control devices that function as subloops in the overall plant control system. For the turbine generator system, valving is the most obvious protective system. However, there are a number of less obvious systems that activate alarms or trips and include:

1. Low-vacuum trip.
2. Low-bearing-oil-pressure trip.
3. Low-main-oil-pressure trip.
4. Rotor and differential expansion trip.
5. Rotor-position alarm (or trip).
6. Rotor-eccentricity meter and alarm.
7. Cylinder-expansion meter.
8. Thrust meter.
9. Overspeed trip.

73.8.1 Throttle Valves

The throttle valve is located at the turbine main steam inlet immediately upstream of the governing valves. The throttle valve is often used for initial roll of the turbine during startup. However, once synchronization occurs, the throttle valve is latched in the wide open position. Subsequent activation of the throttle valve by any of the emergency trip systems is a nonrecoverable action that closes the throttle valve (and reheat stop valve and opens the generator circuit breaker). The turbine will coast down to lower speed before the throttle valve can be relatched and steam readmitted to the turbine.

The throttle valve assembly can be incorporated in a separate valve body that is joined to the governor valve body as in Fig. 73.31. The throttle valve assembly may be part of a combined throttle valve and steam chest as in Fig. 73.30 or the throttle and governor valves may be in close proximity in a single valve body.

73.8.2 Reheat Stop and Interceptor Valves

In reheat plants additional protective valves are needed because of the large steam volume present in the reheater and reheat piping. Two valves—interceptor valve and reheat stop valve—are provided for double protection, Fig. 73.32. The interceptor valve of a fossil reheat turbine is similar to a throttle valve in appearance and is wide open during normal operation. The interceptor valve closes and then modulates during overspeed provided that a trip has not occurred. Under these circumstances the modulating action of the interceptor valve involves periodic opening and closing to discharge the steam trapped in the reheater while maintaining turbine speed below the trip setting.

A second line of protection, the reheat stop valve (a clapper type of valve), provides backup in case the interceptor valve malfunctions. When the reheat stop valve closes, it is a nonrecoverable action like the tripping of the throttle valve.

On some LWR turbines the interceptor valves are similar to their fossil counterparts. On other LWR turbines, both the reheat stop and interceptor valves are butterfly valves.

73.8.3 Antimotoring Protection

If the turbine steam flow is terminated by tripping of the throttle and reheat stop valves and if motoring of the generator maintains turbine speed, the turbine exhaust can overheat. Consequently, the occurrence and duration of motoring is controlled by antimotoring devices that interrupt the supply of electrical power to the generator.

73.8.4 Fast Valving

In the preceding section it was noted that tripping of the throttle and reheat stop valves is a nonrecoverable event. If the governor or interceptor valves could be closed before trip speed is reached in the time interval needed to clear faults or to successfully withstand transient disturbances at or near the power plant, the unit could be kept on line. Among the more recent innovations that maintain transient stability during a disturbance in the electrical system is fast valving or early valve activation. In order to reduce the turbine power output during a transient load disturbance, the interceptor valves are closed rapidly, kept closed for a short period of time, and then are opened rapidly.[35,36] This has sometimes resulted in lifting of the reheater safety valves. An improvement on this procedure involves simultaneous partial closure of the governor valves and a revised reopening rate for the interceptor valves.[37,38]

73.8.5 Load and Flow Control—Main Steam Governor Valves

The main steam governor valves may operate in unison or in a predetermined sequence to control speed and steam flow or load. These valves are fully modulating, being capable of operating anywhere between the closed to the wide-open position. For the case of full-arc admission first stages in the turbine, all of the first-stage nozzles are active at all levels of load and all of the governor valves are positioned simultaneously. With partial-arc admission first stages, the number of active first-stage nozzles is varied in response to load changes. The governor valves are individually positioned, controlling flow to specific groups of nozzles.

73.8.6 Controlled Extraction Valves

In order to maintain steam pressure for process steam on automatic extraction turbines, additional control valves are located immediately downstream of the controlled extraction point. These control valves may be located in a steam chest as shown in Fig. 73.55. When the controlled pressure is low and the volume of steam is large, a grid valve is used instead of a steam chest as shown in Fig. 73.56. The steam chest or grid valve discharges to a partial-arc admission stage to improve efficiency when load and extraction flow demand are variable.

Figure 73.57 is a longitudinal section of a double automatic extraction turbine (two controlled-extraction pressure locations). The rightmost steam chest at the turbine inlet contains the governor valves, which control speed and load. The two steam chests at the middle and left side of the turbine control steam pressure at these locations. Although the control valves at the controlled-extraction points regulate pressure, they affect the turbine output because of changes in the turbine flow through the downstream blading. Therefore, control signals from these extraction pressure control valves are integrated into the operation of the governor valves at the turbine inlet when controlling speed and load.

In any speed-controlling governor, one part always rotates with the shaft speed and produces a

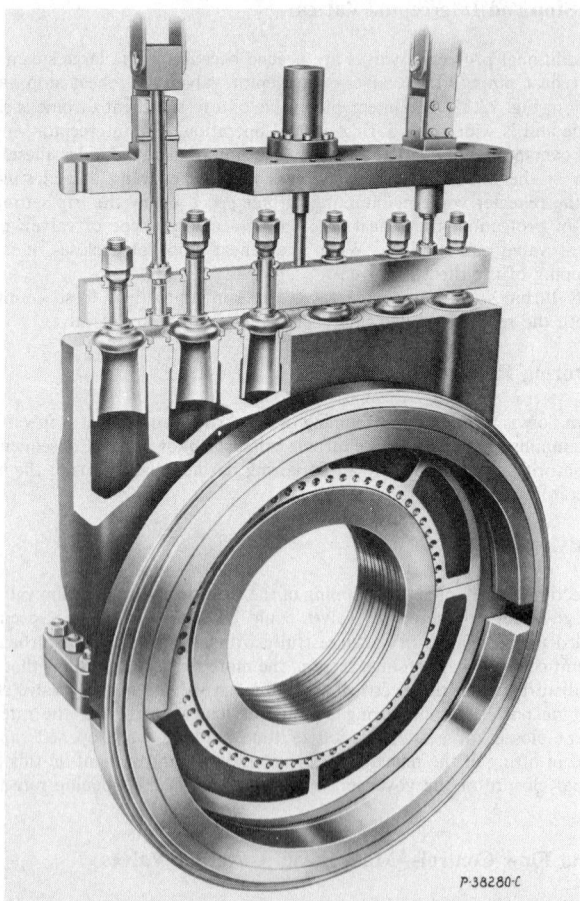

P-38280-C

Fig. 73.55 Steam chest at automatic extraction point.

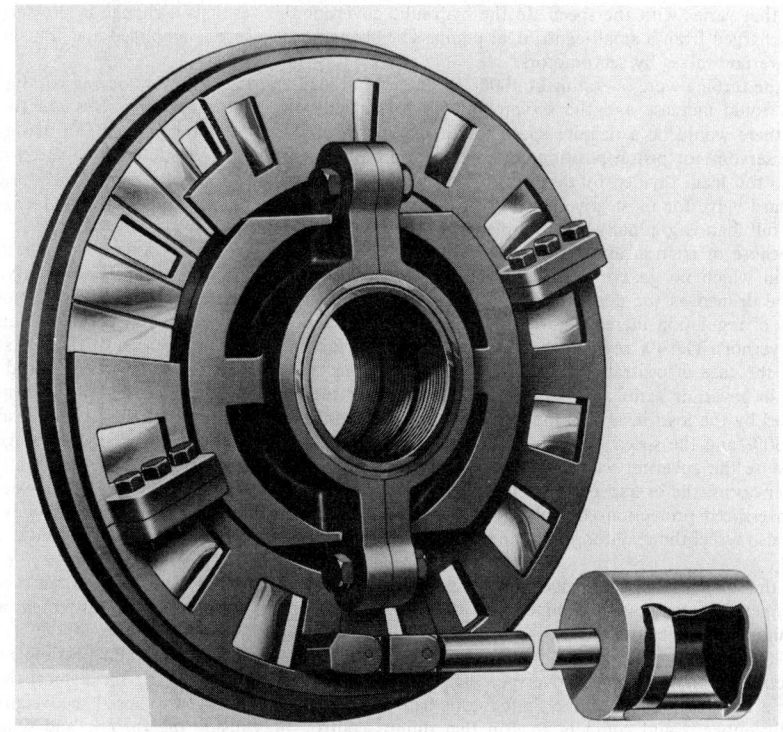

Fig. 73.56 Grid valve.

30000 KW STEAM TURBINE
DOUBLE EXTRACTION
3600 R.P.M. NON CONDENSING
1250 $^{LB}/_{IN}{}^2$ 825° F.T.T. 25 $^{LB}/_{IN}{}^2$ B.P.
AUTOMATIC EXTRACTION
AT 425 $^{LB}/_{IN}{}^2$ & 165 $^{LB}/_{IN}{}^2$

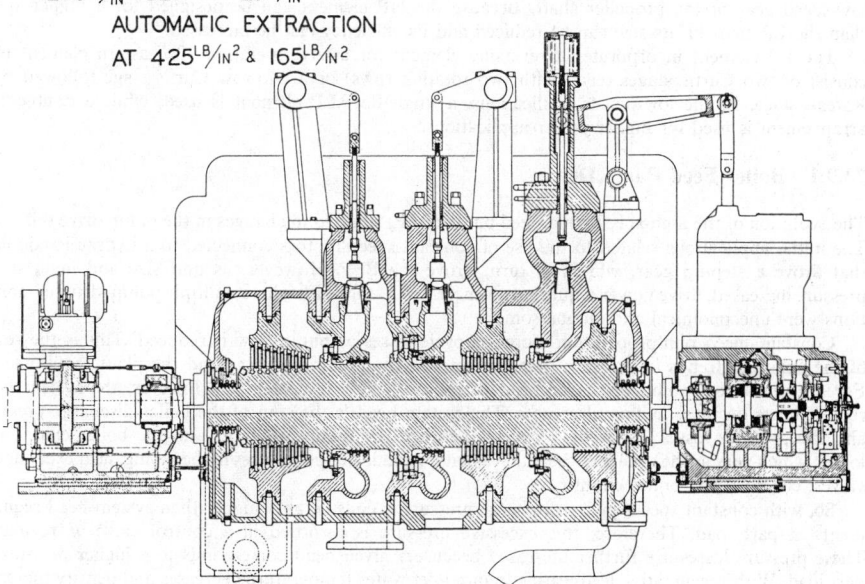

Fig. 73.57 Double automatic extraction turbine.

result that varies with the speed. In the hydraulic governor this result is a change in pressure of the oil discharged from a small centrifugal pump. The change in pressure is amplified and transmitted to the governor valves by servomotors.

If the turbine were operating at 100% speed and full load and the load were decreased, the turbine speed would increase and the servomotor would partially close the governor valves. As the valves close there would be a definite speed for each position of the servomotor piston. The two extremes of the servomotor piston position correspond to no load and full load. The change in speed from no load to full load, divided by the full load or rated speed, is called the regulation. For 4% regulation (standard valve for most governors) the speed of the turbine might increase from 3600 to 3744 rpm when full load is gradually and completely removed.

Because of friction and fluid leakages in the governor system, there is a speed change or dead band in which no governor motion occurs. This is a measure of the governor's sensitivity, which may be defined as the percentage speed change required to produce a corrective valve action. Low values of regulation increase governor sensitivity but reduce stability (and may result in hunting of the governor). The 4% regulation represents an acceptable compromise between sensitivity and stability.

In the case of central station units, speed deviations of 3 rpm or less from design speed do not result in governor action. For speed deviations greater than 3 rpm the governing system response is dictated by the load level and the position of the generator circuit breaker. If the level of load is less than 30% and the speed has not reached 103% of rated speed, the regulation of the speed governor will cause the governor valves to close. If the unit load is more than 30% and the generator circuit breaker opens, the overspeed protection control (sometimes called the auxiliary governor) is activated. The overspeed protection control (OPC) closes the governor valves and the interceptor valves. The OPC also resets the main or speed governor so that it will assume control when rated speed is reestablished.

If the final turbine speed does not exceed the trip speed, 110–111% of rated speed, as a result of OPC operation, the governor valves are closed and the interceptor valves are closed and then reopened to maintain speed below the trip speed.

If OPC action is unsuccessful in keeping speed below the trip point, one and possibly two trips are activated: the mechanical overspeed trip and the electrical speed trip. The mechanical overspeed trip is a movable weight on the shaft that is held in place by springs. At trip speed the weight moves radially outward and contacts an arm that dumps control oil, causing the throttle and reheat stop valves to close. The electrical speed trip, which is a supplemental trip on some control systems, measures rpm and activates a solenoid dump valve in the control oil circuit when speed reaches the trip point.

73.9 NONGENERATOR DRIVES

Modern marine propulsion turbines are disk and diaphragm designs with low-reaction stages except for the last few stages at the exhaust. By using separate HP, Fig. 73.58, and LP, Fig. 73.59, elements, greater flexibility is realized in selecting shaft speeds. Each turbine element is geared through a common low-speed gear to the propeller shaft. Because the HP element can be designed for a higher speed than the LP element, its size can be reduced and its efficiency can be increased.

The LP element incorporates a reversing element for astern operation. The astern element may consist of two Curtis stages (each with two rotating rows) or a two-row Curtis stage followed by a Rateau stage. In the low-power applications, a single-flow LP element is used, while a double-flow arrangement is used for higher power applications.

73.9.1 Boiler Feed Pump Drives

The evolution of the high-speed boiler feed pump (BFP) resulted in changes in the pump drive selection. The initial applications related to the use of constant-speed motors connected to a hydraulic coupling that drove a step-up gear, which, in turn, drove the BFP. However, as unit size and main steam pressure increased, drive power requirements increased to the point where multiple pump–drive combinations were uneconomical and cumbersome.

Constant-speed pump operation increases cycle losses as pump flow is reduced. This is the result of the opposing trends in pressure that characterize the pump discharge and the plant requirements. For a given load, pump discharge head increases as flow is decreased. However, the system head requirements decrease as flow decreases. The system head is the sum of the boiler discharge pressure and the pressure losses in the fluid circuits between the pump discharge and the boiler exit. These losses, which vary as the square of the fluid velocity, include pressure losses in the piping, heat exchangers, valves, controls, and boiler circuitry.

So, with constant speed, pump discharge pressure is considerably higher than system head requirements at part load. Therefore, this excessive pressure is throttled in a control valve or regulator. These pressure losses are further increased because a given head corresponds to a higher pressure at part load. With regenerative feedheating, pump inlet water temperature decreases and density increases as plant load decreases. In addition, pump efficiency decreases as pump flow is decreased.

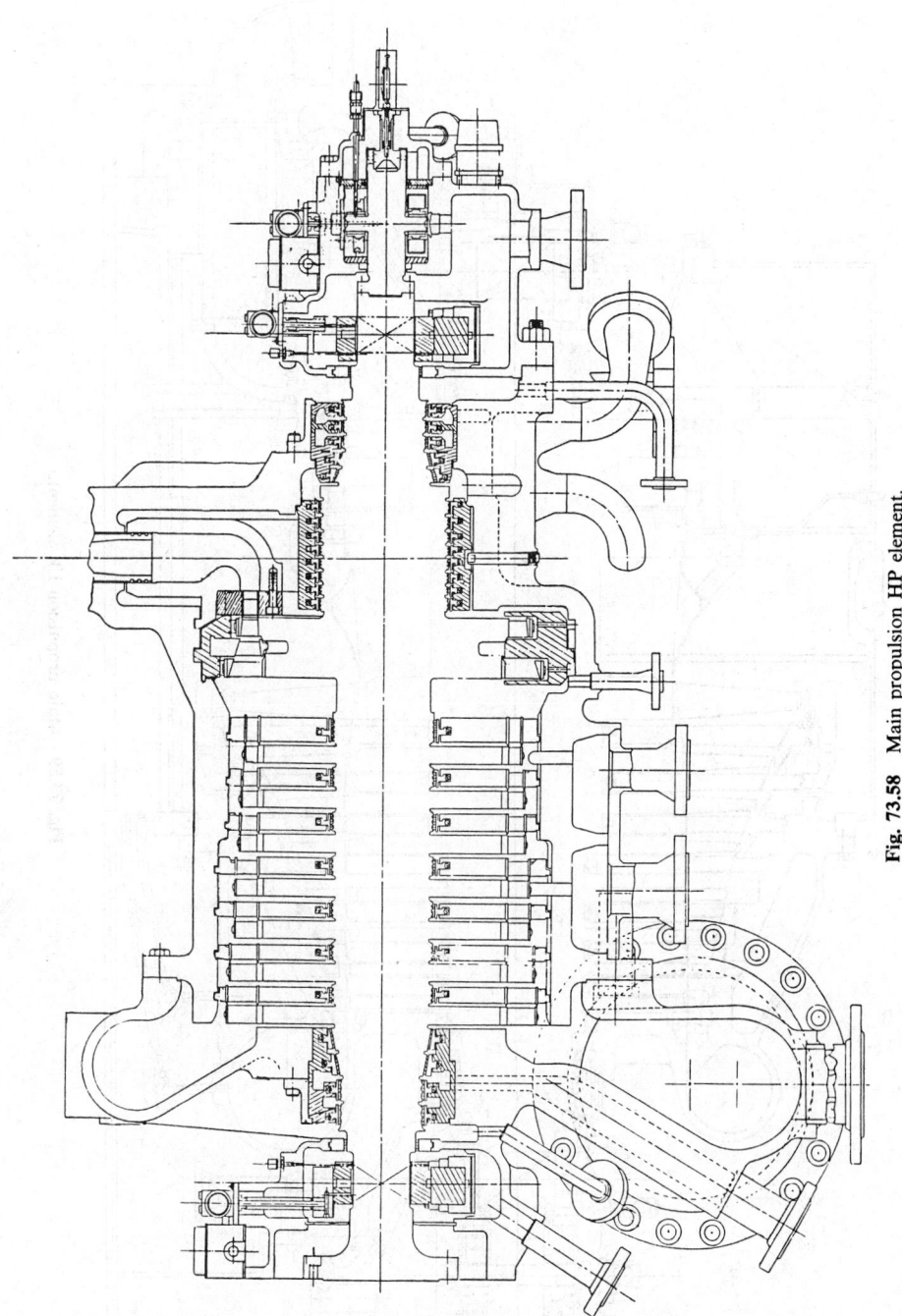

Fig. 73.58 Main propulsion HP element.

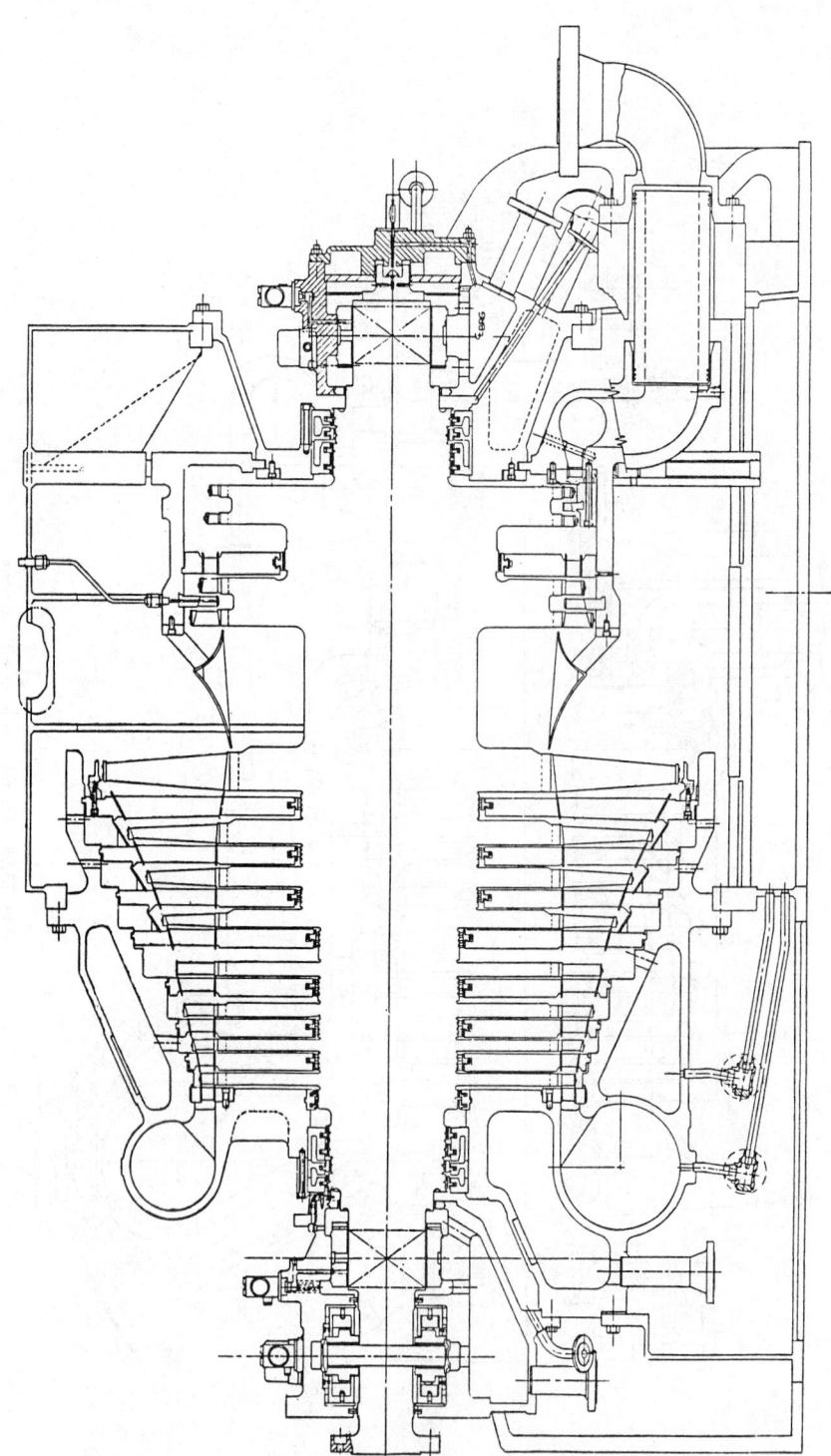

Fig. 73.59 Main propulsion LP element.

With a variable-speed pump drive, it would be possible to vary pump speed so that the pump discharge head more closely matches the system head requirements, minimizing the pressure losses in the feedwater pressure regulator. A secondary benefit of such a strategy is that pump efficiency is generally higher with variable-speed operation than with constant-speed operation.

Public Service Electric & Gas Co. (PSE&G) pioneered in the use of the main generator shaft for driving the boiler feed pump (BFP) through a fluid coupling and step-up gear.[39] This afforded the advantages of variable-speed operation as well as simplifying the plant arrangement. The shaft-driven BFP reduced pumping power and heat rate at part load as compared to a motor drive and was less expensive.

The use of a separate high-speed turbine, which was directly connected to the pump, soon proved to have superior thermal performance over the shaft drive. Initial applications were noncondensing drives that received steam from the cold reheat line with the drive turbine exhausting to the extraction line at the main-unit IP turbine exhaust. Moreover, the drive turbine often supplied steam to one or more feedheaters.[40,41] This was, in turn, succeeded by the condensing turbine drive that, most often, was supplied from the IP turbine exhaust. This variation has been the most widely applied, on both fossil and nuclear plants.[42,43]

The power output of the drive turbines is quite large, reaching about 4.0% of plant output on supercritical units. Single drives as large as 47 MW have been built.[44] Even the pump power requirements of nuclear-fueled plants can be quite large because of the large power ratings that typify these plants. It is important to note that a 50-MW turbine generator was a large unit in the early 1950s and, less than 30 years later, plant auxiliaries exceeded that power level.

73.9.2 Boiler Fans and Compressors

The success of the turbine-driven BFP has prompted designers to use turbine drives for the boiler fans. The use of turbine drives makes variable-fan-speed operation both possible and practical. The benefits are higher combined fan-drive efficiency and increased net plant output at both full load and part load.[45]

In like manner the variable and high speed of a turbine make it possible to improve the operating efficiency of compressors.

73.10 COMPARATIVE TURBINE CYCLE PERFORMANCE

Practically all central station fossil units use the reheat cycle. The incorporation of reheat in a practical Rankine regenerative cycle improves heat rate or efficiency by 4–4.5% depending on the throttle steam conditions. The application of reheat allowed the use of higher throttle pressures than were practical with Rankine-regenerative cycles by reducing the amount of moisture that is formed during the turbine expansion. For example, the highest practical throttle pressure on nonreheat cycles has been 12.4 MPa (1800 psig) at a throttle temperature of 565°C (1050°F). In contrast, throttle pressures up to 24.14 MPa (3500 psig) have been adopted with 538°C (1000°F) throttle and reheat steam temperatures. The exhaust wetness on these reheat cycles is lower than on the 1800 psig nonreheat cycle.

Typical reductions in heat rate for changes in throttle pressure, throttle temperature, reheat temperature, and other cycle variables in the 1450–3500-psig range are presented in Table 73.3. Representative values of final feedwater temperature and high-pressure turbine exhaust pressure are given in Table 73.4.

Turbine generator ratings were often based on operation at rated steam conditions and 3.5 in. Hg A exhaust pressure with 3% feedwater makeup. The turbine also has 5% flow margin to allow for manufacturing tolerances and other variations. In addition, the turbine could operate at 5% overpressure at the throttle. The flow margin and overpressure capability would increase output by about 10%.

The heat rate level at a given set of steam conditions is also affected by the condenser pressure, the turbine rating, and the last stage annulus area. Representative heat rate data for typical fossil applications, reflecting these factors, are presented in Ref. 46. While turbine performance can be estimated from procedures typified by those used by Craig and Cox or Kacker and Okapuu, they are too complex for most users.[47,48] The simplified turbine performance method presented in Ref. 49 is more suited to most users' needs.

Simplified turbine performance methods for nuclear steam turbines with present day cycles have also been published.[50,51] Reference 30 also presents typical performance levels and cycle arrangements for nuclear steam turbines.

73.10.1 Stage Pressure Prediction

The turbine design point involves the selection of extraction pressures for regenerative feedwater heating, determination of reheat pressure level, and selection of locations for turbine interstage moisture removal

Table 73.3

	Change in Net Heat Rate (%)			
Throttle pressure (psig)	1800	2400	3500	3500
Number of reheats	1	1	1	2
Throttle pressure (change from preceding column)	1.6–1.7[a]	1.8–2.0	1.8–2.0	1.5–2.0
50° Δ throttle temperature	0.7	0.8	0.9	0.7
50° Δ first-reheat temperature	0.8	0.8	0.8	0.4
50° Δ second-reheat temperature	—	—	—	0.6
One point in %Δ reheater pressure drop	0.1	0.1	0.1	0.1
Heater above reheat point	0.6–0.7	0.5–0.6	0.5–0.7	—

[a] From 1450 psig, single-reheat.

on LWR applications, for example. It is often necessary to determine performance at other levels of load and with cycle variations.

Field experience has demonstrated that turbine stage pressures vary almost linearly with the mass flow to the downstream stage. In the case of reheat plants, the pressures at all locations downstream of the first reheat point are linear with downstream flow. For the locations in the HP element and for nonreheat cycles, the pressure flow curve is bowed with the midpoint being 3–4% below a straight line joining the pressure-flow point at the design point and the origin of the curve.

There is a theoretical basis for these empirical correlations. For an isentropic expansion through a turbine stage or group of stages, when the theoretical velocity is combined with the continuity equation, the result is an equation that relates the flow G, inlet pressure P_1, exit pressure P_2, the inlet temperature T_1, the exit area A_2, and the isentropic exponent γ:

$$G = \frac{A_2}{v_2} \sqrt{\frac{2g_c \gamma}{\gamma - 1} RT_1 \left[1 - \left(\frac{P_2}{P_1} \right)^{\gamma/(\gamma-1)} \right]} \tag{73.15}$$

since

$$P_1 v_1^\gamma = P_2 v_2^\gamma \tag{73.16}$$

$$v_2 = v_1 \left(\frac{P_1}{P_2} \right)^{1/\gamma} \tag{73.17}$$

also

$$RT_1 = P_1 v_1 \tag{73.18}$$

The use of $P_1 v_1$ rather than RT_1 results in vapors exhibiting more idealized behavior similar to perfect gases. Therefore,

$$G = \frac{A_2}{A_1} \left(\frac{P_2}{P_1} \right)^{1/\gamma} \sqrt{\frac{2g_c \gamma}{\gamma - 1} (P_1 v_1) \left[1 - \left(\frac{P_2}{P_1} \right)^{\gamma/(\gamma-1)} \right]} \tag{73.19}$$

$$= A_2 \sqrt{\frac{2g_c \gamma}{\gamma - 1} \frac{P_1}{v_1} \left(\frac{P_2}{P_1} \right)^{2/\gamma} \left[1 - \left(\frac{P_2}{P_1} \right)^{\gamma/(\gamma-1)} \right]} \tag{73.20}$$

$$= A_2 \sqrt{\frac{2g_c \gamma}{\gamma - 1} \frac{P_1}{v_1} \left[\left(\frac{P_2}{P_1} \right)^{2/\gamma} - \left(\frac{P_2}{P_1} \right)^{2/(\gamma-1)} \right]} \tag{73.21}$$

If the stage or blade group pressure ratios remain constant when the flow is varied, then

$$G = C \sqrt{\frac{P_1}{v_1}} = \frac{CP_1}{\sqrt{P_1 v_1}} \tag{73.22}$$

where C is a proportionality constant replacing the other terms in Eq. (73.21).

In the superheated region Pv has strong enthalpy dependence and very weak pressure dependence. In the wet-steam region strong enthalpy dependence is present but the pressure dependence is much greater than in the superheated region.

At the extraction locations downstream of the first reheat, the enthalpy is practically constant across the load range, so Pv is practically constant. Therefore the flow G and the pressure P vary linearly. In the HP element and in nonreheat turbines, the enthalpy and, consequently, Pv vary somewhat. The end result is a bowing of the pressure-flow curve as was described in the initial paragraphs of this section.

Table 73.4 Maximum Flow at Rated Throttle Pressure[a]

Steam Conditions	1800 (1890) psig, 1000/1000°F	2400 (2520) psig, 1000/1000°F	3500 (3675) psig, 1000/1000°F	3500 (3675) psig, 1000/1000/1000°F
Final feedwater temperature (°F)	(464) (468)	481 (486)	510 (515)	545 (550)
Average cold reheat enthalpy (Btu/lb)	1342 (1340)	1312 (1308)	1277 (1272)	1306 (1301)
				1387 (1386)
High-pressure turbine exhaust pressure (psia)	502 (527)	585 (610)	745 (783)	1013 (1066)
First-reheat turbine exhaust pressure (psia)	—	—	—	355 (373)

[a] Values in parentheses relate to 5% overpressure.

Equation (73.22) does not account for the cases where stage pressure ratio does vary with flow. For example, the last stage may discharge at a fixed condenser pressure while flow varies or the flow may be held constant while the condenser pressure varies. A relationship described as Stodola's steam cone can handle this case as well as the one when the pressure ratio is constant. The equation is

$$G = C \sqrt{\frac{P_1^2 - P_2^2}{P_1 v_1}} \tag{73.23}$$

This equation bears a great similarity to Martin's formula, Eq. (73.11).

73.11 OTHER CYCLE VARIATIONS AND OPERATIONAL FACTORS

While changes in main steam pressure, main steam temperature, reheat temperature, and number of reheats affect plant efficiency and heat rate, other cycle variations can also affect performance. Among these operational factors are the variation in throttle pressure with load, single versus miltipressure condensers, mass velocity of the last stage and exhaust (condenser) pressure.

73.11.1 Constant and Sliding Throttle Pressure Operation

Main steam flow and load can be controlled either by varying the active nozzle area of the first turbine stage or by varying the steam supply pressure to the first-stage nozzles. The first mode of operation is often called partial-arc admission (sequential or multivalve point operation) as the first-stage periphery is divided into discrete segments or arcs of admission. With full-arc admission all of the governor valves operate in unison. This is also known as single-valve-point operation. The steam flow can be controlled by either throttling on the valves with constant throttle pressure (thereby varying the first-stage nozzle inlet pressure) or by varying throttle pressure. Refer to Section 73.4.1. Full-arc admission with sliding throttle pressure results in higher efficiency at high (95–100%) load as compared to partial-arc-admission constant-throttle-pressure operation. In contrast the throttling losses associated with constant-throttle-pressure full-arc-admission operation impairs part load performance. Operation with sliding throttle pressure eliminates the valve throttling losses and reduces boiler feed pump power, thereby improving efficiency.

In the past, most U.S. units have operated at constant throttle pressure with partial-arc admission. However, constant-throttle-pressure operation increases the potential for low-cycle thermal fatigue in the turbine during load change as compared to sliding-throttle-pressure operation. Offsetting this is the higher efficiency that is realizable with constant-throttle-pressure partial-arc admission over the entire load range with the exception of a small portion in the vicinity of maximum load.

Using a judicious combination of partial-arc admission and sliding-throttle-pressure operation, the cycle efficiency at low load would be improved over constant-throttle-pressure operation while reducing low-cycle thermal fatigue considerably. This combined mode of operation is sometimes referred to as hybrid operation.[52] Moreover, hybrid operation is superior to full-arc-admission sliding-throttle-pressure operation over most of the load range.

Starting from full load, load reductions are initially achieved by closing governor valves sequentially and maintaining constant throttle pressure. When a particular valve point is reached, further load reductions are achieved by holding valve position constant and varying throttle pressure. The optimum point for the transition from constant- to sliding-throttle-pressure operation occurs at 50% admission.[53] At 50% admission one-half of the first-stage nozzles are active (supplying steam) and one-half of the nozzles are inactive. Turbine steam flow at 50% admission is about 65% of maximum throttle flow. Figure 73.60 compares the performance of a number of operating alternatives. The heat rate differentials between alternatives illustrate that hybrid operation is the most efficient. The poorest performance results from constant-throttle-pressure single-valve-point or throttling control. Some improvement is obtained on the single-valve-point design by adopting sliding-throttle-pressure operation. The curves also show the advantages of sliding pressure at 50% admission as compared to sliding at some other value of admission.

Figure 73.61 illustrates the change in first-stage exit temperature (a measure of the potential for low-cycle thermal fatigue) that results from the various modes of operation.

For cycles with inlet steam conditions varying from slightly wet to modest amounts of superheat (as typified by LWRs), sliding-throttle-pressure operation results in an increase in heat rate on cycles using throttle steam for reheating. This is because the reheater outlet temperature is related to the saturation temperature of the throttle steam. A reduction in throttle pressure results in a reduction in LP inlet temperature with a commensurate increase in moisture losses. Moreover, since the first-stage exit would be in the wet region, its steam temperature would vary with the stage exit pressure. Therefore, the temperature would be independent of the mode of operation, constant or sliding throttle pressure.

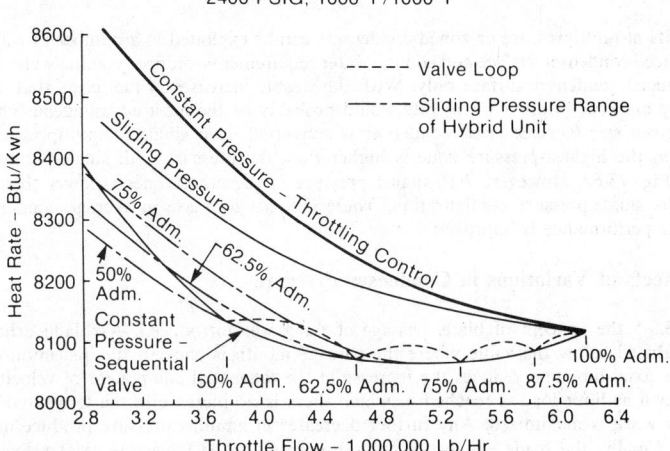

Fig. 73.60   Sliding- versus constant-throttle-pressure performance comparison.

73.11.2   Zoned or Multipressure Condensers

The heat rejection per unit length of a condenser is not uniform. The condenser inlet removes more heat per unit length of tubing than the outlet because the temperature difference between the steam and cooling water is greater at the inlet than at the outlet. If the turbine has multiple LP elements and each LP element exhausts to a different and segregated zone in the condenser, we have a zoned or multipressure condenser.

There are two general reasons for the improvement in cycle performance resulting from the use of multipressure condensers. The first is the lowering of the average exhaust pressure and the second is the possibility of warmer condensate leaving the condenser. Multipressure condensers give lower average exhaust pressures because the heat rejection per unit length of condenser becomes more uniform. As more zones are added, the area weighted average circulating water temperature approaches the ultimate minimum limit of an arithmetic average between the inlet and outlet temperatures that would occur with an infinite number of zones. The magnitude of the additional work obtained from the

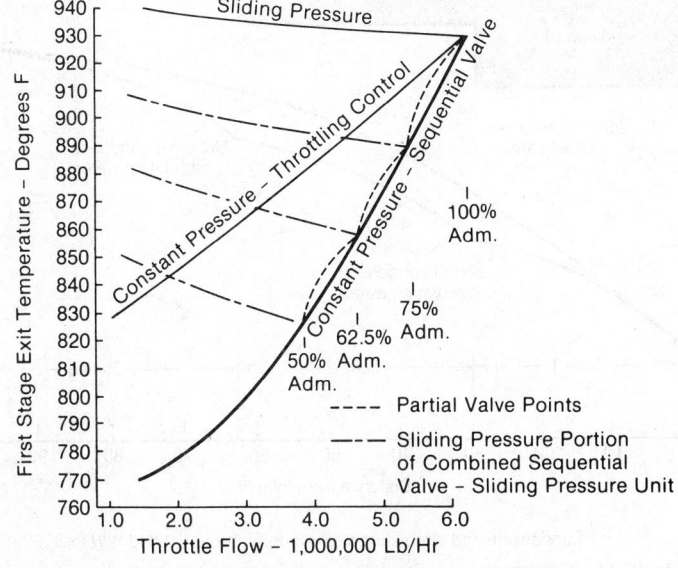

Fig. 73.61   First-stage exit temperature variation.

turbine because of the lower exhaust pressure is largely dependent on the level of loading of the LP turbines.[54]

The benefits of multipressure or zoned condensers can be exploited in a number of ways: improved heat rate, reduced condenser surface and cooling water requirements, reduced cooling water requirements only, and reduced condenser surface only. With the sizable increases in fuel costs that are typical of today's energy markets, improved heat rate would probably be the most advantageous choice.[55]

When a given size (surface area) condenser is converted from single to multipressure operation, the pressure in the highest-pressure zone is higher than the pressure with single-pressure operation as shown in Fig. 73.62. However, the exhaust pressure in the other zones is lower than the exhaust pressure in the single-pressure configuration. There is a net decrease in average condenser pressure and so turbine performance is improved.

73.11.3 Effects of Variations in Condenser Pressure

In Section 73.3.5 the concept of blade passage choking was introduced. As blade exhaust annulus pressure is reduced below the value where the passage mouth is choked, the deflection of the steam jet toward the axial direction reduces the increase in the tangential component of velocity. Finally, a condition known as limit load is reached in which there is no increase in the tangential velocity and the blade row work is maximized. Any further decreases in annulus pressure produce no changes in work output. Finally, the blade exhaust annulus becomes choked when the axial velocity reaches a critical value of Mach number. Because of losses in the exhaust hood connecting the turbine last row exhaust annulus and the condenser, the exhaust hood chokes at inlet axial Mach numbers in the 0.87–0.90 range on modern high-efficiency units. In a well matched last stage–exhaust hood combination, load limit and hood choking occur at about the same value of exhaust annulus axial Mach number.

As condenser pressure is raised above the choking value, last-stage work decreases. At an axial Mach number of about 0.27, last-stage net work is zero. Increases in exhaust pressure above this point result in rapid increases in exhaust temperature because of the work being done on the steam. In addition, stall flutter is a concern.[56] Stall flutter or flow-induced vibration may occur when there is appreciable flow incidence (flow at an off-design inlet angle) on the turbine blade. This differs from the more normal occurrences of vibration that are associated with resonant frequency.

With the range of exhaust pressures that are normally present with conventional types of cooling systems, the probability of stall flutter is small. However, with cooling systems that result in a wider variation of exhaust pressure (dry or dry–wet cooling systems), both choking and flutter may occur. One possible solution is to limit the maximum allowable exhaust pressure. Another is to increase the last-stage mass velocity or end loading, lb/hr/ft² of exhaust annulus area. This increases the choking

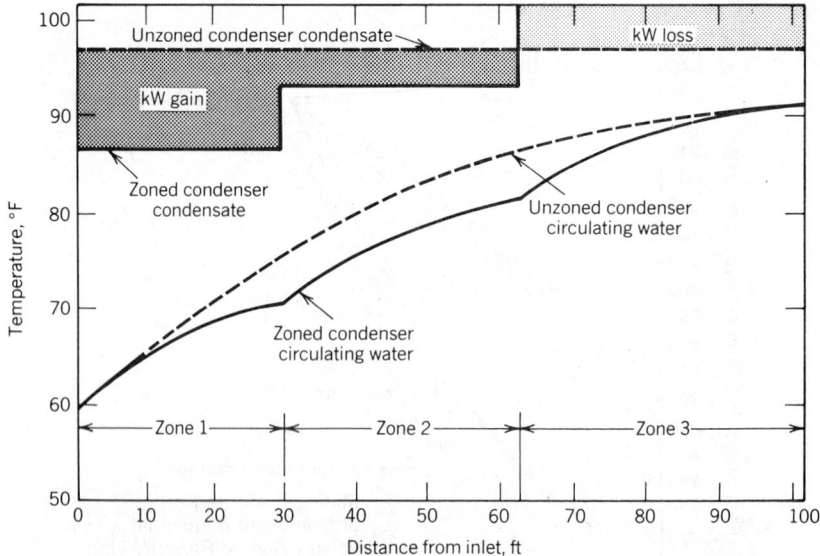

Condensate and circulating-water temperatures entitled "kW loss"

Fig. 73.62 Condensate and circulating water temperature, zoned condenser.

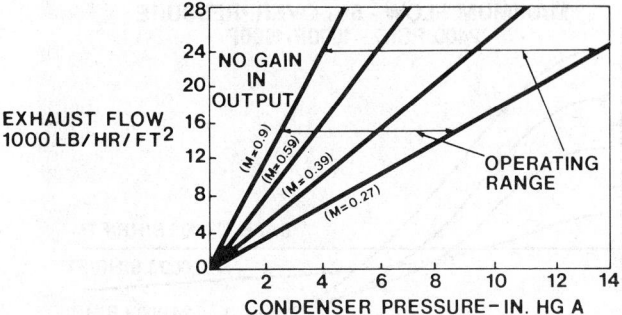

Fig. 73.63   Exhaust-stage operating range.

pressure with an attendant loss in output during periods of low or moderate ambient temperature as well as increasing the maximum allowable exhaust pressure. This is shown in Fig. 73.63 that shows choking exhaust pressure ($M = 0.90$) and the zero work locus ($M = 0.27$) at end loadings of 15,000 and 24,000 lb/hr/ft², respectively.

For many years maximum allowable exhaust pressures were in the 5.0–5.5 in. Hg A range. Recently, LP element designs have become available that substantially increase the level of maximum allowable exhaust pressure at a given level of end loading as compared to the predecessor designs.[57-59] Test data have demonstrated the ability of these designs to operate at considerably higher exhaust pressures than the design values.[57-59]

The variation in turbine performance with exhaust pressure is shown in Fig. 73.64 for a given level of end loading (maximum flow value). A comparison at different levels of ending loading is shown in Fig. 73.65. Figure 73.66 shows the reduced sensitivity of a fossil-fueled reheat turbine to exhaust pressure variations as compared to an LWR unit, both designs having a mass velocity of 15,000 lb/hr/ft². The reason for the reduced sensitivity to exhaust pressure by the fossil-fueled turbine is that it expands the steam through a greater pressure and temperature range (more available energy, Btu/lb) than the LWR turbine. Consequently, the percentage reduction in available energy is lower for a given increase in exhaust pressures than the LWR plant.

For dry or wet–dry cooling applications the fuel cost has a large effect on the magnitude of the exhaust end loading. With low fuel costs, designs with end loadings up to 30,000 lb/hr/ft² may be the economic choice by reducing the size of the heat rejection system.[60] With more typical fuel costs, the optimum end loading selection is in the range of applications with conventional heat rejection systems, if the turbines can be operated efficiently and reliably over the wide exhaust pressure range identified in Refs. 57–59.

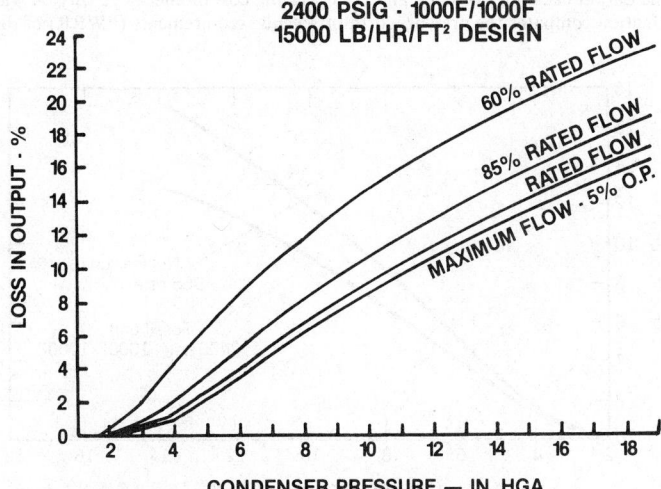

Fig. 73.64   Loss in output versus condenser pressure.

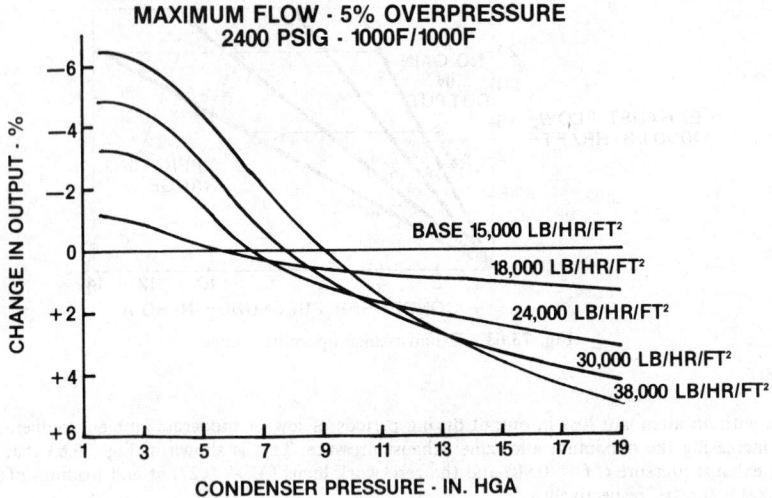

Fig. 73.65 Change in output versus condenser pressure and end loading.

73.11.4 Operating Variations

During operation there may be variations from the design values of cycle parameters such as throttle temperature, reheat temperature, reheater pressure drop, and condenser pressure. The changes in heat rate related to the first three factors are somewhat larger than those tabulated in Table 73.4 because of accompanying changes in turbine blading efficiency. The changes in Table 73.4 relate to turbines designed for those specific cycle parameters.

The variations in steam temperature are the result of the steam temperature–steam flow characteristic (temperature droop) of the steam generator. Sliding-throttle-pressure operation extends the range of load over which design steam temperatures can be maintained. Moreover, the variations in steam supply temperature and the changes in first-stage exit temperature that are related to load changes and first-stage operation, Fig. 73.61, result in low-cycle thermal fatigue during startup and load changes.[61,62] To reduce low-cycle thermal fatigue, manufacturers have recommended rates of load change for specific changes in first-stage exit temperature.

73.12 ECONOMIC EVALUATION

In evaluating the total cost of equipment, purchasers must consider three major penalties: the capital cost penalty, the output capability penalty, and the operating cost or energy penalty. A widely accepted method of evaluation compares the present worth of revenue requirements (PWRR) of these penalties.

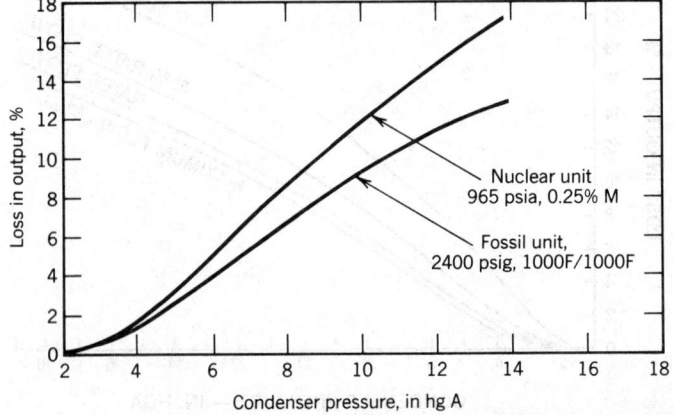

Fig. 73.66 Change in output versus condenser pressure (15,000 lb/hr/ft³).

PWRR has distinct advantages in that it brings all future revenues (costs and expenses) throughout the entire economic life of the project to a present day sum for comparison.[63]

The capital cost penalty is the present worth of the additional capital cost of an alternative plant compared to a base design. The capability penalty relates to any difference in the plant's firm power rating that must be replaced by additional generating capacity. The energy penalty relates to the difference in the energy costs required to generate electricity over the life of the plant when comparing the base and the alternative plants.

73.12.1 Capability Penalty

The capability penalty is evaluated for the time period when the system peak load occurs. The selection of the replacement capacity can range from peaking units such as gas turbines or cycling steam turbines to base load units such as nuclear plants or high efficiency-low fuel cost steam plants. Each selection will involve capital cost-efficiency-fuel cost trade-offs.

73.12.2 Capacity Factor

The unit load or capacity factor is very important. The capacity or load factor is the ratio of the average load on a machine, plant, or system for a given period of time to the capacity of the machine or equipment.[64] Base load units typically have high capacity factors during the early years of their life with the load factor declining over time. The reason for this is that subsequently purchased units may have higher efficiency and/or lower fuel costs or because of equipment deterioration with age. To perform the economic evaluation the useful lifetime average capacity factor must be determined.

73.12.3 Levelized Fuel Costs

Since fuel costs change with time, an average or levelized fuel cost is often used. This levelized cost reflects the expected escalation rate of the fuel cost, the expected interest rate, and the plant life. The levelized fuel cost (LFC) is

$$\text{LFC} = \text{PFC}\left(\frac{1}{1+i}\right)\frac{\left[\dfrac{1-\left(\dfrac{1+e}{1+i}\right)^n}{1-\left(\dfrac{1+e}{1+i}\right)}\right]}{\left[\dfrac{(1+i)^n-1}{i(1+i)^n}\right]} \tag{73.24}$$

where

$$\begin{aligned}
\text{PFC} &= \text{fuel cost in the first year of commercial operation} \\
i &= \text{cost of money (interest rate)} \\
e &= \text{fuel cost escalation rate} \\
n &= \text{plant life, years}
\end{aligned}$$

In reality, the LFC is determined by dividing the sum of the present worth fuel costs by the sum of the present worth factors.

The difficulty in predicting the rate of escalation is illustrated by the rapid increases in oil prices in the 1970–1981 time period and the reversal in the trend that has occurred since then. Neither was there stability in interest rates in the same time period.

In spite of the difficulties in predicting future fuel costs and interest rates, equipment selection must be made from alternatives that have different capital, capability, and energy penalties. Some specific types of comparisons will be identified in the following sections.

73.12.4 Fixed Charge Rate

To establish the minimum fixed charge rate that must be earned on the capital investment, it is necessary to identify the sum of the interest charges on the long-term loans, the dividends, the depreciation charges on equipment, federal and local taxes, and insurance. Fixed charge rates of 18–20% per year are not uncommon for private (investor-owned) electric utilities. The fixed charge rate is most sensitive to variations in the return on investment.[65] Return on investment includes interest charges, dividends, and retained earnings, ranging between 14% and 16% for private utilities. The fixed charge rate for public utilities is 10–12%, reflecting their lower interest rates (tax exempt bonds) and the absence of dividend payments.

73.12.5 Selected Types of Economic Comparisons

When equipment alternatives are compared, the evaluation can be performed for a number of turbine and or steam-supply system (fossil-fueled steam generator or nuclear-fueled steam-supply system) variations. For example, the alternatives can relate to a fixed steam supply rating as is the case for a nuclear-fueled plant, cases 1 and 3. They can relate to a fixed steam supply rating and identical turbine ratings for operation at normal exhaust conditions, case 2.

In other instances the choice of the fixed parameters is at the discretion of the investigator. Suppose that the economic feasibility of adding a feedheater that extracts at a pressure higher than the cold reheat point is being studied. One choice would be to hold the turbine rating constant. In this instance, the throttle flow would vary, affecting the steam-supply system, the feed train, and the heat rejection system. Another choice would be to keep throttle flow constant. This would have a comparatively minor effect on the steam supply. The turbine and plant rating would decrease for the cycle with the additional heater, resulting in a capacity debit. However, the size of the feed train and heat rejection system would decrease, reducing the capital cost and the plant efficiency would be higher.

Still another choice would be based on a fixed value of turbine exhaust flow. The heat rejection system and most of the feed train would be the same for both alternatives except for the highest-pressure feedheater. The reheater flow would be practically the same for both cycles. However, the throttle flow would be greater with the additional heater, and this cycle would have a greater output as well as higher efficiency. The turbine HP element, electric generator, and the steam-supply system would increase in size.

Case 1—Partial-Arc versus Full-Arc Admission Nuclear Turbine

This is a simple comparison in that there is no difference in performance at 100% load; therefore, the balance of plant equipment is not affected. Figure 73.67 gives the reduction in heat rate versus percent load that results from the selection of a partial-arc admission turbine as compared to a full-arc admission turbine. The partial-arc design has valve points at 50%, 75%, and 100% admission. The wiggly shape of the curve results from the presence of the valve loops.

The load duration curve for a typical utility is represented by Fig. 73.68. If the unit runs at 100% load during the first 5 years of operation with increasing amounts of part load operation during each successive period thereafter, the heat rate savings and the corresponding time periods would be as shown in Table 73.5.[66]

The present worth of all of the future fuel savings would establish the increased value of the partial-arc design to the user. Assume that the levelized nuclear cost is $0.60 per 10^6 Btu (MBtu) and 10% is the minimum acceptable rate of return. These quantities are more current values than those used in the analysis in Ref. 66.

The worth W_n of an increment of unit heat rate for a given year n is given by the formula

$$W_n = (L_n)(H_n)(C)(\Delta E) \tag{73.25}$$

where L_n = load on the unit for a given time period in year n
H_n = number of operating hours at load L_n in year n
C = fuel cost in $/MBtu
ΔE = heat rate increment in Btu/kWhr at load L_n

With the given fuel cost and the time duration, the total fuel cost savings for years 6–10 is $287,500. The present worth factors (PWFs) for uniform annual payments at 10% interest for 10 years and 5

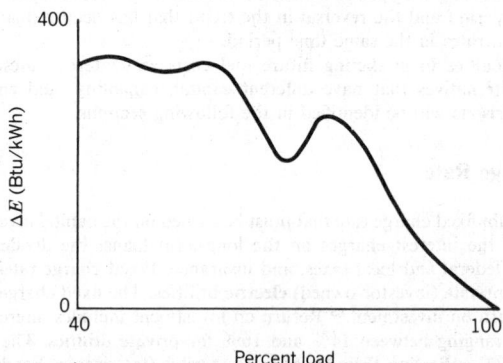

Fig. 73.67 Difference in heat rate—full-arc versus partial-arc admission design.

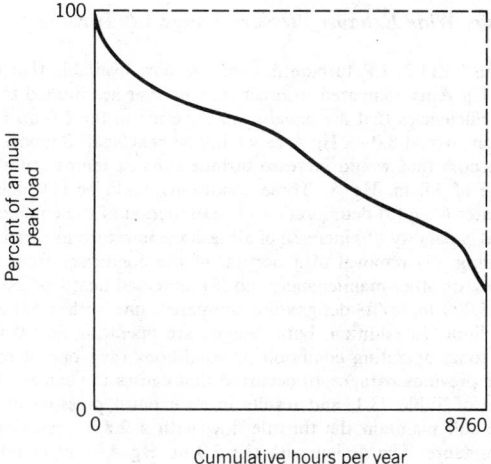

Fig. 73.68 Load duration curve.

years are 6.144 and 4.355, respectively. The present worth (PW) of the fuel savings for the 6–10 year period is

$$PW = (\$287,500)(PWF_{10} - PWF_6)$$
$$= (\$287,500)(6.144 - 4.355)$$
$$= \$426,677$$

For years 11–20, W_n is $354,500 and the difference in the present worth factors is 2.019, resulting in a present worth fuel savings of $765,836. For years 21–40, W_n is $452,100 and the net present worth factor is 1.130, which results in a present worth of $510,873.

Therefore, the partial-arc nuclear turbine has a net present worth of $1,703,386 over a unit with full-arc admission. If the fixed charge rate were increased to 15%, the net present worth of the partial-arc design would be $867,170 greater than the full-arc admission design.

Table 73.5

Year	Load (kW)	Hours at Load	ΔE (Btu)
0–5	1,000,000	7900	0
6–10	1,000,000	4500	0
	900,000	1000	75
	800,000	1000	150
	700,000	1000	200
	500,000	400	350
11–20	1,000,000	3500	0
	900,000	1000	75
	800,000	1000	150
	700,000	1000	200
	600,000	1000	325
	500,000	400	350
21–40	1,000,000	2500	0
	900,000	1000	75
	800,000	1000	150
	700,000	1400	200
	600,000	1000	325
	500,000	1000	350

Case 2—Limited versus Wide Exhaust Pressure Range LP Turbine

As was noted in Section 73.11.3, LP turbine designs are now available that can operate at exhaust pressures up to 8.0 in. Hg A as compared to other designs that are limited to 5.0 in. Hg A. The 8.0 in. Hg A designs have efficiencies that are equal to or superior to the 5.0 in. Hg A counterparts.

Under what conditions would 8.0 in. Hg A capability be beneficial? Suppose that some combination of adverse conditions occurs that would increase turbine exhaust temperature and pressure above its normal maximum value of 5.0 in. Hg A. These conditions could be (1) high ambient temperature; (2) restricted cooling water flow; (3) deterioration or malfunction of the cooling tower; (4) degradation of condenser air removal capability; (5) increase of air leakage into turbine generator–feedheater system; (6) condenser tube fouling; (7) removal of a portion of the condenser from operation for cleaning, plugging of leaking tubes, or other maintenance; and (8) increased heat load because of plant uprating.

Suppose that two 15,000 lb/hr/ft² designs are compared, one with a 5.0 in. Hg A limit and one with an 8.0 in. Hg A limit. In addition, both designs are operating at 5.0 in. Hg A at maximum throttle flow. Suppose some operating condition or conditions (any one of or a combination of the eight enumerated in the previous paragraph) occurred that causes the exhaust temperature to increase by 13.1°F, first column of Table 73.1, and results in an exhaust pressure of 7.0 in. Hg A. The 8.0 in. Hg A design could still maintain the throttle flow with a 2.8% reduction in output because of the higher condenser pressure. The design with the 5.0 in. Hg A limit could not allow the exhaust steam temperature and the condenser pressure to increase; therefore, the load (and, consequently, the throttle flow) would have to be reduced by 62% in order to limit the exhaust pressure to 5.0 in. Hg A. The loss in output for the two designs is shown in Table 73.6 for other combinations of exhaust steam temperature increase and the associated condenser pressures.

With 8.0 in. Hg A capability, on-line condenser cleaning, by removing a portion of the condenser from service, can be done more often with minimal effect on plant output and with an improvement in plant performance after the cleaning. One observer noted that on a conventional 400-MW unit the load had to be reduced by 60 MW to hold condenser pressure to acceptable limits when cleaning one-quarter of the condenser while the unit was still in operation.[67]

There are two aspects to the economic evaluation. First, with both designs the increase in system heat rate during the condenser cleaning operation is compared to the improved performance following the cleaning. The lost output capability during cleaning must be replaced by older, less efficient, units. Therefore, the 5.0 in. Hg A design would result in a larger increase in system heat rate. The lower loss in output of the 8.0 in. Hg A design, coupled with the more frequent cleaning, could improve yearly unit and system heat rate. The yearly incremental heat rate improvement of the 8.0 in. Hg A design over the 5.0 in. Hg A design would be evaluated using a present worth revenue evaluation similar to case 1.

The second aspect of the economic evaluation is the loss of load probability. This refers to the probability that the system load requirements could not be met at some time.[68] A typical acceptable probability is one day in a 10-year period.

Since condenser cleaning is deferrable, it would not be done during a peak load period. However, many of the conditions resulting in high exhaust pressure relate to equipment malfunction or degradation. In this instance, equipment reliability data are needed (forced outage rates, mean time between failures, mean time to repair, etc.). The data should consist not only of the actual plant information but also data from suppliers and industry associations such as the Edison Electric Institute.

Table 73.6 Load Reduction Comparison for 15,000 lb/hr/ft² LP Turbine Annulus Area Designs

Increase in Exhaust Steam Temperature (°F)	5.0 in. Hg A Design		8.0 in. Hg A Design	
	Condenser Pressure (in. Hg A)	Load Reduction (%)	Condenser Pressure (in. Hg A)	Load Reduction (%)
0[a]	5.0	0	5.0	0
1.0	5.0	4.8	5.17	0.3
3.4	5.0	14.3	5.48	0.8
5.5	5.0	23.8	5.78	1.2
7.5	5.0	33.4	6.10	1.7
9.4	5.0	42.9	6.40	2.1
11.1	5.0	52.4	6.69	2.4
13.1	5.0	62.0	7.0	2.8

[a] Reference.

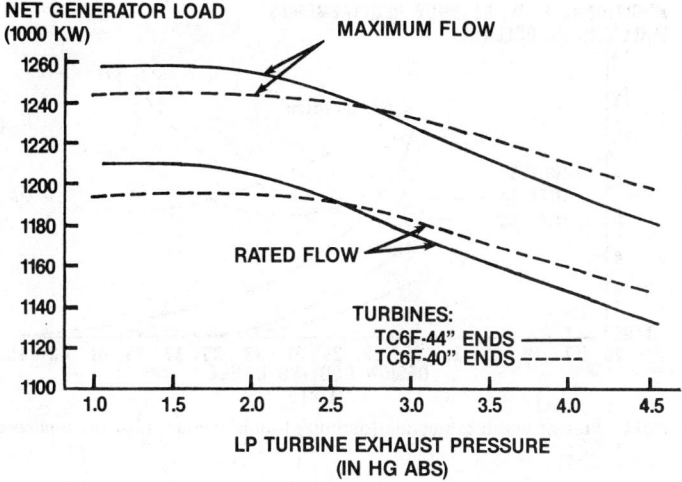

Fig. 73.69 Turbine performance (44-in. versus 40-in. last-row blades).

The level of system load as compared to the system capacity affects the loss of load probability (LOLP), so does the size of the generating units. If there is considerable reserve capacity, there is less probability that loss of a unit would result in the inability to meet system load demand with the attendant loss in revenue. Smaller units reduce the adverse impact of the loss of an individual unit.

The probability of various events or combination of events, that can increase condenser pressure, is combined, based on the actual or expected reliability of the plant components. Then the expected LOLP is compared to the acceptable LOLP to determine if the higher exhaust pressure capability is beneficial.

Case 3—LP Exhaust Size–Cooling Tower Selection

To more accurately choose between LP turbine last-row-blade sizes (and areas) the economic evaluation should also include the influence of condenser-cooling tower design and cost. The turbine, condenser, and cooling tower should be evaluated as a single system. The comparative performance of a nuclear turbine with 40-in. and 44-in. last-row blades is shown in Fig. 73.69. The turbine exhaust area will influence the cooling range (temperature change of cooling water) and the approach temperature of the cooling tower. The approach temperature is the difference in temperature between the cooled water and the ambient temperature. Moreover, zoned or multipressure condensers will influence the economic evaluation as indicated by Figs. 73.70 and 73.71.[69]

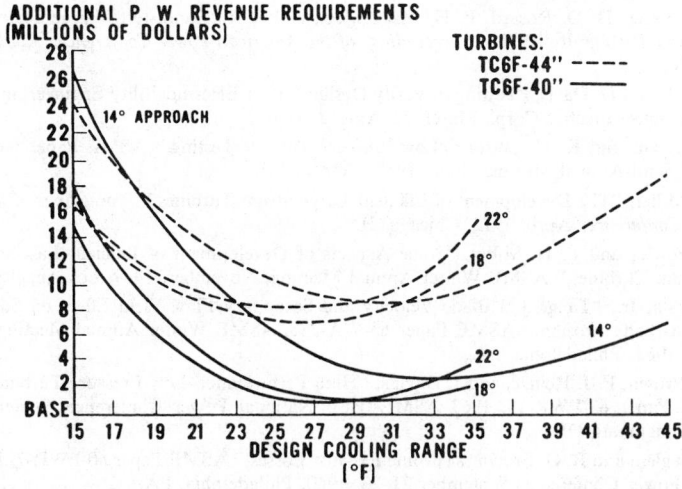

Fig. 73.70 Present worth evaluation [optimized single-pressure (2-shell) condensers].

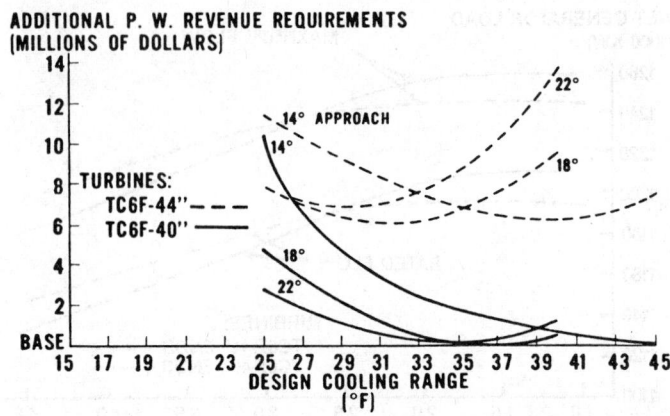

Fig. 73.71 Present worth evaluation [optimized multipressure (3-shell) condensers].

73.12.6 Time Factors in Planning Studies

The selection of the type and size of generating unit should consider the selection of future generating units as well as the unit under study. Moreover, the length of the look-ahead period will significantly affect the selection. When short look-ahead periods are used, the optimization process tends to favor units that are small in size and low in capital costs, like peaking units. Moreover, as the look-ahead period increases, the economics of base load units becomes more favorable.[70] This is illustrated by Tables 73.7 and 73.8 from Ref. 70.

REFERENCES

1. J. H. Keenan, "Reaction Tests of Turbine Nozzles for Supersonic Velocities," *Trans. ASME,* **71,** 773–780 (1949).

2. H. Kraft, "Reaction Tests of Turbine Nozzles for Subsonic Velocities," *Trans. ASME,* **71,** 781–787 (1949).

3. H. Kraft and T. M. Berry, "Automatic Integrating Pressure-Traverse Recorder for Study of Flow Phenomena in Steam Turbine Nozzles and Buckets," *Trans. ASME,* **62,** 479–488 (1940).

4. W. R. New, "An Investigation of Energy Losses in Steam Turbine Elements by Impact-Traverse Static Test with Air at Subacoustic Velocities," *Trans. ASME,* **62,** 489–502 (1940).

5. E. F. Church, *Steam Turbines,* 3rd ed., McGraw-Hill, New York, 1950.

6. C. A. Meyer, C. E. Seglem, and J. T. Wagner, "A Turbine-Testing Facility," ASME Paper 59-A-88, ASME Winter Annual Meeting, December, 1959, Atlantic City, NJ.

7. W. G. Steltz, D. D. Rosard, P. H. Maedel, Jr., and R. L. Bannister, "Large Scale Testing for Improved Turbine Reliability," *Proceedings of the American Power Conference,* April, 1977, Chicago, IL.

8. J. Davids and D. Davis, "Testing to Verify Design," 1981 Electric Utility Engineering Conference, Westinghouse Electric Corp., March 22–April 3, 1981.

9. J. E. Downs and K. C. Cotton, "Low-Pressure Turbine Testing," ASME Paper No. 58-SA-38, ASME Semi-Annual Meeting, June, 1958, Detroit, MI.

10. E. H. Miller, "The Development of Efficient Large Steam Turbines," *Proceedings of the American Power Conference,* April, 1972, Chicago, IL.

11. J. E. Fowler and E. H. Miller, "Some Aspects of Development of Efficient Last-Stage Buckets for Steam Turbines," ASME Winter Annual Meeting, November, 1969, Los Angeles, CA.

12. T. Vuksta, Jr., "Tangential Blade Velocity and Secondary Flow Field Effect on Steam-Turbine Exhaust-Blade Erosion," ASME Paper 63-WA-238, ASME Winter Annual Meeting, November 17–22, 1963, Philadelphia, PA.

13. R. O. Brown, F. J. Heinze, and J. Davids, "High Performance–Low Pressure Turbine Elements," ASME Paper 63-PWR-15, 1963 ASME-IEEE National Power Conference, September 22–25, 1963, Cincinnati, OH.

14. C. E. Seglem and R. O. Brown, "Turbine Exhaust Losses," ASME Paper 60-PWR-7, 1960 ASME-AIEE Power Conference, September 21–23, 1960, Philadelphia, PA.

Table 73.7 Optimal Installation Schedules for Utility E for Five Different Look-Ahead Periods

							Number of Years in Look-Ahead Period								
	One			Two			Three			Four			Twenty		
Year	50 MW CT	600 MW Coal	200 MW Coal	50 MW CT	600 MW Coal	200 MW Coal	50 MW CT	600 MW Coal	200 MW Coal	50 MW CT	600 MW Coal	200 MW Coal	50 MW CT	600 MW Coal	200 MW Coal
1980	0	0	0	0	0	0	0	0	0	0	0	0	0	0	0
1981	0	0	0	0	0	0	0	0	0	0	0	0	0	0	0
1982	0	1	0	0	1	0	0	1	0	0	1	0	0	1	0
1983	0	0	0	0	0	0	0	0	0	0	0	0	0	0	0
1984	0	0	0	0	0	0	0	0	0	0	0	0	0	1	0
1985	0	1	0	0	1	0	0	1	0	0	1	0	0	1	0
1986	0	0	1	2	0	0	0	0	0	0	0	0	0	1	0
1987	1	0	1	3	0	0	0	1	0	0	0	0	0	1	0
1988	1	0	1	0	1	0	0	1	0	0	1	0	0	0	0
1989	3	0	1	0	1	0	0	1	0	0	0	0	0	1	0
1990	2	0	1	0	1	0	0	1	0	0	1	0	0	1	0
1991	2	0	0	0	0	0	0	0	0	0	0	0	0	0	0
1992	3	0	1	2	1	0	0	1	0	0	1	0	0	0	0
1993	0	1	0	0	0	0	0	0	0	0	0	0	0	1	0
1994	2	0	0	2	1	0	2	1	0	2	1	0	0	0	0
1995	0	1	0	0	0	0	0	0	0	0	0	0	0	0	0
1996	3	0	0	0	1	0	0	1	0	0	1	0	0	0	0
1997	0	1	0	0	0	0	0	0	0	0	0	0	0	0	0
1998	0	0	0	0	0	0	0	0	0	0	1	0	1	0	0
1999	0	1	0	0	1	0	3	0	0	3	1	0	0	1	1
Totals	17	6	6	9	9	0	5	9	0	5	9	0	1	9	1

Table 73.8   Costs of Optimal Expansion Plans for Different
Look-Ahead Periods

Number of Years in Look-Ahead Period	System Costs (10^6$, 1980 Present Worth Dollars)		
	Fixed	Variable	Total
1	4168	14,871	19,039
2	4248	14,451	18,699
3	4528	13,996	18,524
4	4678	13,807	18,485
20	5391	12,937	18,328

15. C. G. Thatcher, "Reheat Factors for Expansion of Superheated and Wet Steam," *Trans. ASME* (1938).

16. R. B. Smith, "Calculation of Steam Turbine Reheat Factors," *Trans. ASME* (1938).

17. C. A. Meyer, G. J. Silvestri, and J. A. Martin, "Availability Balance of Steam Power Plants," *Trans. ASME,* **81,** Series A (January, 1959).

18. C. E. Seglem and R. O. Brown, "Turbine Exhaust Losses," ASME Paper 60-PWR-7, ASME-AIEE Power Conference, September 21–23, 1960, Philadelphia, PA.

19. H. M. Martin, "Steam Leakages in Dummies of the Ljungstrom Type," *Engineering* (January 3, 1919).

20. A. Egli, "The Leakage of Gases Through Narrow Channels," *Journal of Applied Mechanics, 7, Trans. ASME,* **59** (June, 1937).

21. V. P. Swaminathan, J. E. Steiner, and R. I. Jaffee, "High Temperature Turbine Rotor Forgings by Advanced Steel Melting Technology," ASME Paper 82-JPGC-Power-24, presented at Joint Power Generating Conference, October, 1982, Denver, CO.

22. R. Viswanathan and R. I. Jaffee, "Toughness of Cr-Mo-V Steels for Steam Turbine Rotors," ASME Paper 82-JPGC-Power-35, presented at Joint Power Generating Conference, October, 1982, Denver, CO.

23. R. M. Curran, D. L. Newhouse, and J. C. Newman, "The Development of Improved Rotor Forgings for Modern Large Steam Turbines," ASME Paper 82-JPGC-Power-25, presented at the Joint Power Generating Conference, October, 1982, Denver, CO.

24. H. Gunter, "Dynamic Stability of Rotor Bearing Systems," NASA SP-113, 1966.

25. C. B. Campbell, "Reheat in Power Plants," presented to Philadelphia Section, ASME, 1952.

26. R. L. Reynolds, "Recent Development of the Reheat Steam Turbine," *Mechanical Engineering* (January, 1952).

27. C. Schabtach and R. Sheppard, "Modern Reheat Turbine," *Mechanical Engineering* (February, 1952).

28. S. N. Fiala, "First Commerical Supercritical-Pressure Steam-Electric Generating Unit for Philo Plant," ASME Annual Meeting, Chicago, IL, November, 1955.

29. J. H. Harlow, "Engineering the Eddystone Plant for 5000-Lb and 1200-Deg. Steam," *Trans. ASME* (1957).

30. F. A. Artusa, "Turbines and Cycles for Nuclear Power Plant Application," *Proceedings of the American Power Conference,* Chicago, IL, April, 1976.

31. J. M. Kovacik, "Impact of High Energy Costs on Turbine System Selection for Industry," *Proceedings of the American Power Conference,* Chicago, IL, April, 1977.

32. "Linden Uses the World's Largest Automatic Extraction Turbine," *Power* (October, 1958).

33. M. A. Eggenberger and P. G. Ipsen, "The Control System of 225,000-KW Double-Automatic Extraction Steam Turbine and Related Reducing Stations," *Trans. ASME, J. Engrg. Power* (April, 1959).

34. P. C. Webb, "Midland Nuclear Cogeneration Plant," ASCE Spring Convention and Exhibition, New York, NY, May, 1981.

35. E. W. Cushing, H. G. Marshall, and H. R. Stewart, "Fast Valving as an Aid to Power System Transient Stability and Prompt Synchronization and Rapid Reload after Full Load Rejection," IEEE Paper 71TP705-PWR.

36. R. H. Park, "Fast Turbine Valving," IEEE Paper T72635-1, presented at the Joint Power Generation Conference, Boston, MA, September, 1972.

37. H. F. Martin, D. N. Tapper, and T. M. Alston, "Sustained Fast Valving Applied to Tennessee Valley Authority's Watts Bar Nuclear Units," ASME Paper 76-JPGC-Pwr-5, presented at the Joint Power Generation Conference, Buffalo, NY, September, 1976.

38. P. L. McGaha and T. L. Dresner, "A Nuclear Turbine Interceptor Valve and Control System for Fast Valving," paper presented at the Joint Power Generation Conference, Long Beach, CA, September, 1977.

39. R. A. Baker, "Evolution of the Boiler Feed-Pump Drive," Joint ASME–AIEE Power Conference, Boston, MA, September, 1958.

40. A. W. Rankin, "Auxiliary Turbine Drives for Boiler Feed Pumps," *Combustion* (January, 1964).

41. J. A. Tillinghast and J. R. Dolan, "The Integration of Single Turbine-Driven Feed Pumps in Large Generating Units," *Electric Light and Power* (Nov. 1, 1958).

42. R. C. Spencer, H. A. Mayor, and S. Styrma, "Heat Rate Gains in Utility Power Plants with Condensing Auxiliary Turbine Drives," *Proceedings of the American Power Conference,* Chicago, IL, April, 1968.

43. M. F. Pierpoline and J. Munnis, "Feed Pump Turbines for Nuclear Plants," *Proceedings of the American Power Conference,* Chicago, IL, April, 1968.

44. T. J. Fritsch, "World's Largest Boiler Feed Pump," *Proceedings of the American Power Conference,* Chicago, IL, April 1974.

45. E. L. Williamson, J. C. Block, A. F. Destribats, W. N. Inliano, and F. A. Reed, "Increased Plant Kilowatts Using Turbine-Driven Boiler Fans," *Proceedings of the American Power Conference,* Chicago, IL, April 1972.

46. "Heat Rates for Fossil Reheat Cycles Using General Electric Steam Turbine Generators 150,000 KW and Larger," General Electric Co. Publication GET-2050C, Schenectady, NY.

47. H. R. M. Craig and H. J. A. Cox, "Performance Estimation of Axial Flow Turbines," *Proceedings of the Institution of Mechanical Engineers,* 1970–1971, Vol. 185 32/71, p. 407.

48. S. C. Kacker and U. Okapuu, "A Mean Line Prediction Method for Axial Flow Turbine Efficiency," *Trans. ASME, J. Enging. Power,* **104,** 111 (1982).

49. R. C. Spencer, K. C. Cotton, and C. N. Cannon, "A Method for Predicting the Performance of Steam Turbine Generators . . . 16500KW and Larger," *Trans. ASME, Series A, J. Enging. Power,* **85,** 249 (1963).

50. F. G. Baily, K. C. Cotton, and R. C. Spencer, "Predicting the Performance of Large Steam Turbine Generators Operating with Saturated and Low Superheat Steam Conditions," *Proceedings of the American Power Conference,* Chicago, IL, April, 1967 (amended in July 1968).

51. F. G. Baily, J. A. Booth, K. C. Cotton, and E. H. Miller, "Predicting the Performance of 1800-RPM Large Steam Turbine-Generators Operating with Light Water-Cooled Reactors," General Electric Company Publication GET-6020.

52. G. J. Silvestri, Jr., O. J. Aanstad, and J. T. Ballantyne, "A Review of Sliding Throttle Pressure for Fossil Fueled Steam Turbine Generators," *Proceedings of the American Power Conference,* Chicago, IL, April, 1972.

53. L. Boyce, H. C. Temme, P. G. Ipsen, and R. E. Lowe, "Design Features for Load Cycling for KCPL's 670 MW Unit 1 at Iatan Station," *Proceedings of the American Power Conference,* Chicago, IL, April, 1978.

54. C. C. Peake, P. C. Holden, E. J. Barsness, and W. K. Price, "Multipressure Condenser Application," paper presented at the ASME Winter Annual Meeting, Chicago, IL, Nov. 1965.

55. W. E. Palmer and E. H. Miller, "Why Multipressure Surface Condenser-Turbine Operation," *Proceedings of the American Power Conference,* Chicago, IL, April 1965.

56. G. J. Silvestri, Jr. and J. Davids, "Effects of High Condenser Pressure on Steam Turbine Design," *Proceedings of the American Power Conference,* Chicago, IL, April, 1971.

57. G. J. Silvestri, Jr. and J. Davids, "Laboratory and Field Tests on Low Pressure Steam Turbines at High Exhaust Pressure," *Proceedings of an EPRI Workshop on Water Conserving Cooling Systems,* Palo Alto, CA, May 1982.

58. G. J. Silvestri, Jr., "Progress in the Development of Reliable and Economical Steam Turbines for Dry Cooling Applications," ASME Century 2-ETC Potpurri Conference, San Francisco, CA, September, 1980 and *Combustion* (March, 1981).

59. G. J. Silvestri, Jr., "High Efficiency and High Exhaust Pressure—Now You Can Have Both," *Power Engineering* (August, 1981).

60. R. J. Gray, W. C. Brauer, and S. C. Leland, "Design and Initial Operation of the Wyodak Plant," *Proceedings of the American Power Conference,* Chicago, IL, April, 1979.

61. L. .B. Podolsky, H. F. Martin, J. L. Barkam, S. A. Kidwell, and J. B. Parkes, "A Study of Cyclic Operation for Fossil-Fueled Power Plants," *Proceedings of the American Power Conference,* Chicago, IL, April, 1979.

62. I. Fruchtman, J. Martin, R. Helt, and J. Kalins, "Considerations for Cyclic Duty Steam Turbine Design with a Review of Some Plant Operating Experience," *Proceedings of the American Power Conference,* Chicago, IL, April, 1979.

63. P. Leung and R. F. Durning, "Power System Economics: On Selection of Engineering Alternatives," ASME Paper 77-JPGC-Pwr-7, presented at the Joint Power Generation Conference, Long Beach, CA, Sept. 1977.

64. R. L. Bartlett, *Steam Turbine Performance and Economics,* McGraw-Hill, New York, 1958.

65. K. A. Gulbrand and P. Leung, "Power System Economics: A Sensitivity Analysis of Annual Fixed Charges," Paper No. 74-WA/Pwr-4 presented at ASME Winter Annual Meeting, New York, NY, Nov. 1974, reprinted in *Transactions of ASME, Journal of Engineering for Power, Series A,* **97** (July, 1975).

66. W. H. Peace, "A Method of Calculating the Economic Present Worth of Multi-Valve Operation of a Steam Turbine in Part-Load Service," Westinghouse Electric Corp., Steam Turbine-Generator Division Publication.

67. R. J. Bell and R. A. Nery, "An Engineering Approach to a Cost Effective Schedule for Condenser Cleaning," Fossil Plant Heat Rate Improvement, 1981 Conference and Workshop, EPRI Publication CS-2180, Charlotte, NC, August 1981.

68. J. T. Day, P. B. Shortley, and J. W. Skooglund, "Expected Value of Generation Deficit: A Supplemental Measure of Power System Reliability," IEEE Power Apparatus and Systems Transactions Paper T72077-1, presented at the 1972 IEEE Winter Power Meeting, New York, NY, January 1972.

69. L. G. Hauser, K. A. Oleson, and R. J. Budenholzer, "An Advanced Optimization Technique for Turbine, Condenser, Cooling System Combinations," *Proceedings of the American Power Conference,* Chicago, IL, April 1971.

70. K. D. Le, "Time Factor in Long-Range Generation Planning Studies," *Proceedings of the American Power Conference,* Chicago, IL, April, 1981.

CHAPTER 74

INTERNAL COMBUSTION ENGINES

RONALD DOUGLAS MATTHEWS

General Motors Foundation Automotive Research
and Combustion Sciences Laboratories
University of Texas at Austin
Austin, Texas

An internal combustion engine is a device that operates on an open thermodynamic cycle and is used to convert the chemical energy of a fuel to rotational mechanical energy. This rotational mechanical energy is most often used directly to provide motive power through an appropriate drivetrain, such as for an automotive application. The rotational mechanical energy may also be used directly to drive a propeller for marine or aircraft applications. Alternatively, the internal combustion engine may be coupled to a generator to provide electric power or may be coupled to a hydraulic pump or a gas compressor. It may be noted that the favorable power-to-weight ratio of the internal combustion engine makes it ideally suited to mobile applications and therefore most internal combustion engines are manufactured for the motor vehicle, rail, marine, and aircraft industries. The high power-to-weight ratio of the internal combustion engine is also responsible for its use in other applications where a lightweight power source is needed, such as for chain saws or lawn mowers.

This chapter is devoted to a discussion of the internal combustion engine, including types, applications, fuels, theory, performance, efficiency, and emissions.

74.1 TYPES AND APPLICATIONS OF INTERNAL COMBUSTION ENGINES

This chapter discusses internal combustion engines that have an intermittent combustion process. Gas turbines, which are internal combustion engines that incorporate a continuous combustion system, are discussed in a separate chapter.

Internal combustion (IC) engines may be most generally classified by the method used to initiate combustion as either spark ignition (SI) or compression ignition (CI or diesel) engines. Another general classification scheme involves whether the rotational mechanical energy is obtained via reciprocating piston motion, as is more common, or directly via the rotational motion of a rotor in a rotary (Wankel) engine (see Fig. 74.1). The physical principles of a rotary engine are equivalent to those of a piston engine if the geometric considerations are properly accounted for, so that the following discussion will focus on the piston engine and the rotary engine will be discussed only briefly. All of these IC engines include five general processes:

1. An intake process, during which air or a fuel–air mixture is inducted into the combustion chamber.

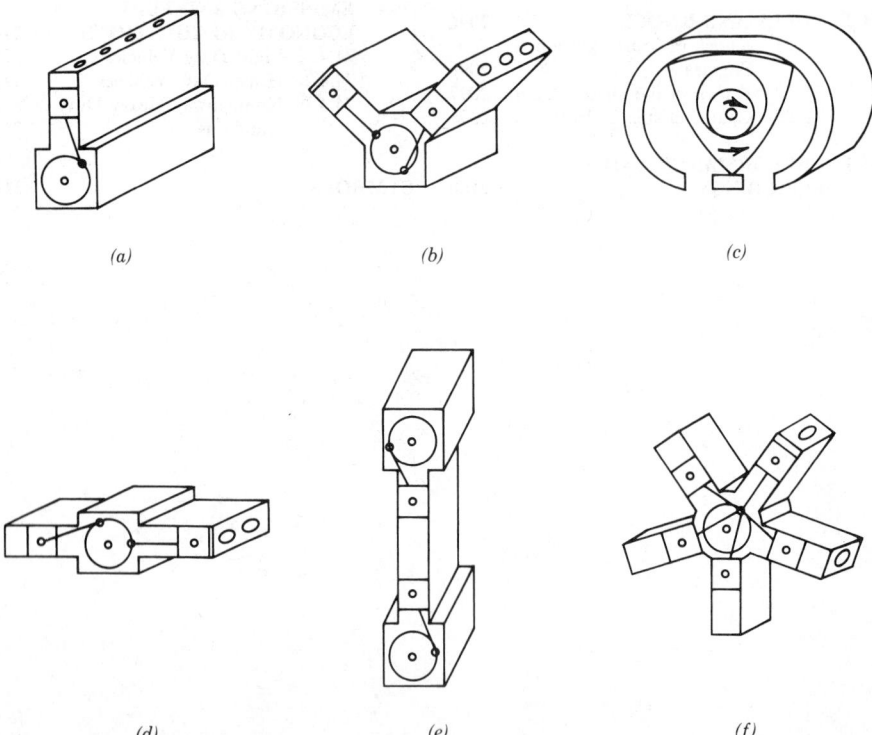

Fig. 74.1 IC engine configurations: (a) inline 4; (b) V6; (c) rotary (Wankel); (d) horizontal, flat, or opposed cylinder; (e) opposed piston; (f) radial.

2. A compression process, during which the air or fuel–air mixture is compressed to higher temperature, pressure, and density.

3. A combustion process, during which the chemical energy of the fuel is converted to thermal energy of the products of combustion.

4. An expansion process, during which a portion of the thermal energy of the working fluid is converted to mechanical energy.

5. An exhaust process, during which most of the products of combustion are expelled from the reaction chamber.

The mechanics of how these five general processes are incorporated in an engine may be used to more specifically classify different types of internal combustion engines.

74.1.1 Spark Ignition Engines

In SI engines, the combustion process is initiated by a precisely timed discharge of a spark across an electrode gap in the combustion chamber. Before ignition, the combustible mixture may be either homogeneous (i.e., the fuel–air mixture ratio may be approximately uniform throughout the combustion chamber) or stratified (i.e., the fuel–air mixture ratio may be more fuel-lean in some regions of the combustion chamber than in other portions). In all SI engines, except the direct injection stratified charge (DISC) SI engine, the power output is controlled by controlling the air flow rate (and thus the volumetric efficiency) through the engine and the fuel–air ratio is approximately constant (and approximately stoichiometric) for all operating conditions. The power output of the DISC engine is controlled by varying the fuel flow rate, and thus the fuel–air ratio is variable while the volumetric efficiency is approximately constant.

In the homogeneous charge SI engine, a mixture of fuel and air is inducted during the intake process. Traditionally, the fuel is mixed with the air in the venturi section of a carburetor. More recently, as more precise control of the fuel–air ratio has become desirable, throttle body fuel injectors have taken the place of carburetors for some automotive applications. Intake port fuel injection is also occasionally used. The five processes mentioned above may be combined in the homogeneous charge SI engine to produce an engine that operates on either a 4-stroke cycle or on a 2-stroke cycle.

In the more common 4-stroke cycle (see Fig. 74.2), the first stroke is the movement of the piston from top dead center (TDC—the closest approach of the piston to the cylinder head, yielding the minimum combustion chamber volume) to bottom dead center (BDC—when the piston is farthest from the cylinder head, yielding the maximum combustion chamber volume) during which the intake valve is open and the fresh fuel–air charge is inducted into the combustion chamber. The second stroke is the compression process, during which the intake and exhaust valves are both in the closed position and the piston moves from BDC back to TDC. The compression process is followed by combustion of the fuel–air mixture. Combustion is a rapid hydrocarbon oxidation process (not an explosion) of finite duration. Because the combustion process requires a finite, though very short, period of time, the spark is timed to initiate combustion slightly before the piston reaches TDC to allow the maximum pressure to occur at or slightly after TDC. The combustion process is essentially complete shortly after the piston has receded away from TDC. However, for the purposes of a simple analysis and because combustion is very rapid, it may be approximated as being instantaneous and occurring while the piston is motionless at TDC. The third stroke is the expansion process or power stroke, during which the piston returns to BDC. The fourth stroke is the exhaust process during which the exhaust valve is open and the piston proceeds from BDC to TDC and expels the products of combustion. Once the piston reaches TDC, the intake valve opens and the exhaust valve closes and the cycle repeats, starting with a new intake stroke. Considerations of spark timing and valve timing are not necessary for this simple explanation of the 4-stroke cycle but do have significant effects on performance and efficiency.

Applications[1]: The 4-stroke SI engine is used in:

Almost all passenger cars and light duty trucks.

Some heavy and medium duty trucks.

Most motorcycles since they have become subject to exhaust pollution control regulations.

Inboard motorboats.

Sometimes for emergency, portable, or remote electrical power generators.

Outboard motorboats.

Aircraft.

Lawn mowers.

Snowblowers.

Light duty tractors.

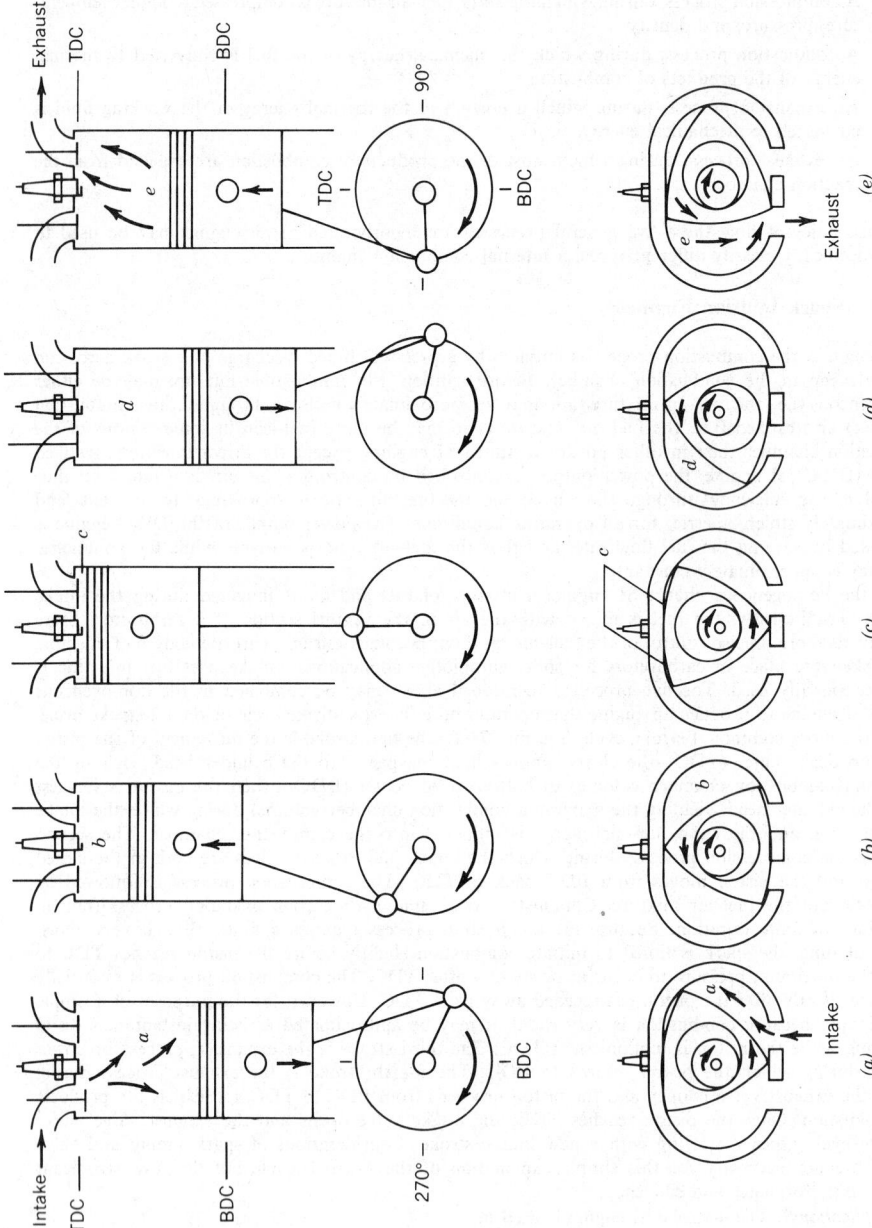

Fig. 74.2 Schematic of processes for 4-stroke SI piston and rotary engines (for 4-stroke CI, replace spark plug with fuel injector): (a) intake, (b) compression, (c) spark ignition, (d) expansion for power stroke, (e) exhaust.

The rotary engine is sometimes used in automobiles, motorcycles, and snowmobiles, and for stationary power generation.

Alternatively, these five processes may be incorporated into a homogeneous charge SI engine that requires only two strokes per cycle (see Fig. 74.3). All 2-stroke SI engines are of the homogeneous charge type. That is, any nonuniformity of the fuel–air ratio within the combustion chamber is unintentional in current 2-stroke SI engines. The 2-stroke SI engine does not have valves, but rather has

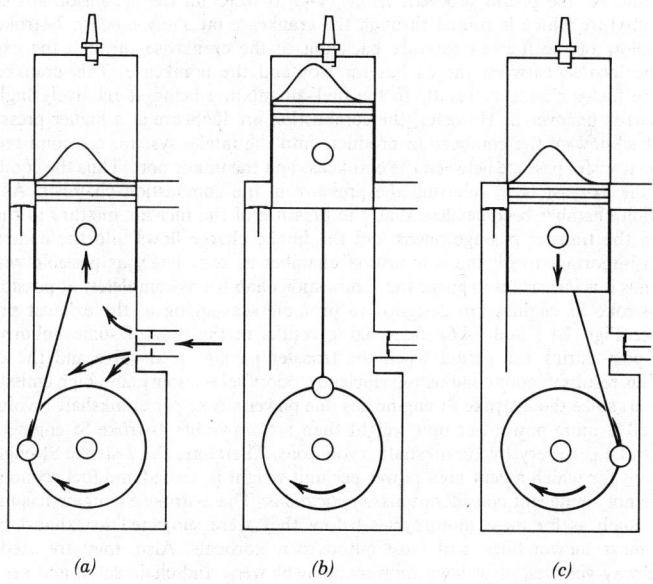

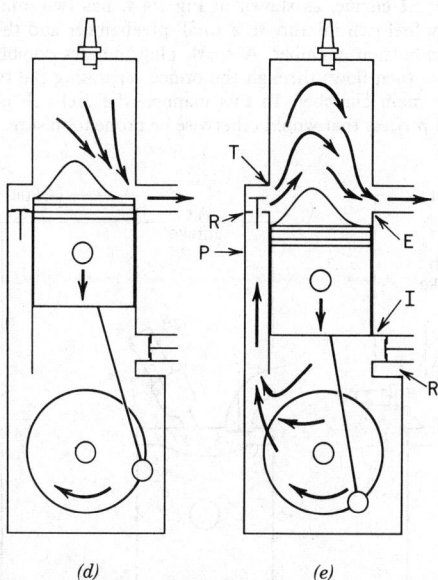

Fig. 74.3 Processes for 2-stroke crankcase compression SI engine (for CI engine, replace spark plug with fuel injector): (*a*) compression of trapped working fluid and simultaneous intake to crankcase, (*b*) spark ignition and combustion (for CI, fuel injection, and autoignition), (*c*) expansion or power, (*d*) beginning of exhaust, (*e*) intake and "loop" scavenging. E, exhaust port; I, intake port; P, transfer passage; R, reed valve; T, transfer port.

intake and exhaust ports which are normally located across from each other near the position of the crown of the piston when the piston is at BDC. When the piston moves toward TDC, it covers the ports and the compression process begins. As previously discussed, for the ideal SI cycle combustion may be perceived to occur instantaneously while the piston is motionless at TDC. The expansion process then occurs as the high pressure resulting from combustion pushes the piston back toward BDC. As the piston approaches BDC, the exhaust and intake ports are uncovered. The exhaust and intake processes occur simultaneously as the piston proceeds to BDC and then returns and covers the ports again. As the piston proceeds from TDC to BDC on the expansion stroke, it compresses the fuel–air mixture which is routed through the crankcase on many modern 2-stroke SI engines. To prevent backflow of the fuel–air mixture back out of the crankcase through the carburetor, a reed valve may be located between the carburetor exit and the crankcase. This crankcase compression process of the fuel–air mixture results in the fuel–air mixture being at relatively high pressure when the intake port is uncovered. However, the combustion products are at a higher pressure at this time. To prevent backflow of the combustion products into the intake system, a second reed valve may be located in the transfer passage between the crankcase and the intake port. Thus the combustion products flow out of the exhaust port, relieving the pressure in the combustion chamber. As the pressure in the combustion chamber becomes less than the pressure of the fuel–air mixture in the crankcase, the reed valve in the transfer passage opens and the intake charge flows into the combustion chamber. Because it is important to fill the combustion chamber as completely as possible with fresh fuel–air charge and thus it is important to purge the combustion chamber as completely as possible of combustion products, 2-stroke SI engines are designed to promote scavenging of the exhaust products via fluid dynamics (see Figs. 74.3 and 74.6). Scavenging results in the flow of some unburned fuel through the exhaust port during the period when the transfer passage reed valve and the exhaust port are both open. This results in poor combustion efficiency, poor fuel economy, and high emissions of hydrocarbons. However, since the 2-stroke SI engine has one power stroke per crankshaft revolution, it develops as much as 80% more power per unit weight than a comparable 4-stroke SI engine, which has only one power stroke per every two crankshaft revolutions. Therefore, the 2-stroke SI engine is best suited for applications for which a very high power per unit weight is needed and fuel economy and pollutant emissions are not significant considerations. *Applications:* The 2-stroke SI engine has many recreational applications, such as for most motorcycles before they were subjected to exhaust pollution control regulations, most snowmobiles, and most outboard motorboats. Also, they are used for many light duty nonhighway vehicles, most lawn mowers, snowblowers, and chain saws, and are sometimes used to power compressors on gas pipe lines.[1]

All stratified charge SI engines operate on the 4-stroke cycle. They may be subclassified as being either divided chamber engines or direct injection stratified charge SI engines.

The divided chamber SI engine, as shown in Fig. 74.4, has two intake systems: one providing a stoichiometric or slightly fuel-rich mixture to a small prechamber and the other providing a fuel-lean mixture to the main combustion chamber. A spark plug initiates combustion in the prechamber. A jet of hot reactive species then flows through the orifice separating the two chambers and ignites the fuel-lean mixture in the main chamber. In this manner, the fuel-rich combustion process stabilizes the fuel-lean combustion process that would otherwise be prone to misfire. This overall fuel-lean system

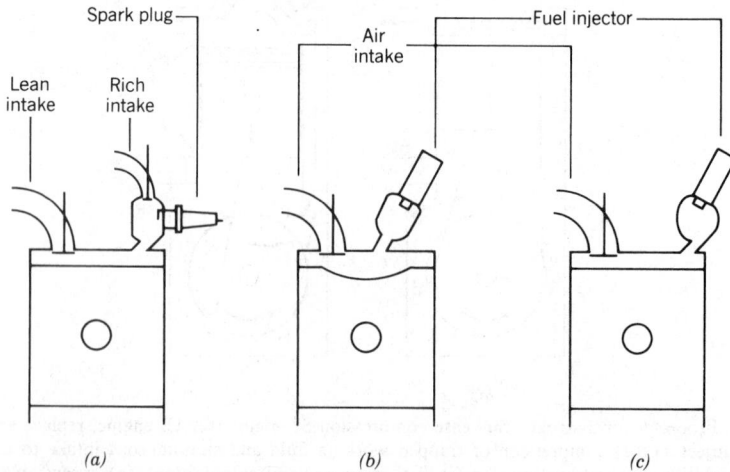

Fig. 74.4 Schematic cross sections of divided chamber engines: (*a*) prechamber SI, (*b*) prechamber IDI diesel, (*c*) swirl chamber IDI diesel.

is desirable since it results in decreased emissions of the regulated pollutants in comparison to the usual, approximately stoichiometric, combustion process. *Applications:* This engine is used in some passenger cars.

In the direct injection stratified charge (DISC) engine, only air is inducted during the intake stroke. As shown in Fig. 74.5, fuel is then injected directly near the center of the combustion chamber and ignited by a spark plug. The DISC engine has three primary advantages: (1) a wide fuel tolerance, that is, the ability to burn fuels with a relatively low octane rating without knock (see Section 74.2); (2) this decreased tendency to knock allows use of a higher compression ratio, which in turn results in higher power per unit displacement and higher efficiency (see Section 74.3); and (3) since the power output is controlled by the amount of fuel injected instead of the amount of air inducted, the DISC engine is not throttled (except at idle) resulting in higher volumetric efficiency and higher power per unit displacement for part load conditions (see Section 74.3). *Applications:* The DISC engine is currently being developed for passenger cars and light and medium duty utility vehicles.

74.1.2 Compression Ignition (Diesel) Engines

CI engines induct only air during the intake process. Late in the compression process, fuel is injected directly into the combustion chamber and mixes with the air that has been compressed to a relatively high temperature. The high temperature of the air serves to ignite the fuel. Like the DISC SI engine, the power output of the diesel is controlled by controlling the fuel flow rate while the volumetric efficiency is approximately constant. Although the fuel–air ratio is variable, the diesel always operates overall fuel-lean, with the maximum allowable fuel–air ratio limited by the production of unacceptable levels of smoke (also called soot or particulates). Diesel engines are inherently stratified because of the nature of the fuel injection process. The fuel–air mixture is fuel-rich near the center of the fuel injection cone and fuel-lean in areas of the combustion chamber that are farther from the fuel injection cone. Unlike the combustion process in the SI engine which occurs at almost constant volume, the combustion process in the diesel engine ideally occurs at constant pressure. That is, the combustion process in the CI engine is relatively slow, and the fuel–air mixture continues to burn during a significant portion of the expansion stroke (fuel continues to be injected during this portion of the expansion stroke) and the high pressure that would normally result from combustion is relieved as the piston recedes. After the combustion process is completed, the expansion process continues until the piston reaches BDC. The diesel may complete the five general engine processes through either a 2-stroke cycle or a 4-stroke cycle. Furthermore, the diesel may be subclassified as either an indirect injection diesel or a direct injection diesel.

Indirect injection (IDI) or divided chamber diesels are geometrically similar to divided chamber stratified charge SI engines. All IDI diesels operate on a 4-stroke cycle. Fuel is injected into the prechamber and combustion is initiated by autoignition. A glow plug is also located in the prechamber, but is only used to alleviate cold start difficulties. As shown in Fig. 74.4, the IDI may be designed so that the jet of hot gases issuing into the main chamber promotes swirl of the reactants in the main chamber. This configuration is called the swirl chamber IDI diesel. If the system is not designed to promote swirl, it is called the prechamber IDI diesel. The divided chamber design allows a relatively inexpensive pintle-type fuel injector to be used on the IDI diesel. *Applications:* The IDI diesel is currently used on all diesel powered passenger cars sold in the United States.

Direct injection (DI) or "open" chamber diesels are similar to DISC SI engines. There is no

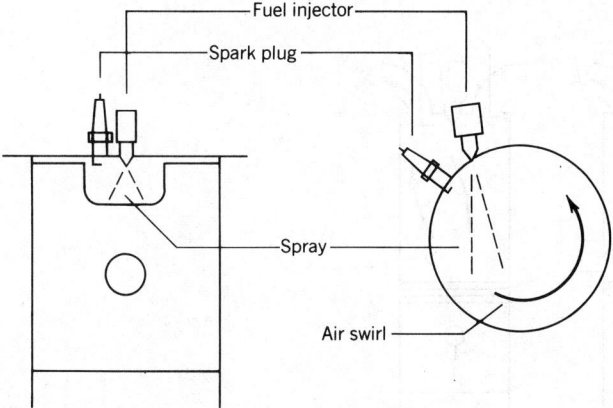

Fig. 74.5 Schematic of DISC SI engine combustion chambers.

prechamber, and fuel is injected directly into the main chamber. Therefore, the characteristics of the fuel injection cone have to be tailored carefully for proper combustion, avoidance of knock, and minimum smoke emissions. This requires the use of a high-pressure close-tolerance fuel injection system that is relatively expensive. The DI diesel may operate on either a 4-stroke or a 2-stroke cycle. Unlike the 2-stroke SI engine, the 2-stroke diesel often uses a mechanically driven blower for supercharging rather than crankcase compression and also may use multiple inlet ports in each cylinder, as shown in Fig. 74.6. Also, an exhaust valve in the top of the cylinder may be used instead of an exhaust port near the bottom of the cylinder, resulting in "through" or "uniflow" scavenging rather than "loop" or "cross" scavenging. *Applications:* The 4-stroke DI diesel is used in most heavy duty and medium duty trucks and for large inboard motorboats. The 2-stroke DI diesel is used in many buses. Both 2-stroke and 4-stroke DI diesels are used for agricultural, off-highway, and military vehicles, locomotives, light naval craft, and ships. They are also commonly used for limited electric power generation.[1]

74.2 FUELS AND KNOCK

Knock is the primary limiting factor in design of most IC engines. Knock is the result of engine design characteristics, engine operating conditions, and fuel properties. The causes of knock are discussed in this section. Fuel characteristics, especially those that affect either knock or performance, are also discussed in this section.

74.2.1 Knock in Spark Ignition Engines

Knock occurs in the SI engine if the fuel–air mixture autoignites too easily. At the end of the compression stroke, the fuel–air mixture exists at a relatively high temperature and pressure, the specific values of which depend primarily on the compression ratio and the intake manifold pressure (which is a function of the load). The spark plug then ignites a flame that travels toward the periphery of the combustion chamber. The increase in temperature and number of moles of the burned gases behind the flame front causes the pressure to rise throughout the combustion chamber. The "end gases" located in the peripheral regions of the combustion chamber are compressed to even higher temperatures by this increase in pressure. If these reactive end gases remain at this condition for a sufficient period of time (i.e., longer than the "ignition delay time"), then the mixture will autoignite. Normal combustion occurs if the flame front passes through the end gases before they autoignite.

For most fuels, autoignition is characterized by two stages. In the initial stage, a "low-temperature" ("low" is used as a relative term) chemical mechanism produces a so-called "cool flame." Enough energy may be released during this stage to initiate a "high-temperature" chemical mechanism.[1,2] Because the flame speed is a function of the thermal diffusivity, the reaction rate, and the turbulence intensity, the secondary flame front will necessarily have a higher flame speed than that of the primary (usual) flame. This is because the secondary flame was started at a higher initial temperature and thus produces a flame with a higher flame temperature and the reaction rate is an exponential function of the flame temperature. Also, the thermal diffusivity of the gases ahead of the secondary flame is higher than that of the gases ahead of the primary flame since the temperature is higher. If the rate of reaction is greater than the rate of expansion, then a strong pressure gradient will result. The pressure wave thus established will travel throughout the combustion chamber, reflect off the walls, and oscillate at the natural frequency characteristic of the combustion chamber geometry. This acoustic vibration results in an audible sound called "knock." It should be noted that the flame speeds associated

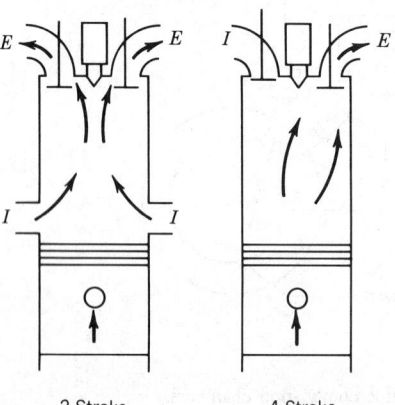

2-Stroke   4-Stroke

Fig. 74.6 Schematic of 2-stroke and 4-stroke DI diesels. The 2-stroke incorporates "uniflow" scavenging.

with knock are generally considered to be lower than the flame speeds associated with detonation (or explosion).[3,4] Nevertheless, the terms "knock" and "detonation" are often used interchangeably in reference to end gas autoignition.

The tendency of the SI engine to knock will be affected by any factors that affect the temperature and pressure of the end gases, the ignition delay time, the end gas residence time (before the normal flame passes through the end gases), and the reactivity of the mixture. The flame speed is a function of the turbulence intensity in the combustion chamber, and the turbulence intensity increases with increasing engine speed. Thus, the end gases will have a shorter residence time at high engine speed and there will be a decreased tendency to knock. As the load on the engine increases, the carburetor throttle plate is opened wider and the pressure in the intake manifold increases, thereby increasing the end gas pressure resulting in a greater tendency to knock. Thus, knock is most likely to be observed for SI engines used in motor vehicles at conditions of high load and low engine speed, such as acceleration from a standing start.

Other factors that increase the knock tendency of an SI engine[1-5] include increased compression ratio, increased inlet air temperature, increased distance between the spark plug and the end gases, location of the hot exhaust valve near the region of the end gases that is farthest from the spark plug, and increased intake manifold temperature and pressure due to supercharging. Factors that decrease the knock tendency of an SI engine[1-5] include retarding the spark timing, operation with either rich or lean mixtures (and thus the ability to operate the DISC SI engine at a high compression ratio since the end gases for this engine are extremely lean and therefore not very reactive), and increased inert levels in the mixture (via exhaust gas recirculation, water injection, etc.). The fuel characteristics that affect knock are quantified using octane rating tests, which are discussed in more detail in Section 74.2.3. A fuel with a higher octane number has a decreased tendency to knock.

74.2.2 Knock in the Diesel Engine

Knock occurs in the diesel engine if the fuel–air mixture does not autoignite easily enough. Knock occurs at the beginning of the combustion process in a diesel engine, whereas it occurs near the end of the combustion process in an SI engine. After the fuel injection process begins, there is an ignition delay time before the combustion process is initiated. This ignition delay time is not caused solely by the chemical delay, as in the SI engine, but is also due to a physical delay. The physical delay results from the need to vaporize and mix the fuel with the air to form a combustible mixture. If the overall ignition delay time is high, then too much fuel may be injected prior to autoignition. This oversupply of fuel will result in an energy release rate which is too high. In turn, this will result in an unacceptably high rate of pressure rise and cause the audible sound called knock.

The factors that will increase the knock tendency of a diesel engine[1,3,5] are those that decrease the rates of atomization, vaporization, mixing, and reaction, and those that increase the rate of fuel injection. The diesel engine is most prone to knock under cold start conditions because: (1) the fuel, air, and combustion chamber walls are initially cold, resulting in high fuel viscosity (poor mixing and therefore a longer physical delay), poor vaporization (longer physical delay), and low initial reaction rates (longer chemical delay); (2) the low engine speed results in low turbulence intensity (poor mixing, yielding a longer physical delay) and may result in low fuel injection pressures (poor atomization and longer physical delay); and (3) the low starting load will lead to low combustion temperatures and thus low reaction rates (longer chemical delay).

After a diesel engine has attained normal operating temperatures, knock will be most liable to occur at high speed and low load (exactly the opposite of the SI engine). The low load results in low combustion temperatures and thus low reaction rates and a longer chemical delay. Since most diesel engines have a gear-driven fuel injection pump, the increased rate of injection at high speed will more than offset the improved atomization and mixing (shorter physical delay).

Because the diesel knocks for essentially the opposite reasons than the SI engine, the factors that increase the knock tendency of an SI engine will decrease the knock tendency of a diesel engine: increased compression ratio, increased inlet air temperature, increased intake manifold temperature and pressure due to supercharging, and decreased concentrations of inert species. The knock tendency of the diesel engine will be increased if the injection timing is advanced or retarded from the optimum value and if the fuel has a low volatility, a high viscosity, and/or a low "cetane number." The cetane rating test and other fuel characteristics are discussed in more detail in the following section.

74.2.3 Characteristics of Fuels

Several properties are of interest for both SI engine fuels and diesel fuels. Many of these properties are presented in Table 74.1 for the primary reference fuels, for various types of gasolines and diesel fuels, and for the alcohols that are currently of interest for blending with regular unleaded gasoline to produce "gasohol" or premium unleaded gasoline.

The relative amount of air and fuel in the combustion chamber is usually specified by the air–fuel mass ratio (AF), the fuel–air mass ratio (FA = 1/AF), or the equivalence ratio (ϕ). Measuring

Table 74.1 **Thermochemical Data for Various Fuels**

Name	Formula	MW	AF_s	LHV_p	$h_f{}^a$	$h_v{}^a$	sg^b	RON	MON	Ref.
Primary Reference Fuels										
Isooctane	C_8H_{18}	114	15.1	44.6	-224.3^c	35.1^c	0.69	100	100	5
Normal cetane	$C_{16}H_{34}$	226	14.9	44.1	-418.3^d	50.9^e	0.77	—	0^f	5
Normal heptane	C_7H_{16}	100	15.1	44.9	-187.9^c	36.6^c	0.68	0	0	5
Alcohols										
Ethanol	C_2H_5OH	46	9.0	27.8	-235.5^g	42.4	0.78	111	92	5, 21
Methanol	CH_3OH	32	6.4	21.2	-201.3^g	37.5	0.79	112	91	5, 21
Tertiary butanol	C_4H_9OH	74	11.1	33.8	-281.3^d	49.5	0.79	113	110	5, 21
Blends										
Gasoline										
"Typical"	C_8H_{17}	113	14.9	44.2	-197.8^d	39.4	0.70	—	—	1
Regular	$C_8H_{15.6}$	112	14.8	43.5	-169.2^d	42.1–56.0	0.72–0.75	—	—	3
Premium	$C_8H_{15.6}$	112	14.7	42.7	-258.7^d	46.8	0.73–0.78	—	—	3
Regular unleaded	C_8H_{15}	111	14.6	42.8–44.6	-100.8^h	33.0–34.4	0.72–0.75	91	82	21
Premium unleaded	C_8H_{15}	111	14.6	42.8–44.6	-100.8^h	33.0–34.4	0.72–0.75	97	87	21
Aviation	C_8H_{17}	113	14.9	41.9–43.1	-390.3^h	—	0.72	—	—	3
Diesel										
Automotive	$C_{12}H_{23.7}$	168	14.7	40.6–44.4	-445.0^h	90.1–131.7	0.81–0.85	—	—	3
No. 1D	$C_{12}H_{26}$	170	15.0	42.4	-641.5^d	45.4	0.88	—	—	1
No. 2D	$C_{13}H_{28}$	184	15.0	41.8	-792.6^d	44.9	0.92	—	—	1
No. 4D	$C_{14}H_{30}$	198	15.0	41.3	-940.6^d	46.0	0.96	—	—	1

[a] At 298 K in MJ/mole.
[b] sg is ρ_{fu} at 4°C/ρ_f at 20°C (1000 kg/m³). For values from Ref. 1, reference temperatures are 15°C.
[c] From Ref. 23.
[d] Calculated.
[e] At 1 atm and boiling temperature.
[f] Estimate from p. 147, Ref. 1.
[g] Ref. 22.
[h] Calculated from avg. LHV_p.

instruments may be used to determine the mass flow rates of air and fuel into an engine so that AF and FA may be easily determined. Alternatively, AF and FA may be calculated if the exhaust product composition is known, using any of several available techniques.[5] The equivalence ratio normalizes the actual fuel–air ratio by the stoichiometric fuel–air ratio (FA_s) where "stoichiometric" refers to the chemically correct mixture with no excess air and no excess fuel. Recognizing that the stoichiometric mixture contains 100% "theoretical air" (by volume or mole) allows the equivalence ratio to be related to the actual percentage of theoretical air (TA, percentage by volume):

$$\phi = FA/FA_s = AF_s/AF = 100/TA \tag{74.1}$$

The equivalence ratio is a convenient parameter because $\phi < 1.0$ refers to a fuel-lean mixture, $\phi > 1.0$ refers to a fuel-rich mixture, and $\phi = 1.0$ refers to a stoichiometric mixture.

The stoichiometric fuel–air and air–fuel ratios can be easily calculated from a reaction balance by assuming "complete combustion" [only water vapor, (H_2O) and carbon dioxide (CO_2) are formed during the combustion process], even though the actual combustion process will almost never be complete. The reaction balance for the complete combustion of a stoichiometric mixture of air with a fuel of the atomic composition $C_x H_y$ is

$$C_x H_y + (x + 0.25y)O_2 + 3.764(x + 0.25y)N_2 \\ = xCO_2 + 0.5yH_2O + 3.764(x + 0.25y)N_2 \tag{74.2}$$

where air is taken to be 79% by volume nitrogen (N_2) and 21% by volume oxygen (O_2) and thus the nitrogen-to-oxygen ratio of air is $0.79/0.21 = 3.764$. Given that the molecular weight (MW) of air is 28.967, the MW of carbon (C) is 12.011, and the MW of hydrogen (H) is 1.008, then AF_s and FA_s for any hydrocarbon fuel may be calculated from

$$AF_s = 1/FA_s = (x + 0.25y) \times 4.764 \times 28.967/(12.011x + 1.008y) \tag{74.3}$$

The stoichiometric air–fuel ratios for a number of fuels of interest are presented in Table 74.1.

The energy content of the fuel is most often specified using the constant-pressure lower heating value (LHV_p). The lower heating value is the maximum energy that can be released during combustion of the fuel if (1) the water in the products remains in the vapor phase, (2) the products are returned to the initial reference temperature of the reactants (298 K), and (3) the combustion process is carried out such that essentially complete combustion is attained. If the water in the products is condensed, then the higher heating value (HHV) is obtained. If the combustion system is a flow calorimeter, then the constant-pressure heating value is measured (and, most usually, this is HHV_p). If the combustion system is a bomb calorimeter, then the constant-volume heating value is measured (usually HHV_v). The constant-pressure heating value is the negative of the standard enthalpy of reaction (ΔH_R^{298}, also known as the heat of combustion) and ΔH_R^{298} is a function of the standard enthalpies of formation (h_F^{298}) of the reactant and product species. For a fuel of composition $C_x H_y$, Eq. (74.4) may be used to calculate the constant-pressure heating value (HV_p) given the enthalpy of formation of the fuel or may be used to calculate h_F^{298} of the fuel given HV_p:

$$HV_p = -\Delta H_R^{298} = [(h_{F, C_x H_y} - \beta h_{v, C_x H_y}) + 393.418x \\ + 0.5y(241.763 + 43.998\alpha)]/(12.011x + 1.008y) \tag{74.4}$$

In Eq. (74.4): (1) $\alpha = 0$ if the water in the products is not condensed (yielding LHV_p) and $\alpha = 1$ if the water is condensed (yielding HHV_p); (2) $\beta = 0$ if the fuel is initially a vapor and $\beta = 1$ if the fuel is initially a liquid; (3) $h_{v, C_x H_y}$ is the enthalpy of vaporization per mole of fuel at 298 K; (4) the standard enthalpies of formation are CO_2: -393.418 MJ/mole,* H_2O: -241.763 MJ/mole, O_2: 0 MJ/mole, N_2: 0 MJ/mole; (5) the enthalpy of vaporization of H_2O at 298 K is 43.998 MJ/mole; and (6) the denominator is simply the molecular weight of the fuel yielding the heating value in MJ/kg of fuel. Also, the relationship between the constant volume heating value (HV_V) and HV_P for a fuel $C_x H_y$ is

$$HV_p = HV_v + (0.25y - 1) \times 2.478/(12.011x + 1.008y) \tag{74.5}$$

Of the several heating values that may be defined, LHV_P is preferred for engine calculations since condensation of water in the combustion chamber is definitely to be avoided and since an engine is essentially a steady-flow device and thus the enthalpy is the relevant thermodynamic property (rather than the internal energy). For diesel fuels HHV_v may be estimated from nomographs given the density and the "mid-boiling-point temperature" or given the "aniline point," the density, and the sulfur content of the fuel.[6]

Values for LHV_p, h_F^{298}, and h_v^{298} for various fuels of interest are presented in Table 74.1.

The density of a fuel (ρ_f) may be easily calculated from the specific gravity of the fuel (sg_f) and the density of water (ρ_w) at the same temperature as the fuel (pressure has an insignificant effect on the density) from the relationship

$$\rho_f = sg_f \cdot \rho_w \tag{74.6}$$

* "Mole" based on a kg, also referred to as a kmole or kg-mole.

In turn, the specific gravity of the fuel can be easily calculated from a simple measurement of the American Petroleum Institute gravity (API):

$$sg_f = 141.5/(API + 131.5) \tag{74.7}$$

Values of sg_f for various fuels are presented in Table 74.1.

The knock tendency of SI engine fuels (gasolines) are rated using an octane number (ON) scale. A higher octane number indicates a higher resistance to knock. Two different octane rating tests are currently used. Both use a single-cylinder variable-compression-ratio SI engine for which all operating conditions are specified (see Table 74.2). The fuel to be tested is run in the engine and the compression ratio is increased until knock of a specified intensity (standard knock) is obtained. Blends of two primary reference fuels are then tested at the same compression ratio until the mixture is found that produces standard knock. The two primary reference fuels are 2,2,4-trimethylpentane (also called isooctane), which is arbitrarily assigned an ON of 100, and n-heptane, which is arbitrarily assigned an ON of 0. The ON of the test fuel is then simply equal to the percentage of isooctane in the blend that produced the same knock intensity at the same compression ratio. However, if the test fuel has an ON above 100, then isooctane is blended with tetraethyllead instead of n-heptane. After the knock tests are completed, the ON is then computed from

$$ON = 100 + 28.28T/[1.0 + 0.736T + (1.0 + 1.472T - 0.035216T^2)^{1/2}] \tag{74.8}$$

where T is the number of milliliters of tetraethyllead per U.S. gallon of isooctane. The two different octane rating tests are called the Motor method (American Society of Testing and Materials, ASTM Standard D2700-82) and the Research method (ASTM D2600-82), and thus a given fuel will have two different octane numbers: a Motor octane number (MON) and a Research octane number (RON). The Motor method produces the lowest octane numbers, primarily because of the high intake manifold temperature for this technique, and thus the Motor method is said to be a more severe test for knock. The "sensitivity" of a fuel is defined as the RON minus the MON of that fuel. The "antiknock index" is the octane rating posted on gasoline pumps at service stations in the United States and is simply the average of RON and MON. Octane numbers for various fuels of interest are presented in Table 74.1.

The standard rating test for the knock tendency of diesel fuels (ASTM D613-82) produces the centane number (CN). Because SI and diesel engines knock for essentially opposite reasons, a fuel with a high ON will have a low CN and therefore would be a poor diesel fuel. A single-cylinder variable-compression-ratio CI engine is used to measure the CN, and all engine operating conditions are specified. The compression ratio is increased until the test fuel exhibits an ignition delay of 13°. Here, it should be noted that ignition delay rather than knock intensity is measured for the CN technique. A blend of two primary reference fuels (n-cetane: CN = 100; heptamethylnonane: CN = 15) are then run in the engine and the compression ratio is varied until a 13° ignition delay is obtained. The CN of this blend is given by

$$CN = \% \ n\text{-cetane} + 0.15 \ (\% \ \text{heptamethylnonane}) \tag{74.9}$$

Various blends are tried until compression ratios are found that bracket the compression ratio of the test fuel. The CN is then obtained from a standard chart. General specifications for diesel fuels are presented in Table 74.3 along with characteristics of "average" diesel fuels for light duty vehicles.

Many other thermochemical properties of fuels may be of interest, such as vapor pressure, volatility, viscosity, cloud point, aniline point, mid-boiling-point temperature, and additives. Discussion of these characteristics is beyond the scope of this chapter but is available in the literature.[1,4-7]

74.3 PERFORMANCE AND EFFICIENCY

The performance of an engine is generally specified through the brake power (bp), the torque (τ), or the brake mean effective pressure (bmep), while the efficiency of an engine is usually specified through

Table 74.2 Test Specifications that Differ for Research and Motor Method Octane Tests—ASTM D2699-82 and D2700-82

Operating Condition	RON	MON
Engine speed (rpm)	600	900
Inlet air temperature (°C)	a	38°C
FA Mixture temperature (°C)	b	149°C
Spark advance	13°BTDC	c

[a] Varies with barometric pressure.

[b] No control of fuel–air mixture temperature.

[c] Varies with compression ratio.

Table 74.3 Diesel Fuel Oil Specifications—ASTM D975-81

Property	Units	Fuel Type 1D[b]	Fuel Type 2D[c]	Fuel Type 4D[d]
Minimum flash paint	°C	38	52	55
Maximum H$_2$O and sediment	Vol. %	0.05	0.05	0.50
Maximum carbon residue	%	0.15	0.35	—
Maximum ash	Wt. %	0.01	0.01	0.10
90% distillation temperature, min/max	°C	—/288	282/338	—/—
Kinematic viscosity^a, min/max	mm^2/sec	1.3/2.4	1.9/4.1	5.5/24.0
Maximum sulfur	Wt.%	0.5	0.5	2.0
Maximum Cu strip corrosion	—	No. 3	No. 3	—
Minimum cetane number	—	40	40	30

^{a} At 40°C.

^{b} Preferred for high-speed diesels, especially for winter use, rarely available. 1976 U.S. average properties[17,24]: API, 42.2; 220°C midboiling point; 0.081 wt. % sulfur; and cetane index^e of 49.5.

^{c} For high-speed diesels (passenger cars and trucks) 1972 U.S. average properties[17,24]: API = 35.7; MBP = 261°C; 0.253 wt. % sulfur; cetane index^e = 48.4.

^{d} Low- and medium-speed diesels.

^{e} The cetane index is an approximation of the CN, calculated from ASTM D976-80 given the API and the midpoint temperature, and accurate within $\pm$ 2 CN for $30 \leq CN \leq 60$ for 75% of distillate fuels tested.

the brake specific fuel consumption (bsfc) or the overall engine efficiency (η_e). Experimental and theoretical determination of important engine parameters is discussed in the following sections.

74.3.1 Experimental Measurements

Engine dynamometer (dyno) measurements can be used to obtain these parameters using the relationships[5,8,9]

$$bp = LRN/9549.3 = LN/K \tag{74.10}$$
$$\tau = LR = 9549.3\ bp/N \tag{74.11}$$
$$bmep = 60{,}000\ bp\ X/DN \tag{74.12}$$
$$bsfc = \dot{m}_f/bp \tag{74.13}$$
$$\eta_e = 3600\ bp/(\dot{m}_f\ LHV_p) = 3600/(bsfc\ LHV_p) \tag{74.14}$$

Definitions and standard units* for the variables in the above equations are presented in the symbols list. The constants in the above equations are simply unit conversion factors. The brake power is the useful power measured at the engine output shaft. Some power is used to overcome frictional losses in the engine and this power (the friction power, fp) is not available at the output shaft. The total rate of energy production by the engine is called the indicated power (ip)

$$ip = bp + fp \tag{74.15}$$

where the friction power can be determined from dyno measurements using:

$$fp = FRN/9549.3 = FN/K \tag{74.16}$$

The efficiency of overcoming frictional losses in the engine is called the mechanical efficiency (η_M) which is defined as

$$\eta_M = bp/ip = 1 - fp/ip \tag{74.17}$$

The definitions of ip and η_M allow determination of the indicated mean effective pressure (imep) and the indicated specific fuel consumption (isfc):

$$imep = bmep/\eta_M = 60{,}000\ ip\ X/DN \tag{74.18}$$
$$isfc = bsfc\ \eta_M = \dot{m}_f/ip \tag{79.19}$$

* Standard units are not in strict compliance with the International System of Units, in order to produce numbers of convenient magnitude.

Three additional efficiencies of interest are the volumetric efficiency (η_v), the combustion efficiency (η_c), and the indicated thermal efficiency (η_{ti}).

The volumetric efficiency is the effectiveness of inducting air into the engine[5,10-13] and is defined as the actual mass flow rate of air ($\dot{m}_a$) divided by the theoretical maximum air mass flow rate ($\rho_a DN/X$):

$$\eta_v = \dot{m}_a/(60\rho_a DN/X) \tag{74.20}$$

The combustion efficiency is the efficiency of converting the chemical energy of the fuel to thermal energy (enthalpy) of the products of combustion.[12-14] Thus,

$$\eta_c = -\Delta H_{R,\,act}^{298}/LHV_p \tag{74.21}$$

The actual enthalpy of reaction ($\Delta H_{R,\,act}^{298}$) may be determined by measuring the mole fractions in the exhaust of CO_2, CO, O_2, and unburned hydrocarbons (expressed as "equivalent propane" in the following) and calculating the mole fractions of H_2O and H_2 from reaction balances. For a fuel of composition $C_x H_y$,

$$-\Delta H_{R,\,act}^{298} = h_{F,\,C_x H_y}^{298} - \{(\Sigma \bar{y}_i h_{F,i}^{298})x/[(\bar{y}_{CO_2} + \bar{y}_{CO} + 3\bar{y}_{C_3 H_8})MW_{C_x H_y}]\} \tag{74.22}$$

where the summation is over the six product species (i) discussed above and $\bar{y}_i$ is the mole fraction of species i in the "wet" exhaust. In Eq. (74.22), a carbon balance was used to convert moles of species i per mole of product mixture to moles of species i per mole of fuel burned. If a significant amount of soot is present in the exhaust (e.g., a diesel under high load), then the carbon balance becomes inaccurate and an oxygen balance would have to be substituted.

The indicated thermal efficiency is the efficiency of the actual thermodynamic cycle. This parameter is difficult to measure directly, but may be calculated from

$$\eta_{ti} = 3600\,ip/(\eta_c\,\dot{m}_f\,LHV_p) \tag{74.23a}$$
$$= 3600/(isfc\,\eta_c\,LHV_p) \tag{74.23b}$$

Because the engine performance depends on the air flow rate through it, the ambient temperature, barometric pressure, and relative humidity can affect the performance parameters and efficiencies. It is often desirable to correct the measured values to standard atmospheric conditions. The use of correction factors is discussed in the literature,[5,8,9] but is beyond the scope of this chapter.

The four fundamental efficiencies η_{ti}, η_c, η_v, and η_M are related to the global performance and efficiency parameters in the following section. Methods for modeling these efficiencies are also discussed in the following section.

74.3.2 Theoretical Considerations and Modeling

A set of *exact* equations relating the fundamental efficiencies to the global engine parameters is[12,13]

$$bp = \eta_{ti}\,\eta_c\,\eta_v\,\eta_M\,\rho_a\,D\,N\,FA\,LHV_p/60X \tag{74.24}$$
$$ip = \eta_{ti}\,\eta_c\,\eta_v\,\rho_a\,D\,N\,FA\,LHV_p/60X \tag{74.25}$$
$$\tau = 1000\,\eta_{ti}\,\eta_c\,\eta_v\,\eta_M\,\rho_a\,D\,FA\,LHV_p/2\pi X \tag{74.26}$$
$$bmep = 1000\,\eta_{ti}\,\eta_v\,\eta_c\,\eta_M\,\rho_a\,FA\,LHV_p \tag{74.27}$$
$$imep = 1000\,\eta_{ti}\,\eta_v\,\eta_c\,\rho_a\,FA\,LHV_p \tag{74.28}$$
$$bsfc = 3600/\eta_{ti}\,\eta_c\,\eta_M\,LHV_p \tag{74.29}$$
$$isfc = 3600/\eta_{ti}\,\eta_c\,LHV_p \tag{74.30}$$
$$\eta_e = \eta_{ti}\,\eta_c\,\eta_M \tag{74.31}$$

where, again, the constants are simply unit conversion factors. Equations (74.24)–(74.31) are of interest because they (1) can be derived solely from physical and thermodynamic considerations,[12] (2) can be used to explain observed engine characteristics,[12] and (3) can be used as a basis for modeling engine performance.[13] For example, Eqs. (74.27) and (74.28) demonstrate that mep is useful for comparing different engines because it is a measure of performance that is independent of displacement (D), engine speed (N), and whether the engine is a 2-stroke ($X = 1$) or a 4-stroke ($X = 2$). Similarly, Eqs. (74.24), (74.27), and (74.29) show that a diesel should have less power, lower bmep, and better bsfc than a comparable SI engine because the diesel has higher η_{ti}, η_c, and η_v but lower η_M and LHV_p and much lower FA.

The performance of an engine may be theoretically predicted by modeling each of the fundamental efficiencies (η_{ti}, η_c, η_v, and η_M) and then combining these models using Eqs. (74.24)–(74.31) to yield the performance parameters. Simplified models for each of these fundamental efficiencies are discussed below. More detailed engine models are available with varying degrees of sophistication and accuracy,[2,5,11,13,15,16] but, because of their length and complexity, are beyond the scope of this chapter.

The combustion efficiency may be most simply modeled by assuming complete combustion. It can be shown[13] that for complete combustion

$$\eta_c = 1.0, \qquad \phi \leq 1 \tag{74.32a}$$
$$\eta_c = 1.0/\phi, \qquad \phi \geq 1 \tag{74.32b}$$

Equation (74.32) implies that η_c is only dependent on FA. It has been shown that for homogeneous charge 4-stroke SI engines using fuels with a carbon-to-hydrogen ratio similar to that of gasoline, η_c is approximately independent of compression ratio, engine speed, ignition timing, and load. Although no data are available, it is expected that this is also true of 4-stroke stratified charge SI engines and diesel engines (at least up to the point of production of appreciable smoke). Such relationships will be less accurate for 2-stroke engines, especially the 2-stroke SI engine. As shown in Fig. 74.7, Eq. (74.32) is accurate within 5–10% for fuel-lean combustion with accuracy decreasing to about 20% for the very fuel-rich equivalence ratio of 1.5 for the 4-stroke SI engine. Also shown in Fig. 74.7 is a quasiequilibrium equation which is slightly more accurate for fuel-lean systems and much more accurate for fuel-rich combustion:

$$\eta_c = 0.959 + 0.129\phi - 0.121\phi^2, \qquad 0.5 \le \phi \le 1 \qquad (74.33a)$$
$$\eta_c = 2.594 - 2.173\phi + 0.546\phi^2, \qquad 1.0 \le \phi \le 1.5 \qquad (74.33b)$$

The indicated thermal efficiency may be most easily modeled using air standard cycles. The values of η_{ti} predicted in this manner will be too high by a factor of about 2 or more, but the trends predicted will be qualitatively correct.

The SI engine ideally operates on the air standard Otto cycle, for which[2,4,5,10,11,14]

$$\eta_{ti} = 1 - (1/CR)^{k-1} \qquad (74.34)$$

where CR is the compression ratio and k is the ratio of specific heats of the working fluid. The assumptions of the air standard Otto cycle are (1) no intake or exhaust processes and thus no exhaust residual and no pumping loss, (2) isentropic compression and expansion, (3) constant-volume heat addition and therefore instantaneous combustion, (4) constant-volume heat rejection replacing the exhaust blowdown process, and (5) air is the sole working fluid and is assumed to have a constant value of k. The errors in this model primarily result from failure to account for (1) a working fluid with variable composition and variable specific heat, (2) finite duration of combustion, (3) heat losses, and (4) fluid mechanics (especially as affecting flow past the intake and exhaust valves). The air standard Otto cycle $P-v$ and $T-s$ diagrams are shown in Fig. 74.8. Equation (74.34) indicates that η_{ti} for the SI engine is only a function of CR. While this is not strictly correct, it has been shown[13] that, for

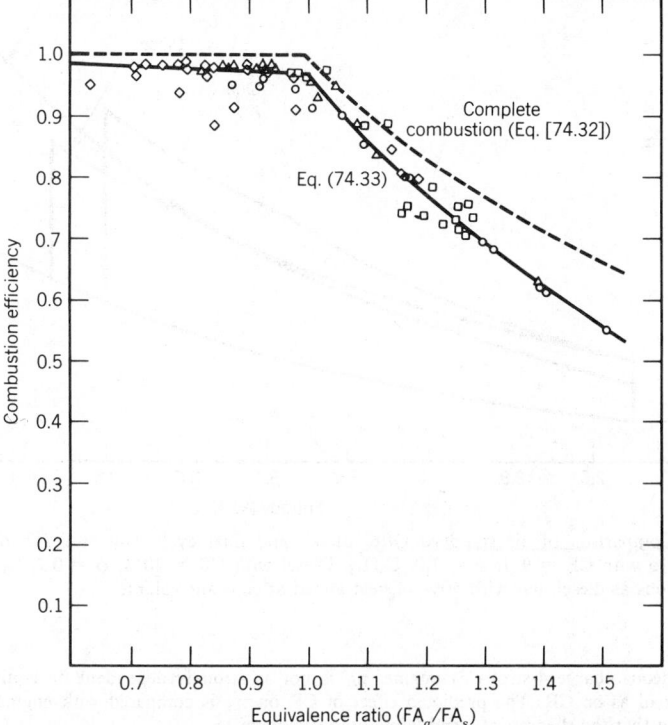

Fig. 74.7 Effect of equivalence ratio (normalized fuel–air ratio) on combustion efficiency for several different 4-stroke SI engines operating on indolene or isooctane.[13] Model predictions of Eqs. (74.32) and (74.33) also shown.

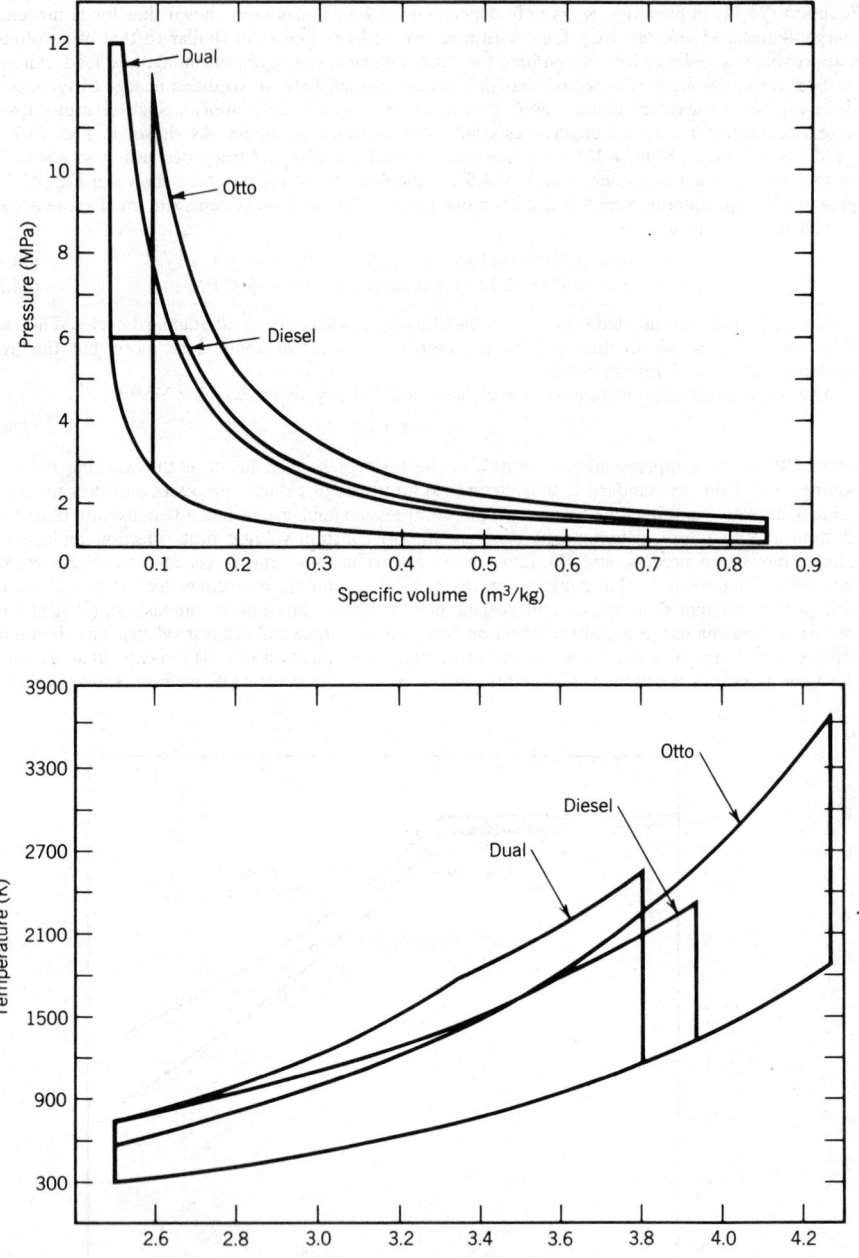

Fig. 74.8 Comparison of air standard Otto, diesel, and dual cycle P–v and T–s diagrams with $k = 1.3$. Otto with CR = 9:1, $\phi = 1.0$, C_8H_{18}. Diesel with CR = 20:1, $\phi = 0.7$, $C_{12}H_{26}$. Dual at same conditions as diesel, but with 50% of heat added at constant volume.

the homogeneous charge 4-stroke SI engine, η_{ti} is not as strongly dependent on equivalence ratio, speed, and load as on CR. The predicted effect of CR on η_{ti} is compared with engine data in Fig. 74.9, showing that the theoretical trend is qualitatively correct.

The traditional simplified model for η_{ti} of the CI engine is the air standard diesel cycle, but the air standard dual cycle is more representative of most modern diesel engines. Figure 74.8 compares

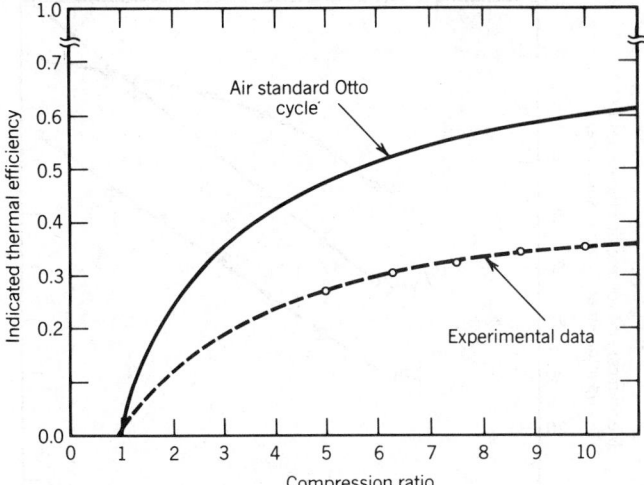

Fig. 74.9 Effect of compression ratio on indicated thermal efficiency of SI engine. Dashed line—experimental data for single cylinder 4-stroke SI engine.[13] Solid line—prediction of air standard Otto cycle [Eq. (74.34)].

the P–v and T–s diagrams for the air standard diesel, dual, and Otto cycles. For the air standard diesel cycle, it can be shown that[2,4,5,11,14]

$$\eta_{ti} = 1 - (r_T^k - 1)/[k\,CR^{k-1}(r_T - 1)] \tag{74.35}$$

where r_T is the ratio of the temperature at the end of combustion to the temperature at the beginning of combustion, and is thus a measure of the load. Assumptions for this cycle are (1) no intake or exhaust process, (2) isentropic compression and expansion, (3) constant-pressure heat addition (combustion), (4) constant-volume heat rejection, and (5) air is the sole working fluid and has constant k. For the air standard dual cycle, it can be shown that[2,4,10,14]

$$\eta_{ti} = 1 - (r_p r_v^k - 1)/\{CR^{k-1}[(r_p - 1) + k r_p (r_v - 1)]\} \tag{74.36}$$

where r_p is the ratio of the maximum pressure to the pressure at the beginning of the combustion process and r_v is the ratio of the volume at the end of the combustion process to the volume at the beginning of the combustion process (TDC). The assumptions are the same as those for the air standard diesel cycle except that combustion is assumed to occur initially at constant volume with the remainder of the combustion process occurring at constant pressure. Equations (74.35) and (74.36) indicate that the η_{ti} of the diesel is a function of both compression ratio and load, predictions that are qualitatively correct.

The volumetric efficiency is the sole efficiency for which a simplified model cannot be developed from thermophysical principles without the need for iterative calculations. Factors affecting η_v include heat transfer, fluid mechanics, and residual exhaust fraction. In fact, a supercharged engine may have greater than 100% volumetric efficiency since Eq. (74.20) is referenced to the air inlet density rather than to the density of the air in the intake manifold. However, for unsupercharged engines, observed engine characteristics allow values to be assumed for η_v for use in Eqs. (74.24)–(74.31). For the DISC SI (except at idle), the diesel, and the SI engine at wide open throttle (full load), η_v is approximately independent of operating conditions. A constant value for η_v of 0.7–1.0 may thus be assumed, recognizing that there is no justification for choosing any particular number unless engine data for that specific engine are available. Because the SI engine (except the DISC) controls power output by varying η_v, the effect of load on η_v must be taken into account for this type of engine. Fortunately, other engine operating conditions have much less effect on η_v than does the load,[13] so that only load need be considered for this simplified approach. As shown in Fig. 74.10 for the 4-stroke SI engine, η_v is linearly related to load (imep). Because choked flow is attained at no load, a value for η_v at zero load at one-half that assumed at full load may be used and a linear relationship between η_v and load (or intake manifold pressure) may then be used.

The mechanical efficiency may be most simply modeled by first determining η_{ti}, η_c, and η_v and then calculating the indicated power using Eq. (74.25). The friction power may then be calculated from an empirical relationship,[4] which has been shown to be very accurate for a variety of production

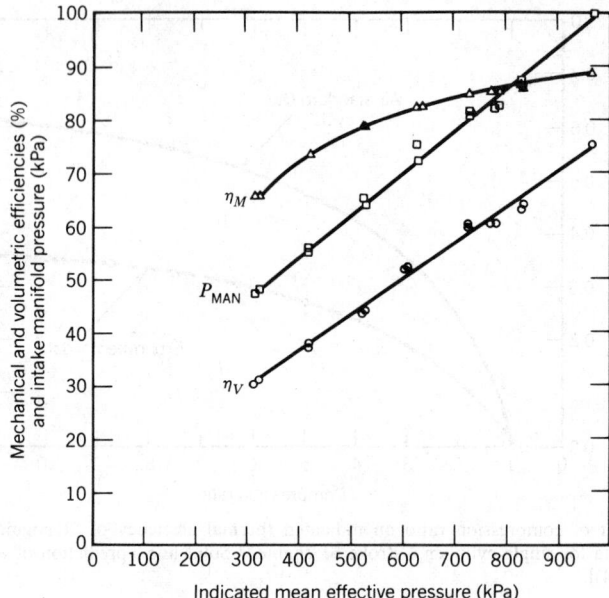

Fig. 74.10 Effect of load (imep) on volumetric efficiency, mechanical efficiency, and intake manifold absolute pressure for 4-stroke V6 SI engine at 2000 rpm.[13]

multicylinder SI engines (and there is little reason to believe that it does not apply equally well to diesels):

$$fp = 1.975 \times 10^{-9} \, D \, S \, CR^{1/2} \, N^2 \qquad (74.37)$$

where S is the stroke in mm. It should be noted that oil viscosity (and therefore oil and coolant temperatures) can significantly affect fp, and that Eq. (74.37) applies for normal operating temperatures. Given fp and ip, η_M may be calculated using Eq. (74.17). Since load strongly affects ip and speed strongly affects fp, then η_M is more strongly dependent on speed and load than on other operating conditions. Figures 74.10 and 74.11 demonstrate the effects of speed and load on η_M for a 4-stroke SI engine. Similar trends would be observed for other types of engines.

The accuracy of the performance predictions discussed above depends on the accuracies of the predictions of η_{ti}, η_c, η_v, and fp. Of these, the models for η_c and fp are generally acceptable. Thus, the primary sources for error are the models for η_{ti} and η_v. As stated earlier, highly accurate models are available if more than qualitative performance predictions are needed.

74.3.3 Engine Comparisons

Table 74.4 presents comparative data for various types of engines.

One of the primary advantages of the SI engine is the wide speed range attainable owing to the short ignition delay, high flame speed, and low rotational mass. This results in high specific weight (bp/W) and high power per unit displacement (bp/D). Additionally, the energy content (LHV$_p$) of gasoline is about 3% greater than that of diesel fuel. This aids both bmep and bp. The low CR results in high η_M and low engine weight. The low weight combined with the relative mechanical simplicity (especially of the fueling system), results in low initial cost. The high η_M, LHV$_p$, and FA result in high bmep and bp. However, the low CR results in low η_{ti}, which in turn causes low η_e, high part load bsfc, and poor low speed τ. Because η_{ti} increases with load, reasonable bsfc is attained near full load. Also, since the product of η_{ti} and N initially increases faster with engine speed than η_M decreases, then good medium speed torque is attained. Another disadvantage of the SI engine is that the near stoichiometric FA used results in high emissions of the gaseous regulated exhaust pollutants: NO$_x$ (oxides of nitrogen), CO (carbon monoxide) and HCs (unburned hydrocarbons). On the other hand, the nature of the premixed combustion process results in particulate emission levels that are almost too low to measure.

One of the primary advantages of the diesel is that the high CR results in high η_{ti} and thus good full load bsfc. The higher η_{ti} and the high η_v produce high η_e and good low speed τ. Additionally, because of the relationship between η_{ti} and load for the diesel, bsfc does not decrease as rapidly

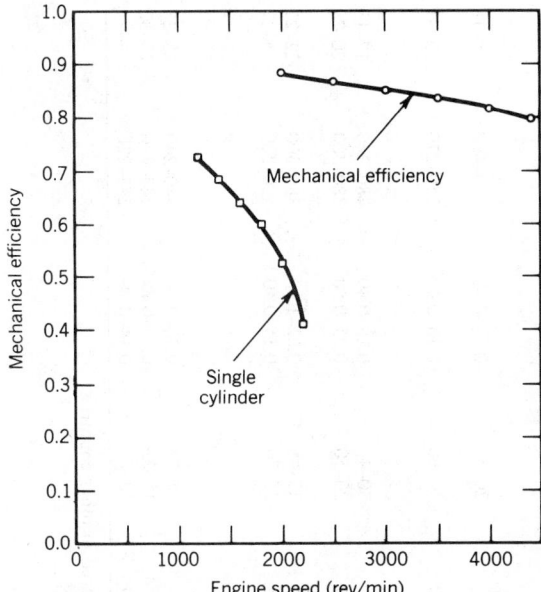

Fig. 74.11 Effect of engine speed on mechanical efficiency of 4-stroke SI V6 engine[13] and of 4-stroke SI single-cylinder research engine.

with decreasing load for the diesel as for the SI engine and thus good part load bsfc is another advantage of the diesel. The low volatility of diesel fuel results in lower evaporative emissions and a lower risk of accidental fire. The high CR also results in relatively low η_M and high engine weight. The high W and the mechanical sophistication (especially of the fuel-injection system) lead to high initial cost. The low η_M, coupled with the somewhat lower LHV_p and the much lower FA in comparison to the SI engine, yield lower bmep. The low bmep and limited range of engine speeds yield low bp and low bp/D. The low bp and high W produce low bp/W. The diesel is ideally suited to supercharging, since supercharging inhibits knock. However, the lower exhaust temperatures of the diesel (500–600°C)[3] means that there is less energy available in diesel exhaust and thus it is somewhat more difficult to turbocharge the diesel. The diffusion flame nature of the combustion process produces high particulate and NO_x levels. The overall fuel-lean nature of the combustion process produces relatively low levels of HCs and very low levels of CO. Diesel engines are much noisier than SI engines.

The IDI diesel is used in all diesel-powered passenger cars sold in the United States, because of the high bsfc of the diesel in comparison to the SI engine. The IDI diesel passenger car currently has approximately 35% better fuel economy on a kilometers per liter of fuel basis (19% better on a per unit energy basis, since the density of diesel fuel is about 16% higher then that of gasoline) than the comparable SI-engine-powered passenger car.[17] The IDI diesel is currently used in passenger cars rather than the DI diesel because of the more limited engine speed range of the DI diesel and because the necessarily more sophisticated fuel-injection system leads to higher initial cost. Also, the IDI diesel is less fuel sensitive (less prone to knock), is generally smaller, runs more quietly, and has less odorous emissions than the DI diesel.[17]

In comparison to the IDI diesel, the DI diesel has 15–20% better fuel economy.[17] This is a result of a higher η_{ti}, which yields better bsfc and η_e. The engine speed range of current-generation DI diesels is more limited than that of the IDI diesel, although efforts are ongoing to improve the speed range of both the DI and the IDI diesel. The DI diesel is easier to start and rejects less heat to the coolant. The somewhat higher exhaust temperature makes the DI diesel more suitable for turbocharging. At full load, the DI emits more smoke and is noisier than the IDI diesel.

The primary advantage of the 2-stroke engine is that it has one power stroke per revolution ($X = 1$) rather than one per every two revolutions ($X = 2$). This results in high τ, bp, and bp/W. However, the scavenging process results in very high HC emissions, and thus low η_e, bmep, bsfc, and η_e. When crankcase supercharging is used to improve scavenging, η_v is still low,[3] being in the range of 0.3–0.7. Blowers may be used to improve scavenging and η_v, but result in decreased η_M. The mechanical simplicity of the valveless crankcase compression design results in high η_M, low weight, and low initial cost. Air-cooled designs have an even lower weight and initial cost, but have limited CR because of engine-cooling considerations. In fact, both air-cooled and water-cooled engines have

Table 74.4 Comparative Data for Motor Vehicle Engines[g]

Engine Type	CR	Peak N (rpm)	bmep (kPa)	bp/D (kW/liter)	bp/W (kW/kg)	bsfc (g/kWhr)	η_0 (%)
Spark Ignition							
Motorcycles							
2-stroke	6.5–11	4500–7000[e]	400–550	20–50	0.17–0.40	600–400	14–18
4-stroke	6–10	5000–7500[e]	700–1000	30–60	0.18–0.40	340–270	25–31
Passenger cars							
2-stroke	6–8	4500–5400	450–600	30–45	0.18–0.40	480–340	17–18
4-stroke	7–11[c]	4000–7500[f]	700–1000	20–50	0.25–0.50	380–300	20–28
Rotary	8–9	6000–8000	950–1050	35–45	0.62–1.1	380–300	22–27
Trucks (4-stroke)	7–8	3600–5000	650–700	25–30	0.15–0.40	400–300	16–27
Diesel							
Passenger cars[a]	12–23[d]	4000–5000	500–750	18–22	0.20–0.40	340–240	23–28
Trucks—NA[b]	16–22	2100–4000	600–900	15–22	0.14–0.25	245–220	23–33
Trucks—TC[b]	15–22	2100–3200	1200–1800	18–26	0.14–0.29	245–220	—

[a] Exclusively IDI in United States. 2-stroke and 4-stroke DI may be used in other countries.
[b] NA: naturally aspirated, TC: turbocharged. Trucks are primarily DI.
[c] U.S. average about 8.5.
[d] U.S. average about 22.
[e] Many modern motorcycle engines have peak speeds exceeding 10,000 rpm.
[f] Production engines capable of exceeding 6000 rpm are rare.
[g] Adopted with permission from Adler and Bazlen.[3] Design trends to be read from left to right.

high thermal loads because of the lack of no-load strokes. For the 2-stroke SI engine, the CR is also limited by the need to inhibit knock.

The primary advantages of the rotary SI engine are the resulting low vibration, high engine speed, and high η_v. The high η_v produces high bmep and good low speed τ. The high bmep and high attainable engine speed produce high bp. The valveless design results in decreased W. The high bp and low W result in very high bp/W. Because η_e and bsfc are independent of η_v, these parameters are approximately equal for the rotary SI and the 4-stroke piston SI engines.

74.4   EMISSIONS AND FUEL ECONOMY REGULATIONS*

Light duty vehicles (less than 8,500 lb) sold in the United States are currently the subject of federal and state regulations concerning exhaust emissions, evaporative emissions, and fuel economy. Heavy duty vehicles are the subject of regulations concerning gaseous exhaust emissions, evaporative emissions, and smoke. Similar regulations are in effect in many foreign countries.

74.4.1   Light Duty Vehicles

U.S. federal and state of California exhaust emissions regulations are presented in Table 74.5. Light-duty-vehicle (LDV) emissions regulations are divided into those applicable to passenger cars and those applicable to light duty trucks [defined as trucks of less than 6000 lb gross vehicle weight (GVW) prior to 1979 and less than 8500 lb after, except those with more than 45 ft² frontal area]. Emissions levels are for operation over the federal test procedure transient driving cycle (FTP, see Fig. 74.12), with 50,000 miles of durability. The FTP cycle is intended to represent average urban driving habits and therefore has an average vehicle speed of only 31.46 km/hr and includes 1.42 stops per kilometer. The total test length is 11.98 km and the duration is 1371 sec with 19% of this time spent with the vehicle idling. The motor vehicle manufacturer will commit at least two vehicles of each type to the certification procedure. The first vehicle will be used to accumulate 50,000 miles so that a deterioration factor (DF) can be determined for each of the regulated species. The DF is determined by periodic FTP testing over the 50,000 miles and is intended to reveal how the emissions levels change with mileage accumulation. The second vehicle is subjected to the FTP test after 4000 miles of accumulation. The emission level of each regulated species is then multiplied by the DF for that species and the product is used to determine compliance with the regulatory standard. All data are submitted to the U.S. Environmental Protection Agency (EPA), which may perform random confirmatory testing, especially of the 4000 mile vehicles.

The 1981 standards represent a 96% reduction in HC and CO and a 76% reduction in NO_x from precontrol levels. Beginning in 1980, the California HC standard was specified as either 0.41 g of total hydrocarbons per mile or 0.39 g of nonmethane hydrocarbons per mile, in recognition that methane is not considered to be photochemically reactive and therefore does not contribute to smog formation.[17,18] Light duty diesel (LDD) vehicles emit much lower levels of methane than light duty gasoline (LDG) vehicles[17,18] and therefore the nonmethane HC standard may be more difficult for the LDD. LDGs emit essentially immeasurable quantities of particulates, and therefore are not required to demonstrate compliance with the particulate standards.

LDVs are also required to meet evaporative emissions standards as determined using a "Shed Test." The precontrol average for evaporative emissions was 50.6 g per test, which is approximately equivalent to 4.3 g HC per mile.[19] The federal regulations are 6 g/test for 1978–1980 and 2 g/test thereafter. The California standard is 2 g/test beginning in 1980. Evaporative emissions from the LDD are essentially immeasurable, and thus they are not required to demonstrate compliance with this regulation. California allows LDDs a 0.16 g/mi credit on exhaust HC in recognition of the negligible evaporative HC emissions of these vehicles.

Beginning in 1982, all LDVs sold in specified high altitude counties are required to meet the following standards at altitude: HC: 0.57 g/mi; CO: 7.8 g/mi; NO_x: 1.0 g/mi (with a possible waiver to 1.5 g/mi for diesel or innovative technology LDVs); evaporative HC: 2.6 g/test. Beginning in 1985, all LDGs will be required to meet normal standards at all altitudes. Separate high altitude standards continue to be in effect for light duty trucks.

Beginning in 1978, the motor vehicle manufacturers were required to meet the corporate average fuel economy (CAFE) standards shown in Table 74.6 (a different, and lower, set of standards is in effect for "low-volume" manufacturers). The CAFE standard is the mandated average fuel economy as determined by averaging over all vehicles sold by a given manufacturer. The CAFE requirement is based on the "composite" fuel economy (MPG_c, in mi/gal), which is a weighted calculation of the fuel economy measured over the FTP cycle (MPG_u) plus the fuel economy measured over the highway fuel economy test (HFET, see Fig. 74.12) transient driving cycle (MPG_h):

$$MPG_c = 1/[(0.55/MPG_u) + (0.45/MPG_h)] \tag{74.38}$$

---

* The regulations are specified in mixed English and metric units and these specified units are used in this section to avoid confusion.

Table 74.5 Exhaust Emission Standards for Light Duty Vehicles in g/mi

Year	Hydrocarbons		Carbon Monoxide		Oxides of Nitrogen		Particulates	
	Federal	California[g]	Federal	California[g]	Federal	California[g]	Federal	California
Passenger Cars								
1960[a]	10.6		84		4		—	—
1975–1977	1.5	0.9	15.0	9.0	3.1	2.0	—	—
1977–1979	1.5	0.41	15.0	9.0	2.0	1.5	—	—
1980	0.41	0.39[b]	7.0	9.0	2.0	1.0	—	—
1981	0.41	0.39[b]	3.4[c]	7.0	1.0[d]	0.7[e]	—	—
1982	0.41	0.39[b]	3.4[c]	7.0	1.0[d]	0.7[e]	0.6	0.6
1983–1984	0.41	0.39[b]	3.4	7.0	1.0[d]	0.4[e]	0.6	0.6
1985	0.41	0.39[b]	3.4	7.0	1.0	0.4	0.2[f]	0.2
Light Duty Trucks								
1975	2.0	2.0	20	20	3.1	3.1	—	—
1976–1978	2.0	0.9	20	17	3.1	2.0	—	—
1979	1.7	0.41/0.50[c]	18	9/9	2.3	1.5/2.0	—	—
1980–1981	1.7	0.39[c]/0.50[c]	18	9/9	2.3	1.5/2.0	—	—
1982–1984	1.7	0.39[c]/0.50[c]	18	9/9	2.3	1.5/2.0	0.6	0.6
1985	1.7	0.39[c]/0.50[c]	18	9/9	2.3	1.5/2.0	0.26[f]	0.26

[a] Average values with no controls.
[b] Nonmethane hydrocarbon standard.
[c] Waiver possible to 7.0 g/mi.
[d] Waiver possible to 1.5 g/mi.
[e] 1 g/mi option with 100,000 miles durability also possible.
[f] Averaging of car and truck particulate emissions may be used beginning in 1985.
[g] For light duty trucks, California standards by inertia weight <4,000 lb/>4,000 lb.

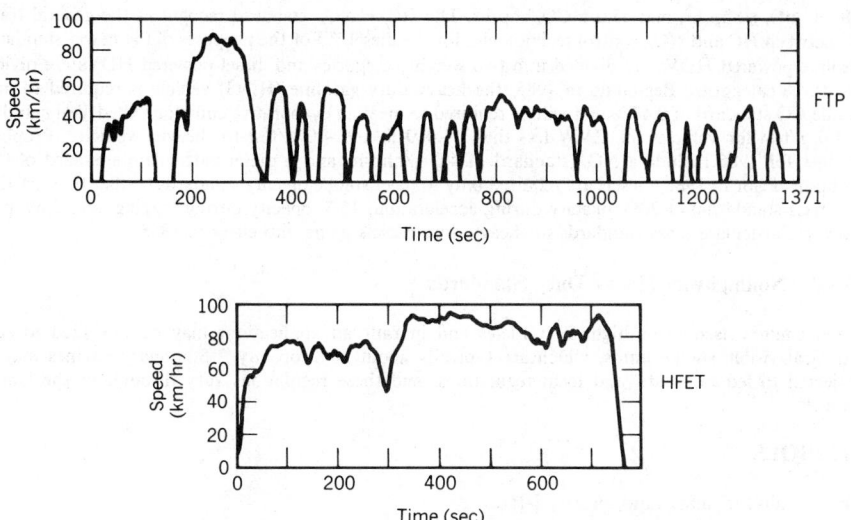

Fig. 74.12   FTP and HFET transient driving cycles.

The manufacturer is assessed a fine of $5 per vehicle produced for every 0.1 mpg below the CAFE standard. The four major American manufacturers of light duty vehicles each received credits during the years 1979–1982 because the fleets sold exceeded the CAFE standards. Beginning in 1985, an adjustment factor is added to the measured CAFE to account for the effects of changes in the test procedure. Additionally, the motor vehicle manufacturer is subjected to a "gas guzzler" tax for each vehicle sold that is below a minimum composite fuel economy, with the minimum level being: 1983, 19.0 mpg; 1984, 19.5 mpg; 1985, 21.0 mpg; 1986 on, 22.5 mpg. For 1986, the tax ranges from $500 for each vehicle with 21.5–22.4 mpg up to a maximum of $3850 for each vehicle with less than 12.5 mpg.

The federal and state regulations are intended to require compliance based on the best available technology. Therefore, future standards may be subjected to postponement, modification, or waiver depending on available technology. For more detailed information regarding the standards and measurement techniques, refer to the U.S. Code of Federal Regulations, Title 40, Part 86, Subpart B and to the Society of Automotive Engineers (SAE) Recommended Practices.[20]

74.4.2   Heavy Duty Vehicles

The test procedure for heavy duty vehicles (HDVs) is different from that for the LDV and is based on grams of pollutant emitted per unit of energy produced (g/bhp-hr, mixed metric and English units, with bhp meaning brake horsepower). The 1980–1983 federal standards are: 1.5 g HC/bhp-hr, 10 g

Table 74.6   CAFE Standards for Passenger Cars Required by the Energy Policy and Conservation Act of 1975

Model Year	Average mpg
1978	18.0
1979	19.0
1980	20.0
1981	22.0[a]
1982	24.0[a]
1983	26.0[a]
1984	27.0[a]
1985	27.5[a]

[a] Set by the Secretary of the U.S. Department of Transportation.

(HC + NO$_x$)/bhp-hr, and 25.0 g CO/bhp-hr. The HC + NO$_x$ standard recognizes the general trade-offs between HC and NO$_x$ control technologies for the diesel.[17] For the purposes of the newer standards, gasoline powered HDVs are divided into two weight categories and diesel powered HDVs are divided into three categories. Beginning in 1985, the heavy duty gasoline (HDG) vehicle is required to meet an idle CO standard of 0.47% and will be required to meet an evaporative emissions Shed Test standard of 3.0 g/test for vehicles for GVW less than 14,000 lb and 4.0 g/test for heavier vehicles. Proposed changes for 1986 include a NO$_x$ standard of 4.0 g/bhp-hr and a diesel particulate standard of 0.25 g/bhp-hr. Prior to 1988, diesels are required only to meet smoke opacity standards (rather than particulate mass standards) of 20% opacity during acceleration, 15% opacity during lugging, and 50% peak opacity. Particulate mass standards for heavy duty diesels come into effect in 1988.

74.4.3 Nonhighway Heavy Duty Standards

Diesel engines used in off-highway vehicles and in railroad applications may be subjected to state and local visible smoke limits, which are typically about 20% opacity.[19] Stationary engines may be subjected to federal, state, and local regulations, and these regulations vary throughout the United States.[19]

SYMBOLS

AF	air–fuel mass ratio, $\dot{m}_a/\dot{m}_f$ [-]*
AF$_s$	stoichiometric air–fuel mass ratio [-]
API	American Petroleum Institute gravity [°]
BDC	bottom dead center
bmep	break mean effective pressure [kPa]
bp	brake power [kW]
bsfc	brake specific fuel consumption [g/kW–hr]
CI	compression ignition (diesel) engine
CN	cetane number [-]
CR	compression ratio (by volume) [-]
C$_x$H$_y$	average fuel molecule, with x atoms of carbon and y atoms of hydrogen
D	engine displacement [liters]
DI	direct injection diesel engine
DISC	direct injection stratified charge SI engine
DF	emissions deterioration factor [-]
F	force on dyno torque arm with engine being motored [N]
FA	fuel–air mass ratio, $\dot{m}_f/\dot{m}_a$ [-]
FA$_s$	stoichiometric fuel–air mass ratio [-]
fp	friction power [kW]
FTP	federal test procedure transient driving cycle
$h_{F,i}^{298}$	standard enthalpy of formation of species i at 298 K [MJ/mole of species i]†
$h_{v,i}^{298}$	enthalpy of vaporization of species i at 298 K [MJ/mole of species i]
ΔH_R^{298}	enthalpy of reaction (heat of combustion) at 298 K [MJ/kg of fuel]
HDV	heavy duty vehicle
HFET	highway fuel economy test transient driving cycle
HV	heating value [MJ/kg of fuel]
HHV$_p$	constant-pressure higher heating value [MJ/kg of fuel]
HHV$_v$	constant-volume higher heating value [MJ/kg of fuel]
HV$_p$	constant-pressure heating value [MJ/kg of fuel]
HV$_v$	constant-volume heating value [MJ/kg of fuel]
IDI	indirect injection diesel engine
imep	indicated mean effective pressure [kPa]
ip	indicated power [kW]
isfc	indicated specific fuel consumption [g/kW–hr]

* Standard units, which do not strictly conform to metric practice in order to produce numbers of convenient magnitude.
† kmole or kg-mole.

k	ratio of specific heats
K	dyno constant [N/kW-min]
L	force on dyno torque arm with engine running [N]
LDD	light duty diesel vehicle
LDG	light duty gasoline vehicle
LDV	light duty vehicle
LHV_p	constant-pressure lower heating value [MJ/kg of fuel]
LHV_v	constant-volume lower heating value [MJ/kg of fuel]
$\dot{m}_a$	air mass flow rate into engine [g/hr]
$\dot{m}_f$	fuel mass flow rate into engine [g/hr]
MON	octane number measured using Motor method [-]
MPG_c	composite fuel economy [mi/gal]
MPG_h	highway (HFET) fuel economy [mi/gal]
MPG_u	urban (FTP) fuel economy [mi/gal]
MW	molecular weight [kg/mole]
N	engine rotational speed [rpm]
ON	octane number of fuel [-]
P	pressure [kPa]
R	dyno torque arm length [m]
RON	octane number measured using Research method [-]
r_p	pressure ratio (end of combustion/end of compression) [-]
r_T	temperature ratio (end of combustion/end of compression) [-]
r_v	volume ratio (end of combustion/end of compression) [-]
s	entropy [$kJ/kg\text{-}K$]
S	piston stroke [mm]
SI	spark ignition engine
sg_f	specific gravity of fuel
T	milliliters of tetraethyllead per gallon of isooctane [ml/gal]
TA	percent theoretical air
T	temperature [K]
TDC	top dead center
v	specific volume [m^3/kg]
W	engine weight [kg]
X	crankshaft revolutions per power stroke
x	atoms of carbon per molecule of fuel
y	atoms of hydrogen per molecule of fuel
$\bar{y}_i$	mole fraction of species i in exhaust products [-]
α	1.0 for HHV, 0.0 for LHV [-]
β	1.0 for liquid fuel, 0.0 for gaseous fuel [-]
η_c	combustion efficiency [-]
η_e	overall engine efficiency [-]
η_M	mechanical efficiency [-]
η_{ti}	indicated thermal efficiency [-]
η_v	volumetric efficiency [-]
ϕ	equivalence ratio, FA/FA_s [-]
ρ_a	density of air at engine inlet [kg/m³]
ρ_f	density of fuel [kg/m³]
ρ_w	density of water [kg/m³]
τ	engine output torque [N-m]

REFERENCES

1. C. F. Taylor, *The Internal Combustion Engine in Theory and Practice*, MIT Press, Cambridge, MA, 1979, Vol. II.

2. R. S. Benson and N. D. Whitehouse, *Internal Combustion Engines,* Pergamon, New York, 1979.

3. U. Adler and W. Bazlen (Eds.), *Automotive Handbook,* Bosch, Stuttgart, 1978.

4. L. C. Lichty, *Combustion Engine Processes,* McGraw-Hill, New York, 1967.

5. E. F. Obert, *Internal Combustion Engines and Air Pollution,* Harper and Row, New York, 1973.

6. Fuels and Lubricants Committee, "Diesel Fuels—SAE J313 APR82," *SAE Handbook,* Vol. 3, pp. 23.34–23.39, 1983.

7. Fuels and Lubricants Committee, "Automotive Gasolines—SAE J312 JUN82," *SAE Handbook,* Vol. 3, pp. 23.25–23.34, 1983.

8. Engine Committee, "Engine Power Test Code—Spark Ignition and Diesel—SAE J1349 DEC80," *SAE Handbook,* Vol. 3, pp. 24.09–24.10, 1983.

9. Engine Committee, "Small Spark Ignition Engine Test Code—SAE J607a," *SAE Handbook,* Vol. 3, pp. 24.14–24.15, 1983.

10. C. F. Taylor, *The Internal Combustion Engine in Theory and Practice,* MIT Press, Cambridge, MA, 1979, Vol. I.

11. A. S. Campbell, *Thermodynamic Analysis of Combustion Engines,* Wiley, New York, 1979.

12. R. D. Matthews, "Relationship of Brake Power to Various Efficiencies and Other Engine Parameters: The Efficiency Rule," *International Journal of Vehicle Design,* 4(5), 491–500 (1983).

13. R. D. Matthews, S. A. Beckel, S. Z. Miao, and J. E. Peters, "A New Technique for Thermodynamic Engine Modeling," *Journal of Energy* 7(6): 667–675 (1983).

14. B. D. Wood, *Applications of Thermodynamics,* Addison-Wesley, Reading, MA, 1982.

15. P. N. Blumberg, G. A. Lavoie, and R. J. Tabaczynski, "Phenomenological Model for Reciprocating Internal Combustion Engines," *Progress in Energy and Combustion Science,* 5, 123–167 (1979).

16. J. N. Mattavi and C. A. Amann (Eds.), *Combustion Modeling in Reciprocating Engines,* Plenum, New York, 1980.

17. Diesel Impacts Study Committee, *Diesel Technology,* National Academy Press, Washington, DC, 1982.

18. R. D. Matthews, "Emission of Unregulated Pollutants from Light Duty Vehicles," *International Journal of Vehicle Design,* 4(5) 5(4): 475–489 (1983).

19. Environmental Activities Staff, *Pocket Reference,* General Motors, Warren, MI, 1983.

20. Automotive Emissions Committee, "Emissions," *SAE Handbook,* Vol. 3, Section 25, 1983.

21. Fuels and Lubricants Committee, "Alternative Automotive Fuels—SAE J1297 APR82," *SAE Handbook,* Vol. 3, pp. 23.39–23.41, 1983.

22. R. C. Weast (Ed.), *CRC Handbook of Chemistry and Physics,* 55th ed., CRC Press, Cleveland, OH, 1974.

23. ASTM Committee D-2, *Physical Constants of Hydrocarbons C_1 to C_{10},* American Society for Testing and Materials, Philadelphia, PA, 1963.

24. E. M. Shelton, "Diesel Fuel Oils, 1976," U.S. Energy Research and Development Administration, Publication No. BERC/PPS-76/5, 1976.

BIBLIOGRAPHY

Combustion and Flame, The Combustion Institute, Pittsburgh, PA, published monthly.

Combustion Science and Technology, Gordon and Breach, New York, published monthly.

International Journal of Vehicle Design, Inderscience, Jersey, C.I., UK, published quarterly.

Journal of Energy, American Institute of Aeronautics and Astronautics, New York, published bimonthly.

Proceedings of the Institution of Mechanical Engineers (Automobile Division), I Mech E, London.

Progress in Energy and Combustion Science, Pergamon, Oxford, UK, published quarterly.

SAE Technical Paper Series, Society of Automotive Engineers, Warrendale, PA.

Symposia (International) on Combustion, The Combustion Institute, Pittsburgh, PA, published biennially.

Transactions of the Society of Automotive Engineers, SAE, Warrendale, PA, published annually.

CHAPTER 75

INDOOR ENVIRONMENTAL CONTROL

JERALD D. PARKER
F. C. McQUISTON

Oklahoma State University
Stillwater, Oklahoma

75.1 MOIST AIR PROPERTIES AND CONDITIONING PROCESSES

75.1.1 Properties of Moist Air

Atmospheric air is a mixture of many gases plus water vapor and countless pollutants. Aside from the pollutants, which may vary considerably from place to place, the composition of the dry air alone is relatively constant, varying slightly with time, location, and altitude. In 1949 a standard composition of dry air was fixed by the International Joint Committee on Psychrometric Data as shown in Table 75.1.[1]

The molecular weight M of dry air is 28.965, and the gas constant R is 53.353 ft · lbf/lbm · R or 287 J/kg · K.

The basic medium in air-conditioning practice is a mixture of dry air and water vapor. The amount of water vapor may vary from zero to a maximum determined by the temperature and pressure of the mixture. The latter case is called saturated air, a state of neutral equilibrium between the moist air and the liquid or solid phases of water.

Moist air up to about 3 atm pressure obeys the perfect gas law with sufficient accuracy for engineering calculations. The Gibb's–Dalton law for a mixture of perfect gases states that the mixture pressure is equal to the sum of the partial pressures of the constituents. Because the various constituents of the dry air may be considered to be one gas, it follows that the total pressure P of moist air is the sum of the partial pressures of the dry air p_a and the water vapor p_v:

$$P = p_a + p_v$$

Humidity ratio W (sometimes called the specific humidity) is the ratio of the mass of the water vapor m_v to the mass of the dry air m_a in the mixture:

$$W = \frac{m_v}{m_a}$$

Relative humidity ϕ is the ratio of the mole fraction of the water vapor x_v in a mixture to the mole fraction x_s of the water vapor in a saturated mixture at the same temperature and pressure:

$$\phi = \left(\frac{x_v}{x_s}\right)_{t,P}$$

For a mixture of perfect gases the mole fraction is equal to the partial pressure ratio of each constituent. The mole fraction of the water vapor is

$$x_v = \frac{p_v}{P}$$

Thus

$$\phi = \frac{p_v/P}{p_s/P} = \frac{p_v}{p_s}$$

Dew point temperature t_d is the temperature of saturated moist air at the same pressure and humidity ratio as the given mixture. It can be shown that

$$\phi = \frac{W p_a}{0.6219 p_s}$$

where p_s is the saturation pressure of the water vapor at the mixture temperature.

The enthalpy i of a mixture of perfect gases is equal to the sum of the enthalpies of each constituent and is usually referenced to a unit mass of dry air:

$$i = i_a + W i_v$$

Each term has the units of energy per unit mass of dry air. With the assumption of perfect-gas behavior the enthalpy is a function of temperature only. If zero Fahrenheit or Celsius is selected as the reference

Table 75.1 Composition of Dry Air[7]

Constituent	Molecular Weight	Volume Fraction
Oxygen	32.000	0.2095
Nitrogen	28.016	0.7809
Argon	39.944	0.0093
Carbon dioxide	44.010	0.0003

state where the enthalpy of dry air is zero, and if the specific heats c_{pa} and c_{pv} are assumed to be constant, simple relations result:

$$i_a = c_{pa}\, t$$

$$i_v = i_g + c_{pv}\, t$$

where the enthalpy of saturated water vapor i_g at 0°F is 1061.2 Btu/lbm and 2501.3 kJ/kg at 0°C.

75.1.2   The Psychrometric Chart

At a given pressure and temperature of an air–water vapor mixture one additional property is required to completely specify the state, except at saturation.

A practical device used to determine the third property is the psychrometer. This apparatus consists of two thermometers, or other temperature-sensing elements, one of which has a wetted cotton wick covering the bulb. The temperatures indicated by the psychrometer are called the wet bulb and the dry bulb temperatures. The wet bulb temperature is the additional property needed to determine the state of moist air.

To facilitate engineering computations, a graphical representation of the properties of moist air has been developed and is known as a psychrometric chart, Fig. 75.1.[2]

In Fig. 75.1 dry bulb temperature is plotted along the horizontal axis in degrees Fahrenheit or Celsius. The dry bulb temperature lines are straight but not exactly parallel and incline slightly to the left. Humidity ratio is plotted along the vertical axis on the right-hand side of the chart in lbm_v/lbm_a or kg_v/kg_a. The scale is uniform with horizontal lines. The saturation curve with values of the wet bulb temperature curves upward from left to right. Dry bulb, wet bulb, and dew point temperatures all coincide on the saturation curve. Relative humidity lines with a shape similar to the saturation curve appear at regular intervals. The enthalpy scale is drawn obliquely on the left of the chart with parallel enthalpy lines inclined downward to the right. Although the wet bulb temperature lines appear to coincide with the enthalpy lines, they diverge gradually in the body of the chart and are not parallel to one another. The spacing of the wet bulb lines is not uniform. Specific volume lines appear inclined from the upper left to the lower right and are not parallel. A protractor with two scales appears at the upper left of the chart. One scale gives the sensible heat ratio and the other the ratio of enthalpy difference to humidity ratio difference. The enthalpy, specific volume, and humidity ratio scales are all based on a unit mass of dry air.

75.1.3   Space Conditioning Processes

When air is heated or cooled without the loss or gain of moisture, the process is a straight horizontal line on the psychrometric chart because the humidity ratio is constant. Such processes can occur when moist air flows through a heat exchanger. In cooling, if the surface temperature is below the dew point temperature of the moist air, dehumidification will occur. This process will be considered later. Figure 75.2 shows a schematic of a device used to heat or cool air. Under steady-flow–steady-state conditions the energy balance becomes

$$\dot{m}_a i_2 + \dot{q} = \dot{m}_a i_1$$

The direction of the heat transfer is implied by the terms heating and cooling, and i_1 and i_2 may be obtained from the psychrometric chart. The convenience of the chart is evident. Figure 75.3 shows heating and cooling processes. The relative humidity decreases when the moist air is heated. The reverse process of cooling results in an increase in relative humidity.

When moist air is cooled to a temperature below its dew point, some of the water vapor will condense and leave the air stream. Figure 75.4 shows a schematic of a cooling and dehumidifying device and Fig. 75.5 shows the process on the psychrometric chart. Although the actual process path will vary considerably depending on the type surface, surface temperature, and flow conditions, the heat and mass transfer can be expressed in terms of the initial and final states. The total amount of heat transfer from the moist air is

$$\dot{q} = \dot{m}_a (i_1 - i_2) - \dot{m}_a (W_1 - W_2) i_w$$

The last term on the right-hand side is usually small compared to the others and is often neglected.

The cooling and dehumidifying process involves both sensible heat transfer, associated with the decrease in dry bulb temperature, and latent heat transfer, associated with the decrease in humidity ratio. We may also express the latent heat transfer as

$$\dot{q}_l = \dot{m}_a (i_1 - i_a)$$

and the sensible heat transfer is given by

$$\dot{q}_s = \dot{m}_a (i_a - i_2)$$

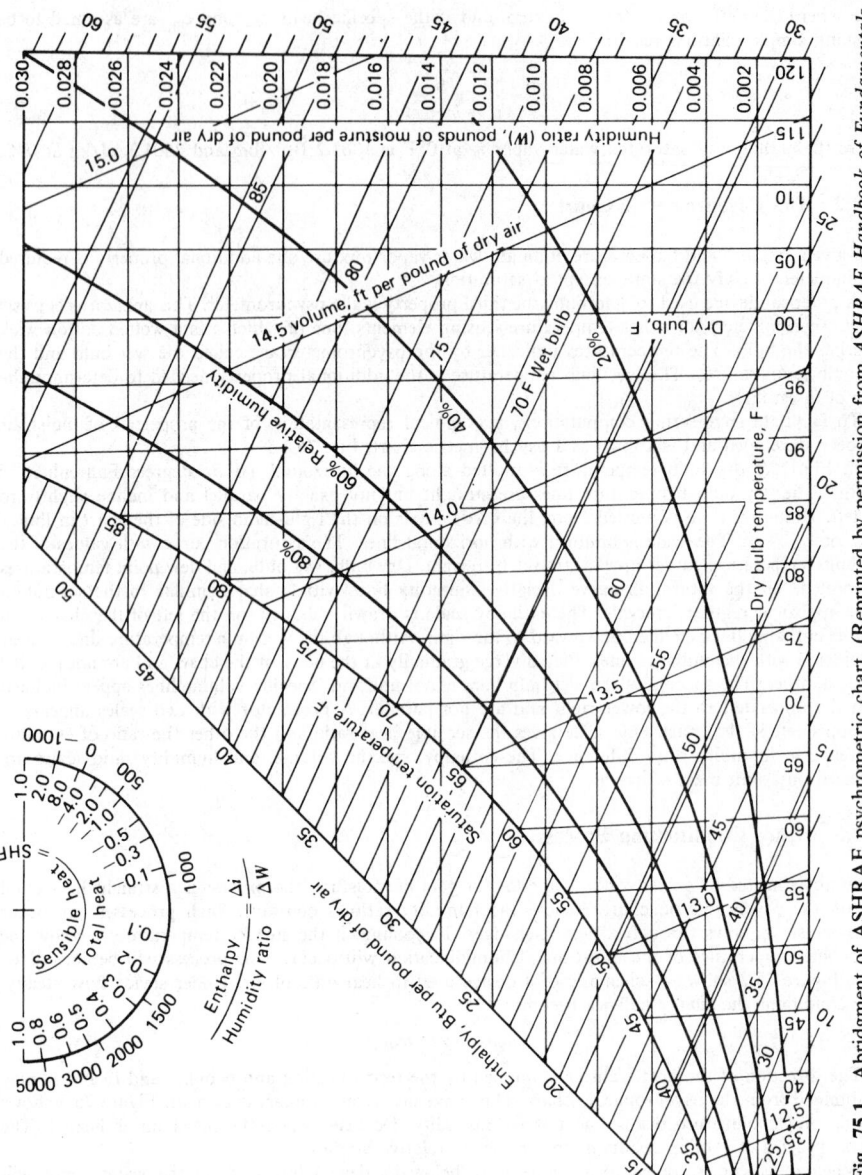

Fig. 75.1 Abridgment of ASHRAE psychrometric chart. (Reprinted by permission from *ASHRAE Handbook of Fundamentals*, 1981.)

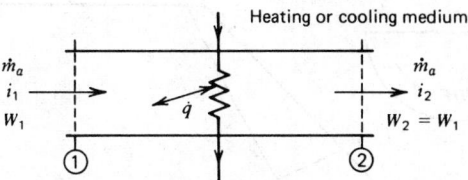

Fig. 75.2 Schematic of a heating or cooling device.[7]

The energy of the condensate has been neglected. Obviously

$$\dot{q} = \dot{q}_s + \dot{q}_l$$

The sensible heat factor (SHF) is defined as $\dot{q}_s / \dot{q}$. This parameter is shown on the semicircular scale of Fig. 75.1.

A device to heat and humidify moist air is shown schematically in Fig. 75.6. An energy balance on the device and a mass balance on the water yields

$$\frac{i_2 - i_1}{W_2 - W_1} = \frac{\dot{q}}{\dot{m}_w} + i_w$$

This gives the direction of a straight line that connects the initial and final states on the psychrometric chart. Figure 75.7 shows a typical combined heating and humidifying process.

A graphical procedure makes use of the circular scale in Fig. 75.1 to solve for state 2. The ratio of enthalpy to humidity ratio $\Delta i / \Delta w$ is defined as

$$\frac{\Delta i}{\Delta W} = \frac{i_2 - i_1}{W_2 - W_1} = \frac{\dot{q}}{\dot{m}_w} + i_w$$

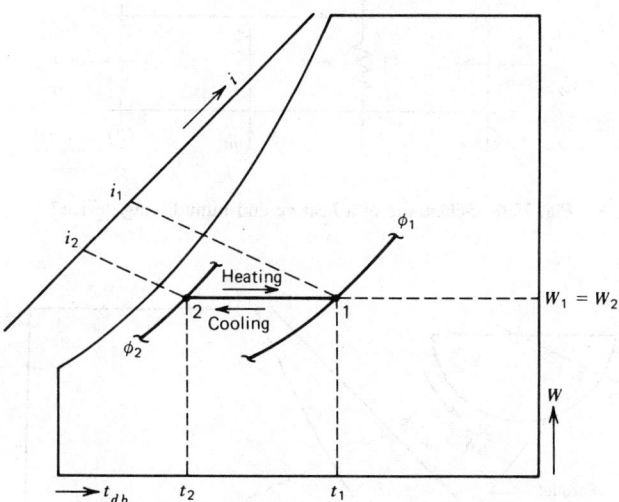

Fig. 75.3 Sensible heating and cooling process.[7]

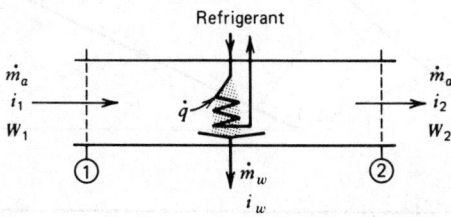

Fig. 75.4 Schematic of a cooling and dehumidifying device.[7]

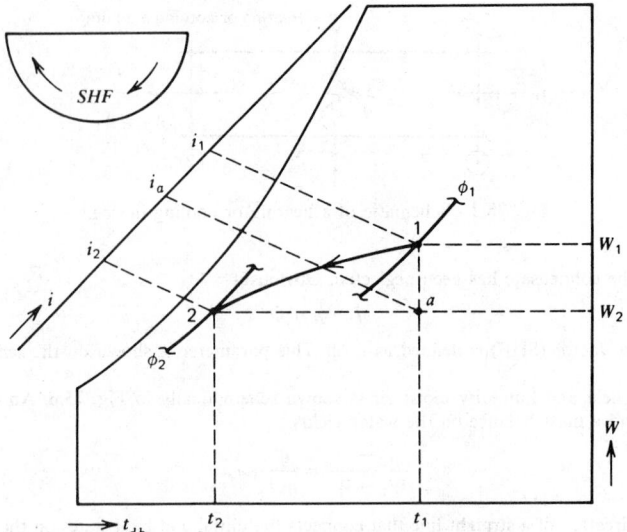

Fig. 75.5 Cooling and dehumidifying process.[7]

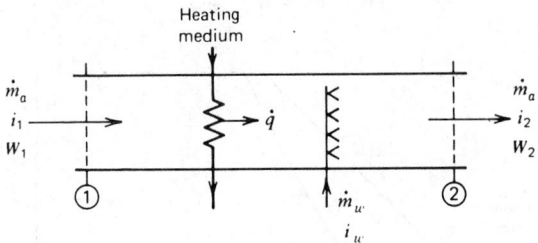

Fig. 75.6 Schematic of a heating and humidifying device.[7]

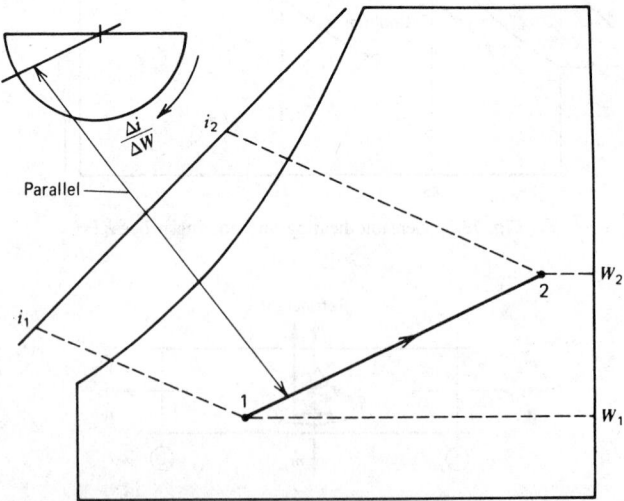

Fig. 75.7 Typical heating and humidifying process.[7]

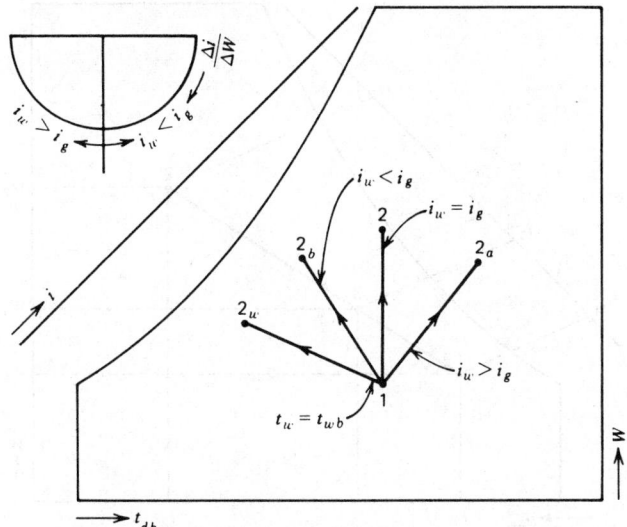

Fig. 75.8 Humidification processes without heat transfer.[7]

Figure 75.7 shows the procedure where a straight line is laid out parallel to the line on the protractor through state point 1. The intersection of this line with the computed value of w_2 determines the final state.

Moisture is frequently added without the addition of heat. In such cases, $q = 0$ and

$$\frac{\Delta i}{\Delta W} = \frac{i_2 - i_1}{W_2 - W_1} = i_w$$

The direction of the process on the psychrometric chart can therefore vary considerably. If the injected water is saturated vapor at the dry bulb temperature, the process will proceed at a constant dry bulb temperature. If the water enthalpy is greater than saturation, the air will be cooled and humidified. Figure 75.8 shows these processes. When liquid water at the wet bulb temperature is injected, the process follows a line of constant wet bulb temperature.

The mixing of air streams is quite common in air-condition systems, usually under adiabatic conditions and with steady flow. Figure 75.9 illustrates the mixing of two air streams. Combined energy and mass balances give

$$\frac{i_2 - i_3}{i_3 - i_1} = \frac{W_2 - W_3}{W_3 - W_1} = \frac{\dot{m}_{a1}}{\dot{m}_{a2}}$$

This shows that the state of the mixed streams must lie on a straight line between states 1 and 2. This is shown in Fig. 75.10. The length of the various line segments are proportional to the masses of dry air mixed. This fact provides a very convenient graphical procedure for solving mixing problems.

The complete air-conditioning system may involve two or more of the processes just considered. In the air conditioning of a space during the summer the air supplied must have a sufficiently low temperature and moisture content to absorb the total heat gain of the space. Therefore, as the air flows through the space, it is heated and humidified. If the system is a closed loop, the air is then returned to the conditioning equipment where it is cooled and dehumidified and supplied to the space again. If fresh air is required in the space, outdoor air may be mixed with the return air before it

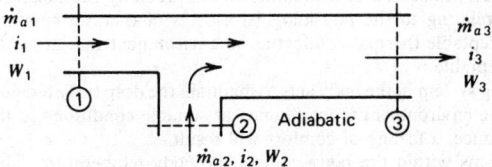

Fig. 75.9 Schematic adiabatic mixing of two air streams.[7]

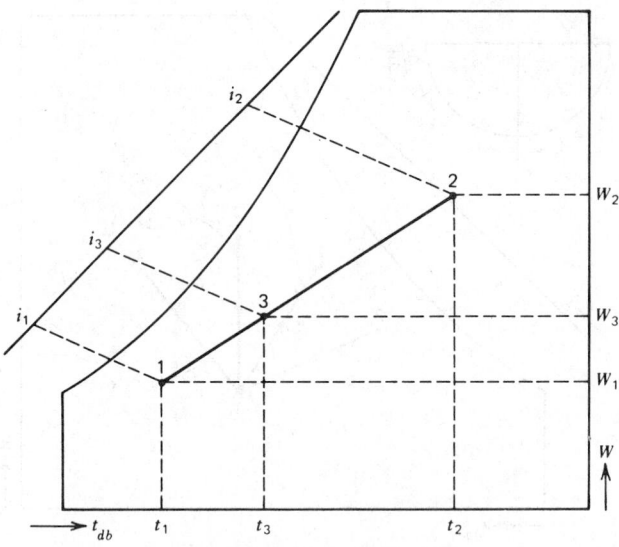

Fig. 75.10 Adiabatic mixing process.[7]

goes to the cooling and dehumidifying equipment. During the winter months the same general processes occur but in reverse. During the summer months the heating and humidifying elements are inactive, and during the winter the cooling and dehumidifying coil is inactive. With appropriate controls, however, all of the elements may be continuously active to maintain precise conditions in the space.

The previous section treated the common-space air-conditioning problem assuming that the system was operating steadily at the design condition. Actually the space requires only a part of the designed capacity of the conditioning equipment most of the time. A control system functions to match the required cooling or heating of the space to the conditioning equipment by varying one or more system parameters. For example, the quantity of air circulated through the coil and to the space may be varied in proportion to the space load. This approach is known as variable air volume (VAV). Another approach is to circulate a constant amount of air to the space, but some of the return air is diverted around the coil and mixed with air coming off the coil to obtain a supply air temperature that is proportional to the space load. This is known as face and bypass control, because face and bypass dampers are used to divert the flow. Another possibility is to vary the coil surface temperature with respect to the required load by changing the temperature or the amount of heating or cooling fluid entering the coil. This technique is usually used in conjunction with VAV and face and bypass systems. However, control of the coolant temperature or quantity may be the only variable in some systems.

75.1.4 Human Comfort

Air conditioning is the simultaneous control of temperature, humidity, cleanliness, odor, and air circulation as required by the occupants of the space. We are concerned with the conditions that actually provide a comfortable and healthful environment. Not everyone within a given space can be made completely comfortable by one set of conditions, owing to a number of factors, many of which cannot be completely explained. However, clothing, age, sex, and the level of activity of each person are considerations. The factors that influence comfort, in their order of importance, are temperature, radiation, humidity, and air motion, and the quality of the air with regard to odor, dust, and bacteria. With a complete air-conditioning system all of these factors may be controlled simultaneously. In most cases a comfortable environment can be maintained when two or three of these factors are controlled. The *ASHRAE Handbook of Fundamentals* is probably the most up-to-date and complete source of information relating to the physiological aspects of thermal comfort.[3] ASHRAE Comfort Standard 55 defines acceptable thermal comfort as an environment that at least 80% of the occupants will find thermally acceptable.[4]

A complex regulating system in the body acts to maintain the deep body temperature at approximately 98.6°F or 36.9°C. If the environment is maintained at suitable conditions so that the body can easily maintain an energy balance, a feeling of comfort will result.

Two basic mechanisms within the body control the body temperature. The first is a decrease or increase in the internal energy production as the body temperature rises or falls, a process called metabolism. The metabolic rate depends on the level of activity such as rest, work, or exercise. The second is the control of the rate of heat dissipation by changing the rate of cutaneous blood circulation (the blood circulation near the surface of the skin). In this way heat transfer from the body can be increased or decreased.

Heat transfer to or from the body is principally by convection and conduction and, therefore, the air motion in the immediate vicinity of the body is a very important factor. Radiation exchange between the body and surrounding surfaces, however, can be important if the surfaces surrounding the body are at a different temperature than the air.

Another very important regulatory function of the body is sweating. Under very warm conditions great quantities of moisture can be released by the body to help cool itself.

There are many parameters to describe the environment in term of comfort. The dry bulb temperature is the single most important index of comfort. This is especially true when the relative humidity is between 40% and 60%. The dry bulb temperature is especially important for comfort in the colder regions. When humidity is high, the significance of the dry bulb temperature is less.

The dew point temperature is a good single measure of the humidity of the environment. The usefulness of the dew point temperature in specifying comfort conditions is, however, limited.

The wet bulb temperature is useful in describing comfort conditions in the regions of high temperature and high humidity where dry bulb temperature has less significance. For example, the upper limit for tolerance of the average individual with normal clothing is a wet bulb of about 86°F or 30°C when the air movement is in the neighborhood of 50–75 ft/min or 0.25–0.38 m/sec.

Relative humidity, although a direct index, has no real meaning in terms of comfort unless the accompanying dry bulb temperature is known. Very high or very low relative humidity is generally associated with discomfort, however.

Air movement is important since the convective heat transfer from the body depends on the velocity of the air moving over it. One is more comfortable in a warm humid environment if the air movement is high. If the temperature is low, one becomes uncomfortable if the air movement is too high. Generally, when air motion is in the neighborhood of 50 ft/min or 0.25 m/sec, the average person will be comfortable.

Clothing, through its insulation properties, is an important modifier of body heat loss and comfort. Clothing insulation can be described in terms of its clo value [1 clo = 0.88 ft² · hr · °F/Btu = 0.155 m² · °C/W]. A heavy two-piece business suit and accessories has an insulation value of about 1 clo, whereas a pair of shorts is about 0.05 clo.

Ventilation. The dominating function of outdoor air is to control air quality, and spaces that are more or less continuously occupied require some outdoor air. The required outdoor air is dependent on the rate of contaminant generation and the maximum acceptable contaminant level. In most cases more outdoor air than necessary is supplied. However, some overzealous attempts to save energy through reduction of outdoor air have caused poor-quality indoor air. Table 75.2, from ASHRAE Standard 62-81 (1980), prescribes the requirements for acceptable air quality.[4] Ventilation air is the combination of outdoor air, of acceptable quality, and of recirculated air from the conditioned space which after passing through the air-conditioning unit becomes supply air. The ventilation air may be 100% outdoor air. The term makeup air may be used synonymously with outdoor air, and the terms return and recirculated air are often used interchangeably. A situation could exist where the supply

Table 75.2   Ambient Air Quality Standards[7,a,*]

Contaminant	Long Term		Short Term	
	Level	Time	Level	Time
Carbon Monoxide			40 mg/m³	1 hr
			10 mg/m³	8 hr
Hydrocarbons			160 µg/m³	3 hr (6–9 AM)
Lead	1.5µg/m³	3 mo.		
Nitrogen Dioxide	100 µg/m³	Year		
Oxidants (Ozone)			240 µg/m³	1 hr
Particulates	75 µg/m³	Year	260 µg/m³	24 hr
Sulfur Dioxide	80 µg/m³	Year	365 µg/m³	24 hr

[a]U.S. Environmental Protection Agency, National Primary and Secondary Ambient Air Quality Standards, Code of Federal Regulations, Title 40 Part 50 (40C.F.R. 50). Pertinent local regulations should also be checked. Some regulations may be more restrictive than those given here, and additional substances may be regulated.
*Reprinted by permission from ASHRAE Standard 62-81, 1981.

air required to match the heating or cooling load is greater than the ventilation air. In that case an increased amount of air would be recirculated to meet this condition.

A minimum supply of outdoor air is necessary to dilute the carbon dioxide produced by metabolism and expired from the lungs. This value, 5 cfm or 2.5 liter/sec per person, allows an adequate factor of safety to account for health variations and some increased activity levels. Therefore, outdoor air requirements should never be less than 5 cfm or 2.5 liter/sec per person regardless of the treatment of the recirculated air. At this time no practical equipment exists to remove carbon dioxide.

75.2 SPACE HEATING

75.2.1 Heat Transmission in Structures

The design of a heating system is dependent on a good estimate of the heat loss in the space to be conditioned. Precise calculation of heat-transfer rates is difficult, but experience and experimental data make reliable estimates possible. Because most of the calculations require a great deal of repetitive work, tables that list coefficients and other data for typical situations are used. Thermal resistance is a very useful concept and is used extensively.

Generally all three modes of heat transfer—conduction, convection, and radiation—are important in building heat gain and loss.

Thermal conduction is heat transfer between parts of a continuum because of the transfer of energy between particles or groups of particles at the atomic level. The Fourier equation expresses steady-state conduction in one dimension:

$$\dot{q} = -kA\frac{dt}{dx}$$

where $\dot{q}$ = heat transfer rate, Btu/hr or W
k = thermal conductivity, Btu/hr · ft · °F or W/m · °C
A = area normal to heat flow, ft or m
dt/dx = temperature gradient, °F/ft or °C/m

A negative sign appears because $\dot{q}$ flows in the positive direction of x when dt/dx is negative.

Consider the flat wall of Fig. 75.11a, where uniform temperatures t_1 and t_2 are assumed to exist on each surface. If the thermal conductivity, the heat-transfer rate, and the area are constant, integration gives

$$\dot{q} = \frac{-kA(t_2 - t_1)}{(x_2 - x_1)}$$

Another very useful form is

$$\dot{q} = \frac{-(t_2 - t_1)}{R'}$$

where R' is the thermal resistance defined by

$$R' = \frac{x_2 - x_1}{kA} = \frac{\Delta x}{kA}$$

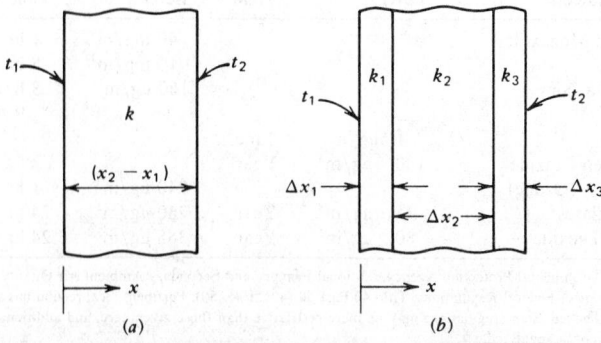

Fig. 75.11 Nomenclature for conduction in plane walls.[7]

The thermal resistance for a unit area of material is very commonly used. This quantity, sometimes called the R-factor, is referred to as the unit thermal resistance or simply the unit resistance, R. For a plane wall the unit resistance is

$$R = \frac{\Delta x}{k}$$

Note that thermal resistance R' is analogous to electrical resistance and q and $t_2 - t_1$ are analogous to current and potential difference in Ohm's law. This analogy provides a very convenient method of analyzing a wall or slab made up of two or more layers of dissimilar material. Figure 75.11b shows a wall constructed of three different materials. The heat transferred by conduction is given by

$$R' = R'_1 + R'_2 + R'_3 = \frac{\Delta x_1}{k_1 A} + \frac{\Delta x_2}{k_2 A} + \frac{\Delta x_3}{k_3 A}$$

Thermal convection is the transport of energy by mixing in addition to conduction. Convection is associated with fluids in motion, generally through a pipe or duct or along a surface. In the very thin layer of fluid next to the surface the transfer of energy is by conduction. In the main body of the fluid mixing is the dominant energy-transfer mechanism. A combination of conduction and mixing exists between these two regions. The transfer mechanism is complex and highly dependent on whether the flow is laminar or turbulent.

The usual, simplified approach in convection is to express the heat-transfer rate as

$$\dot{q} = hA(t - t_w)$$

where $\dot{q}$ = heat transfer rate from fluid to wall, Btu/hr or W
h = film coefficient, Btu/hr · ft² · °F or W/m² · sec
t = bulk temperature of the fluid, °F or °C
t_w = wall temperature, °F or °C

The film coefficient h, sometimes called the unit surface conductance or alternatively the convective heat transfer coefficient, may also be expressed in terms of thermal resistance:

$$\dot{q} = \frac{t - t_w}{R'}$$

where

$$R' = \frac{1}{hA} \qquad (\text{hr} \cdot \text{°F/Btu or °C/W})$$

or

$$R = \frac{1}{h} = \frac{1}{C}$$

where C is the unit thermal conductance. The thermal resistance for convection may be summed with the thermal resistances arising from pure conduction.

The film coefficient h depends on the fluid, the fluid velocity, the flow channel, and the degree of development of the flow field. Many correlations exist for predicting the film coefficient under various conditions. Correlations for forced convection are given in Chapter 2 of the *ASHRAE Handbook.*[2,5]

When the bulk of the fluid is moving relative to the heat-transfer surface, the mechanism is called forced convection, because such motion is usually caused by a blower, fan, or pump, which is forcing the flow. In forced convection buoyancy forces are negligible. In free convection, on the other hand, the motion of the fluid is due entirely to buoyancy forces, usually confined to a layer near the heated or cooled surface. Free convection is often referred to as natural convection.

Natural or free convection is an important part of HVAC applications. Various empirical relations for natural convection film coefficients can be found in the *ASHRAE Handbook of Fundamentals* (1981).[2]

Most building structures have forced convection along outer walls or roofs, and natural convection in inside air spaces and on the inner walls. There is considerable variation in surface conditions, and both the direction and magnitude of the air motion on outdoor surfaces are very unpredictable. The film coefficient for these situations usually ranges from about 1.0 Btu/hr · ft² · °F or 6 W/m² · °C for free convection up to about 6 Btu/hr · ft² · °F or 35 W/m² · °C for forced convection with an air velocity of about 15 miles per hour, 20 ft/sec, or 6 m/sec. Because of the low film coefficients the amount of heat transferred by thermal radiation may be equal to or larger than that transferred by free convection.

Thermal radiation, the transfer of thermal energy by electromagnetic waves, can occur in a perfect vacuum and is actually impeded by an intervening medium. The direct net transfer of energy by radiation between two surfaces which see only each other and which are separated by a nonabsorbing medium is given by

$$\dot{q}_{1-2} = \frac{\sigma(T_1^4 - T_2^4)}{\dfrac{1-\epsilon_1}{A_1\epsilon_1} + \dfrac{1}{A_1F_{12}} + \dfrac{1-\epsilon_2}{A_2\epsilon_2}}$$

where σ = Boltzmann constant, 0.1713×10^{-8} Btu/hr $\cdot$ ft² $\cdot$ °R⁴ or 5.673×10^{-8} W/m $\cdot$ K⁴
$\quad T$ = absolute temperature, °R or K
$\quad \epsilon$ = emittance
$\quad A$ = surface area, ft² or m²
$\quad F$ = configuration factor, a function of geometry only

It has been assumed that both surfaces are "gray" (where the emittance ϵ equals the absorptance α).[6] Figure 75.12 shows situations where radiation may be a significant factor. For the wall,

$$\dot{q}_i = \dot{q}_w = \dot{q}_r + \dot{q}_o$$

and for the air space,

$$\dot{q}_i = \dot{q}_r + \dot{q}_c = \dot{q}_o$$

The resistances can be combined to obtain an equivalent overall resistance R' with which the heat-transfer rate can be computed using

$$\dot{q} = \frac{-(t_o - t_i)}{R'}$$

The thermal resistance for radiation is not easily computed, however, because of the fourth power temperature relationship.

Tables are available that give conductances and resistances for air spaces as a function of position, direction of heat flow, air temperature, and the effective emittance of the space.[5] The effective emittance E is given by

$$\frac{1}{E} = \frac{1}{\epsilon_1} + \frac{1}{\epsilon_2} - 1$$

where ϵ_1 and ϵ_2 are for each surface of the air space. Resistors connected in series may be replaced by an equivalent resistor equal to the sum of the series resistors; it will have an equivalent effect on the circuit:

$$R_e' = R_1' + R_2' + R_3' + \cdots + R_n'$$

Figure 75.13 is an example of a wall being heated or cooled by a combination of convection and radiation on each surface and having five different resistances through which the heat must be conducted. The equivalent thermal resistance R_e' for the wall is given by

$$R_e' = R_i' + R_1' + R_2' + R_3' + R_o'$$

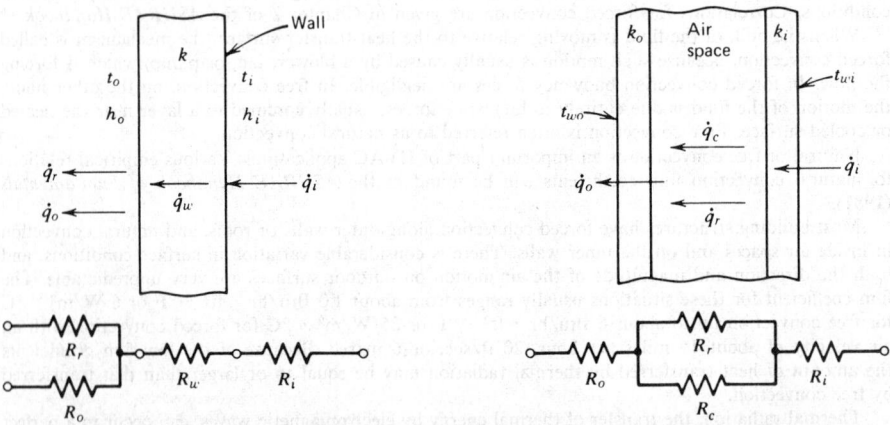

Fig. 75.12 Wall and air space illustrating thermal radiation effects.[7]

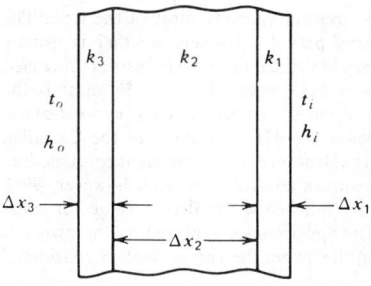

Fig. 75.13   Wall with thermal resistances in series.[7]

Each of the resistances may be expressed in terms of fundamental variables giving

$$R'_e = \frac{1}{h_i A_i} + \frac{\Delta x_1}{k_1 A_1} + \frac{\Delta x_2}{k_2 A_2} + \frac{\Delta x_3}{k_3 A_3} + \frac{1}{h_o A_o}$$

The film coefficients and the thermal conductivities may be obtained from tables. For a plane wall, the areas are all equal and cancel.

The concept of thermal resistance is very useful and convenient in the analysis of complex arrangements of building materials. After the equivalent thermal resistance has been determined for a configuration, however, the overall unit thermal conductance, usually called the overall heat transfer coefficient U, is frequently used:

$$U = \frac{1}{R'A} = \frac{1}{R} \quad \text{(Btu/hr} \cdot \text{ft}^2 \cdot °\text{F or W/m}^2 \cdot °\text{C)}$$

The heat transfer rate is then given by

$$\dot{q} = UA \, \Delta t$$

where UA = conductance, Btu/hr $\cdot$ °F or W/°C
       A = surface area, ft² or m²
       Δt = overall temperature difference, °F or °C

Tabulated Overall Heat Transfer Coefficients.   For convenience of the designer, tables have been constructed that give overall coefficients for many common building sections including walls and floors, doors, windows, and skylights. The tables in the *ASHRAE Handbook* have a great deal of flexibility and are widely used.[2]

75.2.2 Design Conditions

Prior to the design of the heating system an estimate must be made of the maximum probable heat loss of each room or space to be heated. During the coldest months, sustained periods of very cold, cloudy, and stormy weather with relatively small variation in outdoor temperature may occur. In this situation heat loss from the space will be relatively constant and in the absence of internal heat gains will peak during the early morning hours. Therefore, for design purposes the heat loss is usually estimated for steady-state heat transfer for some reasonable design temperature. Transient analyses are often used to study the actual energy requirements of a structure in simulation studies. In such cases solar effects and internal heat gains are taken into account.

Here is the general procedure for calculation of design heat losses of a structure[7]:

1. Select the outdoor design conditions: temperature, humidity, and wind direction and speed.
2. Select the indoor design conditions to be maintained.
3. Estimate the temperature in any adjacent unheated spaces.
4. Select the transmission coefficients and compute the heat losses for walls, floors, ceilings, windows, doors, and floor slabs.
5. Compute the heat load due to infiltration.
6. Compute the heat load due to outdoor ventilation air. This may be done as part of the air quantity calculation.
7. Sum the losses due to transmission and infiltration.

The ideal heating system would provide enough heat to match the heat loss from the structure. However, weather conditions vary considerably from year to year, and heating systems designed for the worst weather conditions on record would have a great excess of capacity most of the time. The failure of a system to maintain design conditions during brief periods of severe weather is usually not critical. However, close regulation of indoor temperature may be critical for some industrial processes.

The outdoor design temperature should generally be the 97½% value. The 97½% value is the temperature equaled or exceeded 97½% of the total hours (2160) in December, January, and February. During a normal winter there would be about 54 hr at or below the 97½% value. For the Canadian stations the 97½% value pertains only to hours in January. If the structure is of lightweight construction, or poorly insulated, has considerable glass, and space temperature control is critical, however, 99% values should be considered. Should the outdoor temperature fall below the design value for some extended period, the indoor temperature may do likewise. The performance expected by the owner is a very important factor, and the designer should make clear to the owner the various factors considered in the design.

The indoor design temperature should be kept relatively low so that the heating equipment will not be oversized. ASHRAE Standard 90 A specifies 72°F or 22°C.[5] Even properly sized equipment operates under partial load, at reduced efficiency, most of the time; therefore, any oversizing aggravates this condition and lowers the overall system efficiency. The indoor design value of relative humidity should be compatible with a healthful environment and the thermal and moisture integrity of the building envelope.

75.2.3 Calculation of Heat Losses

The heat transferred through walls, ceiling, roof, window glass, floors, and doors is all sensible heat transfer, referred to as transmission heat loss and computed from

$$\dot{q} = UA\,(t_i - t_o)$$

A separate calculation is made for each different surface in each room of the structure. To ensure a thorough job in estimating the heat losses, a worksheet should be used to provide a convenient and orderly way of recording all the coefficients and areas. Summations are conveniently made by room and for the complete structure.

All structures have some air leakage or infiltration. This means a heat loss because the cold dry outdoor air must be heated to the inside design temperature and moisture must be added to increase the humidity to the design value. The heat required is given by

$$\dot{q}_s = \dot{m}_o c_p\,(t_i - t_o)$$

where $\dot{m}_o$ = mass flow rate of the infiltrating air, lbm/hr or kg/sec
c_p = specific heat capacity of the moist air, Btu/lbm · °F or J/kg · °C

Infiltration is usually estimated on the basis of volume flow rate at outdoor conditions:

$$\dot{q}_s = \frac{\dot{Q}c_p\,(t_i - t_o)}{v_o}$$

where $\dot{Q}$ = volume flow rate, ft³/hr or m³/sec
v_o = specific volume, ft³/lbm or m³/sec

The latent heat required to humidify the air is given by

$$\dot{q}_l = \dot{m}\,(W_i - W_o)i_{fg}$$

where $W_i - W_o$ = difference in design humidity ratio, lbm$_v$/lbm$_a$ or kg$_v$/kg$_a$
i_{fg} = latent heat of vaporization at indoor conditions, Btu/lbm$_v$ or J/kg$_v$

In terms of volume flow rate,

$$\dot{q}_l = \frac{Q}{v_o}\,(W_i - W_o)i_{fg}$$

Infiltration can account for a large portion of the heating load.

Two methods are used in estimating air infiltration in building structures. In one method the estimate is based on the characteristics of the windows and doors and the pressure difference between inside and outside. This is known as the crack method, because of the cracks around window sash and doors. The other approach is the air-change method, which is based on an assumed number of air changes per hour for each room depending on the number of windows and doors. The crack method is generally considered to be the most accurate when the window and pressure characteristics can be properly evaluated. However, the accuracy of predicting air infiltration is restricted by the limited information on the air-leakage characteristics of the many components that make up a structure. The pressure differences are also difficult to predict because of variable wind conditions and stack effect in tall buildings.

75.2.4 Air Requirements

There are many cases, especially in residential and light commercial applications, when the latent heat loss is quite small and may be neglected. The air quantity is then computed from

$$\dot{q} = \dot{m}c_p \, (t_s - t_r)$$

or

$$\dot{q} = \frac{\dot{Q}c_p}{v_s} \, (t_s - t_r)$$

where v_s = specific volume of supplied air, ft³/lbm or m³/kg
t_s = temperature of supplied air, °F or °C
t_r = room temperature, °F or °C

Residential and light commercial equipment operates with a temperature rise of 60–80°F or 33–44°C, whereas commercial applications will allow higher temperatures. The temperature of the air to be supplied must not be high enough to cause discomfort to occupants before it becomes mixed with room air.

In the unit-type equipment typically used for residences and small commercial buildings each size is able to circulate a relatively fixed quantity of air. Therefore, the air quantity is fixed within a narrow range when the heating equipment is selected. A slightly oversized unit is usually selected with the capacity to circulate a larger quantity of air than theoretically needed. Another condition that leads to greater quantities of circulated air for heating than needed is the greater air quantity sometimes required for cooling and dehumidifying. The same fan is used throughout the year and must therefore be large enough for the maximum air quantity required. Some units have different fan speeds for heating and for cooling.

After the total air flow rate required for the complete structure has been determined, the next step is to allocate the correct portion of the air to each room or space. This is necessary for design of the duct system. Obviously, the air quantity for each room should be apportioned according to the heating load for that space; therefore,

$$\dot{Q}_{rn} = \dot{Q} \, (\dot{q}_{rn} / \dot{q})$$

where $\dot{Q}_{rn}$ = volume flow rate of air supplied to room n, ft³/min or m³/sec
$\dot{q}_{rn}$ = total heat loss rate of room n, Btu/hr or W

The worksheet should have provisions for recording the air quantity for the structure and for each room.

75.2.5 Fuel Requirements

It is often desirable to estimate the quantity of energy necessary to heat the structure under typical weather conditions and with typical imputs from internal heat sources. This is a distinct procedure from design heat load calculations, which are usually made for one set of design conditions neglecting solar effects and internal heat sources. Simulation usually requires a digital computer.

In some cases where computer simulation is not possible or cannot be justified, such as residential buildings, reasonable results can be obtained using hand calculation methods such as the degree-day or bin method.

The degree-day procedure for computing fuel requirements is based on the assumption that, on a long-term basis, solar and internal gains will offset heat loss when the mean daily outdoor temperature is 65°F or 18°C. It is further assumed that fuel consumption will be proportional to the difference between the mean daily temperature and 65°F or 18°C. Degree days are defined by the relationship

$$DD = \frac{(t - t_a) \, N}{24}$$

where N is the number of hours for which the average temperature t_a is computed and t is 65°F or 18°C. The general relation for fuel calculations using this procedure is

$$F = \frac{24 \, DD \, \dot{q} \, C_D}{\eta(t_i - t_o) \, H}$$

where F = the quantity of fuel required for the period desired; the units depend on H
DD = the degree days for period desired, °F-day or °C-day
$\dot{q}$ = the total calculated heat loss based on design condition, t_i and t_o, Btu/hr or W
η = an efficiency factor, which includes the effects of rated full load efficiency, part load performance, oversizing, and energy conservation devices
H = the heating value of fuel, Btu or kWhr per unit volume or mass
C_D = the interim correction factor for degree days based on 65°F or 18°C, Fig. 75.14

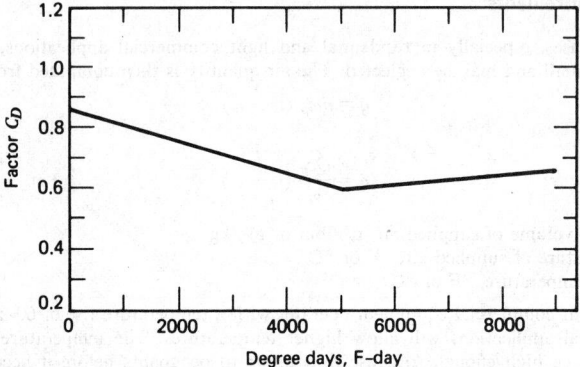

Fig. 75.14 Correction factor. (Reprinted by permission from *ASHRAE Handbook and Product Directory-Systems,* 1980.)

75.3 SPACE COOLING

75.3.1 Heat Gain, Cooling Load, and Heat Extraction Rate

A larger number of variables are considered in making cooling load calculations than in heating load calculations. In design for cooling, transient analysis must be used if satisfactory results are to be obtained. This is because the instantaneous heat gain into a conditioned space is quite variable with time primarily because of the strong transient effect created by the hourly variation in solar radiation. There may be an appreciable difference between the heat gain of the structure and the heat removed by the cooling equipment at a particular time. This difference is caused by the storage and subsequent transfer of energy from the structure and contents to the circulated air. If this is not taken into account, the cooling and dehumidifying equipment will usually be grossly oversized and estimates of energy requirements meaningless.

Heat gain is the rate at which energy is transferred to or generated within a space. It has two components, sensible heat and latent heat, which must be computed and tabulated separately. Heat gains usually occur in the following forms:

1. Solar radiation through openings.
2. Heat conduction through boundaries with convection and radiation from the inner surface into the space.
3. Sensible heat convection and radiation from internal objects.
4. Ventilation (outside) and infiltration air.
5. Latent heat gains generated within the space.

The cooling load is the rate at which energy must be removed from a space to maintain the temperature and humidity at the design values. The cooling load will generally differ from the heat gain at any instant of time, because the radiation from the inside surface of walls and interior objects as well as the solar radiation coming directly into the space through openings does not heat the air within the space directly. This radiant energy is mostly absorbed by floors, interior walls, and furniture, which are then cooled primarily by convection as they attain temperatures higher than that of the room air. Only when the room air receives the energy by convection does this energy become part of the cooling load. The heat-storage characteristics of the structure and interior objects determine the thermal lag and therefore the relationship between heat gain and cooling load. For this reason the thermal mass (product of mass and specific heat) of the structure and its contents must be considered in such cases. The reduction in peak cooling load because of the thermal lag can be quite important in sizing the cooling equipment.

The heat-extraction rate is the rate at which energy is removed from the space by the cooling and dehumidifying equipment. This rate may be equal to the cooling load. However, this is rarely true and some fluctuation in room temperature occurs. Because the cooling load is below the peak or design value most of the time, intermittent or variable operation of the cooling equipment is required.

75.3.2 Design Conditions

The problem of selecting outdoor design conditions for calculation of heat gain is similar to that for heat loss. Again it is not reasonable to design for the worst conditions on record because a great excess of capacity will result. The heat-storage capacity of the structure also plays an important role in this regard. A massive structure will reduce the effect of overload from short intervals of outdoor temperature above the design value. The *ASHRAE Handbook of Fundamentals* gives extensive outdoor design data.[2] Tabulation of dry bulb and mean coincident wet bulb temperatures that are equaled or exceeded 1%, 2½%, and 5% of the total hours during June through September (2928 hr) are given. The 2½% values are recommended for design purposes by ASHRAE Standard 90 A.[5] The daily range of temperature is the difference between the average maximum and average minimum for the warmest month. The daily range is usually larger for the higher elevations where temperatures may be quite low late at night and during the early morning hours. The daily range has an effect on the energy stored by the structure. The variation in dry bulb temperature for a typical design day may be computed using the peak outdoor dry bulb temperature and the daily range, assuming a cosine relation with a maximum temperature at 3 PM and a minimum at 5 AM.

The local wind velocity for summer conditions is usually taken to be about one-half the winter design value but not less than about 7½ mph or 3.4 m/sec.

The indoor design conditions for the average job in the United States and Canada is 75°F or 24°C dry bulb and a relative humidity of 50%, when activity and dress of the occupants are light. The designer should be alert for unusual circumstances that may lead to uncomfortable conditions. Certain activities may require occupants to engage in active work or require heavy protective clothing, both of which would require lower design temperatures.

75.3.3 Calculation of Heat Gains

Design cooling loads for one day as well as long-term energy calculations may be done using the transfer-function approach. Details of this method are discussed in the *ASHRAE Handbook of Fundamentals.* [2]

It is not always practical to compute the cooling load using the transfer-function method; therefore, a hand calculation method has been developed from the transfer-function procedure and is referred to as the cooling-load-temperature-difference (CLTD) method. The method involves extensive use of tables and charts and various factors to express the dynamic nature of the problem and predicts cooling loads within about 5% of the values given by the transfer-function method.[5]

The CLTD method makes use of a temperature difference in the case of walls and roofs and cooling load factors (CLF) in the case of solar gain through windows and internal heat sources. The CLTD and CLF vary with time and are a function of environmental conditions and building parameters. They have been derived from computer solutions using the transfer-function procedure. A great deal of care has been taken to sample a wide variety of conditions in order to obtain reasonable accuracy. These factors have been derived for a fixed set of surface and environmental conditions; therefore, correction factors must often be applied. In general, calculations proceed as follows.

For walls and roofs

$$\dot{q}_\theta = (U)(A)(\text{CLTD})_\theta$$

where U = overall heat transfer coefficient, Btu/hr · ft² · °F or W/m² · °C

A = area, ft² or m²

$(\text{CLTD})_\theta$ = temperature difference which gives the cooling load at time θ, °F or °C

The CLTD accounts for the thermal response (lag) in the heat transfer through the wall or roof, as well as the response (lag) due to radiation of part of the energy from the interior surface of the wall to objects within the space.

For solar gain through glass

$$\dot{q}_\theta = (A)(\text{SC})(\text{SHGF})(\text{CLF})_\theta$$

where A = area, ft² or m²

SC = shading coefficient (internal shade)

SHGF = solar heat gain factor, Btu/hr · ft² or W/m²

$(\text{CLF})_\theta$ = cooling load factor for time θ

The SHGF is the maximum for a particular month, orientation, and latitude. The CLF accounts for the variation of the SHGF with time, the massiveness of the structure, and internal shade. Again the CLF accounts for the thermal response (lag) of the radiant part of the solar input.

For internal heat sources

$$\dot{q}_\theta = (\dot{q}_i)(\text{CLF})_\theta$$

where $\dot{q}_i$ = instantaneous heat gain from lights, people, and equipment, Btu/hr or W

$(\text{CLF})_\theta$ = cooling load factor for time θ

The CLF accounts for the thermal response of the space to the various internal heat gains and is slightly different for each.

The time of day when the peak cooling load will occur must be estimated. In fact, two different types of peaks need to be determined. First, the time of the peak load for each room is needed in order to compute the air quantity for that room. Second, the time of the peak load for a zone served by a central unit is required to size the unit. It is at these peak times that cooling load calculations should be made. The estimated times when the peak load will occur are determined from the tables of CLTD and CLF values together with the orientation and physical characteristics of the room or space. The times of the peak cooling load for walls, roofs, windows, and so on, is obvious in the tables and the most dominant cooling load components will then determine the peak time for the entire room or zone. For example, rooms facing west with no exposed roof will experience a peak load in the late afternoon or early evening. East-facing rooms tend to peak during the morning hours. A zone made up of east and west rooms with no exposed roofs will tend to peak when the west rooms peak. If there is a roof, the zone will tend to peak when the roof peaks. High internal loads may dominate the cooling load in some cases and cause an almost uniform load throughout the day.

The details of computing the various cooling load components are discussed in ASHRAE Manual GRP158.[5]

It is emphasized that the total space cooling load does not generally equal the load imposed on the central cooling unit or cooling coil. The outdoor ventilation air is usually mixed with return air and conditioned before it is supplied to the space. The air circulating fan may be upstream of the coil, in which case the fan power input is a load on the coil. In the case of vented light fixtures, the heat absorbed by the return air is imposed on the coil and not the room.

The next steps are to determine the air quantities and to select the equipment. These steps may be reversed depending on the type of equipment to be used.

75.3.4 Air Requirements

Computing air quantity for cooling and dehumidification requires the use of psychrometric charts. The cooling and dehumidifying coil is designed to match the sensible and latent heat requirements of a particular job and the fan is sized to handle the required volume of air. The fan, cooling coil, control dampers, and the enclosure for these components, referred to as an air handler, are assembled at the factory in a wide variety of coil and fan models to suit almost any requirement. The design engineer usually specifies the entering and leaving moist air conditions, the volume flow rate of the air, and the total pressure the fan must produce.

Specifically constructed equipment cannot be justified for small commercial and residential applications. Furthermore, these applications generally have a higher sensible heat factor, and dehumidification is not as critical as it is in large commercial buildings. Therefore, the equipment is manufactured to operate at or near one particular set of conditions. For example, typical residential and light commercial cooling equipment operates with a coil SHF of 0.75–0.8 with the air entering the coil at about 80°F or 27°C dry bulb and 67°F or 19°C wet bulb temperature. This equipment usually has a capacity of less than 10 tons or 35 kW. When the peak cooling load and latent heat requirements are appropriate, this less expensive type of equipment is used. In this case the air quantity is determined in a different way. The peak cooling load is first computed as 1.3 times the peak sensible cooling load for the structure to match the coil SHF. The equipment is then selected to match the peak cooling load as closely as possible. The air quantity is specified by the manufacturer for each unit and is about 400 cfm/ton or 0.0537 m³/sec · kW. The total air quantity is then divided among the various rooms according to the cooling load of each room.

75.3.5 Fuel Requirements

The only reliable methods available for estimating cooling equipment energy requirements require hour by hour predictions of the cooling load and must be done using a computer and representative weather data. This is mainly because of the great importance of thermal energy storage in the structure and the complexity of the equipment used. This approach is becoming much easier due to the development of desktop minicomputers. This complex problem is discussed in Ref. 3.

There has been recent work related to residential and light commercial applications that is adaptable to hand calculations. The analysis assumes a correctly sized system. Figure 75.15 summarizes the results of the study of compressor operating time for all locations inside the contiguous 48 states. With the compressor operating time it is possible to make an estimate of the energy consumed by the equipment for an average cooling season. The Air-Conditioning and Refrigeration Institute (ARI) publishes data concerning the power requirements of cooling and dehumidifying equipment and most manufacturers can furnish the same data. For residential systems it is generally best to cycle the circulating fan with the compressor. In this case fans and compressors operate at the same time. However, for light commercial applications the circulating fan will probably operate continuously, and this should be taken into account.

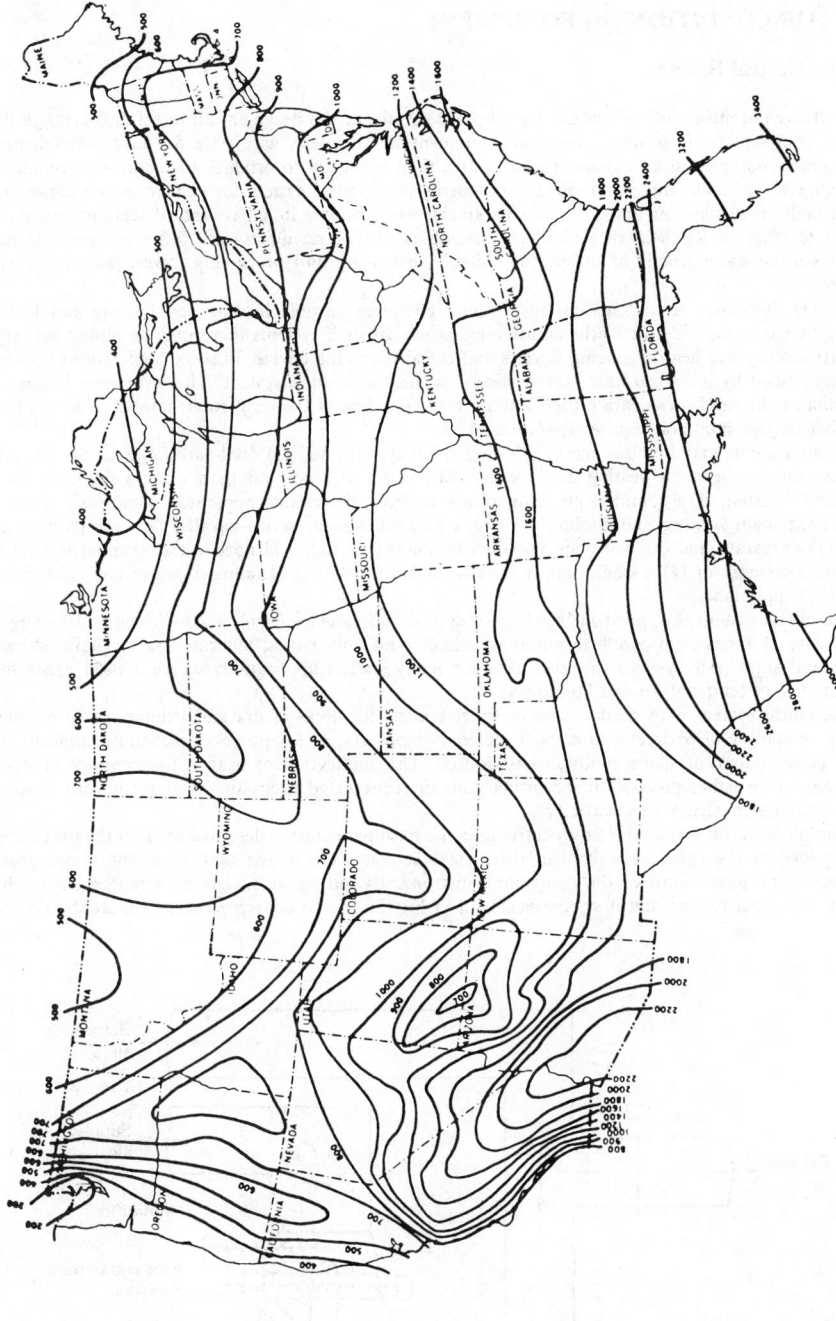

Fig. 75.15 Hours of compressor operation for residential systems. [Reprinted by permission from *ASHRAE Journal*, **16** (Part I, No. 2) (1974).]

75.4   AIR-CONDITIONING EQUIPMENT

75.4.1   Central Systems

When the requirements of the system have been determined, the designer can select and arrange the various components. It is important that equipment be adequate, accessible for easy maintenance, and no more complex in arrangement and control than necessary to produce the conditions required.

Figure 75.16 shows the air-handling components of a central system for year-round conditioning. It is a built-up system, but most of the components are available in subassembled sections ready for bolting together in the field or completely assembled by the manufacturer. Other components not shown are the water heater or boiler, the chiller, condensing unit or cooling tower, pumps, piping, and controls.

All-Air Systems. An all-air system provides complete sensible heating and cooling and latent cooling by supplying only air to the conditioned space. In such systems there may be piping between the refrigerating and heat-producing devices and the air-handling device. In some applications heating is accomplished by a separate air, water, steam, or electric heating system. The term zone implies a provision or the need for separate thermostatic control, whereas the term room implies a partitioned area that may or may not require separate control.

All-air systems may be classified as (1) single-path systems and (2) dual-path systems. Single-path systems contain the main heating and cooling coils in a series flow air path using a common duct distribution system at a common air temperature to feed all terminal apparatus. Dual-path systems contain the main heating and cooling coils in a parallel flow or series–parallel flow air path using either (1) a separate cold and warm air duct distribution system that is blended at the terminal apparatus (dual-duct system), or (2) a single supply duct to each zone with a blending of warm and cold air at the main supply fan.

The all-air system is applied in buildings requiring individual control of conditions and having a multiplicity of zones such as office buildings, schools and universities, laboratories, hospitals, stores, hotels, and ships. Air systems are also used for many special applications where a need exists for close control of temperature and humidity.

The reheat system is to permit zone or space control for areas of unequal loading, or to provide heating or cooling of perimeter areas with different exposures, or for process or comfort applications where close control of space conditions is desired. The application of heat is a secondary process, being applied to either preconditioned primary air or recirculated room air. The medium for heating may be hot water, steam, or electricity.

Conditioned air is supplied from a central unit at a fixed temperature designed to offset the maximum cooling load in the space. The control thermostat activates the reheat unit when the temperature falls below the upper limit of the controlling instrument's setting. A schematic arrangement of the components for a typical reheat system is shown in Fig. 75.17. To conserve energy reheat should not

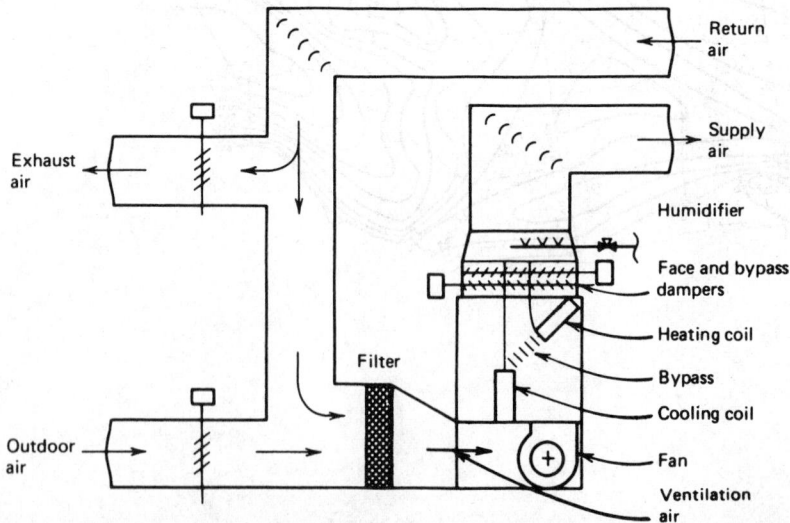

Fig. 75.16   Typical central air system.[7]

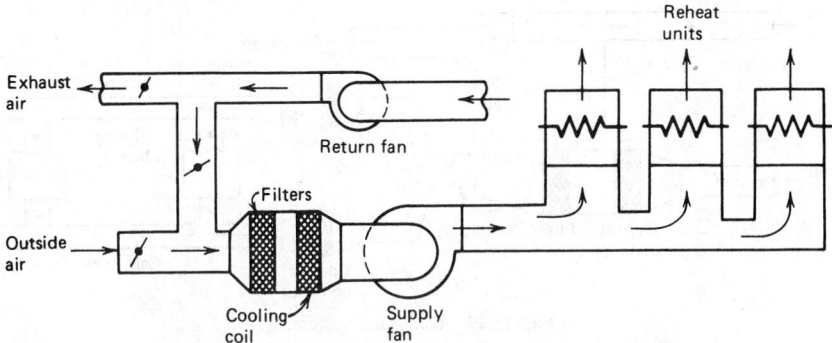

Fig. 75.17 Arrangement of components for a reheat system.[7]

be used unless absolutely necessary. At the very least, reset control should be provided to maintain the cold air at the highest possible temperature to satisfy the space cooling requirement.

The variable-volume system compensates for varying load by regulating the volume of air supplied through a single duct. Special zoning is not required because each space supplied by a controlled outlet is a separate zone. Figure 75.18 is a schematic of a true variable-air-volume (VAV) system.

Significant advantages are low initial cost and low operating costs. The first cost of the system is low because it requires only single runs of duct and a simple control at the air terminal. Where diversity of loading occurs, smaller equipment can be used and operating costs are generally the lowest among all the air systems. Because the volume of air is reduced with a reduction in load, the refrigeration and fan horsepower follow closely the actual air-conditioning load of the building. During intermediate and cold seasons, outdoor air can be used for economy in cooling. In addition, the system is virtually self-balancing.

Until recently there were two reasons why variable-volume systems were not recommended for applications with loads varying more than 20%. First, throttling of conventional outlets down to 50–60% of their maximum design volume flow might result in the loss of control of room air motion with noticeable drafts resulting. Second, the use of mechanical throttling dampers produces noise, which increases proportionally with the amount of throttling.

With improvements in volume-throttling devices and aerodynamically designed outlets, this system can now handle interior areas as well as building perimeter areas where load variations are greatest, and where throttling to 10% of design volume flow is often necessary. It is primarily a cooling system and should be applied only where cooling is required the major part of the year. Buildings with internal spaces with large internal loads are the best candidates. A secondary heating system should be provided for boundary surfaces. Baseboard perimeter heat is often used. During the heating season, the VAV system simply provides tempered ventilation air to the exterior spaces.

An important aspect of VAV system design is fan control. There are significant fan power savings where fan speed is reduced in relation to the volume of air being circulated.

In the dual-duct system the central station equipment supplies warm air through one duct run and cold air through the other. The temperature in an individual space is controlled by a thermostat that mixes the warm and cool air in proper proportions. One form is shown in Fig. 75.19.

From the energy-conservation viewpoint the dual-duct system has the same disadvantage as reheat. Although many of these systems are in operation, few are now being designed and installed.

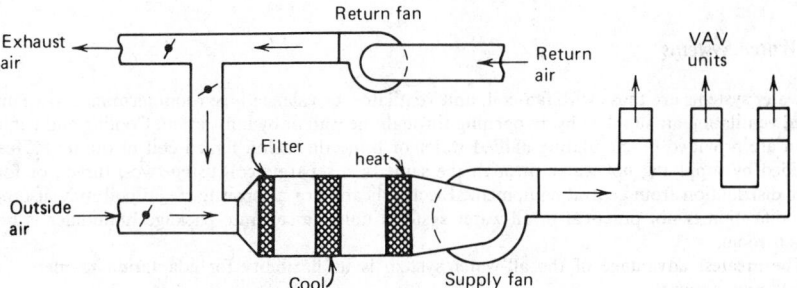

Fig. 75.18 Variable-air-volume system.[7]

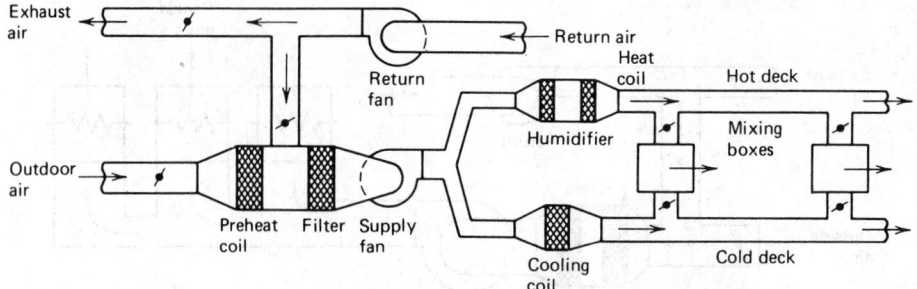

Fig. 75.19   Dual-duct system.[7]

The multizone central station units provide a single supply duct for each zone, and obtain zone control by mixing hot and cold air at the central unit in response to room or zone thermostats. For a comparable number of zones this system provides greater flexibility than the single-duct and involves lower cost than the dual-duct system, but it is physically limited by the number of zones that may be provided at each central unit.

The multizone, blow-through system is applicable to locations and areas having high sensible heat loads and limited ventilation requirements. The use of many duct runs and control systems can make initial costs of this system high compared to other all-air systems. To obtain very fine control this system might require larger refrigeration and air-handling equipment.

The use of these systems with simultaneous heating and cooling is now discouraged for energy conservation.

Air and Water Systems

In an air and water system both air and water are distributed to each space to perform the cooling function. In virtually all air–water systems both cooling and heating functions are carried out by changing the air or water temperatures (or both) to permit control of space temperature during all seasons of the year.

The quantity of air supplied can be low compared to an all-air system, and less building space need be allocated for the cooling distribution system.

The reduced quantity of air is usually combined with a high-velocity method of air distribution to minimize the space required. If the system is designed so that the air supply is equal to the air needed to meet outside air requirements or that required to balance exhaust (including exfiltration) or both, the return air system can be eliminated for the areas conditioned in this manner.

The pumping power necessary to circulate the water throughout the building is usually significantly less than the fan power to deliver and return the air. Thus not only space but also operating cost savings can be realized.

Systems of this type have been commonly applied to office buildings, hospitals, hotels, schools, better apartment houses, research laboratories, and other buildings. Space saving has made these systems beneficial in high-rise structures.

Air and water systems are categorized as two-pipe, three-pipe, and four-pipe systems. They are basically similar in function, and all incorporate both cooling and heating capabilities for all-season air conditioning. However, arrangements of the secondary water circuits and control systems differ greatly.

All-Water Systems

All-water systems are those with fan-coil, unit ventilator, or valance-type room terminals, with unconditioned ventilation air supplied by an opening through the wall or by infiltration. Cooling and dehumidification are provided by circulating chilled water or brine through a finned coil in the unit. Heating is provided by supplying hot water through the same or a separate coil using two-, three-, or four-pipe water distribution from central equipment. Electric heating or a separate steam coil may also be used. Humidification is not practical in all-water systems unless a separate package humidifier is provided in each room.

The greatest advantage of the all-water system is its flexibility for adaptation to many building module requirements.

75.4.2 Unitary Systems

Unitary Air Conditioners

Unitary air-conditioning equipment consists of factory-matched refrigerant cycle components for inclusion in air-conditioning systems that are field designed to meet the needs of the user. They may vary in:

1. Arrangement: single or split (evaporator connected in the field).
2. Heat rejection: air cooled, evaporative condenser, water cooled.
3. Unit exterior: decorative for in-space application, functional for equipment room and ducts, weatherproofed for outdoors.
4. Placement: floor standing, wall mounted, ceiling suspended.
5. Indoor air: vertical upflow, counterflow, horizontal, 90° and 180° turns, with fan, or for use with forced air furnace.
6. Locations: indoor—exposed with plenums or furred in ductwork, concealed in closets, attics, crawl spaces, basements, garages, utility rooms, or equipment rooms; wall—built in, window, transom; outdoor—rooftop, wall mounted, or on ground.
7. Heat: intended for use with upflow, horizontal, or counterflow forced air furnace, combined with furnace, combined with electrical heat, combined with hot water or steam coil.

Unitary air conditioners as contrasted to room air conditioners are designed with fan capability for ductwork, although some units may be applied with plenums.

Heat pumps are also offered in many of the same types and capacities as unitary air conditioners.

Packaged reciprocating and centrifugal water chillers can be considered as unitary air conditioners particularly when applied with unitary-type chilled water blower coil units. Consequently, a higher level of design ingenuity and performance is required to develop superior system performance using unitary equipment than for central systems, since only a finite number of unitary models is available. Unitary equipment tends to fall automatically into a zoned system with each zone served by its own unit.

For large single spaces where central systems work best, the use of multiple units is often an advantage because of the movement of load sources within the larger space, giving flexibility to many smaller independent systems instead of one large central system.

A room air conditioner is an encased assembly designed as a unit primarily for mounting in a window, through a wall, or as a console. The basic function of a room air conditioner is to provide comfort by cooling, dehumidifying, filtering or cleaning, and circulating the room air. It may also provide ventilation by introducing outdoor air into the room, and by exhausting the room air to the outside. The conditioner may also be designed to provide heating by reverse cycle (heat pump) operation or by electric resistance elements.

75.4.3 Heat Pump Systems

The heat pump is a system in which refrigeration equipment is used such that heat is taken from a heat source and given up to the conditioned space when heating service is wanted and is removed from the space and discharged to a heat sink when cooling and dehumidification are desired. The thermal cycle is identical with that of ordinary refrigeration, but the application is equally concerned with the cooling effect produced at the evaporator and the heating effect produced at the condenser. In some applications both the heating and cooling effects obtained in the cycle are utilized.

Unitary heat pumps are shipped from the factory as a complete preassembled unit including internal wiring, controls, and piping. Only the ductwork, external power wiring, and condensate piping are required to complete the installation. For the split unit it is also necessary to connect the refrigerant piping between the indoor and outdoor sections. In appearance and dimensions, casings of unitary heat pumps closely resemble those of conventional air-conditioning units having equal capacity.

Heat Pump Types

The air-to-air heat pump is the most common type. It is particularly suitable for factory-built unitary heat pumps and has been widely used for residential and commercial applications. Outdoor air offers a universal heat-source, heat-sink medium for the heat pump. Extended-surface, forced-convection heat-transfer coils are normally used to transfer the heat between the air and the refrigerant.

Figure 75.20 shows typical curves of heat pump capacity versus outdoor dry bulb temperature. Imposed on the figure are approximate heating and cooling load curves for a building. In the heating

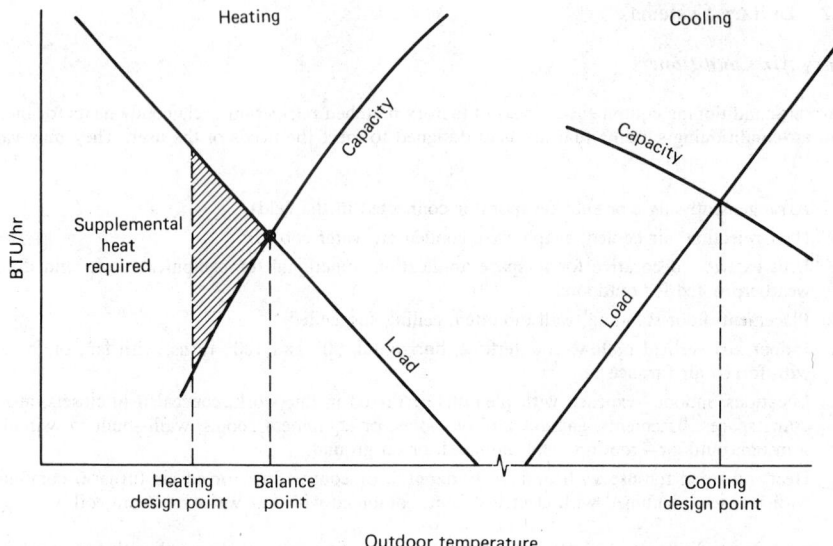

Fig. 75.20 Comparison of building heat loads with heat pump capacities.[7]

mode it can be seen that the heat pump capacity decreases and the building load increases as the temperature drops. In the cooling mode the opposite trends are apparent. If the cooling load and heat pump capacity are matched at the cooling design temperature, then the balance point, where heating load and capacity match, is then fixed. This balance point will quite often be above the heating design temperature. In such cases supplemental heat must be furnished to maintain the desired indoor condition.

The most common type of supplemental heat for heat pumps in the United States is electrical-resistance heat. This is usually installed in the air-handler unit and is designed to turn on automatically, sometimes in stages, as the indoor temperature drops. In some systems the supplemental heat is turned on when the outdoor temperature drops below some preset value. Heat pumps which have fossil-fuel-fired supplemental heat are referred to as hybrid or bivalent heat pumps.

If the heat pump capacity is sized to match the heating load, care must be taken that there is not excessive cooling capacity for summer operation, which could lead to poor summer performance, particularly in dehumidification of the air.

Air-to-water heat pumps are commonly used in large buildings where zone control is necessary and are also sometimes used for the production of hot or cold water in industrial applications as well as heat reclaiming. Heat pumps for hot water heating are commercially available in residential sizes.

A water-to-air heat pump uses water as a heat source and sink and uses air to transmit heat to or from the conditioned space.

A water-to-water heat pump uses water as the heat source and sink for both cooling and heating operation. Heating–cooling changeover may be accomplished in the refrigerant circuit, but in many cases it is more convenient to perform the switching in the water circuits.

Water may represent a satisfactory and in many cases an ideal heat source. Well water is particularly attractive because of its relatively high and nearly constant temperature, generally about 50°F or 10°C in northern areas and 60°F or 16°C and higher in the south. However, abundant sources of suitable water are not always available, and the application of this type of system is limited. Frequently, sufficient water may be available from wells, but the condition of the water may cause corrosion in heat exchangers or it may induce scale formation. Other considerations to be made are the costs of drilling, piping, and pumping, and the means for disposing of used water.

Surface or stream water may be used, but under reduced winter temperatures the cooling spread between inlet and outlet must be limited to prevent freeze-up in the water chiller, which is absorbing the heat.

Under certain industrial circumstances waste process water such as spent warm water in laundries and warm condenser water may be a source for specialized heat pump operations.

A building may require cooling in interior zones while needing heat in exterior zones. The needs of the north zones of a building may also be different from those of the south. In many cases a closed-loop heat pump system is a good choice. Closed-loop systems may be solar assisted. A closed-loop system is shown in Fig. 75.21.

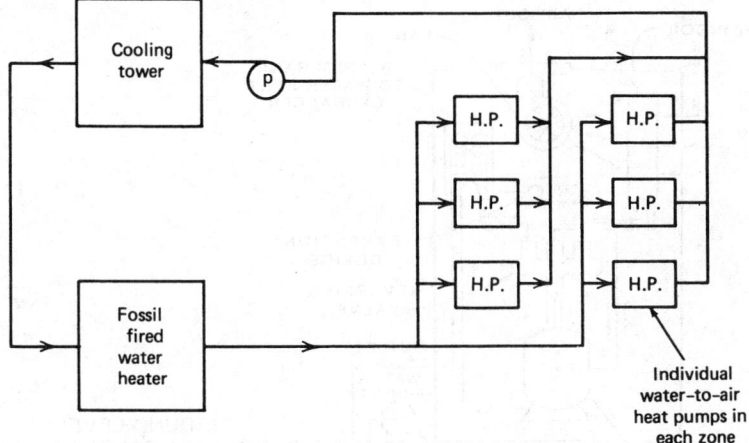

Fig. 75.21 Schematic of a closed-loop heat pump system.[7]

Individual water-to-air heat pumps in each room or zone accept energy from or reject energy to a common water loop, depending on whether that area has a call for heating or for cooling. In the ideal case the loads will balance, and there will be no surplus or deficiency of energy in the loop. If cooling demand is such that more energy is rejected to the loop than is required for heating, the surplus is rejected to the atmosphere by a cooling tower. In the other case, an auxiliary furnace furnishes any deficiency.

The ground has been used successfully as a source–sink for heat pumps with both vertical and horizontal pipe installation. Water from the heat pump is pumped through plastic pipe and exchanges heat with the surrounding earth before being returned back to the heat pump, Fig. 75.22. Tests and analyses have shown rapid recovery in earth temperature around the pipe after the heat pump cycles off. Proper sizing depends on the nature of the earth surrounding the pipe, the water table level, and the efficiency of the heat pump.

Although still largely in the research stage, the use of solar energy as a heat source either on a primary basis or in combination with other sources is attracting increasing interest. Heat pumps may be used with solar systems in either a series or a parallel arrangement, or a combination of both.

75.5 ROOM AIR DISTRIBUTION

75.5.1 Basic Considerations

The object of air distribution in warm air heating, ventilating, and air-conditioning systems is to create the proper combination of temperature, humidity, and air motion in the occupied portion of the conditioned room. To obtain comfort conditions with this space, standard limits for an acceptable effective draft temperature have been established. This term comprises air temperature, air motion, relative humidity, and their physiological effect on the human body, any variation from accepted standards of one of these elements may result in discomfort to the occupants. Discomfort also may be caused by lack of uniform conditions within the space or by excessive fluctuation of conditions in the same part of the space. Such discomfort may arise owing to excessive room air temperature variations (horizontally, vertically, or both), excessive air motion (draft), failure to deliver or distribute the air according to the load requirements at the different locations, or rapid fluctuation of room temperature or air motion (gusts).

75.5.2 Jet and Diffuser Behavior

Conditioned air is normally supplied to air outlets at velocities much higher than would be acceptable in the occupied space. The conditioned air temperature may be above, below, or equal to the temperature of the air in the occupied space. Proper air distribution therefore causes entrainment of room air by the primary air stream and reduces the temperature differences to acceptable limits before the air enters the occupied space. It also counteracts the natural convection and radiation effects within the room.

When a jet is projected parallel to and within a few inches of a surface, the induction or entrainment is limited on the surface side of the jet. A low-pressure region is created between the surface and the jet, and the jet attaches itself to the surface. This phenomenon results if the angle of discharge between

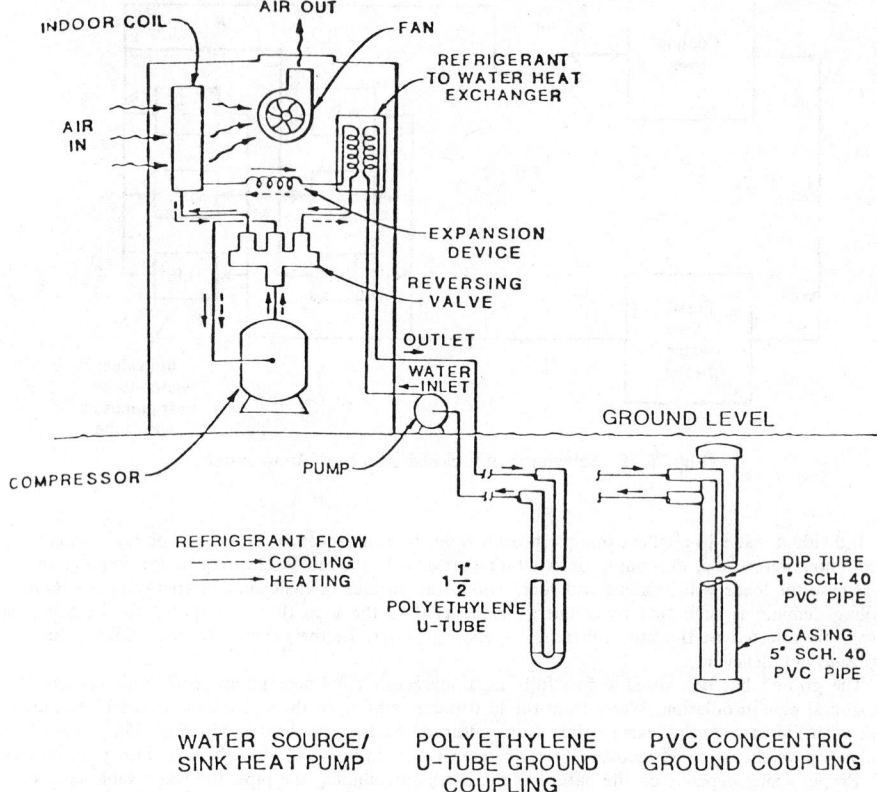

Fig. 75.22 Schematic of a ground-coupled heat pump system.[7]

the jet and the surface is less than about 40° and if the jet is within about 1 ft of the surface. The jet from a floor outlet is drawn to the wall, and the jet from a ceiling outlet is drawn to the ceiling.

Room air near the jet is entrained and must then be replaced by other room air into motion. Whenever the average room air velocity is less than about 50 ft/min or 0.25 m/sec, buoyancy effects may be significant. In general, about 8–10 air changes per hour are required to prevent stagnant regions (velocity less than 15 ft/min or 0.08 m/sec). However, stagnant regions are not necessarily a serious condition. The general approach is to supply air in such a way that the high-velocity air from the outlet does not enter the occupied space. The region within 1 ft of the wall and above about 6 ft from the floor is out of the occupied space for practical purposes.[7]

Perimeter-type outlets are generally regarded as superior for heating applications. This is particularly true when the floor is over an unheated space or a slab and where considerable glass area exists in the wall. Diffusers with a wide spread are usually best for heating because buoyancy tends to increase the throw. For the same reason the spreading jet is not as good for cooling applications because the throw may not be adequate to mix the room air thoroughly. However, the perimeter outlet with a nonspreading jet is quite satisfactory for cooling. Diffusers are available that may be changed from the spreading to nonspreading type according to the season.

The high sidewall type of register is often used in mild climates and on the second and succeeding floors of multistory buildings. This type of outlet is not recommended for cold climates or with unheated floors. A considerable temperature gradient may exist between floor and ceiling when heating; however, this type outlet gives good air motion and uniform temperatures in the occupied zone for cooling application. These registers are generally selected to project air from about three-fourths to full room width.

The ceiling diffuser is very popular in commercial applications and many variations of it are available. Because the primary air is projected radially in all directions, the rate of entrainment is large, causing the high momentum jet to diffuse quickly. This feature enables the ceiling diffuser to handle larger quantities of air at higher velocities than most other types. The ceiling diffuser is quite effective for cooling applications but generally poor for heating. However, satisfactory results may be obtained in commercial structures when the floor is heated.

The return air intake generally has very little effect on the room air motion. But the location may have a considerable effect on the performance of the heating and cooling equipment. Because it is desirable to return the coolest air to the furnace and the warmest air to the cooling coil, the return air intake should be located in a stagnant region.

Noise produced by the air diffuser and air can be annoying to the occupants of the conditioned space. Noise criteria (NC) curves are used to describe the noise in HVAC systems.[5]

The selection and placement of the air outlets is ideally done purely on the basis of comfort. However, the architectural design and the functional requirements of the building often override comfort. When the designer is free to select the type of air-distribution system based on comfort, the perimeter type of system with vertical discharge of the supply air is to be preferred for exterior spaces when the heating requirements exceed 2000 degree (F) days. This type system is excellent for heating and satisfactory for cooling when adequate throw is provided. When the floors are warmed and the degree (F) day requirement is between about 3500 and 2000, the high sidewall outlet with horizontal discharge toward the exterior wall is acceptable for heating and quite effective for cooling. When the heating requirement falls below about 2000 degree (F) days, the overhead ceiling outlet or high sidewall diffuser is recommended because cooling is the predominant mode. Interior spaces in commercial structures are usually provided with overhead systems because cooling is required most of the time.

Commercial structures often are constructed in such a way that ducts cannot be installed to serve the desired air-distribution system. Floor space is very valuable and the floor area required for outlets may be covered by shelving or other fixtures, making a perimeter system impractical. In this case an overhead system must be used. In some cases the system may be a mixture of the perimeter and overhead type.

The Air Distribution Performance Index (ADPI) is defined as the percentage of measurements taken at many locations in the occupied zone of a space which meet a -3 to $2°F$ effective draft temperature criteria. The objective is to select and place the air diffusers so that an ADPI approaching 100% is achieved. ADPI is based only on air velocity and effective draft temperature, a local temperature difference from the room average, and is not directly related to the level of dry bulb temperature or relative humidity. These effects and other factors such as mean radiant temperature must be accounted for. The ADPI provides a means of selecting air diffusers in a rational way. There are no specific criteria for selection of a particular type of diffuser except as discussed above, but within a given type the ADPI is the basis for selecting the throw. The space cooling load per unit area is an important consideration. Heavy loading tends to lower the ADPI. However, loading does not influence design of the diffuser system significantly. Each type of diffuser has a characteristic room length. Table 75.3, the ADPI selection guide, gives the recommended ratio of throw to characteristic length that should maximize the ADPI. A range of throw-to-length ratios are also shown that should give a minimum ADPI. Note that the throw is based on a terminal velocity of 50 ft/min for all diffusers except the ceiling slot type. The general procedure for use of Table 75.3 is as follows:

1. Determine the airflow requirements and the room size.
2. Select the type of diffuser to be used.
3. Determine the room characteristic length.
4. Select the recommended throw-to-length ratio from Table 75.3.
5. Calculate the throw.
6. Select the appropriate diffuser from catalog data.
7. Make sure any other specifications are met (noise, total pressure, etc.).

75.6 BUILDING AIR DISTRIBUTION

This section discusses the details of distributing the air to the various spaces in the structure. Proper design of the duct system and the selection of appropriate fans and accessories are essential. A poorly designed system may be noisy, inefficient, and lead to discomfort of occupants. Correction of faulty design is expensive and sometimes practically impossible.

75.6.1 Fans

The fan is an essential component of almost all heating and air-conditioning systems. Except in those cases where free convection creates air motion, a fan is used to move air through ducts and to induce air motion in the space. An understanding of the fan and its performance is necessary if one is to design a satisfactory duct system.

The centrifugal fan is the most widely used because it can effectively move large or small quantities of air over a wide range of pressures. The principle of operation is similar to the centrifugal pump in that rotating impeller mounted inside a scroll type of housing imparts energy to the air or gas being moved.

Table 75.3 ADPI Selection Guide[a]

Terminal Device	Room Load		$T_{0.25}/L(T_{50}/L)$ for Max. ADPI	Maximum ADPI	For ADPI Greater Than	Range of $T_{0.25}/L(T_{50}/L)$
	W/m²	Btu/hr · ft²				
High sidewall	250	80	1.8	68	—	—
grilles	190	60	1.8	72	70	1.5–2.2
	125	40	1.6	78	70	1.2–2.3
	65	20	1.5	85	80	1.0–1.9
Circular ceiling	250	80	0.8	76	70	0.7–1.3
diffusers	190	60	0.8	83	80	0.7–1.2
	125	40	0.8	88	80	0.5–1.5
	65	20	0.8	93	90	0.7–1.3
Sill grille straight	250	80	1.7	61	60	1.5–1.7
vanes	190	60	1.7	72	70	1.4–1.7
	125	40	1.3	86	80	1.2–1.8
	65	20	0.9	95	90	0.8–1.3
Sill grille spread	250	80	0.7	94	90	0.8–1.5
vanes	190	60	0.7	94	80	0.6–1.7
	125	40	0.7	94	—	—
	65	20	0.7	94	—	—
Ceiling slot dif-	250	80	0.3[b]	85	80	0.3–0.7
fusers[b]	190	60	0.3[b]	88	80	0.3–0.8
	125	40	0.3[b]	91	80	0.3–1.1
	65	20	0.3[b]	92	80	0.3–1.5
Light troffer dif-	190	60	2.5	86	80	<3.8
fusers	125	40	1.0	92	90	<3.0
	65	20	1.0	95	90	<4.5
Perforated and louvered ceiling diffusers	35–160	11–51	2.0	96	90	1.4–2.7
					80	1.0–3.4

Characteristic Room Length for Several Diffuser Types

Diffuser Type	Characteristic Length, L
High sidewall grille	Distance to wall perpendicular to jet
Circular ceiling diffuser	Distance to closest wall or intersecting air jet
Sill grille	Length of room in the direction of the jet flow
Ceiling slot diffuser	Distance to wall or midplane between outlets
Light troffer diffusers	Distance to midplane between outlets plus distance from ceiling to top of occupied zone
Perforated, louvered ceiling diffusers	Distance to wall or midplane between outlets

[a] Reprinted by permission from *ASHRAE Handbook of Fundamentals*, 1981.

[b] Given for $T_{0.50}/L(T_{100}/L)$.

The vaneaxial fan is mounted on the centerline of the duct and produces an axial flow of the air. Guide vanes are provided before and after the wheel to reduce rotation of the air stream.

The tubeaxial fan is quite similar to the vaneaxial fan but does not have the guide vanes.

Axial flow fans are not capable of producing pressures as high as those of the centrifugal fan but can move large quantities of air at low pressure. Axial flow fans generally produce higher noise levels than centrifugal fans.

Fan efficiency may be expressed in two ways. The total fan efficiency is the ratio of total air power to the shaft power input:

$$\eta_t = \frac{\dot{W}_t}{\dot{W}_{sh}}$$

where Q = volume flow rate, ft³/min or m³/sec
$P_{01} - P_{02}$ = change in total pressure, lbf/ft² or Pa
$\dot{W}_{sh}$ = shaft power, ft · lbf/min or W

It has been common practice in the United States for Q to be in ft³/min, $P_{01} - P_{02}$ to be in inches of water, and for $\dot{W}_{sh}$ to be in horsepower. In this special case,

$$\eta_t = \frac{\dot{Q}(P_{01} - P_{02})}{6350 \dot{W}_{sh}}$$

The static fan efficiency is

$$\eta_s = \frac{\dot{Q}(P_1 - P_2)}{6350 \dot{W}_{sh}}$$

Figure 75.23 illustrates typical performance curves for centrifugal fans. Note the differences in the pressure characteristics and in the point of maximum efficiency with respect to the point of maximum pressure.

Table 75.4 compares some of the more important characteristics of centrifugal fans.

The noise emitted by a fan is important in many applications. For a given pressure the noise level is proportional to the tip speed of the impeller and to the air velocity leaving the wheel. Fan noise is roughly proportional to the pressure developed regardless of the blade type; however, backward-curved fan blades generally have the better (lower) noise characteristics.

The pressure developed by a fan is limited by the maximum allowable speed. If noise is not a factor, the straight radial blade is superior. Fans may be operated in series to develop higher pressures, and multistage fans are also constructed. When fans are used in parallel, surging back and forth between fans may develop, particularly if the system demand is changing. Forward-curved blades are particularly unstable when operated at the point of maximum efficiency.

Combining both the system and fan characteristics on a plot is very useful in matching a fan to a system and to ensure fan operation at the desired conditions. Figure 75.24 illustrates the desired operating range for a forward-curved blade fan. The range is to the right of the point of maximum efficiency. The backward-curved blade fan has a selection range that brackets the range of maximum efficiency and is not so critical to the point of operation; however, this type should always be operated to the right of the point of maximum pressure. For a given system the efficiency does not change with speed; however capacity, total pressure, and power all depend on the speed. Changing the fan speed will not change the relative point of intersection between the system and fan characteristics. This can only be done by changing fans.

There are several simple relationships between fan capacity, pressure, speed, and power, which are referred to as the fan laws. The most useful fan laws are:

1. Capacity is directly proportional to fan speed.
2. Pressure (static, total, or velocity) is proportional to the square of the fan speed.
3. Power required is proportional to the cube of fan speed.

Three other fan laws are useful.

4. Pressure and power are proportional to the density of the air at constant speed and capacity.
5. Speed, capacity, and power are inversely proportional to the square root of the density at constant pressure.
6. Capacity, speed, and pressure are inversely proportional to the density, and the power is inversely proportional to the square of the density at a constant mass flow rate.

In a variable-air-volume system it is desirable to reduce fan speed as air-volume flow rate is reduced under part load conditions to reduce the fan power.

Fan Selection

To select a fan it is necessary to know the capacity and total pressure requirement of the system. The type and arrangement of the prime mover, the possibility of fans in parallel or series, nature of the load (variable or steady), and the noise constraints must also be considered. After the system characteristics have been determined, the main considerations in the actual fan selection are efficiency, reliability, size and weight, speed, noise, and cost.

To assist in the actual fan selection, manufacturers furnish graphs with the areas of preferred operation shown. In many cases manufacturers present their fan performance data in the form of

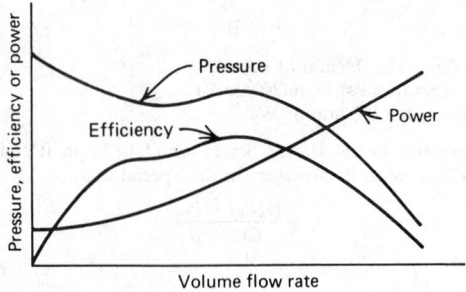

Forward-tip fan characteristics.

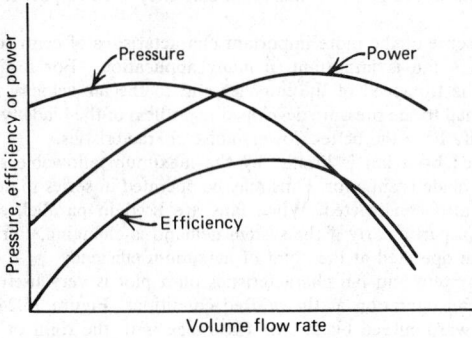

Backward-tip fan characteristics.

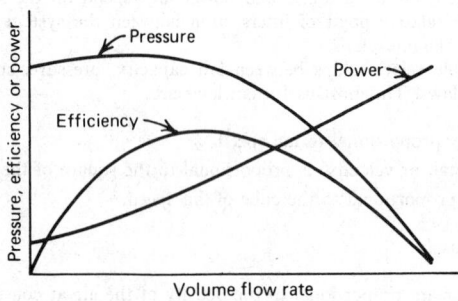

Radial-tip fan characteristics.

Fig. 75.23 Performance curves for centrifugal fans.[7]

tables. The static pressure is often given but not the total pressure. The total pressure may be computed from the capacity and the fan outlet dimensions. Data pertaining to noise are also available from most manufacturers.

It is important that the fan be efficient and quiet. Generally, a fan will generate the least noise when operated near the peak efficiency. Operation considerably beyond the point of maximum efficiency will be noisy. Forward-curved blades operated at high speeds will be noisy and straight blades are generally noisy, especially at high speed. Backward-curved blades may be operated on both sides of the peak efficiency at relatively high speeds with less noise than the other types of fans.

Fan Installation

The performance of a fan can be drastically reduced by improper connection to the duct system. In general, the duct connections should be such that the air may enter and leave the fan as uniformly

Table 75.4 Comparison of Centrifugal Fan Types[7]

Item	Forward-Curved Blades	Radial Blades	Backward-Curved Blades
Efficiency	Medium	Medium	High
Space required	Small	Medium	Medium
Speed for given pressure rise	Low	Medium	High
Noise	Poor	Fair	Good

as possible with no abrupt changes in direction or velocity. The designer must rely on good judgment and ingenuity in laying out the system. Space is often limited for the fan installation, and a less than optimum connection may have to be used. In this case the designer must be aware of the penalties (loss in total pressure and efficiency). Some manufacturers furnish application factors from which a modified fan curve can be computed.

The Air Movement and Control Association, Inc. (AMCA) has published system effect factors in their *Fan Applications Manual* that express the effect of various fan connections on system performance.[8]

75.6.2 Variable-Volume Systems

In variable-air-volume systems the total amount of circulated air may vary between some minimum and the full load design quantity. Normally, the minimum is about 20–25% of the maximum. The volume flow rate of the air is controlled independent of the fan by the terminal boxes, and the fan must respond to the system. The fan speed should be decreased as volume flow rate decreases. Variable speed electric motors have very low efficiency that offsets the benefit of lowering fan speed. Fan drives that make use of magnetic couplings have been developed and are referred to as eddy current drives. These are excellent devices with almost infinite adjustment of fan speed. Their only disadvantage is high cost. A change may be made in the fan speed by changing the diameter of the V-belt drive pulley by adjusting the pulley shives. This requires a mechanism that will operate while the drive is turning. The main disadvantage of this approach is maintenance. The eddy current and variable pulley drives appear to be the most practical at present.

Another approach to control of the fan is to throttle and introduce a swirling component to the air entering the fan that alters the fan characteristic in such a way that less power is required at the lower flow rates. This is done with variable inlet vanes that are a radial damper system located at the inlet to the fan. Gradual closing of the vanes reduces the volume flow rate of air and changes the fan characteristic. This approach is not as effective in reducing fan power as fan speed reduction, but the cost and maintenance are low.

Airflow in Ducts

The steady-flow energy equation applies to the flow of air in a duct. Neglecting the elevation head terms, assuming that the flow is adiabatic, and no fan is present,

$$\frac{g_c}{g}\frac{P_1}{\rho}+\frac{V_1^2}{2g}=\frac{g_c}{g}\frac{P_2}{\rho}+\frac{V_2^2}{2g}+l_f$$

and in terms of the total head

$$\frac{g_c}{g}\frac{P_{01}}{\rho}=\frac{g_c}{g}\frac{P_{02}}{\rho}+l_f$$

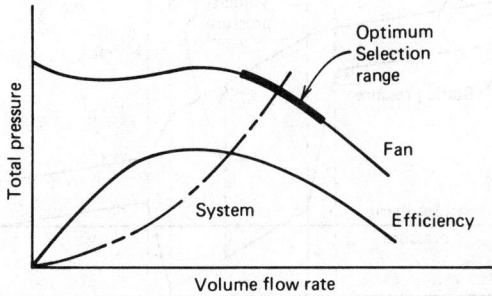

Fig. 75.24 Optimum match between system and forward-curved blade fan.[7]

where V = average air velocity at a duct cross section, ft/min or m/sec

  l_f = lost head due to friction, ft or m

The static and velocity head terms are interchangeable and may increase or decrease in the direction of flow depending on the duct cross-sectional area. Because the lost head must be positive, the total pressure always decreases in the direction of flow, as in Fig. 75.25.

For duct flow the units of each term are usually inches of water because of their small size. The equations may be written

$$H_{s1} + H_{v1} = H_{s2} + H_{v2} + l_f$$

and

$$H_{01} = H_{02} + l_f$$

For air at standard conditions

$$H_v = \left(\frac{V}{4005}\right)^2 \quad \text{in. } H_2O$$

where V is in ft/min,

$$P_v = \left(\frac{V}{1.29}\right)^2 \quad \text{Pa}$$

where V is in m/sec.

The lost head due to friction, l_f, in a straight, constant area duct may be determined by use of a friction factor. Because this approach becomes tedious when designing ducts, special charts have been prepared. Figure 75.26 is such a chart for air flowing in ducts. The chart is based on standard air and fully developed flow. For the temperature range of 50°F or 10°C to about 100°F or 38°C there is no need to correct for viscosity and density changes. Above 100°F or 38°C, however, a correction should be made. The density correction is also small for moderate pressure changes. For elevations below about 2000 ft or 610 m the correction is small. The correction for density and viscosity will normally be less than 1. The effect of roughness is an important consideration and difficult to assess.

A common problem to designers is determination of the roughness effect of fibrous glass duct liners and fibrous ducts. This material is manufactured in several grades with various degrees of absolute roughness. The usual approach to account for this roughness effect is to use a correction factor that is applied to the pressure loss obtained for galvanized metal duct.

The head loss due to friction is greater for a rectangular duct than for a circular duct of the same cross-sectional area and capacity. For most practical purposes ducts of aspect ratio not exceeding 8:1 will have the same lost head for equal length and mean velocity of flow as a circular duct of the same hydraulic diameter. When the duct sizes are expressed in terms of hydraulic diameter D and when the equations for friction loss in round and rectangular ducts are equated for equal length and capacity, an equation for the circular equivalent of a rectangular duct is obtained:

$$D_e = 1.3 \frac{(ab)^{5/8}}{(a+b)^{1/4}}$$

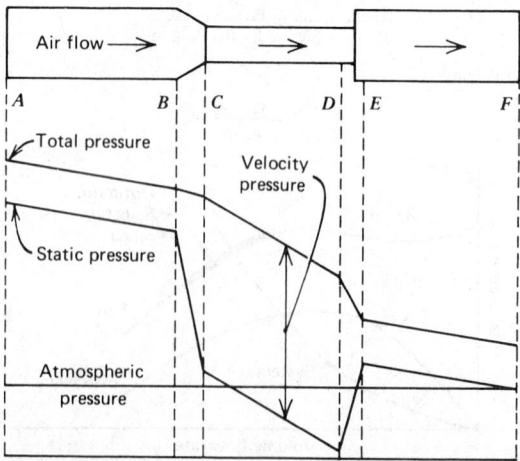

Fig. 75.25 Pressure changes during flow in ducts.[7]

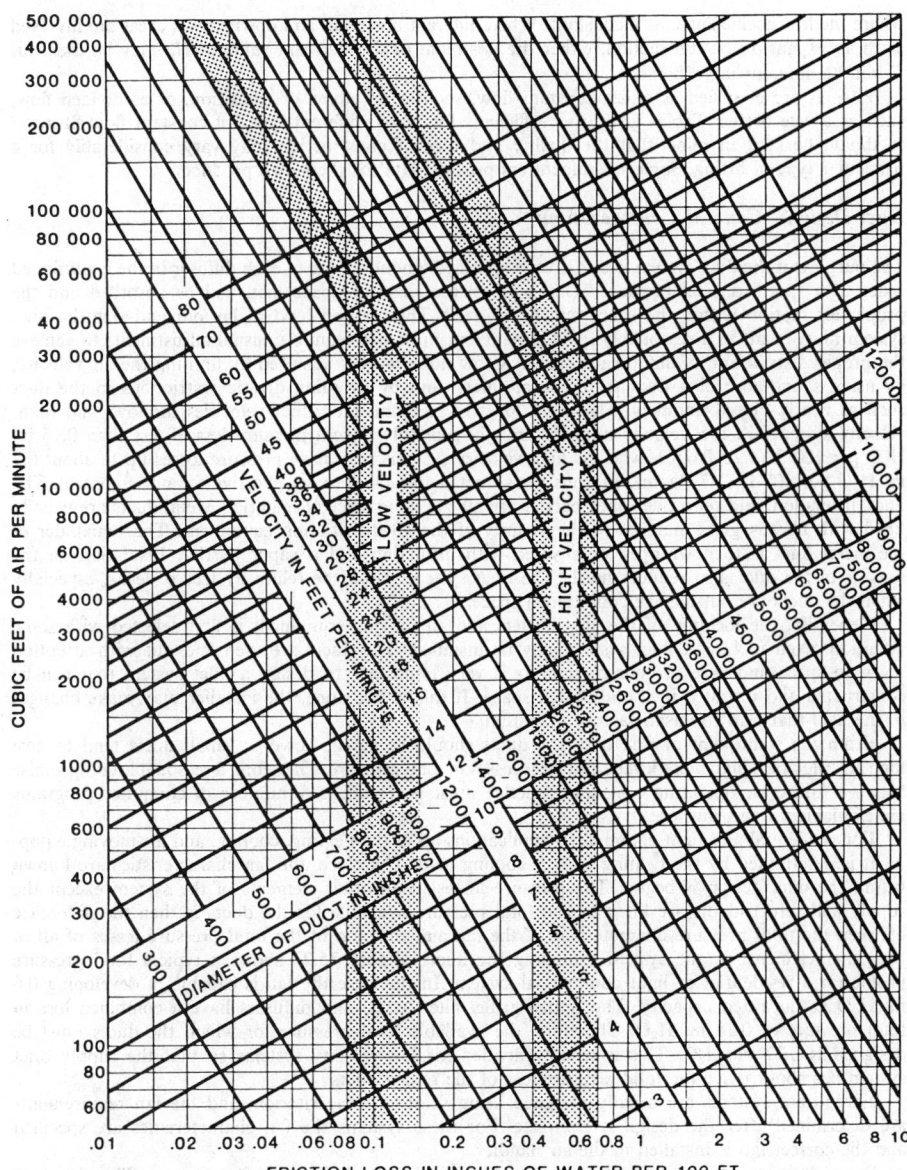

Fig. 75.26 Lost head due to friction for air flowing in ducts. (Reprinted by permission from *ASHRAE Handbook of Fundamentals*, 1981.)

where a and b are the rectangular duct dimensions in any consistent units and D_e is the equivalent diameter. A table of equivalent diameters is given in the *ASHRAE Handbook*.[2]

Air Flow in Fittings

Whenever a change in area or direction occurs in a duct or when the flow is divided and diverted into a branch, substantial losses in total pressure may occur. These losses are usually of greater magnitude than the losses in the straight pipe and are referred to as dynamic losses.

Dynamic losses vary as the square of the velocity and are conveniently represented by

$$H_0 = (C)(H_v)$$

where the loss coefficient C is a constant. When different upstream and downstream areas are involved as in an expansion or contraction, either the upstream or downstream value of H_v may be used but C will be different in each case.

Fittings are classified as either constant flow, such as an elbow or transition, or as divided flow, such as a wye or tee. Tables give loss coefficients for many different types of constant flow fittings.[2] It should be kept in mind that the quality and type of construction may vary considerably for a particular type of fitting. Some manufacturers provide data for their own products.

Duct Design—General Considerations

The purpose of the duct system is to deliver a specified amount of air to each diffuser in the conditioned space at a specified total pressure. This is to ensure that the space load will be absorbed and the proper air motion within the space will be realized. The method used to lay out and size the duct system must result in a reasonably quiet system and must not require unusual adjustments to achieve the proper distribution of air to each space. A low noise level is achieved by limiting the air velocity, by using sound-absorbing duct materials or liners, and by avoiding drastic restrictions in the duct such as nearly closed dampers. Figure 75.26 gives recommended duct velocities for low- and high-velocity systems. A low-velocity duct system will generally have a pressure loss of less than 0.15 in. H_2O per 100 ft (1.23 Pa/m), whereas high-velocity systems may have pressure losses up to about 0.7 in. H_2O per 100 ft (5.7 Pa/m). Fibrous glass duct materials are very effective for noise control. The duct, insulation, and reflective vapor barrier are all the same piece of material. Metal ducts are usually lined with fibrous glass material in the vicinity of the air-distribution equipment. The remainder of the metal duct is then wrapped or covered with insulation and a vapor barrier. Insulation on the outside of the duct also reduces noise. The duct system should be relatively free of leaks, especially when the ducts are outside the conditioned space.

Generally, the location of the air diffusers and air-moving equipment is first selected with some attention given to how a duct system may be installed. The ducts are then laid out with attention given to space and ease of construction. It is very important to design a duct system that can be constructed and installed in the allocated space. If this is not done, the installer may make changes in the field that lead to unsatisfactory operation.

From the standpoint of first cost, the ducts should be small; however, small ducts tend to give high air velocities, high noise levels, and large losses in total pressure. Therefore, a reasonable compromise between first cost, operating cost, and practice must be reached. A number of computer programs are available for this purpose.

For residential and light commercial applications all of the heating, cooling, and air-moving equipment is determined by the heating and/or cooling load. Therefore, the fan characteristics are known before the duct design is begun. The pressure losses in all other elements of the system except the supply and return ducts are known. The total pressure available for the ducts is then the difference between the total pressure characteristic of the fan and the sum of the total pressure losses of all of the other elements in the system excluding the ducts. Figure 75.27 shows a typical total pressure profile for a residential or light commercial system. In this case the fan is capable of developing 0.6 in. H_2O at the rated capacity. The return grille, filter, coils, and diffusers have a combined loss in total pressure of 0.38 in. H_2O. Therefore, the available total pressure for which the ducts must be designed is 0.22 in. H_2O. This is usually divided for low-velocity systems so that the supply duct system has about twice the total pressure loss of the return ducts.

Large duct systems are usually designed using velocity as a criterion, and the fan requirements are determined after the design is complete. For these systems the fan characteristics are specified and the correct fan is installed in the air handler.

Some designers neglect velocity pressure when dealing with low-velocity systems. This does not simplify the duct design procedure and is unnecessary. When the air velocities are high, the velocity pressure must be considered to achieve reasonable accuracy. If static and velocity pressure are computed separately, the problem becomes very complex. It is best to use total pressure in duct design because it is simpler and accounts for all of the flow energy.

Design of Low-Velocity Duct Systems

The methods described in this section pertain to low-velocity systems where the average velocity is less than about 1000 ft/min or 5 m/sec. These methods can be used for high-velocity system design, but the results will not be satisfactory in most cases.

Equal Friction Method

This method makes the pressure loss per foot of length the same for the entire system. If all runs from fan to diffuser are about the same length, this method will produce a well balanced design.

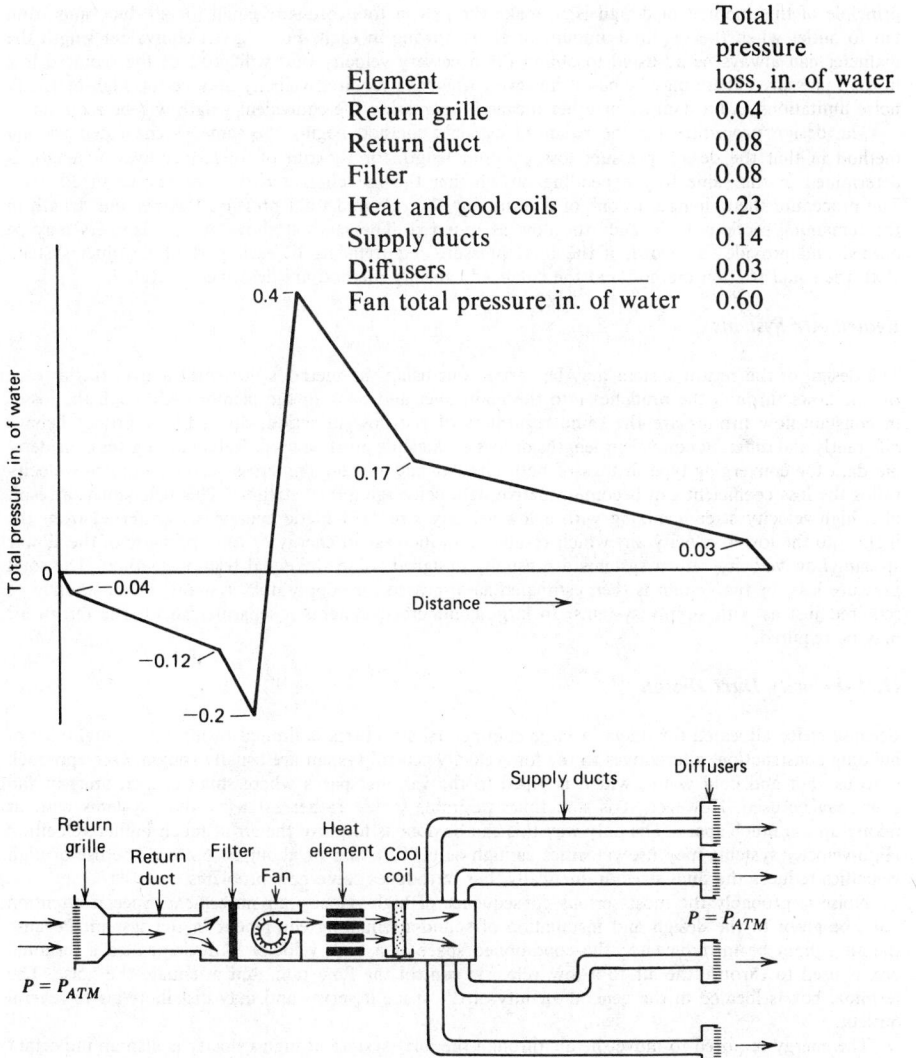

Element	Total pressure loss, in. of water
Return grille	0.04
Return duct	0.08
Filter	0.08
Heat and cool coils	0.23
Supply ducts	0.14
Diffusers	0.03
Fan total pressure in. of water	0.60

Fig. 75.27 Total pressure profile for a typical residential or light commercial system.[8]

However, most duct systems have duct runs ranging from long to short. The short runs will have to be dampered, which can cause considerable noise.

The usual procedure is to select the velocity in the main duct adjacent to the fan and to provide a satisfactory noise level for the particular application. The known flow rate then establishes the duct size and the lost pressure per unit of length. This same pressure loss per unit length is then used throughout the system. A desirable feature of this method is the gradual reduction of air velocity from fan to outlet, thereby reducing noise problems. After sizing the system the designer must compute the total pressure loss of the longest run (largest flow resistance), taking care to include all fittings and transitions. When the total pressure available for the system is known in advance, the design loss value may be established by estimating the equivalent length of the longest run and computing the low pressure per unit length.

Balanced Capacity Method

This method of duct design has been referred to as the "balanced pressure loss method." However, it is the flow rate or capacity of each outlet that is balanced and not the pressure. As previously discussed, the loss in total pressure automatically balances regardless of the duct sizes. The basic principle of this method of design is to make the loss in total pressure equal for all duct runs from fan to outlet when the required amount of air is flowing in each. For a given equivalent length the diameter can always be adjusted to obtain the necessary velocity that will produce the required loss in total pressure. There may be cases, however, when the required velocity may be too high to satisfy noise limitations and a damper or other means of increasing the equivalent length will be required.

The design procedure for the balanced capacity method begins the same as the equal friction method in that the design pressure loss per unit length for the run of longest equivalent length is determined in the same way depending on whether the fan characteristics are known in advance. The procedure then changes to one of determining the required total pressure loss per unit length in the remaining sections to balance the flow as required. The method shows where dampers may be needed and provides a record of the total pressure requirements of each part of the duct system. Both the equal friction method and the balanced capacity method are described in Ref. 7.

Return Air Systems

The design of the return system may be carried out using the methods described above. In this case the air flows through the branches into the main duct and back to the plenum. Although the losses in constant-flow fittings are the same regardless of the flow direction, divided-flow fittings behave differently and different equivalent lengths or loss coefficients must be used. Reference 2 gives considerable data for converging-type fittings of both circular and rectangular cross section. For low-velocity ratios the loss coefficient can become negative with converging-flow streams. This behavior is a result of a high-velocity stream mixing with a low-velocity stream. Kinetic energy is transferred from the higher- to the lower-velocity air, which results in an increase in energy or total pressure of the slower stream. Low-velocity return systems are usually designed using the equal friction method. The total pressure loss for the system is then estimated as discussed for supply duct systems. Dampers may be required just as with supply systems. In large commercial systems a separate fan for the return air may be required.

High-Velocity Duct Design

Because space allocated for ducts in large commercial structures is limited owing to the high cost of building construction, alternatives to the low-velocity central system are usually sought. One approach is to use hot and cold water, which is piped to the various spaces where small central units or fan coils may be used. However, it is sometimes desirable to use rather extensive duct systems without taking up too much space. The only way this can be done is to move the air at much higher velocities. High-velocity systems may use velocities as high as 6000 ft/min or about 30 m/sec. The use of high velocities reduces the duct sizes dramatically, but introduces some new problems.

Noise is probably the most serious consequence of high-velocity air movement. Special attention must be given to the design and installation of sound-attenuating equipment in the system. Because the air cannot be introduced to the conditioned space at a high velocity, a device called a terminal box is used to throttle the air to a low velocity, control the flow rate, and attenuate the noise. The terminal box is located in the general vicinity of the space it serves and may distribute air to several outlets.

The energy required to move the air through the duct system at high velocity is also an important consideration. The total pressure requirement is typically on the order of several inches of water. To partially offset the high fan power requirements, variable-speed fans are sometimes used.

Because the higher static and total pressures required by high-velocity systems aggravate the duct leakage problem, an improved duct fabrication system has been developed for high-velocity systems. The duct is generally referred to as spiral duct and has either a round or oval cross section. The fittings are machine formed and are especially designed to have low pressure losses and close fitting joints to prevent leakage.

The criterion for designing the high-velocity duct systems is somewhat different from that used for low-velocity systems. Emphasis is shifted from a self-balancing system to one that has minimum losses in total pressure.

REFERENCES

1. J. A. Goff, "Standardization of Thermodynamic Properties of Moist Air," *Transactions ASHVE,* **55** (1949).

2. *ASHRAE Handbook of Fundamentals,* American Society of Heating, Refrigerating and Air-Conditioning Engineers, New York, 1981.

3. ASHRAE Standard 55-81, *Thermal Environmental Condition for Human Occupancy,* American Society of Heating, Refrigerating and Air-Conditioning Engineers, New York, 1981.

4. ASHRAE Standard 62-81, *Standards for Ventilation Required for Minimum Acceptable Indoor Air Quality*, American Society of Heating, Refrigerating and Air-Conditioning Engineers, New York, 1981.

5. Heating and Cooling Load Calculation Manual GRP158, ASHRAE, 1979.

6. J. D. Parker, J. H. Boggs, and E. F. Blick, *Introduction to Fluid Mechanics and Heat Transfer*, Addison Wesley, Reading, MA, 1959.

7. F. C. McQuiston and J. D. Parker, *Heating, Ventilating, and Air Conditioning*, 2nd ed., Wiley, New York, 1982.

8. *AMCA Fan Applications Manual*, Parts 1, 2, and 3, Air Movement and Control Association, Inc., 30 West University Drive, Arlington Heights, IL.

CHAPTER 76

REFRIGERATION

K. W. COOPER

K. E. HICKMAN

Borg Warner Corporation
York, Pennsylvania

Reprinted with additions from *Kirk-Othmer Encyclopedia of Chemical Technology,* 3rd ed., Volume 20, Wiley, New York, 1982, by permission of the publisher.

76.1 INTRODUCTION

Refrigeration is the use of mechanical or heat-activated machinery for cooling purposes. The production of extremely low temperatures, $<-150°C$,[1] is usually thought of as cryogenics, and the use of refrigeration equipment to provide human comfort is known as air conditioning. This chapter deals only with refrigeration, which, in its broadest sense, covers such diverse uses as food processing and storage, supermarket display cases, skating rinks, ice manufacture, and biomedical applications such as blood and tissue storage or hypothermia used in surgery. The first commercial machine used for refrigeration was built in 1844.[2]

Refrigeration is used in installations covering a broad range of cooling capacities and temperature levels. The variety of applications results in a diversity of mechanical specifications and equipment requirements. Nevertheless, the methods for producing refrigeration are well standardized.

76.2 BASIC PRINCIPLES

Thermodynamic principles govern all refrigeration processes. Refrigeration is accomplished in the reverse Carnot cycle by using a fluid that evaporates and condenses at suitable pressures for practical equipment designs. The vapor cycle is illustrated by a pressure–enthalpy diagram, Fig. 76.1.

The compressor raises the pressure of the refrigerant vapor so that its saturation temperature is slightly above the temperature of an available cooling medium, for example, air or water. This difference in temperature allows transfer of heat from the vapor to the cooling medium so that the vapor can condense. The liquid next expands to a pressure such that its saturation temperature is slightly below the temperature of the product to be cooled. This difference in temperature allows transfer of heat

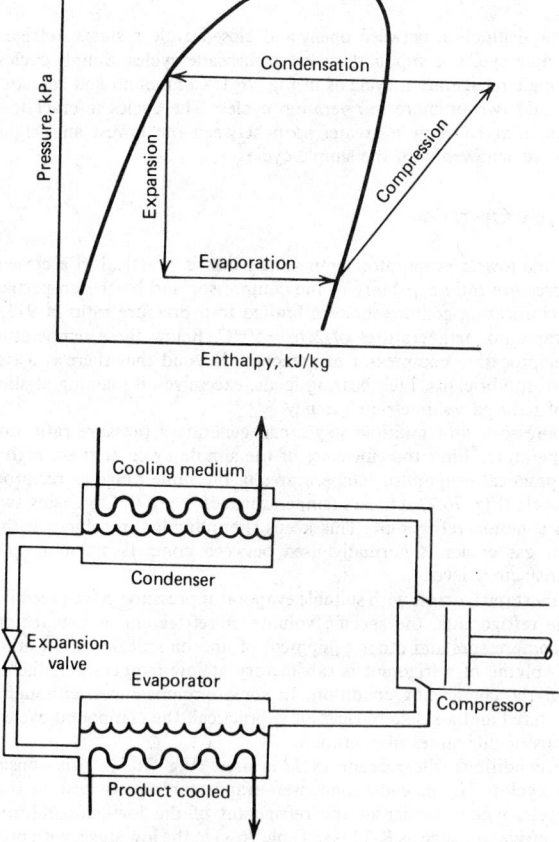

Fig. 76.1 Basic vapor cycle. To convert kPa to atm, divide by 101.3. To convert kJ/kg to Btu/lb, multiply by 0.4302.[9]

from the product to the refrigerant, causing the refrigerant to evaporate. The vapor formed must be removed by the compressor at a sufficient rate to maintain the low pressure in the evaporator and keep the cycle operating.

Pumped recirculation of refrigerant rather than direct evaporation of refrigerant is often used to service remotely located or specially designed heat exchangers. This technique provides the users with wide flexibility in applying refrigeration to complex processes and greatly simplifies operation. Secondary refrigerants or brines are also commonly used today for simple control and operation. Direct application of ice and brine storage tanks may be used to level off batch cooling loads and reduce equipment size. This approach provides stored refrigeration where temperature control is vital as a safety consideration to prevent runaway reactions or pressure buildup.

All mechanical cooling results in the simultaneous production, somewhere else, of a greater amount of heat. It is not always realized that this heat can be put to good use by applying the heat-pump principle. This requires provision to recover the heat normally rejected to cooling water or air in the refrigeration condenser. Recovery of this waste heat at temperatures up to 65°C is frequently used in modern plants to achieve improved heat balance and operating economy.

Historically, capacities of mechanical refrigeration systems have been stated in tons of refrigeration, a unit of measure related to the ability of an ice plant to freeze one short ton (907 kg) of ice in 24 hr. Its value is 3.51 kW$_t$ (12,000 Btu/hr). Often a kilowatt of refrigeration capacity is identified as kW$_t$ to distinguish it from the amount of electricity (kW$_e$) required to produce the refrigeration.

76.3   REFRIGERATION CYCLES AND SYSTEM OVERVIEW

Refrigeration can be accomplished in either closed-cycle or open-cycle systems. In a closed cycle, the refrigerant fluid is confined within the system and recirculates through the processes in the cycle. The system shown at the bottom of Fig. 76.1 is a closed cycle. In an open cycle, the fluid used as the refrigerant passes through the system once on its way to be used as a product or feedstock outside the refrigeration process. An example is the cooling of natural gas to separate and condense heavier components.

In addition to the distinction between open- and closed-cycle systems, refrigeration processes are also described as simple cycles, compound cycles, or cascade cycles. Simple cycles employ one set of components and a single refrigeration cycle as in Fig. 76.1. Compound and cascade cycles use multiple sets of components and two or more refrigeration cycles. The cycles interact to accomplish cooling at several temperatures or to allow a greater span between the lowest and highest temperatures in the system than can be achieved with the simple cycle.

76.3.1   Closed-Cycle Operation

For a simple cycle, the lowest evaporator temperature that is practical in a closed-cycle system (Fig. 76.1) is set by the pressure-ratio capability of the compressor and by the properties of the refrigerant. Most high-speed reciprocating compressors are limited to a pressure ratio of 9:1, so that the simple cycle is used for evaporator temperatures of 2 to −50°C. Below these temperatures, the application limits of a single reciprocating compressor are reached. Beyond that there is a risk of excessive heat which may break down lubricants, high bearing loads, excessive oil foaming at startup, and inefficient operation because of reduced volumetric efficiency.

Centrifugal compressors with multiple stages can generate a pressure ratio up to 18:1, but their high discharge temperatures limit the efficiency of the simple cycle at these high pressure ratios. As a result, they have practical evaporator temperatures in the same range as reciprocating compressors.

The compound cycle (Fig. 76.2) achieves temperatures of ∼ −100°C by using two or three compressors in series and a common refrigerant. This keeps the individual machines within their application limits. A refrigerant gas cooler is normally used between compressors to keep the final discharge temperature at a satisfactory level.

Below −100°C, most refrigerants with suitable evaporator pressures have excessively high condensing pressures. For some refrigerants, the specific volume of refrigerant at low temperatures may be so great as to require compressors and other equipment of uneconomical size. In other refrigerants, even though the specific volume of refrigerant is satisfactory at low temperature, the specific volume may become too small at the condensing condition. In some circumstances, although none of the above limitations is encountered and a single refrigerant is practical, the compound cycle is not used because of oil-return problems or difficulties of operation.

To satisfy these conditions, the cascade cycle is used (Fig. 76.3). This consists of two or more separate refrigerant cycles. The cascade condenser–evaporator rejects heat to the evaporator of the high-temperature cycle, which condenses the refrigerant of the low-temperature cycle. This makes possible the use of a refrigerant such as R-13 (see Table 76.1) in the low stage, with pressure–temperature–volume characteristics well suited to the low temperature. Refrigerants with pressure–temperature–volume characteristics more favorable at higher temperatures, for example, R-12 or R-22, are used

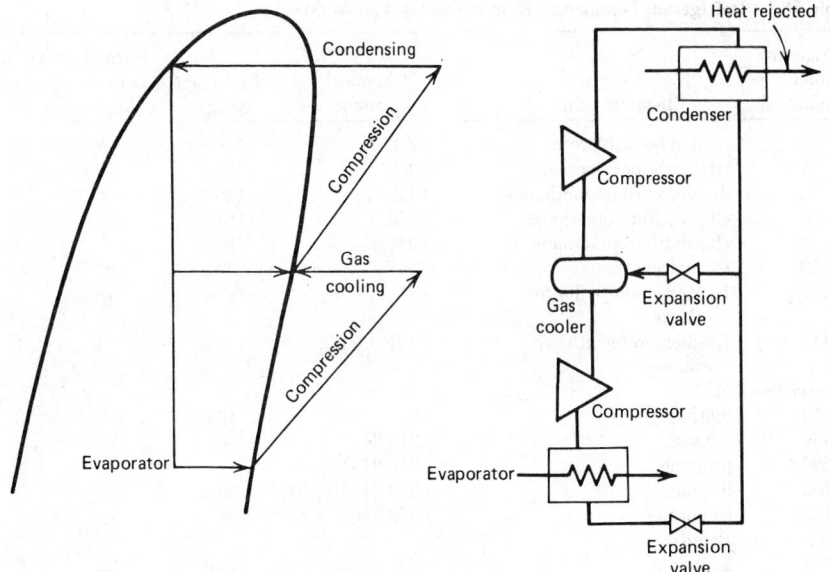

Fig. 76.2 Compound cycle.[9]

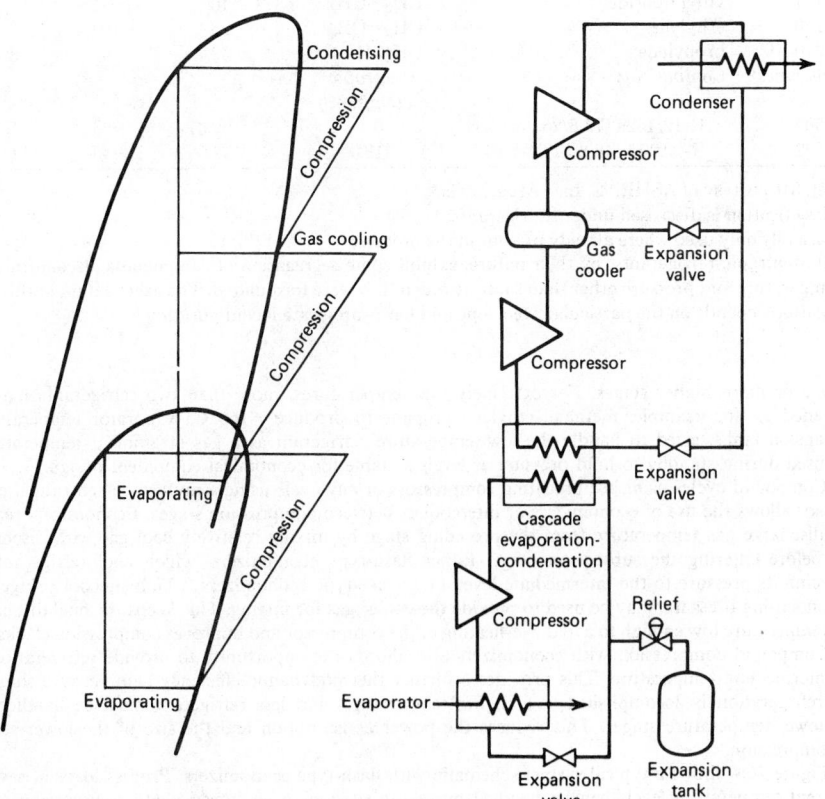

Fig. 76.3 Cascade cycle.[9]

Table 76.1   Refrigerant Numbering System (ASHRAE 34-78)[a]

Refrigerant number designation	Chemical name	Chemical formula	Molecular weight	Normal boiling point, °C	Group classification[b]
10	carbon tetrachloride	CCl_4	153.8	77	2
11	trichlorofluoromethane	CCl_3F	137.4	24	1
12	dichlorodifluoromethane	CCl_2F_2	120.9	−30	1
13	chlorotrifluoromethane	$CClF_3$	104.5	−81	1
22	chlorodifluoromethane	$CHClF_2$	86.5	−41	1
23	trifluoromethane	CHF_3	70.0	−82	
113	1,1,2-trichlorotrifluoro-ethane	CCl_2FCClF_2	187.4	48	1
114	1,2-dichlorotetrafluoro-ethane	$CClF_2CClF_2$	170.9	4	1
Hydrocarbons					
50	methane	CH_4	16.0	−161	3
170	ethane	CH_3CH_3	30	−89	3
290	propane	$CH_3CH_2CH_3$	44	−42	3
600	butane	$CH_3CH_2CH_2CH_3$	58.1	0	3
600a	isobutane	$CH(CH_3)_3$	58.1	−12	
Inorganic compounds					
717	ammonia	NH_3	17.0	−33	2
732	oxygen	O_2	32.0	−183	
744	carbon dioxide	CO_2	44.0	−78	1
764	sulfur dioxide	SO_2	64.1	−10	2
Unsaturated organic compounds					
1140	vinyl chloride	$CH_2{=}CHCl$	62.5	−14	c
1141	vinyl fluoride	$CH_2{=}CHF$	46	−72	
1150	ethylene	$CH_2{=}CH_2$	28.1	−104	3
1270	propylene	$CH_3CH{=}CH_2$	42.1	−48	3
Azeotropes[d]	*Composition*	*Azeotropic temp, °C*			
500	R-12/152a (73.8/26.2 wt %)	0	99.31	−33	1
502	R-22/115 (48.8/51.2 wt %)	19	112	−45	1

[a] Ref. 3. Courtesy of ASHRAE, Inc., Atlanta, Ga.
[b] Classification is discussed under Refrigerants.
[c] Normally only used where already present in the process.
[d] All azeotropic refrigerants, by their nature, exhibit some segregation of components at conditions of temperature and pressure other than those at which they were formulated. The exact extent of this segregation depends on the particular azeotrope and hardware-system configuration.

in one or more higher stages. For extremely low temperatures, more than two refrigerants may be cascaded as, for example, methane–ethylene–propane to produce −160°C evaporator temperatures. Expansion tanks, sized to handle the low-temperature refrigerant as a gas at ambient temperatures, are used during standby to hold pressure at levels suitable for economical equipment design.

Compound cycles using reciprocating compressors or any cycle using a multistage centrifugal compressor allows the use of economizers or intercoolers between compression stages. Economizers reduce the discharge gas temperature from the preceding stage by mixing relatively cool gas with discharge gas before entering the subsequent stage. Either flash-type economizers, which cool refrigerant by reducing its pressure to the intermediate level, or surface-type economizers, which subcool refrigerant at condensing pressure, may be used to provide the cooler gas for mixing. This keeps the final discharge gas temperature low enough to avoid overheating of the compressor and improves compression efficiency.

Compound compression with economizers also affords the opportunity to provide refrigeration at an intermediate temperature. This provides a further thermodynamic efficiency gain because some of the refrigeration is accomplished at a higher temperature, and less refrigerant must be handled by the lower-temperature stages. This reduces the power consumption and the size of the lower stages of compression.

Figure 76.4 shows a typical system schematic with flash-type economizers. Process loads at several different temperature levels can be handled by taking suction to an intermediate compression stage as shown. The corresponding pressure–enthalpy (*P–H*) diagram illustrates the thermodynamic cycle.

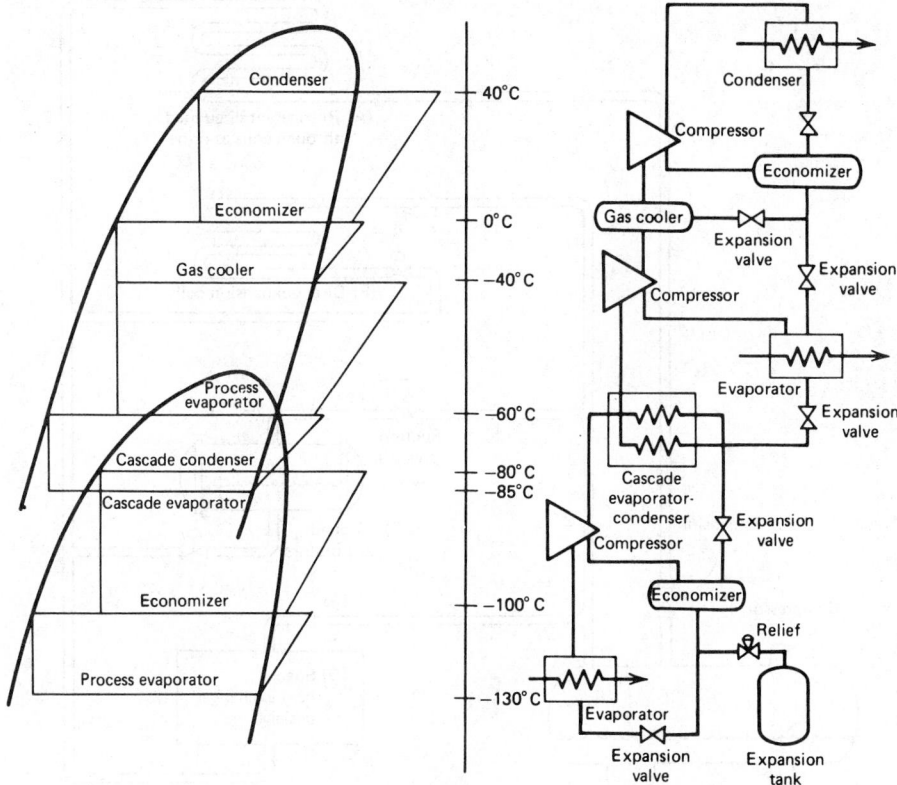

Fig. 76.4 System with flash economizers.[9]

Flooded refrigeration systems are a version of the closed cycle that may reduce design problems in some applications. In flooded systems, the refrigerant is circulated to heat exchangers or evaporators by a pump. Figure 76.5 shows the flooded cycle, which can use any of the simple or compound closed-refrigeration cycles.

The refrigerant-recirculating pump pressurizes the refrigerant liquid and moves it to one or more evaporators or heat exchangers which may be remote from the receiver. The low-pressure refrigerant may be used as a single-phase heat-transfer fluid as in (a) of Fig. 76.5, which eliminates the extra heat-exchange step and increased temperature difference encountered in a conventional system that uses a secondary refrigerant or brine. This approach may simplify the design of process heat exchangers, where the large specific volumes of evaporating refrigerant vapor would be troublesome. Alternatively, the pumped refrigerant in the flooded system may be routed through conventional evaporators as in (b) and (c), or special heat exchangers as in (d).

The flooded refrigeration system is helpful when special heat exchangers are necessary for process reasons, or where multiple or remote exchangers are required.

76.3.2 Refrigerant Selection for the Closed Cycle

In any closed cycle, the choice of the operating fluid is unrestricted and is based on the one whose properties are best suited to the operating conditions. The choice depends on a variety of factors, some of which may not be directly related to the refrigerant's ability to remove heat. For example, flammability, toxicity, density, viscosity, availability, and similar characteristics are often deciding factors. The suitability of a refrigerant also depends on factors such as the kind of compressor to be used (i.e., centrifugal, rotary, or reciprocating), safety in application, heat-exchanger design, application of codes, size of the job, and temperature ranges. The factors below should be taken into account when selecting a refrigerant.

Discharge (condensing) pressure should be low enough to suit the design pressure of commercially available pressure vessels, compressor casings, etc. However, discharge pressure, that is, condenser

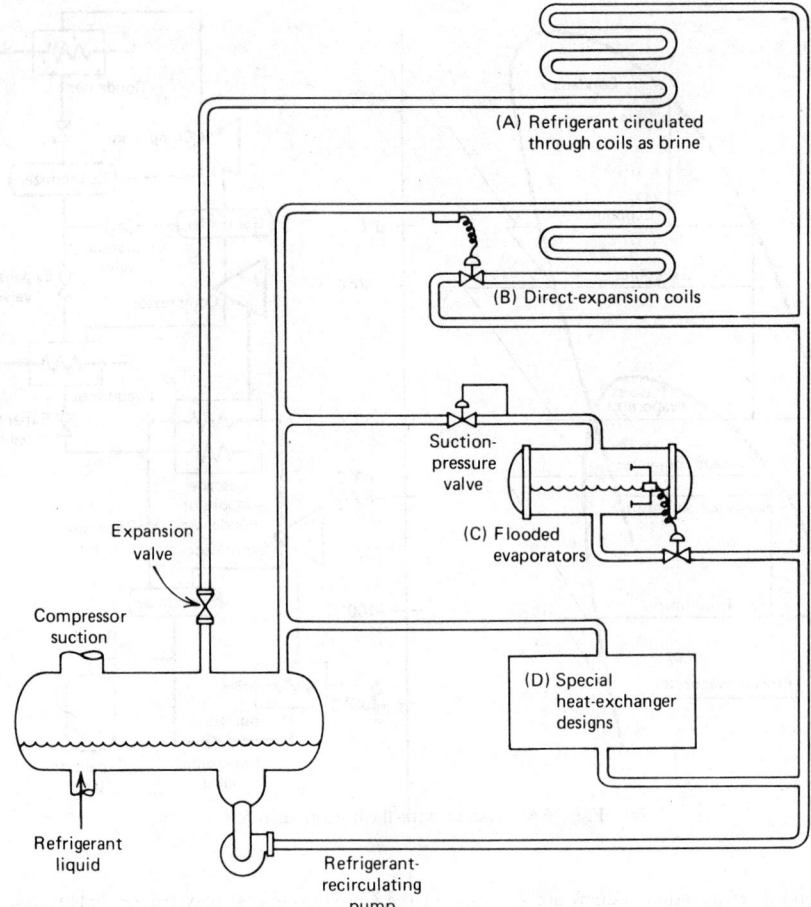

Fig. 76.5 Liquid recirculation.[9]

liquid pressure, should be high enough to feed liquid refrigerant to all the parts of the system that require it.

Suction (evaporating) pressure should be above approximately 3.45 kPa (0.5 psia) for a practical compressor selection. When possible, it is preferable to have the suction pressure above atmospheric to prevent leakage of air and moisture into the system. Positive pressure normally is considered a necessity when dealing with hydrocarbons because of the explosion hazard presented by any air leakage into the system.

Standby pressure (saturation at ambient temperature) should be low enough to suit equipment design pressure unless there are other provisions in the system for handling the refrigerant during shutdown, for example, inclusion of expansion tanks.

Critical temperature and pressure should be well above the operating level. As the critical pressure is approached, less heat is rejected as latent heat compared to the sensible heat from desuperheating the compressor discharge gas, and cycle efficiency is reduced. Methane (R-50) and chlorotrifluoromethane (R-13) usually are cascaded with other refrigerants because of their low critical points.

Suction volume sets the size of the compressor. High suction volumes require centrifugal or screw compressors and low suction volumes dictate the use of reciprocating compressors. Suction volumes also may influence evaporator design, particularly at low temperatures, since they must include adequate space for gas–liquid separation.

Freezing point should be lower than minimum operating temperature. This generally is no problem unless the refrigerant is used as a brine.

Theoretical power required for adiabatic compression of the gas is slightly less with some refrigerants than others. However, this is usually a secondary consideration offset by the effects of particular equipment selections, for example, line-pressure drops, etc., on system power consumption.

Vapor density (or molecular weight) is an important characteristic when the compressor is centrifugal because the lighter gases require more impellers for a given pressure rise, that is, head, or temperature lift. On the other hand, centrifugal compressors have a limitation connected with the acoustic velocity in the gas, and this velocity decreases with the increasing molecular weight.

Low vapor densities are desirable to minimize pressure drop in long suction and discharge lines.

Liquid density should be taken into account. Liquid velocities are comparatively low so that pressure drop is usually no problem. However, static head may affect evaporator temperatures, and should be considered when liquid must be fed to elevated parts of the system.

Latent heat should be high because it reduces the quantity of refrigerant that needs to be circulated. However, large flow quantities are more easily controlled because they allow use of larger, less sensitive throttling devices and apertures.

Refrigerant cost depends on the size of the installation and must be considered both from the standpoint of initial charge and of composition owing to losses during service. Although a domestic refrigerator contains only a few dollars worth of refrigerant, the charge for a typical chemical plant may cost thousands of dollars.

Other desirable properties. Refrigerants should be stable and noncorrosive. For heat-transfer considerations, a refrigerant should have low viscosity, high thermal conductivity, and high specific heat. For safety to life or property, a refrigerant should be nontoxic and nonflammable, should not contaminate products in case of a leak, and should have a low leakage tendency through normal materials of construction.

76.3.3 Refrigerants

No one refrigerant meets all these requirements for the wide range of temperature and the multitude of applications found in modern refrigeration systems. Most systems use one of the nontoxic, nonflammable refrigerants (see Table 76.1) or ammonia. The chemical industry uses low cost fluids such as propane and butane whenever they are available in the process. These hydrocarbon refrigerants, often thought of as too hazardous because of flammability, are entirely suitable for use in modern compressors, and frequently add no more hazard than already exists in the oil refinery or petrochemical field. These low cost refrigerants are used in simple, compound, and cascade systems, depending on operating temperatures.

With a flammable refrigerant, extra precautions may have to be taken in the engineering design if it is required to meet the explosion-proof classification. It may be more economical to use a higher cost, but nonflammable, refrigerant.

A standard numbering system, shown in Table 76.1, has been devised to identify refrigerants without the use of the cumbersome chemical name. Numbers assigned to the hydrocarbons and halohydrocarbons of the methane, ethane, propane, and cyclobutane series are such that the number uniquely specifies the refrigerant compound. ASHRAE Standard 34-78 describes the method of coding.[3]

The American National Standards Institute (ANSI)[4] groups refrigerants in three classes, depending on toxicity and flammability. Group 1 refrigerants are nontoxic and nonflammable; Group 2 are slightly toxic or flammable; Group 3 are highly toxic or flammable. The most commonly used refrigerants are shown in Table 76.2.

The Group 1 refrigerants generally fulfill the basic requirements for an ideal refrigerant with considerable flexibility as to refrigeration capacity. They are ideal for comfort air conditioning since they are nontoxic and nonflammable. These refrigerants are used selectively in terms of compressor displacement and operating temperature levels desired.

Refrigerant 12, dichlorodifluoromethane, is at present the most widely known and used refrigerant. It is ideal for close-coupled or remote systems ranging from small reciprocating to large centrifugal units. It has been used for temperatures as low as $-90°C$ although $-85°C$ is a more practical lower limit because of the high gas volumes necessary for attaining these temperatures. It is suited for single-stage or compound cycles using reciprocating and centrifugal compressors.

Refrigerant 22, chlorodifluoromethane, is used in many of the same applications, as R-12, but its lower boiling point and higher latent heat permit the use of smaller compressors and refrigerant lines than R-12. The higher-pressure characteristics also extend its use to lower temperatures in the range of $-100°C$.

Refrigerant 11, trichlorofluoromethane, has a low-pressure–high-volume characteristic suitable for use in close-coupled centrifugal compressor systems for water or brine cooling. Its temperature range extends no lower than $\sim -7°C$.

Refrigerant 114, dichlorotetrafluoroethane, is similar to R-11 but its slightly higher pressure and lower volume characteristic than R-11 extend its use to $\sim -17°C$ and higher capacities.

Refrigerant 13, chlorotrifluoromethane, is used in low-temperature applications to $\sim -126°C$. Because of its low volume, high condensing pressure, or both, and because of its low critical pressure and temperature, R-13 is usually cascaded with other refrigerants at a discharge pressure corresponding to a condensing temperature in the range of -56 to $-23°C$.

Of the Group 2 refrigerants, only ammonia is still used to any extent. Although toxic and flammable

Table 76.2 Properties of Common Refrigerants[a]

Refrigerant number[a]	Chemical formula	Normal temp range, °C	System capacity	Cycle type	Compressor type	Relative cost	Recommended equipment design pressure, kPa[b]	
							High side	Low side
Group 1, Nontoxic and nonflammable								
R-11	CCl_3F	−5 to 60	low	simple	centrifugal	medium	210	210
R-114	$CClF_2CClF_2$ CCl_2FCF_3	−20 to 65	low	simple	centrifugal	medium	450	450
R-12	CCl_2F_2	−90 to 60	medium	simple and compound	centrifugal and reciprocating	medium	1650	1140
R-22	$CHClF_2$	−100 to 60	high	simple and compound	centrifugal and reciprocating	medium	2170	1140
R-13	$CClF_3$	−125 to −20	high	cascade	centrifugal and reciprocating	high	2170[c]	2170[c]
Group 2, Slightly toxic and flammable								
R-717	NH_3	−30 to 40	high		reciprocating	medium	2170	1140
Group 3, Highly toxic or flammable								
R-600	$CH_3CH_2CH_2CH_3$	2 to 60	low	simple	centrifugal	low	760	450
R-290	$CH_3CH_2CH_3$	−40 to 60	medium	simple	centrifugal and reciprocating	low	2170	1140
R-1270	$CH_3CH{=}CH_2$	−45 to 60	high	simple	centrifugal and reciprocating	low	2170	1140
R-170	CH_3CH_3	−90 to −5	medium	cascade	centrifugal and reciprocating	low	2170[c]	2170[c]
R-1150	$CH_2{=}CH_2$	−105 to −30	high	cascade	centrifugal and reciprocating	low	2170[c]	2170[c]
R-50	CH_4	−160 to −110	high	cascade	centrifugal and reciprocating	low	2170[c]	2170[c]

[a] For corresponding chemical names see Table 1.
[b] To convert kPa to psi, multiply by 0.145.
[c] Expansion tank to allow all liquid to vaporize is required when design pressure as shown is used.

within specific limits, ammonia is one of the best and most widely used refrigerants. Ammonia liquid has a high specific heat, an acceptable density and viscosity, and high conductivity. This makes it an ideal heat-transfer fluid with reasonable pumping costs, pressure drop, and flow rates. As a refrigerant, ammonia provides high heat transfer except when affected by oil at temperatures below ∼ −29°C, where oil films become viscous. To limit the ammonia-discharge-gas temperature to safe values, its normal maximum condensing temperature is 38°C. Generally, ammonia is used with reciprocating compressors, although relatively large centrifugal compressors (≥3.5 MW$_t$ or 1.2 × 10^6 Btu/hr) with 8–12 impeller stages required by its low molecular weight are in use today. Materials that are to contain ammonia should contain no copper (with the exception of Monel metal).

Refrigerants in Group 3 are generally applicable where a flammability of explosion hazard is already present and their use does not add to the hazard. These refrigerants have the advantage of low cost and, although they have fairly low molecular weight, they are suitable for centrifugal compressors of larger sizes. Because of the high acoustic velocity in Group 3 refrigerants, centrifugal compressors may be operated at high impeller tip speeds, which partly compensates for the higher head requirements than Group 1 refrigerants.

These refrigerants should be used at pressures greater than atmospheric to avoid increasing the explosion hazard by the admission of air in case of leaks. In designing the system, it also must be recognized that these refrigerants are likely to be impure in refrigerant applications. For example, commercial propane liquid may contain about 2 wt % ethane, which in the vapor phase might represent as much as 16–20 vol %. Thus, ethane may appear as a noncondensable. Either this gas must be purged or the compressor displacement must be increased ∼ 20% if it is recycled from the condenser; otherwise, the condensing pressure will be higher than required for pure propane and the power requirement will be increased.

Propane is the most commonly used Group 3 refrigerant. It is well suited for use with reciprocating and centrifugal compressors in close-coupled or remote systems. Its temperature range extends to ∼ −40°C (see Table 76.2).

Propylene is similar to propane but has slightly higher pressure characteristics.

Butane occasionally is used for close-coupled systems in the medium temperature range of 2°C. It has a low-pressure and high-volume characteristic suitable for centrifugal compressors where the capacity is too small for propane and the temperature is within range.

Ethane normally is used for close-coupled or remote systems at −87 to −7°C. It must be used in a cascade cycle because of the high-pressure characteristics.

Ethylene is similar to ethane but has a slightly higher-pressure, lower-volume characteristic which extends its use to −104 to −29°C. Like ethane it must be used in the cascade cycle.

Methane is used in an ultralow range of −160 to −110°C. It is limited to cascade cycles. Methane condensed by ethylene, which is in turn condensed by propane, is a cascade cycle commonly employed to liquefy natural gas.

Refrigerant Mixtures

Because the bubble point and dew point temperatures are not the same for a given pressure, nonazeotropic mixtures may be used to help control the temperature differences in low-temperature evaporators. This may be seen in the lowest stage of some LNG plants[5]; although undesirable mixtures or impurities such as are commonly found in hydrocarbon processing can have the opposite effect and raise the condensing pressure or lower the evaporating pressure if the impurities are not properly accounted for.

76.3.4 Open-Cycle Operation

In many chemical processes, the product to be cooled can itself be used as the refrigerating liquid. An important example of this is in the gathering plants for natural gas. Gas from the wells is cooled, usually after compression and after some of the heavier components are removed as liquid. This liquid may be expanded in a refrigeration cycle to further cool the compressed gas, which causes more of the heavier components to condense. Excess liquid not used for refrigeration is drawn off as product. In the transportation of liquefied petroleum gas and of ammonia in ships and barges, the LPG or ammonia is compressed, cooled, and expanded. The liquid portion after expansion is passed on as product until the ship is loaded.

Open-cycle is similar to closed-cycle operation, except that one or more parts of the closed cycle may be omitted. For example, the compressor suction may be taken directly from gas wells, rather than from an evaporator. A condenser may be used, and the liquefied gas may be drained to storage tanks.

Compressors may be installed in series or parallel for operating flexibility or for partial standby protection. With multiple reciprocating compressors, or with a centrifugal compressor, gas streams may be picked up or discharged at several pressures (see Fig. 76.6) if there is refrigerating duty to

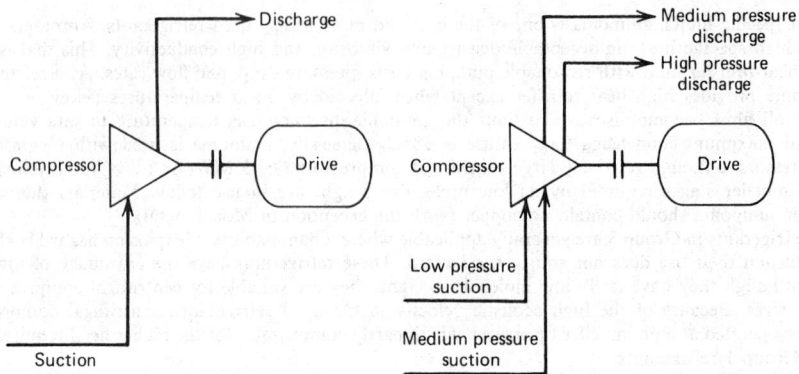

Fig. 76.6 Open-cycle compressors.[9]

be performed at intermediate temperatures. It always is more economical to refrigerate at the highest temperature possible.

Principal concerns in the open cycle involve dirt and contaminants, wet gas, compatibility of materials and lubrication circuits, and piping to and from the compressor. The possibility of gas condensing under various ambient temperatures either during operation or during standby must be considered. Beyond these considerations, the open-cycle design and its operation are governed primarily by the process requirements. The open system can use standard refrigeration hardware. It is not necessary to go to the expensive specialty field for the equipment.

76.3.5 Open-Cycle Refrigerant Selection

Process gases used in the open cycle include chlorine, ammonia, and mixed hydrocarbons. These create a wide variety of operating conditions and corrosion problems. Gas characteristics affect both heat exchangers and compressors, but their impact is far more critical on compressor operation. All gas properties and conditions should be clearly specified to obtain the most economical and reliable compressor design. If the installation is greatly overspecified, design features result that not only add significant cost but also complicate the operation of the system and are difficult to maintain. Specifications should consider the following:

Composition. Molecular weight, enthalpy–entropy relationship, compressibility factor, and operating pressures and temperatures influence the selection and performance of compressors. If process streams are subject to periodic or gradual changes in composition, the range of variations must be indicated.

Corrosion. Special materials of construction and types of shaft seals may be necessary for some gases. Gases that are not compatible with lubricating oils or that must remain oil-free may necessitate reciprocating compressors designed with carbon rings or otherwise made oilless, or the use of centrifugal compressors designed with isolation seals. However, these features are unnecessary on most installations. Standard designs usually can be used to provide savings in first cost, simpler operation, and reduced maintenance.

Dirt and liquid carryover. Generally, the carryover of dirt and liquids can be controlled more effectively by suction scrubbers than by costly compressor design features. Where this is not possible, all anticipated operating conditions should be stated clearly so that suitable materials and shaft seals can be provided.

Polymerization. Gases that tend to polymerize may require cooling to keep the gas temperature low throughout compression. This can be handled by liquid injection or by providing external cooling between stages of compression. Provision may be necessary for internal cleaning with steam.

These factors are typical of those encountered in open-cycle gas compression. Each job should be thoroughly reviewed to avoid unnecessary cost and obtain the simplest possible compressor design for ease of operation and maintenance. Direct coordination between the design engineer and manufacturer during final stages of system design is strongly recommended.

76.4 ABSORPTION SYSTEMS

If a steady supply of waste heat is available, an absorption machine can be chosen for continuous cooling duty. For water chilling service, absorption systems generally use water as the refrigerant and lithium bromide as the absorbent solution. For process applications requiring chilled fluid $< 7°C$, the ammonia–water pair is used with ammonia serving as the refrigerant.

A typical arrangement for a lithium bromide absorption system is shown schematically in Fig. 76.7. The absorbent, lithium bromide may be thought of as a carrier fluid bringing spent refrigerant from the low-pressure side of the cycle (the absorber) to the high-pressure side (the generator). There, the waste heat, steam, or hot water, which drives the system separates the water from the absorbent by a distillation process. The regenerated absorbent returns to the absorber where it is cooled so it will absorb the refrigerant (water) vapor produced in the evaporator and thereby establish the low-pressure level which controls the evaporator temperature. Thermal energy released during the absorption process is transferred to the cooling water flowing through tubes in the absorber shell.

The external heat exchanger shown saves energy by heating the strong liquid flowing to the generator as it cools the hot absorbent flowing from the generator to the absorber. If the weak solution that passes through the regenerator to the absorber does not contain enough refrigerant and is cooled too much, crystallization can occur. Leaks or process upsets that cause the generator to overconcentrate the solution are indicated when this occurs. The slushy mixture formed does not harm the machine, but it interferes with continued operation. External heat and added water may be required to redissolve the mixture.

The generator alone is sufficient for distillation when the absorbent material is nonvolatile (lithium bromide). In ammonia absorption systems, the absorbent (water) is volatile and tends to carry over into the evaporator where it interferes with vaporization. This problem is overcome by adding a rectifier to purify the ammonia vapor flowing from the generator to the condenser.

Single-stage absorption systems are most common with generator heat input < 95°C. The coefficient of performance (COP) of a system is the cooling achieved in the evaporator divided by the heat input to the generator. The COP of a lithium bromide machine generally is 0.65–0.70 for water chilling duty. The heat rejected by the cooling tower from both the condenser and the absorber is the sum of the waste heat supplied plus the cooling produced, requiring larger cooling towers and cooling water flows than for vapor compression systems.

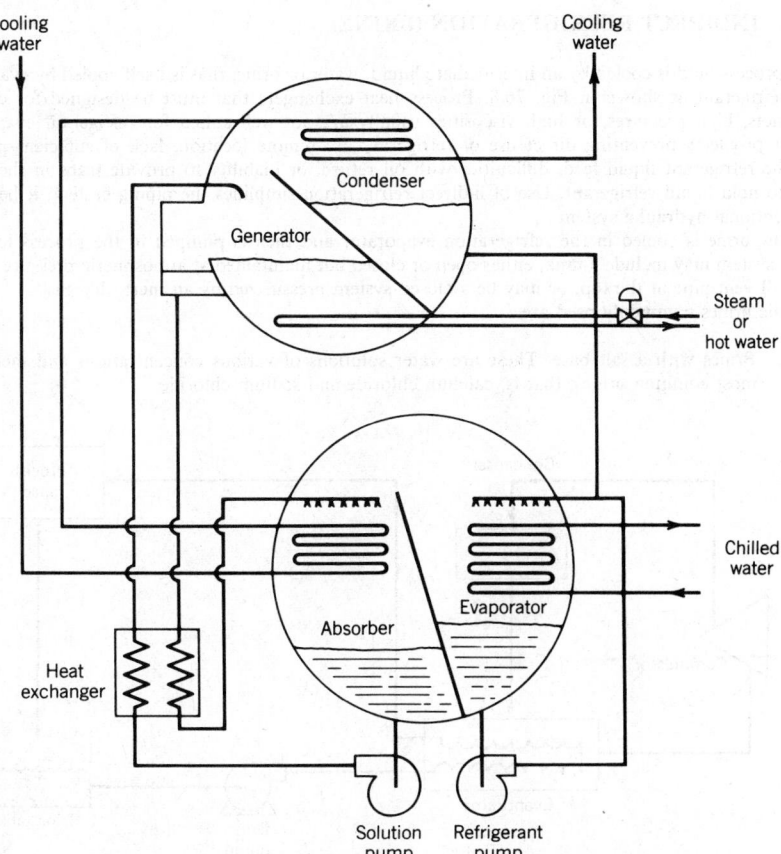

Fig. 76.7 Two-shell lithium bromide absorption system.

Absorption machines can be built with a two-stage generator for heat input temperatures $> 150°C$. Such machines are called dual-effect machines. Coefficients of performance above 1.0 can be obtained, but at the expense of increased design and operating complexity.

76.5 STEAM JET REFRIGERATION

Steam jet refrigeration represents yet another variation of the standard vapor compression cycle. Water is the refrigerant, so very large volumes of low-pressure (~ 1 kPa absolute) vapor must be compressed. A steam jet ejector offers a simple, inexpensive, but inefficient alternative to large centrifugal compressors required for systems of even moderate cooling capacity: 54 liter/sec of water vapor must be handled per kW of refrigeration at evaporator temperatures of 7°C.

The evaporator vessel should have a large surface area to enhance evaporative cooling. Sprays or cascades of water in sheets may be used. Because condenser pressure is subatmospheric (~ 7.6 kPa absolute), leakage of air into the system can cause poor condenser performance, so a small two-stage ejector is commonly used to remove the noncondensable vapors from the condenser. The condenser must condense not only the water vapor generated by the evaporator cooling load, but also the steam from the ejector primary flow nozzle. The condenser rejects two to three times the amount of heat that a mechanical vapor compression cycle would require.

Steam jet refrigeration systems are available in 35–3500-kW capacities. Steam jet refrigeration can be used in process applications where direct vaporization can be used for concentration or drying of foods and chemicals. The cooling produced by the vaporization reduces the processing temperature and helps to preserve the product. No heat exchanger or indirect heat-transfer process is required, making the steam jet system more suitable than mechanical refrigeration for these applications. Examples are concentration of fruit juices, freeze-drying of foods, dehydration of pharmaceuticals, and chilling of leafy vegetables. When applied to process or batch applications such as these, the noncondensables ejector for the condenser must be large enough to obtain the system evacuation rate desired. Detailed information on the performance of steam jet systems can be found in the *ASHRAE Handbook.* [6]

76.6 INDIRECT REFRIGERATION (BRINE)

The process fluid is cooled by an intermediate liquid, water or brine, that is itself cooled by evaporating the refrigerant as shown in Fig. 76.8. Process heat exchangers that must be designed for corrosive products, high pressures, or high viscosities usually are not well suited for refrigerant evaporators. Other problems preventing direct use of refrigerant are remote location, lack of sufficient pressures for the refrigerant liquid feed, difficulties with oil return, or inability to provide traps in the suction line to hold liquid refrigerant. Use of indirect refrigeration simplifies the piping system; it becomes a conventional hydraulic system.

The brine is cooled in the refrigeration evaporator and then is pumped to the process load. The brine system may include a tank, either open or closed but maintained at atmospheric pressure through a small vent pipe at the top, or may be a closed system pressurized by an inert, dry gas.

The brines commonly used are:

1. **Brines with a salt base.** These are water solutions of various concentrations and include the most common brines, that is, calcium chloride and sodium chloride.

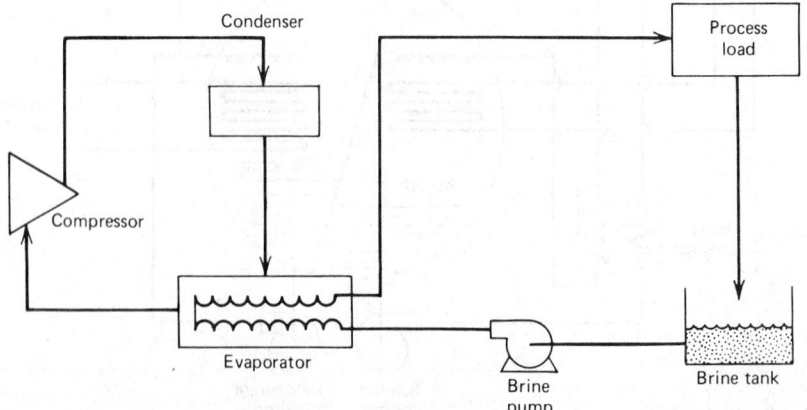

Fig. 76.8 Secondary brine system. [9]

2. Brines with a glycol base. These are water solutions of various concentrations, most commonly ethylene glycol or propylene glycol.

3. Brines with an alcohol base. Where low temperatures are not required, the alcohols are occasionally used in alcohol–water solutions.

4. Brines for low-temperature heat transfer. These usually are pure substances such as methylene chloride, trichloroethylene, R-11, acetone, methanol, and ethanol.

Table 76.3 shows the general areas of application for the commonly used brines. Criteria for selection are discussed in the following paragraphs. The order of importance depends on the specific application.

Costs of brines vary widely. Even considering the labor cost of mixing and the cost of water treating and inhibitors at temperatures where calcium or sodium chloride solutions can be used, these solutions are much less expensive than brines such as R-11.

Corrosion problems with sodium chloride and calcium chloride brines limit their use. Nevertheless, when properly maintained in a neutral condition and protected with inhibitors, they will give 20–30 years of service without corrosive destruction of a closed system. Glycol solutions and alcohol–water solutions are generally less corrosive than salt brines, but they require inhibitors to suit the specific application for maximum corrosion protection. Methylene chloride, trichloroethylene, and R-11 do not show general corrosive tendencies unless they become contaminated with impurities such as moisture. However, methylene chloride and trichloroethylene must not be used with aluminum or zinc; they also attack most rubber compounds and plastics. Alcohol in high concentrations will attack aluminum. Reaction with aluminum is of concern because, in the event of leakage into the refrigeration compressor system, aluminum compressor parts will be attacked.

Toxicity is an important consideration in connection with exposure to some products and to operating personnel. Where brine liquid, droplets, or vapor may contact food products, as in an open spray-type system, sodium chloride and propylene glycol solutions are acceptable because of low toxicity. All other brines are toxic to some extent or produce odors which require that they be used only inside of pipe coils or a similar pressure-tight barrier.

Flash-point and explosive-mixture properties of some brines require precautions against fire or explosion. Acetone, methanol, and ethanol are in this category but are less dangerous when used in closed systems.

Specific heat of a brine determines the mass rate of flow that must be pumped to handle the cooling load for a given temperature rise. The low-temperature brines, such as trichloroethylene, methylene chloride, and R-11 have specific heats approximately one-third to one-fourth those of the water-soluble brines. Consequently, a significantly greater mass of the low temperature brines must be pumped to achieve the same temperature change.

Stability at high temperatures is important where a brine may be heated as well as cooled. Above 60°C, methylene chloride may break down to form acid products. Trichloroethylene can reach 120°C before breakdown begins.

Viscosities of brines vary greatly. The viscosity of propylene glycol solutions, for example, makes them impractical for use below −7°C because of the high pumping costs and the low heat-transfer coefficient at the concentration required to prevent freezing. Mixtures of ethanol and water can become

Table 76.3 Common Brines[9]

Brines	Minimum practical temperature, °C	Toxic	Explosive	Corrosive	Cost
Salts					
calcium chloride	−40	no	no	yes	low
sodium chloride	−10	no	no	yes	low
Glycols					
propylene	−10	no	no	some	medium
ethylene	−15	yes	no	some	medium
Alcohols					
methanol	−35	yes	yes	some	high
ethanol	−30	yes	yes	some	high
Low temperature brines					
methylene chloride	−85	no	no	no	high
trichloroethylene	−75	no	no	no	high
trichlorofluoromethane	−100	no	no	no	high
acetone	−100	no	yes	no	high

highly viscous at temperatures near their freezing points, but 190-proof ethyl alcohol has a low viscosity at all temperatures down to near the freezing point. Similarly, methylene chloride and R-11 have low viscosities down to −73°C. In this region, the viscosity of acetone is even more favorable.

Since a brine cannot be used below its freezing point, certain brines are not applicable at the lower temperatures. Sodium chloride's eutectic freezing point of −20°C limits its use to ∼ −12°C, even though the viscosity is not unreasonable. The eutectic freezing point of calcium chloride brine is −53°C, but achieving this limit requires such an accuracy of mixture that −40°C is a practical low limit of usage.

Water solubility in any open or semiopen system can be important. The dilution of a salt or glycol brine, or of alcohol by entering moisture, merely necessitates strengthening of the brine. But for a brine that is not water-soluble, such as trichloroethylene or methylene chloride, precautions must be taken to prevent free water from freezing on the surfaces of the heat exchanger. This may require provision for dehydration or periodic mechanical removal of ice, perhaps accompanied by replacement with fresh brine.

Vapor pressure is an important consideration for brines that will be used in open systems, especially where the brine may be allowed to warm to room temperature between periods of operation. It may be necessary to pressurize such systems during periods of moderate temperature operation. For example, at 0°C the vapor pressure of R-11 is 39.9 kPa (299 mm Hg); that of a 22% solution of calcium chloride is only 0.49 kPa (3.7 mm Hg). The cost of vapor losses, the toxicity of the escaping vapors, and their flammability should be carefully considered in the design of the semiclosed or open system.

Energy requirements of brine systems may be greater because of the power required to circulate the brine and because of the extra heat-transfer process which necessitates the maintenance of a lower evaporator temperature. Figure 76.9 shows the approximate efficiency of the system using various brines.

Use of Ice

Where water is not harmful to a product or process, ice may be used to provide refrigeration. Direct application of ice or of ice and water is a rapid way to control a chemical reaction or remove heat from a process. The rapid melting of ice furnishes large amounts of refrigeration in a short time and allows leveling out of the refrigeration capacity required for batch processes. This stored refrigeration also is desirable in some processes where cooling is critical from the standpoint of safety or serious product spoilage.

Large ice plants such as the block-ice plants built during the 1930s are not being built today. However, ice still is used extensively, and equipment to make flake or cube ice at the point of use is commonly employed. This method avoids the loss of crushing and minimizes transportation costs.

76.7 SYSTEM COMPONENTS, CONSTRUCTION, AND OPERATION

76.7.1 Compressors

Reciprocating, screw, or centrifugal compressors can be used singly or in parallel and series combinations for process compression and refrigeration applications.

Modern high-speed reciprocating compressors with displacements up to 0.283–0.472 m³/sec (600–1000 cfm) generally are limited to a pressure ratio of about 9. The reciprocating compressor is basically a constant-volume variable-head machine. It handles various discharge pressures with relatively small changes in inlet-volume flow rate as shown by the heavy line in Fig. 76.10.

Open systems and many processes require nearly fixed compressor suction and discharge pressure levels. This load characteristic is represented by the horizontal typical open-system line in Fig. 76.10. In contrast, condenser operation in many closed systems is related to ambient conditions, for example, through cooling towers, so that on cooler days the condenser pressure can be reduced. When the refrigeration load is lower, less refrigerant circulation is required. The resulting load characteristic is represented by the typical closed-system line in Fig. 76.10.

The compressor must be capable of matching the pressure and flow requirements imposed upon it by the system in which it operates. The reciprocating compressor matches the imposed discharge pressure at any level up to its limiting pressure ratio. Varying flow requirements can be met by providing devices that unload individual or multiple cylinders. This unloading is accomplished by blocking the suction or discharge valves that open either manually or automatically. Speed control also can be used to effect changes in capacity.

Most reciprocating compressors have a lubricated design. Oil is pumped into the refrigeration system during operation. Thus, systems must be designed carefully to return oil to the compressor crankcase to provide for continuous lubrication and also to avoid contaminating heat-exchanger surfaces. At very low temperatures (∼ −50°C or lower, depending on refrigerants used) oil becomes too viscous to return, and provision must be made for periodic plant shutdown and warmup to allow manual transfer of the oil.

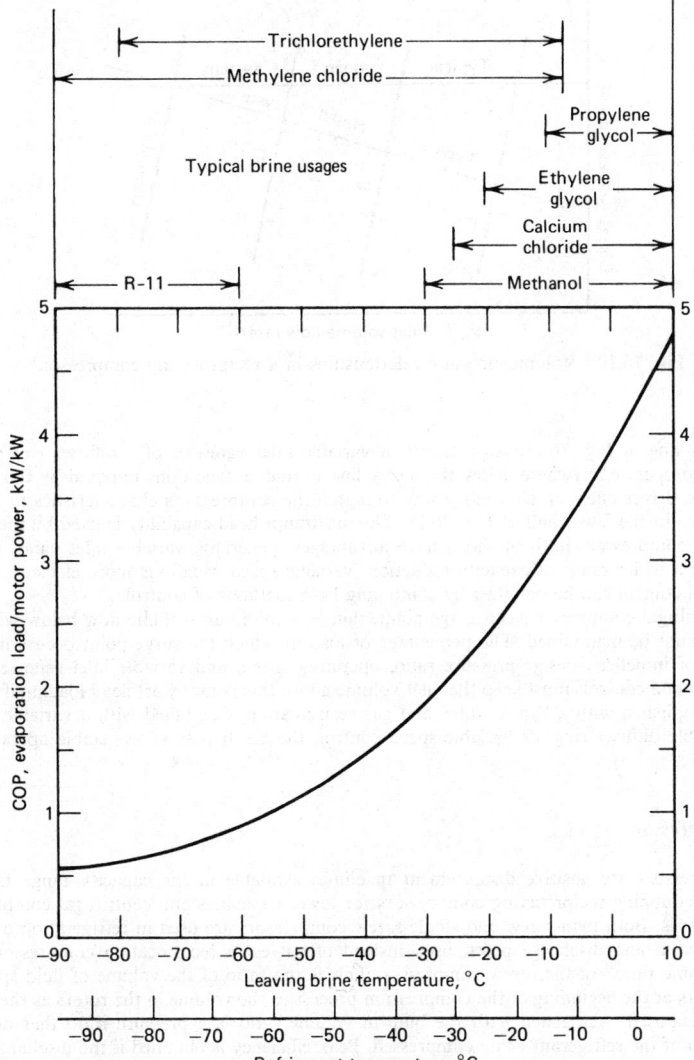

Fig. 76.9 Power requirements at various temperature levels. Centrifugal brine chilling systems based on 29.5°C condenser water. Includes gear losses.[9]

Compressors usually are arranged to start unloaded so that normal torque motors are adequate for starting. When gas engines are used for reciprocating compressor drives, careful torsional analysis is essential.

The centrifugal compressor is preferred whenever the gas volume is high enough to allow its use, because it offers better control, simpler hookup, minimal lubrication problems, and lower maintenance. Single-impeller designs are directly connected to high-speed drives or driven through an internal speed increaser. These machines are ideally suited for clean, noncorrosive gases in moderate-pressure process or refrigeration cycles in the range of 0.236–1.89 m³/sec (500–4000 cfm). Multistage centrifugal compressors are built for direct connection to high-speed drives or for use with an external speed increaser. Designs available from suppliers generally provide for two to eight impellers per casing covering the range of 0.236–11.8 m³/sec (500–25,000 cfm) depending on the operating speed. A wide choice of materials and shaft seals to suit any gas composition, including dirty or corrosive process streams, is available.

The centrifugal compressor has a more complex head–volume characteristic than reciprocating machines. Changing discharge pressure may cause relatively large changes in inlet volume, as shown

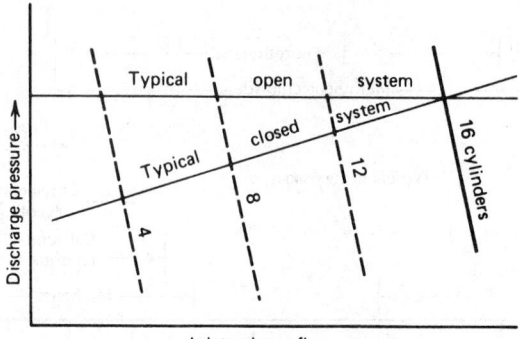

Fig. 76.10   Volume–pressure relationships in a reciprocating compressor.[9]

by the heavy line in Fig. 76.11. Adjustment of variable inlet vanes or of a diffuser ring allows the compressor to operate anywhere below the heavy line to match conditions imposed by the system. A variable-speed driver offers an alternative way to match the compressor's characteristics to the system load, as shown in the lower half of Fig. 76.11. The maximum head capability is fixed by the operating speed of the compressor. Both methods have advantages: generally, variable inlet vanes or diffuser rings provide a wider range of capacity reduction; variable speed usually is more efficient. Maximum efficiency and control can be obtained by combining both methods of control.

The centrifugal compressor has a surge point, that is, a minimum-volume flow below which stable operation cannot be maintained. The percentage of load at which the surge point occurs depends on the number of impellers, design-pressure ratio, operating speed, and variable inlet-vane setting. The system design and controls must keep the inlet volume above this point by artificial loading if necessary. This is accomplished with a bypass valve and gas recirculation. Combined with a variable inlet-vane setting, variable diffuser ring, or variable speed control, the gas bypass allows stable operation down to zero load.

Screw Compressors

Screw compressors are positive displacement machines available in the capacity range from 15 to 1100 kW, overlapping reciprocating compressors for lower capacities and centrifugal compressors for higher capacities. Both twin-screw and single-screw compressors are used in refrigeration duty.

Fixed suction and discharge ports, used instead of valves in reciprocating compressors, set the "built-in volume ratio" of the screw compressor. This is the ratio of the volume of fluid space in the meshing rotors at the beginning of the compression process to the volume in the rotors as the discharge port is first exposed. Associated with the built-in volume ratio is a pressure ratio that depends on the properties of the refrigerant being compressed. Peak efficiency is obtained if the discharge pressure imposed by the system matches the pressure developed by the rotors when the discharge port is exposed. If the interlobe pressure is greater or less than discharge pressure, energy losses occur but no harm is done to the compressor.

Capacity modulation is accomplished by slide valves that are used to provide a variable suction bypass or delayed suction port closing, reducing the volume of refrigerant actually compressed. Continuously variable capacity control is most common, but stepped capacity control is offered in some manufacturers' machines. Variable discharge porting is available on a few machines to allow control of the built-in volume ratio during operation.

Oil is used in screw compressors to seal the extensive clearance spaces between the rotors, to cool the machines, to provide lubrication, and to serve as hydraulic fluid for the capacity controls. An oil separator is required for the compressor discharge flow to remove the oil from the high-pressure refrigerant so that performance of system heat exchangers will not be penalized and the oil can be returned for reinjection in the compressor.

Screw compressors can be direct driven at two-pole motor speeds (50 or 60 Hz). Their rotary motion makes these machines smooth running and quiet. Reliability is high when the machines are applied properly. Screw compressors are compact so they can be changed out readily for replacement or maintenance. The efficiency of the best screw compressors matches that of reciprocating compressors at full load today. Part load performance characteristics are not as competitive, but design improvements in screw compressors are narrowing the differences.

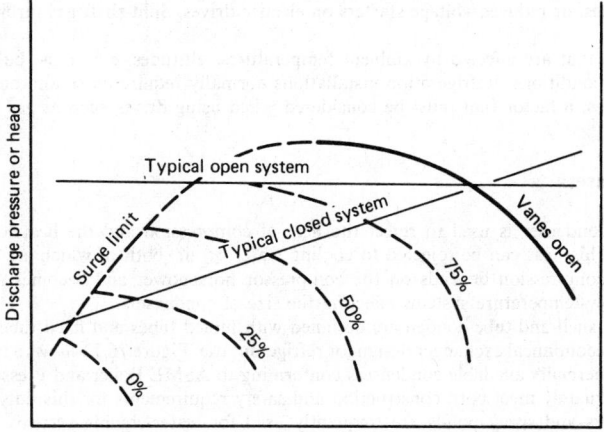

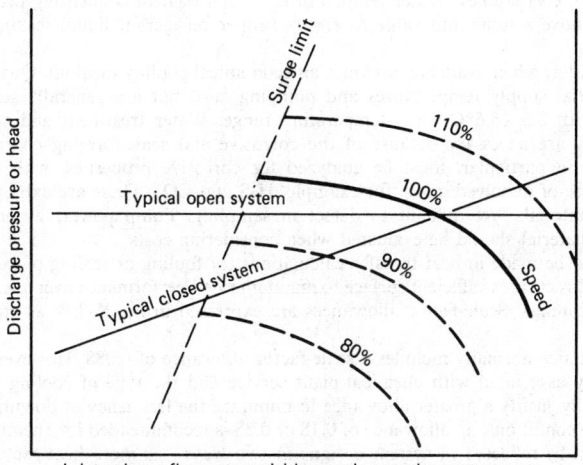

Fig. 76.11 Volume–pressure relationships in a centrifugal compressor.[9]

76.8 SYSTEM OPERATION

Provision for minimum load operation is strongly recommended for all installations because there will be fluctuations in plant load. For chemical plants this permits the refrigeration system to be started up and thoroughly checked out independently of the chemical process.

Contrast between the operating characteristics of the positive displacement compressor and the centrifugal compressor are important considerations in plant design to achieve satisfactory performance. Unlike positive displacement compressors, the centrifugal compressor will not rebalance abnormally high system heads. The drive arrangement for the centrifugal compressor must be selected with sufficient speed to meet the maximum head anticipated. The relatively flat head characteristics of the centrifugal compressor necessitates different control approaches than for positive displacement machines, particularly when parallel compressors are utilized. These differences, which account for most of the troubles experienced in centrifugal-compressor systems, cannot be overlooked in the design of a refrigeration system.

A system that uses centrifugal compressors designed for high pressure ratios and that requires the compressors to start with high suction density existing during standby will have high starting torque. If the driver does not have sufficient starting torque, the system must have provisions to reduce the suction pressure at startup. This problem is particularly important when using single-shaft gas turbine engines, or reduced-voltage starters on electric drives. Split-shaft gas turbines are preferred for this reason.

Drive ratings that are affected by ambient temperatures, altitudes, etc., must be evaluated at the actual operating conditions. Refrigeration installations normally require maximum output at high ambient temperatures, a factor that must be considered when using drives such as gas turbines and gas engines.

76.8.1   Condensers

The refrigerant condenser is used to reject the heat of compression and the heat load picked up in the evaporator. This heat can be rejected to cooling water or air, both of which are commonly used.

The heat of compression depends on the compressor horsepower and becomes a significant part of the load on low-temperature systems affecting the size of condensers.

Water-cooled shell-and-tube condensers designed with finned tubes and fixed tube sheets generally provide the most economical exchanger design for refrigerant use. Figure 76.12 shows a typical refrigerant condenser. Commercially available condensers conforming to ASME Boiler and Pressure Vessel Code[7] construction adequately meet both construction and safety requirements for this duty.

Cooling towers and spray ponds are frequently used for water-cooling systems. These generally are sized to provide 29°C supply water at design load conditions. Circulation rates typically are specified so that design cooling loads are handled with a 5.6°C cooling-water temperature rise. Pump power, tower fans, makeup water (about 3% of the flow rate), and water treatment should be taken into account in operating cost studies. Water temperatures, which control condensing pressure, may have to be maintained above a minimum value to ensure proper refrigerant liquid feeding to all parts of the system.

River or well water, when available, provides an economical cooling medium. Quantities circulated will depend on initial supply temperatures and pumping cost, but are generally selected to handle the cooling load with 8.3–16.6°C water-temperature range. Water treatment and special exchanger materials frequently are necessary because of the corrosive and scale-forming characteristics of the water. Well water, in particular, must be analyzed for corrosive properties, with special attention given to the presence of dissolved gases, for example, H_2S and CO_2. These are extremely corrosive to many exchanger materials, yet difficult to detect in sampling. Pump power, water treatment, and special condenser material should be evaluated when considering costs.

Allowances must be made in heat-transfer calculations for fouling or scaling of exchanger surfaces during operation. This ensures sufficient surface to maintain rated performance over a reasonable interval of time between cleanings. Scale-factor allowances are expressed in $m^2 \cdot K/kW$ as additional thermal resistance.

Commercial practice normally includes a scale-factor allowance of 0.088. However, the long hours of operation usually associated with chemical-plant service and the type of cooling water frequently encountered generally justify a greater allowance to minimize the frequency of downtime for cleaning. Depending on these conditions, an allowance of 0.18 or 0.35 is recommended for chemical-plant service. Scale allowance can be reflected in system designs in two ways—as more heat-exchanger surface or as higher design condensing temperatures with attendant increase in compressor power. This is illustrated in Fig. 76.13. Generally, a compromise between these two approaches is most economical. For extremely bad water, parallel condensers, each with 60–100% capacity, may provide a more economical selection and permit cleaning one exchanger while the system is operating.

Use of air-cooled condensing equipment is on the increase. With tighter restrictions on the use of water, air-cooled equipment is used even on larger centrifugal-type refrigeration plants, although it

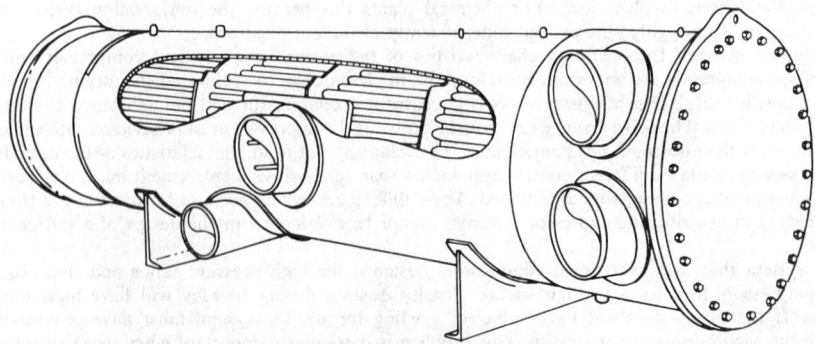

Fig. 76.12   Typical refrigerant condenser.[9]

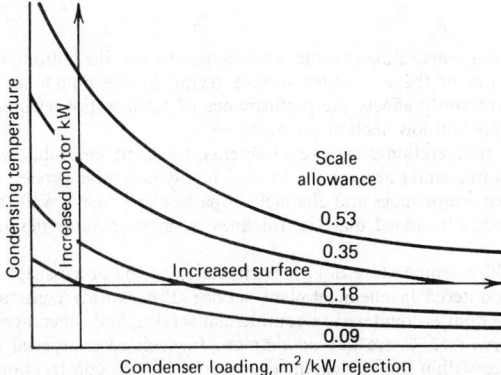

Fig. 76.13 Effect of different scale factors.[9]

requires more physical space than cooling towers. A battery of air-cooled condensers, with propeller fans located at the top, pull air over the condensing coil. Circulating fans and exchanger surface are usually selected to provide design condensing temperatures of 49–60°C with 35–38°C ambient dry bulb temperature.

The design dry bulb temperature should be carefully considered since most weather data reflect an average or mean maximum temperature. If full load operation must be maintained at all times, care should be taken to provide sufficient condenser capacity for the maximum recorded temperature. This is particularly important when the compressor is centrifugal because of its flat-head characteristics and the need for adequate speed. Multiple-circuit or parallel air-cooled condensers must be provided with traps to prevent liquid backup into the idle circuit at light load. Pressure drop through the condenser coil must also be considered in establishing the compressor discharge pressure.

In comparing water-cooled and air-cooled condensers, the compression horsepower at design conditions is invariably higher with air-cooled condensing. However, ambient air temperatures are considerably below the design temperature most of the time, and operating costs frequently compare favorably over a full year. In addition, air-cooled condensers usually require less maintenance, although dirty or dusty atmospheres may affect this.

76.8.2 Evaporators

There are special requirements for evaporators in refrigeration service that are not always present in other types of heat-exchanger design. These include problems of oil return, flash-gas distribution, gas–liquid separation, and submergence effects.

Oil Return

When the evaporator is used with reciprocating-compression equipment, it is necessary to ensure adequate oil return from the evaporator. If oil will not return in the refrigerant flow, it is necessary to provide an oil reservoir for the compression equipment and to remove oil mechanically from the low side of the system on a regular basis. Evaporators used with centrifugal compressors do not normally require oil return from the evaporator since centrifugal compressors pump very little oil into the system. However, even with centrifugal equipment, low-temperature evaporators eventually may become contaminated with oil which must be reclaimed.

Flash-Gas Distribution

As a general rule, refrigerants are introduced into the evaporator by expanding liquid from a higher pressure. In the expansion process, a significant amount of refrigerant flashes off into gas. This flash gas must be introduced properly into the evaporator for satisfactory performance. Improper distribution of this gas can result in liquid carryover to the compressor and in damage to the exchanger tubes from erosion or from vibrations.

Gas–Liquid Separation

The suction gas leaving the evaporator must be dry to avoid compressor damage. The design should provide adequate separation space or include mist eliminators. Liquid carryover is one of the most common sources of trouble with refrigeration systems.

Submergence Effect

In flooded evaporators the evaporating pressure and temperature at the bottom of the exchanger surface is higher than at the top of the exchanger surface owing to the liquid head. This static head or submergence effect significantly affects the performance of refrigeration evaporators operating at extremely low temperatures and low suction pressures.

Beyond these basic refrigeration-design requirements, the chemical industry imposes many special conditions. Exchangers frequently are applied to cool highly corrosive process streams; consequently, special materials for evaporator tubes and channels of particularly heavy wall thicknesses are dictated. Corrosion allowances, that is, added material thicknesses, in evaporator design may be necessary in chemical service.

High-pressure and high-temperature design, particularly on the process side of refrigerant evaporators, is frequently encountered in chemical-plant service. Process-side construction may have to be suitable for pressures seldom encountered in commercial service, and differences between process inlet and leaving temperatures $\geq 55°C$ are not uncommon. In such cases, special consideration must be given to thermal stresses within the refrigerant evaporator. U-tube construction or floating-tube-sheet construction may be necessary. Minor process-side modifications may permit use of less expensive standard commercial fixed-tube-sheet designs. However, coordination between the equipment supplier and chemical-plant designer is necessary to tailor the evaporator to the intended duty. Relief devices and safety precautions common to the refrigeration field normally meet chemical-plant needs but should be reviewed against individual plant standards. Here again, it must be the mutual responsibility of the refrigeration equipment supplier and the chemical-plant designer to evaluate what special features, if any, must be applied to modify commercial equipment for chemical-plant service.

Refrigeration evaporators are usually designed to meet the ASME Boiler and Pressure Vessel Code,[7] which provides for a safe reliable exchanger at economical cost. In refrigeration systems, these exchangers generally operate with relatively small temperature differentials for which fixed-tube-sheet construction is preferred. Refrigerant evaporators also operate with simultaneous reduction in pressure as temperatures are reduced. This relationship results in extremely high factors of safety on pressure stresses, eliminating the need for expensive nickel steels from -59 to $-29°C$. Most designs are readily modified to provide suitable materials for corrosion problems on the process side.

The basic shell-and-tube exchanger with fixed tube sheets (Fig. 76.14) is most widely used for refrigeration evaporators. Most designs are suitable for fluids up to ~ 2170 kPa (300 psig) and for operation with up to $38°C$ temperature differences. Above these limits, specialized heat exchangers generally are used to suit individual requirements.

With the fluid on the tube side, the shell side is flooded with refrigerant for efficient wetting of the tubes (see Fig. 76.15). Designs must provide for distribution of flash gas and liquid refrigerant entering the shell and for separation of liquid from the gas leaving the shell before it reaches the compressor.

In low-temperature applications and large evaporators, the exchanger surface may be sprayed rather than flooded. This eliminates the submergence effect or static-head penalty, which can be significant in large exchangers, particularly at low temperatures. The spray cooler (Fig. 76.16) is recommended for some large coolers to offset the cost of refrigerant inventory or charge which would be necessary for flooding.

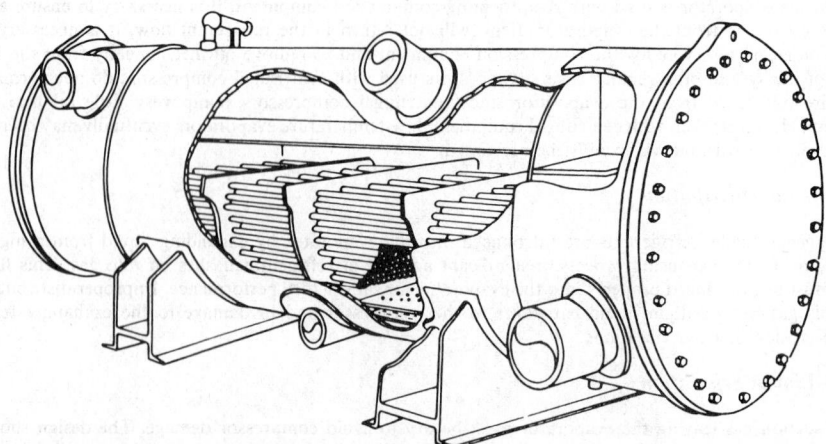

Fig. 76.14 Typical fixed-tube-sheet evaporator.[9]

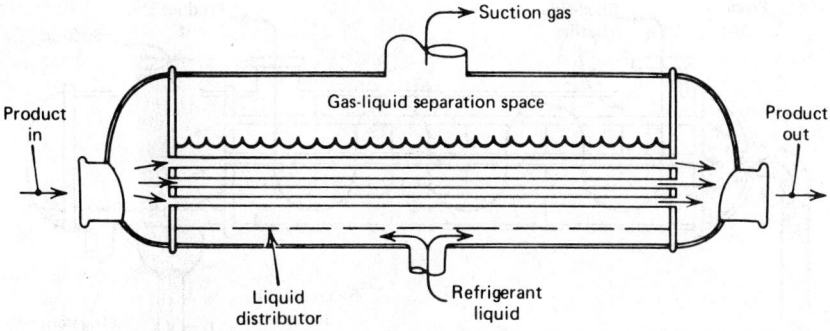

Fig. 76.15 Typical flooded shell-and-tube evaporator.[9]

Where the Reynolds number in the process fluid is low, as for a viscous or heavy brine, it may be desirable to handle the fluid on the shell side to obtain better heat transfer. In these cases, the refrigerant must be evaporated in the tubes. On small exchangers, commonly referred to as direct-expansion coolers, refrigerant feeding is generally handled with a thermal-expansion valve.

On large exchangers, this can best be handled by a small circulating pump to ensure adequate wetting of all tubes (see Fig. 76.17). An oversize channel box on one end provides space for a liquid reservoir and for effective liquid–gas separation.

76.9 REFRIGERATED WAREHOUSE DESIGN

The location of private refrigerated space within a plant is dictated by other operations of the owner. The selected location must fit in with long-range planning of production operations, traffic, and general storage and plant expansion. Public space should be located to meet a definite need to serve a producing area, a transit storage point, a large consuming area, or various combinations which will develop a good average occupancy.

Many factors in addition to location must be considered when designing refrigerated warehouses, such as overall size, stacking arrangement, single or multiple story configuration, and utility spaces such as the general offices, locker rooms, warmup rooms, and checkers' office. Insulation and vapor barrier requirements along with fire protection systems must also be considered. A more detailed discussion of these factors may be found in the *ASHRAE Handbook.*[8]

Because of material-handling costs and certain long-term requirements for product storage, special warehouse designs such as bulk bin storage with pneumatic conveying systems, drive-in and/or drive-through rack systems as well as automated warehouses have been developed. Controlled-atmosphere storage rooms may be required for specialized storage, such as those handling apples. These storages include special gas-tight seals to facilitate the maintenance of an atmosphere that is lower in oxygen and higher in nitrogen and carbon dioxide than normal atmosphere. Optimum atmospheres for most products have not been determined; therefore, the desired atmosphere must be experimentally determined for the commodity produced in the local area.

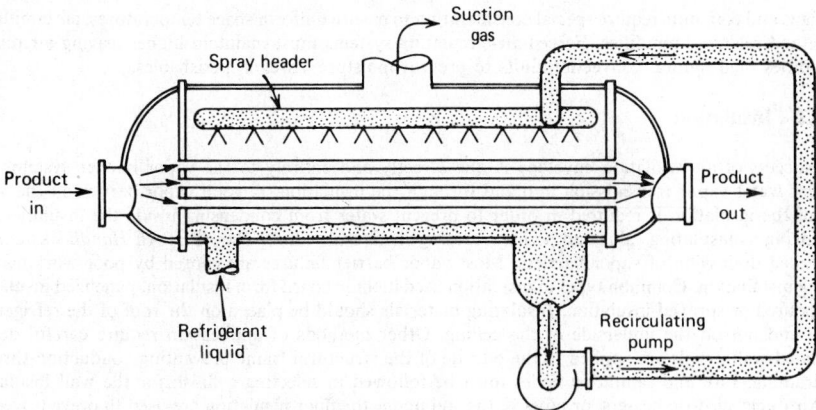

Fig. 76.16 Typical spray-type evaporator.[9]

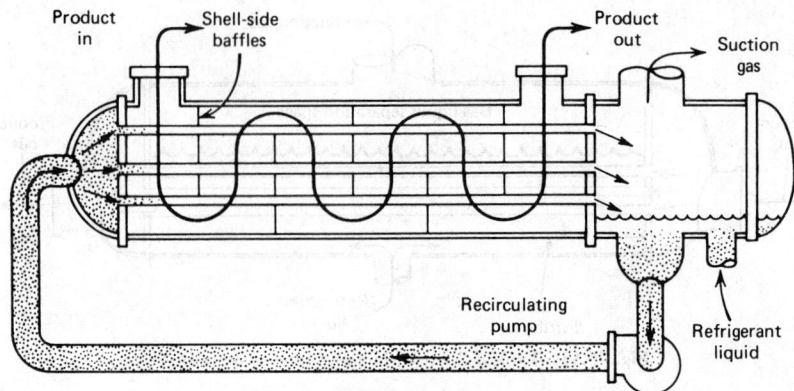

Fig. 76.17 Typical baffled-shell evaporator.[9]

76.9.1 Refrigeration System Selection

The selection of the most practical and economic system must be established in the early stages of planning. For single-purpose, low-temperature storage, almost any type of system can be applied. However, if stored commodities require different temperatures and humidities, the refrigeration system will have to accommodate several isolated rooms and conditions. Factory packaged equipment may be used for small structures or several unrelated storage rooms. The central compressor room is still accepted as the standard for larger installations. Screw and reciprocating compressors are used. Direct refrigeration systems, rather than brine systems, are used almost exclusively in modern designs. Ammonia is still used extensively, although halocarbon refrigerants may be used.

Refrigeration loads of warehouses of the same capacity will vary widely, therefore, no rule of thumb can be applied and a detailed analysis should always be made. Cooling equipment should be designed for maximum daily requirements, which will be well above any monthly average. Factors that should be considered include:

1. Heat transmission.
2. Infiltration.
3. Pumps, fans, equipment, lights, and people working.
4. Heat removed from product in reducing it to storage temperature.
5. Heat produced by goods in storage.
6. Heat removed if product must be frozen.
7. Other loads, that is, office air conditioning, car precooling, or special operations.
8. Refrigerated shipping docks.
9. Heat from automatic defrost units.

Fans and coil units require special consideration to ensure uniform space temperatures, air circulation rates, and relative humidities. Forced-air-circulating systems must maintain higher leaving air relative humidities than natural convection units to prevent moisture losses in perishables.

76.9.2 Insulation

The success of an insulation envelope is due directly and entirely to the vapor barrier systems that prevent water vapor transmission into and through the insulation. A good vapor barrier on the warm side of the insulation is required in order to prevent water from condensing inside the insulation and destroying its insulating value and perhaps causing structural failure. The *ASHRAE Handbooks* contain a detailed discussion of vapor barriers. Most vapor barrier failures are caused by poor workmanship during installation. Common types of insulation used include board form insulation, panelized insulation, and poured or sprayed insulation. Insulating materials should be placed on the roof of the refrigerated space and not on the underside of the ceiling. Other methods of application require careful design. Insulated wall panels are applied to the outside of the structural frame preventing conduction through the framing. Fire and sanitation codes must be followed in selecting a finish for the wall insulation.

Air ducts, electric heaters, or pipes in the soil under the floor insulation are used to prevent freezing and heavage of the floors and columns. A pipe grid that circulates a nonfreeze liquid has usually been found to be the most practical. With this method, heat rejected from the refrigeration system can be used to warm the fluid.

The trend in insulated doors for refrigerated plants is to have fewer doors. Newer door designs are generally efficient and highly refined for the application. There are four basic types: swinging, horizontal sliding, vertical sliding, and double acting. Proper door selection is mandatory to provide maximum traffic capacity with minimum loss of refrigeration and minimum maintenance. Automatic doors are often selected.

76.9.3 Commodity Storage

Precooling of products prior to refrigerated storage is recommended because rapid removal of field heat will substantially increase the storage life. Fresh fruits and vegetables in storage are alive and thus heat of respiration must always be considered as part of the refrigeration load. Table 76.4 gives the approximate rate of heat of evolution for various commodities. Storage requirements for specific commodities are given in Table 76.5. Requirements are so varied that generalizations to other products cannot be made, however. Further information can be obtained from the *ASHRAE Handbook*. [8]

76.10 ICE RINKS

The freezing of an ice sheet for hockey, curling, figure skating, speed skating, or recreational skating is usually accomplished by the circulation of a heat-transfer fluid through a network of pipes or tubes located below the surface of the ice. The heat-transfer fluid is predominantly a brine such as glycol, methanol, or calcium chloride. The amount of refrigeration capacity required is difficult to calculate on a purely theoretical basis. In addition to conduction and convection, the heat load factors considered include type of service, length of season, usage, type of enclosure, radiant load from roof and lights, and the geographic location with associated wet and dry bulb temperatures. For outdoor rinks, the sun must also be considered. From instrumented rinks, the major loads were found to be radiant (30–35%), system pump work (12–15%), ambient temperature (15%), humidity (15%), and ice resurfacing (9–12%).

The open sand fill floor is the least expensive and may be used where there is no other use for the space. Pipe corrosion due to the wet sand can be a problem. Most indoor rinks have a permanent, general-purpose concrete floor so that the structure can be used for other purposes. Subfloor insulation should be installed when quick changeovers are desired, when there is high subsoil moisture, when the floor is elevated, or when the rink is used continuously for more than nine months. Subfloor heating by either electrical cables or a pipe recirculating system using warm brine should be installed to prevent the danger of heaving. Waste heat from the condenser can be used for this purpose as well as to melt the snow and ice in the snow pit that is scraped off during resurfacing.

During mild weather, condensation may drip from the roofs and roof supports and fog may form because there is insufficient internal heat load. To prevent this, moisture may have to be removed from the air by refrigeration, adding to the total load of the plant.

76.11 SYSTEM DESIGN CONSIDERATIONS

Associated with continuous operation are refrigeration startup and shutdown conditions that invariably differ, sometimes widely, from those of the process itself. These conditions, although they occupy very little time in the life of the installation, must be properly accommodated in the design of the refrigeration system. Consideration must be given to the amount of time required to achieve design operating conditions, the need for standby equipment, etc.

In batch processing, operating conditions are expected to change with time, usually in a repetitive pattern. The refrigeration system must be designed for all extremes. Use of brine storage or ice banks can reduce equipment sizes for batch processes.

Closed-cycle operation involves both liquid and gas phases. System designs must take into account liquid-flow problems in addition to gas-flow requirements and must provide for effective separation of the liquid and gas phases in different parts of the system. These factors require careful design of all components and influence the arrangement or elevation of certain components in the cycle.

Liquid pressures must be high enough to feed liquid to the evaporators at all times, especially when evaporators are elevated or remotely located. In some cases, a pump must be used to suit the process requirements. The possibility of operation with reduced pressures caused by colder condensing temperatures than the specified design conditions must also be considered. Depending on the types of liquid valves and relative elevation of various parts of the system, it may be necessary to maintain condensing pressures above some minimum level, even if doing so increases the compression power.

Provision must be made to handle any refrigerant liquid that can drain to low spots in the system upon loss of operating pressure during shutdown. It must not be allowed to return as liquid to the compressor upon startup.

Table 76.4 Approximate Heat Evolution Rates of Fresh Fruits and Vegetables When Stored at Temperatures Shown[a]

Commodity	Watts per Megagram (W/Mg)			
	0°C	5°C	10°C	15°C
Apples	10–12	15–21	41–61	41–92
Asparagus	81–237	161–403	269–902	471–970
Avocados		59–89		183–464
Bananas, ripening			65–116	87–164
Beans, green or snap		101–103	161–172	251–276
Beets, red (roots)	16–21	27–28	35–40	50–69
Blueberries	7–31	27–36	69–104	101–183
Broccoli, sprouting	55–63	102–474		514–1000
Brussels sprouts	46–71	95–143	186–250	282–316
Cabbage	12–40	28–63	36–86	66–169
Carrots, topped	46	58	93	117
Cauliflower	53–71	61–81	100–144	136–242
Celery	21	32	58–81	110
Cherries, sweet	12–16	28–42		74–133
Corn, sweet	125	230	331	482
Cucumbers			68–86	71–98
Garlic	9–32	17–29	27–29	32–81
Gooseberries	20–26	36–40		64–95
Grapefruit			20–27	35–38
Grapes, American	8	16	23	47
Lemons	9	15	33	47
Lettuce, head	27–50	39–59	64–118	114–121
Mushrooms	83–129	210	297	
Onions	7–9	10–20	21	33
Oranges	9	14–19	35–40	38–67
Peaches	11–19	19–27	46	98–125
Pears	8–20	15–46	23–63	45–159
Peas, green (in pod)	90–138	163–226		529–599
Peppers, sweet			43	68
Plums, Wickson	6–9	12–27	27–34	35–37
Potatoes, mature		17–20	20–30	20–35
Radishes, topped	16–17	23–24	45–47	82–97
Raspberries	52–74	92–114	82–164	243–300
Squash, yellow	35–38	42–55	103–108	222–269
Strawberries	36–52	48–98	145–280	210–273
Tomatoes		21	45	61
Turnips, roots	26	28–30		63–71

[a] Abridged with permission, ASHRAE, Inc., Atlanta, GA.

The operating charge in various system components fluctuates depending on the load. For example, the operating charge in an air-cooled condenser is quite high at full load but is low, that is, essentially dry, at light load. A storage volume such as a liquid receiver must be provided at some point in the system to accommodate this variation. If the liquid controls permit the evaporator to act as the variable storage, the level may become too high, resulting in liquid carryover to the compressor.

Abnormally high process temperatures may occur either during startup or process upsets. Provision must be made for this possibility, for it can cause damaging thermal stresses on refrigeration components and excessive boiling rates in evaporators, forcing liquid to carry over and damage the compressor.

Factory-designed and built packages, which provide cooling as a service or utility, can require several thousand kilowatts of power to operate, but in most cases, they require no more installation

Table 76.5 Storage Requirements and Properties of Perishable Products[f]

Commodity	Storage Temperature (°C)	Relative Humidity (%)	Approximate Storage Life[e]	Water Content (%)	Highest Freezing (°C)	Specific Heat Above Freezing[c] (J/kg · °C)	Specific Heat Below Freezing[c] (J/kg · °C)	Latent Heat[d] (J/kg)
Vegetables[a]								
Asparagus	0–2	95	2–3 weeks	93	−0.6	3.952	2.005	310.20
Beans, snap or green	4–7	90–95	7–10 days	89	−0.7	3.818	1.955	296.86
Beets, roots	0	95–100	4–6 months	88	−0.9	3.785	1.943	293.52
Broccoli	0	95	10–14 days	90	−0.6	3.852	1.968	300.20
Brussels sprouts	0	95	3–5 weeks	85	−0.8	3.684	1.905	283.52
Cabbage, late	0	95–100	5–6 months	92	−0.9	3.919	1.993	306.87
Carrots, topped—mature	0	98–100	5–9 months	88	−1.4	3.785	1.943	293.52
Cauliflower	0	95	2–4 weeks	92	−0.8	3.919	1.993	306.87
Celery	0	95	1–2 months	94	−0.5	3.986	2.018	313.54
Corn, sweet	0	95	4–8 days	74	−0.6	3.316	1.767	246.83
Cucumbers	10–13	90–95	10–14 days	96	−0.5	4.053	2.043	320.21
Lettuce, head	0–1	95–100	2–3 weeks	95	−0.2	4.019	2.031	316.87
Mushrooms	0	90	3–4 days	91	−0.9	3.885	1.980	303.53
Dry and onion sets	0	65–75	1–8 months	88	−0.8	3.785	1.943	293.52
Peas, green	0	95	1–3 weeks	74	−0.6	3.316	1.767	246.83
Sweet peppers	7–10	90–95	2–3 weeks	92	−0.7	3.919	1.993	306.87
Main crop potatoes	3–10	90–95	5–8 months	78	−0.7	3.450	1.817	260.17
Sweet potatoes	13–16	85–90	4–7 months	69	−1.3	3.148	1.704	230.15
Winter radishes	0	95–100	2–4 months	95	−0.7	4.019	2.031	316.87
Summer squash	0–10	85–95	5–14 days	94	−0.5	3.986	2.018	313.54
Tomatoes								
Mature green	13–21	85–90	1–3 weeks	93	−0.6	3.952	2.005	310.20
Firm, ripe	7–10	85–90	4–7 days	94	−0.5	3.986	2.018	313.54
Turnips, roots	0	95	4–5 months	92	−1.1	3.919	1.993	306.87

Table 76.5 (Continued)

Commodity	Storage Temperature (°C)	Relative Humidity (%)	Approximate Storage Life[e]	Water Content (%)	Highest Freezing (°C)	Specific Heat Above Freezing[c] (J/kg·°C)	Specific Heat Below Freezing[c] (J/kg·°C)	Latent Heat[d] (J/kg)
Fruits and Melons[a]								
Apples	−1–4	90	3–8 months	84	−1.1	3.651	1.892	280.18
Avocados	4–13	85–90	2–4 weeks	65	−0.3	3.014	1.654	216.81
Bananas	a	85–95	a	75	−0.8	3.349	1.779	250.16
Blueberries	−1–0	90–95	2 weeks	82	−1.6	3.584	1.867	273.51
Sweet cherries	−1	90–95	2–3 weeks	80	−1.8	3.517	1.842	266.84
Cranberries	2–4	90–95	2–4 months	87	−0.9	3.751	1.930	290.19
Dates, cured	−18 or 0	75 or less	6–12 months	20	−15.7	1.507	1.089	66.71
Grapefruit	10–16	85–90	4–6 weeks	89	−1.1	3.818	1.955	296.86
Grapes, American	−1–0	85–90	2–8 weeks	82	−1.6	3.584	1.867	273.51
Lemons	0 or 10	85–90	1–6 months	89	−1.4	3.818	1.955	296.86
Oranges	0–9	85–90	3–12 weeks	87	−0.8	3.751	1.930	290.19
Peaches	−0.5–0	90	2–4 weeks	89	−0.9	3.818	1.955	296.86
Pears	−1.6–−0.5	90–95	2–7 months	83	−1.6	3.617	1.880	276.85
Pineapples, ripe	7	85–90	2–4 weeks	85	−1.0	3.684	1.905	283.52
Plums	−1–0	90–95	2–4 weeks	86	−0.8	3.718	1.918	286.85
Red raspberries	−0.5–0	90–95	2–3 days	84	−0.6	3.651	1.892	280.18
Strawberries	−0.5–0	90–95	5–7 days	90	−0.8	3.852	1.968	300.20
Seafood (Fish)[a]								
Haddock, cod, perch	−1–1	95–100	12 days	81	−2.2	3.550	1.855	270.17
Halibut	−1–1	95–100	18 days	75	−2.2	3.349	1.779	250.16
Herring, smoked	0–2	80–90	10 days	64	−2.2	2.981	1.641	213.47
Frozen fish	−29–−18	90–95	6–12 months					

Seafood (Shellfish)[a]

Scallop meat	0–1	95–100	12 days	80	‑2.2	3.517	1.842	266.84
Shrimp	‑1–1	95–100	12–14 days	76	‑2.2	3.383	1.792	253.50
Lobster, American	5–10	In seawater	Indefinitely in seawater	79	‑2.2	3.483	1.830	263.50
Oysters, clams (meat & liquor)	0–2	100	5–8 days	87	‑2.2	3.751	1.930	290.19
Meat								
Beef, fresh, average	0–1	88–92	1–6 weeks	62–77	‑2.2—‑1.7[b]	2.914–3.426	1.616–1.804	206.80–256.83
Carcass choice, 60% lean	0–4	85–90	1–3 weeks	49	‑1.7	2.478	1.453	163.44
Liver	0–1	90	1–5 days	70	‑1.7	3.182	1.717	233.48
Pork, fresh, average	0–1	85–90	3–7 days	32–44	‑2.2—‑2.7[b]	1.909–2.311	1.239–1.390	106.74–146.76
Ham								
74% lean	0–1	80–85	3–5 days	56	‑1.7[b]	2.713	1.541	186.79
Light cure	3–5	80–85	1–2 weeks	57		2.746	1.553	190.12
Lamb, fresh, average	0–1	85–90	5–12 days	60–70	‑2.2—‑1.7[b]	2.847–3.182	1.591–1.717	200.01–233.48
Poultry, fresh, average	0	85–90	1 week	74	‑2.8	3.316	1.767	246.83
Dairy Products[a]								
Butter, frozen	‑23	70–85	12 months	37	‑13.3	2.077	1.302	123.41
Cheese, cheddar, long storage	‑1–1	65–70	18 months	39	‑7.2	2.143	1.327	130.08
processed	4.4	65–70	12 months					
Ice cream, 10% fat	‑29—‑26		3–23 months	63	‑5.6	2.948	1.629	210.14
Milk, whole, pasteurized, Grade A	0–1.1		2–4 months	87	‑0.56	3.751	1.930	290.19
Eggs, shell	‑2–0	80–85	5–6 months	66	‑2.2[b]	3.048	1.666	220.14
Miscellaneous[a]								
Bread[a]	‑18	45–55	3–13 weeks	32–37		1.993	1.271	106.74–123.41
Fur and fabrics	1–4		Several years					

Table 76.5 (Continued)

Commodity	Storage Temperature (°C)	Relative Humidity (%)	Approximate Storage Life[e]	Water Content (%)	Highest Freezing (°C)	Specific Heat Above Freezing[c] (J/kg · °C)	Specific Heat Below Freezing[c] (J/kg · °C)	Latent Heat[d] (J/kg)
Honey	Below 10		1 year plus	17		1.407	1.051	56.70
Milk chocolate	−18–1.1	40	6–12 months	1		0.871	0.850	3.34
Nuts	0–10	65–75	8–12 months	3–6		0.938–1.038	0.875–0.913	10.01–20.01
Oil, vegetable, salad	21		1 year plus	0				
Oleo margarine	2	60–70	1 year plus	16		1.372	1.038	53.37
Orange juice	−1–2		3–6 weeks	89		3.818	1.955	296.86
Popcorn, unpopped	0–4	85	4–6 weeks	10		1.172	0.963	33.36

[a] *ASHRAE Handbook*, Applications, 1982,[8] for more specific information about many items listed.

[b] Average freezing point.

[c] Calculated by Siebel's formula and converted to SI metric units. For values above freezing, S in kJ/kg · °C (Btu/lb · °F) = 0.0355a + 0.8374 (0.008a + 0.20). For values below freezing, S in kJ/kg · °C (Btu/lb · °F) = 0.0126a + 0.8374 (0.003a + 0.20). Value a = percent of water. Recent work by H. E. Staph, B. E. Short, and others at the University of Texas has shown that Siebel formula is not particularly accurate in the frozen region, because foods are not simple mixtures of solids and liquids, they are not completely frozen even at −29°C (−20°F).

[d] Values for latent heat (latent heat of fusion) in kilojoules per kilogram (Btu/lb), calculated by multiplying the percentages of water content by the latent heat of fusion of water, 333.548 kJ/kg (143.4 Btu).

[e] Not based on maintaining nutritional value.

[f] Abridged with permission, ASHRAE, Inc., Atlanta, GA.

than connection of power, utilities, and process lines. As a result, there is a single source of responsibility for all aspects of the refrigeration cycle involving the transfer and handling of both saturated liquids and saturated vapors throughout the cycle, oil return, and other design requirements. These packages are custom engineered, including selection of components, piping, controls, base designs, torsional and critical speed analysis, and individual chemical process requirements. Large packages are designed in sections for shipment but are readily interconnected in the field.

As a general rule, field-erected refrigeration systems should be close-coupled to minimize problems of oil return and refrigerant condensation in suction lines. Where process loads are remotely located, pumped recirculation or brine systems are recommended. Piping and controls should be reviewed with suppliers to assure satisfactory operation under all conditions.

76.12   REFRIGERATION SYSTEM SPECIFICATIONS

To minimize costly and time-consuming alterations owing to unexpected requirements, the refrigeration specialist who is to do the final design must have as much information as possible before the design is started. Usually, it is best to provide more information than thought necessary, and it is always wise to note where information may be sketchy, missing, or uncertain. Carefully spelling out the allowable margins in the most critical process variables and pointing out portions of the refrigeration cycle that are of least concern is always helpful to the designer.

A checklist of minimum information (Table 76.6) needed by a refrigeration specialist to design a cooling system for a particular application may be helpful.

Table 76.6   Necessary Information for Design of a Cooling System

Process flow sheets and thermal specifications
  type of process
    batch
    continuous
  normal heat balances
  normal material balances
  normal material composition
  normal operating pressures and temperatures
  normal refrigeration loads
  energy recovery possibilities
  manner of supplying refrigeration; that is, pri-
      mary or secondary refrigerant

Basic specifications
  mechanical system details
    construction standards
      industry
      company
      local plant
    insulation requirements
    special corrosion prevention requirements
    special sealing requirements
      process streams to the environment
      process stream to refrigerant
  operating environment
    indoor or outdoor location
    extremes
    special requirements
  special safety considerations
    known hazards of process
    toxicity and flammability constraints
    maintenance limitations
  reliability requirements
    effect of loss of cooling on process safety
    maintenance intervals and types that may
      be performed
  redundancy requirement
  acceptance test requirements

Instrumentation and control requirements
  safety interlocks
  process interlocks

Table 76.6 *(Continued)*

special control requirements
  at equipment
  central control room
special or plant standard instruments
degree of automation: interface requirements
industry and company control standards

Off-design operation
process startup sequence
  degree of automation
  refrigeration loads vs time
  time needed to bring process onstream
  frequency of startup
  process pressure, temperature, and compo-
      sition changes seen by refrigeration
      equipment during startup
  special safety requirements
minimum load
need for standby capability
peak-load pressures and temperatures
composition extremes
process shutdown sequence
  degree of automation
  refrigeration load vs time
  shutdown time span
  process pressure, temperature, and compo-
      sition changes
  special safety requirements

Process flow sheets. For chemical process designs, seeing the process flow sheets is the best overall means for the refrigeration engineer to become familiar with the chemical process for which the refrigeration equipment is to be designed. In addition to providing all of the information shown in Table 76.6, they give the engineer a feeling for how the chemical plant will operate as a system and how the refrigeration equipment fits into the process.

Basic specifications. This portion of Table 76.6 fills in the detailed mechanical information that tells the refrigeration engineer how the equipment should be built, where it will be located, and specific safety requirements. This determines which standard equipment can be used and what special modifications need to be made.

Instrumentation and control requirements. These tell the refrigeration engineer how the system will be controlled by the plant operators. Particular controller types as well as control sequencing and operation must be spelled out to avoid misunderstandings and costly redesign. The refrigeration engineer needs to be aware of the degree of control required for the refrigeration system; for example, the process may require remote starting and stopping of the refrigeration system from the central control room. This could influence the way in which the refrigeration safeties and interlocks are designed.

Off-design operation. It is likely that the most severe operation of the refrigeration system will occur during startup or shutdown. The rapidly changing pressures, temperatures, and loads experienced by the refrigeration equipment can cause motor overloads, compressor surging, or loss of control if they are not anticipated during design.

REFERENCES

1. *ASHRAE Handbook,* American Society of Heating, Refrigerating, and Air Conditioning Engineers, Inc., Publication Department, Atlanta, GA, 1978, Chap. 50.

2. W. R. Woolrich, *The Men Who Created Cold,* Exposition Press, Inc., New York, 1967, p. 23.

3. *ASHRAE Standard 34-78, Number Designation of Refrigerants,* ASHRAE, Inc., Publications Department, Atlanta, GA, 1978.

4. *ASHRAE Standard 15-78, Safety Code for Mechanical Refrigeration,* ASHRAE, Inc., 1978.

5. G. G. Haselden, *Mech. Eng.,* **44** (Mar. 1981).

6. *ASHRAE Handbook,* American Society of Heating, Refrigerating, and Air Conditioning Engineers, Atlanta, GA, 1983.

7. *ASME Boiler and Pressure Vessel Code,* Sect. VIII, Div. 1, The American Society of Mechanical Engineers, New York, 1980.

8. *ASHRAE Handbook,* American Society of Heating, Refrigerating, and Air Conditioning Engineers, Atlanta, GA, 1983, Vol. 40.

9. K. W. Cooper and K. E. Hickman, "Refrigeration" in *Encyclopedia of Chemical Technology,* 3rd ed., Wiley, New York, 1983, Vol. 20, pp. 78–107.

CHAPTER 77
CRYOGENIC SYSTEMS

LEONARD A. WENZEL

Lehigh University
Bethlehem, Pennsylvania

77.1 CRYOGENICS AND CRYOFLUID PROPERTIES

The science and technology of deep refrigeration processing occurring at temperatures lower than about 150 K is the field of cryogenics (from the Greek *kryos,* icy cold). This area has developed as a special discipline because it is characterized by special techniques, requirements imposed by physical limitations, and economic needs, and unique phenomena associated with low-thermal-energy levels.

Compounds that are processed within the cryogenic temperature region are sometimes called cryogens. There are only a few of these materials; they are generally small, relatively simple molecules, and they seldom react chemically within the cryogenic region. Table 77.1 lists the major cryogens along with their major properties, and with a reference giving more complete thermodynamic data.

All of the cryogens except hydrogen and helium have conventional thermodynamic and transport properties. If specific data are unavailable, the reduced properties correlation can be used with all the cryogens and their mixtures with at least as much confidence as the correlations generally allow. Qualitatively *T–S* and *P–H* diagrams such as those of Figs. 77.1 and 77.2 differ among cryogens only by the location of the critical point and freezing point relative to ambient conditions.

Air, ammonia synthesis gas, and some inert atmospheres are considered as single materials although they are actually gas mixtures. The composition of air is shown in Table 77.13. If a thermodynamic diagram for air has the lines drawn between liquid and vapor boundaries where the pressures are equal for the two phases, these lines will not be at constant temperature, as would be the case for a pure component. Moreover, these liquid and vapor states are not at equilibrium, for the equilibrium states have equal Ts and Ps, but differ in composition. That being so, one or both of these equilibrium mixtures is not air. Except for this difference the properties of air are also conventional.

Hydrogen and helium differ in that their molecular mass is small in relation to zero-point-energy

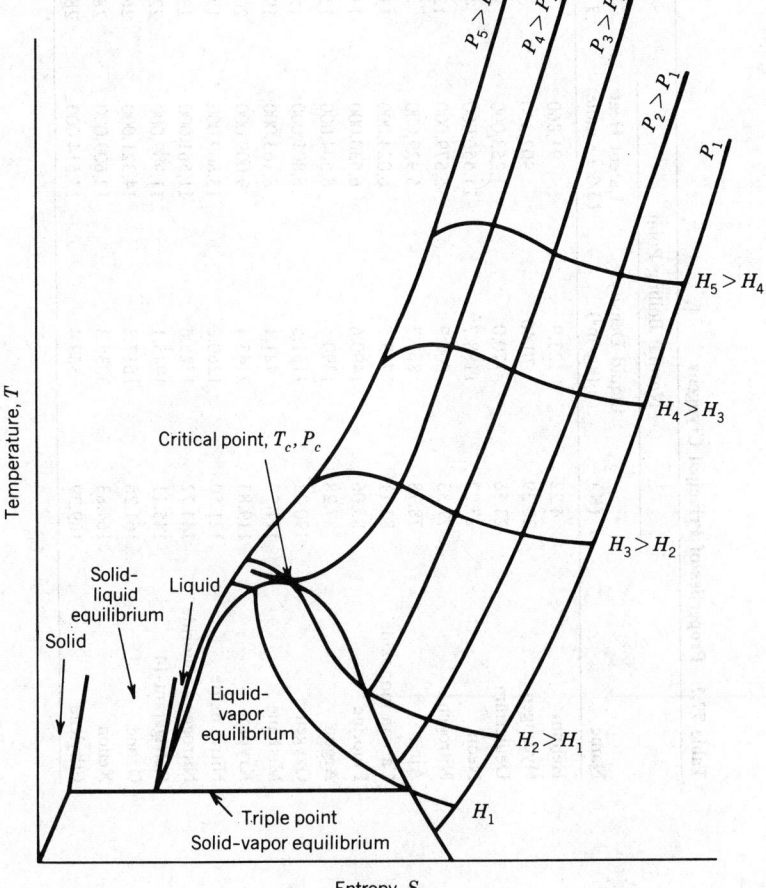

Fig. 77.1 Skeletal *T–S* diagram.

Table 77.1 Properties of Principal Cryogens

Name	Normal Boiling Point			Critical Point		Triple Point		Reference
	T (K)	Liquid Density (kg/m³)	Latent Heat (J/kg · mole)	T (K)	P (kPa)	T (K)	P (kPa)	
Helium	4.22	123.9	91,860	5.28	227			1
Hydrogen	20.39	70.40	902,300	33.28	1296	14.00	7.20	2, 3
Deuterium	23.56	170.0	1,253,000	38.28	1648	18.72	17.10	4
Neon	27.22	1188.7	21,556,000	44.44	2723	26.28	43.23	5
Nitrogen	77.33	800.9	5,579,000	126.17	3385	63.22	12.55	6
Air	78.78	867.7	5,929,000					7, 8
Carbon monoxide	82.11	783.5	6,024,000	132.9	3502	68.11	15.38	9
Fluorine	85.06	1490.6	6,530,000	144.2	5571			10
Argon	87.28	1390.5	6,504,000	151.2	4861	83.78		11, 12, 13
Oxygen	90.22	1131.5	6,801,000	154.8	5081	54.39	0.14	6
Methane	111.72	421.1	8,163,000	190.61	4619	90.67	11.65	14
Krypton	119.83	2145.4	9,009,000	209.4	5488	116.00	73.22	15
Nitric oxide	121.50	1260.2	13,809,000	179.2	6516	108.94		
Nitrogen trifluoride	144.72	1525.6	11,561,000	233.9	4530			
Refrigerant-14	145.11	1945.1	11,969,000	227.7	3737	89.17	0.12	16
Ozone	161.28	1617.8	14,321,000	261.1	5454			
Xenon	164.83	3035.3	12,609,000	289.8	5840	161.39	81.50	17
Ethylene	169.39	559.4	13,514,000	282.7	5068	104.00	0.12	18

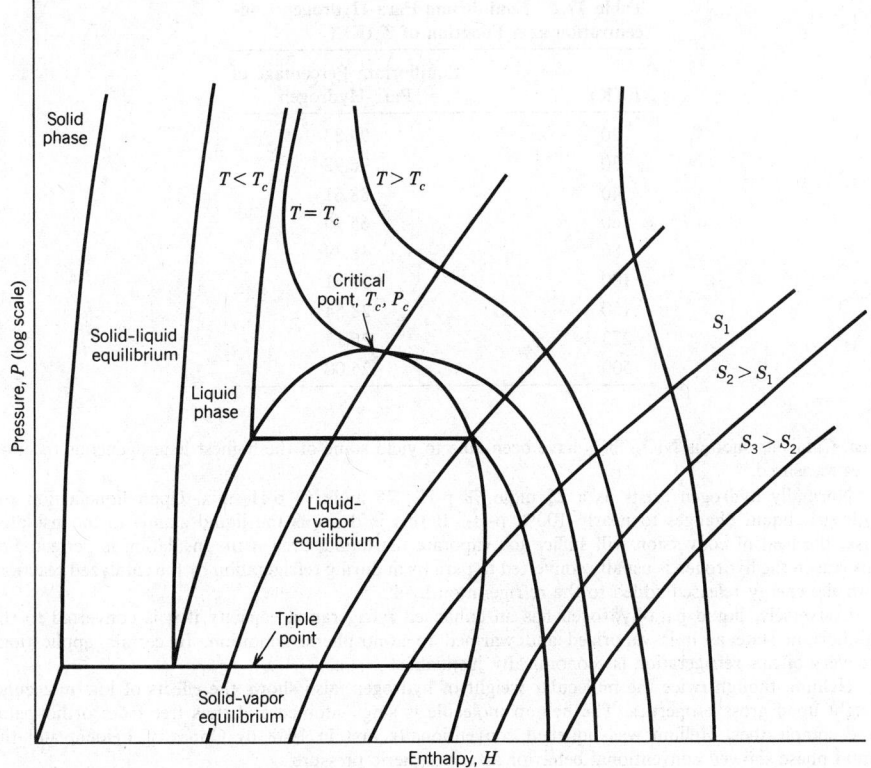

Fig. 77.2 Skeletal *P–H* diagram.

levels. Thus quantum differences are large enough to produce measurable changes in gross thermodynamic properties.

Hydrogen and its isotropes behave abnormally because the small molecular weight allows quantum differences stemming from different molecular configurations to affect total thermodynamic properties. The hydrogen molecule consists of two atoms, each containing a single proton and a single electron. The electrons rotate in opposite directions as required by molecular theory. The protons, however, may rotate in opposed or parallel directions. Figure 77.3 shows a sketch of the two possibilities, the parallel rotating nuclei identifying ortho-hydrogen and the opposite rotating nuclei identifying para-hydrogen. The quantum mechanics exhibited by these two molecule forms are different, and produce different thermodynamic properties. Ortho- and para-hydrogen each have conventional thermodynamic properties. However, ortho- and para-hydrogen are interconvertible with the equilibrium fraction of pure H_2 existing in para form dependent on temperature, as shown in Table 77.2. The natural ortho- and para-hydrogen reaction is a relatively slow one and of second order[19]:

$$\frac{dx}{d\theta} = 0.0114x^2 \quad \text{at} \quad 20 \text{ K}$$

where θ is time in hours and x is the mole fraction of ortho-hydrogen. The reaction rate can be greatly accelerated by a catalyst that interrupts the molecular magnetic field and possesses high surface

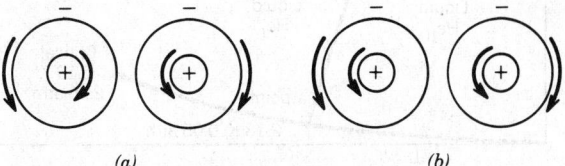

(a) (b)

Fig. 77.3 Molecular configurations of (a) para- and (b) ortho-hydrogen.

Table 77.2 Equilibrium Para-Hydrogen Concentration as a Function of T (K)

T (K)	Equilibrium Percentage of Para-Hydrogen
20	99.82
30	96.98
40	88.61
60	65.39
80	48.39
100	38.51
150	28.54
273	25.13
500	25.00

area. Catalysts such as NiO_2/SiO_2 have been able to yield some of the highest heterogeneous reaction rates measured.[20]

Normally hydrogen exists as a 25 mole % p-H_2, 75 mole % o-H_2 mix. Upon liquefaction the hydrogen liquid changes to nearly 100% p-H_2. If this is done as the liquid stands in an insulated flask, the heat of conversion will suffice to evaporate the liquid, even if the insulation is perfect. For this reason the hydrogen is usually converted to para form during refrigeration by the catalyzed reaction, with the energy released added to the refrigeration load.

Conversely, liquid para-hydrogen has an enhanced refrigeration capacity if it is converted to the equilibrium state as it is vaporized and warmed to atmospheric condition. In certain applications recovery of this refrigeration is economically justifiable.

Helium, though twice the molecular weight of hydrogen, also shows the effects of low molecular weight upon gross properties. The helium molecule is single-atomed and thus free from ortho–para-type complexities. Helium was liquefied conventionally first in 1908 by Onnes of Leiden, and the liquid phase showed conventional behavior at atmospheric pressure.

As temperature is lowered, however, a second-order phase change occurs at 2.18 K (0.05 atm) to produce a liquid called HeII. At no point does solidification occur just by evacuating the liquid. This results from the fact that the relationship between molecular volume, thermal energy (especially zero-point energy), and van der Waals attractive forces is such that the atoms cannot be trapped into a close-knit array by temperature reduction alone. Eventually, it was found that helium could be solidified if an adequate pressure is applied, but that the normal liquid helium (HeI)–HeII phase transition occurs at all pressures up to that of solidification. The phase diagram for helium is shown in Fig. 77.4. The HeI–HeII phase change has been called the lambda curve from the shape of the heat capacity curve for saturated liquid He, as shown in Fig. 77.5. The peculiar shape of the heat

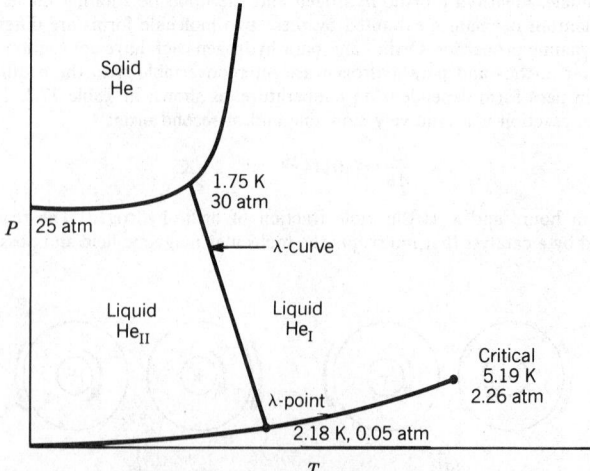

Fig. 77.4 Phase diagram for helium.

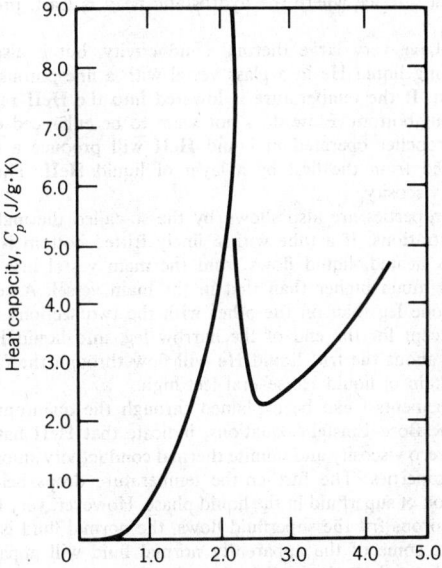

Fig. 77.5   Heat capacity of saturated liquid ⁴He.

capacity curve produces a break in the curve for enthalpy of saturated liquid He as shown in Fig. 77.6.

HeII is a unique liquid exhibiting properties that were not well explained until after 1945. As liquid helium is evacuated to increasingly lower pressures, the temperature also drops along the vapor-pressure curve. If this is done in a glass vacuum-insulated flask, heat leaks into the liquid He causing boiling and bubble formation. As the temperature approaches 2.18 K, boiling gets more violent, but then suddenly stops. The liquid He is completely quiescent. This has been found to occur because the thermal conductivity of HeII is extremely large. Thus the temperature is basically constant and

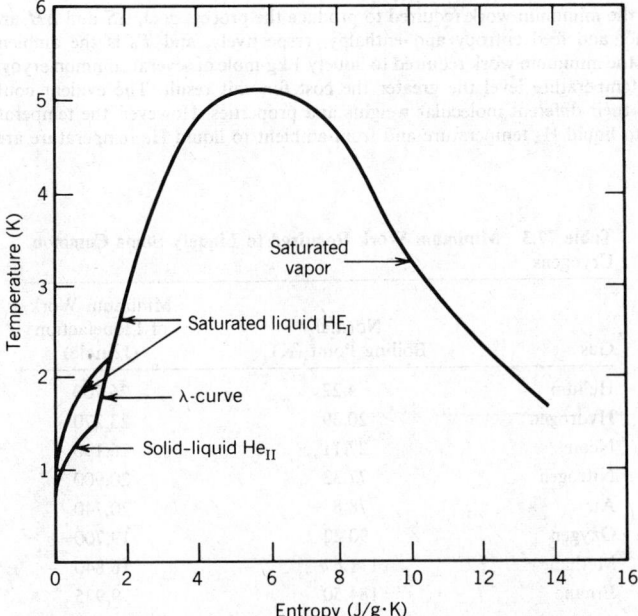

Fig. 77.6   Temperature–entropy diagram for saturation region of ⁴He.

all boiling occurs from the surface where the hydrostatic head is least, producing the lowest boiling point.

Not only does HeII have very large thermal conductivity, but it also has near zero viscosity. This can be seen by holding liquid He in a glass vessel with a fine porous bottom such that normal He does not flow through. If the temperature is lowered into the HeII region, the helium will flow rapidly through the porous bottom. Flow does not seem to be enhanced or hindered by the size of the frit. Conversely, a propeller operated in liquid HeII will produce a secondary movement in a parallel propeller separated from the first by a layer of liquid HeII. Thus HeII has properties of finite and of infinitesimal viscosity.

These peculiar flow properties are also shown by the so-called thermal–gravimetric effect. There are two common demonstrations. If a tube with a finely fritted bottom is put into liquid HeII and the helium in the tube is heated, liquid flows from the main vessel into the fritted tube until the liquid level in the tube is much higher than that in the main vessel. A second, related, experiment uses a U-tube, larger on one leg than on the other with the two sections separated by a fine frit. If this tube is immersed, except for the end of the narrow leg, into liquid HeII and a strong light is focused on the liquid He about the frit, liquid He will flow through the frit and out the small tube opening producing a fountain of liquid He several feet high.

These and other experiments[21] can be explained through the quantum mechanics of HeII. The pertinent relationships, the Bose–Einstein equations, indicate that HeII has a dual nature: it is both a "superfluid," which has zero viscosity and infinite thermal conductivity among other special properties, and a fluid of normal properties. The further the temperature drops below the lambda point the greater the apparent fraction of superfluid in the liquid phase. However, very little superfluid is required. In the flow through the porous frit the superfluid flows, the normal fluid is retained. However, if the temperature does not rise, some of the apparently normal fluid will apparently become superfluid. Although the superfluid flows through the frit, there is no depletion of superfluid in the liquid He left behind. In the thermogravimetric experiments the superfluid flows through the frit but is then changed to normal He. Thus there is no tendency for reverse flow.

At this point applications have not developed for HeII. Still, the peculiar phase relationships and energy effects may influence the design of helium processes, and do affect the shape of thermodynamic diagrams for helium.

77.2 CRYOGENIC REFRIGERATION AND LIQUEFACTION CYCLES

One characteristic aspect of cryogenic processing has been its early and continued emphasis on process efficiency, that is, on energy conservation. This has been forced on the field by the very high cost of deep refrigeration. For any process the minimum work required to produce the process goal is

$$W_{min} = T_0 \Delta S - \Delta H \tag{77.1}$$

where W_{min} is the minimum work required to produce the process goal, ΔS and ΔH are the difference between product and feed entropy and enthalpy, respectively, and T_0 is the ambient temperature. Table 77.3 lists the minimum work required to liquefy 1 kg-mole of several common cryogens. Obviously, the lower the temperature level the greater the cost for unit result. The evident conflict in H_2 and He arises from their different molecular weights and properties. However, the temperature differences from ambient to liquid H_2 temperature and from ambient to liquid He temperature are similar.

Table 77.3 Minimum Work Required to Liquefy Some Common Cryogens

Gas	Normal Boiling Point (K)	Minimum Work of Liquefaction (J/mole)
Helium	4.22	26,700
Hydrogen	20.39	23,270
Neon	27.11	26,190
Nitrogen	77.33	20,900
Air	78.8	20,740
Oxygen	90.22	19,700
Methane	111.67	16,840
Ethane	184.50	9,935
Ammonia	239.78	3,961

A refrigeration cycle that would approach the minimum work calculated as above would include ideal process steps as, for instance, in a Carnot refrigeration cycle. The cryogenic engineer aims for this goal while satisfying practical processing and capital cost limitations.

77.2.1 Cascade Refrigeration

The cascade refrigeration cycle was the first means used to liquefy air in the United States.[22] It uses conveniently chosen refrigeration cycles, each using the evaporator of the previous fluid cycle as condenser, which will produce the desired temperature. Figures 77.7 and 77.8 show a schematic T–S diagram of such a cycle and the required arrangement of equipment.

Obviously, this cycle is mechanically complex. After its early use it was largely replaced by other cryogenic cycles because of its mechanical unreliability, seal leaks, and poor mechanical efficiency. However, the improved reliability and efficiency of modern compressors has fostered a revival in the cascade cycle. Cascade cycles are used today in some base-load natural gas liquefaction (LNG) plants[23] and in some peak-shaving LNG plants. They are also used in a variety of intermediate refrigeration processes. The cascade cycle is potentially the most efficient of cryogenic processes because the major heat-transfer steps are liquefaction–vaporization exchanges with each stream at a constant temperature. Thus heat transfer coefficients are high and ΔTs can be kept very small.

77.2.2 The Linde or Joule–Thomson Cycle

The Linde cycle was used in the earliest European efforts at gas liquefaction and is conceptually the simplest of cryogenic cycles. A simple flow sheet is shown in Fig. 77.9. Here the gas to be liquefied or used as refrigerant is compressed through several stages each with its aftercooler. It then enters the main countercurrent heat exchanger where it is cooled by returning low-pressure gas. The gas is

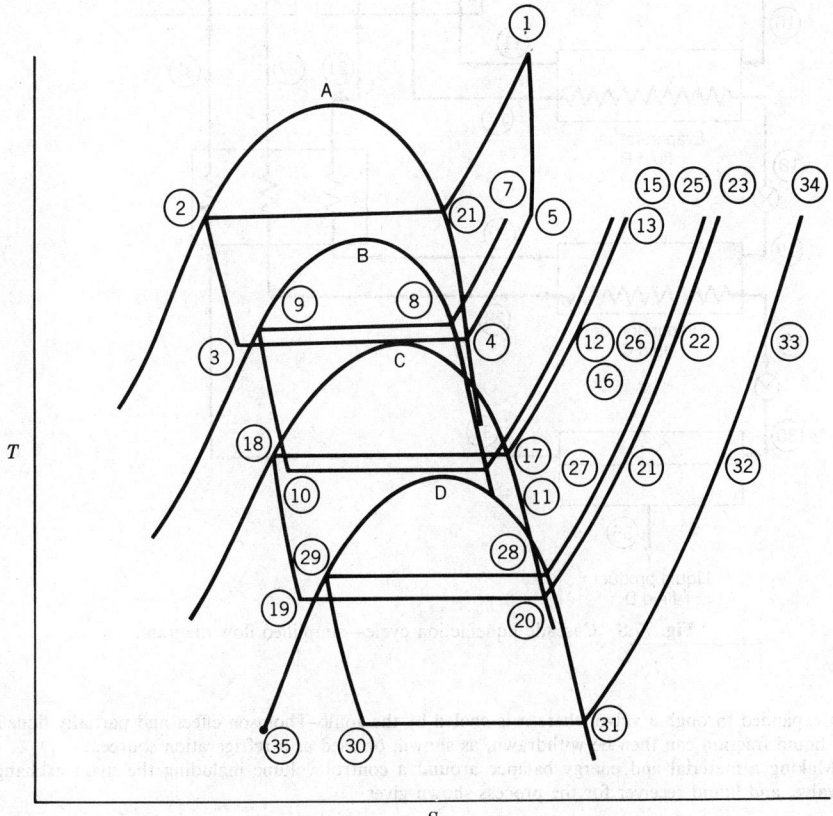

Fig. 77.7 Cascade refrigeration system on T–S coordinates. Note that T–S diagram for fluids A, B, C, and D are here superimposed. Numbers here refer to Fig. 77.8 flow points.

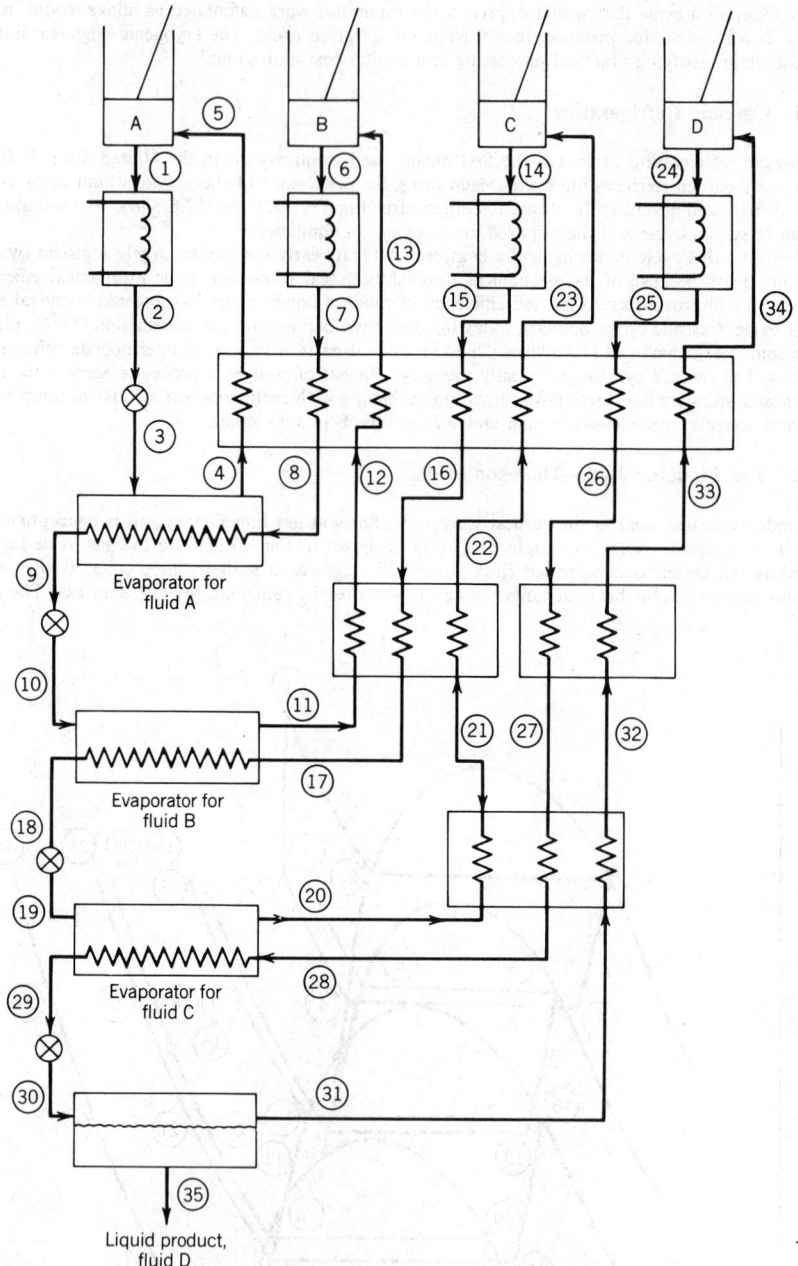

Fig. 77.8 Cascade liquefaction cycle—simplified flow diagram.

then expanded through a valve where it is cooled by the Joule–Thomson effect and partially liquefied. The liquid fraction can then be withdrawn, as shown, or used as a refrigeration source.

Making a material and energy balance around a control volume including the main exchanger, JT valve, and liquid receiver for the process shown gives

$$X_5 = \frac{H_7 - H_2}{H_5 - H_7} \qquad (77.2)$$

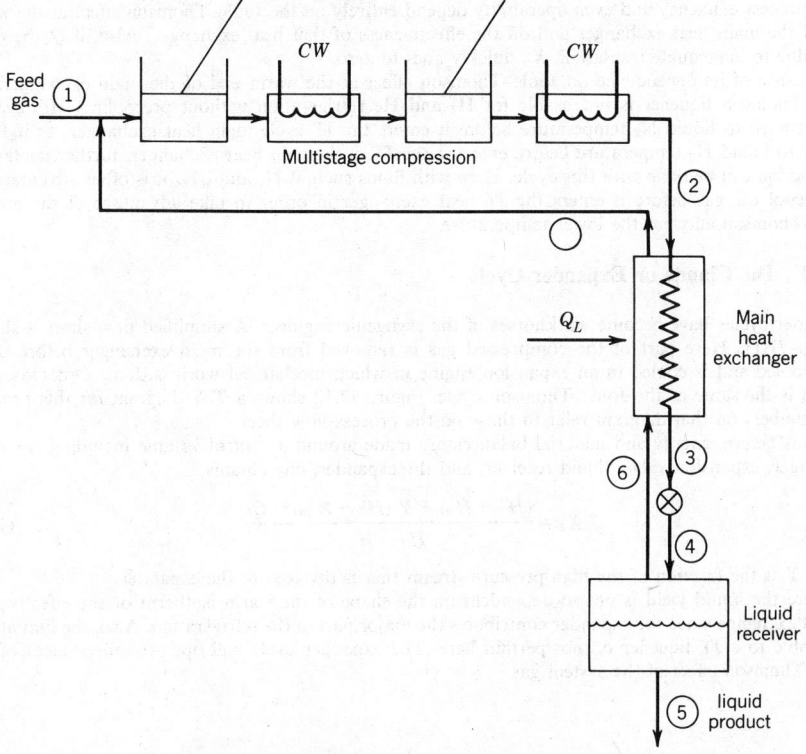

Fig. 77.9 Simplified Joule–Thomson liquefaction cycle flow diagram.

where X_5 is mass fraction liquid at point 5 in Fig. 77.9 and H_7, H_5, H_2 are enthalpies at indicated points on the flow sheet. If heat leaks from ambient into the control volume, Eq. (77.2) becomes

$$X_5 = \frac{H_7 - H_2 - Q_L}{H_5 - H_7} \tag{77.3}$$

The process is shown on a ln P–H diagram in Fig. 77.10. Note that the fraction of the process stream liquefied depends on the difference $(H_7 - H_5)$ that results from the curvature of the *high*-temperature process isotherm and the stream-to-stream ΔT at the warm end of the main exchanger.

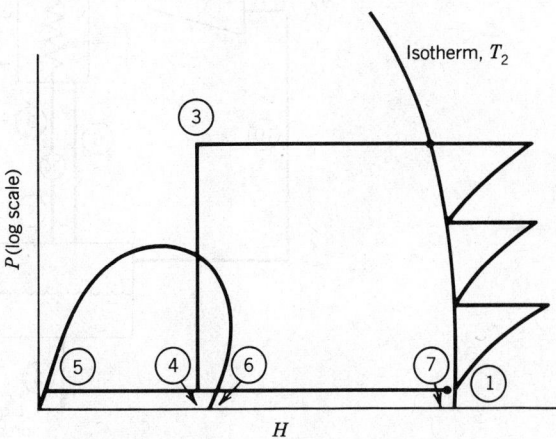

Fig. 77.10 Joule–Thomson liquefaction cycle shown on a pressure–enthalpy diagram.

Thus process efficiency and even operability depend entirely on the Joule–Thomson effect at the warm end of the main heat exchanger and on the effectiveness of that heat exchanger. Also, if Q_2 becomes large due to inadequate insulation, X_5 quickly goes to zero.

Because of its dependence on Joule–Thomson effect at the warm end of the main exchanger, the Joule–Thomson liquefier is not usable for H_2 and He refrigeration without precooling. However, if H_2 is cooled to liquid N_2 temperature before it enters the JT cycle main heat exchanger, or if He is cooled to liquid H_2 temperature before entering the JT cycle main heat exchanger, further cooling to liquefaction can be done with this cycle. Even with fluids such as N_2 and CH_4 it is often advantageous to precool the gas before it enters the JT heat exchanger in order to take advantage of the greater Joule–Thomson effect at the lower temperature.

77.2.3 The Claude or Expander Cycle

Expander cycles have become workhorses of the cryogenic engineer. A simplified flow sheet is shown in Fig. 77.11. Here part of the compressed gas is removed from the main exchanger before being fully cooled and is cooled in an expansion engine in which mechanical work is done. Otherwise, the system is the same as the Joule–Thomson cycle. Figure 77.12 shows a T–S diagram for this process. The numbers on that diagram refer to those on the process flow sheet.

If, as before, energy and material balances are made around a control volume including the main exchanger, expansion valve, liquid receiver, and the expander, one obtains

$$X_5 = \frac{(H_7 - H_2) + Y (H_9 - H_{10}) - Q_L}{H_7 - H_5} \tag{77.4}$$

where Y is the fraction of the high-pressure stream that is diverted to the expander.

Here the liquid yield is not so dependent on the shape of the warm isotherm or the effectiveness of heat exchange since the expander contributes the major part of the refrigeration. Also, the limitations applicable to a JT liquefier do not pertain here. The expander cycle will operate independent of the Joule–Thomson effect of the system gas.

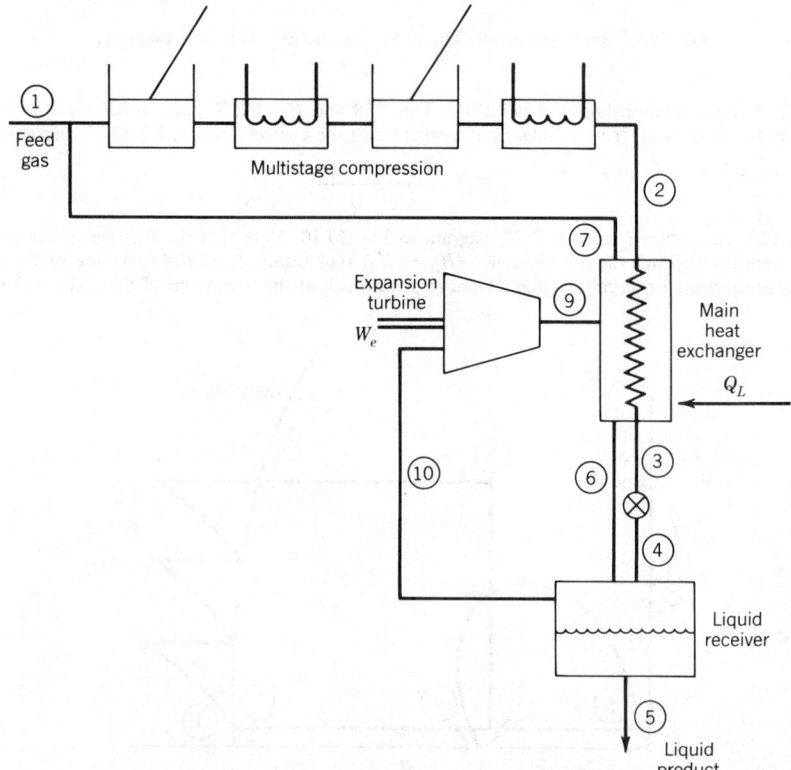

Fig. 77.11 Expander cycle simplified flow diagram.

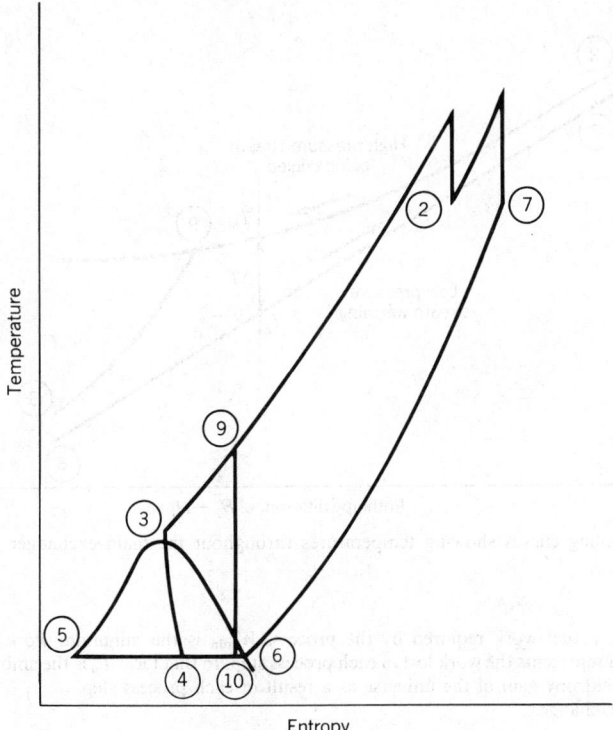

Fig. 77.12   Expander cycle shown on a *T–S* diagram.

The expansion step, line 9–10 on the *T–S* diagram, is ideally a constant entropy path. However, practical expanders operate at 60–90% efficiency and hence the path is one of modestly increasing entropy. In Fig. 77.12 the expander discharges a two-phase mixture. The process may be designed to discharge a saturated or a superheated vapor. Most expanders will tolerate a small amount of liquid in the discharge stream. However, this should be checked carefully with the manufacturer, for liquid can rapidly erode some expanders and can markedly reduce the efficiency of others.

Any cryogenic process design requires careful consideration of conditions in the main heat exchanger. The cooling curve plotted in Fig. 77.13 shows the temperature of the process stream being considered, T_i, as a function of the enthalpy difference $(H_o - H_i)$, where H_o is the enthalpy for the process stream as it enters or leaves the warm end of the exchanger, and H_i is the enthalpy of that same stream at any point within the main heat exchanger. The enthalpy difference is the product of the ΔH obtainable from a thermodynamic diagram and the mass flow rate of the process stream. If the mass flow rate changes, as it does at point 9 in the high-pressure stream, the slope will change. $H_o - H_i$ below such a point would be obtained from $H_o - H_i = (H_o - H_i) \cdot (1 - y)$ if the calculation is made on the basis of unit mass of high-pressure gas.

It is conventional practice to design cryogenic heat exchangers to obtain equal temperatures at a single cross section of the exchanger. The temperature difference between the high- and low-pressure streams $(T_h - T_c)$ at that point is the ΔT available for heat transfer. Obviously, the simple ΔT_{lm} approach to calculation of heat-exchanger area will not be satisfactory here, for that method depends on linear cooling curves. The usual approach here is to divide the exchanger into segments of ΔH such that the cooling curves are linear for the section chosen and to calculate the exchanger area for each section. It is especially important to examine cryogenic heat exchangers in this detail because temperature ranges are likely to be large, thus producing heat-transfer coefficients that vary over the length of the exchanger, and because the curvature of the cooling curves well may produce regions of very small ΔT. In extreme cases the designer may even find that ΔT at some point in the exchanger reaches zero, or becomes negative, thus violating the second law. No exchanger designed in ignorance of this situation would operate as designed.

Minimization of cryogenic process power requirements, and hence operating costs, can be done using classical considerations of entropy gain. For any process

$$W = W_{\min} + T_0 \Delta S_T \tag{77.5}$$

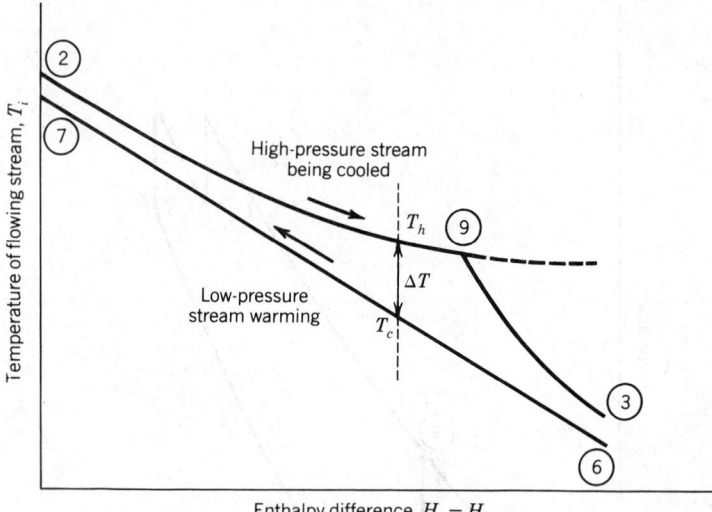

Fig. 77.13 Cooling curves showing temperatures throughout the main exchanger for the expander cycle.

where W is the actual work required by the process, W_{min} is the minimum work [see Eq. (77.1)], and the last term represents the work lost in each process step. In that term T_0 is the ambient temperature, and ΔS_T is the entropy gain of the universe as a result of each process step.

For a heat exchanger

$$T_0 \Delta S_T = W_L = T_0 \int \frac{T_h - T_c}{T_h T_c} dH_i \qquad (77.6)$$

where T_h and T_c represent temperatures of the hot and cold streams and the integration is carried out from end to end of the heat exchanger.

A comparison of the Claude cycle (so named because Georges Claude first developed a practical expander cycle for air liquefaction in 1902) with the Joule–Thomson cycle can thus be made by considering the W_L in the comparable process steps. In the cooling curve diagram, Fig. 77.13, the dotted line represents the high-pressure stream cooling curve of a Joule–Thomson cycle operating at the same pressure as does the Claude cycle. In comparison, the Claude cycle produces much smaller ΔTs at the cold end of the heat exchanger. If this is translated into lost work as done in Fig. 77.14, there is considerable reduction. The Claude cycle also reduces lost work by passing only a part of the high-pressure gas through a valve, which is a completely irreversible pressure reduction step. The rest of the high-pressure gas is expanded in a machine where most of the pressure lost produces usable work.

There are other ways to reduce the ΔT, and hence the W_L, in cryogenic heat exchangers. These methods can be used by the engineer as process conditions warrant. Figure 77.15 shows the effect of (a) intermediate refrigeration, (b) varying the amount of low-pressure gas in the exchanger, and (c) adding a third cold stream to the exchanger.

77.2.4 Low-Temperature Engine Cycles

The possibility that Carnot cycle efficiency could be approached by a refrigeration cycle in which all pressure change occurs by compression and expansion has encouraged the development of several cycles that essentially occur within an engine. In general, these have proven useful for small-scale cryogenic refrigeration under unusual conditions as in space vehicles. However, the Stirling cycle, discussed below, has been used for industrial-scale production situations.

The Stirling Cycle

In this cycle a noncondensable gas, usually helium, is compressed, cooled both by heat transfer to cooling water and by heat transfer to a cold solid matrix in a regenerator (see Section 77.3.3), and expanded to produce minimum temperature. The cold gas is warmed by heat transfer to the fluid or space being refrigerated and to the now-warm matrix in the regenerator and returned to the compressor.

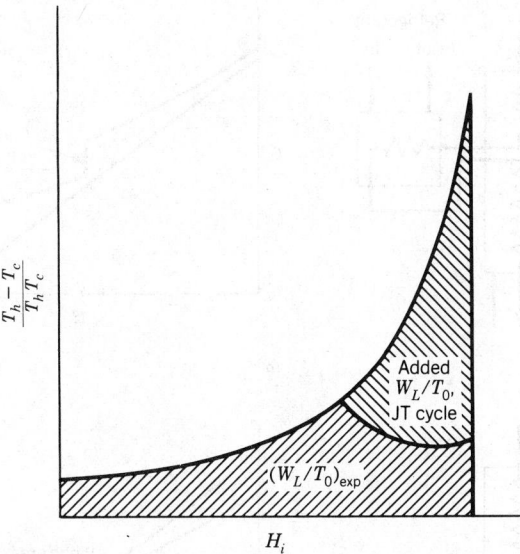

Fig. 77.14 Calculation of W_L in the main heat exchanger using Eq. (77.6) and showing the comparison between JT and Claude cycles.

Figure 77.16 shows the process path on a T–S diagram. The process efficiency of this idealized cycle is identical to a Carnot cycle efficiency.

In application the Stirling cycle is usually operated in an engine where compression and expansion both occur rapidly with compression being nearly adiabatic. Figure 77.17 shows such a machine. The compressor piston (1) and a displacer piston (16 with cap 17) operate off the same crankshaft. Their motion is displaced by 90°. The result is that the compressor position is near the top or bottom of its cycle as the displacer is in most rapid vertical movement. Thus the cycle can be traced as follows:

1. With the displacer near the top of its stroke the compressor moves up compressing the gas in space 4.

2. The displacer moves down, and the gas moves through the annular water-cooled heat exchanger (13) and the annular regenerator (14) reaching the upper space (5) in a cooled, compressed state. The regenerator packing, fine copper wool, is warmed by this flow.

3. Displacer and compressor pistons move down together. Thus the gas in (5) is expanded and cooled further. This cold gas receives heat from the chamber walls (18) and interior fins (15) thus refrigerating these solid parts and their external finning.

4. The displacer moves up, thus moving the gas from space (5) to space (4). In flowing through the annular passages the gas recools the regenerator packing.

The device shown in Fig. 77.17 is arranged for air liquefaction. Room air enters at (23), passes through the finned structure where water and then CO_2 freeze out, and is then liquefied as it is further cooled as it flows over the finned surface (18) of the cylinder. The working fluid, usually He, is supplied as needed from the tank (27).

Other Engine Cycles

The Stirling cycle can be operated as a heat engine instead of as a refrigerator and, in fact, that was the original intent. In 1918 Vuilleumier patented a device that combines these two forms of Stirling cycle to produce a refrigerator that operated on a high-temperature heat source rather than on a source of work. This process has received recent attention[24] and is useful in situations where a heat source is readily available but where power is inaccessible or costly.

The Gifford–McMahon cycles[25] have proven useful for operations requiring a light-weight, compact refrigeration source. Two cycles exist: one with a displacer piston that produces little or no work; the other with a work-producing expander piston. Figure 77.18 shows the two cycles.

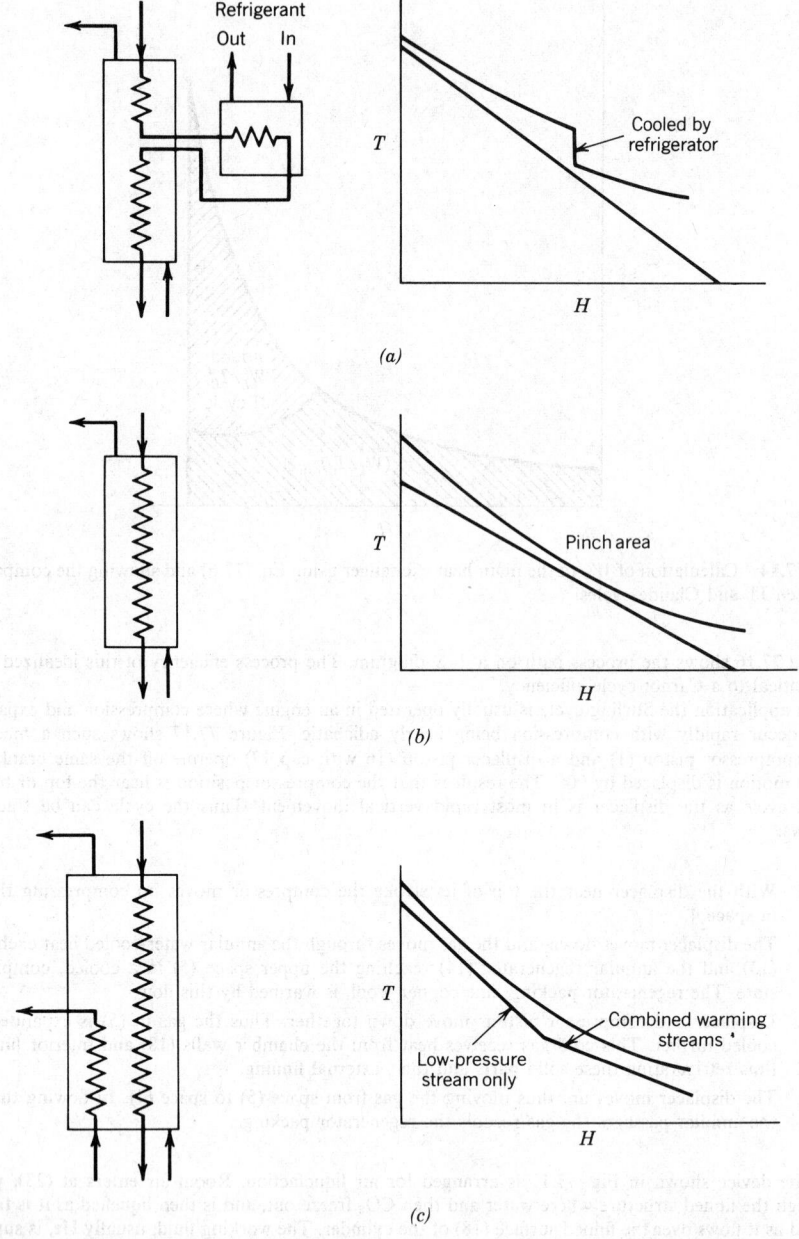

Fig. 77.15 Various cooling curve configurations to reduce W_L: (a) Cooling curve for intermediate refrigerator case. (b) Use of reduced warming stream to control ΔTs. (c) Use of an additional warming stream.

In both these cycles the compressor operates continuously maintaining the high-pressure surge volume of P_1, T_1. The sequence of steps for the system with work-producing piston are:

1. Inlet valve opens filling system to P_2.
2. Gas enters the cold space below the piston as the piston moves up doing work and thus cooling the gas. The piston continues, reducing the gas pressure to P_1.
3. The piston moves down pushing the gas through the heat load area and the regenerator to the storage vessel at P_1.

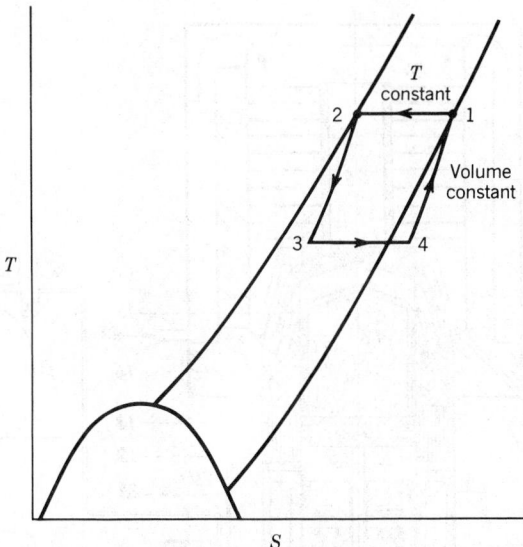

Fig. 77.16 The idealized Stirling cycle represented on a *T–S* diagram.

The sequence of steps for the system with the displacer is similar except that gas initially enters the warm end of the cylinder, is cooled by the heat exchanger, and then is displaced by the piston so that it moves through the regenerator for further cooling before entering the cold space. Final cooling is done by "blowing down" this gas so that it enters the low-pressure surge volume at P_1.

If the working fluid is assumed to be an ideal gas, all process steps are ideal, and compression is isothermal, the COPs for the two cycles are:

$$\text{COP (work producing)} = 1 \left/ \frac{RT_1 \ln P_2/P_1}{C_{P_0}T_3[1 - (P_4/P_3)^{(k-1)/k}]^{-1}} \right.$$

$$\text{COP (displacer)} = \frac{P_3 - P_1}{P_3(T_1/T_3 - P_1/P_3)\ln P_2/P_1}$$

In these equations states 1 and 2 are those immediately before and after the compressor. State 3 is after the cooling step but before expansion, and state 4 is after the expansion at the lowest temperature.

77.3 CRYOGENIC HEAT-TRANSFER METHODS

In dealing with heat-transfer requirements the cryogenic engineer must effect large quantities of heat transfer over small ΔTs through wide temperature ranges. Commonly heat capacities and/or mass flows change along the length of the heat-transfer path, and often condensation or evaporation takes place. To minimize heat leak these complexities must be handled using exchangers with as large a heat-transfer surface area per exchanger volume as possible. Compact heat-exchanger designs of many sorts have been used, but only the most common types will be discussed here.

77.3.1 Coiled-Tube-in-Shell Exchangers

The traditional heat exchanger for cryogenic service is the Hampson or coiled-tube-in-shell exchanger as shown in Fig. 77.19. The exchanger is built by turning a mandrel on a specially built lathe, and wrapping it with successive layers of tubing and spacer wires. Since longitudinal heat transfer down the mandrel is not desired, the mandrel is usually made of a poorly conducting material such as stainless steel, and its interior is packed with an insulating material to prevent gas circulation. Copper or aluminum tubing is generally used. To prevent uneven flow distribution from tube to tube, tube winding is planned so that the total length of each tube is constant independent of the layer on which the tube is wound. This results in a constant winding angle, as shown in Fig. 77.20. For example, the tube layer next to the mandrel might have five parallel tubes, whereas the layer next to the shell might have 20 parallel tubes. Spacer wires may be laid longitudinally on each layer of tubes, or they may be wound counter to the tube winding direction, or omitted. Their presence and size depends

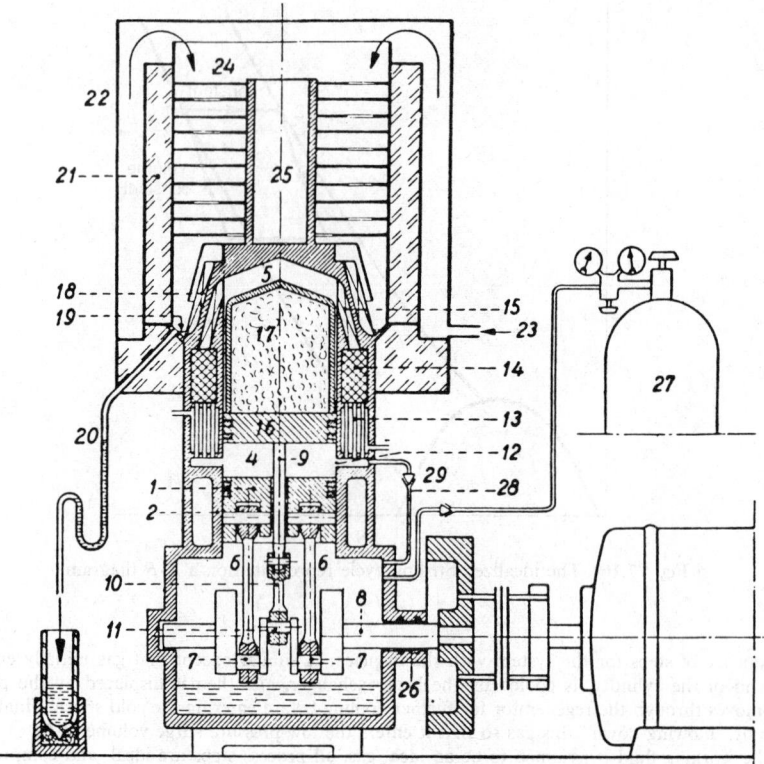

Fig. 77.17 Stirling cycle arranged for air liquefaction reference points have the following meanings: 1, compressor; 2, compression cylinder; 4, working fluid in space between compressor and displacer; 5, working fluid in the cold head region of the machine; 6, two parallel connecting rods with cranks, 7, of the main piston; 8, crankshaft; 9, displacer rod, linked to connecting rod, 10, and crank, 11, of the displacer; 12, ports; 13, cooler; 14, regenerator; 15, freezer; 16, displacer piston, and 17, cap; 18, condenser for the air to be liquefied, with annular channel, 19, tapping pipe (gooseneck) 20, insulating screening cover, 21, and mantel 22; 23, aperture for entry of air; 24, plates of the ice separator, joined by the tubular structure, 25, to the freezer (15); 26, gas-tight shaft seal; 27, gas cylinder supplying refrigerant; 28, supply pipe with one-way valve, 29. (Courtesy U.S. Philips Corporation.)

on the flow requirements for fluid in the exchanger shell. Successive tube layers may be wound in opposite or in the same direction.

After the tubes are wound on the mandrel they are fed into manifolds at each end of the tube bundle. The mandrel itself may be used for this purpose, or hook-shaped manifolds of large diameter tubing can be looped around the mandrel and connected to each tube in the bundle. Finally, the exchanger is closed by wrapping a shell, usually thin-walled stainless steel, over the bundle and welding on the required heads and nozzles.

In application the low-pressure fluid flows through the exchanger shell, and high-pressure fluids flow through the tubes. This exchanger is easily adapted for use by three or more fluids by putting in a pair of mandrels for each tube-side fluid to be carried. However, tube arrangement must be carefully engineered so that the temperatures of all the cooling streams (or all the warming streams) will be identical at any given exchanger cross section. The exchanger is typically mounted vertically so that condensation and gravity effects will not result in uneven flow distribution. Most often the cold end is located at the top so that any liquids not carried by the process stream will move toward warmer temperatures and be evaporated.

Heat-transfer coefficients in these exchangers will usually vary from end to end of the exchangers because of the wide temperature range experienced. For this reason, and because of the nonlinear ΔT variations, the exchanger area must be determined by sections, the section lengths chosen so that linear ΔTs can be used and so that temperature ranges are not excessive. For inside tube heat-transfer coefficients with single-phase flow the Dittus–Boelter equation is used altered to account for the spiral flow:

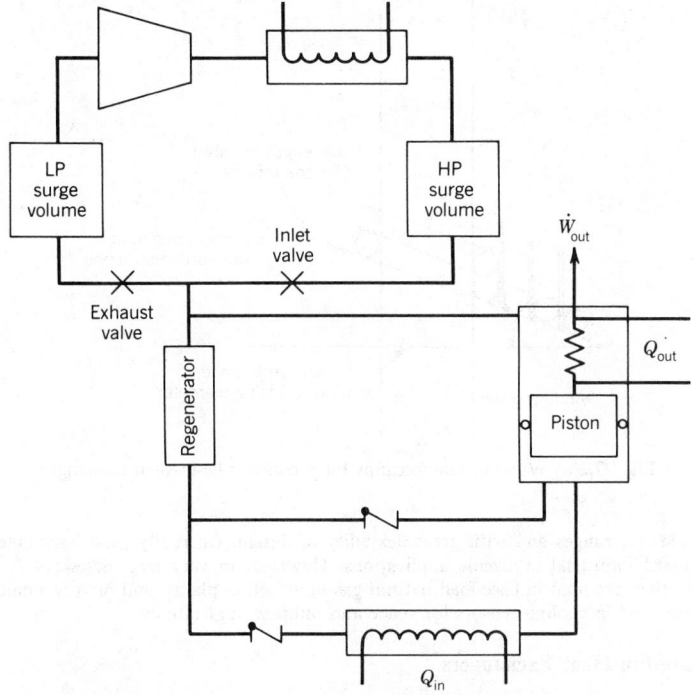

Fig. 77.18 Gifford–McMahon refrigerator. The dashed line and the cooler are present only when the piston is to be used as a displacer with negligible work production.

$$\frac{hD}{k} = 0.023 \, N_{\text{Re}}^{0.8} N_{\text{Pn}}^{0.32} \left(1 + 3.5 \frac{d}{D} \right) \tag{77.7}$$

where D is the diameter of the helix and d the inside tube diameter.

For outside heat-transfer coefficients the standard design methods for heat transfer for flow across tube banks with in-line tubes are used. Usually the metal wall resistance is negligible. In some cases adjacent tubes are brazed or soldered together to promote heat transfer from one to the other. Even here wall resistance is usually a very small part of the total heat-transfer resistance.

Pressure drop calculations are made using equivalent design tools. Usually the low-pressure-side ΔP is critical in designing a usable exchanger.

The coiled-tube-in-shell exchanger is expensive, requiring a large amount of hand labor. Its advantages are that it can be operated at any pressure the tubes can withstand, and that it can be built

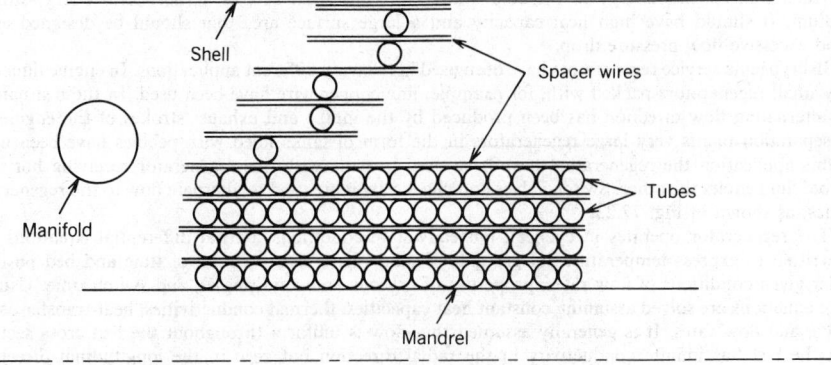

Fig. 77.19 Section of a coiled-tube-in-shell heat exchanger.

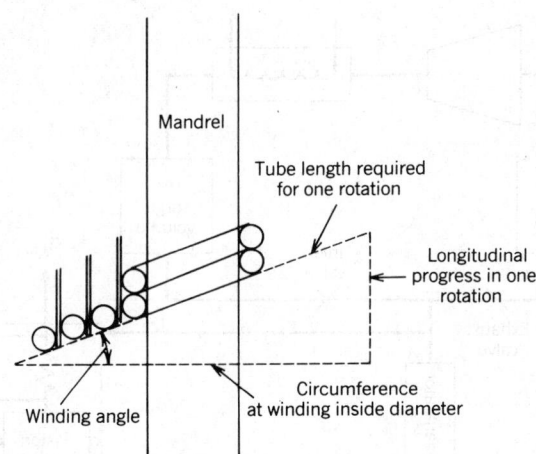

Fig. 77.20 Winding relationships for a coiled-tube-in-shell exchanger.

over very wide size ranges and with great flexibility of design. Currently these exchangers are little used in standard industrial cryogenic applications. However, in very large sizes (14 ft diameter × 120 ft length) they are used in base-load natural gas liquefaction plants, and in very small size (finger sized) they are used in cooling sensors for space and military applications.

77.3.2 Plate-Fin Heat Exchangers

The plate-fin exchanger has become the most common type used for cryogenic service. This results from its relatively low cost and high concentration of surface area per cubic foot of exchanger volume. It is made by clamping together a stack of alternate flat plates and corrugated sheets of aluminum coated with brazing flux. This assembly is then immersed in molten salt where the aluminum brazes together at points of contact. After removal from the bath the salt and flux are washed from the exchanger parts, and the assembly is enclosed in end plates and nozzles designed to give the desired flow arrangement. Usually the exchanger is roughly cubic, and is limited in size by the size of the available salt bath and the ability to make good braze seals in the center of the core. The core can be arranged for countercurrent flow or for cross flow. Figure 77.21 shows the construction of a typical plate-fin exchanger.

Procedures for calculating heat-transfer and pressure loss characteristics for plate-fin exchangers have been developed and published by the exchanger manufacturers. Table 77.4 and Fig. 77.22 present one set of these.

77.3.3 Regenerators

A regenerator is essentially a storage vessel filled with particulate solids through which hot and cold fluid flow alternately in opposite directions. The solids absorb energy as the hot fluid flows through, and then transfer this energy to the cold fluid. Thus this solid acts as a short-term energy-storage medium. It should have high heat capacity and a large surface area, but should be designed as to avoid excessive flow pressure drop.

In cryogenic service regenerators have been used in two very different applications. In engine liquefiers very small regenerators packed with, for example, fine copper wire have been used. In these situations the alternating flow direction has been produced by the intake and exhaust strokes of the engine. In air separation plants very large regenerators in the form of tanks filled with pebbles have been used. In this application the regenerators have been used in pairs with one regenerator receiving hot fluid as cold fluid enters the other. Switch valves and check valves are used to alternate flow to the regenerator bodies, as shown in Fig. 77.23.

The regenerator operates in cyclical, unsteady-state conditions. Partial differential equations can be written to express temperatures of gas and of solid phase as a function of time and bed position under given conditions of flow rates, properties of gaseous and solid phases, and switch time. Usually these equations are solved assuming constant heat capacities, thermal conductivities, heat-transfer coefficients, and flow rates. It is generally assumed that flow is uniform throughout the bed cross section, that the bed has infinite conductivity in the radial direction but zero in the longitudinal direction, and that there is no condensation or vaporization occurring. Thermal gradients through the solid

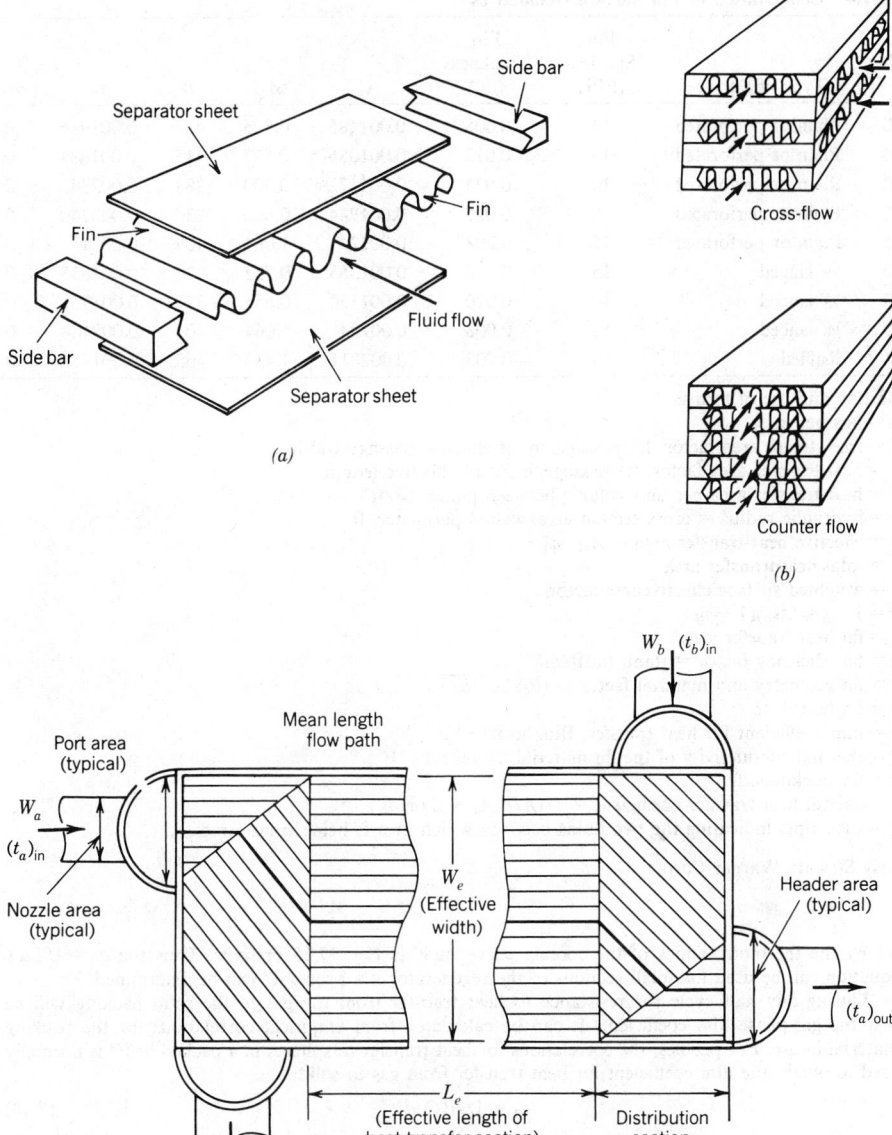

Fig. 77.21 Construction features of a plate-fin heat exchanger. (*a*) Detail of plate and fin. (*b*) Flow arrangements. (*c*) Total assembly arrangement.

particles are usually ignored. These equations can then be solved by computer approximation. The results are often expressed graphically.[26]

An alternative approach compares the regenerator with a steady-state heat exchanger and uses exchanger design methods for calculating regenerator size.[27] Figure 77.24 shows the temperature–time relationship at several points in a regenerator body. In the central part of the regenerator ΔTs are nearly constant throughout the cycle. Folding the figure at the switch points superimposes the temperature data for this central section as shown in Fig. 77.25. It is clear that the solid plays only a time-delaying function as energy flows from the hot stream to the cold one. Temperature levels are

Table 77.4 Computation of Fin Surface Geometrics[a]

Fin Height (in.)	Type of Surface	Fin Spacing (FPI)	Fin Thickness (in.)	A_c'	A_{ht}''	B	r_h	A_r/A_{ht}
0.200	Plain or perforated	14	0.008	0.001185	0.596	437	0.001986	0.751
0.200	Plain or perforated	14	0.012	0.001086	0.577	415	0.001884	0.760
0.250	Plain or perforated	10	0.025	0.001172	0.500	288	0.00234	0.750
0.375	Plain or perforated	8	0.025	0.001944	0.600	230	0.003240	0.778
0.375	Plain or perforated	15	0.008	0.00224	1.064	409	0.00211	0.862
0.250	⅛ lanced	15	0.012	0.001355	0.732	420	0.001855	0.813
0.250	⅛ lanced	14	0.020	0.001150	0.655	378	0.001751	0.817
0.375	⅛ lanced	15	0.008	0.00224	1.064	409	0.002108	0.862
0.455	Ruffled	16	0.005	0.002811	1.437	465	0.001956	0.893

[a] Definition and use of terms:
FPI = fins per inch
A_c' = free stream area factor, ft²/passage/in. of effective passage width
A_{ht}'' = heat-transfer area factor, ft²/passage/in./ft of effective length
B = heat-transfer area per unit volume between plates, ft²/ft³
r_h = hydraulic radius = cross section area/wetted perimeter, ft
A_r = effective heat-transfer area = $A_{ht} \cdot \eta_0$
A_{ht} = total heat-transfer area
η_0 = weighted surface effectiveness factor
   $= 1 - (A_r/A_{ht})(1 - \eta_f)$
A_f = fin heat-transfer area
η_f = fin efficiency factor = $[\tanh (ml)]/ml$
ml = fin geometry and material factor = $(b/s)\sqrt{2h/k}$
b = fin height, ft
h = film coefficient for heat transfer, Btu/hr·ft² · °F
k = thermal conductivity of the fin material, Btu/hr·ft·°F
s = fin thickness, ft
U = overall heat transfer coefficient = $1/(A/h_a A_a + A/h_b A_b)$
a,b = subscripts indicating the two fluids between which heat is being transferred

Courtesy Stewart–Warner Corp.

set by the thermodynamics of the cooling curve such as Fig. 77.25 presents. Thus the $q = UA\Delta T$ equation can be used for small sections of the regenerator if a proper U can be determined.

During any half cycle the resistance to heat transfer from the gas to the solid packing will be just the gas-phase film coefficient. It can be calculated from empirical correlations for the packing material in use. For pebbles, the correlations for heat transfer to spheres in a packed bed[28] is normally used to obtain the film coefficient for heat transfer from gas to solid:

$$h_{gs} = 1.31(G/d)^{0.93} \tag{77.8}$$

where h_{gs} = heat transfer from gas to regenerator packing or reverse, J/hr·m²·K
   G = mass flow of gas, kg/hr·m²
   d = particle diameter, m

The heat that flows to the packing surface diffuses into the packing by a conductive mode. Usually this transfer is fast relative to the transfer from the gas phase, but it may be necessary to calculate solid surface temperatures as a function of heat-transfer rate and adjust the overall ΔT accordingly. The heat-transfer mechanisms are typically symmetrical and hence the design equation becomes

$$A = \frac{q}{U\Delta T} = \frac{q}{h/2 \times \Delta T/2} = \frac{4q}{h_{gs}\Delta T}$$

This calculation can be done for each section of the cooling curve until the entire regenerator area is calculated. However, at the ends of the regenerator temperatures are not symmetrical nor is the ΔT constant throughout the cycle. Figure 77.26 gives a correction factor that must be used to adjust the calculated area for these end effects. Usually a 10–20% increase in area results.

The cyclical nature of regenerator operation allows their use as trapping media for contaminants simultaneously with their heat-transfer function. If the contaminant is condensable, it will condense

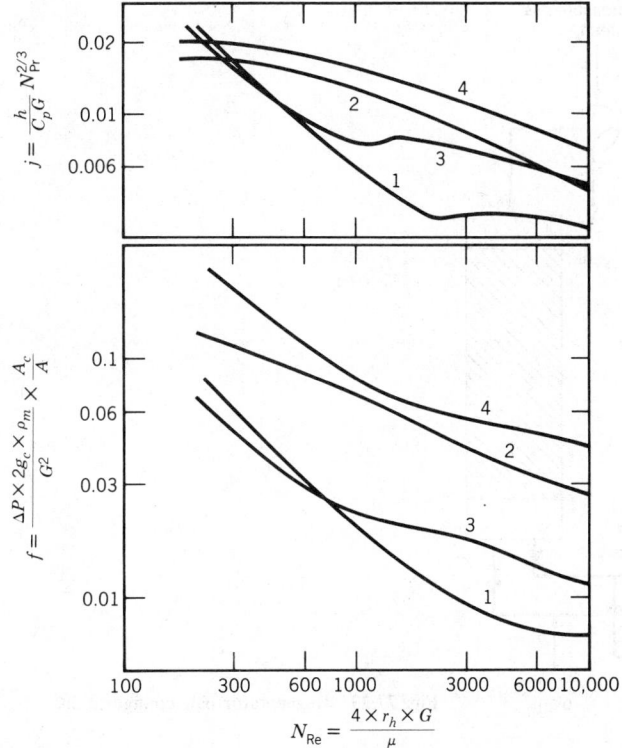

Fig. 77.22 Heat-transfer and flow friction factors in plate and fin heat exchangers. Curves 1: plain fin (0.200 in. height, 14 fins/in.—0.008 in. thick). Curves 2: ruffled fin (0.445 in. height, 16 fins/in.—0.005 in. thick). Curves 3: perforated fin (0.375 in. height, 8 fins/in.—0.025 in. thick). (Courtesy, Stewart-Warner Corp.)

and solidify on the solid surfaces as the cooling phase flows through the regenerator. During the warming phase flow, this deposited condensed phase will evaporate flowing out with the return media.

Consider an air-separation process in which crude air at a moderate pressure is cooled by flow through a regenerator pair. The warmed regenerator is then used to warm up the returning nitrogen at low pressure. The water and CO_2 in the air deposit on the regenerator surfaces and then reevaporate into the nitrogen. If deposition occurs at thermodynamic equilibrium, and assuming Raoult's law,

$$y_{H_2O} \text{ or } y_{CO_2} = \frac{P^\circ_{H_2O \text{ or } CO_2}}{P} \qquad (77.9)$$

where y = mole fraction of H_2O or CO_2 in the gas phase
  P° = saturation vapor pressure of H_2O or CO_2
  P = total pressure of flowing stream

This equation can be applied to both the depositing, incoming situation and the reevaporating, outgoing situation. If the contaminant is completely removed in the regenerator, and the return gas is pure as it enters the regenerator, the moles of incoming gas times the mole fraction of contaminant must equal that same product for the outgoing stream if the contaminant does not accumulate in the regenerator. Since the vapor pressure is a function of temperature, and the returning stream pressure is lower than the incoming stream pressure, these relations can be combined to give the maximum stream-to-stream ΔT that may exist at any location in the regenerator. Figure 77.27 shows the results for one regenerator design condition. Also plotted on Fig. 77.27 is a cooling curve for these same design conditions. At the conditions given H_2O will be removed down to very low concentrations, but CO_2 solids may accumulate in the bottom of the regenerator. To prevent this it would be necessary to remove some of the air stream in the middle of the regenerator for further purification and cooling elsewhere.

Cryogenic heat exchangers often are called on to condense or evaporate and two-phase heat-transfer commonly occurs, sometimes on both sides of a given heat exchanger. Heat-transfer coefficients and

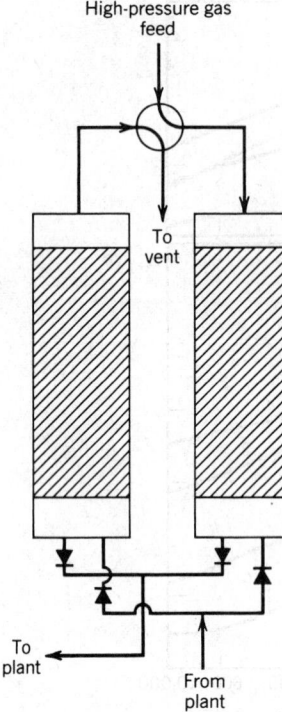

High-pressure gas
feed

To
vent

To
plant

From
plant

Fig. 77.23 Regenerator pair configuration.

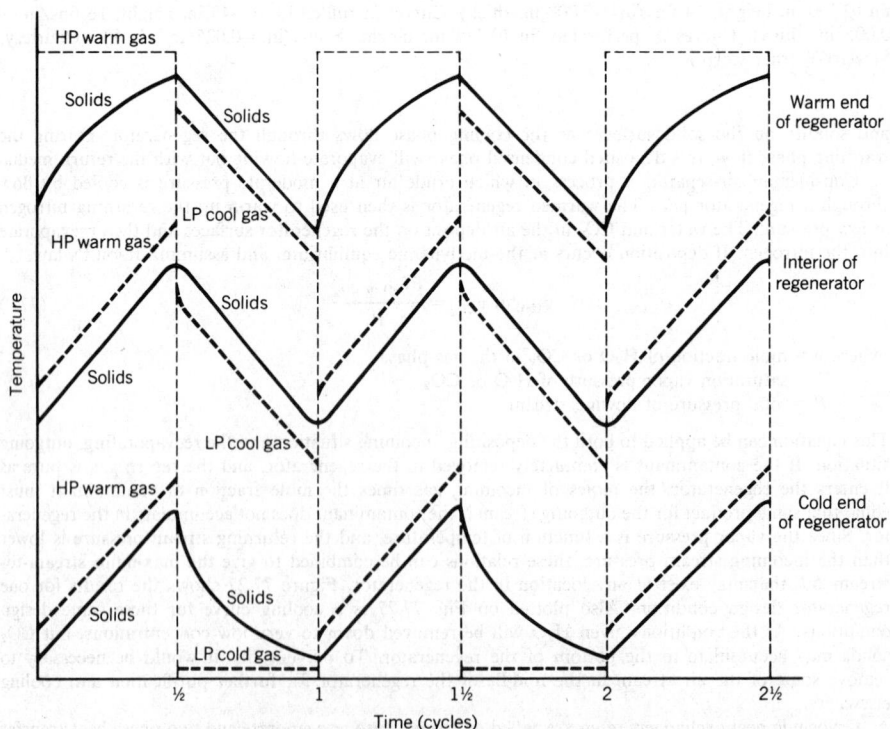

Fig. 77.24 Time–temperature histories in a regenerator.

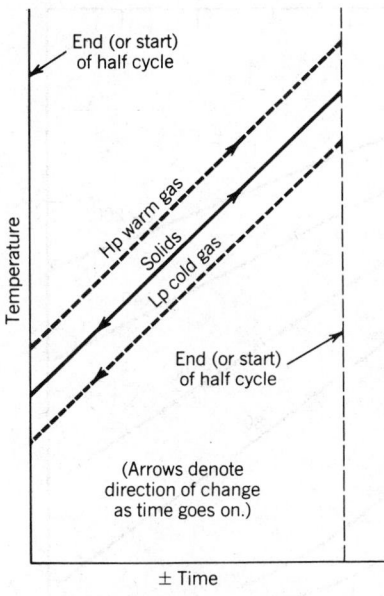

Fig. 77.25 Time–temperature history for a central slice through a regenerator.

flow pressure losses are calculated using correlations taken from high-temperature data.[29] The distribution of multiphase processing streams into parallel channels is, however, a common and severe problem in cryogenic processing. In heat exchangers thousands of parallel paths may exist. Thus the designer must ensure that all possible paths offer the same flow resistance and that the two phases are well distributed in the flow stream approaching the distribution point. Streams that cool during passage through an exchanger are likely to be modestly self-compensating in that the viscosity of a cold gas is lower than that of a warmer gas. Thus a stream that is relatively high in temperature (as would be the case if that passage received more than its share of fluid) will have a greater flow resistance than a cooler system, so flow will be reduced. The opposite effect occurs for streams being warmed, so that these streams must be carefully balanced at the exchanger entrance.

77.4 INSULATION SYSTEMS

Successful cryogenic processing requires high-efficiency insulation. Sometimes this is a processing necessity, as in the Joule–Thomson liquefier, and sometimes it is primarily an economic requirement, as in the storage and transportation of cryogens. For large-scale cryogenic processes, especially those operating at liquid nitrogen temperatures and above, thick blankets of fiber or powder insulation, air or N_2 filled, have generally been used. For lower temperatures and for smaller units, vacuum insulation has been enhanced by adding one or many radiation shields, sometimes in the form of fibers or pellets, but often as reflective metal barriers. The use of many radiation barriers in the form of metal-coated plastic sheets wrapped around the processing vessel within the vacuum space has been used for most applications at temperatures approaching absolute zero.

77.4.1 Vacuum Insulation

Heat transfer occurs by convection, conduction, and radiation mechanisms. A vacuum space ideally eliminates convective and conductive heat transfer but does not interrupt radiative transfer. Thus heat transfer through a vacuum space can be calculated from the classic equation:

$$q = \sigma A F_{12}(T_1^4 - T_2^4) \qquad (77.10)$$

where q = rate of heat transfer, J/sec
σ = Stefan–Boltzmann constant, 5.73×10^{-8} J/sec·m²·K
F_{12} = combined emissivity and geometry factor
T_1, T_2 = temperature (K) of radiating and receiving body, respectively

In this formulation of the Stefan–Boltzmann equation it is assumed that both radiator and receiver are gray bodies, that is, emissivity ϵ and absorptivity are equal and independent of temperature. It is also assumed that the radiating body loses energy to a totally uniform surroundings and receives energy from this same environment.

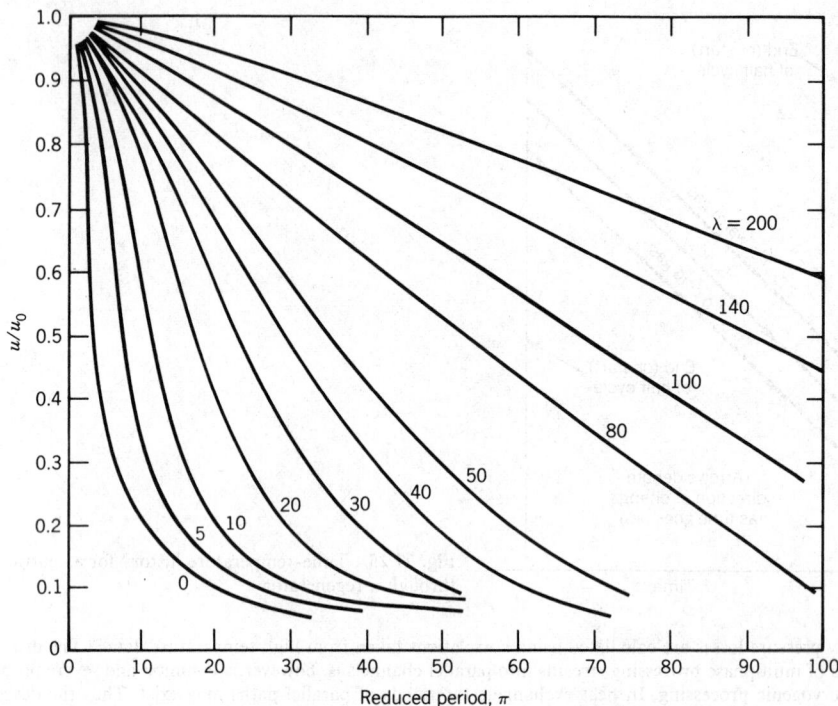

Fig. 77.26 End correction for regenerator heat transfer calculation using symmetrical cycle theory[27] (Courtesy Plenum Press):

$$\lambda = \frac{4H_0 S(T_c + T_w)}{C_c T_c + C_w T_w} = \text{reduced length}$$

$$\pi = \frac{12 H_0 (T_c + T_w)}{c \rho_s d} = \text{reduced period}$$

$$U_0 = \frac{1}{4}\left[\frac{1}{h} + \frac{0.1d}{k}\right]$$

where T_w, T_c = switching times of warm and cold streams, respectively, hr
$\quad\quad S$ = regenerator surface area, m²
$\quad\quad U_0$ = overall heat transfer coefficient uncorrected for hysteresis, kcal/m² · hr · °C
$\quad\quad U$ = overall heat transfer coefficient
$\quad\quad C_w$, C_c = heat capacity of warm and cold stream, respectively, kcal/hr · °C
$\quad\quad c$ = specific heat of packing, kcal/kg · °C
$\quad\quad d$ = particle diameter, m
$\quad\quad \rho_s$ = density of solid, kg/m³

For two infinite parallel plates or concentric cylinders or spheres with diffuse radiation transfer from one to the other,

$$F_{12} = 1 \left/ \left[\frac{1}{\epsilon_1} + \frac{A_1}{A_2}\left(\frac{1}{\epsilon_2} - 1\right)\right]\right. \tag{77.11}$$

If A_1 is a small body in a large enclosure, $F_{12} = \epsilon_1$. If radiator or receiver has an emissivity that varies with temperature, or if radiation is spectral, F_{12} must be found from a detailed statistical analysis of the various possible radiant beams.[30]

Table 77.5 lists emissivities for several surfaces of low emissivity that are useful in vacuum insulation.[31]

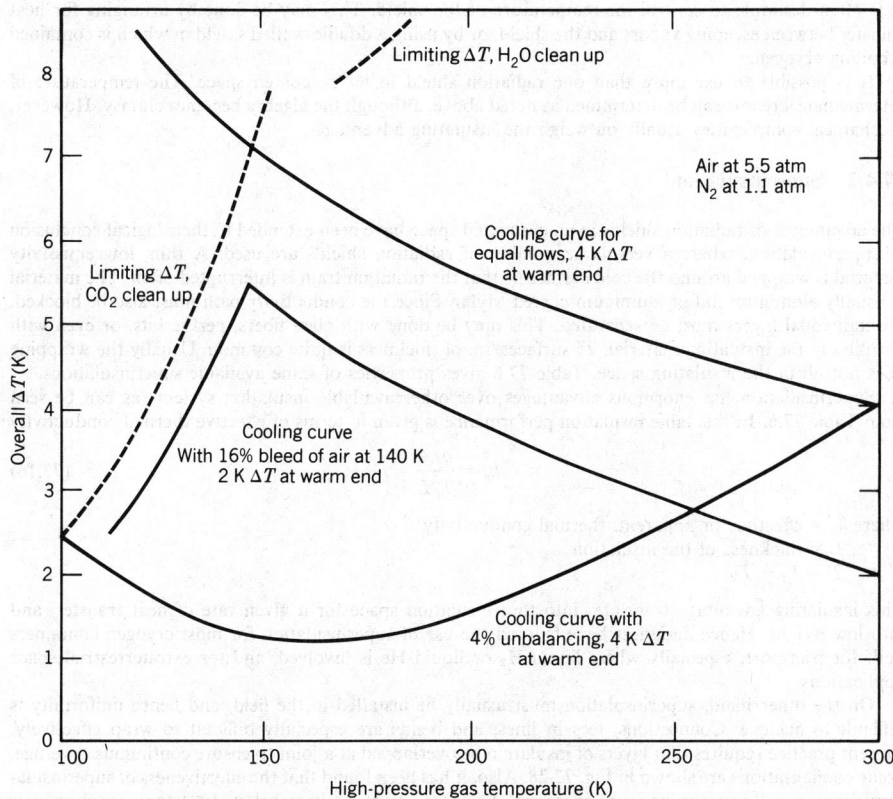

Fig. 77.27 ΔT limitation for contaminant cleanup in a regenerator.

The form of the Stefan–Boltzmann equation shows that the rate of radiant energy transfer is controlled by the temperature of the hot surface. If the vacuum space is interrupted by a shielding surface, the temperature of that surface will become T_s, so that

$$q/A = F_{1s}(T_1^4 - T_s^4) = F_{s2}(T_s^4 - T_2^4) \tag{77.12}$$

Since q/A will be the same through each region of this vacuum space, and assuming $F_{1s} = F_{s2} = F_{12}$

$$T_s = \sqrt[4]{\frac{T_1^4 + I_2^4}{2}} \tag{77.13}$$

Table 77.5 Emissivities of Materials Used for Cryogenic Radiation Shields

Material	Emissivity at		
	300 K	77.8 K	4.33 K
Aluminum plate	0.08	0.03	
Aluminum foil (bright finish)	0.03	0.018	0.011
Copper (commercial polish)	0.03	0.019	0.015
Monel	0.17	0.11	
304 stainless steel	0.15	0.061	
Silver	0.022		
Titanium	0.11		

It is often desirable to control the temperature of the shield. This may be done by arranging for heat transfer between escaping vapors and the shield, or by using a double-walled shield in which is contained a boiling cryogen.

It is possible to use more than one radiation shield in an evacuated space. The temperature of intermediate streams can be determined as noted above, although the algebra becomes clumsy. However, mechanical complexities usually outweigh the insulating advantages.

77.4.2 Superinsulation

The advantages of radiation shields in an evacuated space have been extended to their logical conclusion in superinsulation, where a very large number of radiation shields are used. A thin, low emissivity material is wrapped around the cold surface so that the radiation train is interrupted often. The material is usually aluminum foil or aluminum-coated Mylar. Since the conductivity path must also be blocked, the individual layers must be separated. This may be done with glass fibers, perlite bits, or even with wrinkles in the insulating material; 25 surfaces/in. of thickness is quite common. Usually the wrapping does not fill in the insulating space. Table 77.6 gives properties of some available superinsulations.

Superinsulation has enormous advantages over other available insulation systems as can be seen from Table 77.6. In this table insulation performance is given in terms of effective thermal conductivity

$$k_e = \frac{q/A}{T/L} \tag{77.14}$$

where k_e = effective, or apparent, thermal conductivity
L = thickness of the insulation
$T = T_1 - T_2$

This insulating advantage translates into thin insulation space for a given rate of heat transfer, and into low weight. Hence designers have favored the use of superinsulation for most cryogen containers built for transport, especially where liquid H_2 or liquid He is involved, and for extraterrestrial space applications.

On the other hand, superinsulation must usually be installed in the field, and hence uniformity is difficult to achieve. Connections, tees in lines, and bends are especially difficult to wrap effectively. Present practice requires that layers of insulation be overlapped at a joint to ensure continuous coverage. Some configurations are shown in Fig. 77.28. Also, it has been found that the effectiveness of superinsulation drops rapidly as the pressure increases. Pressures must be kept below 10^{-3} torr; evacuation is slow; a getter is required in the evacuated space; and all joints must be absolutely vacuum tight. Thus the total system cost is high.

77.4.3 Insulating Powders and Fibers

Fibers and powders have been used as insulating materials since the earliest of insulation needs. They retain the enormous advantage of ease of installation, especially when used in air, and low cost. Table 77.7 lists common insulating powders and fibers along with values of effective thermal conductivity.[32]

Table 77.6 Properties of Various Multilayer Insulations (Warm Wall at 300 K)

Sample Thickness (cm)	Shields per Centimeter	Density (g/cm³)	Cold Wall T (K)	Conductivity (μW/cm · K)	Material[a]
3.7	26	0.12	76	0.7	1
3.7	26	0.12	20	0.5	1
2.5	24	0.09	76	2.3	2
1.5	76	0.76	76	5.2	3
4.5	6	0.03	76	3.9	4
2.2	6	0.03	76	3.0	5
3.2	24	0.045	76	0.85	5
1.3	47	0.09	76	1.8	5

[a] 1. Al foil with glass fiber mat separator.
  2. Al foil with nylon net spacer.
  3. Al foil with glass fabric spacer.
  4. Al foil with glass fiber, unbonded spacer.
  5. Aluminized Mylar, no spacer.

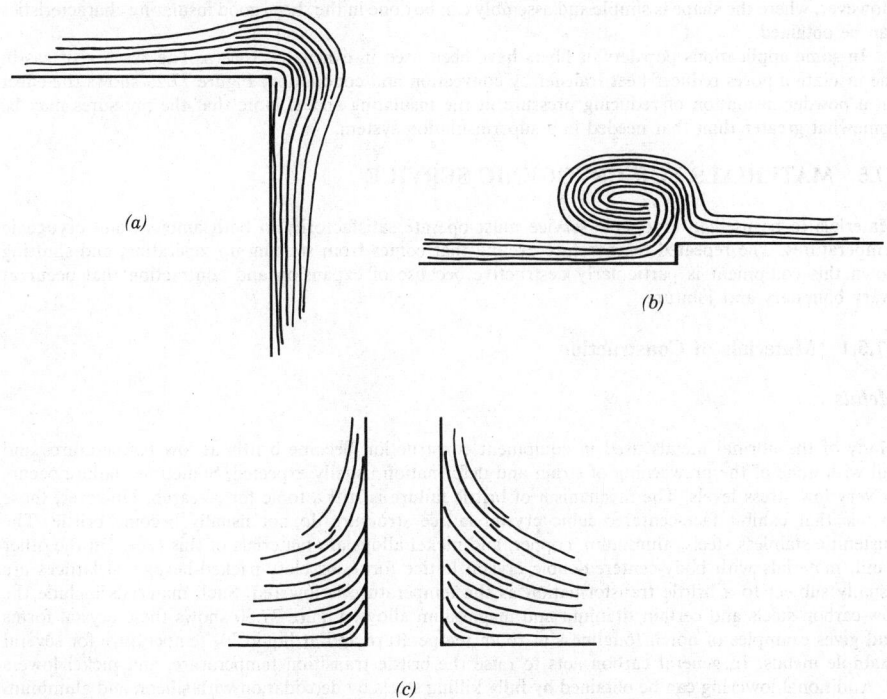

Fig. 77.28 Superinsulation coverage at joints and nozzles: (*a*) Lapped joint at corner. Also usable for nozzle or for pipe bend. (*b*) Rolled joint used at surface discontinuity, diameter change, or for jointure of insulation sections. (*c*) Multilayer insulation at a nozzle.

Since the actual thermal conductivity is a function of temperature, these values may only be used for the temperature ranges shown.

For cryogenic processes of modest size and at temperatures down to liquid nitrogen temperature, it is usual practice to immerse the process equipment to be insulated in a cold box, a box filled with powder or fiber insulation. Insulation thickness must be large, and the coldest units must have the thickest insulation layer. This determines the placing of the process units within the cold box. Such a cold box may be assembled in the plant and shipped as a unit, or it can be constructed in the field. It is important to prevent moisture from migrating into the insulation and forming ice layers. Hence the box is usually operated at a positive gauge pressure using a dry gas, such as dry nitrogen. If rock wool or another such fiber is used, repairs can be made by tunneling through the insulation to the process unit. If an equivalent insulating powder, perlite, is used, the insulation will flow from the box through an opening into a retaining bag. After repairs are made, the insulation may be poured back into the box.

Polymer foams have also been used as cryogenic insulators. Foam-in-place insulations have proven difficult to use because as the foaming takes place cavities are likely to develop behind process units.

Table 77.7 Effective Thermal Conductivity of Various Common Cryogenic Insulating Materials (300 to 76 K)

Material	Gas Pressure (mm Hg)	P (g/cm^2)	K (W/cm·K)
Silica aerogel (250A)	$<10^{-4}$	0.096	20.8×10^{-6}
	N$_2$ at 628	0.096	195.5×10^{-6}
Perlite (+30 mesh)	$<10^{-5}$	0.096	18.2×10^{-6}
	N$_2$ at 628	0.096	334×10^{-6}
Polystyrene foam	Air, 1 atm	0.046	259×10^{-6}
Polyurethane foam	Air, 1 atm	0.128	328×10^{-6}
Foamglas	Air, 1 atm	0.144	346×10^{-6}

However, where the shape is simple and assembly can be done in the shop, good insulating characteristics can be obtained.

In some applications powders or fibers have been used in evacuated spaces. The absence of gas in the insulation pores reduces heat transfer by convection and conduction. Figure 77.29 shows the effect on a powder insulation of reducing pressure in the insulating space. Note that the pressures may be somewhat greater than that needed in a superinsulation system.

77.5 MATERIALS FOR CRYOGENIC SERVICE

Materials to be used in cryogenic service must operate satisfactorily in both ambient and cryogenic temperatures. The repeated temperature cycling that comes from starting up, operating, and shutting down this equipment is particularly destructive because of expansion and contraction that occur at every boundary and jointure.

77.5.1 Materials of Construction

Metals

Many of the normal metals used in equipment construction become brittle at low temperatures and fail with none of the prewarning of strain and deformation usually expected. Sometimes failure occurs at very low stress levels. The mechanism of brittle failure is still a topic for research. However, those metals that exhibit face-centered-cubic crystal lattice structure do not usually become brittle. The austenitic stainless steels, aluminum, copper, and nickel alloys are materials of this type. On the other hand, materials with body-centered-cubic crystal lattice forms or close-packed-hexagonal lattices are usually subject to a brittle transformation as the temperature is lowered. Such materials include the low-carbon steels and certain titanium and magnesium alloys. Figure 77.30 shows these crystal forms and gives examples of notch toughness at room temperature and at liquid N_2 temperature for several example metals. In general carbon acts to raise the brittle transition temperature, and nickel lowers it. Additional lowering can be obtained by fully killing steels by deoxidation with silicon and aluminum and by effecting a fine grain structure through normalizing by addition of selected elements.

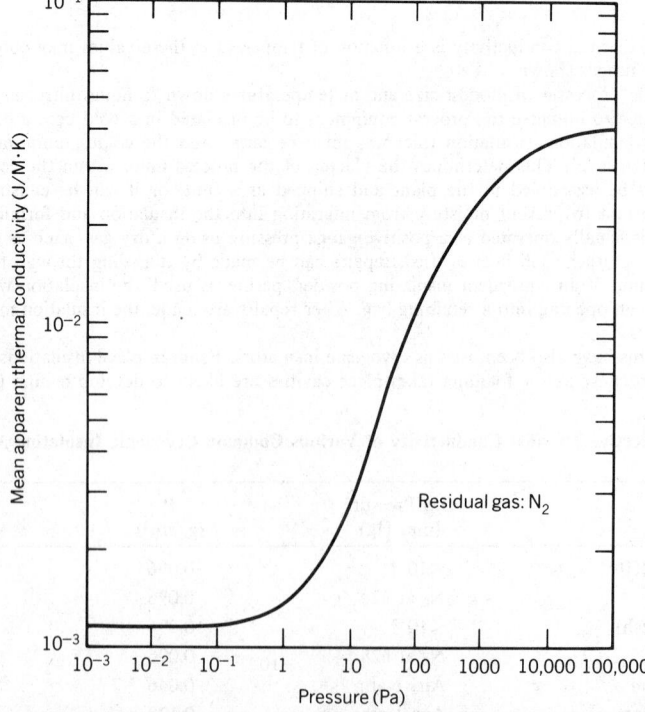

Fig. 77.29 Effect of residual gas pressure on the effective thermal conductivity of a powder insulation—perlite, 30–80 mesh, 300 to 78 K.

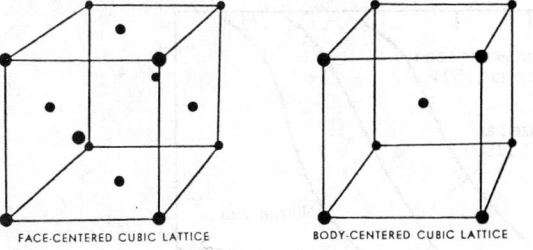
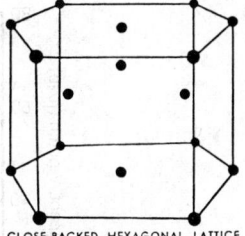

FACE-CENTERED CUBIC LATTICE   BODY-CENTERED CUBIC LATTICE   CLOSE-PACKED HEXAGONAL LATTICE

| Metal | Crystal Lattice | Energy to Break, Foot-pounds Keyhole | |
		Room Temperature	−320°F
Austenitic Stainless Steel	Face-centered Cubic	43	50
Aluminum	Face-centered Cubic	19	27
Copper	Face-centered Cubic	43	50
Nickel	Face-centered Cubic	89	99
Iron	Body-centered Cubic	78	1.5
Titanium	Close-packed Hexagonal	14.5	6.6
Magnesium	Close-packed Hexagonal	4	(3 at −105° F)

(Courtesy—American Society for Metals)

Fig. 77.30 Effect of crystal structure on brittle impact strengths of some metals. (Courtesy, American Society for Metals.)

In selecting a material for cryogenic service, several significant properties should be considered. The toughness or ductibility is of prime importance. Actually, these are distinctively different properties. A material that is ductile, as measured by elongation, may have poor toughness as measured by a notch impact test, particularly at cryogenic temperatures. Thus both these properties should be examined. Figures 77.31 and 77.32 show the effect of nickel content and heat treatment on Charpy impact values for steels. Figure 77.33 shows the tensile elongation before rupture of several materials used in cryogenic service.

Tensile and yield strength generally increase as temperature decreases. However, this is not always true, and the behavior of the particular material of interest should be examined. Obviously if the material becomes brittle, it is unusable regardless of tensile strength. Figure 77.34 shows the tensile and yield strength for several stainless steels.

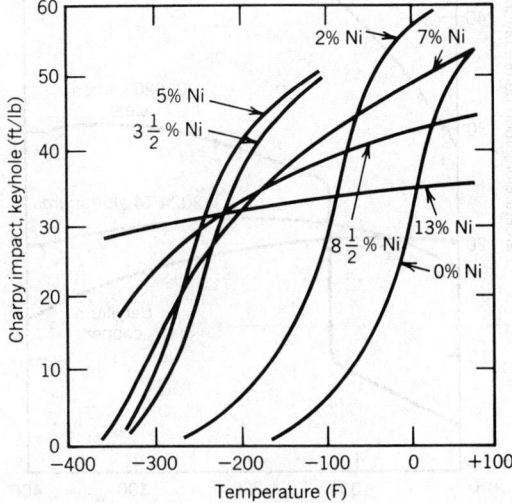

Fig. 77.31 Effect of nickel content in steels on Charpy impact values. (Courtesy, American Iron and Steel Institute.)

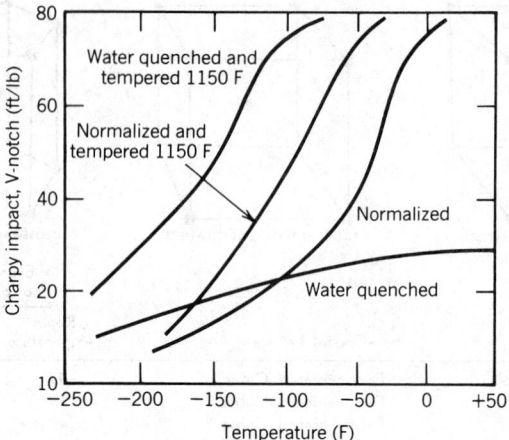

Fig. 77.32 Effect of heat treatment on Charpy impact values of steel. (Courtesy, American Iron and Steel Institute.)

Fatigue strength is especially important where temperature cycles from ambient to cryogenic are frequent, especially if stresses also vary. In cryogenic vessels maximum stress cycles for design are about 10,000–20,000 rather than the millions of cycles used for higher-temperature machinery design. Because fatigue strength data for low-temperature applications are scarce, steels used in cryogenic rotating equipment are commonly designed using standard room-temperature fatigue values. This allows a factor of safety because fatigue strength usually increases as temperature decreases.

Coefficient of expansion information is critical because of the stress that can be set up as temperatures are reduced to cryogenic or raised to ambient. This is particularly important where dissimilar materials

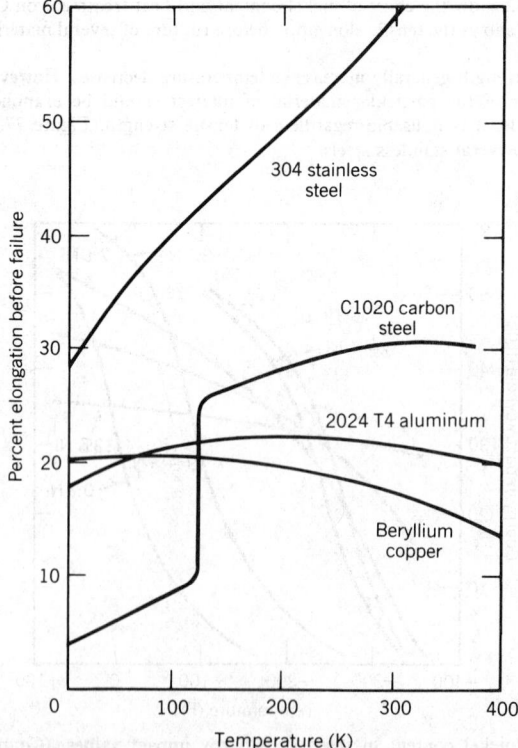

Fig. 77.33 Percent elongation before rupture of some materials used in cryogenic service.[33]

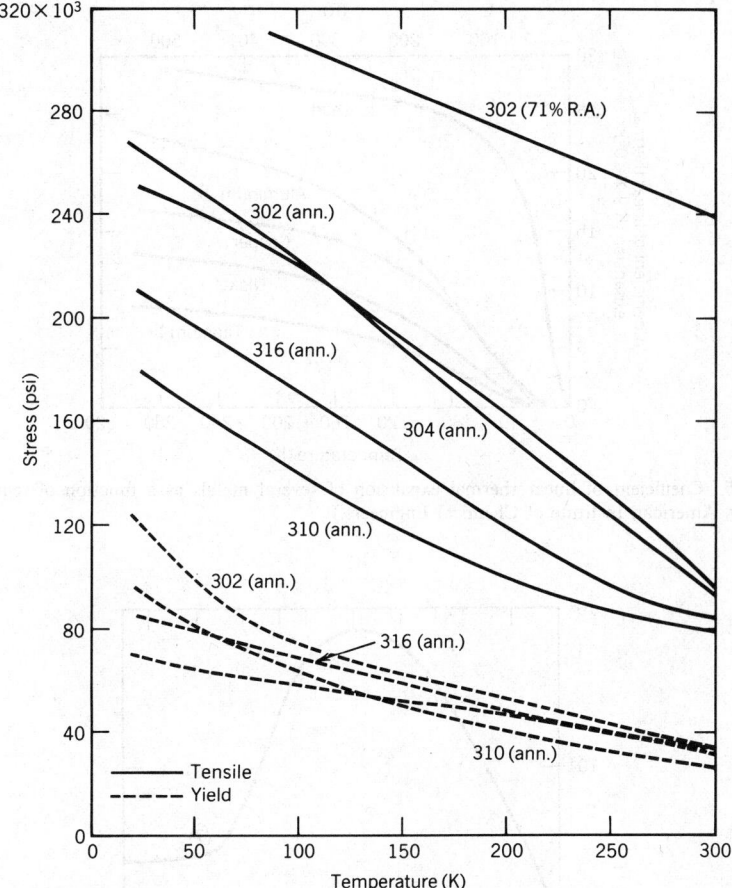

Fig. 77.34 Yield and tensile strength of several AISI 300 series stainless steels.[33] (Courtesy, American Iron and Steel Institute.)

are joined. For example, a 36 ft long piece of 18-8 stainless will contract more than an inch in cooling from ambient to the boiling point of liquid H_2. And stainless steel has a coefficient of linear expansion much lower than that of copper or aluminum. This is seen in Fig. 77.35.

Thermal conductivity is an important property because of the economic impact of heat leaks into a cryogenic space. Figure 77.36 shows the thermal conductivity of some metals in the cryogenic temperature range. Note that pure copper shows a maximum at very low temperatures, but most alloys show only modest effect of temperature on thermal conductivity. One measure of the suitability of a material for cryogenic service is the ratio of tensile strength to thermal conductivity. On this basis stainless steel looks very attractive and copper much less so.

The most common materials used in cryogenic service have been the austenitic stainless steels, aluminum alloys, copper alloys, and aluminum-alloyed steels. Fine grained carbon–manganese steel and aluminum-killed steel and the 2.5% Ni steels can be used to temperatures as low as −50°C. A 3.5% Ni steel may be used roughly to −100°C; 5% Ni steels have been developed especially for applications in liquefied natural gas processing, that is, for temperatures down to about −170°C. Austenitic stainless steels with about 9% Ni such as the common 304 and 316 types are usable well into the liquid He range (−252°C). Aluminum and copper alloys have been used throughout the cryogenic temperature range. However, in selecting a particular alloy for a given application the engineer should consider carefully all of the properties of the material as they apply to that application.

Stainless steel may be joined by welding. However, the welding rod chosen and the joint design must both be selected for the material being welded and the expected service. For example, 9% nickel steel can be welded using nickel-based electrodes and a 60–80° single V joint design. Inert gas welding using Inconel-type electrodes is also acceptable. Where stress levels will not be high types 309 and

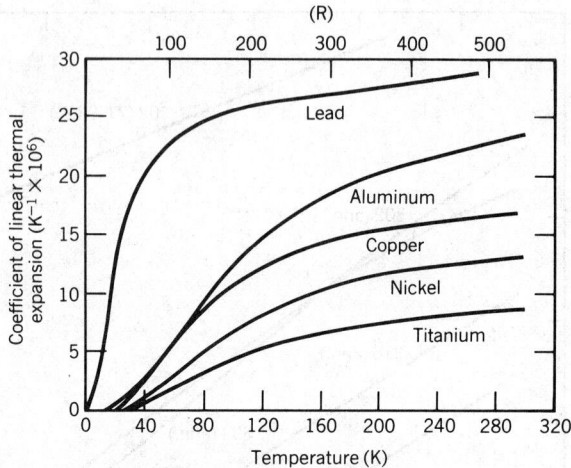

Fig. 77.35 Coefficient of linear thermal expansion of several metals as a function of temperature. (Courtesy, American Institute of Chemical Engineers.)

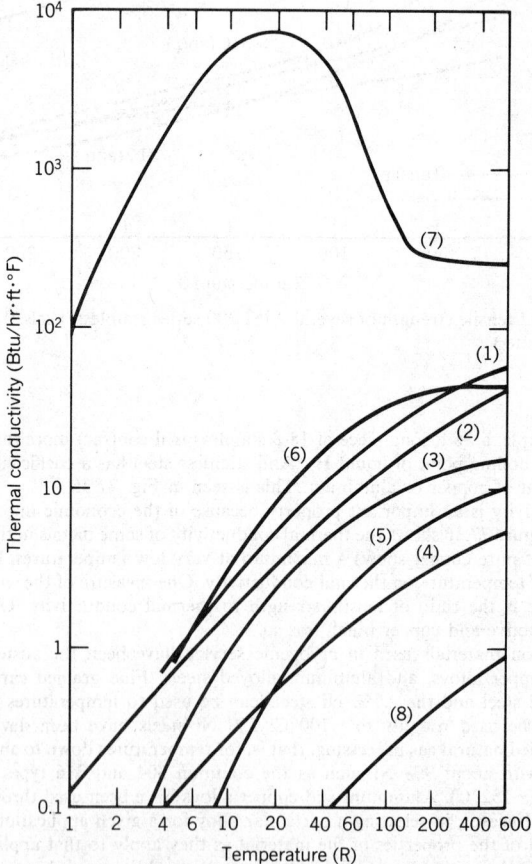

Fig. 77.36 Thermal conductivity of materials useful in low temperature service. (1) 2024TA aluminum; (2) beryllium copper; (3) K-Monel; (4) titanium; (5) 304 stainless steel; (6) C1020 carbon steel; (7) pure copper; (8) Teflon.[35]

310 austenitic-stainless-steel electrodes can be used despite large differences in thermal expansion between the weld and the base metal.

Dissimilar metals can be joined for cryogenic service by soft soldering, silver brazing, or welding. For copper-to-copper joints a 50% tin/50% lead solder can be used. However, these joints have little ductility and so cannot stand high stress levels. Soft solder should not be used with aluminum, silicon-bronze, or stainless steel. Silver soldering is preferred for aluminum and silicon bronze and may also be used with copper and stainless steel.

Polymers

Polymers are frequently used as structural materials in research apparatus, as windows into cryogenic spaces, and for gaskets, O-rings, and other seals. Their suitability for the intended service should be as carefully considered as metals. At this point there is little accumulated, correlated data on polymer properties because of the wide variation in these materials from source to source. Hence properties should be obtained from the manufacturer and suitability for cryogenic service determined case by case.

Tables 77.8 and 77.9 list properties of some common polymeric materials. These are not all the available suitable polymers, but have been chosen especially for their compatibility with liquid O_2. For this service chemical inertness and resistance to flammability are particularly important. In addition to these, nylon is often used in cryogenic service because of its machinability and relative strength. Teflon and similar materials have the peculiar property of losing some of their dimensional stability at low temperatures, thus they should be used in confined spaces or at low stress levels.

Glass

Glasses, especially Pyrex and quartz, have proven satisfactory for cryogenic service because of their amorphous structure and very small coefficient of thermal expansion. They are commonly used in laboratory equipment, even down to the lowest cryogenic temperatures. They have also successfully been used as windows into devices such as hydrogen bubble chambers that are built primarily of metal.

77.5.2 Seals and Gaskets

In addition to careful selection of materials, seals must be specially designed for cryogenic service. Gaskets and O-rings are particularly subject to failure during thermal cycling. Thus they are best if confined and/or constructed of a metal–polymer combination. Such seals would be in the form of metal rings with C or wedge cross sections coated with a sealant such as Kel-F, Teflon, or soft metal. Various designs are available with complex cross sections for varying degrees of deflection. The surfaces against which these seal should be ground to specified finish. Elastomers such as neoprene and Viton-A have proven to be excellent sealants if captured in a space where they are subjected to 80% linear compression. This is true despite the fact that they are both extremely brittle at cryogenic temperatures without this stress.

Adhesive use at low temperatures is strictly done on an empirical basis. Still, adhesives have been used successfully to join insulating and vapor barrier blankets to metal surfaces. In every case the criteria are that the adhesive must not become crystalline at the operating temperature, must be resistant to aging, and must have a coefficient of contraction close to that of the base surface. Polyurethane, silicone, and various epoxy compounds have been used successfully in various cryogenic applications.

77.5.3 Lubricants

The lubrication of cryogenic machinery such as valves, pumps, and expanders is a problem that has generally been solved by avoidance. Valves usually have a long extention between the seat and the packing gland. This extension is gas filled so that the packing gland temperature stays close to ambient. For low-speed bearings babbitting is usually acceptable, as is graphite and molybdenum sulfide. For high-speed bearings, such as those in turboexpanders, gas bearings are generally used. In these devices some of the gas is leaked into the rotating bearing and forms a cushion for rotation. If out-leakage of the contained gas is undesirable, N_2 can be fed to the bearing and controlled so that leakage of N_2 goes to the room and not into the cryogenic system. Bearings of this sort have been operated at speeds up to 100,000 rpm.

77.6 SPECIAL PROBLEMS IN LOW-TEMPERATURE INSTRUMENTATION

Cryogenic systems usually are relatively clean and free flowing, and they often exist at a phase boundary where the degrees of freedom are reduced by one. Although these factors ease measurement problems,

Table 77.8 Properties of Polymers Used in Cryogenic Service

Elastomer Type	Silicone Rubber	Vinylidene Fluoride Hexa-fluoropropylene	Fluorosilicone	Polytrifluoro-chloroethylene
Trade Name	Silastic[a] Silicone Rubber[e,f]	Viton[b] Fluorel[c]	Silastic LS-53[a]	Kel-F[c,d]
Physical and Mechanical Properties				
Durometer range (shore A)	45–60	55–90	50–60	55–90
Specific gravity (base elastomer)	1.17–1.46	1.4–1.85	1.41–1.46	1.4–1.85
Density, lb/in.^{3} (base elastomer)	0.045	0.051–0.067	0.051	0.051–0.067
Tensile strength, psi:				
Pure gum	Under 400	>2000	1000	350–600
Reinforced	600–1500	—	—	—
Elongation, percent:				
Pure gum	Under 200	>350	200	500–800
Reinforced	200–800	—	—	—
Thermal conductivity, g, Btu/hr/ft^2/(°F/ft)	0.13	—	0.13	—
Coefficient of thermal expansion, cubical, in.^{3}/in.^{3}/°F	45×10^{-5}	27×10^{-5}	45×10^{-5}	—
Electrical insulation	Excellent	Excellent	Good	Excellent
Rebound:				
Cold	Very good	Good	Very good	—
Hot	Very good	Excellent	Very good	—
Compression set	Good to excellent	Good to excellent	Good	Good to excellent
Resistance Properties				
Temperature:				
Tensile strength at 250°F, psi	850	300–800	—	300–800
Tensile strength at 400°F, psi	400	150–300	—	150–300
Elongation at 250°F, percent	350	100–350	—	100–350
Elongation at 400°F, percent	200	50–160	—	50–160
Low temperature brittle point, °F	−90 to −200	10 to −60	−90	10 to −60

Property	-60 to -120 [a]	20 to -30 [c]	— [d]	-20 to -30
Low temperature range of rapid stiffening, °F	-60 to -120	20 to -30	—	-20 to -30
Drift, room temperature	Poor to excellent	Good	—	Good
Drift, elevated temperature (158° to 212°F)	Excellent	Good to excellent	Excellent	Good to excellent
Heat aging (212°F)	Excellent	Excellent	Excellent	Excellent
Maximum recommended continuous service temperature, °F	480	450	500	400
Minimum recommended service temperature, °F	-178	-50	-90	-60
Mechanical:				
Tear resistance	Poor	Poor to good	—	Poor to good
Abrasion resistance	Poor	Good	Poor	Good
Impact resistance (fatigue)	Poor	Poor to good	Poor	Poor to good
Chemical:				
Sunlight aging	Excellent	Excellent	Excellent	Excellent
Weather resistance	Excellent	Excellent	Excellent	→
Oxidation	Excellent	Excellent	Excellent	Poor to fair
Acids:				
Dilute	Very good	Excellent	Excellent	Excellent
Concentrated	Good	Good	Very good	Good to excellent
Alkali	Fair to excellent	Poor to fair	Fair to excellent	Excellent
Alcohol	Good	Excellent	Poor to good	Good
Petroleum products, resistance	Poor to fair	Good to excellent	Excellent	
Coal tar derivatives, resistance	Poor	Excellent	—	
Chlorinated solvents, resistance	Good	Good	—	
Hydraulic oils:				
Silicates	Poor	Good	—	Good
Phosphates	Good	Poor	—	Poor
Water swell resistance	Good	Good	Excellent	Good
Permeability to gases	Good	Excellent	Fair	Excellent

[a] Dow Corning Corp.

[b] E. I. duPont de Nemours.

[c] Minnesota Mining and Manufacturing Co.

[d] CTFE compounded with vinylidine fluoride.

[e] General Electric.

[f] Union Carbon and Carbide.

Table 77.9 Properties of Polymers Used in Cryogenic Service

Common Name	Fluorinated Ethylene Propylene	Polychlorotrifluoroethylene	Polyvinylidene fluoride	Polytetrafluoroethylene	Polyimide
Trade Name	Teflon FEP[a]	Kel-F[b]	Kynar[c]	Fluorosint[d,e] Teflon TFE[a] Halon TFE[f]	Kapton H[a] Kapton F[a] Vespel[a] Polymer SP-1[a]
Physical and Mechanical Properties					
Specific gravity	2.14–2.17	2.1–2.2	1.76–1.77	2.13–2.22	1.42
Tensile strength, psi	2700–3100	4500–6000	7000	2000–4500	25,000[g]; 10,500
Elongation, percent	250–330	30–250	100–300	200–400	70[g]; 6–8
Tensile modulus, psi	0.5×10^5	$1.5 \times 10^5 – 3 \times 10^5$	1.2×10^5	0.58×10^5	4.3×10^5
Compressive strength, psi	2200	32,000–80,000	10,000	1700	24,400
Flexural strength, psi	—	7400–9300	—	—	14,000
Impact strength, ft-lb/in. of notch	No break	0.8–5.0	3.5	3.0	0.9
Rockwell hardness	R25	R110–R115	D80 (Shore)	D50–D65 (Shore)	H85–H95
Thermal conductivity, Btu/hr/ft²/(°F/in.)	1.75	0.9	0.9	1.75	2.2
Specific heat, Btu/lbm/°F	0.28	0.22	0.33	0.25	0.27
Coefficient of linear expansion, in./in./°F $\times 10^{-5}$	4.7×10^{-5} to 5.8×10^{-5}	5×10^{-5} to 15×10^{-5}	6.7×10^{-5}	5.5×10^{-5}	28×10^{-5} to 35×10^{-5}
Volume resistivity, ohm-cm	$>2 \times 10^{18}$	1.2×10^{18}	2×10^{14}	$>10^{18}$	10^{18}
Clarity	Transparent to translucent	Transparent to translucent	Transparent to translucent	Opaque	Opaque
Processing Properties					
Molding qualities	Excellent	Excellent	Excellent	—	—
Injection molding temperature, °F	625–760	440–600	450–550	—	—

Mold shrinkage, in./in.	0.03–0.06	0.005–0.010	0.030	—	—
Machining qualities	Excellent	Excellent	Excellent	Excellent	—
Resistance Properties					
Mechanical abrasion and wear Tabor CS 17 wheel mg, loss/1000 cycles	—	0.01	17.6	—	—
Temperature:					
Flammability	None	None	Self-extinguishing	None	—
Low temperature brittle point, °F	−420	−400	−80	−420	—
Resistance to heat, °F (continuous)	400	350–390	300	550	500
Deflection temperature under load, °F	—	258 (66 psi)	300 (66 psi), 195 (264 psi)	250 (66 psi)	—
Chemical:					
Effect of sunlight	None	None	Slight bleaching on long exposure	None	Degrades after prolonged exposure
Effect of weak acids	None →	Halogenated compounds cause slight swelling →	None	None →	None
Effect of strong acids			Attacked by fuming sulfuric		—
Effect of weak alkalies			None		—
Effect of strong alkalies			None		Attacked
Effect of organic solvents			Resists most solvents		Resistant to most organic solvents

a E. I. duPont de Nemours.
b Minnesota Mining and Manufacturing Co.
c Pennsalt Chemicals Corp.
d Polymer Corp. of Pennsylvania.
e Polypenco, Inc.
f Allied Chemical Corp.
g Film.

Table 77.10   Defining Fixed Points of the International Practical Temperature Scale, 1968

Equilibrium Point	T (K)
Triple point of equilibrium H_2	13.81
Boiling point of equilibrium H_2 ($P = 33330.6$ N/m²)	17.042
Boiling point of equilibrium H_2 ($P = 1$ atm)	20.28
Boiling point of neon ($P = 1$ atm)	27.102
Triple point of O_2	54.361
Boiling point of O_2 ($P = 1$ atm)	90.188
Triple point of H_2O ($P = 1$ atm)	273.16
Freezing point of Zn ($P = 1$ atm)	692.73
Freezing point of Ag ($P = 1$ atm)	1235.08
Freezing point of Au ($P = 1$ atm)	1337.58

the fact that the system is immersed in insulation and therefore not easily accessible, the desire to limit thermal leaks to the system, and the likelihood that vaporization or condensation will occur in instrument lines all add difficulties.

Despite these differences all of the standard measurement techniques are used with low-temperature systems, often with ingenious changes to adapt the device to low-temperature use.

77.6.1   Temperature Measurement

Temperature may be measured using liquid-in-glass thermometers down to about −40°C, using thermocouples down to about liquid H_2 temperature, and using resistance thermometers and thermistors down to about 1 K. Although these are the usual devices of engineering measurement laboratory measurements have been done at all temperatures using gas thermometers and vapor pressure thermometers.

Table 77.10 lists the defining fixed points of the International Practical Temperature Scale of 1968. This scale does not define fixed points below the triple point of equilibrium He.[36] Below that range the NBS has defined a temperature scale to 1 K using gas thermometry.[37] At still lower temperatures measurement must be based on the fundamental theories of solids such as paramagnetic and superconducting phenomena.[38]

The usefulness of vapor pressure thermometry is limited by the properties of available fluids. This is evident from Table 77.11. For example, in the temperature range from 20.4 to 24.5 K there is no usable material. Despite this, vapor pressure thermometers are accurate and convenient. The major problem in their use in that the hydraulic head represented by the vapor line between point of measurement and the readout point must be taken into account. Also, the measurement point must be the coldest point experienced by the device. If not, pockets of liquid will form in the line between the point of measurement and the readout point greatly affecting the reading accuracy.

Standard thermocouples may be used through most of the cryogenic range, but, as shown in Fig. 77.37 for copper–constantan, the sensitivity with which they measure temperature drops as the temperature decreases. At low temperatures heat transfer down the thermocouple wire may markedly affect

Table 77.11   Properties of Cryogens Useful in Vapor Pressure Thermometers

Substance	Triple Point (K)	Boiling Point (K)	Critical Point (K)	dP/dT (mm/K)	Hydraulic Head at Boiling Point (K/cm²)
³He	—	3.19	3.32	790	0.000054
⁴He	—	4.215	5.20	715	0.00013
p-H_2 (20.4 K equilibrium)	13.80	20.27	32.98	224	0.00023
Ne	24.54	27.09	44.40	230	0.0039
N_2	63.15	77.36	126.26	89	0.0067
Ar	83.81	87.30	150.70	80	0.013
O_2	54.35	90.18	154.80	79	0.011

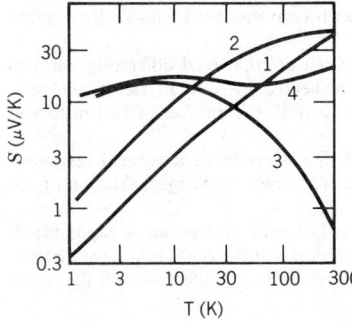

Fig. 77.37 Thermoelectric power of some thermocouples useful for cryogenic temperature measurement (courtesy, Plenum Press): (1) Copper versus constantan; (2) Au + z at % Co versus silver normal (Ag + 0.37 at % Au); (3) Au + 0.03 at % Fe versus silver normal; (4) Au + 0.03 at % Fe versus Chromel.

the junction temperature. This is especially dangerous with copper wires, as can be seen from Fig. 77.36. Also, some thermocouple materials, for example, iron, become brittle as temperature decreases. To overcome these difficulties special thermocouple pairs have been used. These usually involve alloys of the noble metals. Figure 77.37 shows the thermoelectric power, and hence sensitivity of three of these thermocouple pairs.

Resistance thermometers are also very commonly used for cryogenic temperature measurement. Metal resistors, especially platinum, can be used from ambient to liquid He temperatures. They are extremely stable and can be read to high accuracy. However, expensive instrumentation is required because resistance differences are small requiring precise bridge circuitry. Resistance as a function of temperature for platinum is well known.[36]

At temperatures below 60 K, carbon resistors have been found to be convenient and sensitive temperature sensors. Since the change in resistance per given temperature difference is large (580 ohms/K would be typical at 4 K) the instrument range is small, and the resistor must be selected and calibrated for use in the narrow temperature range required.

Germanium resistors that are single crystals of germanium doped with minute quantities of impurities are also used throughout the cryogenic range. Their resistance varies approximately logarithmically with temperature, but the shape of this relation depends on the amount and type of dopant. Again, the germanium semiconductor must be selected and calibrated for the desired service.

Thermistors, that is, mixed, multicrystal semiconductors, like carbon and germanium resistors, give exponential resistance calibrations. They may be selected for order-of-magnitude resistance changes over very short temperature ranges or for service over wide temperature ranges. Calibration is necessary and may change with successive temperature cycling. For this reason they should be temperature-cycled several times before use. These sensors are cheap, extremely sensitive, easily read, and available in many forms. Thus they are excellent indicators of modest accuracy but of high sensitivity, such as sensors for control action. They do not, however, have the stability required for high accuracy.

77.6.2 Flow Measurement

Measurement of flow in cryogenic systems is often made difficult because of the need to deal with a liquid at its boiling point. Thus any significant pressure drop causes vaporization, which disrupts the measurement. This may be avoided by subcooling the liquid before measurement. Where this is possible, most measurement problems disappear, for cryogenic fluids are clean, low-viscosity liquids. Where subcooling is not possible, flow is most often measured using turbine flow meters or momentum meters.

A turbine meter has a rotor mounted axially in the flow stream and moved by the passing fluid. The rate of rotation, which is directly proportional to the volumetric flow rate, is sensed by an electronic counter that senses the passage of each rotor blade. There are two problems in the use of turbine meters in cryogenic fluids. First, these fluids are nonlubricating. Hence the meter rotor must be self-lubricated. Second, during cool-down or warm-up slugs of vapor are likely to flow past the rotor. These can flow rapidly enough to overspeed and damage the rotor. This can be avoided by locating a bypass around the turbine meter shutting off the meter during unsteady operation.

Momentum meters have a bob located in the flow stream to the support of which a strain gage is attached. The strain gage measures the force on the bob, which can be related through drag calculations or correlation to the rate of fluid flow past the bob. These meters are flexible and can be wide of range. They are sensitive to cavitation problems and to overstrain during upsets. Generally, each instrument must be calibrated.

77.6.3 Tank Inventory Measurement

The measurement of liquid level in a tank is made difficult by the cryogenic insulation requirements. This is true of stationary tanks, but even more so when the tank is in motion, as on a truck or space-ship, and the liquid is sloshing.

The simplest inventory measurement is by weight, either with conventional scales or by a strain gage applied to a support structure.

The sensing of level itself can be done using a succession of sensors that read differently when in liquid than they do in vapor. For instance, thermistors can be heated by a small electric current. Such devices cool quickly in liquid, and a resistance meter can "count" the number of thermistors in its circuit that are submerged.

A similar device that gives a continuous reading of liquid depth would be a vertical resistance wire, gently heated, while the total wire resistance is measured. The cold, submerged, fraction of the wire can be easily determined.

Other continuous reading devices include pressure gages, either with or without a vapor bleed, that read hydrostatic head, capacitance probes that indicate the fraction of their length that is submerged, ultrasonic systems that sense the time required for a wave to return from its reflectance off the liquid level, and light-reflecting devices.

77.7 EXAMPLES OF CRYOGENIC PROCESSING

Here three common, but greatly different, cryogenic technologies are described so that the interaction of the cryogenic techniques discussed above can be shown.

77.7.1 Air Separation

Oxygen, nitrogen, and argon are all major items of commerce used in many applications. In 1977 oxygen production was 410 billion standard cubic feet (SCF). Steel making was the largest consumer with the chemical industry second with oxygen being used in such processes as ethylene and propylene oxide production and with heavy uses proposed in coal gasification projects. Other uses include waste water treatment, welding and cutting, and hospital and aviation breathing devices. These uses combine to make oxygen the third or fourth largest production chemical in the United States. A maximum consumption of 430 billion SCF was consumed in 1981, but consumption fell sharply and in 1983 stood at 360 billion SCF. Nitrogen is widely used for inert atmosphere generation in the metals, electronics, and chemical industries and as a source of deep refrigeration, especially for food freezing and transportation. Its 1977 production was 330 billion SCF placing it sixth among the chemicals produced in the United States. Production rose steadily to 530 billion SCF in 1983. Argon, mainly used in welding and in the making of stainless steel by the argon–oxygen decarburization process, has a demand only about 1.5% of that of oxygen, but represents about 20% of the value of oxygen shipments.

Since all of the industrial gases are expensive to ship long distances, the industry was developed by locating a large number of plants close to markets and sized to meet nearby market demand. Maximum oxygen plant size has now grown into the 1000 ton/day range, but these plants are also located close to the consumer with the product delivered by pipe line. Use contracts are often long-term take-or-pay rental arrangements.

Air is a mixture of about the composition shown in Table 77.12. In an air separation plant O_2 is typically removed and distilled from liquefied air. N_2 may also be recovered. In large plants argon may be recovered in a supplemental distillation operation. In such a plant the minor constituents (H_2–Xe) would have to be removed in bleed streams, but they are rarely collected. When this is done the Ne, Kr, Xe are usually adsorbed onto activated carbon at low temperature and separated by laboratory distillation.

Table 77.12 Approximate Composition of Dry Air

Component	Composition (mole %)
N_2	78.03
O_2	20.99
Ar	0.93
CO_2	0.03
H_2	0.01
Ne	0.0015
He	0.0005
Kr	0.00011
Xe	0.000008

Figure 77.38 is a simplified flow sheet of a typical merchant oxygen plant meeting a variety of O_2 needs. Argon is not separated, and no use is made of the effluent N_2. Inlet air is filtered and compressed in the first of four compression stages. It is then sent to an air purifier where the CO_2 is removed by reaction with a recycling NaOH solution in a countercurrent packed tower. Usually the caustic solution inventory is changed daily. The CO_2-free gas is returned to the compressor for the final three stages after each of which the gas is cooled and water is separated from it. The compressed gas then goes to an adsorbent drier where the remaining water is removed onto silica gel or alumina. Driers are usually switched each shift and regenerated by using a slip stream of dry, hot N_2 and cooled to operating temperature with unheated N_2 flow.

The compressed, purified air is then cooled in the main exchanger (here a coiled tube type, but more usually of the plate-fin type) by transferring heat to both the returning N_2 and O_2. The process is basically a variation of that invented by Georges Claude where part of the high-pressure stream is withdrawn to the expansion engine (or turbine). The remainder of the air is further cooled in the main exchanger and expanded through a valve.

The combined air stream, nearly saturated or partly liquefied, enters the bottom of the high-pressure column. This distillation column condenses nearby pure N_2 at its top using boiling O_2 in the low-pressure column as heat sink. If the low-pressure column operates at about 140 kN/m^2 (20 psia), the high-pressure column must operate at about 690 kN/m^2 (100 psia). The bottom product, called crude O_2, is about 65 mole % N_2. The top product from the high-pressure column, nearly pure N_2, is used as N_2 reflux in the low-pressure column.

The crude O_2 is fed to an activated carbon bed where hydrocarbons are removed, is expanded to low-pressure column pressure, goes through a subcooler in which it supplies refrigeration to the liquid O_2 product, and is fed to the low-pressure column. The hydrocarbons removed in the adsorber may come in as impurities in the feed or may be generated by decomposition of the compressor oil. If they are not fully removed, they are likely to precipitate in the liquid O_2 at the bottom of the low-pressure column. They accumulate there and can form an explosive mixture with oxygen whenever the plant is warmed up. Acetylene is especially dangerous in this regard because it is so little soluble in liquid oxygen.

The separation of O_2 and N_2 is completed in the low-pressure column. In the column, argon accumulates below the crude O_2 feed and may be withdrawn at about 10 mole % for further distillation. If it is not so removed, it leaves as impurity in the N_2 product. Light contaminants (H_2 and He) must be removed periodically from the top of the condenser/reboiler. Heavy contaminants are likely to leave as part of the O_2 product.

This plant produces O_2 in three forms: liquid, high-pressure O_2 for cylinder filling, and lower-pressure O_2 gas for pipe line distribution. The liquid O_2 goes directly from the low-pressure column to the storage tank. The rest of the liquid O_2 product is pumped to high pressure in a plunger pump after it is subcooled so as to avoid cavitation. This high-pressure liquid is vaporized and heated to ambient in the main heat exchanger. An alternate approach would be to warm the O_2 to ambient at high-pressure column pressure and then compress it as a gas. Cylinder pressure is usually too great for a plate-and-fin exchanger, so if the option shown in this flow sheet is used, the main exchanger must be of the coiled tube sort.

The nitrogen product, after supplying some refrigeration to the N_2 reflux, is warmed to ambient in the shell of the main exchanger. Here the N_2 product is shown as being vented to atmospheric. However, some of it would be required to regenerate the adsorbers and to pressurize the cold box in which the distillation columns, condenser/reboiler, main exchanger, hydrocarbon adsorber, subcoolers, throttling valves, and the liquid end of the liquid oxygen pump are probably contained.

This process is self-cooling. At startup refrigeration needed to cool the unit to operating temperatures is supplied by the expansion engine and the three throttling valves. During that time the unit is probably run at maximum pressure. During routine operation that pressure may be reduced. The lower the liquid O_2 demand, the less refrigeration is required and the lower the operating pressure may be.

77.7.2 Liquefaction of Natural Gas

Natural gas liquefaction has been commercially done in two very different situations. Companies that distribute and market natural gas have to meet a demand curve with a sharp maximum in midwinter. It has been found to be much more economic to maintain a local supply of natural gas liquid that can be vaporized and distributed at peak demand time than to build the gas pipe line big enough to meet this demand and to contract with the supplier for this quantity of gas. Thus the gas company liquefies part of its supply all year. The liquid is stored locally until demand rises high enough to require augmenting the incoming gas. Then the stored liquid is vaporized and added to the network. These "peak-shaving" plants consist of a small liquefier, an immense storage capacity, and a large capacity vaporizer. They can be found in most large metropolitan areas where winters are cold, especially in the northern United States, Canada, and Europe.

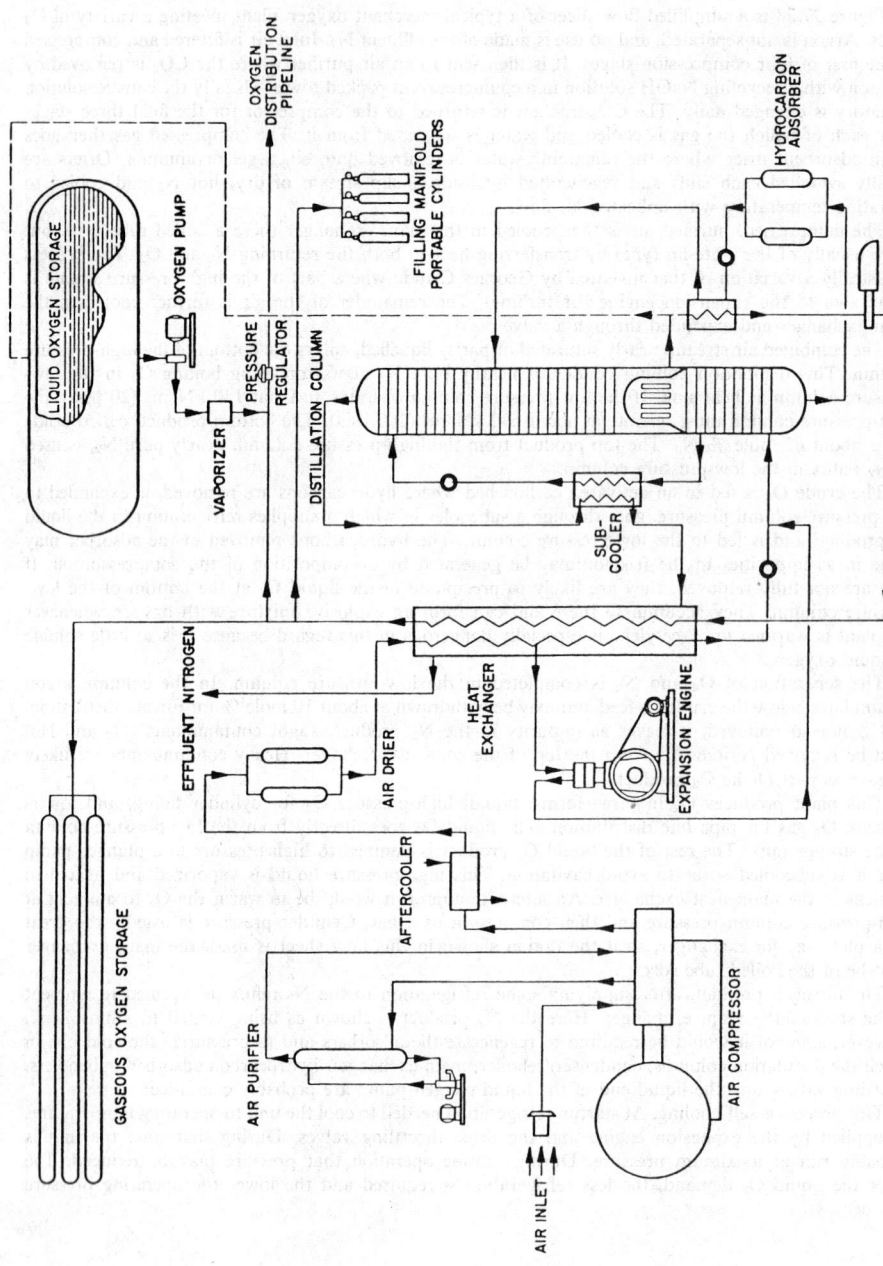

Fig. 77.38 Flow sheet of a merchant oxygen plant. (Courtesy, Air Product and Chemicals, Inc.)

OXYGEN DISTRIBUTION PIPELINE

OXYGEN PUMP

LIQUID OXYGEN STORAGE

HYDROCARBON ADSORBER

LIQUID OXYGEN PUMP

FILLING MANIFOLD PORTABLE CYLINDERS

VAPORIZER

PRESSURE REGULATOR

DISTILLATION COLUMN

SUB-COOLER

EFFLUENT NITROGEN

HEAT EXCHANGER

AIR DRIER

EXPANSION ENGINE

AFTERCOOLER

GASEOUS OXYGEN STORAGE

AIR PURIFIER

AIR COMPRESSOR

AIR INLET

The second situation is that of the oil/gas field itself. These fields are likely to be at long distances from the market. Oil can be readily transported, since it is in a relatively concentrated form. Gas is not. This concentration is done by liquefaction prior to shipment, thus reducing the volume about 600-fold. Subsequently, revaporization occurs at the port near the market. These "base-load" LNG systems consist of a large liquefaction plant, relatively modest storage facilities near the source field, a train of ships moving the liquid from the field to the port near the market, another storage facility near the market, and a large capacity vaporizer. Such a system is a very large project. Because of the large required investment, world political and economic instability, and safety and environmental concerns in some developed nations, especially the United States, only a few such systems are now in operation or actively in progress.

Peak-Shaving Plants

The liquefaction process in a peak-shaving installation is relatively small capacity, since it will be operating over the bulk of the year to produce the gas required in excess of normal capacity for two to six weeks of the year. It usually operates in a region of high energy cost but also of readily available mechanical service and spare parts, and it liquefies relatively pure methane. Finally, operating reliability is not usually critical because the plant has capacity to liquefy the required gas in less time than is the maximum available.

For these reasons efficiency is more important than system reliability and simplicity. Cascade and various expander cycles are generally used, although a wide variety of processes have been used including the Stirling cycle.

Figure 77.39 shows a process in which an N_2 expander cycle is used for low-temperature refrigeration, whereas the methane itself is expanded to supply intermediate refrigeration. This is done because of the higher efficiency of N_2 expanders at low temperature and the reduced need for methane purification. The feed natural gas is purified and filtered and then split into two streams. The larger is cooled in part of the main exchanger, expanded in a turboexpander, and rewarmed to supply much of the warm end refrigeration, after which it is sent to the distribution system. The smaller fraction is cooled both by methane and by N_2 refrigeration until it is largely liquid, whereupon it goes to storage. Heavier liquids are removed by phase separation along the cooling path. Low-temperature refrigeration is supplied by a two-stage Claude cycle using N_2 as working fluid.

The LNG is stored in very large, insulated storage tanks. Typically such a tank might be 300 ft in diameter and 300 ft high. The height is made possible by the low density of LNG compared to other hydrocarbon liquids. LNG tanks have been built in ground as well as aboveground and of concrete as well as steel. However, the vast majority are aboveground steel tanks.

In designing and building LNG tanks the structural and thermal requirements added to the large size lead to many special design features. A strong foundation is necessary, and so the tank is often set on a concrete pad placed on piles. At the same time the earth underneath must be kept from freezing and later thawing and heaving. Thus electric cables or steam pipes are buried in the concrete to keep the soil above freezing. Over this pad a structurally sound layer of insulation, such as foam glass, is put to reduce heat leak to the LNG. The vertical tank walls are erected onto the concrete pad. The inner one is of stainless steel, the outer one is usually of carbon steel, and the interwall distance would be about 4 ft. The walls are field erected with welders carried in a tram attached to the top of the wall and lifted as the wall proceeds. The wall thickness is, of course, greater at the bottom than it is higher up.

The floor of the tank is steel laid over the foam glass and attached to the inner wall with a flexible joint. This is necessary because the tank walls will shrink upon cooling and expand when reheated. The dish roof is usually built within the walls over the floor. When the walls are completed, a flexible insulating blanket is put on the inside wall and the rest of the interwall space is filled with perlite. The blanket is necessary to counter the wall movement and prevent settling and crushing the perlite. At the end of construction the roof is lifted into position with slight air pressure. Usually this roof has hanging from it an insulated subroof that also rises and protects the LNG from heat leak to the roof. When this structure is in place, it is welded in and cover plates are put over the insulated wall spaces.

For safety considerations these tanks are usually surrounded by a berm designed to confine any LNG that escapes. LNG fire studies have shown such a fire to be less dangerous than a fire in an equivalent volume of gasoline. Still, the mass of LNG is so large that opportunities for disaster are seen as equally large. The fire danger will be reduced if the spill is more closely confined, and hence these berms tend to be high rather than large in diameter. In fact, a concrete tank berm built by the Philadelphia Gas Works is integral with the outside tank wall. That berm is of prestressed concrete thick enough to withstand the impact of a major commercial airliner crash.

Revaporization of LNG is done in large heat exchangers using air or water as heat sink. Shell-in-tube exchangers, radiators with fan-driven air for warming, and cascading liquid exchangers have all been used successfully, although the air-blown radiators tend to be noisy and subject to icing.

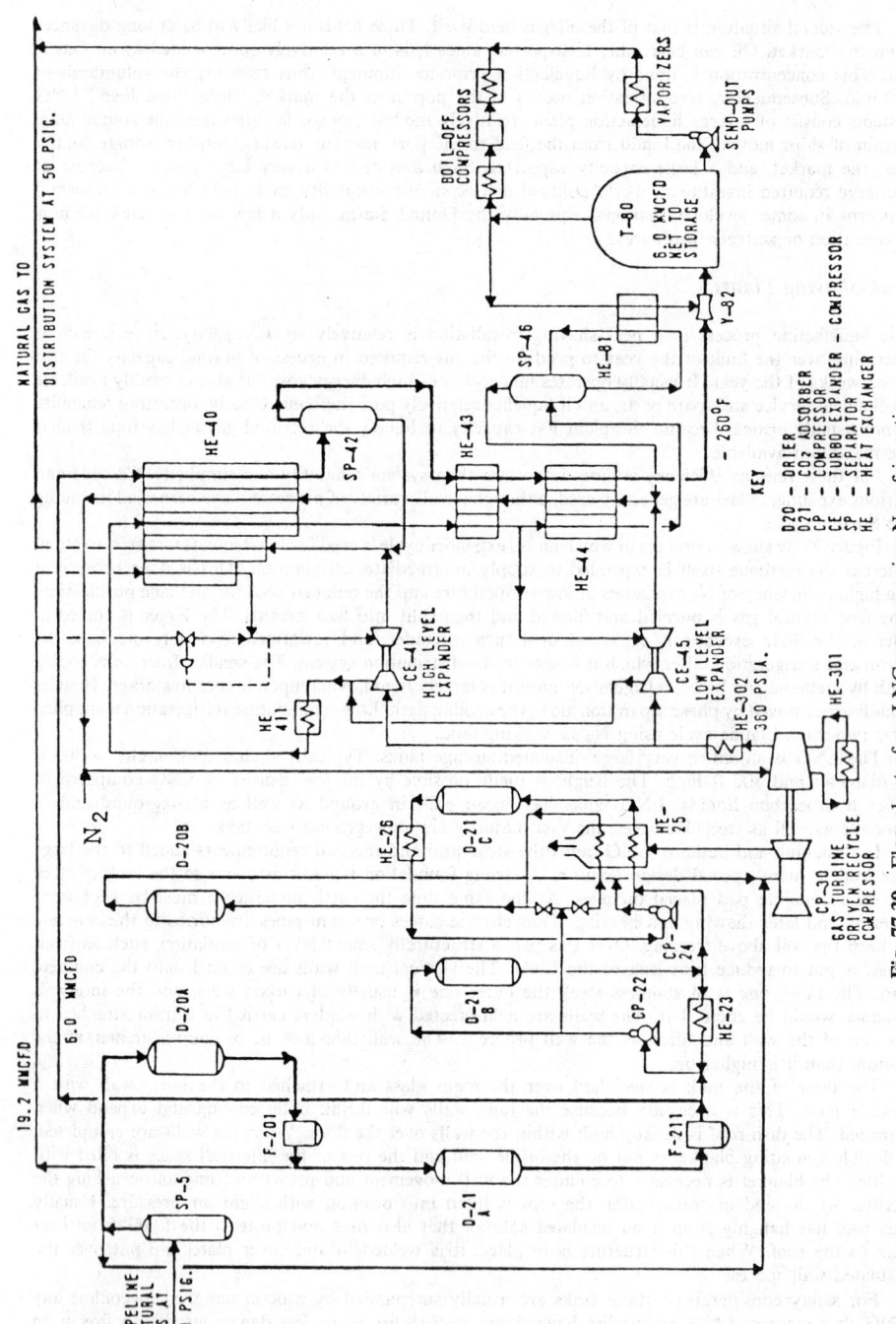

Fig. 77.39 Flow sheet of an LNG process using N_2 refrigeration.

KEY

D20 — DRIER
D21 — CO2 ADSORBER
CP — COMPRESSOR
CE — TURBO EXPANDER – COMPRESSOR
SP — SEPARATOR
HE — HEAT EXCHANGER

Base-Load LNG Plant

Table 77.13 lists the base load LNG plants in operation in 1982. Products from these plants produce much of the natural gas used in Europe and in Japan, but United States use has been low, primarily because of the availability of large domestic gas fields.

In contrast to peak-shaving plants, liquefiers for these projects are large, primarily limited by the size of compressors and heat exchangers available in international trade. Also, these plants are located in remote areas where energy is cheap but repair facilities expensive or nonexistent. Thus, only two types of processes have been used: the classic cascade cycle and the mixed refrigerant cascade. Of these the mixed refrigerant cascade has gradually become dominant because of its mechanical simplicity and reliability.

Figure 77.40 shows a simplified process flow sheet of a mixed refrigerant cascade liquefier for natural gas. Here the natural gas passes through a succession of heat exchangers, or of bundles in a single heat exchanger, until liquefied. The necessary refrigeration is supplied by a multicomponent refrigeration loop, which is essentially a Joule–Thomson cycle with successive phase separators to remove liquids as they are formed. These liquid streams are subcooled, expanded to low pressure, and used to supply the refrigeration required both by the natural gas and by the refrigerant mixture.

The success of this process depends on a selection of refrigerant composition that gives a cooling curve with shape closely matching the shape of the natural gas cooling curve. Thus all heat transfer will be across small ΔTs. This is shown in Fig. 77.41, a cooling curve for a mixed refrigerant cycle.

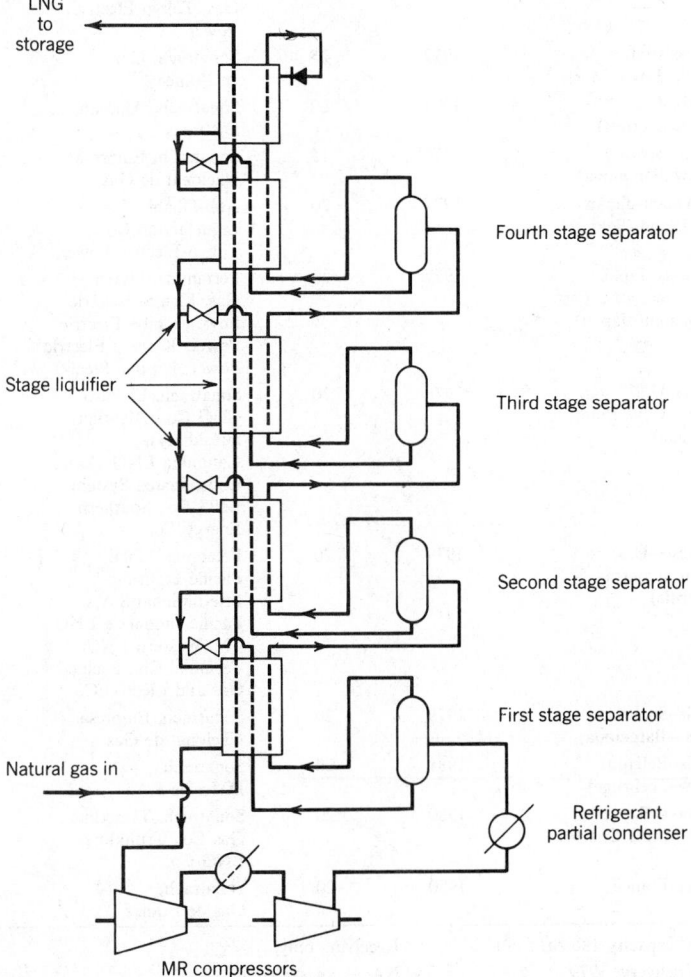

Fig. 77.40 Mixed refrigerant LNG process flow sheet. (Courtesy Plenum Press.)

Table 77.13 Base-Load LNG Plants in Operation, 1982

Trade	Contract Initial Delivery	Contract Term (yr)	Companies Involved	Daily Gas Volume (Million cf.)
Algeria–United Kingdom (Arzew–Canvey Island)	1964	15	Camel, British Gas Corp., Conch Int.	100
Algeria–France (Arzew–Le Harve)	1964	25	Camel, Gaz de France	50
Total cost				
Alaska–Japan (Kenai–Negishi)	1969	15	Phillips Pet. Co., Marathon Oil Co., Tokyo Electric, Tokyo Gas	135[a]
Libya–Italy (Marsa el Brega–La Spezia)	1969	20	Esso Standard Libya, SNAM	235
Libya–Spain (Marsa el Brega–Barcelona)	1969	15	Esso Standard Libya, Gas Natural	110
Total cost				
Brunei–Japan (Lumut–Sodegura Negishi, Semboku)	1972	20	Brunei LNG, Colgas Trading, Osaka Gas, Tokyo Gas, Tokyo Electric Power	737
Algeria–France (Skikda–Fos s/Mer)	1972	25	Sonatrach, Gaz de France	350
Algeria–U.S. (Skikda–Everett)	1971	20	Sonatrach, Alocean, Distrigas	115
Algeria–Spain (Arzew–Barcelona)	1974[b]	15	Sonatrach, Empresa Nacional de Gas	50[c]
Abu-Dhabi–Japan (Das Island–Tokyo)	1977	20	Abu-Dhabi Liquefaction Co., Tokyo Electric Power	450
Indonesia–Japan (North Sumatra, East Kalimantan–Japan)	1977	20	Pertamina, Osaka Gas, Konsai Electric Power, Chiba Electric Power, Kyushu Electric Power, Nippon Steel	1050
Algeria–U.S. (Arzew–Cove Point, Savannah)	1978	20	Sonatrach, El Paso LNG Co. (Algerian Subsidiary), Columbia LNG Corp., Consolidated System LNG Co., Southern Energy Co.	1000
Indonesia–U.S. (North Sumatra–California)	1979	20	Petamina–Mobil, Pacific Lighting, International S.A., Pacific Indonesia LNG Co., Western LNG Terminal Co., Pacific Gas and Electric Co.	550
Algeria–Spain (Skikda–Barcelona)	1979	20	Sonatrach, Empresa Nacional de Gas	436
Algeria–Belgium (Arzew–Zeebruge)	1981	20	Sonatrach, Distrigas S.A.	340[d]
Algeria–U.S. (Arzew–Lake Charles, LA)	1980	20	Sonatrach, Trunkline Gas Co., Trunkline LNG Co.	420
Algeria–France	1980	20	Sonatrach, Gas de France	575

[a] Plant capacity 186 MM cfd.
[b] Full delivery, 1979.
[c] Full delivery 150 MM cfd.
[d] With option for 485 MM cfd.
[e] Receiving only.
[f] NA = not available.
[g] Liquefaction only.
[h] Costs include export facilities for Pacific Lighting Intl., S.A.

Number of Ships and Country of Construction	Capacity of Ships (m³)	Estimated Investment ($ Million)		
		Facilities	Tankers	Total
2, United Kingdom	27,400		28	
1, France	25,500		15	
		205	43	248
2, Sweden	71,500	125	85	210
3, Italy	40,000		78	
1, Spain	40,000		26	
		229	104	333
7, France	75,000	445	325	770
2, France	40,000	190	60	250
1, France	1–129,500	49^e	105	73
1, NA^f	30,000	30^e	NA^f	NA^f
3, Norway	125,000	450^g	650	1100
1, Norway	87,600			
1, Germany	125,000			
7, U.S.	125,000	1365^h	735	2100
3, France	125,000	1320	765	2085
6, U.S.	125,000			
2, France	122,300	260	1150	1410
1, France	125,300			
6, U.S.	125,000			
1, France	129,500	340^g	NA^f	NA^f
2, NA^f	120,000	NA^f	270	NA^f
2, U.S.	125,000	164^e	670	834
3, NA^f				
4, NA^f	40,000	NA^f	600	NA^f

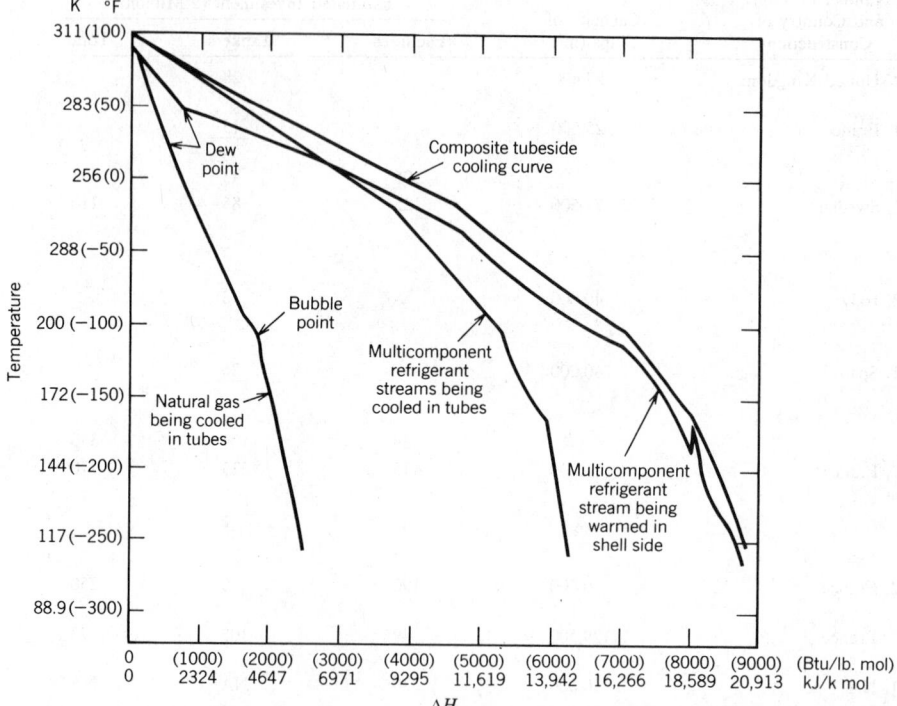

Fig. 77.41 Mixed refrigerant process cooling curve. (Courtesy Plenum Press.)

The need to deal with a mixed refrigerant and to control the composition of the refrigerant mixture are the major difficulties with these processes. They complicate design, control, and general operation. For instance, a second process plant, nearly as large as the LNG plant, must be at hand to separate refrigerant components and supply makeup as needed by the liquefier.

Also not shown in this flow sheet is the initial cleanup of the feed natural gas. This stream must be filtered, dried, purified of CO_2 before it enters the process shown here.

As noted above, both compressors and heat exchangers will be at the commercial maximum. The heat exchanger is of the coiled-tube-in-shell sort. Typically it would have ¾ in. aluminum tubes wrapped on a 2–3 ft diameter mandrel to a maximum 14 ft diameter. The exchanger is probably in two sections totaling about 120 ft in length. Shipping these exchanger bundles across the world challenges rail and ship capacities.

Ships used to transport LNG from the terminal by the plant to the receiving site are essentially supertankers with insulated storage tanks. These tanks are usually built to fit the ship hull. There may be four or five of them along the ship's length. Usually they are constructed at the shipyard, but in one design they are built in a separate facility, shipped by barge to the shipyard, and hoisted into position. Boiloff from these tanks is used as fuel for the ship. On long ocean hauls 6–10% of the LNG will be so consumed. In port the evaporated LNG must be reliquefied, for which purpose a small liquefier circuit is available onboard.

77.7.3 Helium Purification and Liquefaction

Helium exists in minute concentrations in air (see Table 77.12), but these amounts are not usually adequate for economic separation. It also exists in a few natural gas deposits in the United States, as shown in Table 77.14, and in like amounts in some deposits in Venezuela and USSR. This fossil material is apparently the total world supply. In the United States natural gas containing He has been routinely sent for distribution throughout metropolitan areas. The vital role that He plays in welding, superconductivity applications, in certain heat transfer and inert environment needs, and in a wide variety of research requirements lead to the demand that helium be conserved. This project was undertaken after World War II by the Bureau of Mines, and a series of helium separation plants was built in the southwest. In general, these produced a 60% He stream that was pumped to storage in abandoned natural gas fields from which it could later be recovered and purified. These plants generally ceased production during the early 1970s because of shifting government policies and budgetary

Table 77.14 Helium in Natural Gases in the United States

A. Composition of Some He-Rich Natural Gases in the United States

Location	Typical Composition (vol %)					
	CH_4	C_2H_6	N_2	CO_2	O_2	He
Colorado (Las Animas Co.)	0		77.6	14.7	0.3	7.4
Kansas (Waubaunsee, Elk, McPherson Cos.)	30	30	66.4	0.2	0	3.4
Michigan (Isabella Co.)	57.9	25.5	14.3	0	0.3	2.0
Montana (Musselshell)			54	30		16
Utah (Grand)	17		1.0	3.5		7.1

B. Estimated Helium Reserves (1966)

Location	Estimated Reserve (SCF)
Rocky Mountain area (Arizona, Colorado, Montana, New Mexico, Utah, Wyoming)	25×10^9
Midcontinent area (Kansas, Oklahoma, Texas)	169×10^9
He stored in the cliffside structure	9×10^9
Total	203×10^9

limitations. As of 1975 the United States had about 280 billion cubic feet of proved recoverable He reserves. At present rates of use extrapolated to 1990 it is estimated that about 15 billion cubic feet will remain. The remaining 200 billion cubic feet will have been dispersed in the atmosphere as the natural gas in which it resides is burned.

Figure 77.42 shows a helium recovery process that had been part of the conservation program. This plant was designed to process 475 million SCF/day of natural gas containing 0.45 mole % He. Natural gas from the pipe line enters the plant after heavier hydrocarbons have been removed. It is cooled countercurrently against returning streams and then enters a column of seven flash stages operating in series. The liquid from each stage is flashed into the next stage, which operate at successively lower pressure. In this way the He loss into the liquid phase is greatly reduced in comparison to the single flash situations. The final output liquid is reheated through the main heat exchanger and returned to the pipe line. The vapor streams from all these flash steps are combined, cooled to almost complete liquefaction, and sent to an additional three flash stages. The combined vapors from these flash steps are compressed, cooled, and phase separated to produce a vapor of about 83 mole % He. This crude He is rewarmed and sent to storage.

This process operates essentially on Joule–Thomson effects of the gases (mostly methane) accompanying the He. Helium liquefaction processes cannot depend only on Joule–Thomson cooling because of the negative Joule–Thomson effect exhibited by He at temperatures greater than that of liquid hydrogen. Figure 77.43 shows the simplified flow diagram of a helium liquefaction process. This process takes pure He and liquefies it using refrigeration supplied by liquid N_2 and by repeated expansions through turbines. The need for liquid N_2 depends on the number of expanders used, and can be zero for four or more expander steps.

77.8 SUPERCONDUCTIVITY AND ITS APPLICATIONS

For normal electrical conductors the resistance decreases sharply as temperature decreases, as shown in Fig. 77.44. For pure materials this decrease tends to level off at very low temperatures. This results from the fact that the resistance to electron flow results from two factors: the collision of electrons with crystal lattice imperfections and electron collisions with the lattice atoms themselves. The former effect is not temperature dependent, but the latter is. This relationship has, itself, proven of interest to engineers, and much thought and development has gone toward the building of power transmission lines operating at cryogenic temperatures and taking advantage of the reduced resistance.

77.8.1 Superconductivity

In 1911 Dr. Onnes of Leiden was investigating the electrical properties of metals at very low temperatures, helium having just been discovered and liquefied. He was measuring the resistance of frozen mercury as the temperature was reduced into the liquid He range. Suddenly the sample showed zero resistance. At first a short circuit was suspected. However, very careful experiments showed that the electrical

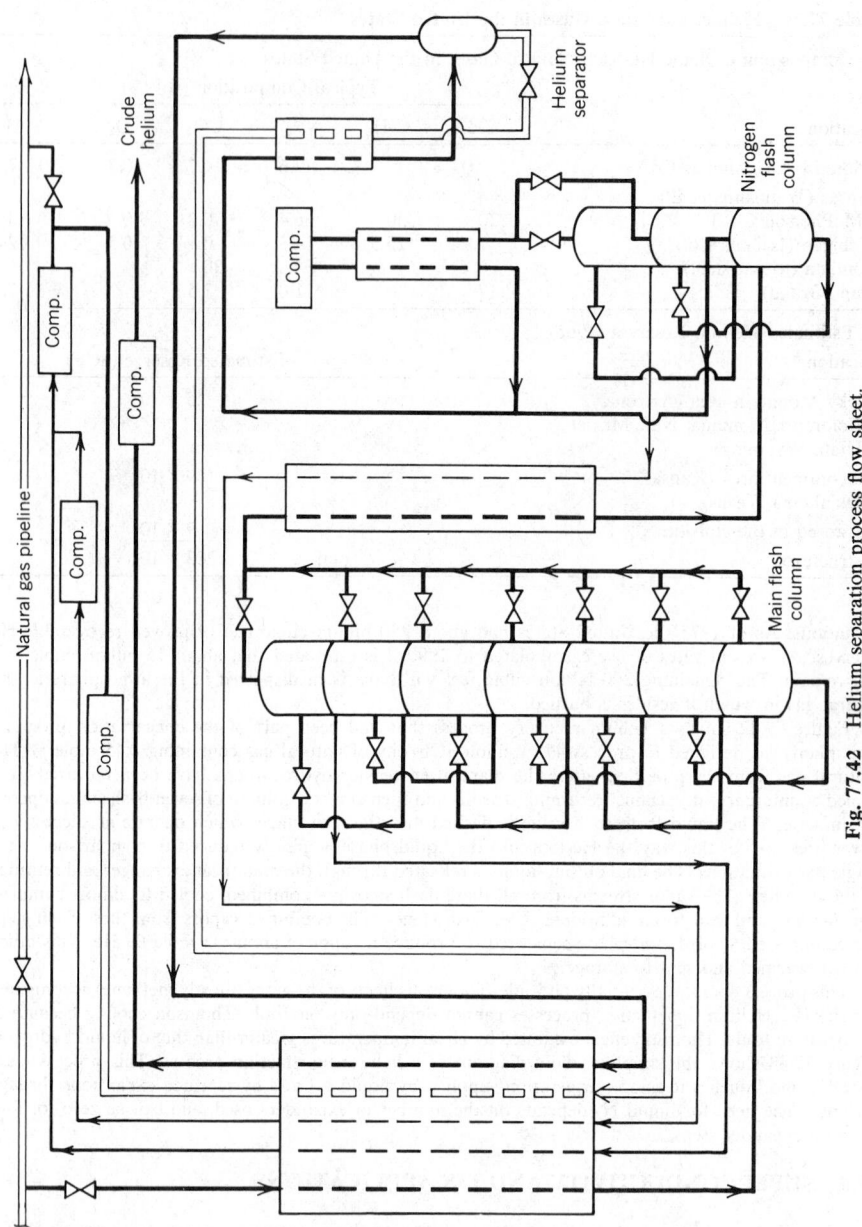

Fig. 77.42 Helium separation process flow sheet.

Natural gas pipeline

Crude helium

Comp.

Helium separator

Nitrogen flash column

Main flash column

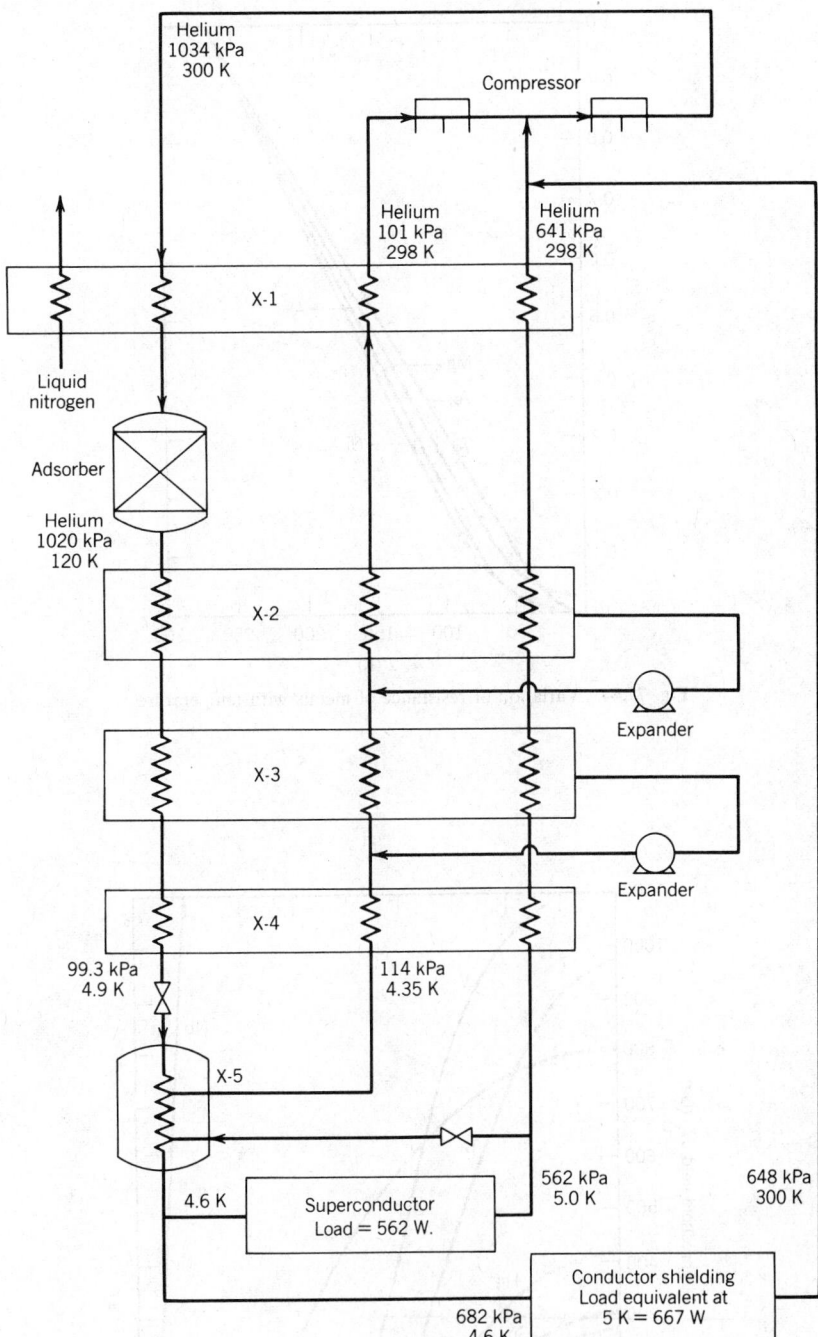

Fig. 77.43 Helium liquefier flow sheet.

conductivity of the sample had dropped discontinuously to a very low value. The phenomenon of superconductivity has since been found to occur in a wide range of metals and alloys. The resistance of a superconductor has been found to be smaller than can be measured by the best instrumentation available. Possibly it is zero. Early on this was demonstrated by initiating a current in a superconducting ring which could then be maintained, undiminished, for months.

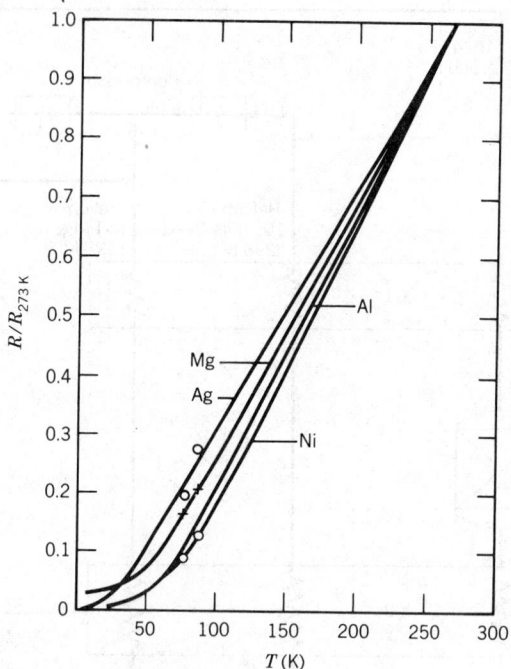

Fig. 77.44 Variation of resistance of metals with temperature.

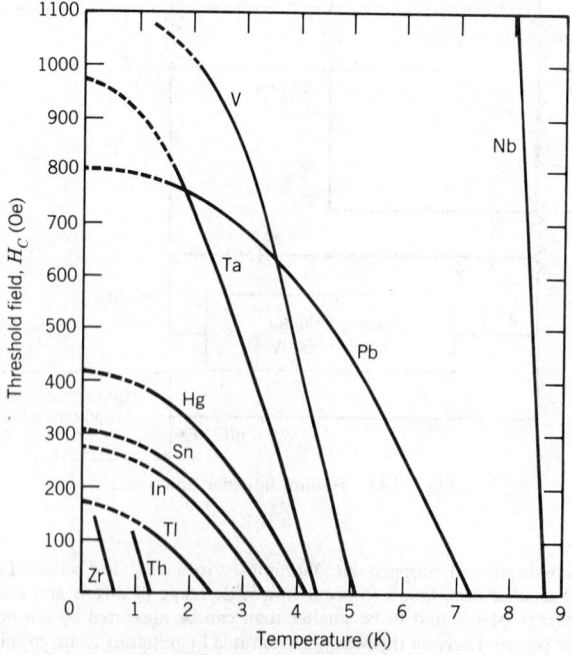

Fig. 77.45 Limits to superconductive behavior.

The phenomenon of superconductivity has been studied ever since in attempts to learn the extent of the phenomena, to develop a theory that will explain the basic mechanism and predict superconductive properties, and to use superconductivity in practical ways.

On an empirical basis it has been found that superconductors are diamagnetic, that is, they exclude a magnetic field, and that they exist within a region bounded by temperature and magnetic field strength. This is shown in Fig. 77.45. In becoming superconductive a material also changes in specific heat and in thermal conductivity.

A great deal of work has been done to determine what metals and alloys are superconductive, and to determine transition curves for materials that do exhibit superconductivity. Even very recent theory has not helped to guide this effort. Generally it has been found that materials with 3, 5, or 7 valence electrons are possible superconductors. It has even been found that alloys of nonsuperconductive elements may show superconductive properties if they are compounded to give an average of 3, 5, or 7 valence electrons. Table 77.15 shows the transition temperature at zero field strength for the superconducting elements. Alloys have been formulated with transition temperatures as great as 21 K, a temperature well within the range of practical application.

The theory of superconductivity developed after the discovery in 1933 by Meissner of the magnetic field exclusion. This led to a qualitative "two-fluid" model analogous to the theory underlying HeII. Since then this theory has been recast in quantum mechanical theory terms, most completely and successfully by J. Bardeen, L. N. Cooper, and R. Schrieffer of the University of Illinois in 1957 (BCS theory).

The BCS theory accounts for the Meissner effect and for other physical behavior phenomena of superconductors. It does not yet allow the prediction of superconductive transition points for new materials. The theory predicts an energy gap between normal and superconductive states existing simultaneously and visualizes the flow of paired electrons through the crystal lattice and the quantization of the magnetic flux. The quantized flux lines are fundamental to the explanation of the difference between type I superconductors that exhibit perfect Meissner effects, and which have relatively low transition temperatures and field tolerance, and type II superconductors that have imperfect Meissner effects, higher transition temperatures, and greater tolerance of magnetic fields. For example, Nb_3Sn, which is a type II superconductor, can be used for generation of magnetic fields of 100,000 gauss. These materials allow the penetration of magnetic field above a lower critical field strength, H_{c1}, but remain superconductive to a much greater field strength, H_{c2}. At fields above H_{c1}, flux enters the material in the form of quantized bundles that are pinned by dislocations so that the flux does not move easily and lead to normalization of the material.

Thus both HeII and superconductors may be considered examples of superfluids. Each exhibits a nondissipative process for mass transfer. In HeII the mass transferred is the fluid itself in inviscid flow; with a superconductor it is the electrons flowing without encountering resistance. In both cases a flow velocity greater than a critical value restores normality. In superconductors circulating currents are set up in the penetration layer at the surface to cancel the applied magnetic field. When this field is increased to a critical value, the critical current is reached and the material reverts to the normal state.

77.8.2 Applications of Superconductivity

Development of applications of superconductivity has proceeded very rapidly over the past decade. However, by far the widest use of superconductivity has been in magnet applications. Superconductive magnets were constructed and tested soon after such practical superconductors as Nb_3Sn and rhodium–zirconium were discovered about 1960. Field strengths as high as 7 T were predicted then. Since then magnets in a wide range of sizes and shapes have been made, tested, and used with field strengths approaching 20 T.

Commercially available superconducting materials are listed in Table 77.16. The ductile materials are readily fabricated by conventional wire-drawing techniques. These form the cheapest finished products, but cannot be used at high field strengths. The Nb_3Sn, which is the most widely used superconductive material, and V_3Ga are formed into tape by chemical vapor deposition. The tape is clad with copper for stability and stainless steel for strength. Material for multifilament conductor formation is produced by the bronze process. In this process filaments of Nb or V are drawn down in a matrix of Sn–Cu or Sn–Ga alloy, the bronze. Heat treatment then produces the Nb_3Sn or V_3Ga. The residual matrix is too resistive for satisfactory stabilization. Hence copper filaments are incorporated in the final conductors.

Multifilament conductors are then made by assembling superconductive filaments in a stabilizing matrix. For example, in one such conductor groups of 241 Ni–Ti filaments are sheathed in copper and cupronickel and packed in a copper matrix to make a 13,255-filament conductor. Such a conductor can be wound into an electromagnet or other large-scale electrical device.

Superconducting magnets have been used, or are planned to be used for particle acceleration in linear accelerators, for producing the magnetic fields in the plasma step of magnetohydrodynamics, for hydrogen bubble chambers, for producing magnetic "bottles" for nuclear fusion reactors such as

Table 77.15 Superconductive Elements with Zero-Field Transition Temperatures

Mg											Al 1.20°	Si	P
Ca	Sc	Ti 0.40°	V 4.9°	Cr	Mn	Fe	Co	Ni	Cu	Zn 0.88°	Ga 1.10°	Ge	As
Sr	Y	Zr 0.75°	Nb 9.5°	Mo 0.95°	Tc 11.2°	Ru 0.47°	Rh	Pd	Ag	Cd 0.56°	In 3.40°	Sn 3.73°	Sb
Ba	La 5.4°	Hf 0.35°	Ta 4.5°	W	Re 1.7°	Os 0.71°	Ir 0.15°	Pt	Au	Hg 4.15°	Tl 2.39°	Pb 7.22°	Bi
Ra	Ac	Th 1.37°	Pa	U-β 1.8°									

the Tokomak, for both levitation and propulsion of ultra high speed trains, for research in solid-state physics, for field windings in motors, and for a host of small uses usually centered on research studies. In fact superconductive magnetics with field strength approaching 10 T are an item of commerce. They are usable where liquid helium temperatures are available and produce magnetic fields more conveniently and cheaply than can be done with a conventional electromagnet. Table 77.17 lists the superconductive magnets in use for various energy related applications.

Perhaps the most interesting of these applications is in high-speed railroads. Studies in Japan, Germany, Canada, and the United States are aimed at developing passenger trains that will operate at 300 mph and above. The trains would be levitated over the track by superconductive magnets, sinking to track level only at start and stop. Propulsion systems vary but are generally motors often with superconductive field windings. Such railroads are proposed for travel from Osaka to Tokyo and from San Diego to Los Angeles. Design criteria for the Japanese train are given in Table 77.18.

Superconductive electrical power transmission has been seriously considered for areas of high density use. Superconductors make it possible to bring the capacity of a single line up to 10,000–30,000 MW at a current density two orders of magnitude greater than conventional practice. The resulting small size and reduced energy losses reduce operating costs of transmission substantially.

The economic attraction of a superconductive transmission line depends on the cost of construction and the demand for power, but also on the cost of refrigeration. Thus a shield is built in and kept at liquid N_2 temperature to conserve on helium. Also, superinsulation is used around the liquid N_2 shield.

Other applications of superconductivity have been found in the microelectronics field. Superconductive switches have been proposed as high-speed, high-density memory devices and switches for computers and other electronic circuits. The ability of the superconductor to revert to normal and again to superconductive in the presence or absence of a magnetic field makes an electric gate or a record of the presence of an electric current. However, these devices have been at least temporarily overshadowed by the rapid development of the electronic chip. Ultimately, of course, these chips will be immersed in a cryogen to reduce resistance and dissipate resistive heat.

77.9 CRYOBIOLOGY AND CRYOSURGERY

Cryogenics has found applications in medicine, food storage and transportation, and agriculture. In these areas the low temperatures can be used to produce rapid tissue freezing and to maintain biological materials free of decay over long periods.

The freezing of food with liquid N_2 has become commonplace. Typically the loose, prepared food material is fed through an insulated chamber on a conveyor belt. Liquid N_2 is sprayed onto the food, and the evaporated N_2 flows countercurrent to the food movement to escape the chamber at the end in which the food enters. The required time of exposure depends on the size of individual food pieces and the characteristics of the food itself. For example, hamburger patties freeze relatively quickly because there is little resistance to nitrogen penetration. Conversely, whole fish may freeze rapidly on the surface, but the enclosing membranes prevent nitrogen penetration, so interval freezing occurs by conductive transfer of heat through the flesh. Usually a refrigerated holding period is required after the liquid N_2 spray chambers to complete the freezing process.

The advantages of liquid N_2 food freezing relative to more conventional refrigeration lie in the speed of freezing that produces less tissue damage and less chance for spoilage, and the inert nature of nitrogen, which causes no health hazard for the freezer plant worker or the consumer.

Liquid N_2 freezing and storage has also been used with parts of living beings such as red blood cells, bull semen, bones, and various other cells. Here the concern is for the survival of the cells upon thawing, for in the freezing process ice crystals form which may rupture cell walls upon freezing and thawing. The rate of survival has been found to depend on the rate of cooling and heating, with

Table 77.16 Some Commercially Available Superconductive Materials[a]

Material	T_c (K)	H_{c2} (T) at 4.2 K	J_c (10^5 A/cm² at 4.2 K)				Fabrication
			2.5 T	5 T	10 T	15 T	
Nb–25 wt% Zr	11	7.0	1.1	0.8	0	0	Fairly ductile
Nb–33 wt% Zr	11.5	8.0	0.9	0.8	0	0	Fairly ductile
Nb–48 wt% Ti	9.5	12.0	2.5	1.5	0.3	0	Ductile
Nb_3Sn	18.0	22.0	17.0	10.0	4.0	0.5	CVD diffusion bronze
V_3Ga	15.0	23.0	5.0	2.5	1.4	0.9	Diffusion bronze

[a] Courtesy Plenum Press.

Table 77.17 General Characteristics of Superconductive Magnets for Energy Conversion and Storage Systems[40,a]

Application Field	Magnet Type	Typical Stored Energy in the Winding (MJ)	Operated			Largest Prototype So Far
			dc	Pulsed	Transients	
MHD generators	Dipole magnet with warm aperature, possibly tapered	500–5000	Yes	No	Yes, from the MHD fluid	60-MJ magnet
Homopolar machines	Solenoid	10–100	Yes	No	No	3-MW generator
Synchronous machines	Rotating dipole or quadrupole winding	Power plant machines, 50–100; airborne systems, 0.5–1	Yes	No	Yes, in case of unbalanced load	5-MVA generator
Fusion magnets Tokamak or similar low-β confinement	Toroidal field coils	$\geq 10^5$	Yes	Yes, in case of fast voltage control	Yes, pulsed field harmonic components of poloidal field	—
	Poloidal field coils	$\geq 10^3$	No	Yes, with rise times of seconds	Yes, dc field components from the toroidal field	—
Mirror confinement	Baseball coils	$\geq 10^5$	Yes	No	No	Baseball coils with 9 MJ
Energy storage; operation of pulsed fusion magnets Theta pinch	No optimal shape defined yet	≥ 100 per unit	No	Yes, transfer time about 30 msec	—	—
Tokamak	No optimal shape defined yet	$\geq 10^4$	No	Yes, transfer time about seconds	—	300 kJ
Load leveling in the grid	No optimal shape defined yet	$\geq 10^8$	No	Yes, transfer time about hours	—	—

[a] Courtesy Plenum Press.

Table 77.18 Design Criteria for Japanese High Speed Train[41,a]

Maximum number of coaches/train	16
Maximum operation speed	550 km/hr
Maximum acceleration and deceleration:	
Acceleration	3 km/hr/sec
Deceleration, normal brake	5 km/hr/sec
Deceleration, emergency brake	10 km/hr/sec
Starting speed of levitation	100 km/hr
Effective levitation height (between coil centers)	250 mm
Accuracy of the track	±10 mm/10 m
Hours of operation	From 6 AM to 12 PM at 15-min intervals
Period of operation without maintenance service	18 hr
Number of superconducting magnets	
Levitation	4 × 2 rows/coach
Guiding and drive	4 × 2 rows/coach
Carriage weight	30 tons
dimensions	25 m × 3.4 m × 3.4 m
Propulsion	Linear synchronous motor

[a] Courtesy Plenum Press.

each class of material showing individual optima. Figure 77.46 shows the survival fractions of several cell types as a function of cooling velocity. Better than half the red blood cells survive at cooling rates of about 3000 K/min. Such a cooling rate would kill all of the yeast cells.

The mechanism of cell death is not clearly understood, and may result from any of several effects. The cell-wall rupture by crystals is the most obvious possibility. Another is the dehydration of the cell by water migration during the freezing process. In any case the use of additives such as glycerol, dimethyl sulfoxide, pyridine n-oxide, and methyl and dimethyl acetamide has greatly reduced cell mortality in various specific cases. The amount and type of additive that is most effective depends upon the specific cell being treated.

Controlled freezing has proven useful in several surgical procedures. In each of these the destruction of carefully selected cells and/or their removal has been the goal of the operation.

In treating Parkinson's disease destruction of some cells in the thalmus can lead to sharp reduction in tremors and muscular rigidity. The operation is done under local anesthetic using a very fine probe consisting of three concentric tubes. Liquid N_2 flows in through the center tube, returning as vapor through the central annulus. The outer annulus is evacuated and insulates all but the probe tip. The surgeon inserts the probe using X-ray pictures for guidance. He or she gently cools the probe tip using temperatures just below freezing. If the patient's tremors subside without other side effects, the right location has been found. Freezing of a quarter inch sphere around the probe tip can proceed.

In ophthalmic surgery cryogenic probes are used to lift cataracts from the lens of the eye. Here

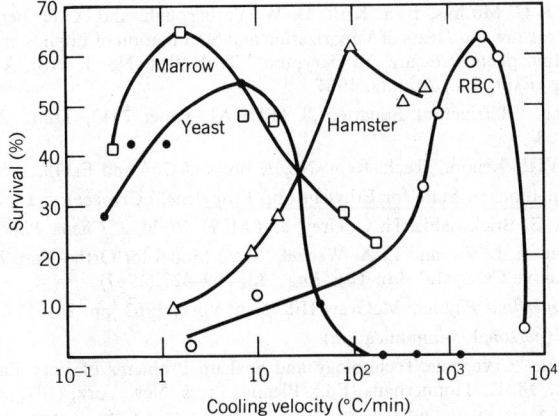

Fig. 77.46 Survival rate for various cells frozen to liquid N_2 temperature.[42] (Courtesy Plenum Press.)

the cataract is frozen to the cryo-probe tip and carefully separated from the eye. Liquid N_2 is not needed and Freons or Joule–Thomson cooling is sufficient.

Malignant or surface tumors can also be removed cyrogenically. The freezing of such a cell mass helps to prevent the escape of some of the cells into the blood stream or the body cavity.

REFERENCES

1. R. D. McCarty, "Thermodynamic Properties of Helium-4 from 2 to 1500 K at Pressures to 10^8 Pa," *J. Chem. Phys. Ref. Data,* **2**(4), 923 (1973); D. B. Mann, "Thermodynamic Properties of Helium from 3 to 300°K Between 0.5 and 100 Atmospheres," NBS Tech. Note 154 (Note 154A for British Units), Jan. 1962.

2. H. M. Roder and R. D. McCarty, "A Modified Benedict–Webb–Rubin Equation of State for Parahydrogen-2," NBS Report NBS1R 75-814, June 1975.

3. J. G. Hurst and R. B. Stewart, "A Compilation of the Property Differences of Ortho- and Para-Hydrogen or Mixtures of Ortho- and Para-Hydrogen," NBS Report 8812, May, 1965.

4. R. Prydz, K. D. Timmerhaus, and R. B. Stewart, "The Thermodynamic Properties of Deuterium," *Adv. Cryo. Eng.,* **13**, 384 (1968).

5. R. D. McCarty and R. B. Stewart, "Thermodynamic Properties of Neon from 25 to 300 K between 0.1 and 200 Atmospheres," Third Symposium on Thermophysical Properties, ASME, 1965, p. 84.

6. R. T. Jacobsen, R. B. Stewart, and A. F. Myers, "An Equation of State of Oxygen and Nitrogen," *Adv. Cryo. Eng.,* **18**, 248 (1972).

7. A. Michels, T. Wassenaar, and G. Wolkers, *Appl. Sci. Res.,* **A5**, 121 (1955).

8. T. R. Strobridge, "The Thermodynamic Properties of Nitrogen from 64 to 300 K between 0.1 and 200 Atmospheres," NBS Tech. Note 129 (Note 129A for British Units), Jan., 1962 and Feb., 1963.

9. J. G. Hust and R. B. Stewart, "Thermodynamic Properties Valves for Gaseous and Liquid Carbon Monoxide from 70 to 300 K with Pressures to 300 Atmospheres," NBS Tech. Note 202, Nov. 30, 1963.

10. R. Prydz, G. C. Straty, and K. D. Timmerhaus, "The Thermodynamic Properties of Fluorine," *Adv. Cryo. Eng.,* **16**, 64 (1971).

11. E. Bendu, "Equations of State Exactly Representing the Phase Behavior of Pure Substances," *Proceedings of the 5th Symposium on Thermophysical Properties,* ASME, 1970, p. 227.

12. A. L. Gosman, R. D. McCarty, and J. G. Hust, "Thermodynamic Properties of Argon from the Triple Point to 300 K at Pressures to 1000 Atmospheres," NBS Reference Data Series (NSRDS-NSB 27), Mar. 1969.

13. L. A. Weber, "Thermodynamic and Related Properties of Oxygen from the Triple Point to 300 K at Pressures to 330 Atmospheres," NBS Rpt. 9710 (Rpt. 9710A for British Units), June and Aug. 1968; L. A. Weber, *NSB J. Res.,* **74A**(1), 93 (1970).

14. R. D. McCarty, "A Modified Benedict–Webb–Rubin Equation of State for Methane Using Recent Experimental Data," *Cryogenics,* **14**, 276 (1974).

15. W. T. Ziegle, J. C. Mullins, B. S. Kirk, D. W. Yarborough, and A. R. Berquist, "Calculation of the Vapor Pressure and Heats of Vaporization and Sublimation of Liquids and Solids, Especially Below One Atmosphere Pressure: VI, Krypton," Tech. Rpt. No. 1, Proj. A-764, Georgia Inst. of Tech. Engrg. Expt. Sta., Atlanta, 1964.

16. R. C. Downing, "Refrigerant Equations," ASHRAE Paper 2313, *Trans. ASHRAE,* **80**, Part III, 1974, p. 158.

17. See Ref. 15, "VIII, Xenon," Tech. Rept. No. 3, Projs. A-764 and E-115, 1966.

18. E. Bendu, "Equations of State for Ethylene and Propylene," *Cryogenics,* **15**, 667 (1975).

19. R. B. Scott, F. G. Brickwedde, H. C. Urey, and M. H. Wahl, *J. Chem. Phys.,* **2**, 454 (1934).

20. A. H. Singleton, A. Lapin, and L. A. Wenzel, "Rate Model for Ortho–Para Hydrogen Reaction on a Highly Active Catalyst," *Adv. Cry. Eng.,* **13**, 409–427 (1967).

21. C. T. Lane, *Superfluid Physics,* McGraw-Hill, New York, 1962, pp. 161–177.

22. L. S. Twomey (personal communication).

23. O. M. Bourguet, "Cryogenic Technology and Scaleup Problems of Very Large LNG Plants," *Adv. Crys. Prg.,* **18**, K. Timmerhaus (Ed.), Plenum Press, New York, (1972), pp. 9–26.

24. T. T. Rule and E. B. Quale, "Steady State Operation of the Idealized Vuillurmier Refrigerator," *Adv. Cryo. Eng.,* **14**, 1968.

25. W. E. Gifford, "The Gifford–McMahon Cycle," *Adv. Cryo. Eng.,* **11,** 1965.

26. H. Hausen, "Warmeubutragung in Gegenstrom, Gluchstrom, und Kiezstrom," Springer-Verlag, Berlin, 1950.

27. D. E. Ward, "Some Aspects of the Design and Operation of Low Temperature Regenerator," *Adv. Cryo. Eng.,* **6,** 525 (1960).

28. G. O. G. Lof and R. W. Hawley, *Ind. Ing. Clem.,* **40,** 1061 (1948).

29. G. E. O'Connor and T. W. Russell, "Heat Transfer in Tubular Fluid–Fluid Systems," *Advances in Chemical Engineering,* T. B. Drew et al. (Eds.), Academic Press, New York, 1978, Vol. 10, pp. 1–56.

30. M. Jacob, *Heat Transfer,* Wiley, New York, 1957, Vol. II, pp. 1–199.

31. W. T. Ziegh and H. Cheung, *Adv. Cryo. Eng.,* **2,** 100 (1960).

32. R. H. Kropschot, "Cryogenic Insulation," *ASHRAE Journal* (1958).

33. T. F. Darham, R. M. McClintock, and R. P. Reed, "Cryogenic Materials Data Book," Office of Tech. Services, Washington, DC, 1962.

34. R. J. Coruccini, *Chem. Eng. Prog.,* 342 (July, 1957).

35. R. B. Stewart and V. J. Johnson (eds.), "A Compendium of Materials at Low Temperatures, Phases I and II," WADD Tech. Rept. 60-56, NBS, Boulder, CO, 1961.

36. C. R. Barber et al., "The International Practical Temperature Scale," *Metrologia,* **5,** 35 (1969).

37. F. G. Brickwedde, H. van Diyk, M. Durieux, J. M. Clement, and J. K. Logan, *J. Res. Natl. Bur. Stds.,* **64A,** 1 (1960).

38. R. P. Reis and D. E. Mapother, *Temperature, Its Measurement in Science and Industry,* H. H. Plumb (ed.), **4,** 885–895 (1972).

39. L. L. Sperikr, R. L. Powell, and W. J. Hall, "Progress in Cryogenic Thermocouples," *Adv. Cryo. Eng.,* **14,** 316 (1968).

40. P. Komacek, "Applications of Superconductive Magnets to Energy with Particular Emphasis on Fusion Power," *Adv. Cryo. Eng.,* **21,** 115 (1975).

41. K. Oshima and Y. Kyotani, "High Speed Transportation Levitated by Superconducting Magnet," *Adv. Cryo. Eng.,* **19,** 154 (1974).

42. E. G. Cravalho, "The Application of Cryogenics to the Reversible Storage of Biomaterials," *Adv. Cryo. Eng.,* **21,** 399 (1975).

CHAPTER 78

THE IMPACT OF CHEMICAL AND THERMAL PROCESSES ON THE ENVIRONMENT

RICHARD M. STEPHENSON

The University of Connecticut
Storrs, Connecticut

78.1 INTRODUCTION

Today, the first problem facing anyone who plans to construct or modify a power plant is to determine how much pollution will result from the proposed new facilities. One of the most difficult aspects of the pollution problem is to develop guidelines defining what actually constitutes pollution. The rock concert that is such a joy to your teenager's ears is simply loud and unpleasant noise to the neighbors. Some people would never under any circumstances live in New York City, while others would never live anywhere else. To the fisherman, anything that alters his or her favorite stream is pollution; to the industrial worker, the most important thing is a job—even if it does cause some contamination of the environment.

In the strictest sense, pollution can be defined as anything that in any way changes the natural environment. With such a definition, human life itself constitutes pollution. We make noise, we produce heat which must be dissipated to the environment, we carry out agriculture, we consume food crops, and we produce waste products which pollute the environment. Of course, most people do not feel that their lives represent pollution and adopt a rather selfish attitude. Typically, we consider pollution to be any change in the environment that we personally find objectionable; changes that give us pleasure are considered desirable.

A commonly accepted definition of pollution is "an undesirable change in the physical, chemical, or biological characteristics of our air, land, or water that may or will harmfully affect human life or that of our desirable species." This really does not help us very much because it leaves open the question of *how much* change we can tolerate in our environment before it becomes "undesirable." Also, there will always be the question of what is or is not a desirable species. For example, when sprays were first used to control the famous New Jersey mosquitoes, there was opposition from certain religious groups who insisted that "God also created the mosquito."

Since any human activity must, of necessity, cause some pollution, it is meaningless to talk about a zero level of pollution. The goal of any pollution-control program must be to determine an *optimal level* of pollution: the point where the benefits to society of the proposed activity outweigh the harmful effects of the inevitable pollution that results. However, in view of the many uncertainties involved, it is perhaps not surprising that any proposed new industrial construction often arouses widely divergent opinions as to the potential for environmental damage, and often routine requests for approval of new construction become highly polarized political battles which delay and add to the cost of doing business.

78.2 ENVIRONMENTAL PROTECTION

All matters of environmental pollution fall under the control of the Environmental Protection Agency (EPA). Their activities are very broad and include the following general categories: (1) control of air pollution and establishment of national air standards; (2) control of water pollution, including chemical wastes, radioactive wastes, and thermal pollution; (3) garbage disposal and the management of land-fill areas; (4) control of pesticides and enforcement of the Federal Insecticide, Fungicide, and Rodenticide Act; (5) building and highway construction; (6) protection of wetlands; (7) noise emission standards for trucks, railroads, and construction equipment; (8) ocean dumping; and (9) discharge of oil and hydrocarbons to the environment.

Since the establishment of EPA, an *environmental impact study* is required before anyone can do anything that materially changes the natural environment. In some cases, the studies themselves are very detailed and costly. There is no question that environmental regulations have become a very serious and, in some instances, an overriding consideration in the location of industrial plants today and sometimes even in the selection of the particular processes that are used. For example, the electric utility industry has essentially abandoned any new construction of nuclear power plants because of the violent opposition by the general public. Environmental considerations have led to the closing of many plants, have greatly increased the cost of many other operations, and have inhibited the growth of much of the chemical, petroleum, mining, and utility industries in recent years.

To illustrate some of the problems encountered in any modification or new construction of power plants, Table 78.1 gives EPA values for maximum emissions from any fossil-fuel-fired boiler having more than 73 megawatts of heat input. In addition to these regulations, there are others governing plant location, thermal discharges, disposal of liquid wastes, disposal of solid wastes such as coal ash, and an entirely separate section for the construction of nuclear plants. And besides satisfying federal authorities, some states and municipalities have their own regulations, which may be even more restrictive than the federal ones!

78.3 HEATING THE GLOBAL ENVIRONMENT

A question frequently asked today concerns whether or not our increased scale of technological activity can adversely affect the entire global environment. We know, for example, that changes in the amount

Table 78.1   EPA Standards of Performance for Fossil-Fuel-Fired Steam Generators Constructed or Modified after September 18, 1978[a]

Material	Maximum Allowable Emission	
	nanograms/J	lb$_m$/Btu
A. Particulate matter		
From solid, liquid, or gaseous fuel	13	0.30
B. Sulfur dioxide		
From liquid or gaseous fuels	340	0.80
From solid fuels	520	1.2
C. Oxides of nitrogen		
Gaseous fuels derived from coal	210	0.50
All other gaseous fuels	86	0.20
Liquid fuels derived from coal	210	0.50
Liquid fuels from shale oil	210	0.50
All other liquid fuels	130	0.30
Solid fuels derived from coal	210	0.50
Subbituminous coal	210	0.50
Bituminous coal	260	0.60
Anthracite coal	260	0.60
All other solid fuels	260	0.60
D. Opacity. Not over 20% opacity except for one six-minute period per hour not over 27% opacity.		

[a] Standards of Performance for New Stationary Sources, *Code of Federal Regulations,* Title 40, Part 60, revised through July 1, 1982.

of carbon dioxide and particulate matter in the atmosphere can affect plant growth and rainfall. There is some evidence that release of chemicals such as the fluorocarbons may affect the earth's ozone layer, thereby increasing ultraviolet radiation and the incidence of skin cancer. It is also well known that even a slight decrease in global temperature could trigger another ice age; similarly, a slight increase in global temperature could cause considerable melting of the polar ice caps, bringing a rise in the level of our oceans and flooding most coastal cities.

In general, the question of changes in our global environment is extremely complex, and there are no easy answers. For example, consider the temperature of the earth. It is known that the earth's climate has changed many times in the past and that the last great Ice Age retreated some 10,000 years ago. Since that time, the earth has experienced periods of warming and cooling where the average temperature changed by one or two degrees over the period of perhaps a century. The first half of the 20th century saw a pronounced warming trend that continued until about 1940. Since then, we have been back on a cooling trend, and there is no way to tell whether or how long this will continue. These natural periodic fluctuations in temperature make it impossible to determine whether or not human activities have had any effect on average global temperature up to this point, and make it very difficult to predict any possible future trends.

The average temperature of the earth obviously depends on a balance between the amount of energy contributed by sunlight and the amount of energy radiated back to space. Theoretically, human activity could change the world's energy balance in several ways. For example, much of the earth's surface has been quite drastically altered—cities and highways cover much of the landscape, forests have been converted to croplands, and deserts have developed and continue to expand in size. Has this been enough to significantly affect the amount of sunlight reflected or absorbed by the earth's surface?

Fortunately, the global energy balance is determined primarily by the polar regions and the oceans (which themselves cover three-quarters of the earth's surface). While changes in surface reflection and absorption resulting from human activity can strongly affect local conditions (cities tend to be pretty warm on a summer day), they are fortunately too small to materially change our overall global temperature.

A second factor that might influence global temperature is the tremendous amount of heat released to the environment through the burning of fossil fuels and the greater use of nuclear energy. This does produce local changes in climate, but on a global scale is not as significant as might be expected. The total amount of heat generated by all human activity is about 1000 times smaller than the heat energy supplied by the sun. Since we know now that it is very doubtful that our consumption of uranium and the fossil fuels will ever be much greater than it is today, it does not appear that this source of heat will ever substantially affect our global temperature.

Another question often asked is whether the hundreds of millions of tons of dust, oil, smog, and acid fumes poured into the atmosphere each year by human activities have any significant influence on global climate. These materials do increase the cloudiness of our atmospheres and act in two ways to reduce global temperatures: they reflect a larger fraction of the sunlight striking the upper atmosphere and they absorb a greater fraction of the sunlight passing down through the atmosphere to the surface of the earth.

There is no question that dust in the atmosphere causes a drop in global temperature, as has been demonstrated vividly by volcanic eruptions. The best example of this is probably the eruption of Tambora in Indonesia in 1815, the greatest eruption in recent history. The following year, 1816, was cold all over the world and became known as "the year without a summer." The average summer temperature in Great Britain was 2–3°C below normal, all of northern Europe suffered, and there were serious food shortages in Ireland and Wales. In New England there was widespread snow between June 6 and June 11. Frosts occurred in every summer month; some crops did not ripen at all, and others rotted in the fields.

But once again the situation is not as serious as might be expected. When we consider the total "loading" of the atmosphere, it turns out that there is a natural injection of material into the atmosphere from forests (organic vapors), oceans (in the form of salt), deserts and other wind-blown dust, forest fires, and periodic volcanic eruptions. Over two-thirds of the pollution of our atmosphere is from natural sources, and less than one-third comes from human activities. Much of the air pollution by industry is now being reduced by filters, precipitators, and scrubbers. Unless we enter a period of violent volcanic activity, it seems unlikely that atmospheric dust will materially affect global temperature.

78.4   THE GREENHOUSE EFFECT

To understand the highly publicized greenhouse* effect, now considered by some to be potentially the most serious consequence of human activity on global climate, we must consider the spectral distribution of sunlight and the radiation properties of air. Sunlight striking the upper atmosphere of the earth consists of all possible wavelengths from ultraviolet through visible and infrared light. Of unique importance in the upper atmosphere is the ozone layer, which absorbs practically all of the short-wavelength ultraviolet radiation—a good thing for us, since we know this radiation causes skin cancer and is lethal to exposed microorganisms. Part of the sunlight that strikes our upper atmosphere is reflected back to space by clouds, ice crystals, and, to a smaller extent, dust; the rest passes downward through the atmosphere and is further reduced in intensity by absorption in clouds, dust, and other materials. On a clear day, sunlight reaching the surface of the earth is about 10% ultraviolet, 45% visible, and 45% infrared light.

When the earth cools at night, the loss of heat is through radiation from the surface of the earth to the atmosphere and eventually to outer space. The spectral distribution of this radiation is given by Planck's law (assuming a black body), and for temperatures equivalent to those at the surface of the earth this radiation is entirely in the infrared region of the spectrum. Simple gases such as oxygen and nitrogen do not absorb infrared radiation; more complex molecules, such as water vapor and carbon dioxide, do readily absorb it and change it to heat energy. (This is why the earth cools easily on a clear, dry night but does little cooling on a damp, cloudy night.)

The greenhouse effect, according to its proponents, can be attributed to an increased concentration of carbon dioxide in the earth's atmosphere as a result of widespread burning of fossil fuels and increased agricultural activity which speeds up the breakdown of biomass. This increased concentration of carbon dioxide means that the earth does not lose quite as much heat at night as it used to, and there is a slight increase in global temperature. There is no question that global concentration of carbon dioxide in the atmosphere is now about 15% higher than it was prior to the industrial revolution, and it is estimated that it will be about 30% higher by the end of the 20th century. Calculations show that—other things being equal—this increased carbon dioxide content of the atmosphere will cause about a 0.6°C rise in global temperature by the end of the 20th century.

Because of other factors that interact, it is very difficult to make a really accurate prediction of this kind. The oceans have about 60 times as much carbon dioxide as the atmosphere, and readily absorb carbon dioxide from the atmosphere and tie it up as a loose bicarbonate solution according to the equation

$$CO_2(g) + Na_2CO_3 + H_2O \rightleftharpoons 2 NaHCO_3 \text{ (soln)}$$

It is estimated that as much as one-half of the carbon dioxide added to the atmosphere by human activity may have dissolved in the oceans; again, however, there is considerable question as to the accuracy of this figure and whether or not there has been time enough to establish an equilibrium between atmospheric carbon dioxide and that contained in the ocean depths.

* This is really a misnomer, since operation of a real greenhouse depends only partly on the radiation property of the glass and mainly on the use of a barrier to keep out the cold air.

A second factor that must be considered is that increased carbon dioxide content of the atmosphere causes an increased rate of photosynthesis. In other words, plants grow faster in an atmosphere that is richer in carbon dioxide, and this acts to remove carbon dioxide more rapidly from the atmosphere and tie more up in the form of biomass.

A third factor is that increased global temperature means increased evaporation of water and hence increased cloudiness. This increases the amount of sunlight reflected in the upper atmosphere, reduces the amount of sunlight reaching the earth's surface, and reduces the amount of heat lost by the earth during nighttime cooling.

Obviously it is very hard to make any kind of precise evaluation of the importance of the greenhouse effect. It is even impossible to say whether or not it even exists, since the maximum global temperature change which could be attributed to it is masked by the natural periodic global temperature variation during the past century or so. Since most of the carbon dioxide has been added in the past 40 years when global temperatures have been cooling, it might be argued that the greenhouse effect is nonexistent. While it now appears that the greenhouse effect is not something we have to worry about for the immediate future, it does bear close watching—particularly if we should get into another period of global warming.

78.5 ENVIRONMENTAL EFFECTS OF POWER GENERATION

All power plants are required to meet certain restrictions as established by EPA. For example, any new construction necessitates the preparation of an environmental impact study that considers such topics as plant effluents; possible effects of the new facility on lakes, streams, and wetlands; thermal pollution; traffic and noise; visual pollution; and the like. Any discharge of solid, liquid, or gaseous materials is subject to strict controls. Public hearings are permitted, during which individual citizens or groups of citizens may file objections. It is often a long, complicated, and expensive procedure.

Since the problems encountered depend to a large extent on the type of facility, let us consider them individually.

78.5.1 Hydropower

Hydro is by far the cheapest form of power and best from an environmental standpoint. Today's hydroelectricity is produced by a conventional water turbine whereby water under high pressure rotates blades mounted on a shaft directly connected to a generator that converts the mechanical energy to electrical energy. A modern hydro plant is situated near the base of a reservoir or waterfall. Water is admitted to the inlet of the turbine either through a tunnel at the bottom of the dam or through a pipe if the plant is located downstream from the dam. Discharge water at atmospheric pressure leaves the turbine and flows downstream. There is no need to worry about exhaust gases, scrubbers, precipitators, radioactive fuel elements, and similar problems encountered in other methods of producing electricity.

Hydro plants do have some adverse social and environmental effects that must be considered. They often deny access to the spawning grounds of important migratory fish such as salmon, although this can be avoided by the use of fish ladders around the dam. Large reservoirs in some cases cover the best farmland and may flood small towns or villages. Stream conditions can be substantially altered downstream from a dam, particularly by the rapid fluctuation of the water level often encountered during intermittent power generation. Underground seepage from reservoirs can raise the surrounding water table, sometimes dissolving subsurface salts that may impair the fertility of the soil. In addition, there is always the potential for a disastrous flood if a dam upstream from a populated area should suddenly give way.

The availability of hydroelectricity can vary considerably over the period of a year, normally being a maximum during the early spring and a minimum during the warm summer months. If a reservoir is also used for irrigation or for maintenance of a navigational channel, not all of the water will be available for power production and there may be limitations on its use. If a reservoir is used for flood control, it must be kept low during periods when floods may occur; this reduces the available head and may also restrict the amount of water that can be used later for power production.

However, despite these problems, hydroelectricity is still the cheapest and the best. In the United States, it represents about 10% of our total electrical production. Unfortunately, we are already using practically all our easily developed sites, further development of hydropower is coming under increasingly vociferous environmental attack by public interest groups, and it is very unlikely that we will be able to materially increase production of hydroelectricity.

78.5.2 Pumped Storage

A major problem for any electric utility is to meet the cyclic demand for electricity that peaks on weekdays during the morning and afternoon rush hours. One way to help meet this demand is through the use of pumped storage.

Pumped storage is used in mountainous areas such as New England and upstate New York. The process involves the construction at the top of a mountain of a large storage reservoir that is connected through turbines to a river at the base of the mountain. When demand for electricity is low, water is pumped from the river to the reservoir and stored there in the form of potential energy until such time as it is needed. During periods of peak demand, the flow is reversed and the pumps operate as turbines to produce electricity.

Pumped storage gives considerable flexibility to a utility, since base-load nuclear power can be used during nights and weekends to pump water when demand is low. There is some loss of power, since only about 60% of the energy used to pump the water is actually recovered again as power, but it eliminates the need for new generation facilities and does not consume critical fuels such as petroleum and natural gas.

The only major environmental problem in connection with pumped storage is the large reservoir that must be provided and the continuous large fluctuations in water level which prohibit any possible recreational use. But these problems are minor in comparison to those encountered in the construction of additional fossil-fuel-fired or nuclear power plants.

78.5.3 Peaking Units

Another way for a utility to handle the increased demand during morning and afternoon rush hours is through the use of peaking units. Two types are used, *gas turbines* (which can be operated on natural gas, but usually use a light fuel oil) and *diesel engines* similar to those used on a large ship. Both of these are directly coupled to a generator to produce electricity. They simply burn oil or natural gas with air under pressure, and pass the exhaust gases directly to a stack.

Although both gas turbines and diesel engines burn very expensive fuel, they only operate for short periods of time and can be easily started and stopped as needed to meet the demand. As in the case of pumped storage, they eliminate the need for more complicated generating facilities. There are EPA regulations for plant emissions of particulate matter, sulfur dioxide, and oxides of nitrogen, but normally there is no problem in meeting these standards as long as low sulfur fuel is burned.

78.5.4 Fossil-Fuel-Fired Plants

About 85% of the electricity generated in the United States comes from the conventional steam turbine. High-pressure steam is produced in a boiler by burning a fuel such as coal, oil, or natural gas or by using a nuclear reactor as a source of heat. The high-pressure steam is run through a steam turbine to generate electricity, and the low-pressure steam exhausted by the turbine is sent to a condenser where it is liquefied and normally recycled back to the boiler or nuclear reactor. Because of the Carnot efficiency for the conversion of heat into work, only about one-third of the heat energy of the fuel is converted to electricity and two-thirds of the heat energy is simply "dumped" as waste heat. Since tremendous quantities of cooling water are needed for the steam condenser, close proximity to a large body of water is the first requirement in the location of any type of thermal power plant.

The following are some of the most important environmental restrictions governing the construction and operation of thermal power plants.

Thermal Discharges

A large modern generating plant will produce some thousand megawatts of electricity, which means that roughly two thousand megawatts (6.8×10^9 Btu/hr) must be dumped as thermal energy. Formerly this was done by locating near a large body of water that served as a "sink" for the plant condenser water. Today, EPA regulations theoretically forbid any type of discharge of heated water to the environment.

One way to eliminate discharge of heated water is by the use of cooling towers. Heated condenser water is recycled to the top of a large tower, where it flows down over suitable packing or grids countercurrent to a stream of air which enters the bottom of the tower. Some of the water is evaporated (this is known as *evaporative cooling*), thereby cooling the remaining water in the same way a swimmer is cooled when he or she steps out of a pool. The lowest temperature which can be reached is the *adiabatic saturation temperature*, which in the case of air and water is identical with the wet bulb temperature.

The chief disadvantage of cooling towers is that they are extremely expensive both to build and to operate, so that the cost of electricity is substantially increased. The degree of cooling depends on the temperature and humidity of the air; on a warm, humid summer day, cooling towers are not very effective. Cooling towers effectively dump the heat to the air, and add large quantities of moisture to the air. Under some conditions they may produce considerable fog and mist, which itself may be a nuisance. They are enormous structures, and may be a visual "eyesore" to those living in the area.

A second option for a utility is to use some kind of holding pond or canal that will furnish sufficient time for the discharged cooling water to dissipate some of its heat energy. In most cases this appears to be environmentally acceptable, and is the preferred method today whenever approval can be obtained from EPA.

At one time, there was considerable question as to possible adverse effects of heated water on the various biological organisms living in our bodies of water. Considerable research has been carried out on this problem, and many years experience now appear to indicate that the effect of a moderate temperature increase is minimal. In fact there is even some evidence that it may actually be beneficial, since the best fishing is sometimes found in the discharge canal from a power plant.

Particulate Matter

Usually there is no problem in meeting EPA regulations for emission of particulate matter from oil- and gas-fired plants. Only small traces of noncombustible materials are present in these fuels. The only other possible source of particulate matter is carbon resulting from incomplete combustion, but this can be controlled by proper regulation of the furnace. Under some conditions an electrostatic precipitator might be used for final cleanup, but often the exhaust gases pass directly to the stack without treatment.

Coal-fired power plants are a different story. Here very large quantities of dirt and dust (fly ash) are carried off with the furnace gases, and elaborate cleaning devices are required to meet EPA emission standards. Two types of system are in common use, *electrostatic precipitators* and *fabric filters*.

An electrostatic precipitator consists essentially of a series of suspended vertical plates containing alternate *discharge* and *collecting* electrodes. The individual electrode bundles have a width of perhaps 4 m and a height of up to 15 m, and may be stacked to a depth of 25–45 m. In the typical coal-fired plant, furnace gases pass in series through a steam boiler, superheater, reheater, economizer, and an air preheater before entering the electrostatic precipitator at a temperature of 150–290°C. A distribution system sends the gases between the electrodes where the corona discharge of the discharge electrode places an electrostatic charge (ionizes) the individual particles of fly ash. An electrical field then causes the particles to be deposited on the surfaces of both electrodes. Perhaps every 10 min a hammer strikes each electrode plate to knock off collected fly ash, which falls down to a hopper from which it can be removed.

The electrostatic precipitator is the system most commonly used today for final cleanup of exhaust gases. Each system must be individually engineered, since performance depends on the moisture, sulfur, sodium, and other chemical elements present in the coal. There are some costs associated with the electricity which must be provided, and in addition the electrodes often require maintenance. Efficiencies as high as 99.9% have been achieved in actual plant operation.

The second type of gas cleaner that is used is the fabric filter, sometimes referred to as the bag filter or baghouse. This consists of an assembly of vertical cylindrical cloth bags which filter the gas in the same manner as a household vacuum cleaner. They are cleaned periodically, either by shaking, by reverse flow of air, or by the use of an air jet; fly ash falls down to a hopper where it is collected and removed.

Fabric filters have the advantage that they are fuel insensitive, so that they do not have to be individually designed. Although fairly new to the utility industry, they have been used in other industries for almost a century. They have a greater pressure drop than the electrostatic precipitator, so that the total power consumption of the two systems is not greatly different. The efficiency of the fabric filter is also about the same, up to 99.9% in commercial operation.

The collected fly ash is normally simply a waste material, although it has been used for landfill and for highway construction. It has also found some use as a raw material for the manufacture of portland cement.

Sulfur Dioxide

The most difficult environmental problem faced by the electric power industry today is the emission of sulfur dioxide. All fossil fuels contain more or less sulfur, which burns according to the equation

$$S + O_2(g) \rightarrow SO_2(g)$$

The sulfur dioxide produced simply leaves with the gases from the furnace.

In addition to the formation of sulfur dioxide, there is always some oxidation to sulfur trioxide (SO_3) which, in the presence of moisture, reacts to form sulfuric acid:

$$SO_3(g) + H_2O(g) \rightarrow H_2SO_4(l)$$

Since sulfuric acid has an extremely low vapor pressure, it condenses to form small colloidal droplets of acid mist that are very hard to remove from the furnace gases. If this acid mist carries on to the stack, it produces a white plume that increases the opacity of the stack gases.

Annual worldwide emissions of sulfur dioxide by human activities now total about 160 million metric tons, with some 70% coming from the burning of coal and 16% from the burning of petroleum products, particularly the high-sulfur residual fuel oil formerly used by many industrial plants. Sulfur dioxide emissions have grown steadily for the past 30 years, but are now expected to level off and perhaps even decrease somewhat as greater restrictions are placed on industrial pollution.

It is of interest to note that, in addition to the sulfur emitted to the atmosphere by human activities, an estimated 90 million metric tons of hydrogen sulfide gas (the familiar "rotten egg" gas that tarnishes silver) are contributed annually by natural biological decay processes occurring on land and in the oceans. It is also estimated that some 40 million metric tons of sulfur per year enter the atmosphere from sea spray, largely in the form of minerals such as calcium sulfate. Thus natural processes account for almost half the total sulfur emissions to the atmosphere.

Sulfur dioxide is recognized as an irritant gas, causing problems for those with asthma, emphysema, or other respiratory diseases. In moderate concentrations, it is very harmful to vegetation. This has been demonstrated near mining towns where sulfide ores are roasted to burn off the sulfur, and the sulfur dioxide gases were at one time simply discharged to the stack. Sudbury, Ontario and some mining towns of the western United States are surrounded by large desolate areas filled with dead and dying trees and other vegetation.

Sulfur dioxide is removed rapidly from the atmosphere. Quite soluble in water, it forms a dilute solution of sulfurous acid according to the equation

$$SO_2(g) + H_2O(l) \rightleftharpoons H_2SO_3(l)$$

This readily removes the sulfur dioxide as an acid rain. If oxides of nitrogen are also present in the atmosphere, these catalytically oxidize the sulfur dioxide–sulfurous acid mixture to form sulfuric acid, thereby increasing the acidity of the rain.

Although acid rain corrodes structural materials, particularly limestone, the chief problem is the increased acidity of water and soil that affects the growth of crops and changes the pH of lakes and streams, thereby killing fish. At one time, utilities "solved" the problem by building taller smokestacks, but this merely exported the problem to someone else. Acid rain is particularly bad in northeastern United States and in southeastern Canada, largely because of sulfur dioxide emissions by our coal-burning midwestern utilities. Acid rain is also a problem in northern Europe, where countries such as Sweden and Norway are being affected by sulfur dioxide emissions that originate in the United Kingdom.

Emissions of sulfur dioxide can be eliminated by removing the sulfur from the fuel before it is burned. (This is usually much easier than trying to remove sulfur dioxide from the tremendous volume of stack gases which are produced.) Removal of sulfur from liquid and gaseous fuels is no real problem, although it does materially increase the cost of fuel oil and prevents the direct use of the cheap high-sulfur residual oil that was formerly used. However, it is not easy to remove sulfur from a solid fuel such as coal. Perhaps half the sulfur present in coal may be in the form of the mineral pyrites, which can be mechanically separated. The remainder is present as complex sulfur compounds that can be removed only if the coal is liquefied or gasified. Processes are now being investigated to "solvent refine" coal, but these increase the cost and it is not always easy to remove enough sulfur to meet EPA emission standards.

When coal is used as a fuel, a second way to meet EPA emission standards is through the use of scrubbers to remove sulfur dioxide from the exhaust gases. This can be done by passing the gases to the bottom of an open tower, and spraying into the top of the tower a slurry of limestone, $CaCO_3$, or slaked lime, $Ca(OH)_2$, which absorb sulfur dioxide according to the typical equation

$$Ca(OH)_2 + SO_2(g) \rightarrow CaSO_3 + H_2O$$

In practice, there is always some oxidation to give a mixture of calcium sulfite, $CaSO_3$, and calcium sulfate, $CaSO_4$.

Two types of system have been developed. The wet process, developed by companies such as Combustion Engineering, places the spray tower directly before the stack and uses sufficient water with the entering scrubbing solution to produce a calcium sulfite–sulfate slurry at the bottom of the tower. This slurry is concentrated in thickener tanks to give a solution that is recycled to the top of the tower with additional lime or limestone. The concentrated solids are vacuum dried to give a filter cake containing about 65% solids. This is mixed with fly ash and lime to give a "fixated" sludge, which in theory does not leach and can be used in landfills.

The other system for sulfur dioxide removal is the dry process developed by Combustion Engineering and Wheelabrator-Frye. Here the spray tower is located just before the electrostatic precipitator or fabric filter. The amount of water in the scrubbing solution is reduced and the gas is admitted to the top of the tower; as the gas and scrubbing solution pass downward through the tower, the water vaporizes immediately to leave a dry powder that is removed with the fly ash in either an electrostatic precipitator or fabric filter.

The wet system removes about 90% of the sulfur dioxide in the exhaust gases. The dry system has the advantage of being simpler and cheaper; however, it is not as efficient, and appears to be most suitable where low-sulfur coals are burned. It must be emphasized that both of these systems are still very much in the developmental stage, and it is entirely possible that they will be modified when more operating experience is obtained.

Oxides of Nitrogen

The third pollutant that must be considered in the design of a power plant is nitrogen oxides. These are obtained in any combustion process where the temperature is high enough to "fix" the nitrogen in the air by causing it to react directly with oxygen according to the equation

$$N_2(g) + O_2(g) \rightleftharpoons 2NO(g)$$

The production of nitric oxide, NO, depends on the temperature, the amount of excess oxygen, and the rate of cooling of the gases (the decomposition rate for nitric oxide is much greater than its rate of formation.)

The initial high-temperature product is nitric oxide, but on cooling this gradually oxidizes to nitrogen dioxide according to the equation

$$2NO(g) + O_2(g) \rightleftharpoons 2NO_2(g)$$

Since there is always a mixture of nitric oxide and nitrogen dioxide in stack gases, this is often referred to as NO_x, where x is somewhere between 1 and 2.

It is estimated that about 55 million metric tons of nitrogen oxides per year are produced by human activities, with roughly half coming from the combustion of coal. Nitrogen is, of course, essential for plant growth. The nitrogen cycle is a critical element in all biological processes, and it is estimated that the amount of nitrogen dioxide released to the atmosphere through natural biological action is some 15 times that contributed by industrial pollution and automobile exhaust. In addition, biological reactions release even greater quantities of nitrogen in the form of ammonia and nitrous oxide, N_2O. Thus, as far as the total quantity of oxides of nitrogen are concerned, pollution sources are not very important.

Nitrogen dioxide reacts readily with moisture to form nitric acid, which is easily removed from the atmosphere in rainfall. This does contribute slightly to acid rain, although sulfur dioxide is far more important. Oxides of nitrogen are an irritant, and may cause problems for those subject to asthma or other respiratory ailments. However, the chief environmental problem associated with oxides of nitrogen is their ability to produce smog.

The word *smog* was originally coined to describe a mixture of smoke and fog. Today it is used more often to categorize the photochemical smog produced by reactions between hydrocarbons, oxides of nitrogen, and ozone that are catalyzed by ultraviolet light from the sun. Smog is a particular problem for cities such as Los Angeles, Denver, and Salt Lake City where there is much sunlight and where there are geographic barriers to prevent the normal movement of air. Smog causes severe eye irritation and difficulty in breathing; it is a hazard to the very young, the elderly, and to anyone affected by respiratory problems.

The most serious consequences of smog have occurred during periods of temperature inversion. Normally the temperature of the atmosphere decreases with altitude, and since hot air rises, the discharge from a chimney simply continues upward until it is gradually dissipated in the atmosphere. A temperature inversion occurs when warm air "overruns" cold air, so that the temperature increases with altitude. The discharged pollutants are now trapped below the inversion level with no place to go, and they can rapidly build up to unpleasant or even dangerous levels.

There have been several well-publicized instances when temperature inversions have had serious consequences. The factory town of Donora, Pennsylvania is located in a steep valley of the Monongahela River; when a smog persisted for five days in 1948, roughly half of its inhabitants became ill, and there were 20 deaths. In London, a combination of fog and coal-smoke which lasted for several days in 1952 caused such a dense smog that people could barely see their own feet, and some even walked off the docks and fell into the Thames River. Mortality figures included some 4000 "excess" deaths that could be attributed to the smog. Los Angeles has periodic smog alerts, and a severe temperature inversion in November of 1966 caused New York City to issue its first dangerous smog warnings.

Despite their contribution to the problem of smog, oxides of nitrogen are not often a serious consideration in the operation of power plants except possibly in areas such as southern California that have more stringent regulations on industrial emissions. Production of oxides of nitrogen can be reduced by proper furnace design and by regulating the amount of excess air used for combustion. Also if an absorption system is used to clean up sulfur dioxide emissions, this will automatically remove a substantial fraction of the oxides of nitrogen which are present in the gases.

Carbon Dioxide

Since carbon dioxide is a natural constituent of air, it is not considered to be a pollutant, and there are no regulations concerning emission of carbon dioxide. As we have seen from our discussion of the greenhouse effect, it is possible that future generations may have to worry about carbon dioxide emissions. However, since carbon dioxide is the inevitable product from the burning of fossil fuels, there are only two ways to reduce production of carbon dioxide: either burn less fuel or place greater emphasis on nuclear power reactors. Neither of these is very acceptable today.

Carbon Monoxide

Carbon monoxide gas is the air pollutant produced in the largest quantities by human activity, and on a global scale exceeds the mass of all other air pollutants combined. Total worldwide production of carbon monoxide is roughly 360 million metric tons a year, with some 55% coming from motor vehicle emissions. The carbon monoxide content of our atmosphere is quite variable, ranging from about 1 to 70 ppm in builtup areas, with brief peak levels as high as 140 ppm. Concentrations over 600 ppm may cause headaches and nausea, and prolonged exposure to concentrations over 1000 ppm may be fatal.

In spite of the large amount of carbon monoxide introduced into the atmosphere by human activity, it is believed that natural sources of carbon monoxide contribute roughly 10 times as much. Carbon monoxide is produced by electrical storms, by forest and prairie fires, and by the decomposition of plant and animal materials. Methane and other hydrocarbons released to the atmosphere can react with oxygen to form carbon monoxide. Large quantities are also produced in the oceans by marine organisms such as jellyfish and by the decomposition of organic material in surface waters.

Even with this tremendous and continuous pollution of the atmosphere by carbon monoxide, the concentration in "clean" air remains extremely low, 0.2 ppm or less, and has been unchanged for a century or more. Obviously there are natural processes that rapidly and effectively remove carbon monoxide from the atmosphere. These are not completely understood, but it is known that sunlight can photochemically activate the oxidation of carbon monoxide to carbon dioxide, particularly in the upper atmosphere or stratosphere. There are also friendly soil bacteria and fungi that effectively remove carbon monoxide from the air, and it is now believed that the capacity of the soils in the United States is sufficient to take care of our entire production of carbon monoxide.

The physiological effects of carbon monoxide result from its strong affinity for hemoglobin, roughly 300 times that of oxygen. However, the absorption of carbon monoxide by the body is entirely reversible; it is not a cumulative poison, and is easily removed from the blood when the individual is exposed to pure air or oxygen. Many individuals, such as traffic officers in vehicular tunnels, have been exposed to high concentrations of carbon monoxide for much of their lives without known harm. Except possibly for those who live in very high traffic areas, carbon monoxide cannot be considered a serious environmental pollutant.

Carbon monoxide emissions are not a problem in the operation of power plants or other industrial facilities. There are no EPA regulations concerning emission of carbon monoxide by the power industry.

Other Volatile Pollutants

In addition to the chief pollutants listed above, there are two others that should be included, hydrocarbons and heavy metals. Since combustion of any fuel is never 100% complete, stack gases from power plants will always contain small traces of unburned volatile hydrocarbons. These can, of course, under proper conditions react with ozone or oxides of nitrogen to form smog. Emission of hydrocarbons is not normally a problem in the operation of power plants except, again, in smog-producing areas such as southern California.

When coal is used as a fuel, there are always many different impurities that may be present in trace amounts. Some of these such as arsenic and mercury are quite volatile and are also biological poisons. They are not removed by the electrostatic precipitator or fabric filter, and leave with the stack gases. There are no present EPA regulations concerning this type of emission from power plants, but it is something that does bear watching for the future.

78.6 ENVIRONMENTAL EFFECTS OF NUCLEAR POWER

This is an extremely broad subject, and the construction of a nuclear power plant is a very long, complicated, and expensive proposition. First, there must be an environmental impact study that describes all possible environmental effects, particularly any emission of radioactive materials. Second, a construction permit must be obtained from the Nuclear Regulatory Commission permitting construction to begin. Third, after the plant is finished, an operating license must be obtained from the Nuclear Regulatory Commission before a single kilowatt-hour of electricity can be generated. All these proceedings are open to the general public, and objections may be filed by individuals or special-interest groups. The time required is about 12 years, and the construction costs are several billions of dollars.

The chief problem in obtaining approval of a nuclear power plant is the question of radiation safety. EPA regulations require the annual whole-body dose to the "general public" to be not over 25 millirems of radiation. Usually this is considered to be the dose received by someone standing at the plant boundary 24 hr a day, 365 days a year. The total quantity of radioactive materials entering the general environment from the whole uranium fuel cycle for each gigawatt-year of electrical energy produced must be less than 50,000 curies of krypton-85, 5 millicuries of iodine-129, and 0.5 millicuries combined of plutonium-239 and other similar transuranium radionuclides having half-lives greater than one year. In addition, the Nuclear Regulatory Commission has established regulations for exposure of people inside the plant boundaries and for release of any radioactive materials in the plant area.

Another problem that must be faced by a nuclear power plant is the question of thermal effluents. Nuclear plants are large facilities, generating 1000 megawatts or more of electricity. In addition, the efficiency for the conversion of heat into work is moderately low because it is not possible to use superheated steam. Thus the thermal discharge from a nuclear plant is higher than that from a fossil-fuel-fired plant, and since there is the possibility that the effluent water could be slightly radioactive, it is easy to understand why this can cause problems.

Nuclear power does have one substantial advantage: there are no releases of sulfur dioxide, oxides of nitrogen, particulate matter, or even carbon dioxide! If the greenhouse effect does turn out to be real, nuclear power and hydro will be the only options for the generation of electricity.

78.7 ENVIRONMENTAL EFFECTS OF COGENERATION

The environmental effects of cogeneration are, in general, the same as those encountered in conventional fossil-fuel-fired plants. One substantial advantage of cogeneration is that the heat in the turbine exhaust is recovered instead of being dumped to the environment. It should also be noted that the EPA regulations given in Table 78.1 apply to fossil-fuel-fired boilers having more than 73 megawatts of heat input. Some plants using cogeneration are quite small, and these would fall under different regulations.

78.8 ENVIRONMENTAL EFFECTS OF THE PROCESS INDUSTRIES

Again, this is a very broad subject and we can only give a brief outline of some of the main problems. EPA regulations now cover emissions from all the process industries. For example, emission of nitric oxides from nitric acid plants cannot be more than 1.5 kg per metric ton of 100% HNO_3 produced. Emission of sulfur dioxide from sulfuric acid plants cannot be over 2 kg per metric ton of 100% H_2SO_4 produced, acid mist cannot be over 0.075 kg per ton of acid produced, and opacity of the exit gases cannot be greater than 10%. Petroleum refineries have regulations for the emission of particulate matter, sulfur dioxide, hydrogen sulfide, and volatile emissions from storage vessels. Very stringent regulations apply to particularly hazardous air pollutants such as asbestos, beryllium, mercury, and vinyl chloride.

Similar regulations apply to the discharge of aqueous effluents such as oil and grease, suspended solids, free available chlorine, and all the thousands of chemicals produced commercially. Special regulations apply to toxic pollutants such as polychlorinated biphenyls (PCB), DDT, and the chlorinated hydrocarbons such as chloroform and carbon tetrachloride.

A particularly serious problem faced by the chemical process industries is the disposal of their waste products. A particularly interesting example is the lime-soda process for the manufacture of soda ash (Na_2CO_3). About a ton of calcium chloride waste is produced for every ton of product, and calcium chloride has no commercial value. Millions of tons have been dumped on the open land or discharged to rivers and quarries where it has contaminated drinking water and killed vegetation. Enforcement of environmental restrictions has now closed down almost the entire industry, and we will have to rely on natural soda ash for our future requirements.

Another example is the discharge of mercury by chemical plants using the mercury cell for the production of chlorine and caustic soda. It was found that microorganisms in river and ocean sediments can convert inorganic mercury into volatile organic mercury compounds, which become concentrated as they move up the food chain, resulting in high concentrations of mercury in some types of fish. During the 1970s, contamination from chemical plants was so bad that some concentrations of mercury in tuna and swordfish were greater than the Food and Drug Administration limit of 0.5 ppm.

An extremely serious environmental problem today is the disposal of the millions of tons of nondegradable synthetic organic polymers we are producing annually such as 5.7 million metric tons of polyethylene and 2.5 million metric tons of polyvinyl chloride. Unlike the natural polymers such as cellulose, there are no microorganisms that will consume materials such as nylon, polyethylene, and the like, so that they persist for very long periods of time once they are placed in the environment. It would certainly be a good idea to at least minimize the use of these materials in our "throw-away" society.

Sometimes unexpected problems are encountered in the use of synthetic organic chemicals such as in the case of the polychlorinated biphenyls (PCBs). These highly chlorinated synthetic chemicals were first produced in 1929 and proved effective as heat-transfer fluids and lubricants, and in the manufacture of paints, varnishes, adhesives, and waxes. Because they were nonflammable and had good electrical properties, they were widely used as heat-transfer fluids for large electrical transformers and capacitators. They were not considered to be particularly hazardous at first, and large amounts escaped to the environment from burning of wastes, vaporization of paints and coatings, and from simply dumping PCB wastes into rivers and lakes.

In the early 1960s, evidence began to accumulate to indicate that PCBs were becoming a hazard. They were extremely persistent chemicals when discharged to the environment and were being distributed rapidly throughout the entire Northern Hemisphere. When ingested, they concentrated in fat, and animal experiments showed them to be carcinogenic. High levels of PCBs were found in fish and other organisms collected in the open sea, miles from land. Traces began to appear in human fat and milk. Because of these adverse environmental effects, U.S. production of PCBs was voluntarily ended in the 1970s, and even the use of PCBs has now been essentially outlawed by the passage of the Toxic Substances Control Act of 1976.

The fluorocarbons are another very interesting example of unexpected environmental problems encountered by the process industries. Developed some 50 years ago for use as refrigerants, these nonflammable and nontoxic fluorocarbons such as CCl_2F_2 and CCl_3F were widely used as propellants for push-button sprays. By the 1950s, they were used to propel roughly 90% of all nonfood aerosols, and represented a substantial market for chemical manufacturers.

About 20 years ago, preliminary plans were made by the United States to develop a supersonic transport plane to compete with those under development in Britain and France. This produced much opposition in the United States because of concern that the injection of water vapor and oxides of nitrogen into the stratosphere might act to destroy the ozone layer that protects us from harmful shortwave ultraviolet radiation. When experiments were carried out to test this theory, it was found that chlorine was even more effective in destroying ozone. It was also pointed out that millions of tons of chlorine-containing fluorocarbons were being released to the atmosphere through the use of push-button sprays, and since these compounds are chemically inert, they can gradually migrate to the stratosphere where the intense sunlight can dissociate them into free chlorine. In 1975, a special Federal Task Force issued a report which concluded that release of fluorocarbons was a legitimate cause for concern. Shortly after, the chemical industry bowed to pressure and voluntarily discontinued the use of fluorocarbons for aerosols and other applications that involved release to the atmosphere.

However, the major environmental problem facing the process industries today is the disposal of chemical wastes. The much-publicized story of Love Canal made the public only too aware of previous abuses by industry. Recently, Congress established a $1.6 billion "superfund" for the cleanup of chemical waste dumps, with 88% of the cost coming from industry. Future disposal of chemical wastes must be in approved impermeable sites where there is no contamination of air or groundwater. The real problem is in finding suitable locations from waste dumps, since the universal attitude of states and municipalities is "not in my backyard!" Federal action may eventually be necessary, since it is impossible to operate the chemical process industries without some way to dispose of unwanted wastes.

BIBLIOGRAPHY

Code of Federal Regulations, Title 40, Part 60, Standards of Performance for New Stationary Sources, revised through July 1, 1982.

National Research Council, *Energy in Transition 1985–2000,* W. H. Freeman, San Francisco, 1980.

S. Fred Singer, *The Changing Global Environment,* D. Reidel, Boston, 1975.

R. Stephenson, *Living with Tomorrow: A Factual Look at America's Resources,* Wiley-Interscience, New York, 1981.

INDEX